STANDARD HANDBOOK FOR ELECTRICAL ENGINEERS

OTHER McGRAW-HILL HANDBOOKS OF INTEREST

INDEX TO SECTIONS FOR QUICK REFERENCE

STANDARD HANDBOOK FOR ELECTRICAL ENGINEERS

DONALD G. FINK, Editor-in-Chief

General Manager, Institute of Electrical and Electronics Engineers;
Formerly Vice President–Research, Philco Corporation,
President of the Institute of Radio Engineers,
Editor of the Proceedings of the IRE;
Fellow of the Institute of Electrical and Electronics Engineers;
Fellow of the Institution of Electrical Engineers (London);
Eminent Member, Eta Kappa Nu

JOHN M. CARROLL, Associate Editor

Associate Professor of Industrial Engineering, Lehigh University;
Senior Member, Institute of Electrical and Electronics Engineers;
Registered Professional Engineer (Pennsylvania)

Tenth Edition

McGRAW-HILL BOOK COMPANY

New York St. Louis San Francisco Düsseldorf Johannesburg
Kuala Lumpur London Mexico Montreal New Delhi
Panama Rio de Janeiro Singapore Sydney Toronto

STANDARD HANDBOOK FOR ELECTRICAL ENGINEERS

Library of Congress Catalog Card Number 56-6964

ISBN 07–020973–1

891011 COCO 7654

Total issue, 1907 to 1968: 310,000

PREFACE TO THE TENTH EDITION

With this Tenth Edition, the Standard Handbook enters its second half-century of service to electrical engineers. In the eleven years that have elapsed since the Ninth Edition was published (an interval lengthened by the untimely death of its distinguished editor, Archer E. Knowlton) many important new techniques, briefly treated in that volume, have reached maturity. Notable among these are nuclear generation of electric power, high-voltage direct-current transmission, the electronic computer, and the associated new methods for the transmission of data in digitalized form. All these are treated in depth in new sections specially written for this edition.

In addition, the direct conversion of thermal and chemical energy into electricity (as contrasted to indirect use to drive generators by steam, water power, or internal combustion) has received concentrated attention in the past ten years. To take account of this work, the section on Conversion of Electric Power has been enlarged to include such topics as thermionic conversion, thermoelectric conversion, fuel cells and magnetohydrodynamics. On the same account the section on Batteries has been greatly enlarged, with particular attention given to the rechargeable dry cell.

Still another development is the semiconductor controlled rectifier, which, preeminiently among new electronic devices, has achieved application in rectification, conversion, inversion, and control at power levels once thought impossible in solid-state materials. The SCR accordingly has a prominent place throughout this edition. Other advances in the development of semiconductor materials and devices are also treated.

The aim of the Tenth Edition carries forward the philosophy of the preceding editions: to contain in a single volume all pertinent data within its scope, to be accurate and comprehensive in technical treatment, to be of use in engineering practice (as well as in study in preparation for such practice), and above all, to be oriented toward application, with economic factors in mind. To cover the latter requirement, the principal sections devoted to the design of equipment and systems include subsections on economic factors.

The scope of this edition, as of its predecessors, includes the generation, transmission, distribution, control, conversion, and application of electric power. Many phases of the electronic arts relating to communications, control, data processing, and industrial applications are also included, but no attempt has been made to provide comprehensive design data on electronic circuits, equipment, and systems (which would require another volume of equal size).

To cover the new developments, it proved necessary to add approximately 160 new pages, despite care in editing to condense material throughout the book. The new edition has been entirely set from new type, making possible consistent use of the new standard symbols and definitions recommended by the IEEE, as listed in Section 1. The statistics are impressive, at least to the editors who have spent over three years with manuscripts and proofs: 2,400 pages, 1,900 illustrations, 575 tables, and 1.4 million words of text matter.

Great care has been taken by the contributors and editors to check and recheck the information and tabular data herein, and it is believed that the new edition is as free from errors as can be expected in a work of its scope and detail. Some errors have no doubt escaped us; the editor would appreciate their being brought to his attention.

The editor gratefully acknowledges the firm foundation provided by his friend and colleague, Archer Knowlton, the painstaking preparation and review by the 116 contributors, and the able assistance of Associate Editor Carroll, in making this volume a fit successor to the previous editions.

DONALD G. FINK
Editor-in-Chief

CONTRIBUTORS TO THE TENTH EDITION

Willis A. Adcock, Vice President, Technical Development, Texas Instruments Incorporated (Sec. 5)

J. G. Anderson, Technical Director, Project EHV, General Electric Company (Sec. 26)

S. Annestrand, Acting Head, High Voltage Group, U.S. Department of the Interior, Bonneville Power Administration (Sec. 14)

R. E. Appleyard, Manager, Product Engineering, Hydrogenerator Department, Allis-Chalmers (Secs. 6 and 12)

G. S. Axelby, Advisory Engineer, Aerospace Division, Westinghouse Electric Corporation (Sec. 20)

Thomas D. Barnes, Engineering Manager, Meter Division, Westinghouse Electric Corporation (Sec. 3)

Bruce B. Barrow, Laboratory Manager, Sylvania Electronics Systems, An Operating Group of Sylvania Electric Products, Inc. (Sec. 1)

Theodore Baumeister, Consulting Engineer; Stevens Professor Emeritus of Mechanical Engineering, Columbia University (Sec. 8)

John Berdy, Applications Engineer, Electric Utility Engineering Operation, General Electric Company (Sec. 15)

Daniel Berg, Insulation and Chemical Technology Department, Westinghouse Research Laboratories (Sec. 4)

R. W. Bergmann, Manager, Electrical Engineering, Marion Power Shovel Company (Sec. 20)

S. Berneryd, Manager of H.V.D C. Valve Design, H.V.D.C. Department, ASEA (Sec. 14)

Julius Bleiweis, Administrative Manager, Northeast Power Coordinating Council (Sec. 16)

Warren B. Boast, Anson Marston Distinguished Professor and Head of Electrical Engineering, Iowa State University (Sec. 2)

Perry A. Borden (Deceased), Formerly Assistant Professor of Electrical Engineering, University of Dayton (Sec. 3)

H. Russell Brownell, Vice President, Hallmark Standards, Inc. (Sec. 3)

Harvey F. Brush, Manager of Engineering, Power and Industrial Division, Bechtel Corporation (Sec. 9)

David Burns, Professional Engineer, Redington Beach, Florida (Sec. 4)

John F. Cachat, Vice President–General Manager, TOCCO Division, Park-Ohio Industries, Inc. (Sec. 22)

John M. Carroll, Associate Professor of Industrial Engineering, Lehigh University (Sec. 27)

V. J. Cissna, Systems Studies Engineer (Retired), Power System Planning Branch, Tennessee Valley Authority (Sec. 13)

William Russell Clark, Staff Assistant to the Senior Vice President, Technical Affairs, Leeds and Northrup Company (Sec. 15)

Nathan Cohn, Senior Vice President, Technical Affairs, Leeds and Northrup Company (Sec. 15)

J. J. Coleman, Vice President, Engineering & Research, Burgess Battery Company, Division of Servel, Inc.

Irvin L. Cooter, Consultant, Formerly Chief, Magnetic Measurements Section, National Bureau of Standards (Sec. 4)

E. J. Croop, Insulation and Chemical Technology Department, Westinghouse Research Laboratories (Sec. 4)

T. W. Dakin, Insulation and Chemical Technology Department, Westinghouse Research Laboratories (Sec. 4)

W. Kenneth Davis, Vice President, Bechtel Corporation (Sec. 9)

M. C. Deibert, Department of Chemical Engineering, Massachusetts Institute of Technology (Sec. 12)

Thomas J. Dolan, Head, Department of Theoretical and Applied Mechanics, College of Engineering, University of Illinois (Sec. 4)

J. F. Donohue, Manager, Product Development, Alkaline Battery Division, Gould-National Batteries, Inc. (Sec. 24)

W. Crawford Dunlap, Assistant Director for Electronics Components Research, Electronics Research Center, National Aeronautics and Space Administration (Sec. 4)

Charles Emerson, *American Machinist* (Sec. 20)

P. G. Engström, Director, Semiconductor and Electronics Division, Allmänna Svenska Elektriska Aktiebolaget, Sweden (Sec. 14)

E. F. Errico, P. R. Mallory & Company, Inc. (Sec. 24)

Stephen Foldes, President, Endicott Coil Company, Incorporated (Sec. 5)

B. Funke, Manager of H.V.D.C. Valve Development, H.V.D.C. Department, ASEA (Sec. 14)

G. C. Gainer, Insulation and Chemical Technology Department, Westinghouse Research Laboratories (Sec. 4)

Hans Gartmann, Chief Engineer (Retired), Pump and Compressor Department, De Laval Turbine, Inc. (Sec. 20)

T. A. Griffin, Jr., Assistant Vice President, Consolidated Edison Company of New York, Inc. (Sec. 10)

J. H. Hagenguth, Manager, High Voltage Research, General Electric Company (Sec. 26)

Harry W. Hale, Professor of Electrical Engineering, Iowa State University (Sec. 2)

R. C. Hardie, Manager, Mine Hoist Department, Process Machinery Division, Nordberg Manufacturing Company (Sec. 20)

Forrest K. Harris, Chief, Absolute Measurements Section, National Bureau of Standards (Sec. 3)

George W. Hedderson, Electrical Engineer (Retired), Exide Industrial Division, The ESB Incorporated (Sec. 24)

C. G. Helmick, Industry Engineering Department, Westinghouse Electric Corporation (Sec. 20)

S. W. Herwald, Vice President, Electronic Specialty Products Group, Westinghouse Electric Corporation (Sec. 20)

F. N. Houser, Mechanical Editor, *Railway Age*; Managing Editor, *Railway Locomotives and Cars* (Sec. 21)

Paul L. Howard, President, P. L. Howard Associates, Inc. (Sec. 24)

Kasson Howe Electrical Engineer (Retired), Ward Leonard Electric Company (Sec. 5)

J. F. Hower, Chief Electrical Engineer, Lehigh Portland Cement Company (Sec. 20)

Daniel Jackson, Jr., Associate Editor, *Coal Age* (Sec. 20)

W. E. Jacobsen, Industrial Sales Division, General Electric Company (Sec. 21)

A. A. Jones, Chief Engineer-Administrative, Anaconda Wire and Cable Company (Sec. 4)

A. S. Judd, Electrical Engineer, Imperial Oil Enterprises Ltd. (Sec. 20)

CONTRIBUTORS TO THE TENTH EDITION

J. A. Keeley, Manager, Operations Research, *The Miami Herald* (Sec. 20)

J. J. Lander, Director of Electrochemical Research, Delco-Remy Division, General Motors Corporation (Sec. 24)

Franklin J. Leerburger, Consulting Engineer (Sec. 8)

I. Lidén, Electrical Engineer, H.V.D.C. Department, ASEA (Sec. 14)

Frederick A. Lowenheim, Research Coordinator, M & T Chemicals, Inc. (Sec. 23)

John Lusti, Chief Engineer, Otis Elevator Company (Sec. 20)

C. T. Main II, Associate, Chas. T. Main, Inc. (Sec. 20)

Clifford Mannal, Consultant, Engineering Services, General Electric Company (Sec. 12)

H. Mårtensson, Chief Engineer, H.V.D.C. Department, ASEA (Sec. 14)

Jacob H. Martin, Technical Assistant to the Vice President, Engineering, Sprague Electric Company (Sec. 5)

Norman W. Mather, Professor of Electrical Engineering, Princeton University (Sec. 12)

W. T. Matzen, Manager, Molecular Electronics Branch, Semiconductor Research and Development Laboratory, Texas Instruments, Incorporated (Sec. 5)

H. P. Meissner, Professor of Chemical Engineering, Massachusetts Institute of Technology (Sec. 12)

H. F. Minter, Insulation and Chemical Technology Department, Westinghouse Research Laboratories (Sec. 4)

J. P. Mize, Member, Technical Staff, Semiconductor Research and Development Laboratory, Texas Instruments, Incorporated (Sec. 5)

Robert E. Moe, P.E., Tube Department, General Electric Company (Retired) (Sec. 5)

E. H. Myers, Manager, D-C Motor and Generator Engineering, Large Rotating Apparatus Division, Westinghouse Electric Corporation (Sec. 7)

A. H. Myles, Chief Engineer, Heavy Industry Division, Square D Company (Sec. 20)

Frederick R. Nelson, Senior Communications Engineer, American Electric Power Service Corporation (Sec. 15)

Leonard M. Olmsted, Senior Editor, *Electrical World* (Sec. 16)

T. B. Owen, Missile & Space Systems Division, Douglas Aircraft Company, Inc. (Sec. 21)

Jack F. Parsons, P.E., Division Illuminating Engineer (Retired), Niagara Mohawk Corporation (Sec. 19)

David B. Peck, Vice President, Engineering, Sprague Electric Company (Sec. 5)

W. E. Phillips, Head, Electric Power Section, Research and Development Department, Leeds and Northrup Company (Sec. 15)

F. B. Pipal, Manager, Battery Engineering Department, Consumer Products Division, Union Carbide Corporation (Sec. 24)

Leslie D. Price, Consultant, Formerly Manager, Engineering and Safety Regulations Department, National Electrical Manufacturers Association (Sec. 29)

L. L. Quinlan, Assistant General Manager, Operating Services, Inland Steel Company (Sec 20)

William J. Rheingans, Consulting Engineer (Sec. 8)

John B. Rice, Chief Development Engineer, Control-Switchgear, General Products Division, Allis-Chalmers (Sec. 12)

Gerald E. Rigsby, Senior Design Engineer, Delco-Remy Division, General Motors Corporation (Sec. 21)

William L. Ringland, Research Division, Allis-Chalmers (Sec. 6)

Arthur N. Robertson, Electronics and Communications Engineer, American Electric Power Service Corporation (Sec. 15)

R. L. Robertson, Senior Staff Engineer, Control Systems Products, General Products Division, Allis-Chalmers (Sec. 12)

E. J. Rogers, Brookhaven National Laboratory (Sec. 26)

Samuel Ruben, Consultant, P. R. Mallory & Company, Inc. (Sec. 24)

R. N. Sampson, Insulation and Chemical Technology Department, Westinghouse Research Laboratories (Sec. 4)

Wesley P. Schiffries, Supervising Engineer, Scientific Development Department, Bechtel Corporation (Sec. 9)

H. E. Seemann, Physicist (Retired), Research Laboratory, Eastman Kodak Company (Sec. 26)

J. F. Sellers, Chief Engineer, D-c Machines, Allis-Chalmers (Sec. 12)

John D. Shepard, Consulting Engineer, Greensboro, North Carolina (Sec. 20)

Claude M. Smith, Process Computer Business Section, Information System Division, General Electric Company (Sec. 27)

E. H. Spreckelmeier, Chief Electrical Engineer, American Laundry Machinery Industries (Sec. 20)

G. R. Sprengling, Insulation and Chemical Technology Department, Westinghouse Research Laboratories (Sec. 4)

H. Ståckegard, Manager of H.V.D.C. Project Planning Department, ASEA (Sec. 14)

Karl A. Staley, Lamp Division, General Electric Company (Sec. 19)

E. C. Starr, Consulting Engineer, U. S. Department of the Interior, Bonneville Power Administration (Sec. 14)

W. T. Stuart, Editor, *Electrical Construction and Maintenance* (Sec. 17)

Walter Sturrock, Illuminating Engineer, Lamp Division, General Electric Company (Sec. 19)

Jack Swiss, Insulation and Chemical Technology Department, Westinghouse Research Laboratories (Sec. 4)

Joseph Teno, Principal Research Engineer, Avco Everett Research Laboratory (Sec. 18)

Douglas W. Turrell, Senior Scientist, Development Division, Leeds and Northrup Company (Sec. 15)

E. Uhlmann, Consulting Engineer, H.V.D.C. Department, ASEA (Sec. 14)

Roland W. Ure, Jr., Fellow Physicist, Westinghouse Research Laboratories (Sec. 12)

R. L. Webb, Associate Inside Plant Engineer, Electrical Engineering Department, Consolidated Edison Company of New York, Inc. (Sec. 10)

L. Wetherill, Consulting Engineer, Power Transformer Department, General Electric Company (Sec. 11)

C. R. Williamson, Assistant Vice President—Engineering, American Telephone and Telegraph Company (Sec. 25)

V. C. Wilson, Research and Development Center, General Electric Company (Sec. 12)

Harold Winograd, Consulting Engineer (Retired), Allis-Chalmers (Sec. 12)

F. T. Wolford, Assistant Professor, Agriculture Machinery Department, Delaware Valley College of Science and Agriculture (Sec. 20)

F. A. Yeoman, Insulation and Chemical Technology Department, Westinghouse Research Laboratories (Sec. 4)

Vin Zeluff, Electronics Consultant (Sec. 28)

CONTENTS

CONTENTS

CONTENTS

STANDARD HANDBOOK FOR ELECTRICAL ENGINEERS

SECTION 1

QUANTITIES, UNITS, AND CONVERSION FACTORS

BY

BRUCE B. BARROW Laboratory Manager, Sylvania Electronic Systems, An Operating Group of Sylvania Electric Products, Inc.; Senior Member, Institute of Electrical and Electronics Engineers; Member, IEEE Standards Committee; Chairman USASI Committee Y10.9 on Letter Symbols for Radio

CONTENTS

Numbers refer to paragraphs

SECTION 1

QUANTITIES, UNITS, AND CONVERSION FACTORS

BASIC CONCEPTS

1. Physical Quantity. Physical quantities are concepts used for the qualitative and quantitative description of physical phenomena. A physical quantity is characterized qualitatively by its *kind* and quantitatively by its *magnitude.* Quantities of the same kind may be compared; e.g., one may ask which of two electric currents is the larger. Comparison of quantities of different kinds, such as a mass and a current, however, is not meaningful.

Quantities of the same kind are subject to measurement, which means that we can experimentally determine a ratio between any two of them. Thus, if one sample quantity from the set of quantities of a given kind is chosen as a reference quantity, called the *physical unit* (or, simply, *unit*), then any other quantity from this set can be expressed as a product of the unit and a number, called the *measure.**

For a quantity symbolized by q, this relationship may be described by the statement: q is $\{q\}$ times as large as q_u. Here q_u is a symbol for the sample quantity chosen to be the unit, and the number $\{q\}$ is the measure of the quantity q.

For the concept of physical quantity to be useful in the quantitative description of physical phenomena, we must require of every particular quantity q that it be independent of the sample quantity q_u chosen as a unit. Thus, if we change our unit from q_{u1} to q_{u2}, where q_{u2} is α times as large as q_{u1}, then the measures of *any* q, relative to the two units, must satisfy the equation

$$\{q\}_2 = \{q\}_1/\alpha \tag{1-1}$$

(Note that here, as above, braces indicate the measure of a quantity, $\{q\}_1$ denoting the measure of q relative to the unit q_{u1}, and similarly for $\{q\}_2$.) Equation (1-1) is a statement of the basic principle that *the measure of a physical quantity is inversely proportional to the unit chosen for its measurement.*

The experimental laws of physics are expressible as proportions among measures, or *measure equations,* such as

$$\{F\} = k\{m\}\{a\} \tag{1-2}$$

Here the proportionality constant k depends upon the units in which the quantities F, m, and a are measured.

2. Symbolic Quantity. To set up the appropriate mathematical structure of theoretical physics, each physical quantity may be represented by a mathematical element called a *symbolic quantity.* The mathematics of group theory has been applied by Page† and others to the development of the *quantity calculus,* which provides a rigorous mathematical framework for handling the symbolic quantities. Symbolic quantities may be manipulated according to the conventional rules of algebra, even when there is no apparent physical significance to a similar manipulation involving the corresponding physical quantities or their measures. Thus we may model physical laws using sym-

* The terms *magnitude, value,* and *numerical value* have also been used in this sense.
† Superscripts refer to the Bibliography (Par. **21**).

bolic quantities in *quantity equations*, such as

$$F = kma \tag{1-3}$$

where k is, as before, a proportionality constant. Here F, m, and a represent not just the measures, but the quantities themselves. What is indicated as a multiplication of m and a is actually a mathematical statement of the fact that to each pair consisting of one m element taken together with one a element there corresponds precisely one F element.

3. Theoretical and Experimental Points of View. The similarity in form between Eqs. (1-2) and (1-3) results from the close correspondence between theoretical model and experimental model that the theoretician strives for. In the scientific literature, moreover, measure equations are usually written exactly as quantity equations [i.e., the braces used in Eq. (1-2) are omitted], and as a result the distinctions become blurred. Usually this causes no difficulty, but when the philosophy of physical measurement underlying the concepts of quantities and units is discussed, it is essential to distinguish carefully between the two contrasting points of view held by those who ordinarily work with measure equations and those who tend to think in terms of quantity equations. The two habits of thought and the apparent paradoxes and controversies that result from mixing them unwittingly have been carefully discussed by Silsbee[2,3] in terms of an idealized experimentalist, the Realist, who "thinks only in terms of physical quantities and units and considers all his equations to be measure equations," and an idealized theoretician, the Synthetiker, who "thinks only in terms of symbolic quantities and units and considers all his equations to be quantity equations."

The letter symbols in the measure equations of the Realist denote only numbers, and hence the symbols can be manipulated algebraically. Since the addition of physical quantities of the same kind (e.g., the addition of one length to another) has a very concrete physical significance, he may write

$$l = 7.5 \text{ m} \tag{1-4}$$

as a sort of shorthand for the statement that "the length l is 7.5 meters." In Eq. (1-4) m is simply an abbreviation for the noun meter, and 7.5 is an adjectival modifier.

If he thinks of Eq. (1-4) as an *equation*, however, he may concede that both sides can be divided by the unit of length; i.e., he may concede a physical significance to the equation

$$l/\text{m} = \{l\}_{\text{m}} = 7.5 \tag{1-5}$$

This suggests that the following operations on physical quantities have significance: (1) two physical quantities of the same kind can be added or subtracted; (2) a physical quantity can be multiplied by a number; and (3) a physical quantity can be divided by another physical quantity *of the same kind*. Beyond this point, however, the experimentalist approach does not go. Without the concepts of quantity calculus it is not permitted to multiply quantities, even quantities of the same kind. Thus the Realist interprets Ohm's law

$$V = IR \tag{1-6}$$

purely as a measure equation, since, as Silsbee has put it, "the Realist dealing with physical quantities finds it meaningless to talk about multiplying a procession of electrons by a property of some alloy which resists such a procession."[3]

For the theoretician, however, the letter symbols appearing in Eq. (1-6) completely represent the symbolic quantities and therefore connote kind as well as magnitude. He interprets Eq. (1-6) as meaning that the product of a point in I space and a point in R space yields a point in V space. Furthermore, he is willing to call his symbolic quantities by the names long used for the corresponding physical quantities—voltage, current, and resistance. Since for a very long time no one has seriously proposed that an arbitrary coefficient of proportionality, like the k of Eqs. (1-2) and (1-3), should be included in the basic statement of Ohm's law, harmony appears to be achieved.

Consider for contrast, however, the expression for the total electric flux Ψ resulting from a static electric charge Q. In an unrationalized system of measurement this expression is

$$\Psi_u = Q \tag{1-7a}$$

while in a rationalized system it is

$$\Psi_r = 4\pi Q \tag{1-7b}$$

The experimentalist, considering both these expressions to be measure equations and knowing that both refer to the same physical situation, concludes that the change in the measure of Ψ results from a change in *unit*. Because he conceives it a fundamental principle of measurement that the size of a unit varies inversely as the measure, he decides that the physical unit of electric flux in an unrationalized system is 4π times as large as that in a rationalized system (the same unit being used for charge in both).

The theoretician, on the other hand, interprets Eqs. (1-7*a*) and (1-7*b*) as quantity equations and passes from the first to the second by a substitution of variables (e.g., "Let $\Psi_r = 4\pi\Psi_u$"). He sees this as a change in the *symbolic quantities*. This change does not imply any change in the symbolic unit for electric flux, for the procedure followed by the theoretician in choosing the mathematical elements that he uses as symbolic units ensures that the size of the unit does not *in general* vary inversely as the measure.[3]

From the above we see that one physical quantity and two symbolic quantities, together with two physical units and one symbolic unit, are used in constructing models to discuss the single aspect of the electrostatic field that we conceive to be a flux. All three quantities are ordinarily called *electric flux*, without distinction, and this is one of the principal sources of confusion. Of course, rationalization affects many quantities other than electric flux, and there are many more than two sets of equations used in modeling the phenomena of electricity and magnetism, so that the need for care in a discussion of quantities and units is obvious.

The habits of thought of the experimental approach are generally shared by electrical engineers and experimental physicists, while those of the quantity calculus are more congenial to theoreticians. Neither point of view is more "correct" in its modeling of the universe; each offers the possibility of constructing a complete and internally consistent system of measurement.

SYSTEMS OF MEASUREMENT

4. Unit Definitions. For each quantity used in a branch of science a unit must be defined. These definitions may be made quite independently of each other, but if this is done, each of the measure equations used to express the relationships of physics will include an experimentally determined coefficient. In most scientific and engineering work, therefore, a small number of quantities are chosen, and units for other related quantities are derived from the basic units.

5. Construction of a System of Measurement. A system of measurement consists of a set of N quantities of different kinds, sufficient to treat the area of science being considered, together with N units, one for each quantity. To construct a system of measurement, we set up M independent equations involving the N different quantities. These M equations represent fundamental facts about our description of the physical universe, including the relations of geometry. The equations involve constants of proportionality appropriate to the geometrical situation involved.

To complete the system of measurement it is necessary to choose $N - M$ quantities that are to be regarded as basic and to select a basic unit corresponding to each one of the basic quantities. The system of equations allows us then to derive the other quantities in terms of the basic quantities and to derive appropriate units as well. Note that the term *basic* is here used in the sense of defining a base on which to construct the system; it should not be inferred that the basic quantities are in any sense more fundamental to the philosophy of science than the others.

As a simple example, consider a part of the field of mechanics, which can be described by the quantities length (l), velocity (v), time (t), acceleration (a), force (F), mass (m), and work (W). Here we have seven quantities ($N = 7$). In constructing our system of measurement we might set up the following equations involving these quantities:

$$l = k_1vt \tag{1-8}$$
$$v = k_2at \tag{1-9}$$
$$a = k_3F/m \tag{1-10}$$
$$W = k_4Fl \tag{1-11}$$

The first of these equations is to be interpreted as meaning that the length l traversed by a point moving at constant velocity v during time t is proportional to the product of v and t, and the other equations have similarly direct interpretations in our model of the physical world.

In the common systems of measurement all four of the coefficients (k's) are set equal to unity. It remains to choose three of the quantities as basic quantities (since $N-M = 3$) and to define basic units corresponding to these basic quantities. A number of reasonable choices could be made, but the experimentalist usually prefers to select mass, length, and time for the construction of a system of measurement. Occasionally force is chosen as basic—it can certainly be argued that force, which appears in electricity, magnetism, gravitation, and dynamics, is as fundamental as any quantity. Mass can be more accurately measured than force, however, and more stable standards of mass can be maintained. For these reasons metrologists have usually preferred to use mass as a basic quantity.

The final step in constructing this system of measurement is to define the basic units, which are normally chosen to be of convenient size for the work at hand. Common choices in modern technology are the meter, kilogram, and second, or the foot, pound, and second.

The experimentalist thinks of Eqs. (1-8) through (1-11) as measure equations and defines derived units for velocity, acceleration, force, and work, with the coefficients of proportionality equal to unity. In other words, he conceives a physical unit of velocity such that a point moving at that uniform velocity will traverse one unit of length in one unit of time. Such an internally consistent set of physical units is called a *germane* set of units.*

It is important to stress the arbitrariness of the steps that are involved in establishing a system of measurement. A total of $2N$ choices must be made—N quantities must be chosen, M equations must be selected (each of which involves a coefficient of proportionality), and $N - M$ basic quantities, with their appropriate basic units, must be selected. Although most systems of measurement used in mechanics, including the example given above, involve three basic quantities, it is not essential that $N - M$ equal 3. For example, the equation expressing the gravitational force of attraction between two masses separated by a distance l may be added to the system of equations given above. Thus we have

$$F = k_5m_1m_2/l^2 \tag{1-12}$$

If this equation, which is certainly of fundamental importance in physics, is used to derive a quantity, then k_5 is assigned an arbitrary numerical value and $N - M$ is equal to 2, so that only two quantities may be given basic status and only two basic units may be chosen. Such a system of measurement is used occasionally in some astronomical work, but it has no wide appeal because it is impossible to measure the gravitational attraction between masses to a satisfactory accuracy for general work. Therefore the coefficient k_5 is ordinarily regarded as a physical constant which must be measured experimentally.

6. Dimensions. It can be shown[3] that any derived quantity Q may be expressed by means of an equation of the form

$$Q = kA^\alpha B^\beta \cdots \tag{1-13}$$

where $A, B, \cdots$ are the basic quantities of the system and k is a constant that is not affected by any change in the basic symbolic units of the system provided coherence

* Some authors use the term *coherent*. Here, however, we reserve the term *coherent* to imply an internally consistent set of symbolic units (Silsbee, reference 3).

is maintained. As a consequence of Eq. (1-13), if the symbolic unit of quantity A is multiplied by the factor a^{-1}, then the measure of quantity Q is multiplied by the factor a^{α} and the coherent symbolic unit in which Q is measured is multiplied by the factor $a^{-\alpha}$. Under these circumstances, it is said that Q has the dimension $A^{\alpha}B^{\beta} \cdots$, which is commonly written

$$[Q] = [A^{\alpha}B^{\beta}\cdots] \tag{1-14}$$

If all the exponents in Eq. (1-14) are equal to zero, Q is said to be "dimensionless," i.e., to have the dimensions of a numeric.

If we denote the dimensions associated with the basic quantities mass, length, and time by M, L, and T, then we have, for the derived quantities used in the example of Par. **5**,

$$[v] = LT^{-1} \tag{1-15}$$

$$[a] = LT^{-2} \tag{1-16}$$

$$[F] = MLT^{-2} \tag{1-17}$$

$$[W] = ML^{2}T^{-2} \tag{1-18}$$

A dimension is thus seen to be a label attached to a quantity which gives some information about the nature of the quantity and about its relationship to other quantities. Dimensions are commonly used in two ways. First, they provide a very useful tool for detecting errors, since most equations used in analytical work are homogeneous, which is to say that all terms have the same dimensions. It is easy to check the work at various stages to see that dimensional homogeneity is preserved.

The second very important application of dimensional analysis uses Buckingham's Π-theorem, which states that, under certain restrictions, any complete physical equation can be put in the form

$$f(\Pi_1, \Pi_2, \cdots, \Pi_i) = 0 \tag{1-19}$$

where each of the Π's is a "dimensionless" product of powers of some of the symbolic quantities involved in the equation, where f indicates a function of the independent arguments $\Pi_1, \Pi_2, \cdots$, and where i, the maximum number of independent arguments of the function f, is equal to the number of quantities involved in the particular problem minus the number of basic quantities in the system of measurement. The smaller the number of Π's, the greater the amount of information yielded by dimensional analysis, and it is partly for this reason that theoreticians have preferred electromagnetic systems with four, or even five, basic dimensions to systems with three.

It would be useful if the dimension label completely defined the quantity, i.e., if no two quantities of different kinds had the same dimensions. This is not achieved in common systems, where, for example, work and torque both have the dimensions of force times length. Page has shown[1] that resolution of such ambiguities requires assigning dimensions other than numeric to plane and solid angles. Torque is quite properly defined as work per unit angle (of rotation), for example, and if the dimension of plane angle is explicitly retained in dimensional analysis, torque and work then have different dimensions.

It is worth remarking, before leaving the subject of dimensions, that one can obtain the coherent symbolic unit for a quantity by replacing each basic dimension in the dimensional expression by the basic unit. Thus, from Eq. (1-18), we obtain (using angle brackets to indicate units)

$$\langle W \rangle = \langle m \rangle \langle l \rangle^{2} \langle t \rangle^{-2} = \text{kg}\cdot\text{m}^{2}\cdot\text{s}^{-2} \tag{1-20}$$

in the MKS system.

7. Rationalization. The process of rationalization may best be understood in terms of assigning coefficients of proportionality, like the k's in Eqs. (1-8) to (1-11). The coefficients are chosen to be appropriate for the geometry concerned, as has been pointed out. Rationalization is not exclusively a problem of electricity and magnetism. Workers in other fields, e.g., photometry, have also experienced some difficulty in agreeing on the coefficients to be used. In electricity and magnetism, however, what has been called "rationalization" has been a chief cause of the bewildering profusion of systems of measurement, and the problem has received a great deal of attention in the literature.

In early systems of equations, as set up by Maxwell and others, the equations for the force between two electric charges and for the force between two magnetic poles do not involve the factor π. As a result, π is absent in the equations describing many cases involving spherical or circular symmetry, while it appears in equations such as that for the capacitance of a parallel-plate capacitor, where the geometry would not lead one to expect it. Heaviside, in the 1880s, called for a more "rational" choice of coefficients, and since then several schemes for rationalization have been proposed.[3] The rationalized MKSA system (see Par. **12**) is most widely used today.

We have already noted in Par. **3** that the theoretical approach sees rationalization as implying a change in quantities, whereas the experimental approach effects rationalization by changing units. The choice is not between right and wrong, but between two alternative systems, each of which is internally logical and consistent.

8. Units and Standards. A *unit* is a defined sample of a quantity. Since it is established by definition, it does not depend upon physical conditions such as temperature or pressure.

A *standard* is the physical embodiment of a unit. In general a standard may depend upon its environment and will then embody the unit only under specified conditions. Thus the International Prototype Meter, the fundamental standard of length used by most countries of the world until 1960, was considered to represent a length of exactly one meter only when at a specified temperature and supported in a specified manner. Now only one basic unit, the kilogram, is defined in terms of a standard, in this case the International Prototype Kilogram. This standard mass thus assumes a fundamental character. All other basic units are now defined with reference to a fundamental physical situation; e.g., the meter is defined in terms of the wavelength of a particular spectral line.

9. The "Fourth-unit" Problem. Although the discussion of systems of measurement in Par. **5** was general, the simple example given there concerned only quantities of mechanics. When a system of measurement is to be expanded to include an additional branch of physics, a defining equation must be introduced to link the new quantities with the old. Thus, in order to include electromagnetism, we must choose an equation that involves both mechanical and electromagnetic quantities. One such equation expresses the force per unit length l between two infinitely long, parallel, current-carrying conductors, separated by a distance r,

$$F = k_6 I_1 I_2 l/r \tag{1-21}$$

If we continue as we have in the past, we assume k_6 to be a numeric. As a consequence we have the dimensional statement

$$[I] = [F]^{1/2} = M^{1/2}L^{1/2}T^{-1} \tag{1-22}$$

We have thus included the electrical branch of physics without increasing the number of basic units or of dimensions, but the results of dimensional analysis in this system are not pleasing. If we proceed along this route, which leads to the classic electromagnetic system based upon three basic quantities, we find that electrical inductance has the dimension of length, and magnetic flux density B has the same dimensions as magnetic field strength H, which makes it difficult to be consistent about keeping these two quantities separate. Furthermore, many corresponding quantities in the classic electromagnetic and electrostatic systems with three basic units (see Par. **10**) have different dimensions.

To avoid difficulties such as these, it has been suggested that a system of measurement for electromagnetism should include a fourth basic unit, drawn from among the electromagnetic quantities. This means that we are allowed to assign dimensions other than numeric to k_6, so that when Eq. (1-21) is written, two new quantities, one of which may be regarded as basic, are introduced with one equation.

We may thus rewrite Eq. (1-21)

$$F = \mu_r \Gamma_m I_1 I_2 l/2\pi r \tag{1-23}$$

where μ_r, the relative magnetic permeability, is a numeric used to describe the medium;

where the 2π in the denominator accounts for the geometry (the equation is in its "rationalized" form); and where Γ_m is given the name *magnetic constant.* Note that $\mu_r\Gamma_m$ is the absolute magnetic permeability. Since μ_r is equal to unity for vacuum, Γ_m may be identified as the magnetic permeability of vacuum. Because Γ_m is a defined constant, however, assigned an arbitrary value when a system of measurement is constructed, a movement to call it the *magnetic constant* has developed. This constant is the fundamental link between mechanical and electromagnetic quantities.

In the MKSA system, the measure of Γ_m is defined to be $4\pi \times 10^{-7}$, and the ampere is chosen as the fourth basic unit. Silsbee[3] has pointed out that it would have been more logical to choose the unit of magnetic permeability as the fourth basic unit and to regard vacuum as the prototype standard of permeability. As things now stand, the ampere, although called a basic unit, is not defined in terms of a prototype standard of current but is defined as the amount of current required to make the measure of Γ_m, as defined in Eq. (1-23), exactly equal to $4\pi \times 10^{-7}$.

10. Terms Used to Describe Systems of Measurement. In describing electromagnetic units or systems of measurement, the term *absolute* is often used to indicate that the electromagnetic units are chosen systematically and are based upon the units of length, mass, and time, rather than on prototype standards of the electromagnetic quantities themselves or on electrical properties of certain alloys, for example.

If, in choosing a link between the mechanical and electromagnetic quantities, a conventional value is chosen for the magnetic constant (permeability of a vacuum), then the system is called *electromagnetic.* In that case, the electric constant (permittivity of vacuum) becomes an experimental physical constant to be determined, since the product of electric constant and magnetic constant is inversely proportional to the square of the velocity of electromagnetic wave propagation in vacuum. Conversely, if the link between mechanical and electromagnetic quantities is established by assigning a conventional value to the electric constant (permittivity of vacuum), the system is called *electrostatic.* In this case the magnetic constant becomes a physical constant to be determined by measurement.

A system is called *rationalized* if the coefficients in the defining equations include explicit factors of π for those cases where spherical or cylindrical symmetry is involved and do not include such factors where the appropriate geometry is rectilinear.

A system is called *comprehensive* or *complete* if it is designed so that it may be extended to cover the whole range of physical quantities, and it is called *partial* or *incomplete* if it is intended for use in only a restricted field of science.

11. The International System of Units. In 1954 the Tenth General Conference on Weights and Measures (CGPM), the international treaty organization set up by the major nations of the world to coordinate matters concerning weights and measures, adopted the following six units "to serve as a basis for the establishment of a practical system of measures for international purposes":

Length	meter	m
Mass	kilogram	kg
Time	second	s
Electric current	ampere	A
Thermodynamic temperature*	kelvin	K
Luminous intensity	candela	cd

* In 1967, the CGPM voted to give the name kelvin to the SI unit of temperature (formerly the "degree Kelvin") and to assign the symbol K, without the symbol °.

In later actions, the CGPM gave the name International System of Units and the international designation SI to this system. It also recognized the following units as belonging to the International System:

Supplementary Units

Plane angle	radian	rad
Solid angle	steradian	sr

Derived Units

Area	square meter	m^2
Volume	cubic meter	m^3
Frequency	hertz	Hz
Density	kilogram per cubic meter	kg/m^3
Velocity	meter per second	m/s
Angular velocity	radian per second	rad/s
Acceleration	meter per second squared	m/s^2
Angular acceleration	radian per second squared	rad/s^2
Force	newton	N
Pressure (stress)	newton per square meter	N/m^2
Kinematic viscosity	square meter per second	m^2/s
Dynamic viscosity	newton-second per square meter	$N \cdot s/m^2$
Work, energy, quantity of heat	joule	J
Power	watt	W
Electric charge	coulomb	C
Voltage, potential difference, electromotive force	volt	V
Electric field strength	volt per meter	V/m
Electric resistance	ohm	Ω
Capacitance	farad	F
Magnetic flux	weber	Wb
Inductance	henry	H
Magnetic flux density	tesla	T
Magnetic field strength	ampere per meter	A/m
Magnetomotive force	ampere	A
Luminous flux	lumen	lm
Luminance	candela per square meter	cd/m^2
Illumination	lux	lx
Wave number	1 per meter	m^{-1}
Entropy	joule per kelvin	J/K
Specific heat capacity	joule per kilogram kelvin	$J/kg \cdot K$
Thermal conductivity	watt per meter kelvin	$W/m \cdot K$
Radiant intensity	watt per steradian	W/sr
Activity (of a radioactive source)	1 per second	s^{-1}

The following prefixes were adopted to indicate multiples and submultiples of units:

	Prefix	*Symbol*
10^{12}	tera	T
10^{9}	giga	G
10^{6}	mega	M
10^{3}	kilo	k
10^{2}	hecto	h
10^{1}	deka*	da
10^{-1}	deci	d
10^{-2}	centi	c
10^{-3}	milli	m
10^{-6}	micro	μ
10^{-9}	nano	n
10^{-12}	pico	p
10^{-15}	femto	f
10^{-18}	atto	a

* This prefix is spelled "déca" in French.

The International System is comprehensive and will be expanded by the addition of other basic and derived units. It includes as a subsystem the MKSA system, the absolute, electromagnetic, rationalized system recommended by the International Electrotechnical Commission.

In the SI, the radian and steradian have been given the special status of "supplementary units." They are clearly not derived from the basic units but are in fact required to be precise in defining some of the derived units. The international scientific organizations have not yet seen fit to give them the status of basic units, presumably because they are still officially thought to be numerics.

The SI is seeing rapid adoption in many fields of work. The National Bureau of Standards and the American Society for Testing and Materials have recommended it for the publication of data, and the IEEE, in a series of recommendations published in 1966 (see Par. **13**), has given it strong preference over all other systems.

12. Other Systems of Units. *CGS Dynamical System.* For a time after the introduction of the metric system, attempts were made to introduce new derived units that were germane to the meter, gram, and second. In 1873, however, a committee established by the British Association for the Advancement of Science recommended basing the definitions of units in both dynamics and electricity on the centimeter, gram, and second, largely because in a CGS system the density of water is essentially unity. This committee also introduced the names *dyne* and *erg* for the CGS units of force and energy, respectively. The work of the BA committee was carefully done and timely, and it has had a wide influence on all fields of science and technology.

CGS Electrical Systems. The most commonly used CGS systems in electricity and magnetism are the *CGS electromagnetic system,* in which the magnetic constant is set equal to unity, and the *CGS electrostatic system,* in which the electric constant is set equal to unity. Most of the work carried on in these systems has been theoretical, and authors have frequently expressed numerical results in *emu* (electromagnetic units) or in *esu* (electrostatic units). The prefix *ab-* has been added to the names of the practical electrical units (volt, ohm, farad, etc.) to designate units in the CGS electromagnetic system, and the prefix *stat-* has similarly been used to designate units in the CGS electrostatic system. These usages did not spread beyond the United States and have never received international sanction. The International Electrotechnical Commission (IEC) did, however, adopt the names gauss, oersted, maxwell, and gilbert for the CGS electromagnetic units of magnetic flux density, magnetic field strength, magnetic flux, and magnetomotive force, respectively. Some confusion has resulted from the fact that gauss was the name given to the unit of "magnetic field intensity" (H) before 1930, when the IEC officially decided to regard H and B as quantities that differed in kind.

A number of other CGS electrical systems are discussed by Silsbee.[3] These include the Gaussian system, in which both electric constant and magnetic constant are set equal to unity and in which, as a result, the velocity of propagation of electromagnetic waves appears explicitly in many of the fundamental equations. Lorentz applied rationalization to the Gaussian system in a manner suggested by Heaviside, and the resultant Heaviside-Lorentz system, which is extremely elegant for theoretical work, has been widely used.

The theoretical advantages of having a fourth basic unit have prompted some authors to suggest expanding the classical three-dimensional CGS systems by adding a fourth basic unit. Thus we have proposals for the CGSB system, an electromagnetic system in which the name biot is given to the unit of current, and the CGSF system, an electrostatic system in which the name franklin is given to the unit of charge. To the experimentalist the names biot and franklin are mere synonyms for abampere and statcoulomb, respectively, but to the theoretician they offer the advantages provided by a fourth basic unit.

The Practical System and the International Units. The various CGS electrical units have never been used very much for practical work, for many of them are of an extremely impractical size. Therefore, six electrical units (volt, ampere, ohm, farad, coulomb, and henry) and two mechanical units, the joule and watt, were defined as exact decimal multiples of units of the CGS electromagnetic system.

These eight units composed what for many years was called the "practical system" of units. In the nineteenth century absolute measurement of the electrical quantities (measurement in terms of the basic units of length, mass, and time) was considerably less accurate than measurement by comparison with prototype standards of resistance, voltage, etc. The IEC, therefore, recommended "for the purposes of electrical measurements and as a basis for legislation" a separate system of "International Electrical Units" in 1908. These international units were intended to represent the absolute units (as they were called) of the practical system as closely as possible. The International Ohm, for example, was defined as the resistance, at 0°C, of a column of mercury 106.300 cm long and weighing 14.4521 grams. As the techniques of absolute measurement increased in accuracy, the need for the International Electrical Units vanished, and in 1948 the United States and other countries passed legally from the international units to the absolute units. At the time of the change, the difference between the absolute units and the international units, as embodied in the prototype standards of the various nations, amounted to several hundred parts per million.*

The MKSA System. The classical CGS systems had several drawbacks for electricity and magnetism—they were not rationalized, the units were of inconvenient size, and three basic quantities are not sufficient to permit a powerful dimensional analysis in electricity and magnetism. Giorgi therefore proposed, in 1901, a system of units, germane to the meter, kilogram, and second, which would have a fourth basic unit drawn from the field of electricity and magnetism, which would include the practical units already so firmly entrenched, and which would have most of the advantages of rationalization as proposed by Heaviside. Unfortunately, Giorgi's proposal involved adopting a measure of $4\pi \times 10^{-7}$ for the magnetic constant (the permeability of vacuum). After decades of discussion, in 1950 the IEC recommended the use of the MKS system with the equations in the rationalized form suggested by Giorgi, and adopted the ampere as a fourth basic unit, thus making the MKSA system.† The MKSA system has gained rapid support throughout the world and has been incorporated in the comprehensive International System of Units. (See Par. **11**.)

Foot-Pound-Second Systems. Systems of electrical units germane to the foot and the pound have never received serious support, thanks largely to the efforts of Lord Kelvin and the British Association for the Advancement of Science, which from the very first chose to define the absolute electrical units in terms of the metric system. Partial systems of units based on the foot, pound, and second have, however, been used in mechanics in the English-speaking countries. If the pound is taken to be a unit of mass, then the derived unit of force is given the name *poundal* (pdl). If the pound is taken to be a unit of force (more properly called the pound-force), then the derived unit of mass is called the *slug*. Other germane units, e.g., the foot poundal as a unit of energy, are frequently defined but are not given special names.

Systems of Photometric Units. All systems of photometric units currently used are based on the candela as a unit of luminous intensity, and all include the lumen as the unit of luminous flux. Systems differ, however, in the unit of length that is used to derive other photometric units. Some are based on the meter, others on the centimeter or foot. There is also a question concerning the position of π in the defining equations, quite analogous to the question of rationalization in electromagnetism. The current trend is toward the photometric units contained in the International System, which are based on the candela and the meter. Definitions of the photometric quantities and of the major units are given in Par. **15**.

13. IEEE Recommendations. The following paragraphs are excerpted from *IEEE Standards Publication* 268, Recommended Practice for Units in Published Scientific and Technical Work, published in 1966.

"Recommendations Regarding Preferred System of Units"

"Technical and scientific data, except in cases such as those cited below, should be given in units of the International System, followed if desired by the equivalent

* The International Electrical Units, which were abandoned in 1948, must not be confused with the electrical units of the International System (the SI units), which are in fact the absolute units of the old practical system. The name International System of Units was not officially applied to a comprehensive system of units until 1960, and Silsbee[3] and other authors writing before that date occasionally used the term International System to apply to the system of units based on the International Electrical Units.

† This system has also been called the Giorgi system.

data in other units given in parentheses. *This recommendation applies only for the publication of data and does not in any way affect the choice of units for manufacturing processes or for industrial standards.*

"Exceptions"

"When a nonmetric industrial standard is referred to, the nonmetric data should be given first place, with the SI equivalent data given in parentheses. Examples include standard inch sizes of nuts and bolts, American Wire Gauge sizes of electrical conductors, and the like.

"Where, in special fields, for reasons related to the field, a unit not in the International System offers significant advantages, such a unit may be used, with the SI equivalent data given in parentheses. Examples include the use of the electronvolt in physics, the nautical mile in navigation, and the astronomical unit in astronomy.

"Experimental data taken in non-SI units may be quoted without change, followed by the SI equivalents in parentheses.

"When a non-SI unit is at present too widely used to be eliminated immediately, its use may be continued with the SI equivalent given in parentheses. Examples include the use of the horsepower to measure mechanical power and the use of the torr to measure air pressure. This exception is not intended as a permanent protection for irrational practices, however, and as future standards are drafted such anachronisms as the horsepower should be eliminated.

"Some nonelectrical quantities are frequently expressed in units that are decimal multiples of SI units. Such units include the liter, the hectare, the bar, the tonne (metric ton), and the angstrom. These units may be used in appropriate fields. The SI units *newton* and *joule* are, however, to be used in place of the *dyne* and *erg*."

"Specific Recommendations"

"The various CGS units of electrical and magnetic quantities are no longer to be used. This includes the various ab- and stat- designations (abvolt, statcoulomb, etc.) and the gilbert, oersted, gauss, and maxwell. It is of course recognized that in some cases it may be desirable to state CGS equivalents in parentheses following data given in the International System.

"The use of prefixes* to express decimal multiples of SI units is permitted. Thus, although the SI unit for current density is the ampere per square meter, use of the ampere per square centimeter where convenient is entirely permissible.

"Use of metric-system units that are not decimal multiples of SI units, such as the calorie and the kilogram-force, is especially to be avoided. Note that the 9th General Conference on Weights and Measures has adopted the joule as the unit of heat, recommending that the calorie be avoided wherever possible.

"Where it is absolutely necessary to use British-American units, the conversion to SI units should be kept as obvious as possible. For example, if an electric generator is built to inch specifications, it may be inconvenient to express magnetic flux density in teslas (webers per square meter). In such a case webers per square inch should be used, not maxwells ("lines") per square inch.

"Recommendation on Unit Conversions"

"Unit conversions are to be handled with careful regard to the implied correspondence between accuracy of data and the number of figures given.

"Remarks"

[Some applicable rules for conversion and rounding are given in USA Standard B48.1.]

"When a dimension or other quantity is specified with tolerances, care must be taken to ensure that the range of values permitted after unit conversion lies within the original tolerances.

"Whenever possible, unit conversions should be made by the author of the document,

* The official list of prefixes is given in Par. **11**. Compound prefixes (e.g., millimicro-) are not to be used. Multiplication beyond the range covered by the prefixes should be handled by using powers of ten.

for he is best able to determine how many figures are significant and should be retained after conversion. For example, a length of 125 feet converts exactly to 38.1 meters. If, however, the 125-foot length had been obtained by rounding to the nearest 5 ft, the conversion should be given as 38 m; and if it had been obtained by rounding to the nearest 25 ft, the conversion should be given as 40 m."

"Recommendations Where a Unit Has More Than One Name"

"General Principles"

"Many of the derived units in the International System have been given special names, which may be used as alternatives to the compound forms. When the special names are formally recognized by the General Conference on Weights and Measures (CGPM), they have a standing which is equal to that of the compound forms. Both names are technically correct, and the choice between them must be made partly on the basis of taste, keeping in mind the particular application.

"*Example*: The unit of electrical resistance is the *ohm,* which is equivalent to the *volt per ampere.* Resistance is almost always expressed in ohms, but there are occasions when the explicit expression in volts per ampere is desirable. Such usage is entirely correct.

"Some names that have been proposed for derived SI units have not yet been recognized by the CGPM. If such a proposed name is used, care must be taken to make sure that the intended meaning is clear.

"*Example*: The name *pascal* has been proposed for the SI unit of pressure, the newton per square meter.

"The use of special names for decimal multiples of SI units is not recommended. Some well-established, nonelectrical units [e.g., the liter, hectare, etc.] are, however, recognized as exceptions.

"The decimal prefixes are not recommended for use with British-American units. An exception is made for the *microinch,* a unit that is frequently used in precision machine work.

"Except for the unit of electrical conductance, the practice of giving special names to reciprocal units is to be discouraged. In particular, use of the name *daraf* for the *reciprocal farad* is not recommended.

"Specific Recommendations"

"*Frequency.* The CGPM has adopted the name *hertz* for the unit of frequency, but *cycle per second* is widely used. Although *cycle per second* is technically correct, the name *hertz* is preferred because of the widespread use of *cycle* alone as a unit of frequency. Use of *cycle* in place of *cycle per second,* of *kilocycle* in place of *kilocycle per second,* etc., is incorrect.

"*Magnetic Flux Density.* The CGPM has adopted the name *tesla* for the SI unit of magnetic flux density. The name *gamma* shall not be used for the unit *nanotesla.*

"*Electrical Conductance.* The CGPM has not yet adopted a short name for the ampere per volt (or reciprocal ohm), the SI unit of electrical conductance. The International Electrotechnical Commission has recommended the name *siemens* for this unit, but the name *mho* has been more widely used. In IEEE publications the name *mho* is preferred.

"*Temperature Scale.* In 1948 the CGPM abandoned *centigrade* as the name of a temperature scale. The corresponding scale is now properly named the *Celsius* scale, and further use of *centigrade* for this purpose is deprecated.

"*Luminous Intensity.* The SI unit of luminous intensity has been given the name *candela,* and further use of the old name *candle* is deprecated. Use of the term *candlepower,* either as the name of a quantity or as the name of a unit, is deprecated.

"*Luminous Flux Density.* The common British-American unit of luminous flux density is the *lumen per square foot.* The name *footcandle,* which has been used for this unit in the U.S., is deprecated.

"*Micrometer and Micron.* Although the name *micron* has been widely used for the *micrometer,* it is not recommended. The name *nanometer* is preferred over *millimicron,* which is deprecated.

"*Gigaelectronvolt.* Because *billion* means a thousand million in the United States but a million million in most other countries, its use should be avoided in technical writing. The term *billion electronvolts* is deprecated; use *gigaelectronvolt* instead."

"Recommendations Concerning British-American Units"

"In principle the number of British-American units in use should be reduced as rapidly as possible.

"As a start toward implementing the recommendation above, the following should be abandoned:

British thermal unit
horsepower
Rankine temperature scale
US dry quart, US liquid quart, and UK (Imperial) quart, together with their various multiples and subdivisions*
footlambert†

"Quantities are not to be expressed in mixed units. For example, a mass should be expressed as 12.75 lb, rather than as 12 lb, 12 oz."

DEFINITIONS OF QUANTITIES AND UNITS

14. Quantities of Electricity and Magnetism. Most of the definitions given in this section are based on the "Proposed Standard Definitions of General (Fundamental and Derived) Electrical and Electronics Terms," *IEEE No.* 270, published 1966.

Admittance (Y). An admittance of a linear constant-parameter system is the ratio of the phasor equivalent of the steady-state sine-wave current or current-like quantity (response) to the phasor equivalent of the corresponding voltage or voltage-like quantity (driving force).

Capacitance (C). Capacitance is that property of a system of conductors and dielectrics which permits the storage of electrically separated charges when potential differences exist between the conductors. Its value is expressed as the ratio of an electric charge to a potential difference.

Coefficient of Coupling (k). Coefficient of coupling (used only in the case of resistive, capacitive, and inductive coupling) is the ratio of the mutual impedance of the coupling to the square root of the product of the self-impedances of similar elements in the two circuit loops considered. Unless otherwise specified, coefficient of coupling refers to inductive coupling, in which case $k = M/(L_1L_2)^{1/2}$, where M is the mutual inductance, L_1 the self-inductance of one loop, and L_2 the self-inductance of the other.

Conductance (G)

1. The conductance of an element, device, branch, network, or system is the factor by which the mean-square voltage must be multiplied to give the corresponding power lost by dissipation as heat or as other permanent radiation or loss of electromagnetic energy from the circuit.

2. Conductance is the real part of admittance.

Conductivity (γ). The conductivity of a material is a factor such that the conduction current density is equal to the electric field strength in the material multiplied by the conductivity.

Current (I). Current is a generic term used when there is no danger of ambiguity to refer to any one or more of the currents described below. (For example, in the expression "the current in a simple series circuit," the word current refers to the conduction current in the wire of the inductor and to the displacement current between the plates of the capacitor.)

Conduction Current. The conduction current through any surface is the integral of the normal component of the conduction current density over that surface.

---

* If it is absolutely necessary to express volume in British-American units, the cubic inch or cubic foot should be used.

† If it is absolutely necessary to express luminance in British-American units, the candela per square foot or lumen per steradian square foot should be used.

Displacement Current. The displacement current through any surface is the integral of the normal component of the displacement current density over that surface.

Current Density (***J***). Current density is a generic term used when there is no danger of ambiguity to refer either to conduction current density or to displacement current density or to both.

Displacement Current Density. The displacement current density at any point in an electric field is (in the International System) the time rate of change of the electric-flux-density vector at that point.

Conduction Current Density. The electric conduction current density at any point at which there is a motion of electric charge is a vector quantity whose direction is that of the flow of positive charge at this point, and whose magnitude is the limit of the time rate of flow of net (positive) charge across a small plane area perpendicular to the motion, divided by this area, as the area taken approaches zero in a macroscopic sense, so as to always include this point. The flow of charge may result from the movement of free electrons or ions but is not in general, except in microscopic studies, taken to include motions of charges resulting from the polarization of the dielectric.

Damping Coefficient (δ). If F is a function of time given by

$$F = A \exp(-\delta t) \sin(2\pi t/T),$$

then δ is the damping coefficient.

Elastance (S). Elastance is the reciprocal of capacitance.

Electric Charge, Quantity of Electricity (Q). Electric charge is a fundamentally assumed concept, required by the existence of forces measurable experimentally. It has two forms, known as positive and negative.

The electric charge on (or in) a body or within a closed surface is the excess of one form of electricity over the other.

Electric Constant, Permittivity of Vacuum (Γ_e). The electric constant pertinent to any system of units is the scalar which in that system relates the electric flux density D, in vacuum, to E, the electric field strength ($D = \Gamma_e E$). It also relates the mechanical force between two charges in vacuum to their magnitudes and separation. Thus in the equation $F = \Gamma_r Q_1 Q_2 / 4\pi\Gamma_e r^2$ for the force F between charges Q_1 and Q_2 separated by a distance r, Γ_e is the electric constant and Γ_r is a dimensionless factor which is unity in a rationalized system and 4π in an unrationalized system.

NOTE: In the CGS electrostatic system Γ_e is assigned measure unity and the dimension "numeric."

In the CGS electromagnetic system the measure of Γ_e is that of $1/c^2$ and the dimension is $[L^{-2}T^2]$.

In the International System the measure of Γ_e is $10^7/4\pi c^2$, and the dimension is $[L^{-3}M^{-1}T^4I^2]$. Here c is the speed of light expressed in the appropriate system of units.

Electric Field Strength (***E***). The electric field strength at a given point in an electric field is the vector limit of the quotient of the force that a small stationary charge at that point will experience, by virtue of its charge, to the charge as the charge approaches zero.

Electric Flux (Ψ). The electric flux through a surface is the surface integral of the normal component of the electric flux density over the surface.

Electric Flux Density, Electric Displacement (***D***). The electric flux density is a quantity related to the charge displaced within a dielectric by application of an electric field. Electric flux density at any point in an isotropic dielectric is a vector which has the same direction as the electric field strength and a magnitude equal to the product of the electric field strength and the permittivity, ϵ. In a nonisotropic medium, ϵ may be represented by a tensor, and ***D*** is not necessarily parallel to ***E***.

Electric Polarization (***P***). The electric polarization is the vector quantity defined by the equation $\boldsymbol{P} = (\boldsymbol{D} - \Gamma_e \boldsymbol{E})/\Gamma_r$, where ***D*** is the electric flux density, Γ_e is the electric constant, ***E*** is the electric field strength, and Γ_r is a coefficient that is set equal to unity in a rationalized system and to 4π in an unrationalized system.

Electric Susceptibility (χ_e). Electric susceptibility is the quantity defined by $\chi_e =$

$(\epsilon_r - 1)/\Gamma_r$, where ϵ_r is the relative permittivity and Γ_r is a coefficient that is set equal to unity in a rationalized system and to 4π in an unrationalized system.

Electrization ($\boldsymbol{E}_i$). The electrization is the electric polarization divided by the electric constant of the system of units used.

Electrostatic Potential (V). The electrostatic potential at any point is the potential difference between that point and an agreed-upon reference point, usually the point at infinity.

Electrostatic Potential Difference (V). The electrostatic potential difference between two points is the scalar-product line integral of the electric field strength along any path from one point to the other in an electric field resulting from a static distribution of electric charge.

Impedance (Z). An impedance of a linear constant-parameter system is the ratio of the phasor equivalent of a steady-state sine-wave voltage or voltage-like quantity (driving force) to the phasor equivalent of a steady-state sine-wave current or current-like quantity (response).

In electromagnetic radiation electric field strength is considered the driving force and magnetic field strength the response. In mechanical systems mechanical force is always considered as a driving force and velocity as a response. In a general sense the dimension (and unit) of impedance in a given application may be whatever results from the ratio of the dimensions of the quantity chosen as the driving force to the dimensions of the quantity chosen as the response. However, in the types of systems cited above any deviation from the usual convention should be noted.

Mutual Impedance. Mutual impedance between two loops (meshes) is the factor by which the phasor equivalent of the steady-state sine-wave current in one loop must be multiplied to give the phasor equivalent of the steady-state sine-wave voltage in the other loop caused by the current in the first loop.

Self-impedance. Self-impedance of a loop (mesh) is the impedance of a passive loop with all other loops of the network open-circuited.

Transfer Impedance. A transfer impedance is the impedance obtained when the response is determined at a point other than that at which the driving force is applied.

NOTE: In the case of an electric circuit the response may be determined in any branch except that which contains the driving force.

Logarithmic Decrement (Λ). If F is a function of time given by

$$F = A \exp(-\delta t) \sin(2\pi t/T),$$

then the logarithmic decrement $\Lambda = T\delta$.

Magnetic Constant, Permeability of Vacuum (Γ_m). The magnetic constant pertinent to any system of units is the scalar which in that system relates the mechanical force between two currents in vacuum to their magnitudes and geometrical configurations. For example, the equation for the force F on a length l of two parallel straight conductors of infinite length and negligible circular cross section, carrying constant currents I_1 and I_2 and separated by a distance r in vacuum, is $F = \Gamma_m \Gamma_r I_1 I_2 l/2\pi r$, where Γ_m is the magnetic constant and Γ_r is a coefficient set equal to unity in a rationalized system and to 4π in an unrationalized system.

NOTE: In the CGS electromagnetic system Γ_m is assigned the magnitude unity and the dimension "numeric."

In the CGS electrostatic system the magnitude of Γ_m is that of $1/c^2$, and the dimension is $[L^{-2}T^2]$.

In the International System Γ_m is assigned the magnitude $4\pi \times 10^{-7}$ and has the dimension $[LMT^{-2}I^{-2}]$.

Magnetic Field Strength ($\boldsymbol{H}$). Magnetic field strength is that vector point function whose curl is the current density, and which is proportional to magnetic flux density in regions free of magnetized matter.

Magnetic Flux (Φ). The magnetic flux through a surface is the surface integral of the normal component of the magnetic flux density over the surface.

Magnetic Flux Density, Magnetic Induction ($\boldsymbol{B}$). Magnetic flux density is that vector quantity producing a torque on a plane current loop in accordance with the relation $\boldsymbol{T} = IA\boldsymbol{n} \times \boldsymbol{B}$, where $\boldsymbol{n}$ is the positive normal to the loop and A is its area.

The concept of flux density is extended to a point inside a solid body by defining the flux density at such a point as that which would be measured in a thin disk-shaped cavity in the body centered at that point, the axis of the cavity being in the direction of the flux density.

Magnetic Moment ($\boldsymbol{m}$). The magnetic moment of a magnetized body is the volume integral of the magnetization. The magnetic moment of a loop carrying current I is $\boldsymbol{m} = (I/2)\int \boldsymbol{r} \times d\boldsymbol{r}$ where $\boldsymbol{r}$ is the radius vector from an arbitrary origin to a point on the loop and where the path of integration is taken around the entire loop.

NOTE: The magnitude of the moment of a plane current loop is IA, where A is the area of the loop.

The reference direction for the current in the loop indicates a clockwise rotation when the observer is looking through the loop in the direction of the positive normal.

Magnetic Polarization, Intrinsic Magnetic Flux Density ($\boldsymbol{J}$, $\boldsymbol{B}_i$). The magnetic polarization is the vector quantity defined by the equation $\boldsymbol{J} = (\boldsymbol{B} - \Gamma_m \boldsymbol{H})/\Gamma_r$, where $\boldsymbol{B}$ is the magnetic flux density, Γ_m is the magnetic constant, $\boldsymbol{H}$ is the magnetic field strength, and Γ_r is a coefficient that is set equal to unity in a rationalized system and to 4π in an unrationalized system.

Magnetic Susceptibility (χ_m). Magnetic susceptibility is the quantity defined by $\chi_m = (\mu_r - 1)/\Gamma_r$, where μ_r is the relative permeability and Γ_r is a coefficient that is set equal to unity in a rationalized system and to 4π in an unrationalized system.

Magnetic Vector Potential ($\boldsymbol{A}$). The magnetic vector potential is a vector point function characterized by the relation that its curl is equal to the magnetic flux density and its divergence vanishes.

Magnetization ($\boldsymbol{M}$, $\boldsymbol{H}_i$). The magnetization is the magnetic polarization divided by the magnetic constant of the system of units used.

Magnetomotive Force (F_m). The magnetomotive force acting in any closed path in a magnetic field is the line integral of the magnetic field strength around the path.

Mutual Inductance (M). The mutual inductance between two loops (meshes) in a circuit is the quotient of the flux linkage produced in one loop divided by the current, in another loop, which induces the flux linkage.

Permeability. Permeability is a general term used to express various relationships between magnetic flux density and magnetic field strength. These relationships are either (1) *absolute permeability* (μ), which in general is the quotient of a change in magnetic flux density divided by the corresponding change in magnetic field strength; or (2) *relative permeability* (μ_r), which is the ratio of the absolute permeability to the magnetic constant.

Permeance (P_m). Permeance is the reciprocal of reluctance.

Permittivity, Capacitivity (ϵ). The permittivity of a homogeneous, isotropic dielectric, in any system of units, is the product of its relative permittivity and the electric constant appropriate to that system of units.

Relative Permittivity, Relative Capacitivity, Dielectric Constant (ϵ_r). The relative permittivity of any homogeneous isotropic material is the ratio of the capacitance of a given configuration of electrodes with the material as a dielectric to the capacitance of the same electrode configuration with a vacuum as the dielectric. Experimentally, vacuum must be replaced by the material at all points where it makes a significant change in the capacitance.

Power (P). Power is the time rate of transferring or transforming energy. *Electrical power* is the time rate of flow of electrical energy. The *instantaneous electrical power* at a single terminal pair is equal to the product of the instantaneous voltage multiplied by the instantaneous current. If both voltage and current are periodic in time, the time average of the instantaneous power, taken over an integral number of periods, is the *active power*, usually called simply the *power* when there is no danger of confusion.

If the voltage and current are sinusoidal functions of time, the product of the rms value of the voltage and the rms value of the current is called the *apparent power*; the product of the rms value of the voltage and the rms value of the in-phase component of the current is the *active power*; and the product of the rms value of the voltage and the rms value of the quadrature component of the current is called the *reactive power*.

The SI unit of instantaneous power and of active power is the watt. The germane unit for apparent power is the voltampere, and for reactive power the var.

Power Factor (F_p). Power factor is the ratio of active power to apparent power.

Q. Q, sometimes called *quality factor,* is that measure of the quality of a component, network, system, or medium considered as an energy storage unit in the steady state with sinusoidal driving force which is given by

$$Q = \frac{2\pi \times (\text{maximum energy in storage})}{\text{energy dissipated per cycle of the driving force}}$$

NOTE: For single components, such as inductors and capacitors, the Q at any frequency is the ratio of the equivalant series reactance to resistance, or of the equivalent shunt susceptance to conductance.

For networks that contain several elements and for distributed parameter systems the Q is generally evaluated at a frequency of resonance.

The *nonloaded* Q of a system is the value of Q obtained when only the incidental dissipation of the system elements is present. The *loaded* Q of a system is the value of Q obtained when the system is coupled to a device that dissipates energy.

The "period" in the expression for Q is that of the driving force, not that of energy storage, which is usually half that of the driving force.

Reactance (X). Reactance is the imaginary part of impedance.

Reluctance (R_m). Reluctance is the ratio of the magnetomotive force in a magnetic circuit to the magnetic flux through any cross section of the magnetic circuit.

Reluctivity (ν). Reluctivity is the reciprocal of permeability.

Resistance (R)

1. The resistance of an element, device, branch, network, or system is the factor by which the mean-square conduction current must be multiplied to give the corresponding power lost by dissipation as heat or as other permanent radiation or loss of electromagnetic energy from the circuit.

2. Resistance is the real part of impedance.

Resistivity (ρ). The resistivity of a material is a factor such that the conduction current density is equal to the electric field strength in the material divided by the resistivity.

Self-inductance (L)

1. Self-inductance is the quotient of the flux linkage of a circuit divided by the current in that same circuit which induces the flux linkage. If v = voltage induced, $v = d(Li)/dt$.

2. Self-inductance is the factor L in $\frac{1}{2}Li^2$ if the latter gives the energy stored in the magnetic field as a result of the current i.

NOTE: Definitions 1 and 2 are not equivalent except when L is constant. In all other cases the definition being used must be specified.

The two definitions are restricted to relatively slow changes in i, i.e., to low frequencies, but by analogy with the definitions equivalent inductances may often be evolved in high-frequency applications such as resonators, waveguide equivalent circuits, etc. Such "inductances," when used, must be specified.

The two definitions are restricted to cases in which the branches are small in physical size compared with a wavelength, whatever the frequency. Thus in the case of a uniform 2-wire transmission line it may be necessary even at low frequencies to consider the parameters as "distributed" rather than to have one inductance for the entire line.

Susceptance (B). Susceptance is the imaginary part of admittance.

Transfer Function (H). A transfer function is that function of frequency which is the ratio of a phasor output to a phasor input in a linear system.

Transfer Ratio (H). A transfer ratio is a dimensionless transfer function.

Voltage, Electromotive Force (V). The voltage along a specified path in an electric field is the dot product line integral of the electric field strength along this path. As here defined, voltage is synonymous with potential difference only in an electrostatic field.

15. Quantities of Radiation and Light. The definitions given in this section are based on USAS Z7.1-1967.

Candlepower. Candlepower is luminous intensity expressed in candelas. (Term deprecated by IEEE.)

Emissivity. Total Emissivity, (ϵ). The total emissivity of an element of surface of a temperature radiator is the ratio of its radiant flux density (radiant exitance) to that of a blackbody at the same temperature.

Spectral Emissivity, $\epsilon(\lambda)$. The spectral emissivity of an element of surface of a temperature radiator at any wavelength is the ratio of its radiant flux density per unit wavelength interval (spectral radiant exitance) at that wavelength to that of a blackbody at the same temperature.

Light. For the purposes of illuminating engineering, light is visually evaluated radiant energy. *Note 1*: Light is psychophysical, neither purely physical nor purely psychological. Light is not synonymous with radiant energy, however restricted, nor is it merely sensation. In a general nonspecialized sense, light is the aspect of radiant energy of which a human observer is aware through the stimulation of the retina of the eye. *Note 2*: Radiant energy outside the visible portion of the spectrum must not be discussed using the quantities and units of light; it is nonsense to refer to "ultra-violet light," or to express infrared flux in lumens.

Luminance (Photometric Brightness) (L). Luminance in a direction, at a point on the surface of a source or of a receiver, or on any other real or virtual surface, is the quotient of the luminous flux (Φ) leaving, passing through, or arriving at a surface element surrounding the point, and propagated in directions defined by an elementary cone containing the given direction, divided by the product of the solid angle of the cone $(d\Omega)$ and the area of the orthogonal projection of the surface element on a plane perpendicular to the given direction $(dA \cos\theta)$. $L = d^2\Phi/d\Omega(dA \cos\theta) = dI/(dA \cos\theta)$. In the defining equation θ is the angle between the direction of observation and the normal to the surface.

In common usage the term *brightness* usually refers to the intensity of sensation which results from viewing surfaces or spaces from which light comes to the eye. This sensation is determined in part by the definitely measurable luminance defined above and in part by conditions of observation such as the state of adaptation of the eye. In much of the literature the term brightness, used alone, refers to both luminance and sensation. The context usually indicates which meaning is intended.

Luminous Efficacy of Radiant Flux. The luminous efficacy of radiant flux is the quotient of the total luminous flux divided by the total radiant flux. It is expressed in lumens per watt.

Spectral Luminous Efficacy of Radiant Flux, $K(\lambda)$. Spectral luminous efficacy of radiant flux is the quotient of the luminous flux at a given wavelength divided by the radiant flux at the wavelength. It is expressed in lumens per watt.

Spectral Luminous Efficiency of Radiant Flux. Spectral luminous efficiency of radiant flux is the ratio of the luminous efficacy for a given wavelength to the value at the wavelength of maximum luminous efficacy. It is a numeric.

NOTE: The term *spectral luminous efficiency* replaces the previously used terms *relative luminosity* and *relative luminosity factor.*

Luminous Flux (Φ). Luminous flux is the time rate of flow of light.

Luminous Flux Density at a Surface. Luminous flux density at a surface is luminous flux per unit area of the surface. In referring to flux incident on a surface, this is called *illumination* (E). The preferred term for luminous flux *leaving* a surface is *luminous exitance* (M), which has been called *luminous emittance.*

Luminous Intensity (I). The luminous intensity of a source of light in a given direction is the luminous flux proceeding from the source per unit solid angle in the direction considered. $(I = d\Phi/d\Omega.)$

Quantity of Light (Q). Quantity of light (luminous energy) is the product of the luminous flux by the time it is maintained; i.e., it is the time integral of luminous flux.

Radiance (L). Radiance in a direction, at a point on the surface of a source or of a receiver, or on any other real or virtual surface, is the quotient of the radiant flux (P) leaving, passing through, or arriving at a surface element surrounding the point, and propagated in directions defined by an elementary cone containing the given direction, divided by the product of the solid angle of the cone $(d\Omega)$ and the area of the orthogonal projection of the surface element on a plane perpendicular to the given direction $(dA \cos\theta)$. $L = d^2P/d\Omega(dA \cos\theta) = dI/(dA \cos\theta)$. In the defining equation θ is the angle between the normal to the element of the source and the direction of observation.

Radiant Density (w). Radiant density is radiant energy per unit volume.

Radiant Energy (W). Radiant energy is energy traveling in the form of electromagnetic waves.

Radiant Flux Density at a Surface. Radiant flux density at a surface is radiant flux per unit area of the surface. When referring to radiant flux incident on a surface, this is called *irradiance* (E). The preferred term for radiant flux *leaving* a surface is *radiant exitance* (M), which has been called *radiant emittance.*

Radiant Intensity (I). The radiant intensity of a source in a given direction is the radiant flux proceeding from the source per unit solid angle in the direction considered. ($I = dP/d\Omega$.)

Radiant Power, Radiant Flux (P). Radiant flux is the time rate of flow of radiant energy.

16. Units. In the following definitions, the designation SI identifies units belonging to the International System (see Par. **11**).

Ab-. A prefix used with the common electrical units (volt, ohm, etc.) to form names for the corresponding units in the electromagnetic CGS system. These units are deprecated by the IEEE.

Ampere (A). Basic SI unit of current. The ampere is the constant current that, if maintained in two straight parallel conductors that are of infinite length and negligible cross section and are separated from each other by a distance of 1 meter in a vacuum, will produce between these conductors a force equal to 2×10^{-7} newton per meter of length.

Ampere per Meter (A/m). The ampere per meter (sometimes called ampere-turn per meter) is the SI unit of magnetic field strength. The ampere per meter is the magnetic field strength in the interior of an elongated uniformly wound solenoid which is excited with a linear current density in its winding of one ampere per meter of axial distance.

Candela (cd). Basic SI unit of luminous intensity. The candela is the luminous intensity, in the perpendicular direction, of a surface of 1/60,000 square meter of a blackbody at the freezing point of platinum under a pressure of 101,325 newtons per square meter.

Coulomb (C). The coulomb is the SI unit of electric charge. It is the quantity of electric charge which passes any cross section of a conductor in one second when the current is maintained constant at one ampere.

Dyne (dyn). The dyne is the unit of force in the CGS systems.

Electromagnetic Units (emu). The general name often given to units, of whatever quantity, belonging to the electromagnetic CGS system. These units are deprecated by the IEEE.

Electrostatic Units (esu). The general name often given to units, of whatever quantity, belonging to the electrostatic CGS system. These units are deprecated by the IEEE.

Erg. The erg is the unit of work and of energy in the CGS systems. The erg is 10^{-7} joule.

Farad (F). The farad is the SI unit of capacitance. It is the capacitance of a capacitor in which a charge of one coulomb produces a potential difference of one volt between the terminals.

Footcandle (fc). The footcandle is the unit of illumination equal to one lumen per square foot. The IEEE has recommended using the explicit name *lumen per square foot* in place of the name *footcandle* for this unit.

Footlambert (fL). The footlambert is a unit of luminance equal to $(1/\pi)$ candela per square foot. This unit is deprecated by the IEEE.

Gauss (G). The gauss is the unit of magnetic flux density in the CGS electromagnetic system. It is deprecated by the IEEE.

Gilbert (Gb). The gilbert is the unit of magnetomotive force in the CGS electromagnetic system. It is deprecated by the IEEE.

Henry (H). The henry is the SI unit of inductance. It is the inductance of a circuit in which a current of one ampere induces a flux linkage of one weber.

Hertz (Hz). The hertz is the SI unit of frequency. It is equal to a frequency of one cycle per second.

Joule (J). The joule is the SI unit of work and energy. It is the work done by a force of one newton acting through a distance of one meter.

Kelvin (K). Basic SI unit of thermodynamic temperature. The kelvin is the fraction 1/273.16 of the thermodynamic temperature of the triple point of water.

Kilogram (kg). Basic SI unit of mass. The kilogram is equal to the mass of the International Prototype Kilogram [a particular cylinder of platinum-iridium alloy preserved in a vault at Sèvres, France, by the International Bureau of Weights and Measures].

Lambert (L). The lambert is a unit of luminance equal to $(1/\pi)$ candela per square centimeter.

Lumen (lm). The lumen is the SI unit of luminous flux. It is equal to the flux through a unit solid angle (steradian) from a uniform point source of one candela.

Lumen-hour (lm·h). The lumen-hour is a unit of quantity of light (luminous energy). lm·h = 3600 lm·s.

Lumen-second (lm·s). The lumen-second is the SI unit of quantity of light. It is the quantity of light that is delivered in one second by one lumen.

Lux (lx). The lux is the SI unit of illumination. It is equal to the lumen per square meter.

Maxwell (Mx). The maxwell (sometimes called line) is the unit of magnetic flux in the CGS electromagnetic system. It is deprecated by the IEEE.

Meter (m). Basic SI unit of length. The meter is the length equal to 1,650,763.73 wavelengths in vacuum of the radiation corresponding to the unperturbed transition between the levels $2p_{10}$ and $5d_5$ of the krypton-86 atom.

Mho (mho). The mho is the SI unit of conductance (and of admittance). The mho is the conductance of a conductor whose resistance is one ohm. The name siemens (S) is also used for this unit.

Newton (N). The newton is the unit of force in the SI. The newton is the force which will impart an acceleration of one meter per second squared to a mass of one kilogram. One newton equals 10^5 dynes.

Nit (nt). The nit is the short name sometimes given to the SI unit of luminance (candela per square meter).

Oersted (Oe). The oersted is the unit of magnetic field strength in the CGS electromagnetic system. It is deprecated by the IEEE.

Ohm (Ω). The ohm is the unit of resistance (and of impedance) in the SI. The ohm is the resistance of a conductor such that a constant current of one ampere in it produces a voltage of one volt between its ends.

Phot (ph). The phot is a unit of illumination that is equal to one lumen per square centimeter.

Second (s). Basic SI unit of time. The second is the duration of 9,192,631,770 periods of the radiation corresponding to the transition between the two hyperfine levels of the ground state of the cesium-133 atom.

Stat-. A prefix used with the common electrical units (volt, ohm, etc.) to form names for the corresponding units in the electrostatic CGS system. These units are deprecated by the IEEE.

Stilb (sb). The stilb is the unit of luminance equal to one candela per square centimeter.

Tesla (T). The tesla is the SI unit of magnetic flux density. It is the magnetic flux density of a uniform field that produces a torque of one newton-meter on a plane current loop carrying one ampere and having a projected area of one square meter on the plane perpendicular to the field. T = N/A·m.

Var (var). The var is the unit of reactive power germane to the SI. The var is the reactive power at the port of entry of a single-phase two-wire circuit when the product of the rms value in amperes of the sinusoidal current by the rms value in volts of the voltage and by the sine of the angular phase difference by which the voltage leads the current is equal to one.

Volt (V). The volt is the unit of voltage or potential difference in the SI. The volt is the voltage between two points of a conducting wire carrying a constant current of one ampere, when the power dissipated between these points is one watt.

Voltampere (VA). The voltampere is the unit of apparent power germane to the

SI. The voltampere is the apparent power at the port of entry of a single-phase two-wire circuit when the product of the rms value in amperes of the current by the rms value in volts of the voltage is equal to one.

Watt (W). The watt is the unit of power in the SI. The watt is equal to one joule per second.

Watthour (Wh). The watthour is a unit of energy equal to 3600 joules.

Weber (Wb). The weber is the SI unit of magnetic flux. It is the magnetic flux passing through an area of one square meter placed normal to a uniform magnetic field of magnetic flux density equal to one tesla. Wb = T·m². If the flux linked by a circuit changes at a uniform rate of one weber per second, a voltage of one volt is induced in the circuit. Wb = V·s.

LETTER SYMBOLS FOR QUANTITIES AND UNITS

17. Letter symbols fall into two categories: symbols for physical quantities (*quantity symbols*) and symbols for the units in which these quantities are measured (*unit symbols*).

A quantity symbol is a single letter (for example, I for electric current), specified as to general form of type, and modified when appropriate by one or more subscripts or superscripts. A unit symbol is a letter or group of letters (for example, cm for centimeter), or in a few cases a special sign, that may be used in the place of the name of the unit.

Symbols for quantities are printed in italic type, while symbols for units are printed in roman type. Subscripts and superscripts that are letter symbols for quantities or for indices are printed in roman type:

C_p	heat capacity at constant pressure p
a_{ij}, a_{45}	matrix elements
I_{i}, I_{o}	input current, output current

For indicating the vector character of a quantity, boldface italic type is used, e.g. $\boldsymbol{F}$ for force. Ordinary italic type is used to represent the magnitude of a vector quantity.

The product of two quantities is indicated by writing ab. The quotient may be indicated by writing

$$\frac{a}{b}, \qquad a/b, \quad \text{or} \quad ab^{-1}$$

If more than one solidus (/) is required in any algebraic term, parentheses must be inserted to remove any ambiguity. Thus one may write $(a/b)/c$ or a/bc, but not $a/b/c$.

Unit symbols are written in lower-case letters, except for the first letter when the name of the unit is derived from a proper name, and except for a very few that are not formed from letters. When a compound unit is formed by multiplication of two or more other units, its symbol consists of the symbols for the separate units joined by a raised dot (for example, N·m for newton meter). The dot may be omitted in the case of familiar compounds such as watthour (symbol Wh) if no confusion would result. Hyphens should not be used in symbols for compound units. Positive and negative exponents may be used with the symbols for units.

When a symbol representing a unit that has a prefix (see Par. **11**) carries an exponent, this indicates that the multiple (or submultiple) unit is raised to the power expressed by the exponent.

Examples:

$$2\ \mathrm{cm}^3 = 2(\mathrm{cm})^3 = 2(10^{-2}\ \mathrm{m})^3 = 2\cdot10^{-6}\ \mathrm{m}^3$$

$$1\ \mathrm{ms}^{-1} = 1(\mathrm{ms})^{-1} = 1(10^{-3}\ \mathrm{s})^{-1} = 10^3\ \mathrm{s}^{-1}$$

Phasor quantities, represented by complex numbers or complex time-varying functions, are extensively used in certain branches of electrical engineering. The following notation and typography are standard:

	Notation	Remarks
Complex quantity	Z	$Z = \|Z\| \exp(j\phi)$ $Z = \mathrm{Re}\,Z + j\,\mathrm{Im}\,Z$
Real part	Re Z, Z'	
Imaginary part	Im Z, Z''	
Conjugate complex quantity	Z^*	$Z^* = \mathrm{Re}\,Z - j\,\mathrm{Im}\,Z$
Modulus of Z	$\|Z\|$	
Phase of Z, Argument of Z	arg Z	arg $Z = \phi$

18. Standard Quantity Symbols. The following quantity symbols are given in *USA Standard Y10.5*, Letter Symbols for Quantities used in Electrical Science and Electrical Engineering. Commas separate symbols of equal standing. Where two symbols are separated by three dots, the second is to be used only when there is specific need to avoid conflict.

Quantity	Quantity symbol	Unit based on International System	Remarks
Space and time:			
Angle, plane	$\alpha, \beta, \gamma, \theta, \phi, \psi$	radian	Other Greek letters are permitted where no conflict results.
Angle, solid	$\Omega \cdots \omega$	steradian	
Length	l	meter	
Breadth, width	b	meter	
Height	h	meter	
Thickness	d, δ	meter	
Radius	r	meter	
Diameter	d	meter	
Length of path line segment	s	meter	
Wavelength	λ	meter	
Wave number	$\sigma \cdots \tilde{\nu}$	reciprocal meter	$\sigma = 1/\lambda$ The symbol $\tilde{\nu}$ is used in spectroscopy.
Circular wave number Angular wave number	k	radian per meter	$k = 2\pi/\lambda$
Area	$A \cdots S$	square meter	
Volume	V, v	cubic meter	
Time	t	second	
Period	T	second	
Time constant	$\tau \cdots T$	second	
Frequency	$f \cdots \nu$	hertz	
Speed of rotation Rotational frequency	n	revolution per second	
Angular frequency	ω	radian per second	$\omega = 2\pi f$
Angular velocity	ω	radian per second	
Complex (angular) frequency Oscillation constant	$p \cdots s$	reciprocal second	$p = -\delta + j\omega$
Angular acceleration	α	radian per second squared	
Velocity	v	meter per second	
Speed of propagation of electromagnetic waves	c	meter per second	In vacuum, c_0
Acceleration (linear)	a	meter per second squared	
Acceleration of free fall Gravitational acceleration	g	meter per second squared	
Damping coefficient	δ	neper per second	
Logarithmic decrement	Λ	(numeric)	
Attenuation coefficient	α	neper per meter	
Phase coefficient	β	radian per meter	
Propagation coefficient	γ	reciprocal meter	$\gamma = \alpha + j\beta$
Mechanics:			
Mass	m	kilogram	
(Mass) density	ρ	kilogram per cubic meter	Mass divided by volume
Momentum	p	kilogram meter per second	
Moment of inertia	I, J	kilogram meter squared	
Force	F	newton	
Weight	W	newton	Varies with acceleration of free fall
Weight density	γ	newton per cubic meter	Weight divided by volume

Quantity	Quantity symbol	Unit based on International System	Remarks
Moment of force	M	newton meter	
Torque	$T \cdots M$	newton meter	
Pressure	p	newton per square meter	The name *pascal* has been suggested for this unit.
Normal stress	σ	newton per square meter	
Shear stress	τ	newton per square meter	
Stress tensor	σ	newton per square meter	
Linear strain	ϵ	(numeric)	
Shear strain	γ	(numeric)	
Strain tensor	ϵ	(numeric)	
Volume strain	θ	(numeric)	
Poisson's ratio	μ, ν	(numeric)	Lateral contraction divided by elongation
Young's modulus Modulus of elasticity	E	newton per square meter	$E = \sigma/\epsilon$
Shear modulus Modulus of rigidity	G	newton per square meter	$G = \tau/\gamma$
Bulk modulus	K	newton per square meter	$K = -p/\theta$
Work	W	joule	
Energy	E, W	joule	U is recommended in thermodynamics for internal energy and for blackbody radiation.
Energy (volume) density	w	joule per cubic meter	
Power	P	watt	
Efficiency	η	(numeric)	
Heat:			
Thermodynamic temperature	$T \cdots \Theta$	kelvin	
Temperature Customary temperature	$t \cdots \theta$	degree Celsius	The word *centigrade* has been abandoned as the name of a temperature scale
Heat	Q	joule	
Internal energy	U	joule	
Heat flow rate	$\Phi \cdots q$	watt	Heat crossing a surface divided by time
Temperature coefficient	α	reciprocal kelvin	
Thermal diffusivity	α	square meter per second	
Thermal conductivity	$\lambda \cdots k$	watt per meter kelvin	
Thermal conductance	G_θ	watt per kelvin	
Thermal resistivity	ρ_θ	meter kelvin per watt	
Thermal resistance	R_θ	kelvin per watt	
Thermal capacitance Heat capacity	C_θ	joule per kelvin	
Thermal impedance	Z_θ	kelvin per watt	
Specific heat capacity	c	joule per kelvin kilogram	Heat capacity divided by mass
Entropy	S	joule per kelvin	
Specific entropy	s	joule per kelvin kilogram	Entropy divided by mass
Enthalpy	H	joule	
Radiation and light:			
Radiant intensity	$I \cdots I_e$	watt per steradian	
Radiant power Radiant flux	$P, \Phi \cdots \Phi_e$	watt	
Radiant energy	$W, Q \cdots Q_e$	joule	The symbol U is used for the special case of blackbody radiant energy
Radiance	$L \cdots L_e$	watt per steradian square meter	
Radiant exitance	$M \cdots M_e$	watt per square meter	
Irradiance	$E \cdots E_e$	watt per square meter	
Luminous intensity	$I \cdots I_v$	candela	
Luminous flux	$\Phi \cdots \Phi_v$	lumen	
Quantity of light	$Q \cdots Q_v$	lumen second	
Luminance	$L \cdots L_v$	candela per square meter	
Luminous exitance	$M \cdots M_v$	lumen per square meter	
Illuminance Illumination	$E \cdots E_v$	lux	
Luminous efficacy†	$K(\lambda)$	lumen per watt	
Total luminous efficacy	K, K_t	lumen per watt	
Refractive index Index of refraction	n	(numeric)	

Quantity	Quantity symbol	Unit based on International System	Remarks
Emissivity†	$\epsilon(\lambda)$	(numeric)	
Total emissivity	ϵ, ϵ_t	(numeric)	
Absorptance†	$\alpha(\lambda)$	(numeric)	
Transmittance†	$\tau(\lambda)$	(numeric)	
Reflectance†	$\rho(\lambda)$	(numeric)	
Fields and circuits:			
Electric charge Quantity of electricity	Q	coulomb	
Linear density of charge	λ	coulomb per meter	
Surface density of charge	σ	coulomb per square meter	
Volume density of charge	ρ	coulomb per cubic meter	
Electric field strength	$E \cdots K$	volt per meter	
Electrostatic potential Potential difference	$V \cdots \phi$	volt	
Retarded scalar potential	V_r	volt	
Voltage Electromotive force	V, $E \cdots U$	volt	
Electric flux	Ψ	coulomb	
Electric flux density (Electric) displacement	D	coulomb per square meter	
Capacitivity Permittivity Absolute permittivity	ϵ	farad per meter	Of vacuum, ϵ_v
Relative capacitivity Relative permittivity Dielectric constant	ϵ_r, κ	(numeric)	
Complex relative capacitivity Complex relative permittivity Complex dielectric constant	ϵ_r^*, κ^*	(numeric)	$\epsilon_r^* = \epsilon_r' - j\epsilon_r''$ ϵ_r'' is positive for lossy materials. The complex absolute permittivity ϵ^* is defined in analogous fashion.
Electric susceptibility	$\chi_e \cdots \epsilon_i$	(numeric)	$\chi_e = \epsilon_r - 1$ MKSA
Electrization	$E_i \cdots K_i$	volt per meter	$E_i = (D/\Gamma_e) - E$ MKSA
Electric polarization	P	coulomb per square meter	$P = D - \Gamma_e E$ MKSA
Electric dipole moment	p	coulomb meter	
(Electric) current	I	ampere	
Current density	$J \cdots S$	ampere per square meter	
Linear current density	$A \cdots \alpha$	ampere per meter	Current divided by the breadth of the conducting sheet
Magnetic field strength	H	ampere per meter	
Magnetic (scalar) potential Magnetic potential difference	U, U_m	ampere	
Magnetomotive force	F, $F_m \cdots \mathfrak{F}$	ampere	
Magnetic flux	Φ	weber	
Magnetic flux density Magnetic induction	B	tesla	
Magnetic flux linkage	Λ	weber	
(Magnetic) vector potential	A	weber per meter	
Retarded (magnetic) vector potential	A_r	weber per meter	
Permeability Absolute permeability	μ	henry per meter	Of vacuum, μ_v
Relative permeability	μ_r	(numeric)	
Initial (relative) permeability	μ_o	(numeric)	
Complex relative permeability	μ_r^*	(numeric)	$\mu_r^* = \mu_r' - j\mu_r''$ μ_r'' is positive for lossy materials. The complex absolute permeability μ^* is defined in analogous fashion.
Magnetic susceptibility	$\chi_m \cdots \mu_i$	(numeric)	$\chi_m = \mu_r - 1$ MKSA
Reluctivity	ν	meter per henry	$\nu = 1/\mu$
Magnetization	H_i, M	ampere per meter	$H_i = (B/\Gamma_m) - H$ MKSA
Magnetic polarization Intrinsic magnetic flux density	J, B_i	tesla	$J = B - \Gamma_m H$ MKSA
Magnetic (area) moment	m	ampere meter squared	The vector product $\boldsymbol{m} \times \boldsymbol{B}$ is equal to the torque
Capacitance	C	farad	
Elastance	S	reciprocal farad	$S = 1/C$
(Self-) inductance	L	henry	
Reciprocal inductance	Γ	reciprocal henry	
Mutual inductance	L_{ij}, M_{ij}	henry	If only a single mutual inductance is involved, M may be used without subscripts
Coupling coefficient	$k \cdots \kappa$	(numeric)	$k = L_{ij}(L_i L_j)^{-1/2}$
Leakage coefficient	σ	(numeric)	$\sigma = 1 - k^2$
Number of turns (in a winding)	N, n	(numeric)	
Number of phases	m	(numeric)	
Turns ratio	$n \cdots n_*$	(numeric)	

Quantity	Quantity symbol	Unit based on International System	Remarks
Transformer ratio	a	(numeric)	Square root of the ratio of secondary to primary self-inductance. Where the coefficient of coupling is high, $a \approx n_*$.
Resistance	R	ohm	
Resistivity Volume resistivity	ρ	ohm meter	
Conductance	G	mho	$G = \text{Re } Y$
Conductivity	γ, σ	mho per meter	$\gamma = 1/\rho$ The symbol σ is used in field theory, as γ is there used for the propagation coefficient.
Reluctance	$R, R_m \cdots \mathcal{R}$	reciprocal henry	Magnetic potential difference divided by magnetic flux
Permeance	$P, P_m \cdots \mathcal{P}$	henry	$P_m = 1/R_m$
Impedance	Z	ohm	
Reactance	X	ohm	
Capacitive reactance	X_C	ohm	For a pure capacitance, $X_C = -1/\omega C$
Inductive reactance	X_L	ohm	For a pure inductance $X_L = \omega L$
Quality factor	Q	(numeric)	
Admittance	Y	mho	$Y = 1/Z = G + jB$
Susceptance	B	mho	$B = \text{Im } Y$
Loss angle	δ	radian	$\delta = (R/\|X\|)$
Active power	P	watt	
Reactive power	$Q \cdots P_q$	var	
Apparent power	$S \cdots P_s$	voltampere	
Power factor	$\cos \phi \cdots F_p$	(numeric)	
Reactive factor	$\sin \phi \cdots F_q$	(numeric)	
Input power	P_i	watt	
Output power	P_o	watt	
Poynting vector	S	watt per square meter	
Characteristic impedance Surge impedance	Z_0	ohm	
Intrinsic impedance of a medium	η	ohm	
Voltage standing-wave ratio	S	(numeric)	
Resonance frequency	f_r	hertz	
Critical frequency Cutoff frequency	f_c	hertz	
Resonance angular frequency	ω_r	radian per second	
Critical angular frequency Cutoff angular frequency	ω_c	radian per second	
Resonance wavelength	λ_r	meter	
Critical wavelength Cutoff wavelength	λ_c	meter	
Wavelength in a guide	λ_g	meter	
Hysteresis coefficient	k_h	(numeric)	
Eddy-current coefficient	k_e	(numeric)	
Phase angle Phase difference	ϕ, θ	radian	

† NOTE: (λ) is not part of the basic symbol but indicates that the quantity is a function of wavelength.

Symbols for Physical Constants

Name of constant	Symbol	Value	Remarks
Speed of propagation of electromagnetic waves in vacuum	c_o	$(2.997\,925 \pm 0.000\,003) \times 10^8$ m/s	See note
Magnetic constant	Γ_m	$4\pi \times 10^{-7}$ H/m	
Electric constant	Γ_e	$(8.854\,185 \pm 0.000\,018) \times 10^{-12}$ F/m	$\Gamma_e = 1/\Gamma_m c_0^2$
Elementary charge Electronic charge	e	$(1.602\,10 \pm 0.000\,07) \times 10^{-19}$ C	See note
Avogadro constant	N_A	$(6.022\,52 \pm 0.000\,28) \times 10^{23}$ mol^{-1}	See note
Faraday constant	F	$(9.648\,70 \pm 0.000\,16) \times 10^4$ C/mol	See note
Planck constant	h	$(6.6256 \pm 0.0005) \times 10^{-34}$ J·s	See note
	$\hbar$	$(1.054\,50 \pm 0.000\,07) \times 10^{-34}$ J·s	$\hbar = h/2\pi$
Boltzmann constant	k	$(1.380\,54 \pm 0.000\,18) \times 10^{-23}$ J/K	See note
Gravitational constant	G	$(6.670 \pm 0.015) \times 10^{-11}$ N·m^2/kg^2	See note
Standard acceleration of free fall	g_n	9.806 65 m/s^2	Defined by the Conférence Générale des Poids et Mesures (CGPM) in 1901

NOTE: These values are taken from *Natl. Bur. Standards Tech. News Bull.*, October, 1963. The estimated error limits shown are based on three standard deviations.

19. Standard Unit Symbols. The following unit symbols are given in USA Standard Y10.19-1967 (IEEE Standard 260):

Unit	Symbol	Remarks
ampere	A	
ampere-hour	Ah	
angstrom	Å	
atmosphere (normal)	atm	1 atm = 101 325 N/m^2
atomic mass unit (unified)	u	The (unified) atomic mass unit is defined as one-twelfth of the mass of an atom of the ^{12}C nuclide.
bar	bar	1 bar = 100,000 N/m^2
barn	b	1 b = 10^{-28} m^2
billion electronvolts	GeV	The name *billion electronvolts* is deprecated; see *gigaelectronvolt*
British thermal unit	Btu	
calorie (International Table calorie)	cal_{IT}	1 cal_{IT} = 4.1868 J
calorie (thermochemical calorie)	cal_{th}	1 cal_{th} = 4.1840 J
candela	cd	
candela per square foot	cd/ft^2	
candela per square meter	cd/m^2	The name *nit* is sometimes used for this unit.
candle	cd	The unit of luminous intensity has been given the name *candela*; use of the word *candle* for this purpose is deprecated.
centimeter	cm	
circular mil	cmil	1 cmil = $(\pi/4) \cdot 10^{-6}$ in^2
coulomb	C	
cubic centimeter	cm^3	
cubic foot per second	ft^3/s	
cubic meter	m^3	
cubic meter per second	m^3/s	
curie	Ci	Unit of activity in the field of radiation dosimetry
cycle per second	c/s	
decibel	dB	
decibel referred to one milliwatt	dBm	
degree (plane angle)	...°	
degree (temperature)		Note that there is no space between the symbol ° and the letter. The use of the word *centigrade* for the Celsius temperature scale was abandoned by the Conférence Générale des Poids et Mesures in 1948.
degree Celsius	°C	
degree Fahrenheit	°F	
kelvin	K	In 1967 the CGPM voted to give the name *kelvin* to the SI unit of temperature, which was formerly called *degree Kelvin*, and to assign it the symbol K.
dyne	dyn	
electronvolt	eV	
farad	F	
foot	ft	
footcandle	fc	The name *lumen per square foot* (lm/ft^2) is preferred for this unit.
footlambert	fL	If luminance is to be measured in English units, the candela per square foot (cd/ft^2) is preferred.
foot per minute	ft/min	
foot per second	ft/s	
foot pound-force	ft·lbf	
gallon	gal	
gauss	G	
gigacycle per second	Gc/s	
gigaelectronvolt	GeV	
gigahertz	GHz	
gram	g	
henry	H	
hertz	Hz	
horsepower	hp	
hour	h	
inch	in	
inch per second	in/s	
joule	J	
joule per kelvin	J/K	
kelvin	K	
kilocycle per second	kc/s	
kiloelectronvolt	keV	
kilogauss	kG	
kilogram	kg	
kilogram-force	kgf	In some countries the name *kilopond* (kp) has been adopted for this unit.
kilohertz	kHz	
kilohm	kΩ	
kilometer	km	
kilometer per hour	km/h	
kilovar	kvar	
kilovolt	kV	
kilovoltampere	kVA	
kilowatt	kW	
kilowatthour	kWh	

Unit	Symbol	Remarks
lambert	L	
liter	l	
lumen	lm	
lux	lx	1 lx = 1 lm/m²
maxwell	Mx	
megacycle per second	Mc/s	
megahertz	MHz	
megawatt	MW	
meter	m	
mho	mho	
microampere	μA	
microfarad	μF	
microhenry	μH	
micrometer	μm	
micromho	μmho	
micron	μm	The name *micrometer* is preferred.
microsecond	μs	
microwatt	μW	
mile (statute)	mi	
nautical mile	nmi	
mile per hour	mi/h	
milliampere	mA	
millibar	mbar	
milligram	mg	
millihenry	mH	
milliliter	ml	
millimeter	mm	
conventional millimeter of mercury	mmHg	1 mmHg = 133.322 N/m²
millimicron	nm	The name *nanometer* is preferred.
millisecond	ms	
millivolt	mV	
milliwatt	mW	
minute (plane angle)	. . .′	
minute (time)	min	Time may also be designated by means of superscripts as in the following example: $9^h46^m30^s$
nanofarad	nF	
nanometer	nm	
nanosecond	ns	
nanowatt	nW	
neper	Np	
newton	N	
newton meter	N·m	
newton per square meter	N/m²	
oersted	Oe	
ohm	Ω	
ounce (avoirdupois)	oz	
picofarad	pF	
picowatt	pW	
pint	pt	
pound	lb	
poundal	pdl	
pound-force	lbf	
pound-force foot	lbf·ft	
pound-force per square inch	lbf/in²	
pound per square inch		Although use of the abbreviation psi is common, it is not recommended. See pound-force per square inch.
quart	qt	
rad	rd	Unit of absorbed dose in the field of radiation dosimetry
radian	rad	
rem	rem	Unit of dose equivalent in the field of radiation dosimetry
revolution per minute	r/min	Although use of the abbreviation rpm is common, it is not recommended
revolution per second	r/s	
roentgen	R	Unit of exposure in the field of radiation dosimetry
second (plane angle)	. . .″	
second (time)	s	
siemens	S	$1\ S = 1\ \Omega^{-1}$
square foot	ft²	
square meter	m²	
steradian	sr	
tesla	T	
tonne	t	1 t = 1,000 kg
var	var	
volt	V	
voltampere	VA	
watt	W	
watthour	Wh	
watt per steradian	W/sr	
watt per steradian square meter	W/(sr·m²)	
weber	Wb	1 Wb = 1 V·s
yard	yd	

20. Conversion Factors

Table 1-1. Length

	Meter	Inch	Foot	Yard	Mile
1 m =	1	39.370	3.2808	1.0936	$0.621\,37 \times 10^{-3}$
1 in =	0.0254†	1	0.083 333	0.027 778	$0.015\,783 \times 10^{-3}$
1 ft =	0.3048†	12†	1	0.333 33	$0.189\,39 \times 10^{-3}$
1 yd =	0.9144†	36†	3†	1	$0.568\,18 \times 10^{-3}$
1 mi =	1,609.3	63,360†	5,280†	1,760†	1

1 angstrom = 10^{-10} meter = 0.1 nanometer†
1 micron = 10^{-6} meter = 1.0 micrometer†
1 nautical mile = 1,852 meters†

Table 1-2. Area

	Square Meter	Square Inch	Square Foot
1 m^2 =	1	1,550.0	10.764
1 in^2 =	6.4516×10^{-4}†	1	6.9444×10^{-3}
1 ft^2 =	0.092 903	144†	1

1 circular mil = 5.0671×10^{-4} square millimeter
1 acre = 4,046.9 square meters = 43,560 square feet†
1 barn = 10^{-28} square meter†
1 hectare = 10,000 square meters†

Table 1-3. Volume

	Cubic Meter	Cubic Inch	Cubic Foot
1 m^3 =	1	61.024×10^{-3}	35.315
1 in^3 =	16.387×10^{-6}	1	$0.578\,70 \times 10^{-3}$
1 ft^3 =	28.317×10^{-3}	1,728†	1

1 fluid ounce (U.K.) = 28.413 cubic centimeters
1 fluid ounce (U.S.) = 29.574 cubic centimeters
1 gallon (U.K.) = 4,546.1 cubic centimeters
1 gallon (U.S.) = 3,785.4 cubic centimeters
1 barrel (U.S.) (for petroleum, etc.) = 0.158 99 cubic meter
1 acre foot = 1233.5 cubic meters
1 liter = 1,000 cubic centimeters†

Table 1-4. Velocity

	Meter per Second	Kilometer per Hour	Foot per Second	Foot per Minute	Mile per Hour
1 m/s =	1	3.6†	3.2808	196.85	2.2369
1 km/h =	0.277 78	1	0.911 34	54.681	0.621 37
1 ft/s =	0.3048†	1.0973	1	60†	0.681 82
1 ft/min =	0.005 08†	0.018 288	0.016 667	1	0.011 364
1 mi/h =	0.447 04	1.6093	1.4667	88†	1

1 knot = 0.514 44 meter per second

Table 1-5. Mass

	Kilogram	Ounce Avoirdupois	Pound	Ton
1 kg =	1	35.274	2.2046	1.1023×10^{-3}
1 oz =	0.028 350	1	0.0625†	$0.031\,25 \times 10^{-3}$†
1 lb =	0.453 59	16†	1	0.5×10^{-3}†
1 ton =	907.18	32,000†	2,000†	1

1 grain = 0.064 799 gram
1 slug = 14.594 kilograms
1 long ton = 1,016.0 kilograms = 2,240 pounds†
1 tonne = 1,000 kilograms†

† Exact conversion.

Table 1-6. Density

1 pound per cubic foot = 16.018 kilograms per cubic meter
1 pound per cubic inch = 27 680 kilograms per cubic meter

Table 1-7. Force

	Newton	Kilogram-force	Pound-force
1 N =	1	0.101 97	0.224 81
1 kgf =	9.806 65†	1	2.2046
1 lbf =	4.4482	0.453 59	1

1 poundal = 0.138 25 newton
1 ounce-force = 0.278 01 newton
1 dyne = 10^{-5} newton†

Table 1-8. Pressure or Stress

	Newton per Square Meter	Kilogram-force per Square Centimeter	Pound-force per Square Inch
1 N/m^2 =	1	0.101 97	$0.145\ 04 \times 10^{-3}$
1 kgf/cm^2 =	98,066.5†	1	14.223
1 lbf/in^2 =	6894.8	0.070 307	1

1 poundal per square foot = 1.4882 newtons per square meter
1 pound-force per square foot = 47.880 newtons per square meter
1 conventional foot of water = 2989.1 newtons per square meter
1 conventional millimeter of mercury = 133.32 newtons per square meter
1 torr = 133.32 newtons per square meter
1 normal atmosphere (760 torr) = 101 325 newtons per square meter†
1 bar = 100,000 newtons per square meter†

Table 1-9. Energy or Work

	Joule	Kilowatthour	Foot Pound-force	British Thermal Unit (I.T.)
1 J =	1	$0.277\ 78 \times 10^{-6}$	0.737 56	0.9478×10^{-3}
1 kWh =	3.6×10^{6}†	1	$2,655.2 \times 10^{3}$	3,412
1 ft·lbf =	1.3558	$0.376\ 62 \times 10^{-6}$	1	1.285×10^{-3}
1 Btu_{IT} =	1055	0.2931×10^{-3}	778.2	1

1 foot poundal = 0.042 140 joule
1 British thermal unit (thermochemical) = 1054 joules
1 British thermal unit (International Table) = 1055 joules
1 calorie (thermochemical) = 4.184 joules†
1 calorie (International Table) = 4.1868 joules†
1 electronvolt = 1.602×10^{-19} joule
1 erg = 10^{-7} joule†

Table 1-10. Power

	Watt	Foot Pound-force per Second	Horsepower (Electrical)
1 W =	1	0.737 56	1.341×10^{-3}
1 ft·lbf/s =	1.3558	1	1.818×10^{-3}
1 hp =	746†	550.2	1

1 horsepower (metric) = 735.50 watts
1 horsepower (British) = 745.70 watts
1 horsepower (electrical) = 746 watts†
1 British thermal unit (I.T.) per hour = 0.2931 watt
1 erg per second = 10^{-7} watt†

Table 1-11. Temperature

	Kelvins	Degrees Celsius	Degrees Fahrenheit
T_K =	—	$t_C + 273.15$†	$5/9(t_C + 459.67)$
t_C =	$T_K - 273.15$†	—	$5/9(t_F - 32)$†
t_F =	$1.8T_K - 459.67$	$1.8t_C + 32$†	

† Exact conversion.

Table 1-12. Quantities of Light

1 footcandle = 10.764 lux (lumens per square meter)
1 footlambert = 3.4263 candelas per square meter
1 candela per square foot = 10.764 candelas per square meter

Table 1-13. Quantities of Electricity and Magnetism*

1 esu of current = 3.3356 $\times 10^{-10}$ ampere
1 emu of current = 10 amperes†
1 esu of voltage = 299.79 volts
1 emu of voltage = 10^{-8} volt†
1 esu of capacitance = 1.1126 $\times 10^{-12}$ farad
1 emu of capacitance = 10^{9} farads†
1 esu of inductance = 8.9876 $\times 10^{11}$ henrys
1 emu of inductance = 10^{-9} henry†
1 esu of resistance = 8.9876 $\times 10^{11}$ ohms
1 emu of resistance = 10^{-9} ohm†
1 gilbert = 0.79577 ampere
1 oersted = 79.577 amperes per meter
1 maxwell = 10^{-8} weber†
1 gauss = 10^{-4} tesla†
1 gamma = 10^{-9} tesla†

* Because corresponding electrical and magnetic quantities have different dimensions in different measurement systems (see Pars. **6** to **13**), the conversion relations are not true equations between units. The "equations" given above may be strictly interpreted as measure conversions, however. For example, the first "equation" indicates that, if the measure of CGS electrostatic current is 1, the measure of current in the International System is 3.3356 $\times 10^{-10}$.

† Exact conversion.

Table 1–14. Resistivity and Conductivity Conversion (see following page)

21. Bibliography on Quantities and Systems of Measurement

1. Page, Chester H. Physical Entities and Mathematical Representation; *J. Res. Natl. Bur. Standards*, October–December, 1961, Vol. 65B, pp. 227–235. *See also* The Mathematical Representation of Physical Entities; *IEEE Trans. on Education*, June, 1967, Vol. E-10, pp. 70–74.

2. Silsbee, F. B. Simplification of Systems of Units, in Systems of Units, C. F. Kayan (ed.), Washington, American Association for the Advancement of Science, 1959, pp. 7–24.

3. Silsbee, F. B. Systems of Electrical Units; *J. Res. Natl. Bur. Standards*, April–June, 1962, Vol. 66C, pp. 137–174.

4. Cornelius, P., de Groot, W., and Vermeulen, R. Quantity Equations, Rationalization and Change of Number of Fundamental Quantities (in three parts); *Appl. Sci. Res.*, 1965, Vol. B12, pp. 1, 235, 248.

Table 1–14. Resistivity and Conductivity Conversion*

Perform indicated operation to obtain ↓ / Given N →	Volume resistivity at 20°C				Mass resistivity at 20°C		Conductivity at 20°C	
	Ohm—circular mil/ft	Ohm—mm²/meter	Microhm—inch	Microhm—cm	Ohm—pound/mile²	Ohm—gram/meter²	Percent IACS (volume basis)	Percent IACS (mass basis)
Volume Resistivity at 20°C								
Ohm—circular mil/ft	....	$N \times 601.53$	$N \times 15.279$	$N \times 6.0153$	$N \times 0.10535 \times \frac{1}{\delta}$	$N \times 601.53 \times \frac{1}{\delta}$	$\frac{1}{N} \times 1037.1$	$\frac{1}{N} \times 9220.0 \times \frac{1}{\delta}$
Ohm—mm²/meter	$N \times 0.0016624$	....	$N \times 0.025400$	$N \times 0.010000$	$N \times 0.00017513 \times \frac{1}{\delta}$	$N \times \frac{1}{\delta}$	$\frac{1}{N} \times 1.7241$	$\frac{1}{N} \times 15.328 \times \frac{1}{\delta}$
Microhm—inch	$N \times 0.065450$	$N \times 39.370$	....	$N \times 0.39370$	$N \times 0.0068950 \times \frac{1}{\delta}$	$N \times 39.370 \times \frac{1}{\delta}$	$\frac{1}{N} \times 67.879$	$\frac{1}{N} \times 603.45 \times \frac{1}{\delta}$
Microhm—cm	$N \times 0.16624$	$N \times 100.00$	$N \times 2.5400$	....	$N \times 0.017513 \times \frac{1}{\delta}$	$N \times 100.00 \times \frac{1}{\delta}$	$\frac{1}{N} \times 172.41$	$\frac{1}{N} \times 1532.8 \times \frac{1}{\delta}$
Mass Resistivity at 20°C								
Ohm—pound/mile²	$N \times 9.4924 \times \delta$	$N \times 5710.0 \times \delta$	$N \times 145.03 \times \delta$	$N \times 57.100 \times \delta$	....	$N \times 5710.0$	$\frac{1}{N} \times 9844.8 \times \delta$	$\frac{1}{N} \times 87520$
Ohm—gram/meter²	$N \times 0.0016624 \times \delta$	$N \times \delta$	$N \times 0.025400 \times \delta$	$N \times 0.010000 \times \delta$	$N \times 0.00017513$	....	$\frac{1}{N} \times 1.7241 \times \delta$	$\frac{1}{N} \times 15.328$
Conductivity at 20°C								
Percent IACS (volume basis)	$\frac{1}{N} \times 1037.1$	$\frac{1}{N} \times 1.7241$	$\frac{1}{N} \times 67.879$	$\frac{1}{N} \times 172.41$	$\frac{1}{N} \times 9844.8 \times \delta$	$\frac{1}{N} \times 1.7241 \times \delta$	....	$N \times 0.11249 \times \delta$
Percent IACS (mass basis)	$\frac{1}{N} \times 9220.0 \times \frac{1}{\delta}$	$\frac{1}{N} \times 15.328 \times \frac{1}{\delta}$	$\frac{1}{N} \times 603.45 \times \frac{1}{\delta}$	$\frac{1}{N} \times 1532.8 \times \frac{1}{\delta}$	$\frac{1}{N} \times 37520$	$\frac{1}{N} \times 15.328$	$N \times 8.89 \times \frac{1}{\delta}$	....

* From Table III, ASTM Standard B-193-65. δ = density in g per cc.
Note: These factors are applicable only to resistivity and conductivity values corrected to 20°C (68°F). They are applicable for any temperature when used to convert between volume units only or between mass units only.

SECTION 2

ELECTRIC AND MAGNETIC CIRCUITS

BY

WARREN B. BOAST Anson Marston Distinguished Professor and Head of Electrical Engineering, Iowa State University; Fellow, Institute of Electrical and Electronics Engineers

HARRY W. HALE Professor of Electrical Engineering, Iowa State University; Senior Member, Institute of Electrical and Electronics Engineers

CONTENTS

Numbers refer to paragraphs

SECTION 2

ELECTRIC AND MAGNETIC CIRCUITS

BY WARREN B. BOAST

ELECTRIC CHARGE

1. Electron Theory. According to the electronic theory of electricity, there is an indivisible particle of negative electricity, called the *electron*. An atom of matter consists of one or more electrons and a nucleus which carries a charge of positive electricity. A negatively charged particle of matter is one in which there are more electrons than necessary to neutralize its positive electricity. A positively electrified body is one which has lost some of the electrons it had in the neutral state. Atoms which have gained or lost electrons are called *ions*.

The unit of electric charge in the mksa system of units is the *coulomb*. The magnitude of the charge associated with an electron is 1.602×10^{-19} C.

ELECTRIC POTENTIAL

2. Sources of Electric-potential Difference. A difference of electric potential, sometimes called *electromotive force*, and abbreviated *emf*, is caused by the separation of opposite charges of electricity. This separation may be forced by physical motion, or it may be initiated or complemented by thermal, chemical, or magnetic causes or by radiation. The various classifications of these causes are called:

a. Electromagnetic induction (see Par. **32**).
b. Contact of dissimilar substances (see Par. **6**).
c. Thermoelectric action (see Pars. **3, 4,** and **5**).
d. Chemical action (Sec. **23**).
e. Friction between dissimilar substances.
f. Photoelectric effect (Sec. **19**).

3. Thomson Effect. A temperature gradient in a metallic conductor is accompanied by a small voltage gradient whose magnitude and direction depend upon the particular metal. When an electric current flows, there is an evolution or absorption of heat due to the presence of the thermoelectric gradient, with the net result that the heat evolved in an interval bounded by different temperatures is slightly greater or less than that accounted for by the resistance of the conductor. In copper, the evolution of heat is greater when the current flows from hot to cold parts and less when the current flows from cold to hot. In iron, the effect is the reverse. Discovery of this phenomenon in 1851 is credited to Sir William Thomson (Lord Kelvin), the English physicist.

4. Peltier Effect. When a current is passed across the junction between two different metals, an evolution or an absorption of heat takes place. This effect is different from the evolution of heat i^2r, owing to the resistance of the junction, and is reversible, heat being evolved when the current passes one way across the junction and absorbed when the current passes in the other direction. The junction is the source of a Peltier electromotive force. When current is forced across the junction against the direction of the emf, a heating action occurs. If the current is forced in the direction of the Peltier emf, the junction is cooled. Refrigerators can be constructed using this principle.

Since the Joule effect (see Par. **20**) produces heat in the conductors leading to the junction, the Peltier cooling must be greater than the Joule effect in that region for refrigeration to be successful. The phenomenon was discovered by Jean Peltier, the French physicist, in 1834.

5. Seebeck Effect. In a closed circuit consisting of two different metals, if the two junctions are maintained at different temperatures, within certain temperature ranges, an electric current flows. Thus, if one junction of a copper-iron circuit is kept in melting ice and the other in boiling water, current passes from copper to iron across the hot junction. The resulting device is usually called a *thermocouple.* The phenomenon was discovered in 1821 by Thomas Johann Seebeck.

6. Volta Effect, or Contact Potential. When pieces of various materials are brought in contact, an emf is developed between them. Thus, in the case of zinc and copper, zinc becomes charged positively and copper negatively. According to the electron theory, different substances possess different tendencies to give up their negatively charged particles. Zinc gives them up very easily; therefore a number of negatively charged particles pass from it to copper. Measurable emfs are observed even between two pieces of the same substance having different structures, for instance, between a piece of cast copper and electrolytic copper. Frictional electricity is explained in a similar way, except that more intimate contact is necessary where the conductivities of the substances are small.

7. Thermocouples and Batteries. To utilize contact emfs, means must be devised to supply energy to the system continuously, for example, by heat or by chemical reaction. The thermocouple is an example of the former and the battery of the latter. For further discussion of thermocouples, see Sec. **3**. Batteries are treated in Sec. **24**.

8. Literature References

Foecke, Harold A. "Introduction to Electrical Engineering Science"; Englewood Cliffs, N.J., Prentice-Hall, Inc., 1961, Chap. 7.

McGraw-Hill Encyclopedia of Science and Technology, "Thermoelectricity," 1st ed.; New York, McGraw-Hill Book Company, 1960, Vol. 13, pp. 577–585.

Ward, Robert P. "Introduction to Electrical Engineering"; Englewood Cliffs, N.J., Prentice-Hall, Inc., 1960, Chap. 13.

CONDUCTORS

9. Conductors and Insulators. The principal materials used in electrical engineering are *conductors* and *insulators.* A conducting material allows a continuous current to pass through it under the action of a continuous emf. An ideal insulator (more correctly called a *dielectric*) allows only a brief transient current which charges it electrostatically. This charge, or *displacement,* of electricity produces a counter emf equal and opposite to the applied emf, and the flow of current ceases. The division into conductors and dielectrics is not strictly correct but is convenient for practical purposes. A substance may have practically no current when the applied voltage is sufficiently low but may be unsuitable as an insulator at high voltages. Some materials which are practically nonconducting at ordinary temperatures become good conductors when sufficiently heated. For numerical data and tables of conducting and insulating properties of the principal materials used in practice, see Sec. **4**.

Metals and other solid conductors possess free (unbound) electrons, in addition to those associated with the molecules. These free electrons move as if they were the particles of a gas dissolved in the metal. When a positive emf is applied, each electron gains a component velocity which, on account of its negative charge, is in the direction opposite to the emf. The drift of electrons constitutes a current. In their motion the electrons collide with the molecules and give up part of their momentum. This loss is supposed to account for the joulean (i^2r) heat set free in the conductor.

10. Gaseous Conduction. A gas may be put in the conducting state by such means as raising its temperature; placing it in the neighborhood of flames, arcs, or glowing metals; or passing an electric discharge through it. The conductivity is due to free electrons. The process by which a gas is made conducting is called *ionization.* The movement of free electrons constitutes the current through the gas.

11. Electrolytes. In liquid chemical compounds (electrolytes), the passage of an electric current is accompanied by chemical decomposition. Atoms of metals and hydrogen travel through the liquid in the direction of the positive current, while oxygen and acid radicals travel against the positive current. Thus, while in solid conductors electricity travels "across" the matter, in electrolytes it travels "with" the matter. For details of electrolytic conduction, see Sec. **23**.

CONTINUOUS-CURRENT CIRCUITS

12. Ohm's Law. When the current in a conductor is steady and there are no emfs within the conductor, the value of the voltage e between the terminals of the conductor is proportional to the current i, or

$$e = ri \tag{2-1}$$

where the coefficient of proportionality r is called the *resistance* of the conductor. The same law may be written in the form

$$i = ge \tag{2-2}$$

where the coefficient of proportionality $g = 1/r$ is called the *conductance* of the conductor. When the current is measured in amperes and the emf in volts, the resistance r is in ohms and g is in mhos.

When there is a counter emf e_c within the conductor, Ohm's law becomes

$$e_t - e_c = ri \tag{2-3}$$

or

$$i = g(e_t - e_c) \tag{2-4}$$

where e_t is the voltage between the terminals of the conductor.

13. Cylindrical Conductors. For current directed along the axis of the cylinder, the resistance r is proportional to the length l and inversely proportional to the cross section A, or

$$r = \rho \frac{l}{A} \tag{2-5}$$

where the coefficient of proportionality ρ (rho) is called the *resistivity* (or *specific resistance*) of the material. For numerical values of ρ for various materials, see Sec. **4**.

The conductance of a cylindrical conductor is

$$g = \sigma \frac{A}{l} \tag{2-6}$$

where σ (sigma) is called the *conductivity* of the material. Since $g = 1/r$, the relation also holds that

$$\sigma = \frac{1}{\rho} \tag{2-7}$$

14. Change of Resistance with Temperature. The resistance of a conductor varies with the temperature. The resistance of metals and most alloys increases with the temperature, while the resistance of carbon and electrolytes decreases with the temperature.

For usual conditions, as for about 100°C change in temperature, the resistance at a temperature t_2 is given by

$$R_{t2} = R_{t1}[1 + \alpha_{t1}(t_2 - t_1)] \tag{2-8}$$

where R_{t1} is the resistance at an initial temperature t_1 and α_{t1} is called the temperature coefficient of resistance of the material for the initial temperature t_1. For copper having a conductivity of 100% of the International Annealed Copper Standard, $\alpha_{20} = 0.00393$, where temperatures are in degrees centigrade (see Sec. **4**).

An equation giving the same results as Eq. (2-8), for copper of 100% conductivity, is

$$\frac{R_{t2}}{R_{t1}} = \frac{234.4 + t_2}{234.4 + t_1} \tag{2-9}$$

where −234.4 is called the "inferred absolute zero" because if the relation held (which it does not over such a large range) the resistance at that temperature would be zero. For hard-drawn copper of 97.3% conductivity, the numerical constant in Eq. (2-9) is changed to 241.5.

See Sec. **4** for values of these numerical constants for copper; and for other metals see Sec. **4** under the metal being considered.

For 100% conductivity copper,

$$\alpha_{t1} = \frac{1}{234.4 + t_1} \tag{2-10}$$

When R_{t1} and R_{t2} have been measured, as at the beginning and end of a heat run, the "temperature rise by resistance" for 100% conductivity copper is given by

$$t_2 - t_1 = \frac{R_{t2} - R_{t1}}{R_{t1}} (234.4 + t_1) \tag{2-11}$$

15. Resistances and Conductances in Series. When two or more resistances are connected in series, the equivalent resistance of the combination is equal to the sum of the resistances of the individual resistors, or

$$r_{eq} = r_1 + r_2 + \cdots \tag{2-12}$$

When conductances are connected in series, the equivalent conductance g_{eq} is determined from the relation

$$\frac{1}{g_{eq}} = \frac{1}{g_1} + \frac{1}{g_2} + \cdots \tag{2-13}$$

that is, the reciprocal of the equivalent conductance is equal to the sum of the reciprocals of the individual conductances.

16. Resistances and Conductances Connected in Parallel. The equivalent resistance r_{eq} of a parallel combination of resistors is determined from the relation

$$\frac{1}{r_{eq}} = \frac{1}{r_1} + \frac{1}{r_2} + \cdots \tag{2-14}$$

or in conductance notation

$$g_{eq} = g_1 + g_2 + \cdots \tag{2-15}$$

When two resistances are connected in parallel, Eq. (2-14) reduces to

$$r_{eq} = \frac{r_1 r_2}{r_1 + r_2} \tag{2-16}$$

17. Series-Parallel Circuits. In a combination like the one shown in Fig. 2-1, where some of the resistances are in series, some in parallel, and it is required to find the equivalent resistance between A and B, the problem is solved step by step, by combining the resistances in series, converting them into conductances, and adding them with other conductances in parallel. For instance, in the case shown in Fig. 2-1, begin by combining the resistance r_2 and R into one, and determine the corresponding conductance

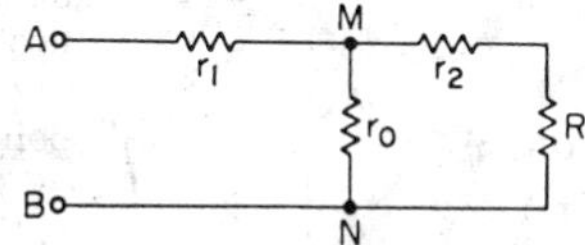

Fig. 2-1. Series-parallel circuit.

$$\frac{1}{R + r_2} \tag{2-17}$$

Then add this conductance to the conductance $1/r_0$. This gives the total conductance between the points M and N, and its reciprocal gives the equivalent resistance between these points. The total resistance between the points A and B is found by adding r_1 to this resistance. When a network of conductors cannot be reduced to a series-parallel combination, the problem is solved by methods as shown in Pars. **21** and **23** to **27**.

18. Power and Energy. When a steady current i exists in a conductor and the voltage across the terminals of the conductor is e, the power, i.e., the energy delivered to the conductor per unit time, is

$$P = ei \tag{2-18}$$

If the current is expressed in amperes and the potential difference in volts, the power P is in watts (joules per second). When the voltage and the current are variable, their instantaneous values being represented by e and i, respectively, the preceding equation gives the instantaneous power, that is, the instantaneous rate at which energy is being delivered to the conductor.

The total energy delivered to the conductor during a time t is

$$W = eit = eQ \tag{2-19}$$

where Q is the total quantity of electricity (coulombs) which passed through the conductor. If Q is in coulombs (ampere-seconds), then W is in joules (watt-seconds). When the voltage and the current are variable, the total energy is expressed by

$$W = \int_{t_1}^{t_2} ei\,dt \tag{2-20}$$

where the time interval is $t_2 - t_1$.

19. Poynting's Law. The power density can be determined from the product of the values of the magnetic field and the component of the electric field which is perpendicular to the magnetic field. The flow of energy at any point is in a direction perpendicular to both the fields. The total power is obtained by integrating the expression for the power density over the appropriate surface.

20. Joule's Law. When the conductor contains ohmic resistance only and no counter emfs, we have $e = ri = i/g$, so that the power

$$P = i^2r = \frac{i^2}{g} = \frac{e^2}{r} = e^2g \tag{2-21}$$

This expression is known as Joule's law.

If the conductor contains a counter emf, e_c, for instance, that developed by a motor or a battery, the power is given by

$$P = e_ti = e_ci + i^2r \tag{2-22}$$

where e_t is the voltage across the terminals of the conductor. In this expression, e_ci is useful power, and i^2r is the heat loss in the conductor (see Par. **12**).

21. Kirchhoff's Laws. In an arbitrary network of conductors, such as Fig. 2-2, with sources of emf connected in one or more places, the distribution of currents is such that two conditions are satisfied, namely:

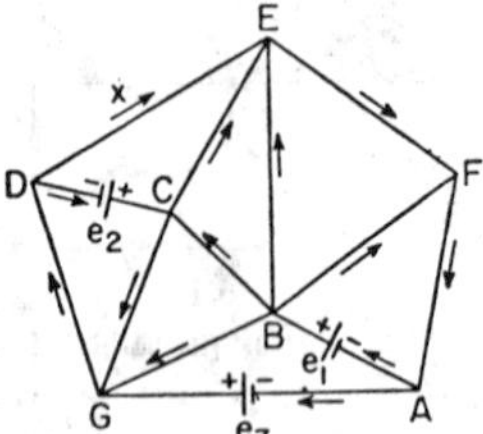

Fig. 2-2. Network of conductors.

1. The algebraic sum of the currents toward any junction point is zero.

2. The algebraic sum of the voltages around any closed path in the network is zero.

The unknown currents in the branches may be denoted by letters, and a direction for each current may be assumed by marking an arrow along the branch. If the direction of the current is not in the assumed direction of the arrow, the computation will show the value of the corresponding current to be negative. The total number of independent equations is equal to the number of unknown quantities, which can therefore be determined by solving the simultaneous equations. For an example of such equations see Par. **23**.

22. Wheatstone Bridge. The combination of six resistances shown in Fig. 2-3 is called the Wheatstone bridge. The resistances are denoted by a, b, c, α, β, γ; the currents, by x, y, z, ξ, η, ζ. An electric battery of emf E is connected in the branch BC, and the value of a includes the internal resistance of the battery. In practice, a galvanometer is usually connected in the branch OA, and α includes its resistance. When the four resistances b, c, β, γ are so adjusted that no current flows through OA, the bridge is said to be balanced and the condition holds that

$$b\beta = c\gamma \tag{2-23}$$

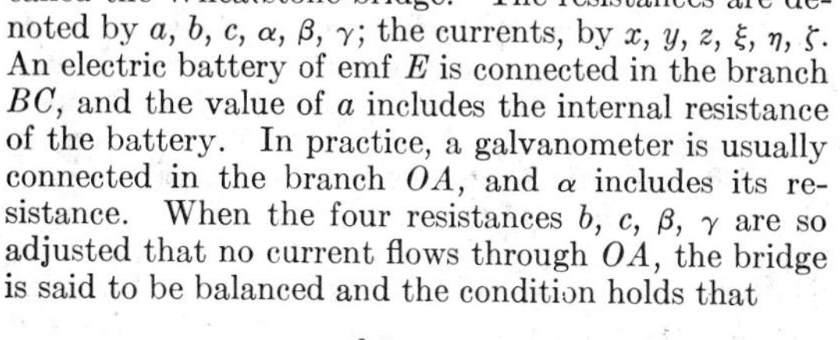

Fig. 2-3. Wheatstone bridge.

23. Unbalanced Bridge. When the Wheatstone bridge is not balanced, Ohm's law and Kirchhoff's laws give the following equations:

$$\begin{array}{lll} ax = C - B + E & \alpha\xi = A & \xi + y - z = 0 \\ by = A - C & \beta\eta = B & \eta + z - x = 0 \\ cz = B - A & \gamma\zeta = C & \zeta + x - y = 0 \end{array}$$

Here E is the battery emf, and A, B, C denote the potentials of these points below that at O. These nine equations contain nine unknown quantities, viz., six currents and three potentials. If these are solved as simultaneous equations, any of the unknown quantities may be determined. For instance, the current in the galvanometer circuit is

$$\xi = \frac{E}{D}(b\beta - c\gamma) \tag{2-24}$$

where the "determinant" D is given by[1]

$$D = abc + bc(\beta + \gamma) + ca(\gamma + \alpha) + ab(\alpha + \beta) + (a + b + c)(\beta\gamma + \gamma\alpha + \alpha\beta)$$

For practical forms of the Wheatstone bridge and its application to the measurement of resistance see Sec. **3**.

24. Networks of Conductors. In a general case (Fig. 2-2), as many Kirchhoff equations (Par. **21**) may be written as there are branches in the network; the unknown quantities may be the currents, the resistances, or the voltages, also any combination of these, provided that the total number of unknown quantities is equal to the number of equations and that at least one known quantity exists in each branch of the network.

25. Loop Currents. In some cases it is convenient to consider, instead of the actual currents, fictitious currents in each loop or mesh of the network. The method was originated by Maxwell (*ibid.*, Art. 282*b*). The actual current in each conductor is equal to the algebraic sum of the fictitious currents. For instance, in Fig. 2-4, the current in conductor f is the difference of the fictitious currents X and Y. The Kirchhoff equations of voltages are written using the loop currents. No current equations enter into the necessary simultaneous equations. Reconstruction of the actual currents then is accomplished after the solution by the proper addition of the components, such as $f = X - Y$, where f in Fig. 2-4 is presumed measured to the right.

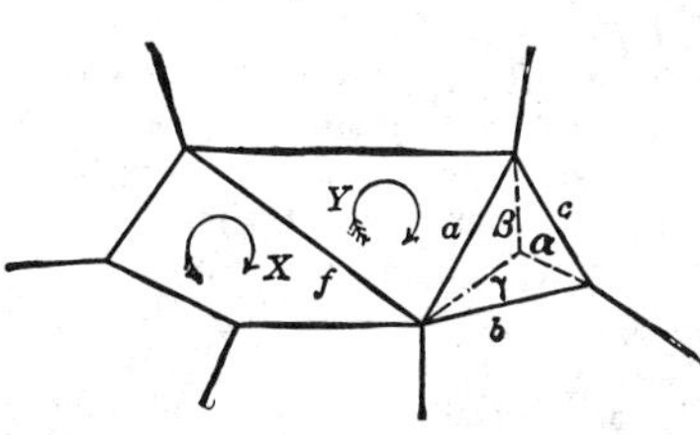

Fig. 2-4. Method of simplifying networks by loop currents or by transformations.

Examples of such solutions will be found in Ward, "Introduction to Electrical Engineering," pp. 79–83; Seshu and Balbanian, "Linear Network Analysis," pp. 25–31;

[1] Maxwell, J. C. "A Treatise on Electricity and Magnetism"; Vol. I, Art. 347.

Cassell, "Linear Electric Circuits," pp. 217–221 and Chap. 11; Brenner and Javid, "Analysis of Electric Circuits," pp. 478–481; and in many other texts (see Par. **31**).

26. Node Voltages. In a similar manner it is sometimes convenient to use a voltage notation in the simultaneous-equation solution. The voltages A, B, and C of Fig. 2-3 constitute three voltages with respect to the reference node O, and these voltages are called *node voltages*. Mathematical procedures using the node voltages as variables of the problem result in establishment of current equations written at the junction nodes. No voltage equations enter into these necessary simultaneous equations. After the results for the node voltages are established, the unknown conditions within each branch of the network can be determined by equations such as $ax = C - B + E$ for the branch between nodes B and C of Fig. 2-3.

Examples of such solutions will be found in the texts referred to above on the following pages: Ward, pp. 87–91; Seshu and Balbanian, pp. 31–35; Cassell, pp. 215–217 and Chap. 11; Brenner and Javid, pp. 471–478 (see Par. **31**).

27. Equivalent Star and Delta. The number of equations required in the solution of a network may be reduced by replacing a triangular connection of branches of the network (called a "delta" or a "mesh"), such as abc (Fig. 2-4), by a star connection $\alpha\beta\gamma$ which is externally equivalent to the delta. It is sometimes advantageous to replace a star by the equivalent delta, and sometimes such successive replacements can reduce a network to a single impedance. This is useful in the calculation of short-circuit currents in power networks, in telephone-cable calculations, and elsewhere.

To replace a delta by a star, put

$$\alpha = \frac{bc}{a+b+c} \qquad \beta = \frac{ca}{a+b+c} \qquad \gamma = \frac{ab}{a+b+c} \tag{2-25}$$

To replace a star by a delta, put

$$a = \beta + \gamma + \frac{\beta\gamma}{\alpha} \qquad b = \gamma + \alpha + \frac{\gamma\alpha}{\beta} \qquad c = \alpha + \beta + \frac{\alpha\beta}{\gamma} \tag{2-26}$$

The six quantities a, b, c, α, β, γ may be complex quantities, representing impedances (see Par. **146**); or they may be the values of resistance when no reactance is involved or the values of reactance when the resistance is negligible.

To replace a star having n branches by a complete mesh joining every pair of the points A, B, C, ... of the star (thus reducing the junction points by one), put

$$AB = \alpha\beta\left(\frac{1}{\alpha} + \frac{1}{\beta} + \frac{1}{\gamma} + \frac{1}{\delta} + \cdots\right)$$

$$BC = \beta\gamma\left(\frac{1}{\alpha} + \frac{1}{\beta} + \frac{1}{\gamma} + \frac{1}{\delta} + \cdots\right) \tag{2-27}$$

and so on, where AB is the impedance of the branch of the mesh between A and B, α is the impedance of the branch of the star leading to A, etc.

No general formula for replacing a mesh of more than three sides by a star exists.

28. Superposition Principle. The response of a network of constant resistances (or impedances for the a-c case) to a number of simultaneously applied excitations is equal to the sum of the responses to those excitations as though taken one at a time.

If in two different locations in a network of constant resistances (impedances), emfs E_1 and E_2 are applied, the currents that will result in the network may be calculated as the result of superposing one set of currents on another as follows: The first set consists of the currents that E_1 acting alone would drive through the branches of the network when E_2 is reduced to zero, but with the connection along the branch originally containing E_2 not broken. The second set consists of currents that E_2 would cause to exist in the various branches of the network when E_1 is similarly reduced to zero.

The principle applies also in networks containing sources that supply fixed currents as well as known voltages. In removing such current sources while the component currents caused by other sources are being investigated, the branch containing the current source must be opened.

29. Thévenin's Theorem. If two terminals exist in a network containing voltage and/or current sources and composed of passive constant linear resistances (impedances for the a-c case), the network may be considered insofar as all external circuit conditions are concerned as composed of a simple series circuit consisting of an emf equal to the open-circuit voltage of the network across the terminals and of an equivalent resistance (impedance) equal to that observed in looking back into the pair of terminals, with all energy sources (voltages and currents) properly removed. For the evaluation of the equivalent resistance (impedance), the removal of voltage sources must be with the branches in which they existed not broken, and the removal of current sources must be made with the branches open.

30. Norton's Theorem. Analogous to Thévenin's theorem is Norton's theorem, which may be stated as follows: If any two terminals emerge from a network containing voltage and/or current sources and with passive elements composed of constant linear resistances (impedances for the a-c case), the network may be considered insofar as all external circuit conditions are concerned as composed of a simple parallel circuit consisting of an ideal current source equal to the short-circuit current of the network and of a shunt conductance (admittance for the a-c case) equal to that observed in looking back into the pair of terminals with all energy sources properly removed. The removal of such sources must be done according to the same principles stated in Par. **29**, Thévenin's Theorem.

31. Literature References

WARD, ROBERT P. "Introduction to Electrical Engineering," 3d ed.; Englewood Cliffs, N.J., Prentice-Hall, Inc., 1960.

SESHU, SUNDARAM, and BALBANIAN, NORMAN. "Linear Network Analysis"; New York, John Wiley & Sons, Inc., 1959.

CASSELL, WALLACE L. "Linear Electric Circuits"; New York, John Wiley & Sons, Inc., 1964.

BRENNER, EGON, and JAVID, MANSOUR. "Analysis of Electric Circuits"; New York, McGraw-Hill Book Company, 1959.

FOECKE, HAROLD A. "Introduction to Electrical Engineering Science"; Englewood Cliffs, N.J., Prentice-Hall, Inc., 1961.

HAMMOND, S. B. "Electrical Engineering"; New York, McGraw-Hill Book Company, 1961.

SABBAGH, ELIAS M. "Circuit Analysis"; New York, The Ronald Press Company, 1961.

MANNING, LAURENCE A. "Electrical Circuits"; New York, McGraw-Hill Book Company, 1965.

NEAL, J. P. "Introduction to Electrical Engineering Theory"; New York, McGraw-Hill Book Company, 1960.

HAYT, WILLIAM H., JR., and KEMMERLY, JACK E. "Engineering Circuit Analysis"; New York, McGraw-Hill Book Company, 1962.

ELECTROMAGNETIC INDUCTION OF EMF

32. Faraday's Law of Induction. According to Faraday's law, in any closed linear path in space, when the magnetic flux ϕ (see Par. **46**) surrounded by the path varies with the time, an emf is induced around the path equal to the negative rate of change of the flux in webers per second,

$$e = -\frac{\partial \phi}{\partial t} \qquad \text{(volts)} \tag{2-28}$$

The minus sign denotes that the direction of the induced emf is such as to produce a current opposing the flux. See the right-handed-screw rule (Par. **60** and Fig. 2-8). If the flux is changing at a constant rate, the emf is numerically equal to the increase or decrease in webers in 1 s.

The closed linear path (or circuit) is the boundary of a surface and is a geometrical line, having length but infinitesimal thickness and not having branches in parallel. It can be changing in shape or position.

If a loop of wire of negligible cross section occupies the same place and has the same motion as the path just considered, the emf e will tend to drive a current of electricity around the wire and this emf can be measured by a galvanometer or voltmeter connected in the loop of wire. As with the path, the loop of wire is not to have branches in parallel; if it has, the problem of calculating the emf shown by an instrument is more complicated and involves the resistances of the branches.

For accurate results, the simple equation (2-28) cannot be applied to metallic circuits having finite cross section. In some cases, the finite conductor can be considered as being divided into a large number of filaments connected in parallel, each having its own induced emf and its own resistance. In other cases, such as the common ones of d-c generators and motors and homopolar generators, where there are sliding and moving contacts between conductors of finite cross section, the induced emf between neighboring points is to be calculated for various parts of the conductors. These can then be summed up or integrated. For methods of computing the induced emf between two points, see Par. **37** and texts on electromagnetic theory (Par. **176**).

In cases such as a d-c machine or a homopolar generator, there may at all times be a conducting path for current to flow, and this may be called a "circuit," but it is not a closed linear circuit without parallel branches and of infinitesimal cross section, and therefore Eq. (2-28) does not strictly apply to such a circuit in its entirety, even though approximately correct numerical results can sometimes be obtained.

If such a practical circuit or current path is made to enclose more magnetic flux by a process of connecting in one parallel branch conductor in place of another, then such a change in enclosed flux does not correspond to an emf according to Eq. (2-28). Although it is possible in some cases to describe a loop of wire having infinitesimal cross section and sliding contacts for which Eq. (2-28) gives correct numerical results, the equation is not reliable, without qualification, for cases of finite cross section and sliding contacts. It is advisable not to use equations involving $\partial\phi/\partial t$ directly on complete circuits where there are sliding or moving contacts.

Where there are no sliding or moving contacts, if a coil has N turns of wire in series closely wound together so that the cross section of the coil is negligible compared with the area enclosed by the coil, or if the flux is so confined within an iron core that it is enclosed by all N turns alike, the emf induced in the coil is

$$e = -N\frac{\partial\phi}{\partial t} \qquad \text{(volts)} \tag{2-29}$$

In such a case, $N\phi$ is called the number of interlinkages of lines of magnetic flux with the coil, or simply the *flux linkage*.

For the above equations, the change in flux may be due to relative motion between the coil and the magnetomotive force (mmf, the agent producing the flux), as in a rotating-field generator; it may be due to change in the reluctance of the magnetic circuit, as in an inductor-type alternator or microphone; it may be due to variations in the primary current producing the flux, as in a transformer; it may be due to variations in the current in the secondary coil itself, as in Eq. (2-30); or it may be due to change in shape or orientation of the loop or coil.

33. Self-induction. When the flux ϕ is produced by a changing current i in the loop or coil itself, the potential difference caused by the change in the flux is called the emf of self-induction. Its direction is related to the current and the emf of resistance by the equation

$$e = ri + L\frac{di}{dt} \tag{2-30}$$

where

$$L = \frac{\partial(N\phi)}{\partial i} = \text{coefficient of self-inductance} \tag{2-31}$$

(see Par. **75**).

The resistance voltage ri tends to oppose the current. For a decreasing current, di/dt is negative, and the inductance voltage $L\,di/dt$ tends to maintain the current. This is the "extra current" flowing in the arc caused by interrupting a current with a switch. The direction assumed positive for e in Eq. (2-30) is not the same as the direction assumed positive in Eq. (2-28).

34. Lenz's Law. When the flux through a secondary circuit is changed because of the relative motion of primary and secondary circuits, the direction of the induced current in the secondary is related to the mechanical force between the circuits according to Lenz's law, which is stated by Maxwell as follows: "If a constant current flows in the primary circuit A, and if, by the motion of A or of the secondary circuit B, a current is induced in B, the direction of this induced current will be such that, by its electromagnetic action on A, it tends to oppose the relative motion of the circuits" ("A Treatise on Electricity and Magnetism," by J. C. Maxwell, Vol. II, Art. 542). A generator is an example of this law; the currents induced by the relative motion of the field and armature tend to oppose the motion; that is, it requires mechanical power to maintain the rotation of the generator.

35. Sinusoidal Flux Variation. When the flux through a coil varies sinusoidally with the time, that is, according to sine law, at a frequency of f c/s, the maximum induced emf is

$$E_m = 2\pi f N \phi_m \qquad \text{(volts)} \tag{2-32}$$

where ϕ_m is the maximum instantaneous value of the flux, in webers. The effective value of the induced emf is

$$E = 0.707E_m = 4.44fN\phi_m \qquad \text{(volts)} \tag{2-33}$$

36. Average Induced EMF. For an interval of time $t_2 - t_1$, no matter what the law of variation of the flux with the time, the average emf is

$$E_{avg} = N\frac{\phi_1 - \phi_2}{t_2 - t_1} \qquad \text{(volts)} \tag{2-34}$$

37. EMF Induced in a Short Length of Straight Conductor. It is sometimes convenient to calculate the emf induced in a short section of the conductor or path. Examples are antennas, skin effect, the path of electrons in a vacuum tube, circuits containing moving contacts, and circuits or parts of circuits made up of straight conductors. A particular solution for the emf is given by the rule for motional emf,

$$e = Blv \qquad \text{(volts)} \tag{2-35}$$

where B is the flux density in webers per square meter at the location of the conductor, l is the length of the conductor in meters, and v is the relative velocity in meters per second between the flux and the conductor. The directions of B, l, and v are assumed mutually perpendicular; if they are not, their projections at right angles to each other are used in Eq. (2-35). The product Blv is equal to the number of magnetic lines of force which cut the conductor per second. This cannot always be the resultant field.

The conductor in Eq. (2-35) is assumed to be of infinitesimal cross section and so short that it can be considered straight. The magnetic field adjacent to it is considered uniform. In computing the value of v, it is customary to consider the magnetic lines of force produced by an electromagnetic coil or magnet as moving with the coil or magnet, as, for example, in the case of the rotating field of an a-c generator. The direction of the emf is given by Fleming's right-handed rule (Par. **61** and Fig. 2-9).

The voltage induced in a short length of conductor by a short element of alternating current (the conductor and the element of current being of infinitesimal cross section) may be determined by calculating the rate at which lines of force cut the conductor. As stated in Par. **41**, the lines of the magnetic field in air due to a short element of current are circles concentric on the straight line in which the short element lies. The rule can be used that, when the current dies to zero, these circular lines of force collapse, each in its own plane, and that in doing so each line cuts any conductors extending through that plane and induces voltages in them.

If the current varies as a sine wave, the total number of lines ϕ_m which cut the conductor in a quarter cycle is computed by integrating the maximum flux densities from the conductor outward. The maximum voltage is given by Eq. (2-32). If effective flux is calculated from effective current, effective volts will be obtained. These rules for calculating induced voltages in metallic conductors and the restrictions to be applied to them are subject to modification in unusual and complicated situations.

The discussion of induced voltages given in the preceding paragraphs does not include voltages produced by radiated fields. For the latter, see texts on modern electromagnetic theory (Par. **176**).

Practical formulas giving the emf induced in a-c and d-c machines are given in Secs. **6** and **7**. A special method of computation is sometimes employed when coils are moving through a pulsating field. The induced emf at any instant is taken equal to the sum of the emf induced by a constant flux in a moving coil and that induced by a pulsating flux in a stationary coil.

38. The Fundamental Circuital Laws.[1] The following summary based on electromagnetic theory has been contributed by Dr. M. S. Vallarta: "Electromagnetic theory is based on two fundamental laws, called the 'circuital laws.' First law (Ampère's law): 'The line integral of the magnetic field strength or intensity taken around any closed path is proportional to the total current flowing across any area bounded by that path.' Second law (Faraday's law): 'The line integral of the total electric field strength taken around any closed path is proportional to the negative rate of change, with respect to time, of the magnetic flux across any area bounded by that path.' In symbols, these laws are

$$\oint H\,ds = I \tag{2-36a}$$

$$\oint \mathcal{E}\,ds = -\frac{d}{dt}\int B_n\,da \tag{2-36b}$$

where B_n is the normal magnetic induction. The Maxwell theory, in addition to the foregoing laws, includes the analytical statements of the following two facts: (1) The nonexistence of separate north and south magnetic poles (the total magnetic flux across a closed surface appropriately chosen in a magnetic field is zero; in symbols $\int B_n\,da = 0$). (2) The electric charge is the source of the electric field (the total dielectric flux across any closed surface enclosing a charge is proportional to that charge; in symbols $\int D_n\,da = Q$, the enclosed charge in coulombs). These four laws constitute the Maxwell theory.

"Equation (2-36*b*) holds without restriction for constant direct currents. For alternating currents, it holds only for those points the distance of which to the element of current is small compared with the wave length. See Abraham's 'Theorie der Elektrizitat,' Vol. 1, p. 325."

39. Electromagnetic Wave Propagation Phenomena.[2] The following was contributed by Dr. E. A. Guillemin: "When the circuital laws mentioned in the foregoing paragraph are applied to nonconducting regions, one of them, namely, Ampère's circuital law [Eq. (2-36*a*)], must be modified to the extent of adding to the total conduction current I the total displacement current $d\psi/dt$ enclosed by the path in question. ψ is the electric flux or displacement and is given by an appropriate surface integral of the flux density $D = \epsilon\mathcal{E}$.

"Differential forms for these circuital-law equations are obtained through dividing both sides by the area which the path encloses and then allowing the path to become smaller and smaller. The limiting forms thus arrived at are written: (*a*) curl $\mathcal{E} = -\partial B/\partial t$ and (*b*) curl $H = J + \partial D/\partial t$, in which J, the conduction current density, is zero if the medium is a nonconductor (dielectric). The two additional relationships $\int D_n\,da = Q$ (Gauss's theorem) and $\int B_n\,da = 0$ may similarly be expressed in equivalent differential forms through dividing both sides of these equations by the enclosed volume and considering the limits of the resulting ratios as the volume becomes smaller and smaller. The limiting forms are written: (*c*) div $D = \rho$ and (*d*) div $B = 0$, in which ρ is the charge density. Equations (*a*), (*b*), (*c*), and (*d*) are Maxwell's equations in differential form.

[1] Units as originally stated by Dr. Vallarta have been converted to mksa units.
[2] Units as originally stated by Dr. Guillemin have been converted to mksa units.

"Physically, the operation denoted by curl, when applied to a vector function, yields a measure of the rotational or swirling intensity (also called 'vortex density') of the field represented by that function; the operation divergence (abbreviated div) yields a measure of the flux produced per unit volume at a point. The former is an axial vector (similar to a mechanical torque); the latter is a scalar. In rectangular coordinates, the x component of curl $\mathcal{E}$ is $\partial \mathcal{E}_z/\partial y - \partial \mathcal{E}_y/\partial z$, from which the y and z components are had through simultaneously advancing, by one and by two letters, each coordinate symbol in the cyclic order $xyzxy \ldots$, while

$$\text{div } D = \partial D_x/\partial x + \partial D_y/\partial y + \partial D_z/\partial z$$

"In a charge-free dielectric, for example in free space, the field equations yield the so-called 'wave' equations governing the behavior of the electric and magnetic field intensities. It is thus found that these fields propagate with a finite velocity

$$v = 1/\sqrt{\epsilon\mu}$$

which, for free space, is the velocity of light. In the case of sinusoidally time-varying fields of frequency f, the distance λ traveled in one period τ is called a 'wavelength.' Thus, this wavelength $\lambda = v\tau = v/f$. The propagational behavior of the fields is, however, not dependent upon the frequency and proceeds essentially unchanged even at zero frequency or for the d-c case.

"The energies stored in the fields travel with them, and this phenomenon is the basic and sole mechanism whereby electric power transmission takes place. Thus, the electrical energy transmitted by means of transmission lines flows through the space surrounding the conductors, the latter acting merely as guides. Hollow pipes, or so-called 'waveguides,' are less conventional forms of transmission lines in which the boundary conditions permitting the propagation of the fields are fulfilled only if λ is less than or of the same order of magnitude as the inside diameter. The energy radiated from a broadcast antenna propagates in the same manner along the earth's surface or into space. The fundamental mechanism of energy propagation is the same in all these applications.

"A significant point about this phenomenon is the fact that electromagnetic energy flows predominantly well through dielectrics (nonconductors). Metals are conductors for current but nonconductors for the flow of energy, while dielectrics are good conductors for the flow of energy.

"Near the surface of a transmission-line conductor, the vector representing energy flow (Poynting's vector) is slightly inclined toward the conductor surface, thus giving rise to a small component of energy flow into the conductor. This component electromagnetic wave causes the conductor current, which in turn causes a loss but does not contribute usefully to the power transmission. Since this component electromagnetic wave is directed normally to the conductor surface and at the higher frequencies attenuates rapidly as it propagates into the conductor, the associated current density is largest at the surface and drops off more or less rapidly with increase of depth—a phenomenon known as the 'skin effect.'

"The usually accepted view that the conductor current produces the magnetic field surrounding it must be displaced by the more appropriate one that the electromagnetic field surrounding the conductor produces, through a small drain on its energy supply, the current in the conductor. Although the value of the latter may be used in computing the transmitted energy, one should clearly recognize that physically this current produces only a loss and in no way has a direct part in the phenomenon of power transmission."

MAGNETIC FIELDS

40. Early Concepts of Magnetic Poles. Substances now called magnetic, such as iron, were observed centuries ago as exhibiting forces upon one another. From this beginning the concept of magnetic poles evolved, and a quantitative theory built upon the concept of these poles, or small regions of magnetic influence, was developed. André Ampère observed forces of a similar nature between conductors carrying currents.

Further developments have shown that all theories of magnetic materials can be developed and explained through the magnetic effects produced by electric charge motions.

41. Ampère's Formula. The magnetic field intensity dH produced at a point A by an element of a conductor ds (in meters) through which there is a current of i A is

$$dH = i\,ds\,(\sin \alpha / 4\pi r^2) \qquad \text{(A/m)} \tag{2-37}$$

where r is the distance between the element ds and the point A, in meters, and α is the angle between the directions of ds and r. The intensity dH is perpendicular to the plane containing ds and r, and its direction is determined by the right-handed-screw rule given in Par. **60** and Fig. 2-8.

The magnetic lines of force due to ds are concentric circles about the straight line in which ds lies. The field intensity produced at A by a closed circuit is obtained by integrating the expression for dH over the whole circuit.

42. The magnetic field due to an indefinitely long, straight conductor carrying a current of i A consists of concentric circles which lie in planes perpendicular to the axis of the conductor and have their centers on this axis. The magnetic field intensity at a distance of r m from the axis of the conductor is

$$H = \frac{i}{2\pi r} \qquad \text{(A/m)} \tag{2-38}$$

its direction being determined by the right-handed-screw rule (Par. **60**).

43. Magnetic Field in Air Due to a Closed Circular Conductor. If the conductor carrying a current of i A is bent in the form of a ring of radius r m (Fig. 2-5), the magnetic field intensity at a point along the axis at a distance b m from the ring is

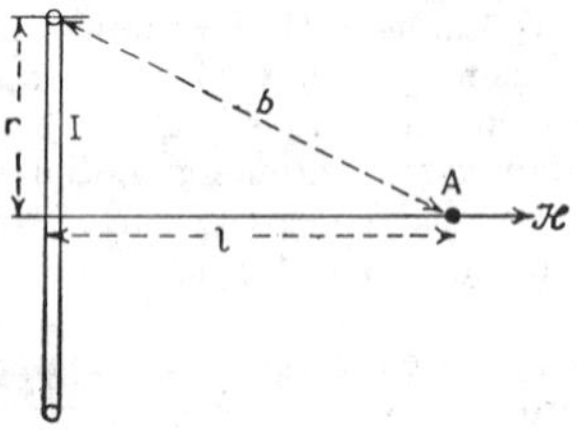

$$H = \frac{r^2 i}{2b^3} = \frac{r^2 i}{2(r^2 + l^2)^{3/2}} \qquad \text{(A/m)} \tag{2-39}$$

When $l = 0$,

$$H = \frac{i}{2r} \tag{2-40}$$

and when l is very great in comparison with r,

$$H = \frac{r^2 i}{2l^3} \tag{2-41}$$

FIG. 2-5. Magnetic field along the axis of a circular conductor.

44. The magnetic field intensity within a solenoid made in the form of a **torus ring,** and also in the middle part of a **long, straight solenoid,** is approximately

$$H = n_1 i \qquad \text{(A/m)} \tag{2-42}$$

where i is the current in amperes and n_1 is the number of turns per meter length.

45. Magnetic Flux Density. The magnetic flux density resulting in free space, or in substances not possessing magnetic behaviors differing from those in free space, is

$$B = \mu_0 H = 4\pi \times 10^{-7} H \tag{2-43}$$

where B is in teslas (or webers per square meter), H in amperes per meter, and the constant $\mu_0 = 4\pi \times 10^{-7}$ is the *permeability* of free space and has units of henrys per meter. In the so-called practical system of units the flux density is frequently expressed in *lines* or *maxwells per square inch* (see Par. **46**). The maxwell per square centimeter is called the *gauss*.

For substances such as iron, and other materials possessing magnetic density effects greater than those of free space, a term μ_r is added to the relationship as

$$B = 4\pi \times 10^{-7} \mu_r H \tag{2-44}$$

where μ_r is the relative permeability of that substance under the conditions existing in it compared with that which would result in free space under the same magnetic-field-intensity condition. μ_r is a dimensionless quantity.

46. Magnetic Flux. The magnetic flux in any cross section of magnetic field is

$$\phi = \int B \cos \alpha \, dA \qquad \text{(webers)} \tag{2-45}$$

where α is the angle between the direction of the magnetic flux density B and the normal at each point to the surface over which A is measured. In the so-called practical system of units the magnetic *line* (or *maxwell*) is frequently used, where 10^8 lines is equivalent to 1 Wb.

47. The density of magnetic energy, or the magnetic energy stored per cubic meter of a magnetic field in free space, is

$$dW/dv = \tfrac{1}{2}\mu_0 H^2 = 2\pi \times 10^{-7} H^2$$

$$= \frac{B^2}{2\mu_0} = \frac{B^2}{8\pi \times 10^{-7}} \qquad (\text{J/m}^3) \tag{2-46}$$

In magnetic materials the energy density stored in a magnetic field as a result of a change from a condition of flux density B_1 to that of B_2 can be expressed as

$$dW/dv = \int_{B_1}^{B_2} H \, dB \tag{2-47}$$

47a. Flux plotting by a graphical process is useful for determining the properties of magnetic and other fields in air. The field of flux required is usually uniform along one dimension, and a cross section of it is drawn. The field is usually required between two essentially equal magnetic potential lines, such as two iron surfaces. The field map consists of lines of force and equipotential lines, which must intersect at right angles. For the graphical method, a field map of curvilinear squares is recommended when the problem is two-dimensional. The squares are of different sizes, but the number of lines of force crossing every square is the same.

In sketching the field map, first draw those lines which can be drawn by symmetry. If parts of the two equipotential lines are straight and parallel to each other, the field map in the space between them will consist of lines which are practically straight, parallel, and equidistant. These can be drawn in. Then extend the series of curvilinear squares into other parts of the field, making sure, first, that all the angles are right angles and, second, that in each square the two diameters are equal, except in regions where the squares are evidently distorted, as near sharp corners of iron or regions occupied by current-carrying conductors. The diameters of a curvilinear square may be taken to be the distances between midpoints of opposite sides.

The magnetic field map near an iron corner is drawn as if the iron had a small fillet; that is, a line issues from an iron angle of 90° at 45° to the surfaces.

Inside a conductor which carries current, the magnetic field map is not made up of curvilinear squares, as in free space on air. In such cases special rules for the spacing of the lines must be used. The equipotential lines converge to a point called the "kernel."

The graphical method described can be used for determining other kinds of flow. Solutions of problems involving current flow and electrostatic flux can be obtained by this method.[1]

FORCE ACTING ON CONDUCTORS

48. Force on a Conductor Carrying a Current in a Magnetic Field. Let a conductor of length l m carrying a current of i A be placed in a magnetic field the density of which is B Wb/m². The force tending to move the conductor across the field is

$$F = Bli \qquad \text{(newtons)} \tag{2-48}$$

This formula presupposes that the direction of the axis of the conductor is at right

[1] See Boast, W. B. "Vector Fields"; New York, Harper & Row Publishers, Inc., 1964, chap. 23.

angles to the direction of the field. If the directions of i and B form an angle α, the expression must be multiplied by $\sin \alpha$.

The force F is perpendicular to both i and B, and its direction is determined by the right-handed-screw rule (Par. **60**). The effect of the magnetic field produced by the conductor itself is to increase the original flux density (B) on one side of the conductor and to reduce it on the other side. The conductor tends to move away from the denser field. A closed metallic circuit carrying current tends to move so as to enclose the greatest possible number of lines of magnetic force.

49. Force between Two Long, Straight Lines of Current. The force upon a unit length of either of two long, straight, parallel conductors carrying currents of i_1 and i_2 A and placed in a nonmagnetic medium (that is, not near masses of iron) is

$$\frac{F}{l} = \frac{2 \times 10^{-7} i_1 i_2}{b} \tag{2-49}$$

where F is in newtons and l (length of the long wires) and b (the spacing between them) are in the same units, such as meters. The force is an attraction or a repulsion according to whether the two currents are flowing in the same or in opposite directions. If the currents are alternating, the force is pulsating. If i_1 and i_2 are effective values, as measured by a-c ammeters, the maximum momentary value of the force may be as much as 100% greater than given by Eq. (2-49). The natural frequency (resonance) of mechanical vibration of the conductors may add still further to the maximum force, so that a factor of safety should be used in connection with Eq. (2-49) for calculating stresses on bus bars.

If the conductors are straps, as is usual in bus bars, the following form of equation results for thin straps placed parallel to each other, b m apart,

$$\frac{F}{l} = \frac{2 \times 10^{-7} i_1 i_2}{s^2}\left(2s \tan^{-1}\frac{s}{b} - b \log_\epsilon \frac{s^2 + b^2}{b^2}\right) \qquad \text{(N/m)} \tag{2-50}$$

where s is the dimension of the strap width in meters and the thickness of the straps placed side by side is presumed small with respect to the distance b between them.

50. Pinch Effect. Mechanical force exerted between the magnetic flux and a current-carrying conductor is also present within the conductor itself and is called *pinch effect*. The force between the infinitesimal filaments of the conductor is an attraction, so that a current in a conductor tends to contract the conductor. This effect is of importance in some types of electric furnaces where it limits the current that can be carried by a molten conductor. This stress also tends to elongate a liquid conductor (see also Sec. **23**).

THE MAGNETIC CIRCUIT

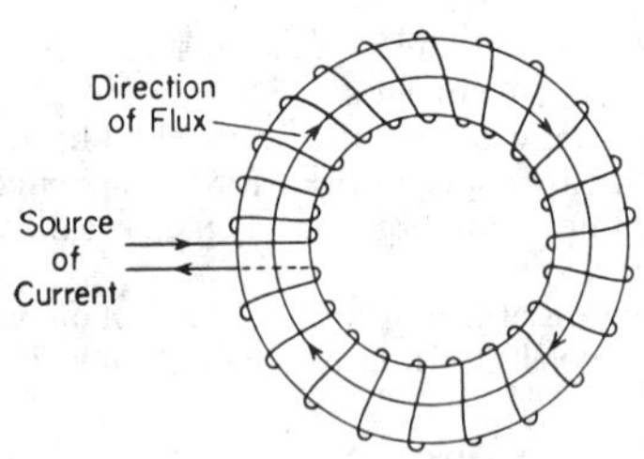

Fig. 2-6. Closed magnetic circuit.

51. The Simple Magnetic Circuit. A simple magnetic circuit is a uniformly wound torus ring (Fig. 2-6). The relation between the mmf $\mathfrak{F}$ and the flux ϕ is similar to Ohm's law (Par. **12**), viz.,

$$\mathfrak{F} = \mathcal{R}\phi \qquad \text{(At)} \tag{2-51}$$

where $\mathcal{R}$ is called the **reluctance** of the magnetic circuit. The relation is sometimes written in the form

$$\phi = \mathcal{P}\mathfrak{F} \qquad \text{(Wb)} \tag{2-52}$$

where $\mathcal{P} = 1/\mathcal{R}$ is called the **permeance** of the magnetic circuit. Reluctance is analogous to resistance, and permeance is analogous to conductance of an electric circuit.

$$\mathfrak{F} = NI \qquad \text{(At)} \tag{2-53}$$

where N is the number of turns of conductor around the magnetic circuit, as in Fig. 2-6, and I is the current in the conductor, in amperes.

52. Permeability and Reluctivity. The reluctance of a uniform magnetic path (Fig. 2-6) is proportional to its length l and inversely proportional to its cross section A,

$$\mathcal{R} = \nu \frac{l}{A} \tag{2-54}$$

and

$$\mathcal{P} = \mu \frac{A}{l} \tag{2-55}$$

In these expressions, ν is called the "reluctivity" and μ the "permeability" of the material of the magnetic path, it being assumed that there is no residual magnetism. The dimensions l and A are in metric units. For a vacuum, air, or other nonmagnetic substance, the reluctivity and permeability are usually written ν_0 and μ_0, and their values are $1/(4\pi \times 10^{-7})$ and $4\pi \times 10^{-7}$, respectively.

53. Magnetic field intensity H is defined as the mmf per unit length of path of the magnetic flux. It is known also as the magnetizing force or the magnetic potential gradient.

In a uniform field,

$$H = \frac{\mathcal{F}}{l} \qquad \text{(At/m)} \tag{2-56}$$

In a nonuniform magnetic circuit,

$$H = \frac{\partial \mathcal{F}}{\partial l} \tag{2-57}$$

Inversely, for a uniform field,

$$\mathcal{F} = Hl \tag{2-58}$$

and for a nonuniform field,

$$\mathcal{F} = \int H\, dl \tag{2-59}$$

By Ampère's law (Pars. **38** and **41**), when this integral is taken around a complete magnetic circuit,

$$\mathcal{F} = \oint H\, dl = I \tag{2-60}$$

where I is the total current, in amperes, surrounded by the magnetic circuit. The circle on the integral sign indicates integration around the complete circuit. In Eqs. (2-58) to (2-60) it is presumed that H is directed along the length of l; otherwise the factor $\cos\theta$ must be added to the product of $H\, dl$ where θ is the angle between H and dl.

54. Flux density B is the magnetic flux per unit area, the area being perpendicular to the direction of the magnetic lines of force. In a uniform field,

$$B = \frac{\phi}{A} \qquad \text{(Wb/m}^2\text{)} \tag{2-61}$$

Flux density is also commonly expressed in lines or maxwells per square inch (see Pars. **45** and **46**). Combinations of metric and practical units are also used as demonstrated in Fig. 2-7.

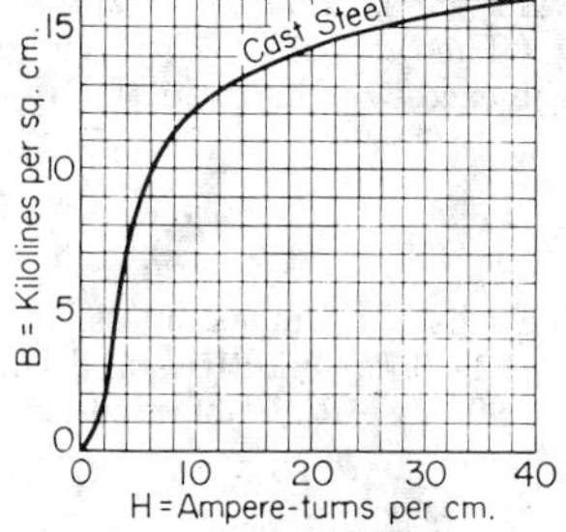

FIG. 2-7. Typical *B-H* curve.

55. Reluctances and Permeances in Series and in Parallel. Reluctances and permeances are added like resistances and conductances (Pars. **15** to **17**), respectively. That is, *reluctances are added when in series, and permeances are added when in parallel.* If several permeances are given connected in series, they are converted into reluctances by taking the reciprocal of each. If reluctances are given in a parallel combination, they are similarly converted into permeances.

56. Magnetization Characteristic or Saturation Curve. The magnetic properties of steel or iron are represented by a saturation or magnetization curve (Fig. 2-7). Magnetic field intensities H in ampere-turns per meter or in ampere-turns per centimeter, per meter, or per inch are plotted as abscissas, and the corresponding flux densities B in teslas (webers per square meter) or in kilolines per square centimeter or per square inch as ordinates.

The practical use of a magnetization curve may be best illustrated by an example. Let it be required to find the number of exciting ampere-turns for magnetizing a steel ring so as to produce in it a flux of 168 kilolines. Let the cross section of the ring be 3 by 4 cm and the mean diameter 46 cm. Let the quality of the material be represented by the curve in Fig. 2-7. The flux density is $168/(3 \times 4) = 14$ kilolines/cm². For this flux density, the corresponding abscissa from the curve is about 18 At/cm. The total required number of ampere-turns is then $18 \times \pi \times 46 = 2{,}600$.

For curves of various grades of steel and iron, see Sec. **4**. The principal methods for experimentally obtaining magnetization curves will be found in Sec. **3**.

57. Ampere-turns for an Air Gap. In a magnetic circuit consisting of iron with one or more small air gaps in series with the iron, the magnetic flux density in each of the air gaps may be considered approximately uniform. If the length across a given air gap in the direction of the flux is l m, the ampere-turns required for that air gap is given by the equation

$$\text{T(Wb/m}^2) = 4\pi \times 10^{-7}\ \text{At/m} \tag{2-62}$$

or

$$\text{At/cm} = 0.7958 \times \text{lines/cm}^2 \tag{2-63}$$

or

$$\text{At/in} = 0.3133 \times \text{lines/in}^2 \tag{2-64}$$

The ampere-turns for each portion of iron, computed from iron magnetization curves such as Fig. 2-7, and the ampere-turns for the air gaps are added together to give the ampere-turns for the complete magnetic circuit.

58. Analysis of Magnetization Curve. Three parts are distinguished in a magnetization curve (Fig. 2-7): the lower, or nearly straight, part; the middle part, called the knee of the curve; and the upper part, which is nearly a straight line. As the magnetic intensity increases, the corresponding flux density increases more and more slowly, and the iron is said to approach saturation (see Sec. **4**).

59. Magnetization per Unit Volume and Susceptibility. If a portion of ferromagnetic material is magnetized by an mmf, H At/m, the resulting flux density in teslas may be written as

$$B = \mu_0(H + M) \tag{2-65}$$

where M is the magnetization per unit volume of the material (see Sec. **4**).

The ratio of M/H is symbolized by χ and is called the *magnetic susceptibility*. It is the excess of the ratio of $B/\mu_0 H$ above unity, that is,

$$\chi = \frac{B}{\mu_0 H} - 1 \tag{2-66}$$

This is a dimensionless quantity. See American Standards Association (ASA) C61.1-1961, Quantities and Units Used in Electricity.

60. The Right-handed-screw Rule. The direction of the flux produced by a given current is determined as shown in Fig. 2-8 (see also Fig. 2-6). If the current is established in the direction of rotation of a right-handed screw, the flux is in the direction of the progressive movement of the screw. If the current in a straight conductor is in the direction of the progressive motion of a right-handed screw, then the flux encircles this conductor in the direction

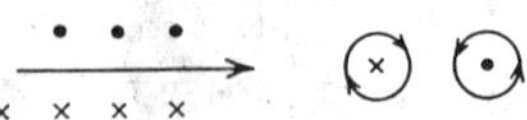

FIG. 2-8. Relation between directions of current and flux.

in which the screw must be rotated in order to produce this motion. The dots in the figure indicate the direction of flux or current toward the reader; the crosses, that away from him.

61. Fleming's Rules. The relative direction of flux, emf, and motion in a revolving-armature generator may be determined with the right hand by placing the thumb, index, and middle fingers so as to form the three axes of a coordinate system and pointing the index finger in the direction of the flux (north to south) and the thumb in the direction of motion; the middle finger will give the direction of the generated emf (Fig. 2-9). In the same way, in a revolving-armature motor, by using the left hand and pointing the index finger in the direction of the flux and the middle finger in the direction of the current in the armature conductor, the thumb will indicate the direction of the force and, therefore, the resulting motion. These two rules, indicated in Fig. 2-9, are known as Fleming's rules.

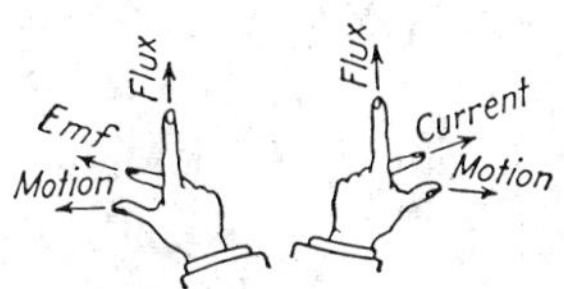

FIG. 2-9. Fleming's generator and motor rules.

62. Magnetic Tractive Force. The carrying weight of a lifting magnet is

$$F = \frac{1}{2}\frac{AB^2}{\mu_0} = \frac{AB^2}{8\pi \times 10^{-7}} \qquad \text{(newtons)} \qquad (2\text{-}67)$$

where B is the flux density in the air gap expressed in teslas (webers per square meter) and A is the total area of the contact between the armature and the core, in square meters.

If the area (A') and flux density (B') are expressed in square centimeters and lines per square centimeter, respectively, the force (F') in pounds is

$$F' = \frac{(A')(B')^2}{11.18} \times 10^{-6} \qquad \text{(lb)} \qquad (2\text{-}68)$$

See also Sec. **5**.

63. Magnetic Force, or Torque. The mechanical force, or the torque, between two parts of a magnetic or electric circuit may in some cases be conveniently calculated by making use of the principle of **virtual displacements.** An infinitesimal displacement between the two parts is assumed. The energy supplied from the source of current is then equal to the mechanical energy for producing the motion, plus the change in the stored magnetic energy, plus the energy for resistance loss.

When the differential motion ds m of a part of a circuit carrying a current I A changes its self-inductance by a differential dL H, the mechanical force on that part of the circuit, in the direction of the motion, is

$$F = \frac{1}{2} I^2 \frac{dL}{ds} \qquad \text{(newtons)} \qquad (2\text{-}69)$$

When the motion of one coil or circuit carrying a current I_1 A changes its mutual inductance by a differential dM H with respect to another coil or circuit carrying a current I_2 A, the mechanical force on each coil or circuit, in the direction of the motion, is

$$F = I_1 I_2 \frac{dM}{ds} \qquad \text{(newtons)} \qquad (2\text{-}70)$$

where ds represents the differential of distance in meters. For a discussion of self- and mutual inductance L and M see Par. **75**.

HYSTERESIS AND EDDY CURRENTS IN IRON

64. The Hysteresis Loop. When a sample of iron or steel is subjected to an alternating magnetization, the relation between B and H is different for increasing and

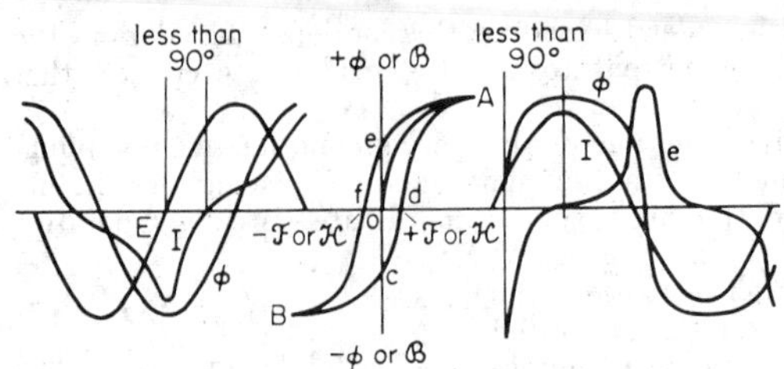

Fig. 2-10. Periodic waves of current, flux, and emf; hysteresis loop.

decreasing values of the magnetic intensity (Fig. 2-10). This phenomenon is due to irreversible processes which result in energy dissipation, producing heat. Each time the current wave completes a cycle, the magnetic flux wave must also complete a cycle, and the elementary magnets are turned. The figure $AefBcdA$ in Fig. 2-10 is called the *hysteresis loop.*

65. Retentivity. If the coil shown in Fig. 2-6 is excited with alternating current, the ampere-turns and consequently the mmf will, at any instant, be proportional to the instantaneous value of the exciting current. Plotting a B-H (or ϕ-$\mathfrak{F}$) curve (Fig. 2-10) for one cycle, the closed loop $AefBcdA$ is obtained. The first time the iron is magnetized, the **virgin,** or **neutral, curve** OA will be produced; but it cannot be produced in the reverse direction AO, because when the mmf drops to zero there will always be some remaining magnetism ($+Oe$ or $-Oc$). This is called **residual magnetism;** to reduce this to zero, an mmf ($-Of$ or $+Od$) of opposite polarity must be applied. This mmf is called the **coercive force.**

66. Wave Distortion. In Fig. 2-10, the instantaneous values of the exciting current I (which is directly proportional to the mmf) and the corresponding values of the flux ϕ and voltage E (or e) are plotted against time as abscissas, beside the hysteresis loop. (*a*) If the voltage applied to the coil is *sinusoidal* (E, to the left), the current wave is distorted and displaced from the corresponding sinusoidal flux wave. The latter wave is in quadrature with the voltage wave. (*b*) If the *current* through the coil is *sinusoidal* (I, to the right), the flux is distorted into a flat-top wave and the induced voltage e is peaked.

67. Components of Exciting Current. The alternating current that flows in the exciting coil (Fig. 2-10) may be considered to consist of two components, one exciting magnetism in the iron, and the other supplying the iron loss. For practical purposes, both components may be replaced by equivalent sine waves and phasors (Fig. 2-11) (see Pars. **141** to **145**). We have

$$I_r = I \cos\theta = \text{power component of current}$$

$$P_h = IE \cos\theta = I_r E = \text{iron loss} \qquad \text{(watts)} \qquad (2\text{-}71)$$

$$I_m = I \sin\theta = \text{magnetizing current}$$

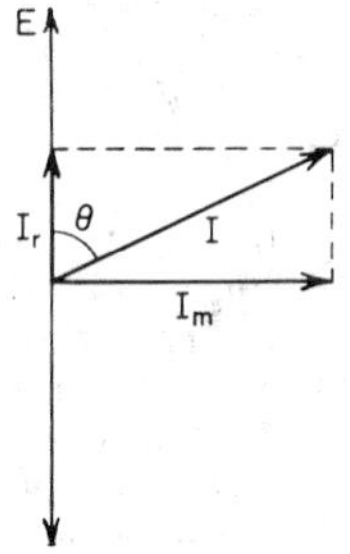

Fig. 2-11. Components of exciting current; hysteretic angle.

where I is the total exciting current and θ the angle of time-phase displacement between current and voltage.

68. Hysteretic Angle. Without iron loss, the current I would be in phase quadrature with E. For this reason, the angle $\alpha = 90 - \theta$ is called the **angle of hysteretic advance of phase.**

$$\sin\alpha = \frac{I_r}{I} = \frac{I_r E}{IE} = \frac{\text{W loss}}{\text{VA}} \qquad (2\text{-}72)$$

In practice, the measured loss usually includes eddy currents (Par. **71**), so that the name "hysteretic" is somewhat of a misnomer.

69. The energy lost per cycle from hysteresis is proportional to the area of the hysteresis loop (Fig. 2-10). This is a consequence of the evaluation over a cycle of Eq. (2-47).

70. Steinmetz's Formula. According to experiments by C. P. Steinmetz, the heat energy due to hysteresis released per cycle per unit volume of iron is approximately

$$W_h = \eta B_{max}^{1.6} \qquad (2\text{-}73)$$

The exponent of B_{max} varies between 1.4 and 1.8 but is generally taken as 1.6. Values of the hysteresis coefficient η are given in Sec. **4**.

71. Eddy-current losses are I^2R losses due to secondary currents (Foucault currents) established in those parts of the circuit which are interlinked with alternating or pulsating flux.

Referring to Fig. 2-12, which shows a cross section of a transformer core, the primary current I produces the alternating flux ϕ, which by its change generates an emf e in the core; this emf then sets up the secondary current i. Now, if the core is divided into two (b), four (c), or n parts, the emf in each circuit is $e/2$, $e/4$, e/n; and the conductance, $g/2$, $g/4$, g/n, respectively. Thus, the loss per lamination will be $1/n^3$ times the loss in the solid core, and the total loss is $1/n^2$ times the loss in the solid core.

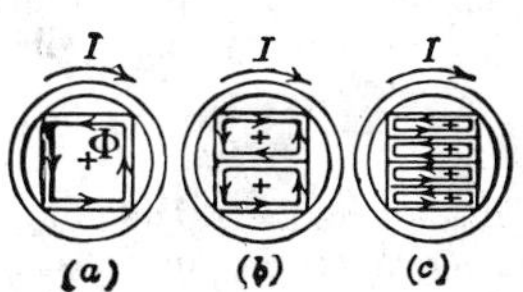

FIG. 2-12. Section of transformer core.

Eddy currents can be greatly reduced by laminating the circuit, i.e., by making it up of thin sheets each insulated from the others. The same purpose is accomplished by using separately insulated strands of conductors or bundles of wires.

A formula for the eddy loss in conductors of circular section, such as wire, is

$$P_e = \frac{(\pi r f B_{max})^2}{4\rho \times 10^{16}} \qquad (\text{W/cm}^3) \qquad (2\text{-}74)$$

where r is the radius of the wire in centimeters; f, the frequency in hertz (cycles per second); B_{max}, the maximum flux density in lines per square centimeter; and ρ, the specific resistance in ohm-centimeters.

A formula for the loss in sheets is

$$P_e = \frac{(\pi t f B_{max})^2}{6\rho \times 10^{16}} \qquad (\text{W/cm}^3) \qquad (2\text{-}75)$$

where t is the thickness in centimeters; f, the frequency in hertz (cycles per second); B_{max}, the maximum flux density in lines per square centimeter; and ρ, the specific resistance.

The specific resistance of various materials is given in Sec. **4**.

72. Effective Resistance and Reactance. When an a-c circuit has appreciable hysteresis, eddy currents, and skin effect (Par. **86**), it can be replaced by a circuit of equivalent resistances and equivalent reactances (Par. **146**) in place of the actual ones. These effective quantities are so chosen that the energy relations are the same in the equivalent circuit as in the actual one. In a series circuit, let the true power lost in ohmic resistance, hysteresis, and eddy currents be P; and the reactive (wattless) volt-amperes, Q. Then the effective resistance and reactance are determined from the relations

$$i^2 r_{eff} = P \qquad i^2 x_{eff} = Q \qquad (2\text{-}76)$$

In a parallel circuit, with a given voltage, the equivalent conductances and susceptance (Par. **147**) are calculated from the relations

$$e^2 g_{eff} = P \qquad e^2 b_{eff} = Q \qquad (2\text{-}77)$$

Such equivalent electric quantities, which replace the core loss, are used in the analytical theory of transformers and induction motors.

73. Core Loss. In practical calculations of electrical machinery, the total core loss is of interest rather than the hysteresis and the eddy currents separately. For such computations, empirical curves are used, obtained from tests on various grades of steel and iron (see Sec. **4**).

74. Separation of Hysteresis Losses from Eddy-current Losses. For a given sample of laminations, the total core loss P, at a constant flux density and at variable frequency f, can be represented in the form

$$P = af + bf^2 \qquad (2\text{-}78)$$

where af represents the hysteresis loss and bf^2 the eddy, or Foucault-current, loss, a and b being constants. The voltage waveform should be very close to a sine wave. If we write this equation for two known frequencies, two simultaneous equations are obtained from which a and b are determined.

It is convenient to divide the foregoing equation by f, because the form

$$P/f = a + bf \qquad (2\text{-}79)$$

represents a straight line relating P/f and f. Known values of P/f are plotted against f as abscissas, and a straight line having the closest approximation to the points is drawn. The intersection of this line with the axis of ordinates gives a; b is calculated from the preceding equation. The separate losses are calculated at any desired frequency from af and bf^2, respectively.

INDUCTANCE AND REACTANCE

75. The electromagnetic inductance, or the **coefficient of self-induction,** L is defined from either of the following fundamental equations:

$$e = L\frac{di}{dt} \qquad (2\text{-}80)$$

$$W = \tfrac{1}{2}i^2L \qquad (2\text{-}81)$$

The first equation expresses the fact that the self-induced voltage e is proportional to the time rate of change of the current i in the circuit, and L, the coefficient of proportionality, is the self-inductance. In the second equation, the magnetic energy W stored in the circuit is proportional to the square of the current i, where L is the coefficient of proportionality. The units in the above equations are e volts, i amperes, t seconds, W joules, and L henrys.

76. Torus Ring or Toroidal Coil of Rectangular Section with Nonmagnetic Core (Fig. 2-6). The inductance of a rectangular toroidal coil, uniformly wound with a single layer of fine wire, is

$$L = 2 \times 10^{-7}N^2b\left(\ln\frac{r_2}{r_1}\right) \quad \text{(henrys)} \qquad (2\text{-}82)$$

where N equals the number of turns of wire on the coil; b, the axial length of the coil, in meters; and r_2 and r_1 are the outer and inner radial distances in meters.

77. Torus Ring or Toroidal Coil of Circular Section with Nonmagnetic Core (Fig. 2-6). A toroidal coil of circular section, uniformly wound with a single layer of fine wire of N turns, has an inductance of

$$L = 4\pi \times 10^{-7}N^2(g - \sqrt{g^2 - a^2}) \quad \text{(henrys)} \qquad (2\text{-}83)$$

where g is the mean radius of the toroidal ring and a is the radius of the circular cross section of the core, both measured in meters.

78. Inductance of a Very Long Solenoid. A solenoid uniformly wound in a single layer of fine wire possesses an inductance of

$$L = \frac{4\pi^2 \times 10^{-7}N^2R^2}{S} \quad \text{(henrys)} \qquad (2\text{-}84)$$

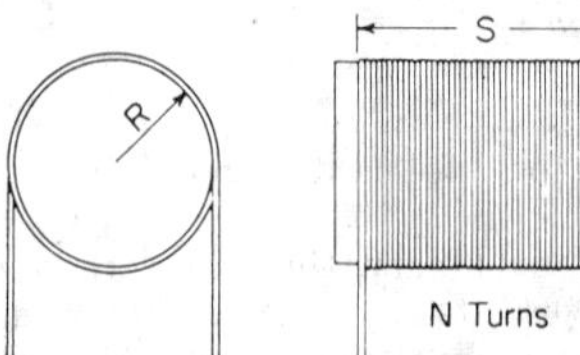

Fig. 2-13. Cylindrical solenoid.

where R and S are the radius and length of the solenoid in meters, as illustrated in Fig. 2-13. The assumption is made that S is very large with respect to R.

79. Inductance of the Finite Solenoid. The inductance of a short solenoid is less than that given by Eq. (2-84), by a factor k (a dimensionless quantity). The inductance relation then is

$$L = k\frac{4\pi^2 \times 10^{-7}N^2R^2}{S} \quad \text{(henrys)} \qquad (2\text{-}85)$$

where the values of k for various ratios of R and S are given in Fig. 2-14.

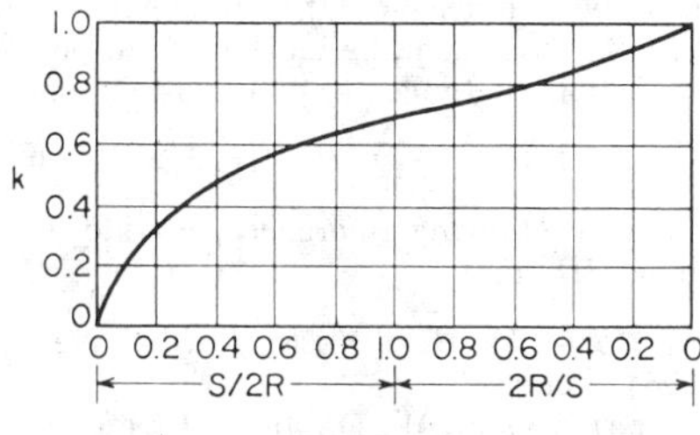

FIG. 2-14. Factor k of Eq. (2-85). (*From W. B. Boast, "Vector Fields," New York, Harper & Row, Publishers, Incorporated, 1964.*)

Inductance relations for other configurations of coils are given in W. B. Boast, "Vector Fields," New York, Harper & Row, Publishers, Incorporated, 1964, Chaps. 17 and 18.

80. Inductance per Unit Length of a Coaxial Cable. For low-frequency applications, where skin effect is not predominant, (uniform current density over nonmagnetic current-carrying cross sections), the inductance per unit length of a coaxial cable is

$$l = \frac{10^{-7}}{2}\left[1 + 4\ln\frac{R_2}{R_1} + \frac{4R_3^4}{(R_3^2 - R_2^2)^2}\ln\frac{R_3}{R_2} - \frac{3R_3^2 - R_2^2}{R_3^2 - R_2^2}\right] \quad \text{(H/m)} \qquad (2\text{-}86)$$

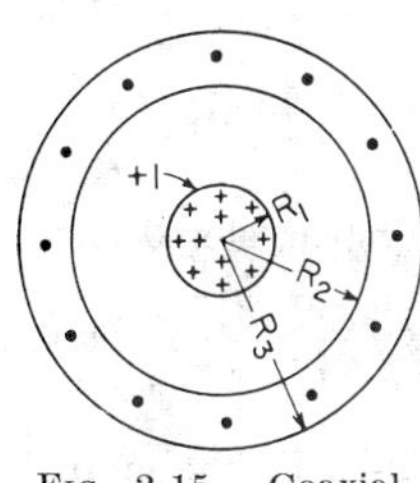

FIG. 2-15. Coaxial cable.

where R_1, R_2, and R_3 are the radii of the inner conductor, the inner radius of the outer conductor, and the outer radius of the outer conductor, in meters respectively, as shown in Fig. 2-15. For very thin outer shells the last two terms drop out of the equation, and for very small inner conductors the first term becomes less important. For high-frequency applications the first, third, and fourth terms are all suppressed, and for the extreme situation where all the current is essentially at the boundaries formed by R_1 and R_2, respectively, the inductance per unit length becomes

$$l = 2 \times 10^{-7}\ln(R_2/R_1) \quad \text{(H/m)} \qquad (2\text{-}87)$$

81. Inductance of Two Long, Cylindrical Conductors, Parallel and External to Each Other. The inductance per unit length of two separate parallel conductors is

$$l = 10^{-7}\left(1 + 4\ln\frac{D}{\sqrt{R_1R_2}}\right) \quad \text{(H/m)} \qquad (2\text{-}88)$$

where D is the distance between centers of the two cylinders and R_1 and R_2 are the radii of the conductor cross sections.

If $R_1 = R_2 = R$ and the skin-effect phenomenon applies as at very high frequencies, the inductance per unit length becomes

$$l = 4 \times 10^{-7}\ln(D/R) \quad \text{(H/m)} \qquad (2\text{-}89)$$

82. Inductance of Transmission Lines. The inductance relationships used in predicting the performance of power-transmission systems often involve the effects of stranded and bundled conductors operating in parallel, as well as configurations of these groups of current-carrying elements of one phase group of the system coordinated with similar groups constituting other phases, in polyphase systems. In such systems the several current-carrying elements of a phase are considered mathematically as a cylindrical shell of current of radius D_s (meters) called the *self-geometric mean radius* of the phase; and the mutual distances [between the current in a particular phase and the other (return) currents in the other phases] are replaced by a distance D_m (meters) called the *mutual geometric mean distance* to the return. The inductance of all phases may be balanced by transposing the conductors over the length of the transmission line, so that each phase occupies all positions equally in the length of the line.

The inductance per phase is then one-half as large as that of Eq. (2-89), that is,

$$l = 2 \times 10^{-7} \ln (D_m/D_s) \qquad \text{(H/m)} \tag{2-90}$$

The following references provide methods for computing the geometric mean distances D_m and D_s:

Boast, W. B. "Vector Fields"; New York, Harper & Row, Publishers, Incorporated, 1964, Chap. 19.

Stevenson, W. D., Jr. "Elements of Power System Analysis," 2d ed.; New York, McGraw-Hill Book Company, 1962, Chap. 2.

Clarke, E. "Circuit Analysis of A-C Power Systems"; New York, John Wiley & Sons, Inc., 1943, Vol. I.

"Electrical Transmission and Distribution Reference Book," 4th ed.; Westinghouse Electric Corporation, 1950.

83. Mutual Inductance. When two independent circuits or coils 1 and 2 are in proximity to each other, a change in current in one is accompanied by a change in its magnetic field, which induces an emf in the other. Thus

$$e_1 = M \frac{di_2}{dt} \qquad \text{and} \qquad e_2 = M \frac{di_1}{dt} \tag{2-91}$$

M is measured in henrys when i is in amperes and e in volts.

When there is no magnetic material, the total energy of the system, stored in the magnetic field, is

$$W = \tfrac{1}{2} L_1 i_1^2 + \tfrac{1}{2} L_2 i_2^2 + M i_1 i_2 \tag{2-92}$$

where L_1 and L_2 are the self-inductances in henrys.

For interference between transmission and telephone lines caused by the mutual-inductance coupling between them, and for transpositions of telephone lines, see Sec. **24**.

For effects of mutual coupling between circuits in power-transmission systems, see Sec. **13**.

84. Leakage Inductance. In electrical apparatus, such as transformers, generators, and motors, in which the greater part of the flux is carried by an iron core, the difference between self-inductance and mutual inductance of the primary and secondary windings is small. This small difference is called "leakage inductance." It is of great importance in the characteristics and operation of the apparatus and is usually calculated or measured separately. The loss in voltage in such apparatus, due to inductance, is associated with the leakage.

85. Magnetizing Current. The mutual inductance of the windings of apparatus with iron cores is not usually stated in henrys, but the effective alternating current required to produce the flux is stated in amperes and is called the **exciting current.** One component of this current supplies the energy corresponding to the core loss (Pars. **67** and **73**). The remaining component is called **magnetizing current.** Solenoids and other coils with only one winding are usually treated in a similar manner when they have iron cores. The exciting current usually does not have a sine-wave form. See Par. **66** and Fig. 2-10.

SKIN EFFECT

86. Real, or ohmic, resistance is the resistance offered by the conductor to the passage of electricity. Although the specific resistance is the same for either alternating or continuous current, the total resistance of a wire is greater for alternating than for continuous current. This is due to the fact that there are induced emfs in a conductor in which there is alternating flux. These emfs are greater at the center than at the circumference, so that the potential difference tends to establish currents that oppose the current at the center and assist it at the circumference. The current is thus forced to the outside of the conductor, thus reducing the effective area of the conductor. This phenomenon is called **skin effect.**

87. Skin-effect Resistance Ratio. The ratio of the a-c resistance to the d-c resistance is a function of the cross-sectional shape of the conductor and its magnetic and electrical properties as well as of the frequency. For cylindrical cross sections with presumed constant values of relative permeability μ_r and resistivity ρ, the function that determines the skin-effect ratio is

$$mr = \sqrt{\frac{8\pi^2 \times 10^{-7} f \mu_r}{\rho}}\, r \qquad (2\text{-}93)$$

where r is the radius of the conductor and f is the frequency of the alternating current. The ratio of R, the a-c resistance, to R_0, the d-c resistance, is shown as a function of mr in Fig. 2-16.

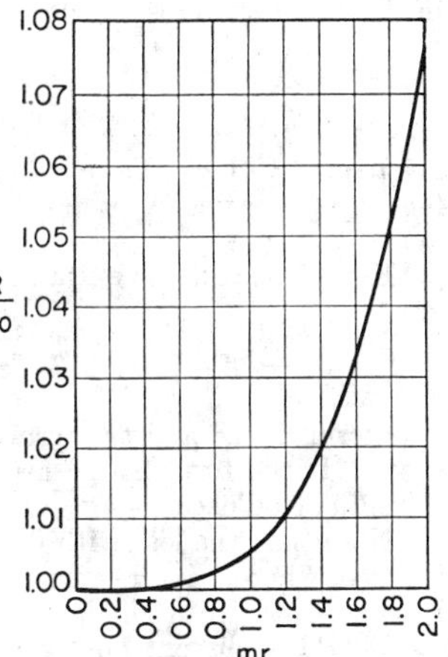

FIG. 2-16. Ratio of a-c to d-c resistance of a cylindrical conductor. (*From W. D. Stevenson, Jr., "Elements of Power System Analysis," New York, McGraw-Hill Book Company*, 1962.)

88. The skin effect of steel wires and cables cannot be accurately calculated by assuming a constant value of the permeability, which varies throughout a large range during every cycle. Therefore curves of measured characteristics should be used. See "Transmission Line Formulas," by H. B. Dwight, p. 77, and Tables 24 and 25, p. 204.

89. Skin Effect of Tubular Conductors. Cables of large size are often made so as to be, in effect, round, tubular conductors. Their effective resistance due to skin effect may be taken from the curves of Sec. **13**; or from Chap. 25, "Electrical Coils and Conductors," by H. B. Dwight; or from the paper by A. W. Ewan, *Gen. Elec. Rev.*, April, 1930, p. 250. The effective resistance may be calculated as described in Chap. 20 of the foregoing book.

The skin-effect ratio of square, tubular bus bars may be obtained from semiempirical formulas in the paper A-C Resistance of Hollow, Square Conductors, by A. H. M. Arnold, *Jour. IEE* (*London*), 1938, Vol. 82, p. 537. These formulas have been compared with tests. The resistance ratio of square tubes is somewhat larger than that of round tubes. Values may be read from the curves of Fig. 4, Chap. 25, of "Electrical Coils and Conductors."

90. Penetration Formula. For wires and tubes (and approximately for other compact shapes) where the resistance ratio is comparatively large, the conductor can be approximately considered to be replaced by its outer shell, of thickness equal to the "penetration depth," given by

$$\delta = \frac{1}{2\pi}\sqrt{\frac{10^7 \rho}{f \mu_r}} \quad \text{(meters)} \qquad (2\text{-}94)$$

where ρ is the resistivity in ohm-meters and μ_r is a presumed constant value of relative permeability. The resistance of the shell is then the effective resistance of the conductor. For Eq. (2-94) to be applicable, δ should be small compared with the dimensions of the cross section. In the case of tubes, δ is, evidently, less than the thickness of the tube. See Eq. (30), Chap. 19, of the book "Electrical Coils and Conductors."

EDDY-CURRENT LOSS IN TRANSFORMER WINDINGS

91. Eddy-current loss in a winding of rectangular wires in series, at power frequencies. The eddy-current loss in the primary winding of a transformer is computed separately from that in the secondary. The main eddy-current loss is that produced by the leakage magnetic flux in the copper winding due to the load current. This is determined from the quantity D, given in Eq. (2-95). The loss thus depends on the length of the leakage flux path s, in inches, the net extent of copper in the direction of s, w, in inches, the number q of the rectangular wires counted across the leakage flux (from zero to maximum flux density), the height of the wires h, in inches, measured in the direction across the leakage flux, and the frequency f, in hertz (cycles per second).

$$D = \left(\frac{h^2fw}{8.23s}\right)^2 \qquad \text{(at 75°C)} \tag{2-95}$$

See paper by H. W. Taylor, *Jour. IEE* (*London*), 1920, Vol. 58, p. 279. The numerical factor 8.23 varies in proportion to $(234.4° + T°)$ for copper of 100% conductivity, at T°C.

Extra loss ratio or **extra resistance ratio**

$$\frac{R_{ac} - R_{dc}}{R_{dc}} = q^2\left(\frac{D}{9} - \frac{17}{3{,}780}D^2 \cdots\right) - \frac{D}{45} + \frac{D^2}{900} \cdots \tag{2-96}$$

The terms in D^2 and the terms that do not involve q are usually negligible in practical cases. If the terms in D^2 are not negligible, the series does not converge rapidly and should not be used.

Note that the loss with direct current is I^2R_{dc} and the loss with alternating current is I^2R_{ac}, where I is the current in amperes in the winding. The ratio of change in loss is equal to the ratio of change in resistance.

92. Eddy-current Loss in Laminated Conductors, Thoroughly Transposed. If the copper laminations are thoroughly transposed so that each lamination can be taken to have the same number of amperes as every other lamination, then the magnetic field and current-distribution conditions are the same as if all the laminations were in series. Equations (2-95) and (2-96) are used, q being the total number of laminations across the direction of leakage magnetic flux, from zero to maximum flux density, in either the primary or the secondary, whichever is being computed. The dimension h is in this case the height of one copper lamination, in the direction across the leakage flux.

EDDY-CURRENT LOSS IN ARMATURE COILS

93. Eddy-current Loss in Coils of Rectangular Wires in Series, at Power Frequencies. The phase angle between the currents of the upper half and the lower half of a slot in a 3-phase machine (except in mesh-wound machines) is usually either 0° or 60°. In a certain proportion of the slots, ϕ is 0°; and in the remainder, 60°. The value of extra loss ratio is to be computed for each case.

Let the length of leakage flux path (that is, the width of the slot) be s in; the net extent of copper in the direction of s be w in; q be the number of rectangular wires one above the other in the half slot, connected in series, each of height h in; and the frequency be f hertz (c/s).

Find D by Eq. (2-95) adjusted for the operating temperature. The extra loss ratio for the lower half of the slot-embedded part of coil is then

$$\frac{R_{ac} - R_{dc}}{R_{dc}} = D\left(\frac{q^2}{9} - \frac{1}{45}\right) \tag{2-97}$$

The extra loss ratio for the upper half of the slot-embedded part of coil is

$$\frac{R_{ac} - R_{dc}}{R_{dc}} = D\left(\frac{7}{9}q^2 - \frac{1}{45} - \frac{2}{3}q^2\sin^2\frac{\phi}{2}\right) \tag{2-98}$$

where ϕ is 0° or 60°, as previously described.

The average of the foregoing gives the extra loss ratio for the complete slot-embedded part of the coil,

$$\frac{R_{ac} - R_{dc}}{R_{dc}} = D\left(\frac{4q^2}{9} - \frac{1}{45} - \frac{q^2}{3}\sin^2\frac{\phi}{2}\right) \tag{2-99}$$

where q is the number of rectangular wires in the *half* slot, one above the other.

Terms in D^2 have been omitted. It is to be noted that the amount of insulation between the wires does not affect the loss ratio, in this case. For wires in series, it is usually assumed that the eddy-current loss is negligible in the part of the coil not embedded in the slot.

94. Eddy-current Loss in Laminated Conductors, Thoroughly Transposed. If the copper laminations are thoroughly transposed as, for instance, in the Punga and Roebel types of transposed conductors illustrated in Sec. **6**, so that each lamination can be taken to have the same number of amperes as every other lamination, then (as stated in Par. **92**) the current-distribution conditions are the same as if all the laminations were in series. Equations (2-97) to (2-99) are used, q being the number of copper laminations, one above the other, in the half slot and h being the height of one lamination. The loss in the nonembedded part is taken to be negligible.

95. Eddy-current Loss in Laminated Conductors, Not Thoroughly Transposed, Laminations Being Soldered Together at Beginning and End of Coil. In this case, circulating currents flow along the upper laminations and return along the lower ones, thus producing extra copper loss in the entire coil, in both the embedded and the nonembedded parts. This extra loss, the **long-path eddy-current loss,** is in addition to the **short-path eddy-current loss** [Eqs. (2-97) to (2-99)], which occurs in the embedded part of the coil.

In calculating the long-path eddy-current loss, any laminated conductor may be called right side up, denoted by D (direct).

Note that, in this discussion, "laminations" are connected in parallel but "conductors" are connected in series in the coil.

For a coil denoted by D, let

$$I_0 = +I_b \tag{2-100}$$

where I_b is the vector sum of all the current below it in the same slot.

For conductors wrong side up, denoted by R (reversed),

$$I_0 = -I - I_b \tag{2-101}$$

that is, (-1) times all the current in the slot below the double line of the conductor considered (see Fig. 2-17), added vectorially, where I is the current in the conductor considered and I_b is the vector sum of all the current below it.

The twisting of the laminated conductors, by which D can be changed to R as desired (Fig. 2-17), is illustrated in Reduction of Armature Copper Losses, by I. H. Summers, *Trans. AIEE*, 1927, p. 102.

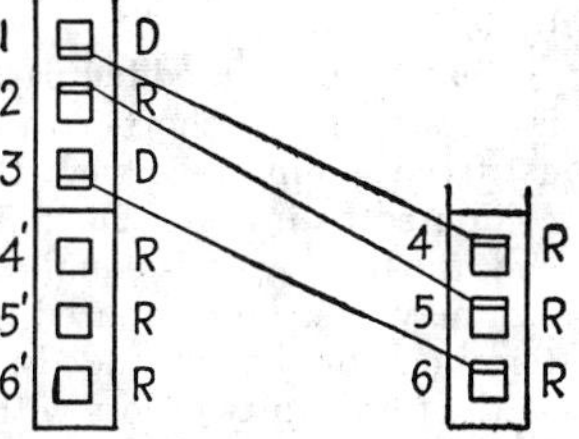

Fig. 2-17. Armature coils with laminated conductors.

The extra loss ratio due to long-path eddy current in a coil with any arrangement of direct and reversed laminated conductors is

$$\frac{R_{ac} - R_{dc}}{R_{dc}} = b^2c^2 \left[M_r - 1 + \left(\left| \frac{I_0}{I} \right|^2 + \left| \frac{I_0}{I} \right| \cos \delta \right) N_r \right] \tag{2-102}$$

where the average value of I_0/I is used. The angle δ is the phase angle between I_0 and I. The ratio of the embedded part to the total is b; that is, b equals the core length divided by one-half the mean length of the coil turns. The ratio of gross height of the laminated conductor to the net copper height of the same conductor is c. The ratio b is less than 1; c is greater than 1.

In Eq. (2-102), M_r and N_r are given by

$$M_r - 1 = \frac{4}{45} Dn^4 - \frac{16}{4{,}725} D^2n^8 \cdots \tag{2-103}$$

$$N_r = \frac{D_n^4}{3} - \frac{17}{1{,}260} D^2n^8 \cdots \tag{2-104}$$

where D is computed for one lamination by Eq. (2-95) and n is the number of laminations per conductor. If the terms in D^2 are not negligible, the series does not converge rapidly and should not be used.

See "Heat Losses in the Conductors of A-C Machines," by Waldo V. Lyon, *Trans. AIEE*, 1921, p. 1378.

96. Normal Diamond Type of Coil without Special Twisting and with Upper and Lower Currents in Phase.

Arrangement $DD\cdots RR\cdots$ (Fig. 2-17).

Extra loss ratio due to long-path eddy current

$$\frac{R_{ac} - R_{dc}}{R_{dc}} = b^2c^2\left[\frac{4}{45}Dn^4 - \frac{16}{4{,}725}D^2n^8\cdots + \frac{q^2-1}{4}\left(\frac{Dn^4}{3} - \frac{17}{1{,}260}D^2n^8\cdots\right)\right] \qquad (2\text{-}105)$$

Above the coil, with upper and lower currents 60° out of phase, use expression (2-105).

97. Diamond Coil with Twisted Conductors, with an Even Number of Turns per Coil (q Even). Use the arrangement $DRDR\cdots RDRD\cdots$ (Fig. 2-17). This can be obtained by twisting each conductor, turning it upside down, as it passes the end of the coil away from the terminals. The extra loss ratio due to long-path eddy current is the same whether ϕ is 0° or 60° and is

$$\frac{R_{ac} - R_{dc}}{R_{dc}} = b^2c^2\left(\frac{Dn^4}{180} - \frac{D^2n^8}{75{,}600}\cdots\right) \qquad (2\text{-}106)$$

This is usually very much less than (2-105).

98. Diamond Coil with Twisted Conductors, with an Odd Number of Turns per Coil (q Odd). To obtain the lowest loss, for three turns per coil, use *DRD, RRR,* starting at the top or open end of the slot, as in Fig. 2-17.

For five turns per coil, use *DRDRR, RDDDD,* starting at the top.

For seven turns per coil, use *DRDRDRR, RDDRDDD,* starting at the top.

For all these coils, when $\phi = 0$, use Eq. (2-106). When $\phi = 60°$, the extra loss ratio due to long-path eddy current is not given by Eq. (2-106) but may be determined by Eq. (2-102).

99. In the **embedded part of the coil,** as stated in Par. **95,** the total extra loss ratio is the sum of the short-path extra loss ratio, given by Eqs. (2-97) to (2-99), computed as if the laminations were thoroughly transposed, and the long-path extra loss ratio, computed as in Pars. **95** to **98**. In the nonembedded part of the coil, that is, the coil end, the total ratio is equal to the long-path extra loss ratio. For an entire winding, the loss may first be computed for direct current; then the watts of extra loss may be determined for the embedded parts and for the coil ends, where $\phi = 0$ and $\phi = 60°$; and so the total watts of loss with alternating current may be found. See also Heat Losses in Stranded Armature Conductors, by W. V. Lyon, *Trans. AIEE,* 1922, p. 199.

100. In designing a coil, if it is desired to reduce the short-path extra loss ratio, a larger number of thinner laminations may be specified. If it is desired to reduce the long-path extra loss ratio, the conductors may be twisted, or a thoroughly transposed type of coil may be specified (see Par. **94**).

ELECTROSTATICS

101. Electrostatic Force. Electrically charged bodies exert forces upon one another according to the following principles:

1. Like charged bodies repel; unlike charged bodies attract one another.
2. The force is proportional to the product of the magnitudes of the charges upon the bodies.
3. The force is inversely proportional to the square of the distance between charges if the material in which the charges are immersed is extensive and possesses the same uniform properties in all directions.
4. The force acts along the line joining the centers of the charges.

Two concentrated charges Q_1 and Q_2 coulombs located R m apart experience a force between them of

$$F = \frac{Q_1Q_2}{4\pi\epsilon_0 R^2} \qquad \text{(newtons)} \tag{2-107}$$

where $\epsilon_0 = 8.85 \times 10^{-12}$ F/m and is the permittivity of free space.

102. Electrostatic Potential. The electric potential resulting from the location of charged bodies in the vicinity is called electrostatic potential. The potential at R m from a concentrated charge Q C is

$$\Phi = \frac{Q}{4\pi\epsilon_0 R} \qquad \text{(volts)} \tag{2-108}$$

where $\epsilon_0 = 8.85 \times 10^{-12}$ F/m. This potential is a scalar quantity.

103. Electric Field Intensity. The electric field intensity is the force per unit charge that would act at a point in the field on a very small test charge placed at that location.

The electric field intensity $\mathcal{E}$ at a distance R m from a concentrated charge Q C is

$$\mathcal{E} = \frac{Q}{4\pi\epsilon_0 R^2} \qquad \text{(N/C)} \tag{2-109}$$

where ϵ_0 is 8.85×10^{-12} F/m.

104. Electric Potential Gradient in Electrostatic Fields. The space rate of change of the electric potential is the electric potential gradient of the field, symbolized by $\nabla\Phi$. The general relationship between the gradient of the electric potential and the electric field intensity is

$$\mathcal{E} = -\nabla\Phi \qquad \text{(V/m)} \tag{2-110}$$

The units for the electric potential gradient, volts per meter, are frequently also used for the electric field intensity since their magnitudes are the same.

105. Electric Flux Density. The density of electric flux (symbol D) in a region where simple dielectric materials exist is determined from the electric field intensity from

$$D = \epsilon\mathcal{E} = \epsilon_0 K\mathcal{E} \qquad \text{(C/m}^2\text{)} \tag{2-111}$$

where $\epsilon_0 = 8.85 \times 10^{-12}$ F/m and K is a dimensionless number called the *dielectric constant*. In free space K is unity. For numerical values of dielectric constant of various dielectrics, see Sec. **4**.

106. Polarization. The polarization is the excess of electric flux density that results in dielectric materials over that which would result at the same electric field intensity if the space were free of material substance. Thus

$$P = D - \epsilon_0\mathcal{E} \qquad \text{(C/m}^2\text{)} \tag{2-112}$$

107. Crystalline Atomic Materials. In simple isotropic materials the directions of the vectors P, D, and $\mathcal{E}$ are the same. For crystalline atomic structures that are not isotropic, Eq. (2-112) is the only relationship which is meaningful and Eq. (2-111) should not be used.

108. Electric Flux. Electric flux and its density are related by

$$\psi = \int D \cos\alpha \, dA \qquad \text{(coulombs)} \tag{2-113}$$

where α is the angle between the direction of the electric flux density D and the normal at each differential surface area dA.

109. Capacitance. The capacitance between two oppositely charged bodies is the ratio of the magnitude of charge on either body to the difference of electric potential between them. Thus

$$C = \frac{Q}{V} \qquad \text{(farads)} \tag{2-114}$$

where Q is in coulombs and V is the voltage between the two equally but oppositely charged bodies, in volts. The unit microfarad (μF), equal to a millionth of a farad, is frequently used.

110. Elastance. The reciprocal of capacitance, called *elastance,* is

$$S = V/Q \qquad \text{(darafs)} \tag{2-115}$$

111. Electric Field Outside an Isolated Sphere in Free Space. The electric field intensity at a distance r m from the center of an isolated charged sphere located in free space is

$$\mathcal{E} = \frac{Q}{4\pi\epsilon_0 r^2} \qquad \text{(V/m)} \tag{2-116}$$

where Q is the total charge (which is distributed uniformly) on the sphere and $\epsilon_0 = 8.85 \times 10^{-12}$ F/m.

112. Spherical Capacitor. The capacitance between two concentric charged spheres is

$$C = \frac{4\pi\epsilon_0 K}{\dfrac{1}{R_1} - \dfrac{1}{R_2}} \qquad \text{(farads)} \tag{2-117}$$

where R_1 is the outside radius of the inner sphere, R_2 is the inside radius of the outer sphere, K is the dielectric constant of the space between them, and $\epsilon_0 = 8.85 \times 10^{-12}$ F/m.

113. Electric Field Intensity Created by an Isolated, Charged, Long Cylindrical Wire in Free Space. The electric field intensity in the vicinity of a long, charged cylinder is

$$\mathcal{E} = \frac{\Lambda}{2\pi\epsilon_0 r} \qquad \text{(V/m)} \tag{2-118}$$

where Λ is the charge per unit of length in coulombs per meter (distributed uniformly over the surface of the isolated cylinder), r is the distance in meters from the center of the cylinder to the point at which the electric field intensity is evaluated, and $\epsilon_0 = 8.85 \times 10^{-12}$ F/m.

114. Coaxial Cable. The capacitance per unit length of a coaxial cable composed of two concentric cylinders is

$$c = \frac{2\pi\epsilon_0 K}{\ln \dfrac{R_2}{R_1}} \qquad \text{(F/m)} \tag{2-119}$$

where R_1 is the outside radius of the inner cylinder, R_2 is the inside radius of the outer cylinder, K is the dielectric constant of the space between the cylinders, and $\epsilon_0 = 8.85 \times 10^{-12}$ F/m.

115. Two-wire Line. The capacitance per unit length between two long, oppositely charged, cylindrical conductors of equal radii, parallel and external to each other, is

$$c = \frac{\pi\epsilon_0 K}{\ln \left[\dfrac{D}{2R} + \sqrt{\left(\dfrac{D}{2R}\right)^2 - 1} \right]} \qquad \text{(F/m)} \tag{2-120}$$

where D is the distance in meters between centers of the two cylindrical wires each with radius R, K is the uniform dielectric constant of all space external to the wires, and $\epsilon_0 = 8.85 \times 10^{-12}$ F/m.

116. Capacitance of Two Flat, Parallel Conductors Separated by a Thin Dielectric. The capacitance is approximately

$$C = \frac{\epsilon_0 K A}{t} \qquad \text{(farads)} \tag{2-121}$$

where A is the area of either of the two conductors, t is the spacing between them, K is the dielectric constant of the space between the conductors, and $\epsilon_0 = 8.85 \times 10^{-12}$ F/m. Strictly, the linear dimensions of the flat conductors should be very large compared with the spacing between them. Good results are obtained from Eq. (2-121) even though the conductors are curved provided that the spacing t is small with respect to the radius of curvature.

117. Induced Charges. The surface of a conducting body, near a charge Q, through which no currents are flowing is an equipotential surface, a condition maintained by the motion of positive and negative charges to the parts of the conductor near Q and distant from it. Hence the potential at any point on the conductor, due to all the charges of the system, is a constant. The charges on the conductors are said to be induced by Q, and the conductor is said to be electrified by induction.

118. Electrostatic Induction on Parallel Wires. Two insulated wires running parallel to a wire carrying a charge Λ C/m display a potential difference (provided that the two wires are not connected to each other or to other conductors) of

$$\phi = \frac{\Lambda}{2\pi\epsilon_0} \ln \frac{b}{a} \qquad \text{(volts)} \qquad (2\text{-}122)$$

where b and a are the distances of the two insulated wires from the charged wire.

If the two wires are connected together, as, for example, through telephone instruments, the current flowing from one wire to the other is that required to equalize their potential difference.

THE DIELECTRIC CIRCUIT

119. Circuit Concepts with Capacitive Elements. When a continuous voltage is applied to the terminals of a capacitor (AB, Fig. 2-18), a positive charge of electricity $+Q$ appears on one plate and a negative charge $-Q$ on the other. A quantity of electricity Q flows through the connecting wires, and this quantity of electricity is said to be displaced through the dielectric. An electrostatic field, as has been described in Par. **104,** then exists between the two charged plates.

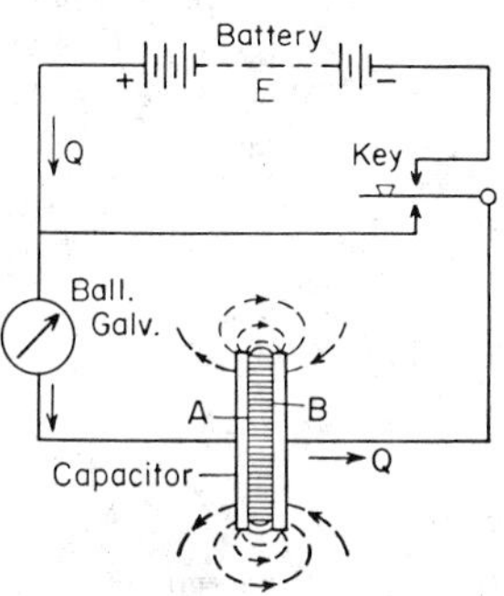

FIG. 2-18. Circuit containing a capacitor.

When an applied voltage e is changing with time the current i through the capacitive circuit is the time derivative of Q in Eq. (2-114),

$$i = \frac{dQ}{dt} = C\frac{de}{dt} \qquad \text{(amperes)} \qquad (2\text{-}122a)$$

120. Electrostatic Flux. The space between the plates of a capacitor can be treated as a dielectric circuit through which passes a dielectric flux ψ, in coulombs. In any dielectric circuit, one coulomb of electrostatic flux passes from each coulomb of positive charge to each coulomb of negative charge, and this is true with any insulating substance or group of substances. That is, electrostatic flux lines end only on charges of electricity. Their number is not affected when they pass from one dielectric to another, unless there is a charge of electricity on the surface of separation (see Par. **39**). Electrostatic flux lines are also called "lines of electrostatic induction."

121. Capacitors in Series and in Parallel. When capacitors are connected in parallel, the equivalent capacitance is equal to the sum of all the capacitances of the component capacitors, or

$$C_{eq} = \sum C \qquad (2\text{-}123)$$

When two or more capacitors are connected in series, the equivalent capacitance is determined from the relation

$$\frac{1}{C_{eq}} = \sum \frac{1}{C} \qquad (2\text{-}124)$$

Analogously, for a series connection of elastances (Par. **110**),

$$S_{eq} = \sum S \qquad (2\text{-}125)$$

and for parallel connection of elastances,

$$\frac{1}{S_{eq}} = \sum \frac{1}{S} \qquad (2\text{-}126)$$

122. As an example, let two capacitances $C_1 = 0.2\ \mu F$ and $C_2 = 0.3\ \mu F$ be connected in parallel with each other and in series with a third capacitor for which $C_3 = 0.4\ \mu F$. The capacitance of the combination is required. The capacitance of the two capacitors in parallel is $C_1 + C_2 = 0.5\ \mu F$, and the elastance of the combination is 2 Mdarafs. The elastance of the third capacitor $1/C_3 = 2.5$ Mdarafs, and the total elastance of the combination is $2 + 2.5 = 4.5$ Mdarafs. The equivalent capacitance is $1/4.5 = 0.222\ \mu F$.

123. The capacitance to neutral of a conductor in an a-c line is defined as the capacitance that, when multiplied by $2\pi f$ and by the voltage to neutral, gives the charging current of the conductor, f being the frequency. This is not the same as the capacitance to a neutral wire measured electrostatically. The voltage to neutral of a single-phase line is one-half the voltage between conductors. The voltage to neutral of a balanced 3-phase line is equal to the voltage between conductors divided by 1.732.

When the conductors are round wires, for either single-phase or 3-phase overhead lines, the capacitance to neutral is

$$c = \frac{2\pi\epsilon_0 K}{\ln \sqrt{\dfrac{s}{d} + \sqrt{\left(\dfrac{s}{d}\right)^2 - 1}}} \qquad \text{(F/m, to neutral)} \qquad (2\text{-}127)$$

or, approximately,

$$c = \frac{0.0388}{\ln \dfrac{2s}{d}} \qquad (\mu\text{F/mi, to neutral}) \qquad (2\text{-}128)$$

where s is the axial spacing and d is the diameter of the conductors, in the same units.

Values of charging kVA for transmission lines are tabulated in Sec. **13**.

The capacitance of a complete single-phase line is one-half the capacitance to neutral of one conductor. The capacitance of stranded conductors may be approximately calculated by using the outside diameter of the conductors. The capacitance of iron or steel conductors is calculated by the same formulas as that of copper conductors.

The above relations assume equilateral spacing for 3-phase systems. If unbalanced spacings are present and the phases are balanced by transposing the conductors over the length of the line, the approximate capacitance per phase can be obtained from the concepts of geometric mean distances D_m and D_s (Par. **82**). Then, approximately,

$$c = \frac{2\pi\epsilon_0 K}{\ln \dfrac{D_m}{D_s'}} \qquad \text{(F/m)} \qquad (2\text{-}129)$$

The self-geometric mean distance D_s' in Eq. (2-129) differs slightly from that for D_s in Eq. (2-90) in that for good conductors the transverse gradient of the electric field is confined principally to the air space about the conductors and D_s' is slightly larger than D_s of Eq. (2-90). In the latter equation, internal flux linkages in the conductor contribute to the meaning of D_s.

The following references may be consulted for the many details needed in computing the geometric mean distances of Eq. (2-129):

Boast, W. B. "Vector Fields"; New York, Harper & Row, Publishers, Inc., 1964, Chap. 8, 10, and 11.

Stevenson, W. D., Jr. "Elements of Power System Analysis"; New York, McGraw-Hill Book Company, 1962, Chap. 3.

Clarke, E. "Circuit Analysis of A-C Power Systems"; New York, John Wiley & Sons, Inc., 1943, Vol. I.

"Electrical Transmission and Distribution Reference Book," 4th ed.; Westinghouse Electric Corporation, 1950.

124. Velocity of Propagation on Long Transmission Lines. The inductance and capacitive parameters per unit length of a transmission line determine the velocity with which such effects as switching surges are propagated along the line. The velocity of propagation is

$$v = \frac{1}{\sqrt{lc}} \quad \text{(m/s)} \tag{2-130}$$

where l is the inductance per unit length from Eq. (2-90) and c is the capacitance per unit length from Eq. (2-129). Substituting these values gives

$$v = 3 \times 10^8 \sqrt{\frac{\ln (D_m/D_s')}{K \ln (D_m/D_s)}} \quad \text{(m/s)} \tag{2-131}$$

The fact that D_s' is slightly larger than D_s produces a velocity of propagation along the transmission line which is slightly less than the velocity of propagation of electromagnetic radiation in free space (3×10^8 m/s). Since the dielectric constant K of the atmosphere surrounding the transmission line is somewhat greater than unity, the velocity of propagation is reduced slightly. Magnetic materials in the conductors tend to increase the inductance in the denominator of Eq. (2-130) and reduce further the velocity of propagation by a small amount.

125. The Energy Stored in a Capacitor. The energy stored in a capacitor is

$$W = \frac{CV^2}{2} = \frac{V^2}{2S} = \frac{VQ}{2} \quad \text{J (W-s)} \tag{2-132}$$

where the voltage V is in volts, the charge Q in coulombs, and the capacitance C in farads. Equivalently if C is in microfarads and Q in microcoulombs, W is in microjoules.

The energy stored per unit volume of the dielectric is

$$W' = \tfrac{1}{2}\epsilon \mathcal{E}^2 \quad \text{(J/m}^3\text{)} \tag{2-133}$$

where ϵ is the permittivity (8.85×10^{-12} F/m multiplied by the dielectric constant) and $\mathcal{E}$ is the voltage gradient (electric field intensity) in volts per meter.

126. The dielectric strength of insulating materials (rupturing voltage gradient) is the maximum voltage per unit thickness that a dielectric can withstand in a uniform field before it breaks down electrically. The dielectric strength is usually measured in kilovolts per millimeter or per inch. It is necessary to define the dielectric strength in terms of a uniform field, for instance, between large parallel plates a short distance apart. If the striking voltage is determined between two spheres or electrodes of other defined shape, this fact must be stated. In designing insulation, a factor of safety is assumed depending upon conditions of operation. For numerical values of rupturing voltage gradients of various insulating materials, see Sec. **4**.

DIELECTRIC LOSS AND CORONA

127. Dielectric Hysteresis and Conductance. When an alternating voltage is applied to the terminals of a capacitor, the dielectric is subjected to periodic stresses and displacements. If the material were perfectly elastic, no energy would be lost during any cycle, because the energy stored during the periods of increased voltage would be given up to the circuit when the voltage is decreased. However, since the electric elasticity of dielectrics is not perfect, the applied voltage has to overcome molecular friction or viscosity, in addition to the elastic forces. The work done against friction is converted into heat and is lost. This phenomenon resembles magnetic hysteresis (Par. **64**) in some respects but differs in others. It has commonly been called "dielectric hysteresis" but is now often called "dielectric loss." The energy lost per cycle is proportional to the square of the applied voltage. Methods of measuring dielectric loss are described in Sec. **3** (see J. B. Whitehead, "Lectures on Dielectric Theory and Insulation," New York, McGraw-Hill Book Company, 1927, pp. 57–59).

An imperfect capacitor does not return on discharge the full amount of energy put into it. Some time after the discharge an additional discharge may be obtained. This phenomenon is known as **dielectric absorption.**

A capacitor that shows such a loss of power can be replaced in some cases for calculation by a perfect capacitor with an ohmic conductance shunted around it. This conductance (or "leakance") is of such value that its I^2R loss is equal to the loss of power from all causes in the imperfect capacitor. The actual current through the capacitor is then considered as consisting of two components—the leading reactive component through the ideal capacitor and the loss component, in phase with the voltage, through the shunted conductance.

128. Electrostatic Corona. When the electrostatic flux density in the air exceeds a certain value, a discharge of pale violet color appears near the adjacent metal surfaces. This discharge is called *electrostatic corona.* In the regions where the corona appears, the air is electrically ionized and is a conductor of electricity. When the voltage is raised further, a brush discharge takes place, until the whole thickness of the dielectric is broken down and a disruptive discharge, or spark, jumps from one electrode to the other.

Corona involves power loss, which may be serious in some cases, as on transmission lines (Secs. **13** and **14**). Corona can form at sharp corners of high-voltage switches, bus bars, etc.; so the radii of such parts are made large enough to prevent this. A voltage of 12 to 25 kV between conductors separated by a fraction of an inch, as between the winding and core of a generator or between sections of the winding of an air-blast transformer, can produce a voltage gradient sufficient to cause corona. A voltage of 100 to 200 kV may be required to produce corona on transmission-line conductors that are separated by several feet. Corona can have an injurious effect on fibrous insulation. For numerical data in application to transmission lines see Secs. **13** and **14.**

TRANSIENT CURRENTS AND VOLTAGES

129. Transient electric phenomena occur, for instance, when a load is suddenly changed and an appreciable time elapses before the generators and the line adapt themselves to the new conditions. The currents and the voltages during the intermediate time are known as transient.

130. Closing a Circuit Containing a Resistance R (Ohms) and an Inductance L (Henrys) in series with a continuous emf. When a deenergized circuit is suddenly connected to a source of continuous voltage E, the current gradually rises to the final value $i_0 = E/R$ according to the law

$$i = i_0(1 - \epsilon^{-tR/L}) \qquad (2\text{-}134)$$

where t is time in seconds and ϵ is the base of natural logarithms. This expression is known as **Helmholtz's law.**

When the source of emf is short-circuited, the current in the remaining circuit decreases to zero according to a similar law

$$i = i_0\epsilon^{-tR/L} \qquad (2\text{-}135)$$

131. Alternating EMF, RL Circuit. When a deenergized circuit containing R and L is suddenly connected at time $t = 0$ to a source of alternating voltage

$$e = E_m \sin (2\pi ft + \alpha)$$

the current in the circuit varies according to the law

$$i = \frac{E_m}{Z} \sin (2\pi ft + \alpha - \phi) - \frac{E_m}{Z} \sin (\alpha - \phi)\epsilon^{-tR/L} \qquad (2\text{-}136)$$

In this equation, $Z = \sqrt{R^2 + (2\pi fL)^2}$ is the impedance of the circuit and ϕ is the phase displacement between the current and the voltage (determined by $\tan \phi = 2\pi fL/R$). The angle α is the phase displacement between the voltage e and the reference wave which passes through zero at the time $t = 0$, and f is the frequency in hertz (cycles per second). The first term in Eq. (2-136) is the current corresponding to the steady-state condition. The second term is the transient, which approaches zero with the time.

132. Closing a Circuit Containing a Resistance R **(Ohms) and a Capacitance** C **(Farads) in Series.** When the capacitor is initially uncharged, the current produced by a constant applied emf E is

$$i = i_0\epsilon^{-t/RC} \tag{2-137}$$

where $i_0 = E/R$ is the initial current. In practice some inductance is always present in the circuit, and this smooths down the initial change in current.

The charge on the capacitor, initially uncharged, is

$$q = EC(1 - \epsilon^{-t/RC}) \tag{2-138}$$

133. Discharging Capacitor. When a capacitor, initially charged to a voltage E_0, is discharged through resistance R, the discharge current at the first instant is $i_0 = -E_0/R$. Thereafter, the current varies according to

$$i = i_0\epsilon^{-t/RC} \tag{2-139}$$

The charge on the capacitor decreases according to

$$q = Q_0\epsilon^{-t/RC} \tag{2-140}$$

where Q_0 is the initial charge.

134. Alternating EMF, RC **Circuit.** When a deenergized circuit containing R Ω and C F is suddenly connected at $t = 0$ to a source of alternating voltage $e = E_m \sin (2\pi ft + \alpha)$, the current in the circuit is

$$i = \frac{E_m}{Z} \sin (2\pi ft + \alpha + \phi) - \frac{E_m \cos (\alpha + \phi)}{2\pi fRCZ} \epsilon^{-t/RC} \tag{2-141}$$

In this equation $Z = \sqrt{R^2 + (1/2\pi fC)^2}$ is the impedance of the circuit, and ϕ is the phase displacement between the current and the voltage, determined by

$$\cot \phi = 2\pi fCR$$

The angle α is the phase displacement between the voltage E and the reference wave which passes through zero at the time $t = 0$; f is the frequency in hertz. The first term in Eq. (2-141) is the current corresponding to the steady-state condition; the second term is a transient, which approaches zero with the time (compare Par. **131**).

For the case described for Eq. (2-141),

$$q = -\frac{E_m}{2\pi fZ} \cos (2\pi ft + \alpha + \phi) + \frac{E_m}{2\pi fZ} \cos (\alpha + \phi)\epsilon^{-t/RC} \tag{2-142}$$

135. Single-energy and Double-energy Transients. The two preceding cases are examples of single-energy transients; that is, the energy is stored in one form only (electromagnetic or electrostatic), and the energy change consists in an increase or a decrease of the stored energy.

When both inductance and capacitance are present, the energy of the circuit is stored in two forms, and there is a periodic transfer of energy from magnetic to dielectric form, and vice versa, producing electric oscillations.

A **triple-energy transient** occurs, for instance, when a synchronous motor is hunting at the end of a long transmission line which possesses inductance and capacitance. In the latter case, the energy of the system is stored in magnetic, dielectric, and mechanical forms.

136. RLC **Circuit.** For a series circuit of resistance R (ohms), inductance L (henrys), capacitance C (farads), and constant emf E (volts), there are three types of current i (amperes).

Nonoscillatory case (overdamped), when $R^2 > 4L/C$.

Oscillatory case (underdamped), when $R^2 < 4L/C$.

Critical case, when $R^2 = 4L/C$.

137. Nonoscillatory Case. When the initial charge on the capacitor and the current are zero,

$$i = \frac{E}{2Lb} (\epsilon^{(-a+b)t} - \epsilon^{(-a-b)t}) \qquad \text{(amperes)} \tag{2-143}$$

where $a = \frac{R}{2L}$ and $b = \sqrt{\frac{R^2}{4L^2} - \frac{1}{LC}}$

Note that $-a + b$ is negative. The charge on the capacitor is

$$q = EC\left(1 - \frac{a+b}{2b} \epsilon^{(-a+b)t} - \frac{a+b}{2b} \epsilon^{(-a-b)t}\right) \qquad \text{(coulombs)} \tag{2-144}$$

When the initial charge on the capacitor is Q_0 and the initial current is zero,

$$i = -\frac{Q_0}{2LCb} (\epsilon^{(-a+b)t} - \epsilon^{(-a-b)t}) \qquad \text{(amperes)} \tag{2-145}$$

$$q = Q_0\left(\frac{a+b}{2b} \epsilon^{(-a+b)t} - \frac{a-b}{2b} \epsilon^{(-a-b)t}\right) \qquad \text{(coulombs)} \tag{2-146}$$

138. Oscillatory Case. $R^2 < L/C$. When the initial charge and current are zero,

$$i = \frac{E}{Lg} \epsilon^{-at} \sin gt \qquad \text{(amperes)} \tag{2-147}$$

where

$$a = \frac{R}{2L}$$

$$g = \sqrt{\frac{1}{LC} - \frac{R^2}{4L^2}}$$

and E is the applied direct voltage.

$$q = EC\left[1 - \epsilon^{-at}\left(\cos gt + \frac{a}{g} \sin gt\right)\right] \qquad \text{(coulombs)} \tag{2-148}$$

When the initial charge is Q_0 and the initial current is zero,

$$i = -\frac{Q_0}{LCg} \epsilon^{-at} \sin gt \qquad \text{(amperes)} \tag{2-149}$$

$$q = Q_0 \epsilon^{-at}\left(\cos gt + \frac{a}{g} \sin gt\right) \qquad \text{(coulombs)} \tag{2-150}$$

The frequency of the oscillations in the oscillatory case of the RLC series circuit is

$$f = \frac{g}{2\pi} = \frac{1}{2\pi}\sqrt{\frac{1}{LC} - \frac{R^2}{4L^2}} \qquad \text{(Hz or c/s)} \tag{2-151}$$

When the resistance is negligible, that is, when R^2 is very small compared with $4L/C$, the frequency of oscillations is very close to

$$f_0 = \frac{1}{2\pi \sqrt{LC}} \qquad \text{(Hz)} \tag{2-152}$$

139. Critical Case. $R^2 = 4L/C$. When the initial charge and current are zero,

$$i = \frac{Et}{L} \epsilon^{-tR/2L} \qquad \text{(amperes)} \tag{2-153}$$

$$q = EC - EC\left(1 + \frac{tR}{2L}\right) \epsilon^{-tR/2L} \qquad \text{(coulombs)} \tag{2-154}$$

where E is the applied direct voltage.

When the initial charge is Q_0 and the initial current is zero,

$$i = -\frac{Q_0 t}{LC}\epsilon^{-tR/2L} \qquad \text{(amperes)} \tag{2-155}$$

$$q = Q_0\left(1 + \frac{tR}{2L}\right)\epsilon^{-tR/2L} \qquad \text{(coulombs)} \tag{2-156}$$

140. Stored Energy. When energy is suddenly changed at some point on a transmission line, for instance, because of an indirect lightning stroke, a wave travels along the line, carrying the energy change to the ends of the line. Part of the wave enters the apparatus at the ends, part is reflected, and the rest is converted into heat.

The total energy stored in a transmission at any instant is

$$W = \tfrac{1}{2}Li^2 + \tfrac{1}{2}Ce^2 \qquad \text{(joules)} \tag{2-157}$$

where L is the inductance of the line in henrys, i instantaneous current, C the capacitance of the line in farads, and e the corresponding instantaneous voltage. The term $\frac{1}{2}Li^2$ represents the electromagnetic energy; the term $\frac{1}{2}Ce^2$, the electrostatic energy. At certain instants, the current is equal to zero; at others, the voltage is zero. Since energy remains constant (no losses are assumed), maximum values of the two energies must be equal and

$$\frac{e_{\max}}{i_{\max}} = \sqrt{\frac{L}{C}} \qquad \text{(ohms)} \tag{2-158}$$

Thus the maximum instantaneous current $i_{\max}$ can be calculated from the maximum voltage $e_{\max}$, and vice versa.

In the case of a lightning stroke, for example, the maximum voltage is limited by the disruptive strength of the insulation, and the maximum current disturbance may be calculated from the preceding equation.

The quantity $\sqrt{L/C}$ is the **surge impedance** of the line; its reciprocal is the **surge admittance.**

ALTERNATING-CURRENT CIRCUITS

141. Phasor Quantities. If e and e' (Fig. 2-19) are the components of a phasor voltage E along two perpendicular axes, E may be represented symbolically as

$$\dot{E} = e + je' \tag{2-159}$$

where

$$j = \sqrt{-1} \tag{2-160}$$

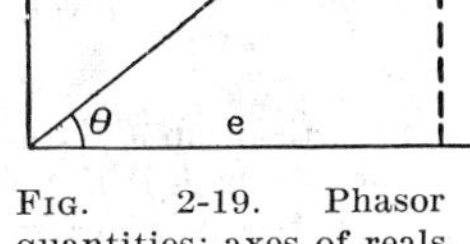

FIG. 2-19. Phasor quantities; axes of reals and imaginaries.

The dot over E signifies that this quantity has a direction as well as a magnitude. (Note that here the dot does not represent the first derivative. The dot over the letter is very frequently omitted, as it is usually obvious when a symbol denotes a complex quantity.)

The quantity E in Eq. (2-159) may denote an alternating voltage of sine-wave form of frequency f Hz (c/s). The effective or root-mean-square (rms) value of the voltage is

$$E_{\text{eff}} = \frac{E_{\max}}{\sqrt{2}} = 0.707E_{\max} \tag{2-161}$$

The instantaneous value of the voltage is $E_{\max} \sin 2\pi ft$, where t is the time in seconds and the angle $2\pi ft$ is in radians.

The effective value of any alternating voltage or current (not necessarily a sine wave) is defined in Par. **162**.

142. Addition and Subtraction of Phasors. When two phasor voltages are represented as

$$\dot{E}_1 = e_1 + je_1'$$

$$\dot{E}_2 = e_2 + je_2'$$

the sum or the difference of these two phasors is

$$\dot{E}_3 = \dot{E}_1 \pm \dot{E}_2 = (e_1 \pm e_2) + j(e_1' \pm e_2') \tag{2-162}$$

143. Rotation of a Phasor. Multiplying a phasor by j rotates it by 90° in the positive direction (counterclockwise). Thus,

$$j\dot{E} = j(e + je') = -e' + je \tag{2-163}$$

Note that $j^2 = -1$. Conversely, multiplying a phasor by $-j$ rotates the phasor by 90° in the negative direction, that is, clockwise.

The phasor $\dot{E}$ may be also represented symbolically (Fig. 2-19) in the polar form as

$$\dot{E} = E\underline{/\theta} = E(\cos\theta + j\sin\theta) \tag{2-164}$$

where E without the dot stands for the magnitude only.

The operator $\underline{/\phi}$ is

$$\underline{/\phi} = \epsilon^{j\phi} = \cos\phi + j\sin\phi \tag{2-165}$$

where ϵ is the base of the natural (naperian) logarithms. It turns a phasor by the angle ϕ in the positive direction. Thus,

$$\begin{aligned}\dot{E}(\cos\phi + j\sin\phi) &= E(\cos\theta + j\sin\theta)(\cos\phi + j\sin\phi)\\ &= E[\cos(\theta+\phi) + j\sin(\theta+\phi)]\end{aligned} \tag{2-166}$$

The operator $\epsilon^{-j\phi} = \cos\phi - j\sin\phi$ turns a phasor by an angle ϕ in the negative direction, that is, clockwise.

The angle ϕ between two alternating quantities (phasors) represents a time-phase difference, such that the period of one complete cycle $(1/f)$ is equal to 2π rad. The phase displacement between alternating quantities is commonly measured in electrical degrees. One electrical degree ($\frac{1}{360}$th part of a complete cycle) is $2\pi/360$ electrical rad.

The horizontal line at the bottom of Fig. 2-19 is called the reference phase. The phase angle θ of E is measured from it.

144. The absolute, or numerical, value of a phasor expressed in the rectangular form $\dot{E} = e + je'$ is equal to

$$|\dot{E}| = \sqrt{e^2 + e'^2} \tag{2-167}$$

The dot and the vertical lines are often omitted.

The rectangular form $e + je'$ may be changed to the polar form $E\underline{/\theta}$ by means of the equations

$$\begin{aligned}E &= |\dot{E}| = \sqrt{e^2 + e'^2}\\ \cos\theta &= \frac{e}{\sqrt{e^2 + e'^2}}\\ \sin\theta &= \frac{e'}{\sqrt{e^2 + e'^2}}\end{aligned} \tag{2-168}$$

The expression $\theta = \tan^{-1}(e'/e)$ indicates two possible angular positions 180° apart, only one of which is the correct angle for a given phasor. For instance, if e' and e are both negative, θ is an angle greater than 180° and the appropriate value of $\tan^{-1}(e'/e)$ must be selected. (See "Tables of Integrals and Other Mathematical Data," by H. B. Dwight, No. 401.2.)

The numerical values of two or more phasors *cannot* be added unless the phasors are in phase. Otherwise the phasors must be added according to Eq. (2-162). Similarly, impedances and admittances must be added as in Eqs. (2-175) and (2-181).

145. Square Root of a Complex Quantity. If a complex quantity is expressed in the polar form $r\underline{/\theta}$, its two square roots are $\sqrt{r}\underline{/\pm\theta/2}$. If the quantity is in the rectangular form $a + jb$ or $a - jb$ (where b is positive),

$$\begin{aligned}\sqrt{a+jb} &= \pm\left(\sqrt{\frac{r+a}{2}} + j\sqrt{\frac{r-a}{2}}\right)\\ \sqrt{a-jb} &= \pm\left(\sqrt{\frac{r+a}{2}} - j\sqrt{\frac{r-a}{2}}\right)\end{aligned} \tag{2-169}$$

where $r = \sqrt{a^2 + b^2}$. The positive square roots of $(r + a)/2$ and $(r - a)/2$ are used.

If desired, Eqs. (2-168) may be used by changing the rectangular form to the polar form. In spite of the simplicity of the polar expression, it is sometimes less work to use Eqs. (2-169) than to change to the polar form and back.

146. Impedance. An impedance consisting of a resistance r in series with a reactance x is represented by the impedance operator

$$Z = r + jx \qquad \text{(ohms)} \tag{2-170}$$

This is a complex quantity but not a rotating phasor.

The letter x has a positive numerical value, equal to $2\pi fL$, for inductive reactance, where f is the frequency in hertz and L is in henrys. It has a negative value $-1/2\pi fC$ for capacitive reactance, where C is in farads.

When inductive and capacitive reactances are in series,

$$x = 2\pi fL - 1/2\pi fC \qquad \text{(ohms)} \tag{2-171}$$

A sine-wave current in an impedance Z is represented by a current phasor I. The voltage drop across the impedance is

$$E = IZ = Ir + jIx \qquad \text{(volts)} \tag{2-172}$$

E is a phasor such that its horizontal component (Fig. 2-19) is equal to Ir and its vertical component is Ix.

The angle θ between the voltage E and the current I is determined by

$$\tan\theta = x/r \tag{2-173}$$

When x is positive, as for inductive reactance, the current comes to a maximum after the voltage and so lags behind the voltage. This is in agreement with the positions of the phasors for voltage and current. When x is negative, as for capacitive reactance, the current leads the voltage.

When inductive and capacitive reactances are in series, the current is lagging or leading with respect to the voltage according to whether x, given by Eq. (2-171), is positive or negative, i.e., according to whether the effect of the inductance or capacitance predominates, respectively.

When a current $I = p + jq$ exists in an impedance $Z = r + jx$, the voltage across the impedance is

$$\begin{aligned} E = IZ &= (p + jq)(r + jx) \\ &= (pr - qx) + j(px + qr) \qquad \text{(volts)} \end{aligned} \tag{2-174}$$

When impedances are connected in series, the complex quantities are added as

$$r + jx = r_1 + jx_1 + r_2 + jx_2 \tag{2-175}$$

If it is not necessary to keep account of phase angles, the numerical value of impedance drop is

$$|E| = |IZ| = |I|\sqrt{r^2 + x^2} \qquad \text{(volts)} \tag{2-176}$$

Similarly, the numerical value of current is

$$|I| = \frac{|E|}{|Z|} = \frac{|E|}{\sqrt{r^2 + x^2}} \qquad \text{(amperes)} \tag{2-177}$$

147. Admittance. An admittance consisting of a conductance g in parallel with a susceptance b is represented by the admittance operator

$$Y = g - jb \qquad \text{(mhos)} \tag{2-178}$$

The quantity b has a positive numerical value $1/2\pi fL$ for inductance, where L is in henrys. It has a negative value $-2\pi fC$ for capacitance, where C is in farads.

When inductive and capacitive susceptances are in parallel,

$$b = 1/2\pi fL - 2\pi fC \qquad \text{(mhos)} \tag{2-179}$$

If the voltage e is represented by a horizontal phasor E, the total current I is obtained by multiplying the expressions for voltage and admittance and is

$$I = EY = Eg - jEb \qquad \text{(amperes)} \tag{2-180}$$

When b is positive, as for inductance, the current lags behind the voltage; and when b is negative, as for capacitance, the current is leading. When inductive and capacitive susceptances are in parallel, the total current lags or leads the voltage according to whether b, given by Eq. (2-179), is positive or negative, i.e., according to whether the effect of the inductance or the capacitance predominates, respectively.

When branches of a network are connected in parallel, the complex quantities representing the admittances are added, as

$$g - jb = g_1 - jb_1 + g_2 - jb_2 \tag{2-181}$$

If it is not desired to keep account of phase angles, the numerical value of current is

$$|I| = |EY| = |E|\sqrt{g^2 + b^2} \qquad \text{(amperes)} \tag{2-182}$$

Similarly the numerical value of voltage is

$$|E| = \frac{|I|}{|Y|} = \frac{|I|}{\sqrt{g^2 + b^2}} \qquad \text{(volts)} \tag{2-183}$$

148. The admittance of a circuit or network whose impedance is known can be obtained directly from the equation

$$Y = 1/Z \tag{2-184}$$

where Y and Z are complex quantities. From this,

$$Y = g - jb = \frac{1}{r + jx} = \frac{r}{r^2 + x^2} - \frac{jx}{r^2 + x^2} \tag{2-185}$$

If the circuit consists of a resistance r in series with a reactance x, it is possible to complete, by Eq. (2-185), the equivalent conductance g and susceptance b of the circuit. These are equivalent to the resistance and reactance, connected in parallel, which together take the same current and power as the actual series circuit, at the same voltage.

In a series-parallel circuit if r and x are the resistance and reactance of the equivalent series circuit, g and b of Eq. (2-185) give the conductance and susceptance of the equivalent parallel circuit.

149. The impedance of a circuit or a network whose admittance is known can be obtained from the equation

$$Z = 1/Y \tag{2-186}$$

where Y and Z are complex quantities. Then

$$Z = r + jx = \frac{1}{g - jb} = \frac{g}{g^2 + b^2} + \frac{jb}{g^2 + b^2} \tag{2-187}$$

The values of r and x are the resistance and reactance in a simple series circuit which is equivalent to the circuit of admittance $g - jb$.

150. Impedance of Circuit Combinations. See Tables 2-1 and 2-2.

151. Power, Reactive Voltamperes, and Power Factor. When the alternating current in a circuit is $I\underline{/\alpha}$ and the voltage is $E\underline{/\beta}$, the power is equal to the product of the effective values of E and I multiplied by the cosine of the phase angle between them and is

$$P = E_{\text{eff}} I_{\text{eff}} \cos(\beta - \alpha) \qquad \text{(watts)} \tag{2-188}$$

This is sometimes called the *average power*.

Table 2-1. Impedance of Series-connected Circuit Elements*

	CIRCUIT	IMPEDANCE $Z = R + jX$ (ohms)	MAGNITUDE OF IMPEDANCE $\lvert Z\rvert = [R^2+X^2]^{1/2}$ (ohms)	PHASE ANGLE $\theta = \tan^{-1}\frac{X}{R}$ (radians)	ADMITTANCE $Y = 1/Z$ (mhos)
1	R	R	R	0	$\frac{1}{R}$
2	L	$j\omega L$	ωL	$+\frac{\pi}{2}$	$-j\frac{1}{\omega L}$
3	C	$-j\frac{1}{\omega C}$	$\frac{1}{\omega C}$	$-\frac{\pi}{2}$	$j\omega C$
4	R_1 R_2	R_1+R_2	R_1+R_2	0	$\frac{1}{R_1+R_2}$
5	M; L_1 L_2	$j\omega(L_1+L_2\pm 2M)$	$\omega(L_1+L_2\pm 2M)$	$+\frac{\pi}{2}$	$-j\frac{1}{\omega(L_1+L_2\pm 2M)}$
6	C_1 C_2	$-j\frac{1}{\omega}\left(\frac{C_1+C_2}{C_1C_2}\right)$	$\frac{1}{\omega}\left(\frac{C_1+C_2}{C_1C_2}\right)$	$-\frac{\pi}{2}$	$j\omega\left(\frac{C_1C_2}{C_1+C_2}\right)$
7	R L	$R+j\omega L$	$[R^2+\omega^2L^2]^{1/2}$	$\tan^{-1}\frac{\omega L}{R}$	$\frac{R-j\omega L}{R^2+\omega^2L^2}$
8	R C	$R-j\frac{1}{\omega C}$	$\left[\frac{\omega^2C^2R^2+1}{\omega^2C^2}\right]^{1/2}$	$\tan^{-1} -\frac{1}{\omega RC}$	$\frac{\omega^2C^2R+j\omega C}{\omega^2C^2R^2+1}$
9	L C	$j\left(\omega L-\frac{1}{\omega C}\right)$	$\left(\omega L-\frac{1}{\omega C}\right)$	$\pm\frac{\pi}{2}$	$-j\frac{\omega C}{\omega^2LC-1}$
10	R L C	$R+j\left(\omega L-\frac{1}{\omega C}\right)$	$\left[R^2+\left(\omega L-\frac{1}{\omega C}\right)^2\right]^{1/2}$	$\tan^{-1}\frac{\left(\omega L-\frac{1}{\omega C}\right)}{R}$	$\frac{R-j\left(\omega L-\frac{1}{\omega C}\right)}{R^2+\left(\omega L-\frac{1}{\omega C}\right)^2}$

* From B. Dudley, *Electronics*, December, 1942.

The reactive voltamperes, or vars, are

$$Q = E_{\text{eff}}I_{\text{eff}}\sin(\beta-\alpha) \qquad \text{(vars)} \tag{2-189}$$

When α is less than β, the current is lagging and expression (2-189) is positive. If, as is customary in the case of power systems, the voltage is taken as a reference, the angle $\beta = 0$ and the relation of the current and voltage is shown in the phasor diagram (Fig. 2-20) for lagging current. The term **power factor** is used for the factor $\cos(\beta-\alpha)$ in Eq. (2-188).

FIG. 2-20. Current and voltage phasor diagram for current lagging voltage.

FIG. 2-21. Complex plane diagram for volt-amperes for current lagging voltage.

Table 2-2. Impedance of Parallel-connected Circuit Elements*

	CIRCUIT	IMPEDANCE $Z = R + jX$ (ohms)	MAGNITUDE OF IMPEDANCE $\lvert Z \rvert = \left[R^2 + X^2\right]^{1/2}$ (ohms)	PHASE ANGLE $\theta = \tan^{-1} \frac{X}{R}$ (radians)	ADMITTANCE $Y = 1/Z$ (mhos)
1	R_1, R_2	$\frac{R_1 R_2}{R_1 + R_2}$	$\frac{R_1 R_2}{R_1 + R_2}$	0	$\frac{R_1 + R_2}{R_1 R_2}$
2	L_1, L_2, M	$+j\omega\left[\frac{L_1 L_2 - M^2}{L_1 + L_2 \mp 2M}\right]$	$\omega\left[\frac{L_1 L_2 - M^2}{L_1 + L_2 \mp 2M}\right]$	$+\frac{\pi}{2}$	$-j\frac{1}{\omega}\left[\frac{L_1 + L_1 \mp 2M}{L_1 L_2 - M^2}\right]$
3	C_1, C_2	$-j\frac{1}{\omega(C_1 + C_2)}$	$\frac{1}{\omega(C_1 + C_2)}$	$-\frac{\pi}{2}$	$+j\omega(C_1 + C_2)$
4	R, L	$\frac{\omega^2 L^2 R + j\omega L R^2}{\omega^2 L^2 + R^2}$	$\frac{\omega L R}{\left[\omega^2 L^2 + R^2\right]^{1/2}}$	$\tan^{-1}\frac{R}{\omega L}$	$\frac{\omega L - jR}{\omega L R}$
5	R, C	$\frac{R - j\omega R^2 C}{1 + \omega^2 R^2 C^2}$	$\frac{R}{\left[1 + \omega^2 R^2 C^2\right]^{1/2}}$	$\tan^{-1} - \omega RC$	$\frac{1}{R} + j\omega C$
6	L, C	$j\frac{\omega L}{1 - \omega^2 LC}$	$\frac{\omega L}{1 - \omega^2 LC}$	$\pm\frac{\pi}{2}$	$-j\left(\frac{1 - \omega^2 LC}{\omega L}\right)$
7	R, L, C	$\frac{\frac{1}{R} - j(\omega C - \frac{1}{\omega L})}{\left[(\frac{1}{R})^2 + (\omega C - \frac{1}{\omega L})^2\right]}$	$\frac{1}{\left[(\frac{1}{R})^2 + (\omega C - \frac{1}{\omega L})^2\right]^{1/2}}$	$\tan^{-1} R\left(\frac{1}{\omega L} - \omega C\right)$	$\frac{1}{R} + j\left(\omega C - \frac{1}{\omega L}\right)$
8	R, L, C	$\frac{\frac{R}{\omega^2 C^2} - j\left[\frac{R^2}{\omega C} + \frac{L}{C}(\omega L - \frac{1}{\omega C})\right]}{R^2 + (\omega L - \frac{1}{\omega C})^2}$	$\frac{\left\{\left(\frac{R}{\omega^2 C^2}\right)^2 + \left[\frac{R^2}{\omega C} + \frac{L}{C}(\omega L - \frac{1}{\omega C})\right]^2\right\}^{\frac{1}{2}}}{R^2 + (\omega L - \frac{1}{\omega C})^2}$	$\tan^{-1} - \frac{\left[\frac{R^2}{\omega C} + \frac{L}{C}(\omega L - \frac{1}{\omega C})\right]}{\left(\frac{R}{\omega^2 C^2}\right)}$	$\frac{R + j\omega\left[R^2 C - L + \omega^2 L^2 C\right]}{R^2 + \omega^2 L^2}$
9	R_1, L, R_2, C	$\frac{R_1 R_2 (R_1 + R_2) + \omega^2 L^2 R_2 + \frac{R_1}{\omega^2 C^2}}{(R_1 + R_2)^2 + (\omega L - \frac{1}{\omega C})^2} + j\frac{\omega R_2^2 L - \frac{R_1^2}{\omega C} - \frac{L}{C}(\omega L - \frac{1}{\omega C})}{(R_1 + R_2)^2 + (\omega L - \frac{1}{\omega C})^2}$	$\left\{\left[\frac{R_1 R_2 (R_1 + R_2) + \omega^2 L^2 R_2 + \frac{R_1}{\omega^2 C^2}}{(R_1 + R_2)^2 + (\omega L - \frac{1}{\omega C})^2}\right]^2 + \left[\frac{\omega L R_2^2 - \frac{R_1^2}{\omega C} - \frac{L}{C}(\omega L - \frac{1}{\omega C})}{(R_1 + R_2)^2 + (\omega L - \frac{1}{\omega C})^2}\right]^2\right\}^{\frac{1}{2}}$	$\tan^{-1}\frac{\left[\omega L R^2 - \frac{R_1^2}{\omega C} - \frac{L}{C}(\omega L - \frac{1}{\omega C})\right]}{\left[R_1 R_2 (R_1 + R_2) + \omega^2 L^2 R_2 + \frac{R_1}{\omega^2 C^2}\right]}$	$\frac{R_1 + \omega^2 R_1 R_2 C^2 (R_1 + R_2) + \omega^4 L^2 C^2 R_2}{(R_1^2 + \omega^2 L^2)(\omega^2 R_2^2 C^2 + 1)} + j\frac{\omega\left[R_1^2 C - L + \omega^2 LC(L - R_2^2 C)\right]}{(R_1^2 + \omega^2 L^2)(\omega^2 R_2^2 C^2 + 1)}$

* From B. Dudley, *Electronics*, December, 1942.

Power and reactive voltamperes can be shown in a complex plane diagram, as in Fig. 2-21. Voltamperes are plotted as

$$E\hat{I}\ (\text{VA}) = P\ (\text{W}) + jQ\ (\text{vars}) \qquad (2\text{-}190)$$

where it is understood that E and I are effective values and the vars are positive when current lags the voltage. $\hat{I}$ is the conjugate of the vector $I\underline{/\alpha}$, that is, it has the same magnitude but its angle is $-\alpha$. When a complex quantity is expressed in the rectangular form, the conjugate has the same form except that j is changed to $-j$, or vice versa.

The voltampere diagram (Fig. 2-21) has the same shape as the current diagram (Fig. 2-20) except that its vertical components are reversed.

The above relations can be expressed in rectangular form (all quantities are effective values),

$$E = A + jB \qquad (2\text{-}191)$$

$$I = C + jD \qquad (2\text{-}192)$$

Then the power is

$$P = \text{Re}\,(E\hat{I}) = \text{Re}\,[(A + jB)(C - jD)] = AC + BD \qquad (\text{watts}) \quad (2\text{-}193)$$

where Re denotes "real part of." The reactive voltamperes, or vars, can be expressed as

$$Q = \text{Im}\,(E\hat{I}) = \text{Im}\,[(A + jB)(C - jD)] = BC - AD \qquad (\text{vars}) \quad (2\text{-}194)$$

where Im denotes "imaginary part of." Q is sometimes called reactive power.

152. Leading current through an inductive line can raise the voltage at the receiving end of the circuit. Referring to Fig. 2-22, let E be the voltage at the generator end of a circuit; e, the voltage at the receiver end; and i, the line current. Let the load be of such a nature that the current is leading with respect to the voltage e. By adding to e the ohmic drop ir in the line (Fig. 2-23) in phase with i and the reactive drop ix in leading quadrature with i, the impressed voltage E is obtained. It will be seen that E may be less than e with a leading current. With a lagging current, E is greater than e (Fig. 2-24).

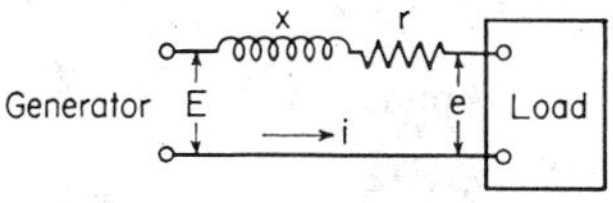

FIG. 2-22. Load connected to inductive line.

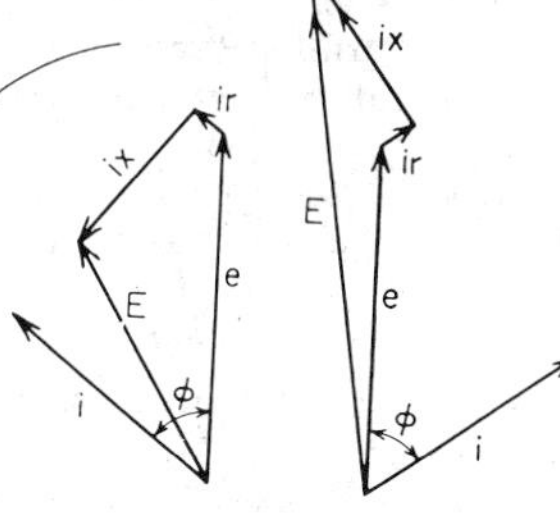

FIG. 2-23. Effect of inductive line with leading current.

FIG. 2-24. Effect of inductive line with lagging current.

Leading current, as described in this paragraph, is usually obtained in practice from static capacitors or overexcited synchronous machines connected in parallel with the load. Leading current from synchronous machines can be controlled by automatic voltage regulators so as to give constant voltage (or rising voltage) from no load to full load.

A static capacitor may be connected in series with the line. This has the effect of reducing the line reactance or even making it negative, according to Eq. (2-171). It does not reduce the line current as parallel capacitance does. Lagging current through a capacitance will tend to raise the voltage. Series capacitance can improve the division of current between unlike lines in parallel.

153. Series Resonance. In a circuit which contains inductive reactance and capacitive reactance in series, it is possible to obtain a very great rise in voltage across the reactances by adjusting them or the frequency. Thus, according to Par. **146**,

$$E = IZ = I\sqrt{r^2 + (x_l - x_c)^2} \qquad (\text{volts}) \qquad (2\text{-}195)$$

The voltage across the capacitive reactance is $e = Ix_c$, so that

$$\frac{e}{E} = \frac{x_c}{\sqrt{r^2 + (x_l - x_c)^2}} = \frac{x_c}{Z} \tag{2-196}$$

Z (the total impedance) may be less than x_c, and in this case e (the drop across the capacitor's terminals) will be greater than E (the total impressed emf). When the frequency is

$$f = \frac{1}{2\pi\sqrt{LC}} \quad \text{(Hz or c/s)} \tag{2-197}$$

the reactances are equal,

$$x_l = x_c \tag{2-198}$$

This condition gives the highest current for voltage E. Moreover, if r (the resistance of the circuit) is assumed to be zero,

$$\frac{e}{E} = \infty \tag{2-199}$$

and we have an extreme case of voltage resonance. Actually, some resistance r is always present. Then

$$\frac{e}{E} = \frac{x_c}{r} \quad \text{when } f = \frac{1}{2\pi\sqrt{LC}} \tag{2-200}$$

where L is the coefficient of self-induction in henrys and C the capacitance in farads of the apparatus connected in series. Resonance at one of the higher harmonics of the applied voltage may also take place when Eq. (2-200) is satisfied by the frequency of the harmonic.

154. Parallel Resonance. When an inductive reactance and a capacitive reactance are in parallel, they can be so adjusted or the frequency can be so chosen that current resonance takes place.

Let the total conductance of the combination be g, the inductive susceptance b_l, and the capacitive susceptance b_c. Then the total current is

$$I = Ey = E\sqrt{g^2 + (b_l - b_c)^2} \tag{2-201}$$

The current through the capacitive susceptance is

$$i_c = Eb_c \tag{2-202}$$

Hence

$$\frac{i_c}{I} = \frac{b_c}{\sqrt{g^2 + (b_l - b_c)^2}} = \frac{b_c}{y} \tag{2-203}$$

When the total admittance y is smaller than b_c, the total line current I is less than one of its components i_c. A similar relation may be proved for i_l. When the frequency is

$$f = \frac{1}{2\pi\sqrt{LC}} \quad \text{(c/s)} \tag{2-204}$$

it follows that

$$b_l = b_c \tag{2-205}$$

and

$$I = Eg \qquad i_l = i_c \tag{2-206}$$

The line current is comparatively small, but there is a large interchange of current between the inductance and the capacitance in parallel.

COMPLEX (NONSINUSOIDAL) WAVEFORMS

155. Examples of Complex Waveforms. Figure 2-25 illustrates the effect of the inductance and capacitance in a circuit to which is applied an alternating emf differing from the simple sine wave. The curves were taken simultaneously with an oscillograph.

E is the impressed emf; I_l, the current taken by an inductance coil; and I_c, that taken by a capacitor. Figure 2-26 shows the circuit.

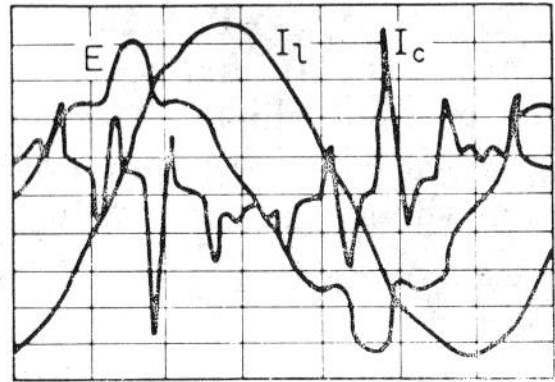

FIG. 2-25. Complex a-c waveforms.

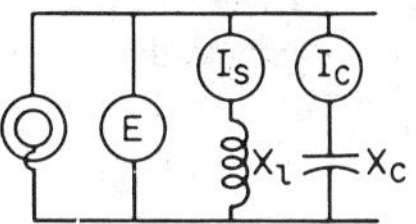

FIG. 2-26. Circuit in which the waveforms of Fig. 2-25 were observed.

156. Waveform of Reactive Emf Due to Inductive Reactance. On the assumption that the reluctance of the iron core in the inductance coil is constant, which is approximately true below the saturation point, the value of the flux is proportional to the current I_l. The instantaneous value of the emf (see Par. **75**) is

$$e = n\frac{d\phi}{dt} = L\frac{di}{dt} \tag{2-207}$$

That is, the curve E will have its maximum amplitude when the curve I passes through zero. This is not precisely true, because the current needed to supply losses in the resistance and in the iron is in phase with the emf E.

156a. Waveform of Current through Capacitive Reactance. The capacitive current is proportional to the rate of change of the emf (see Par. **119**); the instantaneous value is

$$i_c = C\frac{de}{dt} \tag{2-208}$$

That is, the curve I_c has its maximum when the rate of change of the curve E is a maximum. When E is a sine curve, I_c is also a sine curve, in quadrature with E. When the curve of emf is not a sine curve, as in Fig. 2-25, the maximum amplitude of the current I_c occurs at the point where the slope of the emf curve is a maximum.

156b. Effects of Inductive and Capacitive Reactance on Waveform. The curves in Fig. 2-25 show the effect upon the current waveform of inductive reactance and capacitive reactance. The waveform E produced by the generator contains several harmonics (see Par. **167**). The inductive reactance tends to damp out the higher harmonics, while the capacitive reactance emphasizes them.

157. Determination of Total Complex Current Waveform. When the applied voltage contains higher harmonics (Par. **167**), the total current through an impedance is found by summing the harmonic currents due to each harmonic of the voltage acting alone. Thus, the reactance at the fundamental frequency f is $x_1 = 2\pi fL$; the reactance to the nth harmonic is $x_n = 2\pi nfL$; and the impedance to the nth harmonic is

$$Z_n = \sqrt{r^2 + (2\pi nfL)^2} \qquad \text{(ohms)} \tag{2-209}$$

158. Power and Energy. The general expression for the energy delivered to an a-c circuit with any waveform of current and voltage is

$$W = \int_{t_1}^{t_2} ei\,dt \qquad \text{(joules)} \tag{2-210}$$

where e is the instantaneous value of the voltage in volts; i is the corresponding instantaneous current in amperes; and $t_2 - t_1 = T$ is the interval of time, in seconds, for which the energy is to be determined. The average power delivered during the interval T is

$$P = \frac{1}{t_2 - t_1}\int_{t_1}^{t_2} ei\,dt \qquad \text{(watts)} \tag{2-211}$$

159. Power and Power Factor, Sine Waveforms. When the current and the voltage vary according to the sine law, the power $P = EI \cos \phi$, where E and I are the effective values of the voltage and the current, respectively, and ϕ is the phase angle between the two, $\cos \phi$ being known as the power factor of the circuit. See also Par. **151**.

160. Power, Complex Waveforms. When e and i are irregular curves, the average power is found as the average ordinate of a curve the ordinates of which are proportional to the product ei. If e and i are resolved into their harmonics, each harmonic contributes its own share of power as if it were acting alone, so that the average power for a large number of cycles is

$$P = E_1 I_1 \cos \phi_1 + E_3 I_3 \cos \phi_3 + \cdots \tag{2-212}$$

where $I_1, I_3, \ldots$ and $E_1, E_3, \ldots$ are the effective values of the harmonic currents and voltages, respectively, and the angles ϕ are the respective phase displacements.

161. Energy Component and Reactive Component of Voltage or Current. In a simple harmonic circuit with the voltage E, current I, and phase displacement ϕ between the two, $E \cos \phi$ is called the energy component of the voltage; and $E \sin \phi$, the reactive component of the voltage. Analogously, $I \cos \phi$ is the energy component of the current, and $I \sin \phi$ is the reactive component of the current. Similar components are used in circuits with nonsinusoidal currents and voltages, provided that these are first replaced by equivalent sine waves.

162. Effective Value of Any Wave. The effective value of a variable current or voltage is defined as that continuous value which gives the same total i^2r loss. That is, if I is the effective value of a periodic current i and T is the time of one cycle,

$$I^2rT = \int_0^T i^2r\,dt \qquad \text{(joules)}$$

from which

$$I = \sqrt{\frac{1}{T}\int_0^T i^2\,dt} \qquad \text{(amp)} \tag{2-213}$$

This may be expressed by saying that the *effective value of a current or voltage is equal to the square root of the mean square (rms) of the variable values taken throughout one cycle.* Hot-wire instruments and electrodynamometer-type instruments indicate directly the effective values of alternating currents and voltages.

163. Effective Value of a Sine Wave. For sine waves, the effective value is given in Par. **141**. In terms of the maximum value, the effective value is

$$E_{\text{eff}} = 0.707E_{\text{max}} \tag{2-214}$$

164. The crest factor is the ratio of the maximum or crest value (peak value) to the effective value; thus,

$$\frac{y_{\text{max}}}{y_{\text{eff}}} = \text{crest factor} \tag{2-215}$$

The crest factor for a sine wave is $\sqrt{2} = 1.414$.

165. The form factor is the ratio of the effective value to the half-period mean value; thus,

$$\frac{y_{\text{eff}}}{y_{\text{mean}}} = \text{form factor} \tag{2-216}$$

The form factor for a sine wave is $\dfrac{\pi}{2\sqrt{2}} = 1.111$.

166. Waveform Analysis; Fourier's Series. In the mathematical treatment of alternating waves, it is most convenient to work with those having sine form. Waves differing from the sine form may be resolved into a fundamental sine wave and its harmonics. The general equation of any alternating wave, as given by Fourier's series, is

$$y = Y_1 \sin(\omega t + \theta_1) + Y_2 \sin(2\omega t + \theta_2) + \cdots + Y_n \sin(n\omega t + \theta_n) \tag{2-217}$$

wherein y is the ordinate of the resultant wave at time t; Y_1, Y_3, ... Y_n, ... are the maximum ordinates or amplitudes of the first, second, ..., nth, ... harmonics; θ_1, θ_2, ..., θ_n, the constant angles which determine the relative time-phase position of the corresponding harmonics; and $\omega = 2\pi f$, the angular velocity of the generating vector of the fundamental wave, corresponding to the fundamental frequency f in hertz.

167. Waveform Analysis; Fischer-Hinnen's Method. Waves having like loops above and below the time axis contain only odd harmonics, whereas waves having unlike loops above and below the axis contain both even and odd harmonics. A direct method of wave analysis given by J. Fischer-Hinnen[1] is based on the following equations:

$$A_n = \frac{1}{n}(y_4 + y_8 + y_{12} + \cdots + y_{2n-2} - y_2 - y_6 - Y_{10} - \cdots - y_{2n-4}) \qquad (2\text{-}218)$$

and

$$B_n = \frac{1}{n}(y_1 + y_5 + y_9 + \cdots + y_{2n-1} - y_3 - y_7 - y_{11} - \cdots - y_{2n-3}) \qquad (2\text{-}219)$$

where y_1, y_2, ..., y_n are ordinates at points along the base of the half wave, which is divided into $2n$ equal parts; and A_n and B_n are the ordinates of two quadrature components of the nth harmonic. The maximum ordinate of the nth harmonic is $\sqrt{A_n^2 + B_n^2}$, and its time-phase displacement from the resultant wave is

$$\theta_n = \tan^{-1}\frac{A_n}{-B_n} \qquad (2\text{-}220)$$

θ_n being measured in terms of the nth harmonic.

The above equations for the nth harmonic do not take into account the harmonics of the harmonics $2n$, $3n$, etc. This correction is practically negligible for all harmonics, except the first or fundamental, and a correction rarely needs to be carried beyond the ninth harmonic.

Since waveforms generated by electric machinery almost never contain even harmonics, they do not enter into the correction. Denoting the corrected values by prime, we have

$$A'_n = A_n - A'_{3n} - A'_{5n} - A'_{7n} - \cdots \qquad (2\text{-}221)$$

and

$$B'_n = B_n + B'_{3n} - B'_{5n} + B'_{7n} - \cdots \qquad (2\text{-}222)$$

In applying this to the first harmonic, A_n is the ordinate of the resultant wave at y_0 (Fig. 2-28), and B_n is the ordinate displaced 90 time deg therefrom, at y_3.

168. Example of Wave Analysis. As an example,[2] assume the wave given in Fig. 2-27, which is split into three harmonics: the first, or fundamental; the third; and the fifth. Figure 2-28 shows the method of determining a given harmonic, in this case the third. The base of the wave is divided into $2n$ or six equal parts, and ordinates are erected. The ordinates are;

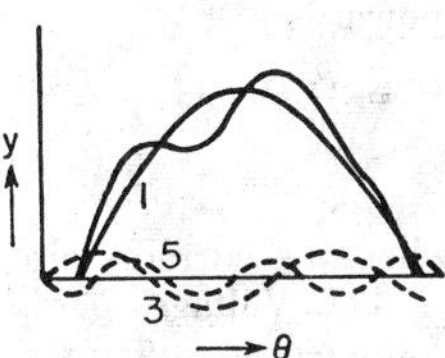

FIG. 2-27. Waveform analysis, Par. **168.**

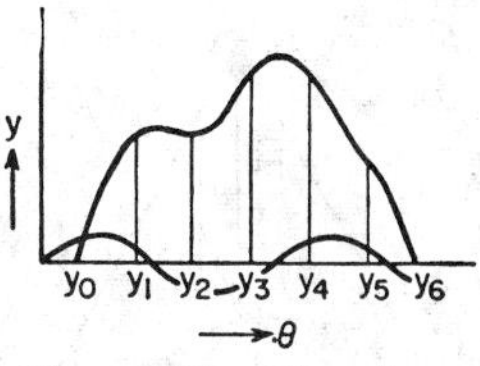

FIG. 2-28. Waveform analysis, Par. **168.**

$$y_1 = 676 \quad y_2 = 660 \quad y_3 = 940 \quad y_4 = 1{,}004 \quad y_5 = 554 \quad y_6 = 0 \qquad (2\text{-}223)$$

Then

$$A_3 = \frac{1}{3}(y_4 - y_2) = \frac{1{,}004 - 660}{3} = 114.7 \qquad (2\text{-}224)$$

[1] FISCHER-HINNEN, J. *Elek. Zt.*, 1901, Vol. 22, p. 396. Also LINCOLN, P. M. *Elec. Jour.*, 1908, vol. 5, p. 386.

[2] *Elec. Jour.*, 1908, Vol. 5, p. 386.

and
$$B_3 - B_2 = \frac{1}{2}(y_1 + y_5 - y_3) = \frac{676 + 554 - 940}{3}$$
$$= 96.7 \tag{2-225}$$

The maximum ordinate is
$$\sqrt{(114.7)^2 + (96.7)^2} = 150 \tag{2-226}$$
and the phase angle is
$$\theta_3 = \tan^{-1}\frac{-114.7}{96.7} = -50°\ * \tag{2-227}$$

In a similar manner, it is found that $A_5 = -92.8$ and $B_5 = 37.4$.

In this example, the wave contains only the third and the fifth harmonics; therefore, the fundamental is determined as follows:

$$A_1 = y_0 - A_3' - A_5' = 0 - 114.7 + 92.8 = -21.9$$
$$B_1 = y_3 + B_3' - B_5' = 940 + 96.7 - 37.4 = 999.3$$
$$\theta_1 = \tan^{-1}\frac{21.9}{999.3} = 1°15' \text{ (approx.)}$$

THREE-PHASE SYSTEMS

169. Three-phase Y and Δ Connections. In a balanced 3-phase system, the star connection is also called the Y connection (Fig. 2-29). The relations of the currents and the voltages are (Fig. 2-31)

$$E_\Delta = E_Y\sqrt{3} = 1.732E_Y \quad \text{and} \quad I_\Delta = \frac{I_Y}{\sqrt{3}} = \frac{I_Y}{1.732} \tag{2-228}$$

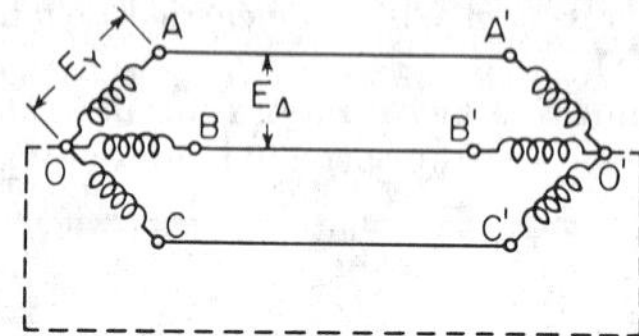

FIG. 2-29. Three-phase Y-(wye) connection.

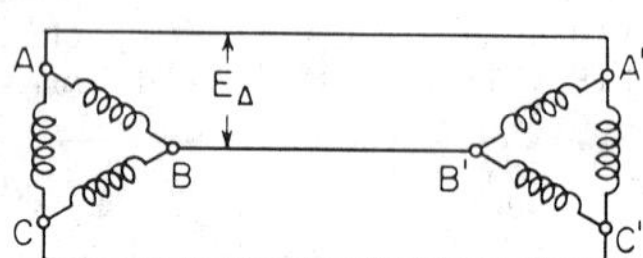

FIG. 2-30. Three-phase Δ-(delta) connection.

170. Three-phase Power. In a balanced 3-phase system (Figs. 2-29 and 2-30), the power is

$$P = 3I_YE_Y\cos\phi = 3I_\Delta E_\Delta\cos\phi = I_YE_\Delta\sqrt{3}\cos\phi \quad \text{(watts)} \tag{2-229}$$

when the currents are in amperes and the voltages in volts.

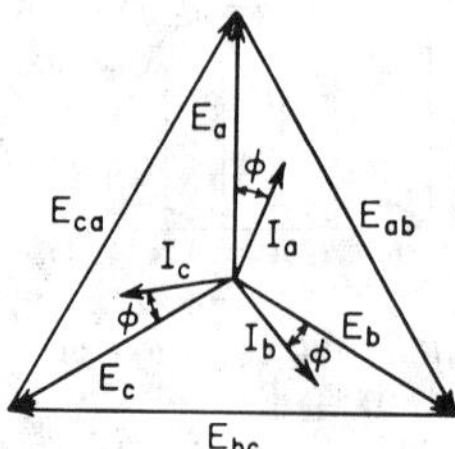

FIG. 2-31. Three-phase voltages and currents.

171. Voltage Drop in Unsymmetrical Circuits. The voltage drop due to resistance and self- and mutual inductance in any conductor of a group of long, parallel, round, nonmagnetic conductors forming a single-phase or polyphase circuit, and with one or more conductors connected electrically in parallel, may be calculated by summing the flux due to each conductor up to a certain large distance u. The vectorial sum of all the currents is zero in a complete system of currents in the steady state, and the quantity u cancels out, so that the result is the same, no matter how large u may be. The currents may be unbalanced, and, in addition, the arrangement of the conductors may be unsymmetrical.

* Fifty degrees in terms of the third harmonic $3f$, or $50/3°$ in terms of the fundamental frequeney f.

The voltage drop in any conductor a of a group of round conductors $a, b, c, \ldots$ is

$$I_aR_a = j0.2794(I_a \log_{10} G_a + I_b \log_{10} S_{ab} + I_c \log_{10} S_{ac} + \cdots) \quad \text{(V/mi at 60 cycles)} \qquad (2\text{-}230)$$

where $I_a + I_b + I_c + \cdots = 0$, the values of the currents being complex quantities; R_a = resistance of conductor a, per mile; G_a = self-geometric mean distance of conductor a; S_{ab} = axial spacing between conductors a and b, etc. The values of G and S should be in the same units (see Par. **82**). For further reference on the evaluation of the self- and mutual geometric mean distances see reference given in Par. **82**.

172. Symmetrical Components—Resolution of an Unbalanced Three-phase System into Balanced Systems. Let the three cube roots of unity, 1, $\epsilon^{j(2\pi/3)}$, $\epsilon^{j(4\pi/3)}$, be 1, a, a^2, where $j = \sqrt{-1}$,

$$a = 1\underline{/120°} = -0.5 + j0.866$$

and

$$a^2 = 1\underline{/240°} = -0.5 - j0.866$$

Any three vectors Q_a, Q_b, Q_c (which may be unsymmetrical or unbalanced, i.e., with unequal magnitudes or with phase differences not equal to 120°) can be resolved into a system of three equal vectors Q_{a0}, Q_{a0}, Q_{a0} and two symmetrical (balanced) 3-phase systems Q_{a1}, a^2Q_{a1}, aQ_{a1} and Q_{a2}, aQ_{a2}, a^2Q_{a2}, the first of which is of positive phase sequence and the second of negative phase sequence. Thus

$$\begin{aligned} Q_a &= Q_{a0} + Q_{a1} + Q_{a2} \\ Q_b &= Q_{a0} + a^2Q_{a1} + aQ_{a2} \qquad (2\text{-}231) \\ Q_c &= Q_{a0} + aQ_{a1} + a^2Q_{a2} \end{aligned}$$

The values of the component vectors are

$$\begin{aligned} Q_{a0} &= \tfrac{1}{3}(Q_a + Q_b + Q_c) \\ Q_{a1} &= \tfrac{1}{3}(Q_a + aQ_b + a^2Q_c) \qquad (2\text{-}232) \\ Q_{a2} &= \tfrac{1}{3}(Q_a + a^2Q_b + aQ_c) \end{aligned}$$

The three equal vectors Q_{a0} are sometimes called the "residual quantities," or the zero-phase, or uniphase, sequence system. Any of the vectors Q_a, Q_b, or Q_c may have the value zero. If two of them are zero, the single-phase system may be resolved into balanced 3-phase systems by the above equations. The symbol Q may denote any vector quantity such as voltage, current, or electric charge.

There are similar relations for n-phase systems. See Method of Symmetrical Co-ordinates Applied to the Solution of Polyphase Networks, by C. L. Fortescue, *Trans. AIEE*, 1918, p. 1027.

173. The calculation of short-circuit currents in 3-phase power networks is a common application of the method of symmetrical components.

The location of a probable short circuit or fault having been selected, three networks are computed in detail from the neutrals to the fault, one for positive, one for negative, and one for zero-phase sequence currents. The three phases are assumed to be identical, in ohms and in mutual effects, except in the connection of the fault itself. Let Z_1, Z_2, and Z_0 be the ohms per phase between the neutrals and the fault in each of the networks, including the impedance of the generators.

Then, for a line-to-ground fault,

$$I_{a1} = I_{a2} = I_{a0} = \frac{I_a}{3} = \frac{E_a}{Z_1 + Z_2 + Z_0} \qquad (2\text{-}233)$$

where E_a is the line-to-neutral voltage and I_{a1} is the positive-phase-sequence current flowing to the fault in phase a and similarly for I_{a2} and I_{a0}. I_a is the total current flowing to the fault in phase a.

The component currents in phases b and c are derived from those in phase a, by means of the relations $I_{b1} = a^2 I_{a1}$, $I_{b2} = aI_{a2}$, $I_{b0} = I_{a0}$, $I_{c1} = aI_{a1}$, $I_{c2} = a^2 I_{a2}$, and $I_{c0} = I_{a0}$.

Each of the component currents divides in the branches of its own network according to the impedance of that network. Thus, each of the component currents, and therefore the total current, at any part of the power system can be determined.

For a line-to-line fault between phases b and c,

$$I_{a1} = -I_{a2} = \frac{E_a}{Z_1 + Z_2} \tag{2-234}$$

and

$$I_{a0} = 0 \tag{2-235}$$

For a double-line-to-ground fault between phases b and c and ground,

$$I_{a1} = \frac{E_a}{Z_1 + \dfrac{Z_2 Z_0}{Z_2 + Z_0}} \tag{2-236}$$

$$I_{a0} = -\frac{I_{a1} Z_2}{Z_2 + Z_0} \tag{2-237}$$

and

$$I_{a2} = -I_{a1} - I_{a0} \tag{2-238}$$

See C. F. Wagner and R. D. Evans, "Symmetrical Components," McGraw-Hill Book Company; W. V. Lyon, "Applications of the Method of Symmetrical Components," McGraw-Hill Book Company; W. D. Stevenson, Jr., "Elements of Power System Analysis," McGraw-Hill Book Company, 1962; and "Electrical Transmission and Distribution Reference Book," 4th ed., Westinghouse Electric Corporation, 1950.

If there is no current in the power system before the fault occurs, the voltage E_a of every generator is the same in magnitude and phase. Such a condition often is assumed in calculated circuit-breaker duty and relay currents although the effects of loads on the system can be included in the analysis.

In calculating power-system stability, however, it must be assumed that current exists in the lines before the fault occurs. The voltage E_a becomes the positive-sequence voltage at the point of fault before the fault occurs. A practical method of computing the positive-sequence current under fault conditions is to leave the positive-sequence network unchanged, with each generator at its own voltage and phase angle. The equivalent Z_1 of the network need not be computed. Certain 3-phase impedances are inserted between line and neutral at the location of the fault. For a single line-to-ground fault $Z_2 + Z_0$ is inserted, for a line-to-line fault Z_2 is inserted, and for a double line-to-ground fault $Z_2 Z_0/(Z_2 + Z_0)$ is inserted. This gives one phase of an equivalent balanced 3-phase circuit for which the positive-sequence currents driven by all the generators in all the branches under fault conditions can be found by means of a network analyzer or computed on a digital computer. The power transmitted after the fault occurs can be determined from these positive-sequence currents.

If it is desired to find the negative-sequence and zero-sequence currents (some relays are operated by the latter), they can be computed from Eqs. (2-233) to (2-238) that do not involve E_a, after finding I_{a1} to the fault.

The impedance Z_f of each arc is mainly resistance. It may be brought into the computation. For single line-to-ground and double line-to-ground faults, Z_f is added to each of Z_1, Z_2, and Z_0. For line-to-line faults, Z_f is added to Z_2 only.

See Chap. 6, "Electrical Elements of Power Transmission Lines," by H. B. Dwight, The Macmillan Company.

174. Load Studies. In calculations relating to the steady-state operation of power systems, in which it is desired to determine the voltage, power, reactive power, etc., at various points, the loads may be designated by kilowatts and kilovars, rather than by impedances. The effect of the impedance of the transmission and distribution lines, transformers, etc., of the network can be computed. The modern method is a process of iterations using a digital computer for the calculations. Alternatively, the system may be simulated on a network analyzer. The division of current in branches, the voltage at various points, and the required ratings of synchronous capacitors can be determined.

Conditions can be estimated at one or two points and a solution for the rest of the network calculated on the basis of these assumptions. If the assumptions are not correct, discrepancies will appear at the end of the work. For instance, two different voltages may be obtained for the same point, one calculated before, and one after, going around a loop of the network. The necessary correction to the first estimates may be based on the discrepancies, thus giving successive approximations which are improvements on the preceding ones.

THREE-PHASE ARMATURE WINDINGS

175. The armature winding of a 3-phase generator or motor is an important type of electric circuit. Windings consisting of diamond-shaped coils, with two coil sides per slot, are connected in groups of coils, three groups or phase belts being opposite each pole. In general, the number of slots per pole per phase is a fraction equal to the average number of coils per phase belt. There are a larger number and a smaller number of coils per phase belt, differing by 1. The winding is usually found to be divided into repeatable sections of several poles each, the sections being duplicates of each other.

The number of poles in a section is found by writing the fraction equal to the number of slots divided by the number of poles and canceling factors to the extent possible. The denominator is the number of poles per section, and the numerator is the number of slots per section.

If the final value of the numerator is not divisible by 3, a balanced 3-phase winding cannot be made, since the windings for phases *a*, *b*, and *c* in a section each require the same number of slots and they must be duplicates except for the phase shift of 120°. This gives rise to the rule for balanced 3-phase windings that the factor 3 must occur at least one more time in the number of slots than in the number of poles.

It can be shown that the slots of a repeatable section have phase angles which, when suitably drawn, are all different and equidistant. They fill the space of 180 elec deg like the blades of a Japanese fan (see "Electrical Coils and Conductors," by H. B. Dwight, Chap. 8, pp. 43 and 44). The angle between the vectors in this fan is

$$\beta = \frac{180}{\text{slots per section}} \qquad \text{(deg)} \tag{2-239}$$

The vectors lying from 0 to 59 deg may be assigned to phase a or $-a$, those from 60 to 119 deg to phase $-c$ or c, and those from 120 to 179 deg to phase b or $-b$.

The phase angles for the upper coil sides of the slots should be tabulated to indicate the proper connections of the winding. Since the diamond coils are all alike, the total resultant voltage developed in the lower coil sides of a phase is a duplicate of that developed in the upper coil sides and can be added on by means of the pitch factor.

The phase angle between two adjacent slots is

$$\frac{\text{Poles per section} \times 180}{\text{Slots per section}} = q\beta \qquad \text{(deg)} \tag{2-240}$$

where q is the number of poles in the repeatable section. From this, the phase angle for every slot in the section can be written. To save numerical work, especially where β is a fractional number of degrees, the angles may be expressed in terms of the angle β, as given in the example below. They may be expressed in degrees and fractions of a degree, but decimal values of degrees should not be used in this part of the work. The required accuracy is obtained by using fractions instead of decimals. Appropriate multiples of 180° should be subtracted to keep the angles less than 180°, thus indicating the relative position of each coil side with respect to the nearest pole. When an odd number times 180° has been subtracted, the coil side is tabulated as $-a$ instead of a, etc., since it will be opposite a south pole when a is opposite a north pole. The terminals of a coil marked $-a$ are reversed with respect to the terminals of a coil marked a with which it is in series.

Example. 21 slots per repeatable section; 5 poles per section; $1\frac{2}{5}$ slots per pole per phase.

$$\beta = \frac{180}{21} = 8\frac{4}{7}\text{ deg} \qquad [\text{by Eq. (2-239)}]$$

It is more convenient in this case to express the angles in terms of β rather than by fractions of degrees. Note that $21\beta = 180°$ and $7\beta = 60°$. The range for coils to be marked $\pm a$ is from 0 to 6β, inclusive; coils marked $\mp c$ from 7β to 13β; and coils marked $\pm b$ from 14β to 20β. Subtract multiples of $21\beta = 180°$. The angle between two adjacent slots is $q\beta = 5\beta$.

Tabulation of phase angles:

1	2	3	4	5	6	7	8	9	10
0	5β	10β	15β	20β	$(25\beta)4\beta$	9β	14β	19β	$(24\beta)3\beta$
a	a	$-c$	b	b	$-a$	c	$-b$	$-b$	a

11	12	13	14	15	16	17	18	19	20	21	[22]
8β	13β	18β	$(23\beta)2\beta$	7β	12β	17β	$(22\beta)\beta$	6β	11β	16β	$[(21\beta)0]$
$-c$	$-c$	b	$-a$	c	c	$-b$	a	a	$-c$	b	$[-a]$

The seven vectors of phase a make a regular fan covering $7\beta = 60°$.

The resultant terminal voltage produced by the coils of phase a is equal to the numerical sum of the voltages in those coils multiplied by the "distribution factor"

$$\frac{\sin\,(n\beta/2)}{n\sin\,(\beta/2)} \tag{2-241}$$

where n is the number of vectors in the regular fan covering 60° and β is the angle between adjacent vectors, given by Eq. (2-238). The number n is large, and the perimeter approaches the arc of a circle. Expression (2-240) is of the same form as the formula for breadth factor, which also is based on a vector diagram that is a regular fan.

The distribution factor for the winding of the foregoing example is

$$\frac{\sin\dfrac{7\times 60°}{2\times 7}}{7\sin\dfrac{60°}{2\times 7}} = \frac{\sin 30°}{7\sin 4\dfrac{2°}{7}} = \frac{0.5}{7\times 0.0746} = 0.956$$

Other possible balanced 3-phase windings for this example could be specified by having some of the vectors of phase a lie outside the 60° range. This would result in a lower distribution factor. The voltage, and hence the rating of the machine, would be lower by 2% or more than in the case described. The canceling of the harmonic voltages would apparently not be improved, and there would be no advantage to compensate for the reduction in kva rating, which would correspond to a loss or waste of 2% or more of the cost of the machine.

GENERAL BIBLIOGRAPHY

176. General Reference Literature

Abraham, M. "The Classical Theory of Electricity and Magnetism," revised by R. Becker, translated into English by J. Dougall; Glasgow, Blackie & Son, Ltd., 1932.

Attwood, S. S. "Electric and Magnetic Fields"; New York, John Wiley & Sons, Inc., 1941.

Bitter, F. "Introduction to Ferromagnetism"; New York, McGraw-Hill Book Company, 1937.

Boast, W. B. "Vector Fields"; New York, Harper & Row, Publishers, Incorporated, 1964.

Booker, H. G. "An Approach to Electrical Science"; New York, McGraw-Hill Book Company, 1959.

Brenner, E., and Javid, M. "Analysis of Electric Circuits"; New York, McGraw-Hill Book Company, 1959.

Cassell, W. L. "Linear Electric Circuits"; New York, John Wiley & Sons, Inc., 1964.

Christie, C. V. "Electrical Engineering"; New York, McGraw-Hill Book Company, 1952.

Clarke, E. "Circuit Analysis of A-C Power Systems"; New York, John Wiley & Sons, Inc., 1943, Vols. I and II.

Dwight, H. B. "Electrical Elements of Power Transmission Lines"; New York, The Macmillan Company, 1954.

"Electrical Transmission and Distribution Reference Book"; Westinghouse Electric Corporation, 1950.

Faraday, M. "Experimental Researches in Electricity," 3 vols.; London, B. Quaritch, 1839–1855.

Fitzgerald, A. E., and Kingsley, C., Jr. "Electric Machinery", 2d ed.; New York, McGraw-Hill Book Company, 1961.

Foecke, H. A. "Introduction to Electrical Engineering Science"; Englewood Cliffs, N.J., Prentice-Hall, Inc., 1961.

Frank, N. H. "Introduction to Electricity and Optics," 2d ed.; New York, McGraw-Hill Book Company, 1950.

Hammond, S. B. "Electrical Engineering"; New York, McGraw-Hill Book Company, 1961.

Harnwell, G. P. "Principles of Electricity and Electromagnetism," 2d ed.; New York, McGraw-Hill Book Company, 1949.

Hayt, W. H., Jr. "Engineering Electromagnetics"; New York, McGraw-Hill Book Company, 1958.

Hayt, W. H., Jr., and Kemmerly, J. E. "Engineering Circuit Analysis"; New York, McGraw-Hill Book Company, 1962.

Jeans, J. H. "Mathematical Theory of Electricity and Magnetism"; New York, Cambridge University Press, 1908.

McGraw-Hill Encyclopedia of Science and Technology, 1960.

Manning, L. A. "Electrical Circuits"; New York, McGraw-Hill Book Company, 1965.

Maxwell, J. C. "A Treatise on Electricity and Magnetism," 2 vols.; New York, Oxford University Press, 1904.

Neal, J. P. "Introduction to Electrical Engineering Theory"; New York, McGraw-Hill Book Company, 1960.

Page, L., and Adams, N. I. "Principles of Electricity"; Princeton, N.J., D. Van Nostrand Company, Inc., 1934.

Peek, F. W., Jr. "Dielectric Phenomena in High-voltage Engineering"; New York, McGraw-Hill Book Company, 1929.

Rosa, E. B., and Grover, F. W. Formulas and Tables for the Calculation of Mutual and Self-inductance, *NBS Sci. Paper* 169, 1916. Published also as Pt. 1 of Vol. 8, *NBS Bull.* Contains also skin-effect tables.

Sabbagh, E. M. "Circuit Analysis"; New York, The Ronald Press Company, 1961.

Sears, F. W. "Principles of Physics"; Reading, Mass., Addison-Wesley Publishing Company, Inc., 1946, Vol. 2, Electricity and Magnetism.

Seshu, S., and Balabanian, N. "Linear Network Analysis"; New York, John Wiley & Sons, Inc., 1959.

Skilling, H. H. "Electromechanics"; New York, John Wiley & Sons, Inc., 1962.

Smythe, W. R. "Static and Dynamic Electricity," 2d ed.; New York, McGraw-Hill Book Company, 1950.

Stevenson, W. D., Jr. "Elements of Power System Analysis," 2d ed.; New York, McGraw-Hill Book Company, 1962.

Ward, R. P. "Introduction to Electrical Engineering"; Englewood Cliffs, N.J., Prentice-Hall, Inc., 1960.

Woodruff, L. F. "Principles of Electric Power Transmission"; New York, John Wiley & Sons, Inc., 1938.

FILTERS

By Harry W. Hale

177. Definition. A filter is a two-terminal-pair network designed so that currents and/or voltages in certain frequency ranges are passed with small (ideally zero) attenuation and those in other ranges are subjected to large attenuation. Frequency ranges of small attenuation are called passbands, and those of large attenuation are called stop bands. The frequency at the boundary between a passband and a stop band is called a cutoff frequency.

Most filters are composed of inductance and capacitance, although certain classes of filters may be composed of resistance and either capacitance or inductance. The filters discussed here are of the first category and are based on the "image-parameter" method of design.

178. Classification by Pass Region. The low-pass filter passes all frequencies below a specified cutoff frequency, whereas the high-pass filter passes all frequencies above a cutoff frequency. The bandpass filter passes all frequencies within given limits, between two or more cutoff frequencies, whereas the band-elimination filter attenuates frequencies within such a band or bands.

The degree of loss in the attenuation regions increases with the number of sections included in the filter and, in addition, varies with frequency. Within the pass regions, there is a loss called the insertion loss of the filter, and this loss may vary with frequency within the pass region.

In general, the losses in the pass region may be minimized by employing inductive and capacitive elements which have a minimum of resistive component (high-Q elements). Losses may also arise, even if the resistive components are small, due to reflections of the signal at frequencies at which there is an impedance mismatch between sections of the filter or between the filter and its termination.

179. Characteristic Image Impedance. Since filter networks are usually terminated in resistive transducers (which make use of the energy thus delivered), it is necessary to design the networks to match the impedance presented by the terminating resistor, at least as closely as possible within the pass region, so as to minimize reflection losses and variations of attenuation with frequency. For each given filter section, composed of ideal inductive and capacitive elements, a characteristic "image" impedance may be computed, which is presented (within stated limits of variation) to the terminals of the network. This characteristic impedance is usually designed to have the same value as the terminating resistance.

180. Attenuation in filters is measured in decibels or nepers. In the absence of reflections (when the impedances are matched at each junction between sections), the total attenuation of a multisection filter is the sum of the attenuations of the individual sections. The shape of the attenuation curve, particularly the sharpness of cutoff between the pass- and stop bands, depends on the type of individual filter sections used.

181. Pi-, L-, and T-sections. The two basic filter configurations, shown in Fig. 2-32, are the pi-section and the T-section. Each may be broken down into half sections which have an inverted L shape. The pi- and T-sections are symmetrical and may offer identical input and output impedances, whereas the L-section is asymmetrical. If the multisection filter (Fig. 2-32) has series elements of one value Z_1 and shunt elements of another value Z_2, then the T-section consists of two series arms of $Z_1/2$ and a shunt arm of Z_2; the pi-section consists of one series element of Z_1 and two shunt elements of $2Z_2$; and the L-section consists of one series element of $Z_1/2$ and one shunt element of $2Z_2$.

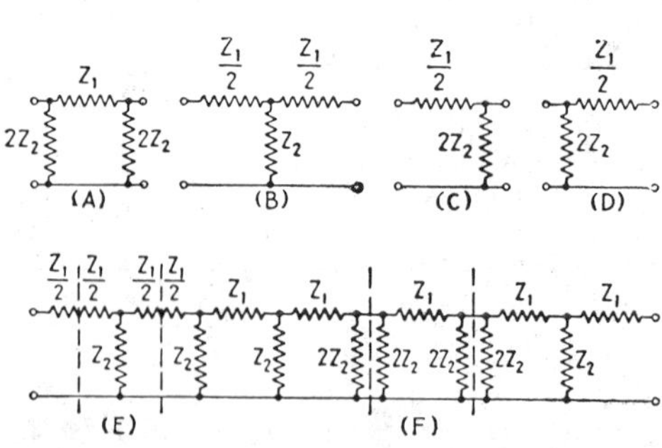

Fig. 2-32. Elementary and combined filter sections; *A*, the pi-section; *B*, the T-section; *C* and *D*, L-sections. The pi- and T-sections are formed by combining L-sections, as shown in *E* and *F*.

182. Constant-K Filter Section. The basic type of filter section is the "constant-K" type, which takes its name from the fact that the product of the impedances of the shunt and series arms is made equal to a constant K^2,

$$Z_1 \times Z_2 = K^2 \tag{2-242}$$

The value of K is taken as the nominal value of the terminal impedance, although when the filter is terminated in a resistance $R = K$, the impedance is matched perfectly only for one frequency in the passband, being to some degree mismatched at all other frequencies. The reflection losses incident to the mismatch cause the cutoff of the constant-K section to be more gradual than may be attained in more involved sections, such as the m-derived section (Par. **183**).

183. The m-derived Filter has sharper cutoff characteristics than the constant-K filter, from which its elements are derived by multiplying with a factor m or some function of m. In general, the m-derived filter has additional shunt and/or series elements which not only allow sharper cutoff and more uniform attenuation within the pass region but also allow the choice of a frequency (or frequencies) outside the passband at which very great attenuation may be obtained. If the extra impedance elements are added as shunt elements, the filter is a shunt m-derived type, whereas the extra impedance is a series element in the series m-derived form. When the multiplying factor m is unity, the m-derived section is the same as the constant-K section.

184. Filter-section Design. In the next four paragraphs, formulas are given for the design of half sections (Fig. 2-32C and D) for the various types of filters. In each case the impedance of the series arm is $Z_1/2$, and that of the shunt arm is $2Z_2$.

185. Formulas for Low-pass Filters. The configurations of filters shown in Fig. 2-33 include the low-pass forms, which are designed as follows: In each case, R is the value of terminating resistance, i.e., the nominal value of the image impedance; f_c is the cutoff frequency; and the filter section is the L-section from which a recurrent multisection filter may be formed.

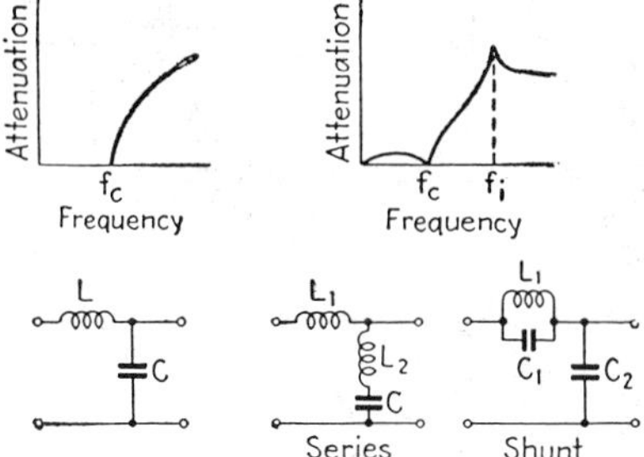

FIG. 2-33. Constant-K and m-derived low-pass half sections.

Constant-K low-pass filter,

$$L = \frac{R}{2\pi f_c} \tag{2-243}$$

$$C = \frac{1}{2\pi f_c R} \tag{2-244}$$

Series m-derived low-pass filter (f_i is the frequency of infinite attenuation),

$$L_1 = \frac{mR}{2\pi f_c} \tag{2-245}$$

$$L_2 = \frac{(1 - m^2)R}{2m\pi f_c} \tag{2-246}$$

$$C = \frac{m}{2\pi f_c R} \tag{2-247}$$

$$m = \sqrt{1 - f_c^2/f_i^2} \tag{2-248}$$

Shunt m-derived low-pass filter,

$$L = \frac{mR}{2\pi f_c} \tag{2-249}$$

$$C_1 = \frac{1 - m^2}{2m\pi f_c R} \tag{2-250}$$

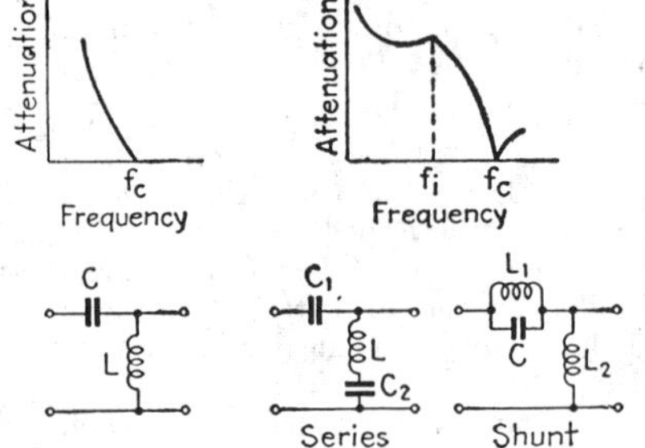

FIG. 2-34. Constant-K and m-derived high-pass half sections.

$$C_2 = \frac{m}{2\pi f_c R} \tag{2-251}$$

$$m = \sqrt{1 - f_c^2/f_i^2} \tag{2-252}$$

186. Formulas for High-pass Filters. The high-pass filters shown in Fig. 2-34 are designed as follows (same symbols as in Par. **185**):

Constant-K high-pass filter,

$$L = \frac{R}{2\pi f_c} \tag{2-253}$$

$$C = \frac{1}{2\pi f_c R} \tag{2-254}$$

Series m-derived high-pass filter (f_i is the frequency of infinite attenuation),

$$L = \frac{R}{2\pi f_c m} \tag{2-255}$$

$$C_1 = \frac{1}{2\pi f_c m R} \tag{2-256}$$

$$C_2 = \frac{m}{2(1 - m^2)\pi f_c R} \tag{2-257}$$

$$m = \sqrt{1 - f_i^2/f_c^2} \tag{2-258}$$

Shunt m-derived high-pass filter,

$$L_1 = \frac{mR}{2(1 - m^2)\pi f_c} \tag{2-259}$$

$$L_2 = \frac{R}{2\pi f_c m} \tag{2-260}$$

$$C = \frac{1}{2\pi f_c m R} \tag{2-261}$$

$$m = \sqrt{1 - f_i^2/f_c^2} \tag{2-262}$$

187. Formulas for Bandpass Filters. The bandpass filters shown in Fig. 2-35 are designed in terms of the following formulas, where R is the nominal-value image impedance; f_{cl}, the low-frequency cutoff frequency; f_{ch}, the high-frequency cutoff frequency; f_{il}, the lower frequency of infinite attenuation; and f_{ih}, the higher frequency of infinite attenuation:

Constant-K bandpass filter,

$$L_1 = \frac{R}{2\pi(f_{ch} - f_{cl})} \tag{2-263}$$

$$L_2 = \frac{(f_{ch} - f_{cl})R}{2\pi f_{ch} f_{cl}} \tag{2-264}$$

$$C_1 = \frac{f_{ch} - f_{cl}}{2\pi f_{ch} f_{cl} R} \tag{2-265}$$

$$C_2 = \frac{1}{2\pi(f_{ch} - f_{cl})R} \tag{2-266}$$

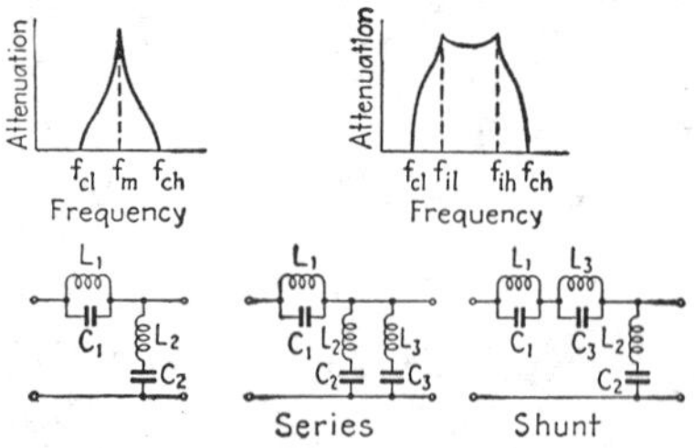

FIG. 2-35. Constant-K and m-derived bandpass half sections.

Series m-derived bandpass filter (Fig. 2-35),

$$L_1 = \frac{mR}{2\pi(f_{ch} - f_{cl})} \tag{2-267}$$

$$L_2 = \frac{aR}{2\pi(f_{ch} - f_{cl})} \tag{2-268}$$

$$L_3 = \frac{bR}{2\pi(f_{ch} - f_{cl})} \tag{2-269}$$

$$C_1 = \frac{f_{ch} - f_{cl}}{2\pi f_{ch} f_{cl} m R} \tag{2-270}$$

$$C_2 = \frac{f_{ch} - f_{cl}}{2\pi f_{ch} f_{cl} b R} \tag{2-271}$$

$$C_3 = \frac{f_{ch} - f_{cl}}{2\pi f_{ch} f_{cl} a R} \tag{2-272}$$

$$m = \frac{h}{1 - f_{ch} f_{cl}/f_{ih}^2} \tag{2-273}$$

$$h = \sqrt{(1 - f_{cl}^2/f_{ih}^2)(1 - f_{ch}^2/f_{ih}^2)} \tag{2-274}$$

$$a = \frac{(1 - m^2)(f_{ch} f_{cl})(1 - f_{il}^2/f_{ih}^2)}{h f_{il}^2} \tag{2-275}$$

$$b = \frac{(1 - m^2)(1 - f_{il}^2/f_{ih}^2)}{h} \tag{2-276}$$

$$f_{ih} f_{il} = f_{ch} f_{cl} \tag{2-277}$$

Shunt m-derived bandpass filter,

$$L_1 = \frac{(f_{ch} - f_{cl})R}{2\pi f_{cl} f_{ch} b} \tag{2-278}$$

$$L_2 = \frac{(f_{ch} - f_{cl})R}{2\pi f_{ch} f_{cl} m} \tag{2-279}$$

$$L_3 = \frac{(f_{ch} - f_{cl})R}{2\pi f_{ch} f_{cl} a} \tag{2-280}$$

$$C_1 = \frac{a}{2\pi(f_{ch} - f_{cl})R} \tag{2-281}$$

$$C_2 = \frac{m}{2\pi(f_{ch} - f_{cl})R} \tag{2-282}$$

$$C_3 = \frac{b}{2\pi(f_{ch} - f_{cl})R} \tag{2-283}$$

NOTE: a, b, and m are given by Eqs. (2-275), (2-276), and (2-273), respectively, the value of h given in Eq. (2-274) being used. The condition in Eq. (2-277) also applies.

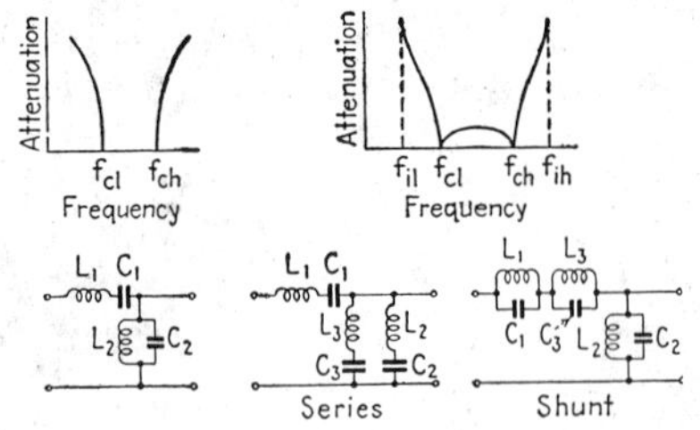

FIG. 2-36. Constant-K and m-derived band-elimination half sections.

188. Formulas for Band-elimination Filters. The band-elimination filters shown in Fig. 2-36 are designed as follows: R is the nominal value of the image impedance; f_{ch}, the higher cutoff frequency; f_{cl}, the lower cutoff frequency; f_{ih}, the higher frequency of infinite attenuation (within the eliminated band); and f_{il}, the lower frequency of infinite attenuation.

Constant-K band-elimination filter,

$$L_1 = \frac{(f_{ch} - f_{cl})R}{2\pi f_{ch} f_{cl}} \tag{2-284}$$

$$L_2 = \frac{R}{2\pi(f_{ch} - f_{cl})} \tag{2-285}$$

$$C_1 = \frac{1}{2\pi(f_{ch} - f_{cl})R} \tag{2-286}$$

$$C_2 = \frac{f_{ch} - f_{cl}}{2\pi R f_{ch} f_{cl}} \tag{2-287}$$

The frequency of maximum attenuation $f_m = \sqrt{f_{ch} f_{cl}}$.

Series m-derived band-elimination filter,

$$L_1 = \frac{mR(f_{ch} - f_{cl})}{2\pi f_{ch} f_{cl}} \tag{2-288}$$

$$L_2 = \frac{aR}{2\pi(f_{ch} - f_{cl})} \tag{2-289}$$

$$L_3 = \frac{bR}{2\pi(f_{ch} - f_{cl})} \tag{2-290}$$

$$C_1 = \frac{1}{2\pi(f_{ch} - f_{cl})mR} \tag{2-291}$$

$$C_2 = \frac{f_{ch} - f_{cl}}{2\pi f_{ch} f_{cl} bR} \tag{2-292}$$

$$C_3 = \frac{f_{ch} - f_{cl}}{2\pi f_{ch} f_{cl} aR} \tag{2-293}$$

$$m = \frac{h}{\sqrt{1 - f_{cl}/f_{ch}}} \tag{2-294}$$

$$h = \sqrt{\left(1 - \frac{f_{cl}^2}{f_{ih}^2}\right)\left(1 - \frac{f_{ih}^2}{f_{ch}^2}\right)} \tag{2-295}$$

$$b = \frac{1}{m} + \frac{f_{ih}^2}{m f_{cl} f_{ch}} \tag{2-296}$$

$$a = \frac{1}{m} + \frac{f_{cl} f_{ch}}{m f_{ih}^2} \tag{2-297}$$

$$f_{ih} f_{il} = f_{ch} f_{cl} \tag{2-298}$$

Shunt m-derived band-elimination filter,

$$L_1 = \frac{(f_{ch} - f_{cl})R}{2\pi f_{ch} f_{cl} b} \quad (2\text{-}299)$$

$$L_2 = \frac{R}{2\pi (f_{ch} - f_{cl}) m} \quad (2\text{-}300)$$

$$L_3 = \frac{(f_{ch} - f_{cl})R}{2\pi f_{ch} f_{cl} a} \quad (2\text{-}301)$$

$$C_1 = \frac{a}{2\pi (f_{ch} - f_{cl}) R} \quad (2\text{-}302)$$

$$C_2 = \frac{m(f_{ch} - f_{cl})}{2\pi f_{ch} f_{cl} R} \quad (2\text{-}303)$$

$$C_3 = \frac{b}{2\pi (f_{ch} - f_{cl}) R} \quad (2\text{-}304)$$

NOTE: The values of m, a, and b are given in Eqs. (2-294), (2-297), and (2-296), respectively, the value of h given in Eq. (2-295) being used. The condition in Eq. (2-298) also applies.

189. Multisection Filters. The half sections of Pars. **185** to **188** can be connected to form T- or pi-sections. Constant-K half sections can be connected to form constant-K T- or pi-sections. Series m-derived half sections can be connected to form a T-section with a characteristic impedance the same as that of a constant-K T-section. Shunt m-derived half sections can be connected to form a pi-section with a characteristic impedance the same as that of a constant-K pi-section.

Multisection filters can be of the uniform type, in which all sections are identical, or of the composite type, in which different types of sections (and their effects) are combined. Composite filters usually consist of one or more internal sections, all being T-sections or all being pi-sections, together with terminating m-derived half sections for connection to resistive terminations. One or more of the internal sections may be m-derived to give increased sharpness of cutoff or large attenuation at specific frequencies. The appropriate value of m for the terminating half sections is 0.6. If the internal sections are T-sections, the terminating half sections should be series m-derived and connected so that their shunt arms are presented to the resistive terminations. If the internal sections are pi-sections, the terminating half sections should be shunt m-derived and connected so that their series arms are presented to the resistive terminations.

190. Lattice and Bridged-T Networks. The constant-K and m-derived filters, discussed in Pars. **185** to **188** inclusive, serve for most practical purposes in communications work. The m-derived form, in particular, allows sufficient choice of design parameters to meet most practical needs. In certain cases, however, it is necessary to design a filter whose image impedance and transfer constant (an index of the time delay and attenuation suffered by the signal in passage through the filter section) must be specified independently. When the attempt is made to satisfy this requirement with ladder-type filters of the constant-K or m-derived types, negative impedances are encountered which cannot be realized in passive circuit elements. It can be shown that the requirement can be met in the most general type of recurrent network, the lattice-filter section shown in Fig. 2-37. The lattice is a form of bridge network with the input terminals across one diagonal and the output terminals across the other. The image impedance of this filter is

$$Z_1 = \sqrt{Z_A Z_B} \quad (2\text{-}305)$$

and the image-transfer constant θ is given by

$$\tanh \frac{\theta}{2} = \sqrt{\frac{Z_A}{Z_B}} \quad (2\text{-}306)$$

Evidently, any combination of Z_1 and θ may be realized by suitable choice of Z_A and Z_B.

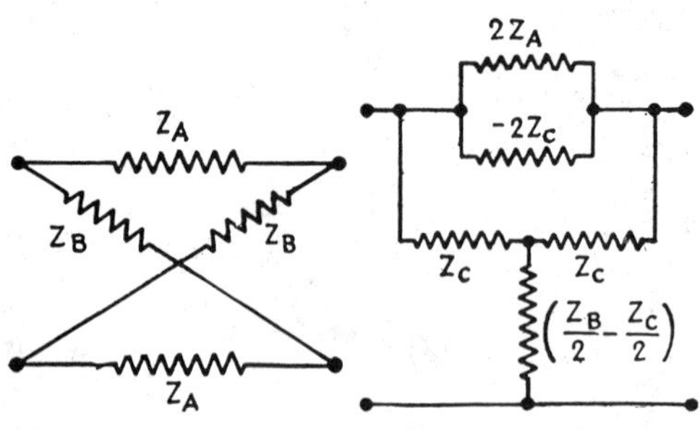

Fig. 2-37. Lattice filter section in bridge-network form.

Fig. 2-38. Unbalanced network in bridged-T form.

The design of lattice networks proceeds from the fact that a pass-band occurs when Z_A and Z_B have opposite signs whereas a stop band occurs when these impedances have the same sign. Hence, at a cutoff frequency, one impedance must change its sign while the other retains it. The position of the cutoff frequencies in a particular case may be determined by plotting the impedances as functions of frequency and noting the regions of frequency within which relative changes in sign occur. The general method of designing lattice filters to meet specific requirements is not so straightforward as the procedure for constant-K and m-derived sections previously listed.

The lattice filter of Fig. 2-37 is a balanced network which cannot be used in grounded circuits. The corresponding unbalanced form is the bridged-T section shown in Fig. 2-38. An additional impedance Z_c, related to Z_A or Z_B, must be inserted, and it must possess certain characteristics that are not always realizable in practice. Hence, the bridged-T network is a form not so general as the lattice. It should be noted that the ladder filters (constant-K and m-derived) are special cases of the lattice network and have correspondingly restricted properties.

BIBLIOGRAPHY ON FILTERS

191. Reference to Literature on Filters

Bode, H. W. "Network Analysis and Feedback Amplifier Design"; Princeton, N.J., D. Van Nostrand Company, Inc., 1945.

Cassell, W. L. "Linear Electric Circuits"; New York, John Wiley & Sons, Inc., 1964.

Cauer, W. New Theory and Design of Wave Filters; *Physics*, 1932, Vol. 2, pp. 242–268.

Darlington, S. Synthesis of Reactance 4-poles Which Produce Prescribed Insertion Loss Characteristics, Including Special Applications to Filter Design; *Jour. Math. Phys.*, 1939, Vol. 18, pp. 257–353.

Dishal, M. Design of Dissipative Band-pass Filters Producing Exact Amplitude-frequency Characteristics; *Proc. IRE*, 1949, Vol. 37, pp. 1050–1069.

Green, E. Exact Amplitude-frequency Characteristics of Ladder Networks; *Marconi Rev.*, 1953, Vol. 16, pp. 25–68.

Guillemin, E. A. "Communication Networks"; New York, John Wiley & Sons, Inc., 1935, Vol. II.

Javid, M., and Brenner, E. "Analysis, Transmission, and Filtering of Signals"; New York, McGraw-Hill Book Company, 1963.

Mole, J. H. "Filter Design Data for Communication Engineers"; New York, John Wiley & Sons, Inc., 1952.

Reed, M. B. "Electric Network Synthesis"; Englewood Cliffs, N.J., Prentice-Hall, Inc., 1955.

Rushton, H., and Bordogna, J. "Electric Networks: Functions, Filters, Analysis"; New York, McGraw-Hill Book Company, 1966.

Shea, T. E. "Transmission Networks and Wave Filters"; Princeton, N.J., D. Van Nostrand Company, Inc., 1929.

Van Valkenburg, M. E. "Introduction to Modern Network Synthesis"; New York, John Wiley & Sons, Inc., 1960.

Weinberg, L. "Network Analysis and Synthesis"; New York, McGraw-Hill Book Company, 1962.

SECTION 3

MEASUREMENTS AND INSTRUMENTS

BY

FOREST K. HARRIS Chief, Absolute Electrical Measurements Section, National Bureau of Standards; Adjunct Professor, School of Engineering and Applied Science, The George Washington University; Fellow, Institute of Electrical and Electronics Engineers

H. RUSSELL BROWNELL Vice President, Hallmark Standards, Inc.; Member, Institute of Electrical and Electronics Engineers

THOMAS D. BARNES Engineering Manager, Meter Division, Westinghouse Electric Corporation; Fellow, Institute of Electrical and Electronics Engineers

PERRY A. BORDEN (Deceased)

CONTENTS

Numbers refer to paragraphs

SECTION 3

MEASUREMENTS AND INSTRUMENTS

ELECTRIC AND MAGNETIC MEASUREMENTS

GENERAL (See also Sec. 1)

1. Measurement of a quantity consists either in its comparison with a unit quantity of the same kind or in its determination as a function of quantities of different kinds whose units are related to it by known physical laws. An example of the first kind of measurement is the evaluation of a resistance (in ohms) with a Wheatstone bridge, in terms of a calibrated resistance and a ratio. An example of the second kind is the calibration of the scale of a wattmeter (in watts) as the product of current (in amperes) in its field coils and the potential difference (in volts) impressed on its potential circuit.

2. Units employed in electrical measurements comprise the absolute MKSA units of the Système International, abbreviated as SI,[1] used throughout the world and based on the prototype mechanical units—meter, kilogram, and second—and the ampere.[2] The MKSA units are identical in value with the practical units—volt, ampere, ohm, coulomb, farad, henry—used by engineers. Certain prefixes have been adopted internationally to indicate decimal multiples and fractions of the basic units. These are given in Sec. **1**, Par. **11**.

3. A reference standard is a concrete representation of a unit or of some fraction or multiple of it. Standard cells and fixed resistors, capacitors, and inductors are regularly used as reference standards. It must be recognized that all physical measurements are subject to uncertainty and hence that the numerical value assigned to any reference standard should be capable of being traced through a chain of measurements to the *National Reference Standards* maintained by the National Bureau of Standards.

4. The National Reference Standards are those maintained by the National Bureau of Standards. The standard of resistance is a group of ten 1-Ω resistors of special construction[3] whose average preserves the *legal* unit of resistance for the country. Similarly, the *legal* unit of emf is preserved by the average of a group of 44 saturated Weston cells[4] kept at constant temperature in a stirred oil bath. Values were assigned to these resistors and cells by international agreement in 1948, and the group averages are assumed not to have changed since that time.[5] Values of resistors and of standard cells tested at NBS are reported in terms of the National Reference Standards.

5. Precision—a measure of the spread of repeated determinations of a particular quantity—depends on various factors. Among these are the resolution of the method used, variations in ambient conditions (such as temperature and humidity) that may influence the value of the quantity or of the reference standard, instability of some element of the measuring system, and many others. In the National laboratory, where every precaution is taken to obtain the best possible value, intercomparisons

---

[1] The absolute units of the Système International must not be confused with the old International Units in use before 1948, based on the "mercury" ohm and "silver" ampere (see also Sec. **1**, Par. **10**).

[2] The SI definition of the *ampere*—the current which, in two infinite parallel conductors spaced 1 m apart, would produce a force of 10^{-7} N/m of length—is based on an assigned value of $4\pi \times 10^{-7}$ as the permeability of free space.

[3] Thomas, J. L. *NBS Jour. Res.*, 1946, Vol. 36, p. 107.

[4] Hamer, W. J. *NBS Mono.* 84, 1965.

[5] The constancy of these standards is checked by regular intercomparisons between members of the group, by periodic comparisons with the national reference standards of other countries, and by occasional absolute measurements at NBS and at the national laboratories of other countries.

may have a precision of a few parts in 10^7. In commercial laboratories, where the objective is to obtain results that are reliable but only to the extent justified by engineering or other requirements, precision ranges from this figure to a part in 10^3 or more, depending on circumstances. For commercial measurements such as the sale of electrical energy, where the cost of measurement is a critical factor, a precision of 1 or 2 percent is considered acceptable in some jurisdictions.

6. Accuracy—a statement of the limits which bound the departure of a measured value from the true value of a quantity—includes the imprecision of the measurement, together with all the accumulated errors in the measurement chain extending downward from the basic reference standards to the specific measurement in question. In engineering measurement practice, accuracies are generally stated in terms of the values assigned to the National Reference Standards—the *legal* units. It is only rarely that one needs also to state accuracy in terms of the *defined* SI unit by taking into account the uncertainty in the assignment of the National Reference Standard.

7. Classification of instruments by accuracy limits has become a universal practice. Table 3-1 shows the accuracy limits specified in the USA Standard for Indicating Instruments.

8. General precautions should be observed in electrical measurements, and sources of error should be avoided, as detailed below:

a. The accuracy limits of the instruments, standards, and methods used should be known, so that appropriate choice of these measuring elements may be made. It should be noted that instrument *accuracy classes* state the "initial" accuracy. Operation of an instrument, with energy applied over a prolonged period, may cause errors due to elastic fatigue of control springs or resistance changes in instrument elements because of heating under load. USAS C39.1 specifies permissible limits of error of portable instruments because of sustained operation.

b. In any other than rough determinations, the *average of several readings* is better than one, particularly if measurement conditions can be altered to avoid or minimize the effect of accidental errors.

c. The *range* of the measuring instrument should be such that the measured quantity produces a reading large enough to yield the desired precision. The deflection of a measuring instrument should preferably exceed half scale. Voltage transformers, wattmeters, and watthour meters should be operated near to rated voltage for best performance. Care should be taken to avoid either momentary or sustained overloads.

d. Magnetic fields, produced by currents in conductors or by various classes of electrical machinery or apparatus, may combine with the fields of portable instruments to produce errors. Alternating or time-varying fields may induce emfs in loops formed in connections or the internal wiring of bridges, potentiometers, etc., to produce an error signal or even "electrical noise" that may obscure the desired reading. The effects of stray alternating fields on a-c indicating instruments may be eliminated generally by using the average of readings taken with direct and reversed connections; with direct fields and d-c instruments the second reading (to be averaged with the first) may be taken after rotating the instrument through 180°. If instruments are to be mounted in magnetic panels, they should be calibrated in a panel of the same material and thickness.

e. In measurements involving high resistances and small currents, *leakage paths* across insulating components of the measuring arrangement should be eliminated if they shunt portions of the measuring circuit. This is done by providing a guard circuit to intercept current in such shunt paths or to keep points at the same potential between which there might otherwise be improper currents.

f. Variations in *ambient temperature* or internal temperature rise from self-heating under load may cause errors in instrument indications. If the temperature coefficient and the instrument temperature are known, readings can be corrected where precision requirements justify it. Where measurements involve extremely small potential differences, thermal emfs resulting from temperature differences between junctions of dissimilar metals may produce errors; heat from the observer's hand or heat generated by the friction of a sliding contact may cause such effects.

g. Phase-defect angles in resistors, inductors, or capacitors and in instruments and instrument transformers must be taken into account in many a-c measurements.

Table 3-1. Accuracy Limits for Electrical Instruments

Panel instruments (Round or rectangular panel instruments mounted in a vertical panel)

Size, in.	Function	Response time, sec	Initial accuracy, % of full-scale value	Min. scale length, in.
1½ (rectangular only)	D-c	2	3	1
2½	D-c	3 for μamp, others 2	2	1.5
	A-c	2.5	2	1.5
	R-f	2.5	2	1.5
	Rectifier	3 for μamp, others 2	5	1.5
3½	D-c	3 for μamp, others 2	2	2
	A-c	2.5	2	2
	R-f	2.5	2	2
	Rectifier	3 for μamp, others 2	5	2
4½	D-c	3 for μamp, others 2	2	2.7
	A-c	2.5	2	2.7
	R-f	2.5	2	2.7
	Rectifier	3 for μamp, others 2	5	2.7

Switchboard instruments (Rectangular switchboard instruments for vertical panel mounting)

Size, in.	Scale angle, deg	Function	Response time, sec	Initial accuracy, % of full-scale value	Min. scale length, in.
4½	90	D-c	2.5	1	3.2
	250	D-c	2.5	1	6.8
4½	90	A-c	2.5	1	5.2
	250	A-c	2.5	1	6.8
4½	90	1- and 2-element wattmeter	2.5	1	3.2
	250	1- and 2-element wattmeter	2.5	1	6.8
4½	90	Single and polyphase power factor	2.5*	0.01 p.f.*	3.2
	250	Single and polyphase power factor	2.5†	0.01 p.f.†	5
4½	90	Frequency	5	3	3.2
	250	Frequency	5	3	5
6		D-c	2.5	1	5
		A-c	2.5	1	5
		1- and 2-element wattmeter	2.5	1	5
		Single and polyphase power factor	2.5‡	0.01 p.f.‡	5
		Frequency	5	3	5

* See restrictions, p. 34, plate 20, USAS C39.1-1951.
† See restrictions, p. 35, plate 21, USAS C39.1-1951.
‡ See restrictions, p. 41, plate 27, USAS C39.1-1951.

Portable instruments

	Response time, sec.	Accuracy, % of full scale	Scale length, in.
D-c ma, amp, mv, volts			
Single and multiple	10	0.1	12
	3.5	0.25	5
	3	0.5	3.2
	2.5	1	2.5
	2.5	2	1.5
A-c ma, amp, volts			
Single and multiple	10	0.1	12
	5	0.25	5
	5	0.5	3.2
	2.5	1	2.5
	2.5	2	1.5
Single-element wattmeter	10	0.1	12
	5	0.25	5
	5	0.5	3.2
	2.5	1	2.5
	2.5	2	1.5

h. Large *potential differences* are to be avoided between the windings of an instrument or between its windings and frame. Electrostatic forces may produce reading errors, and very large potential difference may result in insulation breakdown. Instruments should be connected in the ground leg of a circuit where feasible. The moving-coil end of the voltage circuit of a wattmeter should be connected to the same line as the current coil. When an instrument must be at a high potential, its case must be adequately insulated from ground and connected to the line in which the instrument circuit is connected, or the instrument should be enclosed in a screen that is connected to the line. Such an arrangement may involve shock hazard to the operator, and proper safety precautions must be taken.

i. Electrostatic charges and consequent disturbance to readings may result from rubbing the insulating case or window of an instrument with a dry dustcloth; such charges can generally be dissipated by breathing on the case or window. Low-level measurements in very dry weather may be seriously affected by charges on the clothing of the observer; some of the synthetic textile fibers—such as nylon and dacron—are particularly strong sources of charge; the only effective remedy is the complete screening of the instrument on which charges are induced.

j. Position influence (resulting from mechanical unbalance) may affect the reading of an indicating instrument if it is used in a position other than that in which it was calibrated. Portable instruments of the better accuracy classes (with antiparallax mirrors) are normally intended to be used with the axis of the moving system vertical, and the calibration is generally made with the instrument in this position.

DETECTORS AND GALVANOMETERS

9. Detectors are used to indicate approach to balance in bridge or potentiometer networks. They are generally responsive to small currents or voltages, and their sensitivity—the value of current or voltage that will produce an observable indication—ultimately limits the resolution of the network as a means for measuring some electrical quantity.

10. Galvanometers are deflecting instruments which are used, mainly, to *detect* the presence of a small electrical quantity—current, voltage, or charge—but which are also used in some instances to measure the quantity through the magnitude of the deflection.

11. The D'Arsonval (moving-coil) galvanometer consists of a coil of fine wire suspended between the poles of a permanent magnet. The coil is usually suspended from a flat metal strip which both conducts current to it and provides control torque directed toward its neutral (zero-current) position. Current may be conducted from the coil by a helix of fine wire which contributes very little to the control torque (pendulous suspension) or by a second flat metal strip which contributes significantly to the control torque (taut-band suspension). An iron core is usually mounted in the central space enclosed by the coil; and the pole pieces of the magnet are shaped to produce a uniform radial field throughout the space in which the coil moves. A mirror attached to the coil is used in conjunction with a lamp and scale or a telescope and scale to indicate coil position. In the former case, an incandescent filament or an illuminated target is reflected by the mirror and focused on a scale; in the latter, the scale image reflected by the mirror is observed in a telescope. Angular motion of the coil is indicated by changes in scale reading.

The *pendulous-suspension* type of galvanometer has the advantage of higher sensitivity (weaker control torque) for a suspension of given dimensions and material and the disadvantage of responsiveness to mechanical disturbances to its supporting platform, which produce anomalous motions of the coil. The *taut-suspension* type is generally less sensitive (stiffer control torque) but may be made much less responsive to mechanical disturbances if it is properly balanced, i.e., if the center of mass of the moving system is in the axis of rotation determined by the taut upper and lower suspensions.

12. Galvanometer sensitivity can be expressed in a number of ways, depending on application:

a. The *current* constant is the current in microamperes that will produce unit deflection on the scale—usually a deflection of 1 mm on a scale 1 m distant from the galvanometer mirror.

b. The *megohm* constant is the number of megohms in series with the galvanometer through which 1 V will produce unit deflection. It is the reciprocal of the current constant.

c. The *voltage* constant is the number of microvolts which, in a critically damped circuit (or another specified damping), will produce unit deflection.

d. The *coulomb* constant is the charge in microcoulombs which, at a specified damping, will produce unit ballistic throw.

e. The *flux-linkage* constant is the product of change of induction and turns of the linking search coil which will produce unit ballistic throw.

Table 3-2. Pointer-type Reflecting Galvanometers—Portable, Fixed-scale, Taut Suspension

Galv. resist., ohms	Period, sec	Sensitivity		Scale factor		Critical damping resistance (external), ohms
		Voltage-scale div. per μv	Current-scale div. per μa	Voltage, μv per scale div	Current, μa per scale div	
10	1.5	.....	0.3	...	3.5	100
10	2	0.01	....	100	.....	8
30	...	.....	0.33	...	3	70
300	...	.....	1	...	1	1,000
1,100	...	.....	2	...	0.5	3,500
12	2.5	0.016	....	64	.....	20
25	3	.....	1	...	1	110
250	...	.....	4	...	0.25	1,800
2,500	...	.....	8	...	0.125	10,000
16	4.5	0.022	....	46	.....	30

All these sensitivities (galvanometer response characteristics) can be expressed in terms of current sensitivity, circuit resistance in which the galvanometer operates, relative damping (see Par. **14**), and period. If we define *current* sensitivity S_i as deflection per unit current, then—in appropriate units—the *voltage* sensitivity (the deflection per unit voltage) is

$$S_e = \frac{S_i}{R},$$

where R is the resistance of the circuit, including the resistance of the galvanometer coil. The *coulomb* sensitivity is

$$\frac{\theta}{Q} = \frac{2\pi}{T_0} S_i \exp\left(\frac{-\gamma}{\sqrt{1-\gamma^2}} \tan^{-1} \frac{\sqrt{1-\gamma^2}}{\gamma}\right)$$

where T_0 is the undamped period and γ is the relative damping in the operating circuit. The *flux-linkage* sensitivity is

$$\frac{\theta}{\int e\,dt} \approx S_i \frac{2\pi}{T_0} \frac{1}{2R_c} \frac{1}{1-\gamma_0}$$

for the case of greatest interest—maximum ballistic response—where the galvanometer is heavily overdamped, γ_0 being the open-circuit relative damping, $\int e\,dt$ the time integral of induced voltage or the change in flux linkages in the circuit, and R_c the circuit resistance (including that of the galvanometer) for which the galvanometer is critically damped.[1]

13. Typical sensitivities of commercial galvanometers are given in Tables 3-2, 3-3 and 3-4.

[1] See Harris, F. K., "Electrical Measurements," p. 316, for the general expression of flux-linkage sensitivity.

Table 3-3. Reflecting Galvanometers—Portable, Mirror-Lamp, Integral Scale, Taut Suspension

Galv. resist., ohms	Period, sec	Sensitivity		Scale factor		Critical damping resistance, ohms
		Voltage-scale div. per μv	Current-scale div. per μa	Voltage, μv per scale div.	Current, μa per scale div.	
13	2	0.4	.....	2.5	......	13
25	...	0.18	.....	5.5	......	50
200	...	....	30	....	0.03	600
350	...	....	80	....	0.012	2,500
1,100	2.5	....	40	....	0.025	8,000
1,100	...	....	140	....	0.007	8,000
23	3	0.04	.....	25	......	80
300	...	....	25	....	0.04	2,000–10,000
500	...	....	250	....	0.004	6,000
1,000	...	....	40	....	0.025	15,000
1,100	...	....	200	....	0.005	10,000
30	4	0.67	.....	1.5	......	40
40	5	....	100	....	0.01	225
60	...	....	200	....	0.005	600
500	...	....	670	....	0.0015	10,000
1,500	6	....	1,700	....	0.0006	40,000

14. Galvanometer motion is described by the differential equation

$$P\ddot{\theta} + \left(K + \frac{G^2}{R}\right)\dot{\theta} + U\theta = \frac{GE}{R},$$

where θ is the angle of deflection in radians, P is the moment of inertia, K is the mechanical damping coefficient, G is the motor constant (G = coil area turns $\times$ air-gap field), R is total circuit resistance (including the galvanometer), and U is the suspension stiff-

Table 3-4. Reflecting Galvanometers—Platform or Wall Type, Mirror-detached Scale, Slack Suspension

Galv. resist., ohms	Period, sec	Sensitivity		Scale factor		Critical damping resistance, ohms
		Voltage, mm per μv	Current, mm per μa	Voltage, μv per mm	Current, μa per mm	
21	1.5	2	....	0.5	....	40
500	....	.....	125	....	0.008	2,500
500	3	.....	330	....	0.003	500
10	5	5	....	0.2	....	25
12	....	2	....	0.5	....	50
15	....	10	....	0.1	....	20
16	....	5	....	0.2	....	40
25	....	10	....	0.1	....	50
40	....	.....	200	....	0.005	300
50	....	.....	400	....	0.0025	500
500	6	.....	2,000	....	0.0005	7,000
650	....	.....	2,000	....	0.0005	10,000
16	7	20	....	0.05	....	10
500	....	.....	3,300	....	0.0003	10,000
17	7.5	10	....	0.1	....	25
12	8	20	....	0.05	....	10
115	....	.....	70	....	0.014	10,000
30	9	2	....	0.5	....	60
12	10	10	....	0.1	....	15
650	11	.....	1,000	....	0.001	3,000
35	12	0.67	....	1.5	....	165
2,500	....	.....	2,000	....	0.0005	30,000
650	13	.....	10,000	....	0.0001	30,000
500	14	.....	10,000	....	0.0001	22,000
1,000	....	.....	1,000	....	0.001	10,000
8,000	18	.....	5,000	....	0.0002	56,000
800	20	.....	25,000	....	0.00004	70,000
800	40	.....	100,000	....	0.00001	100,000

ness. If the viscous and circuital damping are combined,

$$K + \frac{G^2}{R} = A,$$

the roots of the auxiliary equation are

$$m = \frac{A}{2P} \pm \sqrt{\frac{A^2}{4P^2} - \frac{U}{P}}.$$

Three types of motion can be distinguished.

a. Critically damped motion occurs when $A^2/4P^2 = U/P$. It is an aperiodic, or dead-beat, motion in which the moving system approaches its equilibrium position without passing through it in the shortest time of any possible aperiodic motion. This motion is described by the equation

$$y = 1 - \left(1 + \frac{2\pi t}{T_o}\right) \exp\left(\frac{-2\pi t}{T_o}\right),$$

where y is the fraction of equilibrium deflection at time t and T_o is the undamped period of the galvanometer—the period that the galvanometer would have if $A = 0$. If the total damping coefficient at critical damping is A_c, we can define relative damping as the ratio of the damping coefficient A for a specific circuit resistance to the value A_c it has for critical damping—$\gamma = A/A_c$, which is unity for critically damped motion.

b. In *overdamped* motion, the moving system approaches its equilibrium position without overshoot and more slowly than in critically damped motion. This occurs when

$$\frac{A^2}{4P^2} > \frac{U}{P}$$

and $\gamma > 1$. For this case, the motion is described by the equation

$$y = 1 - \left(\frac{\gamma}{\sqrt{\gamma^2 - 1}} \sinh \frac{2\pi t}{T_o} \sqrt{\gamma^2 - 1} + \cosh \frac{2\pi t}{T_o} \sqrt{\gamma^2 - 1}\right) \exp\left(\frac{-2\pi t}{T_o} \gamma\right)$$

c. In *underdamped* motion, the equilibrium position is approached through a series of diminishing oscillations, their decay being exponential. This occurs when $A^2/4P^2 < U/P$ and $\gamma < 1$. For this case, the motion is described by the equation

$$y = 1 - \frac{1}{\sqrt{1 - \gamma^2}} \left[\sin\left(\frac{2\pi t}{T_o} \sqrt{1 - \gamma^2} + \sin^{-1} \sqrt{1 - \gamma^2}\right)\right] \exp\left(\frac{-2\pi t}{T_o} \gamma\right)$$

15. Damping factor is the ratio of deviations of the moving system from its equilibrium position in successive swings. More conveniently, it is the ratio of the equilibrium deflection to the "overshoot" of the first swing past the equilibrium position, or

$$F = \frac{\theta_1 - \theta_F}{\theta_F - \theta_2} = \frac{\theta_F}{\theta_1 - \theta_F},$$

where θ_F is the equilibrium deflection and θ_1 and θ_2 are the first maximum and minimum deflections of the damped system. It can be shown that damping factor is connected to relative damping by the equation

$$F = \exp\left(\frac{\pi\gamma}{\sqrt{1 - \gamma^2}}\right).$$

16. The logarithmic decrement of a damped harmonic motion is the naperian logarithm of the ratio of successive swings of the oscillating system. It is expressed by the equation

$$\ln \frac{\theta_1 - \theta_F}{\theta_F - \theta_2} = \ln \frac{\theta_F}{\theta_1 - \theta_F} = \lambda$$

and in terms of relative damping

$$\lambda = \frac{\pi\gamma}{\sqrt{1-\gamma^2}}.$$

17. The period of a galvanometer (and, generally, of any damped harmonic oscillator) can be stated in terms of its undamped period T_o and its relative damping γ as $T = T_o/\sqrt{1-\gamma^2}$.

18. Reading time is the time required, after a change in the quantity measured, for the indication to come and remain within a specified percentage of its final value. Minimum reading time depends on the relative damping and on the required accuracy (see Table 3-5).

Table 3-5. Minimum Reading Time for Various Accuracies

Accuracy, percent	Relative damping	Reading time/free period
10	0.6	0.37
1	0.83	0.67
0.1	0.91	1.0

Thus, for a reading within 1% of equilibrium value, minimum time will be required at a relative damping of $\gamma = 0.83$. Generally in indicating instruments, this is known as *response time* when the specified accuracy is the stated accuracy limit of the instrument. The values of response time in Table 3-1 are those which are permissible for electrical indicating instruments under USAS C39.1.

19. External critical damping resistance (***CDRX***) is the external resistance connected across the galvanometer terminals that produces critical damping ($\gamma = 1$).

20. Measurement of damping and its relation to circuit resistance can be accomplished by a simple procedure in the circuit of Fig. 3-1. Let R_a be very large (say, 150 kΩ) and R_b small (say, 1 Ω), so that, when E is a 1.5-V dry cell, the driving voltage in the local galvanometer loop is a few microvolts (say, 10 μV). Since circuital damping is related to *total* circuit resistance ($R_c + R_b + R_g$), the galvanometer resistance R_g must be determined first. Under the stated circuit conditions, if $R_g + R_c \geqq 10R_b$, the driving voltage in the local galvanometer loop may be considered constant. R_c is adjusted to a value that gives a convenient deflection and then to a new value R_c' for which the deflection is cut in half. Then, since deflection is proportional to current, the following relation may be written: $2(R_g + R_b + R_c) = R_g + R_b + R_c'$, whence $R_g = R_c' - 2R_c - R_b$. Now let R_c be set at such a value that, when the switch is closed, the overshoot is readily observed. After noting the open-circuit deflection θ_o, the switch is closed and the peak value θ_1 of the first overswing, and the final deflection θ_F are noted. Then

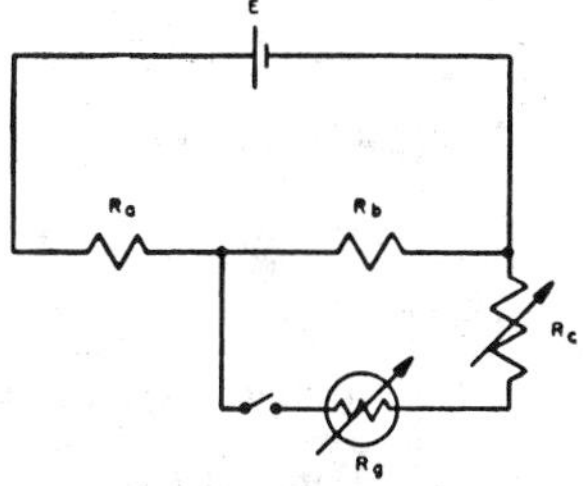

Fig. 3-1. Determination of relative damping.

$$\ln\frac{\theta_F - \theta_o}{\theta_1 - \theta_F} = \frac{\pi\gamma_1}{\sqrt{1-\gamma_1^2}},$$

γ_1 being the relative damping corresponding to the circuit resistance $R_1 = R_g + R_b + R_c$. The switch is now opened, and the first overswing θ_2 past the open-circuit equilibrium position θ_o is noted. Then

$$\ln\frac{\theta_F - \theta_o}{\theta_2 - \theta_o} = \frac{\pi\gamma_o}{\sqrt{1-\gamma_o^2}},$$

γ_o being the open-circuit relative damping. The relative damping γ_x for any circuit resistance R_x is given by the relation

$$\frac{R_x}{R_1} = \frac{\gamma_1 - \gamma_o}{\gamma_x - \gamma_o},$$

where it should be noted that the galvanometer resistance R_g is included in both R_x and R_1. For critical damping R_d can be computed by setting $\gamma_x = 1$, and the external critical damping resistance $CDRX = R_d - R_g$.

21. Galvanometer shunts are used to reduce the response of the galvanometer to a signal. However, in any sensitivity-reduction network, it is important that relative damping be preserved for proper operation. This can always be achieved by a suitable combination of series and parallel resistance. In Fig. 3-2, let r be the external circuit resistance and R_g the galvanometer resistance such that $r + R_g$ gives an acceptable damping (e.g., $\gamma = 0.8$) at maximum sensitivity. This damping will be preserved when the sensitivity-reduction network (S, P) is inserted, if $S = (n - 1)r$ and $P = nr/(n - 1)$, n being the factor by which response is to be reduced. The Ayrton-Mather shunt, shown in Fig. 3-3, may be used where the circuit resistance r is so high

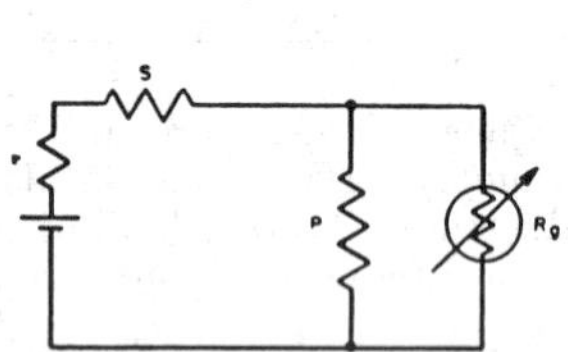

FIG. 3-2. Galvanometer shunt.

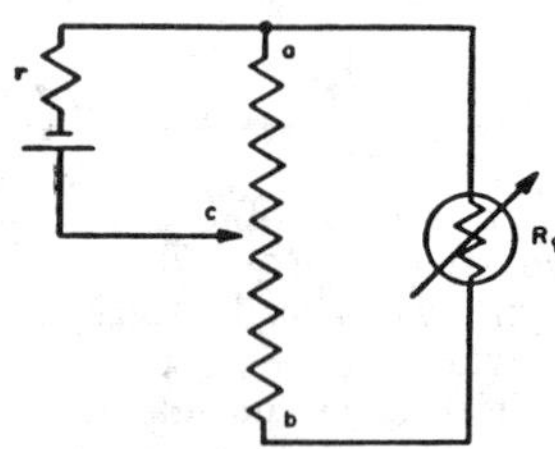

FIG. 3-3. Ayrton-Mather universal shunt.

that it exerts no appreciable damping on the galvanometer. R_{ab} should be such that correct damping is achieved by $R_{ab} + R_g$. In this network, sensitivity reduction is

$$n = \frac{R_{ac}}{R_{ab}},$$

and the ratio of galvanometer current I_g to line current I is

$$\frac{I_g}{I} = \frac{R_{ab}}{n(R_g + R_{ab})}.$$

22. The ultimate resolution of a detection system is the magnitude of the signal it can discriminate against the noise background present. In the absence of other noise sources, this limit is set by the *Johnson noise* generated by electron thermal agitation in the resistance of the circuit. This is expressed by the formula $e = \sqrt{4k\theta Rf}$, where e is the rms noise voltage developed across the resistance $\boldsymbol{R}$, K is Boltzmann's constant 1.4×10^{-23} J/°K, θ is the absolute temperature of the resistor in degrees Kelvin, and f is the bandwidth over which the noise voltage is observed. At room temperature (300°K), and with the assumption that the peak-to-peak voltage is $5 \times$ rms value, the peak-to-peak Johnson noise voltage is $6.5 \times 10^{-10}\sqrt{Rf}$ V. If, in a d-c system, we use the approximation that $f = 1/3t$, where t is the system's response time, the Johnson voltage is $4 \times 10^{-10}\sqrt{R/t}$ V (peak to peak).

By using reasonable approximations, it can be shown that the random brownian-motion deflections of the moving system of a galvanometer, arising from impulses by the molecules in the air around it, are equivalent to a voltage indication $e = 5 \times 10^{-10}\sqrt{R/T}$ V (peak to peak), where R is circuit resistance and T is the galvanometer period in seconds. If the galvanometer damping is such that its response time is $t = 2T/3$ (for $\gamma \approx 0.8$), the Johnson noise voltage to which it responds is about $5 \times 10^{-10}\sqrt{R/T}$ V (peak to peak). This value represents the limiting resolution of a galvanometer, since its response to smaller signals would be obscured by the random excursions of its moving system. Thus, a galvanometer with a 4-s period would have a limiting resolution of about 2 nV in a 100-Ω circuit and 1 nV in a 25-Ω circuit.

It is not surprising that one arrives at the same value from considerations either of random electron motions in the conductors of the measuring circuit or of molecular motions in the fluid that surrounds the system. The resulting figure rests on the premise that the law of equipartition of energy applies to the measuring system and that the galvanometer coil—a body with one degree of freedom—is statistically in thermal equilibrium with its surroundings.

23. Optical systems used with galvanometers and other indicating instruments avoid the necessity for a mechanical pointer and thus permit smaller, simpler balancing arrangements since the mirror attached to the moving system can be symmetrically disposed close to the axis of rotation. In portable instruments, the entire system—source, lenses, mirror, scale—is generally integral with the instrument; and the optical "pointer" may be folded one or more times by fixed mirrors so that it is actually much longer than the mechanical dimensions of the instrument case. In some instances, the angular displacement may be magnified by use of a cylindrical lens or mirror. For a wall- or bracket-mounted galvanometer, the lamp and scale arrangement is external, and the length of the light-beam pointer can be controlled. Whatever the arrangement, the pointer length cannot be indefinitely extended with consequent increase in resolution at the scale. The optical resolution of such a system is, in any event, limited by image diffraction, and this limit—for a system limited by a circular aperture—is $\alpha \approx 1.2\lambda/nd$, where α is the angle subtended by resolvable points, λ is the wavelength of the light, n is the index of refraction of the image space, and d is the aperture diameter. In this case, d is the diameter of the moving-system mirror, and $n = 1$ for air. If we assume that points 0.1 mm apart can just be resolved by the eye at normal reading distance, the resolution limit is reached at a scale distance of about 2 m in a system with a 1-cm mirror, which uses no optical magnification. Thus, for the usual galvanometer, there is no profit in using a mirror-scale separation greater than 2 m. Since resolution is a matter of subtended angle, the corresponding scale distance is proportionately less for systems that make use of magnification.

24. The photoelectric galvanometer amplifier is a detector system in which the light beam from the moving-system mirror is split between two photovoltaic cells connected in opposition as shown in Fig. 3-4. As the mirror of the primary galvanometer turns in

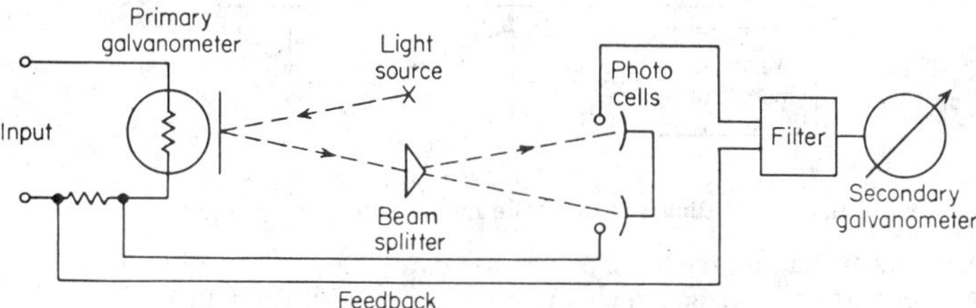

FIG. 3-4. Photoelectric galvanometer amplifier.

response to an input signal, the light flux is increased on one of the photocells and decreased on the other, resulting in a current and thence an enhanced signal in the circuit of the secondary (reading) galvanometer. Since the photocells respond to the total light flux on their sensitive elements, the system is not subject to resolution limitation by diffraction as is the human eye; and the ultimate resolution of the primary instrument—limited only by its brownian motion and the Johnson noise of the input circuit—may be realized.

A feedback loop, energized by the photocell output, is frequently used to stabilize the primary galvanometer and to control its sensitivity. A low-pass filter is sometimes inserted before the secondary galvanometer to increase the response time of the output circuit and thus to enhance the ultimate resolution of the system. Such a detector system must be isolated from mechanical disturbances or made unresponsive to them. The moving system of the primary galvanometer should be very carefully balanced about its axis of rotation. As a further precaution, the moving system is, in some instances, floated with zero buoyancy in a closed, liquid-filled compartment.

25. Chopper amplifier systems are also used effectively as voltage detectors in low-

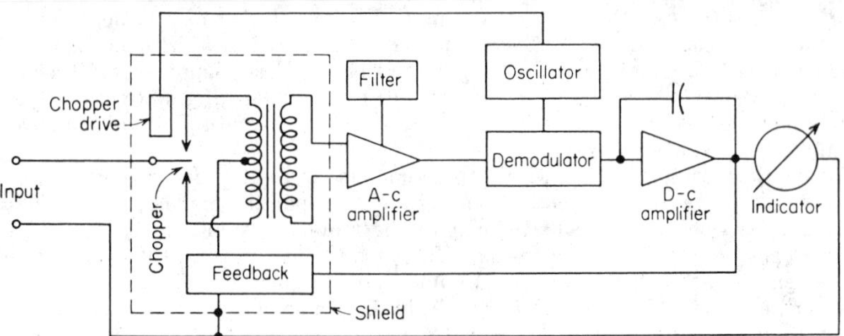

FIG. 3-5. Block diagram of Keithley model 148 nanovoltmeter.

resistance circuits. The typical elements of such a system are shown in Fig. 3-5. The chopper converts the d-c input signal to alternating current; it is amplified, demodulated, and further amplified before reaching the d-c indicating instrument. In the instrument diagrammed, the chopper frequency is 94 Hz. This frequency is not harmonically related to either 50 Hz or 60 Hz, simplifying the rejection of line-frequency interference. The a-c amplifier is sharply tuned to the chopper frequency, and the capacitor in the loop around the final d-c amplifier acts as a low-pass second filter. The feedback loop from the d-c amplifier to the input circuit stabilizes the gain of the system and is also used as a sensitivity control to alter the instrument range. The input circuit has magnetic and electrostatic shielding and is designed to minimize thermal emfs. This instrument is capable of resolution down to the Johnson-noise level.

26. Magnetic modulator amplifier systems use saturable cores to convert the d-c input signals to alternating current, which is then amplified, demodulated, and presented on a center-zero d-c indicating instrument, as shown schematically in Fig. 3-6.

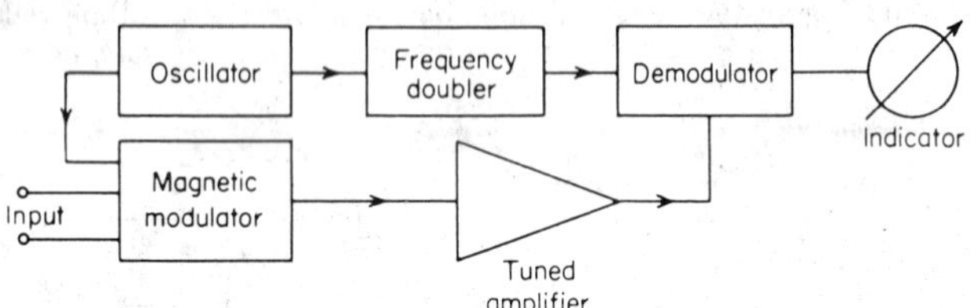

FIG. 3-6. Block diagram of Fluke model 840 electronic galvanometer.

In the instrument diagrammed, a 2.5-kHz oscillator provides excitation to two identical but oppositely wound cores, which are carried to saturation in each a-c half cycle. Since the two cores carry identical exciting windings, there is no net flux in the input winding which surrounds both. A d-c signal in this input winding results in earlier saturation in one core section and later saturation in the other during each excitation half cycle. Under this condition, there will be a net flux at double the excitation frequency in a winding that surrounds both core sections. The double-frequency voltage induced in this winding is proportional to the d-c input signal and reversed in phase if the polarity of the input signal is reversed. The a-c signal goes to a sharply tuned a-c amplifier and thence to a ring demodulator whose phase discrimination is provided by a signal from the oscillator via the frequency doubler. Thus, the zero-center d-c instrument receiving the demodulated signal indicates the polarity as well as the magnitude of the input signal. The sensitivity of the device is selected by means of an amplifier gain-control network. The instrument described is essentially a current-sensitive device with a resolution of 1 to 2 nA on its most sensitive range; voltage resolution is also good in a low-resistance circuit. Devices of this type (and that described in Par. **25**), which do not depend on a mechanically balanced primary sensing element, are insensitive to the mechanical disturbances that interfere with the proper operation of galvanometer detectors.

27. The a-c detectors used for balancing bridge networks are usually low-level tuned amplifiers coupled to an appropriate display device. The narrower the passband of the amplifier, the better is its signal resolution in general, since the narrow passband discriminates most effectively against noise in the input circuit. It is also advantageous to introduce tuning at the earliest possible stage of amplification in order that the desired signal be amplified selectively by early suppression of input noise, and the performance of many commercial a-c amplifier-detector systems is improved if used in conjunction with a narrow-passband preamplifier. In many cases, the amplifier output is rectified and displayed on a d-c indicating instrument, and added resolution is gained by introducing phase selection at the demodulator, since the wanted signal is regular in phase, while interfering noise is generally random. Further improvement can result from the use of a low-pass filter inserted between the demodulator and the d-c indicator such that the signal of selected phase is integrated over an appreciable time interval, say, a response time of a few seconds.

An alternative type of display is the screen of a cathode-ray oscillograph where the a-c signal to be observed is connected to the y terminals of the oscillograph and a synchronous signal of adjustable phase and magnitude is connected to the x terminals. The resulting Lissajous figure on the screen is an ellipse; and, by proper phase adjustment of the x signal, the in-phase and quadrature components of an unbalance signal from a bridge correspond, respectively, to the opening of the ellipse and the angle which its major axis makes with the horizontal. In this way, the two components of the bridge balance can be independently adjusted.

CONTINUOUS EMF MEASUREMENTS

28. A standard of emf may be either an electrochemical system or a Zener diode operated under precisely specified conditions of temperature and current.

29. The Weston standard cell has a positive electrode of metallic mercury and a negative electrode of cadmium-mercury amalgam (usually about 10% Cd). The electrolyte is a saturated solution of cadmium sulfate with an excess of $Cd \cdot SO_4 \cdot \frac{8}{3}H_2O$ crystals, usually acidified with sulfuric acid (0.04 to 0.08 normal). A paste of mercurous sulfate and cadmium sulfate crystals over the mercury electrode is used as a depolarizer. The saturated cell has a substantial temperature coefficient of emf. Vigoureux and Watts[1] of the National Physical Laboratory have given the following formula, applicable to cells with a 10% amalgam:

$$E_t = E_{20} - 39.39 \times 10^{-6}(t - 20) - 0.903 \times 10^{-6}(t - 20)^2 + 0.00660 \times 10^{-6}(t - 20)^3 - 0.000150 \times 10^{-6}(t - 20)^4$$

where t is the temperature in degrees Celsius (centigrade). Since cells are frequently maintained at 28°C, the following equivalent formula by Hamer[2] is useful:

$$E_t = E_{28} - 52.899 \times 10^{-6}(t - 28) - 0.80265 \times 10^{-6}(t - 28)^2 + 0.001813 \times 10^{-6}(t - 28)^3 - 0.0001497 \times 10^{-6}(t - 28)^4$$

A group of saturated Weston cells, maintained at a constant temperature in an air bath or a stirred oil bath, is quite generally used as a laboratory reference standard of emf. The bath temperature must be constant within a few thousandths of a degree if the reference emf is to be reliable to a microvolt. It is even more important that temperature gradients in the bath be avoided, since the individual limbs of the cell have very large temperature coefficients (about +315 μV/°C for the positive limb and −379 μV/°C for the negative limb—more than −50 μV/°C for the complete cell—at 28°C). Frequently, two or three groups of cells are used, one as a reference standard which never leaves the laboratory, the others as transport groups which are used for interlaboratory comparisons with a standards laboratory. The use of mechanical retainers to hold the solid materials of the cell in place are usually avoided in saturated

[1] VIGOUREUX and WATTS *Proc. Phys. Soc. (London)*, 1933, Vol. 45, p. 172.
[2] HAMER, W. J. *NBS Mono.* 84, 1965.

cells, and they cannot be inverted without damage. Thus, they cannot be shipped, and the usual practice is to hand-carry them for interlaboratory comparisons.

30. The Weston unsaturated cell uses the same electrode system as the saturated cell, but its electrolyte is not saturated at ordinary room temperatures. Its concentration is such that it reaches saturation at 3-4°C. Its emf at room temperature is about 0.05% higher than that of the saturated cell, and its temperature coefficient of emf is much lower (less than 10 μV/°C). Here, also, the temperature coefficients of the individual limbs are large, and a copper case is used to ensure their temperature equality. Unsaturated cells are widely used for the reference emf on potentiometers or wherever the accuracy of the voltage reference need not be better than 0.005%. To ensure this accuracy, the cell should be checked annually against the reference voltage from a properly maintained bank of saturated cells, since the emf of the unsaturated cell is expected to decrease slowly with time. The unsaturated cell is usually built with retaining barriers over the electrodes to hold them in place, and they can be shipped by express or even by parcel post.

31. Precautions in Using Standard Cells. *a.* The cell should not be exposed to extreme temperatures—below 4°C or above 40°C.

b. Temperature gradients (differences between the cell limbs) should be avoided.

c. Abrupt temperature changes should be avoided—the recovery period after a sudden temperature change may be as much as several days and depends on the magnitude of the shock, the age of the cell, and probably other parameters.

d. Current in excess of 100 μA should never be passed through the cell in either direction, and it is better to limit current to 10 μA or less and to maintain this for as short a time as is feasible in using the cell as a reference. A cell which has been short-circuited is likely to suffer a permanent change in emf.

32. Zener diodes are being used in increasing numbers to provide the reference voltage for recording potentiometers and digital voltmeters, and in other measurement applications where requirements are not so severe as to demand the use of standard cells. Commercial devices using Zener-diode control are available with guaranteed accuracies of 0.01% or better, and further improvements may reasonably be anticipated. The reverse current of a silicon diode rectifier is very small up to a particular voltage—the Zener voltage—at which point there is a sharp current increase; the resistance suddenly falls to a small value. This dynamic resistance is to some extent a function of current, but the voltage-current relationship in the Zener region is quite repeatable so long as the permissible power dissipation of the diode is not exceeded. Thus, the Zener voltage may be used as a reliable reference; and this critical voltage may be controlled within wide limits through processing the diode. The establishment of a reference voltage may be illustrated as follows: In Fig. 3-7, let R_1 be a large resistance in series with the source voltage E; R_2 the resistance across which the reference voltage E_r is to be developed; R_d the dynamic resistance of the diode at the current level selected for operation. For small changes in source voltage, it can be shown that

$$\frac{\Delta E_r}{\Delta E} = \frac{1}{1 + (R_1/R_d) + (R_1/R_2)}$$

Thus, if R_1 is large compared with R_d, the variation of E_r is much less than the variation of the supply voltage. Now, if the supply voltage is closely regulated, the variation of the reference voltage can be very small. The voltage across only a portion of E_r can be used as reference, and if the tap point is adjustable, reference voltages can be selected within the limit set by the Zener voltage.

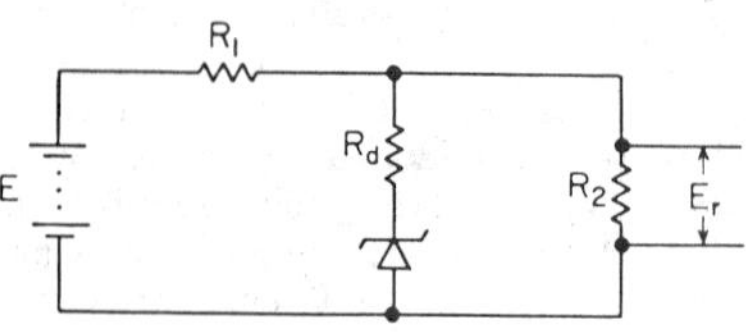

FIG. 3-7. Establishment of reference voltage with Zener diode.

The temperature coefficient of this critical voltage is strongly dependent on the Zener voltage itself, being negative for low-voltage diodes and positive for those whose Zener voltage is greater than about 5 V. Thus, it is possible to select diodes whose critical voltage is nearly independent of temperature; or, alternatively, a low temperature coefficient can be achieved by combining diodes with appropriate positive and negative coefficients.

33. Potentiometers are used for the precise measurement of emf in the range below 1.5 V. This is accomplished by opposing to the unknown emf an equal *IR* drop. There are two possibilities: either the current is held constant while the resistance across which the *IR* drop is opposed to the unknown is varied; or current is varied in a fixed resistance to achieve the desired *IR* drop. Most general-purpose laboratory potentiometers are of the constant-current type. Use of the constant-resistance type is confined to the millivolt and microvolt range or, in some designs, to the lower decades of a potentiometer whose upper decades are of the constant-current type.

Figure 3-8 shows schematically most of the essential features of a general-purpose constant-current instrument. With the standard-cell dial set to read the emf of the reference standard cell, the potentiometer current I is adjusted until the IR drop across 10 of the coarse-dial steps plus the drop to the set point on the standard-cell dial balances the emf of the reference cell. The correct value of current is indicated by a null reading of the galvanometer in position G_1. This adjustment permits the potentiometer to be read directly in volts. With the galvanometer in position G_2, the unknown emf is balanced by varying the opposing IR drop. Resistances used from the *coarse* and *intermediate* dials and the *slide wire* are adjusted until the galvanometer again reads null, and the unknown emf can be read directly from the dial settings. The ratio of the unknown and reference emfs is precisely the ratio of the resistances for the two null adjustments, provided that the current is the same.

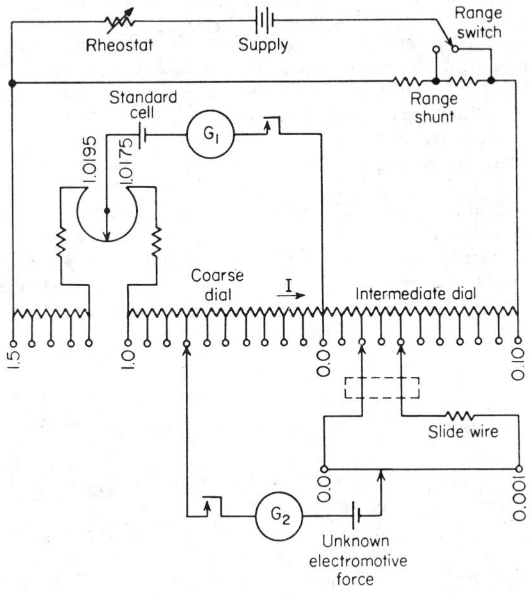

FIG. 3-8. General-purpose constant-current potentiometer.

The switching arrangement is usually such that the galvanometer can be shifted quickly between the G_2 and G_1 positions to check that the current has not drifted from the value at which it was standardized. It will be noted that the contacts of the *coarse-dial* switch and *slide wire* are in the galvanometer branch of the circuit. At balance, they carry no current, and their contact resistance does not contribute to the measurement. However, there can be only two noncontributing contact resistances in the network shown; the switch contacts for adjusting the intermediate-dial position do carry current, and their resistance does enter the measurement. Care is taken in construction that the resistances of such current-carrying contacts are low and repeatable; and frequently, as in the example illustrated, the circuit is arranged so that these contributing contacts carry only a fraction of the reference current, and the contribution of their *IR* drop to the measurement is correspondingly reduced.

Another feature of many general-purpose potentiometers, illustrated in the diagram, is the availability of a reduced range. The resistances of the range shunts have such values that, at the 0.1 position of the range-selection switch, only a tenth of the reference current goes through the measuring branch of the circuit, and the range of the potentiometer is correspondingly reduced. Frequently, a $\times 0.01$ range is also available.

In addition to the effect of *IR* drops at contacts in the measuring circuit, accuracy limits are also imposed by thermal emfs generated at circuit junctions. These limiting factors are increasingly important as potentiometer range is reduced. Thus, in low-range or microvolt potentiometers, special care is taken to keep circuit junctions and contact resistances out of the direct measuring circuit as much as possible, to use thermal shielding, and to arrange the circuit and galvanometer keys so that temperature differences will be minimized between junction points that are directly in the measuring circuit. Generally also, in microvolt potentiometers, the galvanometer is connected to the circuit through a special *thermofree* reversing key so that thermal emfs in the galvanometer can be eliminated from the measurement—the balance point being that which produces zero change in galvanometer deflection on reversal.

A typical operating circuit for a microvolt potentiometer is shown in Fig. 3-9. Here the measuring circuit consists of two constant-current decades followed by a constant-resistance element (known as a Lindeck element). The milliammeter of the latter, with 100 divisions, is nearly the equivalent of three more decades, since its indications can be estimated to one- or two-tenths of a division. The upper dials of the illustrative example are Diesselhorst rings, frequently used in low-range potentiometers. The current entering and leaving the rings is constant at the standardized value and divides between the parallel paths which constitute the ring so that the IR drop in the measuring element increases in 10 equal increments as the dial switch is moved from the 0 to the 10 position. It should be noted that all switch contacts are in the supply circuits where neither their IR drops nor their thermal emfs have any significant effect on the measurement. The dashed line indicates the circuit elements that must be included within an isothermal shield. Such a potentiometer may have a meaningful least count of 0.01 μV.

Nanovolt potentiometers are now also commercially available. It is of the utmost importance that parasitic emfs and IR drops be avoided in such instruments.

34. Voltage dividers are used to reduce voltages by a known factor. *Volt boxes,* used to extend the range of potentiometers, are resistive voltage dividers with a limited number of fixed-position taps, as shown in Fig. 3-10. The unknown voltage is connected

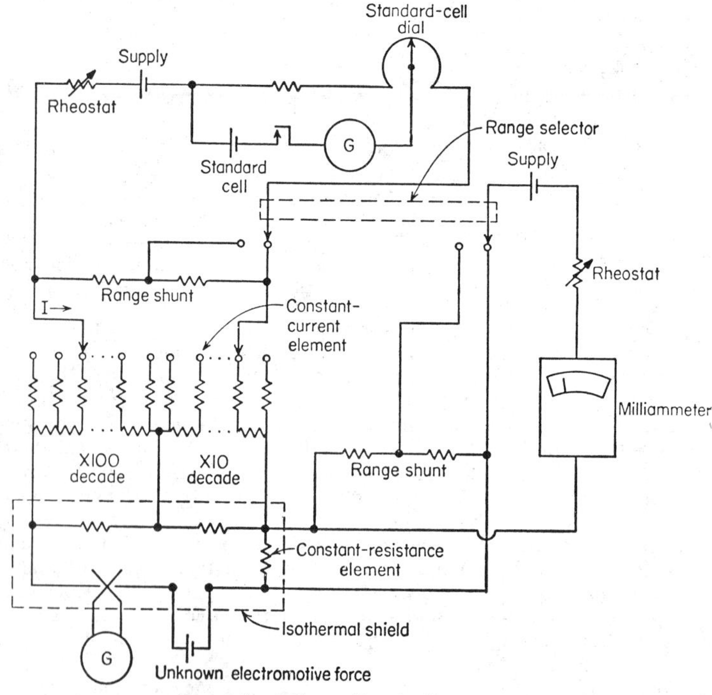

FIG. 3-9. Microvolt potentiometer.

between appropriately marked input terminals (say, 0 and 150), and the potentiometer is connected between the 0 and 1.5 output terminals. When balance is achieved on the potentiometer in the usual way, the input voltage is obtained by multiplying the potentiometer reading by the indicated factor (×100). While no current is taken by the potentiometer at balance, current is supplied by the source to the divider through its input terminals. Hence, the voltage drop between the input terminals is measured rather than the open-circuit emf of the source. The resistances of working volt boxes range from about 200 to 750 Ω/V, and even higher—to 1 kΩ/V—in reference-standard types.

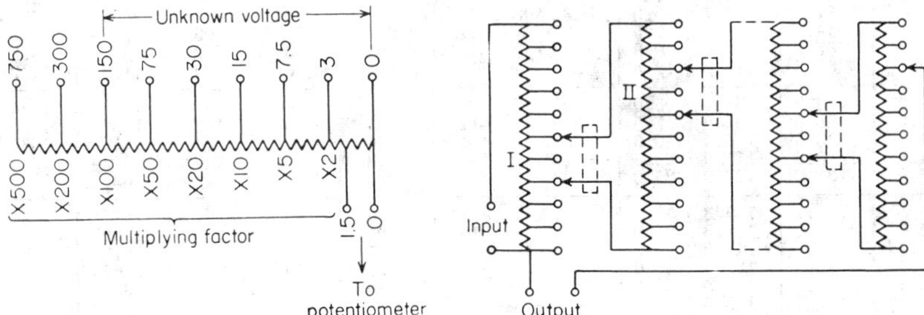

Fig. 3-10. Volt box.

Fig. 3-11. Decade voltage divider.

Since changes in ratio as a result of self-heating are much larger in low- than in high-resistance circuits, particularly in the higher voltage ranges, the present trend is toward the higher ohms per volt figure. However, the quality of insulation becomes more critical in high-resistance boxes, since insulating structures which support the resistance elements are themselves high-resistance leakage paths. Thus, the high-accuracy reference-standard boxes of 1,000 Ω/V resistance are provided with a guard circuit which maintains shields at appropriate potentials along the insulating structure of the box and supplies the leakage currents across the insulators.

Decade voltage dividers generally use the Kelvin-Varley circuit arrangement shown in Fig. 3-11. It will be seen that two elements of the first decade are shunted by the entire second decade, whose total resistance equals the combined resistance of the shunted steps of decade I. The two sliders of decade I are mechanically coupled and move together, keeping the shunted resistance constant regardless of switch position. Thus, the current divides equally between decade II and the shunted elements of decade I, and the voltage drop in decade II equals the drop in one unshunted step of decade I. The effect of contact resistance at the switch points is somewhat diminished because of the division of current. The Kelvin-Varley principle is used in succeeding decades except the final one, which has only a single switch contact. Such voltage dividers may have as many as six decades and have ratio accuracies approaching 1 part in 10^6 of input.

SPARK GAPS FOR VOLTAGE MEASUREMENT

35. General. The sparking distance between two electrodes in air is a standard method of measuring high voltages. The maximum length of gap which a given voltage will break down depends on the air density, the crest value of the voltage, the gap geometry, and other factors (see also Sec. **4,** Par. **308**).

36. Sphere gaps constitute a recognized means for the measurement of crest values of alternating voltages. Many investigators have carried out calibrations of gaps with sphere diameters ranging from a few centimeters to 2 m. The accepted values for spark-over crest voltages of various sizes of spheres set at various gap spacings are given in Table 3-6. These tables are taken from AIEE Standard 4, 1953, Standards for Measurement of Test Voltages in Dielectric Tests.

a. Dimensions. Spheres used for a standard spark gap must be of equal diameter, highly polished, and the curvature of the opposing surfaces should not vary by more than 1% from that of a true sphere of the required diameter. No extraneous conducting body other than the supporting shanks should be nearer the gap than twice sphere

Table 3-6. Sphere-gap Spark-over Crest Voltages in Kilovolts

(At 760 mm and 25°C. One sphere grounded)

Sec. A.—Values for 60 Hz and for negative impulses

Gap, cm	Diameter of spheres, cm		Gap, cm	Diameter of spheres, cm			Gap, cm	Diameter of spheres, cm		
	6.25	12.5		25	50	75		100	150	200
0.5	17.0	.....	2.5	72	...	....	10	261	261	261
1.0	31.3	31.7	5.0	136	136	136	20	504	505	506
1.5	44.5	44.9	7.5	192	197	200	30	700	736	746
2.0	57.0	58.0	10.0	241	260	261	40	862	947	973
2.5	68.8	70.8	12.5	278	317	324	50	985	1120	1172
3.0	78.8	83.5	15.0	309	367	380	60	1084	1254	1346
3.5	86.6	95.0	17.5	338	411	433	70	1163	1360	1505
4.0	93.6	106.0	20.0	362	451	484	80	1234	1458	1635
4.5	99.8	117.0	22.5	379	486	528	90	1295	1552	1752
5.0	105.5	127.0	25.0	393	519	573	100	1338	1628	1857
5.5	.....	135.3	30.0	...	573	653	110	....	1695	
6.0	.....	143.5	35.0	...	615	721	120	....	1760	
6.25	.....	147.5	40.0	...	651	777	130	....	1815	
7.0	.....	157.7	45.0	...	681	827	140	....	1865	
8.0	.....	170.5	50.0	...	707	870	150	....	1900	
9.0	.....	182.0								
			55.0	...	...	910				
			60.0	...	...	945				
			65.0	...	...	977				
			70.0	...	...	1003				
			75.0	...	...	1025				

Sec. B.—Values for positive impulses

Gap, cm	6.25	12.5	Gap, cm	25	50	75	Gap, cm	100	150	200
0.5	17.0	.....	2.5	72	...	.....	10	261	261	261
1.0	31.3	31.7	5.0	136	136	136	20	504	505	506
1.5	44.8	44.9	7.5	196	197	200	30	715	736	746
2.0	57.4	58.0	10.0	252	260	261	40	888	955	973
2.5	69.3	70.8	12.5	296	319	324	50	1024	1140	1178
3.0	79.4	83.5	15.0	334	374	380	60	1124	1293	1364
3.5	88.0	95.3	17.5	364	426	443	70	1209	1400	1533
4.0	....	108.0	20.0	390	474	499	80	1284	1502	1671
4.5	....	120.0	22.5	409	511	548	90	1344	1597	1788
5.0	....	132.3	25.0	426	547	597	100	1390	1678	1896
5.5	....	142.5	30.0	...	605	687	110	....	1755	
6.0	....	153.8	35.0	...	655	755	120	....	1824	
6.25	....	158.0	40.0	...	698	816	130	....	1880	
7.0	....	171.0	45.0	...	732	870	140	....	1920	
			50.0	...	758	917	150	....	1944	
			55.0	...	...	960				
			60.0	...	...	999				
			65.0	...	...	1031				
			70.0	...	...	1058				
			75.0	...	...	1081				

diameter; insulating bodies must be at least a sphere diameter away; shanks should be no greater in diameter than one-fifth sphere diameter.

b. Air density, determined by temperature and pressure, affects spark-over voltage. For large spheres, the effect is proportional to density, and Table 3-7*a* gives the density of dry air at various temperatures and pressures relative to its density at the reference condition—760 mm and 25°C—and the spark-over voltages of Table 3-6 must be multiplied by the appropriate relative air density to give correct voltage for spheres of 100, 150, and 200 cm diameter. For smaller spheres, whose voltage-density relation is not linear, the relative air density must be found and the corresponding correction factor obtained from Table 3-7*b*. The effect of humidity is negligible unless condensation occurs on the spheres.

c. Polarity. Section A of Table 3-6 is for use at 60 Hz and for impulse voltages when the ungrounded sphere is negative. Section B is for use on impulse voltages when the ungrounded sphere is positive. In a vertical arrangement, the lower sphere is grounded.

Table 3-7. Air-density Correction Factors for Sphere Spark Gaps

Sec. A.—Correction factors for 100-, 150-, and 200-cm spheres

(Air density relative to dry air at 760 mm and 25°C)

Pressure, mm	Temperature, deg Centigrade														
	−40	−30	−20	−10	−5	0	5	10	15	20	25	30	35	40	45
660	1.111	1.065	1.023	0.984	0.966	0.948	0.931	0.914	0.899	0.883	0.868	0.854	0.840	0.827	0.814
670	1.128	1.081	1.038	0.999	0.980	0.962	0.945	0.928	0.912	0.897	0.882	0.867	0.853	0.839	0.826
680	1.144	1.097	1.054	1.014	0.995	0.977	0.959	0.942	0.926	0.910	0.895	0.880	0.866	0.852	0.838
690	1.161	1.113	1.069	1.029	1.010	0.991	0.973	0.956	0.939	0.923	0.908	0.893	0.878	0.864	0.851
700	1.178	1.130	1.085	1.044	1.024	1.005	0.987	0.970	0.953	0.937	0.921	0.906	0.891	0.877	0.863
710	1.195	1.146	1.100	1.059	1.039	1.020	1.001	0.984	0.967	0.950	0.934	0.919	0.904	0.889	0.875
720	1.212	1.162	1.116	1.074	1.053	1.034	1.016	0.998	0.980	0.964	0.947	0.932	0.917	0.902	0.888
730	1.229	1.178	1.131	1.088	1.068	1.048	1.030	1.011	0.994	0.977	0.961	0.945	0.929	0.914	0.900
740	1.245	1.194	1.147	1.103	1.083	1.063	1.044	1.025	1.007	0.990	0.974	0.958	0.942	0.927	0.912
750	1.262	1.210	1.162	1.118	1.097	1.077	1.058	1.039	1.021	1.004	0.987	0.971	0.955	0.940	0.925
760	1.279	1.226	1.178	1.133	1.112	1.092	1.072	1.053	1.035	1.017	1.000	0.984	0.968	0.952	0.937
770	1.296	1.242	1.193	1.148	1.127	1.105	1.086	1.067	1.048	1.030	1.013	0.996	0.980	0.965	0.949
780	1.313	1.259	1.209	1.163	1.140	1.120	1.100	1.081	1.062	1.044	1.026	1.009	0.993	0.977	0.962
790	1.330	1.275	1.224	1.178	1.156	1.135	1.114	1.095	1.076	1.057	1.039	1.022	1.006	0.990	0.974
800	1.346	1.291	1.240	1.194	1.170	1.149	1.128	1.108	1.089	1.071	1.053	1.035	1.018	1.002	0.986

Sec. B.—Correction factors for the smaller spheres

(Based on standard conditions of 760 mm and 25°C)

Relative air density	Diameter of standard spheres, cm				
	6.25	12.5	25	50	75
0.50	0.547	0.535	0.527	0.519	0.517
0.55	0.595	0.583	0.575	0.567	0.565
0.60	0.640	0.630	0.623	0.615	0.613
0.65	0.686	0.677	0.670	0.663	0.661
0.70	0.732	0.724	0.718	0.711	0.709
0.75	0.777	0.771	0.766	0.759	0.757
0.80	0.821	0.816	0.812	0.807	0.805
0.85	0.866	0.862	0.859	0.855	0.854
0.90	0.910	0.908	0.906	0.904	0.903
0.95	0.956	0.955	0.954	0.952	0.951
1.00	1.000	1.000	1.000	1.000	1.000
1.05	1.044	1.045	1.046	1.048	1.049
1.10	1.090	1.092	1.094	1.096	1.097

Calibration values are stated to apply also to gaps with one-third sphere diameter for spheres 25 cm and smaller and no more than one-fourth sphere diameter for larger spheres. The tabulated values apply to impulses in which the time to crest is more than 1 μsec; for shorter times, the true crest value is higher by an amount that depends strongly on voltage rise time.

d. Current at spark-over should be limited by a noninductive resistance of at least 1 Ω/V, connected in series on the ungrounded side of the gap; if neither sphere is grounded, half the resistance should be connected in series with each. Resistors should be of a type whose resistance is not affected by voltage; carbon resistors should not be used.

e. Irradiation of the gap with ultraviolet light will considerably reduce the scattering of values of breakdown voltage, particularly for the smaller spheres; a quartz-tube mercury-vapor lamp, operated on direct current and placed five gap lengths from the spheres, is recommended.

CONTINUOUS CURRENT MEASUREMENTS

36a. Absolute current measurement relates the value of the current unit—the ampere—to the prototype mechanical units of length, mass, and time—the meter, the kilogram, and the second—through force measurements in an instrument called a *current balance*. Such instruments are to be found generally only in national standards laboratories, which have the responsibility of establishing and maintaining the electrical units. In a current balance, the force between fixed and movable coils is opposed by the gravitational force on a known mass, the balance equation being $I^2(\partial M/\partial X) = mg$. The construction of the coil system is such that the rate of change with displacement of mutual inductance between fixed and moving coils can be computed from measured coil dimensions. Absolute current determinations are used to assign the emf of reference standard cells. A 1-Ω resistance standard is connected in series with the fixed- and moving-coil system, and its drop is compared with the emf of a cell during the force measurement. Thus, the unit of current is, in effect, stored in terms of the ratio of the reference volt to the ohm. The stated uncertainty associated with the most recent absolute-ampere determination reported by the National Bureau of Standards was 5 ppm.

37. The potentiometer method of measuring continuous currents is commonly used where a value must be more accurate than can be obtained from the reading of an indicating instrument. The current to be measured is passed through a four-terminal resistor (shunt) of known value, and the voltage developed between its potential terminals is measured with a potentiometer. If the current is small, so that there is no significant temperature rise in the shunt, the measurement accuracy can be 0.01% or better. In general, the accuracy of potentiometer measurements of continuous currents is limited by how well the shunt resistance is known under operating conditions.

38. Measurement of very small continuous currents, down to 10^{-17} A, have been accomplished by means of "electrometer" tubes—vacuum tubes designed so that the grid has practically no leakage current either over its insulating supports or to the cathode. The current to be measured flows through a very high resistance (up to 10^{12} Ω), and the voltage drop is impressed on the grid of an electrometer tube. The plate current is observed, and the voltage drop is duplicated by producing the plate current with a known adjustable voltage. The current can then be calculated from the voltage and resistance.

INDICATING INSTRUMENTS

39. Classification of electrical indicating instruments into groups is possible according to two broad criteria. Grouped as to style and accuracy, they may be designated as *panel, portable,* and *laboratory-standard* instruments. To these could be added a special class of panel instruments known as *switchboard* types. An alternative method of grouping instruments is on the basis of operating principle. These groups are *permanent-magnet moving-coil, dynamometer, moving-iron,* and *electrostatic* instruments, with the combination of *converters* operating with *permanent-magnet moving-coil* instruments as a separate class.

40. Panel instruments, as the name implies, are designed for vertical panel mounting. They are calibrated and balanced for operation with the axis of rotation of the moving system horizontal. Scale lengths are 1 to 7 in and accuracies 1 to 5% of full-scale value. Accuracy classification, definitions of terminology, and allowable errors for various operating conditions are given in USAS C-39.1, The American Standard for Indicating Instruments, for these and other classes of instruments—portable and laboratory-standard. This standard also specifies mounting dimensions recommended for panel and switchboard instruments.

41. Portable instruments, as a recognized class, have scale lengths of 4 to 8 in and accuracies of 0.2 to 2% of full-scale value. This category is not generally understood to include the small, low-accuracy instruments used primarily for service work. Although the latter are portable, they are classed as panel instruments mounted in portable cases. Portable instruments in accuracy classes ½% or better are generally made with a hand-calibrated scale and provided with an antiparallax mirror for accurate reading. They are calibrated for operation in a horizontal position—with the axis of the moving element vertical. Use in this position results in better reading accuracy and repeatability, because it eliminates pivot roll and reduces friction in pivot- and jewel-mounted elements and element sag in taut-band types. Because of the advantages of horizontal operation (with the axis of the moving element vertical), some makers furnish portable-type instruments in an edgewise type of case for vertical panel mounting, preserving the accuracy of horizontal operation in a panel instrument.

42. Laboratory-standard instruments are long-scale, high-accuracy instruments used primarily for the calibration of other instruments by direct comparison. Scale lengths are 8 to 13 in, and accuracies are 0.1 to 0.2%. (In some instances, 0.05% accuracy instruments are available.) These instruments represent the best that can

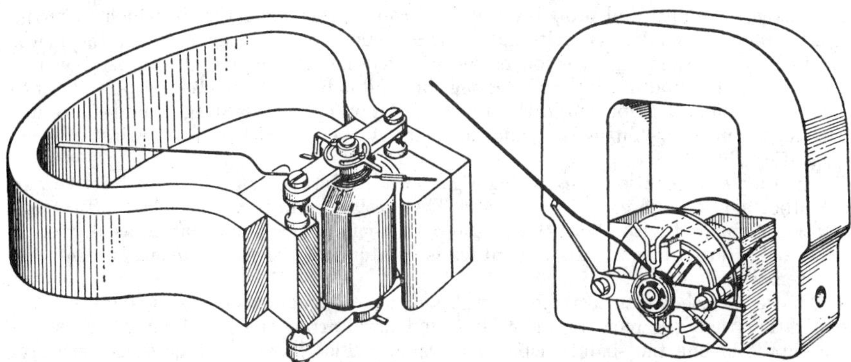

Fig. 3-12. Permanent-magnet moving-coil mechanism (external magnet).

be achieved in indicating instruments. All instruments in this group have antiparallax mirrors. In general, they may be classed as high-burden rather than high-sensitivity instruments, since, with the objective of high-accuracy performance, the operating torque is large. However, as they are intended specifically for calibration of other instruments rather than for general measurement use, the high-burden characteristic usually presents no problems. Highly stable, accurate power supplies combined with special calibration consoles are now coming into use and may eventually replace laboratory-standard instruments for calibration purposes.

43. Switchboard instruments are special panel-mounted instruments designed for long-term continuous operation in the presence of vibration and shock and occasional severe transients.

44. Permanent-magnet-coil instruments are the most common type in use. The moving element is a coil of fine wire suspended in such a manner that it can rotate in an annular gap which has an axial magnetic field. The torque caused by the reaction between the coil and the magnetic field is opposed by some form of spring restraint, and the resulting coil position is indicated by the deflection of a pointer over a scale. To the extent that the field is uniform, the spring restraint linear, and the mechanical positioning symmetrical, the pointer deflection will be linear with respect to the coil current. Figure 3-12 shows one of the possible arrangements of the mechanism. The coil is held between top and bottom pivot and jewel bearings. Helical restraining springs are used, and they also serve to conduct current to and from the coil. The coil moves in a double annular gap between a cylindrical core and two pole pieces, with the field provided by an external magnet.

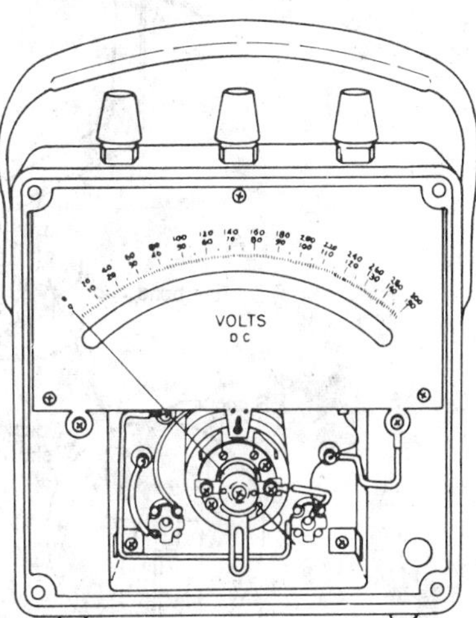

Fig. 3-14. Internal construction of portable voltmeter.

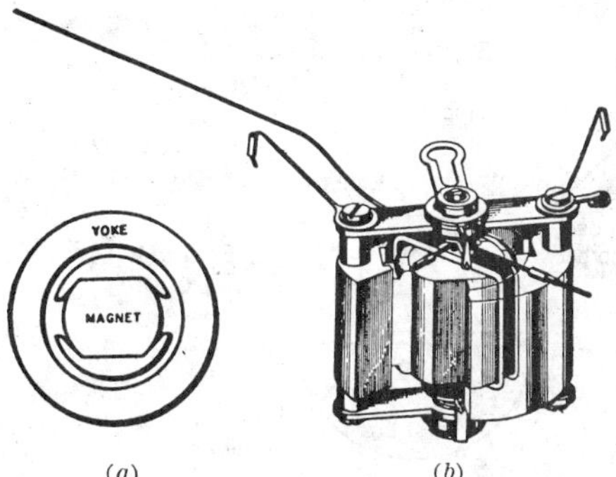

(a) (b)

Fig. 3-13. (a) Core magnet construction; (b) internal-magnet design.

This is the oldest and probably the most common configuration in which this type of instrument is made. An alternative arrangement, which makes use of the newer magnet materials, is an inversion of the external-magnet design. Here, as shown in Fig. 3-13, an internal magnet replaces the core and is fitted with soft-iron pole shoes to give uniform flux distribution in the air gap. An outer soft-iron ring provides a return path for the magnetic flux and, in addition, acts as a very efficient shield against external magnetic fields.

Another configuration uses a single gap, with only one side of the coil active and the other side on the axis of rotation. This design permits as much as 240° of coil rotation compared with about 90° in the double-gap arrangement and is used in some switchboard instruments. Its advantage is greater scale length in an instrument case of given size.

Figure 3-14 shows an instrument with an external magnet and a taut-band suspension rather than the more familiar pivot and jewel arrangement. Figure 3-15 shows the details of one taut-band suspension system. Flat bands of high-tensile-strength material are stretched between end tension springs with a force of 50 to 150 g. The material (frequently a 90% platinum–10% nickel alloy) is rolled to a 10:1 width-to-thickness ratio, and the twisting moment supplies the restraining torque as well as being the support mechanism.

The taut-band construction avoids one of the most common causes of friction found in the pivot and jewel system. Pivots tend to be distorted and become dull with shock and cause excessive friction. Restraining forces in taut bands can also operate at a low level at which helical springs could not satisfactorily hold their shape under

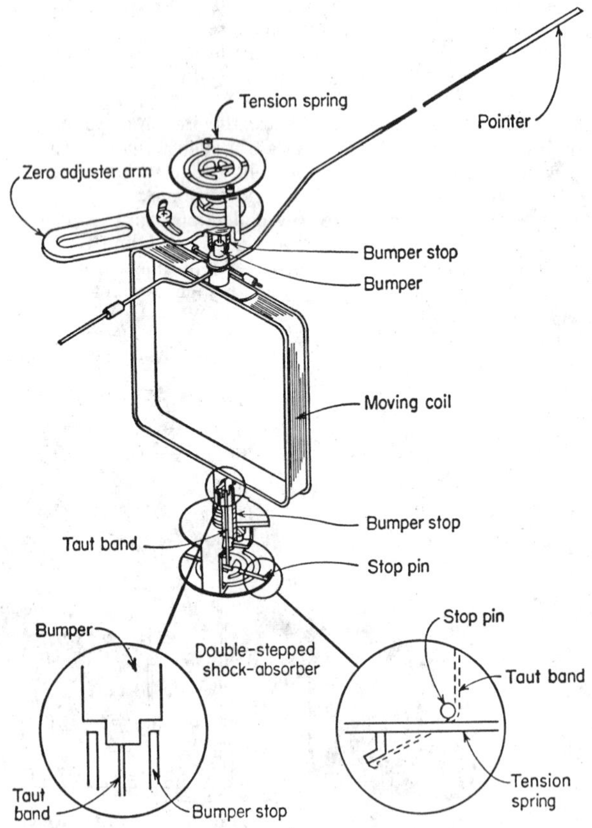

FIG. 3-15. Construction of taut-band suspension.

normal shock. This allows appreciably greater sensitivity to be obtained with the same coil and magnet system.

Permanent-magnet moving-coil instruments are used with coil currents from 1 μA to about 50 mA. For higher currents and for multirange instruments, the coil is shunted to a suitable value; the shunt can be tapped to provide a number of related ranges higher than the base range of the coil itself. Figure 3-16 shows a multirange "ring" shunt. A ring shunt is better than a group of individual shunts because it maintains substantially constant damping for all ranges. Shunt damping can be used either in place of or in addition to coil damping, which is accomplished in one or both of two ways. The moving coil may be wound on an aluminum former which constitutes a short-circuited turn, or a certain portion of the coil turns may be short-circuited. Current induced in the short-circuited turns by coil motion in the magnetic field dissipates the energy of motion as I^2R heat and thus brings the coil to rest. Damping is usually held within one to three overswings past a final rest point when a current is instantaneously applied, but damping characteristics can be varied widely for special applications. The deflection is proportional to coil current, and if a resistor is connected in series, the scale can be marked with the voltage required to produce the current in the combination. This is the method used in voltmeters, multiple ranges being obtained by switching a series of resistors as shown in Fig. 3-17. Voltmeter sensitivity is usually stated in ohms per volt, the reciprocal of full-scale current in amperes. Commercial instruments range from 50 Ω/V to as high as 1 MΩ/V.

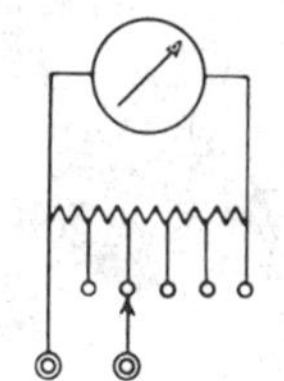

FIG. 3-16. Multirange ring shunt.

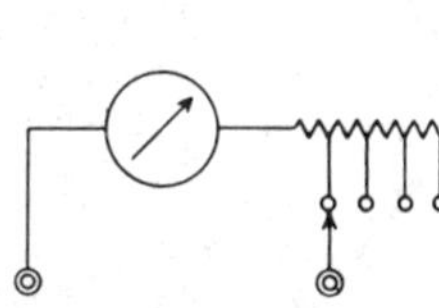

FIG. 3-17. Multirange circuit for voltmeter.

45. Permanent-magnet moving-coil instruments with converters are used for a-c measurements. A half-wave or full-wave rectifier with a d-c movement will allow the reading of a-c values in terms of the direct current through the moving coil. A perfect rectifier would allow the instrument to be read directly in terms of the *average* value of the applied signal (a selected half wave in the case of a half-wave rectifier). However, the *rms* (or effective) value of current or voltage is usually desired. For sinusoidal current or voltage, this is 1.11 times the average; and it is customary to draw the scale of a rectifier instrument to include this 1.11 factor. This can lead to large errors in measuring nonsinusoidal waveforms. The actual scale reading will be 1.11 × average regardless of waveform but is the correct rms value only for sine-wave signals. The scales of commercially available rectifier instruments show marked nonlinearities in low-voltage ranges, and their temperature coefficients are also high. Usually separate scales are drawn for each low-voltage range. Rectifier instruments have advantages in the form of high sensitivity, ruggedness, and low cost.

Thermoelements are also used to convert permanent-magnet moving-coil instruments to a-c operation. The thermoelement consists of a heater wire which carries the current to be measured and a thermocouple junction of two dissimilar metals in intimate thermal contact with the heater wire. Heat in the element is converted to thermoelectric output voltage, which is applied to a d-c millivoltmeter for indication. The heat dissipated in the heater is proportional to I^2R and hence affords a true *rms* indication of the current through it. The scales of thermocouple instruments have a square-law characteristic with marked crowding at the low end of the scale. For low currents, the thermoelement may be enclosed in a vacuum bulb to avoid heat loss by convection and thus improve efficiency. Vacuum elements range from 1 mA to 1 A and exposed elements from about 100 mA to several hundred amperes. Thermoelements have very good high-frequency characteristics, being used in some instances to 10^3 MHz. Most commercial instruments, particularly multirange instruments, are limited in frequency span by resistors and switch wiring rather than by any characteristic of the thermoelement itself. Thermoelements have very limited overload capacity. At rated current, the heater normally operates at a rather high temperature in order that the thermocouple output may be sufficient to operate the indicating meter. The

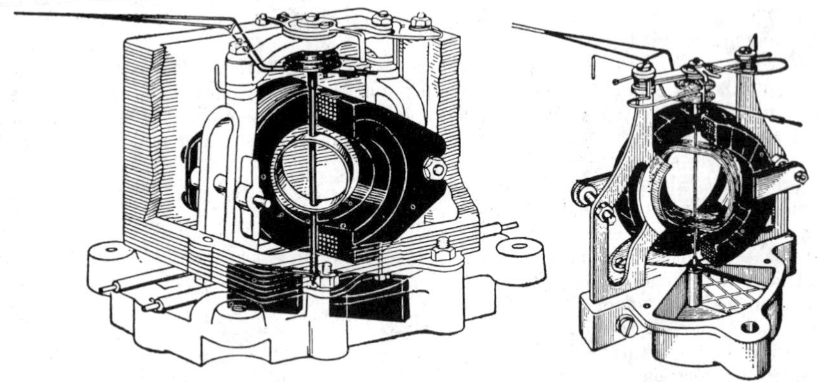

FIG. 3-18. Weston dynamometer instrument.

heater temperature increases with the square of the current and, at moderate overloading, may reach a temperature at which it suffers permanent damage or even burns out.

46. Dynamometer instruments have a moving coil positioned so that it develops torque by reacting with the magnetic field of a fixed-coil system carrying the same or a related current. Figure 3-18 shows the construction of such an instrument, with the moving coil rotating inside a pair of field coils which supply the magnetic field. Suspension systems similar to d-c instruments are used, either pivot and jewel or taut bands. Damping cannot be incorporated in the moving coil of a dynamometer instrument but must be separately accomplished. In the figure, a pair of light vanes on the lower end of the shaft move with small clearance in an enclosure to provide the required damping. Damping can also be accomplished by allowing an aluminum vane to move between the poles of a strong magnet. In this case, the energy of motion is absorbed in the eddy-current circuits of the moving vane. Figure 3-19 shows a magnetic damping-vane assembly. Dynamometer instruments are used to measure current, voltage, or power, depending on the connections. If the moving coil and field coils are connected in series, the instrument deflection will be in the same direction regardless of polarity, so that they operate equally on alternating or direct current. Since moving-coil torque is proportional to the product of fixed and moving currents, response is to I^2 for the series connection, and the *rms* value of the operating current is indicated.

Dynamometer instruments are used as current meters, with coils in series, from 10 to 100 mA. For higher ranges, the moving coil is shunted and carries only a fraction of the measured current. Voltmeters are low-current instruments (with fixed and moving coils in series) together with a series-connected resistor of appropriate magnitude, with ranges available from 1 to 1,000 V in self-contained instruments. Low-voltage instruments draw heavy currents, and instruments in ranges below 10 or 15 V are unusual.

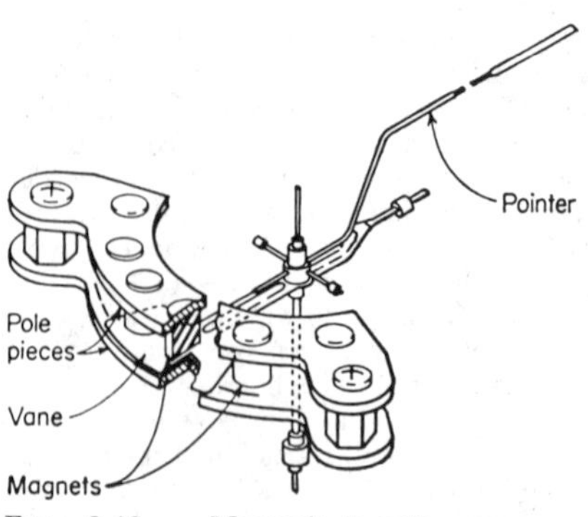

FIG. 3-19. Magnetic-damping vane assembly.

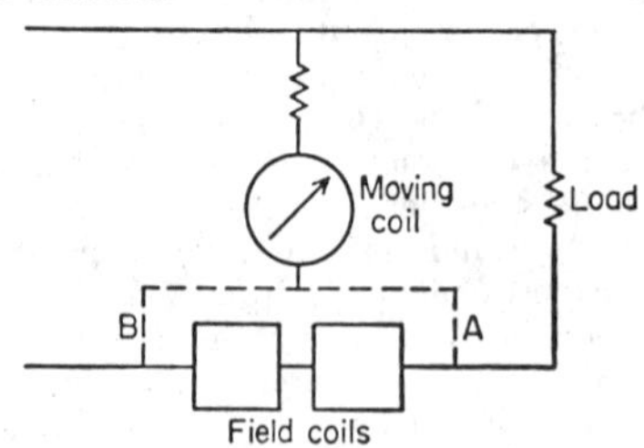

FIG. 3-20. Alternative wattmeter connections.

Wattmeters require separate connections for field and moving-coil circuits. The moving coil in series with a suitable resistance is connected in parallel with the load, and the moving-coil current is proportional to load voltage. The field coils are connected in series with the load and carry the load current. The instantaneous torque is thus proportional to the instantaneous product of load voltage and current, and this is instantaneous load power. Over a complete alternation, the instrument response is to average power, $EI \cos \theta$, where θ is the angular displacement between voltage and current; that is, $\cos \theta$ is load power factor.

Figure 3-20 shows two connections for a wattmeter measuring power delivered to a load. If the voltage circuit is connected as shown in A, the total indicated power is that supplied to the load plus the power taken by the voltage circuit of the wattmeter itself. If the circuit is connected at B, the indication is of load power plus the power taken by the field coils of the wattmeter. In any case, the wattmeter reading includes the losses in one or the other of its own circuits unless it carries a special compensating arrangement. However, the correction is small for internal losses and, except in low-range instruments, can frequently be neglected. Most wattmeters are for use below 125 Hz unless compensated for operation at higher frequencies. Instruments are available whose performance is acceptable up to 10 kHz. Polyphase wattmeters consist actually of two matched single-phase elements on a common shaft, shielded from one another so that they do not interact and with a single damping element. The torques of the two elements are additive, and the total power is indicated by the pointer position.

47. Moving iron instruments are of two basic types: those in which an iron vane is attracted into a coil, and those in which two vanes in a coil show a mutual repulsion. The attracted-vane type provides an inexpensive construction and is frequently used where price is a prime consideration. The repulsion type consists of a movable and a fixed vane within the same coil. Current in the coil magnetizes the two vanes with the same polarity so that they repel, and the movable vane tends to rotate away from the fixed vane. The usual restraining springs or taut bands are needed but are not required to carry current to the moving element. While this type is also an *rms* indicating instrument whose torque is proportional to I^2 in the exciting coil, the scale distribution can be controlled within wide limits and can even be made nearly uniform (except near its zero end) over 80% or more of its total length by properly shaping and positioning the fixed and moving irons.

Figure 3-21 shows the construction of a repulsion-type instrument. Damping is either by an air vane in a closed chamber or by a vane moving in the gap of a permanent magnet. Moving-iron instruments can have current ranges as low as 5 mA and as high as 100 A and voltage ranges of 1 to 1,000 V. In theory such instruments should indicate correctly over only a small frequency range and, because of hysteresis in the irons, should show appreciable directional differences on direct current; but with modern magnetic materials, both hysteresis and remanence effects can be very small, and d-c indication can be satisfactory. Frequencies of 50 to 400 Hz cover the usual range of construction, but, with special design and compensation, good performance can be achieved well into the audio-frequency range.

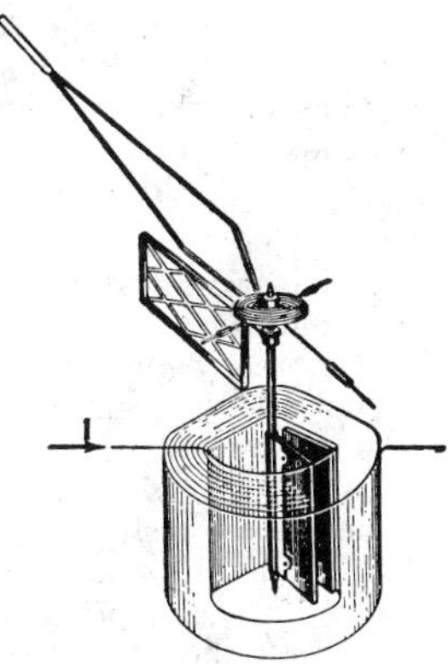

FIG. 3-21. Radial vane mechanism.

48. Electrostatic voltmeters are the only type that do not derive their operating torque from current but are truly voltage-operated; any current is merely incidental to operation. The instrument consists of two vanes (or sets of vanes), one fixed and one movable. A voltage between them will produce a mutual attraction, and the movable vane will rotate toward the fixed one, i.e., in a direction to increase the capacitance of the combination. Low-voltage instruments have multiple rotor and stator vanes and resemble adjustable capacitors of the type used in tuning radio circuits. Full-scale voltages can be as low as 100 V. High-voltage units use single vanes and, in many cases, have a movable electrode which can give several ranges by means of accurate positioning at various spacings. All single-range instruments and those multirange instruments which change range by stator position may be used on direct or alternating voltage. Instruments that employ a capacitive voltage divider for range changing can be used only on alternating current. Instruments in the range 3 to 100 kV can be expected to have d-c resistances of $10^{15}\ \Omega$ or more and capacitances of 10 pF or less. Accuracies ½ to 2% are commercially available.

Electrostatic instruments have rms response and are not subject to waveform error. The upper limit of frequency response is determined by the capacitance current that can be passed safely through restraining springs or taut-band suspensions or the frequency at which mechanical dimensions become an appreciable fraction of a wavelength. The upper frequency limit of electrostatic voltmeters varies widely with construction and may be as low as 25 kHz or as high as 5 MHz.

49. Electronic voltmeters vary widely in performance characteristics and frequency range covered, depending on the circuitry used. A common type uses an initial diode to charge a capacitor. This may be followed by a stabilized amplifier with a microammeter as indicator. Range may be selected by appropriate cathode resistors in the amplifier section. Such instruments normally have very high input impedance (a few picofarads), respond to peak voltage, and are suitable for use to very high frequencies (100 MHz or more). While the response is to *peak* voltage, the scale of the indicating element may be marked in terms of rms for a sine-wave input, that is, 0.707 × *peak* voltage. Thus, for a nonsinusoidal input, the scale (read as rms volts) may include a serious waveform error; but if the scale reading is multiplied by 1.41, the result is the value of the *peak* voltage.

An alternative network, used in some electronic voltmeters, is an attenuator for range selection, followed by an a-c amplifier and finally a rectifier and microammeter. This system has substantially lower input impedance, and limits of frequency range are fixed by the characteristics of the amplifier. The response in this arrangement may be to *average* value of the input signal, but, again, the scale marking may be in terms of rms value for a sine wave. In this case also, the waveform error for nonsinusoidal input must be borne in mind; but if the scale reading is divided by 1.11, the *average* value is obtained. Within these limitations, accuracy may be as good as 1% of full-scale indication in some types of electronic voltmeter, although in many cases a 2 to 5% accuracy may be anticipated.

D-C TO A-C TRANSFER

50. General transfer capability is essential to the measurement of voltage, current, power, and energy. The standard cell, the unit of voltage which it preserves, and the unit of current derived from it in combination with a standard of resistance are applicable only to the measurement of d-c quantities, while the problems of measurement in the power and communication fields involve alternating voltages and currents. It is only by means of transfer devices that one can assign the values of a-c quantities or calibrate a-c instruments in terms of the basic d-c reference standards. In most instances, the rms value of a voltage or current is required, since the transformation of electrical energy to other forms involves the square of voltages or currents; and the transfer from direct to alternating quantities is made with devices that respond to the square of current or voltage. Three general types of transfer instruments are capable of high-accuracy rms measurements: (1) electrodynamic instruments—which depend on the force between current-carrying conductors; (2) electrothermic instruments—which depend on the heating effect of current; (3) electrostatic instruments—which depend on the force between electrodes at different potentials. While two of these depend on current and the third on voltage, the use of series and shunt resistors makes all three types available for current or voltage transfer. Traditional American practice has been to use electrodynamic instruments for current and voltage transfer as well as power transfer from direct to alternating current; but recent developments in thermoelements[1] have improved their transfer characteristics until they are now the preferred

[1] Hermach, F. L. *Trans. IEEE*, December, 1965, Vol. 1, p. 14.

means for *current* and *voltage* transfer, although the electrodynamic wattmeter is still the instrument of choice for *power* transfer up to 1 kHz.

51. Electrothermic transfer standards for current and voltage use a thermoelement consisting of a heater and a thermocouple. In its usual form, the heater is a short, straight wire suspended by two supporting lead-in wires in an evacuated glass bulb. One junction of a thermocouple is fastened to its midpoint and is electrically insulated from it with a small bead. The thermal emf—5 to 10 mV at rated current in a conventional element—is a measure of heater current. For voltage measurement, a resistor is connected in series with the heater so that output emf is a measure of input voltage. For high-frequency applications, the series element consists of one or more metal-film resistors in a coaxial metal guard cylinder with the thermoelement mounted in a separate cylinder. The guard cylinders are so arranged and connected to the circuit as to minimize the phase-defect angle of the combination. These elements are useful into the radio-frequency range—to 100 MHz or more—and at audio frequencies, their transfer accuracy may be a few parts in 10^6.

For current measurements, an a-c shunt (preferably of coaxial design to minimize its time constant) is used with a low-range thermoelement connected across its potential terminals. Here, also, transfer accuracies of a few parts per million are possible in the audio-frequency range. Equal direct and alternating quantities should give the same thermocouple output (since their heating effect is the same), and the magnitude of the d-c quantity can be measured by potentiometer techniques with resistive voltage dividers and standard resistors. Thus, the magnitude of the a-c quantity is referred to the basic d-c standards of voltage and resistance through the transfer device, avoiding the limitations of ordinary indicating instruments. It is generally the a-c–d-c differences of the shunt or series resistor, rather than the characteristics of the thermoelement, that determine the upper frequency limit of applicability of a transfer standard.

52. The electrodynamic wattmeter is still the basic a-c–d-c transfer device for power and energy measurements at low frequencies. For this purpose, the wattmeter must be carefully designed to be as free as possible from sources of a-c error, such as eddy currents, capacitance, and inductance. The best electrodynamic wattmeters are capable of a-c–d-c power transfer to an accuracy within a few hundredths of a percent of the applied voltamperes at any power factor at low frequencies and are usable with fair accuracy well into the audio-frequency range, to 10 kHz in some instances.

INSTRUMENT TRANSFORMERS (See also Sec. 11)

53. A-c range extension beyond the reasonable capability of indicating instruments is accomplished with instrument transformers, since the use of heavy-current shunts and high-voltage multipliers would be prohibitive both in cost and in power consumption. The current circuits of instruments and meters normally have very low impedance, and current transformers must be designed for operation into such a low-impedance secondary burden. The insulation from the primary to secondary of the transformer must be adequate to withstand line-to-ground voltage, since the connected instruments are usually at ground potential. Normal design is for operation with a rated secondary current of 5 A, and the input current may range upward to many thousand amperes. The potential circuits of instruments are of high impedance, and voltage transformers are designed for operation into a high-impedance secondary burden. In the usual design, the rated secondary voltage is 115 V, and instrument transformers have been built for rated primary voltages up to 350 kV.

54. Current transformers, whose primary winding is series connected in the line, serve the double purpose of (1) convenient measurement of large currents and (2) insulation of instruments, meters, and relays from high-voltage circuits. Such a transformer has a high-permeability core of relatively small cross section operated normally at a very low flux density. The secondary winding is usually in excess of 100 turns, and the primary is of few turns and may even be a single turn or a section of a bus bar threading the core. The nominal current ratio of such a transformer is the inverse of the turns ratio, but for accurate current measurement, the actual ratio must be determined under loading corresponding to use conditions. Insulation of primary from secondary and core must be sufficient to withstand, with a reasonable safety factor, the voltage to ground of the circuit into which it is connected; secondary insulation is much less, as the connected instrument burden is at ground potential or nearly so.

The overload capacity of station-type current transformers and the mechanical strength of the winding and core structure must be high to withstand possible short circuits on the line. Various compensation schemes are used in many transformers to retain ratio accuracy up to several times rated current. The secondary circuit—the current elements of connected instruments or relays—*must never be opened* while the transformer is excited by primary current. The flux density in the core—normally very low—will rise to saturation in the absence of the counter mmf of a secondary current and will induce a secondary voltage that might be hazardous to insulation and to personnel; and the core may be left with a magnetic bias that will alter the ratio and phase angle when the secondary burden is reconnected.

55. Stated accuracy limits of current transformers are shown in the following table,

Accuracy class	Accuracy-deviation limits, %		Limits of power factor for this accuracy (lagging)	Corresponding phase-angle limits, min	
	100% rated current	10% rated current		100% rated current	10% rated current
2.4	±2.4	±4.8	1.0–0.6	62	124
1.2	±1.2	±2.4	1.0–0.6	31	62
0.6	±0.6	±1.2	1.0–0.6	15	31
0.3	±0.3	±0.6	1.0–0.6	7	15
0.5	±0.5*	±0.5	1.0–0.6	13	13

* This limit also applies to 150% rated current.

taken from USAS C57.13 for instrument transformers, and the standard secondary burdens used for accuracy rating purposes are as follows:

B-0.1(X), 2.5 VA at 0.9 pf current lagging; B-0.2, 5 VA at 0.9 pf current lagging; B-0.5(Y), 12.5 VA at 0.9 pf current lagging; B-2(Z), 50 VA at 0.9 pf current lagging.

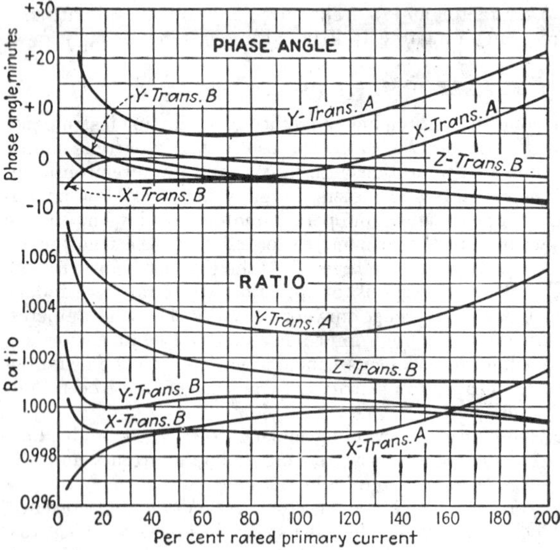

FIG. 3-22. Characteristics of current transformers with standard burdens, 60-cps performance.

The performance of an instrument transformer is specified in terms of accuracy class and burden. Figure 3-22 shows typical ratio and phase-angle curves for two current transformers which would be rated 0.3X, 0.6Y (transformer *A*) and 0.3X, 0.3Y, 0.3Z (transformer *B*), respectively. While these accuracy classes specify error limits when the transformer is operating with a burden equal to or less than that stated, for best accuracy the errors must be determined with a burden whose impedance is identical with that of the connected instrument circuits.

56. Measurement of ratio and phase angle of current transformers is accomplished either in a circuit in which the magnitudes and phase relation of primary and secondary currents can be directly compared or by means of a previously calibrated "standard" transformer in a circuit which compares the secondary currents of the "standard" and "unknown" transformers while their primaries are connected in series and carry identical input currents. A typical null method of direct comparison is shown schematically in Fig. 3-23. Here the voltage drop in a noninductive resistor R_1, carrying the primary current, is made equal to the voltage drop in an adjustable noninductive resistor R_2, which carries the secondary current. The phase angle between primary and secondary currents is compensated by an induced quadrature voltage injected into the measuring circuit from the secondary of an adjustable mutual inductor M. The resistance R_2 and the primary winding of the mutual inductor must be counted as part of the secondary burden. The current ratio is

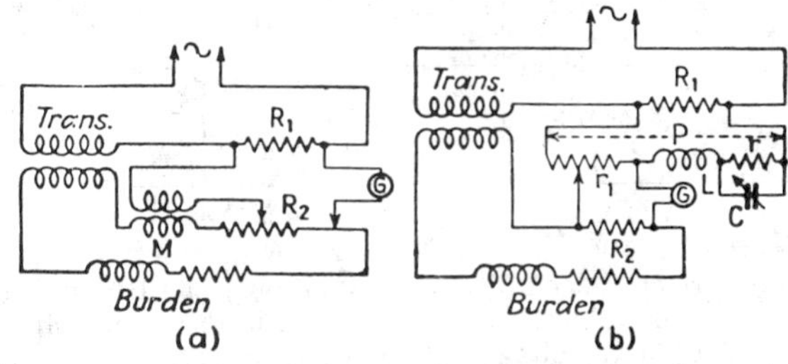

FIG. 3-23. Ratio and phase angle of current transformer by resistance method.

$$\frac{I_p}{I_S} = \frac{R_2}{R_1}\sqrt{1 + \tan^2 \delta},$$

and the phase angle between primary and secondary currents is $\tan \delta = (WM/R_2)$. Usually δ is small enough so that the correction term in the ratio expression can be neglected.

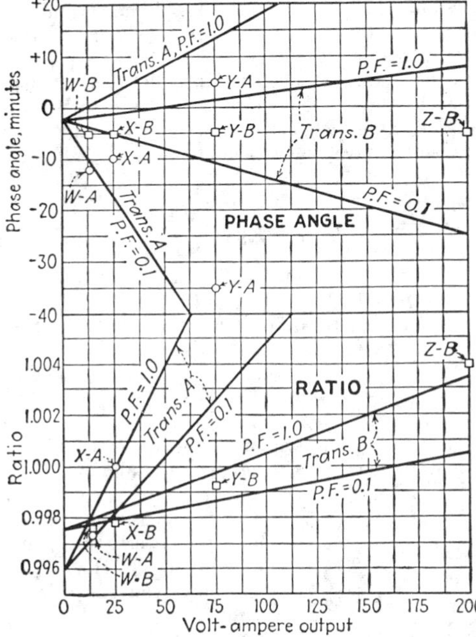

FIG. 3-24. Characteristics of potential transformers, performance with standard burdens indicated.

57. Voltage transformers (potential transformers) are connected between the lines whose potential difference is to be determined and are used to step the voltage down (usually to 115 V) and to supply the voltage circuits of the connected instrument burden. Their basic construction is similar to that of a power transformer operating at the same input voltage, except that they are designed for optimum performance with the high-impedance secondary loads of the connected instruments. The core is operated at high flux density, and the insulation must be appropriate to the line-to-ground voltage.

58. Standard accuracy limits of voltage transformers are shown in the following table taken from USAS C57.13 for instrument transformers, and the secondary burdens used for accuracy rating purposes are:

W, 12.5 VA at 0.10 pf current lagging; X, 25 VA at 0.70 pf current lagging; Y, 75 VA at 0.85 pf current lagging; Z, 200 VA at 0.85 pf current lagging.

In applying these ratings, a transformer is rated 0.6X if its ratio and phase angle lie within the limits for class 0.6 with X as the connected burden. One transformer might be rated 0.6W, 0.6X, 1.2Y; another, of larger capacity and better quality, might

Accuracy class	Accuracy-deviation limits, %	Limits of power factor for this accuracy (lagging)	Corresponding phase-angle limits, min
1.2	±1.2	1.0–0.6	31
0.6	±0.6	1.0–0.6	15
0.3	±0.3	1.0–0.6	7

be rated 0.3W, 0.3X, 0.3Y, 0.6Z. Figure 3-24 shows typical performance for two transformers, *A* and *B*, with varying unity and 10% power-factor loads. Performances with standard burdens are also shown in the figure. It will be seen that the performance, at a fixed voltage and power factor, varies in a strictly linear fashion with burden magnitude and that the performance characteristics may be completely described if the open-circuit corrections and those at a single burden (at a fixed power factor) are known.

59. Measurement of ratio and phase angle of voltage transformers is accomplished either in a circuit in which the magnitudes and phase relation of primary and secondary terminal voltages can be compared directly or by means of a previously calibrated standard transformer in a circuit which compares the secondary voltages of the standard and unknown transformers while their primaries are connected in parallel to the same voltage source. Voltage dividers are used as the reference elements in a direct-comparison network. A typical resistive divider and a capacitance divider arrangement are shown in Fig. 3-25. The ratio of transformation, when the detector null indicates balance, is

$$\text{Ratio} = \frac{V_p}{V_s} = \left(1 + \frac{R}{r}\right)\sqrt{1 + \tan^2 \delta} \qquad \text{and} \qquad \tan \delta = \frac{WM}{S}.$$

Where the phase angle θ is small, the expression under the radical in the ratio expression can usually be neglected. Similar expressions can be derived for the network that makes use of a capacitive voltage divider.

POWER MEASUREMENT

60. Laboratory-standard wattmeters use an electrodynamic type of mechanism, have scales about 12 in long, and are in the 0.1% accuracy class for direct and for alternating current up to 133 Hz. This accuracy can be maintained up to 1 kHz or more. Such instruments are shielded from the effects of external magnetic fields by enclosing the coil system in a laminated iron cylinder. Current ranges are usually double—0.5/1.0, 2.5/5, 5/10, 10/20 A—using series and parallel connections of the field coils. Voltage ranges—generally 75, 150, 300—have resistances of about 25 Ω/V. Beyond these currents and voltages, ranges are extended with precision instrument transformers.

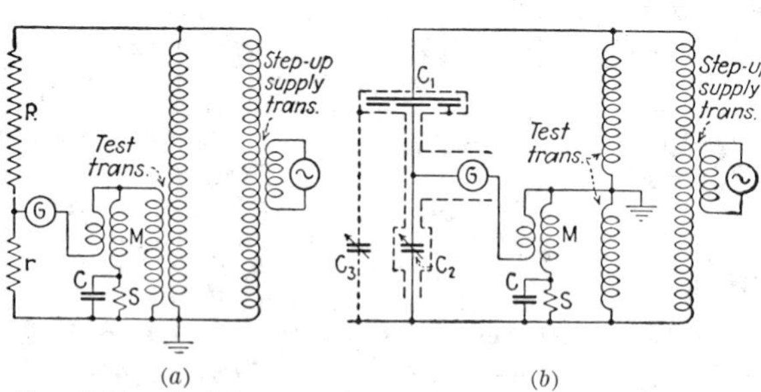

Fig. 3-25. Resistance and capacitance methods for determining ratio of potential transformers.

61. Portable wattmeters are generally of the electrodynamic type. The current element consists of two fixed coils wound with heavy wire or strip, connected in series with each other and with the load to be measured. The voltage element is a moving coil supported on jewel bearings or suspended by taut bands between the fixed field coils. The moving coil has a large number of fine turns, as in a voltmeter, and is connected in series with a relatively large noninductive resistor across the load circuit. The coils are mounted in a laminated iron shield to minimize coupling with external magnetic fields. Switchboard wattmeters have the same coil structure but are of broader accuracy class and do not have the temperature compensation, knife-edge pointers, and antiparallax mirrors required for the better-class portable instruments.

62. Line connections should be such that the moving-coil end of the voltage circuit and the current coils are on the same side of the circuit being measured to minimize potential differences between the fixed and moving coils. When used with instrument transformers, the moving-coil end of the voltage circuit should be connected to the ground terminal of the voltage transformer, and an electrostatic tie (a resistance of a few thousand ohms) should be connected between this terminal and one of the current terminals. Otherwise, there may be sufficient electrostatic forces between the fixed and moving coils to cause an error, or if their voltage difference is large, insulation between the windings may be broken down.

63. Correction for wattmeter power consumption may be important when the power measured is small. When the wattmeter is connected directly to the circuit (without the interposition of instrument transformers), the instrument reading will include the power consumed in the element connected next to the load being measured. If the instrument loss cannot be neglected, it is better to connect the voltage circuit next to the load and include its power consumption rather than that of the current circuit, since it is generally more nearly constant and is more easily calculated. In some low-range wattmeters, designed for use at low power factors, the loss in the voltage circuit is automatically compensated by carrying the current of the voltage circuit through compensating coils wound over the field coils of the current circuit. In this case, the voltage circuit must be connected next to the load to obtain compensation.

64. The inductance error of a wattmeter may be important at low power factor. At power factors near unity, the noninductive series resistance in the voltage circuit is large enough to make the effect of the moving-coil inductance negligible at power frequencies; but with low power factor, the phase angle of the voltage circuit may have to be considered. This may be computed as $\alpha = 21.6fL/R$, where α is the phase angle in minutes, f is the frequency in hertz, L is the moving-coil inductance in millihenrys, and R is the total resistance of the voltage circuit in ohms (see Par. **73** for correcting the effect of inductance).

65. Characteristics of wattmeters, as stated in makers' literature, are given in Table 3-8.

66. Wattmeter calibration can best be checked on direct current by using normal potentiometer techniques to measure current supplied to the field coils and voltage supplied to the voltage circuit from independent sources, but with an electrostatic tie (a high resistance) between one current terminal and the terminal at the moving-coil end of the voltage circuit to avoid errors from electrostatic forces between fixed and moving coils. The product of the measured current and voltage should equal the indicated power without correction for power consumption in either voltage or current circuit. For this test, a "compensated" wattmeter should be connected with its selector switch in the "uncompensated" position. A polyphase meter, with two independent measuring circuits connected by a common moving-element shaft, can be checked as though it were a single-phase meter, by connecting the voltage circuits in parallel to the voltage supply and the current circuits in series to the current supply. However, the following two tests should first be made: (1) independence can be checked by exciting only the current circuit of one element and the voltage circuit of the other—zero scale indication verifies that there is no interaction between the systems; (2) electrodynamic balance between the systems can be checked by connecting both voltage circuits in parallel to the voltage source and both current circuits in series opposition to the current source—here again at rated current and voltage, there should be no deflection of the pointer.

67. A 2-phase 4-wire circuit (not interconnected) may be treated as equivalent to two single-phase circuits. Two wattmeters are connected as shown in Fig. 3-26; total power is the arithmetical sum of the two instrument readings.

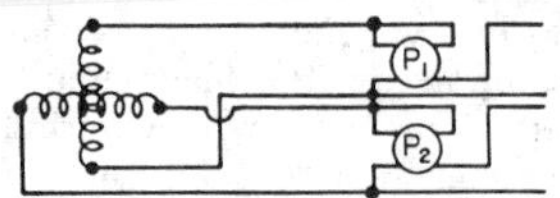

FIG. 3-26. Power in 2-phase, 4-wire circuit (not interconnected).

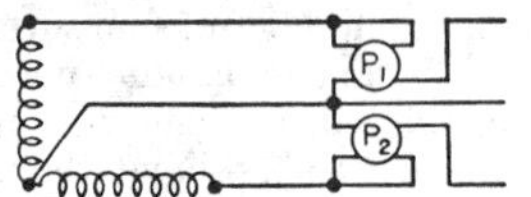

FIG. 3-27. Power in 2-phase, 3-wire circuit.

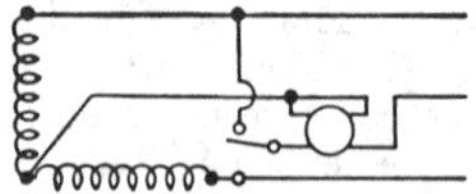

FIG. 3-28. Power in 2-phase, 3-wire circuit, one wattmeter.

68. A 2-phase 3-wire circuit requires two wattmeters connected as shown in Fig. 3-27; total power is the algebraic sum of the two readings. This connection is correct for any condition of load and power factor. One wattmeter may be used, connected as shown in Fig. 3-28, if there is no load across the outer conductors and the phases are balanced as to load and power factor; readings are summed for the two switch positions.

Table 3-8. Characteristics of Wattmeters

Manufacturer and model number of wattmeters	Full-scale accuracy, per cent	Scale length, in.	Resistance of potential circuit, ohms*	Impedance of current circuit, ohms†	Type of damping	Magnetic shield	Weight, lb.
Portable wattmeters							
Weston, Model 310	0.25	5.25	4,500	0.04	Air	Yes	12
Weston, Model 310, p.f. = 0.20	0.25	5.25	3,000	0.16	Air	Yes	12
Sensitive Research, Model DW	0.25	6.4	4,500	0.20	Air	Yes	6
Sens. Res., Model DLW, p.f. = 0.20	0.5	6.4	68,000	1.80	Air	Yes	6
General Electric, Type P-3	0.2	6.5	5,400	0.09	E.M.	Yes	11.5
Weston, Model 432	0.5	4.0	11,000	0.04	Air	Yes	3.3
Westinghouse, Type PY-5	0.5	5.0	6,500	0.04	Air	Yes	6
General Electric, Type AP-9	0.75	4.1	18,500	0.03	E.M.	Yes	2.5
Switchboard wattmeters							
Weston, Models 343 and 498	1.0	5.1	3,500	0.03	Air	Case	7
General Electric, Type AD	1.0	5.1	3,700	0.07	E.M.	Case	
General Electric, Type AH-12	1.0	6.0	2,200	0.16	E.M.	Case	
Westinghouse, Type KY	1.0	3.5	4,000	0.08	Air	Case	4
Westinghouse, Type HY	1.0	5.0	4,000	0.08	Air	Case	4
Weston, Model 610	1.0	3.5	3,500	0.03	Air	Case	2

* For 150-volt (max.) range.
† For 5-amp. coil at 60 cps.

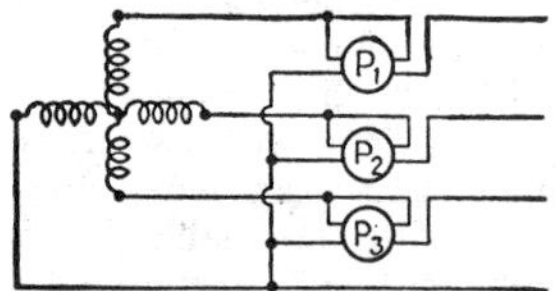

FIG. 3-29. Power in 2-phase, 4-wire interconnected circuit.

69. A 2-phase 4-wire interconnected circuit requires three wattmeters, connected as shown in Fig. 3-29; total power is the algebraic sum of the three readings. This connection is correct under all conditions of load and power factor. It will be noted that the voltage impressed on P_3 is 1.414 times the voltage on P_1 and P_2. Two wattmeters, one in each phase, will give the power only when the load is balanced in all four legs.

70. A 3-phase 3-wire circuit requires two wattmeters connected as shown in Fig. 3-30; total power is the algebraic sum of the two readings under all conditions of load and power factor. If the load is balanced, at unity power factor, each instrument will read half the load; at 50% power factor one instrument reads all the load and the other reading is zero; at less than 50% power factor one reading will be negative. When the load is balanced, power may be measured with one wattmeter, using a Y-box as shown in Fig. 3-31. This arrangement, which creates an artificial neutral, has two branches which have the same impedance and power factor as the wattmeter's voltage circuit, which is the third branch of the Y. Total power is three times the reading of the wattmeter.

71. The 3-phase 4-wire circuits require three wattmeters as shown in Fig. 3-32. Total power is the algebraic sum of the three readings under all conditions of load and power factor. A 3-phase "Y"-system with a grounded neutral is the equivalent of a 4-wire system and requires the use of three wattmeters. If the load is balanced, one wattmeter can be used with its current coil in series with one conductor and the voltage circuit connected between that conductor and the neutral. Total power is three times the wattmeter reading in this instance.

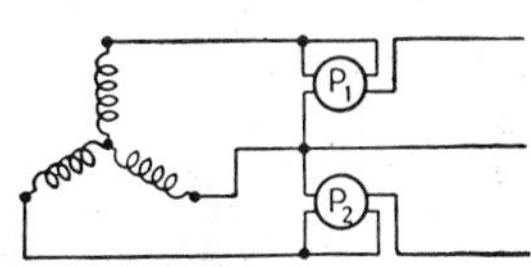

FIG. 3-30. Power in 3-phase, 3-wire circuit, two wattmeters.

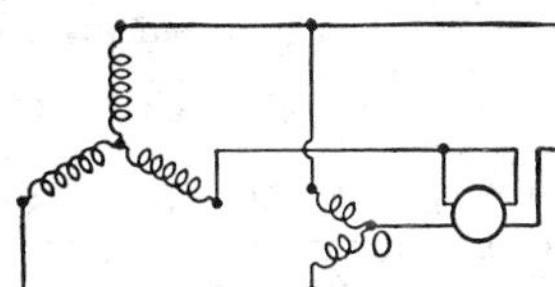

FIG. 3-31. Power in 3-phase, 3-wire circuit, one wattmeter with "Y" box.

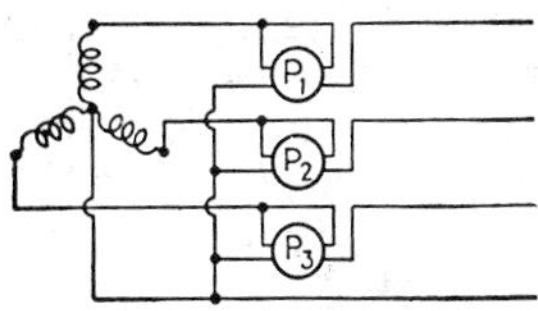

FIG. 3-32. Power in 3-phase, 4-wire circuit, using three wattmeters.

72. Reactive power (reactive voltamperes, or vars) is measured by a wattmeter with its current coils in series with the circuit and the current in its voltage element in quadrature with the circuit voltage.

73. Corrections for instrument transformers are of two kinds. *Ratio* errors, resulting from deviations of the actual ratio from its nominal, may be obtained from a calibration curve showing true ratio at the instrument burden imposed on the transformer and for the current or voltage of the measurement. The effect of *phase-angle* changes introduced by instrument transformers is to modify the angle between the current in the field coils and that in the moving coil of the wattmeter; the resulting error depends on the power factor of the circuit and may be positive or negative depending on phase relations as shown in the following table. If $\cos\theta$ is the *true* power factor in the circuit and $\cos\theta_2$ is the

Line power factor	Sign to be used for phase angle					
	α wattmeter		β current transf.		γ voltage transf.	
	Lead[1]	Lag	Lead	Lag	Lead	Lag
Lead	+	−	−	+	+	−
Lag	−	+	+	−	−	+

[1] In general, α will be leading only when the inductance of the potential coil has been overcompensated with capacitance.

apparent power factor (i.e., as determined from the wattmeter reading and the secondary voltamperes), and if K_c and K_v are the true ratios of the current and voltage transformers, respectively, then

$$\text{Main-circuit watts} = K_c K_v \frac{\cos\theta}{\cos\theta_2} \times \text{wattmeter watts.}$$

The line power factor $\cos\theta = \cos(\theta_2 \pm \alpha \pm \beta \pm \gamma)$, where θ_2 is the phase angle of the secondary circuit, α is the angle of the wattmeter's voltage circuit (see Par. **64**), β is the phase angle of the current transformer, and γ is the phase angle of the voltage transformer. These angles—α, β, and γ—are given positive signs when they act to decrease and negative when they act to increase the phase angle between instrument current and voltage with respect to that of the circuit. This is so because a decreased phase angle gives too large a reading and requires a negative correction (and vice versa), as shown in the preceding table of signs. The correction formula may be written, with sufficient precision for convenient use, as follows:

$$\text{Main-circuit watts} = CV(1 + c + v + p) \text{ wattmeter watts}$$

where C and V are the nominal ratios of current and voltage transformers, c and v their relative deviations from the nominal ratios, and p the phase-angle correction factor. Figures 3-33 and 3-34 give the value of the correction factor p in percent for various

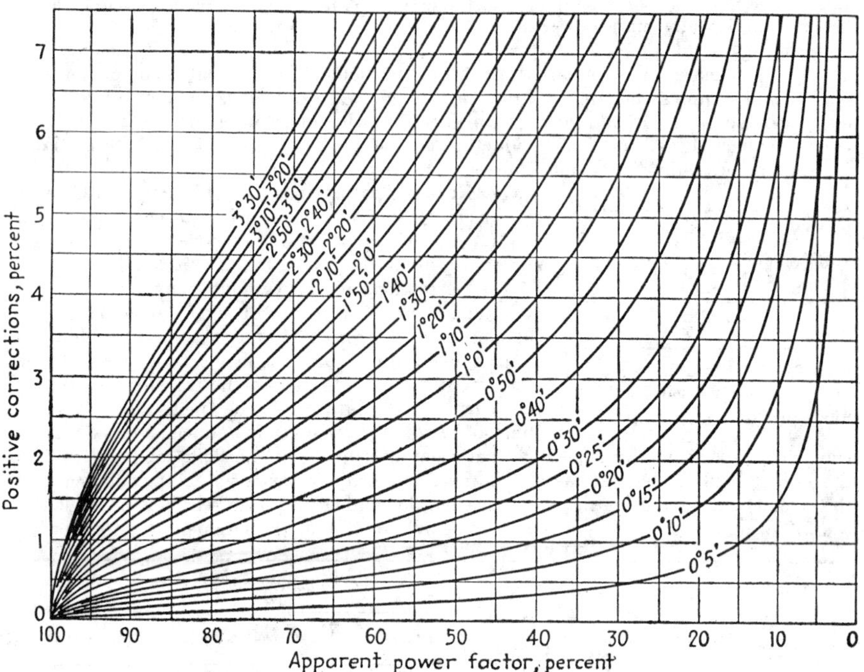

Fig. 3-33. Instrument-transformer correction factors: for *leading* current when $(\alpha + \beta + \gamma)$ is *positive*, for *lagging* current when $(\alpha + \beta + \gamma)$ is *negative*.

values of power factor and net phase angle $\alpha + \beta + \gamma$. Figure 3-33 gives the positive correction to be used when $\alpha + \beta + \gamma$ is negative and the current is lagging, or vice versa; Fig. 3-34 gives the negative correction to be used when $\alpha + \beta + \gamma$ is positive and current is lagging, or vice versa. The following example illustrates the application of the various corrections:

Observed power, 300 W. Observed current, 4.0 A. Observed voltage, 120.0 V. Current lagging.

Nominal current-transformer ratio 20:1. Nominal voltage-transformer ratio 60:1. *CT* ratio at 40 A, 19.95/1, $c = -0.25\%$. *VT* ratio at 120 V, 59.95/1, $v = -0.1\%$. Phase angle of wattmeter voltage circuit, 5 min lag, $\alpha = +5$. Phase angle of *CT* at 4.0 A, 10 min lead, $\beta = +10$. Phase angle of *VT* at 120.0 V, 3 min lead, $\gamma = -3$. $\alpha + \beta + \gamma = +5 + 10 - 3 = +12$ min. Apparent power factor $(\cos\ \theta_2) = 300/(4 \times 120) = 62.5\%$ lag. Correction factor from Fig. 3-34, $p = 0.45\%$. Total correction factor $c + v + p = -0.25 - 0.1 - 0.45 = -0.8\%$.

$$\text{Main circuit watts} = 20 \times 60 \times 300 \text{ decreased by } 0.8\%$$

$$= 360 \text{ kW} - 2.9 \text{ kW} = 357.1 \text{ kW}$$

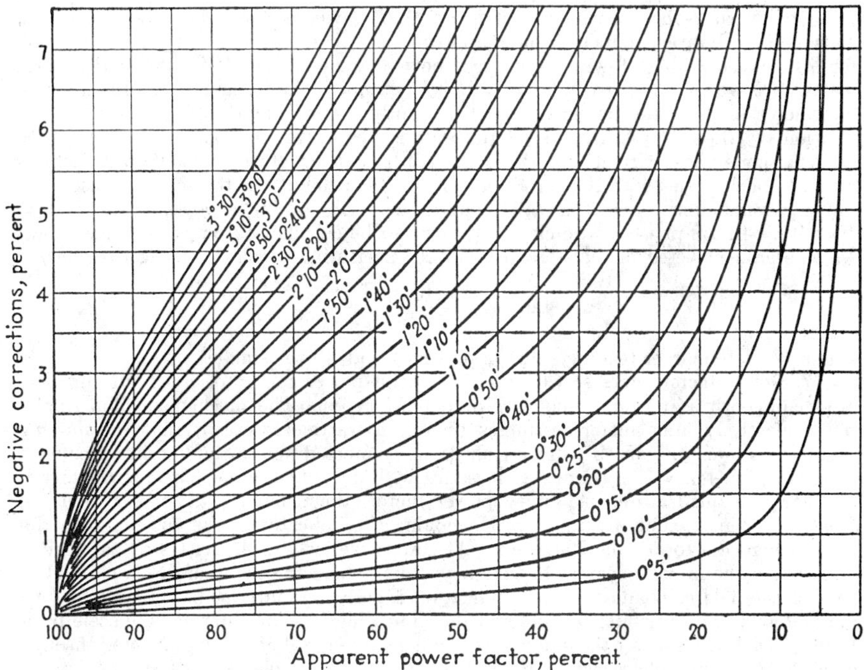

FIG. 3-34. Instrument-transformer correction factors: for *lagging* current when $(\alpha + \beta + \gamma)$ is *positive*, for *leading* current when $(\alpha + \beta + \gamma)$ is *negative*.

74. Dielectric loss, which occurs in cables and insulating bushings used at high voltages, represents an undesirable absorption of available energy and, more importantly, a restriction on the capacity of cables and insulating structures used in high-voltage power transmission. The problem of measuring the power consumed in these insulators is quite special, since their power factor is extremely low and the usual wattmeter techniques of power measurement are not applicable. While many methods have been devised over the past half century for the measurement of such losses, the Schering bridge is almost universally the method of choice at the present time. Figure 3-35 shows the basic circuit of the bridge, as described by Schering and Semm in 1920. The balance equations are $C_x = C_sR_1/R_2$ and $\omega r_xC_x = \tan \delta_x = \omega R_1P$, where C_x is the cable or bushing whose losses are to be determined, C_s is a high-voltage loss-free air capacitor, R_1 and R_2 are noninductive resistors one of which must be an adjustable low-voltage capacitor having negligible losses. For details of screening, shielding, and operation, the reader should refer to a text on a-c bridge techniques, for example, Hague, "Alternating Current Bridge Methods," 5th ed., Pitman, revised 1957.

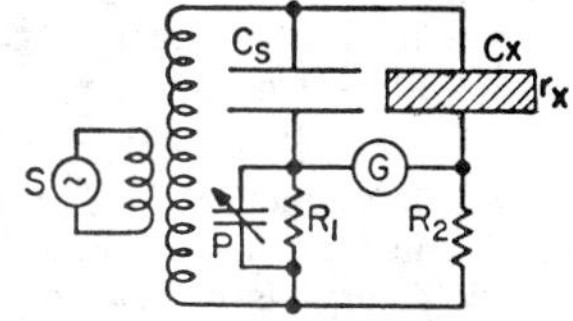

FIG. 3-35. Schering and Semm's bridge for measuring dielectric loss.

POWER-FACTOR MEASUREMENT

75. The power factor of a single-phase circuit is the ratio of the true power in watts, as measured with a wattmeter, to the apparent power in voltamperes, obtained as the product of the voltage and current. When the waveform is sinusoidal (and only then), the power factor is also equal to the cosine of the phase angle.

76. The power factor of a polyphase circuit which is balanced is the same as that of the individual phases. When the phases are not balanced, the true power factor is indeterminate. In the wattmeter-voltmeter-ammeter method, the power factor for a balanced 2-phase 3-wire circuit is $P/(\sqrt{2}EI)$, where P is total power in watts, E is voltage between outside conductors, and I is current in an outside conductor; for a balanced 3-phase 3-wire circuit, the power factor is $P/(\sqrt{3}EI)$, where P is total watts, E is volts between conductors, and I is amperes in a conductor. In the two-wattmeter method, the power factor of a 2-phase 3-wire circuit is obtained from the relation $W_2/W_1 = \tan\theta$, where W_1 is the reading of a wattmeter connected in one phase as in a single-phase circuit and W_2 is the reading of a wattmeter connected with its current coil in series with that of W_1 and its voltage coil across the second phase. At unity power factor $W_2 = 0$; at 0.707 power factor $W_2 = W_1$; at lower power factors $W_2 > W_1$. In a 3-phase 3-wire circuit, power can be calculated from the reading of two wattmeters connected in the standard way for measuring power, by using the relation

$$\tan\theta = \frac{\sqrt{3}(W_1 - W_2)}{W_1 + W_2},$$

where W_1 is the larger reading (always positive) and W_2 the smaller.

77. Power-factor meters, which indicate the power factor of a circuit directly, are made both as portable and as switchboard types. The mechanism of a single-phase electrodynamic meter resembles that of a wattmeter except that the moving system has two coils M, M'. One coil, M, is connected across the line in series with a resistor, whereas M' is connected in series with an inductance. Their currents will be nearly in quadrature. At unity power factor, the reaction with the current-coil field results in maximum torque on M, moving the indicator to the 100 mark on the scale, where the torque on M is zero. At zero power factor, M' exerts all the torque and causes the moving system to take a position where the plane of M' is parallel to that of the field coils and the scale indication is zero. At intermediate power factors, both M and M' contribute torque, and the indication is at an intermediate scale position. In a 2-phase meter the inductance is not required, coil M being connected through a resistance to one phase, while M' with a resistance is connected to the other phase; the current coil may go in the middle conductor of a 3-wire system. Readings are correct only on a balanced load. In one form of polyphase meter, for balanced circuits, there are three coils in the moving system, connected one across each phase. The moving system takes a position where the resultant of the three torques is minimum, and this is dependent on power factor. In another form, three stationary coils produce a field which reacts on a moving voltage coil. When the load is unbalanced, neither form is correct.

ENERGY MEASUREMENTS

78. The practical unit of electrical energy is the watthour, which is the energy expended in 1 h when the power (or rate of expenditure) is 1 W.

79. Energy is measured in watthours (or kilowatthours) by means of a watthour meter. A watthour meter is a motor mechanism in which a rotor element revolves at a speed proportional to power flow and drives a registering device on which energy consumption is integrated. Meters for continuous current are usually of the mercury-motor type, whereas those for alternating current utilize the principle of the induction motor.

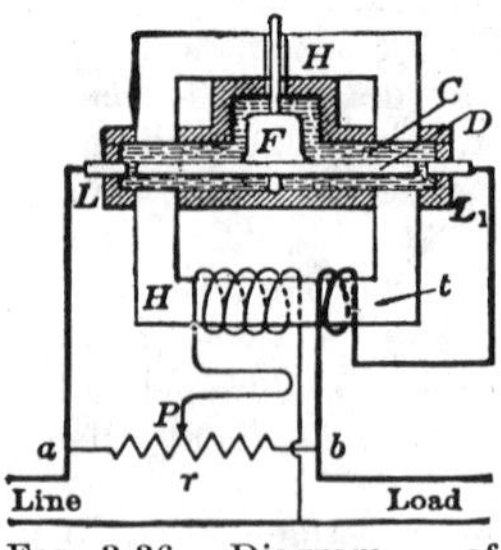

Fig. 3-36. Diagram of Sangamo mercury-motor-type watthour meter.

80. Continuous-current meters have been historically of two types: the commutator type, which has not been built since the early 1920s; and the mercury-motor type, now used in d-c reduction processing, etc. The d-c meter made by the Sangamo Electric Company is representative of the mercury-motor type. Figure 3-36 shows the circuits and scheme of operation. D is a copper disk, having a number of radial slots and floating in a pool of mercury; F is a float which supports the shaft, producing an upward thrust which is taken by a jewel bearing at the top of the shaft; H is a laminated iron core; and C is a chamber filled with mercury. A ring-jewel guide bearing at the bottom of the shaft holds the rotor on a fixed axis. The flux, produced in the core H by the shunt (voltage) coil, cuts the disk at two diametrically opposite points. The line current passes from L to L_1 into the mercury, entering and leaving the copper disk in a flow pattern that is directed along its diameter by the slots. Since this disk is cut by magnetic flux, a torque is produced that is proportional to the product of current and voltage. The usual drag (braking) disk and magnets, gear train, and register are provided at the top of the shaft. The friction of the disk rotating in the mercury may be compensated with the Ayrton shunt ab of Fig. 3-36 across the current circuit, allowing a small adjustable current to be supplied by the voltage circuit.

In an alternative compensation method, a small current is introduced at L, L_1 from a thermocouple whose junction is heated by a coil in series with the voltage circuit. This current is adjusted by a slide-wire resistor in the thermocouple circuit. The increase, with speed, of friction between the mercury and disk is compensated by a series turn t on the core H. The rotor speed, proportional to load, is adjusted by shifting the drag magnets radially with respect to the meter shaft, thus altering the retarding torque. This torque, which is also proportional to rotor speed, is minimum with the magnets close to the shaft and maximum with the magnets near the edge of the disk.

In some types, a micrometer screw is arranged to move the magnets in or out; in others, a soft-iron disk, which acts as a magnetic shunt to alter the flux cutting the braking disk, is mounted on a micrometer screw and can be moved toward or away from the air gap of the drag magnets. The light-load speed is adjusted independently by altering the position of P on the Ayrton shunt. Figure 3-37 shows typical load

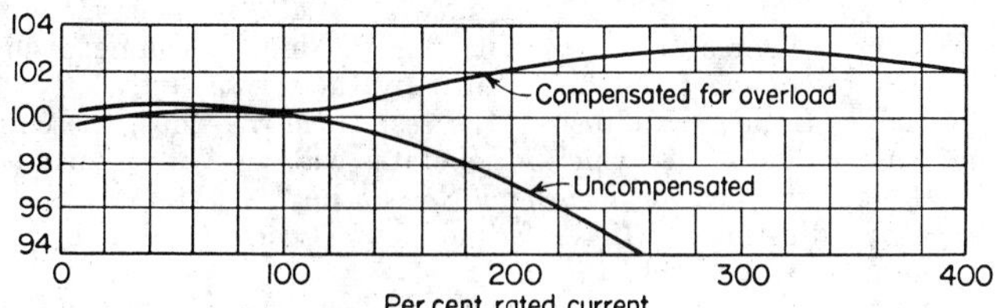

FIG. 3-37. Typical accuracy curves for d-c watthour meters.

accuracy curves both for older types of meters and for the more recent types, which are temperature-compensated for large sustained overloads.

81. The a-c watthour meters measure energy by using the principle of the induction motor. The essential features are shown in Fig. 3-38. P is the voltage coil, S the series (current) coils, and c a compensating coil. An aluminum disk is free to revolve between the poles. The alternating fluxes from these poles will establish currents in the disk as indicated by the arrows in the sketch to the right, which shows the poles and a portion of the disk. The voltage winding P has many turns and is highly inductive so that the flux from its pole tip lags the applied voltage by nearly 90°. The fluxes in the current poles, set up by the line current in the series coils S, are in phase with that current. Now the emf induced in a conductor by an alternating flux which cuts it is in time quadrature with that flux, and the eddy currents in the disk produced by the flux from P will be maximum at almost the same time as the flux from S is maximum from the component of load current in phase with line voltage.

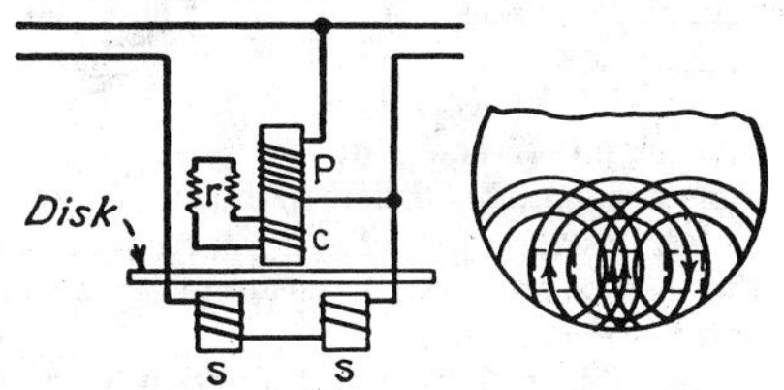

FIG. 3-38. Diagram of induction-type watthour meter.

Similarly, the eddy currents set up in the disk by fluxes from the inphase component of current in S (a quarter period later) will be maximum at almost the instant that the flux from P is maximum. Thus, a torque will be produced which is porportional to the instantaneous product of the eddy currents in the disk and the flux from the pole around which the eddy current flows. This torque will be proportional to power in the load circuit if the fluxes from poles P and S are exactly in quadrature at unity power factor in the load. To provide a rotor speed proportional to this driving torque, a retarding (braking) torque must be exerted on the disk that is also proportional to speed. This is accomplished by a permanent-magnet field through which the disk moves, producing eddy currents proportional to its speed.

The interaction between these eddy currents and the magnet flux that produced them results in a counter torque (or braking action) which is proportional to rotor speed. Consequently, the speed of the rotor is proportional to driving torque and thus to power in the circuit. The above argument assumes for simplicity that the current in P and the flux from its pole are exactly in quadrature with the line voltage. But because of the ohmic resistance of this coil, the current lag is somewhat less than 90°, and the flux from this pole tip must be brought into quadrature with line voltage by the compensating coil c. As shown in Fig. 3-39, the flux E_f produced by the current in the voltage coil lags the applied voltage E by less than 90° and is, in consequence, slightly ahead of the eddy current I_e (resulting from line current) rather than in phase with it.

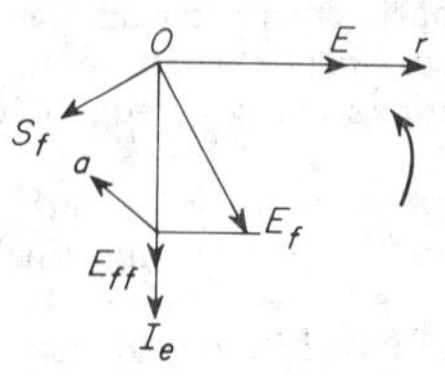

Fig. 3-39. Theory of lag adjustment, induction watthour meter.

The resulting torque, proportional to the product of I_e and the component oa of E_f in phase with it, is less than the product I_eE_f, and the meter will run slow. As a practical matter, this error is negligibly small for a unity-power-factor load but increases as the power factor decreases. A compensating coil (lag coil) is used to eliminate this error at low power factors. This is the coil c on pole P (Fig. 38), short-circuited through the resistance r. The current in the compensating coil will lag the emf induced in it by the flux E_f by nearly 90°, and its magnitude can be controlled by adjusting the resistance r. Thus, the current in the compensating coil and the flux S_f which it generates will lag E_f by nearly 90°, and the resultant flux E_{ff}, the vector sum of E_f and S_f, can be brought into quadrature with the line voltage by adjusting r. If this adjustment is correct, the meter registration will be correct at any load power factor.

With no load current in the current coils, any lack of symmetry in the flux from the voltage pole could produce a forward or reverse torque. Also, there is necessarily friction in the rotor bearings, gear train, and register of the meter which tends to make the disk rotation slower than it should be at very small load currents. To compensate for these tendencies, a controlled driving torque is added by means of a shading-pole loop or plate in the disk gap, to react with the voltage flux. The position of this plate can be adjusted tangentially to cover more or less of the voltage pole. Since the driving torque it produces by its shading action is constant as long as the line voltage is constant, its effect on meter registration is inversely proportional to line current and is negligible at large loads, whereas it counteracts the effect of friction and brings the registration to its correct value for light loads.

Even with correct calibration with this adjustment at light load, there may still be a slight "creep" torque (generally forward) on voltage alone. To prevent continuous creep, two anticreep holes or slots are cut into the disk diametrically opposite. The holes restrict the eddy currents induced in the disk by the voltage flux and cause the disk to stop at a position of least coupling between the conductive path in the disk and the flux field.

82. Adjustments of induction watthour meters are normally made only at *light* and *full* load. The position of the light-load compensation plate can be changed with conveniently placed screws and the light-load speed thus adjusted to be correct. Speed adjustment at full load (but affecting the speed at all loads) is made by shifting the drag magnets relative to the disk axis or by shunting flux by means of a movable soft-iron strip bridging the air gap. The lag or power-factor adjustment is made at the meter factory and, if properly made, should never require correction. Temperature compensation for sustained overloads and for variations of ambient temperature has been secured by using a piece of temperature-sensitive iron-nickel-copper alloy[1] as a

[1] Kinnard and Faus *Trans. AIEE*, 1925, p. 275.

shunt across the air gap of the drag magnets. Figure 3-40 shows typical performance characteristics of modern watthour meters.

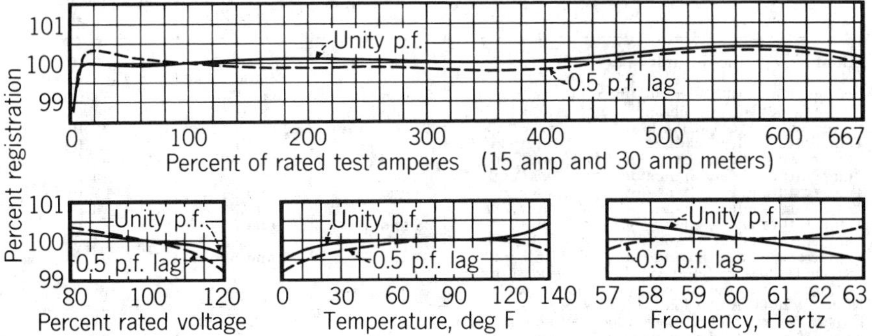

FIG. 3-40. Characteristic performance curves of watthour meters.

83. Meter Ratings. Watthour meters were formerly rated only at full or rated load and at light load, generally one-tenth of full-load value. Maximum load capacities, however, had little relationship to the rated loads as the overload ranges were extended to higher and higher loads. Consequently, in 1960 the industry rerated watthour meters into classes on the basis of their maximum capacities. The class designation of a watthour meter denotes the maximum of the load range in amperes. Generally used classes are 10 (for use with current transformers), 100, and 200. The previous full-load rating, now known as TA or test ampere rating, is retained simply for purposes of calculating test constants and determining the percent registration of a watthour meter at heavy and light loads.

84. Typical characteristics of the generally used single-phase Class 200, 240-V, 3-wire, 60-c watthour meter are shown in Table 3-9.

Table 3-9. Induction Watthour-meter Data*

Item	Duncan MQS	General Electric I60S	Sangamo J3S	Westinghouse D3S
Speed at rated TA load	10	16⅔	10	16⅔
Torque at rated load, mmg	23	40	45	38
Weight of rotor, g	20	17	24	18
Loss in potential circuit, at rated voltage, W	1.1	1.1	1.0	1.2
Power factor of potential circuits, %	16	14.7	15	14
Loss in current circuit at TA current, W	0.21	0.28	0.28	0.22
Starting watts	24	22	24	22

* Data supplied by manufacturers.

85. Measurement of energy in a-c circuits is effected with watthour meters connected in exactly the same manner as are wattmeters for the measurement of power (see Figs. 3-26 to 3-32). In 3-wire 2-phase or 3-phase systems, polyphase meters may be used. Such meters comprise merely two single-phase meters in one case, with a common shaft, and connected to the main circuit in the same manner as two single-phase meters.

Four-wire systems, unless balanced, require three single-phase meters. A 3-phase system with grounded neutral should be considered a 4-wire system requiring three meters, unless it is completely and continually balanced. Three-element meters are made with three driving-braking disks on one shaft and may be conveniently used instead of the three single-phase meters.

Condition at 25°C ±5°C	Percent of class designations		Maximum deviation in % from normal accuracy
	Class 10	Class 100 Class 200	
Initial performance at unity pf	1.5	1.0	±3.0
	5 to 100	3 to 100	±2.0
Voltage effect, 10% change, unity pf	2.5 and 25	1.5 and 15	±1.0
Frequency effect, 5% change, unity pf	2.5 and 25	1.5 and 15	±1.0
Power-factor effect at 0.5 lag	5, 50, and 100	3, 50, and 100	±2.0
Temperature effect, change of +25°C unity pf	2.5	1.5	±2.0
Temperature effect, change of +25°C unity pf	25 and 50	15 and 50	±1.0
Temperature effect, change of +25°C 0.5 lag pf	5	3	±3.0
Temperature effect, change of +25°C 0.5 lag pf	25 and 50	15 and 50	±2.0
Temperature effect, change of −25°C unity pf	2.5	1.5	±3.0
Temperature effect, change of −25°C unity pf	25 and 50	15 and 50	±2.0
Temperature effect, change of −25°C 0.5 lag pf	5	3	±4.0
Temperature effect, change of −25°C 0.5 lag pf	25 and 50	15 and 50	±3.0
External field effect (100 At) through conductors arranged as specified in ASA C12-1965, Sec. 5.18.10	2.5	1.5	±1.0
Effect of 2,000 A overload for 0.1 s	...	1.5	−1.5, +0.5
Equality of current circuits, relative to combined	2.5, 5, 25, 50	1.5, 3, 15, 30	±1.0
Effect of tilt at 4° from vertical, forward, backward, left, and right	2.5	1.5	±1.0
	25	15	±0.5
Independence of stators on multistator meters	5, 10, 15	3, 6, 9	±1.0
	25, 50, 75	15, 30, 45	
20,000 A (20 × 50 μsec) surge through vertical conductor 1.5 in from meter	15	15	±1.0

86. Acceptance Accuracy Limits (from the Code for Electricity Metering, USAS C12-1965)

87. Total energy in a 3-phase circuit is the algebraic sum of the indications of two single-phase meters, just as the total power is the algebraic sum of the readings of two wattmeters. If a polyphase meter is used, the summation is automatically performed, and when one element tends to run backward (power factor less than 50%), it simply reduces the torque of the other one, so that the actual speed is still proportional to the total power in the circuit. Totalizing meters are used to add together the energy outputs of several generators, to combine the energy values of several feeders, etc. Such meters have one or more driving disks mounted on a common shaft and rotated by the torques contributed by the pairs of ordinary watthour-meter elements provided for each generator or feeder; the total energy is then indicated on a single register.

88. Polyphase Meter Connections. Obviously, it is extremely important that the various circuits of a polyphase meter be properly connected. If, for example, the current-coil connections of Fig. 3-30 are interchanged and the line power factor is 50%, the meter will run at the normal 100% power-factor speed, thus giving an error of 100%.

A test for correct connections of the arrangement of Fig. 3-30 is as follows: If the line power factor is over 50%, rotation will always be forward when the potential or the current circuit of either element is disconnected but in one case the speed will be less than in the other. If the power factor is less than 50%, the rotation in one case will be backward.

When it is not known whether the power factor is less or greater than 50%, this may be determined by disconnecting one element and noting the speed produced by the remaining element. Then change the voltage connection of the remaining element from the middle wire to the other outside wire, and again note the speed. If the power factor is over 50%, the speed will be different in the two cases, but in the same direction. If the power factor is less than 50%, the rotation will be in opposite directions in the two cases.

When instrument transformers are used, care must be exercised in determining correct connections; if terminals of similar instantaneous polarity have been marked on both current and voltage transformers, these connections can be verified and the usual test made to determine power factor. If the polarities have not been marked, or if the identities of instrument transformer leads have been lost in a conduit, the correct connections can still be established but the procedure is more lengthy.

89. Use of Instrument Transformers with Watthour Meters. When the capacity of the circuit is over 200 A, instrument current transformers are generally used to step

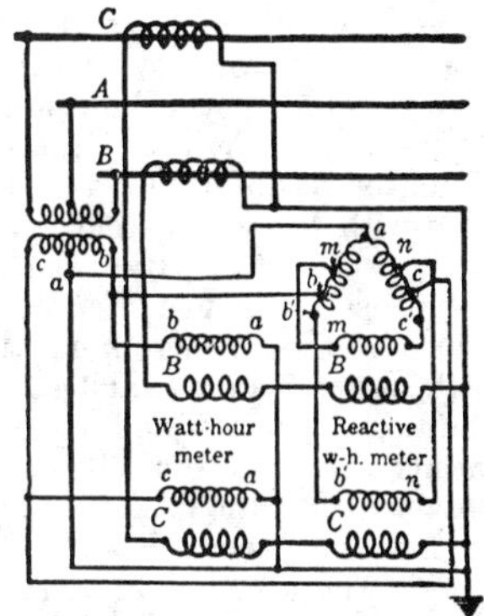

FIG. 3-41. Connections for measuring active and reactive energy in 3-phase 3-wire circuit.

down the current to 5 A. If the voltage is over 440 V, current transformers are almost invariably employed, irrespective of the magnitude of the current, in order to insulate the meter from the line; in such cases, voltage transformers are also used to reduce the voltage to 110 V. The ratio and phase-angle errors of these transformers must be taken into account where high accuracy is important, as in the case of a large installation. These errors can be largely compensated for by adjusting the meter speed. The percentage of error corresponding to various phase angles is given in Figs. 3-33 and 3-34. For polyphase meters, it is usually sufficiently accurate to use average values of phase angles and ratios. Space limits permit showing a diagram of connections only for the metering of a 3-phase 3-wire system with two watthour-meter elements and current and voltage instrument transformers (Fig. 3-41). Diagrams to cover most conditions will be found in Sec. 7 of the "Electrical Metermen's Handbook," Edison Electric Institute, 1965.

90. Reactive voltampere-hour (var-hour) meters are generally ordinary watthour meters in which the current coil is inserted in series with the load in the usual manner while the voltage coil is arranged to receive a voltage in quadrature with the load voltage. In 2-phase circuits, this is easily accomplished by using two meters as in power measurements, with the current coils connected directly in series with those of the "active" meters but with the voltage coils connected across the quadrature phases. Evidently, if the meters are connected to rotate forward for an inductive load, they will rotate backward for capacitive loads.

The simplest arrangement for 3-phase 3-wire circuits is the counterpart of the two-wattmeter method for the measurement of power (Fig. 3-30), but with potential coils interchanged. With this arrangement, the voltage in each element is in quadrature with the current in that element at unity-power-factor load. Watthour meters of standard construction can be used, but the sum of the two readings is equal to 1.1547 times the total reactive voltamperes, with balanced load.

To eliminate this correction factor, autotransformers are utilized with midpoint taps m and n (see Fig. 3-41) and step-up taps b' and c', so that voltages of the correct magnitude in quadrature with those from BA and CA, respectively, are applied to make the meter direct-reading.

91. Errors of Var-hour Meters. The 2- and 3-phase arrangements described above give correct values of reactive energy when the voltages and currents are balanced. The 2-phase arrangement still gives correct values for unbalanced currents but will be in error if the voltages are also unbalanced. Both 3-phase arrangements give erroneous readings for unbalanced currents or voltages; the autotransformer arrangement will usually show less error for a given condition of unbalance than the simple arrangement with interchanged potential coils. To register correctly, the reactive energy in a 3-phase 3-wire system under all conditions of power factor, unbalanced current, and voltages requires three watthour meters, connected in the usual manner for energy measurements and provided with voltage circuits so designed that the torque developed in the rotating element is proportional to $EI \sin \theta$. Such a meter is not yet commercially available; however, the comparatively small errors which will occur on the average system, especially with present-day rate schedules, are more than offset by the ability to use standard-type watthour meters for the measurement of reactive energy.

92. Total voltampere-hours, or "apparent energy" expended in a load, is of interest to engineers because it determines the heating of generating, transmitting, and distributing equipment, and hence their rating and investment cost. The apparent energy may be computed if the power factor is constant, from the observed watthours P and the observed reactive voltampere-hours Q; thus, voltampere hours $= \sqrt{P^2 + Q^2}$. This method may be greatly in error when the power factor is not constant; the computed value is always too small.

A number of devices have been offered for the direct measurement of the apparent energy. In one class (*a*) are those in which the meter power factor is made more or less equal to the line power factor. This is accomplished automatically (in the Angus meter) by inserting a movable member in the voltage-coil pole structure which shifts the resulting flux as line power factor changes. In others, autotransformers are used with the voltage elements to give a power factor in the meter close to expected line power factor. By using three such pairs of autotransformers and three complete polyphase watthour-meter elements operating on a single register, with the record determined by the meter running at the highest speed, an accuracy of about 1% is achieved, with power factors ranging from unity down to 40%. In the other class (*b*), vector addition of active and reactive energies is accomplished either by electromagnetic means or by electromechanical means, many of them very ingenious. But the results obtained with the use of modern watthour and var-hour meters are generally adequate for most purposes.

93. The accuracy of a watthour meter is the percentage of the total energy passed through a meter which is registered by the dials. The watthours indicated by the meter in a given time are noted, while the actual watts are simultaneously measured with standard instruments. Because of the time required to get an accurate reading from the register, it is customary to count revolutions of the rotating element instead of the register. The accuracy of the gear-train ratio between the rotating element and the first dial of the register can be determined by count. Since the energy represented by one revolution, or the watthour constant, has been assigned by the manufacturer and marked on the meter, the indicated watthours will be $K_h \times R$, where K_h is the watthour constant and R the number of revolutions.

94. Power-factor variation, in meter testing, can be obtained by several methods. In the two-alternator method, two generators are mounted on a common base with a common shaft. The stationary members (armature or field) are made movable about the shaft with respect to the base and to each other. Thus, with the voltage coil of the meter connected to one machine and the current coil to the other, any phase relation can be obtained by adjusting one movable stator with respect to the other.

In the autotransformer method, a variable-ratio autotransformer is connected across one phase of a 3-phase circuit, and the potential coil of the meter is connected in such a manner that any phase relation can be obtained. Thus, referring to Fig. 3-42, the current coil of the meter is connected in series with conductor *a* of a 3-phase circuit, and the voltage coil is connected to *o* and to *p*, the latter being a tap on a transformer connected across phase *bc*. It is apparent that any phase angle between the current and the voltage can be obtained in a range from 0 to 60° by moving the connection point *p* along the transformer winding. Angles of 60 to 90°, lead or lag, can be obtained by changing the transformer to either of the other two phases and the meter voltage-coil connection from *o* to *x* or *y*.

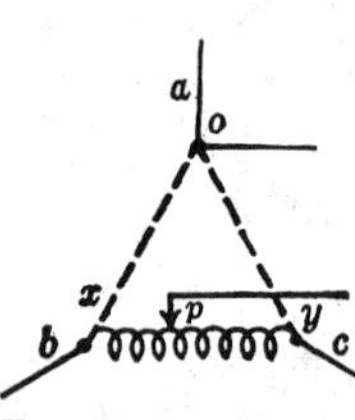

Fig. 3-42. Power-factor variation, 3-phase transformer method.

These changes can be instantly made with suitable switching arrangements. But with these connections, the voltage varies in magnitude as well as in phase, so that it is convenient to introduce a second autotransformer between the taps *o*, *p* and the meter for the purpose of compensating for the variations in the voltage between *o* and *p* and keeping the voltage constant at the meter. Two variable-ratio autotransformers arranged for automatic operation in this manner make a convenient phase shifter.

The most satisfactory phase shifter is similar to a 3-phase induction motor with a wire-wound rotor (see Sec. **7**) which may be moved to any position around the circle by means of a worm gear and handwheel. In this way, a single-phase voltage or the 3-phase voltages induced in the rotor with 3-phase voltage applied to the stator may be made to take any desired phase. The voltage coils of the meter under test are connected to the secondary low-voltage coils of step-down transformers excited from the rotor coils. Such phase shifters are also made for 2-phase operation. Manifestly either the 2-phase shifter or the 3-phase shifter can be used in tests of single-phase meters.

95. Reference standards for d-c meter tests in the laboratory may be ammeters and voltmeters, in portable or laboratory-standard types, or potentiometers; in a-c meter tests, use is made of indicating wattmeters and a time reference standard such as a stopwatch or contact clock. A more common reference is a standard watthour meter, which is started and stopped automatically by light pulsing through the anti-creep holes of the meter under test.

96. The rotating-standard method of watthour meter testing is that most often used, because only one observer is required and it is more accurate with fluctuating loads. Rotating standards are watthour meters similar to standard house-type service meters, except that they are made with extra care, are usually provided with more than one current and one voltage range, and are portable. A pointer, attached directly to the shaft, moves over a dial divided into 100 parts, so that fractions of a revolution are easily read. Such a standard meter is used by connecting it to measure the same energy as is being measured by the meter to be tested; the comparison is made by the "switch" method, in which the register only (in d-c standards) or the entire moving element (in a-c standards) is started at the beginning of a revolution of the meter under test, by means of a suitable switch, and stopped at the end of a given number of revolutions. The accuracy is determined by direct comparison of the number of whole revolutions of the meter under test with the whole revolutions and a fraction of the standard. In the laboratory, use is sometimes made of a special rotating standard having a notched disk through which a beam of light shines intermittently and flashes a neon lamp. A target pattern marked on the disk of the meter under test is then viewed stroboscopically under this illumination, and the difference in speeds is counted and used in the computation of accuracy. Another method of measuring speed of rotation in the laboratory is to use a tiny mirror on the rotating member which reflects a beam of light into a photoelectric cell; the resulting impulses may be recorded on a chronograph or used to define the period of operation of a synchronous electric clock, etc.

97. Watthour meters used with instrument transformers are usually checked as secondary meters; i.e., the meter is removed from the transformer secondary circuits (current transformers must first be short-circuited) and checked as a 5-A 110-V meter in the usual manner. The meter accuracy is adjusted so that, when the known corrections for ratio and phase-angle errors of the current and potential transformers have been applied, the combined accuracy will be as close to 100% as possible, at all load currents and power factors. An overall check is seldom required, both because of the difficulty and because of the decreased accuracy as compared with the secondary check. Standardized current and potential transformers must be provided and instrument-transformer corrections applied to corrected readings of the standard watthour meter.

98. General precautions to be observed in testing watthour meters are as follows: (*a*) The test period should always be sufficiently long and a sufficiently large number of independent readings should be taken to ensure the desired accuracy. In service tests, the period preferably should be not less than 30 s and the number of readings not less than three. In laboratory tests, 100-s periods and five readings are preferable. (*b*) Capacity of the standards should be so chosen that readings will be taken at reasonably high percentages of their capacity in order to make observational or scale errors as small as possible. (*c*) Where indicating instruments are used on a fluctuating load, their average deflections should be estimated in such a manner as to include the time of duration of each deflection as well as the magnitude. (*d*) Instruments should be so connected that neither the standards nor the meter being tested are measuring the voltage-circuit loss of the other, that the same voltage is impressed on both, and that the same load current passes through both. (*e*) When the meter under test has not been previously in circuit, sufficient time should be allowed for the temperature of the voltage circuit to become constant, preferably not less than 10 min; this is important with d-c meters, especially in the case of rotating standards. In some types of the latter, special provision is made for rapid heating. (*f*) Guard against the effect of stray fields by locating the standards and arranging the temporary test wiring in a judicious manner.

99. Meter Constants. The following definitions of various meter constants are taken from the Code for Electricity Metering, 5th edition, USAS C12, 1965.

100. Register constant K_r is the factor by which the register reading must be multiplied in order to provide proper consideration of the register or gear ratio and of the instrument-transformer ratios to obtain the registration in the desired units.

101. Register ratio R_r is the number of revolutions of the first gear of the register, for one revolution of the first dial pointer.

102. Gear ratio R_g is the number of revolutions of the rotor for one revolution of the first dial pointer.

103. Watthour constant K_h is the registration expressed in watthours corresponding to one revolution of the rotor. (When a meter is used with instrument transformers, the watthour constant is expressed in terms of primary watthours. For a secondary test of such a meter, the constant is the primary watthour constant, divided by the product of the nominal ratios of transformation.)

104. Test current of a watthour meter is the current marked on the nameplate by the manufacturer (identified as TA on meters manufactured since 1960) and is the current in amperes which is used as the basis for adjusting and determining the percent registration of a watthour meter at heavy and light loads.

105. Percentage registration of a meter is the ratio of the actual registration of the meter to the true value of the quantity measured in a given time, expressed as a percentage. Percentage registration is also sometimes referred to as the accuracy or percentage accuracy of a meter. The value of one revolution having been established by the manufacturer in the design of the meter, meter watthours $= K_h \times R$, where $K_h =$ watthour const and $R =$ number of revolutions of rotor in S s. The corresponding power in meter watts is $P_m = (3{,}600 \times R \times K_h)/S$. Hence, multiplying by 100 to convert to terms of percentage registration (accuracy):

$$\text{Percentage registration} = \frac{K_h \times R \times 3{,}600 \times 100}{PS},$$

where $P =$ true watts. This is the basic formula for watthour meters in terms of true watt reference.

106. Average Percentage Registration (Accuracy) of Watthour Meters. The Code for Electricity Metering makes the following statement under the heading. Methods of Determination:

The percentage registration of a watthour meter is, in general, different at light load than at heavy load, and may have still other values at intermediate loads. The determination of the average percentage registration of a watthour meter is not a simple matter as it involves the characteristics of the meter and the loading. Various methods are used to determine one figure which represents the average percentage registration for all practical purposes, the method being prescribed by commissions in many cases. Common usage has, in general, caused the adoption of "average accuracy" or "final average accuracy" in place of "average percentage registration." Two methods of determining the average percentage registration are in common use:

Method 1. Average percentage registration is the weighted average of the percentage registration at light load (LL) and at heavy load (HL), giving the heavy-load registration a weight of four. By this method:

$$\text{Weighted average percentage registration} = \frac{LL + 4HL}{5}.$$

Method 2. Average percentage registration is the average of the percentage registration at light load (LL) and at heavy load (HL). By this method:

$$\text{Average percentage registration} = \frac{LL + HL}{2}.$$

107. In-service performance tests, as specified in the Code for Electricity Metering, USAS C12-1965, shall be made in accordance with a *periodic test schedule,* except that self-contained single-phase meters and 3-wire network meters may also be tested under either of two other systems, provided that all meters are tested under the same system. These systems are the *variable interval plan* and the *statistical sampling plan.*

108. Periodic Interval System. The chief characteristic of this system is that a fixed percentage of the meters in service shall be tested annually.

In the test intervals specified below, the word "years" means calendar years. The periods stated are recommended test intervals. There may be instances where individual meters or groups of meters should be tested more or less frequently.

In general, periodic test schedules should be as follows:

1. Meters used with instrument transformers:
   a. Polyphase meters—at least once in 4 years.
   b. Single-phase meters—at least once in 8 years.
2. Self-contained polyphase meters—at least once in 6 years.
3. Self-contained single-phase meters and 3-wire network meters—at least once in 8 years.

The chief weaknesses of the above periodic test schedule are that it fails to recognize the differences in accuracy characteristics of various types of meters as a result of technical advance in meter design and construction and fails to provide incentives for maintenance and modernization programs.

109. Variable Interval Plan. This plan differs from the periodic interval system in that the percentage of meters to be tested in any given year depends upon the test data obtained during the preceding year or years up to a maximum of three. It also provides for classifying meters into homogeneous groups by types. The number of meters to be tested in each group is a function of the percentage of meters outside of acceptable limits. The relationship used to determine the test rate from the test data provides for increasing the test rate with increasing percentages of meters outside acceptable limits. A recommended formula provides for a test rate of 12.5% when the percentage of meters outside of limits equals 3.0% and a maximum test rate of 25% when the percentage of meters outside acceptable limits equals or exceeds 6.0%. It is also recommended that the minimum number of meters tested in each group should be 200 or 12.5%, whichever is less.

110. Statistical Sampling Plans. Statistical sample testing is a method of estimating the condition of a large population of meters by the annual selection and testing of a random sample of meters from each homogeneous group and evaluation of the test results. There are two basic types of sampling plans: the method of attributes and the method of variables. The required size of the sample and the test criteria for making the decision as to whether or not a group of meters is acceptable are a mathematical calculation determined by the parameters of (1) proportion defective in an acceptable group, (2) proportion defective in an unacceptable group, (3) the utility's risk of rejecting a good group, and (4) the consumer's risk of passing an unacceptable group.

111. Method of Attributes. In this method, the decision criterion for acceptability is the number of permissible defectives in an acceptable sample. If more than this critical number are defective, the entire lot is rejected; or in some double sampling plans, a new sample is drawn and tested with different accept and reject numbers. If the test criteria for the second sample are met, the lot is accepted. USAS C12 specifies a minimum sample of 300 meters from each homogeneous group.

112. Method of Variables. The second type of statistical sampling plan is known as the method of variables. These plans require more processing of sample-test data than is required by attributes plans. The bar X, or average accuracy of the sample lot, and the sigma, or average deviation from the mean, must be calculated. The extra processing required to obtain such data is compensated for, however, by the smaller test sample permitted. USAS C12 specifies a minimum test sample of 100 meters. Although variables plans permit smaller test samples, there is a danger that, if the distribution of meter accuracies within the population is not normal or gaussian, the conclusions drawn from the test data about the percent defective may be erroneous. Therefore, users of these plans should be reasonably assured, by prior testing for normality before using such plans, that their meter populations have normal distributions.

113. Ampere-hour meters measure only electrical quantity, i.e., coulombs or ampere-hours; and, therefore, where they are used in the measurement of electrical energy, the potential is assumed to remain constant at a "declared" value and the meter is calibrated or adjusted accordingly. The outstanding example in the United States of ampere-hour meters is the Sangamo mercury meter, which is practically the same as the Sangamo d-c watthour meter (Par. **80**) except that the electromagnet is replaced by permanent magnets. Models are available especially designed for use with storage batteries, electric railways, etc.

114. Ampere-hour or volt-hour meters for alternating current are not practical but ampere-squared-hour or volt-squared-hour meters are readily built in the form of the induction watthour meter. Ampere-hours or volt-hours are then obtainable by extracting the square root of the registered quantities.

115. Maximum-demand Meters. Some methods of selling energy involve the maximum amount which is taken by the customer in any period of a prescribed length, i.e., the maximum demand. Many types of meters for measuring this demand have been developed, but space permits only a brief description of a few. There are two general classes of demand meters in common use: (1) integrated-demand meters and (2) thermal, logarithmic, or lagged-demand meters. Both have the same function, which is to meter energy in such a way that the registered value is a measure of the load as it affects the heating (and therefore the load-carrying capacity) of the electrical equipment.

116. Integrated-demand meters consist of an integrating meter element (kWh or kvarh) driving a mechanism in which a timing device returns the demand actuator to zero at the end of each timing interval, leaving the maximum demand indicated on a passive pointer, display, or chart, which in turn is manually reset to zero at each reading period, generally 1 month. Such demand mechanisms operate on what is known as the block-interval principle. There are three types of block-interval registers: (1) the indicating type, in which the maximum demand obtained between each reading period is indicated on a scale or numeric display; (2) the cumulative type, in which the accumulated total of maximum demand during the preceding periods is indicated during the period after the device has been reset and before it is again reset, i.e., the maximum demand for any one period is equal or proportional to the difference between the accumulated readings before and after reset; (3) the recording type, in which the demand is transferred as a permanent record onto a tape by printing, punching, or magnetic means or onto a circular or strip chart.

117. Thermal, logarithmic, or lagged-demand meters are devices in which the indication of the maximum demand is subject to a characteristic time lag by either mechanical or thermal means. The indication is often designed to follow the exponential heating curve of electrical equipment. Such a response, inherent in thermal meters, averages on a logarithmic and continuous basis, which means that more recent loads are heavily weighted but that, as time passes, their effect decreases. The demand interval for the lagged meter is defined as the time required to indicate 90% of the full value of a constant load suddenly applied.

118. Concordance of Demand Meters and Registers. The measurement of demand may be obtained with meters and registers having various operating principles and employing various means of recording or indicating the demand. On a constant load of sufficient duration, accurate demand meters and registers of both classifications will give the same value of maximum demand, within the limits of tolerance specified. On varying loads, the values given by accurate meters and registers of different classifications may differ because of the different underlying principles of the meters themselves. In commercial practice, the demand of an installation or a system is given with acceptable accuracy by the record or indication of any accurate demand meter or register of acceptable type.

ELECTRICAL RECORDING INSTRUMENTS

119. Recording instruments are, in many instances, essentially high-torque indicating instruments arranged so that a permanent, continuous record of the indication is made on a chart. They are made for recording all electrical quantities that can be measured with indicating instruments—current, voltage, power, frequency, etc. In general, the same type of electrical mechanism is used—permanent-magnet moving-coil for direct current and moving-iron or dynamometer for alternating current. The indicator is an inking pen or stylus that makes a record on a chart moving under it at constant speed. This requires a higher torque to overcome friction, so that the operating power required for a recording instrument is greater than for a simple indicating instrument. Overshoot is generally undesirable, and recording instruments are slightly overdamped, whereas indicating instruments are usually somewhat underdamped. Some recorders use strip charts; graduations along the length of the chart are usually of time intervals, and the graduations across the chart represent the instrument scale. Alternatively, the chart may be circular, with radial graduations for the instrument scale and time markers around the circumference. The chart paper should be well made and glazed to minimize dimensional changes from temperature and humidity. The ink should be in accordance with the maker's specification for the particular paper used so that it is accepted readily and does not run or blot the paper. Chart drives may be electrical or clockwork. In strip charts, perforations along the edges of the paper are engaged by a drive pinion; circular charts are rotated from a central hub.

120. Potentiometric self-balancing recorders are systems incorporating d-c potentiometers, used either alone or with a transducer to measure various quantities.

121. Transducers include those for voltage, current, power, power factor, frequency, temperature, humidity, steam or water flow, gas velocity, neutron density, and many other applications.

122. Types of systems are classified according to the means of detecting and correcting electrical unbalance in the potentiometer circuit.

123. The Leeds & Northrup Micromax recorder senses unbalance of a D'Arsonval galvanometer every 2.4 s. When unbalance occurs, a motor-driven cam clamps the galvanometer in its unbalanced position and metal feelers indicate this out of balance and cause a clutch arm to engage the slide-wire disk onto a constantly running motor so that the slide wires rotate until a balance condition is reached. The sensing time of 2.4 s gives a standard across the chart time of 24 s.

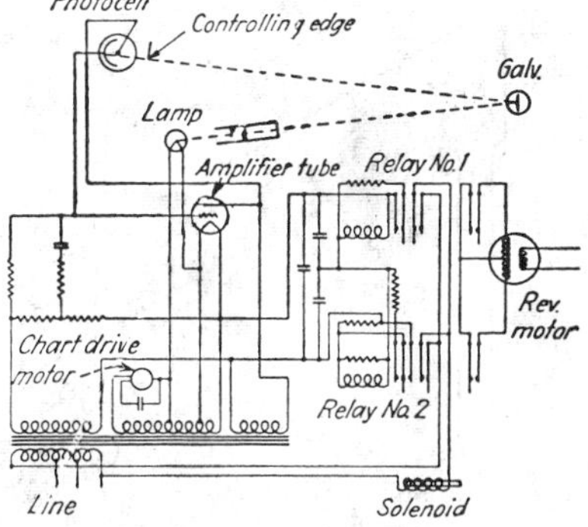

Fig. 3-43. Tagliabue Celectray recorder.

124. The Tag Celectray recorder (see Fig. 3-43), also typical of the galvanometer-operated type, contains the light-beam and phototube combination applied to the D'Arsonval galvanometer. Deflection of the galvanometer changes the intensity of light falling on the vacuum tube, and the output of this phototube is electrically amplified to operate two relays which control the motion of the reversible balancing motor connected to the slide wire.

125. Electrical self-balancing systems employ an amplifier to amplify the d-c unbalance in the potentiometer circuit and a reversible motor to produce the required rebalancing action. The *drift* which occurs in d-c amplifiers is equivalent to a spurious input signal, so that it has become general to employ d-c to a-c conversion to obtain a stable amplifier, free from zero drift. The a-c control signal is amplified by the a-c coupled amplifier and applied to the control winding of a 2-phase induction motor, whose power phase is fed from the power line. Polarity reversal of

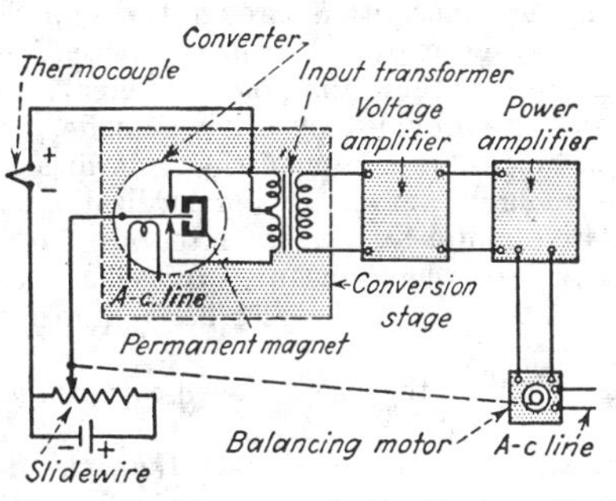

Fig. 3-44. Brown Electronik recorder.

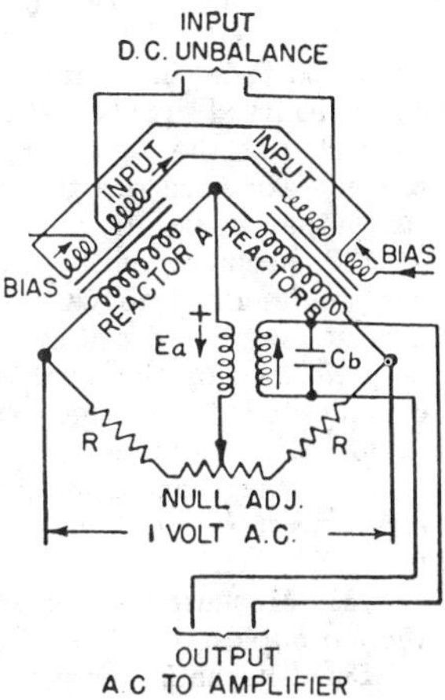

Fig. 3-45. Bailey Pyrotron recorder.

the d-c unbalance signal causes phase reversal of the a-c signal into the control winding of the motor, so that the resulting unbalance drives the motor in the correct direction to rebalance the potentiometer.

126. The Brown Electronik, shown schematically in Fig. 3-44, uses a vibrating reed to change the d-c unbalance signal to alternating current, which is then amplified and passed to the reversible induction motor. The reed converter is a metal reed driven in synchronism with the power frequency and oscillating between two contacts connected to opposite ends of the primary winding of the input transformer of the a-c amplifier. The d-c signal is thus applied in turn to alternate sides of the transformer, inducing in it an alternating voltage of power frequency whose magnitude and phase depend on the magnitude and polarity of the d-c signal.

127. The Bailey Pyrotron (see Fig. 3-45) employs a saturable core reactor in place of the vibrating reed to give d-c to a-c conversion, amplifiers and the 2-phase motor operation being as in the figure.

128. Leeds & Northrup Speedomax G, whose circuits are shown in Fig. 3-46 in

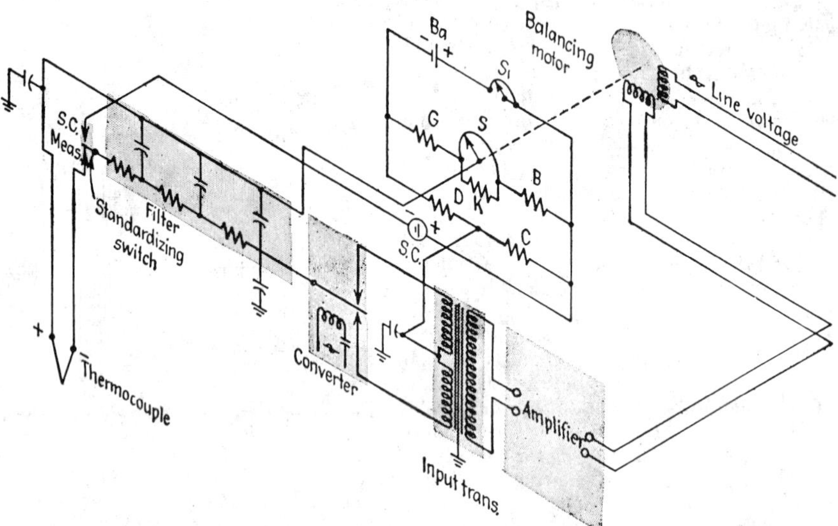

FIG. 3-46. Leeds & Northrup Speedomax G recorder.

simplified form, illustrates the complete system. The amplifiers are prone to pick up stray electrical signals that could give false indications. To avoid this, a filter network connected to the input serves the double purpose of acting as a damping network to prevent excessive pointer overshoot and, at the same time, screening out electrical strays that would otherwise be picked up and transmitted through the amplifier.

129. Accuracy of the order of ¼% may be expected from the potentiometer recorders described above. To maintain this accuracy, the potentiometer is referenced against a standard cell or a reference voltage provided by a Zener diode. This may be performed by the operator pressing a button to give manual standardization whenever desired. A further refinement is to have automatic standardization, in which the operation is initiated by the chart-drive motor at specified intervals.

130. Range extension of potentiometric recorders upward is by means of shunt or series resistors. Extension below the basic range of the recorder requires preamplifiers.

131. Measurement of a-c quantities requires the use of a-c–d-c transducers, e.g., thermocouples, rectifiers, etc.

132. Alternating-current potentiometer recorders are simpler than the d-c types, as they require no standardization against a standard cell or Zener reference voltage, and d-c–a-c conversion is not required, eliminating the requirement for a vibrator or saturable reactor. The amplifier and motor-control circuits can be the same as in the d-c recorder. By far the greatest application is with a-c bridges, where the a-c amplifier acts as an unbalance detector. Strain-gage bridges and bridges which employ platinum or nickel resistive elements for narrow-range temperature measurements frequently employ recorders of this type.

133. Proximity-type recorders use a high-frequency oscillator whose operation is started or stopped by the insertion of a metal vane into a pair of coils. If the vane is mounted on the pointer of an indicating instrument, the oscillator can sense movement between the pointer and a pair of coils fitted to the oscillator. Servo motion of the coils on displacement of the instrument pointer is accomplished by coupling the oscillator output to the input of a servo amplifier which drives the control motor. This gives a graphic record that follows but does not constrain movement of the instrument pointer. In this way, quantities which can operate an indicating instrument can be recorded without using a transducer.

134. Telemetering is the indicating or recording of a quantity at a distant point. Telemetering is employed in power measurements to show at a central point the power loads at a number of distant stations, and often to indicate total power on a single meter, but practically any electrical quantity which is measured can be transmitted, together with a large number of nonelectrical quantities such as levels, positions, pressures, etc. Telemetering systems may be classified by type: current, voltage, frequency, position, and impulse.

a. In *current* systems, the movement of the primary measuring element calls for a current in the attached control member to balance the torque created by the quantity measured. This balancing current (usually d-c) is sent over the transmitting circuit to be indicated and recorded. Totalizing is possible by the addition of such currents from several sources in a common indicator. The receiver may be as much as 50 mi from the transmitter.

b. In *voltage* systems, a voltage balance may be produced through a control-member voltmeter, or a voltage may be generated by thermocouples heated by the quantity to be measured, or produced as an IR drop as a result of a current torque balance, or generated by a generator driven at a speed proportional to the measured quantity. These voltages, however produced, are recorded at a distance by a potentiometer recorder. Here, also, the recorder may be 50 mi from the transmitter.

c. A *variable frequency* may be produced for telemetering by causing the primary element to move a capacitor plate in an r-f oscillator or to change the speed of a small d-c motor driving an alternator. High-frequency systems cannot be used for transmission over many miles.

d. In *position* systems, the movement of the primary element or of a pilot controlled by the primary element is duplicated at a distance. The pilot may be a bridge balancing resistance or reactance, a variable mutual inductance, or a selsyn motor where the position of a rotor relative to a 3-phase stator is reproduced at the receiver end. Satisfactory operation is usually limited to a few miles.

e. The *impulse* type of transmission of measured quantities is represented by the largest number of devices. The number of impulses transmitted in a given time may represent the magnitude of the quantity being measured, and these may be integrated by a notching device or by a clutch, or the duration of the pulse may be governed by the primary element and interpreted at the receiver. If the impulses are transmitted at high frequency, inductance and capacitance effects in the transmitting line limit the distance of satisfactory transmission; systems using d-c impulses operate over 50 to 250 mi.

RESISTANCE MEASUREMENTS

135. The unit of resistance, the ohm, has been determined directly in terms of the mechanical units by *absolute-ohm* experiments performed at the National Bureau of Standards and at national laboratories in other countries. The reactance of an inductor or capacitor of special construction whose value can be computed from its measured dimensions is compared with a resistance at a known frequency. The value of this resistance can then be assigned in absolute units, in terms of length and time—the dimensions of the inductor or capacitor and the time interval corresponding to the comparison frequency. These measurements are made with high precision, and it is believed that the assigned value of the National Reference Standards of resistance, maintained at the National Bureau of Standards, differs from its intended *absolute* value by not more than 1 part in 10^6.

136. The National Reference Standard of resistance is a group of ten 1-Ω resistors of special construction, sealed in double-walled enclosures away from contact with the air, that are kept at the National Bureau of Standards. To ensure that their values are *constant,* they are intercompared regularly, are compared every 3 years with the reference standards maintained by other countries, and their value is checked at intervals by repeated *absolute-ohm* determinations. This reference group serves as the basis for all resistance measurements made in the country.

137. Resistance standards, used in precise measurements, are made with high-resistivity metal, in the form of wire or strip. Manganin—a copper-nickel-manganese alloy— is generally used in resistance standards because, when properly treated and protected from air and moisture, it has a number of desirable characteristics, including stable value, low temperature coefficient, low thermal emf at junctions with copper, and relatively high resistivity. A copper-nickel-chromium-aluminum alloy, Evanohm, has recently been used for high-resistance standards, since it has the same desirable characteristics as manganin and a much higher resistivity. Two forms of standard are in general use: the Reichsanstalt form developed in the German National Laboratory is illustrated in Fig. 3-47; the NBS form developed at the National Bureau of Standards is shown in Fig. 3-48. Both are intended to be used with their terminal lugs in mercury cups and are suspended in an oil bath to dissipate heat and to hold the temperature constant at a known value during measurements. For highest precision, power dissipation must be kept below 0.1 W, although as much as 1 W can be dissipated in stirred oil with very small changes in value. The maker's recommendations should be followed regarding safe operating-current levels. High-resistance and low-resistance standards use different terminal arrangements. In standards of 1 Ω or lower value, the current and voltage terminals are separated, whereas in higher-value standards they may not be separated. The four-terminal construction is required to make the value of a low-resistance standard determinate. Connections to the current-carrying circuit range from a few microhms upward and, in a two-terminal construction, would make the resistance value uncertain to the extent that the connection resistance varies. With four-terminal construction, the resistance of the standard can be *exactly* defined as the voltage drop between the voltage terminals for unit current in and out at the current terminals.

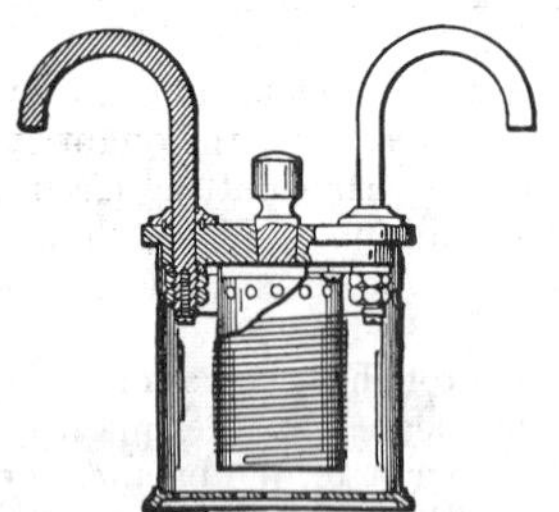

Fig. 3-47. Standard resistor, Reichsanstalt form, showing internal construction.

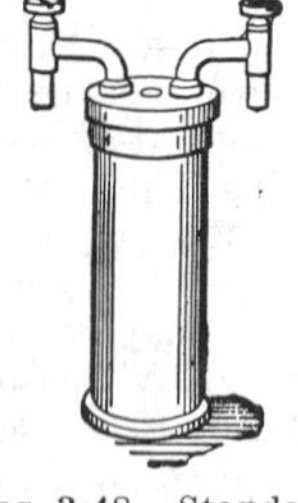

Fig. 3-48. Standard resistor, NBS form.

138. Current standards are precision four-terminal resistors used to measure current by measuring the voltage drop between the voltage terminals with the current introduced at the current terminals. These standards, designed for use with potentiometers for precision current measurement, correspond in structure to the shunts used with millivoltmeters for current measurement with indicating instruments. Current standards must be designed to dissipate the heat they develop at rated current, with only a small temperature rise. They may be oil-cooled or air-cooled, the latter design having much greater surface, since heat transfer to still air is much less efficient than to oil. Figure 3-49 shows an air-cooled current standard of 20 μΩ resistance and 2,000 A capacity, with an accuracy of 0.04%. Very-low-resist-

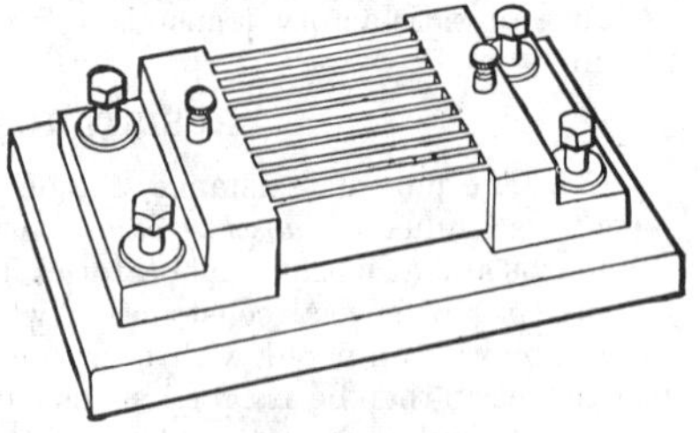

Fig. 3-49. Leeds & Northrup air-cooled standard resistor or shunt.

ance oil-cooled standards are mounted in individual oil-filled containers provided with copper coils through which cooling water is circulated and with propellers to provide continuous oil motion.

139. Alternating-current resistors for current measurement require further design consideration. For example, if the resistor is to be used for current-transformer calibration, its a-c resistance must be identical with its d-c resistance within $\frac{1}{100}\%$ or better, and the voltage difference between its voltage terminals must be in phase with the current through it within a few tenths of a minute. Thin strips or tubes of resistance material are used to limit eddy currents and minimize "skin" effect, the current circuit must be arranged to have small self-inductance, and the leads from the voltage taps to the potential terminals should be arranged so that, as nearly as possible, the mutual inductance between the voltage and current circuits opposes and cancels the effect of the self-inductance of the current circuit. Figure 3-50 shows three types of construction. In (*a*) a metal strip has been folded into a very narrow U; in (*b*) the current circuit consists of coaxial tubes soldered together at one end and to terminal blocks at the other end; in (*c*) a straight tube is used as the current circuit, and the potential leads are snugly fitting coaxial tubes soldered to the resistor tube at the desired separation and terminating at the center.

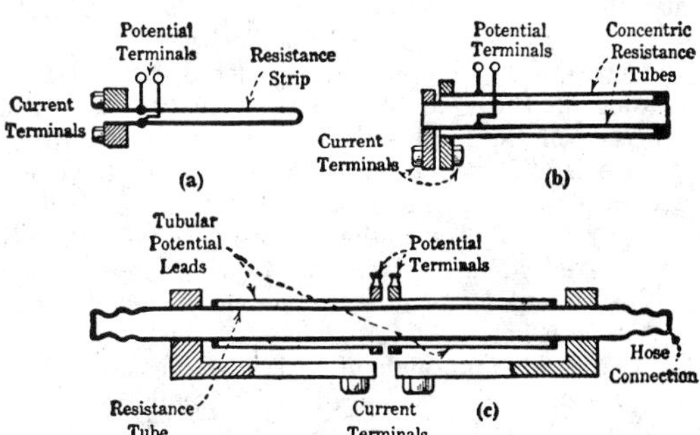

FIG. 3-50. Types of low-inductance standard resistors.

140. Resistance coils consist of insulated resistance wire wound on a bobbin or winding form, hard-soldered at the ends to copper terminal wires. Metal tubes are widely used as winding form for d-c resistors, as they dissipate heat more readily than insulating bobbins; but if the resistor is to be used in a-c measurements, a ceramic winding form is greatly to be preferred, as it contributes less to the phase-defect angle of the resistor. The resistance wire ordinarily is folded into a narrow loop and wound bifilar onto the form, to minimize inductance. This construction results in considerable associated capacitance of high-resistance coils, for which the wire is quite long; and an alternative construction is to wind the coil inductively on a thin mica or plastic card. The capacitive effect is greatly reduced, and the inductance is still quite small if the card is thin.

Resistors in which the wire forms the warp of a woven ribbon have lower time constants than either the simple bifilar- or card-wound types. Manganin is the resistance material most generally employed, but Evanohm and similar alloys are beginning to be extensively used for very-high-resistance coils. Enamel or silk is used to insulate the wire; and the finished coil is ordinarily coated with shellac or varnish to protect the wire from the atmosphere. Such coatings do not completely exclude moisture, and dimensional changes of insulation with humidity will result in small resistance changes, particularly in high resistances where fine wire is used.

141. Resistance boxes usually have two to four decades of resistance so that with reasonable precision they cover a considerable range of resistance, adjustable in small steps. For convenience of connection, terminals of the individual resistors are brought to copper blocks or studs, which are connected into the circuit by means of plugs or of dial switches using rotary laminated brushes; clean, well-fitted plugs probably have lower resistance than dial switches but are much less convenient to use. The residual inductance of decade groups of coils due to switch wiring, and the capacitance of connected but inactive coils, will probably exceed the residuals of the coils themselves, and it is to be expected that the time constant of an assembly of coils in a decade box will be considerably greater than that of the individual coils.

142. Measurement of resistance makes use of many methods whose applicability depends on the magnitude of the resistance and the accuracy required. Most methods can be broadly classified as voltage-drop or bridge methods.

143. Voltage-drop (or fall-of-potential) methods of measuring resistance consist, in their elementary form, of noting the voltage drop across the resistor when it carries a known current and calculating the resistance by Ohm's law. The shunting effect of the voltmeter must be considered, particularly if the resistance is high, since a milliammeter in series with the combination will indicate the current taken by the voltmeter as well as by the resistor. The formula for the resistance becomes $R = E/(I - E/V)$, where V is the resistance of the voltmeter. The possibility should not be overlooked when use of this method is considered that the current required to give an acceptable reading on the voltmeter may cause overheating in the resistor. In any event, the precision of the method is limited by the resolution of the indicating instruments and its use is dictated by considerations of convenience rather than accuracy. Better accuracy can be achieved by a deflection method in which voltage drops are compared across a standard and the unknown resistor in series carrying the same current, and will be best when the two deflections are nearly equal. Still higher accuracy is possible if the current is constant and a potentiometer can be substituted for the indicating instrument.

144. Bridge methods of resistance measurement are the most accurate, because they are null methods and because the comparison is made directly with standardized resistors that can be accurately known. The common types of bridges for d-c resistance measurements are (1) Wheatstone bridges for two-terminal resistors and (2) Kelvin bridges for four-terminal resistors.

145. The Wheatstone bridge is generally used for two-terminal resistors. In the low-resistance range where four-terminal construction is normal, the resistance of connections into the network may be a significant fraction of the total resistance to be measured, and the Wheatstone network is not applicable. Figure 3-51 shows the arrangement of a Wheatstone bridge, where A, B, and C are known resistances and D is the resistance to be measured. One or more of the known arms is adjusted until the galvanometer G indicates a null; then $D = B(C/A)$. In case D is inductive, the battery switch S_1 should be closed before the galvanometer key S_2 to protect the galvanometer from the initial transient current. In a common form of bridge, B is a decade resistance, adjustable in small steps, while C and A (the ratio arms of the bridge) can be altered to select ratios in powers of ten, from $C/A = 10^{-3}$ to 10^3. If the value of the unknown resistor is not very different from that of a known resistor, accuracy may be improved by substituting the known and unknown in turn into arm D and noting the difference in balance readings of the adjustable arm B. Since there has been no change in the ratio arms, any errors they may have do not affect the difference measurement, and only those errors in arm B which were involved in the difference between the settings affect the difference value; in effect, the unknown is measured in terms of a known resistor by a substitution procedure. An alternative form of Wheatstone bridge is frequently assembled from standards and a ratio box of limited range called a "direct-reading ratio set." This latter has a nominal ratio of unity, with ratio adjustments ranging from 1.005000 to 0.995000, that is, four decades of adjustment of which the largest has steps of 0.1%. If a balance is made with the two standards in arms B and D and a second balance with the standards interchanged, their difference is half the difference between the balance readings. A similar technique can be used wherever small resistance differences are involved, for example, in the determination of temperature coefficients.

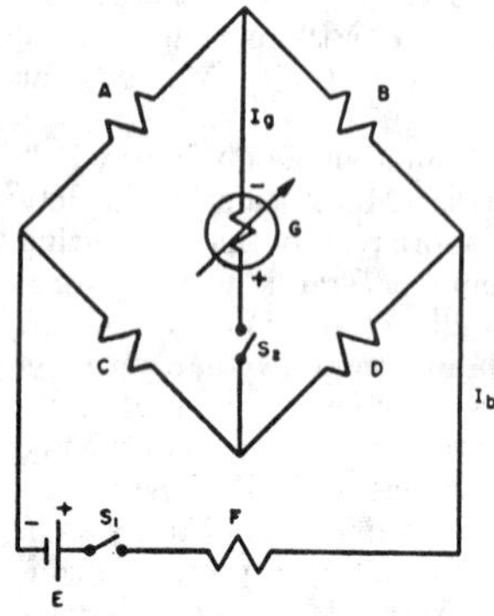

FIG. 3-51. Wheatstone bridge.

146. Bridge sensitivity can be determined in the following way. The voltage that would appear in the galvanometer branch of the circuit with switch S_2 open is

$$e = \frac{EBD\,\Delta B}{(B + D)^2},$$

where E is the supply voltage and ΔB is the amount in proportional parts by which B departs from balance. If, now, the voltage sensitivity of the galvanometer is known for operation in a circuit whose external resistance is that of the bridge as seen from the galvanometer terminals, its response for the unbalance ΔB can be computed. The current in the galvanometer with S_2 closed is

$$I_g = \frac{e}{G + BD/(B + D) + AC/(A + C)},$$

where G is the resistance of the galvanometer. If there is a large current-limiting resistance F in the battery branch of the bridge, the terminal voltage at the AC and BD junction points should be used rather than the supply voltage E in computing e. In connecting and operating a bridge, the allowable power dissipation of its components should first be checked to ensure that these limits are not exceeded, either in any element of the bridge itself or in the resistance to be measured.

147. The Kelvin double bridge is used for the measurement of low resistances of four-terminal construction, i.e., whose current and voltage terminals are separate. Figure 3-52 shows the network. The balance equation is

$$\frac{X}{S} = \frac{A}{B} + \frac{l}{S}\frac{\beta}{\alpha + \beta + l}\left(\frac{A}{B} - \frac{\alpha}{\beta}\right).$$

If the resistances X and S being compared are small, so that the resistance of the link l connecting them is comparable, the term of the balance equation involving l could be significant; but if the ratio A/B is equal to the ratio α/β, the correction term vanishes. This equality can be demonstrated, after the bridge is balanced, by opening the link l; if the inner and outer ratios are equal, the bridge will remain balanced. It should be noted that the resistance of the leads r_1, r_2, r_3, r_4 between the bridge terminals and the voltage terminals of the resistors may contribute to a ratio unbalance; these lead resistances should be in the same ratio as the arms to which they are connected. In some Kelvin bridges, small adjustable resistors are provided for balancing leads; another technique is to shunt the α or β arm with a high resistance until $A/B = \alpha/\beta$ with the link removed. When this balance is achieved, the link l is replaced and the main bridge balance is readjusted. In some bridges, the outer and inner ratio arms are adjustable only in decimal steps, and the main balance is secured by means of an adjustable standard resistor consisting of a Manganin strip with nine voltage taps of 0.01 or 0.001 Ω each and a Manganin slide wire. Portable bridges may use slide-wire ratio arms and reference resistors to cover a range from 10 μΩ to 10 Ω.

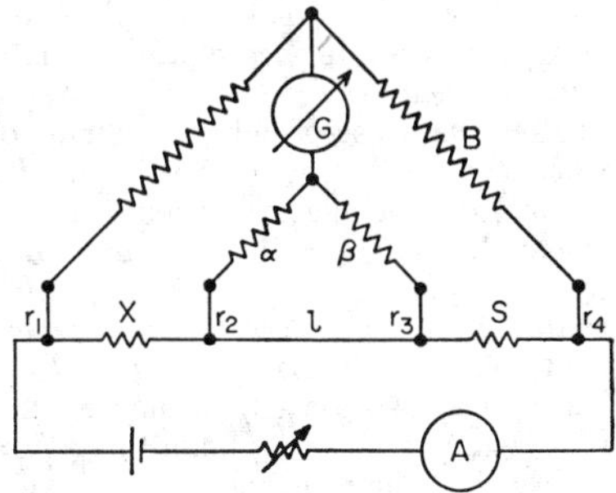

Fig. 3-52. Kelvin double bridge.

148. Insulation resistance is generally measured by deflection methods. In the case of resistances of the order of a few megohms, a Wheatstone bridge may be used with low to moderate accuracy. A portable megohm bridge is made by General Radio Company. It operates as a Wheatstone bridge with an amplifier and d-c indicating instrument as the detector system. A choice of high-resistance ratio arms gives ranges of 0.1 to 10^4 MΩ. On the highest range, the resolution limit is about 10^6 MΩ. The deflection methods fall in two general classes: (1) direct-deflection methods and (2) loss-of-charge methods.

149. Direct-deflection methods (insulation resistance) involve the simple application of Ohm's law. When the resistance is of the order of 1 MΩ, an ordinary voltmeter can give results that are good enough for most purposes. Two readings are taken, one with the voltmeter directly across the source of voltage, the other with the resistance to be measured connected in series with the voltmeter. The resistance is $R =$

$V(d_1 - d_2)/d_2$, where V is the resistance of the voltmeter, d_1 is the voltmeter deflection on the first reading, and d_2 is the deflection on the second reading. The greater the resistance of the voltmeter per volt, the higher the resistance that can be measured. For higher resistances, a reflecting galvanometer with high current sensitivity is used. Figure 3-53 is a diagram of the arrangement for measuring the insulation resistance of a cable.

The measurement is made as follows: The galvanometer shunt S is set at the highest shunting value, and the circuit is closed. The shunt is decreased until a large, readable deflection is obtained. The reading is taken 1 min after closing the main switch. This procedure is repeated with only the standard resistor r_s (usually 0.1 or 1 MΩ) in the circuit, the specimen being short-circuited. The resistance of the specimen in megohms is $R = (G/d_1s_1) - r_s$, where d_1 is the first reading and s_1 the multiplier corresponding to the shunt setting. G, the galvanometer megohm constant, is obtained from the second reading, $G = dr_ss$, where d is deflection, r_s the value of the standard resistor in megohms, and s the shunt multiplier. The conductor is preferably negative to the sheath or water. The standard resistor r_s is left in the circuit as a protection to the galvanometer against accidental short circuit in the sample. The guard for the cable ends is shown by the broken line. Removing braid for several inches at the ends of the sample and dipping the ends in hot paraffin tend to reduce leakage across the face of the insulation from sheath to central conductor, especially in damp weather.

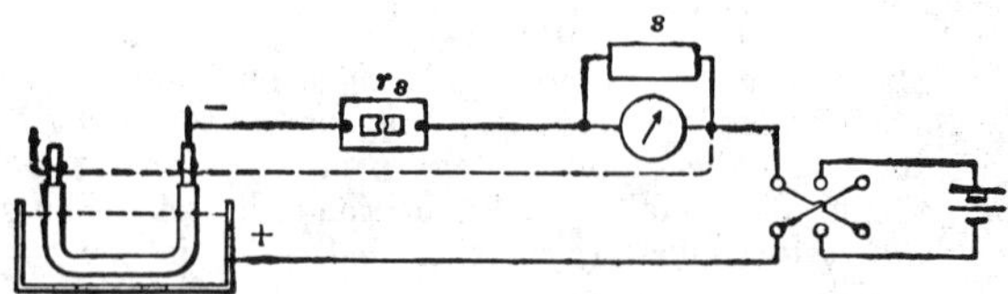

FIG. 3-53. Diagram for insulation resistance of cable.

150. The loss-of-charge method of measuring insulation resistance may be used when the resistance is very high, such as the resistance of porcelain and glass, and the surface leakage resistance of line insulators. The principle is shown in Fig. 3-54, where the resistance r to be measured is connected in parallel with a capacitor C. Key a is closed and immediately opened, charging the capacitor. Key b is closed immediately after a is opened and the ballistic throw d_1 of the galvanometer noted. The process is repeated, but now a time t s is allowed to pass from the instant of charging before key b is closed and a deflection d_2 observed. The resistance is

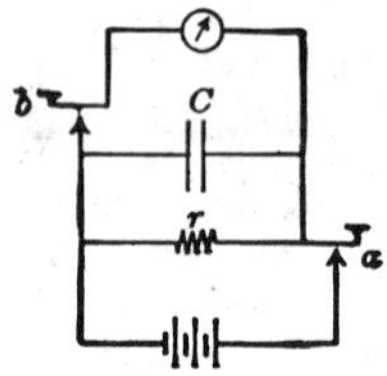

$$r = \frac{t}{2.303C \log_{10} (d_1/d_2)} \text{ M}\Omega,$$

FIG. 3-54. Leakage method of measuring insulation resistance.

where C is the capacitance in microfarads.

The insulation resistance of the capacitor is not infinite and should be measured in a similar manner with r removed. The two resistances are in parallel, and the corrected value is

$$r = \frac{r_1 r_2}{r_2 - r_1},$$

where r_1 is the resistance value obtained in the first measurement and r_2 is the resistance of the capacitor. For even higher resistance, a growth-of-charge method may be used. In this case, the resistance to be measured is connected in series with a capacitor (preferably an air capacitor), and a known voltage E is applied for t s, the voltage on the capacitor being e at the end of this time. This value, e, is best measured with an electrostatic voltmeter connected continuously across C. The resistance is

$$r = \frac{t}{2.303C \log_{10} [E/(E - e)]} \text{ M}\Omega.$$

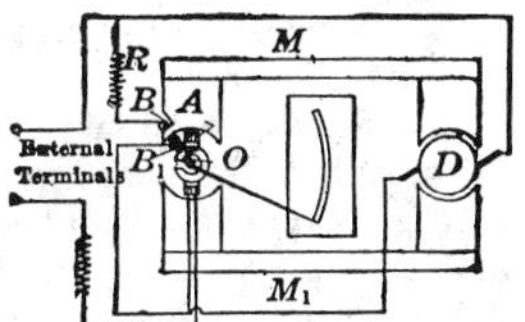

Fig. 3-55. Evershed Megger.

151. The Evershed Megger[2] is an ohmmeter that uses the principle of the ratio meter, as shown in Fig. 3-55, where A is a coil in series with the resistance to be measured and B, B_1 are coils which, with the resistance R, are connected to a hand-driven generator D. All three coils are rigidly coupled together, mounted for rotation about an axis O, and connected to the circuit by fine copper strips that exert no control force. If the external circuit is open, B and B_1 are deflected to the position where they will intercept the least flux from the permanent magnets M, M_1, that is, opposite the gap in the C-shaped iron piece about which coils A and B_1 move. The pointer then stands at "infinity" on the scale. If a finite resistance is connected across the terminals, the current in A will produce a torque and, as the system moves, the coils B, B_1 exert an opposing torque whose magnitude increases with displacement from the infinity position. The system will come to rest at a position where the torques are equal, depending on the external resistance. This position is practically independent of the voltage generated at D. In some Meggers, there is a slip clutch which enables the generator to turn at constant speed when hand-driven. This applies a constant voltage, 500 or 1,000 V, to the external circuit and permits a voltage-withstand test of the insulation connected to the external circuit as well as a measure of insulation resistance.

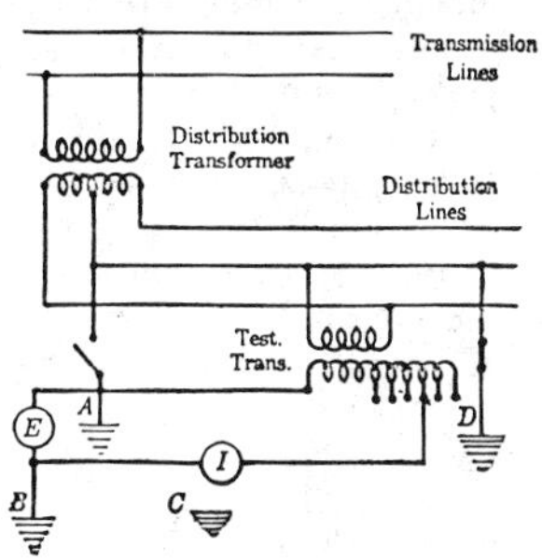

Fig. 3-56. Resistance of earth connections.

152. The resistance of earth connections may be measured by a three-electrode method. In Fig. 3-56, A is the connection whose resistance to earth is to be measured; it is temporarily disconnected from the distribution system while ground connection is preserved through a connection at D, either temporary or permanent. Two additional "grounds," B and C, are established, separated from each other and from A by not less than 15 ft. These auxiliary grounds may be pieces of metal buried in the earth, such as a guy wire or a steel pole, making sufficient contact with ground for a good current reading. Resistances between the three electrodes taken in pairs are measured by a voltmeter-ammeter method. These resistances are r_{ab}, r_{bc}, r_{ac}. Then the resistances are as follows:

At A: $$R_a = \frac{r_{ab} - r_{bc} + r_{ac}}{2}$$

At B: $$R_b = \frac{r_{ab} - r_{bc} - r_{ac}}{2}$$

At C: $$R_c = \frac{r_{bc} - r_{ab} + r_{ac}}{2}$$

The measurement should be made with alternating current, which can be taken from the distribution system through an isolating transformer with secondary taps as shown. A low-range voltmeter is usually required. An Evershed ratio instrument, similar to that described in the previous paragraph, is used for the measurement of ground resistance. One of the moving coils is traversed by the current sent through the ground from the attached hand-operated generator; the other is energized by the voltage drop to an auxiliary, driven electrode.

153. Location of Line Faults.[1] Faults in electric lines for the transmission and distribution of power, speech, etc., may be divided into two classes, closed-circuit faults and open-circuit faults. **Closed-circuit faults** consist of **shorts,** where the insu-

[1] Henneberger, T. C., and Edwards, P. G. Bridge Methods for Locating Resistance Faults in Cable Wires; *Bell System Tech. Jour.*, 1931, Vol. 10, p. 382.

[2] A registered trademark of Evershed and Vignoles, Ltd.

lation between conductors becomes faulty, and **grounds,** where the faulty insulation permits the conductor to make more or less perfect contact with the earth. **Open-circuit faults,** or **opens,** are produced by breaks in the conductors.

a. When the short is a low-resistance union of the two conductors, such as at M in Fig. 3-57, the resistance should be measured between the ends AB; from this value and the resistance per foot of conductor, the distance to the fault can be computed. A measurement of resistance between the other ends $A'B'$ will confirm the first computation or will permit the elimination of the resistance in the fault, if this is not negligible.

b. The location of a ground, as at N in Fig. 3-57, or of a high-resistance short is made by either of the two classical "loop" methods, provided that a good conductor remains.

Figure 3-58 shows the arrangement of the **Murray loop test,** which is suitable for low-resistance grounds. The faulty conductor and a good conductor are joined together at the far end, and a Wheatstone-bridge arrangement is set up at the near ends with two arms a and b comprised in resistance boxes which can be varied at will; the two segments of line x and $y + l$ constitute the other two arms; the battery current flows through the ground; the galvanometer is across the near ends of the conductors. At balance,

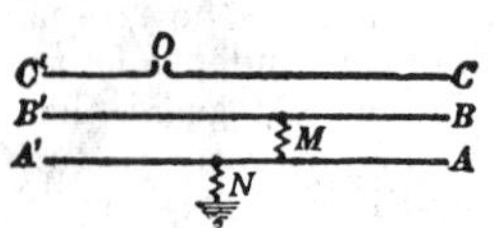

Fig. 3-57. Line faults.

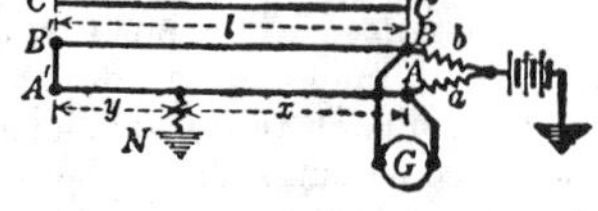

Fig. 3-58. Murray loop.

$$\frac{a}{b} = \frac{x}{y + l} \quad \text{or} \quad \frac{a + b}{b} = \frac{x + y + l}{y + l} \quad \text{(ohms)}$$

The sum $x + y + l$ may be measured or known. If the conductors are uniform and alike and x and l are expressed as lengths, say, in feet,

$$x = \frac{2al}{a + b} \quad \text{(ft)}$$

If the ground is of high resistance, very little current will flow through the bridge with the arrangement of Fig. 3-58. In that case, battery and galvanometer should be interchanged, and the galvanometer used should have a high resistance. If ratio arms a and b consist of a slide wire (preferably with extension coils), the sum $a + b$ is constant and the computation is facilitated. Observations should be taken with direct and reversed currents, especially in work with underground cables.

In the **Varley loop,** shown in Fig. 3-59, fixed ratio coils, equal in value, are employed, and the bridge is balanced by adding a resistance r to the near leg of the faulty conductor.

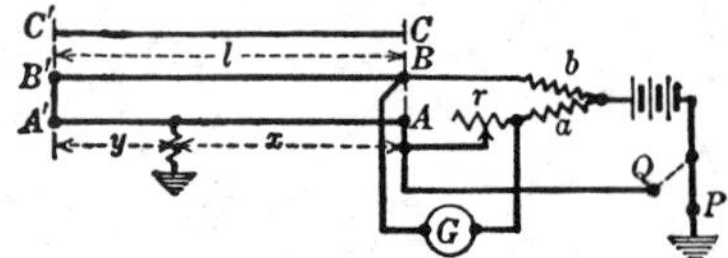

Fig. 3-59. Varley loop.

$$\frac{a}{b} = \frac{r + x}{y + l} \quad \text{or} \quad \frac{a + b}{b} = \frac{x + y + l + r}{y + l} \quad \text{(ohms)}$$

If $a = b$,

$$x = y + l - r \quad \text{or} \quad x = \frac{1}{2(x + y + l - r)} \quad \text{(ohms)}$$

The total line resistance $x + y + l$ is conveniently determined by shifting the battery connection from P to Q and making a new balance, r'. The equation then becomes $x = 1/2(r' - r)$. When a and b are slightly unequal, a second set of readings should be taken with a and b interchanged and the average values of r and r' substituted in the foregoing equations.

c. **Opens,** such as O in Fig. 3-57, are located by measuring the electrostatic capacitance to ground (or to a good conductor) of the faulty conductor and of an identical good conductor; the position of the fault is determined from the ratio of the capacitances.

d. **Shorts** and **grounds** may be detected by sending through the defective conductor an alternating current of audible frequency, say 1,000 c/s. A pickup coil connected to a telephone receiver worn on the head of the tester is then carried along the line; the note in the receiver will cease when the fault has been passed.

INDUCTANCE MEASUREMENTS

154. General. The self-inductance, or coefficient of self-induction, of a circuit is the constant by which the time rate of change of the current in the circuit must be multiplied to give the self-induced counter emf. Similarly, the **mutual inductance** between two circuits is the constant by which the time rate of change of current in either circuit must be multiplied to give the emf thereby induced in the other circuit. Self-inductance and mutual inductance depend upon the shape and dimensions of the circuits, the number of turns, and the nature of the surrounding medium.

155. Computable standards of self- or mutual inductance have been used traditionally in *absolute-ohm* determinations; but they are not suitable for use in assigning the values of other inductors—they are bulky and have relatively large capacitance to ground and considerable coupling to other circuits, their ratio of inductance to resistance is relatively low, and they exhibit appreciable skin effect even at moderately high frequencies, since they must be wound with rather heavy wire. All these undesirable features inevitably follow from the requirement that their values be computable from measured dimensions. Computable self-inductors and the primaries of computable mutual inductors are wound as single-layer solenoids on a dimensionally stable nonmagnetic form. The best of them are on cylinders of fused silica, and the winding is laid in a groove lapped into the form to ensure uniform winding pitch. The primary winding of a computable mutual inductor is in two or three sections spaced at such intervals that there is a region outside and in its central plane in which its field gradients are very small. The secondary—a multilayer winding—is located in this position so that its position and dimensions will be less critical.

156. Working standards of inductance are usually multilayer coils wound on nonmagnetic forms of bakelite, marble, or ceramic to ensure reasonable dimensional stability. A toroidal core gives a coil that is practically immune to external magnetic fields. Approximate astaticism is also achieved by using two equal coils, connected in series and so located with respect to each other that their coupling with external fields tends to cancel each other. Since there is always capacitance associated with a winding, the effective value of an inductor will always be a function of frequency to a greater or lesser extent and an accurate statement of value must necessarily include the frequency with which the value is associated. Inductance standards for radio frequencies are wound on open frames with copper or other binders. Single-layer winding or "loose basket weave" is essential to reduce the distributed capacitance and the consequent change of effective inductance with frequency. Insulating material is kept to a minimum to reduce dielectric loss.

157. Inductometers are continuously adjustable inductance standards. The Ayrton-Perry inductometer uses pairs of coaxial coils wound on zones of spheres; the outer pair is fixed, and the inner pair can be rotated about a vertical axis. The coils are so proportioned that the scale is uniform over most of its length. This inductometer is not astatic, and its coupling with external fields can cause significant measurement errors. The Brooks inductometer, a better design from several viewpoints, consists of six link-shaped coils. The four stator coils are mounted in pairs above and below the rotor coils, which are located diametrically opposite one another in a flat disk. These two fixed- and moving-coil combinations are so connected that their coupling with external fields tends to cancel. The shape of the link coils gives a scale that is completely uniform except at its extreme ends, and the time constant of the inductometer is much higher than in the Ayrton-Perry arrangement. Ratio of maximum to minimum inductance is about 8:1; and change of calibration with wear in the bearings is negligibly small. Terminals of the fixed and movable coils are usually brought out separately so that inductometers can be used as either adjustable self-inductors or adjustable mutual inductors.

158. Measurement methods at power and audio frequency are (1) null methods employing bridges if accurate values are required or (2) deflection methods in which the inductance is computed from measured values of impedance and power factor, the measurements being made with indicating instruments—ammeter, voltmeter, wattmeter. At radio frequencies, resonance methods are used.

159. Bridges for inductance measurements can assume a variety of forms, depending on available components and reference standards, magnitude and time constant of the inductance to be measured, and a variety of other factors. In a four-arm bridge similar to the Wheatstone network, an inductance can be (1) compared with another inductance in an adjacent arm with two resistors forming the "ratio" arms or (2) measured in terms of a combination of resistance and capacitance in the opposite arm with two resistors as the "product" arms. It is generally better, where possible, to measure inductance in terms of capacitance and resistance rather than by comparison with another inductance, because the problems of stray fields and coupling between bridge components are more easily avoided. The basic circuits will be described for a few bridges which can be used to measure inductance, but the reader must refer to a text on bridges (e.g., Hague, "A-C Bridge Methods," 5th edition, Pitman, 1957) for discussion of shielding, physical arrangement of components, effects of residuals, etc. In the balance equations which will be stated below, the inductance, L or M, will be expressed in henrys, the resistance R in ohms, capacitance in farads, and ω is $2\pi \times$ frequency in hertz (cycles per second). The time constant of an inductor is L/R; its storage factor Q is $\omega L/R$.

160. Inductance comparison is accomplished in the simple Wheatstone network shown in Fig. 3-60, in which A and B are resistive ratio arms, L_x and r_x represent the inductor being measured, and L_s and r_s are the reference inductor and the associated resistance (including that of the inductor itself) required to make the time constants of the two inductive arms equal. At balance,

$$\frac{A}{B} = \frac{L_x}{L_s} = \frac{r_x}{r_s}.$$

An inductometer may be used to achieve balance, together with an adjustable resistance in the same bridge arm, as indicated in the diagram. If only a fixed-value standard inductor is available, balance can be secured by varying one of the ratio arms but there must also be an adjustable resistance in series with L_x or L_s to balance the time constants of the inductive arms. Care must be taken to ensure that there is no inductive coupling between L_s and L_x, as this would lead to a measurement error.

161. The Maxwell-Wien bridge for the determination of inductance in terms of capacitance and resistance is shown in Fig. 3-61. The balance equations are $L_x = ASC$

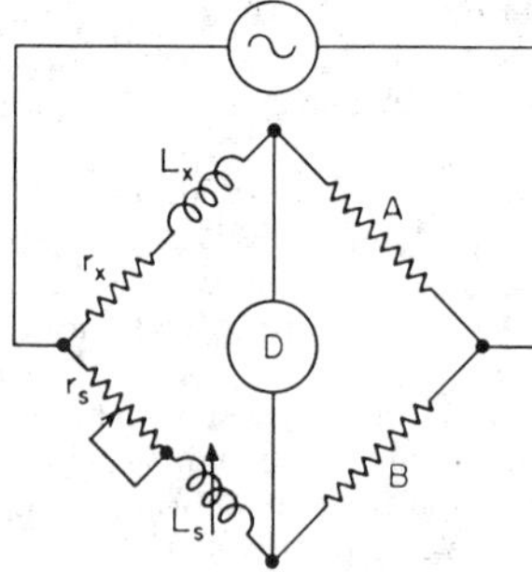

Fig. 3-60. Inductance bridge.

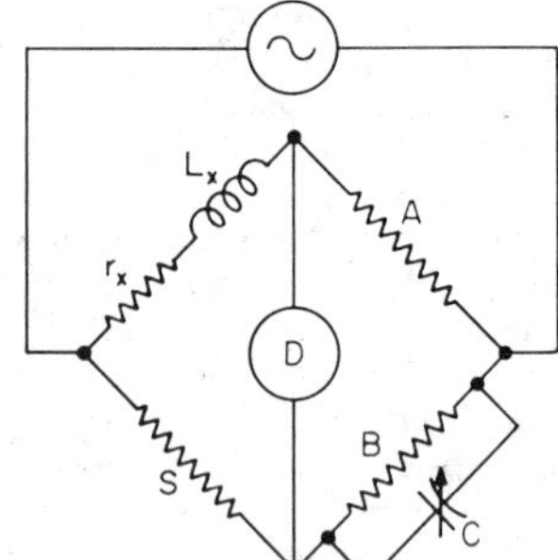

Fig. 3-61. Wien-Maxwell inductance-capacitance bridge.

and $r_x = AS/B$. This bridge is widely used for accurate inductance measurements. It is most easily balanced by adjustments of capacitor C and resistor B; these elements are in quadrature, and therefore their adjustments do not interact.

162. Anderson's bridge, shown in Fig. 3-62, can be used for measurement over a wide range of inductances with reasonable values of R and C. Its balance equations are $L_x = CAS(1 + R/S + R/B)$ and $r_x = AS/B$. Balance adjustments are best made with R and r_x. This bridge has also been used to measure the residuals of resistors, a substitution method being employed in which the unknown and a loop of resistance wire with calculable residuals are substituted in turn into the L arm. If A and B are equal and if the resistances of the unknown and the calculable loop are matched, the residuals in the various bridge arms do not enter the final calculation, except the residual of Δr_x, the change in r_x between balances. The elimination of bridge-arm residuals from the *exact* balance equations is characteristic of substitution methods; and quite generally, residuals or corrections to the arms that are unchanged between the balances do not have to be taken into account in the final calculation when the difference is small between the substituted quantities.

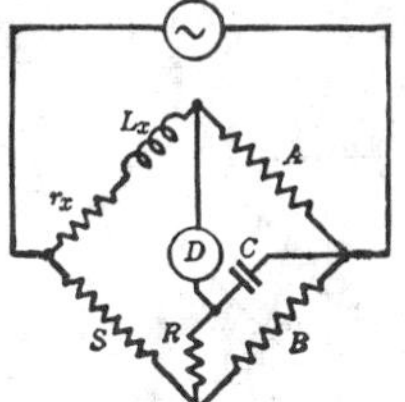

FIG. 3-62. Anderson's bridge.

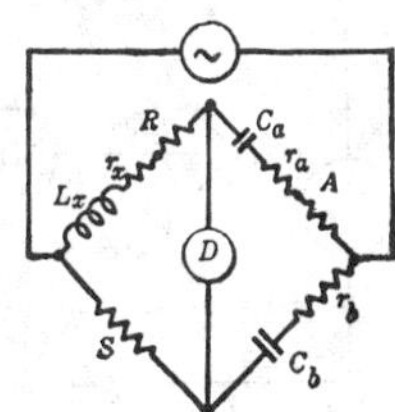

FIG. 3-63. Owen's inductance-capacitance bridge.

163. Owen's bridge, shown in Fig. 3-63, can be used to measure a wide range of inductance with a standard capacitor C_b of fixed value, by varying the resistance arms S and A. In operation, the resistance S and capacitor $C_b(r_b)$ are usually fixed, balance being secured by successive adjustments of A and R. At balance

$$r_x + R = (C_b/C_a)S + \omega L_x \omega C_b r_b,$$

and $L_x(1 + \tan \delta_b \tan \delta_x) = C_b S(A + r_a)$. If $C_b(r_b)$ is a loss-free air capacitor, so that $r_b = 0$ and $\tan \delta_b = 0$, $r_x = (C_b/C_a)S' - R$ and $L_x = C_b S(A + r_a)$. This is a bridge which is much used for examining the properties of magnetic materials; inductance may be measured with direct current superposed. With a low-reactance blocking capacitor in series with the detector and another in series with the source, a d-c supply may be connected across the test inductance without current resulting in any other branch of the network; a high-reactance, low-resistance "choke" coil should be connected in series with the d-c source (see Ferguson, *Bell System Tech. Jour.*, 1927, Vol. 6, p. 375, for details of shielding, etc.).

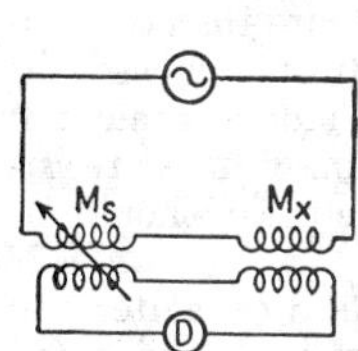

FIG. 3-64. Comparison of mutual inductometers.

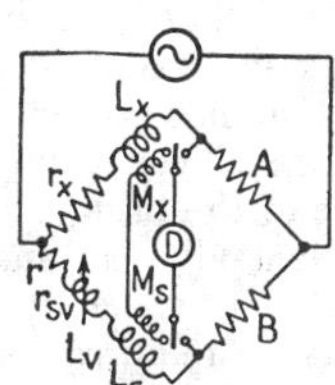

FIG. 3-65. Campbell's bridge for comparing mutual inductances.

164. Mutual inductance can be measured readily if an adjustable standard of proper range is available. Connections are made as in Fig. 3-64. At balance, $M_x = M_s$, so that the range of measurement is limited to values that can be read on M_s with the desired precision. Care should be taken in arranging the circuit to avoid coupling between the mutual inductors. Campbell's *mutual-inductance bridge* (Fig. 3-65) makes possible the comparison of mutual inductors of quite different value. The resistances and self-inductances of their primaries must also be balanced. This is accomplished by adjusting L_v and r in the self-inductive bridge with the switches in the detector branch to the right. The mutual-inductance balance is then made with the switches to the left. At balance,

$$\frac{A}{B} = \frac{M_x}{M_s} = \frac{L_x}{L_v + L_s} = \frac{r_x}{r + r_{sv}}.$$

The *Maxwell-Wien bridge* can also be adapted to the measurement of mutual inductance as shown in Fig. 3-66. Balancing adjustments are to be made with r_x and C. With the selector switch of the supply branch in position 1, the self-inductance L_x of the primary winding is determined in the normal manner. The balance equation is $L_x = ASC_1$. With the selector switch in position 2, a capacitance value C_2 is required for balance. The mutual inductance is given by the equation

$$M = (C_1 - C_2)\frac{ABS}{B + S}.$$

The values of resistors A, B, and S must not be changed between the two balances, and the connections should be such that the emf induced in L_x from the winding of M which is in the supply circuit will oppose the emf of self-inductance in L_x (see Harris, "Electrical Measurements," John Wiley & Sons, Inc., 1952, p. 721, for an analysis of the effect of bridge residuals in this measurement). Mutual inductance may also be measured in any self-inductance bridge by the method illustrated in Fig. 3-67. The coils L_1 and L_2 are connected in series (a) so that the emf of mutual inductance aids the emf of self-inductance. The measured value of total inductance is $\lambda_a = L_1 + L_2 + 2M$. A second measurement is made with one coil reversed in the series connection of (b), so that the emfs of self- and mutual inductance are opposed. The measured value is now $\lambda_b = L_1 + L_2 - 2M$. The mutual inductance is calculated from the expression

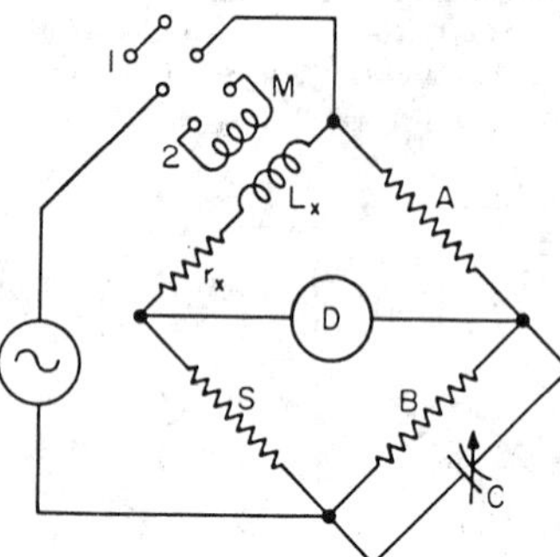

Fig. 3-66. Maxwell-Wien bridge for mutual inductance.

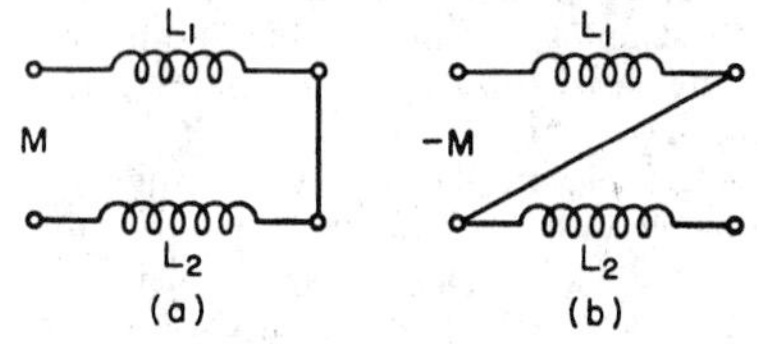

Fig. 3-67. Mutual inductance connected for measurement of self-inductance.

$$M = \frac{\lambda_a - \lambda_b}{4}.$$

165. Iron-cored inductors vary in value with frequency and with current, so that measurements must be made at known current and frequency; bridge methods can, of course, be adapted to this measurement, care being exercised to ensure that the current capacities of the various bridge components are not exceeded. In such a case, the waveform of the voltage drop across the circuit branch containing the inductor may not be sinusoidal, whereas that across the other side of the bridge, containing linear resistances and reactances, may be undistorted. Generally, a tuned detector should be used.

166. Ballistic measurement of mutual inductance requires a standard capacitor and a ballistic galvanometer. The secondary circuit of the mutual inductance is connected to the ballistic galvanometer and to a series resistance whose value is adjusted so that the total circuit resistance is R, including the resistance of the galvanometer and the secondary coil. When a current I (amperes) is reversed in the primary circuit, the charge Q (coulombs) flowing through the secondary circuit is $Q = 2MI/R$. From the ballistic throw of the galvanometer and its coulomb constant K, Q may be determined. K is determined in turn by observing the deflection s, when a capacitor C (microfarads) charged to E (volts) is discharged through the galvanometer. From these data, $M = RECd/(2Is \times 10^6)$.

167. Deflection methods for measuring inductance, in the simplest case, require measurement of the current I in the inductor when the voltage across it is E. The d-c resistance r_x of the coil and the frequency f must be known. Then $2\pi fL = \sqrt{(E/I)^2 - r_x^2}$. The voltmeter current should not be included in the measured current (i.e., the voltmeter should be across both the inductor and the ammeter). The d-c resistance and inductance of the ammeter must be negligible, or allowance must be made for them.

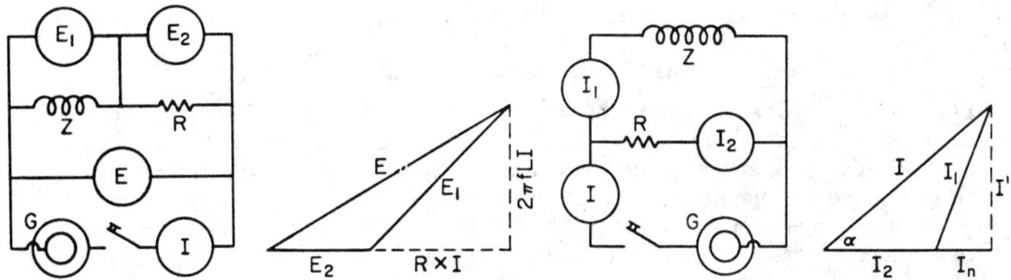

FIG. 3-68. Three-voltmeter method of measuring inductance.
FIG. 3-69. Vector diagram for three-voltmeter method of measuring inductance.
FIG. 3-70. Three-ammeter method of measuring inductance.
FIG. 3-71. Vector diagram for three-ammeter inductance measurement.

In the *three-voltmeter method,* the inductance to be measured is connected in series with a noninductive resistance as shown in Fig. 3-68. The total current I is measured and the total voltage as well as the voltage across Z and R. From these voltage readings, the triangle of Fig. 3-69 is constructed. I and f being known, L is obtained by calculation. The resistance of the voltmeter must be large in comparison with R and Z.

The *three-ammeter method* is similar, and the connections are shown in Fig. 3-70. From the three measured currents, the triangle of Fig. 3-71 is constructed. The voltage E can be measured, or if R is known, $E = I_2R$ can be calculated. Then $r_x = (EI''/I_1^2)$ and $L_x = (EI'/I^2)$. The resistances and reactances of the ammeters must be negligible.

168. Resonance methods can be used to measure inductance at radio frequencies. A suitable source is used to establish an r-f field whose wavelength is λ m. The inductance L_x (microhenrys) to be measured is placed in this field and connected to a calibrated variable capacitor through a thermocouple ammeter (i.e., a current-indicating instrument without reactance). The capacitor is adjusted to resonance at a value of C (picofarads). Then $L_x = 0.2815\lambda^2/C$. If a calibrated inductor L_s of the same order as L_x is available, the wavelength need not be known and a substitution method can be used. The resonance settings are C_s and C_x, with L_s and L_x, respectively, in the circuit. Then, $L_x = L_sC_s/C_x$. The value of L_x is the effective inductance at the frequency of measurement and includes the effect of associated coil capacitance. The frequency of the source must not be affected by the substitution of L_x for L_s.

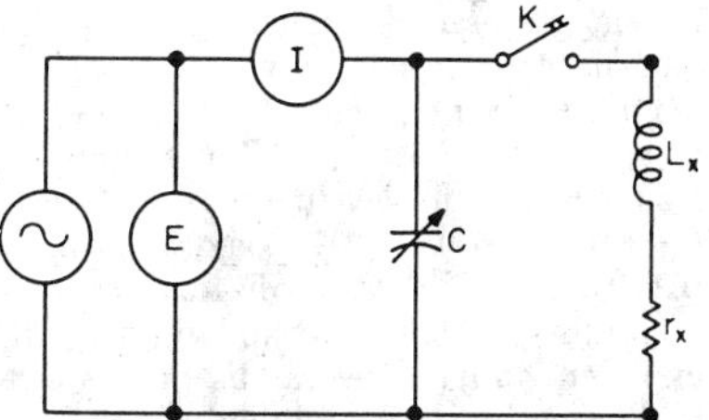

FIG. 3-72. Resonance-impedance method of measuring inductance.

A *resonance-impedance method,* suitable for high frequencies, is indicated in Fig. 3-72. The capacitor C is adjusted until the same current is indicated by the ammeter with switch K open or closed. (The applied voltage must be constant.) Then $L_x = (1/2\omega^2C)$ H, if C is in farads and the frequency is $f = \omega/2\pi$. The waveform must be practically sinusoidal and the ammeter of negligible impedance. This method may be used to measure the effective inductance of choke coils with superposed direct current (see Turner, *Proc. IRE*, 1928, Vol. 16, p. 1559, for details).

169. The residual inductance of a resistor or a length of cable at high frequency can often be determined by connecting the resistor in series with a fixed air capacitor and measuring its effective capacitance in an appropriate bridge with and without the series resistor S. If C_1 and C_2 are the measured capacitances in farads, without and with the series resistor, then

$$L = \frac{C_2 - C_1}{\omega^2 C_1 C_2} \qquad \text{and} \qquad S = \frac{1 - \omega^2 LC_1}{\omega C_1} \tan \delta,$$

S is the effective resistance in ohms, L is the residual inductance in henrys, and δ is the loss angle of the capacitor-resistor combination computed from the second bridge balance.

CAPACITANCE MEASUREMENTS

170. General. The *capacitance* between two electrodes may be defined for measurement purposes as the charge stored per unit potential difference between them. It depends on their area, spacing, and the character of the dielectric material or materials, which is affected by the electric field between them. The value of a capacitor, measured in farads or a convenient submultiple of this unit, will be influenced quite generally by temperature, pressure, or any ambient condition that changes the dimensions or spacing of the electrodes or the characteristics of the dielectric. The *dielectric constant* of a material is defined as the ratio of the capacitance of a pair of electrodes, with the material occupying all the space affected by the field between them, to the capacitance of the same electrode configuration in vacuum.

171. Computable capacitors known to a part in 10^6 or better have recently been constructed at the National Bureau of Standards and at certain other national laboratories as a basis for their *absolute-ohm* determinations. Such capacitors now serve as the "base" unit in assigning values to standard capacitors. The electrode arrangement of these conputable capacitors conforms to the geometry prescribed in the recently discovered *Thompson-Lampard theorem*: If four cylindrical conductors of arbitrary sections are assembled with their generators parallel, to form a completely enclosed cylinder in such a way that the internal cross capacitances per unit length are equal, then in vacuum these cross capacitances are each

$$\frac{\log_e 2}{4\pi^2 \mu_0 V^2}.$$

In the MKSA system of electrical units, where μ_0 has the assigned value $4\pi \times 10^{-7}$ and V is the speed of light in vacuum in meters per second, this capacitance is in farads per meter. The capacitance of such a cross capacitor is about 2 pF/M. A practical realization of such a capacitor consists of four equal closely spaced cylindrical rods with their axes parallel and at the corners of a square. Arranged as a three-terminal capacitor and with end effect eliminated, its value can be computed as accurately as its effective length can be measured.

The capacitance of vacuum capacitors with electrodes of simple geometry can be computed approximately in a few cases: (1) Flat, parallel plates with guard ring, $C = 0.08854A/t$ pF, where A is area of the guarded plate in square centimeters and t is spacing in centimeters between electrodes; if dimensions are in inch units, $C = 0.2249A/t$. (2) Coaxial cylinders with guard cylinders at both ends, $C = 0.24161L/\log(R_2/R_1)$ pF for centimeter units, or $C = 0.6137L/\log(R_2/R_1)$ pF for inch units, where L is the length of the guarded cylinder, R_1 is the radius of the inner cylinder, and R_2 the radius of the outer cylinder. (3) Concentric spheres, where R_1 is the radius of the inner sphere and R_2 is the radius of the outer sphere, $C = 1.1127R_1R_2/(R_2 - R_1)$ pF for centimeter units, or $C = 2.8262R_1R_2/(R_2 - R_1)$ pF for inch units. These formulas give only approximate values because they assume no contributing field beyond the edges of the bounding surfaces, and take no account of possible eccentricity, lack of parallelism of surfaces, finite width of gap between guard and working electrode, etc., all of which would require small correction terms.

172. Standard capacitors at levels up to 10^3 pF are generally of a multiple-parallel-plate variety with dry gas (air or nitrogen) as dielectric. Low temperature coefficient is secured by use of invar—a low-expansion alloy—as the electrode material and a good degree of stability achieved by careful, strain-free mounting of fully annealed components and by hermetically sealing the unit. A very high degree of stability has been achieved in a solid-dielectric construction at the 10-pF level in which a disk of fused silica is provided with fired-on silver electrodes. Direct capacitance is through the interior of the disk between its parallel faces, and a silver coating on the cylindrical face acts as guard electrode and confines the field. Very narrow gaps at the edges of the disk between the guard and active electrodes, together with continuation of the shielding in the mounting arrangement, eliminate the possibility of any portion of the measured capacitance being through an outside path between the parallel-plate electrodes. The assembly is hermetically sealed in dry nitrogen, in a shock-resistant, resilient mounting together with a resistance thermometer so that temperature corrections can be accurately applied. Standards of this type have shown variations less than 1 part in 10^7 over a year interval. From 10^3 pF to 1 μF, standard capacitors generally have clear mica as dielectric. The electrodes may be metal foils laid out between the mica sheets, the assembly impregnated with paraffin, and the excess wax squeezed out under high pressure. In an alternative construction, the mica sheets are silvered, assembled under pressure, and the assembly hermetically sealed. Neither construction is as stable with time as the lower-value air-dielectric units, and the mica units are characterized by low but appreciable loss angles, whereas the loss angle of the air-dielectric standards is negligible in almost all applications. Continuously adjustable air capacitors have two stacks of interleaved parallel metal plates, one stack being mounted to rotate on an axis. The maximum capacitance occurs when the fixed and movable plates completely overlap; the minimum, a small value but not zero, occurs 180° from this position.

173. A three-terminal construction is required if the value of the capacitor is to be definite and independent of its proximity to other objects. In a nominally two-terminal arrangement, each of the electrodes has some capacitance to surrounding objects or to ground which may depend on spacing and which actually forms a second capacitance circuit in parallel with the capacitor of interest, as will be seen from Fig. 3-73*a* and *b*. It is only in case *c*, where there is an actual third electrode which com-

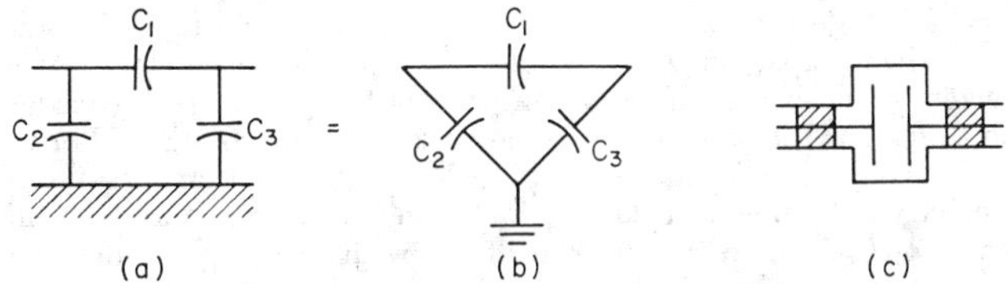

Fig. 3-73. Two-terminal and three-terminal capacitors.

pletely encloses the other two, that the value can be made definite and completely independent of any object or field outside the assembly. A second advantage of the three-terminal construction is that the direct capacitance between the two enclosed electrodes can be made loss-free, since the solid insulation required to support them mechanically can be in the auxiliary capacitances between the enclosing shield electrode and the shielded electrodes.

174. Methods of measuring capacitance can be classified as *null* methods, which quite generally involve the use of bridges, and *deflection* methods, in which some characteristic, usually impedance, is measured with the aid of indicating instruments. In the equations that follow, the capacitance C will be expressed in farads and resistance A, B, S in ohms. δ will be the *loss angle,* the amount which the current lacks of a true quadrature relation with voltage. The *power factor* of a capacitor is then $\cos(\pi/2 - \delta) = \sin \delta$. The dissipation factor D is the name given to $\tan \delta$. It is convenient to represent a capacitor as consisting of a capacitance C (farads) in series with a resistance r (ohms), such that $\tan \delta = 2\pi f C r$ at a frequency f. The *power loss,* for an impressed voltage E (volts), is $P = 2\pi f C E^2 \sin \delta$. Since most bridges yield $\tan \delta$, the power loss can be expressed conveniently as $P = 2\pi f C E^2 \tan \delta$, where δ is small, or $P = \omega C E^2 D$.

175. Bridge methods for the comparison of capacitors are to be preferred over methods in which capacitance is determined in terms of inductance, since it is simple to shield capacitors so that their values are completely independent of neighboring objects and their electric fields are completely confined, whereas the magnetic fields of inductors cannot be so confined. Error voltages can enter bridges through coupling of an inductor with an external field, through mutual coupling with eddy-current circuits induced by the inductor in neighboring metal objects, etc. The reader should refer to a text on bridges (e.g., Hague, "A-C Bridge Methods," 5th ed., Pitman, 1957) for discussion of shielding, physical arrangements of components, effects of residuals, etc.

DeSauty's bridge, shown in Fig. 3-74, is a simple Wheatstone network in which capacitors may be compared in terms of a resistance ratio. It should be noted that the loss angles of the two capacitance arms must be equal, so that a series resistor is inserted in the branch with the smaller loss angle. In the case illustrated, the resistance S is in series with the reference capacitor C_s. At balance $C_x = C_s(B/A)$, and $\tan \delta_x = \omega C_x r_x = \omega C_s(r_s + S) = \tan \delta_s + \omega C_s S$.

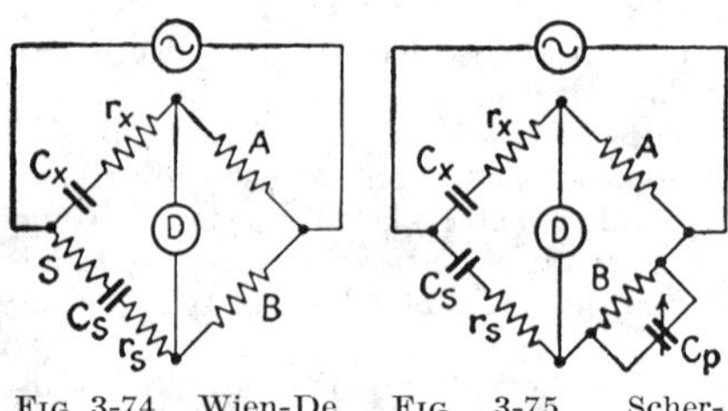

Fig. 3-74. Wien-De Sauty bridge.

Fig. 3-75. Schering's bridge.

Schering's bridge, shown in Fig. 3-75, has found wide application in measuring the loss angles of high-voltage power cables and high-voltage insulators. For this purpose, the supply voltage is connected as shown, and a ground connection is made at the junction of branches A and B, so that the balance adjustments may be made close to ground potential. The adjustable components are generally A and C_p. It is also customary to enclose the A, B, and detector branches in a grounded screen and to protect this low-voltage section against possible breakdown of the test specimen by an air gap paralleling branch A. Such a gap can be set to spark over at 100 V or so and provides a low-resistance path to ground for breakdown current from the specimen. The balance equations are $C_x = C_s(B/A)(1 + \tan \delta_s \tan \delta_p)$ and $\tan \delta_x = \omega C_p B + \tan \delta_s$. Usually the reference capacitor C_s is a high-voltage air or compressed-gas capacitor with a negligible phase-defect angle, in which case the correction terms to the balancing equation drop out. The Schering bridge is also an excellent one to use for the comparison of capacitors at low voltage. For this purpose, it is used in its conjugate form with supply and detector branches interchanged to increase sensitivity. C_p must, of course, be connected across branch A instead of B if the loss angle of C_s is greater than that of C_x, with a corresponding modification of the balance equations. When the loss angles of C_s and C_x are both very small, adjustable capacitors must be connected across both A and B arms and the difference in the phase-defect angles they introduce into the bridge must equal the difference in loss angles of C_s and C_x. This modification of the bridge is made necessary by the fact that the capacitance of an adjustable capacitor cannot be reduced to zero in the usual construction.

The *transformer bridge* has been developed recently into the most precise tool available for the comparison of capacitors, especially for three-terminal capacitors with complete shielding. A three-winding transformer is used so that the bridge ratio is the ratio of the two secondary windings of the transformer which are of low resistance and uniformly distributed around a toroidal core to minimize leakage reactance. A stable ratio, known to better than 1 part in 10^7, can be achieved in this way. For details of such a transformer construction and the construction and shielding of bridge components, the reader should consult the following papers:

Thompson, A. M. Precise Measurement of Small Capacitances; *Trans. IRE,* Instrumentation, December, 1958, Vol. I-7.

McGregor, M. C., et al. New Apparatus at NBS for Absolute Capacitance Measurement; *Trans. IRE,* Instrumentation, December, 1958, Vol. I-7.

Cutkosky, R. D., and Shields, J. Q. Precise Measurement of Transformer Ratios; *Trans. IRE,* December, 1960, Vol. I-9.

A variety of schemes for balancing adjustment have been used successfully. One of these, employing *inductive voltage dividers,* is shown schematically in Fig. 3-76, but simplified by omitting the necessary shielding. Current in phase with the main current is injected at the junction between the capacitors being compared, C_1 and C_2, to balance their inequality in magnitude. This current, through capacitor C_5, is controlled by adjusting the tap position on inductive voltage divider B, supplied from an appropriate tap point on the main transformer-ratio arm. Quadrature current, to balance the phase difference between C_1 and C_2, is similarly injected through R and the current divider $C_3/(C_3 + C_4)$, controlled by adjusting the tap point on divider A. The current divider is used so that R may have a reasonable value, a few megohms at most. In the illustrated network, it is assumed that $C_1 > C_2$ and that $\delta_1 < \delta_2$. The balance equations are $C_2 = C_1 + N_B C_5$ and $\delta_2 = \delta_1 + \omega R C_1 \cdot N_A \cdot C_3/(C_3 + C_4)$, where N_B is the fraction of the voltage across C_2 which is impressed on C_5, that is, the product of the tap-point ratios of the main transformer and divider B_5 and N_A is the corresponding fraction of the voltage across C_1 which is impressed on R. The reactance of C_3 and C_4 in parallel must be small compared with the resistance of R.

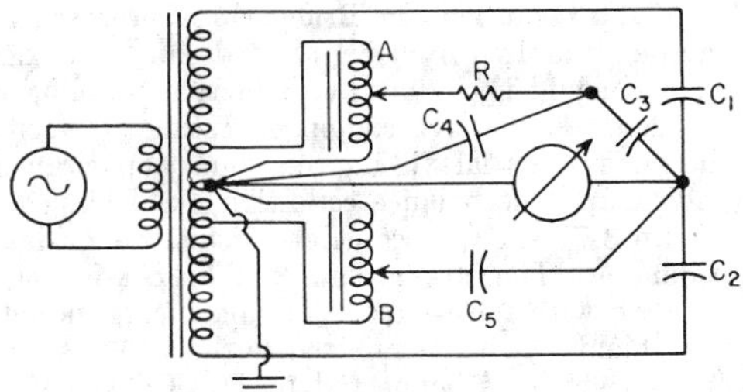

FIG. 3-76. Transformer bridge for capacitor comparison.

176. Detectors used in bridge measurements are selected with regard to frequency and impedance. *Vibration galvanometers* can be used at power frequencies in low-impedance circuits; they discriminate well against harmonics and have high sensitivity, but they must be tuned sharply to the use frequency.

Wave analyzers, which are commercially available with internal crystal control, also have a narrow passband and a high rejection of frequencies on either side. They can be used with a preamplifier when maximum sensitivity is required, and it is desirable that the preamplifier itself be sharply tuned in its first stage to improve noise rejection. This system can be used at any frequency throughout the audio region.

Cathode-ray oscilloscopes of adequate sensitivity (or used with tuned preamplifiers) make particularly good null detectors. If a phase-adjustable voltage from the bridge supply is impressed on the horizontal plates and the unbalance signal in the detector branch impressed on the vertical plates, the resulting Lissajous figure is an ellipse which, with proper phase adjustment, will change its opening with magnitude adjustment and the slope of its major axis with quadrature adjustment in the bridge. Balance is indicated by a straight horizontal trace on the screen. It is essential in this system that the initial stages of amplification be sharply tuned or that the bridge input be sinusoidal, for otherwise the pattern on the screen is confused and difficult to interpret. Phase discrimination of this type in the null detector is of considerable value in achieving balance, as it informs the operator of the individual magnitudes of inphase and quadrature unbalance.

Telephone receivers may be used at audio frequencies (maximum sensitivity being at about 1 kHz), but their response is usually quite broad, and the balance point may be masked by the presence of harmonics.

In *resonance methods* at radio frequencies, a thermocouple ammeter can be employed

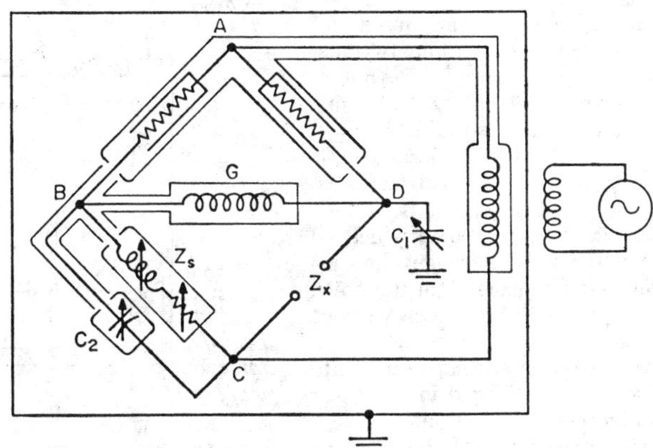

FIG. 3-77. Completely shielded bridge.

to show the current maximum. A crystal rectifier with an electronic voltmeter is used at ultrahigh frequencies.

177. Precautions in Bridge Measurements. The effect of stray magnetic fields can be minimized by using twisted-pair or coaxial leads and by avoiding loops in which an emf could be induced. Inductive coupling between bridge components should be avoided. Capacitive coupling existing between parts of the bridge which are not at the same potential will impress shunt capacitance across one or more of the bridge arms and modify the balance condition. Shielding must be used to minimize these effects. Figure 3-77 shows a completely shielded bridge suitable for comparing inductive impedances. The ratio arms AB and AD are enclosed in metal shields connected to their junction with the source. The balancing inductance and resistance of Z_s are enclosed in a shield which is connected to the source at terminal C. The detector G is shielded, and its shield is connected to D. Definite fixed shunting capacitances are thereby established in these arms. The capacitance between ratio-arm shield and reference-standard shield is eliminated by enclosing the former in a second shield which is extended over the secondary winding of the supply transformer and connected to C. Thus, the entire bridge is encased in a grounded shield; D is connected to ground through an adjustable capacitor C_1, and a shielded balancing capacitor is connected across arm BC. It is often required that the terminals B and D shall not be connected directly to ground, but that they must be maintained at ground potential. *Auxiliary Wagner arms* are used for this purpose. In Fig. 3-78 impedances M and N are connected in series across the source, with their junction point a grounded. If their ratio is such that $M/N = A/B$, the junction points b and c will also be at ground potential and admittances to ground from the other two bridge corners will shunt the auxiliary Wagner arms rather than the working arms of the bridge. The ratio M/N must be balanced for phase as well as magnitude with the main bridge arms. This balance can best be achieved by connecting the detector temporarily between a and b or c.

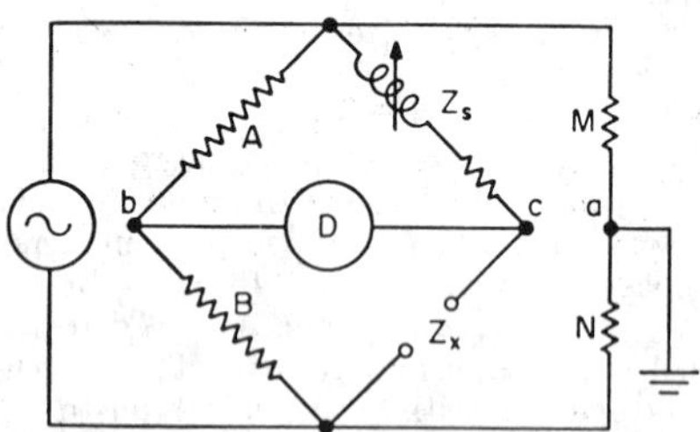

Fig. 3-78. Illustrating the Wagner arms.

Substitution methods should be used where possible to eliminate systematic errors in ratio arms and adjustable balancing components, as well as the effects of stray admittances and the residuals in arms that are unchanged by the substitution. The technique used is essentially a difference measurement, and substitution methods are particularly effective when the known standard that is substituted has very nearly the same value as the unknown.

178. Resonance methods are used for capacitance measurements at radio frequencies, a coil of known inductance L (microhenrys) at a known wavelength λ (m) being employed. Resonance is produced by varying λ and is detected with a thermocouple ammeter. At resonance $C_x = 0.2815\lambda^2/L$ in picofarads. λ and L need not be known if a substitution method is used in which an adjustable capacitor with a range that includes C_x is connected in place of the unknown capacitor and adjusted to resonance without altering the frequency, so that $C_x = C_s$. The leads used to connect the capacitors into the circuit must not be changed in length or position in making the substitution.

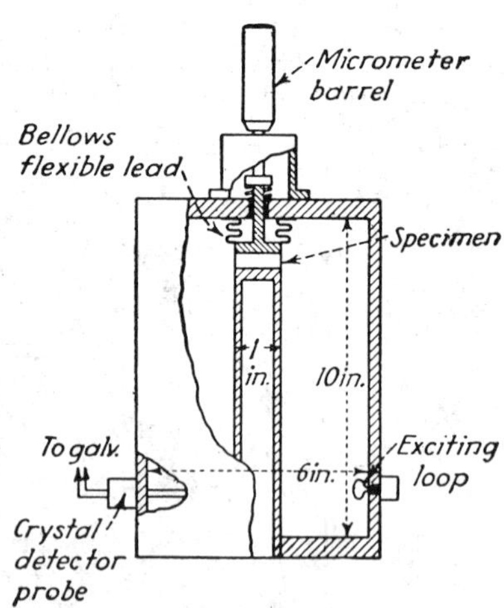

Fig. 3-79. Diagram of a cavity resonator for about 200 Mc.

179. A cavity resonator can be used at

frequencies of the order of 200 to 1,000 MHz for measuring the characteristics of insulating material placed between electrodes within the cavity. The arrangement is shown in Fig. 3-79. Resonance is established with excitation of a small loop of wire within the cavity by connection to an oscillator, and resonance is shown by a crystal-rectifier probe connected to an electronic voltmeter.

180. The impedance method may be used for determining capacitance if proper precautions are observed. Figure 3-80 shows the circuit. The voltmeter must be an electrostatic or electronic voltmeter whose capacitance is negligible compared with the capacitance to be measured; the ammeter should be of the thermocouple type to avoid resonance errors that might be present if an electrodynamic instrument were used; the waveform must be sinusoidal, or the value is only an "apparent" capacitance; the power factor of the capacitor should be low. The capacitance is

$$C = \frac{10^6 I}{2\pi f E}.$$

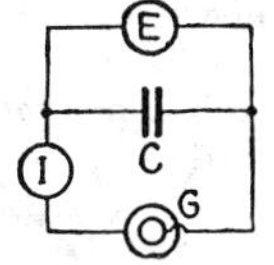

Fig. 3-80. Impedance method for measuring capacitance.

181. The ballistic galvanometer method is perhaps the simplest way of measuring capacitance. The capacitor is charged to a known voltage E_x and discharged through the galvanometer. The operation is repeated with a known capacitor C_s charged to the same voltage or another voltage E_s. The deflections are d_x and d_s, respectively, and should be nearly equal for best precision. Then, the unknown capacitance is $C_x = C_s E_s d_x / E_x d_s$. The method is best suited to relatively large capacitances.

INDUCTIVE DIVIDERS

182. General. Inductive dividers are employed in precise voltage- and current-ratio applications. Their ratios are used for comparing impedances and for calibrating devices with known nominal ratios such as other dividers, synchros, and resolvers. A divider usually consists of an autotransformer adjustable in decade steps. Such transformers, with high ratio accuracy for voltage or current comparison, have been made by using high-permeability magnetic-core materials and ingenious winding and connection techniques. Such a transformer can be represented electrically by the equivalent circuit of Fig. 3-81. The components of this circuit can be measured directly and will predict the performance of the divider. D is a perfect divider with infinite input impedance and zero output impedance. A' is the transfer ratio of D, the ratio of the voltage between the open-circuited *tap point* and the *low* end to the voltage between the *high* and *low* ends. A' is also the ratio of the short-circuit current between the high and low ends to the current into the tap point and out of the low point. Z_{oc} is impedance between high and low points, with the tap point open-circuited. This impedance is quite high and is a function of input voltage and frequency primarily. Its major components are the winding capacitance, charging inductance, and leakage reactance in parallel. Z_{sc} is the impedance between tap and low points, with the high and low points short-circuited together. This impedance is quite low and is a function of frequency and setting. Its major components are winding and contact resistances and the leakage inductance in series. The autotransformer configuration can produce voltage and current ratios of very high accuracy.

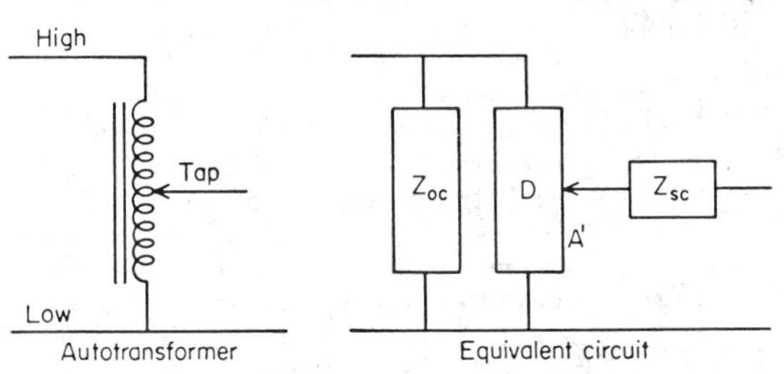

Fig. 3-81. Autotransformer and equivalent circuit.

183. Voltage-divider Operation. As a voltage divider, the circuit can be represented by a Thévenin equivalent consisting of a zero-impedance generator, with a voltage which is the product of the input voltage times the transfer ratio, and an output impedance equal to Z_{sc}. This low-output impedance provides high accuracy even with appreciable load admittance. For example, a 5,000-Ω load will change the output voltage by only 0.1% if the output impedance is 5 Ω.

184. Current-divider Operation. As a current divider, the circuit can be represented by a Norton equivalent consisting of an infinite-impedance generator, with a current which is the product of the transfer ratio A' and the input current and an impedance equal to Z_{oc} in shunt across the output. This high-shunt output impedance provides high accuracy even with appreciable load impedance. For example, a 500-Ω load will receive a current only 0.1% less than short-circuit current if the output impedance Z_{oc} is 500,000 Ω.

185. Impedance comparison, using the comparison ratio $A'/(1 - A')$, can be accomplished in either the voltage mode of operation or the current mode of operation, as shown in Fig. 3-82. For impedance comparison, the divider impedances Z_{oc} and Z_{sc}

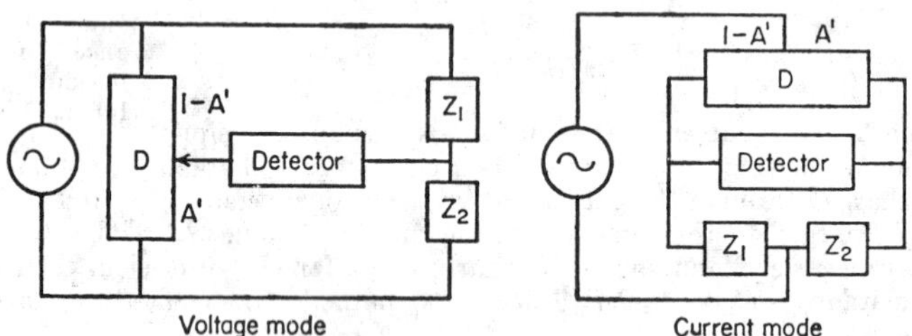

Fig. 3-82. Voltage mode (left) and current mode for impedance comparison.

are of no consequence. In the voltage-ratio mode, Z_{oc} is outside the bridge circuit, and at null no current is drawn through Z_{sc}. In the current-ratio mode, Z_{sc} is outside the bridge circuit, and at null there is no voltage across Z_{oc}. In either mode, the balance equation is $Z_2 = Z_1A'/(1 - A')$.

186. Quadrature Transfer Ratio. A well-designed autotransformer type of inductive divider has very little phase difference between its input and output signals in most applications. Impedances being compared and other circuits being measured do, however, form divider circuits which may have appreciable phase difference between input and output signals. For such bridge circuits, a quadrature component B must be added to the transfer ratio to obtain a balance, as indicated in Fig. 3-83. The balance equation is $Z_2 = Z_1(A' + jB)/(1 - A' - jB)$. This quadrature voltage may be developed

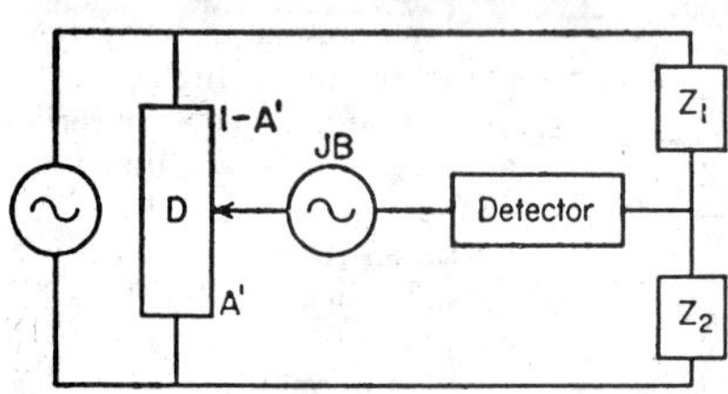

Fig. 3-83. Quadrature transfer ratio.

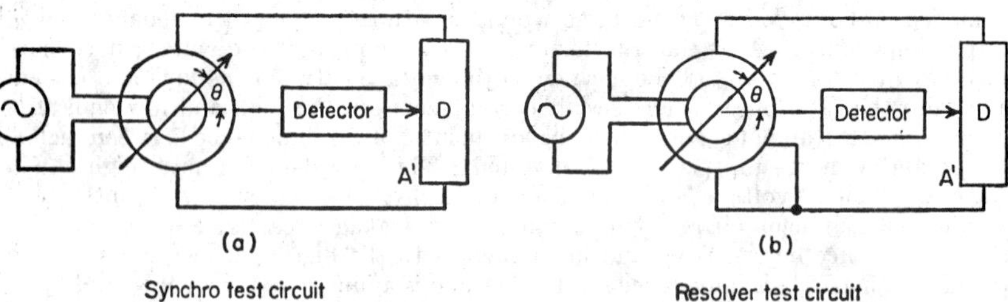

Fig. 3-84. Synchro and resolver calibration.

through an appropriate phase-shifting network or injected from a calibrated-divider source through a suitably shielded transformer.

187. Synchro and resolver calibration can be accomplished by supplying a signal to the rotor and measuring the ratio of the resulting stator voltages. The synchro test circuit is shown in Fig. 3-84*a* and of the resolver in Fig. 3-84*b*. For these applications, the divider impedances can usually be ignored, but for measurements of the highest accuracy and at high frequencies, Z_{oc} may cause measurable loading on the stator circuit supplying it.

188. Calibration of other dividers may be checked in terms of a reference divider by using the circuit of Fig. 3-85. A calibrated inductive divider D and a calibrated source of quadrature voltage B are required to give a null for the desired setting on the unknown divider. The balance equation is $Az = A + a + j(b + B)$. Corrections for the inductive divider are given in terms of the linearity deviation a, the quadrature deviation b, and the nominal ratio A (usually the dial setting); thus, $A' = A + a + jb$.

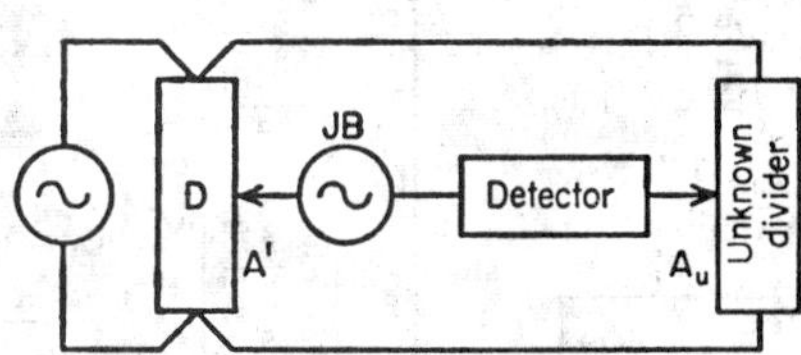

FIG. 3-85. Divider calibration.

189. Specifications of the performance of an inductive voltage divider are stated in terms of the characteristics of its equivalent circuit.

a. Ratio accuracy is $A' = A + a + jb$, where A is the nominal ratio given by the dial setting, usually corresponding to the turns ratio of the autotransformer.

b. Voltage limits are imposed by (1) the input voltage at which the core saturates and the divider output becomes distorted and (2) the ability of the winding insulation to withstand breakdown.

c. Operating Frequency. The voltage-frequency limit and the Z_{oc} and Z_{sc} impedances can be varied by core and winding choices so that dividers can be designed for optimum operation in different frequency bands and at different excitation levels. Typical values of the various characteristics are given in the table for two designs which give their best performance at the frequency stated.

Frequency	Accuracy		Z_{oc}			Z_{sc}		Volt-frequency limit
	a_{max}	b_{max}	R	L	C	R_{max}	L_{max}	
60 Hz	10^{-6}	......	1 MΩ	2 kH	10 nF	5 Ω	500 mH	2.5 × *f*
1 kHz	10^{-6}	10^{-5}	500 kΩ	200 H	1 nF	3 Ω	100 mH	0.35 × *f*

WAVEFORM MEASUREMENTS

190. Methods. The instantaneous variations of current and voltage in a circuit can be measured by oscillographs, whose basic operating principle may be either that of a D'Arsonval galvanometer whose inertia is low enough to permit it to follow the variations or that of an electron beam which has no sensible inertia and whose deflection is governed by electric or magnetic fields. In addition to tracing waveforms, oscillographs are used for measurements of transient phenomena, such as those which occur in switching operations or in the impulse-voltage testing of insulating structures and disturbances resulting from lightning discharges.

191. The galvanometer oscillograph may have a light, low-inertia coil or, for higher frequency response, a pair of thin metal ribbons tightly stretched across insulating bridges and tied together by a small mirror at their midpoints, mounted in the field of a permanent magnet. A light beam from the galvanometer mirror traces its response to varying current on a moving photographic film or, by means of an intermediate rotating mirror, on a stationary viewing screen. Galvanometer elements have been built with natural response frequencies as high as 8 kHz (a more common construction has

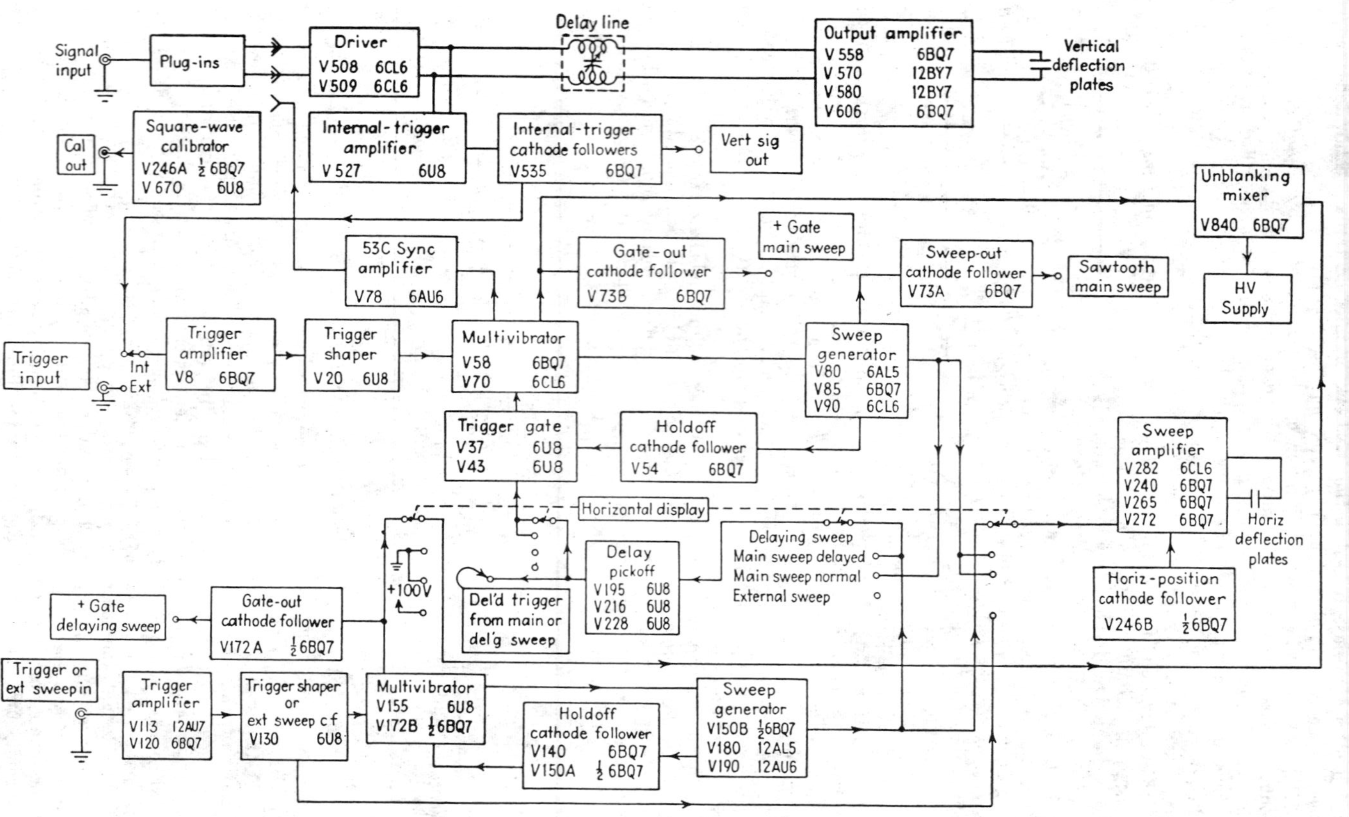

FIG. 3-87. Schematic diagram of tektronic oscilloscope.

a resonance frequency of about 3 kHz) and, if damped at about 0.7 of critical, have a response to signals which is practically free from distortion up to about half their resonant frequency; at resonant frequency, the deflection sensitivity has decreased to about 70% of their d-c sensitivity for this damping.

192. Cathode-ray-oscillograph tube elements are shown in Fig. 3-86, and Fig. 3-87 is a block diagram of the circuitry of a modern oscilloscope. Electrons emitted by a heated cathode are accelerated toward an anode which has a small central aperture. A grid system serves to concentrate the electrons into a narrow beam and focus it into a small spot on a fluorescent screen at the far end of the tube. The grid nearest the cathode controls the beam current and the brightness of the spot on the fluorescent screen. After going through the focusing electrode system, the beam passes between two plate pairs set at right angles to one another and is deflected by the electric fields impressed on the plates. If a saw-tooth voltage is impressed on the plate pair controlling horizontal deflection and is synchronized with an a-c signal impressed on the vertical-deflection plate pair, the waveshape of the signal is traced out on the fluorescent screen. Since the deflection sensitivity of this system is relatively low, the plate pairs are preceded by amplifiers which control the actual magnitude of deflections on the screen. A variety of circuits are required to control the screen displays for repetitive or nonrepetitive signals, to initiate, form, and synchronize a variety of sweep-signal voltages, etc. Modern oscilloscopes are very flexible tools for studying the nature of both continuous and transient signals and provide uniform response well into the high-frequency region, capable of reliable calibration in both voltage and frequency or time interval.

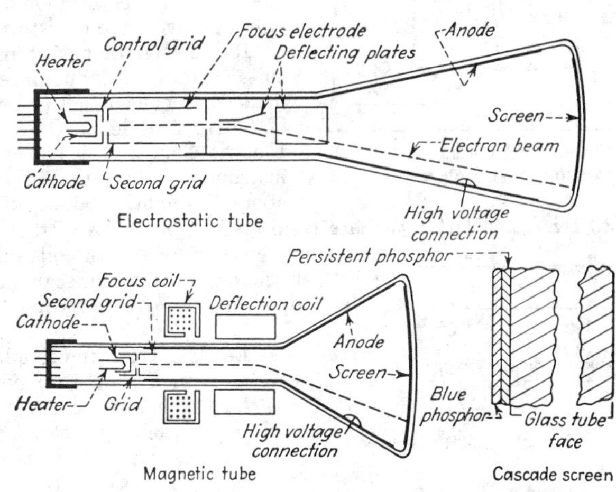

FIG. 3-86. Elements of cathode-ray oscillograph.

FREQUENCY MEASUREMENTS

193. Reed-type frequency meters have a number of steel strips rigidly fastened to a bar at one end and free to vibrate at the other. These strips are located in the field of an electromagnet which is energized from the circuit whose frequency is to be measured, as shown in Fig. 3-88. The strips have been accurately adjusted by solder weights to resonant vibration frequencies that differ by ¼ or ½ c/s, and the one with a period corresponding to the alternations of the voltage will be set into vibration. The free ends of the strips or reeds are turned up and painted white so that the reed which is vibrating will be indicated by an extended white band or blur.

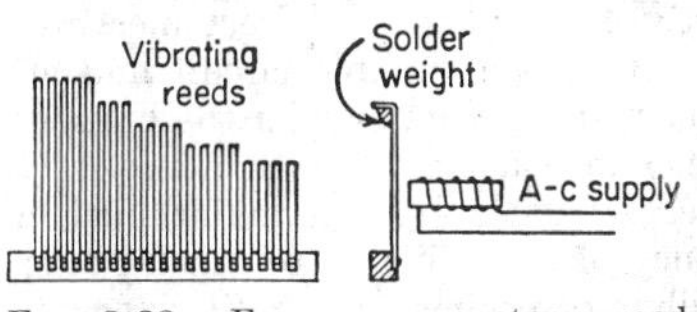

FIG. 3-88. Frequency meter, reed type.

194. The Weston frequency meter is shown in Fig. 3-89, where 1, 1 and 2, 2 are fixed coils, 90° apart, and c, c is the movable element consisting of a simple, soft-iron core mounted on a shaft, with no control of any kind. Coil 2, 2 is connected in series with a noninductive resistance R_2 and coil 1, 1 in series with an inductance X_1. A second noninductive resistance R_1 is connected in parallel with 1, 1 and X_1. A second inductance X_2 is connected in parallel with 2, 2 and R_2. The soft-iron core takes up the position of the resultant field produced by the two coils. When the frequency increases, the current decreases in 1, 1 and increases in 2, 2, thus shifting the direction of the resultant field and the position of *c*, *c* to which the pointer is attached. The opposite effect takes place when the frequency is decreased. The series inductance *X* serves merely to damp the higher harmonics which are present if the voltage waveshape is distorted.

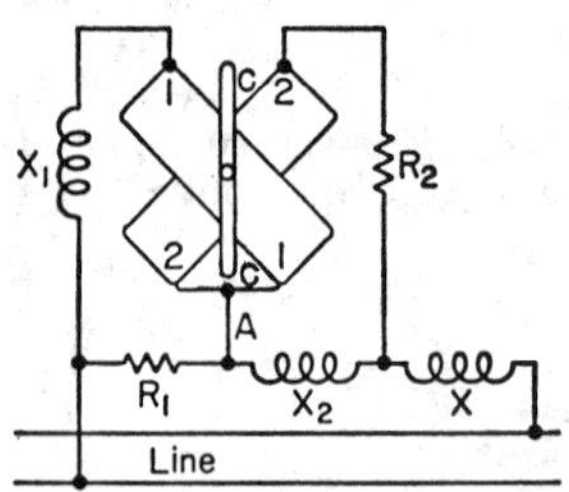

Fig. 3-89. Circuits of the Weston frequency meter.

195. Resonant circuit meters, operating from circuits containing inductance and capacitance, can be made sensitive enough to indicate frequency variations of 0.01 Hz or less. One form, used by the General Electric Company, is shown in Fig. 3-90. In a 60-Hz instrument, one circuit is adjusted for resonance at about 70, another at about 58, and a third at about 36 Hz. The latter two are connected in parallel and then to coil *A*; the first circuit is in series with coil *A′*, both coils being in series with the field *F*. With the center of a 6-in scale marked for 60 Hz, end-scale deflection is obtained for variations of 5 Hz from the central value.

196. The Leeds & Northrup frequency recorder uses a Wien-bridge network with a slide wire between the resistance ratio arms *A* and *B*, as shown in Fig. 3-91. The other two arms have capacitors C_1 and C_2, one with series and the other with parallel resistance. With a change in frequency, the reactance of one arm is increased and the other is decreased; the movement of the detector *G* along the slide wires S_1 and S_2 required to restore balance brings the recording pen to a new position corresponding to the changed frequency.

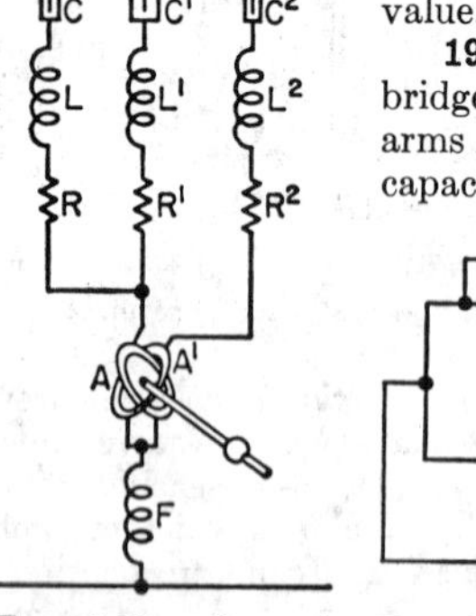

Fig. 3-90. General Electric resonating frequency indicator.

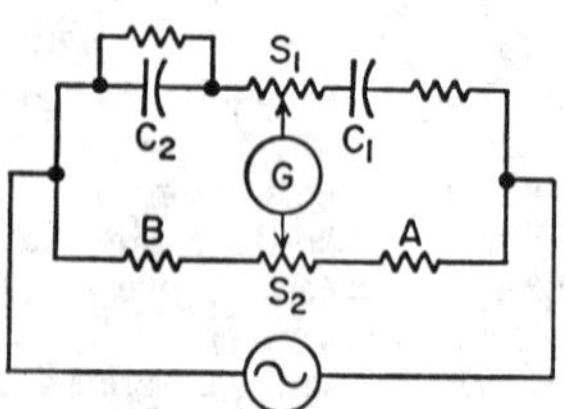

Fig. 3-91. Circuits of Leeds & Northrup frequency recorder.

197. Precise frequency control is also accomplished with resonance techniques. Small-range indicators or recorders can be built as relays to monitor the frequency of a power system or generator, injecting an appropriate signal into a control system to restore frequency to a particular value. Such control may be made precise enough for use of the system frequency for electric-clock operation. Any tendency to frequency drift may be detected and corrected at the source by comparing an electric clock with a precise pendulum clock or one driven by a quartz-crystal oscillator.

198. Radio frequencies may be measured directly or indirectly. Direct measurement may be made with a wavemeter, an instrument with a tunable circuit and an ammeter to indicate the resonance frequency by a current maximum. In the indirect method, the unknown-frequency signal is introduced into a circuit with a precisely known frequency, and the beat frequency is counted. Quartz crystals maintained in temperature-regulated ovens will control the frequency of an oscillator to much better than 1 part in 10^6. Such a crystal-controlled oscillator, serving as a local reference standard of frequency, can be monitored against the very precise *standard* frequencies continuously broadcast by the National Bureau of Standards from its low-frequency station WWVL, operated at 60 kHz or its high-frequency stations WWV and WWVH, which broadcast at a large number of higher frequencies. These broadcast frequencies are controlled by crystals operating under conditions that are most favorable to stability and are, in turn, monitored against the frequency of an atomic-beam resonator. The transmitted frequencies, as sent from the Bureau stations, are accurate to about 1 part in 10^{12}. Frequencies from these broadcasts are modified somewhat in transmission by diurnal and moment-to-moment variations in the ionosphere, and their accuracy as received may be reduced by more than an order of magnitude.

199. Audio frequencies can be measured with a frequency-sensitive bridge, such as the Wien bridge with parallel- and series-connected capacitance-resistance arms; or they can be conveniently observed with a cathode-ray oscilloscope, if a known reference frequency is available. One set of plates of the oscilloscope is excited by the known- and the other set by the unknown-frequency signals. If the two frequencies have an exact, simple fractional relation, the Lissajous figure formed on the screen is stationary. For a 1/1 relationship, the pattern is an ellipse; for other fractional relationships, the pattern is more complicated, the relationship being determined from the number of loops. If the relationship cannot be represented by a simple fraction, the pattern will change continuously and a count of the beat frequency is made over a measured time interval.

SLIP MEASUREMENTS

200. The **slip** of a rotating a-c machine is the difference between its speed and the synchronous speed, divided by the synchronous speed; slip is usually expressed in percent. It may be computed from the measured speed of the machine and the synchronous speed, but direct methods are more accurate.

201. Millivoltmeter Method. If sufficient stray field is produced by the current in the secondary of an induction motor, a d-c millivoltmeter connected to an adjacent coil of wire or across the motor shaft or frame will oscillate at slip frequency, each swing being one pole slip. In motors with wire-wound rotors the millivoltmeter may be connected across the rotor slip rings.

202. Dooley's Method of Measuring Slip. One form of device for indicating the slip of an induction motor is shown diagrammatically in Fig. 3-92. A small cylinder made of conducting material and in two parts, each insulated from the other, is mounted in a frame. Four small brushes, 1, 2, 3, and 4, bear upon the cylinder as shown. Brushes 3, 4 are connected through resistance r across one phase of the supply circuit, and brushes 1, 2 are connected to a low-reading d-c ammeter I. Each time the brushes 1, 2 bridge the insulating strip as the cylinder rotates, the circuit is completed in alternate directions through the ammeter. The cylinder should have as many segments as the motor has poles. The ammeter will indicate a constant current at synchronous speed and an oscillating current for any speed above or below synchronism, because the impulses of current through brushes 1, 2 will occur at the same point on the wave at synchronous speed and at constantly advancing or retarding points for other speeds. Thus, the ammeter will be reversed each time the motor loses one-half cycle and will reach a maximum positive value each time the motor loses one complete cycle. If the motor loses n c/min, then the slip in percent $= 100n/60f$, where f = frequency of the system in cycles per second.

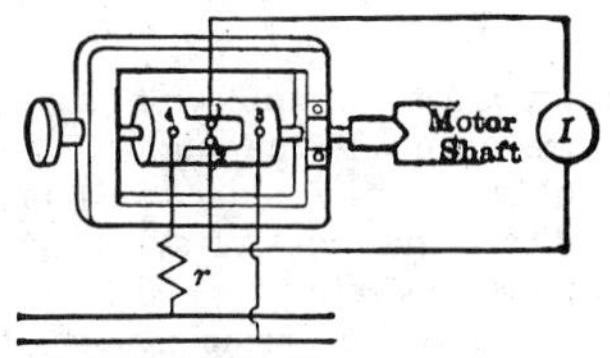

Fig. 3-92. Slip-measuring device.

203. Stroboscopic Method. In Fig. 3-93 a black disk with white sectors, equal in number to the number of poles of the induction motor, is attached to the induction-motor shaft. It is observed through another disk having an equal number of sector-shaped slits and carried on the shaft of a small self-starting synchronous motor, in turn fitted with a revolution counter which can be thrown in and out of gear at will. If n is the number of passages of the sectors, then $100n/n_sn_r$ = slip in percent, where n_s = number of sectors and n_r = number of revolutions recorded by the counter during the interval of observation. For large values of slip the observations can be simplified by using only one sector ($n_s = 1$); then n = slip in revolutions.

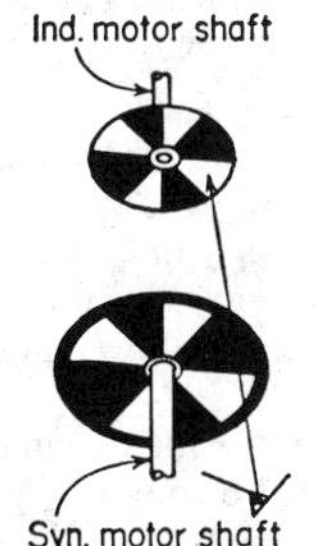

Fig. 3-93. Slip measurement by stroboscopic method.

With a **synchronous light source** to illuminate the target on the induction-motor shaft, the synchronous motor is no longer necessary. An arc lamp connected across the a-c supply may be used, but the carbons must be readjusted from time to time. A neon lamp makes a satisfactory source of light when the general illumination is not too bright.

204. Synchronizing. In order to connect any synchronous machine in parallel with another machine or system, the two voltages must be made equal and the machines must be synchronized, i.e., the speed so adjusted that corresponding instantaneous values on the two waves are reached at the same instant, when they will be in exact phase. Furthermore, with polyphase machines, the direction of phase rotation must assuredly be the same. This, however, is usually made right once and for all when the machines are installed, the phases being so connected to the switches that the phase rotation will always be correct.

205. The lamp method of synchronizing is the simplest. The principle of lamp synchronizers is shown in Fig. 3-94 where a, a_1 are the sources being connected in parallel and t, t_1 are transformers, the secondaries of which are connected in opposition through incandescent lamps l, l'. When the two sources are in synchronism, the secondary emfs neutralize each other, and the lamps will be "dark." As the phase difference increases, the current through the lamps will increase, reaching a maximum at 180° of phase difference. If the machines run at different speeds, the lamps will "flicker." If the secondary of one transformer is reversed, the lamps will be brightest at synchronism and dark at 180° of phase difference. The former connection is preferable, because the point of total "darkness" is more easily detected than the point of maximum "brightness." A voltmeter may be substituted for the lamps by connecting it so that synchronism is indicated when the reading is a maximum. The disadvantage of this method is that it does not indicate which frequency is the higher. **Synchronism indicators** are instruments which not only overcome this objection but indicate the point of synchronism more accurately.

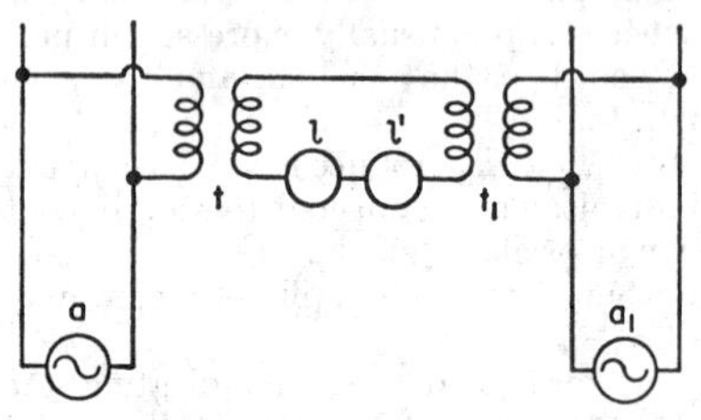

FIG. 3-94. Connections for synchronizing with lamps.

206. The principle of the Westinghouse synchroscope is shown in Fig. 3-95, where a rotating field is produced by the coils M and N connected to the buses through the reactance P and the resistance Q, respectively. An iron vane A, free to rotate, is mounted in this rotating field and magnetized by the coil C, which in turn is connected across the incoming machine. As the vane is attracted or repelled by the rotating field from M and N, it will take up a position where this field is zero at the same instant that the field from C is zero. Hence the position at any instant indicates the difference in phase. When the two frequencies are different, this position is constantly changing, and the pointer will rotate in the "fast" or "slow" direction, coming to rest again at the zero-field position when the frequencies are equal. In a larger type, the split-phase winding is placed on the movable member, similar to the arrangement shown in Fig. 3-96, which represents the General Electric synchroscope.

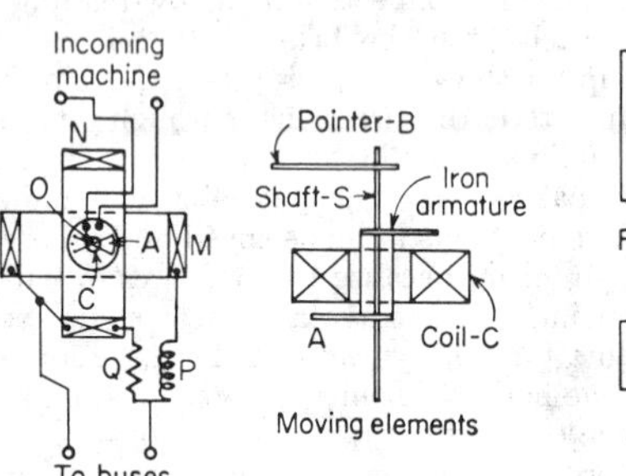

FIG. 3-95. Circuits of Westinghouse synchroscope.

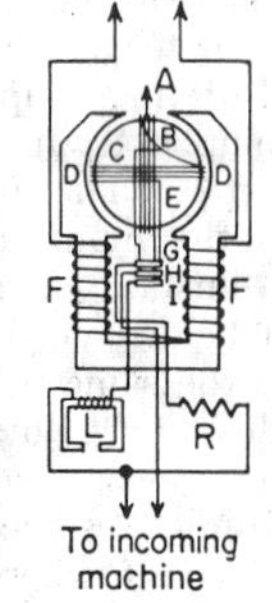

FIG. 3-96. Circuits of General Electric synchroscope.

207. In the Weston synchroscope there is no iron in the instrument, and the moving element is not allowed to rotate. The elements are practically the same as in an electrodynamometer wattmeter. The fixed coils are connected in series with a resistance and to the buses. The moving coil is connected in series with a capacitor and the incoming machine. The two circuits are adjusted to exactly 90° difference in phase.

At synchronism there is no torque, and the moving coil is held at the zero position by the control spring. If the frequencies are the same but there is a phase difference, a torque will be exerted and the coil will move to a position of balance at the right or left ("fast" or "slow"). If the frequencies are different, the torque will continually vary and the pointer will oscillate over the dial. A synchronizing lamp illuminates the scale simultaneously.

208. The phase sequence of a 3-phase system is often desired. Figure 3-97 shows two lamp methods. In I, two lamps and a highly reactive coil, such as the potential coil of a watthour meter, are used. The bright lamp indicates the particular phase sequence. In arrangement II, a noninductive resistance and a reactive coil of equal impedance are used in conjunction with a lamp, the brightness of which indicates the sequence.

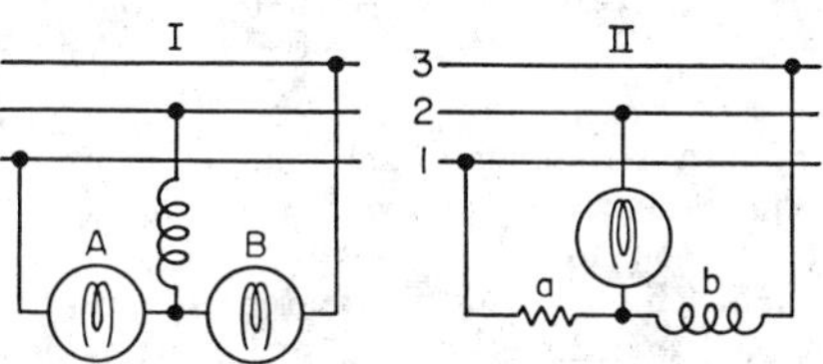

FIG. 3-97. Phase-sequence indicators.

MAGNETIC MEASUREMENTS

209. The intensity of a magnetic field H at a point is equal to the magnetic potential gradient at that point (see Sec. **2**); the **average** intensity of field between two points may be considered to be the average fall of magnetic potential along the path between the points and is expressed in oersteds or in gilberts per centimeter. In a long, straight solenoid $H = 4\pi NI/10l$, where I = current in amperes and N/l = turns per centimeter length. **Magnetic induction,** or **flux density,** B is measured in lines of magnetic induction per square centimeter (gausses). When the substance in which the field exists is nonmagnetic, $B = H$, and the ratio B/H, or **permeability,** is $\mu = 1$. When the substance is magnetic, B becomes much greater than H, owing to the decreased magnetic reluctance; in general, B varies with H, but the relationship is seldom a linear one and must be determined experimentally for a given material.

210. The normal induction, or **B-H, curve** of a magnetic material is the curve plotted between the strength of the magnetic field H existing in the material and the magnetic induction B produced by that field when the material is in a neutral or normal condition. A **permeability curve** is plotted between the permeability μ and B or between μ and H.

211. The hysteresis curve is a curve plotted between B and H for various values of H from a maximum value in the positive direction to a maximum value in the negative direction and back again, or through a complete cycle of values. The ends of a hysteresis loop will lie in the normal induction curve.

212. Magnetic measurements may be divided into two classes: (1) those in which the strength of a magnetic field is determined (such as the earth's field, the field due to a conductor carrying a current, the field in the air gap of a magnet), and (2) those made to determine the properties of magnetic material.

213. Field-strength measurements may be made by induction methods, with an oscillating bar magnet (Par. **216**), with a bismuth spiral (Par. **217**), or with the magnetron (Par. **218**). In the induction method a coil of known turns and area is so arranged that it can be made to cut the field in the desired region in a direction perpendicular to the field. The emf generated in the coil, and hence the field producing it, is determined from the quantity of electricity discharged through a ballistic galvanometer connected to the coil terminals.

In measuring the field strength in air gaps, the coil is so arranged that one side can be moved quickly across the entire gap, or through a definite distance, by means of a spring or weight suddenly released. When there is sufficient space, a more accurate and convenient method is to rotate the coil through 180°. Very weak fields, such as the earth's field and that due to conductors, may be measured in this manner by using a sufficiently large coil of many turns.

214. Formula for Field Strength by Induction Method. The flux density in lines per square centimeter = B; by definition $B = H$ in air; hence

$$H \text{ (field strength)} = \frac{10^8 dkR}{xan} \qquad \text{(Oe)}$$

where d = deflection, k = constant of galvanometer in coulombs per scale division, R = total resistance in galvanometer circuit, n = turns in coil, a = mean area of coil in square centimeters, $x = 1$ when coil is removed from field, $x = 2$ when coil is rotated through 180° in field.

215. The Grassot fluxmeter is a portable instrument with which magnetic flux may be read directly on a scale. It is essentially a ballistic galvanometer with the moving element suspended by a torsionless suspension so that it remains deflected. The instrument is connected to an exploring coil which is placed in the air gap to be measured. The flux is measured by noting the deflection when the magnetizing circuit is either opened or closed or when the coil is moved as described in Par. **213.**

216. In the oscillating-magnet method, a small, simple bar magnet is suspended by untwisted silk fibers. The magnet is set to vibrating through an angle of about 5°, and the period of oscillation determined. The average of at least three observations should be taken. The field strength is

$$H = \frac{4\pi^2 K}{MT^2} \qquad \text{(Oe)}$$

where K = moment of inertia computed from the mass and dimensions, M = magnetic moment, and T = period of oscillation. M may be determined with a magnetometer or by calibration in a known field (Helmholtz coil). This method is suitable only for weak fields, such as the earth's field. If mounted in a wooden box with a glass front, the apparatus will be protected from air currents and can be conveniently moved about when making magnetic surveys.

217. Bismuth-spiral Method. The resistance of bismuth wire increases when placed in a magnetic field. This property is utilized by noting the increase in resistance of a flat spiral coil of bismuth wire when placed in the field to be measured, the leading-in wires being arranged noninductively. The device is calibrated with known field strengths. It is particularly suitable for exploring small air gaps, such as those in motors and generators. Bismuth has a considerable resistance-temperature coefficient which must be taken into account (see the Smithsonian Tables for values).

218. Magnetron Method. A magnetron is a vacuum tube with a cylindrical anode and an axial wire cathode. When the magnetron is placed in a magnetic field of which the direction is parallel to the axial cathode, the paths of the electrons become curvilinear, and if the field is of sufficient intensity, the radius of curvature of the paths becomes so small that the electrons miss the anode altogether and return to the cathode. The plate current, which has hitherto remained practically constant as the field intensity increased, now falls abruptly to zero. The abruptness of the drop depends on the degree of vacuum attained and on the approach to perfect symmetry in the electrodes; a change in H of about 10% is usually required to reduce the current from maximum value to zero. The critical value of magnetic field is given by the following equation:

$$H = \frac{6.72 E^{1/2}}{R} \qquad \text{(Oe)}$$

where E = voltage between anode and cathode and R = radius of anode cylinder in centimeters. The plate current, with no field, is given by

$$I = 14.7 E^{3/2} \frac{L}{R} \qquad (\mu\text{A})$$

where E = voltage between anode and cathode, L = length, R = radius of anode in centimeters or inches. The useful range is approximately 20 to 200 Gb/cm. For weaker fields the tube is placed within a coil carrying a constant current which produces a known field intensity; the weak field is then measured by difference. In this way a field of $H = 0.01$ Oe may be detected without difficulty (see Albert W. Hull, Measurement of Magnetic Fields by Means of a Magnetron; *Phys. Rev.*, 1923, Vol. 22, p. 279).

219. Measurement of Magnetic Properties. The magnetic properties of iron and steel which are of most commercial importance are **normal induction,** or **permeability, hysteresis loss,** and **total losses** (**core loss**) with alternating magnetizing forces of commercial frequencies.

220. Normal-induction Data. The various methods are distinguished principally by the method employed to measure B, for in most methods H is determined from the magnetizing coil. B can be measured directly by ballistic methods as in the **fluxmeter permeameters** or indirectly as in the **traction permeameters.** The fluxmeter method is usually employed in the more accurate measurements. The better known are the **ring method, Burrows' permeameter,** and **Fahy's permeameter.** In all these methods the flux is measured with a ballistic galvanometer connected to a test coil which is cut by the flux when the exciting current is reversed.

221. The ring method, devised by Rowland, is one of the earliest methods of measuring the magnetic properties of iron and is still used for small specimens which cannot be machined into rods or punched into strips and for specimens to be tested at audio frequencies (see Par. **232**). The connections are shown diagrammatically in Fig. 3-98, where T is the test specimen. The latter is an annular ring, either solid or built up of punchings of sheet metal, with a diameter preferably 8 or 10 times the radial thickness. After covering with a thin layer of insulation, a test coil of very fine double-silk-covered wire is wound on a portion of the ring. The magnetizing coil is wound uniformly over the test coil and the entire ring; it is usually of magnet wire, of sufficient size to carry the maximum current without raising the temperature of the iron above 30°C. The formulas for H and B are

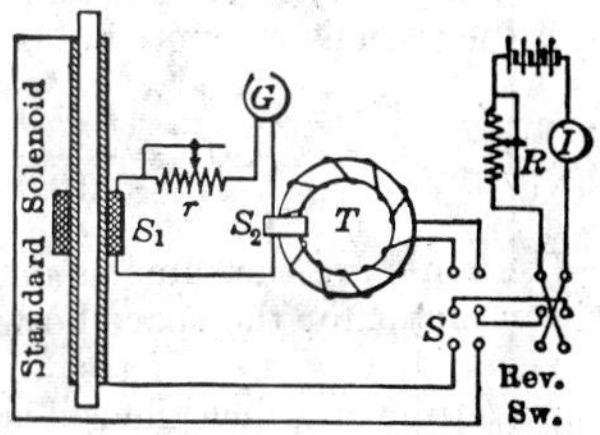

FIG. 3-98. Permeability tests, ring method.

$$(a)\quad H = \frac{4\pi NI}{10l} \qquad \text{and} \qquad (b)\quad B = \frac{10^8 dkR}{2an}$$

where N = total turns of exciting coil, I = exciting current in amperes, l = mean length of magnetic path in centimeters (NI/l = ampere-turns per centimeter), d = deflection when current is reversed, R = total resistance of test-coil circuit, k = galvanometer constant, a = area of specimen in square centimeters, n = turns in test coil. There is an appreciable error with wide rings due to differences in field intensity between outer and inner periphery.[1] If strains are likely to be set up in the specimens when prepared by punching, they must be removed by an acceptable annealing method.

222. The Burrows permeameter is a double-bar double-yoke apparatus devised by C. W. Burrows;[2] it is recognized as a permeameter of high precision for tests at magnetizing forces up to 300 Oe. The essential feature is the distribution of the magnetizing winding in sections, which permits the independent adjustment of the magnetizing force in various parts of the magnetic circuit. Thus the effect of nonuniform reluctance at joints, etc., can be overcome, and the induction made uniform throughout the entire magnetic circuit. Exploring coils are placed at various positions so that the uniformity of the induction can be tested.

The scheme for rod or bar specimens (similar also for specimens of sheets) is shown in Fig. 3-99, where b, b are two bars (standard size, 1 cm diameter and about 30 cm long), one to be tested and the other an auxiliary bar of similar material; y, y are yokes of soft iron about 15 cm long and more than 10 cm² in area, into which the bars are clamped. The mmf is applied in three sections: one coil N_t is placed over the test specimen, another N_a over the auxiliary rod, and the third N_j is divided into four parts, one near each joint. The corresponding exploring coils are n_t, n_a, and n_j. These coils

[1] LLOYD, M. G. Errors in Magnetic Testing with Ring Specimens; *NBS Bull.*, 1908, Vol. 5, p. 435.
[2] BURROWS, C. W. Determination of Magnetic Induction in Straight Bars; *NBS Bull.*, 1909, Vol. 6, p. 31.

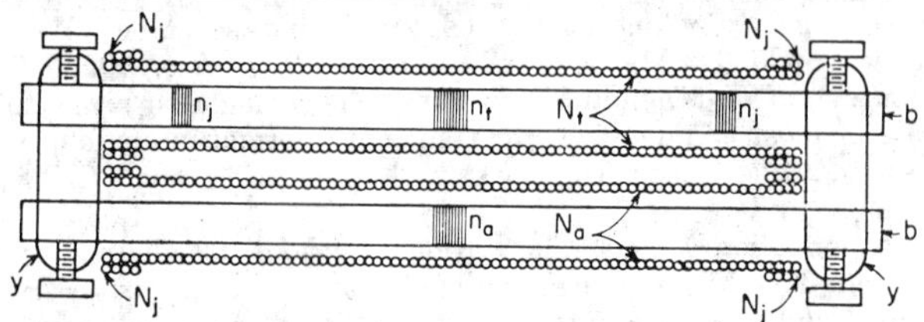

FIG. 3-99. Permeability tests, double-yoke method.

all have the same number of turns; that is, $n_t = n_a =$ the two parts of n_j in series. They are so connected to a switch that n_a or n_j may be connected through the ballistic galvanometer in opposition to n_t, thus providing a zero method of determining uniformity of flux. The magnetizing coils are connected to special reversing switches so made that they can be operated simultaneously. When uniform flux has been established, the galvanometer is connected to n_t and the deflection noted when the current in the various magnetizing coils is reversed simultaneously. Then

$$(a)\quad H = \frac{4\pi NI}{10l} \qquad \text{and} \qquad (b)\quad B = \frac{10^8 dkR}{2an} - \left(\frac{A - a}{a}\right) H$$

The units are the same as in Par. **221**. The quantity in the parentheses is the correction factor for the space between the surface of the bar and the test coil; a = area of bar, and A = area of test coil. Ordinarily, this correction is very small, because the brass tube on which the coil is wound is made very thin and the test coil is wound under the magnetizing coil.

223. The procedure of the ASTM[1] permits the main magnetizing coils N_t and N_a and the main test coils n_t and n_a to be connected in series, so that average values for the two specimens are thus obtained. Manifestly, the two specimens must be of the same material under this procedure. Four "compensating" test coils n_j are then required for opposition connection to n_t and n_a.

224. The Fahy permeameter, devised by F. P. Fahy, determines H as the difference of magnetic potential between the ends of the specimen. The Simplex form is represented in Fig. 3-100. Y, Y is a yoke of high-grade laminated steel with a magnetizing coil M to give an H up to 200 Oe. X, X is the specimen (strips of sheet material or a rod in adapters) surrounded by the test coil T of 100 turns and pressed against Y, Y by two iron pieces P, P, which are moved by jackscrews. P, P carry between them a nonmagnetic bar encased throughout its entire length by a coil H consisting of several thousand turns of fine wire. The current in M is adjusted to the approximate value of H desired (1 A is equivalent to H = 100, nearly) and then reversed, the deflection of the ballistic galvanometer connected to T being noted. The galvanometer is next connected to H and a second deflection noted upon reversal of current. B is determined from the first reading, using the equations in Par. **222**. H is determined from the second reading, using the same equation without the area-correction term. In both cases the area turns an may be determined in a standard solenoid (Par. **235**) but are usually furnished by the maker. The method is no longer accurate within 3% for materials having permeability above 8,000.

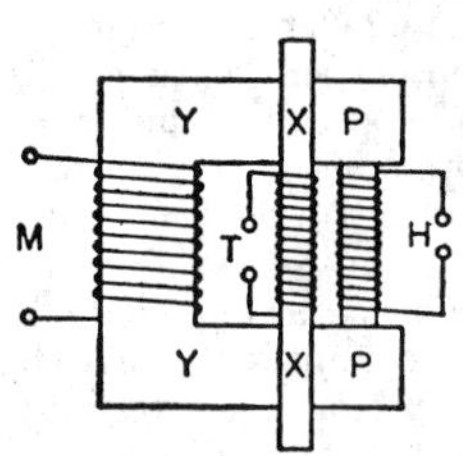

FIG. 3-100. Fahy Simplex permeameter.

225. For measurements at high field intensities (order of H = 2,000), a modification of **Ewing's isthmus method** has been found satisfactory. An apparatus used at the National Bureau of Standards[2] applies to the specimen the magnetizing force generated in two conjugate yokes, each shaped like a square U, of laminated steel, each with its magnetizing coil and provided with rectangular pole pieces which, applied

[1] ASTM Standard Method of Testing Magnetic Materials (A34-47).

[2] SANFORD, R. L., and BENNETT, E. G. Apparatus for Magnetic Testing at Magnetizing Forces Up to 5000 Oersteds; *NBS Jour. Res.*, 1939, Vol. 23, p. 415.

to the test bar, give an effective length of specimen of 5.5 cm (2.2 in). This length was found most satisfactory by trial. Field intensity is measured by a search coil close to the specimen, flux by the usual search coil surrounding the specimen, by using the ballistic method with reversal of the magnetizing current.

226. The principal advantages of traction permeameters are their ruggedness and simplicity, which are important features in shop testing where rapidity is essential and only comparative data are required. **S. P. Thompson's permeameter** is of the traction type and is shown schematically in Fig. 3-101, where AB is the test specimen in the form of a rod which passes through a hole in the top of a heavy yoke F. The surfaces of the end of the rod and the yoke at A are carefully machined. The force necessary to move the rod is measured with a spring balance at S. The induction is

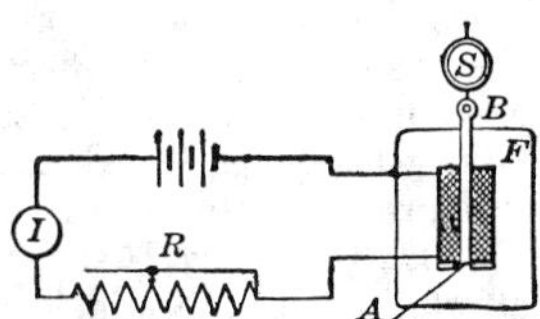

FIG. 3-101. Thompson permeameter.

$$B = 156.9\sqrt{\frac{P}{a}} + H \qquad \text{(gausses)}$$

where B = induction in lines per square centimeter (gausses), P = pull in grams, a = area of specimen in square centimeters.

227. Ballistic Step-by-step Method of Determining Hysteresis. This method is carried out as follows: The current in the magnetizing circuit is adjusted to a value corresponding to maximum B. It is then reversed a few times to get the specimen in a cyclic state. By means of suitable switching arrangements, some resistance is suddenly cut into the circuit, thus reducing H to a new value, which is carefully noted. The corresponding change in B is determined with a ballistic galvanometer in the usual manner. This process is continued, step by step, until zero current is reached, after which the circuit is reversed and the resistance is cut out, step by step, until the same maximum value of H is reached in the opposite direction. The whole process is then repeated and the other side of the loop obtained. With this method, the errors will be cumulative, but this is overcome in a second method in which the magnetizing current is brought back each time to the original maximum H; furthermore, it is then a simple matter to ensure that the conditions remain constant, by occasionally checking the value of B corresponding to maximum H.

With the value of B known at the start, the values corresponding to the various steps in the step-by-step method are obtained by adding algebraically the various changes in B to this initial value. The resulting values of B and H suitably plotted on cross-section paper form the **hysteresis loop.**

228. Other means of determining a hysteresis curve are available, but that described in the preceding paragraph is the one usually employed for quantitative measurement. The cathode-ray oscillograph will furnish **hysteresis cyclograms,** with a deflecting coil carrying the magnetizing current, and the deflecting plates are furnished with a voltage proportional to B. The hysteresis curve of sheet steel may be computed from waveforms of induced voltage and magnetizing current of a transformer in which the core is the material under test.

229. Hysteresis-loss Measurements. The area of the hysteresis loop, measured in the units of the ordinates of the curve (by planimeter or otherwise) and divided by 4π, gives the hysteresis loss in ergs per cubic centimeter per cycle, between flux densities $+B_{max}$ and $-B_{max}$. This method of measuring hysteresis loss is much too slow and expensive for commercial purposes, and several other methods have been devised by means of which hysteresis losses can be measured directly, by electrical or mechanical means.

Robinson's Method of Measuring Hysteresis Loss. An example of an electrical method of measuring hysteresis loss is one used by L. T. Robinson.[1] It was designed primarily to reduce to a minimum the time and expense in preparing the sample, as well as the time spent in testing. The specimen is a bundle of strips 0.5 in (1.27 cm) by 10 in (25.4 cm), weighing 1 lb (0.45 kg). It is placed in a simple, straight solenoid,

[1] ROBINSON, L. T. Commercial Testing of Sheet Iron for Hysteresis Loss; *Trans. AIEE*, 1911, p. 741.

with a sensitive wattmeter (reflecting electrodynamometer) in series with the magnetizing coil. A separate winding is provided for the wattmeter potential coil and the correction for the wattmeter loss thus simplified. The induction is determined from the indication of a voltmeter, which is also connected to a separate winding at the center of the specimen. The flux which it indicates is much higher than that at the ends, but experiment has shown that the ratio of the maximum to the average is 1.3:1. Due allowance is therefore made in adjusting the magnetizing current to such a value that the voltmeter deflection will correspond to the required average flux density. Measurements are made at 10 c, or less, in order to reduce the eddy-current loss to a point where it can be eliminated by means of empirical corrections without serious error. The precision obtained is about ±5%.

230. Core-loss Measurements. The total loss in iron or steel subjected to an alternating magnetic field is most accurately measured with a wattmeter. In making precision measurements, the Rowland ring specimen can be used, but the **Epstein** apparatus is more convenient and has been adopted by the ASTM.[1] Figure 3-102 shows the scheme diagrammatically.

The specimen is arranged in the form of a rectangle. The magnetizing winding is divided into four solenoids, each being wound on a form into which one side of the rectangular specimen is placed. The form is nonmagnetic and nonconducting and has the following dimensions: inside cross section 4 cm (1.57 in) by 4 cm; thickness of wall not over 0.3 cm (0.12 in); winding length 42 cm (16.5 in). Each limb of the specimen consists of 2.5 kg (5.5 lb) of strips 3 cm (1.18 in) wide and 50 cm (19.7 in) long. Two of the bundles are made up of strips cut in the direction of rolling and two at right angles to the direction of rolling. The strips may be held with tape wound loosely around the bundle. The bundles form butt joints at the corners with tough paper 0.01 cm (0.004 in) thick between. They are held firmly in position by clamps placed at the corners after the magnetizing current has been reduced to a minimum by tapping the joints together.

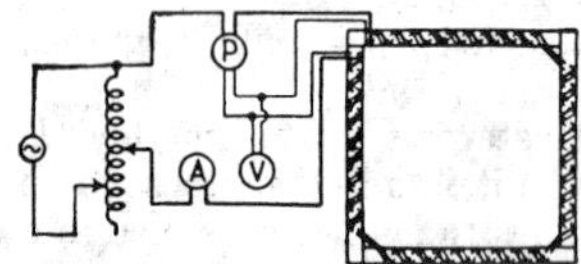

FIG. 3-102. Core-loss measurements, Epstein method.

231. The magnetizing winding on each solenoid consists of 150 turns uniformly distributed over the 42-cm (16.5-in) winding length and has a resistance of between 0.075 and 0.125 ohm. A secondary winding is uniformly wound underneath the first; it also contains 150 turns in each solenoid and energizes the potential circuit of the wattmeter and also the voltmeter with which the induction is measured. The resistance should not exceed 0.25 Ω/solenoid. With a sine-wave emf impressed on the magnetizing winding, the maximum induction is

$$B = \frac{E4lD10^8}{4fNnM} \quad \text{(gausses)}$$

where E = rms volts indicated by voltmeter, l = length of specimen, D = specific gravity (see ASTM Method A34-47 for assumed values where density is not determined), f = form factor of magnetizing emf (1.11 for sine wave), N = total secondary turns, n = cycles per second, and M = total mass in grams. The wattmeter gives the total loss in the iron, plus that in the secondary circuit. The latter is calculated from the resistance and the voltage and deducted from the wattmeter reading.

In the foregoing equation E/f is the *average* value of voltage. If the emf is not sinusoidal, a voltmeter should be used which reads the average value of the alternating emf such as the suitable type of vacuum-tube voltmeter (Par. **49**) or a rectifier type of voltmeter with avoidance of temperature errors. A **core-loss voltmeter** may be used for the same purpose; this is really a wattmeter measuring the loss in a small "standard" iron core of similar material.

232. Magnetic tests at low inductions and audio frequencies are usually made on Epstein apparatus specimens by employing an a-c bridge method. The simple inductance bridge (Par. **160**) may be used, or a modification of Owen's bridge (Par. **163**), or the induced emf may be measured with a coordinate a-c potentiometer. The max-

[1] ASTM Standard Method of Testing Magnetic Materials; ASTM Standard A34-47, Pars. 14–17.

imum induction B is computed from the number of turns in the magnetizing coil and the voltage across the coil or from the measured inductance and magnetizing current. Permeability may be obtained from the inductance and the dimensions of the specimen. Core loss is computed from the effective a-c resistance of the specimen [see ASTM Standard Method of Testing Magnetic Materials (A34-47), Pars. **24-32**].

233. Separation of eddy-current and hysteresis losses in the core is usually accomplished by taking advantage of the fact that with a given value of B_{max} the hysteresis loss varies directly with the frequency and the eddy-current loss with the square of the frequency. The specimen is arranged as in a core-loss test, and the loss noted at two frequencies, the induction being kept the same in the two cases. By means of two simultaneous equations, with two unknown quantities, both losses can be calculated.

234. Precautions in Magnetic Testing. Where the induction is measured by means of a stationary test coil surrounding the specimen, the space between the coil and the test specimen should be as small as possible. However, this space cannot be reduced to zero, and for precision work the necessary correction must be made [see Eq. (b) of Par. **222**]. The test-coil area A can be determined by noting the deflection of a ballistic galvanometer when the test coil is placed in a standard solenoid (Par. **235**) and the current in the solenoid is reversed.

Before induction measurements are made, the specimen should be carefully demagnetized. This is best done by first magnetizing to a value well above the maximum at which measurements will be made; the current is then gradually reduced to zero, being rapidly reversed meanwhile. Alternating current of 25-c frequency is convenient where much work is to be done.

In loss measurements, the temperature of the specimen should be carefully noted, because this affects the eddy-current loss. The exciting winding should therefore be sufficiently large to avoid heating of the specimen. In precision work, the apparatus is placed in an oil bath.

In tests of sheet materials, the strips should not be too narrow, because of the hardening at the edges due to cutting. This effect is negligible with a width of 2 in (5 cm). Care should be taken that burrs are removed from the edges and that the only insulation between sheets is the natural scale or oxide. The test specimen should be composed of strips cut from the sheets in both directions.

The readings of a ballistic galvanometer should be kept at about the same magnitude by varying the resistance in series with it. The observational error is thus kept about constant.

235. Standard Solenoid. A ballistic galvanometer is easily calibrated with a mutual inductor (see Par. **157**). Standard mutual inductors of either the fixed or the variable type are produced by several manufacturers and are usually preferable to standard solenoids in which the mutual inductance must be computed from carefully determined dimensions. Standard solenoids are often required, however, for the determination of the area turns of test coils; such solenoids are coils in which the field within the coil is exceptionally uniform, and the value of the field intensity is readily computed from the current and the constants of the coil. In a long, straight solenoid with a length 25 times its diameter, the field intensity at the center is

$$H = \frac{1.257NI}{L} \qquad \text{(oersteds)}$$

where N/L = turns per centimeter in winding and I = current in amperes. The error in this equation is less than 0.1% for a single-layer coil wound with small wire.

In the **Helmholtz-coil** arrangement, two equal coils are set coaxially with their mean planes separated by a distance equal to the mean radius of a coil; the field strength midway between the coils is uniform over a considerable area. If N is the number of turns per coil, R (centimeters) is the mean radius, and if the depth of the winding is not more than one-eighth of this radius, the field at a point on the axis halfway between the coils is given with an accuracy better than 0.1% by

$$H = \frac{0.8992NI}{R} \qquad \text{(oersteds)}$$

BIBLIOGRAPHY

236. Selected List of Reference Literature on Electric and Magnetic Measurements. The following are some of the "leading" texts:

NORTHRUP, E. F. "Methods of Measuring Electrical Resistance"; New York, McGraw-Hill Book Company, 1912.

LAWS, F. A. "Electrical Measurements," 2d ed.; New York, McGraw-Hill Book Company, 1938.

FARMER, F. MALCOLM "Electrical Measurements in Practice"; New York, McGraw-Hill Book Company, 1917.

GLAZEBROOK, T. "Dictionary of Applied Physics"; London, Macmillan & Co., Ltd., 1922, Vol. 2, Electricity.

KARAPETOFF, V. (rev. by Boyd C. Dennison) "Experimental Electrical Engineering," 4th ed.; New York, John Wiley & Sons, Inc., Vol. 1, 1923; Vol. 2, 1941.

Radio Instruments and Measurements, 2d ed., *NBS Circ.* 74, 1924.

HARRIS, FOREST K. "Electrical Measurements"; New York, John Wiley & Sons, Inc., 1952.

DRYSDALE, C. V., JOLLEY, A. C., and TAGG, G. F. "Electrical Measuring Instruments"; New York, John Wiley & Sons, Inc., 1952.

BANNER, E. H. W. "Electronic Measuring Instruments"; London, Chapman & Hall, Ltd., 1954.

MOULLIN, E. B. "Radio-frequency Measurement," 2d ed.; London, Charles Griffin & Co., Ltd., 1931.

HARTSHORN, L. "Radio-frequency Measurements by Bridge and Resonance Methods"; New York, John Wiley & Sons, Inc., 1940.

BATCHER, RALPH R., and MOULIC, WILLIAM "Electronic Engineering Handbook"; New York, Electronic Development Associates, 1944.

SPOONER, THOMAS "Properties and Testing of Magnetic Materials"; New York, McGraw-Hill Book Company, 1927. (Includes excellent bibliography.)

Code for Electricity Meters, 5th ed.; New York, Edison Electric Institute, 1965.

KEINATH, G. "Die Technik elektrischer Messgeraete," 3d ed.; Munich, R. Oldenbourg KG, 1929.

STUBBINS, G. W. "Commercial A-C Measurements"; Princeton, N.J., D. Van Nostrand Company, Inc., 1930.

HUND, AUGUST "High-frequency Measurements," 2d ed.; New York, McGraw-Hill Book Company, 1951.

KNOWLTON, A. E. "Electric Power Metering"; New York, McGraw-Hill Book Company, 1934.

TERMAN, F. E. "Measurements in Radio Engineering"; New York, McGraw-Hill Book Company, 1935.

HAGUE, B. "Instrument Transformers"; London, Sir Isaac Pitman & Sons, Ltd., 1936.

HAGUE, B. "Alternating-current Bridge Methods," 4th ed.; New York, Pitman Publishing Corporation, 1938.

CANFIELD, DONALD T. "Measurement of Alternating-current Energy"; New York, McGraw-Hill Book Company, 1939.

"Electrical Metermen's Handbook," 5th ed.; New York, Edison Electric Institute, 1940.

Descriptions of new methods of measurement and new measuring apparatus usually appear first in the following technical periodicals [generally, articles sought can be located most quickly with the aid of *Science Abstracts*, which appears annually in two volumes: (*A*) Physics and (*B*) Electrical Engineering]:

Scientific Papers and *Journal of Research*, National Bureau of Standards, Washington; *Transactions*, American Institute of Electrical Engineers, New York; *Electrical Engineering* (formerly *Journal AIEE*) monthly; *Proceedings*, Institute of Radio Engineers, New York; *Journal*, Optical Society of America, Menasha, Wis.; *Journal*, Franklin Institute, Philadelphia; *Physical Review*, Lancaster, Pa.; *Science Abstracts*, Spon & Chamberlain, New York; *Electrical World*, New York; *General Electric Review*, Schenectady, N.Y.; *Bell System Technical Journal*, New York; *Instruments*, Pittsburgh; *Review of Scientific Instruments*, Lancaster, Pa.; *RCA Review*, New York.

Journal, Institution of Electrical Engineers, London; *Proceedings*, Physical Society, London; *Proceedings*, Royal Society, London; *Philosophical Transactions*, Royal Society, London; *Philosophical Magazine and Journal of Science*, London; *Journal of Scientific Instruments*, London; *Archiv für Elektrotechnik*, Berlin; *Annalen der Physik*, Leipzig; *Elektrotechnische Zeitschrift für Instrumentenkunde*, Berlin; *Archiv für technisches Messen*, Munich; *Comptes rendus*, Paris; *Annales de physique*, Paris; *Revue générale de l'électricité*, Paris; *L'Elettrotecnica*, Milan.

MECHANICAL POWER MEASUREMENTS

TORQUE MEASUREMENTS

237. Torque is best measured with dynamometers, of which there are two classes: absorption and transmission. Absorption dynamometers absorb the total power delivered by the machine being tested, whereas transmission dynamometers absorb only that part represented by friction in the dynamometer itself. Made in a wide variety of forms, typical forms are described in the following paragraphs.

238. The Prony brake is the most common type of **absorption dynamometer.** The torque developed by the machine to overcome the friction is determined from the product of force required to prevent rotation of the brake and the lever arm. The load is applied by tightening the brake band or adding weights.

239. Dissipation of Heat in Friction Brakes. The energy dissipated in the brake appears in the form of heat. In small brakes, natural cooling is sufficient, but in large brakes, special provisions have to be made to dissipate the heat. Water cooling is the usual method, one common scheme employing a pulley with flanges at the edges of the rim which project inward. Water from a hose is played on the inside surface of the pulley and collected again by means of a suitable scooping arrangement. About 100 in^2 of rubbing surface of brake should be allowed with air cooling or about 25 to 50 in^2 with water cooling per horsepower.

240. The Westinghouse turbine brake employs the principle of the water turbine and is capable of absorbing several thousand horsepower at very high speeds.

In the **magnetic brake,** a metallic disk on the shaft of the machine being tested is rotated between the poles of magnets mounted on a yoke which is free to move. The pull due to the eddy currents induced in the disk is measured in the usual manner by counteracting the tendency of the yoke to revolve. This form of brake can be made in very small sizes and is, therefore, convenient for very small motors.

241. The principal forms of transmission dynamometers are the torsion and the cradle types.

In torsion dynamometers, the deflection of a shaft or spiral spring, which mechanically connects the driving and driven machines, is used to measure the torque. The spring or shaft can be calibrated statically by noting the angular twist corresponding to a known weight at the end of a known lever arm perpendicular to the axis. When in use, the angle can be measured by various electrical and optical methods.

The cradle dynamometer is a convenient and accurate device which is extensively used for routine measurements of the order of 100 hp or less. An electric generator is mounted on a "cradle" supported on trunnions and mechanically connected to the machine being tested. The pull exerted between the armature and field tends to rotate the field. This torque is counterbalanced and measured with weights moved along an arm in the usual manner.

SPEED MEASUREMENTS[1]

242. Tachometers, or **speed indicators,** indicate the speed directly and thus include the time element. The principal types are centrifugal, liquid, reed, and electrical. In the **centrifugal type,** a revolving weight on the end of a lever moves under the action of centrifugal force in proportion to the speed, as in a flyball governor. This movement is indicated by a pointer which moves over a graduated scale. In the portable or hand

[1] A comprehensive discussion of speed measurements will be found in ASME Power Test Code, Instruments and Apparatus, Pt. 13, entitled Speed Measurements, issued September, 1930.

type, the tachometer shaft is held in contact with the end of the shaft being measured, and in the stationary type, the instrument is either geared or belted. In the **liquid tachometer** of the Veeder type, a small centrifugal pump is driven by a belt consisting of a light cord or string. This pump discharges a colored liquid into a vertical tube, the height of the column being a measure of the speed.

Reed tachometers are similar to reed-type frequency indicators (Par. **193**), the reeds being set in resonant vibration corresponding to the speed of the machine. The instrument may be set on the bed frame of the machine, where any slight vibration due to the unbalancing of the reciprocating or revolving member will set the corresponding reed in vibration. Some forms are belted to the revolving shaft and the vibrations imparted by a mechanical device. **Electrical tachometers** may be either reed instruments operated electrically from small alternators geared or belted to the machine being measured or ordinary voltmeters connected to small permanent-magnet d-c generators driven by the machine being tested.

243. Chronographs are speed-recording instruments in which a graphic record of speed is made. In the usual forms, the record paper is placed on the surface of a drum which is driven at a certain definite and exact speed by clockwork or weights, combined with a speed-control device so that 1 in on the paper represents a definite time. The pens which make the record are attached to the armatures of electromagnets. With the pens in contact with the paper and making a straight line, an impulse of current causes the pen to make a slight lateral motion and, therefore, a sharp indication in the record. This impulse can be sent automatically by a suitable contact mechanism on the shaft of the machine or by a key operated by hand. The time per revolution is then determined directly from the distance between marks.

244. Stroboscopic methods are especially suitable for measuring the speed of small-power rotating machines where even the small power required to drive an ordinary speed counter or tachometer would change the speed, also for determining the speed of machine parts which are not readily accessible or where it is not practicable to use mechanical methods or where the speed is variable (see Par. **203** for description of stroboscopic principle).

One convenient form of stroboscopic tachometer employs a neon lamp connected to an oscillating circuit supplied from a 60-Hz circuit, which is adjusted to "flash" the neon lamp at the frequency necessary to make the moving part which the lamp illuminates appear to stand still. Speeds from a few hundred to many thousands of rpm can be very conveniently measured.

PYROMETRY

245. Thermoelectric Pyrometry. In a pyrometer of this type temperatures are measured by means of the change with temperature of the emf of a thermocouple. Such a pyrometer ordinarily consists of a thermocouple, an indicator for measuring the emf, and connecting leads. A thermocouple is fundamentally a pair of electrical conductors of dissimilar materials so joined as to produce a thermal emf when the junctions are at different temperatures. The conductors are usually joined at the end which is exposed to the temperature to be measured, the other ends being free to connect to an instrument for measuring the emf.

246. Metals Used for Thermocouples. The combinations of metals and alloys in Table 3-10 are extensively employed for measuring temperatures in the ranges indicated.

247. Temperature-EMF Relations for Various Couples. Table 3-11 gives representative temperature-emf relations for the thermocouples listed in Par. **246**. The temperature-emf relations for thermocouples employing constantan as one element vary with different manufacturers, and the value for these couples issued by some manufacturers may differ from those given in the table by as much as 20°C. However, most manufacturers maintain a uniform product and the calibrations of individual couples supplied by reputable pyrometer manufacturers can be depended upon to agree with the curve issued by the manufacturer for that type of couple to ±10°C and usually to ±5.

Table 3-10. Metals and Alloys for Thermocouples

Type of thermocouple	For long-time service	For short-time service*
Platinum to platinum—10% rhodium	0–1450 C	1700 C
Platinum to platinum—13% rhodium	0–1450 C	1700 C
Chromel to alumel	0–1100 C	1350 C
Iron to constantan	0– 900 C	1100 C
Chromel to constantan	0– 900 C	1100 C
Copper to constantan	−190– 350 C	600 C

* Thermocouples which have been subjected to the temperatures given in this column cannot be depended upon to retain their calibrations and should not be used for accurate work subsequently without being recalibrated.

For fuller details, including emf for smaller intervals of temperature than 100°C, the Temperature-Emf Tables for Thermocouples, ISA *R.P.* I-6 of the Instrument Society of America, should be consulted.

248. Indicating Instruments. The indicating instruments employed to measure the emfs of thermocouples operate on the galvanometric principle, potentiometric principle, or a combination of these.

Galvanometer Method. Since the emfs are small, the galvanometers, which are essentially ammeters graduated to read emfs, must be very sensitive to voltage. This could be accomplished by using an instrument of low resistance; but on account of the variable line resistance of a thermoelectric circuit, it is necessary to make the resistance of the galvanometer as high as possible, consistent with substantial construction. If the indicator is graduated to read emf at its terminals, the relation between scale reading e_0 and true emf of the couple e becomes

$$e_0 = \frac{R_g e}{R_g + R_c + R_L}$$

where R_g, R_c, and R_L = resistances of galvanometer, couple, and lead wires, respectively. If R_g is large compared with $R_c + R_L$, the foregoing equation reduces approximately to $e_0 = e$ and the indicator reads correctly. When R_g is low, it is still possible to graduate the scale to read true emf for a definite line resistance, but if this changes on account of deterioration of the couple, etc., serious error may result. Thus a 300-Ω indicator designed for use with a 2-Ω line resistance will be in error by only 7° at 1000°C if the line resistance changes to 4 Ω, while a 100-Ω indicator under similar conditions

Table 3-11. Calibration Data of Representative Couples (Millivolts)

(Cold-junction temperature 0°C)

Temp., deg C	Type J iron-constantan	Type K chromel-alumel	Type S platinum, 10% rhodium-platinum	Type R platinum, 13% rhodium-platinum	Type T copper-constantan
−100	−4.63	−3.49	......	......	−3.349
0	0	0	0	0	0
100	5.27	4.10	.643	.645	4.277
200	10.78	8.13	1.436	1.465	9.288
300	16.33	12.21	2.316	2.395	14.864
400	21.85	16.40	3.251	3.399	20.874
500	27.30	20.65	4.221	4.455	
600	33.11	24.91	5.224	5.563	
700	39.15	29.14	6.260	6.720	
800	45.53	33.30	7.329	7.924	
900	......	37.36	8.432	9.175	
1000	......	41.31	9.570	10.471	
1100	......	45.16	10.741	11.817	
1200	......	48.89	11.935	13.193	
1300	......	52.46	13.138	14.582	
1400	......	......	14.337	15.969	
1500	......	......	15.530	17.355	
1600	......	......	16.716	18.727	
1700	......	......	17.891		

would read 140°C too low. This emphasizes the importance of the use of galvanometers of high resistance.

The advantages of this method are quick reading, simplicity, and moderate initial cost. The disadvantages lie largely in the effects due to thermocouple and lead resistance cited above.

249. Potentiometer Method. The most accurate instrument for measuring thermoelectric emfs is the potentiometer. Special low-range instruments adapted to pyrometry and portable potentiometers for either rare-metal or base-metal couples or for both with a double scale are available. There are several important advantages in the potentiometer method. The emf or temperature scale is easily made very open, thus permitting accurate readings. The calibration of the scale is in no way dependent upon the constancy of magnets, springs, or jewel bearings or upon the level of the instrument. From the pyrometric standpoint, however, the greatest advantage is the complete elimination of any error due to ordinary changes in the resistance of the couple or of the lead wires. Thus, if a potentiometer is correctly balanced and the resistance of the thermocouple is then greatly increased, the balance is unchanged. However, excessive resistance in the couple circuit reduces the sensitivity of a setting. The objections to the potentiometer are initial cost and the fact that usually a manual adjustment must be made to obtain a setting.

250. Cold-junction Corrections. If a couple is calibrated with a cold-junction temperature of t_0°C and used with a cold-junction temperature of t_0' °C, it is necessary to add to the observed emf the value of the emf developed when the hot junction is at t_0' and the cold junction at t_0 to obtain correct temperature from the calibration data. If the indicator is graduated in temperature, add to the observed reading $(t_0' - t_0)K$, where K = factor depending upon couple and temperature. Values of K for typical couples are as in Table 3-12.

Cold-junction correction figures for representative samples in 1 °F and 1 °C intervals may be found in the Instrument Society of America publication ISA *R.P.* I-6.

Table 3-12. Cold-junction Correction Factors

Platinum to platinum—10 % rhodium		Platinum to platinum—13 % rhodium		Copper-constantan		Iron-constantan		Chromel-alumel	
Deg C	K*	Deg C	K*	Deg C	K*	Deg C	K*	Deg C	K*
265–450	0.65	250–400	0.60	0–50	1.00	0–100	1.00	0–800	1.00
450–650	0.60	400–550	0.55	50–80	0.95	100–600	0.95	800–1100	1.05
650–1000	0.55	550–900	0.50	80–110	0.90	600–1000	0.85		
1000–1450	0.50	900–1450	0.45	110–150	0.85				
				150–200	0.80				
				200–270	0.75				
				270–350	0.70				

* Based on calibration with t_0 = 0°C.

251. Compensators. Various methods are employed for avoiding the necessity of making these corrections. Simple indicators may be set on open circuit to read the cold-junction temperature. An indicator operating on the galvanometric principle is usually equipped with an automatic cold-junction compensator, which consists of a bimetallic spring attached to the springs of the moving coil in such a way that on open circuit the pointer will indicate the temperature of the instrument and cold junctions in case the cold junctions are located at the instrument. Potentiometric instruments may be equipped with either manually or automatically operated compensators. Even with such devices, however, it is desirable to control the cold-junction temperature as much as feasible.

252. Electrical-resistance pyrometry makes use of the variation in the electrical resistance of platinum and is capable of great sensitivity. In one of its simplest forms the pyrometer consists of a coil of platinum wire wound on mica and encased in a protecting tube of porcelain. Owing to permanent changes in resistance when exposed to high temperatures, probably the result of contamination or distillation of platinum, high-resistance coils of small wire are not used much above 900°C. However, coils constructed of 0.6-mm wire may serve satisfactorily to 1200°C for investigational work.

253. Three-lead Wheatstone-bridge Method. Figure 3-103 illustrates the wiring diagram for a simple Wheatstone bridge and thermometer with the Siemens three-lead compensation. The platinum or gold lead wires *C'd* and *T'e* in the thermometer are constructed of as nearly the same resistance as possible, and the copper lead wires *CC'* and *TT'* must also be of equal resistance. The battery *B* is connected between the ratio arms r_1 and r_2 of the bridge and to the compensating lead wire *cd*. A sensitive galvanometer *G* is connected to the points *f* and *g*, as illustrated. By changing r_3 the bridge is balanced until the galvanometer shows zero deflection. From the principle of the balanced Wheatstone bridge is obtained the following relation:

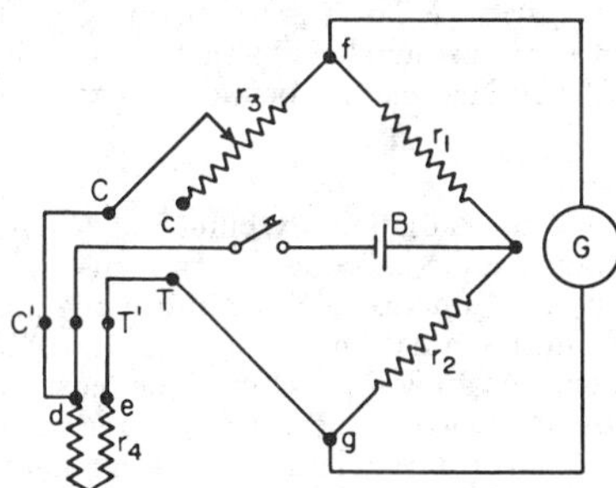

Fig. 3-103. Wheatstone bridge and thermometer with Siemens three-lead compensation.

$$\frac{r_3 + Cd}{r_1} = \frac{r_4 + Te}{r_2}$$

Now, r_1 is always constructed equal to r_2, and since $Cd = Te$, it follows that $r_3 = r_4$. Hence the measured resistance is independent of the resistance of the lead wires.

254. Compensating Lead Wires of Callendar or Siemens Form. The Callendar form of compensation requires four lead wires. Two of the lead wires joined to the ends of the coil are connected in one arm of the bridge, and the other lead wires, which are "dummy" leads formed of a single loop of wire extending to the top of the coil, are connected in the corresponding arm of the bridge. The two sets of lead wires are alike, so that variations in temperature affect each set similarly. The Siemens method of compensation, described in Par. **253**, requires three leads and is more often employed industrially in this country.

255. Radiation. The temperature of bodies may be estimated from the radiant energy, which they send out in the form of visible light, or the longer, infrared rays, which may be detected by their thermal effects. Since the intensity of radiation increases very rapidly with a rise in temperature, it would appear that a system of pyrometry based on the intensity of the light or total radiation from a hot body would be an ideal and simple one. However, different substances at the same temperature show vastly different intensities at a given wavelength; in other words, the **absorbing or emissive powers** may vary with the substance, with the wavelength, and also with the temperature.

256. Blackbody Radiation. A substance which absorbs all the radiation of any wavelength falling upon it is known as a **blackbody.** Such a body will emit the maximum intensity of radiation for any given temperature and wavelength. No such material exists, but a very close approximation is obtained by heating the walls of a hollow opaque enclosure as uniformly as possible and observing the radiation coming from the inside through a very small opening in the wall.

257. Stefan-Boltzmann Law. The relation between the total energy radiated by a blackbody and its temperature is expressed by the equation

$$J = \sigma(T^4 - T_0^4)$$

where J = energy of all wavelengths emitted per second per square centimeter of surface, T and T_0 = absolute temperatures of radiator and surroundings, respectively, σ = a constant of the value 5.7×10^{-12} W/(cm²)(°C⁴). In general T_0^4 is negligible in comparison with T^4, so that the above relation becomes $J = \sigma T^4$. Although the total energy emitted by any substance is not that emitted by a blackbody at the same temperature, it may be considered as some fractional part of that from the ideal radiator, this fraction E being known as the **total emissivity.** If S denotes the apparent absolute temperature, i.e., the temperature on the blackbody scale corresponding to an amount of energy equivalent to that emitted by the nonblack substances at a true temperature $T°$ abs, the relation between its total emissivity E and the quantities S and T is

$$\log E = 4(\log S - \log T)$$

258. Radiation Pyrometry. In the radiation pyrometer, the radiant energy of all wavelengths is brought to a focus by means of a quartz lens or a concave mirror upon the hot junctions of a minute thermopile. The cold junctions are suitably screened from the radiation of the hot body. The radiant energy focused upon the hot junctions develops an emf which may be measured by a potentiometer or automatically recorded. In practice the indicating or recording instrument is calibrated to read temperature directly. The relation between the emf e and the absolute temperature T may be expressed by the equation

$$e = aT^b \qquad \text{or} \qquad \log e = k + b \log T$$

where a (or k) and b = empirical const. The constant b is usually between 4 and 5.

Leeds & Northrup Rayotube. This radiation pyrometer is made either with a quartz lens or with a mirror. The mirror type is suitable for the lower temperatures at which the energy radiated is largely in the long wavelengths which would be too greatly absorbed by a lens.

259. Optical Pyrometry. Optical pyrometers are based upon the photometric principle of comparing the intensity of visible monochromatic radiation emitted by a body with that of the same wavelength or color from a constant and reproducible comparison source, such as an electric lamp. The light from the comparison source must be calibrated in terms of the formula

$$\log_e \frac{J_2}{J_1} = \frac{1.432}{\lambda}\left(\frac{1}{1{,}336} - \frac{1}{t + 273}\right)$$

specified in the International Temperature Scale either by direct observations on a crucible of freezing gold and by use of sector disks to determine ratios of brightness or by checking against a standard pyrometer which has been previously so calibrated.

The above formula comes directly from applying Wien's law

$$J_\lambda = C_1\lambda^{-5} \exp\left(-\frac{1.432}{\lambda(t + 273)}\right)$$

to the ratio of monochromatic radiation of wavelength λ from a blackbody at two different temperatures. Practically all optical pyrometers now in use are of the disappearing-filament type.

Table 3-13. Monochromatic Emissivity for Red Light ($\lambda = 0.65\ \mu$)

Material	E_λ	Material	E_λ
Silver	0.07	Nichrome (oxidized)	0.90
Gold, solid	0.14	Cuprous oxide	0.70
Gold, liquid	0.22	Iron oxide	0.80
Platinum, solid	0.30	Nickel oxide, 800 C	0.96
Platinum, liquid	0.38	Nickel oxide, 1300 C	0.85
Palladium, solid	0.33	Nickel, solid and liquid	0.36
Palladium, liquid	0.37	Iridium	0.30
Copper, solid	0.10	Rhodium	0.24
Copper, liquid	0.15	Graphite powder (estimated)	0.95
Tantalum, 1100 C	0.60	Carbon	0.85
Tantalum, 2600 C	0.48	Porcelain (? ?)	0.25–0.50
Tungsten, 1000 C	0.46		
Tungsten, 2000 C	0.43		
Tungsten, 3000 C	0.41		

260. Disappearing-filament Pyrometers. The filament of a small electric lamp F (Fig. 3-104) is placed at the focal point of an objective L and ocular, forming an ordinary telescope which superposes upon the lamp the image of the source viewed. Red glass, such as Corning "high-transmission red," is mounted at the ocular to produce approximately monochromatic light. In making a setting, the current through the lamp is adjusted by rheostat until the tip or some definite part of the filament is of the same brightness as the source viewed.

The lamps should be operated at temperatures no higher than 1500°C, on account of deterioration of the tungsten filament. If this temperature is not exceeded, the calibration of the lamp is good for hundreds of hours of ordinary use. For higher temperatures, absorption glasses S (Fig. 3-104) are placed between the lamp and the objective, or in front of the objective, to diminish the observed intensity of the source. The relation between the temperature of the source, $T°$ abs, and the observed temperature, $T_0°$ abs, measured with the absorption glass interposed, is as follows: $1/T - 1/T_0 = A$, where A = for most practical purposes a constant. Sometimes an instrument is calibrated to read the lamp current, and a table is furnished showing the relation between the current and the temperature both with and without the absorption glass. Often, however, the instrument reads directly in temperature and is provided with a double scale to take care of the high and low ranges.

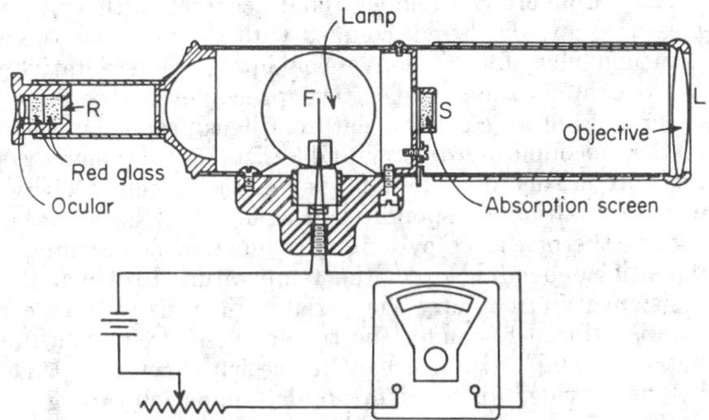

FIG. 3-104. Disappearing-filament optical pyrometer.

Instruments may be obtained with an additional special glass for use when sighting upon iron or steel. This special glass differs from the regular absorption glass in such a way that its substitution for the regular screen automatically compensates for the emissivity of the metal. The instrument will thus read on the same scale either the true temperature of the metal (with the special glass in place) or the temperature of a blackbody (with the regular glass in place) (see Table 3-13).

Table 3-14. Corrections to Optical Pyrometer in Degrees C When Used on Molten Cast Iron ($\lambda = 0.65\ \mu$)

Apparent temperature, deg C	Below transition point		Above transition point	
	True temperature, deg C	Correction, deg C	True temperature, deg C	Correction, deg C
1160	1194	34		
1180	1215	35		
1200	1236	36		
1220	1257	37		
1240	1278	38		
1260	1299	39	1365	105
1280	1320	40	1387	107
1300	1341	41	1410	110
1320	1362	42	1433	113
1340	1383	43	1456	116
1360	1404	44	1479	119
1380	....	..	1502	122
1400	....	..	1526	126
1420	....	..	1549	129
1440	....	..	1572	132
1460	....	..	1595	135

261. Emissivity Corrections for Optical Pyrometers. Optical pyrometers will indicate true temperatures when sighted upon a blackbody. Blackbody conditions are approximated in practice by a peephole in the side of a furnace or kiln or a closed porcelain tube thrust into molten metals or salts. When sighting upon objects in the open, certain corrections must be applied. Table 3-13 gives the emissivity of various substances for red light ($\lambda = 0.65\ \mu$).

262. Emissivity Correction for Molten Cast Iron. Measurements of the true temperature with a thermocouple and apparent temperature with an optical pyrometer at the National Bureau of Standards show that the character of the surface of molten cast iron undergoes a change in the neighborhood of 1375°C true temperature. Below this transition point the emissivity is 0.7, while above this point it is about 0.4 (see Table 3-14).

ELECTRICAL MEASUREMENT OF NONELECTRICAL QUANTITIES

263. Converter. The ease and accuracy with which electrical magnitudes may be measured and displayed, coupled with the fact that there is virtually no variable phenomenon incapable of conversion into a representative electrical signal, have led to a broad application of electrical principles to the measurement of quantities not in themselves of an electrical nature. This practice is well exemplified in the techniques of thermocouple pyrometry and resistance thermometry (Pars. **245** and **252**). Chief requisite in such measurement is a device which will serve as a **converter** to provide an electrical variable responding to changes in the value of a quantity to be measured; e.g., in thermocouple **pyrometry** a junction of dissimilar metals produces an electrical potential depending upon the temperature to which the junction is exposed, and in **resistance thermometry** the variation in resistivity of certain materials with temperature is utilized to render the resistance of a circuit a function of temperature. Since both emf and resistance can be measured with extreme precision, a practical means is thus provided for **temperature** determination through a wide range.

264. Transducer. Any device wherein variations in energy magnitude of one form produce corresponding variations in energy of another form is known as a **transducer.** In common usage, either the input or the output of a transducer is electrical in its nature. Thermocouples and temperature-sensitive resistors (**thermistors**) both fall into that category. Transducers include also **thermal converters,** wherein a readily measurable electrical output effect (d-c millivolts) is derived from a thermal effect in turn representative of an electrical measure (a-c watts, vars, volts, or amperes), differing in nature from the output.

A concise discussion of this branch of measurement is rendered difficult by the two conditions that (*a*) several different electrical methods are generally available for the measurement of any specified variable and (*b*) practically every form and principle of transducer is applicable to a variety of measurements. "Frequently the principle of operation will dictate the selection of a specific transducer; for instance, where long cables are necessary, piezoelectric transducers may not perform satisfactorily, or where temperatures are high, **crystals** may disintegrate with heat. Capacitive devices, though relatively sensitive, usually require intervening electronic circuitry, whereas magnetic transducers should not be used in the presence of strong magnetic fields."[1]

265. Mechanical displacement may be converted into an electrical variable by the simple expedient of adjusting resistance in an electrical circuit. A slide-wire resistor, having a movable contact attached to the part whose displacement is to be measured, may be connected through a 2-conductor circuit to a steady-voltage source in series with an ammeter (or milliammeter) calibrated in terms of the displacement. If the resistor is connected as a voltage divider, the need for a regulated supply is eliminated, and with a 3-conductor circuit the display instrument may be a ratio meter or a potentiometer. Such combinations are common and are available for both d-c and a-c

[1] From "Transducers," by David B. Kret. Published 1953 by Allen B. DuMont Laboratories, Inc., Clifton, N.J. This work is a comprehensive treatment of transducers, especially, but not wholly, for use with oscillographic display devices. It includes a list of more than 800 transducers and their applications, as well as a tabulation of over 200 accessory devices and a bibliography of 200 references.

operation. Where deflections are small—of the order of 0.1 in—measurement may be made by use of a differential transformer.

266. Strain Gage. In the **strain gage,** microscopic relative displacements are electrically magnified and are displayed on an indicating or a recording meter or on an oscillograph. Modern resistance-type strain gages comprise fine wire windings arranged to be more or less elongated when subjected to deformation. The units may be used singly, in pairs, or in sets of four constituting a complete Wheatstone bridge. There are two main classes of wire-wound strain gages, (*a*) "bonded" and (*b*) "unbonded."

a. The *bonded strain gage* is composed of fine wire, wound and cemented on a resilient insulating support, usually a wafer unit. Such units may be mounted upon or incorporated in mechanical elements or structures whose deformations under stress are to be determined. While there are no limits to the basic values which may be selected for strain-gage resistances, a typical example may be taken as of the order of 100 to 500 Ω.

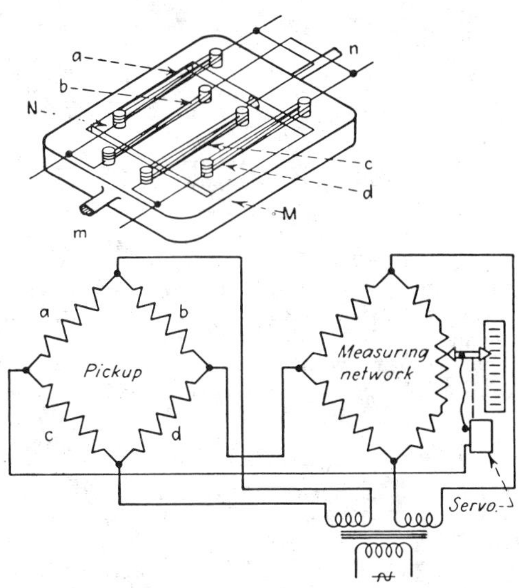

FIG. 3-105. Diagram of unbonded wire strain gage. Supports M and N are attached by rods m and n, respectively, to points between which displacement is to be measured. Pickup and measuring networks are energized from similar but isolated sources. Unbalance originating in pickup is detected and balanced by servo-actuated measuring network, providing a reading of strain on graduated scale.

b. In the *unbonded strain gage,*[1] the resistance structure comprises a fine wire winding stretched between insulating supports mounted alternately on the two members between which displacement is to be measured (see Fig. 3-105). These wires comprise the four arms of a Wheatstone-bridge network of which two opposite arms are tightened and the other two slackened by the displacement. While a bonded gage tends to respond to the average strain in the surface to which it is cemented, the unbonded form measures displacement between the two points to which the respective supports are attached. Unbonded wire strain gages are usually operated on input potentials ranging up to 35 V direct or alternating current. Under conditions of extreme unbalance, corresponding to full operating range, the open-circuit emf may be of the order of 8 to 10 mV and the closed-circuit current up to 100 μA.

Recently developed types of conductive rubber show considerable promise in the design of nonlinear resistive transducers capable of wide ranges of deformation. Also, the National Bureau of Standards has published information concerning springs which change their resistances with elongation. Strain gages for use on a-c circuits are supplied in both capacitive and inductive forms, wherein the corresponding characteristics of a-c circuit components are varied by the displacements to be measured.

267. Small Displacements. A popular means for measuring **small displacements** of the order of 0.1 in is the *linear,* or *differential, transformer.*[2] This device (Fig. 3-106) is generally produced with a single primary winding and two secondaries, all disposed along a common axis and having in the common magnetic circuit a movable iron core longitudinally displaceable with the motion to be measured. The secondaries may be connected additively or differentially and may be included in the circuit of a null-type instrument balanced either by shifting the core of a similar transformer excited from

[1] HELFAND, B. B. Developments in Unbonded Strain Gage Transducers; *Rept. AIEE-IRE Conf. Telemetering,* April, 1953, p. 27.

[2] Differential transformers may be obtained for direct displacements as great as 10 in.

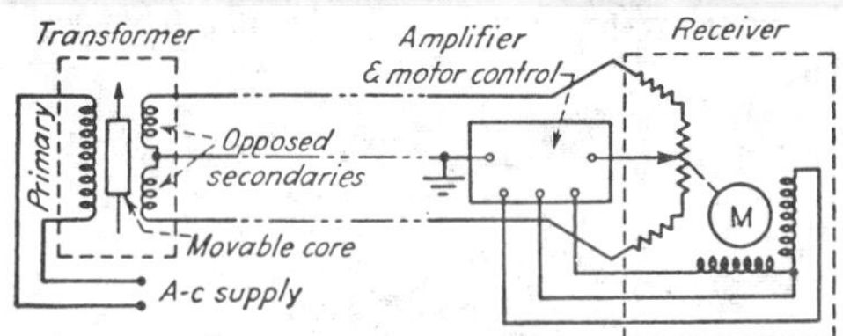

Fig. 3-106. Schematic diagram of differential-transformer transducer with servo-actuated receiver.

the same source or by the use of a slide-wire potentiometer. Linear transformers are regularly supplied for operation at all frequencies up to 30,000 Hz. The sensitivity, of course, increases with the frequency. Linear transformers[1] may be interconnected in a great variety of arrangements to perform computations or to express desired mathematical functions of measured variables.

268. Pressure Gage. Strain gages permanently attached to diaphragms, tubes, and other pressure-sensitive elements find a wide application as components of **electrically actuated pressure gages.** By electrically combining simultaneous measurements of torque and velocity, continuous determination of mechanical power may be obtained, the combination becoming an electrical-transmission dynamometer.

269. Vibration may be determined by a strain gage, but the fact that this magnitude involves motion renders it generally preferable to utilize alternating potentials developed by periodic change in the geometry of the measuring circuit. This may be embodied in either a *capacitor* or an *inductive device.* In a recently developed apparatus there are no moving parts except the object being shaken, and the vibration displacement is sensed by its effect on an electrostatic field between the pickup and the moving part. **Piezoelectric crystals** are particularly adapted to the measurement of vibration. The emf so obtained is proportional to the amplitude of deflection multiplied by the frequency squared.

270. Air velocities and the **flow of gases** in general may be measured by the **hot-wire anemometer.** In its simplest form, this device utilizes the cooling effect of the gas stream to establish a temperature difference between exposed and protected bridge arms. Where the flow is in an enclosed conduit, a heating element may be introduced and the volume of flow determined by the amount of heat transferred between the heater and the temperature-sensitive bridge wires.

271. Flow of an electrically conducting liquid may be determined by measuring the emf developed between a pair of electrodes set in opposite sides of an insulating conduit due to the movement of the liquid through a magnetic field established transversely of the conduit and perpendicular to both the flow and the line joining the electrodes (Fig. 3-107). By using an alternating field the effects of electrode polarization may be eliminated. Null measurement of the generated voltage renders the apparatus independent of the resistance of the liquid.

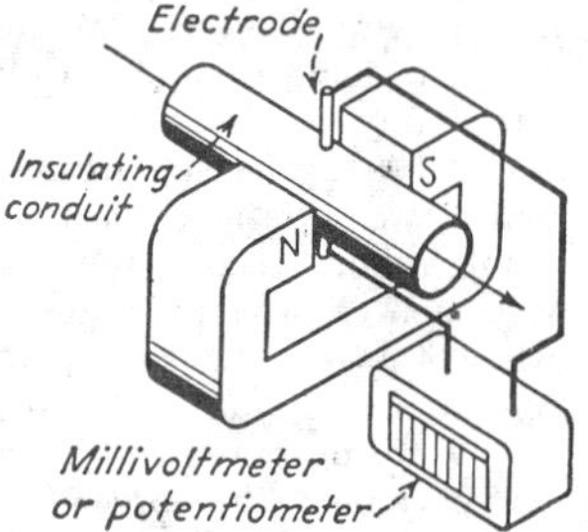

Fig. 3-107. Electromagnetic flow meter.

272. Liquid level may be expressed electrically by the use of a transducer responsive to the vertical position of a float or by a pressure-sensitive strain gage immersed in the liquid below its lowest level. Variation in resistance of an immersed conductor is a widely accepted principle, especially in fuel tanks. If the liquid is an electrical insulator and of constant characteristics, its depth may be determined by its dielectric effect between a pair of vertically disposed capacitor plates. On the other hand, if the liquid is a conductor and very small changes in level are to be detected or regulated, the liquid may be made one electrode of a capacitor whose other electrode is a horizontal plate positioned above the surface.

273. Levels of corrosive liquids or those operating under extreme pressures, temperatures, or other conditions rendering them inaccessible for measurement by conventional means may be determined by the use of **gamma radiation.** Several gamma-ray sources are spaced at equal vertical intervals in the tank or reactor containing the

[1] Hornfeck, A. J. Computing Circuits and Devices for Industrial Process Functions; *Tech. Paper* 52-191, *Trans. AIEE*, Communication and Electronics, July, 1952, p. 183.

liquid to be measured but are positioned so that none of them obstructs the line of sight of a Geiger counter tube placed at the top of the container. The response of the Geiger tube depends upon the depth of the process material, and the output is measured on a null-type recording instrument.

274. Vacuum may be measured by determining either *energy dissipation* or *electron emission* in the space under test. The former principle provides the basis of the **Pirani gage,** wherein two similar heated filaments forming arms of a bridge are located, respectively, in a reference bulb and a bulb connected to the evacuated space. Heat dissipation will vary with the degree of evacuation, while conditions in the reference bulb remain constant. The electrical condition of the bridge then provides a continuous measure of the vacuum. The normal range of operation of the Pirani gage is from 10^{-7} to 5 mm Hg. As the performance of a thermionic tube is highly responsive to the degree of vacuum, its action under controlled electrical conditions is a criterion of internal atmosphere. This principle forms the basis of a number of **electronic vacuum gages.** The normal range of operation lies between 10^{-7} and 10^{-3} mm Hg.

275. Gas Analysis. Electrical methods for **analyzing gases,** while essentially thermal in their nature, are made practicable only by the application of electrical principles in determining thermal relationships. In the *thermal-conductivity method,* as best exemplified in the CO_2 recorder, two cells or sections of conduit containing, respectively, a standard sample and the gas under test have in them adjacent arms of a bridge network composed of wires having known resistance variation with temperature and carrying sufficient current to raise their temperatures appreciably above their surroundings. As more or less heat is dissipated in the test cell as compared with the reference cell, the relative resistance of the bridge arms varies, providing an electrical basis for measurement of the gas composition.

276. Detection of Flammable Gases. The **catalytic-combustion method** is especially adapted to detection of flammable gases or determination of explosibility. The arrangement of cells and bridge wires may be similar to that of the thermal-conductivity type, but the filament is composed usually of activated platinum and is operated at a temperature sufficient to ignite the gas when a critical proportion is attained (Fig. 3-108). The increased heating of the bridge wire due to combustion abruptly disturbs the balance and provides a positive indication of explosibility. In some forms of this instrument the temperature rise is determined by thermocouples. The catalytic-combustion method is useful in determining mixtures containing such gases as propane, acetone vapor, carbon disulfide, and carbon monoxide. The equipment finds use in (1) solvent-recovery processes, (2) solvent-evaporating ovens, (3) combustible-gas storage rooms, (4) storage vaults, (5) gas-generating plants, (6) refineries, and (7) mines.

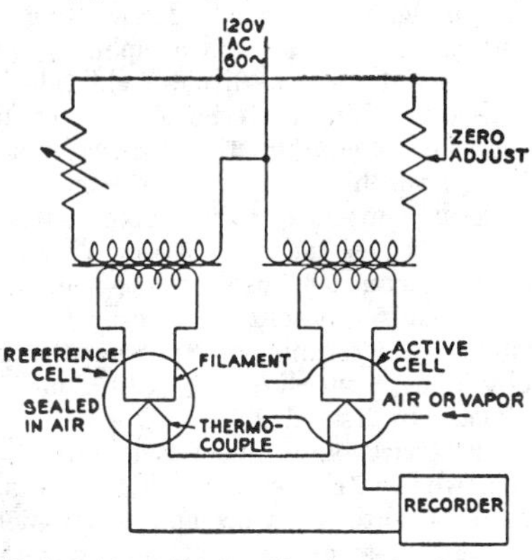

Fig. 3-108. Basic circuit of thermocouple-type gas analyzer.

277. Oxygen Content of Gases. Both the conventional thermal-conductivity method and the catalytic combustion method are applicable. In addition to these, use is made of the **magnetic susceptibility** of oxygen as a basis of operation. In one such instrument, a hot-wire bridge similar to that of a CO_2 recorder is employed, one of the gas chambers being placed in a strong magnetic field. This stimulates the flow of oxygen-containing gas through that chamber, thereby unbalancing the bridge by a measurable amount. In the other magnetic analyzer, a test chamber contains a small magnetic member rotatable in a distorted field whose conformation depends upon the amount of oxygen present. The resultant angular displacement of the test member may be used similarly to that of a galvanometer in either a direct-deflecting or a null-type instrument.[1]

[1] Rigg, O. W. Oxygen Recorders; *Instruments,* February, 1953, p. 248.
Seffern, O. K. Oxygen Measurement; *Instruments,* September, 1953, p. 1210.

278. Toxic Gases. An analyzer especially suited to measurement of **toxic ionizable gases** or vapors to and beyond the toxic limits utilizes the electrical conductivity of an aqueous solution of the gas. The vapor under test is bubbled through distilled water at a fixed rate, and the conductivity of the solution becomes a measure of gas concentration. A typical use is the continuous recording of small quantities of substances like sulfur dioxide, hydrogen sulfide, chlorine, and carbon disulfide in the air.

279. Atmospheric contamination may be determined by an **electronic leak detector,** utilizing emission of positive ions from an incandescent filament exposed to the air. The filament is enclosed in an open inner cylinder and heated by alternating current. The atmosphere under test is forced through the annular space between the inner and an outer cylinder at a predetermined rate, and the electron flow due to a d-c potential maintained between the cylinders is measured as an index of the amount of contaminant. Presence of extremely small proportions of halogen vapor compounds, of which Freon, chloroform, and carbon tetrachloride are good examples, greatly increases the emission. At room temperatures the device does not respond to Pyranol, but if this material is heated sufficiently to give off vapor, a response is obtained. It also responds to solid particles of the halogens and therefore will detect **smoke** from burning materials containing these elements. The instrument is also available as a recorder and/or a controller.[1]

280. Relative humidity is determinable electrically by methods involving either of two basic principles: (1) variation of electrical conductivity or of dielectric constant of a hydrophilic element and (2) computation based on "dry-bulb" and "wet-bulb" temperatures of the atmosphere whose moisture content is to be determined. The most common embodiment of the former method consists in an insulating card, plate, or cylinder carrying a bifilar winding of conductive wire and having a relatively large surface exposed to the atmosphere. The two strands of wire are bridged by a coating of material, such as lithium chloride or colloidal graphite, having a high affinity for moisture. This material quickly assumes a water content corresponding to that of the atmosphere, and the electrical resistance between the conductors becomes a function of the humidity to be measured. A similar principle is used in determining the moisture content of hygroscopic materials, such as wood, grain, or pulp. In such applications a resistance-measuring circuit terminates in electrodes or probes which are pressed against or inserted into the material to be tested.

Moisture content of material in a web or sheet form, such as paper, may be continuously determined by passing the web between the plates of a capacitor and thus obtaining a measurement determined by the dielectric constant of the material as affected by its water content.

Electrical determination of humidity by the "wet-and-dry-bulb" method requires somewhat intricate computing circuits, which for accurate results must take account of absolute temperature and of barometric pressure.[2]

281. Determination of **dew point,** or the temperature at which condensation takes place on a polished surface, as a function of absolute humidity, employs essentially a thermal and optical method of measuring, but such a system may be rendered continuous and automatic by photoelectrically observing the condition of a polished surface in the tested atmosphere, utilizing a servo system to regulate its temperature, and thus obtaining an indication or a record of the dew point.

282. Electric Micrometers. The two most popular types are (1) that utilizing the magnified output of a strain gage and (2) that based upon precise determination of capacitance between two electrodes whose spacing corresponds to the measured dimension.

283. Ultrasonic thickness gages may be used to measure steel walls ranging in thickness from ⅛ in to 1 ft, utilizing the fact that sound vibrations tend to establish standing waves within the mass of the material upon which they are impressed. This device combines a variable-frequency oscillator with a piezoelectric crystal which is pressed against the wall to be tested. The circuit is tuned until the metal oscillates,

---

[1] WHITE, W. C., and HICKEY, J. J. Electronics Simulates Sense of Smell; *Electronics*, March, 1948, Vol. 21, p. 100.

[2] BEHR, LEO A New Relative Humidity Recorder; *Jour. Opt. Soc. Am.*, 1925, Vol. 12, p. 623.

causing a sharp increase in the loading. The frequency of this resonance indicates the thickness of the material.[1]

Selection of a method for determining the **thickness of sheet material** in process will depend primarily upon the inherent electrical conductivity of that material. If it is essentially a nonconductor, as rubber, plastic, or paper, measurement may be continuously performed by passing the sheet or web between the plates of a capacitor. (In such measurements on hygroscopic materials, moisture content may become a dominating factor.)

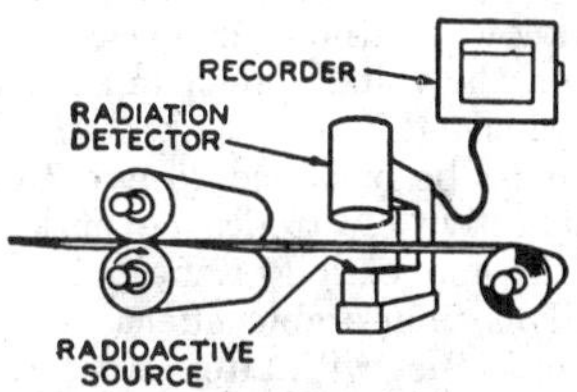

FIG. 3-109. Simplified diagram of beta-ray gage.

284. Sheet Thickness. Sheet materials, whether conducting or insulating, may be measured by the **beta-ray gage.** In this device a stream of beta rays passes through the sheet to a pickup head whose response is amplified and continuously recorded and, if desired, made the controlling influence in automatic regulation. Provision is made for the combined radiation source and pickup to traverse the strip of material and scan its whole width.

285. Thickness of coatings,[2] such as varnish or lacquer on conducting materials, may be determined by a continuous measurement of capacitance between the base and a reference electrode, the coating being included as a dielectric. With a magnetic base, such measurement may be performed effectively by determining the effect of the coating upon the gap in a magnetic circuit.

286. Surface roughness may be determined either on an absolute basis or by comparison with a "standard" surface. A common method involves passing a small stylus systematically over the surface, similarly to a phonograph needle, and measuring the resulting vibration. The stylus may be attached to a strain gage, a piezoelectric crystal, or a magnetic pickup. The resulting alternating emf may be amplified and displayed on an oscillograph, or it may be rectified and measured with a millivoltmeter. A basis for quantitative determination of surface roughness is found in USAS B46.2.

An absolute method of determining roughness utilizes the electrical capacitance of the tested surface in contact with an electrolyte as compared with that of an ideal (mercury) surface. On the assumption that the capacitance varies as the surface area, the comparison provides a figure representing the ratio of the tested surface to one of perfect smoothness.

287. Radiant energy may be measured either by a **thermocouple** or by a **photocell.** Response of the former, restricted almost wholly to the longer waves, is manifested in a thermoelectric effect. Photocells may be either photovoltaic, relying upon a barrier layer to develop an emf depending upon light intensity, or photoresistive, wherein the resistance of the cell varies with the light. Since photocells of both classes are selective with respect to wavelength and are not essentially linear in response, quantitative measurement of radiation by such means becomes complicated and subject to a number of variables. Thus, in precision photometry, the photocell is usually employed essentially as a detector in a null-balance system.

288. Transparency (or opacity) determination of materials and continuous monitoring of smoke density involve passing the substance to be examined through the path of a light beam directed upon a photocell. Uninterrupted measurement is made by means of a potentiometer or a bridge, according to the class of cell employed.

289. Viscosity measurement is essentially mechanical in its nature, and the application of electrical methods consists in determination of stress or displacement set up in the measuring apparatus owing to the characteristic of the fluid. One method involves measuring the electrical input to a small motor driving an impeller or stirrer in the fluid. Another method is based on electrical determination of the angle of lag (torque measurement) in a resilient mechanism through which an impeller is driven.

[1] BRANSON, N. G. Portable Ultrasonic Thickness Gage; *Electronics*, January, 1948, p. 88.

[2] Three Electronic Thickness Gages for Metallic Coatings; *NBS Tech. News Bull.* 38, September, 1941, No. 9.

CLARKE, E., CARLIN, J. R., and BARBOUR, W. E. Measuring the Thickness of Thin Coatings with Radiation Backscattering; *Elec. Eng.*, January, 1951, p. 35.

Thickness Meters, "The Instrument Manual"; London, United Trade Press, 1953, p. 12.

A further method utilizes **magnetostriction** to produce longitudinal oscillations in a steel rod carrying a diaphragm immersed in the liquid. Determination of the electrical loading on the exciting circuit provides a measure of viscosity.

290. Chemical Magnitudes. Electrical measurement has superseded many of the older methods of quantitative determination of chemical magnitudes. The two best-known methods are based, respectively, on the electrical **conductivity** of solutions and on the **voltaic effect** in specific cells. The basic principles of these measurements are wholly different, as are their applications. In the **conductivity cell** every precaution must be observed to avoid electrolytic effects, the prime requisite being that the respective electrodes be of identical material. Even then, the passage of current or the application of the potential tends to produce internal polarization emf in the cell. This undesirable effect may be almost wholly eliminated by measuring electrolytic resistance with alternating current, and the highly sensitive a-c detectors now available enable such tests to be made with precision. Outstanding among the uses of the resistance cell is determination of the **purity of water** for domestic and industrial purposes. Conductivity of water solutions usually increases in proportion to the amount of dissolved electrolytic material. Perfectly pure water has a specific resistance of 18 to 20 million Ω/cm^3, but in practice such values are virtually unobtainable. Only by careful distillation or deionization is it possible to obtain water of 400,000 to 800,000 specific Ω at a reference temperature of 70°F. Continuously operating water-conductivity recorders are supplied for use with commercial a-c power supply, and a typical range is 100,000 specific Ω to infinity.

291. Electrolytic cells utilize measurement of emf developed between a standard combination of electrodes by the solution under test. Development of the principle has reached its highest refinement in the measurement of pH, or hydrogen-ion concentration, which is a criterion of the activity with which the solution will enter as an acid into a chemical reaction. The pH value is a logarithmic function of the emf developed with a given strength of the solution in a specified cell. For pure water, which is "neutral" in its reaction, lying midway between the acids and the bases, the pH value is 7. pH measurement is essential in practically every industry involving any chemical process, as well as in waterworks, sewage systems, biological laboratories, and agricultural experiment stations.

BIBLIOGRAPHY

292. Electrical Measurement of Nonelectrical Quantities

Borden, P. A. Electrical Measurement of Physical Values; *Trans. AIEE*, February, 1925, Vol. 44, pp. 238–263. (330 refs.)

Supplementary bibliographies to above:

Trans. AIEE, 1927, Vol. 46, pp. 709–712.

Trans. AIEE, 1928, Vol. 47, pp. 1168–1171.

Pflier, P. M. "Elektrische Messung mechanischer Grossen"; Berlin, Springer-Verlag OHG, 1943. (Extensive bibliography.)

Roberts, H. C. "Mechanical Measurements by Electrical Methods"; Pittsburgh, Instruments Publishing Co., 1946. (470 refs.)

TELEMETERING

293. Telemetering is measurement with the aid of intermediate means which permit the measurement to be interpreted at a distance from the primary detector. Note: The distinctive feature of telemetering is the nature of the translating means, which includes provision for converting the measurand into a representative quantity of another kind that can be transmitted conveniently for measurement at a distance. The actual distance is irrelevant.

294. Electric telemetering is telemetering performed by deriving from the measurand or from an end device a quantitatively related separate electrical quantity or quantities as a translating means.

A **measurand** is a physical quantity, property, or condition which is to be measured.

Telemetering has been practiced many years in the central station industry and in the transmission and distribution of electric power but until lately only to a limited extent in the nonelectrical fields. With the phenomenal expansion of pipe lines for gas and for oil, the need has vastly increased, and electric telemetering installations have become indispensable in the remote measurement, totalization, regulation, and dispatching of these utilities. Telemetering has also found wide application in extensive industrial plants, such as refineries, steel mills, and large chemical plants, and in these installations it often forms an essential part of remote regulating apparatus.

295. Field of Application. The past decade has seen a rapidly increasing use of telemetering in aircraft, meteorology, ordnance, and guided missiles. This has led to a sharp demarcation of telemetering philosophies and techniques into two classes, "mobile" and "stationary." In the former, the apparatus is expected to operate for a very short period of time—often only a matter of seconds. The transmitting unit at least must be considered as expendable, and the combination is generally subject to an overall calibration for each isolated test in which it is used. Obviously, there can be no interconnecting physical circuit, and a radio link is an essential part of the system.[1]

296. Stationary systems in general involve transmitting and receiving units at fixed locations. These are usually of a permanent nature and are intended for operation over extended periods of time. Signal transmission between the stations usually involves a physical circuit, and even where radio principles are utilized, the most common practices require guiding of the signal by means of a more or less continuous conducting path.

297. A telemetering system incorporates the same three essential elements as are required in a system for measurement of nonelectrical quantities by electrical means, viz., a **transmitting unit** (transducer or pickup), a **receiving unit** (an instrument for measuring an electrical variable), and an **interconnecting circuit** or channel by which the electrical variable (signal) originating at the transmitter is carried to and impressed upon the receiver.

298. Channel. In transmission of measurement over considerable distances the **circuit or channel** may become the predominating factor in the system. In the ideal telemetering system the terminal apparatus would be inherently self-compensating, so that variations in circuit conditions would not adversely modify the signal. Merit of a telemetering system is directly related to the degree to which it approaches this ideal. Distance criterion of a telemetering system is not so much the number of feet or miles over which it will operate as it is the nature and magnitude of circuit impedance through which its signals will maintain their identity and proportionality. Since the data have been determined for specific types of circuits and channels, such magnitudes may generally be expressed in units of distance. A continually increasing proportion of telemetering is being carried out over circuits and channels leased from communication companies. With information available respecting the type of signal to be transmitted, the telephone or telegraph company provides a suitable circuit and assumes responsibility for its operation. Where privately owned circuits are used for telemetering, their maintenance and protection correspond to those for comparable communication circuits.

299. Classification of Telemeters. In classifying telemetering systems the USASI has adopted a grouping recommended by the AIEE and based on the nature of the electrical variable transmitted through the interconnecting circuit or channel. The names of the five classes are more or less self-explanatory and are as follows: *current, voltage, frequency, position,* and *impulse* types. In each of the first three of these classes the corresponding characteristic of the electrical output of transducer comprising the transmitting unit is varied with variations in the measurand. In the *position* system, the quantitative ratio, or the phase relationship, between two electrical voltages or currents determines the nature of the transmitted signal, usually requiring a circuit of three or more conductors. There are several *impulse* systems, in all of which the transmitting instrument acts to "key" a signal impressed upon the circuit, producing a series of successive pulses which, according to their nature, are interpreted by the receiving instrument and expressed in terms of the measurand.

Telemetering systems are not always mutually exclusive. A single installation may represent a combination of several of the named systems. In some instances it

[1] Treatment of mobile telemetering systems lies beyond the scope of this handbook, and until the appearance of an authoritative textbook on the subject, reference may be made to the *Reports* of the Annual Conferences on Telemetering, sponsored by the AIEE in cooperation with other technical organizations.

becomes difficult to decide into which of the specified classes a particular method of telemetering may fall.

300. Telemetering of electrical quantities, such as volts, watts, vars, etc., presents a problem owing to the inherently low torque of direct-deflecting instruments, whereas devices for measuring such magnitudes as position, flow, liquid level, etc., are not subject to such restrictions. Accordingly, where measurements of electric units are to be transmitted, practice favors those systems which place a minimum of burden upon the primary measuring instruments and preferably those adapted to transmitters having no moving parts. Thus, photoelectric, thermoelectric, and capacitive transmitters have found considerable favor in the electric industry.

301. Integrated Quantities. In transmitting measurements originating in integrating meters, such as watthour or varhour meters, the mechanism of the meter, either by photoelectric or electronic means or by a contact arrangement, is caused to develop a series of electrical pulses whose frequency of occurrence is proportional to the instantaneous value of the measured load. By a simple electronic network including capacitors charged and discharged at the frequency of the pulses, there is produced a direct current whose value is proportional to that frequency, the telemetering system being thus placed in the *current* class. On the other hand, the pulses may be directly impressed on the communication channel, whereupon the system falls into the *frequency* group.

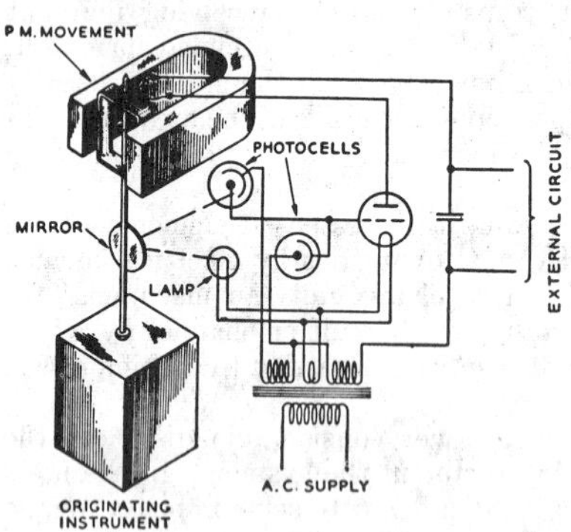

FIG. 3-110. Photoelectric torque balance.

302. Torque Balance. Where the basic measurement is performed by a low-torque instrument of the direct-deflecting class, such as a wattmeter, common telemetering practice involves either **balancing the torque** or matching the deflection of the instrument by the effect of an automatically regulated direct current in the winding of a permanent-magnet moving-coil mechanism (Fig. 3-110). This current, remaining proportional to the instrument torque, is transmitted through a metallic circuit for measurement at the receiving station and, if desired, may be included with other and similar currents in a load totalization.

303. Thermal Converter. A most flexible method for the transmission and totalization of electric power measurements involves the use of a **thermal converter.** The several commercially available forms of this device operate on a long-known but only recently applied principle combining the circuit of the thermal wattmeter with that of the thermocouple. In the former, the temperatures of two resistors are caused to assume values differing by an amount proportional to the power in the measured circuit. In the latter, there is developed an emf proportional to the temperature difference or to the *power* in the measured circuit, irrespective of power factor, frequency, or waveform. Thermal converters are supplied in single-element, two-element, and three-element forms, and the a-c input circuits may be wired into the instrument-transformer secondaries on any conventional polyphase power system. The output from the d-c terminals is either directly measured or interconnected with that of other converters to provide totals of measured loads. The full-load potentials are usually rated at 50 or 100 mV, according to make and type, and measurement is preferably made with a self-balancing potentiometer. For best results, thermal-converter output circuits, which, of course, must be wholly metallic, should be well shielded from parasitic electrical effects and should preferably be in a sheathed cable. An advantage of thermal-converter installations, even for relatively short distances within the plant, is that the 7 or 8 conductors necessary for connecting instrument-transformer secondaries to wattmeters or varmeters are replaced

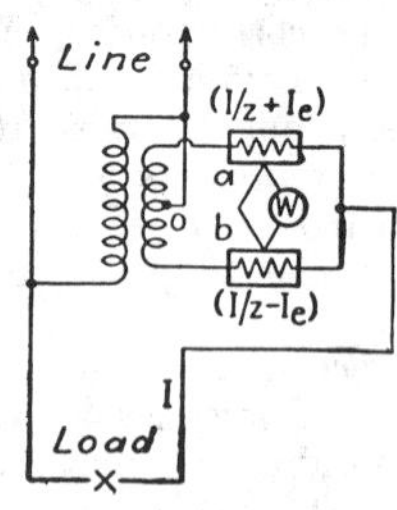

FIG. 3-111. Thermal watt converter.

by two small wires operating at a negligible power level. Furthermore, physical damage to the output wiring, whether in the nature of an open circuit or a short circuit, is not hazardous to equipment or personnel, and upon restoration of the circuit, normal operation will be resumed without loss of accuracy.

304. Electrical impulses may be used as signals for telemetering in a number of ways, the most important in stationary installations being that based on **frequency** and that based on **duration** of successive impulses. Impulse systems are to telemetering what telegraphy is to other forms of communication. The function of the transmitting instrument is essentially one of "keying" a circuit. Since the significance of the transmitted signal is based on time only, it follows that the method is most nearly immune to circuit conditions, such as voltage variation, impedance changes, attenuation, poor connections, and pickup from adjacent disturbing influences. Impulses whose frequency represents the measured variable may be transmitted as such, then falling into the category of the **frequency system of telemetering,** or they may be converted into a proportional direct current and be classified with the **current** systems.

305. Impulse-duration Telemetering. Signals recur at uniform intervals, and each has a duration corresponding to the then existing value of the measured magnitude. The transmitting instrument includes a constantly running cam or scroll plate having a spiral trailing edge and operating in the plane of the pointer but perpendicular to the line of excursion (Fig. 3-112). At a fixed point in each revolution of the cam, the

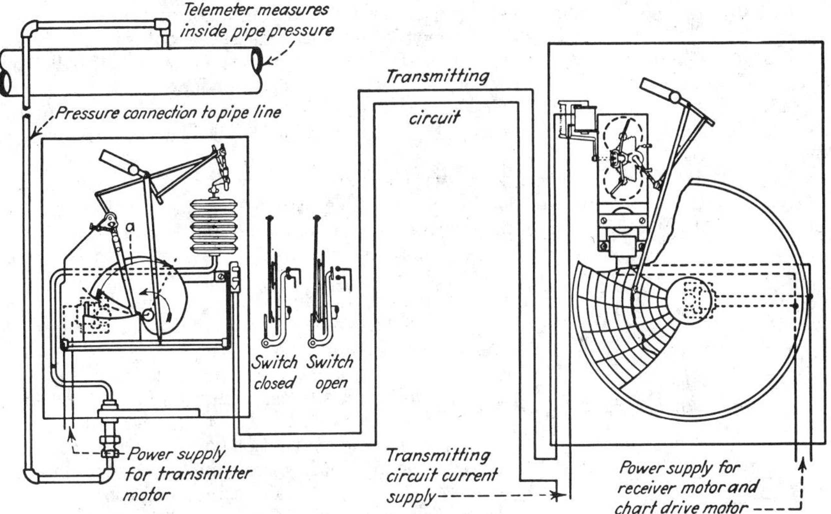

FIG. 3-112. Impulse-duration telemetering.

pointer is engaged and brought against the cam face until subsequently released by the trailing edge. With engagement and disengagement, the pointer is slightly deflected perpendicular to its line of travel and actuates a contact in a signal circuit. Because of the spiral form of the trailing edge, the length of the signal depends upon the position of the pointer and thus represents the measured variable.

The receiving unit includes the equivalent of a pair of electromagnetic clutches continuously driven by a constant-speed motor. These clutches are actuated by the incoming signals, one in an "up-scale" and the other in a "down-scale" sense, according to whether the transmitter pointer is on or off the cam. The receiver pointer or pen is frictionally retained in position and is "nudged" alternately toward one end or the other of its range by impellers or dogs carried by the clutches and respectively reset to zero as the corresponding clutch is released. Thus, with each signal, the receiver pointer finds or maintains a position corresponding to that of the transmitter pointer.

If the measuring element follows a linear law, the cam is shaped to an arithmetical spiral. If nonlinear, the trailing edge may be made to provide the needed compensation. For example, in a flow meter, wherein the deflection is basically proportional to the square of the rate of flow, correction for the quadratic law may be incorporated in the cam contour, with resulting linear signals. Not only does this produce a uniform scale in the receiving instrument, but it enables integration to be accomplished by driving a timing train in response to the up-scale signals. The system being inherently "sampling," the transmitter pointer is free to deflect during those intervals when it is not in contact with the cam. Thus, the impulse-duration system may be applied to even a delicate electrical instrument mechanism without loss of accuracy. Since each operating cycle of the receiver is established by the incoming impulses from the transmitter, such synchronism as is necessary is inherent in the principle, it being required only that both units run at constant and properly related speeds. Furthermore, upon circuit trouble or power interruption the system is self-restoring as soon as normal service is resumed. The signals, being in the nature of discrete impulses, may, of course, be transmitted by carrier, by radio, or by microwave.

306. Position Telemetering. In the **position system** of telemetering, the characteristic signal involves the relationship between two electrical quantities of a similar nature, i.e., two voltages or two currents. Unless carrier is used, position systems (with one exception[1]) require an interconnecting circuit of three or more conductors. The simplest position-telemetering arrangements are those of the rheostatic, or bridge, type, either direct or alternating current. Mechanical attachment of the measuring element to a voltage-dividing resistor provides a transmitting unit wherein the relative value of two voltages may be made proportional to the measured quantity. The receiving instrument may take the form of a ratio meter or may be a self-balancing bridge. The accuracy of such systems is affected by the impedance of the interconnecting circuit, but by maintaining this value small in comparison with that of the terminal instruments, the error may be made negligible for considerable lengths of line.

307. Selsyn. The inductive type of position system is best exemplified in the "selsyn" position motor or any one of its several equivalents. The transmitting and receiving units (Fig. 3-113) may be identical in structure. Each involves a stator and a rotor, one being provided with a single-phase and the other with a polyphase winding. The single-phase windings are excited from a common a-c source, and the polyphase windings are interconnected. The rotors of the two units will tend to assume duplicate angular positions, so that if one is attached to a measuring element, the other will provide a remote indication of its position. This system requires three line conductors in addition to the pair comprising the common power supply. The versatility and flexibility of the **differential transformer** (Par. **267**) render it particularly adaptable to telemetering of mechanical displacements.

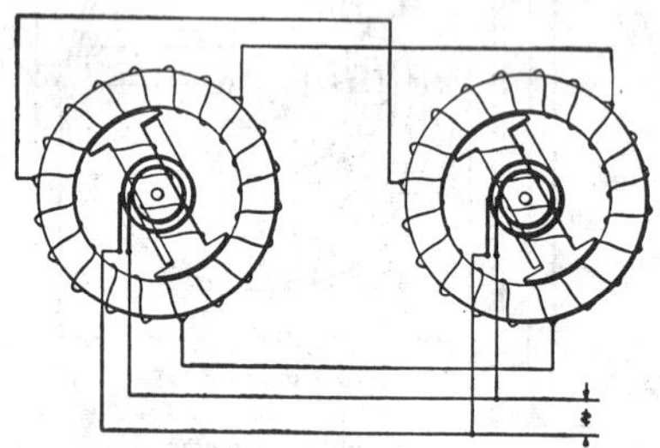

Fig. 3-113. Position motor.

308. Totalization of power loads and of other measured quantities may readily be effected in the current or voltage systems by connecting the outputs of the respective transmitters in parallel or in series, as the case may be. Subtotals and other mathematical functions also may be obtained. Telemetering, especially totalization and retransmission, is greatly facilitated by the power and flexibility of servo-actuated potentiometers and bridges. With these instruments available, there is practically no limit to the possibilities of telemetering, not only in the electrical-utility field but in association with pipe lines and large industrial plants. By **multiplexing**[2] the circuits it is possible for several telemetering transmitters and receivers to share a common communication channel. The most common systems of multiplexing are those based on *frequency* and those based on *time*. The frequency method transmits the signals

[1] The *rectifier* system: see Borden and Thynell (p. 70) in Bibliography (Par. **309**).

[2] Jacobson, A. W. A Time-multiplex System for Impulse-duration Telemetering; *Proc. Natl. Telemetering Conf.*, Chicago, 1953, p. 86.

Table 3-15. Typical Telemetering Systems
(Data selected from AIEE *Report* on Telemetering, 1948)

Designation	ASA class	Nature of transmitted current: Kind	Volts (max.)	Current, ma	No. of wires	Variables for which especially suited
Torque balance	Current	D-c	125	8	2	Electric power, or other quantities with specially adapted measuring elements
Photoelectric	Current	D-c	250	5	2	(same)
Current balance	Current	D-c	250	25	2	(same)
Thermal converter	Voltage	D-c	1	50	2	Electric power
Position motor	Position	A-c	22	10	3 or 5	Variables as measured
Bridge	Position	D-c or a-c	6	5	3	Variables as measured
Electronic	Position	A-c	48	65	3	Variables as measured
Impulse duration	Impulse	D-c or a-c	115	50	2*	Variables as measured
			Frequency range, cycles			
"Frequency"	Frequency	A-c	6–27 20–25 80–100		2†	Electric power

* Values given represent unrelayed signal. May be amplified or converted to any type of signal suited to channel.
† Signal voltage and current values for frequency systems not given. These are normally adapted to the specific installation.

on carriers having a specific frequency allotted to each transmitter and receiver combination. Time multiplexing involves the use of a multiple-point switch at each end of the circuit. These switches are progressively advanced at definite intervals, providing connection successively between each receiver and its corresponding transmitter. After a predetermined number of operations a distinct synchronizing signal checks and, if necessary, adjusts the relative position of the switches at the transmitting and the receiving stations.

BIBLIOGRAPHY

309. Publications Containing Telemetering Bibliographies

STABELIN, W. "Die Technik der Fernwirkanlagen"; Munich, R. Oldenbourg KG, 1934. (530 refs.)

Telemetering, Supervisory Control and Associated Circuits: Joint Subcommittee Report, AIEE, 1948. (43 refs.) (Under revision, 1956.)

BORDEN, P. A., and THYNELL, G. M. "Principles and Methods of Telemetering"; New York, Reinhold Publishing Corporation, 1948. (80 refs. and extensive list of patents.)

MABEY, C. A. Bibliography on Telemetering; *AIEE Publ.* S68, December, 1954. (850 refs.)

PERRY, C. C., and LISSNER, H. R. "The Strain Gage Primer"; New York, McGraw-Hill Book Company, 1955.

MEASUREMENT ERRORS[1]

310. General. The universal presence of uncertainty in physical measurements must be recognized as a starting point in a discussion of errors in measuring systems. These errors arise in (1) the measuring system itself and (2) the standards used for calibration of the system. This statement applies equally to control systems.

This subsection describes (1) sources of error, (2) evaluation of errors, including

[1] Reproduced by permission from "Handbook of Applied Instrumentation"; edited by D. M. Considine and S. D. Ross, New York, McGraw-Hill Book Company, 1964.

systematic and random errors, and (3) rules regarding the expression of data, especially significant figures.

311. Definition of Error. In making any physical measurement, the essential purpose is to assign a value, consisting of an appropriately chosen unit and an associated numeric, which will express the magnitude of the physical quantity being measured. For example, in the measurement of a temperature, the chosen unit may be degrees Fahrenheit and the associated numeric may be 110. Thus, 110°F. The extent of failure in exactly specifying this magnitude, and hence the departure of the stated value from the true value of the quantity, constitutes the error of measurement.

312. Types of Measurement. In considering and evaluating measurement errors, it is helpful to keep in mind the scheme of measurement employed. Some of the more common types of measurement are described as follows.

Direct Comparison. A measurement may consist of comparing the quantity being measured with a standard of the same physical nature. In such cases, the ratio between, or the difference of the standard from the unknown magnitude, is determined. The use of a Wheatstone bridge to determine a resistance in terms of a known resistance and a ratio is an example of this technique.

Adjustment to Equality. A calibrated magnitude is adjusted by known amounts until it is equal to the unknown. This is done in the determination of mass with a chemical balance, or in the measurement of voltage with a potentiometer. This type of measurement may be considered as a special case of direct comparison.

Direct Actuation of a Physical System. Some property of the measurand is used to operate a suitable indicating device, and its magnitude is read from an appropriate scale. For example, the magnitude of an electric current may be measured by the torque which it produces on the moving system of an ammeter, and the value is read from the angular deflection of the instrument pointer. Or a temperature may be measured by the expansion of liquid in a thermometer, and its value read from the height of the liquid column in the capillary.

313. Universality of Error. Whatever scheme of measurement is used, the value of the numeric assigned as a result of the measurement to describe the magnitude of the measurand will be in error to a greater or lesser extent, i.e., will depart somewhat from the true value of the quantity. No measurement, however elaborate or precise, or how often repeated, can ever completely eliminate this uncertainty. Thus, the true value of a measured physical quantity can never be stated with complete exactness. One of the most important phases of the art of measurement consists in the reduction of measurement errors to limits that can be tolerated for the purpose at hand.

Errors are unavoidable in the comparison of an unknown with a reference standard, or in the calibration and use of a measuring system. The value of the reference standard itself is also uncertain by an amount that depends on the whole chain of measurements extending back to the primary standards designed for maintaining all measurements on a common basis.

The established system of electrical units may be taken as an example. By virtue of certain relations between electrical quantities, values of all of them can be established in terms of the ohm and the volt. Values for these latter units are physically maintained by means of a group of wire-wound resistors and a group of saturated cadmium cells, which serve as primary standards at the National Bureau of Standards.

The values assigned to the primary standards are the experimental realization, through an elaborate series of measurements, of their legal definition, in terms of the fundamental mechanical units of length, mass, and time. Uncertainties are present at every step of this measurement procedure, extending back to the basic standards—the prototype meter bar and kilogram preserved in the vaults of the International Bureau of Weights and Measures at Sèvres, France.[1]

The skill of the various scientists who carried out these measurements was such that the uncertainties in values assigned to the primary standards, accumulated in the process of transferring from the mechanical prototypes, are probably less than 1 ppm

[1] In addition to the physical standards of length and mass, the prototype meter and kilogram, an absolute measurement of the electrical units requires that a standard of time measurement be set up at the time of the absolute measurement and that a value of permeability be assigned at the place of the measurement.

for the ohm and about 10 ppm for the volt. Values assigned to laboratory reference standards, by comparison with these primary standards and with other standards derived from them, are somewhat more uncertain, because of the further measurement procedures involved. Also, the laboratory reference standards themselves are of a lower quality, and are inherently less stable than the primary standards.

Reference standards for laboratory use are readily available for which uncertainties of value are well within 1/100%. Some of the best laboratory-standard deflecting instruments (ammeters, voltmeters, wattmeters) have errors of not more than 1/10%, while the errors of the usual grade of working laboratory instruments may amount to 1/4 or 1/2%, and the better grades of switchboard instruments are in the 1% accuracy class. Thus, in the case of indicating instruments, the uncertainty of value which must be accepted may be much more a result of inherent limitations in their design, construction, operation, and stability than of uncertainty in the working standards used for their calibration.

314. Sources of Error. In addition to the errors which necessarily result from faulty calibration of a measuring system, there are a number of sources of error that should be examined. These include: (1) noise, (2) response time, (3) design limitations, (4) energy gained or lost by interaction, (5) transmission, (6) deterioration of the measuring system, (7) ambient influences on the system, and (8) incorrect interpretation by the observer.

315. Noise in Measuring Systems. Noise may be defined broadly as any signal that does not convey useful information. Extraneous disturbances, generated in the measuring system itself or coming from the outside, frequently constitute a background against which the desired signal must be read. The background noise in a radio receiver is a familiar example; likewise, vibration or sudden displacements of the moving system of an indicating instrument may be classed as noise when they result from vibration pickup or shock excitation.

Sources of Noise. Noise may originate (1) at the primary sensing device, (2) in a communication channel or other intermediate link, or (3) in the indicating element of the system. In amplifying systems, the signal-to-noise ratio sets an upper limit to the useful amplification and, therefore, a lower limit to the magnitude of the wanted signal that can be observed against the background noise. Such disturbances are particularly a problem in electronic amplifiers, where they may arise from fluctuations in the resistance of a circuit element or contact point or from microphonics in a vacuum tube. Disturbing signals are, of course, more serious at the input end, where they undergo full amplification, than at an intermediate or final stage, where their amplification is less, relative to the input signal.

Noise signals may also be picked up by an electrical or mechanical coupling between an external source and an element or communication channel of the system. Another type of noise is the electrical signal produced by static charges generated between the conducting and insulating structure of a cable, as a result of flexing or other mechanical motion of the cable. Wherever signal disturbances are present, they contribute to the uncertainty of the measurement.

Fundamental Character of Noise. Under the most favorable circumstances, where noise signals are reduced to a minimum through filtering, careful selection of components, shielding, and isolation of the entire measuring system, there still exist certain noise levels that are always present. These random disturbances result from the fact that, in the final analysis, energy is always transferred in discrete, finite amounts (by molecular action in a gas, electrons as carriers of electric charge, and so on). The structure of natural phenomena is not infinitely fine-grained. Although the magnitude of fluctuations in the rate of energy transfer is usually small compared with the total transfer involved in a measurement, nevertheless these fluctuations supply a noise background and limit the ultimate sensitivity to which a measurement can be carried. Examples of sensitivity-limiting mechanisms are Brownian motion in a mechanical system, the shot effect in the heated cathode of a vacuum tube, the Johnson effect in conductors or resistance elements, and the Barkhausen effect in magnetic elements.

Figure 3-114 shows the voltage background resulting from Johnson noise in a low-resistance network. The figure was traced originally from an actual voltage record. The voltage from the network was fed through an amplifier to a recorder. The background voltage is approximately 10^{-3} μV, and the signal voltage is 20 times as great (2×10^{-2} μV) in each instance. The gain of the amplifier is different in the two records, but the amplitude of the background voltage (noise) is about 5% of the deflection produced by the signal in each record. With either low or high gain in the amplifier, the minimum signal that could be determined against the existing background would be about the same.

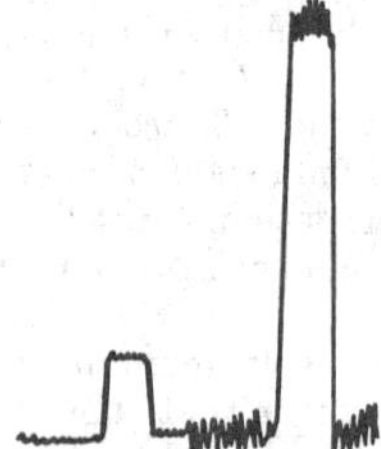

FIG. 3-114. Reproduced from a tracing of voltage record, showing a 0.02-μV signal impressed on a noise background of 0.001 μV. The amplification was increased by a factor of 5 for the second record. In each instance, the height of the signal is 20 times the height of the random background. (*From Considine and Ross, "Handbook of Applied Instrumentation."*)

316. Response Time. The time of response of a measuring system to an impressed signal may also contribute to the uncertainty of the measurement. If the signal is not constant in value, lag, or delay, in the system response results in an indication whose value depends on a sequence of values of the stimulus over an interval of time.

As an illustration of the effect of response time on indication, consider a simple system such as a mercury thermometer in air or a pressure gage consisting of a bellows connected to a source of pressure by a tube of small diameter. Heat must flow through the wall of the thermometer bulb and be absorbed by the mercury, raising its temperature and expanding it, or gas must flow through the tube to increase the pressure in the bellows and expand it. In either instance, the measuring system has capacitance and is connected through a resistance to the system under measurement. Such a measuring system is known as a first-order system, since its dynamic response can be expressed by a first-order differential equation.

This response is shown in Fig. 3-115*a* for a step change in the measurand and in Fig. 3-115*b* for a linear change. It will be observed that, for both types of change, the

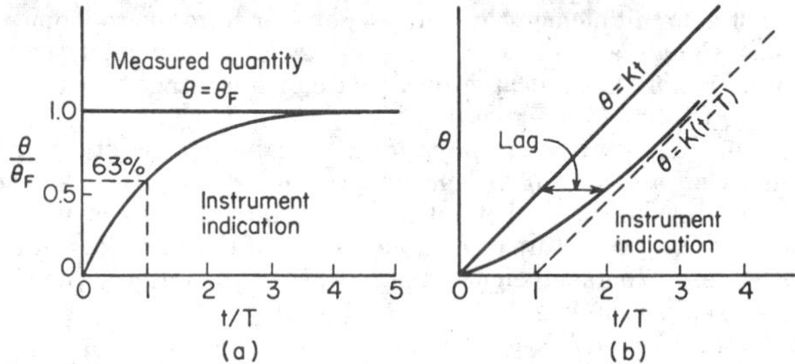

FIG. 3-115. Response of a measuring system: (*a*) to a step change; (*b*) to a linear change. (*From Considine and Ross, "Handbook of Applied Instrumentation."*)

response equation has a transient term that describes the initial response and a steady-state term (the asymptotic value of the response curve) that describes the behavior of the system after an interval which is long compared with the response time T. It will be seen that, in these simple examples, the indication of the measuring system is a function of its response time as well as of changes in the measurand.

More complicated systems or situations can, in many instances, be analyzed only approximately. But it can be generally stated that, for any system having a finite response time, the indication at any instant is the result of the events which happened over a previous time interval; that the magnitude of the indication not only depends on the variation of the signal over an interval of time previous to the observation but may also depend in a more or less complicated way on the response characteristics of the system itself.

317. Design Limitations. Limitations and defects in the design and construction of measuring systems are also factors in the uncertainty of measurements.

Friction. In moving parts, not only does friction contribute an uncertain amount to the damping of the system, but, because a certain minimum force is required to overcome the friction and to initiate motion, there results an uncertainty in the rest position of the indicator. This uncertainty can frequently be reduced by gently tapping the instrument, so that the forces transmitted to the moving system are just enough to overcome the frictional forces present that prevent the system from assuming its natural rest position. However, vigorous tapping may defeat its purpose by supplying enough force to produce deflection, thus moving the indicator to a new position which is still a function of friction. Also, if the bearings are delicate, they may be injured by vigorous tapping and, thus, increase the friction even more.

Resolving Power. Broadly speaking, this is the ability of the observer to distinguish between nearly equal quantities. In an optical system, such as that of a microscope, resolving power may be stated in terms of the smallest angle at which points in the field of view can be distinguished as separate. If the lenses in the optical train were perfect, this angle would be fixed by the effective aperture of the objective lens and the wavelength of the light used.

For a measuring system in which a scale reading is used to determine magnitudes, resolution would be limited by the smallest fraction of a scale division that could be read with certainty. Since, without a vernier, observers may attempt to read to one-tenth scale division but will not generally agree on a reading to better than perhaps two-tenths division, this latter figure probably represents about the limiting resolution of the scale.

In these instances, the limiting (or theoretical) resolving power is an ideal that is never attained because of imperfections or defects in the measuring system. For example, the scale of a good electrical instrument is usually constructed by accurately locating a series of cardinal points (or major scale divisions) by electrical measurements. The scale subdivision between cardinal points is then done by some sort of dividing mechanism. Thus, the actual location of a particular division is uncertain, both by the error of the measurements made to locate the adjacent cardinal points and by the imperfections of the mechanical device used to subdivide the scale.

318. Energy Exchanged by Interaction. Wherever the energy required for operating the measuring system is extracted from the measurand, the value of the latter is altered to a greater or lesser extent.

Low Energy Levels. The preceding statement is always true in some degree but is particularly significant where the available energy is limited in amount. The heat capacity of the thermometer used in a calorimetric experiment must be taken into account. A voltmeter, used to measure the voltage of a high-resistance source, draws current so that its reading is in error by the amount of the internal impedance drop in the source.

Coupling and Feedback Errors. In instances where energy is supplied to the measuring system from an auxiliary source, the value of the measurand may be altered by coupling to the measuring system and consequent feedback of energy. This may occur in electric circuits, for example, through resistive, inductive, or capacitive coupling and in mechanical circuits through elastic or viscous coupling. Actually, of course, the flow of energy may be in either direction under appropriate circumstances.

Interference. A third type of interaction is interference of an element of the measuring system with the action of the measurand. A simple example is the alteration of flow resulting from the introduction of an orifice plate into a pipe. As another example, a milliammeter, in measuring current in a low-resistance circuit, introduces additional resistance, and may alter the current by a significant amount.

319. Transmission. In the transmission of information from sensing element to indicator, any (or all) of three types of error may arise: (1) the signal may be attenuated by being absorbed or otherwise consumed in the communication channel; (2) it may be distorted by attenuation, resonance, or delay phenomena whose actions are selective on various signal components; or (3) it may suffer loss through leakage. In any of these circumstances, the signal reaching the indicator will differ in some respects from that at the primary sensing element.

320. Deterioration of Measuring System. Changes in the measuring system itself constitute a source of error in measurement. Physical or chemical deterioration or other alterations in characteristics of measuring elements may change their response and their indication. The oxidation of weights used in a laboratory balance, the change in resistance of a circuit element through strain relief, and the weakening of a permanent magnet, or a bellows, through aging, are examples. The alteration of thermocouple characteristics in an oxidizing or a reducing atmosphere is another example.

321. Ambient Influences on Measuring Systems. Of the various ambient conditions that may alter the calibration of an instrument, temperature influence almost always affects the measurement in one way or another. As examples, a change in temperature may alter the elastic constant of a spring, change the dimensions of a measuring element or linkage in the system, or alter the resistance of an electric-circuit element or the flux density in a magnetic element.

Other influences, not so universally active but often more important, are: humidity, barometric pressure; the presence of smoke or other foreign constituents in the air; the effect of the departure of an unbalanced mechanical system from its proper operating position; and several other influences which come from outside the instrument but which tend to affect its calibration.

322. Errors of Observation and Interpretation. Personal errors in the observation, interpretation, and recording of data must also be considered among the sources of uncertainty in measurements.

Parallax Errors. If the observer's line of sight to the pointer of an indicating instrument is not perpendicular to the scale face of the instrument, the reading will be high or low, depending on whether the observer's eye is to the left or right of this perpendicular. The amount of the parallax error will depend on the height of the pointer above the scale and on the sine of the angle between the line of sight and the perpendicular.

Linear Interpolation of Scales. In estimating the correction to be applied to readings of an indicating instrument, it is common practice to determine by measurement the corrections to be applied at the cardinal points of the scale and then to determine by linear interpolation the corrections to be applied at intermediate scale points. Such a procedure is simple and convenient but can be completely justified only if the scale is uniform and if the subdivisions between the cardinal points are correctly located. If the scale is not uniform or if errors are present in its subdivision, the determination of corrections by linear interpolation can result in error.

Personal Bias of Observer. Another reader error results from conscious or unconscious bias on the part of the observer. Some individuals will tend to favor even (or odd) tenths in estimating indicator position between marked scale divisions or perhaps will tend to read high in the lower half of the interval and low in the upper half. These bias patterns, of which the observer is not usually aware, vary considerably from one individual to another. They are consistently followed over long periods of time and generally tend to symmetry about the midpoint of the interval. The prevalence of observer bias patterns is such that readings made by a single observer are always questionable by one-tenth division and are frequently in error by two-tenths, even when the observer is consistent in his readings.

Mistakes. Faulty logging, either through writing down an incorrect digit or through transposition of digits, obviously must be classed as an observer error.

323. Classification of Errors. In estimating the magnitude of the uncertainty or error in the value assigned to a quantity as the result of measurement, a distinction must be made between two general classes of error: (1) systematic and (2) random. Apart from calibration errors which result in the consistent use of incorrectly assigned values and which are, therefore, systematic, the various sources of error give rise in some cases to systematic errors and in others to random errors.

324. Systematic Errors. Systematic errors are those which are consistently repeated with repetition of the experiment. Faulty calibration of the measuring system or a change in the system that causes its indication to depart consistently from the value assigned in calibration is of this type. Examples might include changes with age of the elastic properties of a spring or diaphragm or the reduction in strength of a magnet through shock or aging. Failure to take into account the energy extracted from a low-level source to operate the measuring system would also result in a systematic error.

As the observer is usually unaware of the presence or magnitude of systematic errors (since otherwise he could make appropriate corrections), this type of error is difficult to evaluate. It cannot be demonstrated by merely repeating the measurement under identical conditions, as this would lead to results that were consistently wrong. The skillful experimenter may be distinguished from the unskilled by his ability to plan and perform measurements in such a way as to minimize or avoid systematic errors.

In searching for systematic errors and in evaluating them, it is generally helpful to make definite known changes in those parameters of the measurement which are under the operator's control and to use different instruments or, if possible, a different method of measurement. In this way, errors that are functions of one of the controlled parameters are changed in magnitude, or those which arise from incorrect instrument calibration or which are inherent in a particular method may be altered. At times, it is possible to measure something whose magnitude is accurately and independently known and which is similar to the measurand. Such a measurement constitutes a check on the calibration of the measuring system and should help in evaluating systematic errors that may be present. Checking the accuracy of a micrometer with calibrated gage blocks or the accuracy of a thermometer at one or more fixed points on the temperature scale is of this nature. Another example is the checking of a potentiometer by measuring with it the emf of the standard cell that was used to standardize its current or, alternatively, the measurement of the emf of another standard cell whose value is independently known.

325. Random Errors. Random errors are those which are accidental, whose magnitude (and sign) fluctuates in a manner that cannot be predicted from a knowledge of the measuring system and the conditions of measurement.

In the measurement of any physical quantity, the observations are influenced by a multitude of contributing factors. These are the parameters of measurement. In an ideal measurement, all the parameters are fixed in value, so that the magnitude of the measurand is completely defined and may be exactly determined. Repeated observations of the magnitude of a quantity differ as a result of the operator's failure to (1) control the parameters closely enough or (2) to apply proper corrections for their influence. These uncontrolled, or unknown, or uncorrected influences disturb the observations and cause the result to depart from the value it would have if all parameters were completely defined and fixed.

If it is assumed that the various parameters beyond the observer's control (and the uncontrollable residue of those he tries to hold fixed) act in a completely random fashion, probability theory can be used to deduce certain helpful results. Now, combinations of circumstances that produce large departures of the observed from the true value of a quantity will occur less frequently than those producing small deviations, and each influence may be presumed equally likely to cause either a positive or a negative departure. Hence, the general tendency of the effects of all influences will be to cancel one another rather than to be additive. If all measurement errors followed the laws of chance, it could be expected that the true value of a quantity would be the average of an infinite number of observations.

There is an important weakness in the foregoing argument. It may well be that some of the parameters influencing the result of a measurement do not act in a completely random fashion. There is never assurance that all systematic errors have been eliminated from a measurement or that proper corrections have been made for those errors present. Under such conditions, the average of a large number of observations would not approach the true value of the quantity but would differ from the truth by the algebraic sum of the uncorrected systematic errors. The laws of probability take no account of an unknown but constant error superposed on deviations that are truly random.

In Table 3-16, a classification is suggested of the various sources of error with respect to the type of error produced. It will be seen that errors from a particular source cannot, in general, be uniquely classified as either systematic or random. For example, noise is inherently random in nature; but if the response of the indicating system is always in the same direction independently of the signal polarity, then, for a weak signal, the noise present may produce a positive indication that is larger on the average than would be produced, in the absence of noise, by the signal being measured. Thus, for some response systems, noise may give rise to a spurious signal that is always added to the measured value, and its effect is systematic. To illustrate this point further, it is suggested that the reader attempt his own classification of errors and then look for the exceptions in the various categories.

Table 3-16. Types and Sources of Measurement Errors

Source of error	Type of error	
	Random	Systematic
Noise	Generally	May be
Response time	Seldom	Almost always
Design limitations	Usually	Sometimes
Energy of interaction	May be	Usually
Transmission	Sometimes	Usually
Deterioration	Seldom	Usually
Ambient influences	Usually	May be

326. Evaluation of Data. Statistical procedures are available that make it possible to state from a limited group of data the most probable value of a quantity, the probable uncertainty of a single observation, and the probable limits of uncertainty of the best value that can be derived from the data. It must be borne in mind here that the objective is precision (or consistency) of values rather than their accuracy (or approach to the truth). That this is necessarily so has already been pointed out; the laws of chance operate only on random errors, not on systematic errors. Some of the important results of the theory of probability, as applied to the treatment of data, are given in the following paragraphs.

327. Arithmetic Average or Mean. The best value that can be obtained from a group of similar measurements of a quantity is usually the arithmetic average or mean.

Suppose, for example, that a series of measurements are made (all with the same care) of the length of a rod wherein a scale divided into millimeters is used. Estimations are made to tenths of a millimeter. To avoid a systematic error that might result from an attempt to align the end of the rod with the end of the scale, the rod is laid along the scale. The scale position corresponding to each end of the rod is read. The rod is shifted along the scale for each new pair of readings, and the difference between the members of the pair is taken as the observed value of the rod length. As an example, the results of 10 such measurements are as follows:

Reading *A*, mm	Reading *B*, mm	Difference, observed length, mm
78.8	37.1	41.7
67.4	25.4	42.0
92.1	50.3	41.8
56.8	14.8	42.0
88.1	46.0	42.1
50.5	8.6	41.9
74.7	32.7	42.0
82.4	40.5	41.9
61.6	19.1	42.5
65.4	23.6	41.8

The arithmetic average of the observed values is

$$X = \frac{\Sigma x}{n} = \frac{419.7}{10} = 41.97$$

328. Standard Deviation. One of the best measures of the dispersion of a set of observations is the root mean square of the deviations of individual observations from the arithmetic average of the set. This is the standard deviation, and can be expressed by

$$\sigma = \sqrt{\frac{\Sigma d_m^2}{n}}$$

where d_m is the deviation of the individual observation from the group mean, and n is the number of observations in the set. Statisticians customarily give as an estimate of σ the quantity

$$s = \sqrt{\frac{\Sigma d_m^2}{n - 1}}$$

because it more accurately describes the dispersion of the set when the number of observations is small. It will be noted that the value of s approaches that of σ as n is increased and differs from it by a significant amount only when n is quite small. If the difference is taken between each of the individual observations of rod length and their arithmetic average in the example given in the preceding paragraph, the standard deviation can be computed as follows:

d_m	d_m^2
−0.27	7.29×10^{-2}
0.03	0.09×10^{-2}
−0.17	2.89×10^{-2}
0.03	0.09×10^{-2}
0.13	1.69×10^{-2}
0.07	0.49×10^{-2}
0.03	0.09×10^{-2}
−0.07	0.49×10^{-2}
0.53	28.09×10^{-2}
−0.17	2.89×10^{-2}

$$\sigma = \sqrt{\frac{\Sigma d_m^2}{n}} = \sqrt{\frac{44.10 \times 10^{-2}}{10}} = 0.21 \text{ mm}$$

$$s = \sqrt{\frac{\Sigma d_m^2}{n - 1}} = \sqrt{\frac{44.10 \times 10^{-2}}{9}} = 0.22 \text{ mm}$$

It will be apparent that, even with as small a number as ten observations, there is little or no significant difference between σ and s.

329. Probable Error of a Single Observation. As an index of the consistency of a set of observations, the probable error of a single observation may be used. It is defined as that deviation from the group mean for which the probability that it will be exceeded is equal to the probability that it will not be exceeded. If the number of observations is large, the probable deviation of a single observation from the mean is given by

$$r = \pm 0.6745\sigma$$

where σ is the standard deviation defined previously. If the number of observations is small, this formula should be modified to

$$r = \pm 0.6745 \sqrt{\frac{\Sigma d_m^2}{n - 1}}$$

While probable error expresses correctly the range in which the chances are equally good that the numerical value will or will not be found, provided that a sufficiently large group of data is available to make this estimate meaningful, it has actually no more significance as a precision index than the standard deviation from which it is derived. Hence, it has largely fallen into disuse, in favor of a statement of standard deviation. It may be found, however, in older literature as an index of the precision of data.

In the foregoing example,

$$r = \pm 0.6745 \sqrt{\frac{\Sigma d_m^2}{n - 1}} = \pm 0.6745 \sqrt{\frac{44.10 \times 10^{-2}}{9}} = \pm 0.15 \text{ mm}$$

330. Probable Error of the Mean. This is defined as the amount R by which the

Table 3-17. Confidence Level and Confidence Interval Values Where Number of Observations Is Large

Confidence level	Confidence interval	Values lying outside confidence interval
0.50	$\mu \pm 0.674\sigma$	1 in 2
0.80	$\mu \pm 1.282\sigma$	1 in 5
0.90	$\mu \pm 1.645\sigma$	1 in 10
0.95	$\mu \pm 1.960\sigma$	1 in 20
0.99	$\mu \pm 2.576\sigma$	1 in 100
0.999	$\mu \pm 3.291\sigma$	1 in 1,000

mean of a group of observations may be expected to differ (with a probability of 50%) from the mean of an infinite set taken under the same conditions of measurement. It can be calculated by the formula

$$R = \frac{r}{\sqrt{n}}$$

or, if the number of observations is large, by its equivalent

$$R = \pm 0.6745 \frac{\sigma}{\sqrt{n}}$$

the symbols having the same meaning as in the foregoing description. Only if the number of observations is large will the value of R, calculated as above, accurately represent the departure of the group mean from the mean of an infinite set with a 50 per cent probability. If the number of observations is small, R becomes an increasingly inaccurate index of precision. The present trend in the statistical treatment of data is toward the abandonment of probable error in favor of a much broader and more useful concept, that of "confidence intervals," which are described in the following paragraphs.

In the previous example, the probable error of the mean of 10 observations is given by

$$R = \frac{r}{\sqrt{n}} = \frac{\pm 0.15}{\sqrt{10}} = \pm 0.05 \text{ mm}$$

331. Confidence Intervals. Probable error, previously defined, is a special case of a much broader concept. It is possible by the statistical analysis of data to state a range of deviation from the mean value within which a certain fraction of all values may be

Table 3-18. Factors for Establishing Confidence Interval Where Number of Observations Is Small

Number of degrees of freedom	Number of observations	Confidence level 0.5	0.9	0.95	0.99
		Confidence interval			
1	2	$\mu \pm 1.00s$	$\mu \pm 6.31s$	$\mu \pm 12.71s$	$\mu \pm 63.66s$
2	3	$\mu \pm 0.82s$	$\mu \pm 2.92s$	$\mu \pm 4.30s$	$\mu \pm 9.92s$
3	4	$\mu \pm 0.77s$	$\mu \pm 2.35s$	$\mu \pm 3.18s$	$\mu \pm 5.84s$
4	5	$\mu \pm 0.74s$	$\mu \pm 2.13s$	$\mu \pm 2.78s$	$\mu \pm 4.60s$
5	6	$\mu \pm 0.73s$	$\mu \pm 2.02s$	$\mu \pm 2.57s$	$\mu \pm 4.03s$
6	7	$\mu \pm 0.72s$	$\mu \pm 1.94s$	$\mu \pm 2.45s$	$\mu \pm 3.71s$
7	8	$\mu \pm 0.71s$	$\mu \pm 1.90s$	$\mu \pm 2.37s$	$\mu \pm 3.50s$
8	9	$\mu \pm 0.71s$	$\mu \pm 1.86s$	$\mu \pm 2.31s$	$\mu \pm 3.36s$
9	10	$\mu \pm 0.70s$	$\mu \pm 1.83s$	$\mu \pm 2.26s$	$\mu \pm 3.25s$
10	11	$\mu \pm 0.70s$	$\mu \pm 1.81s$	$\mu \pm 2.23s$	$\mu \pm 3.17s$
15	16	$\mu \pm 0.69s$	$\mu \pm 1.75s$	$\mu \pm 2.13s$	$\mu \pm 2.95s$
∞	∞	$\mu \pm 0.67s$	$\mu \pm 1.64s$	$\mu \pm 1.96s$	$\mu \pm 2.58s$

NOTE: This table is a modification and abridgment of Table IV in Fisher and Yates, "Statistical Tables for Biological, Agricultural, and Medical Research," Edinburgh and London, Oliver & Boyd, Ltd.

expected to lie. Such a range is called a "confidence interval," and the probability that the value of a randomly selected observation will lie within this range is called the "confidence level." If the number of observations is large and their errors are random (the normal distribution of errors), various confidence intervals about the mean value μ are as stated in Table 3-17.

If the number of observations is small and the standard deviation σ is, therefore, not accurately known, these intervals must be broadened. Here one would compute $s = d_m^2/(n - 1)$ and multiply it by an appropriate factor, as shown in Table 3-18, to establish the confidence interval, i.e., the interval within which one would expect to find a randomly selected observation, with a particular level of confidence.

It will be noted that the factors in the table corresponding to an infinite number of observations are the same as the factors multiplying σ in Table 3-18.

To obtain the confidence intervals for the group mean from the corresponding intervals for an individual observation, the latter is divided by $\sqrt{n}$. Thus, the expectation that the mean of a group of observations will differ by not more than a certain amount from the theoretical mean of an infinite set can also be expressed in terms of a confidence interval and a confidence level.

In the previous example for the measurement of rod length, $s = 0.22$ mm, so that the confidence intervals for an individual observation corresponding to various confidence levels are:

Confidence level.	0.50	0.90	0.95	0.99
Confidence interval.	±0.15	±0.40	±0.50	±0.71

For the group average (41.97), the corresponding confidence intervals are ±0.05, ±0.13, ±0.16, ±0.23. Thus, the average value (41.97) found from the 10 observations made would be expected, with a probability of 50%, to be within 0.05 mm of the mean of an infinite number of measurements. At the 99% level of confidence, it may be stated that the observed and theoretical means differ by not more than 0.23 mm.

332. Rejection of Data. Occasionally, one individual value in a set of observations is noticeably different from the others, and a decision must be made to either use or discard it. If the observer knows at the time the data are taken that the system is not behaving properly or that a blunder has been made, the observer should immediately discard the data. If the observer is not able to assign a valid reason for discarding it at the time the observation is made, it is questionable that it should be eliminated at all. The cause should be sought when a single observation is outstandingly different from the others of the set. It may be that the quantity being measured has changed temporarily or that there has been some other significant change in the conditions of measurement. To eliminate an observation simply because it differs from the others by more than normally would be expected is not sound practice.

A criterion sometimes used for discarding an observation is that its deviation from the mean is greater than four times the probable error of a single observation. This corresponds to discarding data that lie outside a confidence interval for a single observation at a level of 0.993. A better criterion, which would avoid the difficulty of estimating probable error when the set is small and the standard deviation σ is not accurately known, would be to discard data that lie outside the interval corresponding to a confidence level of 0.99 for a single observation. On this basis, not more than 1 in 100 observations would lie outside this range if only random influences were operating to produce dispersion in the data. However, rather than to use such a criterion arbitrarily as a reason for discarding an observation, it is perhaps better to use it as a criterion for thoroughly examining the conditions of the measurement in order to find its cause. Better yet, the interval corresponding to a confidence level of 0.95 may be used as a criterion for the need to scrutinize the measurement procedure and the quantity measured.

In the example of the measurement of rod length, the ninth observation in the set differed from the mean by 0.53 mm. This is outside the interval (μ0.50) corresponding to a confidence level of 0.95, but inside the interval (μ0.71) corresponding to the 0.99

level. On the basis of the rejection criterion just stated, the observation should be retained.

333. Comparison of Averages. If two sets of measurements, taken under different conditions, yield different average values (A and B) of a quantity, with different standard deviations σ_A and σ_B, the question of their consistency arises. Is the difference $A - B$ consistent with the assumption that random errors alone operate? Or does the difference result from the presence of systematic errors that operate differently in sets A and B? The sets are inconsistent if the difference of their averages is more than twice the sum of their standard deviations. This is a rough but simple test to apply. Thus, A and B are inconsistent (show the operation of systematic errors) if $A - B > 2(\sigma_A + \sigma_B)$.

A much more sensitive test is based on the confidence intervals previously described. Furthermore, this test (the t test) is equally applicable to large and small sets. Let the individual values in the first set be $x_1, x_2, \cdots, x_m$ (m observations) with individual differences from the average $d_i = x_i - A$. Let the values in the second set be $y_1, y_2, \cdots, y_N$ (n observations) with the individual differences from the average $f_i = y_i - B$. Then

$$s^2 = \frac{\Sigma d_i^2 + \Sigma f_i^2}{(m-1) + (n-1)}$$

The estimated standard deviation of an individual measurement is s, and that of the difference between the average is

$$\frac{s}{\sqrt{mn/(m+n)}}$$

The ratio of the difference of the averages to the standard deviation of the difference is

$$t = \frac{A-B}{s}\sqrt{\frac{mn}{m+n}}$$

This computed t ratio is a measure of the consistency of the two sets of measurements, and should be compared with the factor previously used to multiply s in establishing confidence intervals. The critical values of t are given in Table 3-19 for confidence levels of 0.99 and 0.95, corresponding to 1 chance in 100 or 1 in 20, respectively, that random errors alone cause a larger difference.

Table 3-19. Values of t Used in Establishing Confidence Levels

Total number of observations $(m + n)$	4	6	8	10	12	17	32	∞
Degrees of freedom $(m - 1) + (n - 1)$	2	4	6	8	10	15	30	∞
Critical value of t_{99}	9.92	4.60	3.71	3.36	3.17	2.95	2.75	2.58
Critical value of t_{95}	4.30	2.78	2.45	2.31	2.23	2.13	2.04	1.96

NOTE: More elaborate tables of t will be found in most modern textbooks on statistical methods, for example, Table 5 in the appendix of W. J. Dixon and F. J. Massey, Jr., "Introduction to Statistical Analysis," 2d ed.; New York, McGraw-Hill Book Company, 1957.

If the computed ratio is significantly larger than the critical value of t, it may be concluded, with a corresponding level of confidence, that there are systematic errors which operate differently in A and B. Of course, the presence of systematic errors that affect both sets of measurements in the same way will not be disclosed by either test.

334. Propagation of Errors. It frequently occurs that independent measurements $x_1, \cdots, x_N$ are obtained, and that some function of them $f(x_1, \cdots, x_N)$ is of interest. If $f(x_1, \cdots, x_N)$ has continuous first and second derivatives, and if each x_i has approximately a normal distribution with a mean μ_i and standard deviation σ_i, where each σ_i is small, then $f(x_1, \cdots, x_N)$ has approximately a normal distribution with mean

$f(\mu_i, \cdots, \mu_N)$ and standard deviation

$$\sigma_f = \sqrt{\sum_{i=1}^{N} \left(\frac{\partial f}{\partial x_i}\right)^2 \Bigg|_{x_i=\mu_i} \cdot \sigma_i^2}$$

For practical purposes, since the mean values μ_i are usually not known, the observed x_i may be used in evaluating the partial derivatives. Subject to the stated restrictions, the standard deviation of $f(x, y, z, \cdots)$, where $x, y, z, \cdots$ are independent, is

$$\sigma_f = \sqrt{\left(\frac{\partial f}{\partial x}\right)^2 \sigma_x^2 + \left(\frac{\partial f}{\partial y}\right)^2 \sigma z^2 + \left(\frac{\partial f}{\partial z}\right)^2 \sigma_z^2 + \cdots}$$

This can be written in the following simple instances as follows:

Let

$$f = ax + by$$

$$f = kxy$$

$$f = k\frac{x}{y}$$

$$f = kx^n$$

$$f = ke^x$$

$$f = k \ln x$$

Then

$$\sigma_f \approx \sqrt{a^2\sigma_x^2 + b^2\sigma_y^2}$$

$$\sigma_f \approx k\sqrt{y^2\sigma_x^2 + x^2\sigma_y^2}$$

$$\sigma_f \approx k\sqrt{\sigma_x^2 + \left(\frac{x}{y}\right)^2 \sigma_y^2}$$

$$\sigma_f \approx nkx^{n-1}\sigma_x$$

$$\sigma_f \approx ke^x - \sigma_x$$

$$\sigma_f \approx \frac{k}{x}\sigma_x$$

335. Normal-distribution Law. The normal pattern of distribution about a mean value is shown in Fig. 3-116. Starting with the assumption that the most probable value of a quantity is the arithmetic mean of a large number of determinations, each

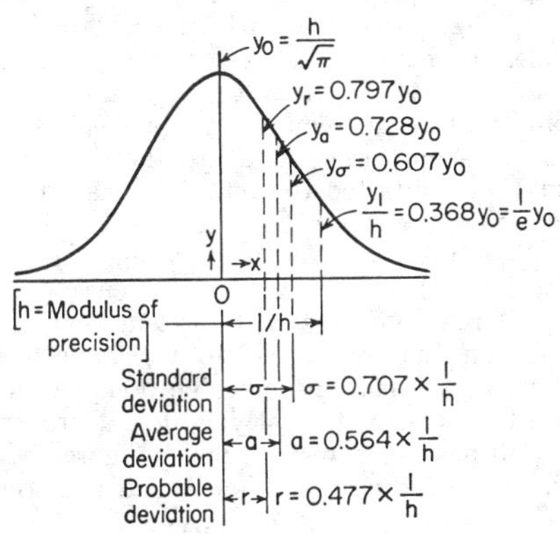

FIG. 3-116. Indices of precision. (*From Considine and Ross, "Handbook of Applied Instrumentation."*)

made with equal care, the normal law of distribution

$$y = \frac{h}{\sqrt{\pi}} e^{-h^2x^2}$$

can be deduced. Here x is the amount that a particular observation deviates from the group mean and y is the frequency of occurrence of such a deviation in the group.

Normal distribution law: $$y = \frac{h}{\sqrt{\pi}} e^{-h^2x^2}$$

Probability of deviation between $-x$ and x: $$P_x = \frac{h}{\sqrt{\pi}} \int_{-x}^{x} e^{-h^2x^2}\, dx$$

The precision (or repeatability) of the observations in the group is indicated by the maximum height $h/\sqrt{\pi}$ and by the deviation $x = 1/h$ at which the height is $1/e$ times the maximum. Thus, the parameter h is an indication of the narrowness or "spread" of the curve and has been called the "modulus of precision."

Certain indices of precision sometimes used (namely, the standard, average, and probable deviation of observations from the mean) are indicated in Fig. 3-116. The frequency of occurrence of each in a normal distribution is also shown. Ordinates of the normal distribution curve can be considered as probability coefficients; the probability of a deviation x from the mean is given by the area under the curve between x and y. Thus

$$P_x = \frac{h}{\sqrt{\pi}} \int_{-x}^{x} e^{-h^2x^2}\, dx$$

The law of normal distribution can also be written in terms of standard deviation as

$$y = \frac{1}{\sigma\sqrt{2\pi}} \exp\left[-\frac{1}{2}\left(\frac{x-\mu}{\sigma}\right)^2\right]$$

where σ is the standard deviation and μ is the mean value.

336. Significant Figures. In computing or stating the results of a measurement, it is important that the recorded figures include all that convey usable information. Frequently, it is also important that figures be excluded that do not convey information of value in the determination. In recording a numerical result, if only those digits which are meaningful are set down, the manipulation of surplus digits is avoided in subsequent computations and opportunities for arithmetical errors are lessened.

Because data are more easily tested in terms of precision (or repeatability) than in terms of accuracy (or approach to the true value of the quantity), the number of digits making up the significant figures is frequently the recorded statement of the experimenter's precision of measurement. Wherever statistical tests of precision are made, or if there is any possibility that they may be made in the future, the recorded data should include the numerical figures which contain information on the variation between determinations. This applies not only to the worker's notebook, but also to published data that may be subjected to statistical analysis by an interested reader. Usually, the experimenter is also the person best qualified to form a judgment concerning the accuracy of his results. No measurement can ever be considered really complete without some kind of evaluation of the accuracy of results. If the worker distrusts the accuracy of his results to such an extent that, in rounding off numerical values, he suppresses the information needed to determine their precision, his work is thereby decreased in value. When this must be done, it should be set forth in an accompanying statement if it is not clearly apparent from the results themselves. There are certain conventions regarding significant figures which have attained wide acceptance, but which should be used judiciously and with the foregoing argument in mind.

Retention of Digits. The last digit in a numerical result should represent the point of uncertainty. Although there is no universal agreement on a rule for deciding how many digits to record in tabulating values of known precision, it is usually considered acceptable to retain the last figure which is uncertain by not more than 10 units. Thus, the value 24.3 would at best lie between 24.2 and 24.4 and at worst between 23.3 and 25.3.

Rounded Numbers. In rounding off a number, the last retained digit should be increased by one unit when the first dropped digit is greater than 5 or is 5 followed by digits other than zero. The last digit should not be changed when the dropped digit is less than 5. Thus, if 24.352 were rounded off to three significant figures, it would be stated as 24.4, whereas 24.349 would be stated as 24.3. When the dropped digit is 5 and no further figures follow it, a practice which is frequently followed is to round off to the even number. Thus, 24.35 would round off to 24.4, whereas 24.25 would be rounded off to 24.2. To a limited extent, such a practice improves a value obtained by averaging, since the rounding will cause an increase to the higher digit about as often as a decrease to the lower digit.

Significant Zeros. To avoid misunderstanding and for convenience in multiplication and division, zeros which are not significant but which serve only to indicate the location of the decimal point should not be used in the stated number. It is better to indicate the location of the decimal point exponentially, using an appropriate power of ten. Thus, 12,500 is ambiguous as stated. One does not know whether three or five significant figures are intended. If there are only three significant figures, the ambiguity is eliminated if it is written 125×10^2 or 1.25×10^4. If five-place significance is intended, there is no ambiguity if it is written 125.00×10^2 or 1.2500×10^4. When numbers are to be added or subtracted, they must, of course, first be reduced to the same units and expressed in terms of a common power of ten.

Multiplication and Division. In multiplication and division, we need retain in each factor only the number of digits which will produce in that factor a percentage uncertainty that is no greater than the uncertainty in the factor having the fewest significant figures. Thus, the product of 103.24 and 8.1 would be written $103 \times 8.1 = 830$. The factor 8.1 being known only to about 1%, the factor 103.24 need be considered only to 1% (or 103).

Addition and Subtraction. In these operations, no digit need be retained in the result whose position with respect to the decimal point is to the right of the last significant figure in any number entering the computation. Thus, 24.3 and 2.102 would be added as 24.3 plus 2.1 = 26.4. No digit in the result farther to the right of the decimal point would have a meaning.

Slide-rule Computation. If the accuracy of the result need not be better than ¼%, a 10-in slide rule is adequate for the computation. If accuracy requirements are better than ¼%, machine methods, longhand, or logarithms should be used.

Computation with Logarithms. In computing with logarithms, no more digits need be retained in the mantissa of the logarithm than are significant in the corresponding numerical factor. Thus, the briggsian logarithm of 103.2 may be written as 2.0137 rather than 2.013679.

Averaging. In taking the average of four or more numbers, an additional digit beyond those of the individual values may be retained as having possible significance.

Precision Index. A number representing a precision index need never be stated to more than two significant figures.

REFERENCES ON MEASUREMENT ERRORS

337. Bibliography

ASTM Manual on Presentation of Data, 1945.

Curtis, H. L. "Electrical Measurements"; New York, McGraw-Hill Book Company, 1937.

Harris, F. K. "Electrical Measurements"; New York, John Wiley & Sons, Inc., 1952.

"Handbuch der experimental Physik"; Leipzig, Akademische Verlagsgesellschaft Geest & Portig KG, 1926, Vol. 1, Messmethoden.

Eckman, D. P. "Industrial Instrumentation"; New York, John Wiley & Sons, Inc., 1950.

WHITEHEAD, T. N. "Instruments and Accurate Mechanism"; New York, Dover Publications, Inc., 1954.

WILSON, E. B., JR. "An Introduction to Scientific Research"; New York, McGraw-Hill Book Company, 1952.

DIXON, W. J., and MASSEY, F. J., JR. "Introduction to Statistical Analysis," 2d ed.; McGraw-Hill Book Company, New York, 1957.

YOUDEN, W. J. "Statistical Methods for Chemists"; New York, John Wiley & Sons, Inc., 1951.

WORTHING, A. G., and GEFFNER, J. "Treatment of Experimental Data"; New York, John Wiley & Sons, Inc., 1943.

BACKSTRÖM, H. E. *Z. Instrumentenk.*, 1930, Vol. 50, pp. 561, 609, 665; 1932, Vol. 52, pp. 105, 260.

CERNI, R. Sources of Instrument Error; *ISA Jour.*, June, 1962, Vol. 9, No. 6, pp. 29–32.

YOUDEN, W. J. Realistic Estimates of Error; *ISA Jour.*, October, 1962, Vol. 9, No. 10, pp. 57–58.

TOPPING, J. "Errors of Observation and Their Treatment"; New York, Reinhold Publishing Corporation, 1957.

CHATTERTON, J. B. The Uncertainty of Measuring Systems; *Trans. IRE*, March, 1958, Vol. I-7, pp. 90–94.

ENTIN, L. P. Is Instrument Zero Output Really Zero? *Control Eng.*, December, 1959, Vol. 6, No. 12, pp. 95–96.

ENTIN, L. P. What about Scale Factor and Resolution? *Control Eng.*, February, 1960, Vol. 7, No. 2, pp. 75–77.

SECTION 4

PROPERTIES OF MATERIALS

BY

A. A. JONES Chief Engineer-Administrative, Anaconda Wire and Cable Company; Fellow, Institute of Electrical and Electronics Engineers; Fellow, Institution of Electrical Engineers (London); Member, American Society for Testing and Materials; Registered Professional Engineer (New York)

IRVIN L. COOTER Consultant, formerly Chief, Magnetic Measurements Section, National Bureau of Standards; Senior Member, Institute of Electrical and Electronics Engineers

T. W. DAKIN Insulation and Chemical Technology Department, Westinghouse Research Laboratories; Senior Member, Institute of Electrical and Electronics Engineers

R. N. SAMPSON Insulation and Chemical Technology Department, Westinghouse Research Laboratories

G. R. SPRENGLING Insulation and Chemical Technology Department, Westinghouse Research Laboratories

DANIEL BERG Insulation and Chemical Technology Department, Westinghouse Research Laboratories

H. F. MINTER Insulation and Chemical Technology Department, Westinghouse Research Laboratories

JACK SWISS Insulation and Chemical Technology Department, Westinghouse Research Laboratories; Member, Institute of Electrical and Electronics Engineers

E. J. CROOP Insulation and Chemical Technology Department, Westinghouse Research Laboratories; Senior Member, Institute of Electrical and Electronics Engineers

G. C. GAINER Insulation and Chemical Technology Department, Westinghouse Research Laboratories

F. A. YEOMAN Insulation and Chemical Technology Department, Westinghouse Research Laboratories

THOMAS J. DOLAN Head, Department of Theoretical and Applied Mechanics, College of Engineering, University of Illinois

DAVID BURNS Professional Engineer, Redington Beach, Florida; Senior Member, Institute of Electrical and Electronics Engineers

W. CRAWFORD DUNLAP Assistant Director for Electronics Components Research, Electronics Research Center, National Aeronautics and Space Administration; Fellow, Institute of Electrical and Electronics Engineers

CONTENTS

Numbers refer to paragraphs

SECTION 4

PROPERTIES OF MATERIALS

CONDUCTOR MATERIALS

By A. A. Jones

GENERAL PROPERTIES

1. Conducting Materials. A conductor of electricity is any substance or material which will afford continuous passage to an electric current when subjected to a difference of electric potential. The greater the density of current for a given potential difference, the more efficient the conductor is said to be. Virtually all substances in solid or liquid state possess the property of electric conductivity in some degree, but certain substances are relatively efficient conductors, while others are almost totally devoid of this property. The metals, for example, are the best conductors, while many other substances, such as metal oxides and salts, minerals, and fibrous materials, are relatively poor conductors, but their conductivity is beneficially affected by the absorption of moisture. Some of the less efficient conducting materials, such as carbon and certain metal alloys, as well as the efficient conductors such as copper and aluminum, have very useful applications in the electrical arts.

Certain other substances possess so little conductivity that they are classed as nonconductors, a better term being insulators or dielectrics. In general, all materials which are used commercially for conducting electricity for any purpose are classed as conductors.

2. Definition of Conductor. A conductor is a body so constructed from conducting material that it may be used as a carrier of electric current.[1]* In ordinary engineering usage, a conductor is a material of relatively high conductivity.

3. Definition of Circuit. An electric circuit is the path of an electric current, or, more specifically, it is a conducting part or a system of parts through which an electric current is intended to flow.[1]

4. General Properties of Conductors. Electric circuits in general possess four fundamental electrical properties, consisting of resistance, inductance, capacitance, and leakance. That portion of a circuit which is represented by its conductors will also possess these four properties, but only two of them are related to the properties of the conductor considered by itself. Capacitance and leakance depend in part upon the external dimensions of the conductors and their distances from one another and from other conducting bodies and in part upon the dielectric properties of the materials employed for insulating purposes. The inductance is a function of the magnetic field established by the current in a conductor, but this field as a whole is divisible into two parts, one being wholly external to the conductor and the other being wholly within the conductor; only the latter portion can be regarded as corresponding to the magnetic properties of the conductor material. The resistance is strictly a property of the conductor itself. Both the resistance and the internal inductance of conductors change in effective values when the current changes with great rapidity, as in the case of high-frequency alternating currents; this is termed the "skin effect."

In certain cases, conductors are subjected to various mechanical stresses. Con-

* For footnote references, see Par. **224**.

sequently their weight, tensile strength, and elastic properties require consideration in all applications of this character. Conductor materials as a class are affected by changes in temperature and by the conditions of mechanical stress to which they are subjected in service. They are also affected by the nature of the mechanical working and the heat treatment which they receive in the course of manufacture or fabrication into finished products.

5. Types of Conductor. In general, a conductor consists of a solid wire or a multiplicity of wires stranded together, made of a conducting material and used either bare or insulated. Only bare conductors are considered in this subsection. Usually the conductor is made of copper or aluminum, but for applications requiring higher strength, such as overhead transmission lines, bronze, steel, and various composite constructions are used. For conductors having very low conductivity and used as resistor materials, a group of special alloys is available.

METAL PROPERTIES

6. Specific Gravity and Density. Specific gravity is the ratio of mass of any material to that of the same volume of water at 4°C. Density is the unit weight of material expressed as pounds per cubic inch, grams per cubic centimeter, etc., at some reference temperature, usually 20°C. For all practical purposes, the numerical values of specific gravity and density are the same, expressed in g/cm³.

7. Density and Weight of Copper. Pure copper, rolled, forged, or drawn and then annealed, has a density[2] of 8.89 g/cm³ at 20°C or 8.90 g/cm³ at 0°C. Samples of high-conductivity copper will vary usually from 8.87 to 8.91 and occasionally from 8.83 to 8.94. Variations in density may be caused by microscopic flaws or seams or the presence of scale or some other defect; the presence of 0.03% oxygen will cause a reduction of about 0.01 in density. Hard-drawn copper has about 0.02% less density than annealed copper, on the average, but for practical purposes the difference is negligible.

The International standard of density, 8.89 at 20°C, corresponds to a weight of 0.32117 lb/in³ or 3.0270×10^{-6} lb/(cmil) (ft) or 15.982×10^{-3} lb/(cmil) (mile). Multiplying either of the last two figures by the square of the diameter of the wire in mils will produce the total weight of wire in pounds per foot or per mile, respectively.

8. Density and weight of copper alloys varies with the composition. For hard-drawn wire covered by ASTM Specification B105-55, the density of alloys 85 to 20 is 8.89 g/cm³ (0.32117 lb/in³) at 20°C; alloy 15 is 8.54 (0.30853); alloys 13 and 8.5 are 8.78 (0.31720).

9. Density and weight of copper-clad steel wire is a mean between the density of copper and the density of steel, which can be calculated readily when the relative volumes or cross sections of copper and steel are known. For practical purposes a value of 8.15 g/cm³ (0.29444 lb/in³) at 20°C is used.

10. Density and weight of aluminum wire[3] (commercially hard-drawn) is 2.703 g/cm³ (0.09765 lb/in³) at 20°C. The density of electrolytically refined aluminum (99.97% Al) and for hard-drawn wire of the same purity is 2.698 at 20°C. With less pure material there is an appreciable decrease in density on cold working. Annealed metal having a density of 2.702 will have a density of about 2.700 when in the hard-drawn or fully cold-worked condition (see *NBS Circ.* 346, pp. 68 and 69).

11. Density and weight of aluminum-clad wire is a mean between the density of aluminum and the density of steel, which can be calculated readily when the relative volumes or cross sections of aluminum and steel are known. For practical purposes a value of 6.59 g/cm³ (0.23808 lb/in³) at 20°C is used.

12. Density and weight of aluminum alloys varies with type and composition. For hard-drawn aluminum alloy wire 5005-H19 and 6201-T81, a value of 2.703 g/cm³ (0.09765 lb/in³) at 20°C is used.

13. Density and weight[4] **of pure iron** is 7.90 g/cm³ [2.690×10^{-6} lb/(cmil) (ft)] at 20°C.

14. Density and weight of galvanized steel wire (EBB, BB, HTL-85, HTL-135, and HTL-195) with Class A weight of zinc coating is 7.83 g/cm³ (0.283 lb/in³) at 20°C, with Class B is 7.80 g/cm³ (0.282 lb/in³), and with Class C is 7.78 g/cm³ (0.281 lb/in³).

15. Percent Conductivity. It is very common to rate the conductivity of a conductor in terms of its percentage ratio to the conductivity of chemically pure metal of the same kind as the conductor is primarily constituted or in ratio to the conductivity of the international copper standard. Both forms of the conductivity ratio are useful for various purposes.

This ratio can also be expressed in two different terms, one where the conductor cross sections are equal and therefore termed the **volume-conductivity ratio** and the other where the conductor masses are equal and therefore termed the **mass-conductivity ratio.**

16. The International Annealed Copper Standard (IACS) is the internationally accepted value[2] for the resistivity of annealed copper of 100% conductivity. This standard is expressed in terms of mass resistivity as **0.15328** Ω·g/m², or the resistance of a uniform round wire **1** m long weighing **1** g at the standard temperature of **20**°C. Equivalent expressions of the annealed copper standard, in various units of mass resistivity and volume resistivity, are as follows:

0.15328	Ω·g/m²	at 20°C
875.20	Ω·lb/mi²	at 20°C
1.7241	μΩ·cm	at 20°C
0.67879	μΩ·in	at 20°C
10.371	Ω·cmil/ft	at 20°C
0.017241	Ω·mm²/m	at 20°C

The above values are the equivalent of 1/58 Ω·mm²/m, so that the volume conductivity can be expressed as 58 mho·mm²/m at 20°C.

17. Conductivity of conductor materials varies with chemical composition and processing. For industry specification values, see Table 4-1.

18. Electrical resistivity is a measure of the resistance of a unit quantity of a given material. It may be expressed in terms of either mass or volume; mathematically,

Mass resistivity:
$$\delta = \frac{Rm}{l^2} \tag{4-1}$$

Volume resistivity:
$$\rho = \frac{RA}{l} \tag{4-2}$$

where R = resistance, m = mass, A = cross-sectional area, l = length.

19. Electrical resistivity of conductor materials varies with chemical composition and processing. For industry specification values, see Tables 4-1 and 4-2.

20. Electrical Conductivity and Resistivity: Nonferrous Conductors. See Table 4-1.

21. Electrical Resistivity: Ferrous Conductors. See Table 4-2.

22. Conversion Factors for Electrical Resistivity and Conductivity. See Table 1-14.

23. Effects of Temperature Changes. Within the temperature ranges of ordinary service there is no appreciable change in the properties of conductor materials, except in electrical resistance and physical dimensions. The change in resistance with change in temperature is sufficient to require consideration in many engineering calculations. The change in physical dimensions with change in temperature is also important in certain cases, such as in overhead spans and in large units of apparatus or equipment.

24. Temperature Coefficient of Resistance. Over moderate ranges of temperature, such as 100°C, the change of resistance is usually proportional to the change of temperature. Resistivity is always expressed at a standard temperature, usually 20°C (68°F). In general if R_{t_1} is the resistance at a temperature t_1, and α_{t_1} is the temperature coefficient at that temperature, the resistance at some other temperature t_2 is expressed by the formula

$$R_{t_2} = R_{t_1}[1 + \alpha_{t_1}(t_2 - t_1)] \tag{4-3}$$

Over wide ranges of temperature the linear relationship of this formula is not usually applicable, and the formula then becomes a series involving higher powers of t, which is unwieldy for ordinary use.

Table 4-1. Electrical Conductivity and Resistivity*—Nonferrous Conductors

Specification†	Temper and shape	Size limits, in	Conductivity at 20°C (68°F), IACS, %	Weight resistivity at 20°C (68°F)		Volume resistivity at 20°C (68°F)			
				Ω·g/m²	Ω·lb/mi²	Ω·cmil/ft	Ω·mm²/m	μΩ·cm	μΩ·in
Copper and copper alloy—specific gravity 8.89:									
....	....	....	100.17	0.15302‡	873.75	10.354	0.017213	1.7213	0.67767
B4-42 B5-43 B170-59 B187-62 B188-61 B298-64 B75-65	Low-resistance Lake wirebar Electrolytic wirebar Oxygen-free wirebar Soft bus bar, rod and shape Soft bus tube Silver-coated soft, round Tube, soft, OF copper	All	100.00	**0.15328**§	875.20	10.371	0.017241	1.7241	0.67879
B49-52 B3-63 B48-65	Hot-rolled rod Soft, round Soft, rectangular	1.375–0.250 All All	100.00	0.15328	**875.20**	10.371	0.017241	1.7241	0.67879
....	....	....	99.50	0.15405	879.60	10.423	0.017328	1.7328	0.68220
....	....	....	99.00	0.15482	884.04	10.476	0.017416	1.7416	0.68565
....	....	....	98.50	0.15561	888.53	10.529	0.017504	1.7504	0.68913
B187-62	Hard bus bar, rod and shape	Over 1 in OD Over 0.375 × 4	98.40	**0.15577**	889.42	10.539	0.017521	1.7521	0.68981
B188-61	Hard bus tube, rectangular or square	Over 6 in							
B75-65	Tube, soft, DLP copper	All	98.16	**0.15614**					
....	....	....	98.00	0.15640	893.06	10.583	0.017593	1.7593	0.69265
B188-61	Hard bus tube, rectangular or square	Up to 6 × 3/16 in wall	97.80	**0.15673**	894.90	10.604	0.017629	1.7629	0.69406
B2-52 B33-63 B189-63	Medium hard, round Tinned soft, round Lead-coated soft, round	0.460–0.325 0.460–0.290 0.460–0.290	97.66	0.15694	**896.15**	10.619	0.017654	1.7654	0.69504
....	....	....	97.50	0.15721	897.64	10.637	0.017683	1.7683	0.69620
B187-62	Hard bus bar and rod	Up to 1 in OD Up to 0.375 × 4	97.40	**0.15737**	898.55	10.648	0.017701	1.7701	0.69690
B188-61	Hard bus tube, round Hard bus pipe, IPS and extra strong Hard bus tube, rectangular or square	Over 1 in OD Over 4 in OD Up to 6 in, over 3/16 in wall							
B75-65 B372-63	Tube, hard, OF copper Waveguide tube, OF copper	All							

. . . .	. . . .	. . . .	97.30	0.15753	899.49	10.659	0.017720	1.7720	0.69763
B1-56	Hard, round	0.460–0.325							
B47-64 and B116-64	Hard trolley wire	All	97.16	0.15775	**900.77**	10.674	0.017745	1.7745	0.69863
B33-63	Tinned soft, round	0.289–0.103							
B189-63	Lead-coated soft, round								
. . . .	. . . .	. . . .	97.00	0.15802	902.27	10.692	0.017775	1.7775	0.69979
B2-52	Medium hard, round	0.324–0.0403	96.66	0.15857	**905.44**	10.729	0.017837	1.7837	0.70224
B188-61	Hard bus pipe, IPS and extra strong	Up to 4 in OD	96.60	**0.15865**	905.86	10.734	0.017845	1.7845	0.70257
. . . .	. . . .	. . . .	96.50	0.15884	906.94	10.747	0.017867	1.7867	0.70341
B1-56	Hard, round	0.324–0.0403	96.16	0.15940	**910.15**	10.785	0.017930	1.7930	0.70590
B33-63	Tinned soft, round	0.102–0.0201							
B189-63	Lead-coated soft, round	0.102–0.0201							
B75-65	Tube, hard, DLP copper	All	96.16	**0.15940**	910.15	10.785	0.017930	1.7930	0.70590
B372-63	Waveguide tube, DLP copper								
B355-65T	Nickel-coated soft, round, Class 2	All	96.00	0.15966	**911.67**	10.803	0.017960	1.7960	0.70708
B33-63	Tinned soft, round	0.0200–0.0111	94.16	0.16279	**929.52**	11.015	0.018312	1.8312	0.72092
B189-63	Lead-coated soft, round								
B355-65T	Nickel-coated soft, round, Class 4	All	94.0	0.16306	**931.06**	11.033	0.018342	1.8342	0.72212
B246-64	Tinned medium hard, round	0.2043–0.103	93.22	0.16443	**938.85**	11.125	0.018495	1.84949	0.72816
B33-63	Tinned soft, round	0.0110–0.0030	93.15	0.16454	**939.51**	11.133	0.018508	1.8508	0.72867
B189-63	Lead-coated soft, round								
B246-64	Tinned hard, round	0.2043–0.103	92.72	0.16532	**943.92**	11.185	0.018595	1.85947	0.73209
B246-64	Tinned medium hard, round	0.103–0.0508	92.51	0.16569	**946.06**	11.211	0.018637	1.86369	0.73375
B246-64	Tinned hard, round	0.103–0.0508	91.96	0.16668	**951.72**	11.278	0.018748	1.87484	0.73814
B355-65T	Nickel-coated soft, round, Class 7	All	91.0	0.16844	**961.76**	11.397	0.018947	1.8947	0.74593
. . . .	. . . .	. . . .	90.00	0.17031	972.45	11.524	0.019157	1.9157	0.75421
B355-65T	Nickel-coated soft, round, Class 10	All	88.0	0.17418	**994.55**	11.785	0.019592	1.9592	0.77136
B105-55	Hard round, alloy 85	All	85.00‖	0.18039	**1,030**	12.206	0.020291	2.0291	0.79885
B105-55	Hard round, alloy 80	All	80.00‖	0.19160	**1,094**	12.964	0.021551	2.1551	0.84849
B9-64	Trolley wire, alloy 80								
B355-65T	Nickel-coated soft, round, Class 27	All	71.0	0.21588	**1,232.7**	14.607	0.024284	2.4284	0.95605
B105-55	Hard round, alloy 65	All	65.00‖	0.23573	**1,346**	15.950	0.026516	2.6516	1.0439
B9-64	Trolley wire, alloy 65								
B105-55	Hard round, alloy 55 }	All	55.00‖	0.27864	1,591	18.854	0.031343	3.1343	1.2340
B9-64	Trolley wire, alloy 55 }								
B105-55	Hard round, alloy 40 }	All	40.00‖	0.38320	2,188	25.928	0.043103	4.3103	1.6970
B9-64	Trolley wire, alloy 40 }								
B105-55	Hard round, alloy 30	All	30.00‖	0.51086	2,917	34.567	0.057465	5.7465	2.2624
B105-55	Hard round, alloy 20	All	20.00‖	0.76638	4,376	51.856	0.086207	8.6207	3.3940
Resistivity temperature constant		. . . .	. . . .	0.000597	3.41	0.0409	0.0000681	0.00681	0.00268

Table 4-1. Electrical Conductivity and Resistivity*—Nonferrous Conductors.—*Concluded*

Specification†	Temper and shape	Size limits, in	Conductivity at 20°C (68°F), IACS, %	Weight resistivity at 20°C (68°F)		Volume resistivity at 20°C (68°F)			
				Ω·g/m²	Ω·lb/mi²	Ω·cmil/ft	Ω·mm²/m	μΩ·cm	μΩ·in
Copper alloy—specific gravity 8.54:									
B105-55	Hard, round, alloy 15	All	15.00‖	0.98162	**5,605**	69.142	0.11494	11.494	4.5253
Resistivity temperature constant				0.000597	3.41	0.0409	0.0000681	0.00681	0.00268
Copper alloy—specific gravity 8.78:									
B105-55	Hard, round, alloy 13	All	13.00‖	1.1645	**6,649**	79.778	0.13263	13.263	5.2215
B105-55	Hard, round, alloy 8.5	All	8.50‖	1.7809	**10,169**	122.01	0.20284	20.284	7.9857
Resistivity temperature constant				0.000597	3.41	0.0409	0.0000681	0.00681	0.00268
Aluminum and aluminum alloy—specific gravity 2.703:									
....	....	....	62.0	0.075167	429.20	16.728	0.027809	2.7809	1.0948
B233-64	Redraw rod EC-O	0.375	61.8	0.075410	**430.59**	16.782	0.027899	2.7899	1.0983
B233-64	Redraw rod EC-H12	0.375	61.5	0.075778	**432.69**	16.864	0.028035	2.8035	1.1037
B233-64	Redraw rod EC-H14	0.375	61.4	0.075901	**433.39**	16.891	0.028080	2.8080	1.1055
B233-64	Redraw rod EC-H16	0.375	61.3	0.076025	**434.10**	16.919	0.028126	2.8126	1.1073
B236-64	Bus bar	All	61	**0.07640**	436.24	17.002	0.028264	2.8624	1.1128
B230-60 B262-61 B323-61 B314-60	Hard, round Three-quarter hard, round Half hard, round Half hard, round	All	61.0	0.076397	436.24	**17.002**	0.028264	2.8624	1.1128
B324-60	All tempers, rectangular	All	61.0	0.076397	**436.24**	17.002	0.028264	2.8624	1.1128
B317-64	T64 temper, extruded alloy 6101	All	59.5	**0.0782**	446.74	17.430	0.028976	2.8976	1.1408
B317-64	H111 temper, extruded alloy 6101	All	59.0	**0.0789**	450.52	17.578	0.029222	2.9222	1.1505
B317-64	T61 temper, extruded alloy 6101	All	57.0	**0.0817**	466.33	18.195	0.030247	3.0247	1.1909
B317-64	T62 & T8 temper, extruded alloy 6101	All	56.0	**0.0831**	474.66	18.520	0.030788	3.0788	1.2121
B317-64	T6 temper, extruded alloy 6101	All	55.0	**0.0846**	483.29	18.856	0.031347	3.1347	1.2342
B396-63T	Hard round, alloy 5005	All	53.5	0.087106	497.38	**19.385**	0.032226	3.2226	1.2687
B398-63T	Hard round, alloy 6201	All	52.5	0.088764	506.85	**19.754**	0.032839	3.2839	1.2929
Resistivity temperature constant				0.000305	1.74	0.0689	0.000115	0.0115	0.00451
Copper-clad steel—specific gravity 8.15:									
B227-65	Hard, round, grades 40 HS and EHS	All	40‖	0.035837	2,046.3	**26.45**	0.043971	4.3971	1.7311
B227-65	Hard, round, grades 30 HS and EHS	All	30‖	0.047773	2,727.8	**35.26**	0.058617	5.8617	2.3078
Resistivity temperature constant (40%)				0.00135	7.68	0.100	0.000167	0.0167	0.00657
Resistivity temperature constant (30%)				0.00179	10.24	0.134	0.000222	0.0222	0.00875
Aluminum-clad steel—specific gravity 6.59:									
B415-64T	Hard, round	0.204–0.080	20‖	0.55886	3,191.0	**51.01**	0.084805	8.4805	3.3384
Resistivity temperature constant				0.00200	11.40	0.184	0.000306	0.0306	0.0121

* The value established as standard is indicated in boldface type; other values given are calculated.
† ASTM unless otherwise noted.
‡ Matthiessen's standard.
§ International Annealed Copper Standard (IACS).
‖ Nominal value.

Table 4-2. Electrical Resistivity—Ferrous Conductors

Material	ASTM specification	Weight resistivity at 20°C (68°F)		Volume resistivity at 20°C (68°F)			
		Maximum	Average	Range	Average	Range	Average
		Ω·lb/mi²		Ω·cmil/ft		μΩ·cm	
Pure iron	....	4,410	....	....	58.83	....	9.78
Contact rails:							
Open hearth	....	....	....	100–135	......	16.6–22.4	
Mild to soft steel	....	....	....	77–95	......	12.8–15.8	
Telephone and telegraph (galvanized):							
Extra Best Best (EBB)	A111–59	5,000					
Best Best (BB)	A111–59	5,600					
Grade 85	A326–59	5,800					
Grade 135 and 195	A326–59	6,500					
Commercial galvanized:							
Siemens-Martin	....	....	7,280	....	97.9*	....	16.3*
High strength	....	....	9,000	....	121*	....	20.1*
Extra-high strength	....	....	9,360	....	126*	....	20.9*

* Calculated from average weight resistivity, with average specific gravity 7.83.

When the temperature of reference t_1 is changed to some other value, the coefficient changes also. Upon assuming the general linear relationship between resistance and temperature previously mentioned, the new coefficient at any temperature t within the linear range is expressed

$$\alpha_t = \frac{1}{(1/\alpha_{t_1}) + (t - t_1)} \tag{4-4}$$

The reciprocal of α is termed the inferred absolute zero of temperature. Equation (4-3) takes no account of the change in dimensions with change in temperature and therefore applies to the case of conductors of constant mass, usually met in engineering work. For a more extended discussion of this subject see J. H. Dellinger, The Temperature Coefficient of Resistance of Copper, *NBS Bull.*, 1911, Vol. 8, pp. 71–101; also see *NBS Handbook* 100, Copper Wire Tables.

The coefficient for copper of less than standard (or 100%) **conductivity** is proportional to the actual conductivity, expressed as a decimal percentage. Thus, if n is the percentage conductivity (95% = 0.95), the temperature coefficient will be $\alpha_t' = n\alpha_t$, where α_t is the coefficient of the annealed copper standard.

The coefficients given in Table 4-3 were computed from the formula

$$\alpha_1 = \frac{1}{[1/n(0.00393)] + (t_1 - 20)} \tag{4-5}$$

The inferred absolute zero of temperature, upon assuming a linear relationship between resistance and temperature, is given as quantity $-T$ in the last column of Table 4-3. At the absolute zero of temperature, the resistance would be zero (see Fig. 4-1).

The coefficient changes with the temperature of reference as shown in Table 4-3.

25. Temperature-resistance Coefficients for Copper. See Table 4-3.

26. Temperature-resistance coefficients for copper alloys usually can be approximated by multiplying the corresponding coefficient for copper (100% IACS) by the alloy conductivity expressed as a decimal. For some complex alloys, however, this relation does not hold even approximately, and suitable values should be obtained from the supplier.

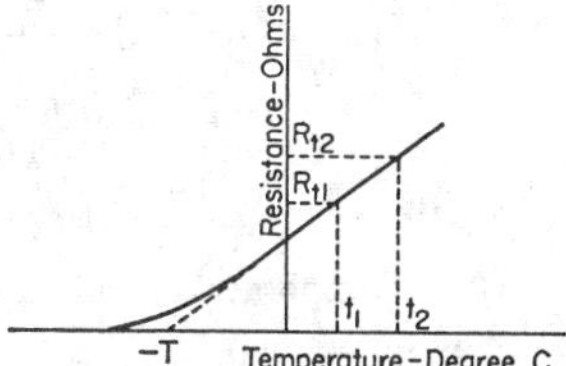

Fig. 4-1. Resistance-temperature relationship.

Table 4-3. Temperature-resistance Coefficients for Aluminum and Copper

Conductivity IACS, %	Temperature, deg C: 0	15	20	25	30	50	Temperature $-T$ for inferred-zero resistance,* deg C
	Temperature coefficient of resistance, α_1 per deg C						
Aluminum							
55	0.00392	0.00370	0.00363	0.00357	0.00351	0.00328	255.2
56	0.00400	0.00377	0.00370	0.00363	0.00357	0.00333	250.3
57	0.00407	0.00384	0.00377	0.00370	0.00363	0.00338	245.6
58	0.00415	0.00391	0.00383	0.00376	0.00369	0.00344	241.0
59	0.00423	0.00398	0.00390	0.00382	0.00375	0.00349	236.6
60	0.00431	0.00404	0.00396	0.00389	0.00381	0.00354	232.3
60.6	0.00435	0.00409	0.00400	0.00393	0.00385	0.00357	229.8
60.97	0.00438	0.00411	0.00403	0.00395	0.00387	0.00359	228.3
61.0	0.00438	0.00411	**0.00403**	0.00395	0.00387	0.00360	228.1
61.3	0.00441	0.00413	0.00405	0.00397	0.00389	0.00361	226.9
61.4	0.00441	0.00414	0.00406	0.00398	0.00390	0.00362	226.5
61.5	0.00442	0.00415	0.00406	0.00398	0.00390	0.00362	226.1
61.8	0.00445	0.00417	0.00408	0.00400	0.00392	0.00364	224.9
62.0	0.00446	0.00418	0.00410	0.00401	0.00393	0.00365	224.1
63	0.00454	0.00425	0.00416	0.00408	0.00400	0.00370	220.3
64	0.00462	0.00432	0.00423	0.00414	0.00406	0.00375	216.5
65	0.00470	0.00439	0.00429	0.00420	0.00412	0.00380	212.9
Copper							
95	0.00403	0.00380	0.00373	0.00367	0.00360	0.00336	247.8
96	0.00408	0.00385	0.00377	0.00370	0.00364	0.00339	245.1
97	0.00413	0.00389	0.00381	0.00374	0.00367	0.00342	242.3
97.5	0.00415	0.00391	0.00383	0.00376	0.00369	0.00344	241.0
98	0.00417	0.00393	0.00385	0.00378	0.00371	0.00345	239.6
99	0.00422	0.00397	0.00389	0.00382	0.00374	0.00348	237.0
100	0.00427	0.00401	**0.00393**	0.00385	0.00378	0.00352	234.5
101	0.00431	0.00405	0.00397	0.00389	0.00382	0.00355	231.9
102	0.00436	0.00409	0.00401	0.00393	0.00385	0.00358	229.5

* See Par. **24**.
Conductivities 95 to 102 from *NBS Handbook* 100.
Coefficient 0.00403 at 20°C for 61.0% conductivity from ASTM Designation B193-65; others calculated on same basis.
Boldface type indicates standard values.

27. Temperature-resistance coefficient for copper-clad steel wire is 0.00378/°C.

28. Temperature-resistance Coefficients for Aluminum. See Table 4-3.

29. Temperature-resistance coefficients for aluminum-alloy wires are: for 5005-H19, 0.00353/°C; for 6201-T81, 0.00347/°C.

30. Temperature-resistance coefficient for aluminum-clad wire is 0.0036/°C.

31. Temperature-resistance coefficient for pure iron is 0.0064/°C at 20°C. The coefficient, determined by extrapolation from tests on galvanized telephone and telegraph wire, for wire of 100% conductivity at 20°C, is 0.0061/°C.

32. Temperature-resistance coefficients for galvanized-steel conductors are as follows:

Commercial Grade	Approximate Temperature Coefficient per °C
Extra Best Best (EBB)	0.0056
Best Best (BB), HTL-85	0.0046
HTL-135, HTL-195	0.0043
High strength	0.0032
Extra-high strength	0.0031

33. Temperature-resistance coefficients for composite conductors are as follows:

Type	Approximate Temperature Coefficient per °C
Copper–copper-clad steel	0.00381
ACSR (aluminum-steel)	0.00403
Aluminum–aluminum alloy	0.00394
Aluminum–aluminum-clad steel	0.00396

34. Reduction of Observations to Standard Temperature. A table of convenient corrections and factors for reducing resistivity and resistance to standard temperature, 20°C, will be found in Copper Wire Tables, *NBS Handbook* 100.

35. Resistivity-temperature Constant. The *change of resistivity per degree* may be readily calculated, taking account of the expansion of the metal with rise of temperature. The proportional relation between temperature coefficient and conductivity may be put in the following convenient form for reducing *resistivity* from one temperature to another: *The change of resistivity of copper per degree centigrade is a constant, independent of the temperature of reference and of the sample of copper. This "resistivity-temperature constant" may be taken, for general purposes, as* 0.00060 Ω *(meter, gram), or* 0.0068 $\mu\Omega \cdot cm$. More exact values for this constant are given in Table 4-1.

Details of the calculation of the resistivity-temperature constant will be found in Copper Wire Tables, *NBS Handbook* 100; also see this reference for expressions for the temperature coefficients of resistivity and their derivation.

36. Calculation of Percent Conductivity. The percent conductivity of a sample of copper is calculated by dividing the resistivity of the International Annealed Copper Standard at 20°C by the resistivity of the sample at 20°C. Either the mass resistivity or the volume resistivity may be used. Inasmuch as the temperature coefficient of copper varies with the conductivity, it is to be noted that a different value will be found if the resistivity at some other temperature is used. This difference is of practical moment in some cases. In order that such differences may not arise, it is best always to use the 20°C value of resistivity in computing the percent conductivity of copper. When the resistivity of the sample is known at some other temperature t, it is very simply reduced to 20°C by adding the quantity $20 - t$ multiplied by the resistivity-temperature constant, given in Table 4-1.

37. Temperature coefficient of expansion (linear) of pure metals over a range of several hundred degrees is not a linear function of the temperature but is well expressed by a quadratic equation,

$$\frac{L_{t_2}}{L_{t_1}} = 1 + [\alpha(t_2 - t_1) + \beta(t_2 - t_1)^2] \tag{4-6}$$

Over the temperature ranges for ordinary engineering work (usually 0 to 100°C), the coefficient can be taken as a constant (assumed linear relationship) and a simplified formula employed,

$$L_{t_2} = L_{t_1}[1 + \alpha_{t_1}(t_2 - t_1)] \tag{4-7}$$

Changes in linear dimensions, superficial area, and volume take place in most materials with changes in temperature. In the case of linear conductors, only the change in length is ordinarily important.

The coefficient for changes in superficial area is approximately twice the coefficient of linear expansion for relatively small changes in temperature. Similarly the volume coefficient is three times the linear coefficient, with similar limitations.

38. Temperature-expansion Coefficients for Conductors. See Table 4-4.

39. Specific heat of electrolytic tough pitch copper is 0.092 cal/(g)(°C) at 20°C[5] (see *NBS Circ.* 73).

40. Specific heat of aluminum is 0.226 cal/(g)(°C) at room temperature (see *NBS Circ.* C447, Mechanical Properties of Metals and Alloys).

41. Specific heat of iron (wrought) or very soft steel from 0 to 100°C is 0.114 cal/(g)(°C); the true specific heat of iron at 0°C is 0.1075 cal/(g)(°C) (see "International Critical Tables," vol. II, p. 518; also ASM "Metals Handbook").

42. Thermal conductivity of electrolytic tough pitch copper at 20°C is 0.934 cal/(cm^2)(cm)(sec)(°C), adjusted to correspond to an electrical conductivity of 101%[5] (see *NBS Circ.* 73).

43. Thermal-electrical Conductivity Relation of Copper. The Wiedemann-Franz-Lorenz law, which states that the ratio of the thermal and electrical conductivities at a given temperature is independent of the nature of the conductor, holds closely for copper. The ratio $K/\lambda T$ (where K = thermal conductivity, λ = electrical conductivity, T = absolute temperature) for copper is 5.45 at 20°C.[7]

44. Thermal Conductivity of Copper Alloys

ASTM alloy (Spec. B105-55)	Thermal conductivity (volumetric) at 20 C*	
	Btu per sq ft per ft per hr per deg F	Cal per sq cm per cm per sec per deg C
8.5	31	0.13
15	50	0.21
30	84	0.35
55	135	0.56
80	199	0.82
85	208	0.86

* The Anaconda American Brass Company.

45. Thermal Conductivity of Aluminum. The determination made by the Bureau of Standards at 50°C for aluminum of 99.66% purity is 0.52 cal/(cm^2) (cm) (s) (°C) (*Circ.* 346; also see "Smithsonian Physical Tables" and "International Critical Tables").

46. Thermal conductivity of iron (mean) from 0 to 100°C is 0.143 cal/(cm^2) (cm) (s) (°C); with increase of carbon and manganese content, it tends to decrease and may reach a figure of approximately 0.095 with about 1% carbon, or only about half that figure if the steel is hardened by water quenching (see "International Critical Tables," Vol. II, p. 518).

47. Influence of Chemical Composition. The resistivity of most metals is very sensitive to changes in chemical composition or constituents. This applies particularly to the case of relatively small amounts of impurities present in metals which approach closely to the chemically pure state. Such impurities may consist of oxides, slag, or traces of various foreign ingredients which escape elimination in the processes of reducing the original ores to the metallic state, refining these intermediate products and fabricating the ingots into the form of conductors by various processes of hot and cold working. The combined effects of such impurities are usually summed up in the percentage of conductivity of the conductor expressed as the ratio to the conductivity of chemically pure metal.

Table 4-4. Temperature-expansion Coefficients for Conductors

Conductor type	Temperature coefficient	
	Per °F	Per °C
Copper	9.4×10^{-6}	16.92×10^{-6}
Copper alloy	9.4×10^{-6}	16.92×10^{-6}
Copper-clad steel	7.2×10^{-6}	12.96×10^{-6}
Aluminum	12.8×10^{-6}	23.0×10^{-6}
Aluminum alloy	12.8×10^{-6}	23.0×10^{-6}
Aluminum-clad steel	7.2×10^{-6}	13.0×10^{-6}
Pure iron	6.72×10^{-6}	12.1×10^{-6}
Galvanized steel:		
Mild steel	6.22×10^{-6}	11.2×10^{-6}
ACSR core wire	6.4×10^{-6}	11.52×10^{-6}
HTL-85, HTL-135, HTL-195	5.7×10^{-6}	10.26×10^{-6}
Composite conductors:		
Copper–copper-clad:		
Type 2A to 6A	8.5×10^{-6}	15.30×10^{-6}
Type F	9.0×10^{-6}	16.20×10^{-6}
Type E	8.4×10^{-6}	15.12×10^{-6}
Type EK	8.8×10^{-6}	15.84×10^{-6}
Copper–copper alloy	9.4×10^{-6}	16.92×10^{-6}
Aluminum–aluminum clad:		
AWAC 5/2	10.0×10^{-6}	18.00×10^{-6}
AWAC 4/3	9.3×10^{-6}	16.74×10^{-6}
AWAC 3/4	8.6×10^{-6}	15.48×10^{-6}
AWAC 2/5	8.0×10^{-6}	14.40×10^{-6}

48. Influence of Mechanical Treatment. The fabrication of conductors, from the ingot to the finished state, normally starts with hot rolling and finishes with cold drawing. Annealing operations may or may not take place at intermediate stages or at the finished state. The cold-working operations tend in general to harden the material, reduce its ductility, increase its tensile strength, and very slightly increase its resistivity. The increase in tensile strength is frequently very useful, and consequently many types of conductor are finished by cold working, in which condition they are usually described as hard-drawn.

49. Copper is a highly malleable and ductile metal, of reddish color. It can be cast, forged, rolled, drawn, and machined. Mechanical working hardens it, but annealing will restore it to the soft state. The density varies slightly with the physical state, 8.9 being an average value. It melts at 1083°C (1981°F) and in the molten state has a sea-green color. When heated to a very high temperature, it vaporizes and burns with a characteristic green flame. Copper readily alloys with many other metals. In ordinary atmospheres it is not subject to appreciable corrosion. Its electrical conductivity is very sensitive to the presence of slight impurities in the metal.

Copper when exposed to ordinary atmospheres becomes oxidized, turning to a black color, but the oxide coating is protective, and the oxidizing process is not progressive. When exposed to moist air containing carbon dioxide, it becomes coated with green basic carbonate, which is also protective. At temperatures above 180°C it oxidizes in dry air. In the presence of ammonia it is readily oxidized in air, and it is also affected by sulfur dioxide. Copper is not readily attacked at high temperatures below the melting point by hydrogen, nitrogen, carbon monoxide, carbon dioxide, or steam. Molten copper readily absorbs oxygen, hydrogen, carbon monoxide, and sulfur dioxide, but on cooling, the occluded gases are liberated to a great extent, tending to produce blowholes or porous castings. Copper in the presence of air does not dissolve in dilute hydrochloric or sulfuric acid but is readily attacked by dilute nitric acid. It is also corroded slowly by saline solutions and sea water.

50. Commercial grades of copper in the United States are electrolytic, oxygen-free, Lake, fire-refined, and casting. **Electrolytic copper** is that which has been electrolytically refined from blister, converter, black, or Lake copper. **Oxygen-free** copper is produced by special manufacturing processes which prevent the absorption of oxygen during the melting and casting operations or by removing the oxygen by reducing agents. It is used for conductors subjected to reducing gases at elevated temperature where reaction with the included oxygen would lead to the development of cracks in the metal. **Lake** copper is electrolytically or fire-refined from Lake Superior native copper ores and is of two grades, low resistance and high resistance. **Fire-refined** copper is a lower purity grade intended for alloying or for fabrication into products for mechanical purposes; it is not intended for electrical purposes. **Casting** copper is the grade of lowest purity and may consist of furnace-refined copper, rejected metal not up to grade, or melted scrap; it is exclusively a foundry copper.

51. Copper Content of Commercial Grades

Commercial grade	ASTM Designation	Copper content, minimum %
Electrolytic	B5-43	99.900
Oxygen-free electrolytic	B170-59	99.95
Lake, low resistance	B4-42	99.900
Lake, high resistance	B4-42	99.900
Fire-refined	B216-49	99.88
Casting	B119-64	98

52. Hardening and Heat-treatment of Copper. There are but two well-recognized methods for hardening copper; one is by mechanically working it, and the other is by the addition of an alloying element. The properties of copper are not affected by a rapid cooling after annealing or rolling, as are those of steel and certain copper alloys.

53. Annealing of Copper. Cold-worked copper is softened by annealing, with decrease of tensile strength and increase of ductility. In the case of pure copper hardened by cold reduction of area to one-third of its initial area, this softening takes place with maximum rapidity between 200 and 325°C. However, this temperature range is affected in general by the extent of previous cold reduction and the presence of impurities. The greater the previous cold reduction, the lower is the range of softening temperatures. The effect of iron, nickel, cobalt, silver, cadmium, tin, antimony, and tellurium is to lower the conductivity and raise the annealing range of pure copper in varying degrees (see Effect of Iron, Cobalt and Nickel on Some Properties of High-purity Copper and Effect of Certain Fifth-period Elements on Some Properties of High-Purity Copper, by J. S. Smart, Jr., and A. A. Smith, Jr., *Trans. AIME*, 1942, Vol. 147, p. 48, and 1943, Vol. 152, p. 103). Oxygen content lowers the annealing range.

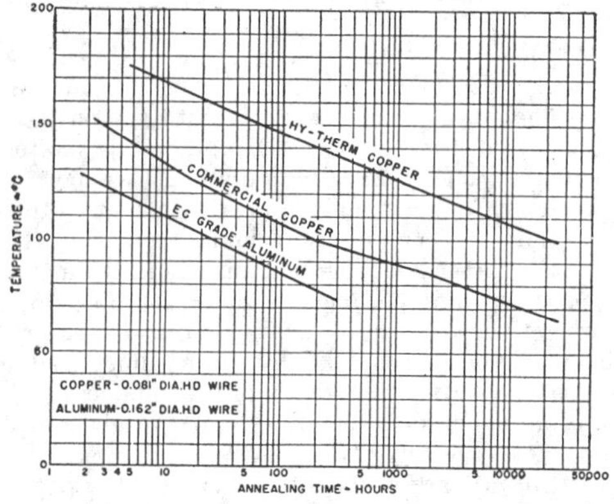

Fig. 4-2. Annealing characteristics of copper and aluminum for 5% loss in strength, no tension.

Trade-named coppers, such as **Hy-Therm, Tensilok,** and **High Thermo,** are produced by adding minute amounts of hardening agents to pure electrolytic copper. These coppers meet industry specifications for purity and conductivity but have higher annealing characteristics than commercial brands, as illustrated in Fig. 4-2. Their higher resistance to annealing permits higher continuous and short-time emergency overload current-carrying capacity of conductors used in overhead lines (see Hy-Therm Copper—An Improved Overhead-line Conductor, by L. F. Hickernell, A. A. Jones, and C. J. Snyder, *Trans. AIEE*, 1949, Vol. 68, Pt. 1, p. 22).

54. Alloying of Copper. Elements that are soluble in moderate amounts in a solid solution of copper, such as manganese, nickel, zinc, tin, and aluminum, generally harden it and diminish its ductility but improve its rolling and working properties. Elements that are but slightly soluble, such as bismuth and lead, do not harden it but diminish both the ductility and the toughness and impair its hot-working properties. See *NBS Circ.* 73, 2d ed., 1922, pp. 53–68, for extended discussion. Small additions (up to 1.5%) of manganese, phosphorus, or tin increase the tensile strength and hardness of cold-rolled copper.

Brass is usually a binary alloy of copper and zinc, but the brasses are seldom employed as electrical conductors, as they have relatively low conductivity though comparatively high tensile strength. In general, brass is not suitable for use where exposed to the weather, owing to the difficulty from stress-corrosion cracking; the higher the zinc content, the more pronounced does this become.

Bronze in its simplest form is a binary alloy of copper and tin, in which the latter element is the hardening and strengthening agent. This material is rather old in the arts and has been used to some extent for electrical conductors for many years past, especially abroad. Modern bronzes are frequently ternary alloys, containing as the third constituent such elements as phosphorus, silicon, manganese, zinc, aluminum, or cadmium; in such cases the third element is usually given in the name of the alloy, as phosphor bronze, silicon bronze. Certain bronzes are quaternary alloys or contain two other elements in addition to copper and tin.

In bronzes for use as electrical conductors the content of tin and other metals is usually less than in bronzes for structural or mechanical applications where physical properties and resistance to corrosion are the governing considerations. High resistance to atmospheric corrosion is always an important consideration in selecting bronze conductors for overhead service.

55. Commercial Grades of Bronze. Various bronzes have been developed for use as conductors, and these are now covered by ASTM Specification B105-55. They all have been designed to provide conductors having high resistance to corrosion and tensile strengths greater than hard-drawn copper conductors. The standard specification covers 10 grades of bronze, designated by numbers according to their conductivities. The types of alloy now commonly used for each of the several grades are listed below.

Alloy	Composition	Alloy	Composition
8.5...	Copper, silicon, iron Copper, silicon, manganese Copper, silicon, zinc Copper, silicon, tin, iron Copper, silicon, tin, zinc	20....	Copper, tin
		30....	Copper, tin Copper, zinc, tin
		40....	Copper, tin Copper, tin, cadmium
13....	Copper, aluminum, tin Copper, aluminum, silicon, tin Copper, silicon, tin	55....	Copper, tin, cadmium
		65....	Copper, tin Copper, tin, cadmium
15....	Copper, aluminum, silicon Copper, aluminum, tin Copper, aluminum, silicon, tin Copper, silicon, tin	80....	Copper, cadmium
		85....	Copper, cadmium

56. Copper-Chromium Alloy. A patented alloy of this type contains from 0.5 to 3.0% of chromium. When cast, rolled, heat-treated, and drawn it may have an electrical conductivity of 80% and a tensile strength of 72,000 to 80,000 lb/in^2.

57. Copper-beryllium alloy containing 0.4% of beryllium may have an electrical conductivity of 48% and a tensile strength (in 0.128-in wire) of 86,000 lb/in^2. A content of 0.9% of beryllium may give a conductivity of 28% and a tensile strength of 122,000 lb/in^2. The effect of this element in strengthening copper is about ten times as great as that of tin.

58. Copper-clad steel wires have been manufactured by a number of different methods. The general object sought in the manufacture of such wires is the combination of the high conductivity of copper with the high strength and toughness of iron or steel.

The principal manufacturing processes now in commercial use are: (*a*) coating a steel billet with a special flux, placing it in a vertical mold closed at the bottom, heating the billet and mold to yellow heat, and then casting molten copper around the billet, after which it is hot-rolled to rods and cold-drawn to wire, and (*b*) electroplating a dense coating of copper on a steel rod and then cold drawing to wire.

59. Aluminum ductile metal, silver-white in color, which can be readily worked by rolling, drawing, spinning, extruding, and forging. Its specific gravity is 2.703. Pure aluminum melts at 660°C (1220°F). Aluminum has relatively high thermal and electrical conductivities. The metal is always covered with a thin, invisible film of oxide which is impermeable and protective in character. Aluminum, therefore, shows stability and long life under ordinary atmospheric exposure.

Exposure to atmospheres high in hydrogen sulfide or sulfur dioxide does not cause severe attack of aluminum at ordinary temperatures, and for this reason aluminum or its alloys can be used in atmospheres which would be rapidly corrosive to many other metals.

Aluminum parts should, as a rule, not be exposed to salt solutions while in electrical contact with copper, brass, nickel, tin, or steel parts, since galvanic attack of the aluminum is likely to occur. Contact with cadmium in such solutions results in no appreciable acceleration in attack on the aluminum, while contact with zinc (or zinc-coated steel as long as the coating is intact) is generally beneficial, since the zinc is attacked selectively and cathodically protects adjacent areas of the aluminum.

Most organic acids and their water solutions have little or no effect on aluminum at room temperature, although oxalic acid is an exception and is corrosive. Concentrated nitric acid (about 80% by weight) and fuming sulfuric acid can be handled in aluminum containers. However, more dilute solutions of these acids are more active. All but the most dilute (less than 0.1%) solutions of hydrochloric and hydrofluoric acids have a rapid etching action on aluminum.

Solutions of the strong alkalies, potassium, or sodium hydroxides dissolve aluminum rapidly. However, ammonium hydroxide and many of the strong organic bases have little action on aluminum and are successfully used in contact with it (see *NBS Circ.* 346).

Aluminum in the presence of water and limited air or oxygen rapidly converts into aluminum hydroxide, a whitish powder.

60. Commercial grades of aluminum in the United States are designated by their purity, such as 99.99, 99.95, 99.90%.

61. Electrical conductor (EC) **grade aluminum,** having a purity of approximately 99.5% and a minimum conductivity of 61.0% IACS, is used for conductor purposes. Specified physical properties are obtained by closely controlling the kind and amount of certain impurities.

62. Annealing of Aluminum. Cold-worked aluminum is softened by annealing, with decrease of tensile strength and increase of ductility. The annealing temperature range is affected in general by the extent of previous cold reduction and the presence of impurities. The greater the previous cold reduction, the lower is the range of softening temperatures. Typical annealing characteristics of grade EC aluminum are shown in Fig. 4-2.

63. Alloying of Aluminum. Aluminum can be alloyed with a variety of other elements, with a consequent increase in strength and hardness. With certain alloys, the strength can be further increased by suitable heat-treatment. The alloying elements most generally used are copper, silicon, manganese, magnesium, chromium, and zinc. Some of the aluminum alloys, particularly those containing one or more of the following elements—copper, magnesium, silicon, and zinc—in various combinations, are susceptible to heat-treatment.

Pure aluminum, even in the hard-worked condition, is a relatively weak metal for construction purposes. Strengthening for castings is obtained by alloying elements. The alloys most suitable for cold-rolling seldom contain less than 90 to 95% aluminum. By alloying, working, and heat-treatment it is possible to produce tensile strengths ranging from 8,500 lb/in^2 for pure annealed aluminum up to 82,000 lb/in^2 for special wrought heat-treated alloy, with densities ranging from 2.65 to 3.00.

64. Electrical conductor alloys of aluminum are principally alloys 5005 and 6201 covered by ASTM Specifications B396-63T and B398-63T.

65. Aluminum-clad steel wires have a relatively heavy layer of aluminum surrounding and bonded to the high-strength steel core. The aluminum layer can be formed by compacting and sintering a layer of aluminum powder over a steel rod or by electroplating a dense coating of aluminum on a steel rod, and then cold drawing to wire.

66. Magnesium is a ductile metal, silver-white in color, which is distinguished by its light weight (sp. gr. 1.74) and ease of machining.

Pure magnesium has relatively low strength, so its uses are limited to applications where strength is of little importance. However, by alloying magnesium with small amounts of other metals, particularly aluminum, zinc, or manganese or combinations of these, alloys have been developed which show excellent mechanical properties and lead to the high strength-weight ratios of the magnesium products. In a general way the mechanical properties compare favorably with those of the commercial aluminum alloys. The various alloys are divided into those suitable for castings and for wrought products. A number of the standard alloys are susceptible to heat-treatment to improve the properties in general or for the improvement of some particular property.

Magnesium alloys are resistant to attack by alkalies and many organic chemicals. They are attacked by acids of any strength with the exception of pure hydrofluoric and chromic acids. Salt solutions in general corrode the metal, and applications involving continuous contact with saline solutions are not recommended. The alloys are commercially stable against ordinary atmospheric conditions, and for many uses the surface needs no protection.

A number of coatings applied by chemical treatment have been developed to provide a surface stability suitable for more difficult situations.

Pure magnesium is used for ingot, powder, shavings, extruded wire and strip, and rolled ribbon. Sand, permanent-mold, and die castings are available in magnesium alloys. Wrought magnesium alloys are available in extruded round, square, hexagonal, and rectangular bar; as special shapes, moldings, and structural sections; as rolled plate and sheet; and as hammer and hot-pressed forgings.

67. Silicon is a light metal having a specific gravity of approximately 2.34. There is lack of accurate data on the pure metal, because its mechanical brittleness bars it from most industrial uses. However, it is very resistant to atmospheric corrosion and to attack by many chemical reagents. Silicon is of fundamental importance in the steel industry, but for this purpose it is obtained in the form of ferrosilicon, which is a coarse granulated or broken product. It is very useful as an alloying element in steel for electrical sheets and substantially increases the electrical resistivity and thereby reduces the core losses. Silicon is peculiar among metals in the respect that its temperature coefficient of resistance may change sign in some temperature range, the exact behavior varying with the impurities.

68. Beryllium is a light metal having a specific gravity of approximately 1.84 or nearly the same as magnesium. It is normally hard and brittle and difficult to fabricate. Copper is materially strengthened by the addition of small amounts of beryllium, without very serious loss of electrical conductivity. The principal uses for this metal appear to be as an alloying element with other metals, such as aluminum and copper.

69. Sodium is a soft, bright, silvery metal obtained commercially by the electrolysis of absolutely dry fused sodium chloride. It is the most abundant of the alkali group of metals, is extremely reactive, and is never found free in nature. It oxidizes readily and rapidly in air. In the presence of water (it is so light that it floats) it may ignite spontaneously, decomposing the water with evolution of hydrogen and formation of sodium hydroxide. This can be explosive. Sodium should be handled with respect, as it can be dangerous when improperly handled. It melts at 97.8°C, below the boiling point of water, and in the same range as many fuse metal alloys. Sodium is approximately one-tenth as heavy as copper and has roughly three-eighths the conductivity; hence 1 lb of sodium is about equal electrically to $3\frac{1}{2}$ lb of copper. Interest in sodium as an electrical conductor recently was renewed with the development of a means of extruding a polyethylene insulating tube, simultaneously filling it with liquid sodium fed through a tube from a closed container, and cooling them together.

70. Iron and Steel. Iron is a hard tenacious metal which has a silvery-white luster and takes a high polish. It is strongly attracted by a magnet but retains practically no magnetism. It softens at a red heat and may be readily welded at a white heat. Its melting point is higher than that of steel or wrought iron. Iron is very reactive chemically and dissolves in most dilute acids with liberation of hydrogen. In dry air it undergoes no change, but when exposed to atmospheres containing moisture it corrodes more or less rapidly and forms rust (hydrated ferric oxide); such corrosion is accelerated by the presence of carbon dioxide and sulfur dioxide. There are three oxides of iron: FeO, or ferrous oxide, is a black powder; Fe_2O_3, or ferric oxide, is known also as hematite and is a steel-gray crystalline substance with considerable luster; Fe_3O_4, or ferroso-ferric oxide, is also known as magnetite and is attracted by a magnet but not always magnetic of itself. Iron forms numerous compounds with carbon, sulfur, phosphorus, oxygen, hydrogen, nitrogen, and other elements; it alloys readily with numerous metallic elements such as manganese, silicon, nickel, cobalt, chromium, and tungsten.

The element iron is the base of all commercial iron and steel products, but in order to define these products in reasonably complete terms it is desirable to state both their constituents and the processes by which they were made. Iron and steel are always different substances in the commercial sense and moreover are manufactured in many different commercial varieties. Even the common varieties of steel are somewhat complicated substances; it is insufficient to describe them as alloys of iron and certain other elements, because the alloys take many different forms and impart certain distinctive properties, depending on both the alloy proportions and the processes of manufacture.

71. Protective Coatings for Iron and Steel. Iron and steel wires for outdoor service as conductors or guys are protected from corrosion by the application of zinc or aluminum coatings. Such coatings may be applied by the hot-dip process or by electroplating. These coatings themselves are not impregnable against atmospheric attack but afford the best protection which has yet been found practicable.

72. Galvanized-iron telephone- and telegraph-line wire is available commercially in two grades, EBB and BB, and with three weights of zinc coating. Characteristics are given in ASTM Specification A111-59. See Tables 4-2 and 4-24.

73. Galvanized high-tensile steel telephone- and telegraph-line wire is available commercially in three grades, 85, 135, and 195 and with three weights of zinc coating. Characteristics are given in ASTM Specification A326-59. See Tables 4-2 and 4-24.

74. Steel Rails. The resistivities of steel rails are given in Table 4-2.

The effective resistance of steel rails to alternating currents is increased on account of skin effect, which in turn is a function of magnetic permeability (see data in "Report of the Electric Railway Test Commission," New York, McGraw-Hill Book Company, 1906; also Experimental Researches on the Skin Effect in Steel Rails, by A. E. Kennelly, F. H. Achard, and A. S. Dana, *Jour. Franklin Inst.*, August, 1916).

CONDUCTOR PROPERTIES

75. Electrical conductors are manufactured in various forms and shapes for various purposes. These may be wires, cables, flat straps, square or rectangular bars, angles, channels, or special designs for particular requirements. The most extensive use of conductors, however, is in the form of round solid wires and stranded conductors. The following terminology[6] describes properly the various terms relating to conductors:

76. Definitions of Electrical Conductors.

Wire. A slender rod or filament of drawn metal.

The definition restricts the term to what would ordinarily be understood by the term "solid wire." In the definition, the word "slender" is used in the sense that the length is great in comparison with the diameter. If a wire is covered with insulation, it is properly called an "insulated wire"; while primarily the term "wire" refers to the metal, nevertheless when the context shows that the wire is insulated, the term "wire" will be understood to include the insulation.

Conductor. A wire or combination of wires not insulated from one another, suitable for carrying a single electric current.

The term "conductor" is not to include a combination of conductors insulated from one another, which would be suitable for carrying several different electric currents.

Rolled conductors (such as bus bars) are, of course, conductors but are not considered under the terminology here given.

Stranded Conductor. A conductor composed of a group of wires or any combination of groups of wires.

The wires in a stranded conductor are usually twisted or braided together.

Cable. A stranded conductor (single-conductor cable) or a combination of conductors insulated from one another (multiple-conductor cable).

The component conductors of the second kind of cable may be either solid or stranded, and this kind of cable may or may not have a common insulating covering. The first kind of cable is a single conductor, while the second kind is a group of several conductors. The term "cable" is applied by some manufacturers to a solid wire heavily insulated and lead covered; this usage arises from the manner of the insulation, but such a conductor is not included under this definition of "cable." The term "cable" is a general one and in practice it is usually applied only to the larger sizes. A small cable is called a "stranded wire" or a "cord," both of which are defined below. Cables may be bare or insulated, and the latter may be armored with lead or with steel wires or bands.

Strand. One of the wires or groups of wires of any stranded conductor.

Stranded Wire. A group of small wires used as a single wire.

A wire has been defined as a slender rod or filament of drawn metal. If such a filament is subdivided into several smaller filaments or strands and is used as a single wire, it is called "stranded wire." There is no sharp dividing line of size between a "stranded wire" and a "cable." If used as a wire, for example, in winding inductance coils or magnets, it is called a stranded wire and not a cable. If it is substantially insulated, it is called a "cord," defined below.

Cord. A small cable, very flexible and substantially insulated to withstand wear.

There is no sharp dividing line in respect to size between a cord and a cable, and likewise no sharp dividing line in respect to the character of insulation between a cord and a stranded wire. Usually the insulation of a cord contains rubber.

Concentric Strand. A strand composed of a central core surrounded by one or more layers of helically laid wires or groups of wires.

Concentric-lay Cable. A single-conductor cable composed of a central core surrounded by one or more layers of helically laid wires.

Rope-lay Cable. A single-conductor cable composed of a central core surrounded by one or more layers of helically laid groups of wires.

This kind of cable differs from the preceding in that the main strands are themselves stranded.

N-conductor Cable. A combination of N conductors insulated from one another.

It is not intended that the name as here given be actually used. One would instead speak of a "3-conductor cable," a "12-conductor cable," etc. In referring to the general case, one may speak of a "multiple-conductor cable."

N-conductor Concentric Cable. A cable composed of an insulated central conducting core with N-1 tubular-stranded conductors laid over it concentrically and separated by layers of insulation.

This kind of cable usually has only two or three conductors. Such cables are used in carrying alternating currents.

The remark on the expression "N conductor" given for the preceding definition applies here also.

77. Wire sizes have been for many years indicated in commercial practice almost entirely by gage numbers, especially in America and England. This practice is accompanied by some confusion because numerous gages are in common use. The most commonly used gage for electrical wires, in America, is the **American wire gage.** The most commonly used gage for steel wires is the **Birmingham wire gage.**

There is no legal standard wire gage in this country, although a gage for sheets was adopted by Congress in 1893. In England there is a legal standard known as the **Standard wire gage.** In Germany, France, Austria, Italy, and other Continental countries practically no wire gage is used, but wire sizes are specified directly in millimeters. This system is sometimes called the **Millimeter wire gage.** The wire sizes used in France, however, are based to some extent on the old Paris gage (*jauge de Paris de* 1857) (for a history of wire gages see *NBS Handbook* 100, Copper Wire Tables; also see *Circ.* 67, Wire Gages, 1918).

There is a tendency to *abandon gage numbers* entirely and specify wire sizes by the **diameter in mils** (thousandths of an inch). This practice holds particularly in writing specifications and has the great advantages of being both simple and explicit. A number of the wire manufacturers also encourage this practice, and it was definitely adopted by the U.S. Navy Department in 1911.

78. Mil is a term universally employed in this country to measure wire diameters and is a unit of length equal to one-thousandth of an inch.

79. Circular mil is a term universally used to define cross-sectional areas, being a unit of area equal to the area of a circle 1 mil in diameter. Such a circle, however, has an area of 0.7854 (or $\pi/4$) **mil²**. Thus a wire 10 mils in diameter has a cross-sectional area of 100 cmils or 78.54 mils². Hence, 1 cmil equals 0.7854 mil².

80. American wire gage, also known as the **Brown & Sharpe gage,** was devised in 1857 by J. R. Brown. It is usually abbreviated AWG. This gage has the property, in common with a number of other gages, that its sizes represent approximately the successive steps in the process of wire drawing. Also, like many other gages, its numbers are retrogressive, a larger number denoting a smaller wire, corresponding to the operations of drawing. These gage numbers are not arbitrarily chosen, as in many gages, but follow the mathematical law upon which the gage is founded.

Basis of the AWG is a simple mathematical law. The gage is formed by the specification of two diameters and the law that a given number of intermediate diameters are formed by geometrical progression. Thus, the diameter of No. 0000 is defined as 0.4600 in and of No. 36 as 0.0050 in. There are 38 sizes between these two; hence the ratio of any diameter to the diameter of the next greater number is given by this expression:

$$\sqrt[39]{\frac{0.4600}{0.0050}} = \sqrt[39]{92} = 1.122\,932\,2 \tag{4-8}$$

The square of this ratio = 1.2610. The sixth power of the ratio, i.e., the ratio of any diameter to the diameter of the sixth greater number, = 2.0050. The fact that this ratio is so nearly 2 is the basis of numerous useful relations or short cuts in wire computations.

There are a number of approximate rules applicable to the AWG which are useful to remember:

1. An increase of three gage numbers (e.g., from No. 10 to 7) doubles the area and weight and consequently halves the d-c resistance.
2. An increase of six gage numbers (e.g., from No. 10 to 4) doubles the diameter.
3. An increase of 10 gage numbers (e.g., from No. 10 to 1/0) multiplies the area and weight by 10 and divides the resistance by 10.
4. A No. 10 wire has a diameter of about 0.10 in, an area of about 10,000 cmils, and (for standard annealed copper at 20°C) a resistance of approximately 1.0 Ω/1,000 ft.
5. The weight of No. 2 copper wire is very close to 200 lb/1,000 ft.

81. Steel wire gage, also known originally as the **Washburn & Moen gage** and later as the **American Steel & Wire Co.'s gage,** was established by Ichabod Washburn about 1830. This gage also, with a number of its sizes rounded off to thousandths of an inch, is known as the **Roebling gage.** It is used exclusively for steel wire and is frequently employed in wire mills.

82. Birmingham wire gage, also known as **Stubs' wire gage** and **Stubs' iron wire gage,** is said to have been established early in the eighteenth century in England, where it was long in use. This gage was used to designate the Stubs soft-wire sizes and should not be confused with Stubs' steel-wire gage. The numbers of the Birmingham gage were based upon the reductions of size made in practice by drawing wire from rolled rod. Thus, a wire rod was called "No. 0," "first drawing No. 1," and so on. The gradations of size in this gage are not regular, as will appear from its graph. This gage is generally in commercial use in the United States for iron and steel wires.

83. Standard wire gage, which more properly should be designated **(British) Standard wire gage,** is the legal standard of Great Britain for all wires, adopted in 1883. It is also known as the **New British Standard gage,** the **English legal standard gage,** and the **Imperial wire gage.** It was constructed by so modifying the Birmingham gage that the differences between consecutive sizes become more regular. This gage is largely used in England but never has been used extensively in America.

84. Old English wire gage, also known as the **London wire gage,** differs very little from the Birmingham gage. It formerly was used to some extent for brass and copper wires but is now nearly obsolete.

85. Millimeter wire gage, also known as the **Metric wire gage,** is based on giving progressive numbers to the progressive sizes, calling 0.1 mm diameter "No. 1," 0.2 mm "No. 2," etc.

86. German wire gage, in which the diameter or thickness is expressed in millimeters, is retrogressive and contains 25 sizes.

German Wire Gage Table
(Diameters in millimeters)

No.	Diam.	No.	Diam.	No.	Diam.	No.	Diam.	No.	Diam.
1	5.50	6	3.75	11	2.50	16	1.375	21	0.750
2	5.00	7	3.50	12	2.25	17	1.250	22	0.625
3	4.50	8	3.25	13	2.00	18	1.125	23	0.562
4	4.25	9	3.00	14	1.75	19	1.000	24	0.500
5	4.00	10	2.75	15	1.50	20	0.875	25	0.438

87. Conductor-size Designation. America uses, for sizes up to 4/0, mil, decimals of an inch, or AWG numbers for solid conductors and AWG numbers or circular mils for stranded conductors; for sizes larger than 4/0, circular mils is used throughout. **U.S. Navy** uses circular mils for all sizes. **Britain** uses square inch area. Other countries ordinarily use square millimeter area.

88. Conductor-size conversion can be accomplished from the following relation:

$$\text{cmils} = \text{in}^2 \times 1{,}273{,}200 = \text{mm}^2 \times 1{,}973.5 \qquad (4\text{-}9)$$

89. Measurement of wire diameters may be accomplished in many ways, but most commonly, by means of a micrometer caliper. Stranded cables usually are measured by means of a circumference tape calibrated directly in diameter readings.

90. Comparison of Wire Gages. See Table 4-10.

91. Stranded conductors are used generally because of their increased flexibility and consequent ease in handling. The greater the number of wires in any given cross section, the greater will be the flexibility of the finished conductor. Most conductors above 4/0 AWG in size are stranded. Generally, in a given concentric-lay stranded conductor, all wires are of the same size and the same material, although special conductors are available embodying wires of different sizes and of different materials. The former will be found in some insulated cables, and the latter in overhead stranded conductors combining high-conductivity and high-strength wires.

The flexibility of any given size of strand obviously increases as the total number of wires increases. It is common practice to increase the total number of wires as the strand diameter increases, in order to provide reasonable flexibility in handling. So-called **flexible concentric strands** for use in insulated cables have about one to two more layers of wires than the standard type of strand for ordinary use.

92. Number of Wires in Stranded Conductors. Each successive layer in a concentrically stranded conductor contains six more wires than the preceding one. The total number of wires in a conductor is:

For 1-wire core constructions (1, 7, 19, etc.),

$$N = 3n(n + 1) + 1 \tag{4-10}$$

For 3-wire core constructions (3, 12, etc.),

$$N = 3n(n + 2) + 3 \tag{4-11}$$

where n is number of layers over core, which is not counted as a layer.

93. Wire size in stranded conductors is

$$d = \sqrt{\frac{A}{N}} \tag{4-12}$$

where A = total conductor area in circular mils and N = total number of wires.

Copper cables are manufactured usually to certain cross-sectional sizes specified in total circular mils or by gage numbers in AWG. This necessarily requires individual wires drawn to certain prescribed diameters, which are different as a rule from normal sizes in AWG (see Table 4-16).

94. Diameter of stranded conductors (circumscribing circle) is

$$D = d(2n + k) \tag{4-13}$$

where d = diameter of individual wire, n = number of layers over core which is not counted as a layer, k = 1 for constructions having 1-wire core (1, 7, 19, etc.), and k = 2.155 for constructions having 3-wire core (3, 12, etc.).

For standard concentric-lay stranded conductors, the following rule gives a simple method of determining the outside diameter of a stranded conductor from the known diameter of a solid wire of the same cross-sectional area.

To obtain the diameter of concentric-lay stranded conductor, multiply the diameter of the solid wire of the same cross-sectional area by the appropriate factor as follows:

Number of wires	Factor	Number of wires	Factor
3	1.244	91	1.153
7	1.134	127	1.154
12	1.199	169	1.154
19	1.147	217	1.154
37	1.151	271	1.154
61	1.152		

95. Area of stranded conductors is

$$A = Nd^2 \text{ cmils} = \tfrac{1}{4}\pi Nd^2 \times 10^{-6} \text{ in}^2 \quad (4\text{-}14)$$

where N = total number of wires and d = individual wire diameter in mils.

96. Effects of Stranding. All wires in a stranded conductor except the core wire form continuous helices, of slightly greater length than the axis or core. This causes slight increase in weight and electrical resistance and slight decrease in tensile strength and sometimes affects the internal inductance, as compared theoretically with a conductor of equal dimensions but composed of straight wires parallel with the axis.

97. Lay, or Pitch. The axial length of one complete turn, or helix, of a wire in a stranded conductor is sometimes termed the lay, or pitch. This is often expressed as the **pitch ratio,** which is the ratio of the length of the helix to its **pitch diameter** (diameter of the helix at the centerline of any individual wire or strand equals the outside diameter of the helix minus the thickness of one wire or strand). If there are several layers, the pitch expressed as an axial length may increase with each additional layer, but when expressed as the ratio of axial length to pitch diameter of helix, it is usually the same for all layers, or nearly so. In commercial practice, the pitch is commonly expressed as the ratio of axial length to outside diameter of helix, but this is an arbitrary designation made for convenience of usage. The **pitch angle** is shown in Fig. 4-3, where ac represents the axis of the stranded conductor and l is the axial length of one complete turn or helix, ab is the length of any individual wire $l + \Delta l$ in one complete turn, and bc is equal to the circumference of a circle corresponding to the pitch diameter d of the helix. The angle bac, or θ, is the pitch angle, and the pitch ratio is expressed by $p = l/d$. There is no standard pitch ratio used by manufacturers generally, since it has been found desirable to vary this depending on the type of service for which the conductor is intended. Applicable lay lengths generally are included in industry specifications covering the various stranded conductors. For bare overhead conductors, a representative commercial value for pitch length is 13.5 times the outside diameter of each layer of strands.

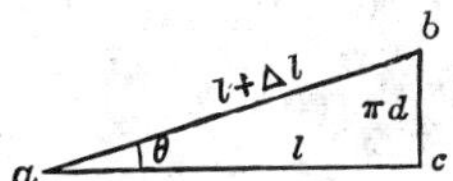

FIG. 4-3. Pitch angle in concentric-lay cable.

98. Direction of Lay. The direction of lay is the lateral direction in which the individual wires of a cable run over the top of the cable as they recede from an observer looking along the axis. **Right-hand lay** recedes from the observer in clockwise rotation or like a right-hand screw thread; **left-hand lay** is the opposite. The outer layer of a cable is ordinarily applied with the left-hand lay, although the opposite lay can be used if desired.

99. Increase in Weight Due to Stranding. Referring to Fig. 4-3, the increase in weight of the spiral members in a cable is proportional to the increase in length,

$$\frac{l + \Delta l}{l} = \sec\theta = \sqrt{1 + \tan^2\theta}$$

$$= \sqrt{1 + \frac{\pi^2}{p^2}} = 1 + \frac{1}{2}\frac{\pi^2}{p^2} - \frac{1}{8}\left(\frac{\pi^2}{p^2}\right)^2 + \cdots \quad (4\text{-}15)$$

As a first approximation this ratio equals $1 + 0.5(\pi^2/p^2)$, and a pitch of 15.7 produces a ratio of 1.02. This correction factor should be computed separately for each layer if the pitch p varies from layer to layer. Practical correction factors for stranded copper conductors are given in Table 4-16.

100. Increase in Resistance Due to Stranding. If it were true that no current flows from wire to wire through their lineal contacts, the proportional increase in the total resistance would be the same as the proportional increase in total weight. If all the wires were in perfect and complete contact with each other, the total resistance would decrease in the same proportion that the total weight increases, owing to the slightly increased normal cross section of the cable as a whole. The contact resistances are normally sufficient to make the actual increase in total resistance nearly as much, proportionately, as the increase in total weight, and for practical purposes they are usually assumed to be the same. See Table 4-16.

101. Decrease in Strength Due to Stranding. When a concentric-lay cable is subjected to mechanical tension, the spiral members tend to tighten around those layers under them and thus produce internal compression, gripping the inner layers and the core. Consequently the individual wires, taken as a whole, do not behave as they would if they were true linear conductors acting independently. Furthermore, the individual wires are never exactly alike in diameter or in strength or in elastic properties. For these reasons there is ordinarily a loss of a few percent in total tensile efficiency, usually estimated at approximately 10%, in comparison with the sum of the tensile strengths of the individual wires. This reduction tends to increase as the pitch ratio decreases. Actual tensile tests on cables furnish the most dependable data on their ultimate strength.

102. Tensile efficiency of a stranded conductor is the ratio of its ultimate strength to the sum of the ultimate strengths of all its individual wires. Concentric-lay cables of 12 to 16 pitch ratio have a normal tensile efficiency of approximately 90%; rope-lay cables, approximately 80%.

103. Preformed Cable. This type of cable is made by preforming each individual wire (except the core) into a spiral of such length and curvature that the wire will fit naturally into its normal position in the cable instead of being forced into that shape under the usual tension in the stranding machine. This method has the advantage in cable made of the stiffer grades of wire that the individual wires do not tend to spread or untwist if the strand is cut in two without first binding the ends on each side of the cut.

104. Weight. A uniform cylindrical conductor of diameter d, length l, and density δ has a total weight expressed by the formula

$$W = \delta l \left(\frac{\pi d^2}{4} \right) \tag{4-16}$$

The weight of any conductor is commonly expressed in pounds per unit of length, such as 1 ft, 1,000 ft, or 1 mi. The weight of stranded conductors can be calculated using Eq. (4-16), but allowance must be made for increase in weight due to stranding (see Par. 99). Rope-lay stranding has greater increase in weight because of the multiple stranding operations. As an example, industry standards for increase in weight of **stranded copper conductors** are as follows:

ASTM stranding classes: AA, A, B, C, D	Increase, %	ASTM stranding classes: G, H	Increase, %	ASTM stranding classes: I, J, K, L, M, O, P, Q	Increase, %
1 to 4 AWG (AA only)	1	49 wires or less....	3	Single bunched strand.	2
Up to 2,000 Mcm.....	2	133 wires..........	4	7 ropes of bunched strand............	4
Over 2,000–3,000 Mcm.	3	259 wires..........	4.5	19 ropes of bunched strand............	5
Over 3,000–4,000 Mcm.	4	427 wires..........	5	7 × 7 ropes of bunched strand............	6
Over 4,000–5,000 Mcm.	5	Over 427 wires....	6	19, 37 or 61 × 7 ropes of bunched strand...	7

105. Ultimate Strength. This term is not synonymous with breaking strength, because with some classes of materials the ultimate strength is the greater of the two and must be reached and passed before rupture occurs. In all cases the ultimate strength is the greatest load which a material will sustain when the load is gradually increased until rupture or failure takes place.

106. Total Elongation at Rupture. When a sample of any material is tested under tension until it ruptures, measurement is usually made of the total elongation in a certain initial test length. In certain kinds of testing, the initial test length has been standardized, but in every case, the total elongation at rupture should be referred to the initial test length of the sample on which it was measured. Such elongation is usually expressed in percentage of original unstressed length and is a general index of the ductility of the material. Elongation is determined on solid conductors or on individual wires before stranding; it is rarely determined on stranded conductors.

107. Elasticity. All materials are deformed in greater or lesser degree under application of mechanical stress. Such deformation may be either of two kinds, known, respectively, as "elastic deformation" and "permanent deformation." When a material is subjected to stress and undergoes deformation but resumes its original shape and dimensions when the stress is removed, the deformation is said to be elastic. If the stress is so great that the material fails to resume its original dimensions when the stress is removed, the permanent change in dimensions is termed permanent deformation or "set." In general the stress at which appreciable permanent deformation begins is termed the "working elastic limit." Below this limit of stress the behavior of the material is said to be elastic, and in general the deformation is proportional to the stress.

108. Stress and Strain. The stress in a material under load, as in simple tension or compression, is defined as the total load divided by the area of cross section normal to the direction of the load, assuming the load to be uniformly distributed over this cross section. It is commonly expressed in pounds per square inch. The strain in a material under load is defined as the total deformation measured in the direction of the stress, divided by the total unstressed length in which the measured deformation occurs, or the deformation per unit length. It is expressed as a decimal ratio or numeric.

In order to show the complete behavior of any given conductor under tension, it is customary to make a graph in terms of loading or stress as the ordinates and elongation or strain as the abscissas. Such graphs or curves are useful in determining the elastic limit and the yield point if the loading is carried to the point of rupture. Graphs showing the relationship between stress and strain in a material tested to failure are termed load-deformation or stress-strain curves.

109. Hooke's law consists of the simple statement that the stress is proportional to the strain. It obviously implies a condition of perfect elasticity, which is true only for stresses less than the elastic limit.

110. Stress-Strain Curves. A typical stress-strain diagram of hard-drawn copper wire is shown in Fig. 4-4, which represents No. 9 AWG. The curve *ae* is the actual stress-strain curve; *ab* represents the portion which corresponds to true elasticity, or for which Hooke's law holds rigorously; *cd* is the tangent to *ae* which fixes the Johnson elastic limit (see Par. **114**); and the curve *af* represents the set, or permanent elongation due to flow of the metal under stress, being the difference between *ab* and *ae*.

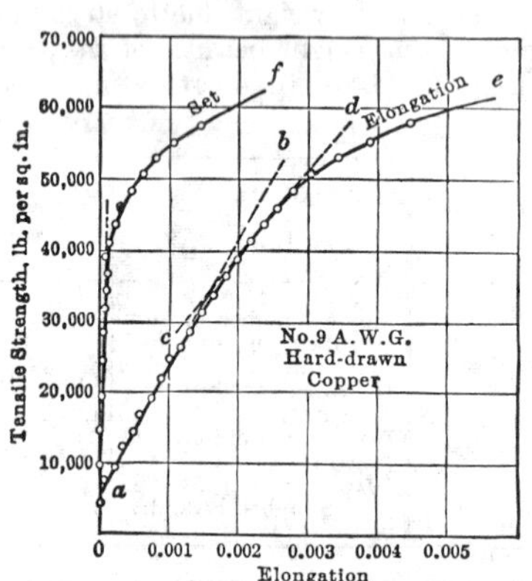

Fig. 4-4. Stress-strain curves of No. 9 AWG hard-drawn copper wire (Watertown Arsenal Test).

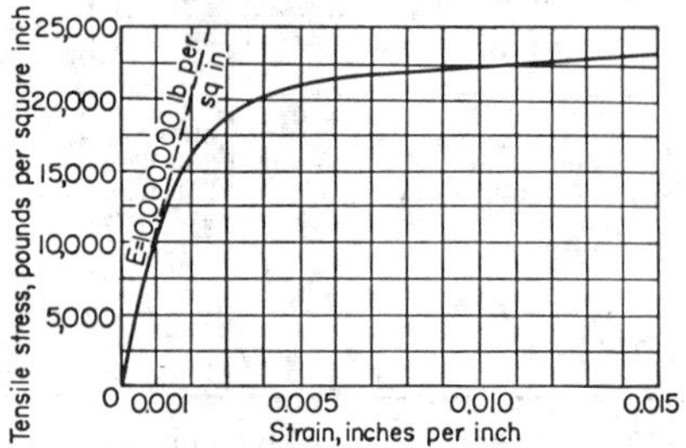

Fig. 4-5. Typical stress-strain curve of hard-drawn aluminum wire.

A typical stress-strain diagram of hard-drawn aluminum wire, based on data furnished by the Aluminum Company of America, is shown in Fig. 4-5.

111. Modulus (or coefficient) of elasticity is the ratio of internal stress to the corresponding strain or deformation. It is a characteristic of each material, form (shape or structure), and type of stressing. For deformations involving changes both in volume and in shape, special coefficients are used. For conductors under axial tension, the ratio of stress to strain is called **Young's modulus.**

If F is the total force or load acting uniformly on the cross section A, the stress is

Table 4-5. Young's Moduli for Conductors

Conductor	Young's modulus,* lb/in²		Reference
	Final†	Virtual initial‡	
Copper wire, hard-drawn	17.0×10^6	14.5×10^6	Copper Wire Engineering Assoc.
Copper wire, medium hard-drawn	16.0×10^6	14.0×10^6	Anaconda Wire and Cable Co.
Copper cable, hard-drawn, 3 and 12 wire	17.0×10^6	14.0×10^6	Copper Wire Engineering Assoc.
Copper cable, hard-drawn, 7 and 19 wire	17.0×10^6	14.5×10^6	Copper Wire Engineering Assoc.
Copper cable, medium hard-drawn	15.5×10^6	14.0×10^6	Anaconda Wire and Cable Co.
Bronze wire, alloy 15	14.0×10^6	13.0×10^6	Anaconda Wire and Cable Co.
Bronze wire, other alloys	16.0×10^6	14.0×10^6	Anaconda Wire and Cable Co.
Bronze cable, alloy 15	13.0×10^6	12.0×10^6	Anaconda Wire and Cable Co.
Bronze cable, other alloys	16.0×10^6	14.0×10^6	Anaconda Wire and Cable Co.
Copper-clad steel wire	24.0×10^6	22.0×10^6	Copperweld Steel Co.
Copper-clad steel cable	23.0×10^6	20.5×10^6	Copperweld Steel Co.
Copper–copper-clad steel cable, type E	19.5×10^6	17.0×10^6	Copperweld Steel Co.
Copper–copper-clad steel cable, type EK	18.5×10^6	16.0×10^6	Copperweld Steel Co.
Copper–copper-clad steel cable, type F	18.0×10^6	15.5×10^6	Copperweld Steel Co.
Copper–copper-clad steel cable, type 2A to 6A	19.0×10^6	16.5×10^6	Copper Wire Engineering Assoc.
Aluminum wire	10.0×10^6	....	Reynolds Metals Co.
Aluminum cable	9.1×10^6	7.3×10^6	Reynolds Metals Co.
Aluminum-alloy wire	10.0×10^6	....	Reynolds Metals Co.
Aluminum-alloy cable	9.1×10^6	7.3×10^6	Reynolds Metals Co.
Aluminum-steel cable, aluminum wire	10.0×10^6	....	Aluminum Co. of America
Aluminum-steel cable, steel wire	29.0×10^6	....	Aluminum Co. of America
Aluminum-clad steel wire	23.5×10^6	22.0×10^6	Copperweld Steel Co.
Aluminum-clad steel cable	23.0×10^6	21.5×10^6	Copperweld Steel Co.
Aluminum-clad steel–aluminum cable:			
AWAC 5/2	13.5×10^6	12.0×10^6	Copperweld Steel Co.
AWAC 4/3	15.5×10^6	14.0×10^6	Copperweld Steel Co.
AWAC 3/4	17.5×10^6	16.0×10^6	Copperweld Steel Co.
AWAC 2/5	19.0×10^6	18.0×10^6	Copperweld Steel Co.
Galvanized-steel wire, Class A coating	28.5×10^6	....	Indiana Steel & Wire Co.
Galvanized-steel cable, Class A coating	27.0×10^6	....	Indiana Steel & Wire Co.

* For stranded cables the moduli are usually less than for solid wire and vary with number and arrangement of strands, tightness of stranding, and length of lay. Also, during initial application of stress, the stress-strain relation follows a curve throughout the upper part of the range of stress commonly used in transmission-line design.

† Final modulus is the ratio of stress to strain (slope of the curve) obtained after fully prestressing the conductor. It is used in calculating design or final sags and tensions.

‡ Virtual initial modulus is the ratio of stress to strain (slope of the curve) obtained during initial sustained loading of new conductor. It is used in calculating initial or stringing sags and tensions.

F/A. If this magnitude of stress causes an elongation e in an original length l, the strain is e/l. Young's modulus is then expressed

$$M = \frac{Fl}{Ae} \tag{4-17}$$

If a material were capable of sustaining an elastic elongation sufficient to make e equal to l, or such that the elongated length is double the original length, the stress required to produce this result would equal the modulus. This modulus is very useful in computing the sags of overhead conductor spans under loads of various kinds. It is usually expressed in pounds per square inch.

Stranding usually lowers the Young's modulus somewhat, rope-lay stranding to a greater extent than concentric-lay stranding.

When a new cable is subjected initially to tension and the loading is carried up to the maximum working stress, there is an apparent elongation which is greater than the subsequent elongation under the same loading. This is apparently due to the removal of a very slight slackness in the individual wires, causing them to fit closely together and adjust themselves to the conditions of tension in the strand. When a new cable is loaded to the working limit, unloaded, and then reloaded, the value of Young's modulus determined on initial loading may be on the order of one-half to two-thirds of its true value on reloading. The latter figure should approach within a few percent of the modulus determined by test on individual straight wires of the same material.

For those applications where elastic stretching under tension needs consideration, the stress-strain curve should be determined by test, with the precaution not to prestress the cable before test unless it will be prestressed when installed in service.

Commercially used values of Young's modulus for conductors are given in Table 4-5.

112. Young's Moduli for Conductors. See Table 4-5.

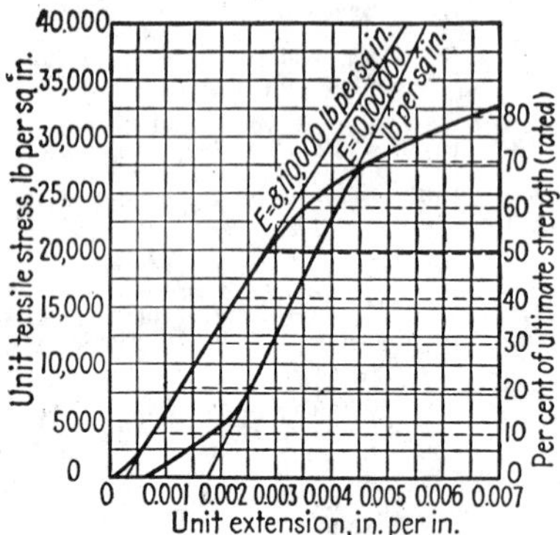

FIG. 4-6. Repeated stress-strain curve, 795,000 cmils ACSR; 54 × 0.1214 in aluminum/7 × 0.1214 in steel.

113. Young's Modulus for ACSR. The permanent modulus of ACSR is dependent upon the proportions of steel and aluminum in the cable and upon the distribution of stress between aluminum and steel. This latter condition is dependent upon temperature, tension, and previous maximum loadings. Because of the interchange of stress between the steel and the aluminum caused by changes of tension and temperature, graphical methods are much superior in accuracy and convenience to analytical methods for sag-tension calculations.

Because ACSR is a composite cable made of aluminum and steel wires, additional phenomena occur which are not found in tests of cable composed of a single material. As shown in Fig. 4-6, the part of the curve obtained in the second stress cycle contains a comparatively large "foot" at its base, which is caused by the difference in extension at the elastic limits of the aluminum and steel.

114. Elastic Limit. This is variously defined as the limit of stress beyond which permanent deformation occurs or the stress limit beyond which Hooke's law ceases to apply or the limit beyond which the stresses are not proportional to the strains or the **proportional limit.** In some materials the elastic limit occurs at a point which is readily determined, but in others it is quite difficult to determine because the stress-strain curve deviates from a straight line but very slightly at first, and the point of departure from true linear relationship between stress and strain is somewhat indeterminate.

The late Dean J. B. Johnson of the University of Wisconsin, well-known authority on materials of construction, proposed the use of an arbitrary determination referred to frequently as the "Johnson definition of elastic limit." This proposal, which has been quite largely used, was that an "apparent elastic limit" be employed defined as that point on the stress-strain curve at which the rate of deformation is 50% greater than at the origin. The apparent elastic limit thus defined is a practical value, which is suitable for engineering purposes, as it involves negligible permanent elongation.

The **Johnson elastic limit** is that point on the stress-strain curve at which the natural tangent is equal to 1.5 times the tangent of the angle of the straight or linear portion of the curve, with respect to the axis of ordinates, or Y axis.[8]

115. Yield Point. In many materials a point is reached on the stress-strain diagram at which there is a marked increase in strain or elongation without an increase in stress or load. The point at which this occurs is termed the yield point. It is usually quite noticeable in ductile materials but may be scarcely perceptible or possibly not present at all in certain hard-drawn materials such as hard-drawn copper.

116. Maximum working stresses of conductor materials must be determined by tests on samples or specimens comparable in size and condition with the shapes or members intended for use under service conditions. The ratio of the safe maximum working stress to the ultimate strength is from 50 to 60% in many classes of materials, but in others it may range as high as 75 to 80%. The maximum working stresses of any material should be determined only from complete knowledge of its properties and the conditions under which it will be used in service.

The working strength of strands is also affected by the mode of attachment to their supports. When gripped in clamping devices, the edges of the clamps should be rounded to prevent injury to the wires, and the grooves in the clamps should be of suitable diameter to fit the strand very closely and of sufficient length to grip all the wire surfaces firmly. If these precautions are not followed, the strength of any strand as a whole may be appreciably impaired at the clamps. In some forms of construction, strands are supported by means of tapered sockets, to which their ends are made fast by having a matrix, such as zinc, cast in the open end of the socket so as to fill it and grip the individual wires.

117. Maximum Working Stress for Annealed-copper Conductors. The stress-strain curves for annealed copper show that it has no definite elastic limit and starts to take a permanent deformation, or set, at comparatively small stresses. It is characteristic of such wire to stretch slowly but permanently under relatively moderate stresses, but in so doing it hardens and tends to increase its own elastic limit. The actual condition of any given wire depends on its previous loading. In overhead spans, where the slack has been pulled up repeatedly, the maximum working stress may approach a high percentage of the original ultimate strength; according to one authority this may become as much as 85%.

118. Maximum Working Stress for Medium-hard-drawn Copper Conductors. The average is 50% of the ultimate tensile strength, but where more exact information is desired, stress-strain curves should be determined by test. If the wire will be prestressed at the time it is installed for service, it should be similarly prestressed before determining the stress-strain curve or tested in appropriate manner to show the stress-strain curve after prestressing.

119. Maximum Working Stress for Hard-drawn Copper Conductors. For wire sizes from 0.460 to 0.325 in, inclusive, the average is 55% of the ultimate tensile strength, with a minimum of 50%. For sizes from 0.324 to 0.040 in, inclusive, the average is 60%, with a minimum of 55%. For more exact information, stress-strain curves should be determined by test.

120. Maximum Working Stress for ACSR. Although the ultimate strength of the steel strands is 190,000 lb/in^2 and that of the aluminum strands is from 23,000 to 30,000 lb/in^2, depending on the size, the ultimate stresses do not occur at the same elongation. Since the aluminum strands have less elongation at rupture than the steel strands, it is evident that the breaking strength of ACSR is the sum of two quantities, the first being the cross-sectional area of aluminum multiplied by its ultimate stress and the second being the cross-sectional area of steel multiplied by its stress at the ultimate elongation of the aluminum wires. The ultimate strength of ACSR thus obtained is the tension at which the aluminum strands fail; the steel core will then stretch, thus reducing the tension. The safe maximum working tension may be defined as that value of tension which can be experienced repeatedly without altering the tensile properties which obtain after the first application of that tension.

121. Maximum Working Stresses for Copper-clad, Aluminum-clad, and Galvanized-steel Conductors. For sizes customarily employed, 60% of the ultimate breaking strength is used. Under some special conditions, this value may be exceeded.

122. Prestressed Conductors. In the case of some materials, especially those of considerable ductility, which tend to show permanent elongation or "drawing" under loads just above the initial elastic limit, it is possible to raise the working elastic limit by loading them to stresses somewhat above the elastic limit as found on initial loading. After such loading, or prestressing, the material will behave according to Hooke's law at all loads less than the new elastic limit. This applies not only to many ductile materials, such as soft or annealed copper wire, but also to cables or stranded conductors, in which there is a slight inherent slack or looseness of the individual wires that can be removed only under actual loading. It is sometimes the practice, when erecting such conductors for service, to prestress them to the working elastic limit or safe maximum working stress and then reduce the stress to the proper value for installation at the stringing temperature without wind or ice.

123. Resistance is the property of an electric circuit or of any body that may be used as part of an electric circuit which determines for a given current the average rate at which electrical energy is converted into heat. The term is properly applied only when the rate of conversion is proportional to the square of the current and is then equal to the power conversion divided by the square of the current.[6] A uniform cylindrical conductor of diameter d, length l, and **volume resistivity** ρ has a total resistance to continuous currents expressed by the formula

$$R = \frac{\rho l}{\pi d^2/4} \tag{4-18}$$

The resistance of any conductor is commonly expressed in ohms per unit of length, such as 1 ft, 1,000 ft, or 1 mi. When used for conducting alternating currents, the effective resistance may be higher than the d-c resistance defined above. In the latter case it is common practice to apply the proper factor, or ratio of effective a-c resistance to d-c resistance, sometimes termed the "skin-effect resistance ratio" (see Par. **127**). This ratio may be determined by test, or it may be calculated if the necessary data are available.

124. Magnetic permeability applies to a field in which the flux is uniformly distributed over a cross section normal to its direction or to a sufficiently small cross section of a nonuniform field so that the distribution can be assumed as substantially uniform. In the case of a cylindrical conductor, the magnetomotive force (mmf) due to the current flowing in the conductor varies from zero at the center or axis to a maximum at the periphery or surface of the conductor and sets up a flux in circular paths concentric with the axis and perpendicular to it but of nonuniform distribution between the axis and the periphery. If the permeability is nonlinear with respect to the mmf, as is usually true with magnetic materials, there is no correct single value of permeability which fits the conditions, although an apparent or equivalent average value can be determined. In the case of other forms of cross section, the distribution is still more complex, and the equivalent permeability may be difficult or impossible to determine except by test.

125. Internal Inductance. A uniform cylindrical conductor of nonmagnetic material, or of unit permeability, has a constant magnitude of internal inductance per unit length, independent of the conductor diameter. This is commonly expressed in microhenrys or millihenrys per unit of length, such as 1 ft, 1,000 ft, or 1 mi. When the conductor material possesses magnetic susceptibility, and when the magnetic permeability μ is constant and therefore independent of the current strength, the internal inductance is expressed in absolute units by the formula

$$L = \frac{\mu l}{2} \tag{4-19}$$

In most cases μ is not constant but is a function of the current strength. When this is true, there is an effective permeability, one-half of which ($\mu/2$) expresses the inductance per centimeter of length, but this figure of permeability is virtually the ratio of the effective inductance of the conductor of susceptible material to the inductance of a conductor of material which has a permeability of unity. When used for conducting alternating currents, the effective inductance may be less than the inductance with direct current; this is also a direct consequence of the same skin effect which results in an increase of effective resistance with alternating currents, but the overall effect is usually included in the figure of effective permeability. It is usually the practice to determine the effective internal inductance by test, but it may be calculated if the necessary data are available.

126. Skin effect is a phenomenon which occurs in conductors carrying currents whose intensity varies rapidly from instant to instant but does not occur with continuous currents. It arises from the fact that elements or filaments of variable current at different points in the cross section of a conductor do not encounter equal components of inductance, but the central or axial filament meets the maximum inductance, and in general the inductance offered to other filaments of current decreases as the distance of the filament from the axis increases, becoming a minimum at the surface or periphery of the conductor. This, in turn, tends to produce unequal current density over the cross section as a whole; the density is a minimum at the axis and a maximum at the periphery. Such distribution of the current density produces an increase in effective resistance and a decrease in effective internal inductance; the former is of more practical importance than the latter. In the case of large copper conductors at commercial power frequencies, and in the case of most conductors at carrier and radio frequencies, the increase in resistance should be considered (see Pars. **131** and **132**).

127. Skin-effect Ratios. If R' is the effective resistance of a linear cylindrical conductor to sinusoidal alternating current of given frequency and R is the true resistance with continuous current, then

$$R' = KR \qquad \text{(ohms)} \tag{4-20}$$

where K is determined from Table 4-6 in terms of x. The value of x is given by

$$x = 2\pi a \sqrt{\frac{2f\mu}{\rho}} \tag{4-21}$$

where a = radius of conductor in centimeters, f = frequency in cycles per second, μ = magnetic permeability of conductor (here assumed to be constant), ρ = resistivity in abohm-centimeters (abohm = $10^{-9}\ \Omega$).

For practical calculation, Eq. (4-21) can be written

$$x = 0.063598 \sqrt{\frac{f\mu}{R}} \tag{4-22}$$

where R = d-c resistance at operating temperature in ohms per mile.

If L' is the effective inductance of a linear conductor to sinusoidal alternating current of a given frequency,

$$L' = L_1 + K'L_2 \tag{4-23}$$

where L_1 = external portion of inductance, L_2 = internal portion (due to the magnetic field within the conductor), and K' is determined from Table 4-6 in terms of x. Thus the total effective inductance per unit length of conductor is

$$L' = 2 \ln \frac{d}{a} + K' \frac{\mu}{2} \tag{4-24}$$

The inductance is here expressed in abhenrys per centimeter of conductor, in a linear circuit; a is the radius of the conductor and d is the separation between the conductor and its return conductor, expressed in the same units.

Values of K and K' in terms of x are shown in Table 4-6 and Figs. 4-7 and 4-8 (see *NBS Circ.* 74, pp. 309–311, for additional tables, and *Sci. Paper* 374).

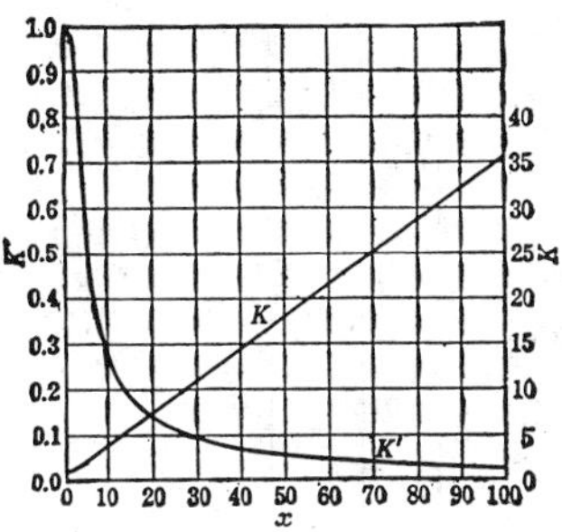

FIG. 4-7. K and K' for values of x from 0 to 100.

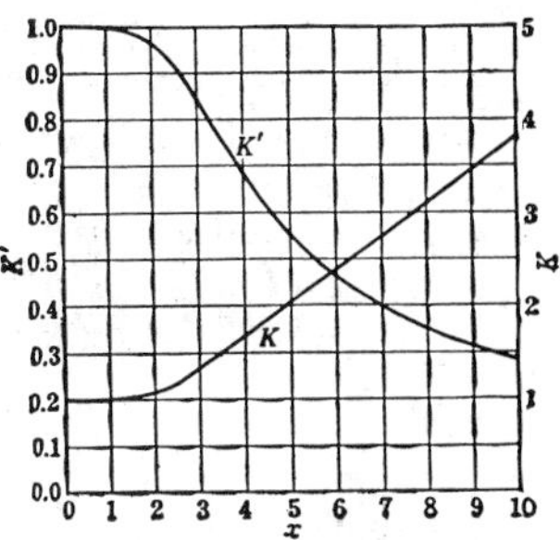

FIG. 4-8. K and K' for values of x from 0 to 10.

Table 4-6. Skin-effect Ratios
(Bur. Std. *Bull.* 169, pp. 226–228)

x	K	K′	x	K	K′	x	K	K′	x	K	K′
0.0	1.00000	1.00000	2.9	1.28644	0.86012	6.6	2.60313	0.42389	17.0	6.26817	0.16614
0.1	1.00000	1.00000	3.0	1.31809	0.84517	6.8	2.67312	0.41171	18.0	6.62129	0.15694
0.2	1.00001	1.00000	3.1	1.35102	0.82975	7.0	2.74319	0.40021	19.0	6.97446	0.14870
0.3	1.00004	0.99998	3.2	1.38504	0.81397	7.2	2.81334	0.38933	20.0	7.32767	0.14128
0.4	1.00013	0.99993	3.3	1.41999	0.79794	7.4	2.88355	0.37902	21.0	7.68091	0.13456
0.5	1.00032	0.99984	3.4	1.45570	0.78175	7.6	2.95380	0.36923	22.0	8.03418	0.12846
0.6	1.00067	0.99966	3.5	1.49202	0.76550	7.8	3.02411	0.35992	23.0	8.38748	0.12288
0.7	1.00124	0.99937	3.6	1.52879	0.74929	8.0	3.09445	0.35107	24.0	8.74079	0.11777
0.8	1.00212	0.99894	3.7	1.56587	0.73320	8.2	3.16480	0.34263	25.0	9.09412	0.11307
0.9	1.00340	0.99830	3.8	1.60314	0.71729	8.4	3.23518	0.33460	26.0	9.44748	0.10872
1.0	1.00519	0.99741	3.9	1.64051	0.70165	8.6	3.30557	0.32692	28.0	10.15422	0.10096
1.1	1.00758	0.99621	4.0	1.67787	0.68632	8.8	3.37597	0.31958	30.0	10.86101	0.09424
1.2	1.01071	0.99465	4.1	1.71516	0.67135	9.0	3.44638	0.31257	32.0	11.56785	0.08835
1.3	1.01470	0.99266	4.2	1.75233	0.65677	9.2	3.51680	0.30585	34.0	12.27471	0.08316
1.4	1.01969	0.99017	4.3	1.78933	0.64262	9.4	3.58723	0.29941	36.0	12.98160	0.07854
1.5	1.02582	0.98711	4.4	1.82614	0.62890	9.6	3.65766	0.29324	38.0	13.68852	0.07441
1.6	1.03323	0.98342	4.5	1.86275	0.61563	9.8	3.72812	0.28731	40.0	14.39545	0.07069
1.7	1.04205	0.97904	4.6	1.89914	0.60281	10.0	3.79857	0.28162	42.0	15.10240	0.06733
1.8	1.05240	0.97390	4.7	1.93533	0.59044	10.5	3.97477	0.26832	44.0	15.80936	0.06427
1.9	1.06440	0.96795	4.8	1.97131	0.57852	11.0	4.15100	0.25622	46.0	16.51634	0.06148
2.0	1.07816	0.96113	4.9	2.00710	0.56703	11.5	4.32727	0.24516	48.0	17.22333	0.05892
2.1	1.09375	0.95343	5.0	2.04272	0.55597	12.0	4.50358	0.23501	50.0	17.93032	0.05656
2.2	1.11126	0.94482	5.2	2.11353	0.53506	12.5	4.67993	0.22567	60.0	21.46541	0.04713
2.3	1.13069	0.93527	5.4	2.18389	0.51566	13.0	4.85631	0.21703	70.0	25.00063	0.04040
2.4	1.15207	0.92482	5.6	2.25393	0.49764	13.5	5.03272	0.20903	80.0	28.53593	0.03535
2.5	1.17538	0.91347	5.8	2.32380	0.48086	14.0	5.20915	0.20160	90.0	32.07127	0.03142
2.6	1.20056	0.90126	6.0	2.39359	0.46521	14.5	5.38560	0.19468	100.0	35.60666	0.02828
2.7	1.22753	0.88825	6.2	2.46338	0.45056	15.0	5.56208	0.18822	∞	∞	0
2.8	1.25620	0.87451	6.4	2.53321	0.43682	16.0	5.91509	0.17649			

Table 4-7. Skin-effect Ratios—Copper Conductors NOT in Close Proximity
(See also Table 4-8; A-C/D-C Resistance Ratios—Copper and Aluminum Conductors IN Close Proximity)

Conductor size, Mcm	Skin-effect ratio **K** at 60 cycles and 65 C (149 F)															
	Inside conductor diameter, in.															
	0*		0.25		0.50		0.75		1.00		1.25		1.50		2.00	
	Outside diameter, in.	**K**	Outside diameter, in.	**K**	Outside diameter, in.	**K**	Outside diameter, in.	**K**	Outside diameter, in.	**K**	Outside diameter, in.	**K**	Outside diameter, in.	**K**	Outside diameter, in.	**K**
3000	1.998	1.439	2.02	1.39	2.08	1.36	2.15	1.29	2.27	1.23	2.39	1.19	2.54	1.15	2.87	1.08
2500	1.825	1.336	1.87	1.28	1.91	1.24	2.00	1.20	2.12	1.16	2.25	1.12	2.40	1.09	2.75	1.05
2000	1.631	1.239	1.67	1.20	1.72	1.17	1.80	1.12	1.94	1.09	2.09	1.06	2.25	1.05	2.61	1.02
1500	1.412	1.145	1.45	1.12	1.52	1.09	1.63	1.06	1.75	1.04	1.91	1.03	2.07	1.02	2.47	1.01
1000	1.152	1.068	1.19	1.05	1.25	1.03	1.39	1.02	1.53	1.01	1.72	1.01				
800	1.031	1.046	1.07	1.04	1.16	1.02	1.28	1.01	1.45	1.01						
600	0.893	1.026	0.94	1.02	1.04	1.01										
500	0.814	1.018	0.86	1.01	0.97	1.01										
400	0.728	1.012	0.78	1.01												
300	0.630	1.006														

* For standard concentric-stranded conductors (i.e., inside diameter = 0).

Value of μ for nonmagnetic materials (copper, aluminum, etc.) is 1; for magnetic materials, it varies widely with composition, processing, current density, etc., and should be determined by test in each case.

128. Skin-effect Ratios. See Table 4-6.

129. Skin-effect Ratios—Copper Conductors NOT in Close Proximity. See Table 4-7.

130. Alternating-current Resistance. For small conductors at power frequencies, the frequency has a negligible effect, and d-c resistance values can be used. For large conductors, frequency must be taken into account in addition to temperature effects. To do this, first calculate the d-c resistance at the operating temperature, then determine the skin-effect ratio K, and finally determine the a-c resistance at operating temperature (see Par. **127**).

131. A-c resistance for copper conductors NOT in close proximity can be obtained from the skin-effect ratios given in Tables 4-6 and 4-7. Practical values for usual conductor sizes are given in Table 4-29.

132. A-c resistance for copper conductors in close proximity or for insulated copper conductors installed in conduit may be calculated from Table 4-8.

133. A-C/D-C Resistance Ratios: Copper Conductors in Close Proximity. See Table 4-8.

134. A-c resistance for copper-clad and aluminum-clad steel conductors is dependent on several variables and should be determined by test. Practical values for usual conductor sizes are given in Tables 4-30, 4-31, and 4-33.

Table 4-8. A-C/D-C Resistance Ratios—Copper and Aluminum Conductors IN Close Proximity
(*IPCEA Publ.* P-34-359)

Conductor size, Mcm or AWG	A-c/d-c resistance ratio at 60 cycles and 65°C (149°F)							
	Single-conductor cable in air or separate **nonmetallic** conduit						5–15-kV nonleaded shielded power cable, 3 single conductor cables in same **metallic** conduit	
	Concentric		Segmental		Annular		Concentric	
	Copper	Aluminum	Copper	Aluminum	Copper	Aluminum	Copper	Aluminum
5,000	1.77	1.42	....	....	1.12	1.05		
4,500	1.69	1.36	....	....	1.12	1.05		
4,000	1.61	1.31	1.18	1.08	1.12	1.05		
3,500	1.52	1.25	1.15	1.06	1.11	1.04		
3,000	1.43	1.20	1.11	1.04	1.11	1.04		
2,500	1.33	1.14	1.08	1.03	1.07	1.03		
2,250	1.28	1.12	1.06	1.03	....	....		
2,000	1.24	1.10	1.05	1.02	1.05	1.02		
1,750	1.19	1.08	1.04	1.02	1.04	1.02		
1,500	1.14	1.06	1.03	1.01	1.04	1.01		
1,250	1.10	1.04	1.02	1.01	1.04	1.01		
1,000	1.07	1.03	1.01	1.01	1.03	1.01	1.36	1.17
900	1.06	1.02	....	....	1.03	1.01	1.30	1.14
800	1.04	1.02	....	....	1.02	1.01	1.24	1.11
750	1.04	1.01	....	....	1.02	1.01	1.22	1.10
700	1.03	1.01	....	....	....	....	1.19	1.09
600	1.03	1.01	....	....	....	....	1.14	1.07
500	1.02	1.01	....	....	....	....	1.10	1.05
400	1.01	1.01*	....	....	....	....	1.07	1.03
350	1.01	....	....	....	....	....	1.05	1.03
300	1.01	....	....	....	....	....	1.04	1.02
250	1.01*	....	....	....	....	....	1.03	1.01
4/0	1.01*	....	....	....	....	....	1.02	1.01
3/0	....	....	....	....	....	....	1.01	
2/0	....	....	....	....	....	....	1.01	

* Conductor skin effect less than 1%.

135. A-C Resistance for Aluminum Conductors. The increase in resistance and decrease in internal inductance of cylindrical aluminum conductors can be determined from the data in Par. **127.** It is not the same as for copper conductors of equal diameter but is slightly less because of the higher volume resistivity of aluminum.

136. A-C/D-C Resistance Ratios: Aluminum Conductors in Close Proximity. See Table 4-8.

137. A-C Resistance for ACSR. In the case of ACSR conductors, the steel core is of relatively high resistivity, and therefore its conductance is usually neglected in computing the total resistance of such strands. The effective permeability of the grade of steel employed in the core is also relatively small. It is approximately correct to assume that such a strand is hollow and consists exclusively of its aluminum wires; in this case the laws of skin effect in tubular conductors will be applicable. Conductors having a single layer of aluminum wires over the steel core have higher a-c/d-c ratios than those having multiple layers of aluminum wires. Practical values for usual conductor sizes are given in Table 4-32.

138. A-C Resistance for Steel Conductors. The increase in effective resistance with alternating currents depends fundamentally upon the circular permeability; there is also an increase, of much smaller proportions, caused by hysteresis. The permeability is very sensitive to variations in composition, heat treatment, and working of the metal; for this reason it is not usually feasible to compute the skin effect except as an approximation. The results of tests show great variations depending upon the factors just mentioned. Typical test results[9] are shown in Figs. 4-9 and 4-10. Practical values for small conductors are given in Table 4-26.

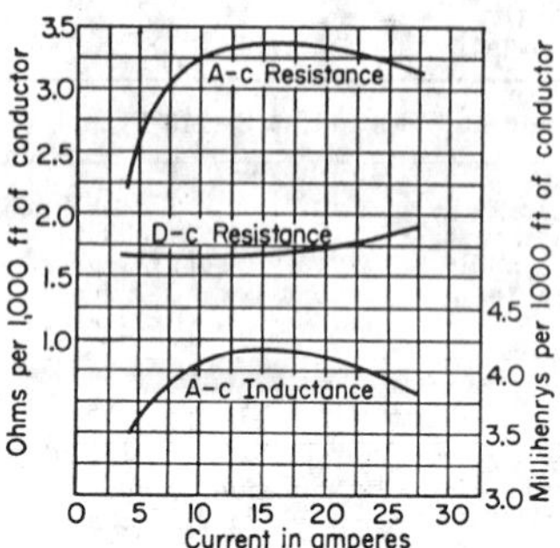

FIG. 4-9. Effective resistance and internal inductance at 60 cycles of No. 6 BWG galvanized BB wire.

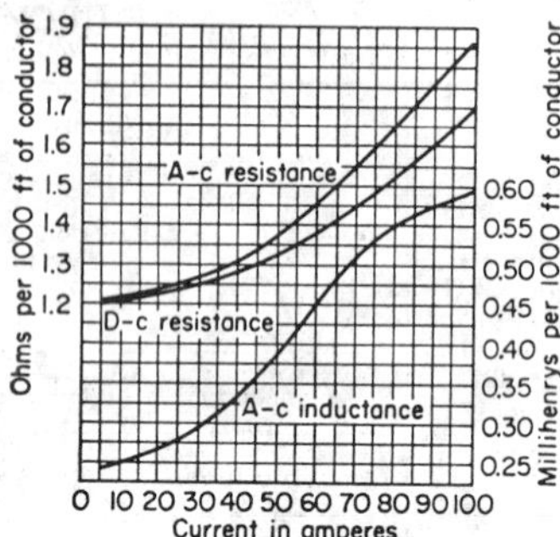

FIG. 4-10. Effective resistance and internal inductance at 60 cycles of ⅜-in 7-wire Class A galvanized high-strength steel strand.

139. Inductive Reactance. Present practice is to consider inductive reactance as split into two components[10]: (1) that due to flux within a radius of 1 ft including the internal reactance within the conductor of radius r and (2) that due to flux between 1 ft radius and the equivalent conductor spacing D_s or geometric mean distance (GMD).

The fundamental inductance formula is

$$L = 2 \ln \frac{D_s}{r} + \frac{\mu}{2} \qquad [\text{abH/(cm) (conductor)}] \tag{4-25}$$

This can be rewritten

$$L = 2 \ln \frac{D_s}{1} + 2 \ln \frac{1}{r} + \frac{\mu}{2} \tag{4-26}$$

where the term $2 \ln (D_s/1)$ represents inductance due to flux between 1 ft radius and the equivalent conductor spacing and $2 \ln (1/r) + (\mu/2)$ represents the inductance due to flux within 1 ft radius [2 ln $(1/r)$ represents inductance due to flux between conductor surface and 1 ft radius, and $\mu/2$ represents internal inductance due to flux within the conductor].

By definition, geometric mean radius (GMR) of a conductor is the radius of an infinitely thin tube having the same internal inductance as the conductor. Therefore,

$$L = 2 \ln \frac{D_s}{1} + 2 \ln \frac{1}{\text{GMR}} \tag{4-27}$$

Since inductive reactance $= 2\pi fL$, for practical calculation Eq. (4-27) can be written

$$X = 0.004657f \log \frac{D_s}{1} + 0.004657f \log \frac{1}{\text{GMR}} \qquad [\Omega/(\text{mi})(\text{conductor})] \tag{4-28}$$

In the conductor tables in this section, inductive reactance is calculated from Eq. (4-28), considering that

$$X = x_a + x_d \tag{4-29}$$

140. Inductive reactance for conductors using steel varies in a manner similar to a-c resistance (see Pars. **134** to **138**). Practical values for usual conductor sizes are given in Tables 4-26, 4-30, 4-31, and 4-32.

141. Capacitive Reactance. By the same reasoning used in Par. **139,** the capacitive reactance can be considered in two parts also, giving

$$X = \frac{4.099}{f} \log \frac{D_s}{1} + \frac{4.099}{f} \log \frac{1}{r} \qquad [\text{M}\Omega/(\text{mi})(\text{conductor})] \tag{4-30}$$

In the conductor tables in this section, capacitive reactance is calculated from Eq. (4-30), it being considered that

$$X' = x_a' + x_d' \tag{4-31}$$

It is important to note that in capacitance calculations the conductor radius used is the actual physical radius of the conductor.

142. Capacitive Susceptance

$$B = \frac{1}{x_a' + x_d'} \qquad [\mu\text{mhos}/(\text{mi})(\text{conductor})] \tag{4-32}$$

143. Charging Current

$$I_C = eB \times 10^{-3} \qquad [\text{A}/(\text{mi})(\text{conductor})] \tag{4-33}$$

where e = voltage to neutral in kilovolts.

144. Current-carrying Capacity of Bare and Weatherproof Conductors. No method has been accepted generally by the industry for the calculation of the current-carrying capacity of conductors for overhead power-transmission lines. For various methods in use, see the following:

Schurig, O. R., and Frick, C. W. Heating and Current-carrying Capacity of Bare Conductors for Outdoor Service; *Gen. Elec. Rev.*, March, 1930, Vol. 33, No. 3, p. 141.

Frick, C. W. Current-carrying Capacity of Bare Cylindrical Conductors for Indoor and Outdoor Service; *Gen. Elec. Rev.*, August, 1934, Vol. 34, No. 8, p. 464.

Kidder, A. H., and Woodward, C. B. Ampere Load Limits for Copper in Overhead Lines; *Trans. AIEE*, 1943, March section, Vol. 62, p. 149.

Enos, H. A. Current-carrying Capacity of Overhead Conductors; *Elec. World*, May 15, 1943, p. 1612.

Zucker, Myron. Thermal Rating of Overhead Line Wire; *Trans. AIEE*, 1943, July section, Vol. 62, p. 501.

Olmstead, L. M. Safe Ratings for Overhead Line Conductors; *Trans. AIEE*, 1943, Vol. 62, p. 845.

Seelye, H. P., and Malmstrom, A. L. Determining Current Ratings of Overhead Conductors; *Elec. Light and Power,* December, 1943, p. 42, and January, 1944, p. 60.

George, E. E. Electrical Heating Characteristics of Overhead Conductors; *Elec. Light and Power,* December, 1944, p. 48; January, 1945, p. 78; and April, 1945, p. 58.

House, H. E., and Tuttle, P. D. Current Carrying Capacity of ACSR; *Trans. AIEE,* 1958, Vol. 77, Pt. III, p. 1169.

145. Current-carrying Capacity of Insulated Conductors. For **power transmission** and distribution cables in air, in enclosed and exposed conduit and in underground ducts, the current-carrying capacities sponsored by the Insulated Power Cable Engineers Association (IPCEA) are in general use. See Power Cable Ampacities, *AIEE Publ.* S-135-1, 1962, Vol. 1, Copper Conductors, Vol. 2, Aluminum Conductors (available from IEEE, 345 E. 47th St., N.Y. 10017, at $15 per set).

For **interior wiring** under jurisdiction of the National Electrical Code, the latest issue of the Standard of the National Board of Fire Underwriters for Electric Wiring and Apparatus, *NBFU Pamphlet* 70, should be consulted (see Sec. **15**).

146. Contact (Trolley) Wires. The special requirements for trolley contact wires have resulted in the development of materials having special characteristics. For general use, hard-drawn copper wire manufactured under particular specifications gives satisfactory service. Under unusual conditions of service, however, harder materials are needed to avoid constant wire replacements.

Fig. 4-11

Fig. 4-12

Fig. 4-13

Fig. 4-11. Grooved section.
Fig. 4-12. Figure 8 section.
Fig. 4-13. Figure 9 deep section.

Trolley contact wires are designed and manufactured for the dual purpose of providing current-carrying capacity and resisting the constant abrasive effect of the trolley wheel, shoe, or pantograph. They must be of good-quality material with a hard-wearing surface, and special care is necessary in manufacture to obtain a surface smooth and free from imperfections.

Because of the need for special methods of supporting trolley contact wires, the use of grooved wires is very common. These as well as other special designs are illustrated in Figs. 4-11 to 4-13.

147. Bus conductors require that greater attention be given to certain physical and electrical characteristics of the metals than is usually necessary in designing line conductors. These characteristics are current-carrying capacity, emissivity, skin effect, expansion, and mechanical deflection. To obtain the most satisfactory and economical designs for bus bars in power stations and substations, where they are used extensively, consideration must be given to choice not only of material but also of shape. Both copper and aluminum are used for bus bars, and in certain outdoor substations, steel has proved satisfactory. The most common bus-bar form for carrying heavy current, especially indoors, is flat copper bar. Bus bars in the form of angles, channels, and tubing have been developed for heavy currents and, because of better distribution of the conducting material, make more efficient use of the metal both electrically and mechanically. All such designs are based upon the need for proper current-carrying capacity without excess bus-bar temperatures and upon the necessity for adequate mechanical strength (see Sec. **12**).

148. Solid conductors (wire) for bare power conductors are usually made of copper or copper alloy, although for particular applications aluminum, aluminum alloy, copper-clad steel, aluminum-clad steel, or steel is used.

149. Copper solid conductors for overhead line applications are supplied normally in hard-drawn or medium-hard-drawn tempers in sizes No. 10 to 4 AWG. Number 2 AWG is occasionally used, but for this size and larger, stranded conductors are preferable. Soft or annealed conductors are available in all sizes for use in insulated conductors, grounding or bonding wires, weather-resisting (weatherproof) wire, etc.

150. Hard-drawn copper wire is drawn through dies from rod to finished product without intermediate annealing. This results in a wire of high strength and low elongation. Both these characteristics can be controlled to some extent by choice of size of rod and by modification of certain details of the drawing process.

151. Medium hard-drawn copper wire is essentially and necessarily a special product, because when wire has once started on its course through the drawing operations, it can finish only as a hard-drawn wire to be used as such or to be annealed and become soft or annealed wire. Medium hard-drawn wire is annealed wire drawn to a slightly smaller diameter.

Medium hard-drawn wire approaches hard-drawn wire in its characteristics, but from the very nature of the product, exact uniformity in tensile strength cannot be obtained; hence the necessity for establishing a range of tensile strength within which standard medium-hard-drawn wire must be expected to be found.

152. Soft or annealed copper wire is drawn by customary operations and annealed, finished by cleaning when necessary to remove scale or oxide. The wire is so soft and ductile that it is easily marred and even stretched by careless handling in the operations of winding or cabling; hence the necessity for confining specifications and inspection to wire in packages as it leaves the manufacturer and before being put through processes incident to its use by the purchaser.

153. Copper-alloy solid conductors for overhead line applications normally are supplied only in hard-drawn temper. The usual range of sizes is No. 10 to 4 AWG. Wires are available in any of the several alloys designated in ASTM B105-55 as alloys 8.5 to 85 in accordance with their conductivities.

154. Copper-clad steel solid conductors are rarely used for purposes other than signal or communication wire (weatherproofed or insulated) or telephone drop wire. Usual sizes are No. 17 to 10 AWG.

Aluminum-clad steel solid conductors are used primarily for signal and communication wire (either bare or insulated). The usual sizes employed are No. 8 through No. 12 AWG.

155. Aluminum solid conductors are practically never used for overhead-line applications because of their relatively poor physical characteristics. Intermediate temper or hard-drawn conductors are used in all permitted sizes of insulated solid conductors.

156. Iron and steel solid conductors are used occasionally for telephone and telegraph lines in Nos. 16, 10, and 8 BWG and Nos. 14, 13, and 12 SWG (see Table 4-10).

157. Concentric-lay stranded conductors are made up of successive layers of helically laid solid members (wires). Usual strandings have 6 wires more in each added layer and start with either a 1- or a 3-wire core.

158. Bunch-stranded conductors are made up of any number of wires grouped together without regard to their accurate geometric arrangement. Usually they are twisted together; if not, they are generally referred to as "parallel-strand."

159. Rope-lay stranded conductors are made up of successive layers of helically laid stranded members. The members may be concentric-lay or bunch-stranded.

160. Special Stranded Conductor Shapes. For use in multiconductor insulated cables whose finished cross section must be round, special shapes such as D shape (hemispherical), sector shape (triangular), and semisector (oval) are commonly used.

Annular conductors are formed by stranding helically laid wires over a central core, which may be (1) rope or fibrous material, (2) copper helix, or (3) twisted copper I-beam. This construction reduces the skin-effect ratio and is desirable in order to obtain economical use of the copper at high currents (see Table 4-8).

Segmental conductors are single conductors composed of either four or three segments which are combined to give a substantially circular cross section. The segments are electrically separated (usually by means of paper tape), and each strand of the individually stranded segments is alternately transposed between the inner and outer positions in the complete conductor due to its concentric lay in its segment. This construction reduces the skin-effect ratio and is desirable where high current-carrying capacity must be combined with small diameter (see Table 4-8).

161. Copper stranded conductors for overhead-line applications are normally supplied in hard-drawn or medium hard-drawn tempers in No. 4 AWG and larger.

Soft or annealed conductors are used in all sizes for insulated conductors and to some extent for weather-resisting (weatherproof) conductors in overhead distribution systems.

162. Copper-alloy stranded conductors are available in the same grades as copper-alloy solid conductors (see Par. **153**). Generally they are used for high strength together with conductance, where corrosion conditions do not permit the use of cheaper constructions; for applications such as overhead ground wires, messengers, railway catenaries; etc.

163. Copper-clad steel stranded conductors are used in the same manner as copper-alloy stranded conductors, where their higher strength and limited range of conductance are suitable.

Aluminum-clad steel stranded conductors are used where high strength, limited conductance, and good corrosion resistance are required. They are widely used for overhead ground wires, neutral messenger and messenger strands, antenna conductors, and guy wires.

164. Aluminum stranded conductors for bare and weatherproof overhead-line applications are normally supplied hard-drawn in No. 6 AWG and larger. Hard-drawn or three-quarter hard-drawn conductors are used in all sizes for insulated conductors.

165. Iron and steel stranded conductors are used for overhead ground wires, messengers, and guy wires.

166. Hollow (expanded) conductors are used on high-voltage transmission lines when, in order to reduce corona loss, it is desirable to increase the outside diameter without increasing the area beyond that needed for maximum line economy. Not only is the initial corona voltage considerably higher than for conventional conductors of equal cross section, but the current-carrying capacity for a given temperature rise is also greater because of the larger surface area available for cooling and the better disposition of the metal with respect to skin effect when carrying alternating currents.

167. Expanded ACSR is formed by interposing aluminum spacer wires and/or impregnated fibrous filler material between a stranded galvanized-steel core and a layer of helically laid hard-drawn aluminum wires (Fig. 4-14).

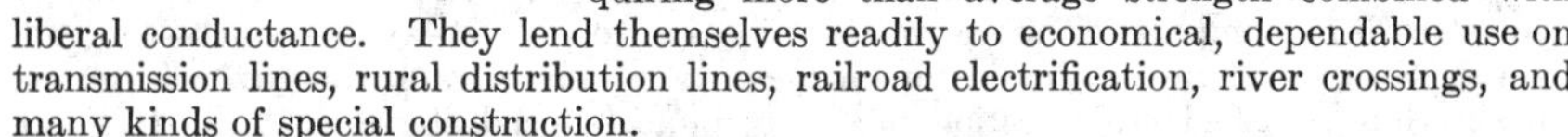

Fig. 4-14. Expanded ACSR conductor.

168. Composite conductors are those made up of usually two different types of wire having differing characteristics. They are generally designed for a ratio of physical and electrical characteristics different from that found in homogeneous materials.

Cables of this type are particularly adaptable to long-span construction or other service conditions requiring more than average strength combined with liberal conductance. They lend themselves readily to economical, dependable use on transmission lines, rural distribution lines, railroad electrification, river crossings, and many kinds of special construction.

169. Commercial composite conductors are available in a range of standard designs and types. Usually they consist of a single wire or stranded core of high-strength material surrounded by one or more layers of high-conductivity material. Special designs to meet unusual requirements can be made readily.

CONDUCTOR DATA TABLES

See Pars. **170** to **197** for identification of tables (p. **4**-56).

Table 4-9. Industry Specifications for Conductors

Sponsoring organization	Specification designation	Date of issue	Specification title	Corresponding ASA specification
General definitions				
AIEE	Book	1957	Definitions of electrical terms	C42-1957
AIEE/IEEE	No. 30	1944	Definitions and general standards for wire and cable	C8.1-1944 (R1953)
ASTM	B354-64T	1964	Definitions of terms relating to uninsulated metallic electrical conductors	
Conductor materials				
ASTM	B258-65	1965	Standard nominal diameters and cross-sectional areas of AWG sizes of solid round wires used as electrical conductors	C7.36-1966
ASTM	B5-43 (1961)	1961†	Electrolytic copper wire, bars, cakes, slabs, billets, ingots, and ingot bars	H17.2-1943
ASTM	B4-42 (1961)	1961†	Lake copper wire, bars, cakes, slabs, billets, ingots, and ingot bars	H17.1-1942
ASTM	B49-52	1952	Hot-rolled copper rods for electrical purposes	C7.7-1953 (R1960)
ASTM	B170-59	1959	Oxygen-free electrolytic copper wire bars, billets, and cakes	
ASTM	B224-58	1958	Classification of coppers	
ASTM	B263-58	1965†	Method for determination of cross-sectional area of stranded conductors	C7.29-1957 (R1962)
ASTM	B233-64	1964	Rolled aluminum rods (EC grade) for electrical purposes	C7.23-1965
ASTM	B296-63	1963	Recommended practice for temper designation of aluminum and magnesium alloys, cast and wrought	
Bare solid copper wire				
ASTM	B1-56	1956	Hard-drawn copper wire	C7.2-1953 (2d ed.) (R1957)
ASTM	B2-52	1952	Medium hard-drawn copper wire	C7.3-1953 (R1960)
ASTM	B3-63	1963	Soft or annealed copper wire	C7.1-1964
ASTM	B48-65	1965	Soft rectangular and square bare copper wire for electrical conductors	C7.9-1966
ASTM	B47-64	1964	Copper trolley wire	C7.6-1965
ASTM	B116-64	1964	ASTM figure 9 deep-section grooved and figure 8 copper trolley wire for industrial haulage	C7.11-1965
Coated-copper wire				
ASTM	B33-63	1963	Tinned soft or annealed copper wire for electrical purposes	C7.4-1964
ASTM	B246-64	1964	Tinned hard-drawn and medium-hard-drawn copper wire for electrical purposes	C7.37-1965
ASTM	B189-63	1963	Lead-coated and lead-alloy-coated soft copper wire for electrical purposes	C7.15-1964
ASTM	B298-64	1964	Silver-coated soft or annealed copper wire	C7.38-1965
ASTM	B355-65T	1965	Nickel-coated soft or annealed copper wire	C7.48-1966
Copper-clad steel wire				
ASTM	B227-65	1965	Hard-drawn copper-clad steel wire	C7.17-1966
Copper-alloy wires				
ASTM	B9-64	1964	Bronze trolley wire	C7.5-1965
ASTM	B105-55	1955	Hard-drawn copper-alloy wires for electrical conductors	C7.10-1956 (R1961)

Table 4-9. Industry Specifications for Conductors.—*Continued*

Sponsoring organization	Specification designation	Date of issue	Specification title	Corresponding ASA specification
Bare solid aluminum wire				
ASTM	B230-60	1961*	Aluminum wire, EC-H19, for electrical purposes	C7.20-1960
ASTM	B262-61	1961	Aluminum wire, EC-H16 or -H26, for electrical purposes	C7.35-1961
ASTM	B314-60	1960	Aluminum wire for communication cable	C7.40-1960
ASTM	B323-61	1961	Aluminum wire, EC-H14 or -H24, for electrical purposes	C7.42-1961
ASTM	B324-60	1960	Aluminum rectangular and square wire for electrical purposes	C7.43-1960
Aluminum-alloy wire				
ASTM	B396-63T	1963	Aluminum alloy 5005-H19 wire for electrical purposes	
ASTM	B398-63T	1963	Aluminum alloy 6201-T81 wire for electrical purposes	
Aluminum-clad steel wire				
ASTM	B415-64T	1964	Hard-drawn aluminum-clad steel wire	
Copper cable				
ASTM	B8-64	1964	Concentric-lay-stranded copper conductors, hard, medium hard, or soft	C7.8-1965
ASTM	B172-64	1964	Rope-lay-stranded copper conductors having bunch-stranded members for electrical conductors	C7.12-1965
ASTM	B173-64	1964	Rope-lay-stranded copper conductors having concentric-stranded members for electrical conductors	C7.13-1965
ASTM	B174-64	1964	Bunch-stranded copper conductors for electrical conductors	C7.14-1965
ASTM	B226-64	1964	Cored, annular, concentric-lay-stranded copper conductors	C7.16-1965
ASTM	B286-65T	1965	Copper conductors for use in hookup wire for electronic equipment	C7.39-1966
Copper-clad steel and composite cables				
ASTM	B228-65	1965	Concentric-lay-stranded copper-clad steel conductors	C7.18-1966
ASTM	B229-65	1965	Concentric-lay-stranded copper and copper-clad steel composite conductors	C7.19-1966
Aluminum and ACSR cables				
ASTM	B231-64	1964	Aluminum conductors, concentric-lay-stranded	C7.21-1965
ASTM	B232-64T	1964	Aluminum conductors, concentric-lay-stranded steel-reinforced (ACSR)	C7.22-1965
ASTM	B245-63	1963	Standard-weight zinc-coated (galvanized) steel-core wire for aluminum conductors, steel-reinforced (ACSR)	C7.28-1964
ASTM	B261-63	1963	Zinc-coated (galvanized) steel-core wire (with coatings heavier than standard weight) for aluminum conductors, steel-reinforced (ACSR)	C7.34-1964
ASTM	B341-63T	1963	Aluminum-coated (aluminized) steel-core wire for aluminum conductors, steel-reinforced (ACSR)	C7.47-1964
ASTM	B400-63T	1963	Compact round concentric-lay-stranded EC-grade aluminum conductors, hard-drawn	
ASTM	B401-63T	1963	Compact round concentric-lay-stranded aluminum conductors, steel-reinforced (ACSR)	
Aluminum-alloy cables				
ASTM	B397-63T	1963	Concentric-lay-stranded 5005-H19 aluminum alloy conductors	
ASTM	B399-63T	1963	Concentric-lay-stranded 6201-T81 aluminum alloy conductors	

Table 4-9. Industry Specifications for Conductors.—*Concluded*

Sponsoring organization	Specification designation	Date of issue	Specification title	Corresponding ASA specification
Aluminum-clad steel cables				
ASTM	B416-64T	1964	Concentric-lay-stranded aluminum-clad steel conductors	
Bus conductors				
ASTM	B187-62	1962	Copper bus bar, rod, and shapes	C7.25
ASTM	B188-61	1961	Seamless copper bus pipe and tube	C7.26-1963
ASTM	B236-64	1965*	Aluminum bars for electrical purposes (bus bars)	C7.27-1964 (2d ed.)
ASTM	B75-65	1965	Seamless copper tube	H23.3-1965
ASTM	B317-64	1965*	Aluminum-alloy extruded bar, rod, pipe, and structural shapes for electrical purposes (bus conductors)	C7.45-1964 (2d ed.)
Waveguide tube				
ASTM	B372-63	1963	Seamless copper and copper-alloy rectangular waveguide tube	H37.1-1963
Stainless-steel strand				
ASTM	A368-55	1955	Stainless-steel wire strand	
Galvanized steel wire and strand				
ASTM	A111-59	1959	Zinc-coated (galvanized) "iron" telephone and telegraph line wire	C7.31-1963
ASTM	A326-59	1959	Zinc-coated (galvanized) high tensile steel telephone- and telegraph-line wire	C7.30-1963
ASTM	A363-65	1965	Zinc-coated (galvanized) steel overhead ground wire strand	
ASTM	A411-65	1965	Zinc-coated (galvanized) low-carbon steel armor wire	
ASTM	A475-62T	1962	Zinc-coated steel wire strand	C7.46-1963
Methods of test				
ASTM	B193-65	1965	Resistivity of electrical conductor materials	C7.24-1966
ASTM	E8-65T	1965	Tension testing of metallic materials	
ASTM	B279-60	1960	Stiffness of bare soft square and rectangular copper wire for magnet wire fabrication	C7.41-1960
ASTM	A90-53	1965†	Weight of coating on zinc-coated (galvanized) iron or steel articles	G8.12-1956
ASTM	A239-41	1965†	Uniformity of coating by the Preece test (copper sulfate dip) on zinc-coated (galvanized) iron or steel articles	
ASTM	B342-63	1963	Electrical conductivity by use of eddy currents	C7.44-1961

* Editorially revised.
† Reapproved without change.

Table 4-10. Wire Gages—Diameter, Area, COPPER Weight

Gage name						Diameter at 20 C (68 F)		Area at 20 C (68 F)			Weight at 20 C (68 F) bare copper wire		
American (B & S) wire gage, AWG	Steel wire gage, Stl. wg	Birmingham (Stubs' iron) wire gage, BWG	Old English (London) wire gage	(British) standard wire gage, SWG	Metric wire gage	Mils	Mm	Sq mils	Cir mils	Sq mm	Lb per 1,000 ft	Lb per mile	Kg per km
Gage No.													
...	...	...	...	7/0	...	500	12.70	196,300	250,000	126.7	756.7	3996	1126
...	7/0	...	...	...	...	490.0	12.45	188,600	240,100	121.7	726.8	3837	1082
...	...	...	...	6/0	...	464	11.79	169,100	215,000	109.1	651.7	3441	969.8
...	6/0	...	...	...	...	461.5	11.72	167,300	213,000	107.9	644.7	3404	959.4
4/0	...	...	...	...	...	460.0	11.68	166,200	211,600	107.2	640.5	3382	953.2
...	...	4/0	4/0	...	...	454	11.53	161,900	206,100	104.4	623.9	3294	928.5
...	...	...	...	5/0	...	432	10.97	146,600	186,600	94.56	564.9	2983	840.7
...	5/0	...	...	...	...	430.5	10.93	145,600	185,300	93.91	561.0	2962	834.8
...	...	3/0	3/0	...	...	425	10.80	141,900	180,600	91.52	546.8	2887	813.7
3/0	...	...	...	...	...	409.6	10.40	131,800	167,800	85.01	507.8	2681	755.7
...	...	...	...	4/0	...	400	10.16	125,700	160,000	81.07	484.3	2557	720.8
...	4/0	...	...	...	...	393.8	10.00	121,800	155,100	78.58	469.4	2479	698.6
...	...	...	...	...	100	393.7	10.00	121,700	155,000	78.54	469.2	2477	698.2
...	...	2/0	2/0	...	...	380	9.652	113,400	144,400	73.17	437.1	2308	650.5
...	...	...	...	3/0	...	372	9.449	108,700	138,400	70.12	418.9	2212	623.4
2/0	...	...	...	...	...	364.8	9.266	104,500	133,100	67.43	402.8	2127	599.5
...	3/0	...	...	...	...	362.5	9.208	103,200	131,400	66.58	397.8	2100	591.9
...	...	...	...	...	90	354.3	9.0	98,589	125,500	63.62	380.0	2007	565.6
...	...	...	...	2/0	...	348	8.839	95,115	121,100	61.36	366.6	1936	545.5
...	...	1/0	1/0	...	...	340	8.636	90,790	115,600	58.58	349.9	1848	520.7
...	2/0	...	...	...	...	331.0	8.407	86,050	109,600	55.52	331.6	1751	493.5
1/0	...	...	...	...	...	324.9	8.252	82,891	105,600	53.49	319.5	1687	477.6
...	...	...	...	1/0	...	324	8.230	82,450	105,000	53.19	317.8	1678	472.9
...	...	...	...	...	80	314.96	8.0	77,931	99,200	50.27	300.3	1585	446.9
...	1/0	...	...	...	...	306.5	7.785	73,780	93,940	47.60	284.4	1501	423.2
...	...	1	1	1	...	300	7.620	70,690	90,000	45.60	272.4	1438	405.4
1	...	...	...	...	...	289.3	7.348	65,730	83,690	42.41	253.3	1338	377.0
...	...	2	2	...	...	284	7.214	63,350	80,660	40.87	244.4	1289	363.3
...	1	...	...	...	...	283.0	7.188	62,900	80,090	40.58	242.4	1280	360.8
...	...	...	...	2	...	276	7.010	59,830	76,180	38.60	230.6	1217	343.1
...	...	...	...	...	70	275.59	7.0	59,650	75,950	38.49	229.9	1214	342.1
...	2	...	...	...	...	262.5	6.668	54,120	68,910	34.92	208.6	1101	310.4
...	...	3	3	...	...	259	6.579	52,690	67,080	33.99	203.1	1072	302.2
2	...	...	...	...	...	257.6	6.543	52,120	66,360	33.62	200.9	1061	298.9
...	...	...	...	3	...	252	6.401	49,880	63,500	32.18	192.2	1015	286.1

...	3	...	...	...	...	243.7	6.190	46,640	59,390	30.09	179.8	949.2	267.5
...	...	4	4	...	...	238	6.045	44,490	56,640	28.70	171.5	905.3	255.2
...	...	...	...	...	60	236.2	6.000	43,830	55,800	28.27	168.9	891.8	251.4
...	...	...	...	4	...	232	5.893	42,270	53,820	27.27	162.9	860.3	242.5
3	...	...	...	...	...	229.4	5.827	41,330	52,620	26.67	159.3	841.1	237.1
...	4	...	...	...	...	225.3	5.723	39,870	50,760	25.72	153.7	811.3	228.7
...	...	5	5	...	...	220	5.588	38,010	48,400	24.52	146.5	773.6	218.0
...	...	...	...	5	...	212	5.385	35,300	44,940	22.77	136.0	718.3	202.5
...	5	...	...	...	...	207.0	5.258	33,605	42,850	21.71	129.7	684.9	193.0
4	...	...	...	...	...	204.3	5.189	32,780	41,740	21.15	126.3	667.1	188.0
...	...	6	6	...	...	203	5.156	32,370	41,210	20.88	124.7	658.6	185.6
...	...	...	...	...	50	196.8	5.0	30,430	38,750	19.63	117.3	619.3	174.6
...	6	...	...	6	...	192.0	4.877	28,950	36,860	18.68	111.6	589.2	166.1
5	...	...	...	...	...	181.9	4.620	25,990	33,090	16.77	100.2	528.8	149.0
...	...	7	7	...	...	180	4.572	25,450	32,400	16.42	98.07	517.8	146.0
...	...	...	...	...	45	177.2	4.500	24,650	31,390	15.90	95.01	501.7	141.4
...	7	...	...	...	...	177.0	4.496	24,610	31,330	15.87	94.83	500.7	141.1
...	...	...	...	7	...	176	4.470	24,330	30,980	15.70	93.76	495.1	139.5
...	...	8	8	...	...	165	4.191	21,380	27,220	13.80	82.41	435.1	122.6
6	8	...	...	...	...	162.0	4.115	20,610	26,240	13.30	79.44	419.4	118.2
...	...	...	...	8	...	160	4.064	20,110	25,600	12.97	77.49	409.2	115.3
...	...	...	...	...	40	157.5	4.000	19,480	24,800	12.57	75.09	396.4	111.7
...	9	...	...	...	...	148.3	3.767	17,270	21,990	11.14	66.57	351.5	99.07
...	...	9	9	...	...	148	3.759	17,200	21,900	11.10	66.30	350.1	98.67
7	...	...	...	...	...	144.3	3.665	16,350	20,820	10.55	63.03	332.8	93.80
...	...	...	...	9	...	144	3.658	16,290	20,740	10.51	62.77	331.4	93.41
...	...	...	...	...	35	137.8	3.5	14,910	18,990	9.622	57.47	303.5	85.53
...	10	...	...	...	...	135.0	3.429	14,310	18,220	9.235	55.17	291.3	82.10
...	...	10	10	...	...	134	3.404	14,100	17,960	9.098	54.35	287.0	80.89
8	...	...	...	...	...	128.5	3.264	12,970	16,510	8.367	49.98	263.9	74.38
...	...	...	...	10	...	128	3.251	12,870	16,380	8.302	49.59	261.9	73.80
...	11	...	...	...	...	120.5	3.061	11,400	14,520	7.358	43.95	232.1	65.41
...	...	11	11	...	...	120	3.048	11,310	14,400	7.297	43.59	230.2	64.87
...	...	...	...	...	30	118.1	3.0	10,960	13,950	7.068	42.22	222.9	62.83
...	...	...	...	11	...	116	2.946	10,570	13,460	6.818	40.73	215.1	60.61
9	...	...	...	...	...	114.4	2.906	10,280	13,090	6.631	39.61	209.2	58.95
...	...	12	12	...	...	109	2.769	9,331	11,880	6.020	35.96	189.9	53.52
...	12	...	...	...	...	105.5	2.680	8,742	11,130	5.640	33.69	177.9	50.14
...	...	...	...	12	...	104	2.642	8,495	10,820	5.481	32.74	172.9	48.72
10	...	...	...	...	...	101.9	2.588	8,155	10,380	5.261	31.43	166.0	46.77
...	...	...	...	...	25	98.42	2.5	7,609	9,687	4.908	29.32	154.8	43.63
...	...	13	13	...	...	95	2.413	7,088	9,025	4.573	27.32	144.2	40.65
...	...	...	...	13	...	92	2.337	6,648	8,464	4.289	25.62	135.3	38.13
...	13	...	...	...	...	91.5	2.324	6,576	8,372	4.242	25.34	133.8	37.71
11	...	...	...	...	...	90.7	2.304	6,460	8,230	4.17	24.90	131	37.1

Table 4-10. Wire Gages—Diameter, Area, COPPER Weight.—*Continued*

Gage name						Diameter at 20 C (68 F)		Area at 20 C (68 F)			Weight at 20 C (68 F) bare copper wire		
American (B & S) wire gage, AWG	Steel wire gage, Stl. wg	Birmingham (Stubs' iron) wire gage, BWG	Old English (London) wire gage	(British) standard wire gage, SWG	Metric wire gage	Mils	Mm	Sq mils	Cir mils	Sq mm	Lb per 1,000 ft	Lb per mile	Kg per km
Gage No.													
...	...	14	14	...	.....	83	2.108	5,411	6,889	3.491	20.85	110.1	31.03
12	...	...	...	...	.....	80	2.05	5,130	6,530	3.310	19.8	104	29.4
...	14	...	...	14	.....	80.0	2.032	5,029	6,400	3.243	19.37	102.3	28.83
...	...	...	...	...	20	78.74	2.0	4,869	6,200	3.142	18.77	99.09	27.93
...	15	15	15	15	.....	72.0	1.829	4,072	5,184	2.627	15.69	82.85	23.35
13	...	...	...	...	.....	72.0	1.83	4,070	5,180	2.63	15.7	82.9	23.4
...	...	...	...	...	18	70.87	1.8	3,944	5,022	2.545	15.20	80.27	22.62
...	...	16	16	...	.....	65	1.651	3,318	4,225	2.141	12.79	67.53	19.03
14	...	...	...	...	.....	64.1	1.63	3,230	4,110	2.08	12.4	65.7	18.5
...	...	...	...	16	.....	64	1.626	3,217	4,096	2.075	12.40	65.46	18.45
...	...	...	...	...	16	62.99	1.6	3,116	3,968	2.011	12.01	63.41	17.87
...	16	...	...	...	.....	62.5	1.588	3,068	3,906	1.979	11.82	62.43	17.60
...	...	17	17	...	.....	58	1.473	2,642	3,364	1.705	10.18	53.77	15.15
15	...	...	...	...	.....	57.1	1.45	2,560	3,260	1.650	9.87	52.1	14.7
...	...	...	...	17	.....	56	1.422	2,463	3,136	1.589	9.493	50.12	14.13
...	...	...	...	...	14	55.12	1.4	2,386	3,038	1.539	9.196	48.56	13.69
...	17	...	...	...	.....	54.0	1.372	2,290	2,916	1.478	8.827	46.60	13.14
16	...	...	...	...	.....	50.8	1.29	2,030	2,580	1.31	7.81	41.2	11.6
...	...	18	18	...	.....	49	1.245	1,886	2,401	1.217	7.268	38.37	10.82
...	...	...	...	18	.....	48	1.219	1,810	2,304	1.167	6.974	36.82	10.38
...	18	...	...	...	.....	47.5	1.207	1,772	2,256	1.143	6.830	36.06	10.16
...	...	...	...	...	12	47.24	1.200	1,753	2,232	1.131	6.756	35.67	10.05
17	...	...	...	...	.....	45.3	1.150	1,610	2,050	1.040	6.21	32.8	9.24
...	...	19	...	...	.....	42	1.067	1,385	1,764	0.8938	5.340	28.19	7.946
...	19	...	...	...	.....	41.0	1.041	1,320	1,681	0.8518	5.088	26.87	7.572
18	...	...	...	...	.....	40.3	1.02	1,280	1,620	0.823	4.92	26.0	7.32
...	...	...	19	19	.....	40	1.016	1,257	1,600	0.8107	4.843	25.57	7.207
...	...	...	...	...	10	39.37	1	1,217	1,550	0.7854	4.692	24.77	6.982
...	...	...	...	20	.....	36	0.9144	1,018	1,296	0.6567	3.923	20.71	5.838
19	...	...	...	...	.....	35.9	0.912	1,010	1,290	0.653	3.90	20.6	5.81

...	...	...	...	...	9	35.43	0.90	986.1	1,255	0.6362	3.800	20.07	5.656
...	...	20	20	...	...	35	0.8890	962.1	1,225	0.6207	3.708	19.58	5.520
...	20	...	...	...	...	34.8	0.8839	951.1	1,211	0.6136	3.666	19.36	5.455
...	...	21	...	21	...	32	0.8128	804.2	1,024	0.5189	3.100	16.37	4.613
20	...	...	...	...	...	32.0	0.813	804.0	1,020	0.519	3.10	16.4	4.61
...	21	...	...	...	...	31.7	0.8052	789.2	1,005	0.5092	3.042	16.06	4.527
...	...	...	21	...	8	31.5	0.80	779.1	992	0.5027	3.003	15.85	4.469
...	...	...	22	...	...	29.50	0.7493	683.3	870	0.4410	2.634	13.91	3.920
...	22	...	...	...	...	28.6	0.7264	642.4	818.0	0.4145	2.476	13.07	3.685
21	...	...	...	...	...	28.5	0.724	638.0	812.0	0.412	2.46	13.0	3.66
...	...	22	...	22	...	28	0.7112	615.8	784.0	0.3973	2.373	12.53	3.532
...	...	...	...	...	7	27.56	0.70	596.5	759.5	0.3848	2.299	12.14	3.421
...	...	...	23	...	...	27.00	0.6858	572.6	729.0	0.3694	2.207	11.65	3.284
...	23	...	...	...	...	25.8	0.6553	522.8	665.6	0.3373	2.015	10.64	2.998
22	...	...	...	...	...	25.3	0.643	503.0	640.0	0.324	1.94	10.2	2.88
...	...	23	24	...	...	25	0.6350	490.9	625.0	0.3167	1.892	9.989	2.815
...	...	...	...	23	...	24	0.6096	452.4	576.0	0.2919	1.744	9.206	2.595
...	...	...	...	...	6	23.62	0.60	438.3	558.0	0.2827	1.689	8.918	2.514
...	24	...	25	...	...	23.0	0.5842	415.5	529.0	0.2680	1.601	8.455	2.383
23	...	...	...	...	...	22.6	0.574	401.0	511.0	0.259	1.55	8.16	2.30
...	...	24	...	24	...	22	0.5588	380.1	484.0	0.2452	1.465	7.736	2.180
...	...	...	26	...	...	20.5	0.5207	330.1	420.3	0.2129	1.272	6.717	1.893
...	25	...	...	...	...	20.4	0.5182	326.9	416.2	0.2109	1.260	6.651	1.875
24	...	...	...	...	...	20.1	0.511	317.0	404.0	0.205	1.22	6.46	1.82
...	...	25	...	25	...	20	0.5080	314.2	400.0	0.2027	1.211	6.393	1.802
...	...	...	...	...	5	19.68	0.50	304.3	387.5	0.1963	1.173	6.193	1.746
...	...	...	27	...	...	18.75	0.4763	276.1	351.6	0.1781	1.064	5.619	1.584
...	26	...	...	...	...	18.1	0.4597	257.3	327.6	0.1660	0.9916	5.236	1.476
...	...	26	...	26	...	18	0.4572	254.5	324.0	0.1642	0.9807	5.178	1.460
25	...	...	...	...	...	17.9	0.455	252.0	320.0	0.162	0.970	5.12	1.44
...	...	...	...	...	4.5	17.72	0.45	246.5	313.9	0.1590	0.9501	5.016	1.414
...	27	...	...	...	...	17.3	0.4394	235.1	299.3	0.1517	0.9059	4.783	1.348
...	...	...	28	...	...	16.50	0.4191	213.8	272.3	0.1380	0.8241	4.351	1.226
...	...	...	...	27	...	16.4	0.4166	211.2	269.0	0.1363	0.8141	4.299	1.212
...	28	...	...	...	...	16.2	0.4115	206.1	262.4	0.1330	0.7944	4.194	1.182
...	...	27	...	...	...	16	0.4064	201.1	256.0	0.1297	0.7749	4.092	1.153
26	...	...	...	...	...	15.9	0.404	199.0	253.0	0.128	0.765	4.04	1.14
...	...	...	...	...	4	15.75	0.40	194.8	248.0	0.1257	0.7507	3.964	1.117
...	...	...	29	...	...	15.50	0.3937	188.7	254.1	0.1217	0.7272	3.840	1.082
...	29	...	...	...	...	15.0	0.3810	176.7	225.0	0.1140	0.6811	3.596	1.014
...	...	...	...	28	...	14.8	0.3759	172.0	219.0	0.1110	0.6630	3.501	0.9867
27	...	...	...	...	...	14.2	0.361	158.0	202.0	0.102	0.610	3.22	0.908
...	30	28	...	...	...	14.0	0.3556	153.9	196.0	0.09932	0.5933	3.133	0.8829
...	...	...	...	...	3.5	13.78	0.35	149.1	189.9	0.09621	0.5747	3.035	0.8553
...	...	...	30	...	...	13.75	0.3493	148.5	189.1	0.09580	0.5723	3.022	0.8517

Table 4-10. Wire Gages—Diameter, Area, COPPER Weight.—*Concluded*

Gage name						Diameter at 20 C (68 F)		Area at 20 C (68 F)			Weight at 20 C (68 F) bare copper wire		
American (B & S) wire gage, AWG	Steel wire gage, Stl. wg	Birmingham (Stubs' iron) wire gage, BWG	Old English (London) wire gage	(British) standard wire gage, SWG	Metric wire gage	Mils	Mm	Sq mils	Cir mils	Sq mm	Lb per 1,000 ft	Lb per mile	Kg per km
Gage No.													
...	...	...	...	29	.....	13.6	0.3454	145.3	185.0	0.09372	0.5599	2.956	0.8332
...	31	...	...	...	.....	13.2	0.3353	136.8	174.2	0.08829	0.5274	2.785	0.7849
...	...	29	...	...	.....	13	0.3302	132.7	169.0	0.08563	0.5116	2.701	0.7613
...	32	...	...	...	.....	12.8	0.3251	128.7	163.8	0.08302	0.4959	2.619	0.7380
28	...	...	...	...	.....	12.6	0.320	125	159.0	0.0804	0.481	2.54	0.715
...	...	...	...	30	.....	12.4	0.3150	120.8	153.8	0.07791	0.4654	2.457	0.6926
...	...	...	31	...	.....	12.25	0.3112	117.9	150.1	0.07604	0.4542	2.398	0.6760
...	...	30	...	...	.....	12	0.3048	113.1	144.0	0.07297	0.4359	2.301	0.6487
...	...	...	...	...	3	11.81	0.30	109.6	139.5	0.07069	0.4223	2.230	0.6284
...	33	...	...	...	.....	11.8	0.2997	109.4	139.2	0.07055	0.4215	2.225	0.6272
...	...	...	...	31	.....	11.6	0.2946	105.7	134.6	0.06818	0.4073	2.151	0.6061
29	...	...	...	...	.....	11.3	0.287	100.0	128.0	0.0647	0.387	2.04	0.575
...	...	...	32	...	.....	11.25	0.2858	99.40	126.6	0.06413	0.3831	2.023	0.5701
...	...	...	...	32	.....	10.8	0.2743	91.61	116.6	0.05910	0.3531	1.864	0.5254
...	34	...	...	...	.....	10.4	0.2642	84.95	108.2	0.05481	0.3274	1.729	0.4872
...	...	...	33	...	.....	10.25	0.2604	82.52	105.1	0.05324	0.3180	1.679	0.4733
30	...	...	...	...	.....	10.0	0.254	78.50	100.0	0.0507	0.303	1.60	0.450
...	...	31	...	33	.....	10.0	0.2540	78.54	100.0	0.05067	0.3027	1.598	0.4505
...	...	...	...	...	2.5	9.842	0.25	76.09	96.87	0.04909	0.2932	1.548	0.4364
...	35	...	34	...	.....	9.5	0.2413	70.88	90.25	0.04573	0.2732	1.442	0.4065
...	...	...	...	34	.....	9.2	0.2337	66.48	84.64	0.04289	0.2562	1.353	0.3813
...	36	32	35	...	.....	9.0	0.2286	63.62	81.00	0.04104	0.2452	1.295	0.3649
31	...	...	...	...	.....	8.9	0.226	62.20	79.2	0.0401	0.240	1.27	0.357
...	37	...	...	...	.....	8.5	0.2159	56.75	72.25	0.03661	0.2187	1.155	0.3255
...	...	...	...	35	.....	8.4	0.2134	55.42	70.56	0.03575	0.2139	1.128	0.3178
...	38	33	...	...	.....	8.0	0.2032	50.27	64.00	0.03243	0.1937	1.023	0.2883
32	...	...	...	...	.....	8.0	0.203	50.30	64.0	0.0324	0.194	1.02	0.288
...	...	...	...	...	2	7.874	0.20	48.69	62.00	0.03142	0.1877	0.9909	0.2793
...	...	...	...	36	.....	7.6	0.1930	45.36	57.76	0.02927	0.1748	0.9231	0.2602
...	39	...	36	...	.....	7.5	0.1905	44.18	56.25	0.02850	0.1703	0.8990	0.2534
...	...	...	...	...	1.8	7.087	0.1800	39.44	50.22	0.02545	0.1520	0.8026	0.2262
33	...	...	...	...	.....	7.1	0.180	39.60	50.40	0.0255	0.153	0.806	0.227
...	40	34	...	...	.....	7.0	0.1778	38.48	49.00	0.02483	0.1483	0.7831	0.2207
...	...	...	...	37	.....	6.8	0.1727	36.32	46.24	0.02343	0.1400	0.7390	0.2083
...	41	...	...	...	.....	6.6	0.1676	34.21	43.56	0.02207	0.1319	0.6962	0.1962

...	...	...	37	...	...	6.50	0.1651	33.18	42.25	0.02141	0.1279	0.6753	0.1903
34	...	...	...	...	...	6.3	0.160	31.20	39.7	0.0201	0.120	0.634	0.179
...	...	...	...	...	1.6	6.299	0.16	31.16	39.68	0.02011	0.1201	0.6342	0.1787
...	42	...	...	...	...	6.2	0.1575	30.19	38.44	0.01948	0.1164	0.6144	0.1732
...	43	...	...	38	...	6.0	0.1524	28.27	36.00	0.01824	0.1090	0.5754	0.1622
...	...	...	...	...	1.5	5.906	0.15	27.39	34.87	0.01767	0.1056	0.5575	0.1571
...	44	...	...	...	...	5.8	0.1473	26.42	33.64	0.01705	0.1018	0.5377	0.1515
...	...	...	38	...	...	5.75	0.1461	25.97	33.06	0.01675	0.1001	0.5284	0.1489
35	...	...	...	...	...	5.6	0.142	24.60	31.4	0.0159	0.0949	0.501	0.141
...	...	...	...	...	1.4	5.512	0.14	23.86	30.38	0.01539	0.09196	0.4855	0.1369
...	45	...	...	...	...	5.5	0.1397	23.76	30.25	0.01533	0.09157	0.4835	0.1363
...	46	...	...	39	...	5.2	0.1321	21.24	27.04	0.01370	0.08185	0.4322	0.1218
36	47	35	39	...	...	5.0	0.127	19.60	25.00	0.0127	0.0757	0.400	0.113
...	48	...	...	40	...	4.8	0.1219	18.10	23.04	0.01167	0.06974	0.3682	0.1038
...	...	...	...	...	1.2	4.724	0.12	17.53	22.32	0.01131	0.06756	0.3567	0.1005
...	49	...	...	...	...	4.6	0.1168	16.62	21.16	0.01072	0.06405	0.3382	0.09532
...	...	...	40	...	...	4.50	0.1143	15.90	20.25	0.01026	0.06130	0.3236	0.09122
37	...	...	...	...	...	4.50	0.114	15.90	20.20	0.0103	0.0613	0.324	0.0912
...	50	...	...	41	...	4.4	0.1118	15.21	19.36	0.009810	0.05860	0.3094	0.08721
...	...	36	...	42	...	4.0	0.1016	12.57	16.00	0.008107	0.04843	0.2557	0.07207
38	...	...	...	...	...	4.0	0.102	12.60	16.0	0.00811	0.0484	0.256	0.0721
...	...	...	...	...	1	3.937	0.10	12.17	15.50	0.007854	0.04692	0.2477	0.06982
...	...	...	...	43	...	3.6	0.09144	10.18	12.96	0.006567	0.03923	0.2071	0.05838
39	...	...	...	...	...	3.5	0.0889	9.62	12.2	0.00621	0.0371	0.196	0.0552
...	...	...	...	44	...	3.2	0.08128	8.042	10.24	0.005189	0.0310	0.1637	0.04613
40	...	...	...	...	...	3.1	0.0787	7.55	9.61	0.00487	0.0291	0.154	0.0433
41	...	...	...	...	...	2.8	0.0711	6.16	7.84	0.00397	0.0237	0.125	0.0353
...	...	...	...	45	...	2.8	0.07112	6.158	7.840	0.003973	0.02373	0.1253	0.03532
42	...	...	...	...	...	2.5	0.0635	4.91	6.25	0.00317	0.0189	0.0999	0.0282
...	...	...	...	46	...	2.4	0.06096	4.524	5.760	0.002919	0.01744	0.09206	0.02595
43	...	...	...	...	...	2.2	0.0559	3.80	4.84	0.00245	0.0147	0.0774	0.0218
...	...	...	...	47	...	2.0	0.05080	3.142	4.000	0.002027	0.01211	0.06393	0.01802
44	...	...	...	...	...	2.0	0.0508	3.14	4.00	0.00203	0.0121	0.0639	0.0180
...	...	...	...	...	0.5	1.969	0.05	3.043	3.875	0.001963	0.01173	0.06193	0.01746
45	...	...	...	...	...	1.8	0.0457	2.54	3.24	0.00164	0.00981	0.0519	0.0146
...	...	...	...	48	...	1.6	0.04064	2.011	2.560	0.001297	0.007749	0.04092	0.01153
46	...	...	...	...	...	1.6	0.0406	2.01	2.56	0.00130	0.00775	0.0409	0.0115
47	...	...	...	...	...	1.4	0.0356	1.54	1.96	0.000993	0.00593	0.0313	0.00883
48	...	...	...	...	...	1.2	0.0305	1.13	1.44	0.000730	0.00436	0.0230	0.00649
...	...	...	...	49	...	1.2	0.03048	1.131	1.440	0.0007297	0.004359	0.02301	0.006487
49	...	...	...	...	...	1.1	0.0279	0.950	1.21	0.000613	0.00366	0.0193	0.00545
...	...	...	...	50	...	1.0	0.02540	0.7854	1.000	0.0005067	0.003027	0.01598	0.004505
50	...	...	...	...	...	1.0	0.0254	0.785	1.00	0.000507	0.00303	0.0160	0.00450

Table 4-11. Copper Wire—Tensile Strength, Elongation
(ASTM Specifications B1-56, B2-52, B3-63)

Size,* AWG	Diameter at 20 C (68 F),† in	Area at 20 C (68 F)		Tensile strength, psi				Elongation, min %				
				Hard	Medium		Soft§	Hard		Medium		Soft
		Cir mils	Sq in	Min	Min	Max		In 10 in	In 60 in	In 10 in	In 60 in	In 10 in
4/0	0.4600	211,600	0.1662	49,000	42,000	49,000	......	3.75	....	3.75	....	35
3/0	0.4096	167,800	0.1318	51,000	43,000	50,000	......	3.25	....	3.60	....	35
2/0	0.3648	133,100	0.1045	52,800	44,000	51,000	......	2.80	....	3.25	....	35
1/0	0.3249	105,600	0.08291	54,500	45,000	52,000	......	2.40	....	3.00	....	35
1	0.2893	83,690	0.06573	56,100	46,000	53,000	......	2.17	....	2.75	....	30
2	0.2576	66,360	0.05212	57,600	47,000	54,000	......	1.98	....	2.50	....	30
3	0.2294	52,620	0.04133	59,000	48,000	55,000	......	1.79	....	2.25	....	30
4	0.2043	41,740	0.03278	60,100	48,330	55,330	......	....	1.24	....	1.25	30
5	0.1819	33,090	0.02599	61,200	48,660	55,660	......	....	1.18	....	1.20	30
...	0.1650‡	27,220	0.02138	62,000	......	......	......	....	1.14			
6	0.1620	26,240	0.02061	62,100	49,000	56,000	......	....	1.14	....	1.15	30
7	0.1443	20,820	0.01635	63,000	49,330	56,330	......	....	1.09	....	1.11	30
...	0.1340‡	17,960	0.01410	63,400	......	......	......	....	1.07			
8	0.1285	16,510	0.01297	63,700	49,660	56,660	......	....	1.06	....	1.08	30
9	0.1144	13,090	0.01028	64,300	50,000	57,000	......	....	1.02	....	1.06	30
...	0.1040‡	10,820	0.008495	64,800	......	......	......	....	1.00			
10	0.1019	10,380	0.008155	64,900	50,330	57,300	......	....	1.00	....	1.04	25
...	0.0920‡	8,460	0.00665	65,400	......	......	......	....	0.97			
11	0.0907	8,230	0.00646	65,400	50,660	57,660	......	....	0.97	....	1.02	25
12	0.0808	6,530	0.00513	65,700	51,000	58,000	......	....	0.95	....	1.00	25
...	0.0800‡	6,400	0.00503	65,700	......	......	......	....	0.94			
13	0.0720	5,180	0.00407	65,900	51,330	58,330	......	....	0.92	....	0.98	25
...	0.0650‡	4,220	0.00332	66,200	......	......	......	....	0.91			
14	0.0641	4,110	0.00323	66,200	51,660	58,660	......	....	0.90	....	0.96	25
15	0.0571	3,260	0.00256	66,400	52,000	59,000	......	....	0.89	....	0.94	25
16	0.0508	2,580	0.00203	66,600	52,330	59,330	......	....	0.87	....	0.92	25
17	0.0453	2,050	0.00161	66,800	52,660	59,660	......	....	0.86	....	0.90	25
18	0.0403	1,620	0.00128	67,000	53,000	60,000	......	....	0.85	....	0.88	25
19	0.0359	1,290	0.00101	......	......	......	......	....	....	....	....	25
20	0.0320	1,020	0.000804	......	......	......	......	....	....	....	....	25
21	0.0285	812	0.000638	......	......	......	......	....	....	....	....	25
22	0.0253	640	0.000503	......	......	......	......	....	....	....	....	25
23	0.0226	511	0.000401	......	......	......	......	....	....	....	....	25
24	0.0201	404	0.000317	......	......	......	......	....	....	....	....	20
25	0.0179	320	0.000252	......	......	......	......	....	....	....	....	20
26	0.0159	253	0.000199	......	......	......	......	....	....	....	....	20
27	0.0142	202	0.000158	......	......	......	......	....	....	....	....	20
28	0.0126	159	0.000125	......	......	......	......	....	....	....	....	20
29	0.0113	128	0.000100	......	......	......	......	....	....	....	....	20
30	0.0100	100	0.0000785	......	......	......	......	....	....	....	....	15
31	0.0089	79.2	0.0000622	......	......	......	......	....	....	....	....	15
32	0.0080	64.0	0.0000503	......	......	......	......	....	....	....	....	15
33	0.0071	50.4	0.0000396	......	......	......	......	....	....	....	....	15
34	0.0063	39.7	0.0000312	......	......	......	......	....	....	....	....	15
35	0.0056	31.4	0.0000246	......	......	......	......	....	....	....	....	15
36	0.0050	25.0	0.0000196	......	......	......	......	....	....	....	....	15
37	0.0045	20.2	0.0000159	......	......	......	......	....	....	....	....	15
38	0.0040	16.0	0.0000126	......	......	......	......	....	....	....	....	15
39	0.0035	12.2	0.00000962	......	......	......	......	....	....	....	....	15
40	0.0031	9.61	0.00000755	......	......	......	......	....	....	....	....	15
ASTM Specification Designation............				B1-56	B2-52		B3-63	B1-56		B2-52		B3-63

* The use of gage numbers to specify wire sizes is not recognized in these specifications, because of the possibility of confusion.

† The value of wire diameters in this table which correspond to gage numbers of the AWG are in agreement with the standard nominal diameters prescribed in ASTM Specification B258-65. For wire whose nominal diameter is more than 0.001 in (1 mil) greater than a size listed in the table and less than that of the next larger size, the requirements of the next larger size shall apply.

‡ Diameters often employed by purchasers for communication lines, but not in the AWG (B. & S. wire gage) series. They correspond to certain of the numbers of the Birmingham wire gage or of the (British) standard wire gage.

§ No requirements for tensile strength are specified.

Table 4-12. Copper Wire—Weight, Breaking Strength, D-c Resistance

(Based on ASTM Specifications B1-56, B2-52, B3-63)

Size, AWG	Diameter, in.	Area		Weight		Hard		Medium		Soft	
		Cir mils	Sq in.	Lb per 1,000 ft	Lb per mile	Breaking strength, minimum,* lb	D-c resistance at 20 C (68 F) maximum,† ohms per 1,000 ft	Breaking strength, minimum,* lb	D-c resistance at 20 C (68 F) maximum,† ohms per 1,000 ft	Breaking strength, maximum,‡ lb	D-c resistance at 20 C (68 F) maximum,† ohms per 1,000 ft
4/0	0.4600	211,600	0.1662	640.5	3382	8143	0.05045	6980	0.05019	5983	0.04901
3/0	0.4096	167,800	0.1318	507.8	2681	6720	0.06362	5666	0.06330	4744	0.06182
2/0	0.3648	133,100	0.1045	402.8	2127	5519	0.08021	4599	0.07980	3763	0.07793
1/0	0.3249	105,600	0.08291	319.5	1687	4518	0.1022	3731	0.1016	2985	0.09825
1	0.2893	83,690	0.06573	253.3	1338	3688	0.1289	3024	0.1282	2432	0.1239
2	0.2576	66,360	0.05212	200.9	1061	3002	0.1625	2450	0.1617	1928	0.1563
3	0.2294	52,620	0.04133	159.3	841.1	2439	0.2050	1984	0.2039	1529	0.1971
4	0.2043	41,740	0.03278	126.3	667.1	1970	0.2584	1584	0.2571	1213	0.2485
5	0.1819	33,090	0.02599	100.2	528.8	1590	0.3260	1265	0.3243	961.5	0.3135
6	0.1620	26,240	0.02061	79.44	419.4	1280	0.4110	1010	0.4088	762.6	0.3952
7	0.1443	20,820	0.01635	63.03	332.8	1030	0.5180	806.7	0.5153	605.1	0.4981
8	0.1285	16,510	0.01297	49.98	263.9	826.1	0.6532	644.0	0.6498	479.8	0.6281
9	0.1144	13,090	0.01028	39.61	209.2	660.9	0.8241	513.9	0.8199	380.3	0.7925
10	0.1019	10,380	0.008155	31.43	166.0	529.3	1.039	410.5	1.033	314.0	0.9988
11	0.0907	8,230	0.00646	24.9	131	423	1.31	327	1.30	249	1.26
12	0.0808	6,530	0.00513	19.8	104	337	1.65	262	1.64	197	1.59
13	0.0720	5,180	0.00407	15.7	82.9	268	2.08	209	2.07	157	2.00
14	0.0641	4,110	0.00323	12.4	65.7	214	2.63	167	2.61	124	2.52
15	0.0571	3,260	0.00256	9.87	52.1	170	3.31	133	3.29	98.6	3.18
16	0.0508	2,580	0.00203	7.81	41.2	135	4.18	106	4.16	78.0	4.02
17	0.0453	2,050	0.00161	6.21	32.8	108	5.26	84.9	5.23	62.1	5.05
18	0.0403	1,620	0.00128	4.92	26.0	85.5	6.64	67.6	6.61	49.1	6.39
19	0.0359	1,290	0.00101	3.90	20.6	68.0	8.37	54.0	8.33	39.0	8.05
20	0.0320	1,020	0.000804	3.10	16.4	54.2	10.5	43.2	10.5	31.0	10.1
21	0.0285	812	0.000638	2.46	13.0	43.2	13.3	34.4	13.2	24.6	12.8
22	0.0253	640	0.000503	1.94	10.2	34.1	16.9	27.3	16.8	19.4	16.2
23	0.0226	511	0.000401	1.55	8.16	27.3	21.1	21.9	21.0	15.4	20.3
24	0.0201	404	0.000317	1.22	6.46	21.7	26.7	17.5	26.6	12.7	25.7
25	0.0179	320	0.000252	0.970	5.12	17.3	33.7	13.9	33.5	10.1	32.4
26	0.0159	253	0.000199	0.765	4.04	13.7	42.7	11.1	42.4	7.94	41.0

Table 4-12. Copper Wire—Weight, Breaking Strength, D-c Resistance.—*Concluded*

Size, AWG	Diameter, in.	Area		Weight		Hard		Medium		Soft	
		Cir mils	Sq in.	Lb per 1,000 ft	Lb per mile	Breaking strength, minimum,* lb	D-c resistance at 20 C (68 F) maximum,† ohms per 1,000 ft	Breaking strength, minimum,* lb	D-c resistance at 20 C (68 F) maximum,† ohms per 1,000 ft	Breaking strength, maximum,‡ lb	D-c resistance at 20 C (68 F) maximum,† ohms per 1,000 ft
27	0.0142	202	0.000158	0.610	3.22	10.9	53.5	8.87	53.2	6.33	51.4
28	0.0126	159	0.000125	0.481	2.54	8.64	67.9	7.02	67.6	4.99	65.3
29	0.0113	128	0.000100	0.387	2.04	6.97	84.5	5.68	84.0	4.01	81.2
30	0.0100	100	0.0000785	0.303	1.60	5.47	108	4.48	107	3.14	104
31	0.0089	79.2	0.0000622	0.240	1.27	4.35	136	3.6	135	2.49	131
32	0.0080	64.0	0.0000503	0.194	1.02	3.53	169	2.90	168	2.01	162
33	0.0071	50.4	0.0000396	0.153	0.806	2.79	214	2.30	213	1.58	206
34	0.0063	39.7	0.0000312	0.120	0.634	2.20	272	1.82	270	1.25	261
35	0.0056	31.4	0.0000246	0.0949	0.501	1.75	344	1.44	342	0.985	331
36	0.0050	25.0	0.0000196	0.0757	0.400	1.40	431	1.16	429	0.785	415
37	0.0045	20.2	0.0000159	0.0613	0.324	1.13	533	0.944	530	0.636	512
38	0.0040	16.0	0.0000126	0.0484	0.256	0.898	674	0.750	671	0.503	648
39	0.0035	12.2	0.00000962	0.0371	0.196	0.691	880	0.577	876	0.385	847
40	0.0031	9.61	0.00000755	0.0291	0.154	0.543	1120	0.455	1120	0.302	1080
41	0.0028	7.84	0.00000616	0.0237	0.125	....	1380	....	1370	0.246	1320
42	0.0025	6.25	0.00000491	0.0189	0.0999	....	1730	....	1720	0.196	1660
43	0.0022	4.84	0.00000380	0.0147	0.0774	....	2230	....	2220	0.152	2140
44	0.0020	4.00	0.00000314	0.0121	0.0639	....	2700	....	2680	0.126	2590
ASTM Specification Designation						B1-56		B2-52		B3-63	

* No. 19 AWG and smaller, based on Anaconda data.
† Based on nominal diameter and ASTM resistivities.
‡ No requirements for tensile strength are specified in ASTM B3-63. Values given here based on Anaconda data.

Table 4-13. Copper Cable—Stranding Classes, Uses

(ASTM Specifications)

ASTM Designation	Construction	Class	Application
B8-64	Concentric lay	AA	For bare conductors usually used in overhead lines
		A	For weather-resistant (weatherproof), slow-burning conductors For bare conductors where greater flexibility than is afforded by Class AA is required
		B	For conductors insulated with various materials such as rubber, paper, varnished-cambric, etc. For the conductors indicated under Class A where greater flexibility is required
		C D	For conductors where greater flexibility is required than is provided by Class B
B173-64	Rope lay with concentric-stranded members	G	Conductor constructions having a range of areas from 5,000,000 cir mils and employing 61 stranded members of 19 wires each down to No. 14 AWG containing 7 stranded members of 7 wires each (Typical uses are for rubber-sheathed conductors, apparatus conductors, portable conductors, and similar applications)
		H	Conductor constructions having a range of areas from 5,000,000 cir mils and employing 91 stranded members of 19 wires each down to No. 9 AWG containing 19 stranded members of 7 wires each (Typical uses are for rubber-sheathed cords and conductors where greater flexibility is required, such as for use on take-up reels over sheaves and extra-flexible apparatus conductors)
B226-64	Annular stranded		For bare conductors, or covered with weather-resistant (weatherproof) materials or insulated with rubber, varnished-cambric or solid-type impregnated paper

ASTM Designation	Construction	Class	Conductor size, AWG	Individual wire size, In.	Individual wire size, AWG	Application
B174-64	Bunch stranded	I	7, 8, 9, 10	0.0201	24	Rubber-covered, varnished-cambric, and paper-insulated conductors
		J	10, 12, 14, 16, 18, 20	0.0126	28	Fixture wire
		K	10, 12, 14, 16, 18, 20	0.0100	30	Fixture wire, flexible cord, and portable cord
		L	10, 12, 14, 16, 18, 20	0.0080	32	Fixture wire and portable cord with greater flexibility than Class K
		M	14, 16, 18, 20	0.0063	34	Heater cord and light portable cord
		O	16, 18, 20	0.0050	36	Heater cord with greater flexibility than Class M
		P	16, 18, 20	0.0040	38	More flexible conductors than provided in preceding classes
		Q	18, 20	0.0031	40	Oscillating fan cord. Very great flexibility

ASTM Designation	Construction	Class	Conductor size, cir mils	Individual wire size, In.	Individual wire size, AWG	Application
B172-64	Rope lay with bunched-stranded members	I	Up to 2,000,000	0.0201	24	Typical use is for special apparatus cable
		K	Up to 1,000,000	0.0100	30	Typical use, special portable cord and conductors
		M	Up to 1,000,000	0.0063	34	Typical use is for welding conductor

Table 4-14. Copper Cable, Concentric Lay—Dimensions, Weight
(ASTM Specification B8-64)

Conductor size, Mcm or Awg	Class AA			Class A			Class B			Class C†		Class D†		Approximate weight, lb per 1000 ft
	No. of wires	Diameter each wire, mils	Nominal conductor diam, in.	No. of wires	Diameter each wire, mils	Nominal conductor diam, in.	No. of wires	Diameter each wire, mils	Nominal conductor diam, in.	No. of wires	Diameter each wire, mils	No. of wires	Diameter each wire, mils	
5,000*	..	.....	.....	169	172.0	2.580	217	151.8	2.581	271	135.8	271	135.8	15,890
4,500	..	.....	.....	169	163.2	2.448	217	144.0	2.448	271	128.9	271	128.9	14,300
4,000	..	.....	.....	169	153.8	2.307	217	135.8	2.309	271	121.5	271	121.5	12,590
3,500	..	.....	.....	127	166.0	2.158	169	143.9	2.159	217	127.0	271	113.6	11,020
3,000*	..	.....	.....	127	153.7	1.998	169	133.2	1.998	217	117.6	271	105.2	9,353
2,500*	..	.....	.....	91	165.7	1.823	127	140.3	1.824	169	121.6	217	107.3	7,794
2,000*	..	.....	.....	91	148.2	1.630	127	125.5	1.632	169	108.8	217	96.0	6,175
1,900	..	.....	.....	91	144.5	1.590	127	122.3	1.590	169	106.0	217	93.6	5,866
1,800	..	.....	.....	91	140.6	1.547	127	119.1	1.548	169	103.2	217	91.1	5,558
1,750*	..	.....	.....	91	138.7	1.526	127	117.4	1.526	169	101.8	217	89.8	5,403
1,700	..	.....	.....	91	136.7	1.504	127	115.7	1.504	169	100.3	217	88.5	5,249
1,600	..	.....	.....	91	132.6	1.459	127	112.2	1.459	169	97.3	217	85.9	4,940
1,500*	..	.....	.....	61	156.8	1.411	91	128.4	1.412	127	108.7	169	94.2	4,631
1,400	..	.....	.....	61	151.5	1.364	91	124.0	1.364	127	105.0	169	91.0	4,323
1,300	..	.....	.....	61	146.0	1.314	91	119.5	1.315	127	101.2	169	87.7	4,014
1,250*	..	.....	.....	61	143.1	1.288	91	117.2	1.289	127	99.2	169	86.0	3,859
1,200	..	.....	.....	61	140.3	1.263	91	114.8	1.263	127	97.2	169	84.3	3,705
1,100	..	.....	.....	61	134.3	1.209	91	109.9	1.209	127	93.1	169	80.7	3,396
1,000*	37	164.4	1.151	61	128.0	1.152	61	128.0	1.152	91	104.8	127	88.7	3,088
900	37	156.0	1.092	61	121.5	1.094	61	121.5	1.094	91	99.4	127	84.2	2,779
800*	37	147.0	1.029	61	114.5	1.031	61	114.5	1.031	91	93.8	127	79.4	2,470
750*	37	142.4	0.997	61	110.9	0.998	61	110.9	0.998	91	90.8	127	76.8	2,316
700*	37	137.5	0.963	61	107.1	0.964	61	107.1	0.964	91	87.7	127	74.2	2,161
650	37	132.5	0.928	61	103.2	0.929	61	103.2	0.929	91	84.5	127	71.5	2,007
600*	37	127.3	0.891	37	127.3	0.891	61	99.2	0.893	91	81.2	127	68.7	1,853
550	37	121.9	0.853	37	121.9	0.853	61	95.0	0.855	91	77.7	127	65.8	1,698
500*	19	162.2	0.811	37	116.2	0.813	37	116.2	0.813	61	90.5	91	74.1	1,544
450	19	153.9	0.770	37	110.3	0.772	37	110.3	0.772	61	85.9	91	70.3	1,389
400*	19	145.1	0.726	19	145.1	0.726	37	104.0	0.728	61	81.0	91	66.3	1,235
350*	12	170.8	0.710	19	135.7	0.679	37	97.3	0.681	61	75.7	91	62.0	1,081

Table 4-14. Copper Cable, Concentric Lay—Dimensions, Weight.—*Concluded*

Conductor size, Mcm or Awg	Class AA			Class A			Class B			Class C†		Class D†		Approximate weight, lb per 1000 ft
	No. of wires	Diameter each wire, mils	Nominal conductor diam, in.	No. of wires	Diameter each wire, mils	Nominal conductor diam, in.	No. of wires	Diameter each wire, mils	Nominal conductor diam, in.	No. of wires	Diameter each wire, mils	No. of wires	Diameter each wire, mils	
300*	12	158.1	0.657	19	125.7	0.629	37	90.0	0.630	61	70.1	91	57.4	926.3
250*	12	144.3	0.600	19	114.7	0.574	37	82.2	0.575	61	64.0	91	52.4	771.9
4/0*	7	173.9	0.522	7	173.9	0.552	19	105.5	0.528	37	75.6	61	58.9	653.3
3/0*	7	154.8	0.464	7	154.8	0.464	19	94.0	0.470	37	67.3	61	52.4	518.1
2/0*	7	137.9	0.414	7	137.9	0.414	19	83.7	0.419	37	60.0	61	46.7	410.9
1/0*	7	122.8	0.368	7	122.8	0.368	19	74.5	0.373	37	53.4	61	41.6	325.8
1	3	167.0	0.360	...	.....	.....	...	.....	.....	...	.....	...	.....	255.9
1*	..	.....	.....	7	109.3	0.328	19	66.4	0.332	37	47.6	61	37.0	258.4
2	3	148.7	0.320	...	.....	.....	...	.....	.....	...	.....	...	.....	202.9
2*	..	.....	.....	7	97.4	0.292	7	97.4	0.292	19	59.1	37	42.4	204.9
3	3	132.5	0.285	...	.....	.....	...	.....	.....	...	.....	...	.....	160.9
3*	..	.....	.....	7	86.7	0.260	7	86.7	0.260	19	52.6	37	37.7	162.5
4	3	118.0	0.254	...	.....	.....	...	.....	.....	...	.....	...	.....	127.6
4*	..	.....	.....	7	77.2	0.232	7	77.2	0.232	19	46.9	37	33.6	128.9
5*	..	.....	.....	...	.....	.....	7	68.8	0.206	19	41.7	37	29.9	102.2
6*	..	.....	.....	...	.....	.....	7	61.2	0.184	19	37.2	37	26.6	81.05
7*	..	.....	.....	...	.....	.....	7	54.5	0.164	19	33.1	37	23.7	64.28
8*	..	.....	.....	...	.....	.....	7	48.6	0.146	19	29.5	37	21.1	50.97
9*	..	.....	.....	...	.....	.....	7	43.2	0.130	19	26.2	37	18.8	40.42
10*	..	.....	.....	...	.....	.....	7	38.5	0.116	19	23.4	37	16.7	32.06
12*	..	.....	.....	...	.....	.....	7	30.5	0.0915	19	18.5	37	13.3	20.16
14*	..	.....	.....	...	.....	.....	7	24.2	0.0726	19	14.7	37	10.5	12.68
16*	..	.....	.....	...	.....	.....	7	19.2	0.0576	19	11.7	...	.....	7.974
18*	..	.....	.....	...	.....	.....	7	15.2	0.0456	19	9.2	...	.....	5.015
20*	..	.....	.....	...	.....	.....	7	12.1	0.0363	19	7.3	...	.....	3.154

* The sizes of conductors which have been marked with an asterisk provide for one or more schedules of preferred series and are commonly used in the industry. Those not marked are given simply as a matter of reference, and it is suggested that their use be discouraged.

† To calculate the nominal diameters of Class C or Class D conductors or of any concentric-lay-stranded conductors made from round wires of uniform diameters, multiply the diameter of an individual wire by that one of the following factors which applies:

Number of wires in conductor	3	7	12	19	37	61	91	127	169	217	271
Diameter calculation factor	2.155	3	4.155	5	7	9	11	13	15	17	19

Table 4-15. Copper Cable, Rope Lay—Dimensions, Weight
(ASTM Specification B173-64)

Conductor size, Mcm or Awg	Class G						Class H					
	No. of wires	No. of members	No. of wires in each member	Diameter each wire, mils	Nominal conductor diameter, in.	Approx weight, lb per 1000 ft	No. of wires	No. of members	No. of wires in each member	Diameter each wire, mils	Nominal conductor diameter, in.	Approx weight, lb per 1000 ft
5000	1159	61	19	65.7	2.957	16,050	1729	91	19	53.8	2.959	16,060
4500	1159	61	19	62.3	2.804	14,435	1729	91	19	51.0	2.805	14,430
4000	1159	61	19	58.7	2.642	12,820	1729	91	19	48.1	2.646	12,840
3500	1159	61	19	55.0	2.475	11,255	1729	91	19	45.0	2.475	11,235
3000	1159	61	19	50.9	2.291	9,635	1729	91	19	41.7	2.294	9,650
2500	703	37	19	59.6	2.086	8,015	1159	61	19	46.4	2.088	8,010
2000	703	37	19	53.3	1.866	6,415	1159	61	19	41.5	1.868	6,400
1900	703	37	19	52.0	1.820	6,100	1159	61	19	40.5	1.823	6,100
1800	703	37	19	50.6	1.771	5,775	1159	61	19	39.4	1.773	5,770
1750	703	37	19	49.9	1.747	5,620	1159	61	19	38.9	1.751	5,625
1700	703	37	19	49.2	1.722	5,460	1159	61	19	38.3	1.724	5,455
1600	703	37	19	47.7	1.670	5,130	1159	61	19	37.2	1.674	5,145
1500	427	61	7	59.3	1.601	4,775	703	37	19	46.2	1.617	4,815
1400	427	61	7	57.3	1.547	4,460	703	37	19	44.6	1.561	4,485
1300	427	61	7	55.2	1.490	4,135	703	37	19	43.0	1.505	4,170
1250	427	61	7	54.1	1.461	3,975	703	37	19	42.2	1.477	4,015
1200	427	61	7	53.0	1.431	3,810	703	37	19	41.3	1.446	3,845
1100	427	61	7	50.8	1.372	3,500	703	37	19	39.6	1.386	3,535
1000	427	61	7	48.4	1.307	3,180	703	37	19	37.7	1.320	3,205
900	427	61	7	45.9	1.239	2,860	703	37	19	35.8	1.253	2,895
800	427	61	7	43.3	1.169	2,545	703	37	19	33.7	1.180	2,560
750	427	61	7	41.9	1.131	2,385	703	37	19	32.7	1.145	2,410
700	427	61	7	40.5	1.094	2,230	703	37	19	31.6	1.106	2,255
650	427	61	7	39.0	1.053	2,070	703	37	19	30.4	1.064	2,085
600	427	61	7	37.5	1.013	1,910	703	37	19	29.2	1.022	1,920
550	427	61	7	35.9	0.969	1,750	703	37	19	28.0	0.980	1,770
500	259	37	7	43.9	0.922	1,585	427	61	7	34.2	0.923	1,590
450	259	37	7	41.7	0.876	1,425	427	61	7	32.5	0.878	1,435
400	259	37	7	39.3	0.825	1,265	427	61	7	30.6	0.826	1,270
350	259	37	7	36.8	0.773	1,110	427	61	7	28.6	0.772	1,110
300	259	37	7	34.0	0.714	945	427	61	7	26.5	0.716	953
250	259	37	7	31.1	0.653	795	427	61	7	24.2	0.653	795
4/0	133	19	7	39.9	0.599	668	259	37	7	28.6	0.601	670
3/0	133	19	7	35.5	0.533	529	259	37	7	25.5	0.536	533
2/0	133	19	7	31.6	0.474	419	259	37	7	22.7	0.477	422
1/0	133	19	7	28.2	0.423	334	259	37	7	20.2	0.424	334
1	133	19	7	25.1	0.377	264	259	37	7	18.0	0.378	266
2	49	7	7	36.8	0.331	207	133	19	7	22.3	0.335	208
3	49	7	7	32.8	0.295	164	133	19	7	19.9	0.299	167
4	49	7	7	29.2	0.263	130	133	19	7	17.7	0.266	132
5	49	7	7	26.0	0.234	103	133	19	7	15.8	0.237	105
6	49	7	7	23.1	0.208	82	133	19	7	14.0	0.210	82
7	49	7	7	20.6	0.185	65	133	19	7	12.5	0.188	65
8	49	7	7	18.4	0.166	51	133	19	7	11.1	0.167	52
9	49	7	7	16.4	0.148	40.8	133	19	7	9.9	0.149	41
10	49	7	7	14.6	0.131	32.3						
12	49	7	7	11.6	0.104	20.3						
14	49	7	7	9.2	0.083	12.8						

Table 4-16. Copper Cable, Classes AA, A, B—Weight, Breaking Strength, D-C Resistance

(ASTM Specifications B1-56, B2-52, B3-63, B8-64)

Conductor size, Mcm or Awg	No. of wires (ASTM stranding class)	Wire diameter, in.	Conductor diameter, in.	Conductor area, sq in.	Conductor weight, lb		Hard		Medium		Soft	
					Per 1000 ft	Per mile	Breaking strength, minimum,* lb	D-c resistance at 20 C (68 F), ohms per 1000 ft	Breaking strength, minimum,* lb	D-c resistance at 20 C (68 F), ohms per 1000 ft	Breaking strength, maximum,† lb	D-c resistance at 20 C (68 F), ohms per 1000 ft
5000	169 (A)	0.1720	2.580	3.927	15,890	83,910	216,300	0.002265	172,000	0.002253	145,300	0.002178
5000	217 (B)	0.1518	2.581	3.927	15,890	83,910	219,500	0.002265	173,200	0.002253	145,300	0.002178
4500	169 (A)	0.1632	2.448	3.534	14,300	75,520	197,200	0.002517	154,800	0.002504	130,800	0.002420
4500	217 (B)	0.1440	2.448	3.534	14,300	75,520	200,400	0.002517	156,900	0.002504	130,800	0.002420
4000	169 (A)	0.1538	2.307	3.142	12,590	66,490	175,600	0.002804	138,500	0.002790	116,200	0.002697
4000	217 (B)	0.1358	2.309	3.142	12,590	66,490	178,100	0.002804	139,500	0.002790	116,200	0.002697
3500	127 (A)	0.1660	2.158	2.749	11,020	58,180	153,400	0.003205	120,400	0.003188	101,700	0.003082
3500	169 (B)	0.1439	2.159	2.749	11,020	58,180	155,900	0.003205	122,000	0.003188	101,700	0.003082
3000	127 (A)	0.1537	1.998	2.356	9,353	49,390	131,700	0.003703	103,900	0.003684	87,180	0.003561
3000	169 (B)	0.1332	1.998	2.356	9,353	49,390	134,400	0.003703	104,600	0.003684	87,180	0.003561
2500	91 (A)	0.1657	1.823	1.963	7,794	41,150	109,600	0.004444	85,990	0.004421	72,650	0.004273
2500	127 (B)	0.1403	1.824	1.963	7,794	41,150	111,300	0.004444	87,170	0.004421	72,650	0.004273
2000	91 (A)	0.1482	1.630	1.571	6,175	32,600	87,790	0.005501	69,270	0.005472	58,120	0.005289
2000	127 (B)	0.1255	1.632	1.571	6,175	32,600	90,050	0.005501	70,210	0.005472	58,120	0.005289
1750	91 (A)	0.1387	1.526	1.374	5,403	28,530	77,930	0.006286	61,020	0.006254	50,850	0.006045
1750	127 (B)	0.1174	1.526	1.374	5,403	28,530	78,800	0.006286	61,430	0.006254	50,850	0.006045
1500	61 (A)	0.1568	1.411	1.178	4,631	24,450	65,840	0.007334	51,950	0.007296	43,590	0.007052
1500	91 (B)	0.1284	1.412	1.178	4,631	24,450	67,540	0.007334	52,650	0.007296	43,590	0.007052
1250	61 (A)	0.1431	1.288	0.9817	3,859	20,380	55,670	0.008801	43,590	0.008755	36,320	0.008463
1250	91 (B)	0.1172	1.289	0.9817	3,859	20,380	56,280	0.008801	43,880	0.008755	36,320	0.008463
1000	37 (AA)	0.1644	1.151	0.7854	3,088	16,300	43,830	0.01100	34,400	0.01094	29,060	0.01058
1000	61 (A-B)	0.1280	1.152	0.7854	3,088	16,300	45,030	0.01100	35,100	0.01094	29,060	0.01058
900	37 (AA)	0.1560	1.092	0.7069	2,779	14,670	39,510	0.01222	31,170	0.01216	26,150	0.01175
900	61 (A-B)	0.1215	1.094	0.7069	2,779	14,670	40,520	0.01222	31,590	0.01216	26,150	0.01175
850	37 (AA)	0.1516	1.061	0.6676	2,624	13,860	37,310	0.01294	29,440	0.01288	24,700	0.01245
850	61 (A-B)	0.1180	1.062	0.6676	2,624	13,860	38,270	0.01294	29,840	0.01288	24,700	0.01245
800	37 (AA)	0.1470	1.029	0.6283	2,470	13,040	35,120	0.01375	27,710	0.01368	23,250	0.01322
800	61 (A-B)	0.1145	1.031	0.6283	2,470	13,040	36,360	0.01375	28,270	0.01368	23,250	0.01322
750	37 (AA)	0.1424	0.997	0.5890	2,316	12,230	33,400	0.01467	26,150	0.01459	21,790	0.01410
750	61 (A-B)	0.1109	0.998	0.5890	2,316	12,230	34,090	0.01467	26,510	0.01459	21,790	0.01410

Table 4-16. Copper Cable, Classes AA, A, B—Weight, Breaking Strength, D-C Resistance.—*Concluded*

Conductor size, Mcm or Awg	No. of wires (ASTM stranding class)	Wire diameter, in.	Conductor diameter, in.	Conductor area, sq in.	Conductor weight, lb		Hard		Medium		Soft	
					Per 1000 ft	Per mile	Breaking strength, minimum,* lb	D-c resistance at 20 C (68 F), ohms per 1000 ft	Breaking strength, minimum,* lb	D-c resistance at 20 C (68 F), ohms per 1000 ft	Breaking strength, maximum,† lb	D-c resistance at 20 C (68 F), ohms per 1000 ft
700	37 (AA)	0.1375	0.963	0.5498	2,161	11,410	31,170	0.01572	24,410	0.01563	20,340	0.01511
700	61 (A-B)	0.1071	0.964	0.5498	2,161	11,410	31,820	0.01572	24,740	0.01563	20,340	0.01511
650	37 (AA)	0.1325	0.928	0.5105	2,007	10,600	29,130	0.01692	22,670	0.01684	18,890	0.01627
650	61 (A-B)	0.1032	0.929	0.5105	2,007	10,600	29,770	0.01692	22,970	0.01684	18,890	0.01627
600	37 (AA-A)	0.1273	0.891	0.4712	1,853	9,781	27,020	0.01834	21,060	0.01824	17,440	0.01763
600	61 (B)	0.0992	0.893	0.4712	1,853	9,781	27,530	0.01834	21,350	0.01824	18,140	0.01763
550	37 (AA-A)	0.1219	0.853	0.4320	1,698	8,966	24,760	0.02000	19,310	0.01990	15,980	0.01923
550	61 (B)	0.0950	0.855	0.4320	1,698	8,966	25,230	0.02000	19,570	0.01990	16,630	0.01923
500	19 (AA)	0.1622	0.811	0.3927	1,544	8,151	21,950	0.02200	17,320	0.02189	14,530	0.02116
500	37 (A-B)	0.1162	0.813	0.3927	1,544	8,151	22,510	0.02200	17,550	0.02189	14,530	0.02116
450	19 (AA)	0.1539	0.770	0.3534	1,389	7,336	19,750	0.02445	15,590	0.02432	13,080	0.02351
450	37 (A-B)	0.1103	0.772	0.3534	1,389	7,336	20,450	0.02445	15,900	0.02432	13,080	0.02351
400	19 (AA-A)	0.1451	0.726	0.3142	1,235	6,521	17,810	0.02750	13,950	0.02736	11,620	0.02645
400	37 (B)	0.1040	0.728	0.3142	1,235	6,521	18,320	0.02750	14,140	0.02736	11,620	0.02645
350	12 (AA)	0.1708	0.710	0.2749	1,081	5,706	15,140	0.03143	12,040	0.03127	10,170	0.03022
350	19 (A)	0.1357	0.679	0.2749	1,081	5,706	15,590	0.03143	12,200	0.03127	10,170	0.03022
350	37 (B)	0.0973	0.681	0.2749	1,081	5,706	16,060	0.03143	12,450	0.03127	10,580	0.03022
300	12 (AA)	0.1581	0.657	0.2356	926.3	4,891	13,170	0.03667	10,390	0.03648	8,718	0.03526
300	19 (A)	0.1257	0.629	0.2356	926.3	4,891	13,510	0.03667	10,530	0.03648	8,718	0.03526
300	37 (B)	0.0900	0.630	0.2356	926.3	4,891	13,870	0.03667	10,740	0.03648	9,071	0.03526
250	12 (AA)	0.1443	0.600	0.1963	771.9	4,076	11,130	0.04400	8,717	0.04378	7,265	0.04231
250	19 (A)	0.1147	0.574	0.1963	771.9	4,076	11,360	0.04400	8,836	0.04378	7,265	0.04231
250	37 (B)	0.0822	0.575	0.1963	771.9	4,076	11,560	0.04400	8,952	0.04378	7,559	0.04231
4/0	7 (AA-A)	0.1739	0.522	0.1662	653.3	3,450	9,154	0.05199	7,278	0.05172	6,149	0.04999
4/0	12 —	0.1328	0.552	0.1662	653.3	3,450	9,483	0.05199	7,378	0.05172	6,149	0.04999
4/0	19 (B)	0.1055	0.528	0.1662	653.3	3,450	9,617	0.05199	7,479	0.05172	6,149	0.04999
3/0	7 (AA-A)	0.1548	0.464	0.1318	518.1	2,736	7,366	0.06556	5,812	0.06522	4,876	0.06304
3/0	12 —	0.1183	0.492	0.1318	518.1	2,736	7,556	0.06556	5,890	0.06522	4,876	0.06304
3/0	19 (B)	0.0940	0.470	0.1318	518.1	2,736	7,698	0.06556	5,970	0.06522	5,074	0.06304
2/0	7 (AA-A)	0.1379	0.414	0.1045	410.9	2,169	5,926	0.08267	4,640	0.08224	3,867	0.07949
2/0	12 —	0.1053	0.438	0.1045	410.9	2,169	6,048	0.08267	4,703	0.08224	3,867	0.07949
2/0	19 (B)	0.0837	0.419	0.1045	410.9	2,169	6,152	0.08267	4,765	0.08224	4,024	0.07949
1/0	7 (AA-A)	0.1228	0.368	0.08289	325.8	1,720	4,752	0.1042	3,705	0.1037	3,067	0.1002
1/0	12 —	0.0938	0.390	0.08289	325.8	1,720	4,841	0.1042	3,755	0.1037	3,191	0.1002
1/0	19 (B)	0.0745	0.373	0.08289	325.8	1,720	4,901	0.1042	3,805	0.1037	3,191	0.1002

1	3 (AA)	0.1670	0.360	0.06573	255.9	1,351	3,621	0.1302	2,879	0.1295	2,432	0.1252
1	7 (A)	0.1093	0.328	0.06573	258.4	1,364	3,804	0.1314	2,958	0.1308	2,432	0.1264
1	19 (B)	0.0664	0.332	0.06573	258.4	1,364	3,899	0.1314	3,037	0.1308	2,531	0.1264
2	3 (AA)	0.1487	0.320	0.05213	202.9	1,071	2,913	0.1641	2,299	0.1633	1,929	0.1578
2	7 (A-B)	0.0974	0.292	0.05213	204.9	1,082	3,045	0.1657	2,361	0.1649	2,007	0.1594
3	3 (AA)	0.1325	0.285	0.04134	160.9	849.6	2,359	0.2070	1,835	0.2059	1,530	0.1990
3	7 (A-B)	0.0867	0.260	0.04134	162.5	858.0	2,433	0.2090	1,885	0.2079	1,592	0.2010
4	3 (AA)	0.1180	0.254	0.03278	127.6	673.8	1,879	0.2610	1,465	0.2596	1,213	0.2509
4	7 (A-B)	0.0772	0.232	0.03278	128.9	680.5	1,938	0.2636	1,505	0.2622	1,262	0.2534
5	7 (B)	0.0688	0.206	0.02600	102.2	539.6	1,542	0.3323	1,201	0.3306	1,001	0.3196
6	7 (B)	0.0612	0.184	0.02062	81.05	427.9	1,288	0.4191	958.6	0.4169	793.8	0.4030
7	7 (B)	0.0545	0.164	0.01635	64.28	339.4	977.1	0.5284	765.2	0.5257	629.5	0.5081
8	7 (B)	0.0486	0.146	0.01297	50.97	269.1	777.2	0.6663	610.7	0.6629	499.2	0.6408
9	7 (B)	0.0432	0.130	0.01028	40.42	213.4	618.2	0.8402	487.4	0.8359	395.9	0.8080
10	7 (B)	0.0385	0.116	0.008155	32.06	169.3	491.7	1.060	388.9	1.054	314.0	1.019
12	7 (B)	0.0305	0.0915	0.005129	20.16	106.5	311.1	1.685	247.7	1.676	197.5	1.620
14	7 (B)	0.0242	0.0726	0.003225	12.68	66.95	197.1	2.679	157.7	2.665	124.2	2.576
16	7 (B)	0.0192	0.0576	0.002028	7.974	42.10	124.7	4.259	100.4	4.237	81.14	4.096
18	7 (B)	0.0152	0.0456	0.001276	5.015	26.48	78.99	6.773	63.91	6.738	51.03	6.513
20	7 (B)	0.0121	0.0363	0.0008023	3.154	16.65	50.04	10.77	40.67	10.71	32.09	10.36
ASTM Designation	B8-64						B1-56 & B8-64		B2-52 & B8-64		B3-63 & B8-64	

* No. 10 AWG and smaller, based on Anaconda data.
† No requirements for tensile strength are specified in ASTM B3-63. Values given here based on Anaconda data.

Weight and Resistance

Stranding class	Conductor size, Mcm or Awg	Increment of resistance and weight, %
AA	4–1	1
	1/0–1000	2
A, B, C, D	2000 and under	2
	Over 2000–3000	3
	Over 3000–4000	4
	Over 4000–5000	5

Resistance
(ASTM requirements)

Temper	Conductivity at 20 C (68 F), IACS, %	Resistivity at 20 C (68 F), ohms (mile, lb)
Hard	96.16	910.15
Medium	96.66	905.44
Soft	100	875.20

The resistance values in this table are trade maximums and are higher than the average values for commercial cable.

CONDUCTOR DATA TABLES

170. Industry specifications for conductors are listed in Table 4-9.

Special designs usually are manufactured to conform with requirements of the purchaser or those prepared by manufacturers to cover the highly technical characteristics of specially designed conductors.

171. Industry Specifications for Conductors. See Table 4-9.

172. Wire Gages: Diameter, Area, Copper Weight. See Table 4-10.

173. Copper Wire: Tensile Strength, Elongation. See Table 4-11.

174. Copper Wire: Weight, Breaking Strength, D-C Resistance. See Table 4-12.

175. Copper Cable: Stranding Classes, Uses. See Table 4-13.

176. Copper Cable, Concentric Lay: Dimensions, Weight. See Table 4-14.

177. Copper Cable, Rope Lay: Dimensions, Weight. See Table 4-15.

178. Copper Cable, Classes AA, A, B: Weight, Breaking Strength, D-C Resistance. See Table 4-16.

179. Copper-clad Steel Wire and Cable: Weight, Breaking Strength, D-C Resistance. See Table 4-17.

180. Copper-clad Steel-copper Cable: Weight, Breaking Strength, D-C Resistance. See Table 4-18.

181. Aluminum Wire: Dimensions, Weight, D-C Resistance. See Table 4-19.

182. Aluminum Cable: Stranding Classes, Uses. See Table 4-20.

183. Aluminum Cable: Weight, Breaking Strength, D-C Resistance. See Table 4-21.

184. Aluminum Steel (ACSR) Cable: Weight, Breaking Strength, D-C Resistance. See Table 4-22.

185. Aluminum-clad Steel Wire and Cable: Weight, Breaking Strength, D-C Resistance. See Table 4-23.

186. Galvanized-steel Wire: Weight, Breaking Strength, D-C Resistance. See Table 4-24.

187. Galvanized-steel Strand: Dimensions, Weight, Breaking Strength. See Table 4-25.

188. Galvanized-steel Conductors: Weight, Breaking Strength, A-C Resistance, Reactance. See Table 4-26.

189. Copper Trolley Wire: Weight, Breaking Strength, D-C Resistance. See Table 4-27.

190. Bronze Trolley Wire: Weight, Breaking Strength, D-C Resistance. See Table 4-28.

191. Copper Wire and Cable: Electrical Characteristics. See Table 4-29.

192. Copper-clad Steel Cable: Electrical Characteristics. See Table 4-30.

193. Copper-clad-steel Copper Cable: Electrical Characteristics. See Table 4-31.

194. Aluminum-steel (ACSR) Cable: Electrical Characteristics. See Table 4-32.

195. Aluminum-clad Steel Cable: Electrical Characteristics. See Table 4-33.

196. Inductive-reactance Spacing Factors. See Table 4-34.

197. Capacitive-reactance Spacing Factors. See Table 4-35.

(*Numbered paragraphs resume on page* 4-81.)

Table 4-17. Copper-clad Steel Wire and Cable—Weight, Breaking Strength, D-C Resistance

(Based on ASTM Specifications B227-65 and B228-65)

Conductor size,* AWG or in.	Conductor stranding		Conductor diam., in.	Conductor area		Conductor weight, lb		Breaking strength, min., lb			D-c resistance at 20 C (68 F), ohms per 1,000 ft	
	No. of wires	Wire size, AWG		Cir mils	Sq in.	Per 1,000 ft	Per mile	High strength		Extra-high strength	Conductivity, IACS	
								Conductivity, IACS				
								40%	30%	30%	40%	30%
Solid (B227-65)												
4	..	..	0.2043	41,740	0.03278	115.8	611.6	3,541	3,934	4,672	0.6337	0.8447
5	..	..	0.1819	33,090	0.02599	91.86	485.0	2,938	3,250	3,913	0.7990	1.065
0.165	..	..	0.1650	27,230	0.02138	75.55	398.9	2,523	2,780	3,368	0.9715	1.295
6	..	..	0.1620	26,240	0.02061	72.85	384.6	2,433	2,680	3,247	1.008	1.343
7	..	..	0.1443	20,820	0.01635	57.77	305.0	2,011	2,207	2,681	1.270	1.694
8	..	..	0.1285	16,510	0.01297	45.81	241.9	1,660	1,815	2,204	1.602	2.136
0.128	..	..	0.1280	16,380	0.01287	45.47	240.1	1,647	1,802	2,188	1.614	2.152
9	..	..	0.1144	13,090	0.01028	36.33	191.8	1,368	1,491	1,790	2.020	2.693
0.104	..	..	0.1040	10,820	0.008495	30.01	158.5	1,177	1,283	1,487	2.445	3.260
10	..	..	0.1019	10,380	0.008155	28.81	152.1	1,130	1,231	1,460	2.547	3.396
12	..	..	0.0808	6,530	0.005129	18.12	95.68	785	......	......	4.051	
0.080	..	..	0.0800	6,400	0.005027	17.76	93.77	770	......	900	4.133	5.509
Stranded (B228-65)												
7/8	19	5	0.910	628,900	0.4940	1770	9344	50,240	55,570	66,910	0.04264	0.05685
13/16	19	6	0.810	498,800	0.3917	1403	7410	41,600	45,830	55,530	0.05377	0.07168
23/32	19	7	0.721	395,500	0.3107	1113	5877	34,390	37,740	45,850	0.06780	0.09039
21/32	19	8	0.642	313,700	0.2464	882.7	4660	28,380	31,040	37,690	0.08550	0.1140
9/16	19	9	0.572	248,800	0.1954	700.0	3696	23,390	25,500	30,610	0.1078	0.1437
5/8	7	4	0.613	292,200	0.2295	818.9	4324	22,310	24,780	29,430	0.09143	0.1219
9/16	7	5	0.546	231,700	0.1820	649.4	3429	18,510	20,470	24,650	0.1153	0.1537
1/2	7	6	0.486	183,800	0.1443	515.0	2719	15,330	16,890	20,460	0.1454	0.1938
7/16	7	7	0.433	145,700	0.1145	408.4	2157	12,670	13,910	16,890	0.1833	0.2444
3/8	7	8	0.385	115,600	0.09077	323.9	1710	10,460	11,440	13,890	0.2312	0.3081
11/32	7	9	0.343	91,650	0.07198	256.9	1356	8,616	9,393	11,280	0.2915	0.3886
5/16	7	10	0.306	72,680	0.05708	203.7	1076	7,121	7,758	9,196	0.3676	0.4900
.....	3	5	0.392	99,310	0.07800	277.8	1467	8,373	9,262	11,860	0.2685	0.3579
.....	3	6	0.349	78,750	0.06185	220.3	1163	6,934	7,639	9,754	0.3385	0.4513
.....	3	7	0.311	62,450	0.04905	174.7	922.4	5,732	6,291	7,922	0.4269	0.5691
.....	3	8	0.277	49,530	0.03890	138.5	731.5	4,730	5,174	6,282	0.5383	0.7176
.....	3	9	0.247	39,280	0.03085	109.9	580.1	3,898	4,250	5,129	0.6788	0.9049
.....	3	10	0.220	31,150	0.02446	87.13	460.0	3,221	3,509	4,160	0.8559	1.141
.....	3	12	0.174	19,590	0.01539	54.80	289.3	2,236	....	......	1.361	

* To determine copper equivalent of copper-clad steel conductor, multiply circular-mil area by percent conductivity expressed as a decimal.

Table 4-18. Copper-clad Steel-copper Cable—Weight, Breaking Strength, D-C Resistance

(ASTM Specification B229-65)

Hard-drawn copper equivalent,* Mcm or AWG	Conductor type	Conductor stranding				Conductor diam., in	Conductor area, in²	Conductor weight, lb		Breaking strength, min., lb	D-c resistance at 20°C (68°F), Ω/1,000 ft
		EHS 30% copper-clad wires		Hard-drawn copper wires				Per 1,000 ft	Per mi		
		No.	Diam., in	No.	Diam., in						
350	E	7	0.1576	12	0.1576	0.788	0.3706	1403	7409	32,420	0.03143
350	EK	4	0.1470	15	0.1470	0.735	0.3225	1238	6536	23,850	0.03143
300	E	7	0.1459	12	0.1459	0.729	0.3177	1203	6351	27,770	0.03667
300	EK	4	0.1361	15	0.1361	0.680	0.2764	1061	5602	20,960	0.03667
250	E	7	0.1332	12	0.1332	0.666	0.2648	1002	5292	23,920	0.04400
250	EK	4	0.1242	15	0.1242	0.621	0.2302	884.2	4669	17,840	0.04400
4/0	E	7	0.1225	12	0.1225	0.613	0.2239	848.3	4479	20,730	0.05199
4/0	EK	4	0.1143	15	0.1143	0.571	0.1950	748.4	3951	15,370	0.05199
4/0	F	1	0.1833	6	0.1833	0.550	0.1847	710.2	3750	12,290	0.05199
3/0	E	7	0.1091	12	0.1091	0.545	0.1776	672.7	3552	16,800	0.06556
3/0	EK	4	0.1018	15	0.1018	0.509	0.1546	593.5	3134	12,370	0.06556
3/0	F	1	0.1632	6	0.1632	0.490	0.1464	563.2	2974	9,980	0.06556
2/0	F	1	0.1454	6	0.1454	0.436	0.1162	446.8	2359	8,094	0.08265
1/0	F	1	0.1294	6	0.1294	0.388	0.09206	354.1	1870	6,536	0.1043
1	F	1	0.1153	6	0.1153	0.346	0.07309	280.9	1483	5,266	0.1315
2†	A	1	0.1699	2	0.1699	0.366	0.06801	256.8	1356	5,876	0.1658
2	F	1	0.1026	6	0.1026	0.308	0.05787	222.8	1176	4,233	0.1658
4†	A	1	0.1347	2	0.1347	0.290	0.04275	161.5	852	3,938	0.2636
6†	A	1	0.1068	2	0.1068	0.230	0.02688	101.6	536.3	2,585	0.4150
8†	A	1	0.1127	2	0.07969	0.199	0.01995	74.27	392.2	2,233	0.6598

* Area of hard-drawn copper cable having the same d-c resistance as that of the composite cable.
† Sizes commonly used for rural distribution.

Table 4-19. Aluminum Wire—Dimensions, Weight, D-C Resistance

(Based on ASTM Specifications B230-60, B262-61, and B323-61)

Conductor size, AWG	Diam. at 20 C (68 F), mils	Area at 20 C (68 F)		D-c resistance at 20 C (68 F),* ohms per 1,000 ft	Weight at 20 C (68 F),† lb		Length at 20 C (68 F), ft per ohm
		Cir mils	Sq in.		Per 1,000 ft	Per ohm	
2	257.6	66,360	0.05212	0.2562	61.07	238.4	3903
3	229.4	52,620	0.04133	0.3231	48.43	149.9	3095
4	204.3	41,740	0.03278	0.4074	38.41	94.30	2455
5	181.9	33,090	0.02599	0.5139	30.45	59.26	1946
6	162.0	26,240	0.02061	0.6479	24.15	37.28	1544
7	144.3	20,820	0.01635	0.8165	19.16	23.47	1225
8	128.5	16,510	0.01297	1.030	15.20	14.76	971.2
9	114.4	13,090	0.01028	1.299	12.04	9.272	769.7
10	101.9	10,380	0.008155	1.637	9.556	5.836	610.7
11	90.7	8,230	0.00646	2.07	7.57	3.66	484
12	80.8	6,530	0.00513	2.60	6.01	2.31	384
13	72.0	5,180	0.00407	3.28	4.77	1.45	305
14	64.1	4,110	0.00323	4.14	3.78	0.914	242
15	57.1	3,260	0.00256	5.21	3.00	0.575	192
16	50.8	2,580	0.00203	6.59	2.38	0.361	152
17	45.3	2,050	0.00161	8.29	1.89	0.228	121
18	40.3	1,620	0.00128	10.5	1.49	0.143	95.5
19	35.9	1,290	0.00101	13.2	1.19	0.0899	75.8
20	32.0	1,020	0.000804	16.6	0.942	0.0568	60.2
21	28.5	812	0.000638	20.9	0.748	0.0357	47.8
22	25.3	640	0.000503	26.6	0.589	0.0222	37.6
23	22.6	511	0.000401	33.3	0.470	0.0141	30.0
24	20.1	404	0.000317	42.1	0.372	0.00884	23.8
25	17.9	320	0.000252	53.1	0.295	0.00556	18.8
26	15.9	253	0.000199	67.3	0.233	0.00346	14.9
27	14.2	202	0.000158	84.3	0.186	0.00220	11.9
28	12.6	159	0.000125	107	0.146	0.00136	9.34
29	11.3	128	0.000100	133	0.118	0.000883	7.51
30	10.0	100	0.0000785	170	0.0920	0.000541	5.88

* Conductivity = 61.0% IACS.
† Density = 2.703 g per cu cm (0.09765 lb per cu in.).

Table 4-20. Aluminum Cable—Stranding Classes, Uses

(ASTM Specification B231-64)

Construction	Class	Application
Concentric lay	AA	For bare conductors usually used in overhead lines
	A	For conductors to be covered with weather-resistant (weatherproof), slow-burning materials and for bare conductors where greater flexibility than is afforded by Class AA is required. Conductors intended for further fabrication into tree wire or to be insulated and laid helically with or around aluminum or ACSR messengers shall be regarded as Class A conductors with respect to direction of lay only
	B	For conductors to be insulated with various materials such as rubber, paper, varnished cloth, etc., and for the conductors indicated under Class A where greater flexibility is required
	C, D	For conductors where greater flexibility is required than is provided by Class B conductors

Table 4-21. Aluminum Cable—Weight, Breaking Strength, D-C Resistance

(Based on ASTM Specifications B231-64, B230-60, and B262-61)

Conductor size		Hard-drawn copper equivalent,* Mcm or AWG	Conductor stranding		Conductor diam., in	Conductor area, in²	Conductor weight, lb		Breaking strength, ultimate, lb		D-c resistance at 20°C (68° F),† 61% conductivity, Ω/1,000 ft
AWG	Cmils		No. of wires	Wire diam., in			Per 1,000 ft	Per mi	Hard-drawn	¾-hard-drawn	
...	4,000,000	2,520	217	0.1358	2.309	3.142	3,829	20,215	....	48,066	0.004421
...	3,500,000	2,200	169	0.1439	2.159	2.749	3,350	17,688	....	42,058	0.005052
...	3,500,000	2,200	127	0.1660	2.158	2.749	3,350	17,688	59,380	....	0.005052
...	3,000,000	1,890	169	0.1332	1.998	2.356	2,844	15,016	....	36,050	0.005838
...	3,000,000	1,890	127	0.1537	1.998	2.356	2,844	15,016	50,890	....	0.005838
...	2,500,000	1,570	127	0.1403	1.824	1.963	2,370	12,514	43,300	30,041	0.007005
...	2,500,000	1,570	91	0.1657	1.823	1.963	2,370	12,514	42,410	....	0.007005
...	2,000,000	1,260	127	0.1255	1.632	1.571	1,878	9,916	35,340	24,033	0.008671
...	2,000,000	1,260	91	0.1482	1.630	1.571	1,878	9,916	34,640	....	0.008671
...	1,900,000	1,195	127	0.1223	1.590	1.492	1,784	9,420	33,580	22,832	0.009127
...	1,900,000	1,195	91	0.1445	1.590	1.492	1,784	9,420	32,900	....	0.009127
...	1,800,000	1,132	127	0.1191	1.548	1.414	1,690	8,923	32,450	21,630	0.009634
...	1,800,000	1,132	91	0.1406	1.547	1.414	1,690	8,923	31,170	....	0.009634
...	1,750,000	1,101	127	0.1174	1.526	1.374	1,643	8,675	31,540	21,029	0.009910
...	1,750,000	1,101	61	0.1694	1.525	1.374	1,643	8,675	29,690	....	0.009910
...	1,750,000	1,101	91	0.1387	1.526	1.374	1,643	8,675	30,920	....	0.009910
...	1,700,000	1,069	127	0.1157	1.504	1.335	1,596	8,427	30,640	20,428	0.01020
...	1,700,000	1,069	61	0.1669	1.502	1.335	1,596	8,427	28,840	....	0.01020
...	1,700,000	1,069	91	0.1367	1.504	1.335	1,596	8,427	30,040	....	0.01020
...	1,600,000	1,006	127	0.1122	1.459	1.257	1,502	7,931	28,840	19,227	0.01084
...	1,600,000	1,006	91	0.1326	1.459	1.257	1,502	7,931	28,270	....	0.01084
...	1,600,000	1,006	61	0.1620	1.458	1.257	1,502	7,931	27,140	....	0.01084
...	1,500,000	943	91	0.1284	1.412	1.178	1,408	7,435	26,510	18,025	0.01156
...	1,500,000	943	61	0.1568	1.411	1.178	1,408	7,435	25,450	....	0.01156
...	1,400,000	880	91	0.1240	1.364	1.100	1,314	6,939	24,740	16,823	0.01239
...	1,400,000	880	61	0.1515	1.364	1.100	1,314	6,939	23,750	....	0.01239
...	1,300,000	818	91	0.1195	1.315	1.021	1,220	6,442	23,430	15,622	0.01334
...	1,300,000	818	61	0.1460	1.314	1.021	1,220	6,442	22,510	....	0.01334
...	1,250,000	786	91	0.1172	1.289	0.9817	1,173	6,196	22,530	15,021	0.01387
...	1,250,000	786	61	0.1431	1.288	0.9817	1,173	6,196	21,650	....	0.01387
...	1,200,000	755	91	0.1148	1.263	0.9425	1,127	5,948	21,630	14,420	0.01445
...	1,200,000	755	61	0.1403	1.263	0.9425	1,127	5,948	20,780	....	0.01445
...	1,100,000	692	91	0.1099	1.209	0.8639	1,033	5,452	20,210	13,218	0.01577
...	1,100,000	692	61	0.1343	1.209	0.8639	1,033	5,452	19,440	....	0.01577
...	1,000,000	629	61	0.1280	1.152	0.7854	938.8	4,957	17,670	12,017	0.01734
...	1,000,000	629	37	0.1644	1.151	0.7854	938.8	4,957	16,960	....	0.01734
...	900,000	566	61	0.1215	1.094	0.7069	844.9	4,461	15,900	10,815	0.01927
...	900,000	566	37	0.1560	1.092	0.7069	844.9	4,461	15,260	....	0.01927
...	800,000	503	61	0.1145	1.031	0.6283	751.0	3,965	14,420	9,613	0.02168
...	800,000	503	37	0.1470	1.029	0.6283	751.0	3,965	13,850	....	0.02168
...	750,000	472	61	0.1109	0.998	0.5890	704.1	3,717	13,520	9,012	0.02312
...	750,000	472	37	0.1424	0.997	0.5890	704.1	3,717	12,990	....	0.02312
...	700,000	440	61	0.1071	0.964	0.5498	657.1	3,470	12,860	8,412	0.02477
...	700,000	440	37	0.1375	0.962	0.5498	657.1	3,470	12,370	....	0.02477
...	650,000	409	61	0.1032	0.929	0.5105	610.2	3,222	11,950	7,811	0.02668
...	650,000	409	37	0.1325	0.928	0.5105	610.2	3,222	11,480	....	0.02668
...	600,000	377	61	0.0992	0.893	0.4712	563.2	2,974	11,450	7,210	0.02890
...	600,000	377	37	0.1273	0.891	0.4712	563.2	2,974	10,600	....	0.02890
...	550,000	346	61	0.0950	0.855	0.4320	516.3	2,726	10,500	6,609	0.03153
...	550,000	346	37	0.1219	0.853	0.4320	516.3	2,726	9,720	....	0.03153
...	500,000	314	37	0.1162	0.813	0.3927	469.4	2,478	9,010	6,008	0.03468
...	500,000	314	19	0.1622	0.811	0.3927	469.4	2,478	8,480	....	0.03468
...	450,000	283	37	0.1103	0.772	0.3534	422.4	2,230	8,110	5,407	0.03854
...	450,000	283	19	0.1539	0.770	0.3534	422.4	2,230	7,635	....	0.03854
...	400,000	252	37	0.1040	0.728	0.3142	375.5	1,983	7,350	4,807	0.04336
...	400,000	252	19	0.1451	0.726	0.3142	375.5	1,983	6,925	....	0.04336
...	350,000	220	37	0.0973	0.681	0.2749	328.6	1,735	6,680	4,206	0.04955
...	350,000	220	19	0.1357	0.678	0.2749	328.6	1,735	6,185	....	0.04955
...	350,000	220	12	0.1708	0.710	0.2749	328.6	1,735	5,940	....	0.04955
...	300,000	188.7	37	0.0900	0.630	0.2356	281.6	1,487	5,830	3,605	0.05781
...	300,000	188.7	19	0.1257	0.628	0.2356	281.6	1,487	5,300	....	0.05781
...	300,000	188.7	12	0.1581	0.657	0.2356	281.6	1,487	5,090	....	0.05781
...	250,000	157.2	37	0.0822	0.575	0.1963	234.7	1,239	4,860	3,004	0.06937
...	250,000	157.2	19	0.1147	0.574	0.1963	234.7	1,239	4,505	....	0.06937
...	250,000	157.2	12	0.1443	0.600	0.1963	234.7	1,239	4,330	....	0.06937

Table 4-21. Aluminum Cable—Weight, Breaking Strength, D-C Resistance.—*Concluded*

Conductor size		Hard-drawn copper equivalent,* Mcm or AWG	Conductor stranding		Conductor diam., in	Conductor area, in²	Conductor weight, lb		Breaking strength, ultimate, lb		D-c resistance at 20°C (68°F),† 61% conductivity, Ω/1,000 ft
AWG	Cmils		No. of wires	Wire diam., in			Per 1,000 ft	Per mi	Hard-drawn	¾-hard-drawn	
4/0	211,600	2/0	19	0.1055	0.528	0.1662	198.6	1,049	3,890	2,543	0.08196
4/0	211,600	2/0	7	0.1739	0.522	0.1662	198.6	1,049	3,590	....	0.08196
3/0	167,800	1/0	19	0.0940	0.470	0.1318	157.5	831.7	3,200	2,016	0.1033
3/0	167,800	1/0	7	0.1548	0.464	0.1318	157.5	831.7	2,845	....	0.1033
2/0	133,100	1	19	0.0837	0.419	0.1045	124.9	659.6	2,587	1,599	0.1303
2/0	133,100	1	7	0.1379	0.414	0.1045	124.9	659.6	2,350	....	0.1303
1/0	105,600	2	19	0.0745	0.372	0.08286	99.07	523.1	2,090	1,269	0.1642
1/0	105,600	2	7	0.1228	0.368	0.08286	99.07	523.1	1,866	....	0.1642
1	83,690	3	19	0.0664	0.332	0.06573	78.57	414.8	1,686	1,006	0.2072
1	83,690	3	7	0.1093	0.328	0.06573	78.57	414.8	1,538	....	0.2072
2	66,360	4	7	0.0974	0.292	0.05213	62.30	329.0	1,266	797	0.2613
3	52,620	5	7	0.0867	0.260	0.04134	49.41	260.9	1,023	632	0.3296
4	41,740	6	7	0.0772	0.232	0.03278	38.18	206.9	826	502	0.4155
6	26,240	8	7	0.0612	0.184	0.02061	24.63	130.1	529	315	0.6609
8	16,510	10	7	0.0486	0.146	0.01297	15.50	81.84	292	198	1.050
10	10,380	12	7	0.0385	0.116	0.008155	9.748	51.47	183	125	1.670
12	6,530	14	7	0.0305	0.092	0.00513	6.129	32.36	115	78.5	2.656
14	4,110	16	7	0.0242	0.073	0.00323	3.857	20.37	72.6	49.4	4.221
16	2,580	18	7	0.0192	0.058	0.00203	2.423	12.79	45.6	31.0	6.720
18	1,620	20	7	0.0152	0.046	0.00128	1.525	8.050	28.7	19.5	10.68
20	1,020	22	7	0.0121	0.036	0.000804	0.961	5.076	18.1	12.3	16.94
Copper equivalent sizes											
...	1,590,000	1000	91	0.1322	1.454	1.2488	1,493	7,881	28,100	....	0.01091
...	1,590,000	1000	61	0.1614	1.453	1.2488	1,493	7,881	27,000	....	0.01091
...	1,431,000	900	61	0.1531	1.378	1.1239	1,343	7,093	24,300	....	0.01212
...	1,272,000	800	61	0.1444	1.300	0.9990	1,194	6,305	22,000	....	0.01363
...	1,192,500	750	61	0.1398	1.258	0.9366	1,119	5,911	21,000	....	0.01454
...	1,113,000	700	61	0.1350	1.215	0.8741	1,045	5,517	19,660	....	0.01558
...	1,033,500	650	61	0.1302	1.172	0.8117	970.2	5,123	18,260	....	0.01678
...	1,033,500	650	37	0.1671	1.170	0.8117	970.2	5,123	17,530	....	0.01678
...	954,000	600	61	0.1250	1.125	0.7493	895.6	4,729	16,860	....	0.01818
...	954,000	600	37	0.1606	1.124	0.7493	895.6	4,729	16,180	....	0.01818
...	874,500	550	61	0.1197	1.077	0.6868	820.9	4,335	15,760	....	0.01983
...	874,500	550	37	0.1537	1.076	0.6868	820.9	4,335	14,830	....	0.01983
...	795,000	500	61	0.1142	1.028	0.6244	746.3	3,941	14,330	....	0.02181
...	795,000	500	37	0.1466	1.026	0.6244	746.3	3,941	13,770	....	0.02181
...	715,500	450	61	0.1083	0.975	0.5620	671.7	3,546	13,150	....	0.02424
...	715,500	450	37	0.1391	0.974	0.5620	671.7	3,546	12,640	....	0.02424
...	636,000	400	37	0.1311	0.918	0.4995	597.1	3,152	11,240	....	0.02727
...	556,500	350	37	0.1226	0.858	0.4371	522.4	2,758	9,834	....	0.03116
...	556,500	350	19	0.1711	0.856	0.4371	522.4	2,758	9,440	....	0.03116
...	477,000	300	37	0.1135	0.795	0.3746	447.8	2,364	8,600	....	0.03636
...	477,000	300	19	0.1585	0.793	0.3746	447.8	2,364	8,090	....	0.03636
...	397,500	250	19	0.1447	0.724	0.3122	373.2	1,970	6,880	....	0.04363
...	336,400	4/0	19	0.1331	0.666	0.2642	315.8	1,667	5,940	....	0.05155
...	266,800	3/0	19	0.1185	0.593	0.2095	250.5	1,322	4,810	....	0.06500
...	266,800	3/0	7	0.1952	0.586	0.2095	250.5	1,322	4,525	....	0.06500

* Area of hard-drawn copper cable (conductivity = 97% IACS) having the same d-c resistance as that of the aluminum cable (conductivity = 61% IACS).

† Increased 2% to allow for stranding.

Table 4-22. Aluminum-steel (ACSR) Cable—Weight, Breaking Strength, D-C Resistance
(Based on ASTM Specification B232-64T)

Hard-drawn copper equivalent,* Mcm or AWG	Conductor stranding: Aluminum, No. of wires	Conductor stranding: Aluminum, Wire diam., in.	Conductor stranding: Steel, No. of wires	Conductor stranding: Steel, Wire diam., in.	Steel core diam., in.	Conductor diam., in.	Aluminum area: Cir mils or AWG	Aluminum area: Sq in.	Conductor area, sq in.	Conductor weight, lb: Per 1,000 ft	Conductor weight, lb: Per mile	Component weights, %: Aluminum	Component weights, %: Steel	Breaking strength, ultimate, lb	D-c resistance at 20 C (68 F) 61% conductivity, ohms per 1,000 ft
1000	54	0.1716	19	0.1030	0.515	1.545	1,590,000	1.249	1.407	2044.5	10,795	73.7	26.3	56,000	0.01101
950	54	0.1673	19	0.1004	0.502	1.506	1,510,500	1.186	1.337	1942.3	10,255	73.7	26.3	53,200	0.01159
900	54	0.1628	19	0.0977	0.489	1.465	1,431,000	1.124	1.266	1839.8	9,714	73.7	26.3	50,400	0.01224
850	54	0.1582	19	0.0949	0.475	1.424	1,351,500	1.062	1.196	1737.2	9,172	73.7	26.3	47,600	0.01296
800	54	0.1535	19	0.0921	0.461	1.382	1,272,000	0.9990	1.126	1635.2	8,634	73.7	26.3	44,800	0.01377
750	54	0.1486	19	0.0892	0.446	1.338	1,192,500	0.9366	1.055	1533.3	8,096	73.7	26.3	43,100	0.01469
700	54	0.1436	19	0.0862	0.431	1.293	1,113,000	0.8741	0.9849	1431.3	7,557	73.7	26.3	40,200	0.01573
650	54	0.1384	7	0.1384	0.415	1.246	1,033,500	0.8117	0.9169	1331.3	7,029	73.3	26.7	37,100	0.01686
650	36	0.1694	1	0.1694	0.169	1.186	1,033,500	0.8117	0.8342	1049.1	5,539	92.8	7.2	21,130	0.01686
600	54	0.1329	7	0.1329	0.399	1.196	954,000	0.7493	0.7464	1228.5	6,486	73.3	26.7	34,200	0.01827
600	36	0.1628	1	0.1628	0.163	1.140	954,000	0.7493	0.7701	968.4	5,113	92.8	7.2	19,510	0.01827
566	54	0.1291	7	0.1291	0.387	1.162	900,000	0.7069	0.7985	1159.0	6,120	73.3	26.7	32,300	0.01936
550	54	0.1273	7	0.1273	0.382	1.146	874,500	0.6868	0.7759	1126.4	5,947	73.3	26.7	31,400	0.01993
500	30	0.1628	19	0.0977	0.489	1.140	795,000	0.6244	0.7668	1235.1	6,521	60.9	39.1	38,400	0.02197
500	26	0.1749	7	0.1360	0.408	1.108	795,000	0.6244	0.7261	1094.3	5,778	68.5	31.5	31,200	0.02192
500	54	0.1214	7	0.1214	0.364	1.093	795,000	0.6244	0.7053	1024.1	5,407	73.3	26.7	28,500	0.02192
500	36	0.1486	1	0.1486	0.149	1.040	795,000	0.6244	0.6417	807.0	4,261	92.8	7.2	16,540	0.02192
450	30	0.1544	19	0.0926	0.463	1.081	715,500	0.5620	0.6901	1110.8	5,865	60.9	39.1	34,600	0.02442
450	26	0.1659	7	0.1290	0.387	1.051	715,500	0.5620	0.6535	984.8	5,200	68.5	31.5	28,100	0.02436
450	54	0.1151	7	0.1151	0.345	1.036	715,500	0.5620	0.6348	921.4	4,865	73.3	26.7	26,300	0.02436
419	54	0.1111	7	0.1111	0.333	1.000	666,600	0.5235	0.5914	858.4	4,532	73.3	26.7	24,500	0.02614
419	24	0.1667	7	0.1111	0.333	1.000	666,600	0.5235	0.5914	858.6	4,533	73.2	26.8	23,770	0.02614
400	36	0.1329	1	0.1329	0.133	0.930	636,000	0.4995	0.5134	645.6	3,409	92.8	7.2	13,460	0.02740
400	30	0.1456	19	0.0874	0.437	1.019	636,000	0.4995	0.6134	988.2	5,218	60.9	39.1	31,500	0.02747
400	26	0.1564	7	0.1216	0.365	0.990	636,000	0.4995	0.5809	875.3	4,622	48.5	31.5	25,000	0.02740
400	54	0.1085	7	0.1085	0.326	0.977	636,000	0.4995	0.5643	819.0	4,324	73.3	26.7	23,600	0.02740
400	24	0.1628	7	0.1085	0.326	0.977	636,000	0.4995	0.5643	819.2	4,325	73.2	26.8	22,600	0.02740
400	18	0.1880	1	0.1880	0.188	0.940	636,000	0.4995	0.5273	691.0	3,648	86.4	13.6	15,840	0.02740
380.5	30	0.1420	19	0.0852	0.426	0.994	605,000	0.4752	0.5835	939.7	4,962	60.9	39.1	30,000	0.02888
380.5	26	0.1525	7	0.1186	0.356	0.966	605,000	0.4752	0.5526	832.6	4,396	68.5	31.5	24,100	0.02881
380.5	54	0.1059	7	0.1059	0.318	0.953	605,000	0.4752	0.5368	779.3	4,115	73.3	26.7	22,500	0.02881
380.5	24	0.1588	7	0.1059	0.318	0.953	605,000	0.4752	0.5368	779.5	4,116	73.2	26.8	21,500	0.02881
350	30	0.1362	7	0.1362	0.409	0.953	556,500	0.4371	0.5391	872.0	4,604	60.4	39.6	27,200	0.03139
350	26	0.1463	7	0.1138	0.341	0.927	565,500	0.4371	0.5083	766.1	4,045	68.5	31.5	22,400	0.03132
350	24	0.1523	7	0.1015	0.3045	0.914	556,500	0.4371	0.4940	716.8	3,785	73.2	26.8	19,860	0.03132
350	18	0.1758	1	0.1758	0.1758	0.879	556,500	0.4371	0.4614	604.2	3,190	86.4	13.6	13,850	0.03132
300	30	0.1261	7	0.1261	0.3783	0.883	477,000	0.3746	0.4620	747.4	3,946	60.4	39.6	23,300	0.03662
300	26	0.1355	7	0.1054	0.3162	0.858	477,000	0.3746	0.4356	656.8	3,468	68.5	31.5	19,430	0.03653

300	24	0.1410	7	0.0940	0.2820	0.846	477,000	0.3746	0.4232	614.5	3,245	73.2	26.8	17,200	0.03653
300	18	0.1628	1	0.1628	0.1628	0.814	477,000	0.3746	0.3954	518.0	2,735	86.4	13.6	11,870	0.03653
250	30	0.1151	7	0.1151	0.3453	0.806	397,500	0.3122	0.3850	622.8	3,288	60.4	39.6	19,980	0.04395
250	26	0.1236	7	0.0961	0.2883	0.783	397,500	0.3122	0.3630	546.9	2,888	68.5	31.5	16,190	0.04384
250	24	0.1287	7	0.0858	0.2574	0.772	397,500	0.3122	0.3527	512.1	2,704	73.2	26.8	14,680	0.04384
250	18	0.1486	1	0.1486	0.1486	0.743	397,500	0.3122	0.3295	431.7	2,279	86.4	13.6	10,030	0.04384
4/0	30	0.1059	7	0.1059	0.3177	0.741	336,400	0.2642	0.3259	527.1	2,783	60.4	39.6	17,040	0.05193
4/0	26	0.1138	7	0.0885	0.2655	0.721	336,400	0.2642	0.3072	463.1	2,445	86.5	31.5	14,050	0.05180
4/0	24	0.1184	7	0.0789	0.2367	0.710	336,400	0.2642	0.2984	433.2	2,287	73.2	26.8	12,550	0.05180
4/0	18	0.1367	1	0.1367	0.1367	0.684	336,400	0.2642	0.2789	365.3	1,931	86.4	13.6	8,630	0.05180
188.7	26	0.1074	7	0.0835	0.2505	0.680	300,000	0.2356	0.2740	412.8	2,179	68.5	31.5	12,650	0.05809
3/0	26	0.1013	7	0.0788	0.2356	0.642	266,800	0.2095	0.2436	367.3	1,939	68.5	31.5	11,250	0.06532
3/0	6	0.2109	7	0.0703	0.2109	0.633	266,800	0.2095	0.2367	342.4	1,808	73.2	26.8	9,645	0.06500
3/0	18	0.1217	1	0.1217	0.1217	0.609	266,800	0.2095	0.2211	289.7	1,531	86.4	13.6	6,830	0.06532
2/0	6	0.1878	1	0.1878	0.1878	0.563	4/0	0.1662	0.1939	291.1	1,537	67.9	32.1	8,420	0.08155
2/0	5	0.2057	1	0.1443	0.1443	0.556	4/0	0.1662	0.1825	252.9	1,335	78.2	21.8	6,600	0.08155
1/0	6	0.1672	1	0.1672	0.1672	0.502	3/0	0.1318	0.1538	230.9	1,219	67.9	32.1	6,675	0.1028
1/0	5	0.1832	1	0.1285	0.1285	0.495	3/0	0.1318	0.1448	200.5	1,059	78.2	21.8	5,240	0.1028
1	6	0.1490	1	0.1490	0.1490	0.447	2/0	0.1045	0.1219	183.1	966	67.9	32.1	5,345	0.1297
1	5	0.1632	1	0.1632	0.1145	0.441	2/0	0.1045	0.1148	159.0	840	78.2	21.8	4,200	0.1297
2	6	0.1327	1	0.1327	0.1327	0.398	1/0	0.0829	0.0967	145.2	766	67.9	32.1	4,280	0.1634
2	5	0.1453	1	0.1019	0.1019	0.393	1/0	0.0829	0.09105	126.1	666	78.2	21.8	3,375	0.1634
3	6	0.1182	1	0.1182	0.1182	0.355	1	0.0657	0.0767	115.2	608	67.9	32.1	3,480	0.0262
4	7	0.0974	1	0.1299	0.1299	0.325	2	0.0521	0.0653	106.7	563	58.2	41.8	3,525	0.2601
4	6	0.1052	1	0.1052	0.1052	0.316	2	0.0521	0.0608	91.3	482	67.9	32.1	2,790	0.2601
5	6	0.0937	1	0.0937	0.0937	0.281	3	0.0413	0.0482	72.5	383	67.9	32.1	2,250	0.3280
6	7	0.0772	1	0.1029	0.1029	0.257	4	0.0328	0.0411	67.0	354	58.2	41.8	2,288	0.4134
6	6	0.0834	1	0.0834	0.0834	0.250	4	0.0328	0.0383	57.4	303	67.9	32.1	1,830	0.4134
7	6	0.0743	1	0.0743	0.0743	0.223	5	0.0260	0.0303	45.5	240	67.9	32.1	1,460	0.5215
8	6	0.0661	1	0.0661	0.0661	0.198	6	0.0206	0.0240	36.1	190	67.9	32.1	1,170	0.6577
High mechanical strength†															
127.8	16	0.1127	19	0.0977	0.4885	0.714	203,200	0.1596	0.3020	676.8	3,574	28.3	71.7	27,500	0.08372
132.9	12	0.1327	7	0.1327	0.3981	0.663	211,300	0.1660	0.2628	527.5	2,785	37.8	62.2	19,640	0.08051
120	12	0.1261	7	0.1261	0.3783	0.631	190,800	0.1499	0.2373	476.3	2,514	37.8	62.2	17,730	0.08916
111.2	12	0.1214	7	0.1214	0.3642	0.607	176,900	0.1389	0.2200	441.4	2,330	37.8	62.2	16,440	0.09616
100	12	0.1151	7	0.1151	0.3453	0.576	159,000	0.1249	0.1977	396.8	2,095	37.8	62.2	15,200	0.1070
84.6	12	0.1059	7	0.1059	0.3177	0.530	134,600	0.1057	0.1674	336.0	1,773	37.8	62.2	12,920	0.1264
69.7	12	0.0961	7	0.0961	0.2883	0.481	110,800	0.0870	0.1378	276.6	1,460	37.8	62.2	10,730	0.1535
64.2	12	0.0921	7	0.0921	0.2763	0.461	101,800	0.0800	0.1266	254.1	1,342	37.8	43.3	9,860	0.1671
50.3	8	0.1000	1	0.1670	0.1670	0.367	80,000	0.0628	0.0847	149.0	787	50.4	71.7	5,200	0.2126

* Area of hard-drawn copper cable (conductivity = 97% IACS) having the same d-c resistance as that of the ACSR (aluminum conductivity = 61% IACS).

† These conductors have been designed to meet special requirements for high mechanical strength and relatively small current-carrying capacity. They are used largely for overhead ground wires, river-crossing spans, and other types of construction in which mechanical strength is of primary importance.

Table 4-23. Aluminum-clad Steel Wire and Cable—Weight, Breaking Strength, D-C Resistance

(Based on ASTM Specifications B415-64T and B416-64T)

Conductor stranding		Conductor diam., in	Conductor area		Conductor weight, lb		Breaking strength, min., lb	D-c resistance at 20°C (68°F) Ω/1,000 ft
No. of wires	Wire size, AWG		Cmils	In²	Per 1,000 ft	Per mi		
Solid (B415-64T)								
1	4	0.2043	41,740	0.03278	93.63	494.3	5,081	1.222
1	5	0.1819	33,100	0.02600	74.25	392.0	4,290	1.541
1	6	0.1620	26,250	0.02062	58.88	310.9	3,608	1.943
1	7	0.1443	20,820	0.01635	46.69	246.6	3,025	2.450
1	8	0.1285	16,510	0.01297	37.03	195.6	2,529	3.089
1	9	0.1144	13,090	0.01028	29.37	155.1	2,005	3.896
1	10	0.1019	10,380	0.008155	23.29	123.0	1,590	4.912
1	11	0.09074	8,234	0.006467	18.47	97.52	1,261	6.194
1	12	0.08081	6,530	0.005129	14.65	77.33	1,000	7.811
Stranded (B416-64T)								
19	5	0.910	628,900	0.4940	1,430	7,552	73,350	0.08224
19	6	0.810	498,800	0.3917	1,134	5,990	61,700	0.1037
19	7	0.721	395,500	0.3107	899.5	4,750	51,730	0.1308
19	8	0.642	313,700	0.2464	713.5	3,767	43,240	0.1649
19	9	0.572	248,800	0.1954	565.8	2,987	34,290	0.2079
19	10	0.509	197,300	0.1549	448.7	2,369	27,190	0.2622
7	5	0.546	231,700	0.1820	524.9	2,772	27,030	0.2264
7	6	0.486	183,800	0.1443	416.3	2,198	22,730	0.2803
7	7	0.433	145,700	0.1145	330.0	1,743	19,060	0.3535
7	8	0.385	115,600	0.09077	261.8	1,382	15,930	0.4458
7	9	0.343	91,650	0.07198	207.6	1,096	12,630	0.5621
7	10	0.306	72,680	0.05708	164.7	869.4	10,020	0.7088
7	11	0.272	57,640	0.04527	130.6	689.4	7,945	0.8938
7	12	0.242	45,710	0.03590	103.6	546.8	6,301	1.127
3	5	0.392	99,310	0.07800	224.5	1,186.0	12,230	0.5177
3	6	0.349	78,750	0.06185	178.1	940.2	10,280	0.6528
3	7	0.311	62,450	0.04905	141.2	745.6	8,621	0.8232
3	8	0.277	49,530	0.03890	112.0	591.3	7,206	1.038
3	9	0.247	39,280	0.03085	88.81	468.9	5,715	1.309
3	10	0.220	31,150	0.02446	70.43	371.8	4,532	1.651

Table 4-24. Galvanized-steel Wire—Weight, Breaking Strength, D-C Resistance

(ASTM Specifications A111-59 and A326-59)

Conductor size, BWG	Conductor diam., in	Conductor area, in²	Weight at 20°C* (68°F), lb/mi	Breaking strength, min., lb					D-c resistance at 20°C (68°F), max., Ω/mi				
				Grade EBB†	Grade BB†	Grade 85	Grade 135	Grade 195	Grade EBB	Grade BB	Grade 85	Grade 135	Grade 195
4	0.238	0.04449	797	2,028	2,270	.....	.....	.....	6.27	7.02			
6	0.203	0.03237	580	1,475	1,650	.....	.....	.....	8.62	9.65			
8	0.165	0.02138	383	975	1,090	.....	.....	.....	13.0	14.6			
9	0.148	0.01720	308	785	880	1,462	.....	.....	16.2	18.2	18.8		
10	0.134	0.01410	253	645	720	1,199	.....	.....	19.8	22.2	22.9		
11	0.120	0.01131	203	515	575	.....	.....	.....	24.7	27.6			
12	0.109	0.009331	167	425	475	793	1,213	1,800	29.9	33.5	34.7	38.9	38.9
14	0.083	0.005411	97.0	247	275	460	.....	.....	51.6	57.7	59.8		

* Density = 7.83 g per cu cm at 20°C.
† ASTM designation: Extra Best Best (EBB), Best Best (BB).

Table 4-25. Galvanized-steel Strand—Dimensions, Weight, Breaking Strength

(ASTM Specifications A363-65, A475-62T)

Strand diameter, in.		Stranding		Strand area, sq in.	Strand weight, lb per 1000 ft	Breaking strength, minimum, lb							
						Utilities grade*							
Nominal	Actual	No. of wires	Diameter of coated wires, in.			1	2	3	4	Common	Siemens-Martin	High strength	Extra-high strength
1¼	1.253	37	0.179	0.9311	3248	....	....	....	......	44,600	73,000	113,600	162,200
1⅛	1.127	37	0.161	0.7533	2691	....	....	....	......	36,000	58,900	91,600	130,800
1	1.001	37	0.143	0.5942	2057	....	....	....	......	28,300	46,200	71,900	102,700
1	1.000	19	0.200	0.5969	2073	....	....	....	......	28,700	47,000	73,200	104,500
⅞	0.885	19	0.177	0.4675	1581	....	....	....	......	21,900	35,900	55,800	79,700
¾	0.750	19	0.150	0.3358	1155	....	....	....	......	16,000	26,200	40,800	58,300
⅝	0.625	19	0.125	0.2332	796	....	....	....	......	11,000	18,100	28,100	40,200
⅝	0.621	7	0.207	0.2356	813	....	....	....	......	11,600	19,100	29,600	42,400
9/16	0.565	19	0.113	0.1905	637	....	....	....	......	9,640	16,100	24,100	33,700
9/16	0.564	7	0.188	0.1943	671	....	....	....	......	9,600	15,700	24,500	35,000
½	0.500	19	0.100	0.1492	504	....	....	....	......	7,620	12,700	19,100	26,700
½	0.495	7	0.165	0.1497	517	....	....	....	**25,000**	**7,400**	**12,100**	**18,800**	26,900
7/16	0.435	7	0.145	0.1156	399	....	....	....	**18,000**	**5,700**	**9,350**	**14,500**	20,800
⅜	0.360	7	0.120	0.07917	273	....	....	....	**11,500**	**4,250**	**6,950**	10,800	15,400
⅜	0.356	3	0.165	0.06415	220.3	....	....	**8500**					
5/16	0.327	7	0.109	0.06532	225	**6000**							
5/16	0.312	7	0.104	0.05946	205	....	....	....	......	**3,200**	5,350	**8,000**	**11,200**
5/16	0.312	3	0.145	0.04954	170.6	....	....	**6500**					
9/32	0.279	7	0.093	0.04755	164	**4600**	....	....	......	2,570	4,250	6,400	8,950
¼	0.240	7	0.080	0.03519	121	....	....	....	......	**1,900**	**3,150**	4,750	6,650
¼	0.259	3	0.120	0.03393	116.7	....	**3150**	**4500**					
7/32	0.216	7	0.072	0.02850	98.3	....	....	....	......	1,540	2,560	3,850	**5,400**
3/16	0.195	7	0.065	0.02323	80.3	**2400**							
3/16	0.186	7	0.062	0.02113	72.9	....	....	....	......	**1,150**	1,900	2,850	**3,990**
5/32	0.156	7	0.052	0.01487	51.3	....	....	....	......	**870**	1,470	2,140	**2,940**
⅛	0.123	7	0.041	0.00924	31.8	....	....	....	......	**540**	910	1,330	**1,830**
Elongation in 24 in.:						%							
....	.....	..	....	......	......	10	8	5	4	10	8	5	4

* Used principally by communication and power and light industries.

NOTE: Sizes and grades in bold-faced type are those most commonly used and readily available.

Table 4-26. Galvanized-steel Conductors—Weight, Breaking Strength, A-C Resistance, Reactance

Conductor size, BWG	Conductor stranding		Conductor diameter, in	Conductor area		Conductor weight, lb		Breaking strength, minimum, lb		Current, A	A-c resistance at 20°C (68°F), 60 cycles, Ω/mi	Inductive reactance (series) at 1-ft spacing, 60 cycles (X_a),* Ω/(conductor)(mi)
	No. of wires	Wire diameter, in		Cmils	In²	Per 1,000 ft	Per mi	Grade 80	Grade 130		Grade 130	Grade 130
4	3	0.138	0.297	57,132	0.04487	156	823	3,624	5,610	1	8.07	1.26
										2.5	8.20	1.27
										5	8.39	1.31
										7.5	8.60	1.34
										10	8.83	1.39
6	3	0.117	0.252	41,067	0.03225	112	590	2,604	4,295	1	11.29	1.28
										2.5	11.31	1.29
										5	11.36	1.33
										7.5	11.43	1.36
										10	11.53	1.41

* See Table 4-34.
Data copyrighted 1943 and 1945 by Indiana Steel & Wire Co.

Table 4-27. Copper Trolley Wire—Weight, Breaking Strength, D-C Resistance
(ASTM Specifications)

Conductor shape (ASTM specification)	Nominal conductor size, Mcm or AWG	Conductor diam.,* in	Conductor area, actual		Conductor weight,† lb		Tensile strength, min., psi	Elongation in 10 in, min., %	Breaking strength, min., lb	D-c resistance at 20°C (68°F),‡ ohms per 1,000 ft
			Cmils	Sq in	Per 1,000 ft	Per mi				
Round (B47-64)	300	0.5477	300,000	0.2356	908.0	4,794	46,400	4.50	10,930	0.03558
	4/0	0.4600	211,600	0.1662	640.5	3,382	49,000	3.75	8,143	0.05045
	3/0	0.4096	167,800	0.1318	507.8	2,681	51,000	3.25	6,720	0.06362
	2/0	0.3648	133,100	0.1045	402.8	2,127	52,800	2.80	5,519	0.08021
	1/0	0.3249	105,600	0.08291	319.5	1,687	54,500	2.40	4,518	0.1011
Grooved (B47-64)	350	0.620	351,200	0.2758	1,063	5,612	42,800	4.50	11,800	0.03040
	300	0.574	299,800	0.2355	907.6	4,792	44,200	4.50	10,410	0.03560
	4/0	0.482	212,000	0.1665	641.9	3,389	46,600	3.75	7,759	0.05035
	3/0	0.430	167,300	0.1314	506.4	2,674	48,500	3.25	6,373	0.06380
	2/0	0.392	137,900	0.1083	417.6	2,205	50,200	2.80	5,437	0.07741
Figure-8 (B116-64)	350	0.754 × .570	350,100	0.2750	1,060	5,597	42,800	4.50	11,770	0.03049
	4/0	0.600 × .450	211,600	0.1662	640.5	3,382	46,600	3.75	7,745	0.05044
	3/0	0.540 × .400	167,800	0.1318	508.0	2,682	48,500	3.25	6,392	0.06361
	2/0	0.480 × .352	133,100	0.1045	402.8	2,127	50,200	2.80	5,246	0.08021
	1/0	0.420 × .312	105,600	0.0829	319.5	1,687	51,800	2.40	4,294	0.1011
Figure-9 deep section (B116-64)	400	0.745 × .552	397,200	0.3120	1,202	6,347	41,300	4.50	12,890	0.02687
	350	0.707 × .496	348,900	0.2740	1,056	5,576	42,800	4.50	11,730	0.03060

* For Figure-8 and Figure-9 conductor, dimensions given are nominal height of entire section and width of lower lobe.
† Density = 8.89 g per cu cm (0.32117 lb per cu in) at 20°C.
‡ Conductivity = 97.16% IACS.

Table 4-28. Bronze Trolley Wire—Weight, Breaking Strength, D-C Resistance
(Based on ASTM Specification B9-64)

Nominal conductor size, Mcm or AWG	Conductor diam.,* in	Conductor area, actual		Conductor weight†		Tensile strength, min., lb/in² (Conductivity, %)			Elongation in 10 in, min., %
		Cmils	In²	Lb/1,000 ft	Lb/mi	40 and 55	65	80	
Round									
300	0.5477	300,000	0.2356	908.0	4,794	64,800	57,800	61,500	4.50
4/0	0.4600	211,600	0.1662	640.5	3,382	69,000	61,000	65,000	3.75
3/0	0.4096	167,800	0.1318	507.8	2,681	71,000	63,000	67,000	3.25
2/0	0.3648	133,100	0.1045	402.8	2,127	74,000	65,000	69,000	2.75
1/0	0.3249	105,600	0.08291	319.5	1,687	76,000	68,000	72,000	2.40
Grooved									
350	0.620	351,200	0.2758	1,063	5,612	62,500	56,200	59,500	4.00
300	0.574	299,800	0.2355	907.6	4,792	64,800	57,800	61,500	4.00
4/0	0.482	212,000	0.1665	641.9	3,389	69,000	61,000	65,000	3.25
3/0	0.430	167,300	0.1314	506.4	2,674	71,000	63,000	67,000	2.75
2/0	0.392	137,900	0.1083	417.6	2,205	73,000	65,000	69,000	2.25
Figure 9 deep section									
335	0.680 × 0.482	336,400	0.2642	1,020	5,386	61,500	54,000	56,800	4.00

Nominal conductor size, Mcm or AWG	Breaking strength, min., lb (Conductivity, %)			D-c resistance at 20°C (68°F), ohms per 1,000 ft (Conductivity, %)			
	40 and 50	65	80	40	55	65	80
Round							
300	15,270	13,620	14,490	0.08644	0.06285	0.05317	0.04322
4/0	11,470	10,140	10,800	0.1225	0.08910	0.07538	0.06127
3/0	9,536	8,301	8,828	0.1545	0.1124	0.09507	0.07727
2/0	7,630	6,794	7,212	0.1948	0.1417	0.1199	0.09742
1/0	6,301	5,638	5,969	0.2456	0.1786	0.1511	0.1228
Grooved							
350	17,240	15,500	16,410	0.07384	0.05370	0.04544	0.03692
300	15,260	13,610	14,480	0.08647	0.06289	0.05321	0.04324
4/0	11,490	10,160	10,820	0.1223	0.08895	0.07526	0.06115
3/0	9,329	8,278	8,804	0.1550	0.1127	0.09537	0.07749
2/0	7,906	7,040	7,473	0.1880	0.1368	0.1157	0.09402
Figure 9 deep section							
335	16,250	14,270	15,010	0.07708	0.05605	0.04742	0.03854

* For Figure 9 conductor, dimensions given are nominal height of entire section and width of lower lobe.
† Density = 8.89 g/cm³ (0.32117 lb/in³) at 20°C.

Table 4-29. Copper Wire and Cable—Electrical Characteristics

(Compiled from tables published by Westinghouse Electric Corp., "Electrical Transmission and Distribution Reference Book")

Conductor size, Awg or Mcm	Stranding		Conductor diameter, in	Breaking strength, lb	Conductor weight, lb per mile	Geometric mean radius at 60 cycles, ft	Resistance at 25°C (77°F)*				Resistance at 50°C (122°F)*				Inductive reactance (series) at 1 ft spacing (xa)†			Capacitive reactance (shunt) at 1 ft spacing (xa')‡			Current-carrying capacity at 60 cycles§ (approx), amp
	No. of wires	Wire diameter, in					D-c	25 cycles	50 cycles	60 cycles	D-c	25 cycles	50 cycles	60 cycles	25 cycles	50 cycles	60 cycles	25 cycles	50 cycles	60 cycles	
							Ohms per conductor per mile								Ohms per conductor per mile			Megohms per conductor per mile			
Solid conductors:																					
2	1	.....	0.258	3,003	1,061	0.00836	0.864	0.864	0.864	0.864	0.945	0.945	0.945	0.945	0.242	0.484	0.581	0.323	0.1614	0.1345	220
3	1	.....	0.229	2,439	841	0.00745	1.090	1.090	1.090	1.090	1.192	1.192	1.192	1.192	0.248	0.496	0.595	0.331	0.1656	0.1380	190
4	1	.....	0.204	1,970	667	0.00663	1.374	1.374	1.374	1.374	1.503	1.503	1.503	1.503	0.254	0.507	0.609	0.339	0.1697	0.1415	170
5	1	.....	0.1819	1,591	529	0.00590	1.733	1.733	1.733	1.733	1.895	1.895	1.895	1.895	0.260	0.519	0.623	0.348	0.1738	0.1449	140
6	1	.....	0.1620	1,280	420	0.00526	2.18	2.18	2.18	2.18	2.39	2.39	2.39	2.39	0.265	0.531	0.637	0.356	0.1779	0.1483	120
7	1	.....	0.1443	1,030	333	0.00468	2.75	2.75	2.75	2.75	3.01	3.01	3.01	3.01	0.271	0.542	0.651	0.364	0.1821	0.1517	110
8	1	.....	0.1285	826	264	0.00417	3.47	3.47	3.47	3.47	3.80	3.80	3.80	3.80	0.277	0.554	0.665	0.372	0.1862	0.1552	90
Stranded conductors:																					
1000	37	0.1644	1.151	43,830	16,300	0.0368	0.0585	0.0594	0.0620	0.0634	0.0640	0.0648	0.0672	0.0685	0.1666	0.333	0.400	0.216	0.1081	0.0901	1300
900	37	0.1560	1.092	39,510	14,670	0.0349	0.0650	0.0658	0.0682	0.0695	0.0711	0.0718	0.0740	0.0752	0.1693	0.339	0.406	0.220	0.1100	0.0916	1220
800	37	0.1470	1.029	35,120	13,040	0.0329	0.0731	0.0739	0.0760	0.0772	0.0800	0.0806	0.0826	0.0837	0.1722	0.344	0.413	0.224	0.1121	0.0934	1130
750	37	0.1424	0.997	33,400	12,230	0.0319	0.0780	0.0787	0.0807	0.0818	0.0853	0.0859	0.0878	0.0888	0.1739	0.348	0.417	0.226	0.1132	0.0943	1090
700	37	0.1375	0.963	31,170	11,410	0.0308	0.0836	0.0842	0.0861	0.0871	0.0914	0.0920	0.0937	0.0947	0.1759	0.352	0.422	0.229	0.1145	0.0954	1040
600	37	0.1273	0.891	27,020	9,781	0.0285	0.0975	0.0981	0.0997	0.1006	0.1066	0.1071	0.1086	0.1095	0.1799	0.360	0.432	0.235	0.1173	0.0977	940
500	37	0.1162	0.814	22,510	8,151	0.0260	0.1170	0.1175	0.1188	0.1196	0.1280	0.1283	0.1296	0.1303	0.1845	0.369	0.443	0.241	0.1205	0.1004	840
500	19	0.1622	0.811	21,590	8,151	0.0256	0.1170	0.1175	0.1188	0.1196	0.1280	0.1283	0.1296	0.1303	0.1853	0.371	0.445	0.241	0.1206	0.1005	840
450	19	0.1539	0.770	19,750	7,336	0.0243	0.1300	0.1304	0.1316	0.1323	0.1422	0.1426	0.1437	0.1443	0.1879	0.376	0.451	0.245	0.1224	0.1020	780
400	19	0.1451	0.726	17,560	6,521	0.0229	0.1462	0.1466	0.1477	0.1484	0.1600	0.1603	0.1613	0.1619	0.1909	0.382	0.458	0.249	0.1245	0.1038	730
350	19	0.1357	0.679	15,590	5,706	0.0214	0.1671	0.1675	0.1684	0.1690	0.1828	0.1831	0.1840	0.1845	0.1943	0.389	0.466	0.254	0.1269	0.1058	670
350	12	0.1708	0.710	15,140	5,706	0.0225	0.1671	0.1675	0.1684	0.1690	0.1828	0.1831	0.1840	0.1845	0.1918	0.384	0.460	0.251	0.1253	0.1044	670
300	19	0.1257	0.629	13,510	4,891	0.01987	0.1950	0.1953	0.1961	0.1966	0.213	0.214	0.214	0.215	0.1982	0.396	0.476	0.259	0.1296	0.1080	610
300	12	0.1581	0.657	13,170	4,891	0.0208	0.1950	0.1953	0.1961	0.1966	0.213	0.214	0.214	0.215	0.1957	0.392	0.470	0.256	0.1281	0.1068	610
250	19	0.1147	0.574	11,360	4,076	0.01813	0.234	0.234	0.235	0.235	0.256	0.256	0.257	0.257	0.203	0.406	0.487	0.266	0.1329	0.1108	540

Table 4-29. Copper Wire and Cable—Electrical Characteristics.—*Concluded*

Conductor size, Awg or Mcm	Stranding		Conductor diameter, in	Breaking strength, lb	Conductor weight, lb per mile	Geometric mean radius at 60 cycles, ft	Resistance at 25°C (77°F)*				Resistance at 50°C (122°F)*				Inductive reactance (series) at 1 ft spacing (x_a)†			Capacitive reactance (shunt) at 1 ft spacing (x_a')‡			Current-carrying capacity at 60 cycles§ (approx), amp
	No. of wires	Wire diameter, in					D-c	25 cycles	50 cycles	60 cycles	D-c	25 cycles	50 cycles	60 cycles	25 cycles	50 cycles	60 cycles	25 cycles	50 cycles	60 cycles	
							Ohms per conductor per mile								Ohms per conductor per mile			Megohms per conductor per mile			
Stranded conductors:—(Concluded)																					
250	12	0.1443	0.600	11,130	4,076	0.01902	0.234	0.234	0.235	0.235	0.256	0.256	0.257	0.257	0.200	0.401	0.481	0.263	0.1313	0.1094	540
4/0	19	0.1055	0.528	9,617	3,450	0.01668	0.276	0.277	0.277	0.278	0.302	0.303	0.303	0.303	0.207	0.414	0.497	0.272	0.1359	0.1132	480
4/0	12	0.1328	0.552	9,483	3,450	0.01750	0.276	0.277	0.277	0.278	0.302	0.303	0.303	0.303	0.205	0.409	0.491	0.269	0.1343	0.1119	490
4/0	7	0.1739	0.522	9,154	3,450	0.01579	0.276	0.277	0.277	0.278	0.302	0.303	0.303	0.303	0.210	0.420	0.503	0.273	0.1363	0.1136	480
3/0	12	0.1183	0.492	7,556	2,736	0.01559	0.349	0.349	0.349	0.350	0.381	0.381	0.382	0.382	0.210	0.421	0.505	0.277	0.1384	0.1153	420
3/0	7	0.1548	0.464	7,366	2,736	0.01404	0.349	0.349	0.349	0.350	0.381	0.381	0.382	0.382	0.216	0.431	0.518	0.281	0.1405	0.1171	420
2/0	7	0.1379	0.414	5,926	2,170	0.01252	0.440	0.440	0.440	0.440	0.481	0.481	0.481	0.481	0.222	0.443	0.532	0.289	0.1445	0.1205	360
1/0	7	0.1228	0.368	4,752	1,720	0.01113	0.555	0.555	0.555	0.555	0.606	0.607	0.607	0.607	0.227	0.455	0.546	0.298	0.1488	0.1240	310
1	7	0.1093	0.328	3,804	1,364	0.00992	0.699	0.699	0.699	0.699	0.765	0.765	0.765	0.765	0.233	0.467	0.560	0.306	0.1528	0.1274	270
1	3	0.1670	0.360	3,620	1,351	0.01016	0.692	0.692	0.692	0.692	0.757	0.757	0.757	0.757	0.232	0.464	0.557	0.299	0.1495	0.1246	270
2	7	0.0974	0.292	3,045	1,082	0.00883	0.881	0.882	0.882	0.882	0.964	0.964	0.964	0.964	0.239	0.478	0.574	0.314	0.1570	0.1308	230
2	3	0.1487	0.320	2,913	1,071	0.00903	0.873	0.873	0.873	0.873	0.955	0.955	0.955	0.955	0.238	0.476	0.571	0.307	0.1537	0.1281	240
3	7	0.0867	0.260	2,433	858	0.00787	1.112	1.112	1.112	1.112	1.216	1.216	1.216	1.216	0.245	0.490	0.588	0.322	0.1611	0.1343	200
3	3	0.1325	0.285	2,359	850	0.00805	1.101	1.101	1.101	1.101	1.204	1.204	1.204	1.204	0.244	0.488	0.585	0.316	0.1578	0.1315	200
4	3	0.1180	0.254	1,879	674	0.00717	1.388	1.388	1.388	1.388	1.518	1.518	1.518	1.518	0.250	0.499	0.599	0.324	0.1619	0.1349	180
5	3	0.1050	0.226	1,505	534	0.00638	1.750	1.750	1.750	1.750	1.914	1.914	1.914	1.914	0.256	0.511	0.613	0.332	0.1661	0.1384	150
6	3	0.0935	0.201	1,205	424	0.00568	2.21	2.21	2.21	2.21	2.41	2.41	2.41	2.41	0.262	0.523	0.628	0.341	0.1703	0.1419	130

* Resistance is based on conductivity = 97.3% IACS.
Resistance is increased to allow for stranding: 3-wire conductors = 1%; all others = 2%.
For resistance temperature conversion see Par. **24**.
† See Table 4-34.
‡ See Table 4-35.
§ For conductor at 75°C, air at 25°C, wind 2 fps (1.4 mph), average tarnished surface.

Table 4-30. Copper-clad Steel Cable—Electrical Characteristics

(Compiled from tables published by Westinghouse Electric Corp., "Electrical Transmission and Distribution Reference Book"; and Copperweld Steel Co.)

Nominal conductor size, in	Conductor stranding		Conductor diam., in	Conductor area, cir mils	Breaking strength, rated, lb		Conductor weight, lb per mile	Geometric mean radius at 60 cycles, average currents, ft	Resistance at 25°C (77°F),* small currents				Inductive reactance (series) at 1 ft spacing, average currents (x_a)†			Capacitive reactance (shunt) at 1 ft spacing (x_a')‡			Current-carrying capacity at 60 cycles§ (approx), amp
	No. of wires	Wire size, AWG			High strength	Extra high strength			D-c	25 cycles	50 cycles	60 cycles	25 cycles	50 cycles	60 cycles	25 cycles	50 cycles	60 cycles	
									Ohms per conductor per mile				Ohms per conductor per mile			Megohms per conductor per mile			
30% conductivity																			
7/8	19	5	0.910	628,900	56,570	66,910	9344	0.00758	0.306	0.316	0.326	0.331	0.261	0.493	0.592	0.233	0.1165	0.0971	620
13/16	19	6	0.810	498,800	45,830	55,530	7410	0.00675	0.386	0.396	0.406	0.411	0.267	0.505	0.606	0.241	0.1206	0.1005	540
23/32	19	7	0.721	395,500	37,740	45,850	5877	0.00601	0.486	0.496	0.506	0.511	0.273	0.517	0.621	0.250	0.1248	0.1040	470
21/32	19	8	0.642	313,700	31,040	37,690	4660	0.00535	0.613	0.623	0.633	0.638	0.279	0.529	0.635	0.258	0.1289	0.1074	410
9/16	19	9	0.572	248,800	25,500	30,610	3696	0.00477	0.773	0.783	0.793	0.798	0.285	0.541	0.649	0.266	0.1330	0.1109	360
5/8	7	4	0.613	292,200	24,780	29,430	4324	0.00511	0.656	0.664	0.672	0.676	0.281	0.533	0.640	0.261	0.1306	0.1088	410
9/16	7	5	0.546	231,700	20,470	24,650	3429	0.00455	0.827	0.835	0.843	0.847	0.287	0.548	0.654	0.269	0.1347	0.1122	360
1/2	7	6	0.486	183,800	16,890	20,460	2719	0.00405	1.043	1.050	1.058	1.062	0.293	0.557	0.668	0.278	0.1388	0.1157	310
7/16	7	7	0.433	145,700	13,910	16,890	2157	0.00361	1.315	1.323	1.331	1.335	0.299	0.569	0.683	0.286	0.1429	0.1191	270
3/8	7	8	0.385	115,600	11,440	13,890	1710	0.00321	1.658	1.666	1.674	1.678	0.305	0.581	0.697	0.294	0.1471	0.1226	230
11/32	7	9	0.343	91,650	9,393	11,280	1356	0.00286	2.09	2.10	2.11	2.11	0.311	0.592	0.711	0.303	0.1512	0.1260	200
5/16	7	10	0.306	72,680	7,758	9,196	1076	0.00255	2.64	2.64	2.65	2.66	0.316	0.604	0.725	0.311	0.1553	0.1294	170
..	3	5	0.392	99,310	9,262	11,860	1467	0.00457	1.926	1.931	1.936	1.938	0.289	0.545	0.654	0.293	0.1465	0.1221	220
..	3	6	0.349	78,750	7,639	9,754	1163	0.00407	2.43	2.43	2.44	2.44	0.295	0.556	0.668	0.301	0.1506	0.1255	190
..	3	7	0.311	62,450	6,291	7,922	922.4	0.00363	3.06	3.07	3.07	3.07	0.301	0.568	0.682	0.310	0.1547	0.1289	160
..	3	8	0.277	49,530	5,174	6,282	731.5	0.00323	3.86	3.87	3.87	3.87	0.307	0.580	0.696	0.318	0.1589	0.1324	140
..	3	9	0.247	39,280	4,250	5,129	580.1	0.00288	4.87	4.87	4.88	4.88	0.313	0.591	0.710	0.326	0.1629	0.1358	120
..	3	10	0.220	31,150	3,509	4,160	460.0	0.00257	6.14	6.14	6.15	6.15	0.319	0.603	0.724	0.334	0.1671	0.1392	100
40% conductivity																			
7/8	19	5	0.910	628,900	50,240	......	9344	0.01175	0.229	0.239	0.249	0.254	0.236	0.449	0.539	0.233	0.1165	0.0971	690
13/16	19	6	0.810	498,800	41,600	......	7410	0.01046	0.289	0.299	0.309	0.314	0.241	0.461	0.553	0.241	0.1206	0.1005	610
23/32	19	7	0.721	395,500	34,390	......	5877	0.00931	0.365	0.375	0.385	0.390	0.247	0.473	0.567	0.250	0.1248	0.1040	530
21/32	19	8	0.642	313,700	28,380	......	4660	0.00829	0.460	0.470	0.480	0.485	0.253	0.485	0.582	0.258	0.1289	0.1074	470
9/16	19	9	0.572	248,800	23,390	......	3696	0.00739	0.580	0.590	0.600	0.605	0.259	0.496	0.595	0.266	0.1330	0.1109	410

Table 4-30. Copper-clad Steel Cable—Electrical Characteristics.—*Concluded*

(Compiled from tables published by Westinghouse Electric Corp., "Electrical Transmission and Distribution Reference Book"; and Copperweld Steel Co.)

Nominal conductor size, in	Conductor stranding: No. of wires	Conductor stranding: Wire size, AWG	Conductor diam., in	Conductor area, cir mils	Breaking strength, rated, lb: High strength	Breaking strength, rated, lb: Extra high strength	Conductor weight, lb per mile	Geometric mean radius at 60 cycles, average currents, ft	Resistance at 25°C (77°F),* small currents (Ohms per conductor per mile): D-c	Resistance: 25 cycles	Resistance: 50 cycles	Resistance: 60 cycles	Inductive reactance (series) at 1 ft spacing, average currents (x_a)† (Ohms per conductor per mile): 25 cycles	Inductive reactance: 50 cycles	Inductive reactance: 60 cycles	Capacitive reactance (shunt) at 1 ft spacing (x_a')‡ (Megohms per conductor per mile): 25 cycles	Capacitive reactance: 50 cycles	Capacitive reactance: 60 cycles	Current-carrying capacity at 60 cycles§ (approx), amp
40 % conductivity—(*Concluded*)																			
5/8	7	4	0.613	292,200	22,310	......	4324	0.00792	0.492	0.500	0.508	0.512	0.255	0.489	0.587	0.261	0.1306	0.1088	**470**
9/16	7	5	0.546	231,700	18,510	......	3429	0.00705	0.620	0.628	0.636	0.640	0.261	0.501	0.601	0.269	0.1347	0.1122	**410**
1/2	7	6	0.486	183,800	15,330	......	2719	0.00628	0.782	0.790	0.798	0.802	0.267	0.513	0.615	0.278	0.1388	0.1157	**350**
7/16	7	7	0.433	145,700	12,670	......	2157	0.00559	0.986	0.994	1.002	1.006	0.273	0.524	0.629	0.286	0.1429	0.1191	**310**
3/8	7	8	0.385	115,600	10,460	......	1710	0.00497	1.244	1.252	1.260	1.264	0.279	0.536	0.644	0.294	0.1471	0.1226	**270**
11/32	7	9	0.343	91,650	8,616	......	1356	0.00443	1.568	1.576	1.584	1.588	0.285	0.548	0.658	0.303	0.1512	0.1260	**230**
5/16	7	10	0.306	72,680	7,121	......	1076	0.00395	1.978	1.986	1.994	1.998	0.291	0.559	0.671	0.311	0.1553	0.1294	**200**
..	3	5	0.392	99,310	8,373	......	1467	0.00621	1.445	1.450	1.455	1.457	0.269	0.514	0.617	0.293	0.1465	0.1221	**250**
..	3	6	0.349	78,750	6,934	......	1163	0.00553	1.821	1.826	1.831	1.833	0.275	0.526	0.631	0.301	0.1506	0.1255	**220**
..	3	7	0.311	62,450	5,732	......	922.4	0.00492	2.30	2.30	2.31	2.31	0.281	0.537	0.645	0.310	0.1547	0.1289	**190**
..	3	8	0.277	49,530	4,730	......	731.5	0.00439	2.90	2.90	2.91	2.91	0.286	0.549	0.659	0.318	0.1589	0.1324	**160**
..	3	9	0.247	39,280	3,898	......	580.1	0.00391	3.65	3.66	3.66	3.66	0.292	0.561	0.673	0.326	0.1629	0.1358	**140**
..	3	10	0.220	31,150	3,221	......	460.0	0.00348	4.61	4.61	4.62	4.62	0.297	0.572	0.687	0.334	0.1671	0.1392	**120**
..	3	12	0.174	19,590	2,236	......	289.5	0.00276	7.32	7.33	7.33	7.34	0.310	0.596	0.715	0.351	0.1754	0.1462	**90**

* For resistance temperature conversion see Pars. **24** and **27**.
† See Table 4-34.
‡ See Table 4-35.
§ For conductor at 125°C, air at 25°C, wind 2 fps (1.4 mph), average tarnished surface.

Table 4-31. Copper-clad-steel Copper Cable—Electrical Characteristics

(Compiled from tables published by Westinghouse Electric Corp., "Electrical Transmission and Distribution Reference Book"; and Copperweld Steel Co.)

Conductor size: Type	Hard-drawn copper equivalent: Mcm or AWG	Conductor stranding: EHS 30% Copper-clad steel wires: No.	EHS 30% Copper-clad steel wires: Diam., in	Hard-drawn copper wires: No.	Hard-drawn copper wires: Diam., in	Conductor diam., in	Breaking strength, rated, lb	Conductor weight, lb/mi	Geometric mean radius at 60 cycles, ft	Resistance at 25°C (77°F),* small currents, Ω/(conductor) (mi): D-c	25 cycles	50 cycles	60 cycles	Resistance at 50°C (122°F),* current approximately 75% capacity, Ω/(conductor) (mi): D-c	25 cycles	50 cycles	60 cycles	Inductive reactance (series) at 1-ft spacing (x_a)†, Ω/(conductor) (mi): 25 cycles	50 cycles	60 cycles	Capacitive reactance (shunt) at 1-ft spacing (x_a')‡, MΩ/(conductor) (mi): 25 cycles	50 cycles	60 cycles	Current-carrying capacity at 60 cycles§ (approx), amp
350 E	350	7	0.1576	12	0.1576	0.788	32,420	7,409	0.0220	0.1658	0.1728	0.1789	0.1812	0.1812	0.1915	0.201	0.204	0.1929	0.386	0.463	0.243	0.1216	0.1014	660
350 EK	350	4	0.1470	15	0.1470	0.735	23,850	6,536	0.0245	0.1658	0.1682	0.1700	0.1705	0.1812	0.1845	0.1873	0.1882	0.1875	0.375	0.450	0.248	0.1241	0.1034	680
300 E	300	7	0.1459	12	0.1459	0.729	27,770	6,351	0.0204	0.1934	0.200	0.207	0.209	0.211	0.222	0.232	0.235	0.1969	0.394	0.473	0.249	0.1244	0.1037	600
300 EK	300	4	0.1361	15	0.1361	0.680	20,960	5,602	0.0227	0.1934	0.1958	0.1976	0.1981	0.211	0.215	0.218	0.219	0.1914	0.383	0.460	0.254	0.1269	0.1057	610
250 E	250	7	0.1332	12	0.1332	0.666	23,920	5,292	0.01859	0.232	0.239	0.245	0.248	0.254	0.265	0.275	0.279	0.202	0.403	0.484	0.255	0.1276	0.1064	540
250 EK	250	4	0.1242	15	C.1242	0.621	17,840	4,669	0.0207	0.232	0.235	0.236	0.237	0.254	0.258	0.261	0.261	0.1960	0.392	0.471	0.260	0.1301	0.1084	540
4/0 E	4/0	7	0.1225	12	0.1225	0.613	20,730	4,479	0.01711	0.274	0.281	0.287	0.290	0.300	0.312	0.323	0.326	0.206	0.411	0.493	0.261	0.1306	0.1088	480
4/0 EK	4/0	4	0.1143	15	0.1143	0.571	15,370	3,951	0.01903	0.274	0.277	0.278	0.279	0.300	0.304	0.307	0.308	0.200	0.401	0.481	0.266	0.1331	0.1109	490
4/0 F	4/0	1	0.1833	6	0.1833	0.550	12,290	3,750	0.01558	0.273	0.280	0.285	0.287	0.299	0.309	0.318	0.322	0.210	0.421	0.505	0.269	0.1344	0.1120	470
3/0 E	3/0	7	0.1091	12	0.1091	0.545	16,800	3,552	0.01521	0.346	0.353	0.359	0.361	0.378	0.391	0.402	0.407	0.212	0.423	0.508	0.270	0.1348	0.1123	420
3/0 EK	3/0	4	0.1018	4	0.1018	0.509	12,370	3,134	0.01697	0.346	0.348	0.350	0.351	0.378	0.382	0.386	0.386	0.206	0.412	0.495	0.274	0.1372	0.1143	420
3/0 F	3/0	1	0.1632	6	0.1632	0.490	9,980	2,974	0.01388	0.344	0.351	0.356	0.358	0.377	0.388	0.397	0.401	0.216	0.432	0.519	0.277	0.1385	0.1155	410
2/0 F	2/0	1	0.1454	6	0.1454	0.436	8,094	2,359	0.01235	0.434	0.441	0.446	0.448	0.475	0.487	0.497	0.501	0.222	0.444	0.533	0.285	0.1427	0.1189	350
1/0 F	1/0	1	0.1294	6	0.1294	0.388	6,536	1,870	0.01099	0.548	0.554	0.559	0.562	0.599	0.612	0.622	0.627	0.228	0.456	0.547	0.294	0.1469	0.1224	310
2 A	2	1	0.1699	2	0.1699	0.366	5,876	1,356	0.00763	0.869	0.875	0.880	0.882	0.950	0.962	0.973	0.979	0.247	0.493	0.592	0.298	0.1489	0.1241	240
2 F	2	1	0.1026	6	0.1026	0.308	4,233	1,176	0.00873	0.871	0.878	0.884	0.885	0.952	0.967	0.979	0.985	0.230	0.479	0.575	0.310	0.1551	0.1292	230
4 A	4	1	0.1347	2	0.1347	0.290	3,938	853	0.00604	1.382	1.388	1.393	1.395	1.511	1.525	1.540	1.545	0.258	0.517	0.620	0.314	0.1572	0.1310	180
6 A	6	1	0.1068	2	0.1068	0.230	2,585	536	0.00479	2.20	2.20	2.21	2.21	2.40	2.42	2.44	2.44	0.270	0.540	0.648	0.331	0.1655	0.1379	140
8 A	8	1	0.1127	2	0.0797	0.199	2,233	392	0.00394	3.49	3.50	3.51	3.51	3.82	3.84	3.86	3.87	0.280	0.560	0.672	0.341	0.1706	0.1422	100

* For resistance-temperature conversion see Pars. **24** and **27**.
Resistance at 50°C total temperature, based on ambient of 25°C plus 25°C rise due to heating effect of current. The approximate magnitude of current necessary to produce the 25°C rise is 75% of the "approximate current-carrying capacity at 60 cycles."
† See Table 4-34.
‡ See Table 4-35.
§ For conductor at 75°C, air at 25°C, wind 2 ft/s (1.4 mi/h), average tarnished surface.

Table 4-32. Aluminum-steel (ACSR)

(Compiled from tables published by Westinghouse Electric Corp., "Electrical

Conductor size, Mcm or Awg	Hard-drawn copper equivalent,* Mcm or Awg	Stranding: Aluminum, No. of wires	Aluminum, No. of layers	Aluminum, Wire diameter, in.	Steel, No. of wires	Steel, Wire diameter, in.	Conductor diameter, in.	Breaking strength, ultimate, lb	Conductor weight, lb per mile	Geometric mean radius at 60 cycles, ft
										Multilayer
1590	1000	54	3	0.1716	19	0.1030	1.545	56,000	10,777	0.0520
1510.5	950	54	3	0.1673	19	0.1004	1.506	53,200	10,237	0.0507
1431	900	54	3	0.1628	19	0.0977	1.465	50,400	9,699	0.0493
1351	850	54	3	0.1582	19	0.0949	1.424	47,600	9,160	0.0479
1272	800	54	3	0.1535	19	0.0921	1.382	44,800	8,621	0.0465
1192.5	750	54	3	0.1486	19	0.0892	1.338	43,100	8,082	0.0450
1113	700	54	3	0.1436	19	0.0862	1.293	40,200	7,544	0.0435
1033.5	650	54	3	0.1384	7	0.1384	1.246	37,100	7,019	0.0420
954	600	54	3	0.1329	7	0.1329	1.196	34,200	6,479	0.0403
900	566	54	3	0.1291	7	0.1291	1.162	32,300	6,112	0.0391
874.5	550	54	3	0.1273	7	0.1273	1.146	31,400	5,940	0.0386
795	500	54	3	0.1214	7	0.1214	1.093	28,500	5,399	0.0368
795	500	26	2	0.1749	7	0.1360	1.108	31,200	5,770	0.0375
795	500	30	2	0.1628	19	0.0977	1.140	38,400	6,517	0.0393
715.5	450	54	3	0.1151	7	0.1151	1.036	26,300	4,859	0.0349
715.5	450	26	2	0.1659	7	0.1290	1.051	28,100	5,193	0.0355
715.5	450	30	2	0.1544	19	0.0926	1.081	34,600	5,865	0.0372
666.6	419	54	3	0.1111	7	0.1111	1.000	24,500	4,527	0.0337
636	400	54	3	0.1085	7	0.1085	0.977	23,600	4,319	0.0329
636	400	26	2	0.1564	7	0.1216	0.990	25,000	4,616	0.0335
636	400	30	2	0.1456	19	0.0874	1.019	31,500	5,213	0.0351
605	380.5	54	3	0.1059	7	0.1059	0.953	22,500	4,109	0.0321
605	380.5	26	2	0.1525	7	0.1186	0.966	24,100	4,391	0.0327
556.5	350	26	2	0.1463	7	0.1138	0.927	22,400	4,039	0.0313
556.5	350	30	2	0.1362	7	0.1362	0.953	27,200	4,588	0.0328
500	314.5	30	2	0.1291	7	0.1291	0.904	24,400	4,122	0.0311
477	300	26	2	0.1355	7	0.1054	0.858	19,430	3,462	0.0290
477	300	30	2	0.1261	7	0.1261	0.883	23,300	3,933	0.0304
397.5	250	26	2	0.1236	7	0.0961	0.783	16,190	2,885	0.0265
397.5	250	30	2	0.1151	7	0.1151	0.806	19,980	3,277	0.0278
336.4	4/0	26	2	0.1138	7	0.0885	0.721	14,050	2,442	0.0244
336.4	4/0	30	2	0.1059	7	0.1059	0.741	17,040	2,774	0.0255
300	188.7	26	2	0.1074	7	0.0835	0.680	12,650	2,178	0.0230
300	188.7	30	2	0.1000	7	0.1000	0.700	15,430	2,473	0.0241
266.8	3/0	26	2	0.1013	7	0.0788	0.642	11,250	1,936	0.0217
										Single layer
										Current approximately 75% capacity
266.8	3/0	6	1	0.2109	7	0.0703	0.633	9,645	1,802	0.00684
4/0	2/0	6	1	0.1878	1	0.1878	0.563	8,420	1,542	0.00814
3/0	1/0	6	1	0.1672	1	0.1672	0.502	6,675	1,223	0.00600
2/0	1	6	1	0.1490	1	0.1490	0.447	5,345	970	0.00510
1/0	2	6	1	0.1327	1	0.1327	0.398	4,280	769	0.00446
1	3	6	1	0.1182	1	0.1182	0.355	3,480	610	0.00418
2	4	6	1	0.1052	1	0.1052	0.316	2,790	484	0.00418
2	4	7	1	0.0974	1	0.1299	0.325	3,525	566	0.00504
3	5	6	1	0.0937	1	0.0937	0.281	2,250	384	0.00430
4	6	6	1	0.0834	1	0.0834	0.250	1,830	304	0.00437
4	6	7	1	0.0772	1	0.1029	0.257	2,288	356	0.00452
5	7	6	1	0.0743	1	0.0743	0.223	1,460	241	0.00416
6	8	6	1	0.0661	1	0.0661	0.198	1,170	191	0.00394

* Area of hard-drawn copper cable (conductivity = 97% IACS) having the same d-c resistance as that of the ACSR aluminum (conductivity = 61% IACS).

† For resistance temperature conversion see Pars. **24** and **28**.

Resistances at 50°C total temperature, based on ambient of 25°C plus 25°C rise due to heating effect of current. The approximate magnitude of current necessary to produce the 25°C rise is 75% of the "approximate current-carrying capacity at 60 cycles."

Cable—Electrical Characteristics

Transmission and Distribution Reference Book"; and Aluminum Co. of America)

Resistance at 25°C (77°F),† small currents				Resistance at 50°C (122°F)† current approximately 75% capacity				Inductive reactance (series) at 1 ft spacing (x_a)‡			Capacitive reactance (shunt) at 1 ft spacing (x_a')§			Current-carrying capacity at 60 cycles‖ (approx), amp
D-c	25 cycles	50 cycles	60 cycles	D-c	25 cycles	50 cycles	60 cycles	25 cycles	50 cycles	60 cycles	25 cycles	50 cycles	60 cycles	
Ohms per conductor per mile								Ohms per conductor per mile			Megohms per conductor per mile			
conductors														
.0587	.0588	.0590	.0591	.0646	.0656	.0675	.0684	.1495	.299	.359	.1953	.0977	.0814	1380
.0618	.0619	.0621	.0622	.0680	.0690	.0710	.0720	.1508	.302	.362	.1971	.0986	.0821	1340
.0652	.0653	.0655	.0656	.0718	.0729	.0749	.0760	.1522	.304	.365	.1991	.0996	.0830	1300
.0691	.0692	.0694	.0695	.0761	.0771	.0792	.0803	.1536	.307	.369	.201	.1006	.0838	1250
.0734	.0735	.0737	.0738	.0808	.0819	.0840	.0851	.1551	.310	.372	.203	.1016	.0847	1200
.0783	.0784	.0786	.0788	.0862	.0872	.0894	.0906	.1568	.314	.376	.206	.1028	.0857	1160
.0839	.0840	.0842	.0844	.0924	.0935	.0957	.0969	.1585	.317	.380	.208	.1040	.0867	1110
.0903	.0905	.0907	.0909	.0994	.1005	.1025	.1035	.1603	.321	.385	.211	.1053	.0878	1060
.0979	.0980	.0981	.0982	.1078	.1088	.1118	.1128	.1624	.325	.390	.214	.1068	.0890	1010
.104	.104	.104	.104	.1145	.1155	.1175	.1185	.1639	.328	.393	.216	.1078	.0898	970
.107	.107	.107	.108	.1178	.1188	.1218	.1228	.1646	.329	.395	.217	.1083	.0903	950
.117	.118	.118	.119	.1288	.1308	.1358	.1378	.1670	.334	.401	.220	.1100	.0917	900
.117	.117	.117	.117	.1288	.1238	.1288	.1288	.1660	.332	.399	.219	.1095	.0912	900
.117	.117	.117	.117	.1288	.1288	.1288	.1288	.1637	.327	.393	.217	.1085	.0904	910
.131	.131	.131	.132	.1442	.1452	.1472	.1482	.1697	.339	.407	.224	.1119	.0932	830
.131	.131	.131	.131	.1442	.1442	.1442	.1442	.1687	.337	.405	.223	.1114	.0928	840
.131	.131	.131	.131	.1442	.1442	.1442	.1442	.1664	.333	.399	.221	.1104	.0920	840
.140	.140	.141	.141	.1541	.1571	.1591	.1601	.1715	.343	.412	.226	.1132	.0943	800
.147	.147	.148	.148	.1618	.1638	.1678	.1688	.1726	.345	.414	.228	.1140	.0950	770
.147	.147	.147	.147	.1618	.1618	.1618	.1618	.1718	.344	.412	.227	.1135	.0946	780
.147	.147	.147	.147	.1618	.1618	.1618	.1618	.1693	.339	.406	.225	.1125	.0937	780
.154	.155	.155	.155	.1695	.1715	.1755	.1775	.1739	.348	.417	.230	.1149	.0957	750
.154	.154	.154	.154	.1700	.1720	.1720	.1720	.1730	.346	.415	.229	.1144	.0953	760
.168	.168	.168	.168	.1849	.1859	.1859	.1859	.1751	.350	.420	.232	.1159	.0965	730
.168	.168	.168	.168	.1849	.1859	.1859	.1859	.1728	.346	.415	.230	.1149	.0957	730
.187	.187	.187	.187	.206	.206	.206	.206	.1754	.351	.421	.234	.1167	.0973	690
.196	.196	.196	.196	.216	.216	.216	.216	.1790	.358	.430	.237	.1186	.0988	670
.196	.196	.196	.196	.216	.216	.216	.216	.1766	.353	.424	.235	.1176	.0980	670
.235	.235	.235	.235	.259	.259	.259	.259	.1836	.367	.441	.244	.1219	.1015	590
.235	.235	.235	.235	.259	.259	.259	.259	.1812	.362	.435	.242	.1208	.1006	600
.278	.278	.278	.278	.306	.306	.306	.306	.1872	.376	.451	.250	.1248	.1039	530
.278	.278	.278	.278	.306	.306	.306	.306	.1855	.371	.445	.248	.1238	.1032	530
.311	.311	.311	.311	.342	.342	.342	.342	.1908	.382	.458	.254	.1269	.1057	490
.311	.311	.311	.311	.342	.342	.342	.342	.1883	.377	.452	.252	.1258	.1049	500
.350	.350	.350	.350	.385	.385	.385	.385	.1936	.387	.465	.258	.1289	.1074	460

conductors

								Small currents, cycles			Current approximately 75% capacity, cycles						
								25	50	60	25	50	60				
0.351	0.351	0.351	0.352	0.386	0.430	0.510	0.552	.194	.388	.466	.252	.504	.605	.259	.1294	.1079	460
0.441	0.442	0.444	0.445	0.485	0.514	0.567	0.592	.218	.437	.524	.242	.484	.581	.267	.1336	.1113	340
0.556	0.557	0.559	0.560	0.612	0.642	0.697	0.723	.225	.450	.540	.259	.517	.621	.275	.1377	.1147	300
0.702	0.702	0.704	0.706	0.773	0.806	0.866	0.895	.231	.462	.554	.267	.534	.641	.284	.1418	.1182	270
0.885	0.885	0.887	0.888	0.974	1.01	1.08	1.12	.237	.473	.568	.273	.547	.656	.292	.1460	.1216	230
1.12	1.12	1.12	1.12	1.23	1.27	1.34	1.38	.242	.483	.580	.277	.554	.665	.300	.1500	.1250	200
1.41	1.41	1.41	1.41	1.55	1.59	1.66	1.69	.247	.493	.592	.277	.554	.665	.308	.1542	.1285	180
1.41	1.41	1.41	1.41	1.55	1.59	1.62	1.65	.247	.493	.592	.267	.535	.642	.306	.1532	.1276	180
1.78	1.78	1.78	1.78	1.95	1.98	2.04	2.07	.252	.503	.604	.275	.551	.661	.317	.1583	.1320	160
2.24	2.24	2.24	2.24	2.47	2.50	2.54	2.57	.257	.514	.611	.274	.549	.659	.325	.1627	.1355	140
2.24	2.24	2.24	2.24	2.47	2.50	2.53	2.55	.257	.515	.618	.273	.545	.655	.323	.1615	.1346	140
2.82	2.82	2.82	2.82	3.10	3.12	3.16	3.18	.262	.525	.630	.279	.557	.665	.333	.1666	.1388	120
3.56	3.56	3.56	3.56	3.92	3.94	3.97	3.98	.268	.536	.643	.281	.561	.673	.342	.1708	.1423	100

‡ See Table 4-34.
§ See Table 4-35.
‖ For conductor at 75°C, air at 25°C, wind 2 fps (1.4 mph), average tarnished surface.

Table 4-33. Aluminum-clad Steel Cable—Electrical Characteristics
(Compiled from tables published by Copperweld Steel Company)

Conductor stranding: No. of wires	Conductor stranding: Wire size, AWG	Geometric mean radius at 60 cycles, average currents, ft	Resistance at 25°C (77°F),* small currents, Ω/(conductor)(mi): D-c	Resistance at 25°C: 50 cycles	Resistance at 25°C: 60 cycles	Resistance at 75°C (167°F), current approx 75% of capacity, Ω/(conductor)(mi): D-c	Resistance at 75°C: 50 cycles	Resistance at 75°C: 60 cycles	Reactance at 1-ft spacing: Inductive (series) (x_a)†, Ω/(conductor)(mi): 50 cycles	Inductive: 60 cycles	Reactance at 1-ft spacing: Capacitive (shunt) (x_a')‡, MΩ/(conductor)(mi): 50 cycles	Capacitive: 60 cycles	Current-carrying capacity at 60 cycles§ (approx), A
19	5	0.004929	0.4420	0.4507	0.4507	0.5202	0.7203	0.7585	0.537	0.645	0.1165	0.0971	485
19	6	0.004387	0.5574	0.5683	0.5683	0.6559	0.8517	0.8886	0.548	0.658	0.1206	0.1005	425
19	7	0.003905	0.7030	0.7171	0.7171	0.8273	1.027	1.064	0.561	0.673	0.1248	0.1040	380
19	8	0.003478	0.8864	0.9038	0.9038	1.043	1.243	1.280	0.572	0.687	0.1289	0.1074	335
19	9	0.003098	1.118	1.140	1.140	1.315	1.518	1.554	0.584	0.701	0.1331	0.1109	295
10	10	0.002757	1.409	1.437	1.437	1.658	1.861	1.896	0.596	0.715	0.1360	0.1133	260
7	5	0.002958	1.217	1.240	1.240	1.432	1.634	1.669	0.589	0.707	0.1346	0.1122	280
7	6	0.002633	1.507	1.536	1.536	1.773	1.977	2.01	0.601	0.721	0.1388	0.1157	250
7	7	0.002345	1.900	1.937	1.937	2.24	2.44	2.47	0.612	0.735	0.1429	0.1191	220
7	8	0.002085	2.40	2.44	2.44	2.82	3.03	3.06	0.624	0.749	0.1471	0.1226	190
7	9	0.001858	3.02	3.08	3.08	3.56	3.77	3.80	0.636	0.763	0.1512	0.1260	160
7	10	0.001658	3.81	3.88	3.88	4.48	4.70	4.73	0.647	0.777	0.1552	0.1294	140
3	5	0.002940	2.78	2.78	2.78	3.27	3.52	3.56	0.589	0.707	0.1465	0.1221	170
3	6	0.002618	3.51	3.51	3.51	4.13	4.36	4.41	0.601	0.721	0.1506	0.1255	150
3	7	0.002333	4.42	4.42	4.42	5.21	5.43	5.47	0.612	0.735	0.1547	0.1289	130
3	8	0.002078	5.58	5.58	5.58	6.57	6.78	6.82	0.624	0.749	0.1589	0.1324	110

* For resistance-temperature conversion see Pars. **24** and **30**.
† See Table 4-34.
‡ See Table 4-35.
§ For conductor at 125°C, air at 25°C, wind 2 ft/s (1.4 mi/h), average tarnished surface.

Table 4-34. Inductive-reactance Spacing Factors (x_d)

Frequency, cycles	Tens \ Units	0	1	2	3	4	5	6	7	8	9
		Equivalent conductor spacing, ft									
		Ohms per conductor per mile									
25	0	......	0	0.0350	0.0555	0.0701	0.0814	0.0906	0.0984	0.1051	0.1111
	1	0.1164	0.1212	0.1256	0.1297	0.1334	0.1369	0.1402	0.1432	0.1461	0.1489
	2	0.1515	0.1539	0.1563	0.1585	0.1607	0.1627	0.1647	0.1666	0.1685	0.1702
	3	0.1720	0.1736	0.1752	0.1768	0.1783	0.1798	0.1812	0.1826	0.1839	0.1852
	4	0.1865	0.1878	0.1890	0.1902	0.1913	0.1925	0.1936	0.1947	0.1957	0.1968
50	0	......	0	0.0701	0.1111	0.1402	0.1627	0.1812	0.1968	0.2103	0.2222
	1	0.2328	0.2425	0.2513	0.2594	0.2669	0.2738	0.2804	0.2865	0.2923	0.2977
	2	0.3029	0.3079	0.3126	0.3170	0.3214	0.3255	0.3294	0.3333	0.3369	0.3405
	3	0.3439	0.3472	0.3504	0.3536	0.3566	0.3595	0.3624	0.3651	0.3678	0.3704
	4	0.3730	0.3755	0.3779	0.3803	0.3826	0.3849	0.3871	0.3893	0.3914	0.3935
60	0	......	0	0.0841	0.1333	0.1682	0.1953	0.2174	0.2361	0.2523	0.2666
	1	0.2794	0.2910	0.3015	0.3112	0.3202	0.3286	0.3364	0.3438	0.3507	0.3573
	2	0.3635	0.3694	0.3751	0.3805	0.3856	0.3906	0.3953	0.3999	0.4043	0.4086
	3	0.4127	0.4167	0.4205	0.4243	0.4279	0.4314	0.4348	0.4382	0.4414	0.4445
	4	0.4476	0.4506	0.4535	0.4564	0.4592	0.4619	0.4646	0.4672	0.4697	0.4722

Total inductive reactance $= x_a + x_d$.
See Par. **139**.

Table 4-35. Capacitive-reactance Spacing Factors (x_d')

Frequency, cycles	Tens \ Units	0	1	2	3	4	5	6	7	8	9
		Equivalent conductor spacing, ft									
		Megohms per conductor per mile									
25	0	......	0	0.0494	0.0782	0.0987	0.1146	0.1276	0.1386	0.1481	0.1565
	1	0.1640	0.1707	0.1769	0.1826	0.1879	0.1928	0.1974	0.2017	0.2058	0.2097
	2	0.2133	0.2168	0.2201	0.2233	0.2263	0.2292	0.2320	0.2347	0.2373	0.2398
	3	0.2422	0.2445	0.2468	0.2490	0.2511	0.2532	0.2552	0.2571	0.2590	0.2609
	4	0.2627	0.2644	0.2661	0.2678	0.2695	0.2711	0.2726	0.2742	0.2756	0.2771
50	0	......	0	0.0247	0.0391	0.0494	0.0573	0.0638	0.0693	0.0740	0.0782
	1	0.0820	0.0854	0.0885	0.0913	0.0940	0.0964	0.0987	0.1009	0.1029	0.1048
	2	0.1067	0.1084	0.1100	0.1116	0.1131	0.1146	0.1160	0.1173	0.1186	0.1199
	3	0.1211	0.1223	0.1234	0.1245	0.1255	0.1266	0.1276	0.1286	0.1295	0.1304
	4	0.1313	0.1322	0.1331	0.1339	0.1347	0.1355	0.1363	0.1371	0.1378	0.1386
60	0	......	0	0.0206	0.0326	0.0411	0.0478	0.0532	0.0577	0.0617	0.0652
	1	0.0683	0.0711	0.0737	0.0761	0.0783	0.0803	0.0823	0.0841	0.0858	0.0874
	2	0.0889	0.0903	0.0917	0.0930	0.0943	0.0955	0.0967	0.0978	0.0989	0.0999
	3	0.1009	0.1019	0.1028	0.1037	0.1046	0.1055	0.1063	0.1071	0.1079	0.1087
	4	0.1094	0.1102	0.1109	0.1116	0.1123	0.1129	0.1136	0.1142	0.1149	0.1155

Total capacitive reactance $= x_a' + x_d'$.
See Pars. **141** and **142**.

Table 4-36. Properties of Resistance Metals and Alloys

Material	Chemical composition	Resistivity at 20°C (68°F)	Resistance		Linear expansion		Melting point, approx, °C	Tensile strength at 20°C (68°F), min, psi	Specific gravity	Weight, lb/in³
		Ω·cmils/ft	Temperature coefficient, per °C	Temperature range, °C	Temperature coefficient, per °C	Temperature range, °C				
Driver-Harris Co., Harrison, N. J.										
Karma*	Ni 73%-Cr 20% + Al + Fe	800	....	−50–105	0.00001	20–100	1400	130,000	8.105	0.292
Nichrome*	Ni 60%-Cr 16%-balance Fe	675	0.00015	20–500	0.000017	20–1000	1350	95,000	8.247	0.2979
Nichrome V*	Ni 80%-Cr 20%	650	0.00011	20–500	0.000017	10–1000	1400	100,000	8.412	0.3039
Chromax*	Ni 35%-Cr 20%-balance Fe	600	0.00036	20–500	0.0000158	20–500	1380	70,000	7.950	0.2872
Nilvar*	Ni 36%-balance Fe	484	0.00135	20–100	0.000001	20–100	1425	70,000	8.08	0.292
Stainless type 304	Cr 18%-Ni 8%-balance Fe	438	0.00094	20–500	0.000020	0–1000	1399	100,000	7.93	0.286
142 alloy	Ni 42%-balance Fe	400	0.0012	20–500	0.0000053	20–400	1425	70,000	8.12	0.293
Advance*	Ni 43%-balance Cu	294	±0.00002	20–100	0.0000149	20–100	1210	60,000	8.9	0.321
Therlo*	Ni 29%-Co 17%-balance Fe	294	0.0038	0–100	0.000006	30–500	1450	75,000	8.36	0.302
Manganin	Mn 13%-balance Cu	290	±0.000015	15–35	0.0000187	15–35	1020	40,000	8.192	0.296
146 alloy	Ni 46%-balance Fe	275	0.0027	20–500	0.000008	25–425	1425	70,000	8.17	0.295
152 alloy (52)	Ni 51%-balance Fe	260	0.0029	20–500	0.0000095	20–500	1425	70,000	8.247	0.2979
Duranickel	Nickel plus additions	260	0.001	20–100	0.000014	20–500	1435	90,000	8.75	0.316
Midohm*	Ni 23%-balance Cu	180	0.00018	−50–150	0.0000175	20–500	1100	50,000	8.9	0.321
R-63 alloy	Mn 4%-Si 1%-balance Ni	130	0.003	20–250	0.0000152	20–500	1425	70,000	8.72	0.315
Hytemco*	Ni 72%-balance Fe	120	0.0042	20–100	0.000015	20–1000	1425	70,000	8.46	0.305
Permanickel	Nickel plus additions	100	0.0036	30–500	0.000014	30–1000	1450	90,000	8.75	0.316
90 alloy	Ni 11%-balance Cu	90	0.00049	−50–150	0.0000175	20–500	1100	35,000	8.9	0.321
Gr. A nickel	Ni 99%	60	0.0050	0–100	0.000015	20–500	1450	60,000	8.9	0.321
Lohm*	Ni 6%-balance Cu	60	0.0008	−50–150	0.000018	20–500	1100	50,000	8.9	0.321
99 alloy	Ni 99.8%	48	0.0060	0–100						
30 Alloy	Ni 2.25%-balance Cu	30	0.0015	−50–150	0.0000175	20–500	1100	30,000	8.9	0.321
Hoskins Manufacturing Co., Detroit, Mich.										
Chromel AA*	Ni 68%-Cr 20%-Fe 8%	700	0.00011	20–500	0.0000135	20–1000	1390	120,000	8.33	0.301
Chromel A*	Ni 80%-Cr 20%	650	0.00011	20–500	0.000017	10–1000	1400	100,000	8.412	0.3039
Chromel C*	Ni 60%-Cr 16%-balance Fe	675	0.00015	20–500	0.000017	20–1000	1350	95,000	8.247	0.2979
Chromel D*	Ni 35%-Cr 20%-balance Fe	600	0.00036	20–500	0.0000158	20–500	1380	70,000	7.950	0.2872
Copel*	Ni 43%-balance Cu	294	±0.00002	20–100	0.0000149	20–100	1210	60,000	8.9	0.321
Alloy 875	Cr 22.5%-Al 5.5%-balance Fe	875	0.00002	20–500	0.0000174	20–1000	1520	110,000	7.10	0.256
Alloy 815	Cr 22.5%-Al 4.6%-balance Fe	815	0.00008	20–500	0.0000159	20–1000	1520	110,000	7.25	0.262
Alloy 750	Cr 15%-Al 4%-balance Fe	750	0.00015	20–500	0.0000150	20–1000	1520	110,000	7.43	0.268

Table 4-36. Properties of Resistance Metals and Alloys.—*Concluded*

Material	Chemical composition	Resistivity at 20°C (68°F)	Resistance		Linear expansion		Melting point, approx, °C	Tensile strength at 20°C (68°F), min, psi	Specific gravity	Weight, lb/in³
		Ω·cmils/ft	Temperature coefficient, per °C	Temperature range, °C	Temperature coefficient, per °C	Temperature range, °C				
The Kanthal Corp., Bethel, Conn.										
Kanthal DR*	Fe 75%-Cr 20%-Al 4.5%-Co 0.5%	812	0.00007	20–150	0.0000119	20–100	1505	100,000	7.2	0.262
Nikrothal L*	Ni 75%-Cr 17%-balance Si + Mn	800	0.000003	20–150	0.0000126	20–100	1410	150,000	8.1	0.292
Nikrothal 6*	Ni 60%-Cr 16%-balance Fe	675	0.000140	20–100	0.000013	20–100	1350	90,000	8.25	0.298
Nikrothal 8*	Ni 80%-Cr 20%	650	0.000080	20–100	0.000014	20–100	1400	95,000	8.41	0.304
Cuprothal 294*	Ni 45%-balance Cu	294	0.00002	20–100	....	....	....	60,000	8.9	0.321
Cuprothal 180*	Ni 22%-balance Cu	180	0.00018	20–100	....	....	....	50,000	8.9	0.321
Cuprothal 90*	Ni 11%-balance Cu	90	0.00045	20–100	....	....	....	35,000	8.9	0.321
Cuprothal 60*	Ni 6%-balance Cu	60	0.0008	20–100	....	....	....	35,000	8.9	0.321
Cuprothal 30*	Ni 2%-balance Cu	30	0.0014	20–100	....	....	....	30,000	8.9	0.321
Pure Metals										
Platinum	....	63.80	0.00300	20	0.0000089	20	1773	....	21.45	0.7750
Iron	....	60.14	0.0050	20	0.0000117	20	1535	50,000	7.86	0.2840
Molybdenum	....	34.27	0.0033	20	0.000005	20	2625	100,000	10.2	0.3685
Tungsten	....	33.22	0.0045	18	0.000004	20	3410 ± 20	490,000	19.3	0.6973
Aluminum	....	16.06	0.00446	....	0.000024	20	660	35,000	2.7	0.0975
Gold	....	14.55	0.0034	20	0.0000142	20	1063	....	19.3	0.6973
Copper	....	10.37	0.00393	20	0.0000166	20	1083	35,000	8.92	0.3223
Silver	....	9.796	0.0038	20	0.0000189	20	960	....	10.5	0.3793

* Trademark.

Table 4-37. Resistance of Chromel and Copel Wire

(Compiled from tables published by Hoskins Manufacturing Co., Detroit)

Wire size, AWG	Wire diameter, in	Resistance at 20°C (68°F), ohms per foot				
		Chromel AA* 68 Ni-20 Cr-8 Fe	Chromel A* 80 Ni-20 Cr	Chromel C* 60 Ni-16 Cr	Chromel D 35 Ni-18.5 Cr	Copel 55 Cu-45 Ni
3/0	0.410	0.00417	0.00387	0.00402	0.00357	0.00175
2/0	0.365	0.00526	0.00488	0.00507	0.00451	0.00221
1/0	0.325	0.00623	0.00615	0.00639	0.00568	0.00278
1	0.289	0.00838	0.00778	0.00808	0.00718	0.00352
2	0.258	0.105	0.00977	0.01014	0.00901	0.00442
3	0.229	0.0133	0.01239	0.01287	0.0144	0.00561
4	0.204	0.0168	0.01562	0.01622	0.01442	0.00761
5	0.182	0.0211	0.01962	0.0204	0.01811	0.00888
6	0.162	0.0267	0.0248	0.0257	0.0299	0.01120
7	0.144	0.0338	0.0314	0.0326	0.0289	0.01418
8	0.128	0.0427	0.0397	0.0412	0.0366	0.01794
9	0.114	0.0539	0.0500	0.0519	0.0462	0.0226
10	0.102	0.0673	0.0625	0.0649	0.0577	0.0283
11	0.091	0.0845	0.0785	0.0815	0.0725	0.0355
12	0.081	0.107	0.0991	0.1029	0.0915	0.0448
13	0.072	0.135	0.1255	0.1302	0.1157	0.0567
14	0.064	0.171	0.1588	0.1648	0.1465	0.0718
15	0.057	0.216	0.200	0.208	0.1847	0.0905
16	0.051	0.269	0.250	0.260	0.231	0.1130
17	0.045	0.346	0.321	0.333	0.296	0.1452
18	0.040	0.438	0.406	0.422	0.375	0.1838
19	0.036	0.540	0.502	0.521	0.463	0.227
20	0.032	0.684	0.635	0.659	0.586	0.287
21	0.0285	0.862	0.800	0.831	0.739	0.362
22	0.0253	1.09	1.017	1.055	0.937	0.459
23	0.0226	1.37	1.272	1.322	1.175	0.576
24	0.0201	1.73	1.609	1.671	1.485	0.728
25	0.0179	2.19	2.03	2.11	1.873	0.918
26	0.0159	2.77	2.57	2.67	2.37	1.163
27	0.0142	3.47	3.23	3.35	2.98	1.458
28	0.0126	4.41	4.09	4.25	3.78	1.852
29	0.0113	5.48	5.09	5.29	4.70	2.30
30	0.0100	7.00	6.50	6.75	6.00	2.94
31	0.0089	8.84	8.21	8.52	7.58	3.71
32	0.0080	10.9	10.16	10.55	9.38	4.59
33	0.0071	13.9	12.90	13.39	11.90	5.83
34	0.0063	17.6	16.37	17.00	15.12	7.41
35	0.0056	22.3	20.7	21.5	19.13	9.38
36	0.0050	28.0	26.0	27.0	24.0	11.76
37	0.0045	34.6	32.1	33.3	29.6	14.52
38	0.0040	43.8	40.6	42.2	37.5	18.38
39	0.0035	57.2	53.1	55.1	49.0	24.0
40	0.0031	72.9	67.6	70.2	62.4	30.6
...	0.00275	....	86.0	89.3	....	38.9
...	0.00250	....	104.0	108.0	....	47.0
...	0.00225	....	128.5	133.4	....	58.1
...	0.00200	....	162.5	168.8	....	73.5
...	0.00175	....	212.4	221	....	96.0
...	0.00150	....	289	300	....	130.7
...	0.00140	....	332	344	....	150.0
...	0.00130	....	385	399	....	174.0
...	0.00120	....	451	469	....	204
...	0.00110	....	537	558	....	243
...	0.00100	....	650	675	....	294

Table 4-37. Resistance of Chromel and Copel Wire.—*Concluded*

°F	°C	Increase in resistance when heated, %				
		Chromel AA* 68 Ni-20 Cr-8 Fe	Chromel A* 80 Ni-20 Cr	Chromel C* 60 Ni-16 Cr	Chromel D 25 Ni-18.5 Cr	Copel 55 Cu-45 Ni
68	20	0	0	0	0	
200	93	1.0	1.6	1.9	2.9	
400	204	2.6	3.7	4.4	6.7	
600	315	5.0	5.4	7.0	10.5	
800	427	6.7	6.6	9.2	13.7	Negligible between 68 and 900°F
1000	538	6.3	7.0	10.8	16.7	
1200	649	5.2	6.4	11.2	18.7	
1400	760	5.1	6.2	11.8	20.6	
1600	871	5.6	6.4	13.0	22.3	
1800	982	6.3	7.2	14.5		
2000	1093	7.0	7.8			

* The change in resistance with increase in temperature varies slightly with the size of the wire and its rate of cooling, following the annealing operations in manufacture. Not all sizes of wire can be treated alike, and hence the cooling rates differ. A wire that has been cooled slowly has a higher resistivity than if it had been cooled quickly. Also its increase in resistance when heated will be less. The data given above and in Fig. 4-15 may be safely used for most design problems. If extreme accuracy is desired, a test unit should be made and tested under operating conditions and the proper length determined by experiment.

ELECTRICAL RESISTANCE ALLOYS

198. Electrical resistance alloys are made in a great variety of compositions and are commonly marketed under trade names by Driver-Harris Co., Harrison, N.J.; Hoskins Manufacturing Co., Detroit, Mich.; W. B. Driver Co., Newark, N.J.; C. O. Jellieff Co., Southport, Conn.; Alloy Metal Wire Div., H. K. Porter Co., Inc., East Orange, N.J.; and the Kanthal Corp., Stamford, Conn. The principal elements employed as constituents in these alloys are carbon, copper, chromium, iron, manganese, nickel, silicon, and zinc. Many of these alloys are also heat-resistant and corrosion-resistant. A wide range of resistivities can be obtained in commercial resistance alloys.

199. Nickel alloys for electrical resistances, rheostats, and electric heating units may be divided into several groups, classified by composition. Nickel-chromium alloys have high electrical resistivity and low temperature coefficient, together with high resistance to oxidation and alteration at high temperatures. Ferro-nickel alloys have somewhat lower resistivity and resistance to oxidation and are considerably lower in cost. Copper-nickel alloys have lower resistivity than nickel-chromium alloys and inferior heat-resisting properties but have practically negligible temperature coefficient of resistivity at ordinary temperatures, which is useful in precision measuring instruments. Copper-nickel-zinc alloys, or nickel silver, are the oldest for electrical purposes but have been largely displaced by the alloys before mentioned. Also see The Selection of Material for Electrical Heating Alloys, *ASM Metals Handbook*, 8th ed., Vol. 1, p. 620, and Electrical Resistance Alloys for Instruments and Controls, *ASM Metals Handbook*, 8th ed., Vol. 1, p. 797.

200. Chromium-nickel alloys are covered by ASTM Specification B344-65, which gives the following requirements:

Composition and properties	80 Ni-20 Cr alloy	60 Ni-16 Cr alloy	35 Ni-20 Cr alloy
Composition, %:			
Nickel	Remainder	57 (min)	34–37
Chromium	19–21	14–18	18–21
Manganese	2.5 (max)	1.0 (max)	1.0 (max)
Carbon	0.15 (max)	0.15 (max)	0.15 (max)
Silicon	0.75–1.5	0.75–1.5	1.0–3.0
Sulfur	0.01 (max)	0.01 (max)	0.01 (max)
Iron	1.0 (max)	Remainder	Remainder
Elongation in 10 in, min:			
No. 39 AWG and larger, %	20	20	20
No. 40 to 44, %	10	10	10
Nominal resistivity at 77°F (25°C):			
Round wire, Ω (cmil, ft)	650	675	610
Round wire, μΩ·cm	108	112.2	101.7
Ribbon, Ω (mil², ft)	510	530	478
Ribbon, μΩ·cm	108	112.2	101.7
Average resistance change with temperature:			
2000°F (1093°C)	1.014	1.102	1.235
1800°F (982°C)	1.007	1.090	1.225
1600°F (871°C)	1.003	1.082	1.214
1400°F (760°C)	1.003	1.073	1.196
1200°F (649°C)	1.007	1.069	1.178
1000°F (538°C)	1.019	1.066	1.156
900°F (482°C)	1.023	1.063	1.144
800°F (427°C)	1.022	1.058	1.130
600°F (315°C)	1.017	1.044	1.100
400°F (204°C)	1.011	1.026	1.065
200°F (93°C)	1.004	1.010	1.026
77°F (25°C)	1.000	1.000	1.000

201. Ferrous alloys for electrical resistances are chiefly of the ferrochromium and ferrosilicon types, with small additional contents of other elements such as carbon, copper, manganese, and nickel. Many of these alloys are of the stainless-steel or rustless-iron types.

202. Commercial Alloys. Tables 4-36 and 4-37 show various commercial resistance metals and alloys listed under their trade names and the physical and electrical characteristics of each. Also included are similar data for pure metals.

203. Properties of Resistance Metals and Alloys. See Table 4-36.

204. Resistance of Chromel and Copel Wire. See Table 4-37 and Fig. 4-15*a*.

205. Resistance of Cr-Al-Fe Resistance Alloys. See Table 4-38 and Fig. 4-15*b*.

206. Current-Temperature Characteristics of Nichrome-V Wire. See Table 4-39.

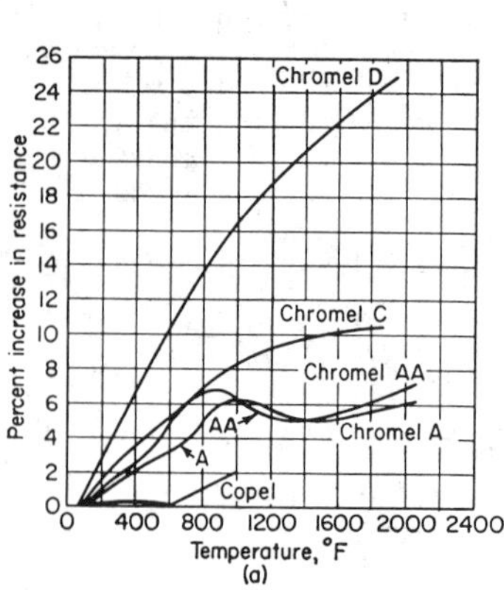

FIG. 4-15*a*. Temperature-resistance relationship for Chromel AA, A, C, D and Copel wire.

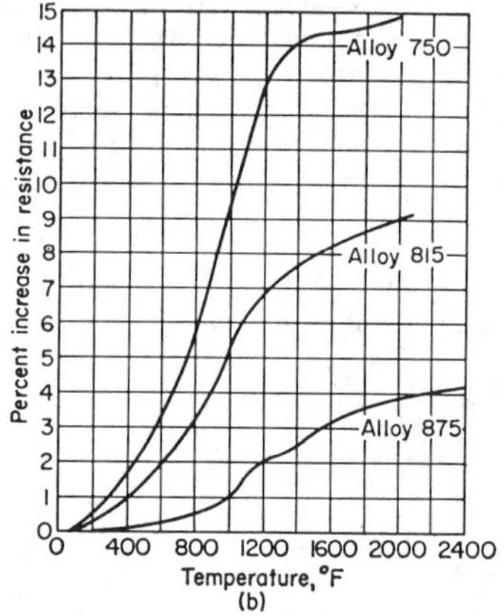

FIG. 4-15*b*. Temperature-resistance relationship for Cr-Al-Fe resistance wire.

Table 4-38. Resistance of Cr-Al-Fe Resistance Alloys
(Compiled from tables published by Hoskins Manufacturing Co., Detroit)

Wire size, AWG	Wire diameter, in	Resistance at 20°C (68°F), Ω/ft		
		Alloy 875 22.5 Cr–5.5 Al–72 Fe	Alloy 815 22.5 Cr–4.6 Al–72.9 Fe	Alloy 750 15 Cr–4 Al–81 Fe
3/0	0.410	0.005206	0.004848	0.004461
2/0	0.365	0.006569	0.006119	0.005631
1/0	0.325	0.008286	0.007718	0.007102
1	0.289	0.010476	0.009758	0.00898
2	0.258	0.01314	0.01224	0.01126
3	0.229	0.01668	0.01554	0.01430
4	0.204	0.02102	0.01958	0.01802
5	0.182	0.02640	0.02460	0.02264
6	0.162	0.03333	0.03104	0.02856
7	0.144	0.04219	0.03930	0.03617
8	0.128	0.05341	0.04974	0.04577
9	0.114	0.06731	0.06270	0.05770
10	0.102	0.08411	0.07834	0.07209
11	0.091	0.1056	0.09841	0.09056
12	0.081	0.1333	0.1242	0.11430
13	0.072	0.1690	0.1574	0.1448
14	0.064	0.2138	0.1991	0.1832
15	0.057	0.2691	0.2507	0.2307
16	0.051	0.3364	0.3133	0.2883
17	0.045	0.4320	0.4024	0.3703
18	0.040	0.5468	0.5093	0.4687
19	0.036	0.6751	0.6288	0.5787
20	0.032	0.8544	0.7958	0.7323
21	0.0285	1.077	1.003	0.9230
22	0.0253	1.369	1.275	1.1730
23	0.0226	1.712	1.595	1.468
24	0.0201	2.165	2.017	1.856
25	0.0179	2.731	2.544	2.341
26	0.0159	3.462	3.224	2.967
27	0.0142	4.346	4.048	3.725
28	0.0126	5.506	5.128	4.719
29	0.0113	6.853	6.383	5.874
30	0.0100	8.751	8.150	7.500
31	0.0089	11.05	10.29	9.469
32	0.0080	13.68	12.74	11.720
33	0.0071	17.38	16.18	14.890
34	0.0063	22.04	20.53	18.89
35	0.0056	27.90	25.98	23.91
36	0.0050	35.00	32.60	30.00
37	0.0045	43.21	40.24	37.03
38	0.0040	54.68	50.93	46.87
39	0.0035	71.43	66.53	61.22
40	0.0031	91.05	84.80	78.04
...	0.00275	....	107.8	99.28
...	0.00250	....	130.4	120.10
...	0.00225	....	161.0	148.30
...	0.00200	....	203.7	187.60
...	0.00175	....	....	245.10
...	0.00150	....	....	333.30

°F	°C	Increase in resistance when heated, %		
		Alloy 875 22.5 Cr–5.5 Al–72 Fe	Alloy 815 22.5 Cr–4.6 Al–72.9 Fe	Alloy 750 15 Cr–4 Al–81 Fe
68	20	0	0	0
200	93	0	0.3	0.6
400	204	0.1	1.0	1.8
600	315	0.2	1.9	3.3
800	427	0.5	3.2	5.8
1000	538	1.0	5.2	9.3
1200	649	2.0	6.9	12.6
1400	760	2.5	7.7	14.1
1600	871	3.1	8.1	14.4
1800	982	3.5	8.6	14.6
2000	1093	3.8	9.0	14.9
2200	1204	4.0		
2400	1316	4.2		

Table 4-39. Current-temperature Characteristics of Nichrome-V Wire
(Compiled from tables published by Driver-Harris Co., Harrison, N.J.)

Wire size, AWG	Wire diam., in.	Arbor diam., in.	Current, amp (approx.), to produce temperature of								
			400 F 204 C	600 F 316 C	800 F 427 C	1000 F 538 C	1200 F 649 C	1400 F 760 C	1600 F 871 C	1800 F 982 C	2000 F 1093 C
Straight wire*											
1	0.289	.....	78.20	109.0	140.0	177.0	219.0	268.0	322.0	378.0	434.0
2	0.258	.....	65.41	91.59	117.5	148.7	184.1	225.1	270.4	317.3	364.5
3	0.229	.....	54.71	76.94	98.57	124.9	154.7	189.1	227.1	266.4	306.2
4	0.204	.....	45.76	64.64	82.71	104.9	130.0	158.9	190.7	223.6	257.3
5	0.182	.....	38.28	54.30	69.40	88.09	109.3	133.4	160.1	187.6	216.0
6	0.162	.....	32.02	45.62	58.23	73.99	91.86	112.1	133.5	157.5	181.4
7	0.144	.....	26.78	38.33	48.85	62.15	77.21	94.16	112.9	132.2	152.4
8	0.128	.....	22.40	32.20	41.00	52.20	64.89	79.10	94.80	111.0	128.0
9	0.114	.....	18.80	27.00	34.60	43.90	54.60	66.90	80.00	94.00	108.0
10	0.102	.....	16.11	22.95	29.35	37.17	46.22	56.66	67.79	79.62	91.41
11	0.091	.....	13.82	19.52	24.90	31.47	39.12	47.99	57.44	67.44	77.36
12	0.081	.....	11.84	16.59	21.12	26.65	33.12	40.65	48.67	57.12	65.48
13	0.072	.....	10.15	14.10	17.92	22.57	28.03	34.64	41.24	48.38	55.42
14	0.064	.....	8.70	11.98	15.20	19.11	23.73	29.16	34.95	40.98	46.91
15	0.057	.....	7.465	10.19	12.89	16.18	20.08	24.70	29.62	34.71	39.70
16	0.051	.....	6.390	8.66	10.94	13.70	17.00	20.90	25.10	29.40	33.60
17	0.045	.....	5.53	7.48	9.51	11.80	14.50	17.60	21.10	24.60	28.10
18	0.040	.....	4.964	6.53	8.23	10.17	12.48	15.11	18.06	21.00	24.03
19	0.036	.....	4.26	5.70	7.12	8.77	10.74	12.97	15.46	17.92	20.55
20	0.032	.....	3.73	4.97	6.17	7.56	9.24	11.13	13.23	15.30	17.57
21	0.0285	.....	3.27	4.34	5.34	6.52	7.95	9.55	11.32	13.06	15.03
22	0.0253	.....	2.87	3.78	4.62	5.62	6.85	8.20	9.69	11.15	12.85
23	0.0226	.....	2.52	3.30	4.00	4.85	5.89	7.04	8.30	9.52	10.99
24	0.020	.....	2.21	2.88	3.46	4.18	5.06	6.04	7.10	8.12	9.40
25	0.0179	.....	1.92	2.52	3.02	3.58	4.32	5.18	6.10	7.25	8.04
26	0.0159	.....	1.66	2.17	2.62	3.12	3.76	4.49	5.27	6.22	6.90
27	0.0142	.....	1.43	1.87	2.28	2.73	3.26	3.89	4.55	5.33	5.92
28	0.0126	.....	1.23	1.62	1.98	2.38	2.84	3.37	3.93	4.57	5.09
29	0.0113	.....	1.06	1.40	1.72	2.08	2.47	2.92	3.39	3.92	4.37
30	0.010	.....	0.915	1.21	1.50	1.81	2.14	2.53	2.93	3.36	3.75
31	0.0089	.....	0.789	1.04	1.30	1.58	1.86	2.19	2.53	2.88	3.22
32	0.008	.....	0.680	0.900	1.13	1.38	1.62	1.90	2.18	2.47	2.76
33	0.0071	.....	0.580	0.780	0.980	1.18	1.40	1.62	1.87	2.12	2.35
34	0.0063	.....	0.511	0.686	0.853	1.02	1.21	1.39	1.60	1.81	2.01
35	0.0056	.....	0.451	0.603	0.742	0.884	1.04	1.20	1.38	1.55	1.72
36	0.005	.....	0.397	0.530	0.646	0.766	0.902	1.04	1.18	1.33	1.47
37	0.0045	.....	0.350	0.467	0.561	0.663	0.777	0.892	1.01	1.14	1.26
38	0.004	.....	0.309	0.410	0.489	0.574	0.671	0.768	0.869	0.971	1.09
39	0.0035	.....	0.272	0.361	0.425	0.497	0.579	0.662	0.746	0.830	0.923
40	0.0031	.....	0.240	0.310	0.370	0.430	0.500	0.570	0.640	0.710	0.790
Helical coils†											
11	0.091	0.190	7.982	12.94	18.22	23.73	29.49	36.31	43.48	50.99	58.77
12	0.081	0.190	6.92	11.05	15.40	19.93	24.41	29.75	35.90	40.57	46.20
13	0.072	0.190	6.222	9.989	13.98	18.14	22.40	26.84	31.40	36.05	40.80
14	0.064	0.190	5.28	8.32	11.49	14.75	18.10	22.29	26.70	31.31	36.10
15	0.057	0.190	4.31	6.96	9.78	12.74	15.80	19.08	22.49	25.96	29.50
16	0.051	0.190	3.48	5.51	7.64	9.84	12.10	15.25	18.56	22.14	25.90
17	0.045	0.190	3.06	4.78	6.58	8.42	10.30	12.77	15.28	18.13	21.00
18	0.040	0.128	2.51	3.93	5.41	6.93	8.48	10.41	12.48	14.53	16.70
19	0.036	0.128	2.20	3.46	4.78	6.14	7.53	9.25	11.06	12.94	14.91
20	0.032	0.128	1.63	2.64	3.72	4.84	6.01	7.44	8.96	10.54	12.20
21	0.0285	0.128	1.35	2.19	3.09	4.03	5.01	6.23	7.52	8.89	10.30
22	0.0253	0.128	1.17	1.82	2.48	3.15	3.84	4.93	6.13	7.42	8.81
23	0.0226	0.128	1.02	1.60	2.19	2.81	3.43	4.32	5.28	6.29	7.37
24	0.020	0.128	0.991	1.52	2.05	2.59	3.13	3.82	4.55	5.30	6.07
25	0.0179	0.128	0.859	1.31	1.77	2.24	2.71	3.32	3.95	4.61	5.29
26	0.0159	0.128	0.686	1.08	1.50	1.93	2.37	2.87	3.39	3.92	4.47
27	0.0142	0.128	0.583	0.896	1.22	1.54	1.87	2.28	2.70	3.05	3.60
28	0.0126	0.128	0.410	0.663	0.933	1.22	1.51	1.87	2.26	2.56	3.09
29	0.0113	0.128	0.328	0.538	0.742	0.965	1.19	1.53	1.86	2.24	2.61
30	0.010	0.128	0.277	0.448	0.631	0.822	1.02	1.32	1.64	1.90	2.38

Table 4-39. Current-temperature Characteristics of Nichrome-V Wire.—*Concluded*

Wire size, AWG	Wire diam., in.	Arbor diam., in.	Current, amp (approx.), to produce temperature of								
			400 F 204 C	600 F 316 C	800 F 427 C	1000 F 538 C	1200 F 649 C	1400 F 760 C	1600 F 871 C	1800 F 982 C	2000 F 1093 C
31	0.0089	0.062	0.205	0.337	0.480	0.632	0.790	1.04	1.31	1.61	1.94
32	0.0080	0.062	0.165	0.286	0.420	0.572	0.733	0.938	1.16	1.40	1.66
33	0.0071	0.062	0.144	0.250	0.373	0.508	0.652	0.830	1.023	1.230	1.45
34	0.0063	0.062	0.139	0.219	0.303	0.389	0.478	0.585	0.696	0.812	0.932
35	0.0056	0.062	0.083	0.148	0.224	0.309	0.401	0.502	0.627	0.763	0.886
36	0.005	0.032	0.070	0.129	0.199	0.278	0.366	0.468	0.579	0.747	0.824
37	0.0045	0.032	0.063	0.115	0.175	0.245	0.327	0.402	0.485	0.580	0.674
38	0.0040	0.032	0.050	0.108	0.163	0.222	0.284	0.350	0.416	0.482	0.568
39	0.0035	0.032	0.044	0.081	0.126	0.176	0.231	0.288	0.340	0.412	0.466
40	0.0031	0.032	0.035	0.065	0.105	0.149	0.196	0.236	0.275	0.310	0.354

NOTE: For further discussion of resistance-coil design, see W. A. Gatward, Practical Heating Element Designs, *Elec. Mfg.*, May, 1940, p. 72.
* Applies only to straight wires stretched horizontally in free air.
† Coiled on arbors, stretched to twice the close-wound lengths, in open air.

FUSIBLE METALS AND ALLOYS

207. Fusible alloys having melting points in the range from about 60 to 200°C are made principally of bismuth, cadmium, lead, and tin in various proportions. Many of these alloys have been known under the names of their inventors (see index of alloys in "International Critical Tables," Vol. 2), but typical compositions and melting points are shown in Table 4-40.

208. Compositions and Melting Points of Fusible Alloys. See Table 4-40.

Table 4-40. Compositions and Melting Points of Fusible Alloys
(International Critical Tables, Vol. 2, p. 391)

Chemical composition, %					Melting point, deg C	Chemical composition, %					Melting point, deg C
Bi	Pb	Sn	Cd	Hg		Bi	Pb	Sn	Cd	Hg	
20	20	..	..	60	20	..	32	50	18	..	145
50	27	13	10	..	72	50	50	..	..	..	160
52	40	..	8	..	92	15	41	44	..	..	164
53	32	15	..	..	96	33	..	67	..	..	166
54	26	..	20	..	103	20	..	80	..	..	200
29	43	28	..	..	132	..	..	..	..	..	

Compositions and melting points are approximate.

209. Fuse metals for electric fuses of the open-link enclosed and expulsion types are ordinarily made of some low-fusible alloy; aluminum also is used to some extent. The resistance of the fuse causes dissipation of energy, liberation of heat, and rise of temperature. Sufficient current will obviously melt the fuse and thus open the circuit if the resultant arc is self-extinguishing. Metals which volatilize readily in the heat of the arc are to be preferred to those which leave a residue of globules of hot metal. The rating of any fuse depends critically upon its shape, dimensions, mounting, enclosure, and any other factors which affect its heat-dissipating capacity.

210. Fusing currents of different kinds of wire were investigated by W. H. Preece,[11] who developed the formula

$$I = ad^{3/2} \qquad (4\text{-}34)$$

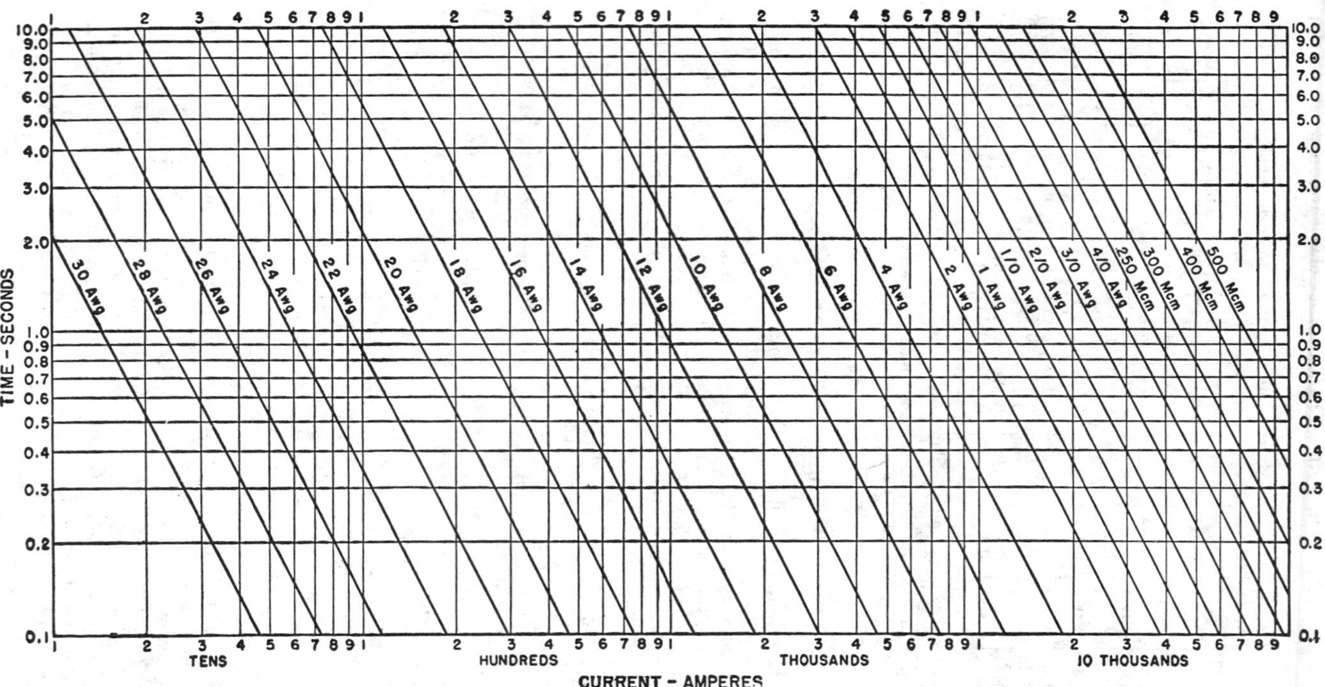

FIG. 4-16. Fusing current time for copper conductors.

where I = fusing current in amperes, d = diameter of the wire in inches, a = a constant depending upon the material. He found the following values for a:

Copper	10,244	Iron	3,148
Aluminum	7,585	Tin	1,642
Platinum	5,172	Alloy (2Pb–1Sn)	1,318
German silver	5,230	Lead	1,379
Platinoid	4,750		

Although this formula has been used to a considerable extent in the past, it gives values that usually are erroneous in practice, because it is based on the assumption that all heat loss is due to radiation. A formula of the general type

$$I = kd^n \tag{4-35}$$

can be used with accuracy if k and n are known for the particular case (material, wire size, installation conditions, etc.).

211. Fusing current-time for copper conductors and connections may be determined by an equation developed by I. M. Onderdonk:[12]

$$33\left(\frac{I}{A}\right)^2 S = \log\left(\frac{T_m - T_a}{234 + T_a} + 1\right) \tag{4-36}$$

$$I = A\sqrt{\frac{\log\left(\dfrac{T_m - T_a}{234 + T_a} + 1\right)}{33S}} \tag{4-37}$$

where I = current in amperes, A = conductor area in circular mils, S = time current applied in seconds, T_m = melting point of copper in degrees centigrade, T_a = ambient temperature in degrees centigrade.

Copper Conductors. E. R. Stauffacher[12] has prepared a chart of the fusing current for sizes from 30 AWG to 500,000 cmils from 0.1 to 10 s (see Fig. 4-16). This chart is based on the assumptions that (1) radiation may be neglected owing to the short time involved, i.e., 10 s; (2) resistance of 1 cm cube of copper at 0°C is 1.589 μΩ; (3) temperature-resistance coefficient of copper at 0°C is 1/234; (4) melting point of copper is 1083°C; and (5) ambient temperature is 40°C.

For most practical purposes, Eq. (4-37) also may be applied where the melting temperature T_m of solder or other materials used in making connections is the determining factor; e.g.:

Soldered Connections. Select a value of T_m corresponding to the melting temperature for the composition of tin-lead alloy used. This may be determined by test or approximated from Fig. 4-17 prepared from the "Smithsonian Physical Tables" (e.g., T_m = 183°C for 70:30 solder).

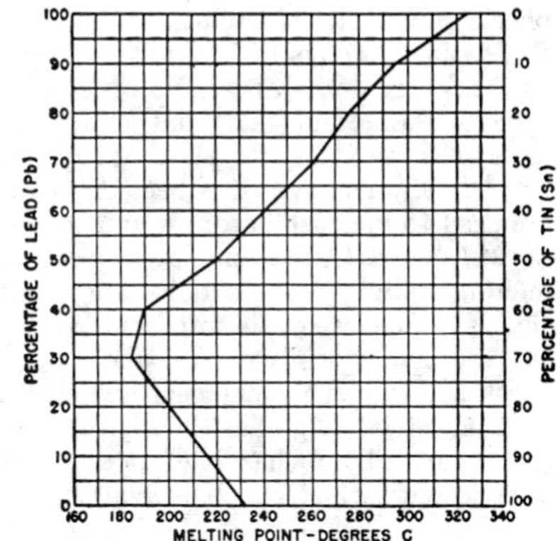

FIG. 4-17. Melting temperatures (T_m) for solder.

Brazed Connections. A reasonable value of T_m is 450°C.

Bolted Connections. Generally accepted value of T_m is 250°C.

MISCELLANEOUS METALS AND ALLOYS

212. Contact metals may be grouped into three general classifications:

Hard metals, which have high melting points, e.g., tungsten and molybdenum. Contacts of these metals are employed usually where operations are continuous or very frequent and current has nominal value of 5 to 10 A. Hardness to withstand mechanical wear and high melting point to resist arc erosion and welding are their outstanding advantages. Tendency to form high-resistance oxides is a disadvantage, but this can be overcome by several methods, such as using high contact-closing force, a hammering or wiping action, and a properly balanced electrical circuit.

Highly conductive metals, of which silver is the best for both electric current and heat. Its disadvantages are softness and a tendency to pit and transfer. In sulfurous atmosphere, a resistant sulfide surface will form on silver, which results in high contact-surface resistance. These disadvantages are overcome usually by alloying.

Noncorroding metals, which for the most part consist of the noble metals, such as gold and the platinum group. Contacts of these metals are used on sensitive devices, employing extremely light pressures or low currents in which clean contact surfaces are essential. Because most of these metals are soft, they are usually alloyed.

The metals commonly used are tungsten, molybdenum, platinum, palladium, gold, silver, and their alloys. Alloying materials are copper, nickel, cadmium, iron, and the rarer metals such as iridium and ruthenium. Some are prepared by powder metallurgy.

Commercial grades are available under trade names or alloy numbers from Baker & Co., Newark, N.J.; Fansteel Metallurgical Corp., North Chicago (Fasaloy, Fastell); North American Phillips Co., New York (Elmet); P. R. Mallory & Co., Indianapolis; and others.

213. Tungsten (W) is a hard, dense, slow-wearing metal, a good thermal and electrical conductor, characterized by its high melting point and freedom from sticking or welding. It is manufactured in several grades having various grain sizes.

214. Molybdenum (Mo) has contact characteristics about midway between tungsten and fine silver. It often replaces either metal where greater wear resistance than that of silver or lower contact-surface resistance than that of tungsten is desired.

215. Platinum (Pt) is one of the most stable of all metals under the combined action of corrosion and electrical erosion. It has a high melting point and does not corrode, and surfaces remain clean and low in resistance under most adverse atmospheric and electrical conditions.

Platinum alloys of iridium (Ir), ruthenium (Ru), silver (Ag), or other metals are used to increase hardness and resistance to wear.

216. Palladium (Pd) has many of the properties of platinum and frequently is used as an alternate for platinum and its alloys.

Palladium alloys of silver (Ag), ruthenium (Ru), nickel (Ni), and other metals are used to increase hardness and resistance to wear.

217. Gold (Au) is similar to platinum in corrosion resistance but has a much lower melting point. Gold and its alloys are ductile and easily formed into a variety of shapes. Because of its softness it is usually alloyed.

Gold alloys of silver (Ag) and other metals are used to impart hardness and improve resistance to mechanical wear and electrical erosion.

218. Silver (Ag) has the highest thermal and electrical conductivity (110%, IACS) of any metal. It has low contact-surface resistance, since its oxide decomposes at approximately 300°F. It is available commercially in three grades:

Grade	Typical composition, %	
	Silver	Copper
Fine silver	99.95+	
Sterling silver	92.5	7.5
Coin silver	90	10

Fine silver is used extensively under low contact pressure where sensitivity and low contact-surface resistance are essential or where the circuit is operated infrequently.

Sterling and coin silvers are harder than fine silver and resist transfer at low voltage (6 to 8 V) better than fine silver. Since their contact-surface resistance is greater than that of fine silver, higher contact-closing forces should be used.

Silver alloys of copper (Cu), nickel (Ni), cadmium (Cd), iron (Fe), carbon (C), tungsten (W), molybdenum (Mo), and other metals are used to improve hardness, resistance to wear and arc erosion, and for special applications.

219. High-melting-point Metals. Table 4-41 shows the melting-point range of all the metals.

Table 4-41. Melting Points of Metals
(Compiled from tables published by Fansteel Metallurgical Corp.)

Melting point, deg C	Metal	Melting point, deg C	Metal
3500–3400		1500–1400	Silicon, nickel, cobalt, yttrium
3400–3300	**Tungsten**	1400–1300	Beryllium
3300–3200		1300–1200	Manganese
3200–3100	Rhenium	1200–1100	
3100–3000		1100–1000	Gold, copper
3000–2900	**Tantalum**	1000– 900	Praseodymium, germanium, radium, silver
2900–2800		900– 800	Cerium, arsenic, neodymium, calcium, lanthanum
2800–2700		800– 700	Barium, strontium
2700–2600	**Molybdenum**	700– 600	Antimony, magnesium, aluminum
2600–2500	Osmium, iridium	600– 500	
2500–2400	**Columbium**	500– 400	Zinc, tellurium
2400–2300	Boron	400– 300	Cadmium, thallium, lead
2300–2200	Hafnium	300– 200	Selenium, tin, bismuth
2200–2100	Zirconium	200– 100	Indium, lithium
2100–2000		100– 0	Cesium, gallium, rubidium, potassium, sodium
2000–1900	Chromium, ruthenium, rhodium	0 to −100	Mercury
1900–1800	Thorium		
1800–1700	Vanadium, titanium, platinum		
1700–1600	Uranium		
1600–1500	Iron, palladium		

Table 4-42 contains the physical properties of four of the metals in the highest melting-point range.

220. Melting Points of Metals. See Table 4-41.

221. Properties of High-melting-point Metals. See Table 4-42.

222. Selenium is a nonmetallic element chemically resembling sulfur and tellurium and occurs in several allotropic forms varying in specific gravity from 4.3 to 4.8. It melts at 217°C and boils at 690°C. At 0°C it has a resistivity of approximately 60,000 Ω·cm. The dielectric constant ranges from 6.1 to 7.4. It has the peculiar property that its resistivity decreases upon exposure to light; the resistivity in darkness may be anywhere from 5 to 200 times the resistivity under exposure to light (see paper by W. J. Hammer, *Trans. AIEE*, 1903, Vol. 21, pp. 372–393).

Table 4-42. Properties of High-melting-point Metals
(Fansteel Metallurgical Corp.)

Property	Tungsten	Tantalum	Molybdenum	Columbium
Specific gravity at 20 C	19.3	16.6	10.2	8.57
Electrical resistivity at 20 C, microhm-cm	5.5	12.4	5.17	13.1
Electrical resistance at 20 C, ohm (mil, ft)	33.1	74.6	31.1	79.0
Temperature coefficient of resistance at 20 C	0.0051	0.0036	0.0047	0.00395
Tensile strength, unannealed wire, lb per sq in.	200,000	130,000	105,000	96,000
Coefficient of linear expansion, per deg C	4.3×10^{-6}	6.5×10^{-6}	4.9×10^{-6}	7.1×10^{-6}
Specific heat at 20 C, cal per g atom per deg C	6.18	6.512	6.24	6.012
Melting point, deg C	3410	2996	2625	2415

BIBLIOGRAPHY

223. General References on Conductor Materials. Many specific references appear in the preceding paragraphs. The following are references on the general subject of conductors which will be found of assistance in solving cable engineering problems:

REEVES, H. J. Copper and Aluminum Cable Fusing Time-current Formulas (letter to editor); *Trans. AIEE (Elec. Eng.)*, October, 1935, Vol. 54, No. 10, p. 1127.

EWAN, A. W. A Set of Curves for Skin Effect in Isolated Tubular Conductors; *Gen. Elec. Rev.*, April, 1930, Vol. 33, No. 4, p. 249.

Central Station Engineers "Electrical Transmission and Distribution Reference Book," 4th ed.; Westinghouse Electric & Manufacturing Co., East Pittsburgh, Pa., 1950.

Current-carrying Capacity. See Par. **144.**

Conductor Specifications. See Table 4-9.

Committee on Insulated Conductors Classified Bibliography on Insulated Conductors; *AIEE Special Publ.* S73, June, 1954.

FOOTNOTE REFERENCES

224. Numbered items following are those to which there are corresponding references by number throughout the preceding text.

[1] American Standard Definitions of Electrical Terms, ASA C42, 1941.

[2] Ratified at the meeting of the International Electrotechnical Commission held in Berlin, Sept. 1 to 6, 1913; see *Trans. AIEE*, 1913, Vol. 32, Pt. 2, p. 2156; also *NBS Handbook* 100.

[3] Specifications for Hard-drawn Aluminum Wire for Electrical Purposes, ASTM Designation B230-60.

[4] "International Critical Tables," Vol. II, pp. 456, 460, 478, and Vol. VI, p. 136.

[5] "Metals Handbook," 8th ed.; American Society for Metals, 1961, Vol. 1, Properties and Selection, p. 1007.

[6] Definitions and General Standards for Wires and Cables; AIEE/IEEE Standard 30-1944 [ASA C8.1-44(R1953)].

[7] SMITH, C. S. Thermal Conductivity of Copper Alloys, I. Copper-Zinc Alloys; *Trans. AIMME, Inst. Metals Div.*, 1930, Vol. 89, p. 84.

[8] JOHNSON, J. B. "Materials of Construction"; New York, John Wiley & Sons, Inc., 1925.

[9] Made by Indiana Steel & Wire Co., Muncie, Ind.

[10] Method proposed by W. A. Lewis, first published in "Symmetrical Components," C. F. Wagner and R. D. Evans, New York, McGraw-Hill Book Company, 1933, Appendix VII.

[11] PREECE, W. H. On the Heating Effects of Electric Currents; *Proc. Roy. Soc. (London)*, April, 1884; December, 1887; April, 1888.

[12] STAUFFACHER, E. R. Short-time Current-carrying Capacity of Copper Wires; *Gen. Elec. Rev.*, June, 1928, p. 326.

CARBON AND GRAPHITE

(Prepared by The Carbon Section, National Electrical Mauufacturers Association)

225. Forms of Carbon. Carbon occurs in two forms, amorphous and crystalline. The crystalline forms include **diamond** and **graphite,** the latter also being known as **plumbago.** The amorphous forms include charcoal, coke, lampblack, bone black; coal is an impure variety of amorphous carbon. The specific gravity of carbon in the diamond state is 3.47 to 3.56; graphite, 2.10 to 2.266; charcoal, 0.28 to 0.57; coke, 1.0 to 1.7; gas carbon, 1.88; lampblack, 1.7 to 1.8.

Most carbon used for electrical purposes is made from a mixture of powdered carbon and/or graphite (such as lampblack and petroleum coke) and binders (such as pitch and resins), which are mixed into a homogeneous mass, extruded or molded, and then baked. When the mixture is baked to approximately 900°C with the air excluded, the volatile part of the binding material is driven off and the remaining binder is carbonized. The resulting product can be converted into electrographite by furnacing it in the absence of oxygen to a temperature of not less than 2200°C.

226. Properties of Carbon. See Table 4-43.

227. Properties of Manufactured Carbon and Graphite. See Table 4-44.

228. Resistivity. See Tables 4-43 to 4-46 inclusive.

229. Temperature Coefficient of Resistance. Carbon exhibits a decreasing electrical resistivity and a decreasing thermal resistivity with rising temperature. Graphite exhibits but little change in electrical resistivity, tending downward with rising temperature, but its thermal resistivity increases slightly with rising temperature (see Fig. 4-18).

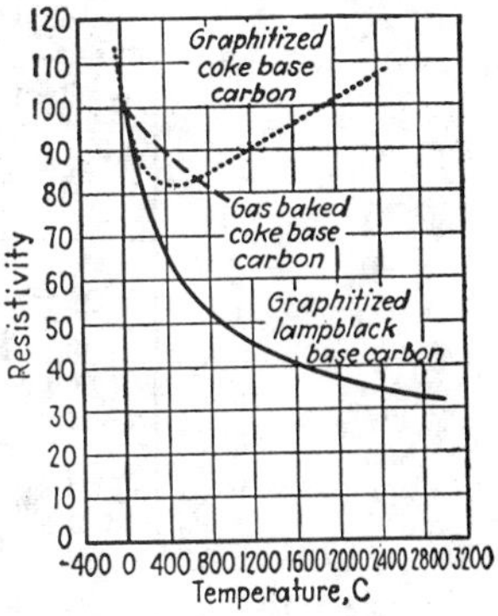

230. Properties of Carbon and Graphite Electrodes. See Table 4-45.

231. Carbon-brush Applications. The term carbon brush is used to designate all types of sliding electrical contacts that contain any appreciable percentage of carbon or graphite in their composition. Other ingredients may be metals and suitable binders.

Carbon and graphite brushes for commutator-type machines and collector rings of a-c machines are made in various grades with appropriate characteristics for the types of service, including atmospheric conditions and load cycles.

Fig. 4-18. Temperature-resistance relationship of carbon. 100 = resistivity at 20°C.

Functions to be performed by carbon brushes on electrical machines are these:

Carbon brushes on a slip-ring machine have only to provide a suitable sliding electrical connection between the line and the rotor.

Carbon brushes on a d-c machine have two entirely distinct functions to perform: (1) They must carry the current into and out of the commutator. (2) They participate in the reversal of all or part of the current in the armature coils during the time they are short-circuited. Performance of both functions simultaneously complicates the problem of design. In order that the current in the short-circuited armature coils may be reversed without arcing or sparking, there must be an appreciable resistance in the contact between the brush and commutator. The amount of the resistance depends upon the magnitude of the reactance voltage.

The carbon brush in its various types and compositions provides the characteristics needed for an ideal sliding electrical connection, namely, good conductivity, high contact resistance, low coefficient of friction and high durability.

Table 4-43. Properties of Carbon

Material	Specific gravity	Electrical resistivity at 20 C (68 F)		Specific heat at temperatures					
		Ohm-in.	Ohm-cm	26–76 C	26–282 C	26–538 C	36–902 C	47–1193 C	56–1450 C
				G-cal per deg C					
Carbon, coke base, gas calcined	1.98–2.10*	0.0014–0.0018	0.0035–0.0046	0.168	0.200	0.199	0.315	0.352	0.387
Carbon, coke base, graphitized	2.20–2.24	0.0003–0.0005	0.0007–0.0012	0.168	0.200	0.199	0.315	0.352	0.387
Carbon, lampblack base, gas calcined	1.80–1.85*	0.0023–0.0032	0.0058–0.0081						
Carbon, lampblack base, graphitized	1.98–2.08	0.0018–0.0026	0.0046–0.0066						
Anthracite, gas calcined	1.79*								
Anthracite, electric calcined	1.90–1.97*								
Graphite, pure	2.25	0.0003–0.0005	0.0008–0.0013	0.165	0.195	0.234	0.324	0.350	0.390
Diamond	3.51	....	....	0.160	0.315	0.415			

* Dependent on source, character of bond, and degree of calcination.

Table 4-44. Properties of Manufactured Carbon and Graphite*

Material	Form	Apparent density		Strength			Elastic modulus, psi $\times 10^5$	Electrical resistivity at 20 C (68 F), ohm-in.	Thermal expansion per deg F $\times 10^{-7}$
		Grams per cu cm	Lb per cu ft	Tensile, psi	Compressive, psi	Transverse, psi			
Carbon rods	½–3 in. diam.	1.55/1.7	96/105	800/2000	7100/10000	2250/4000	9.0/15.0	0.0013/.0018	12/16
	3–8 in. diam.	1.54/1.65	96/100	650/1100	2800/5000	1200/2200	5.5/8.0	0.0013/.0016	12/15
	9–15 in. diam.	1.52/1.54	94/97	400/500	2000/2600	900/1200	5.0/6.0	0.0012/.0015	12/15
	16–20 in. diam.	1.52/1.54	94/97	400/500	2000/2600	700/1000	5.0/6.0	0.0012/.0016	12/15
	30–44 in. diam.	1.52/1.54	94/97	400/500	1800/2400	700/1000	4.0/5.0	0.0020/.0030	11/14
Carbon blocks	3 × 3–6 × 6	1.5/1.7	96/105	800/1600	6000/10000	1700/4000	7.5/15.0	0.0014/.0020	12/16
	6 × 6–20 × 20	1.5/1.7	95/100	500/800	2000/4000	1200/2000	6.0/7.0	0.0014/.0017	13/15
	Larger blocks	1.5/1.6	94/96	350/450	1700/2200	800/1500	4.0/5.0	0.0020/.0030	11/13
Graphite rods	½–3 in. diam.	1.53/1.64	96/105	800/1400	4000/6000	2000/3500	8.0/10.0	0.00030/.00040	4/12
	3–8 in. diam.	1.54/1.62	95/100	800/1200	4000/5000	1800/3000	7.5/9.5	0.00030/.00040	5/12
	9–15 in. diam.	1.52/1.58	94/98	550/800	3000/4000	1500/2000	6.5/8.0	0.00035/.00040	7/12
	16–20 in. diam.	1.52/1.58	94/97	450/600	3000/3600	1400/1800	6.5/7.5	0.00037/.00041	8/12
Graphite blocks	3 × 3–6 × 6	1.55/1.7	96/100	700/1200	3000/4500	1700/4000	8.0/9.5	0.00035/.00041	5/12
	6 × 6–20 × 20	1.54/1.6	96/100	700/1200	3000/4500	1500/3500	7.0/8.5	0.00035/.00041	6/12
	Larger blocks	1.52/1.55	94/97	550/800	3000/3800	1500/3000	6.5/7.5	0.00036/.00042	8/12

* Characteristics shown are typical values and will differ with individual grades and manufacturers.

For discussion on electric brushes and commutation, see the bibliography under that title issued by the National Electrical Manufacturers Association. See also NEMA Standards CB1, 1961.

232. Brush Characteristics. See Table 4-46.

233. Resistance of Arc-lamp Carbons. The resistance of ½-in diameter by 12-in enclosed-arc carbons varies from 0.012 to 0.015 Ω/lin in. Other sizes down to ⅜ in diameter vary according to their cross-sectional areas. The resistance of a ⅝-in-diameter projector carbon varies from 0.009 to 0.011 Ω/lin in. The ½- and 9/16-in-diameter carbons vary according to their cross-sectional areas.

All high-grade forms of carbon, such as those used in the manufacture of searchlight carbons and also enclosed-arc carbons, may be given the value of about 0.002 Ω·in. Flame-arc carbon material such as is used in the homogeneous electrodes varies from 0.004 to 0.006 Ω·in. All the above values are for ordinary room temperatures.

Table 4-45. Properties of Carbon and Graphite Electrodes

Material	Electrical resistivity 20 C (68 F), ohm-in.	Comparative section area for same voltage drop	Weight, lb per cu in.	Tensile strength lengthwise, psi	Temp. of oxidation in air, deg C
Graphite electrodes	0.00032	1.0	0.0574	450/1200	640
Small high-quality electrically baked carbon electrodes	0.00124	3.8	0.0564	1000/1500	500
Small gas-baked carbon electrodes	0.00161	4.4	0.0560	1000/1500	500
Large carbon electrodes	0.0022	6.8	0.0580	700/1200	500

234. Resistances of carbon contacts vary with pressure, current, and time (see results of investigation published by A. L. Clark in *Phys. Rev.*, January, 1913). The property of variable contact resistance with varying contact pressure is very useful in carbon-pile resistors, which can be varied over a wide range, from practically open circuit to very low resistance with manipulation of the pressure on the pile.

Carbon contacts for relays have the same characteristics as low-resistance carbon brushes (Par. **231** and Table 4-46).

235. Properties of Coke-base Carbon Materials. The apparent density ranges from 1.53 to 1.64 g/cm³ or approximately 100 lb/ft³, the true density ranges from 1.98 to 2.10 g/cm³, and the porosity is approximately 20 to 30%. The approximate tensile strength is 400 to 2,500 lb/in², and the crushing strength is 2,000 to 10,000 lb/in². The thermal conductivity is 0.00786 cal/(cm)(°C). The coefficient of thermal expansion from 20 to 1000°C is 7.3×10^{-6}/°C. The specific heat between 26 and 282°C is 0.200 gcal/°C. The electrical resistivity is 0.0014 to 0.0030 Ω·in. The sublimation point is approximately 3500°C.

236. Properties of Coke-base Electrographitic Materials. The scleroscope hardness ranges from 12 to 75, and the materials can be easily machined. The resistance in general first decreases as the temperature increases and becomes a minimum at 500°C, or about 80% of its value at room temperature; then as the temperature increases further, the resistance also increases; and at 1200°C the resistance becomes the same as at room temperature; from 1200 to about 1800°C the resistance continues to rise, becoming about 9% above normal at 1800°C; above this temperature it is practically constant. The specific heat ranges from 0.166 at 76°C to 0.390 at 1450°C. There is no substantial oxidation at temperatures below 500°C; at a bright-red heat in an oxidizing atmosphere the material is rapidly consumed. The coefficient of thermal longitudinal expansion from 20 to 600°C is 2.7×10^{-6}/°C, and the transverse expansion is 3.7×10^{-6}/°C. The apparent density is 1.56 to 1.70 g/cm³ or approximately 97.5 lb/ft³. Its true density is 2.20 to 2.24 g/cm³, and the porosity ranges from 23 to 30%. The tensile strength ranges from 450 to 1,400 lb/in², and the crushing strength ranges from 3,000 to 6,000 lb/in².

237. Thermal expansion of coke-base graphite is different in the longitudinal and transverse directions and increases somewhat with temperature. The average longitudinal coefficient is 1.9×10^{-6}/°C in the range from 20 to 100°C and 2.7×10^{-6}

Table 4-46. Range of Properties and Characteristics of Typical Brush Materials*

Physical properties and characteristics	Types of brush materials					
	Carbon	Carbon graphite	Graphite	Graphite (resin bonded)	Electrographitic	Metal graphite
Specific resistance, ohm-in	0.0010–0.0035	0.0005–0.0025	0.0003–0.0018	0.0005–0.050	0.0004–0.0035	0.000002–0.0001
Scleroscope hardness	45–85	40–75	10–30	10–35	12–75	7–35
Transverse strength, psi	3,500–10,000	3,000–7,500	1,000–4,500	1,000–5,000	1,500–10,000	2,500–10,000
Current-carrying capacity, amp per sq in	35–45	45–50	40–60	10–60	35–70	75–150
Contact drop†	Medium–high	Low–medium	Low–medium	Medium–very high	Medium–high	Very low–low
Coefficient of friction‡	Medium–high	Low–medium	Low–medium	Low–medium	Low–medium	Very low–medium
Abrasiveness (cleaning action)	Pronounced	Medium–pronounced	Slight	Slight	Slight	Slight–medium
Peripheral speed, fpm	2,000–4,000	3,000–5,000	9,000–12,000	9,000–12,000	3,000–9,000	3,000–6,000

* See NEMA Standard CB1, 1955.
† These terms have the following meanings: high, over 2.5 volts; medium, 1.8 to 2.5 volts; low, 1.0 to 1.8 volts; very low, below 1.0 volt.
‡ These terms have the following meanings: high, over 0.26; medium, 0.20 to 0.26; low, 0.15 to 0.20; very low, 0.15 and below.

in the range from 20 to 600°C; the corresponding average transverse coefficients are 2.9×10^{-6} and 3.7×10^{-6}/°C (see *NBS Tech. Paper* 335).

238. Therapeutic carbons contain various salts or compounds of metals such as cerium, calcium, cobalt, and strontium and are employed in arc lamps which produce ultraviolet radiation. Such types of radiation have certain sterilizing properties and can also be made to simulate the properties of sunlight.

239. Photographic carbons are made in forms specially adapted for arc-lamp projectors of various types and have important applications in the field of motion pictures.

240. Silicon carbide or carborundum is a crystalline compound of silicon and carbon, SiC, which is very hard and abrasive. Resistance rods made principally of this material are employed as electric heating units in various types of furnaces, bakers, ovens, and heaters.

MAGNETIC MATERIALS

By Irvin L. Cooter

241. Terms, Symbols, and Definitions. The following definitions of terms and symbols relating to magnetics have been selected from ASTM Standards A340:

Aging Coefficient. The percentage change in a specific magnetic property resulting from a specified aging treatment.

Note. The aging treatments usually specified are:

a. 100 h at 150°C, or

b. 600 h at 100°C.

Aging, Magnetic. The change in the magnetic properties of a material resulting from metallurgical change. This term applies whether the change results from a continued normal or a specified accelerated aging condition.

Note. This term implies a deterioration of magnetic materials for electronic and electrical applications, unless otherwise specified.

Ampere-turn. Unit of mmf in the rationalized mksa system. One ampere-turn equals $4\pi/10$ or 1.257 Gb.

Ampere-turn per Meter. Unit of magnetizing force (magnetic field strength) in the rationalized mksa system. One ampere-turn per meter is $4\pi \times 10^{-3}$ or 0.01257 Oe.

Coercive Force, H_c. The (d-c) magnetizing force at which the magnetic induction is zero when the material is in a symmetrically cyclically magnetized condition.

Coercive Force, Intrinsic, H_{ci}. The (d-c) magnetizing force at which the intrinsic induction is zero when the material is in a symmetrically cyclically magnetized condition.

Coercivity, H_{cs}. The maximum value of coercive force.

Core Loss, Incremental, P_Δ. The core loss in a magnetic material when subjected simultaneously to a d-c biasing magnetizing force and an alternating magnetizing force.

Core Loss, Specific, $P_{B;f}$. The power expended per unit mass of magnetic material in which there is a symmetrical, sinusoidally varying induction of a specified maximum value, B, at a specified frequency f.

Core Loss (Total), P_c. The power expended in a magnetic specimen in which there is a cyclically alternating induction.

Note. Measurements of core loss are normally made with sinusoidally alternating induction, or the results are corrected for deviations from the sinusoidal condition.

Curie Temperature, T_c. The temperature above which a ferromagnetic material becomes paramagnetic.

Cyclically Magnetized Condition, CM. A magnetic material is in a cyclically magnetized condition when, after having been subjected to a sufficient number of identical cycles of magnetizing force, it follows identical hysteresis loops on successive cycles which are not necessarily symmetrical with respect to the origin of the axes.

Demagnetization Curve. That portion of a normal (d-c) hysteresis loop which lies in the second or fourth quadrant, that is, between the residual induction point B_r and the coercive force point H_c. Points on this curve are designated by the co-ordinates B_d and H_d.

Demagnetizing Force, H_d. A magnetizing force (on the demagnetization curve) applied in such a direction as to reduce the induction in a magnetized body (see Demagnetization Curve).

Diamagnetic Material. A material whose relative permeability is less than unity.

Eddy-current Loss, Normal, P_e. That portion of the core loss which is due to induced currents circulating in the magnetic material subject to an SCM excitation.

Electrical Sheet or Strip. A term used commercially to designate a flat-rolled iron-silicon alloy used for its magnetic properties.

Energy-product Curve, Magnetic. The curve obtained by plotting the product of the corresponding coordinates, B_d and H_d, of points on the demagnetization curve as abscissa against the induction B_d as ordinates.

NOTE 1. The maximum value of the energy product, $(B_dH_d)_m$, corresponds to the maximum value of the external energy.

NOTE 2. The demagnetization curve is plotted to the left of the vertical axis and usually the energy-product curve to the right.

Ferromagnetic Material. A material which, in general, exhibits the phenomena of hysteresis and saturation, and of which the permeability is dependent on the magnetizing force.

Flux-current Loop, Normal. A dynamic loop of flux ϕ versus current I, or induction B versus magnetizing force H, that is obtained by using a symmetrically alternating magnetizing force to produce an SCM excitation in the core material.

NOTE. The area of the loop is proportional to the sum of the static hysteresis loss and all dynamic losses.

Form Factor, ff. The ratio of the rms value of a periodically alternating quantity to its average absolute value.

NOTE. For a sinusoidal variation the form factor is

$$\frac{\pi}{2\sqrt{2}} = 1.1107$$

Gauss (plural gausses). The unit of magnetic induction in the cgs electromagnetic system. The gauss is equal to 1 Mx/cm² or 10^{-4} T [see Magnetic Induction (Flux Density)].

Gilbert. The unit of mmf in the cgs electromagnetic system. The gilbert is an mmf of $10/4\pi$ At (see Magnetomotive Force).

Hysteresis Loop, Normal. A closed curve obtained with a ferromagnetic material by plotting (usually to rectangular coordinates) corresponding d-c values of magnetic induction B for ordinates and magnetizing force H for abscissa when the material is passing through a complete cycle between equal definite limits of either magnetizing force $\pm H_m$ or magnetic induction $\pm B_m$. In general the normal hysteresis loop has inversion symmetry with respect to the origin of the B and H axes, but this may not be true for special materials.

Hysteresis Loss, Incremental, $P_{\Delta h}$. The hysteresis loss occurring in a ferromagnetic material when subjected simultaneously to a constant biasing magnetic force and a cyclic excitation.

Hysteresis Loss, Normal, P_h. (1) The power expended in a ferromagnetic material, as a result of hysteresis, when the material is subjected to an SCM excitation. (2) The energy loss per cycle in a magnetic material as a result of magnetic hysteresis when the induction is cyclic (but not necessarily periodic).

Hysteresis, Magnetic. The property of a ferromagnetic material revealed by the lack of correspondence between the changes in induction resulting from increasing magnetizing force and from decreasing magnetizing force.

Induction, Incremental, B_Δ. One-half the algebraic difference of the extreme values of the magnetic induction during a cycle in a magnetic material that is subjected simultaneously to a biasing magnetizing force and a symmetrically cyclically varying magnetizing force. Twice the incremental induction is indicated by the symbol ΔB, thus,

$$B_\Delta = \frac{\Delta B}{2} \qquad (4\text{-}38)$$

Induction, Intrinsic, B_i. The vector difference between the magnetic induction in a magnetic material and the magnetic induction that would exist in a vacuum under the influence of the same magnetizing force. This is expressed by the equation

$$B_i = B - \Gamma_m H \tag{4-39}$$

Γ_m is the magnetic constant.

NOTE. In the cgs-em system $B_i/4\pi$ is often called magnetic polarization.

Induction, Normal, B. The maximum induction, in a magnetic material that is in a symmetrically cyclically magnetized condition.

NOTE. Normal induction is a magnetostatic parameter usually measured by ballistic methods.

Induction, Remanent, B_d. The magnetic induction that remains in a magnetic circuit after the removal of an applied mmf.

NOTE. If there are no air gaps or other inhomogeneities in the magnetic circuit the remanent induction B_d will equal the residual induction B_r; if air gaps or other inhomogeneities are present, B_d will be less than B_r.

Induction, Residual, B_r. The magnetic induction corresponding to zero magnetizing force in a magnetic material that is in a symmetrically cyclically magnetized condition.

Induction, Saturation, B_s. The maximum intrinsic induction possible in a material.

Induction Curve, Normal. A curve of a previously demagnetized specimen depicting the relation between normal induction and corresponding ascending values of magnetizing force. This curve starts at the origin of the B and H axes.

Lamination Factor (Space Factor, Stacking Factor) S. The ratio of the calculated volume of a stack of laminations to the measured volume under a given pressure. It is usually expressed as a percentage.

NOTE. The calculated volume is the equivalent solid volume based on the weight of the material and on assumed or measured density.

Magnetic Constant (Permeability of Space) Γ_m. The dimensional scalar factor that relates the mechanical force between two currents to their intensities and geometrical configurations. That is,

$$\mathbf{dF} = \mathbf{\Gamma}_m \mathbf{I}_1 \mathbf{I}_2 \, \mathbf{dl}_1 \times (\mathbf{dl}_2 \times \mathbf{r}_1)/nr^2 \tag{4-40}$$

where dF is the element of force of a current element $I_1\, dl_1$ on another current element $I_2\, dl_2$ at a distance r; r_1 = unit vector in the direction from dl_1 to dl_2; and n = dimensionless factor. The symbol n is unity in unrationalized systems and 4π in rationalized systems.

NOTE 1. The numerical value of Γ_m depends upon the system of units employed. In the cgs-em system $\Gamma_m = 1$; in the rationalized mksa system $\Gamma_m = 4\pi \times 10^{-7}$ H/m.

NOTE 2. The magnetic constant expresses the ratio of magnetic induction to the corresponding magnetizing force at any point in a vacuum and therefore is sometimes called the permeability of space μ_v.

NOTE 3. The magnetic constant times the relative permeability is equal to the absolute permeability.

$$\mu_{abs} = \Gamma_m \mu_r \tag{4-41}$$

Magnetic Field of Induction. A state of a region such that a conductor carrying a current in the region would be subjected to a mechanical force and an emf would be induced in an elementary loop rotated with respect to the field in such a manner as to change the flux linkage.

Magnetic Flux ϕ. The product of the area of a surface (or cross section), A, and the component of the magnetic induction, B, normal to the plane of the surface, when B is uniformly distributed.

$$\phi = \mathbf{B} \cdot \mathbf{A} \tag{4-42}$$

where ϕ = magnetic flux, B = magnetic induction, and A = area of the surface.

NOTE 1. If the magnetic induction is not uniformly distributed over the surface, the flux ϕ is the surface integral of the normal component of B over the area.

$$\phi = \iint_s \mathbf{B} \cdot \mathbf{dA} \tag{4-43}$$

NOTE 2. Magnetic flux is a scalar and has no direction.

Magnetic Flux Density B. See Magnetic Induction (Flux Density).

Magnetic Induction (Flux Density) B. That magnetic vector quantity which at any point in a magnetic field is measured either by the mechanical force experienced by an element of electric current at the point or by the emf induced in an elementary loop during any change in flux linkage with the loop at the point.

NOTE 1. If the magnetic induction B is uniformly distributed and normal to a surface or cross section, then the magnetic induction is

$$B = \frac{\phi}{A} \tag{4-44}$$

where B = magnetic induction, ϕ = total flux, and A = area.

NOTE 2. B_{in} is the instantaneous value of the magnetic induction and B_m is the maximum value of the magnetic induction.

Magnetic Polarization, J. In the cgs-em system, the intrinsic induction divided by 4π is sometimes called magnetic polarization or magnetic dipole moment per unit volume.

Magnetizing Force (Magnetic Field Strength) H. That magnetic vector quantity at a point in a magnetic field which measures the ability of electric currents or magnetized bodies to produce magnetic induction at the given point.

NOTE 1. The magnetizing force H may be calculated from the current and the geometry of certain magnetizing circuits. For example, in the center of a uniformly wound, long solenoid

$$H = C\frac{NI}{l} \tag{4-45}$$

where H = magnetizing force, C = constant whose value depends on the system of units, N = number of turns, I = current, and l = axial length of the coil.,

If I is expressed in amperes and l is expressed in centimeters, then $C = 4\pi/10$ in order to obtain H in the cgs-em units, oersteds. If I is expressed in amperes and l is expressed in meters, then $C = 1$ in order to obtain H in the mksa units, ampere-turns per meter.

NOTE 2. The magnetizing force H at a point in air may be calculated from the measured value of induction at the point by dividing this value by the magnetic constant Γ_m.

Magnetizing Force, Incremental, H_Δ. One-half the algebraic difference of the maximum and minimum values of the magnetizing force during a cycle in a magnetic material that is subjected simultaneously to a biasing magnetizing force and a symmetrical, periodically varying magnetizing force.

Twice the incremental magnetizing force is indicated by the symbol ΔH. Thus

$$H_\Delta = \frac{\Delta H}{2} \tag{4-46}$$

Magnetomotive Force $\mathfrak{F}$. The line integral of the magnetizing force around any flux loop in space.

$$\mathfrak{F} = \oint H \cdot dl \tag{4-47}$$

where $\mathfrak{F}$ = mmf, H = magnetizing force, and dl = unit length along the loop.

NOTE. The mmf is proportional to the net current linked with any closed loop of flux or closed path.

$$\mathfrak{F} = CNI \tag{4-48}$$

where $\mathfrak{F}$ = mmf, N = number of turns linked with the loop, I = current in amperes, and C = constant whose value depends on the system of units. In the cgs system $C = 4\pi/10$. In the mksa system $C = 1$.

Maxwell. The unit of magnetic flux in the cgs electromagnetic system. One maxwell equals 10^{-8} Wb (see Magnetic Flux).

NOTE

$$e = -N\, d\phi/dt \times 10^{-8} \tag{4-49}$$

where e = induced instantaneous emf in volts, $d\phi/dt$ = time of change of flux in maxwells per second, and N = number of turns surrounding the flux, it being assumed that each turn is linked with all the flux.

Mksa (Giorgi) Rationalized System of Units. System for measuring physical quantities in which the basic units are the meter, kilogram, and second, and the ampere is a derived unit defined by assigning the magnitude $4\pi \times 10^{-7}$ to the rationalized magnetic constant (sometimes called the permeability of space).

NOTE 1. The electrical units of this system were formerly called the "practical" electrical units.

NOTE 2. In this system dimensional analysis is customarily used with the four independent (basic) dimensions: length, mass, time, current.

Oersted. The unit of magnetizing force (magnetic field strength) in the cgs electromagnetic system. One oersted equals an mmf of 1 Gb/cm of flux path. One oersted equals $1{,}000/4\pi$ or 79.58 At/m [see Magnetizing Force (Magnetic Field Strength)].

Paramagnetic Material. A material having a relative permeability which is greater than unity, and which is practically independent of the magnetizing force.

Permeability. Permeability is a general term used to express various relationships between changes in magnetic induction B and changes in magnetizing force H. These relationships are either (1) absolute permeability, which in general is the change in magnetic induction divided by the corresponding change in magnetizing force, or (2) relative permeability, which is the ratio of the absolute permeability to the magnetic constant Γ_m.

NOTE 1. The magnetic constant Γ_m is a fixed constant, a particular value for which is chosen conventionally as a basis for each electromagnetic system. In the unrationalized cgs system Γ_m is 1, and in the mksa rationalized system $\Gamma_m = 4\pi \times 10^{-7}$ H.

NOTE 2. Relative permeability is a pure number which is the same in all unit systems; the value and dimension of absolute permeability depend on the system of units employed.

NOTE 3. For any ferromagnetic material permeability is a function of the magnetization. However, initial permeability μ_0 and maximum permeability μ_m are specific values for a given material under specified conditions.

Permeability, A-C. A general dynamic term used to express various relationships between magnetic induction B and magnetizing force H for magnetic material subjected to an SCM excitation by alternating current at commercial or higher frequencies. The values of a-c permeability obtained for a given material depend on many factors, including resistivity, thickness of laminations, frequency, and methods and conditions of measurements. Alternating-current permeabilities often used are:

Peak permeability μ_p. The ratio of the measured peak value of magnetic induction to the measured peak value of the magnetizing force for a material in the SCM condition.

NOTE. The peak value for magnetizing force is obtained from measurements of peak current made by either the mutual-inductance method or the peak-reading-voltmeter method.

Impedance (rms) permeability, μ_z. The ratio of the calculated maximum value of magnetic induction to the calculated maximum value of the magnetizing force for a material in the SCM condition.

NOTE. The maximum value of magnetizing current is obtained from multiplying the measured value of rms current by 1.414. This implies that deviations from sinusoidal currents are ignored.

Inductance permeability μ_L. The permeability evaluated in terms of core and circuit geometry (assumed uniform and homogeneous) and the measured parallel inductance of the path that is considered to carry the magnetizing current exclusively.

NOTE. This is the permeability value which is measured by ASTM bridge methods in terms of the core inductance L_1.

Permeability, Differential, μ_d. The absolute value of the slope of the hysteresis loop at any point.

Permeability, Incremental, μ_Δ. The ratio of a cyclic change in magnetic induction to the corresponding cyclic change in magnetizing force when the mean induction differs from zero.

NOTE 1. When the cyclic change is reduced to zero, incremental permeability μ_Δ becomes the reversible permeability μ_{rev}.

NOTE 2. When the mean induction approaches zero, the incremental permeability μ_Δ approaches the normal permeability μ.

Permeability, Initial, μ_0. The limiting value approached by the normal permeability as the applied magnetizing force H is reduced to zero.

Permeability, Maximum, μ_m. The maximum value of normal permeability for a given material.

Permeability, Normal, μ. The ratio of the normal induction to the corresponding magnetizing force.

Permeability, Relative, μ_r. The ratio of the absolute permeability of a material to the magnetic constant Γ_m.

NOTE. In the cgs-em system of units the relative permeability is numerically the same as the absolute permeability.

Permeability, Space, Γ_m. See Magnetic Constant (Permeability of Space).

Permeance $\mathcal{P}$. The reciprocal of the reluctance.

Reluctance $\mathcal{R}$. That quantity which determined the magnetic flux ϕ resulting from a given mmf $\mathcal{F}$ around a magnetic circuit.

$$\mathcal{R} = \frac{\mathcal{F}}{\phi} \tag{4-50}$$

where $\mathcal{R}$ = magnetic reluctance, $\mathcal{F}$ = mmf, and ϕ = flux. The reluctance is measured in gilberts per maxwell (magnetic ohms) in the cgs-em system and in ampere-turns per weber in the mksa system.

Remanence B_{dm}. The maximum value of the remanent induction for a given geometry of the magnetic circuit.

NOTE. If there are no air gaps or other inhomogeneities in the magnetic circuit, the remanence B_{dm} is equal to the retentivity B_{rs}; if air gaps or other inhomogeneities are present, B_{dm} will be less than B_{rs}.

Retentivity B_{rs}. That property of a magnetic material which is measured by its maximun value of the residual induction.

Specific Exciting Power P_z/m_1. The product of the rms voltage induced in the primary winding of a reactance coil divided by the effective mass of the magnetic core of the coil.

NOTE. When a reactance coil has a secondary winding, the induced primary voltage is obtained from the measured open-circuit secondary voltage multiplied by the appropriate turns ratio.

Susceptibility κ. The ratio of the intrinsic induction B_i due to the magnetization of a material to the induction in space due to the influence of the corresponding magnetizing force H.

$$\kappa = \frac{B_i}{\Gamma_m H} = \mu_r - 1 \tag{4-51}$$

where Γ_m = magnetic constant and μ_r = relative permeability.

NOTE 1. The above equations apply to an isotropic material if the mksa system of units is used.

NOTE 2. In the classical cgs-em system of units

$$\kappa = \frac{B_i}{4\pi\Gamma_m H} = \frac{\mu_{r-1}}{4\pi} \tag{4-52}$$

NOTE 3. This susceptibility divided by the density of a body is called the susceptibility per unit mass, χ, or simply the mass susceptibility.

$$\chi = \frac{\kappa}{\delta} \tag{4-53}$$

where δ = density.

Symmetrically Cyclically Magnetized Condition SCM. A magnetic material is in an SCM condition when, under the influence of a magnetizing force that varies cyclically between two equal positive and negative limits, its successive hysteresis loops are both identical and symmetrical with respect to the origin of the axes.

Tesla. The unit of magnetic induction in the mksa (Giorgi) system. The tesla is equal to 1 Wb/m^2 or 10^4 G.

Weber. The unit of magnetic flux in the mksa and in the practical system. The weber is the magnetic flux whose decrease to zero when linked with a single turn induces in the turn a voltage whose time integral is 1 V·s. One weber equals 10^8 Mx. (See Magnetic Flux.)

242. Standard Forms of Magnetic Data. The magnetic properties of ferromagnetic materials are customarily displayed in various standard curves and/or tables.

Magnetic Hysteresis and Normal Induction Curves. If a demagnetized specimen is subjected to the influence of a magnetizing force H which is increased from zero to higher and higher values, the magnetic induction B also increases, but not linearly with H. This is shown by a curve *oabcd* (see Fig. 4-19). If the increase in H is stopped at a point such as b and then decreased, the induction does not retrace the original curve in reverse order but lags behind it as indicated by the curves b, B_r, H_c, etc. This lag is called *magnetic hysteresis*. The point where the magnetizing force is zero is called the residual induction B_r. The negative magnetizing force at which the induction becomes zero is called the *coercive force* H_c. The closed curve starting from b through B_r, H_c, etc., back again to b is called a *hysteresis loop*. The loop does not always close at the first reversal of the magnetizing force but will close after enough reversals have been made. If the limits of H in each direction are equal, the limits of B in the two directions will also be equal and the material is said to be in a *symmetrically cyclic condition*. The induction at the tip of such a loop is called the normal induction. The ratio of B to H at this point is called the *normal permeability* $\mu = B/H$.

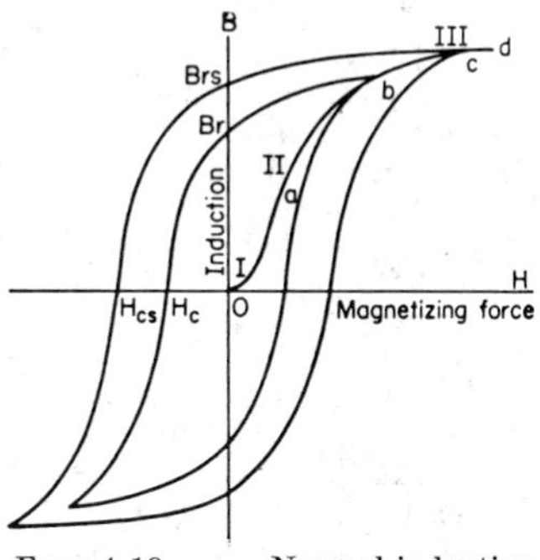

FIG. 4-19. Normal-induction curve and hysteresis loops.

It is easy to see that the size of a hysteresis loop for a given material depends upon the value of induction at its tip. The *normal induction curve oabcd* is the locus of the tips of a family of cyclically symmetrical hysteresis loops.

The normal induction curve usually consists of three distinct stages. In stage I the rate of increase of B as H is increased is comparatively small. The steep part of the curve represents stage II of the magnetization. In this stage, a small increase in H produces a relatively large increase in B. In stage III the rate of increase of B is again small. In this stage the intrinsic induction B_i asymptotically approaches a limiting value which is called the *saturation induction* B_s. For this reason, stage III is sometimes called the saturation range. Magnetic saturation is another of the distinguishing characteristics of ferromagnetic materials.

The magnetizing force and induction at the tip of a hysteresis loop are denoted by the symbols H_m and B_m, respectively. As B_m is carried higher and higher in the saturation range there comes a time when further increase does not produce any further increase in B_r and H_c. The maximum values of these quantities are called *retentivity* B_{rs} and *coercivity* H_{cs}, respectively.

For permanent-magnet materials, the important part of the hysteresis loop is that portion in the second quadrant between the residual-induction point B_r and the coercive-force point H_c. This is called the *demagnetization curve*. Points on this curve are designated by the coordinates B_d and H_d. The product of B_d and H_d for any point on the demagnetization curve is related to the energy external to the magnet which could be maintained under ideal conditions. A curve obtained by plotting the products of the corresponding coordinates B_d and H_d as abscissas against the induction B_d as ordinates is called the *energy-product curve*. The maximum value of the energy-product, $(B_dH_d)_m$, is a good criterion of the relative quality of permanent-magnet materials.

Permeability curves are obtained by plotting the permeability (a-c or d-c) against other variables such as induction, magnetizing force, frequency, thickness of lamination, etc. These curves are often prepared on log-log or semilog graph paper (see Figs. 4-20, 4-25, and 4-42).

Incremental-permeability curves. Since many applications of magnetic materials require that both d-c and a-c components of flux be present, it is also customary to use graphs of incremental permeability plotted against incremental induction.

Core-loss curves or tables are usually prepared to show the relation between the summation of the hysteresis and eddy-current components of the core loss for a specified induction and frequency, when the induction is varying harmonically.

Exciting RMS voltamperes per unit weight (exciting power) curves. This quantity, which is very useful in design calculation of electrical apparatus, is usually plotted against induction.

Energy-product curves are frequently used as the basis of the selection of permanent-magnet material. The maximum energy product is related to the maximum amount of energy that a magnet would impart to a keeper as it is inserted in the gap.

Hysteresis loss. A complete cycle of magnetization (hysteresis loop) causes a loss of energy in the material, which appears as heat. The loss per cycle is proportional to the area of the loop. Steinmetz developed an empirical equation to express this loss in the following terms:

$$W_h = \eta f B_m{}^k \qquad (4\text{-}54)$$

In this expression W_h is the hysteresis loss in ergs per cubic centimeter per second, f is the frequency in cycles per second (hertz), B_m is the maximum value of the induction, η is the hysteresis coefficient, and k is an exponent. The values of both η and k are dependent on the material and its geometry. Consequently they may not be constants for any given material if the grade or dimensions are changed.

Measurements of η, the hysteresis coefficient, have given values ranging from 0.012 for annealed iron down to 0.0002 for Permalloy. Intermediate values of 0.003 for annealed low-silicon sheet and 0.00046 for the best grades of electrical sheet have been obtained.

For the specimens of various irons and steels commonly used in the construction of electromagnetic apparatus at the time of Steinmetz's measurements, a value of approximately 1.6 was found for k. However, for modern materials the value of this coefficient may vary between 1.5 and 2.5, and of course may not be constant for any given material.

Eddy-current loss. The effect of varying flux density, as in a complete cycle of magnetization, is to induce internal emfs in the core materials and produce local circulating currents termed "eddy currents," which dissipate energy as heat. If the penetration of the sample by the field is complete and uniform, then the loss can be expressed as follows:

$$W_e = \epsilon[(ff)ftB_m]^2 \qquad (4\text{-}55)$$

In this expression ϵ is the eddy-current coefficient, ff is the form factor of the flux wave, f is the frequency in cycles per second (hertz), t is the thickness of sheets or laminations in centimeters. W_e is the eddy-current loss in ergs per cubic centimeter per second.

Theoretically for sheet material $\epsilon = \pi^2/6_\rho$, where ρ is the volume resistivity of the magnetic specimen. However, the geometry of the specimen, low resistance between laminations, and air gaps within the core make the eddy-current calculations more accurate if ϵ is determined from total core-loss measurements.

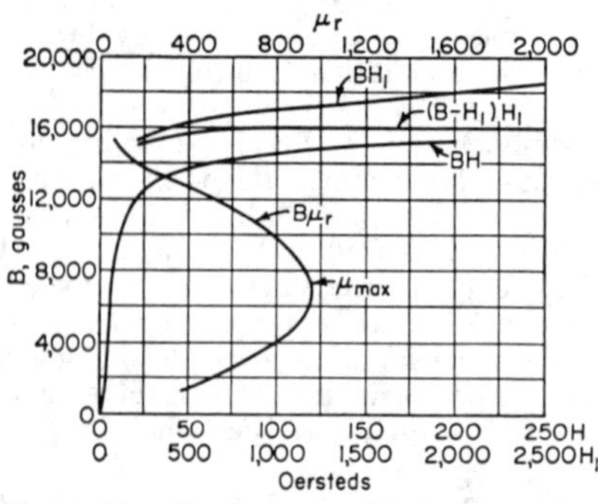

FIG. 4-20. Typical normal-induction, magnetization, and saturation curves.

243. Application of Magnetic Data. The proper selection of magnetic data and the calculation of their performance in actual apparatus are often difficult and empirical processes. Factors, usually gained from experience, must be used in the design calculations to allow for nonsymmetrical circuits, nonuniformity of materials, and flux leakage. The data presented in the standard forms are the properties of the material obtained under "ideal" conditions such as sinusoidal waveform, uniform flux in the magnetic path, closed magnetic circuit, test specimen free of any mechanical stress, specified flux path, etc. Therefore the results obtained in practice with any given materials are functions not only of the properties of the material but also of the inherent characteristics of the magnetic circuit as a whole.

The following effects are often serious enough to require the application of "correcting" factors:

Leakage. The principal difficulty in the design of many magnetic circuits is due to the lack of a material which can act as an insulator with respect to magnetic flux. This results in the total magnetic flux seldom being confined uniformly to the desired magnetic path. Furthermore, except in very special and unusual cases, it is not feasible to calculate the leakage with high accuracy. However, a reasonable estimate is needed to serve as a guide in designing the circuit. This estimate is usually based on experience with similar designs or on an experimental determination for the particular design.

Waveform. Even when a sinusoidal voltage is applied to an electromagnetic circuit, the resulting current and magnetic flux are usually not sinusoidal. The harmonic components of the flux will cause increases in core loss and exciting current which may become comparatively large, especially at moderate or high inductions. This should be taken into account in using data on magnetic material obtained under sinusoidal flux conditions.

Accurate measurement of core loss when the flux is distorted requires that the maximum flux density be determined by a meter responsive to the average voltage.[1]

Fabrication. Stresses introduced in magnetic materials by shearing, machining, forming, riveting, or other fabricating operations often change the properties of the materials remarkably. This is particularily true of materials having high permeability. These effects can be eliminated by a suitable stress-relieving anneal after fabrication to the final shape. When stress relief is impractical or when some stresses are unavoidable in assembly, a compensating factor must be used in the design.

Joints and Gaps. Joints or gaps in a magnetic circuit can cause a large increase in the total excitation requirements. This results in the effective permeability of the magnetic circuit at a given induction being considerably less than that of the magnetic material.

Nonuniform Flux Distribution. Calculations of magnetic circuits having curved portions, such as rings, are sometimes in error because several magnetic paths around the circuit exist, with unequal lengths. Thus the normal component of the induction may not be uniform across a cross section of the circuit, and the real flux distribution may not be the same as the idealized one used for the calculation. Both these effects lead to errors which can seldom be calculated and are difficult to determine experimentally. When possible, the geometry of the circuit should be chosen to make this effect negligible.

Effect of Unsymmetrical Periodic Cycles on Hysteresis Loss. When both direct and alternating magnetizing forces are applied simultaneously, the resulting hysteresis loop is unsymmetrical and different for different values of the d-c excitation. Superimposed d-c excitation decreases the hysteresis loss for any given alternating component of flux density in the range above a certain value which may be called the critical value. In the range of alternating flux density below this value, the hysteresis loss tends first to rise, then to reach a maximum, and finally to decrease as d-c excitation is superimposed and the alternating component of flux density is held unchanged.

Critical value of alternating flux density for silicon sheet steel lies in the range 13,200 to 13,800 G (85 to 89 kilolines/in^2). The a-c excitation required for a given alternating flux density is increased when d-c excitation is superimposed. The d-c excitation for a given direct flux density in the low or moderate density range is decreased when a small amount of a-c excitation is superimposed but is increased when larger amounts of a-c excitation are added.[2]

---

[1] Smith, B. M., and Concordia, C. Core-loss Measurements at High Flux Densities; *Elec. Eng.*, January, 1932, Vol. 51, p. 36. Also Camilli, G. A Flux Voltmeter for Magnetic Tests; *Jour. AIEE*, October, 1929, Vol. 45, p. 989.

[2] Edgar, R. F. Loss Characteristics of Silicon Steel at 60 Cycles with D-C Excitations; *Trans. AIEE*, 1933, Vol. 52, p. 721.

Core Loss in High-permeability Materials. The determination of hysteresis and eddy-current losses by the measurement of total core loss at two frequencies and extrapolation to $f = 0$ is valid only at moderate flux densities and for materials of permeability less than 10,000.[1] In such materials as oriented-grain 3% silicon steel, and the nickel-iron alloys, eddy-current shielding causes the flux near the middle of the specimen to lag behind the flux at the surface and to produce a nonhomogeneous induction. At a given frequency, such a lag will cause the effective permeability to be substantially below that measured ballistically. The total core loss as measured under such conditions is markedly higher than that calculated from the ballistic hysteresis loop and the "eddy-current loss," as in Eq. (4-55).

244. Commercial Magnetic Materials. Commercial magnetic materials are generally divided into two main groups:

a. The magnetically "soft," or nonretentive, materials.

b. The magnetically "hard," or retentive, materials.

a. The "soft" materials are often referred to as core materials and include both the common materials used in most electromagnetic circuits and the special materials having magnetic and physical properties suitable for special applications.

The common materials used in core applications have a wide range of magnetic and other physical properties. The selection of a material for a specific application is usually based on a compromise between the requirements for physical properties and economy. If superior magnetic properties are required and the design permits, then grain-oriented electrical steel is often selected. Its magnetic properties approach the best characteristics of single crystals of iron. Cast iron has inferior magnetic properties but is frequently used in certain applications because of its low cost.

For the special core materials the properties of interest may require high permeabilities of various kinds, special shapes of hysteresis loops, special temperature-permeability relations, unusually good magnetic properties at low or high induction, or high frequencies, etc. Although on a tonnage basis these materials represent only a small fraction of the commercial output, they are very important and in some cases are indispensable for the operation of certain equipment.

b. The magnetically "hard," or permanent-magnet, materials are used in many applications, such as meters, relays, loudspeakers, television, latches, and other equipment where it is not convenient or economical to obtain a magnetic field by electromagnetic means. Materials used in the production of permanent magnets range from almost pure iron and ferrous metallic alloys to chemical compounds having ceramic-like properties. The desirable magnetic properties for permanent-magnet materials include high coercivity and high maximum energy product.

245. "Soft" Magnetic Materials. The principal use of these core materials is in magnetic circuits of electromagnetic apparatus and equipment. The required magnetic and physical properties as well as economics govern the selection and adaptation of a material for a particular purpose. Occasionally, for certain applications, one of the basic ferromagnetic elements, iron, nickel, or cobalt, may be used by itself because it has the desired magnetic and/or physical properties. This was particularly true in the early development of electromagnetic apparatus, when iron was extensively used. In present-day practice, however, metallic magnetic materials are generally alloys of the basic ferromagnetic elements and other elements.

Alloying elements frequently used with the basic ferromagnetic elements include aluminum, arsenic, cerium, chromium, copper, manganese, molybdenum, silicon, thorium, titanium, tungsten, and vanadium.

These metallic, magnetically "soft" alloys employed in commercial practice may be considered under the three classifications: (1) solid-core materials, (2) electrical sheet or strip, and (3) special alloys. In addition to the metallic alloys, in recent years nonmetallic materials having various composition have become important for application as a core material. These are the chemical compounds known as "soft" ferrites or ceramic core materials.

---

[1] BOZORTH, R. M. "Ferromagnetism"; Princeton, N.J., D. Van Nostrand Company, Inc., 1951, p. 782.

246. Basic Ferromagnetic Elements. The only elements which are strongly ferromagnetic at normal temperatures are iron, cobalt, and nickel. In Fig. 4-21 are shown typical normal-induction curves of annealed samples of iron, nickel, and cobalt of comparatively high purity. These curves are given only for the purpose of general comparison and should not be considered as representing critical values. Small variations in the degree of purity or in the annealing procedure lead to substantial differences in normal induction. It may be noted that iron has the greatest permeability, and this factor coupled with its low cost makes it the only element of commercial importance by itself.

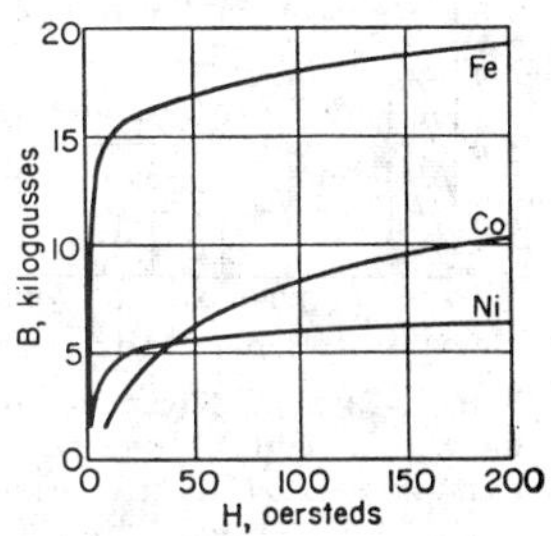

FIG. 4-21. Typical normal-induction curves of annealed samples of iron, nickel, and cobalt.

247. Solid-core Materials. These materials are used for the cores of d-c electromagnets, relays, field frames of d-c machines, and other applications which use essentially d-c excitation. The principal requirement is high permeability at relatively high inductions. For the majority of uses it is desirable that the coercive force and hysteresis be low.

248. Vacuum-fused Electrolytic Iron. Very pure iron may be made by remelting electrolytic iron in vacuo and subjecting it to annealing in special atmospheres. Cioffi[1] by means of hydrogen treatment has achieved a maximum permeability of about 250,000. Curve I in Fig. 4-22 shows a normal-induction curve for vacuum-fused iron. The striking characteristics of high-purity electrolytic iron are high induction at low magnetizing forces, high maximum permeability, and low hysteresis loss; the saturation induction is not materially influenced by small amounts of impurity.

249. Ingot iron is made by the open-hearth process and is refined to the highest degree commercially attainable in large-scale production. A typical analysis of Armco ingot iron shows the following percentages of impurities present; carbon 0.012, manganese 0.018, phosphorus 0.004, sulfur 0.023, silicon 0.003, and copper about 0.05 or less. Its properties are stated by the manufacturer to be as follows at 20°C: density 7.85; resistivity 10.7 $\mu\Omega\cdot$cm; temperature coefficient of resistance 0.0055/°C. Its magnetic properties are affected by mechanical working and heat-treatment.

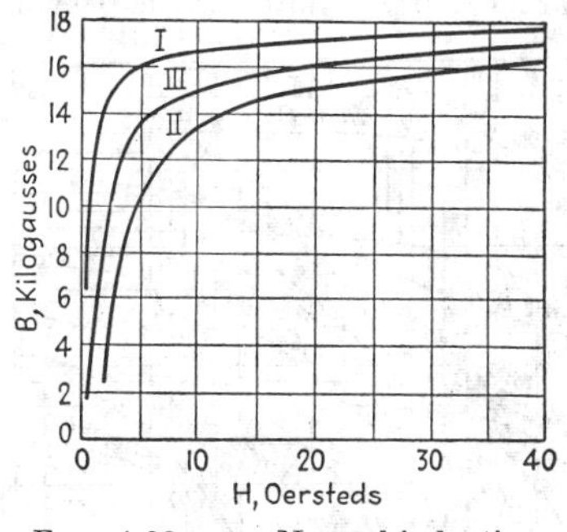

FIG. 4-22. Normal-induction curves of (I) vacuum-fused electrolytic iron, (II) mild open-hearth steel unannealed, and (III) Armco ingot iron.

A typical normal-induction curve of an annealed Armco ingot-iron bar is shown in curve III (Fig. 4-22).

In order to obtain the best results with this material it should be annealed thoroughly after forging or machining. In general, annealing to a maximum soaking temperature of 760°C and slowly cooling will give good results. A long-time anneal in hydrogen at 1100°C or higher will give the best magnetic results and prevent aging.

250. Wrought iron is a ferrous material, aggregated from a solidifying mass of pasty particles of highly refined metallic iron, into which is incorporated, without subsequent fusion, a minutely and uniformly distributed quantity of slag. The minute slag inclusions distinguish wrought iron from ingot iron or steel made by the fusion process. Norway iron and Swedish iron are names frequently applied to the better grades of wrought iron. In bar form it is widely used as cores for relays. For this purpose it must be well annealed, preferably in hydrogen to reduce hysteresis and prevent aging. Annealing, as in the case of all materials used for magnetic circuits, should be done

[1] *Phys. Rev.*, May, 1934, Vol. 45, p. 472.

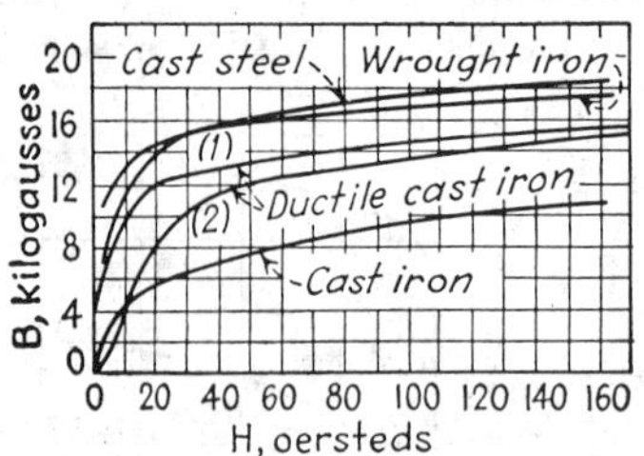

Fig. 4-23. Normal-induction curves of typical wrought iron, cast steel, and ductile cast iron (3% Si) as (1) cast and (2) annealed.

after all shaping, forming, or machining operations.

A normal induction curve of wrought iron is shown in Fig. 4-23, with similar curves of other materials for comparison.

251. Mild, or Low-carbon, Steel. The effect of combined carbon on the magnetic properties of steel is adverse from the standpoint of obtaining high permeability and low hysteresis loss. Open-hearth steel of very low cabon content has magnetic properties comparable with those of wrought iron and cast steel (Figs. 4-22 and 4-24). It is used to a considerable extent for laminated and solid magnetic parts but is inferior to silicon steel as regards aging and core losses.

252. Gray cast iron is magnetically inferior to wrought iron or steel but is used to a limited extent because of the ease of casting complex shapes. Its permeability is decreased by carbon somewhat in proportion to the amount of combined carbon and increased by silicon. Cast iron usually contains 2.7 to 3.6% total carbon, of which 0.2 to 0.8% may be in the combined form, and 2.0 to 2.7% silicon. Phosphorus ranges from 0.15 to 0.50%, and sulfur under 0.15% (see Figs. 4-23 and 4-25).

253. Ductile (nodular) cast iron is a low-phosphorus, low-sulfur cast iron which has been treated with magnesium, thereby changing the form of the graphite from the flake of gray iron to crystalline spheroids. It has the good fluidity, castability, and machinability of gray iron and many of the engineering advantages of steel. Magnetically it is greatly superior to gray cast iron and, depending upon the silicon and combined carbon contents, it approaches mild cast steel. Figures 4-23 and 4-25 show curves for ductile iron of nromal composition and with 3% silicon.

254. Malleable cast iron is magnetically superior to cast iron, being lower in combined carbon and improved by heat treatment which it receives. Curve 5 in Fig. 4-25 is for malleable cast iron containing 0.83 combined carbon, 2.02 graphite carbon, 0.93 Si,

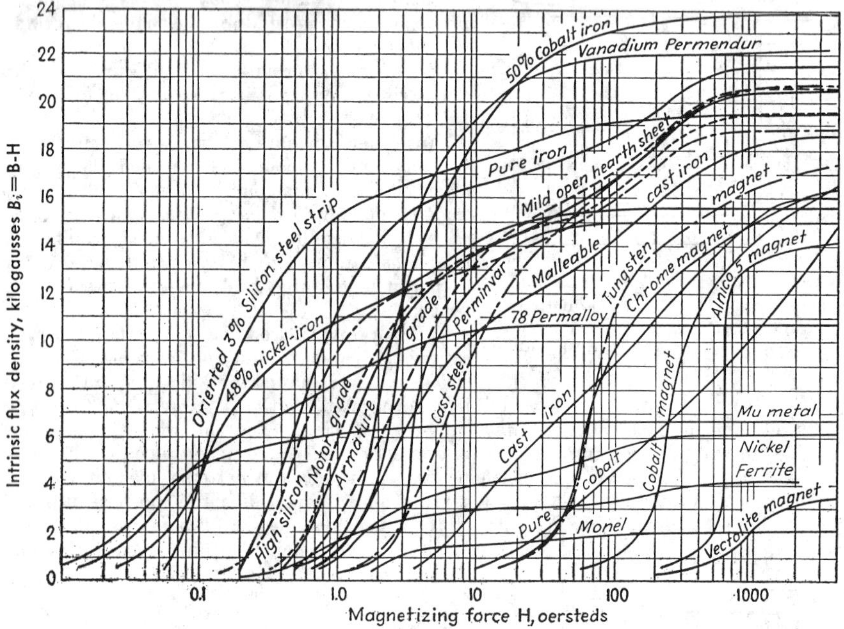

Fig. 4-24. Direct-current magnetization curves for a wide range of ferromagnetic materials.

0.12 Mn, 0.039 P, and 0.08 S. Cast steel is extensively used for those portions of magnetic circuits which carry uniform or continuous flux and need superior mechanical strength. Cast steel of good magnetic qualities should be limited in its composition as follows: combined carbon 0.25, Si 0.20, Mn 0.50, P 0.08, S 0.05. The lower the impurities, the better the magnetic properties. Annealing improves the quality of both cast iron and cast steel. Rolled and welded frames of soft steel plate are now widely used in place of cast steel.

255. Iron-Cobalt Alloys. The addition of cobalt to iron has the effect of raising the saturation intensity of iron up to about 36% cobalt (Fe_2Co). This alloy is useful

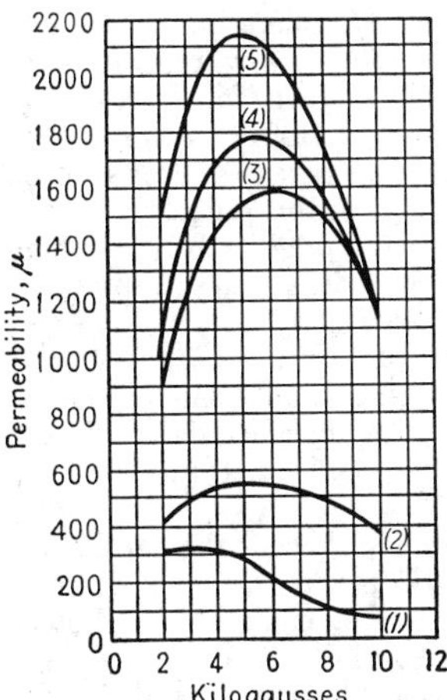

FIG. 4-25. Permeability induction curves of (1) cast iron, (2) ductile cast iron (normal composition) as cast and (3) annealed, (4) ductile iron (3% Si) annealed, (5) malleable cast iron.

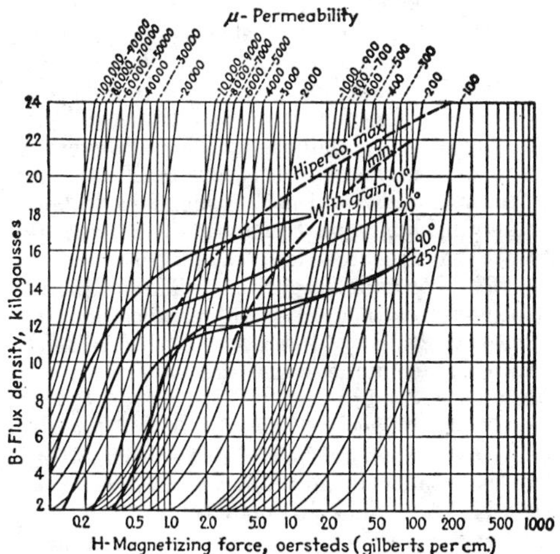

FIG. 4-26. Highly oriented silicon strip showing effect of angle of magnetization with rolling direction on permeability. Broken lines show range of values for Hyperco, a high-saturation Co-Fe alloy.

for pole pieces of electromagnets and for any application where high magnetic intensity is desired. It is workable hot but quite brittle cold. Hyperco contains ⅓ Co, ⅔ Fe plus 1 to 2% "added element" (see Fig. 4-26). Total core loss is about 2.5 W/lb at 15 kG for 0.010-in-thick material. It is available as hot-rolled sheet, cold-rolled strip, plates, and forgings. The 50% cobalt-iron alloy Permendur has a high permeability for magnetizing forces up to 50 Oe and, with about 2% vanadium added, it can be cold-rolled.

256. Effects of Temperature. Iron is ferromagnetic up to about 790°C and above that point is paramagnetic. This critical temperature is termed the Curie point. With weak magnetizing forces, the permeability increases with temperatures up to near this point. Under moderate magnetizing forces, the permeability increases slightly in the lower range of rising temperature, and then a point is reached above which it decreases rapidly toward the critical point. Under strong magnetizing forces, the permeability is not at first affected by rising temperature, but after a certain point is passed, it decreases rapidly as the critical point is approached. In hard carbon steel the permeability rises continuously with increasing temperature up to a point below the critical temperature and then decreases very abruptly as the critical point is approached. This magnetic-change point is profoundly influenced by alloying—with 30% nickel or 13% manganese, the alloys are nonmagnetic at room temperature.

257. Aging. Iron subjected to elevated temperatures over long periods of time, as in transformer cores and rotors in continuous operation, develops magnetic fatigue or aging characteristics manifested by decreased permeability and increased hysteresis loss. Prolonged heating at even so low a temperature as 50°C, if continued for several weeks, will produce an appreciable effect. The effect increases with the temperature, and the increase in hysteresis loss may exceed 110%. Silicon steel has substantially nonaging characteristics, obtained by the addition of silicon as an alloying element. The ASTM standard for aging is the percentage change in standard core loss after continued heating at 150°C for 100 h or at 100°C for 600 h. From a practical point of view, aging is negligible for silicon contents above about 1%.

258. Effects of stress on magnetization are discussed in Chap. 13 of Bozorth, "Ferromagnetism," D. Van Nostrand, 1951, in Chap. II of Williams, "Magnetic Phenomena," New York, McGraw-Hill Book Company, 1931, and in Chap. III of Honda, "Magnetic Properties of Matter," 1928. These phenomena come generally under the head of magnetostriction, which relates to the stresses and dimension changes attending magnetization. The magnetostriction of randomly oriented polycrystalline iron is positive until its intrinsic flux density approaches 18 kG, and then it becomes negative for higher flux densities. Its maximum expansion is approximately 6×10^{-6} cm/cm in the region of 16 kG. Nickel contracts continuously and reaches a maximum of about 35×10^{-6} cm/cm at saturation induction. Conversely, magnetization increases with tensile stress in iron and decreases with tensile stress in nickel for fields in this range. A 6¼% silicon-iron alloy and a 75% nickel-iron alloy (Permalloy) have practically no volume or length change with magnetization. Single crystals show marked anisotropy with respect to magnetostriction.

259. Effects of Impurities. The magnetic properties of all ferromagnetic materials depend upon their chemical composition, mechanical working, and heat-treatment. The general effect of impurities is to decrease the permeability and increase the hysteresis loss. The common impurities in iron are carbon, manganese, silicon, copper, sulfur, phosphorus, and oxygen.

Carbon increases the resistivity, decreases the permeability, lowers the saturation point, and increases the coercive force and the retentivity. Concurrently the hysteresis loop is broadened, and its area increased. These effects are greater in hardened steel than in soft or annealed materials.

Very small proportions of manganese are not injurious in any substantial degree, but it is customary to limit the proportion of manganese as much as practicable. When the manganese content reaches 12%, the steel becomes practically nonmagnetic.

The presence of silicon, up to contents of 6½%, is beneficial to the magnetic properties but decreases the ductility appreciably when present in amounts above 3%.

The addition of aluminum has a similar effect as silicon but is more expensive. Arsenic and tin also have similar effects.

Copper is frequently added in very small proportions to iron and mild steel to increase the corrosion resistance. The effect on the magnetic properties is inappreciable up to 0.5%.

Sulfur, phosphorus, and oxygen are injurious in their effects and should be reduced to the lowest attainable limits.

260. Heat-treatment. The effects of cold working, rapid cooling, and machining are to leave strains in the metal which impair its magnetic properties. These effects can be overcome only by thorough annealing of the finished parts in order to relieve strains. The anneal recommended for low-carbon steels consists of heating to 760°C and holding at that temperature for a period of 4 h, followed by cooling not faster than 50°/h to 500°C. An oxidizing atmosphere is generally used to accomplish the removal of carbon (see Fig. 4-27).

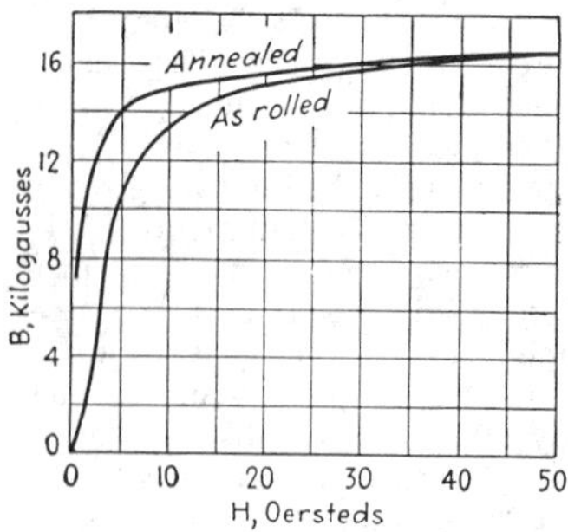

Fig. 4-27. Effect of annealing on normal induction of low-carbon steel sheet.

261. Time lag in magnetization, or magnetic viscosity, may be very noticeable in soft iron members of large cross section. The phenomenon depends much upon the size of the specimen, and while noticeable in an iron rod 4 mm in diameter, it could not be detected in fine annealed iron wire. In the case of large, bulky sections, such as a solid field frame, the effect is very pronounced; it occurs both upon magnetization and upon demagnetization. The phenomenon occurs to a slight degree in hard iron and steel but is scarcely observable. In general, the thinner the cross section, the less the effect is observed. Magnetic drift or decrease in permeability with time occurs at low magnetizing forces and is considerable only at densities below 1,000 G. It is a factor affecting the ratio and phase-angle error in current transformers. Drift is greatest in silicon steels and very low in Nicaloi and Mumetal.

262. Magnetic analysis for detection of flaws in ferromagnetic metals has been applied in various ways to all kinds of important structural members which must be sound and free from cracks. All the methods used for this purpose are based on the common property of such materials of being very sensitive magnetically to changes in composition and structure. The presence of obscure cracks[1] or seams or other abrupt discontinuities in metal otherwise of homogeneous structure can frequently be detected by these methods.

263. Electrical Sheet, or Strip (Flat-rolled Electrical Steel).[2] These terms are commercially used to designate low-carbon steels in which electrical characteristics,

Table 4-47. Maximum-core-loss Tables, Fully Processed—Conventional Types
(Hot-rolled and cold-rolled flat-rolled electrical steel)

Maximum core losses, W/lb at 60 c/s					
Electrical Steel Standard Gage No.			29	26	24
Thickness, in.			0.0140	0.0185	0.0250
Grade name	Type No.	Normal silicon content	15 kG		
	M-14	4.0–5.0	1.30		
Transformer	M-15	2.8–5.0	1.45	1.70	
	M-19	2.5–3.8	1.64	1.86	2.10
Dynamo	M-22	2.5–3.5	1.77	1.97	2.23
Motor	M-27	1.7–3.0	1.92	2.16	2.45
Electrical	M-36	1.4–2.2	2.15	2.47	2.90
Armature	M-43	0.6–1.3	....	2.92	3.50
			10 kG		
	M-14	4.0–5.0	0.52		
Transformer	M-15	2.8–5.0	0.58	0.68	
	M-19	2.5–3.8	0.67	0.76	0.91
Dynamo	M-22	2.5–3.5	0.73	0.83	0.98
Motor	M-27	1.7–3.0	0.81	0.93	1.10
Electrical	M-36	1.4–2.2	0.92	1.06	1.27
Armature	M-43	0.6–1.3	....	1.25	1.52

Losses determined according to ASTM A34 and A343.*

NOTE 1. Core losses are commonly specified at inductions of either 15 kG or 10 kG, but not at both, on specimens one-half parallel and one-half transverse to direction of rolling.
NOTE 2. To convert the above values to watts per pound at 50 c/s, multiply values shown by 0.79.
NOTE 3. To convert the above values to watts per kilogram at 60 c/s, multiply values shown by 2.204.
* Samples not annealed after shearing (AS).

mainly low core loss and high permeability, are achieved either by the use of silicon or by special processing or both. Other elements are sometimes used in place of, or in addition to, silicon. These materials are generally produced in sheet or strip form and used as core materials in a-c apparatus such as transformers, motors, electromagnets, or relays. The several grades differ mainly with respect to their silicon content, which ranges up to approximately 6%. Although the chemical compositions (carbon, manganese, phosphorus, sulfur, silicon, and aluminum) may be similar, magnetic characteristics may be widely dissimilar depending on the processing employed.

The magnetic characteristics of electrical sheet or strip may vary with respect

[1] Symposium on the Role of Non-destructive Testing; *ASTM Spec. Tech. Publ.* 112, June, 1950.
[2] This section is based on the tables and discussions contained in the *Steel Products Manual*, Flat Rolled Electrical Steel, published by the AISI.

to the rolling direction. However, if this material is processed in such a manner that it possesses a minimum amount of directionality in magnetic properties, it is known as the *conventional type* of electrical steel.

By a suitable combination of cold rolling and heat-treatment, electrical sheet or strip may be produced in which better magnetic properties are obtained in the preferred direction than in ordinary grades. Their maximum permeability is approximately twice as high, and they have much lower core losses combined with higher permeability at high induction than the conventional type. Material processed in this way is known as the *grain-oriented type*.

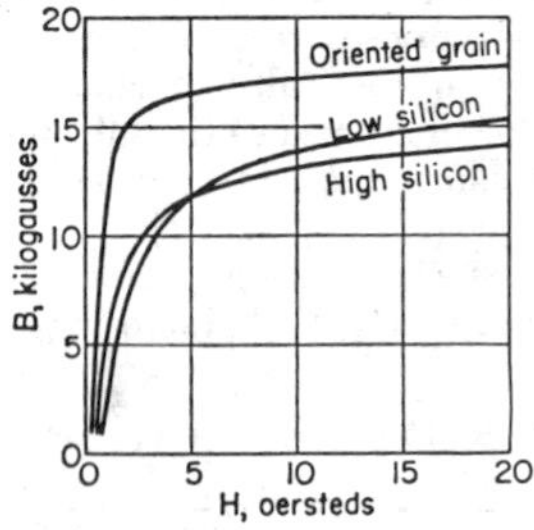

FIG. 4-28. Typical normal-induction curves for electrical sheet.

Typical normal-induction curves for two grades of electrical sheet and grain-oriented material are shown in Fig. 4-28. The improvement in the grain-oriented material is easily noticed.

264. Conventional Type. The conventional type of electrical steel is available in four specific classes:

1. Hot-rolled fully processed.
2. Cold-rolled fully processed.
3. Hot-rolled semiprocessed.
4. Cold-rolled semiprocessed.

The magnetic properties of fully processed electrical steel have been fully developed by the producer to meet maximum core-loss values (see Table 4-47)

Table 4-48. Maximum Core Losses, Semiprocessed—Conventional Types
(Hot-rolled and cold-rolled flat-rolled electrical steel losses determined according to ASTM A34 and A343*)

	Maximum core losses, W/lb, at 60 c/s			
Electrical Steel Standard Gage No.	26	24	26	24
Thickness, in.	0.0185	0.0250	0.0185	0.0250
	15 kG		10 kG	
Type No.:				
M-22	1.90	2.10	0.80	0.93
M-27	2.03	2.25	0.90	1.05
M-36	2.35	2.70	1.02	1.21
M-43	2.72	3.35	1.16	1.35
M-45	....	4.00		

NOTE 1. Core losses are commonly specified at inductions of either 15 kG or 10 kG, but not at both, on specimens one-half parallel and one-half transverse to direction of rolling.
NOTE 2. To convert the above values to watts per pound at 50 c/s, multiply values shown by 0.79.
NOTE 3. To convert the above values to watts per kilogram at 60 c/s, multiply values shown by 2.204.
* Sample is annealed at 1550°F for approximately 1 h, except M-45, where 1450°F for approximately 1 h applies.

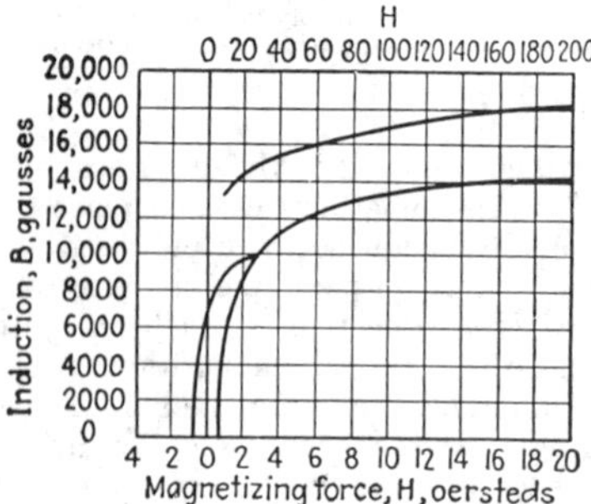

FIG. 4-29. Normal-induction curve and hysteresis loop of dynamo sheet, No. 29 gage.

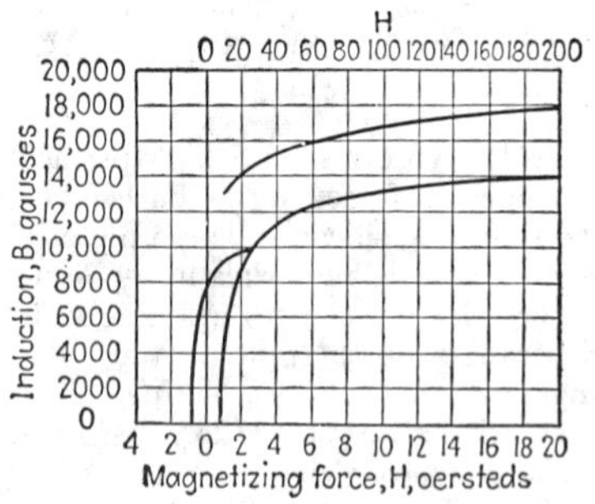

FIG. 4-30. Normal-induction curve and hysteresis loop of motor sheet, No. 29 gage.

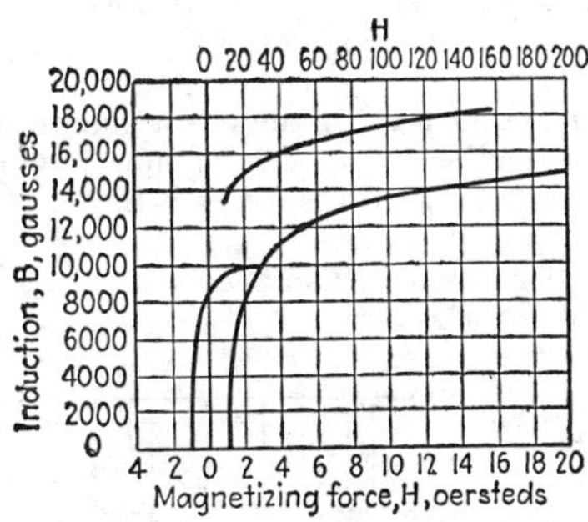

FIG. 4-31. Normal-induction curve and hysteresis loop of electrical sheet, No. 22 gage.

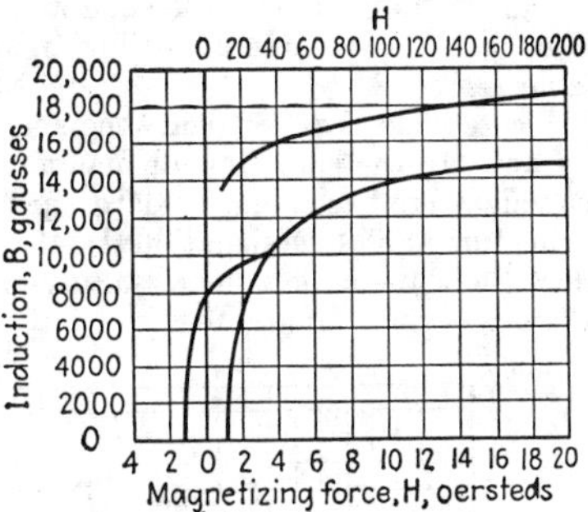

FIG. 4-32. Normal-induction curve and hysteresis loop of armature, No. 26 gage.

Table 4-49. Some Characteristics and Typical Applications for Specific Types of Flat-rolled Electrical Steel

AISI Type No.	Nominal silicon %	Some characteristics	Typical applications
M-4 M-5 M-6	2.8–3.5	Highly directional magnetic properties due to grain orientation. Very low core loss and high permeability in rolling direction	Highest-efficiency power and distribution transformers with lower weight per kVa
M-7 M-8	2.8–3.5	Grain-oriented, but properties are less directional than M-6	Large generators and power transformers
M-14	4.0–5.0	Lowest-core-loss conventional grade. High permeability at low induction, but it is low at high induction. Comparatively brittle	Distribution and power transformers and rotating machines of high efficiency
M-15	2.8–5.0	Characteristics depend upon which of two annealing processes is used. The open-annealed lower-silicon product is quite ductile	Punchings requiring low core loss and excellent permeability at low and moderate induction
M-19	2.5–3.8	Usually more ductile than types of lower core loss. Moderately high permeability at all inductions	Communication transformers and reactors. High-efficiency fractional-horsepower motors
M-22	2.5–3.5	Ductile. Relatively good properties	Cores of high reactance, stators of high-efficiency rotating electrical equipment, intermittent-duty transformers, and magnetos
M-27	1.7–3.0	Ductile. Relatively good punching properties	Continuous-duty high-efficiency motors. Small transformers operating at moderate induction
M-36	1.4–2.2	Ductile. Excellent permeability at high induction	Used extensively for rotating machines including a-c and d-c motors
M-43	0.6–1.3	Ductile. High core loss but good permeability at high induction	Suitable for fractional-horsepower motors, pole pieces, and relays
M-45	0–0.6	Ductile. High core loss but good permeability at high induction	Suitable for fractional-horsepower motors, pole pieces, and relays
M-50	0–0.6	See Note	Used for intermittently operating electrical apparatus and pole pieces. This grade is not commonly subject to magnetic-test requirements

NOTE: Core loss is considered the principal measure of quality for flat-rolled electrical steel, except for type M-50. Tests for other properties, such as permeability, lamination factor, electrical resistivity, interlamination resistance, aging, ductility, tensile strength, and hardness are sometimes required. Methods for performing tests are described in ASTM Designation A34. For test purposes the individual supplier should be consulted regarding which ASTM density corresponds most closely with the actual density, since aluminum or other elements may also be present in significant amounts.

The magnetic properties of the semiprocessed electrical steel must be developed by the consumer with a suitable annealing treatment. Maximum core-loss values are given in Table 4-48.

The core losses for electrical sheets in Tables 4-47 and 4-48 are the latest guaranteed values and are the same for all manufacturers. They are revised from time to time. These values are based on standard Epstein tests made in accordance with the method prescribed in ASTM Designation A-343.

Because of the almost universal use of electrical sheets at higher than 10 kG, manufacturers now make guarantees at 15 kG as in Table 4-47.

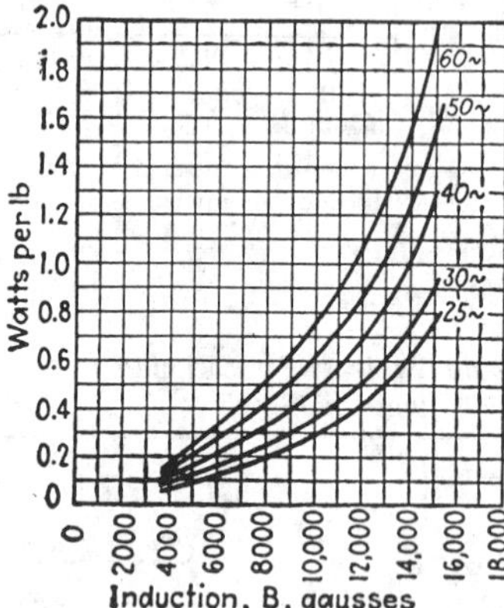

FIG. 4-33. Typical core-loss curves for dynamo grade, No. 29 gage at various frequencies.

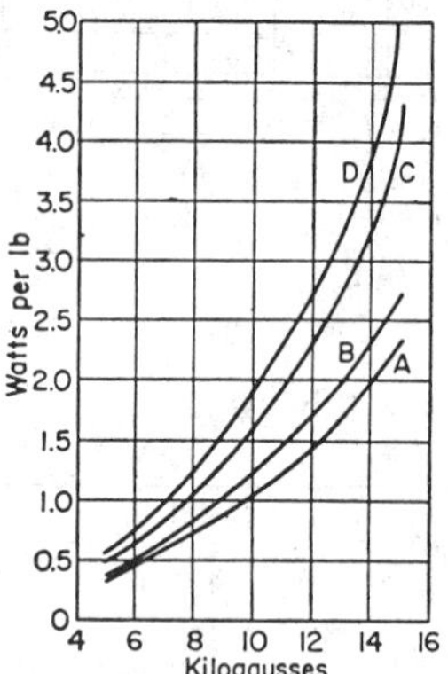

FIG. 4-34. Typical loss curves 24-gage (0.025 in) electrical sheets, 60 cycles. *A.* Dynamo grade. *B.* Motor grade. *C.* Electrical grade. *D.* Armature grade.

The AISI type number, nominal silicon content, and some characteristics and typical applications for the conventional types are given in Table 4-49 beginning with M-14.

Composition. Silicon alloys are the best materials for commercial power apparatus, such as transformers, generators, and motors, which are designed for continuous duty. They are superior to iron or low-carbon steel because they are nonaging and lower in total core loss. High-grade silicon sheets, usually made by the basic open-hearth process, will be low in impurities—carbon under 0.07, sulfur and phosphorus under 0.025, and manganese usually about 0.25% or less.

Table 4-50. Maximum Core Losses, Grain-oriented Types, Flat-rolled Electrical Steel

(Losses determined according to ASTM A34 and A343*)

	Maximum core losses, W/lb at 60 c/s		
Thickness, in.	0.011	0.012	0.014
Thickness, mm	0.28	0.30	0.36
		15 kG	
Type No.:			
M-4	0.53		
M-5	0.57	0.58	0.60
M-6	0.63	0.64	0.66
M-7	....	0.71	0.73
M-8	....	0.78	0.80

NOTE 1. To convert the above values to watts per pound at 50 c/s, multiply values shown by 0.76.
NOTE 2. To convert the above values to watts per kilogram at 60 c/s, multiply values shown by 2.204.
NOTE 3. To convert the above values to watts per kilogram at 50 c/s, multiply values shown by 1.675.
* Samples cut parallel to direction of rolling and stress relief annealed at 1450°F for approximately 1 h.

Sheets of the highest quality have the largest silicon content, approximately 4 to 5%. The silicon increases the hardness of the material so that the percentage present is determined by the use to which the sheets are to be put. For rotating machines the upper limit is about 4%, on account of the danger of brittleness. Lower-silicon sheets are also to be preferred where the operating densities are high or where high heat conductivity is desirable. It is commercial practice to grade sheets according to silicon content, ranging from 5 and sometimes 6¼% down to ordinary mild open-hearth steel containing only traces of silicon. The core loss and the aging coefficient increase as the silicon content decreases.

Commercial Grades. The manufacture of electrical sheets has become standardized to a considerable extent, and the magnetic data published by the various manufacturers of this material do not differ, grade for grade, very materially; most manufactures make about five standard grades for power apparatus and a few grades having lower core losses to meet special requirements. The commercial grades given in Table 4-47 are representative.

Magnetic Characteristics

a. Normal induction and hysteresis curves (see Figs. 4-29 to 4-32).

b. Typical core-loss curves as a function of grade, thickness, flux density, frequency, etc. (see Figs. 4-33 to 4-36).

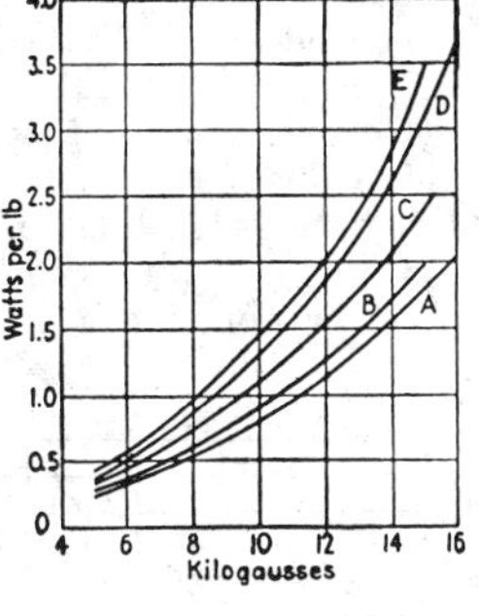

FIG. 4-35. Typical loss curves for 26-gage (0.185 in) electrical sheets, 60 cycles. *A.* Transformer grade. *B.* Dynamo grade. C. Motor grade. *D.* Electrical grade. *E.* Armature grade.

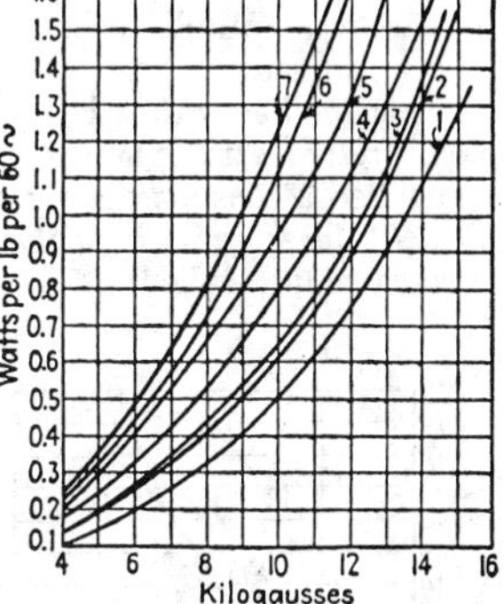

FIG. 4-36. Typical 60-cycle loss curves for electrical-sheet grades, No. 29 gage (0.014 in). 1. Transformer 52. 2. Transformer 65. 3. Transformer 72. 4. Dynamo. 5. Motor. 6. Electrical. 7. Armature.

265. Grain-oriented Type. This type of material was specifically developed for low core loss and high permeability in the rolling direction. Through special control of the hot-rolling practice, the amount of cold reduction between annealing steps, and the use of suitabie high-temperature anneal it has been found possible to align most of the crystal lattices so that their axes of easiest magnetization are nearly parallel to the direction of rolling. However, this technique can produce the desired grain orientation only in silicon steel having approximately 2.8 to 3.5% silicon.

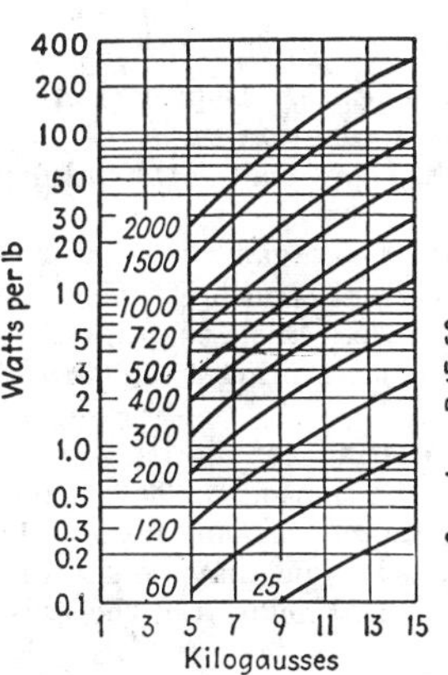

FIG. 4-37. Core loss vs. induction at frequencies 25 to 2,000 c/s for highly oriented 3% silicon strip steel, 0.014 in thick, parallel-grain specimen, annealed.

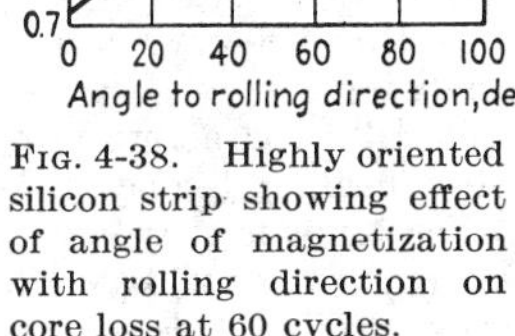

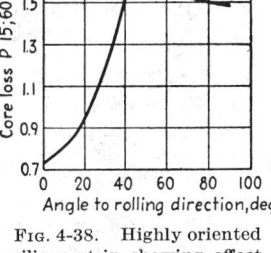

FIG. 4-38. Highly oriented silicon strip showing effect of angle of magnetization with rolling direction on core loss at 60 cycles.

Table 4-49 gives some characteristics and typical applications for the grain-oriented type of flat-rolled electrical steel. Table 4-50 gives the maximum core losses for this material. Figures 4-26, 4-37, and 4-39 show in graphical form some of the magnetic properties of grain-oriented material.

266. Annealing. A stress-relief anneal is usually required in order to remove the stresses introduced by fabrication and to restore the magnetic properties of fully processed electrical steel. Semiprocessed electrical steel must be annealed both to relieve stresses and to develop the inherent magnetic properties that were not fully developed by the producer. The most common methods of annealing flat-rolled electrical steels are:

a. Continuous annealing.

b. Box annealing.

c. Open-coil annealing.

Continuous annealing consists in passing the steel continuously through a furnace similar to the "normalizing furnace" used in the processing of certain types of plain carbon steel. This type of annealing is ordinarily used for strands of cold-reduced or welded sheets but may be utilized for the annealing of individual sheets.

Box annealing is the process of heating steel in a closed container through which a controlled atmosphere may be circulated for the protection of the charge. This method is readily adaptable to hot-rolled or cold-reduced electrical sheets either as cut lengths or in coil form.

Open-coil annealing is a process of heating coils of steel in a closed furnace containing a controlled atmosphere. The individual laps of the coil are separated so that the controlled atmosphere can circulate between each lap of the coil.

267. Interlaminar insulation is produced by forming a natural oxide layer on the surface of flat-rolled electrical steel or applying a surface coating. This treatment,

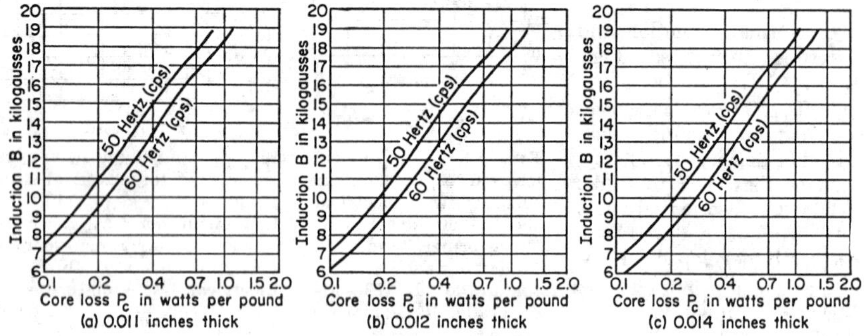

FIG. 4-39. Core loss of grain-oriented material at high inductions.

of course, does not reduce the eddy currents within the laminations. Improved interlamination resistance is usually obtained by annealing the lamination under slightly oxidizing conditions, which increase the surface oxide, and then core-plating the finished laminations.

The insulating coatings or finishes may be broadly classified as organic or inorganic.

268. Organic insulation consists, in general, of enamels or varnishes which are applied to the steel surfaces to provide interlaminar resistance. Such coatings may also improve the punchability of steel used for laminations not requiring a stress-relief anneal.

Flat-rolled electrical steel having the organic type of coating cannot be given a stress-relief anneal without impairing the insulation value of the coating. The coating, however, will withstand normal operating temperatures. Some organic insulations are suitable only in air-cooled cores, whereas other organic insulations may be suitable for cores in both air-cooled and oil-immersed transformers. The thickness of this type of insulation is approximately 0.0001 in/side.

269. Inorganic insulation is generally characterized by high resistance and by the ability to withstand stress-relief annealing temperatures. It is intended for cores in air-cooled or oil-immersed transformers and may also contribute to improved punchability. Descriptions of various categories of flat-rolled electrical steel insulation (sometimes called core plates) are shown in Table 4-51.

Table 4-51. Descriptions of Flat-rolled Electrical Steel Insulations or Core Plates

Suggested Identification	Description
C-0	This identification is merely for the purpose of describing the natural oxide surface which occurs on flat-rolled silicon steel, which gives a slight but effective insulating layer sufficient for most small cores, and which will withstand normal stress-relief annealing temperatures. This oxidized surface condition may be enhanced in the stress-relief anneal of finished cores by controlling the atmosphere to be more or less oxidizing to the surface.
C-1	This insulation consists of an enamel or varnish coating intended for cores not immersed in oil. It is most often used to enhance punchability rather than for its insulating quality. This type of organic core plate is resistant to normal operating temperatures but will not withstand stress-relief annealing.
C-2	This identification is for the purpose of describing an inorganic insulation which consists of a glasslike film which forms during high-temperature hydrogen anneal of grain-oriented silicon steel as the result of the reaction of an applied coating of MgO and silicates in the surface of the steel. This insulation is intended for air-cooled or oil-immersed cores. It will withstand stress-relief annealing temperatures and has sufficient interlamination resistance for wound cores of narrow-width strip such as is used in distribution transformer cores. It is not intended for stamped laminations because of the abrasive nature of the coating.
C-3	This insulation consists of an enamel or varnish coating intended for air-cooled or oil-immersed cores. The interlamination resistance provided by this coating is superior to the C-1 type of coating, which is primarily utilized as a die lubricant. The C-3 coating will also enhance punchability and is resistant to normal operating temperatures but will not withstand stress relief annealing.
C-4	This insulation consists of a chemically treated or phosphated surface intended for air-cooled or oil-immersed cores. It will withstand stress-relief annealing and serves to promote punchability. It is often applied over the C-2 coating on grain-oriented silicon steel. This combination of coatings develops excellent interlamination resistance.
C-5	This is an inorganic insulation similar to the chemically treated C-4 finish, but with ceramic fillers added to enhance the surface resistance. It is often applied over the C-2 coating on grain-oriented silicon steel. It is intended for air-cooled or oil-immersed cores and will withstand stress-relief annealing.

270. Special-purpose Materials. For certain applications of soft, or nonretentive, materials, special alloys and other materials have been developed which, after proper fabrication and heat-treatment, have superior properties in certain ranges of magnetization. Several of these alloys and materials will be described.

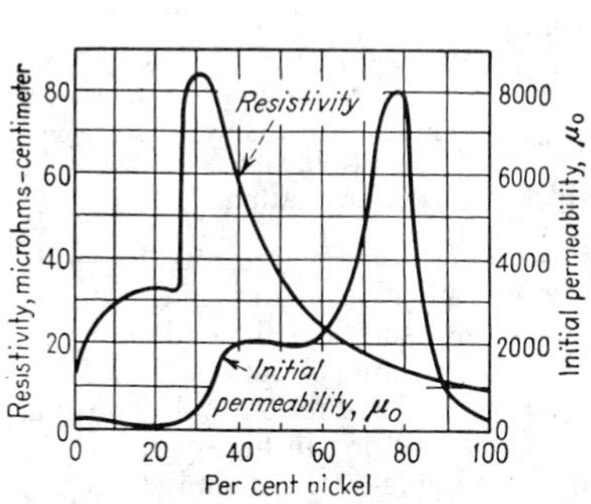

FIG. 4-40. Electrical resistivity and initial permeability of iron-nickel alloys with various nickel contents.

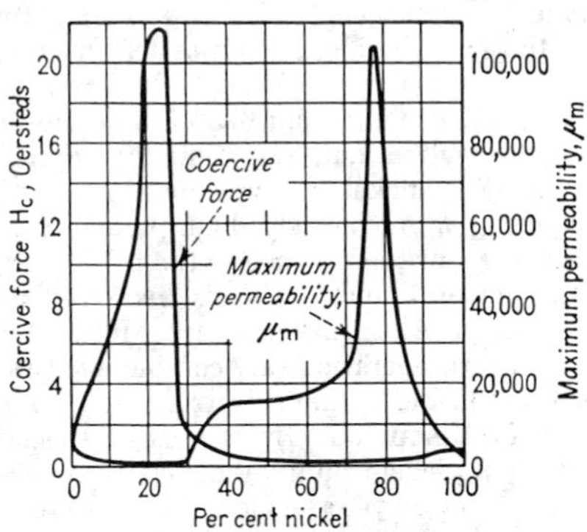

FIG. 4-41. Maximum permeability and coercive force of iron-nickel alloys with various nickel contents.

271. Nickel-Iron Alloys. Nickel alloyed with iron in various proportions produces a series of alloys with a wide range of magnetic properties. With 30% nickel, the alloy is practically nonmagnetic and has a resistivity of 86 $\mu\Omega$/cm. With 78% nickel the

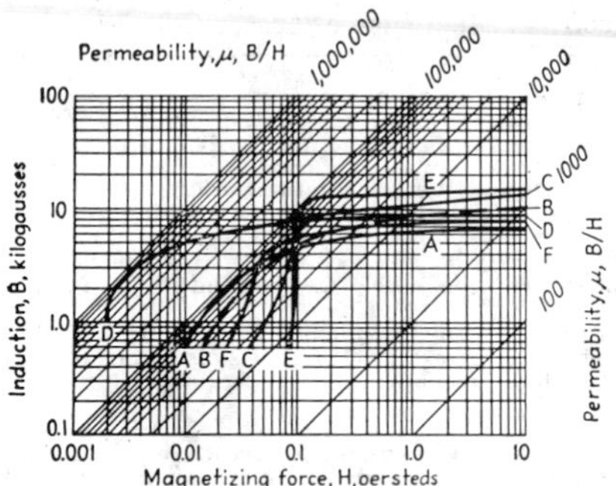

FIG. 4-42. Induction-permeability curves of some high-nickel alloy strip (0.014 in). *A.* Mumetal. *B.* Permalloy. *C.* 48% nickel-iron. *D.* Supermalloy. *E.* Deltamax. *F.* MoPermalloy.

alloy properly heat-treated has very high permeability. These effects are shown in Figs. 4-40 and 4-41. Many variations of this series have been developed for special purposes. Table 4-52 lists some of the more important commercial type of nickel-iron alloys, with their approximate properties. These alloys are all very sensitive to heat-treatment; so their properties are largely influenced thereby. A comparison of their normal-induction curves is given in the curves of Fig. 4-42 and Fig. 4-24.

272. Permalloy[1] is a term applied to a number of nickel-iron alloys developed by the Bell Laboratories, each specified by a prefix number indicating the nickel content. The term is usually associated with the 78.5% nickel-iron alloys, the important properties of which are high permeability and low hysteresis loss in relatively low magnetizing fields. These properties are obtained by a unique heat-treatment consisting of a high-temperature anneal, preferably in hydrogen, with slow cooling followed by rapid cooling from about 625°C. The alloy is very sensitive to mechanical strain; so it is desirable to heat-treat the alloy in its final form. The addition of 3.8% chromium or molybdenum increases the resistivity from 16 to 65 and 55 $\mu\Omega\cdot$cm, respectively, without seriously impairing the magnetic quality. In fact, low-density permeabilities are better with these additions. These alloys have found their principal application as a material for the continuous loading of submarine cables and in loading coils for land lines (see Figs. 4-42 and 4-43 and Table 4-52).

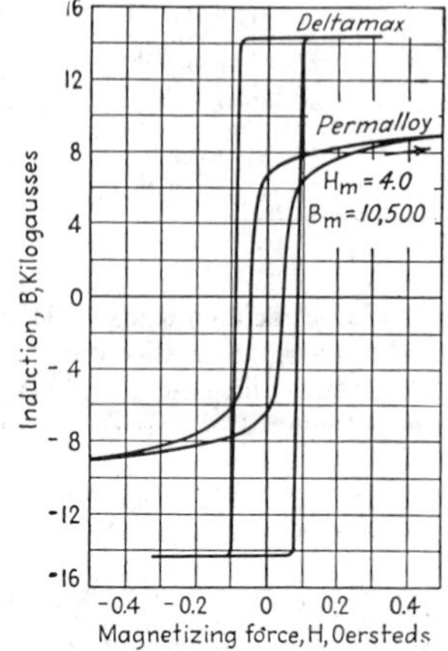

FIG. 4-43. Hysteresis curves of 78 Permalloy and Deltamax from saturation.

By special long-time high-temperature treatments, maximum permeability values greater than 1 million have been obtained. The double treatment required by the 78% Permalloy is most effective when the strip is thin, say under 10 mils. For greater thicknesses, the quick cooling from 625°C is not uniform throughout the section, and loss of quality results.

A 48% nickel-iron was developed for applications requiring a moderately high-permeability alloy with higher saturation density than 78 Permalloy. The same general composition is marketed under many names, as Hyperm 50, Hipernik, Audiolloy, Allegheny Electric Metal, 4750, Carpenter 49 alloy. Annealing after all mechanical operations are completed is recommended. These alloys have found extensive use in radio, radar, instrument, and magnetic-amplifier components (see Fig. 4-42).

273. Deltamax. By the use of special techniques of cold reduction and annealing, the 48% nickel-iron alloy develops directional properties resulting in high permeability and a square hysteresis loop in the rolling direction (see Fig. 4-43). A similar product is sold under the name of Orthonic. For optimum properties, these materials are rapidly cooled after a 2-h anneal in pure hydrogen at 1100°C. They are generally used in wound cores of thin tape for applications such as pulse transformers and magnetic amplifiers.

274. Iron-Nickel-Copper-Chromium. The addition of copper and chromium to high nickel-iron alloys has the effect of raising the permeability at low flux density.

[1] *Elem. Bell System Tech. Jour.*, 1936, Vol. 15, p. 113.

Table 4-52. Special-purpose Materials

Name	Approximate composition, %	Saturation, G	Maximum permeability	Coercivity (from saturation), Oe	Initial permeability	Resistivity, microhm-cm
78 Permalloy	78.5 Ni	10,500	70,000	−0.05	8,000	16
MoPermalloy	79 Ni, 4.0 Mo	8,000	90,000	−0.05	20,000	55
Supermalloy	79 Ni, 5 Mo	7,900	900,000	−0.002	100,000	60
48% nickel-iron	48 Ni	16,000	60,000	−0.06	5,000	45
Monimax	47 Ni, 3 Mo	14,500	35,000	−0.10	2,000	80
Sinimax	43 Ni, 3 Si	11,000	35,000	−0.10	3,000	85
Mumetal	77 Ni, 5 Cu, 2 Cr	6,500	85,000	−0.05	20,000	60
Deltamax	50 Ni	15,500	85,000	−0.10	....	45

Alloys of this type are marketed under the names of Mumetal, 1040 alloy, Hymu 80. A typical induction characteristic is curve *A* in Fig. 4-42; for optimum properties they are annealed after cutting and forming for 4 h at 1100°C in pure hydrogen and cooled slowly. Important applications are as magnetic shielding for instruments and electronic equipment and as cores in magnetic amplifiers.

275. Constant-permeability alloys having a moderate permeability which is quite constant over a considerable range of flux densities are desirable for use in circuits in which waveform distortion must be kept at a minimum. Isoperm and Conpernik are two alloys of this type. They are nickel-iron alloys containing 40 to 55% nickel which have been severely cold-worked. Perminvar is the name given to a series of cobalt-nickel-iron alloys (for example, 50% nickel, 25% cobalt, 25% iron) which also exhibit this characteristic of constant permeability over a low ($\cong$800 G) density range. When magnetized to higher flux densities, they give a double loop constricted at the origin so as to give no measurable remanence or coercive force. The characteristics of the alloys in this group vary greatly with the chemical content and the heat-treatment. A sample containing approximately 45 Ni, 25 Co, and 30 Fe, baked for 24 h at 425°C and slowly cooled, had hysteresis losses as follows: At 100 G, 214×10^{-4} erg/(cm cube) (cycle); at 1,003 G, 15.27 ergs; at 1,604 G, 163 ergs; at 4,950 G, 1,736 ergs; and at 13,810 G, 4,430 ergs. Over the range of flux densities in which the permeability is constant (from 0 to 600 G), the hysteresis loss is very small, or on the order of the foregoing figure for 100 G. The resistivity of the sample was 19.63/microhm-cm.

276. Monel metal is an alloy of 67% nickel, 28% copper, and 5% other metals. It is slightly magnetic below 95°C (see Fig. 4-24).

277. Iron-Cobalt Alloys. The addition of cobalt to iron has the effect of raising the saturation intensity of iron up to about 36% cobalt (Fe_2Co). This alloy is useful for pole pieces of electromagnets and for any application where high magnetic intensity is desired. It is workable hot but quite brittle cold. **Hyperco** contains approximately ⅓ Co, ⅔ Fe, plus 1 to 2% "added element" (see Fig. 4-26). Total case loss is about 2.5 W/lb at 15 kG and 0.010 in thick. It is available as hot-rolled sheet, cold-rolled strip, plates, and forgings. The 50% cobalt-iron alloy Permendur has a high permeability in fields up to 50 Oe and, with about 2% vanadium added, can be cold-rolled (see Fig. 4-24).

278. Iron-Silicon Aluminum Alloys. Aluminum in small percentages, usually under 0.5%, is a valuable addition to the iron-silicon alloy. Its principal function appears to be as a deoxidizer. Masumoto[1] has investigated soft magnetic alloys containing much higher percentages of aluminum and found several that have high permeabilities and low hysteresis losses. Certain compositions have very low magnetostriction and anisotropy, high initial permeability, and high electrical resistivity. An alloy of 9.6% silicon and 6% aluminum with iron has better low-flux-density properties than the Permalloys. However, poor ductility has limited these alloys to d-c applications in cast configurations or in insulated pressed-powder cores for high-frequency uses. These

[1] Masumoto. On a New Alloy "Sendust" and Its Magnetic and Electrical Properties; *Tohoku Imp. Univ. Rept.*, 1936, Anniversary Vol., Ser. I, p. 388.

alloys are commonly known as Sendust. The material has recently been prepared in sheet form[1] by special processes.

279. Temperature-sensitive Alloys.[2] Inasmuch as the Curie point of metal may be moved up or down the temperature scale by the addition of other elements, it is possible to select alloys which lose their ferromagnetism at almost any desired temperature up to 1115°C, the change point in cobalt. Iron-base alloys are ordinarily used to obtain the highest possible permeability at points below the Curie temperature. Nickel, manganese, chromium, and silicon are the most effective alloy elements for this purpose;[1,3] and most alloys made for temperature-control application, such as instruments, reactors, and transformers, use one or more of these. Figure 4-44 shows the magnetization-temperature characteristics of a group of these alloys. The Carpenter Temperature Compensator 30 is a nickel-copper-iron alloy which loses its magnetism at 55°C and is used for temperature compensation in meters.[4]

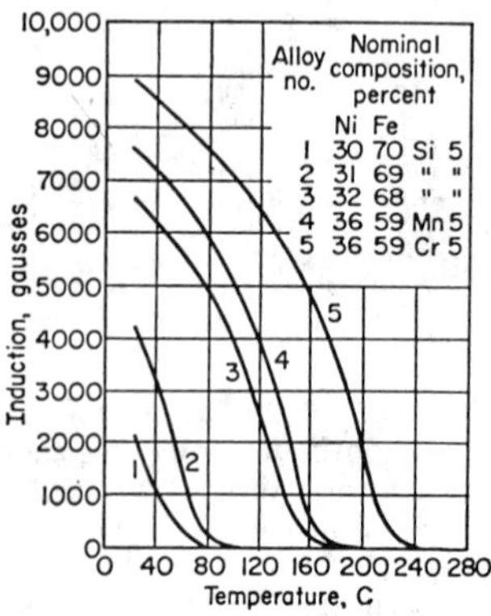

FIG. 4-44. Thermosensitive alloys. Temperature-induction for $H = 20$.

280. Heusler's alloys are ferromagnetic alloys composed of "nonmagnetic" elements. Copper, manganese, and aluminum are frequently used as the alloying elements. The saturation induction is about one-third that of pure iron.

281. High-frequency Materials Applications.[5] The ideal core material for small reactors and transformers employed in communication equipment should possess the characteristics of constant permeability, small hysteresis loss, and small eddy-current loss within the range of small magnetizing forces and over the wide range of frequencies met in such applications. At the higher frequencies, the control of eddy currents becomes of primary importance, not only to reduce losses but also to minimize skin effect produced by eddy-current shielding. This is accomplished by the use of high-permeability alloys in the form of wound cores of thin tape, by the use of compressed powdered iron-alloy cores, or by sintered ferrites. Comparisons of these types of materials are given in Table 4-53.

Table 4-53. Properties of High-frequency Core Materials

Core material	Permeability		Saturation, kg	Resistivity $\mu\Omega\cdot$cm	Curie point, °C
	Maximum	Initial			
Oriented 3% silicon tape 0.004 in.	35,000	1,250	20	50	750
Monimax tape 0.004 in.	38,000	2,000	15	80	390
4-79 MoPermalloy tape 0.004 in.	75,000	20,000	8	55	580
Ferrites	1,000-5,000	400-1,000	2-5	10^3-10^8	100-330
Powdered iron	50-160	25-125	8	......	700
Sendust	120,000	30,000	10	80	

Thin electrical steels are insulated 3% iron-silicon alloys designed for applications involving frequencies of 400 to 20,000 c/s and for pulse components. They are made in strip 0.001 to 0.007 in thick for wound cores, with high effective permeability and low losses at relatively high flux densities.

[1] HELMS and ADAMS. Sendust Sheet-processing Techniques and Magnetic Properties; *Jour. Appl. Phys.*, March, 1964, Vol. 35.
[2] JACKSON and RUSSELL. Temperature Sensitive Magnetic Alloys and Their Uses; *Instruments*, November, 1938, p. 280.
[3] SHAW, J. L. Curie Temperature Alloys; *Product Eng.*, June, 1948.
[4] EBERLY, W. S. Temperature-compensator Alloys; *Machine Design*, May, 1954.
[5] BOZORTH, R. M. "Ferromagnetism"; Princeton, N.J., D. Van Nostrand Company, Inc., 1951.

282. Nickel-alloy tape of high permeability is used in thicknesses of 0.001 to 0.010 in for wound cores designed for the frequency range 100 to 100,000 c/s. Commonly used alloys for this purpose are the Permalloys, MoPermalloy, Deltamax, and Supermalloy, the thickness being chosen to provide the desired permeability at the application frequency. Figure 4-45 shows the effect of tape thickness on the initial permeability of these types of materials (see Table 4-53).

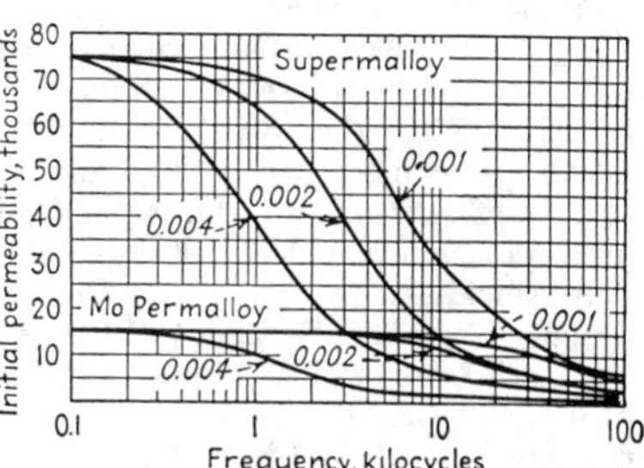

FIG. 4-45. Effect of tape thickness on the initial permeability of Supermalloy and MoPermalloy at various frequencies.

283. Ferrite cores are molded from a mixture of metallic oxide powders such that certain iron atoms in the cubic crystal of magnetite (ferrous ferrite) are replaced by other metal atoms, such as Mn and Zn, to form manganese zinc ferrite, or by Ni and Zn to form nickel zinc ferrite.[1] They resemble ceramic materials in production processes and physical properties. The d-c resistivities correspond to those of semiconductors, being at least 1 million times those of metals. Magnetic permeabilities may be as high as 5,000 and apparent dielectric constants in excess of 100,000.[2] The Curie point is quite low, however, in the range 100 to 300°C. Saturation flux density is generally below 5,000 G. Ferrite cores provide design advantages over strip and power cores for such uses as filter cores up to 200 kc, as deflection transformers and yokes, and in antenna rods, pulse transformers, delay lines, and waveguide elements (see Table 4-53, Fig. 4-46, and Fig. 4-47).

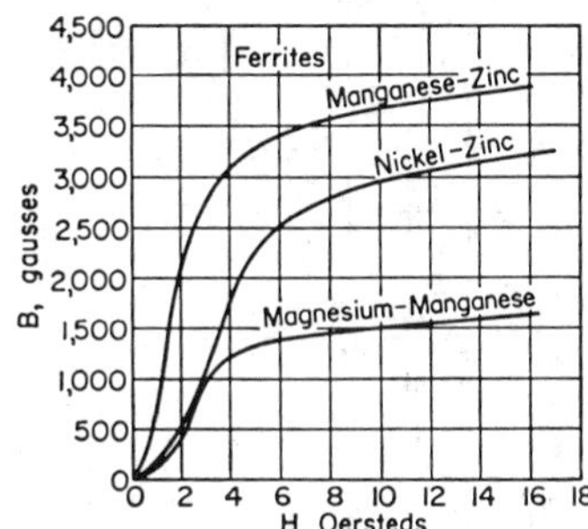

FIG. 4-46. Typical normal-induction curves for "soft" ferrites.

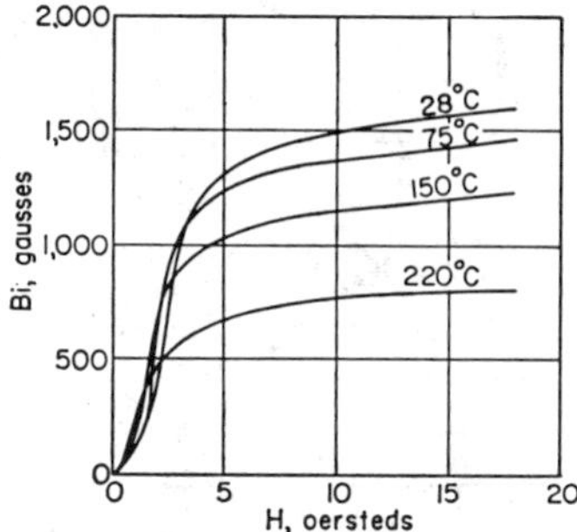

FIG. 4-47. Temperature dependence of a magnesium-manganese ferrite.

284. Powdered-iron Cores. Iron powders having a grain size of the order of 10 microns (or 10^{-3} cm) are manufactured by a chemical process and coated to a thickness of 1 micron with a special insulating material. The powder is then mixed with phenol resin binder, compressed at high pressure, and baked. The product is a chemically stable magnetic body containing 90% pure iron by weight; it can be worked mechanically in the same manner as soft metals. This material has been developed for applications in r-f apparatus. It has a core loss, at 500 kc/s and a flux density of 1 G, of 0.012 erg/(cm)/(cycle). For additional data see W. J. Polydoroff, "High-frequency Magnetic Materials," John Wiley & Sons, Inc., 1960.

285. Compressed powdered alloy, such as Permalloy, has also been developed for applications similar to those of compressed powdered iron and has superseded the

[1] SNOEK, J. L. Nonmetallic Magnetic Materials for High Frequencies; *Philips Tech. Rev.*, December, 1946, Vol. 8, pp. 353–360.
[2] HOH, S. R. *WADC Tech. Rept.* 52–287, November, 1952.

latter for certain purposes, such as loading coils for long telephone-cable circuits. The superior magnetic properties permit the use of smaller cores, with considerable saving in the overall dimensions of a coil of given inductance. It is necessary, however, to apply the proper heat-treatment in order to develop the desired properties of Permalloy. *Compressed powdered molybdenum Permalloy*,[1] having lower hysteresis losses and higher resistivity than the 78 or 81% nickel-iron alloy, is considered the best of this class of alloys. It contains about 2% molybdenum, 81% nickel, and 17% iron. Much of the quality of all these powder cores depends upon the use of specially developed particle insulation which will withstand the desired pressure and temperature and be of minimum thickness.

286. Permanent-magnet (Retentive) Materials. Although it is possible to make permanent magnets of almost any kind of steel that is capable of being hardened by heat-treatment, it is best to use materials especially produced for this purpose. Before the development of the special magnet steels, magnets were generally made of plain high-carbon tool steel. This type of steel is relatively inexpensive, but its magnetic properties are greatly inferior to those of the special steels.

Permanent-magnet materials may be grouped in five classes as follows:

a. Precipitation-hardened alloys.
b. Quench-hardened alloys.
c. Ceramic.
d. Iron-powder compacts.
e. Work-hardened materials.

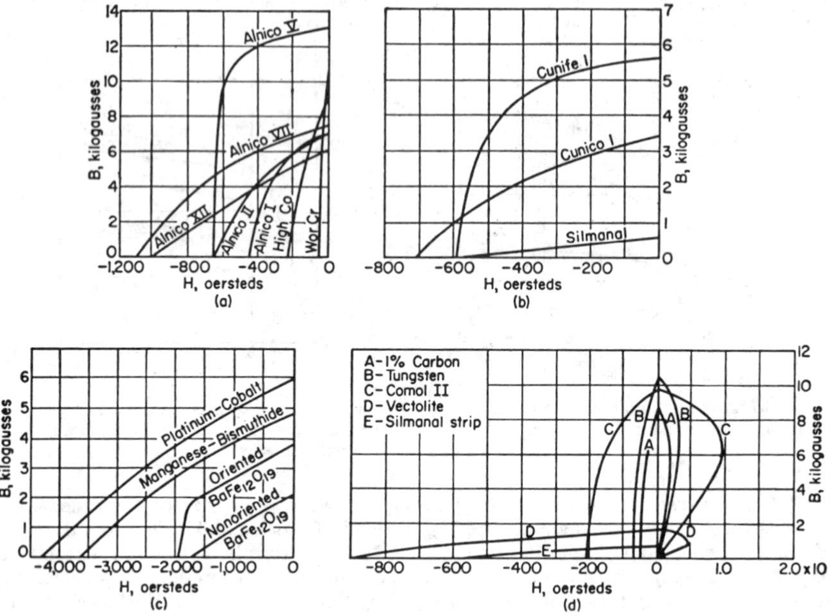

FIG. 4-48. Typical demagnetization curves for permanent-magnet materials.

Figure 4-48 and Table 4-54 show typical demagnetization curves and information for several permanent magnet materials.

287. Precipitation-hardening Alloys. Alnico[2] magnet alloys have the highest energy per unit of cost or volume of any permanent-magnet material commercially available.

[1] LEGG, V. E. and GIVEN, F. J. *Trans. AIEE*, 1940.
[2] NEUMANN, H. Materials for Permanent Magnets; *Arch. Tech. Messen*, March, 1937, Vol. 69, p. T38-43.

Table 4-54. Permanent-magnet Materials

Material	Retentivity B_{rs}, G	Coercivity H_{cs}, Oe	Maximum energy product $(B_dH_d)_m$, G·Oe × 10^6
Alnico I	7,200	450	1.35
Alnico II	7,400	540	1.65
Alnico III	6,900	470	1.35
Alnico IV	5,500	730	1.30
Alnico V	12,500	630	5.5
Alnico Vcc	13,000	750	7.0
Alnico VI	10,500	750	3.7
Alnico VII	7,500	1,100	3.0
Alnico VIII	8,500	1,300	4.5
Alnico XII	5,500	950	1.6
Remalloy	9,700	250	1.0
Cunife	5,400	550	1.7
Cunico	3,400	700	0.85
Vicalloy	7,500	250	0.80
Silmanal	550	550	0.08
1% carbon	9,000	50	0.2
3.5% chromium	9,500	65	0.29
6% tungsten	10,000	70	0.32
17% cobalt	7,500	200	0.60
36% cobalt	9,600	250	0.95
Platinum-cobalt	6,000	4,000	7.0
Platinum-iron	5,400	1,550	3.1
Barium ferrite (nonoriented)	2,000	1,800	1.0
Barium ferrite (oriented)	3,900	2,100	3.5
Iron compacts	7,900	560	2.2
Fe-Co compacts	9,000	850	3.6
Mn Bi	4,800	3,650	5.3

They are characterized by a higher coercivity, a higher energy, and (except for Alnico V and VI) a lower retentivity than the magnet-steel types. They are formed only by casting or sintering, are relatively weak and brittle, cannot be readily machined except by grinding, and require a field of 2,000 to 3,000 Oe for complete magnetization. The standard grades differ in properties and somewhat in cost, and their choice depends upon the use and design of the magnet (see Fig. 4-48*a*). Alnico V, which has the highest external energy of the group, gets its superior properties by cooling in a magnetic field. It is therefore strongly directional in its characteristics and must be magnetized in the direction of the field applied during heat-treatment. The large grain size of Alnico V makes it possible to produce a structure of essentially columnar grains[1] by differential cooling during the casting process. Alnico V with this columnar structure has a much fuller hysteresis loop, a larger coercivity, and a greater maximum energy product. This material is designated as Alnico Vcc in Table 4-54.

Koch[2] and his coworkers have demonstrated improved magnetic properties in a titanium Alnico alloy, using an isothermal magnetic field treatment. The commercial version of this development is called Alnico VIII. It has the highest coercivity of any Alnico alloy. By using the titanium alloys it has been possible in the laboratory to increase the energy product to 12 × 10^6 G·Oe.

Cobalt-molybdenum-iron (known as Remalloy or Comol) is a cast and hot-rolled magnet material, preferably containing 12% Co and 17% Mo, of the precipitation-hardening type. After quenching in air or oil at 1200 to 1300°C, it can be formed and machined and is then aged for 1 h at 650 to 700°C, after which it is not sensitive to further aging.

Cunife[3] is a copper-nickel-iron alloy that is malleable, ductile, and machinable, even in the age-hardened form. It has directional properties and should be magnetized in the direction in which it was drawn. In small sizes, Cunife has a tensile strength of approximately 10,000 lb/in². Remanence decreases markedly at elevated temperatures, about 50% at 325°C, and is nonmagnetic above 400°C (see Fig. 4-48*b* to *d*).

---

[1] Ebeling, D. G. U.S. Patent 2,578,407.
McCaig, M. *Proc. Phys. Soc.* (*London*), 1949, Vol. B62, pp. 652–656.

[2] Koch, A. J., v. Stag, M. G., and de Vos, K. J. Conference on Magnetism and Magnetic Materials, February, 1957.

[3] Neumann, H., Buchner, A., and Reinboth, H. *Metalkunde*, 1937, Vol. 29, pp. 173–185. Mechanically soft iron-nickel-copper permanent-magnet alloys.

Cunico[1] is a copper-nickel-cobalt alloy which is ductile prior to its final heat-treatment but which cannot be readily machined thereafter. Magnets can be made from rods, strips, and wire and can also be cast. The magnetic properties are independent of the direction of cold working or heat-treatment (see Fig. 4-48*b*).

Vicalloy[2] is the trade name for permanent-magnet alloys of iron, cobalt, and vanadium (see Table 4-54 and Fig. 4-48*b*). Vicalloy I contains 9.5% vanadium and has a much higher energy product, which is obtained by heavy cold working. It is therefore strongly directional, having its best properties in the direction of cold working. It is aged at 600°C, after which it is no longer ductile. It is used in tape form for magnetic recording.

Silmanal[3] is a ternary alloy of silver, manganese, and aluminum, having the highest intrinsic coercive force of any of the magnet materials. It will withstand severe demagnetizing effects.[4] The alloy is ductile enough to be drawn to a fine wire and can be machined as readily as soft steel. Magnets are made in a wide variety of shapes from wire, swaged rods, or rolled sheets. Silmanal requires a magnetizing force of 20,000 Oe and is therefore commonly used only in lengths less than 1½ in. Care must be taken not to heat the material above 200°C, after which it is no longer ductile. It is used in tape form for magnetic recording.

288. Quench-hardened Alloys. *Carbon Magnet Steel.* The coercivity and retentivity of quenched carbon steel increase with the carbon content up to the eutectoid point, or about 0.85% of carbon; with still higher carbon content the retentivity decreases (see Fig. 4-48*d*).

Tungsten magnet steel contains approximately 5 to 6% tungsten, 0.60 to 0.80% carbon, and about 0.50% manganese. There are two general types, viz., oil hardening and water hardening. In designs subject to breakage in water quenching, the oil-hardening type should be used (see Fig. 4-48*d*).

Chrome magnet steels contain approximately 2½ to 6% chromium and about the same proportions of carbon and manganese as tungsten magnet steel. They are nearly as efficient magnetically as tungsten steel and less expensive (see Fig. 4-48*a*).

Cobalt-chrome magnet steel is an alloy steel containing about 11% cobalt and 9% chromium. It has the advantage of being readily machinable under production conditions. Magnetically it is not quite so efficient as cobalt steel but is less expensive.

Cobalt magnet steel, which is also known as K. S. Steel, contains approximately 36% cobalt, 4% tungsten, and 6% chromium. It forms and punches well when hot but is not well suited for magnets requiring considerable machining, although it can be drilled. Because of its high cobalt content it is expensive (see Table 4-54 for data on its magnetic properties; also see P. H. Brace, Cobalt Magnet Steel, *Elec. Jour.*, March, 1929, pp. 111–121. See Fig. 4-48*a*).

Platinum-iron alloys containing 60 to 90% platinum develop high values of coercive force depending on heat-treatment. Highest values are obtained by quenching from 1200°C. The Curie points for these alloys are high (near 1100°C), making them of value in high-temperature applications. The preferred alloy contains about 78% platinum (see Table 4-54).

Platinum-cobalt alloys have the highest energy product of any of the alloys with noble metals. The high coercivity is attained by quenching at 1200°C, after which they may be machined to shape, followed by aging at 700°C. Machining after aging is not possible because of their hardness (see Table 4-54).

289. Ceramic Magnet Materials. Commercial developments in permanent-magnet materials which are increasing in importance each year are the barium ferrite ceramic permanent-magnet materials. These are chemical compounds with mechanical characteristics similar to those of other ceramics. These materials are hard and brittle and have a lower density than metals and extremely high electrical resistivity. The basic ingredients are barium carbonate and iron oxide. The materials in powdered form

[1] Dannohl, W., and Neumann, H. *Metalkunde*, 1938, Vol. 30, pp. 217–231. On permanent-magnet alloys of Co, Cu, and Ni.

[2] Nesbitt. Vicalloy, a Workable Permanent Magnet Alloy; *Metals Technol.*, 13, No. 1973, pp. 1-11.

[3] Potter, H. H. *Phil. Mag.*, Vol. 12, No. 7, pp. 225–264.

[4] Neumann, H., Buchner, A., and Reinboth, H. *Metalkunde*, 1937, Vol. 29, pp. 173–185. Mechanically soft iron-nickel permanent-magnet alloys.

are compressed in dies under high pressure to the required shape. This compacted material is then sintered at a high temperature. This process produces a material which has an H_{cs} of approximately 1,800 Oe, a B_{rs} of approximately 2,000 G, and a demagnetization curve which is practically a straight line. Further improvements in ceramic materials have resulted in a highly oriented barium-iron oxide whose magnetic properties, on a weight basis, are almost equal to those of Alnico V. The coercivity is approximately 2,000 Oe, the retentivity 4,000 G, and the energy product 3.5 times that of the unoriented variety. At right angles to the direction of grain orientation, however, this material exhibits negligible permanent-magnet properties and has relative permeability of only approximately 1.0 (see Fig. 4-48*c*).

290. Powder Magnets. Although pure iron is usually regarded as a high-permeability, or "magnetically soft," material, yet theory has predicted and experiments have proved that compacts of pure iron powders may produce very good permanent magnets. Powder magnets have been produced of iron and iron alloys (such as 70% iron and 30% cobalt) with particle size of about 10^{-5} cm diameter with H_{cs} up to 850 Oe, B_{rs} of 9,000 G, and energy product of 3.5×10^6 G·Oe. The permanent-magnet properties result from the discrete small particles of a single phase instead of from the presence of two or more phases, as in most other metallic permanent-magnet material. Further experimental work with particle size and shape and processes of manufacture has produced, in the laboratory, magnets with energy products comparable with those of Alnico V, and theoretical considerations predict even higher values.

Manganese-bismuth permanent magnets also belong to this group. This material is an anisotropic aggregate of crystals of the intermetallic compound manganese bismuthide (MnBi) and is a product of powder metallurgy. Manganese bismuthide is prepared from the chemical action between molten bismuth and powdered manganese when heated to approximately 700°C in an inert atmosphere of argon or helium. Cooling is accomplished in such a manner as to produce crystallization of the compound. Laboratory-produced material may have a retentivity of approximately 4,800 G and a coercivity of 3,600 Oe, with energy-product values as high as 5×10^6 G·Oe.

Powder metallurgy has also produced sintered Alnico magnets. These magnets have greater mechanical strength and more uniform magnetic properties than the cast variety, at the expense of a slight decrease in the magnetic properties.

Magnet materials prepared from metal oxides such as cobalt ferrites and Vectolite have been made and used for many years; however, they have been practically superseded by the barium ferrites.

291. Work-hardened Alloys. Several ordinarily "nonmagnetic" alloys of iron may become ferromagnetic after cold working owing to a phase change in the material. Stainless steel (18% chromium, 8% nickel) is "nonmagnetic" at room temperature after being rapidly cooled from 1200°C in the usual process of manufacture. However, if it is hardened by cold working such as drawing through a reducing die, it may develop properties such that it makes an acceptable permanent-magnet material at room temperature. If this work-hardened alloy is then reheated to a high temperature and cooled slowly, it regains its original nonmagnetic condition at room temperature. Another alloy that shows this property contains 45% iron, 15% Ni, and 40% Cu. Nesbitt[1] has measured a coercivity of 240 Oe and a retentivity of 4,400 G in wire of this composition, which, after quenching from 1000°C, was then cold-drawn from 0.026 to 0.006 in. Increasing the percentage of iron to 60%, decreasing the percentage of copper to 25%, with 15% nickel, produced an alloy that after similar treatment as above resulted in a coercivity of 170 Oe and retentivity of 11,000 G.

292. Bibliography. Information on the properties of commercial magnetic materials may be obtained from technical bulletins and handbooks issued by the various producers of magnetic materials. Other sources of information include:

Sanford, R. L., and Cooter, I. L. *NBS Monograph* 47, Basic Magnetic Quantities and the Measurement of the Magnetic Properties of Materials.

AISI Steel Products Manual, Flat-rolled Electrical Steel.

Bozorth, R. M. "Ferromagnetism"; Princeton, N.J., D. Van Nostrand Company, Inc., 1951.

[1] Bozorth, R. M. "Ferromagnetism"; Princeton, N.J., D. Van Nostrand Company, Inc., 1951, p. 401.

ASTM standards relating to magnetic properties: Specifications for Flat-rolled Electrical Steel, A345; Methods of Test for Magnetic Materials, A34, Normal Induction and Hysteresis of Magnetic Materials, A341, Permeability of Feebly Magnetic Materials, A342, Alternating Current Magnetic Materials Using Epstein Specimens, A343, Electrical and Mechanical Properties of Magnetic Materials, A344, Alternating Current Magnetic Properties of Laminated Core Specimens, A346; Definitions of Terms, Symbols and Conversion Factors Relating to Magnetic Testing, A340.

PARKER and STUDDERS. "Permanent Magnets and Their Applications"; New York, John Wiley & Sons, Inc., 1962.

INSULATING MATERIALS

GENERAL PROPERTIES

By T. W. DAKIN

293. Electrical Insulation and Dielectric Defined. Electrical insulation is a medium or a material which, when placed between conductors at different potentials, permits only a small or negligible current in phase with the applied voltage to flow through it. The term dielectric is almost synonymous with electrical insulation, which can be considered the applied form of the dielectric. The perfect dielectric would pass no current between conductors. A perfect vacuum (at low stresses between uncontaminated pure metal surfaces) is the only perfect dielectric.

The range of resistivities of substances which can be considered insulators is from greater than 10^{20} Ω·cm downward to the vicinity of 10^6 Ω·cm, depending on the application and voltage stress. There is no sharp boundary defined between low-resistance insulators and semiconductors. If the voltage stress is low and there is little concern about the level of current flow (other than that which would heat and destroy the insulation), relatively low-resistance insulation can be tolerated.

294. Circuit Analogy of a Dielectric or Insulation. Any dielectric or electrical insulation can be considered as equivalent to a combination of capacitors and resistors which will duplicate the current-voltage behavior at a particular frequency or time of voltage application. In the case of some dielectrics, simple linear capacitors and resistors do not adequately represent the behavior. Rather, resistors and capacitors with particular nonlinear voltage-current or voltage-charge relations must be postulated to duplicate the dielectric current-voltage characteristic.

The simplest circuit representation of a dielectric is a parallel capacitor and resistor, as shown in Fig. 4-49 for $R_S = 0$. The perfect dielectric would be simply a capacitor. Another representation of a dielectric is a series-connected capacitor and resistor as in Fig. 4-49 for $R_p = \infty$, while still another involves both R_S and R_p.

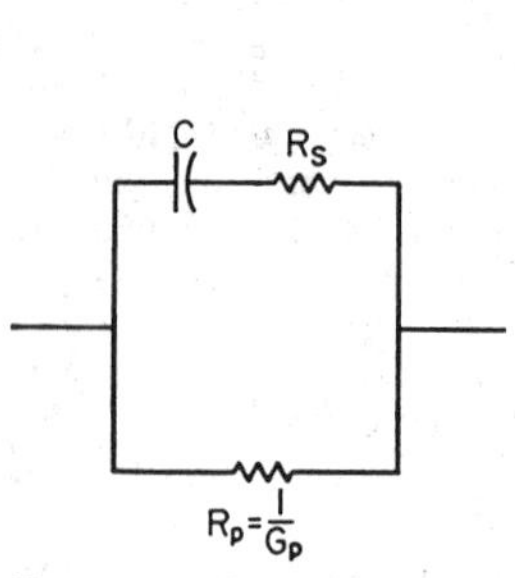

FIG. 4-49. Equivalent circuit of a dielectric.

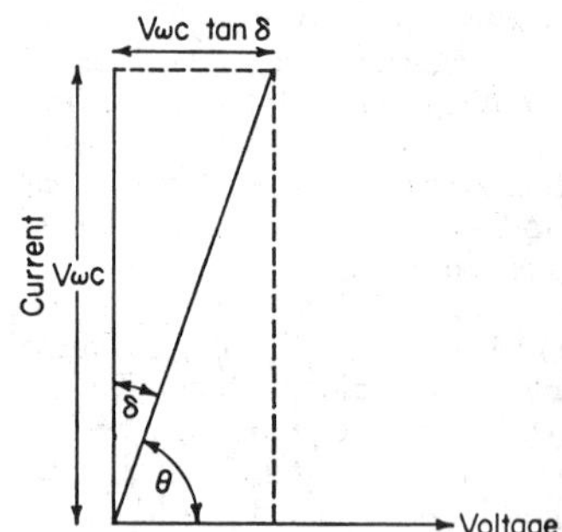

FIG. 4-50. Current-voltage phase relation in a dielectric.

The a-c dielectric behavior is indicated by the phase diagram (Fig. 4-50). The perfect dielectric capacitor has a current which leads the voltage by 90°, but the imperfect dielectric has a current which leads the voltage by less than 90°. The dielectric phase angle is θ, and the difference, $90° - \theta = \delta$, is the loss angle. Most measurements of dielectrics give directly the tangent of the loss angle $\tan \delta$ (known as the *dissipation factor*) and the capacitance C. In Fig. 4-49, $\tan \delta = 2\pi f C R_s$ for $R_p = \infty$, and for $R_s = 0$, $\tan \delta = 1/2\pi f C R_p$.

The a-c power or heat loss in the dielectric is $V^2 2\pi f C \tan \delta$ W, or $VI \sin \delta$ W, where $\sin \delta$ is known as the power factor, V is the applied voltage, I is the total current through the dielectric, and f is the frequency. From this it can be seen that the equivalent parallel conductance of the dielectric, σ (the inverse of the equivalent parallel resistance ρ), is $2\pi f C \tan \delta$. The a-c conductivity is

$$\sigma = (5/9) f k \tan \delta \times 10^{-12}\ \Omega^{-1}\cdot\text{cm}^{-1} = 1/\rho \qquad (4\text{-}56)$$

where k is the permittivity (or relative dielectric constant) and f is the frequency. While the a-c conductivity theoretically increases in proportion to the frequency, in practice, it will depart from this proportionality insofar as k and $\tan \delta$ change with frequency.

295. Capacitance and Permittivity or Dielectric Constant. The capacitance between plane electrodes in a vacuum (with fringing neglected) is

$$C = k_0 A/t = 0.0884 \times 10^{-12} A/t \qquad \text{farads} \qquad (4\text{-}57)$$

where k_0 is the dielectric constant of a vacuum, A the area in square centimeters, and t the spacing of the plates in centimeters. k_0 is 0.225×10^{-12} F/in when A and t are expressed in inch units.

When a dielectric material fills the volume between the electrodes, the capacitance is higher by virtue of the charges within the molecules and atoms of the material, which attract more charge to the capacitor plates for the same applied voltage. The capacitance with the dielectric between the electrodes is

$$C = k k_0 A/t \qquad (4\text{-}58)$$

where k is the relative dielectric constant of the material. The capacitance relations for several other commonly occurring situations are:

Coaxial conductors: $$C = \frac{2\pi k_0 k L}{\ln (r_2/r_1)} \qquad \text{farads} \qquad (4\text{-}59)$$

Concentric spheres: $$C = \frac{4\pi k_0 k r_1 r_2}{r_2 - r_1} \qquad \text{farads} \qquad (4\text{-}60)$$

Parallel cylindrical conductors: $$C = \frac{\pi k_0 k L}{\cosh^{-1}(D/2r)} \qquad \text{farads} \qquad (4\text{-}61)$$

In these equations, L is the length of the conductors, r_2 and r_1 are the outer and inner radii, and D is the separation between centers of the parallel conductors with radii r. For dimensions in centimeters, k_0 is 0.0884 F/cm.

The value of k depends on the number of atoms or molecules per unit volume and the ability of each to be polarized (that is, to have a net displacement of their charge in the direction of the applied voltage stress). Values of k range from unity for vacuum to slightly greater than unity for gases at atmospheric pressure (see Par. **306** on Gases), 2 to 8 for common insulating solids and liquids, 35 for ethyl alcohol and 91 for pure water, and 1,000 to 10,000 for titanate ceramics (see Table 4-55 for typical values).

The relative dielectric constant of materials is not constant with temperature, frequency, and many other conditions and is more appropriately called the dielectric permittivity. Refer to the volume by Smyth[1]* for a discussion of the relation of k to molecular structure and to von Hippel[2,3] and other tables of dielectric materials[4] from the MIT Laboratory for Insulation Research. The permittivity of many liquids has been tabulated in *NBS Circ.* 514. The "Handbook of Chemistry and Physics" (Chemical Rubber Publishing Co.) also lists values for a number of plastics and other materials.

The permittivity of many plastics, ceramics, and glasses varies with the composi-

* Superscripts refer to bibliography, Par. **305**.

Table 4-55. Dielectric Permittivity (Relative Dielectric Constant)

	k
Inorganic crystalline:	
NaCl, dry crystal	5.5
$CaCO_3$ (av)	9.15
Al_2O_3	10.0
MgO	8.2
BN	4.15
TiO_2 (av)	100
$BaTiO_3$ crystal	4,100
Muscovite mica	7.0–7.3
Fluorophlogopite (synthetic mica)	6.3
Ceramics:	
Alumina	8.1–9.5
Steatite	5.5–7.0
Forsterite	6.2–6.3
Aluminum silicate	4.8
Typical high-tension porcelain	6.0–8.0
Titanates	50–10,000
Beryl	4.5
Zirconia	8.0–10.5
Magnesia	8.2
Glass-bonded mica	6.4–9.2
Glasses:	
Fused silica	3.8
Corning 7740 (common laboratory Pyrex)	5.1
Polymer resins:	
Nonpolar resins:	
Polyethylene	2.3
Polystyrene	2.5–2.6
Polypropylene	2.2
Polytetrafluoroethylene	2.0
Polar resins:	
Polyvinyl chloride (rigid)	3.2–3.6
Polyvinyl acetate	3.2
Polyvinyl fluoride	8.5
Nylon	4.0–4.6
Polyethylene terephthalate	3.25
Cellulose cotton fiber (dry)	5.4
Cellulose Kraft fiber (dry)	5.9
Cellulose cellophane (dry)	6.6
Cellulose triacetate	4.7
Tricyanoethyl cellulose	15.2
Epoxy resins unfilled	3.0–4.5
Methylmethacrylate	3.6
Polyvinyl acetate	3.7–3.8
Polycarbonate	2.9–3.0
Phenolics (cellulose-filled)	4–15
Phenolica (glass-filled)	5–7
Phenolics (mica-filled)	4.7–7.5
Silicones (glass-filled)	3.1–4.5

tion, which is frequently variable in nominally identical materials. In the case of some plastics, it varies with degree of cure and in the case of ceramics with the firing conditions. Plasticizers often have a profound effect in raising the permittivity of plastic compositions.

There is a force of attraction between the plates of a capacitor having an applied voltage. The stored energy is $\frac{1}{2}CV^2$ J. The force equals the derivative of this energy with respect to the plate separation: $\frac{1}{2}kk_0E^2 \times 10^7$ dyn/cm², where E is the electric field in volts per centimeter. The force is greater for the same voltage if the capacitance or permittivity is larger. This leads to a force of attraction of dielectrics into an electric field, that is, a net force which tends to move them toward a region of high field. If two dielectrics are present, the one with higher permittivity will displace the one with lower permittivity in the higher-field region. For example, air bubbles in a liquid are repelled from high-field regions. Correspondingly, elongated dielectric bodies are rotated into the direction of the electric field. In general, if the voltage on a dielectric system is maintained constant, the dielectrics move (if they are able) to create a higher capacitance.

296. Resistance and Resistivity of Dielectrics and Insulation. The measured resistance R of insulation depends upon the geometry of the specimen or system measured, which for a parallel-plate arrangement is

$$R = \rho t/A \qquad \text{ohms} \tag{4-62}$$

where t is the insulation thickness in centimeters, A is the area in square centimeters, and ρ is the dielectric resistivity in ohm-centimeters. If t and A vary from place to place, the effective "insulation resistance" will be determined by the effective integral of the t/A ratio over all the area under stress, on the assumption that the material resistivity ρ does not change. If the material is not homogeneous and materials of different resistivities appear in parallel, the system can be treated as parallel resistors: $R = R_aR_b/(R_a + R_b)$. In this case, the lower-resistivity material usually controls the overall behavior. But if materials of different resistivities appear in series in the electric field, the higher-resistivity material will generally control the current and a majority of the voltage will appear across it, as in the case of series resistors.

The resistance of dielectrics and insulation is usually time-dependent and (for the same reason) frequency-dependent. The d-c behavior of dielectrics under stress is an extension of the low-frequency behavior. The a-c and d-c resistance and permittivity can, in principle, be related for comparable times and frequencies.

Current flow in dielectrics can be divided into parts: (*a*) the true d-c current, which is constant with time and would flow indefinitely, is associated with a transport of charge from one electrode into the dielectric, through the dielectric, and out into the other electrode; and (*b*) the polarization or absorption current, which involves, not charge flow through the interface between the dielectric and the electrode, but rather the displacement of charge within the dielectric. This is illustrated in Fig. 4-51, where it is shown that the displaced or absorbed charge is responsible for a reverse current when the voltage is removed.

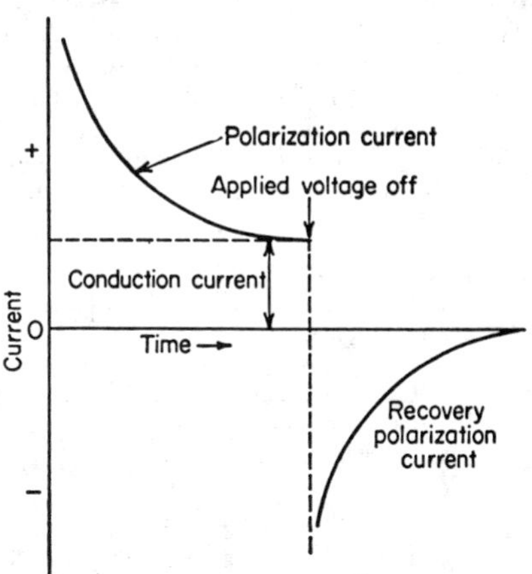

Fig. 4-51. Typical d-c dielectric current behavior.

Polarization current results from any of the various forms of limited charge displacement which can occur in the dielectric. The displacement occurring first (within less than nanoseconds) is the electronic and intramolecular charged atom displacement responsible for the very-high-frequency permittivity. The next slower displacement is the rotation of dipolar molecules and groups which are relatively free to move. The displacement most commonly observed in d-c measurements, i.e., currents changing in times of the order of seconds and minutes, is due to the very slow rotation of dipolar molecules and ions moving up to internal barriers.

In composite dielectrics (material with relatively lower resistance intermingled with a material of relatively higher resistance) a large interfacial or "Maxwell-Wagner" type of polarization can occur.[5] A circuit model of such a situation can be represented by placing two of the circuits of Fig. 4-49 in series and making the parallel resistance of one much lower than the other. To get the effect, it is necessary that the time constant R_pC be different for each material. This model assumes $R_S = 0$.

A simple model of the polarization current predicts an exponential decline of the current with time: $I_p = Ae^{-\alpha t}$, similar to the charging of a capacitor through a resistor. Composite materials are likely to have many different time constants, $\alpha = 1/RC$, superimposed. It is found empirically that the polarization or absorption current decreases inversely as a simple negative exponent of the time,

$$I = At^{-n} \tag{4-63}$$

The ratio of the current at 1 min to that at 10 min has been called the polarization index and is used to indicate the quality of composite machine insulation. A low polarization index associated with a low resistance sometimes indicates parallel current leakage paths through or over the surface of insulation (for example, in adsorbed water films).

The level of the conduction current which flows essentially continuously through insulation is an indication of the level of the ionic concentration and mobility in the material. Frequently, as with salt in water, the ions are provided by dissolved, adsorbed, or included impurity electrolytes in the material, rather than by the material itself. Purifying the material will therefore often raise the resistivity. If it is a liquid, purification can be done with adsorbent clays or ion-exchange resins.

The conductivity is given by the equation[6]

$$\sigma = \mu ec \text{ ohm}^{-1} \text{ cm}^{-1} \tag{4-64}$$

where μ is the ion mobility, e is the ionic charge in coulombs, and c is the ionic concentration per cubic centimeter. The mobility, expressed in centimeters per second-volt per centimeter, decreases inversely with the effective internal viscosity and is very low for hard resins, but it increases with temperature and with softness of the resin.

The ionic conductivity of insulating materials varies widely, depending on their purity. Among the polymers and resins, nonpolar resins such as polyethylene are likely to have high resistivities, of the order of 10^{16} or greater, since they do not readily dissolve or dissociate ionic impurities. Harder or crystalline polar resins have higher resistivity than do similar softer resins. Ceramics and glasses are lower if they contain alkali ions (sodium and potassium), since these ions are highly mobile.

Water is particularly effective in decreasing the resistivity by increasing the ionic concentration and mobility of materials, on the surface as well as internally. Water associates with impurity ions or ionizable constituents within or on the surface or interfaces. It helps to dissociate the ions by virtue of its high dielectric constant and provides a local environment of greater mobility, particularly as surface water films. Electrolyte conduction is discussed in Ref. 6, Par. **305.** Table 4-56 indicates the effect of water on the resistivity of some insulating materials.

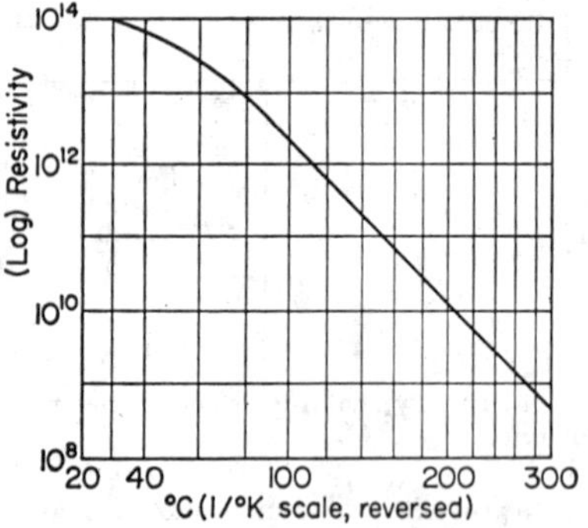

FIG. 4-52. Typical dielectric resistivity-temperature dependence (Corning Glass 7740).

The ionic conductivity, σ, exclusive of polarization effects, can be expected to increase exponentially with temperature according to the relation

$$\sigma = \sigma_0 e^{-B/T} \tag{4-65}$$

where T is the Kelvin temperature and σ_0 and B are constants. This relation, log σ versus $1/K$, is shown in Fig. 4-52. It is often observed that at lower temperatures, where the resistivity is higher, the resistivity trends lower than the extrapolated higher temperature line would predict. There are at least two possible reasons for this: the effect of adsorbed moisture and the contribution of a very slowly decaying polarization current.

297. Variation of Dielectric Properties with Frequency. The permittivity of dielectrics invariably tends downward with increasing frequency, owing to the inability of the polarizing charges to move with sufficient speed to follow the increasing rate of alternations of the electric field. This is indicated in Fig. 4-53. The sharper decline in permittivity is known as a dispersion region. At the lower frequencies the ionic-interface polarization declines first; next the molecular dipolar polarizations decline. With some polar polymers two or more dipolar dispersion regions may occur owing to different parts of the molecular rotation.

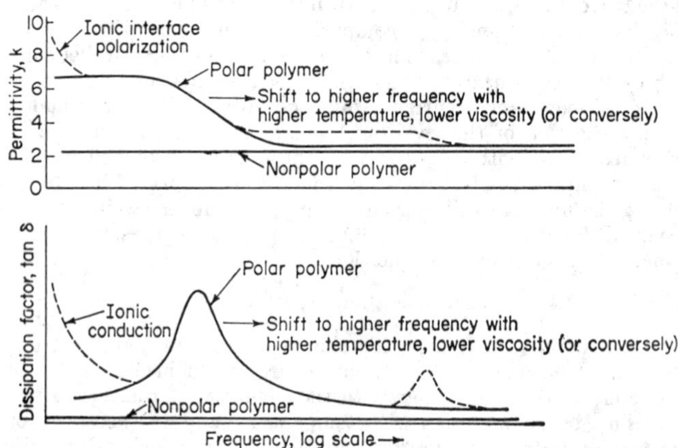

FIG. 4-53. Typical variation in dielectric properties with frequency.

Table 4-56. Surface Resistivities of Insulation and Effect of Humidity
(From R. F. Field)

Material	Ω/square	% *RH*/decade decrease of Ω/square
Hydrocarbon wax, modified	$>20 \times 10^{12}$	
Cellulose acetate butyrate	$>20 \times 10^{12}$	
Silicone rubber	10×10^{12}	
Polytetrafluoroethylene	3.6×10^{12}	
Polystyrene (sheet)	840×10^{9}	
Polydichlorostyrene 2–5	29×10^{9}	7
Hydrocarbon wax	20×10^{9}	13
Ethyl cellulose	13×10^{9}	9
Cellulose acetate	7.0×10^{9}	6
Polyvinyl chloride acetate	5.7×10^{9}	12
Polystyrene (plasticized)	5.0×10^{9}	4
Phenolic, mica-filled	5.0×10^{9}	9
Aniline formaldehyde	4.2×10^{9}	4
Polyamide	3.8×10^{9}	14
Porcelain, glazed	3.7×10^{9}	15
Glass (high *K*)	3.4×10^{9}	10
Mica	3.0×10^{9}	12
Polystyrene (molded)	2.4×10^{9}	10
Polystyrene (plasticized)	2.4×10^{9}	8
Steatite (L-3)	1.6×10^{9}	
Quartz	1.4×10^{9}	
Polyethylene	1.3×10^{9}	9
Phenolic, XX	1.3×10^{9}	16
Phenolic, asbestos-filled	1.2×10^{9}	9
Phenolic, XXXP	660×10^{6}	15
Steatite (L-4)	640×10^{6}	
Phenolic, LE	500×10^{6}	18
Phenolic, mica-filled	320×10^{6}	8
Steatite (L-4)	280×10^{6}	
Polydichlorostyrene 3–4	240×10^{6}	6
Phenolic, cellulose-filled	240×10^{6}	10
Aniline formaldehyde, glass matte	240×10^{6}	9
Phenolic, C	220×10^{6}	16
Vulcanized fiber	220×10^{6}	
Aniline formaldehyde, glass cloth	200×10^{6}	12
Quartz	190×10^{6}	
Phenol formaldehyde (plasticized)	100×10^{6}	12
Glass (sintered)	90×10^{6}	
Glass-bonded mica	64×10^{6}	18
Melamine, glass cloth	38×10^{6}	14
Phenolic, mica-filled	30×10^{6}	11

Figure 4-53 is typical of polymers and liquids, but not of glasses and ceramics. Glasses, ceramics, and inorganic crystals usually have much flatter permittivity-frequency curves similar to that shown for the nonpolar polymer, but at a higher level, owing to their atom-ion displacement polarization, which can follow the electric field usually up to infrared frequencies.

The dissipation factor–frequency curve in Fig. 4-53 indicates the effect of ionic migration conduction at low frequency. It shows a maximum at a frequency corresponding to the permittivity dispersion region. This maximum is usually associated with a molecular dipolar rotation and occurs when the rotational mobility is such that the molecule rotation can just keep up with frequency of the applied field. Here it has its maximum movement in phase with the voltage, thus contributing to conduction current. At lower frequencies the molecule dipole can rotate faster than the field and contributes more to permittivity. At higher frequencies it cannot move fast enough. Such a dispersion region can also occur because of ionic migration and interface polarization if the interfaces are closely spaced, and if the frequency and mobility have the required values.

The frequency region where the dipolar dispersion occurs depends on the rotational mobility. In mobile, low-viscosity liquids it is in the 100- to 10,000-Mc range. In viscous liquids it occurs in the region of 1 to 100 Mc. In soft polymers it may occur in the audio-frequency range, and with hard polymers it is likely to be at very low frequency (indistinguishable from d-c properties). Since the viscosity is affected by the temperature, increased temperature shifts the dispersion to higher temperatures.

298. Variation of Dielectric Properties with Temperature. The trend in a-c permittivity and conductivity, as measured by the dissipation factor, is controlled by

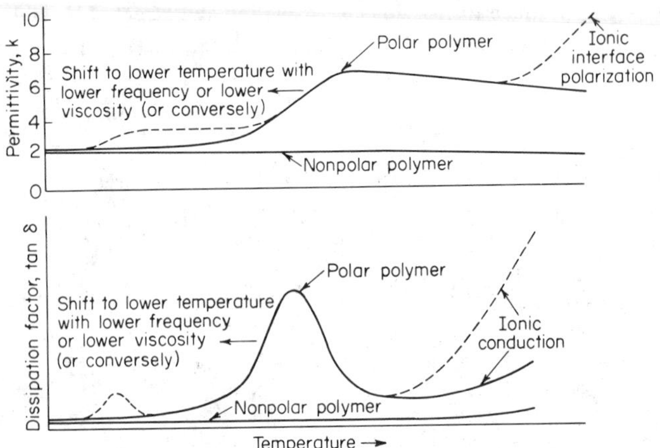

Fig. 4-54. Typical variation in dielectric properties with temperature.

the increasing ionic migrational and dipolar molecular rotational mobility with increasing temperature. This curve, which is indicated in Fig. 4-54, is in most respects a mirror image of the frequency trend shown in Fig. 4-53, since the two effects are interrelated.

The permittivity-dispersion and dissipation-factor maximum region occurs below room temperature for viscous liquids, and still lower for mobile liquids. In fact mobile liquids may crystallize before they would show dispersion, except at high frequencies. With polymers the dissipation-factor maximum is likely to occur, at power frequencies, at a temperature close to a softening-point temperature or internal second-order transition-point temperature. Dielectric dispersion and mechanical modulus dispersion can usually be correlated at the same temperature for comparable frequencies.

299. Composite Dielectrics. The dielectric properties of composite dielectrics are generally a weighted average of the individual component properties, unless there is interaction, such as dissolving (as opposed to intermixing) of one material in another, or chemical reaction of one with another. Interfaces created by the mixing present a special factor, which can often lead to higher dissipation factor and lower resistivity as a result of moisture and/or impurity concentration at the interface.

The a-c properties of sheets of two dielectrics of dielectric constant k_1 and k_2 and of thickness t_1 and t_2 placed in series are related to the properties of the individual materials by the series capacitance and impedance relations,

$$C = \frac{k_0 k_1 k_2 A}{k_1 t_2 + k_2 t_1} \tag{4-66}$$

$$\tan \delta = \frac{(t_1/t_2) k_2 \tan \delta_1 + k_1 \tan \delta_2}{k_1 + k_2 (t_1/t_2)} \tag{4-67}$$

Similarly the properties of two dielectrics in parallel are

$$C = k_0 \left(\frac{k_1 A_1}{t_1} + \frac{k_2 A_2}{t_2} \right) \tag{4-68}$$

$$\tan \delta = \frac{t_2 k_1 A_1 \tan \delta_1 + t_1 k_2 A_2 \tan \delta_2}{t_2 k_1 A_1 + t_1 k_2 A_2} \tag{4-69}$$

With steady d-c voltages, the resistivities control the current. With equal-area layer dielectrics in series,

$$R = R_1 + R_2 = \frac{1}{A} (\rho_1 t_1 + \rho_2 t_2) \tag{4-70}$$

When the dielectrics are in parallel and of equal thickness t,

$$R = \frac{R_1R_2}{R_1 + R_2} = \frac{\rho_1\rho_2 t}{\rho_1 A_2 + \rho_2 A_1} \tag{4-71}$$

300. Potential Distribution in Dielectrics. The maximum potential gradient in dielectrics is of critical significance insofar as the breakdown is concerned, since breakdown or corona is usually initiated at the region of highest gradient. In a uniform-field arrangement of conductors or electrodes, the maximum gradient is simply the applied voltage divided by the minimum spacing. In divergent fields the gradient must be obtained by calculation (which is possible for some simple arrangements) or by field mapping.

A common situation is the coaxial geometry with inner and outer radii R_1 and R_2. The gradient at radius r (centimeters) with V V applied is given by the equation

$$E = \frac{V}{r \ln (R_2/R_1)} \quad \text{V/cm} \tag{4-72}$$

The gradient is a maximum at $r = R_1$. Reference books which consider other geometries are Schwaiger and Sorensen,[7] Stratton,[8] and Ollendorff.[9]

When different dielectrics appear in series, the greater stress with a-c fields is on the material having the lower dielectric constant. This material will frequently break down first unless its dielectric strength is much higher,

$$\frac{E_1}{E_2} = \frac{k_2}{k_1} \quad \text{and} \quad E_1 = \frac{V}{t_1 + t_2k_1/k_2} \tag{4-73}$$

With d-c fields the stress distributes according to the resistivities of the materials, the higher stress being on the higher-resistivity material.

301. Dielectric Strength. Dielectric strength of an insulating material is defined by the ASA as the maximum potential gradient that the material can withstand without rupture. It is obtained, for practical purposes, by dividing the breakdown voltage by the thickness of the material between the test electrodes, regardless of the actual maximum voltage gradient. The value obtained will depend on the method and conditions of test.

Breakdown appears to require not only sufficient electric stress but also a certain minimum amount of energy. It is a property which varies with many factors such as thickness of the specimen size and shape of electrodes used in applying stress, form or distribution of the field of electric stress in the material, frequency of the applied voltage, rate and duration of voltage application, fatigue with repeated voltage applications, temperature, moisture content, and possible chemical changes under stress.

To state the dielectric strength correctly, the size and shape of specimen, method of test, temperature, manner of applying voltage, and other attendant conditions should be particularized as definitely as possible.

The dielectric strength varies as the time and manner of voltage application as indicated in Fig. 4-55. With unidirectional pulses of voltage, having rise times of less than a few microseconds, there is a time lag of breakdown, which results in an apparent higher strength for very short pulses. In testing sheet insulation in mineral oil there is usually observed a higher strength for pulses of slow rise time and a still somewhat higher strength for d-c voltages.

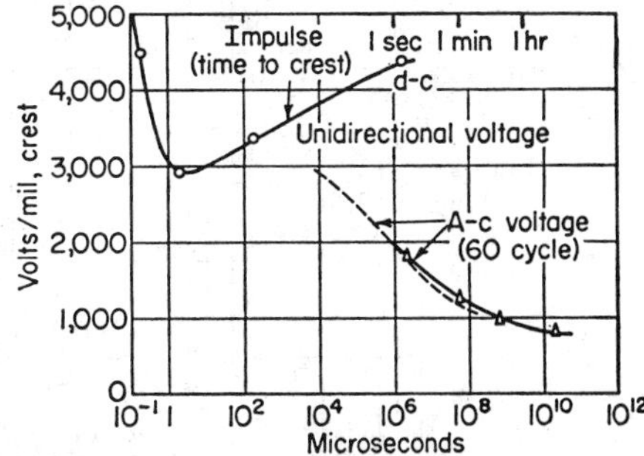

FIG. 4-55. Dielectric strength of 0.032-in pressboard in oil as a function of time of voltage application.

With a-c voltages the apparent strength declines steadily with time as a result of corona discharges[17] (in the ambient medium at the conductor or electrode edge), which penetrate the solid insulation. The discharges result from breakdown of the gas or liquid prior to the immediate breakdown of the solid. The long-time strength with a-c voltage declines, as shown in Fig. 4-57, and levels off at the corona-discharge threshold (usually

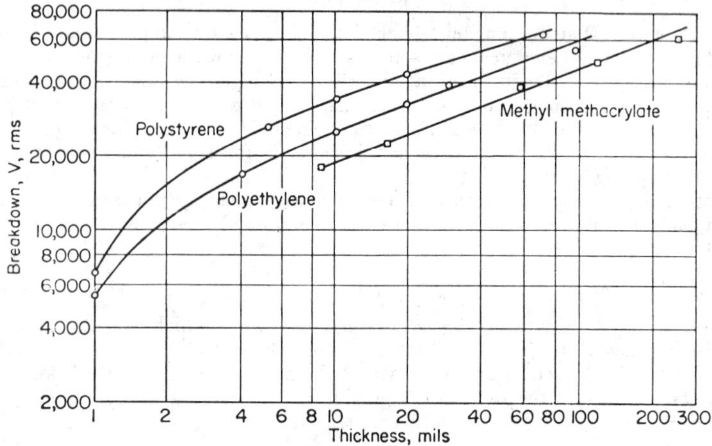

FIG. 4-56. Variation in breakdown strength with thickness, 60 c/s a-c voltage, 2 kV/sec rate of rise.

offset) voltage. Mica in particular, as well as other inorganic materials, is more resistant to such discharges. Organic resins should be used with caution where the a-c voltage gradient is high and corona discharge may be present. If discharges erode the insulation, the time to failure varies inversely as the applied frequency, since the number of discharges per unit time increases in almost direct proportion to the frequency. At higher frequencies corona and dielectric heating may also reduce the failure time.

With a-c voltage, under corona conditions, an inverse-power-law dependency of failure time t on gradient E, $t = AE^{-n}$, has been found empirically to apply over limited ranges of stress. A similar dependence has been noted for the long-time life of d-c insulation at much higher gradients, such as is used in d-c capacitors.

In practical tests of the dielectric strength, the measured strength usually declines with increasing thickness of material. The breakdown gradient decreases approximately as the inverse half power of the thickness, or, conversely, the breakdown voltage increases as the half power of the thickness. This is illustrated in Fig. 4-56. The value of the exponent may vary somewhat with conditions. An exception to this behavior is noted with sheet or wrapper insulation having defects, which are covered by increased thickness. Figure 4-56 illustrates this effect for 1-mil film, which has small defects, giving it a somewhat reduced strength.

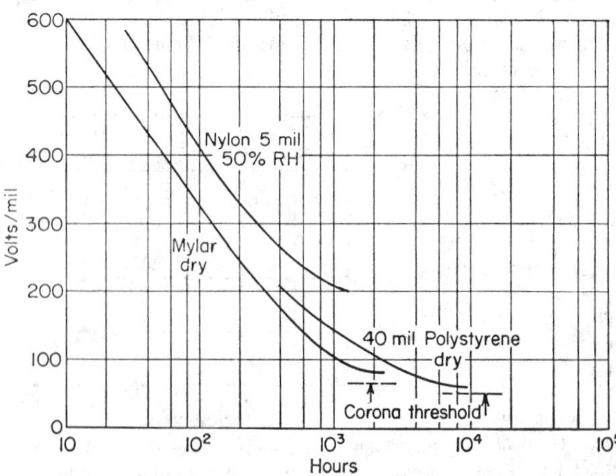

FIG. 4-57. Alternating-current dielectric strength with surface conductor edge corona.

The d-c strength of solid sheet insulation is usually higher and declines less with time than the a-c strength, since corona discharges are minimized.

The dielectric strength is much higher where surface discharges are avoided and when the electric field is uniform. This can be achieved with solid materials by recessing spherical cavities into the material and using conducting paint electrodes.

The "intrinsic" electric strength of solid materials measured in uniform fields, avoiding surface discharges, ranges from levels of the order of 0.5 to 1 MV/cm for alkali halide crystals, which are about the lowest, upward to somewhat more than 10 MV/cm. Polymers and some inorganic materials such as mica, aluminum oxide, etc., have strengths of 2 to 20 MV/cm for thin films. The strength decreases with increasing thickness and with temperature above a critical temperature (which is usually from 1 to 100°C), below which the strength has a level value or a moderate increase with increasing temperature. Below the critical temperature, the breakdown is believed to be strictly electronic in nature. Above this temperature, it is believed to be influenced by thermal heating.

The breakdown voltage of thin insulating materials containing defects, which give the minimum breakdown voltage, declines as the area under stress increases. The effect of area on the strength can be estimated from the standard deviation S of tests on smaller areas by applying minimum value statistics:[10] $V_1 - V_2 = 1.497S \log (A_1/A_2)$, where V_1 and V_2 are the breakdown voltages of areas A_1 and A_2.

If the a-c or d-c conductivity of a dielectric is high, or the frequency is high, breakdown can occur as a result of dielectric heating, which raises the temperature of the material sufficiently to cause melting or decomposition, formation of gas, etc. This effect can be detected by measuring the conductivity as a function of applied electric stress. If the conductivity rises with time, with constant voltage, and at constant ambient temperature, this is evidence of an internal dielectric heating. If the heat transfer to the electrodes and ambient surroundings is adequate, the internal temperature may eventually stabilize, but if this heat transfer is inadequate, the temperature will rise until breakdown occurs. The criterion of this sort of breakdown is the heat balance between dielectric heat input and loss to the surroundings.

The dielectric heat input is given by the equation

$$\sigma E^2 = (5/9 \epsilon' f \tan \delta \times 10^{-12}) E^2 \qquad \text{watts/cm}^3 \tag{4-74}$$

where E is the field in volts per centimeter. When this quantity is of the order of 0.1 or greater, dielectric heating can be a problem. It is much more likely to occur with thick insulation and at elevated temperatures. Dielectric breakdown is covered by Whitehead,[11] Peek,[12] and Roth[13] and in a review chapter by Mason.[14]

302. Arc Tracking of Insulation. High current arc discharges between conductors across the surface of organic resin insulation may carbonize the material and produce a conducting track. In the presence of surface water films, formed from rain or condensation, etc., small arc discharges form between interrupted parts of the water film, which is fairly conducting, and conducting tracks grow progressively across the surface, eventually bridging between conductors and causing complete breakdown. Materials vary widely in their resistance to tracking, and there are a variety of dry and wet tests for this property. Table 4-57, from a survey paper by Mandelcorn and Sommerman,[15] indicates the difference between materials and the correlation or lack of correlation between the tests. With proper fillers, some organic resins can be made essentially nontracking. Some resins such as polymethyl methacrylate and polymethylene oxide burst into flame under arcing conditions.

Review references are by Mandelcorn and Sommerman[15] and Olyphant.[16]

303. Thermal Aging. Organic resinous insulating materials in particular are subject in varying degrees to deterioration due to thermal aging, which is a chemical process involving decomposition or modification of the material to such an extent that it may no longer function adequately as the intended insulation. The aging effects are usually accelerated by increased temperature, and this characteristic is utilized to make accelerated tests to failure or to an extent of deterioration considered dangerous. Such tests are made at appreciably higher than normal operating temperatures, if the expected life is to be several years or more, since useful accelerated tests should reasonably be completed in less than a year.

Table 4-57. Comparison of Tracking Resistance of Various Materials Measured with Seven Test Procedures

Column heading (reference)	A(15)	B(25, 26)	C(9, 28)	D(9, 27)	E(2)	F(11)	G(30)
Test method designation	D495-61	IEC 113; VDE	D2132-62T	Lin.-Accel.	Differential	Inclined	Inclined
Units	Equiv s/10	Drops, 0.9 kV	Std Dust-Fog	Dust-Fog	Wet Track	plane I	plane II
		Nekal	h, 1.5 kV	H, 1.5 kV	W·min	V, kV	H, 2.5 kV
Symbol (data reference)	Liii(9)	L . . (27)	L. . . (9)	L. . . (27)	W. . . . (2, 32)	V . (11)	L . (32, 33)
Polyvinyl chloride	0.5 Tr		0.5 Tr	0.5 Tr			
Phenolic laminate, paper base	0.5 Tr	1	0.5 Tr		0.2* Tr	1.5 Tr	
Epoxy resin, unfilled	1.7 Tr	60 + No Tr	0.5 Tr		1.6 Tr		
Polyamide resin	58 + Er	5 Tr	0.5 Tr		1.3 Int		
Silicone resin, glass cloth	54 Tr	10 Tr	1.0 Tr	1.0 Tr	1.8 Tr	1.5 Tr	
Melamine resin, glass cloth	47 Tr	6 Tr	3.5 Tr	2.5 Tr	2.3 Tr	2.3 Tr	0.2 Tr
Polyethylene	13 Tr	60 + No Tr	27 Tr	10 Er + Tr			
Polyester, glass mat, h-m-f, 1	25 Tr	No Tr	50 Tr	12 Tr	3.7† Tr	2 Tr	1.1 Tr
Polymethylmethacrylate	100 + Er	No Tr	90 Er	33 Tr	8.1 + Er	6 F	
Polypropylene	310 Er	No Tr	180 Er	40 Tr	8.1 + Er	3.8 Tr	
Epoxy resin, h-m-f	100 + Er	No Tr	200 Er				
Polyester, glass mat, h-m-f, 2	51 Tr	No Tr	350 Tr	90 Tr	6.4 Tr	3 Tr	11 Tr
Butyl rubber, h-m-f	100 + Er	No Tr	450 Er	100 Er + Tr	8.1 + Er	6 F	
Silicone rubber, n-m-f	5 Tr		750 Er	120 Er + Tr		3.7 Tr	
Polytetrafluoroethylene	310 + Er	No Tr	2,700 Er	330 Tr	8.1 + Er	7 F	

h-m-f = hydrated-mineral-filled; n-m-f = nonhydrated-mineral-filled; Tr = tracked; No Tr = no tracking; Er = eroded; Int = internal; C = carbonized; F = flame.
* Failed 1.3 W, 1 s.
† Failed 5.5 W, 18 s.

Frequently other environmental factors influence the life, in addition to the temperature. These include presence or absence of oxygen, moisture, and electrolysis. Mechanical and electrical stress may reduce the life by setting a required level of performance at which the insulation must perform. If this level is high, less deterioration of the insulation is required to fall below it.

Sometimes complete apparatus is life-tested, as well as smaller specimens involving only one insulation material or simple combination of these in a simple model. New tests are being devised continually,[18] but there has been some standardization of tests by the IEEE and ASTM and internationally by the IEC.

It is important to note that frequently materials are assigned temperature ratings based on tests of the material alone. Often that material, combined with others in an apparatus or system, will perform satisfactorily at appreciably higher temperatures. Conversely, because of incompatibility with other materials, it may not perform at as high a temperature as it would alone. For this reason it is considered desirable to make functional operating tests on complete systems. These can also be accelerated at elevated temperatures and environmental exposure conditions such as humidification, vibration, cold-temperature cycling, etc., introduced intermittently.

The basis for temperature rating of apparatus and materials is discussed thoroughly in *IEEE Standards Publ.* 1. Tests for determining ratings are described in *IEEE Publs.* 1C, 1D, and 1E.

304. Application of Electrical Insulation. In applying an insulating material it is necessary to consider not only the electrical requirements but also the mechanical and environmental conditions of the application. Mechanical failure often leads to electrical failure, and mechanical failure is frequently the primary cause for failure of an aged insulation.

The initial properties of an insulation are frequently more than adequate for the application, but the effects of aging and environment may degrade the insulation rapidly to the point of failure. Thus, the thermal and environmental stability should be considered of equal importance. The effects of moisture and surface dirt contamination should be particularly considered, if these are likely to occur.

The application of insulation to shipboard insulation and rotating machinery generally is reviewed by Moses.[19] Application to a variety of apparatus is covered by Jackson.[20] A reference book[21] by Clark surveys the properties of a wide variety of materials and their application characteristics.

References which should be consulted for further details include the following: The ASTM Standards on Insulating Materials[22] are continually revised, and a single-volume collection of the electrical tests and specifications is published every two years. "Progress in Dielectrics" reviews the literature of the field.[23] The annual "Digest of Literature on Dielectrics"[24] is a reference source for each year's published papers in classified form. The "Reports of the Annual National Research Council Conference on Electrical Insulation" and the reports of the approximately biennial NEMA-IEEE "Electrical Insulation Conference" offer collections of papers covering the recent developments.

305. General References on Insulating Materials

1. Smyth, C. P. "Dielectric Behavior and Structure"; New York, McGraw-Hill Book Company, 1955.

2. von Hippel, A. "Dielectric Materials and Applications"; Cambridge, Mass., and New York, The Technology Press of the Massachusetts Institute of Technology and John Wiley & Sons, Inc., 1954.

3. von Hippel, A. "Dielectrics and Waves"; New York, John Wiley & Sons, Inc., 1954.

4. Tables of Dielectric Materials, Vols. I–VI, *MIT Lab. Insulation Research Tech. Repts.*, 1944–1958.

5. Miner, D. F. "Insulation of Electrical Apparatus"; New York, McGraw-Hill Book Company, 1941.

6. MacInnes, D. A. "The Principles of Electrochemistry"; New York, Reinhold Publishing Corporation, 1939.

7. Schwaiger, A., and Sorensen, R. W. "The Theory of Dielectrics"; New York, John Wiley & Sons, Inc., 1932.

8. STRATTON, J. A. "Electromagnetic Theory"; New York, McGraw-Hill Book Company, 1941.

9. OLLENDORFF, F. "Potential Felder der Elektrotechnik"; Berlin, Springer-Verlag, OHG, 1932.

10. WEBER, K. H., and ENDICOTT, H. S. Area Effect and Its Extremal Basis for the Electric Breakdown of Transformer Oil; *Trans. AIEE*, 1956, Vol. 5–III, p. 371.

11. WHITEHEAD, S. "Dielectric Breakdown of Solids"; New York, Oxford University Press, 1951

12. PEEK, F. W. "Dielectric Phenomena in High Voltage Engineering"; New York, McGraw-Hill Book Company, 1929.

13. ROTH, A. "Hochspannungstechnik"; Vienna, Springer-Verlag, OHG, 1959.

14. MASON, J. H. "Progress in Dielectrics"; London, Heywood & Co., 1959, Vol. 1, Chap. 1, Breakdown of Solid Dielectrics.

15. SOMMERMAN, G. M. L. Electrical Tracking Resistance of Polymers; *Trans. AIEE*, 1960, Vol. 79–III, pp. 69–74. MANDELCORN, L., and SOMMERMAN, G. M. L. "Collected Papers of the 1963 NEMA-IEEE Electrical Insulation Conference"; Chicago, Ill.

16. OLYPHANT, M. "Arc Resistance I & II"; *ASTM Bull.*, 1952, Vol. 181, p. 60, Vol. 185, p. 31.

17. DAKIN, T. W., PHILOFSKY, H. M., and DIVENS, W. C. Effect of Electrical Discharges on the Breakdown of Solid Insulation; *Trans. AIEE*, 1954, Vol. 73–I, pp. 155–162.

18. A Bibliography on Testing of Insulating Materials and Systems for Thermal Degradation, *AIEE Spec. Publ.* S–87.

19. MOSES, G. L. "Electrical Insulation, Its Application to Shipboard Electrical Equipment"; New York, McGraw-Hill Book Company, 1951.

20. JACKSON, W. "The Insulation of Electrical Equipment"; London, Chapman & Hall, Ltd., 1954.

21. CLARK, FRANK M. "Insulating Materials for Design and Engineering Practice"; New York, John Wiley & Sons, Inc., 1962.

22. ASTM Standards, Pt. 29, Electrical Insulating Materials, 1964 and succeeding even years.

23. "Progress in Dielectrics"; 1959–1962, London and New York, Heywood & Co. and Academic Press, Inc., Vols. 1 to 4 issued.

24. "Digest of Literature on Dielectrics"; National Academy of Sciences—National Research Council, Conference on Electrical Insulation, Washington, D.C.

25. DAKIN, T. W. Electrical Insulation Deterioration Treated as a Chemical Rate Phenomenon; *Trans. AIEE*, 1948, Vol. 67, p. 113.

26. FIELD, R. F. The Formation of Ionized Water Films on Dielectrics under Conditions of High Humidity; *J. Appl. Phys.*, 1946, Vol. 17, p. 318.

INSULATING GASES

BY T. W. DAKIN

306. General Properties of Gases. A gas is a highly compressible dielectric medium, usually of low conductivity and with dielectric constant only a little greater than unity, except at high pressures. In high electric fields the gas may become conducting as a result of impact ionization of the gas molecules by electrons accelerated by the field, and by secondary processes, which produce partial breakdown (corona) or complete breakdown. Conditions which ionize the gas molecules, such as very high temperatures and ionizing radiation (ultraviolet rays, x-rays, gamma rays, high-velocity electrons, and ions, such as alpha particles), will also produce some conduction in a gas.

The gas density d (grams per liter) increases with pressure p (torrs or millimeters of mercury) and gram-molecular weight M and decreases inversely with the absolute temperature (degrees centigrade + 273) T, according to the relation

$$d = \frac{M}{22.4} \cdot \frac{p}{760} \cdot \frac{273}{T} \qquad \text{g/l} \tag{4-75}$$

The above relation applies exactly only to ideal gases but is approximately correct for most common gases. For exact values, tables should be consulted and more exact equations such as the Van der Waals equation.[1*]

If the gas is a vapor in equilibrium with a liquid or solid, the pressure will be the vapor pressure of the liquid or solid. The logarithm of the pressure varies as $-\Delta H/RT$, where ΔH is the heat of vaporization in calories per mole and R is the model gas constant, 1.98 cal/(mole) (°C).[1] This relation also applies to all common atmospheric gases at low temperatures, below the points where they liquefy.

307. Dielectric Properties at Low Electric Fields. *Dielectric Constant.* The dielectric constant k of gases is a function of the molecular electrical polarizability and the gas density. It is independent of magnetic and electric fields except when a significant number of ions is present. Values of the dielectric constant of some common gases are given in Table 4-58 at atmospheric pressure and 20°C. The increment above unity $(k - 1)$ may be estimated[2] approximately by assuming that it varies proportionally to the pressure and inversely to the Kelvin temperature.

Table 4-58. Dielectric Constant of Gases, 20°C, 1 atm gas

Air*	1.000536	He	1.000065
N_2	1.000547	A	1.000517
O_2	1.000495	SF_6	1.002084
CO_2	1.000921	H_2	1.000254

* Dry, CO_2 free.

Conduction. The conductivity of a pure molecular gas at moderate electric stress and moderate temperature can be assumed, in the absence of any ionizing effect such as radiation, to be practically zero.

Ionizing radiation induces conduction in the gas to a significant extent, depending on the amount absorbed and the volume of gas under stress.[3] The energy of the radiation must exceed, directly or indirectly, the ionization energy of the gas molecules and thus to produce an ion pair (usually an electron and positive ion). The threshold ionization energy is of the order of 10 to 25 electronvolts (eV)/molecule for common gases (10.86 eV for methyl alcohol, 12.2 for oxygen, 15.5 for nitrogen, and 24.5 for helium). Only very-short-wavelength ultraviolet light is effective directly in photoionization, since 10 eV corresponds to a photon of ultraviolet with a wavelength of 1,240 Å. Since the photoelectric work function of metal surfaces is much lower (2 to 6 eV, e.g., copper about 4 eV) the longer-wavelength ultraviolet commonly present is effective in ejecting electrons from a negative conductor surface. Such cathode-ejected electrons give the gas apparent conductivity.

High-energy radiation from nuclear disintegration is a common source of ionization in gases. Nuclear sources usually produce gamma rays of the order of 10^6 eV energy. Only a small amount is absorbed in passing through a low-density gas. A flux of 1 R/h produces ion pairs corresponding to a saturation current (segment *a-b* of Fig. 4-58) of 0.925×10^{-13} A/cm³ of air at 1 atm pressure, if all the ions formed are collected at the electrodes. The effect is proportional to the flux and the gas density.

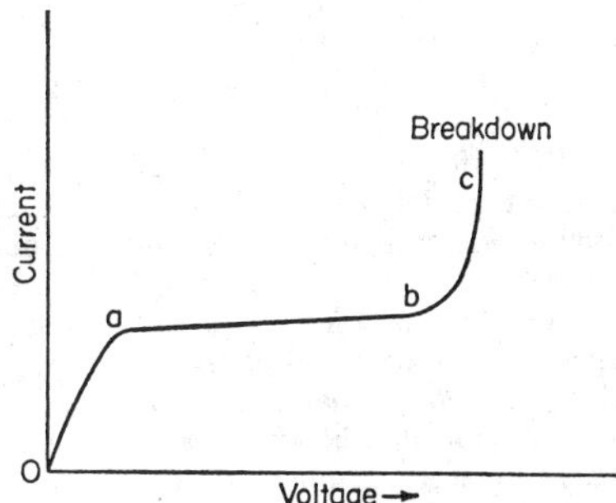

FIG. 4-58. Current-voltage behavior of a slightly ionized gas.

At a voltage stress below about 100 V/cm, some of the ions formed will recombine before being collected and the current will be correspondingly less (segment *o-a* of Fig. 4-58). Higher stresses do not increase the current if all the ions formed are collected.

* Superscripts refer to bibliography, Par. **315**.

A very small current, of the order of 10^{-21} A/cm^3 of air, is attributable to cosmic rays and residual natural radioactivity.

Electrons (beta rays) produce much more ionization per path length than gamma rays, because they are slowed down by collisions and lose their energy more quickly. Correspondingly, the slower alpha particles (positive helium nuclei) produce a very dense ionization in air over a short range. For example, a 3-million-eV (MeV) alpha particle has a range in air of 1.7 cm and creates a total of 6.8×10^5 ion pairs. A beta particle of the same energy creates only 40 ion pairs/cm and has a range of 13 m.

Thermally induced conductivity occurs in gases, at very high temperatures, as a result of impact ionization by the very-high-velocity molecules in the gas. This ionization can be calculated from the Saha equation[4] if the ionization energy is known. Such conductivity, in air, becomes significant only above 2000°C. Introduction of quantities of "seed" atoms such as sodium and potassium, which have low ionization energies, has been used in MHD generators to increase the gas conductivity substantially at high temperatures. The chemical reactions in flames also produce significant quantities of ions, and these can carry currents.

At temperatures increasing above 600°C, it has been shown that thermionic electron emission from negative conductor surfaces produces significant currents compared with levels typical of electrical insulation. The order of magnitude of this effect can be estimated from the Richardson thermionic-emission equation.[4]

Since the rate of production of ions by the various sources mentioned above is limited, the current in the gas does not follow Ohm's law, unless the rate of collection of the ions at the electrodes is small compared with the rate of production of these ions, as in the initial part of segment *o-a* in Fig. 4-58.

308. Dielectric Breakdown. *Uniform Fields.* The dielectric breakdown of gases is a result of an exponential multiplication of free electrons induced by the applied electric field. It is indicated by segment *b-c* of Fig. 4-58. It is generally assumed that the initiation of breakdown requires only one electron. However, if only a few electrons are present prior to breakdown, it is not easily possible to measure the trend of current shown in Fig. 4-58. If the breakdown is completed between metal electrodes, the spark develops extremely rapidly into an arc, involving copious emission of electrons from the cathode metal and, if the necessary current flow is permitted, vaporization of metal from the electrodes. Table 4-59 gives the dielectric strength of typical gases.

Table 4-59. Relative Dielectric Strengths of Gases

(0.1 in gap)

Gas	Strength	Gas	Strength
Air	0.95	CF_4	1.1
N_2	1.0	C_2F_6	1.9
CO_2	0.90	C_3F_8	2.3
H_2	0.57	C_4F_8 cyclic	2.8
A	0.28	CF_2Cl_2	2.4
Ne	0.13	C_2F_5Cl	2.6
He	0.14	$C_2F_4Cl_2$	3.3
SF_6	2.3–2.5		

In uniform electric fields breakdown occurs at a critical voltage which is a function of the product of the pressure p and spacing d (Paschen's law), as illustrated in Fig. 4-59 for several gases. The relation shown between voltage and $p \cdot d$ applies at constant temperature (20°C in this case).

It would be more accurate to consider the gas density–spacing product, since the dielectric strength varies with the temperature only as the latter affects the gas density. It will be noted that the electric field at breakdown decreases as the spacing increases. This is typical of all gases and is due to the fact that a minimum amount of multiplication of electrons must occur before breakdown occurs. A single electron accelerated by the field creates an avalanche which grows exponentially as $e^{\alpha d}$, where d is the spacing and α is the Townsend ionization coefficient (number of electrons formed by collision per centimeter), which increases rapidly with increasing electric field. At smaller spacings, α and the field must reach higher levels to produce sufficient multiplication. Actually, it is found that the value of αd at breakdown in air at atmospheric pressure increases from about 1 for spacings of the order of 10^{-3} cm up to 45 at spacings of 10 cm.

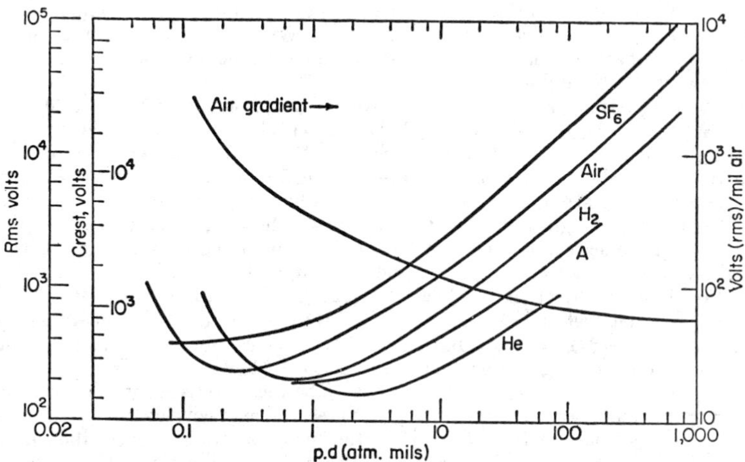

FIG. 4-59. Pressure-spacing dependence of the dielectric strength of gases (Paschen's curves).

Table 4-60. Sphere-gap Spark-over Crest Voltages
(Condensed from AIEE Standard No. 4, 1953)
At 25°C and 760 mm barometric pressure, one sphere grounded. For corrections at other air densities see Table 4-61

Sphere gap spacing, cm	Sphere diameter, cm					
	6.25		12.5		25	
	60-cycle and negative-impulse kv crest	Positive-impulse kv crest	60-cycle and negative-impulse kv crest	Positive-impulse kv crest	60-cycle and negative-impulse kv crest	Positive impulse kv crest
0.5	17.0	17.0				
1.0	31.3	31.3	31.7	31.7		
1.5	44.5	44.8	44.9	44.9		
2.0	57.0	57.4	58.0	58.0		
2.5	68.8	69.3	70.8	70.8	72	72
3.0	78.8	79.4	83.5	83.5		
3.5	86.6	88.0	95.0	95.3		
4.0	93.6	....	106.0	108.0		
4.5	99.8	....	117.0	120.0		
5.0	105.5	....	127.0	132.3	136	136
5.5	.....	....	135.3	142.5		
6.0	.....	....	143.5	153.8		
6.25	.....	....	147.5	158.0		
7.0	.....	....	157.7	171.0		
7.5	.....	....	......	.....	192	196
8.0	.....	....	170.5			
9.0	.....	....	182.0			
10.0	.....	....	192.0*	.....	241	252
11.0	.....	....	200.0*			
12.0	.....	....	208.0*			
12.5	.....	....	211.0*	.....	278	296
15.0	.....	....	......	.....	309	334
17.5	.....	....	......	.....	338	364
20.0	.....	....	......	.....	362	390
22.5	.....	....	......	.....	379	409
25.0	.....	....	......	.....	393	426

* These values are to be used with power frequencies only. When using these values, the error may not be within the 3% specified.

The degree of multiplication required depends on the number of secondary electrons formed per avalanche, which in turn start other avalanches. With larger spacings, the avalanches become so large as to cause space-charge field distortion and the onset of streamer type of breakdown.

The dielectric strength of air for larger sphere gaps is given in Table 4-60, selected from IEEE Standard 4 (revision of AIEE Standard 4). These values are used as voltage standards, but they indicate also the trend of the breakdown stress downward with increasing spacing. They also indicate that the impulse strength is almost identical with the crest 60-c/s strength for smaller spacings but is a little greater for large spacings. The positive (high-terminal) d-c strength is the same as the positive-impulse strength. The higher values than the crest 60-c/s on the larger spacings are because of a slight asymmetry of the field due to the ground plane.

309. Air-density Effect on Dielectric Strength. Pressure and (moderate) temperature affect the dielectric strength only as they affect the gas density according to Paschen's law (Fig. 4-59). IEEE Standard 4 gives correction factors for the relative pressure effect on sphere gap breakdown (see Table 4-61).

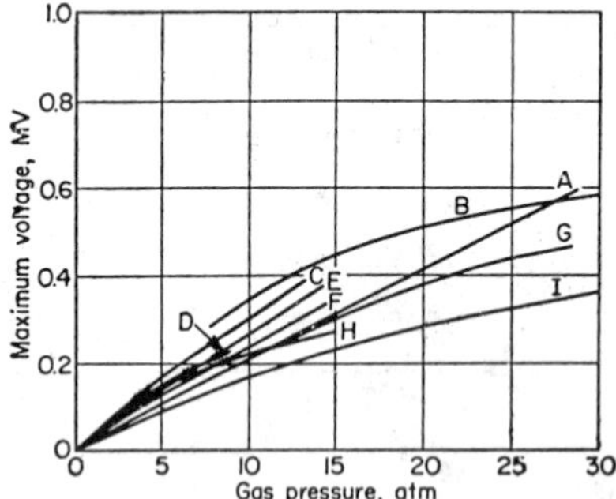

Fig. 4-60. Breakdown of N_2, CO_2, and air at high pressures, 12.7 mm gap. *Philp* (*Trans. AIEE*, 1963, *Vol.* 82, *p.* 356): 64-mm sphere facing negative high-voltage terminal in $N_2 + CO_2$ (50%) (*A*); *Kusko*: Uniform field gap in $N_2 + CO_2$ (*B*); *Trump*, *Safford*, and *Cloud*: Uniform field gap in air (*C*); *Gänger*: 50-mm-diameter sphere-to-plane gap in N_2 (*D*); *Finkelmann*: Uniform field gap in N_2 (*E*); *Finkelmann*: Uniform field gap in CO_2 (*F*); *Trump, Cloud, Mann,* and *Hanson* (*Elec. Eng.*, 1950, *Vol.* 69, *p.* 961): Uniform field gap in $N_2 + CO_2$ (*G*); *Palm*: Uniform field gap in N_2 (*H*); *Howell*: Uniform field gap in air (*I*).

The dielectric strength of gases can be increased very considerably by increasing the pressure (and hence the density). At moderate pressures the increase in strength is slightly less than proportional to the pressure. At higher pressures the increase becomes appreciably less than proportional to the pressure, as indicated[5] in Fig. 4-60. In several cases such as CO_2, N_2, SF_6, and hexane vapor, the compressed gas strength[6] has been shown to approach that of the pure liquid. Cathode field emission of electrons is believed to be an important contributing factor in very-high-pressure breakdown. The breakdown voltage is affected by the cathode metal as well as by surface imperfections and particles.

With nonuniform fields when corona occurs before breakdown, maxima in the breakdown voltage-pressure curve are observed with electronegative gases, such as SF_6.

310. Relative Dielectric Strengths of Gases. The relative dielectric strength, with few exceptions, tends upward with increasing molecular weight. There are a number of factors other than molecular or atomic size which influence the retarding effect on electrons. These include ability to absorb electron energy on collision and trap electrons to form negative ions. The noble atomic gases (helium, argon, neon, etc.) are poorest in these respects and have the lowest dielectric strengths. Table 4-59 gives the relative dielectric strengths of a variety of gases at 1 atm pressure at a $p \cdot d$ value of 1 atm $\times$ 0.25 cm. The relative strengths vary with the $p \cdot d$ value, as well as gap geometry, and particularly in divergent fields where corona begins before breakdown. It is best to consult specific references with regard to divergent field breakdown values.

311. Corona and Breakdown in Nonuniform Fields between Conductors. In nonuniform fields, when the ratio of spacing to conductor radius of curvature is about 3 or less, breakdown occurs without prior corona. The breakdown voltage is controlled by the integral of the Townsend ionization coefficient, α, across the gap.[6] At larger ratios of spacing to radius of curvature, corona discharge occurs at voltage levels below complete gap breakdown.

Table 4-61. Air-density Correction Factors for Sphere Gaps

Relative air density	Sphere diameter, cm		
	6.25	12.5	25
0.50	**0.547**	**0.535**	**0.527**
0.55	**0.595**	**0.583**	**0.575**
0.60	**0.640**	**0.630**	**0.623**
0.65	**0.686**	**0.677**	**0.670**
0.70	**0.732**	**0.724**	**0.718**
0.75	**0.777**	**0.771**	**0.766**
0.80	**0.821**	**0.816**	**0.812**
0.85	**0.866**	**0.862**	**0.859**
0.90	**0.910**	**0.908**	**0.906**
0.95	**0.956**	**0.955**	**0.954**
1.00	**1.000**	**1.000**	**1.000**
1.05	**1.044**	**1.045**	**1.046**
1.10	**1.090**	**1.092**	**1.094**

According to Peek,[7] corona in air at atmospheric pressure occurs before breakdown when the ratio of outer to inner radius of coaxial electrodes exceeds 2.72, or where the ratio of gap to sphere radius between spheres exceeds 2.04. These discharges project some distance from the small-radii conductor but do not continue out into the weaker electric field region, until a higher voltage level is reached.

Such partial breakdowns are often characterized by rapid pulses of current and radio noise. With some conductors at intermediate voltages between onset and complete breakdown, they blend into a pulseless glow discharge around the conductor. When corona occurs before breakdown, it creates an ion space charge around the conductor, which modifies the electric field, reducing the stress at sharp conductor points in the intermediate voltage range. At higher voltages, streamers break out of the space-charge region and cross the gap.

The surface voltage stress at which corona begins increases above that for uniform field breakdown stress, since the field to initiate breakdown must extend over a finite distance. An empirical relation developed by Peek[7] is useful for expressing the maximum surface stress for corona onset for concentric cylinders of radius r cm,

$$E = 31\delta(1 + 0.308/\sqrt{\delta r}) \quad \text{kV/cm} \tag{4-76}$$

for parallel wires,

$$E = 29.8\delta(1 + 0.301/\sqrt{\delta r}) \quad \text{kV/cm} \tag{4-77}$$

for spheres

$$E = 27.2\delta(1 + 0.54/\sqrt{\delta r}) \quad \text{kV/cm} \tag{4-78}$$

where δ is the density of air relative to that at 25°C and 1 atm pressure.

312. Corona Discharges on Insulator Surfaces. It has been shown by a number of investigators that the discharge-threshold voltage stress on or between insulator surfaces is the same as between metal electrodes.[8] Thus, the threshold voltage for such discharges can be calculated from the series dielectric-capacitance relations for internal gaps of simple shapes, such as plane and coaxial gaps, insulated conductor surfaces, hollow spherical cavities, etc.

The corona-initiating voltage at a conductor edge on a solid barrier depends on the electric stress concentration and generally on the ratio of the barrier thickness to its dielectric constant, but not on the barrier material,[8] except when it is capable of adsorbing a water film. Any adsorbed water or conducting film raises the corona threshold voltage by reducing stress concentration at the conductor edge on the surface.

It is sometimes possible to overvolt such gaps considerably prior to the first discharge, and the offset voltage may be below the proper voltage due to surface-charge concentration. With a-c voltages, pulse discharges occur regularly back and forth each half cycle, but with d-c voltage the first discharge deposits a surface charge on the insulator surface which must leak away, before another discharge can occur. Thus, corona on or between insulator surfaces is very intermittent with steady d-c voltages, but discharges occur when the voltage is raised or lowered.

313. Breakdown at High Frequency. The a-c dielectric strength of gases declines only slightly (6 to 10% at 600 kc) as the frequency is increased, until the time of a half cycle is about the same as the transit time, first of positive ions, and then of electrons across the gap.[10] At these critical frequencies (10^5 to 10^7 c/s) small maxima have been observed. Above the critical frequency, cumulative ionization occurs in the gap, and there is a sharp drop in breakdown voltage. At these high frequencies, the breakdown voltage is set by the equilibrium between production of electrons by electron impact ionization and loss by diffusion to the walls or electrodes.

314. Vacuum Breakdown. When the pressure and gas density in a system are so low that the electron mean free path is much larger than the spacing of conductors, electron multiplication by impact ionization of the gas molecules cannot take place. This occurs at pressures well below the Paschen's minima shown in Fig. 4-59.

In the absence of direct gaseous ionization, breakdown can occur, at high stresses, from electrode effects. While the exact mechanism of vacuum breakdown has not been determined, there are several phenomena which can lead to breakdown. One of these is cathode field emission, which may be enhanced by imperfections, in or on the cathode surface, which increase the local stress, or even heating by the high current density. Steady cathode emission currents, which can lead to breakdown at elevated stresses, have been observed.

Another process which also seems likely to occur is a cathode-anode regeneration process of elementary particles. Electrons strike the anode with enough energy to create photons and positive ions which return to the cathode to generate more electrons and ions, etc.

At larger spacings, breakdown seems to be controlled by the total voltage rather than the gradient. The breakdown voltage increases approximately as the square root of the spacing. One mechanism which can account for this, together with other aspects of vacuum breakdown, is the Cranberg clump hypothesis, which presumes that a microscopic particle of many atoms is accelerated by the field from one electrode to the other, gaining enough kinetic energy to vaporize itself and some atoms from the electrode when it strikes the electrode. The vapor formed by this impact then leads to breakdown by a gas discharge process. Figure 4-61 indicates the range of breakdown voltages of vacuum, from a review paper by Hawley.[11] Breakdown in vacuum is very sensitive to residual particulate matter on the electrodes or in the system. Frequently, initial breakdown values on fresh systems are quite low, and electrodes can be "conditioned" to higher breakdown levels by repeatedly breaking down the system with limited current discharges.

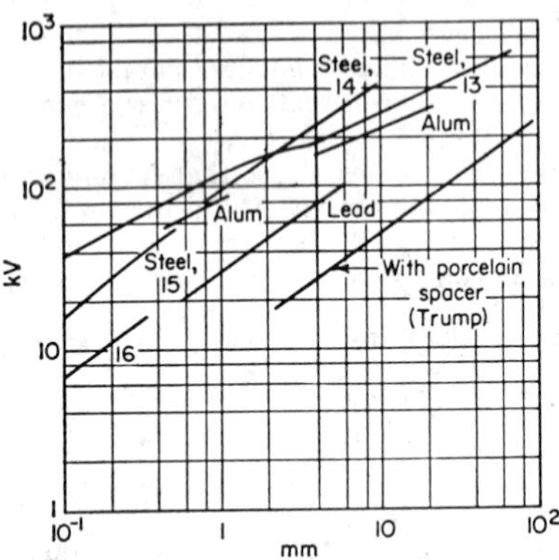

FIG. 4-61. Breakdown voltage of vacuum gaps. (Numbers correspond with those in bibliography, Par. **315**.)

Supporting insulators between electrodes in vacuum may reduce the breakdown voltage drastically below the level of breakdown in clear gaps.[12] It has been shown that flashover of such insulators in vacuum is initiated at the contact between the insulator and the cathode. If this region is shielded from the field and the insulator properly shaped, much higher breakdown voltage values can be obtained.

315. References on Insulating Gases

1. MACDOUGALL, F. H. "Physical Chemistry"; New York, The Macmillan Company, 1952.

2. SMYTH, C. P. "Dielectric Behavior and Structure"; New York, McGraw-Hill Book Company, 1955.

3. CURRAN, S. C., and CRAGGS, J. D. "Counting Tubes"; London, Butterworth Scientific Publications, 1949.

4. Cobine, J. D. "Gaseous Conductors"; New York, Dover Publications, Inc., 1958.

5. Trump, J. G., Cloud, R. W., Mann, J. G., and Hanson, E. P. Influence of Electrodes on D-C Breakdown in Gases at High Pressures; *Elec. Eng.*, 1950, Vol. 69, p. 961. Philp, S. F. Compressed Gas Insulation in the Million Volt Range, SF_6, N_2, CO_2; *Trans. IEEE*, 1963, Vol. 82, p. 356.

6. Loeb, L. B. Electrical Breakdown of Gases, "Encyclopedia of Physics"; Berlin, Springer-Verlag, OHG, 1956, Vol. XXII.

7. Peek, F. W. "Dielectric Phenomena in High Voltage Engineering"; New York, McGraw-Hill Book Company, 1929.

8. Hall, H. C., and Russek, R. M. Discharge Inception and Extinction in Dielectric Voids; 1954, *Proc. IEE*, Vol. 101–2, p. 47.

9. Dakin, T. W., Philofsky, H., and Divens, W. Effect of Electrical Discharges on the Breakdown of Solid Insulation; *Trans. AIEE*, 1954, Vol. 73–I, pp. 155–162.

10. Brown, S. C. Breakdown in Gases, Alternating and High Frequency Fields, "Encyclopedia of Physics"; Berlin, Springer-Verlag, OHG, 1956, Vol. XXII.

11. Hawley, R. Vacuum as an Insulator; *Vacuum*, 1960, Vol. 10, p. 310.

12. Kofoid, M. J. Effect of Metal-Dielectric Junction Phenomena on High Voltage Breakdown over Insulators in Vacuum; 1960, *Trans. AIEE*, Vol. 79–III, p. 999.

13. Trump, J. G., and Van de Graaf, R. The Insulation of High Voltages in Vacuum; *J. Appl. Phys.*, 1947, Vol. 18, p. 327.

14. Slivkov, I. N. Mechanism for Electrical Discharge in Vacuum; *Soviet Phys. Tech. Phys.*, 1957, Vol. 2, p. 1919.

15. Denholm, A. S. The Electrical Breakdown of Small Gaps in Vacuum; *Can. J. Phys.*, 1958, Vol. 36, p. 476.

16. Leader, D. Electrical Breakdown in Vacuum; *Proc. IEE*, 1953, Vol. 100–2A, p. 138.

Reference Books

17. Meek, J. M., and Craggs, J. D. "Electrical Breakdown of Gases"; New York, Oxford University Press, 1953.

18. Llewellyn Jones, F. "Ionization and Breakdown in Gases"; London, Methuen & Co., Ltd.

19. Ganger, B. "Der Elektrische Durchschlag von Gasen"; Berlin, Springer-Verlag, OHG, 1953.

20. Gas Discharges I, "Encyclopedia of Physics"; Berlin, Springer-Verlag, OHG, 1956, Vols. XXI and XXII.

21. Dakin, T. W., and Berg, D. Theory of Gas Breakdown, chapter in "Progress in Dielectrics"; London and New York, Heywood & Co., Ltd., and Academic Press, Inc., 1962, Vol. 4.

22. Roth, A. "Hochspannungstechnik," 4th ed.; Berlin, Springer-Verlag, OHG, 1959.

INSULATING OILS AND LIQUIDS

By T. W. Dakin

316. General Considerations. Typical insulating liquids are natural or synthetic organic compounds and frequently consist of mixtures of essentially isomeric compounds with some range of molecular weight. The mixture of very similar but not exactly the same molecules, with a range of molecular size and with chain and branched hydrocarbons, prevents crystallization and results in a low freezing point, together with a relatively high boiling point. Typical insulating liquids have permittivities (dielectric constants) of 2 to 7 and a wide range of conductivities, depending upon their purity. The d-c conductivity in these liquids is usually due to dissolved impurities, which are ionized by dissociation. Higher ionized impurity and conductivity levels occur in liquids having higher permittivities and lower viscosities.

The function of insulating liquids is to provide electrical insulation and heat transfer. As insulation, the liquid is used to displace air in the system and provide a medium of high electric strength to fill pores, cracks, and gaps in insulation systems. It is usually necessary to fill and impregnate systems with liquid under vacuum, so that all air bubbles are eliminated. If air is completely displaced in all high electric field regions, the corona threshold voltage and breakdown voltage for the system are greatly increased. The viscosity selected for a liquid insulation is often a compromise to provide the best balance between electrical insulation and heat transfer and other limitations such as flammability, solidification at low temperatures, and pressure development at high temperatures in sealed systems.

The most commonly used insulating liquids are natural hydrocarbon mineral oils refined to give low conductivity and selected viscosity and vapor-pressure levels for transformer, circuit-breaker, and cable applications. Other common liquids are the chlorinated aromatic (benzene and diphenyl) liquids, used for their nonflammability characteristics and their higher permittivity (for capacitors). Less common, but of critical importance for higher-temperature applications, are the silicone fluids. Completely fluorinated (perfluoro) compound liquids have found use in applications to provide high heat-transfer rates (by boiling and convection), together with good dielectric properties. Organic ester fluids, both natural and synthetic, are used in special applications, usually in capacitors.

317. Mineral Insulating Oils. Mineral insulating oils are hydrocarbons (compounds of hydrogen and carbon) refined from crude petroleum deposits from the ground.[1]* They consist partly of aliphatic compounds with the general formula C_nH_{2n+2} and C_nH_{2n}, comprising a mixture of straight and branched chain and cyclic or partially cyclic compounds. Many oils also contain a sizable fraction of aromatic compounds related to benzene, naphthalene, and derivatives of these with aliphatic side chains. The ratio of aromatic to aliphatic components depends on the source of the oil and its refining treatment. The percent aromatics is of importance to the gas-absorption or evaluation characteristics under electrical discharges[2] and to the oxidation characteristics.[3]

The important physical properties of a mineral oil (as for other insulating liquids as well) are listed in Table 4-62 for three types of mineral oils. In addition to these properties, mineral oils which are exposed to air in their application have distinctive oxidation characteristics which vary with type of oil and additives and associated materials.[4]

Table 4-62. Characteristic Properties of Insulating Liquids

Type of liquid	Mineral oil			Transformer Askarel
	Transformer	Cable and capacitor	Solid cable	
Specific gravity	0.88	0.885	0.93	1.56–1.57
Viscosity, Saybolt sec at 37.8°C	57–59	0.100	100	52–56† 41–45‡
Flash point, °C	135	165	235	
Fire point, °C	148	185	280	None
Pour point, °C	−45	−45	−5	< −32°C† < −44°C‡
Specific heat	0.425	0.412	....	0.251
Coefficient of expansion	0.00070	....	0.00075	0.0007
Thermal conductivity, cal/(cm)(s)(°C)	0.39	....	....	0.30
Dielectric strength,* kV~	30	....	....	>35
Permittivity at 25°C	2.2	....	....	4.5–4.7
Resistivity, Ω·cm × 10[12]	1–10	50–100	1–10	0.1

* ASTM D877.
† is a mixture of 60% hexachlorobiphenyl and 40% trichlorbenzene.
‡ is a mixture of 45% hexachlorobiphenyl and 55% trichloro- and tetrachlorobenzenes.

Many manufacturers now approve the use of any of several brands of mineral insulating oil in their apparatus provided that they meet their specifications which are similar to ASTM D1040, values from which are tabulated in Table 4-62. Low values of dielectric strength may indicate water or dirt contamination. A high neutralization

* Superscripts refer to bibliography, Par. **326**.

number will indicate acidity, developed very possibly from oxidation, particularly if the oil has already been used. Presence of sulfur is likely to lead to corrosion of metals in the oil.

Table 4-63. Chlorinated Diphenyl Properties

% chlorination	Viscosity, Saybolt sec, 210°F	Boiling range, °C	*K*, 25°C
32	31–32	290–235	5.7
42	34–35	325–360	5.8
54	44–48	365–390	5.0
60	72–78	385–420	4.3

The solubility of gases and water in mineral oil is of importance in regard to its function in apparatus. Solubility is proportional to the partial pressure of the gas above the oil,

$$S = S_0(p/p_0) \tag{4-79}$$

where S is the amount dissolved at pressure p if the solubility is expressed as the amount S_0 dissolved at pressure p_0.

The solubility is frequently expressed in volume percent of the oil. Values for solubility of some common gases in transformer oil[5] at atmospheric pressure (760 torrs) and 25°C are air 10.8%; nitrogen 9.0%; oxygen 14.5%; carbon dioxide 99.0%; hydrogen 7%; methane 30% by volume. The solubilities of all the gases, except CO_2, increase slightly with increasing temperature. Water is dissolved in new transformer oil to the extent of about 60 to 80 ppm at 100% relative humidity and 25°C. The amount dissolved is proportional to the relative humidity. Solubility of water increases with oxidation of the oil and the addition of polar impurities, with which the water becomes associated. Larger quantities of water can be suspended in the oil as fine droplets.

318. Dielectric Properties of Mineral Oils. The permittivity of mineral insulating oils is low, since they are essentially nonpolar, containing only a few molecules with electric dipole moments. Some oils possess a minor fraction of polar constituents, which have not been identified. These contribute a dipolar character to the dielectric properties at low temperature and/or high frequency, similar to the trends shown in Figs. 4-53 and 4-54. A typical permittivity for American transformer oil at 60 cycles is 2.19 at 25°C, declining almost linearly to 2.11 at 100°C. At low temperatures and high frequencies, values of permittivity as high as 2.85 have been noted in oils with a relatively high level of polar constituents.

The d-c conductivity levels of mineral oils range from about 10^{-15} ohm^{-1} cm^{-1} for pure new oils up to 10^{-12} ohm^{-1} cm^{-1} for contaminated used oils.[6] This conductivity is due to dissociated impurity ions or ions developed by oil oxidation.[7] It increases approximately exponentially with temperature about 1 decade in 80°C.

Alternating-current dissipation-factor values are nearly proportional to the d-c conductivity, 10^{-13} ohm^{-1} cm^{-1}, corresponding to a tan δ of 0.008. If no electrode polarization effects are present, the d-c conductivity σ should be related to the a-c conductivity (tan δ) by $\sigma = \frac{5}{9}\epsilon' f \tan\delta \times 10^{-12}$, where ϵ' is the dielectric permittivity (Table 4-55) and f is the frequency.

319. Dielectric Strength of Mineral Oils. The dielectric strength of mineral oils, as with all liquids, varies considerably with the state of purity, particularly with respect to particulate matter and moisture. Typical values (ASTM D877, D1816 standard test gaps) are shown in Fig. 4-62 as a function of moisture content.[8]

The dielectric strength of mineral oil has been shown by Weber and Endicott,[10] Par. **305**, to decrease with increasing area under stress according to the relation

$$V_1 - V_2 = 1.497S \log (A_1/A_2) \tag{4-80}$$

where S is the standard deviation of tests with the smaller area. This relation is derived by application of minimum value statistics, assuming that the largest defect controls the breakdown.

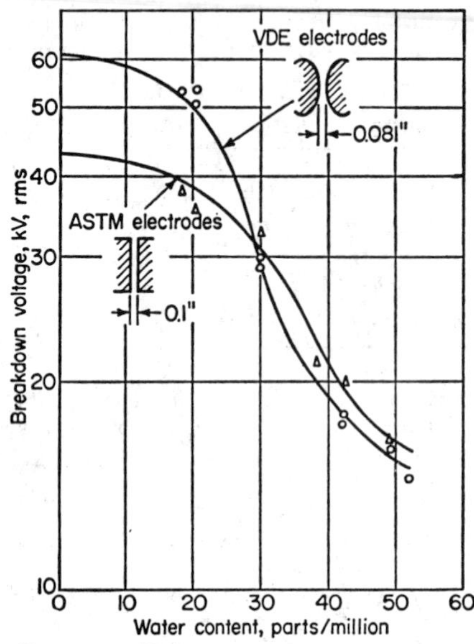

FIG. 4-62. Electric strength of transformer oil vs. water content with ASTM and VDE electrodes (rate of voltage rise, 2 kV rms/sec).

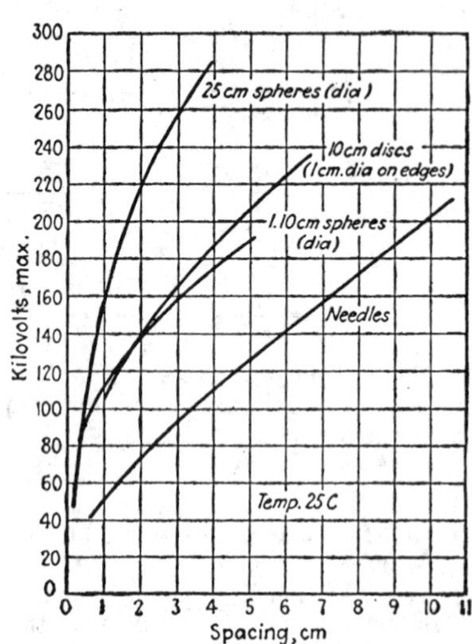

FIG. 4-63. Spark-over of various shaped electrodes in oil at 60 c/s.

Typical a-c values of the dielectric strength vs. spacing and electrode geometry, which affects the maximum stress, are shown[9] in Fig. 4-63. It must not be assumed that the levels shown can be maintained indefinitely, since particulate matter may move into the field and reduce the strength. The dielectric strength is thus dependent upon the time of voltage application. Typical impulse breakdown voltages of transformer oil are shown in Fig. 4-64. Usually the impulse strength is about two to three times the crest 60-c/s 1-min strength. The difference decreases as the oil purity increases.

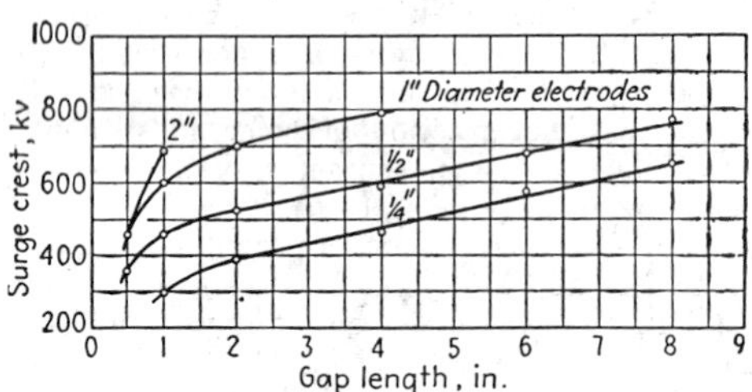

FIG. 4-64. Curve showing relation of gap length to minimum surge crest voltage required for breakdown between cylindrical electrodes with hemispherical ends immersed in oil, 1½ × 40-μs wave.

The covering of metal conductor surfaces has been known to increase the a-c strength of oil gaps.[11]

Corona or partial breakdown can occur in mineral oil, as with any liquid or gas, when the electric stress is locally very high and complete breakdown is limited by a solid barrier or large oil gap (as with a needle point in a large gap). Such discharges produce hydrogen and methane gas, and sometimes carbon with larger discharges. Dissolved air is also sometimes released by the discharge. If the gas bubbles formed are not ejected away from the high field, they will reduce the subsequent discharge threshold voltage as much as 80%.

320. Deterioration of Oil. Oil in service is subject to oxidation, which leads to acidity and sludge. There is no correlation between the amount of acid and the likelihood of sludging or the amount of sludge. Sludge clogs the ducts, reduces the heat transfer, and accelerates the rate of deterioration. ASTM tests for oxidation of oils are D1904, D1934, D1313, and D1314. Copper and lead and certain other metals accelerate the oxidation of mineral oils. Oils are considerably more stable in nitrogen atmospheres.

Inhibitors are now commonly added both to new and to used oils to delay the oxidation. Ditertiary butyl paracresol (DBPC) is the inhibitor most commonly used at present.

321. Servicing, Filtering, and Treating. Oil in service is usually maintained by testing for dielectric strength, neutralization number, and color. Proposals are being made to use power factor, resistivity, and interfacial tension. The ASTM D117 test states that all these properties indicate contaminating influences. It has been suggested that interfacial tension below a certain value indicates that sludging is imminent or has started (see *ASTM Spec. Tech. Publ.* 135, 1952).

Depending on the voltage rating of the apparatus, the oil is maintained above 16 to 22 kV (ASTM Test D877). The usual contaminants are water, sludge, acids, and, in circuit-breaker oils, carbon. The centrifuge is best suited for removing large quantities of water, heavier solid particles, etc. The blotter filter press is used for the removal of minute quantities of water, fine carbon, etc. In another method, after removing the larger particles the oil is heated and sprayed into a vacuum chamber, where the water and volatile acids are removed. Sludge and very fine solids are then taken out by a blotter filter press. All units are assembled together so that the process is completed in a single pass. Some work has been done on reclaiming oil by treating it to reduce acidity. One process is similar to the later stages in refining. Another treatment uses activated alumina, fuller's earth, or silica gel. The IEEE guide for Maintenance of Insulating Oil is published as *IEEE Standards Publ.* 64.

322. Chlorinated Aromatic—Askarel Liquids. These liquids are commonly of two types: (*a*) chlorinated diphenyl liquids containing 2 to 6 chlorine atoms/diphenyl molecule and (*b*) mixtures of the more highly (60%) chlorinated diphenyl with trichlorobenzene or tetrachlorobenzene. The former liquids, as shown in Table 4-63, are more viscous and are used mostly as capacitor impregnants. The latter liquids usually have a viscosity similar to that of transformer oils and are used in transformers.

These liquids have somewhat higher dielectric strengths than mineral oil. When subjected to discharges or arcing, they produce hydrogen chloride. To avoid the effects of this acidic hydrogen chloride, scavengers such as tin tetraphenyl and epoxy compounds, which react rapidly with the hydrogen chloride, have been added, usually less than 0.25%. The use of these liquids at high frequency is not recommended. The impulse strength in pressboard is lower than with mineral oil.

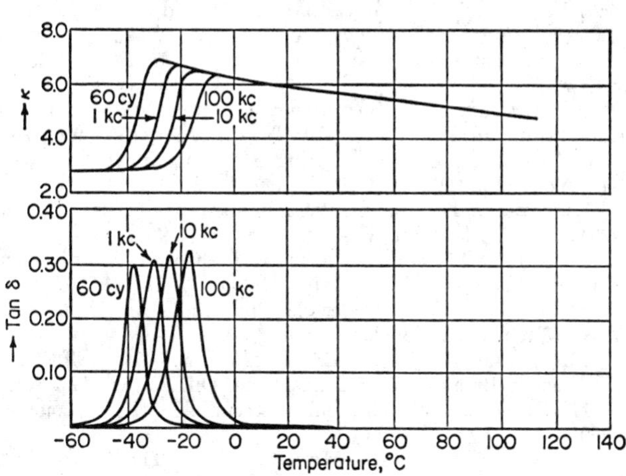

FIG. 4-65. Dielectric properties of trichlorodiphenyl (Aroclor 1242).

These liquids show a large dispersion of permittivity and a loss tangent maximum at low temperatures[12] as illustrated by Fig. 4-65. The dispersion region for higher-viscosity liquids is moved to higher temperatures. The dissipation factor and conductivity above room temperature decrease inversely with the viscosity and increase with the ionic contamination.

323. Fluorocarbon Liquids. A variety of nonpolar nonflammable perfluorinated aliphatic compounds, in which the hydrogen has been completely replaced by fluorine, is available with different ranges of viscosity and boiling point from below room temperature to more than 200°C. These compounds have low permittivities (near 2.0) and very low conductivity. They are inert chemically and have low solubilities for most other materials. The chemical formula for these compounds is one of the following: C_nF_{2n}, C_nF_{2n+2}, and $C_nF_{2n}O$. The presence of the oxygen in the latter formula does not seem to reduce the stability. These compounds have been used for filling electronic apparatus[13] and large transformers to give high heat-transfer rates together with high dielectric strength. The vapors of these liquids also have high dielectric strengths.[14]

324. Silicone Fluids. These fluids, chemically formed from Si—O chains with organic (usually methyl) side groups, have a high thermal stability, low temperature coefficient of viscosity, low dielectric losses, and high dielectric strength. They can be obtained with various levels of viscosity and correlated vapor pressures. Rated service temperatures extend from −65 to 200°C, some having short time capability up to 300°C. Their permittivity is about 2.6 to 2.7, declining with increasing temperature. These fluids have a tendency to form heavier carbon tracks than other insulating liquids when breakdown occurs. They cannot be considered fireproof.

325. Ester Fluids. There are a few applications, mostly for capacitors, where organic ester compounds are used. These liquids have a somewhat higher permittivity, in the range of about 4 to 7, depending on the ratio of ester groups to hydrocarbon chain lengths. Their conductivities are generally somewhat higher than those of the other insulating liquids discussed here. The compounds are easily subject to hydrolysis with water to form acids and alcohols and should be kept dry, particularly if the temperature is raised. Their thermal stability is poor. Specifically dibutyl sebacate has been used in high-frequency capacitors and castor oil in energy-storage capacitors.

326. References on Insulating Oils and Liquids

1. Gruse, W. A., and Stevens, D. R. "Chemical Technology of Petroleum"; 3d ed.; New York, McGraw-Hill Book Company, 1960.

2. Berberich, L. J. Influence of Gaseous Electric Discharge on Hydrocarbon Oils; *Ind. Eng. Chem.*, 1938, Vol. 30, p. 280. Blodgett, R. B., and Bartlett, S. C. *Trans. AIEE*, 1961, Vol. 80, p. 528. Olds, W. F., Feich, G., and Eich, E. *Ann. Rept. NRC* 1960 *Conf. on Electrical Insulation*, p. 93.

3. Berberich, L. J. Oxidation Inhibitors in Electrical Insulating Oils; *ASTM Bull.* 149, pp. 65–73, 1947. Ford, J. G., and Sloat, T. K. Inhibitors Lengthen Life of Transformer Oil; *Westinghouse Eng.*, 1950, p. 250.

4. Symposium on Insulating Oils; *ASTM Bulls.* 146 and 149, 1947. (Several authors.)

5. Kaufman, R. B., Shimanski, E. J., and MacFadyen, K. W. Gas and Moisture Equilibriums in Transformer Oil; *Trans. AIEE*, 1955, Vol. 74–III, p. 312.

6. Clark, F. M. "Insulating Materials for Design and Engineering Practice"; New York, John Wiley & Sons, Inc., 1962.

7. Piper, J. D. Chapter in Dielectric Materials and Applications, A. von Hippel (ed.); New York, John Wiley & Sons, Inc., 1954.

8. Rohlfs, A. F., and Turner, F. J. Correlation between the Breakdown Strength of Large Oil Gaps and Oil Quality Gauges; *Trans. AIEE*, 1956, Vol. 75–III.

9. Peek, F. W. "Dielectric Phenomena in High Voltage Engineering"; New York, McGraw-Hill Book Company, 1929.

10. Weber, K. H., and Endicott, H. S. Area Effect and Its Extremal Basis for the Electric Breakdown of Transformer Oil; *Trans. AIEE*, 1956, Vol. 75–III, p. 371.

11. Roth, A. "Hochspannungstechnik," 4th ed.; Vienna, Springer-Verlag, OHG, 1959.

12. White, A. H., and Morgan, S. O. The Dielectric Properties of Chlorinated Diphenyl; *J. Franklin Inst.*, 1933, Vol. 216, p. 635.

13. Kilham, L. F., and Ursch, P. R. *Proc. Natl. Electronics Computer Conf.*, Los Angeles, May, 1955.

14. Berberich, L. J., Works, C. N., and Lindsay, E. W. *Trans. AIEE*, 1955, Vol. 74–I, p. 660.

PLASTIC INSULATING MATERIALS

By R. N. Sampson

327. Definitions. A plastic is a material which contains an organic substance of high molecular weight, which is a solid in its finished state, and which at some stage in its manufacture can be shaped by flow. The organic substances are called resins or polymers and are derived from oil, coal, or in some cases natural materials. A special case occurs with silicone plastics wherein the silicon atom replaces the carbon in parts of the polymer. The polymers are usually mixed with other materials to improve or modify their properties. For example, fillers are added to improve physical properties, fluids are added to modify the flow characteristics, reinforcements are included to increase the strength, and antioxidants are added to improve the oxidation stability. Plastics are available in a number of forms. Those of concern in this section are molding materials, films, and laminates.

328. Molded Materials. Molded materials are formed by causing the plastic to flow in a mold with the application of heat and pressure. The plastic materials may be pure polymer or heterogeneous mixtures of polymer, fillers, reinforcements, and additives. Little or no orientation of the ingredients occurs. The most common processes are compression molding, injection molding, extrusion, and modifications of these processes. Molding compounds can be divided into three categories: thermoplastic materials, thermosetting materials, and cold-molded materials.

329. Thermoplastic Materials. Thermoplastic molding compounds are capable of being repeatedly softened or melted by increasing temperatures. In these polymers, the synthesis reactions are fully complete, and changes in state are physical rather than chemical. A typical thermoplastic material is paraffin. Thermoplastic materials can be molded quickly, are generally inexpensive, but are heat-sensitive.

Standards. The SPI (Society of the Plastics Industry) classifies thermoplastic materials based on electrical, optical, thermal, mechanical, and aging properties, derived from other sources and standards such as ASTM, National Bureau of Standards, Manufacturing Chemists Association, and military standards. Classification is based on three properties, i.e., heat-distortion temperature, impact strength, and tensile strength. The classification consists of prefix letters serving as a generic description defining the plastic family and five numbers defining the other three properties. The plastics classified are:

ABS—copolymers of acrylonitrile, butadiene, and styrene.
CA—cellulose acetate.
CAB—cellulose acetate butyrate.
CP—cellulose propionate.
HH—halogenated hydrocarbons.
MM—methylmethacrylate.
PA—polyamide.
PE—polyethylene.
PS—polystyrene.
VC—polyvinylidene chloride.
VCA—polyvinylchloride-acetate.

For more complete definitions, see "Plastics Engineering Handbook," 3d ed., SPI, Reinhold Publishing Company, New York, 1960.

Many other thermoplastic materials are in common usage but have not yet been classified. Examples of such plastics are polypropylene, polyethers, polycarbonate, polysulfone, polyphenylene oxide, styrene-acrylonitrile copolymers (SAN), and others.

Properties. Typical properties for thermoplastic materials are given in Table 4-64. A wide variety of compounds exists in each category, and no effort has been made in the table to identify any one material. The manufacturers should be consulted for exact property data. The data supplied are descriptive of molded, extruded, vacuum formed, and blown parts made from thermoplastics without fillers. Thermoplastic materials with fibrous reinforcements have higher physical properties, being superior in impact, tensile and flexural strength.[1*]

* Superscripts refer to bibliography, Par. **337**.

Table 4-64. Thermoplastic Molding Materials

Material		Acetal	ABS	Acrylic	Cellulose acetate	Cellulose acetate butyrate	Cellulose propionate	Chlorinated polyether	Chlorotri-fluoro-ethylene	Nylon (polyamide)	Poly-carbonate
SPI prefix	ASTM No.		ABS	MM	CA	CAB	CP		HH	PA	
Electrical properties:											
Arc resistance, sec	D495	129	90	No track	200	....	180	....	>360	140	120
Dielectric constant	D150										
60 c/s	....	3.8	3.0	4.0	7.5	6.4	4.0	3.1	2.8	5.5	3.2
10^6 c/s	....	3.8	3.0	3.5	7.0	6.3	4.0	3.0	2.7	4.9	3.0
10^9 c/s	....	3.8	3.0	3.2	7.0	6.2	3.6	2.9	2.5	4.7	3.0
Dissipation factor	D150										
60 c/s	....	0.004	0.003	0.04	0.01	0.02	0.01	0.01	0.001	0.01	0.0009
10^6 c/s	....	0.004	0.003	0.03	0.01	0.02	0.01	0.01	0.023	0.01	0.0021
10^9 c/s	....	0.004	0.005	0.02	0.01	0.05	0.01	0.01	0.009	0.03	0.01
Dielectric strength, V/mil, step by step	D149	400	350	350	200	250	300	400	450	320	364
Volume resistivity, Ω·cm	D257	10^{14}	10^{16}	10^{14}	10^{13}	10^{14}	10^{15}	10^{15}	10^{18}	10^{15}	10^{16}
Mechanical properties:											
Tensile strength, lb/in²	D651, D638	9,000	7,000	11,000	8,500	6,900	7,800	6,000	6,000	14,000	9,500
Tensile modulus, lb/in², $\times 10^5$	....	4.1	3.5	4.5	4.0	2.0	2.2	1.6	3.0	3.8	3.5
Elongation, %	....	15	....	10	70	88	100	160	250	320	100
Compressive strength, lb/in²	D695	18,000	7,000	18,000	36,000	22,000	22,000	....	7,400	13,000	12,500
Flexural strength, lb/in²	D790	14,000	10,500	17,000	16,000	9,300	11,400	5,000	9,300	....	13,500
Impact strength, ft·lb/in of notch	D256	1.4	7.0	0.5	5.2	6.3	11.5	0.4	2.7	4.0	16
Hardness, Rockwell	D785	R120	M110	M105	R125	R115	R122	R100	R95	R118	R118
Thermal properties:											
Heat-distortion temp. at 264 lb/in²	D648	255	200	210	190	202	228	285	258	167	280
Maximum-use temp., °F	....	185	210	190	220	220	220	290	390	250	250
Coefficient of thermal expansion, in/(in)(°C) $\times 10^{-5}$	D696	8.1	11	9	16	17	17	8	7	8	7
Thermal conductivity, cal/(cm²)(sec)(°C/cm) $\times 10^{-4}$	C177	5.5	3.0	0.0	8.0	8.0	8.0	3.1	5.3	5.9	4.6
Flammability, in/min	D635	1.1	....	1.2	....	....	1.3	No burn	No burn	No burn	No burn
Chemical properties:											
Water absorption, % in 24 h	D570	0.25	0.45	0.4	6.5	2.2	2.8	0.01	0.00	1.88	0.15
Not resistant to	D543	Strong acids	Oxidizing acids, ketones, esters, chlorinated solvents	Ketones, esters, aromatic and chlorinated solvents	....	Strong bases, ketones, esters, aromatic and chlorinated solvents	....	Oxidizing acids	Chlorinated solvents	Strong acids, phenol	Alkalies, aromatic and chlorinated solvents

Trade names											
Trade names	....	Betalux Celcon Delrin Dielux Formaldafil Thermo- comp	Abson Cycolac Cyclon Kralastic Lustran Royalite Sullvac Triform Tybrene	Acrapon Acryglass Acrylite Acrylux Acrysol Araset Bovick Glopaque Implex Interpole Kydex Lucite Luctrelite Oraglos Plexiglas Stypol XT Poly- mer Zerlon	Cal-Stix Celanese Joda Kodacel Plasticel Tenite	Acelon Cabulite Joda Tenite Uvex	Forticel Tenite	Penton	Kel-F Halon	Capran Catalin Firestone Fosta- Nylon Glastil Moleculoy Monocast Nylafil Nylux Plaskon Spencer Thermo- comp. X-Tal Zytel	Dupilon Lexan Merlon Penntube IV Polycarbafil Thermo- comp. Zelux

Table 4-64. Thermoplastic Molding Materials.—*Concluded*

Material		Poly-ethylene, low-density	Poly-ethylene, med.-density	Poly-ethylene, high-density	Poly-propylene	Poly-styrene	Poly-sulfone	Poly-phenylene oxide	Phenoxy	Poly-vinyl chloride	SAN	Tetra-fluoro-ethylene
SPI prefix	ASTM No.	PE	PE	PE		PS				VC		HH
Electrical properties:												
Arc resistance, sec	D495	140	200	200	185	100	122	75	....	80	150	>200
Dielectric constant	D150											
60 c/s	....	2.4	2.4	2.4	2.6	3.4	3.1	2.6	4.1	3.6	3.4	2.1
10^6 c/s	....	2.4	2.4	2.4	2.6	3.2	3.1	2.6	4.1	3.3	2.5	2.1
10^9 c/s	....	2.4	2.4	2.4	2.6	3.1	3.1	2.6	3.8	3.1	3.1	2.1
Dissipation factor	D150											
60 c/s	....	<0.0005	<0.0005	<0.0005	<0.0005	0.0004	0.0008	0.0004	0.001	0.007	0.004	<0.0002
10^6 c/s	....	<0.0005	<0.0005	<0.0005	<0.0005	0.0004	0.001	....	0.002	0.009	0.007	<0.0002
10^9 c/s	....	<0.0005	<0.0005	<0.0005	<0.0005	0.0004	0.005	0.0009	0.03	0.006	0.007	<0.0002
Dielectric strength, V/mil, step by step	D149	420	500	550	450	300	400	400	400	375	300	430
Volume resistivity, Ω·cm	D257	10^{16}	10^{16}	10^{16}	10^{16}	10^{16}	10^{17}	10^{13}	10^{13}	10^{16}	10^{16}	10^{18}
Mechanical properties:												
Tensile strength, lb/in²	D651, D638	2,300	3,500	5,500	5,500	6,800	10,200	11,000	9,500	9,000	12,000	4,500
Tensile modulus, lb/in², × 10^5	....	0.35	0.55	1.5	2.3	4.5	3.6	3.8	3.8	6.0	5.6	0.58
Elongation, %	....	800	600	100	700	80	100	80	100	40	3.5	400
Compressive strength, lb/in²	D695	....	....	3,200	8,000	9,000	14,000	13,000	12,000	13,000	17,000	1,700
Flexural strength, lb/in²	D790	....	7,000	1,000	8,000	10,000	15,400	15,000	14,000	16,000	19,000	
Impact strength, ft·lb/in of notch	D256	No break	>16	20	1.5	8	1.3	1.9	12	20	0.5	3.0
Hardness, Rockwell	D785	....	....	....	R110	R100	R120	R123	R123	....	M90	
Thermal properties:												
Heat-distortion temp. at 264 lb/in²	D648	105	120	120	145	205	....	375	....	164	215	>250
Maximum-use temp., °F	....	212	250	250	320	175	345	....	300	175	205	550
Coefficient of thermal expansion, in/(in)(°C) × 10^{-5}	D696	18	16	13	10	21	6	3	6	18	8	10
Thermal conductivity, cal/(cm²)(sec)(°C/cm) × 10^{-4}	C177	8.0	10.0	12.4	2.8	3.0	6.2	4.5	8.4	7.0	2.9	6
Flammability, in/min	D635	1.04	1.04	1.04	1.04	1.0	No burn	No burn	No burn	No burn	1.0	No burn
Chemical properties:												
Water absorption, % in 24 h	D570	<0.01	<0.01	<0.01	0.03	0.6	0.2	0.06	0.13	0.4	0.3	0.00
Not resistant to	D543	Oxidizing acids, aromatic and chlorinated solvents			....	Oxidizing acids, aromatic chlorinated solvents	Aromatic solvents	Aromatic and chlorinated solvents		Ketones, esters, aromatic solvents	Acids, ketones, esters, chlorinated solvents	

Trade names........................	....	Agilene, Althon, Bakelite, Cao X-L, Dylan, El-Rex, Epolene, Ethylux, Excelite, Fortiflex, Hi-Fax, Marlex, Modulene, Petrothene, Poly-Eth	Aerotuf Avisun Bakelite Chevron Escon Marlex Petrothene Pro-Fax	Bakelite Biax Dylene El-Rex Evenglo E-Z Flow Fostarene Gilco Grace Hypac Kardel Lustrex Shell Solar Styrafil Styroflex Styrolux Styron	Bakelite	PPO	Bakelite	Bakelite Blacar Dacovin Durelene Escambia Ethyl Excelon Exon Geon Insular Kenron Kohinor Marvinol Nalgon Opalon Pliovic Resinite Rucoblend Secron Trulon Tygon Vicoa Vygen Vyran Xyran	Acrylafil Bakelite Catalin Kralac Lustran Plaxacrin Tyril	Halon Teflon

Table 4-65. Thermosetting Molding Materials

Properties		Diallyl phthalate			Epoxy		Melamine		
SPI prefix	ASTM No.	DAP					MF		
Filler		Glass fiber	Mineral	Synthetic fiber	Glass fiber	Mineral	Alpha cellulose	Asbestos	Glass fiber
Electrical properties:									
Volume resistivity, Ω·cm	D257	10^{16}	10^{13}	10^{16}	10^{14}	10^{14}	10^{14}	10^{12}	10^{11}
Dielectric str., short-time, V/mil	D149	450	420	400	400	400	400	430	300
Dielectric str., step-by-step, V/mil	D149	400	400	410	400	400	300	320	240
Dielectric constant, 60 c/s	D150	4.3	5.2	5.0	5.0	5.0	9.5	10.2	11.1
Dielectric constant, 10^3 c/s	D150	4.4	5.3	3.9	5.0	5.0	9.2	9.0	...
Dielectric constant, 10^6 c/s	D150	4.5	4.0	3.6	5.0	5.0	8.4	6.7	7.5
Dissipation factor, 60 c/s	D150	0.01	0.03	0.026	0.01	0.01	0.030	0.07	0.14
Dissipation factor, 10^3 c/s	D150	0.004	0.03	0.004	0.01	0.01	0.015	0.07	...
Dissipation factor, 10^6 c/s	D150	0.009	0.02	0.012	0.01	0.01	0.027	0.041	0.013
Arc resistance, sec	D495	180	190	130	180	190	180	180	180
Mechanical properties:									
Specific gravity	D792	1.78	1.68	1.39	2.0	2.0	1.52	2.0	2.0
Specific volume, in^3/lb	D792	17.2	16.8	20.7	15.4	14.2	18.2	13.8	13.8
Tensile strength, lb/in^2	D638, D651	11,000	8,700	6,800	30,000	15,000	13,000	7,000	10,000
Elongation, %	D638	...	...	...	4	...	0.9	0.45	...
Tensile modulus, $lb/in^2 \times 10^5$	D638	22	22	6.0	30.4	...	14	19.5	24
Compressive strength, lb/in^2	D695	35,000	32,000	30,000	40,000	40,000	45,000	30,000	35,000
Flexural strength, lb/in^2	D790	19,000	11,000	19,000	60,000	15,000	16,000	11,000	23,000
Impact strength, ft·lb/in of notch	D256	10.0	0.45	8.0	10.0	0.4	0.35	0.4	6.0
Hardness, Rockwell	D785	M108–110	M100–103	M108–115	M100–110	M100–110	M110–125	M110	...
Thermal properties:									
Thermal conductivity, cal/(sec)(cm^2) 10^{-4} (1°C/cm)	C177	5–10	7–25	5–6	4–10	4–30	7–10	13–17	11–0
Thermal expansion, in/in $\times 10^{-5}$/°C	D696	3.6	4.2	6.0	3.5	5.0	4.0	2.0–4.5	1.5–1.7
Maximum-use temp., °F	...	350	350	300	300	300	210	250	300
Heat-distortion temp., °F at 264 lb/in^2	D648	350	320	300	250	250	360	265	400
Chemical properties:									
Water absorption, % in 24 h	D570	0.35	0.5	0.2	0.2	0.04	0.6	0.14	0.21
Burning rate	D635	Self-exting. to nonburning	Self-exting. to nonburning	Self-exting.	Self-exting.	Self-exting.	Self-exting.	Self-exting.	Self-exting.
Effect of sunlight	...	None	None	None	Slight	Slight	Slight color change	Slight color change	Slight
Effect of weak acids	D543	None	None	None	None	None	None	None to slight	None
Effect of strong acids	D543	Slight	Slight	Slight	Negligible	None	Decomposes	Decomposes	Decomposes
Effect of weak alkalies	D543	None to slight	None to slight	None	None	None	None	Very slight attack	None

Effect of strong alkalies	D543	Slight	Slight	Slight	None	Slight	Attacked	Slight attack	None to slight
Effect of organic solvents	D543	None	None	None	None	None	None	None	None
Trade names	....	Acme, Cosmic, Dapon, Diall, Poly-Dap, RX			Bakelite, Dri-Coat, Eccomold, EMC, Epiall, Eposet, Fibercore, Formitt, High Strength, Hysol, Plenco, Polyset, Scotchply, Smooth-on, Trevarno, Unipoxy		Amres, Cymel, Diaron, Melmac, Permelite, Plenco, Resimene, Resloom, Syr-U-Tex		

Table 4-65. Thermosetting Molding Materials.—*Concluded*

Properties		Phenolic			Polyester		Silicone		Urea formaldehyde
SPI prefix	ASTM No.	PF			EA		S		UF
Filler		Woodflour and cotton flock	Asbestos	Glass fiber	Glass fiber	Mineral	Glass fiber	Mineral	Alpha cellulose
Electrical properties:									
Volume resistivity, Ω·cm	D257	10^{13}	10^{13}	10^{12}	10^{15}	10^{14}	10^{14}	10^{14}	10^{13}
Dielectric str., short-time, V/mil	D149	400	350	400	420	450	400	400	400
Dielectric str., step-by-step, V/mil	D149	375	300	270	390	350	300	380	300
Dielectric constant, 60 c/s	D150	13	50	7.1	7.3	7.5	5.2	3.6	9.5
Dielectric constant, 10^3 c/s	D150	9.0	30	6.9	4.68	6.2	5.0	...	7.5
Dielectric constant, 10^6 c/s	D150	6.0	10	6.6	6.4	5.5	4.7	6.3	6.8
Dissipation factor, 60 c/s	D150	0.05	0.1	0.05	0.011	0.009	0.004	0.004	0.035
Dissipation factor, 10^3 c/s	D150	0.04	0.1	0.02	...	0.01	0.0035	...	0.025
Dissipation factor, 10^6 c/s	D150	0.03	0.4	0.012–0.026	0.008	0.015	0.002	0.002	0.25
Arc resistance, sec	D495	Tracks	120	120	180	150	250	420	150
Mechanical properties:									
Specific gravity	D792	1.45	1.9	1.95	2.3	2.30	2.0	2.82	1.52
Specific volume, in³/lb	D792	17.8	11.9	14.1	...	5.4	13.8	...	18.2
Tensile strength, lb/in²	D638, D651	10,000	7,500	18,000	10,000	8,000	5,000	3,500	13,000
Elongation, %	D638	0.8	0.50	0.2	...	...	...	...	1.0
Tensile modulus, lb/in² × 10^5	D638	17	30	33	25	26	...	...	15
Compressive strength, lb/in²	D695	36,000	35,000	70,000	30,000	25,000	15,000	18,000	45,000
Flexural strength, lb/in²	D790	12,000	14,000	60,000	20,000	10,000	14,000	8,000	18,000
Impact strength ft·lb/in of notch	D256	0.60	3.5	18	16.0	0.50	15	0.35	0.40
Hardness, Rockwell	D785	M96–120	M95–115	M95–100	...	...	M84	M71–95	M110–120
Thermal properties:									
Thermal conductivity, cal/(sec)(cm²)(°C/cm) × 10^{-5}	C177	4.7	8–22	9–14.5	10–16	15–25	7.51–7.54	11–13	7–10
Thermal expansion, in/in × 10^{-5}/°C	D696	3.0–4.5	0.8–4	0.8–1.6	2.5–3.3	3.5–5	0.8	2–4	2.2–3.6
Maximum-use temp., °F	...	350	350	350	300	300	>600	>600	170
Heat-distortion temp., °F at 264 lb/in²	D648	260	300	300	>400	350	>900	>900	260
Chemical properties:									
Water absorption, % in 24 h	D570	0.7	0.5	1.2	0.28	0.5	0.2	0.13	0.8
Burning rate	D635	Very low	None	None	Slow to self-exting.	Slow to non-burning	None to slow	Self-exting.	Self-exting.
Effect of sunlight	...		Darkens		Depends on pigmentation	None	None to slight	None to slight	Pastels gray
Effect of weak acids	D543		None		Slight	None	None to slight	None	None to slight
Effect of strong acids	D543		Oxidizing acids		Attacked	Attacked	Slight	Slight	Decomposed if surface attacked
Effect of weak alkalies	D543		Slight		Slight to attacked	Attacked	None to slight	None to slight	Slight to marked

Effect of strong alkalies	D543	Attacked	Slight to attacked	Decomposes	Slight to marked	Slight to marked	Decomposes
Effect of organic solvents	D543	None	None	None	Attacked by some	Attacked by some	None to slight
Trade names	....	Amres, Arochem, Bakelite, Catalin, Celeron, Durez, Fiberite, Haveg, Mouldrite, Nestorite, Permelite, Plenco, Plyophen, Resinox, Snap-Cure, Synvaren, Synvorite, Tetra-Flex, Tybon, Varcum	Alpon, Co-Rezyn, Durez, Fibercore, Formadall, Glasdramatic, Glaskyd, Glasrin, Glastic, Haysite, Insulstruct, Mobaloy, Parr, Plaskon, Politen, Polyglas, Premix, Resistrac, Rosite, Trevarno		Dow Corning		Amres, Arodure, Beetle, Mouldrite, Nestorite, Plaskon, Resimene, Resloom, Sylplast, Syn-U-Flex, Synvarol, Tetra-Ria, Tybon

330. Thermosetting Materials. Thermosetting materials are partly polymerized materials which are cured in a mold with heat and pressure, thereby completing the chemical reaction, the material being converted into an infusible, insoluble product. Since the change in state is chemical rather than physical, thermosetting materials cannot be softened to the flow point by heat after curing. In general, thermosetting resins are combined with fillers and reinforcements to produce molding compounds. Fillers may be wood flour or any of a wide variety of mineral materials such as calcium carbonate. Reinforcements may be fiber glass, asbestos, or synthetic fibers.

Standards. Many ASTM and military standards apply to thermoset materials. The SPI has classified these molding compounds, as follows:

DAP—diallyl phthalate.
EA—ester alkyd.
MF—melamine formaldehyde.
PF—phenol formaldehyde.
UF—urea formaldehyde.
S—silicone.

Other molding materials are in common usage but have not yet been included in the SPI classification. The epoxies furnish an example of these.

Properties. Typical properties for thermosetting materials are given in Table 4-65. These properties apply to compression and transfer molded materials.[2]

331. Cold-molded Compounds. Cold-molded materials are formed in molds at room temperature under high pressure and are subsequently hardened in ovens. Cold-molding materials are either refractory (inorganic) or nonrefractory (organic), depending on the nature of the binder. Refractory cold-molded materials consist of cement, lime, water glass, or silica mixed with water and a filler such as asbestos. These materials are more heat-resistant than the nonrefractory materials and have higher arc resistances. Nonrefractory materials consist of asphalts, coal tar pitches, oils, gums, resins, or waxes filled with such materials as asbestos, silica, or magnesia compounds. The cold-molded materials are frequently used in electrical insulation where thermal stability and arc resistance are important. These materials are inexpensive, lack a high luster, and are available in a limited color range. More information is given in the "Plastics Engineers Handbook," 3d ed., SPI, Reinhold Publishing Company, New York, 1960.

Properties. Typical properties for a refractory compound and a nonrefractory material are given in Table 4-66.

332. Films. Films are made from plastics by extrusion, calendering, or solvent casting. The polymers used for films are generally thermoplastic. Films are available in thicknesses of ¼ to 200 mils and in a wide variety of widths. One of the most useful properties of films to an electrical designer is the high dielectric strength usually associated with these materials. Physical properties are also high. The strength properties of some films can be dramatically increased by orientation. Orientation is the process of selectively stretching the film, thereby reducing its thickness and causing changes in the crystallinity of the polymer. Usually, it is accomplished under conditions of elevated temperature during processing, and the benefits are lost if the processing temperature is thereafter exceeded during service.

Most films can be bonded to other substrates with a variety of adhesives. Those films which do not readily accept adhesives can be surface-treated for bonding by several chemical means. Some films are combined to obtain bondable surfaces. Examples of these composites are olefins laminated to polyester films and fluorocarbons laminated to polyimide films.

Films are used extensively in wire and cable, fractional-horsepower motors, capacitors, coils, transformers, and batteries.

Properties. Typical properties for films[3] are given in Table 4-67.

333. Laminated Sheet. Laminated products are formed by bonding together two or more layers of material. The layers generally comprise reinforcing fibers such as cellulose, glass fibers, or asbestos in the form of fabrics, papers, or mats and impregnated with a thermosetting resin. The laminates may be parallel-laminated, in which case the reinforcements are oriented parallel to the strongest direction in tension, or cross-

Table 4-66. Cold-molded Materials

Properties	ASTM test method	Cold-molded compounds	
		Nonrefractory (organic)	Refractory (inorganic)
Electrical properties:			
Volume resistivity, Ω·cm	D257	10^{12}	
Dielectric strength, short-time, V/mil	D149	115	45
Dielectric strength, step-by-step, V/mil	D149	75	80
Dielectric constant, 60 c/s	D150	28	
Dielectric constant, 10^3 c/s	D150	18	
Dielectric constant, 10^6 c/s	D150	6.0	
Dissipation factor, 60 c/s	D150	0.20	
Dissipation factor, 10^6 c/s	D150	0.07	
Arc resistance, sec	D495	200	500
Mechanical properties:			
Specific gravity	D792	2.15	2.2
Specific volume, in³/lb	D792	12.9	12.7
Tensile strength, lb/in²	D638	3,000	2,200
Compressive strength, lb/in²	D695	15,000	18,000
Flexural strength, lb/in²	D790	10,000	7,500
Impact strength, ft·lb/in of notch	D256	0.4	0.4
Hardness, Rockwell	D785	M80–90	M75–95
Thermal properties:			
Resistance to heat, °F	......	500	1,300
Deflection temp., °F at 264 lb/in²	D648	>800	>400
Chemical properties:			
Water absorption, % in 24 h	D570	0.55	0.5
Burning rate	D635	Nil	Nil
Effect of sunlight	......	....	Nil
Effect of weak acids	D543	Slight	Slight
Effect of strong acids	D543	Decomposes	Decomposes
Effect of weak alkalies	D543	None	None
Effect of strong alkalies	D543	Decomposes	None
Effect of organic solvents	D543	Attacked by some	None
Trade names	......	Aico, Blak-Stretchy, Cetic, Ebrok, Gra-Tufy, Gumman, Okon, Rosite, Thermoplax	Aico, Alphide, Blue-sil, Coltstone, Ebony Asbestos, Hemit, Howelex, Tegit

laminated, wherein the reinforcements are at 90° to one another. Laminates with other orientations are also manufactured. Laminates are frequently used for structural applications in electrical devices because of their superior physical properties.

Standards. Standards and specifications for thermosetting laminates have been prepared by ASTM (D709-62T) and NEMA (LI 1-1965). Five types have been established; paper, cellulose fabric, asbestos, glass, and nylon-based laminates for sheets, rods and tubes. The grades consist of the following laminates and contain phenolic resin unless otherwise stated:

Grade X—paper-based sheet and tubes intended for mechanical applications where electrical properties are of little importance. Not equal to fabric-based grade in impact strength.

Grade XP—paper-based sheet intended for hot punching. More flexible and not so strong as grade X.

Grade XPC—paper-based sheet for cold punching. More flexible, lower flexural strength, higher cold flow than grade XP.

Grade XX—paper-based sheet, tube, and rods suitable for electrical applications. Good machinability.

Grade XXP—paper-based sheet with better electrical and moisture properties than grade XX. More suitable for hot punching.

Grade XXX—paper-based sheet, tube, and rod for r-f work. High humidity and cold-flow resistance.

Grade XXXP—paper-based sheet with high insulation resistance and low losses at high humidities. More suited for hot punching than grade XXX.

Table 4-67. Insulating Films

Film base	ASTM No.	Cellulose acetate	Cellulose triacetate	Polytrifluoro-chloroethylene copolymers	Polyurethane elastomer	FEP-fluorocarbon	Polyethylene	Polyvinyl chloride	Ionomer	Polyimide
Method of processing	...	Cast, extruded	Cast	Extruded	Cast, calendered extruded	Extruded	Extruded	Cast, calendered extruded	Extruded	...
Forms available	...	Rolls, sheets, tapes	Rolls, sheets	Rolls	Rolls, sheets	Rolls, sheets, tapes	Rolls, sheets, tapes	Rolls, sheets, tapes	Rolls, sheets, tapes	Rolls
Thickness range, in	...	0.0005–0.250	0.0008–0.020	0.005–0.030	0.005–0.030	0.0005–0.020	0.00075–0.010	0.00075–0.08	0.001–0.010	0.001–0.005
Maximum width, in	...	60	46	54	52	48	240	84	60	18
Area factor, $in^2/(lb)(mil)$	...	22,000	21,400	13,000	22,000	12,900	29,500	23,000	29,400	19,400
Specific gravity	D792-60T	1.31	1.31	2.15	1.26	2.15	0.940	1.50	0.94	1.42
Tensile strength, lb/in^2	D882-61T	16,400	16,000	8,000	9,000	3,000	3,500	5,600	5,000	25,000
Elongation, %	D882-61T	70	40	150	650	300	650	500	450	70
Bursting strength, 1 mil thickness, Mullen points	D774-63T	40	70	42	...	11	...	20	...	75
Tearing strength, g	D689-62	10	10	26	...	125	300	1,400	70	8
Tearing strength, lb/in	D689-62	415	395	900	600	600	...	490	...	...
Folding endurance, cycles	D643-63T	2,000	4,000	...	...	4,000	>100,000	...	...	10,000
Water absorption, % in 24 h	D570-63	9	2–4.5	0.00	0.55–0.77	<0.01	<0.01	...	0.4	2.9
Dielectric constant, 10^3 c/s	D1531-62	3.6	3.2	2.7	7.5	2.0	2.2	6.0	2.4	3.5
Dielectric constant, 10^6 c/s	D1531-62	3.2	3.3	2.4	7.1	2.0	2.2	4.0	2.4	3.4
Dielectric constant, 10^9 c/s	D1531-62	3.2	3.2	2.3	...	...	2.2	...	2.4	...
Dissipation factor, 10^3 c/s	D1531-62	0.10	0.10	0.027	0.060	0.0002	0.0003	0.16	0.007	0.003
Dissipation factor, 10^6 c/s	D1531-62	0.10	0.10	0.017	...	0.0007	0.0003	0.14	0.007	0.010
Dissipation factor, 10^9 c/s	D1531-62	0.094	0.094	0.004	...	...	0.0003	...	0.007	...
Dielectric strength, V/mil	...	5,000	3,700	...	500	5,000	500	600	1,000	7,000
Volume resistivity, Ω·cm	...	10^{13}	10^{15}	10^{18}	10^{11}	10^{19}	10^{16}	10^{14}	10^{16}	10^{18}
Orientation possible	...	No	No	No	Yes	Yes	Yes	Yes	Yes	No
Trade names	...	Auburn Acelon Bexfilm Bexoid Cadco Campco Inceloid Kodacel Lumarith Midlon Saplasco Tenite Vuepak	Bexfilm Celanese Kodacel	Kel-F Polyfluoron	Adiprene Estane Perflex Texin	Teflon	Arnar Amerifilm Ampacet Auburn Boltaron Clysar Conolean Crown-Seal Dynafilm Ger-Pac Hypac Katharon Midlon Plicose Polython Prohi Seilon Vis-Queen Zee	Agilide Amerifilm Arnar Boltaflex Cadco Clopane Colovin Dorn Fabray Fabtex Ger-Pac Katharon Koroseal Krene Monosol Pantex Randfilm Resinite Reynolon	Surlyn	Kapton

Zendel	Rucoam Saplasco Touchstone Ultron Velon Vylene Wataseal

Table 4-67. Insulating Films.—*Concluded*

Film base	ASTM No.	Polypropylene	Polyvinyl fluoride	Polyester	TFE tetra-fluoroethylene	Vinylidene chloride	Fluoro-halocarbon	Vinylidene fluoride	Polyamide
Method of processing		Extruded	Extruded	Cast	Skived, extruded	Extruded, cast	Extruded	Extruded	Extruded, cast
Forms available		Rolls, sheets, tapes	Rolls	Rolls, sheets, tapes	Sheets, tapes	Rolls, tapes	Rolls	Rolls	Rolls
Thickness range, in		0.0005–0.020	0.0005–0.02	0.00015–0.014	0.0005–0.01	0.0004–0.02	0.005–0.02	0.001–0.010	0.0005–0.02
Maximum width, in		60	138	120	38	40	54	16	54
Area factor, in²/(lb)(mil)		31,300	17,200	22,600	12,800	16,800	13,000	15,000	24,500
Specific gravity	D792-60T	0.9	1.38	1.4	2.2	1.64	2.2	1.76	1.13
Tensile strength, lb/in²	D882-61T	10,000	18,000	30,000	4,000	20,000	11,000	6,500	11,500
Elongation, %	D882-61T	1,000	250	120	350	80	150	300	400
Bursting strength, 1 mil thickness, Mullen points	D774-63T		70	66			31	20	
Tearing strength, g	D689-62	45	40	15	100		150	60	90
Tearing strength, lb/in	D689-62		1,400	1,300			29		1,200
Folding endurance, cycles	D643-63T	>100,000	47,000	14,000		50,000	>100,000	75,000	>250,000
Water absorption, % in 24 h	D570-63	0.005	0.5	0.8	0.00	0.00	0.00	0.04	9.5
Dielectric constant, 10³ c/s	D1531-62	2.1	8.5	3.1	2.1	6.0	2.6	7.72	3.8
Dielectric constant, 10⁶ c/s	D1531-62	2.1	7.4	3.0	2.1	5.0	2.3	6.43	3.7
Dielectric constant, 10⁹ c/s	D1531-62	2.1		2.8	2.1	4.0		2.98	3.4
Dissipation factor, 10³ c/s	D1531-62	0.0003	0.009	0.0047	0.0002	0.045	0.039	0.019	0.010
Dissipation factor, 10⁶ c/s	D1531-62	0.0003		0.016	0.0002	0.075	0.037	0.159	0.016
Dissipation factor, 10⁹ c/s	D1531-62	0.0003		0.003	0.0002	0.08		0.11	0.025
Dielectric strength, V/mil		3,000		7,000	430	5,000	3,000	1,280	1,700
Volume resistivity, Ω·cm		10^{16}	10^{13}	10^{18}	10^{18}	10^{16}	10^{17}	10^{14}	
Orientation possible		Yes	No	Yes	No	Yes	No	No	No
Trade names		Bexphone Clysar Dynafilm Electro-film Hypac Midlon Olefane Profax Propylux Udel Vypro	Tedlar	Celenar Kodar Mylar Scotchpar Scotchpack Videne	Halon Teflon	Cryovac Saran	Aclar	Kynar	Cadco Califilm Capran Plastex Sapalaso

Grade XXXPC—paper-based sheet similar to grade XXXP but with lower punching temperatures.

Grade FR-3—paper-based sheet with epoxy resin binder. Self-extinguishing, low dielectric losses at high humidity, high flexural strength, and punchable.

Grades ES-1, ES-2, ES-3—paper-based sheets for name-plate engraving.

Grade C—cotton-fabric (>4 oz/yd^2) sheet, tube, and rod. Mechanical grade with high impact strength.

Grade MC—cotton-fabric (>4 oz/yd^2) sheet and rod with melamine resin binder. Good alkali and arc resistance.

Grade CE—sheet, tube, and rod made from same fabric as grade C. For electrical applications requiring greater toughness than grade XX. More moisture-resistant than grade C. Not recommended for primary insulation at voltages over 600 V.

Grade L—cotton-fabric (<4 oz/yd^2) sheet, tube, and rod. Mechanical grade not so tough as grade C and suitable for gears.

Grade LE—sheet tubes and rods similar to grade L. Not recommended for primary insulation at voltages over 600 V.

Grade A—asbestos paper-based sheet and tube. More resistant to flame and heat than cellulose-based material. Limited to voltages lower than 250 V.

Grade AA—asbestos fabric sheet and tube. More heat-resistant and stronger than grade A. Not recommended for primary insulation.

Grade G-2—sheet made of staple fiber-type glass cloth. Electrical and heat-resistant. Lowest in dielectric loss except silicone grade.

Grade G-3—sheet, tube, and rod made of continuous filament-type glass cloth. General-purpose grade, high impact strength, low bond strength.

Grade G-5—sheet, tube, and rod made of continuous filament glass fabric with melamine resin binder. Highest mechanical strength and hardest grade. Good heat and arc resistance. Excellent electrically when dry. Good dimensional stability.

Grade G-6—sheet made of staple fiber-type glass cloth with silicone resin binder. Excellent electrically when dry and good at high humidities. Dielectric strength perpendicular to laminations is low. Heat- and arc-resistant. Meets IEEE Class H insulation requirements.

Grade G-7—sheet made of continuous filament-type glass cloth with silicone resin binder. Similar to grade G-6, but bond strength and flame resistance slightly lower. Higher electric strength.

Grade G-9—sheet, tube, and rod made of continuous filament-type glass cloth with heat-resistant melamine resin binder. Similar to grade G-6, except more heat-resistant.

Grade G-10—sheet, tube, and rod made of continuous filament-type glass cloth with epoxy resin binder. Extremely high mechanical strength. Good electrical-loss properties both wet and dry.

Grade G-11—sheet, tube, and rod made of continuous filament-type glass cloth with heat-resistant epoxy resin. Similar to grade G-10, but retains 50% of its initial flexural strength at 150°C after 1 h at 150°C.

Grade GPO-1—sheet made of random glass mat with a polyester resin binder, mechanical and electrical grade.

Grade GPO-2—sheet made of random glass mat with a low-flammability polyester resin. Similar to grade GPO-1, but is self-extinguishing.

Grade FR-4—sheet and rod of continuous filament-type glass cloth with epoxy resin binder. Similar to grade G-10, but is self-extinguishing.

Grade FR-5—sheet and rod of continuous filament-type glass cloth with heat-resistant epoxy resin binder. Similar to grade G-11, but is self-extinguishing.

Grade N-1—sheet with nylon cloth base. Excellent electrical properties at high humidity. Good impact strength, but subject to creep at elevated temperatures.

Grade CF—sheet made from cotton (>4 oz/yd^2). Postforming grade. Postforming refers to the process of forming sheets into shapes by applying heat in a forming die.

Properties. Representative properties of laminates are shown in Table 4-68. These properties are typical and should not be used as specification requirements.[4]

334. Tubes, Rods, and Special Shapes. The properties given in Table 4-68 relate

Table 4-68. ASTM-NEMA Laminated Thermosetting Products

ASTM and NEMA grade designations	X	XX	XXX	XP	XPC	XXP	XXXP	C	CE	L	LE
Tensile strength, lb/in²:											
LW*	20,000	16,000	15,000	12,000	10,500	11,000	12,400	11,200	12,000	14,000	13,500
CW†	16,000	13,000	12,000	9,000	8,500	8,500	9,500	9,500	9,000	10,000	9,500
Modulus of elasticity in tension, lb/in²:											
LW	1,900,000	1,500,000	1,300,000	1,200,000	1,000,000	900,000	1,000,000	1,000,000	900,000	1,200,000	1,000,000
CW	1,400,000	1,200,000	1,000,000	900,000	800,000	700,000	800,000	900,000	800,000	900,000	850,000
Modulus of elasticity in flexure, lb/in²:											
LW	1,800,000	1,400,000	1,300,000	1,200,000	1,000,000	900,000	1,000,000	1,000,000	900,000	1,100,000	1,000,000
CW	1,300,000	1,100,000	1,000,000	900,000	800,000	700,000	700,000	900,000	800,000	850,000	850,000
Compressive strength, lb/in²:											
Flatwise	36,000	34,000	32,000	25,000	22,000	25,000	25,000	37,000	39,000	35,000	37,000
Edgewise	19,000	23,000	25,500	....	....	....	....	23,500	24,500	23,500	25,000
Rockwell hardness (M scale)	M-110	M-105	M-110	M-95	M-75	M-100	M-105	M-103	M-105	M-105	M-105
Dielectric strength perpendicular to laminations, V/mil:	....	....	....	....	....	....	....	....	....	....	....
Short-time test											
1/32 in.	950	950	900	900	850	950	900		....		700
1/16 in.	700	700	650	650	600	700	650		500		500
1/8 in.	500	500	470	470	425	500	470	No values recommended	360	No values recommended	360
Step-by-step test											
1/32 in.	700	700	650	650	625	700	650		....		450
1/16 in.	500	500	450	450	425	500	450		300		300
1/8 in.	360	360	320	320	290	360	320		220		220
Insulation resistance (condition C-96/35/90), MΩ	....	60	1,000	....	....	500	20,000	....	....	....	30
Specific gravity	1.36	1.34	1.32	1.33	1.34	1.32	1.30	1.36	1.33	1.35	1.33
Specific volume, in³/lb	20.4	20.6	21.0	20.8	20.6	21.0	21.3	20.4	20.8	20.5	20.8
Thermal expansion, cm/(cm)(°C)						0.000020					
Thermal conductivity, cal/sec cm²(°C/cm)						0.00070					
IEEE insulation class	A	A	A	A	A	A	A	A	A	A	A
Specific heat, Btu/(lb)(°F)						0.35–0.40					

* LW means lengthwise, or tested in the direction of orientation.
† CW means crosswise, or tested 90° to the direction of orientation.

Table 4-68. ASTM-NEMA Laminated Thermosetting Products.—*Concluded*

ASTM and NEMA grade designations	A	AA	G-2	G-3	G-5	G-6	G-7	N-1	ES-1	ES-2	ES-3	GPO-1	GPO-2
Tensile strength, lb/in²:													
LW*	10,000	12,000	16,000	23,000	37,000	...	23,000	8,500	12,000	13,000	15,000	12,000	10,000
CW†	8,000	10,000	11,000	20,000	30,000	...	18,500	8,000	8,500	9,000	12,000	10,000	9,000
Modulus of elasticity in tension, lb/in²:													
LW	2,500,000	1,700,000	1,800,000	2,000,000	2,300,000	...	1,800,000	400,000	...	...	...	1,000,000	1,000,000
CW	1,600,000	1,500,000	1,200,000	1,700,000	2,000,000	...	1,800,000	400,000	...	...	...	900,000	900,000
Modulus of elasticity in flexure, lb/in²:													
LW	2,300,000	1,600,000	1,300,000	1,500,000	1,700,000	...	1,400,000	600,000	...	...	...	1,200,000	1,200,000
CW	1,400,000	1,400,000	1,000,000	1,200,000	1,500,000	...	1,200,000	500,000	...	...	...	1,000,000	1,000,000
Compressive strength, lb/in²:													
Flatwise	40,000	38,000	38,000	50,000	70,000	40,000	45,000	...	...	...	...	30,000	
Edgewise	17,000	21,000	15,000	17,500	25,000	9,000	14,000	...	...	...	...	20,000	
Rockwell hardness (M scale)	M-111	M-103	M-105	M-100	M-120	M-95	M-100	M-105	M-118	M-118	M-120	M-100	M-100
Dielectric strength perpendicular to laminations, V/mil:	...	...	0.40	0.30	0.30	...	0.30						
Short-time test													
1/32 in.	...		...	750	...	...	450	850					
1/16 in.	225		500	700	350	250	400	600	750	...	...	400	
1/8 in.	160	No values recommended	425	600	260	185	350	450	...	550	...	350	
Step-by-step test													
1/32 in.	...		...	550	...	...	400	650					
1/16 in.	135		360	500	220	220	350	450	550				
1/8 in.	95		300	450	160	160	250	300	...	400			
Insulation resistance (condition C-96/35/90), MΩ	...	...	5,000	...	100	...	2,500	50,000					
Specific gravity	1.72	1.70	1.50	1.65	1.90	1.65	1.68	1.15	1.45	1.40	1.38		
Specific volume, in³/lb	16.1	16.3	18.5	16.8	14.6	16.8	16.5	24.1	19.1	19.8	20.1		
Thermal expansion, cm/(cm)(°C)	0.000015 (A–AA)			0.000018 (G-2–G-5)		0.000010 (G-6–G-7)		...		0.000020 (ES-1–ES-3)			
Thermal conductivity, cal/sec cm²(°C/cm)	...	...	...	...	0.00120	0.00070	0.00070	...	0.00070	0.00070	0.00070		
IEEE insulation class	B	B	B	B	B	H	H	A	A	A	A		
Specific heat, Btu/(lb)(°F)	0.30 (A–AA)		0.30 (G-2–G-3)		0.26	0.25 (G-6–G-7)		0.35–0.40 (N-1–ES-3)					

* LW means lengthwise, or tested in the direction of orientation.
† CW means crosswise, or tested 90° to the direction of orientation.

Table 4-69. Filament-wound Products

Specific gravity	2.0
Density, lb/in³	0.072
Thermal conductivity, Btu/(h)(ft²)(°F/in)	2.2
Thermal expansion, in/in $\times 10^{-6}$/°F	7.0
Specific heat, Btu/(lb)(°F)	0.227
Maximum-use temp., °F	400
Hoop strength:	
Unidirectional windings, lb/in²	230,000
Helical windings, lb/in²	135,000
Compressive strength, lb/in²	70,000
Flexural strength, lb/in²	100,000
Bearing strength, lb/in²	35,000
Shear strength:	
Interlaminar, lb/in²	6,000
Cross, lb/in²	18,000
Modulus of elasticity, tension, lb/in²	6.0×10^6
Modulus of rigidity, torsion, lb/in²	2×10^6
Dielectric strength, step-by-step, V/mil	400

to sheet products. Since the processes of manufacturing tubes and rods differ from that of laminating, the properties also differ and Table 4-68 should not be used for tubes and rods. Laminated products are also available in angles, channels, beams, and other special shapes.

335. Factor of Safety. Based on the standard properties of the various grades of sheets, tubes, and rods a minimum factor of safety of 4 for mechanical strength and 6 for dielectric strength is recommended. This statement is taken from *NEMA Standards Publ.*, Vol. LI 1-1965, Industrial Laminated Thermosetting Products.

336. Filament-wound Materials. Filament winding refers to the process of wrapping resin-impregnated continuous filaments of glass around a mandrel, forming a surface of revolution. When enough layers have been wound, the part is cured and the mandrel is removed. Filament-wound parts have extremely high hoop strengths and have been used in the electrical industry as fuse tubes, lightning-arrester tubes, bushings, and binding bands on motors and generators. Typical properties[5] for filament-wound parts are given in Table 4-69.

337. References on Plastics

1. "Modern Plastics Encyclopedia"; New York, McGraw-Hill Book Company, September, 1965, Vol. 43, No. 1A.

2. "Modern Plastics Encyclopedia"; New York, McGraw-Hill Book Company, September, 1965, Vol. 43, No. 1A.

3. *Plastics World,* October, 1965, Vol. 23, No. 10.

4. *NEMA Standards Publ.* LD 1–1965, 1964.

5. Rosato, D. V., and Grove, C. S. "Filament Winding"; New York, John Wiley & Sons, Inc., 1964.

PAPER, FIBER, WOOD, AND INSULATING FABRICS

By G. R. Sprengling

338. Cellulose. Cellulose is one of the oldest of electrical insulating materials, and it remains today the most widely used. It is employed in the form of papers, fabrics, and pressboards. For the chief commercial varieties, see Pars. **339, 340,** and **345.** Pure cellulose, commonly termed alpha cellulose, is essentially a polymer of glucose and is thus a carbohydrate of the formula $(C_6H_{10}O_5)_n$, in which *n,* the number of units in the polymer, varies from 1,100 to 8,000, depending on the source. This molecule contains three hydroxyl groups, which cause cellulose to have both a high dielectric constant (6.0 for kraft cellulose, at 60 cycles, decreasing with the degree of crystallinity) and a high affinity for water.[1]* The latter of these properties is the salient cause of difficulty in dealing with cellulose as insulation.

Common insulating papers are made chiefly from coniferous woods but also from cotton and linen rags, rope, and other materials. These are first treated by a chemical process, most commonly the sulfite or the kraft (sulfate) process, and then comminuted to a dispersed pulp by "beating." The pulp suspension is formed into a loose sheet

* Superscripts refer to bibliography, Par. **344.**

by filtering on a moving wire screen (the Fourdrinier) and compacted into paper by calendering with heated rolls. Accordingly, paper has different properties, chiefly mechanical, in the machine direction (MD) and that transverse to it (CMD). Thus the MD strength is 1.5 to 5 times higher but the elongation 2 to 3 times smaller than the CMD properties.

Table 4-70. Properties of Some Polymer Fibers of Use in Insulation

Material	Examples of trade names	Tenacity dry, g/denier	Density	Moisture absorption, %	Max. overload temp.,°C*	Flammability
Silk	....	4–5	1.25	11	130°D	Burns
Cotton	....	3–5	1.54	7	110°D	Burns
Viscose rayon	....	1.5–2.4	1.53	12–13	110°D	Burns
Acetate rayon	....	1.3–1.5	1.32	3–6	120°D	Slow burn
Polyamide	Nylon 6	6.8–8.6	1.14	4	100°D	Self-exting.
Polyvinyl chloride	Vinyon HH, PCU	0.7–3.8	1.34	0.1–0.5	75°S	Self-exting.
Polyvinylidene chloride–vinyl chloride	Saran	1.5–2.0	1.70	0.1	100°D	Self-exting.
Polyacrylonitrile	Orlon, Creslan, Acrilan	2.2–2.6	1.14	1.5	135°S	Slow burn
Acrylonitrile–vinyl chloride	Dynel, Vinyon N	2.5–3.3	1.30	0.3–0.4	130°S	Self-exting.
Polyethylene terephthalate	Dacron, Terylene	4.4–4.5	1.38	0.4	150°D	Slow burn
Polyethylene, LD	....	1–3	0.925	Nil	75°S	Slow burn
Polyethylene, HD	Velon LP	5–8 (4–6)	0.955	Nil	100°D	Slow burn
Polypropylene	....	5.5–7.0	0.905	Nil	100°D	Slow burn
Polytetrafluoroethylene	Teflon	1.6–1.7	2.3	Nil	180°S	Incombustible
Aromatic polyamide	Nomex	5.5	1.38	8	200°D	Self-exting.

* These temperatures give a rough guide to the conditions a material can withstand for approximately 100 h. The suffixes indicate the chief reason for the limit: D = degrades; S = softens. These are not service temperatures. See text.

References:

1. Cook, J. G. "Handbook of Textile Fibres"; Watford, Herts, England, Merrow Publishing Co., 1960.
2. Hill, R. "Fibres from Synthetic Polymers"; Amsterdam, Elsevier Publishing Co., 1953.

The paper- and pulp-making process does not produce pure cellulose. Even a rather pure grade of electrical paper will contain about 4% of pentosans, 3% of lignin, and 0.35 to 1.0% of ash impurities. The processing, especially the degree of beating, also determines the proportion of the cellulose present in the form of crystallites or in the much more hygroscopic amorphous form. With modern deionized water washing methods the various processes can produce paper of equivalent electrical quality.

339. Unimpregnated cellulose paper is used for a variety of purposes in insulation,

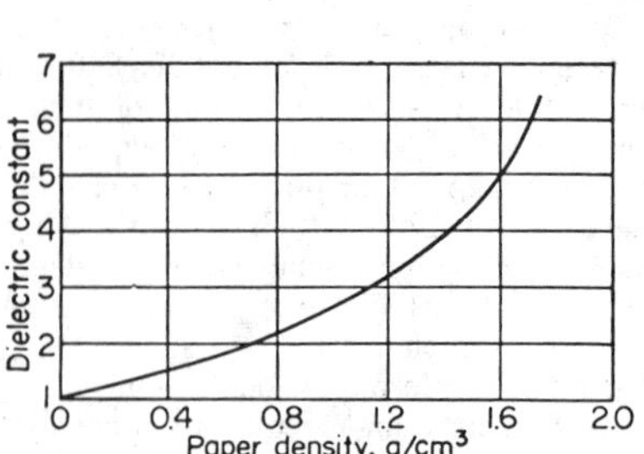

Fig. 4-66. Effect of density on the dielectric constant of cellulose paper. [*Sakamoto and Yoshida, ETJ (Japan)*, 1956.]

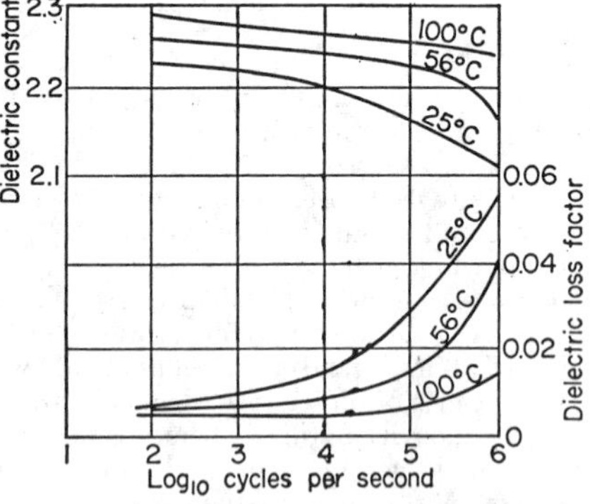

Fig. 4-67. Effect of frequency on dielectric constant and loss in cable paper. (*Race, Hemphill, Endicott, Gen. Elec. Rev.*, 1940, *Vol.* 43, *p.* 492.)

such as telephone-cable insulation, household electric cable, small transformers and capacitors, and the like. So-called fish paper used in motor-slot cell insulation is a type of vulcanized fiber (which see). The electrical properties of such papers depend chiefly on their pore volume, which may vary from 10 to 60%, and on their moisture content, which is normally 7 to 12% in contact with air of average humidity. The apparent dielectric constant for pure, dry paper varies with porosity, as shown[2] in Fig. 4-66. The dielectric constant and loss factor vary with frequency, as shown[3] in Fig. 4-67.

The dielectric loss factor (tan δ) of dry paper may be as low as 0.0009 to 0.004 at 40°C. Both the loss factor and the dielectric constant change sharply in papers exposed to humid air, as shown[4] in Fig. 4-68. At the same time the volume resistivity of the paper changes in a straight line from 5×10^{15} Ω·cm in equilibrium with dry air to 10^{10} Ω·cm at 84% relative humidity. The loss factor of moist paper also rises with increasing test voltage gradient.

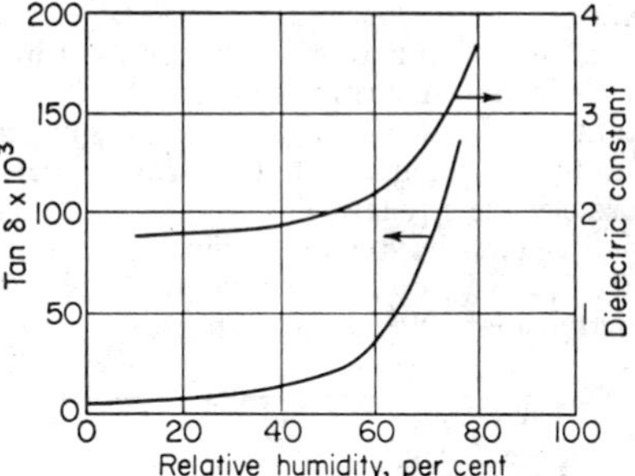

FIG. 4–68. Effect of ambient humidity on dielectric constant and loss in cable paper. (*Dieterle, Bull. Swiss Electrotech. Inst.*, 1955, *Vol.* 22, *p.* 3.)

These figures demonstrate the importance of using dry paper for insulation. The drying of cellulose is a difficult operation, since at least a portion of the contained water is so tightly bound that it can be removed only at elevated temperatures where the process overlaps with thermal degradation of the cellulose itself.[5] Such degradation is accelerated by the water present, especially if trapped in equipment, and also by the presence of air, some metallic oxides, alkali, and light. Commercial drying of paper at temperatures above 115 to 120°C is usually impractical. Insofar as the electrical properties achievable by drying are of major importance, as, for example, in telephone cable, paper dryness must be carefully maintained by sealing the insulation in an impervious sheath (of lead), sometimes even pressurized with inert, dry gas. Otherwise moisture is rapidly regained in contact with air. Because of the above factors the IEEE and NEMA accepted use of unimpregnated paper insulation extends only to apparatus operating at temperatures not above 90°C.

The electric strength of untreated kraft insulating paper may be 160 volts/mil of thickness. It is characteristic of paper that this electric strength is a function not only of paper density but also of the distribution of defects in the electrical sense. A study on single-thickness paper has shown these to be invariably either "holes" (low-density areas) or opaque inclusions, which are usually electrically conducting. When two or more layers of paper are tested together, the defects are usually unassignable.[6]

340. Pressboards and fiber are other forms in which cellulose is used for insulation. Pressboard is essentially a thicker form of paper, made in thicknesses up to 0.5 in, which is used chiefly impregnated with oil in transformers. Its properties are treated under Pars. **339** and **341.** Vulcanized fiber is made by chemical treatment of cellulose (usually with zinc chloride) followed by lamination to the desired thickness. Standard grades of fiber (NEMA Standard VU-1-1954) are: (1) Electrical insulation grade, gray in color and made to ⅛ in thickness. Thinner samples of this grade are known as "fish paper" and are used as slot liners in motors. (2) Commercial grade, thicker than the above and red, black, or gray in color. (3) Bone grade, harder than the above, gray, furnished in tubes and rods. (4) Trunk grade, used for suitcases. In general, fiber is of low electric strength (50 to 300 V/mil) and high water absorption. It is used for its toughness and abrasion resistance in low-voltage applications. Fiber is also used in arc-interrupting devices, since it leaves no carbon residue when burned and evolves large quantities of gas when heated, which helps extinguish the arc.

341. Impregnated papers are usually impregnated with mineral oils or chlorinated hydrocarbons (Askarels) or with various synthetic resins. So impregnated paper becomes capable of high-voltage applications. Thermal stability is also improved, permitting application in apparatus operating at 105°C. As a general rule of thumb the life of such insulation can be expected to change by a factor of 2 for each 8°C change in the operating temperature. One system of paper additives has been found which will enable a rise in the operating temperature limit for oil-impregnated paper by up to 30°C or which will yield a corresponding eightfold increase in life at equal temperature.[7]

In general, the same factors affecting the electrical properties and thermal life of cellulose as noted above also dictate the properties of impregnated papers, notably the ingress of moisture. These effects are retarded but not prevented by the impregnants. Where optimum properties are needed, as in capacitors and transformers, such insulation is therefore often used in sealed enclosures. In dry, impregnated paper the effective dielectric constant of the composite can be approximated by calculation from the known constants of the impregnant and cellulose and the paper density. Impregnated paper is used at higher voltages, and the major desired property is therefore electric strength.

Vacuum-dried and oil-impregnated papers have a short-time electric strength of 300 to 900% of the untreated value. The electric strength-thickness relationship of such insulation is usually shown as $V = AT^n$, where A is the strength per unit thickness and T is the thickness. The exponent n is commonly determined at about 0.8 but for design purposes is better set at 0.5. This is because the values of both A and n vary with degree of impregnation, temperature, voltage gradient, electrode configuration, duration of voltage stress, and other factors. The effect of voltage stress duration is shown[8] in Fig. 4-69. This effect is roughly cumulative for repeated stressing. Because of the many variables involved in its performance, impregnated paper is used at voltages corresponding to only 25 to 35% of its "inherent" electric strength.

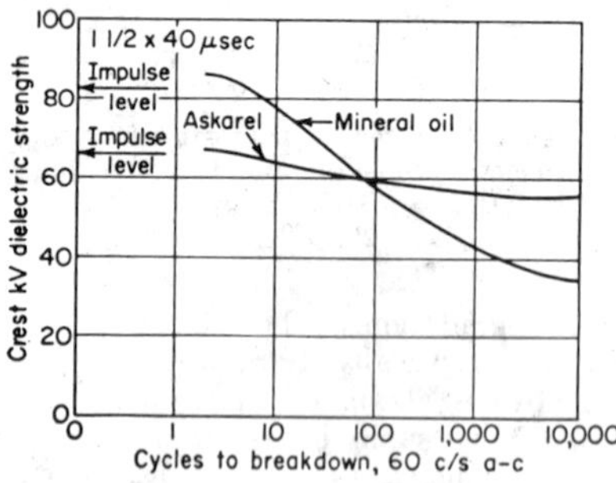

FIG. 4-69. Time dependence of electric strength in impregnated paper. (*Dakin and Works, Transactions AIEE, 1952, Vol. F1-I, p. 321.*)

Paper is also impregnated with natural oils, synthetic ester fluids, silicones, and a variety of other materials. Though promising, these have not found wide use so far.

342. Altered cellulose and synthetic fiber papers have also found some use in insulation. The cellulose molecule has been chemically reacted to form cyanoethylated and acetylated papers,[9] which generally have greater thermal and moisture resistance than cellulose. Papers or "felts" are also made from polyester, acrylic, nylon, glass, and other fibers. These are generally of low electric strength as such and are therefore used as a base for resin-impregnated and laminated structures rather than by themselves.

Nomex paper (E. I. du Pont de Nemours & Co.), made from a synthetic, aromatic polyamide fiber, is used for insulation as such. This paper has capability for use at operating temperatures of 180°C or over. This paper has a dielectric constant of 2.6, an electric-loss factor of 0.011 at 1,000 c/s, and an electric strength of 600 V/mil, all at 5-mil thickness. Mechanical properties are equal or superior to those of the best rope papers.

343. Wood. Wood, once extensively used as electrical insulation, today has been largely replaced by synthetic materials for general use. The dielectric constant of wood varies in the range 2.5 to 7.7; the dry resistivity is in the order of 10^{10} to 10^{13} Ω·cm. Dry wood will withstand a potential gradient of 10 kV/in in service. These properties vary over a wide range, depending on the type of wood, season of cutting, grain direction, and especially the water content. The chief remaining use of wood in electrical apparatus is in operating rods and tie rods, where it is used in resin-impregnated form. Here its mechanical stiffness and impact resistance are of value.

By far the chief use of wood for insulation is in poles and other structural insulating members for electrical transmission lines.[10]

Wood contributes to the impulse strength of such a structure at the rate of 10 to

130 kV/ft when dry. In green wood or wood wet by even a short rain, this value falls to 50 to 70% of the above for longer lengths and may approach zero in very short, wet lengths. For design purposes, these values apply with sufficient accuracy to common hard or soft woods.[11] Other electrical virtues of wood in such use lie in its arc-quenching properties, which can prevent formation of a continued power-follow arc after a flashover caused by lightning, and in the fact that the impulse strength of wood is not seriously impaired even by repeated flashover. The arc-quenching properties of wood depend on the fact that the voltage across a flashover arc in wood or on its surface does not drop to zero, as it would in air or over porcelain, but falls only to a minimum arc gradient averaging 1.1 kV rms/in. Thus, if the gradient imposed by line voltage on the wood is less than 0.5 kV rms/in, there is very low probability that a flashover caused by a surge will be sustained as an arc.[12]

344. References on Paper, Fiber, and Wood

1. Dakin, T. W. The Absolute Dielectric Constant of Cellulose Fibers; *Ann. Rept., NRC Conf. on Electrical Insulation,* 1950.

2. Sakamoto, T., and Yoshida, Y. Research on Dielectric Properties of Impregnated Paper as Composite Dielectrics; *E.T.J.* (*Japan*), March, 1956, p. 3.

3. Race, H. H., Hemphill, R. J., and Endicott, H. S. Important Properties of Electrical Insulating Papers; *Gen. Elec. Rev.*, 1940, Vol. 43, p. 492.

4. Dieterle, W. Acetylated Paper as an Electrical Insulating Material; *Bull. Swiss Electrotech. Inst.*, 1955, Vol. 22, p. 3.

5. Flowers, L. C. Moisture Evolution Rates Calculated for Cellulose Paper Undergoing Thermal Decomposition; *Insulation,* September, 1964, p. 23.

6. Bullwinkel, E. P. *NAS-NRC Publ.* 1238, *Conf. on Electrical Insulation,* 1964, p. 101.

7. Ford, J. G., Lockie, A. M., and Leonard, M. G. INSULDUR—A New and Improved Heat-stabilized Insulation; *AIEE Conf. Paper* 60–936, June, 1960.

8. Dakin, T. W., and Works, C. N. Impulse Dielectric Strength Characteristics of Liquid-impregnated Pressboard; *Trans. AIEE Paper* 52–228, 1952.

9. Dieterle, W. Insulation Features of Acetylated Paper; *Insulation,* June, 1962, Vol. 8, No. 7, p. 19.

10. Dean, P. S. Insulation Tests for the Design and Uprating of Wood-pole Transmission Lines, 1966 *IEEE Trans.*, Vol. PAS-85, p. 1258.

11. Lusignan, J. T., and Miller, C. F. *Trans. AIEE,* 1940, Vol. 59, pp. 534–540.

12. Darveniza, M., Limbourn, G. J., and Prentice, S. A. *AIEE* 31 TP 66–94.

345. Insulating Fabrics. Base materials for insulating fabrics include natural fibers such as cellulose, cotton, and silk; synthetic organic fibers of, for example, cellulose derivatives, polyamides (nylon), polyethylene terephthalates (e.g., Dacron), polyacrylonitrile, and many others; and inorganic fibers, chiefly glass and asbestos. Nonwoven fabrics of natural fibers overlap in properties with the papers, fiberboards, and asbestos composites treated in Pars. **338** to **344** and **356** to **359.** Nonwoven synthetic organic fibers are usually bonded into a fabric by use of a bonding resin or by fusion. They find electrical use chiefly as a base for resin-impregnated insulation.

Unimpregnated woven fabrics find some limited use in electrical insulation. The electric strength of such fabrics generally does not exceed the breakdown strength of an equivalent air gap and may indeed be less. Their chief use, therefore, is to confer mechanical strength, abrasion resistance, and mechanical spacing of conductors in low-voltage applications. The effects of ambient moisture and temperature on the properties of cellulosic fabrics are similar to those observed in cellulose papers (see Par. **339**). Often the properties of such fabrics are upgraded by impregnation of a varnish after application. Better results are generally obtained if the fabric is impregnated prior to application.

Nonwoven synthetic organic mats may also be composed of two types of fibrous material: fibrids, which are interspersed with and have a lower fusion point than the fibrils, which compose the bulk of the material. On hot calendering the fibrids fuse to bond the mat. Depending on the proportion of fibrids present and the calendering process, rather impervious nonwoven fabrics of fairly high electric strength may be obtained. One such product is Nomex (refer to Par. **342**).

Table 4-70 gives some of the properties of the more common organic fibers used in insulation. It should be understood that the long-time service-temperature rating applicable to a fabric material cannot be assessed by its chemical type alone. It is also a function of the particular filament or fiber, the weave, and the treatment given in manufacture. Service temperature also varies widely with the nature of the application. It must therefore be separately determined for a specific case. The overload temperatures given in the table are intended only as a quick guide to danger areas in possible misapplication. Service temperatures of a given fabric or composites made therefrom may be lower or might even be higher.

346. Impregnated fabrics and those coated with elastomers are much more widely used for insulation than unimpregnated ones. The base fabric is generally woven. At least some degree of flexibility is required of all such products, ranging from smooth, rather stiff and elastic materials (e.g., for slot cells) to soft and rubbery tapes. Stretchability of such materials to allow their conforming to irregular surfaces is obtained by using base fabrics with a bias weave.

Electric strength is the chief electrical property demanded of impregnated fabrics, since other electrical properties, such as loss tangent, are relatively less important in the applications of which these are chiefly used. The electric strength of such fabrics is conferred upon them by the impregnation. It is a function of the completeness of impregnation primarily and secondarily of the type of resin used. The type of resin is of primary importance in determining whether electric strength is maintained after mechanical or thermal stress.

The breakdown voltage of such materials varies with thickness according to the usual logarithmic expression $V_B = AT^n$, where A is the electric strength of unit thickness and T is the thickness, expressed in the same units. The exponent n in this expression varies from about 0.66 to nearly 1.0 for impregnated fabrics. The values for both A and n for a given material vary with the test (or use) conditions approximately as follows: With increasing duration of voltage stress A decreases sharply, often by a factor of 2 or more (or ultimately to the corona-starting voltage gradient), whereas n may go up or down; with increasing electrode area A decreases, and n usually increases. The value of A also usually decreases with rising test temperature to a degree depending on the material. Methods for measuring electric strength and a discussion of the significance of the results are given in ASTM Standards, Pt. 29, Method D149-64.

347. Thermal endurance of varnish-impregnated fabrics may be somewhat higher than normally warranted by the base fabric used, since an impregnating resin which is itself of appropriate thermal stability will protect the fabric to some degree. In general, cellulosic fabric-based materials are limited to operating temperatures not to exceed 105°C and synthetic fiber-base materials to a maximum of 155°C (with the exception of aromatic polyimide and polytetrafluoroethylene fabrics). Above the latter temperatures glass and asbestos fabrics find use.

The maximum-use temperature is commonly determined by following the change in electric strength of an impregnated fabric with exposure time to a series of temperatures indefinitely higher than the intended service temperature. Such testing is described by ASTM Method D1830-63, in which curved electrodes imposing a 2% elongation on the test-sample outer surface are used for electric-strength test. This takes account of the fact that use conditions of impregnated-fabric insulation may be such as to stress the material mechanically. The logarithm of the life of the tested material to, for example, 50% decrease in electric strength may then be plotted against the reciprocal of the absolute (Kelvin) test temperature and the points extrapolated to yield the service temperature for a desired life.

Other tests on impregnated and coated fabrics for electrical insulation are described in ASTM Methods D295-61, D1000-64, D2148-64T, D902-61, D350-61T, D1458-62T, D69-63. Specifications are set forth in ASTM D373-64T, D372-61T, D1459-59, D1931-62T, and various NEMA, AIEE, and military specifications, such as MIL-1-17205C.

It should be remembered that the service voltage stresses that any material of this class may be expected to withstand in extended service are only a small fraction of the breakdown stresses determined by electric-strength tests.

348. Uses of impregnated and coated fabrics fall into several categories:

Tapes for wrapping of conductors, cables, leads, bus bars, connections, and splices.

These are made both with straight-weave substrate and with bias cut, the latter having more stretch for application to irregular surfaces. They may be varnish-impregnated, rubber- or silicone-rubber-coated, or supplied with a fusible coating to seal on heating or with a contact adhesive.

Cloth for coil wrapping or slot insulation. Often supplied as combinations of varnished fabric with fish paper, synthetic polymer films, or mica composites.

Sleeving for lead insulation. Generally this is a continuous braid tubular fabric impregnated with a very flexible resin or rubber.

Table 4-71. Varnish-impregnated and -coated Fabrics

Material	Thicknesses, in	Tensile strength MD, lb/in of width	Electric strength, V/mil (short time)	Max.-use temp., °C	Application
1. Black varnished cambric tape	0.008-0.012	40-45	1,400-1,450	105	Flexible cable tape
2. Black bias tape and cloth.	0.007-0.015	40	1,100-1,150	...	Wrapper insulation on armatures, cable joints, irregular shapes
3. Black varnished canvas...	0.016-0.031	50-100	380-500	105	Coil supports, padding for end windings, bonding cushion
4. Tan varnished silk tape...	0.005	16	1,600	105	Layer and wrapper insulation
5. Rag paper-cambric varnished composite	0.005-0.015	......	650-750	105	Slot cells and other applications requiring high toughness
6. Epoxy varnished glass-polyester film	0.013-0.017	400	1,400	130	Slot and phase insulation
7. Black varnished asbestos fabric	0.015-0.050	......	....	...	Padding
8. Epoxy varnished polyester mat tape	0.0025-0.007	14-68	1,600-2,370	...	Dry-type transformer layer insulation
9. Class F varnished glass cloth	0.003-0.015	60-200	1,900	155	Layer and wrapper insulation
10. Self-sealing varnished glass-polyester tape	0.012	40	800	155	Encapsulation of wrapped coils
11. Silicone varnished glass cloth	0.0035-0.013	70-250	450-1,500	180	Flexible layer, wrapper and tape
12. Silicone-rubber-coated glass cloth	0.007-0.010	125	900-1,000	180	Layer insulation and cable tape
13. Silicone-rubber-coated bias weave glass cloth	0.007	100	585	180	Taping irregular surfaces and shapes

Padding, used for spacing of coil ends and the like. This is often a resin-impregnated nonwoven fabric such as a felt or asbestos. Electrical requirements are low.

An illustrative selection of materials typical of those available in the class of coated and impregnated fabrics is given in Table 4-71, together with appropriate properties.

MICA AND MICA PRODUCTS

By T. W. Dakin

349. Mica Sources. Mica insulation is derived from a class of minerals of finely laminated structure and very easy cleavage, the flakes of which are very flexible and tough and extremely resistant to heat. Reference 1 discusses the structure, properties, and sources in detail.[1*]

The two classes most commonly used for electrical purposes are (1) the ferromagnesia mica phlogopite, also known as "amber" or "silver mica," and (2) the potassium mica muscovite, known as "India," "white," "ruby," or potash mica. The phlogopite mica is produced principally in Madagascar and Canada. The muscovite is produced principally in India, with quantities also obtained in the United States, Africa, and South America. Mica for electrical purposes must be carefully selected owing to mineral and vegetable inclusions occurring in the slabs. ASTM Designation D351 describes the standard methods for grading and classification of natural muscovite mica in blocks and films. Blocks range from 0.007 in thickness; films range from 0.0008 to 0.0040 in. NEMA specifies the thickness of 10 splittings at 0.006 to 0.0011 in. NEMA phlogopite splittings range from 0.007 to 0.012 in for 10 splittings.

* Superscripts refer to bibliography, Par. **355**.

350. Composition and Physical Properties. The micas are complex and variable in composition. Muscovite mica is translucent; is white, ruby, green, or brown; and is harder and less flexible than the phlogopite, which is opaque and ranges from pale yellow (amber) and silver to brown and green.

The idealized formula for muscovite mica is $K_2Al_4Al_2Si_6O_{20}(OH)_4$ and for phlogopite mica $K_2Mg_6Al_2Si_6O_{20}(OH)_4$. The structure consists of firmly bonded double sheets of aluminum silicate (muscovite) or magnesium aluminum silicate (phlogopite). The individual layers of the double sheets are internally held together with hydroxyl groups and other atoms of aluminum and magnesium. The cleavage plane occurs between these double sheets, which are more loosely bonded by potassium atoms, which lie in the cleavage plane. If clear, thin sheets of muscovite or phlogopite mica are heated to a temperature of 400 to 600°C, no perceptible alteration occurs, and their clearness and elasticity are retained, but muscovite starts to lose H_2O at about 500°C, phlogopite at 1000°C. At higher temperatures, between 900 and 1000°C, muscovite becomes silver-white, with a pronounced metallic appearance, loses considerably in clearness, and becomes rather brittle, so that it can be pulverized into a thin white dust. Phlogopite withstands this temperature much better; it loses but little in transparency and does not become so brittle. The melting points are in the range of 1200 to 1300°C. The safe maximum temperature is about 500°C for muscovite and 1000°C for phlogopite. (Also see results of heating tests reported[3] in Ref. 3.) The specific heat varies from 0.206 to 0.209. Mica will withstand great mechanical pressure perpendicular to the plane of lamination, but the laminations have very weak cleavage and are easily split into very thin flakes or leaves. It resists to a high degree the attack of gases such as combustion products but is attacked by warm hydrofluoric acid, molten potassium hydrate, warm alkaline carbonates, and water containing carbon dioxide. Magnesia mica is attacked by concentrated sulfuric acid, but the potassium micas are not attacked by either hydrochloric acid or sulfuric acid. Contact with oil is injurious to mica, as it penetrates the laminae and seriously impairs their cohesion.

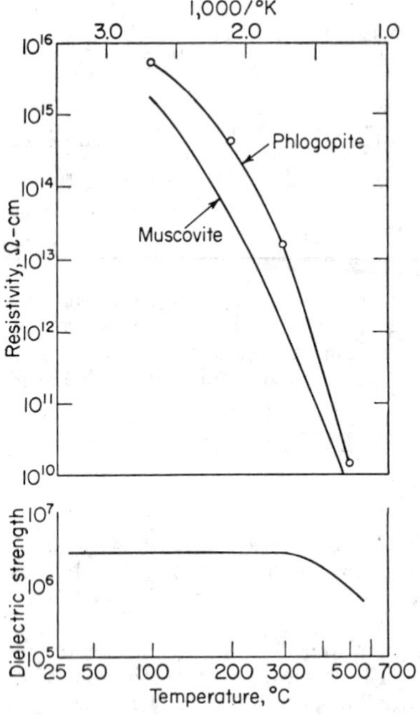

FIG. 4-70. Resistivity (*Dakin et al.*) and apparent dielectric strength (*Hackett and Thomas*) of mica splittings at elevated temperature.

351. Dielectric Properties of Mica. The resistivity of mica at 25°C ranges from about 10^{12} to 10^{16} Ω·cm depending on inclusions, etc., muscovite usually having a higher resistivity. The dissipation factors (tan δ) of muscovite splittings range from 0.0001 to 0.0008 and phlogopite from 0.003 to 0.09, over the frequency range from 60 cycles to 1 Mc. Changes in permittivity and tan δ with temperature are illustrated in Fig. 4-71 from Ref. 2. The high-temperature resistivity of muscovite and phlogopite splittings is illustrated in Fig. 2 from data of Dakin and coworkers.

Reference 3 gives data[3] for frequencies of 100 to 1,000 kc, at a test temperature of 25°C. All samples were muscovite, except two phlogopite samples from Madagascar.

The presence of stains or inclusions so seriously affects the tan δ as to render such stained micas unsuited for radio purposes. The tan δ of phlogopite mica was found to be so high as to render it also unsuitable for radio purposes. See ASTM D748 for measurement methods and acceptable values. Table 4-72 lists the properties of mica of different origins.

352. Dielectric Strength of Mica. The dielectric strength of India ruby mica was investigated by Moon and Norcross with d-c potentials, and the breakdown was found

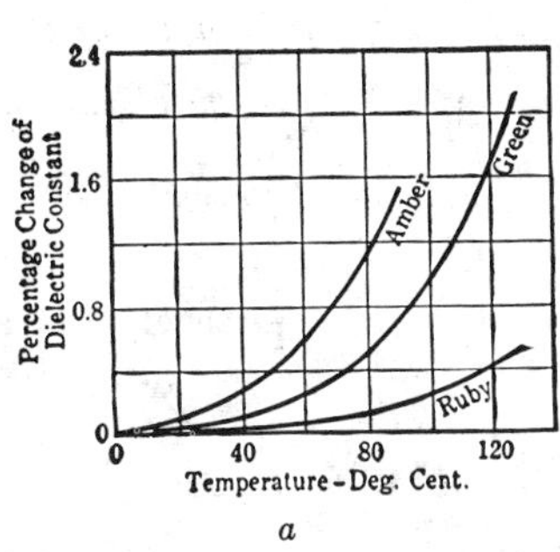

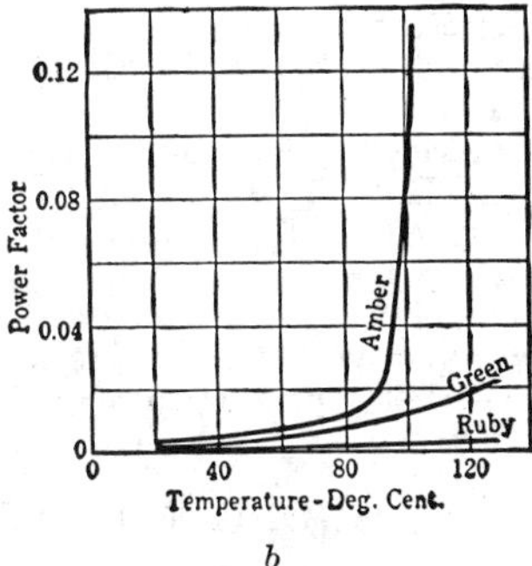

FIG. 4-71. Characteristics of mica. (a) Change in dielectric constant with temperature; (b) change in power factor with temperature.

to have a linear relationship to thickness, independent of temperature up to at least 100°C. The breakdown gradients ranged from 3,520 kV/cm in air to 6,960 kV/cm in transformer oil and 10,600 kV/cm in a mixture of xylene and aniline, at 20°C. Higher temperature-strength values are shown in Fig. 4-70.

In these tests the edge effect of the test electrodes was not eliminated except in the bath of xylene and aniline. Dielectric strength tests[3] made in transformer oil, at 60 cycles, gave results on 19 lots of domestic clear ruby samples of 5.7 kV/cm for 1-mil thickness, 9.9 kV for 5-mil thickness, and 13.2 kV for 9-mil thickness. Reference 3 also gives similar results for mica from other sources. Mica is particularly noted for its resistance to corona discharges, thus maintaining its a-c dielectric strength much better than any other thin dielectric material, when exposed to discharges either on the surface or within internal gas spaces in built-up insulation.

Table 4-72

Origin	Kind	No. of samples	Permittivity		tan δ	
			Average	Spread	Average	Spread
India	Clear	4	7.32	7.90–7.07	0.02	0.02–0.01
United States except North Carolina	Clear	4	7.02	7.36–6.62	0.02	0.03–0.01
North Carolina	Clear	8	7.22	8.69–6.57	0.02	0.04–0.01
Haywood County, N.C.	Clear	4	7.31	8.69–6.57	0.02	0.04–0.01
United States and India	Stained	17	....	9.64–5.83	....	8.36–0.06
Madagascar	....	2	....	6.07–5.41	....	7.12–0.38

353. Synthetic Mica. A modified mica is now made synthetically. Its structure is similar to phlogopite mica with the hydroxyl groups (OH) replaced by fluorine. Its permittivity is 6.5 and its tan δ and resistivity similar to those of muscovite. Its principal advantage in properties is its thermal stability, particularly in the absence of air, where it will withstand heating to 1100 to 1200°C. In the presence of atmospheric moisture at elevated temperatures it will very slowly change, losing fluorine. It is not available in sizable splittings, but only as a compacted aggregate.

354. Mica Paper. Several methods have been developed for disintegrating rough mica into very thin, small-area splittings and re-forming it into a continuous paperlike sheet. Such sheets are of themselves weak mechanically and must usually be supported by resin impregnation or backing with plastic films or fibrous material. The properties of these composite sheets, which are also available as tapes, vary widely depending on the impregnating and supporting material. The sheets are now used widely in many applications where mica splittings were formerly employed. Some forms, when properly applied, have very high dielectric strength, in excess of 1,000 V/mil, which is maintained under corona discharges much better than that of organic resins. Mica paper is also available with inorganic (commonly phosphate) impregnation which can be cured up hard by heating.

355. References on Mica

1. Skow, M. L. Mica, A Material Survey; *U.S. Dept. Interior Information Circ.* 8125, 1962.

2. Dannatt, C., and Goodall, S. E. The Permittivity and Power Factor of Micas; *Jour. IEEE*, 1931, Vol. 69, pp. 490, 1034.

3. Lewis, A. B., Hall, E. L., and Caldwell, F. R. *NBS Jour. Res.*, 1931, Vol. 7, p. 403.

4. Moon, P. H., and Norcross, A. S. *Trans. AIEE*, 1930, Vol. 49, p. 755.

5. Hackett, W., and Thomas, A. M. The Electric Strength of Mica and Its Variation with Temperature; *Proc. IEE*, 1940, Vol. 88, p. 295.

INORGANIC INSULATING MATERIALS

By Daniel Berg

356. Asbestos Sources. Asbestos is a fibrous crystalline material, and chrysotile is the principal source for commercial use. It is a hydrated silicate of magnesium, whose composition formula is $3MgO$, $2SiO_2$, $2H_2O$. However, the composition does vary from approximately 37 to 44% silicon oxide, 39 to 44% magnesium oxide, 12 to 15% H_2O, and up to 6% iron oxide. It is a fibrous mineral having a specific gravity of 2.2 to 2.6, a hardness of 2.5 to 4.0 on a Mohs scale. Individual fibers are very strong, with tensile strengths up to 400,000 lb/in^2. The mineral loses its water of crystallization at approximately 400°C and becomes weaker. At about 700°C all the water has been removed, and at 775°C the crystal structure changes. The melting temperature of asbestos is reported to be 1525°C. The principal sources are Canada, United States, Turkey, Rhodesia, and China. Most asbestos has magnetite and other iron oxides as impurities. Processes have been developed for removing the impurities in order to obtain good electrical insulation properties.

357. Fabrication of Asbestos. Asbestos fibers are extremely fine, ranging from 850,000 to 1,400,000/lin in. Under a high-powered microscope they appear to be very smooth, as if polished. The fiber lengths generally are found up to 1 in in the commercial varieties. Often other fibers such as cotton are mixed in, in order to facilitate spinning the smooth asbestos fibers. For nonelectrical products, metal wire is often used.

358. Asbestos Insulation. Asbestos is not a good insulator unless it is totally dry. It absorbs moisture too readily for many electrical applications. It is dried and impregnated with varnishes or inorganic binders like phosphate solutions. Refining processes have been developed which remove water-soluble electrolytes and iron oxide. The fibers are then dispersed and formed into thin sheets, as in papermaking. These products are known as Quinterra and Novabestos. The materials are then generally used in a treated manner, either by impregnation or in laminates.

359. Untreated Asbestos. These materials are often used for low-voltage insulation or for barriers to provide separation where heat-resisting or arc-resisting properties are required. Asbestos cloth is used for electrical safety blankets. Magnet coils, mill and railway motors, and transformer coils are often insulated with asbestos-woven and paper tape. An asbestos cord is available which uses long-fiber asbestos. Occasionally, woven asbestos sleeving is used for fireproof insulation and coil leads.

360. Glass. *Constitution.* Glass is an amorphous substance, often consisting of silicates though in some cases borates, phosphates, etc., are used. Most glass is made by fusing together some form of silica such as sand, an alkali oxide such as is found in potash or soda, and some other base such as lime or lead oxide. It is a hard and brittle material when cold, generally breaking with a conchoidal fracture, but on heating sufficiently it softens, becoming plastic and, finally, very fluid. It may be blown, pressed, cast, and cut to a great variety of shapes. Colors are imparted by the addition of metallic oxides which absorb light of the appropriate wavelength. The total number of formulas for glassmaking is exceedingly large, but in general the chief constituent is SiO_2 (silica), which generally may be 50 to 90% of the total content.

Glass generally possesses high resistivity and dielectric strength at ordinary temperatures. However, there are glasses which can be made semiconducting by addition of material such as vanadium oxide (V_2O_5). The resistivity of glasses typically ranges from 10^{11} to 10^{18} Ω·cm at room temperature, while for a high-quality pure fused silica the resistivity may be as high as 10^{19} Ω·cm. The temperature coefficient is negative and very large. For most glasses the conductivity is ionic, and either the alkaline metal or other cations do most of the conducting. At high temperatures glasses become more conducting and eventually when molten have very high conductivities, as do most molten ionic salts. Figure 4-72 shows some d-c volume resistivities as a function of temperature for some commercial glasses. The mechanical properties of glass in general are rather poor, with tensile strengths very low, in the order of 2,000 to 12,000 lb/in². However, the compressive strengths are much higher, varying from 20,000 to 50,000 lb/in². The density of ordinary glass varies from 2.2 to 2.8; however, flint glass, which contains lead silicate, can have a density as high as 5.9. Coefficients of thermal expansion vary considerably with composition and range from 0.8×10^{-6} to 13×10^{-6}/°C. Specific heat is approximately 0.2. Thermal conductivity generally is approximately 0.0025 cgs unit and increases with temperature. The dielectric constant generally varies from 3.8 for fused silica to as high as 10 for lead silicate glasses having high lead content.

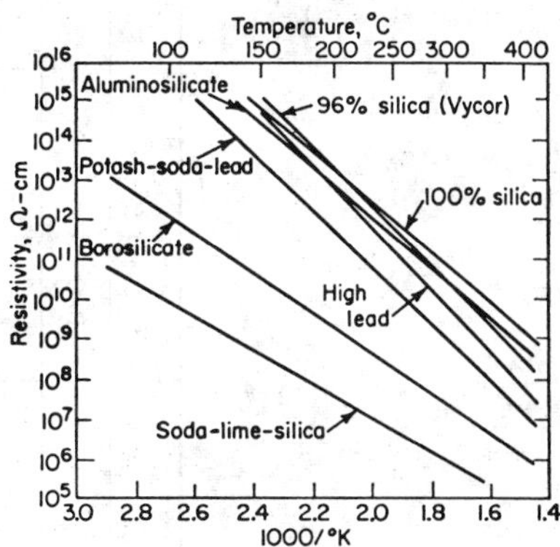

FIG. 4-72. Direct-current volume resistivity of commercial glasses. (*Kingery, "Introduction to Ceramics"; New York, John Wiley & Sons, Inc.*, 1960.)

361. Dielectric Strength of Glass. The dielectric strength of glass, as with all solids, varies with the test conditions. For typical use conditions the dielectric strength is governed by the thermal breakdown process, in which the heat dissipated in the conductivity process increases the temperature of the glass, leading to an increased conductivity, which in turns leads to further temperature increase of the glass. This typical process eventually leads to a runaway condition in which the conductivity of the glass increases to such a point that the glass is no longer an insulating material. Dielectric strength for glasses ranges anywhere from 50 to 3,000 kV/cm. The impulse strength of glass is generally close to the d-c or 60-cycle peak value. One advantage of glass, especially in the form of glass flake, is that its corona-withstand voltage or equivalently its life under corona-discharge conditions is very much higher than that of organic materials. However, the corona resistance of mica flakes is even higher.

362. Typical Commercial Glasses. Data supplied by the Corning Glass Works are included in Table 4-73 and show the approximate analysis, designation, and product number of typical glasses (see "Engineering Materials Handbook"; New York, McGraw-Hill Book Company).

363. Electrical and Mechanical Properties of Commercial Glasses. Electrical and mechanical properties of commercial glasses are included in Table 4-74 for the glasses listed in Table 4-73. The changes of dielectric properties with temperature are shown in Figs. 4-73 and 4-74 based on data supplied by the Corning Glass Works.

364. Glass Fiber and Ribbon. Glass in thin sheets or ribbon drawn in thicknesses down to 0.0025 cm is now used in capacitors. Also, processes have been developed for producing glass fibers of 0.0005 cm diameter, which are then spun into yarns and woven. The fibers are made by drawing molten glass through small orifices. These fibers are made into glass cloth and tape and used either untreated or treated with organic resins as a bonding agent. There is a wide variety of glass fibers available, and new ones are continually being developed. Often the glass cloth is combined with mica into a lamination for coil and slot insulation.

365. Fused Silica and Quartz. Quartz is a form of silica occurring in hexagonal crystals which are commonly colorless and transparent. The density is about 2.65. Fused silica is made by melting sand in an electric furnace and pressing it in carbon molds. It is an insulating material of excellent properties and is very heat-resistant. Crystalline quartz is an anisotropic material and on heating has a coefficient of thermal expansion on its main axis of 7.8×10^{-6}/°C and along the transverse axis of 14×10^{-6}.

Table 4-73. Approximate Compositions of Commercial Glasses
(Corning Glass Works)

Designation	Corning Glass No.	Percent								
		SiO_2	Na_2O	K_2O	CaO	MgO	BaO	PbO	B_2O_3	Al_2O_3
Silica glass, fused silica	....	99.5+								
90% silica glass*	7900, 7910, 7911	96.3	<0.2	<0.2	....	....	....	..	2.9	0.4
Soda-lime, plate glass	....	71–73	12–14	.....	0–12	1–4	....	..	....	0.5–1.5
Soda-lime, electric-lamp bulbs	0080	73.6	16	0.6	5.2	3.6	....	..	....	1
Lead silicate, electrical	0010	63	7.6	6	0.3	0.2	....	21	0.2	0.6
Lead silicate, high lead	8870	35	.....	7.2	....	....	....	58		
Borosilicate, low expansion†	7740	80.5	3.8	0.4	....	....	....	..	12.9	2.2
Borosilicate, low electrical loss†	7070	70.6	2.1	0.4	0.1	0.2	....	..	25.2	1.1
Borosilicate, tungsten sealing	7050	67.3	4.6	1.0	....	0.2	....	..	24.6	1.7
Aluminosilicate	1710 1720	57	1.0	.....	5.5	12	....	..	4	20.5

* Vycor brand.
† Pyrex brand.

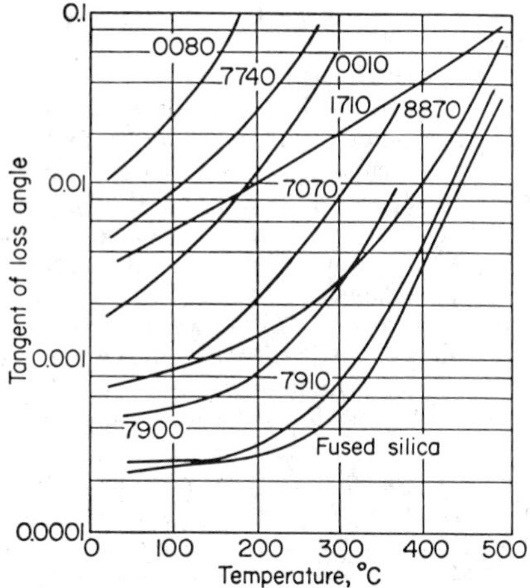

Fig. 4-73. Dielectric loss of electrical glass at 1 Mc/s and various temperatures.

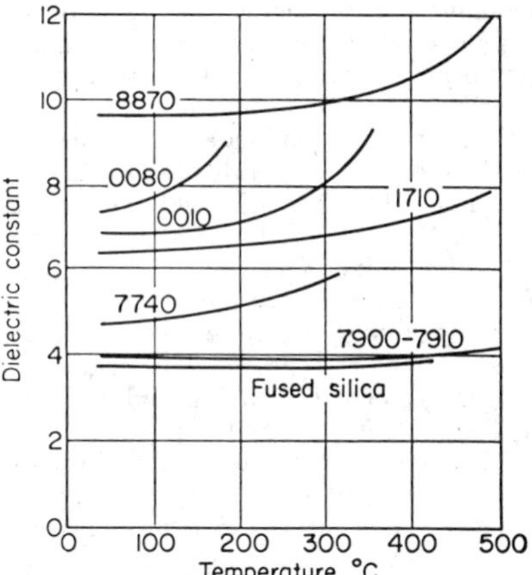

Fig. 4-74. Dielectric constant of electrical glasses at 1 Mc/s and various temperatures.

In the temperature range from 1100 to 1400°C fused quartz undergoes gradual devitrification. The vitreous nature, acid resistivity, and very low coefficient of expansion of fused quartz give it exceptional properties as an electrical insulator. Fused silica has a density of 2.0 to 2.2; softens at 1400 to 1500°C; has specific heat of 0.168, coefficient of thermal expansion of 0.5×10^{-6}, dielectric constant of 3.8, and resistivity of 10^{14} to 10^{17} Ω·cm at room temperature. Fused quartz has a coefficient of expansion of 0.5×10^{-6}/°C and a specific heat of 0.185 to 100°C.

366. Molded Mica and Glass. This is a composition comprised of ground or crushed mica, a fusible vitreous binder of soft glass or lead borate in the correct proportions of 2 parts mica to 1 part soft glass. The mixture is heated to a temperature sufficient to soften the glass, and then the product is molded, under hydraulic pressure, while hot. It is a vitreous material which may be machined, sawed, drilled, and tapped. Low-melting metals may be cast around the composite, or metal inserts may be used. The material has been used for vacuum-tube faces, insulated pipe joints, etc. Its use temperature is to about 250°C. A trade name is Mycalex.

367. Porcelain. *Constitution.* A typical porcelain used for electrical insulation consists of approximately 28% china clay, 10% ball clay, 35% feldspar, 25% flint, and 2% talc. The feldspar, which has a composition K_2O, $Al_2O_3 \cdot 6SiO_2$, serves as the fluxing material which assists in dissolving the more insoluble consituents, clay and quartz. The porcelain is usually fused at 1200°C. During the process there is about a 12% shrinkage. The raw materials used in making the porcelain are finely ground and intimately mixed in a liquid suspension. The mixed material is then converted to a plastic state by filtering under pressure; it is next molded into the desired shape, dried, dipped in a glaze, and finally fired at a slowly rising temperature level, which fuses the flux in the body of the material and also converts the glaze into a smooth glass coating. There are three different processes for making porcelain insulators.

368. Wet-process Porcelain. Pieces of the plastic material about the consistency of putty are worked into a convenient size and shape and placed in plaster-of-paris molds which determine the shape of the outer surface. The surface now in contact with the mold is worked into the desired shape by machine forming or pressing. After the pieces have been partly dried and have become stiff enough to handle, they are removed from the mold. The surfaces which were not in contact with the mold are finished to accurate dimensions. The pieces are then dried, glazed, and fired. Another method of making wet-processed porcelain is by extruding the plastic material into tubes or bars. In some cases finishing operations are performed on lathes where the desired shapes are obtained by turning. This process is being tried extensively in making high-voltage porcelain. Disks for suspension insulators are made by hot pressing. The mold is plaster of paris into which the soft clay is placed. The heated plunger presses the material into the mold and also forms the other surface.

369. Casting Process for Porcelain. Porcelains of high electric strength and of complicated form can be made by pouring the liquid material into multipart plaster-of-paris forms. The cast pieces are removed from the molds after they have stiffened sufficiently to permit handling. Plaster of paris is particularly useful because it absorbs water and accelerates the drying. The molded shapes are dried, finished, glazed, and fired. The large high-voltage insulators commonly used are made by casting.

370. Dry-process Porcelain. The preparation of the body consists in taking the pressed material and drying, crushing, dampening, and pulverizing it. The result of pulverization of the dampened mass is to produce granules. These are pressed in a die, where they adhere to each other. The molded shapes are dried, finished to dimensions, glazed, and fired. The general result is a nonuniform porous mass which is not capable of withstanding high voltages but is suitable only for low-voltage applications and for use in dry locations.

371. Porcelain Glazes. The impervious character of porcelain is not obtained by the glaze but is the result of the vitrification of the primary body. The glaze is a smooth glass coating which provides a finish and prevents the accumulation of dirt. In order to avoid cracks, it is important to employ a glaze which has the same thermal coefficient of expansion as the underlying porcelain.

372. Steatite Ceramics. These are products formed by firing steatite or talc, which is a magnesium silicate in the form of $Mg_3(OH)_2Si_4O_{10}$. The natural material is a soft creamy white and has a soapy feel. The material is machined to dimension and then fired, resulting in a very hard product with small dimensional changes. The water absorption in the naturally fired stone is high. These ceramics are used for low-tension and high-frequency insulators and for use at elevated temperatures for the thermal shock resistances desired. Trade names of steatite insulators include Lava, Lava Rock, Lavite, and AlSiMag.

373. Titania Bodies. These are formed from rutile crystals of titanium dioxide. The mineral rutile (TiO_2) has a dielectric constant of 183 parallel to the crystalline axis and 83 perpendicular to the crystalline axis. Titania bodies are used for high-frequency low-loss insulators and dielectrics.

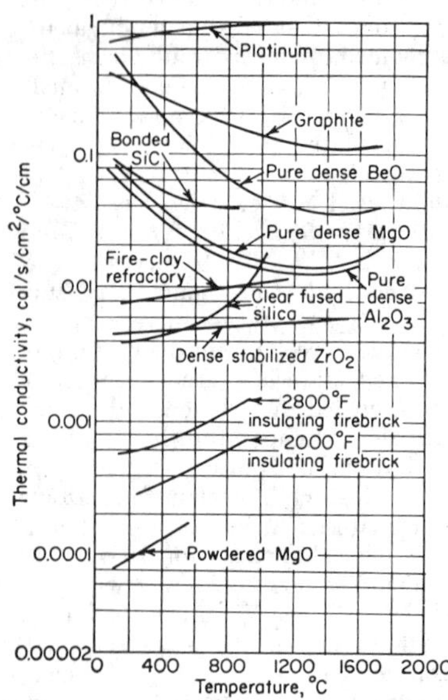

FIG. 4-75. Conductivity of ceramic materials over a wide temperature range. (*Kingery, "Introduction to Ceramics"; New York, John Wiley & Sons, Inc.*, 1960.)

374. Other Ceramic Insulators. Many other minerals are used to produce ceramics having special characteristics such as low thermal coefficient of expansion, low loss, etc. Some of these materials are: alumina (Al_2O_3), forsterite (Mg_2SiO_4), mullite ($Al_6Si_2O_{13}$), sillimanite ($AlSiO_5$), spodumene ($Li_2Al_2Si_4O_{12}$), zircon ($ZrSiO_4$), and wollastonite ($CaSiO_3$). Spark-plug porcelains are generally high-alumina materials, the remainder consisting of oxides of silicon, zirconium, magnesium, and manganese. A comprehensive listing of the mechanical and electrical properties is given in Table 4-75. Figure 4-75 shows the thermal conductivity of a variety of ceramics over a wide temperature range.

It should be noted that the properties of ceramic materials generally exhibit a strong dependence on impurity level, crystalline imperfection, stoichiometry, and microstructure. In addition, the method of measurement is important, particularly at elevated temperatures. Thus, the values given in the literature by different investigators may be inconsistent, although an estimate of consistency may be obtained from density or purity values, which should be given.

375. References on Inorganic Insulating Materials

1. NORTON, F. H. "Refractories," 3d ed.; New York, McGraw-Hill Book Company, 1949.

2. GREEN, A. T., and STEWART, G. H. (eds.) "Ceramics, A Symposium"; British Ceramics Society, 1953.

3. RYSCHKEWITSCH, E. "Oxydkeramik der Einstuffsystemme"; Berlin, Springer-Verlag, OHG, 1949.

4. LEBEAN, P., and TROMBE, F. (eds.) "Les Hautes températures et leur utilization en chimie"; Paris, Masson et Cie, 1950, Vol. II.

5. CAMPBELL, I. E. (ed.) "High-temperature Technology"; New York, John Wiley & Sons, Inc., 1956.

6. DODD, A. E. "Dictionary of Ceramics"; London, George Newnes Ltd., 1964.

7. KINGERY, W. D. "Introduction to Ceramics"; New York, John Wiley & Sons, Inc., 1960.

8. POPPER, P. "Special Ceramics"; New York, Academic Press Inc., 1960.

9. KINGERY, W. D. "Property Measurements at High Temperature"; New York, John Wiley & Sons, Inc., 1959.

10. RYSHKEWITCH, E. "Oxide Ceramics"; New York, Academic Press Inc., 1960.

11. CLARK, F. M. "Insulating Materials for Design and Engineering Practice"; New York, John Wiley & Sons, Inc., 1962.

12. "Standard Handbook for Electrical Engineers," 9th ed.; New York, McGraw-Hill Book Company, 1957.

376. Inorganic Insulating Systems. A relatively new concept in electrical insulation is complete systems for insulating electrical devices composed only of inorganic materials. Such materials as mica, glass, asbestos, cements, and ceramics have been used for years, but mostly in conjunction with organic resins or insulation. With the advent of metallic phosphates, alkaline silicates, and glasses as a replacement for the organic resins a complete insulation system has evolved. Some of the materials most commonly used in inorganic insulation systems are described below.

377. Conductor Insulation. The electrical insulation on a conductor is primarily used to direct the current flow in the conductor in the desired direction. Some examples are:

a. Fibrous Insulation. Covering conductors with fibrous glass and asbestos is the most usual insulating process. The glass fibers have the lower space factor (3 to 7 mils) and lower temperature resistance (400°C).

b. Enamel and Refractory Glass. A combination of finely divided, low-melting lead glass and a refractory such as silica or alumina are dispersed in an organic wire enamel. This enamel composition is applied to a conductor, cured, and the conductor used in an electrical device. The conductor, which is held in place by an inorganic encapsulant, is then fired to about 650 to 700°C to burn out the organic enamel and fuse the glass-refractory combination. The insulated conductor has a conductor-to-conductor electric strength of 450 V and a temperature-use limit of 500°C.

c. Glass Coatings. Glass coatings fused onto a conductor have provided electrical insulation capabilities up to 400°C and 400 V electric strength between conductors. The coating produces very small cracks on bending and may fracture and separate on small radius bends. A very thin organic coating such as Teflon may be applied over the glass to improve handling characteristics.

d. Chemical Coatings. Anodized aluminum wire has found some uses at temperatures up to 500°C. The outer layer of aluminum on the conductor has been converted electrochemically to aluminum oxide, an electrical insulator. The conductor-to-conductor electric strength averages 400 V despite the fact that microscopic cracks are found when the conductor is wound into electrical devices.

378. Flexible Sheet Insulation. Flexible sheet insulation is primarily used on curved or uneven surfaces to prevent grounding of electrical conductors to metal components and to prevent electrical breakdown between voltage differentials.

a. Mica. The only positive barrier to electrical breakdown is provided by mica. Mica splittings or paper can be combined with an inorganic binder to provide a semi-flexible sheet material having an electrical strength of 400 V/min. The temperature limit of use is 550°C.

b. Synthetic Mica Paper. A highly flexible synthetic mica paper is physically

Table 4-74. Electrical and Mechanical Properties of Commercial Glasses

	Volume resistivity, Ω·cm		Dielectric properties		Specific gravity	Young's modulus, lb/in²	Coefficient of expansion, °C	Viscosity data, °C		
	250°C	350°C	Power factor	Dielectric constant				Strain point*	Annealing point	Softening point
Silica glass, fused silica	10^{12}	10^{9}	0.0002	3.78	2.20	10×10^{6}	5.5×10^{-7}	1070	1140	1667
99% silica glass (7900)	10^{9}	10^{8}	0.0005	3.8	2.18	9.7×10^{6}	8×10^{-7}	820	910	1500
99% silica glass (7911)	10^{11}	10^{9}	0.0002	3.8	2.18	9.7×10^{6}	8×10^{-7}	820	910	1500
Soda-lime, plate glass	10^{6}–10^{7}	10^{5}	0.004–0.010	7.0–7.6	2.46–2.49	10×10^{6}	87×10^{-7}	515	550	735
Soda-lime, electric-lamp bulbs	10^{6}	10^{5}	0.009	7.2	2.47	9.8×10^{6}	92×10^{-7}	478	510	696
Lead-silicate, electrical	10^{8}	10^{7}	0.0016	6.6	2.85	9.0×10^{6}	91×10^{-7}	397	428	626
Lead-silicate, high lead	10^{11}	10^{9}	0.0009	9.5	4.28	7.6×10^{6}	91×10^{-7}	398	429	580
Borosilicate, low expansion	10^{8}	10^{6}	0.0046	4.6	2.23	9.8×10^{6}	32×10^{-7}	515	555	820
Borosilicate, low electrical loss	10^{11}	10^{9}	0.0006	4.0	2.13	6.8×10^{6}	32×10^{-7}	455	490	
Borosilicate, tungsten sealing	10^{8}	10^{7}	0.0033	4.9	2.25	....	46×10^{-7}	461	496	703
Aluminosilicate	10^{11}	10^{9}	0.0037	6.3	2.63	12.7×10^{6}	42×10^{-7}	672	712	915

* Safe operating temperature 10 to 20°C below strain point, on the assumption that temperature differences within the body are not severe. Mechanical strength retained up to this temperature.

Notes apply to **Table 4-75** (page 4-184).

NOTES: (1) Single crystals available; superior high-temperature strength. (2) Best mechanical properties of sintered oxides. (3) Highest thermal conductivity of oxides; powder highly toxic. (4) Properties strongly directional; oxidizes in air above red heat. (5) Very hard, oxidizes readily at red heat. (6) Hydrates in damp air; unstable in strongly reducing atmosphere. (7) Strength increased to max. 7,000 lb/in (TS) at 2500°C; oxidizes air; reacts $H_2 > 2500$°C. (8) Dissociated in vacuo and reducing atmospheres >1700°C. (9) Vapor pressure 2 microns at 2150°C. (10) Brittle; forms protective SiO_2 coating in air. (11) Good load-bearing capacity up to fusion temperature. (12) Good for some electrical applications and for relatively high expansion coeff. (13) Few advantages over alumina. (14) Tensile strength 2,000 lb/in² at 1600°C; 5×10^{-5} %/h creep at 500 lb/in², 750°C. (15) Tensile strength 8,000 lb/in² at 1600°C; 1×10^{-1} %/h creep at 500 lb/in², 750°C. (16) Excellent load resistance at high temperatures. (17) Good machinability; embrittled in hydrogen. (18) Oxidizes rapidly at temperatures above red heat. (19) Loses oxygen in nonoxidizing atm; samples weak unless fine-grained. (20) Single-crystal tensile strength of 3,400 lb/in² at 2250°C reported. (21) Excellent chemical and mechanical stability. (22) Other rare-earth oxides similar. (23) Gains oxygen and shatters on heating in air. (24) Good deformation resistance at very high temperatures; good slag resistance. (25) Dissociates into $ZrO_2 + SiO_2$ at high temperatures. (26) Devitrifies at temperatures above 1100°C. (27) Can be used in place of fused silica in many applications. (28) Most useful laboratory glass. (29) Good for laboratory ware. (30) Best porcelain available. (31) Mainly for electrical applications. (32) Not much used in laboratory. (33) Good resistance to basic slags. (34) Good thermal conductivity and load-bearing capacity. (35) Light weight, good thermal insulation, easy to shape and use. (36) Light weight, good thermal insulation, easy to shape and use. (37) Light weight, good thermal insulation, easy to shape and use.

From KINGERY, W. D. "Property Measurements at High Temperature"; New York, John Wiley & Sons, Inc., 1959.

Table 4-75. Mechanical, Thermal, and Electrical Properties of Refractories

(Mainly from References 1–5)

	Composition	Porosity (vol. %)	Fusion temp., °C	Max. normal-use temp., °C	Density, bulk (*b*), true (*t*), g/cm³	Specific heat capacity, cal/(g)(°C), 20–1000°C	Linear-expansion coeff. (10^{-6} in)(in)(°C), 20–1000°C
Special refractories:							
Sapphire crystal	99.9 Al_2O_3	0	2030	1950	3.97(*t*)	0.26	8.6
Sintered alumina	99.8 Al_2O_3	3–7	2030	1900	3.97(*t*)	0.26	8.6
Sintered beryllia	99.8 BeO	3–7	2570	1900	3.03(*t*)	0.50	8.9
Hot-pressed boron nitride	98 BN, 1.5 B_2O_3	3–7	2730	1700	2.25(*t*)	0.39	0.77–7.51
Hot-pressed boron carbide	99.5 B_4C	2–5	2450	1900	2.52(*t*)	0.36	4.5
Sintered calcia	99.8 CaO	5–10	2600	2000	3.32(*t*)	0.23	13.0
Graphite	99.9 C	20–30	3700	2600	2.22(*t*)	0.34	1.5–2.5
Sintered magnesia	99.8 MgO	3–7	2800	1900	3.58(*t*)	0.25	13.5
Molybdenum	99.8 Mo	0	2625	2200	10.2(*t*)	0.065	5.45
Sintered molybdenum silicide	99.8 $MoSi_2$	0–10	2030	1700(air)	6.2(*t*)	0.11	9.2
Sintered mullite	72 Al_2O_3, 28 SiO_2	3–10	1810	1750	3.03(*t*)	0.25	5.3
Sintered forsterite	99.5 Mg_2SiO_4	4–12	1885	1750	3.22(*t*)	0.23	10.6
Sintered spinel	99.8 Mg Al_2O_4	3–10	2135	1850	3.58(*t*)	0.25	8.8
Platinum	99.9 Pt	0	1774	1550	21.45(*t*)	0.035	10.1
Platinum-20 rhodium	80 Pt-20 Rh	0	1900	1650	18.74(*t*)	0.048	10.3
Dense silicon carbide	98 SiC, 1–2 Si, <1C	2–5	>2700	1600(air)	3.22(*t*)	0.20	4.0
Tantalum	99.8 Ta	0	3000	2000	16.6(*t*)	0.036	6.5
Sintered titanium carbide	98 TiC, <1C, <1O	3–10	3140	2500	4.25(*t*)	0.18	7.4
Sintered titania	99.5 TiO_2	3–7	1840	1600	4.24(*t*)	0.20	8.7
Tungsten	99.8 W	0	3410	3000	19.3(*t*)	0.034	4.0
Sintered thoria	99.8 ThO_2	3–7	3050	2500	10.00(*t*)	0.06	9.0
Sintered yttria	99.8 Y_2O_3	2–5	2410	2000	4.50(*t*)	0.13	9.3
Sintered urania	99.8 UO_2	3–10	2800	2200	10.96(*t*)	0.06	10.0
Sintered stabilized zirconia	92 ZrO_2, 4 HfO_2, 4 CaO	3–10	2550	2200	5.6(*t*)	0.14	10.0
Sintered zircon	99.5 $ZrSiO_4$	5–15	2420	1800	4.7(*t*)	0.16	4.2
Commercial refractories:							
Silica glass	99.8 SiO_2	0	1710	1100	2.20(*t*)	0.18	0.5
Vycor glass	96 SiO_2, 4B_2O_3	0	......	950	2.18(*t*)	0.19	0.7
Pyrex glass	81 SiO_2, 13 B_2O_3, 2 Al_2O_3, 4 M_2O	0	......	650	2.23(*t*)	0.20	3.2
Mullite porcelain	70 Al_2O_3, 27 SiO_2, 3 MO + M_2O	2–10	1750	1400	2.8(*b*)	0.25	5.5
High-alumina porcelain	90–95 Al_2O_3, 4–7 SiO_2, 1–4 MO + M_2O	2–5	1800	1500	3.75(*b*)	0.26	7.8
Steatite porcelain	35 MgO, 60 SiO_2, 5 Al_2O_3	2–5	1450	1200	2.7(*b*)	0.26	10.2
Superduty fireclay brick	40–45 Al_2O_3, 55–50 SiO_2	10–20	1750	1650	2.1(*b*)	0.26	5.3
Magnesite brick	83–93 MgO, 2–7 FeO	10–20	2100	1750	3.2(*b*)	0.25	14
Bonded silicon carbide	90 SiC, 6–9 SiO_2, 2–4 Al_2O_3	20	2200	1450	2.95(*b*)	0.20	4.5
2000°F I.F.B	Fireclay or diatomaceous	80–85	1600	1090	0.52(*b*)	0.25	5.3
2600°F I.F.B	Fireclay	72–77	1750	1430	0.75(*b*)	0.25	5.3
3000°F I.F.B	Aluminous fireclay	60–65	1800	1650	1.04(*b*)	0.25	5.3

Table 4-75. Mechanical, Thermal, and Electrical Properties of Refractories.—*Concluded*

	Thermal conductivity, cal/(cm²)(cm)(sec)(°C)		Modulus of rupture (MR) or tensile strength (TS), lb/in²		Modulus of elasticity, 10^6 lb/in²	Thermal stress resistance	Electrical resistivity, Ω·cm		Notes
	At 100°C	At 1000°C	At 20°C	At 1000°C			20°C	1000°C	
Special refractories:									
Sapphire crystal	0.072	0.019	40,000–150,000 (MR)	30,000–100,000 (MR)	55	Very good	$>10^{14}$	10^8	1
Sintered alumina	0.069	0.014	30,000 (MR)	22,000 (MR)	53	Good	$>10^{14}$	5×10^7	2
Sintered beryllia	0.500	0.046	20,000 (MR)	10,000 (MR)	45	Excellent	$>10^{14}$	10^8	3
Hot-pressed boron nitride	0.04–0.07	0.03–0.06	7,000–15,000 (MR)	1,000–2,150 (MR)	12	Good	10^{10}	10^4	4
Hot-pressed boron carbide	0.07	0.05	50,000 (MR)	40,000 (MR)	42	Good	0.5	...	5
Sintered calcia	0.033	0.017	...	...	...	Fair–poor	$>10^{14}$	10^6	6
Graphite	0.30	0.10	3,500 (TS)	4,000 (TS)	1.3	Excellent	10^{-3}	10^{-3}	7
Sintered magnesia	0.082	0.016	14,000 (MR)	12,000 (MR)	30.5	Fair–poor	$>10^{14}$	10^7	8
Molybdenum	0.35	0.28	90,000–250,000 (TS)	30,000 (TS)	45	Excellent	5.2×10^{-6}	24×10^{-6}	9
Sintered molybdenum silicide	0.075	0.03	100,000 (TS)	40,000 (TS)	59	Good	22×10^{-6}	...	10
Sintered mullite	0.013	0.008	12,000 (MR)	7,000 (MR)	21	Good	$>10^{14}$	...	11
Sintered forsterite	0.010	0.005	10,000 (MR)	...	...	Fair–poor	$>10^{14}$	10^6	12
Sintered spinel	0.033	0.013	12,300 (MR)	11,000 (MR)	34.5	Fair	$>10^{14}$	10^6	13
Platinum	0.166	0.22	24,000 (TS)	8,000 (TS)	21.3	Excellent	11.4×10^{-6}	45×10^{-6}	14
Platinum-20 rhodium	...	...	70,000 (TS)	30,000 (TS)	28	Excellent	20.8×10^{-6}	32×10^{-6}	15
Dense silicon carbide	0.133	0.05	24,000 (MR)	24,000 (MR)	68	Excellent	10	4	16
Tantalum	0.130	0.12	50,000–180,000 (TS)	25,000 (TS)	27	Excellent	12.4×10^{-6}	54×10^{-6}	17
Sintered titanium carbide	0.08	0.02	160,000 (TS)	140,000 (TS)	45	Very good	1×10^{-4}	...	18
Sintered titania	0.015	0.008	8,000 (MR)	6,000 (MR)	...	Excellent	5.48×10^{-6}	25×10^{-6}	19
Tungsten	0.40	0.30	100,000–500,000 (TS)	60,000 (TS)	60	Fair–poor	$>10^{14}$	10^4	20
Sintered thoria	0.022	0.007	12,000 (MR)	7,000 (MR)	21	Fair–poor	$>10^{14}$	10^5	21
Sintered yttria	(0.02)	...	...	...	...	Fair–poor	...	...	22
Sintered urania	0.020	0.007	12,000 (MR)	18,000 (MR)	25	Fair–poor	...	...	23
Sintered stabilized zirconia	0.005	0.005	20,000 (MR)	15,000 (MR)	22	Fair–good	10^8	500	24
Sintered zircon	0.015	0.008	12,000 (MR)	6,000 (MR)	30	Good	$>10^{14}$	10^5	25
Commercial refractories:									
Silica glass	0.004	0.012	15,500 (MR)	...	10.5	Excellent	$>10^{14}$	10^6	26
Vycor glass	0.004	...	10,000 (MR)	...	10.5	Excellent	$>10^{14}$	...	27
Pyrex glass	0.004	...	10,000 (MR)	...	10	Very good	$>10^{14}$	...	28
Mullite porcelain	0.007	0.006	10,000 (MR)	6,000 (MR)	10	Good	$>10^{14}$	10^4	29
High-alumina porcelain	0.05	0.015	50,000 (MR)	...	53	Very good	$>10^{14}$	10^4	30
Steatite porcelain	0.008	0.006	20,000 (MR)	...	10	Fair–poor	$>10^{14}$	10^5	31
Superduty fireclay brick	0.004	0.005	750 (MR)	700 (MR)	14	Poor	10^8	10^4	32
Magnesite brick	0.040	0.009	4,000 (MR)	4,000 (MR)	25	Fair	10^8	10^5	33
Bonded silicon carbide	0.080	0.030	2,000 (MR)	2,000 (MR)	50	Excellent	10^5	10^3	34
2000°F I.F.B	0.00015	...	40 (MR)	...	...	Fair–good	...	...	35
2600°F I.F.B	0.0005	0.0011	170 (MR)	...	...	Fair–good	...	...	36
3000°F I.F.B	0.0007	0.0014	290 (MR)	...	...	Fair–good	...	...	37

See notes on page 4-182.

strong and thermally stable to 980°C. The electric strength is 140 V/mil at room temperature, with little loss up to 600°C. A reversible dehydration amounting to 4.5% by weight occurs at 110°C and may cause delamination of the paper.

c. Other Fibrous Papers. A number of other fibrous papers can be used as thin electrical insulating sheets, but all have low (about 35 to 50 V/mil) electric strength. Electrical-grade glass cloth has high initial strength but loses most of its physical strength after exposure to 450°C. The new S-type glass cloth retains more strength at elevated temperatures and can be used to 500°C. Synthetic aluminum silicate fiber mats are electrically useful to 750°C. Asbestos in paper or tape form has fair to good physical strength, especially when combined with organic or glass fibers. Dehydration at about 500°C substantially weakens it, but electrical properties are maintained. A great number of other fibers have appeared on the market in paper, mat, and fabric form for use in specialized applications.

379. Rigid Sheet Insulation. Rigid sheet is generally used where both electrical insulation and structural strength are required.

a. Mica Sheet. Both mica flakes and mica paper are made into rigid sheets using glass and metal phosphate as a bond. Electric strengths up to 2,500 V/mil can be obtained and flexural strengths up to 20,000 lb/in². Some grades can be used up to 800°C.

b. Asbestos Sheet. A number of grades of asbestos boards are made using cement or alkaline silicate bonds. The length of the asbestos fiber, from almost a powder to 2 in, greatly influences the structural strength of the sheet. The low-density short-fiber sheets have flexural strengths of about 500 lb/in², while the long-fiber sheets can achieve a strength of 32,000 lb/in². All are somewhat porous, thus giving a low electrical strength of 35 to 60 V/mil. The asbestos fiber generally used is chrysotile, which dehydrates and loses strength rapidly above 500°C, thus limiting its use temperature to this value.

c. Ceramic Sheet. A number of ceramics can be made into sheets for insulation and structural purposes.

d. Glass-Mica Sheet. Glass-bonded mica sheet material is made from both natural and synthetic mica. The natural mica composition is usable to 450°C and the synthetic mica to 620°C. Maximum electric strength is 500 V/mil, and flexural strength is 15,000 lb/in².

380. Encapsulants. An encapsulant is primarily used to provide structural integrity and environmental protection to an electrical device. The inorganic encapsulants are cements made from various inorganics, usually metal oxides as fillers and a bonding material. The bonding material is generally a hydraulic or water-setting compound or a chemical-setting compound such as an alkaline silicate or a metal phosphate. All cements are porous and thus have low electric strength (35 to 60 V/mil).

a. Alkaline Silicate Cements. These hard, strong, acid-resistant cements provide good electrical insulation to 500°C.

b. Metal Phosphate Cements. These alkali-resistant cements are the best for both structural and electrical properties. They can be used at temperatures up to 1000°C.

c. Hydraulic, or Water-setting, Cements. Plaster of paris and portland cement are examples of this type of cement. The presence of water gives them poorer electrical properties.

Dehydration of the cement above 100°C considerably reduces their already weakened physical strength.

d. Other Cements. One of the most important of these cements is a magnesium oxysulfate bonded magnesium oxide. This finds extensive use around heating elements because of its high thermal conductivity. Use temperature is 1000°C. Several other cements are made for specialized electrical applications which utilize both hydraulic and chemical setting.

381. References on Insulating Systems

1. GOLDSMITH, A., et al. "Handbook of Thermophysical Properties of Solid Materials"; New York, The Macmillan Company, 1961, Vol. III.

2. Hughes Aircraft Co., Electronic Properties Information Center, data sheets on electrical insulating materials.

3. Kueser, P. E., et al. Electrical Conductors and Electrical Insulation Materials Topical Report, NASA–CR–54092.

4. Vondracek, C. H. Inorganic Potting Compounds for High Temperatures; *Materials in Design Eng.*, December, 1964, p. 100.

382. Ferroelectrics and Ferroelectricity. A ferroelectric material possesses a reversible polarization as shown by a dielectric hysteresis loop (Fig. 4-76). It lacks a center of symmetry in the structure and possesses a polar axis. Furthermore the hysteresis loop disappears above a certain temperature (Curie point), and a transition occurs in the crystal structure to a form of higher symmetry. A more workable definition would be that:

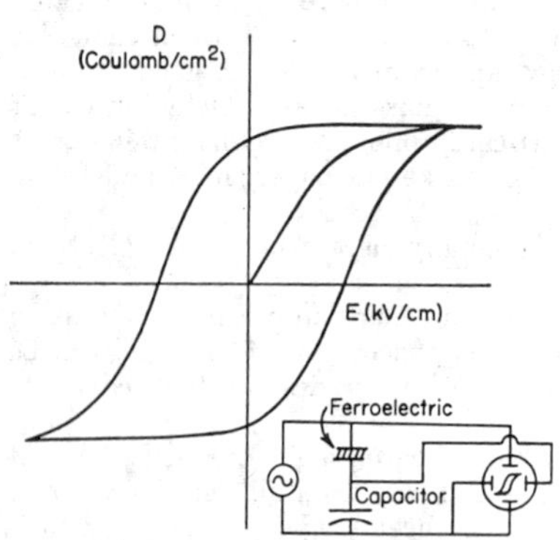

Fig. 4-76. Typical ferroelectric ceramic hysteresis loop and Sawyer-Tower circuit.

a. Upon applying an electric field E, a displacement field D is generated containing frequency components not present in E.

b. A linear combination of two electric fields does not produce a corresponding linear combination of the two displacement fields. A number of ferroelectric materials such as rochelle salt, alkali-metal dihydrogen phosphates and arsenates, and also a group represented by guanidine aluminum sulfate hexahydrate have been discovered but have not proved useful in practical application because of such difficulties as chemical instability and the incompatibility of the material's Curie temperature and the temperature of practical working devices. The only group that has found widespread application is that possessing the perovskite or oxygen octahedra structure. Members of this group include solid solutions or mixtures of titanates, zirconates, niobates, tantalates, and stannates. The most widely used member of this group is $BaTiO_3$, which has been successfully used in capacitor, transducer, memory, and thermistor applications.

The polar axis caused by the small ionic displacement from symmetry is believed to contribute to the polarization of the structure. At the Curie temperature of $BaTiO_3$ (120°C) the thermal energy of the ions is sufficient to overcome the electrostatic forces holding the ions at these positions. When this occurs, the structure changes from tetragonal to cubic and the ferroelectric effect disappears. The Curie temperature can be increased by substituting lead at a rate of approximately 4°C/(mole) (%) lead sub-

Table 4-76. Piezoelectric Properties of Several Typical Ferroelectrics

Material	Dielectric constant	Curie temp., °C	Electromechanical coupling, k	Strain constant, C/N, d
Rochelle salts	500	−18, +24	$k_{14} = 0.78$	$d_{14} = 870$ $d_{25} = -53$ $d_{36} = 12$
$BaTiO_3$	1,500	120	$k_{31} = 0.20$ $k_{33} = 0.50$ $k_{15} = 0.50$	$d_{31} = -56$ $d_{33} = 150$
Lead:	1,500	350	$k_{31} = 0.32$	$d_{31} = -140$
Zirconate	.....	....	$k_{33} = 0.675$	$d_{33} = 320$
Titanate	.....	....	$k_{radial} = 0.54$	
$BaTiO_3$ (88%) $PbTiO_3$ (12%)	850	150	$k_{31} = 0.125$ $k_{33} = 0.365$ $k_{radial} = 0.210$	$d_{31} = 30$ $d_{33} = 90$
$PbNbO_2$ (80%) $BaNbO_3$ (20%)	400	425	$k_{radial} = 0.20$	$d_{31} = 25$
$PbNbO_3$ (60%) $SrNbO_3$ (40%)	755	310	$k_{radial} = 0.26$	$d_{31} = -53$

stitution. A strontium substitution will lower the Curie temperature at a rate of approximately 3°C/(mole)(%) strontium. Other additions can affect the Curie temperature of $BaTiO_3$.

383. Ferroelectric Capacitors. An advantage of ferroelectric material is its large dielectric constant, which permits the use of physically small capacitors. Several other characteristics that should be taken into consideration are the materials temperature coefficient, voltage or bias sensitivity, and breakdown characteristics. The temperature variance of the dielectric constant of several typical types of high-dielectric-constant dielectrics is shown in Fig. 4-77. In general, the higher the dielectric constant, the more pronounced the effect of temperature. The curves may be shifted or flattened by changes in composition or process treatment.

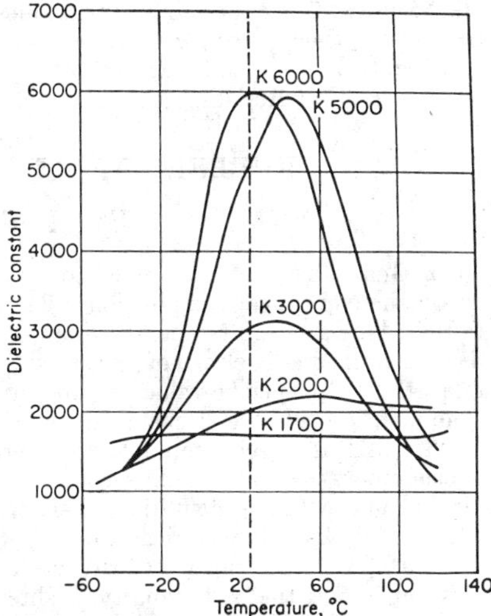

FIG. 4-77. Temperature dependence of dielectric constant for typical ferroelectric ceramics.

The voltage or bias sensitivity shown in Fig. 4-78 indicates that the higher-dielectric-constant materials are more sensitive to increasing bias fields. The K_β/K_0 ratio is the incremental change in the dielectric constant, where K_β is the value of dielectric constant with d-c bias voltage and K_0 is without the d-c bias voltage. In ferroelectrics a high electric field applied to a device can cause voltage breakdown. The general rule is that, the higher the dielectric constant, the lower the dielectric strength. A figure of 2.5 kV/mm at room temperature for high-dielectric-constant materials (4,000 to 6,000) is a nominally accepted value, and this can be increased to 15 kV/mm for low-dielectric-constant (~30) dielectrics. These values would have to be derated for elevated temperatures.

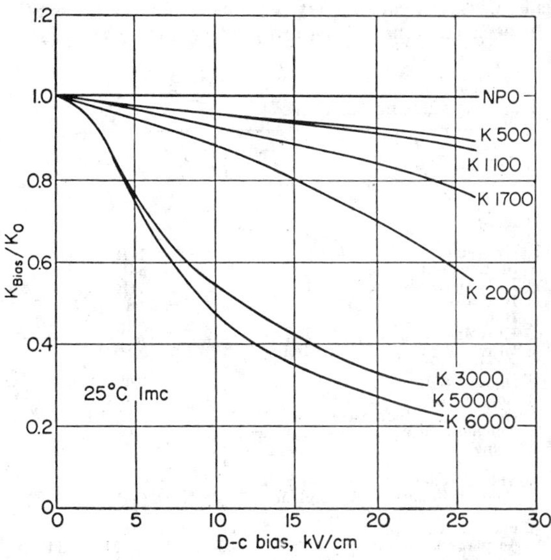

FIG. 4-78. Effect of d-c bias on dielectric constant.

Capacitors with very high dielectric constants, in the order of 100,000 and higher, have been developed by treating $BaTiO_3$ in a reducing atmosphere. These capacitors are semiconductor in nature and rather lossy and are therefore limited to very-low-voltage applications such as those in transistor circuits. Other ferroelectrics and their properties are shown in Table 4-76.

384. References on Ferroelectrics

1. MEGAW, H. O. "Ferroelectricity in Crystals"; London, Methuen & Co., Ltd., 1957.

2. FOTLAND, R. A. Ferroelectrics as Solid-state Devices; *Elec. Mfg.*, March, 1958.

3. SCHICKE, H. M. "Essentials of Dielectromagnetic Engineering"; New York, John Wiley & Sons, Inc., 1961.

4. IRE Standards on Piezoelectric Crystals; *Proc. IRE*, July, 1961, Vol. 49, No. 7.

5. WEDLOCK, B. D. "Properties of Piezoelectric Materials"; (AD-611212) Materials Technology Division, U.S. Army Materials Research Agency, Watertown, Mass.

6. SABURI, O. Processing Techniques and Applications of PTC Thermistors; *Trans. IEEE Component Parts*, June, 1963, Vol. CP-10, No. 2.

7. ICHIKAWA, Y., and CARLSON, W. G. Yttrium-doped Ferroelectric Solid Solutions with PTC of Resistance; *Am. Ceram. Soc. Bull.*, May, 1963, Vol. 42, No. 5.

RUBBER—NATURAL AND SYNTHETIC

BY H. F. MINTER

385. General. Rubbers are elastomers which are better defined by their properties than by their chemical composition. They are described by ASTM as: "A polymeric material which at room temperature can be stretched to at least twice its original length and upon immediate release of the stress will return quickly to approximately its original length." No specific values of time, strain, or stress are indicated in the description, and no mention is made of chemical composition. The broad definition is appropriate, since "rubber" is loosely employed to describe at least 14 chemically different vulcanizable elastomers in commercial use.

In keeping with the definition of elastomers such materials as hard rubber and the plastics widely used in electrical applications are not discussed in this section.

The ASTM classifications of rubbers are based upon service requirements such as oil resistance and thermal stability. Three general types are specified: Class *R*, Non-oil-resistant; Class *S*, Resistant to Petroleum Chemicals; and Class *T*, Temperature-resistant. In Table 4-77, these types are listed with the subdivisions as outlined in ASTM D-735 and SAE Standard J-14.*

The type S designation is subdivided into three classes. Class SA has very low volume swell in petroleum chemicals, Class SB calls for low volume swell, and Class SC permits medium volume swell. In each case the values are relative to the very high swell of the non-oil-resistant types.

Table 4-77. Insulating Rubbers

ASTM-SAE		Popular name	Chemical composition	Standard symbol (ASTM)
Type	Class			
R		Natural rubber	Isoprene	NR
		SBR	Styrene/butadiene	SBR
		Butyl	Isobutylene/isoprene	IIR
		Polyisoprene	Isoprene	IR
		Polybutadiene	Butadiene	BR
		EPR	Ethylene/propylene	EPM
		EPT	Ethylene/propylene terpolymer	EPDM
S	SA	Thiokol*	Organic polysulfide	ET, EOT
	SB	Nitrile	Acrylonitrile/butadiene	NBR
	SB	Polyurethane	Diisocyanate/polyester or polyether	AU, EU
	SB	Hydrin†	Epichlorohydrin or copolymer with ethylene oxide	CO
	SC	Neoprene	Chloroprene	CR
	SC	Hypalon‡	Chlorosulfonated polyethylene	CSM
T	TA	Silicone	Polysiloxane	Si, FVSi
	TB	Polyacrylate	Acrylate copolymers	ACM, ANM
	TB	Fluorocarbon	Vinylidene fluoride/perfluoropropylene	FPM
			Vinylidene fluoride/chlorotrifluoroethylene	FM

* Trade name of Thiokol Chemical Corporation.
† Trade name of B. F. Goodrich Chemical Company.
‡ Trade name of E. I. du Pont de Nemours & Co., Inc.

* A more recent system is ASTM D2000-64T, SAE J200. ASTM D735 is presented here because of its greater simplicity.

Type T rubbers are designated as Class TA, which is suitable for both low and high temperatures, and Class TB, which is resistant to hot air and oil.

386. Vulcanization. Dry rubbers, as distinguished from latex, are highly viscous solids which are seldom used as such. They are thermoplastic until vulcanized. In the course of vulcanization the long molecular chains which are characteristic of rubbery materials are linked together, classically with sulfur, into a three-dimensional network which retains the elasticity but has lost the plasticity of the original rubber. The vulcanized rubber displays resistance to solvents and stability of physical properties over a wide temperature range, in contrast to the unvulcanized material. Other vulcanizing agents than sulfur may be used and indeed are required for many of the synthetic elastomers. To facilitate processing and to achieve uniform properties, the primary vulcanizing agent is normally supplemented with accelerators, which both hasten and complete the curing of the rubber.

387. Compounding of Rubbers. The properties of rubber parts are only partly due to their chemical nature. Commercial rubber compounds are vulcanized mixtures of the rubber with modifiers such as fillers, plasticizers, and protective agents which are added to enhance the properties of the basic gum. Such properties as tensile strength, elongation, resistance to solvents including water, abrasion resistance, stability on weathering, resistance to oxygen and ozone, and low temperature embrittlement may be improved by the proper choice of materials added to the rubber. Dielectric properties, particularly dissipation factor, high temperature resistance, resilience, permeability to gases, and resistance to oils, are inherent properties subject to only slight modification by compounding. Interesting and useful exceptions are conducting rubbers, which are used for stress grading and dissipation of static charges.

Cost of rubber compounds is a variable which may be considerably modified by skillful compounding. Because the price of a rubber item reflects not only the costs of the raw materials but the time in process and the equipment required, apparently similar products may differ in price because different methods of production are required. Ease of processing, minimum raw-materials cost, and rapid vulcanization are areas in which skillful compounding contributes to lower cost.

Throughout the processing and forming steps, the rubber must remain in the thermoplastic state; yet it must cure (or vulcanize) rapidly. For this reason, vulcanizing systems are often complex mixtures of accelerators, "ultra" accelerators, and, at times, retarders in order to provide safe processing and fast curing.

388. Processing of Rubber. Raw rubber is a tough, viscous gum into which the ingredients are incorporated by mixing in internal mixers, on rubber rolls, or in the development stage, the addition of all the components, with the rubber in crumb form, to an extruder-mixer. The internal mixers are typified by the Banbury mixer, in which the rubber is masticated between large rotors and smeared against the walls. In mixing, a ram confines the materials to the mixing chamber, from which they are discharged to a mill, on which sheets are formed preliminary to further processing. A rubber mill is comprised of two (generally steel) counterrotating rolls with a speed differential on which, as in the internal mixer, the principal mode of mixing is shear. Because the mixing is intensive, the heat buildup is often sufficient so that a cooling period (by storage and/or water spray or immersion) is necessary before the vulcanizing agents are added in a second mixing operation. The end product is then formed by compression, transfer or injection molding, by extrusion, by calendering into thin sheet, or by other means. Cloth reinforcement may be added as in sheet and conveyor belting, hose, and V-belts. Metal inserts may be included during molding or rubber-metal composites achieved by cementing. A rubber grommet or bumper pad, for example, may be formed by compression, transfer, or injection molding; hose, weather stripping, and wire-cable covering produced by extrusion; and flat circular gaskets made by wrapping calendered sheet on a mandrel, curing, and slitting.

For description of process equipment, see Chap. 20, "Introduction to Rubber Technology," by Morton, Reinhold, 1959.

389. Natural and Synthetic Rubbers. The rubbers below are listed in groups of similar cost, based on the prices of the base elastomers (see also Table 4-78). The ASTM designation is given in parentheses when assigned. Variations within groups and between adjacent groups are to be expected, as compounding materials and processing costs vary widely with the required performance of the product. Roughly, the first three groups include materials in the range of 20 cents to $1.25 per pound. The silicones are intermediate at about $2 to $6 per pound, and the fluoroelastomers at present cost more than $7 per pound.

390. Group I Rubbers. *Natural Rubber (NR).* As obtained from the tropical tree *Hevea brasiliensis,* natural rubber is present in a latex, or emulsion, at about 40% solids. The latex may be concentrated, stabilized, and used in the preparation of "proofed" goods, such as sheeting, and for articles prepared by dipping of forms, such as gloves. When coagulated, washed, and dried, the rubber is compressed into bales, which are the raw material for the rubber manufacturer.

Until about 1930 natural rubber had no competitors. The first synthetics were polysulfide rubber (Thiokol) and neoprene, and the advent of World War II forced further developments when the rubber-producing areas of the world became inaccessible. The success of the synthetics is indicated by 1964 figures (*Rubber Age,* April, 1965), which showed natural rubber usage to be only one-fourth the total world consumption.

The outstanding properties of natural rubber are its high resilience and low hysteresis, which gives low heat buildup in rapid flexing applications such as automotive tires. Chemically, natural rubber is a polymer of isoprene [$CH_2{=}C(CH_3){-}CH{=}CH_2$], as is gutta-percha, a hard, tough, plastic material, with the difference attributable to their molecular structures. Maximum continuous-service temperature is 70°C. Good low-temperature flexibility is characteristic of natural rubber compounds, but extended static exposure at low temperatures results in crystallization and loss of rubbery properties. The crystallization rate is increased by imposed strain, but it is completely reversible by warming and has no permanent effect.

Polyisoprene (IR) is a synthetic counterpart of natural rubber. Though some processing characteristics vary, finished products of the two rubbers are indistinguishable. As with natural rubber continuous-service temperature is limited to 70°C, and flexibility to −50°C or lower is attainable.

Styrene-butadiene rubber (SBR) is the major general-purpose non-oil-resistant synthetic elastomer. Many types are manufactured, with the preponderance having a chemical make-up of about 75% butadiene [($CH_2{=}CH{-}CH{=}CH_2$)] and 25% styrene [($C_6H_5CH{=}CH_2$)]. Both water resistance and electrical properties are better than those of natural rubber. On aging, SBR hardens, unlike natural rubber, which softens. Continuous-service temperature is limited to 70°C.

Butyl rubber (IIR) is a copolymer of isobutylene [$CH_2{=}C(CH_3){-}CH_3$] with a small (less than 5%) amount of isoprene. The electrical properties and resistance to ozone are very good. An outstanding property of butyl rubber is impermeability to gases, used to advantage in gaskets for gas-filled equipment and in gas retainers such as bladders and inner tubes for automobile tires. Heat resistance is superior to that of natural rubber and SBR, and 90 to 100°C applications are possible when special heat-resistant grades are specified. Degradation is evidenced by softening for most grades.

Polybutadiene rubber (BR) is most commonly used in blends with natural rubber or SBR to improve physical properties. It is more resilient than natural and imparts toughness, cut resistance, and wear properties to a blend as well as improving low-temperature flexibility. The thermal stability of blends is generally equivalent to that of the base rubber used.

Ethylene-propylene rubbers (EPR and EPDM) are general-purpose rubbers with very good thermal stability and superior resistance to ozone and weathering. Electrical properties are inherently good, as with butyl. Gas-barrier properties are similar to those of natural rubber and SBR. Resistance to steam and to hot water is an outstanding property. EPR differs from EPDM in that a third component is included in EPDM which permits sulfur vulcanization, whereas EPR must be cured by peroxides, sometimes a limiting factor in processing.

391. Group II Rubbers. *Neoprene* (CR) is a general-purpose oil-resistant rubber which has very good resistance to thermal aging. The physical properties are generally equivalent to those of natural rubber, but resistance to oxidation and impermeability to gases are superior. It is suitable for low-frequency insulation and is often used for cable covers and low-voltage conductors. It is rated as self-extinguishing in flame tests. The many types available are principally polymers of chloroprene [$CH_2{=}C(Cl){-}CH{=}CH_2$], though copolymers are used in special grades. Like natural rubber, neoprene crystallizes at low temperatures, causing loss of flexibility. The phenomenon is reversible by warming and has no permanent effect but should be considered in low-temperature sealing applications. The crystallization is strain-enhanced and is a function of both time of exposure and temperature, with the maximum rate achieved at −10°C. Crystallization-resistant grades are available for continuous low-temperature applications. Maximum continuous-service temperature is 70°C, and compounds serviceable at −40°C and lower are available.

Hypalon (CSM) is chlorosulfonated polyethylene elastomer, with outstanding weathering and abrasion resistance. It is more heat-resistant than neoprene and has similar oil resistance. Colored materials are readily made and show good color retention on exposure to weather so that color-coded wiring for commercial and military vehicles is a growing use. Electrical properties in general are similar to those of neoprene. At low temperatures (below −30°C) Hypalon stiffens and loses its elasticity. Like neoprene, it is rated as self-extinguishing. Maximum continuous-service temperature is 70°C.

Nitrile rubbers (NBR) are copolymers of acrylonitrile [$(CH_2{=}CH{-}CN)$] and butadiene [$(CH_2{=}CH{-}CH{=}CH_2)$], with acrylonitrile contents of about 18 to 50%. As the acrylonitrile content is increased, resistance to oils and solvents and hardness increase, while low-temperature flexibility decreases. Both thermal stability and abrasion resistance are very good. Mixtures of the rubber with polyvinyl chloride give very good ozone and weather resistance, which the base rubber lacks. Electrically, dissipation factor is intrinsically high. Cable covers for excessively oily locations are a major electrical use, as are gaskets for oil-filled equipment. Recommended maximum continuous-service temperature is 70°C, though oil-immersed parts may be satisfactory at a higher temperature because exposure to oxygen and light are minimized.

392. Group III Rubbers. *Polysulfide rubbers* (Thiokol) are condensation products of organic dichlorides such as ethylene dichloride ($Cl{-}CH_2{-}CH_2{-}Cl$) and sodium polysulfide, NaS_x. They are the most outstanding of the rubbers in resistance to swelling by oils, greases, and organic solvents. On storage and on aging, the polysulfide elastomers harden. Tensile and elongation values are low, and the rubbers have an objectionable odor. In diaphragms for gas meters, printer's rolls, solvent-carrying hose, and linings for chemical equipment and in sealing and caulking compounds advantage is taken of their outstanding chemical resistance. Maximum continuous-service temperature is 50°C.

Polyurethane rubbers are reaction products of polyethers or polyesters with polyisocyanates. They offer a wide range of physical properties and have the highest tensile strength (to 6,000 lb/in^2) available in elastomers. Very high hardness, resistance to abrasion, the ability to sustain high compressive loading with little deflection, and good oil resistance are among the assets of these rubbers. Steam, hot water, acids, and alkalies will cause the polyurethanes to deteriorate, but they may be used in dry applications to 120°C. Ozone has little effect. Typical applications are metal-handling rolls, die-forming pads, oil seals, and vibration-isolation and support pieces.

Acrylic rubbers (ANM) are special-purpose elastomers of which there are several types. Their service temperature of 130 to 140°C is intermediate between the general-purpose rubbers and the silicones and fluoroelastomers. Outstanding properties are resistance to sunlight, ozone, and extreme-pressure lubricants. Continued exposure to steam, hot water, acids, and alkalies is not recommended, nor is use at temperatures below −10 to −20°C. A major use is in automatic transmission seals, where high temperatures and sulfur-bearing oils rapidly cause deterioration in other rubbers, particularly those vulcanized with sulfur.

Epichlorohydrin rubbers (CHR and CHC) are a new elastomer type. Formed by polymerization of epichlorohydrin

$$(Cl{-}CH_2{-}\overset{\;}{CH}{-}CH_2) \quad \text{(epoxide O bridging CH—CH}_2\text{)}$$

or modified by copolymerization of ethylene oxide

$$(CH_2{-}CH_2) \quad \text{(epoxide O bridging CH}_2\text{—CH}_2\text{)}$$

they are polyethers as $(O—CHX—CH_2)_n$, where X may represent hydrogen or CH_2Cl groups. They offer oil and solvent resistance equal or superior to that of neoprene, very good low-temperature flexibility (in the copolymer), and gas-barrier properties superior to those of butyl rubber. They may be used continuously to 90°C. The CHR grade is rated as self-extinguishing, the CHC as slow-burning. Because these elastomers are not fully commercial at the time of writing, they are not included in the table of general properties (Table 4-78).

393. Group IV Rubbers. *Silicone rubbers* are unique in that the basic chain structure is made up of alternating silicon and oxygen atoms

$$(\overset{|}{\underset{|}{Si}}—O—\overset{|}{\underset{|}{Si}}—O)_n$$

with methyl $(CH_3—)$, phenyl $(C_6H_5—)$, or vinyl $(CH_2{=}CH—)$ groups attached to the silicon atoms. The choice of the attached groups and their ratios permits variations in properties and processing. The silicones are flexible at very low temperatures (−60 to −100°C) and are stable at 180°C, giving them the widest usable temperature range of the elastomers. Dielectric properties, ozone resistance, and solvent resistance are good. Physical properties of the silicones are generally lower than for other elastomers at room temperatures, but because they are less affected by high temperatures, they are superior above 120 to 150°C. Silicone rubbers may be processed and fabricated much like other elastomers. RTV (room-temperature-vulcanizing) types increase the utility of the silicones. Fully cured pieces may be prepared from free-flowing liquids without the need for high temperatures or pressures, making in-place preparation of gaskets and seals practicable. A modification which includes fluorinated groups attached to the silicon atoms confers improved resistance to hydraulic fluids and solvents.

394. Group V Rubbers. *Fluoro rubbers* are outstanding for both high-temperature stability and solvent resistance. They may be used continuously at 180°C and find many applications as hose liners, gaskets, O-rings, rolls, and immersed service parts, where their solvent resistance is a primary requirement. Chemically, they are similar to carbon-based elastomers except that the majority of the hydrogen has been replaced by fluorine, which confers the heat- and solvent-resistant properties. Typical components of the polymers are chlorotrifluoroethylene $[C(Cl)F{=}CF_2]$, vinylidene fluoride $[(CH_2{=}CF_2)]$, and perfluoropropylene $[(CF_2{=}CF—CF_3)]$. Low-temperature utility is limited to about −10°C, where rubbery characteristics are compromised. These elastomers are costly and will probably remain so as their manufacture requires extremes in corrosion-resistant equipment.

395. General Comments on Rubber Insulation.

a. Rubber terminology is deceptive when comparisons are made with metals. Tensile strength, hardness, elongation, and modulus values are not necessarily indicative of utility, as explained below.

b. Few applications use *rubber in tension* at more than 100 lb/in²; thus ultimate tensiles of 1,000 or 3,500 lb/in² hardly affect the rubber's ability to perform a function. Consequently, the tensile strength is more useful as a measure of quality control than as a factor in design applications. Ultimate elongation is subject to similar considerations.

c. The *hardness,* as applied to rubber products, is defined as the relative resistance of the surface to indentation by a standard indenter. Although the hardness is determined by deformation, it should be considered only as a relative measure of stiffness, which is a mass deformation rather than a surface deformation.

d. Since *stress and strain* are not proportional in rubber, the modulus values represent, not a ratio, as does Young's modulus, but the coordinates of a point on a stress-strain curve.

e. Compression-set values are useful in comparing a new material with one proved in service and can eliminate unsuitable material in short-term tests. Compression set is a measure of the retention of imposed deflection or the lack of recovery of a standard sample. The test procedure is defined in ASTM D395. The limited duration of the test as compared with expected service life, the probable different conditions of temperature, stress, and strain in service, indicates that small differences are of little significance.

f. Thermal properties of rubber include the usual thermal conductivity and the coefficient of expansion, but are complicated by the Joule effect. Thermal conductivities generally fall within the Btu/(hr) (ft^2) (in) (°F) range of 1.15 to 1.70. Coefficient of expansion varies from about 50 to 250 $\times 10^{-6}$ in/(in) (°C) and exerts strong influence on practical tolerances of molded products.

g. Stiffness of rubber increases with an increase in temperature. Under constant load rubber will contract when heated; under constant strain an increase in temperature results in higher stress. This is known as the Joule effect.

h. Low-temperature properties are both time- and temperature-dependent. Stiffening is progressive as the temperature is lowered until an abrupt change occurs known as "second-order transition," which usually begins at −20 to −40°F. The lowering of the temperature by 20°F may multiply the stiffness a hundredfold. Little effect is found on further cooling. Natural rubber, neoprene, and butyl rubber are subject to crystallization on prolonged exposure to low temperature. This effect is accentuated by strain. Crystallization results in increased stiffness but does not necessarily result in brittleness. All the low-temperature effects mentioned above are reversible by bringing the rubber to room temperature.

i. Temperature limitations of the various rubbers can be given in the terms applicable to the IEEE insulation classifications. The silicone and fluoro rubbers fall within Class 180, the acrylics within Classes 130 to 155, and the remainder of the rubbers within Class 90. Neither natural rubber nor SBR rubber should be used at temperatures exceeding 70°C continuously. However, neoprene, Hypalon, butyl, and nitrile rubbers are suitable for applications up to Class 105 temperature if heat-resistant compounds are specified.

j. Deterioration is a primary factor to be considered in selecting a rubber. The effects of sunlight, ozone, weathering, heat, oils, and chemicals should be included in the evaluation of rubber products. Rubber is inherently susceptible to degradation, but the variety of rubbers available and their differing resistance properties aid in making an optimum choice. Rubber products should be stored in a cool, dark area, away from sources of heat or ozone-generating equipment, and protected from solvents and oils.

k. The effects of *radiation* on rubbers are variable and are reviewed in Par. **396**.

l. Fungicidal agents may be incorporated in the rubber or applied as surface coatings. Rubber parts to be used in tropical climates should be protected, as fungus growth may deteriorate the rubber and reduce surface resistivity.

m. Sponge products tend to fail more rapidly than their solid counterparts because of the greater exposed area. Solvent attack, oil swelling, etc., will be more rapid and severe because of the thin cross sections of the cell walls.

n. The susceptibility of surfaces to *staining* in contact with rubber products varies with the finish, the exposure conditions, and the rubber composition. Where staining, which is migration of rubber-compound components to a surface, resulting in discoloration, is a problem, cooperative work with the supplier will usually give satisfactory results. Samples of the parts with the finish to be used as they are applied in production should be furnished to the supplier for development evaluation. In many cases, the nonstaining requirement can be met without a premium price if sufficient latitude is allowed on other properties.

o. Avoid use of high-*sulfur-content* rubbers near silver contacts to prevent tarnishing, pitting, and sticking.

p. Care should be exercised in using *silicone rubbers* near current-carrying brushes or other contacts, as even extremely low concentrations of silicone vapors around contact interfaces may lead to high contact drop, contact pitting, and excessive wear.

q. Low-friction surfaces can be imparted to rubber. While not all formulations or all polymers can be treated, many grades can be. Surface coatings of polyurethane resins have been adopted by the automotive industry for glass-to-rubber sliding seals. Also, surface treatments of chlorination and fluorination have given dramatic results. High-speed frictional-wear life of 117 h versus 4 h for untreated butyl rubber and over 100 h versus 15 sec for neoprene has been reported. For applications requiring low sliding friction such as O-rings, seals, and packings, or where a nonseize closure is required, such finishes or treatments should be considered. The insulation classification of these coatings should not be considered more than Class 105.

r. Though *water resistance* of rubber is normally assumed to be excellent, regular production of specially water-resistant synthetic rubbers indicates it to be relative. Therefore, for applications requiring extended water exposure, the specification should emphasize this property.

s. The *swelling* of rubber by solvents, water, oil, steam, etc., does not always degrade the rubber. In many gasket applications, advantage is taken of the swelling action to effect a tighter seal.

t. Aging effects vary with the different rubbers. Natural and most butyl rubber compounds degrade by softening, while most of the other types degrade by stiffening. Oxygen attack is the most common cause of failure.

u. A *surface finish* on sheet and molded items may be specified. Most expensive is a high-gloss surface free from defects. Cost decreases as slight imperfections, such as toolmarks and lack of gloss, are permitted. A dull mat finish on sheet or hose is obtained by molding against cloth, which is later removed.

v. Ring-shaped items such as gaskets are commonly made by molding or by cutting from a tube formed on a mandrel. Better uniformity and quality can usually be obtained from a molded gasket, because the tube method requires that thin sheet rubber

Table 4-78. Properties of Rubbers

Characteristic	Natural	Styrene-butadiene	Butyl	Poly-butadiene	Synthetic natural	Ethylene propylenes	Neoprene
Physical and mechanical properties:							
Tensile strength, filler reinforced	High	Medium	Medium	Medium	High	Medium	High
Elongation	Excellent	Good to excellent	Good to excellent	Good	Excellent	Good to excellent	Excellent
Hardness—Shore A	30–90	40–90	40–75	40–90	30–90	20–90	40–95
Rebound:							
Room temperature	Excellent	Good	Poor	Excellent	Excellent	Good	Good
Hot	Excellent	Good	Excellent	Excellent	Excellent	Good	Good
Low-temperature flexibility	Excellent	Good	Fair	Very good	Excellent	Excellent	Fair
Continuous operating temperature, °C, max.	70	70–90	70	70	70	90	70
Tear resistance	Good	Fair	Good	Very good	Good	Excellent	Fair
Abrasion resistance	Excellent	Good to excellent	Good	Excellent	Excellent	Good	Good to excellent
Flame resistance	Poor	Poor	Poor	Poor	Poor	Poor	Good
Gas impermeability	Fair	Fair	Excellent	Fair	Fair	Good	Good
Cut-growth resistance	Excellent	Excellent	Good	Excellent	Excellent	Good	Excellent
Electrical properties:							
Dielectric strength	High	Medium	High	High	High	High	Medium
Power factor	Low	Low-medium	Low	Low	Low	Low	Medium
Chemical stability:							
Taste impart	Medium	Medium	Medium	Medium	Medium	Medium	Medium
Relative odor	Medium	Medium	Medium	Medium	Medium	Medium	Medium
Resistance to:							
Acid	Fair to good	Fair to good	Excellent	Fair to good	Fair to good	Excellent	Good
Alkali	Good	Good	Excellent	Good	Good	Excellent	Fair to good
Water	Good	Good to excellent	Excellent	Excellent	Excellent	Excellent	Fair to good
Weather	Poor	Fair	Good to excellent	Fair	Fair	Good	Excellent
Ozone	Poor	Poor	Good to excellent	Poor	Poor	Excellent	Excellent
Oxidation	Good	Good	Good to excellent	Good	Good	Good	Good to excellent
Radiation	Good	Good	Poor	Poor	Good	Poor	Good
Aliphatic hydrocarbons	Poor	Poor	Poor	Poor	Poor	Poor	Good
Aromatic hydrocarbons	Poor	Poor	Poor	Poor	Poor	Poor	Poor
Oxygenated solvents	Good	Good	Good	Good	Good	Good	Fair
Animal and vegetable oil	Poor to good	Poor to good	Excellent	Fair	Fair	Good	Good
Swelling in ASTM lubricating oil No. 3	High	High	High	High	High	High	Medium
Swelling in transformer oil	High	High	High	High	High	High	Medium
Relative cost	Low-medium	Low	Low	Low	Low-medium	Low	Medium

Table 4-78. Properties of Rubbers.—*Concluded*

Characteristic	Hypalon	Nitrile	Thiokol	Poly-urethane	Acrylic	Silicone	Fluoro
Physical and mechanical properties:							
Tensile strength, filler reinforced	High	Medium	Low	Very high	Medium	Low	Medium
Elongation	Good	Good	Fair	Good to excellent	Fair	Fair to good	Fair
Hardness—Shore A	40–95	40–95	35–80	40–95	40–90	40–85	60–90
Rebound:							
Room temperature	Good	Good	Fair	Poor	Poor	Excellent	Fair
Hot	Good	Good	Fair	Good	Fair	Excellent	Good
Low-temperature flexibility	Fair	Fair	Good to excellent	Good to excellent	Poor	Excellent	Fair
Continuous operating temperature, °C, max.	70	70	50	120	130–140	180	220
Tear resistance	Fair	Fair	Poor	Excellent	Fair	Poor to fair	Fair to good
Abrasion resistance	Good to excellent	Good to excellent	Poor	Excellent	Good	Poor to fair	Good
Flame resistance	Good	Poor	Poor	Poor	Poor	Fair	Good
Gas impermeability	Good	Excellent	Excellent	Good	Good	Poor	Excellent
Cut-growth resistance	Good	Good	Poor	Good	Good	Poor	Poor
Electrical properties:							
Dielectric strength	High	Medium	Low	High	High	High	Medium
Power factor	Medium	High	High	Medium	Medium	Low	Medium
Chemical stability:							
Taste impart	Medium	Medium	High	Medium	Medium	Medium	Medium
Relative odor	Medium	Medium	High	Medium	Medium	Medium	Medium
Resistance to:							
Acid	Excellent	Good	Fair	Poor to fair	Poor	Fair	Good to excellent
Alkali	Excellent	Good	Fair	Poor	Poor	Poor	Poor
Water	Good to excellent	Excellent	Fair	Fair	Fair	Good to excellent	Good to excellent
Weather	Excellent	Poor to fair	Excellent	Excellent	Excellent	Excellent	Excellent
Ozone	Excellent	Fair	Excellent	Excellent	Excellent	Excellent	Excellent
Oxidation	Excellent	Fair to good	Excellent	Excellent	Excellent	Excellent	Excellent
Radiation	Good	Good	Good	Good	Poor	Good	Good
Aliphatic hydrocarbons	Good	Excellent	Excellent	Excellent	Good	Poor	Excellent
Aromatic hydrocarbons	Fair	Good	Excellent	Good	Poor	Poor	Excellent
Oxygenated solvents	Poor	Fair	Good	Poor	Poor	Fair	Poor
Animal and vegetable oil	Good	Excellent	Excellent	Excellent	Excellent	Fair to excellent	Good to excellent
Swelling in ASTM oil No. 3	Medium	Low	Low	Low	Low	Low	Low
Swelling in transformer oil	Medium	Low	Low	Low	Low	Medium	Low
Relative cost	Medium	Medium	Medium to medium plus	Medium plus	Medium plus	High	Very high

be wrapped to final thickness, confined, cured, and cut or sliced. In the wrapping, poor adhesion or air entrapment may cause ply separation.

396. Radiation Effects of Elastomers. The effects of radiation on elastomers are variable because of the influence of the various compounding agents. Several methods can be used for improving the radiation resistance of elastomers. These include the use of fillers, addition of radiation-resistant resins, and the use of organic additives called antirads. In general, carbon black fillers are superior to mineral fillers. Resins have been used to improve the radiation resistance of gum stocks but appear to have little effect on black stocks.

Natural Rubber. Irradiation of natural rubber induces cross linking and tends to decrease the elastic properties and increase the hardness. This is similar to the effects of overvulcanization, whereby natural rubber acquires a rigidity comparable with that of glass. Natural rubber is unaffected by radiation up to about 2×10^8 ergs/(g)(C). Tensile strength is not affected until the rubber is exposed to 2.4×10^9 ergs/(g)(C). Elongation and set at break are not affected up to about 5.5×10^8 ergs/(g)(C).

Polyurethane Rubber. Polyurethane rubber is not affected up to 9×10^7 ergs/(g)(C). In general, the urethane elastomers tend to soften at exposures up to 4×10^{10} ergs/(g)(C) and then become increasingly harder. The materials also tend to decrease both in tensile strength and in elongation after irradiation. Compounding appears to have little effect on the radiation resistance for most types.

Styrene-Butadiene Rubber (SBR). Styrene-butadiene rubber (SBR) resists radiation better than any of the synthetic hydrocarbon rubbers but is not equal to natural rubber in radiation resistance. Threshold damage is reached at 2×10^{8} ergs/(g)(C). The tensile strength of the material changes less rapidly than that of natural rubber. The rate of property changes for both hot SBR (polymerized at 122°F or higher) and cold SBR (polymerized at 41°F) is about the same under irradiation. Cold SBR has better initial physical properties than hot SBR, and this superiority is evident after irradiation.

Nitrile Rubber (NBR). Nitrile rubber is a copolymer of acrylonitrile and butadiene. Formulations with a high acrylonitrile content have about average radiation stability compared with other elastomers. Compression set degrades by about 25% at 7×10^{8} ergs/(g)(C). Tensile strength is variable after irradiation and increases by 25% at 1.5×10^{10} ergs/(g)(C).

Neoprene Rubber (CR). In general, the properties of neoprene (polychloroprene) rubber are similar to those of nitrile rubber after irradiation.

Hypalon (Chlorosulfonated Polyethylene) Rubber (CSM). It is difficult to predict the tensile strength of Hypalon, because the material exhibits different trends during irradiation. In some cases, the material retains nearly its original value up to 9×10^{9} ergs/(g)(C), after which it starts to increase with higher dosages. In other cases, the tensile strength increases at low doses, drops considerably at about 4.5×10^{9} ergs/(g)(C), and then starts to increase with continued exposure. Continued exposure tends to increase the hardness and decrease the elongation of Hypalon. There is evidence that stability can be improved by adding aromatic plasticizers.

Acrylic Rubber (ANM). Acrylic, or polyacrylate, rubbers are based on polymers of butyl or ethylacrylate. Both types of polymers behave similarly when irradiated. They undergo slight amounts of cross linking and chain cleavage when exposed to 9×10^{9} ergs/(g)(C). Their hardness increases with increased exposure. However, their tensile strength behaves erratically. It increases or decreases after short exposure, remains relatively unchanged for intermediate exposure, and drops and eventually increases with prolonged exposure.

Silicone Rubber. In general, the radiation resistance of silicone rubbers is below the average of other elastomers. Physical properties are damaged by 25% at exposures less than 10^{9} ergs/(g)(C).

Fluorocarbon Rubber. Fluorocarbon rubbers undergo both cross linking and chain cleavage when irradiated. Fluorocarbon rubbers exposed to 9×10^{9} ergs/(g)(C) increase about 25% in hardness and lose about 80% of their elongation. The tensile strength also varies. The stability of the fluorocarbon rubbers is quite dependent on the environment. Thus, the stability of the rubbers in air cannot be used to predict their stability in other media.

Table 4-79. Standard ASTM Tests for Elastomers

Property	ASTM test
Mechanical properties:	
Hardness	D314, D676, D2240
Tensile, elongation, set in tension	D412, D1414
Compression set	D395
Modulus	D575, D797, D945
Resistance	D945, D1054
Abrasion resistance	D394
Tear strength	D624
Electrical properties:	
Surface, volume resistivity	D257, D991
A-c loss, dielectric constant	D150
Dielectric strength	D149
Environmental properties:	
Heat	D454, D573, D865
Low temperature	D746, D797, D1053, D1229
Weathering	D518, D1171
Aging	D454, D572, D573, D865
Ozone and corona	D470, D1149
Fluid resistance	D471, D814
Radiation	D1671, D1672, D2309

Polysulfide Rubber. The hardness of polysulfide rubber does not change significantly up to 2.5×10^{10} ergs/(g)(C). However, tests conducted at 4.5×10^{10} ergs/(g)(C) have caused such great damage that the hardness could not be measured. Also, at the higher exposure, both elongation and tensile strength were reduced to zero, suggesting that the material may undergo chain cleavage. No stress cracking was observed at either exposure.

Butyl Rubber (IIR). In general, butyl rubber appears suitable for use only at relatively low radiation doses. The tensile strength of the material decreases with increasing irradiation. A damage level of 25% is reached for hardnesses at about 5×10^9 ergs/(g)(C) and for tensile strength and elongation at about 10^9 ergs/(g)(C). The material shows no evidence of stress cracking after irradiation.

397. References on Rubber Insulation

1. BATEMAN, L. "The Chemistry and Physics of Rubber-like Substances"; New York, John Wiley & Sons, Inc., 1963.
2. MORTON, M. "Introduction to Rubber Technology"; New York, Reinhold Publishing Corporation, 1959.
3. BUECHE, I. "Physical Properties of Polymers"; New York, Interscience Publishers, Inc., 1962.
4. NAUNTON, W. J. S. "The Applied Science of Rubber"; London, Edward Arnold (Publishers), Ltd., 1961.
5. WHITBY, G. S. "Synthetic Rubber"; New York, John Wiley & Sons, Inc., 1954.
6. ASTM Tests for Elastomers (see Table 4–79).

INSULATED CONDUCTORS

BY JACK SWISS

398. Magnet-wire Insulation. The term magnet wire includes an extremely broad range of sizes of both round and rectangular conductors used in electrical apparatus. Common round sizes for copper are AWG No. 42 (0.0025 in) to AWG No. 8 (0.1285 in). Ultrafine sizes of round wire, used in very small devices, range as low as AWG No. 60 for copper and AWG No. 52 for aluminum.

Approximately 26 different "enamels" are used commercially at present in insulating magnet wire. Enamel insulations generally are lowest in cost and best in space factor. The most widely used materials are based on polyvinyl acetals, polyesters, and epoxy resins. The polyvinyl acetal and polyester materials possess good mechanical properties and good flexibility and perform well in automatic winding machines. Where low cost is important and winding conditions are not too severe, oleoresinous types and modified oleoresinous types are used. Polyurethanes are employed where ease of solderability, without solvent or mechanical stripping, is required. These do not have high cut-through resistance, however. Epoxy enamels are used where resistance to chemicals and to moisture is important. Polyimide and other aromatic polymer types are employed for operation in the 200 to 220°C range.

Table 4-80 lists some of the commonly used enameled wires by temperature class. It should be understood that this temperature rating is based on a thermal test[1]* and does not include other environmental factors such as exposure to high humidity or use with a varnish which may impair its thermal stability. Cycling tests which include humidification greatly reduce the lives at temperature of many insulation systems capable of undergoing hydrolytic as well as thermal and oxidative degradation.

Table 4-81 lists some fibrous insulations commonly used for insulating magnet wire conductors.[2] Fibrous insulations are employed where positive separation and high reliability are required. These are generally higher in cost and poorer in space factor than enamel insulations.

399. Magnet-strip and foil Conductors. Magnet strip is a term generally employed to describe conductors, both copper and aluminum, with a width-to-thickness ratio greater than 50:1, while smaller ratios place the conductors in the category of rectangular magnet wire. If the thickness of the strip is less than 0.008 in, it is often referred to

* Superscripts refer to bibliography, Par. **401**.

as "foil." Strip conductors are used in many electromagnetic devices including transformers,[3] choke coils, welders, motor and generator fields, lift magnet coils, and electric clutches and brakes. Some of the advantages[4,5] of strip conductors are more uniform

Table 4-80. Some Typical Enamel-insulated Wires

Temperature class, °C	Type	NEMA Std.	Mil Spec.	Advantages
105	Acrylic	MW-4	....	Resists refrigerants, low cost
105	Nylon	MW-6	Mil-W-583C, types T_1, T_2, T_3, T_4	Excellent windability. Solderable
105	Oleoresinous (plain enamel)	MW-1	Mil-W-583C, E and E-2	Low cost
105	Polyvinyl formal (Formvar)	MW-15, MW-18	Mil-W-583C, types T_1, T_2, T_3, T_4	Excellent windability
105	Polyvinyl formal, isocyanate modified, for hermetic use (Formetic)	None	Same as above	Excellent resistance to R-22
105	Polyvinyl formal with nylon overcoat	MW-17	Same as above	Improved windability over Formvar and resistance to hot solvents
105	Polyvinyl formal with polyvinyl butyral overcoat	MW-19	None	Can be self-bonded by heat or solvent
105	Cellulose lacquer	None	None	Can be applied in thin coatings to very fine wires. Bonds by solvent activation
105	Polyurethane	MW-2	None	Solderable. Can be coated at high speeds
130	Epoxy	MW-14	Mil-W-583C, types B, B2, B3, B4	Resistance to solvents, chemicals, hydrolysis
130	Epoxy with self-bonding overcoat	None	None	Used in making coils self-supporting. Eliminates varnish dip
155–180	Polyester	MW-5, MW-13, MW-25	Mil-W-583C, Classes 155, 180, 200, depending on type and overcoat (none, polyester, nylon)	Good windability, heat shock, thermal stability
>180	Polytetrafluoroethylene	MW-10	Mil-W-583C	Good thermal stability to 250°C. Solvent resistance
220	Polyimide	MW-16	Mil-W-583C, Class 220, types M, M2, M3, M4	Excellent thermal stability, solvent resistance, flexibility, scrape resistance, cut-through
220	Polyamide-imide	None	Same as above	Somewhat lower cost than polyimide at some sacrifice in properties
220	Ceramic with polytetrafluoroethylene overcoat	MW-8	None	High cut-through resistance
180	Ceramic with silicone overcoat	MW-7	None	High thermal stability. Radiation resistant
>220	Ceramic with polyimide overcoat	None	None	High thermal stability
650	Ceramic	None	None	High thermal stability. Radiation resistant

Table 4-81. Some Typical Fibrous-covered Wires

Temperature class, °C	Type	NEMA Std.	Mil Spec.	Advantages
90–105	Paper	MW-31, MW-33	Mil-W-583C, Class 90, types P, P2	High electric strength when impregnated with oil (oil-filled transformers)
	Cotton yarn	MW-11, MW-12	Mil-W-583C, Class 90, types C, C2	Positive separation of conductors, good varnish absorption and bonding, good abrasion resistance
	Cellulose-acetate fiber	None	Mil-W-583C, Class 90, types F, F_2	Can be self-bonded by solvent activation
130	Asbestos with organic bond	None	Mil-W-583C, Class 130, type AV	High compressive strength
155	Glass fibers, organic bond	MW-41, MW-42	Mil-W-583C, Class 130, types GV, G2V	Positive separation of conductors
	Glass and polyester fibers, organic bond	MW-45, MW-46	Mil-W-583C, Class 130, types DG, DG2	Positive separation of conductors, greatly improved abrasion resistance over glass alone
180	Glass and polyester fibers, silicone bond	None	Mil-W-583C, Class 200, types GH, G2H	Positive separation. High-temperature capability
180	Asbestos with silicone bond	None	Mil-W-583C, Class 130, type AV	High compressive strength
650	Glass fibers, organic bond with dispersed ceramic filler	None	None	Windability and high thermal stability

voltage distribution under surge or impulse conditions, better heat transfer, improved space factor, and stronger coil structure. For insulation, paper and polyester film (0.0005 in or less in thickness) have been used as interleaving materials. The width of the interleave is usually about 0.125 in wider than the strip. Other available interleaving materials are asbestos, polytetrafluoroethylene film, mica, and glass cloth. The most widely used insulation is enamel, which provides the best space factor and lowest cost. Many of the enamels used for insulating wire can be used also for strip, but the most widely used are epoxy and a modified polyester type. The enamel thickness generally ranges from 0.00025 to 0.0005 in on each side, or a build of 0.0005 to 0.0010 in.

400. Wire and Cable Insulation. Many materials are used in wire and cable insulations. Some general references which review the recent state of the art are included at the end of this section (Refs. 6, 7, 8, 9, Par. **401**).

Polyvinyl chloride (PVC) is widely used for primary insulation or jacketing on communication wires, control cable, bell wire, building wire, hookup wire, fixture wire, appliance cords, power cables, motor leads, etc. Many formulations are available, including those with flame resistance. Dielectric strength is excellent, and flexibility is very good. PVC is one of the most versatile of the lower-cost conventional insulations. A conductive PVC can be used for both shielding and jacketing.

Butyl rubber, when properly compounded, is characterized by excellent resistance to oxidation and aging, exceptional ozone resistance, and very good electrical properties. Resistance to moisture and chemicals is also very good. Applications include low and high power cables, apparatus leads, and control cables. Ethylene-propylene terpolymer rubbers (EPT) are replacing butyl in some applications.

Neoprene has been used as a cable-jacketing material for more than thirty years. Its application over lead-sheathed and rubber-insulated cables has grown rapidly

Table 4-82. Some Properties of Common Wire-insulation Materials

Physical properties	Unplasticized PVC	Plasticized PVC	Silicone rubber	Nylon	TFE fluorocarbon	FEP fluorocarbon	Polyethylene	Irradiated polyolefin
Specific gravity*	1.40	1.2–1.5	1.9	1.13	2.15	2.15	0.930	1.2
Tensile strength, lb/in²	6,000–9,000	1,000–3,000	4,200	4,000–7,000	2,500	2,500	1,900–2,600	2,500
Elongation, %	2–40	200–400	....	300–600	200–300	250–330	200	250
Abrasion resistance*	Good	Good	Poor	Excellent	Good	Fair	Good	Good
Maximum continuous operating temperature, °C*	105	105	200	150	260	260	80	135
Melting point, °C	200	200	>375	300	327	275	120	Not thermoplastic
Flexibility at −180°C*	Cracks	Cracks	Cracks	....	Good	Good	Cracks	Fair
Cut-through resistance*	Good	Fair	Fair	Excellent	Fair	Fair	Good	Good
Flammability, in/min	Self-extinguishing	Self-extinguishing	10–78	Self-extinguishing	Nonflammable	Nonflammable	1.0	Self-extinguishing
Dielectric strength, V/mil (short time)	425–1,300	1,000	375	385	480	550	480	1,000
Dielectric constant, 1,000 c/s	5–10	2–4	4.2	4–10	2.0	2.1	2.3	2.6
Volume resistivity, Ω·cm	2×10^{12}	2×10^{14}	$>3 \times 10^{13}$	4.5×10^{13}	Approx. 10^{19}	$>2 \times 10^{13}$	10^{16}	$>10^{16}$

NOTE: Data compiled by Hughes Aircraft Company.
From ADAMS, H. S. *Electrotechnol.*, 1963, Vol. 72, No. 3, p. 133.
* These properties are of particular importance in aerospace applications.

during this time. Although the electrical properties of neoprene are inferior to many other insulations, they are adequate for low-voltage work.

Nitrile-butadiene rubber (NBR) offers excellent resistance to oils and solvents but has low electrical resistivity.

Polyethylene (PE) is used in wires and cables in very large amounts. Polyethylene has excellent electrical properties plus good abrasion resistance and solvent resistance (at temperatures below 50°C). It is employed for hookup wire, coaxial cable, communication cable, line wire, lead wire, high-voltage cable, etc. Chemically cross-linked filled polyethylene is growing in usage for hookup and lead wire. Properties are similar to those of conventional PE except for a marked improvement in heat resistance, mechanical properties, aging characteristics, and freedom from environmental stress cracking. Flame resistance can be provided by proper compounding. Uses include building wire, control cable, automotive wiring, and lead wire for motors and appliances. Polyethylene can also be cross-linked by irradiating the insulation on the wire. Advantages are similar to those of the chemically cross-linked material, but the process is generally limited to thin wall insulations, such as hookup wire (5 to 12 mils wall thickness). Foamed or cellular polyethylene represents a small but important part of the wire and cable insulation field. Dielectric constants of the order of 1.5 can be attained in this manner. In coaxial cables for community antenna television and closed-circuit television, the trend has been away from solid polyethylene to foamed polyethylene cable. Coaxial cables for military applications have either a solid low-density polyethylene insulation or polytetrafluoroethylene (TFE) in solid, semisolid, or tape-wrap form. PE and TFE have dielectric constants and dissipation factors which vary little over wide frequency and temperature ranges.

Polypropylene is the lightest of all plastics. It is similar to polyethylene in electrical properties but offers better heat resistance, tensile strength, and abrasion resistance. The material may be extruded, foamed, and made into cast and biaxially oriented films. Polypropylene film is being used as a cable wrap.

Fluorinated ethylene propylene (FEP) and polytetrafluoroethylene (TFE) are used in critical applications where heat resistance, solvent resistance, and reliability are important, for example, wiring in jet aircraft, military electronic equipment, and supervisory wiring for steam-turbine generators.

Polyimide film laminated to FEP film (HF film) is a heat-sealable material which offers possibilities of savings in weight and space for wire insulation. It is rated for continuous use at 200°C.

Some properties of typical wire insulation materials are shown in Table 4-82 (from Ref. 9).

For a review of insulation for integrated microelectronic circuits see Ref. 10.

401. References on Insulated Conductors

1 Tentative Method of Test for Relative Thermal Endurance of Film Insulated Round Magnet Wire, ASTM D2307–64T (based on IEEE No. 57).

2. Saums, H. L. Magnet Wire, Strip, Hollow Conductors and Superconductors; *Insulation Directory/Encyclopedia Issue,* May, 1965, pp. 332–352, Lake Publishing Corp., Libertyville, Ill.

3. Book, H. W. A New Approach to Distribution Transformer Design; *Westinghouse Engr.,* July, 1964, Vol. 24, No. 4, p. 110.

4. Magnet Strip Conductor; *Publ.* EB 38, 1965, Anaconda Wire and Cable Co, New York.

5. Edge Conditioned Aluminum Strip Conductor, *Publ.* 731–1–8 (5–665), 1965, Reynolds Metals Co., Richmond, Va.

6. Staff Report, "Wire, Cable and Assemblies," Ref. 2, pp. 353–371.

7. Staff Report, Coaxial Cables: 1966; *Electrotechnol.,* January, 1966, Vol. 77, No. 1, p. 71.

8. Noble, M. G., and Savage, R. M. A Status Report on Silicone Rubber for Wire and Cable Insulation; *Insulation,* November, 1965, Vol. 11, No. 12, p. 51.

9. Adams, H. S. Problems in Insulated Wire and Cable in Space-vehicle Systems; *Electrotechnol.,* 1963, Vol. 72, No. 3, pp. 133–135.

10. Staff Report, "Where Does Insulation Technology Stand Today for Integrated Microelectronic Circuits?"; *Insulation,* September, 1965, Vol. 11, No. 10, p. 108.

INSULATING VARNISHES

By E. J. Croop

402. Definition. Varnish is usually defined as a liquid composition, normally a solution of resinous matter in an oil or volatile solvent, which, after application to a surface, dries by either evaporation or chemical action to form a hard, lustrous coating which is more or less resistant to air, moisture, and various chemical agents. Varnishes are sometimes classified by color as "clear" or "black" and/or by functions, as coil-bonding varnish, mica-sticking varnish, or coil-impregnating varnish. Varnish does not contain pigments like paint. The dried film is usually transparent or translucent. When pigments are added to varnish, the mixture is called an "enamel."

Varnishes have been traditionally divided into two classes:

a. Oil or oleoresinous varnishes, which are essentially solutions of natural or synthetic resins or asphalts, in drying oils (especially linseed, tung, or soybean oil).

b. Spirit varnishes, which are solutions of natural or synthetic resins or asphalts in volatile solvents such as alcohol, acetone, and turpentine. They do not contain a drying oil.

Drying of varnishes takes place by (1) evaporation of the solvent, (2) oxidation of the oils, (3) polymerization of the resin and oil. It is usual to classify varnishes as either air drying or baking, depending on the method of drying.

403. Functions of Insulating Varnishes. The chief functions of insulating varnishes are the protection of fibrous insulation (such as fabrics, glass, asbestos, and wood) from mechanical damage, chemical contamination and moisture penetration, improvement of insulating properties, bonding the layers of laminated materials or conductors, imparting a surface finish, and increasing heat transfer. For properties, see Tables 4-85 and 4-86.

404. Manufacture of Varnishes. The details of actual processes used in making varnishes are in large degree held as trade secrets by the various manufacturers, although such processes are relatively simple and easily acquired by those practicing the art.

The oil varnishes are made by treatment and bodying of drying oils at elevated temperatures, then by cooking in varnish kettles a mixture of the oils and resins to attain specific chemical properties, with subsequent addition of required solvents and driers. The spirit varnishes are made by dissolving natural or synthetic resins in appropriate solvents, usually with some heating and agitation. Recently synthetic resins have been developed which may be used as varnishes without oils or solvents, drying either in air or by baking (polyesters, epoxies, etc.).

The raw materials used in manufacture of insulating varnishes consist of drying oils, resins, solvents, and driers. They are briefly described in the following paragraphs.

405. Resins Used in Varnish. A resin is a solid, semisolid, or liquid organic compound, noncrystalline in nature, with color varying from water-white to yellow or brown (usually), transparent or translucent, and soluble in ether or alcohol, etc., but insoluble in water. Chemically, resins differ widely, but all are rich in carbon and hydrogen and also contain some oxygen. Some of the more thermally stable synthetic resins may also contain some nitrogen, sulfur, etc. Resins may be divided into four groups: animal, vegetable, mineral, and synthetic.

Animal Resins. Shellac is the most important animal resin. It is a purified resin made from lac, which consists of excretions from certain insects found on trees in India and the Far East. The crude lac is known as "stick lac"; it is crushed and sifted to produce so-called "grain lac," which is then melted and strained and when cold is marketed in various shapes, such as flakes, buttons, and cakes, known respectively as "shellac," "button lac," and "garnet lac." Shellac is extensively used in making spirit varnishes. It has a resistivity of the order of 10^{15} to 10^{16} $\Omega\cdot$cm and a dielectric constant of 2.9 to 3.7 approximately. For information concerning the testing of shellac see ASTM Designations D29 and D411.

Vegetable resins are obtained from various trees. These resins have been called "copals." The group that is found in fossilized deposits is called "hard copal." Their melting points vary from 75 to 450°C. The second group, from living trees of many species, have melting points from 75 to 100°C. These are called "soft copal" (see Table 4-83).

Mineral Resins. Asphalts are sometimes called mineral resins. They are soluble in carbon disulfide and benzene, are soft at 70°C, and melt at about 100°C. Natural and mineral tars and pitches are in this group. Gilsonite is one of the best for black varnishes.

Synthetic Resins. These are formed directly from organic compounds. They are largely replacing the natural resins because of their uniformity of composition and, in comparison with older varnishes, their superior performance in imparting chemical resistance, thermal stability, and mechanical durability. The types having widest application in oleoresinous varnishes are the modified or unmodified phenolic resins

Table 4-83. Typical Commercial Vegetable Resins

Name	Type and source	Usefulness
Amber	Fossil resin, Germany	Oil varnishes
Kauri	Fossil copal, New Zealand	Oil varnishes
Pale East India	Damar, East Indies	Oil and spirit varnishes
Black East India	Damar, East Indies	Oil varnishes
Congo copal	Fossil resin, Africa	Oil varnishes
Manila	Soft copal, East Indies, Philippines	Oil and spirit varnishes
Mastic	India, North Africa	Spirit varnishes
Sandarac	Africa, Australia	Spirit varnishes
Pine resin gum "thus"	Southern United States, France, India	Oil varnishes, etc.

and the oil-modified glycerol phthalate resins or alkyds. Synthetic resins are also used in spirit varnishes and in some cases are sufficiently fluid for use alone, drying in air or by baking. Some of the more important commercial resins and trade names are: (*a*) alkyd (Amberlac, Beckosol, Dulux, Durez, Glyptal, Plaskon, Rezyl, Teglac); (*b*) ethyl cellulose; (*c*) phenolic (Amberal, Bakelite, Durez, Harvel, Super Beckacite); (*d*) urea (Beckamine, Beetle, Uformite); (*e*) vinyl (Vinylite, Formvar, Geon, Exon); (*f*) melamine formaldehyde (Melmac, Uformite, Super Beckamine); (*g*) silicone; (*h*) epoxy (Epon, Araldite, Epotuf, Epirez, DER); (*i*) urethane (Spenkel, Mondur, Nacconate); (*j*) polyesters (Alkanex, Isonel); (*k*) aromatic polymer (ML); (*l*) acrylic (Elvacite, Acryloid).

Some of the synthetic resins may be coreacted with each other to achieve a resin with final properties that are roughly intermediate between those of the original coreactants, i.e., phenolic and alkyd. See "Modern Plastics Encyclopedia" for complete listings and descriptions of commercial resins.

406. Drying oils used in varnish manufacture are classified according to their drying properties. They are naturally occurring, largely triglycerides of long-chain saturated and unsaturated fatty acids. The drying ability is a function of the amount of unsaturation and conjugated double bonds. The drying mechanism is considered a combination of oxidation and polymerization at the double bonds to form an insoluble and infusible gel. The drying oils include linseed oil and tung oil (china wood oil). The semidrying oils include cottonseed and soybean oil. The nondrying oils are castor, olive, and rosin oils.

Linseed oil is the drying oil used in by far the greatest quantity. It is derived from flaxseed and has a density of 0.932 to 0.936 at 15°C. It dries largely by oxidation to a hard gum and is a good insulator. The action is hastened by using bodied linseed oil or by adding driers or by baking. For specifications, see ASTM D234 and D260. Tung oil is obtained from the nuts of the oriental tung tree, which is now being grown in the southeastern U.S. The density is 0.938 to 0.942 at 15°C. Drying is by polymerization as well as oxidation. See ASTM Specification D12. Rosin oil is also used for impregnating paper insulation in power cables because of its penetrating power and high electric strength, although it is giving way to the use of chlorinated hydrocarbons, such as Askarel or Inerteen, for this purpose.

407. Solvents and thinners are organic liquids used to reduce the viscosity of varnish and facilitate its application. They dissolve the varnish ingredients, are present in a balanced ratio of nonvolatile solids to solvents, then disappear by evaporation after the wet film is applied. Those commonly used in varnish manufacture and application are listed in Table 4-84.

408. Driers. The drying properties of certain oils (see Drying Oils) can be increased by the addition of metallic catalysts which on heating with the oil as in a varnish promote faster polymerization or directly liberate oxygen and thus promote drying. These materials are usually metal-organic compounds of cobalt, lead, or manganese dioxide, etc. However, there are objections from the electrical point of view to the use of any of these metallic salts, even in small quantities. Fortunately, they are effective in concentrations less than 1% addition by weight. The unnecessary use of driers should be avoided.

409. Types of Insulating Varnishes for Various Applications. Catalogues of insulating-varnish manufacturers list the following general types (see Tables 4-85 and 4-86):

a. Clear baking varnish for armatures, field coils, and instruments.

b. Black baking varnish for armatures, field coils, and transformers when higher electric strength and resistance to moisture, acids, and alkalies are wanted. Less oil resistance than that of the clear.

Table 4-84. Typical Solvents and Thinners Used in Varnishes
(For an extensive listing with properties see T. H. Durrans, "Solvents"; 1957. ASTM has specifications for a number of solvents)

Solvent	ASTM Specification	Specific gravity	Flash point of closed cup	Toxicity
Acetone	D329-33	0.791–0.799	2°	Slight
Amyl acetate	D554-39	0.860–0.865	77°	Slight
Amyl alcohol	D319-40	0.812–0.820	97°	Slight
Benzene, benzol	D361-36	0.868–0.882	5°	Considerable
Benzine, petroleum spirits	....	0.69–0.77	70°	Slight
Carbon tetrachloride	....	1.600–1.608	None	Considerable
Chloroform	....	1.49–1.50	None	Considerable
DC naphtha	....	0.745–0.750	25°	Slight
Dichloroethylene	....	1.250–1.278	None	Medium
Ethyl acetate	D302-33	0.883–0.888	25°	Slight
Ethyl alcohol	....	0.790	57°	Slight
Ethylene glycol monoethyl ether	D331-35	0.927–0.933	107°	In question
Methyl alcohol, methanol	D1152	0.796	32°	Considerable
Methylated spirits, denatured alcohol	....	0.790	57°	Slight
MEK, methyl ethyl ketone	....	0.806	24°	Medium
MIBK (methyl isobutyl ketone)	....	0.805	75°	Medium
Petroleum spirits, mineral spirits	D235-39	....	100°	Slight
Toluene	D-362	0.860–0.870	41°	Considerable
Triacetin (plasticizer)	....	1.16–1.17	270°	Slight
Trichloroethylene	....	1.461–1.468	None	Considerable
Turpentine	D-13	0.860–0.875	90°	Slight
VM&P naphtha	....	0.750–0.765	53°	Slight
White spirit, mineral turpentine	....	....	78°	Slight

c. Clear or black baking varnish (internal-curing type) in which the varnish thermosets throughout the depth of the coil during the baking operation and assists in bonding the components of the coil.

d. Clear or black air-drying varnish used where baking is not convenient.

e. Clear red or black finishing varnish, usually contains shellac or synthetic resins, used for producing a harder surface over baking varnishes, protection against oil, moisture, dirt, and metal dust, and improving appearance.

f. Sticking varnish used in cementing cloth, paper, mica, etc.

g. Core-plate varnish (air drying, baking, and flashing) for insulating armature and transformer laminations. The air drying is not suitable for oil-immersed operation.

h. Epoxy resin varnish (baking) for all coil impregnation, internal curing, where superior durability and chemical and moisture resistance are required.

i. Silicone resin varnish (air drying and baking) for motor stators and rotors, transformers, coils; for high-temperature and high-humidity service.

j. Polyester resin varnishes (baking) for motor stators and rotors, transformers, coils, for high-temperature service not so severe as to require silicones.

k. Phenolic varnishes (baking) for hermetic motor coils and bonding of form-wound coils.

Table 4-85. Properties of Insulating Varnishes

(Irvington Varnish and Insulating Division of Minnesota Mining & Manufacturing Co.)

Number	Type	Specific gravity at 30°C	Percent solids	Percent solvent	Drying temp., °C	Drying time, h*	Electric strength, V/min†	Solvents
612C	Internal curing Impregnation	0.8750	55	45	121	8	2,600	VM&P Naphtha
123	Internal curing Impregnation	0.8805	50	50	121	5	1,800	VM&P Naphtha
140	Internal curing Impregnation	0.9210	50	50	121	5	2,700	VM&P Naphtha
133-F	Internal curing Impregnation	0.8805	44	56	121	3	2,200	VM&P Naphtha
9	Clear air-drying impreg. coating	0.8695	50	50	30	4	2,000	VM&P Naphtha
1201	Black air-drying impreg. coating	0.8484	41	59	30	2	800	VM&P Naphtha
30	Red air-drying sealing coating	1.0067	52	48	30	1	1,000	Xylene
2505	Clear air-drying sticking coating	1.0069	64	36	30	½	500	Denatured alcohol

* Drying time refers to internal mass drying in a deep layer coil and not to film drying time.

† Electric strength for internal curing is determined with ¼-in electrodes on varnish which is baked on 2½-mil bond paper. The air-drying-varnish electric strength is determined by test method from MIL-V-1137.

Table 4-86. Properties of Silicone Insulating Varnishes
(From *Dow Corning Corp. Tech. Data Sheets*)

Number	Type	Specific gravity at 25°C	Percent solids	Percent solvents	Drying temp., °C	Drying time, h	Electric strength, V/min	Solvent
991	Air-drying water-repellent film, heat-stable	1.03	50	50	25	2	1,500–2,000	Xylene
994	Internal-curing coating varnish	1.00–1.02	50	50	250	1	2,000–2,500	Xylene
997	Internal-curing impregnating varnish	1.00–1.03	50	50	200	3	1,500–2,000	Xylene
Sylgard 1377	Internal-curing coating impregnating varnish	1.07	45	55	200	5	1,700–2,000	Xylene
40-C	Internal-curing adhesive varnish	1.12–1.15	80	20	100	1	1,000–2,000	Xylene

410. Methods of Testing Varnish. The following specifications and methods of testing have been developed by ASTM (in referring to these standards, always consult the latest annual issue of the ASTM Index).

a. Methods of Testing Varnishes Used for Electrical Insulation (D115).

b. Method of Test for Compatibility of Glass Yarn with Insulating Varnish (D886).

c. Sampling and Analysis of Shellac (D29).

d. Methods of Testing Shellac Used for Electrical Insulation (D411).

e. Specifications for Orange Shellac and Other Indian Lacs for Electrical Insulation (D784).

f. Specifications for Shellac Varnishes (D360).

g. Methods of Testing Silicone Insulating Varnishes (D1346).

h. Thermal Endurance of Flexible Insulating Varnish (D1932).

Other test methods for varnish cover density, viscosity, flash point, time of drying, electric strength as liquid and when solid at two temperatures and after water immersion, heat flexibility, oilproof test, draining or working viscosity, nonvolatile and volatile matter. Other properties which should be determined depend on the application but may include some of the following: impregnating properties, internal drying of saturating varnish, hardness, elasticity, toughness, bonding strength, flex life, resistance to chemicals and corrosive actions, and appearance.

411. Thermal Endurance of Varnishes. It is known that the thermal stability of varnishes determined by simple test methods as well as by field experience varies considerably. ASTM D1932 and IEEE 57 test methods are generally used to determine thermal stability in the laboratory, but experience with present tests does not show that any are reliable for rating varnishes for a particular service, although they do give comparative data. Operating data are the best guide for the selection of a varnish.

412. Methods of Applying Varnishes. Two general methods are in use for treating coils, windings, and insulating parts with insulating varnishes: (1) by vacuum impregnation, and (2) by hot dipping. Finishing varnishes are usually applied by brush or spray. Mica sticking varnishes are applied by brush or sometimes by machine (as by passing a roller which dips in the varnish). Synthetic varnishes are frequently used for impregnation by dipping and require baking to develop their properties fully. Baking ovens of the continuous type are frequently used where the class of production is suited to them. Infrared lamps are used with considerable success for baking varnish-treated coils; generally they reduce the time of bake.

413. Specification for Shellac Varnish. See ASTM Standard Specifications for Shellac Varnishes (D360).

414. References on Insulating Varnishes

1. Parker, D. H. "Principles of Surface Coating Technology"; New York, Interscience Publishers, Inc., 1965.

2. Gardner, H. A. "Physical Examination of Paints, Varnishes, and Colors," 12th ed.; Washington, D.C., 1962.

3. Moses, G. L., Lee, R., and Hillen, R. J. "Insulation Engineering Fundamentals"; Lake Forest, Ill., Lake Publishing Co., 1958, pp. 22–25 and 74–78.

4. Durrans, T. H. "Solvents"; Princeton, N.J., D. Van Nostrand Company, Inc., 1957.

5. Chatfield, H. W. "Paint & Varnish Manufacture"; London, George Newnes, Ltd., 1955.

6. Gordon, P. L., and Gordon, R. "Paint and Varnish Manual"; New York, Interscience Publishers, Inc., 1955.

7. Bidlack, V. C., and Fasig, E. W. "Paint and Varnish Production Manual"; New York, John Wiley & Sons, Inc., 1951.

8. Moses, G. L. "Electrical Insulation: Its Application to Shipboard Electrical Equipment"; New York, McGraw-Hill Book Company, 1951.

9. Von Fischer, W. "Paint and Varnish Technology"; New York, Reinhold Publishing Corporation, 1948.

10. *ASTM Publ.* 310, Solvents.

11. Paint, Varnish, Lacquer, and Related Materials; Methods of Inspection, Sampling and Testing, Federal Specifications TT-P-141-b.

COATING POWDERS

By E. J. Croop

415. General. If one attempts to insulate a sharp edge or corner, as in the bare slots of a small motor stator, with a varnish or a paint, it is impossible to insulate the edges to withstand more than a few volts because surface tension causes the liquid to draw away from the sharp edges. This problem has been overcome by the development of solid coating powders, formulated to have a thixotropic character when molten, which are applied to the preheated object to be coated.

416. Coating Processes. In one method, called the fluidized-bed method, the solid coating powder, which may be thermoplastic or thermosetting, is suspended in air by an air stream blown in from the bottom of a container through a fritted glass filter or other means which distributes the air uniformly. The mixture behaves like a liquid in which air may be considered as the solvent or suspending medium. The object to be coated, usually a metallic material of relatively high heat capacity, is preheated to a temperature above the fusion point of the powdered coating material, then dipped in the fluidized bed, and withdrawn. The heat stored in the object to be coated is great enough to cause the solid particles to melt, coalesce, and flow to form a smooth continuous coating. If the powder is properly formulated and the treating conditions are properly controlled, the coating does not pull away from the edges and uniform high electric strengths are obtained.

Another technique is to apply the powders to the preheated object to be coated by electrostatic spraying. Automatic machines are used in applying coating powders to insulate the slots of small motor stators by this method. Flocking techniques are also used to apply such powders.

Table 4-87. Materials Currently Used in Coating Powders

Material	Minimum Coating Temperature, °C
Epoxy	120–150
Polyethylene	200
Polyvinyl chloride	185
Polypropylene	250
Cellulose acetate butyrate	260
Nylon-6	300
Penton (chlorinated polyether)	290
Teflon T.F.E	400
Teflon F.E.P	340
Polychlorotrifluoroethylene, P.C.T.F.E	250

417. Coating Materials. The technique is applicable to a wide variety of materials, including epoxy, polyamide, and cellulose derivatives (see Table 4-87). Epoxy coatings are used in low-voltage automotive, aircraft, and some fractional-horsepower appliance motors to replace paper and polyester film slot cells. The advantages of these coatings are lower cost and higher temperature capability over the previously used insulating materials.

418. References on Coating Powders

1. German patent No. 933,019, Process and Apparatus for the Preparation of Protective Coatings from Pulverulent, Synthetic, Thermoplastic Materials, Erwin Gemmer (Knapsack-Griesheim Aktiengesellschaft), Sept. 15, 1955.

2. Stott, L. F.: Fluidized Bed Method of Coating; *Organic Finishing*, June, 1956, Vol. 17, pp. 16–17.

3. Anon. Now: Fluidized Coatings; *Chem. Eng.*, January, 1956, Vol. 63, pp. 236–237.

4. Newman, I. A., and Bockhoff, F. J. Fluidized Plastic Coating for Corrosion Resistance; *Product Eng.*, January, 1957, Vol. 28, pp. 140–143.

5. Checkel, R. L. Fluidized Polymer Deposition; *Mod. Plastics*, October, 1958, Vol. 36, pp. 125–132.

6. U.S. patent 2,844,489, Fluidized Bed Coating Process, Erwin Gemmer (Knapsack-Griesheim Atkiengesellschaft), July 22, 1958.

7. Thielking, R. H., and McClenahan, D. L. Fluidized Coating—A Method of Slot Insulation; *Conf. Paper* CP59-471, presented at the winter general meeting, New York, Feb. 4, 1959.

8. Gemmer, E. Das Wiebelsintern-Entwicklungen and neuere Erkenntnisse; *Kunstoffe*, 1957, Vol. 47, pp. 510–512.

9. Armstrong, C. W. Fluidized-bed Processing Equipment, conference paper presented at the regional technical conference, Society of Plastics Engineers, May 22, 1959, Fort Wayne, Ind.

10. Parent, J. D., Yagol, N., and Steiner, C. S. Fluidizing Processes; *Chem. Eng. Progr.*, 1947, Vol. 43, No. 8, pp. 429–436.

11. U.S. patent 2,711,387, Treating Subdivided Solids, G. L. Matheson and H. J. Hall, June 21, 1955.

12. Elbling, I. N. Powdered Insulating Finishes; *Official Digest*, 1959, Vol. 31, No. 419, pp. 1625–1639.

13. Hate, F. L. Powder Coating; *Resin News*, 1964, Vol. 4, No. 12.

14. Sprackling, J. M. Why Use Epoxy Powder Coatings?; *Resin News*, 1964, Vol. 4, No. 12.

IMPREGNATING AND FILLING COMPOUNDS

By G. C. Gainer

419. Nature and Purpose. Two widely different classes of materials are dealt with here, viz. (1) the older impregnating and filling compounds, primarily the bitumens and waxes, which are generally melted in place, and which remain permanently heat-softening; and (2) the newer, important synthetic products of modern polymer chemistry—the solvent-reactive or so-called "solventless" resins. The latter, when properly applied (usually with resort to vacuum and pressure), can be subsequently induced to react (polymerize) *in situ* and thus provide solid, more nearly void-free insulation. The impregnation and sealing of all forms of porous insulating materials, windings, wire coverings, joints, etc., are highly essential in securing satisfactory overall insulation where appreciable voltage stresses will be encountered in service. It is the purpose in utilizing either class of materials to endeavor to provide, as nearly as possible, a void-free insulating structure to seal out moisture, chemicals, and other electrically destructive contamination throughout the useful life of the equipment. A wide variety of commercially available materials is thoroughly delineated[1]* in Ref. 1. To achieve intelligent selection from the myriad materials which are available today, attention should be directed to many properties other than initial electrical and physical properties. Some of these considerations involve such important electrical properties as long-time resistance to the destructive effects of corona. Closely allied is the property of voltage endurance (i.e., ability of the insulation system to withstand required voltage stress throughout the life expectancy of the electrical equipment). No less important are considerations involving the thermal endurance of electrical insulation systems, which frequently reflect a combination of thermal and oxidative degradation of the organic materials which make up the impregnant, whether solid or liquid. Of vital importance in all these considerations is the concept of functional evaluation, where all the affected insulating materials, combined with the active elements in an operating system, are functionally tested to destruction, under specified (usually accelerated) conditions. Thus the materials interactions which occur can reveal major flaws and shortcomings of an operating insulation system. For further information on these considerations, reference should be made to G. L. Moses et al.[2] and to F. M. Clark.[3]

420. Bitumens. The term "bitumen" includes a large number of inflammable mineral substances, consisting mainly of hydrocarbons and including the hard, solid, brittle varieties termed "asphalt"; the semisolid naphtha and mineral tars; the oily petroleum; and even the light volatile naphthas. For discussion of the evolution and meaning of this term see Ref. 3, Chap. X.

421. Bituminous Substances. The following are listed as some of the common bitumens: asphalt, asphaltite (gilsonite), asphaltic pyrobitumen (elaterite), petroleum, mineral wax (ozokerite, montan), tar (coal gas, oil gas, wood), pitch.

The natural *asphalts* follow: soft asphalts, from lakes and springs in Trinidad, etc.;

* Superscripts refer to bibliography, Par. **431**.

hard, rubbery, insoluble asphalts from mineral veins, as elaterite; hard, brittle, oil-soluble asphalts, as gilsonite; petroleum asphalts, from petroleum distillation often modified by cracking, blowing with air, or blending (see Ref. 4).

Wax is defined as any of a class of natural substances composed of carbon, hydrogen, and oxygen and consisting chiefly of esters other than those of glycerin or of free fatty acids. In this class are included beeswax, spermaceti, Chinese wax, carnauba wax, etc. (see Ref. 5). The mineral waxes, such as ozokerite, differ chemically from the true waxes in containing no oxygen. Waxes when melted are good impregnating agents, being waterproof but usually not oilproof.

Table 4-88 lists some properties of the more commonly used bitumens and waxes.

Table 4-88. Properties of Common Bitumens and Waxes

	Density, 25°C	Fusing or dropping point, °C	Elec. strength, V/mil (Monkhouse)	Dielectric constant
Gilsonite	1.05–1.10	130–190	100	
Asphalt	1.04–1.40	80–100	100–400	2.7
Ozokerite	0.85–1.00	70–100	100–140	2.1 (liquid)
Ceresine wax (purified ozokerite)	0.89	70–100	100–140	2.1
Montan wax	0.90–1.00	80–100		
Paraffin	0.85–0.95	45–80	300	1.9–2.3
Beeswax	0.96–0.97 (15°C)	62–64	250	
Rosin oil and pitch	1.08–1.15	60–100		
Halowax (Koppers)	1.19–1.80	Liquid–139		4.9–5.5

422. Bituminous insulating compounds are prepared from a multitude of different formulas, but in many, if not most, cases, their compositions are guarded as manufacturing secrets, and they are marketed under trade names. Semisolid to solid compounds capable of melting under heat are combined in many ways, often with the addition of other substances, including resins, rubber, and animal and vegetable oils and fats; animal, vegetable, and mineral waxes; mineral fillers, sulfur; etc. Such compounds are resistant to moisture, acids, alkalies, changes in temperature, and, in many cases, exposure to weather. They also have reasonably high dielectric strength ranging from 200 to 1,200 V (rms)/mil at 60 cycles.

Vacuum-impregnating bituminous compounds are used for impregnating coils and windings. The conductors are wound with muslin or, in some cases, with asbestos and then impregnated with a melted bituminous or oleoresinous composition. Electrical apparatus treated in this manner should never be immersed in insulating oil, as it will soften and dissolve the impregnating materials.

423. Filling and sealing compounds used to fill voids and seal enclosures containing live parts against the entrance of moisture should have low temperature coefficient of contraction and expansion, high flash and fire points, low dielectric loss and power factor, high dielectric strength, freedom from volatile matter, very high moisture resistance, and chemical inertness. These compounds should be tested for softening point, evaporation, melting or pouring point, flash point, burning point, characteristic at extremes of working temperature; cubical expansion, chemical activity, effect of moisture, and electrical properties (see Ref. 3; also see Plastic and Hard Filling Compounds).

424. Insulating Grease—Petrolatum. The greaselike substance left after the distillation of paraffin-base petroleum is termed "petrolatum." Physical characteristics of unadulterated petrolatum are as follows: density 0.867 to 0.880; melting point 110 to 125°F; flash point 385 to 485°F; fire point 440 to 550°F; Saybolt viscosity (seconds) 52 to 100 at 210°F. The melting point is within the normal range of cable operating temperatures, and therefore voids may occur. The best field of usefulness is in the wrapping process in making cable joints rather than as filling material (see Ref. 6, p. 141). Petrolatum is also known as "Vaseline," "petroleum jelly," and "liquid paraffin." Greases made from silicone fluids, silica, and other thickening agents have been used as potting compounds in special applications.

425. Plastic and hard filling compounds are usually made of asphalt or of pitch derived from asphaltic-base petroleum. Almost all of them are ductile rather than brittle at operating temperatures. Most hard compounds form a seal against the admission of moisture or the emission of oil, while the plastic compounds are effective only against moisture and then only partly so. The hard compounds are therefore well adapted for filling low-voltage potheads when it is necessary to seal the end of the cable; they have been used with cable voltages as high as 26 kV, but owing to the danger of voids, they are generally used only for lower voltages. The volumetric coefficient of expansion varies considerably, ranging from 0.0003 to 0.001/°C; consequently care should be exercised to prevent pocket formations during cooling. These compounds are also difficult to remove when once in place. Their physical characteristics range as follows: density 0.90 to 1.25; pouring temperature 250 to 450°F; flash point 400 to 650°F; fire point 430 to 780°F. Detailed properties are available from the Minerallac Electric Corporation (Chicago).

426. Ideal Filling Compound for "Solid" Cable. The required characteristics are: (*a*) very low coefficient of expansion; (*b*) low viscosity at impregnating temperature, increasing at normal temperatures just sufficiently to prevent bleeding during the leading (sheath-formation) process and during installations and to prevent migration to points of low elevation; solidifying point below the minimum temperature in service; (*c*) sticky and adhesive, forming a good, strong film; (*d*) low dielectric loss, high insulation resistance, and low corresponding temperature coefficients; (*e*) high dielectric strength; (*f*) chemically stable and free of all adsorbed gases and other impurities.

Formation of Voids. Characteristics in items *a* and *b* are specified for high voltages, primarily to reduce the formation of voids. During operating cycles the compound expands and contracts and may migrate. When it is cooled, voids are formed which give rise to ionization and failure.

427. Solventless Impregnating and Filling Resins. The so-called solventless resins are generally mixtures of reactive liquid monomers or solutions of viscous or solid polymers, in liquid reactive monomers. They are characterized by being essentially 100% reactive. A wide variety of commercial resins are available, which provide, when cured, thermoset (nonremeltable) products which range from rigid to flexible, depending on selection. They are produced in a wide range of viscosities. The low-viscosity materials are generally used for impregnation, the higher-viscosity products for embedment, encapsulation, and filling. During cure, polymerization occurs, with volume shrinkage and increase in density, as the polymer converts from a liquid of increasing viscosity, through a soft gel state, finally to solid. For this reason, if it is not properly processed, cracking of the substrate can occur, and mechanical stresses are set up in the embedded structure. A wide variety of mineral fillers is frequently incorporated in the solventless resin, which results, among other changes, in reduced shrinkage, reduced coefficient of thermal expansion, increased thermal conductivity, and change in electrical and physical properties of the cured product. Several widely differing classes of solventless resins are available, viz., the polyesters, epoxies, and solventless silicone resins, gels, and potting materials. (For information on the physical and electrical properties and the numerous manufacturing and design considerations in electrical application, see Harper[7] and Clark,[3] pp. 677, 749, and 716ff.)

428. Polyester Resins.[8] These resins frequently consist of a styrene solution of reactive polymer, the latter having the ester group as the recurring functional unit. They are thermosetting, usually utilize peroxide catalysts to initiate "vinyl" polymerization, and can be formulated to provide a wide degree of flexibility in the cured product. While they are low in cost, perhaps the greatest shortcoming is the high degree of shrinkage on curing (as high as 10%), which can create cracking problems. This phenomenon can usually be overcome by appropriate design modifications. Typical properties are shown in Table 4-89.

429. Epoxy Resins.[9] Epoxy resins are a class of materials which contain usually more than one epoxide (oxirane) group. They are capable of polymerizing with a number of multifunctional compounds, which are called hardeners. The various types of epoxy resins which are available commercially differ essentially in: (*a*) the source and degree of polymerization of the epoxy groups; (*b*) the type of catalyst and hardener(s) employed; and (*c*) the presence or absence of flexibilizing or modifying agents.

The most common hardeners are di- or polyamines, organic acids and anhydrides, and certain polymers such as the polysulfides and polyamides. Some epoxies can be cured at room temperature; others must be cured at some elevated temperature, commonly in the range of 100 to 150°C. Polymerization shrinkage in the unfilled resin is

Table 4-89. Properties of Typical Polyester and Epoxy Resins
(Taken from Harper, "Electronic Packaging with Resins"; New York, McGraw-Hill Book Company, 1961)

Property	Polyester resins		Epoxy resins	
	Rigid	Flexible	Unfilled resin	Silica-filled resin
Specific gravity	1.10–1.46	1.10–1.20	1.11–1.23	1.6–2.0
Tensile strength, lb/in²	6,000–10,000	800–1,800	4,000–13,000	5,000–8,000
Elongation, %	5	40–310		
Modulus of elasticity in tension, 10^5 lb/in²	3.0–6.4	....	4.5	
Compressive strength, lb/in²	13,000–36,500	....	15,000–18,000	17,000–28,000
Flexural strength, lb/in²	8,500–18,300	....	14,000–21,000	8,000–14,000
Impact strength, Izod ft·lb/in notch (½- by ½-in notched bar)	0.2–0.4	7.0	0.2–0.6	0.3–0.45
Hardness	M70–M115 (Rockwell)	84–94 (Shore)	M80–M100	M85–M120
Thermal conductivity, 10^4 cal/(s)(cm²)(°C)(cm)	4	....	4–5	10–20
Thermal expansion, 10^{-5}/°C	5.5–10	....	4.5–6.5	2.0–4.0
Resistance to heat (continuous), °F	250	250	250–600	250–600
Heat-distortion temperature, °F	140–400	....	115–550	160–550
Volume resistivity (at 50% relative humidity and 23°C), Ω·cm	10^{14}	....	10^{12}–10^{17}	10^{13}–10^{16}
Electric strength, ⅛-in thickness, V/mil:				
Short-time	380–500	250–400	400–500	400–550
Step-by-step	280–420	170	380	
Dielectric constant, 60 cycles	3.0–4.36	4.4–8.1	3.5–5.0	3.2–4.5
Dissipation (power) factor, 60 cycles	0.003–0.028	0.026–0.31	0.002–0.010	0.008–0.03
Arc resistance, sec	125	135	45–120	150–300
Water absorption (24 h, ⅛-in thickness), %	0.15–0.60	0.50–2.5	0.08–0.13	0.04–0.10

lower (typically about 4%) than with the polyesters. Adhesion to substrates is usually excellent, which assists greatly in excluding moisture. Chemical resistance is excellent. The cost is somewhat higher than for the polyesters. For a review on these resins, see Ref. 9. Typical properties are shown in Table 4-89.

430. Solventless Silicone Potting, Encapsulating, and Filling Resins. Silicone polymers are characterized by the presence of the recurring siloxane group made up of alternating silicon and oxygen atoms, to which organic groups are attached, through silicon. These resins, therefore, partake somewhat of the nature of inorganic materials and are very stable to heat (for review, see Rochow[10]). An extremely versatile line

Table 4-90. Properties of Some Solventless Silicone Insulating Resins
(Data from Dow Corning Corp.)

Property	Rigid unfilled resin	Rigid filled resin		Rubbery (RTV) resin	Dielectric gel
		Silica	Zircon		
Specific gravity	1.11	1.70	3.3	1.12	0.97
Thermal expansion (coeff.), 10^{-5}/°C	12.5 (lin.)	8.0 (lin.)	1.3 (lin.)	77.0 (vol.)	96.0 (vol.)
Thermal conductivity, cal/(s)(cm²)(°C)(cm) ×10^4	3.6	20.0	27.0	5.0	7.0
Electric strength, V/mil, ASTM D149	350–0.125 in (¼ in elec.)	350–0.125 in (¼ in elec.)	240–0.125 in (¼ in elec.)	400–0.063 in (¼ in elec.)	1,000–0.10 in (½ in ball elec.)
Volume resistivity, Ω·cm	5 × 10^{15}	4 × 10^{15}	4 × 10^{15}	0.5 × 10^{14}	5 × 10^{14}
Dielectric constant (ASTM D150), 23°C, 400 cycles	2.82	3.62	7.30	3.14	3.00
Dissipation factor (ASTM D150), 23°C, 400 cycles	0.002	0.007	0.008	0.01	0.00008

of thermally stable solventless silicone polymer systems is available. The cured products vary in consistency from the so-called dielectric gels, through rubberlike products, to rigid analogs of the polyesters and epoxies. Properties of a few of these are listed in Table 4-90.

431. References on Impregnating and Filling Compounds

1. *Insulation Directory/Encyclopedia Issue,* prepared annually by the Lake Publishing Corp., Libertyville, Ill.

2. Moses, G. L., et al. "Insulation Engineering Fundamentals"; Libertyville, Ill., Lake Publishing Corp., 1958.

3. Clark, F. M. "Insulating Materials for Design and Engineering Practice"; New York, John Wiley & Sons, Inc., 1962.

4. Neppe, S. L. The Chemistry and Rheology of Asphalt Bitumen; *Petrol. Refiner,* 1952, Vol. 31, No. 2, pp. 137–142.

5. Warth, A. H. "The Chemistry and Technology of Waxes," 2d ed.; New York, Reinhold Publishing Corporation, 1956.

6. "Underground Systems Reference Book"; National Electric Light Association, 1931.

7. Harper, C. A. "Electronic Packaging with Resins"; New York, McGraw-Hill Book Company, 1961.

8. Boenig, H. V. "Unsaturated Polyesters, Structure and Properties"; New York, American Elsevier Publishing Company, 1964.

9. Lee, A., and Neville, K. "Epoxy Resins: Their Applications and Technology"; New York, McGraw-Hill Book Company, 1957.

10. Rochow, E. G. "An Introduction to the Chemistry of the Silicones," 2d ed.; New York, John Wiley & Sons, Inc., 1951.

THERMAL CONDUCTIVITY OF INSULATION

By F. A. Yeoman

432. General. The materials used for electric insulation in general are also good thermal insulators. It follows that insulation of electric machines or electronic components which develop heat during their operation results in higher operating temperatures because the insulation hinders the escape of heat. It is very often important to utilize insulation having as high a thermal conductivity as possible owing first to the consideration that higher operating temperatures substantially reduce the life expectancy of most insulating materials and second to the fact that high operating temperatures tend to reduce the efficiency of the machine.

433. Stationary air is a very high-quality thermal insulator. Consequently the presence of very thin layers of air due to poor contact between the elements of a laminated or fibrous insulating structure profoundly reduces the overall thermal conductivity. It is largely due to this effect that taped or laminated insulations based on such components as cloth, paper, mica splittings, etc., tend to exhibit far better thermal conductivity longitudinally than transversely. Impregnation of this type of insulation with oil or resin will accomplish varying degrees of improvement, depending upon the completeness with which the air is displaced.

434. Insulating resins generally have thermal conductivities of the same order of magnitude. Any major improvements in thermal conductivity of cast or molded insulation must come from fillers used with the resin. Those fillers which have the highest thermal conductivities, such as powdered metals, silicon carbide, or graphite, are also electrically conducting in varying degrees, and their use yields compositions of low electric strength. Substantial improvement in thermal conductivity can be achieved, however, through incorporation of such electrically insulating mineral fillers as beryllia (toxic!), zircon, alumina, or silica into the resins used.

C. A. Harper[1*] has pointed out that maximum thermal conductivity in a filled resin system results from use of a filler having the highest possible thermal conductivity and at the same time permitting maximum possible filler content by volume. Fillers

* Superscripts refer to bibliography, Par. **436**.

Table 4-91. Thermal Conductivities of Insulations Based upon Natural Products

Ref.	Material	Thickness of sample, in	Direction of heat flow	Temp. range of measurement, °C	(Cal/(cm) (°C)(s) ×10⁻⁴	W/(in) (°C) ×10⁻³
4	Fish paper:					
	0.010	0.212	Trans.	20–85	4.33	4.62
	0.010	0.748	Long.	20–80	12.15	12.90
4	Paraffined fish paper, 0.007	0.211	Trans.	20–80	4.83	5.13
4	Untreated fuller board:					
	0.030	0.216	Trans.	20–80	6.28	6.67
	0.030	0.500	Long.	20–80	15.8	16.8
4	Varnished cambric (tacky):					
	0.009	0.263	Trans.	20–95	5.44	5.78
	0.009	0.694	Long.	20–100	10.46	11.13
4	Kraft paper and mica:					
	No. 312	0.220	Trans.	20–100	5.45	5.79
	No. 312	0.520	Long.	20–100	28.4	30.2
4	Pressed mica plate, 0.041 in. (white)	0.201	Trans.	20–100	6.23	6.63
4	White pine	0.519	Across grain	20–120	2.55	2.71
4	White pine	0.732	With grain	30–80	6.13	6.52
4	White oak	0.516	Across grain	20–80	4.55	4.84
4	White oak	0.754	With grain	40–70	9.44	10.03
4	Asbestos:					
	0.025-in paper	0.306	Trans.	20–100	3.75	3.99
	0.035-in cloth board	0.356	Trans.	20–80	6.85	7.28
		0.507	Trans.	20–90	19.5	20.8
4	Soapstone	0.715	Trans.	70–130	80.0	85.0
5	Electrical ceramics	.....	....	......	36–64	38–68
5	Alumina ceramics	.....	....	7–100	248–428	263–454
5	Silica glasses	.....	....	7–100	32	34
6	Air (not in motion):					
	100°C	.....	....	100	0.72	0.76
	20°C	.....	....	20	0.60	0.64

Table 4-92. Thermal Conductivities of Insulations Based upon Synthetic Resins

Ref.	Resin	Filler	Filler content		Cal/(cm) (°C)(s) ×10⁻⁴	W/(in) (°C) ×10⁻³
			Wgt. %	Vol. %		
7	Epoxy	None	..	..	1.3–3.8	1.4–4.0
2	Polyester	None	..	..	4	4
2	Solventless silicone	None	..	..	3	3
8	Polystyrene	None	..	..	2.4–3.3	2.5–3.5
8	Polyethylene	None	..	..	8	8
8	Polytetrafluoroethylene	None	..	..	6	6
8	Nylon-66	None	..	..	5.8	6.1
9	Urethane foam, 5 lb/ft³	None	..	..	1	1
9	Polyester	Mica (325-mesh)	45	24	12	13
2	Solventless silicone	Levigated alumina (3-6 microns)	70	41	18	19
2	Solventless silicone	Powdered glass	74	58	8.8	9.3
3	Epoxy	Silica (325-mesh)	60	39	18.3	19.4
2	Solventless silicone	Flint sand (40-mesh)	83	69	27	28
3	Epoxy	Alumina (325-mesh)	77	48	34.0	36.0
9	Epoxy	Tabular alumina	80	53	24.5	26.0
3	Epoxy	Tabular alumina (20 to 30-mesh)	60	45	58.8	62.4
		325-mesh alumina	26	20		
9	Polyester	30-mesh aluminum*	80	62	60	64

* Included for comparison only. Low electric strength would preclude use of this material as electrical insulation.

whose individual particles are in the form of fibers or leaflets yield high-viscosity systems at low-filler-volume contents, with the consequence that high-filler-volume-content systems are not usable. Incorporation of filler systems involving a considerable range of particle sizes generally permits higher filler content by volume within a workable viscosity range. Obviously acceptable particle size and density of fillers in resin-filler systems are limited by tendency of the filler to separate from the resin during storage.

An alternative technique for securing very high volume-filler content involves filling the volume to be insulated with a very coarse filler such as 40-mesh sand and subsequently impregnating the sand with unfilled resin.[2] Not only are very high-filler-volume contents accessible by this technique, but the problem of filler separation is circumvented. Still higher filler contents have been obtained by impregnation of a coarse, dry filler with a resin containing a finely divided suspended filler,[3] but a drawback is the long time required for impregnation with the relatively viscous filled resin. When a dry filler aggregate of large particle size is to be impregnated, best results can be obtained by introducing the resin at the bottom of the system while evacuating from the topside. The resin is introduced slowly in order to prevent channeling as the resin level rises and residual air is displaced toward the top. Impregnation from the topside provides no means for controlling channeling of the resin, with the result that residual air will be trapped in the system.

435. Thermal-conductivity Values. Tables 4-91 and 4-92 present thermal-conductivity values for some representative materials.

436. References on Thermal Conductivities

1. Harper, C. A. Thermally Conductive Cast Resin Compounds for Heat Dissipation; *Electrotechnol.*, April, 1961, Vol. 67, No. 4, pp. 148–152.

2. Nelson, M. E., Christensen, D. F., Yeoman, F. A., and Nixon, D. R. Potting Compound for Canned Motor Pumps; *AIEE Conf. Paper* 62–438, 1962.

3. Colletti, W., and Rebori, L. High Thermal Conductivity Casting Compounds Developed; *Insulation,* January, 1965, Vol. 11, No. 1, pp. 27–30.

4. Taylor, T. S. The Thermal Conductivity of Insulating and Other Materials; *Elec. Jour.*, December, 1919, pp. 526–532.

5. Comparisons of Materials; Thermal Conductivity; *Mater. Design Eng.*, Mid-October, 1965, Vol. 62, No. 5, p. 20.

6. Moses, G. L. "Electrical Insulation"; New York, McGraw-Hill Book Company, 1951, Chap. 1.

7. Hirsch, H., and Koved, F. Thermal Conductivity of Epoxy Resin Systems; *Mod. Plastics*, October, 1964, Vol. 42, No. 2, p. 134.

8. "Technical Data on Plastics"; Washington, D.C., Manufacturing Chemists' Association, Inc., 1957.

9. Harper, C. A. "Electronic Packaging with Resins"; New York, McGraw-Hill Book Company, 1961, Chap. 8.

STRUCTURAL MATERIALS

By Thomas J. Dolan

DEFINITIONS OF PROPERTIES OF STRUCTURAL MATERIALS

437. Stress is the intensity at a point in a body of the internal forces or components of force that act on a given plane through the point. Stress is expressed in force per unit of area (pounds per square inch, kilograms per square millimeter, etc.). There are three kinds of stress: tensile, compressive, and shearing. Flexure involves a combination of tensile and compressive stress. Torsion involves shearing stress. It is customary to compute stress on the basis of the original dimensions of the cross section of the body, though "true stress" in tension or compression is sometimes calculated from the area at the time a given stress exists, rather than from the original area.

438. Strain is a measure of the change, due to force, in the size or shape of a body referred to its original size or shape. Strain is a nondimensional quantity but is frequently expressed in inches per inch, etc. Under tensile or compressive stress, strain is measured along the dimension under consideration. Shear strain is defined as the tangent of the angular change between two lines originally perpendicular to each other.

439. A stress-strain diagram is a diagram plotted with values of stress as ordinates and values of strain as abscissas. Diagrams plotted with values of applied load, moment, or torque as ordinates and with values of deformation, deflection, or angle of twist as abscissas are sometimes referred to as stress-strain diagrams but are more correctly called "load-deformation diagrams." Six stress-strain diagrams are shown in Fig. 4-79, where curve I is typical of normalized high-carbon steel; curve II is typical of low-carbon ductile steels, which have a yield point shown at Y; and curve III is typical of some of the nonferrous alloys; curve IV represents a heat-treated alloy steel; curve V is typical for a gray cast iron; and curve VI shows approximately the type of curve obtained for timber. The stress-strain diagram for some materials is affected by the rate of application of the load, by cycles of previous loading, and again by the time during which the load is held constant at specified values; for precise testing, these conditions should be stated definitely in order that the complete significance of any particular diagram may be clearly understood.

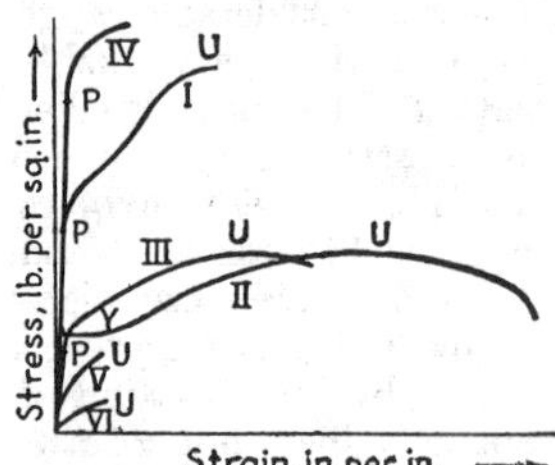

Fig. 4-79. Typical stress-strain diagrams for tensile stress.

440. The modulus of elasticity is the ratio of stress to corresponding strain below the proportional limit. For many materials the stress-strain diagram is approximately a straight line below a more or less well-defined stress known as the "proportional limit." As there are three kinds of stress, there are three moduli of elasticity for a material, i.e., the modulus in tension, the modulus in compression, and the modulus in shear. The value in tension is practically the same, for most ductile metals, as the modulus in compression; the modulus in shear is only about 0.36 to 0.42 of the modulus in tension. The modulus is expressed in pounds per square inch (or kilograms per square millimeter) and measures the elastic *stiffness* (the ability to resist elastic deformation under stress) of the material.

441. Elastic Strength. To the user and the designer of machines or structures one significant value to be determined is a *limiting stress below which the permanent distortion of the material is so small that the structural damage is negligible and above which it is not negligible.* The amount of plastic distortion which may be regarded as negligible varies widely for different materials and for different structural or machine parts. In connection with this limiting stress for elastic action a number of technical terms are in use; some of them are:

a. Elastic Limit. The greatest stress which a material is capable of withstanding without a permanent deformation remaining upon release of stress. Determination of the elastic limit involves repeated application and release of a series of increasing loads until a set is observed upon release of load. Since the elastic limit of many materials is fairly close to the proportional limit, the latter is sometimes accepted as equivalent to the elastic limit for certain materials. There is, however, no fundamental relation between elastic limit and proportional limit. Obviously the value of elastic limit determined will be affected by the sensitivity of apparatus used.

b. Proportional Limit. The greatest stress which a material is capable of withstanding without a deviation from proportionality of stress to strain. The statement that the stresses are proportional to strains below the proportional limit is known as **Hooke's law.** Proportional limits for the metals in Fig. 4-79 are located at the points P; however, the numerical values of the proportional limit are influenced by methods and instruments used in testing and the scales used for plotting diagrams.

c. Yield point is the lowest stress at which marked increase in strain of the material occurs *without increase in load.* It is indicated at point Y on the stress-strain curve II in Fig. 4-79. If the stress-strain curve shows no abrupt or sudden yielding of this nature, then there is no yield point; e.g., curve I in Fig. 4-79 exhibits no yield point. Wrought iron and low-carbon steels have yield points, but most metals do not, es-

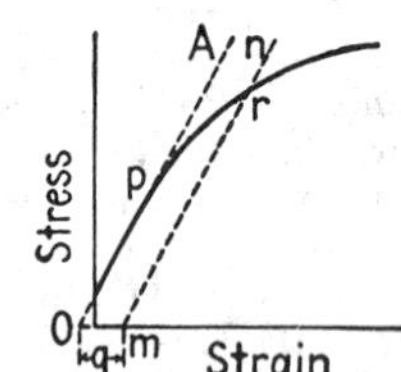

Fig. 4-80. Yield strength of a material having no well-defined yield point.

pecially those steels containing more carbon and those which have been plastically deformed at temperatures below the critical range (cold-worked).

d. Yield strength is the stress at which a material exhibits a specified limiting permanent set. Its determination involves the selection of an amount of permanent set that is considered the maximum amount of plastic yielding which the material can exhibit, in the particular service condition for which the material is intended, without appreciable structural damage. A set of 0.2% has been used for several ductile metals, and values of yield strength for various metals are for 0.2% set unless otherwise stated. On the stress-strain diagram for the material (see Fig. 4-80) this arbitrary set is laid off as q along the strain axis, and the line mn drawn parallel to OA, the straight portion of the diagram. Since the stress-strain diagram for release of load is approximately parallel to OA, the intersection r may be regarded as determining the stress at the yield strength. The yield strength is generally used to determine the elastic strength for materials whose stress-strain curve in the region pr is a smooth curve of gradual curvature. See discussion in ASTM Designation E6.

442. Ultimate strength (**tensile strength or compressive strength**) is the maximum stress which a material will sustain when slowly loaded to rupture. Ultimate strength is computed from the maximum load carried during a test and the original cross-sectional area of the specimen. For materials that fail in compression with a shattering fracture, the compressive strength has a definite value, but for materials that do not fracture, the compressive strength is an arbitrary value depending on the degree of distortion which is regarded as indicating complete failure of the material. In tensile tests, the nominal stress at rupture for many materials, especially those having appreciable plasticity, is less than the ultimate strength. For such materials the ultimate strength corresponds to the point of maximum stress on the stress-strain curve (see points U, Fig. 4-79); beyond that point the stress may decrease somewhat with increasing strain until rupture occurs.

443. Shearing strength is the maximum shearing stress which a material is capable of developing. The general remarks in Par. **442** regarding methods of failure are also applicable to failures in shear. Owing to experimental difficulties of obtaining true shearing strength, the values of modulus of rupture in torsion are usually reported as indicative of the shearing strength.

444. Modulus of rupture in flexure (or torsion) is the term applied to the computed stress, in the extreme fiber of a specimen tested to failure under flexure (or torsion), when computed by the arbitrary application of the formula for stress with disregard of the fact that the stresses exceed the proportional limit. Hence the modulus of rupture does not give the true stress in the member but is useful only as a basis of comparison of relative strengths of materials.

445. Ductility is that property of a material which enables it to acquire large permanent deformation and at the same time develop relatively large stresses (as drawing into a wire). Though ductility is a highly desirable property required by almost all specifications for metals, the quantitative amount needed for structural applications is not entirely clear but probably does not exceed about 3% elongation after the structure is fabricated (see *Proc. ASTM*, Vol. 40, p. 551). The commonly used measures of ductility are:

a. Elongation is the ratio of the increase of length of a specimen, after rupture under tensile stress, to the original gage length; it is usually expressed in percent. The percentage of elongation for any given material depends upon the gage length, which should always be specified.

b. Reduction of area or contraction of area is the ratio of the difference between the original and the fractured cross section to the original cross-sectional area; it is usually expressed in percent.

c. Bend test measures the angle through which a given specimen of material can be bent, at a specified temperature, without cracking. In some cases the maximum angle through which the specimen can be bent around a certain diameter or the number of bendings back and forth through a stated angle are measured. In other cases the elongation in a given gage length across the crack on the tension side of the bend specimen is measured. See ASTM Standard E16.

446. Plasticity is that quality of a material which permits it to be molded or to assume permanent deformations under loads without recovery of the strain when the loads are removed. Plastic materials deform instead of fracturing under load.

447. Brittleness is defined as the ability of a material to fracture under stress with little or no plastic deformation. Brittleness implies a lack of plasticity.

448. Resilience is the amount of strain energy (or work) which may be recovered from a stressed body when the loads causing the stresses are removed. Within the elastic limit the work done in deforming the bar is completely recovered upon removal of the loads; the total amount of work done in stressing a unit volume of the material to the elastic limit is called the **modulus of resilience.**

449. Toughness is the ability to withstand large stresses accompanied by large strains before fracture. The toughness is usually measured by the total work done in stressing a unit volume of the material to complete fracture and may be interpreted as the total area under the stress-strain curve (Fig. 4-79). Ductility differs from toughness in that it deals only with the ability of the material to deform, whereas toughness is measured by the energy-absorbing capacity of the material.

450. Impact Resistance. The ability of a material to resist impact or energy loads without permanent distortion is measured by the modulus of resilience. The ultimate resistance to impact before fracture is measured by the toughness of the material. For members with abrupt changes of section (holes, keyways, fillets, etc.), the resistance to a rapidly applied load depends greatly on the "notch sensitivity" (the resistance to the formation and spread of a crack); above certain critical velocities of loading and below certain critical temperatures, the impact strength is greatly reduced. Relative notch sensitivity under repeated loads is not the same as that in a single-blow notched-bar test. Impact values are influenced by speed of straining, shape and size of specimen, and type of testing machine. Impact tests are sometimes applied to steel rails to detect brittleness.

Charpy or **Izod** impact bend tests measure the energy required to fracture small notched specimens (1 cm square) under a single blow. These tests are sometimes used to detect brittleness induced by improper heat-treatment, excessive cold working, or low-temperature service conditions. Test values are markedly influenced by speed of straining, testing temperature, and shape of notch. See Symposium on Impact Testing, *ASTM STP*176, 1956, and *ASTM* Standard E23.

451. Hardness is the resistance which a material offers to small, localized plastic deformations developed by specific operations such as scratching, abrasion, cutting, or penetration of the surface. Hardness does not imply brittleness, as a hard steel may be tough and ductile. The standard Brinell hardness test is made by pressing a hardened steel ball against a smooth, flat surface under certain standard conditions; the Brinell hardness number is the quotient of the applied load divided by the area of the surface of the impression. A different method of test is employed in the Shore scleroscope, in which a small, pointed hammer is allowed to fall from a definite height onto the material, and the hardness is measured by the height of the rebound, which is automatically indicated on a scale. The Rockwell hardness machine measures the depth of penetration in the metal produced by a definite load on a small indenter of spherical or conical shape. The Vickers hardness test is similar to the Brinell except that a pyramid-shaped diamond indenter is used. (For further data see ASTM Standards E10, E18, and E92; also Metals Handbook, ASM, 1948, pp. 93–105.)

452. Fatigue strength (**fatigue limit**) is a limiting stress below which no evidence of failure by progressive fracture can be detected after the completion of a very large number of repetitions of a definite cycle of stress. The fatigue limits usually reported are those for completely reversed cycles of flexural stress in polished specimens. For stress cycles in which an alternating stress is superimposed on a steady stress the endurance limit (based on the maximum stress in the cycle) is somewhat higher. Most ferrous metals have well-defined limits, whereas the fatigue strength of many nonferrous metals is arbitrarily listed as the maximum stress that is just insufficient to cause fracture after some definite number of cycles of stress, which should always be stated. The fatigue strength of actual members containing notches (holes, fillets, surface scratches, etc.) is greatly reduced and depends entirely on the "stress-raising" effect of these discontinuities and the sensitivity of the material to the localized stresses at the notch.

Severe damage is caused by the simultaneous action of repeated stress and corrosion even with mild corrosive agents. When tested in reversed bending under a stream of fresh water, with a speed of 1,450 c/min of stress, the **corrosion-fatigue** limits of nearly all steels and alloy steels tested generally fall within the limits of 10,000 to 25,000 lb/in^2, even though their fatigue limits in air range from 22,000 to 108,000 lb/in^2. These values are based on about 20,000,000 cycles of stress; for a larger number of stress cycles, a more active corroding medium, or a slower frequency, the corrosion-fatigue limits would be distinctly less. Most of the commonly used protective coatings offer only slight or temporary protection against corrosion fatigue, and metals that are resistant to stressless corrosion may be greatly damaged by corrosion fatigue. (See ASME Handbook, "Metals Engineering—Design," and "Metal Fatigue," by Sines and Waisman, Par. **494.**)

453. Composition and Structure. Chemical analysis of a material is employed to determine whether or not certain useful constituents are present in sufficient quantities and certain undesirable elements are kept below the specified limits. The mere presence of certain elements in a material does not reveal the nature of its structure or indicate how such constituents affect its properties. Microscopy and photography are employed for the investigation of structure and require carefully polished and etched surfaces for examination. X-ray photography is useful for detecting defects such as blowholes, faults, and slag inclusions in metal articles and for investigating crystalline structure. Magnetic analysis is another method of detecting irregularities in iron and steel. Macroscopy is useful for full-scale investigation of structure and for detection of local areas which should be subjected to more detailed microscopic examination. Sections of completed members are often deep-etched by boiling in acids to reveal cracks, inclusions, surface defects, and general grain flow in wrought steels.

454. Permanence of Mechanical Properties. The strength properties of most structural materials subjected to service conditions at ordinary temperatures do not vary with time. However, some nonferrous materials creep at ordinary temperatures, and metals operating at elevated temperatures may undergo changes in structure that may materially alter their physical properties. Most metals (particularly ductile steels) tend to develop increased strength and hardness and corresponding loss of ductility with time after they have been plastically deformed. This process is referred to as **aging** after **cold working** or overstressing. Cold working in one direction, i.e., tension, usually lowers the elastic strength of a steel when tested in the reverse direction (compression).

455. Aging is a spontaneous change in properties of a metal after a heat-treatment or a cold-working operation. Aging tends to restore the material to an equilibrium condition and to remove the unstable condition induced by the prior operation and usually results in increased strength of the metal with corresponding loss of ductility. The fundamental action involved is generally one of precipitation of hardening elements from the solid solution, and the process can usually be hastened by slight increase in temperature.

456. Corrosion Resistance. There is no universal method of determining corrosion resistance, because different types of exposure ordinarily produce entirely dissimilar results on the same material. In general the subject of corrosion is rather complicated; in some cases corrosive attack appears to be chiefly chemical in its nature, while in others the attack is by electrolysis. Owing to the great diversity of materials exposed to corrosive influences in service and the wide range of service conditions, it is impracticable to formulate any universal measure of corrosion resistance. If the service life is likely to be determined by corrosion resistance, the degree of impairment which marks the end of usefulness will ordinarily be established by considerations of safety and reliability or perhaps of appearance. Corrosion testing is conducted in general by two methods: (*a*) normal exposure in service with periodic observations of corrosive action as it progresses under such conditions; (*b*) some type of artificially accelerated test, which may serve merely to obtain comparative results or, again, may simulate the conditions of service exposure. For specific information see H. H. Uhlig, "Corrosion and Corrosion Control," New York, John Wiley & Sons, Inc., 1963, and ASTM STP290, Twenty Year Atmospheric Corrosion Investigation of Zinc-coated and Uncoated Wire and Wire Products, 1961.

457. Powder Metallurgy. Many alloys and metallic aggregates having unusual and very valuable properties are being produced commercially by mixing metal powders, pressing in dies to desired shapes, and sintering at high temperatures. Parts may be produced to close dimensional tolerance, and the process enables the mixing of dissimilar materials which will not normally alloy or which cannot be cast because of insolubility of the constituents. Wide use of powder metallurgy is made in producing copper-molybdenum alloys for contact electrodes for spot welding, extremely hard cemented tungsten carbide tips for use in metal cutting tools, and copper-base alloys containing either graphite particles or a controlled dispersion of porosity for bearings of the "oilless" or oil-retaining types. Silver-nickel and silver-molybdenum alloys (tungsten or graphite may be added) for contact materials having high conductivity but good resistance to fusing can be produced by the method. Powdered iron is being used to manufacture gears and small complex parts where the savings in weight of metal and machining costs are able to offset the additional cost of metal and processing in the powdered form. Small Alnico magnets of involved shape which are exceedingly difficult to cast or machine can be produced efficiently from metallic powders and require little or no finishing. Solid mixtures of metals and nonmetals, such as asbestos, can be produced to meet special requirements. The size and shape of powder particles, pressing temperature and pressure, sintering temperature and time, all affect the final density, structure, and physical properties. For further details see C. G. Goetzel, "Applied and Physical Powder Metallurgy," New York, John Wiley & Sons, Inc., 1950.

STRUCTURAL IRON AND STEEL

458. Classification of Iron. The kinds of iron employed for structural purposes are wrought iron, ingot iron, cast iron, and malleable iron.

459. Wrought iron is a ferrous material, aggregated from a solidifying mass of pasty particles of highly refined metallic iron, with which is incorporated, without subsequent fusion, a minutely and uniformly distributed quantity of slag (ASTM Designation A81). Refined wrought-iron bars under $\frac{5}{8}$ in diameter (ASTM Designation A189) have a minimum tensile strength of 48,000 lb/in^2, minimum yield point of 0.6 of the tensile strength, and minimum elongation in 8 in of 25% (also see ASTM specifications for other wrought-iron products). The normal range of tensile strength is from about 45,000 to 55,000 lb/in^2; Young's modulus, from 25×10^6 to 29×10^6 lb/in^2. The structural uses for wrought iron include pipe, boiler tubes, stay bolts, engine bolts, chain, and applications where toughness, ductility, and resistance to fracture under shock are important. It is preferred by some users for forging and welding (see bibliography, Par. **494**).

460. Ingot iron is a highly refined *steel* made in the basic open-hearth furnace, consisting of almost pure iron and containing a maximum of 0.15% of total impurities. Commercially, however, it goes by the name of iron, and its properties closely resemble those of wrought iron and the mildest carbon steel. It is distinguished from wrought iron by the absence of the slag inclusions, which constitute the chief difference between wrought iron and very mild steel. The average tensile properties of Armco ingot-iron plates are as follows: tensile strength 46,000 lb/in^2; yield point 32,000 lb/in^2; elongation in 8 in, 30%; Young's modulus 29×10^6 lb/in^2. The structural uses include plates, pipe, tubing, sheets, rivets, etc. (see also ASTM Designation A345).

461. Cast Iron. The first product obtained in the reduction of iron ore to metal is known as "pig iron." Cast iron is made by remelting pig iron in a cupola, air furnace, or electric furnace and casting the molten metal in molds of the desired shape of the finished article. Cast iron derives its characteristic qualities from the nonferrous constituents present, which aggregate from about 5 to 8%. These constituents are the same as those in steel, except for graphite, which is one form of carbon. The carbon is always present in two forms: (1) combined carbon, in the form of cementite (Fe_3C); (2) uncombined carbon, or graphite. The total carbon seldom exceeds 3.75% or falls below 2.5%. The larger the proportion of carbon in the combined state, the harder and more brittle will be the metal. Silicon is usually a desirable constituent, because it tends to precipitate carbon in the graphite form; the majority of gray-iron castings contain 1.25 to 2.75% silicon, whereas white and malleable cast irons usually contain less than 1.25% silicon. Sulfur has the opposite effect of silicon and is undesirable.

Manganese increases the total carbon and also the proportion of combined carbon but tends to neutralize the similar effect of sulfur. Phosphorus, if present in sufficient proportions to be chemically active, tends to hold the carbon in combined form and also to weaken the metal. Rapid cooling of the casting tends to cause combined carbon, but slow cooling tends to increase the graphite. When the composition is properly adjusted, it is possible by rapid cooling to retain the carbon as cementite or, by slow cooling, to precipitate it as graphite. When the carbon is largely precipitated, the iron is soft and presents a dull gray fracture, from which it derives the name of "gray cast iron." When the carbon is retained in combined form, the cast iron is very hard and brittle and presents a silvery-white fracture, from which it derives the name of "white cast iron." When these two characteristics are intermingled, the metal is termed "mottled cast iron." (See Withey and Washa, Chap. 28, Par. **494** of this section; and Cast Metals Handbook.)

462. Gray cast iron usually contains 2% or more graphite and less than 1.5% combined carbon. The graphite is mechanically intermingled throughout the body of the metal; although the percentage by weight may be no more than 4%, the percentage by volume may be as much as 14%. The smaller the proportion of combined carbon, the larger will be the proportion of graphite and the softer and more workable the metal. At the same time this produces a metal of minimum strength and greater machinability, because of the free graphite, which acts in a degree as a lubricant in machining operations. The tensile strength ranges from 12,000 lb/in^2 for soft coarse-grained irons to 35,000 lb/in^2 for hard, close-grained irons. The crushing strength ranges from 35,000 to 200,000 lb/in^2; machinery gray iron of good quality will have a crushing strength of 90,000 to 150,000 lb/in^2. The modulus of rupture of a bar of 1.25 in diameter and 15 in long, on a 12-in span, will range from 1.5 to 2.25 times the tensile strength in solid rectangular sections. The stress-strain diagram is not linear, but at a stress of 10,000 lb/in^2 the value of Young's modulus for good gray iron ranges from 12×10^6 to 15×10^6 and occasionally as high as 20×10^6 lb/in^2. The strength in shear is not definitely known but is greater than the tensile strength. Tests in torsion produce tensile failures at stresses ranging from about 11,000 to 33,000 lb/in^2, with a value of at least 25,600 lb/in^2 for good machinery iron. Cast iron under torsion fails through weakness in tension, with a range of modulus of rupture (in torsion) of about 30,000 to 45,000 lb/in^2 and a shearing modulus of elasticity of about 6.4×10^6 to 8.2×10^6 lb/in^2. (See Withey and Washa, Chap. 28; also see ASTM Specifications A48 and A126.)

463. White Cast Iron. The constituents of white cast iron are pearlite, graphite, and cementite. Practically all the carbon is retained in the combined form (cementite), and this imparts the characteristic hard and brittle qualities. It is difficult to machine and resistant to abrasion and has few direct industrial applications, except where it is employed to give a hard coating or surface, as in chilled castings. White cast iron of suitable composition can, however, be rendered useful by the malleable process described in the following paragraph.

464. Malleable Cast Iron. White cast iron of suitable composition can be rendered somewhat malleable and ductile, and greatly toughened, by annealing at a bright red heat for 3 to 6 days in a suitable packing material such as mill scale and siliceous slag. The iron before annealing should have all its carbon in combined form. This heat-treatment transforms the combined carbon into a special type of graphitic carbon known as "temper" carbon. Malleable cast iron is considerably superior to gray cast iron in ductility and strength but inferior to steel castings. The average composition and properties of air-furnace malleable iron are as follows: carbon 1.75 to 2.30%; silicon 0.85 to 1.20%; manganese, under 0.40%; phosphorus, under 0.20%; sulfur 0.06 to 0.15%; density 7.15 to 7.45; coefficient of expansion 12×10^{-6}/°C; specific heat, average 0.122 cal/(g)(°C) between 20 and 100°C; thermal conductivity at 50°C, 0.145 g·cal/(sec)(cm cube)(°C); tensile strength 54,000 lb/in^2; yield point in tension 36,000 lb/in^2; elongation in 2 in, 18%; tension modulus 25×10^6 lb/in^2; shearing strength 48,000 lb/in^2; yield point in shear 23,000 lb/in^2; shear modulus 12.5×10^6 lb/in^2; modulus of rupture in torsion 58,000 lb/in^2; Brinell hardness range 110 to 145; Charpy impact 16.5 ft·lb. The so-called "higher strength malleable iron" has an average tensile strength of 57,600 lb/in^2, yield point of 37,900 lb/in^2, and elongation in 2 in of 23%. Cupola malleable iron has average properties, in three grades as listed in Table 4-93.

Table 4-93. Average Properties of Three Grades of Cupola Malleable Iron

Properties	No. 1	No. 2	No. 3
Tensile strength (lb/sq in.)	49,700	43,000	43,000
Yield strength (lb/sq in.)	41,000	33,000	31,000
Elongation in 2 in.	8.1	7.0	6.5

See "American Malleable Iron, A Handbook," Malleable Founders' Society, Cleveland, 1944; Withey and Washa (Par. **494**); ASTM Specification A47; Cast Metals Handbook.

465. Nodular cast iron has the same carbon content as gray iron; however, the addition of a few hundreds of 1% of either magnesium or cerium causes the uncombined carbon to form spheroidal particles during solidification instead of graphite flakes. Strength properties comparable with those of steel (130,000 lb/in^2 tensile strength, 90,000 lb/in^2 yield strength) may be achieved in the pearlitic iron. The softer, ferritic, and pearlitic as-cast irons, with yield strengths of 60,000 to 80,000 lb/in^2 and ultimate strengths of 80,000 to 100,000 lb/in^2, exhibit considerable ductility, 10% elongation or more. As the hardness and strength are increased by appropriate heat-treatment or the thickness of the casting decreased below approximately ¼ in, the ductility decreases. An austenitic form of nodular iron may be obtained by adding various amounts of silicon, nickel, manganese, and chromium. For many purposes nodular iron exhibits properties superior to those of either gray or malleable cast iron. For more complete information see Metals Handbook (Par. **494**), 1954 supplement.

466. Chilled cast iron is made by pouring cast iron into a metallic mold which cools it rapidly near the surfaces of the casting, thus forming a wear-resisting skin of harder material than the body of the metal. The rapid cooling decreases the proportion of graphite and increases the combined carbon, resulting in the formation of white cast iron.

467. Centrifugally cast pipe is made by several processes. A rapidly rotating, water-cooled, steel mold receives molten metal from a horizontal spout, and the metal freezes in the mold under rotation, which imparts a dense structure. It is also possible to obtain a laminated structure, consisting of an outer layer similar to malleable iron, a central layer of a steel matrix, and an inner layer of gray cast iron.

468. Alloy cast iron contains specially added elements in sufficient amount to produce measurable modification of the physical properties. Silicon, manganese, sulfur, and phosphorus, in quantities normally obtained from raw materials, are not considered alloy additions. Up to about 4% silicon increases the strength of pure iron; greater content produces a matrix of dissolved silicon that is weak, hard, and brittle. Cast irons with 7 to 8% silicon are used for heat-resisting purposes and with 13 to 17% silicon form acid- and corrosion-resistant alloys, which, however, are extremely brittle. Manganese up to 1% has little effect on mechanical properties but tends to inhibit the harmful effects of sulfur. Nickel, chromium, molybdenum, vanadium, copper, and titanium are commonly used alloying elements. The methods of processing or of making the alloy additions to the iron influence the final properties of the metal; hence a specified chemical analysis is not sufficient to obtain required qualities. Heat-treatment is also employed on alloy irons to enhance the physical properties. (See "Alloy Cast Irons," 2d ed., Chicago, American Foundrymen's Association, 1944; Cast Metals Handbook, American Foundrymen's Association, 1944, pp. 362–371, 515–539.)

469. Nickel Cast Iron. The presence of nickel in cast iron causes increased precipitation of carbon in graphitic form but also tends to harden the iron by sorbitizing the pearlite matrix. Grades of gray iron carrying 0.50 to 0.85% combined carbon will in general be hardened, strengthened, and toughened by additions of 1 to 5% nickel. Nickel also tends to refine the grain, increase the hardness, and improve the resistance to wear. (See bulletins of International Nickel Co. on nickel cast iron.)

470. Nickel-copper-chromium cast iron is a corrosion-resistant and heat-resistant material having a range of composition as follows: nickel 12 to 15%; copper 5 to 7%; chromium 1.5 to 4.0%; carbon 2.75 to 3.10%; silicon 1.25 to 2.00%; manganese 1.0 to 1.5%. The alloy of normal composition (14 Ni, 6 Cu, 2 Si) is nonmagnetic and has an electrical resistivity of about 80 times that of copper. Its character with respect to electrical properties is essentially that of the austenitic, high-nickel alloys. By special heat-treatments or manipulation of its composition, it may be made magnetic or nonmagnetic, without changing its basic property of resistance to corrosion. Its tensile strength ranges from about 18,000 to 23,000 lb/in^2; coefficient of expansion, about $19 \times 10^{-6}/°C$; Brinell hardness 110 to 140. The uses for it include pipe, valves, fittings, pumps, compressors, boiler specialties, impellers, marine castings, etc. (See also *Bull.* 208, International Nickel Co.)

471. High-strength Cast Irons. This term is sometimes employed to indicate cast irons ranging above about 35,000 lb/in^2 tensile strength. Rapid progress in the manufacture of cast irons has enabled the production of tensile strengths of 60,000 to be obtained in the as-cast condition. The percentage composition of one cupola iron having a tensile strength of about 60,000 lb/in^2 is: total C, 2.60; Si, 2.15; S, 0.08; P, 0.08; Mn, 0.70; Ni, 1.10. For irons having strengths above 50,000 lb/in^2 the total carbon is usually kept below 3%, and alloying elements such as nickel, chromium, and molybdenum are usually added to increase strength and to help offset the decreased machinability of the higher-strength irons. Foundry difficulties are encountered in producing the higher-strength irons, and generally only relatively small, thin castings are produced with such high strengths. High-strength irons require extremely close metallurgical control. For detailed classification of gray-iron castings on a basis of tensile strengths ranging from 20,000 to 60,000 lb/in^2 see ASTM Specification A48 and "Metals Properties"; ASME (Par. **494**).

472. Density of Cast Iron. The density of cast iron varies over a considerable range and tends to decrease as the proportion of graphitic carbon increases. Approximate values are: pure iron 7.86; white cast iron 7.60; mottled cast iron 7.35; light gray cast iron 7.20; dark gray cast iron 6.80.

473. Thermal Properties of Cast Iron. Thermal properties vary somewhat with the composition and the proportions of graphitic carbon. The average specific heat from 20 to 110°C is 0.119; thermal conductivity, 0.40 watt/(cm^3) per (°C); coefficient of linear expansion, 0.0000106/(°C) at 40°C.

474. Classification of Steel. Steels are usually broadly classified as either plain carbon steels or alloy steels, though recently there has been developed a new group commonly called "low-alloy high-strength" steels. Plain or simple carbon steel is a binary alloy of iron and carbon, in which carbon is the chief element employed to control the mechanical properties, although a little manganese and silicon are usually present and also small amounts of phosphorus and sulfur as impurities. An alloy steel contains added elements (other than carbon) in amounts sufficient to produce marked changes in its mechanical properties. The low-alloy structural steels each contain combinations of small amounts of several alloying elements which are intended to produce a slight increase of strength of the steel in the as-rolled condition without the necessity of further heat-treatment. They are designed to replace structural grades of plain carbon steel in applications where an increased strength is desirable at small additional cost. Steels are further classified by methods of processing as cast, hot-rolled, cold-drawn or cold-rolled, quenched, tempered, etc. Widely divergent mechanical properties are obtained by these different methods of processing, and further variations result from the effect of size (thickness) of section and of its shape, i.e., plate, bar, rod, rolled shapes. The most widely used systems of designating chemical analyses of steels are the American Iron and Steel Institute (AISI) and the Society of Automotive Engineers (SAE) classifications, in which identifying numbers are assigned to each group of alloys.

475. Classification of Steels by SAE or AISI Numbers. See Table 4-94.

476. Plain carbon steels cover a wide range of composition within the following approximate limits: carbon 0.03 to 1.20%; manganese 0.10 to 1.50%; sulfur 0.005 to 0.12%; phosphorus 0.005 to 0.10%; silicon 0.05 to 0.50%; iron, remainder. Additions of about 0.10 to 0.30% copper are being made to some steels to enhance the cor-

Table 4-94. Classification of Steels*

Number	Name	Carbon (C)	Manganese (Mn)	Phosphorus (P) max.	Sulfur (S) max.	Silicon (Si)	Nickel (Ni)	Chromium (Cr)	Molybdenum (Mo)	Vanadium (V)
C1010	Plain carbon steel	0.08–0.13	0.30–0.60	0.040	0.050	†				
C1020		0.18–0.23	0.30–0.60	0.040	0.050	†				
C1045		0.43–0.50	0.60–0.90	0.040	0.050	†				
C1095		0.90–1.03	0.30–0.50	0.040	0.050	†				
C1137	Free-cutting steel	0.32–0.39	1.35–1.65	0.040	0.08–0.13	†				
B1113		0.13 max.	0.70–1.00	0.07–0.12	0.24–0.33	†				
A1330	Manganese steel	0.28–0.33	1.60–1.90	0.040	0.040	0.20–0.35				
A2317		0.15–0.20	0.40–0.60	0.040	0.040	0.20–0.35	3.25–3.75			
A3130	Nickel-chromium steel	0.29–0.33	0.60–0.80	0.040	0.040	0.20–0.35	1.10–1.40	0.55–0.75		
E3316		0.14–0.19	0.45–0.60	0.025	0.025	0.20–0.35	3.25–3.75	1.40–1.75		
A4037	Molybdenum steel	0.35–0.40	0.70–0.90	0.040	0.040	0.20–0.35	....	....	0.20–0.30	
A4130		0.28–0.33	0.40–0.60	0.040	0.040	0.20–0.35	....	0.80–1.10	0.15–0.25	
A4340		0.38–0.43	0.60–0.80	0.040	0.040	0.20–0.35	1.65–2.00	0.70–0.90	0.20–0.30	
A4620		0.17–0.22	0.45–0.65	0.040	0.040	0.20–0.35	1.65–2.00	....	0.20–0.30	
A4817		0.15–0.20	0.40–0.60	0.040	0.040	0.20–0.35	3.25–3.75	....	0.20–0.30	
A5120	Chromium steel	0.17–0.22	0.70–0.90	0.040	0.040	0.20–0.35	....	0.70–0.90		
A5150		0.48–0.53	0.70–0.90	0.040	0.040	0.20–0.35	....	0.70–0.90		
E52100		0.95–1.10	0.25–0.45	0.025	0.025	0.20–0.35	....	1.30–1.60		
A6120	Chromium-vanadium steel	0.17–0.22	0.70–0.90	0.040	0.040	0.20–0.35	....	0.70–0.90	....	0.10 min
A6150		0.48–0.53	0.70–0.90	0.040	0.040	0.20–0.35	....	0.80–1.10	....	0.15 min
A9260	Silicon-manganese steel	0.55–0.65	0.70–1.00	0.040	0.040	1.80–2.20				
A8650	Nickel-chromium molybdenum steel	0.48–0.53	0.75–1.00	0.040	0.040	0.20–0.35	0.40–0.70	0.40–0.60	0.15–0.25	
A8720		0.18–0.23	0.70–0.90	0.040	0.040	0.20–0.35	0.40–0.70	0.40–0.60	0.20–0.30	
A8740		0.38–0.43	0.75–1.00	0.040	0.040	0.20–0.35	0.40–0.70	0.40–0.60	0.20–0.30	
A86B45‡		0.43–0.48	0.75–1.00	0.040	0.040	0.20–0.35	0.40–0.70	0.40–0.60	0.15–0.25	
E9310		0.08–0.13	0.45–0.65	0.025	0.025	0.20–0.35	3.00–3.50	1.00–1.40	0.08–0.15	
A9840		0.38–0.43	0.70–0.90	0.040	0.040	0.20–0.35	0.85–1.15	0.70–0.90	0.20–0.30	

* Scheme of Numbering. The first digit of the identifying number denotes the characteristic alloying element (or elements) as follows: 1, plain carbon steels; 2, nickel; 3, chromium and nickel; 4, molybdenum; 5, chromium; 6, chromium and vanadium; 7, tungsten; 8, nickel-chromium-molybdenum; 9, silicon and manganese. The last two digits show the approximate percentage of carbon (in hundredths of 1%). The remaining intermediate digits show the approximate content of the characteristic alloying element. In addition to the numerals, the AISI standard employs the following prefixes to each designation to indicate the manufacturing process employed in producing the steel: A denotes a basic open-hearth alloy steel; B is acid Bessemer carbon steel; C is basic open-hearth carbon steel; D is acid open-hearth carbon steel; E is electric-furnace steel. The letter B between the first and last two digits indicates that boron is present.

† Silicon content not specified; it ranges from 0.10 to 0.30%.

‡ Boron content, 0.0005% min.

rosion-resistant properties of the plain carbon steel without affecting its other physical properties, and similarly about 0.2% lead is being added to some steels to improve their machinability. The total number of commercial steels within this range is quite large, and the mechanical properties of the steels are influenced by heat-treatment as well as by chemical content. In general, the tensile strength and hardness of a carbon steel increase in certain proportions to the contents of carbon, manganese, and phosphorus, whereas the ductility decreases as the strength is increased. The tensile properties of carbon steel are greatly modified by cold working or heat-treatment, and the range of possible modification generally increases with increased carbon content. Average strengths of some typical carbon steels are shown in Table 4-97; for more complete information on other steels the bibliography in Par. **494** should be consulted.

477. Alloy Steels. When alloying ingredients (in addition to carbon) are added to iron to improve its mechanical properties, the product is known as an alloy steel. Heat-treatment is a necessary part of the manufacture and use of alloy steels; only through proper quenching and tempering can the full beneficial effects of the alloys be obtained. The chief advantages obtained from the addition of alloys to steel are (*a*) to alter the critical temperatures, thus making it possible to produce more uniform properties throughout thick sections with a minimum of distortion in quenching; (*b*) to form chemical compounds which when properly distributed develop desirable properties in the steel, i.e., extreme hardness, corrosion or heat resistance, high strength without excessive brittleness.

Common alloying elements and the lowest content at which they produce appreciable effect are:

Element	%	Element	%
Manganese	1.00	Molybdenum	0.25
Silicon	0.30	Tungsten	1.00
Nickel	0.50	Vanadium	0.10
Chromium	0.50	Aluminum	0.50

The general effects and uses of various alloying elements are shown in Table 4-95.

Mechanical properties of the alloy steels vary over a wide range depending upon size, composition, and heat-treatment. Typical chemical analyses of some commonly used alloy steels are shown in Table 4-94; and approximate strengths of several alloy steels after specific heat-treatments are given in Table 4-97. (See also E. C. Bain, "Functions of the Alloying Elements in Steel," American Society for Metals, 1939; and J. Mitchell, Relative Effects of Elements on Alloy Steels, *Metal Progr.*, July, 1942.)

478. Low-alloy high-strength steels include 30 or more steels that have been developed because they exhibit somewhat greater strengths than the commonly used structural grades of steel without the necessity of heat treatment or the addition of large amounts of expensive alloying elements. Each steel contains small amounts of two to six of the alloying elements listed in Table 4-95 and usually relatively low carbon content. For these steels the range of percentage of each alloying element is approximately as follows: C, 0.08 to 0.40; Mn, 0.10 to 1.75; Si, 0.05 to 1.00; Cu, 0.10 to 1.50; Ni, 0 to 2.00; Mo, 0 to 0.6; P, 0.01 to 0.20; Cr, 0 to 1.50; V, 0 to 0.14; the total alloy content of these steels is usually less than 3 or 4%. Most of the low-alloy steels have tensile strengths (in the hot-rolled condition) ranging from about 70,000 to 95,000 lb/in^2, the yield points are approximately 0.6 to 0.8 of the tensile strength, and the ductility is about the same as or slightly less than that of the structural grades of steel. Many of the low-alloy steels are being extensively used to lighten the necessary dead load in movable structures (such as railroad cars, buses, mine equipment, and shovel buckets). They offer increased strength and a slight improvement in corrosion resistance at a moderate increase in cost over that of the plain carbon steels. (See J. B. Austin, "Trends in the Metallurgy of Low-alloy, High-yield strength Steels," ASTM, 1963; see also ASTM Designation A242.)

479. Values of modulus of elasticity for ferrous metals may be assumed approximately as shown in Table 4-96. The values for all steels are fairly constant, whereas for cast irons the modulus increases somewhat with increased strength of material.

Table 4-95. Effect of Alloying Elements in Steel
(Based on article by R. H. Aborn in *Iron Age*, Mar. 5, 1925)

Alloying element	General influence	Definite effect	Uses and beneficial effects	Disadvantages
Aluminum (Al)	Deoxidizes	"Quiets" molten steel, facilitates escape of gases	Up to content of 0.05% standard remover of gases	Excess of Al tends to cause formation of graphite
Chromium (Cr)	Forms a very stable carbide	Gives great hardness; makes quenching effective to considerable depth	Cutting tools, ball bearings, rolls. A content of 13–18% Cr gives "stainless" iron and steel, and steels serviceable at high temperatures. With Ni, Cr is used for general purpose alloy steels. Adds hardness and strength	With certain heat treatments danger of brittle steel
Manganese (Mn)	Gives fine-grained crystalline structure. Deoxidizes, desulphurizes	Removes oxygen and sulphur from steel. A content of 10–12 % Mn gives an "austenitic" amorphous structure with great resistance to wear	Up to 2% Mn as deoxidizer and desulphurizer, and to increase strength. 10–12% Mn for crusher jaws and other mining and milling equipment, safes, frogs, switches, special rails	Steel with high Mn content extremely difficult to machine
Molybdenum (Mo)	Similar to Cr and Ni	Next to carbon the most effective hardening element	General purpose alloy steels (with Ni, Cr or V). Allows effective control of heat treatment	Expensive, variable quality, volatilizes from surface layers on rolling. Tendency to seams and brittleness
Nickel (Ni)	Gives fine-grained crystalline structure	Up to 4% increases strength with little loss of ductility. Makes quenching effective to greater depths in plain carbon steel	General-purpose alloy steels. Makes possible heat treatment of fairly large pieces. Adds strength with little loss of ductility	Scale formed in rolling; rough surface
Silicon (Si)	Deoxidizes. Coarsens crystalline grain	Up to 1.75% Si elastic strength is increased with little loss of ductility	Structural steel, spring steel (with Mn), standard deoxidizer	Silico-manganese spring steel lacking in toughness
Tungsten (W)	Forms extremely stable carbide. Gives fine-grained crystalline structure	Increased strength accompanied by brittleness. Quenching effect not so deep as for Cr	Cutting tools, permanent magnets. Allows very effective control of heat treatment	Brittleness of tungsten steel unfits it for use as structural steel
Vanadium (V)	Deoxidizer. Gives fine-grained crystalline structure	Deoxidizing and hardening agent. Gives high elastic and tensile strength	Spring steel and fine alloy steel (with Cr). Forms fluid slag giving "clean" metal. Checks grain growth	Expensive

Table 4-96. Approximate Modulus of Elasticity for Ferrous Metals

Metal	Modulus in tension-compression, millions of lb/in.²	Modulus in shear, millions of lb/in.²
All steels	30.0	12.0
Wrought iron	27.0	10.8
Malleable cast iron	23.0	9.2
Gray cast iron, ASTM No. 20	15.0	6.0
Gray cast iron, ASTM No. 60	20.0	8.0

Alloy steels have practically the same modulus as plain carbon steels unless large amounts, say 10%, of alloying material are added; for large percentages of alloying elements the modulus decreases slightly.

480. Heat-treatment of steel is an operation, or combination of operations, involving heating and cooling of a metal or an alloy in the solid state for the purpose of obtaining certain desirable conditions or properties. The operations usually consist of heating the steel to a predetermined temperature for a time interval sufficient to change the crystalline structure of the steel and then cooling in a medium (such as water, oil, air, molten salts, or metals) found suitable to produce the rate of cooling which readjusts the crystalline structure to the desired form. **Hardening** consists in heating and quenching certain iron-base alloys from a temperature either within or above the critical temperature range, within which the pattern of atomic structure is changed. For carbon steels the lower critical temperature is 1330°F, and the upper critical temperature ranges from 1670°F for low-carbon steel to 1330°F for a steel of about 0.8% carbon. **Quenching** implies a relatively rapid rate of cooling. **Normalizing** consists in heating above the critical range followed by cooling in still air. **Annealing** involves heating followed by a very *slow* cooling. Annealing at high temperatures may be used to soften a steel preparatory to machining, but *stress relieving* may be achieved at temperatures below the critical range for steel. Many carbon steels are commonly used in the normalized condition or in a cold-drawn state, whereas alloy steels are usually given a heat-treatment consisting of a hardening and tempering. The tempering temperature is used to control the final mechanical properties, and Fig. 4-81 shows the approximate variation of properties of a typical alloy steel for various tempering temperatures. It has been found (Metals Handbook, 1954 supplement, p. 12; see also Par. **494**) that many of the SAE structural alloy steels are substantially equivalent on the basis of mechanical properties as determined by a tensile test, provided that (1) the cross sections are so chosen that the steels are hardened throughout on quenching; (2) the various steels are tempered to the same tensile strength, irrespective of the tempering temperature necessary to produce that tensile strength; and (3) a tensile strength of 200,000 lb/in² is not exceeded.

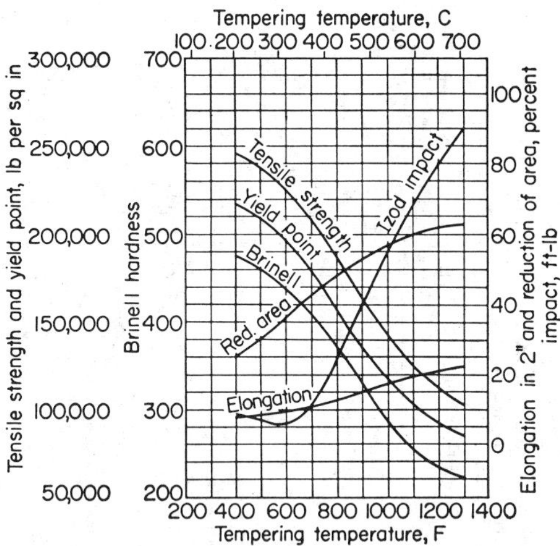

FIG. 4-81. Mechanical properties of SAE 3140 (in small sizes, ½- to 1½-in diameter or thickness). Quenched from 1475 to 1525°F, in oil. Tempered as indicated.

481. Molybdenum steels usually contain manganese and chromium or nickel and are characterized by the following properties: toughness, good depth hardening on quenching, good machinability, minimum susceptibility of temper brittleness, maintenance of mechanical properties at elevated temperatures better than other steels of low-alloy content. Chromium-molybdenum steels of the SAE 4100 series (see Table 4-97) and nickel-molybdenum steels of the SAE 4600 series are widely used in airplane construction and in highly stressed machine parts. Carbon-molybdenum steels, both rolled and in the form of castings, are used where necessity for welding or some other condition precludes the presence of chromium and have excellent creep-resisting properties at elevated temperatures. (For detailed information see "Molybdenum in Steel," New York, Climax Molybdenum Co.)

482. Mechanical Properties of Iron and Steel. See Table 4-97.

483. Manganese Steels. Structural or pearlitic manganese steels contain slightly

Table 4-97. Approximate Mechanical Properties of Iron and Steel

(Based on test data from various materials testing laboratories)

Metal	Strength in tension, psi		Strength in compression, psi		Yield[a] strength in shear, psi[b]	Endurance limit for reversed bending stress, psi	Brinell hardness No.[c]	Elongation in 2 in., %
	Yield[a]	Ultimate	Yield[a]	Ultimate				
Gray cast iron:								
ASTM 20	[d]	20,000	[d]	80,000	[e]	9,000	[h]	Less than 1
ASTM 35	[d]	35,000	[d]	125,000	[e]	15,000	[h]	Less than 1
ASTM 60	[d]	60,000	[d]	145,000	[e]	24,000	[h]	Less than 1
Gray cast iron with 1.15% nickel	[d]	50,000	[d]	156,000	[e]	20,000	[h]	Less than 1
Malleable cast iron	30,000	50,000	30,000	[f]	16,000	25,000	110	10
Commercial pure iron, annealed	19,000	42,000	19,000	[f]	12,000	26,000	69	48
Commercial wrought iron, as rolled	30,000	50,000	30,000	[f]	18,000	25,000	100	35
Structural steel, as rolled, and SAE 1020 steel, as rolled	35,000	60,000	35,000	[f]	21,000	30,000	120	35
SAE 1040 steel, water quenched, drawn at 1050 F	87,000	102,000	87,000	[f]	52,000	57,000	210	23
SAE 1095 steel, oil quenched, drawn at 850 F	97,000	188,000	97,000	[f]	58,000	98,000	380	10
SAE 2340 steel, oil quenched, drawn at 1200 F	91,000	112,000	91,000	[f]	54,000	67,000	248	24
Oil quenched, drawn at 400 F	174,000	282,000	174,000	[f]	96,000	112,000	488	8
SAE 3325 steel, oil quenched, drawn at 700 F	128,000	139,000	128,000	[f]	70,000	68,000	291	18
SAE 4140 steel, water quenched, drawn at 1100 F	116,000	140,000	116,000	[f]	63,000	64,000	250	16
SAE 5150 steel, oil quenched, drawn at 800 F	210,000	235,000	210,000	[f]	115,000	90,000	455	13
SAE 6120 steel, water quenched, drawn at 1100 F	130,000	164,000	130,000	[f]	72,000	92,000	350	16
SAE 9260 steel, oil quenched	100,000	158,000	100,000	[f]	60,000	62,000	240	16
AISI 8650, 1 in. diam., oil quenched, drawn at 1000 F	158,000	170,000	158,000	[f]	[h]	[h]	350	14
AISI E 8740, 1 in. diam., quenched from 1525 F in oil tower, drawn at 1100 F	134,000	149,000	134,000	[f]	[h]	[h]	302	18.3
AISI E 9310, 1 in. diam., oil quenched from 1425 F, drawn at 300 F	118,000	145,000	118,000	[f]	[h]	[h]	302	15.5
AISI 9840, 1¼ in. diam. oil quenched from 1525 F, drawn at 800 F	205,000	220,000	205,000	[f]	[h]	[h]	430	12
18-8 stainless steel, 18% chromium, 8% nickel, water quenched	33,000	75,000	33,000	[f]	18,000	35,000	140	55
Steel casting, 0.35% C, 1.71% Mn, annealed	60,000	104,000	60,000	[f]	35,000	45,000	188	22
Steel casting, 0.25% C, 0.68% Mn, annealed	43,000	77,000	43,000	[f]	24,000	35,000	136	30
Cold-drawn steel rod, 0.20% C	60,000	80,000	60,000	[f]	36,000	38,000	150	18
Drawn wire, iron or soft steel	70,000	85,000	[g]	[g]	40,000	[h]	[h]	[h]
High-carbon steel wire	150,000	275,000	[g]	[g]	80,000	[h]	[h]	[h]

[a] Yield strength taken as yield point, or at 0.2% nominal set, see Par. **441**.
[b] Accurate data on ultimate strength in shear not available.
[c] See Par. **451** for description of Brinell hardness test.
[d] No well-defined yield strength.
[e] Shearing yield strength greater than tensile yield strength.
[f] For ductile metal the ultimate in compression is only slightly greater than yield strength.
[g] Wire can offer resistance only in tension and in shear.
[h] Data lacking.

more manganese than the plain carbon steels (usually about 1.00 to 2.00% Mn) and are made in carbon contents of 0.10 to 0.50%. The additional manganese produces a direct increase in hardness and strength of the steel and an indirect beneficial effect due to its action as a scavenger in counteracting the harmful effects of oxides, gases, and sulfur in the steel. Hadfield found that 2.5 to 6% manganese made a steel very brittle but that steel containing about 13 to 14% manganese (called "austenitic manganese steel") was practically nonmagnetic, very tough, and exceedingly hard and wear-resistant upon cold working. It is almost impossible to form this steel with machine tools, since the slight cold working of the surface produces extreme hardness (450 to 550 Brinell). It is usually cast in the approximate form in which it is to be used; heating to 1800 to 2000°F and quenching in water increase the toughness and work-hardening capacity. Austenitic manganese steel is often used where both heavy pressure and severe abrasion are encountered in service; the kneading effect produces a hardening of the surface that is resistant to wear or abrasion. Typical applications of cast high-manganese steel include railway frogs and crossings, shovel dippers and teeth, sprockets, and cover plates for electric lifting magnets. Its electrical resistance is about seven times that of pure ion; thermal conductivity is about one-sixth that of iron; coefficient of thermal expansion, about 1½ times that of iron. (See ASTM Designation A128. Also see E. E. Thum, The New Manganese Alloy Steels, *Proc. ASTM*, 1930, Vol. 30, Pt. II, pp. 215–240, for steels with manganese content up to about 1.50%.)

484. Vanadium steels usually contain less than about 0.2% vanadium, an addition that acts as a powerful deoxidizer and alloying element promoting a fine-grained, strong, tough steel with good deep-hardening characteristics upon quenching. Carbon-vanadium steel is used to a considerable extent for large castings and forgings of machines such as crankshafts, springs, locomotive parts, and gears, which are said to maintain their dimensions without warping caused by internal stresses during machining operations. Vanadium accentuates the benefits derived from other alloying elements such as manganese, chromium, or nickel, and it is used in a variety of quaternary alloys containing these elements. (See SAE 6120, Table 4-97.) Vanadium in amounts of 0.15 to 2.50% is an important element in a large number of tool steels. (For detailed data see "Vanadium Steels and Irons," New York, Vanadium Corporation of America, 1950.)

485. Tungsten steels are used for tools, dies, and applications requiring extreme hardness and wear resistance. Formerly a magnet steel containing about 5% tungsten was widely used, but it has been superseded by cheaper chromium steels and the magnetically stronger cobalt steels and other complex alloys. Some high-carbon tool steels may contain as little as 0.5% tungsten, whereas up to 20% tungsten is used in some high-speed tool steels. Some tungsten is used in steels to promote high strength at elevated temperatures, and nearly all tungsten steels also contain chromium. Owing to brittleness, tungsten steels are not widely used for stress-carrying parts of machines and structures.

486. Silicon Steel. Silicon is normally present in basic open-hearth steels in relatively insignificant amounts, but appreciable proportions of it are present in acid open-hearth steels, and it may also be added as an alloying element. When added in small percentages, silicon increases the tensile strength and elongation; however, more than 2% silicon decreases ductility, and more than 4% causes the tensile strength to fall off. A steel containing 3.40% silicon, 0.21% carbon, and 0.29% manganese has a tensile strength of 106,000 lb/in²; elastic ratio 0.74; elongation 11%; reduction of area 14%; annealed strength 87,000 lb/in². The silicon content in structural steel is not usually in excess of about 2% as a maximum, but in electrical sheets it is present in proportions up to about 4 to 5%; above the latter figure it causes serious difficulty in hot rolling. Silicon is used in electrical sheets, because it substantially increases the electrical resistivity and thus reduces the core loss; such sheets are much harder than ordinary dynamo sheets of soft steel (see Magnetic Materials in this section). The thermal conductivity of 4% silicon steel, between 20 and 250°C, is 0.32 W/(cm cube) (°C). Iron-silicon alloy containing 12 to 14% of silicon is resistant to special forms of corrosion and is attacked extremely slowly by mineral acids. This alloy is allied to cast iron, comparatively brittle, so hard as to be machinable with difficulty if at all, and cannot be forged or rolled.

487. High-silicon Structural Steel. The influence of silicon on the mechanical properties of steel is similar to that of carbon but proportionately less intensive; its distinctive effect is the tendency to increase the elastic ratio. A steel containing about 0.12% carbon, 0.40 to 0.70% manganese, and about 0.95% silicon as rolled has a tensile strength of about 70,000 lb/in², yield point of about 50,000 lb/in², elongation of about 25% in 8 in, and 60% reduction of area. This elastic ratio of about 70% compares with 60% for ordinary structural (carbon) steel. The silicon content of silicon structural steel will range from a minimum of 0.20% to a maximum of slightly more than 1%, with carbon not in excess of 0.40% and manganese not usually over 0.70%. For additional information see Johnson's "Materials of Construction"; ASTM Specification A94.

488. Nickel Steel. The average effect of adding nickel to steel, in amounts up to 8%, is as follows for each 0.01% of nickel: increase of 40 lb/in² in yield strength; increase of 42 lb/in² in tensile strength; increase in reduction of area 0.005%; and decrease in elongation 0.010%. Heat-treatment by quenching from 850°C increases tensile strength greatly without appreciably lowering ductility or toughness. Nickel depresses the critical temperature range, allowing quenching from lower temperatures, thus resulting in smaller shrinkage strains than plain carbon steels of the same carbon content. In the structural nickel steels the nickel content ranges from about 3 to 4%. Steels containing 0.7 to 0.9% carbon and about 25% nickel have great resistance to oxidation at high temperatures. Nickel steel has a higher resistance to wear or abrasion than carbon steel and greater resistance to corrosion. Special alloys very high in nickel are known as Invar and Platinite. **Invar** contains 36% nickel, and its coefficient of thermal expansion is about one-sixth that of ordinary steel. **Platinite** (42% Ni) has the same coefficient of expansion as glass.

489. Chromium steel containing from 0.5 to 2.0% of chromium and 0.2 to 1.5% of carbon has great hardness, high strength, and a fair degree of toughness. The properties depend upon the heat-treatment and range from about 75,000 lb/in² in annealed condition to as high as 250,000 lb/in² in treated condition. These steels are used for tools, dies, ball bearings, forgings, projectiles, and other special purposes. Chromium is the only alloying element which has been found to produce in iron alloys a condition approaching complete resistance to atmospheric corrosion. In order to attain this result with a pure iron-chrome alloy, more than 11% chromium is necessary and still more if carbon is present as in steel. Alloy steels with corrosion-resistant properties are made with various chromium contents, as 4 to 6%, 11 to 14%, 16 to 20%, and 25 to 30%. An annealed alloy of 17% chromium, maximum manganese and silicon 0.50% each, and maximum carbon 0.10% has the following properties: ultimate strength 75,000 lb/in²; elongation in 2 in, 27%; reduction of area, 55%; Brinell hardness 175; density 7.7; coefficient of linear expansion $11 \times 10^{-6}/(°C)$; elastic modulus 29×10^6 lb/in². The thermal expansion of stainless iron alloys, containing 12 to 16% of chromium and 0.09 to 0.13% of carbon, has an average value of $10 \times 10^{-6}/(°C)$ in the range from 20 to 100°C and $12.5 \times 10^{-6}/(°C)$ in the range from 20 to 800°C. For data on other compositions, see A. B. Kinzel and R. Franks, "Alloys of Iron and Chromium," New York, McGraw-Hill Book Company, 1940, Vol. II, High-chromium Alloys.

490. Chrome-nickel steels in the range of about 0.5 to 1.8% chromium, 1.0 to 3.5% nickel, 0.20 to 0.50% carbon, and 0.30 to 0.80% manganese have very high strength, elastic limit, and hardness combined with good ductility and great toughness, if developed by proper heat-treatment. The tensile strengths range from about 90,000 lb/in² as rolled up to 190,000 lb/in² when heat-treated; the elastic ratios range as high as 90% (see Fig. 4-81 for typical variation of properties when heat-treated). They are used for shafts, axles, gears, armor plate, projectiles, and special structural purposes. The steels which are much higher in chromium and nickel content are described in Par. **491.**

491. Stainless and Heat-resisting Alloy Steels. Stainless steels are of three principal types. Austenitic steels, AISI 300 series, contain 16 to 20% chromium, 6 to 22% nickel, maximum manganese 2.0% and silicon 1.0% each, and 0.08 to 0.20% carbon.

Martensitic stainless steels, AISI 400 and 500 series, contain 4 to 18% chromium, maximum manganese 1% and silicon 1%, and 0.05 to 0.20% carbon. Several alloys in the series (AISI 440A, B, and C) contain a higher percentage of carbon and also molybdenum, 0.75% maximum. Ferritic stainless steels, AISI 400 series, contain 11.50 to 27.00% chromium, maximum manganese 1.5% and silicon 1%, and 0.08% to 0.35% carbon. These alloys are heat-resistant and corrosion-resistant; however, these properties depend upon the alloy and corrosive media. The resistance to oxidation (scaling) depends primarily upon the amount of chromium present. The austenitic steels are practically nonmagnetic, whereas the martensitic and ferritic stainless steels are ferromagnetic. The austenitic steels can be satisfactorily welded by all production methods, whereas welds made with martensitic and ferritic steels exhibit low ductility and must be annealed. The martensitic steels may be hardened by quenching, whereas the austenitic and ferritic steels may be hardened primarily by cold working. For detailed data regarding composition, corrosion resistance, and mechanical properties of stainless steels see Metals Handbook (Par. **494**) and ASTM STP369, 1965.

492. Properties of Typical Stainless Steels. See Table 4-98.

493. Creep Strength. Metals subjected to static loading at elevated temperatures continue to elongate (creep) with time. After an initial period of adjustment to a fairly constant velocity of flow the time rate of deformation under constant stress and temperature (expressed as percentage elongation per hour) is called the **creep rate.** Short-time tensile-test values are not reliable design criteria for metals used at elevated temperatures. The useful strength is limited to the stress that will not produce a damaging amount of deformation during the normal life of the structure. The **creep limit** for a material is the stress that will not produce more than a specified elongation (usually 1%) in a definite time interval (often taken as 10,000 or 100,000 h) at the given temperature. Determination of the creep limit involves long-time testing of a series of specimens to determine initial deformations and creep rates for various stresses. The data are plotted and arbitrarily extrapolated to obtain approximate total creep at future times. Some nonferrous materials such as lead and zinc creep at room temperatures (see Par. **535**), whereas, except for stresses nearly up to the ultimate, no appreciable creep has been observed for steels until temperatures above about 500°F are exceeded. Variations in data reported on steels have led to the conclusion that creep characteristics are too sensitive an index of strength to permit exact duplication either in different laboratories or in duplicate tests in the same laboratory. Figure 4-82 shows the approximate variation of creep stress with temperature for several steels. For boilers, piping, etc., operating at temperatures above 1000°F, the maximum working stresses that can be used (without excessive creep) are so small as to make it difficult to produce economical and safe designs until better creep-resistant materials are available. Eleven grades of alloy steels for service at temperatures from 750 to 1100°F are covered by ASTM Specifications A351, A335, and A193 for specific applications to castings, bolting materials, and seamless pipe. These range from ordinary carbon-molybdenum steels to 18% Cr, 8% Ni austenitic steels. In general, 0.4 to 1.5% molybdenum is used in steels for high-temperature service, since it is the only element thus far proved to be effective in increasing creep resistance when present in only small amounts. (For a comprehensive tabulation of creep data see "High Temperature Strength Data of

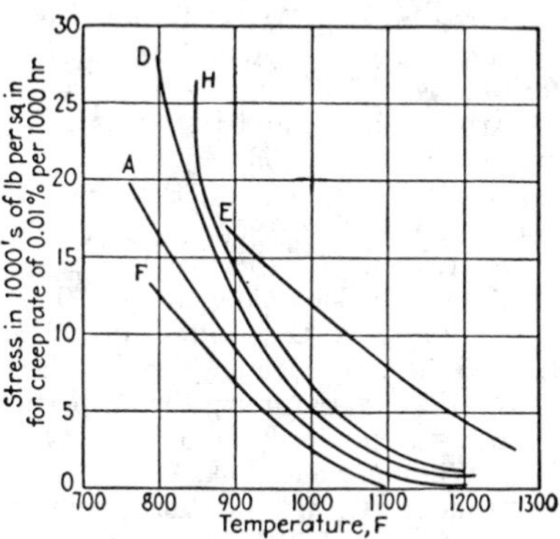

FIG. 4-82. Creep stress for a creep rate of 0.01% per 1,000 hr for several steels. *A.* Wrought 0.10 to 0.20% carbon steels. *D.* Wrought chromium-molybdenum bolt steel (0.40% carbon). *E.* Wrought 18% chromium, 8% nickel steel. *F.* Wrought carbon steels (carbon above 0.20%). *H.* Wrought 1.0 to 2.5% chromium, 0.50 = molybdenum steel (0.20% carbon max.).

Table 4-98. Properties of Typical Stainless Steels
(From "Metals Handbook")

Property	Austenitic		Martensitic	Ferritic
	AISI 309 Cr. 22–24% Ni. 12–15% annealed	AISI 321 Cr. 17–19% Ni. 8–11% annealed	AISI 410 Cr. 11.5–13.5% quenched and tempered at 1000 F	AISI 430 Cr. 14–18% annealed
Ultimate strength, psi	95,000	85,000	145,000	75,000
Yield strength, psi, 0.2% offset	40,000	30,000	115,000	40,000
Elongation in 2 in., %	45	55	20	30
Reduction of area, %	50–65	55–65	65	40–55
Hardness:				
Rockwell	B78–90	B75–90	C31	B79–90
Brinell	140–185	135–185	300	145–185
Density	7.9	7.9	7.7	7.7
Weight, lb per cu in	0.29	0.29	0.28	0.28
Thermal conductivity at 212 F, Btu per hr per sq ft per deg F	8.0	9.3	14.4	15.1
Coefficient of expansion per deg F (mean value from 32 to 1000 F)	9.6×10^{-6}	10.3×10^{-6}	7.2×10^{-6}	6.3×10^{-6}
Elastic modulus, psi	29×10^6	28×10^6	29×10^6	29×10^6
Sealing temp, deg F	2000	1650	1250	1550

Metals and Alloys," Punched Cards, ASTM, 1965; and annual reports of the Joint Research Committee on Effect of Temperature on the Properties of Metals in *Proc. ASTM*.)

494. General Reference Literature on Iron and Steel

Young, J. F. "Materials and Processes," 2d ed.; New York, John Wiley & Sons, Inc., 1954.

Withey, M. O., and Washa, G. W. "Materials of Construction"; New York, John Wiley & Sons, Inc., 1954.

Hoyt, S. L. "Metal Data"; New York, Reinhold Publishing Corporation, 1952.

Grossman, M. A. "Elements of Hardenability"; Cleveland, American Society for Metals, 1952.

Aitchison, L., and Pumphrey, W. I. "Engineering Steels"; London, MacDonald and Evans, Ltd., 1953.

Dolan, T. J., Lazan, B. J., and Horger, O. J. "Fatigue"; Cleveland, American Society for Metals, 1954.

Smith, G. V. "Properties of Metals at Elevated Temperatures"; New York, McGraw-Hill Book Company, 1950.

Simmons, W. F., and Cross, H. C. "The Elevated Temperature Properties of Chromium-Molybdenum Steels," Philadelphia, ASTM–ASME Joint Committee on Effect of Temperature on the Properties of Metals, ASTM STP 151, 1953.

Hehemann, R. F., and Ault, G. M. "High Temperature Materials"; New York, John Wiley & Sons, Inc., 1959.

Horger, O. J. "Metals Engineering—Design," ASME Handbook, 2d ed.; New York, McGraw-Hill Book Company, 1965.

Sines, G., and Waisman, J. L. "Metal Fatigue"; New York, McGraw-Hill Book Company, 1959.

Miner, D. F., and Seastone, J. B. "Handbook of Engineering Materials"; New York, John Wiley & Sons, Inc., 1955.

Samans, C. H. "Metallic Materials in Engineering"; New York, The Macmillan Company, 1963.

Cast Metals Handbook; Chicago, American Foundrymen's Association.

"American Malleable Iron, A Handbook"; Cleveland, Malleable Founders' Society.

Metals Handbook; Cleveland, American Society for Metals, 1961, Vol. I, Properties and Selection of Materials.

Battelle Staff. "Prevention of the Failure of Metals under Repeated Stress"; New York, John Wiley & Sons, Inc., 1941.

Munitions Board Aircraft Committee, Strength of Metal Aircraft Elements, *Document* ANC–5a, U.S. Government Printing Office. (Revised periodically.)

"Fracture of Engineering Materials"; Metals Park, Ohio, American Society for Metals, 1964.

Structure and Properties of Ultrahigh-strength Steels, ASTM STP 370, 1965.

Air Weapons Materials Application Handbook, Metals and Alloys, *ARDC Tech. Rept.* 59–66, December, 1959.

Advances in Technology of Stainless Steel and Related Alloys, ASTM STP 369, 1965.

Symposium on Fracture Toughness Testing and Its Applications, ASTM STP 381, 1965.

STEEL STRAND AND ROPE

495. Iron and Steel Wire. The manufacture and properties of iron and steel wire are covered in Pars. **72** to **73** and Table 4-28, including the physical properties pertinent to structural uses. Annealed wire of iron or very mild steel has a tensile strength in the range of 45,000 to 60,000 lb/in^2; with increased carbon content, varying amounts of cold drawing, and various heat-treatments the tensile strength ranges all the way from the latter figures up to about 500,000 lb/in^2, but a figure of about 250,000 lb/in^2 represents the ordinary limit for wire for important structural purposes. For example, see the following paragraph on bridge wire. Wires of high carbon content can be tempered for special applications such as spring wire. The yield strength of cold-drawn steel wire is 65 to 80% of its ultimate strength. For examples showing the effects of drawing and carbon content on wire, see Hoyt, "Metal Data," p. 55 (see Par. **494**). See Tables 4-30 and 4-31 for description of Copperweld wire.

496. Galvanized-steel bridge wire is well typified by the wire used for the cables and hangers of suspension bridges such as the George Washington Bridge and the San Francisco–Oakland Bay Bridge. The wire is a hot-galvanized, acid open-hearth steel, of 0.196 in diameter, and is cold-drawn to obtain a tensile strength of about 225,000 lb/in^2 and a yield strength of approximately 182,000 lb/in^2. Joints or splices are made with cold-pressed sleeves which develop practically the full strength of the wire. Fatigue tests of galvanized bridge wire in reversed bending indicate that the endurance limit of the coated wire is only about 50,000 to 60,000 lb/in^2. (See L. S. Moisseiff, Investigation of Cold-drawn Bridge Wire, *Proc. ASTM*, 1930, Vol. 30, Pt. II, pp. 313–349; and detailed discussion of properties of wire in W. H. Swanger and G. F. Wohlgemuth, Failure of Heat-treated Steel Wire in Cables of the Mt. Hope, R.I., Suspension Bridge, *Proc. ASTM*, 1936, Vol. 36, Pt. II, pp. 21–84.)

497. Wire rope is made of wires twisted together in certain typical constructions and may be either flat or round. Flat ropes consist of a number of strands of alternately right and left lay, sewed together with soft iron wire to form a band or belt; they are sometimes of advantage in mine hoists. Round ropes are composed of a number of wire strands twisted around a hemp core or around a wire strand or wire rope. The standard wire rope is made of six strands twisted around a hemp core, but for special purposes, four, five, seven, eight, nine, or any reasonable number of strands may be used. The hemp is usually saturated with a lubricant, which should be free from acids or corrosive substances; this provides little additional strength, but acts as a cushion to preserve the shape of the rope and helps to lubricate the wires. The numbers of wires commonly used in the strands are 4, 7, 12, 19, 24, and 37, depending upon the service for which the ropes are intended. When extra flexibility is required, the strands of a rope sometimes consist of ropes, which in turn are made of strands around a hemp core. Ordinarily the wires are twisted into strands in the opposite direction to the twist of the strands into rope. The make-up of standard hoisting rope is 6×19; extra-pliable hoisting rope is 8×19 or 6×37; transmission or haulage rope is 6×7; hawsers and mooring lines are 6×12 or 6×19 or 6×24 or 6×37, etc.; tiller or hand rope is $6 \times 6 \times 7$; highway guard-rail strand is 3×7; galvanized mast-arm rope is 9×4 with a cotton center. The tensile strength of the wire ranges, in different grades, from 60,000 to 350,000 lb/in^2, depending on the material, diameter, and treatment. The maximum tensile efficiency of wire rope is 90%; the average is about 82.5%, being higher for 6×7 rope and lower for 6×37 construction. The apparent modulus of elasticity for steel cables in service may be assumed to be about 9×10^6 to 12×10^6 lb/in^2 of cable section. The various strength grades of steel wire used in rope construction are known by various trade names, such as "iron,"

Table 4-99. Diameters of Sheaves and Drums for Wire Rope

Rope	Sheave diameter ÷ rope diameter		
	Average	Minimum	Larger installations
6 × 7	72	42	96
6 × 19	45	30	90
6 × 37	27	18	
8 × 19	31	21	

"cast steel," "extra-strong cast steel," "crucible steel," "plow steel," etc. Wire rope in service should be lubricated and kept free from moisture unless it is galvanized. It should not be handled like hemp rope, but all twisting or untwisting and kinking must be avoided. Also see wire-rope data books published by various manufacturers. Federal Specification RR-R-571a, May 8, 1945, lists extensive data for a wide variety of constructions.

498. Diameter of Sheaves and Drums for Wire Rope. The average and minimum tread diameters, in accordance with the practice recommended by the American Steel & Wire Co., are shown in Table 4-99; also higher values should be used for larger hoisting installations. Diameters larger than those listed as minimum will give increased rope life.

499. Galvanized-steel strand is made of 7 wires in concentric lay in sizes from ⅛- to ⅝-in diameter and of 19 wires in concentric lay in diameters from ½ to 1 in; it can also be made in larger sizes with 37 or 61 wires, up to 1⅜ in. diameter. This strand is made in four standard grades of strength known as "common," "Siemens-Martin," "high strength," and "extra-high strength." Galvanized strand is extensively used for guying and for aerial-cable messengers. Detailed requirements for strengths,

Table 4-100. Properties of Standard Steel Hoisting Rope (6 × 19)
(From Metals Handbook)

Diameter, in.	Approximate weight, lb per ft.	Approximate strength, tons of 2,000 lb			
		Cast steel	Extra strong cast or mild plow steel	Plow steel	Improved plow steel
2¾	12.10	212.0	234.0	256.0	294.0
2½	10.00	176.0	195.0	214.0	246.0
2¼	8.10	144.0	160.0	176.0	202.0
2⅛	7.23	128.0	143.0	157.0	181.0
2	6.40	114.0	127.0	140.0	161.0
1⅞	5.63	100.0	112.0	123.0	142.0
1¾	4.90	88.0	98.0	108.0	124.0
1⅝	4.23	76.0	85.0	94.0	108.0
1½	3.60	65.0	72.5	80.5	92.5
1⅜	3.03	55.0	61.5	68.0	78.5
1¼	2.50	46.0	51.0	56.5	65.0
1⅛	2.03	37.0	41.5	46.0	53.0
1	1.60	29.5	33.0	36.5	42.0
⅞	1.23	22.8	25.4	28.0	32.2
¾	0.90	16.8	18.7	20.6	23.7
⅝	0.63	11.8	13.1	14.4	16.6
$\frac{9}{16}$	0.51	9.6	10.6	11.7	13.5
½	0.40	7.7	8.5	9.4	10.8
$\frac{7}{16}$	0.31	6.0	6.6	7.3	8.4
⅜	0.23	4.5	5.0	5.5	6.3
$\frac{5}{16}$	0.16	3.2	3.5	3.9	4.5
¼	0.10	2.1	2.3	2.5	2.9

The recommended proper working loads are ⅕ to ⅛ of the above strengths.

sizes, types, and coating tolerances for several varieties of wire and cable products are given in ASTM specifications as follows: A326, Zinc-coated (Galvanized) High Tensile Steel Telephone and Telegraph Line Wire; A363, Zinc-coated Steel Overhead Ground Wire Strand; A474, Aluminum Coated Steel Wire Strand; A475, Zinc-coated Steel Wire Strand.

500. Properties of Standard Steel Hoisting Rope. See Table 4-100.

PROTECTIVE COATINGS FOR IRON AND STEEL

501. Theory of Corrosion. The electrochemical theory is now generally accepted as the one which best explains the established facts regarding the corrosion of iron and steel. When any metal forms the anodic or electronegative element of an electrolytic cell (Sec. **23**), it is attacked or corroded and passes into solution. It is also possible for both anodic and cathodic conditions to exist at the same time at different parts of the same metal, as sometimes between stressed and unstressed portions. The anodic condition explains in general the cause of corrosion under exposure to moist atmospheres, industrial gases, acids, wastes, etc. The presence of both oxygen and water is essential for the corrosion of steel to proceed at ordinary temperatures. For complete exposition see the bibliography in Par. **518.**

502. Protection against corrosion is obtained in general by the use of coatings of various kinds which are either noncorrosive or much less susceptible to corrosion than the base metal. Such coatings are divisible broadly into four classes: (1) paints, (2) metal coatings, (3) chemical coatings, and (4) greases. Painting is very extensively used for the protection of structural iron and steel but must be maintained by periodic renewal. Metal coatings take various ranks in protective efficiency, depending on the metal used and its characteristics as a coating material. The most extensively used metal coating is zinc, which is electronegative to iron. Zinc itself is not immune to corrosive attack, but as long as any zinc is present, it will protect the iron in its immediate vicinity. The electrode potentials of various metals are shown in the following paragraph. Only the metals which are electronegative to iron have this property of protection through anodic action.

503. Electrode Potentials of the Metals. The accompanying table shows the difference of potential at the interface between a metal and its solution of normal ionic concentration in which it is immersed. (For additional information see Sec. **23,** W. M. Latimer, "Oxidation Potentials," Englewood Cliffs, N.J., Prentice-Hall, Inc., 1952.)

Metal (electrode)	Ion	Potential difference, volts	Metal (electrode)	Ion	Potential difference, volts
Gold	Au^{+}	1.7	Cobalt	Co^{++}	−0.277
	Au^{+++}	1.4	Cadmium	Cd^{++}	−0.402
Silver	Ag^{+}	0.799	Iron	Fe^{++}	−0.441
Mercury	Hg^{++}	0.799	Chromium	Cr^{++}	−0.56
Copper	Cu^{+}	0.52		Cr^{+++}	−0.7
	Cu^{++}	0.34	Zinc	Zn^{++}	−0.762
Hydrogen	H^{+}	0	Aluminum	Al^{+++}	−1.67
Lead	Pb^{++}	−0.126	Magnesium	Mg^{++}	−2.37
Tin	Sn^{++}	−0.136	Sodium	Na^{+}	−2.715
Nickel	Ni^{++}	−0.250	Potassium	K^{+}	−2.924

Any metal in this table is electropositive to any element following it and electronegative to any element preceding it. The electropositive element constitutes the cathode, and the electronegative element constitutes the anode in an electrolytic cell or galvanic couple.

504. Protective metallic coatings of various nonferrous metals are applied to iron and steel to prevent or postpone corrosion from various causes. Such coatings are made of zinc, tin, copper, nickel, chromium, cobalt, lead, cadmium, and aluminum; coatings of gold and silver are also used for ornamental purposes. In addition to these, there are black nickel plating and coatings formed by cementation, as with zinc or aluminum, or by carburization or nitriding. There are four principal methods of applying metal coatings, viz., hot dipping, cementation, spraying, and electroplating.

505. Zinc coatings are more widely used for the protection of structural iron and steel than coatings of any other type. They are applied by all four methods mentioned in the last paragraph. The hot-dip process is the earliest type known and is very extensively used at the present time; two improvements, the Crapo process and the Herman, or "galvannealed," process, are used in galvanizing wire. The cementation, or sherardizing, process consists in heating the articles for several hours in a packing or zinc dust in a slowly rotating container. Electroplating is also employed, and heavier coatings can be obtained than are usual with the hot-dip process but require longer time for deposition (see Sec. **23**). (Also see ASTM specifications for zinc-coated iron and steel products.)

506. Aluminum coatings are applied by a cementation process which is commercially known as "calorizing." The articles to be coated are packed in a drum in a mixture of powdered aluminum, aluminum oxide, and a small amount of ammonium chloride. The articles are then slowly rotated and heated in an inert atmosphere, usually of hydrogen. Such coatings are very resistant to oxidation and sulfur attack at high temperatures. Aluminum coatings can also be applied by the hot-dipping method and then are heat-treated to improve the alloy bond. Aluminum can also be applied by spraying.

507. Lead and tin coatings are applied by the hot-dip process. Tin affords excellent protection as long as the coating is intact but is electropositive to iron and therefore lacking in protective properties as soon as any iron is exposed. Tin-coated steel, or so-called "tin plate," is very extensively used for canning foodstuffs. Tin coatings are also applied to electrical conductors which are to be covered with any insulating compound containing traces of free sulfur, e.g., rubber compounds. The best results are obtained in applying lead coatings if a bonding metal, such as tin or antimony, is employed to form an alloy film. Lead and lead-rich alloys are resistant to corrosion by most acid fumes but when used as coatings are sometimes subject to the formation of pinholes and are not resistant to abrasion.

508. Lohmannizing. In this process, iron and steel sheets are first immersed in a bath containing an amalgamating salt and then pickled and dipped in two different baths of molten alloys. The finished surface is in a clean condition and is coated with the protecting metal. The metals used in this process are alloys of zinc, lead, and tin, the proportions of which are varied to suit requirements. Terneplate, or roofing tin, consists of sheet steel coated with an alloy of approximately 75 parts of lead and 25 parts of tin.

509. Metal-spray coatings are applied by passing metal wire through a specially constructed spray gun which melts and atomizes the metal to be used as coating. The surface to be sprayed must be roughened to afford good adhesion of the deposited metal. Nearly all the commonly used protective metals can be applied by spraying, and the process is especially useful for coating large members of repairing coatings on articles already in place. Sprayed coatings can also be applied that will resist wear and can be used to build up worn parts such as armature shafts and bearing surfaces or to apply copper coatings to carbon brushes and resistors.

510. Chromium coatings can be applied by cementation or electroplating. In electroplating, the best results are secured by first plating on a base coating of nickel or nickel copper to receive the chromium. The great hardness of chromium gives it important applications for protection against wear or abrasion; it will also take and retain a high polish. Very thin coatings have a tendency to be inefficient as a result of the presence of minute pinholes.

511. Electroplated Coatings. Electroplating is employed in the application of coatings of nickel, brass, copper, chromium, cadmium, cobalt, lead, and zinc. Only cadmium, chromium, and zinc are electronegative to iron. The other metals mentioned are employed because of their own corrosion-resistant properties and because they afford surface finishes having certain desirable characteristics.

512. Protective paints are extensively employed to protect heavy exposed structures of iron and steel, such as bridges, tanks, and towers. The protection is not permanent but gradually wears away under weather exposure and must be periodically renewed. Various specially prepared paints are used for protecting the surface from dampness, oxidizing gases, and smoke. No one paint is suitable for all purposes, but the choice depends on the nature of the corrosive influence present. Asphaltum and tar protect the surface by formation of an impervious film. A chemical protective action is exerted by paints containing linseed oil as the vehicle and red lead as the pigment; linseed oil absorbs oxygen from the atmosphere and forms a thick elastic covering, a formation hastened by adding salts of manganese or lead to the oil. All dryers, vehicles, and pigments used in paint must be inert to the steel; otherwise corrosion will be hastened instead of prevented. Graphite- and aluminum-flake pigments give very impermeable films but do not show the inhibitive action of red lead or zinc when the films are scratched. Aluminum has the advantage of reflecting both infrared and ultraviolet rays of the sun; hence it protects the vehicle from a source of deterioration and is used to paint gasoline-storage tanks to prevent excessive heating due to the sun's rays. A large number of new protective coatings have recently been developed from synthetic materials, such as silicones, artificial rubbers, and phenolic plastics. Many of these are tightly adhering compounds in the form of paints or varnishes which offer rather good protection against a wide variety of chemical attacks. The majority of these new coatings, however, are sensitive to abrasion, and many of them must be baked on to secure full effectiveness. (See Reports of Committee D1 in *ASTM Proc.;* also bibliography, Par. **518.**)

513. Chemical coatings which are corrosion-resistant include Parkerizing and Bower-Barff finish. **Parkerizing** consists in immersing the steel in a solution of manganese dihydrogen phosphate, then heating to about the boiling point. The pieces are allowed to remain in the bath until effervescence ceases. After oiling, the surface produced has the appearance of gunmetal. These phosphate coatings are used mainly as bases for finishing enamels and paints. **Bower-Barff finish** consists in heating the pieces in a closed retort to a temperature of 1600°F. Alternate injections of superheated steam and carbon monoxide then reduce the oxides formed on the surface to Fe_3O_4. The operations may be repeated several times until a sufficient depth of oxide is obtained. Other types of chemical surface which develop increased surface hardness and resistance to corrosion are produced commercially by treating a steel with nitrogen or silicon for prolonged periods of time at high temperature.

514. Greases and oils of various grades are used to protect the surface by applying a thin film to parts of machines, tools, bearings, and steels which are to be put in storage or are to be shipped. These slushing compounds are usually mineral oils, fats and waxes, lanolin, or greases; oil-soluble chromates are sometimes added to aid in preventing corrosion. Silicone oils and greases are particularly effective because of their imperviousness and repellency of moisture.

515. Corrosion-resistant ferrous alloys such as rustless or stainless iron and steel have come into use for both structural and ornamental purposes but on account of their chromium and nickel contents are relatively expensive in comparison with the ordinary structural steels (see Par. **491**). Copper-bearing iron and steel, containing about 0.15 to 0.25% copper, are used extensively; the copper content tends to retard corrosion slightly but does not prevent it, and some protective coating is usually necessary. Some structural uses have been made of these steels without applying special protective coatings. A tightly adherent brown oxide surface film forms from weathering to serve as the future "protective coating."

516. Stainless-clad Sheets. Plates and sheets of steel faced with stainless steels are widely used in chemical processing, paper mills, nuclear reactors, and transportation or storage of corrosive liquids, food products, etc. The stainless-steel layer usually constitutes 10 to 20% of the thickness of the plate, and either one or both sides may be coated, but the surfacing must be strongly bonded to the base metal. One method of production consists in casting two stainless-steel sheets in the center of a steel ingot with a separating material between. The composite "sandwich" ingot is hot-rolled to the desired size and then separated into two sheets each clad on one face with stainless steel. Another method of production is to bond stainless sheet to a mild-steel plate by means of a great number of resistance welds closely spaced all over the sheet. These methods provide a stainless-clad surface on either thick or thin plates without the added expense of a solid stainless-steel sheet.

517. Copperweld. A series of steel products, including wire, wire rope, bars, clamps, ground rods, and nails, that contain a copper-clad surface are made by the Copperweld process. The copper coating is intimately bonded to the steel by pouring a ring of molten copper about a heated steel billet fastened in the center of a refractory mold.

The solidified composite ingot is then hot-rolled to bar stock and subsequently cold-drawn to the various wire sizes. The thickness of the copper coating on wire is 10 to 12½% of the wire radius and produces a high-strength steel wire with a resistance to corrosion similar to that of a solid copper wire. Their increased electrical conductivity over that of a solid steel wire or rod makes the Copperweld products suitable for high-strength conductors, ground rods, aerial cable messengers, etc.

518. References to Technical Literature on Corrosion and Protective Coatings

UHLIG, H. H. (ed.) "The Corrosion Handbook"; New York, John Wiley & Sons, Inc., 1948.

Reports of Committee A5, Corrosion of Iron and Steel, *Ann. Proc. ASTM.*

VON FISCHER, W., and BOBALEK, E. G. (eds.) "Organic Protective Coatings"; New York, Reinhold Publishing Corporation, 1953.

SPELLER, F. N. "Corrosion, Causes and Prevention"; New York, McGraw-Hill Book Company, 1951.

PAYNE, H. F. "Organic Coating Technology"; New York, John Wiley & Sons, Inc., 1961.

NONFERROUS METALS AND ALLOYS

519. Copper is a very important industrial metal because of its great resistance to corrosion, high electrical and thermal conductivities, and the readiness with which it alloys with zinc, tin, and many other metals. The mechanical properties of wrought and drawn copper are stated in Pars. **6** and **7,** and the thermal properties in Par. **37.** Hard-drawn copper may be softened by annealing, and its physical properties are then about the same as those of cast copper. The ultimate strength of cast copper is about 25,000 lb/in²; it has a yield strength of about 8,000 lb/in². So-called "casting copper" contains minor impurities which greatly diminish its electrical conductivity but do not impair its mechanical properties. Copper castings are improved by using special deoxidizers such as Boroflux and silico-calcium copper alloy. Boroflux is a mixture of boron suboxide, boric anhydride, magnesia, and possibly some magnesium; for data on its use see publications of General Electric Co. By the use of these deoxidizers the castings are improved structurally, and the electrical conductivity can be increased to about 80 to 90% of standard annealed copper. When copper containing oxygen is heated in a slightly reducing atmosphere at temperatures above 400°C, it becomes severely embrittled. Copper deoxidized in various ways differs considerably in its susceptibility to embrittlement, and it is important to avoid prolonged alternate treatment in oxidizing and reducing gases. The industrial uses for copper include plates, bars, rods, bolts, sheets, tubes, pipes, and castings. [See ASTM Standards for specifications for copper in various forms and products. Also see O. W. Ellis, "Copper and Copper Alloys," Cleveland, American Society for Metals, 1948, and Metals Handbook, 1948 (Par. **494**).]

520. Nickel is a brilliant metal which approaches silver in color. It is more malleable than soft steel and when rolled and annealed is somewhat stronger and almost as ductile. The tensile strength ranges from 60,000 lb/in² for cast nickel to 115,000 lb/in² for cold-rolled full-hard strip; yield strength 20,000 to 105,000 lb/in²; elongation in 2 in, 2% when full hard to about 50% when annealed; modulus in tension about 30×10^6 lb/in². Nickel takes a good polish and does not tarnish or corrode in dry air at ordinary temperatures. It has various industrial uses in sheets, pipes, tubes, rods, containers, and the like, where its corrosion resistance makes it especially suitable; it is used in nickel steel and is an important constituent of many heat-resistant and corrosion-resistant alloys. (For complete physical data see publications of the International Nickel Co. and *NBS Circ.* C447, Mechanical Properties of Metals and Alloys, Washington, D.C., 1943.)

521. Aluminum is a white metal of high metallic luster. It is an important industrial metal because it is malleable, ductile, very light, and very strong in proportion to its weight and has a high electrical and thermal conductivity. Refined (99.97%) aluminum in the annealed wrought form has a tensile strength of about 9,000 lb/in², yield strength of 3,000 lb/in², elongation 60%, reduction of area 95%, Brinell hardness 15, Young's modulus 10×10^6 lb/in², and shearing strength of 7,000 lb/in². Commercially pure, annealed wrought aluminum has a tensile strength of about 13,000 lb/in², yield strength of about 5,000 lb/in², and shearing strength of 9,500 lb/in². Cold working increases the tensile strength, increases the ratio of the yield strength to the tensile strength, and reduces the elongation and reduction of area; hard-rolled or drawn aluminum has a tensile strength of about 22,000 to 33,000 lb/in². The physical properties of aluminum conductors are stated in Pars. **120** and **181** to **185**. Aluminum is the base for many alloys, both wrought and cast, and in relatively pure form is used extensively for electrical conductors, for cooking utensils, and in the chemical industry. The physical properties are greatly improved by alloying it with certain other metals. (For very comprehensive data on aluminum and its alloys see "Aluminum," edited by Kent R. Van Horn, American Society for Metals, 1966.)

522. Zinc is a bluish-white metal, which has a metallic luster on a new fracture. The density of cast zinc ranges from 7.04 to 7.16. At ordinary temperature it is brittle, but in the range of about 100 to 150°C it becomes malleable and can be rolled into sheets and drawn into wire. At 200°C it becomes so brittle that it can be pulverized. The tensile strength of cast zinc ranges from about 8,000 to 14,000 lb/in² in an ordinary testing-machine test, and that of drawn zinc from about 22,000 to 30,000 lb/in²; it has a poorly defined proportional limit of about 5,000 lb/in² and exhibits a certain amount of creep at room temperatures; hence it may fracture in service under constant stresses below its testing-machine strength. It strongly resists atmospheric corrosion but is readily attacked by acids. The principal industrial uses for it are for galvanizing iron and steel, for plates and sheets for roofing and other applications, for castings, and for alloying with copper, tin, and other metals; very large quantities are used in the various types of brass. (See Metals Handbook and Withey and Washa, Par. **494.**)

523. Tin is a silvery-white, lustrous metal, very soft and malleable and of very low tensile strength. It has a density of about 7.3 and melts at 232°C. In ductility it equals soft steel. The tensile strength varies with the speed of testing. As a metal it has few uses except in sheets, but large quantities of it are used in various industrial alloys. Its chief uses are in tin- and terneplate, solder, babbitt and other bearing metals, brass, bronze, and foil. Tin is very resistant to atmospheric corrosion, and water hardly affects it at all; however, it is electronegative to iron and therefore is not an efficient protective coating under atmospheric exposures. (For systematic data on tin see C. L. Mantell, "Tin: Its Mining, Production, Technology, and Applications," New York, Reinhold Publishing Corporation, 1929, and Symposium on Tin, *ASTM Spec. Tech. Publ.* 141, 1953.)

524. Lead is a heavy, soft, malleable metal with a blue-gray color; it shows a metallic luster when freshly cut, but the surface is rapidly oxidized in moist air. It can be easily rolled into thin sheets and foil or extruded into pipes and cable sheaths but cannot be drawn into fine wire. Although in an ordinary tensile test lead may develop a tensile strength of 2,400 lb/in², it may creep at ordinary room temperatures at stresses as low as 50 lb/in². Owing to this tendency to creep, it may fracture under long-continued load at stresses as low as 800 lb/in², and the ordinary static tensile properties do not have much significance. (See Moore, Betty, and Dollins, Creep and Fracture of Lead and Lead Alloys for Cable Sheathing, *Univ. Ill. Eng. Expt. Sta. Bull.* 347, 1943, and *Bull.* 440, 1956.) The resistance of pure lead to corrosion makes it useful in the form of sheets, pipes, and cable coverings, and large quantities of lead are used in the manufacture of various alloys, particularly in alloys for bearings. Common alloys of lead for cable sheathings contain (approximately) 0.04% Cu, 0.75% Sb, or 0.03% Ca. (See "Lead in Modern Industry," New York, Lead Industries Association, 1952.)

525. Monel metal is a silvery-white alloy containing approximately 66 to 68% nickel, 2 to 4% iron, 2% manganese, and the remainder copper. It can be cast, forged, rolled, drawn, welded, and brazed and is easily machined. It melts at 1360°C and has a density of 8.80, coefficient of expansion of 14×10^{-6}/°C, thermal conductivity of 0.06 cgs unit, specific heat of 0.127 cal/(g)(°C), and modulus of 25×10^6 lb/in². The tensile strength ranges from 65,000 lb/in² for cast monel metal to 125,000 lb/in² in cold-rolled full-hard strip; yield strength 25,000 to 115,000 lb/in². It is highly resistant to corrosion and the action of sea water or mine waters. The industrial uses for it include many applications where its combination of physical properties and corrosion resistance gives it special advantages. (See technological data published by the International Nickel Co.)

526. Brass, or Copper-Zinc Alloys. Brass is an alloy consisting principally of copper and zinc and sometimes small proportions of other metals such as lead, tin, iron, aluminum, and manganese. The more useful brasses contain from 60 to 90% copper and 10 to 40% zinc and range in color from silver-white with low copper content to copper-red with low zinc content. Common, or standard, brass contains about 2 parts of copper and 1 part of zinc. Brass may be either cast or wrought. The physical properties cover a wide range depending upon composition, working, and heat-treatment.

Muntz metal is an alloy of about 57 to 63% copper and 37 to 43% zinc; it is also known as "malleable brass" and "yellow metal." When slowly cooled from a cherry-red heat, this alloy has a tensile strength of 55,000 to 65,000 lb/in^2 and about 50 to 60% elongation in 2 in; the strength, hardness, and ductility can be somewhat increased by quenching. It is used for bolts, rods, tubes, and extruded shapes where high resistance to corrosion is required.

Yellow brass sand castings should have a minimum tensile strength of 20,000 lb/in^2, and it may be as high as about 50,000 lb/in^2. (For data on the mechanical properties throughout the range of copper and zinc contents see Withey and Washa, Chap. 30, Fig. 9.) Cold working of annealed brass may increase its tensile strength to as high as 80,000 lb/in^2 and reduce the elongation from 50 to 5%. The complex brasses contain one or more other metals in addition to copper and zinc; among them are manganese bronze, naval brass, sterro metal, delta metal, Tobin bronze. (See specifications of the ASTM.)

527. Bronze, or Copper-Tin Alloys. Bronze is an alloy consisting principally of copper and tin and sometimes small proportions of zinc, phosphorus, lead, manganese, silicon, aluminum, magnesium, etc. The useful range of composition is from 3 to 25% tin and 95 to 75% copper. Bronze castings have a tensile strength of 28,000 to 50,000 lb/in^2, with a maximum at about 18% of tin content. The crushing strength ranges from about 42,000 lb/in^2 for pure copper to 150,000 lb/in^2 with 25% tin content. Cast bronzes containing about 4 to 5% tin are the most ductile, elongating about 14% in 5 in. Gunmetal contains about 10% tin and is one of the strongest bronzes. Bell metal contains about 20% tin. Copper-tin-zinc alloy castings containing 75 to 85% copper, 17 to 5% zinc, and 8 to 10% tin have a tensile strength of 35,000 to 40,000 lb/in^2, with 20 to 30% elongation. Government bronze, or Admiralty metal, contains 88% copper, 10% tin, and 2% zinc; it has a tensile strength of 30,000 to 35,000 lb/in^2, yield strength of about 50% of the ultimate, and about 14 to 16% elongation in 2 in; the ductility is much increased by annealing for ½ h at 700 to 800°C, but the tensile strength is not materially affected. Phosphor bronze is made with phosphorus as a deoxidizer; for malleable products, such as wire, the tin should not exceed 4 or 5%, and the phosphorus should not exceed 0.1%. United States Navy bronze contains 85 to 90% copper, 6 to 11% tin, and less than 4% zinc, 0.06% iron, 0.2% lead, and 0.5% phosphorus; the minimum tensile strength is 45,000 lb/in^2, and elongation at least 20% in 2 in. Lead bronzes are used for bearing metals for heavy duty; an ordinary composition is 80% copper, 10% tin, and 10% lead, with less than 1% phosphorus. Steam or valve bronze contains approximately 86% copper, 6.5% tin, 1.5% lead, and 4% zinc; the tensile strength is 34,000 lb/in^2 minimum, and elongation 22% minimum in 2 in (ASTM Specification B61). The bronzes have a great many industrial applications where their combination of tensile properties and corrosion resistance is especially useful. (See A. H. White, "Engineering Materials," New York, McGraw-Hill Book Company, 1948, Chap. XIII; also see ASTM specifications for various kinds of bronze. Also see Pars. **53** to **57**.) Brass and bronze under light service or while in storage occasionally develop cracks which are sometimes referred to as "season" cracking. This cracking is confined to metal that has been cold-worked and is caused by the initial residual stresses developed during cold working. The cracking is accelerated by the presence of moisture or corrosive liquids which develop surface corrosion or by sudden changes of temperature; such cracking can be prevented by annealing to remove the initial stresses.

528. Beryllium-copper alloys containing up to 2.75% beryllium can be produced in the form of sheet, rod, wire, and tube. The alloys are hardenable by a heat-treatment consisting in quenching from a dull red heat, followed by reheating to a low temperature to hasten the precipitation of the hardening constituents. Depending somewhat on the heat-treatment, the alloy of 2.0 to 2.25% beryllium has a tensile strength of

60,000 to 193,000 lb/in², elongation 2.0 to 10.0% in 2 in, modulus of elasticity 18×10^6 lb/in², and endurance limit of about 35,000 to 44,000 lb/in². An outstanding quality of this alloy is its high endurance limit and corrosion resistance; it can be hardened by heat-treatment to give great wear resistance and has high electrical conductivity. Typical applications include nonsparking tools for use where serious fire or explosion hazards exist and many electrical accessories such as contact clips and springs or instrument and relay parts. (See J. Delmonte, Beryllium and Its Alloys, *Metals & Alloys*, August and September, 1936.)

529. Bearing Metals. The white-metal bearing alloys known commercially as "babbitt metal" consist of tin, antimony, lead, and copper, in various proportions. Twelve typical babbitt metals are covered by ASTM Specifications B23, with an appendix showing composition and physical properties. The SAE has also published specifications for five babbitts which meet most ordinary needs. The constitution of babbitt metals appears to be that of a mass of soft tin with hard crystals of a copper-antimony and a tin-antimony compound scattered through it. The hard particles carry the load and resist wear, while the soft mass allows the bearing to adjust itself to the shaft and equalize the bearing pressure. Special brasses and bronzes, including the lead bronzes, are also employed for bearing metals; the copper-tin-zinc alloys may be classed as bronze or brass, depending upon the predominance of tin or zinc, respectively. The development of aluminum-alloy bearings has progressed to the stage where such alloys are now being used for many bearing applications. (Also see Metals Handbook and ASTM Specification B67.) Graphite bronze consisting of a bronze base with about 40% graphite by volume distributed throughout its mass is also used as a bearing metal. The material is porous, and capillarity can be employed to carry oil to the bearing surfaces. (See D. F. Wilcock and E. R. Booser, "Bearing Design and Application," New York, McGraw-Hill Book Company, 1957.)

530. Properties of Miscellaneous Ferrous and Nonferrous Metals and Alloys are listed in Table 4-101.

Table 4-101. Properties of Some Familiar Metals and Alloys
(International Nickel Co.)

Metal	Density	Melting point, deg C	Specific heat	Thermal expansion per deg C	Thermal conductivity*	Electrical conductivity*	Coef. of elect. res. per deg C	Modulus of elasticity, psi
Monel metal	8.80	1315	0.127	0.000014	6.6	4	0.0019	26,000,000
Nickel	8.85	1440	0.130	0.000013	15.5	16	0.0041	30,000,000
Copper	8.89	1083	0.093	0.000017	100	100	0.0040	16,000,000
Brass	8.46	900	0.088	0.000020	28	28	0.0015	13,800,000
Phosphor bronze	8.66	....	0.104	0.000018	.....	36	0.0039	16,000,000
Everdur	8.30	1050	.....	0.000017	30	6	......	15,000,000
Nickel silver	8.75	....	0.095	0.000018	7.6	5.2	0.0003	17,000,000
Iron	7.7	1535	0.110	0.000013	15	15	0.0062	25,000,000
Steel	7.9	1400	.....	0.000013	6–12	3–15	......	30,000,000
Cast iron	7.2	1000–1200	.....	0.000010	10–12	2–12	......	12–27,000,000
Duriron	7.0	1260	.....	0.000028	17.4	2.5		
14% Cr steel	7.7	1490	.....	0.000011	5	2.8	......	30,000,000
17% Cr steel	7.6	1400	.....	0.000010	5	.....	0.0015	
18–8 Cr-Ni steel	7.9	1400	0.118	0.000017	3.6	2.8	......	28,600,000
Zinc	7.14	420	0.094	0.000029	29	28.2	0.0040	†
Lead	11.38	327	0.031	0.000029	9	7.8	0.0041	†
Aluminum								
99.4%‡	2.7	660	0.229	0.000024	54	61	0.0040	10,000,000
6061-T6	2.70	582–652	0.23	0.000023	40	40	0.0026	10,000,000
2014-T6	2.80	510–638	0.23	0.000023	40	40	0.0026	10,600,000
Silver	10.51	960	0.056	0.000019	110	106	0.0040	9,000,000
Platinum	21.5	1755	0.032	0.000008	18	15	0.0036	23,000,000

* Expressed in percent of same properties of copper.
† Modulus of elasticity depends on rate of straining.
‡ Electrical conductor metal.

531. Magnesium Alloys. The important commercial alloys of magnesium usually contain 4 to 12% aluminum and 0.10 to 0.30% manganese, may be used for die casting or sand casting, and can be obtained in the form of rolled sheets, extruded shapes, and press forgings. Their outstanding feature is their light weight (specific gravity about 1.80), which makes them especially useful in the aircraft industry. Their thermal coefficient of expansion is about 0.000029/°C, and melting point is about 620°C. Tensile strengths of castings range from 21,000 to 34,000 lb/in², yield strengths from 9,000 to 22,000 lb/in², and elongation from 1 to 10% in 2 in. Forged or extruded alloys have tensile strengths of 33,000 to 43,000 lb/in², yield strengths 18,000 to 30,000 lb/in², and elongation of 5 to 17% in 2 in. The Brinell hardness ranges from 35 to 78, and the endurance limit from 6,000 to 17,000 lb/in² for the various alloys and heat-treatments. Magnesium alloys in a finely divided state will burn, but large sections are not easily ignited. (See J. Alico, "Introduction to Magnesium and Its Alloys," Chicago, Ziff-Davis Publishing Company, 1946; and publications of the Dow Chemical Co., Midland, Mich.)

532. Mechanical Properties of Nonferrous Metals and Alloys. See Table 4-102.

533. Stainless and Heat-resistant Alloys. Stainless steels are described in Pars. **491** and **492.** In addition, four classes of special heat-resistant alloys have been developed: (1) iron-chromium-nickel alloys containing 12.5 to 23.2% chromium, 9 to 32% nickel, small amounts of carbon and manganese, and in some cases small amounts of molybdenum, tungsten, columbium, titanium, and aluminum; (2) cobalt-nickel-chromium-iron alloys containing 20 to 43% cobalt, 20 to 43% nickel, 18 to 25% chromium, 3 to 33% iron, 0 to 10% molybdenum, small amounts of carbon, manganese, and silicon, and in some cases small amounts of tungsten, columbium, titanium, and aluminum; (3) nickel-base alloys containing 45 to 77% nickel, 15 to 22% chromium, 0 to 28% cobalt, 0 to 10% molybdenum, small amounts of carbon, manganese, and silicon, and in some cases small amounts of columbium, titanium, aluminum, and iron; (4) cobalt-base alloys containing 51 to 62% cobalt, 3 to 17% nickel, 19 to 77% chromium, 1% iron, either 5 to 6% molybdenum or 8 to 15% tungsten, and small amounts of carbon, manganese, and silicon. (Detailed data may be found in Metals Handbook, Par. **494,** and Report on Elevated-temperature Properties of Selected Super-strength Alloys, *ASTM Spec. Tech. Publ.* 160, 1954. See also Symposium on Metallic Materials for Service at Temperatures above 1600°F, ASTM STP 174, 1956; Relaxation Properties of Steels and Super-strength Alloys at Elevated Temperatures, ASTM STP 187, 1961.)

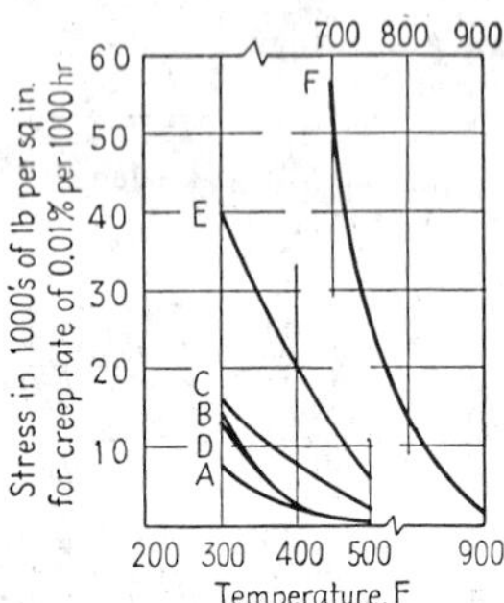

FIG. 4-83. Creep stress for a creep rate of 0.01% per 1,000 hr for several nonferrous metals. *A.* Copper, deoxidized and annealed, grain size 0.013 mm. *B.* Aluminum brass (76% Cu, 22% Zn, and 2% Al), annealed grain size 0.015 mm. *C.* 70–30 brass, annealed, grain size 0.085 mm. *D.* 70–30 brass, annealed, grain size 0.016 mm. *E.* Copper-nickel-phosphorus-tellurium alloy, quenched and aged (98.1% Cu, 1.11% Ni, 0.28% P, 0.51% Te). *F.* Monel metal, cold-drawn 40%.

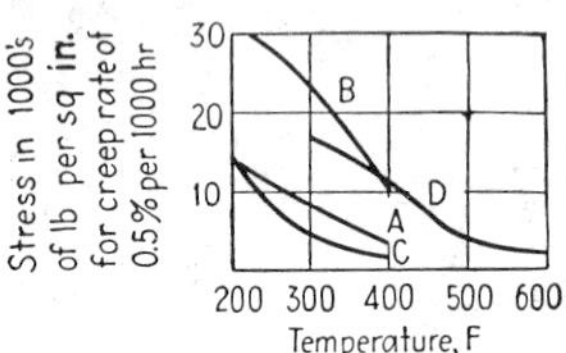

FIG. 4-84. Creep stress for a creep rate of 0.5% per 1,000 hr for several aluminum and magnesium alloys. *A.* 3003-H14 aluminum alloy. *B.* 6061-T6 aluminum alloy (see Par. **535**). *C.* AZ-63 HTS magnesium alloy (5.3 to 6.7% Al, 2.5 to 3.5% Zn, 0.15% Mn). *D.* EM 51 HTA magnesium alloy (1.2% Mn, 3.8 to 6.2% Ce).

534. Titanium and Titanium Alloys. Titanium alloys are important industrially because of their high strength-weight ratio, particularly at temperatures up to 800°F. The density of the commercial titanium alloys ranges from 4.50 to 4.85 g/cm³, or approximately 70% greater than aluminum alloy and 40% less than steel. The purest titanium currently produced (99.9% Ti) is a soft, white metal. The mechanical strength improves rapidly, however, with an increase of the impurities present, particularly carbon, nitrogen, and oxygen. The commercially important titanium alloys, in addition to these impurities, contain

Table 4-102. Approximate Mechanical Properties of Miscellaneous Nonferrous Metals and Alloys

(Compiled from various authorities)

Metal	Condition	Approximate composition, %	Strength in tension,[a] 1,000 lb/in²		Endurance limit,[c] 1,000 lb/in²	Brinell hardness No.	Elongation, 2 in, %	Weight, lb/in³
			Yield[b]	Ultimate				
Aluminum:								
EC-O	Wrought, annealed	99.60% min. Al	4	10	*d*	*d*	23[k]	0.098
EC-H19	Wrought, extra hard	99.60% min. Al	24	27	7	*d*	1.5[k]	0.098
1199-O	Wrought, annealed	99.99 Al	1.5	6.5	*d*	*d*	50	0.098
1100-H14	Wrought, half hard	99.0% min. Al	17	18	7	32	20	0.098
2024-T4	Quenched	Al 93, Cu 4.5, Mn 0.6, Mg 1.5	47[i]	68[i]	20	120	19	0.100
5052-H34	Wrought, half hard	Al 97, Mg 2.5, Cr 0.25	31	38	18	68	16	0.097
6061-T6	Quenched and aged	Al 98, Mg 1, Si 0.6, Cu 0.25, Cr 0.25	40	45	14	95	17	0.098
6063-T5	Quenched and aged	Al 99, Mg 0.7, Si 0.4	21	27	17	60	22	0.097
6101-T6	Quenched and aged	Al 99, Mg 0.6, Si 0.5	28	32	9	71	20	0.097
7075-T6[j]	Quenched and aged	Al 90, Zn 5.6, Mg 2.5, Cu 1.6, Cr 0.3	73	83	22	150	11	0.101
A356-T61	Permanent mold, quenched and aged	Al 9, Si 7, Mg 0.3	30	41	13	90	10	0.097
Brass	Annealed,	Cu 60; Zn 40	18	54	22	72	56	0.30
	cold-drawn	Cu 60; Zn 40	49	97	26	179	13	0.30
Bronze	Annealed,	Cu 95; Sn 5	13	46	23	74	67	0.32
	cold-drawn	Cu 95; Sn 5	59	85	27	166	12	0.32
Bronze, phosphor	Rolled sheet	Sn 8; Zn 0.2; P 0.1, Cu 91	*d*	50–100	*d*	35–90[g]	5–50	*d*
Bronze, aluminum	As rolled	Cu 88; Al 9; Fe 3	28	70	*d*	125	30	0.29
Bronze, manganese	Cast	Cu 57, Zn 41, +Mn, Al, Fe & Sn	30	70	17	115	33	0.30
Copper	Annealed	Commercially pure	5	32	10	47	56	0.32
Copper alloy	Wire	Cu 95; +Si, Sn, Al, Fe	*d*	68–120	*d*	*d*	0.8–3.8[e]	0.32
Copper-silicon alloy	Half hard	Cu 94, Si 3, +Mn, Zn, & Fe	40–50	71–81	*d*	80–90[g]	10–20	0.32
Copper, beryllium	(Heat-treat, half hard)	Be 2, Ni 0.5, Fe 0.2, Cu 97	132	173	35	340	4.8	0.297
Gunmetal	Cast	Cu 88, Sn 10, Zn 2–4	14–20	30–45	*d*	50–80	15–40	0.314
Magnesium alloy:								
FS1-H24, AMC52S-H	Sheet, hard	Mg 96, Al 3, Zn 1, Mn 0.2	32	42	14	73	15	0.064
01-T5, AMC58S-T5	Extrusions, aged	Mg 91, Al 8.5, Zn 0.5, Mn 0.2	39	54	19	82	7	0.065
ZK60A, AMA76S-T5	Extrusions, aged	Mg 94, Zn 5.5, Zr 0.6	43	52	16	82	12	0.066
H-T4, AM265-T4	Sand cast, quenched	Mg 91, Al 6, Zn 3, Mn 0.2	14	40	14	55	12	0.067
C-T6, AM260-T6	Sand cast, quenched and aged	Mg 89, Al 9, Zn 2, Mn 0.1	23	40	13	84	2	0.066
Monel metal (B)	Cold-drawn	Ni 67, Cu 30, +Fe & Mn	60–95	85–125	*d*	160–220	35–15	0.318
	Annealed		30–40	70–85		120–160	50–35	
Nickel	Cast	Ni 97, Si 1.2, +Mn & Fe	21	55	*d*	120	22	0.319
	Hot-rolled	Ni 99, +Fe, Cu, Mn, C, & Co	27	73	33	109	46	0.319
Zinc	Die-cast	Zn 95, Al 4, Cu 1	*f*	45[h]	*d*	70–85	3	0.242
	Rolled sheet	Zn 99.9	*f*	16–19[h]	*d*	*d*	40–70	0.242

[a] The yield strength in compression may be usually assumed approximately equal to the yield strength in tension; the ultimate in compression may be considered slightly above the yield strength; the strength in shear may be considered about 0.6 the strength in tension.
[b] Yield strength determined as stress corresponding to approximately 0.2% permanent set.
[c] For completely reversed cycles of flexural stress; polished specimens.
[d] Test data lacking.
[e] In 60 in.
[f] Not clearly defined owing to creep at low loads.
[g] Rockwell B scale hardness.
[h] Values for ordinary testing machine tests; much lower values probable for long-time tests under steady load.
[i] The strengths of extrusions more than about ¾ in thick will be 15 to 20% higher.
[j] These values are for other products than extrusions, which will have strengths 8 to 10% higher.
[k] Value for wire in 10-in gage length.

small percentages (1 to 7%) of (1) chromium and iron, (2) manganese, and (3) combinations of aluminum, chromium, iron, manganese, molybdenum, tin, or vanadium. The thermal conductivity of the titanium alloys is low, 8 to 10 Btu/(h) (ft²) (ft) (°F), and the electrical resistivity is high, ranging from 54 $\mu\Omega\cdot$cm for the purest titanium to approximately 150 $\mu\Omega\cdot$cm for some of the alloys. The coefficient of thermal expansion of the titanium alloys varies from 5 to 6.4 $\times 10^{-6}$/°F, and the melting-point range is from 2500 to 3100°F for the purest titanium. The tensile modulus of elasticity varies between 15 and 17 $\times 10^{-6}$ lb/in². The mechanical properties, at room temperature, for annealed commercial alloys range approximately as follows: yield strength 110,000 to 140,000 lb/in²; ultimate strength 116,000 to 160,000 lb/in²; elongation 5 to 18%; hardness 300 to 370 Brinell. On the basis of the strength-weight ratio many of the titanium alloys exhibit superior short-time tensile properties as compared with many of the stainless and heat-resistant alloys up to approximately 800°F. However, at the same stress and elevated temperature, the creep rate of the titanium alloys is generally higher than that of the heat-resistant alloys. Above about 900°F the strength properties of titanium alloys decrease rapidly. The corrosion resistance of the titanium alloys in many media is good. (Additional information may be obtained from J. L. Everhart, "Titanium and Titanium Alloys," New York, Reinhold Publishing Corporation, 1954; Metals Handbook, Par. **494;** and "Increased Production, Reduced Costs through Better Understanding of the Machining Process and Control of Materials, Tools and Machines. Titanium," Curtiss-Wright Corp. for U.S. Air Force, 1954, Vol. 3.)

535. Creep Stress for Nonferrous Metals. Figures 4-83 and 4-84 show for several metals the approximate creep strengths (pounds per square inch) to produce a given creep rate per 1,000 h. Creep characteristics are influenced by many factors, such as melting practice and grain size as well as chemical composition. Materials of the same composition will not necessarily have the same creep rate. Test data should therefore be regarded only as qualitative. (For other data see S. L. Hoyt, "Metal Data," New York, Reinhold Publishing Corporation, 1952.)

STRUCTURAL ALUMINUM ALLOYS

536. Structural aluminum alloys are alloys of aluminum with relatively small proportions of several other elements, including copper, manganese, magnesium, silicon, and zinc (see ASTM Specifications B209 and B211). The approximate compositions of alloys frequently used structurally are listed in Table 4-103.

Table 4-103. Approximate Compositions of Structural Aluminum Alloys
(Aluminum Company of America)

Alloy designation	Other elements added to aluminum, %					
	Copper	Manganese	Magnesium	Silicon	Chromium	Zinc
Wrought:						
1100	(Minimum 99.0 aluminum; Si + Fe = 1.0%)					
3003	....	1.2				
3004	....	1.2	1.0			
2014	4.4	0.8	0.4	0.8		
2024	4.5	0.6	1.5			
5052	....	....	2.5	...	0.25	
6061	0.25	....	1.0	0.6	0.25	
6063	....	....	0.7	0.4		
6070	0.25	0.7	0.8	1.3		
7075	1.6	....	2.5	...	0.3	5.6
Cast:						
43	....	....	....	5.0		
A356	....	....	0.30	7.0		
380	3.5	....	....	9.0		

537. Typical Properties of Structural Aluminum Alloys. See Table 4-104.

538. Hardening and Heat-treatment. Alloys 1100, 3003, 3004, and 5052 are produced in annealed and cold-worked tempers. The different tempers depend upon the amount of strain hardening or cold working which the metal receives during fabrication (rolling, drawing, etc.). The annealed temper is designated O, whereas the cold-worked tempers are designated by the letter H followed by two digits. The second

Table 4-104. Typical Properties of Aluminum Alloys in Forms Most Generally Used for Structural Purposes

Alloy designation	Weight, lb/in³	Yield strength, lb/in³	Tensile strength, lb/in²	Elongation in 2 in, %	Shear strength, lb/in²	Fatigue limit, lb/in²	Electrical conductivity
EC-O	0.098	4,000	10,000	23*	8,000	......	63
EC-H19	0.098	24,000	27,000	1.5*	15,000	7,000	62.5
1100-O	0.098	5,000	13,000	40	9,000	5,000	59
1100-H14	0.098	17,000	18,000	20	11,000	7,000	59
1100-H18	0.098	22,000	24,000	15	13,000	9,000	59
3003-O	0.099	6,000	16,000	40	11,000	7,000	46
3003-H14	0.099	21,000	22,000	16	14,000	9,000	46
3003-H18	0.099	27,000	29,000	10	16,000	10,000	46
3004-O	0.098	10,000	26,000	25	16,000	14,000	42
3004-H34	0.098	29,000	35,000	12	18,000	16,000	42
3004-H38	0.098	36,000	41,000	6	21,000	18,000	42
2014-T6	0.101	60,000	70,000	13	42,000	18,000	40
2024-T4	0.100	47,000†	68,000†	19	41,000	20,000	30
5052-O	0.097	13,000	28,000	30	18,000	16,000	35
5052-H34	0.097	31,000	38,000	16	21,000	18,000	35
5052-H38	0.097	37,000	42,000	14	24,000	20,000	35
6061-O	0.098	8,000	18,000	30	12,000	9,000	47
6061-T4	0.098	21,000	35,000	25	24,000	13,000	40
6061-T6	0.098	40,000	45,000	17	30,000	14,000	43
6063-T6	0.098	31,000	35,000	18	22,000	10,000	53
6101-T61	0.098	20,000	25,000	22	17,000	9,000	59
6070-O	0.098	10,000	21,000	20	14,000	9,000	52
6070-T4	0.098	30,000	49,000	20	30,000	13,000	40
6070-T6	0.098	52,000	57,000	12	34,000	14,000	44
7075-T6	0.101	73,000	83,000	11	48,000	22,000	33
43-F	0.097	8,000	19,000	8	14,000	8,000	37
356-T51	0.097	20,000	25,000	2	20,000	8,000	43
A356-T61	0.097	30,000	41,000	10	28,000	13,000	39
380	0.098	24,000	48,000	3	31,000	21,000	23

* Value for wire in 10-in gage length.
† The strengths of extrusions more than about ¾ in thick will be 15 to 20% higher.
Coefficient of thermal expansion per degree centigrade for alloys 3003, 3004, 2014, 5052, and 6061 is 0.000023. For additional data see Alcoa Structural Handbook, issued by Aluminum Company of America.

digit of the temper indicates the amount of cold working after the process anneal, 4 signifying half-hard and 8 signifying hard. Alloy 5052 can be annealed at 650°F, and alloys 3003 and 3004 at 750°F. For the solution heat-treated alloys 2014, 2024, 6061, 6063, and 7075, the high mechanical properties are obtained by solution heat-treatment followed by proper aging. Alloy 2024 ages at room temperature, whereas alloys 2014, 6061, 6063, and 7075 are artificially aged at a higher temperature. The temper of a heat-treatable alloy is designated by the letter T, followed by one or more digits, which indicate the particular treatment given the material. The heat-treatment practice used with these alloys is indicated in Table 4-105.

Table 4-105. Heat-treatment of Aluminum Alloys

Alloy designation	Heat-treating temperature, deg F	Approx duration of heating,* min	Quench†	Aging temperature, deg F	Time of aging
2014	930–950	15–60	Water	340	10 hr
2024	910–930	15–60	Cold water	Room	4 days‡
6061	960–980	15–60	Cold water	315–325	18 hr
7075	860–880	15–60	Cold water	250	24 hr

* This depends on the size and amount of material. In some cases, even longer times may be needed.
† The quench should be made with minimum time loss in transfer from furnace.
‡ More than 90% of the maximum properties are obtained during the first day of aging.

STONE, BRICK, CONCRETE, AND GLASS BRICK

539. Building Stone. Stone is any natural rock deposit or formation of igneous, sedimentary, and/or metamorphic origin, in either its original or its altered form. Building stone is the quarried product of such deposit or formation which is suitable for structural and ornamental purposes. Igneous or volcanic rock, such as granite or basalt, is rock of plutonic or volcanic origin, formed from a fused condition and crystalline in structure. Sedimentary rock, such as limestone, dolomite, and sandstone, is formed by the deposition of particles from water and laminated in structure. Metamorphic rock, such as gneiss, marble, and slate, is rock formation which, in the natural ledge, has undergone marked change in microstructure or character due to heat, pressure, or moisture and therefore exists in form different from the original. (See ASTM Designation C119 for additional definitions.)

540. Weight and Strength of Stone. The properties of stone from different quarries vary over a considerable range. Average values, from Watertown Arsenal Tests in 1894–1895, are listed in Table 4-106. (See M. O. Withey and G. W. Washa, "Materials of Construction," J. Wiley & Sons, Inc., New York, 1954.)

Table 4-106. Weight and Strength of Stone

Stone	Weight, lb per cu ft (Eckel)	Ultimate strength, lb per sq in.			Poisson's ratio
		Compression	Shearing	Modulus of rupture in flexure	
Granite	168	20,000	2250	1600	0.172–0.25
Sandstone	157	12,500	1685	1450	0.091–0.333
Limestone	165	9,000	1400	1240	0.27
Marble	175	12,600	1300	1500	0.222–0.345
Slate	174	......	....	7000	

541. Building brick made from clay or shale is required to have certain physical properties (see Table 4-107) under ASTM Specification C62-58: for sand-lime building brick see ASTM Specification C73-51. (Also see D. F. Miner and J. B. Seastone, "Handbook of Engineering Materials," New York, J. Wiley & Sons, Inc., 1955.)

Table 4-107. Physical Properties of Building Brick

Designation	Minimum compressive strength (brick flatwise), lb per sq in., average, gross area		Maximum water absorption by 5-hr boiling, %		Maximum saturation coefficient*	
	Average of 5 bricks	Individual	Average of 5 bricks	Individual	Average of 5 bricks	Individual
Grade SW	3000	2500	17	20	0.78	0.80
Grade MW	2500	2200	22	25	0.88	0.90
Grade NW	1400	1250	No limit	No limit	No limit	No limit

* The saturation coefficient is the ratio of absorption by 24-hr submersion in cold water to that after 5-hr submersion in boiling water.

542. Clay firebrick for stationary and marine boiler service is covered by ASTM Standard Specifications C64-61. **Refractory brick** for resisting high temperatures and the effects of very hot gases and molten slag and clinker is made of various special compositions known as "firebrick," "silica brick," "magnesia brick," "bauxite brick," "chromite brick," etc. (See ASTM specifications for brick and clay products; also see S. H. White, "Engineering Materials," 2d ed., New York, McGraw-Hill Book Company, 1948, Chap. 19.)

543. Structural gypsum products include gypsum tile, plaster board, wallboard, and plain or reinforced members cast in place. (For data on physical properties and specifications see *Bur. Standards Circ.* 108, Gypsum—Properties, Definitions and Uses, 1921; ASTM specifications for gypsum and various gypsum products; and Withey and Washa, Chap. 11.)

544. Cement is a material which in a plastic, amorphous, or liquid form fills the voids of a mass of particles or unites two adjacent surfaces and, by subsequent hardening, binds the whole together. The structure of the hardened cement may be crystalline or colloidal, depending upon the nature of the base. The cements used in engineering construction consist of hydraulic cements, limes, gypsum plasters, and bitumens. The most important are the calcareous cements, which contain lime as a base, made from some form of limestone. (See Withey and Washa, Chaps. 9, 10, and 12.)

545. Portland cement is the product obtained by finely pulverizing clinker produced by calcining to incipient fusion an intimate and properly proportioned mixture of argillaceous and calcareous materials, with no additions subsequent to calcination excepting water and calcined or uncalcined gypsum. It should not develop initial set in less than 45 min and should attain final set within 10 h. The average tensile strength of three standard briquets composed of 1 part cement and 3 parts standard sand (by weight) should equal or exceed 275 lb/in^2 after 7 days and 350 lb/in^2 after 28 days. (See ASTM Specification C150.)

546. Natural cement is the finely pulverized product resulting from the calcination of an argillaceous limestone at a temperature only sufficient to drive off the carbonic acid gas. It sets more rapidly but has less strength than portland cement. (See ASTM Specification C10.)

547. Mortar is a mixture of sand, screenings, or similar inert particles, with cement and water, which has the capacity of hardening into a rocklike mass. The inert particles are usually less than $\frac{1}{4}$ in in size. The proportions of cement to sand range all the way from 1:0 to 1:4 for various purposes.

548. Concrete is a mixture of crushed stone, gravel, or similar inert material with a mortar. The maximum size of inert particles is variable but usually less than 2 in. These inert constituents of mortar and concrete are known as the "aggregate." In making good concrete, the properties of the aggregate are as important as those of the cement. The fine aggregates consist of sand, screenings, mine tailings, pulverized slag, etc., with particle sizes less than $\frac{1}{4}$ in; the coarse aggregates consist of crushed stone, gravel, cinders, slag, etc. Rubble concrete is made by embedding a considerable proportion of boulders or stone blocks in concrete. The proportions by volume of cement, sand, and coarse aggregate range all the way from 1:1:2 for high compressive strength to 1:4:8 for structures requiring mass more than strength. In general, the strength of concrete increases with the density and richness of mix (proportion of cement) but is decreased in proportion to the amount of mixing water that is added beyond that required to produce a plastic workable mixture. In controlling the quality of a concrete of given mix, the ratio of the volume of mixing water to the volume of cement (water-cement ratio) is often used as a criterion of the strength. For proper curing the concrete should be kept moist for at least a week after placing, and care should be taken to prevent its freezing in cold weather during the early stages of curing. Freshly poured concrete gains strength very slowly in cold weather. (For systematic information see publications of the Portland Cement Association and ASTM Standards.)

549. Compressive Strength of Concrete. For concrete of common proportions cured under good conditions, 25 to 40% of the 2-year strength is developed in 7 days, 50 to 65% in 1 month, and 70 to 90% in 6 months. The tensile strength of concrete is very low (about one-tenth its compressive strength), and hence in structural members the concrete is usually designed to resist the compressive stresses only, the tensile strength of the concrete being considered negligible. For flexural members, steel reinforcing bars are usually inserted on the tensile side of the beam to resist the tensile stresses. The curves in Fig. 4-85 show typical variations of the compressive strength of concrete with the mix, density, and amount of mixing water used. The mix in each case is given as the proportion of cement to the total volume of aggregate used, and the water-cement ratio is plotted as the number of gallons of water per sack of cement. (One sack is assumed to be one cubic foot of cement.)

550. Flexural Strength of Concrete. Slabs and pavements are often designed on the basis of the flexural strength of the concrete. Typical variation of the modulus of rupture with the compressive strength of concrete for a number of different mixes and water-cement ratios is shown in Fig. 4-86. This curve is adapted from data in the 1928 report of the Director of Research of the Portland Cement Association on beams 7 in deep and 10 in wide, loaded with equal loads at the ⅓ points of a 36-in span. (See also Withey and Washa, Chap. 15.)

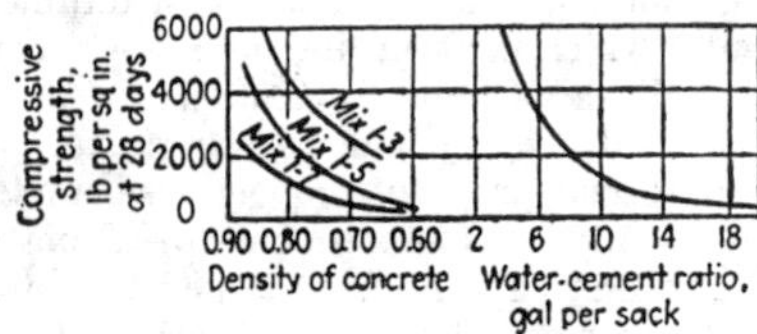

FIG. 4-85. Typical variations of strength of concrete with density and water-cement ratio.

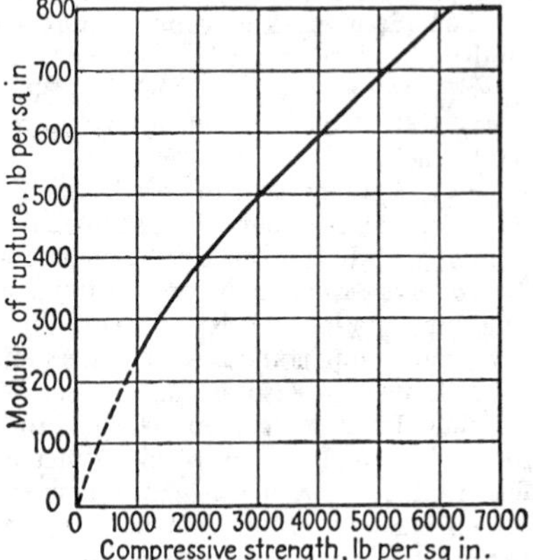

FIG. 4-86. Average relation of modulus of rupture to compressive strength of concrete. Mix 1 to 4 by volume. Age of test, grading, and source of aggregate variable.

551. Glass bricks are not primarily intended as load-bearing structural units but are used mainly for partition walls and outside walls of buildings in which the main loads are carried by other structural units. Safe working stresses in compression (obtained from data published by the manufacturers) would probably be less than about 200 lb/in². The blocks are ribbed in various ways to transmit a large proportion of the incident light with almost complete diffusion to eliminate glare. The interior of these blocks contains a thoroughly dry air under partial vacuum, and they are therefore excellent thermal insulating materials for wall construction.

LUMBER AND TIMBER

(See also Pars. **573** to **587**)

552. Wood consists structurally of cellulose, lignin, and ash-forming minerals. Extractives in wood are not part of its structure but contribute such properties as color, odor, taste, and resistance to decay. Cellulose comprises about 70% of wood; lignin, which cements the structural units of wood together, comprises 18 to 28% of wood; and the ash-forming minerals which are left when the lignin and cellulose are burned make up 0.2 to 1% of the wood.

553. The cross section of a log shows several well-defined features from the outside to the center: bark; wood, which in most species is clearly differentiated into sapwood and heartwood; and the pith, a small central core. With many species in temperate climates there is sufficient difference in the wood formed early (earlywood) and late (latewood) in the growing season to produce well-marked annual rings which represent 1 year's growth. Growth rings are most readily seen in ring-porous hardwoods such as ash and oak and softwoods like southern yellow pine and Douglas fir. In species such as sweetgum and soft maple, differentiation of earlywood and latewood is slight, and annual growth rings are difficult to recognize.

554. Sapwood and Heartwood. The sapwood layer is located next to the bark and is frequently lighter in color than the remaining portion of the log, the heartwood. Sapwood commonly ranges from 1½ to 2 in of radial thickness, although it may be wider in young, vigorously growing trees. Heartwood, with the infiltrations deposited in the cells, is usually more durable than sapwood, but sapwood is more easily treated with preservatives. There is no consistent difference in strength between sapwood and heartwood.

555. Moisture in Wood. Moisture in wood occurs in three forms: water vapor in air spaces in the cell cavities, capillary water in the cell cavities, and water molecules bound to the hydroxyl groups of the cellulose in the cell wall. In most end-use conditions, when wood is not in contact with water, nearly all the moisture present is bound water and is usually between 3 and 30% of the dry weight of the wood. Since this bound water tends to be at equilibrium with the vapor pressure of the surrounding atmosphere, the maximum amount of bound water in wood occurs in a saturated atmosphere. Any increase in moisture content above this maximum is due to capillary water, acquired from contact with liquid water.

556. Effect of Moisture on Strength. Clear wood increases in strength as it dries. The increase, however, does not begin until the wood has dried to the fiber saturation point, generally at about 30% moisture content. On sawed structural lumber the increase in strength of the clear wood from drying below the fiber saturation point is partly offset by the development of seasoning defects such as checks or splits. Design stress values generally are the same for large timbers whether green or dry. Increases ranging up to 25% in stress in the outer fiber in bending, 20% in modulus of elasticity, and 37½% in compression parallel to grain are recognized in 2-in lumber that is at 15% or lower moisture content when manufactured and is fabricated and used under conditions where that moisture content is not exceeded. (For a more complete discussion of this subject see ASTM D245, Methods for Establishing Structural Grades of Lumber.)

557. Shrinkage occurs in wood as a result of the loss of the water molecules which are bound in the cell wall and begins at 20 to 30% moisture content, depending upon the species. The anisotropic nature of wood results in unequal shrinkage in the three principal grain directions. Shrinkage is greatest in the transverse grain direction parallel to the growth rings (tangential), and the total shrinkage from green to the ovendry condition ranges from 4 to 13%, depending upon the species and density. In general, shrinkage increases with density. The transverse shrinkage perpendicular to the growth rings (radial) is usually about one-half that parallel to the growth rings and ranges from 2 to 8%. Total shrinkage along the grain ranges from 0.1 to 0.3%. Some representative shrinkage values are given in Table 4-108. In poles, the tangential shrinkage results in seasoning checks, while radial shrinkage is the dominant factor in reducing pole circumference. Circumferential shrinkage is about 1% for poles dried to approximately 20% moisture content.

558. Specific Gravity and Fiber Saturation Point. The specific gravity of wood is the weight of a given volume of wood divided by the weight of an equal volume of

Table 4-108. Specific Gravity and Shrinkage of Common Woods

Species	Average specific gravity*	Shrinkage, %	
		Radial	Tangential
Douglas fir, various regions	0.43–0.48	3.6–5.0	6.2–7.8
White ash	0.60	4.8	7.8
Aspen	0.38	3.3	7.9
Yellow birch	0.55	7.2	9.2
White fir	0.37	3.2	7.1
Western red cedar	0.33	2.4	5.0
Northern white cedar	0.31	2.2	4.9
Western hemlock	0.42	4.3	7.9
Southern yellow pines	0.51–0.61	4.4–5.5	7.4–7.8
Tamarack	0.53	3.7	7.4
White oak species	0.63–0.68	4.1–5.5	7.2–10.8
Red oak species	0.59–0.69	4.0–5.5	8.2–10.6
Hickory species	0.69–0.75	7.0–7.8	10.0–12.6

* Weight ovendry and volume at 12% moisture content.
† From green to ovendry condition, based on green dimension.

Table 4-109. Mechanical Properties of Various Woods in the Green Condition Grown in the United States

Species	Moisture content, %	Specific gravity*	Static bending		Compression parallel to grain maximum crushing strength, lb/in²	Compression perpendicular to grain stress at proportional limit, lb/in²	Tension perpendicular to grain maximum tensile strength, lb/in²	Hardness†		Maximum shearing strength parallel to grain, lb/in²
			Modulus of rupture, lb/in²	Modulus of elasticity, 1,000 lb/in²				End, lb	Side, lb	
Ash, black	85	0.45	6,000	1,040	2,300	350	490	590	520	860
Ash, white	42	0.55	9,600	1,460	3,990	670	590	1,010	960	670
Aspen	94	0.35	5,100	860	2,140	180	230	280	300	660
Basswood	105	0.32	5,000	1,040	2,220	170	280	290	250	600
Beech	54	0.56	8,600	1,380	3,550	540	720	970	850	1,290
Birch, yellow	67	0.55	8,300	1,500	3,380	430	430	810	780	1,110
Cottonwood, eastern	111	0.37	5,300	1,010	2,280	200	410	380	340	680
Elm, American	89	0.46	7,200	1,110	2,910	360	590	680	620	1,000
Elm, slippery	85	0.48	8,000	1,230	3,320	420	640	750	660	1,110
Hickory, shagbark	60	0.64	11,000	1,570	4,580	840	.....	.....	.....	1,520
Locust, black	40	0.66	13,800	1,850	6,800	1,160	770	1,640	1,570	1,760
Maple, silver	66	0.44	5,800	940	2,490	370	560	670	590	1,050
Maple, sugar	58	0.56	9,400	1,550	4,020	640	.....	1,070	970	1,460
Oak, red	80	0.56	8,300	1,350	3,440	610	750	1,060	1,000	1,210
Oak, white	68	0.60	8,300	1,250	3,560	670	770	1,120	1,060	1,250
Sweetgum	115	0.46	7,100	1,200	3,040	370	540	670	600	990
Sycamore	83	0.46	6,500	1,060	2,920	360	630	700	610	1,000
Yellow poplar	83	0.40	6,000	1,220	2,660	270	510	480	440	790
Baldcypress	91	0.42	6,600	1,180	3,580	400	300	440	390	810
Cedar, northern white	55	0.29	4,200	640	1,990	230	240	320	230	620
Cedar, Port Orford	43	0.40	6,200	1,420	3,130	280	180	460	400	830
Cedar, western red	37	0.31	5,100	920	2,750	270	230	430	270	710
Douglas fir, coast‡	38	0.45	7,700	1,560	3,780	380	300	570	500	900
Fir, white	110	0.37	5,900	1,160	2,900	280	300	410	340	760
Hemlock, western	77	0.42	6,600	1,310	3,360	280	290	500	410	860
Larch, western	58	0.48	7,700	1,460	3,760	400	330	580	510	870
Pine, lodgepole	65	0.38	5,500	1,080	2,610	250	220	320	330	680
Pine, ponderosa	91	0.38	5,100	1,000	1,940	280	310	310	320	700
Pine, loblolly	81	0.47	7,300	1,410	3,490	390	260	420	450	850
Pine, longleaf	62	0.54	8,700	1,600	4,300	480	330	550	590	1,040
Pine, shortleaf	81	0.46	7,300	1,390	3,430	350	320	410	440	850
Pine, western white	54	0.36	5,200	1,170	2,650	240	260	310	310	640
Spruce, Engelmann	80	0.32	4,500	960	2,190	220	240	310	260	590
Spruce, Sitka	42	0.37	5,700	1,230	2,670	280	250	430	350	760

* Specific gravity based on green volume and ovendry weight.
† Load required to embed a 0.444-in ball to half its diameter.
‡ Coast Douglas fir is defined as that coming from counties in Oregon and Washington west of the summit of the Cascade Mountains. For Douglas fir from other sources, see Western Wood Density Survey, *U.S. Forest Service Res. Paper* FPL 27.

Table 4-110. Mechanical Properties of Various Woods in the Air-dry Condition Grown in the United States

Species	Moisture content, %	Specific gravity*	Static bending: Modulus of rupture, lb/in²	Static bending: Modulus of elasticity, 1,000 lb/in²	Compression parallel to grain maximum crushing strength, lb/in²	Compression perpendicular to grain stress at proportional limit, lb/in²	Tension perpendicular to grain maximum tensile strength, lb/in²	Hardness†: End, lb	Hardness†: Side, lb	Maximum shearing strength parallel to grain, lb/in²
Ash, black	12	0.49	12,600	1,600	5,970	760	700	1,150	850	1,570
Ash, white	12	0.60	15,400	1,770	7,410	1,160	940	1,720	1,320	1,160
Aspen	12	0.38	8,400	1,180	4,250	370	260	510	350	850
Basswood	12	0.37	8,700	1,460	4,730	370	350	520	410	990
Beech	12	0.64	14,900	1,720	7,300	1,010	1,010	1,590	1,300	2,010
Birch, yellow	12	0.62	16,600	2,010	8,170	970	920	1,480	1,260	1,880
Cottonwood, eastern	12	0.40	8,500	1,370	4,910	380	580	580	430	930
Elm, American	12	0.50	11,800	1,340	5,520	690	660	1,110	830	1,510
Elm, slippery	12	0.53	13,000	1,490	6,360	820	530	1,120	860	1,630
Hickory, shagbark	12	0.72	20,200	2,160	9,210	1,760	.....	.....	.....	2,430
Locust, black	12	0.69	19,400	2,050	10,180	1,830	640	1,580	1,700	2,480
Maple, silver	12	0.47	8,900	1,140	5,220	740	500	1,140	700	1,480
Maple, sugar	12	0.63	15,800	1,830	7,830	1,470	.....	1,840	1,450	2,330
Oak, red	12	0.63	14,300	1,820	6,760	1,010	800	1,580	1,290	1,780
Oak, white	12	0.68	15,200	1,780	7,440	1,070	800	1,520	1,360	2,000
Sweetgum	12	0.52	12,500	1,640	6,320	620	760	1,080	850	1,600
Sycamore	12	0.49	10,000	1,420	5,380	700	720	920	770	1,470
Yellow poplar	12	0.42	10,100	1,580	5,540	500	540	670	540	1,190
Baldcypress	12	0.46	10,600	1,440	6,360	730	270	660	510	1,000
Cedar, northern white	12	0.31	6,500	800	3,960	310	240	450	320	850
Cedar, Port Orford	12	0.42	11,300	1,730	6,470	620	400	730	560	1,080
Cedar, western red	12	0.33	7,700	1,120	5,020	490	220	660	350	860
Douglas fir, coast‡	12	0.48	12,400	1,950	7,240	800	340	900	710	1,130
Fir, white	12	0.39	9,800	1,490	5,810	530	300	780	480	1,100
Hemlock, western	12	0.45	11,300	1,640	7,110	550	340	900	540	1,250
Larch, western	12	0.52	13,100	1,870	7,640	930	430	1,120	830	1,360
Pine, lodgepole	12	0.41	9,400	1,340	5,370	610	290	530	480	880
Pine, ponderosa	12	0.40	9,400	1,290	5,320	580	420	570	460	1,130
Pine, loblolly	12	0.51	12,800	1,800	7,080	800	470	750	690	1,370
Pine, longleaf	12	0.58	14,700	1,990	8,440	960	470	920	870	1,500
Pine, shortleaf	12	0.51	12,800	1,760	7,070	810	470	750	690	1,310
Pine, western white	12	0.38	9,500	1,510	5,620	440	.....	440	370	850
Spruce, Engelmann	12	0.34	8,700	1,280	4,770	470	350	560	350	1,030
Spruce, Sitka	12	0.40	10,200	1,570	5,610	580	370	760	510	1,150

* Specific gravity based on green volume and ovendry weight.

[2] Load required to embed a 0.444-in ball to half its diameter.

‡ Coast Douglas fir is defined as that coming from counties in Oregon and Washington west of the summit of the Cascade Mountains. For Douglas fir from other sources, see Western Wood Density Survey, *U.S. Forest Service Res. Paper* FPL 27.

water. As both the weight and volume of wood vary with the moisture content, specific gravity of wood is an indefinite quantity unless the conditions under which it is obtained are clearly specified. Most commonly, specific gravity of wood is based on weight ovendry and volume green, ovendry, or some intermediate moisture content. As wood dries, most of the liquid water held in the capillaries evaporates before the bound water molecules begin to leave the cell wall. The fiber saturation point is defined as the moisture content of the wood at this transition point and is the moisture content at which shrinkage begins. The radial and tangential shrinkage and specific gravity of a number of woods are listed in Table 4-108, which was abstracted from *U.S. Dept. Agr.*, 72, Wood Handbook, Forest Products Laboratory, Forest Service.

559. The mechanical properties of the commercially important woods of the United States have been evaluated by the U.S. Forest Products Laboratory. These tests were conducted in accordance with ASTM Standard D143, which specifies small, clear specimens to eliminate the influence of naturally occurring physical defects in the wood.

Tables 4-109 and 4-110 show some mechanical properties for wood in the green condition and at 12% moisture content, respectively. The green properties are obtained from specimens at essentially the same moisture content as in the living tree, well above the fiber saturation point (see Par. **558**).

The increase in strength normally associated with the loss of moisture from wood fibers in small specimens during drying is generally not applicable to large structural members, where the increase is generally offset by drying defects and the loss in section modulus due to shrinkage during drying. The properties in the green condition therefore are generally used as the base for the development of design stresses in most engineering applications. However, in many other instances, where the geometry of a member is roughly comparable in size with the small, clear specimens evaluated and the atmosphere to which the member is to be exposed is relatively dry, the air-dry properties may be used.

Tables 4-109 and 4-110 have been abstracted to include several of the more important hardwood and softwood species and the mechanical properties of each which are likely to be uniquely important for specific uses encountered in electrical engineering applications.

For additional data on other strength properties and other species, refer to the Wood Handbook, *U.S. Dept. Agr., Agr. Handbook* 72.

560. Decay and Its Prevention. At ordinary temperatures, wood is very stable and unless attacked by living organisms remains the same for centuries, either in air or under water. Fungi are the chief enemies of wood, and they thrive best with warmth and abundance of moisture and air, e.g., in contact with the ground. Higher temperatures near the surface of the ground, together with adequate air and a greater prevalence of fungi, cause decay to progress faster near the ground line than at several feet below. Perfect seasoning, together with protection against the entrance of moisture and impregnating with fungus-inhibiting compounds (see Par. **561**) which prevent fungi from feeding on the wood, is the best means of preservation. Only the heartwood is resistant, however; consequently, the sapwood should be preservative-treated, irrespective of the species of wood, if decay resistance is needed. Various species differ materially in their natural resistance to decay. For systematic presentation of the subject of wood preservation, see the references in Par. **572** and the following:

HUNT, G. M., and GARRATT, G. A. "Wood Preservation"; New York, McGraw-Hill Book Company, 1966.

MACLEAN, J. D. Preservative Treatment of Wood by Pressure Methods, *U.S. Dept. Agr., Agr. Handbook* 40, 1960.

American Wood-Preservers' Association "Manual of Recommended Practice."

U.S. Federal Supply Service Wood Preservation: Treating Practices, Specification TT-W-571, 1961.

561. Wood Preservatives. Wood preservatives fall into two main classes: (1) oil-type preservatives and (2) water-borne metallic salts. The former may be further subdivided into (*a*) coal tar creosote with and without the mixture of cheaper materials such as petroleum or coal tar and (*b*) solutions of toxic organic chemicals such as pentachlorophenol dissolved in petroleum oils. Oil-type preservatives are used extensively for products that are exposed to ground contact whereby resistance to leaching is an important requirement of the preservative. These products include poles, crossties, piling, bridge timbers, and fence posts. Water-borne preservatives are used mainly for the treatment of lumber. Wood treated with a water-borne preservative is clean, paintable, and odorless.

Creosote is a distillate of coal tar formed during the coking of coal. On the basis of the quantity of wood treated, it is the most important preservative. Much of the treated wood is used in ground contact; appreciable amounts are also used in coastal waters infested with marine organisms that bore into and destroy untreated wood.

Pentachlorophenol dissolved in petroleum oils of varied nature has come into wide use. As a general rule, the effectiveness is highest when high-boiling oils are used as solvents, but relatively low-boiling oils are sometimes used in the treatment of products, such as millwork, having high cleanliness requirements. Water-repellent materials are generally added to the preservative in millwork treatments to minimize the dimensional changes that accompany fluctuations in the moisture content of the wood. A relatively new type of treatment comprises the solution of pentachlorophenol in a liquefied petroleum gas which is subject to practically complete removal by evaporation, leaving the treated wood very clean and readily paintable.

Water-borne preservatives are generally mixtures of several inorganic salts, the most important of which are salts of copper, chromium, arsenic, and zinc. Sodium fluoride is an ingredient of two widely used commercial preservatives. Traditionally, the use of water-borne preservatives has been restricted to situations where resistance to leaching is not required; however, several formulations now available comprise mixtures of salts that undergo chemical reaction within the wood with the formation of relatively insoluble toxic compounds. Such preservatives give good protection to wood exposed to wet conditions.

Paints, varnishes, and stains are used for decorative effects, but they also afford surface protection by retarding moisture changes and thus decreasing checking, warping, and weathering. Such protection is only superficial, however, and internal decay may be expected unless the wood is kept dry.

Fire-retardant chemicals such as ammonium phosphate and sulfate and salts of zinc and boron are used to decrease the flammability of wood. Some fire-retardant formulations also give protection against decay. For additional information see bibliography in Pars. **560** and **572**.

562. Methods of Treating Wood. The methods of preservative treatment may be divided into two classes, pressure and nonpressure. In pressure methods the wood is enclosed in a vessel, and the liquid preservative is forced into the wood under considerable hydrostatic pressure. Nonpressure methods do not utilize artificial pressure, the preservative being applied by dipping, soaking, brushing, or spraying. A third method, somewhat distinct from the others and called the thermal method, may be mentioned. It consists in heating the wood to expel air and then allowing the wood to cool in the liquid, whereby a partial vacuum forms in the internal spaces. Although movement of the liquid into the wood is due to atmospheric pressure, the process is not classed among pressure processes.

There are several modifications of the pressure process. In the full-cell process, the wood is subjected to a vacuum in order to evacuate the internal cavities before the liquid is injected. This process is generally used for the treatment of marine piling, which requires high retention of creosote for protection against wood-boring animals. The process is also used commonly in treatments of lumber with water-borne preservatives. Much wood for land use is treated with oil-type preservatives by one of the so-called empty-cell methods, whereby it is possible to increase the depth of penetration obtained with a limited retention of preservative. In the Rueping process, air is first injected to create within the wood a pressure greater than atmospheric. The cylinder is filled with preservative in such a way that the injected air is trapped in the wood. The pressure is then increased to force preservative into the wood. After the pressure is released and the cylinder drained, the compressed air in the wood expands to expel some of the preservative. The recovered preservative is called kickback. The Lowry process differs from the Rueping process in that no initial air pressure is applied. The air normally present is compressed during the pressure cycle and produces some kickback when pressure is released.

The conditioning of the wood prior to treatment is an important step. Air seasoning, kiln drying, and various processes of cylinder conditioning are employed. The latter include steaming plus vacuum, boiling in oil under vacuum, and vapor drying, in which green wood is surrounded by hot vapors of distillates of coal tar or petroleum.

When oil preservatives are applied by simple soaking methods, the wood should be well seasoned in order to provide air spaces into which the oil may move. Oil preservatives of low viscosity are preferable. The results attainable vary greatly with the species of wood.

Diffusion methods depend upon the diffusion of water-soluble chemicals into the moisture present in green wood. Here again, the species of wood is an important factor, but the results are affected by other factors such as the nature of the chemical, the concentration of the solution, and the duration of the soaking period.

563. Applications of Preservative Treatment. Preservative treatments are applied to many wood products, the most important being poles, crossties, lumber and structural timbers, fence posts, piling, and crossarms. Approximately 85% of all crossties treated in 1964 were of hardwood species, with oak accounting for 53%. The coniferous species dominated the treatment of other wood items, with southern pine being the most important, followed by Douglas fir.

564. Advantages of Preservative Treatment. In addition to the conservation of a natural resource, preservative treatment results in economic savings due to increased service life and reduced maintenance costs. This has been recognized for many years by railroad companies, utility companies, and other large users of wood products. Because of demonstrated savings, practically all crossties and poles are now given a preservative treatment before installation. There has been a gradual increase in the volume of lumber treated annually, due to more widespread knowledge of the need for such treatment when the wood is to be used under conditions favorable to attack by decay or insects. For best performance it is desirable that all machining operations be completed before treatment.

565. Strength of Treated Lumber. (See also Pars. **575** and **578**.) The effect of a preservative such as creosote or pentachlorophenol, in and of itself, on the strength of treated lumber appears to be negligible. It may be necessary, however, in establishing design stress values, to take into account possible reductions in strength that may result from temperatures or pressures used in the conditioning or treating processes. Results of tests of treated wood show reductions of stress in extreme fiber in bending and in compression perpendicular to grain, ranging from a few percent up to 25%, depending on the processes used. Compression parallel to grain is affected less and modulus of elasticity very little. The effect on resistance to horizontal shear can be estimated by inspection for shakes and checks after treatment. Strength reductions for wood poles agreed upon in formulating fiber-stress recommendations in American Standard Specifications and Dimensions for Wood Poles, ASA Designation 05.1-1963, range from 0 to 15% in various species, depending upon the conditioning and treating processes. Treating conditions specified by the American Wood Preservers' Association should never be exceeded. Reductions of strength can be minimized by restricting temperatures, heating periods, and pressures as much as is consistent with obtaining the absorption and penetration required for proper treatment.

566. Effect of Preservative Treatment on Electrical Resistivity. The electrical resistivity of wood depends on its moisture content to a much greater degree than any other single variable. Ovendry wood is an excellent insulator, but as the wood absorbs moisture, its resistivity decreases rapidly. Wood in normal use, however, where its moisture content may range from about 6 to 14%, is still a good enough insulator for many electrical applications.

When wood has been treated with salts for preservative or fire-retardant purposes, its electrical resistivity may be markedly reduced. The effect of such salt treatment is small when the wood moisture is below about 8% but increases rapidly as the moisture content exceeds about 10%. Treatment with creosote or pentachlorophenol has practically no effect on the resistivity of wood.

The resistivity of wood decreases by about a factor of 2 for each increase of 10°C in the temperature and is about half as great for current flow along the grain as across the grain.

567. American Lumber Standards. Simplified Practice Recommendation 16-53, American Lumber Standards for Softwood Lumber, is a voluntary standard of manufacturers, distributors, and users, promulgated in cooperation with the U.S. Department of Commerce. It provides for use classifications of (1) yard lumber, (2) structural lumber, and (3) factory and shop lumber. Different grading rules apply to each class. Size standards and generalized grade descriptions are a part of SPR 16-53, but details of grading rules are left to the organized regional agencies of the lumber manufacturing industry. The grades and working stresses for structural lumber are referred by SPR 16-53 to the authority of ASTM D245, Methods for Establishing Structural Grades of Lumber, or D2018, Recommended Practice for Determining Design Stresses for Load-sharing Lumber Members.

568. Standard Commercial Names. Standard commercial names of the most commonly used structural softwoods from ASTM D1165, Standard Nomenclature of Domestic Hardwoods and Softwoods, are as follows:

Cedar:
- Alaska cedar
- Port Orford cedar
- Western red cedar

Fir:
- Douglas fir
- White fir

Hemlock:
- Eastern hemlock
- West Coast hemlock

Larch, Western

Pine:
- Jack pine
- Lodgepole pine
- Norway pine
- Ponderosa pine
- Southern yellow pine

Redwood

Spruce:
- Eastern spruce
- Engelmann spruce
- Sitka spruce

569. Standard Structural Grades. Detailed descriptions of the standard structural grades are published in the grading rule books of the organized regional agencies of the lumber manufacturing industry. These are subject to review for compliance with the general requirements of SPR 16-53, American Lumber Standards for Softwood Lumber (see Par. **567**). The principal-use classes of structural lumber are: (1) *joists and planks,* pieces of rectangular cross section 2 to 4 in thick and 4 or more in wide (nominal dimensions), graded primarily for bending strength edgewise or flatwise; (2) *beams and stringers,* pieces of rectangular cross section 5 by 8 in (nominal dimensions) and up, graded for strength in bending when loaded on the narrow face; and (3) *posts and timbers,* pieces of square or nearly square cross section, 5 by 5 in (nominal dimensions) and larger, graded primarily for use as posts and columns.

570. Working Stresses. (See also Pars. **575** and **578.**) Working stresses recommended by the lumber industry for their structural grades are found with the detailed grade descriptions in the grading rule books of the organized regional agencies of the industry. A complete listing of all structural grades and their working stresses is found in the "National Design Specification for Stress-grade Lumber and Its Fastenings," published by the National Forest Products Association. Values for a few typical grades are shown in Table 4-111. Working stresses vary according to the grades and sizes of lumber and their condition with respect to moisture content. Stresses are adjustable also for duration of load and for special conditions such as extreme temperature. Stress increases are provided for "load-sharing members" in which the safety of the structure depends upon the strength of the assemblage of members rather than upon the lowest strength value for any single member. These stress modifications are described in ASTM standards.

Allowable working stresses for the structural grades of lumber are also a part of certain use specifications, such as the Minimum Property Standards of the Federal Housing Administration, the American Railway Engineering Association Manual, and various local or regional building codes. These allowable values may or may not coincide with the lumber industry stress recommendations for the same species and grade.

571. Wood-base Panel Materials. Included in this category are plywood, insulating board, hardboard, particle board, and the medium-density building fiberboards. Plywood, normally fabricated by bonding an odd number of layers of veneers together with the grain direction in adjacent plies at right angles to each other, is more dimensionally stable and more uniform in strength in the plane of the sheet than wood. Qualities of glue line and veneer permitted are set by the various commercial standards for plywood and determine the grades under which plywood is sold. In general, glue-line quality determines whether plywood is classed as being suitable for interior or exterior use.

Commercial Standard CS 35 governs qualities of hardwood plywood; CS 45 and CS 122 do the same for Douglas fir and western softwood plywood, and a southern pine standard is being developed. Certain grades and species of plywood have design working stresses assigned. For further information on methods of designing with plywood, see Forest Products Laboratory, *U.S. Dept. Agr., Agr. Handbook* 72, 1955; Nelson S. Perkins, "Plywood, Properties, Design and Construction," Tacoma, Wash., American Plywood Association, 1962.

Insulation board, hardboard, and medium-density building fiberboard are panel

Table 4-111. Typical Stress Grades and Working Stresses for Structural Lumber*
(Normal duration of load and dry conditions of use)

Species	Grade	Allowable working stress				
		Bending or tension parallel, lb/in²	Horizontal shear, lb/in²	Compression perpendicular to grain, lb/in²	Compression parallel to grain, lb/in²	Modulus of elasticity, lb/in²
Douglas fir	Select structural beams and stringers	1,900	120	415	1,400	1,760,000
	Construction joists and planks	1,500	120	390	1,200	1,760,000
	Standard joists and planks	1,200	95	390	1,000	1,760,000
	Construction posts and timbers	1,200	120	390	1,200	1,760,000
West Coast hemlock	Construction joists and planks	1,500	100	365	1,100	1,540,000
	Construction, MC joists and planks	1,650	105	365	1,250	1,540,000
	Standard joists and planks	1,200	80	365	1,000	1,540,000
Western larch	Standard joists and planks	1,200	95	390	1,000	1,760,000
Southern pine	Dense structural 72 beams and stringers	2,000	135	455	1,550	1,760,000
	Dense structural 58 beams and stringers	1,600	105	455	1,300	1,760,000
	No. 1 dimension, 2-in	1,500	120	390	1,350	1,760,000
	No. 2 dimension, 2-in	1,200	105	390	900	1,760,000
Norway pine	Common structural joists and planks	1,100	75	360	775	1,320,000
Redwood	Heart structural joists and planks	1,300	95	320	1,100	1,320,000
Eastern spruce	1300f structural joists and planks	1,300	95	300	975	1,320,000

* Compiled from "National Design Specification for Stress-grade Lumber and Its Fastenings"; Washington, D.C., National Forest Products Association, 1962.

products made by reducing wood substance to fiber and reconstituting the fiber into stiff panels 4 by 8 ft in area or larger. Insulation board is of either interior or water-resistant quality and is usually manufactured for use where combinations of thermal and sound-insulating properties and stiffness and strength are desired. Hardboard with a density of 50 lb/ft³ or more is used in many applications where a relatively thin, hard, uniform panel material is required. Of great importance in the electrical field are special high-density hardboard products expressly manufactured with high dielectric properties. Medium-density fiberboards with a density between that of insulation board and hardboard are new products.

Particle boards are panel products made by gluing small pieces of wood in a form such as flakes and shavings into relatively thick, rigid panels. Thermosetting resins, usually urea or phenolformaldehyde, are used to provide bonds of either interior or water-resistant quality. Commercial Standards CS 42, CS 236, and CS 251 govern minimum qualities of regular insulation board, particle board, and hardboard, although because of the many special products not covered by commercial standards, individual manufacturers should be consulted. The important physical and strength properties of the various board products are indicated by Table 4-112.

Table 4-112. Strength and Mechanical Properties of Wood-base Fiber and Particle Panel Materials*

Material	Density, lb/ft	Specific gravity	Modulus of rupture, lb/in²	Modulus of elasticity (bending), 1,000 lb/in²	Tensile strength parallel to surface, lb/in²	Tensile strength perpendicular to surface, lb/in²	Compression strength parallel to surface, lb/in²	24-hr water absorption		Thickness swelling, 24-h soak, %	Maximum linear expansion,† %	Thermal conductivity, Btu/(ft²)(h)(°F)(in thickness)
								% by volume	% by weight			
Fibrous-felted boards:												
1. Structural insulating board	10–26	0.16–0.42	200–800	25–125	200–500	10–25	...	1–10	...	...	0.5	0.27–0.45
2. Medium-density building fiberboard	26–50	0.42–0.80	400–4,000	90–700	800–2,000	...	500–3,400	...	6–150	...	0.2–1.30‡	0.50–0.60
3. Hardboard:												
a. Untempered	50–80	0.80–1.28	3,000–7,000	400–800	3,000–6,000	...	1,800–6,000	...	3–30	10–25	0.6	0.80–1.40
b. Tempered	60–80	0.96–1.28	6,500–10,000	800–1,000	4,000–7,800	...	4,200–6,000	...	3–20	8–15	0.4	1.10–1.50
4. Super hardboard	85–90	1.36–1.44	10,000–12,500	1,250	7,800	500	26,500	...	0.3–1.2	...	...	1.85
Particle boards:												
1. Insulating type	10–26	0.16–0.42	700	...	...	...	...	...	...	...	...	0.36
2. Medium-density type	26–50	0.42–0.80										
a. Extrusion	...	...	Values not presented because extruded boards are always used and tested with facings applied									
b. Flat-platen pressed	...	...	1,500–8,000	150–700	500–4,000	40–400	1,400–2,800	...	20–75	20–75	0.6	0.40–1.00
3. Hard-pressed type	50–80	0.80–1.28	3,000–7,500	400–1,000	1,000–5,000	275–400	3,500–4,000	...	15–40	15–40	0.85	1.10–1.50

* The data presented are general round-figure values, accumulated from numerous sources; for more exact figures on a specific product, individual manufacturers should be consulted or actual tests made. Values are for general laboratory conditions of temperature and relative humidity.

† Expansion resulting from a change in moisture content from equilibrium at 50% relative humidity to equilibrium at 90% relative humidity.

‡ For homogeneous and laminated boards, respectively.

572. References on Lumber and Timber. Many references to technical literature are given in the preceding paragraphs of this subsection and also under Wood Poles and Crossarms. The following references may be consulted for more general treatment of the subject:

Wood Handbook, *U.S. Dept. Agr., Agr. Handbook* 72, Forest Products Laboratory, 1955.

"Timber Design and Construction Handbook"; New York, Timber Engineering Company, F. W. Dodge Corporation, 1956.

Scofield and O'Brien. "Modern Timber Engineering"; New Orleans, Southern Pine Association, 1963.

Appropriate standards of the ASTM.

Publications of the U.S. Forest Products Laboratory, Madison, Wis.

Brown, Panshin, A. J., and Forsaith. "Textbook of Wood Technology"; New York, McGraw-Hill Book Company, 1952.

Hunt, G. M., and Garratt, G. A. "Wood Preservation," 2d ed.; New York, McGraw-Hill Book Company, 1953.

Wangaard. "The Mechanical Properties of Wood"; New York, John Wiley & Sons, Inc., 1950.

WOOD POLES AND CROSSARMS

By David Burns

573. Wood Species. (See also Pars. **552** to **572**.) Western red cedar and southern yellow pine are two species most commonly used in the United States for poles to support electric supply and communication equipment. In the northeast part of the country there are still a number of chestnut and northern white cedar poles in service, but these species are no longer available for purchase. Douglas fir, lodgepole pine, and western larch are used in considerable numbers, particularly in the western states. Other species that can be used for poles but not considered so desirable as western cedar or southern yellow pine are eastern hemlock, eastern larch, jack pine (large usage in Canada), northern white pine, ponderosa pine, red (Norway) pine, southern white cedar, spruce, sugar pine, western hemlock, western white pine, and white fir.

574. Standards for Wood Poles. The ASA specifications for wood poles serve as a basis for purchasing and use. The ASA specifications cover fiber stresses, dimensions, defect limitations, and manufacturing requirements. These specifications are also the basis for standards and specifications of using organizations such as Edison Electric Institute and American Telephone and Telegraph Co. EEI Specification TD-100 for non-pressure-treated cedar poles and AT&T Specification AT-7312 are good examples of users' pole-purchasing specifications.

575. Ultimate Fiber Stresses. The ultimate fiber stresses approved by the ASA and contained in its Standard 05.1-1963 are as shown in Tables 4-113, 4-114, and 4-115.

These tables cover all species of poles normally used in communication and electrical power construction.

576. Pole Dimensions. The circumference at "6 ft from butt" in Standard 05.1-1963 is based on the following principles:

a. The classes from the lowest to the highest were arranged in approximate geometric progression, the increments in breaking load between classes being about 25%.

b. The dimensions were specified in terms of circumference in inches at the top and circumference in inches at 6 ft from the butt for poles of the respective classes and lengths, except for three classes having no requirement for butt circumference.

c. All poles of the same class and length were to have, when new, approximately equal strength or, in more precise terms, equal moments of resistance at the ground line.

d. All poles of different lengths within the same class were of sizes suitable to withstand approximately the same breaking load, on the assumption that the load is applied 2 ft from the top and that the break (failure) would occur at the ground line.

The breaking loads referred to in (*d*) above for the classes for which "6 ft from butt" circumferences are given are as follows: Class 1, 4,500 lb; Class 2, 3,700 lb; Class

Table 4-113. Dimensions of Northern White Cedar and Engelmann Spruce Poles

	Class		1	2	3	4	5	6	7	9	10
	Minimum circumference at top, in		27	25	23	21	19	17	15	15	12
	Length of pole, ft	Ground-line distance from butt,* ft	Minimum circumference at 6 ft from butt, in								
Northern white cedar poles (based on a fiber stress of 4,000 lb/in²)	20	4	38.0	35.5	33.0	30.5	28.0	**26.0**	**24.0**	**22.0**	**17.5**
	25	5	42.0	39.5	36.5	34.0	31.5	**29.0**	**27.0**	**24.0**	**19.5**
	30	5½	45.5	**43.0**	**40.0**	**37.0**	**34.5**	**32.0**	**29.5**	**26.0**	
	35	6	49.0	**46.0**	**42.5**	**39.5**	**37.0**	**34.0**	**31.5**		
	40	6	**51.5**	**48.5**	**45.0**	**42.0**	**39.0**	**36.0**			
	45	6½	**54.5**	**51.0**	**47.5**	**44.0**	**41.0**				
	50	7	**57.0**	**53.5**	**49.5**	46.0	43.0				
	55	7½	**59.0**	**55.5**	**51.5**	48.0	44.5				
	60	8	**61.0**	**57.5**	**53.5**	50.0					
Engelmann spruce poles (based on a fiber stress of 5,600 lb/in²)	20	4	34.5	32.0	30.0	28.0	25.5	**23.5**	**22.0**	**19.0**	**15.0**
	25	5	38.0	35.5	33.0	30.5	28.5	**26.0**	**24.5**	**21.0**	**16.5**
	30	5½	41.0	**38.5**	**35.0**	**33.0**	**30.5**	**28.5**	**26.5**	**22.5**	
	35	6	43.5	**41.0**	**38.0**	**35.5**	**32.5**	**30.5**	**28.0**		
	40	6	**46.0**	**43.5**	**40.5**	**37.5**	**34.5**	**32.0**			
	45	6½	**48.5**	**45.5**	**42.5**	**39.5**	**36.5**				
	50	7	**50.5**	**47.5**	**44.5**	41.0	38.0				
	55	7½	**52.5**	**49.5**	**46.0**	42.5	39.5				
	60	8	**54.5**	**51.0**	**47.5**	44.0					
	65	8½	**56.0**	**52.5**	**49.0**	45.5					
	70	9	**57.5**	**54.0**	**50.5**	47.0					
	75	9½	**59.5**	**55.5**	**52.0**	48.5					
	80	10	**61.0**	**57.0**	53.5	49.5					
	85	10½	**62.5**	**58.5**	54.5						
	90	11	**63.5**	**60.0**	56.0						
	95	11	**65.0**	**61.0**	57.0						
	100	11	**66.0**	**62.0**	58.0						

NOTE: Classes and lengths for which circumferences at 6 ft from the butt are listed in boldface type are the preferred standard sizes. Those shown in light type are included for engineering purposes only.

* The figures in this column are intended for use only when a definition of ground line is necessary in order to apply requirements relating to scars, straightness, etc.

3, 3,000 lb; Class 4, 2,400 lb; Class 5, 1,900 lb; Class 6, 1,500 lb; Class 7, 1,200 lb; Class 9, 740 lb; Class 10, 370 lb.

Minimum top circumferences and minimum circumferences at 6 ft from butt are given in Tables 4-113, 4-114, and 4-115.

Length. Poles under 50 ft in length shall not be more than 3 in shorter or 6 in longer than nominal length. Poles 50 ft or over in length shall not be more than 6 in shorter or 12 in longer than nominal length.

Length shall be measured between the extreme ends of the pole.

Circumference. The minimum circumference at 6 ft from the butt and at the top, for each length and class of pole, is listed in the tables of dimensions. The circumference at 6 ft from the butt of poles shall be not more than 7 in or 20% larger than the specified minimum, whichever is greater.

The top dimensional requirement shall apply at a point corresponding to the minimum length permitted for the pole.

Classification. The true circumference class shall be determined as follows: Measure the circumference at 6 ft from the butt. This dimension will determine the true class of the pole, provided that its top (measured at the minimum length point) is large enough. Otherwise the circumference at the top will determine the true class, provided that the circumference at 6 ft from the butt does not exceed the specified minimum by more than 7 in or 20%, whichever is greater.

The above information relating to the pole standards approved by the ASA does not constitute the complete standards. For further information, consult the standards, which may be obtained at a nominal charge from the ASA, 10 East 40th Street, New York, N.Y. 10016.

Table 4-114. Western Red Cedar, Ponderosa Pine, Douglas Fir, and Southern Pine

	Class		1	2	3	4	5	6	7	9	10
	Minimum circumference at top, in		27	25	23	21	19	17	15	15	12
	Length of pole, ft	Ground-line distance from butt,* ft	Minimum circumference at 6 ft from butt, in								
Western red cedar and ponderosa pine poles (based on a fiber stress of 6,000 lb/in^2)	20	4	33.5	31.5	29.5	27.0	25.0	**23.0**	**21.5**	**18.5**	**15.0**
	25	5	37.0	34.5	32.5	30.0	28.0	**25.5**	**24.0**	**20.5**	**16.5**
	30	5½	40.0	**37.5**	**35.0**	**32.5**	**30.0**	**28.0**	**26.0**	**22.0**	
	35	6	42.5	**40.0**	**37.5**	**34.5**	**32.0**	**30.0**	**27.5**		
	40	6	**45.0**	**42.5**	**39.5**	**36.5**	**34.0**	**31.5**	29.5		
	45	6½	**47.5**	**44.5**	**41.5**	**38.5**	**36.0**	33.0	31.0		
	50	7	**49.5**	**46.5**	**43.5**	40.0	37.5	34.5	32.0		
	55	7½	**51.5**	**48.5**	**45.0**	42.0	39.0	36.0			
	60	8	**53.5**	**50.0**	**46.5**	43.5	40.0	37.0			
	65	8½	**55.0**	**51.5**	**48.0**	45.0	41.5				
	70	9	**56.5**	**53.0**	**49.5**	46.0	42.5				
	75	9½	**58.0**	**54.5**	**51.0**	47.5					
	80	10	**59.5**	**56.0**	52.0	48.5					
	85	10½	**61.0**	**57.0**	53.5						
	90	11	**62.5**	**58.5**	54.5						
	95	11	**63.5**	**59.5**	56.0						
	100	11	**65.0**	**61.0**	57.0						
	105	12	**66.0**	**62.0**	58.0						
	110	12	**67.5**	**63.0**	59.0						
	115	12	**68.5**	**64.0**							
	120	12	**69.5**	**65.0**							
	125	12	**70.5**	**66.0**							
Douglas fir and southern pine poles (based on a fiber stress of 8,000 lb/in)^2	20	4	31.0	29.0	27.0	25.0	23.0	**21.0**	**19.5**	**17.5**	**14.0**
	25	5	33.5	31.5	29.5	27.5	25.5	**23.0**	**21.5**	**19.5**	**15.0**
	30	5½	36.5	**34.0**	**32.0**	**29.5**	**27.5**	**25.0**	**23.5**	**20.5**	
	35	6	39.0	**36.5**	**34.0**	**31.5**	**29.0**	**27.0**	**25.0**		
	40	6	**41.0**	**38.5**	**36.0**	**33.5**	**31.0**	**28.5**	26.5		
	45	6½	**43.0**	**40.5**	**37.5**	**35.0**	**32.5**	30.0	28.0		
	50	7	**45.0**	**42.0**	**39.0**	36.5	34.0	31.5	29.0		
	55	7½	**46.5**	**43.5**	**40.5**	38.0	35.0	32.5			
	60	8	**48.0**	**45.0**	**42.0**	39.0	36.0	33.5			
	65	8½	**49.5**	**46.5**	**43.5**	40.5	37.5				
	70	9	**51.0**	**48.0**	**45.0**	41.5	38.5				
	75	9½	**52.5**	**49.0**	**46.0**	43.0					
	80	10	**54.0**	**50.5**	47.0	44.0					
	85	10½	**55.0**	**51.5**	48.0						
	90	11	**56.0**	**53.0**	49.0						
	95	11	**57.0**	**54.0**	50.0						
	100	11	**58.5**	**55.0**	51.0						
	105	12	**59.5**	**56.0**	52.0						
	110	12	**60.5**	**57.0**	53.0						
	115	12	**61.5**	**58.0**							
	120	12	**62.5**	**59.0**							
	125	12	**63.5**	**59.5**							

NOTE: Classes and lengths for which circumferences at 6 ft from the butt are listed in boldface type are the preferred standard sizes. Those shown in light type are included for engineering purposes only.

* The figures in this column are intended for use only when a definition of ground line is necessary in order to apply requirements relating to scars, straightness, etc.

577. Machine shaving of poles has increased as a practice of producers. Approximately 85% of present production is so shaved. Some producers also turn the pole down in the process, thereby obtaining a straighter pole with a specific taper. The machine-processed poles season more rapidly, which is particularly important with species like southern yellow pine which are susceptible to fungus attack before treatment. Machine shaving makes for easier detection of defects and provides a pole of improved appearance. If poles having normally thin sapwood (such as western red cedar and larch) are to be full-length treated with preservative, it is undesirable to reduce the thickness of the sapwood more than necessary to obtain a dressed pole.

578. Preservative Treatment. For a nominal cost the service life of wood can be greatly increased by the use of preservative treatment. Creosote and pentachlorophenol are extensively used for the protection of poles and crossarms. Southern yellow pine because of its thick sapwood requires a pressure treatment. Species with inter-

Table 4-115. Additional Wood Species

	Class		1	2	3	4	5	6	7	9	10
	Minimum circumference at top, in		27	25	23	21	19	17	15	15	12
	Length of pole, ft	Ground-line distance from butt,* ft	Minimum circumference at 6 ft from butt, in								
Jack pine, lodgepole pine, red pine, redwood, Sitka spruce, western fir, and white spruce poles (based on a fiber stress of 6,600 lb/in²)	20	4	32.5	30.5	28.5	26.5	24.5	**22.5**	**21.0**	**18.0**	**14.5**
	25	5	36.0	33.5	31.0	29.0	27.0	**25.0**	**23.0**	**20.0**	**15.5**
	30	5½	39.0	**36.5**	**34.0**	**31.5**	**29.0**	**27.0**	**25.0**	**21.0**	
	35	6	41.5	**38.5**	**36.0**	**33.5**	**31.0**	**28.5**	**26.5**		
	40	6	**44.0**	**41.0**	**38.0**	**35.5**	**33.0**	**30.5**	28.0		
	45	6½	**46.0**	**43.0**	**40.0**	**37.0**	**34.5**	32.0	29.5		
	50	7	**48.0**	**45.0**	**42.0**	39.0	36.0	33.5	31.0		
	55	7½	**49.5**	**46.5**	**43.5**	40.5	37.5	34.5			
	60	8	**51.5**	**48.0**	**45.0**	42.0	38.5	36.0			
	65	8½	**53.0**	**49.5**	**46.0**	43.0	40.0				
	70	9	**54.5**	**51.0**	**47.5**	44.5	41.0				
	75	9½	**56.0**	**52.5**	**49.0**	45.5					
	80	10	**57.5**	**54.0**	50.5	47.0					
	85	10½	**58.5**	**55.0**	51.5						
	90	11	**60.0**	**56.5**	52.5						
	95	11	**61.5**	**57.5**	54.0						
	100	11	**62.5**	**58.5**	55.0						
	105	12	**63.5**	**60.0**	56.0						
	110	12	**65.0**	**61.0**	57.0						
	115	12	**66.0**	**62.0**							
	120	12	**67.0**	**63.0**							
	125	12	**68.0**	**64.0**							
Alaska yellow cedar and western hemlock poles (based on a fiber stress of 7,400 lb/in²)	20	4	31.5	29.5	27.5	25.5	23.5	**22.0**	**20.0**	**17.5**	**14.0**
	25	5	34.5	32.5	30.0	28.0	26.0	**24.0**	**22.0**	**19.5**	**15.0**
	30	5½	37.5	**35.0**	**32.5**	**30.0**	**28.0**	**26.0**	**24.0**	**20.5**	
	35	6	40.0	**37.5**	**35.0**	**32.0**	**30.0**	**27.5**	**25.5**		
	40	6	**42.0**	**39.5**	**37.0**	**34.0**	**31.5**	**29.0**	27.0		
	45	6½	**44.0**	**41.5**	**38.5**	**36.0**	**33.0**	30.5	28.5		
	50	7	**46.0**	**43.0**	**40.0**	37.5	34.5	32.0	29.5		
	55	7½	**47.5**	**44.5**	**41.5**	39.0	36.0	33.5			
	60	8	**49.5**	**46.0**	**43.0**	40.0	37.0	34.5			
	65	8½	**51.0**	**47.5**	**44.5**	41.5	38.5				
	70	9	**52.5**	**49.0**	**46.0**	42.5	39.5				
	75	9½	**54.0**	**50.5**	**47.0**	44.0					
	80	10	**55.0**	**51.5**	48.5	45.0					
	85	10½	**56.5**	**53.0**	49.5						
	90	11	**57.5**	**54.0**	50.5						
	95	11	**58.5**	**55.0**	51.5						
	100	11	**60.0**	**56.0**	52.5						
	105	12	**61.0**	**57.0**	53.5						
	110	12	**62.0**	**58.0**	54.5						
	115	12	**63.0**	**59.0**							
	120	12	**64.0**	**60.0**							
	125	12	**65.0**	**61.0**							
Western larch poles (based on a fiber stress of 8,400 lb/in²)	20	4	30.0	28.5	26.5	24.5	22.5	**21.0**	**19.0**	**17.0**	**13.5**
	25	5	33.0	31.0	29.0	26.5	24.5	**23.0**	**21.0**	**18.5**	**14.5**
	30	5½	35.5	**33.5**	**31.0**	**29.0**	**26.5**	**24.5**	**23.0**	**19.5**	
	35	6	38.0	**35.5**	**33.0**	**31.0**	**28.5**	**26.5**	**24.5**		
	40	6	**40.0**	**37.5**	**35.0**	**32.5**	**30.0**	**28.0**	26.0		
	45	6½	**42.0**	**39.5**	**37.0**	**34.0**	**31.5**	29.0	27.0		
	50	7	**44.0**	**41.0**	**38.5**	35.5	33.0	30.5	28.5		
	55	7½	**45.5**	**42.5**	**40.0**	37.0	34.5	31.5			
	60	8	**47.0**	**44.0**	**41.0**	38.5	35.5	33.0			
	65	8½	**48.5**	**46.0**	**42.5**	39.5	36.5				
	70	9	**50.0**	**47.0**	**44.0**	41.0	38.0				
	75	9½	**51.5**	**48.0**	**45.0**	42.0					
	80	10	**52.5**	**49.5**	46.0	43.0					
	85	10½	**54.0**	**50.5**	47.0						
	90	11	**55.0**	**51.5**	48.5						
	95	11	**56.5**	**53.0**	49.5						
	100	11	**57.5**	**54.0**	50.5						
	105	12	**58.5**	**55.0**	51.5						
	110	12	**59.5**	**56.0**	52.5						
	115	12	**60.5**	**57.0**							
	120	12	**61.5**	**58.0**							
	125	12	**62.5**	**58.5**							

NOTE: Classes and lengths for which circumferences at 6 ft from the butt are listed in boldface type are the preferred standard sizes. Those shown in light type are included for engineering purposes only.

* The figures in this column are intended for use only when a definition of ground line is necessary in order to apply requirements relating to scars, straightness, etc.

mediate sapwood thickness such as Douglas fir, lodgepole pine, and jack pine are treated by either pressure or nonpressure processes. Thin sapwood species such as western red cedar and larch are generally treated by nonpressure processes. Specifications for treatment are American Wood Preservers' Association C-1, C-4, C-7, C-8; Edison Electric Institute TD-100, TD-101 (full-length nonpressure treatment); and American Telephone & Telegraph Company AT-7313 and AT-7336.

Treatment of standing poles with a solution of pentachlorophenol is widely practiced. The economy is greatest on power-line cedar poles. There are contractors specializing in this work.

579. Inspection. Poles are inspected prior to treatment for physical defects and decay and after treatment for penetration and retention of preservative and for cleanliness. Inspection is most effective when made at vendors' plants, because defects that may be hidden by preservative are detected and freight is saved on rejects. Commercial inspection agencies are available at most producing locations, and it is normally economical to utilize their services. This service costs about 3 cents per lineal foot. Quantity users may have their own trained inspectors. **Crossarm inspection** is important because safety of linemen is a consideration in addition to quality of timber. As with poles, inspection should be made before treatment for defects and after treatment for penetration, retention, and cleanliness. Inspection should be done by qualified timber specialists. Inspection cost is about 5 cents per arm.

580. Conductivity is of concern to many electric-utility companies. Pole resistance varies greatly with moisture content. Dry wood of all species exhibits high resistance. Surface absorption of rain water by untreated wood may vary the resistance over a wide range. Full-length-treated poles thoroughly dried before treatment generally show only moderate reduction in resistance following a rain. A rough correlation between resistance and moisture will show that 500,000 Ω over a 20-ft length of pole between contacts driven 3½ in deep corresponds to a moisture content of about 25%. Other average points on the curve band are 50,000,000 Ω 15% moisture and 20,000 Ω 40% moisture.

581. Depth of Pole Setting. The values in the column headed "Ground-line distance from butt" in Tables 4-113 to 4-115 may be accepted as a guide for a satisfactory depth of pole settings in ordinary firm soil. In marshy soil and at unguyed angles in lines, setting depths should be increased 1 to 2 ft. In rock, the indicated settings may be reduced one-half for that part of the pole set in rock. Rock backfill in ordinary earth locations is not considered as set in rock.

582. Pole stubbing can frequently be employed to effect substantial money saving. An otherwise good pole that is decayed at or below the ground line is fastened securely to a new preservative-treated stub set in the ground alongside it. The major part of the savings results from avoidance of transferring wires and equipment.

583. Salvaging. Poles removed for any reason can frequently be salvaged for future use. Users of large quantities can economically do this. One or more of the following operations may be employed: cut off top, cut off butt, remove old hardware, shave, reframe, re-treat.

584. Kinds of Timber for Crossarms. Two kinds of timber are in general use for crossarms, Douglas fir and southern yellow pine. All pine crossarms are treated with creosote or pentachlorophenol. The practice of treating fir crossarms is increasing rapidly as users recognize the need for arm life to match pole life.

Most Douglas fir arms used for communication and power-distribution lines are manufactured from timber selected for the purpose. Dense and close-grain lumber is used. *Publication* 14 of the West Coast Lumberman's Association, dated Aug. 1, 1947, sets forth grading and dressing requirements.

There is no grade of southern pine timber designated as crossarm stock, and crossarm users depend on the limitations set forth in their specifications to obtain a satisfactory quality of product. Pine arms are usually small boxed heart timbers.

Laminated arms are coming into use, but a buying specification is not available. Large transmission-line arms may be laminated structures or framed, treated round poles.

585. Crossarm Specifications. The most widely used specifications for power-distribution crossarms have been prepared by the Transmission and Distribution Committee of the Edison Electric Institute. For fir crossarms: Specification TD-90, which combines both dense and close-grain grades; Specification TD-92, Heavy-duty Douglas Fir Crossarms; Specification TD-93, Heavy-duty Douglas Fir Braces. For pine crossarms: Specification TD-91, Dense Southern Pine Crossarms Preservative Treated. Widely used specifications for communication crossarms are American Telephone and Telegraph Co. Specification AT-7298, Crossarms.

586. Strength of Crossarms. The most reliable source of information on the strength of crossarms comes from tests made under conditions to simulate crossarms in service. Some tests have been made, and others are under consideration. Theoretical considerations, treating a crossarm as a beam, are valuable if those factors which control the actual strength are taken into account. Tests made several years ago on 84 six-pin, 3¼ by 4¼-in by 6-ft crossarms, with a uniformly distributed vertical load, gave average results shown in Table 4-116 (*U.S. Forest Service Circ.* 204, by T. R. C. Wilson). The maximum bending load shown in Table 4-116 is the total distributed vertical load. The maximum crushing strength is under compression parallel to the grain. Methods of tests are covered by ASTM specifications.

587. Ice and wind loading assumptions for use in pole and crossarm design are given in Sec. **13**.

Table 4-116. Bending Load and Crushing Strength of Crossarms

Species	Rings per in.	Per cent			Density (dry)	Max. bending load, lb	Maximum crushing strength lb per sq in.
		Summer wood	Sap wood	Moisture			
Douglas fir	20	40	0	11.5	0.48	7,590	7,080
Longleaf pine (50% heart)	18	44	55	13.4	0.54	8,984	5,425
Longleaf pine (75% heart)	19	53	32	13.5	0.63	10,180	8,950
Longleaf pine (100% heart)	16	44	1	12.8	0.63	9,782	8,940
Shortleaf pine	11	46	79	13.3	0.52	9,260	7,300
Shortleaf pine, creosoted	11	49	..	....	....	7,649	5,770
White cedar	12	45	2	14.3	0.36	5,200	4,700

588. References on Poles and Crossarms

Hunt, George M., and Garratt, George A. "Wood Preservation"; New York, McGraw-Hill Book Company, 1953.

"Manual of Recommended Practice"; Washington, D.C., American Wood Preservers' Association.

Ostman, H. F. Crossarm Loading Studies; *Bull. EEI*, June, 1945.

Eggleston, Richard C. Evaluating the Relative Bending Strength of Crossarms; *Bell System Tech. Jour.*, January, 1945.

Vaughan, J. A. Timber Preservation Practices for Southern Pine; *Elec. Light & Power*, March, 1946.

SEMICONDUCTOR MATERIALS

By W. Crawford Dunlap

589. Introduction. The discussion presents sufficient background theory of the solid state and of semiconductors for the reader to understand technical discussions of the properties of devices such as transistors and diodes without further reference to theory. The properties of the semiconductor devices as such are covered in other parts of the book (Sec. **5**).

It is usual to define a semiconductor as a material whose electrical conductivity is intermediate between that of a metal (10^{-6} Ω·cm resistivity) and that of an insulator (over 10^{10} Ω·cm resistivity). This valid definition is not sufficiently precise, however. Semiconductors have resistivities that are highly sensitive to temperature and to impurity content. In fact, in some semiconductors the conductivity is wholly dependent on impurities. The latter are the so-called *extrinsic* semiconductors. Other semiconductors (*intrinsic* semiconductors) have conductivity which is characteristic of the pure material. All semiconductors may be classed as extrinsic in one temperature range, intrinsic in another (usually higher) temperature range.

Semiconductors are *electronic* conductors; that is, they conduct because of the presence of free or nearly free electrons in the material. Ionic semiconductors also exist, in which ions are mobile, but we shall not deal with them here. Electronic conductors are of two types, *n*-type and *p*-type. The *n*-type materials have negative charge carriers, as if free electrons were conducting; *p*-type semiconductors, however, are those in which there are apparently positive electrons present. There is no simple way to understand clearly how such positive charges, called *positive holes*, can be present. Suffice it to say that the material acts as though positive charges with mass roughly that of the electron are present. They are distinguished by the fact that the Hall effect, the thermoelectric effects, rectification properties, and a number of other properties are reversed in sign for *n*- and *p*-type materials.

Semiconductors are in general *crystalline* materials; the best semiconductors are those which have the most nearly perfect crystalline structures. Silicon and germanium, for example, the best-known transistor materials, have structures which are almost perfect. This is necessary, since the properties of both the material and devices made from it are very sensitive to the presence of crystal imperfections, such as vacancies, dislocations, etc.

590. Basic Theory of the Solid State. *Theory of the Perfect Crystal.* The theory of semiconductors may be resolved into three parts: the quantum theory of the perfect lattice, the theory of the imperfections in the lattice, and the theory of transport of carriers in a crystal as a function of the impurity content, temperature, electric and magnetic fields, and other parameters.

The theory of the perfect lattice is based upon the quantum theory of individual atoms. Unfortunately when a large number of atoms are collected in a crystal, even if it is perfect in structure, the theory becomes so complicated that exact solutions of the basic equations have not been made. Approximations are required, and these limit the validity of the results. For example, we still cannot predict exactly what the crystal structure will be when atoms of a given species, say phosphorus, are brought together to form a lattice.

One of the important theories of perfect lattices is the *band theory*. According to this theory, the individual atoms have discrete energy levels as predicted by the atomic quantum theory (see, for example, Ref. 1, Par. **597**). These sharp energy levels are broadened into bands of allowed energy for electrons, separated by forbidden energy regions in which no electrons of these intermediate energies are allowed. Two classes of semiconductor may be distinguished. The *broad-band semiconductors* such as germanium and indium antimonide have a narrow band for forbidden energies and broad bands of allowed energies. They tend to be highly conductive, to have carriers which move easily through the lattice, and to have low melting points. The second class is the *narrow-band semiconductors*, of which the refractory oxides are good examples. The gap between the bands is large, the width of the bands is small, and the materials tend to be highly refractory, brittle, to have high resistivity, and to have low carrier mobility in electric fields. The two types of band structure are illustrated in Figs. 4-87 and 4-88.

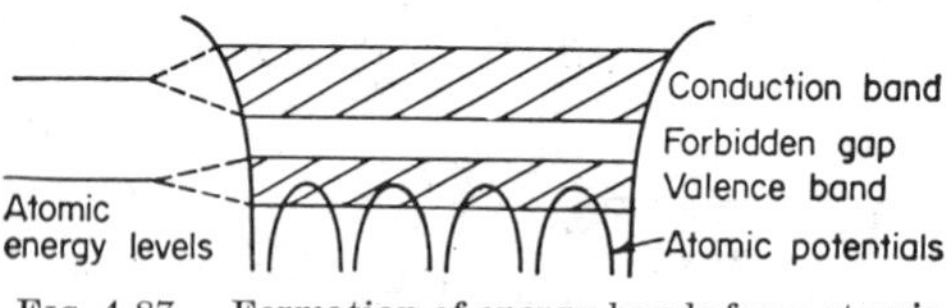

FIG. 4-87. Formation of energy bands from atomic levels for broad-band materials.

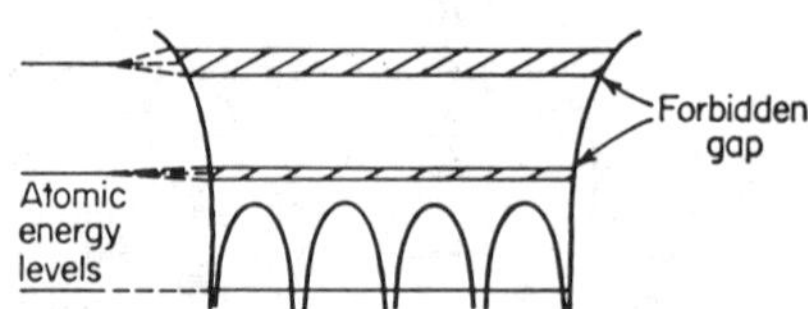

FIG. 4-88. Energy bands of narrow-band semiconductors.

The energy bands of most importance in semiconductors correspond to the highest filled shell of the isolated atom, and the first unfilled shell. The former shell is represented by a completely filled energy band in the solid, called the *valence band*, the latter to the first unfilled energy band, called usually the *conduction band*.

Extrinsic semiconductors of the n type have electrons in the normally empty conduction band, arising from impurities (*donors*) which eject electrons into the band.

Intrinsic semiconductors have carriers of both signs: negative electrons moving about in the conduction band, and positive "holes." The holes take the form of "bubbles" in the valence band, remaining behind when electrons are excited into the conduction band. The holes act as the positively conducting charges previously mentioned. Extrinsic semiconductors of the p type exist by virtue of the impurity atoms (*acceptors*) absorbing electrons from the valence band, immobilizing them, and leaving behind the mobile hole to act as a positive conducting particle.

It might be thought that the "unbound electrons" released, say, from a group of impurity atoms would move in the lattice as if they were free electrons in a vacuum. They do not, although there is some resemblance to the motion of free electrons. The quantum-mechanical conditions modify the behavior of the electrons, the main effect being a change in the mass of the electron. In some materials, the *effective mass* is many times that of the free electron. In other materials, such as germanium, gallium arsenide, and indium antimonide, the carriers act as though the effective mass were much *less* than the free-electron mass (in InSb only about 0.01 the mass m_e of the free electron). Holes also have an effective mass, which may be less than the mass of the free electron.

A second important property of the unbound electrons in a semiconductor crystal is that they interact with the lattice imperfections, and with each other, so that they "scatter"; that is, their motions are irregular. Since most semiconductors have large densities of electronic charge carriers, in the range 10^{10} to $10^{20}/\text{cm}^3$, the carrier motions are calculated by statistical equations derived either from Fermi (quantum) statistics or from Maxwell-Boltzmann (classical) statistics.

Each type of semiconductor provides a "unique" medium for the motion of electronic carriers. Hence there is an individual set of properties for each semiconductor. The rich variety of properties, and applications, in the field of semiconductors results.

The simplest diagram for semiconductor properties is the energy band of Fig. 4-89, which is a rough extraction of the information contained in Figs. 4-87 and 4-88 plus an indicator of the nature of donor and acceptor action. This diagram gives only a small part of the essential information about the energy-band structure, however, and more general diagrams are used in modern theoretical treatments. Corresponding to the simple energy-band picture are the "E versus k" curves, and the Fermi surfaces for the simplest model of the carriers in a semiconductor. This is the "free-electron" model in which electrons behave as though they were free, modified only by their having an effective mass different from that of a free electron. This may be expressed by the parabolic curve of Fig. 4-90, which shows the dependence of electron energy on the electron wave vector k in the quantum-theory representation. Some familiarity with this theory is essential for engineers interested in the basic theory of devices. The relationship is parabolic since $E = \frac{1}{2}mv^2$ for a free particle, which translates to $E = \frac{1}{2}h^2k^2/m^*$ in wave-mechanical language. The curvature of the parabola indicates the effective mass m^*.

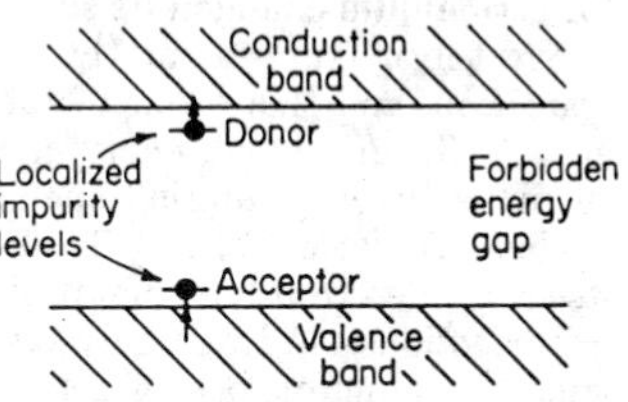

FIG. 4-89. Simple energy-band picture for a semiconductor.

To express the anisotropy of the electron-wave properties, E versus k curves may be plotted for different directions, or surfaces of constant energy may be shown in k space (Fermi surfaces). For isotropic materials, and for crystals of cubic symmetry, these surfaces are often spheres, as in Fig. 4-91. For most anisotropic crystals, and for many cubic crystals, they are ellipsoids or other more complicated types of surface. The theoretical calculation of these E versus k curves forms one of the most important applications of the quantum theory of solids, and a great deal is being done today in developing advanced methods of calculation and computer programs for their solution.

Germanium and silicon represent examples of the importance of proceeding from the simple theory of parabolic energy bands and spherical Fermi surfaces to the more accurate forms. These materials were presumed during the early days of the transistor to conform to the simple theory to a good approximation. However, so many dis-

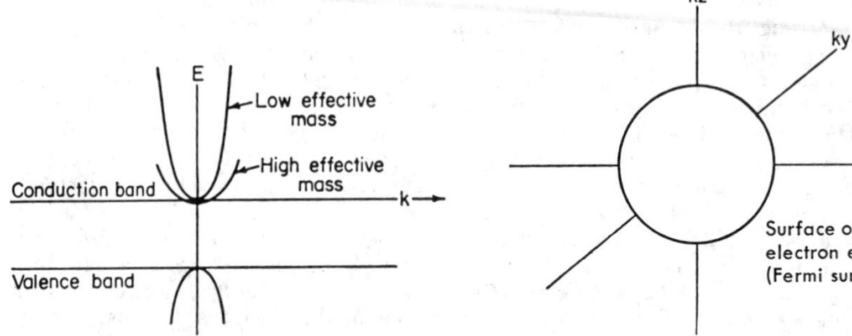

FIG. 4-90. Parabolic E versus k curves for simple model of semiconductor.

FIG. 4-91. Fermi surface for simple model of semiconductor.

crepancies between theory and experiment developed that more precise models had to be made. These showed that the energy-band picture is more like that shown in Fig. 4-92 than that shown in Fig. 4-90. It is to such more complicated models that we must look for explanations of many phenomena such as details of tunnel-diode behavior, the Gunn effect, and semiconductor laser phenomena.

The diagrams show that both the conduction and valence bands are "degenerate," in different ways. This means that there may be several different allowed energy states for a given energy value. The corresponding Fermi surfaces for germanium are shown in Fig. 4-93.

An indication of the important information in these diagrams can be obtained from Fig. 4-92. Figure 4-90 shows the parabolas for the valence band and the conduction band to be directly one above the other. In Fig. 4-92 not only has the conduction band for germanium changed its shape from a parabola, but the minimum is no longer directly above the maximum for the valence band. This makes difficult the "direct" optical transition, and this is one of the reasons semiconductor lasers have not been made with either silicon or germanium, so-called "indirect" materials. Semiconductor lasers require "direct" materials such as gallium arsenide or indium antimonide.

591. Imperfections in Solids. Imperfections are almost as important as the lattice itself in determining the properties of semiconductors. Impurities are themselves imperfections, and not only do they introduce the new energy states which make extrinsic conduction possible, but they also disrupt the lattice perfection and cause *scattering* of carriers. If impurities are present in too great quantity, they exceed the solubility limit, and segregation of second-phase material degrades the material.

Besides impurities, local lattice imperfections have important effects on the properties of semiconductors. A simple missing atom, a so-called *Schottky defect,* may introduce new energy levels as an impurity atom does, and these may have the character of either a donor or an acceptor. Such defects arise naturally during the growth of a crystal, or they may be introduced by mechanical stress and deformation, by thermal stress, or by radiation damage.

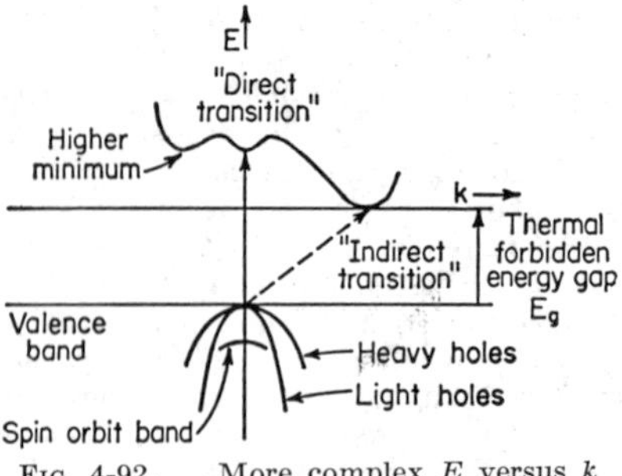

FIG. 4-92. More complex E versus k curves for germanium.

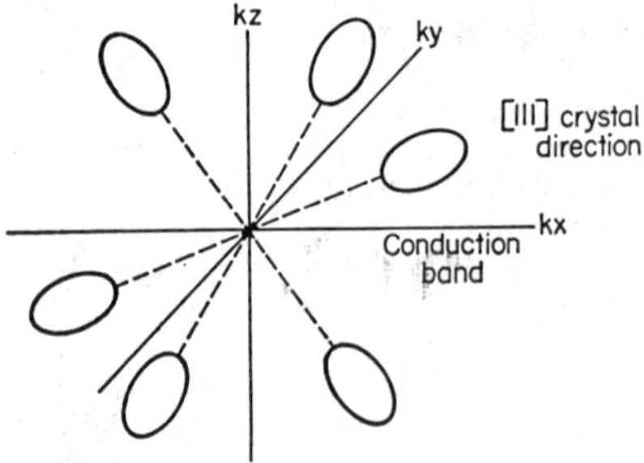

FIG. 4-93. Fermi surfaces for germanium.

Mechanical stress, as well as the crystal-growth process, also may introduce another type of imperfection, the *dislocation*—which is a linear structure as contrasted with the point defect just described. There are several types of dislocations, including *edge, screw,* and *partial* (or *mixed*) dislocations. Like point defects, dislocations can introduce carriers through new energy states, and they can scatter the carriers by specific interactions with them.

Another type of imperfection is the *lattice vibration,* or *phonon.* The phonon is the localized particle representing the strain introduced in the lattice by thermally generated waves, both acoustic and optical in nature, which are present at all temperatures, including absolute zero. These lattice vibrations account for most of the resistivity (*lattice scattering*) in semiconductors at high purity levels, i.e., at levels above those at which *impurity scattering* sets in. Interference by phonons with each other causes *resistivity* to the flow of heat. Every material is characterized by its *phonon spectrum.*

An important aspect of semiconductors is the calculation of the distribution of carriers. In any complicated situation involving simultaneously intrinsic conductivity, impurity contents of various types and magnitudes, changes of temperature, etc., *distribution functions* must be used to determine how many of the available carriers are free in the conduction band, how many free holes there are, how many electrons are bound on donor or acceptor centers, and so on. The fundamental distribution function based upon the Fermi-Dirac quantum statistics has the form $1/(1 + e^{E/kT})$, whereas the classical distribution function (Maxwell-Boltzmann) has the form $e^{-E/kT}$. These two functions are approximately the same, particularly for high temperatures. Hence classical statistics apply to semiconductors for many situations.

592. Conduction Properties of Semiconductors. *Transport Properties.* The energy-band picture of a semiconductor, the density of free carriers, both electrons and holes, the density of scattering centers, and the nature and type of the lattice imperfections characteristic of the material provide most of the data required to predict performance. A basic question remains, however, concerning the conduction properties of the material as a function of the applied parameters such as electric field, magnetic field, light intensity, pressure, temperature, etc.

The *basic properties of a semiconductor* are the density of carriers n and p, the mobilities μ_n and μ_p, and such properties as the conductivity σ and resistivity ρ, the Hall coefficient R, the thermoelectric power (Seebeck coefficient S), and the magnetoresistance $\Delta\rho/\rho_0$.

The conductivity and resistivity are related to carrier density n by the simple relation

$$\sigma = 1/\rho = ne\mu \tag{4-81}$$

where e is the charge on the electron (assumed the same as in free space).

When both holes and electrons are present simultaneously, they contribute independently to the conductivity,

$$\sigma = e(n\mu_n + p\mu_p) \tag{4-82}$$

Conductivity is usually measured by the arrangement in Fig. 4-94. By itself this measurement gives no indication of the type of semiconductor. To get such information, the extra two leads are used (Hall probes), and a magnetic field is applied. This serves to distinguish holes and electrons, since the magnetic field tends to direct holes in one direction, the electrons in the opposite, when they are moving in the same direction in the field.

The *Hall coefficient* gives both the magnitude and the sign of the dominant carrier. If only one carrier is present,

$$|R| = 1/ne \tag{4-83}$$

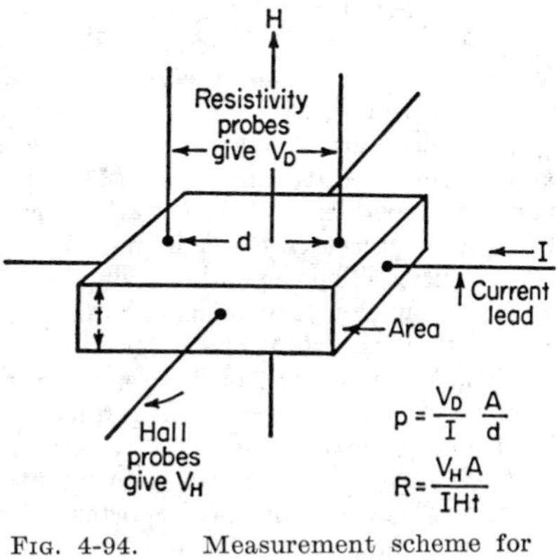

Fig. 4-94. Measurement scheme for conductivity and Hall coefficient.

or $R = 6.25 \times 10^{18}/n$ in cubic centimeters per coulomb when the carrier density is in particles per cubic centimeter. For mixtures of holes and electrons, the Hall coefficient is determined by the difference in carrier densities of the two types, weighted according to mobilities as follows:

$$|R| = (1/e)(n\mu_n^2 - p\mu_p^2)/(n\mu_n + p\mu_p)^2 \tag{4-84}$$

For homogeneous, one-carrier semiconductor samples, the *Hall mobility* is found from the equation $\mu = R\sigma$. Thus conductivity and Hall measurements form basic tools in the analysis of semiconductor properties.

The thermoelectric power, or Seebeck coefficient, is also used for analyzing semiconductors for carrier type and density. Bars of material, with standard copper or platinum reference junctions, are placed in a temperature gradient, and the voltage S produced per unit temperature difference is measured. Theory states that

$$S = -(k/e) \ln (N_c/n) \tag{4-85}$$

where $N_c = 2(2\pi m^* kT/h^2)^{3/2}$. The sign reverses for p-type samples. This equation is of limited applicability. For more general equations, the bibliography, Par. **597**, should be consulted.

The temperature dependence of the resistivity of a semiconductor can be very complicated. In general, however, the resistivity drops with increasing temperature. The classical expression for the carrier density of an extrinsic semiconductor is

$$n = (2N_d)^{1/2} \frac{(2\pi m^* kT)^{3/4}}{h^{3/2}} e^{-E_d/2kT} \tag{4-86}$$

This is applicable to a simple situation only, namely, materials possessing only a single impurity (very few of whose atoms have been ionized by the release of an electron) and negligible intrinsic conduction. It goes over to the equation $n = N_d$ (the saturation region) when the temperature is high.

Intrinsic conduction is determined by the forbidden energy gap of the material and by the numbers of energy states available in the conduction band and the valence band. From the simple parabolic model

$$n_i^2 = \frac{4(2\pi mkT)^3}{h^6} \frac{(m_e m_h)^{3/2}}{m^3} e^{-E_g/kT} = np \tag{4-87}$$

The equation $n_i^2 = np$ means that the product of the electron density and the hole density is independent of the donor and acceptor density and of the contribution of intrinsic conductivity. The intrinsic resistivity of germanium at 290°K is about 47 Ω·cm and of silicon, about 100,000 Ω·cm, corresponding to band gaps of 0.65 and 1.11 eV, respectively. *Thermistors* are important temperature-control and -compensation devices which rely on the large temperature coefficients of resistivity of intrinsic semiconductors.

The calculation of the mobility values for semiconductors is importantly affected by the *scattering mechanisms*. The main agencies for scattering are, as indicated, *lattice scattering, impurity scattering*, and *imperfection scattering*. Scattering from surfaces and from dislocations are also known.

The elementary models give a lattice-scattering mobility with a $T^{-3/2}$ dependence and impurity scattering with $T^{3/2}$ dependence. Imperfection scattering is less temperature-dependent. The overall situation is far more complicated, however, and the basic characteristics of each material determine the specific temperature dependence of mobility. The dependence of scattering of carriers on their velocity is a variable which is important in the calculation of practically all the conduction properties.

Another important characteristic is the dependence of mobility on electric field. As the field increases, the particle velocity reaches a constant limiting value; i.e. the effective mobility decreases. For germanium the limit is about 10^7 cm/sec. Under these conditions, electrons are no longer in a state of quasi equilibrium. Their "temperature" is much higher than the value attributable to the lattice; hence they are called "hot electrons." Many of the important devices now being studied, such as Gunn-effect devices, several types of thin-film amplifiers, and various types of infrared and microwave detectors, depend on hot-electron phenomena.

Another area where the scattering dependence on velocity is important is the dependence of Hall coefficient and resistivity on magnetic field (magnetoresistance). Such effects have now been studied and experiments made on a variety of semiconductors at fields well over 100,000 G.

Photoconductivity and Lifetime in Semiconductors. Thermal equilibrium, the steady-state condition in the case of current-carrying specimens, is usually achieved in times of the order 10^{-13} sec. Only in unusual cases, such as in extremely high-frequency phenomena (for example, cyclotron resonance) and at very low temperatures, can this *relaxation time* of electron distributions be observed. In the case of photoconductivity, however, another lifetime appears, which has to do with the recombination of electrons and holes.

If light of such frequency that its photon energy is greater than the energy gap strikes a semiconductor, absorption generally leads to the formation of free electron-hole pairs. If the light is turned off, and if the number of free pairs exceeds the equilibrium values, then equilibrium is restored according to an equation of the form

$$dp/dt = (p - p_0)/\tau \tag{4-88}$$

where τ is the lifetime and p_0 the equilibrium hole density.

Likewise, if light is steadily and uniformly incident upon a sample, the lifetime determines the steady-state increase in conductivity according to the equation

$$\Delta\sigma = e\mu\,\Delta n = e\mu G\tau \tag{4-89}$$

where G is the rate of optical electron-hole pair production.

Junction and Surface Phenomena. An extremely important set of phenomena appears when n-type material is in contact with a p-type portion of the same material, particularly when the two types are present in the same perfect single crystal with no imperfections in the contact region. Such a structure, called the p-n junction, is the basic element in most modern solid-state rectifier and transistor devices (see Sec. **5**).

One important property of a p-n junction is the existence of an electrostatic potential between the n and p regions when no current flows, arising from the transition from positively charged impurity ions to negatively charged ones. The mobile carriers tend to diffuse away to equalize their concentrations across the junctions, thereby creating the electrostatic field. When external voltages are applied across the junction, in the one direction large injection currents may flow, consisting of holes flowing into the n region, electrons in the p region, where they are minority carriers. They recombine with much the same lifetime as previously described for photoconductivity. In the opposite polarity, little or no current flows.

Of increasing importance is the *heterojunction,* which is the boundary between two different materials, usually with a negligible crystal discontinuity. It may have an n-n, a p-p, or a p-n character and generally is a rectifying junction.

The contact and surface properties of semiconductors are of great importance. Surface properties determine the properties of the materials in many situations, and most devices have weaknesses determined more by surface characteristics than by bulk ones. The contacts between semiconductors and metals are also of great importance from both the theoretical and practical points of view.

593. Properties of Specific Semiconductors. As indicated previously, each semiconductor is a unique electronic medium in itself, so that cataloguing all the properties of all the possible semiconductors is a large task. Table 4-117 summarizes many of the basic properties of the important semiconductors.

There are literally hundreds of semiconductors known, including a number of the elements, many inorganic compounds, and some organic compounds. Most of these are highly crystalline, although a few liquid and amorphous semiconductors are known, of which selenium is probably the most important. Semiconductors are usually metallic in appearance but (unlike metals) are generally hard and brittle. Their resistivities may range from 0.001 Ω·cm or less to over 10^{10} Ω·cm.

In addition to the elements germanium, silicon, tellurium, selenium, boron, phos-

Table 4-117. Room-temperature Properties of Some Semiconductors
(Prepared by 1964 International Semiconductor Conference, Paris)

Material	Band gap, eV	Electron mobility, $cm^2/V \cdot s$	Hole mobility, $cm^2/V \cdot s$	Dielectric constant, $\kappa = n^2$	Lattice constant, Å	Density, g/cm^3	Melting pt., °C
C (diamond)	5.4	1,800	1,200	5.5	3.567	3.51	3550
Si	1.15	1,900	480	11.8	5.42	2.4	1412
Ge	0.65	3,800	1,800	16.0	5.646	5.36	938
Sn	0.08	2,500	2,400	.....	6.47	6.0	232
Se	1.6	......	0.6	8.5	4.35 4.95	4.8	220
Te	0.33	1,100	560	5.0	4.447 5.915	6.24	452
BP	6	......	110	11.6	4.537	2.97	3000
AlP	2.5	3,500	....	11.6	5.43	2.85	1500
AlAs	2.3	1,200	200	.....	5.63	3.81	1600
AlSb	1.52	400	150	10.3	6.13	4.22	1060
GaP	2.25	80	17	8.4	5.44	4.13	1350
GaAs	1.35	8,500	400	13.5	5.65	5.31	1280
GaSb	0.69	4,000	650	15.2	6.095	5.62	728
InP	1.27	4,600	700	10.6	5.869	4.78	1055
InAs	0.35	30,000	240	11.5	6.058	5.66	942
InSb	0.17	70,000	1,000	16.8	6.48	5.775	525
SiC	3.0	60	8	10.2	4.35	3.21	2700
PbS	0.37	800	1,000	17.9	7.5	7.61	1114
PbSe	0.26	1,500	1,500	.....	6.14	8.15	1062
PbTe	0.25	1,620	750	.....	6.45	8.16	904
Bi_2Te_3	0.15	1,250	515	.....	10.48	7.7	580
Cd_3As_2	0.13	15,000	......	.....	8.76	6.21	721
CdSb	0.48	300	300	.....	6.471	6.66	456
ZnO	3.2	190	....	8.5	3.24 5.18	5.60	1975
ZnS	3.65	100	....	8.3	5.423	4.80	
ZnSe	2.6	100	16	5.75	5.667	5.42	685
ZnTe	2.15	......	50	18.6	6.101	5.54	1239
CdS	2.4	200	....	5.9	5.83	4.82	685
CdSe	1.74	500	....	4.30	6.05	5.81	1350
CdTe	1.50	650	45	11.0	6.48	6.20	1098
HgSe	0.3	18,500	....	14.0	6.08	7.1	798
HgTe	0.2	22,000	160	.....	6.429	8.42	670

phorus, tin, carbon, sulfur, and iodine, there are important semiconductor compounds in the 3-5 and the 2-6 groups (the numbers refer to the columns in the periodic table of the elements). In the former class gallium arsenide has proved to be by far the most important, although indium antimonide and gallium antimonide are also widely used. In the 2-6 group, cadmium sulfide is probably the most important, although zinc sulfide, zinc oxide, cadmium selenide, and cadmium telluride are also interesting.

594. Germanium. Germanium, the material in which transistor action was first observed, is one of the most interesting and most important semiconductors. It is a gray metallic-looking material, brittle and glasslike in its mechanical properties. It crystallizes in the diamond cubic lattice. It has an intrinsic resistivity of 47 $\Omega \cdot$cm but may be doped with antimony or arsenic to give *n*-type resistivities of 0.01 or less $\Omega \cdot$cm and with boron, gallium, or aluminum to give *p*-type resistivities of 0.001 $\Omega \cdot$cm or less.

The energy-band structure for germanium is shown in Fig. 4-92. The conduction-band minimum lies at a value of $k \neq 0$, so that germanium is an "indirect" semiconductor. The electrons in the conduction-band minimum have ellipsoidal energy surfaces as indicated in Fig. 4-93. The mobility of the electrons is high, about 3,900 cm/sec (velocity) per volt/centimeter (field) [cm^2/(volt) (sec)]. The effective mass is about 0.20 that of the free electron. An interesting detail, of importance in the theory of tunnel diodes and hot electrons, is the higher minimum of the conduction band in Fig. 4-92, to which electrons can be transferred under some conditions.

The valence band of germanium is also of interest, since it is degenerate, so that there are two types of positive-hole conducting current, namely, "light" and "heavy" holes, the former having 0.044, the latter having 0.28, the mass of the free electron. The third degenerate "spin-orbit split" band is of minor significance.

Germanium is of interest because of the large number of elements which have electrical activity when added to it by diffusion or mixture on melting. The elements in the third and fifth columns (donors and acceptors, respectively) have essentially the same action, called "hydrogenic." These impurity centers release electrons when the energy E_i is supplied. E_i is given by

$$E_i = \frac{13.5}{\kappa^2}\frac{m^*}{m} \quad \text{eV} \tag{4-90}$$

where κ = dielectric const. This is the "ionization energy" for impurity action. Many elements have other ionization energies, and many are characterized by a complex impurity energy-band structure. Gold, for example, produces three acceptor levels spread through the forbidden gap, and a donor (acceptor compensating) level close to the valence band. Germanium doped with these impurities is used in infrared detectors. Figure 4-95 shows the energy levels for various impurities.

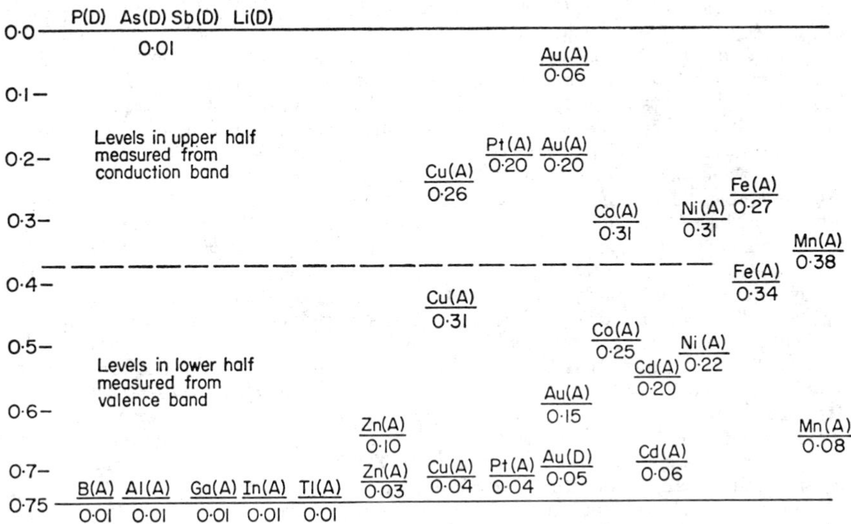

FIG. 4-95. Energy levels of various impurities in germanium.

Germanium is usually prepared in high-purity single-crystal form for electronics use by pulling from the melt, either vertically (Czochralski method) or horizontally (zone refining, or leveling). Because the melt-solid interface tends to exclude most impurities from the solid, the crystal-growing process is also the most important technique of purifying the material. To obtain intrinsic material ($n = p = 2 \times 10^{13}$ carriers/cm^3), impurities of active elements must be in the range 1 part/billion or less. Edge-dislocation densities of 100/cm^2 or less are fairly easy to obtain.

An increasingly important technique for obtaining germanium for transistors or integrated circuits is the *epitaxial* technique, which is also useful for silicon and the 3-5 compounds. This is the growth of very thin layers on a single-crystal substrate by vapor deposition (disproportionation, or quasi-equilibrium decomposition of an iodide, is usually preferred to pyrolytic deposition, which is nonequilibrium and irreversible). The epitaxial technique has now been developed to permit production of controlled resistivities of any type, with perfection rivaling that of Czochralski (vertical-pulled) material.

Junctions of the *p-n* type have been made in germanium, not only by counterdoping from the melt and by epitaxy, but also by diffusion and alloying. In these processes, impurity atoms are driven into the crystal by heat up to 900°C (germanium melts at 937°C). The diffusion profiles produced may be steep or graded, depending on the time and temperature. Alloyed junctions involve greater amounts of impurity, are produced at lower temperatures, and are made directly from metals, which form a eutectic with germanium. Such alloyed junctions tend to be abrupt in transition and usually involve recrystallization phenomena.

The diffusion of various elements in germanium has been an important aspect of germanium research, partly from the impetus given by the use of diffusion as a technical process, partly from the fact that diffusion studies give promise of providing new insight into the nature of the process itself.

The ordinarily active impurities such as gallium, indium, arsenic, and antimony tend to diffuse rather slowly in germanium. The donors diffuse roughly an order of magnitude faster than the acceptors. The typical values are given by the equation

$$D = D_0 e^{-E/kT} \tag{4-91}$$

where D_0 is usually of the order 10 cm²/sec and E has values of the order 2.5 eV. The sensitivity of germanium to small concentrations of impurities makes the electrical method of detecting diffusion an even more powerful tool for studying diffusion than the usual technique of radioactive tracers. Etching, staining, and thermoelectric techniques make possible measurement of diffusion depths as small as a few microns.

Germanium is notable, not only for the number of impurity elements with which it may be used to create deep impurity energy levels, but also for the sensitivity it has to other elements (notably copper, lithium, nickel, and gold) which can move about in the crystal at high speeds at low temperature. Lithium, which seems to create a fast-moving donor with E_i of 0.01 eV, probably is the fastest—it moves with an almost zero activation energy for diffusion, and measurable movements at room temperature are detectable. This diffusion, which is aided by electric fields, is widely used for the creation of broad *p-n* junctions useful for "lithium-drifted" radiation counters for detecting single nuclear particles. Germanium for many years was subject to peculiar "heat-treatment" effects, which later were identified as being due to fast-moving copper picked up from even fairly pure water to which the material was exposed. Now, "gettering" with indium and related techniques have removed this as a problem in the handling of germanium for technical purposes.

595. Silicon. Silicon has become increasingly important as a transistor material, to such an extent that its properties are now being studied more than those of germanium. Its higher band gap offers a greater temperature range than that of germanium, but silicon is handicapped by the lower carrier mobilities, 1,300 and 450 for electrons and holes, respectively, compared with values of 3,900 and 1,800 for germanium. This drawback limits applications of silicon in high-frequency transistors.

Silicon is somewhat more difficult than germanium to produce and to purify because of its higher melting point (1420°C). Alloyed, diffused, and epitaxial junctions are somewhat more difficult to prepare. Not only does the higher temperature make quartz, the usual crucible material, more difficult to work with, but several impurities, notably boron, have segregation coefficients close to 1 so that the crystal-growth process is ineffective in screening them out.

In spite of this, nearly intrinsic silicon with resistivity of 20,000 Ω·cm has been made, and lifetimes of many microseconds are commonplace. Dislocation-free silicon has been prepared through use of small seeds and careful technique, but generally dislocation counts of 100 to 1,000 per cm² are common. Thus, silicon prepared for semiconductor devices is one of the purest and most perfectly structured materials ever made.

The properties of silicon are notably sensitive to the presence of oxygen, which is usually present to a level of 10^{17} to 10^{18} atoms/cm³, unless special care is taken to exclude it. Freedom from contact with oxides, such as silica, is maintained in the so-called "floating-zone" process, in which an r-f-heated liquid band, held by surface tension, is moved mechanically through the bar.

Oxygen in silicon tends to introduce instabilities when the material is subjected to heat-treatments at high temperatures. The mechanism is not well understood, although it is likely that interstitial oxygen acts as a donor. It may also be combined with other impurity atoms to produce neutral complexes.

Silicon is one of the more sensitive elements to nuclear radiations, stemming from its low atomic weight and the high resistivity of the material as generally used. As few as 10^{10} neutrons incident per square centimeter can produce measurable effects on the bulk material. Even with special techniques, such as the use of heavier dopings and very thin base regions, silicon transistors cannot withstand more than perhaps 10^{14} neutrons/cm^2 without damage. The bulk properties of silicon under radiation are particularly sensitive to the presence of oxygen, since the defects produced seem to combine readily with oxygen to produce active centers.

Silicon behaves much like germanium in its sensitivity to a wide variety of impurities. The ionization energy for 3-5 column doping elements is higher than for germanium, however, being generally in the range 0.04 to 0.05 eV. This results from both silicon's smaller dielectric constant, 12 rather than 16, and higher effective mass for both holes and electrons. Silicon also is different in that indium tends to produce deep levels, with ionization energy of 0.15 eV rather than the normal value. The theory of the impurity energy states is still not complete, although much work has been done on it during recent years.

Silicon, like germanium, is also characterized by the number of impurities which, besides having deep levels, move rather rapidly through the lattice at moderate to high temperatures. Among these are copper, iron, manganese, nickel, and cobalt.

Because of the importance of the planar technology (see Sec. **5**), which has developed since 1960, the surface and oxidation properties of silicon are of particular importance. Measurements of recombination velocity, transistor leakage, and related properties indicate the presence of a variety of surface states, some of which appeared to be at the boundary of the silicon surface and the silicon dioxide layer which forms naturally on all silicon exposed to air for any appreciable period of time. These were the so-called "fast states." Other "slow states" accounting for slow changes in the properties of experimental samples or device structures were also observed. Research on the nature and origin of these states continues, and much controversy exists on the interpretation of the results.

Silicon dioxide layers, formed by heavy thermal oxidation at 1000°C or higher for several hours, have proved to have enormous practical significance—partly because they made possible masking of diffusion steps in the transistor process, and partly because they made possible passivation of the device surface. This greatly arrests or even eliminates changes in properties due to variations in ambient conditions. The study of the diffusion of impurities through oxides, the nature of the leakage phenomena in the oxide, and the segregation phenomena of impurities between oxide and bulk silicon during the oxidation process are active and continuing aspects of silicon research at the present time.

596. Compound Semiconductors. *Gallium Arsenide.* Even before germanium and silicon became important, compound semiconductors were the objects of much interest. The foundations of semiconductor science, as a matter of fact, were laid on the basis of studies with copper oxide, zinc sulfide, silicon carbide, and zinc oxide, among others. Beginning with German work on indium antimonide about 1950, attention shifted from these compounds to the so-called 3-5 compounds, of which gallium arsenide has proved by far the most interesting and important.

Gallium arsenide is a material whose characteristics are very closely related to those of germanium. This is understandable, since gallium is the third-column neighbor and arsenic the fifth-column neighbor of germanium in the periodic table. With its large band gap of 1.35 eV and an electron mobility of over 8,000 cm^2/V·sec, gallium arsenide has evident potential advantages over both silicon and germanium for device use. Its band gap, for example, should make possible an almost ideal match to the solar spectrum for production of photovoltaic cells for space and other uses (solar batteries). Unlike germanium, gallium arsenide has a "direct"-type band structure.

There are a number of properties of GaAs, however, which make it inferior to silicon and germanium. The high melting point (1300°C), combined with the high vapor pressure of arsenic at 1200 to 1300°C, makes the production of gallium arsenide of electronic grade an extremely difficult one. Applications of Czochralski technique using high-pressure chambers have been unsatisfactory. Horizontal furnaces using zone refining techniques have proved the best, but the resulting material still is not competitive with Si and Ge in purity or structural perfection. Resistivities of 1 to 10 Ω·cm are generally obtained, and mobilities of 5,000 to 6,000 are common. However, lifetime values are low (10^{-8} sec has been typical until recently).

The problem is complicated by the fact that not only do impurity atoms from neighboring columns of the periodic table dope gallium arsenide, but there is a much greater variety possible of lattice imperfections; gallium and arsenic vacancies may be important, as well as gallium and arsenic interstitials, and elements such as silicon, which is inert as an impurity in germanium, have strong electrical effects. Zinc is ordinarily used as the p dopant in gallium arsenide, and sulfur or tellurium as the n dopant.

Gallium arsenide has proved to be useful for a number of very important devices, including some varieties of switching and parametric diodes, Esaki or tunnel diodes, semiconductor lasers, and most recently, hot-electron "Gunn-effect" diodes. Material-production problems remain one of the chief difficulties in the field, since good samples of material are obtained by selection rather than by planning.

Gallium arsenide impurity research has proceeded along the lines of impurity research on silicon and germanium, but the process is more complicated. The ionization energies of many elements, as well as their diffusion coefficients, solubility, and segregation coefficients, have been measured. Because of the tremendous numbers of experiments involved, this research is not complete.

Recent work on semiconductor lasers and recombination diodes indicates the importance of complex defects, particularly coupled pairs of donors and acceptors, which tend to make an extremely complicated spectrum of light emitted by recombination of electrons in certain levels with holes on other, nearby sites.

Indium Antimonide. Indium antimonide is of interest because it has the highest room-temperature electron mobility of any known material, namely, about 70,000 $cm^2/V \cdot sec$. Because of this, it is of interest for Hall-effect and magnetoresistive devices. Its small band gap, about 0.18 eV, also makes it of some interest for infrared detectors.

Because of the low melting point, 525°C, InSb is much easier to prepare in single-crystal form than GaAs. It has also been prepared in purer form (carrier density $10^{13}/cm^3$ and less) than any other 3-5 compound. Because of its simple band structure (direct type, with nearly parabolic energy bands) it has proved a useful subject for the study of the applications of solid-state theory to the 3-5 compounds.

Cadmium Sulfide. Of the 2-6 compounds, CdS is probably the most interesting and important. It has been used commercially as a photoconductor for many years, as well as a constituent of cathode-ray phosphors. Cadmium sulfide is a high-band-gap material ($E_g = 2.4$ eV) which has been produced mainly by vapor-phase reactions and sublimation techniques. It melts only under high pressures.

Cadmium sulfide is notable for the complex set of trapping states spread through the forbidden energy gap, which tend to produce, on the one hand, extremely high photoconductivities and, on the other, very slow response following changes in light intensity. The exact nature of these traps is not understood, although they are probably lattice vacancies.

Cadmium sulfide can be prepared in the resistivity range from 10 to about 10^{12} $\Omega \cdot cm$, depending on the defects present and the impurities. It is extremely difficult, if not impossible, to produce p-type CdS. This apparently is due to the fact that p-type impurities tend to foster the presence of compensating n-type lattice vacancies.

Silicon Carbide. One of the semiconductors first used to detect radio signals in the 1920s was silicon carbide, a material widely used also for lightning arresters (thyrite). Silicon carbide is extremely refractory, subliming in the region of 2800°C. Because of this and because of the large band gap of 3.0 eV, it has long been hoped that silicon carbide would be useful for very-high-temperature rectifiers and transistors. While such devices have been made, the materials properties have not been good enough to make possible the preparation of devices competitive with silicon or germanium. In addition, the problems of making junctions and low-resistance contacts to silicon carbide have not yet been adequately solved.

597. References on Semiconductor Materials

1. SMITH, R. A. "Semiconductors"; New York, Cambridge University Press, 1959.

2. DUNLAP, W. C. "An Introduction to Semiconductors"; New York, John Wiley & Sons, Inc., 1957.

3. MADELUNG, O. "Physics of III–V Compounds"; New York, John Wiley & Sons, Inc., 1964.

4. BEAM, W. R. "Electronics of Solids"; New York, McGraw-Hill Book Company, 1965.

5. HANNAY, N. B. (ed.) "Semiconductors"; New York, Reinhold Publishing Corporation, 1959.

6. SHOCKLEY, W. "Electrons and Holes in Semiconductors"; Princeton, N.J., D. Van Nostrand Company, Inc., 1950.

7. "American Institute of Physics Handbook"; New York, Institute of Physics, 1964.

8. KROGER, F. A. "The Chemistry of Imperfect Crystals"; Amsterdam and New York, North Holland Publishing Company and John Wiley & Sons, Inc., 1964.

9. HULTGREN, ORR, ANDERSON, and KELLEY "Selected Values of Thermodynamic Properties of Metals and Alloys"; New York, John Wiley & Sons, Inc., 1963.

SECTION 5

CIRCUIT ELEMENTS

BY

STEPHEN FOLDES President, Endicott Coil Company, Incorporated; Member, Institute of Electrical and Electronics Engineers

DAVID B. PECK Vice President, Engineering, Sprague Electric Company; Member, Institute of Electrical and Electronics Engineers

JACOB H. MARTIN Technical Assistant to the Vice President, Engineering, Sprague Electric Company

KASSON HOWE Retired; Former Electrical Engineer, Ward Leonard Electric Company

WILLIS A. ADCOCK Vice President, Technical Development, Texas Instruments Incorporated; Fellow, Institute of Electrical and Electronics Engineers

W. T. MATZEN Manager, Molecular Electronics Branch, Semiconductor Research and Development Laboratory, Texas Instruments, Incorporated; Member, Institute of Electrical and Electronics Engineers

J. P. MIZE Member, Technical Staff, Semiconductor Research and Development Laboratory, Texas Instruments, Incorporated

ROBERT E. MOE, P.E. General Electric Tube Department (retired); Fellow, Institute of Electrical and Electronics Engineers; Chairman, 1966–1967, Joint Electron Tube Council of the Electronic Industries Association; U.S. Technical Advisor 1964–1966, Technical Committee No. 39, Electron Tubes, International Electro-Technical Commission

CONTENTS

Numbers refer to paragraphs

SECTION 5

CIRCUIT ELEMENTS

MAGNETS

By Stephen Foldes

GENERAL

1. A magnet is a body which possesses the property of attracting magnetic substances. Magnets are of two classes: permanent magnets and electromagnets.

2. A permanent magnet has the property of storing magnetic energy after the external magnetizing force which produced it is removed. The stored magnetic energy of such a magnet may be essentially constant for an indefinite period.

3. Electromagnets are magnets which possess magnetic qualities only in the presence of a magnetizing force produced by an electric current. The induced magnetic energy is a function of the strength, direction, and, in some cases, duration of the current.

PERMANENT MAGNETS

4. Permanent Magnets. See also Sec. **4**, Pars. **286** to **291**. Permanent magnets derive their use from residual, or remanent, magnetism, a quality termed retentiveness. Quantitatively the following measures apply:

a. Remanence is the structural flux density, also called residual induction. It is denoted by the symbol B_r and is expressed in units of flux density, gauss (lines per square centimeter) or lines per square inch.

b. Coercive force is equal to the reversed magnetizing force required to reduce the remanence to zero is denoted by the symbol H_c, and is expressed in units of oersteds (ampere-turns per centimeter) or ampere-turns per inch.

The coercive force is also a measure of the ability of a permanent magnet to withstand mechanical shock, vibration, temperature changes, and the passage of time without loss of its properties.

c. Energy product is the product of B_d and H_d of a given magnet at specific points on the magnetizing curve. The peak value of the energy product is often used to compare various magnetic materials for suitability in given applications.

5. Permanent-magnet Materials. The earliest forms of magnetic materials known to man were bits of magnetic oxides and native iron to which mystic properties were attributed and which found their first practical application in the magnetic compass.

With the growth of electrical technology needs developed for stored magnetic energy in reproducible and predictable configuration. This led to the development of various hard alloys of steel, containing carbon, chrome, tungsten, and cobalt, which qualified for permanent-magnet applications owing to their stability and controllable characteristics.

The search for improved materials and for the upgrading of weight and cost ratios to the available magnetic force and energy gave rise to the family of Alnico alloys during the 1930s, whose improved characteristics and unprecedented energy products opened new opportunities for design and application of permanent-magnet devices.

The materials technology of the following decades has made available the family of ceramic magnets. These are nonmetallic, refractory materials characterized by high coercive force and relatively low cost and weight, making them economically applicable in mass-use fields formerly dominated by mechanical devices. Their basic constituents are magnetic oxides and ferrites.

6. Comparative Properties of Permanent-magnet Materials. In general, high remanence (B_r) is associated with low coercive force (H_c) in a permanent magnet. High values of remanence are also generally associated with high-permeability materials, which usually are lower in coercive force. The magnetization curve (hysteresis loop) of a given material is an excellent indicator of its properties as a permanent magnet. Broad magnetization curves are a requirement, since the loss normally associated with materials of this nature is the basis of their ability to store the magnetic energy imparted by the inducing field. Typical magnetization curves of the three types of materials are shown in Fig. 5-1. Values of B_r, H_c, peak energy product, and permeability are tabulated in Table 5-1 for the same materials.

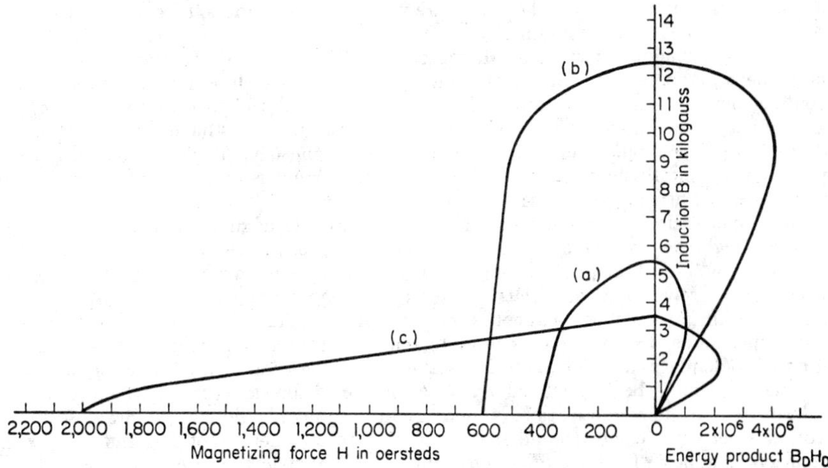

FIG. 5-1. Normal magnetization curves, second quadrant: (*a*) alloy steel; (*b*) Alnico; (*c*) ceramic magnet. Curves shown are typical; many variations in each class exist.

Since the field strength (B) at the pole of a magnet is a product of the unit field strength and the *area* of the magnet at this plane, and since the coercive force of the magnet (H) is the product of the unit coercive force (at the same unit field strength) and the *length* of the magnet, it follows that the energy content, or BH product, is proportional to the *volume* of the magnet. It is therefore theoretically possible to employ suitable materials to provide any desired energy storage, and the selection of one material in preference to the others for a given device is governed by other considerations. These involve the required geometrical configuration, the allowable size and weight, end use, or application, producibility, and economic factors. Of the above, producibility is the only one subject to further general discussion.

The alloy steels can be produced by casting, forging, rolling or forming, and machining in the annealed state to any desired configuration. Development of optimum properties requires subsequent hardening and heat-treating, or "curing," cycles to provide uniform and stable grain structure in the metal. Since critical dimensions

Table 5-1. Typical Magnetic Properties

Magnet material	B_r	H_c	B_dH_d	Permeability
Alloy steel	5,500	400	1.2×10^6	Approx. 15
Alnico	12,500	600	4.5×10^6	Approx. 12.5
Ceramic magnet	3,500	2,000	2.2×10^6	Approx. 1

are not usually maintained during the curing, grinding of the hard material to the desired finished dimensions is frequently required.

The Alnico family of aluminum, nickel, and cobalt alloys is either cast or sintered from powdered metals. Hardening and curing cycles are required, with finishing by grinding, since the materials are too hard for normal machining. It is possible, but costly, to put holes or complex configurations into the material by the EDM process after curing. Sintered parts can usually be held to tolerances of ±0.010 to ±0.005, depending on size.

The ceramic family of materials is universally made by processes resembling clay or porcelain techniques from a wet mix, but it can also include compacting under high pressure to develop suitable densities, similar to sintering. Once completed, these materials are very hard and fragile, with high internal stresses. They respond poorly to finishing operations of any kind. For this reason close mechanical tolerances are not usually achievable, and when they are required, the manufacturers should be consulted to establish practical limits.

7. Magnetizing of Permanent Magnets. The most common method of magnetizing permanent magnets is by electrical fields of suitable strength and configuration. Depending on the material of the permanent magnet and the saturation required, the magnetizing force may vary from 600 to 3000 At/lin in of the magnetic structure. Since the power-input requirements to produce magnetizing fields of this magnitude may be substantial, the time of application of the current is usually very short, to prevent excessive heating and to save power.

8. Conservation of Magnetization. In the case of magnets with short air gaps, or when high-coercive-force materials are used, no special precautions are required to conserve the initial magnetization. However, when long air gaps are involved, or when low-coercive-force materials are used, a "keeper" of soft iron should be kept across the air gap until the magnet is put into service. This is accomplished by sliding the keeper across the pole pieces prior to removal from the electromagnet and sliding it off the pole pieces as the magnet is installed. For best results on critical applications, the magnet should be magnetized after assembly into the structure in which it is used.

To prevent demagnetization, a permanent magnet should not be drawn over an iron or steel surface, nor should it be subjected to any demagnetizing influence, such as heavy magnetic fields or other permanent magnet fields, severe mechanical shock, or extreme temperature variations.

9. The mechanical force produced between a permanent magnet and a magnetic material with flat and parallel surfaces is expressed by

$$F = KB^2S/72.13 \tag{5-1}$$

where F = force in pounds, B = mean flux density in kilolines per square inch, S = mean cross-sectional area of the air gap in square inches, and K = factor depending upon leakage. Since the leakage is determined by the geometric configuration of the assembly, including the length of the air gap, it is not readily definable and empirical determination is necessary. In general, a short, compact structure will yield a higher K than a long structure. For air gaps which are short in relation to the cross section, values of K between 0.2 and 0.8 are usual.

10. Applications of Permanent Magnets. The two basic classifications of permanent-magnet applications are:

Electromagnetic. The interaction of the field of a permanent magnet with a moving conductor results from its interaction with the field.

Mechanical. The magnetic attraction between a permanent magnet and a magnetic material causes a pull, push, holding, or deflection of an object.

Electromagnetic applications include uses in measuring instruments, motors, generators, magnetos, and pickup devices.

ELECTROMAGNETS

11. Electromagnets. Electromagnets derive their use from the control of their magnetic state by the flow and interruption of electric current. This permits positive and rapid control of apparatus however remotely situated.

12. The basic form of an electromagnet is a conducting coil, termed a solenoid. The conductor is usually round magnet wire, but the conductor cross section may be square or rectangular. Ribbon foils may also be used. The length of the coil, as well as the shape of its cross section, is determined by the application. The large majority of applications utilize a coil of round cross section, and this configuration will be assumed in the following discussion unless otherwise noted.

13. The mechanical forces of all forms of electromagnets are due to the tendency of the electromagnet to move so as to possess the greatest linkage of electric current and magnetic flux, i.e., the maximum self-inductance.

Two solenoids mutually attract or repel one another, depending on their relative polarities. This type of operation is very inefficient and is seldom used in practice. Magnetic materials (iron or steel), introduced into the field created by the energized coil, improve performance by several orders of magnitude.

14. Solenoid and Plunger. Figure 5-2 shows a suitably mounted solenoid provided with a movable bar of soft iron or steel, called a plunger. When the coil is energized, the plunger becomes magnetized and mutual attraction then takes place between the coil and the plunger. The solenoid and plunger are a differential device, the net opposing mechanical forces just balancing each other when the neutral polarities of the solenoid and the plunger coincide. When the plunger, or core, is rigidly fastened within the coil, the combination is called a *bar electromagnet.*

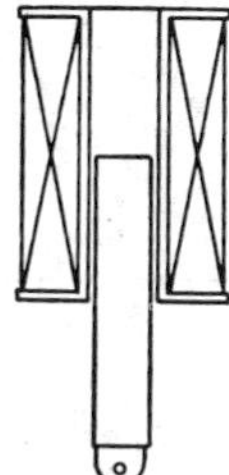

FIG. 5-2. Solenoid and plunger.

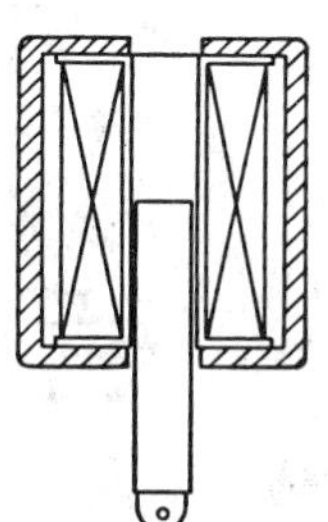

FIG. 5-3. Iron-clad solenoid and plunger.

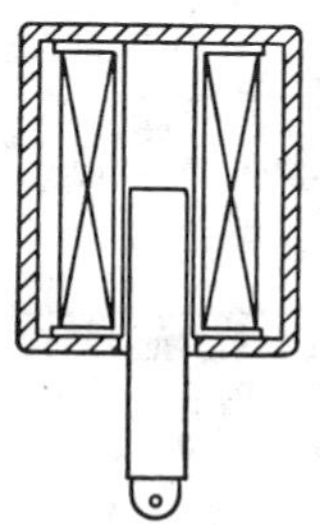

FIG. 5-4. Iron-clad solenoid, closed end.

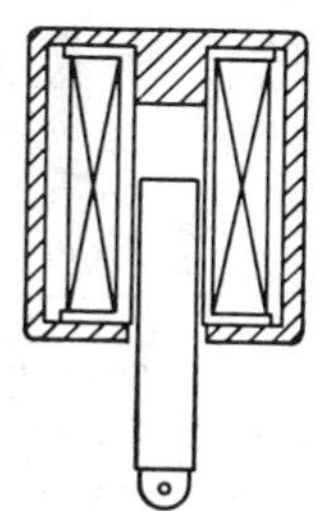

FIG. 5-5. Iron-clad solenoid with fixed core, to increase pull.

Solenoids with plungers only, i.e., without an iron frame, are seldom used in practice except in cases where it is desired to take advantage of their peculiar pull characteristics. However, it is important to understand their actions, since the design of all movable-core electromagnets is influenced by the characteristics of the simple coil and plunger.

15. An iron-clad or steel-clad solenoid is a solenoid and plunger provided with an iron or steel frame or jacket. The effect of the frame is to increase the mechanical force only slightly at the initial position, but greatly at or near the final position of the plunger. In the magnetic-cushion type (Fig. 5-3) the plunger cannot strike the oppo-

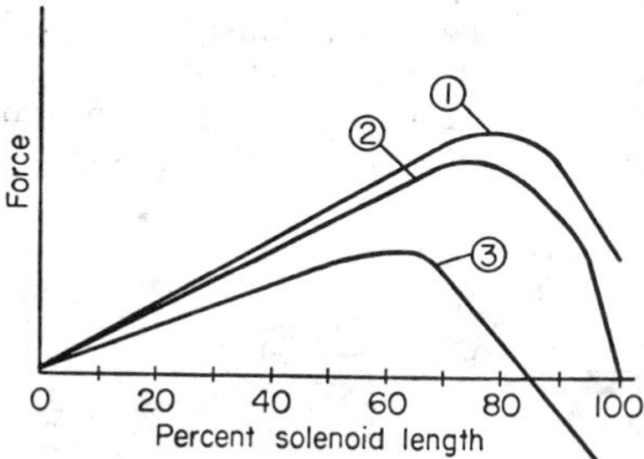

FIG. 5-6. Static-force curves, simple solenoid and plunger.

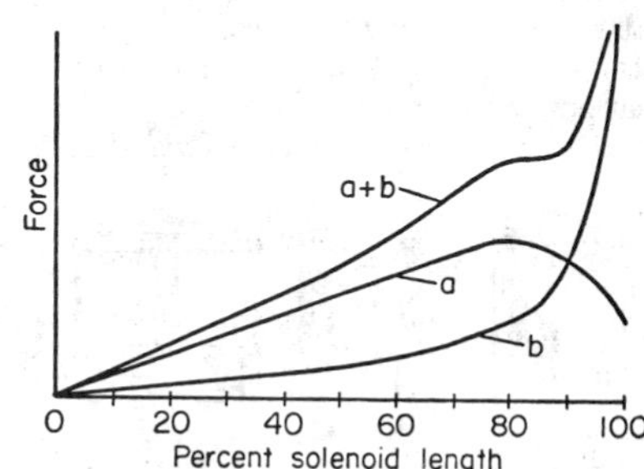

FIG. 5-7. Static-force curves, closed-end solenoid.

site end of the frame as it passes into the coil, thereby eliminating the hammer blow (Par. **27**) which occurs in the iron-clad solenoids shown in Figs. 5-4 and 5-5. The types illustrated in Figs. 5-4 and 5-5 are more efficient than the one in Fig. 5-3, because the iron plug, or stop, extending toward the plunger increases the pull near the final position and also increases the sealing, or holding, pull.

16. The static force of the several forms of electromagnets with constant current is shown in Figs. 5-6 through 5-9. In Fig. 5-6, curve 1 is for a plunger longer than the solenoid; curve 2 is for a plunger of the same length as the solenoid; curve 3 is for a plunger shorter than the solenoid. The pull between the solenoid and the plunger is marked *a* in Figs. 5-7 and 5-8. Curve *b* is for the air-gap pull only between the plunger and the stop. The total pull is the sum of these two pulls, as shown. Substantially, only the pull *b* is available in the clapper-type electromagnet of Fig. 5-9.

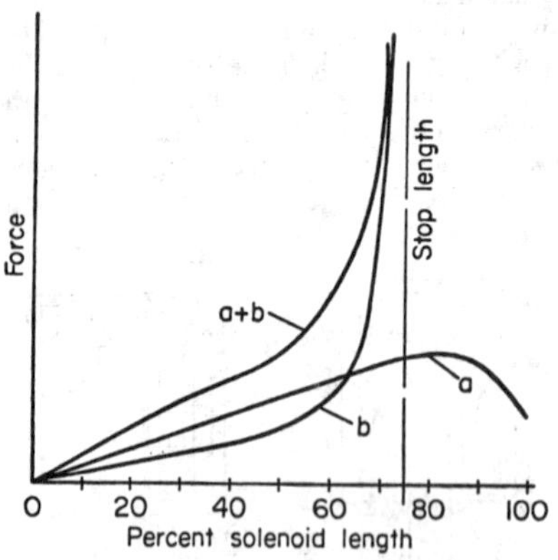

FIG. 5-8. Static-force curves, solenoid with stop.

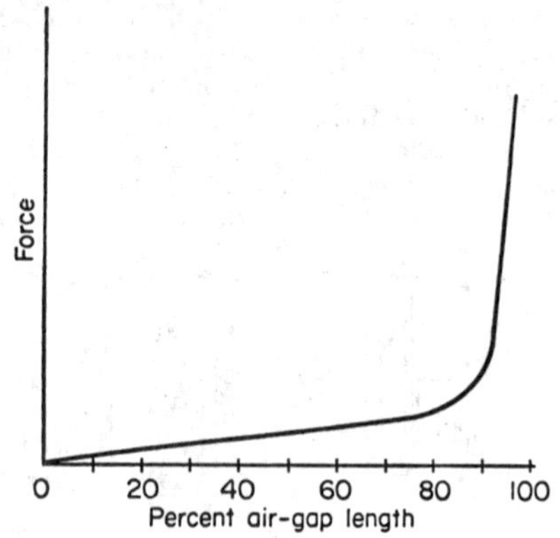

FIG. 5-9. Static-force curve, clapper-type electromagnet.

17. Modification of Characteristics. In some cases, toggles, cams, levers, or bell cranks are employed in connection with electromagnets to obtain force-distance characteristics different from those exhibited by the electromagnet alone. Such mechanical devices (1) increase the mechanical force but correspondingly decrease the effective range of travel; (2) increase the range but correspondingly decrease the mechanical force; or (3) are employed to make the force approximately uniform throughout the range of motion. Any auxiliary device introduces mechanical losses such as friction or binding which reduce the effectiveness of the solenoid.

18. Increasing the Initial Pull. Several methods, illustrated in Figs. 5-10 through 5-13, employ the principle of the magnetic shunt. In cases where it is desired to obtain a greater force at the initial position at the expense of a smaller force in the final position, portions of the armature or core are so arranged that the distance between the coacting parts shall be relatively small at the initial position of the armature, as the latter moves toward the core end, or stop (Figs. 5-12 and 5-13). Enlarged core ends also are employed, as shown in Figs. 5-14 and 5-15, the former being of the general clapper type and the latter of the general horseshoe type. The enlarged core ends tend to reduce the leakage flux, thereby increasing the initial effective field strength.

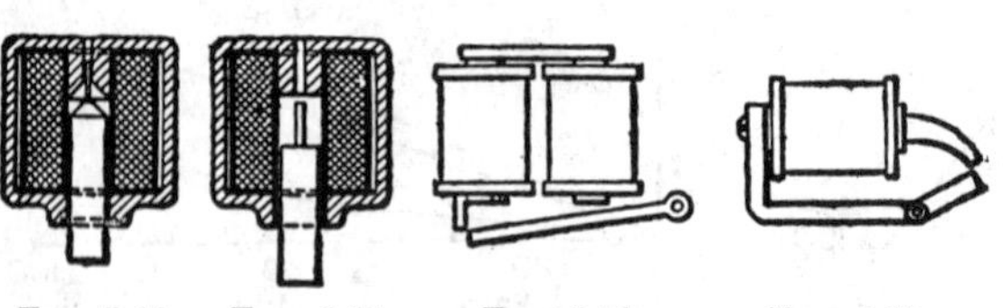

FIG. 5-10. FIG. 5-11. FIG. 5-12. FIG. 5-13.

FIGS. 5-10 to 5-13. Methods for increasing the initial pull.

FIG. 5-14. FIG. 5-15.

FIGS. 5-14 and 5-15. Use of enlarged core ends for increasing initial pull.

Coned plungers and stops, shown in Fig. 5-10, are successfully employed for increasing the initial pull. In a magnetic structure where the air gap is initially relatively short, the saturation characteristic rises very rapidly as the plunger begins to move and further close the air gap (see Par. **25**). When a conical shape is imparted to the plunger and the stop, on the assumption of the same total travel of the plunger, the air-gap distance for the flux is reduced as the sine of the half angle of the cone. Although the surface area of the cone is increased inversely as the sine of the half angle of the cone, the drop in flux density over the larger area is more than compensated by the increase in flux density due to the reduced air gap. Gains up to 50% in initial pull can be realized by the proper cone angle compared with a flat-faced plunger. The minimum practical included angle of the cone is about 30°. With angles smaller than this, mechanical seizure of the plunger and cone can occur on seating.

19. Sealing Pull. This term refers to the properties of an electromagnet after the armature has completed its stroke and is to hold the load against the counterforce of a spring, gravity, compressed air, etc. This pull may be made a maximum by reducing at least one of the attracting areas, as shown in Fig. 5-16. The pull of an electromagnet when the armature is closed is proportional to ϕ^2/S. Hence making the area of contact S small increases the pull for the same magnitude flux ϕ. About the same amount of magnetic flux exists in both core ends, but the area of contact at the rounded end is less than at the flat end; therefore the holding force is greater at the rounded end. It is this effect which permits lifting magnets to hold skull-cracker balls weighing 10 or more tons. In a plunger-type solenoid, the sealing pull, or hold, can similarly be increased by rounding or reducing the contacting surfaces of the plunger and stop. This, of course, will be at the expense of initial pull. The equivalent of a strong sealing pull may also be obtained with toggles and by other mechanical means.

Fig. 5-16. Rounded core end for high sealing pull.

20. The magnetizing force of a solenoid is a maximum at its center, falling off to about one-half at each end of the winding. When one end of the plunger is placed in or near the end of the solenoid, it becomes magnetized and mutual attraction results between the flux in the plunger and the magnetizing force due to the ampere-turns in the winding of the solenoid. When the plunger is not much longer than the solenoid, all tractive effort ceases when the respective magnetic centers of the solenoid and the plunger coincide (see also Par. **16**). As a result, the maximum pull always occurs after the inserted end of the plunger has passed the middle of the solenoid. In long solenoids, with the magnetizing force high compared with the cross-sectional area for the plunger, the maximum pull may occur over a considerable distance on each side of the middle of the solenoid. On the other hand, with very short solenoids the maximum pull may occur at or near the end of the solenoid opposite to where the plunger enters.

21. Characteristic Pull of Solenoid and Plunger. Figure 5-17 shows the approximate pull diagram for different positions of the plunger. The expression for the maximum uniform pull in pounds is

$$F = CsNI/l \tag{5-2}$$

where I = current in amperes, N = number of turns, s = cross-sectional area of the core or plunger in square inches, l = length of the solenoid in inches, and C = pull in pounds per square inch per ampere-turn per inch. C depends upon the proportions of the coil, the degree of saturation, and the length, physical character, and chemical composition of the plunger. This expression is valid when the plunger is inserted between 0.4 and 0.8 of the coil length. Experience indicates that little is gained in maximum pull by making the plunger considerably longer than the coil. When the diameter of the coil is about three times that of the plunger and the length is two to three times the coil diameter, the value of C is about 0.01 lb/(in²)(At)(in) of coil length.

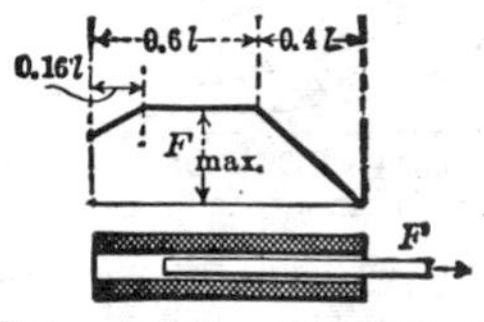

Fig. 5-17. Diagram of plunger pull for simple solenoid and plunger.

22. The characteristic air-gap pull of a solenoid or

Table 5-2. Operate and Release Characteristics

	Operate	Release
Series resistance	Fast (see TC)*	No effect
Series inductance	Slow	No effect
Parallel resistance	Slow	High resistance—slow release
Parallel capacitance	Slow	Slow
Shunt-reversed diode	No effect	Fast
Inductive coupling	Slow	Slow
Hysteresis loss	Slow	Slow
Coil time constant	Small TC, fast operate	No effect
Magnetic shunt	Slow	Fast
Mechanical inertia	Slow	Slow
Developed force	Fast, with more force	No effect
Opposing force	Slow, with more force	Fast, with more force, except friction

*Operating voltage must be increased to compensate for voltage drop across resistance.

electromagnet with flat-faced parallel attracting surfaces is governed largely by the leakage factor c and is expressed by

$$F = s(NI/l_a c)^2 = sKB^2 \tag{5-3}$$

This is based on Maxwell's equation for the mechanical force required to pull apart an energized long-bar electromagnet hypothetically separated at its middle. The symbol l_a stands for the air-gap length in inches.

23. The pull in an iron-clad solenoid is composed of two components: the pull between the end of the stop or frame and the plunger, and the pull between the winding and the plunger. The latter is all-important at the beginning of the stroke, but near the end the air-gap pull becomes predominant. The complete expression for the pull F with plunger inserted 0.4 or more into the coil is the sum of Eq. (5-1) and Eq. (5-2), namely,

$$F = sN[I(NI/l_a^2c^2 + C/l)] \tag{5-4}$$

Approximate values for ordinary soft-iron cores are $c = 2{,}600$ and $C = 0.01$.

24. Pull per Unit Cross Section. A typical curve plotted between pull per mean unit cross section of air gap and mean flux density therein is shown in Fig. 5-18. The pull for a given air gap is seen to obey the square law with variation of flux density [Eq. (5-2)]. Since in magnets with long air gaps the path through the iron can be considered as having no reluctance, the saturation characteristic becomes a straight line and the flux density for the air gap is directly proportional to the ampere-turns. Relationships between pull and mean length of air gap for various ampere-turn values are plotted in Fig. 5-19. This relationship is essentially hyperbolic. These two sets of curves explain the gain in initial pull experienced with a cone-shaped plunger described in Par. **18**.

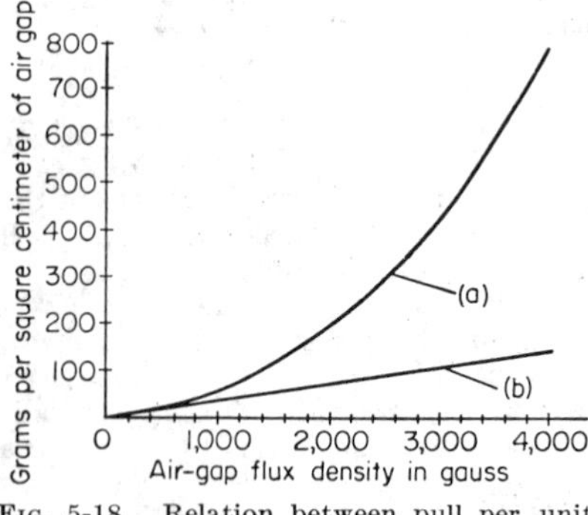

Fig. 5-18. Relation between pull per unit area of air gap and flux density. (*a*) Short air gap; (*b*) long air gap.

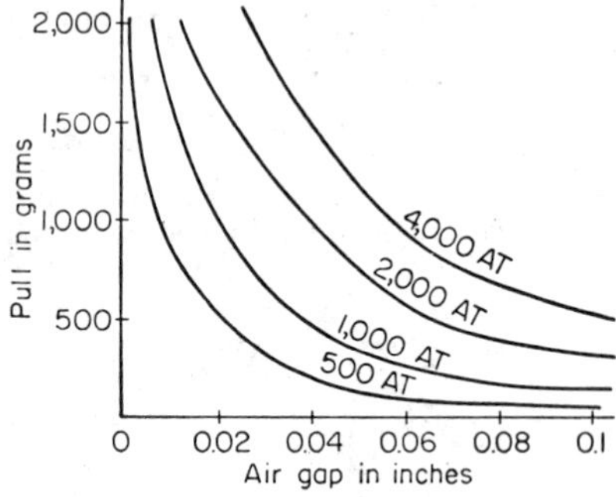

Fig. 5-19. Pull vs. air gap for various ampere-turns in coil.

25. Test Data on Pull Characteristics. Figure 5-20 shows curves from actual test. Curve (1) is that of a single coil and plunger; (2) that of a coil and plunger with stop; (3) that of an iron-clad coil and plunger; (4) shows the same iron-clad coil with a different length of stop. These curves are normalized for a coil with the same excitation in all cases.

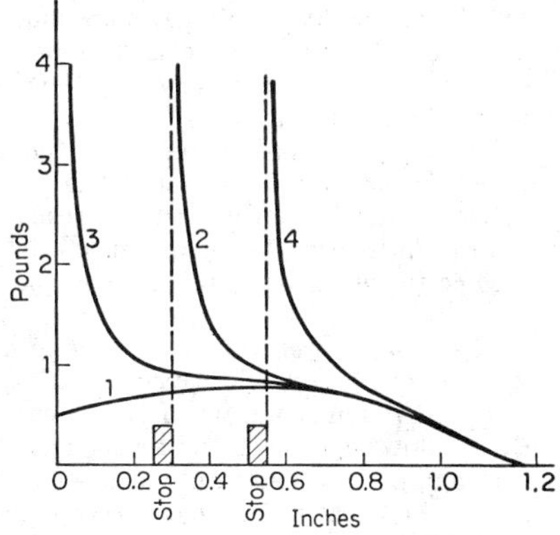

FIG. 5-20. Test curves on pull of solenoid and plunger.

26. Energy Considerations. Barring heat losses, electrical energy in a plunger-type electromagnet is converted into mechanical energy through magnetic energy. An amount of energy equal to that expended by the moving plunger is also stored in the coil as magnetic energy. Therefore, the useful work performed by an electromagnet is equal to the product of the mean net mechanical force and the distance of travel, including the kinetic energy of the moving parts which tends to produce the hammer blow. At least twice as much electrical energy must be supplied to a d-c electromagnet as is required for the mechanical work to be accomplished. Hence the conventional efficiency of a d-c electromagnet does not exceed 50%. The actual overall efficiency is very much less.

27. The Hammer-blow Effect. Electromagnets produce hammer blows when the movements of their elements are arrested by a stop. This constant hammering in electromagnets often is detrimental to the apparatus, and the noise produced is objectionable. In such applications as high-speed character or line printers used as digital-computer outputs, and in electric hammers or riveters, efforts are made to maximize the effect to obtain the desired result. When it is desired to reduce or eliminate the hammer blow, it may be done by the use of the electromagnets shown in Fig. 5-3 and Fig. 5-21, by the use of springs, or by reducing the plunger armature velocities with time-delay methods.

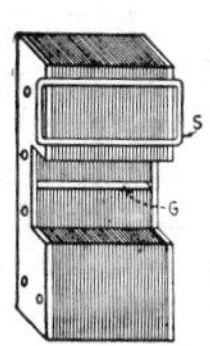

FIG. 5-21. Shading coil and auxiliary air gap at neutral-polarity plane.

28. Operating and Release Characteristics of Electromagnets. The operating characteristic refers to the time interval required for the electromagnet to start its load and perform its work by completing its stroke after voltage has been impressed upon its coil, including the time of response. The release characteristic refers to the time interval required for the electromagnet to release its load or for its armature to return to its initial position after the current has been discontinued or reduced to a predetermined strength.

29. Major factors affecting the operating and release characteristics of electromagnets are:

a. Resistance in series with the coil.
b. Inductance in series with the coil.
c. A resistor connected across the coil.
d. A capacitor connected across the coil.
e. A diode reverse-connected across the coil.
f. Induced currents in magnetically coupled closed electrical circuits.
g. Hysteresis losses in the magnetic structure.
h. The time constant of the coil.
i. Magnetic leakage of shunts.
j. The inertia of the moving parts.
k. The force developed by the electromagnet.
l. The opposing load, or force.

Table 5-2 correlates the desired characteristics with the listed factors.

30. Undesirable behavior may result from three other factors without affecting the inherent dynamic characteristics of the electromagnet:

a. Excessive heating due to the applied current or due to environmental factors will cause an increase in resistance with an accompanying reduction of available ampere-turns input for fixed-voltage operation. At the same time the permeability of the heated iron structure can also diminish, causing the induced total flux to decrease. The reduction of force produced by the electromagnet due to these causes is cumulative and may be significant.

b. Nonuniform and excessive friction in the path of motion may resist in variable manner the operation characteristics, since normally the initial force, or pull, of the magnet is minimum at the start, whereas the friction is maximum under static conditions. Similarly, on release, when the magnetizing force is removed, the restoring force must overcome the static friction. The magnet may fail to close or may stick when it should release.

c. The coercive force of the magnetic material may be sufficiently great so that the remanent magnetism will cause "sticking" of the plunger when the magnetizing force is removed. This condition may be corrected by selecting "soft" magnetic materials with very low remanence and by the introduction of small air gaps in the closed position of the plunger or armature which will minimize the force due to the remanent magnetism. At times the momentary application of a reversed field may be required to secure release.

A-C-OPERATED ELECTROMAGNETS

31. A-c-operated electromagnets have certain inherent characteristics which distinguish them from those operated on direct current. Although the principles heretofore discussed apply to them in general, notable exceptions must be made. A single-phase a-c electromagnet is by its nature a fast-operating device. Since the flux and the resulting force follow the variation of the current magnitude through the cycle, it peaks twice each cycle. If the plunger is unable to accelerate the load in one half cycle or less, it cannot move it at all. For 60-c operation the plunger and load must accelerate and attain sufficient momentum in $1/120$ s or less to sustain motion during the absence of force as the flux passes through zero. For higher frequencies, for example, 400 c, the conditions for operation are more severe. To overcome this condition, either larger solenoids or more power input must be provided; or, as is often the case, the available alternating current must first be rectified, then used as direct current, with a d-c solenoid. In some instances viscous damping, such as an oil dashpot, may be used to slow down the action of an a-c solenoid.

32. Polyphase electromagnets do not have this restriction. However, they are considered to be torque motors, or "linear motors." Here the principles of induction motors apply, and this class of device is not treated in this section.

33. Eddy Currents. Among the characteristic effects of a-c electromagnets are the secondary (eddy) currents induced in their magnetic materials and in other closed electric circuits linked with their magnetic circuits. They behave much as the primary windings of transformers, hence easily are overloaded owing to short-circuited turns. Characteristics that tend to make d-c electromagnets slow-operating have little similar effects on a-c electromagnets, which merely consume more current in consequence, much as the current in the primary of a transformer increases with loads on its secondary without materially changing the magnitude of the magnetic flux. For this reason in higher-power applications laminated-iron structures are used to reduce eddy-current losses.

34. The static-pull characteristics of a constant-voltage or shunt a-c electromagnet are different from those of the same electromagnet operated with direct current for similar relative positions of plungers and stops, and with the best construction for a-c operation. This is due to the fact that the current strength decreases as the plunger moves into the coil because the self-inductance, and hence the inductive reactance, of the electromagnet is increased. The ampere-turns of the magnet are then a function of plunger position, and the complexity of the analysis defies the prediction of typical pull characteristics. Experience has shown that the stops of a-c plunger electromagnets should be about one-third as long as the coil, so that the total pull will be reasonably linear over a portion of the plunger travel and still provide a strong holding pull at the end of the stroke. The effective static mechanical force of a constant-current a-c electromagnet is about 20% less than that due to the same electromagnet when operated with direct current for a given number of ampere-turns. This is due largely to hysteresis and eddy-current losses in the iron or steel parts. It also is affected by the form factor of the current.

35. Shading coils, consisting of a single closed turn of proper resistance such that the main-coil current and the secondary current in the shading coil shall be substantially in time quadrature, are used in the ends of cores, stops, plungers, or armatures to split the phase in order to produce a 2-phase effect after the air gap has tightly closed to give the required sealing pull. This is also the most effective device to overcome the annoying noise caused by the chattering of the closed plunger or armature during alternations of the cycles as the load attempts to break away the plunger or armature when the flux passes through low amplitudes. An improperly seated or overloaded a-c electromagnet will chatter owing to the failure of the shading ring to provide sufficient pull when a minute air gap increases the reluctance of the magnetic path.

36. Plunger ends and stops are precisely ground on high-quality a-c electromagnets to provide good seating contact and quiet operation. However, this can cause slow or faulty release due to remanent magnetism acting through the minute gap. Auxiliary air gaps at the neutral-polarity plane, small leaf springs, or both are used to overcome sticking. Figure 5-21 illustrates the features of a well-designed a-c solenoid.

37. The efficiency of a-c electromagnet of the best design has been estimated to be of the order of 85%. It is at least double that for a d-c electromagnet because most of the magnetic energy stored in the magnetic circuit during the first half of an alternation is returned to the electric circuit during the second half. Also, the copper losses are smaller. Iron losses, as well as pull, increase with flux density; hence the maximum flux density does not give the most efficient results and causes undue heating. The major portion of the total heat loss occurs in the iron rather than in the copper.

38. Polarized electromagnets are employed where relatively strong mechanical forces are desired with relatively weak electric currents in an operating coil due to the fact that a polarizing field already is produced through a magnetizing coil or a permanent magnet. Polarized mechanisms comprising a magnetizing coil and an operating coil require considerable magnetizing current (as in the dynamic loudspeaker). To eliminate magnetizing currents, polarized electromagnets employing permanent magnets are used. They require only operating current for their functions; therefore they are efficient, strong, and sensitive to weak currents. Permanent-magnet loudspeakers, telephone receivers, and polarized relays are examples of this type of application.

Table 5-3. Use of Electromagnets

1. Portative—for holding magnetic material:
   - Lifting magnets.
   - Magnetic chucks.
   - Magnetic clutches and couplings.
   - Magnetic brakes.
   - Magnetic separators.
2. Tractive—for doing work:
   - *A*. Solenoid type.
     - Mechanical actuators.
     - Electric valves.
     - Circuit breakers.
     - Magnetic brakes.
     - Time-delay relays.
   - *B*. Horseshoe type.
     - Bells and buzzers.
     - Telegraph apparatus.
     - Short-stroke actuators.
     - Relays.
   - *C*. Clapper type.
     - Relays.
     - Magnetic switches.
   - *D*. Special types.
     - Vibrators.
     - Latching and unlatching devices.
     - Reed relays.
     - Electric pendulums.

Directional effects are easily obtained through current reversal, and the armatures may be either neutral or biased in position.

39. Uses of electromagnets may be classified by the major operating functions performed. An immense variety of individual applications exist and will continue to be developed, limited only by the imagination of the designer as it is stimulated by the required end result. Table 5-3 lists the major classes, with some typical applications for each.

HEATING OF ELECTROMAGNETS

40. Heating of electromagnets due to the power loss in the winding and, in the case of alternating current, in the iron structure limits the capacity of the electromagnet to a power input (hence output) which will not cause a temperature rise in excess of a safe limit. The safe limit is determined by the internal temperatures which the insulation within the coil can withstand, by the external environment and location of the electromagnet (i.e., ambient temperature, proximity to flammable materials, external cooling by means of convection or heat sinks, etc.), and by the resistance rise of the winding, which may limit the current, therefore the pull, or holding, capability of the electromagnet.

41. Time Rates of Heat Dissipation of Coils. Coils dissipate heat at a time rate directly proportional to the difference in temperature between the coil and its surroundings and inversely proportional to the coil-wall thickness and the amount of electrical insulation (which also is thermal insulation) between the layers of copper wire in the coil. As a result, the time rate of heat dissipation increases and the time rate of heating decreases as the temperature of the coil increases. When these two time rates are numerically equal, no further increase in temperature results. Thus, if a test shows that a coil can dissipate 10 W at a given safe temperature, it is known that a constant 10-W I^2R input may be applied to the coil without causing overheating and that less input will result in proportionally lower temperature rise.

42. The required dimensions of an electromagnet are proportional to the load, the range or stroke, the duration of excitation, and the time interval between excitations. Since the foregoing determine the volume and weight requirements of the coil, whereas the surface area of the electromagnet determines its rate of heat dissipation, it is advantageous to maximize the area with respect to volume of the coil to allow either larger power output or smaller temperature rise. Generally, the surface area increases rapidly with increase in radius and a corresponding decrease in length and much more gradually with an increase in length and a decrease in radius.

43. Electromagnets designed for continuous service must have windings of such volume and cooling surface and wires of such cross sections that the coils will not become overheated and the required pull will be maintained even after the coil has been in the circuit for many hours. The designer should allow too much, rather than too little, space for the winding and the insulation of the coil. A practical rule in the design of heavy-duty coils for continuous service is to calculate the size of wire to use at an ambient of 20° and then use the next larger size because of the increase in the coil's resistance with the rise in temperature. In a-c electromagnets the magnetic materials as well as the coils are heated in service. Whereas heat may flow from coils to cores in d-c electromagnets, this heat flow may be greatly retarded in a-c electromagnets. The final information as to the safe maximum power input to a coil can be obtained only when the coil is mounted in the apparatus in which it is to be used. It is safe to assume that a coil of identical configuration in the same installation will successfully perform its function regardless of the value of the applied voltage or, when the operation is intermittent, provided that the average power input does not exceed the established maximum.

44. The thermal time constant of a coil may be readily established by the cooling method. In this test, the coil should be mounted in the apparatus where it is to be used and under conditions of maximum ambient temperature. Heat the coil with direct current to any safe temperature, as determined by the increase in the coil's resistance, and maintain it at this temperature to assure stabilization. Turn off the power, and measure the time interval required for the coil to cool to one-half the difference between its maximum temperature and the ambient temperature. Multiplying this measured time interval by 1.44 gives the thermal time constant of the coil. The resistance rise of a coil can be calculated from the following formula,

$$R_t - R_c = R_c \alpha t \tag{5-5}$$

where R_c = resistance of coil at the initial temperature, R_t = resistance of coil at the final temperature, α = temperature coefficient of resistance, and t = temperature rise. The value of α for copper at room temperature is 0.00381/°C, so that approximately 1% resistance change results from a temperature change of 2.6°C.

MAGNET WIRE

45. Magnet wires are insulated conductors in round, square, rectangular, or ribbon configurations. The conductor material is generally electrolytic-refined oxygen-free annealed copper. High-purity aluminum has gained some prominence in recent years. Specifications of the USASI and the ASTM define the bare-conductor requirements, whereas the insulated wire product is manufactured to standards prescribed by the NEMA and the IEEE.

46. Specification of magnet wire, both round and square, is by AWG size and by insulation class, reflecting the temperature and dielectric requirements of the application. The temperature class designates the highest continuous internal coil temperature in degrees centigrade at which the wires may be operated for extended periods without severe degradation and failure. The classes are given in Table 5-4.

Table 5-4. Magnet-wire Classes

Class 90	(90°C)	(restricted to unimpregnated natural textile or paper insulations over bare conductor)
Class 105	(105°C)	(UL Class *A*)
Class 130	(130°C)	(UL Class *B*)
Class 155	(155°C)	(UL Class *F*)
Class 180	(180°C)	(UL Class *H*)
Class 220	(220°C)	

Within each class, the dielectric strength is defined by the thickness of insulating film, described as "single," "heavy," "triple," or "quadruple."

A number of different varnishlike insulating compounds are available for each temperature class, the selection being based on other comparative properties, such as mechanical wear and abrasion, resistance to cracking or peeling, solderability, thermal flow, and compatibility with impregnants, oils, tapes, and insulations used within the coils. Table 5-5 lists a commonly used range of wire sizes, with nominal diameters for selected examples of every temperature class and single and heavy insulation thickness. Table 5-6 lists the NEMA MW standards applicable to the various temperature classes.

47. Textile or paper insulation is used where mechanical spacing of the conductors is necessary to improve the dielectric performance, where additional protection is required in winding against mechanical stresses and abrasion, or where very heavy conductors of rectangular shape have to be hammered into place after winding. The insulation is of itself of little value but, being porous in nature, forms a matrix which readily absorbs and holds insulating varnishes or oils to provide the intended insulation. The textiles generally used are purified cotton, acetate yarn, polyester webs or yarn, and glass cloth and yarn. Papers based on both natural and synthetic fibers are available. Fabricators can manufacture a large variety of such materials on demand.

48. The space factor, or activity coefficient, of a coil is the ratio of the space occupied by an insulated conductor wound therein to the volume of solid copper that would occupy the same coil space; hence a thin insulation of high dielectric strength is desirable to achieve a high space factor. The relative space factors of magnet wires are expressed by the numbers of turns per square inch (reciprocals of the squares of the relative diameters of the wires).

A coil having a high space factor contains copper of greater cross section for a given

Table 5-5. Magnet-wire Dimensions, Sizes 14 to 44 AWG

AWG	Bare wire diameter, in			Single		Heavy	
	Minimum	Nominal	Maximum	Minimum increase in diameter, in	Maximum overall diameter, in	Minimum increase in diameter, in	Maximum overall diameter, in
14	0.0635	0.0641	0.0644	0.0016	0.0666	0.0032	0.0682
15	0.0565	0.0571	0.0574	0.0015	0.0594	0.0030	0.0609
16	0.0503	0.0508	0.0511	0.0014	0.0531	0.0029	0.0545
17	0.0448	0.0453	0.0455	0.0014	0.0475	0.0028	0.0488
18	0.0399	0.0403	0.0405	0.0013	0.0424	0.0026	0.0437
19	0.0355	0.0359	0.0361	0.0012	0.0379	0.0025	0.0391
20	0.0317	0.0320	0.0322	0.0012	0.0339	0.0023	0.0351
21	0.0282	0.0285	0.0286	0.0011	0.0303	0.0022	0.0314
22	0.0250	0.0253	0.0254	0.0011	0.0270	0.0021	0.0281
23	0.0224	0.0226	0.0227	0.0010	0.0243	0.0020	0.0253
24	0.0199	0.0201	0.0202	0.0010	0.0217	0.0019	0.0227
25	0.0177	0.0179	0.0180	0.0009	0.0194	0.0018	0.0203
26	0.0157	0.0159	0.0160	0.0009	0.0173	0.0017	0.0182
27	0.0141	0.0142	0.0143	0.0008	0.0156	0.0016	0.0164
28	0.0125	0.0126	0.0127	0.0008	0.0140	0.0016	0.0147
29	0.0112	0.0113	0.0114	0.0007	0.0126	0.0015	0.0133
30	0.0099	0.0100	0.0101	0.0007	0.0112	0.0014	0.0119
31	0.0088	0.0089	0.0090	0.0006	0.0100	0.0013	0.0108
32	0.0079	0.0080	0.0081	0.0006	0.0091	0.0012	0.0098
33	0.0070	0.0071	0.0072	0.0005	0.0081	0.0011	0.0088
34	0.0062	0.0063	0.0064	0.0005	0.0072	0.0010	0.0078
35	0.0055	0.0056	0.0057	0.0004	0.0064	0.0009	0.0070
36	0.0049	0.0050	0.0051	0.0004	0.0058	0.0008	0.0063
37	0.0044	0.0045	0.0046	0.0003	0.0052	0.0008	0.0057
38	0.0039	0.0040	0.0041	0.0003	0.0047	0.0007	0.0051
39	0.0034	0.0035	0.0036	0.0002	0.0041	0.0006	0.0045
40	0.0030	0.0031	0.0032	0.0002	0.0037	0.0006	0.0040
41	0.0027	0.0028	0.0029	0.0002	0.0033	0.0005	0.0036
42	0.0024	0.0025	0.0026	0.0002	0.0030	0.0004	0.0032
43	0.0021	0.0022	0.0023	0.0002	0.0026	0.0004	0.0029
44	0.0019	0.0020	0.0021	0.0001	0.0024	0.0004	0.0027

number of turns than a similar coil having a lower space factor. Hence, the higher the space factor of a coil, the greater the coil conductance, resulting in an increased number of ampere-turns per volt or per watt of power input.

49. Space Factor and Operating Temperature. A coil consisting of solid copper from inside to outside, as an edge-wound coil, conducts heat from the interior of the copper to the inner and outer cooling surfaces in direct proportion to the difference in temperature between the interior and exterior of the copper. Thus a coil of high space factor will carry the maximum current with a minimum temperature rise. The

Table 5-6. NEMA Product Standards, Film-coated Round Magnet Wire

MW1-C	Oleoresinous-enamel-coated round copper, Class 105
MW2-C	Polyurethane-coated round copper, Class 105
MW4-C	Acrylic-coated round copper, Class 105
MW5-C	Polyester-coated round copper, Class 155
MW6-C	Nylon-coated round copper, Class 105
MW7-C	Ceramic-silicone-coated round copper, Class 180
MW9-C	Epoxy-coated round copper, Class 130
MW10-C	Polytetrafluoroethylene-coated round copper, Class 180
MW15-C	Polyvinyl formal-coated round copper, Class 105
MW16-C	Aromatic polyimide-coated round copper, Class 220
MW17-C	Polyvinyl formal-nylon-coated round copper, Class 105
MW19-C	Polyvinyl formal self-bonding round copper, Class 105
MW25-A	Polyester-coated round aluminum, Class 180
MW25-C	Polyester-coated round copper, Class 180
MW27-C	Polyvinyl formal-urethane-coated round copper, Class 105
MW28-C	Polyurethane-nylon-coated round copper, Class 130

higher the temperature, the greater should be the space factor or the cross section of copper conductor to offset the reduction in the electrical conductivity of the copper due to higher temperature. Furthermore, the greater the space factor, the greater the thermal conductivity of the coil wall, because heat can be conducted through copper much better than through the insulation. Time and temperature combined cause deterioration of insulating materials. High-heat coils are made with insulating materials that will stand up well at temperatures of 200°C or more; however, to prevent oxidation of the copper at these temperatures, it is frequently silver- or nickel-plated prior to insulating.

CONSTRUCTION OF MAGNET COILS

50. Methods of coil winding fall into four general classes:

Self-supporting or form-wound coils, where the wire is wound on a mandrel, or form, from which it can be removed to be taped, tied, or secured by other means for finishing and use.

Coils wound in layers on an insulating tube, with thin sheets of insulating material, such as paper applied between layers of the wire, and with the paper extending beyond the ends of the wire winding to provide high insulation values between successive layers of wire and end turns.

Coils wound on an insulating form or bobbin, with the flanges of the bobbin retaining the wire. This method generally requires a minimum of finishing operations, since many accommodations can be incorporated into the bobbin design to simplify finishing.

A method combining some of the features above, where wire is wound on an insulating tube in uniform layers and an insulating thread or yarn such as acetate or cotton is wound simultaneously but traversed much faster than the wire and allowed to dwell beyond the wire layer on each end to build a shoulder or flange which serves to retain the wire and adds additional internal insulation.

All the above methods use winding machines, which are basically similar to lathes in that they employ a rotating arbor to turn the coil form and a traverse mechanism which is synchronized with the arbor to provide the desired turns per layer. Here the similarity ends, since the traverse is arranged to be reversible between limits which correspond to the length of the winding. A wire guide attached to the traverse feeds the wire onto the rotating arbor from the wire spool through an unreeling device and an adjustable tensioning means to provide a tight coil, without undue stress or stretching of the copper.

51. Finishing of Coils. To provide external insulation, mountability, and accessibility of connections to the circuits in which coils are used, some forms of finishing operations are required. Coils wound by the first two methods of Par. **50** almost always require impregnation by varnish or synthetic resins either after winding or after assembly into their core structures. Bobbin-wound coils do not generally require similar fill or bonding, since the form prevents movement of the wire. If increased internal insulation is required, or if the relative movement of conductors must be completely eliminated owing to the application requirements, these coils may also be impregnated. When the fourth (combination) method of Par. **50** is used, impregnation is required for all yarns except acetate to prevent the loosening of the wound coil. When acetate is used, application of moderate quantities of acetone by brush, spray, or vapor will bond the surface layers of the yarn instantaneously, and to sufficient depth to yield firm, self-supporting assemblies. The unbonded inner layers comprise a cushion to shield the coil from mechanical shocks and abuse.

Terminals or lead wires are attached to the magnet wire by soldering, brazing, or welding. The joints are insulated by sleeves, tapes, or film-type insulations, and the coil is then covered with tape or yarn to lock in the leads and to protect them against pullout in handling. The outer covering also protects the magnet wire against damage and adds insulation to preclude high-voltage breakdown to metal housings or assemblies in installation.

Complete hermetic encapsulation of coils may be needed in some instances for environmental protection. This may be done by vacuum impregnation or casting into suitable outline forms or by transfer molding in dies. The materials generally used are epoxies, polyesters, and acrylics, but techniques have been developed for injection molding in nylon or for acetate-based thermoplastic resins.

When the size of the magnet wire used in the coil is sufficiently large to prevent handling damage or breakage in use or installation, lead wire and terminals may be omitted, and external connections can be made directly to the coil wires.

52. Effect of Construction on Space Factor. The best space factor of a coil maximizes the ratio of the volume of conductor to the total volume of the completed coil. The higher the space factor, the more efficient the coil becomes for its size, since the ampere-turns are maximized and heat dissipation is improved. Coils wound by the first method of Par. **50** are generally of the highest space factor, with the third method being a close second when properly designed. Method 4 is next, while the second method has the smallest space factor owing to the inherent volume of the added interlayer insulation. For this reason these coils are generally used only in high-voltage applications, as induction coils, for example, where efficiency must be sacrificed to achieve the desired insulation values.

WINDING CALCULATIONS

53. Winding Formulas. In what follows, for round magnet wires only, let

d_1 = diameter of insulated wire in inches
R_1 = ohms per linear inch
t'' = turns per linear inch
n' = turns per layer
n'' = layers per inch
N_a = turns per square inch
R_v = ohms per cubic inch
T = thickness of wall of winding in inches
L = actual length of winding in inches
p_a = average perimeter or mean length of turn in inches
S = longitudinal cross-sectional area of winding in square inches
V = volume of winding in cubic inches

For layer windings with no paper between layers

$$n'' = t'' \times 1.07 \tag{5-6}$$

$$n' = Lt''/1.05 \tag{5-7}$$

The factors 1.05 and 1.07 represent practical considerations to correct for possible variations in wire dimensions from the nominal and allow for the less than perfect packing of round wires when placed layer upon layer.

When paper is used,

$$n'' = \frac{1}{d_1 + t_p} \tag{5-8}$$

where t_p = thickness of the paper layer. The turns per square inch and the ohms per cubic inch are given by

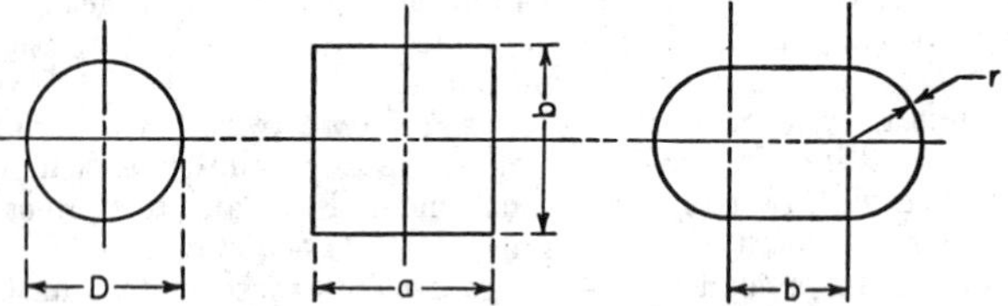

FIG. 5-22. Core shapes and dimensions for windings.

$$N_a = t''n''/1.05 \tag{5-9}$$

$$R_v = N_aR_1 \tag{5-10}$$

The winding cross section and the winding volume are

$$S = TL \tag{5-11}$$

$$V = Sp_a \tag{5-12}$$

In all usual shapes of coils, the thickness of the coil wall is $T = n/n''$, where n = number of layers. For coils wound on cores, or forms, of the shapes and dimensions shown in Fig. 5-22, the equations

$$p_a = \pi(D + T) \qquad \text{(round core)} \tag{5-13}$$

$$p_a = 2(a + b + 2T) \qquad \text{(square or rectangular core)} \tag{5-14}$$

$$p_a = 2(b + \pi r + 1.571T) \qquad \text{(oval core)} \tag{5-15}$$

give the mean length of turn. In all cases, the dimensions of the insulated cores are assumed, so far as the inside dimensions of the winding are concerned.

The total number of turns in a winding is given by

$$N = TLN_a \qquad \text{or} \qquad N = n''t''TL/1.05 \tag{5-16}$$

The resistance of a winding is expressed by

$$R = VR_v \qquad \text{or} \qquad R = p_aNR_1 \tag{5-17}$$

54. Formula for Ampere-turns. The calculations of windings for electromagnets are generally controlled by the ampere-turns to be developed and the energy loss in watts which can be safely allowed. This reduces to the problem of finding the size and kind of insulated wire which will give the proper proportions of turns and resistance. For d-c electromagnets, the equation for the ampere-turns NI for a given emf is

$$NI = E/p_aR_1 \tag{5-18}$$

Conversely, the size of wire is found from the table opposite the nearest value for $R_1 = E/NIp_a$. The proper size of wire having been found, it is customary to make preliminary calculations for turns, resistance, etc., and then to make the final calculations after the turns per layer and number of layers have been determined from the preliminary data.

For a round winding, with given core, length, size of wire, and resistance, the outside diameter is

$$D = [4/\pi L(R/R_v) + Di^2]^{1/2} \tag{5-19}$$

The approximate thickness of wall of winding is

$$T = (D - Di)/2 \tag{5-20}$$

The number of layers is $n = Tn''$; and a new value for T is found for an exact number of layers by $T = n/n''$. The mean length of turn is

$$p_a = \pi(Di + T) \tag{5-21}$$

and the outside diameter is then $D = Di + 2T$. The number of turns is $N = nt''L$; and the resistance $R = Np_aR_1$. For round windings, the radiating surface is $A = \pi DL$, where A is the cylindrical surface.

55. Trends in Magnet Coils. The growth of basic research into the nature of matter and into the investigation of subatomic particles has created a number of requirements for powerful magnetic fields within which some advanced experiments may be carried out. Superstrength fields of hundreds of thousands of gauss have been generated by using superconducting materials and cryogenic techniques at temperatures approaching absolute zero. The containment of ionic plasmas in "magnetic bottles" is one special problem of eventual space propulsion systems, where the future design of supermagnets remains a challenge to engineers and scientists specializing in this field. On the other end of the scale, the known existence of weak magnetic fields of vast proportions in interstellar space stimulates the imagination of some to foresee vehicles propelled through these fields by external magnetic fields generated within the vehicles and achieving high speeds by continuous acceleration of relatively small magnitude.

56. Bibliography

1. Parker, R. J. Analytical Methods for Permanent Magnet Design; *Electro-Technology,* September, 1960, October, 1960.

2. Greenwood, D. "Manual of Electromechanical Devices"; New York, McGraw-Hill Book Company, 1965.

3. "Engineers Relay Handbook"; New York, Hayden Book Company, Inc., 1966.

4. Moskowitz, L. R. Handling Materials with Magnetic Devices; *Automation,* April, 1965.

5. Ludwig, J. T. A Method for the Design of Holding Electromagnets; *AIEE Communications and Electronics,* July, 1960, pp. 300–310.

6. Reitan, D. K., and Szews, A. P. A Simplified Electromagnetic Apparatus for Use in Undergraduate Teaching; *Elec. Eng.,* January, 1963, Vol. 82, pp. 24–29.

7. Montgomery, D. B. High-strength Conductors for Supermagnets; *IEEE Spectrum,* August, 1966, Vol. 3, pp. 111–114.

8. Saums, Harry L. Magnet Wire—State of the Art; *Insulation,* September, 1965.

CAPACITORS

By David B. Peck and Jacob H. Martin

FUNDAMENTALS

57. Capacitance is the property of any system of dielectrics and conductors to store electrical energy when a potential difference exists between the conductors. Its value is expressed as the ratio of electrical charge to the potential difference,

$$C = \frac{Q}{V} \tag{5-22}$$

where Q equals the charge in coulombs or ampere-seconds, V equals the voltage in volts, and C equals the capacitance in farads.

58. A capacitor is a system of conductors and dielectrics so arranged that a large electrical charge is stored in a small volume.

59. Calculation of Capacitance. When the conductors are two flat parallel plates or electrodes, the capacitance is given by

$$C = \frac{0.225AK}{t \times 10^{12}} \tag{5-23}$$

where A equals the area in square inches of the electrodes; K equals the relative dielectric constant; and t equals the space in inches separating the electrodes.

Since the electrode field bends outward at the edges of the electrodes, there is an increase in the capacitance over the value presented in Eq. (5-23). The increase is dependent not only on the geometry and spacing of the electrodes but on the amount of dielectric extension beyond the electrodes. The increase is quite small for electrodes with a large area-to-thickness ratio, such as a rolled-paper dielectric capacitor. Conversely, the increase can be quite significant in ceramic-disk capacitors. This effect has been used advantageously by applying a second dielectric in the fringing area in ceramic capacitors to create a capacitor with a very low temperature coefficient of capacitance.

For nonplanar or nonparallel electrode geometries such as the coaxial cable, the capacitance value is calculated from formulas different from Eq. (5-23).

60. Calculation of Energy Storage. This is determined by the formula

$$W = \tfrac{1}{2}CV^2 \tag{5-24}$$

where W equals the energy in joules or watt-seconds, V equals the potential difference between the electrodes in volts, and C is the capacitance.

61. Relative Dielectric Constant. The relative dielectric constant K of a material is the ratio of the capacitance of a capacitor using the material as a dielectric to separate the electrodes to the capacitance of the same capacitor using a vacuum as a dielectric. By definition, the relative dielectric constant K for a vacuum is 1. The relative dielectric constant of dry air is slightly greater than 1, while most solid and liquid dielectric materials have relative dielectric constants in excess of 2 and as high as 14,000.

62. Equivalent Circuit. All practical capacitors possess not only capacitance but certain amounts of inductance and resistance. A practical equivalent circuit, useful in the determination of various performance characteristics and the fundamentals of a capacitor type, is shown in Fig. 5-23.

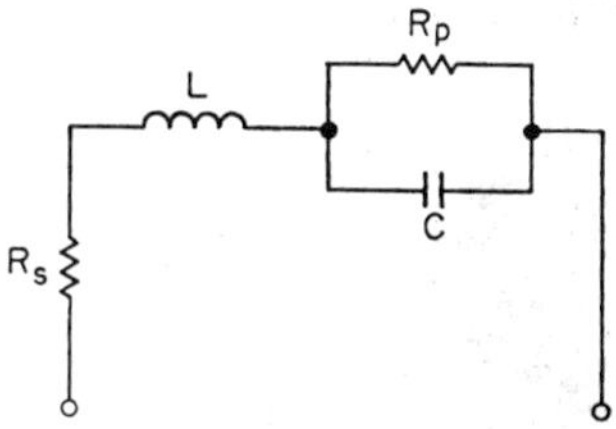

FIG. 5-23. Simple equivalent circuit for a capacitor component.

In this circuit, R_s is the series resistance presented by lead wires, electrodes, and contact terminals, R_p is the shunt resistance of the dielectric material and related losses, and L is the inductance of the electrodes and capacitor terminals. An understanding of this circuit is helpful in the comparison of the basic characteristics of different types of capacitors.

63. Power Factor. The power factor is an expression for the losses in a capacitor under a-c voltage. In a perfect capacitor, the current flowing through the capacitor will lead the voltage by 90°, but in actuality this angle is slightly less. The power factor is the cosine of the angle or the sine of its complement. The complement angle is usually small and is almost equal to its sine. Refer to Fig. 5-24. The complement angle is termed the loss angle. Losses occur owing to dielectric bulk and surface leakage, electronic and atomic dielectric effects, and I^2R losses in the electrodes and terminals.

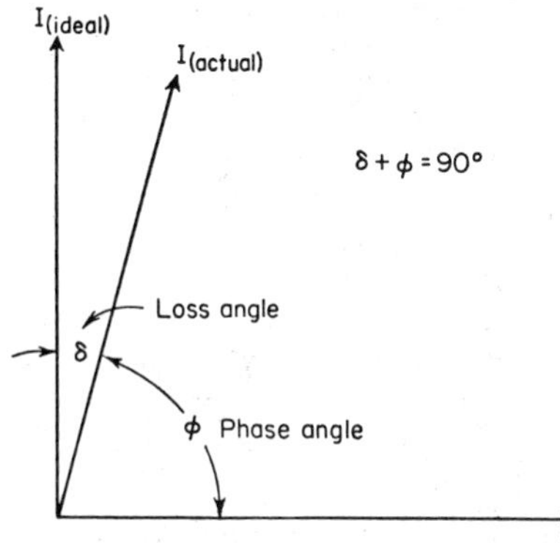

FIG. 5-24. Vector diagram showing the voltage-current phase for an ideal and an actual capacitor.

The power factor is related to the losses in the capacitor in the following manner,

$$\text{pf} = \cos \phi = \frac{R}{Z} = \frac{W}{IV} \tag{5-25}$$

where W equals the power dissipated in the capacitor, R equals the equivalent series resistance, Z equals the impedance, and I and V equal the rms current and voltage.

64. Dissipation Factor. This is an alternative measure of the losses occurring when an alternating current flows through a capacitor. For power-factor values of less than 10%, the power factor and the dissipation factor are essentially equal. The formula for dissipation factor is

$$\text{df} = \tan \delta = 2\pi fRC \tag{5-26}$$

where f equals the frequency in hertz.

65. The Q of a capacitor is a reference normally employed for high-frequency capacitors. The Q is the reciprocal of a dissipation factor and represents the ratio of the pure reactance to the equivalent series resistance.

66. Equivalent Series Resistance. The equivalent series resistance ESR combines R_s and R_p at a particular frequency so that losses in these elements may be expressed as a loss in a single resistor R in the equivalent circuit in series with C.

67. Capacitive Reactance. The reactance of a perfect capacitor is given by the formula

$$X_c = -1/2\pi fC \tag{5-27}$$

Table 5-7. Properties of Some Capacitor Dielectric Material

Dielectric material	Typical thickness range, in	Approx. K	Approx. operating temp., °C	Approx. temp. coefficient, ppm/°C (0–85)
Mica	0.0005 -0.004	7	−66–260	±200
Vitreous ceramic	0.005 -0.005	7	200 max.	±25 and up
Ceramic:				
Class I	0.001 -0.030	6–500	−55–125	+120–5,600
Class II	0.001 -0.030	500–10,000	−55–125	Very high
Paper (Kraft) with:				
Mineral oil	0.00015–0.002	2.2	−55–105	+300
Castor oil	0.00015–0.002	4.7	−25–65	−600
Chlorinated diphenyl	0.00015–0.002	4.9	−55–85	−600
Silicone oil	0.00015–0.002	2.6	−60–125	+150
Polyethylene teraphthalate	0.00015–0.002	3	150 max.	−150 ± 75
Polystyrene	0.00015–0.002	2.6	−70–85	−150 ± 50
Polytetrafluoroethylene	0.00015–0.002	2	200 max.	−150 ± 75
Polycarbonate	0.00015–0.002	3	−70–125	±150

and is entirely negative. In practice however, the inductance of the leads enters into the reactance of the device, especially at high frequencies. At the resonant frequency, the inductive and capacitive reactances are equal.

68. Impedance. Impedance can be expressed by the formula

$$Z = \sqrt{R^2 + X^2} \tag{5-28}$$

where R is the equivalent series resistance and X is the total reactance of the capacitor. X combines the effects of the inductive reactance and the capacitive reactance. At the resonant frequency, the latter two cancel each other, and the impedance of the device is determined entirely by the equivalent series resistance.

69. Leakage Resistance. This is the property shown in the equivalent circuit of Fig. 5-23 as R_p for d-c operation and covers the d-c dielectric losses. Capacitor leakage current I_L measured under specified conditions is often used as a reference characteristic instead of leakage resistance, particularly for electrolytic capacitors.

70. Dielectric Absorption. All practical dielectric materials, other than a perfect vacuum, possess to a greater or lesser degree the property of dielectric absorption. During the period when a voltage is applied to the capacitor, part of the leakage current is trapped in the dielectric. After the capacitor is discharged, the trapped dielectric charge is released and migrates to the electrodes. Whether this charge is electronic, ionic, or both is not clear. Many other details of the physical behavior are also lacking.

It is common practice to measure the dielectric absorption by determining the "reappearing voltage" which appears across a capacitor at some point in time, after it has been fully discharged under short-circuit conditions. Values range from about 0.1 to 4% at room temperature. Polar dielectrics normally have relatively high dielectric absorption.

71. Dielectric Strength. This is the breakdown strength of an insulating material when subjected to voltage stress. It is usually expressed in volts per mil. It is important to note that this is rarely linear with separation, and it is necessary to state the conditions (such as sample thickness, the rate at which the voltage is applied, whether a-c or d-c, etc.) under which any particular value is specified or determined.

72. Corona is the ionization of air or other vapors which causes them to conduct current. Corona is produced by voltage gradients which are sufficiently strong in localized spots to cause ionization. Voids in solids, sharp electrode points, nonhomogeneous insulation, bubbles, etc., can promote the formation of corona. All organic insulations are degraded through corona, and, in time, the corona will lead to insulation failure. Corona can occur with a-c voltages as low as 250 V in air.

In the case of oils, corona leads to the formation of wax and can cause voids and ultimate failure. In chlorinated aromatic dielectrics, corona causes formation of carbon, with breakdown and failure. Corona can be avoided by proper overall design and by the elimination of sharp points, edges, voids, etc.

CAPACITOR TYPES

Electrostatic Capacitors

73. Types of Capacitors. Capacitors may be conveniently classified into two generic categories, electrostatic and electrolytic. Each category is subdivided into various individual types according to the particular dielectric or electrode system employed.

The first category includes capacitors in which the dielectric may be a gaseous material, a liquid material, a solid material, or a combination of any of these. Electrodes are either deposited on the dielectric or provided in the form of plates or foil. The properties of some of the dielectric materials used to make capacitors in this category are shown in Table 5-7.

The second category is the electrolytic capacitor, usually characterized by a very thin metallic oxide dielectric film formed on the surface of one or more electrodes. This oxide film is "direction-dependent" in that the insulating effectiveness depends upon the direction of application of the voltage.

74. Vacuum Capacitors. These capacitors have a relative dielectric constant of 1 and are limited in capacitance to values on the order of 1,000 pF, but the voltage ratings range up to the high-kilovolt range of 50 kV. High current-carrying ability is typical owing to the low dielectric losses. Current can range up to 100 A at radio frequencies.

The r-f power factor is normally about 0.02% and results only from eddy current and I^2R losses in the plates and connectors. With good hermeticity of the glass envelope and terminal seal, the life of these capacitors is indefinite. An air leak, partly destroying the vacuum, will lead to a significant reduction of voltage capability and destruction through arcing. Conventionally, the electrodes are concentric cylinders made of heavy-gage copper to minimize losses while maintaining structural integrity for uniform capacitance. Vacuum variable capacitors are also provided with one electrode mounted on a flexible metal bellows to permit adjustment of the electrode overlap by a 2:1 factor or more.

75. Air Capacitors. The relative dielectric constant of air is very nearly unity and is characterized by very low dielectric losses and extremely low change in dielectric constant with temperature. For this reason, air capacitors are often used for higher-frequency tuned circuits. The most common application is the variable tuning capacitor. These are often ganged into two or three individual capacitors on a single shaft. Voltage ratings are most commonly less than 1 kV but can be made as high as 50 kV when the environment is characterized by low humidity and freedom from suspended contaminants. The insulation resistance is normally extremely high, up to the corona starting voltage.

76. Ceramic Capacitors. Ceramic capacitor materials may be divided into two classes, as suggested by the Electronic Industries Association.

Class I capacitors are usually made from titanium dioxide (TiO_2) or materials containing a large percentage of TiO_2, such as $Ba_2Ti_9O_{20}$, $CaTiO_3$, $MgTiO_3$, and $SrTiO_3$. They are characterized by a nominal temperature coefficient in the range of 25 to 85°C. The relative dielectric constant of these materials ranges from 6 to 500, and the power factor is 0.4% or less.

Class II capacitors use barium titanate ($BaTiO_3$) as the basic dielectric material. A wide variety of compounds such as $SrTiO_3$, $CaTiO_3$, $CaZrO_3$, $Bi_2Sn_3O_9$, Nb_2O_5, etc., are reacted with $BaTiO_3$ to modify its properties. Processing of these materials to achieve the desired properties must be closely controlled.

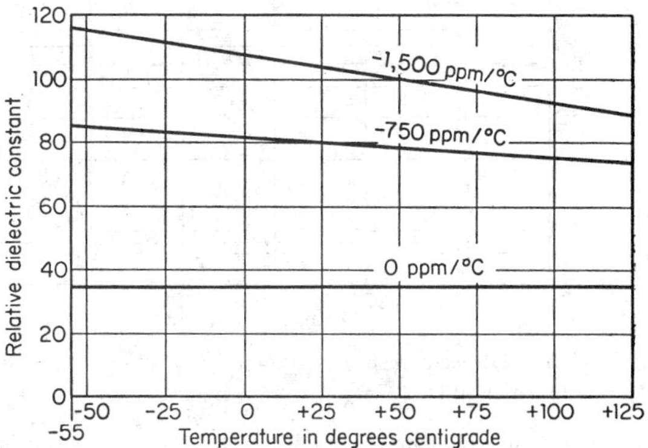

FIG. 5-25. Relative dielectric constant vs. temperature—Class I Ceramic capacitors.

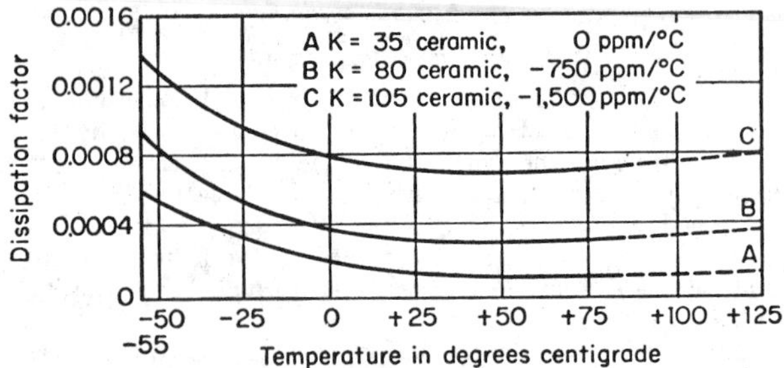

FIG. 5-26. Dissipation factor vs. temperature—Class I ceramic capacitors.

Barium titanate-based dielectrics are nonlinear (i.e., the charge accumulation is not a linear function of the applied voltage.) The relative dielectric constant of Class II materials is typically in the range of 500 to 10,000. The relative dielectric constant decreases upon aging at the rate of ½ to 8% per decade of time, depending upon the dielectric material. The power factor ranges from 0.4% up.

Class I capacitors are used for coupling, fixed tuning at high frequencies, and bypass and temperature-compensating applications. Class II capacitors are used for bypassing at radio frequencies, interstage coupling, and filtering.

Ceramic-capacitor construction is of three general types. In the first type, a disk or hollow cylinder of ceramic is metallized on both sides to form electrodes. The dielectric thickness may be 0.005 to as much as 1.00 in. The disk-style capacitor is popular in the United States, whereas the cylindrical style is common in Europe.

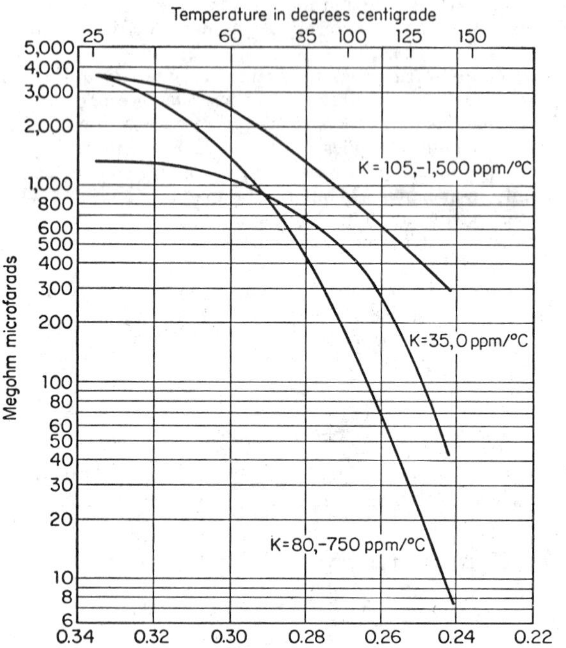

FIG. 5-27. Direct-current resistance vs. temperature—Class I ceramic capacitors.

In another type of ceramic capacitor, thin layers (0.001 to 0.005 in) of unfired, metallized ceramic are formed and fired to produce a monolithic mass with many dielectric layers electrically in parallel. This construction permits very high capacitance per unit volume.

The third type of ceramic capacitor is the barrier-layer type, in which an insulating ceramic disk is heat-treated in a reducing atmosphere so that resistivity decreases to about 10 Ω·cm. A thin layer on the surface of this body is then reconverted to the insulating state by firing in an oxidizing atmosphere, and an electrode is applied to the surface. A capacitor is thereby formed between the electrode and the semiconductive body. In an actual device, electrodes are often applied to both sides of the disk, and the finished capacitor is actually made up of two capacitors in series.

Class I disk and monolithic ceramic capacitors are produced in a wide variety of controlled-temperature coefficients ranging from +100 ppm/°C capacitance change to about −2,200 ppm/°C. Temperature coefficients as great as −5,600 ppm/°C are possible, but such materials have a tendency toward nonlinearity in their dielectric properties.

The properties of commercial disk and monolithic capacitors having temperature coefficients of 0 ppm capacitance change/°C, −750 ppm/°C, and −1,500 ppm/°C are shown in Figs. 5-25, 5-26, and 5-27. The properties presented are dielectric constant, dissipation factor, and resistivity as a function of temperature.

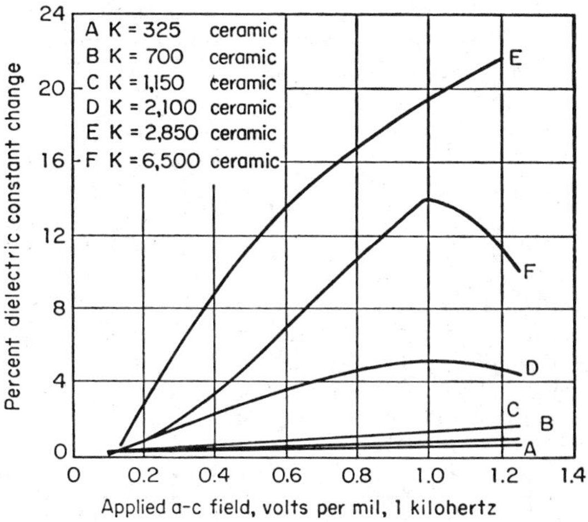

Fig. 5-28. Percent dielectric change vs. 1 kHz applied field—Class II ceramic capacitors.

Class II disk- and monolithic-capacitor nonlinearities are shown in Fig. 5-28. Note that the dielectric constant is a function of the applied a-c measuring field. The nonlinearity generally increases

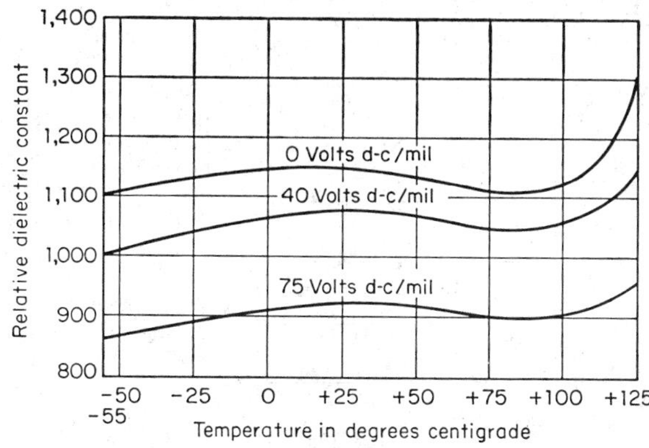

Fig. 5-29. Dielectric constant vs. temperature for a Class II ceramic capacitor at different levels of applied d-c bias.

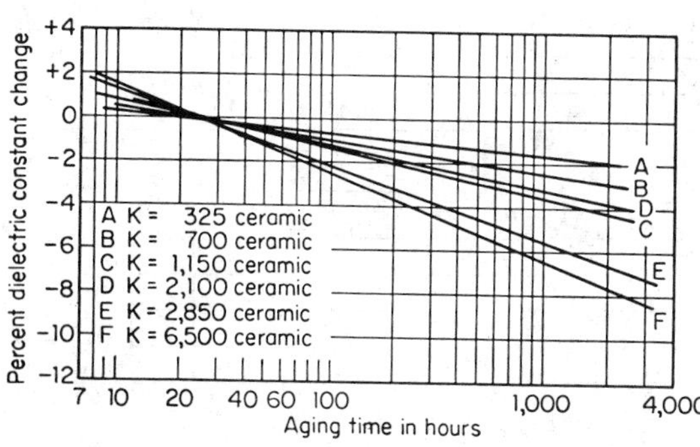

Fig. 5-30. Aging of dielectric constant in ceramic capacitors at 25°C.

Table 5-8

Voltage Ratings, V	Specific Capacitance of Completed Capacitors, μF/in²
3	6–7
10, 12	1–1.5
16	0.5–0.8
25, 30	0.3–0.6

Table 5-9. Capacitance Loss (in Percent) from Room-temperature Capacity at Various Ambient Temperatures

Voltage rating, V	−55°C, %	+10°C, %	+85°C, %
3	−15	−3	+13
10	−23	−5	+17
10	−50	−7	−14
10	−85	−8	−60
12	−71	−17	−50
25	−22	−4	+4
25	−45	−7	−25
25	−75	−8	−30
30	−29	−11	+0.5

as the small-signal dielectric constant increases. The effect of applied d-c voltage on dielectric constant as a function of temperature for several disk ceramic capacitors is shown in Fig. 5-29. The most dramatic changes in dielectric constant as a function of applied d-c voltage occur in the higher dielectric constant materials.

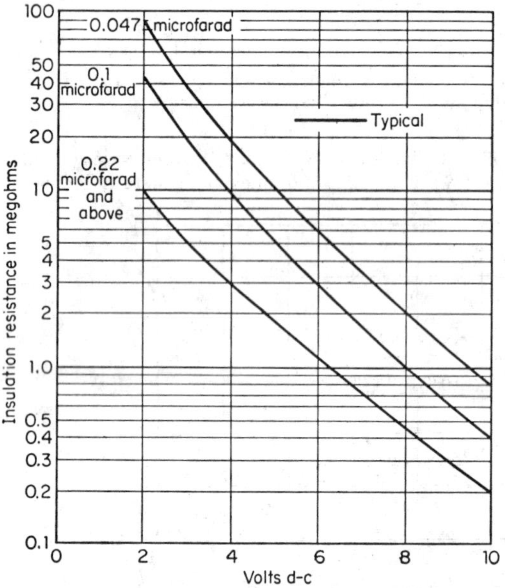

Another feature of Class II ceramic materials is the aging effect, which produces a decrease in capacitance with time. The change of capacitance with time is shown in Fig. 5-30.

The ceramic barrier-layer capacitors are normally catalogued by their voltage ratings. These are related to the thickness of the dielectric under the electrodes, which, in turn, determines the capacitance per square inch. Table 5-8 lists some industry ratings and the specific capacitances of completed capacitors. These parts are normally rated to withstand life testing of 250 to 1,000 h at 85°C and at voltages that range from the rated voltage up to twice the rated voltage.

FIG. 5-31. Insulation resistance vs. applied voltage for barrier-layer capacitors.

The permissible leakage resistances at end of life go up with higher voltage ratings and smaller capacitance values. They range from 5,000 Ω for 2.2-pF 3-V units to 50 MΩ

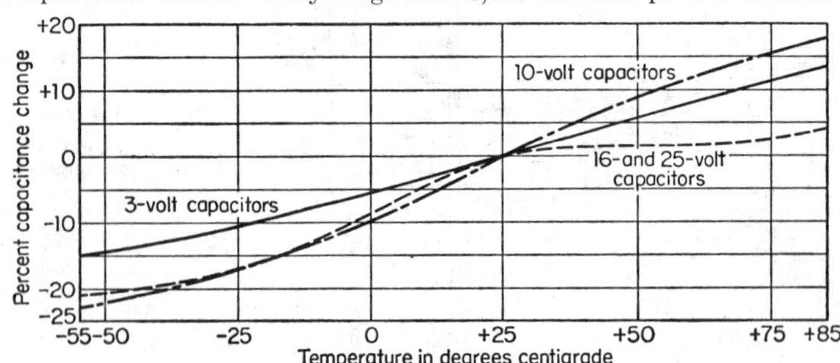

FIG. 5-32. Capacitance vs. temperature at various voltage levels for barrier-layer capacitors.

for 0.01-pF 25-V units. Figure 5-31 shows how the insulation resistance of a 10-V capacitor rises logarithmically when the applied voltage is reduced.

Barrier-layer capacitors are made in a wide range of temperature coefficients, and Table 5-9 presents a listing of the capacitance change with temperature as observed in parts of various voltage ratings made by different manufacturers. Figure 5-32 shows the shape of these curves for one manufacturer at various voltage ratings. Figure 5-33 shows how the insulation resistance of units made to these three voltage ratings varies with temperature. Note that the characteristic curves can change appreciably, depending on the materials and processing cycle.

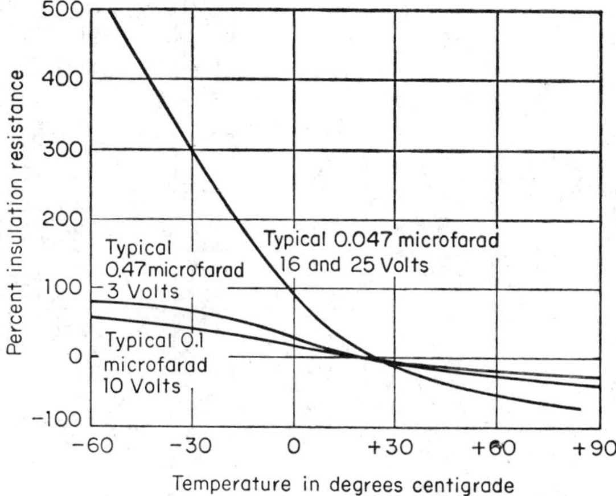

FIG. 5-33. Insulation resistance vs. temperature at various voltage levels for barrier-layer capacitors.

The semiconducting interior of the barrier-layer capacitor presents a low, but finite, series resistance. The dissipation factors of these units increase 3 to 6% between audio and radio frequencies. Figure 5-34 shows the actual variation of capacitance and dissipation factor with frequency. Figure 5-35 shows the impedance vs. frequency for several 10-V barrier-layer capacitors.

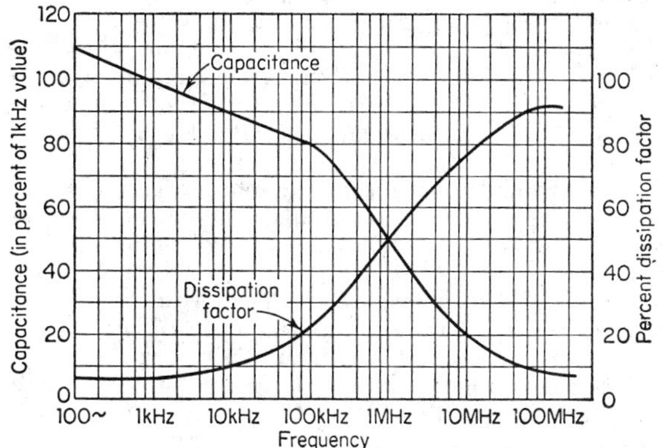

FIG. 5-34. Actual variation of capacitance and dissipation factor with frequency.

77. Vitreous (Glass) Ceramic Capacitors. Electrical-grade glasses exhibit high bulk resistivity. Conduction in glass takes place by ion migration through the glass. Therefore, alkali-metal ions, which are quite mobile in many glasses, must be eliminated or immobilized through the "mixed alkali effect," or the use of additives such as calcium or lead. Glasses have dielectric constants k in the range of 4 to 50. High-dielectric-constant glass contains large percentages of heavy metals such as lead, cadmium, and bismuth. Glass capacitors are quite similar to mica capacitors and can be used in similar applications.

Glass capacitors are made either by stacking alternate layers of glass ribbon and foil and then fusing or by spraying or silk-screening alternate layers of glass frit and powdered metal, followed by a fusion step.

The properties of glass capacitors vary widely depending upon glass composition. The temperature coefficient can be as low as ± 25 ppm/°C. Room-temperature insulation resistance is quite variable, but can be in excess of 1,000 $M\Omega \cdot \mu F$, and the Q is in the 2,000 range. Operating temperature with derating can be as high as 200°C. Life-test changes are on the order of 1% maximum for the better-quality units.

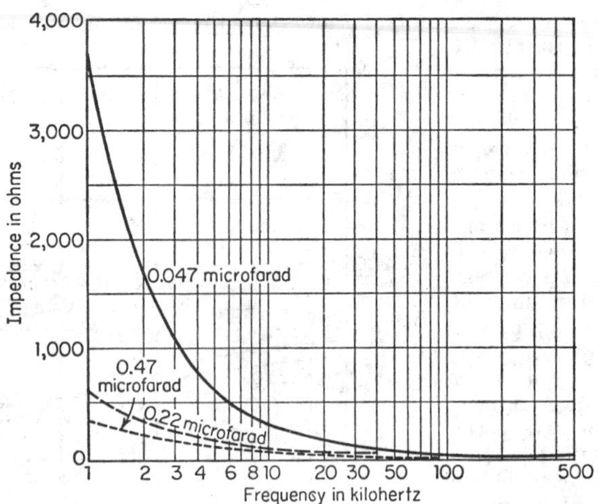

FIG. 5-35. Impedance vs. frequency for several 10-V barrier-layer capacitors.

78. Mica Capacitors. Of the several varieties of mica, two are used in the manufacture of capacitors. Muscovite mica is the most common form. It has great mechanical strength and can be used up to 500°C. One form of muscovite, ruby mica, so called because of its coloration, has the best dielectric properties. Phlogopite mica is softer than muscovite and does not have as good dielectric properties, but it can be used up to 900°C. The dielectric constant of capacitor-grade mica varies from 6.5 to 8.5.

Chemically, muscovite is a hydrogen potassium aluminum silicate, and phlogopite is a hydrogen potassium magnesium aluminum silicate. Each has varying amounts of iron, manganese, copper, and chromium. Mica crystallizes in the monoclinic system and can be cleaved into sheets as thin as 0.00001 in, although 0.001 in is common in capacitor construction.

Mica capacitors are noted for their ability to handle high voltages, high currents, and high frequencies and for their low power factor and low capacitance drift. They are used in high-frequency filtering, bypass, blocking, buffering, coupling, and fixed tuning applications.

Mica capacitors are available from 1 pF to several microfarads, with tolerances of ±20 to ±1%. Special tolerances to ±0.5% are also manufactured. The temperature coefficient of capacitance varies between ±200 ppm, depending upon the construction of the capacitor. The dissipation factor at room temperature and 1 kHz is 0.1% and decreases to 0.02% at 1 MHz. Increased temperature or moisture penetration increases the dissipation factor, while insulation resistance decreases.

The insulation resistance of mica is very high, but if the capacitor is encased in plastic, the plastic may greatly lower the resistance of the assembly. Voltage ratings may range up to 75,000 V d-c. Mica capacitors exhibit very little dielectric absorption.

Mica capacitors are built up of sheets in the range of 0.001 to 0.005 in thick. Foil electrodes of brass, tin, lead, or copper are interleaved between the mica sheets. The stacked assembly is then clamped together to reduce voids between the electrodes and the dielectric. Lead wires are attached to the foils, and the unit may be encapsulated in plastic. Wax impregnation may also be used separately or in conjunction with the plastic impregnation.

Capacitors can likewise be made by firing coats of silver on the faces of the mica sheets. Assembly then proceeds as before, with foil tabs used as contacts for the silver electrodes. This construction offers improved stability.

Another variation in construction involves the use of a type of paper made from shredded mica. The paper in some cases may be impregnated with materials such as silicone or polystyrene. Since the paper can be made in large, flexible sheets, capacitors can be of rolled construction as well as the stacked-plate construction.

79. Paper Capacitors—Liquid Impregnant. Paper capacitors are constructed with two or more layers of kraft cellulose paper ranging in thickness from 0.00015 to about 0.002 in, convolutely wound to separate two long electrode foils, usually aluminum, 0.0002 to 0.0003 in thick. The set of foils and paper layers are wound on removable mandrels of various diameters, according to the physical dimension required. At times the capacitor element is left round, whereas in other cases it is flattened. When the foils are narrower than the paper and the edges do not extend beyond the paper, the construction is referred to as "inductive," with terminals consisting of tinned copper or aluminum tabs placed over the foils usually about the middle of the length of the electrodes.

Extended foil capacitors are constructed so that one of the edges of the foil extends beyond one of the edges of the paper. The connections are made by soldering directly between the leads and the exposed foil surface. This construction considerably reduces the inductance and effective series resistance of the capacitor and is normally employed in all high-frequency constructions. Both the inductive and noninductive constructions are shown in Fig. 5-36.

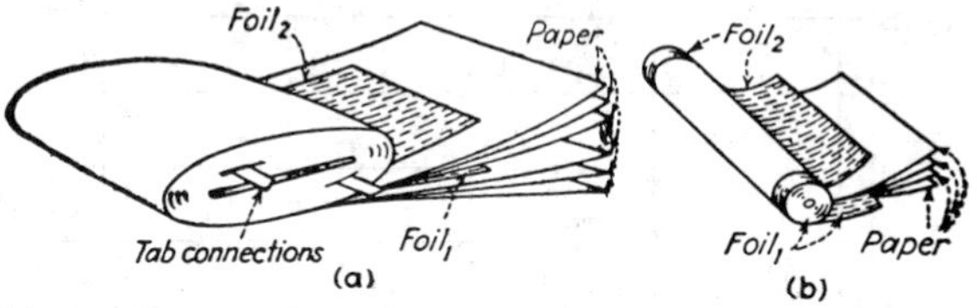

FIG. 5-36. Inductive and noninductive rolled capacitor construction.

Even highly calendered kraft capacitor papers are somewhat porous and, additionally, contain 5 to 10% water. The water is removed by vacuum-drying the capacitor at an elevated temperature, which increases the porosity. To replace the air-filled spaces with a dielectric, vacuum impregnation immediately follows the vacuum drying. The interstices of the paper are filled with a liquid or solid dielectric to increase the overall electrical capacitance and the dielectric breakdown of the combined separator.

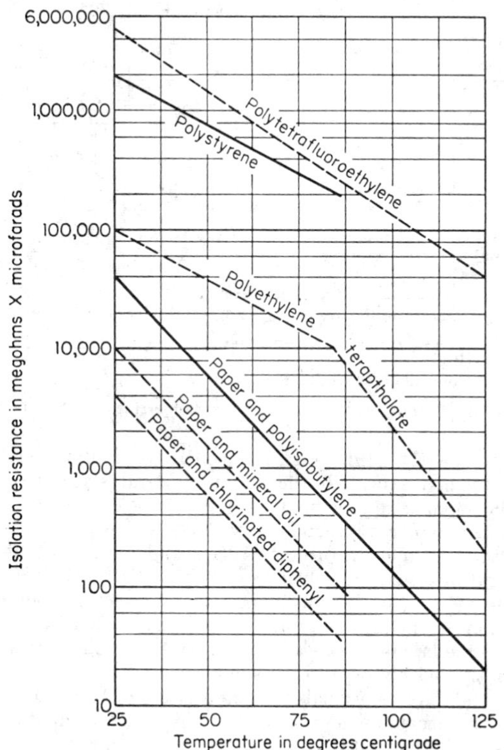

FIG. 5-37. Insulation resistance vs. temperature for various dielectrics.

Many liquid oil impregnants are used, most common of which are mineral oil, chlorinated diphenyl isomers, and polyisobutylene. The chlorinated diphenyl isomer impregnant is polar in nature and has a high dielectric constant. The mineral oil and polyisobutylene are hydrocarbons, possessing a low dielectric constant with superior temperature-coefficient and power-factor characteristics. Capacitance, dissipation factor, and insulation resistance as a function of temperature are presented for both types of dielectrics in Figs. 5-37, 5-38, and 5-39.

Oil-impregnated paper capacitors are available in a wide capacitance range and in voltage ratings to 100 kV and higher. They also find application in d-c and lower-frequency a-c circuits. The hydrocarbon-impregnated paper capacitors can be operated at higher ambient temperatures but do not possess the volumetric efficiency of the higher-effective-dielectric-constant chlorinated-diphenyl–paper combination.

Higher voltage ratings are attained while corona and arcing deterioration are minimized, by combining series and parallel capacitor sections in a single oil-filled container.

80. Paper Capacitors—Solid Impregnant. The general construction of solid-impregnant paper capacitors corresponds to that just described for the liquid-impregnant capacitors except for the physical nature of the impregnant. At normal operating temperatures, the impregnant is either a wax or a resinous material. This simplifies the sealing and packaging of the capacitor. The wax most commonly used is chlorinated naphthalene, with a dielectric constant of 4.5 and a melting point of 90°C.

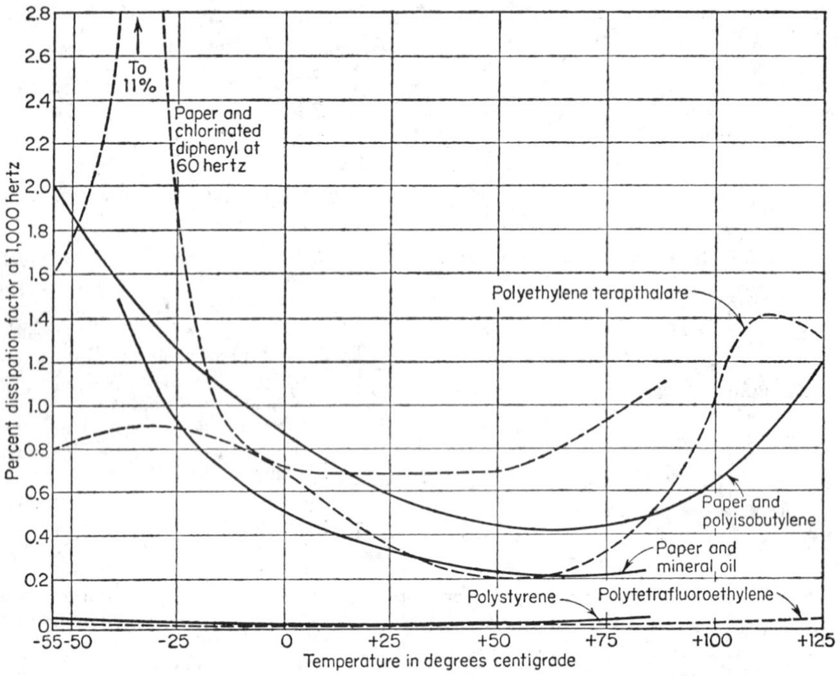

Fig. 5-38. Dissipation factor vs. temperature for various dielectrics.

Polymerizable impregnants are frequently used, and these consist of one or more monomers or low polymers (liquid) which, through the action of catalysts and/or cross-linking agents and heat, further polymerize to a rigid state.

To minimize electrical losses, such polymer impregnants are predominantly hydrocarbon in nature and can be operated at temperatures of 125°C and higher. This contrasts with the 85°C maximum rating usually applied to chlorinated naphthalene.

81. Film Capacitors. The ability to cast or extrude thermoplastic materials, both polarized and nonpolarized in nature, into very thin and uniform films has made possible the widespread use of single and multiple layers of plastic film for capacitor dielectrics, usually without subsequent impregnation with any liquid or solid.

The films are essentially nonporous and are available in thicknesses of 0.00015 to 0.001 in. They exhibit superior resistance to penetration by moisture as compared with paper-based dielectric systems. The capacitor construction corresponds to that described for paper capacitors, but the normal range of voltage ratings is limited to less than 1,000 V d-c or 300 V a-c. For higher operating gradients, it is necessary to thoroughly impregnate the capacitor with a liquid dielectric which will not attack the plastic

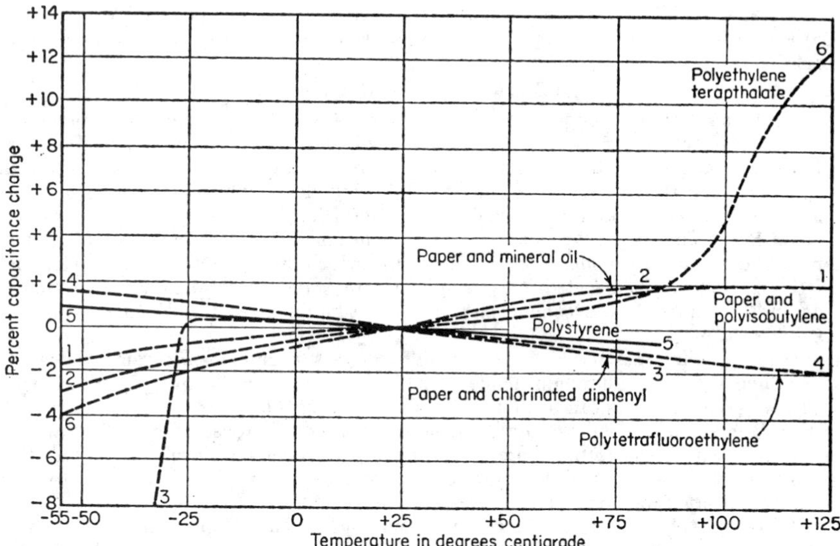

FIG. 5-39. Capacitance change vs. temperature for various dielectrics.

film. When so protected against corona formation at the foil edges, voltage ratings can exceed 1,000 V.

The choice of dielectric films is broad. In the nonpolar group, polystyrene, polyethylene, and polytetrafluoroethylene (Teflon[1]) find wide usage where very low electrical losses are desired coupled with a small and essentially linear temperature coefficient of dielectric constant. As shown in Fig. 5-39, polystyrene has a limited temperature range, while, at the other end of the range, polytetrafluoroethylene is outstanding in its thermal stability.

In polarized film dielectrics, polyethylene terapthalate (Mylar[1]), cellulose esters, and polycarbonate are extensively used, the higher dielectric constants being favorably combined with film thicknesses down to 0.00015 in. Capacitance, dissipation factor, and insulation resistance of some film dielectrics are shown in Figs. 5-37, 5-38, and 5-39.

Rigid control of the production of thin dielectric films makes it possible to avoid holes or conducting particles, thus permitting single-layer capacitor structures which, for low-voltage purposes, are adequate in ratings up to about 0.1 μF. It is possible to combine different dielectric films or a dielectric film with a paper dielectric to secure unique and controlled temperature coefficients of capacitance.

82. Metallized Paper and Film Dielectric Capacitors. Metallized paper or film may be used instead of the layered paper or film with foil electrodes. Metallization is performed in a vacuum system. Deposited metal thickness is on the order of 1,000 Å. Zinc and aluminum are the metals most frequently deposited. This type of capacitor is always made in the extended electrode configuration (noninductive) to reduce the effects of the high electrode sheet resistivity.

The chief advantages of this capacitor are its smaller size and weight and its self-healing properties. A momentary short circuit through the dielectric is eliminated, because the localized heat generated is sufficient to vaporize the thin electrode in the area of the breakdown. This property permits higher voltage ratings for a given dielectric thickness than in the foil construction. It also permits the use of single-layer dielectric construction using paper or certain films, as the local faults in these films can subsequently be cleared.

The dissipation factor is about 0.04% higher than a comparable foil unit. Leakage-resistance values are comparable with, or slightly lower than, those of a similar foil capacitor.

[1] Trademark, Du Pont Company.

The metallizing process, coupled with the formation of lacquer dielectrics down to 0.00004 in thick on a film carrier, is now leading to even more volumetrically efficient capacitors.

ELECTROLYTIC CAPACITORS

83. Electrolytic Capacitors. Although electrolytic capacitors are used interchangeably with electrostatic capacitors in many applications, they are characterized by very different construction and operating parameters. To properly employ electrolytic capacitors with their high *CV* product per unit volume, it is essential that the nature and behavior of these thin-film devices be understood.

84. Basic Theory. The dielectric material in an electrolytic capacitor consists of an anodically formed oxide of the anode metal, which serves as the positive electrode of the capacitor. The metals most employed are aluminum and tantalum, although titanium, niobium, and zirconium and others can be anodized to form dielectric films of limited value.

When aluminum is used, the dielectric oxide is a combination of amorphous and crystalline (gamma) aluminum oxide with the basic formula Al_2O_3. Tantalum oxide is Ta_2O_5. These oxides can be "formed" by connecting the anode metal as the positive electrode in an electrolytic cell containing a phosphate or borate solution. The cathodic electrode used in the forming process normally is copper or nickel. By applying a voltage to the system, the aluminum or tantalum is oxidized at its surface, to form an adherent layer of the oxide of the parent metal.

This film is characterized by a relatively high insulation resistance in the "forward" direction. Passage of current in the reverse direction takes place, presumably, through micropores. Reverse current has little permanent effect on the bulk dielectric oxide unless it causes internal heating, which can destroy the capacitor.

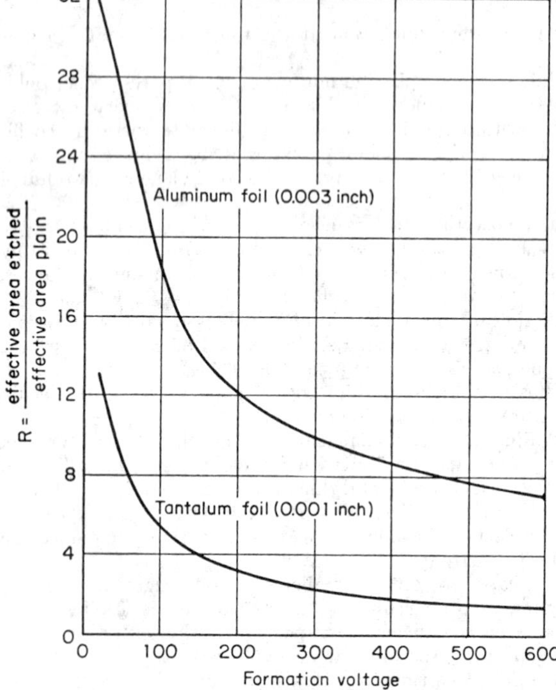

FIG. 5-40. Gain of effective surface area, *R*, vs. oxide-formation voltage.

The thickness of the oxide film depends upon the voltage to which the formation cell is raised. In the case of aluminum, this may be as high as 750 V d-c. Other factors are time, temperature, and the type of formation bath. The thickness of the oxide produced ranges between 11 and 15 Å/V applied. When tantalum is the parent metal, formation voltages rarely exceed 500 V d-c. The oxide thickness ranges from 16 to 20 Å/V applied.

The effective dielectric constant of pure aluminum oxide is 8.4, and that of tantalum oxide is 27. In capacitors, the dielectric constant can be slightly lower for aluminum oxide and appreciably lower for tantalum oxide because of impurities in the oxide. Since the thickness of the film increases linearly with voltage, the capacitance per unit area of electrode decreases inversely according to Eq. (5-23).

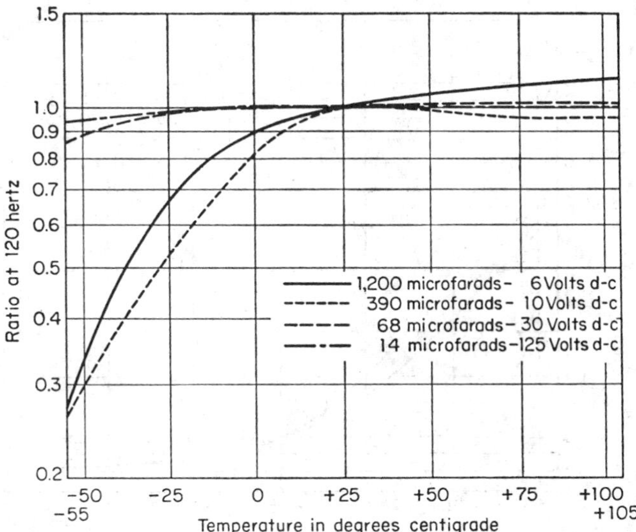

FIG. 5-41. Typical curves of capacitance vs. temperature for sintered tantalum liquid electrolyte capacitors.

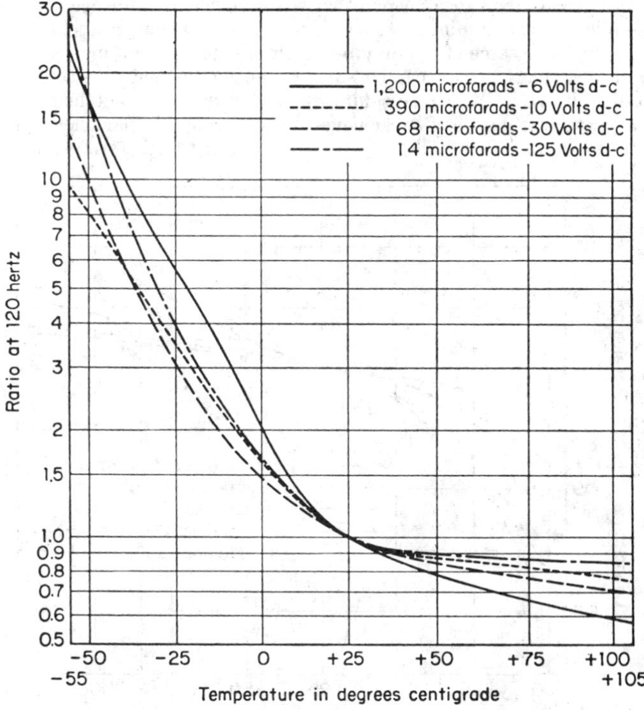

FIG. 5-42. Typical curves of equivalent series resistance vs. temperature for sintered tantalum liquid electrolyte capacitors.

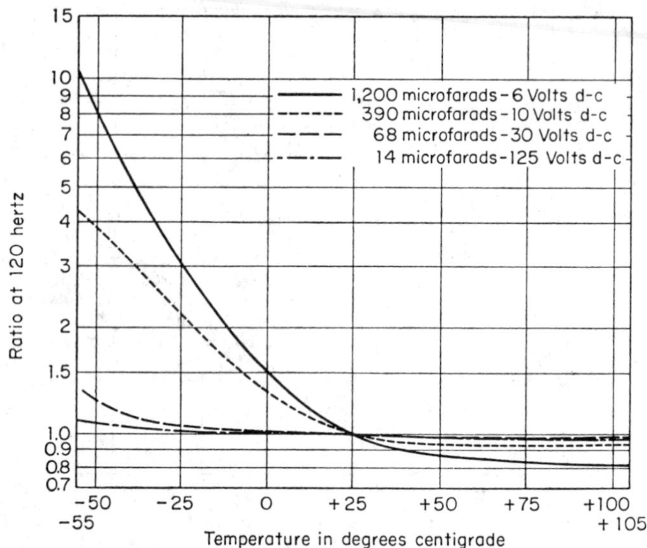

FIG. 5-43. Typical curves of impedance vs. temperature for sintered tantalum liquid electrolyte capacitors.

85. Electrode Surface Area. To increase the capacitance per unit volume of a capacitor, it is customary to etch or otherwise roughen the surface of an electrolytic capacitor anode. With aluminum, this is done by chemical or electrochemical treatment in an etch bath. The same effect may be accomplished by spraying particles of molten aluminum onto an inert porous carrier such as gauze or paper, resulting in a large surface area per unit area. While tantalum foil and wire are also etched to increase surface area, a very large gain can be achieved by sintering fine particles of the tantalum powder into a pellet. This pellet can be impregnated with forming electrolyte and

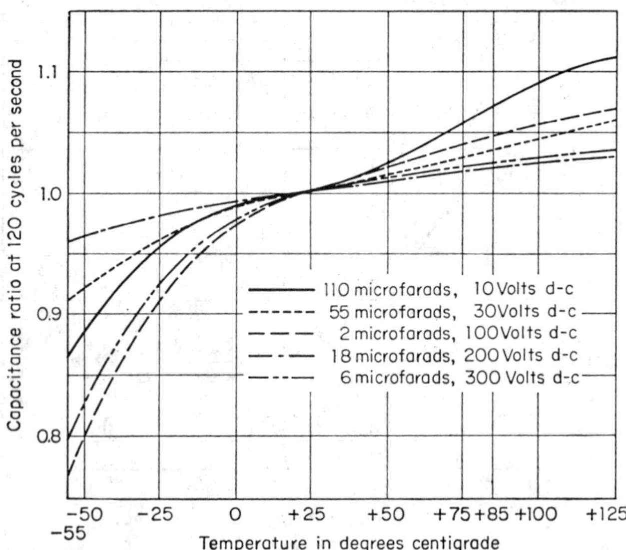

FIG. 5-44. Typical curves of capacitance vs. temperature for tantalum plain-foil capacitors.

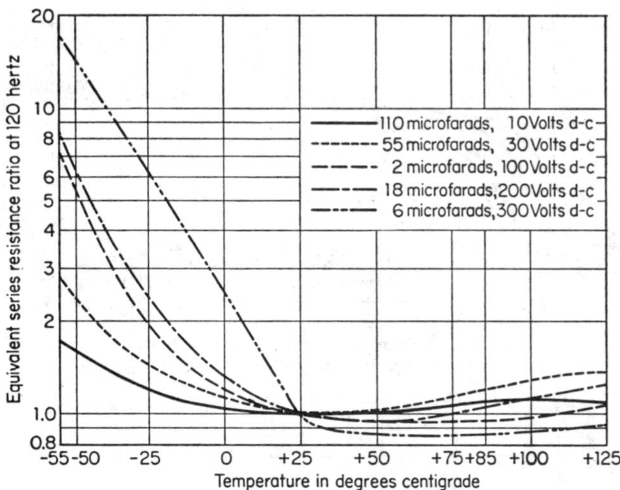

FIG. 5-45. Typical curves of equivalent series resistance vs. temperature for tantalum plain-foil capacitors.

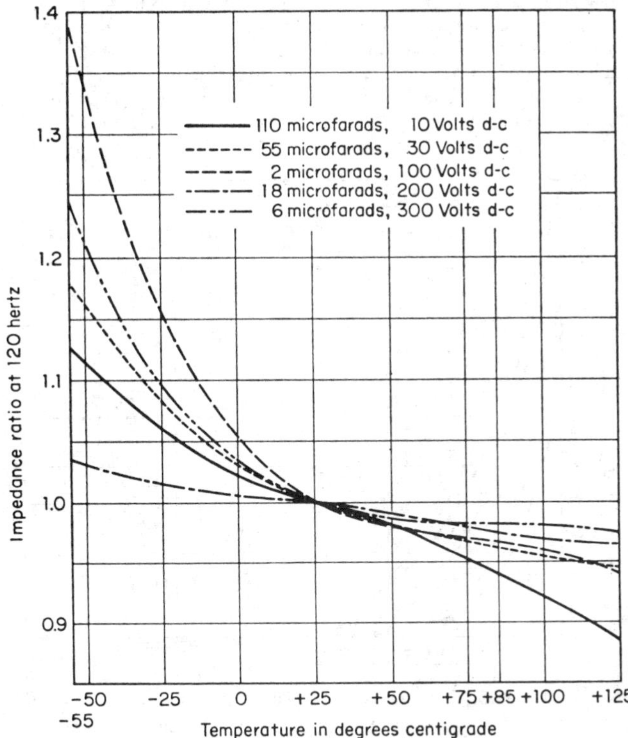

FIG. 5-46. Typical curves of impedance vs. temperature for tantalum plain-foil capacitors.

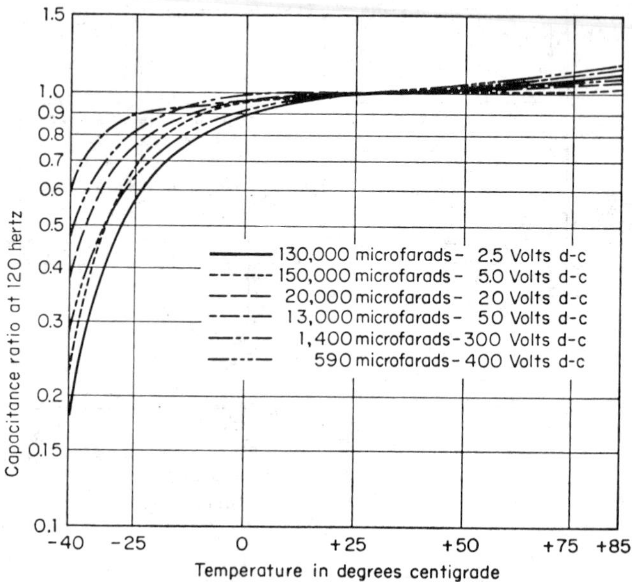

FIG. 5-47. Typical curves of capacitance ratio vs. temperature for aluminum-foil capacitors.

thereafter with an operating electrolyte. Figure 5-40 shows the approximate gain of effective surface area of aluminum and tantalum as accomplished by etching, as a function of the oxide-formation voltage.

86. Capacitor Construction. Several basic and individually useful electrolytic capacitor structures are in use. The more important of these are (1) rolled aluminum foil, (2) rolled etched tantalum foil, (3) wet tantalum pellet, and (4) solid electrolyte tantalum.

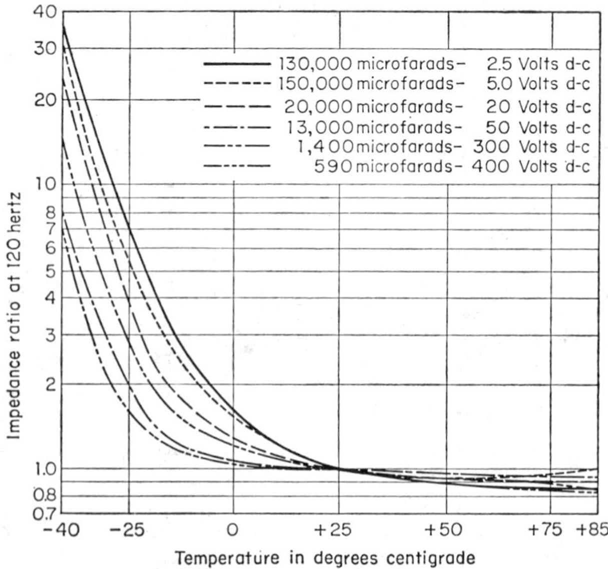

FIG. 5-48. Typical curves of impedance ratio vs. temperature for aluminum-foil capacitors.

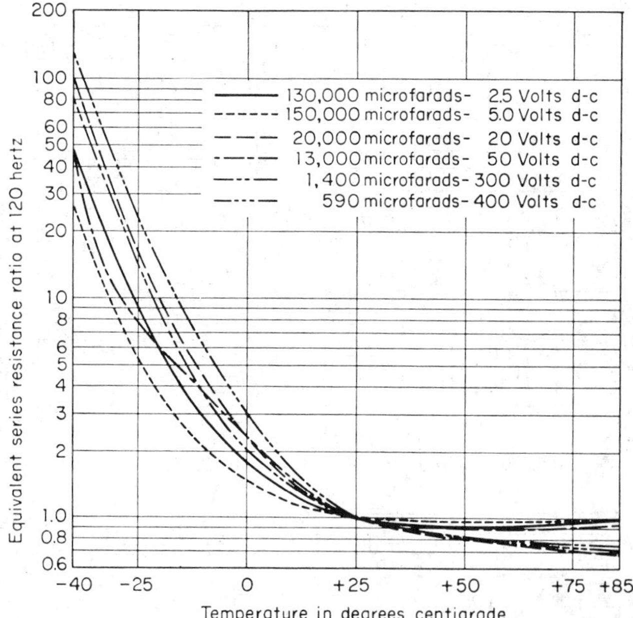

FIG. 5-49. Typical curves of equivalent series resistance ratio vs. temperature for aluminum-foil capacitors.

87. Operating Electrolytes. A variety of electrolyte families are employed in these capacitors. In the aluminum electrolytic capacitors, the electrolyte-separator combination normally consists of an ionogen such as boric acid dissolved in and reacted with glycol, to form a pastelike mass of medium resistivity. This is normally supported in a carrier of a high-purity paper such as Kraft or hemp. Because of the sensitivity of the aluminum oxide film to chloride ions, extreme care must be taken to assure freedom from this type of contamination.

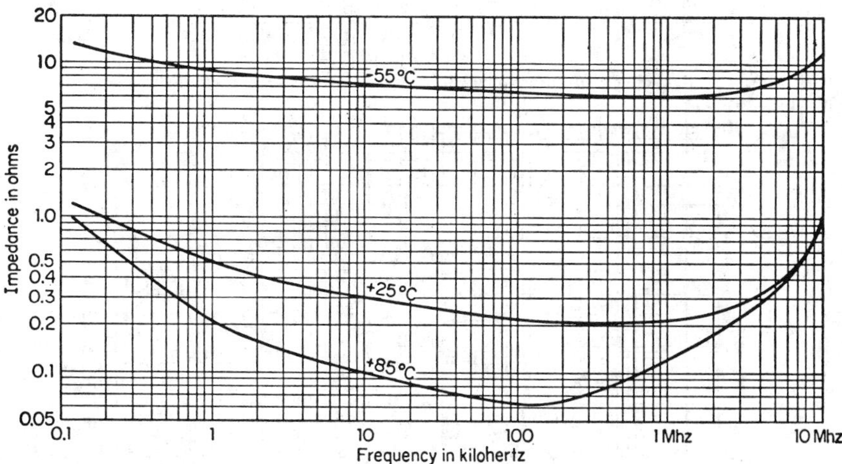

FIG. 5-50. Typical curves of impedance vs. frequency at various temperatures for sintered tantalum liquid electrolyte capacitors.

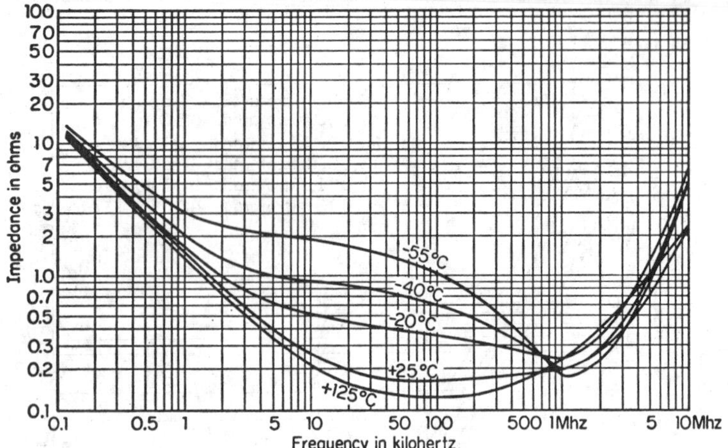

Fig. 5-51. Typical curves of impedance vs. frequency for polarized etched-foil tantalum capacitors.

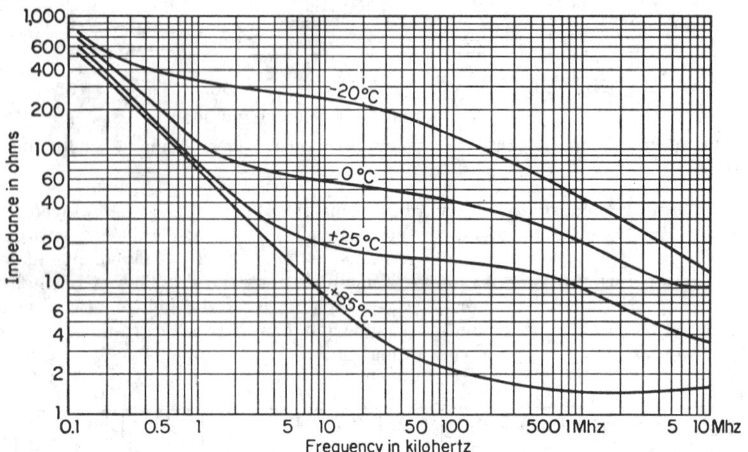

Fig. 5-52. Typical curves of impedance vs. frequency at various temperatures for aluminum-foil capacitors.

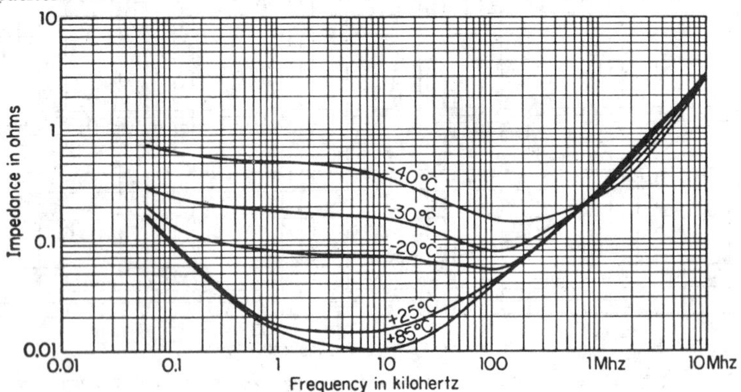

Fig. 5-53. Typical curves of impedance vs. frequency at various temperatures for etched aluminum-foil capacitors.

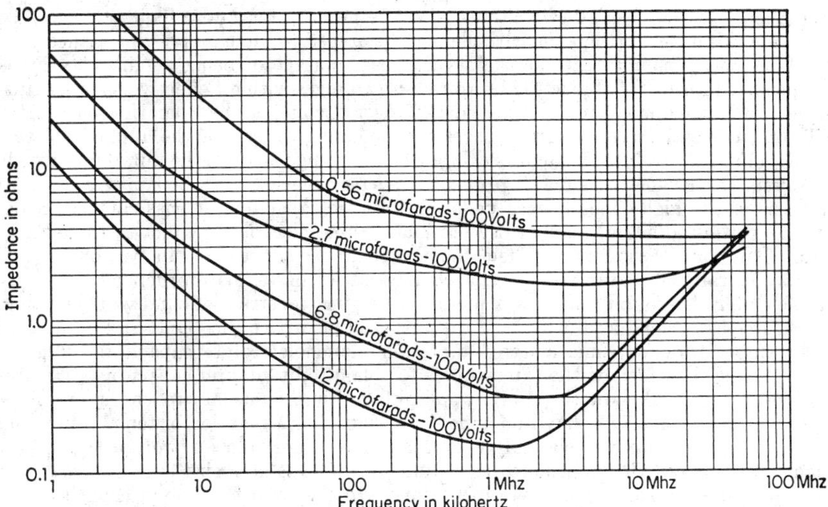

FIG. 5-54. Typical curves of impedance vs. frequency at 25°C for sintered tantalum solid electrolyte capacitors.

Because of the extreme chemical stability and resistance of tantalum oxide film, strong low-resistivity electrolytes may be employed with safety in wet tantalum electrolytic capacitors. Sulfuric acid and lithium chloride electrolytes are popular for sintered pellets. Inorganic salts dissolved in dimethyl formamide are used as electrolytes in foil tantalum capacitors. Where an electrolyte carrier or electrode separator is required, paper or glass fiber may be used.

Solid electrolyte capacitors employ a solid semiconductor such as manganese dioxide or lead peroxide to conduct current from the dielectric film to the cathode of the capacitor. Conventionally, such electrolytes are used without a carrier or structural supporting member, being encased by a metallic cathode.

Cathode construction in electrolytic capacitors requires an electronic conductor terminal element. These generally fall into one of the two categories which follow.

In aluminum electrolytic capacitors, the cathode is normally aluminum of a purity comparable with or slightly less than that of the anode material. Because aluminum anodizes so readily, it is desirable to form the cathode to a low voltage to prevent change

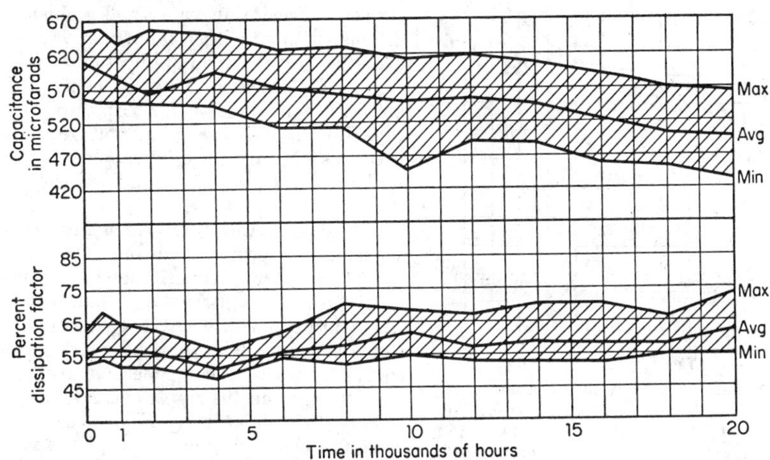

FIG. 5-55. Typical range of capacitance, dissipation factor, and d-c leakage current on 20,000-h life test at 85°C and rated voltage for sintered tantalum liquid electrolyte capacitors.

of capacitance by a series capacitance effect, especially for low-voltage ratings. To minimize this effect further, the cathode may be etched to increase its capacitance.

While the aluminum cathode suffers the disadvantage of being anodizable, this is compensated for by its ability to absorb reverse voltages without effective short-circuiting of the capacitor. Upon application of reverse voltage, the operating electrolyte will form the cathode to a voltage level corresponding to that of the applied voltage.

Tantalum electrolytic capacitors exhibit the following features: the tantalum cathode foil offers the protective advantage during reverse polarity described previously. In the case of the high-effective-area pellet electrolytic capacitors, a passivated cathode, such as platinized silver, may be used to make direct electrical contact to the electrolyte. This construction avoids the cathode capacitance effect noted in Par. **88**.

88. Operational Parameters. The characteristics of capacitors depend, not only on the material used, but on the specifics of processing, construction, and voltage rating.

Figures 5-41, 5-42, and 5-43 show the variation of capacitance ESR and impedance with temperature for some sintered tantalum liquid electrolyte capacitors. Figures 5-44, 5-45, and 5-46 show the same curves for plain-foil tantalum capacitors. Figures 5-47, 5-48, and 5-49 show the curves applicable to aluminum-foil capacitors, and Figs. 5-50, 5-51, 5-52, and 5-53 show the impedance plotted as a function of temperature and frequency for various wet-electrolyte tantalum and aluminum electrolytic capacitors. Figure 5-54 shows the impedance as a function of capacitor size and frequency for sintered tantalum solid electrolyte capacitors. In the solid electrolyte capacitor, the impedance is relatively independent of temperature. Of course, the impedance curves provide adequate data for the calculation of ESR and dissipation factor at a given temperature and frequency.

89. Performance (Life). Properly manufactured electrolytic capacitors show excellent performance on life at rated voltage and rated temperature. Life increases correspondingly as the temperature is lowered from the maximum. In telephone systems, electrolytic capacitors are designed for 20-year service and have proved suitable for this extended period.

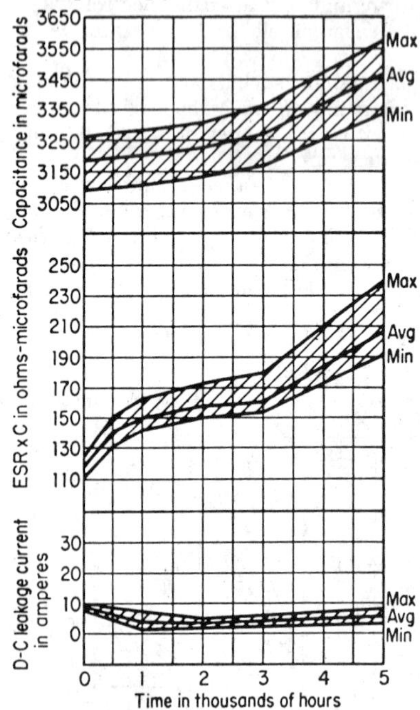

FIG. 5-56. Typical range of capacitance, $R \times C$, and d-c leakage current on 5,000-h life test at 85°C and rated voltage for aluminum electrolytic capacitors.

Figures 5-55 and 5-56 show life-test results for sintered tantalum and aluminum-foil capacitors with liquid electrolyte.

90. Performance (Shelf). In contrast to the wet electrolytic capacitors, and the early dry electrolytic capacitors, all properly manufactured electrolytic capacitors possess stable oxide films and electrolyte systems, whereby a capacitor may be stored "on the shelf" for several years under normal ambient conditions, without either failure or drawing of excessive current upon application of voltage.

Simple accelerated tests, using a high ambient temperature for exposure of the capacitor in the absence of voltage, may be used to determine the adequacy of the capacitor for storage under normal ambient conditions.

91. Ripple Current. In many applications, it is expected that the capacitor will be subjected to a superimposed a-c current on the d-c operating voltage and that some heat will be generated by the passage of this ripple current. The total heat dissipation is equivalent to the heat generated by the d-c bias leakage and the ripple current, the latter usually predominating. Knowledge of the level and the frequency of the ripple voltage makes possible the design of units which will handle substantial ripple currents, including as much as 20 A rms at 120 c. Such units would be used as filters in large, heavy-duty power supplies.

It is essential that ripple-current conditions be exactly specified in order that the capacitor may be designed to have minimum power factor plus the ability to dissipate the heat generated.

92. A-C Capacitors. If both electrodes in an electrolytic capacitor are formed to the same voltage, the capacitor may be operated under alternating polarity conditions.

In most designs, the power factor is high enough to cause excessive heating, and standard construction units are employed only in intermittent applications, such as motor-starting circuits for air conditioners and refrigerators. It is possible to design continuous-duty electrolytic capacitors for lower voltages and, in some special instances, for high voltage. It is imperative that all operating conditions be provided and defined before employing an electrolytic capacitor in a continuous-duty application.

93. Multiple Capacitors. It is possible to design and produce several individual capacitance ratings in a single structure, especially in aluminum electrolytic capacitors. Usually, such capacitors employ a common cathode or ground connection with two to five individual anodes. This has the special advantage of permitting several different anode-formation voltages to be employed if the circuit in which the capacitor is to be used does not call for identical operating conditions. Also, one or more sections can be specially designed to handle substantial ripple currents, such as those found in color-television-set power supplies.

94. Operating Temperature Limits. Solid tantalum and foil capacitors may be operated at rated voltage up to 85°C. They can typically be operated at 125°C using a derating factor of 0.65% of the rated voltage. Rated operating-temperature maximums for commercial aluminum electrolytic capacitors are either 65°C or 85°C. Special constructions can be used to 125°C.

95. Leakage Current. Referring to Figs. 5-55, 5-56, and 5-57, note that the leakage current of an electrolytic capacitor can be relatively low. In wet electrolytic capacitors, the leakage current is dependent on the age of the capacitor and the purity of the materials used in its construction. The resistance-capacitance product is often used to generalize leakage data for various sizes of a single

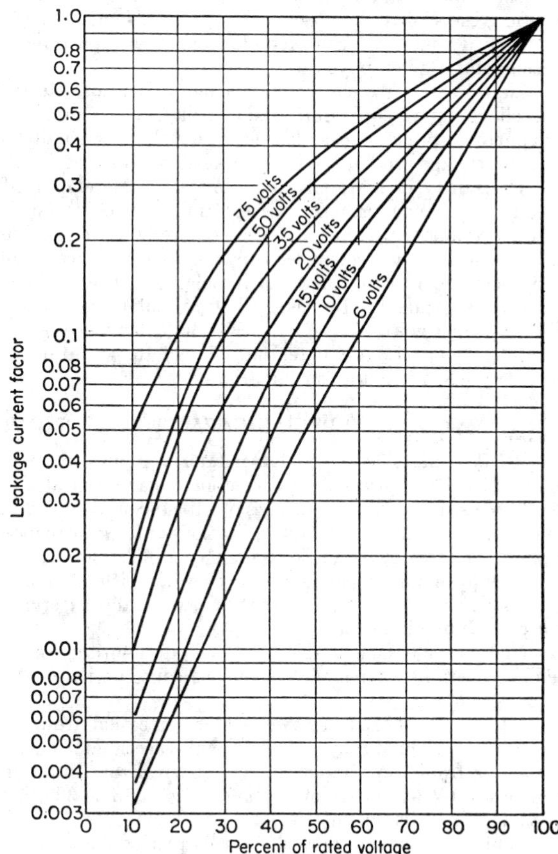

FIG. 5-57. Leakage current factor vs. percent of rated voltage.

$$\text{Leakage current factor} = \frac{\text{leakage at given voltage}}{\text{leakage at rated voltage}}$$

capacitor type. In the case of a typical large power-supply filter capacitor, after it is employed on voltage for a considerable period of time, this product often exceeds 50,000 $M\Omega \cdot \mu F$.

In miniature capacitors, such as aluminum electrolytic styles, the leakage current primarily occurs at imperfections in the junction of the anode foil to the terminal element.

96. Low-temperature Performance. As shown in Figs. 5-42, 5-45, and 5-49, the equivalent series resistance (ESR) of electrolytic capacitors, especially the aluminum type, increases significantly at low temperatures, especially below $-20°C$.

In actual usage, and where sufficient power is available in the circuit, internal heating will occur to increase the capacitor temperature to a point where the power factor and impedance reach tolerable values. In the case of equipment which must be operative immediately, when turned on at a very low temperature, it is advisable to use a tantalum electrolytic capacitor or one of the special grades of aluminum capacitor.

97. Structural Considerations. For low-voltage applications, especially an aluminum electrolytic capacitor construction, it is desirable that all connections between the elements of the capacitor, i.e., the anode-to-terminal and the cathode-to-terminal, have a true electronic bond as opposed to a pressure contact. Otherwise, oxidation or corrosion of the anode or cathode material may lead to an open or intermittent connection, especially when used in low-voltage semiconductor circuits.

Etched foil offers a considerable increase in capacitance, particularly at lower voltage ratings, for a given physical case size.

Liquid and paste electrolyte capacitors can lose electrolyte through vents provided to relieve internal pressure and by vapor transmission through the seal. Modern advances in producing nonhermetic seals have virtually removed the chances of electrolyte leakage around the seal and lead wires as a failure mode.

The purity of materials and the cleanliness employed in the manufacture of wet electrolytic capacitors are of great importance in the quality and reliability of the finished component. For instance, it is well known that chloride ion at parts per million levels can destroy electrolytic capacitor performance. Solid electrolyte capacitors are less susceptible to the effects of impurities. Their properties are also much less temperature-dependent. However, they cannot be operated at more than about 50% of their forming voltage, compared with 90% for wet electrolytes. The lifetime of solid electrolyte capacitors is potentially longer than that of wet electrolyte types, because electrolyte loss is not a consideration.

SPECIAL CAPACITOR APPLICATIONS

98. Capacitors for Power-factor Improvement. Standard sizes for individual closed-terminal indoor power-factor-correction capacitors range from ½ to 50 kvars for single- and 3-phase operation on voltages of 240, 480, and 600 and higher, to match the voltages currently used in industrial applications. Large capacitor banks are made by combining a number of individual units in parallel. For industrial application, these banks are usually mounted in a steel framework with a completely enclosed wiring cubicle attached for housing the internal bus work and external connections, plus the fusing for each individual capacitor.

High-voltage capacitors for use in public-utility service are generally of the non-enclosed terminal type, of single-phase construction, and range in size from 25 to 100 kvars per unit.

All capacitors used in power-factor correction are provided with internal discharge resistors as required by the National Electrical Code. After disconnection of the capacitor from the voltage source, the resistors drain the charge on the capacitors to below 50 V within 1 min for capacitors rated at 600 V or lower and within 5 min for capacitors rated above 600 V.

Public-utility-type power-factor-correction capacitors have standard supporting brackets for pole mounting. Capacitor fuses and circuit breakers are available with time-current characteristics so designed as to prevent rupture of the capacitor case due to internal dielectric breakdown.

All capacitors for power-factor improvement are designed to withstand continuous

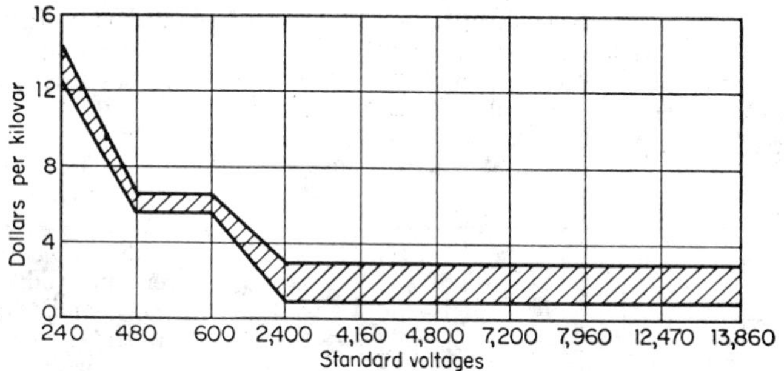

FIG. 5-58. Average cost, dollars per kvar, of medium-sized capacitors at standard operating voltages.

10% overvoltage without impairing the long life of the capacitor. For approximate costs for these units, refer to Fig. 5-58.

99. Energy-storage Capacitors. Many industrial applications utilize substantial sources of energy in the form of a fully charged capacitor. This includes welding equipment, photoflash units, pulse-forming networks, and laser beam and particle accelerators.

When lower-voltage power supplies can be used, the aluminum electrolytic capacitor, with special precautions and design to minimize inductance and series resistance, offers the best source of joules per unit cost. When higher-voltage power sources are required, especially in the area of 1,000 V, electrostatic capacitors are most economical. Of these, the paper-oil dielectrics find widespread use.

100. Induction Heating. Capacitors employed in induction heating installations must exhibit very low I^2R losses to minimize internal heating. Such systems subject the capacitors to frequencies ranging form 1 to 300 kHz at voltages up to 3,000 V.

Larger capacitors may be provided with external cooling fins, or with internal cooling coils that carry water or another coolant under pressure. It is now general practice to employ extremely low-loss dielectrics such as polyethylene and polytetrafluoroethylene to minimize electrical losses in the dielectric. For less rigid duty cycles, paper impregnated with mineral oil or chlorinated diphenyl may be used.

Particular attention is given to all electrode and terminal losses in this type of capacitor. As the frequency of operation increases, it is essential that inductive effects be kept at a minimum, especially in the higher capacitance ratings.

101. Silicon-controlled Rectifier (SCR) Capacitors. The widespread use of SCR devices has led to the design of a family of capacitors used in the triggering circuit of the SCR, and such capacitors are subjected to very fast rise-time pulses. With a typical 60-c power application, the high-frequency content of the waveshape and the normally continuous-duty operation require a combination of low-loss capacitors with the added feature of withstanding high peak voltages and transients, which usually occur in such circuits.

Special designs, normally based on oil-impregnated dielectric systems, are essential for reliable performance of the capacitor in SCR triggering applications. A mineral-oil impregnation of paper-polyethyleneterapthalate spaced capacitor section is generally used, although straight plastic-film dielectrics are also used.

102. Feed-through Capacitors. The conventional type of capacitor is not effective for filtering at high radio frequencies owing to the inherent series inductance of the leads and the internal terminal connections. This inductance resonates with the capacitance of the unit at relatively low frequencies and greatly affects its use as a filter at frequencies above resonance. Even the best conventional-type capacitors resonate at frequencies below 20 MHz.

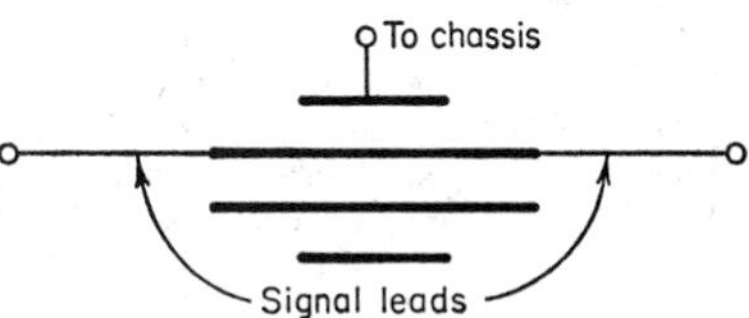

FIG. 5-59. Schematic representation of a single-turn rolled feed-through capacitor, showing the capacitor electrodes only.

Electrically, the feed-through capacitor is a three-terminal device which does not exhibit the series-resonant characteristic of the conventional capacitor. Its physical design lends itself to an extremely low inductance connection to ground, while inductance in the other connections enhances the filtering effect.

Figure 5-59 illustrates the schematic diagram of a feed-through capacitor, and Fig. 5-60 shows the difference in attenuation between an ideal capacitor and a ceramic feed-through type. Feed-through capacitors are made using paper, film, ceramic, and mica dielectrics, and, additionally, electrolytic types are now being manufactured.

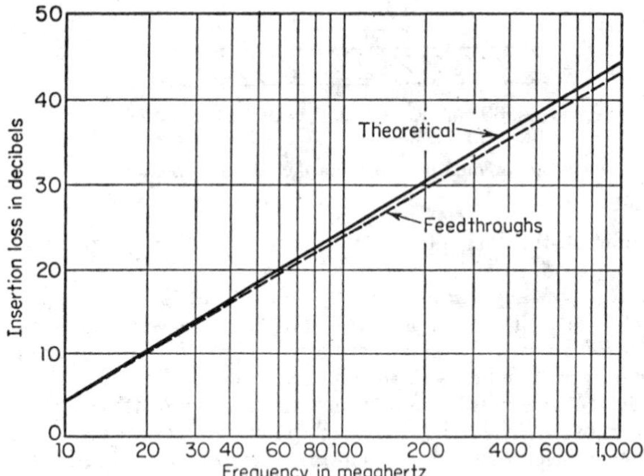

FIG. 5-60. Attenuation differences between an ideal capacitor and a ceramic feed-through capacitor.

Feed-through capacitors are effective in the suppression of radio-frequency interference over a wide frequency range and are especially valuable in filtering power-supply and control-circuit wiring in shielded high-frequency equipment.

103. Importance of Proper Specifications. In order to provide the optimum capacitor for any given circuit or system application, it is essential that the user specify in as much detail as possible all conditions to which the unit will be subjected in actual operation as well as on stand-by.

Actual waveforms to which the capacitor will be exposed, including all transients, should be specified. Environmental conditions under which the capacitors will be stored and operated should be given in detail. The complete electrical performance requirements over the range of environmental operating conditions are also necessary data for proper capacitor selection. Physical size, weight, and outline may also be important, as well as the type of terminations with which the unit is fitted.

Complete and accurate specifications are essential to ensure that an acceptable level of reliability can be achieved at the lowest cost.

104. Military Specifications. The United States Military Triservice coordinated specifications for capacitors are listed in the List of Military Specifications, Standards and Related Documents. This list is prepared semiannually by the Engineering Standardization Directorate, Defense Electronics Supply Center, 1507 Wilmington Pike, Dayton, Ohio. A list of specifications in effect Jan. 31, 1966, follows:

MIL-C-5C, Capacitors, Fixed, Mica-Dielectric

MIL-C-20D, Capacitors, Fixed, Ceramic Dielectric (Temperature Compensating)

MIL-C-25C, Capacitors, Fixed, Paper-Dielectric, Direct Current (Hermetically Sealed in Metallic Cases)

MIL-C-62C, Capacitors, Fixed, Electrolytic (D-C, Aluminum, Dry Electrolyte, Polarized)

MIL-C-81A, Capacitors, Variable, Ceramic Dielectric

MIL-C-92A, Capacitors, Variable, Air Dielectric (Trimmer)

MIL-C-3871, Capacitors, Fixed, Electrolytic (A-C, Dry-Electrolytic, Nonpolarized)

MIL-C-3965C, Capacitors, Fixed, Nonsolid, Electrolytic, (Tantalum, Foil and Sintered-Slug)

MIL-C-10950B, Capacitors, Fixed, Mica Dielectric, Button Style

MIL-C-11015C, Capacitors, Fixed, Ceramic-Dielectric (General Purpose)

MIL-C-11272B, Capacitors, Fixed, Glass Dielectric

MIL-C-11693B, Capacitors, Feed Through, Radio-Interference Reduction, A-C and D-C, (Hermetically Sealed in Metallic Cases)

MIL-C-12889A, Capacitors, By-Pass, Radio-Interference Reduction, Paper Dielectric, A-C and D-C, (Hermetically Sealed in Metallic Cases)
MIL-C-14157D, Capacitors, Fixed, Paper, (Paper-Plastic) or Plastic Dielectric, Direct Current, (Hermetically Sealed in Metal Cases), Established Reliability
MIL-C-14409B, Capacitors, Variable, (Piston Type, Tubular Trimmer)
MIL-C-18312C, Capacitors, Fixed, Metallized Paper (or Polyester Film), Dielectric, Direct Current, (Hermetically Sealed in Metallic Cases)
MIL-C-19978B, Capacitors, Fixed, Plastic (or Paper-Plastic) Dielectric, (Hermetically Sealed in Metallic, Ceramic or Glass Cases)
MIL-C-23183A, Capacitors, Fixed or Variable, Vacuum Dielectric
MIL-C-23269A, Capacitors, Fixed, Glass Dielectric, Established Reliability
MIL-C-26655B, Capacitors, Fixed, Solid Electrolyte, Tantalum
MIL-C-39001, Capacitors, Fixed, Mica Dielectric, Established Reliability
MIL-C-39003, Capacitors, Fixed, Solid-Electrolyte, Tantalum, Established Reliability
MIL-C-39006, Capacitors, Fixed, Nonsolid, Electrolytic, Tantalum (Foil and Sintered Slug), Established Reliability
MIL-C-39011, Capacitors, Feed Through, Radio-Interference Reduction, A-C and D-C, (Hermetically Sealed in Metallic Cases), Established Reliability
MIL-C-39014, Capacitors, Fixed, Ceramic Dielectric (General Purpose)
MIL-C-39018, Capacitors, Fixed, Electrolytic (Aluminum Oxide)
MIL-C-39022, Capacitors, Fixed, Metallized Paper (or Polyester Film), Dielectric, Direct Current, (Hermetically Sealed in Metal Cases), Established Reliability
MIL-C-39028, Capacitors, Packaging of

105. EIA Specifications. Many specifications for commercial electronic parts have been prepared by the Electronic Industries Association, 2001 Eye Street, N.W., Washington, D.C. The EIA capacitor specifications are listed below:

RS-171, Capacitors, ceramic, high voltage, class 2
RS-154-B, Capacitors, dry aluminum electrolytic, for general use
RS-205, Capacitors, dry electrolytic, special quality
RS-198, Capacitors, fixed, ceramic dielectric, classes 1 and 2, up to 500 V
RS-165-A, Capacitors, fixed, ceramic dielectric, classes 1 and 2, 1,000-7,500-volt rating
RS-153, Capacitors, fixed, mica dielectric, molded types
TR-109, Capacitors, fixed, mica dielectric, potted types
RS-218, Capacitors, fixed, paper dielectric, metal encased, d-c applications
RS-154-B, Capacitors, polarized dry aluminum electrolytic
RS-228, Capacitors, tantalum electrolytic (polarized)
RS-182, Capacitors, variable, air, class A
REC-101, Capacitors, variable, air, class B

105a. Other Specification Sources. Specifications are also prepared by the Aircraft Industries Association and the International Electro-Technical Commission.

REFERENCES

106. Reference Books. Although the technology in capacitor theory, design, and construction is constantly changing, there are a number of references containing valuable details. Among these are the following:

Bloomquist, W. "Capacitors for Industry"; New York, John Wiley & Sons, Inc., 1950.

Brotherton, M. "Capacitors"; Princeton, N.J., D. Van Nostrand Company, Inc., 1946.

Coursey, P. R. "Electrical Condensers"; London, Sir Isaac Pitman, 1927.

Coursey, P. R. "Electrolytic Condensers"; New York, John F. Rider & Sons, Ltd., 1938.

Deeley, P. M. "Electrolytic Capacitors"; South Plainfield, N.J., Cornell-Dubilier Electric Corporation, 1938.

Dummer, G. W. A., and Nordenberg, H. M. "Fixed and Variable Capacitors"; New York, McGraw-Hill Book Co., 1960.

Frungel, F. B. A. "High Speed Pulse Technology"; New York, Academic Press, Inc., 1965, Vol. I. (Chapters on capacitor discharges.)

Georgiev, A. M. "The Electrolytic Capacitor"; New York, Murray Hill Books, Inc., 1945.

Guntherschulze, A., and Betz, H. "Elektrolyt-Kondensatoren"; 2d ed.; Berlin, Technischer Verlag Herbert Cram, 1952.

Marbury, R. E. "Power Capacitors"; New York, McGraw-Hill Book Co., 1949.

Young, L. "Anodic Oxide Films"; New York, Academic Press, Inc., 1961.

Von Hippel, A. R. "Dielectric Materials and Applications"; New York, John Wiley & Sons, Inc., 1954.

RESISTORS AND RHEOSTATS

By Kasson Howe

GENERAL

107. Resistance is the (scalar) property of an electric circuit or of any body which may be used as part of an electric circuit which determines for a given current the rate at which electric energy is converted into heat or radiant energy and which has a value such that the product of the resistance and the square of the current gives the rate of conversion of energy.

In general, resistance is a function of the current, but the term is most commonly used in connection with circuits where the resistance is independent of the current.[1]

108. A resistive conductor is a conductor used primarily because it possesses the property of electric resistance. A **resistor unit** is an assembly comprising a resistor element, resistor core, and resistor housing and parts for mechanical mounting and electric connections. The **resistor core** is the insulating support on which the resistor element is wound. The **resistor element** is the material possessing the property of electric resistance. The **resistor housing** is an enclosing member which surrounds the resistor element and the core. A **resistor support** is an assembly of base, insulators, and necessary fittings for mounting the resistor unit.[1]

109. A resistor is a device the primary purpose of which is to introduce resistance into an electric circuit.[1]

110. A constant-torque resistor is a resistor which is intended for use in the armature or rotor circuit of a motor in which the current remains practically constant throughout the entire speed range.[1]

111. A fan-duty resistor is a resistor which is intended for use in the armature or rotor circuit of a motor in which the current is approximately proportional to the speed of the motor.[1]

112. A current-limiting resistor is a resistor inserted in an electric circuit to limit the flow of current to some predetermined value. A current-limiting resistor, used in series with a fuse or circuit breaker, may be employed to limit the flow of circuit or system energy at the time of a fault or short circuit. In a lightning arrester it is installed in series with the characteristic element for the purpose of limiting the flow of current. A **load-indicating resistor** is one used in conjunction with suitable relays and/or meters, for the purpose of determining the value of the connected load. A **load-shifting resistor** is one used in an electric circuit to shift load from one circuit to another.[1]

113. An adjustable resistor is a resistor so constructed that its resistance can be changed by moving and setting one or more movable contact elements.

114. A rheostat is an adjustable resistor so constructed that its resistance may be changed without opening the circuit in which it may be connected. A **faceplate rheostat** is a rheostat consisting of a tapped resistor and a panel with fixed contact members connected to the taps and a lever carrying a contact rider over the fixed members for adjustment of the resistance. A **field rheostat** is a rheostat designed to control the exciting current of an electric machine. A **load rheostat** is a rheostat whose purpose is to dissipate electric energy. A **starting rheostat** is a rheostat designed to control the current taken by a motor during the period of starting and acceleration, but not to control the speed when the motor is running normally.[1]

[1] USAS C42, 1941.

115. Temperature Rise. All the electrical energy dissipated by a resistor or rheostat is converted into heat energy. Suitable means of ventilation must therefore be provided. Air-convection currents are generally used. The ultimate temperature rise of **intermittent-duty resistors** is governed largely by the time cycle and by the mass and specific heat of the resistive element. The smaller the percentage of time in circuit, the more important are the mass and specific heat. With **continuous-duty resistors** the ultimate temperature rise is obtained when the heat radiated, plus that carried off by conduction and convection, equals that generated by the I^2R loss.

When a **temperature test** is made on a **starting- or intermittent-duty resistor** without its motor, the resistor shall be connected to a voltage that will give the initial inrush of current specified. The steps shall be cut out at equal intervals of time in the "time-on" period of the cycle specified, and the current shall be maintained at 125% of the full-load current for those steps through which 125% of full-load current can flow. The specified cycle shall be repeated for 1 hr.[1]

When a **temperature test** is made on a **continuous-duty resistor** without its motor, any tested step shall be subjected to 100% of the current for which it is designed, and this value of current shall be maintained until the maximum temperatures are reached.[1]

When a **temperature test** is made on a **rheostatic dimmer,** it shall be made on a single plate, operated at rated voltage and connected in series with lamps totaling the rated lamp load. Single plates meeting this test will operate at safe temperatures when assembled into switchboard groups, because of the diversity of loading of the various plates. The maximum temperature rise may be approximated by setting the lever so that the dimmer is dissipating maximum wattage.[1]

When a **temperature test** is made on a **resistor, rheostat,** or **dimmer** at the current values, duty cycle and elapsed time specified, the temperature rise above the ambient temperature and the methods of temperature measurement shall be in accordance with the following:

a. For **bare resistive conductors,** the **temperature rise** shall not exceed 375°C as measured by a thermocouple in contact with the resistive conductor.

b. For **resistor units, rheostats, and wall-mounted rheostatic dimmers** which have an embedded resistive conductor, the **temperature rise** shall not exceed 300°C as measured by a thermocouple in contact with the surface of the embedding material.

c. For **rheostatic dimmers** which have embedded resistive conductors and which are arranged for mounting on switchboards or in noncombustible frames, the **temperature rise** shall not exceed 350°C as measured by a thermocouple in contact with the surface of the embedding material.

d. The **temperature rise of the issuing air** shall not exceed 175°C as measured by a mercury thermometer at a distance of 1 in from the enclosure.[1]

116. Temperature Rise of Plate-type Rheostats. For each watt dissipated per square inch of free radiating surface, the temperature of the rheostat will rise about 45°C. Frequently but one side of a plate-type rheostat can radiate freely. Therefore, where A = area, in square inches, of one side of the rheostat; I = current, in amperes, flowing; and R = resistance, ohms, of the rheostat in circuit, the temperature rise will be $45(I^2R/A)$ in degrees centigrade (approximately).

117. The resistive conductors in more common use consist of wire, suitably protected either by a fused vitreous enamel or by a cementlike covering, or suitably supported in air or sand; ribbon so formed as to give it mechanical rigidity; cast grids; carbon plates, disks, or rods; and liquids in suitable containers with electrodes.

118. The specific resistance and the temperature coefficient of resistance of the various conductors used in resistor construction are given in Sec. **4**, Pars. **198** to **206**.

119. The supports for resistive conductors are tubes or bushings of porcelain, pottery, lava, mica, asbestos, or other insulating material when the resistive conductor is wire or ribbon. When cast grids are used, they are ordinarily mounted in a frame made from steel stampings and tie rods suitably insulated. Pottery tubes are used when the resistance wire is to be hermetically sealed in a fused vitreous enamel. *Military Specifications* in the AN-, MIL-R-, and MIL-STD- categories cover numerous types of resistors and rheostats.

---

[1] NEMA Standards for Industrial Control, *Publ.* IC1-1965.

RESISTORS

120. Resistor units are used wherever a resistor that is not readily adjustable is desired. These units are made in many forms and types. **Wire-wound resistors** are usually made by winding a resistance wire or ribbon on a core of ceramic material. The resistance element is held in place by, or embedded in, a coating of cement or vitreous enamel. These units have relatively high wattage ratings. In smaller wattage ratings composition resistors are used extensively, especially in electronic applications. To meet the present trend toward miniaturization requiring high accuracy, stability, and reliability, resistors of the deposited-film type are available.

The colored bands around the body of a color-coded composition resistor represent its value in ohms. These bands are grouped toward one end of the resistor body. Starting with this end of the resistor, the first band represents the first digit of the resistance value; the second band represents the second digit; the third band represents the number by which the first two digits are multiplied. A fourth band of gold or silver represents a tolerance of ±5 or ±10%, respectively. The absence of a fourth band indicates a tolerance of ±20%. The physical size of a composition resistor is related to its wattage rating. Size increases progressively as the wattage rating is increased. The diameters of ½-, 1-, and 2-W resistors are approximately ⅛, ¼, and 5/16 in, respectively. The color-code chart provides the information required to identify the resistors.

Table 5-10. Resistor Color Code

Color	1st digit	2d digit	Multiplier	Tolerance, %*
Black	0	0	1	Gold ±5
Brown	1	1	10	Silver ±10
Red	2	2	100	No band ±20
Orange	3	3	1,000	
Yellow	4	4	10,000	
Green	5	5	100,000	
Blue	6	6	1,000,000	
Violet	7	7	10,000,000	
Gray	8	8	100,000,000	
White	9	9	1,000,000,000	
Gold	....	....	0.1	
Silver	....	....	0.01	

* Tolerance band is in addition to color bands indicating resistance values.

121. A discharge resistor is a resistor used for the protection of field windings of generators and motors and of other large magnet windings. It permits dissipation of the field energy and limits the surge voltage appearing at its terminals when current flow in the field circuit is interrupted. The value of resistance usually employed is equal to that of the field and rarely exceeds twice the resistance value of the field.[1]

122. Power-limiting resistors are large intermittent-duty resistors connected between the neutral of a polyphase transmission system and ground. They limit the flow of current to ground and the induced-voltage strains when an accidental ground occurs elsewhere on the system. Except during the interval of trouble they carry no current.

123. Resistors are often named by the function they perform in the circuit. **Voltage dividers** are potentiometer-connected resistors, placed across a voltage supply, with taps to furnish various lower voltages. **Grid leaks** are resistors, having a high value of resistance, used to provide a path to prevent the accumulation of a charge on the grid of a vacuum tube. **Grid suppressors** are resistors used in series with the grid of a vacuum tube to prevent oscillation. There are numerous other special types of resistor used in electronic circuits.

124. A standard resistor (resistance standard) is a resistor which is adjusted with high accuracy to a specified value, is but slightly affected by variations in temperature, and is substantially constant over long periods of time.[2]

[1] USAS C42-1941 ed.
[2] NEMA Standard for Industrial Control, *Publ.* IC1-1965.

125. Large resistors are made by mounting a suitable number of units on a faceplate or in a framework or enclosure.

126. Mounted resistor units are manufactured in various designs for placing readily in special apparatus or for giving various combinations, etc. Units designed to replace resistance lamps in order that a resistor may be secured that does not change with time or on account of temperature rise are so made as to screw into standard receptacles (see *A*, *B*, *C*, *D*, Fig. 5-61). Other units are made with ferrules on the ends, so that they may be inserted readily in the conventional type of fuse clips (see *E*); still others are mounted on brackets (see *F*) or on porcelain bases (see *G*); etc.

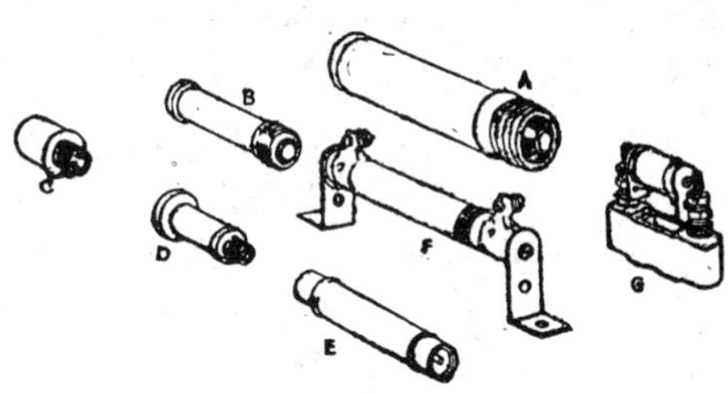

Fig. 5-61. Typical resistor units.

127. Negative temperature coefficient resistors have wide application in precision equipment where it is necessary to compensate for the increase in resistance of coils, etc., due to the positive temperature coefficient of the material with which they are wound. One of the most widely used negative temperature coefficient materials is a special form of carbon. Information on this material will be found in Sec. **4**.

128. Resistors used primarily for heating are not within the scope of this section. They are commonly known as **heating elements.**

129. The temperature rise of resistor units is discussed in Par. **115**.

TYPES OF RHEOSTAT

130. Plate-type rheostats have a resistance element of formed metal wire or ribbon attached to a plate of vitreous-enamel-covered iron or to a plate of insulating material such as soapstone by means of a coating of fused vitreous enamel or by cement. They are used almost universally for field rheostats, theater dimmers, motor-speed controllers, battery-charging rheostats, etc.

131. Box-type rheostats have a resistance element in the form of wire-wound enamel- or cement-covered pottery tubes, cast grids, coils of bare wire, metallic ribbon, carbon disks, or a conducting liquid. They are used for large field rheostats, motor starters, large motor-speed controllers, battery-charging rheostats of the larger sizes, etc.

FIELD RHEOSTATS

132. Field rheostats are used in series with the fields of generators for regulating the field strength, which in turn regulates the voltage of the generator, or with the fields of motors for adjusting the field strength, which in turn regulates the speed or power factor.

133. Generator field rheostats for separately excited d-c machines usually are provided with a value of total resistance about equal to that of the field to be regulated, thus giving an adjustment of the field current from maximum to one-half maximum with a constant excitation voltage.

134. Alternators have field rheostats whose resistance value is usually about twice that of the field, thus giving an adjustment of the field amperes from maximum to one-third maximum at constant excitation voltage.

135. Direct-current motor field rheostats have values of resistance determined by the control desired. The resistance is about equal to two-thirds of that of the field when 25% speed increase is wanted and may be 10 or 15 times that of the field when about 400% increase is wanted. For most machine-tool applications it is desirable to have approximately equal speed increments rather than equal percentage speed increments. To accomplish this, when the required rheostat resistance value is more than 1.5 times the resistance value of the motor field, it is necessary to use a rheostat with a much flatter resistance taper than that dictated by the watt-dissipating capacity of the rheostat. The resistance taper of a rheostat is the ratio of the resistance value of the highest resistance step to that of the lowest resistance step. For the above machine-tool applications it is usual to limit the resistance taper to a value not exceeding 5.

136. The current capacity of a field rheostat should change by approximately equal amounts as each additional step of resistance is inserted. If I_{max} = current when the rheostat is short-circuited, I_{min} = current when all resistance is in circuit, N = total number of steps of resistance, and I_N = change in current per step, then

$$I_N = \frac{I_{max} - I_{min}}{N} \qquad \text{(amperes)} \qquad (5\text{-}29)$$

and the current I_n when n steps are in circuit is

$$I_n = I_{max} - nI_N \qquad \text{(amperes)} \qquad (5\text{-}30)$$

If E is the emf in volts at the terminals of the field circuit, the total resistance value of the rheostat and field when n steps are in circuit is

$$R_n = \frac{E}{I_{max} - nI_N} \qquad \text{(ohms)} \qquad (5\text{-}31)$$

and when $n - 1$ steps are in circuit, the total resistance value of the rheostat and field is

$$R_{n-1} = \frac{E}{I_{max} - (n-1)I_N} \qquad \text{(ohms)} \qquad (5\text{-}32)$$

The resistance value of the nth step is therefore $R_n - R_{n-1}$ Ω.

137. The watt capacity of a field rheostat is the sum of the watt capacities of the individual steps. Upon assuming a fixed value of excitation voltage, a fixed value of field resistance, and an infinite number of steps in the rheostat gives Σ watts or summation watts,

$$\Sigma_w = I_{max} \times I_{min} \times R \qquad \text{(watts)} \qquad (5\text{-}33)$$

where Σ_w = summation of the watt capacities of the individual steps and R = resistance value of the entire rheostat. In practice, the smaller the number of steps and the larger the taper (I_{max}/I_{min}), the more does Σ_w calculated by the above formula exceed the actual sum of the watt capacities of the individual steps. Also,

$$E = \frac{I_{max} \times I_{min} \times R}{I_{max} - I_{min}} \qquad \text{(volts)} \qquad (5\text{-}34)$$

138. The maximum number of watts to be dissipated by any one step will be approximately constant if the change in current per step of resistance inserted is constant and the resistance value of each step is figured as above.

139. An automatic-release feature is sometimes provided on d-c motor field rheostats. This usually consists of a magnetic device, whose solenoid is connected across the armature terminals, which short-circuits the rheostat resistance when the motor stops, thus preventing starting under weakened field.

140. The temperature rise of field rheostats is discussed in Par. **115**.

DIRECT-CURRENT MOTOR-STARTING RHEOSTATS

141. The motor-starting resistors should be so proportioned that the total resistance and the resistance per step admit sufficient current to start the motor but not to exceed certain limits.

142. The normal current is that necessary to keep the motor armature turning over after it is once started.

143. The starting current is equal to the normal current plus a current necessary to overcome the friction of rest at starting, overcome inertia, and produce the desired rate of acceleration. In general, the starting current for d-c motors is assumed to be one and one-half times the normal current.

144. The apportionment of resistance in d-c starters is covered in Sec. **18**.

145. Rating of Resistors for Motor Starters. The resistors for motor starters

Table 5-11. Periodic Ratings of Resistors*

Approximate per cent of full-load current on first point starting from rest	Class numbers						
	Duty cycles						
	5 sec on 75 sec off	10 sec on 70 sec off	15 sec on 75 sec off	15 sec on 45 sec off	15 sec on 30 sec off	15 sec on 15 sec off	Continuous duty
25	111	131	141	151	161	171	91
50	112	132	142	152	162	172	92
75	113	133	143	153	163	173	93
100	114	134	144	154	164	174	94
150	115	135	145	155	165	175	95
200 or over	116	136	146	156	166	176	96

* NEMA Standards for Industrial Control, *Publ.* IC1-1965.

usually are rated by the time cycle and by the approximate percentage of full-load current which they allow to flow when the entire resistor is in circuit (see Table 5-11).

146. The temperature rise shall not exceed that given in Par. **115** when the cycles of operation are continued for 1 h.

147. Automatic features on motor-starting rheostats consist of no-voltage protection, overload protection, overload circuit breaker, vibrating field relay, and field-failure relay. No-voltage protection and overload protection are secured by the use of contact arms which are held in the operating position against the action of a spring. On voltage failure or overload the spring is magnetically released and moves the arm to the open-circuit position.

148. An overload circuit breaker on motor-starting rheostats consists of a separate switch arm or arms held closed by a latch. On overload this latch is tripped by a magnet, and the switch is opened by a spring and breaks the main-line circuit irrespective of the movement of the resistance-controlling arm.

DIRECT-CURRENT MOTOR-SPEED-REGULATING RHEOSTATS

149. Speed-regulating rheostats for main armature circuits have varying amounts of resistance for varying conditions.

150. The required resistance with a shunt-wound motor for a certain percentage of speed reduction is determined as follows: Let S be the normal speed in revolutions per minute; S_r the reduced speed in revolutions per minute; E the line voltage; I the armature current taken at reduced speed; R the rheostat resistance in ohms required to secure the desired speed reduction; then

$$\frac{S}{S_r} = \frac{E}{E - IR} \tag{5-35}$$

or

$$E - IR = \frac{ES_r}{S} \tag{5-36}$$

and

$$R = \frac{E}{I}\left(1 - \frac{S_r}{S}\right) \tag{5-37}$$

151. Motor Current Requirements at Reduced Speed. For machine tools, positive-pressure blowers, compressors, printing presses, etc., the current taken at reduced speeds is approximately that at normal speed. For large ventilating fans, centrifugal pumps on constant-discharge head, etc., the torque or armature current varies approximately as the square of the speed, but for small ventilating fans the current will vary more nearly as the speed.

152. Speed Regulators for Series-wound Motors. For series motors the calculation of the exact resistance value for a definite speed change requires a curve of the motor characteristics. Without this it will be found that a controller designed as above for a shunt machine (Par. **150**) will give approximately the same regulation for a series machine running under constant-torque conditions.

153. Other motor-control applications of resistors and rheostats are covered in Sec. **20**.

154. The maximum allowable temperature rise for speed-regulating rheostats is discussed in Par. **115**.

155. Automatic features on motor-speed-regulating rheostats are the same as for motor-starting rheostats and are discussed in Pars. **147** and **148**.

DIRECT-CURRENT BATTERY-CHARGING RHEOSTATS

156. The resistance value of a battery-charging rheostat for charging storage batteries from a constant-potential d-c source may be determined as follows: If E, in volts, is the emf of the charging circuit; E_{min}, in volts, the lowest emf of the battery during charge; E_{max}, in volts, the highest emf of the battery during charge; I_{min}, in amperes, the lowest value of charging current; then the total rheostat resistance value R_1, in ohms, will be

$$R_1 = \frac{E - E_{min}}{I_{min}} \quad \text{(ohms)} \quad (5\text{-}38)$$

A certain amount of this resistance must carry the maximum charging current I_{max}; this amount is

$$R_2 = \frac{E - E_{min}}{I_{max}} \quad \text{(ohms)} \quad (5\text{-}39)$$

Of this we may arrange so as to be in circuit permanently an amount equal to

$$R_3 = \frac{E - E_{max}}{I_{max}} \quad \text{(ohms)} \quad (5\text{-}40)$$

The balance of the resistance will have a current-carrying capacity tapering from I_{max} to I_{min}.

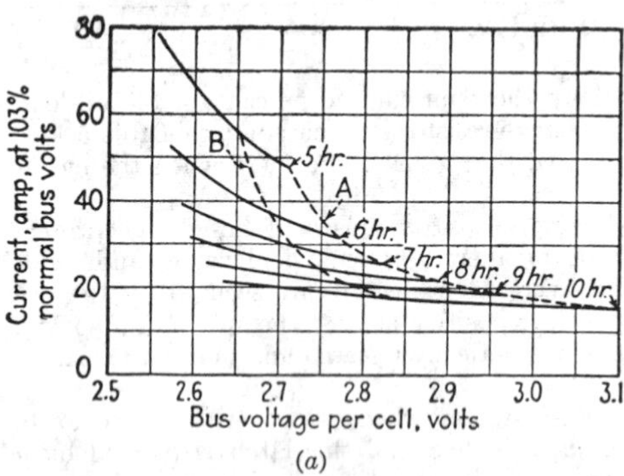

(a)

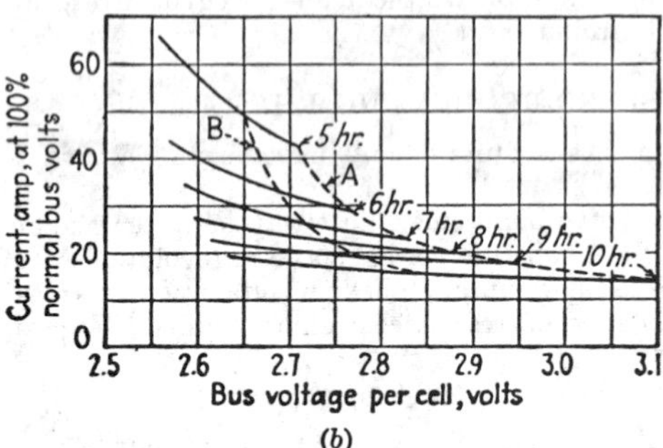

(b)

Fig. 5-62. Tubular-type lead-acid batteries; initial charging rate for 100-amp-h batteries. Curve *A*. Maximum value of bus voltage per cell for types MVM, KXK, ML, and ME. Curve *B*. Maximum values of bus voltage per cell for types TLM, MEH, and FLM. For batteries with cells of other capacity, the initial current will be directly proportional to cell capacity. (*a*) For determining ampere rating of rheostats, meters, switchgear, etc. (*b*) For determining ampere ratings of generators. (*Standard Specifications of Automatic Battery-charging Motor-Generators and Panels; adopted unanimously by the Industrial Truck Statistical Association at Cleveland, Ohio, Mar. 7, 1939.*)

157. Lead-acid batteries normally have a voltage per cell of at least 2.0 V as soon as charging is started, so that in calculating rheostats for lead-acid cells, E_{min} referred to in Par. **156** is equal to the number of cells multiplied by 2.0. At the end of charge the voltage per cell will be 2.6 (2.65) V, and E_{max} referred to in Par. **156** may be taken as equal to the number of cells multiplied by 2.6.

158. Nickel-alkaline batteries normally have a voltage of at least 1.4 V per cell as soon as charging is started, so that in calculating rheostats for nickel-alkaline cells E_{min} referred to in Par. **156** is equal to the number of cells multiplied by 1.4 and E_{max} is equal to the number of cells multiplied by 1.8.

159. The modified constant-voltage method of charging batteries is often employed when automatic charging is desirable. A fixed resistor is connected in series with the line and the battery. The resistance value and current capacity of the resistor depend upon the size, type, and number of cells to be charged; the line voltage; and the time available for the desired charge.

159a. Modified Constant-voltage Rates. The taper charging rates obtained under this system are produced by the combination of a constant-voltage source of power with a calibrated fixed resistance in series with each battery on charge.

This system is perfectly adapted to all types of **lead-acid battery,** for it provides a starting rate as high as is necessary to permit the charge to be completed in the desired time yet tapering the rate to the low values required at the completion of the charge. Values of the starting rates of the tubular positive-plate lead-acid type of battery under different permissible conditions, as affecting generator, panel, and resistor capacities, are determined from the graphs (Figs. 5-62*a* and 5-62*b*). The ohmic values of the resistors are similarly determined from Fig. 5-63.

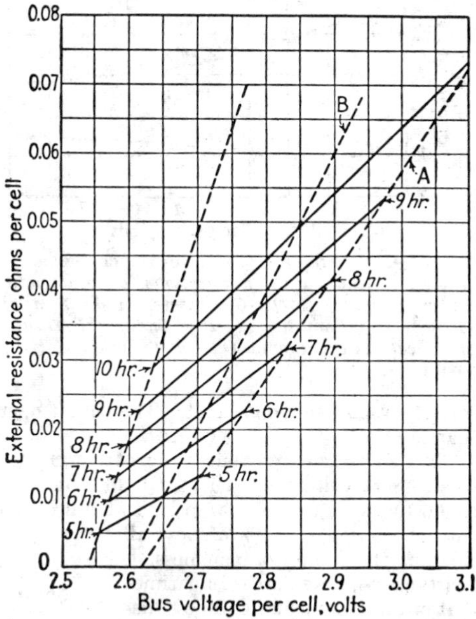

FIG. 5-63. Tubular-type lead-acid batteries; external resistance value for 100-amp-h batteries. Curve *A*. Maximum value of bus voltage per cell for types MVM, KXK, ML, and ME. Curve *B*. Maximum value of bus voltage per cell for types TLM, MEH, and FLM. For batteries with cells of other capacity, the external resistance will be inversely proportional to cell capacity. (*Standard Specifications of Automatic Battery-charging Motor-Generators and Panels; adopted unanimously by the Industrial Truck Statistical Association at Cleveland, Ohio, Mar.* 7, 1939.)

For types of lead-acid battery other than those shown in the charts, similar information will be provided by the battery manufacturers.

The **nickel-alkaline battery** may be charged under the modified constant-voltage system. The generator voltage may be as low as 1.84 times the number of cells in series, but a voltage of 2 V/cell is recommended. At this latter value, the starting current will be about 140% of the normal 5-h discharge rate of the battery, and the finishing

Table 5-12. Normal Rates for Modified Constant-voltage Charging of Nickel-Alkaline Batteries

Cell type				Normal rate, amp	Cell type				Normal rate, amp
...	B1	...	....	3.75	A10	...	...	....	75
...	B2	...	....	7.5	...	...	C7	....	78.75
...	B4	...	....	15	...	...	...	G11	82.5
...	B6	...	....	22.5	A12	...	C8	....	90
A4	...	...	G4	30	A14	...	...	G14	105
A5	...	...	....	37.5	...	...	C10	....	112.5
A6	...	C4	G6	45	A16	...	...	....	120
A7	...	...	G7	52.5	...	...	C12	G18	135
...	...	C5	....	56.25	A20	...	...	....	150
A8	...	...	...	60	...	...	...	G22	165
...	...	C6	G9	67.5	A24	...	...	....	180

current will be about 78% of the same normal value, averaging about 100% of normal during the entire charging period. The charging rates and the fixed calibrated resistance values for different voltages will be found in Figs. 5-64 and 5-65, respectively. The normal rates for various types of cell are given in Table 5-12.

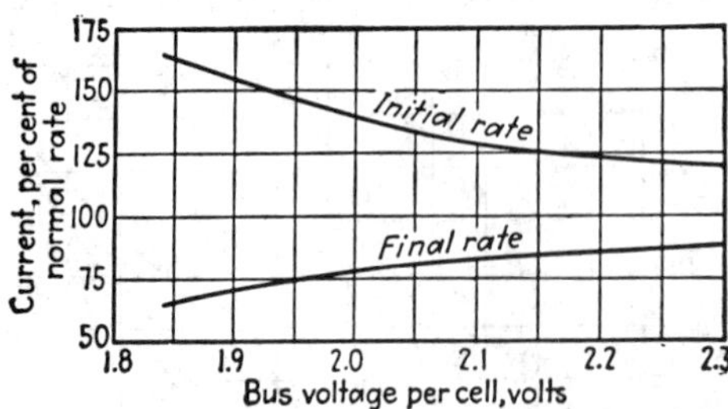

FIG. 5-64. Charging rates for nickel-alkaline batteries. (*Standard Specifications of Automatic Battery-charging Motor-Generators and Panels; adopted unanimously by the Industrial Truck Statistical Association at Cleveland, Ohio, Mar. 7, 1939.*)

THEATER DIMMERS

160. Theater dimmers are made of a number of plate-type rheostats mounted in a bank with each plate controlled by its own **interlocking lever.** These levers interlock through cams into a **master lever** for group control. The master levers are arranged so that any number of them may be operated in the same or in opposite directions by means of either a **grand-master lever** or a **slow-motion wheel.** The circuit operated by one plate seldom carries more than 30 A.

161. For dimming general-purpose tungsten-filament gas-filled lamps to a point where there will be no reflected light from concealed lamps, a resistance value equal to 3.6 times that of the lamp load at full candle power is required. This reduces the current flowing to 22% of normal. To dim the lamp to blackness, a resistance value equal to 4.6 times that of the lamp load at full candlepower is required. This reduces the current to 12% of normal. In order to dim the lamps smoothly the dimmer must have at least 110 steps.

The characteristic lamp curves in Fig. 5-66 show the relationship between the dimmer and the load when the load consists of Mazda "C" lamps. Smoothest dimming is obtained when the dimmer is designed to give approximately equal voltage changes per step.

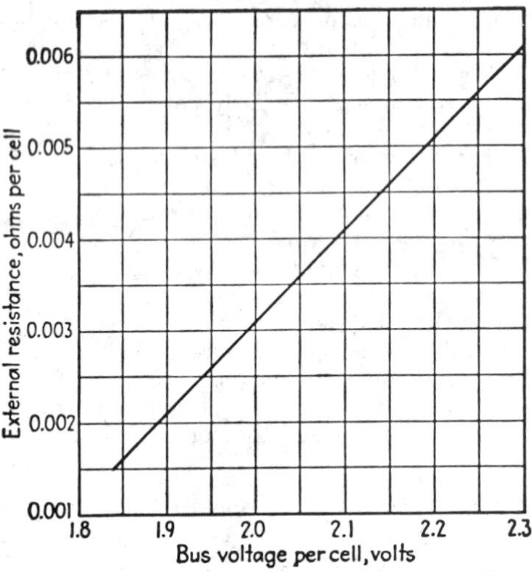

FIG. 5-65. Nickel-alkaline batteries; external resistance value for batteries with 100-amp-h normal rate. For batteries with cells of other capacity, the external resistance value per cell will be inversely proportional to the normal rate. (*Standard Specifications of Automatic Battery-charging Motor-Generators and Panels; adopted unanimously by the Industrial Truck Statistical Association at Cleveland, Ohio, Mar.* 7, 1939.)

162. Reactance dimmers are often used when the wattages per circuit are very large. If the resistance type were used, the dimmers would easily become unmanageable owing to size. Furthermore, the space for the installation of the dimmer on the stage is frequently inadequate. The use of reactance dimmers requires that only the dimmer plates be located on the stage, as the reactors may be located near the centers of their respective lighting loads.

The reactance dimmer consists of a magnetic element and a control rheostat in the form of a dimmer plate. The control plate is operated the same as any dimmer plate, as described in Par. **160**. A single plate can control up to 30 kW of lamps. The elementary idea of the construction is shown in Fig. 5-67. The magnetic element has three paths *A*, *B*, and *C*, as shown. Coils wound on paths *A* and *B*, having equal values of reactance, control the lamp current. The coil on path *C* connects to the control dimmer plate and is operated on a d-c circuit. The lamp circuit must be connected to an a-c supply. When the magnetization of the d-c coil is at its maximum, the paths *A*, *B*, and *C* are saturated. The alternating current, therefore, can induce but a small voltage in windings *A* and *B*. Since the lamp voltage is equal to the line voltage minus the reactor voltage, the lamps are bright. When the d-c magnetization is at a minimum, the paths *A*, *B*, and *C* are practically unsaturated. This allows the alternating current to saturate paths *A* and *B* and induce a voltage in its own windings *A* and *B*. Hence the reactor voltage is at a maximum, and the lamps are dim.

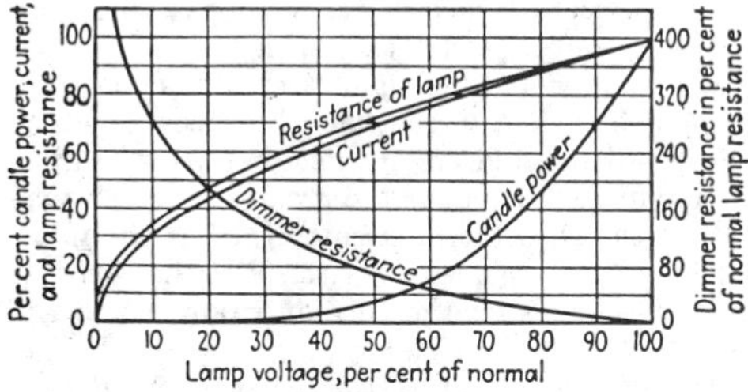

FIG. 5-66. Characteristic lamp curves for Mazda "C" lamps.

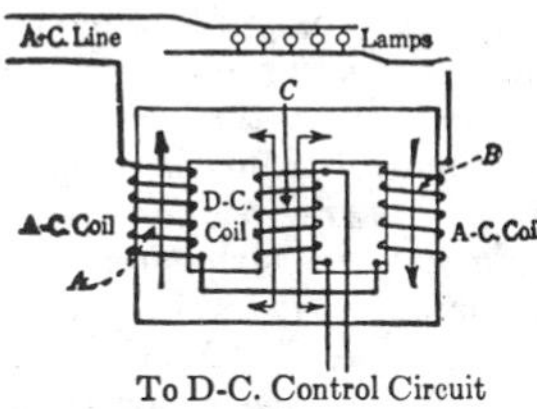

FIG. 5-67. Principle of the reactance dimmer.

The use of individual **controlled rectifiers** to provide the saturating current for each **reactance-type dimmer** permits the use of miniature devices as dimmer controls for large-capacity circuits. Semiconductor controlled rectifier circuits, such as are described in Sec. **12**, have been adapted to dimmer-control applications. Control may also be obtained through the grid of a triode rectifying tube or by means of magnetic elements in the plate circuit of a full-wave rectifying tube. In either case the controlling currents are of very small magnitude and usually at low voltage. Therefore, small-diameter conductors may be used in the circuit between the remote reactance dimmer and the pilot stage switchboard. This represents a considerable saving in copper in an installation containing a large number of dimmer circuits. Because of the low current carried by the pilot control circuit, the pilot control devices may be miniature in size and occupy very little space on the switchboard. Together with miniature switching control and pilot indicating lights, the switchboard and dimmer-control equipment can be made very compact. Additional features are easily incorporated with this type of control circuit. These features include proportional dimming through masters, preselection of intensities for as many as 10 scenes of lighting effect, fading from one preselected scene to a second scene, and extended mastering to remote locations. The control circuit usually incorporates a regulating or a "feedback" characteristic, which permits variable loading on the reactance dimmers while still providing the same dimming characteristics. This involves a slight modification in the design of the reactance dimmer to provide sufficient impedance to dim the minimum expected load to "blackout" and at the same time have sufficient capacity to carry the maximum expected load on the dimmer. Within this range, the same dimming characteristics are obtained with the pilot control circuit, regardless of the connected load.

163. An autrastat dimmer is essentially an adjustable ratio autotransformer (see Sec. **11**) and can be used on alternating current with any lamp load up to its maximum rating. The autrastat winding is connected across the line. The movable contacts make electrical contact with each turn of the winding. The voltage between the movable contacts and one end of the winding is applied to the lamps. This fulfills the conditions for smoothest dimming, giving equal voltage changes per step. The light intensity of each lamp will remain at that intensity, regardless of any change in the number of lamps connected in the circuit. These dimmers require relatively small space per unit of load and can be assembled in dimmer banks. To avoid circulating currents, dimmers must not be connected in parallel.

164. Free or artificial ventilation of theater dimmers is essential, as the amount of heat dissipated by a complete equipment is often very large. **Overloads** caused by using too many lamps or lamps of larger wattage than that for which the dimmer is designed must be carefully avoided.

LABORATORY RHEOSTATS

165. Tubular sliding-contact rheostats are designed for the accurate control of relatively small potentials and currents. Essentially, these rheostats consist of an insulated metal or a porcelain tube upon which is wound resistance wire of appropriate characteristics for the resistance value and watt dissipation required. One or two sliding contacts, which may be either manually or mechanically adjusted, are supported by metal bars running parallel to the axis of the tube. Provision is made so that these rheostats may be potentiometer-connected; each end of the resistance wire forms one terminal connection, and the sliding contact forms the third.

Usually only one size of wire is used to wind any specified value and size of rheostat. In the selection of these rheostats care must be taken to see that the rheostat is not overloaded at the short-circuit end. Some manufacturers list maximum and minimum currents for these rheostats. The maximum current is the greatest current that may be carried safely with but one-tenth of the rheostat in circuit, whereas the minimum current is the greatest current that may be carried safely with all of the rheostat in circuit. If, for a specified resistance value and size, the current values listed are not adequate, a larger size rheostat must be selected.

166. Loading rheostats are used as artificial loads for testing meters, small generators, resistors, rheostats, and similar apparatus that is to be given a load test. The controlling switches, in this apparatus, are connected each in series with a resistor across the line voltage. As each resistor will carry the maximum current passing through it at full-line voltage, overloads are impossible unless the rated voltage is exceeded. The load is regulated by closing as many switches as required to give the proper number of resistors in parallel.

167. Resistance boxes or plug boxes are used in educational and industrial laboratories in measuring values of resistance, current, and emf. Comparatively high values of resistance are usually employed in this work. The boxes usually consist of resistors connected in series and each shunted by a suitable short-circuiting device. Thus, any value of resistance from the minimum to the total of the box may be inserted in the circuit. Most resistance boxes are made for currents not exceeding a few hundredths of an ampere.

168. Carbon rheostats of the compression type have the advantage of fine adjustment. These rheostats are generally made up of plates held in a frame or disks in a tube. The resistance value is adjusted by regulating the pressure between the plates or disks with a screw clamp. Carbon has a negative temperature coefficient of resistance. Carbon-filament incandescent lamps, if heated appreciably, do not make satisfactory rheostats because of the rapid change of resistance with current changes.

169. The resistance value of a carbon rheostat varies inversely as the pressure, so that any desired resistance value within limits can be secured if the pressure is lessened or increased sufficiently. Such rheostats are used for laboratory tests and similar experimental work where the currents to be handled are not large and where it is not necessary to hold the resistance value constant during changes in temperature of the rheostat.

170. Wire Rheostats. Wire can be wound in coils or stretched over insulated

Table 5-13. Dimensions of Wire Coils for Rheostats

Size AWG	Maximum mandrel, inches	Feet per turn	Turns per inch	Max. coil length, inches
6–8	1.25	0.38	4.0	18
9–11	1.00	0.30	4.5	12
12–14	0.75	0.23	7.0	12
15–18	0.50	0.16	9.0	12
19–21	0.25	0.082	14.0	6
22–30	Must be wound on insulated core			

Note: The maximum diameter of mandrel given in the table corresponds to the length given therein; if a stiffer coil is desired a smaller mandrel must be used.

frames. Wires larger than No. 6 AWG are difficult to wind in spiral form, and wires smaller than No. 21 AWG must be wound upon an insulated core. When it is desired to increase the current capacity of a coil resistor beyond that of No. 6 wire, several coils may be connected in parallel. Table 5-13 gives the mechanical dimensions of coils made of different sized wires.

171. Mounting of Wire Coils. The finished coils are generally mounted in an iron frame. They are suspended vertically from insulated supports and interconnected so as to give the proper current capacity and resistance value. The total resistance should be divided into units containing a sufficient number of coils in multiple to give the required current capacity and then enough of these units in series to give the required resistance value. The terminals of each unit should be brought out and so arranged that the unit can be cut out or short-circuited when it is desired to adjust the resistance value.

Only fireproof material should be used in construction of the rheostat. Resistors which are not used continuously, such as starters, can have their heat capacity considerably increased by using plenty of heat-conducting material, which will delay the attainment of the final temperature. The heat capacity and effective radiating surface of a coil resistor can be increased by placing the coils in an iron box filled with sand, oil, or some other insulator which is a fair conductor of heat. Coils may be submerged, temporarily, in water when it is desired to increase their capacity for some special purpose. Wire under these conditions can be worked at about four times its normal capacity (see Par. **178**).

172. Radiation. When coils are to be enclosed in a weatherproof box, 0.5 to 1 in^2 of box surface should be allowed per watt dissipated by the resistor. With good ventilation and about 1 in^2/W the temperature rise will be about 100°F on the surface of the enclosure.

Wires which contain zinc should not be used where the resistor is likely to be overheated, because they will become brittle, or "hot short."

173. Potentiometer-connected rheostats (potential dividers) are often used in laboratory work. They consist of rheostats whose resistance is shunted across the source of potential. The load is connected to the low-current-capacity end of the rheostat resistance and to the moving contact. This arrangement permits the smooth reduction to zero of the potential applied to the load. It is particularly applicable to tubular sliding-contact rheostats (see Par. **165**).

MISCELLANEOUS

174. Enclosures for resistors and rheostats are made in various forms to meet special requirements. They may be acid-, drip-, dust-, explosion-, fume-, gas-, moisture-, oil-, sleet-, splash-, water-, or weatherproof, -resistant, or -tight.

175. Motor drives may be had for almost all types of rheostat, the particular drive depending upon the application.

176. Liquid rheostats are especially adapted to the dissipation of large amounts of power and are often used as artificial loads in testing generators or as starting rheostats for large motors starting under load. The adjustment is perfectly continuous, but unless there is a provision for short-circuiting the electrodes outside the solution, it is impossible to cut out the resistance entirely.

177. Electrodes for Liquid Rheostats. The electrode material is not important so long as it is a good conductor and is not attacked by the liquid. Lead or carbon plates are used with sulfuric acid, copper with copper sulfate, and iron in most other cases. The **current density** should not exceed 1 A/in^2 of surface per electrode.

The solution depends upon the voltage and upon the quantity necessary to radiate the heat. Water alone is seldom used for voltages under 1,000 V. For voltages below this, sulfuric acid or a salt is added to the water to increase its conductivity.

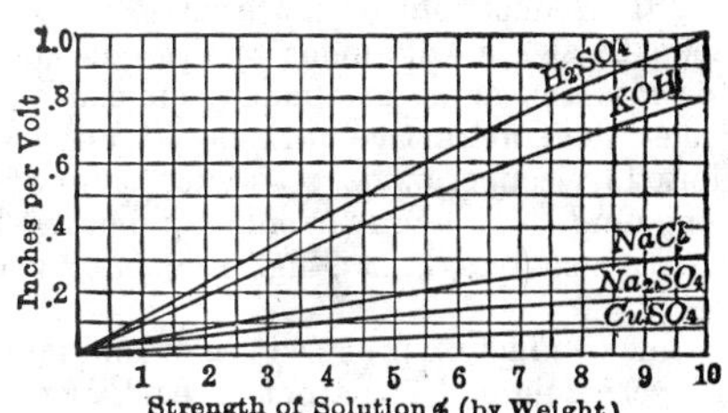

FIG. 5-68. Conductivity of solutions.

Figure 5-68 shows the relative conductivity of various solutions expressed in inches between the plates with a current density of 1 A/in². Ordinary water gives a drop of 2,500 to 3,000 V/in gap at this current density.

178. Water-cooled rheostats are often used for the same applications as liquid rheostats. Rheostats made up of galvanized iron wire mounted on wooden frames and submerged in running water are often used to dissipate energy when making acceptance tests of large apparatus. In this case the power dissipated can be assumed as directly proportional to the surface of the resistor (see Par. **172**).

BIBLIOGRAPHY

179. References on Resistors and Rheostats

BECK, E. New Resistor for Use in Measurements of Impulse Voltages and Currents; *Elec. Jour.*, 1928, p. 1017.

BOULTON, B. K. Water Rheostat as Emergency Load; *Elec. World,* May 14, 1927, p. 1017.

DAVIS, B. L. Printed Circuit Techniques; An Adhesive-tape Resistor System; *NBS Circ.* 530.

EARNHEART, R. L. Bonneville Uses Water Rheostat; *Elec. World,* Jan. 14, 1939, p. 118.

JAMES, H. D." Controllers for Electric Motors"; Princeton, N. J., D. Van Nostrand Company, Inc., 1926.

RAGAZZINI, JOHN Calculation of Resistance Steps in Starting Rheostats; *Elec. Eng.*, July, 1939, Vol. 58, p. 318.

SOLID-STATE DEVICES

BY WILLIS A. ADCOCK, W. T. MATZEN, AND J. P. MIZE

INTRODUCTION

180. Solid-state electronics has had a truly revolutionary effect on the electronics industry, stemming from the commercial introduction of transistors in 1952. Although solid-state devices were used in the early days of radio as detectors, they were superseded by vacuum tubes. They came back into prominence during World War II for microwave detection. It was this resurgence in the use of silicon and germanium that led to the invention of the transistor and the widespread development of solid-state electronics.

From its initial impact in portable radios, the transistor through extensions in performance has gained major importance in computer, military, and space applications. Subsequent progress in technology has led to the development of a wide variety of devices and circuits based on semiconductor principles. Application of these products has been extended into the industrial and consumer fields, where they are rapidly replacing vacuum tubes in conventional electronic equipment and are having widespread effect on functions previously accomplished through mechanical or electromechanical means.

For information on basic semiconductor principles, materials, and technologies, see Sec. **4**, Pars. **589** to **597**. Sources to meet individual requirements and areas of interest are listed in the Bibliography, Par. **197**. These should be referred to for detailed information.

181. Semiconductor Materials. Silicon and germanium are by far the most commercially significant among semiconducting materials. However, other semiconducting elements and many compound semiconductor materials have electrical properties that make them attractive for special applications. Semiconducting properties are exhibited by copper oxide, selenium, cadmium sulfide, gallium arsenide, and some organic compounds. Gallium arsenide junction devices, for example, are used as infrared radiation sources, lasers, high-frequency varactors for parametric amplifiers, and tunnel diodes.

The conduction process in semiconductors is most easily visualized in terms of silicon and germanium. The atoms of each of these elements have four electrons in the outer shell (valence shell). These electrons are normally bound in the crystalline lattice structure. Some of these electrons are free at room temperature and hence can move through the crystal; the higher the temperature, the more electrons are free to move. Each vacancy, or hole, left in the lattice can be filled by an adjacent valence electron. Since a hole moves in a direction opposite to that of an electron, a hole may be considered as a positive-charge carrier. Electrical conduction is due to the motion of holes and electrons under the influence of an applied field.

Intrinsic (pure) semiconductors exhibit a negative coefficient of resistivity, since the number of carriers increases with temperature. Conduction due to thermally generated carriers, however, is usually an undesirable effect, since it limits the operating temperature of the semiconductor device.

At a given temperature, the concentration of thermally generated carriers is related to the **energy gap** of the material. This is the minimum energy (stated in electron volts) required to free a valence electron (1.1 eV for silicon and 0.7 eV for germanium). Silicon devices perform at higher temperatures because of the wider energy gap.

The conductivity of the semiconductor material can be altered radically by **doping** with minute quantities of **donor or acceptor impurities.** Donor (*n*-type) impurity atoms have five valence electrons, whereas only four are accommodated in the lattice structure of the semiconductor. The extra electron is free to conduct at normal operating temperatures. Commonly used donor impurities include phosphorus, arsenic, and antimony. Conversely, acceptor (*p*-type) atoms have three valence electrons; a surplus of holes is created when a semiconductor is doped with them. Typical acceptor dopants include boron, gallium, and indium.

In an **extrinsic (doped) semiconductor,** the current-carrier type introduced by doping predominates. These carriers, electrons in *n*-type material and holes in *p*-type, are called **majority carriers.** Thermally generated carriers of the opposite type are also present in small quantities and are referred to as **minority carriers.** Resistivity is determined by the concentration of majority carriers.

Transistor amplification depends upon injecting a surplus of minority carriers into the base region of the semiconductor device. These carriers propagate across the base by a process of diffusion, a process similar to the spreading of molecules in a gas from a region of high concentration to a region of low concentration. Transistor amplification is therefore dependent upon two properties of the minority carriers, lifetime and mobility.

Lifetime is the average time required for excess minority carriers to recombine with majority carriers. Recombination occurs at centers caused by impurities and imperfections in the semiconductor crystal. Semiconductor junctions are formed in material grown as a single, continuous crystal to obtain the lattice perfection required, and extreme precautions are taken to ensure exclusion of impurities during processing. However, in some applications short lifetime is desired, and in such cases gold doping is used to achieve this.

Carrier mobility is the property of a charge carrier which determines its velocity in an electrical field. High mobility yields a short base-transit time and good frequency response in a transistor. Long lifetime and high mobility are desirable to reduce unwanted recombination during the transit of minority carriers across the base region.

SEMICONDUCTOR JUNCTIONS

182. *p-n* Junctions. The operation of most semiconductor devices is dependent upon the *p-n* junction—the boundary formed between a *p* region and an *n* region in a monocrystalline piece of semiconductor material. Each region contains a high concentration of its majority carriers and an equal number of oppositely charged impurity ions (Fig. 5-69). Since electrons are more highly concentrated in the *n* region

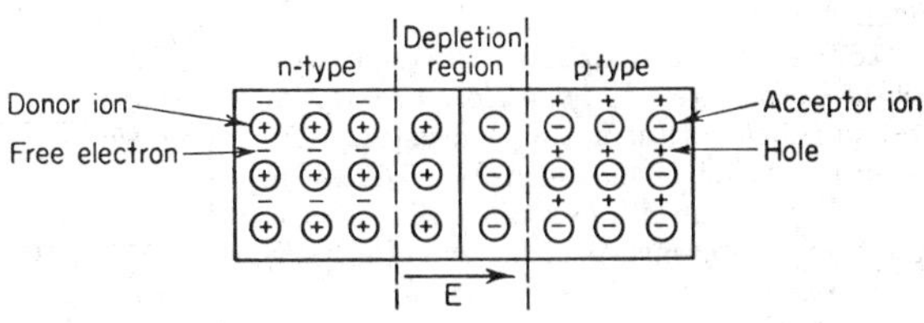

FIG. 5-69. Distribution of ions and carriers near *p-n* junction.

than in the p region, they diffuse across the junction. Similarly, holes diffuse from the p region into the n region. This flow leaves a space-charge region of ionized impurities, depleted of free carriers, on both sides of the junction. This **depletion region** is positively charged (donor ions) on the n side and negatively charged (acceptor ions) on the p side of the junction. This electric field opposes the diffusion of majority carriers across the junction. It also produces a flow of thermally generated minority carriers across the junction in the opposite direction. When no external voltage is applied, the net flow of holes and electrons across the junction is zero and the junction is in equilibrium.

When an external voltage is applied, if its polarity makes the p region positive with respect to the n region, the diffusion of majority carriers across the junction is increased. This process, called injection, raises the concentration of minority carriers on both sides of the junction. These excess carriers diffuse toward the contacts opposite the junction, causing current flow in the external circuit. Small values of applied voltage control this considerable forward current because of the large concentration of majority carriers available.

Conversely, an applied voltage of reverse polarity opposes the diffusion of majority carriers across the junction. The resulting reverse current is small, since it is produced by the flow of minority carriers across the junction.

Ordinarily, junctions are fabricated so that one region is more heavily doped than the other. The forward current is then largely due to the injection of carriers from the heavily doped region into the lightly doped region.

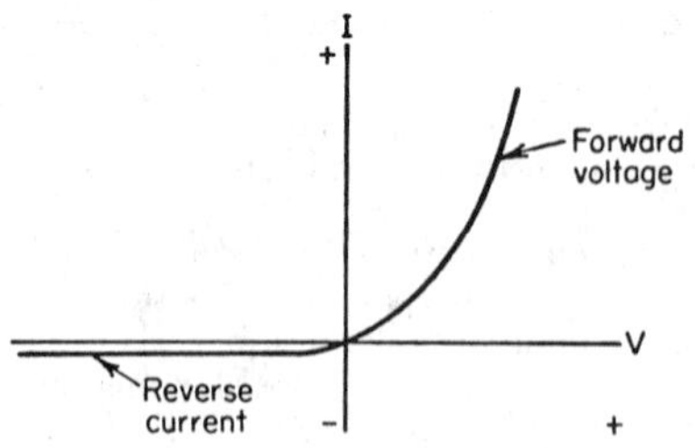

Fig. 5-70. Voltage-current characteristic for p-n junction diode.

The **voltampere characteristic** for a p-n **junction diode** is shown in Fig. 5-70. For typical junction concentrations and current densities, at a temperature of 300°K, forward voltage ranges between 0.2 and 0.3 V in germanium and between 0.5 and 0.75 V in silicon.

The reverse current is related to minority-carrier concentration, which depends upon temperature and the energy gap of the material. Reverse current increases exponentially with temperature. It is a limiting factor in the high-temperature operation of semiconductor junction devices.

The high-frequency response of a semiconductor diode may be seriously limited by charge stored in the depletion region. This charge gives a capacitive effect since it changes with voltage. The value of the stored charge is that of the ionized impurity atoms in the depletion regions on either side of the junction. The width of the depletion region increases with higher reverse voltage and lighter doping. The result is lower capacitance, as in the case of a parallel-plate capacitor with wider spacing between plates.

The **maximum reverse voltage** of a p-n **junction** is limited by the field in the depletion region. This field accelerates carriers, which may gain enough energy to create new hole-electron pairs by colliding with atoms of the lattice structure. Each of these carriers may also create a hole-electron pair. As reverse voltage is increased, an avalanche breakdown point is reached at which this multiplicative action causes the current to increase abruptly. Avalanche breakdown voltage is higher in lightly doped regions, since the depletion region is wider, making the internal electric field smaller for any given voltage.

183. Transistor Action. A transistor consists of two junctions in close proximity and parallel within a single crystal. An n-p-n **transistor** is shown in Fig. 5-71. In normal bias conditions, the emitter-base junction is forward-biased, and the collector-base junction is reverse-biased.

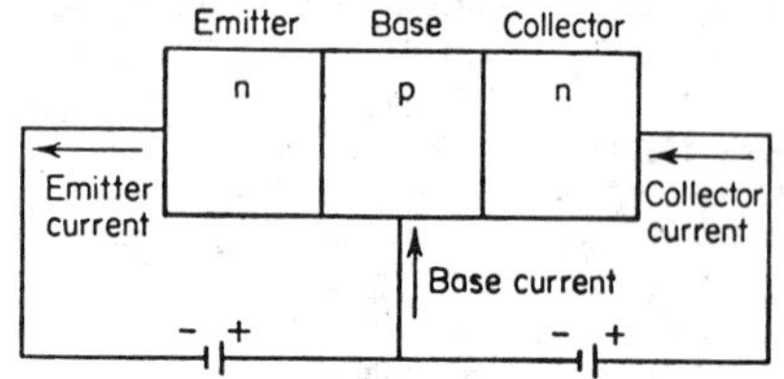

Fig. 5-71. n-p-n junction transistor.

Forward bias of the emitter-base junction causes electrons to be injected into the base region, producing an excess concentration of minority carriers there. These carriers move by diffusion to the collector junction. Here, they are accelerated into the collector region by the field in the depletion region of the reverse-biased collector junction. Some of the electrons recombine before reaching the collector. Current flows from the base terminal to supply the holes for this recombination process. Another component of current flows in the emitter-base circuit because of the injection of holes from the base into the emitter.

Practical transistors have narrow bases and high lifetimes in the base to minimize recombination. Injection of holes from the base into the emitter is made negligible by doping the emitter much more heavily than the base. Thus, the collector current is less than, but almost equal to, the emitter current.

In terms of the emitter current I_E, the collector current I_C is

$$I_C = \alpha I_E + I_{CBO} \tag{5-41}$$

where α is the fraction of the emitter current that is collected and I_{CBO} is due to the reverse current characteristic of the collector-base junction. Increase of I_{CBO} with temperature sets the maximum temperature of operation.

High-frequency transistors are fabricated with very narrow bases to minimize the transit time of minority carriers across the base region. Germanium transistors have slightly better high-frequency response than silicon devices because of the higher mobility of germanium.

TRANSISTOR FABRICATION

184. Transistor-fabrication Processes. Although many techniques are used to fabricate transistors, the major portion of present transistor production employs either the alloy process or the diffusion process.

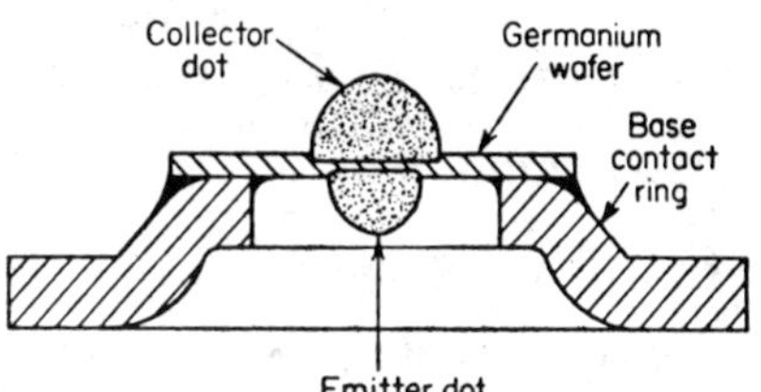

FIG. 5-72. Cross section of germanium-alloy transistor.

The cross section of a **germanium-alloy transistor** is shown in Fig. 5-72. The transistor is formed by placing a thin n-type germanium wafer between two indium pellets and heating it in a furnace. The indium melts and forms a solution with the germanium. When the assembly is cooled, the germanium recrystallizes. The germanium closest to the pellets is doped heavily p type by the indium, forming a p-n-p structure. Since junction areas and the width of the n region cannot be controlled precisely in this process, these devices have large areas and wide base regions. This limits them to low-frequency applications.

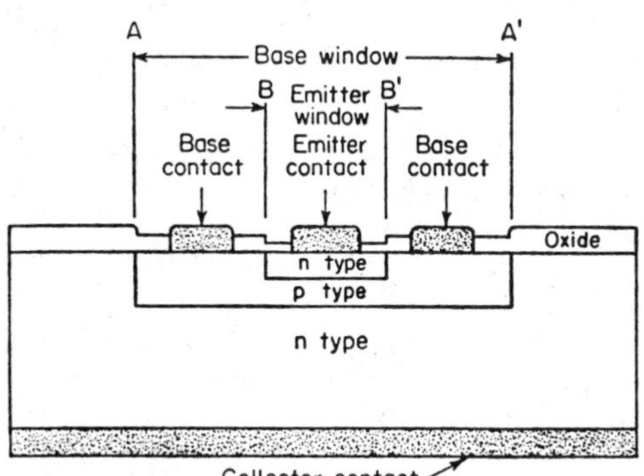

FIG. 5-73. Cross section of n-p-n silicon planar transistor.

Alloy transistors were among the first types manufactured and are still widely used, since the process is inexpensive and performance is good at low frequencies. Germanium-alloy transistors, for example, are used in the audio stages of transistor radios. Both silicon and germanium-alloy transistors are used for low-frequency high-power applications.

Diffusion technology, on the other hand, permits precise control of junction areas and layer widths, yielding high-performance devices having reproducible characteristics. The process for fabricating a **silicon diffused *n-p-n* transistor,** shown in cross section in Fig. 5-73, illustrates the technologies involved. An oxide is formed on an n-type silicon slice. A window (A-A') is cut in the oxide by a photolithographic process. Impurity vapors of the p type, passed over the surface at high temperature, diffuse into the silicon where the surface is exposed. Additional oxide is formed, and a smaller window (B-B') is cut in the oxide. Impurities of the n type are diffused through this opening. Metal contacts are formed by evaporation and photolithographic processing.

Before these diffusions occur, a heavily doped $n+$ region may be diffused into the opposite face of the slice to reduce the ohmic resistance in series with the collector contact (Fig. 5-74). Base and emitter are then diffused into the lightly doped n region, giving low collector capacitance and high breakdown. Alternatively, the $n+$-n structure may be formed by an epitaxial process in which a lightly doped n region is deposited onto an $n+$ substrate from the vapor phase.

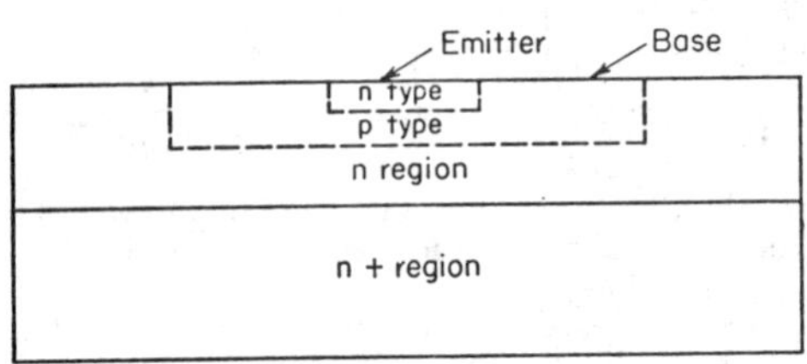

Fig. 5-74. Epitaxial or diffused material for reducing collector series resistance. Subsequent diffusions are indicated by dotted lines.

Junction size, junction shape, and impurity profiles may be varied to optimize performance depending on the application. Variations of the process are used to make **p-n-p transistors** and complex semiconductor devices such as controlled rectifiers and integrated circuits.

TRANSISTOR CIRCUIT MODELS

185. Circuit Model of the Transistor. Performance of the transistor as an active circuit element is analyzed in terms of various small-signal equivalent circuits. The low-frequency T-equivalent circuit of Fig. 5-75 is closely related to the physical structure. This circuit model is used here to illustrate the principle of transistor action. Carriers are injected into the base region by forward current through the emitter-base junction. A fraction α (near unity) of this current is collected. The incremental change in collector current is determined essentially by the current generator αI_e, where I_e is the incremental change of emitter current. The **collector resistance** r_c in parallel with the current generator accounts for the finite resistance of the reverse-biased collector-base junction. The **input impedance** is due to the dynamic resistance r_e of the forward-biased emitter-base junction and the ohmic resistance r_b of the base region.

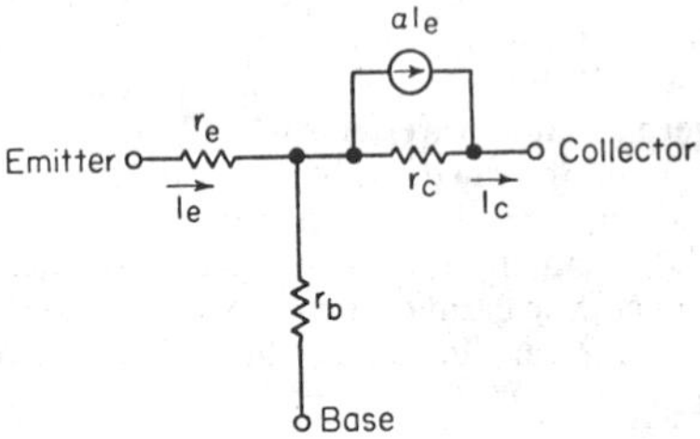

Fig. 5-75. Common-base T-equivalent circuit.

The room-temperature value of r_e is about $26/I_E\ \Omega$, where I_E is the d-c value of the emitter current in milliamperes. Typical ranges of other parameters are as follows: r_b varies from tens of ohms to several hundred ohms; α varies from 0.9 to 0.999; and r_c ranges from a few hundred ohms to several megohms.

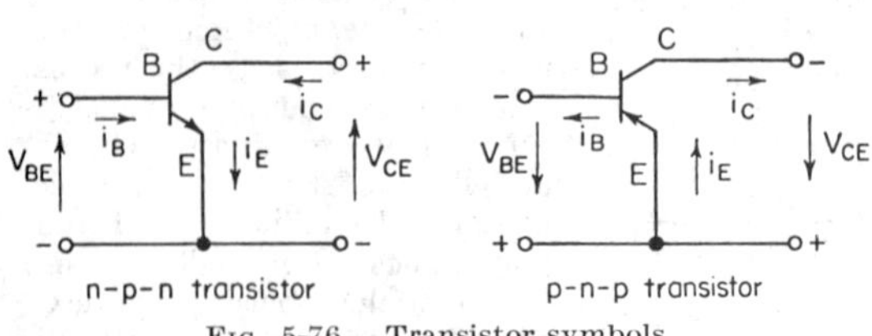

Fig. 5-76. Transistor symbols.

The symbolic representations of an n-p-n and a p-n-p transistor are shown in Fig. 5-76. Direction of conventional current flow and terminal voltage for normal operation as an active device are indicated for each. The voltage polarities and current for the *p-n-p* are reversed from those of the *n-p-n*, since the conductivity types are interchanged.

Transistors may be operated with any of the three terminals as the common, or grounded, element, i.e., common-base, common-emitter, and common-collector. These configurations are shown in Fig. 5-77 for an *n-p-n* transistor.

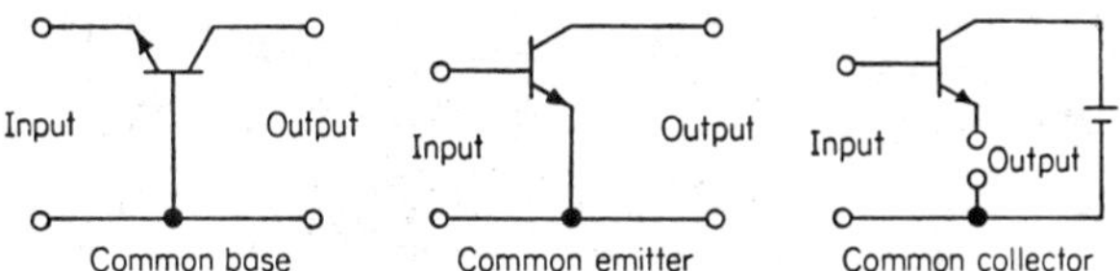

FIG. 5-77. Circuit connections for *n-p-n* transistor.

The transistor action shown in Fig. 5-75 is that of the **common-base connection,** whose current gain ($\simeq\alpha$) is slightly less than 1. Even with less than unity current gain, voltage and power amplification can be achieved, since the output impedance is much higher than the input impedance.

For the **common-emitter connection,** only base current is supplied by the source. Base current is the difference between emitter and collector currents and is much smaller than either; hence, current gain I_c/I_b is high. Input impedance of the common-emitter stage is correspondingly higher than it is in the common-base connection.

In the **common-collector connection,** the source voltage and the output voltage are in series and have opposing polarities. This is a negative-feedback arrangement, which gives a high input impedance and approximately unity voltage gain. Current gain is about the same as that of the common-emitter connection. The common-base, common-emitter, and common-collector connections are roughly analogous to the grounded-grid, grounded-cathode, and grounded-plate (cathode-follower) connections, respectively, of the vacuum tube.

Low-frequency performance of transistors is commonly specified in terms of the small-signal *h* parameters listed in Table 5-14. In the notation system used, the second subscript designates the circuit connection (*b* for common-base and *e* for common-emitter). The forward transfer parameters (h_{fb} and h_{fe}) are current gains measured with the output short-circuited. The current gains for practical load conditions are not greatly different. The input parameters h_{ib} and h_{ie}, although measured for short-circuit load, approximate the input impedances of practical circuits. The output parameters h_{ob} and h_{oe} are the output admittances. Values of these parameters are estimated in Table 5-14 in terms of the T-equivalent circuit elements for which approximate values were given earlier.

Table 5-14. Transistor Small-signal *h* Parameters

	Input parameter	Transfer parameter	Output parameter
Common-base...........	$h_{ib} \cong r_e + (1 - \alpha)r_b$	$h_{fb} \cong \alpha$	$h_{ob} \cong 1/r_c$
Common-emitter.........	$h_{ie} \cong r_e/(1 - \alpha) + r_b$	$h_{fe} \cong \alpha/(1 - \alpha)$	$h_{oe} \cong 1/r_c(1 - \alpha)$

The **current gain** of the common-base stage is slightly less than unity; common-emitter current gains may vary from 10 to several hundred. Input impedance and output admittance of the common-emitter stage are higher than those of the common-base circuit by approximately h_{fe}.

Although matched power gain of the common-base and common-emitter connections is about the same, the higher input impedance and lower output impedance of the common-emitter stage are desirable for most applications. For these reasons, the common-emitter stage is more commonly used. For example, the voltage gain of cascaded common-base stages cannot exceed unity unless transformer coupling is used.

Table 5-15. Transistor Parameter Values (Typical)

Device type	Material	Fabrication method	Application	BV_{CEO}, V	I_{max}	Max. power (case temp. = 25° C)	h_{FE} (d-c) at I		High-frequency response $\|h_{fe}\|$ at f		Switching times	
											t_{on}	t_{off}
2N1893	*n-p-n* silicon	Double-diffused planar	General purpose	80	0.5 A	3W	≧35	10 mA	≧2.5	20 MHz		
2N404	*p-n-p* germanium	Alloy junction	General purpose	24	100 mA	150 mW (25° C free air temp.)	≧30	12 mA	f_{hfb} ≧	4 MHz		
2N930	*n-p-n* silicon	Planar	Low-level amplifier	45	30 mA	600 mW	≧150	0.5 mA	1.0	30 MHz		
2N918	*n-p-n* silicon	Epitaxial planar	U-h-f amplifier	15	50 mA	300 mW	≧20	3 mA	6.0	100 MHz		
2N5043	*p-n-p* germanium	Epitaxial planar	U-h-f amplifier	7	30 mA	30 mW (100° C free air temp.)	≧15	3 mA	3.75	400 MHz		
2N960	*p-n-p* germanium	Epitaxial diffused-base mesa	Switch	7	100 mA	300 mW	≧20	10 mA	3.0	100 MHz	≦50 ns	≦85 ns
2N3011	*n-p-n* silicon	Epitaxial planar	Switch	12	200 mA	1.2 W	≧30	10 mA	4.0	100 MHz	≦15 ns	≦20 ns
2N1724	*n-p-n* silicon	Triple-diffused mesa	Power	80	5 A	50 W (case temp. = 100° C)	≧12	2 A	1.0	10 MHz		
2N4002	*n-p-n* silicon	Epitaxial planar	Power	80	30 A	100 W (case temp. = 100° C)	≧10	30 A	3.0	10 MHz	≦1 μs	≦3 μs
2N3055	*n-p-n* silicon	Single-diffused mesa	Power	70	15 A	115 W	≧20	4 A	f_{hfe} ≧	20 kHz		
2N456B	*p-n-p* germanium	Alloy junction	Power	30	7 A	150 W	≧22	7 A	f_T =	200 kHz		

The common-collector circuit has a higher input impedance and lower output impedance than either of the other connections. It is used primarily for impedance transformation.

The current gain of a transistor decreases with frequency, principally because of the transit time of minority carriers across the base region. The frequency f_T at which h_{fe} decreases to unity is a measure of high-frequency performance. **Parasitic capacitances** of junctions and leads also limit high-frequency capabilities. These high-frequency effects are shown in the modified equivalent circuit of Fig. 5-78. The maximum frequency f_{max} at which the device can amplify power is limited by f_T and the time constant $r_b'C_c$, where r_b' is the ohmic base resistance and C_c is that portion of the collector-base junction capacitance which is under the emitter stripe (see Fig. 5-78). Values of f_T greater than 3 GHz and f_{max} exceeding 10 GHz are obtained by maintaining very thin bases (<0.01 mil) and narrow emitters (<0.1 mil).

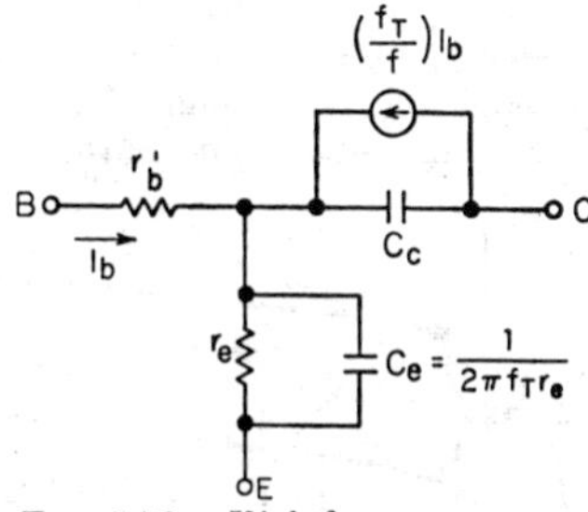

FIG. 5-78. High-frequency common-emitter equivalent circuit.

186. Transistor Voltampere Characteristics. The performance of a transistor over wide ranges of current and voltage is determined from static characteristic curves, such as the common-emitter output characteristics of Fig. 5-79. Collector current I_c is plotted as a function of collector-to-emitter voltage V_{CE} for constant values of base current I_B.

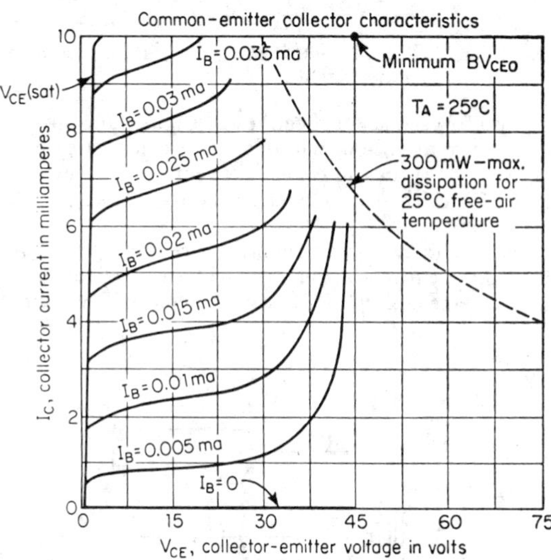

FIG. 5-79. Common-emitter collector characteristics—2N929 *n-p-n* silicon transistor.

Avalanche breakdown limits the maximum voltage that can be applied without risking destruction; this may be as low as 10 V in very-high-frequency transistors or greater than 200 V in power transistors. The minimum voltage $V_{CE(sat)}$ at which the collector-base junction becomes forward-biased is typically a few tenths of a volt. Maximum currents range from a few milliamperes to greater than 50 A.

Average power dissipation must be limited to prevent excessive junction temperature. The major effect of temperature rise is to increase the collector current that flows with the base terminal open.

The leakage current I_{CBO} in the common-base connection is the reverse current of the collector-base junction; common-emitter leakage is higher by the factor $1/(1 - \alpha)$ because of transistor amplification. In either case, the leakage current increases exponentially with temperature. Maximum junction temperatures are limited to about 100°C in germanium and 250°C in silicon. The locus of maximum power dissipation is a hyperbola on the voltampere characteristic curve. Power dissipation must be decreased when higher ambient temperatures exist. Large-area devices and physical heat sinks of high thermal dissipation are used to extend power ratings.

Dynamic variations of voltage and current are analyzed by a **load line** on the characteristic curves, as in vacuum tubes. For a linear transistor amplifier with load resistance R_L, the output varies along a load line of slope $-1/R_L$ about the d-c operating point (Fig. 5-80). Since the minimum voltage $V_{CE(sat)}$ is quite low, good efficiencies can be obtained with low values of supply voltage. The operating point on the V_{CE}-I_C coordinates is established by a d-c bias current in the input circuit. Transistor circuits should be biased for a fixed *emitter* current rather than a fixed *base* current to maintain a stable operating point, since the lines of constant base current are variable between devices of a given type and with temperature.

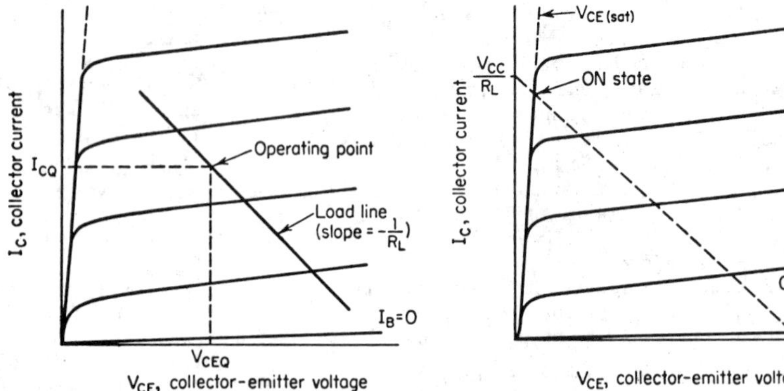

Fig. 5-80. Load line for linear-transistor amplifier circuit.

Fig. 5-81. Switching states for common-emitter circuit.

The common-emitter circuit can also be used as an effective switch, as shown by the load line of Fig. 5-81. When the base current is zero, the collector circuit is effectively open-circuited and only leakage current flows in the collector circuit. The device is turned on by applying base current I_{b1}, which decreases the collector voltage to the saturation value.

187. Transistor Applications. Representative circuit functions are discussed below.

***RC*-coupled amplifiers** similar to the circuit shown in Fig. 5-82 are used for audio and video applications. Bandwidths in excess of 250 MHz can be achieved with high-frequency transistors.

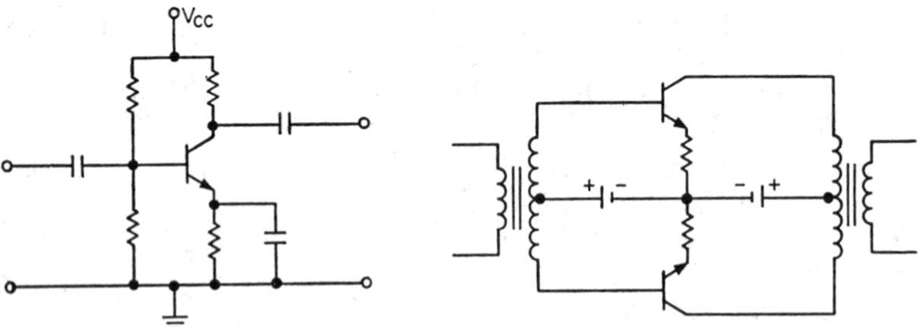

Fig. 5-82. *RC*-coupled amplifier.

Fig. 5-83. Class B push-pull amplifier.

Transistors are used in **Class B push-pull amplifiers** (Fig. 5-83) for high-power linear applications. Since the transistor is a high-current low-voltage device, low impedance loads such as speakers and servomotors can be driven without a matching transformer. Use of complementary stages as in Fig. 5-84 provides versatility of source and load connections not possible with vacuum tubes. The *n-p-n* and *p-n-p* transistors conduct during alternate half cycles, since they are forward-biased for opposite polarities of the input signal.

Direct-coupled transistor circuits are excellent for amplifying low-level signals, particularly from low-impedance sources. Offset voltage of the input can be limited to a few microvolts per degree centigrade of ambient temperature change in differential

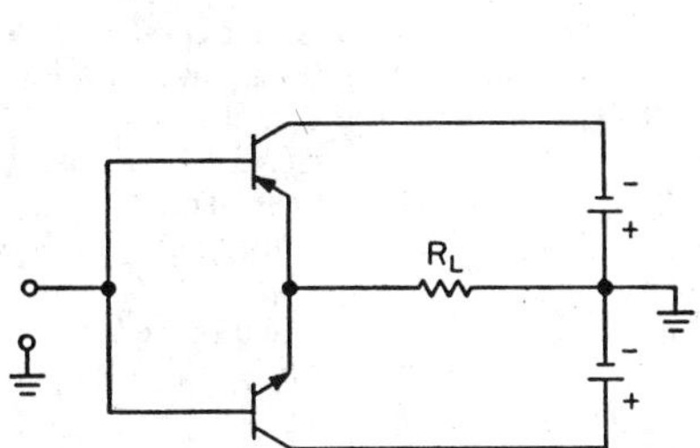

FIG. 5-84. Complementary Class B push-pull amplifier.

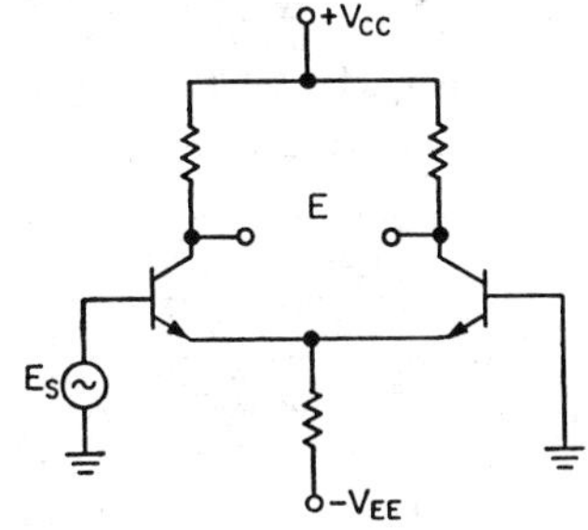

FIG. 5-85. Direct-coupled differential amplifier.

circuits like that of Fig. 5-85. Drift due to changes of V_{BE} with temperature are canceled owing to the balanced circuit; shifts due to I_{CBO} are minimized by using low-leakage silicon transistors. Typical applications are strain-gage amplifiers and operational amplifiers for analog computers.

High-frequency applications of transistors include amplifiers, oscillators, and mixers in communications systems. They provide useful power gains at frequencies as high as 5 GHz (5×10^9 Hz), with noise performance superior to that of a vacuum tube. A typical low-level amplifier used as the i-f stage of an FM receiver is shown in Fig. 5-86. High-frequency power transistors used in mobile transmitters supply as much as 20 W at 500 MHz.

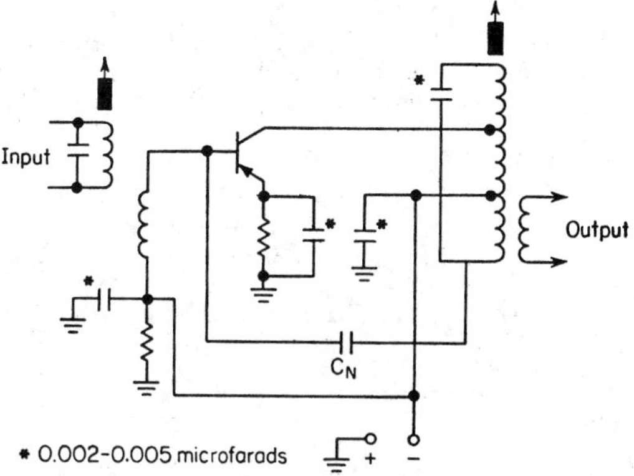

FIG. 5-86. Ten-megacycle single-tuned high-frequency amplifier for intermediate-frequency stage of FM receiver.

Because of their efficiency, reliability, and speed, transistors are used for a variety of **switching functions.** Power transistors have replaced mechanical relays in many applications. High-speed low-level switching transistors are ideal devices for gates and flip-flops (Fig. 5-87) and are the basic logic elements for many digital computers.

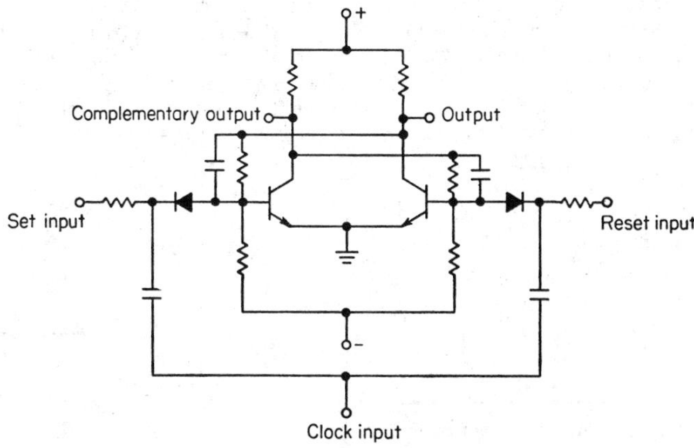

FIG. 5-87. Set-reset flip-flop.

OTHER SEMICONDUCTOR DEVICES AND ASSEMBLIES

188. Field-effect (Unipolar) Transistors. There are two general types of field-effect transistors: junction FETs and metal oxide semiconductor (MOS) transistors.

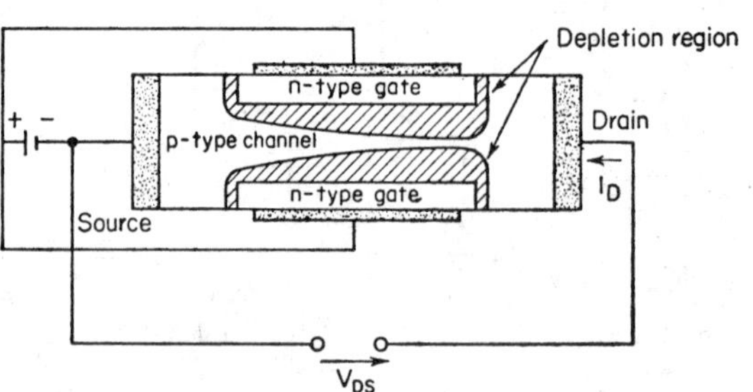

Fig. 5-88. *p*-channel junction field-effect transistor.

The cross section of a *p*-channel junction FET is shown in Fig. 5-88. Channel current is controlled by reverse-biasing the gate-to-channel junction so that the depletion region reduces the effective channel width.

The input impedance of these devices is high because of the reverse-biased diode in the input circuit. In fact, the voltampere characteristics (Fig. 5-89) are quite similar to those of a vacuum tube. Another important feature of the junction FET is the excellent low-frequency noise characteristics, which surpass those of either the vacuum tube or conventional (bipolar) transistor for high impedance sources.

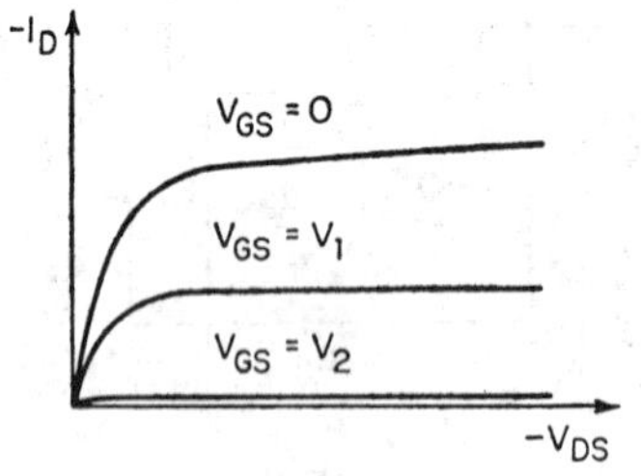

Fig. 5-89. Voltampere characteristics of *p*-channel junction FET.

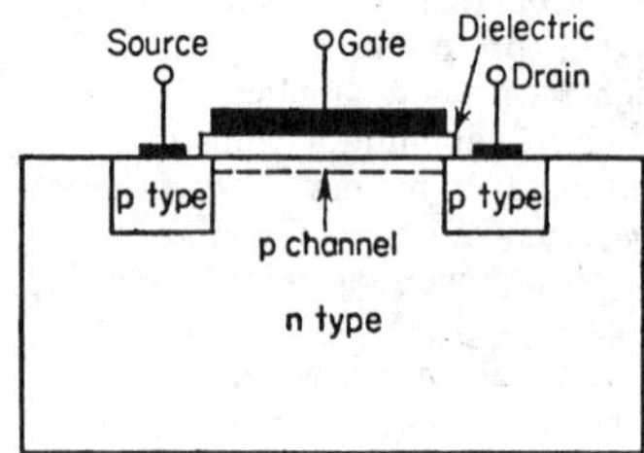

Fig. 5-90. *p*-channel MOS transistor.

The cross section of a *p*-channel MOS transistor is shown in Fig. 5-90. This device operates in the enhancement mode. For zero gate voltage there is no channel, and the drain current is very small. A negative voltage on the gate repels the electrons from the surface and produces a *p*-type conduction region under the gate.

Compared with the junction FET, the MOS transistor has a higher input impedance ($> 10^{11}\ \Omega$). Its simple structure is particularly suitable for integrated logic circuits and arrays.

189. Unijunction Transistors. A unijunction transistor is shown in Fig. 5-91. The input diode is reverse-biased at low voltages owing to *IR* drop in the bulk resistance of the *n*-type region. When V_E exceeds this drop, carriers are injected and the resistance is lowered. As a result, the *IR* drop and V_E decrease abruptly. The voltampere characteristics are shown in Fig. 5-92. The negative resistance characteristic is useful in such applications as oscillators and as trigger devices for silicon-controlled rectifiers.

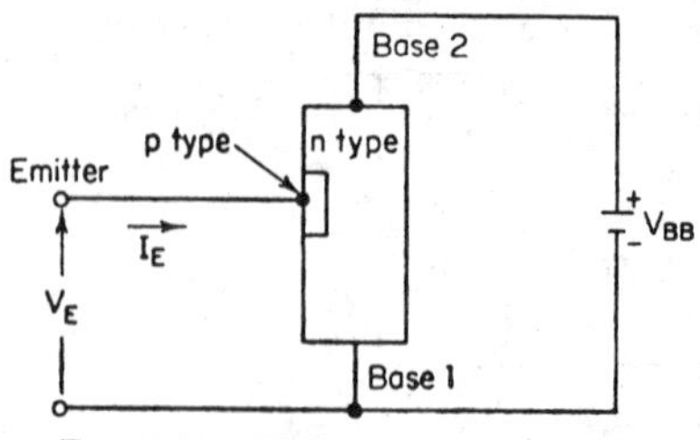

Fig. 5-91. Unijunction transistor.

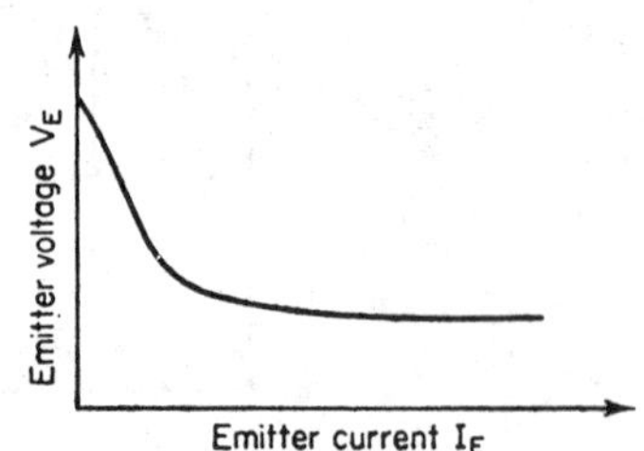

Fig. 5-92. Emitter characteristics for unijunction transistor.

190. Semiconductor-junction Light Sensors. When a semiconductor junction is exposed to light, photons generate hole-electron pairs which diffuse across the junction and constitute a photocurrent. Junction light sensors are normally operated in series with a load resistance and a battery which reverse-biases the junction, as shown in Fig. 5-93. The device acts as a source of current which increases with light intensity.

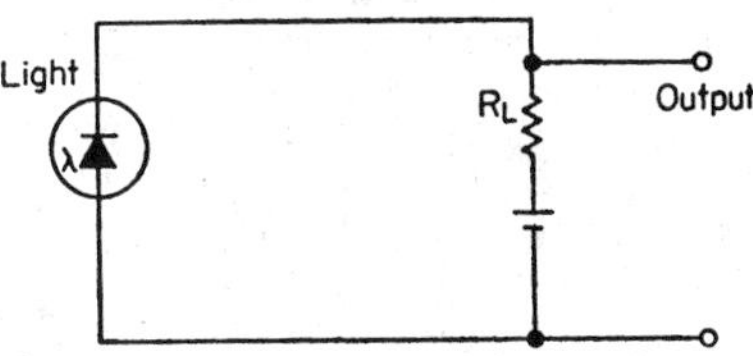

FIG. 5-93. Circuit connection for light sensor.

Silicon sensors are used for sensing light in the visible and near-infrared spectra. These may be fabricated as phototransistors, in which the collector-base junction is light-sensitive. Phototransistors are more sensitive than photodiodes because the photon-generated current is amplified by the h_{FE} of the transistor. Junction sensors have the advantage of very small size (less than 0.1 in in diameter).

191. Integrated Circuits. By using the diffusion processes discussed earlier, it is possible to fabricate transistors, diodes, resistors, and capacitors within a single wafer of silicon, as shown in Fig. 5-94. These circuit elements can be isolated from the silicon substrate by a reverse-biased junction and interconnected to form a complete circuit by evaporating leads across the insulating oxide. A typical digital network contains 16 transistors, 4 diodes, and 16 resistors (totaling 50 kΩ) on a 50- by 50-mil wafer. Circuits are commercially available in flat packs or dual in-line packages.

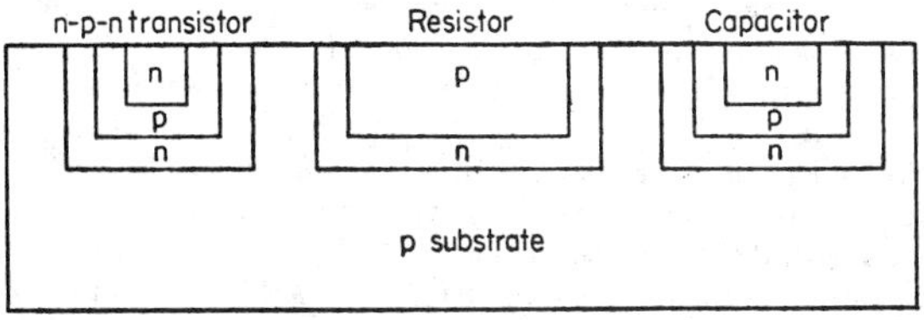

FIG. 5-94. Cross section of typical integrated circuit element.

Integrated circuits have the obvious advantages of small size and weight and provide the convenience of a preassembled circuit. Because of the minimization of external connections, they are more reliable than discrete device circuits. Performance of state-of-the-art integrated switching circuits is superior to that of discrete circuits, because interconnection parasitics and propagation delays between logic stages on the same wafer can be minimized.

Digital logic circuits are readily adapted to integrated circuit technology. Production availability has developed rapidly, since the computer market is large and a small number of circuit types can perform the required logic functions. In contrast, large-volume production of linear integrated circuits has been confined thus far to a few circuit types particularly suited to integrated circuit technology. For example, integrated d-c differential amplifiers are superior in performance to discrete circuits (drift $<5\ \mu v/°C$).

Some commercial integrated circuits and their characteristics are given in Table 5-16. These devices are typical of the functions available at present.

Table 5-16. Monolithic Integrated Circuit Parameters (Typical)

Device type	Description	Characteristics	
SN5420	Dual 4-input NAND gate, TTL logic	Propagation delay = 13 ns	Power dissipation = 10 mW/gate
SN52709	General-purpose operational amplifier	Voltage gain = 45,000	Input offset = 2 mv

192. Signal Diodes. One of the first important applications of semiconductors in electronics was the use of solid-state diodes for rectification of low-frequency alternating current. Selenium was used for this purpose as early as 1886, and the copper-oxide diode was introduced in 1927. Solid-state rectifiers for high-frequency currents initially employed a "cat's whisker" (a fine wire in contact with the semiconductor crystal). Electrical and mechanical instability of these devices led to the introduction of *p-n* junction diodes. Initially, germanium was used extensively to produce diodes because it was relatively easy to fabricate. More recently, however, advanced technology has led to the almost universal acceptance of silicon as the preferred material for *p-n* junction signal diodes.

In practice, an advanced form of the *p-n* junction, the *p-i-n* structure, is employed in most diodes and rectifiers. The *p-i-n* diode consists of three regions: (1) a heavily doped *p* region, (2) the *i* region (intrinsic region) of lightly doped *p*- or *n*-type material, and (3) a heavily doped *n* region. The *i* region, which separates the heavily doped *p* and *n* regions, accommodates most of the depletion layer when the diode is reverse-biased. The resistivity and thickness of the *i* layer essentially determine the avalanche breakdown voltage of the device. Fortunately, the effect of conductivity modulation significantly reduces the contribution that the highly resistive *i* layer would otherwise make to the forward voltage drop of the device. Conductivity modulation, arising from the effect of minority-carrier injection into the *i* region from the heavily doped *p* and *n* layers of the diode, effectively increases the carrier concentration in the *i* layer and thereby lowers its resistivity.

The diode characteristics of primary interest are forward voltage drop, reverse breakdown voltage, reverse leakage current, junction capacitance, and recovery time. The *p-i-n* configuration offers great latitude for the optimization of these design parameters. Diode structures of the *p-i-n* type can be formed by the alloy, mesa-diffused, or planar-diffused processes.

The capabilities of the general-purpose diode as a switch, demodulator, rectifier, limiter, capacitor, and nonlinear resistor suit it to many applications. One of the highest-volume uses of the solid-state diode is in computers. In this application, the computer diode has evolved from the need for diodes exhibiting low forward resistance, high counter resistance, low self-capacitance, low self-inductance, and short recovery time. The computer diode is manufactured specifically for digital applications; it is usually gold-doped and is as small as practical. A typical example of a computer-diode application is shown in the *OR* logic circuit (Fig. 5-95) and includes two inputs against a single output. However, the arrangement is not limited to two; the scheme may be repeated by adding one diode and one resistor for each new input. The circuit will produce a positive output pulse across R_3 if there is a positive pulse at any of the inputs; i.e., if input 1 or 2 is energized, the input pulse forward-biases the diode, and the resulting pulse current develops a pulse across R_3.

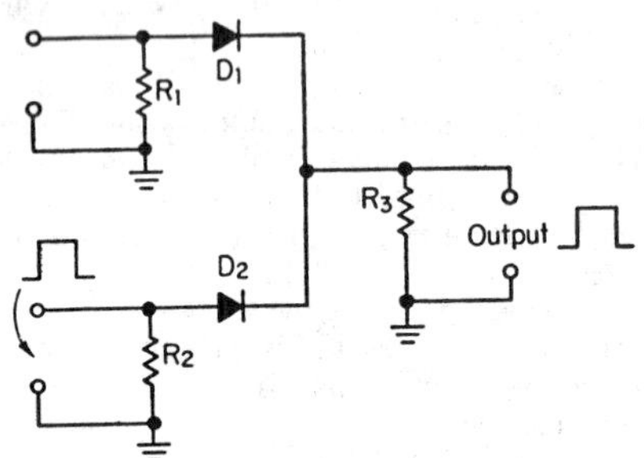

FIG. 5-95. Diode *OR* circuit.

The computer diode is frequently utilized in large matrix arrays. A matrix of this type consists of a lattice of conductors insulated from each other by the diodes. Information in the form of electrical pulses may be switched electronically at rates up to several million times per second from one to another of a large number of paths by means of the diode matrix.

The varactor diode is a *p-n* junction diode that has useful nonlinear voltage-dependent variable-capacitance characteristics. Because of these nonlinear characteristics, a voltage spectrum rich in harmonics is produced when the device is swept with a time-varying voltage signal. By employing the proper filter and impedance matching circuits, the desired output frequency is used to deliver power to the load. If the varactor circuit losses are small, the efficiency of frequency-power conversion will approach 100%. Varactor diodes are also used for sensitive microwave amplification in parametric amplifiers. These amplifiers derive their name from the ability of the *p-n* junction to adjust one of its important parameters (capacitance) in response to changes in bias. This voltage-dependent capacitance effect in the diode permits its use as an electrically controlled tuning capacitor in a radio receiver, to replace the conventional mechanical variable capacitor.

The Schottky-barrier diode (also known as the surface-barrier diode, metal-semiconductor diode, and hot-carrier diode) consists of a rectifying metal-semiconductor junction in which majority carriers carry the current flow. When the diode is forward-biased, the carriers are injected into the metal side of the junction, where they remain majority carriers at some energy greater than the Fermi energy in the metal; i.e., they are "hot" with respect to the metal lattice. This gives rise to the name "hot carrier." The diode can be switched to the off state in picoseconds, and no stored minority-carrier charge exists. The current-voltage characteristic of Schottky-barrier diodes can be described very closely by the ideal diode equation

$$I = I_s[\exp(qV/KT) - 1] \tag{5-42}$$

The reverse d-c current-voltage characteristics of the device are very similar to those of conventional *p-n* junction diodes. The reverse leakage current increases with reverse voltage gradually, until avalanche breakdown is reached.

Schottky-barrier diodes usually consist of silicon semiconductor material onto which gold, platinum, palladium, or silver is deposited by evaporation. By changing the internal barrier voltage through selection of different metals, the forward conduction, or knee, of the diode characteristic can be varied over a range from 0.2 to 0.8 V. Both hot-electron (on *n*-type silicon) and hot-hole (on *p*-type silicon) devices are produced. The hot-electron type is generally preferable because the higher electron mobility gives better high-frequency performance.

Schottky-barrier diodes utilized in detector applications have several advantages over conventional *p-n* junction diodes. Having lower noise and better conversion efficiency, they have greater overall detection sensitivity. Minority carrier storage in the Schottky-barrier diode is so low that turn-on and turn-off delays present in the *p-n* junction diodes are essentially eliminated. Accordingly, Schottky-barrier diodes can be used effectively in pulsed and high-frequency applications such as detecting, mixing, and limiting at microwave frequencies, as well as rapid clamping and gating applications of fractional nanosecond duration.

193. Voltage-regulator (Zener) Diode. The voltage-regulator diode is commonly called a "zener" diode. It is a voltage-limiting diode that has some applications in common with the older voltage-regulator gas tubes but serves a much wider field of application, because the devices cover a wide spectrum of voltages and power levels.

The electrical performance of the zener diode is based on the avalanche characteristics of the *p-n* junction. When a source of voltage is applied to a diode in the reverse direction (negative to anode), a reverse current I_r is observed. As the reverse potential is increased beyond the "zener knee," avalanche breakdown becomes well developed at the zener voltage V_z. At voltage V_z, the high counter resistance drops to a low value and the junction current increases rapidly. The current must of necessity be limited by an external resistance, since the voltage V_z developed across the zener diode remains essentially constant. Avalanche breakdown of the operating zener diode is not destructive as long as the rated power dissipation of the junction is not exceeded.

Externally, the zener diode looks much like other silicon rectifying devices, and electrically it is capable of rectifying alternating current. It serves in a variety of applications; the primary use is as a voltage reference or regulator element. Figure 5-96 shows the fundamental circuit for the zener diode employed as a shunt regulator. In the circuit, the diode element and load R_L draw current through the series resistance R_S. If E_{in} increases, the current through the zener element will increase and thus maintain an essentially fixed voltage across R_L. This ability to maintain the desired

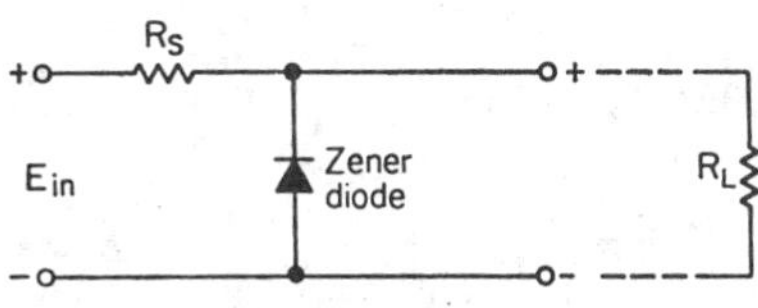

Fig. 5-96. Basic zener-diode regulator circuit.

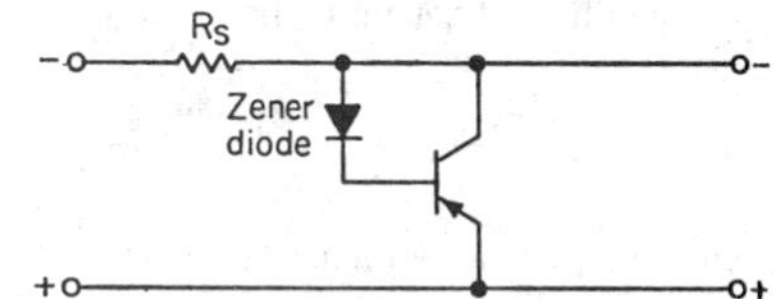

Fig. 5-97. Shunt transistor regulator.

voltage is determined by the temperature coefficient and the diode impedance of the zener device.

The zener diode may also be used to control the reference voltage of a transistor-regulated power supply. An example of this application in a shunt transistor regulator is shown in Fig. 5-97, where the zener element is used to control the operating point of the transistor. The advantages of this circuit over that shown in Fig. 5-96 are increased power-handling capability and a regulating factor improved by utilizing the current gain of the transistor.

The zener diode also finds use in audio or r-f applications where a source of stable reference voltage is required, as in bias supplies. Frequently, zener diodes are connected in series within a single package, with, for example, one junction operating in the reverse direction and possessing a positive temperature V_z coefficient; the remaining diodes are connected to operate in the forward direction and exhibit negative temperature V_z coefficient characteristics. The net result is close neutralization of V_z drift vs. temperature change; such reference units are frequently used to replace standard voltage cells. Zener diodes also find use in computer circuits designed for switching about the avalanche voltage of the diode. Design of the zener diode permits it to absorb overload surges and thereby serves the function of protecting delicate circuitry from overvoltage.

The usual voltage specifications V_z on zener diodes are 3.3 to 200 V, with ±1, 2, 5, 10, or 20% tolerances. Typical power dissipation ratings are 500 mW, 1, 10, and 50 W. The temperature coefficient range on V_z is as low as 0.001%/°C.

194. The Tunnel Diode. The tunnel diode is a semiconductor device whose primary use arises from its negative conductance ($-g_d$) *I-V* characteristic (Fig. 5-98). In a *p-n* junction, a "tunnel" effect is obtained when the depletion layer is made extremely thin ($\simeq 10^{-6}$ in). A depletion layer of this width is obtained by heavily doping the *p* and *n* regions of the device. In this situation it is possible for an electron in the conduction band on the *n* side to "tunnel" into the valence band on the *p* side. This gives rise to an additional current in the diode at a very small forward bias, which disappears when the bias is increased. It is this additional current that produces the negative resistance region in the *I-V* characteristic of the tunnel diode.

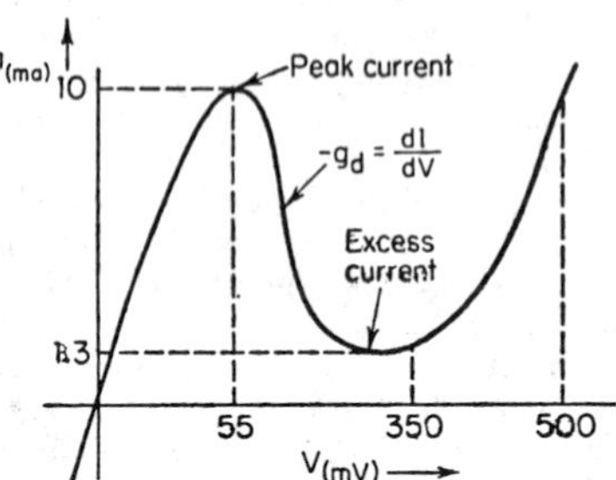

FIG. 5-98. Static characteristics of a germanium tunnel diode.

Commercial tunnel diodes are fabricated from germanium, silicon, and gallium arsenide. They can be made with peak forward current values of a few microamperes to several amperes. Peak-to-valley current ratios are approximately 10:1. A typical *I-V* characteristic of a commercial germanium tunnel diode is shown in Fig. 5-98.

Two significant frequency figures of merit can be assigned to the tunnel diode:

a. Resistive cutoff frequency, above which the tunnel diode cannot amplify,

$$f_{ro} = \frac{|g_d|}{2\pi C}\sqrt{\frac{1}{R_s|g_d|} - 1} \tag{5-43}$$

where g_d is the negative conductance of the device, C is the junction and stray capacitance, and R_s is the series resistance.

b. The self-resonant frequency,

$$f_{xo} = \frac{1}{2\pi}\sqrt{\frac{1}{L_sC} - \left(\frac{g_d}{C}\right)^2} \tag{5-44}$$

where L_s is the parasitic series inductance.

Both the above frequencies are reduced by external circuit components, and the highest possible operating frequency is circuit-dependent. In a transistor package (TO-18), the tunnel diode is limited to frequencies below 1 GHz. Microstrip or microwave packaging, owing to its inherently lower inductance, can raise the frequency limitations to ≃100 GHz.

Since a positive conductance dissipates energy, it follows that a negative conductance generates energy. Thus, through use of its negative conductance $-g_d$ characteristic, the tunnel diode finds application as an amplifier and in oscillator and switching circuitry.

195. The Silicon Power Rectifier. Silicon power rectifiers are *p-n* junction devices that have up to several hundred amperes forward current-carrying capability. The silicon rectifier is a unilateral device in that it passes current in one direction only. An ideal rectifier would have an infinite reverse resistance, infinite breakdown voltage, and zero forward resistance. The silicon rectifier approaches these ideal specifications in that the forward resistance is only a few tenths of an ohm, while the reverse resistance is in the megohm range.

The silicon rectifier exhibits several advantages over rectifiers of copper oxide, selenium, and germanium. For example, a silicon rectifier can operate at junction temperatures as high as 175°C, while the others are limited to temperatures below 100°C. Superior characteristics include high reverse breakdown voltage, low reverse leakage current, and high forward current-handling capabilities. With carefully controlled processes, silicon power rectifiers have been made with breakdown voltages up to 2,000 V. The high current-handling capability of silicon rectifiers is due to the ability of the silicon *p-n* junction to withstand a direct-current flow density of approximately 1,000 A/in² of silicon wafer.

Since silicon rectifiers are primarily used in power supplies, it is evident that thermal considerations are quite important. To avoid excessive heating of the junction, heat sinks must be capable of keeping the junction temperature below the maximum allowable of ≃175°C. The heat generated at the *p-n* junction must be transferred efficiently to the heat sink. The relative efficiency of this heat transfer is expressed in terms of the thermal impedance of the device. For stud-mounted silicon power rectifiers, the thermal impedance range is typically 0.1 to 1°C/W dissipation.

Devices that are required to carry large amounts of current pose particularly difficult problems of electrical contact to the silicon wafer. These high-current devices require the use of large pieces of silicon, which in turn demand special fabrication techniques to ensure good mechanical and thermal stability. A useful technique is that of alloying the silicon wafer directly to a material with a similar coefficient of linear expansion, such as molybdenum or tungsten. This assembly is then soldered to a copper stud and counter electrode.

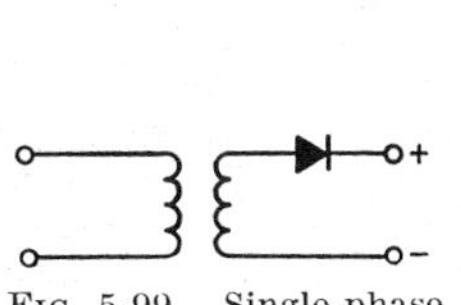

FIG. 5-99. Single-phase half-wave rectification.

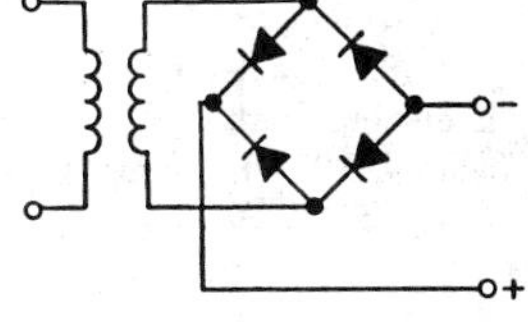

FIG. 5-100. Single-phase bridge full-wave rectification.

Silicon rectifiers are primarily used to convert a-c power to d-c power. Single-phase half-wave, bridge, and center-tap operation are shown in Figs. 5-99, 5-100, and 5-101, respectively. Three-phase half-wave, bridge, and double-Y operation are shown in Figs. 5-102, 5-103, and 5-104, respectively.

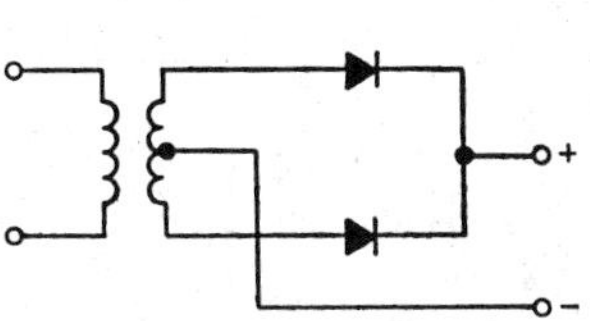

FIG. 5-101. Single-phase center-tap full-wave rectification.

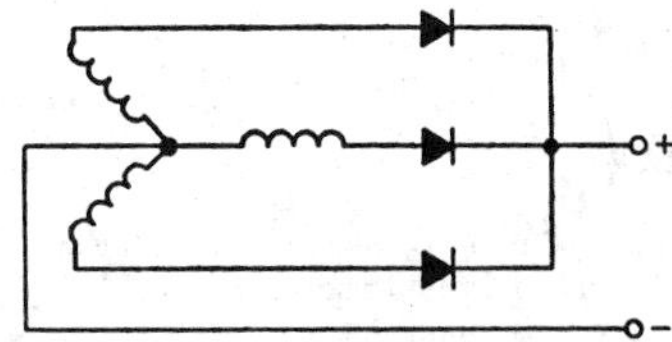

FIG. 5-102. Three-phase half-wave rectification.

In some applications, a single rectifier cannot safely provide for the high peak inverse voltage (PIV) required by the circuit. To distribute a large PIV among several rectifiers, it becomes necessary to connect the devices in series. Series connection

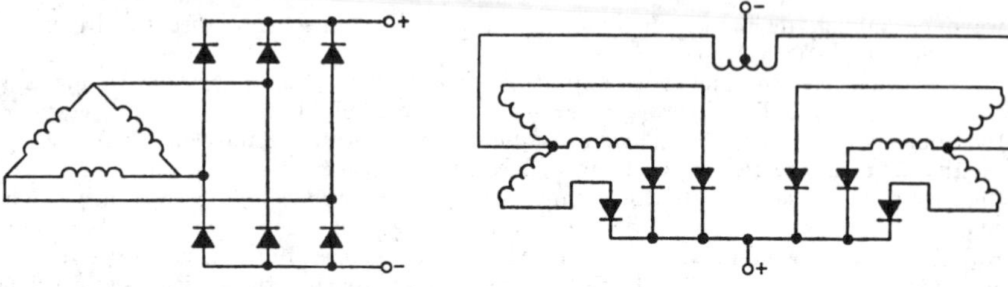

FIG. 5-103. Three-phase full-wave bridge rectification.

FIG. 5-104. Three-phase double-Y full-wave rectification.

("stacking") of rectifiers provides the designer with added flexibility. Precautions must be taken to avoid device failure due to unequal reverse voltage division among the units, caused by unequal reverse resistance among units of the same type. A solution is to shunt each rectifier with a resistance that forces the reverse resistances to be nearly equal. Shunting each rectifier with a capacitor eliminates voltage-distribution problems across the stack of rectifiers arising from transients associated with *p-n* junction storage time.

Since a finite time is taken for rectifiers to revert from forward bias to reverse bias, it is important that the switching, transient, and frequency limitations of these devices be considered in circuit design. For example, silicon rectifiers with recovery times of 10 to 20 μs will have good rectification characteristics up to 1 kHz. Beyond this frequency, the rectification ratio will decrease, and power will be lost as heat in the rectifiers. The recovery time of the rectifier depends on the minority-carrier lifetime within the device and also on both the forward current and reverse voltage applied. The recovery time in most silicon rectifiers ranges from about 0.1 to 50 μs.

Rectifiers are available with PIV ratings of 100 to 1,000 V, in 100-V intervals. Devices with forward current-carrying capabilities are available with ratings of 0.75, 1.6, 5, 10, 12, 15, 18, 20, 35, 70, 120, 160, 240, and 275 A.

SILICON-CONTROLLED RECTIFIER

196. The Silicon-controlled Rectifier. The silicon-controlled rectifier (SCR) is a three-terminal four-layer *p-n-p-n* semiconductor device with electrical characteristics (Fig. 5-105) analogous to those of a thyratron. The forward, or "on," voltage drop of the SCR is, however, only about one-tenth that of a thyratron, making it a much more efficient device. The turn-on time of an SCR is only a few microseconds, and its turn-off time is several orders of magnitude faster than the deionization time of the gas thyratron. The SCR also has all the inherent advantages of a solid-state device, including long life, large power-handling capability per unit volume, and ruggedness.

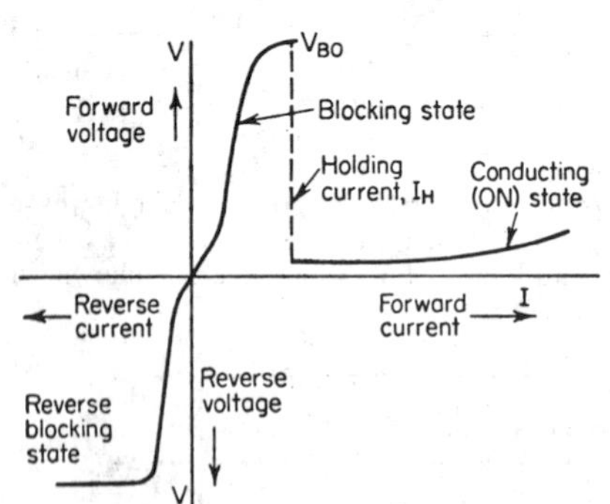

FIG. 5-105. Current-voltage characteristics of the silicon-controlled rectifier.

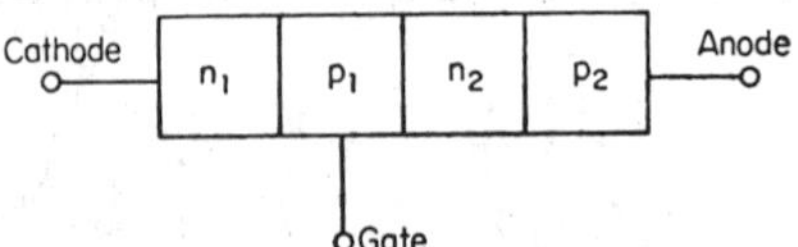

FIG. 5-106. SCR configuration.

An SCR is composed of four successive layers of alternate conductivity-type semiconductor material (Fig. 5-106). When a positive potential is applied to the cathode, the outer two junctions (n_1, p_1, and n_2, p_2) are in a reverse blocking state, as shown in Fig. 5-105. When a negative potential is applied to the cathode (forward bias condition), the middle junction p_1, n_2 is reverse-biased and the device exhibits a high-voltage low-current conduction state. As the voltage is increased, a level is finally reached at which the device switches to a low-voltage high-current conduction state (Fig. 5-105).

The switching action becomes clear if we consider that the two component "transistors" of the SCR share a common collector. In Fig. 5-107 consider the gate current I_g flowing into the base region I_{b1} of the *n-p-n* transistor. This base current causes the emitter n_1 to inject current so that $I_{c1} \sim \alpha I_{e1}$. However, I_{c1} is also the base current I_{b_2} for the *p-n-p* transistor. This base current causes the emitter p_2 of the *p-n-p* transistor to inject current so that I_{c2} is $\sim \alpha I_{e2}$. Thus I_{c2} is added to I_g to form I_{b1}, and the feedback is positive. If α_1 and α_2 are sufficient to cause the loop gain to reach unity, then base currents will be increased rapidly so that both transistors will be driven into saturation even after the base current is removed. The collector junction J_2 will then be biased in the positive direction, and therefore high current will flow from anode to cathode with very low forward voltage drop across the device. Once conduction has been established, the gate terminal no longer has any controlling effect; i.e., the unit is latched "on" in the conducting state. In order to turn the SCR off, the anode current must be reduced to a value less than the holding current, I_H.

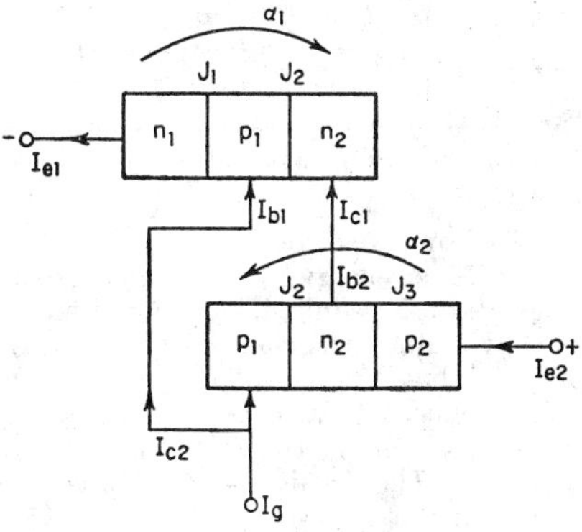

FIG. 5-107. Two-transistor analog of the SCR.

In the conventional SCR, the n_1 and p_2 outer regions are highly doped, while the inner p_1 and n_2 regions are lightly doped. At high current densities, the heavy injection from n_1 and p_2 overrides the doping level of p_1 and n_2, and thus the center region J_2 is highly conductivity-modulated. The structure and its electrical characteristics are then similar to the *p-i-n* power rectifier described previously. In a sense, the *p-n-p-n* structure possesses rectifier characteristics when in the on conducting state and is a "controlled rectifier" in that it can be turned on by the mechanism described above.

Present-day SCRs are made by the diffusion process. Frequently for SCRs of current-carrying capability less than 1 A, the planar process is used. For the larger-power SCRs, the mesa-diffusion process is employed, and the junctions are beveled or surface-contoured (Fig. 5-108). Surface contouring is performed to ensure that the electric field at the surface of the device is lower than in the bulk material of the *p-n* junctions. Avalanche is then confined to the inner bulk region of the material and does not take place in localized areas on the surface of the *p-n* junctions. Voltage blocking stability is thereby enhanced, and local hot-spot formation leading to device failure is eliminated.

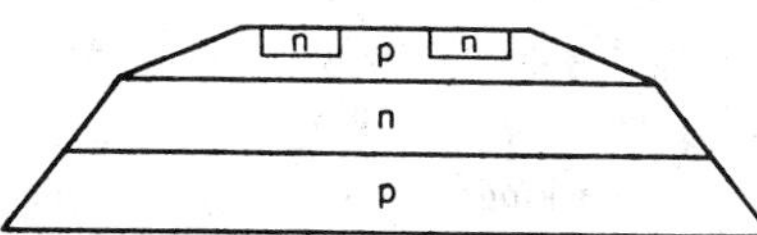

FIG. 5-108. Contoured SCR structure.

Applications for SCRs are experiencing rapid growth, and several classical applications of the device are outlined below:

A. Static contactors.

Static contactor design requires that the contactor be capable of load-current switching and continued operation in the presence of limited overload and transient conditions. The SCR is ideally suited for this application in that high-voltage capability is available and the basic latching nature of the SCR as a switch implies a low power requirement to maintain contact closure. Figure 5-109 shows a simple full-wave a-c power contactor employing SCRs.

B. Alternating-current power control system.

High-power light-dimming and heater control circuits operating from an a-c line represent two typical examples of SCR switching-mode pulse modulation systems. Figure 5-110 is an example of an SCR light control circuit.

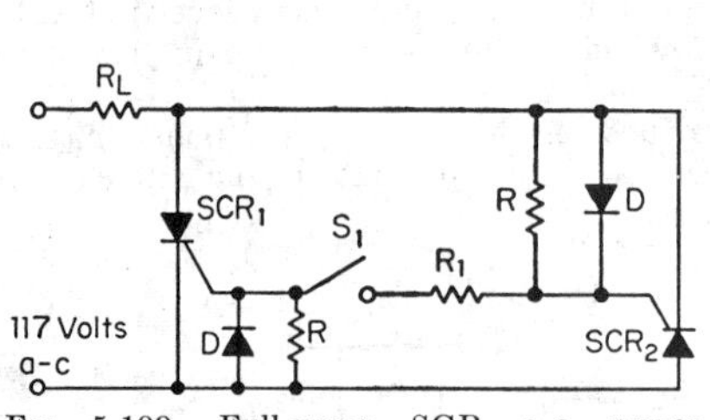

FIG. 5-109. Full-wave SCR a-c power contactor.

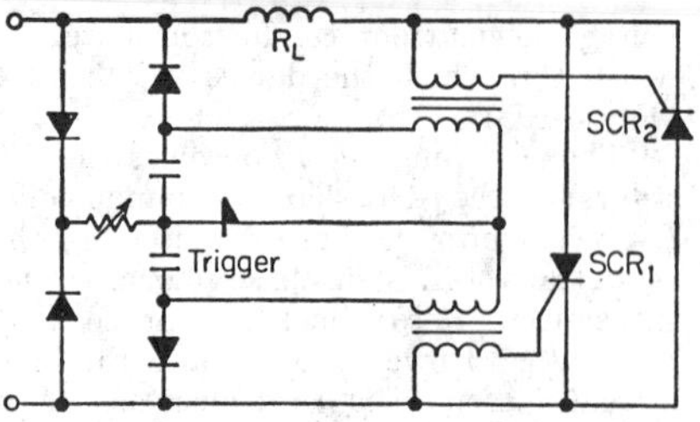

FIG. 5-110. Full-wave SCR light-dimmer circuit.

C. Power inverter systems.

Most energy storage elements store d-c energy; therefore, if power transformation is required, the energy must be converted into an a-c source. Conversion of d-c energy into a-c energy is accomplished by an inverter system. Until solid-state SCR inverters became available, most high-power conversion was done with a mechanical motor-generator system. An example of a basic SCR parallel inverter circuit is shown in Fig. 5-111.

The circuit can be operated as a d-c–d-c converter by rectifying the a-c output of the d-c–a-c inverter. An a-c–a-c frequency converter is possible by phasing the SCR gate pulse to the a-c input frequency.

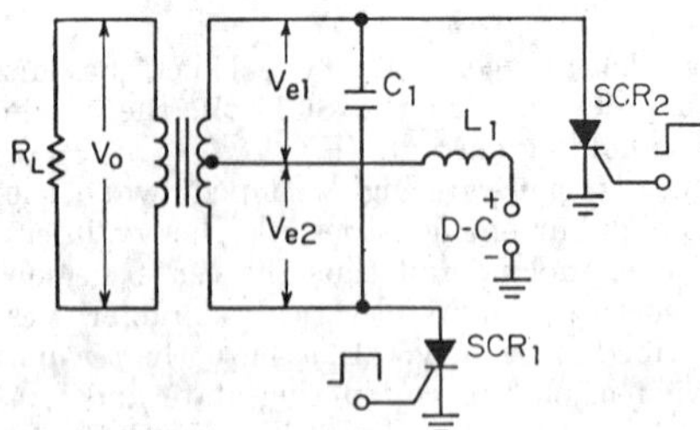

FIG. 5-111. Basic single-phase parallel inverter.

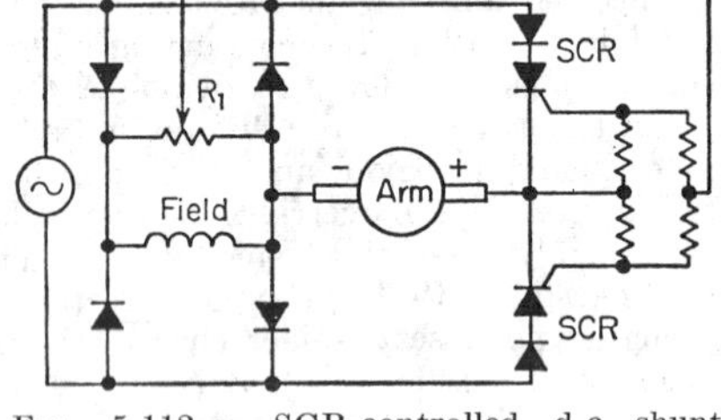

FIG. 5-112. SCR-controlled d-c shunt motor.

D. Motor control system.

An example of SCR chopped voltage control of a shunt d-c motor is shown in Fig. 5-112.

Power SCRs are available with 4.7-, 10-, 16-, 35-, 70-, 150-, and 250-A half-wave average current capabilities. Forward and reverse blocking voltages in the higher current units (>16 A) can be obtained, up to 1,500 V. Forward voltage drop across the SCR at rated average forward current is approximately 2.0 V, and holding current I_H is nominally 100 mA. The turn-off time of an SCR is much longer than turn-on time and is circuit-dependent. Most SCRs are rated through 400-Hz sinusoidal- and square-wave operation. The newer medium-current SCRs are operable to 10 kHz. Thermal-impedance values of power SCRs are approximately 0.1 to 1°C/W. Since the SCR is often used for controlled rectification, in which the device is triggered to start conduction at a particular phase angle, the units are graphically rated so that permissible average forward current is specified as a function of conduction angle. Most SCRs are rated for operation up to 125°C.

197. Bibliography. The following list is suggested to the reader for a more detailed discussion of semiconductor principles, materials, products, and applications within his individual area of interest:

SHOCKLEY, WILLIAM "Electrons and Holes in Semiconductors"; Princeton, N.J., D. Van Nostrand Company, Inc., 1950.

HANNAY, N. B. "Semiconductors"; New York, Reinhold Publishing Corporation, 1959.

Dunlap, W. Crawford "An Introduction to Semiconductors"; New York, John Wiley & Sons, Inc., 1957.

Runyan, W. R. "Silicon Semiconductor Technology"; New York, McGraw-Hill Book Company, 1965.

Semiconductor-Electronics Education Committee, "Elementary Circuit Properties of Transistors"; New York, John Wiley & Sons, Inc., 1964.

Shea, Richard F. "Transistor Circuit Engineering"; New York, John Wiley & Sons, Inc., 1959.

Walston, Joseph A., and Miller, John R. "Transistor Circuit Design"; New York, McGraw-Hill Book Company, 1963.

Shea, Richard F. "Transistor Applications"; New York, John Wiley & Sons, Inc., 1964.

Semiconductor-Electronics Education Committee, "Multistage Transistor Circuits"; New York, John Wiley & Sons, Inc., 1965.

Sevin, L. J. "Field-effect Transistors"; New York, McGraw-Hill Book Company, 1965.

Warner, Raymond M., Jr., and Fordemwalt, James N. "Integrated Circuits"; New York, McGraw-Hill Book Company, 1965.

Gentile, S. P. "Basic Theory and Application of Tunnel Diodes"; Princeton, N.J., D. Van Nostrand Company, Inc., 1962.

Gentry, F. E., et al. "Semiconductor Controlled Rectifiers; Principles and Applications of *p-n-p-n* Devices"; Englewood Cliffs, N.J., Prentice-Hall, Inc., 1964.

Penfield, P., Jr., and Rafuse, Robert P. "Varactor Applications"; Cambridge, Mass., The M.I.T. Press, 1962.

Jonscher, A. K. "Principles of Semiconductor Device Operation"; New York, John Wiley & Sons, Inc., 1960.

"Diode Circuits Workshop Lecture Series"; Boston Section, Institute of Radio Engineers, Fall, 1959.

ELECTRON TUBES

By Robert E. Moe

198. Definition. Electron tubes may be defined as active devices (as contrasted with passive electronic components such as resistors, capacitors, or inductors) which depend on the controlled flow of electrons in a vacuum or a rarefied gas, in a suitable glass, metal, or ceramic envelope.

199. Basic Construction. With a few exceptions, electron tubes require a thermionic source of electrons, such as a heated filament or oxide-coated cathode, and a collector electrode, called the anode, or "plate." The control of the electron flow may be accomplished by various means such as a wound grid, a deflection plate, or a resonant cavity, to influence the electric and/or magnetic field between cathode and anode.

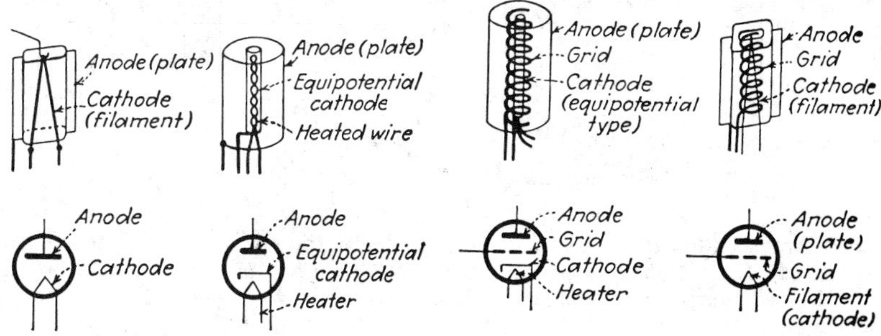

Fig. 5-113. Filamentary and cathode-type vacuum diodes; symbol shown below each structure.

Fig. 5-114. Triode structures and symbols for cathode (left) and filament type (right).

200. Tube Characteristics. The simplest tube, a diode (Fig. 5-113), is characterized by electron flow from cathode to anode only when the anode is positive. It is used as a rectifier to change alternating current to pulsating direct current, since it conducts only on the forward half cycle. The triode (Fig. 5-114) has a negative control grid inserted between cathode and anode to vary the electron flow when the anode is positive. Operating characteristics over a wide range of applied voltages are described by families of curves of anode current vs. anode voltage and, in the case of a triode, as functions of both anode and grid voltages.

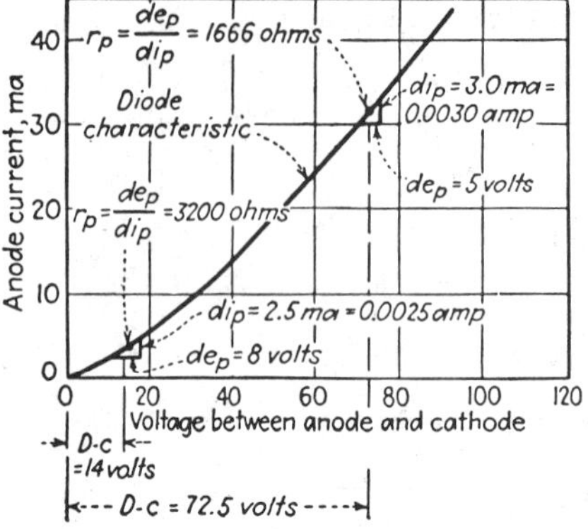

FIG. 5-115. Value of dynamic plate resistance (d_{ep}/d_{ip}), with value of direct voltage applied between cathode and anode.

201. Plate Resistance (Fig. 5-115). Since electron flow follows the Langmuir-Childs law ($I_p \propto E_p{}^{3/2}$) the "dynamic plate resistance" is defined as the slope de_p/di_p of the anode current–anode voltage curve at a given point.

202. Amplification Factor (Fig. 5-116). One of the basic characteristics of a given tube construction, almost entirely a function of element spacings and grid turns per inch, is the amplification factor, or μ, defined as the ratio of the change in plate voltage to obtain a given change in plate current, to the change in grid voltage necessary to accomplish an equal and opposite change in plate current. Thus $\mu = -de_p/de_g$.

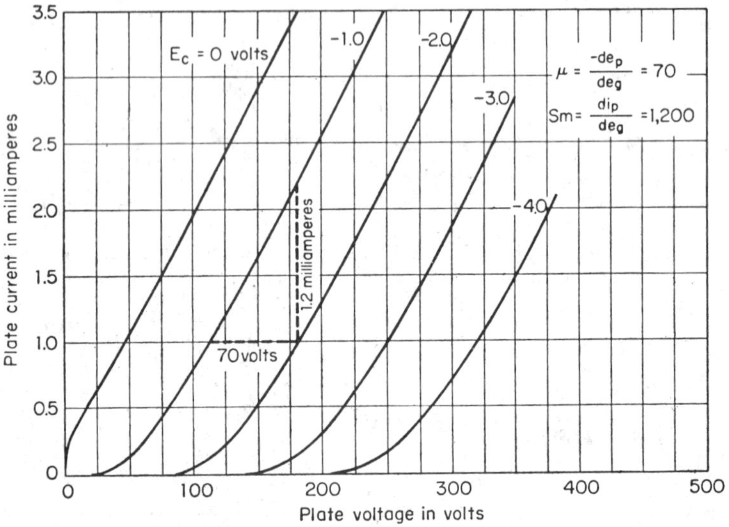

FIG. 5-116. 6T8A-triode plate characteristics, showing derivation of μ and *Sm*.

203. Transconductance (Fig. 5-116). Similarly, the ability of the grid to control the plate current is defined as *gm* (or sometimes *Sm*) $= di_p/de_g$. It is usually expressed in the United States in micromhos (microamperes per volt) and in Europe in milliamperes per volt.

204. Other Characteristics. *Interelectrode Capacitance.* At high frequencies, the grid-cathode (*Cgk*), grid-plate (*Cgp*), and plate-cathode (*Cpk*) capacitances must be considered as part of the total circuit capacitance. Owing to the Miller effect, the total input capacitance is $Cin = Cgk + Cgp(1 + G)$, where G is the gain of the amplifier.

DESCRIPTION AND RATING

The 6T8-A is a miniature triple-diode, high-mu triode intended primarily for use as a combined AM detector, FM detector, and audio-frequency voltage amplifier in radio and television receivers. For flexibility in circuit design, one of the three high-perveance diodes is provided with a separate cathode connection.

GENERAL

ELECTRICAL

Cathode—Coated Unipotential
Heater Characteristics and Ratings

Heater Voltage, AC or DC*	6.3	Volts
Heater Current‡	0.45 ± 0.03	Amperes
Heater Warm-up Time, average§	11	Seconds

Direct Interelectrode Capacitances

	With Shield¶	Without Shield	
Triode Grip to Triode Plate: (Tg to Tp)	1.7	1.7	pf
Triode Input: Tg to (h + Tk + i.s.)	1.7	1.6	pf
Triode Output: Tp to (h + Tk + i.s.)	2.4	1.2	pf
Diode—Number 1 Input: 1Dp to (h + Tk + i.s.)	3.8	3.8	pf
Diode—Number 2 Input: 2Dp to (h + 2Dk + i.s.)	3.8 ⋇	3.8	pf
Diode—Number 3 Input: 3Dp to (h + Tk + i.s.)	3.6	3.4	pf
Diode—Number 2 Cathode to All: 2Dk to (h + Tk + Tg + Tp + 1Dp + 2Dp + 3Dp + i.s.)	8.5 Δ	7.5	pf
Triode Grid to Any Diode Plate: Tg to (1Dp or 2Dp or 3Dp), maximum	0.034	0.034	pf

MECHANICAL

Operating Position—Any
Envelope—T-6 1/2, Glass
Base—E9-1, Small Button 9-Pin
Outline Drawing—EIA 6-2

Maximum Diameter	0.875	Inches
Maximum Over-all Length	2.188	Inches
Maximum Seated Height	1.938	Inches

Physical Dimensions

Terminal Connections

Pin 1 - Diode number 3 plate
Pin 2 - Diode number 2 plate
Pin 3 - Diode number 2 cathode and shield
Pin 4 - Heater
Pin 5 - Heater
Pin 6 - Diode number 1 plate
Pin 7 - Cathode and Shield
Pin 8 - Triode grid
Pin 9 - Triode plate

Basing Diagram

MAXIMUM RATINGS

DESIGN—MAXIMUM VALUES

Plate Voltage	330	Volts
Positive DC Grid Voltage	0	Volts
Plate Dissipation	1.1	Watts
Heater-Cathode Voltage:		
Heater Positive with Respect to Cathode	100	Volts
Heater Negative with Respect to Cathode	100	Volts
Diode Current for Continuous Operation, Each Diode	5.5	Milliamperes

CHARACTERISTICS AND TYPICAL OPERATION

CLASS A_1 AMPLIFIER

Plate Voltage	100	250	Volts
Grid Voltage	−1	−3	Volts
Amplification Factor	70	70	
Plate Resistance, approximate	54,000	58,000	Ohms
Transconductance	1,300	1,200	Micromhos
Plate Current	0.8	1.0	Milliamperes
Average Diode Current, Each Diode With 5.0 Volts DC Applied	20	20	Milliamperes

NOTES

* Heater voltage for a bogey tube at If = 0.45 amperes.

‡ The equipment designer should design the equipment so that heater current is centered at the specified bogey value, with heater supply variations restricted to maintain heater current within the specified tolerance.

§ The time required for the voltage across the heater to reach 80 percent of the bogey value after applying 4 times the bogey heater voltage to a circuit consisting of the tube heater in series with a resistance equal to 3 times the bogey heater voltage divided by the bogey heater current.

¶ With external shield (EIA 315) connected to pin 7, except as noted.

⋇ With external shield (EIA 315) connected to pin 3.

Δ With external shield (EIA 315) connected to pins 4 and 5.

Fig. 5-117. Typical published data sheet—type 6T8A, triple-diode triode.

Voltage Ratings. Element spacings and insulating supports limit the applied voltages, and these are usually expressed as "design maximum"[1] ratings, d-c, a-c, or pulse at a defined duty cycle.

Power Dissipation. The elements carrying electron current must dissipate a certain amount of power, depending on the product of voltage and current, and are rated in accordance with their heat-conducting and -radiating ability.

Mechanical Dimensions. To a large extent envelopes, bases, and socket connections have been standardized for interchangeability between manufacturers. The standard dimensions are available in published form from the Electronic Industries Association (for example, RS-209A). Standard bulb outlines have been assigned numbers like EIA 5-4, where the first digit refers to bulb size (T-5 ½, T-9, T-12, etc.), denoting the diameter in eighths of an inch. Bases are numbered according to the number of socket pins (for example, E9-1, small button 9 pin). Socket connection diagrams have also been standardized for a large range of types and sizes and are denoted by the number of pins plus one or more code letters (for example, EIA 9DX). These code designations are referenced on the manufacturer's published data sheet under the particular outline and socket diagram.

205. Typical Tube Data Sheet. A typical published data sheet (type 6T8A) is shown in Fig. 5-117. Some of the accompanying curves were given in Fig. 5-116. In the EIA type-numbering system, the first digits denote the heater voltage, followed by a code letter or letters and the number of active elements. The suffix denotes an interchangeable later version.

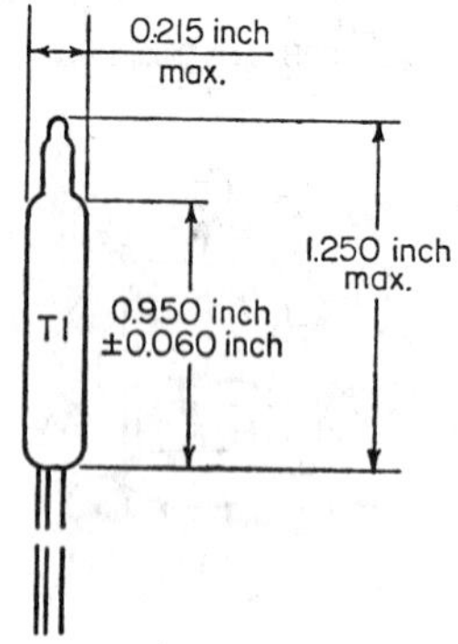

Fig. 5-118. T-1 subminiature, type 5647.

206. Range of Sizes. Electron tubes range in size from the T-1 subminiature (Fig. 5-118) to an 8-ft superpower klystron (Fig. 5-119), with powers ranging from milliwatts to megawatts. Many of the large transmitting types require water-cooled jackets or forced-air-cooled radiators.

207. Low-power (Receiving) Tubes. *Diodes.* Filamentary diodes have been used largely as power-supply rectifiers (e.g., type 5U4G), but they are gradually being replaced by solid-state silicon rectifiers because of the latter's smaller size and lower voltage drop. Cathode-type diodes like the 6AL5 are used as detectors in low-frequency low-power applications because of their high counter resistance, and large diodes like the 6AX3 are used in television in the horizontal-deflection "damper" circuit

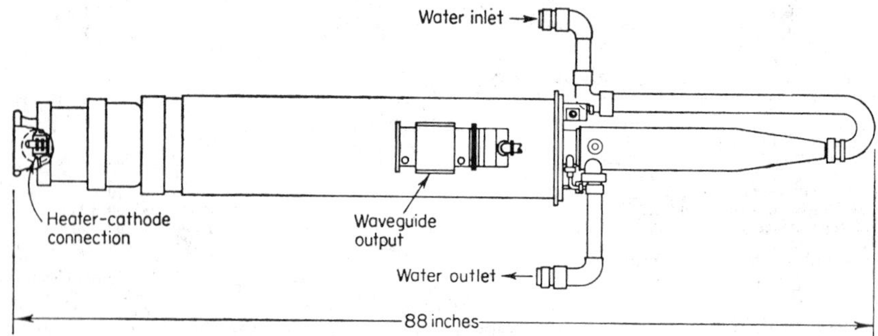

Fig. 5-119. Large high-power klystron. (*Courtesy of General Electric Company.*)

[1] "Design maximum ratings are limiting values of operating and environmental conditions applicable to a bogey electron tube of a specified type as defined by its published data, and should not be exceeded under the worst probable conditions."—E.I.A. Standard.

because of their high counter-voltage ratings. In the latter case, special high-voltage insulation is incorporated between heater and cathode, as the cathode pulse voltage may reach 5 kV.

Triodes and Twin Triodes. Single triodes are being replaced by twin triodes like the 12AX7 (a high-μ audio amplifier) or the 12AT7 (a high-gain medium-μ general-purpose type) and more recently by other combinations of diodes, triodes, and pentodes, as indicated in *Multifunction Tubes,* below.

A family of metal-ceramic triodes with cylindrical envelopes and plane electrodes has extended the frequency capability of triodes used as amplifiers to over 1 Gc/s and as oscillators to over 10 Gc/s (see Fig. 5-120).

Tetrodes. The addition of a positive second grid, wound outside the control grid of a triode, serves to accelerate the electron flow to the plate and to shield the cathode-grid region from the influence of changes in plate voltage, resulting in much higher μ and plate resistance. Secondary electrons ejected from the plate by the primary electron stream flow back to the screen at low plate voltages, resulting in an unstable negative-resistance region. Because of this, few low-power tetrodes are in common use. The introduction of the screen, however, reduces greatly the grid-plate capacitance, permitting much higher gain without oscillation (see Fig. 5-121).

Pentodes. The addition of a third grid, at or near cathode potential, prevents the secondary electrons from reaching the screen and returns them to the plate. This results in a rounded "knee" at the lower voltage end of the plate characteristic. Low-power pentodes like the "frame-frid" 6EJ7 are used for high-gain r-f or i-f amplifiers (see Fig. 5-122).

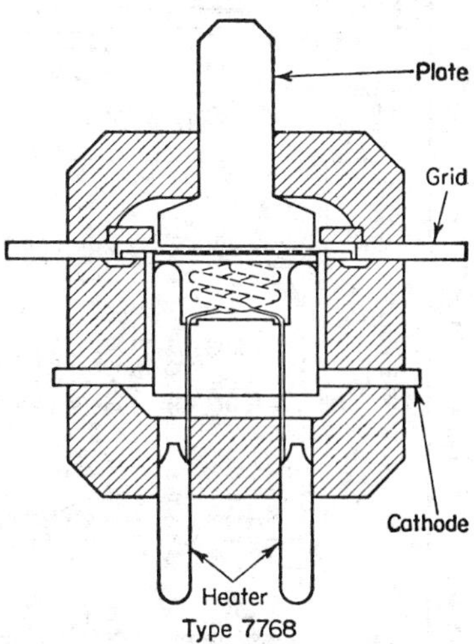

FIG. 5-120. Typical ceramic-metal triodes, types 7077 and 7768. (*Courtesy of General Electric Company.*)

Beam Tetrodes. If, instead of a third grid, a grounded set of beam-forming plates is interposed between screen grid and anode, a zero-

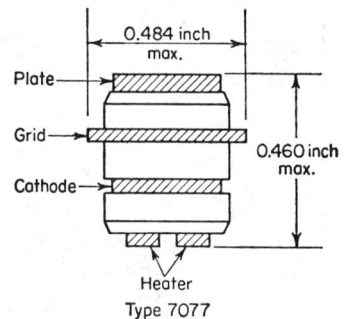

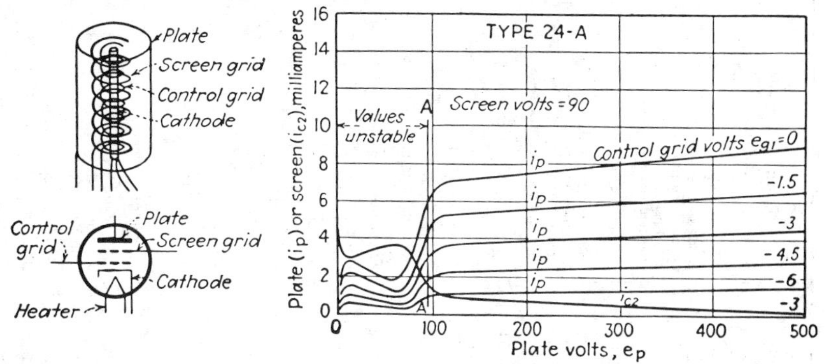

FIG. 5-121. Structure, symbol, and typical static operating characteristics of a tetrode high-vacuum tube.

Table 5-17. Typical Receiving Tubes

Type	Function description	Base pins†	Bulb	Heater V/A	Max. plate/ screen, V	Plate current, mA	Grid voltage, or Rk	Load res., Ω, or plate res., MΩ	Transcon- ductance, μmhos	Output, W	Max. plate diss., W
1B3GT	Diode H-V rect.	8 + TC	T-9	1.25/0.2	22 kV	0.5					
1X2B	Diode H-V rect.	7 + TC	T-5 1/2	1.25/0.2	16 kV	0.5					
3AT2	Diode H-V rect.	12 + TC	T-9	3.15/0.22	30 kV	1.7					
6AG7	Video pent.	8	MT-8	6.3/0.65	300/180	30	−3	0.13	11,000	3.0	9.0
6AF11	Duotriode-pentode	12	T-9	6.3/1.05	330/180	P24	100	0.068	11,000	....	P-5
						T7, 9	220	0.010	5,500–4,400	....	T-1, 2
6AR11	I-F twin pentode	12	T-9	6.3/0.8	330/180	11	56	0.2	10,500	....	3.1
						11	56	0.2	10,500	....	3.1
6AX3	TV horiz. damper diode	12	T-9	6.3/1.2	5 kV	165	......	....	......	....	5.3
6BQ5(EL84)	Audio pentode	9	T-6 1/2	6.3/0.76	300/300	48	135	5,200	11,300	6.0	12
6CW4	Nuvistor triode	5 + S	MT-4	6.3/0.135	110	7	130	0.0066	9,800	....	1.5
6DQ6B	TV hor. sweep-beam tetrode	8 + TC	T-12	6.3/1.2	6.5 kV	65	−22.5	0.018	7,300	....	18
6EH7(EF183*)	Remote C.O. frame grid pentode	9	T-6 1/2	6.3/0.3	250/250	12	−2	0.500	12,500	....	2.5
6EJ7(EF184*)	Sharp C.O. frame grid pentode	9	T-6 1/2	6.3/0.3	250/250	10	−2.5	0.350	15,000	....	2.5
6HF5	TV hor. sweep-beam tetrode	12 + TC	T-12	6.3/2.25	7.5 kV	125	−25	0.0056	11,300	....	28
6JE6A	TV hor. sweep-beam pentode	Novar + TC	T-12	6.3/2.5	7.0 kV	115	−25	0.0055	10,500	....	24
6L6GC	Audio-beam tetrode	8	T-12	6.3/0.9	500/300	54	−18	5,600	6,000	10.8	30
6SN7GTB	Twin triode	8	T-9	6.3/0.6	450	9	−8	0.0077	2,600	....	5.0 either
											7.5 total
6T10	Audio det. and out. pentode-tetrode	12	T-9	6.3/0.95	275/275	35	−8	0.10	6,500	....	T-10
						1.3	560	0.15	1,000	....	P-1.7
6U8A	Triode pentode	9	T-6 1/2	6.3/0.45	330/180	9.5	−1	0.200	5,000	....	P-3.0
						13.5	−1	0.005	7,500	....	T-2.5
12AT7	Twin triode	9	T-6 1/2	6.3/0.3	300	10	200	0.0011	5,500	....	2.2
				12.6/0.15	300	10	200	0.0011	5,500	....	2.2
12AU7	Twin triode	9	T-6 1/2	6.3/0.3	330	10.5	−8.5	0.0077	2,200	....	2.75
				12.6/0.15	330	10.5	−8.5	0.0077	2,200	....	2.75
12AX7A	Twin triode	9	T-6 1/2	6.3/0.3	330	1.2	−2	0.0625	1,600	....	1.2
(ECC83*)				12.6/0.15	330	1.2	−2	0.0625	1,600	....	1.2
12AV6	Duo-diode triode	7	T-5 1/2	12.6/0.15	330	1.2	−2	0.0625	1,600	....	1.2
								0.0625	1,600	....	0.55
12BA6	R-f pentode	7	T-5 1/2	12.6/0.15	330	11	68	1.0 M	4,400	....	3.4
12BE6	Mixer heptode	7	T-5 1/2	12.6/0.15	330/110	2.9	−1.5	1.0 M	475 Sc	....	1.1
12HG7	Video pentode	9	T-9	12.6/0.26	400/165	31	47	0.06	32,000	....	10
				6.3/0.56							
35W4	Diode rect.	7	T-6 1/2	35/0.15	360	110					
38HE7	TV horizontal diode-tetrode	12	T-12	38/0.45	5 kV	60	−22	0.0062	8,800	....	10-T
					4.2 kV	200	......	....	......	....	4-D
50C5	Audio-tetrode	7	T-5 1/2	50/0.15	150	49	−8	2,500	7,500	2.3	7
7077	Grounded grid cer. tri.	2 + 3	0.3 in	6.3/0.24	250	6.5	82	0.010	10,000	(450 Mc/s)	1.1
7587	Nuvistor tetrode	5 + S	MT-4	6.3/0.15	125/50	10	68	0.2	10,600	....	2.2
7768	Cer. triode	2 + 3	0.55 in	6.3/0.4	330	22	22	0.0045	50,000	....	5.5
7868	Audio pentode	Novar	T-9	6.3/0.8	300/300	75	−10	3,000	10,200	11	19
7984	R-f P.O.	12	T-12	13.5/0.58	315/165	150	−74	Radio frequency	13,500	26.5	20

* European equivalents. † TC = top cap; S = metal shield.

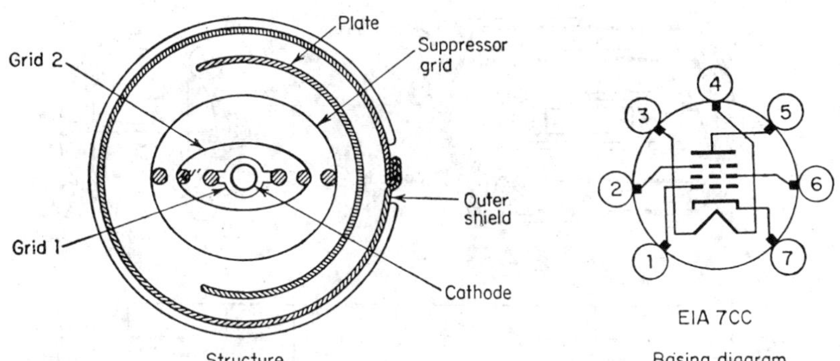

FIG. 5-122. 12BA6-pentode electrode structure and diagram.

potential region is produced, but the "knee" is sharper, allowing a greater swing in plate voltage and more power output. Examples are the 6DQ6 and 6HF5, used in the horizontal output stage of television receivers at powers of 15 to 30 W (see Figs. 5-123 and 5-124).

Hexodes and Heptodes. The addition of more grids allows additional control of the electron stream, as in oscillator mixers. In the pentagrid converter (Fig. 5-125), the first and second grids are used in an oscillator circuit, and the third, or signal, grid superimposes the signal modulation on the oscillating electron stream. The plate current thus carries sum-and-difference components, and the difference component is used in superheterodyne receivers as the intermediate frequency, followed by fixed-tuned stages at that frequency.

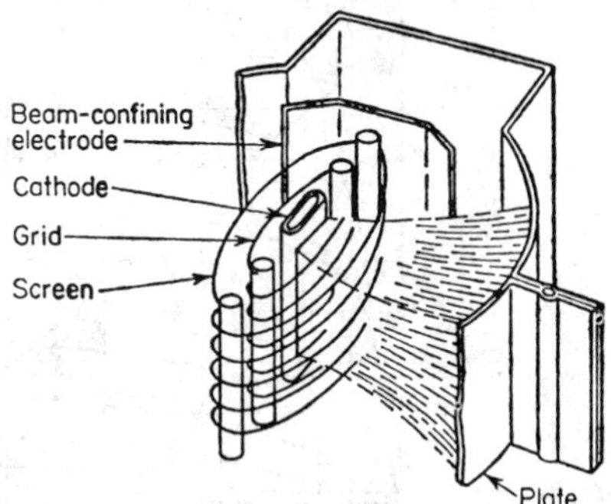

FIG. 5-123. Internal structure of a typical beam-power tetrode. (*From "Electronic Designers' Handbook" by Landee, Davis, and Albrecht, New York, McGraw-Hill Book Company,* 1957.)

Multifunction Tubes. In the interests of increased efficiency and size reduction, diodes, triodes, pentodes, and multielement sections can be combined for specific applications. Duodiode triodes, double triodes, and triode pentodes are available in 9-pin miniature (T-6½) bulbs and duotriode pentodes, triple triodes, and other combinations in T-9 novar and T-9 and T-12 bulbs with a 12-pin base, called compactrons. An example of the latter is the 38HE7, a combination horizontal-output beam tetrode and damper diode, used in small portable television receivers.

Industrial and Military Types. Receiving tubes designed for difficult environmental conditions (e.g., for airline communication and navigation), long-life computer appli-

Table 5-18. Typical High-voltage Diodes

Type	Description	Filament V/A	Plate voltage peak inverse, kV	Output current, A	
				Peak	Avg.
3B24W	T-12 4-pin + top cap	5/3	20	0.3	0.06
5575	6 × 25 in rad.-cooled	20/24	150	1.0	0.2
5973	6 × 10 in rad.-cooled	16/19	75	5.0	1.25
KR9 (x-ray)	3 × 7 in air-cooled	6/7.5	150	0.6	0.05
KR10 (x-ray)	3 × 7 in oil-cooled	6/7.5	150	0.6	0.05

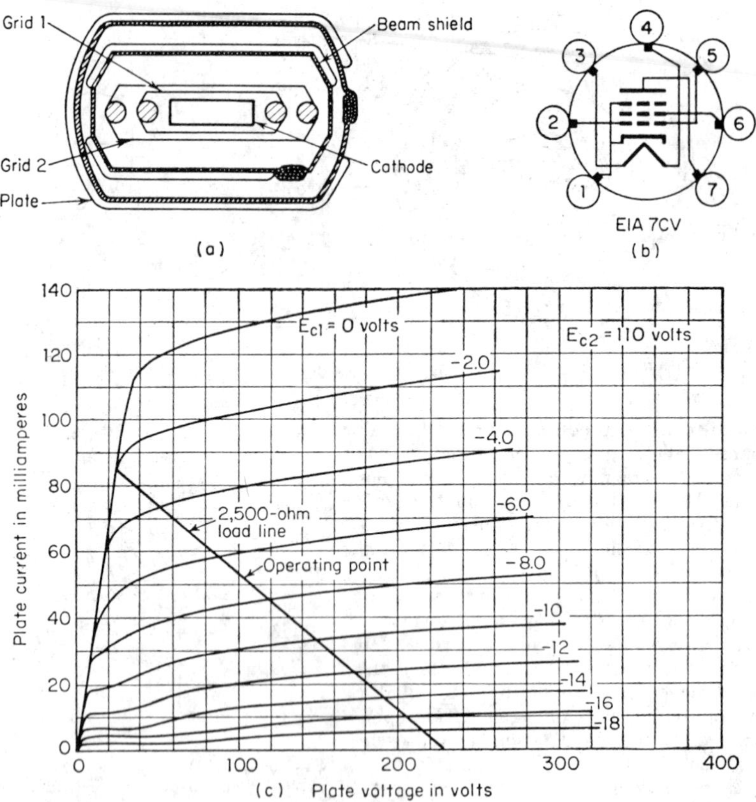

Fig. 5-124. 50C5 beam tetrode electrode structure (*a*), diagram (*b*), and plate characteristics (*c*).

cations, or military specifications are available. These meet more elaborate testing requirements and display a narrower spread of characteristics. They carry a four-digit type number (such as the 5814-A, a rugged, general-purpose twin triode), with specified survival ratings under military specification conditions.

Table 5-17 shows some typical tubes used in radio, television, and low-power applications.

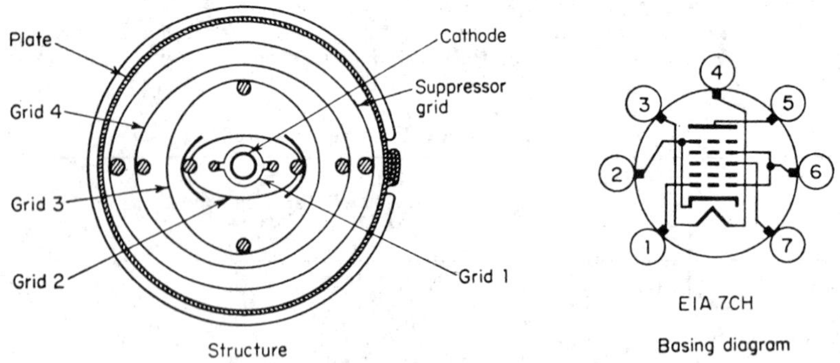

Fig. 5-125. 12BE6 heptode structure and diagram.

More complex and specialized types are required in color-television receivers, such as high-voltage regulator tubes and color subcarrier demodulators. Basically, however, these types perform the function of control of the output current by one or more grids, with frequency components determined by the product of the various control signals.[1]

The boundary between high-power receiving tubes and low-power transmitting tubes is indistinct and overlapping. At low frequencies (50 Mc/s and below), a plate dissipation of 30 W is about maximum for a receiving type, owing to the use of soft glass envelopes with a bulb temperature limitation of around 225°C maximum.

208. High-power Tubes. *Diodes.* High-voltage vacuum rectifiers, used in radio and radar transmitters and electrostatic precipitators, have a high peak inverse voltage rating and carry currents up to several amperes at voltages up to 75 kV and at less than 1A up to 150 kV (see Table 5-18).

Triodes. The original hard-glass transmitting tubes of 50 W and up are giving way to glass-metal and ceramic-metal structures with forced-air-cooled, water-cooled, or vapor-cooled anodes. Some of these use a pure tungsten filament, which offers resistance to residual gas-ion bombardment, but most employ a thoriated-tungsten filament. The latter has higher efficiency at high current densities and a slightly lower operating temperature. Some of the newer microwave types use a sintered tungsten and nickel powder, or "matrix," cathode, but these tend to evaporate a metallic film on the other parts of the tube (see Fig. 5-126 and Table 5-19).

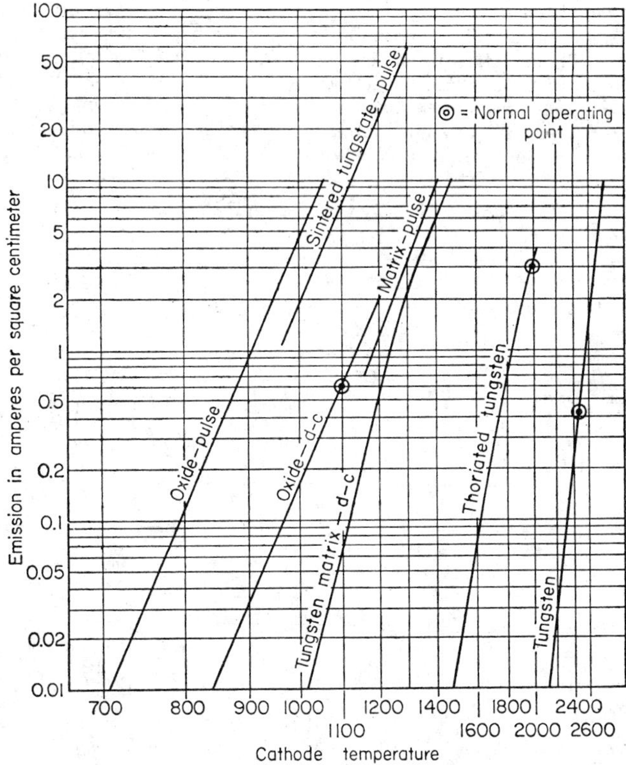

Fig. 5-126. Cathode emission current vs. temperature for typical base materials and emission surfaces.

[1] Fink, D. G. "Television Engineering Handbook"; New York, McGraw-Hill Book Company, 1957, Sec. 13.702.

Table 5-19. Typical High-power Transmitting Tubes

Type	Description	Cathode V	Cathode A	Plate voltage, kV	Plate current, A	Plate diss.	Max. power Class C	Max. freq. full ratings, Mc/s
2C39B	Ceramic disk seal triode	6.3	1.0	1	0.1	100 W	17 W	2,500
7D21	Forced-air-cooled tetrode	6.3	30	4	1	1,200 W	1,600 W	110
8D21	Water-cooled tetrode	3.2	125	6	1.9	6 kW	5.3 kW	300
813	Beam pentode	10	5	2.5	0.29	125 W	490 W	30
862A	Water-cooled triode	33	207	20	5	100 kW	45 kW	1.6
880	Water-cooled triode	12.6	320	10.5	7	20 kW	46 kW	25
6251	Water-cooled ceramic tet.	5.5	19	7	7.5	25 kW	25 kW	220
8513	Air-cooled ceramic tet.	6.7	14	7	1	4 kW	3.5 kW	900
8549	Water-cooled beam triode	7.6	1,900	25	115	500 kW	2.5 MW	30

Tetrodes and Pentodes. These types operate at higher efficiency than triodes and are typically used in v-h-f television transmitters in Class B amplifier service up to 25 kW output.[1] Most of them are of ceramic-metal construction with forced-air-cooled anode radiators. They display low output capacitance.

209. Gas Tubes. *Diodes.* Gas-filled rectifier tubes, which have lower voltage drop and higher current capacity than vacuum diodes, are gradually being replaced by solid-state rectifiers. The most popular types are filled with mercury vapor. All types require a filament heating time before anode voltage is applied to prevent destructive bombardment of the cathode by gas ions. The ionized gases emit a characteristic glow, e.g., blue for mercury, pink for argon, and violet for krypton.

Triodes and Tetrodes. Gas-filled triodes, called "thyratrons," are used for control of high currents at low frequencies. When the grid voltage is sufficiently positive

Table 5-20. Typical Gas and Pool Tubes

Diodes

Type	Description	Heater V	Heater A	Peak inverse plate volts	Peak current, A	Avg. current, A	Ignitor firing V	Ignitor firing A
857B	Large glass	5.0	30	22,000	40	10		
866A	4-pin + top cap	2.5	5.0	10,000	1.0	0.25		
872A	Large 4-pin + top cap	5.0	7.5	10,000	5.0	1.25		
5550	2 × 9-in metal	Ignitron		600 rms	1,400	22.4	200	30
7042	8 × 27-in metal double grid	Ignitron		4,000	2,000	275	300	30

Thyratrons

Type	Description	Heater V	Heater A	Peak inverse plate volts	Peak current, A	Avg. current, A	Critical grid and anode volts	Deionization time, μs
2D21	7-pin miniature	6.3	0.6	1,300	0.5	0.1	−4.2 at 450	1,000
3C23	4-pin + top cap	2.5	7	1,250	6	1.5	−5.5 at 1,000	1,000
FG27A	4-pin + top cap	5.0	4.5	1,000	10	2.5	−2.2 at 100	1,000
672A	Large 4-pin	5	5	2,500	40	3.2	−10 at 1,000	1,000
FG172	2 × 10-in metal	5	10	2,000	40	6.4	−9 at 1,000	1,000
5830	5 × 17-in glass	5	20	10,000	75	12.5	−7 at 10,000	1,000
7890	6 × 13-in ceramic (hydrogen)	6.3	33	40,000	2,400	4	−650	25

[1] Transmitting tubes are rated by the "absolute-maximum" system, ratings which represent the limiting value above which the operating life may be seriously affected. Line-voltage variations, supply regulation, and component tolerances must be considered by the designer, and ratings should not be exceeded on any tube under any combination of conditions.

to permit ionization, the grid loses control until the anode voltage goes below the ionization potential. The gas tetrode utilizes a shield around the grid structure, permitting a smaller grid to be used and allowing some control of firing characteristics by adjusting the shield voltage.

Pool triodes, or "ignitrons," use a pool of mercury as the cathode for very high current applications, and require an "ignitor" to start the luminous, emitting spot on the mercury surface. These devices are gradually being replaced by single or multiple solid-state units called silicon-controlled rectifiers, or SCRs (see Table 5-20).

210. Cold-cathode Tubes. *Diodes.* Cold-cathode gas-filled diodes are used primarily as voltage reference devices or voltage regulators, because of the extensive current range over which the voltage drop is almost constant, once the gas has been ionized ("breakdown voltage"). Simple neon lamps like the NE-2 are often used for this purpose, and others cover voltages of 75 to 150 V. For higher voltages, low-current corona regulators are available. A series resistor must be used to prevent destructive overload currents.

Triodes. A so-called "starter" electrode can be used in a cold-cathode triode to initiate the discharge and then transfer it to the main electrode. These are called "trigger" tubes and are typically used in standby applications, since they draw no power until energized.

Another triode type is the cold-cathode counter and indicator family, in which the control electrode initiates the discharge and another, similar electrode transfers it to the next gap. In the case of the decade counter for example, a series of pulses cycles the discharge from the first anode to the tenth and then back to the first. Because of the visible glow of the discharge, the count can be indicated by its position, as on a dial.

Fig. 5-127. Typical numerical-indicator glow tube.

Another type of cold-cathode gas-filled tube is the numeric indicator, in which a set of superimposed electrodes is shaped in the form of numbers or letters, which can be lit in sequence by energizing the proper socket terminal (see Table 5-21 and Fig. 5-127).

TR, or Duplexer, Tubes. Duplexing devices are used primarily in pulsed radar systems to enable a single antenna to perform both the transmitting and the receiving function.

TR (for "transmit-receive") tubes are gas-filled waveguide cavities, which are used to protect the detector crystal in a radar system. The main pulse of the transmitter breaks down the gas, which is preionized by a keep-alive electrode, and the ionization forms an effective short circuit in the waveguide branch leading from the antenna to the receiver. Extremely high attenuation is required, as the main pulse power may be several megawatts peak, and the crystal can withstand less than 0.2 W without being damaged.

Table 5-21. Typical Cold-cathode Tubes

Type	Description	Base	Bulb	Current, mA	Voltage	Maximum starting voltage	Trigger volts
5641A	Reference	7-pin	T-5 1/2	1.5–3.5	83–87 ± 3	107	
OA2	Regulator	7-pin	T-5 1/2	5–30	151 ± 6	185	
OB2	Regulator	7-pin	T-5 1/2	5–30	108 ± 4	133	
OA3	Regulator	8-pin	ST-12	5–40	75 ± 6.5	105	
1C21	Trigger	8-pin	T-9	25	125–145	. . .	66–80
5823	Trigger	7-pin	T-5 1/2	25	117 rms	. . .	105
6910	Decade counter	13-pin	T-11	0.8	250	400	Guide bias +45 Pulse −85
6844A	Numerical indicator	13-pin	T-9	2	140	170	

The ATR tube is a gas-filled device in the transmitter waveguide which passes the transmitter pulse but prevents received signal power from the antenna from being dissipated in the transmitter branch.

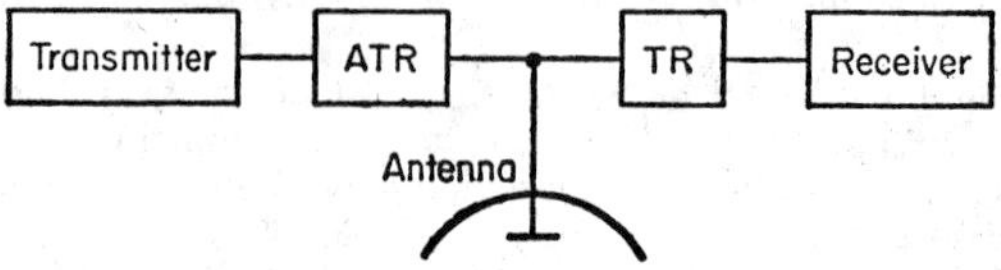

Fig. 5-128. Radar-system diagram showing TR and ATR tubes.

Duplexing circuits can be divided into two classes: branched and balanced circuits. The usual branched circuit consists of two T junctions in the transmission line, with the ATR tube mounted on the T nearest the transmitter and the TR tube (and pre-TR tube, if used) mounted on a T placed an odd number of quarter wavelengths from the ATR tube. A common type of balanced circuit consists of a dual TR tube mounted between two 3-dB hybrid directional couplers (see Fig. 5-128 and Table 5-22).

Table 5-22. Typical TR and ATR Tubes

TR Tubes

Type	Description	Frequency range, Mc/s	Ignitor V	Ignitor A	Max. peak power, kW	Peak leakage, mW	VSWR	Insertion loss, dB
1B24	Circular	8,490–9,600	450	200	. . .	30	. . .	1.5
6602	Rect. waveguide with shutter	3,070–3,530	400	200	750	50	1.4	0.7
8623	Rect. waveguide	2,900–3,100	400	100	50	50	1.2	0.3

ATR Tubes

Type	Description	Frequency range, Mc/s	Min. peak power, kW	Max. peak power, kW	VSWR	Insertion loss, dB
1B51	Rect. waveguide	6,225–6,625	100	. .	11	0.5
6081	Rect. waveguide	5,640	10	. .	. .	0.8
6890	Rect. waveguide	9,000–9,600	0.5	30	. .	1.6

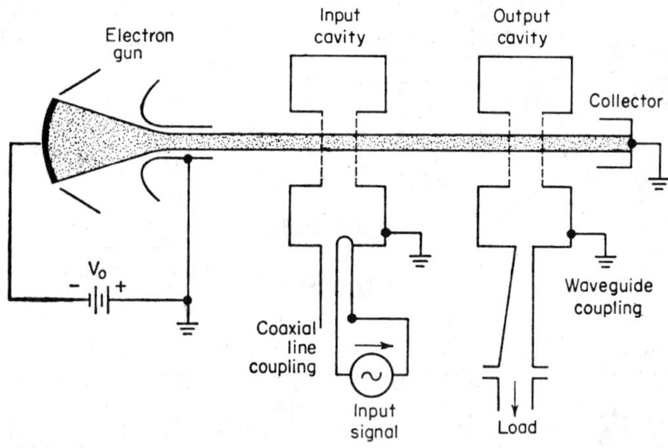

Fig. 5-129. Diagram of a two-cavity klystron amplifier. (*From "Principles of Electron Tubes" by J. W. Gewartowski and H. A. Watson, D. Van Nostrand Company, Inc., Princeton, N. J.*, 1965.)

Gas Noise Sources. An ionized column of gas is an excellent noise generator and, when used under carefully controlled current conditions, can furnish a calibrated noise signal for measuring signal-to-noise ratios in microwave receivers. Type 8402 is an example. Saturated diodes like the 5722 are used at lower frequencies.

211. Electron-beam Amplifiers and Oscillators. This class of tubes is distinguished from the types using a negative control grid by the fact that a beam of electrons from a cathode is focused and controlled by electric and magnetic fields to derive power gain, primarily at microwave frequencies.

Velocity-modulation, or "Klystron," Tubes. In this type, a focused beam of electrons is passed through a gap between two electrodes (or alternatively through an r-f cavity) across which the input signal has imposed an r-f field. This field alternately accelerates and decelerates the passing electrons, so that after a short distance ("drift space") the fast electrons have moved forward and the slow ones back, producing "bunching." The passing of these "bunches" through the gap of an output electrode or cavity induces an amplified field in the output circuit or cavity coupling loop. The beam is then dissipated on a collector electrode, which carries the d-c accelerating potential (see Fig. 5-129).

Reflex Klystron. If in a single-gap tube the collector electrode is given a negative voltage, making it a repeller, the beam turns around and returns through the gap and the distance to and from the repeller permits the bunches to form. This makes the amplified energy excite the input gap, and continuous oscillation takes place at a frequency determined by the cavity resonance, the drift space, and the repeller voltage (see Fig. 5-130).

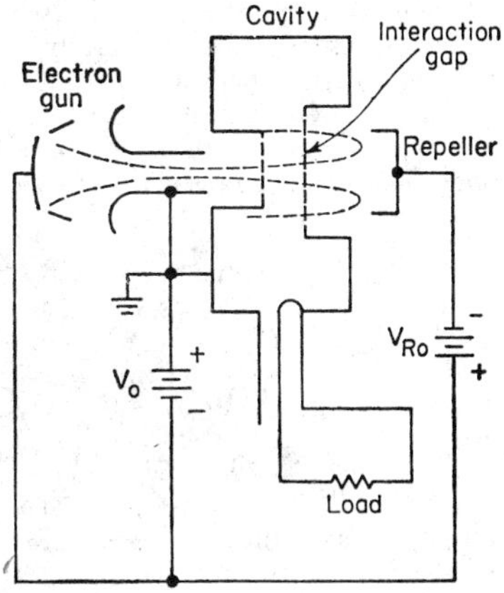

Fig. 5-130. Diagram of a reflex klystron. (*From "Principles of Electron Tubes," by J. W. Gewartowski and H. A. Watson, Princeton, N.J., D. Van Nostrand Company, Inc.*, 1965.)

Traveling-wave Tubes. A traveling-wave tube is a beam-type amplifier tube in which the magnetically focused pencil beam is surrounded by a long helix, with the signal introduced at the cathode end by a coaxial connector. The signal interacts with the beam, propagating a growing wave down the beam, which then transfers power to the output circuit through a coaxial connector at the collector end of the helix. These types are characterized by their low noise (<10 dB), wide bandwidth (for example, 4 to 8 Gc/s), low power (milliwatts to a few watts), and high gain (20 to 30 dB) (see Fig. 5-131).

Backward-wave Oscillator, or "Carcinotron."[1] If the collector end of the helix of a

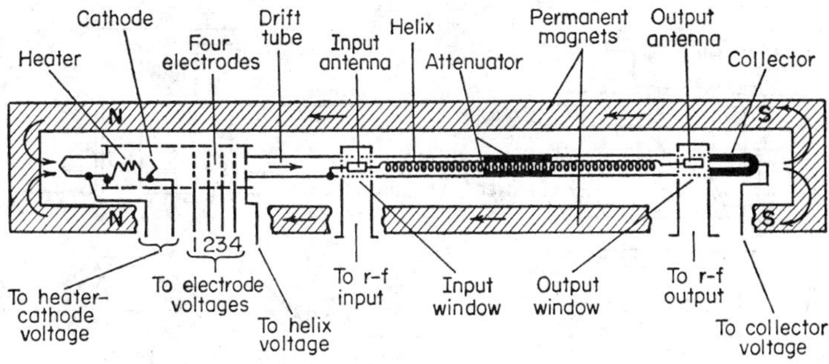

Fig. 5-131. Diagram of a permanent-magnet traveling-wave tube. (*Courtesy of General Electric Company.*)

[1] A trademark of Compagnie Télégraphie Sans Fils.

traveling-wave tube is terminated by a matched attenuator, under proper voltage and field conditions, random noise in the beam will excite a wave traveling toward the cathode, which grows as it progresses backward, and is taken off by a connector at the cathode end of the helix or delay line. Any r-f field in the line at this point induces a signal in the beam which interacts with the line as the electrons flow from cathode to collector, thus maintaining the oscillation. This type of tube is characterized by a wide tuning range (2 to 1) and an extremely high upper frequency limit (>500 Gc/s) (see Fig. 5-132).

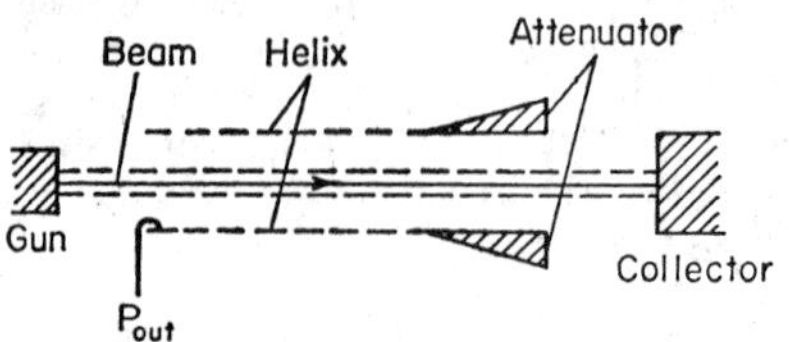

Fig. 5-132. Diagram of a backward-wave oscillator tube.

Magnetrons. In the simplest form of magnetron (Fig. 5-133) with alternate strip anodes surrounding a cathode, and an axial magnetic field, electrons flow from cathode to anode in a spiral path and bunch together like the rotating spokes of a wheel. As these spokes pass the alternately connected anodes, power from the beam is given up to the resonant output circuit.

In the cavity magnetron, the anode looks like the punchings of a motor stator; the open slots form the resonant circuit. Output power is delivered through a coupling loop (Fig. 5-134). Frequency is determined largely by the anode slot dimensions and spacings, the magnetic field, and the anode voltage. High-power magnetrons (up to 30 MW pulse) are used in radar transmitters.

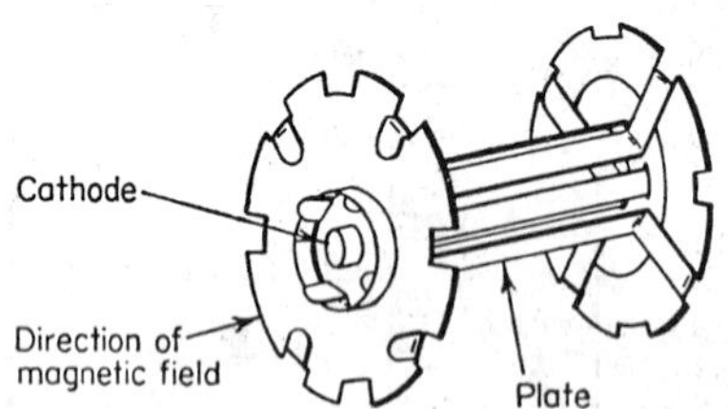

Fig. 5-133. Electrode structure of an elementary low-frequency interdigital magnetron.

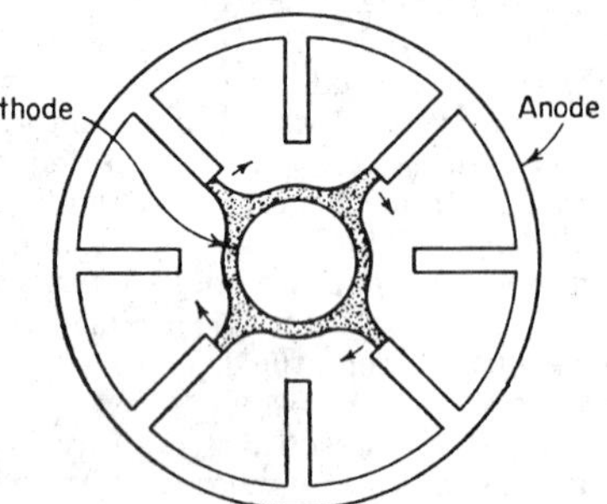

Fig. 5-134. Diagram of electron flow in a cavity magnetron. (*From "Principles of Electron Tubes" by J. W. Gewartowski and H. A. Watson, Princeton, N. J., D. Van Nostrand Company, Inc.,* 1965.)

Voltage-tunable Magnetrons. By injecting electrons from an auxiliary cathode at one end of the cavity structure, the frequency of oscillation can be made to vary over

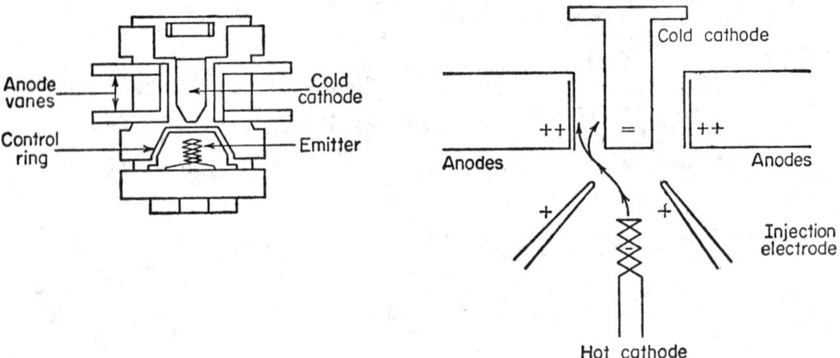

Fig. 5-135. Cross section and electron-flow diagram of a typical voltage-tuned magnetron.

Table 5-23. Typical Microwave Tubes

Type	Description	Frequency range, Mc/s	Power output, W	Anode voltage, V	Cathode current, A	Gain, dB	Noise figure, dB	Duty cycle	Tuning, Mc/(s)(V)
ZM6047	VTM	2,900–3,200	75	3,100	0.07	....	...	CW	0.8
ZM6215	VTM	2,500–3,500	10	1,850	0.025	....	...	CW	1.8
ZM6222	VTM	2,000–4,000	1	1,350	0.010	....	...	CW	2.3
Z3028	TWT	4,000–8,000	0.005	750	0.001	30	8	CW	
Z3031	TWT	14–18 Gc/s	0.005	1,400	0.001	30	14	CW	
Z3040	TWT	35–40 Gc/s	0.003	2,750	0.001	20	13	CW	
VA651	TWT	675	275	2,100	0.8	27	...	CW	
VA136	TWT	2,650–3,050	5,000	42 kV	20	30	...	0.05	
L3919	Klystron	200	100 kW	38 kV	8	30	...	CW	
5777	Reflex klystron	600–2,350	0.160	400	......	....	...	CW	
L3890	Klystron	2,856	24 MW	250 kV	250	50	...	0.001	
QKL1066	Klystron	10.2–10.6 Gc/s	2,500	10.4 kV	1.1	....	...	CW	
VC7153	Reflex klystron	139–141 Gc/s	0.1	2,500	0.025	....	...	CW	
QKA1103	Backward-wave oscillator	1,000–1,400	600	4,000	0.5	....	...	CW	
SFD214	Backward-wave amplifier	9,000–9,600	350 kW	30 kV	30	18.5	...	0.001	
CO-04	Backward-wave oscillator	760–790 Gc/s	0.002						
7547	Magnetron	406–450	2.0 MW	55 kV	97	....	...	0.0018	
6410	Magnetron	2,750–2,860	5.0 MW	76 kV	135	....	...	0.001	
8366	Magnetron	33.0–33.4 Gc/s	50 kW	13 kV	25	....	...	0.001	

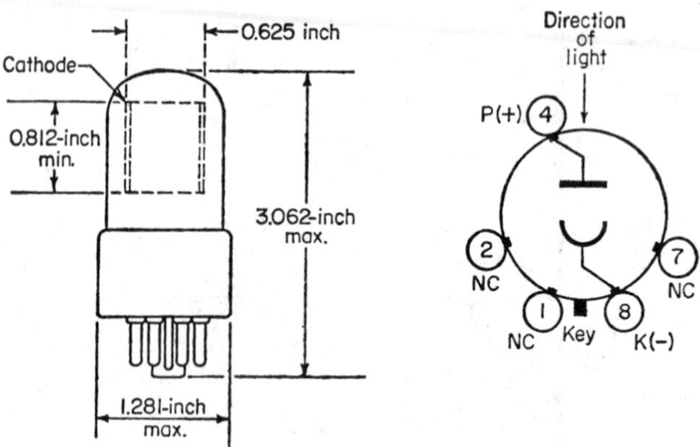

Fig. 5-136. Outline and diagram of a diode phototube—type 1P39.

about a 2-to-1 range. Low-power tunable magnetrons (100 MW) are used as local oscillators for receivers or sweep-signal generators and higher-power (up to 100 W average, 180 kW peak) for radar-altimeter and telemetry applications. See Fig. 5-135 and Table 5-23.

212. Phototubes. A phototube is a vacuum or gas-filled device whose output current is a function of electrons ejected from the photocathode by incident photons of radiant energy.

Vacuum Phototubes. Simple vacuum phototubes like the 1P39 (see Fig. 5-136) use a relatively large semicylindrical photocathode with a cesium-antimony alloy and a central wire anode. The current output, in the microampere region, is linear over a wide range of incident light levels and proportional to the three-halves power of the applied voltage up to the saturation level. The spectral response depends on cathode and envelope materials, the usual soft-glass envelope cutting off most of the ultraviolet response. Special glass windows or fused quartz envelopes are necessary for phototubes sensitive to ultraviolet radiation. Luminous sensitivity is expressed in microamperes per lumen, with typical values of 5 to 40.

Gas Phototubes. Introduction of an inert gas like argon at around 100 μ pressure provides a multiplying factor of 5 to 10 in output current by ionization of the gas molecules by the emitted photoelectrons. The resulting characteristics are nonlinear, but this is unimportant in many applications where sensitivity is the prime factor. Some noise is introduced and frequency response is limited by the ionization process. Care must be taken not to exceed maximum voltage ratings because of ionization breakdown. A typical application is in home motion-picture projectors as the sound track sensor, because of the high output obtainable at moderate cost. Typical sensitivity ranges from 40 to 135 μA/lm.

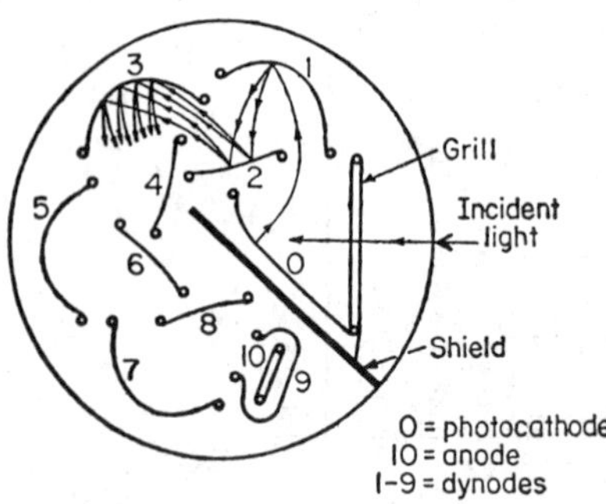

Fig. 5-137. Diagram of a multiplier phototube. (*Courtesy of Radio Corporation of America.*)

Multiplier Phototubes. For low light levels, where photocathode output is extremely small, secondary emission of electrons from additional cathodes (called "dynodes") is employed. Several dynodes, at successively higher voltage, are shaped to collect, reflect, and focus secondary electron beams so as to derive the maximum secondary emission current at the final anode. With a 10-stage unit, a total multiplication of 10^6 is common, with typical sensitivities from 1 to 80 A/lm. A typical application is the count-

Table 5-24. Typical Phototubes

Type	Description	Avg. sensitivity, μA/lm	Max. anode voltage, V	Max. anode current, μA	Cathode window area, in²	Spectral limits, Å	Wavelength of max. sensitivity, Å	Max. dark current, μA
				Vacuum phototubes				
917	4-pin T-8	20	500	10	0.9	5,000–12,000	7,900	0.005
922	Cartridge T-7	20	500	5	0.4	4,000–11,000	8,000	0.005
929	4-pin T-9	45	250	5	0.6	3,000–6,400	4,000	0.0125
1P39	8-pin T-9	45	250	5	0.5	3,000–6,400	4,000	0.005
1P42	Cartridge T-2	37	180	0.4	0.03	3,000–4,800	4,800	0.005
				Gas-filled				
918	4-pin T-8	150	90	10	0.9	4,000–12,000	8,000	0.1
1P40	8-pin T-9	135	90	6	0.5	5,000–12,000	8,000	0.1
1P41	3-pin T-7	90	90	3	0.25	5,000–12,000	8,000	0.1
				Multiplier types				
								Equivalent dark current, μA
931A	11-pin T-9	24×10^6	1,250	1,000	0.25	3,000–6,400	4,000	0.050
1P28	11-pin T-9	50×10^6	1,250	500	0.25	2,300–6,400	3,400	0.025
7029	12-pin T-12 dormer window	40×10^6	1,250	20	0.325	3,000–6,500	4,900	0.004
7046	21-pin 5-in-dia. 14-stage	1.2×10^9	3,400	2,000	15.5	2,500–6,400	4,200	0.040
7850	20-pin T-16 12-stage	6.0×10^9	2,600	2,000	2.22	3,000–6,500	4,400	3
8053	14-pin T-16 10-stage	120×10^6	2,000	2,000	2.22	3,000–6,500	4,400	0.0004

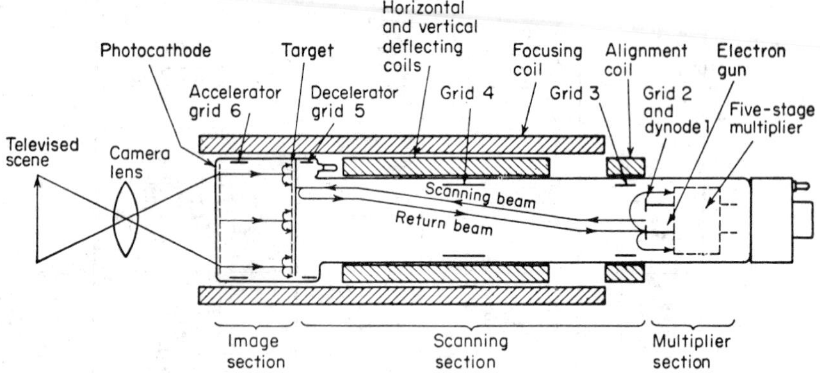

FIG. 5-138. Schematic arrangement of the image orthicon.

ing of bursts of light emitted from a crystal by radioactive emanations, called a "scintillation" counter (see Fig. 5-137 and Table 5-24).

Camera Tubes—Image Orthicon Tube. Early television cameras used a tube known as the "iconoscope," in which photocathode elements were sputtered onto a thin insulating plate and the charge pattern produced by the emitted photoelectrons was scanned by a moving beam of electrons. The signal was taken off the backing plate of the target, as the scanning beam "refilled" the "holes" left by the emitted photoelectrons. Because of its low sensitivity, it was soon found necessary to add many stages of secondary-emission multiplication, resulting in what is called the "image orthicon." In this tube, the photocathode is semitransparent, on an extremely thin membrane, and a low-velocity scanning beam is directed at the rear surface and reflected back to the multiplier stages in proportion to the charges on each spot of the photocathode surface (see Fig. 5-138).

Vidicon Tube. In the vidicon camera tube, the photocathode is a semiconductor, and the scanning beam produces an output proportional to the photoconductivity of the cathode film at each point on the scanned area. Definition of early tubes was not so good as in the image orthicon, and decay time was quite long, so that moving images left a "trail." Recent developments using new materials have improved this fault.

Vidicons are much simpler and less expensive than image orthicons and require less complicated circuitry. A typical application is closed-circuit TV, for remote monitoring and burglar protection (see Fig. 5-139).

Image Intensifiers and Converters. Special photocathodes have been developed which permit the formation of an "electron image" from incident near-infrared, low-

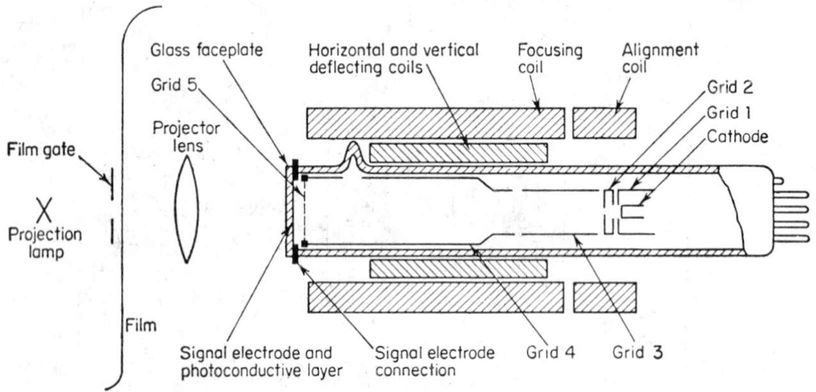

FIG. 5-139. Schematic arrangement of type 6326 vidicon.

Table 5-25. Typical Camera Tubes and Image*

Type	Description	Cathode V	Cathode A	Photo-cathode, volts	Anode, volts	Max. target, volts	Max. signal current peak to peak, μA	Min. S/N ratio	Max. illum. on photo-cathode, F.C.	Sens. satur., F.C.	Sens., A/F.C. at 0.001 F.C.	Peak spectral response, Å
						Image orthicons						
5820A	14-pin 3-in dia.	6.3	0.6	−550	1,350	10	24	35	50	0.02	270	4,400
7629A	14-pin 3-in dia.	6.3	0.6	−550	1,350	4	4	48	50	0.01	800	4,400
8092A	14-pin 3-in dia. with field mesh for color match	6.3	0.6	−550	1,350	6	4	37	50	0.01	800	4,400
						Vidicons						
								Max. dark current, μA			Max. sensitivity at 0.01 F.C., A/F.C.	
6198A	8-pin T-8	6.3	0.6	.....	350	125	0.2	...	.....	10	0.06	4,500
7262A	8-pin T-8	6.3	0.095	.....	750	100	0.3	0.2	1,000	60	0.7	4,600
7735B	8-pin T-8	6.3	0.6	.....	1,000	100	0.4	0.1	1,000	15	1.0	5,500

* F.C. = footcandles.

level visible, or x-ray images, accelerating the electrons in straight lines and focusing them on a fluorescent screen. A typical application is the "snooper scope," a small infrared converter with a transistorized high-voltage supply, for aiming a gun at night. A similar unit sensitive in the x-ray region makes possible "flash" x-rays and x-ray motion pictures and reduces many-fold the necessary radiation dosage (see Fig. 5-140 and Table 5-25).

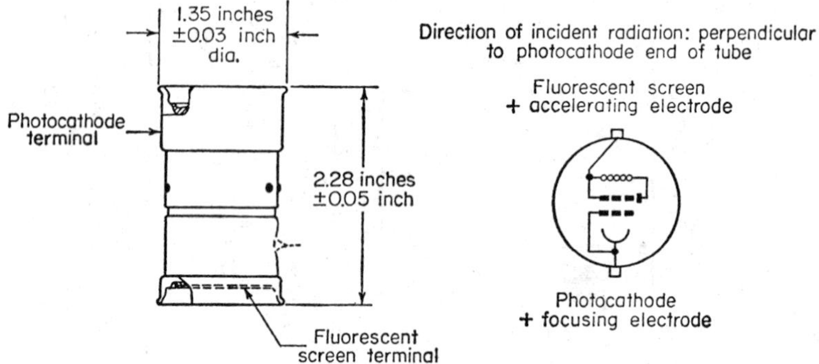

Fig. 5-140. Outline and diagram of an image intensifier.

213. Cathode-ray Tubes. *Electrostatic Deflection.* An electrostatic-deflection cathode-ray tube consists of an electron gun, generating a focused beam of electrons, a control grid to vary the beam intensity, and two sets of parallel plates at right angles to deflect the beam horizontally and vertically. The large end of the bulb is coated with a phosphor to convert the incident beam to visible light. Various color phosphors

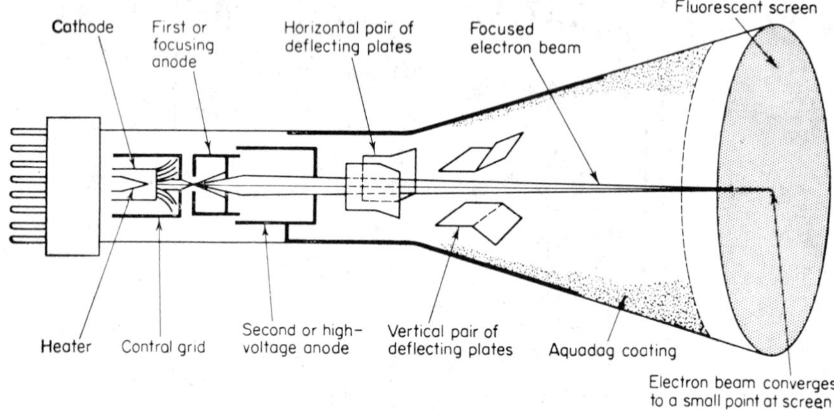

Fig. 5-141. Schematic arrangement of an electrostatic-deflection cathode-ray tube.

are available, with a wide range of persistence and resulting decay times. The phosphors are designated by a suffix letter P followed by digits 1 to 35, indicating the standardized color and decay characteristics. A typical use is in the familiar oscilloscope, where a sawtooth waveshape is used for the horizontal-deflection voltage, producing the "linear time base," and the test voltage under investigation is applied to the vertical plates, to portray its waveshape (see Fig. 5-141).

Magnetic Deflection (Fig. 5-142). Television picture tubes utilize pairs of external coils at right angles to generate horizontal- and vertical-magnetic-deflection fields, to deflect the electron beam across and down the phosphor screen in the standard scanning pattern. The electron gun may use either electrostatic or magnetic focusing, but the former is simpler and used in the majority of TV tubes. Early tubes incorporated a "bent gun" and an ion-trap magnet to prevent negative gas ions from hitting the center of the screen and causing a dark spot. A majority of tubes now use an evaporated layer of aluminum over the phosphor, which not only protects it against ionic bombardment but increases its light output in the forward direction by reflection. Most tubes now incorporate a darkened faceplate to reduce ambient light reflection from the phosphor and a layer of plastic or laminated glass to protect the user in the unlikely event of an implosion. Alternatively, the faceplate is put under compression by a band or frame around the perimeter, which also has mounting ears for fastening to the chassis or cabinet.

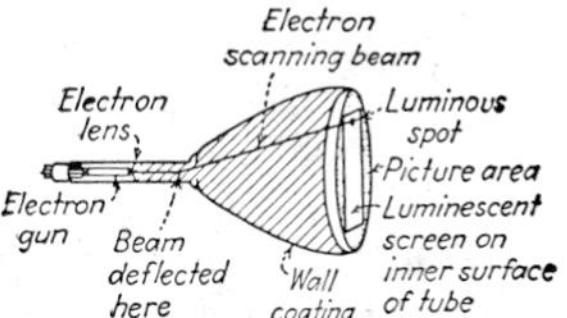

FIG. 5-142. Schematic arrangement of a magnetic-deflection cathode-ray tube.

Color-picture tubes use three guns and focus the three beams through an "aperture-mask" anode in front of the fluorescent screen. The three basic color phosphors, red blue, and green, are deposited in sets of three dots in front of each hole in the aperture mask.

Second anode voltages of 10 to 22 kV are used on black-and-white tubes and 15 to 30 kV on color tubes. In addition to the high voltage itself, precautions should be taken to protect the user from soft x-rays generated by the high-velocity electrons striking the screen and bulb. In the type numbering system for cathode-ray tubes, the initial digits represent the diameter or diagonal of the screen, followed by one or two code letters and the phosphor designation. Successive improvements carry suffix letters A, B, etc. (see Table 5-26).

Storage Tubes (see Fig. 5-143). More properly called "charge-storage tubes," this class of cathode-ray tube utilizes a mosaic or mesh target with an insulating coating capable of retaining individual point charges on every surface element, which can then be "read" off by a scanning electron beam. The charges may be induced by a signal on the backing plate of the mesh ("barrier-grid" type), or by a "writing" gun beam

Table 5-26. Typical Cathode-ray Tubes

Type	Description	Deflection angle and type degrees*	Heater V	Heater A	Anode voltage, kV	Focus voltage, V	Grid 2 volts	Max. grid 1 volts to cutoff
2AP1	Round oscilloscope	30 E	6.3	0.6	1.1	550	550	−60
2EP4	Round TV	30 M	6.0	0.145	9.0	−50/350	300	−25
5AXP4	Round TV	53 M	6.3	0.6	14.0	M	300	−72
5CP1	Round oscilloscope	30 E	6.3	0.6	4	575	2,000	−60
9QP4A	Rect. TV	70 M	4.7	0.3	5.5	0/400	200	−52
11HP4	Rect. TV	110 M	6.3	0.45	11	0	150	−49
14AW4P4	Rect. TV	90 M	6.3	0.45	12	−50/350	50	−47
16BUP4	Rect. TV	114 M	6.3	0.45	12	M	250	−58
19DZP4	Rect. TV	114 M	6.3	0.45	13	−250/150	150	−54
21ALP4A	Rect. TV	90 M	6.3	0.6	14	−55/300	300	−72
23DYP4	Rect. TV	110 M	6.3	0.6	18	0/500	300	−54
30BP4	Round metal	90 M	6.3	0.6	22	M	300	−72
Color—3-gun—shadow mask								
11SP22	Rect. glass	72 M	13.8	0.58	15	−250/500	250	−40
21AXP22A	Round metal	70 M	16.3	1.8	25	3.8/5.3 kV	200	−100
23EGP22	Rect. glass	92 M	6.3	1.35	25	6 kV	650	−175
25AP22A	Rect. glass	90 M	6.3	0.8	25	4.2/5.0 kV	685	−190

Phosphors, P-1 green, P-4 white, P-22 color. * E = electrostatic; M = magnetic.

Table 5-27. Typical Storage Tubes

Type	Description	Focus/ deflection*	Heater		Screen voltage	Writing gun cathode, V	Anode voltage	Screen current, μA	Writing current, μA	Reading current, μA	Resolution, lines/in	Writing speed, in/s
			V	A								
6498	5-in display	E/E	6.3	0.6	5,000	−3,000	.....	300	1,000	3,000	50	35,000
			6.3	0.6								
6866	5-in display	E/E	6.3	0.6	11,000	−2,400	.....	600	4,000	4,000	.....	300,000
			6.3	0.6								
6896	Double-ended	E/M	6.3	0.6	−10	−13,000	0	5	235	15		
			6.3	0.6								
7383	1-in pickup	M/M	6.3	0.6	15	....	.....	1.0	.....	.....	600	
7702	Double-ended	M/M	6.3	0.6	200	0	3,500	.....	.....	.....	1,200	250,000
			6.3	0.6								

* E = electrostatic; M = magnetic.

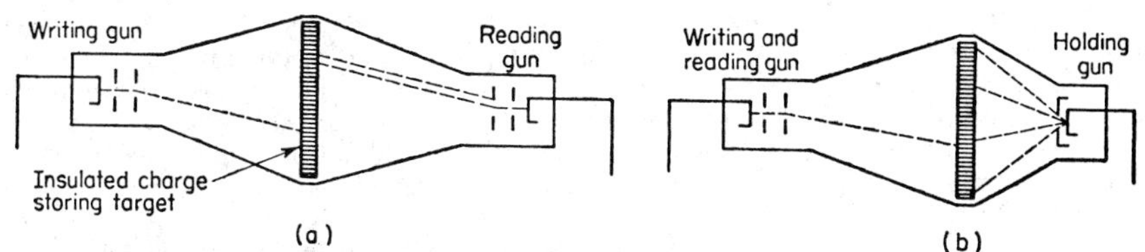

FIG. 5-143. Diagram of two types of storage tubes.

which scans the storage surface. In the latter type, the beam can be switched to go through the storage surface to a fluorescent screen, or a separate "flood" gun can be used, to display the stored charge information until it is erased by a reverse pulse on the mesh. In the double-ended configuration, both "read" and "write" guns scan the target, and simultaneous writing and nondestructive readout are possible. This latter type is commonly used for radar PPI-scan to TV-scan conversion, or conversion from one television scanning-rate picture to another, e.g., European to United States (625 to 525 lines). (It is also possible to accomplish this latter in two separate envelopes by use of a fiber optics faceplate between the phosphor of the writing portion and the photoconductive surface of the reading end.)

Table 5-27 lists typical type numbers and characteristics.

214. General Precautions. *Applied Voltages.* While tubes are able to withstand a wide range of input voltages, life can be greatly improved by close (±1 percent) control of heater voltage. Extensive tests show that cathode life varies inversely as the seventh or eighth power of the heater voltage above its normal rating.

Envelope Temperature. Tubes with soft glass envelopes can be rated up to 225°C envelope temperature but exhibit much longer life if derated to 175°C or if forced-air-cooled to this temperature, because of eventual electrolysis of the glass under electron bombardment and release of surface gases inside the bulb.

Grid Current. Tube manufacturers' rating sheets usually specify a maximum allowable value of external grid resistance, because under some conditions a few microamperes of grid emission current may be generated or a small amount of gas may cause ions to be attracted to the grid. Either of these may result in a runaway condition unless the grid resistor is low enough to limit the effect or a cathode bias resistor provides the necessary degeneration.

Power-amplifier Tuning. The tuning procedure and gradual application of power to large transmitting tubes must closely follow the manufacturer's recommended practice to avoid tank-circuit flashover or internal arcing caused by excessive r-f voltages.

R-F Burns. Accidental exposure to high-power microwave fields in testing tubes or equipment can produce serious external and internal burns. Leakage from high-power output loads and waveguides should be checked and operating personnel adequately protected.

215. Conditions of Use. The designer of electronic equipment has available for most of his applications a choice between tubes and solid-state devices and must make his decision based on a number of factors, such as the following.

Cost. Where cost is a prime objective, tubes usually come out ahead on a per function basis, because of the availability of multifunction tubes like compactrons and the need for fewer passive circuit components.

Size. There is no doubt that semiconductors, and especially integrated circuits, provide at least an order-of-magnitude reduction in volume, where this is important. Electronic computers are a good example.

Temperature. Tubes can withstand a wider range of ambient temperatures (−50 to +350°C) than semiconductors, but on the other hand they dissipate more heat and require more total cooling. However, semiconductors are sensitive to small changes in temperature and require large heat sinks or radiators to avoid exceeding their fairly low maximum junction temperature ratings (100°C).

Reliability. While tubes have come a long way in improved survival ratings and in conservative designs do better than 30,000 h, semiconductors have a potential for almost unlimited life under controlled conditions. As mentioned above, this means well-controlled temperature conditions and avoidance of high-voltage pulses or spikes, such as often accompany line-voltage transients. Semiconductor devices are more easily burned out by overload or accidentally applied potentials, especially in the reverse direction. As an example, a simple ohmmeter should not be used in checking out a circuit, for fear of applying excessive potentials to transistors, whereas this process is not detrimental to a tube circuit.

High Impedance. Tubes provide much higher input and output impedances than semiconductors, except for the "field-effect" transistors. Vacuum-tube voltmeters, for example, provide a long, linear range scale with several thousand megohms input resistance.

Radiation. While some of the newer alloy transistors have been improved to withstand 10^{11} to 10^{13} *NVT* (neutron radiation), most tubes will withstand 10^{15} and special ceramic tubes 10^{18} *NVT*.

High Voltage. Semiconductors are naturally low-voltage high-current devices, and tubes have an inherent advantage for high-voltage rectifiers, high-voltage pulse applications, and high-voltage drivers, as in the horizontal-output stage of a television receiver.

Power Consumption. Tubes consume far more power than transistors because of the need for a heated electron source, so that low-power circuits using battery power are ideal applications for semiconductors.

High Frequency and Power. Tubes at present can generate more power and more gain at u-h-f and microwave frequencies than solid-state devices, because of the inherent transit time and capacitance in the semiconductor. The gap is closing, however, and in some cases it is possible to use varactor frequency multipliers to achieve equivalent output in a size smaller than that of a tube unit. In this case, other factors must be taken into account, such as tuning range and overload tolerance, before deciding on the tube vs. the transistor (see Fig. 5-144).

Special Characteristics. Tubes have an advantage over semiconductors in that they can be designed to provide special input-output transfer functions, as, for example, type 6BN6, a gated-beam limiter and discriminator used as a frequency-modulation detector, and type 6AR8, a beam-deflection color subcarrier demodulator for color television.

In summary, tubes present unique advantages in many of their characteristics, particularly in high-power, high-frequency, and high-voltage applications. At low power and low frequencies, the choice between tubes and solid-state devices is often conditioned by the overall cost of the active device (transistor or tube), circuit, and power supply.

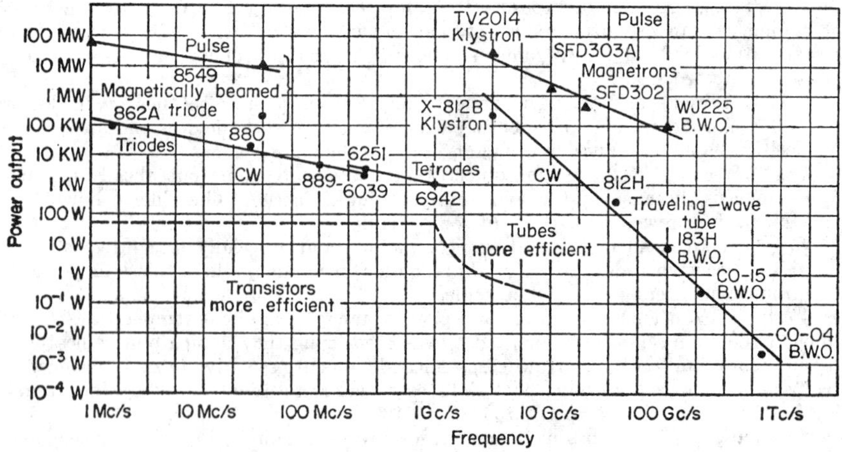

Fig. 5-144. Power output vs. frequency for various classes of tubes.

216. Bibliography

Fink, D. G. "Television Engineering Handbook"; New York, McGraw-Hill Book Company, 1957.

Gewartowski, J. W., and Watson, H. A. "Principles of Electron Tubes"; Princeton, N.J., D. Van Nostrand Company, Inc., 1965.

Landee, R. W., Davis, D. C., and Albrecht, A. P. "Electronic Designers' Handbook"; New York, McGraw-Hill Book Company, 1957.

Reich, H. J. "Theory and Application of Electron Tubes," 2d ed.; New York, McGraw-Hill Book Company, 1944.

Zworykin, V. K., and Morton, G. A. "Television"; New York, John Wiley & Sons, Inc., 1954.

RCA Tech. Manual TT-5, Transmitting Tubes; Harrison, N.J., Radio Corporation of America, October, 1962.

RCA Tech. Manual PT-60, Phototubes and Photocells; Harrison, N.J., Radio Corporation of America, 1963.

"Essential Characteristics Handbook" (receiving, special-purpose, and picture tubes); Owensboro, Ky., General Electric Company, 1964.

"American Standard Dimensional Characteristics of Electron Tubes, Bases, Caps, and Terminals (including Gauges)"; 2001 Eye St. NW, Washington, D.C., Electronic Industries Association, EIA RS-209A, 1964.

"D.A.T.A. Microwave Characteristics Tabulation," 13th ed.; 43 South Day St., Orange, N.J., Derivation and Tabulation Associates, Inc., November, 1964, through April, 1965.

Shea, R. F. "Amplifier Handbook"; New York, McGraw-Hill Book Company, 1966.

SECTION 6

ALTERNATING-CURRENT GENERATORS

BY

R. E. APPLEYARD Manager, Product Engineering, Hydrogenerator Department, Allis Chalmers

WILLIAM L. RINGLAND Research Division, Allis Chalmers

CONTENTS

Numbers refer to paragraphs

SECTION 6

ALTERNATING-CURRENT GENERATORS

BY

R. E. Appleyard and W. L. Ringland

GENERAL

1. General Construction. An alternating-current generator consists principally of a magnetic circuit, d-c field winding, a-c armature winding, and mechanical structure, including cooling and lubricating systems. The magnetic circuit and field windings are arranged so that, as the machine rotates, the magnetic flux linking the armature winding changes cyclically, thereby inducing alternating voltage in the armature winding.

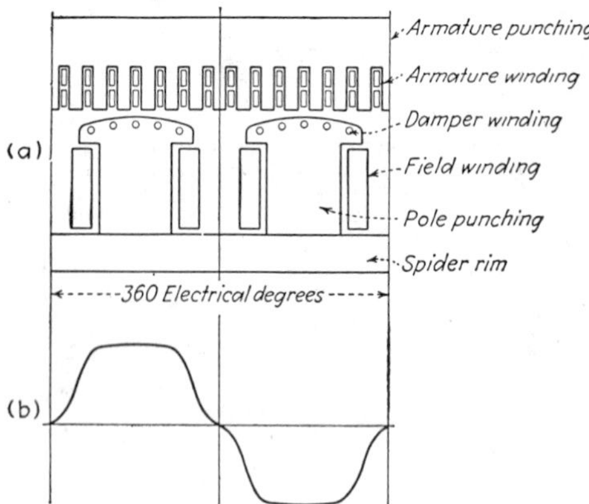

Fig. 6-1. (*a*) Generalized sketch of one pair of poles for a salient-pole machine. (*b*) Flux form for a typical pair of poles (on the basis of current in field winding only).

Many different geometrical arrangements of these elements are possible, and each has its own economical field of application. For high power generation, the most common types are: (1) the salient-pole construction illustrated in Fig. 6-1 and characteristic of hydraulic turbine or large diesel-engine drives and (2) the non-salient-pole or round-rotor machine illustrated in Fig. 6-2 and characteristic of steam-turbine drives. These machines normally have rotating fields and stationary armatures. Automotive alternators and high-frequency generators are frequently inductor-type machines which have a variety of forms, as discussed in Ref. 10.

Most of the principles to be dealt with in the following articles apply equally well to all these types, but the discussion will be related primarily to salient-pole machines illustrated in Fig. 6-1 in the size range of 100 to 500,000 kVA unless otherwise indicated.

2. Synchronous Speed. Two poles must pass a point on the stator to complete 1 c so that

$$\text{r/min} = 60\,\frac{\text{c/s}}{\text{pairs of poles}}$$

$$= \frac{7{,}200}{\text{poles}} \qquad \text{(for 60 c/s)} \tag{6-1}$$

3. Electrical Degrees. Because 1 c of a sine wave is 360°, it is convenient to measure distance around the machine periphery in electrical degrees with two poles spanning 360 elec deg. Thus, elec deg = mech deg × pairs of poles.

Electrical degrees are also commonly used as a measure of time, with

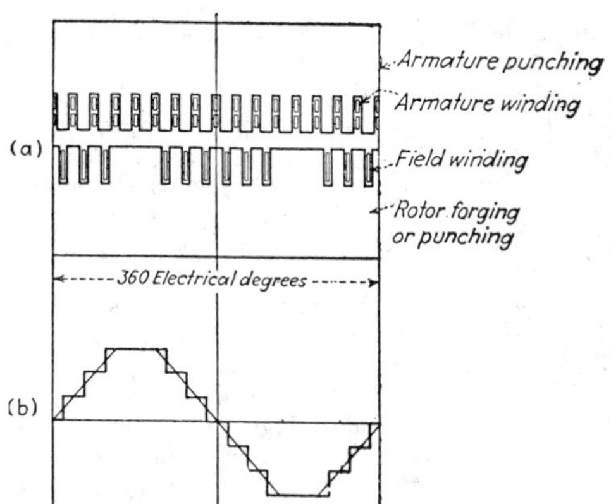

FIG. 6-2. (*a*) Generalized sketch of one pair of poles for a non-salient-pole machine. (*b*) Flux form for a typical pair of poles (on the basis of current in field winding only).

360 elec deg corresponding to the time period of 1 c as illustrated in Fig. 6-3.

4. Air Gap. The clearance between the stator and rotor is commonly called the air gap even though the machine may normally operate in an atmosphere of hydrogen or other gas, rather than air.

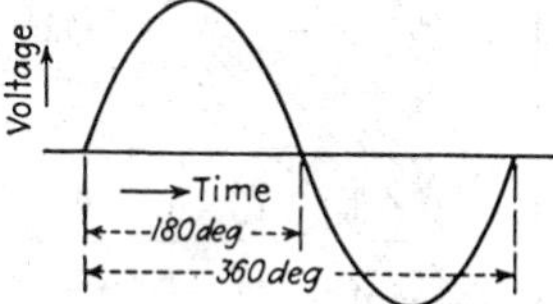

FIG. 6-3. One cycle of the voltage wave of a synchronous machine.

5. Major standards pertaining to a-c generators are now under the jurisdiction of the United States of America Standards Institute. New standards approved by USASI will be designated as USA Standards. This organization replaces the American Standards Association (ASA). The existing ASA Standards remain valid until they are replaced by USA Standards.

A similar situation exists with respect to American Institute of Electrical Engineers (AIEE) Standards, this organization having been replaced by the Institute of Electrical and Electronic Engineers (IEEE).

MAGNETIC CIRCUIT

6. Functions. The magnetic circuit determines to a large extent the output ratings and performance characteristics that are possible for a particular machine. Because output results from the interaction of the current-carrying armature conductors and the air-gap flux and is proportional to their product, the magnetic circuit must provide space for the windings as well as a path for the magnetic flux. The objective of magnetic circuit design is to provide an optimum division of machine volume between the flux-carrying and current-carrying parts.

7. Flux Paths. The paths for the principal components of flux with load are indicated in Fig. 6-4. Additional paths, mainly for the leakage fluxes, exist at the ends of the machine.

8. The proportions of the magnetic circuit are usually consistently related to the pole pitch for a particular number of poles. As the number of poles decreases, the restrictions in space available in the rotor re-

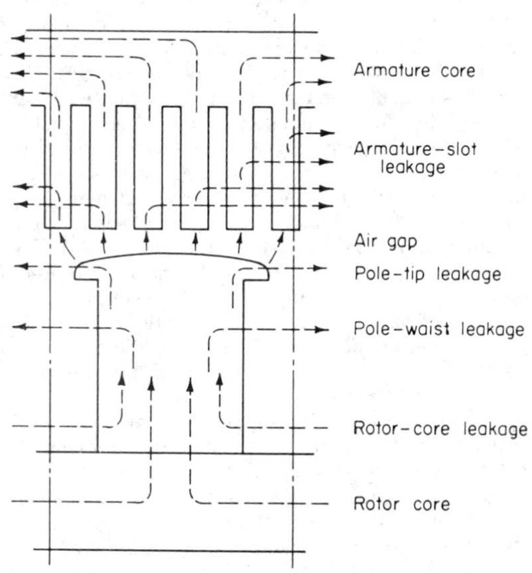

FIG. 6-4. Flux paths for a salient-pole machine with load.

sult in most of the magnetic circuit dimensions being a smaller proportion of the pole pitch. This is illustrated in Fig. 6-5.

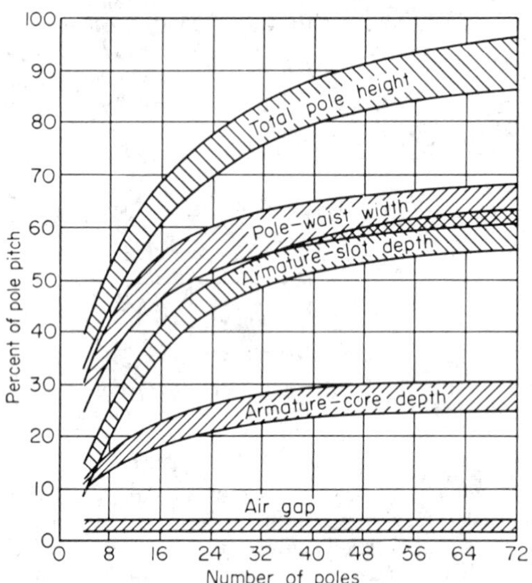

Fig. 6-5. Typical proportions of salient-pole machines.

The armature slot width is determined principally by the insulation thickness required for the machine voltage and is commonly such that the resulting total copper width per slot is 40 to 60% of the slot width.

9. Materials and Losses. The magnetic flux in the rotor is essentially unidirectional and varies only slightly with changes in load or terminal voltage during normal operation. This allows the rotor magnetic circuit to be made of solid steel, which is commonly done on turbine-generators and occasionally on some highly stressed salient-pole machines. However, the armature slots result in local variations of the air-gap flux density which cause eddy currents and losses in the rotor pole faces. Where solid pole faces are used, the air gap is usually relatively large, reducing these losses to acceptable values. Most salient-pole machines, on the other hand, have smaller air gaps relative to the armature slot width, and laminated poles are necessary to reduce the eddy-current losses at the pole faces. Pole laminations are commonly made of low-carbon steel, $\frac{1}{16}$ in thick, with magnetization characteristics specified and controlled in production. Thinner steel, sometimes with some silicon content, may be used where further reduction of eddy-current losses is necessary.

The armature magnetic circuit carries alternating flux and is always laminated, either with complete ring laminations or with segmental laminations depending on the machine size and the available widths of electrical sheet steel. The material most commonly used is about 3.5% silicon electrical sheet steel in 0.014-, 0.018-, or 0.025-in thickness for 60-c machines. Grain-oriented steel, with reduced losses and improved permeability in the direction of rolling, is frequently used for the armature laminations of large turbine-generators. Orientation in the circumferential direction is advantageous in such machines because of the large proportion of steel and moderate flux densities in the ring portion of the armature. At high flux densities characteristic of the armature teeth, the advantages of grain orientation become less pronounced. Typical magnetization characteristics are illustrated in Fig. 6-6.

10. Performance. The field current required for a particular load condition is determined by the magnetic circuit, in conjunction with its armature and field windings. This is calculated in design by evaluating the flux densities and the corresponding ampere-turns in all parts of the magnetic circuit. After the machine is built, the magnetic characteristics are represented by the test performance shown in the article *Characteristics and Vector Diagrams.*

Magnetic saturation significantly affects machine characteristics and performance under normal operating conditions. Unfortunately, much of the mathematical treatment of synchronous machine circuit theory is based on the assumption of negligible magnetic saturation because of the more manageable equations which then result. Equations or analogs intended to represent the machine generally must include the effects of saturation for acceptable accuracy.

11. Size and Output. A commonly used relationship between the rated machine

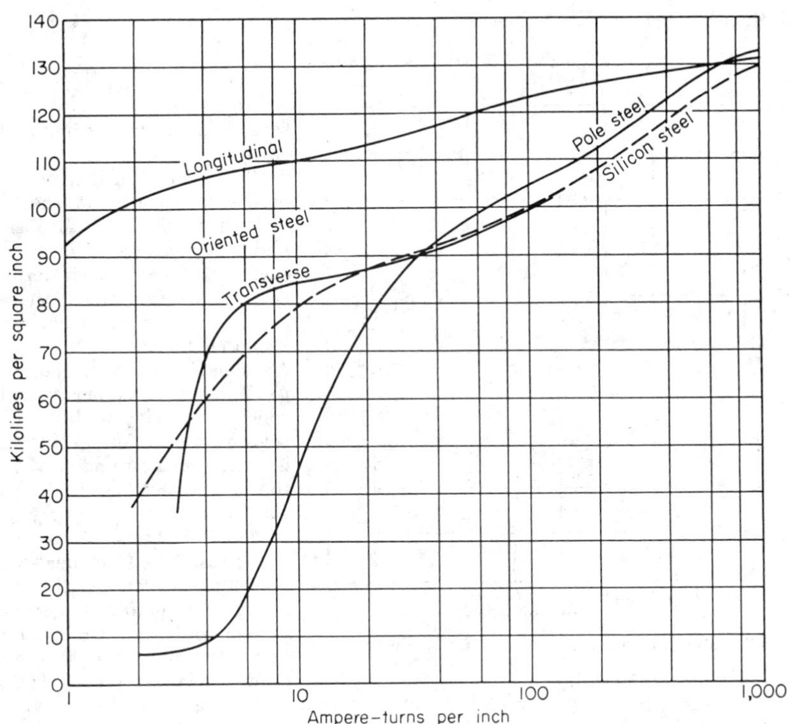

Fig. 6.6. **Magnetization curves for commonly used steels.**

output and the armature inside diameter D and gross core length l is the output factor,

$$\text{Output factor} = \frac{\text{kVA} \times 10^5}{\text{r/min} \times D^2 l} \tag{6-2}$$

The quantities determining the output factor are more clearly indicated in the equation

$$\text{Output factor} = \frac{B_{gf}}{8{,}600} \times kac/\text{in} \tag{6-3}$$

where B_{gf} = peak of fundamental component of air-gap flux density, lines per square inch, and kac/in = rms effective armature kiloampere conductors per in of armature periphery at inside diameter D.

The output factor normally increases with pole pitch for a particular voltage and number of poles because the deeper armature slots and greater field coil space allow more kiloampere conductors per in. However, the output factor normally decreases as the number of poles decreases because of the dimensional restrictions resulting from the mechanical angle between poles. Both B_{gf} and the kiloampere conductors per in are then reduced. Typical output factors for salient-pole generators are shown in Fig. 6-7.

VOLTAGE GENERATION, WAVESHAPE, AND HARMONICS

12. Voltage Generation. Voltage is generated in the armature winding as a result of relative motion between the field and armature. The magnetic flux linking each armature coil changes as the machine rotates, causing induced voltages in accordance with the basic relationship

$$E/N = d\phi/dt \times 10^{-8} \qquad \text{(instantaneous V/turn)} \qquad (6\text{-}4)$$

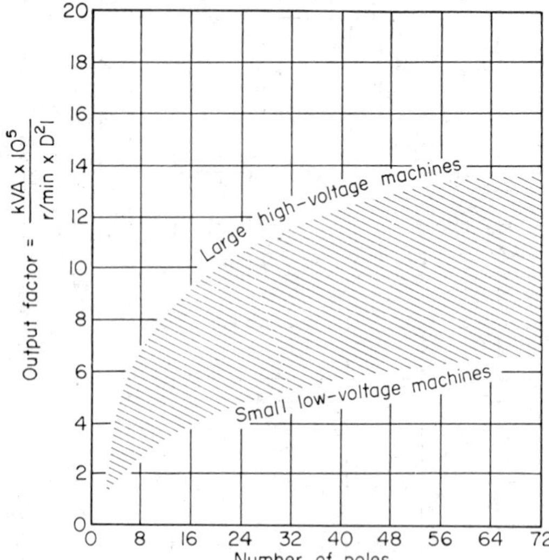

FIG. 6-7. Typical output factors for salient-pole generators.

where $d\phi/dt$ = maxwells per second change in flux linking the turn.

The change in flux per turn occurs principally at the conductors in the armature slots, and it is convenient to consider each conductor separately as though it were cutting the air-gap flux. At a particular rotating speed, the instantaneous volts per conductor are proportional to the air-gap flux density at the conductor. The waveshape of the conductor voltage vs. time is therefore the same as that of the air-gap flux density vs. distance around the periphery.

13. Flux Forms. Typical flux forms resulting from field-winding magnetization and illustrated in Figs. 6-1 and 6-2 are generally not sinusoidal but rather are designed to provide a suitable compromise relationship between the fundamental component of flux and maximum flux density, average flux density, harmonic content, and iron losses in the pole faces.

The ratio of the peak value of the fundamental component of flux density to the actual maximum flux density is designated as C_1. Corresponding values for the nth harmonic are designated as C_n. Flat and wide flux forms may have a large value of C_1, approaching 1.27 as a maximum. Such waveforms result in lower air-gap and armature-teeth flux densities for a given fundamental component, but at the expense of greater total flux, higher harmonic content, and greater pole-face losses. Opposite effects with lower values of C_1 may result from narrow, peaked flux forms.

Reference 11 evaluates C_1 and other quantities for commonly used pole-face shapes and field-winding distributions. A more general approach is to apply Fourier analysis to the flux form determined from the machine dimensions. Even harmonics are not present in symmetrical flux forms.

Although flux forms are commonly evaluated from the air-gap permeance, magnetic saturation in the pole tips, armature teeth, or elsewhere may substantially change the flux form at high flux densities.

Harmonics in the flux form generally do not result in harmonic voltages of corresponding magnitude at the armature-winding terminals because of the factors discussed in the following paragraphs.

14. Pitch Factor. When the coil pitch differs from the pole pitch, the voltage developed in the two sides of a single coil will differ in phase by an angle B which is the angle in electrical degrees by which the coil pitch differs from the pole pitch. This reduces the coil voltage, as compared with a full-pitch coil, by the pitch factor K_p, which is

$$K_p = \cos (B/2) \qquad (6\text{-}5)$$

For the nth harmonic,

$$K_{pn} = \cos (nB/2) \qquad (6\text{-}6)$$

Values for K_{pn} are shown in Fig. 6-8. It is evident that coil pitch can be chosen to reduce at least some harmonics much more than the fundamental.

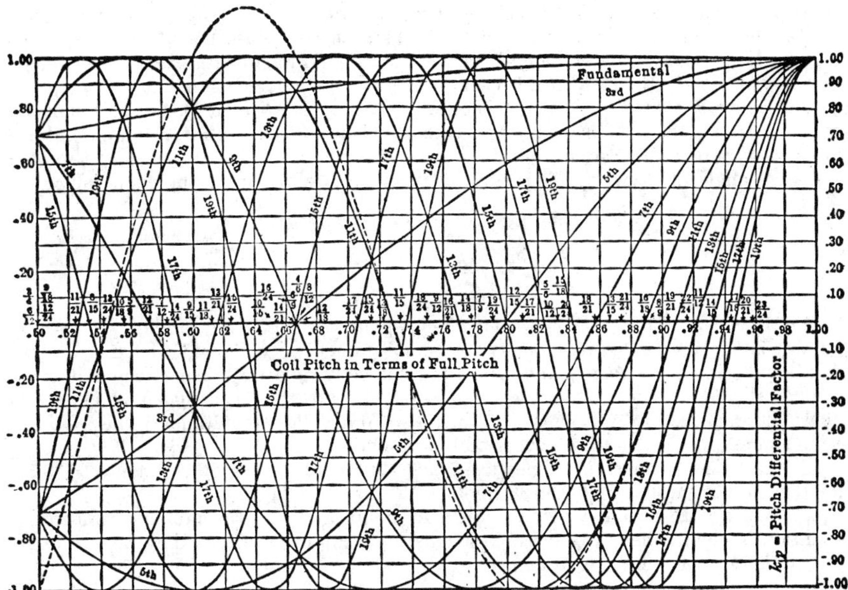

FIG. 6-8. Curves of pitch factors, K_{pn}. NOTE: The dotted curve shows the ratio of K_{pn} to K_p for the eleventh harmonic. This is obtained by dividing the ordinates of the curve for the eleventh harmonic by the corresponding ordinates of the fundamental curve.

15. Distribution Factor K_d. The conductors of one phase are generally distributed in more than one slot per pole. The group of conductors for one phase at a pole is referred to as a phase belt, and the corresponding group of whole coils as a coil group.

For **integral slot windings** where the number of slots per phase per pole is a whole number, all coil groups are identical, but the voltages of the coils in one group differ in phase by an angle corresponding to the slot pitch. For example, with six slots per pole the fundamental components of the two coil voltages of one group are 30° out of phase and the third harmonics are 90° out of phase, etc. The vector sum of the coil voltages is less than their arithmetic sum because of the phase differences. The factor by which these phase differences reduce the total voltage is the distribution factor K_d and is given in Table 6-1 for integral numbers of slots per pole.

For **fractional slot windings,** where the number of slots per phase per pole is not a whole number, the number of coils per group is not the same at all poles. Differences in voltage phase angle exist among coil groups as well as among the coils of a group.

Table 6-1. Distribution Factors K_{dn} for 3-phase Machines with 60° Phase Belts

N_{sp} = number of slots per pole

N_{sp}	3	6	9	12	15	18	21	24	∞
$n = 1$	1.00	+0.966	+0.960	+0.958	+0.957	+0.956	+0.956	+0.956	+0.955
$n = 3$	1.00	+0.707	+0.667	+0.653	+0.647	+0.644	+0.642	+0.641	+0.637
$n = 5$	1.00	+0.259	+0.218	+0.205	+0.200	+0.197	+0.196	+0.194	+0.191
$n = 7$	1.00	−0.259	−0.177	−0.157	−0.149	−0.145	−0.143	−0.141	−0.136
$n = 9$	1.00	−0.707	−0.333	−0.270	−0.248	−0.236	−0.229	−0.225	−0.212
$n = 11$	1.00	−0.966	−0.177	−0.128	−0.109	−0.102	−0.097	−0.095	−0.087
$n = 13$	1.00	−0.966	+0.218	+0.128	+0.102	+0.091	+0.086	+0.083	+0.073
$n = 15$	1.00	−0.707	+0.667	+0.270	+0.200	+0.173	+0.159	+0.149	+0.127
$n = 17$	1.00	−0.259	+0.960	+0.157	+0.102	+0.084	+0.075	+0.070	+0.056
$n = 19$	1.00	+0.259	+0.960	−0.205	−0.109	−0.084	−0.072	−0.066	−0.050
$n = 21$	1.00	+0.707	+0.667	−0.653	−0.248	−0.173	−0.147	−0.127	−0.091
$n = 23$	1.00	+0.966	+0.218	−0.958	−0.149	−0.091	−0.072	−0.063	−0.041
$n = 25$	1.00	+0.966	−0.177	−0.958	+0.200	+0.102	+0.075	+0.063	+0.038
$n = 27$	1.00	+0.707	−0.333	−0.653	+0.647	+0.236	+0.159	+0.127	+0.071

Distribution factors for fractional slot windings are tabulated for 3-phase windings in Ref. 12. For most fractional slot windings with 60° phase belts, the coils of one phase occupy many different angular positions. The fundamental distribution factor is then very nearly the ratio of a 60° chord to a 60° arc, and $K_d = 0.955$.

16. Skew Factor K_s. When a skewed relationship exists between the armature conductors and the axis of symmetry of the field flux form, the total voltage induced in an armature conductor is the integral of the incremental voltages over the total electrical angle of skew λ. The resultant voltage is reduced by the skew factor K_s,

$$K_{sn} = \text{skew factor for } n\text{th harmonic}$$

$$= \frac{\sin (n\lambda/2)}{n\lambda/2} \qquad (6\text{-}7)$$

17. Phase Factor K_ϕ. For Y-connected 3-phase machines the terminal-to-terminal voltage is the vector difference of two terminal-to-neutral voltages which are 120 elec deg out of phase. The resultant fundamental voltage is 0.866 times the arithmetic sum of the two phase voltages, and $K_\phi = 0.866$. For the harmonic components, the phase difference is $n \times 120°$, resulting in $K_{\phi n} = 0$ for all odd multiples of the third harmonic and $K_{\phi n} = 0.866$ for all other odd harmonics.

Similarly, for 2-phase machines the phase difference is $n \times 90°$, and $K_{\phi n} = 0.707$ for the fundamental and all odd harmonics.

18. Voltage Equation, Open Circuit. A modification of the familiar transformer equation is one of the many possible means for including the preceding factors in the calculation of open-circuit voltage,

$$V_{tt} = \text{fundamental rms terminal-to-terminal voltage}$$

$$= 4.44\phi f N(K_p K_d K_s K_\phi) \times 10^{-8} \qquad (6\text{-}8)$$

where ϕ = fundamental flux per pole, maxwells, f = line frequency, cycles per second, N = total number of turns in series between terminals (equals 2 × series turns per phase for 3-phase Y-connection), and $K_p K_d K_s K_\phi$ = factors from preceding paragraphs.

For the nth harmonic, the fundamental voltage is multiplied by the reduction factor $C_n K_{pn} K_{dn} K_{sn} K_{\phi n}/C_1 K_p K_d K_s K_\phi$.

19. Tooth Ripples. The flux forms illustrated in Figs. 6-1 and 6-2, and the preceding analysis, do not include the effects of the armature teeth and slots on the air-gap flux. The resulting harmonic fluxes do not directly generate voltages in the armature conductors but rather cause currents in the field and damper windings. These currents produce flux components which do generate voltages in the armature conductors at frequencies which are multiples of the number of slots per pair of poles plus or minus 1. For example, with 12 slots per pair of poles (6 slots per pole) the harmonics will be the eleventh, thirteenth, twenty-third, twenty-fifth, etc.

20. Voltage Waveform Standards. The harmonic content of the voltage wave is generally specified in terms of (1) the maximum deviation of the voltage waveform from a pure sine wave and (2) the weighted average of all harmonics.

The **deviation factor** is specified and defined by the USA C50 Standards as follows: "The deviation factor of the open-circuit terminal voltage wave of synchronous machines shall not exceed 0.1, unless otherwise specified. The deviation factor of a wave is the ratio of the maximum difference between corresponding ordinates of the wave and of the equivalent sine wave to the maximum ordinate of the equivalent sine wave when the waves are superimposed in such a way as to make this maximum difference as small as possible."

The **telephone influence factor** of the open-circuit terminal voltage waveform is defined in the USA C42 Standards, and standard values are specified in the USA C50 Standards. This factor is the weighted sum of all harmonics in the voltage wave. The weighting of each harmonic is intended to reflect the relative objectionable effect of inductive coupling at the harmonic frequency with telephone communication.

21. Harmonics under Load Conditions. Load on the machine affects the harmonic content principally in the following ways: (1) Increased field current tends to increase all internal voltages, including harmonics. (2) Armature reaction generally reduces the fundamental more than the harmonics and may introduce additional harmonics.

(3) Magnetic saturation changes the harmonic magnitudes. (4) The portion of the harmonic internal voltage appearing at the terminals depends on the relationship of the load impedance and the machine internal impedance at the harmonic frequency.

Generally, these effects tend to offset each other, but it is possible for series resonance to occur between machine internal inductance and load capacitance at some generated harmonic frequency, resulting in a very large harmonic voltage at the machine terminals. Also, the possible malfunction of some external device because of generator harmonics depends as much on the nature of the device and the associated electrical system as it does on the generator characteristics. For these reasons, generator voltage-waveform specifications are usually limited to open-circuit conditions where the generator is isolated from the influences of the load circuit.

ARMATURE REACTION

22. General Effect. Current in the armature conductors produces an mmf which has the same number of poles as the field structure and rotates at synchronous speed. This mmf may add to or subtract from the field mmf depending on its angular displacement from the pole axis. For generators supplying reactive current to an inductive load, the net effect of the armature reaction is to oppose the field mmf, requiring additional field current to offset the armature reaction and sustain the flux and voltage.

23. Armature Ampere-turns. The effective number of turns producing armature reaction is reduced by the same factors as apply for the generated voltage. The effects of the pitch and distribution factors are illustrated in Fig. 6-9 for 1 phase at one pole. The mmfs of the two adjacent phase belts are displaced +60 and −60° in space and time, the effect being to increase the mmf in the axis of the phase illustrated by a factor of 1.5. Then

$$AT_d = \text{peak of fundamental component of armature At/pole}$$

$$= 1.5\sqrt{2}IN_{pp}(K_pK_dK_s)4/\pi \tag{6-9}$$

where I = rms current, N_{pp} = turns per phase per pole, and $K_pK_dK_s$ are from Eq. (6-8) and Pars. **14, 15,** and **16.**

24. Harmonic MMFs and Losses. The distribution of the armature ampere-turns also produces space harmonic mmfs and, in the case of fractional slot windings, subharmonic mmfs,

$$AT_{dn} = \text{peak of } n\text{th harmonic of armature At/pole}$$

$$= (1.5/n)\sqrt{2}IN_{pp}(K_{pn}K_{dn}K_{sn})4/\pi \tag{6-10}$$

where n = order of harmonic or subharmonic (n is less than 1.0 for subharmonics).

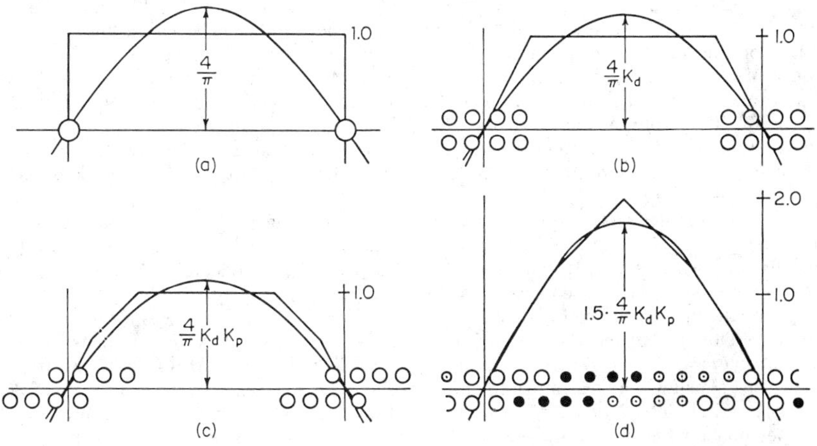

Fig. 6-9. Effect of coil pitch and phase-belt distribution on fundamental armature-reaction mmf. (*a*) Single full-pitch coil. (*b*) Full-pitch coils, 60° phase belts. (*c*) Short-pitch coils, 60° phase belts. (*d*) Three-phase mmf.

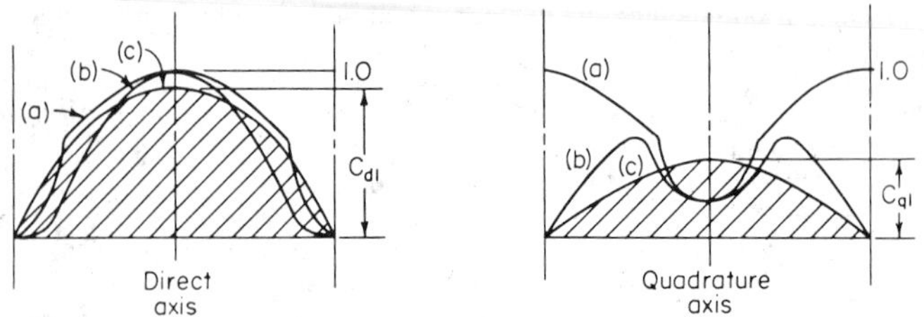

Fig. 6-10. Direct and quadrature axis fluxes with sinusoidal armature mmf. (*a*) Air-gap permeance. (*b*) Actual air-gap flux. (*c*) Fundamental component of air-gap flux.

All these mmfs result from fundamental current and rotate at the synchronous speed corresponding to their order *n*. The corresponding fluxes induce voltages of fundamental frequency in the armature winding. However, these fluxes rotate forward or backward with respect to the rotor and cause losses in the pole faces.

The fifth- and seventh-harmonic mmfs are often sources of substantial loss unless they are reduced by a suitable coil pitch such as ⅚. Another common source of loss is the steps in mmf from one slot to the next. The resulting harmonics are not reduced by K_{pn} or K_{dn}, but more slots per pole or a larger air gap reduces the resulting flux and loss.

25. Direct and Quadrature Axes. The fundamental armature mmf is commonly resolved into two components 90° displaced from each other for purposes of analysis. One of these is the direct-axis component acting in line with the poles, and the other is the quadrature-axis component acting in line with the axis of symmetry midway between poles. These axes are the direct and quadrature axes.

Because the permeance distributions are different in the two axes, the same mmf produces different fluxes in the two cases. Figure 6-10 illustrates these conditions. C_{d1} and C_{q1} are the peak values of the fundamental components of the air-gap flux density in the two axes relative to the flux density corresponding to the peak mmf acting on the minimum effective air gap. These coefficients are evaluated in Ref. 11 for a wide range of pole-face proportions.

ARMATURE WINDINGS

26. Winding Types. A wide variety of winding types are possible to produce a desired voltage in the proper number of phases and with a suitable waveshape. Practical considerations, mainly economic, limit the usual alternator winding to a double-layer 3-phase lap winding, arranged in 60° phase belts in open slots. The number of coils, the number of turns per coil, the coil pitch, the number of circuits, and the connection of the phases are selected to give the desired voltage and waveform.

Double-layer windings in open slots permit the use of form-wound coils which are all alike in a given machine. These coils have the characteristic diamond shape in the end area. Three-phase windings may be either Δ- or Y-connected; Y-connected machines are much more common, particularly in the larger sizes. The winding may be arranged to be connected either Y or Δ, with leads brought out from both ends of each phase to make this possible.

27. Number of Coils. For integral slot windings, where the number of slots per phase per pole is a whole number, all coil groups are identical, and the phase voltages are balanced with respect to magnitude and angle. Fractional slot windings, where the number of slots per phase per pole is not an integer, have unequal coil groups. These can be arranged to produce balanced voltages if the number of phases is a factor of the number of slots per phase at least as many times as it is a factor of the number of poles. For example, a balanced 3-phase winding for a 36-pole alternator will require a number of slots per phase which is a multiple of 3 times 3, since 3 appears twice as a factor of 36. Similarly, a balanced 2-phase winding for a 16-pole alternator will require a number of slots per phase which is a multiple of 16 (2 times 2 times 2 times 2).

Other numbers of slots than those which satisfy the above criterion may be used by making coils which do not all have the same number of turns. Special groupings of the coils may also permit a balanced winding when the number of slots does not conform to this criterion. These expedients may be used because of limitations in available punching dies.

28. Conductor Design. Conductors are stranded except in the smallest ratings to enable easier shaping of the coil and to limit eddy-current losses which result from the flux that crosses the slot. The effect of this flux is to produce a voltage within the strand which results in circulating currents. The thinner the strand, the less the voltage and the resulting eddy currents. Since the strands ordinarily are all connected together at the joints between coils, there will also be eddy currents circulating between strands because of the difference in flux linked by the various strands. This source of loss can often be reduced sufficiently by the use of more and thinner conductors per coil, with a corresponding increase in the number of parallel circuits. As an alternative, some kind of transposition may be used to control eddy currents between strands. The purpose of the transposition is to arrange the strands in one part of the coil with respect to another so that the induced voltages cancel each other.

29. Types of Transpositions. The Roebel type of transposition is often used where conductor strands are arranged two in width by several strands deep. This is illustrated in Fig. 6-11. The effect of the Roebel transposition is the same as twisting the bundle of strands 180 or 360°. This is accomplished without substantially increasing the width of the bundle of strands. For a multiturn coil, a 180° transposition usually suffices. For a single-turn coil, because of its greater depth, a 360° transposition (illustrated) may be necessary in which each strand occupies all possible positions in the slot.

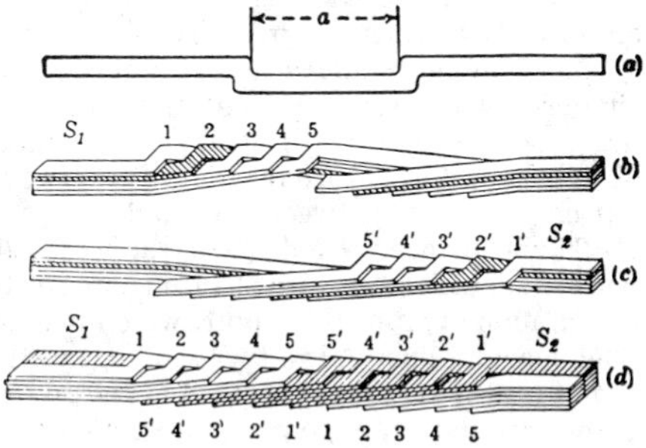

FIG. 6-11. Roebel type of transposition. (*a*) Typical offset conductor strand; (*b*) group of conductor strands comprising half the conductor; (*c*) complementary group of strands; (*d*) completely assembled conductor.

Another type of transposition provides a 180° twist of the strand bundle in the end-turn portion of the winding, where the increased width dimension can be tolerated.

Sometimes it is satisfactory to make a transposition by groups of strands rather than by individual strands. This may be done in the end turns of the winding, by connecting upper groups in one coil in series with lower groups of adjacent coils. The groups of strands must be insulated from adjacent groups throughout the coils included in the transposition.

These transpositions are effective in limiting only eddy currents produced by flux crossing the slots in the core. Eddy currents resulting from voltages induced in the ends of the winding may be controlled by extending a transposition such as the Roebel type into the coil ends. This presents manufacturing difficulties. A convenient means of essentially eliminating eddy currents between strands due to coil-end flux is a 540° Roebel-type transposition entirely in the straight portion of the coils. A 180° transposition occupies the middle half of the coil, and an additional 180° occupies each outer quarter of the coil.

30. Skewing. Sometimes slots are skewed to minimize the effects of voltages resulting from the ripple in air-gap flux produced by the stator slots. Skewing a slot essentially a full slot pitch in the length of the core will eliminate voltages due to slot ripple. The effect of a number of slots which produces undesirable ripple voltages can be limited by some degree of skew.

31. Dead Coils. Coils may be left unconnected in the armature circuit during manufacture to obtain operating characteristics not readily obtainable from adjustment of number of coils and turns. They may also be cut out of machines in service in order to permit operation after one coil or more has been damaged. Where the winding has parallel circuits, the number of coils cut out must be the same in each circuit in order to avoid damaging circulating currents. If more than a few percent of the coils are cut out, it may also be necessary to cut out a like number of coils in the other phases of the machine. The controlling consideration is the voltage balance and the circulating ground currents. This situation would be most critical in a Δ-connected winding.

32. Coil Pitch. Most economical use of the flux in the magnetic circuit results when the stator coil is full pitch, or as near to full pitch as a fractional slot winding will permit. In a fractional slot winding, the maximum pitch ordinarily used is full pitch for the integer; i.e., for a $7\frac{5}{7}$ slots per pole winding, a coil pitch of 1 to 8 would be the maximum ordinarily used. Less than full-pitch coils are used to obtain adjustments in the voltage generated or to limit harmonics.

33. Single-phase windings are usually 3-phase windings with 1 phase not used. Sometimes the coils for the third phase are omitted, but most often they are wound in the machine and might be considered as spares. In small sizes a special concentric winding may be used. A low-resistance damper winding is generally necessary on single-phase machines to reduce the flux pulsations that are set up by the single-phase armature reaction and to reduce the effective armature reactance. Single-phase machines have an inherent torque pulsation at twice rated frequency. The resulting noise and vibration are noticeable on small machines and may require special construction on large machines.

34. Two-phase windings differ from 3-phase windings only in the grouping of the stator coils. Ninety-degree phase belts are ordinarily used.

35. Double windings are sometimes used to reduce short-circuit currents and to simplify switchgear and bus structure problems. The windings have standard coil design with special end connections. The electrical designs must allow for the effects of unbalanced armature reaction if the two windings are not equally loaded and for the mutual reactance between the windings.

36. Multispeed Windings. For some applications a winding is required which will permit operation at more than one speed, but at the same frequency. An example is a hydraulic pump-turbine unit, which may generate at one speed and pump at another. Such machines may have two windings in the same slots or in adjacent slots, or a single armature winding may be reconnected by means of suitable switches to serve this purpose. Special rotor construction and field-winding reconnections are also necessary.

INSULATION

37. Classes of Insulation. The electrical insulation of the alternator windings is designed to operate satisfactorily at the specified voltages and temperature and to retain its dielectric and mechanical strength and dimensional stability over many years of operation. The USASI has defined various classes of insulating systems based on the maximum steady-state operating temperatures and has established voltage proof tests to demonstrate the dielectric capability of the insulation system. Other nondestructive test techniques have been developed to evaluate the dielectric capability and condition of the insulation systems.

Of the several insulation-system classes, four are most applicable to large rotating machines. These are Classes 105, 130, 155, and 180, where the numbers signify the hot-spot temperature in degrees centigrade. These classes were formerly known as A, B, F, and H, respectively. Class 105 and Class 130 are most frequently applied to alternators; most of the discussion in this article will refer to these two.

Class 105 insulation systems are comprised of organic materials such as cotton, silk, paper, and certain synthetic films. Varnishes and synthetic resins are used as binders. **Class 130 systems** are comprised of inorganic materials such as mica, glass fibers, asbestos, and synthetic films, with suitable binders. **Class 155 systems** are comprised of generally similar materials to those of Class 130, with binders selected for suitable life at the higher temperatures. **Class 180 systems** include the silicone elastomers as well as mica, glass fibers, asbestos, etc., and high-temperature binders. Any of these systems may include other materials or combinations of materials in limited quantities if by experience or accepted test they can be shown to have acceptable thermal life at the specified temperature.

The rapid growth of the field of synthetic chemistry has presented a continual flow of materials suitable for electrical insulation. The IEEE has prepared guides for test procedures for the thermal evaluation of electrical insulating materials and systems (see Ref. 5 in Bibliography, Par. **105**).

Table 6-2a. Limiting Observable Temperature Rise of Air-cooled Salient-pole Alternators

(From Ref. 4)

Item	Machine part	Method of temperature determination	Generators other than large and high-voltage hydraulic-turbine-driven generators				Large and high-voltage hydraulic-turbine-driven generators*
			Temperature rise, °C†				
			Class A insulation	Class B insulation	Class F insulation	Class H insulation	Class B insulation
1	Armature winding: *a.* Machines of 1,500 kVA and less‡ *b.* Machines of more than 1,500 kVA	Thermometer Resistance Embedded detector	50 60 60	70 80 80	. . . . . .	110 125 125	60
2	Field windings	Resistance	60	80	. .	125	60
3	Cores and mechanical parts in contact with or adjacent to insulation	Thermometer	50	70	. .	110	55
4	Collector rings	Thermometer	65	85	. .	85	65
5	Amortisseur windings and miscellaneous parts (such as brushholders, brushes, pole tips, etc.)	May attain such temperatures as will not injure the machine in any respect					

* Large and high-voltage machines are those having voltage ratings above 5,000 V and kVA ratings as follows:
Hydraulic turbine-driven generators, 6,250 kVA and above.

Large and high-voltage hydraulic-turbine-driven generators and reversible generator motor units may be operated at 115% load at rated power factor, frequency, and voltage with temperature rises in excess of normal standards for these machines. When operated at loads above rated load, it is recommended that the ambient be kept as low as possible to assure maximum usable life.

† For machines which operate under prevailing barometric pressure and which are designed not to exceed the standard temperature rise at altitudes of 3,300 ft (1,000 m) to 13,000 ft (4,000 m), the temperature rises, as checked by test at low altitude, shall be less than those specified in this standard by 1% of the specified temperature rise for each 330 ft (100 m) of altitude in excess of 3,300 ft (1,000 m).

‡ The method of temperature determination to be used shall be optional with the manufacturer unless otherwise agreed upon. Only one method of temperature determination shall be required in any particular case.

38. Hot-spot Allowance. It is ordinarily not practicable to measure maximum hot-spot temperatures on alternator windings; so machine ratings are based on observable temperatures, and a suitable insulation class is used, based on the expected difference between the observed temperature and the maximum temperature. This temperature difference is called the hot-spot allowance.

Conventional hot-spot allowances have been established depending on the method of determining observed temperature. Although in a specific instance a manufacturer may deviate from these, they provide guidance for selecting suitable insulation systems and for establishing standard temperature rises.

39. Measuring Winding Temperatures. Several methods of observing winding temperatures are defined in the USAS Standards. The **thermometer method** consists in determining the temperature with a mercury thermometer or other suitable temperature-measuring device applied to the hottest parts ordinarily accessible to mercury thermometers. This method is commonly applied to armature windings of smaller alternators. Of the several methods listed here, the thermometer method gives readings furthest from the maximum winding temperature and has the largest hot-spot allowance associated with it. The **resistance method** consists in determining temperature from a comparison of the winding resistance at the operating temperature with its resistance at a known temperature. For the **embedded-detector method,** thermocouples or resistance temperature detectors are built into the machine in locations inaccessible to mercury thermometers. In the commonest application they are placed between the coil sides in a two-layer armature winding near the middle of the core. The resistance and embedded-detector methods give temperature readings closer to the maximum temperature and ordinarily have the same hot-spot allowance associated with them. The **applied-thermocouple method** uses a thermocouple in direct contact with the conductor or separated from it by only the integrally applied insulation of the conductor itself. Applied thermocouples may be located on or near the hottest parts of the winding. This method is not ordinarily associated with machine temperature ratings but is used for experimental testing to determine hot-spot allowances, etc.

40. Temperature Ratings. Temperature ratings of alternator armature windings are usually based on the thermometer method for smaller ratings and on the embedded-detector method for larger ratings. Temperature ratings of alternator fields are usually based on the resistance method. Standards have been established for rating machines, using these temperature-measuring methods and related to the class of insulation system provided. These are summarized in Table 6-2.

41. Armature Winding Insulation. Standard armature voltages range from 220 to

Table 6-2b. Limiting Observable Temperature Rise of Air-cooled Round-rotor Alternators*

(From Ref. 4)

Item	Machine part	Method of temperature determination	Class B insulation	Class H insulation†
1	Armature windings of machines 500 kVA and above	Embedded detector	60	100
2	Field windings of machines 500 kVA and above	Resistance	85	125
3	Cores and mechanical parts in contact with or adjacent to insulation	Thermometer	70	110
4	Collector rings‡	Thermometer	85	85
5	Miscellaneous parts such as brushholders, brushes, etc.	May attain such temperatures as will not injure the machine in any respect.		

* The temperature rises in degrees centigrade in this table apply to machines having the class of insulation shown at the heads of the respective columns.

† The temperature rise values given for Class H insulation are based only on consideration of the insulation. Successful operation of machines at these temperatures may require that special consideration be given to other parts of the machine such as bearings, lubricants, brushes, etc.

‡ The class of insulation refers to the insulation affected by the heat from the collector rings. The limiting observable temperature rise for Class H insulation shall be the same as listed for Class B insulation.

18,000 V or more. Appropriate amounts of turn and ground insulation are provided to withstand normal operating voltages under both steady-state and transient conditions. In low-voltage systems the turn insulation may be applied directly to the conductor as a film or serving. In the higher voltage ranges (generally above 5,000 V) special construction is used to control corona. This is an electrostatic discharge due to the voltage gradient within or at the surface of the coils exceeding the dielectric strength of the air. In the presence of moisture this discharge produces nitrous acid, which decomposes organic materials associated with the insulation system. Insulation for high-voltage coils is applied in such a way that internal voids are minimized. The outer surfaces of the slot portion of the highest-voltage windings are coated with a semiconducting medium to lower the voltage gradient between coil and core. Spacing between coils in the end windings is controlled to limit discharge which would be damaging to the cording and blocking of the coil-support structure.

42. Field Winding Insulation. Field insulation systems, because of the lower voltages involved, present fewer design problems than armature systems. Operating voltages of field windings are in the range of 125 to 375 V, occasionally somewhat higher. Transient conditions, for example, the interruption of full-load field current, may impose voltages several times rated volts for a short time. The same insulation-system classes are applied to field insulation systems. Temperatures are usually measured by the resistance method and are shown in Table 6-2.

Field windings are subjected to the centrifugal forces of rotation, and the assembly must be dimensionally stable so that the turns in the coil do not separate from each other and the coil does not become loose on the pole.

43. Tests of Insulation Systems. The high-potential dielectric test and the insulation-resistance test are the principal methods for evaluating insulation capability and condition. The high-potential test prescribed by the USASI is twice rated terminal-to-terminal voltage, plus 1,000 V for armature windings, and ten times rated voltage, but not less than 1,500 V, for alternator field windings. This is an a-c test, of specified waveshape, applied under controlled conditions, and maintained at its top value for 1 min. It is a severe test and is applied only on new windings after ascertaining that they are dry and otherwise in good condition. The design breakdown level of the insulation may be several times the test voltage, to provide a suitable factor of safety over process control.

The **insulation-resistance test** is often used as a measure of the condition of the winding. Insulation resistance is the ratio of the applied voltage to the current at some specified time after the voltage is applied. Direct, rather than alternating, voltages are used for measuring insulation resistance.

The principal currents affecting insulation resistance after 1 min or 10 min of application of the test potential are (1) leakage current over the winding surface, (2) conduction through the insulation material, and (3) absorption currents in the insulation. The first two currents are essentially steady with time, but the last current decays approximately exponentially from an initial high value. Such insulation-resistance measurements are affected by surface condition (dirt or moisture on the winding surfaces), moisture within the insulation wall, and insulation temperature. The magnitude of the test potential may also affect the insulation value, especially if the insulation is not in good condition. Therefore, it is desirable in using insulation resistance as a measure of winding condition over a period of years to make readings under similar conditions each time.

The **dielectric-absorption characteristic,** which is displayed in the shape of the curve of insulation resistance against time when a constant test potential is applied for 10 or 15 min, is also frequently used as an indication of the winding condition. The resistance of a clean, dry winding will continue to rise as the test potential is maintained, becoming fairly steady after 10 or 15 min. A wet or dirty winding will reach its steady value much sooner.

The ratio of the 10-min reading to the 1-min reading is called the **polarization index.** Other things being equal, then, a high polarization index indicates good winding condition.

A common device for measuring insulation resistance is a direct-indicating ohmmeter with a self-contained hand- or power-driven generator. The megger is a typical device

Table 6-2c. Limiting Observable Temperature Rise of Hydrogen-cooled Round-rotor Alternators*
(From Ref. 4)

Item	Machine part	Conventionally cooled windings		Conductor cooled windings		
		Method of temperature determination	Class B insulation (notes 1 and 2)	Method of temperature determination	Class B insulation (notes 1 and 2): Type of coolant: Liquid	Class B insulation (notes 1 and 2): Type of coolant: Gas
1	Temperature of cold coolant:					
	a. Generators rated at 30 lb/in² gage hydrogen press.	Detector or thermometer (note 3)	46			
	b. Generators rated at 30, 45, 60, or 75 lb/in² gage hydrogen pressure	....	....	Detector or thermometer (note 3)	45–50 (note 4)	45–50 (note 4)
2	Temperature rise of armature windings:					
	a. Generators rated at 30 lb/in² gage hydrogen pressure	Embedded detector	45 (note 5)			
	b. Generators rated at 30, 45, 60, or 75 lb/in² gage hydrogen pressure	....	....	Coolant (notes 6 and 7)	55–50 (notes 4 and 5)	65–60 (notes 4 and 5)
3	Temperature rise of field windings:					
	a. Generators rated at 30 lb/in² gage hydrogen pressure (below 100,000 kVA)	Resistance	79 (note 5)			
	b. Generators rated at 30 lb/in² gage hydrogen pressure (100,000 kVA and above)	Resistance	74 (note 5)			
	c. Generators rated at 30, 45, 60, or 75 lb/in² gage hydrogen pressure	....	....	Resistance	....	65–60 (notes 4 and 5)
4	Temperature rise of cores and mechanical parts in contact with or adjacent to insulation as determined by the type of armature winding affected (note 8):					
	a. Generators rated at 30 lb/in² gage hydrogen pressure (below 100,000 kVA)	Thermometer	64			
	b. Generators rated at 30 lb/in² gage hydrogen pressure (100,000 kVA and above)	Thermometer	49 (note 9)			
	c. Generators rated at 30, 45, 60, or 75 lb/in² gage hydrogen pressure	....	....	Detector or thermometer	85–80 (notes 5 and 10)	85–80 (notes 5 and 10)
5	Temperature rise of collector rings all generators (note 11)	Thermometer	85	Thermometer	....	85
6	Temperature of miscellaneous parts such as amortisseur windings, rotor surface, brushholders, brushes, etc., may attain such levels as will not injure the machine in any respect					

* The temperature and temperature rises in degrees centigrade in this table apply to the types of windings as given at the heads of the respective columns.

Note 1: Because of the large thermal gradient between hottest-spot and observed temperatures of large high-voltage generators and because of mechanical considerations of thermal expansion, it is often desirable to design for lower temperatures than shown in Table 2 on large or high-voltage machines or machines intended for operation with highly variable loads.

Note 2: Hydrogen-cooled generators which operate under controlled pressure do not require a correction for temperature rise at altitude if the pressure of the cooling medium is maintained at the absolute pressure corresponding to the rated value.

Note 3: The method of coolant temperature measurement shall be optional with the manufacturer unless otherwise agreed upon. Only one method of temperature measurement shall be required in any particular case.

Note 4: Cold-coolant temperatures may be provided within the range of 45 to 50°C, at the manufacturers' option, so long as compensating adjustments are made in the rise of the respective parts so that the sum of the cold-coolant temperature and respective part rise does not exceed:

1. 100°C for liquid-cooled and 100°C for gas-cooled armature windings listed in items 2 and 4.
2. 110°C for gas-cooled field windings listed in item 3.

Note 5: In designing to meet the temperature rises of Table 2, it is intended that the hottest-spot total temperature of 130°C will not be exceeded. Such design concepts would be demonstrated by:

1. For armature winding—direct measurement or recognized methods of calculation correlated to special factory tests on a basically similar machine.
2. For field winding—recognized methods of calculation.

Note 6: The temperature rise of the coolant at the outlet of the hottest coil shall be considered the observable temperature rise of conductor-cooled armature winding.

Note 7: The embedded detector may be used in place of the coolant method if the embedded detector reads essentially the same as the hot coolant.

Note 8: The temperature of the core and mechanical parts in contact with or adjacent to insulating material (including that of the winding and of the core laminations) shall not exceed the values in the table. The temperature of all metal parts, including structural members and shielding devices in the end region, is not required to be within the limiting temperature, provided that these parts do not appreciably influence the temperature of insulating material either by contact or radiation. These parts may be operated at temperatures which are considered safe for the particular metals used.

Note 9: Higher localized values of temperature rise on the core end sections and the end structure are permissible during generator operation near unity or leading power factor. Instead of 49°C, the limiting temperature rise may be 64°C. For near unity or leading power factor operation, the highest temperature is usually found at or near the ends of the core on the teeth. The temperature may be obtained by thermocouples or equivalent detectors, if specified.

Note 10: For conductor-cooled armature windings, the values shown for item 4 are limiting regardless of the operating power factor.

Note 11: The class of insulation refers to the insulation affected by the heat from the collector rings.

of this type. Measurements are made at voltages in the range of 500 to 2,500 V, higher voltages sometimes being used in checking very high voltage windings.

Because of the many factors affecting insulation resistance it is impracticable to establish rigid standards for minimum values of either insulation resistance or polarization index. It has been recommended by the IEEE, however, that the minimum insulation-resistance value for alternator windings be 1 MΩ/1,000 V rated voltage + 1 MΩ, measured with the winding at 40°C with a 500-V test potential. It is recommended also that the minimum value of polarization index be 1.5 for Class 105 insulation systems and 2.0 for Class 130 insulation systems.

Reference 6 in the Bibliography provides more information for testing and interpreting insulation condition.

CONVENTIONAL COOLING

44. Definition. Conventionally cooled machines dissipate their losses to a cooling medium which is entirely outside the coil insulation. All air-cooled machines, with rare exceptions, are cooled in this manner, as well as most hydrogen-cooled machines except very large turbine-generators. The latter machines use conductor cooling, as described in the next article.

45. Cooling Mediums. The comparative characteristics of various gases that might be used for cooling are shown in Table 6-3. Air is most commonly used, for obvious reasons. Hydrogen provides better heat transfer with much less windage loss, which is nearly proportional to density. Frequently hydrogen is used at higher than atmospheric pressure, which further improves its heat-transfer capabilities, but with greater windage loss. Other advantages of hydrogen are the reduction of insulation oxidation and fire hazard. Hydrogen purity is normally maintained in the range of 95 to 99% where the mixture is nonexplosive and will not support combustion. The explosive range of hydrogen-air mixtures is 5 to 75% hydrogen.

The effects of different cooling gases and hydrogen pressure variation on temperature rise are illustrated in Figs. 6-12 and 6-13.

46. Ventilating Paths. Stators are usu-

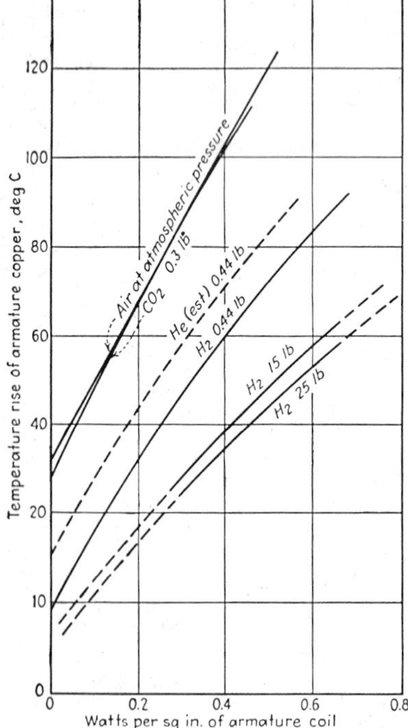

FIG. 6-12. Armature heating of a 6,250-kVA alternator with different cooling gases. (*From Ref.* 14.)

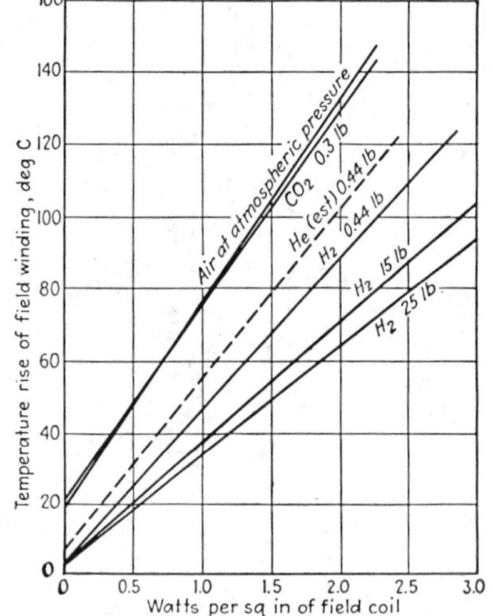

FIG. 6-13. Field heating of a 6,250-kVA alternator with different cooling gases. (*From Ref.* 14.)

Table 6-3. Properties of Cooling Gases
(From Ref. 13)

Characteristics	Air	N_2	CO_2	NH_3	H_2	He	Methane
Thermal conductivity	1	1.08	0.638	0.868	6.69	6.40	1.29
Density	1	0.966	1.52	0.588	0.0696	0.1378	0.554
Specific heat (const. press)	1	1.046	0.848	2.185	14.35	5.25	2.495
Heat capacity	1	1.02	1.29	1.232	0.996	0.72	1.38
Heat transfer	1	1.03	1.132	1.228	1.51	1.18	1.43

ally cooled by blowing air or hydrogen over the coil ends and also through radial ducts in the armature core. The ducts are normally ⅜ in wide, with a spacing of about 2 in, but they may be omitted entirely on machines with short core lengths. In salient-pole machines, the gas normally enters at the ends of the rotor in the space between poles and flows radially outward through the stator ducts. In turbine-generators the cross section for axial flow into the machine is greatly restricted, and the cooling gas is usually ducted to the outer diameter of the stator core, where it is directed radially inward through some of the stator ducts to the air gap. Other radial ducts in the stator allow the gas to discharge radially back into the cooling system.

The field coils of salient-pole machines are normally cooled by the axial flow between poles and by the relative velocities at the ends resulting from rotation. For large salient-pole machines, the field coils may have strap copper conductors wound on edge with some turns extended, or the exposed edge of the strap may be beveled to increase the exposed surface and improve the cooling. Turbine-generator rotors normally have axial gas flow through ventilating slots or other passages to supplement the cooling at the exposed cylindrical surface.

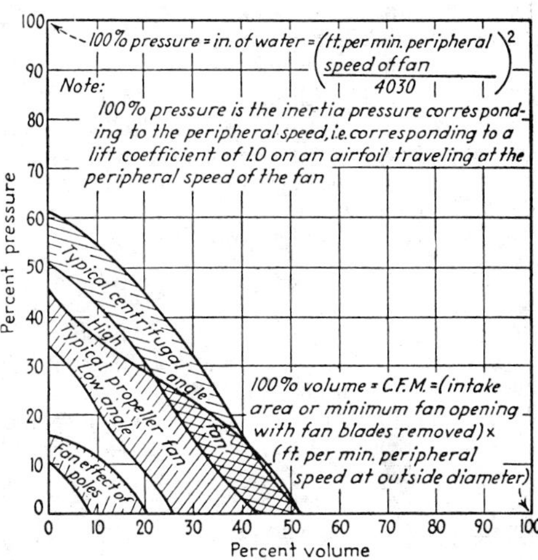

FIG. 6-14. Typical performance of shrouded fans as used without air guides or diffusers on air-cooled machines.

47. Fans and Enclosures. Machine enclosures have a significant effect on the fans and ventilating system. Small slow-speed salient-pole machines often have open end shields and are cooled largely by the unconfined air circulation at the ends produced by simple fan blades on the rotor. Air flow through the center of these machines may be produced largely by the centrifugal effects acting on the air columns between poles. For larger machines, solid end shields are generally used, providing static pressure chambers at the ends and more definite gas-flow paths. More effective fans are also characteristic of larger machines and may be either axial-flow propeller type or radial-flow centrifugal. Typical machine-fan characteristics are shown in Figs. 6-14 and 6-15.

48. Coolers. Most large machines have closed recirculating systems with fin-tube coolers to transfer the heat absorbed by the gas to the cooling water. For hydrogen-cooled machines the coolers usually consist of several units arranged either vertically or horizontally within the stator frame. For air-cooled machines, the closed recirculating system provides clean, cooled air independent of the environment. Horizontal air-cooled machines commonly have a single cooler located in a horizontal position below the machine, while vertical water-wheel generators usually have several coolers distributed around the outer periphery of the stator within an air housing.

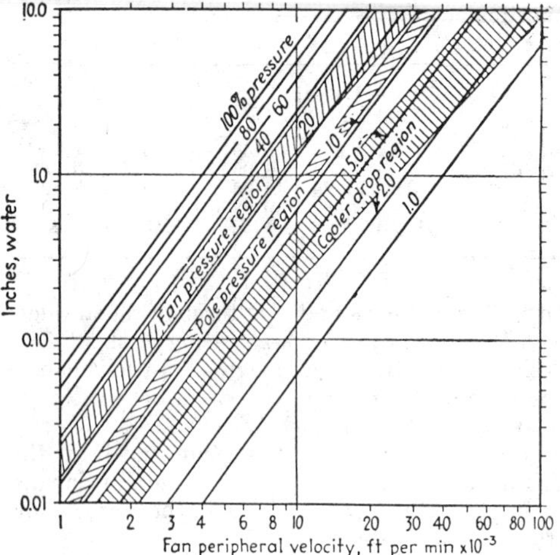

FIG. 6-15. Curve for estimating permissible cooler drops and approximate pressure developed by fans and by poles for air-cooled machines.

Coolers are usually selected to have a gas pressure drop which is low enough so that the gas flow through the machine is not unduly restricted. Typical cooler drops for air-cooled machines are shown in Fig. 6-15.

49. Gas Quantities. A major factor determining the required gas flow through the machine is the temperature rise of the gas resulting from the heat loss it absorbs. Air and hydrogen *at atmospheric pressure* are essentially the same in this respect, and

$$°\text{C gas rise} = 1{,}880 \frac{\text{kW loss absorbed}}{\text{gas, ft}^3/\text{min}} \tag{6-11}$$

A typical volume for operation at atmospheric pressure is 100 ft³/min per kW of loss absorbed, resulting in a gas temperature rise of 18.8°C in passing through the machine. Machines designed to operate at more than atmospheric pressure may require less volumetric flow because the degrees centigrade gas temperature rise is inversely proportional to gas density for a given cubic foot per minute per kilowatt of loss absorbed.

50. Cooling Water. The temperature rise of the cooling water in absorbing the machine losses is usually quite small and is

$$°\text{C water rise} = 3.78 \frac{\text{kW loss absorbed}}{\text{water, gal/min}} \tag{6-12}$$

The water quantity normally used is roughly 1.0 gal/min per kW loss, resulting in a water temperature rise of 3.78°C. Because the required cooler-gas outlet temperature is normally 40°C, the economical water temperature rise may vary considerably depending on the available water temperature, which may be as low as 20°C (68°F) or as high as 35°C (95°F). Also, the required heat-transfer rates in the cooler along with the water-flow pattern and velocity may influence the gallons per minute required. Generally, water requirements are the result of an economic compromise with cooler cost.

51. Hydrogen Seals. Shaft seals for hydrogen-cooled turbine-generators maintain an oil film under pressure in a small clearance between the rotating shaft and a stationary member. The construction may be similar to a journal bearing with a cylindrical oil film or similar to a spring-loaded thrust bearing with the oil film in a plane at right angles to the shaft axis. The oil film is maintained by a supply pressure higher than the hydrogen pressure.

Oil can absorb about 10% by volume of either hydrogen or air. It is important that the seal oil flow toward the hydrogen side be minimized in order to reduce the amount of air carried into the machine and the amount of hydrogen carried out. The oil is sometimes vacuum-treated to minimize its air content before it is supplied to the seal. Any air entering at the seal becomes part of the hydrogen-air mixture within the machine and requires exhausting a mixture volume of 20 to 100 times the air volume and replacing it with pure hydrogen to maintain the hydrogen purity.

52. Scavenging. When hydrogen is removed from the machine to allow inspection or servicing, the scavenging process must be safe from explosion hazard. The usual method is to admit carbon dioxide and exhaust hydrogen from the highest point of the enclosure with the machine deenergized and either at standstill or on the turning gear. For filling with hydrogen, the gas flows are reversed.

CONDUCTOR COOLING

53. Definition. Conductor cooling is the process of dissipating the armature and field coil losses to a cooling medium within the coil insulation wall. Machines cooled in this manner are also called "supercharged," "inner-cooled," and "direct-cooled" by various manufacturers. The cooling medium either is in direct contact with the conductor copper or is separated only by thin materials having little thermal resistance. Conductor cooling eliminates the temperature differential resulting from heat flow through the coil insulation, providing greater current-carrying capability for the same hot-spot temperature rise.

54. Cooling media normally used are hydrogen, oil, and water. Conductor-cooled turbine-generators operate in a hydrogen atmosphere which provides cooling for all the machine except, in some instances, the armature coils. These coils may be cooled from the common hydrogen system or by a separate oil, water, or high-pressure hydrogen system. Because of the limited cross section available within the coils for cooling-medium flow, the temperature rise of the cooling medium in absorbing the heat loss is usually the most important factor determining the hot-spot temperature rise. Hydrogen-cooled machines are often operated at 60 lbf/in^2 or more to increase the mass flow and reduce the hydrogen temperature rise. The surface heat-transfer rates are also improved, of course.

Electrical insulation associated with hydrogen conductor cooling must allow for adequate creepage distances at the coil hydrogen inlets and outlets. Oil and water systems require, in addition, insulated piping. Water is deionized to maintain low electrical conductivity.

55. Coil Construction and Flow Paths. Typical armature coil construction for hydrogen cooling is illustrated at the top of Fig. 6-16. The hydrogen passages are usually thin-wall high-resistance material, such as stainless steel, to reduce eddy-current losses. Liquid-cooled machines may have similar internal tubes, or some or all of the conductor strands may be hollow to convey the liquid. The flow path may have the two sides of a coil in parallel or in series.

Rotor coil constructions and rotor cooling systems in use are more varied. Some salient-pole machines have hollow-conductor liquid-cooled field windings, but hydrogen is almost universally used for turbine-generators. The lower part of Fig. 6-16 shows an edge-cooled copper construction. For larger machines, hollow conductors are common. The flow path may be from end to end, from the ends to discharge openings near the center, or between alternate high- and low-pressure zones along the rotor length. These zones are separated by baffles in the air gap, with radial ducts in the stator arranged to allow the blower to circulate the hydrogen between the zones. The length of the machine is the major factor determining the optimum number of parallel paths required. Another means of providing these parallel-flow paths is the "air-gap pickup" construction, where passages inclined in the direction of rotation scoop hydrogen out of the air gap and convey it to the coil passages. Similar passages with openings inclined backward with respect to rotation provide for the hydrogen discharge.

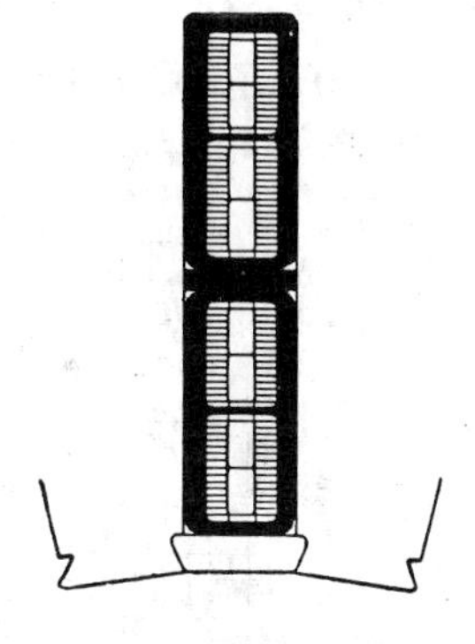

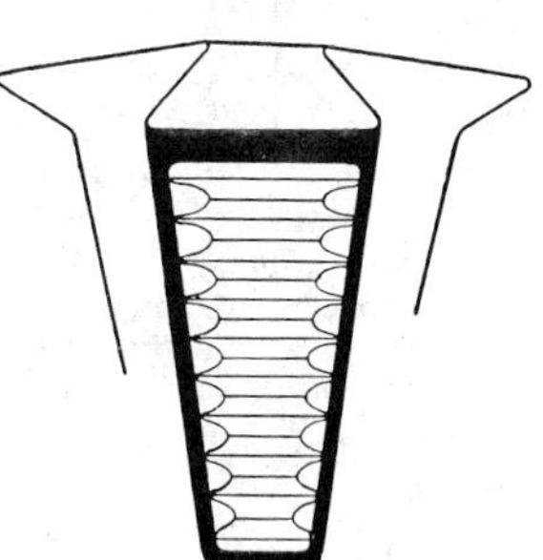

Fig. 6-16. Cross section of stator and rotor slots of conductor-cooled generator. (*See Ref.* 15.)

56. Hydrogen Blowers. Blower requirements for circulating hydrogen in conductor-cooled machines are more critical than for conventional cooling, for two reasons: the required pressure differential is greater, and the increased fan power output requires higher blower efficiency to avoid excessive fan power loss. The blowers are usually mounted on the rotor within the hydrogen enclosure. Multistage axial blowers are most common. Multi- or single-stage centrifugal blowers are also used, sometimes with an impeller diameter considerably exceeding that of the rotor.

57. Characteristics. The increased current-carrying capability of conductor-cooled coils is illustrated in Fig. 6-17. Liquid-cooled armature coils are comparable in performance with hydrogen-cooled coils at the higher hydrogen pressures.

Conductor-cooled machines are now built mainly for larger ratings than are possible with conventional cooling. The increase in available ratings in recent years shown in Figs. 6-18 and 6-19 has come about primarily because of this improved cooling method. Machine size has increased somewhat, but mechanical stresses, shipping clearances, and available forgings are severe limitations for further increases.

The increased armature coil loading results in conductor-cooled machines having higher reactances than conventionally cooled machines. This increased loading also results in more leakage flux in the end regions, often requiring low-resistance nonmagnetic or laminated magnetic shielding to minimize eddy-current losses in the structural parts.

MECHANICAL CONSTRUCTION

58. Basic Types of Constructions. The magnetic circuit and the windings of the alternator are designed to function satisfactorily mechanically as well as electrically and are provided with a suitable

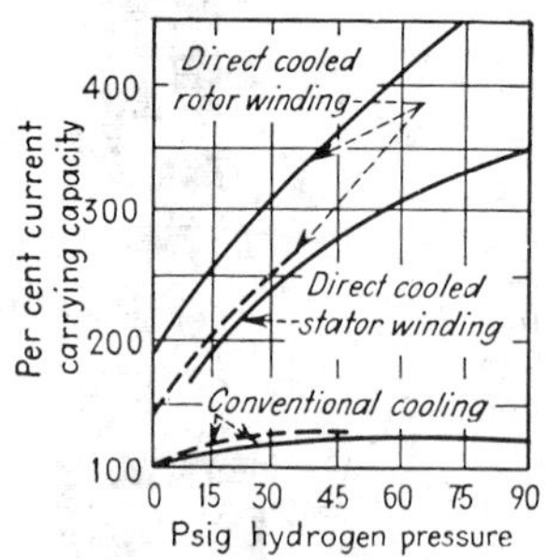

Fig. 6-17. Current-carrying capacity of rotor and stator coils as a function of hydrogen pressure. (*Solid curves from Ref.* 16; *dotted curves from Ref.* 17.)

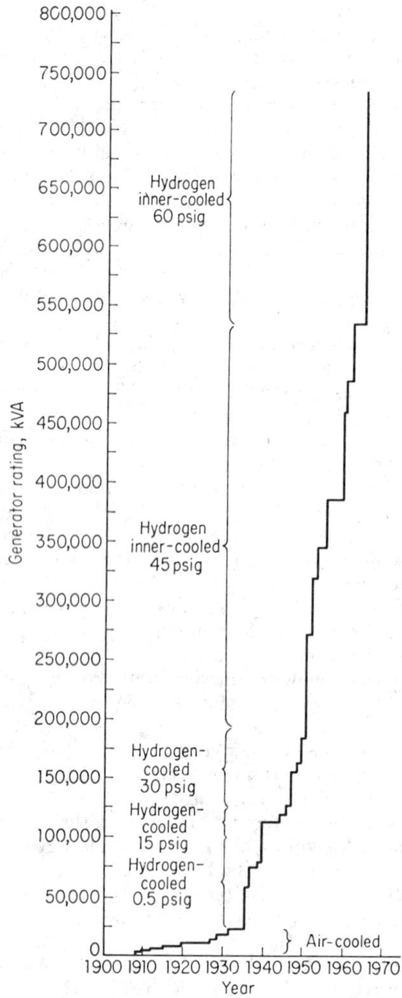

Fig. 6-18. Maximum kVA ratings for 3,600 r/min turbine-generators. (*See Ref.* 18.)

supporting structure consisting of a stator yoke or frame, rotor body, shaft, bearings, and bearing supports. The design criteria are based not only on normal operating conditions but also on abnormal conditions such as short circuits and overspeeds.

Two fundamental variations in mechanical construction are distinguished by the field configuration: salient-pole and round-rotor. In all but the smallest modern alter-

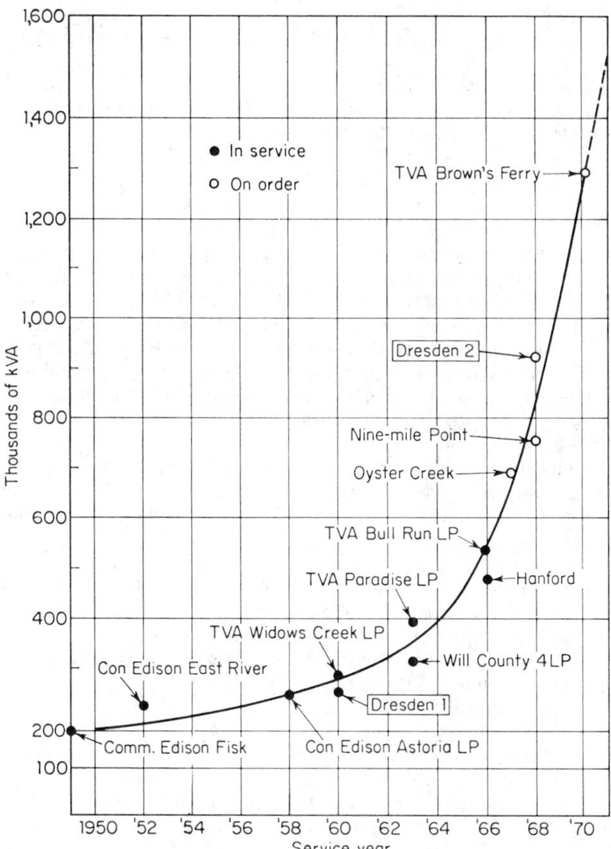

FIG. 6-19. Kilovoltampere ratings for 1,800 r/min turbine-generators. (*See Ref.* 19.)

nators the field is the rotating element (rotor), and the armature is the stationary element (stator). Salient-pole construction, where the field windings are on pole pieces attached to a rotor body, is used on slower-speed machines, 1,200 r/min and less, because of its relatively lower cost. Round-rotor construction, where the field windings are inserted in axial slots in a cylindrical rotor body, is used on essentially all 3,600 r/min machines and on the larger 1,800 r/min machines, because solutions to the problems of attaching salient poles to the rotor bodies of such high-speed machines become impracticable. The fundamentals of stator core and winding construction for these two types of designs are the same.

Two additional variations of construction are characterized by whether the shaft is horizontal or vertical. The prime-mover design ordinarily determines whether the machine will be horizontal or vertical, although special problems in generator design may affect the choice. Most steam-turbine-driven and engine-driven alternators have horizontal shafts; hydraulic-turbine-driven alternators may be either horizontal or vertical, the larger slower-speed units usually being vertical. The magnetic circuits and windings of horizontal and vertical machines are similar; the differences between these two types are in their structural members.

59. Stator Construction. Armature cores are built up of thin laminations, produced as segments or rings, depending on size. Successive layers of the segmented laminations are staggered to minimize the effect of the joints in the magnetic circuit. The core is clamped between pressure plates and fingers to support it with sufficient pressure to prevent undue vibration of the laminations. Especially in long cores, the clamping arrangement may include some provision to compensate for compacting of the core after initial assembly.

The armature windings are fitted tightly in the slots and secured radially by slot sticks, or wedges, driven into suitable notches at the air-gap end of the slots. It is necessary that the stator coil ends be able to resist the abnormal forces associated with short circuits. A supporting structure may be employed for this purpose. There are many variations of support design; most of them provide filler blocks between the coil sides, strategically located to transmit the circumferential forces from coil to coil, and additional structure to counteract the radial forces.

Coil supports ordinarily are designed to suit the need of a particular machine. Large two-pole machines will require a quite elaborate structure; the combination of large short-circuit currents and coil ends inherently flexible because of their long length makes these machines particularly susceptible to coil-end movement. Low-speed machines with stiffer coil ends require less support; in the smallest ratings the coils may be capable of withstanding the short-circuit forces without any additional support.

Stator frames, or yokes, commonly are fabricated from structural steel, designed to support the core in proper alignment with the rotor and to suit the ventilating scheme used.

60. Rotor Construction. The pole pieces of salient-pole alternators may be built up of steel laminations, both as a manufacturing convenience and as a means of limiting the loss in their air-gap surfaces due to pulsations in air-gap flux. The field coils, wound directly on the poles or preformed and then mounted on the poles, are suitably insulated from the poles for the voltages associated with normal and transient operation. The pole and coil assembly is bolted, dovetailed, or otherwise attached to the rotor body. It is the limitation of this attachment which usually dictates when round-rotor construction must be used rather than salient-pole construction.

The rotor body for a salient-pole machine may be a solid forging or assembly of heavy steel plates, for high-speed designs, or a spider-and-rim assembly, for low-speed designs. The shaft may be integral with the body, as in the case of a forging, or may be bolted to or inserted into the body.

When the spider-and-rim construction is used, the entire assembly may be an integral weldment or casting or the rim may be separate from the spider, as in the case of large waterwheel-driven generators. A common construction for this latter case is a rim built up on thin steel laminations, assembled around a cast or fabricated spider, bolted together between steel end plates, and keyed to the spider.

The rotor of a round-rotor machine is cylindrical in shape, with axial slots provided in its body for the field coils. The body is usually a steel forging with the shaft ends integral. In special applications, other constructions may be used, with this same general configuration. The field coils are wound in axial slots in the rotor body, held in place by heavy slot wedges and by retaining rings over the coil ends.

Rotors are designed for operation at overspeeds which depend on the characteristics of the prime mover. Overspeeds (the speed above rated at which the unit must be capable of safe operation) may be as low as 20% for a steam-turbine-driven unit or as high as 125% for some adjustable-blade axial-flow hydraulic-turbine-driven units.

61. Critical Speeds. The shaft system of the entire unit (generator and prime mover) must be designed with regard to critical speeds. Both lateral and torsional critical speeds are considered. Lateral critical speed is the speed corresponding to the natural frequency of the shaft system in response to lateral or transverse forces such as residual unbalance forces. Torsional critical speed relates to the response of the shaft system to torsional forces. A lateral critical speed is associated with each mode of lateral vibration, the first critical speed corresponding to the lowest frequency mode. Critical speeds are affected by shaft support, including foundation (particularly lateral critical speeds), and by internal and external damping.

It is preferable that the operating speed be at least 20% away from the nearest critical speed. Low-speed rotors ordinarily operate below the first critical speed. High-speed rotors, especially two- and four-pole turbogenerator rotors, often operate above the first critical speed and sometimes, in the largest ratings, above the second critical speed. It is particularly important that such rotors be carefully balanced so that the forces, and resultant stresses, while passing through the critical speeds on start-up and shutdown are not excessive.

Torsional critical speeds are excited by external forces, such as a sudden load change or a short circuit or cyclic variations in prime-mover torque, e.g., an internal-combustion engine. Impulses from the buckets of a hydraulic turbine or from the gear teeth in a unit driven through a gear may also excite torsional vibrations.

62. Bearings. Although antifriction bearings are occasionally used on alternators of the smaller ratings, the great majority are furnished with oil-lubricated babbitted bearings. For horizontal shafts these will be self-contained ring-oiled bearings wherever design conditions permit. At higher shaft peripheral speeds and higher bearing loadings ring oiling is supplemented with recirculation of externally cooled oil. The rings may be eliminated, or they may be retained to afford some degree of emergency oil supply in the event of a failure of the external system. Lead-base babbitts are commonly used for journal bearings, although tin-base babbitt may be employed for some heavy-duty applications. While bearing supports may be designed to afford some degree of self-alignment for the bearing bushing or shell, they must be sufficiently rigid so as to not affect unduly the lateral critical speeds of the shaft system.

Two principal types of **thrust bearings** are used on vertical alternators: the pivoted-shoe type and the spring type. The adjustable pivoted-shoe type, introduced in this country by Albert Kingsbury, consists of a flat rotating collar or runner of steel or fine-grained cast iron resting on a stationary member consisting of several babbitted segmental shoes pivoted near their center on adjusting screws, which, by changing the elevation of the shoes, can provide equal loading on them. The screws also permit small adjustments in rotor elevation to correct generator and turbine clearances. The spring-type bearing manufactured in this country by the General Electric Company consists also of a flat rotating thrust collar, resting on a series of stationary babbitted segments supported on a number of precompressed springs.

The bearings are immersed in oil. In operation, a thin, wedge-shaped film of oil is formed between the runner and shoe. The oil is continuously circulated by the rotation of the runner and is cooled either by radiation or by water cooling, usually within the oil bath but occasionally by an external system.

The spring-type bearing is inherently self-equalizing; i.e., each shoe carries very nearly the same amount of load. A variation of the pivoted-shoe bearing, in which the shoes are supported on a system of interconnected levers, provides the same self-equalizing feature.

The **spherical bearing** is another variation of the pivoted-shoe thrust bearing, in which the runner is a part of a sphere and the shoes of corresponding shape. This type of bearing restrains lateral movement of the shaft, serving the dual function of thrust and guide bearing.

Horizontal-shaft alternators occasionally require thrust bearings, as, for example, a single-impeller reaction turbine having unbalanced hydraulic thrust which must be restrained by the bearing. Thrust bearing designs for this application are generally of the pivoted-shoe type, either adjustable or equalizing.

Some thrust bearings, particularly of the adjustable pivoted-shoe type, may be provided with load cells for measuring and equalizing the thrust on the shoes. These may be of the hydraulic or the strain-gage type, the latter being more common in modern applications. In addition to providing a check on the adjustment of the shoe loadings, these devices provide information about the hydraulic-thrust characteristics of the turbine.

Guide bearings for vertical alternators (Fig. 6-20) are oil-lubricated babbitted rings. These are frequently segmental to facilitate assembly and may be composed of individual shoes which are radially adjustable. Guide bearings usually are partly immersed in an oil bath, with oil circulated by the pumping action of sloping grooves in the babbitt surface. Occasionally a separate lubrication system is provided which introduces oil at the top clearance of the bearing, collects it at the bottom, and recirculates it. It is common practice to place a guide bearing closely above the thrust bearing, in the same oil pot. In some instances the guide bearing is on the outside periphery of the thrust runner.

Guide-bearing clearances are on the order of 0.001 to 0.0005 in (diametral)/in of

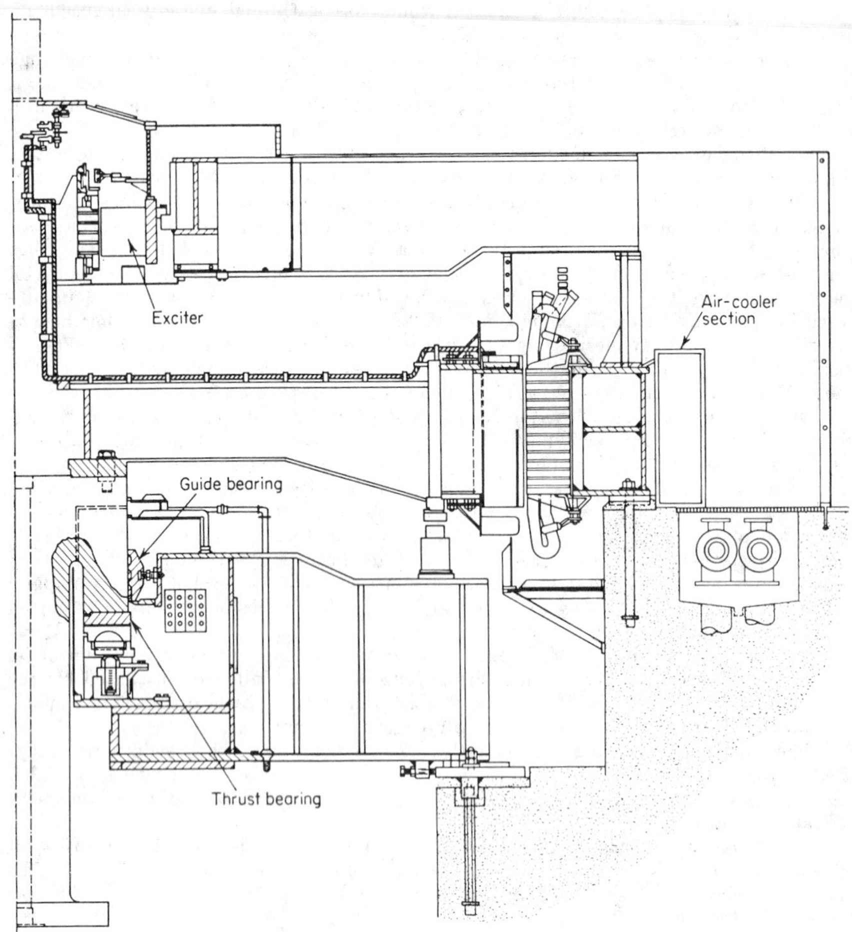

FIG. 6-20. Cross section of typical vertical alternator.

bearing-journal diameter, with this figure decreasing as diameter increases, and with a maximum on the order of 0.025 in.

63. Bearing Arrangements. Vertical alternators are classified by their bearing arrangement. A suspended unit has the thrust bearing above the rotor and is provided with two guide bearings, one above the rotor and one below. The upper guide bearing is frequently placed just above and in the same oil pot as the thrust bearing. An umbrella unit has the thrust bearing below the rotor and one guide bearing also below the rotor, usually in the same oil pot and just above the thrust bearing. When umbrella arrangements are used, careful consideration must be given to the stability of the unit with respect to over-turning moments. Large-diameter slow-speed units, in which the ratio of rotor diameter to the height of rotor center of gravity above the thrust surface is relatively large, lend themselves to umbrella construction. A principal advantage of umbrella construction is a substantial reduction in required powerhouse headroom, since the relatively high bearing support structure is below the rotor in the turbine pit. Furthermore, this structure itself usually spans a shorter distance when placed below the rotor and is not so high because of this, and shaft length may be reduced. A modified umbrella arrangement, having a guide bearing above the rotor, may be used when mechanical stability precludes classic umbrella construction; this retains some of the advantage of reduction in headroom requirement.

64. High-pressure Systems. Horizontal journal bearings and thrust bearings are sometimes provided with high-pressure oil to reduce starting torque and minimize bearing wear on start-up. Oil at pressures on the order of 1,500 lbf/in² is introduced at the bottom of the journal bearing—or at the center of each shoe or segment of a thrust bearing—to lift the rotor and introduce an oil film in the bearing clearance before the shaft rotates.

Steam turbine-generator units frequently are operated at very low speeds when they are disconnected from the system, to prevent the shaft system from sagging while cooling or while at standstill for extended periods. This "turning-gear" operation may be at 5 to 10 r/min, much below the speed at which the bearing would maintain an oil film. High-pressure or flood lubrication may be applied during this operation.

65. Oil Specifications. The oil used in journal and thrust bearings is selected to suit the requirements of the particular application. Table 6-4 gives typical lubricating-oil specifications.

Table 6-4. Typical Lubricating-oil Specifications

Application	Viscosity (SSU)		Flash point (min.), °F	Pour point (max.), °F	Specific gravity at 60°F (max.)	Neutral No. (max.)	Vis. index (min.)
	At 100°F	Min. at 210°F					
Pivoted—shoe thrust bearings, and sleeve bearings on horizontal machines with speeds 1,800 r/min and below	275–375	50	400	+10	0.89	0.2	90
Sleeve bearings on horizontal machines with speeds above 1,800 r/min to and including 3,600 r/min	140–160	43	400	+10	0.89	0.2	90

NOTE: Lubricants are high-quality petroleum oils having rust and oxidation inhibitors.

66. Temperature Limitations. Safe operating temperatures of babbitted bearings are dependent on the ability of the babbitt to withstand plastic deformation. Babbitt temperatures of 95°C are safe, but most bearings are designed for operation at somewhat lower temperatures. Bearing temperature detectors are placed in the bearing shoe material as close as practicable to the babbitt surface, or in the oil discharging from the bearing. The trend in bearing temperature is more significant than the temperature itself; a bearing which suddenly rises from 50 to 85°C, for no apparent reason, is much more a matter for concern than one which operates consistently at any point within this range.

REACTANCES AND TIME CONSTANTS

67. General. To facilitate mathematical analysis of alternators operating alone or in conjunction with other machines and systems, the reactances and time constants of synchronous machines are defined under various operating conditions. Where variations in reactance with rotor position occur because of asymmetry, direct- and quadrature-axis values of reactance are required.

Reactances of synchronous machines generally are taken to be equal to their corresponding impedances. However, resistance has a major influence on time constants.

68. Per Unit System. Voltages and currents are commonly expressed in percent or per unit of rated values. The corresponding base for expressing per unit reactances is the ohms which would produce a voltage drop of rated volts per phase when rated current flows through it. A principal advantage of this system of notation is the ease in comparing similar machines. As an example, if 1.0 per unit voltage is applied to 0.5 per unit reactance, the current will be 2.0 per unit regardless of the actual machine rating. Time constants are expressed in seconds.

69. Principal Reactances. Each of the commonly used reactances, expressed in per unit terms, is equal to the fundamental voltage induced in the armature winding by the flux resulting from rated armature current acting on a particular combination of permeances.

Stator leakage reactance x_l is a portion of all other machine reactances and is a result of flux produced across the armature slots and in the coil end region by armature current. Figure 6-4 shows the slot portion of this flux path. This component of flux is essentially independent of rotor position, but it is drastically affected by coil pitch in the case of zero-sequence reactance x_0, discussed later.

Synchronous reactances are applicable when the rotor is moving in synchronism with the mmf produced by steady-state armature current. The principal flux paths are then as shown in Fig. 6-21*a*. For the sake of simplification the stator slots and

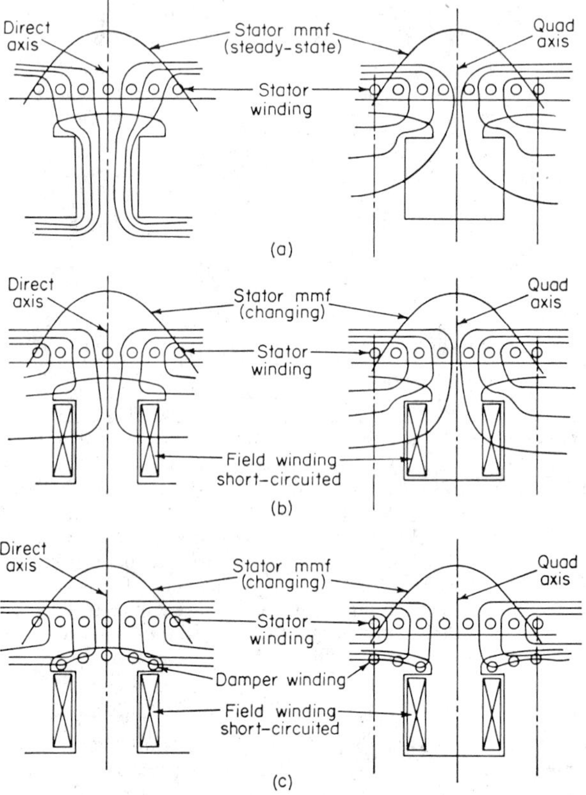

Fig. 6-21. Principal flux paths for direct and quadrature axes. (*a*) Synchronous reactance; (*b*) transient reactance; (*c*) subtransient reactance.

stator leakage fluxes are not included in this figure. The **direct-axis synchronous reactance** x_d corresponds to the condition shown at the left in Fig. 6-21*a*, where the rotor pole axis is in line with the maximum value of the armature mmf. The **quadrature-axis synchronous reactance** x_q corresponds to the condition shown at the right in Fig. 6-21*a*, where the axis between the rotor poles is in line with the peak value of the armature mmf. The difference in air-gap flux between these two conditions is more clearly illustrated in Fig. 6-10. The considerably lower air-gap permeance in the interpole space results in much less actual and fundamental flux in the quadrature axis. The reactances corresponding to these two air-gap permeances are commonly called the magnetizing reactances, or reactances of armature reaction, and are designated as x_{ad} and x_{aq}. Since the synchronous reactances include stator leakage,

$$x_d = x_l + x_{ad} \qquad \text{and} \qquad x_q = x_l + x_{aq}$$

Transient and subtransient reactances are applicable when the armature mmf is changing with respect to time. Currents will then be induced in the rotor windings which will affect the air-gap flux. If the direct-axis mmf is established suddenly and the field circuit is closed, current will be induced in the field winding which will oppose the sudden establishment of flux linkages with it and will force the air-gap flux to cross the space between the poles and not penetrate to the rotor spider. This flux path is shown in Fig. 6-21*b*. The total permeance of this flux path corresponds to the **direct-axis transient reactance** x'_d.

Furthermore, if there is a damper winding in the pole faces, current induced in it will oppose the sudden establishment of flux linkages with it, forcing the air-gap flux into a path above the damper winding. This flux path is shown in Fig. 6-21*c*, corresponding to the **direct-axis subtransient reactance** x''_d.

The currents induced in the damper winding and the field winding decay proportionately to their respective time constants, the damper-winding currents decaying first, permitting flux linkages with these windings, until the steady-state flux pattern of Fig. 6-21*a* is established.

The quadrature-axis transient flux paths shown at the right in Fig. 6-21*b* are the same as the quadrature-axis synchronous flux paths at the right in Fig. 6-21*a*. This is because there are no flux linkages with the field winding in the quadrature axis and no induced field current to affect the air-gap flux. Therefore the **quadrature-axis transient reactance** x'_q is the same as the **quadrature-axis synchronous reactance** x_q.

The flux paths at the right in Fig. 6-21*c* correspond to the **quadrature-axis subtransient reactance** x''_q. x''_q normally will be somewhat higher than x''_d unless the damper winding extends well into the pole tips and has a low impedance connection between poles.

In a round-rotor machine, currents in the surface of the solid rotor are the equivalent of damper-winding currents in a salient-pole machine.

The foregoing reactances are positive-sequence reactances applicable when the armature mmf and the rotor structure are rotating in synchronism. The **negative-sequence reactance** x_2 is applicable with an armature mmf rotating backward at synchronous speed while the rotor is rotating forward at synchronous speed. Currents of twice normal frequency induced in the damper and field windings limit the air-gap flux to the same rotor paths as shown in Fig. 6-21*c*. The permeance varies between the direct- and quadrature-axis values, and negative-sequence reactance is usually taken to be equal to the average of the direct- and quadrature-axis subtransient reactances.

Zero-sequence reactance x_0 is evaluated with the 3 phases of the armature winding connected in parallel, with a single-phase rated frequency voltage applied across them. The air-gap flux is then greatly reduced as compared with the positive- and negative-sequence cases. The armature leakage fluxes and the corresponding reactance are also less, particularly with short-pitch armature coils. x_0 is less than the other reactances normally used.

70. Equation of Short-circuit Current. The direct-axis reactances and their corresponding time constants describe the current of a synchronous machine on sudden short circuit in accordance with the following equation,

$$I = \frac{E}{x_d} + \left(\frac{E}{x'_d} - \frac{E}{x_d}\right)\epsilon^{-t/T'_d} + \left(\frac{E}{x''_d} - \frac{E}{x'_d}\right)\epsilon^{-t/T''_d} \tag{6-13}$$

where I = the a-c rms component of current following a 3-phase symmetrical short circuit from no load, per unit; E = a-c rms voltage prior to short circuit, per unit; t = time after instant of short circuit, seconds; T'_d = direct-axis transient time constant, seconds; and T''_d = direct-axis subtransient time constant, seconds. Armature circuit resistances are neglected, and excitation is assumed to be constant.

Typical oscillograms of the currents and voltages of a sudden 3-phase short circuit are shown in Fig. 6-22. Detailed methods for analyzing the oscillograms of short-circuit currents to obtain values for reactances and time constants are given in Ref. 9.

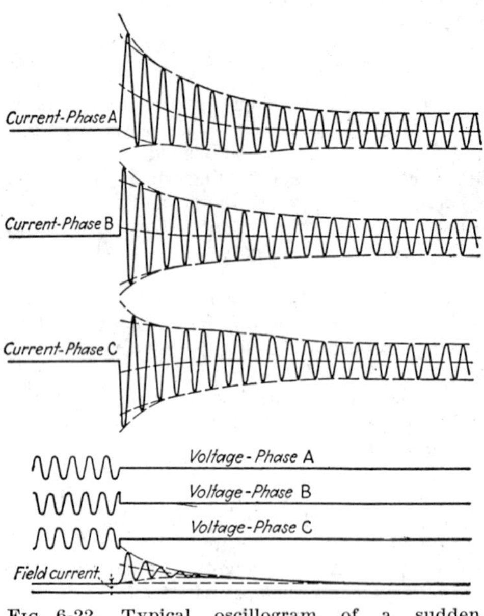

FIG. 6-22. Typical oscillogram of a sudden 3-phase short circuit.

71. Offset Currents. At the instant of a short circuit from rated-voltage no load, each of the 3-phase windings will be linked by some portion of the air-gap flux. To maintain these flux linkages constant, each armature current will have an added d-c component proportional to them. The d-c components of current produce a stationary air-gap mmf which causes a fundamental-frequency current in the field winding, in order that the field flux linkages remain constant. The **armature time constant** T_a is the time constant of the exponential rate of decay of the d-c components.

The d-c components of armature current, which cause the currents to be asymmetrical with respect to the zero axis, are shown in Fig. 6-22. A phase could have zero initial flux linkages, in which case the short-circuit current would be symmetrical. The maximum initial flux linkages would occur if the phase was directly in line with the rotor pole at the instant of short circuit. The d-c component of armature current in that phase would then equal the peak value of the initial a-c component. The total rms asymmetrical current over the first cycle would then be the square root of the sum of the squares of the d-c component and an equal rms a-c component.

CHARACTERISTICS AND VECTOR DIAGRAMS

72. Vector Diagrams. The terminology and the conventions of phase relationships used in this article are in common use in the power field, although others are also used in the literature on this subject.

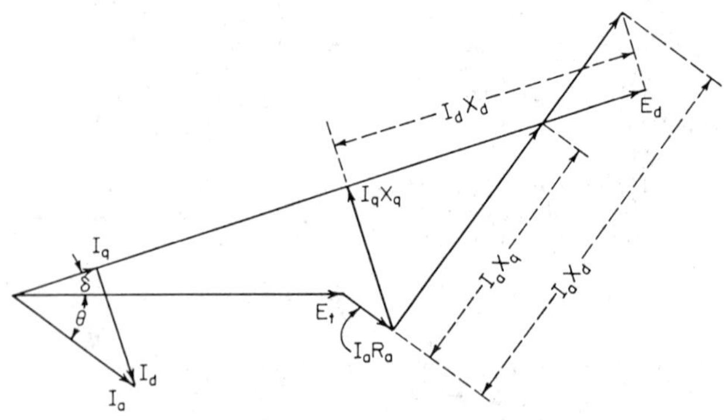

FIG. 6-23. Vector diagram of alternator, with saturation neglected.

Vector diagrams (also known as phasor diagrams) of salient-pole machines are based on the two-reaction theory, which treats the asymmetry of the rotor circuits by resolving voltage and current vectors into direct- and quadrature-axis components. The diagrams are valid only for unsaturated conditions, since saturation in the two axes is not the same and combining them by superposition is therefore invalid.

Saturation is here evaluated only in calculating load excitation, and then only in the direct axis. The method described in Par. **74** is generally accepted because of its good conformance with actual test data.

The vector diagram of an alternator, disregarding saturation, is given in Fig. 6-23, where E_t is the terminal voltage, I_a is the line current, displaced from E_t by power-factor angle θ, I_aR_a is the armature resistance drop drawn parallel to I_a, and I_ax_q and I_ax_d are the quadrature- and direct-axis synchronous reactance drops drawn perpendicular to I_a. The line from the origin through the vector sum of E_t, I_aR_a, and I_ax_q establishes the reference axis for direct-axis voltages. The current may then be resolved into its quadrature- and direct-axis components and the corresponding voltage drops I_qx_q and I_dx_d added to I_aR_a and E_t to determine the magnitude of E_d, which is the internal voltage produced by the field current acting alone. The angle between the internal voltage E_d and the terminal voltage E_t is the displacement angle δ.

If x_q equals x_d, as in a round-rotor machine, the diagram is simplified to Fig. 6-24. It will be noted that, in so far as determining the magnitude of E_d is concerned, the refinement of considering both direct and quadrature axis has a very minor effect. On this basis the USASI has stipulated a method for calculating excitation requirements based on round-rotor theory and using data from the open- and short-circuit saturation curves and the zero-power-factor saturation curve at rated armature current.

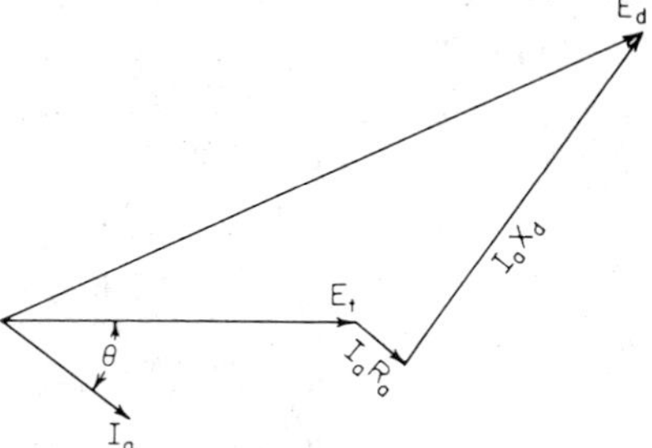

FIG. 6-24. Vector diagram for round-rotor alternator, with saturation neglected.

73. Characteristic Curves. Typical characteristic curves of a salient-pole alternator are shown in Fig. 6-25. The USASI method for calculating load excitation requires calculation of the Potier reactance x_p from these characteristic curves in the following manner: The intersection of the rated-current zero-power-factor saturation curve with the rated-voltage line locates the point **d**. To the left of **d** on the rated-voltage line, the length **ad** is laid off equal to the field current (I_{FSI}) for zero voltage on the rated-current zero-power-factor saturation curve. This is the field current to produce rated-armature-current short circuit. Through **a** the line **ab** is drawn parallel to the air-gap line. The intersection of this line with the no-load saturation curve locates the point **b**. The vertical distance **bc** from the point **b** is the Potier reactance. If the zero-power-factor saturation curve is obtained for some current slightly different from rated current, the distance **bc** is divided by the per unit armature current at which the zero-power-factor saturation curve was actually obtained, to give the Potier reactance.

74. Calculating Excitation Requirement. The rated-load excitation requirement I_{FL} is then determined as shown in Fig. 6-26. I_{FG} is the excitation current corresponding to rated armature voltage on the air-gap line. I_{FSI} is the excitation current to produce rated armature current, short circuit. θ is the rated power-factor angle. θ is laid off clockwise for lagging power factor, i.e., with the generator supplying an inductive current.

Armature resistance is normally neglected. I_{FS} is the field current corresponding to saturation and is measured between the air-gap line and the actual saturation curve at the voltage, E_{xp}, behind Potier reactance. This voltage is the vector sum of rated armature voltage and the voltage drop corresponding to Potier reactance, I_ax_p, added at the power-factor angle θ (see Fig. 6-27). It will be noted that Fig. 6-26 corresponds to the vector diagram for a round-rotor machine (Fig. 6-24), with resistance neglected and the effect of saturation added.

For loads other than rated load, and at voltages other than rated voltage, appropriate values of I_{FG}, I_{FSI}, θ, and I_{FS} are used.

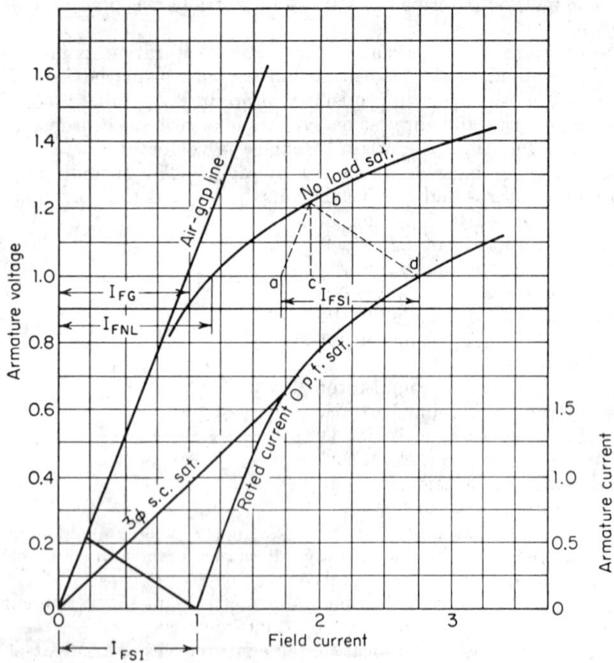

FIG. 6-25. Typical characteristic curves of a salient-pole alternator, showing graphic determination of Potier reactance. Quantities are in per unit.

75. Regulation. The voltage regulation of an alternator is defined as the rise in voltage with constant rated-load field current when the load is reduced from rated load to zero, expressed as a percent of rated voltage. It is the difference between the voltage on the no-load saturation curve at rated-load field current and rated voltage, divided by rated voltage and multiplied by 100.

76. Effect of Sudden Load Application. The voltage drop on sudden application of load is influenced by (1) the applicable alternator reactance, (2) the alternator

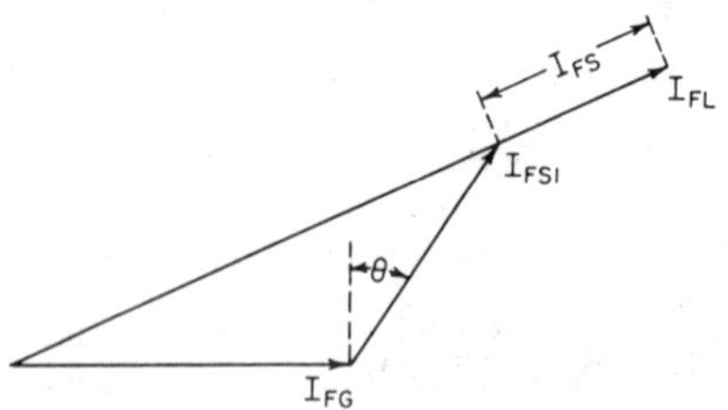

FIG. 6-26. ASA method for calculating load excitation.

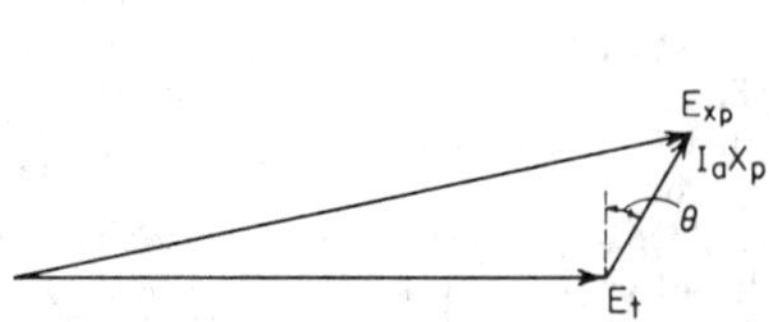

FIG. 6-27. Calculation of voltage behind Potier reactance.

open-circuit time constant, (3) the exciter-system response, and (4) the magnitude and nature of the load. In most cases the initial drop is taken to be the product of the applied load current and the alternator transient reactance X'_d. It is followed by a further decrease in voltage before the exciter system begins to restore the voltage to normal. Figure 6-28 represents a typical voltage-time curve following sudden application of load. References 20 and 21 give detailed discussions of these phenomena.

FIG. 6-28. Voltage vs. time for suddenly applied load.

77. Displacement Angle. The angle δ between E_d and E_t in Fig. 6-23 represents the physical angle measured in electrical degrees between the pole center and the flux corresponding to the terminal voltage. As load on the alternator is increased, this angle increases; i.e., the position of the pole center moves farther ahead of this flux, although remaining in synchronism with it.

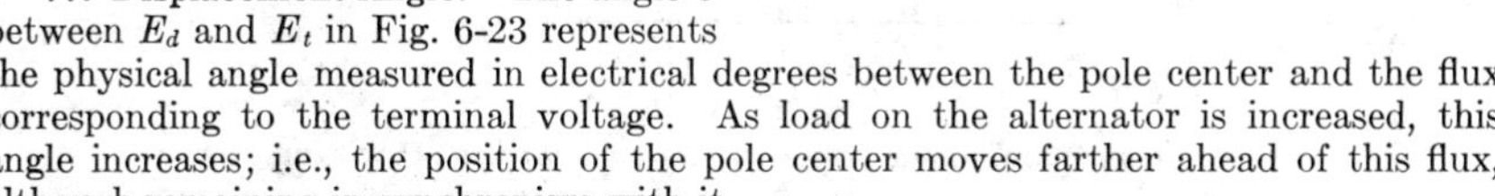

78. Power Output. From the vector diagram (Fig. 6-23) it can be shown that the power delivered by the alternator, saturation and losses being neglected, is

$$\frac{E_d E_t}{X_d} \sin \delta + E_t^2 \frac{X_d - X_q}{2X_d X_q} \sin 2\delta$$

For a round-rotor machine in which X_q equals X_d, the second term drops out and $P = (E_d E_t / X_d) \sin \delta$. The maximum power then is $E_d E_t / X_d$ when δ equals 90°. This is the stability limit under steady-state conditions.

79. Stability. The relationship between power, or torques, and displacement angle represented by the preceding equation is shown in Fig. 6-29. If the machine is operating at a load corresponding to torque T_0 and displacement δ_0 and the torque is suddenly changed to T_1, the rotor will start to increase its displacement toward a new value of δ_1. But the rotor will overshoot with an energy equal to area A_1 which represents the energy that produced the change in displacement and will continue until the area A_2 equals A_1. If area A_1 is larger than the maximum area under the torque-displacement curve above the line corresponding to T_1, the alternator will pull out of step.

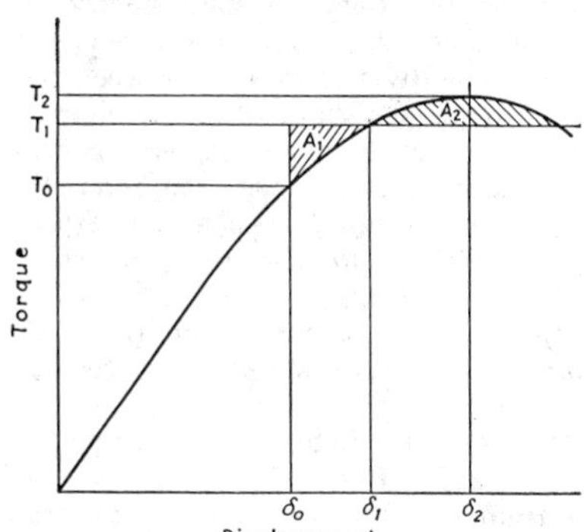

FIG. 6-29. Torque-displacement angle curve.

Torque-displacement angle characteristics are affected by other factors such as saturation and transient currents. Figure 6-29 indicates only the general nature of the phenomenon. Consult the Bibliography for references giving detailed treatment of the subject.

PARALLEL OPERATION

80. Conditions for Paralleling. In connecting two alternators in parallel it is desirable, in order to minimize current surges, that voltages and speeds be equal and corresponding voltages be in phase. Synchroscopes measure the difference between corresponding voltages of alternators to show when they are in synchronism and may be connected together. Various schemes have been devised to sense synchronism and initiate switching sequences automatically. If a synchroscope or automatic synchronizing device is not available, synchronism may be indicated with a system of lamps

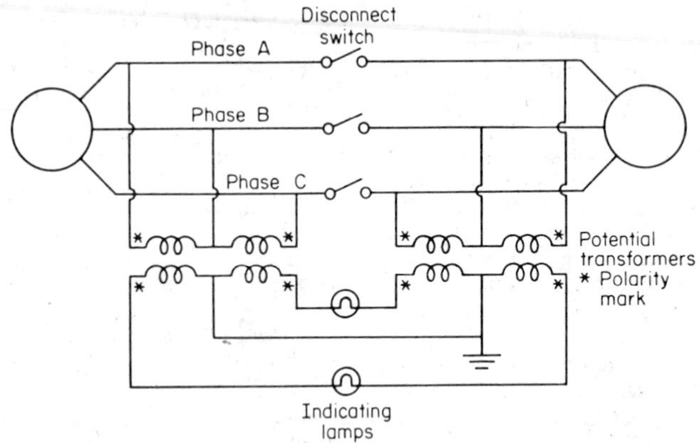

Fig. 6-30. Connection diagram for indicating lamps to show synchronism.

as shown in Fig. 6-30. Indicating lamps connected across the disconnect switch between the two machines measure the difference between two of the terminal voltages. This difference ranges between zero (when the voltages are in phase) and twice rated voltage (when they are 180° out of phase). The lamps are alternately dark and bright as the phase position changes. The two sets of lamps go dark and bright together if the sequence of the terminal voltages is the same; if the sequence is opposite, they go dark and bright alternately.

81. Division of Power. As load is increased on two alternators connected in parallel, there is a reduction in their speed which is sensed by the prime-mover speed-governing systems. The governors act to restore speed to normal. The division of load between the two alternators is determined by the characteristics of their prime-mover governing systems. If one system has speed characteristic *a* in Fig. 6-31 and another has characteristic *b*, they will divide the load in the proportion P_a and P_b when operated at speed *S*. Control of the load on a unit is obtained by adjusting the governor speed characteristic up or down.

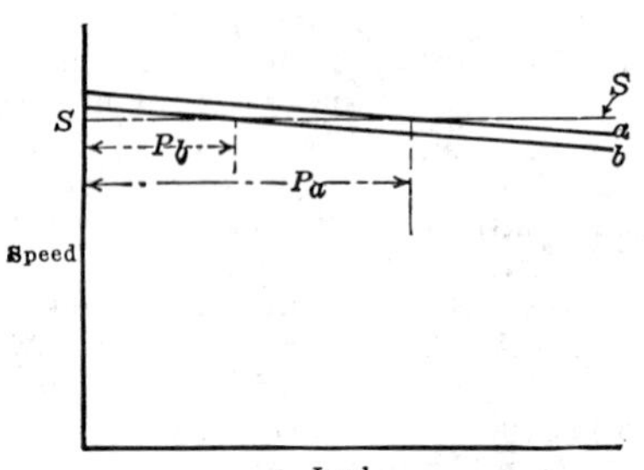

Fig. 6-31. Diagram of division of load between two alternators in parallel.

82. Division of Reactive Voltamperes. The voltage applied to a load connected to two alternators is determined by the total excitation on the alternators. Identical alternators with identical prime-mover speed-governing characteristics will share the load equally and, with equal excitation, will share the reactive voltamperes equally. Each alternator will operate at the same power factor as the load power factor. An increase in excitation on one will increase the system voltage, and that alternator will supply a greater share of the reactive volt amperes. A decrease in excitation of the other alternator will restore the terminal voltage to the original value but will increase the difference in the division of the reactive voltamperes. Adjustment of alternator excitation, then, determines not only the voltage applied to the load but also the division of reactive voltamperes between the alternators.

Automatic voltage regulators used on alternators operating in parallel may be provided with cross-current compensation to adjust the excitation on machines having different saturation characteristics so that they will operate at the same power factor. A voltage regulator applied to an alternator connected to a relatively very large system controls the power factor of the alternator rather than its voltage.

OSCILLATION AND TRANSIENT SPEED CHANGES

83. Oscillatory Characteristics. Although an alternator normally operates at a constant speed corresponding to the frequency and number of poles, variations in the angular velocity of the rotor about its average velocity can occur because of variations in driving torque, load, field excitation, or terminal voltage. These changes in angular velocity are associated with changes in the displacement angle δ, which is discussed in Par. **77** and illustrated in Figs. 6-23 and 6-29. The relationship between torque and displacement angle along with the mechanical inertia of the rotating parts and the damping torque resulting from the rate of change of the displacement angle provide the elements of an oscillatory system. These machine characteristics are discussed in the following paragraphs.

84. Synchronizing torque is commonly evaluated from the **synchronizing power coefficient** P_r, which is the change in shaft power, expressed in synchronous kilowatts, per electrical radian of change in displacement angle. Synchronous kilowatts is the power corresponding to the product of torque and synchronous speed, while elec rad = elec deg $\times$ $(2\pi/360)$.

With slow and small variations in displacement angle, P_r approaches the slope of the steady-state torque-angle curve (Fig. 6-29) at the average operating point, with due regard for the units used. Since the torque-angle curve is determined in part by the field excitation, P_r is influenced by this variable as well as by the average load. More rapid variations in the displacement angle increase P_r because of currents induced in the rotor circuits.

As an approximation, P_r at rated load is commonly taken as rated kilowatts divided by rated-load displacement angle. From Fig. 6-23, with $E_t = 1.0$, $I = 1.0$, and X_q in per unit terms,

$$\text{Rated-load } \delta = \tan^{-1} \frac{X_q \cos \theta}{X_q \sin \theta + 1.0} \quad \text{(deg)} \tag{6-14}$$

$$\text{Rated-load } P_r = \frac{\text{rated kW}}{\text{rated-load } \delta \times (2\pi/360)} \tag{6-15}$$

The use of a **per unit synchronizing torque coefficient** T_s is frequently more convenient, and $T_s = P_r/\text{rated kVA}$. Typical values of T_s are in the range of 1.2 to 2.0 for slow oscillations but may be much greater for rapid oscillations.

85. Damping torque can be evaluated from the **per unit damping torque coefficient** T_d, which is the per unit change in shaft torque per electrical radian per second rate of change in displacement angle. This coefficient T_d depends greatly on machine damper winding construction and is affected by frequency of oscillation, field excitation, and average load. An order-of-magnitude figure for T_d is 0.02 per unit at rated load, but large deviations from this value are common.

86. Inertia torque is evaluated from the mechanical rotational inertia or flywheel effect of the unit, commonly expressed as WK^2, where W is the weight of the rotating parts in pounds and K is the effective radius of gyration in feet. The rotational stored energy is also commonly expressed in per unit terms as the **inertia constant** H, which is the kilowatt-seconds of mechanical stored energy at synchronous speed divided by rated kVA. Then

$$H = \frac{0.231 \times 10^{-6}\,(WK^2)\,(\text{r/min})^2}{\text{rated kVA}} \tag{6-16}$$

Typical values of H range from 0.5 to 4.0, the higher values being for larger, lower-speed machines.

87. Basic Equations. The electromechanical oscillation and transient speed changes of an alternator can be expressed in the same mathematical forms common to other oscillatory systems, including electrical circuits and mechanical spring-mass systems. In this case, the applicable differential equations of motion are based on the essential condition that the torque applied to the shaft must at all times equal the algebraic sum of the synchronizing, damping, and inertia torques.

As an example of the application of these principles, for the simple case of sustained oscillation of a single synchronous machine operating on an infinite system with a component of applied shaft torque varying sinusoidally with time, the equilibrium of torques is shown by the equation

$$T_0 = Y_0[T_s + j\omega_0 T_d - \omega_0^2(H/\pi f)] \qquad (6\text{-}17)$$

where T_0 = per unit applied shaft torque (vector quantity), Y_0 = electrical radians displacement from the average displacement angle (vector quantity), $\omega_0 = 2\pi \times$ frequency of oscillation in cycles per second, and f = line frequency in cycles per second.

Practical problems of sustained and transient oscillation often involve additional machines and other elements interconnected in a system of sufficient complexity to justify computer solutions. In such instances, the alternators may also be represented by analogs to allow for variation of T_s and T_d, which are not always constant.

88. The natural frequency of oscillation f_{on}, with damping neglected, occurs when $T_s = \omega_0^2(H/\pi f)$;

$$f_{on} = \omega_{on}/2\pi$$

$$= \sqrt{\frac{fT_s}{4\pi H}} \quad \text{(c/s)} \qquad (6\text{-}18)$$

Note that f is line frequency in cycles per second. Typical values of f_{on} range from about 1.2 to 4.5 c/s, although added mechanical inertia in the alternator or the prime mover may result in a lower natural frequency.

The USA Standards express the natural frequency of oscillation in different, but equivalent, terms, as follows:

$$F = \frac{35{,}200}{\text{r/min}}\sqrt{\frac{fP_r}{WK^2}} \quad \text{(c/min)} \qquad (6\text{-}19)$$

89. A reciprocating engine drive applies pulsating torque to the alternator shaft. The principal alternating component of the torque has a frequency equal to the product of r/min × (no. cylinders normally fired)/r. Torque components of lesser magnitude and higher frequency are also produced, and inequalities in the torque per cylinder may introduce lower-frequency components.

An alternator driven by such an engine and connected to a large system oscillates in accordance with the equation in Par. **87**. The per unit power variation is $\pm Y_0 T_s$, and the resulting current pulsation could cause objectionable voltage variation in the system, usually evident as light flicker. Since voltage variation caused by one unit can produce oscillation of other units, the analysis of possible oscillations in a system with several reciprocating engine-driven alternators may be quite complex.

One suitable means for limiting the oscillation is to provide flywheel effect in the new unit such that its natural frequency of oscillation for rated-load conditions is well below, perhaps by 20%, the lowest possible forced frequency produced by any of the engines, including that of the new unit. Where this is not practical, oscillation at resonance may be sufficiently limited under some conditions if suitable damping torque is provided, or system operating procedures may be altered to avoid load conditions on individual units resulting in excessive oscillation.

A single reciprocating-engine-driven alternator supplying an isolated load may have little or no synchronizing or damping torque because the line frequency and voltage may both vary with the oscillation in speed resulting from the engine torque pulsations. This speed variation, and the resulting effects on voltage and frequency, is then limited almost entirely by the flywheel effect provided.

90. Governors and voltage regulators have oscillatory and transient characteristics of their own which sometimes influence the specification of alternator characteristics. Increased flywheel effect is usually beneficial from the standpoint of allowing these regulating elements more time to act in the event of load changes before the corresponding speed change becomes too great. Transient stability, as discussed in Par. **79,** is improved. Flywheel effect in a **hydraulic-turbine-driven alternator** is often particularly important because the response of the speed-governing system to loss of load may be deliberately delayed to avoid excessive pressure rise in the penstock as the gates close. Added flywheel effect reduces the overspeed resulting from these conditions.

LOSSES AND EFFICIENCIES

91. Conventional efficiencies used to evaluate alternator performance are based on losses which can be readily measured by test procedures described in the following article. The efficiency is given by the following equation, where the losses are the sum of the individual losses described in the succeeding paragraphs.

$$\text{Efficiency in per unit} = \frac{\text{output}}{\text{input}} = 1 - \frac{\text{losses}}{\text{input}} \qquad (6\text{-}20)$$

92. Losses. Five losses are usually evaluated for alternators. Two of them are considered to be fixed losses; i.e., they are assumed to be fixed with respect to load. The other three are variable losses; they are variable with respect to load. The fixed losses are (1) windage and friction and (2) core loss. The variable losses are (3) field copper loss, (4) armature copper loss, and (5) stray loss or load loss. In addition to these, exciter and/or rheostat losses may be evaluated, but in determining operating efficiency these ordinarily are charged, not to the machine, but to the plant.

Windage and friction loss is affected by the size and shape of the rotating parts, fan design, bearing design, and enclosure arrangements. An approximate combined windage and friction loss is given by the equation

$$\text{Windage and friction} = K\left(\frac{V}{10{,}000}\right)^{2.5} D\sqrt{L} \qquad \text{(kW)} \qquad (6\text{-}21)$$

where V = rotor peripheral velocity, feet per minute, D = rotor diameter, inches, L = rotor effective length, inches, and K = 0.08 to 0.11 for slow-speed salient-pole machines, 0.06 to 0.08 for higher-speed salient-pole machines, and 0.06 to 0.07 for air-cooled turbogenerators.

Core loss is caused by the main flux of the machine and occurs primarily in the stator teeth, in the portion of the stator core behind the teeth, and in the surface of the rotor poles, but it includes also the losses in structural parts of the machine which are exposed to stray alternating magnetic fields.

Stator cores ordinarily are constructed of thin laminations of a silicon steel, insulated from each other, to limit the hysteresis and eddy-current losses in the steel. The spacers in the ventilating ducts and the clamping structure at the ends of the cores may be of nonmagnetic materials to limit losses in these components. The end packets of the core may be stepped back from the air gap to decrease fringing flux which enters the core axially, or they may be provided with radial slits in the teeth to decrease eddy-current losses due to flux entering the ends of the cores axially. Since the stator slots introduce a variation in the air-gap flux density which will produce eddy-current losses in the surfaces of the rotor poles, the poles on salient-pole machines are ordinarily built of relatively thin laminations, or the surface of a round-rotor machine may be grooved, to reduce these losses.

Field copper loss is calculated from the field current and the d-c resistance of the field winding at 75°C. Field current may be measured under actual operating conditions or calculated by the method described in Par. **74.** The voltage drop across the collector ring brushes ordinarily is neglected but may be included in the exciter loss.

Armature copper loss is calculated from the d-c resistance of the armature winding at 75°C. For a 3-phase machine the loss is $1.5RI^2$ or $3R'I^2$, where R is the armature winding resistance at 75°C measured terminal to terminal, I is the line current, and R' is the armature resistance at 75°C measured terminal to neutral.

Stray loss or load loss is caused by the flux produced by armature current and includes eddy-current loss in the armature conductors, core losses set up by the armature current, losses in the core supporting structure and end housings due to armature current, and field surface or damper-winding losses due to armature current. Stray loss ordinarily varies nearly as the square of the armature current and may be expressed as a percentage of the armature copper loss.

93. Typical Efficiencies. Trends in full-load efficiencies of 60-c 80% power-factor synchronous machines, taken from published price book data of large manufacturers, are shown in Fig. 6-32. These curves are representative for normal machines. If voltage, short-circuit ratio,

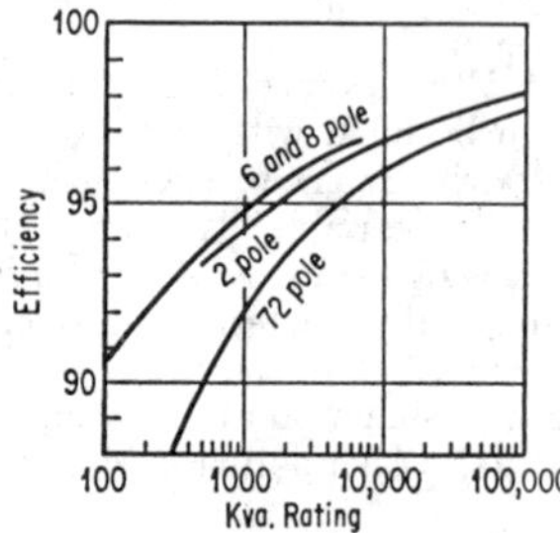

FIG. 6-32. Trends in full-load efficiencies for 60-c 80% power-factor alternators.

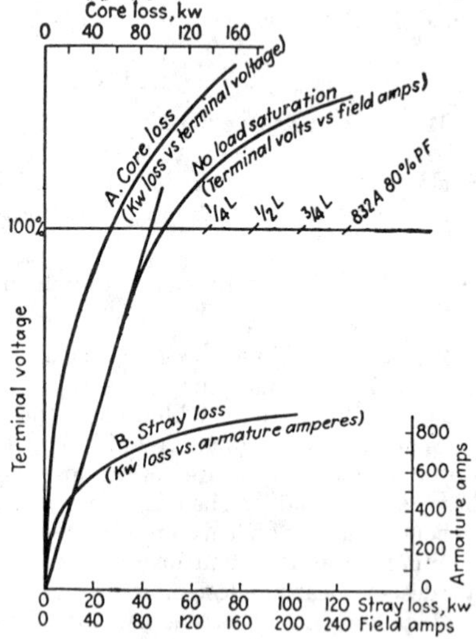

FIG. 6-33. Typical 14-pole 80% power-factor salient-pole alternator losses.

WK^2, or temperature ratings are higher than normal, or if reactances are lower than normal, the tendency will be to reduce these efficiencies. Special designs with higher-cost materials may be employed to increase them.

Losses may be furnished in the form of curves such as Fig. 6-33. A calculation of efficiency from these curves is given in Table 6-5.

Table 6-5. Efficiency Calculation for the Typical Generator in Fig. 6-33

(Rating, 10,000 kVA, 80% pf, 6,900 V, 3-phase, 60 c, 832 A, 514 r/min. Stator resistance, 0.060 Ω at 75°C. Field resistance, 1.82 Ω at 75°C.)

Load, kW	8,000	6,000	4,000	2,000
Power factor, %	80	80	80	80
Field current, A	250	206	164	125
Core loss (see curve *A*, Fig. 6-33)	48	48	48	48
Field and winding loss	45	45	45	45
Field I^2R	52	35	22	13
Stator I^2R	60	34	15	4
Stray loss (see curve *B*, Fig. 6-33)	59	33	15	4
Exciter loss	5	4	3	3
Pilot exciter loss	0.3	0.2	0.2	0.2
Total loss	269.3	199.2	148.2	117.2
Output, kW	8,000	6,000	4,000	2,000
Output and loss, kW	8,269.3	6,119.2	4,148.2	2,117.2
Loss, %	3.25	3.21	3.58	5.52
Efficiency, %	96.75	96.75	96.42	94.48

TESTS ON ALTERNATORS

94. Requirements. Tests are performed on alternators, either at the factory or in the field, to demonstrate that they have met their required performance. The schedule of tests may vary from commercial tests, which may require only the measurement of winding resistance, excitation current at rated-voltage no load, and dielectric tests, to a complete schedule which evaluates essentially all the operating characteristics of the machine. Details of all the tests ordinarily performed are given in Ref. 9, *IEEE Publ.* 115, Test Procedures for Synchronous Machines.

95. Winding resistances are usually measured with a Kelvin bridge or similar equipment. It is important that the temperature of the winding be accurately determined at the time the resistance readings are made. The resistance may be corrected to another temperature from that at which the test was made by the equation

$$R_1 = R_2 \frac{K + T_1}{K + T_2} \tag{6-22}$$

where R_2 is the known resistance at temperature T_2°C, T_1 is the temperature in degrees centigrade at which resistance is desired, and K is a constant depending on conductor material (234.5 for copper, 225 for electrical-conductivity aluminum).

When a Kelvin bridge is not available, the drop-of-potential method may be employed. It is important in using this method that the current be as small as possible so that it does not affect the resistance measurement by heating the winding. The current is applied to the winding being tested through cables clamped to the terminals; collector ring brushes should not be used to carry current at standstill.

96. Saturation curves may be made under various conditions: open-circuit, short-circuit, and with load current. These tests are made with the alternator operated at rated speed and the excitation varied to produce data for a curve of voltage or current vs. excitation. If the open-circuit and short-circuit saturation curves are obtained by a method which enables measuring power input, core loss and stray loss may be obtained at the same time.

97. The separate driving-motor method is commonly used to obtain saturation curves, particularly when losses are also desired. The alternator being tested is mechanically coupled to a driving motor and driven at rated speed. For an **open-circuit saturation curve** the alternator terminals are open-circuited, and its excitation is varied to produce terminal voltage over a range of about 30 to 150% voltage. Simultaneous readings of excitation current and terminal voltage are taken to provide data for plotting the open-circuit saturation curve. If the power input to the driving motor is measured and its losses are known, the input to the alternator can be determined. This input represents windage and friction and core loss. Data taken with zero excitation give the windage and friction, which may be subtracted from the alternator input at the various values of excitation to determine core loss.

A **short-circuit saturation curve** is obtained in much the same way, but with the terminals of the alternator short-circuited. Excitation is varied from a value which will give about 30% stator current to one which will give about 150% stator current. The input to the alternator may be determined from measurements of driving-motor input power and is the sum of windage and friction, armature copper loss, and stray loss. The windage and friction are determined from power-input data taken with zero excitation on the alternator, and the armature copper loss may be calculated from measured values of winding resistance and corrected to the temperature of the armature winding measured during the test. The stray loss is calculated as the alternator input minus the windage and friction and armature copper losses.

The **load saturation curve** most frequently determined by test is with rated armature current at zero power factor. It is obtained by operating the alternator as a motor without connected load, i.e., as a synchronous condenser. Its excitation and the excitation of the supply generator are adjusted to obtain a curve of terminal voltage vs. excitation, with rated armature current. This saturation curve, in conjunction with the open-circuit and short-circuit saturation curves, provides data from which the excitation requirement for any other value of load may be estimated as discussed in Par. **74.** Since the power requirement for this saturation curve is limited to the losses in the two machines, it can be obtained at the factory even for quite large machines, provided that a test machine is available capable of absorbing the reactive kVA from the machine being tested.

98. The deceleration method of obtaining saturation curves and losses may also be used, particularly with large vertical generators. The machine is brought up to a speed above rated speed and allowed to decelerate with the terminal condition and excitation established at the desired values. Excitation is best supplied from a separate source. For an open-circuit saturation-curve and core-loss test, the armature terminals are open-circuited, and the excitation is adjusted to the value desired and held constant while the generator decelerates. The curve of voltage vs. speed will be a straight line, intercepting rated speed at the point on the saturation curve corresponding to the excitation used. The rate of deceleration is proportional to the windage and friction and core loss corresponding to the value of excitation used. By measuring the rate of deceleration, the loss at rated speed may be calculated from the following equation. A deceleration run with zero excitation will give a value of windage and friction.

$$\text{Loss} = 0.462WK^2N\frac{dN}{dt} \times 10^{-6} \qquad \text{(kW)} \tag{6-23}$$

where WK^2 is moment of inertia of rotating parts, pound-feet squared, N is the speed at which loss is being evaluated, revolutions per minute, and dN/dt is the slope of the speed-time curve at N, revolutions per minute per second.

A short-circuit saturation curve and stray-loss measurement is made by deceleration in a similar manner, except that the armature terminals are short-circuited during the deceleration. Points on the short-circuit saturation curve are determined from measurements of the short-circuit current at rated speed for the various values of constant excitation. It will be noted that there is very little variation in short-circuit current with speed. The total loss may also be calculated from the rate of deceleration and the WK^2 of the unit. This total loss is the sum of windage and friction, armature copper loss, and stray loss. A test with zero excitation will give the windage and friction loss; the armature copper loss may be calculated from the current, the measured winding resistance, and the measured winding temperature during the deceleration run.

99. Tests as a Motor. An open-circuit saturation curve may also be obtained by operating the machine as a motor from a variable-voltage source. At a particular voltage, the excitation of the machine is adjusted so that armature current is minimum. The armature current of an unloaded motor may be as low as 2 or 3% of rated current; low-range current-measuring equipment will be required to determine this minimum point. At this point the power factor is unity, and the excitation is very nearly that for open-circuit conditions at the same voltage. A more sensitive indication of unity power factor can be obtained by connecting a single-phase wattmeter with its current coil in one line and voltage coil across the other two. The meter reading is zero at unity power factor. Similar tests at other values of voltage will produce data from which an open-circuit saturation curve is plotted. Measurements of input power may be made to determine open-circuit losses.

A very close approximation of the short-circuit saturation curve may be obtained in a similar manner. The machine is operated as a synchronous motor at a fixed voltage of the lowest value at which stable operation can be obtained. This will be on the order of 30% of rated voltage. Field current is varied from a value which will produce about 150% of rated armature current, down to a value which will produce near zero armature current, recording data at several points. Lower values of field current will produce increasing armature current. This curve of armature current vs. field current is plotted. The short-circuit saturation curve is a curve parallel to it, but passing through the origin. Measurements of input power may be made to determine short-circuit losses, and stray loss can be separated from copper loss if winding temperatures are observed during the test.

These test methods based on operating the machine as a motor are often used on small- and medium-sized units to simplify the test equipment needed and minimize the cost of the testing.

100. Heat runs are made to establish the capability of the machine to operate at its rated load without exceeding the guaranteed temperature rises. The types of heat runs most commonly made in the factory are open-circuit, short-circuit, and zero-power-factor. Heat runs under rated load conditions may be made in the field after the equipment has been installed but ordinarily are not conducted in the factory because of the large power requirements.

A heat run at zero power factor, with the machine operated at no load as a synchronous condenser and with appropriate conditions of armature current, voltage, and frequency maintained until the machine reaches constant temperature, can give temperature data most nearly approximating that corresponding to a rated-load test. When the test is made at rated armature current, it may be desirable to run at less than rated armature voltage to compensate for the higher internal voltage associated with zero-power-factor overexcited operation or to avoid excessive field current and temperature. The field temperature rise may be corrected to the field current corresponding to rated load, by using the following equation from Ref. 9,

$$T_r = T_f + (I_f/I_t)^2(T_t - T_f)\frac{k + T_a + T_f}{k + T_t + T_{at} - (I_f/I_t)^2(T_t - T_f)} \tag{6-24}$$

where T_r = temperature rise corrected to the desired field current, degrees centigrade, I_f = desired field current, amperes, I_t = test field current, amperes, T_t = temperature rise at test field current, degrees centigrade, T_f = temperature rise through the fan or blower, degrees centigrade (usually neglected for salient-pole machines), T_a = ambient temperature at which corrected temperature rise is desired, degrees centigrade, T_{at} = ambient temperature during test, degrees centigrade, and k = constant of field winding material, 234.5 for copper and 225 for electrical-conductivity aluminum.

A heat run at rated voltage with the terminals open-circuited and one at rated armature current with the terminals short-circuited provide an alternative way to approximate rated-load temperatures. The armature temperature rise is nearly the sum of the temperature rises for the two tests. Since these tests are conducted with relatively low values of excitation, it may be desirable to run another open-circuit heat run with excitation corresponding to rated load to approximate the field temperature rise under rated-load conditions.

Heat runs are conducted until the temperatures become essentially constant. For ambient temperatures within the range of 20 to 50°C temperature rises above the ambient ordinarily are not corrected for ambient temperature but are used as observed to evaluate the performance of the machine compared with the guarantees.

101. Reactances. Test values for most of the commonly used reactances can be determined. Synchronous reactances can be measured by applying a positive-sequence voltage to the armature windings with no field current, while the machine is rotated at slightly above or below rated speed, and measuring the maximum and minimum values of the varying current. Direct-axis synchronous reactance is the ratio of the voltage to the minimum value of the current; and quadrature-axis synchronous reactance is the ratio of the voltage to the maximum value of the current. Direct-axis synchronous reactance may also be calculated as the ratio of I_{FSI} to I_{FG} from Fig. 6-25. Direct-axis transient and subtransient reactances and their respective time constants can be calculated from oscillograms taken during symmetrical sudden short circuits, by using the equation given in Par. **70.** Negative-sequence reactance can be measured by applying a negative-sequence voltage to the armature windings while the machine is rotated at rated speed in the positive direction with its field winding short-circuited. Zero-sequence reactance can be measured by applying a single-phase voltage from the machine neutral to the three armature terminals in parallel, with the machine rotated at rated speed and with its field winding short-circuited. Reference 9 describes detailed procedures for making these and other reactance tests.

102. Dielectric Tests. The principal dielectric tests are described in Par. **43.**

BIBLIOGRAPHY

103. Note: Because of the large amount of available literature on synchronous machinery, this bibliography is limited to general sources and references specifically cited in the text.

104. General Bibliographies

1. Bibliography of Rotating Electric Machinery 1886–1947; *AIEE Publ.* S-32, January, 1950.

2. Bibliography of Rotating Electric Machinery for 1948–1961; *Trans. IEEE*, Power Apparatus and Systems, June, 1964, Vol. 83, No. 6, pp. 589–606, 650.

105. Standards

3. Definitions of Electrical Terms, Group 10 (Rotating Machinery); USAS C42.10, 1957.

4. Synchronous Machines (General); USAS C50.10, 1965. Synchronous Motors; USAS C50.11, 1965. Salient Pole Synchronous Generators and Condensers; USAS C50.12, 1965. Cylindrical Rotor Synchronous Generators; USAS C50.13, 1965.

5. Guide for the Preparation of Test Procedures for the Thermal Evaluation of Electrical Insulation Materials; *AIEE Publ.* 1-D, 1957.

6. Recommended Guide for Testing Insulation Resistance of Rotating Machinery; *AIEE Publ.* 43, 1961.

7. Guide for Insulation Maintenance for Large Alternating-current Rotating Machinery; *AIEE Publ.* 56, 1958.

8. Recommended Guide for Making Dielectric Measurements in the Field; *AIEE Publ.* 62, 1958.

9. Test Procedures for Synchronous Machines; *IEEE Publ.* 115, 1965.

106. Articles

10. Pohl, R. Theory of Pulsating Field Machines; *Jour. IEE*, 1946, Vol. 93, Pt. 2, pp. 67–80.

11. Wieseman, R. W. Graphical Determination of Magnetic Fields—Practical Applications to Salient-pole Synchronous Machine Design; *Trans. AIEE*, 1927, Vol. 46, pp. 141–154.

12. Calvert, J. F. Amplitudes of Mmf Harmonics for Fractional Slot Windings; *Trans. AIEE*, 1938, Vol. 57, pp. 777–784. See also *Univ. Iowa Bull.* 142, Ames, Iowa.

13. Fechheimer, C. J. Hydrogen, a Successor to Air; *Elec. Jour.*, September, 1929, p. 405.

14. Knowlton, Rice, and Frieburghouse Hydrogen as a Cooling Medium for Electrical Machinery; *Trans. AIEE*, 1925, Vol. 44, pp. 922–932.

15. Beckwith and Rosenberg A New Fully Supercharged Generator; *Trans. AIEE*, 1954, Vol. 73, Pt. 111-A, pp. 477–483.

16. Baudry and Heller Ventilation of Inner-cooled Generators; *Trans. AIEE*, 1954, Vol. 73, Pt. 111-A, pp. 500–508.

17. Holley and Taylor Direct Cooling of Turbine-Generator Field Windings; *Trans. AIEE*, 1954, Vol. 73, Pt. 111-A, pp. 542–550.

18. Baudry and King Improved Cooling for Generators of Large Rating; *Trans. IEEE*, Power Apparatus and Systems, February, 1965, Vol. 84, No. 2, pp. 106–114.

19. Downs and Holley Progress in the Design of Nuclear Turbine-Generators; *ASME-IEEE Joint Power Generation Conf.*, Denver, Sept. 19, 1966.

20. Harder, E. L., and Cheek, R. C. Regulation of Alternating-current Generators, with Suddenly Applied Loads; *Trans. AIEE*, 1944, Vol. 63, pp. 310–318.

21. Harder, E. L., Cheek, R. C., and Clayton, J. M. Regulation of Alternating Current Generators with Suddenly Applied Loads, Part II; *Trans. AIEE*, 1950, Vol. 69, pp. 395–405.

SECTION 7

DIRECT-CURRENT GENERATORS

BY

E. H. MYERS Manager, D-C Motor and Generator Engineering, Large Rotating Apparatus Division, Westinghouse Electric Corporation; Senior Member, Institute of Electrical and Electronics Engineers

CONTENTS

Numbers refer to paragraphs

SECTION 7

DIRECT-CURRENT GENERATORS

BY E. H. MYERS

THE D-C MACHINE

1. Applications. The most important role played by the d-c generator is the power supply for the important d-c motor. It supplies essentially ripple-free power and precisely held voltage at any desired value from zero to rated. This is truly d-c power, and it permits the best possible commutation on the motor because it is free of the severe

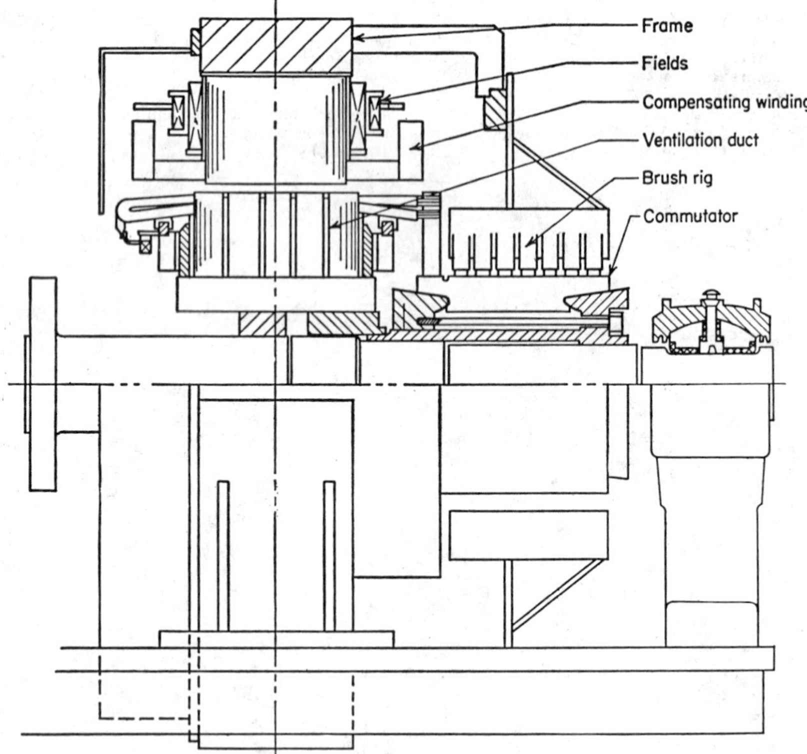

FIG. 7-1. The d-c machine.

waveshapes of "d-c" power from rectifiers. It has excellent response and is particularly suitable for precise output control by feedback control regulators. It is also well suited for supplying accurately controlled and responsive excitation power for both a-c and d-c machines.

The d-c motor plays an ever-increasing vital part in modern industry, because it can operate at and maintain accurately any speed from zero to its top rating. For example, high-speed multistand steel mills for thin steel would not be possible without d-c motors. Each stand must be held precisely at an exact speed which is higher than that of the preceding stand to suit the reduction in thickness of the steel in that stand and to maintain the proper tension in the steel between stands.

2. General Construction. Figure 7-1 shows the parts of a medium or large d-c generator. All sizes differ from a-c machines in having a commutator and the armature on the rotor. They also have salient poles on the stator, and, except for a few small ones, they have commutating poles between the main poles.

3. Construction and Size. Small d-c machines have large surface-to-volume ratios and short paths for heat to reach dissipating surfaces. Cooling requires little more than means to blow air over the rotor and between the poles. Rotor punchings are mounted solidly on the shaft, with no air passages through them.

Larger units, with longer, deeper cores, use the same construction, but with longitudinal holes through the core punchings for cooling air.

Medium and large machines must have large heat-dissipation surfaces and effectively placed cooling air, or "hot spots" will develop. Their core punchings are mounted on arms to permit large volumes of cool air to reach the many core ventilation ducts and also the ventilation spaces between the coil end extensions.

4. Armature-core punchings are usually of high-permeability electrical sheet steel, 0.017 to 0.025 in thick, and have an insulating film between them. Small and medium units use "doughnut" circular punchings, but large units, above about 45 in in diameter, use segmental punchings shaped as in Fig. 7-2, which also shows the fingers used to form the ventilating ducts.

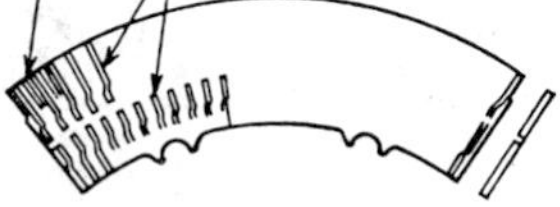

Fig. 7-2. Armature segment for a d-c generator, showing vent fingers applied.

5. Main- and commutating-pole punchings are usually thicker than rotor punchings because only the pole faces are subjected to high-frequency flux changes. These range from 0.062 to 0.125 in thick, and they are normally riveted.

6. The frame yoke is usually made from rolled mild steel plate, but, on high-demand large generators for rapidly changing loads, laminations may be used. The solid frame has a magnetic time constant of ½ s or more depending on the frame thickness. The laminated frame ranges from 0.05 to 0.005 s.

7. The commutator is truly the heart of the d-c machine. It must operate with temperature variations of at least 55°C and with peripheral speeds that may reach 7,000 ft/min. Yet it must remain smooth concentrically within 0.002 to 0.003 in and true, bar to bar, within about 0.0001 in.

The commutator is made up of hard copper bars drawn accurately in a wedge shape. These are separated from each other by mica plate segments, whose thicknesses must be held accurately for nearly perfect indexing of the bars and for no skew. This thickness is 0.020 to 0.050 in depending on the size of the generator and on the maximum voltage that can be expected between bars during operation. The mica segments and bars are clamped between two metal V-rings and insulated from them by cones of mica. On very-high-speed commutators of about 10,000 ft/min shrink rings of steel are used to hold the bars. Mica is used under the rings.

8. Carbon brushes ride on the commutator bars and carry the load current from the rotor coils to the external circuit. The brush holders hold the brushes against the commutator surface by springs to maintain a fairly constant pressure and smooth riding.

GENERAL PRINCIPLES

9. Electromagnetic Induction. A magnetic field is represented by continuous lines of flux considered to emerge from a north pole and to enter a south pole. When the

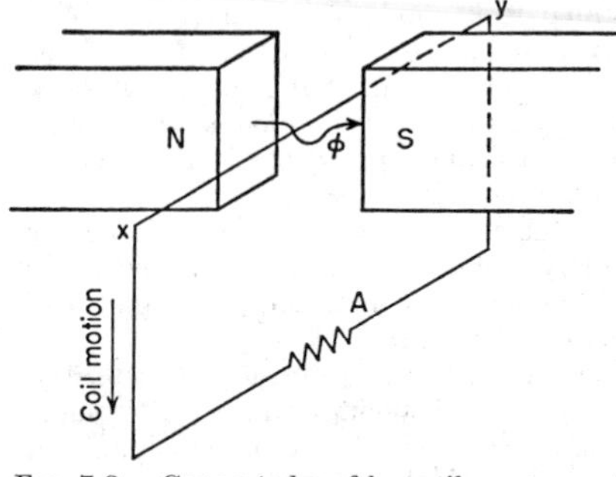

FIG. 7-3. Generated emf by coil movement in a magnetic field.

number of such lines linked by a coil is changed (Fig. 7-3), a voltage is induced in the coil equal to 1 V for a change of 10^8 linkages/s (Mx/s) for each turn of the coil, or $E = (\Delta\phi T \times 10^{-8})/t$ V.

If the flux lines are deformed by the motion of the coil conductor before they are broken, the direction of the induced voltage is considered to be into the conductor if the arrows for the distorted flux are shown to be pointing clockwise and outward if counterclockwise. This is generator action (Fig. 7-4).

10. Force on Current-carrying Conductors in a Magnetic Field. If a conductor carries current, loops of flux are produced around it (Fig. 7-5). The direction of the flux is clockwise if the current flows away from the viewer into the conductor and counterclockwise if the current in the conductor flows toward the viewer.

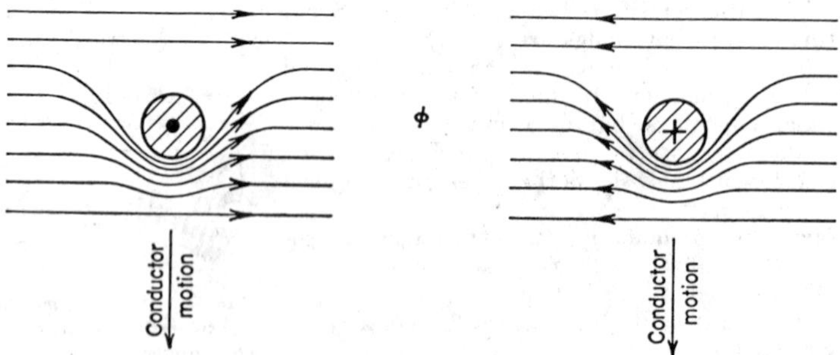

FIG. 7-4. Direction of induced emf by conductor movement in a magnetic field.

If this conductor is in a magnetic field, the combination of the flux of the field and the flux produced by the conductor may be considered to cause a flux concentration on the side of the conductor where the two fluxes are additive and a diminution on the side where they oppose. A force on the conductor results that tends to move it toward the side with reduced flux (Fig. 7-6). This is motor action.

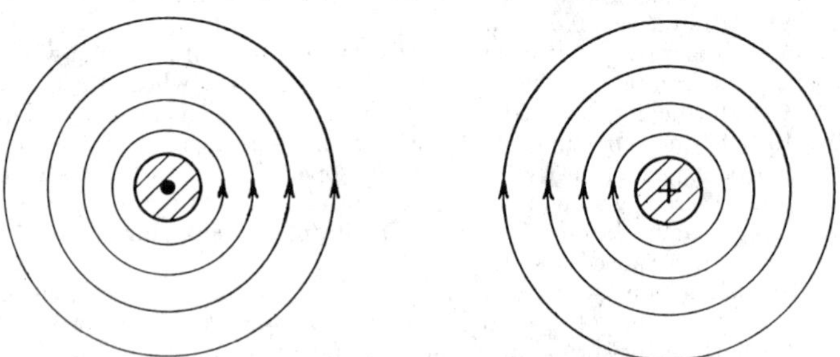

FIG. 7-5. Magnetic fields caused by current-carrying conductors.

11. Generator and Motor Reactions. It is evident that a d-c generator will have its useful voltage induced by the reactions described above and an external driving means must be supplied to rotate the armature so that the conductor loops will move through the flux lines from the stationary poles. However, these conductors must carry current for the generator to be useful, and this will cause retarding forces on them, as described in Par. **10.** The prime mover must overcome these forces.

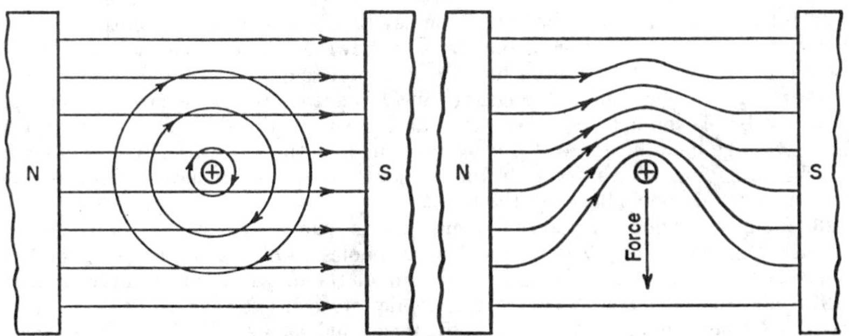

FIG. 7-6. Force on a current-carrying conductor in a magnetic field.

In the case of the d-c motor, the conductor loops will move through the flux, and voltages will be induced in them as described in Par. **9.** These induced voltages are called the "counter emf," and they oppose the flow of currents which produce the forces that rotate the armature. Therefore, this emf must be overcome by an excess voltage applied to the coils by the external voltage source.

12. Direct-current Features. Direct-current machines require many conductors and two or more stationary flux-producing poles to provide the needed generated voltage or the necessary torque. The direction of current flow in the armature conductors under each particular pole must always be correct for the desired results (Fig. 7-7). Therefore, the current in the conductors must reverse at some time while the conductors pass through the space between adjacent north and south poles.

This is accomplished by carbon brushes connected to the external circuit. The brushes make contact with the conductors by means of the commutator.

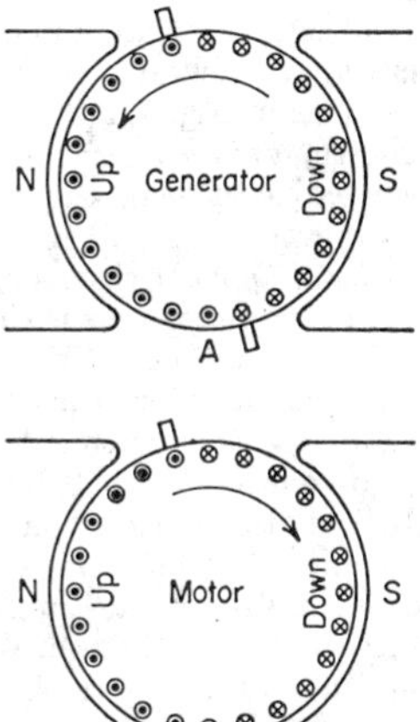

FIG. 7-7. Direction of current in generator and motor.

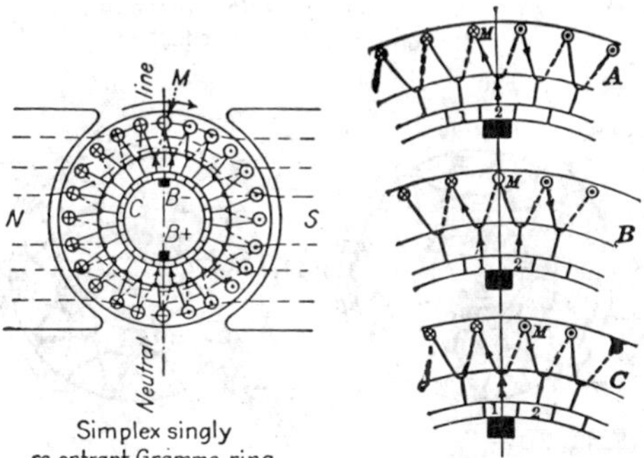

FIG. 7-8. Principle of commutation.

To describe commutation, the Gramme-ring armature winding (which is not used in actual machines) is shown in Fig. 7-8. All the conductors are connected in series and are wound around a steel ring. The ring provides a path for the flux from the north to the south pole. Note that only the outer portions of the conductors cut the flux as the ring rotates. Voltages are induced as shown. With no external circuit, no currents flow, because the voltages induced in the two halves are in opposition. However, if the coils are connected at a commutator C made up of copper blocks insulated from each other, brushes $B-$ and $B+$ may be used to connect the two halves in parallel with respect to an external circuit and currents will flow in the proper direction in the conductors beneath the poles.

As the armature rotates, the coil *M* passes from one side of the neutral line to the other and the direction of the current in it is shown at three successive instants at *A*, *B*, and *C* in Fig. 7-8. As the armature moves from *A* to *C* and the brush changes contact from segment 2 to segment 1, the current in *M* is automatically reversed. For a short period the brush contacts both segments and short-circuits the coil. It is important that no voltage be induced in *M* during that time, or the resulting circulating currents could be damaging. This accounts for the location of the brushes so that *M* will be at the neutral flux point between the poles.

13. Field Excitation. Because current-carrying conductors produce flux that links them as described in Par. **10,** flux from the main poles is obtained by winding conductors around the pole bodies and passing current through them. This current may be supplied in different ways. When a generator supplies its own exciting current, it is "self-excited." When current is supplied from an external source, it is "separately excited." When excited by the load current of the machine, it is "series excited."

ARMATURE WINDINGS

14. Terms. The Gramme-ring winding is not used, because half the conductors (those on the inside of the ring) cut no flux and are wasted. Figures 7-8, 7-10, and 7-11 show such windings only because they illustrate types of connections so well.

A singly reentrant winding closes on itself only after including all the conductors, as shown in Figs. 7-8 and 7-10.

A doubly reentrant winding closes on itself after including half the conductors, as in Fig. 7-11.

As shown, *a simplex winding* has only two paths through the armature from each brush (Fig. 7-8). *A duplex winding* has twice as many paths from each brush and is shown in Figs. 7-10 and 7-11. Note that each brush should cover at least two commutator segments with a duplex winding, or one circuit will be disconnected at times from the external circuit. Although it is possible to use multiplex and multiple reentrant windings, they are uncommon in the United States. They are used in Europe in some large machines.

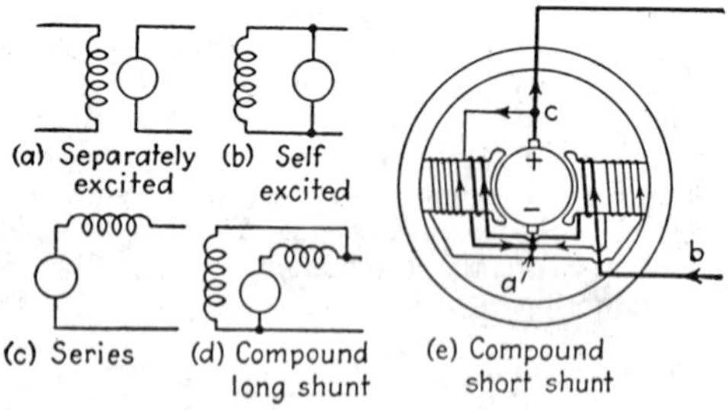

(a) Separately excited (b) Self excited (c) Series (d) Compound long shunt (e) Compound short shunt

Fig. 7-9. Methods of excitation.

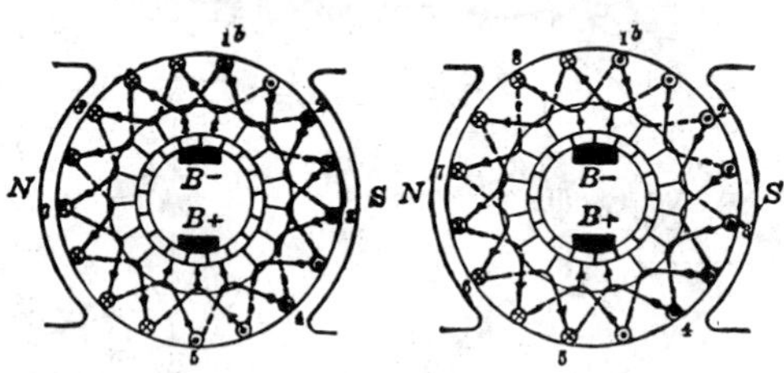

Fig. 7-10. Singly reentrant duplex winding. Fig. 7-11. Doubly reentrant duplex winding.

Modern d-c machines have the armature coils in radial slots in the rotor. Nonmetallic wedges restrain the coils normally but some wedgeless rotors use nonmetallic banding around the core, such as glass fibers in polyester resin. This permits shallower slots and helps to reduce commutation sparking. However, the top conductors are near the pole faces and may have high eddy losses. The coil ends outside the slots are held down on coil supports by glass polyester bands for both types.

15. Multiple, or Lap, Windings. Figure 7-12 shows a *lap-winding coil.* The conductors shown on the left side lie in the top side of the rotor slot. Those on the right side lie in the bottom half of another slot approximately one pole pitch away. At any instant the sides are under adjacent poles, and voltages induced in the two sides are additive. Other coil sides fill the remaining portions of the slots. The coil leads are connected to the commutator segments, and this also connects the coils to form the armature winding. This is shown in Fig. 7-13. The pole faces are slightly shorter than the rotor core.

Almost all medium and large d-c machines use simplex lap windings in which the number of parallel paths in the armature winding equals the number of main poles.

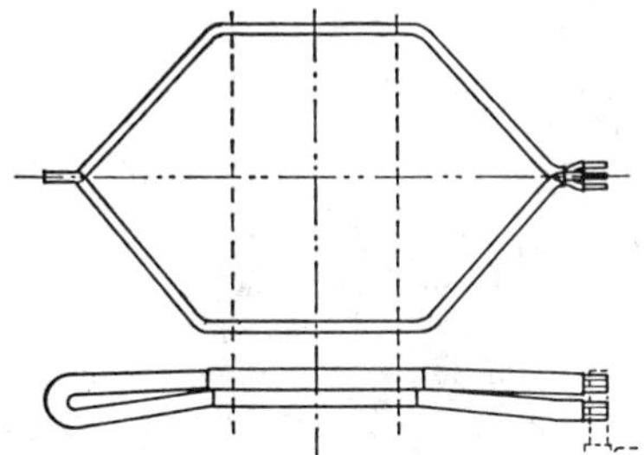

FIG. 7-12. Coil for one-turn lap winding.

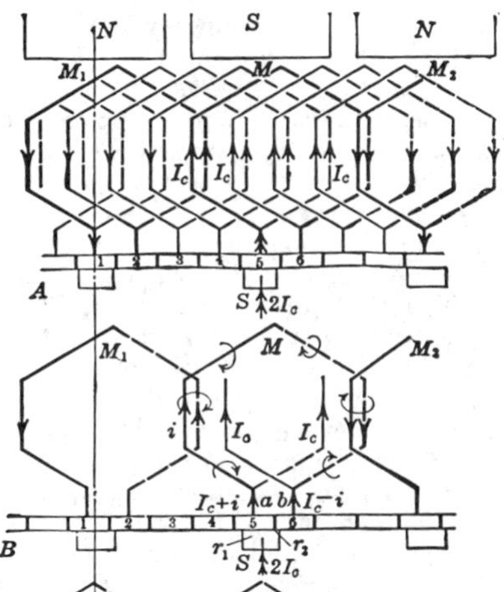

FIG. 7-13. Multiple, or lap, winding.

This permits the current per path to be low enough to allow reasonable-sized conductors in the coils.

Windings. Representations of d-c windings are necessarily complicated. Figure 7-14 shows the lap winding corresponding to the Gramme-ring winding of Fig. 7-8. Unfortunately, the nonproductive end portions are emphasized in such diagrams, and the long, useful portions of the coils in the core slots are shown as radial lines. Conductors in the upper layers are shown as full lines, and those in the lower layers as dotted lines. The inside end connections are those connected to the commutator bars. For convenience, the brushes are shown inside the commutator.

Note that both windings have the same number of useful conductors but that the Gramme-ring winding requires twice the number of actual conductors and twice the number of commutator bars.

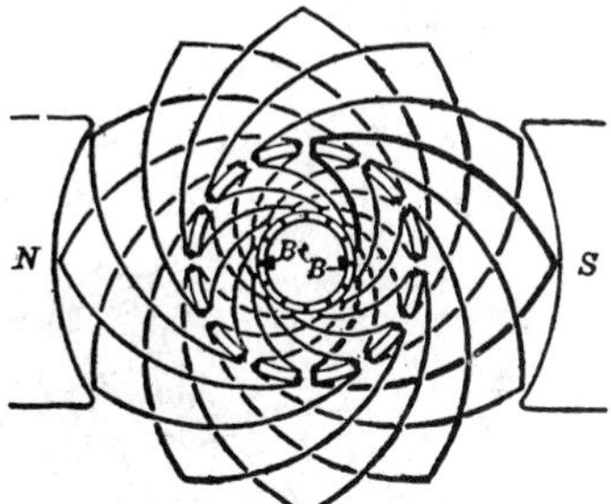

FIG. 7-14. Simplex lap winding.

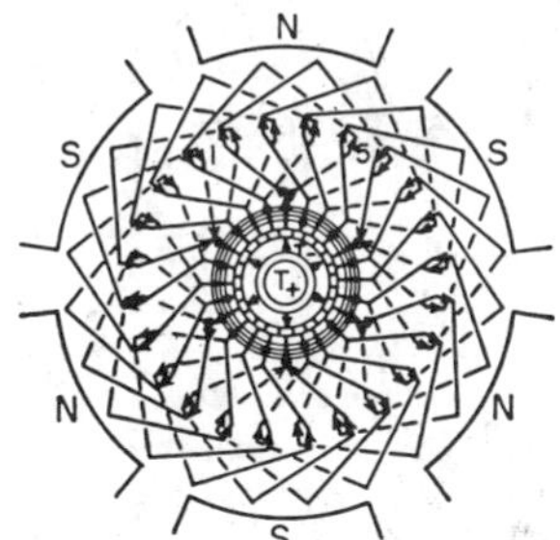

FIG. 7-15. Simplex singly reentrant full-pitch multiple winding with equalizers.

Figure 7-15 shows a six-pole simplex lap winding. Study of this reveals the six parallel paths between the positive and negative terminals. The three positive brushes are connected outside the machine by a copper ring $T+$ and the negative brushes by $T-$.

The two sides of a lap coil may be full pitch (exactly a pole pitch apart), but most machines use a short pitch (less than a pole pitch apart), with the coil throw one-half slot pitch less than a pole pitch. This is done to improve commutation.

Equalizers. As shown in Fig. 7-15, the parallel paths of the armature circuit lie under different poles, and any differences in flux from the poles cause different voltages to be generated in the various paths. Flux differences can be caused by unequal air gaps, by a different number of turns on the main-pole field coils, or by different reluctances in the iron circuits.

With different voltages in the paths paralleled by the brushes, currents will flow to equalize the voltages. These currents must pass through the brushes and may cause sparking, additional losses, and heating. The variation in pole flux is minimized by careful manufacture but cannot be entirely avoided.

To reduce such currents to a minimum, copper connections are used to short-circuit points on the paralleled paths that are supposed to be at the same voltage. Such points would be exactly two pole pitches apart in a lap winding. So in a six-pole simplex lap winding each point in the armature circuit will have two other points that should be at its exact potential. For these points to be accessible, the number of commutator bars and the number of slots must be a multiple of the number of poles divided by 2.

These short-circuiting rings are called "equalizers." Alternating currents flow through them instead of the brushes. The direction of flow is such that the weak poles are magnetized and the strong poles are weakened. Usually one coil in about 30% of the slots is equalized. The cross-sectional area of an equalizer is 20 to 40% of that of the armature conductor.

Involute necks, or connections, to each commutator bar from conductors two pole pitches apart give 100% equalization but are troublesome because of inertia and creepage insulation problems. They are seldom used.

Figure 7-15 shows the equalizing connections behind the commutator connections. Normally they are located at the rear coil extensions; so they are more accessible and less subject to carbon-brush dust problems.

16. Two-circuit, or Wave, Windings. Figure 7-16 shows a wave type of coil. Figure 7-17 gives a six-pole wave winding. Study reveals that it has only two parallel paths between the positive and negative terminals. Thus, only two sets of brushes are needed. Each brush shorts $p/2$ coils in series. Because points a, b, and c are at the same potential (and, also, points d, e, and f), brushes can be placed at each of these points to allow a commutator one-third as long.

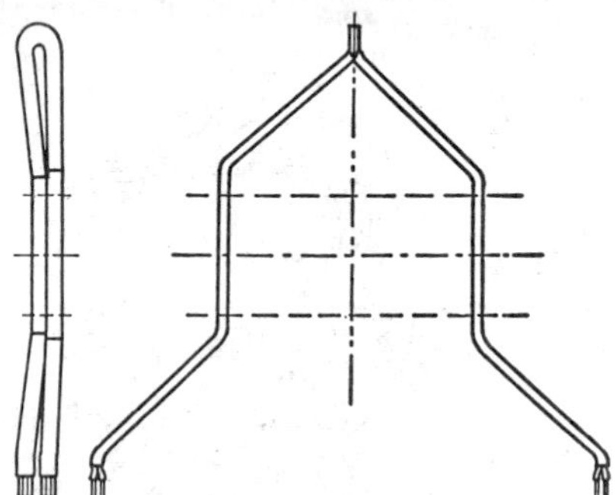

Fig. 7-16. One-turn wave winding.

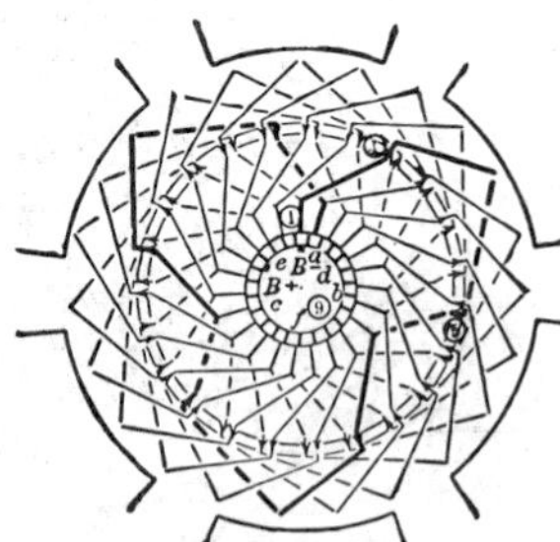

Fig. 7-17. Two-circuit progressive winding.

The winding must progress or retrogress by one commutator bar each time it passes around the armature for it to be singly reentrant. Thus, the number of bars must equal $(kp/2) \pm 1$, where k is a whole number and p is the number of poles. The winding needs no equalizers because all conductors pass under all poles.

While most wave windings are 2-circuit, they can be multicircuit, as 4 or 16 circuits on a 4-pole machine or 6, 12, or 24 circuits on a 12-pole machine. Multicircuit wave windings with the same number of circuits as poles can be made by using the same slot and bar combinations as on a lap winding. For example, with an 8-pole machine with 100 slots and 200 commutator bars, the bar throw for a simplex lap winding would be from bar 1 to bar 2 and then from bar 2 to bar 3, etc. For an 8-circuit wave winding the winding must fail to close by circuits/2 bars, or 4. Thus, the throw would be bar 1 to 50, to bar 99, to bar 148, etc. The throw is (bars ± circuits/2)/(p/2), in this case, $(200 - 4)/4 = 49$. Theoretically such windings require no equalizers, but better results are obtained if they are used.

Since both lap and multiple wave windings can be wound in the same slot and bar combination simultaneously, this is done by making each winding of half-size conductors. This combination resembles a *frog's leg* and is called by that name. It needs no equalizers but requires more insulation space in the slots and is seldom used.

Some wave windings require *dead coils*. For instance, a large 10-pole machine may have a circle of rotor punchings made of 5 segments to avoid variation in reluctance as the rotor passes under the 5 pairs of poles. To avoid dissimilar slot arrangements in the segments, the total number of slots must be divisible by the number of segments, or 5 in this case. This requires the number of commutator bars to be also a multiple k of 5. However, the bar throw for a simplex wave winding must be an integer and equal to (bars ± 1)/(P/2). Obviously $(5k \pm 1)/5$ cannot meet this requirement. So one coil, called a *dead coil*, will not be connected into the winding, and its ends will be taped up to insulate it completely. No bar will be provided for it, and thus the bar throw will be an integer. Dead coils should be avoided because they impair commutation.

ARMATURE REACTIONS

17. Cross-magnetizing Effect. Diagram a of Fig. 7-18 represents the magnetic field produced in the air gap of a two-pole machine by the mmf of the main exciting coils, and part b represents the magnetic field produced by the mmf of the armature winding alone when it carries a load current. If each of the Z armature conductors carries I_c A, then the mmf between a and b is equal to ZI_c/p At. That between c and d (across the pole tips) is $\psi ZI_c/p$ At, where ψ = ratio of pole arc to pole pitch. On the assumption that all the reluctance is in the air gap, half the mmf acts at ce and half at fd; so the cross-magnetizing effect at each pole tip is

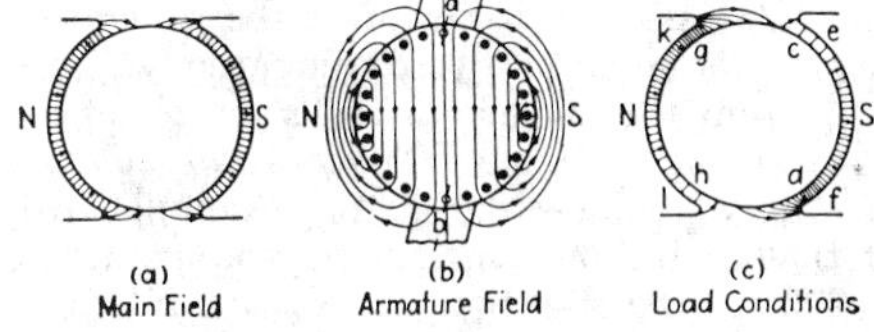

FIG. 7-18. Flux distribution.

$$\psi ZI_c/2p \qquad \text{(At)} \tag{7-1}$$

for any number of poles.

18. Field Distortion. Diagram c of Fig. 7-18 shows the resultant magnetic field when both armature and main exciting mmfs exist together; the flux density is increased at pole tips d and g and is decreased at tips c and h.

19. Flux Reduction Due to Cross Magnetization. Figure 7-19 shows part of a large machine with p poles. Curve D shows the flux distribution in the air gap due to the main exciting mmf acting alone, with flux density plotted vertically. Curve G shows the distribution of the armature mmf, and curve F shows the resultant flux distribution with both acting. Since the armature teeth are saturated at normal flux densities, the increase in density at f is less than the decrease at e, so that the total flux per pole is diminished by the cross-magnetizing effect of the armature.

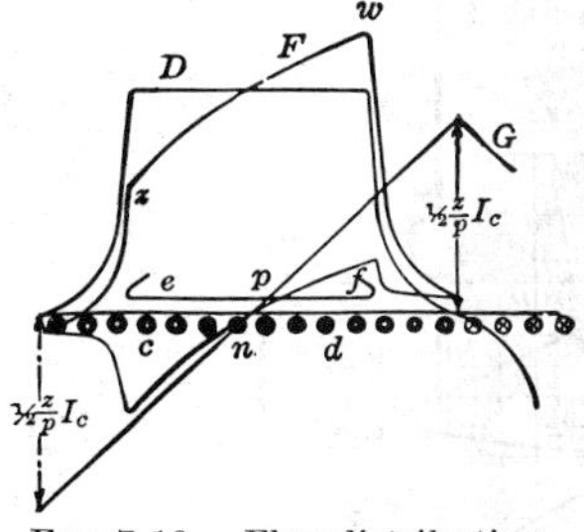

FIG. 7-19. Flux distribution.

20. Demagnetizing Effect of Brush Shift. Figure 7-20 shows the magnetic field produced by the armature mmf with the brushes shifted through an angle θ to improve commutation. The armature field is no longer at right angles to the main field but may be considered as the resultant of two components, one in the direction OY, called the "cross-magnetizing component," discussed in Par. **19** and the other in the direction OX, which is called the "demagnetizing component" because it directly opposes the main field. Figure 7-21 gives the armature divided to show the two components, and it is seen that the demagnetizing ampere-turns per pair of poles are

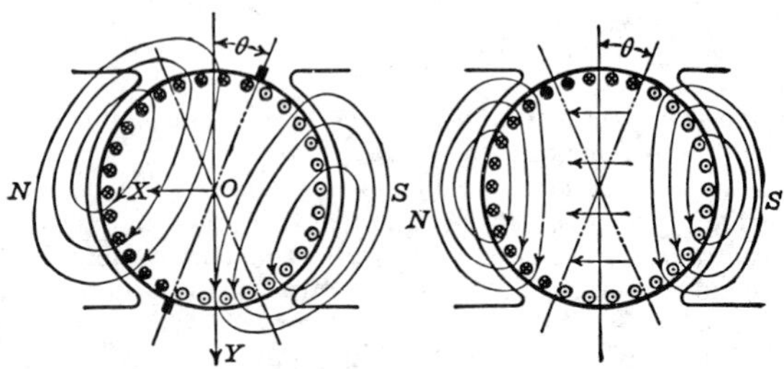

FIGS. 7-20 and 7-21. Demagnetizing and cross-magnetizing effect.

$$\frac{ZI_c}{p} \times \frac{2\theta}{180} \qquad \text{(At)} \qquad \textbf{(7-2)}$$

where $2\theta/180$ is about 0.2 for small noncommutating pole machines where brush shift is used. So the demagnetizing ampere-turns per pole would be

$$0.1ZI_c/p \qquad \text{(At)} \qquad (7\text{-}3)$$

21. No-load and Full-load Saturation Curves. Curve 1 of Fig. 7-22 is the no-load saturation curve of a d-c generator. When full-load current is applied, there is a decrease in useful flux, and therefore a drop in voltage ab due to the armature cross-magnetizing effect (see Par. **19**). A further voltage drop from brush shift is counterbalanced by an increase in excitation $bc = 0.1ZI_c/p$; also a portion cd of the generated emf is required in overcoming the voltage drop from the current in the internal resistance of the machine. The no-load voltage of 240 V requires 8,000 At. At full load at that excitation the terminal voltage drops to 220 V. To have both no-load and full-load voltages equal to 240 volts, a series field of 10,700 − 8,000 = 2,700 At would be required.

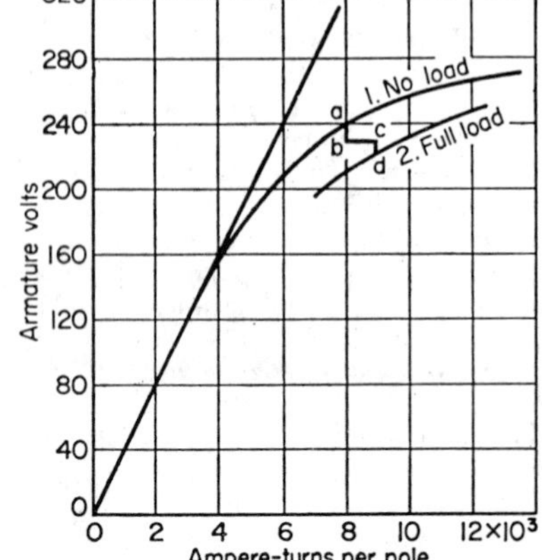

FIG. 7-22. Saturation curves—d-c generator.

COMMUTATION

22. Commutation. The voltages generated in all conductors under a north pole of a d-c generator are in the same direction, and those generated in the conductors under a south pole are all in the opposite direction (Fig. 7-23). Currents will flow in the same directions as induced voltages in generators

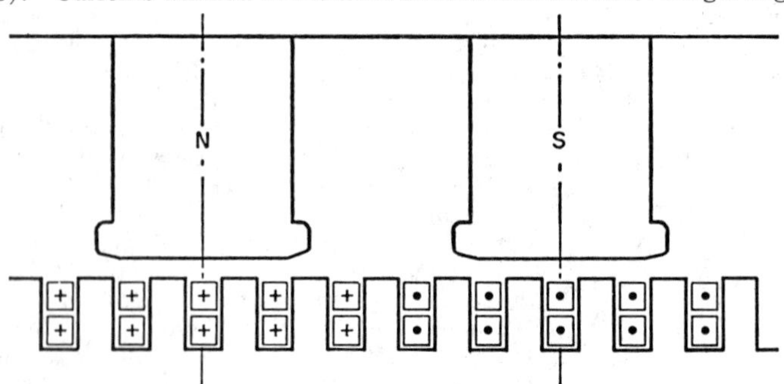

FIG. 7-23. Conductor currents.

and in the opposite direction in motors. Thus, as a conductor of the armature passes under a brush, its current must reverse from a given value in one direction to the same value in the opposite direction. This is called "commutation."

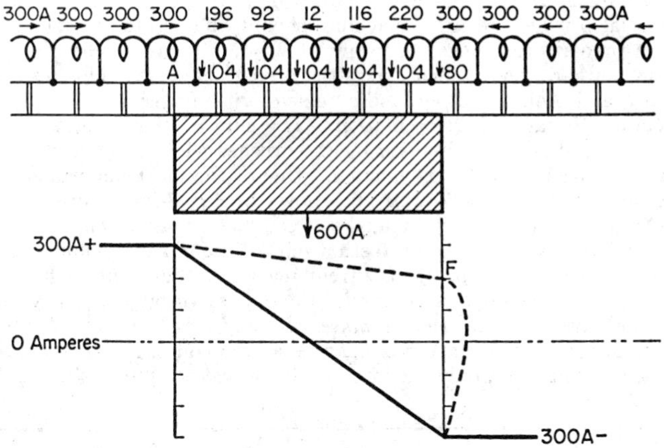

Fig. 7-24. Commutation.

23. Conductor Current Reversal. If commutation is "perfect," the change of the current in a coil will be linear, as shown by the solid line in Fig. 7-24. Unfortunately, the conductors lie in steel slots, and self- and mutual inductances indicated in Fig. 7-25 cause voltages in the coils short-circuited by the brushes. These result in circulating currents that tend to prevent the initial current change, delaying the reversal. In extreme cases, the delay may be as severe as indicated by the dotted line of Fig. 7-24. Because the current must be reversed by the time the coil leaves the brush (when there is no longer any path for circulating currents), the current remaining to be reversed at *F* must discharge its energy in an electric arc from the commutator bar to the heel of the brush. This is commutation sparking. It can burn the edges of the commutator bars and the brushes. However, most large and heavy-duty d-c machines have some nondamaging sparking, and "sparkless" commutation is not required by accepted standards. However, commutation must not require undue maintenance.

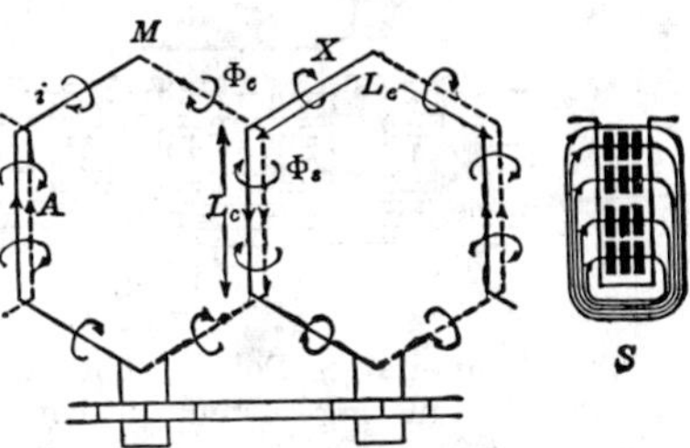

Fig. 7-25. Magnetic field surrounding short-circuited coils.

The undesired voltages causing the circulating currents result from interpolar fluxes from armature reaction (see Par. **18**), leakage fluxes of the current-carrying armature conductors, and, in some cases, main-pole-tip spray flux. Beneficial factors reducing the circulating currents include the resistance of the short-circuited coil, the resistance of the commutator risers, and that of the brush body to transverse currents. However, the most important factor is the voltage drop at the sliding contact between the brush face and the copper commutator surface.

24. Commutator Brushes. Most d-c machines use electrographitic brushes with about 60 A/in^2 current density at full load. These have an essentially constant contact voltage drop at the commutator surface of about 1 V for loads above one third. This effective resistance to circulating currents is important to good operation of d-c machines.

The cross resistance of the brush body to circulating currents can be increased by splitting the brush into two wafers and making the cross currents cross the air gap between the two pieces. This has increased the good commutation range on some machines by 7%. The use of double brush holders which have metal dividers between two brushes in the holder is even more effective and has increased the good commutation range as much as 15% over single solid brushes.

Unless special brushes are used, machines should be operated for not more than a few hours at a time at brush densities below 30 A/in². If this is done, the commutator surface develops a hard glaze which makes the brushes chatter. This results in frayed shunts, chipped and broken brushes, and excessive brush-finger wear.

25. Reactance Voltage of Commutation. The sum of the voltages induced in the armature coil while it is short-circuited by the brushes while undergoing commutation is called the *reactance voltage of commutation*. One of the most important of the fluxes causing this voltage is the *slot leakage flux* shown in Fig. 7-26. This is the resultant flux leakage from current in the individual slot conductors, as shown in Fig. 7-25. Because the radial fluxes in the rotor teeth from adjacent slot conductors essentially cancel except at point *C* (the point of current reversal), the resultant flux is as shown in Fig. 7-26. As the conductors commutate and pass through *C*, they cut the flux shown there and this generates the *reactance voltage of commutation*. Actually part of this voltage is also due to leakage flux changes at the coil ends, to armature reaction flux, etc., but, for simplicity, only the important slot leakage flux is shown.

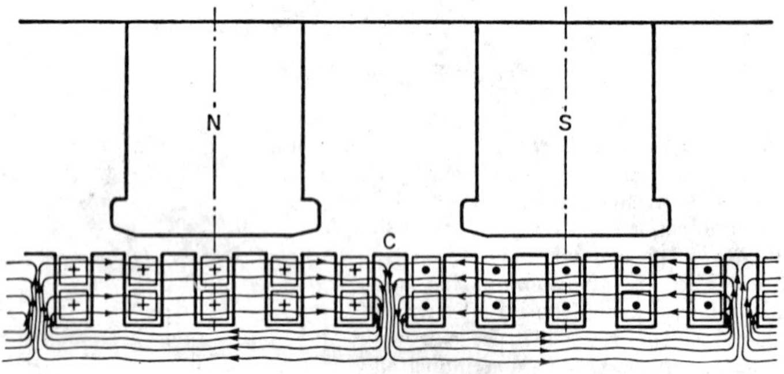

Fig. 7-26. Slot-leakage flux.

26. Commutating Poles. The beneficial factors described in Pars. **24** and **25** that limit the circulating currents in coils being commutated are not adequate to prevent serious delays in current reversal. Other means must be taken to prevent sparking.

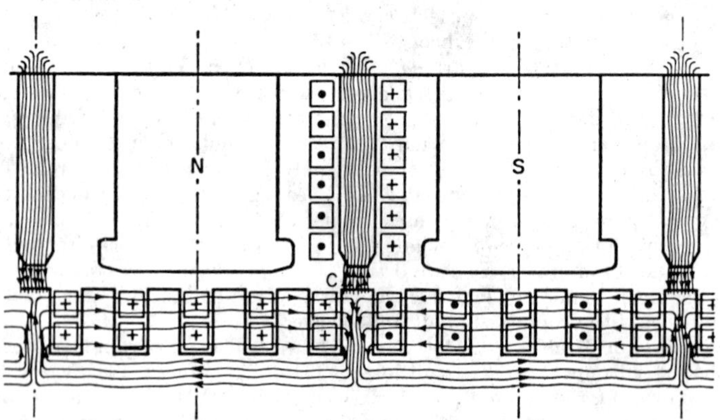

Fig. 7-27. Slot-leakage flux and commutating-pole flux.

If the flux at C (Fig. 7-26) could be nullified by an equal flux in the opposite direction, the circulating currents due to the slot leakage flux would be prevented.

The location of C is fixed by the location of the brushes. If the brushes were shifted toward the south main pole, a position could be found where the main flux upward into the south pole would cancel the downward flux due to slot leakage at C.

This method was used early in the history of d-c machines. Unfortunately, the slot leakage flux at C is proportional to conductor load current, whereas the flux into the south pole is not. Thus, a new brush position is needed for every change in load current.

A better solution is to provide stationary poles midway between the main poles, as shown in Fig. 7-27. Windings on these *commutating poles* carry the load current. Thus, the flux into the pole at C is proportional to the rotor conductor currents and, theoretically, can cancel the voltages induced in the coils being commutated by the slot leakage flux. In the case of the d-c motor, the current reverses in both the armature and the commutating field, and proper canceling is maintained.

Note that the strength of the commutating-pole winding must be greater than the armature-winding ampere-turns per pole by the amount required to carry the needed flux across the commutating-pole air gap.

Almost all modern d-c machines use commutating poles, although some small machines have only half as many as main poles.

The commutating-pole tip is usually shaped with tapered sides, to approximate the shape of the reactance voltage of commutation form (see Figs. 7-27 and 7-28).

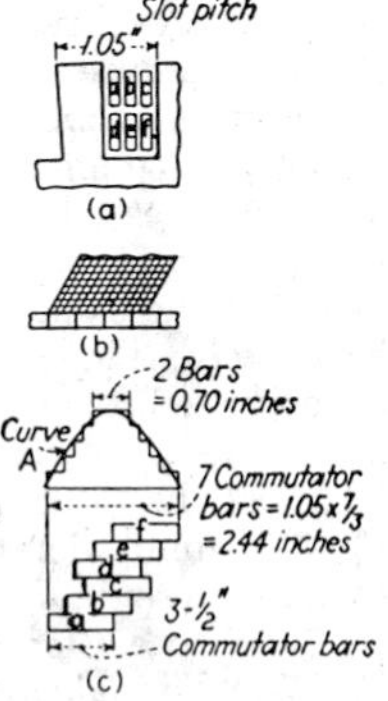

FIG. 7-28. Commutating zone.

27. Reactance Voltage of Commutation Formula. To determine the useful flux needed across the commutating-pole air gap, it is useful to calculate the reactance voltage of commutation (the total of the voltages induced in the armature coil as it undergoes commutation). The approximate value of this voltage may be calculated by the use of the following formula,

$$E_c = \left(\frac{\text{poles}}{\text{paths}}\right)(I_c ZT)(\text{r/min})(5.3)(10^{-10}) \times \left[K_1 + K_2 L_r + \frac{(0.4L_r)(SP)(d_s + 1)}{b_s}\right] \quad \text{(volts)} \quad (7\text{-}4)$$

where
I_c = current per armature conductor, in A
Z = total no. armature conductors
T = no. of turns/coil between commutator bars
L_r = gross armature-core length, in
K_1 = 23 for machines using nonmagnetic bands
= 29 for machines using magnetic bands
K_2 = 4.35 for noncommutating-pole machines
= 0 for machines with commutating-pole length same as gross armature-core length
b_s = width of slot, in
d_s = depth of slot, in
SP = the slot pitch, in

The above formula is based on work by Lamme. (See Theory of Commutation by B. G. Lamme, *Trans. AIEE,* October, 1911, Vol. 30.)

28. The Commutating Zone. This is defined as that space on the armature periphery through which a given slot moves while all the conductors lying in the slot commutate. In chorded windings, it is extended to include the coil edges in the chorded slots. The commutating zone thus depends on the number of commutating bars covered per brush.

The zone may be calculated by the following formula,

$$CZ = \frac{SP[(B/S) + (B/S \times Ch) + (B/Br) - (Cir/p)]}{B/S} \qquad (7\text{-}5)$$

where CZ is the commutating zone in inches, SP the rotor slot pitch in inches, B/S the number of commutator bars per slot, Ch the slot chording as a fraction of the slot pitch, B/Br the number of commutator bars spanned per brush, Cir the number of paralleled circuits in the armature, and p the number of main poles.

Consider an eight-pole simplex lap winding with three bars per slot, chording of ½ slot, 3½ bars per brush, and slot pitch of 1.05 in.

$$CZ = \frac{1.05 \times (3 + 1\tfrac{1}{2} + 3\tfrac{1}{2} - 8/8)}{3} = 2.44 \text{ in}$$

In this machine, all the conductors in a slot are commutated while the armature periphery moves 2.44 in.

This can be seen graphically in Fig. 7-28, where (*a*) shows a slot with six conductors, (*b*) shows a brush covering 3½ bars, and (*c*) shows the graphical solution. In (*c*) the rectangle *a* represents as abscissa the space of 3½ commutator bars if they were at the armature surface. This is the length to commutate coil *a*. The ordinate represents to a convenient scale the commutation voltage induced in this conductor while it is being commutated. Rectangles *b* and *c* are the same for coils *b* and *c*. Since *b* commutates 1 bar later than *a*, it is shown one bar space to the right of *a*, etc. In a similar manner *d*, *e*, and *f* are shown. Normally *d* would be expected to start commutation at the same time as *a*, but, because of chording, it starts later, in this case 1½ bars later. Thus, the commutating zone starts with the beginning of rectangle *a* and is completed at the end of rectangle *f*. Upon adding the spaces of the parts, this is 3½ bars for *f*, 2 bars for the steps of *e* and *d*, and 1½ bars for chording, or a total of 7 bars at the rotor surface, which is 1.05 × 7/3, or 2.44 in.

The summation of the individual rectangles as smoothed off by curve *A* of (*c*) is a rough representation of the reactance voltages induced in the coils during commutation.

29. Single Clearance. The center line of the commutating zone and curve *A* of Fig. 7-28 lie midway between the adjacent main-pole tips if the brushes are not shifted off neutral. The arc on the rotor surface between the tips of adjacent main poles is called the *neutral zone*. If the commutating zone is centered in this arc, the spaces left at each end are called the *single clearance*. Thus, the single clearance is

$$SC = (\text{neutral zone} - \text{commutating zone})/2 \qquad (7\text{-}6)$$

The single clearance is an indication of the probability that spray flux from the main-pole tips might flow into the commutating zone. Such flux would not vary with load and would distort the form of the useful flux from the commutating pole. The commutating-pole useful flux form should closely resemble that of curve *A* in Fig. 7-28.

Noncompensated d-c machines usually have main-pole tips with short radial dimensions and have limited spray flux into the neutral zone. The minimum single clearance for these should be not less than 0.6 in and not less than 0.9, with commutation voltages above 3 or 4 V.

Compensated-machine main poles usually have tips 2 to 3 in deep to accommodate the compensating slots and are more likely to spray flux into the commutating zone. These require single-clearance minimums of 1.2 to 1.4 in.

If there is any question about tip flux reaching the commutating zone, flux plots should be made.

30. Commutating-pole Excitation. Figures 7-18*b* and 7-19 show that flux should normally be expected in the commutation area. It is caused by the armature-winding ampere-turns per pole. It could be reduced to zero if the commutating pole had ampere-turns equal and opposite to those of the armature winding. This is $ZI_c/2p$ At/pole, as explained in Par. **17.**

However, it is necessary that the commutating winding also produce useful flux across the commutating-pole gap to counteract the reactance voltage of commutation, as shown in Fig. 7-27. For this reason, the strength of the commutating field is usually 20 to 30% greater than the armature ampere-turns per pole. This difference is called the *excess ampere-turns*. These must be added to the dotted-line bar diagram of Fig. 7-29. The actual flux across the gap is set accurately during the factory test by adjusting the number of sheet-steel shims behind the commutating poles to set the reluctance of the gap for the exact flux needed.

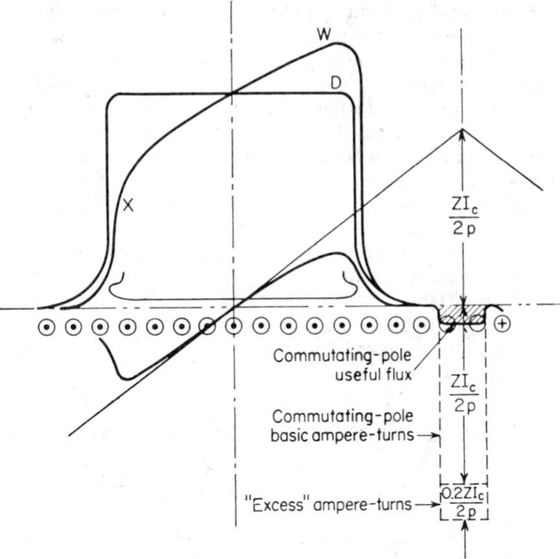

FIG. 7-29. Commutating-pole ampere-turns.

31. Calculation of Commutating-pole Air Gaps. With fixed *excess ampere-turns* on the commutating-pole winding and a certain commutation voltage at rated current and speed, only one particular commutating-pole air gap will result in the most favorable compensation of the commutation voltage. The shape of the pole tip will determine the form of the flux density under it, but the length of the air gap will determine the magnitude of the density.

To counteract the reactance voltage of commutation, E_c, the approximate maximum flux density needed in the commutating-pole air gap is

$$B_m = \frac{E_c \times 23 \times 10^8}{Z/\text{bars} \times DL_c \times \text{r/min}} \tag{7-7}$$

where E_c is the full-load reactance voltage of commutation at speed r/min, Z is the total number of armature conductors, bars is the total number of commutator bars, D is the armature diameter in inches, L_c is the axial length of the commutating poles in inches, and r/min is the revolutions per minute for which E_c was calculated.

The approximate length of the needed commutating-pole single air gap may be calculated by the following formula:

$$\text{Gap} = \frac{3.19 \times \text{excess ampere-turns}}{B_m} \tag{7-8}$$

When the machine is on factory test, the excess ampere-turns can be adjusted to obtain the best commutation possible by placing another d-c generator or a battery across the commutating winding to add to the load current flowing in it or to lower the excess by shunting out some of the load current. This is known as a "boost or buck" test. Afterward the commutating-pole air gap is changed to produce the "best" gap flux density with the actual excess ampere-turns. The new gap will be

$$\text{Gap}_2 = \frac{\text{excess At}_1}{\text{excess At}_2} \times \text{gap}_1 \tag{7-9}$$

32. Dimensions of Commutating Poles. If the useful flux across a commutating-pole air gap is not proportional to the machine load current, the compensation of the reactance voltage of commutation will not be correct for all loads and sparking may damage the brushes and commutator. Thus, the commutating pole must not saturate at the highest load currents to be accommodated. The base of the pole must carry not only the useful air-gap flux but also leakage fluxes from the commutating and main field coils which are near. These leakage fluxes are relatively large and must be determined with care by flux plotting if the danger of commutating-pole saturation exists.

The amount of leakage flux through the base of the pole depends upon the length of the leakage paths, the number of coil ampere-turns, and the location of the commutating field. The leakage paths should be made as long as feasible, the coil ampere-turns as few as reasonable, and the commutating coil located as close to the pole tip as possible. Also, all sections of the commutating pole should be large enough to accommodate their flux.

For a normal compensated machine the leakage flux will be about 75% of the commutating-pole useful flux, or about 140% of the useful flux in a noncompensated machine (see Par. **33**).

The approximate useful flux can be calculated by using the maximum commutating-pole air-gap flux density from Eq. (7-7). The average flux density of the commutating zone will be approximately

$$B_a = 0.83B_m \tag{7-10}$$

The flux density at overload in the base of the pole is

$$B_{cp} = \frac{K_3 \times K_4 \times B_a \times CZ}{L_c \times W_c} \tag{7-11}$$

where K_3 is 1.75 for compensated machines and 2.40 for noncompensated machines, K_4 is the ratio of overload current to rated current, B_a is the average flux density in the commutating zone, CZ is the width of the commutating zone, L_c is the axial length of the commutating pole, and W_c is the circumferential width of the pole at its base. B_{cp} should not exceed 80,000 to 90,000 lines/in² for good commutation.

33. Compensating Windings. While the commutating pole is a good solution for commutation, it does not prevent distortion of the main-pole flux by armature reaction, as explained in Par. **18.** The flux set up across the main-pole face by the armature mmf is shown in Fig. 7-30. If the pole face is provided with another winding, as shown in Fig. 7-31, and connected in series with the load, it can set up an mmf equal and opposite to that of the armature. This would tend to prevent distortion of the air-gap field by armature reaction. Such windings are called *compensating windings* and are usually provided on medium-sized and large d-c machines to obtain the best possible characteristics. They are also often needed to make machines less susceptible to flashovers (see Par. **34**).

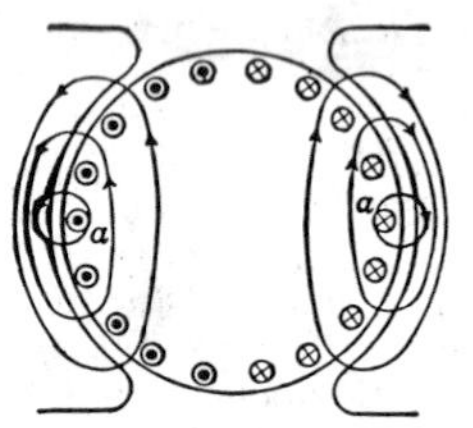

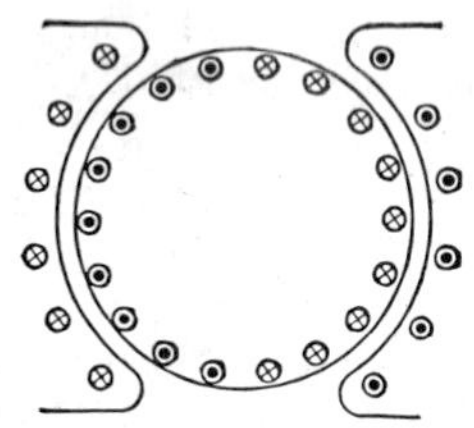

FIGS. 7-30 and 7-31. Armature field without and with compensating windings.

The use of compensating windings reduces the number of turns required on the commutating-pole fields, and this materially reduces the leakage fluxes of the field and, in turn, the pole saturations at high currents. The ampere-turns on the commutating field are reduced by about 50% with the use of a compensating field. This new winding may be considered to be some of the turns taken off the commutating-pole winding and relocated in slots in the main-pole faces.

The number and location of the compensating slots must be carefully chosen to match, as closely as possible, the rotor ampere-turns per inch. However, the slot spacing must not correspond closely to that of the rotor. This would cause a major change in reluctance to the main-pole useful flux every time the rotor moved from a position where the rotor and stator slots all coincided to where the rotor slots coincided with the stator teeth. This would occur once for every slot-pitch movement. The resulting rapid changes in useful flux would cause ripples in the output voltage and also serious magnetic noise. If too few slots are used, local flux distortions occur and the compensating winding loses some of its effectiveness (see Fig. 7-33).

Compensation of armature reaction effectively reduces the armature circuit inductance. This makes the machine less susceptible to the bad effects of $L(di/dt)$ voltages caused by very fast load-current changes.

During manufacture it is possible to locate the compensating winding nonsymmetrically about the center line of the main pole. This causes a direct-axis flux, which will give a series field effect (Fig. 7-32). For generator cumulative compounding, the slots must be shifted in the direction of the machine rotation. This shift gives a motor differential compounding. The effect cannot be adjusted after manufacture. It seldom exceeds ½ in, and this does not materially reduce the effectiveness of the compensation.

34. Volts per Bar. The mica thickness between the commutator segments depends on the machine design and varies from 0.020 in on small machines to 0.050 in on large units. While several hundred volts would normally be required to jump these distances, the presence of ionized air from sparking and the presence of conducting carbon dust make it necessary that the voltage between segments be held to low values. If a low-resistance arc does jump between segments, it raises the voltages across the remaining bars. It also tends to ionize some air to form conducting paths across the rest of the bars. If this progresses until all the segments between brush arms of opposite polarity are bridged, then a *flashover* occurs and severe damage may result to the commutator, brushes, and brushholders.

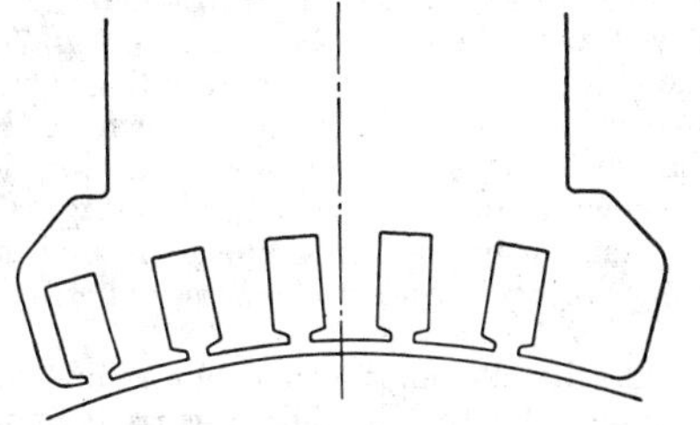

FIG. 7-32. Offset compensating winding.

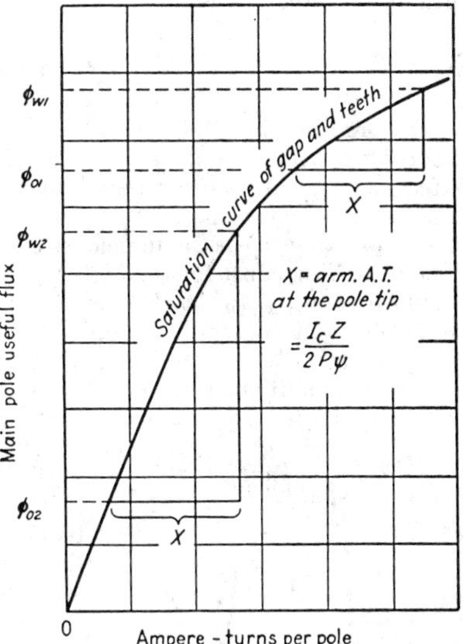

FIG. 7-33. Effect of flux-distorting armature ampere-turns at normal and low saturation.

Because the highest voltage between bars is the "trigger" that starts the flash, this is an important limit. The "average" volts per bar has little significance. Figure 7-29 shows that the maximum volts per bar depends on the field form. For the noncompensated machine shown, the maximum volts between segments exists at w. The segments connected to conductors at x have much less voltage between them, and those beyond the edge of the pole have almost none.

The relation between maximum volts per bar and the average depends on the armature ampere-turns per pole and the saturation curve of the gap and teeth at the pole tips. Upon neglecting the small voltage drop in the series and commutating windings, the voltage between brush arms is the machine voltage V, and the number of bars between arms is B/p. So

$$\text{Avg. volts/bar} = \frac{V \times p}{B} \qquad (7\text{-}12)$$

where B is the total number of commutator bars and p the number of main poles.

Even if no distortion exists, only the conductors under the pole faces generate voltage; so the corrected average volts per bar should be

$$\frac{V \times p}{B \times \Psi} \quad \text{(volts)}$$

where Ψ is the ratio of pole arc to pole pitch, about 0.65. This is represented by D in Fig. 7-29. However, the maximum volts per bar at w is greater than this, as the height w is greater than D, or

$$\text{Max. volts per bar} = \frac{V \times p}{B} \times \frac{w}{D \times \Psi} \tag{7-13}$$

In practice the value of w/D for a noncompensated machine at full-strength main field varies from about 1.7 to 1.9. However, any reduction in saturation causes the effects of the armature ampere-turns (which cause the distortion) to be magnified. The designer must check the actual value of w/D, since it may be as high as 4.5 for a d-c motor at a weak main-field strength (high speed). This is evident in Fig. 7-33. The distorting effect for the high-speed (low-average-flux) condition ϕ_{02} raises the maximum flux to ϕ_{w2}, which is over three times the change for the saturated (low-speed) condition ϕ_{01} to ϕ_{w1} with the same distorting ampere-turns X.

The use of a compensating winding tends to eliminate the flux distortion, and for saturated conditions the flux curve coincides well with the no-load curve D of Fig. 7-29. However, under low saturation conditions the stationary compensating windings permit localized flux distortions. These are shown in Fig. 7-34. Similar distortions occur at low main flux densities on d-c generators, but the output voltage V is reduced in the same proportion as the main flux, and the maximum voltage between bars is not affected seriously.

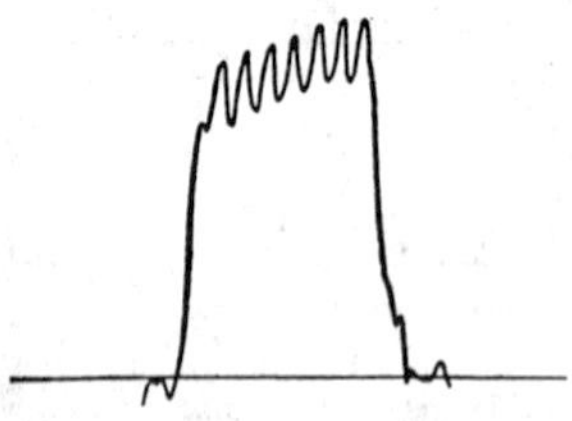

Fig. 7-34. Main-pole flux distortion on a compensated motor at full load and 2½ times base speed.

At full field on well-compensated motors or generators w/D is about 1.4 to 1.5. Direct-current motors at weak field may have ratios of 2.0 or more. On any questionable machine the designer should check this value carefully.

Approximate safe limits of maximum volts per bar are 40 V for motors and 30 V for generators on machines having 0.040-in-thick mica between segments.

35. Brush Potential Curves. When a d-c machine develops some commutation sparking, the user may suspect that the commutating-pole air gap is not set correctly. He often takes "brush potential curves" to prove or disprove his suspicions.

These are taken by measuring the voltage drops between the brush and commutator surface at four points while the machine is operating at constant speed and load current (see Fig. 7-35). The voltages at 1, 2, 3, and 4 are taken by touching the pointed lead of a wooden pencil to the commutator surface. The circuit is completed with leads and a low-reading voltmeter as shown.

The voltages are then plotted. A curve such as A of Fig. 7-35 may indicate undercompensation due to a too large commutating-pole gap. Curve C may indicate overcompensation with too much flux density in the commutating-pole air gap. Curve B is typical of good compensation.

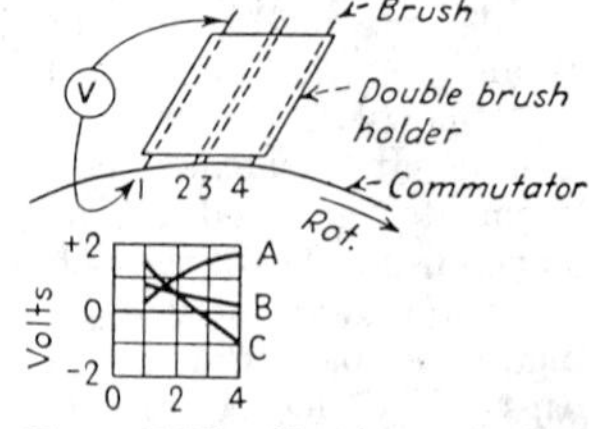

Fig. 7-35. Brush potential curves.

Justification for such conclusions is based on the theory that best commutation (coil current reversal) will be linear while the coil passes under the brush. This is possible only if there are no circulating currents. Undercompensation should cause circulating currents that would crowd the current to the leaving edge of the brush and cause a high voltage at point 4. Overcompensation would reverse the current too soon and would actually reverse the voltage drop at point 4.

Even to an expert, this test is only an indicator that more definitive tests, such as a buck-boost test, are needed [see Par. **31** and Eq. (7-8)]. Many other factors, including brush riding, commutator surface conditions, sparking, etc., influence the readings. Where machine changes may be required, the manufacturer should be consulted.

ARMATURE DESIGN

36. EMF Equation. If 10^8 lines (Mx) of flux are cut by one conductor in 1 s, 1 V is induced in it (see Par. **9**). Therefore, the induced voltage of a d-c machine is

$$E = \phi_t \times \frac{Z}{C} \times \frac{\text{r/min}}{60} \times 10^{-8}\ \text{V} \tag{7-14}$$

where ϕ_t is the total flux in maxwells across the main air gaps and Z/C is the number of conductors in series per circuit (C).

37. Output Equation. Equation (7-14) is converted to watts output if both sides are multiplied by the load current I_L, $I_c \times C$. The formula can then be rearranged,

$$D^2L = \frac{\text{watts} \times 6.08 \times 10^8}{\text{r/min} \times B_g \times \psi \times q} \tag{7-15}$$

where D is the armature diameter and L is the armature gross core length, B_g is the main-pole air-gap density in maxwells (lines), ψ is the ratio of pole arc to pole pitch [see Eq. (7-18)], q is $ZI_c/\pi D$ (a useful loading factor), and ϕ_t is the total air-gap flux equal to

$$B_g\psi\pi DL \qquad \text{[see Eq. (7-13)]} \tag{7-16}$$

38. Rotor Speeds. Standards list d-c generator speeds as high as are reasonable to reduce their size and cost. This relation is seen from (7-15). The speeds may be limited by commutation, maximum volts per bar, or the peripheral speeds of the rotor or commutator. Generator commutators seldom exceed 5,000 ft/min, although motor commutators may exceed 7,500 ft/min at high speeds. Generator rotors seldom exceed 9,500 ft/min. Figure 7-36 shows typical standard speeds. If the prime mover requires lower speeds than these, generators can be designed for them but larger machines result.

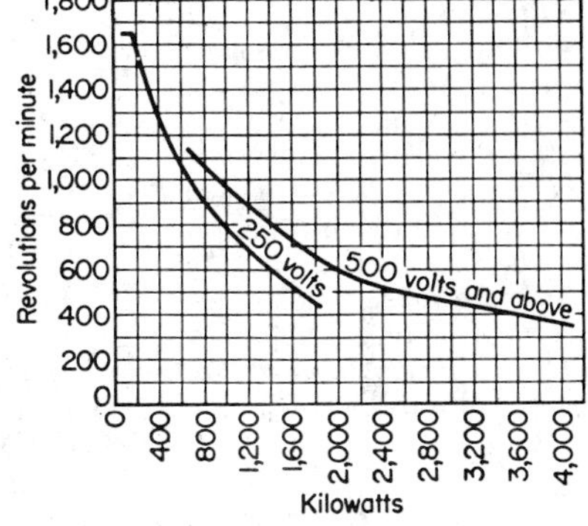

Fig. 7-36. Standard speeds of d-c generators.

39. Rotor Diameters. Difficult commutating generators benefit from the use of large rotor diameters, but diameters are limited by the same factors as rotor speeds listed above. The resultant armature length should be not less than 60% of the pole pitch, because such a small portion of the armature coil would be used to generate voltage. Typical generator diameters are shown in Fig. 7-37.

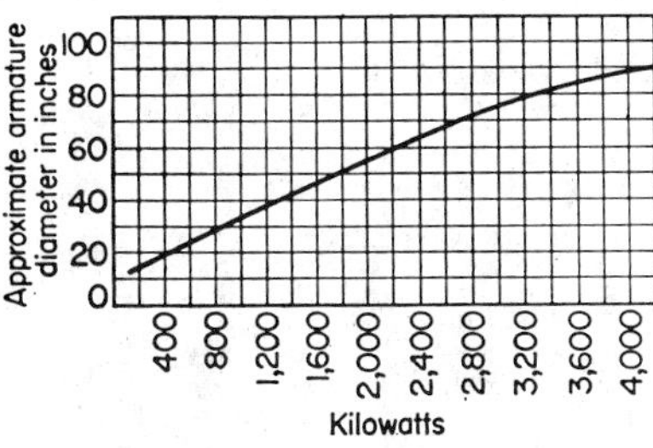

Fig. 7-37. Approximate rotor diameters for standard speeds of d-c generators.

Direct-current motor speeds must suit the application, and often the rotor diameter is selected to meet the inertia requirements of the application. Core lengths may be as long as the diameter. Such motors are usually force-ventilated.

40. Number of Poles. The rotor diameter usually fixes the number of main poles. Typical pole pitches range from 17.5 to 20.5 in on medium and large machines. When a choice is possible, high-voltage generators use fewer poles to allow more voltage space on the commutator between the brush arms.

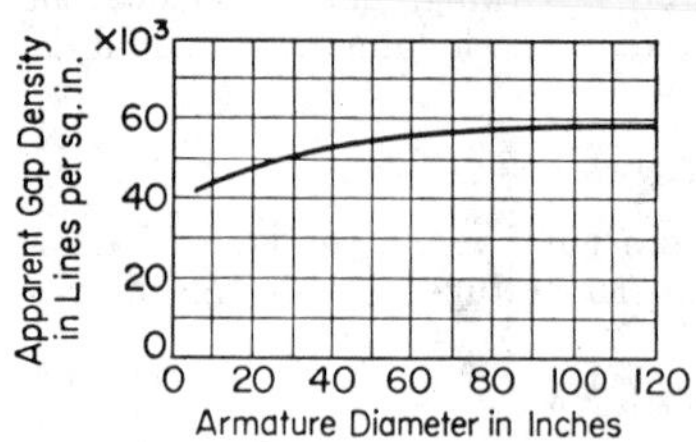

FIG. 7-38. Curve of apparent gap density against armature diameter.

However, high-current generators need many poles to permit more current-carrying brush arms and shorter commutators. Commutators for 1,000 to 1,250 A/(brush arm) (polarity) are costly, and lower values should be used where existing dies will permit.

41. The main-pole air-gap flux density B_g is limited by the density at the bottom of the rotor teeth. The reduced taper in the teeth of large rotors permits the higher gap densities, as shown in Fig. 7-38.

42. Ampere conductors per inch of rotor circumference (q) is limited by rotor heating, commutation, and, at times, saturation of commutating poles. Approximate acceptable values of q are shown in Fig. 7-39.

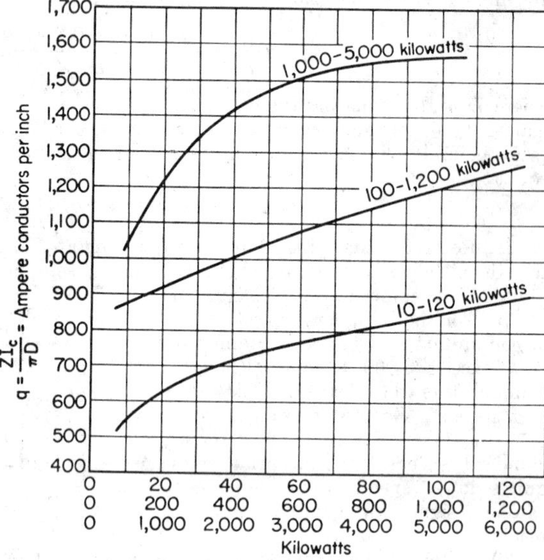

FIG. 7-39. Ampere conductors per inch of armature circumference.

43. The commutator diameter is usually about 55 to 85% of the rotor diameter, depending on the sizes available to the designer, the peripheral speed, and the resulting single clearances. Heating may also limit his choice (see Pars. **28** and **29**).

44. Brushes and brush hoiders are chosen from designs available to limit the brush current density to 60 to 70 A/in² at full load, to obtain the needed single clearance, and to obtain acceptable commutator heating.

45. Selection of an Approximate Design. Consider a generator rated 2,500 kW, 700 V, 3,571 A, and 514 r/min. From Figs. 7-39 and 7-40

Approx. dia	$D = 62$ in
Available dia	$D = 56$ in
No. of poles	10
Pole pitch	17.59 in
Pole arc	12.0 in
(Arc/pitch)	0.687
Neutral zone	6.04 in
B_g gap density @ 721 V	58,500 lines/in²
Approx q (Fig. 7-39)	1,480 A cond./in
D^2L [Eq. (7-15)]	50,200 in³
L (gross core)	16 in
No. ⅜-in vents in core	5
Net core length	14.125 in
ϕ [Eq. (7-14)]	1.12×10^8
Approx. total cond. Z	752 (use 750)
Actual q	1,520 A cond./in
No. commutating bars (1-turn lap)	375
No. slots	125
Slot pitch	1.407 in
Slot throw	12½ (use 12)
Chording	½-slot pitches

Examination of the data indicates that the design appears feasible, and so we may continue.

Commutating dia. (Par. **43**)	39 in
Brush size (35°)	2(0.500 × 1.75) in
CZ (commutating zone) [Eq. (7-5)]	3.53 in
Brushes/arm (Par. **44**)	7*b*/*a*
Commutating speed	5,250 ft/min
Commutating bar pitch	0.327
Brush arc	1.315 in
Bars/arc	4.02 bars
SC [Eq. (7-6)]	1.26 in
Brush density	58.3 A/in²
Brush I^2R (Par. **69**)	7,142 W
Length of commutating face	7(1.75 + 0.063) + 1 = 14.56 in
Brush friction loss [Eq. (7-33)]	6,760 W
Watts/in² of commutating surface	7.8 W/in²

Examination of these data also indicates that the proposed design is reasonable.

46. Armature Slots and Coils. The depth of an armature slot is limited by several factors, including the tooth density, eddy losses in the armature conductors, available core depths, and commutation. For reasonable frequencies (up to 50 c/s on medium and large d-c machines), slots of about 2 in deep can ordinarily be used.

Acceptable slot pitches range from 0.75 to 1.5 in. Small machines have shallower slots and a lower range of slot pitches. For medium and large machines, a reasonable tooth density usually results if the ratio of slot width to slot pitch is about 0.4.

Eddy losses in the conductors can be large compared with their load I^2R losses. Sometimes these must be reduced by making each armature conductor from several strands of insulated copper wire. The number of strands and their size depend on the frequency and the total depth of the conductor. An approximate formula for reasonable eddy losses is

$$\text{No. of strands} = (0.75)(f)(d_c)^2 \qquad (7\text{-}17)$$

where f is the frequency in cycles per second, (r/min × poles)/120, and d_c is the total depth of a conductor.

The insulation space required depends on the type used. Typical conductor strands have about 0.018 in of glass strands and varnish total. Mica wrappers, binding tapes, and varnish and slot finish allowance (0.010 in) total about 0.085 in on the coil width. If the space for the wedge and its retainer is included, the two coils depthwise total about 0.315 in (see Fig. 7-40).

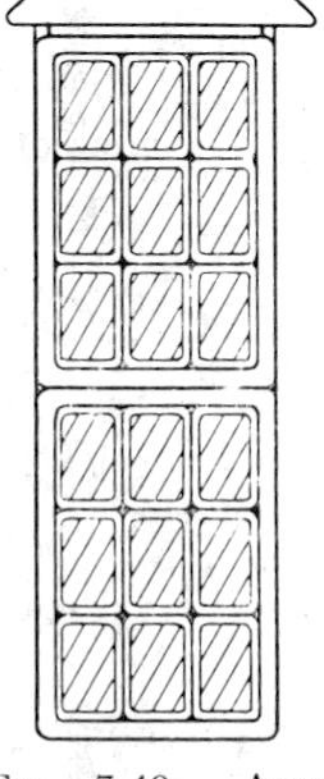

FIG. 7-40. Armature slot cross section.

47. Approximate Slot Design

Width (see Par. **46**) 0.4 × 1.407	0.563 in
Depth	2.0 in
Approx. total cond. depth	0.875 in
Frequency	42.8 c/s
No. strands/conductor [Eq. (7-17)]	3

	Slot width	Depth
Approx.		
Size	0.563	2.000 in
Insulation	0.139 (0.085 + 0.054)	0.423 (0.315 + 0.108)
Bare copper	0.424	1.577 in
Strand size	0.141	0.263 in
Use	3(0.144	0.289) in strands/conductor
Use available slot	0.570	2.250 in

COMPENSATING AND COMMUTATING FIELDS

48. Compensating Winding Data. See Par. **33.** The compensating winding should closely match the armature ampere-turns per inch, should avoid causing magnetic noise, and should result in an acceptable maximum volts per bar (see Par. **34**). Machines for 40°C temperature rise will have compensating bar densities of about 2,500 to 3,000 A/in². The pole tip section will limit the maximum depth of the compensating bar. Localized areas of high flux density must be avoided where flux must funnel between the pole "shoe" surface and the bottom of the compensating slot.

For single compensating bar per slot designs, the typical width required for insulation, varnish, and stacking factor is about 0.140 in. With the wedge space included, the insulation-depth requirement is about 0.400 in.

49. Compensating Winding Calculations

q (armature)	1,520 A cond./in
Pole arc of 12.1 in covers	18,400 A cond.
Approx. compensating At	9,200 At
Load current	3,571 A
Approx. turns/pole 2.68	use 2.5 turns/pole
Consider 5 slots/pole	1 bar/slot = 2.5 turns/pole
Size of compensating bar	0.688 × 2.0 in
Bar density	2,590 A/in²
Compensating slot width 0.828	use 0.830 in
Compensating slot depth 2.400	use 2.400 in
Compensating slot pitch (layout)	2.25 in
Rotor slot pitch	1.407 in
So magnetic noise	improbable
Maximum volts/bar	see Par. **58**

50. Commutating Winding Calculations. See Par. **30**. The total of the commutating and compensating ampere-turns per pole should be about 120 to 130% of those on the rotor.

Armature At/pole = $ZI_c/2p$ = (750 × 357)/(2 × 10)	13,400 At/pole
Equiv. armature-turns/pole on line ampere basis	3.75 turns/pole
Approx. commutating + compensating At/pole (1.2 × 13,400)	16,100 At/pole
Commutating + compensating turns/pole 16,100/3,571	4.5 turns/pole
Less compensating turns/pole	−2.5 turns/pole
Requires commutating winding of	2.0 turns/pole
Excess At/pole, 16,100–13,400	2,700 At/pole

Well-ventilated commutating coils may have densities of 2,000 to 2,500 A/in² (see Fig. 7-49).

51. Commutation Calculations

E_c = reactance voltage of commutation	5.42 V [see Eq. (7-4)]
Commutating-pole gap density B_m	13,550 L/in² [Eq. (7-7)]
Excess At	2,700 At/pole
Commutating-pole air gap	0.609 in [Eq. (7-8)]

MAGNETIC CALCULATIONS

52. Flux Paths. Figure 7-41 shows the paths of the main-pole flux for a typical medium-sized machine. The commutating poles and the compensating slots are not shown. Saturation calculations involve only half the length of a complete flux loop, because that is all that one field coil accommodates. Except for the main-pole air gap and the rotor teeth ampere-turns, the calculations are simple. They require (1) the determination of flux densities by dividing the flux in a section by its cross-sectional area, $\beta = \Phi/\text{area}$, (2) reading a magnetization curve for the material involved to find the ampere-turns per inch needed for the density, and (3) finding the total ampere-turns for the part by multiplying the length of the portion of the path by those ampere-turns per inch. Typical magnetization curves are shown in Fig. 7-42.

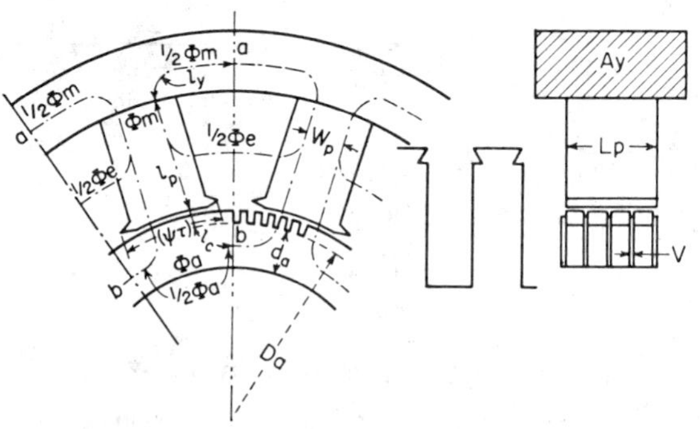

Fig. 7-41. The paths of the main and of the leakage fluxes.

The rotor core is usually built up of sheet steel laminations 0.017 to 0.025 in thick. Because of burrs and surface coatings, a stacking factor of 93% is common. The main poles use thicker laminations, and a factor of 95% is common. If the frame is also made

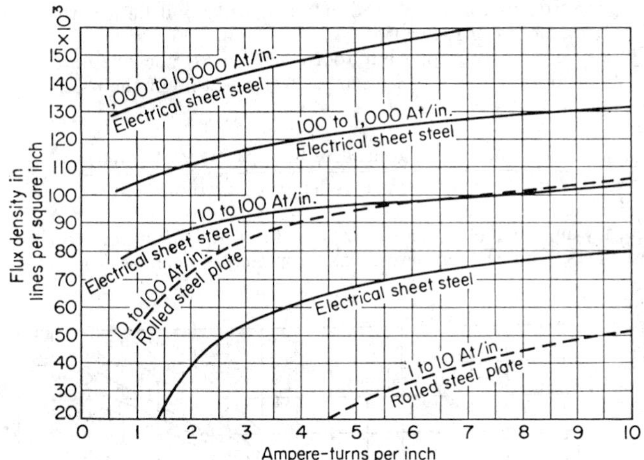

FIG. 7-42. Magnetization curves.

up of laminations, a similar factor is necessary. Of course, a solid frame uses its full area.

The leakage flux ($\frac{1}{2}\phi_e$) in Fig. 7-41 from the main field coils must be included with the useful flux in the frame yoke and the pole body. Calculations will depend on the actual machine dimensions and on the main field ampere-turns. However, the ampere-turns in these parts represent only a small part of the total required for the entire path, and it is usually accurate enough to estimate this leakage to be 12% of the useful flux normally and 20% at high saturations. For accurate calculations the actual leakage can be plotted. No leakage fluxes are considered in computing the gap, teeth, or core densities.

53. The Carter Coefficient and Gap Ampere-turns. The presence of rotor slots, compensating slots, and vent ducts in the generator causes the actual densities in the main-pole air gap to be greater than for a smooth, solid core. Also, the average lengths of the flux paths are longer (see Fig. 7-43). The two effects may be lumped by assuming that the air gap is larger than measured mechanically. Upon considering the three factors (rotor slots, compensating slots, and vents) in succession, the formula

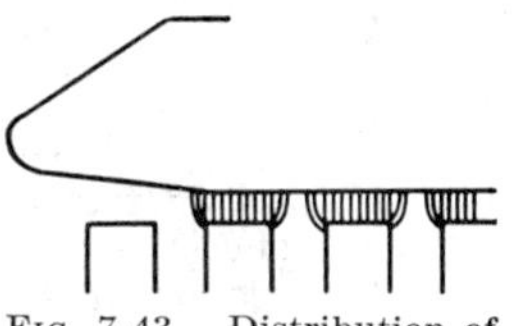

FIG. 7-43. Distribution of flux in the air gap.

$$G_1 = G \times \frac{G + (\text{slot width}/5)}{G + (\text{slot width}/5)\,(1 - \text{slot width}/\text{slot pitch})} \tag{7-18}$$

gives the first corrected air gap G_1; this will closely approximate the effective air gap.

The ampere-turns across the gap will be

$$At_g = \beta_g \times 0.313 \times G_1 \tag{7-19}$$

54. The Rotor Teeth Ampere-turns. For tooth densities below 100,000 lines/in², the ampere-turn drops in a tooth are so low that practically no flux will pass down the adjacent slot because the reluctance of air is so great. However, as tooth flux densities become larger, they produce very high ampere-turn drops from the top of the tooth to its bottom owing to saturation. Because these ampere-turns are also across the parallel flux path in the adjacent slot, when they are large enough, some useful flux will pass down the slot, relieve the tooth of some of its flux, and lower its actual density. If the tooth apparent density is calculated by assuming that all the flux across a slot pitch passes down the tooth, the actual density will be less than the apparent, depending on the amount of saturation.

The relation between the apparent tooth density β_{ta} and the actual tooth density β_t for different ratios of air area to iron area at any section of the tooth is shown in Fig. 7-44. The K of these areas is

$$K = \frac{\text{air area}}{\text{iron area}} = \frac{(\text{gross core length}) \times (\text{slot pitch})}{(\text{eff. core length}) \times (\text{tooth width})} - 1 \qquad (7\text{-}20)$$

For accuracy in calculating tooth ampere-turns, it is desirable to divide the tooth into several parts, find the ampere-turns drop across each section, and total them.

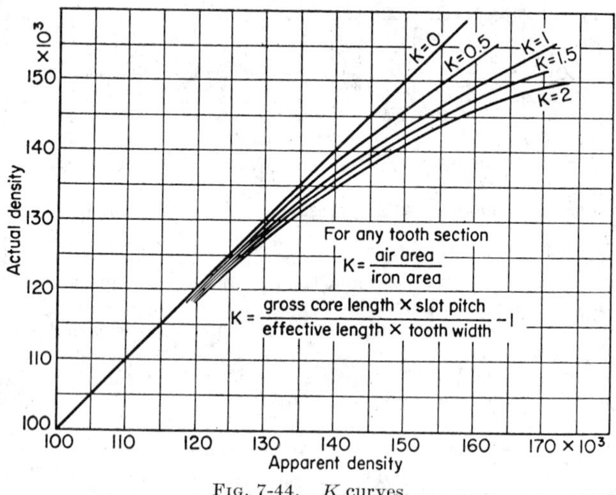

Fig. 7-44. *K* curves.

The flux density is found at the middle of each section, and the K ratio is calculated at the middle of each section.

55. Calculation of No-load Saturation Data. Considering the 2,500-kW 700-V 3,571-A 514-r/min generator of Par. **45,** we have the values shown in Table 7-1. Using

Table 7-1. Magnetic Dimensions

Section	K	Net area, in²	Eff. length, in
Frame yoke 6 × 17	....	102	13.85
Pole body 9½ × 15½	....	140	10.35
Comp. pole teeth (layout)	....	100	2.40
Eff. air gap	....	.....	0.268
Tooth 1 (upper ⅓)	0.92	10.8	0.75
Tooth 2 (middle ⅓)	0.96	10.3	0.75
Tooth 3 (bottom ⅓)	1.00	9.8	0.75
Core	....	79.3	7.15

the magnetization curves of Fig. 7-42 and these data, the no-load saturation curve is calculated for several voltages. Note that 721 volts is chosen in Table 7-2 on the

Table 7-2. Calculated Ampere-turns per Pole

Volts	ϕ_t	Gap, $L = 0.268$	Tooth 1, $L = 0.75$, $K = 0.92$	Tooth 2, $L = 0.75$, $K = 2.96$	Tooth 3, $L = 0.75$, $K = 1.0$	Core, $L = 7.15$	Frame, $L = 13.85$	Pole, $L = 10.35$	C. tooth, $L = 2.40$	Total ampere-turns
		β At	Apparent β Actual β At/in At	Apparent β Actual β At/in At	Apparent β Actual β At/in At	β At/in At	β At/in At	β At/in At	β At/in At	
420	65.5×10^6	33,800 2,850	70,500 70,500 5.4 5	74,600 74,600 7.2 5	78,000 78,000 9.0 5	41,300 2.1 15	36,000 6.5 90	52,400 2.8 30	65,500 4.2 10	3,018
630	98.3×10^6	50,800 4,280	106,000 106,000 130 100	112,000 111,300 210 160	117,000 116,200 320 240	62,000 3.7 25	54,000 11 150	78,600 9.2 95	98,300 6.2 15	5,065
721	112.5×10^6	58,200 4,900	121,200 120,500 440 330	128,100 126,500 660 495	134,000 132,300 1,000 750	71,000 5.7 40	61,800 14 195	90,000 26 270	112,500 225 540	7,520
770	120×10^6	62,100 5,230	129,500 128,000 710 535	137,000 134,600 1,200 900	143,000 139,000 1,850 1,390	75,800 7.9 55	66,000 16 220	96,000 46 475	120,000 410 985	9,790

L = length of flux path, in; K = air area/iron area at particular position on tooth; Apparent β = apparent flux density, lines/in^2; Actual β = actual flux density, lines/in^2; At/in = ampere-turns/in; At = ampere-turns.

assumption that the IR drop in the generator will not exceed 3%, or 21 volts in this case. The generator must have this additional voltage induced in it for a 700-V terminal voltage. In the case of a motor, the induced voltage would be lower by the amount of the IR drop, or 679 V.

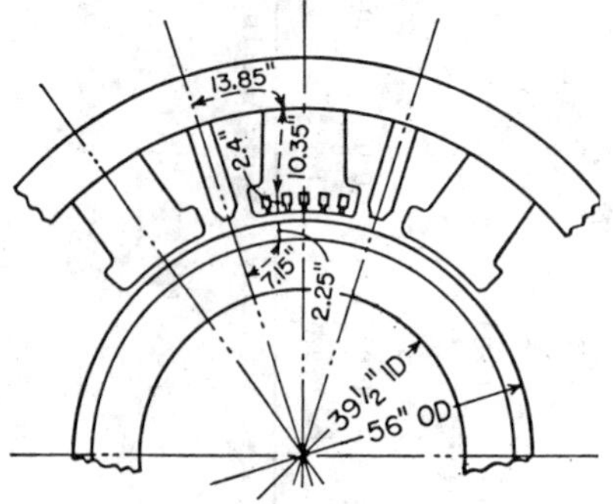

Fig. 7-45. Cross section of a 2,400-kW generator.

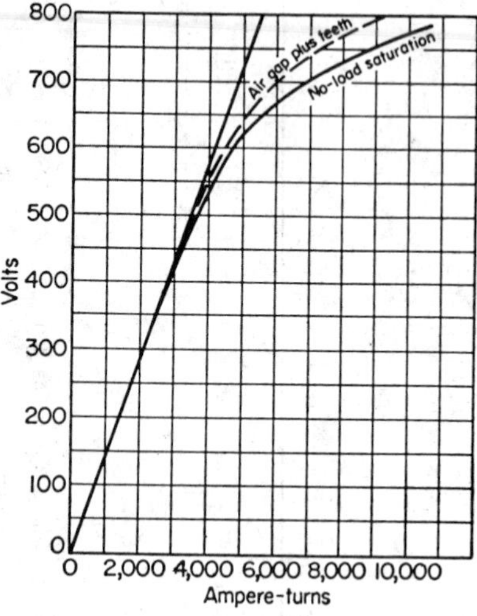

Fig. 7-46. No-load saturation curves.

56. Full-load Saturation Curve for a Compensated Machine. Figure 7-46 shows the calculated no-load saturation curve. For a well-compensated machine, the brushes will have little or no shift, and essentially no useful flux will be lost because of armature reaction (see Par. **17**). Only the armature-circuit-resistance IR drop need be considered, and the full-load excitation ampere-turns required can be read directly from the no-load saturation curve at the induced voltage.

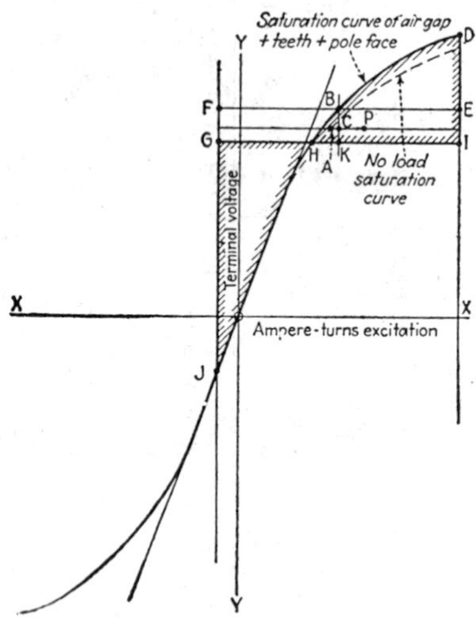

Fig. 7-47. Calculation of load saturation curve.

For the 2,500-kW generator the excitation required at 721 V is 7,520 At at full load.

57. Full-load Saturation Curve for a Noncompensated Machine. With commutating poles, there is no need for brush shift, but the uncompensated armature reaction (see Par. **17**) will result in loss of useful flux as the load is increased. Figure 7-47 shows a method of calculating the additional ampere-turns excitation to replace this lost flux.

OBD = saturation curve of air gap plus teeth and pole face

BC = IR drop in armature circuit plus the brush drop. B = any point chosen on curve OBD

FB = BE = full-load-armature At/pole arc, or $\text{At}/p \times \psi$, laid off on a horizontal line through B. Through E and F draw vertical lines of indefinite length. Move line GI vertically upward or downward parallel to FBE to a position $GHKI$, so that area $JGHOJ$ = area $HABDIKH$. Through B draw a vertical line BCK. Then HK = distortion ampere-turns for the load current considered for point B

Through C draw a horizontal line of indefinite length cutting the no-load saturation curve at A

CP = HK, to be extended from right at C
AP = total ampere-turns required at load current considered to maintain voltage under load at same value as at no load

By choosing several points, such as B, along the saturation curve and making the same calculations for each point, a full-load or any load saturation curve can be produced.

58. Maximum Volts per Bar Calculations. See Par. **34.** The distorting ampere-turns resulting from imperfect compensation of the armature ampere-turns by the compensating winding are found by plotting the two and noting the maximum difference. This is done at the maximum-overload-current point.

The distortion factor (see Par. **34,** Figs. 7-33 and 7-46) is determined from the gap and teeth saturation curve (Fig. 7-46). At double load the induced voltage is considered to be 740 V.

Volts between arms	700 V	
No. poles p	10	
No. commutating bars	375	
Pole arc/pole pitch ψ	0.687	
Distorting ampere-turns	1,600 At	
Distortion factor w/D of Fig. 7-33	1.06	
Max V/bar	$\frac{V \times p}{B} \times \frac{w}{D \times \psi}$	[Eq. (7-13)]
Max V/bar	28.8 V/bar	

This value is acceptable according to Par. **34.**

MAIN FIELDS

59. Main Field and Main Field Heating. Figures 7-48 and 7-49 show three types of d-c main fields. Small machines commonly use those of Fig. 7-48. They are wound on molds and then slipped on the poles. Type A is wound on an insulating spool, and type B uses an insulated steel spool for better heat transfer and mechanical protection.

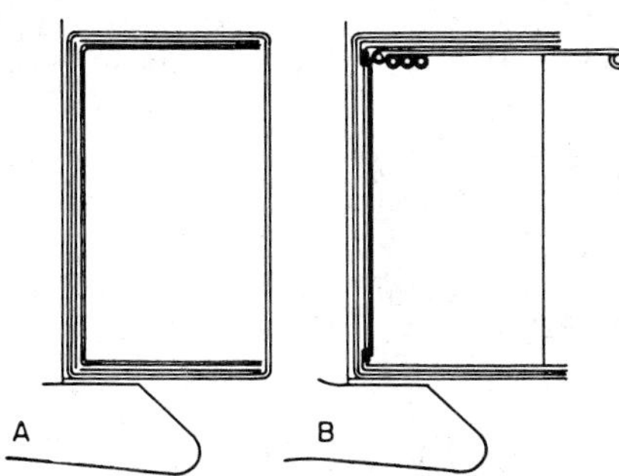

FIG. 7-48. Two types of field-coil insulation, combined with fiber and metal spools, respectively.

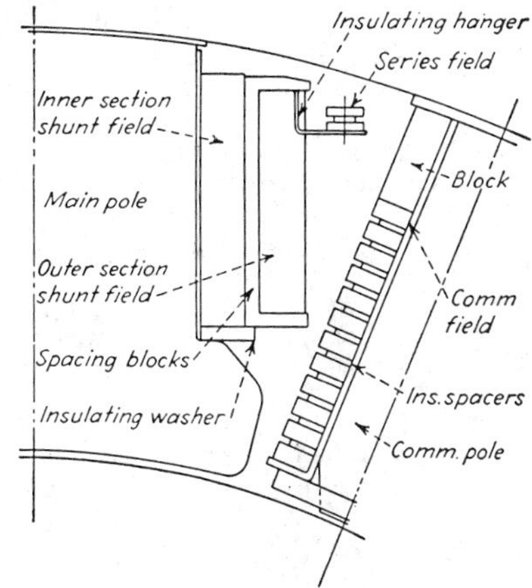

FIG. 7-49. Ventilated field coils.

The arrangement of Fig. 7-49 is common on large and medium-sized d-c machines. The turns of the inner section are wound tightly on the insulated pole body to avoid air spaces between the pole and the coil. This permits maximum heat transfer. The second section is spaced away from the inner coil to permit the cooling air to flow over the maximum surface area possible. The thickness of a coil section is limited to about $1\frac{1}{4}$ to $1\frac{3}{4}$ in for a small temperature gradient within the coil.

All three types may use wire insulated with varnish, double cotton covering, or glass slivers in varnish. Air pockets which act as barriers to transfer of heat must be avoided, and so rectangular wire is common. Also, varnish or resin is liberally applied during winding or applied by vacuum impregnation after the coil is wound.

Design criteria suitable for all d-c machines cannot be established, because the field cooling depends on air pressures from the armature rotation, the air-passage areas through the fields, and the radiation of heat from adjacent parts. These factors vary with machine design. However, on medium and large self-ventilated d-c generators (built as in Fig. 7-49) empirical data are useful.

The main fields receive heat, not only from their own I^2R losses, but from heat radiated from the hot armature and the commutation coils. Also, the air cooling the coils is already heated by the rotor. This lowers the temperature gradient for cooling the coils. The temperature rise of the fields must be calculated, not on the basis of the actual air temperature, but on the basis of the cool ambient-air temperature outside the machine. Figure 7-50 shows empirical data for such typical self-ventilated medium and large machines, built as shown in Fig. 7-49.

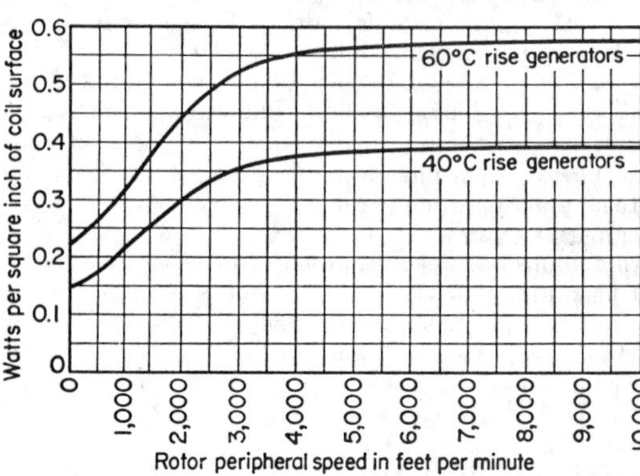

FIG. 7-50. Main field loss per surface area.

The "surface area" for these curves includes the entire periphery of the coil, because the heat transfer to the pole body is as effective as that to the air-cooled surfaces.

Little gain is made in cooling with increase in rotor velocities above 5,000 ft/min, because most of the armature air must pass through the limited field structure area. At high rotor speeds the air is throttled owing to the high-velocity pressure drops.

60. Main Field Calculations. These are made by making a layout similar to that shown in Fig. 7-49. This permits the estimate of approximate mean length of turns (MLT) for the sections.

The means of excitation and the particular application usually determine the IR drop of the main field. This is met in design by selection of the field wire cross-sectional area. This is calculated by Eq. (7-21).

$$\text{Conductor sectional area} = \frac{At/p \times MLT \times p \times 8.25 \times 10^{-7}}{IR} \tag{7-21}$$

where At/p is the number of ampere-turns per pole needed, MLT is the mean length of turns, p is the number of coils in series, and IR is the required voltage drop.

Typical field calculations are:

At/p	7,520 At
Approx. MLT	55 in
IR drop needed	90 V
Conductor area [Eq. (7-21)]	0.038
Insulation conductor	0.018 in
Section of coil	6.78 × 1.6 in
Actual IR	86.5 V
Watts ($IR \times I$)	3,380 W
W/in²	0.362 W/in²
W/in² allowed	0.388 W/in²
Res. 75°C	2.21 Ω
Coils in series	10
Copper	0.162 × 0.258 = 0.04 in²
Coil	24T high × 8lay. 192T/coil
Layout MLT	55.65 in
I = At/t)	39.1 A
Surface $2(H + tk)(MLT)p$	9,350 in²
Rotor velocity	7350 ft/min
Current density	977 A/in²

These data indicate an acceptable field.

COOLING AND VENTILATION

61. Cause of Temperature Rise. The losses in a d-c machine cause the temperature of the parts to rise until the difference in temperature between their surfaces and the cooling air is great enough to dissipate the heat generated.

62. Permissible measured temperature rises of the parts are limited by the maximum "hot-spot" temperature that the insulation can withstand and still have reasonable life. The maximum surface temperatures are fixed by the temperature gradient through the insulation from the hot spot to the surface.

The *IEEE Insulation Standards* have established the limiting hot-spot temperatures for systems of insulation. The *United States of America Standards Institute* Standard C50.4 for d-c machines has determined typical gradients for those systems, listing acceptable surface and average copper temperature rises above specified ambient-air temperatures for various machine enclosures and duty cycles. Typical values are 40°C for Class A systems, 60°C for Class B, and 80°C rise for Class F systems on armature coils. Class H systems usually contain silicones and are seldom used on medium and large d-c machines. Silicone vapors can cause greatly accelerated brush wear at the commutator and severe sparking, particularly on enclosed machines.

63. Temperature Gradients in Rotor Coils. Figure 7-51 represents a current-carrying conductor insulated from the core slot in which it is embedded. The hot spot is probably at the core center line and near the center of the conductor. Heat will probably travel along the conductor to the end turn and also through the insulation to the iron. The amount of heat flowing in each direction is difficult to calculate. Also, variations in the coils, such as resin fill and tightness in the slots, make heat conductivity factors difficult to predict.

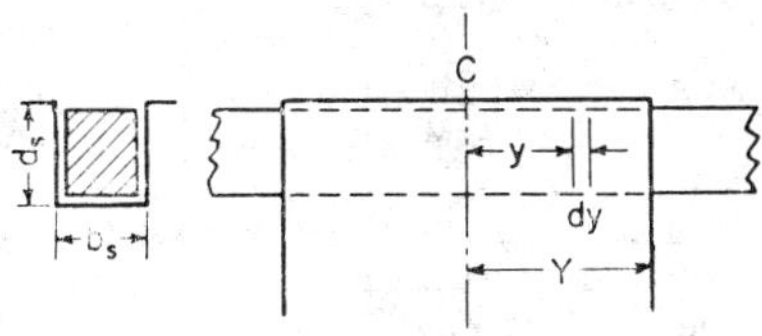

Fig. 7-51. Heat paths in an armature conductor.

a. Assume that all the heat must travel down the conductor to the end turn. What will be the temperature difference in the conductor between the center of the core and its edge?

$$\text{Resistivity of copper at } 75°\text{C} = 8.25 \times 10^{-7}\ \Omega/\text{in}^3$$

$$\text{Thermal cond. copper} = 9.75\ \text{W/(in)}\,(°\text{C}) \text{ for } 1\ \text{in}^2 \text{ section}$$

Therefore, the energy crossing dy of Fig. 7-51 is

$$\text{Watts} = (I_c)^2R_y = \frac{(I_c)^2(y)(8.25 \times 10^{-7})}{A} \tag{7-22}$$

where I_c is the conductor amperes, R_y the resistance of length y, and A the conductor cross-section area.

The difference in temperature between two faces dy apart is

$$°\text{C} = \frac{(I_c)^2(y)(8.25 \times 10^{-7})}{A} \times \frac{dy}{A} \times \frac{1}{9.75} \tag{7-23}$$

and the difference in temperature between the center C and any point y is

$$°\text{C} = \frac{(I_c)^2(8.25 \times 10^{-7})}{A} \int_0^y \frac{y\,dy}{A} \times \frac{1}{9.75}$$

or

$$°\text{C} = 4.22 \left(\frac{I_c}{A}\right)^2 (y)^2(10)^{-8} \tag{7-24}$$

Consider a current density of 2,920 A/in² and a total core length of 16 in. Then the coil temperature gradient from the core center, with no ventilating ducts, to the edge is 28.8°C. This assumes that no heat passes through the insulation to the iron. So medium and large machines normally use ventilating core ducts every few inches.

b. Assume that the end turns are so hot that no heat flows longitudinally down the coil. The I^2R loss of each inch of conductor length is

$$\text{Watts} = \frac{(I_c)^2(8.25 \times 10^7)}{A}$$

If the slot contains several conductors

$$\text{Watts} = (\text{ampere conductors})\,(\text{A/in}^2)\,(8.25 \times 10^{-7})$$

and the temperature difference between the bare conductor and the steel across the insulation is

$$°\text{C} = (\text{amp conductors})\,(\text{A/in}^2) \times \frac{\text{insulation thickness}}{2d_s + b_s} \times \frac{8.25 \times 10^{-7}}{0.003} \qquad (7\text{-}25)$$

The factor 0.003 is the thermal conductivity of the insulation in watts per cubic inch per degree centigrade difference.

Thus, for 2,142 ampere conductors per slot, 2,920 A/in², a surface of two slot depths plus a slot width (times 1 in) = 5.07 in², and an insulation thickness of 0.051 in (data for the 2,500-kW generator of Par. **45**), the temperature drop across the insulation is 17.65°C.

This figure cannot be considered precise because the thermal conductivity can vary widely with the insulation used and the presence of varying amounts of air in it. The conductivity figure for air is 0.0007, while that of mica is 0.007 W/(in³) (°C). Also, heat moves along the coil. Because of these difficulties, empirical data from actual machines are more reliable and easier to use.

64. Heating of End Connections of Armature Windings. Small machines often have "solid" end windings banded down on insulated "shelf"-type coil supports. Larger machines are more heavily loaded per unit volume and usually have narrow coil supports, air spaces between the end turns, and ventilating air scouring both the top and bottom surfaces of the coil extensions.

With this construction the approximate allowable product of ampere conductors per inch of outer circumference times the amperes per square inch for various rotor velocities is shown in Fig. 7-52 for a 40°C rise on the end turns.

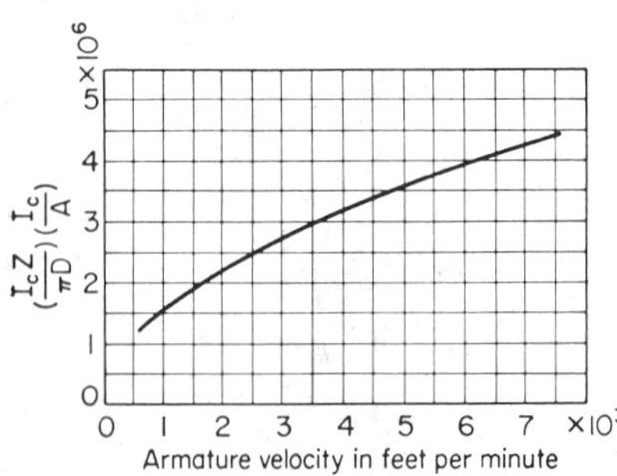

Fig. 7-52. End-winding cooling.

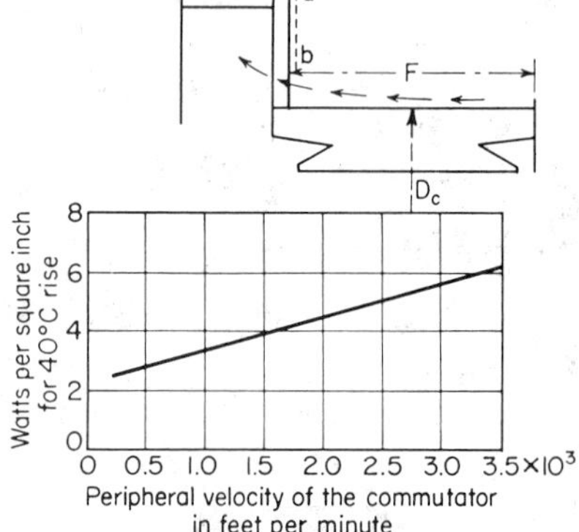

Fig. 7-53. Temperature rise of commutator.

65. Commutator Heating. A modern d-c armature is shown in Fig. 7-53. The commutator diameter ranges from 55 to 85% of the rotor core, and the commutator necks joining the bars with the rotor winding extensions are usually separated from one another by air spaces, so that, when the armature revolves, air circulation is set up as shown by the arrows.

A typical relation between permissible watts per square inch of commutator surface and its peripheral velocity is shown in Fig. 7-53. The radiating surface is the commutator circumference times its face length. Neck area is not included.

The heat to be dissipated is that due to brush friction and the brush contact I^2R losses. There may be other losses due to poor commutation, brush chattering, and commutator surface, and, if so, the rise will be greater than indicated in Fig. 7-53. If commutation is very good and brush riding excellent, the temperature will be lower.

66. Application of Heating Constants. The paragraphs covering the design of the armature, main fields, compensating windings, and commutating windings included typical loading data such as ampere conductors per inch, amperes per square inch, flux densities and watts per square inch of cooling surface. More accurate data depend on the exact arrangements used in a particular design. If possible, new design should be compared with similar machines which have already been tested. Any machine enclosure variation that restricts or increases the ventilation will affect the temperature rises.

LOSSES AND EFFICIENCY

67. Armature Copper I^2R Loss. At 75°C the resistivity of copper is 8.25×10^{-7} Ω/in^3. Thus, for an armature winding of Z conductors, each with a length of $MLT/2$ (half the mean length turn of the coil), each with a cross-sectional area of A and arranged in several parallel circuits, the resistance is

$$R_a = Z \frac{MLT}{2A} \frac{8.25 \times 10^{-7}}{(\text{circuits})^2} \qquad \text{(ohms)} \tag{7-26}$$

The MLT is best found by layout, but an approximate value is

$$MLT = 2[(1.35)(\text{pole pitch}) + (\text{rotor length}) + 3] \tag{7-27}$$

There are also eddy-current losses in the rotor coils, but these may be held to a minimum by conductor stranding in accordance with Eq. (7-17). Some allowance for these is included in the load loss.

68. Compensating, Commutating, and Series Field I^2R Losses. These fields also carry the line current, and the I^2R losses are easily found when the resistance of the coils is known. Their MLT is found from sketch layouts. At 75°C

$$R = T \frac{MLT}{A} \frac{8.25 \times 10^{-7}}{(\text{circuits})^2} p \qquad \text{(ohms)} \tag{7-28}$$

where R is the field resistance in ohms, T the number of turns per coil, p the number of poles, MLT the mean length of turn, and A the area of the conductor. The total of these losses ranges from 60 to 100% of the armature I^2R for compensated machines and is less than 50% for noncompensated machines.

69. The brush I^2R loss is caused by the load current passing through the contact voltage drop between the brushes and the commutator. The contact drop is assumed to be 1 V.

$$\text{Brush } I^2R \text{ loss} = 2(\text{line amperes}) \qquad \text{(watts)} \tag{7-29}$$

70. Load Loss. The presence of load current in the armature conductors results in flux distortions around the slots, in the air gap, and at the pole faces. These cause losses in the conductors and iron that are difficult to calculate and measure. A standard value has been set at 1% of the machine output.

$$\text{Load loss} = 0.01(\text{machine output}) \tag{7-30}$$

71. Shunt Field Loss. Heating calculations are concerned only with the field copper I^2R loss. It is customary, however, to charge the machine with any rheostat losses in determining efficiency. Thus,

$$\text{Shunt field and rheostat loss} = I_f V_{ex} \qquad \text{(watts)} \qquad (7\text{-}31)$$

where I_f is the total field current and V_{ex} is the excitation voltage.

72. Core Loss. As seen from Fig. 7-54, the flux in any portion of the armature passes through $p/2$ c/r (cycles per revolution) or through $(p/2)[(\text{r/min})/60]$ c/s.

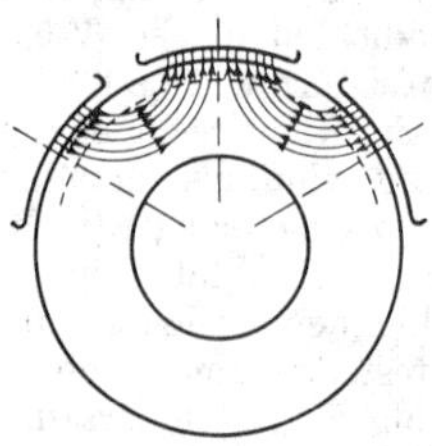

Fig. 7-54. Distribution of flux in the armature.

The *iron losses* consist of the *hysteresis loss,* which equals $K\beta^{1.6}fw$ W, and the *eddy-current loss,* which equals $K_e(\beta ft)^2w$ W. K is the hysteresis constant of the iron used, K_e is a constant inversely proportional to the electrical resistance of the iron, β is the maximum flux density in lines per square inch, f is the frequency in cycles per second, w is the weight in pounds, and t is the thickness of the core laminations in inches.

The *eddy loss* is reduced by using iron with as high an electrical resistance as is feasible. Very-high-resistance iron has a tendency to have low flux permeability and to be mechanically brittle and expensive. It is seldom justified in d-c machines. The loss is kept to an acceptable value by the use of thin core laminations, 0.017 to 0.025 in in thickness.

Another significant loss is the *pole-face loss.* Figure 7-43 shows the distribution of flux in the air gap of a d-c machine. As the armature rotates and the teeth move past the pole face, emfs are induced which tend to cause currents to flow across the pole face. These losses are included in the core loss.

Unfortunately, there are other losses in the core that may differ widely even on duplicate machines and that do not lend themselves to calculation. These include:

a. Loss due to filing of slots. When the laminations have been assembled, it will be found in some cases that the slots are rough and must be filed to avoid cutting the coil insulation. This burrs the laminations and tends to short-circuit the interlaminar resistance.

b. Losses in the solid spider, core end plates, and coil supports from leakage fluxes may be appreciable.

c. Losses due to nonuniform distribution of flux in the rotor core are difficult to anticipate. In calculating core density, it is customary to assume uniform distribution over the core section. However, flux takes the path of least resistance and crowds behind the teeth until saturation forces it into the less used, longer paths below. As a result of the concentration, the core loss, which is about proportional to the square of the density, is greater than calculated.

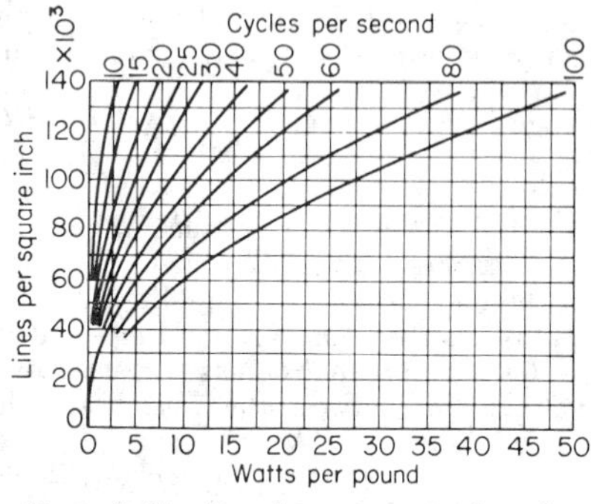

Fig. 7-55. Iron-loss curves for d-c machine.

Thus, it is not possible to predetermine the total core loss by the use of fundamental formulas. Consequently, core-loss calculations for new designs are usually based on the results from tests on similar machines built under the same conditions. Such test results are plotted in Fig. 7-55 for machines using ordinary laminations 0.017 in thick and a limited amount of filing. They do not include the pole-face losses, which would increase the values about 30%.

73. Brush Friction Loss. This loss varies with the condition of the commutator surface and the grade of carbon brush used. A typical machine has about 8 W loss/(in² of brush contact surface) (1,000 ft/min) of peripheral speed when normal brush pressure of 2½ lb/in² is used.

$$\text{Brush friction} = (8)(\text{contact area})(\text{peripheral velocity}/1{,}000) \qquad (7\text{-}32)$$

74. Friction and Windage. Most large d-c machines use babbitt bearings and many small machines use ball or roller bearings, although both types of bearing may be used in machines of any size. The bearing friction losses depend on the speed, the bearing load, and the lubrication. The windage losses depend on the construction of the rotor, its peripheral velocity, and the machine restrictions to air movement. The two losses are lumped in most estimates because it is not practical to separate them during machine testing.

Figure 7-56 shows typical values of friction and windage losses for various rotor diameters referred to rotor velocities.

75. Efficiency. The efficiency of a generator is the ratio of the output to its input. The prime mover must supply the output and, in addition, the sum of the losses listed in Pars. **67** to **74**. This is the input.

$$\text{Eff.} = \frac{\text{output}}{\text{input}} = \frac{\text{output}}{\text{output} + \text{losses}} \qquad (7\text{-}33)$$

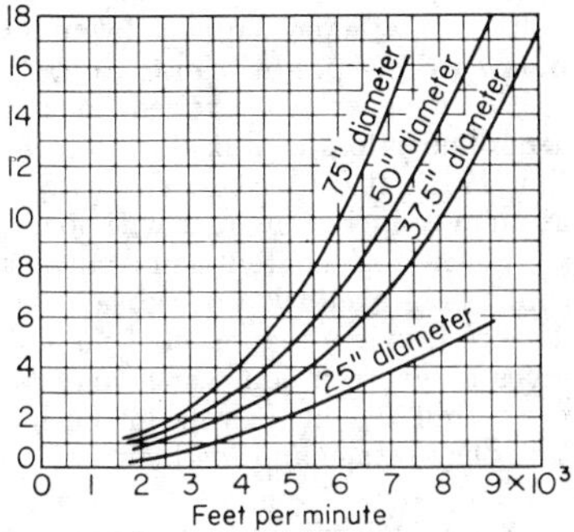

FIG. 7-56. Friction and windage vs. rotor velocity.

GENERATOR CHARACTERISTICS

76. The voltage regulation of a d-c generator is the ratio of the difference between the voltage at no load and that at full load to the rated-load voltage. The characteristic is normally drooping as the load is increased, but it can rise because of series field effects or the action of circulating currents of commutation at very-low-voltage operation.

For a d-c generator, the terminal-voltage equation is

$$TV = E - IR = [K(\phi_t)\,(\text{r/min}) - IR] \qquad (7\text{-}34)$$

where E is the induced emf, IR is the armature circuit drop, K is a constant depending on the machine design, and ϕ_t is the total main-pole flux of the generator.

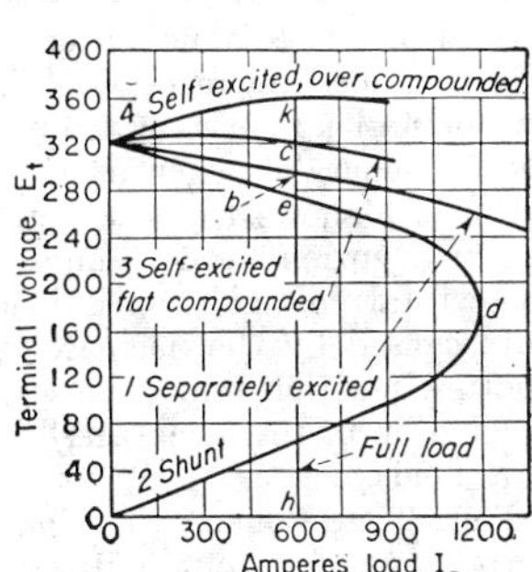

FIG. 7-57. External characteristics.

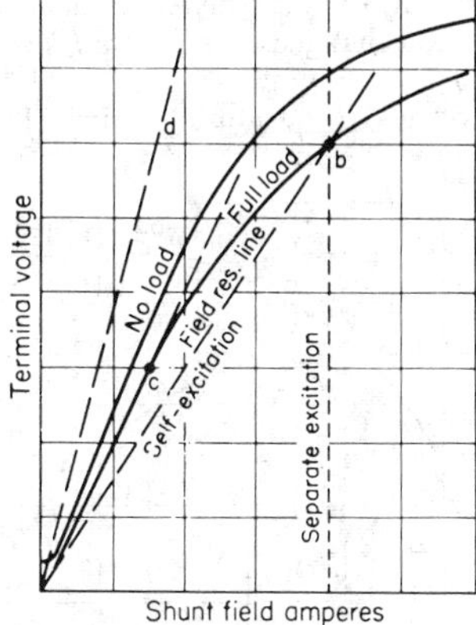

FIG. 7-58. No-load and field-load saturation curves.

The regulation curves are easily calculated by using the no-load and full-load saturation curves determined in Pars. **56** and **57** and shown in Fig. 7-58. The effect of the excitation method is found by the use of the field and rheostat IR line for self-excited machines and by the constant-ampere-turn line for separate excitation.

77. A separately excited compensated generator which is shunt-wound will have a voltage-load characteristic which will approach a straight line; it droops to full load an amount equal to the percent IR drop. There is little or no flux loss due to armature reaction or brush shift.

At voltages 10% or less of rated, the main field strength is so weak that currents circulating in the coils short-circuited by the brushes at commutation may cause an increase in main-pole flux with load that causes a rising characteristic. These armature coils loop the main poles and their ampere-turns produce direct axis flux. A rising voltage characteristic can be undesirable, particularly if the generator supplies a d-c motor whose speed is caused to rise with load, since this causes instability.

78. A separately excited noncompensated d-c generator which is shunt-wound has a nonlinear loss of flux due to armature reaction as the load current is increased (see Par. **19**). It can be seen from Eq. (7-34) that this causes a characteristic which droops at an ever-increasing rate with load increase, giving a curve which is concave downward.

79. A self-excited noncompensated d-c generator which is shunt-wound has its shunt field excitation decreased as the terminal voltage drops, as described in Par. **78.** This results in a reduction of main field ampere-turns and a loss of still more flux. This gives a severe droop which may be so great that, above a certain peak-load current, the terminal voltage will not be high enough to provide enough field current to maintain the voltage and load current and the voltage will collapse, as shown in (*d*) of Fig. 7-57.

80. Instability of Self-excited Generators. A self-excited d-c generator is unstable if the rheostat line does not make a definite intersection with the load-saturation curve (see Fig. 7-58). The shunt field current is fixed by the terminal voltage, and the resistance is in the shunt field circuit. Instability will exist if the slope of the rheostat line is nearly equal to or greater than the slope of a line tangent to the operating point on the saturation curve. In the figure, point *b* is a stable operating condition, but point *c* is not, because a decrease in voltage decreases the shunt field ampere-turns and this produces a further decrease in voltage.

If the field circuit resistance were set at *d*, the self-excited generator would never build up beyond residual voltage. Another cause of failure to build up may be the connection of the shunt field. If the current flow due to residual voltage is such that it tends to kill the flux producing the residual voltage, no build-up occurs.

81. Compound-wound D-C Generators. The generators described above can be compounded by adding series fields excited by the load current. However, the resulting field strength of these fields is linear with load and the shape of the voltage-regulation curve is not changed thereby but is merely rotated upward or downward with the zero-load point as a pivot.

82. Series Generators. Curve 1 of Fig. 7-59 shows the relation between voltage and current if there is no armature resistance or armature reaction. This is actually the no-load curve of the machine obtained by separately exciting the series field. Curve 2 shows the actual relation between load current and terminal voltage. The total voltage drop is made up of a part caused by the decrease in flux by armature reaction and a part caused by the *IR* drop of the armature, brushes, and series fields.

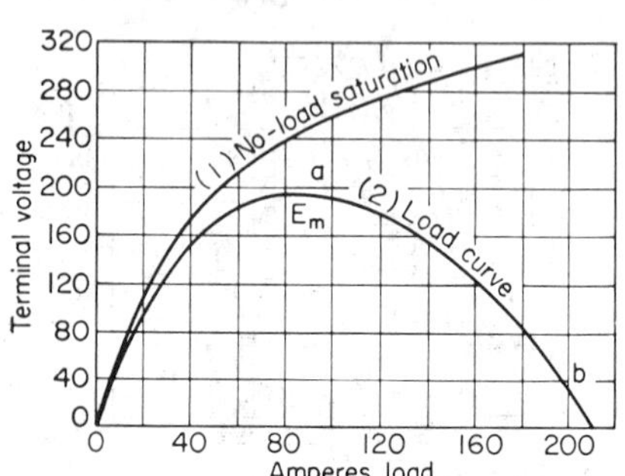

Fig. 7-59. Characteristic curves of a series generator.

83. Field Time Constants. The major delay in change of output voltage by an excitation change is caused by the inductance of the main fields. The time constant of the shunt field is the ratio of its inductance in henrys to its resistance in ohms, and this ratio represents the time in seconds required for 63% of a field current change to occur when the excitation voltage is suddenly changed. In the case of the 2,500-kW generator whose armature was designed in Par. **45** and whose fields were developed in Par. **60,** a mean main field inductance over the voltage range from zero to rated is 6.20 H. The main field resistance is 2.21 Ω. The field time constant is therefore 2.8 s.

The inductance *L* of a coil is the incremental change of flux linkages per incremental change in field current times 10^{-8}. This is proportional to the slope of the saturation curve and is constant over the air-gap line. It is therefore a decreasing variable after the curve leaves the air-gap line (see Fig. 7-46). The overall inductance, as the voltage builds up from zero, is not so high as that of the air-gap portion or as low as at the rated-voltage point. A common compromise is the slope of a straight line drawn from zero voltage through the full-load point at rated voltage. For the 2,500-kW generator the total flux at this point is 112.5×10^6 lines. With a leakage flux of 12%, each coil has a flux of 12.6×10^6 lines (see Table 7-2). As indicated in Par. **60** each coil has 192 turns and there are 10 coils in series. The field current is 39.1 A.

$$L = \frac{\phi T}{I_f} \times 10^{-8} = \frac{(12.6 \times 10^6)(192)(10)}{39.1} \times 10^{-8} = 6.2 \text{ H} \tag{7-35}$$

$$\text{Time constant} = L/R = 6.2/2.21 = 2.8 \text{ s} \tag{7-36}$$

This value is typical for large machines. Smaller generators have less copper in their fields and lower time constants. In cases where drive systems must have very rapid voltage adjustments, it is common to provide large forcing voltages on the field to overcome the inductive lag. These sudden excitation changes may be 4 to 10 times the IR drop of the field. This effectively reduces the time constant to one-fourth or one-tenth its normal value.

84. Armature-circuit Time Constants. Compensating windings effectively lower the inductances of the armature circuit. The 2,500-kW generator developed in this section has an armature-circuit inductance of 0.0001929 H and a circuit resistance of 0.00398 Ω for a time constant of 0.048 s. This value is typical for large d-c machines. Smaller noncompensated units have longer time constants.

TESTING

85. Factory Tests. These depend on the size, application, and design of the d-c generator.

The *United States of America Standards Institute* USAS C50.4 for d-c machines includes lists of recommended tests for d-c generators and motors. The *IEEE Test Code* for d-c machines covers recommended methods to be used for these tests.

GENERATOR OPERATION AND MAINTENANCE

86. General. Despite its rugged construction, a d-c machine is a delicate device. Factory tests on large units may cost thousands of dollars and must be performed carefully to adjust the generator to obtain the best possible characteristics and commutation. Owing to shipping requirements, the generator may then have to be disassembled and shipped in several pieces. If the final assembly is not correctly accomplished, not only have the factory tests been wasted but the machine may be damaged.

The manufacturer's instruction book should be studied carefully.

87. Before Installation. Upon arrival, the generator should be inspected for damage and to be sure it is dry. If it is wet, consult the manufacturer. Drying out with heat should be done only by slowly raising the generator temperature to 100°C so that moisture can escape without forming gas pockets within the insulation.

If the generator is dry and clean, the windings should be checked with a megger for insulation resistance to ground measurements. If any readings less than 1 MΩ are found, check with the manufacturer.

88. Alignment. After the machines are installed and grouted to the foundations, all couplings should be opened and alignments of all shafts finally checked. Regardless of whether solid or flexible couplings are used, the alignment should be as accurate as possible. The difference between the bottom and the top openings should not exceed 0.002 in for 12 in of flange diameter, and the large opening should be at the top. Regardless of the size of coupling the difference should not exceed 0.004 in. Differences at the side should not exceed 0.001 in. Shafts should be rotated 180° and rechecked.

The frame should be set on the *magnetic center* of the core. This position can be located by setting the armature in rotation and forcing it to oscillate longitudinally the full end play of the bearing by pushing on the end of the shaft. While the rotor is coasting and oscillating freely, excite the main field. The stator can then be shifted so that the rotor position with excitation coincides with the center of bearing end play.

Air gaps between the rotor and poles should be uniform. A typical limit of variation is 0.010 in. The brushes should ride properly on the commutator surface at both extremes of bearing end play.

89. Prerunning Checks. The circumferential position of the brushes on the commutator is important for commutation and also to provide the voltage characteristics set at the factory. Brushes should be on the factory test setting. The toes of the brushes should be aligned and should have no skew. The spacing between adjacent arms of brushes should be identical within 0.032 in. The brushes should move freely in their holders and should have a pressure against the commutator of 2 to 3 lb/in^2 on the basis of brush cross section. The faces of the brushes should accurately match the curvature of the commutator surface.

The polarity of the main fields may be checked by tracing the wiring around the frame or by lightly exciting the fields and using a compass around the frame behind the poles.

The oiling system for the bearings should be checked and the oil rings tested for freedom. The entire machine, particularly its air gaps, should be inspected for foreign material.

90. Running Checks. Note any unusual noises as the unit is brought up to speed. Bearing temperatures should level out at acceptable values within a few hours.

The voltage should be slowly raised at no load and commutation observed. If satisfactory, the voltage should be raised to 110% of rated and then reduced.

The generator may then be loaded gradually while commutation is observed, until rated current is reached. If commutation remains satisfactory until stable temperatures are achieved, the generator is ready for work.

91. Shunt-wound Generators in Parallel. *A* and *B* of Fig. 7-60 are two similar generators feeding the same bus bars *C* and *D*. If *A* tends to take more than its share of the total load, its voltage falls and more load is automatically thrown on *B*. Also, if the driver of one of the generators slows down to stop, the emf of the machine falls until the other generator starts to drive it as a motor. This continues until its driver takes over again.

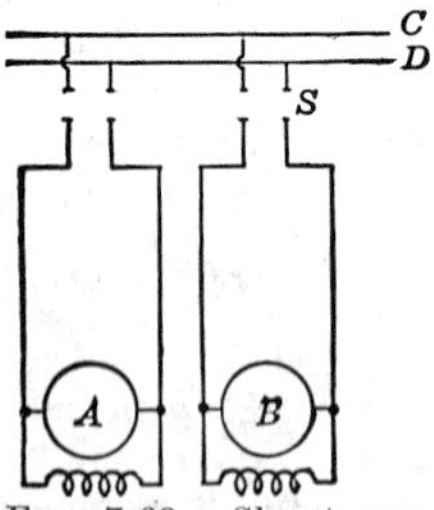

Fig. 7-60. Shunt generators in parallel.

The external characteristics of the two machines are shown in Fig. 7-61. At voltage *E*, the currents in the generators are I_a and I_b, and the line current is $I_a + I_b$. To make machine *A* take more of the load, its excitation must be increased to raise its characteristic curve. If a 1,000-kW generator and a 500-kW machine have the same regulation curves, the machines will divide the load according to their respective capacities, as shown in Fig. 7-62.

92. Compound-wound Generators in Parallel. *A* and *B* of Fig. 7-63 are two compound-wound machines. If *A* tends to take more than its share of the load, the series excitation of *A* increases, its voltage rises, and it takes still more of the load. Thus, the operation is unstable. If this continues until *A* takes all the load and the voltage of *B* drops to the point that *A* reverses the current in *B*, *B* will be driven as a motor. With the reversed current in the series field of *B* it becomes a differentially compounded motor, and the series weakens the flux to speed up the motor. This may progress to a point at which the unit may be damaged mechanically and electrically.

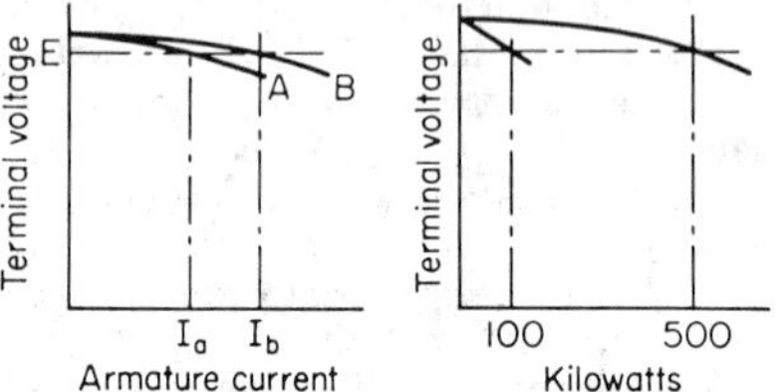

Figs. 7-61 and 7-62. Division of load between two shunt generators in parallel.

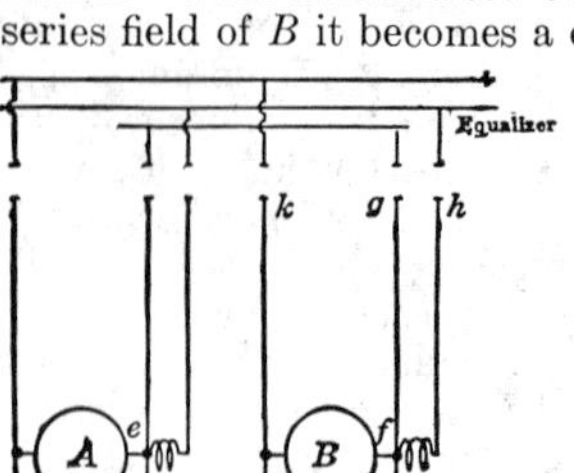

Fig. 7-63. Compound generators in parallel.

To prevent this, a bus bar of large section and of negligible resistance, called an *equalizer bus*, is connected from *e* to *f* (Fig. 7-63). Points *e* and *f* are then practically at the same potential. Therefore, the current in each series coil is independent of the current in its particular generator, is inversely proportional to the resistance of the coils, and is always in the same direction.

When a single compound generator has too much compounding, a shunt in parallel with the series field coils will reduce the current in these coils and so reduce the compounding. When compounded generators are operating in parallel using an equalizer bus, the current in the series field coils depends only on the resistance of the coils and a shunt connected across one of them is actually across all of them, reducing the compounding of all but not disturbing the relative compounding between the machines. To reduce the compounding of a single machine, it is necessary to place a resistance in series with the coils. This may require a large resistor to handle the large load current it must carry.

93. Maintenance. Except for the commutator and its brushes, maintenance of d-c machines differs little from that of other rotating electrical machines. Proper lubrication must be provided for the bearings, and the machine must be kept clean and dry. In addition, the brushes should be checked periodically for commutation, riding ability, freedom of motion in the holders, pressure, and length.

Because the commutator necks are not insulated and receive full voltage, conducting dust from brush wear or from ventilating air can cause creepage currents between the risers and ground over insulated surfaces. To avoid this, the d-c generator must be cleaned and blown out with clean, dry air at regular intervals. Air pressures above 25 lb/in^2 should not be used because of the danger of lifting the edges of insulating tape. The effectiveness of the cleaning program should be verified occasionally by megger readings.

94. Poor Commutation. Sparking and bar burning are usually due to one or more of the following causes:

a. Brushes not in the proper position.

b. Incorrect spacing of brushes. This may be checked by marking an adding-machine tape around the commutator.

c. Projecting-bar-edge mica. Mica between bars should be undercut about 0.063 in below the commutating surface, but occasionally slivers of mica are left inadvertently along the bar.

d. Rough or burned commutator. The commutator should be ground according to the manufacturer's instruction book.

e. Grooved commutator. This may be prevented by properly staggering the brush sets so that the spaces between the brushes of an arm are covered by brushes of the same polarity of other arms.

f. Poor brush contact due to improper fitting of the brushes to the commutator surface. To seat the brushes, sandpaper should be moved between the commutator and the brush face. Emery cloth should not be used, because its abrasive is conducting.

g. Worn brushes replaced by others of wrong size or grade.

h. Sticking brushes, which do not move freely in their holders so that they can follow the irregularities of the commutator.

i. Chattering of the brushes. This is usually due to operation at current densities below 35 A/in^2 and must be corrected by lifting brushes to raise the density or by using a special grade of brush.

j. Vibration. This may be due to poor line-up, inadequate foundations, or poor balance of the rotor.

k. Short-circuited turns on the commutating or compensating fields. These may be obvious on inspection but usually must be found by passing a-c current through them for voltage-drop comparisons.

l. Open or very-high-resistance joints between the commutator neck and the coil leads. In this case the bar at the bad joint will usually be burned.

m. An open armature coil. A broken coil conductor produces an effect similar to that produced by the poor joints described in (*l*). For emergency operation, the open coil may be opened at both ends, insulated from the circuit, and a jumper placed across the two affected necks. Since some sparking will probably result, operation should be limited.

n. Short-circuited main field coils. With the resulting unbalanced air-gap fluxes under the poles, large circulating currents must be expected even with good armature cross connections. The offending coil may be found by comparing voltage drops across the individual coils.

o. Reversed main field coil. This is an extreme case of (*n*).

p. Overloading.

SPECIAL GENERATORS

95. General. The adaptability of the d-c generator for specific uses has led to the development of many special generators. These machines over the years made a significant contribution to industrial progress. However, most of these special applications have disappeared or are now being met with other devices such as silicon-controlled rectifiers or programmed control of field currents to the main d-c generator. In some cases these new devices are used because they have faster response, but in others the solid-state device is preferred to the old rotating device because it is "up to date."

96. Shovel and Dragline Excavator Generators. With electric excavators using dippers in excess of 100 yd^3, the voltage characteristics of the generators which supply power for the hoist, drag, and swing motions are of paramount importance. They must permit very fast accelerations and decelerations but must limit the power available and bucket movement before an unacceptable "bite" bends the boom or snaps the cables. An excellent voltage-load characteristic, shown in Fig. 7-64, results in very high voltage and speed up to high loads but also results in a "stall" of the motors when twice normal torque is reached.

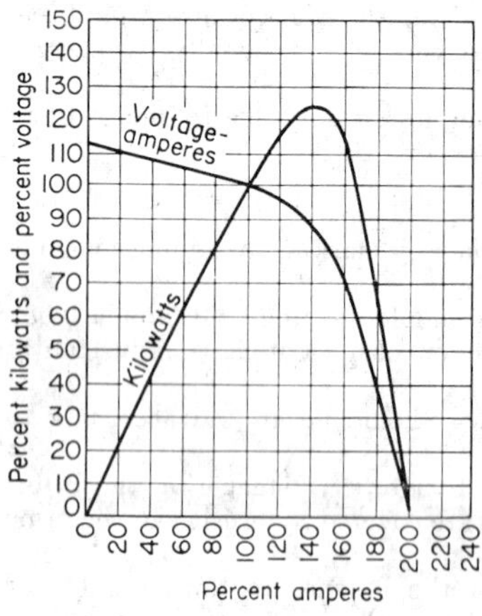

FIG. 7-64. Shovel generator characteristics.

In the past this characteristic has been obtained inherently in the generator by using three main fields. The proportions between a self-excited shunt field, a separately excited shunt field, and a differential series field accomplish this. More recent excavator generators use a single field with controlled input to obtain the same characteristic. A light series field is used to ensure an inherent droop.

97. Exciters. Direct-current generators in excess of 2,250 kW are used as exciters for turbine-driven a-c generators for electric utilities. Those up to 350 kW may be direct-connected at 3,600 r/min. These require special "shrink-ring" commutators to withstand the peripheral speed of approximately 10,000 ft/min.

Above 350 kW the exciters are driven either by speed-reducing gears from the generator shaft or by an a-c motor on a motor-generator set. The latter arrangement uses a flywheel to maintain speed during system disturbances which might lower the a-c voltage to the driving motor. Regardless of the drive, these exciters must have fast voltage response, low saturation, and low-voltage main fields for strong forcing.

A new d-c exciter is now being introduced that may become predominant for turbo-exciters. It eliminates all brushes, commutators, and main-generator slip rings by using a rotating-armature 3,600-r/min a-c synchronous generator feeding fused diodes on the same shaft. The rectified output is carried directly into the rotating field of the main turbine-driven a-c generator. The exciter output voltage is controlled by the field current to the stationary fields of the a-c exciter. Ratings of this type exceed 3,500 kW.

98. Rotating Regulators—Rototrol. Special generators and exciters used as rotating regulators can be employed to regulate industrial power equipment governing such functions as voltage, current, speed, power factor, position, limit, light, heat, and sound or a combination of these functions, without the use of any separate regulating devices. Such machines are known under trade names like Rototrol, Regulex, and Amplidyne. They employ the principle of a separately excited control or pattern field and an opposing differential field excited from the source to be controlled. Figure 7-65 shows a d-c rotating regulator used as an exciter for supplying the excitation to a d-c machine. This arrangement will hold the voltage output of the d-c machine essentially constant. Increased sensitivity can be obtained and a much smaller amount of power is required for the control field if means for self-energizing are incorporated in the rotating controller.

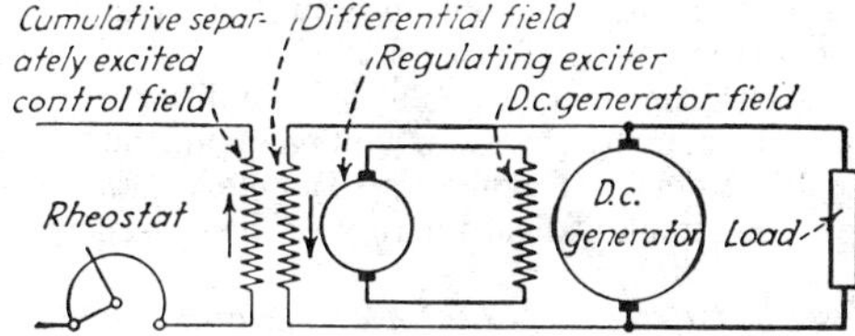

FIG. 7-65. Elementary type of rotating regulator.

Figure 7-66 shows the circuit of a Rototrol used to supply the excitation for and to control the voltage of a d-c generator. This Rototrol has a shunt-wound pattern field A, separately excited from a constant-voltage source; a differential shunt field B, excited from the output of the regulated generator; and a series self-energizing field C, which is excited by and proportional to the output current of the Rototrol exciter.

The saturation curve OYD of the Rototrol is shown in Fig. 7-67. For normal operation the voltage range of the Rototrol is held within the straight portion of the saturation curve, OY. The resistance line of the self-energizing series field C and its circuit is initially

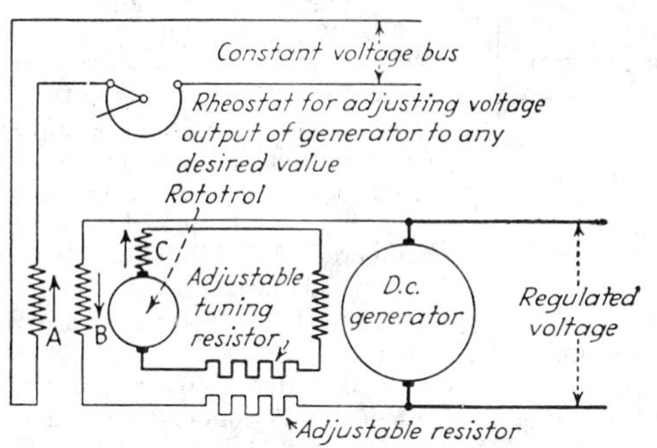

FIG. 7-66. Rototrol exciter regulating to a constant value the output voltage of a d-c generator.

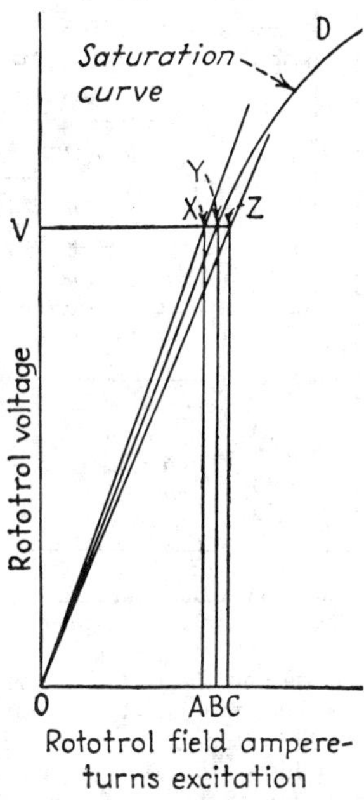

FIG. 7-67. Saturation curve and self-energizing field lines of a Rototrol.

adjusted by the tuning resistor to lie close to the straight portion of the saturation curve. Excitation from a constant-voltage bus is applied to pattern field A, and the voltage of the system builds up. The voltage-adjusting rheostat of pattern field A is set for the desired voltage output of the generator. As this occurs, the self-energizing series field C adds to the total excitation by the load current flowing through it. As the generator voltage increases, the differential field B lowers the total excitation and a stable point is reached when the excitations supplied by fields A and B are exactly equal, if the self-energizing field line coincides with the straight portion of the saturation curve. Sufficient accuracy is obtained if the field line is within $\pm 5\%$ of the straight-line portion of the saturation curve. Under this condition, A and B do not exactly balance. If it is less than the curve as OX of Fig. 7-67, the small excitation XY must be supplied by the pattern field exceeding the differential field B. For the condition of OZ, the differential field will exceed the pattern field.

When the voltage tends to change, as with a change in load, a drop will weaken the differential field ampere-turns and the resultant increase in total ampere-turns causes the voltage of the Rototrol to rise. This, in turn, gives additional excitation from the self-energizing field C. Settled conditions occur when field B again equals A, which occurs when the generator voltage again returns to its correct value.

99. Rotating Regulators—Regulex. Figure 7-68 shows a connection for the Regulex using a shunt field for tuning the self-energizing field to the air-gap line of the saturation curve. The diagram shows the Regulex as an exciter for an a-c generator. The constant-voltage bus is replaced by a static network which provides a source of constant a-c voltage even though the a-c generator output varies. The varying a-c voltage of the generator supplies the voltage of the pilot field. Rectifiers convert these two voltages to direct current for exciting the control field of the Regulex. In the figure the rectified constant-reference output voltage of the generator and the rectified variable output voltage of the generator are combined differentially to supply the control field.

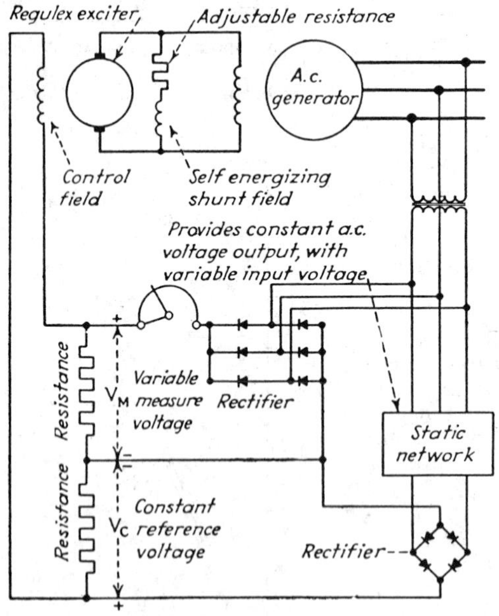

FIG. 7-68. Diagram of a Regulex exciter controlling the voltage of a d-c generator.

The principal difference between the series self-energizing field and the shunt type used here is that the excitation of the series type is in phase with the load current of the regulator. With the shunt self-energizing field, the response of the circuit is more rapid, on the assumption that its time constant is short compared with that of the load circuit. However, the faster response and the fact that the flux due to the self-energizing field is out of phase with the output circuit flux require more careful stabilization.

Both the Rototrol and Regulex regulators can be considered to be single-stage amplifiers with regenerative feedback, using a series or shunt connection or a combination of the two. Amplifications of 25,000 to 1 to 50,000 to 1 are possible. These regulators use any brush polarity for carrying output current, are not sensitive to brush position or brush fit, and are not adversely affected by abnormal commutator surface roughness.

Multipolar Rototrols have been built using two stages of amplification in a single unit. This type has been built up to 350 kW capacity, with amplifications of 1,500,000 to 1. A change in output can be obtained with a control energy of less than 0.01 W.

100. The Amplidyne principle is shown in Fig. 7-69, which shows a two-pole stator with four brush arms. This device has two stages of amplification. It does not use a conductively connected self-energizing field as do the Rototrol and Regulex. The first stage of amplification is from the direct-axis control field F_1F_2 to the quadrature short-circuit axis. Control flux ϕ_1 plus armature rotation generates a speed voltage E in brush circuits XA_1, XA_2. These brushes are short-circuited on themselves. A resulting current I_2 flows through this external short-circuit connection and produces within the machine an armature field ϕ_2 in the quadrature axis. The second stage of amplification is from the short-circuited quadrature axis to the direct output axis. Rotation of the armature in this field ϕ_2 generates an output speed voltage E_3, and the resulting current I_5 flows in the load circuit. This load current flows through a compensating winding C_1C_2 located on the direct axis and completely neutralizes the load armature reaction in a manner similar to that of a compensating field on a conventional d-c machine.

The total amplification is the product of the first- and second-

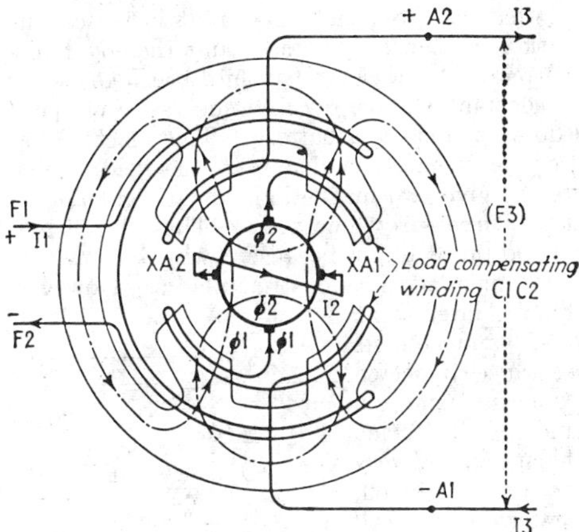

FIG. 7-69. Diagram of a two-pole four-brush-arm Amplidyne with control and quadrature fields indicated.

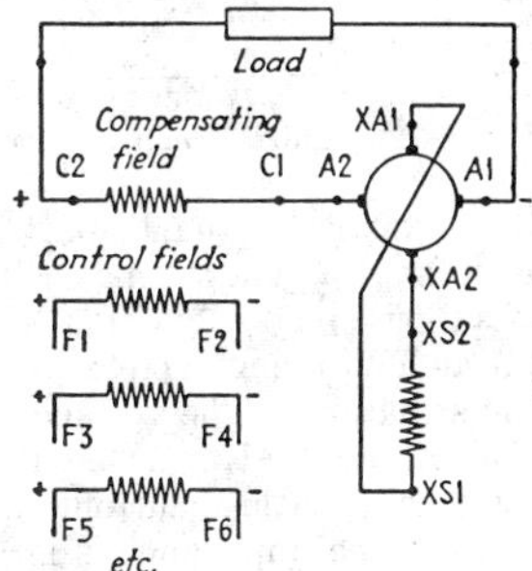

FIG. 7-70. Schematic diagram of an Amplidyne, showing compensating field, quadrature series field, and control fields.

stage amplifications. Because the control flux, and therefore the control field excitation watts, necessary to create the short-circuit current I_2 is very low, full-voltage power amplification of 10,000 to 1 to 250,000 to 1 is obtained. Machine sizes range from 1 to 50 kW. Accurate control is obtained with a field input of less than ½ W. The Amplidyne has a fast response with a time constant of 0.05 to 0.25 s. Field forcing lowers these values markedly, with low power input. Additional quadrature- and direct-axis shunt and series fields may be added to obtain extremely high amplifications, reduce the short-circuit current, and provide other special characteristics. These are shown in Fig. 7-70.

The rotating regulator is still being used in many applications. However, solid-state regulators, such as the magnetic amplifier and more recently the silicon-controlled rectifier regulator, are gradually becoming dominant.

101. The Rosenberg variable-speed generator is self-regulating. It is shown diagrammatically in Fig. 7-71. The battery voltage is approximately constant, and the shunt field produces a flux ϕ_1 which is also about constant. The armature cuts this flux, and an emf is generated between brushes *bb*, but none between brushes *BB*. The former are short-circuited, so that a current flows in the armature and sets up an armature field ϕ_2. Since the armature cuts this field, there is an emf between brushes *BB* and a current in the external circuit. This current *I* sets up a flux ϕ_3 which opposes ϕ_1 but which can never exceed ϕ_1, so that *I* cannot exceed that value at which the armature ampere-turns per pole equal the shunt field ampere-turns per pole. The relationship between current and speed for different shunt field excitations is shown in the figure. The direction of the current *I* is independent of the direction of rotation of the machine.

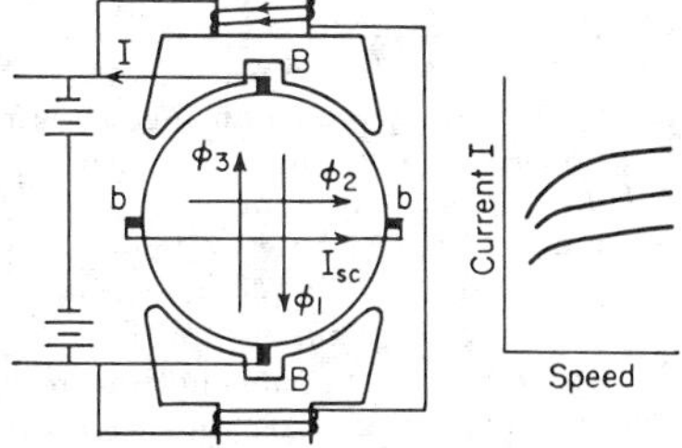

FIG. 7-71. Rosenberg generator.

102. Third-brush Generators. This type of machine is shown diagrammatically in Fig. 7-72. At no load the excitation voltage *ab* is half the terminal voltage *ac*, but the armature reaction weakens the leading pole tip so that, as the armature current increases, the excitation voltage *ab* decreases and thereby limits the current. Another type of third-brush generator gives constant voltage over a wide range of speed and does not require a battery. This machine has a two-pole armature and a four-pole field system excited to give two magnetic fields at right angles.

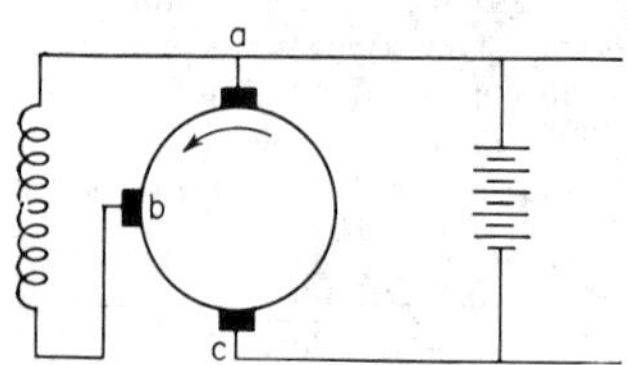

Fig. 7-72. Third-brush generator.

103. Three-wire Generators. These are of limited importance today because 3-wire systems are no longer common in industrial plants. However, the principle of the Dobrowolsky machine is interesting. It is shown diagrammatically in Fig. 7-73 and consists of a 2-wire machine with a coil of high reactance connected through slip rings across diametrically opposite points on the armature. The voltage between *a* and *b* is alternating even at no load or when the load is balanced. A very small alternating current flows in the reactance coil, because the impedance is large. The center *o* of the coil is always midway in potential between the brushes *c* and *d* and is connected to the neutral of the system. With unbalanced currents in the lines, the current difference flows in the neutral wire and through the reactance coil. This coil offers only a small resistance to direct current. The unbalanced current is usually limited to 25% of line amperes.

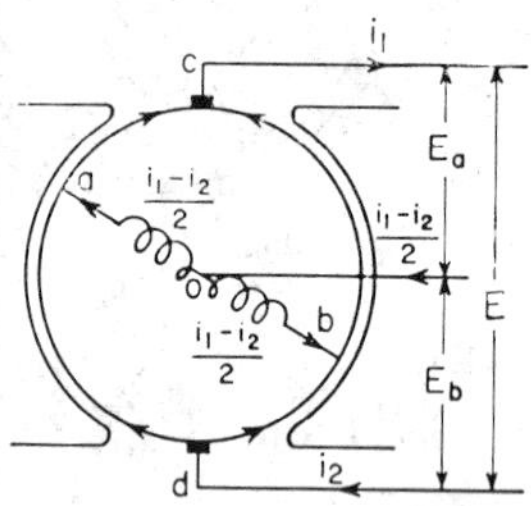

Fig. 7-73. Unbalanced currents in a 3-wire generator.

104. Balancer sets consist of two like generators coupled together, each wound for half voltage and connected in series as shown in Fig. 7-74. With balanced loads they both run light as d-c motors. With unbalanced loads, the currents flow as shown in the figure. Machine *M* acts as a motor and drives *G* as a generator. The current in *M* is greater than that in *G* by the amount required to supply the armature losses. The motor speed drops with the increase in load, and the generator voltage drops because of the decrease of speed and the reduction in excitation. Thus E_g is less than $\frac{1}{2}E$. By crossing the shunt fields, the regulation is improved because the decreased excitation on the motor increases its speed, and the excitation for the generator increases due to the higher voltage across the motor so that its voltage is increased as a result.

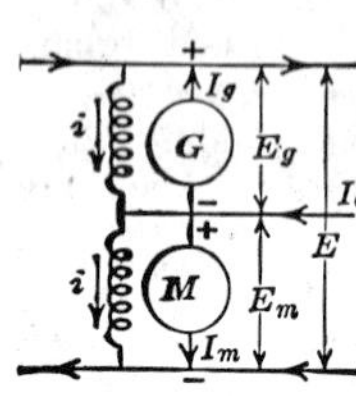

Fig. 7-74. Balancer set.

105. Synchronous converters were used by the hundreds for power supplies for street cars, for interurban lines, and in chemical plants. Their use has declined, and many are now available on the second-hand market. The synchronous converter remains an ingenious machine. It combines a synchronous motor and a d-c generator in one machine using common windings for the a-c and d-c currents. A commutator is provided on one end of the rotor winding and collector rings on the other end. Alternating voltage is supplied to the collector rings, and at no load the units operate as a simple synchronous motor. Under load, d-c current is taken from the commutator, and a-c current is supplied at the collector rings. Both currents flow in opposition in the single armature winding. In off-unity power-factor operation the currents become partially additive so that the amount of deviation from unity power factor must be limited to avoid burning out the tap coils connected to the rings.

As in a synchronous motor, any excess or deficiency of excitation is corrected by current drawn from the line by the armature. Thus within narrow limits the synchronous converter may be used to improve the power factor of the line. Overexcitation of the main fields produces leading currents and underexcitation lagging currents.

The use of a common winding and a common main-pole flux results in definite ratios between the a-c and d-c voltages and the respective currents. Compounding can be achieved on the d-c side only by some means of increasing the voltage to the a-c rings with load increase. This is done by the use of a series field which provides a total flux greater than the synchronous machine can tolerate. A leading a-c current results. The supplying transformer is designed to have a large reactance, and its resultant output voltage to the collector rings is raised by the action of the leading current through the reactance. This in turn causes the d-c voltage to increase with load. Approximate ratios are given in the tabulation:

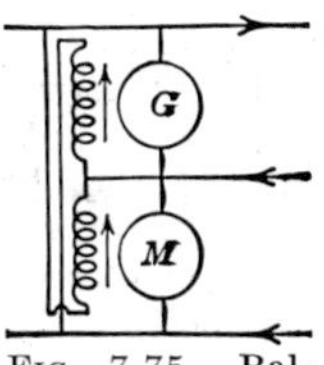

FIG. 7-75. Balancer set.

Phases	Volts d-c to a-c	Amperes a-c to d-c
Single	0.71	1.50
Two	0.71	0.75
Three	0.61	1.00
Six	0.71	0.50

106. Acyclic, or Homopolar, Generators. If a single conductor were continuously moved at a constant speed through a constant magnetic field, a constant d-c voltage would be induced in it. No commutator would be required. This is homopolar operation.

The Conventional Homopolar Generator. Figure 7-76 shows the design used for a 150,000-A 7.5-V 514-r/min generator of this type employed for making longitudinal welds in the manufacture of large steel pipe from rounded plates.

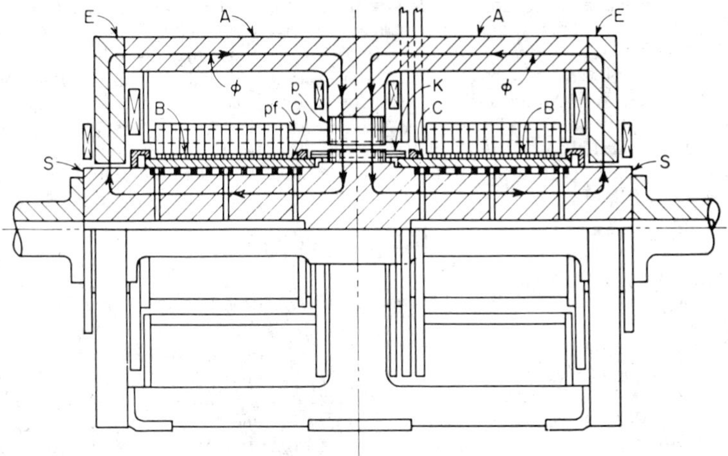

FIG. 7-76. Brush-type homopolar generator.

As shown, the constant magnetic field flux ϕ passes across the main air gap at the center of the machine from a circular north pole p which completely envelops the armature. The single conductor k is divided into many insulated straps located in slots around the rotor. These are short-circuited to form a single conductor by the large collectors C at each end. The flux loops are completed by passing each way through the steel shaft S under the collectors—across the end air gaps, into the end plates E, and finally into the yoke arms A and back to the circular pole.

On this type of homopolar generator there are over 1,400 brushes, and the collector peripheral speed must be low to provide acceptable brush friction losses, brush wear, heating, and maintenance costs. This is the bottleneck of the conventional unipolar unit. The peripheral speed limits the diameter of the collectors and the revolutions per minute of the rotor. The former limits the shaft diameter under the collectors, and this automatically limits the total flux of the machine. The limit on the revolutions per minute limits the voltage that can be obtained from the flux that can be put through the shaft. Even with water-cooled collectors, as shown in Fig. 7-76, and air-blast ventilation on the collector surfaces, the maximum continuous voltage is about 10 to 12 V for acceptable brush wear.

The brushes B are metal graphite, with about 75% copper to obtain low contact resistances and losses. Because the collectors are also large single conductors, any flux leaking through them to the yoke arms or elsewhere as they rotate generates unipolar voltages across them. Such voltages cause poor current division among the many brushes. To prevent as much as possible such leakage flux, two "doughnut" field coils are used on each end of the generator. One is located near the main air gap at the center, and the other is located at the end air gap. These provide the necessary ampere-turns near the points where they are used; thus very few ampere-turns exist between the arms of the stator and the shaft under the collectors. Bucking coils are provided beyond the end plates to keep leakage flux out of the bearing journals.

The huge currents collected by the brushes are carried through the pole-face windings pf to counteract the same currents passing through the rotor conductors.

Liquid-metal Homopolars. The space age brought the need for short-term d-c currents to create high-velocity gases by means of electric arcs. In particular, wind tunnels for space-vehicle nose-cone tests require up to a million and a quarter d-c amperes at about 65 volts. The voltage-limiting features of the conventional unipolar (collector peripheral speeds and metal-graphite brushes) have been overcome by modifying the generator to use a liquid metal to collect the current from the rotor rather than the usual brushes. Sodium potassium (NaK) is used between the rotor collectors and the stator current collector. The liquid metal is continuously pumped and cooled and is kept in the collector regions by seals and inert gases. This gives very low friction losses. Because the contact voltage is almost zero, very high speeds are acceptable and some units have been built for 7,200 r/min. Normally the generators operate at 3,600 r/min, and voltages of 65 volts are usual. The continuous ratings for the wind-tunnel application are about 150,000 A, with short peak loads of 300,000 A. Four units in parallel meet the short-time requirements of 1,250,000 A.

Smaller units have been built for 60,000 A, 30 V at 7,200 r/min and for 80,000 A, 75 V at 3,600 r/min.

BIBLIOGRAPHY

107. General

LAMME, B. G. "Engineering Papers"; East Pittsburgh, Westinghouse Electric Corporation, 1919.

GRAY, A., and LINCOLN, P. M. "Electrical Machine Design," 2d ed.; New York, McGraw-Hill Book Company, 1926.

KLOEFFLER, R. G., BRENNEMAN, J. L., and KERCHNER, R. M. "Direct Current Machinery"; New York, The Macmillan Company, 1950.

LEWSCHITZ and GARRICK. "Fundamentals of D. C. Machines"; Princeton, N.J., D. Van Nostrand Company, Inc., 1951.

MOSES, G. L. "Electric Insulation: Its Application to Shipboard Electrical Equipment"; New York, McGraw-Hill Book Company, 1951.

LINVILLE, T. M., and STRANG, D. P. Current in Equalizer Connections of D-C Machine Armature Windings; *Trans. AIEE*, 1950, Vol. 69, Pt. II, pp. 1219–1227.

SNIVELY, H. D., and ROBINSON, P. B. Measurement and Calculation of D-C Machine Armature Inductance; *Trans. AIEE*, 1950, Vol. 69, Pt. II, pp. 1228–1237.

SHERON, R. L., and GRANT, D. A. Determination of Stray Load Loss in D-C Machines; *Trans. AIEE*, 1956, Vol. 75, Pt. III, pp. 1332–1334.

PASCULLE, M. J. Armature Tooth Pulsations Eddy Currents; *Trans. AIEE*, 1960, Vol. 79, Pt. III, pp. 612–618.

ERDELYI, E. Calculation of Stray Load Losses in D-C Machinery; *Trans. AIEE*, 1960, Vol. 79, Pt. III, pp. 129–138.

Lyon, W. V., Sayne, E., and Henderson, M. L. Heat Losses in D-C Armature Conductors; *Trans. AIEE,* 1928, p. 589.

Darling, A. G., and Linville, T. M. Rate of Rise of Short Circuit Current of D-C Motors and Generators; *Trans. AIEE,* 1952, Vol. 71, Pt. III, pp. 314–325.

O'Connor, J. P., and Cybulski, J. D-C Machines—Short-circuit Calculations and Test Results; *Trans. AIEE,* 1955, Vol. 74, Pt. III, pp. 222–238.

Cybulski, J., Broncato, E. L., and O'Connor, J. P. Transient Performance of D-C Machinery; *Trans. AIEE,* 1953, Vol. 72, Pt. III, pp. 45–52.

Brackman, J. J., and Linkous, C. E. D-C Machines: Response to Impact Excitation; *Trans. AIEE,* 1955, Vol. 74, Pt. III, pp. 500–505.

Dunaiski, R. M. The Effect of Rectifier Power Supply on Large D-C Motors; *Trans. AIEE,* 1960, Vol. 79, Pt. III, pp. 253–259.

Reiter, C. R., and Ammersman, C. R. Increased Losses in a D-C Motor when Operated from Grid-controlled Rectifiers; *Trans. AIEE,* 1952, Vol. 71, Pt. II, pp. 77–82.

Schmidt, J., and Smith, W. P. Operation of Large D-C Motors from Controlled Rectifiers; *Trans. AIEE,* 1948, Vol. 67, Pt. I, pp. 679–683.

108. Commutation

Lynn, C., and Elsey, H. M. Effects of Commutator Surface Film Conditions on Commutation; *Trans. AIEE,* 1949, Vol. 68, Pt. I, pp. 106–112.

Linville, T. M., and Rosenberry, G. M. Jr. Commutation of Large D-C Motors and Generators; *Trans. AIEE,* 1952, Vol. 71, Pt. III, pp. 326–336.

Alger, J. R. M., and Bewley, D. F. An Analysis of D-C Machine Commutation; *Trans. AIEE,* 1957, Vol. 76, Pt. III, pp. 399–416.

McLean, H. J., and Coho, O. C. D-C Motor Flashover Torque; *Trans. AIEE,* 1961, Vol. 80, Pt. III, pp. 850–853.

109. Rotating Regulators

Kopylov, I. P. Direct-current Motor Amplifier; *Electric Technol. (USSR)*, March, 1960, Vol. I, pp. 27–33.

Fisher, A. Design Characteristics of Amplidyne Generators; *Trans. AIEE,* December Supplement, 1940, Vol. 59, pp. 939–944.

Arc Furnace Control by Regulex Exciters; *Trans. AIEE,* June, 1943.

Formhals, W. H. Rototrol—A Versatile Electrical Regulator; *Westinghouse Eng.*, May, 1942.

110. Homopolar Generators

Myers, E. H. The Unipolar Generator; *Westinghouse Eng.*, March, 1956, Vol. 16, pp. 59–61.

Gigot, E. N. Applying Unipolar Generators; *Allis-Chalmers Elec. Rev.*, Second Quarter, 1962.

SECTION 8

POWER PLANTS

BY

THEODORE BAUMEISTER Consulting Engineer; Stevens Professor Emeritus of Mechanical Engineering, Columbia University; Editor-in-chief, Marks' "Standard Handbook for Mechanical Engineers"; Fellow, American Society of Mechanical Engineers

WILLIAM J. RHEINGANS Consulting Engineer; Fellow, American Society of Mechanical Engineers

FRANKLIN J. LEERBURGER Consulting Engineer; Fellow, Institute of Electrical and Electronics Engineers

CONTENTS

Numbers refer to paragraphs

SECTION 8

POWER PLANTS

THERMAL POWER PLANTS, PRIME MOVERS, AND ASSOCIATED EQUIPMENT

By Theodore Baumeister

1. Systems and Sources of Power. Many estimates are available on the magnitudes of the world's raw-energy sources and resources. The primary irreplaceable sources include petroleum, natural gas, coal, shale oil, tar sands, and nuclear fuels. Petroleum, natural gas, and coal are the major sources in today's economy, with the reserves of coal eclipsing the other two by a wide margin.

Nuclear fuels are of increasing significance (see Sec. **9**). The fission of heavy-atomic-weight elements, e.g., uranium and thorium, and the fusion of light-atomic-weight elements, e.g., deuterium, offer almost limitless reserves. It is too early to assess the magnitude of nuclear-energy sources, but they are vast compared with fossil-fuel reserves.

Table 8-1. Heating Values and Air Requirements for Selected Fuel Elements and Compounds

Substance	1 lb of substance requires air		1 cu ft of substance requires air		Heating value of fuel				Heating value fuel-air mixture, Btu per lb
					Btu per lb		Btu per cu ft		
	lb	cu ft	lb	cu ft	High	Low	High	Low	
Carbon to CO_2	11.53	150.6	.....	.....	14,093	14,093	....	....	1125
Carbon to CO	5.76	75.2	.....	.....	4,340	4,340	....	....	642
Hydrogen, H_2	34.34	448.5	0.182	2.38	61,100	51,623	325	275	1729
Carbon monoxide, CO	2.47	32.3	0.182	2.38	4,347	4,347	322	322	1253
Methane, CH_4	17.27	225.6	0.730	9.53	23,880	21,520	1013	913	1307
Ethane, C_2H_6	16.12	210.6	1.277	16.68	22,320	20,430	1792	1641	1304
Ethylene, C_2H_4	14.81	193.4	1.094	14.29	21,640	20,300	1614	1513	1369
Acetylene, C_2H_2	13.30	173.7	0.912	11.91	21,500	20,780	1499	1448	1503
Propane, C_3H_8	15.70	205.0	1.824	23.82	21,660	19,940	2590	2385	1297
Octane, C_8H_{18}	15.10	197.2	4.556	59.5	20,550	19,080	6210	5770	1276
Sulfur, S	4.29	56.0	.....	.....	3,983	3,983	....	....	753

Volumetric values are based on normal pressure and temperature, *i.e.*, 30 in. Hg and 60°F.
Specific volume of air = 13.06 cu ft per lb.
Air analysis: gravimetric, oxygen = 23.2%, nitrogen = 76.8%; volumetric, oxygen = 21.0%, nitrogen = 79.0%.

The primary replaceable sources of energy include elevated water, vegetation, tides, waves, winds, and solar energy. The only practical utilization of replaceable sources is water power, which constitutes about 20% of the electric-utility capacity of the United States. Solar-energy magnitudes, practically unharnessed, are reflected in a radiation level of some 400 Btu/(h) (ft²) at the outside of the earth's atmospheric envelope. The overwhelming capacity of United States power plants is reflected in the automotive industry (approximately 90,000,000 units), constituting some 97%

of the nation's entire power-generating capacity. They are all internal-combustion plants. Central stations, for electric-power generation, aggregate some 250 million kw capacity, 80% of them steam plants. The approximate installed capacity for all power plants of the United States—both stationary and transportation—is in excess of 10,000 million kw. Heat power is of controlling practical significance. Internal-combustion systems require a special, usually fabricated, fuel, e.g., gasoline, diesel fuel, or natural gas, but steam systems can use any source of heat, e.g., coal, oil, gas, nuclear, waste heat, waste fuel, or geothermal. Reserves of some fuels, particularly petroleum, are expected to feel the pinch of demand before the end of the century.

2. Fuels. The elements carbon and hydrogen, or their compounds, are the heat sources of all fossil fuels. Sulfur contributes some heating value, but its presence is both minor and objectionable. Table 8-1 gives heating values and air requirements of fuel elements and compounds, while Table 8-2 gives similar data for representative commercial fuels. Heating values and air requirement may be estimated by the following equations:

$$\text{Coal, Btu per lb} = 14{,}544\ \text{C} + 62{,}028\left(\text{H}_2 - \frac{\text{O}_2}{8}\right) + 4{,}050\ \text{S} \tag{8-1}$$

where C, H_2, O_2, S are the proportional weight parts of carbon, hydrogen, oxygen, and sulfur.

Residual fuel oil (from 10 to 20° API):

$$\text{Btu/lb} = 18{,}300 + 40\ (°\text{API} - 10) \tag{8-2}$$

$$\text{Btu/gal} = 152{,}500 - 700\ (°\text{API} - 10) \tag{8-3}$$

Distillate fuel oil (16 to 36°API):

$$\text{Btu/lb} = 18{,}900 + 40\ (°\text{API} - 16) \tag{8-4}$$

$$\text{Btu/gal} = 151{,}000 - 600\ (°\text{API} - 16) \tag{8-5}$$

where

$$°\text{API} = \frac{141.5}{\text{sp. gr. } 60/60°\text{F}} - 131.5$$

The high heating value is used in American practice for the specification and purchase of fuels. Low heating values are frequently used in Europe and are often convenient in detailed analyses of combustion phenomena. Air requirements can be estimated approximately for most commercial fuels by

$$\text{Chemically necessary air, lb/lb of fuel} = \frac{\text{high heating value, Btu/lb}}{1{,}300} \tag{8-6}$$

Figure 8-1 facilitates conversion and comparison of customary fuel charges on the basis of cost in cents per million Btu. Bulk and weight comparisons are given in Table 8-3 for fuel-storage and transportation-service problems.

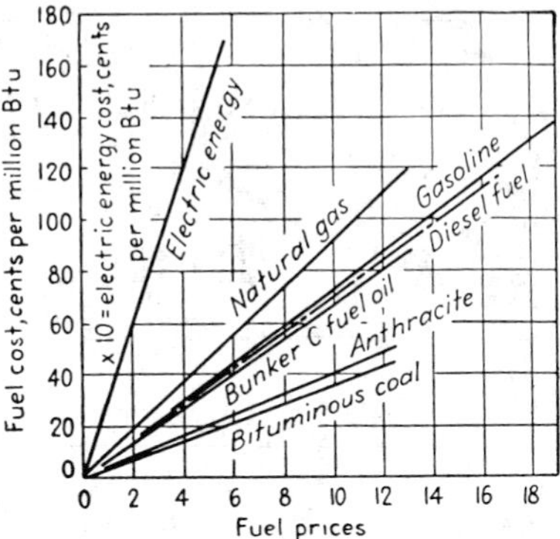

FIG. 8-1. Comparative fuel costs and conversion chart. Coals, dollars per short ton; liquids, cents per gallon; gases, (× 10) cents per thousand cubic feet; electric energy, cents per kilowatthour.

3. Basic Cycles and Performance. Rudimentary arrangements of internal-combustion and steam-power plants are illustrated, schematically, in Fig. 8-2. Modifications, elaborations, and details of components of these plants and cycles are given below. Fuel consumption and fuel cost are of controlling economic importance in the performance of steam and internal-combustion

Table 8-2. Analysis of Some Selected Fuels

A. Coals—Gravimetric basis, as fired*

Fuel	Volatile matter	Fixed carbon	Moisture	Ash	H_2	C	$O_2 + N_2 + S$	High heating value, Btu per lb	Chemically necessary air, lb per lb of fuel
	← Proximate analysis			→					
			← Ultimate analysis				→		
Lignite	17	35	39	9	2	36	14	6,500	4.8
Subbituminous	31	38	22	9	3	54	12	8,800	7.2
Low-rank bituminous	37	42	11	10	4	65	10	11,600	8.8
Middle-rank bituminous	37	50	5	8	4	75	8	12,800	10.0
High-rank bituminous	29	60	3	7	5	77	8	14,000	10.6
Low-rank semibituminous	22	69	3	6	5	79	7	14,600	10.8
High-rank semibituminous	13	76	5	6	4	79	5	14,500	10.5
Semianthracite	9	77	6	8	3	79	3	13,700	10.2
Anthracite	2	85	3	10	2	82	3	13,000	10.2
Coke	..	..	..	13	..	86	1	12,500	10.0

* Many commercial coals, as found in present markets, are poorer than the values listed by as much as 2,000 or 2,500 Btu per lb, because of proportionately high moisture and ash.

B. Liquids—Gravimetric basis, as fired

Fuel	H_2	C	$O_2 + N_2 + S$	High heating value, Btu per lb	Sp. gr. deg API	High heating value, Btu per gal	Chemically necessary air, lb per lb fuel
Bunker C (No. 6)	11.5	85	3.5	18,500	10–18	150,000	13.7
Diesel fuel	13.0	85	2.0	19,200	22–28	145,000	14.3
Furnace oil	13.5	85	1.5	19,500	30–35	140,000	14.4
Kerosine	14.0	85	1.0	19,800	45	133,000	14.6
Gasoline	15.0	85	0.0	20,300	55	130,000	14.9

C. Gases—Volumetric basis

Fuel	Sp. gr. air = unity	H_2	CO	CH_4	Heavy hydrocarbons	$O_2 + N_2 + CO_2$	High heating value, Btu per cu ft	Low heating value, Btu per cu ft	Chemically necessary air, cu ft per cu ft gas
Natural gas	0.6	..	..	96	2	3	1,100	1,000	10
Coke-oven gas	0.4	50	6	32	5	7	600	540	5.2
Producer gas	0.85	15	26	2	1	56	150	140	1.2
Blast-furnace gas	1.0	5	23	..	..	72	90	88	0.67

Table 8-3. Weight and Space Requirements of Some Selected Fuels

Fuel	Density, lb per cu ft	High heating value, Btu per lb or per cu ft	Weight, lb per million Btu	Space cu ft per million Btu
Bituminous coal	52	14,000	71	1.4
Anthracite	57	12,500	80	1.4
Coke	30	12,500	80	2.7
Bunker C fuel oil	60	18,500	54	0.9
Diesel fuel	55	19,500	51	0.93
Gasoline	45	20,000	50	1.1
Natural gas	0.04 NTP	1,100	28	910 NTP
Manufactured gas	0.04 NTP	550	56	1820 NTP

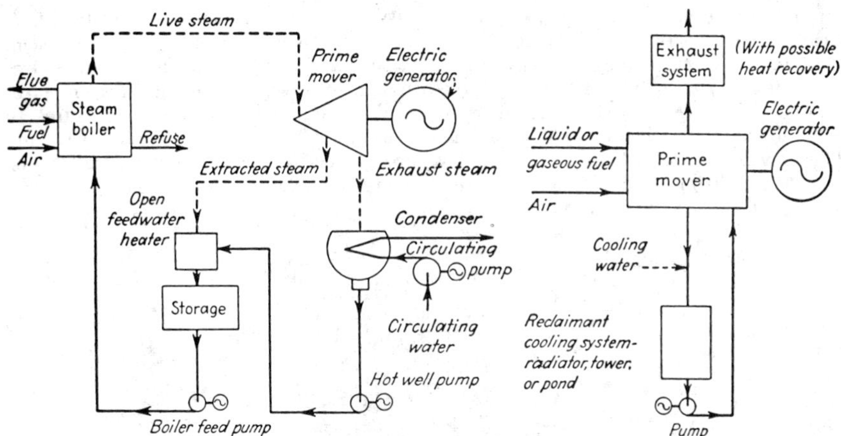

FIG. 8-2a. Basic flow diagram for a condensing-steam power plant with a single stage of extraction feed-water heating.

FIG. 8-2b. Basic flow diagram for an internal-combustion power plant.

systems, where thermal efficiency and heat rate are defined as

$$\text{Thermal efficiency, \%} = \frac{\text{work output}}{\text{heat supplied}} \times 100 \qquad (8\text{-}7)$$

$$\text{Heat rate, Btu/kWh} = \frac{\text{heat supplied in fuel, Btu}}{\text{energy generated, kWh}} \qquad (8\text{-}8)$$

or

$$\text{Thermal efficiency, \%} = \frac{3{,}412.75}{\text{heat rate}} \qquad (8\text{-}9)$$

Table 8-4 gives thermal performance of representative power plants.

Table 8-4. Approximate Capacity and Heat Rate of Selected Thermal Power Plants

Plant type	Unit size, kW	Overall heat rate, Btu/kWh
Recent central station, fossil-fuel-fired steam plants	500,000–1,000,000	8,000–9,000
Recent central station, nuclear-fuel-fired, steam plants	500,000–1,000,000	10,000–12,000
Average steam central station	. . . .	11,000 ±
"By-product" power, steam plant	3,000–50,000	4,500–5,000
Stationary diesel engine	1,000–25,000	10,000–12,000
Gas turbine	5,000–100,000	15,000–20,000
Steam locomotive	1,000–3,000	50,000–75,000
Diesel-electric locomotive	1,000–3,000	12,000–15,000

STEAM GENERATION

4. Steam-generating Equipment. Fire tube boilers, in which hot gases pass through tubes surrounded by water, are found only in some low-cost small-sized (<10,000 lb/h), low-pressure (<150 lb/in²) applications. They have been largely displaced by water tube boilers in which the hot gases sweep over the outside of the tubular surfaces connected to one or more boiler drums. Design variety is dictated by service factors, i.e., safety, weight, bulk, height, portability, character of operating labor, life, efficiency, and cost. Tubes, straight or bent, range from 1 to 4 in. in diameter. Header boilers, using straight tubes, have been superseded by curved-tube drum-type designs (cf. Fig. 8-3a, b, c). Safe metal temperature is maintained by keeping surfaces wet through ample circulation on the water side. Natural circulation (by density difference) is used with pressures up to 2,500 lb/in². Pumps, giving controlled circulation, are often preferred on designs delivering more than 700,000 lb of steam/h at pressures of 1,400 to 2,700 lb/in². The once-through boiler, operating at supercritical pressures (above 3,200 lb/in²), uses no drums. Central-station boilers range in pressures to 5,000 lb/in², capacities to 6,000,000 lb steam/h, and firing rates of 6 tons of coal/min. The prevalent supercritical pressure is 3,500 lb/in², with unit sizes in excess of 500,000 kW, and utilizing a double reheat cycle. Figure 8-3*d* shows a steam generator for a pressurized-water nuclear power plant.

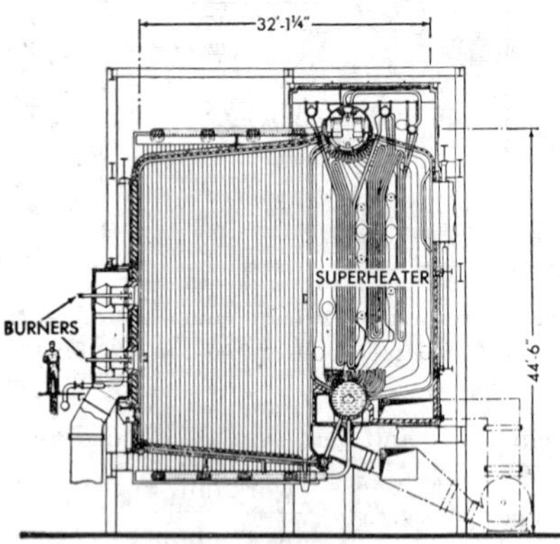

Fig. 8-3*a*. Section through an oil- or gas-fired steam boiler. (*Babcock and Wilcox Company*.) Capacity 310,000 lb/h at 875 lb/in², 910°F; feed temperature 405°F.

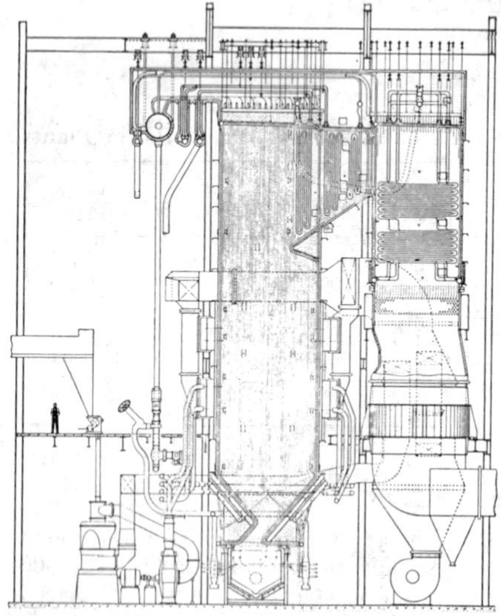

Fig. 8-3*b*. Section through a pulverized-coal-fired controlled-circulation reheat steam boiler. (*Combustion Engineering, Inc.*) Capacity 750,000 lb/h at 1,500 lb/in², 1000°F/1000°F.

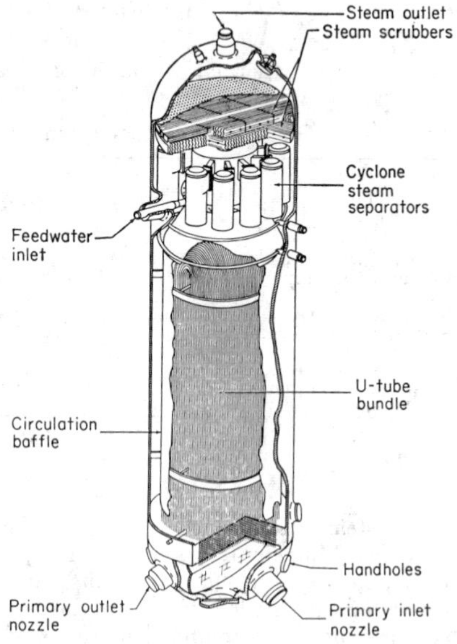

Fig. 8-3*d*. Section through a steam generator for a pressurized-water nuclear power plant.

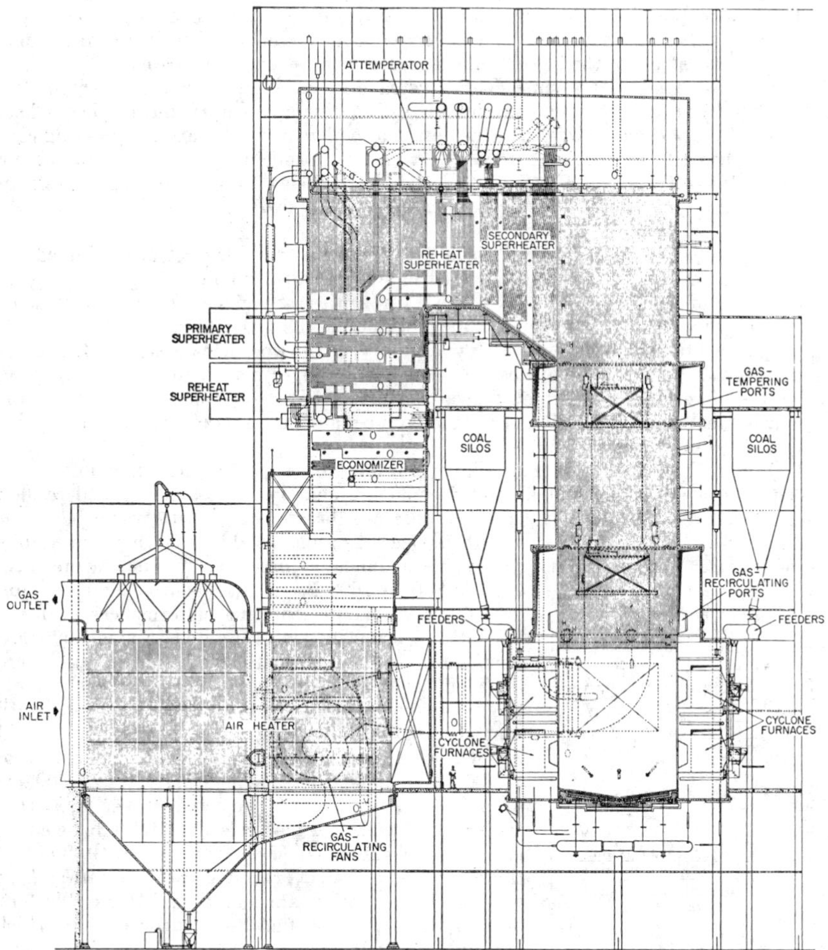

FIG. 8-3*c*. Section through a cyclone-fired forced-circulation supercritical-pressure steam boiler. (*Babcock and Wilcox Company.*)

Table 8-5. Boiler and Furnace Performance Factors

Item	Heat release rate, Btu per hr per cu ft	Excess air, %	Furnace size, 10^6 Btu per hr	Associated boiler efficiency, %	Stack temperature, F
Natural gas	20,000–40,000	10–25	10–2,000	85–88	300–600
Fuel oil	20,000–50,000	15–30	10–2,000	85–89	300–600
Pulverized coal	15,000–35,000	15–30	25–2,000	85–91	250–500
Cyclone furnace		10–20	200–2,000	88–91	250–400
Stokers, large	20,000–40,000	20–40	50–200	80–86	400–700
Stokers, small	15,000–25,000	30–60	2–50	75–82	400–700
Hand firing	5,000–10,000	50–150	0.5–5	50–60	500–700

5. Construction. The construction and operation of boilers are rigorously controlled by law. The ASME Boiler Construction Code is the prevalent standard specifying such items as drum thickness, tube gage, ligament strength, and welding practices. Fusion-welded drums have superseded riveted designs. Large boilers (over 100,000 to 200,000 lb/h) are generally designed for the specific application, but smaller boilers are offered in "package" designs. The design of boilers or modification of existing constructions constitutes a specialized talent which is found only in the manufacturers' organizations. The trend of reliability in boilers is such that spare boilers are seldom provided; unit construction (a single boiler supplying a single turbine) prevails; annual availability factors range to 95%.

6. Capacity. Boiler capacity is usually expressed as pounds of steam delivered per hour and sometimes as the heat delivered in steam, Btu per hour. The latter is particularly significant with reheat boilers. Capacities are given on the continuous, 1- or 4-h-peak bases. The boiler horsepower unit (33,479 Btu/h added in steam) has long been obsolete in the engineer's vocabulary, as the word horsepower is a misnomer as applied to a boiler. The nuclear scientist has, however, reintroduced the concept in the rating of reactors as thermal kilowatts, where 1 kW(th) = 3412.75 Btu/h. It is imperative in the interpretation of atomic-power-plant performance data to distinguish "thermal" from "electrical" kilowatts.

7. Boiler Furnaces. The most important part of a steam-generating unit is the furnace. A good design and a good fireman are equally necessary for good results. Boiler fuels are coal, oil, and gas (cf. Table 8-2 for analyses), with waste heat and waste fuels used industrially and with nuclear reactions as the heat source in atomic power plants. Waste heat is generally economically justifiable if temperature level exceeds 800 to 1000°F. Most steam generation is by fossil fuels with representative performances in Table 8-5. Hand firing of coal is limited to small or transient services (less than 10,000 lb steam/h), where the inefficiency of poor combustion control is acceptable. Stokers give mechanical feeding of coal, with control and proportioning of air, and mechanical removal of refuse. Stokers are built in sizes from 10,000 to 250,000 lb steam/h, firing rates of 300,000 to 500,000 Btu/(ft²) of projected stoker area/(h). Underfeed stokers are particularly suitable for coking-type (semibituminous) coals and run with fuel-bed thicknesses up to 2 ft. Traveling or chain-grate stokers are used for burning anthracite or Middle Western bituminous coals with fuel beds less than 6 in. thick and grate-travel rates of 1 to 2 ft/min. Spreader stokers, with dumping or traveling grates, are suitable for a wide variety of coals using thin fires (2 to 3 in.).

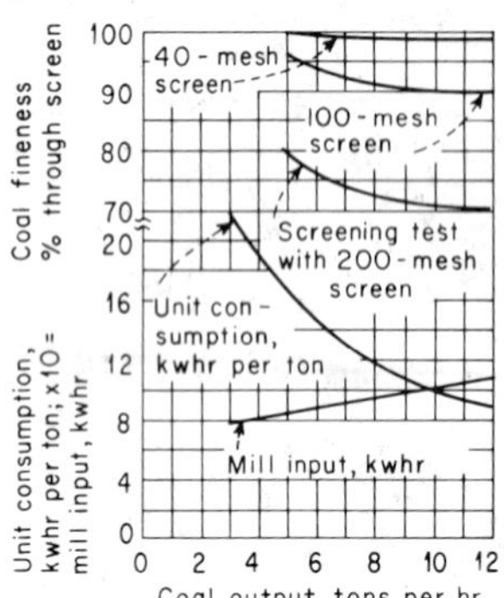

Fig. 8-4. Coal pulverizer performance. Coal analysis: moisture 2%; volatile 23%; fixed carbon 69%; ash 6%. Heating value 14,300 Btu/lb. Grindability 103.

Pulverized coal prevails in the largest coal-fired installations, above 250,000 lb steam/h, and can be justified for units as small as 25,000 lb/h. Coal is dried and pulverized to a fineness specification of 80% through a 200 mesh. Pulverizer performance is reflected in Fig. 8-4. Grindability index (Hardgrove) refers to a standard coal which has an index value of 100. Harder coals show an index less than 100, and more readily pulverizable coals show an index greater than 100. Grindability confirms the preference for bituminous rather than anthracite with pulverized firing. The unit mill system with coal ground immediately before introduction to the furnace is preferred today to the bin-and-storage system. Pulverized coal, when aerated, is easily transferred to the burners, where mixing with air and control of combustion are accomplished. Oil or gas is used for ignition, and coal burns in suspension. Flames may be very long (100 ft) in order to complete combustion of particles. Ash is removed from furnace in molten (wet-bottom) or solid (dry-bottom) condition. Furnace must be designed for one method or the other, which requires knowledge of fusion temperature of the ash and variation of the furnace temperature with boiler load. Fusion temperatures below 1800 to 2200°F favor wet bottom. The cyclone furnace (Fig. 8-3*c*) uses crushed coal (less than ½ in) and burns it by prolonged swirling in a high-heat-intensity zone, gases escaping to a secondary furnace and ash being tapped in a molten condition.

Fuel oil (residue from refinery stills) of 18,500 Btu/lb; 145,000 to 155,000 Btu/gal; and specific gravity ranging from 1.05 (greater than water) to 10 or 15°API (see Table 8-6*a* for ASTM specifications) is common boiler fuel. Pumps, strainers, and heaters

Table 8-6a. Summary Specification (Abbreviated) for Fuel Oils, ASTM D396-48T

Detailed requirements for fuel oils[a]

Grade of fuel oil[b]	Flash point, deg F	Pour point, deg F	Water and sediment, % by volume	Carbon residue on 10% bottoms, %	Ash, % by weight	Distillation temperatures, deg F: 10% point	90% point	End point	Saybolt viscosity, sec: Universal 100 F		Furol At 122 F		Kinematic viscosity, centistokes: At 100 F		At 122 F		Gravity, deg API	Corrosion at 122 F (50 C)[d]
	Min.	Max.	Max.	Max.	Max.	Max.	Max.	Max.	Max.	Min.	Max.	Min.	Max.	Min.	Max.	Min.	Min.	
No. 1: A distillate oil intended for vaporizing pot-type burners and other burners requiring this grade of fuel	100 or legal	0	trace	0.15	....	420	...	625	...	...	...	..	2.2	1.4	.....	....	35	pass
No. 2: A distillate oil for general-purpose domestic heating for use in burners not requiring No. 1 fuel oil	100 or legal	20[c]	0.10	0.35	....	[e]	675	...	40	...	...	..	(4.3)	......	.....	....	26	
No. 4: An oil for burner installations not equipped with preheating facilities	130 or legal	20	0.50	....	0.10	...	...	...	125	45	...	..	(26.4)	(5.8)				
No. 5: A residual-type oil for burner installations equipped with preheating facilities	130 or legal	....	1.00	....	0.10	...	...	...	...	150	40	..	......	(32.1)	(81)			
No. 6: An oil for use in burners equipped with preheaters permitting a high-viscosity fuel	150	....	2.00[f]	...	...	...	...	...	...	...	300	45	......	......	(638)	(92)		

[a] Recognizing the necessity for low-sulfur fuel oils used in connection with heat-treatment, nonferrous metal, glass, and ceramic furnaces and other special uses, a sulfur requirement may be specified in accordance with the following table:

Grade of Fuel Oil	Sulfur, Max., %
No. 1	0.5
No. 2	1.0
No. 4	No limit
No. 5	No limit
No. 6	No limit

Other sulfur limits may be specified only by mutual agreement between the purchaser and the seller.

[b] It is the intent of these classifications that failure to meet any requirement of a given grade does not automatically place an oil in the next lower grade unless in fact it meets all requirements of the lower grade.

[c] Lower or higher pour points may be specified whenever required by conditions of storage or use. However, these specifications shall not require a pour point lower than 0°F under any conditions.

[d] Report as passing when the copper test strip shows no gray or black deposit.

[e] The 10% may be specified at 440°F maximum for use in other than atomizing burners.

[f] The amount of water by distillation plus the sediment by extraction shall not exceed 2.00%. The amount of sediment by extraction shall not exceed 0.50%. A deduction in quantity shall be made for all water and sediment in excess of 1.0%.

deliver clean fuel to burner nozzles, where atomization is effected by the use of steam (less than 300 lb/in^2) or mechanically by pumps with pressures to 1,000 lb/in^2. Steam atomizing gives wider capacity range, but water consumption (as much as 1% of boiler production) tends to limit its preference to stationary installations with mechanical burners prevailing in marine practice.

Table 8-6b. Tentative Summary Classification (Abbreviated) of Diesel Fuel Oils, ASTM Designation D975-52T

Limiting requirements for Diesel fuel oils*

Grade of Diesel fuel oil	Flash point, deg F	Pour point, deg F	Water and sediment, % by volume	Carbon residue on 10% residuum, %	Ash, % by weight	Distillation temperatures, deg F		Viscosity at 100 F		Sulfur, % by weight	Copper strip corrosion	Cetane number
						90% point	End point	Kinematic, centistokes (or Saybolt Universal, sec)				
	Min.	Max.	Max.	Max.	Max.	Max.	Max.	Min.	Max.	Max.	Max.	Min.
No. 1-D: A volatile distillate fuel oil for engines in service requiring frequent speed and load changes	100 or legal	20†	Trace	0.15	0.01	...	625	1.4	.....	0.50	No. 3‡	40¶
No. 2-D: A distillate fuel oil of low volatility for engines in industrial and heavy mobile service	125 or legal	†	0.10	0.35	0.02	675	...	1.8 (32.0)	5.8 (45)	1.0	No. 3‡	40¶
No. 4-D: A fuel oil for low- and medium-speed engines	130 or legal	†	0.50	....	0.10	...	...	5.8 (45)	26.4 (125)	2.0	......	30¶

* To meet special operating conditions, modifications of individual limiting requirements may be agreed upon between purchaser, seller, and supplier.

† For cold-weather operation the pour point should be specified 10°F below the ambient temperature at which the engine is to be operated, except where fuel-oil heating facilities are provided.

‡ A No. 4 ASTM copper strip or darker constitutes grounds for rejection of the fuel.

¶ Low-atmospheric temperatures as well as engine operation at high altitudes may require use of fuels with higher cetane ratings.

Furnace designs, capable of using alternative fuels, take advantage of changing fuel availability and price. Storage facilities may range as high as 6 months, with 100 days as common. Within the plant there are usually (1) a bunker, with coal firing, to use the fuel-handling system on a single-shift basis, and (2) a day tank, with oil firing.

Furnace construction varies from plain refractory walls to completely water-cooled walls, depending upon fuels, firing methods, and firing rates. Water walls are built of tubes of 1 to 4 in diameter, variously spaced, with or without fins or studs, and bare or with different thicknesses of moldable refractory on the inner face. The walls are often of panel construction to minimize field labor. Heat-transfer rates run to 50,000 to 150,000 Btu/(h)(ft^2) of surface. To meet these heat-transmission conditions, circulation on the water side must be adequate, obtained by convection or by pumps. Air-cooled walls are no longer favored. Surrounding the furnace and the rest of the boiler is the insulated casing, which may be finished in plastic or with a sheet-metal sheath. Casings must be built with adequate strength for positive-pressure furnace operation. The boiler is supported on its own structural steel, with provision for expansion by suspension from above in stationary practice and by bottom support in marine practice.

8. Superheaters and reheaters are built in 1- to 2.5-in.-diameter tubes, carbon steel to 950°F steam temperatures, carbon-molybdenum (ferritic) steel to 1050°F, and stainless (austenitic) steel to 1200°F. Steam temperatures in excess of 1100°F are not looked upon with favor in today's central-station practice. In reheat applications primary steam temperatures may be higher than reheat temperatures because of metal economy with the smaller (high-pressure) diameters. Tube-bundle location and arrangement, with countercurrent and/or parallel flow, are dictated by type of firing, required steam temperature, and steam-temperature characteristic (Fig. 8-5). Radiant superheaters (placed in the furnace) give falling characteristics, while convection superheaters (placed well back in the boiler tube bank) give rising characteristics. Desired control of characteristic is obtained by (1) proportioning and placing surfaces in series; (2) use of internal dampers on boiler gas side; (3) attemperating by water; or (4) supplementary burners. Heat-transfer rates of 10 to 12 Btu/(h) (ft²) (deg temperature difference) are representative.

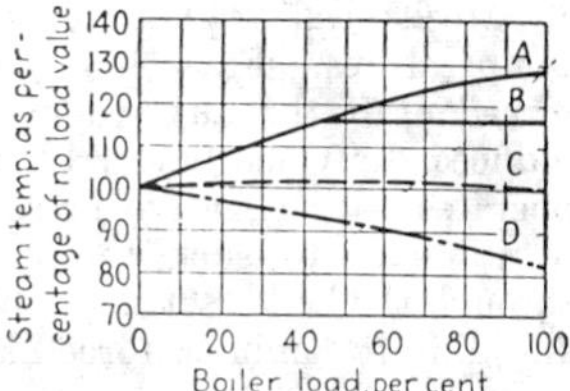

FIG. 8-5. Selected superheater performance characteristics. (A) Convection type. (B) Attemperating or internal-baffle type. (C) Series convection and radiation type, interdeck type. (D) Radiant type.

9. Economizers and Air Heaters. Economizers and air heaters are heat traps which raise boiler efficiency by lowering the stack temperature. Either or both may be used, or they may be omitted in their entirety. Economizers recover heat from the flue gases by adding it to the feed water on its way to the boiler, thus raising the temperature of the water, usually without evaporation. The water must be under boiler pressure. Steel tubes arranged in multiple countercurrent circuits prevail. Surfaces are generally bare steel tubes about 2 in. in diameter but may have gills or rings to increase the ratio of external to internal heating surface. Heat-transfer coefficients of 6 to 9 Btu/(h) (ft²) (deg) obtain. Degasified feed water is needed to reduce corrosion.

Air heaters recover the heat from the flue gases by adding it to the air supplied for combustion. This raises the temperature of the furnace gases, improves combustion rates and efficiency, and lowers the stack temperature, thus improving the overall economy of steam making. Air heaters are of the recuperative or regenerative type. The former employs a bank of straight light-gage steel tubes 1 to 3 in. in diameter, usually with the flue gases inside the tubes and air outside. The regenerative type transfers the heat by the use of a slowly revolving drum (1 to 2 rpm) of corrugated metal which moves alternately through the hot gas and cold air streams. Heat-transfer rates on tubular heaters are of the order of 2.5 to 5 Btu/(h) (ft²) (deg).

Economizers cost more per square foot than air heaters because they are pressure vessels. Heat-transfer rates are higher so that there is an offsetting element which makes both types of heat traps acceptable. Flue gases should not be cooled below the dew point (200 to 300°F), and combustion air temperatures should not be raised too high (500 to 700°F).

10. Draft Systems. The purpose of the draft system is (1) to provide the air for combustion and (2) to remove the products of combustion after they have given up their heat to the boiler surfaces. The air supplied is always in excess of that chemically required with practice represented by the data of Table 8-5. Natural draft is calculated by the equation

$$\text{Theoretical draft, inches of water} = \frac{H(W_{CA} - W_{HG}) \times 12}{W_W} \tag{8-10}$$

where H = height of stack in feet, W_{CA} = density of cold air in pounds per cubic foot, W_{HG} = density of hot gas in stack in pounds per cubic foot, W_W = density of water, 62.4 lb/ft³. The effective draft is less by the amount of the velocity head and the losses due to friction, fittings, and changes of section. Stack velocities range from 30 to 50 ft/sec.

Stack heights, for draft purposes, usually do not exceed 200 ft, but the need for dissipation of flue gases at a suitable atmospheric elevation has led to stacks as tall as 1,000 ft. High issuing jet velocities help in dispersion. Adjacent topography and structures will drastically influence the diffusion pattern of the released flue gases.

Mechanical draft is necessary with heavy fires, high firing rates, and low flue-gas temperatures. Fans and blowers are used in conjunction with a short stack. In forced draft the air is delivered for combustion under positive pressure, while with induced draft the fan or blower operates with negative pressure at its inlet to suck the gases through the boiler setting. Balanced draft combines forced and induced draft to give atmospheric pressure, minus 0.1 in of water maximum, in the furnace. Many large modern central-station-type boilers operate with forced draft only (e.g., 60 in of water), which requires a gastight casing but saves on auxiliary power (approximately 1% of the gross output of an electric central station). Draft requirements for a steam generator vary with the gas-flow quantity (Fig. 8-6*a*). Steam jet blowers are used on small installations where inefficiency is insignificant and low investment, no moving parts, simplicity, and reliability are the objectives. Centrifugal fans prevail in all larger installations, with performance characteristics shown in Fig. 8-6*b*.

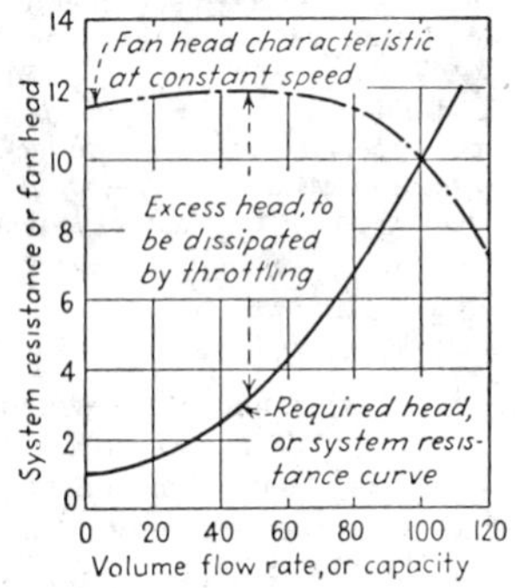

FIG. 8-6*a*. System resistance and fan head characteristics.

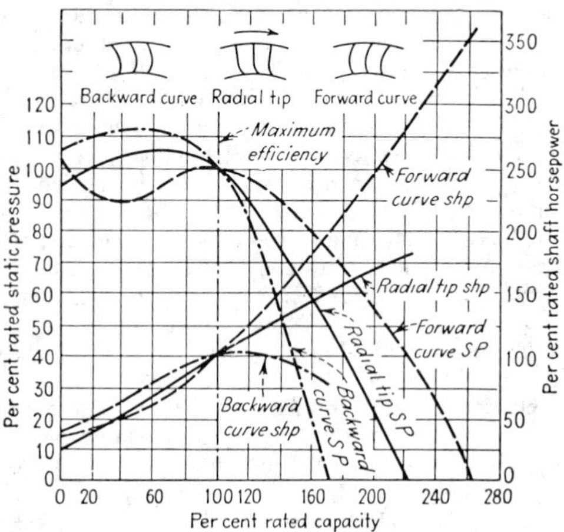

FIG. 8-6*b*. Comparative characteristics for centrifugal fans—percentage basis.

Scale Multiplied by 25 000: 0.001 Inch; 0.01 Inch; 0.1 Inch; 1 Inch; 10 Inches; 8⅓ Feet; 83⅓ Feet; 333⅓ Feet

Tyler Standard Screen Scale: Meshes per Inch 325 200 150 100 65 48 35 28 20 14 10 8 6

Diameter of Particles in Microns: 0.001 0.01 0.02 0.04 0.06 0.08 0.1 0.2 0.4 0.6 0.8 1 2 4 6 8 10 20 40 60 80 100 200 400 600 800 1000 2000 4000

Scale of Atmospheric Impurities:
Fumes
Dusts
Smokes
Fog
Mist
Drizzle
Rain
Permanent Atmospheric Impurities
Temporary Atmospheric Impurities
Heavy Industrial Dust
Bacteria
Plant Spores
Average Size of Smoke Particle
Dust Causing Lung Damage
Pollens Causing Hay Fever
1/128″ 1/64″ 1/32″ 1/16″ 1/8″
Mean Free Space Between Gas Molecules
Ultra Microscope
Microscope
Particles Larger Than 10 Microns Seen with Naked Eye
Electrical Precipitators
Air Filters
Dust Arresters
Centrifugal Cleaners
X-rays
Visible
Infrared
Hertzian Waves

Rate of Settling in Fpm for Spheres of Density 1 at 70°F		0	0	0	0	0.00007 = 3/64″	0.002 = 1.4″ per hr	0.007 = 5″ per hr	0.148	0.592	14.8	59.2	555	790
Number of Particles in 1 Cu Ft	Air Containing 0.0006 Grains of Impurities per Cu Ft (Density = 1)	75×10^{15}	60×10^{13}	75×10^{12}	60×10^{10}	75×10^{9}	60×10^{7}	75×10^{6}	600 000	75 000	600	75	0.6	0.075
Surface Area in Square Inches		365 ≅ 2.53 sq ft	73.0	36.5 ≅ 1/4 sq ft	7.3	3.65 ≅ 1.9 in. sq	0.73	0.365 ≅ 5/8 in. sq	0.073	0.0365 ≅ 3/16 in. sq	0.0073	0.00365 ≅ 1/16 in. sq	0.00073	0.000365

Based, in part, on information from W. G. Frank, and from *Chemical and Metallurgical Engineering*, Vol. 45, p. 133.

FIG. 8-7. The dust spectrum. (*Westinghouse Electric Corporation.*)

For a given fan the performance at some other speed can be determined for a chosen point on the efficiency curve by the rules

$$\text{Capacity, ft}^3/\text{min} \propto \text{rpm} \tag{8-11}$$

$$\text{Head, in of water or ft of air} \propto (\text{rpm})^2 \tag{8-12}$$

$$\text{Horsepower, air or shaft} \propto (\text{rpm})^3 \tag{8-13}$$

Draft control is obtained (1) by throttling excess pressure as in Fig. 8-6*a*, (2) by variable speed, utilizing the principles of Eqs. (8-11) to (8-13), or (3) by inlet vane control. The last two methods save on power but are more costly with the prevalent motor drives. Hydraulic and magnetic couplings find favor in many cases.

11. Dust Catchers and Cinder Traps. The wanton contamination of the atmosphere by the discharge of cinders, fly ash, smoke, and noxious gases is no longer tolerated in enlightened communities. Proper control of combustion will reduce smoke. Low-sulfur fuels will reduce the toxic effects of SO_2 and SO_3. With coal firing, as much as 90% of the refuse contained in the fuel may appear in the stack gases as fly ash and cinders (Fig. 8-7). Particles range in size from ⅛ in across to less than 10 microns (limit of visibility with the naked eye). Mechanical separators which operate on the inertia principle are most effective with coarse particles and heavy dust loadings. Electrostatic precipitators are especially helpful with finer particles (less than 40 microns). Equipment is often large, bulky, and expensive, especially with high separation efficiency (greater than 90%). Combination mechanical and electrostatic designs, with equipment in series, often serve to give a good economic solution. A residual dust loading below 500 grains/1,000 ft³ can be obtained—a loading which is acceptable under many air-pollution control codes.

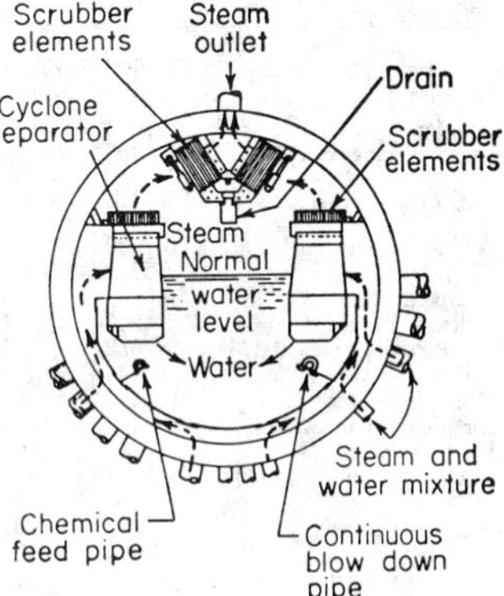

Fig. 8-8. Drum internal arrangement with cyclone steam separators and scrubber elements.

Increasingly severe legal air-pollution standards are to be expected in all phases of purity control of the atmosphere.

12. Boiler Accessories. Accessories and instruments are required by law or by practice to give safer or improved operation of a boiler. These include (1) pressure gages of the Bourdon-tube type; (2) water columns, with or without alarms, indicating drum level directly or by television; (3) safety valves with tamperproof adjustable settings for relieving pressure and for limiting extent of the blowdown; (4) feed-water piping system containing stop valves, check valves, and level regulators; (5) steam take-off system containing scrubbers and driers which can reduce the solid contents below 3 ppm (see Fig. 8-8); (6) stop and nonreturn valves for manual and automatic sectionalizing of the boiler plant; (7) blowdown system of the intermittent or continuous type to limit the solids content of the boiler water (1,000 to 2,000 ppm maximum); (8) soot blowers used periodically to clean the gas side surfaces by high-velocity air or steam jets (100 to 300 lb/in²); (9) water side cleaning by mechanical or chemical means; (10) operating instruments and automatic controls.

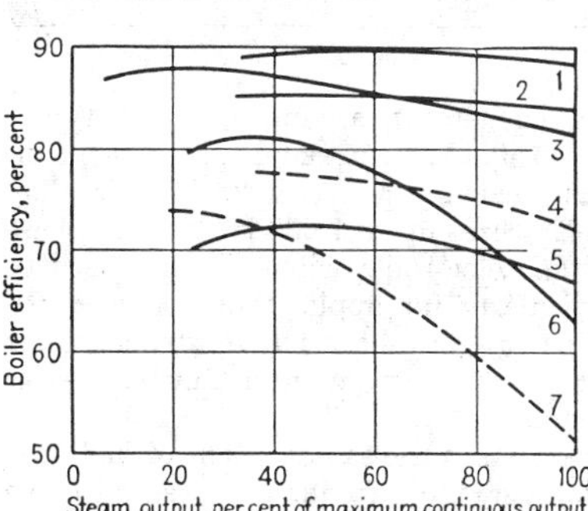

Fig. 8-9. Selected boiler efficiency curves. (1) Coal-fired drum boiler, economizer, and air heater. (2) Gas-fired drum boiler, economizer, and air heater. (3) Oil-fired naval boiler (destroyer). (4) Coal-fired industrial boiler, no air heater. (5) No. 3 buckwheat-fired header boiler, no economizer or air heater. (6) Bituminous-fired four-drum boiler, economizer, no air heater. (7) Coal-fired locomotive boiler.

12a. Boiler Performance. Boiler tests comply with the ASME Power Test Code. Some selected results of tests are shown in Table 8-7. Efficiency variation with load is shown in Fig. 8-9. The efficiency of boilers is subject to wide variation, even under test conditions, because of (1) fuel, (2) method of firing, (3) presence of heat-recovery equipment, (4) condition of the equipment, and (5) skill of the operator. The range can be less than 50 to more than 90%, depending essentially upon the economics, bulk, weight, and reliability limitations. Actual boiler efficiency may be one to three points less than test efficiency because of starting, stopping, and banking losses. Availability of boilers is high (95%) as reflected in the unit construction system (one boiler to one turbine) and the omission of spare boilers in current installations.

Table 8-7. Selected Boiler Performances

	(1)	(2)	(3)	(4)	(5)	(6)	(7)		(8)	
Boiler type	Open pass	Radiant	Radiant	Radiant-controlled circulation	Cyclone	Cyclone	Industrial		Industrial	
Size, lb per hr	930,000	1,150,000	500,000	1,450,000	990,000	400,000	140,000		265,000	
Pressure, psi	2,080	1,550	1,550	1,875	2,050	1,250	420		685	
Temperature, F	1,050	1,005	1,005	1,000	1,050	830	620		730	750
Reheat temperature, F	1,000	1,005	1,005	1,000	1,000					
Fuel type	Bit. Ill.	Gas	Oil	Bit. WVa	Bit, WVa	Bit, WVa	Coal	Oil	Oil	Gas
Heating value, Btu	10,900	1,062	18,270	13,700	13,000	12,960	14,500	18,400	18,500	2,300
Firing method	PC	Tan	Tan	PC	Cyclone	Cyclone	PC	Horiz.		
Temperatures:										
Stack, F	246	303	311	250	295	300	370	370	345	330
Air leaving air heater, F	552	600	461	555	695	715	425	415	513	485
Feed, F	441	462	454	500	547	255	425	415	320	320
Draft, in. water	11.7	9.40	10.35	9.90	56.2	54.4	7.60	4.45	13.2	14.3
Air pressure, in. water	11.0	9.30	7.25	8.80			5.60	6.05		
Heat balance:										
Absorbed in steam, %	89.1	84.05	87.04	90.72	89.9	90.2	86.80	85.30	86.2	80.2
Dry-gas loss, %	3.6	3.93	4.95	3.46	3.8	3.3	7.05	6.47	5.8	4.9
Hydrogen loss, %		10.21	6.03	3.50			2.60	5.94		
Moisture in air, %	5.1	0.11	0.13	0.08	4.4	4.3	0.18	0.17	6.0	12.9
Surface moisture, %		....	....	0.31			0.25			
Incomplete combustion, %	0.4	....	....	0.30	0.1	0.3	1.00			
Combustible in refuse, %										
Radiation and unaccounted for, %	1.8	1.7	1.85	1.63	1.8	1.9	2.12	2.12	2.0	2.0

13. Electric Steam Generators.[1] Electric steam generators are made in three types: (*a*) Immersion-heater type, made in sizes up to 750 kW, with multiple immersion-heater units rated up to 75 kW, 575 V, 3-phase, each. (*b*) With 80Ni-20Cr alloy resistors mounted in the tubes of a fire-tube boiler. Sizes up to 1,000 kW, single-phase or 3-phase, standard low voltage. (*c*) Low-voltage electrode boilers are competitive with boilers of the immersion-heater type, particularly for applications, such as for intermittent services, in which their overload capacity may be used to advantage for raising steam in a minimum of time. They also have an advantage in having greater mobility for mobile units such as steam-jet cleaners.

The auxiliary equipment of electric steam generators is the same as that of fuel-fired steam generators. Full automatic control is, however, more frequently used with the electric steam generator.

The main use of low-voltage electric steam generators is for the supply of steam at points which cannot conveniently be reached by steam pipelines and in locations where a fuel-fired unit is not permissible.

Electrode steam or HTHW (high-temperature hot-water) boilers are installed in industrial plants, and in northern mining districts, in sizes of 10 to 40 MW and operated at 6.6 to 13.2 kV for the utilization of surplus hydroelectric power for process or space

[1] By M. Eaton, Consulting Electrical Engineer, Shawinigan, Que.

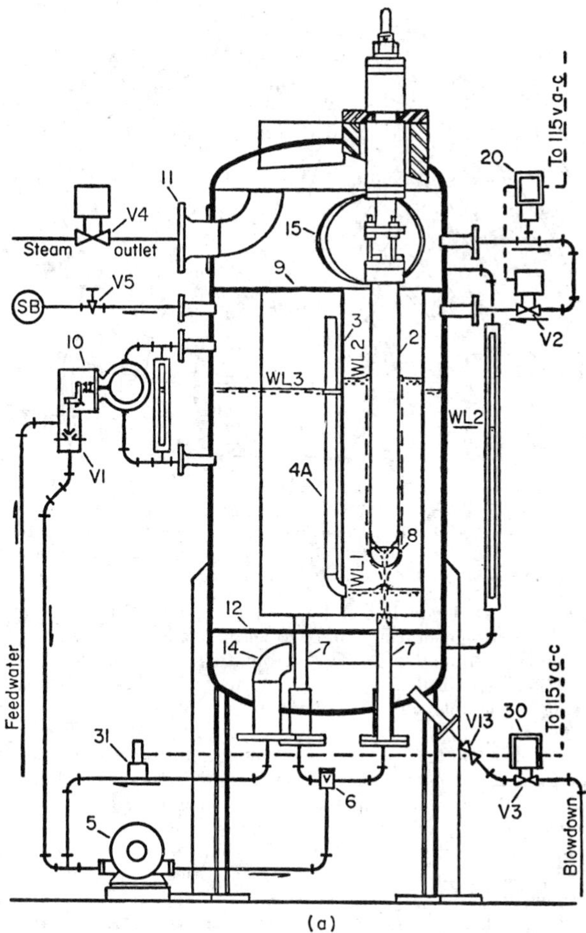

FIG. 8-10*a*. Sectional view of Eaton automatic electrode boiler. (*Canadian General Electric Company*.)

heating. Smaller electrode boilers (down to 500 kW, 2.3 kV) have wider economical applications, such as for the use of off-peak power, for intermittent or standby services, and for installations in urban districts having air-pollution regulations. Their operating efficiency depends on the heat losses in the blowdown required for boiler-water conductivity control (no blowdown is required if all the condensate is returned to the boiler) and is within the range of 92 to 98% (3.0 to 3.2 lb of steam/kWh).

Figure 8-10*a* shows schematically an automatic high-voltage electrode boiler designed to meet adverse operating conditions, viz., (*a*) the high and negative temperature coefficient of water resistivity (about 4.5%/°C from 25°C) causes operation on fluctuating loads to be inherently unstable and difficult to regulate; (*b*) if the water level falls below the electrodes, a flashover occurs (unless means is provided to prevent it), making the safe minimum load about 15% of full load; (*c*) the energy concentration in the water in contact with the electrode tips is about three times greater than along their vertical sides. If a critical value is exceeded, arcing contact through steam causes accelerated electrode wastage. This condition limits either the operating voltage or the boiler-water conductivity or both. Three round, tip-shielded electrodes, 2-8, at 120° spacing, are concentric with cylindrical neutrals 3 opening into a diaphragm plate 9. The load

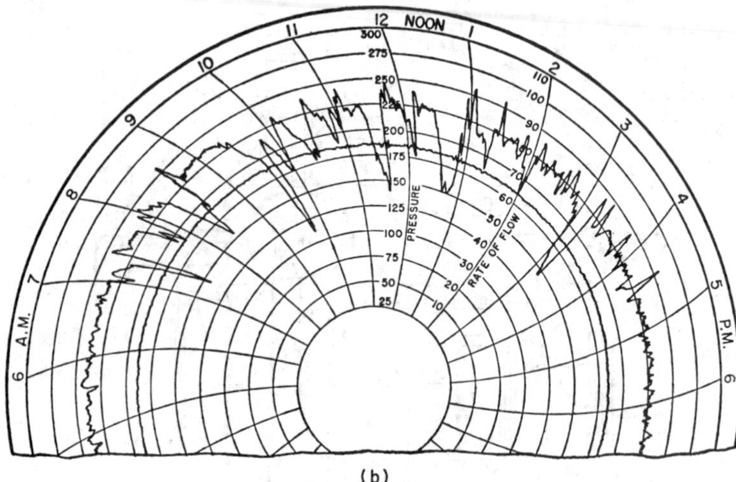

FIG. 8-10*b*. Meter chart from a 30-MW electrode boiler supplying steam for a paper mill. The outer curve is the boiler load in steam flow; the inner curve shows automatically controlled steam pressure (185 lb/in² gage). The fluctuations in demand are caused by interruptions in steam flow to the dryers, caused by paper breaks.

is determined by the electrode immersion and is regulated by automatic transfer of water between the steam-generating compartments, inside the neutrals, and the control compartment, outside the neutrals. The automatic steam pressure-control equipment provides proportional-speed floating-controller action, with the water level on the electrodes serving as the final control element, plus self-regulation inherent in the boiler design. The control performance is illustrated by Fig. 8-10*b*. The boiler functions as a source of steam or HTHW at constant pressure and/or boiler-water temperature. When it is operated as a steam boiler, off-peak power control is obtained by a power controller having single-speed floating action (and neutral zone), with valve V4 serving as the final control element. When it is used as a source of HTHW, the load is limited by bypassing water returned from the heating system to an associated fuel-fired boiler.

A pump 5 takes water from the boiler through outlet 14 and feed water, controlled by regulator 10, which is discharged through nozzles 7 toward the tips of the electrodes. The electrode-tip shields 8 function to maintain effective electrical contact between the electrodes and the jets, which at low water levels would bounce off unprotected electrode tips, with resultant sparking contact and electrode wastage. They also make the only vulnerable parts of the electrodes conveniently replaceable. This water-jet action functions (*a*) to maintain the electric circuit as the water level falls below the electrodes, thereby obtaining a load range down to 2% of full load, and (*b*) to increase the permissible energy concentration at the electrode tips by about 250%. Since the energy concentration is proportional to I^2R, the water-jet action permits either an increase in the boiler voltage of nearly 60% or an increase in the boiler-water conductivity of about 250%. The permissible increase in voltage (depending also on the size of the boiler) may make it possible to operate the boiler at the power-distribution voltage, thus saving the installed cost of a boiler transformer. If the boiler-water conductivity is increased, the rate of blowdown for conductivity control is proportionally decreased, with a corresponding increase in efficiency. A conductivity controller 30, with conductivity cell 31, functions to maintain the boiler-water conductivity at the value for which the boiler is designed. Electrode boilers in sizes up to 2,000 kW are equipped with the type of control gear shown. Larger boilers are supplied with pneumatic (or equivalent) controllers. For more detailed description see patents:

United States 2,611,852, 2,729,738, 2,986,623; Canadian 518,831, 550,072, 626,610.

STEAM PRIME MOVERS

14. Steam Engines and Steam Turbines. Steam prime movers are either reciprocating engines or turbines, the former being the older, dominant type until 1900. Slow speed (100 to 400 r/min), high efficiency in small sizes (less than 500 hp), starting torque, and foolproofness are the advantages of engines. Turbines offer high speed (1,800 to 25,000 r/min); maximum size (1,000,000 kW); minimum floor space, bulk, and weight; maximum efficiencies (80% +) in larger sizes; suitability for highest steam pressures (5,000 lb/in²), steam temperatures (1200°F), vacua (0.5 in Hg abs). All large units (above 1,000 kW) are turbines, but each type has utility, e.g., engines to drive reciprocating pumps and compressors, turbines for electric generators, centrifugal pumps, and turboblowers.

15. Engine Types and Applications. The former great diversity in engine types is today reduced so that (1) simple D-slide engines (less than 100 hp) are used for auxiliary drive; (2) single-cylinder counterflow and uniflow engines (less than 1,000 hp) with Corliss or poppet-type valve gear are used for generator or equipment drive in factories, office buildings, paper mills, hospitals, laundries, and process applications where noncondensing by-product power operations prevail. Multiple-expansion multicylinder constructions are largely obsolete except for some marine applications. While engines as large as 7,500 kW have been built and are still found in service, the field is generally limited today to engines less than 500 kW in size. Engine governing is by flyball or flywheel types to (1) throttle steam supply or (2) vary cutoff.

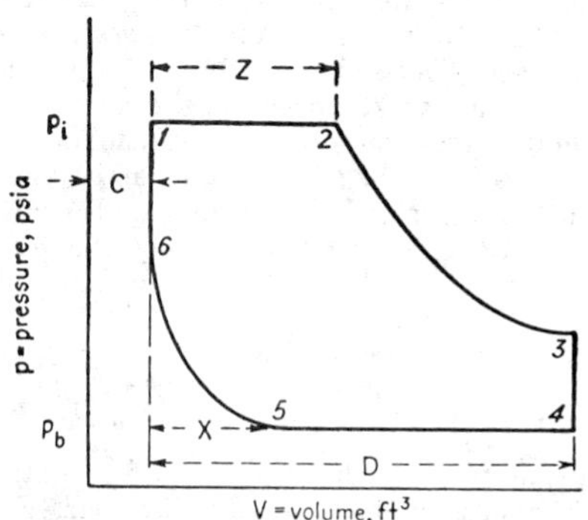

Fig. 8-11. Pressure-volume diagram for steam-engine cycle. Phase 1-2, constant-pressure admission at p_i; 2-3, expansion, $pv = C$; 3-4, release; 4-5, constant-pressure exhaust at p_b; 5-6, compression, $pv = C$; 6-1, constant-volume admission.

Where	Usual Value
D = displacement	0.05–20 ft³
C = clearance, fraction of D	0.03–0.2
Z = cutoff, fraction of D	0.1–0.6
X = compression, fraction of D	0.1–0.8
p_i = initial pressure	100–300 psia
p_b = back pressure	2–30 psia
p_m = mean effective pressure	50–125 psi

16. Engine Performance. The basic thermodynamic cycle is shown in Fig. 8-11. The net work of the cycle is represented by the area enclosed within the diagram and is conveniently represented by the mean effective pressure (mep), i.e., the net work (area) divided by the length of the diagram. The power output is computed by the "plan" equation, viz.,

$$\text{hp} = p_m Lan/33{,}000 \tag{8-14}$$

where hp = horsepower, p_m = mep, pounds per square inch, L = length of stroke, feet, a = net piston area, square inches, and n = number of cycles completed per minute.

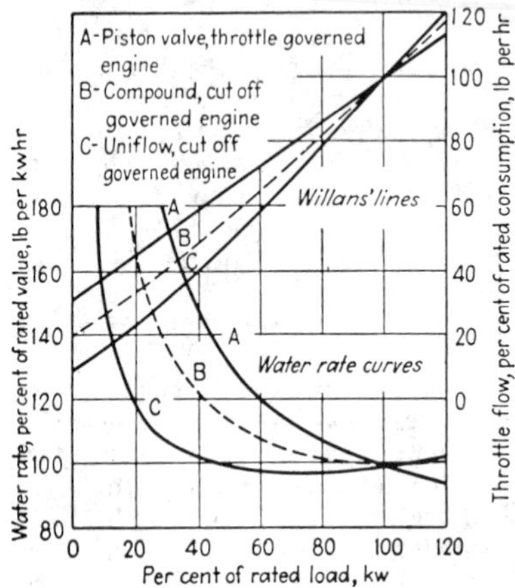

Fig. 8-12. Water-rate curves and Willans' lines for typical steam engines.

The theoretical mep and horsepower are larger than the actual indicated values and are customarily related by a diagram factor ranging between 0.5 and 0.95. The shaft or brake mep and horsepower are lower still, with mechanical efficiency ranging between 0.8 and 0.95. Representative actual water rates (pounds per kilowatthour) and Willans' lines (pounds steam per hour) are shown in Fig. 8-12.

17. Steam Turbines, General. The steam turbine is a product of the twentieth century, built in sizes of 20 hp to 1,000,000 kW. It completely dominates the field of power generation in large blocks. Steam turbines are essentially a series of calibrated nozzles in which kinetic energy is generated, transferred to disks or drums, and delivered at the end of the rotating shaft as usable power. Turbines are built in two distinct types, (1) impulse and (2) reaction. With impulse turbines the nozzles are stationary, but with reaction turbines moving nozzles are also present. Vanes or buckets are arranged about the periphery of a drum or disks to form annular passages as illustrated in Fig. 8-13. If the passage between adjacent blades shows a throat (minimum) section (Fig. 8-13*D*), the assembly is a nozzle. If such nozzles occur on the rotor and in the stationary rows, the machine operates on the reaction principle. Reaction turbines must have full circumferential admission because of the pressure drop across the row of moving buckets. Partial circumferential admission may be used on impulse turbines. As blades must be not less than ⅜ or ½ in in height for good efficiency, the reaction principle is limited to large turbines (more than 5,000 kW) or to the vacuum end where steam volumes are large.

The velocity of steam through a nozzle is calculated by

$$\text{Jet velocity, ft/sec} = 223.7\sqrt{\Delta h} \qquad (8\text{-}15)$$

where Δh is the enthalpy drop, Btu per pound, between initial and final conditions. Δh is most conveniently evaluated by use of the Mollier chart (Fig. 8-14), with isentropic expansion as the ideal limit. Jet velocities may range to 5,000 ft/sec. Bucket

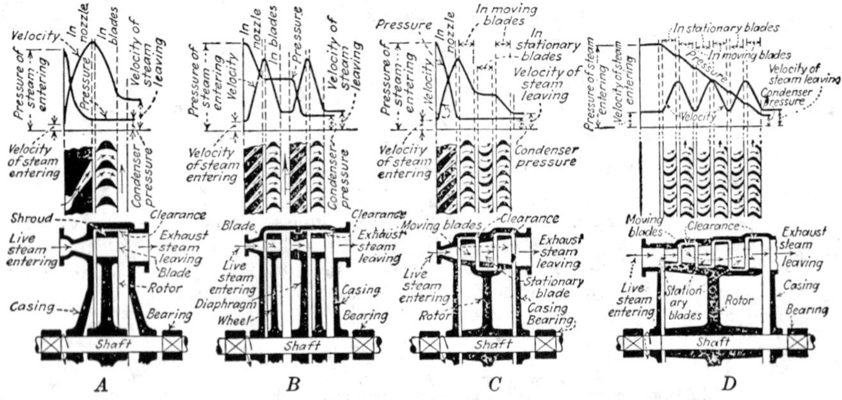

Fig. 8-13. Basic steam-turbine types. (*A*) Simple impulse turbine. (*B*) Pressure staged (Rateau) impulse turbine. (*C*) Velocity staged (Curtis) impulse turbine. (*D*) Reaction (Parsons) turbine.

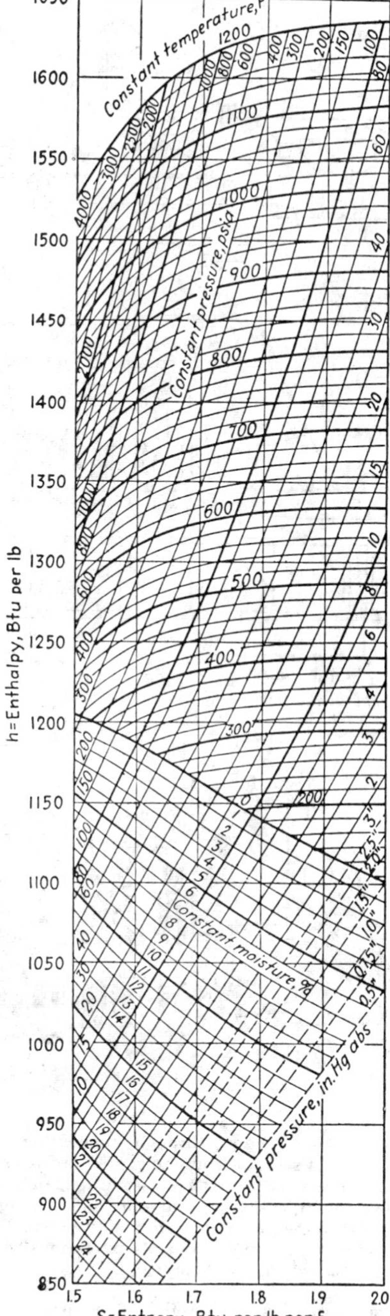

Fig. 8-14. Mollier chart for steam. (*Adapted from Power Handbook, July*, 1951, *p*. 93. *Recommended publication for complete properties of steam, Keenan and Keyes,* "*Thermodynamic Properties of Steam,*" *Wiley.*)

P-V-T Values for Dry Saturated Steam

Pressure, in. Hg abs	Temp., F	Sp. vol., cu ft per lb
0.5	58.8	1,256
1.0	79.0	652
1.5	91.7	445
2.0	101.1	339
3.0	115.1	232
Pressure, psia		
14.7	212.0	26.8
50	281.0	8.52
100	327.8	4.43
200	381.8	2.29
300	417.3	1.543
400	444.6	1.161
600	486.2	0.770
900	532.0	0.501
1,200	567.2	0.362
1,800	621.0	0.218
2,400	662.0	0.141
3,206	705.4	0.050

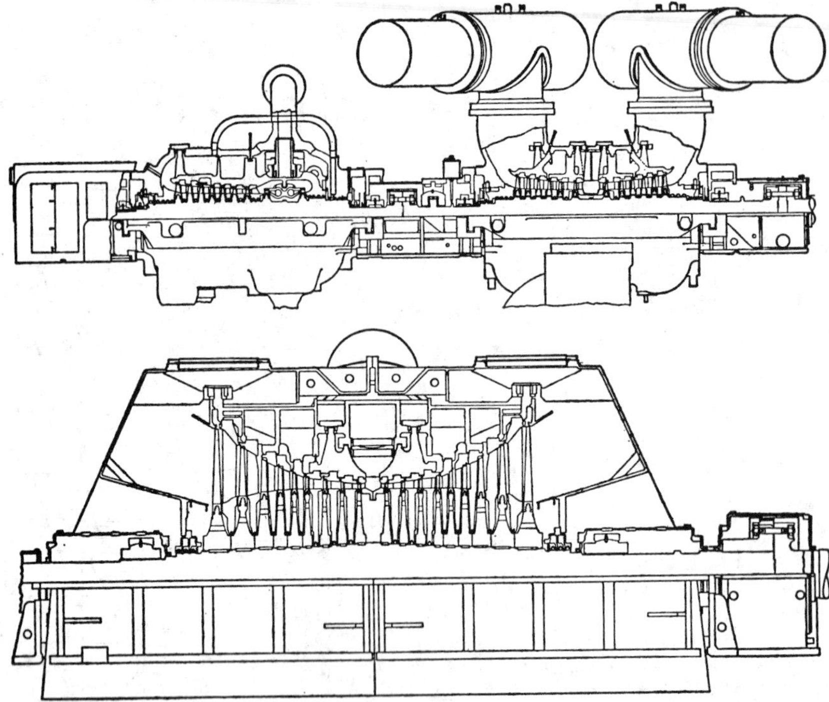

FIG. 8-15. Section through a large high-pressure high-temperature reheat steam turbine. (*General Electric Company*.)

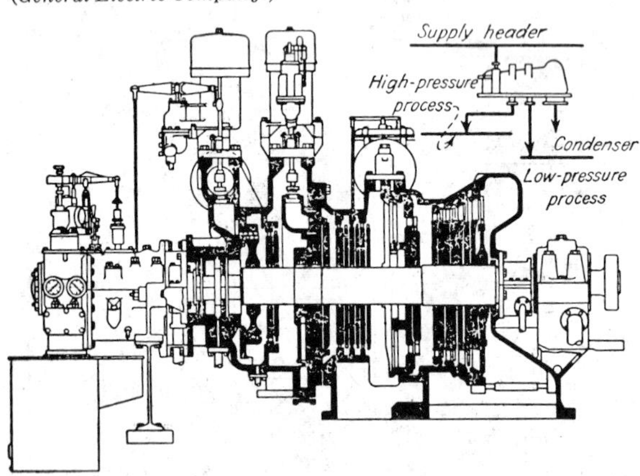

FIG. 8-16. Section through condensing double-automatic-extraction steam turbine to supply process steam at two pressure levels. Steam and electric demands can be varied independently. Poppet valve at high-pressure and grid valve at low-pressure extraction points, respectively. Three-arm governor mechanism controls extractions and speed. (*Power*, *and General Electric Company*.)

(peripheral) speeds are usually 500 to 600 ft/sec, with 1,200 to 1,500 ft/sec as maximum. Single velocity stage (Fig. 8-13*A*) impulse blading runs at 0.45 jet speed; double-row velocity staging (Curtis) (Fig. 8-13*C*) runs at 0.2 to 0.22 jet speed; and reaction (Parsons) (Fig. 8-13*D*) staging runs at 0.9 jet speed for the usual 10° to 20° exit-vane angles. Impulse turbines may also be pressure-staged (Rateau) (Fig. 8-13*B*) with a single wheel in a cellular construction. Reentry velocity staging on a single wheel is used on some small units, usually less than 100 hp. Multiple-row staging, as illustrated in Figs. 8-15 and 8-16, is used in all larger units where efficiency and speed are prime considerations.

18. Turbine Types. Vertical turbines, favored prior to 1914, are no longer built. Horizontal shaft machines may be (1) condensing or noncondensing, (2) single shaft or multiple shaft, (3) single or multiple cylinder, (4) single or double flow, (5) extracting or nonextracting, (6) controlled (automatic) or uncontrolled extraction, (7) reheat or nonreheat, (8) throttle- or multiple-nozzle-governed, (9) constant or variable speed, (10) single or double shell (see Figs. 8-15, 8-16, and 8-17).

19. Turbine Construction. A turbine consists essentially of three parts: (1) a rotor with blading, (2) a bottom half, and (3) a top half of the stationary element each containing blading. Blade-material selection is determined by operating conditions: carbon steel to 850°F; ferritic to 1050°F; austenitic to 1200°F. Alloy steel is preferred even for lower initial temperatures. At the exhaust end of condensing turbines, alloy steel or stellite inserts may be used on the entering edge to reduce moisture erosion. Blades are generally machined from solid stock. The rotor is (1) solid forging or (2) built-up shaft and wheel construction (Figs. 8-15 and 8-16). The casing is commonly of cast steel at the high-pressure end, with cast iron and fabricated construction for the exhaust section. With high temperatures a double- rather than single-shell construction is preferred. The rotor is connected to the generator through a solid or a flexible coupling.

20. Turbine Governors. Most turbines run at constant revolutions per minute so that governors are speed-responsive. Small-capacity turbines (less than 250 kW) use direct-acting governors (Fig. 8-18). Centrifugal action moves the governor weights outward, compressing the axial spring and displacing the crank lever and throttle valve in the steam line. For large machines the power and sensitivity of this type of governor are inadequate, and a relay type of hydraulic governor (Fig. 8-19) is substituted. The primary speed-change impulse from a centrifugal governor positions

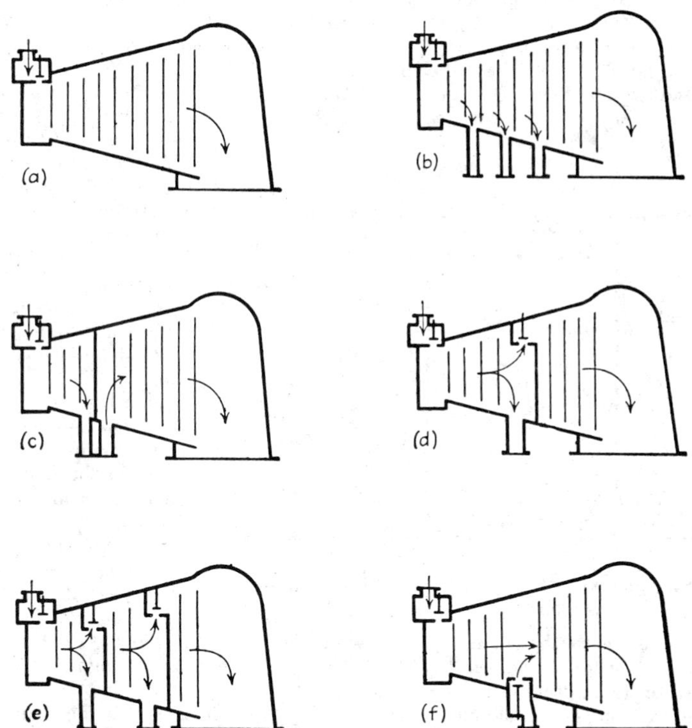

Fig. 8-17. Schematic drawings of important basic steam-turbine types. Similar designs are built for noncondensing service. Types *d*, *e*, and *f* are particularly suited to industrial applications calling for process steam and "by-product" power. (*a*) Straight condensing. (*b*) Condensing bleeding or extracting (three-stage). (*c*) Reheat condensing. (*d*) Single automatic extraction, condensing. (*e*) Double automatic extraction, condensing. (*f*) Mixed pressure, condensing.

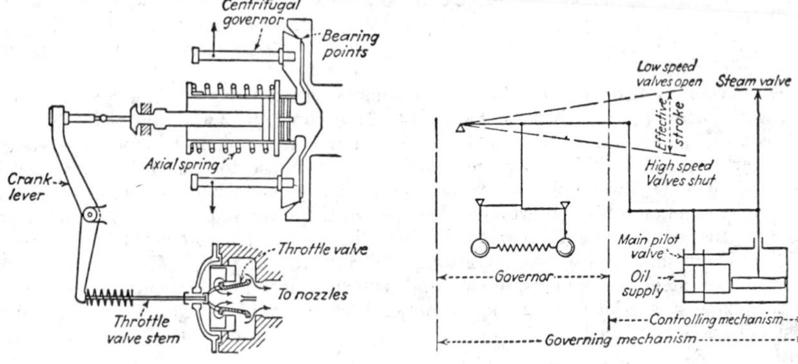

FIG. 8-18. Mechanical governor for small turbines.

FIG. 8-19. Governor with hydraulic power amplifier.

the pilot valve. Oil under pressure is admitted to either side of a large-diameter power piston which opens or closes the steam-admission valves. Large turbine manufacturers recommend that (1) governor sensitivity shall permit governor response with not more than 0.02 of 1% speed change; (2) regulation shall be adjustable between 2 and 6% of normal speed; (3) turbine speed shall not exceed 110% upon loss of full-rated load; and (4) adjustment shall be possible between 98 and 102% full-load speed. An overspeed trip which automatically closes the steam-admission valves at 110% normal speed customarily supplements the speed governor.

21. Lubrication System. Forced-feed lubrication is usual on turbines above 200 hp in size. Figure 8-20 illustrates such a system schematically. The system also supplies pressure oil for the governor mechanism. The auxiliary pump supplies oil, in starting and stopping, for bearings and for operation of throttle or stop valve. Auxiliary oil pumps are motor-driven on reheat units and usually steam-driven on other units. The shaft-driven (main) oil pump is cut in and out by the action of oil pressure to carry the load in the region of normal operating speed. Oil coolers transfer the heat from the bearings with entering oil temperatures between 120 and 160°F and should be proportioned for use with the highest cooling-water temperatures (80 to 100°F). Typical summary oil specifications are given in Table 8. In modern high-pressure, high-temperature steam practice the fire hazard is greatly increased so that seamless tubing, with welded joints, throughout, is justified. Supply lines should be routed inside the return line, and oil tanks should be protected by foam, gas, or fog systems to reduce fire hazards.

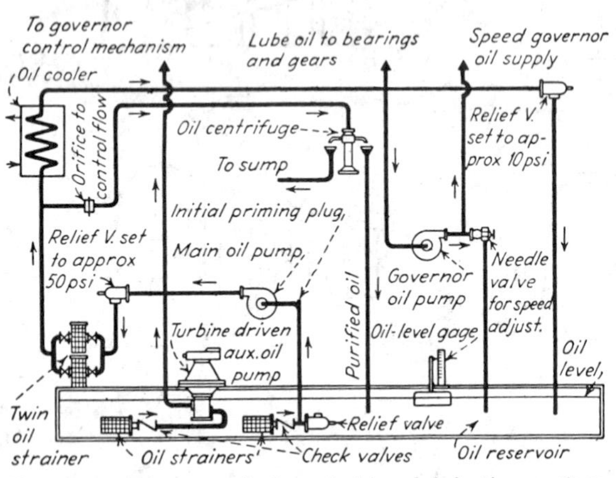

FIG. 8-20. Diagram of steam-turbine lubricating system. (*Power*.)

22. Generator Cooling. Generators are cooled by air or hydrogen circulated by fans mounted on the generator rotor. Hydrogen cooling is used for units larger than 15,000 kW. Liquid-cooling and hollow conductors are usual on the largest machines. Closed systems are preferred with any fluid, with the gas recirculated to windings

Table 8-8. Typical Summary Lubricating-oil Specifications for Steam Turbines (*Power*)

A. New Turbine-oil Analyses

Test	Light	Medium	Heavy
Gravity, API	30	27	27
SAE number	.....	20	30
Saybolt viscosity at 100 F	150	300	400
Saybolt viscosity at 210 F	43	51	58
Flash point, F	350	420	430
Fire point, F	455	495	500
Pour point, F	30	30	30
Carbon residue (Conradson)	0.01	0.08	0.25
Demulsibility at 130 F	1,620	1,620	
Demulsibility at 180 F	.....	.....	1,620
Steam-emulsion number	25	40	80
Neutralization number	0.05	0.05	0.05

B. Lubricating-oil Recommendations

Ring-oiled bearings

Temperature	Saybolt viscosity, 100 F	SAE number
Normal	150–300	Below 10–20
Above 130 F	800–2,000	50–70

Circulating systems

Turbine type	Saybolt viscosity, 100 F	SAE number
Direct-connected	150–200	Below 10–20
Geared (single reduction)	300–350	20
Geared (double reduction)	400–550	30–40

after cooling in surface coolers, where condensate or service water absorbs heat amounting to 75 to 90% of the generator electrical losses. Hydrogen is particularly effective with decreased windage losses and higher heat-transfer rates. Pressures run to 45 lb/in². The system is purged with CO_2 to avoid explosion hazards. Leakage through seals is small, and oil must be separately purified. Care must be exercised in the handling, storage, and maintenance of hydrogen, but hazards are practically eliminated in well-designed and well-maintained systems.

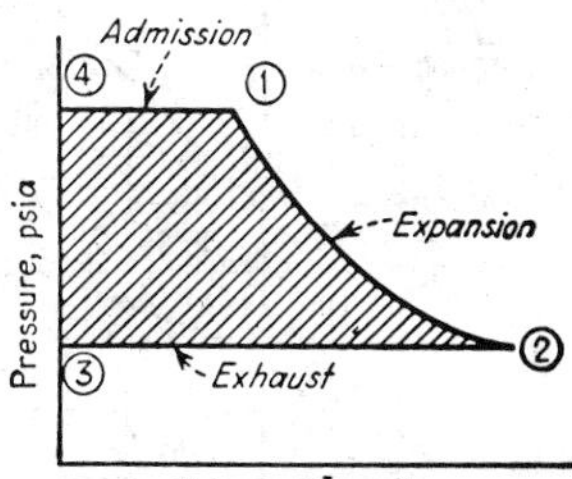

FIG. 8-21. Pressure-volume diagram for the Rankine cycle. Phase 4-1, constant-pressure admission; 1-2, complete isentropic expansion; 2-3, constant-pressure exhaust. Cross-hatched area represents the work of the cycle.

23. Engine Efficiency. The Rankine cycle (Fig. 8-21) defines the ideal performance of a steam turbine. The work is most conveniently evaluated by the use of the Mollier diagram (Fig. 8-14), where

$$\Delta W = h_1 - h_2 \tag{8-16}$$

and ΔW = Rankine-cycle work in Btu per pound' h_1 = steam enthalpy at throttle in Btu per pound, h_2 = steam enthalpy at exhaust in Btu per pound, and h_1 and h_2 are at the same entropy (vertical line). The actual work of a real turbine is less than the ideal Rankine-cycle work, with engine efficiency defined as

$$\text{Engine efficiency} = \frac{\text{actual work, Btu/lb}}{\text{Rankine-cycle work, Btu/lb}} \tag{8-17}$$

The steam rate or water rate, *WR*, in pounds per kilowatthour, follows from these two relations as

$$WR = \frac{3{,}412.75}{\Delta W} \tag{8-18}$$

and ΔW is either the Rankine-cycle work [Eq. (8-16)] or the actual work [Eq. (8-17)].

Engine efficiency is variously expressed as determined by the allocation of losses, such as nozzle friction, blade friction, windage, throttle, leakage, mechanical, and exit (leaving) losses. Representative engine efficiencies are shown in Fig. 8-22 for small turbines and Fig. 8-23 for large turbines. Efficiency generally increases with size, but correction factors for items such as steam pressure, steam temperature, vacuum, reheat, speed, and load should be obtained from turbine manufacturers.

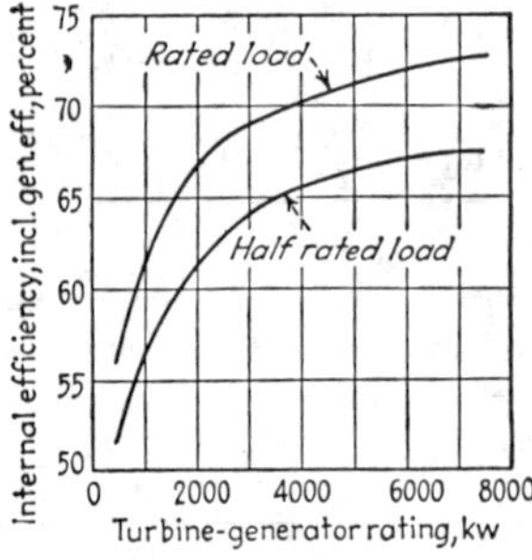

Fig. 8-22. Internal-engine efficiency of small-capacity condensing turbine-generators (3,600 r/min).

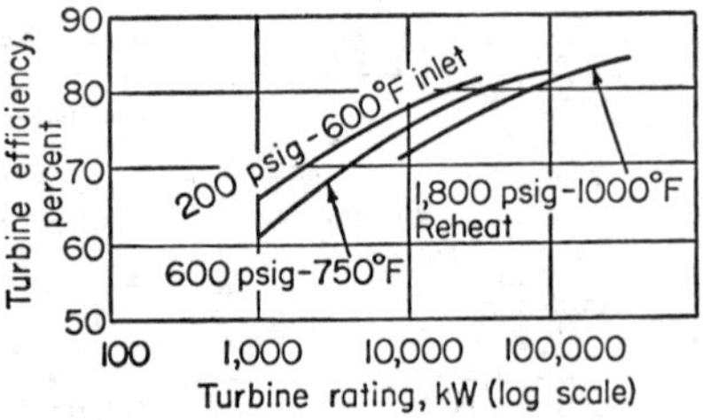

Fig. 8-23. Turbine efficiencies vs. capacity.

24. Willans' Lines and Water-rate Curves. Turbine performance, like reciprocating-engine performance (Fig. 8-12), may be shown as water rate, pounds per kilowatthour, or steam consumed, pounds per hour. The latter, Willans'-line basis, is usually preferable, with the lines approximately straight, or with slight concavity upward, or with discontinuities for each valve opening. For estimating purposes the no-load intercept is 2 to 5% of full load with present-day condensing turbines. It may run to 8 or 10% on older condensing units and varies from 20 to 25% on back-pressure units. With automatic extraction turbines, maintaining constant pressure on a predetermined bleed point for the supply of process or heating steam, a nest of steam-flow curves best shows turbine performance (Fig. 8-24). Performance is limited by various elements as shown.

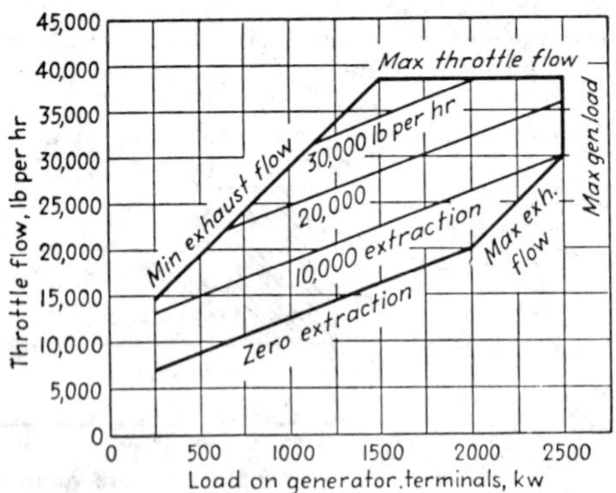

Fig. 8-24. Willans' lines for condensing industrial-type turbine-generator set with single automatic extraction point.

25. Steam-turbine Standards. The largest turbine-generators (ranging to units of a million kilowatts capacity) are of primary interest to the central station production of electric power. They are customarily tailored to the specific installation, with the consequent minimum opportunity for standardization. Small sizes, viz., 25 to 5,000 capacity, are essentially in the hardware category and subject to real standardization. Marine turbines are built to different structural and performance standards from shore turbines because of differences in service requirements, e.g., bulk, weight, speed, flexibility, reversibility, and reliability. Turbines for nuclear-power services incorporate principles for the accommodation of prevalent low steam pressures and temperatures and the avoidance of high moisture content in the lower turbine stages. Turbines supplied with steam from fossil-fuel-fired boilers can utilize maximum pressures (5,000± lb/in²), maximum temperatures (1,100°±F), maximum reheat, and minimum turbine exhaust losses, all for maximum overall conversion efficiency of heat into work. Manufacturers are the best source of factual information.

26. Miscellaneous Turbine Elements and Accessories. Steam leakage between stages and at the ends of shafts is reduced by the use of carbon packing rings in units smaller than 1,500 kW and by labyrinths with steam or water seals in larger sizes. Gland leak-off may be condensed by a heater in the feed cycle or reintroduced at a lower pressure stage in the turbine. Turning gears are installed on units in excess of 10,000 kW capacity to prevent rotor distortion and to shorten the starting period by turning the spindle slowly (1 to 2 r/min) when the unit is out of service. A turbine room crane is necessary for installation and maintenance, with capacity determined by the weight of the largest piece. Foundations are of structural steel or reinforced concrete and of such rigidity as to limit deformation, elastic or otherwise, to a value which will not cause misalignment of the turbine parts. Some designs use the condenser structure to support the turbine, but conventional practice uses a separate turbine foundation. Clearances must be provided for the condenser, generator coolers, oil tanks, oil coolers, and all piping connections.

CONDENSING EQUIPMENT

27. Steam Condensing System—General. Jet or barometric condensers in which steam and circulating water mix are limited to small industrial applications (1,000 kW) where best vacuums are not required (2 to 5 in Hg abs). Surface condensers (Fig. 8-25) are usual in all larger installations (1,000 mW max.) for best vacuums (0.5 to 2.0 in Hg abs.) and where condensate must be conserved. Jet (mixture) condensers have been applied in some large modern installations in conjunction with dry cooling towers.

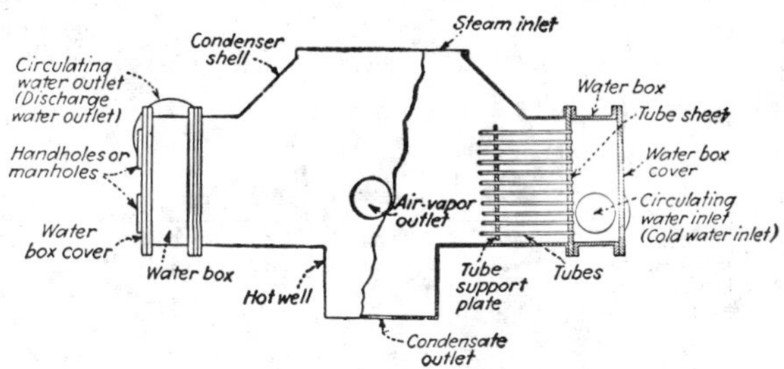

Fig. 8-25. Surface condenser—longitudinal view.

28. Construction. Surface condensers employ welded steel plate for the shell with cast-iron water boxes. Horizontal tube banks using 1±-in-diameter nonferrous tubes and tube sheets are usual. Tubes are generally rolled at both ends, with a slight bowing or with a joint in the shell for expansion. Straight tubes, rolled at one end and with a packing ferrule at the other end, may be used. For fresh water, tube sheets are commonly Muntz or admiralty metal, while tubes are brass, copper, arsenical copper, aluminum brass, or stainless steel. For salt water, admiralty or cupronickel tube sheets are usual with admiralty, aluminum brass, or cupronickel tubes. Tube support plates are of copper-bearing steel. Divided water boxes and water-flow reversal arrangements facilitate maintenance and cleaning.

29. Condenser Support. Attachment of the condenser to the turbine is by (1) direct bolting to the turbine exhaust flange (up to 20,000 kW); (2) direct bolting to the turbine exhaust flange with spring supports under the condenser; (3) solid support of the condenser with the expansion joint between the turbine and the condenser. Condenser weights, when flooded, will run to millions of pounds on large units.

30. Heat Transfer and Condenser Performance. Steam delivered to a condenser usually contains less than 10 to 12% moisture. Heat removal can be taken as 950 Btu/lb when more accurate data are lacking. The circulating-water temperature rise is 5 to 20°F, with terminal temperature differences of the order of 10°. Entering steam velocities range from 25,000 to 50,000 ft/min. Water velocities should not exceed 7 to 8 ft/sec with fresh water and 5 ft/sec with sea water. Steam loadings range from

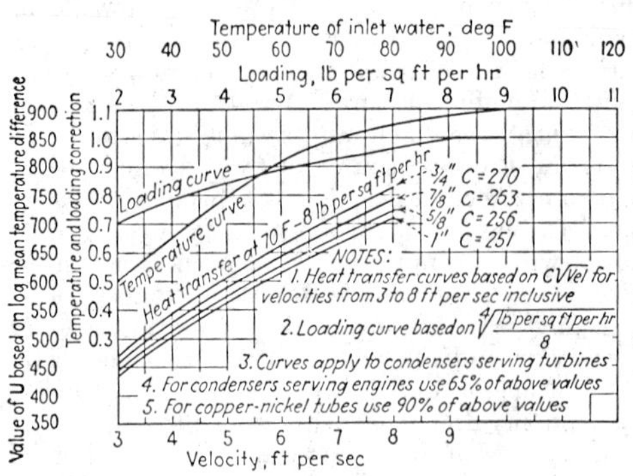

Fig. 8-26. Condenser heat-transfer rates.

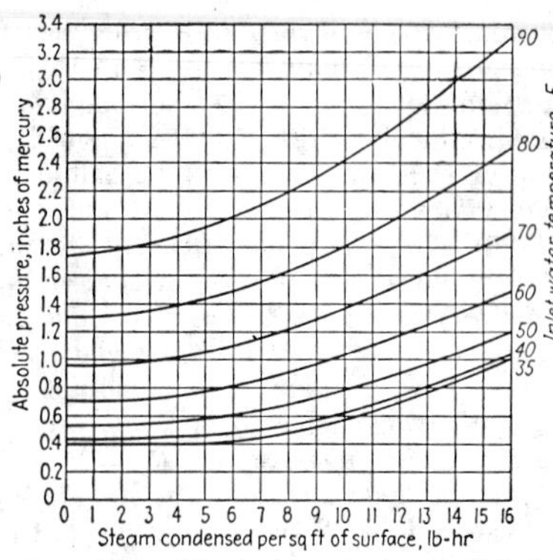

Fig. 8-27. Typical performance of single-pass surface condenser. Curves show performance with a constant quantity of circulating water.

8 to 12 lb/(ft²)(hr). Basic heat-transfer calculations employ the data of Fig. 8-26, using appropriate values of water velocity, tube gage, steam loading, and inlet temperature. Conservative selection usually incorporates a cleanliness factor of 0.8 or 0.9. The heat transferred is given by

$$H = AUt_m \tag{8-19}$$

where H = total heat transferred, Btu per hour, A = condenser heat-exchange surface in square feet, U = heat-transfer coefficient in Btu per hour per square foot per deg (Fig. 8-26), t_m = logarithmic mean temperature difference, defined by Eq. (8-20).

$$t_m = \frac{t_{wo} - t_{wi}}{\log_e \dfrac{t_s - t_{wi}}{t_s - t_{wo}}} \tag{8-20}$$

where t_s = saturated steam temperature, equivalent to the steam pressure in degrees Fahrenheit, t_{wi} = inlet circulating-water temperature in degrees Fahrenheit, t_{wo} = outlet circulating-water temperature in degrees Fahrenheit. A representative set of performance curves is given in Fig. 8-27.

31. Circulating-water Pumps. Water requirements range from 2 gal/(min)(kW) on small units to less than 1 on the largest turbines. Water quantities are so large that the pumping head must be kept to a minimum—approximately 10 ft for velocity and friction head. A siphon loop is essential, and the condenser must be placed to retain its advantages regardless of river stages or tidal fluctuations. These conditions favor the selection of single-stage centrifugal or axial-flow pumps (Fig. 8-28) with constant-

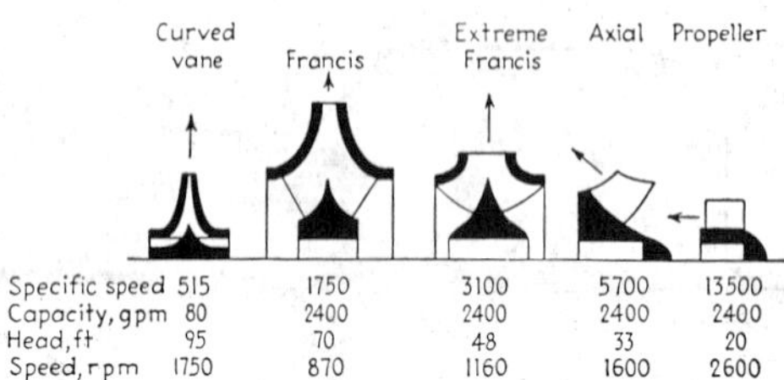

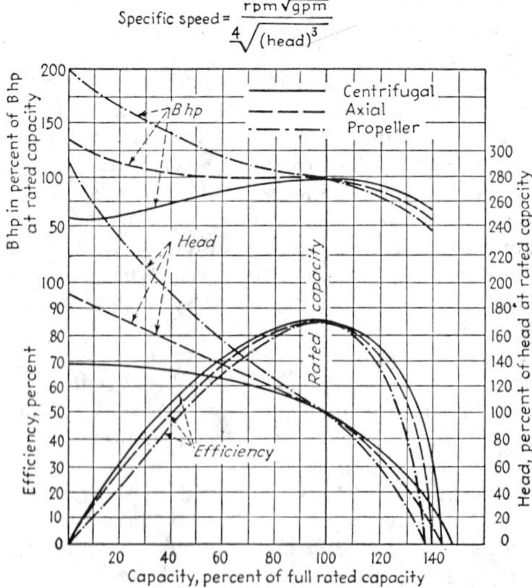

Fig. 8-28. Characteristic curves of centrifugal, axial-flow, and propeller-type pumps.

speed motor drive and attaining efficiencies of 85 to 90% in the larger sizes (100,000 gal/min).

32. Condensate or Hot-well Pumps. Condensate pumps, for removal of the distilled water at the saturation temperature of the condenser vacuum, must have a positive suction head to avoid cavitation (Fig. 8-29) and be of a balanced design—if centrifugal, with positive pressure on exterior stuffing boxes. Their capacity is 0.01 to 0.02 that of the circulating pump, and the total head ranges from 100 to 300 ft, depending upon the ultimate delivery point for the water. Duplicate full-capacity motor-driven constant-speed multistage centrifugal designs prevail.

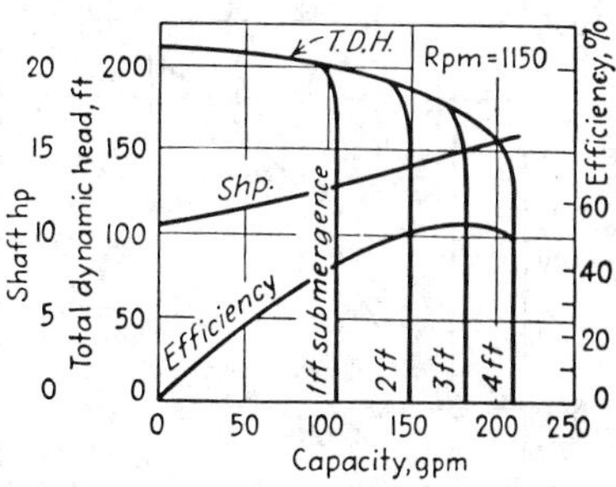

FIG. 8-29. Two-stage hot-well-pump characteristics showing influence of suction-head submergence.

33. Air-removal Pumps. Air and other noncondensable gases will accumulate in the condenser and destroy the vacuum unless removed by a pump. Steam-jet air pumps, in two- or three-stage designs, using 200 to 400 lb per hr of steam per stage at 200± lb per square inch, are standard equipment. Inter- and after-condensers recover heat, returning it to the main feed. Some installations use motor-driven rotary vacuum pumps. Reciprocating dry vacuum and hydraulic types are seldom used in current practice. The wet-vacuum pump is confined to some small low-cost installations. Air-pump suction should be at a cool point in the condenser to reduce capacity, which is of the order of 2 to 5 ft^3 of air/min at NTP.

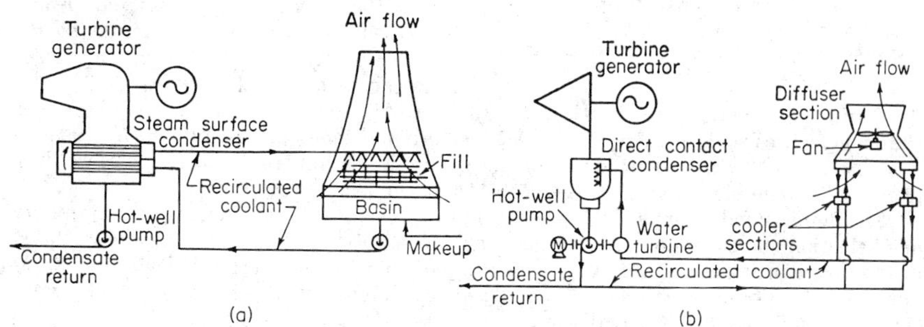

FIG. 8-30. Cycles for (*a*) natural-draft evaporative (wet) and (*b*) mechanical-draft (dry) cooling towers. (*After Fiehn, Combustion, July,* 1966.)

34. Circulating-water Reclamation Systems. The inadequacy of water in many regions and the desire to reduce the extent of so-called temperature pollution of streams and bodies of water are prompting the increasing use of circulating-water reclamation systems. Storage reservoirs, spray ponds, and cooling towers are used, with the last preferred because of smaller ground area, less drift loss, and closer approach to the ambient wet-bulb temperature. Natural-, forced-, and induced-draft types with evaporative cooling are variously used. Induced- and natural-draft types reduce recirculation and icing hazards. In the largest installations the natural-draft hyperbolic tower (Fig. 8-30*a*) with heights to 300± ft finds increasing favor. Performance generally gives a 10 to 15°F approach to the ambient wet-bulb temperature of the warmest season, with a range of 10 to 15°F on the extent of water cooling. Pumping heads approximate 20± ft. When towers employ a wooden cribwork, with intermittent or seasonal operation, it is essential to provide ample fire protection. Dry-cooling towers, coupled with direct-contact condensers (Fig. 8-30*b*), may be selected as an alternative to the evaporative-type towers, where the finned heat-exchange surface is incorporated in the tower and the evaporation loss (somewhat larger than the turbine-exhaust steam rate) is avoided.

35. Miscellaneous Condenser Equipment and Auxiliaries. Intake systems should be designed to receive the coldest water with protecting trash racks and screens. Discharge systems should avoid mixing warm water with the intake supply. Piping should be insulated to avoid sweating. Atmospheric relief valves or rupturing diaphragms are needed to avoid bursting damage on the condenser shell with loss of circulating water. Cleaning systems using wire brushes, nylon brushes, water lances, and rubber slugs are variously available. Organic or algae growth can be controlled by the use of chemical (e.g., Cl or $CuSO_4$) treatment of water.

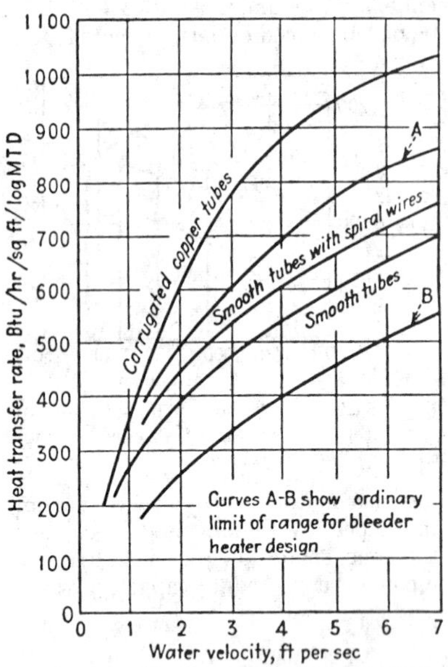

Fig. 8-31. Heat-transfer rates for feed-water heaters, ⅝- and ¾-in OD No. 18 or 20 BWG tubes.

STEAM-PLANT AUXILIARIES

36. Feed-water Heaters—General. Steam may be extracted from as many as 8 or 10 stages of a turbine successively to raise the temperature of the feed water from the condenser hot well to values as high as 500 or 550°F, with consequent improvement in plant heat rate. Closed heaters with water inside bundles of 1±-in-diameter) nonferrous tubes arranged in vertical or horizontal steel shells or open heaters with steam and water mixing may be used. Open heaters can serve a deaerating function simultaneously by degasifying the feed water of dissolved O_2 and CO_2. Open heaters require a pump at the feed-water outlet of each heater, but closed heaters can be arranged in series.

Figure 8-31 can be used with Eqs. (8-19) and (8-20) to estimate surface requirements. Water velocities seldom exceed 8 ft/sec; the terminal temperature difference is usually less than 5 or 10°F; pressure drop on the water side is limited to 25 ft; pressure drop from the turbine bleed flange to the heater is 4 or 5% of the stage pressure.

37. Boiler Feed Pumps. The feed pump is the most vital auxiliary in the plant, and selection should assure a continuing reliable supply of water at a pressure in excess of the boiler pressure (e.g., 400 lb/in² with 2,000 lb/in² operating), with suitable allowances for friction through heaters and piping, differences in elevation, safety-valve settings, regulating valves, and the like (Fig. 8-32). Multiple pumps with different drives are generally chosen for reliability. Multistage low-specific-speed centrifugal pumps (Fig. 8-28) for larger plants (above 1,000 kW) are standard practice, while reciprocating pumps (generally steam-driven) are limited to smaller plants and to some marine service. Constant-speed induction motors, for full-voltage starting, prevail. Steam turbines are used for emergency drive, but high pressure precludes good turbine efficiency in small horsepower ranges despite excellent reliability. On the largest units, e.g., 1,000 mW, the main turbine-generator is effectively used to drive the feed pump. Figure 8-33 shows

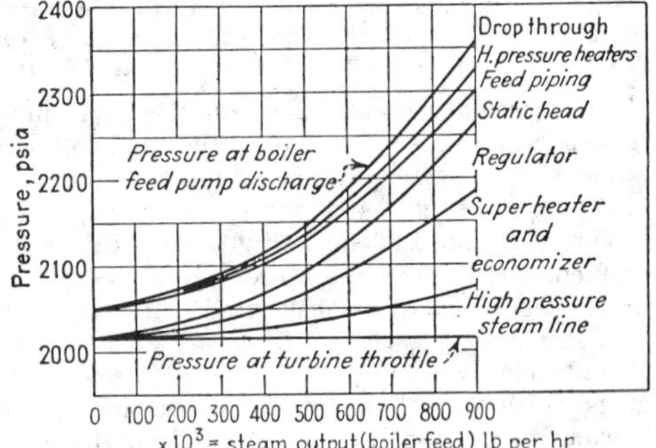

Fig. 8-32. System resistance curve for boiler feed pump.

comparative feed-pump power requirements of large units. Extensive operation at part load favors variable-speed drive, such as steam turbines, slip-ring motors, and hydraulic or magnetic couplings with about 15% speed change. Pump performance is sensitive to water temperature so that allowable suction lift and minimum positive suction head are specified in Fig. 8-34. Superpressures (above 2,000 lb/in²) dictate the use of speed-increasing gears with motor drive to avoid the limit of 3,600 r/min and to reduce the number of pump stages required.

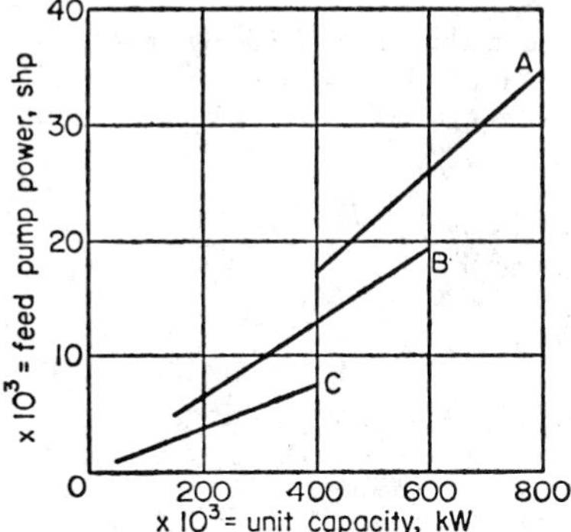

FIG. 8-33. Feed-pump power vs. unit size for various throttle pressures. A = 3,500 lb/in². B = 2,400 lb/in². C = 1,450 lb/in². (*After Fiehn, Combustion, July,* 1966.)

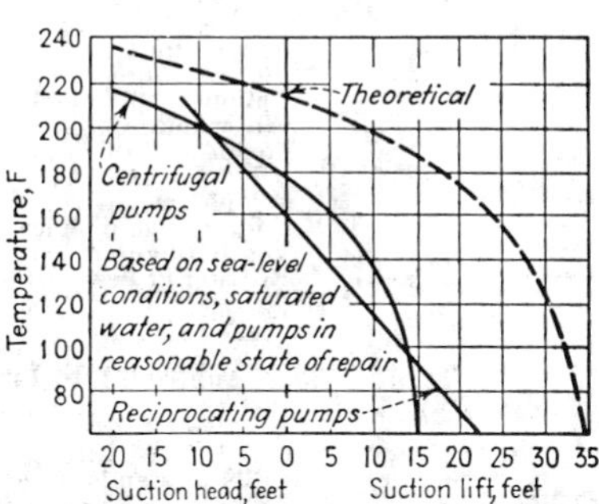

FIG. 8-34. Theoretical and practical suction lifts and heads for pumps handling water at various temperatures.

38. Piping and Insulation. Piping systems include live steam, reheat steam, feed water, bleed steam, condensate, circulating water, process steam, vacuum, compressed air, fuel, lubricant, service water, vent, safety valve, blowdown, and sewage. Materials must be suitable for the service and include black, gray, and galvanized iron; brass; bronze; copper; vitreous; clay; and wood stave. Carbon steel is used for steam lines up to 750°F; carbon molybdenum for 800 to 900°F; ferritic to 1050°F; and austenitic (stainless) above 1050°F (see ASTM specifications for details). Pipe may be bored, seamless, lap-welded, butt-welded, spiral-welded, spiral-riveted, or cast. Most joints on power piping are welded and stress-relieved, with threaded, flanged, bell, and spigot or compression joints used for some services and sizes. ASA standards prevail on piping sizes. Selection of type and size depends on service, operating pressure, temperature, velocity, investment, and cost of friction losses. Friction may be estimated by

$$\text{Head loss due to friction, ft of fluid} = f\frac{2Lv^2}{gD} \qquad (8\text{-}21)$$

where L = length of pipe in feet, v = fluid velocity in feet per second, g = gravitational acceleration, 32.2 ft per second per second, D = diameter of pipe in feet, f = friction factor, depending upon fluid velocity, density, viscosity, pipe size, and surface roughness (Reynolds number), with a safe average value of 0.005 to 0.006 for preliminary estimating purposes.

Table 8-9 shows allowable fluid velocities established by practice. Friction losses for some valves and fittings are estimated by friction factor F from Table 8-10 in

$$\text{Head loss due to friction, ft of fluid} = F\frac{v^2}{2g} \qquad (8\text{-}22)$$

where v and g are the same as above.

Hangers, supports, and cradles hold pipe in place with anchorage and expansion to avoid damage to machinery. Slip-type expansion joints are suitable for low pressures (100 lb/in² and 250°F), but otherwise loops and bends must be provided, requiring elaborate and careful stress analysis. Insulation is a matter of balancing investment cost against heat saving. Heat-loss reduction by 90 to 95% is generally justifiable. Diatomaceous earth and magnesia, canvas- or sheet-metal-covered, are used for high-temperature steam lines; air-cell covering for low-pressure steam lines (5 lb/in²); ground cork, in waterproof paint, for subatmospheric temperature lines, subject to sweating. Valve accessibility, trained personnel, and stenciled directions are better than unmarked elaborate color schemes for pipe identification and operation.

Table 8-9. Usual Allowable Fluid Velocities in Duct and Piping Systems

	Velocity, Ft per Min
Forced draft ducts	2,500–3,500
Flues and breeching	2,000–3,000
Chimneys and stacks	2,000–3,000
Water lines	500–600
Steam lines, high pressure	10,000
Low pressure	12,000–15,000
Vacuum	25,000–40,000
Compressed-air lines	2,000
Ventilating-duct systems	1,200–3,000
Register grilles	500

Table 8-10. Approximate Friction Factors for Valves and Fittings in Piping Systems

Fitting	Factor, F
Elbow, standard, 90 deg	0.75–1.00
Elbow, long-sweep, 90 deg	0.10–0.25
Elbow, blade	0.25
Elbow, 45 deg	0.25
Gate valve	0.20
Globe valve	2.00
Orifice, sharp-edge	0.04
Orifice, bell-mouth	0.04
Re-entrant nozzle, $L > 3$ diam	1.00
Tee, straight-run	0.25
Tee, turn	1.00–2.00

39. Feed-water Treatment. Feed water introduced into the boiler must be pure enough to reduce or eliminate (1) scaling, (2) corrosion, and (3) carry-over of solids into steam lines. Screens, filtration, sedimentation, coagulation, chemicals, evaporators, demineralizers, deaerators, residual scavengers, steam separators, and scrubbers are all used to remove various impurities and to deliver clean steam (approximately 1 ppm of solids). Condensing steam plants operate with 1 to 2% makeup, but industrial-process steam plants may range to 100% makeup. The latter is severe service and calls for feed water with maximum total solids of 300 ppm and pH of 9. Boiler-water pH is generally maintained at 10 to 11, and solids concentration by blowdown at 2,000 ppm with 600 lb/in² service and less than 500 ppm with 2,000 lb/in² service.

Deaerators are of the steam-scrubbing type and deliver water with less than 0.005 cm³/l oxygen when water-temperature rise is more than 30°F. Open heaters and condenser hot wells are limited to 0.03 cm³/l final oxygen content. Residual scavenging with chemicals, such as sulfite or hydrogen, and depending upon boiler conditions, may be used. Removal of ammonia and carbon dioxide is more troublesome than removal of oxygen. Operation of modern high-pressure high-temperature steam plants requires the closest supervision by personnel skilled in chemistry.

40. Miscellaneous Equipment. Service pumps, usually constant-speed and motor-driven, deliver water at 100-ft head for use in bearing, oil, and jacket cooling. Many other constant-speed motor-driven pumps are provided for specific services such as heater drainage, ash sluicing, screen washing, and evaporator feeding. Motor-driven compressors provide air at 100 lb/in². Lubricant-purifying, reclaiming, and cooling systems assure reliability of running machinery.

41. Instruments and Controls. The modern power station with its small operating labor force (1,000 to 5,000 kW/man) requires complete instrumentation and automatic controls for maximum effectiveness and economy. Instruments may be indicating, recording, or integrating and are grouped in a central control room, fully air-conditioned. Commercial, not laboratory, accuracy is to be expected. Automatic combustion and computerized controls should be superimposed on manual control, with operators appropriately trained.

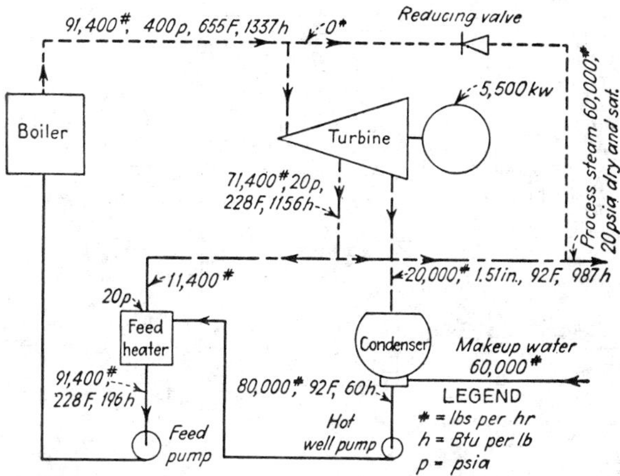

FIG. 8-35. Heat-balance diagram and summary calculations for an industrial-type "by-product" steam power plant delivering 60,000 lb/h of process steam at 5 lb/in² and 5,200 kW.

Condenser	
Steam flow, lb per hr	20,000
Pressure, in. Hg abs	1.51
Hot-well temperature, F	92
Hot-well enthalpy, Btu per lb	60
Make-up water flow, lb per hr	60,000
Feed Heater	
Heater pressure, psia	20
Heater saturation temperature, F	228
Enthalpy of water out, Btu per lb	196
Heat added to water, Btu per lb	136
Heat added to water, 10^6 Btu per hr	10.9
Extraction steam enthalpy, Btu per lb	1,156
Heat available from extraction, Btu per lb	960
Extraction required, lb per hr	11,400
Boiler feed, lb per hr	91,400
Turbine Generator	
Enthalpy at throttle, Btu per lb	1,337
Extraction enthalpy, Btu per lb	1,156
Enthalpy drop (work) to extraction point, Btu per lb	181
Extraction, lb per hr	71,400
Internal generation by extracted steam, kw	3,730
Enthalpy at exhaust, Btu per lb	987
Enthalpy drop (work) to exhaust, Btu per lb	350
Internal generation by condenser flow, kw	2,050
Total internal generation, kw	5,780
Generator efficiency	0.96
Generator output, kw	5,550
Boiler	
Enthalpy at superheater outlet, Btu per lb	1,337
Enthalpy of feed, Btu per lb	196
Heat added to steam, Btu per lb	1,141
Heat added to steam, 10^6 Btu per hr	104.6
Boiler efficiency	0.82
Heat supplied in fuel, 10^6 Btu per hr	127.5
Plant Performance	
Electric auxiliary use, %	6.4
Electric auxiliary use, kw	350
Net plant send-out, kw	5,200
Plant realization ratio	0.95
Net plant heat supply, 10^6 Btu per hr	134

42. Heat Balance and Station Performance. Heat-balance or cycle diagrams, like Figs. 8-35 to 8-37 show the schematic arrangement of all essential equipment assembled in a plant, together with actual performance data on power, heat, steam quantity, water quantity, pressure, temperature, enthalpy, and efficiency. Figure 8-35 contains

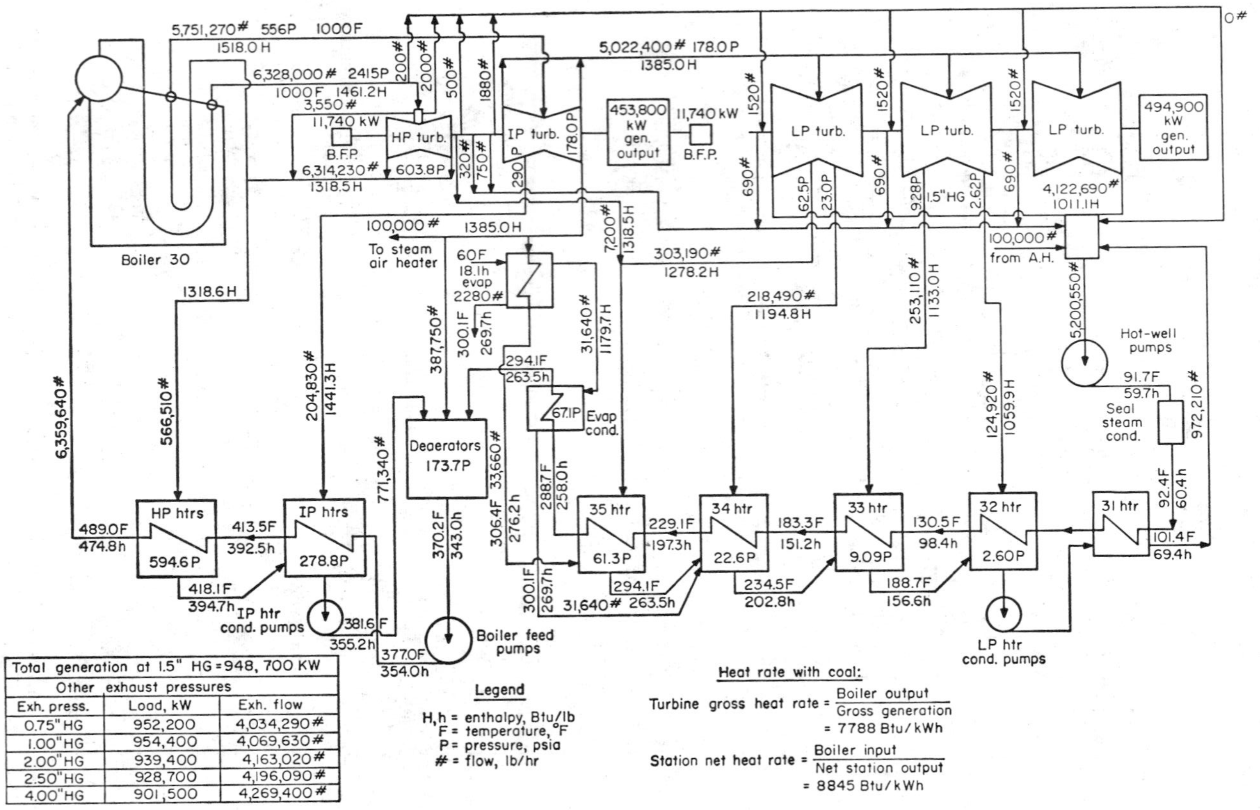

Total generation at 1.5" HG = 948, 700 KW		
Other exhaust pressures		
Exh. press.	Load, kW	Exh. flow
0.75"HG	952,200	4,034,290#
1.00"HG	954,400	4,069,630#
2.00"HG	939,400	4,163,020#
2.50"HG	928,700	4,196,090#
4.00"HG	901,500	4,269,400#

Fig. 8-36. Heat-balance diagram for a steam-electric power station, fossil-fuel-fired (Ravenswood Station, Consolidated Edison Company of New York).

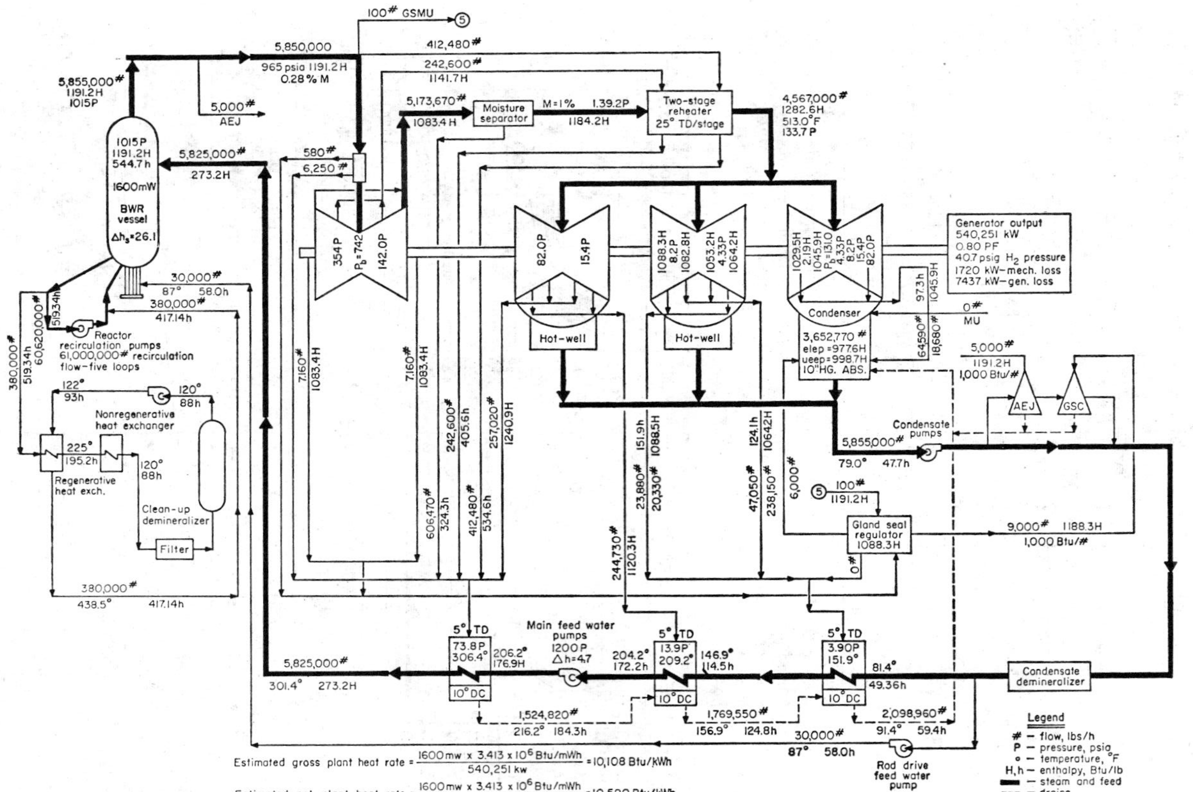

Fig. 8-37. Heat-balance diagram for a steam-electric power station, nuclear-fuel-fired (Oyster Creek Station, Jersey Central Power and Light Company; General Electric Company).

a tabulation of the successive steps in the heat-balance calculations for a simple by-product-type power plant. Figure 8-38 gives the results of a study on a multipurpose by-product steam plant. Figures 8-36 and

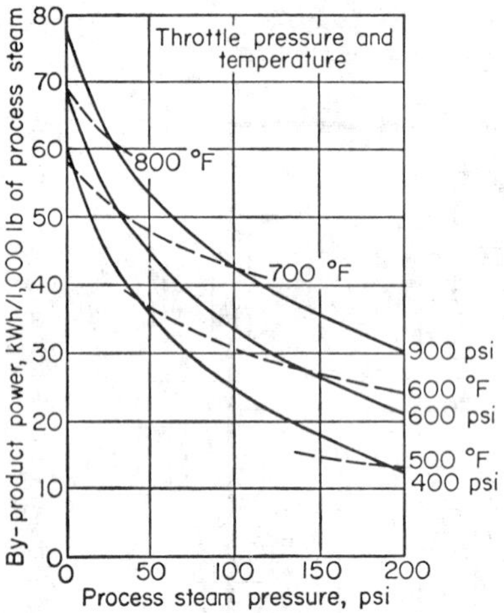

FIG. 8-38. "By-product" power as a function of process steam pressure for various throttle conditions (selected from a multipurpose power-plant study).

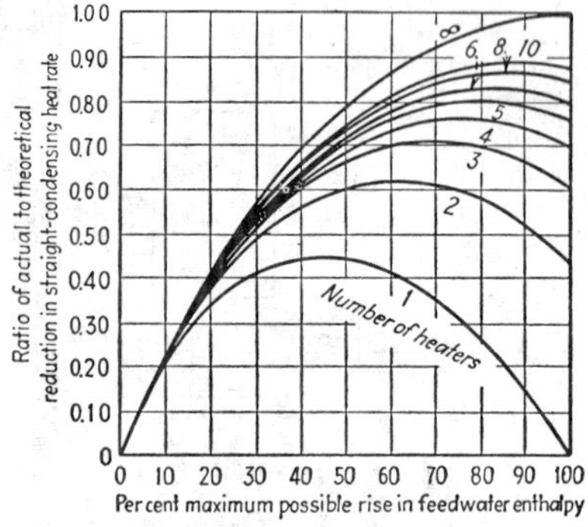

FIG. 8-39. Improvement in heat rate with regenerative feed-water heating. (*Salisbury.*)

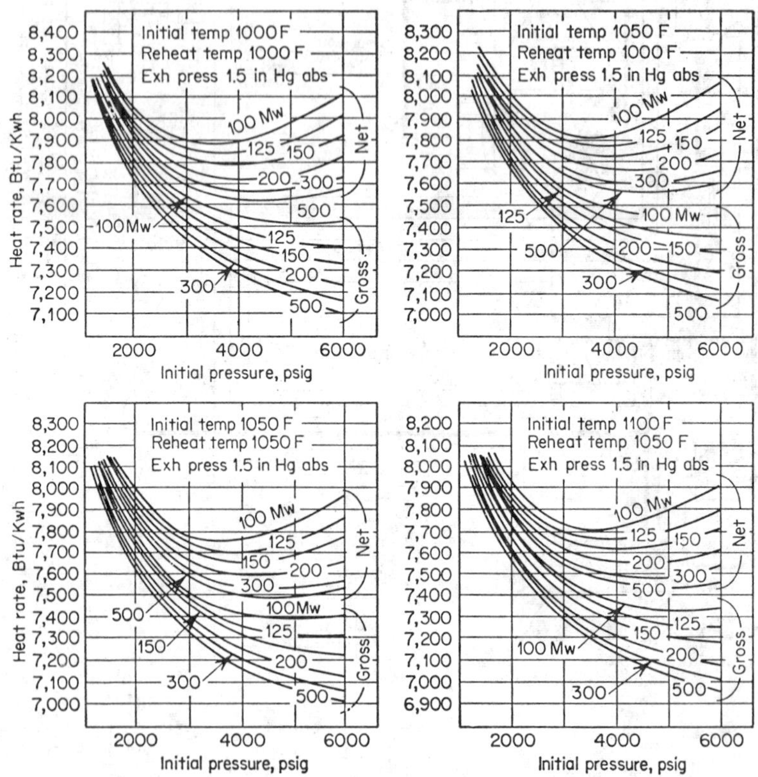

FIG. 8-40. Turbine heat rates as functions of steam pressure, steam temperature, and unit size for large, single reheat units operating on a six-stage regenerative feed cycle. (*Knowlton in "Standard Handbook for Mechanical Engineers."*)

8-37 give, respectively, the data for a modern representative (1) fossil-fuel-fired and (2) nuclear-fuel-fired steam central station. In the selection of initial steam, reheat steam, and feed-water thermal conditions the data of Figs. 8-39 and 8-40 are helpful.

INTERNAL COMBUSTION

43. Internal-combustion Power—General. Internal-combustion power plants are predominantly of the reciprocating-engine type, with some recent emphasis on the gas turbine. Because of self-contained features, low weight, good efficiency, and reasonable price, it is the prevalent transportation power plant, e.g., some 90 million in the United States. It requires a special (usually high-priced) fuel—gas, gasoline (volatile), or distillate (diesel) fuel (see Tables 8-1, 8-2, 8-6*b*, and Fig. 8-1). It will burn these fuels

Table 8-11. Performance of Selected Internal-combustion Engine Plants

Type	Fuel	Shaft horse-power	Com-pression ratio, R_v	Brake mep, p_m, psi	Piston speed, ft per min	Weight, lb per cu in piston displace-ment	Weight, lb per shp	Overall heat rate, Btu per kWh
Mixture engines:								
Automotive	Gasoline	2-300	4.5-10	75-125	800-2000	2-3.5	5-20	14,000-15,000
Stationary								
Low compression	Natural gas	100-1000	4-6	40-75	600-1200	3-5	50-150	13,000-18,000
High compression	Natural gas	300-1500	8-14	80-90	600-1200	3-6	15-100	9,000-11,000
Injection engines:								
Solid injection, compression ignition								
Automotive	Diesel fuel	20-300	12-15	75-100	800-1800	2.5-4	7-25	11,000-14,000
Railroad	Diesel fuel	200-2500	12-15	60-90	800-1800	2.5-4	10-40	9,500-11,000
Stationary								
Unsupercharged	Diesel fuel	50-2500	12-15	70-80	600-1500	2.5-5	10-100	9,500-11,000
Supercharged	Diesel fuel	60-4000	10-13	110-125	600-1500	2.5-5	7.5-75	8,500-10,500
Dual fuel, stationary:								
Unsupercharged	Diesel fuel and natural gas	50-2500	12-15	80-90	600-1500	2.5-5	10-100	9,500-11,000
Supercharged	Diesel fuel and natural gas	60-4000	10-13	120-135	600-1500	2.5-5	7.5-75	8,500-10,500

with high inherent and high actual efficiency (20 to 40%). The trend in the art is toward the development of the high-speed (1,000 to 5,000 r/min) automotive types, with the low-speed engine (less than 300 r/min) confined to a few stationary services and motorships, and with the medium speed (500 to 1,000 r/min) preferred for locomotive application. Internal-combustion engines as judged by steam and hydro central-station practice are small (see Table 8-11) with upper limits of 5,000 hp/unit. The gas-turbine power plant is available in units of 500 to 150,000 kW.

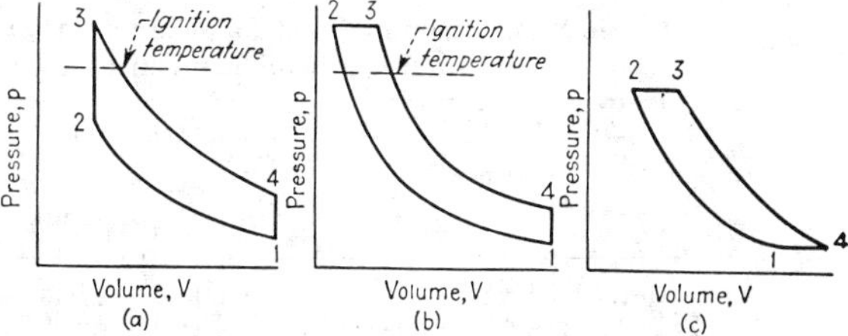

FIG. 8-41. Ideal indicator cards (pressure-volume diagrams) for internal-combustion cycles. (*a*) Otto cycle. (*b*) Diesel cycle. (*c*) Brayton cycle. Phase 1–2, isentropic compression (compression ratio, $R_v = V_1/V_2$); 2–3, heat addition at constant pressure or volume; 3–4, isentropic expansion; 4–1, heat rejection at constant pressure or volume.

44. Cycles. Mixture engines operate on the Otto cycle, injection engines on the diesel cycle, and gas turbines on the Brayton cycle (Fig. 8-41). Thermodynamically, these cycles all show improved thermal efficiency with higher compression ratios, $R_v = V_1/V_2$ (phase 1–2). For the Otto and Brayton cycles

$$\text{Thermal efficiency} = 1 - \frac{1}{R_v^{k-1}} \tag{8-23}$$

where k = ratio of specific heat at constant pressure, C_P, to specific heat at constant volume, C_V. $k = 1.4$ for the cold-air standard often used for purposes of cycle evaluation. For the diesel cycle the thermal efficiency is also improved by shortening the heat-addition line (phase 2–3) so that

$$\text{Thermal efficiency} = 1 - \frac{1}{R_v^{k-1}}\left[\frac{R_c^k - 1}{k(R_c - 1)}\right] \tag{8-24}$$

where R_c = cutoff ratio, V_3/V_2.

Mean effective pressure (p_m, lb/in²) is most useful in evaluating performance and is found by

$$\text{Ideal } p_m = \frac{\Delta W}{144(V_2 - V_1)} \tag{8-25}$$

where ΔW = work of cycle in foot-pounds and V_1 and V_2 are the volumes at the beginning and end of compression (phase 1–2, Fig. 8-41*a* and *b*). The work may be calculated by various thermodynamic methods, such as the cold-air standard, in which all values of work and volume in Eq. (8-25) are based on 1 lb of working substance. ΔW can be related to the heat supplied, ΔQ, in Btu per pound of working gas, by

$$\Delta W,\ \text{ft-lb/lb of working substance} = \Delta Q \times (\text{th. eff.}) \times 778 \tag{8-26}$$

ΔQ ranges from 700 to 1,400 Btu/lb of fuel-air mixture, with 1,200 as representative of many hydrocarbons (Table 8-1).

Horsepower is calculated by Eq. (8-14), where p_m is determined (1) by theoretical methods, using Eq. (8-25), or (2) by actual indicator card, wherein

$$\text{Ind. } p_m = \frac{\text{area of ind. card, in}^2}{\text{length of ind. card, in}} \times \text{spring scale} \tag{8-27}$$

or (3) if the engine speed is too high for use of an indicator, p_m may be determined by measuring the horsepower at the engine shaft (brake horsepower), as by dynamometer, and solving for equivalent brake p_m by Eq. (8-14). Some representative values of p_m and other engine-performance data are given in Table 8-11. Indicated mep at the engine cylinder is related to brake mep through mechanical efficiency, which ranges from 85 to 95% or with a friction mean pressure of 10± lb/in².

The heat supplied in fuel can be accounted for by a heat balance, as shown in Table 8-12, where the bulk of the heat appears in the jacket system and in the exhaust gases.

Table 8-12. Sample Heat Balance and Allocation of Losses in Internal-combustion Electric Power Plants

Item	Btu per kwhr		Percentage	
	Injection	Mixture	Injection	Mixture
Output	3,413	3,413	30	25
Exhaust gases	3,982	5,460	35	40
Cooling water	3,754	4,505	33	33
Generator losses	114	137	1	1
Auxiliary use	114	137	1	1
Input	11,377	13,652	100	100

45. Engine Types. Engines operating on the Otto cycle use an explosive air-fuel (15 to 1) mixture made in a carburetor or mixing valve. Compression (Fig. 8-41) is limited to a value less than the fuel-ignition temperature to avoid preignition. Ignition is by electric spark. Diesel engines are of the injection type in which air alone is compressed (phase 1–2) and fuel is sprayed into the combustion chamber after compression. The temperature at (2) must exceed the ignition temperature to initiate combustion. Fuel burns at substantially constant pressure, with 100±% excess air at full load.

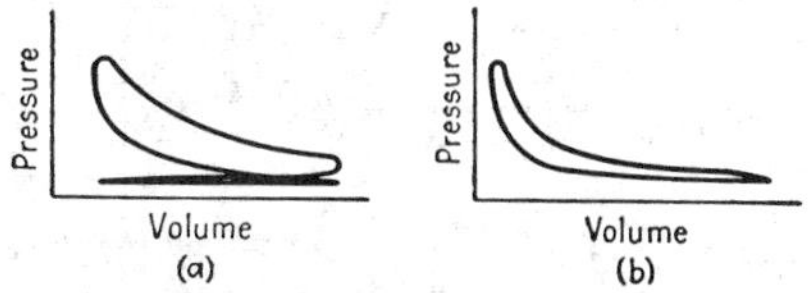

Control of output in the mixture engine is accomplished by throttling the supply of mixture. In an injection engine the air supply is not varied, but the fuel supply is adjusted to the load requirements, the excess air consequently rising as load drops.

Fig. 8-42. Actual internal-combustion-engine indicator cards. (*a*) Four-cycle mixture engine, Otto cycle. (*b*) Two-cycle injection engine, diesel cycle.

Engines may be (1) 4-cycle, using four strokes (two revolutions) to complete a cycle, or (2) 2-cycle, using two strokes (one revolution) to complete a cycle as shown in Fig. 8-42 for an actual (*a*) 4-cycle Otto engine and (*b*) 2-cycle diesel engine. The four-stroke principle is favored with multicylinder automotive-mixture engines; the two-stroke principle, with larger and lower-speed injection engines. The 2-cycle principle gives theoretically twice the power for a given bore, stroke, and speed, thus reducing engine weight and cost. Scavenging requires approximately 1.2 times engine displacement. Two-cycle mixture engines accordingly show poorer economy than 4-cycle engines. These limitations do not apply to the injection engine, as air alone is used for scavenging. Scavenging on 2-cycle diesels is with air at 5 ± lb/in^2 obtained by crankcase compression, front-end compression, or separate rotary, reciprocating, or centrifugal blowers. Supercharging to positive pressures, as high as 10 to 15 lb/in^2, increases the weight charge admitted to the power cylinder and consequently increases the mean pressure and the output of a given machine size, with resultant reduction in engine weight per horsepower.

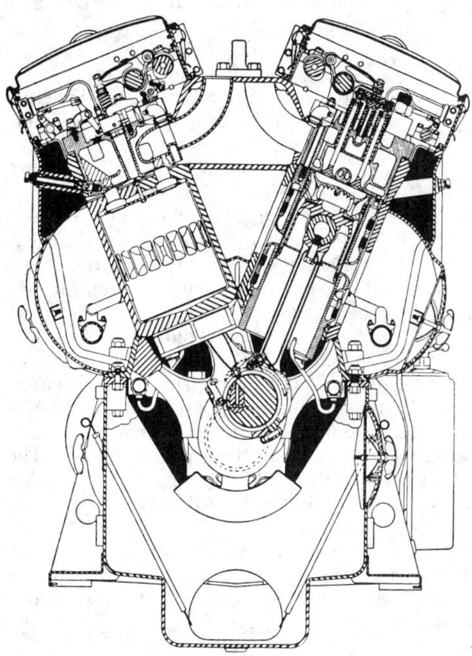

Fig. 8-43. Section through General Motors 567-C diesel engine (Electro Motive Division): 2-cycle, 8½ × 10 in, 6, 8, 12, and 16 cylinders; speeds up to 835 r/min; 600 to 1,750 hp; 350 to 1,000 kW. Application: diesel-electric locomotives, mobile power plants, stationary installations.

45a. Commercial Engines. Engines using liquid fuels dominate the transportation field, with high-priced gasoline burned in automobile, truck, tractor, bus, and aircraft services and distillate (diesel) fuel used for larger truck, tractor, bus, motorboat, and locomotive services where the higher investment can be offset by use of less expensive fuel. In stationary services the gaseous fuels (especially natural gas) and the heavier diesel fuels predominate. The gas producer, rendering coals suitable as fuel for internal-combustion engines, is no longer a contender in American practice. Stationary engines show an evolution using 2- and 4-cycle principles and single- and double-acting constructions and horizontal and vertical arrangements, with practice favoring the vertical or V-multicylinder (Figs. 8-43 and 8-44) engine, and some horizontal radials. One of

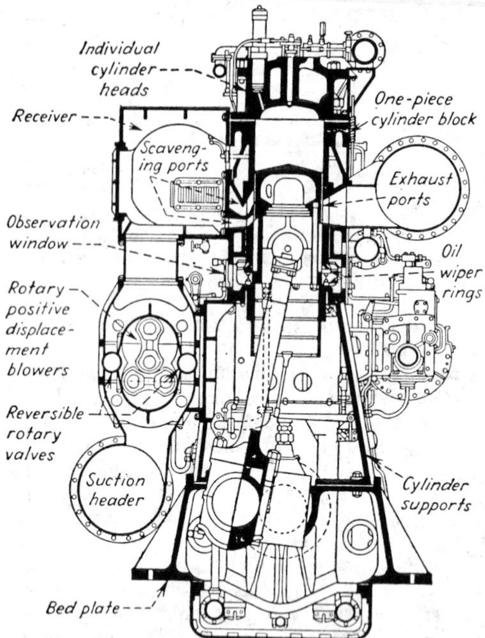

FIG. 8-44. Two-cycle diesel-engine section. (*Busch-Sulzer Bros., and Power.*)

the most notable trends has been toward the use of alternative fuels, as in the dual fuel engines and the gas diesels, to take advantage of market prices for liquid and gaseous fuels. Various designs are offered to utilize alternative fuels such as the gas diesel in which gas under pressure (1,000± lb/in²) and auxiliary pilot oil are injected toward the end of the compression stroke. The oil, which controls burning and stabilizes ignition, can be varied from 100% to as little as 5 or 10% of the total fuel. In locations where oil is much higher in price than gas, high-compression spark-ignition engines operating on a lean mixture may be used. Or the one engine can be changed over from a spark-ignition mixture system burning gas to a compression-ignition system burning oil. Such change-over necessitates shutdown and alteration of the compression ratio.

45b. Starting. Starting automotive and small-sized (less than 250-hp) engines is generally by electric motor with storage battery. Larger engines use compressed air (200 lb/in²) introduced through the cylinder head. With engine-driven generator sets starting may be accomplished by motoring the generator.

45c. Cooling Systems. Air is used to cool many aircraft engines and small (10±-hp) engines. Otherwise water cooling systems predominate. Water should be clean and pure to eliminate corrosion, scaling, and plugging. Temperature should be between 120 and 180°F. Reclamation with a cooling tower or spray pond is common with limited water supply. If natural water is poor, a double circuit, with heat exchanger, instead of a single circuit, is used to keep impurities away from engine parts. The heat-balance data of Table 8-12 define the extent of the cooling system. Evaporation loss is 3 to 4 lb/kWh with injection engines and 5 to 6 lb/kWh with mixture engines. The water-circulation rate is 5 to 10 gal/hph.

45d. Waste Heat. Waste-heat utilization for industrial power plants with demands for both heat and electric energy can be estimated from Table 8-12, where most of the energy supplied in the fuel appears as jacket loss and exhaust loss. The former is at a temperature less than 180°F, which limits its use to minor warming operations. The latter is at a higher temperature level, with estimating data given in Fig. 8-45. Heat is generally available at lower temperature levels with injection engines than with mixture engines because of higher thermal efficiency and higher excess air (100% compared with 10% at full load, typically). These effects are more pronounced at light loads and limit waste-heat boilers to lowest steam pressures. An equation for estimating waste-heat opportunities with diesel engines is

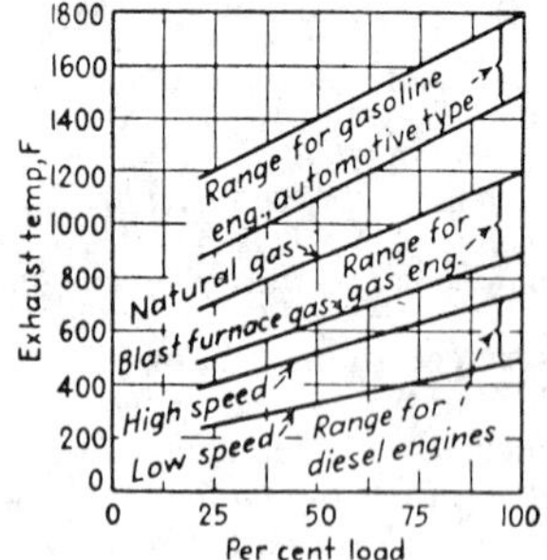

FIG. 8-45. Typical exhaust-gas temperatures for internal-combustion engines.

$$\Delta Q = \text{shp} \times C \times \frac{\Delta T}{4} \qquad (8\text{-}28)$$

where ΔQ = heat recoverable from exhaust gases in

Btu per hour, shp = shaft horsepower at load, ΔT = temperature drop of exhaust gases through heater, degrees Fahrenheit (usually 50 to 100°F), C = 10 for 4-cycle engines, 24 for 2-cycle engines.

46. Miscellaneous Items. Foundations must be designed to handle the stress loading from reciprocating engines and to reduce vibration. Exhaust systems should be equipped with wave-trap silencers to eliminate the nuisance of noise, particularly in the low-frequency ranges. Filter systems to supply both clean fuel and clean air are among the best forms of insuring engine reliability. In fact it has been amply demonstrated that economical engine operation justifies the rigid purchase of clean, dry fuel and the necessary apparatus to assure delivery of that clean fuel to the engine. Instruments should include injection-nozzle testers, jacket-system thermometers, lubrication pressure gages and thermometers, smoke indicators, and exhaust-gas pyrometers, in addition to the generally recognized devices for satisfactory supervision of station operation.

47. Overall Performance, Diesel-Electric Plants. The best source of information on stationary internal-combustion plant performance and costs is the annual ASME Oil and Gas Engine Power Reports. These show typically that heat rates vary between 10,000 and 13,000 Btu/kWh for most of the diesel plants, while lubricating oil consumption may range from 500 to 3,000 gross kWh/gal. Labor and maintenance experience are also detailed. Figure 8-49 shows some comparative data on heat rates of various types of power plants as functions of load.

GAS TURBINES

48. Gas Turbines. Gas turbines operate on the Brayton cycle (Fig. 8-41*c*) or on modifications using regeneration, intercooling during compression, and reheating during expansion. Basically (Fig. 8-46*a*), there is (1) a centrifugal, axial, or rotary compressor which delivers the working gas (air) to (2) a combustion chamber where temperature is raised and (3) an expander which is a single or multistage axial-flow turbine. The work output of the expander exceeds the work input to the compressor, by the amount given in the Brayton cycle. Open-cycle turbines (Fig. 8-46*a*) use air as the fluid, and the fuel (gas or selected oil) is burned directly in the compressed air at 75 to 100 lb/in² with the mixture at 1,500°±F delivered to the expander. Closed-cycle (Fig. 8-46*b*) turbines place a heat exchanger in the combustor which can be direct-fired with any type of fuel. The working fluid can be chosen as to chemical, physical, and nuclear properties and operated at selected pressure levels. The heat exchanger increases investment, bulk, and weight. Both the open and closed cycles can be modified for higher thermal efficiency by introduction of regeneration, reheating, and intercooling, i.e., from 15 or 20% to values as high as 30 or 35% (Fig. 8-46*c*).

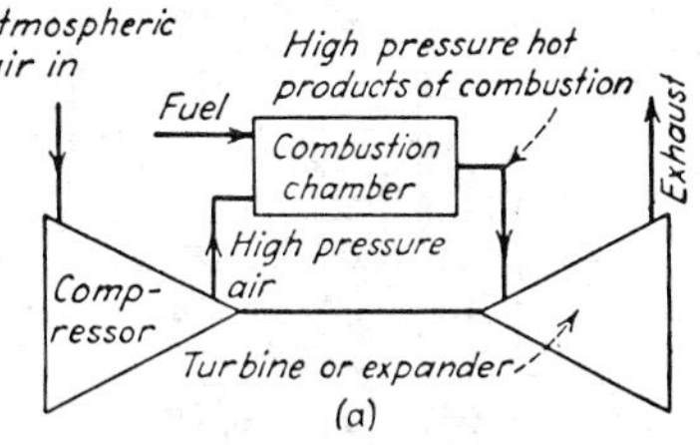

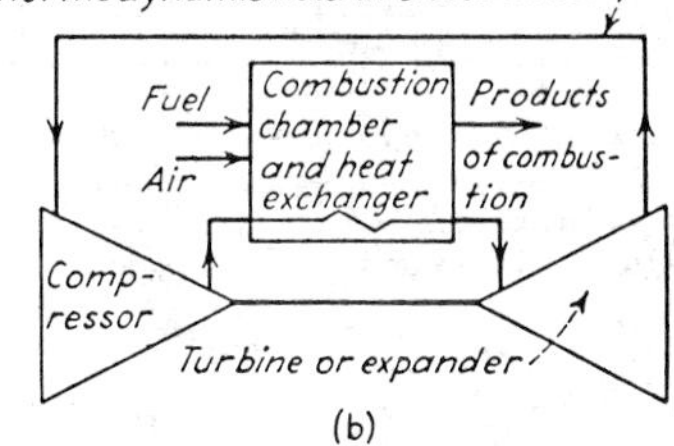

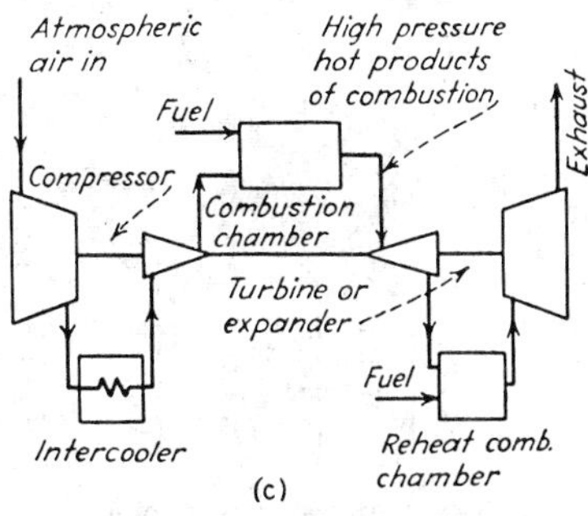

FIG. 8-46. Schematic arrangements—gas-turbine power plants. (*a*) Simple, open cycle; (*b*) simple, closed cycle; (*c*) open cycle with intercooling and reheat.

49. Theory and Performance. Most economical arrangements require that the length of the combustion line (phase 2–3 of Fig. 8-41*c*) be long and the temperature high in order to reduce the compressor size and to increase (1) the work output of the expander and (2) the thermal efficiency of the assembly. The higher the turbine throttle temperature, the better the efficiency of the real unit. Actual net output is given by

$$\Delta W_T \times e_T - \frac{\Delta W_c}{e_c} = \text{net work of unit} \tag{8-29}$$

where ΔW_T = ideal work of the expander, ΔW_c = ideal work of the compressor, e_T = engine efficiency of the expander, e_c = compression efficiency of the compressor. The ideal values of work are found (1) by thermal methods, using Mollier charts, or (2) P-V methods with fixed gases. As both engine and compression efficiencies are less than 1.0, inspection of Eq. (8-29) shows that net work of the unit can rapidly approach zero unless the numerical values of efficiency are high. Values ranging from 0.8 to 0.88 are possible with well-designed and well-maintained axial-flow compressors and turbines. The influence of compression ratio and combustion temperature is given by Fig. 8-47.

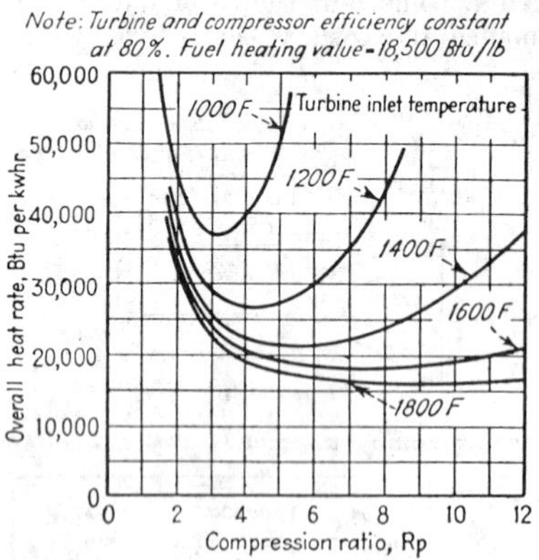

FIG. 8-47. Gas-turbine–electric-plant performance. Open cycle, no regeneration or intercooling. (*Salisbury.*)

50. Fields of Application. Gas turbines have found their greatest market in the propulsion of airplanes, one manufacturer having sold more than a billion horsepower since the initial development during World War II. The features of light weight, compactness, high speed, low lubricant consumption, minimum water requirements, high starting torque, quick starting, low maintenance, and low labor are merits variously usable in other applications, viz., locomotives, gas-line pumping, ship propulsion, automotive, and stationary electric service. Single gas-turbine elements are generally less than 10,000 or 20,000 kW in size and are not too well suited for capacities below 500 or 1,000 kW. Among the smaller-sized power plants the automotive gasoline and diesel engines dominate the markets, and in sizes ranging upward from 20,000-kW units the steam-turbine plants grow increasingly important. In the stationary-power area gas turbines have been used (1) to modernize older steam plants with improved overall heat rates and capacity by better feed-water heating; (2) to provide necessary gas compression for chemical manu-

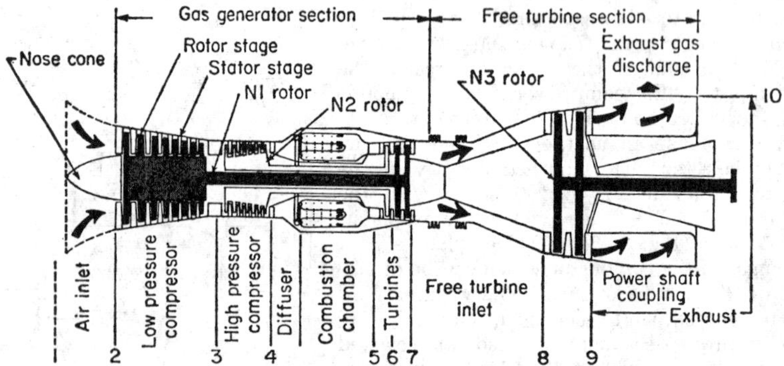

FIG. 8-48. Aircraft gas turbine adapted as a gas generator, for driving an electric generator, compressor, or pump (United Aircraft–Pratt and Whitney).

facturing operations at elevated pressure and temperature; (3) to provide by-product steam and power for process heat, air-conditioning, distillation, desalination, and other multipurpose power operations; and (4) to provide straight electric generation without the inducements of supplementary objectives. In the last category they are finding increasing favor as quick-starting units for peak-load, emergency, and cranking service in electric utility plants and systems. Aircraft propulsion elements can be applied as gas generators to drive electric generator units in sizes ranging to 300,000 kW (see Fig. 8-48). These units generally burn gaseous or distillate fuels. Residue fuel oils, when treated for vanadium and sodium contaminants, can be burned successfully. Coal burning is still much in the future. Adaptation to nuclear service waits for the development of reactors with high operating gas temperatures—a fundamental requirement for an acceptable gas-turbine plant. Figure 8-49 gives some comparative performance data for several selected thermal power plants.

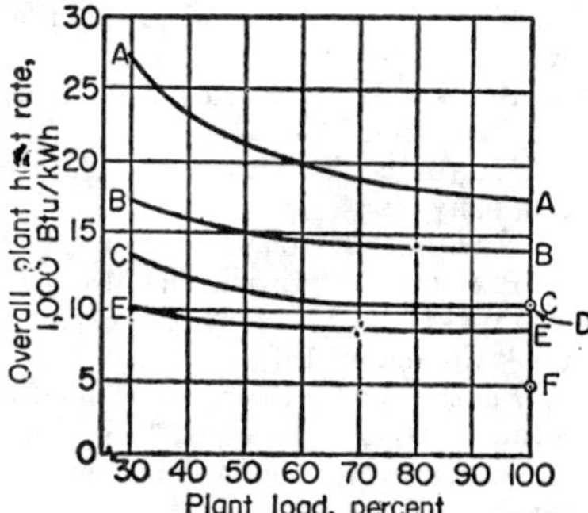

FIG. 8-49. Comparative overall plant heat rates for selected thermal-electric power plants:

A-A, gas turbine, 3,000 kW. *B-B*, steam turbine, 5,000 kW. *C-C*, diesel, 1,000 kW. Point *D*, steam turbine, 500,000 kW, nuclear-fired, at 100% load.

E-E, steam turbine, 300,000 kW, coal-fired.

Point *F*, "by-product" steam plant, 5,000 kW.

51. References

BAUMEISTER, T., and MARKS "Standard Handbook for Mechanical Engineers"; New York, McGraw-Hill Book Company, 1967.

ASME, Boiler and Pressure Vessel Code, Piping Code, Power Test Codes, Fluid Meters Committee Report, Bibliography on Gas Turbines, Annual Reports on Oil Engine Power Costs.

Automotive Industries, Annual Specification Number.

AYRES and SCARLOTT "Energy Sources: The Wealth of the World," McGraw-Hill.

Babcock and Wilcox Company, Steam, Its Generation and Use.

BAUMEISTER, T. "Fans"; New York, McGraw-Hill Book Company (University Microfilm).

Combustion Engineering, Inc., *Combustion Engineering*.

Diesel Engine Manufacturers Association, Standard Practices.

DOWNS and HOLLEY Progress in the Design of Large Steam-turbine Generators, *Natl. Power Conf., IEEE and ASME*, 1965.

Edison Electric Institute, Prime Movers Committee Reports.

Electrical World, Annual Steam Station Design Survey.

ELSTON and KNOWLTON Comparative Efficiencies of Central Station Reheat and Non-reheat Steam Turbine-generator Units, *Gen. Elec. Co.* GER-482.

FAIRES "Applied Thermodynamics"; New York, The Macmillan Company.

FIEHN Major Influences of Large Unit Size on Steam-electric Station Design, *Combustion*, July, 1966.

GAFFERT, G. A. "Steam Power Stations," 4th ed.; New York, McGraw-Hill Book Company, 1952.

General Electric Company, Turbine Generator Foundations, GET-1749.

Heat Exchange Institute, Standards.

HEYBURN and BRANDFON Application of Low-cost Thermal Power for Peaking, 64-PWR-12, *Trans. ASME*.

JENNINGS, B. H., and ROGERS, W. L. "Gas Turbines"; New York, McGraw-Hill Book Company, 1953.

JUSTIN and MERVINE "Power Supply Economics"; New York, John Wiley & Sons, Inc.

KEENAN and KEYES "Thermodynamic Properties of Steam"; New York, John Wiley & Sons, Inc.

KIEFER "A Practical Evaluation of Railroad Motive Power"; Steam Locomotive Research Institute.

LANDSBERG "Resources in America's Future, 1960–2000"; Baltimore, The Johns Hopkins Press.

MCADAMS, W. H. "Heat Transmission," 3d ed.; New York, McGraw-Hill Book Company, 1954.

NEWMAN "Modern Turbines"; New York, John Wiley & Sons, Inc.

PALMER Power Turbine Modules; *Elec. World*, Nov. 29, 1965.

POWELL, S. T. "Water Conditioning for Industry"; New York, McGraw-Hill Book Company, 1954.

Power, Annual Energy Systems Design Survey, Power Handbook, Pump Handbook, Steam Turbine Handbook.

SALISBURY "Steam Turbines and Their Cycles"; New York, John Wiley & Sons, Inc.

SEWARD "Marine Engineering," SNA and ME.

SHELDON and HOVEKE Tandem Steam-Gas Cycle Adds Flexibility, Cuts Costs; *Elec. Light and Power*, November, 1965.

SKROTZKI, B. G. A., and VOPAT, W. A. "Applied Energy Conversion," 1st ed.; New York, McGraw-Hill Book Company.

TAYLOR and TAYLOR "The Internal-combustion Engine"; New York, International Publishers Company, Inc.

ZUCROW "Principles of Jet Propulsion and Gas Turbines"; New York, John Wiley & Sons, Inc.

HYDROELECTRIC POWER PLANTS

BY WILLIAM J. RHEINGANS

GENERAL

52. Nomenclature. The nomenclature used throughout this section is based on *National Electrical Manufacturers Association (NEMA) Publication* HT1-1957, Hydraulic Turbines, Governors and Accessory Equipment, which contains illustrated terms, definitions, and purchase specifications.

A hydroelectric power plant converts the inherent energy of water under pressure into electrical energy. Its main elements are:

An *upper*, or *high-level, reservoir*, usually formed by building a diversion dam across a river.

An *intake*, consisting of a canal or concrete passageway to carry the water directly to low-head turbines or to the pressure conduit used for medium- and high-head turbines.

A *pressure conduit*, consisting of a tunnel, pipeline, or penstock, or any combination thereof, to carry the water under pressure to medium- and high-head turbines.

A *surge tank*, to prevent excessive pressure rises and drops during sudden load changes, installed somewhere along the pressure conduit when this conduit is quite long.

Trash racks at the inlet to the intake or pressure conduit.

Intake and *draft-tube gates*.

A *penstock shutoff valve*, located near the downstream end of the penstock.

A *hydraulic turbine* consisting primarily of a runner, connected to a shaft, for producing prime motive power from the inherent energy of the water under pressure, a mechanism for controlling the quantity of water flowing to the runner, and water passages leading to and away from the runner.

A *governor* for operating the hydraulic-turbine control mechanism.

An *electric generator* connected to the hydraulic-turbine shaft to convert the prime motive power of the turbine to electric power.

A *pressure regulator*, sometimes used instead of a surge tank, to prevent excessive pressure rises and drops during sudden load changes in plants with long pressure conduits.

A *powerhouse* to enclose and support the hydraulic turbine, generator, governor, pressure regulator (if used), and auxiliaries.

A *draft tube*, usually a part of the powerhouse structure to carry the water away from the turbine runner.

A *tailrace,* sometimes used to carry the water away from the draft tube to the tailrace reservoir.

A *tail-water reservoir* which receives the water discharged from the draft tube or tailrace and is usually part of the original river at an elevation lower than the upper reservoir.

The difference in elevation between the water level in the upper reservoir and the level of the water in the tailrace or tail-water reservoir is called the *head* on the plant.

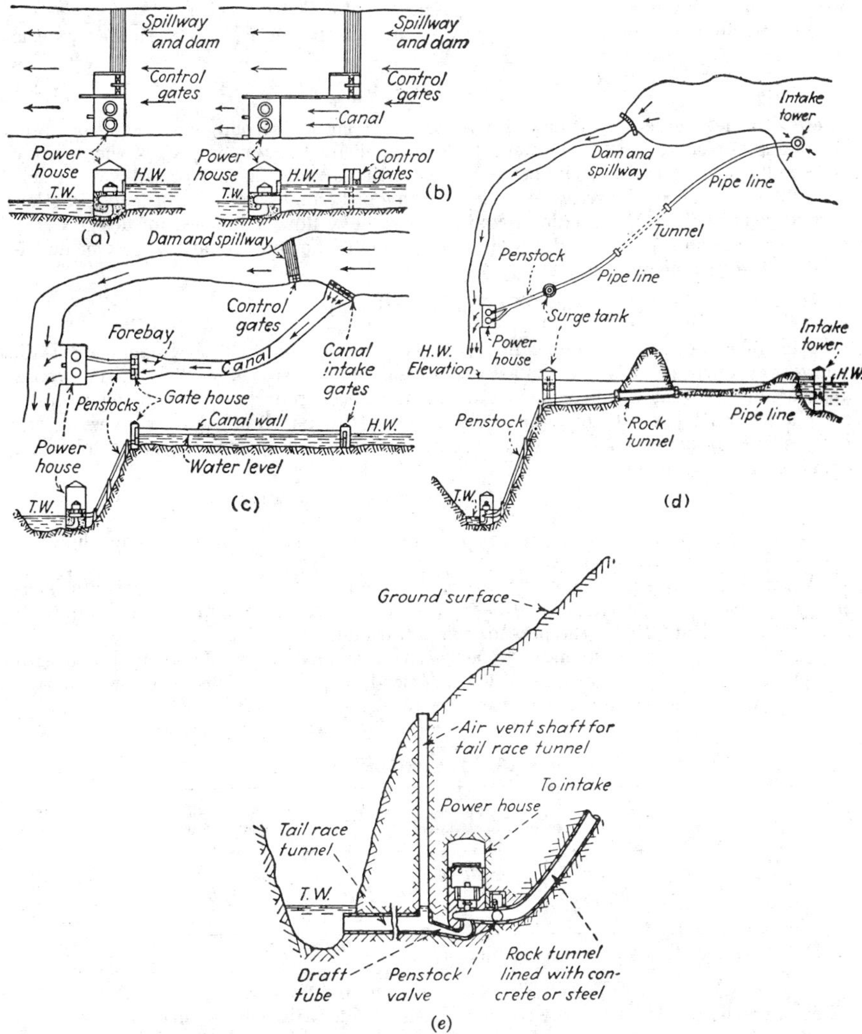

Fig. 8-50. Outline sketches of several typical hydropower developments. (*a*) Low-head development with dam, spillway, and powerhouse as an integral unit. (*b*) Low-head development with short intake canal and powerhouse separate from dam. (*c*) Medium-head development with long intake canal, gatehouse, and penstocks connecting the forebay with the powerhouse. (*d*) High-head development with large storage reservoir, pipeline, and tunnel leading to surge tank at upper end of penstocks. Powerhouse at lower end of penstocks is a considerable distance from the dam and spillway. (*e*) Outline sketch of underground power plant, showing penstock and tailrace tunnels.

The size, location, and type of power plant depend upon the topography, the geological conditions, and the amount of water and head available. Hydropower developments can be classified as low-head, medium-head, or high-head. Figure 8-50 shows in outline the most common arrangement and features of some of the elements listed above for the various developments.

HYDRAULIC TURBINES

53. Turbine Characteristics. Hydraulic turbines derive power from the pressure or force exerted by water falling through a given distance (the head).

The theoretical power, usually expressed in horsepower P_t, is determined by the equation

$$P_t = HQw/550 = HQ/8.82 \tag{8-30}$$

where H = head in feet, Q = flow of water in cubic feet per second, w = weight of water in pounds per cubic foot. The head is established by the topography of the country and the location of the dam, intake works, powerhouse, and tailrace or tail-water reservoir. An analysis of the river-flow records, type of turbine, and type of load (whether base or peak) will fix the maximum and mean value of flow to be used for design.

The actual horsepower P of a hydraulic turbine is the theoretical horsepower multiplied by the turbine efficiency e,

$$P = P_t e = HQe/8.82 \tag{8-31}$$

The efficiency will vary depending upon type of turbine, load, and operating head. For general purposes it is usual to assume a mean efficiency of 85 to 90%. Maximums approaching 95% at the peak of the curve have been reported, based on field tests.

The kilowatt capacity of a hydroelectric unit can be obtained by converting the turbine output in actual horsepower to generator output kilowatts by the following equation,

$$kW = 0.746Pe_g \tag{8-32}$$

where e_g = generator efficiency, which will range from 92% in the smaller machines to over 98% in the larger machines.

A combined efficiency of both turbine and generator of 82% is conservative for general purposes, although cases have been reported where the combined maximum efficiency has reached 93% at the peak of the performance curve.

54. The laws of proportionality (the variation of power, speed and discharge with runner size and head) for turbines of varying size but with the same basic conformation and design (also called homologous turbines) are shown in Table 8-13.

Table 8-13. Proportionality Laws

For constant runner dia.	For constant head	For variable dia. and head
$P \propto H^{3/2}$ $n \propto H^{1/2}$ $Q \propto H^{1/2}$	$P \propto D^2$ $n \propto 1/D$ $Q \propto D^2$	$P \propto H^{3/2}D^2$ $n \propto H^{1/2}/D$ $Q \propto H^{1/2}D^2$

D = nominal diameter of the turbine runner.

55. Specific speed (n_s) is the common basis of comparison between turbine runners of different types and between runners of the same type but different design and performance characteristics. It is the constant relationship between the speed of a runner at the point of highest efficiency and the maximum power output at this speed, regardless of size. However, since both power and speed vary with head, specific speed is defined as the relationship between the speed n_1 and power P_1 at 1 ft head. Subscript 1 denotes that the value is reduced by the proportionality law to 1 ft head basis. Since $n \propto 1/D$ and $P \propto D^2$, the product $n_1\sqrt{P_1}$ remains a constant for a given design runner regardless of its size and is designated the specific speed (n_s) of that particular design runner.

The term specific speed for this relationship stems from the fact that $n_1\sqrt{P_1}$ also is the value of the speed in revolutions per minute at the best efficiency which the runner would have if operated under 1 ft head, the runner being of such size as to develop 1 hp ($P_1 = 1$).

Since $n \propto H^{1/2}$ and $P \propto H^{3/2}$, for homologous runners, $n_1 = n/H^{1/2}$ and $P_1 = P/H^{3/2}$, where n is the speed and P the power output of any size or type of runner operating under head H. Substituting these values for n_1 and P_1 in the formula $n_s = n_1\sqrt{P_1}$, we have specific speed $n_s = (n/H^{1/2})(\sqrt{P/H^{3/2}}) = n\sqrt{P}/H^{5/4}$ for any size and type of runner operating at speed n, with a power output of P under head H.

ELEMENTS OF A HYDROELECTRIC PLANT

56. Principal Elements. *Dams* can be of two types, (1) impounding, or nonoverflow, and (2) spillway, or overflow. If impounding dams are used, means must be provided to release excess flow, by a separate spillway section, by regulating valves, or by large spillway gates. Earth dams, rock-fill dams, and high-reservoir concrete-arch dams are examples of nonoverflow types. Careful control of the reservoir elevation is needed to prevent overtopping these dams and causing damage or even failure. Spillway dams are always concrete, and for low head installations the powerhouse usually forms part of the dam.

Intakes may consist of canals, flumes, or concrete passageways.

Conduits may consist of concrete or rock tunnels, steel pipelines, steel penstocks, or any combination thereof.

Trash racks are provided at the inlet to the intake or the conduit to protect the turbine against floating or other material. Cleaning devices such as rakes, either manual or motor-operated, are provided to remove debris from the racks.

Head gates or *stop logs* are provided at the inlet to the intake or conduit and at the outlet of the draft tube for shutting off the flow to the turbine for safety and for ease of maintenance. The head gates are usually of steel. The head gates or stop logs are lowered and raised by a motor-operated crane.

Pipelines and *tunnels* are the closed conduits connecting the upper reservoir to the surge tank or penstock.

Penstocks are the closed conduits connecting the upper reservoir, tunnel, or surge tank with the turbine casing. In medium-head installations, each turbine usually has its own penstock. In the case of high heads, a single penstock is frequently used and branch connections provided at the lower end to supply two or more turbines. In recent years welded-plate-steel penstocks have nearly superseded riveted penstocks.

Penstock valves located at the intake to the turbine spiral case are usually provided when the conduit is of considerable length. This permits shutting off the flow to each turbine, for safety and maintenance and to reduce leakage losses during long turbine shutdowns, without having to drain and refill a long conduit. Penstock valves are also a necessity where more than one turbine is connected to a single conduit so that the flow can be shut off to each turbine individually.

The butterfly type of valve is suitable for medium- and high-head turbines. In the past, gate valves have been used for high heads in connection with impulse turbines. However, in recent years these have been entirely superseded by rotary or plug valves, which are also sometimes used for medium heads where the loss through the butterfly valve is considered to be excessive owing to the obstruction of the valve wicket to the flow of water.

57. Powerhouse Structure. The powerhouse foundation and superstructure (Fig. 8-60) contain the hydraulic turbine, the generator, the governor system, water passages, draft tube, basements, passageways for access to the turbine casing and draft tube, and sometimes the penstock valve. The electrical apparatus is usually housed in the superstructure.

Transformers and oil circuit breakers are located within the superstructure, on the roof, or, frequently, on a deck built over the draft-tube extension. The transformers and switchgear are sometimes located outdoors adjacent to the powerhouse and are not an integral part of it.

Cranes are provided in the powerhouse to handle the heaviest pieces of the turbine and generator and sometimes extend over the penstock valves.

58. Outdoor Powerhouse. A design developed in recent years is the outdoor powerhouse, where the operating floor is placed adjacent to the turbine pits, with the generator located outdoors on the roof of a one-story structure. All superstructure is omitted and a watertight removable cover placed over the generator. A gantry crane can be used for erection and servicing of the equipment. This outdoor design reduces overall costs of the power plant.

59. Powerhouse Auxiliaries. The hydroelectric powerhouse requires some basic *auxiliaries* such as controls, switchboards, exciters, cranes, circuit breakers, and transformers. In addition, some special apparatus is usually included.

1. *Service Units.* Usually a small hydraulic turbine and generator used for supplying power for internal plant use and as a source of independent power supply in case the power plant is electrically separated from the main system.
2. *Casing drain valves* for draining the turbine.
3. *Strainers,* or *filters,* for bearing or cooling-water supply.
4. *Air compressors* for charging governor oil systems, generator brakes, tail-water depression systems, etc.

TYPES OF HYDRAULIC TURBINES

60. The hydraulic turbine, sometimes referred to as a **water wheel,** is the most important element in a hydroelectric power plant. There are two general groups of hydraulic turbines: (1) *reaction,* where the water under pressure is only partly converted into velocity before it enters the turbine runner (Fig. 8-51) and (2) *impulse,* where the water under pressure is entirely converted into velocity before it enters the turbine runner (Fig. 8-56). A further classification of reaction turbines is: (1) *Francis* (Fig. 8-51) and (2) *propeller,* which can be further subdivided into:

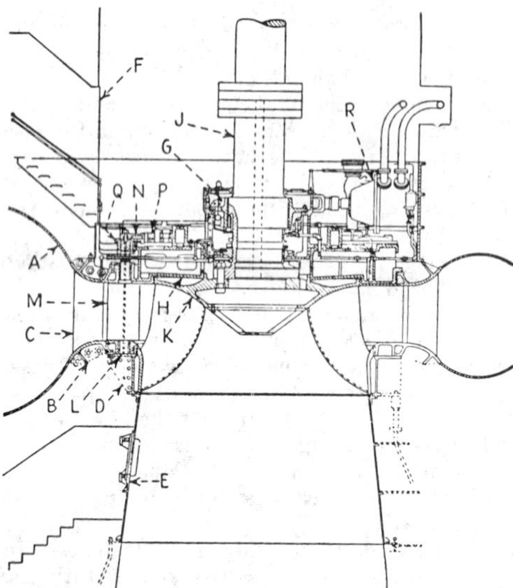

FIG. 8-51. Sectional elevation of a Francis reaction turbine. (*A*) Spiral case; (*B*) stay ring; (*C*) stay vane; (*D*) discharge ring; (*E*) draft tube liner; (*F*) pit liner; (*G*) main shaft bearing; (*H*) head cover; (*J*) main shaft; (*K*) runner; (*L*) bottom ring; (*M*) wicket gates; (*N*) shear pins; (*P*) links; (*Q*) gate levers; (*R*) servomotors.

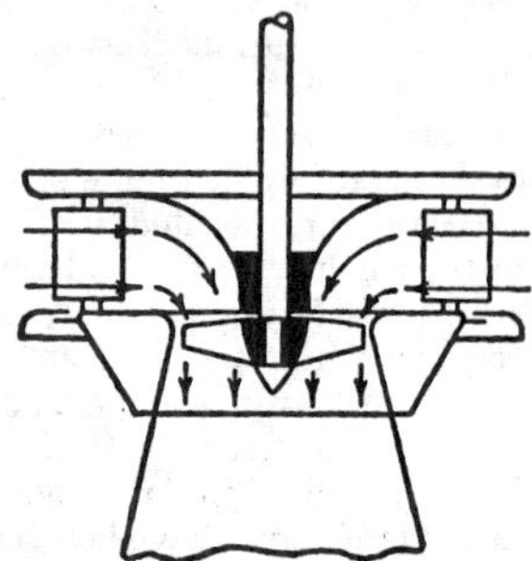

FIG. 8-52. Vertical-shaft reaction turbine, fixed-blade propeller type.

1. *Fixed-blade propeller turbines* (Fig. 8-52).
2. *Adjustable-blade propeller turbines* (Figs. 8-53 and 8-54).
3. *Axial-flow propeller turbines* (Fig. 8-59).
4. *Diagonal-flow turbines* (Fig. 8-55).

Reaction and impulse turbines have in common a stationary guide case and a revolving part, the runner. In the guide case of the reaction turbine, the water under pressure is only partly converted into velocity, leaving a pressure head at the entrance to the runner. This pressure head causes an acceleration of the relative velocity of flow through the runner,

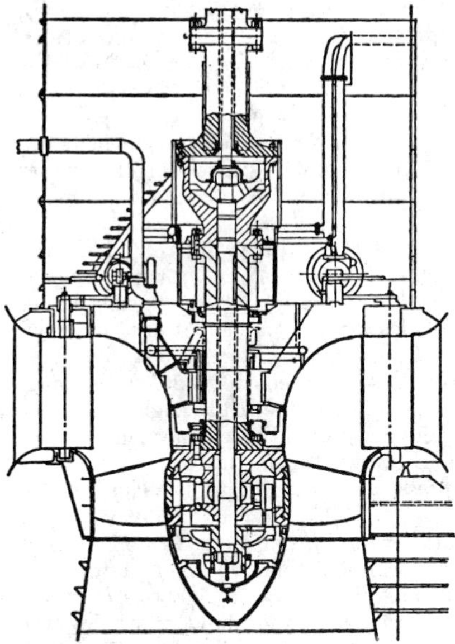

FIG. 8-53. Sectional elevation of a Kaplan turbine.

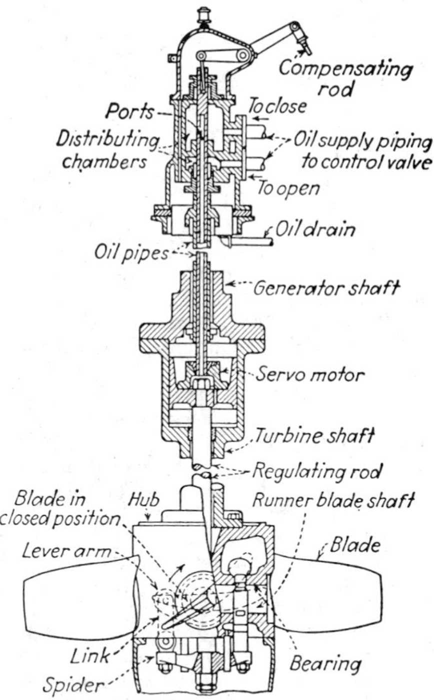

FIG. 8-54. Sectional elevation of rotating element (runner and operating mechanism) of a Kaplan turbine.

the discharge area of which is smaller than the entrance area. The reaction turbine thus utilizes the pressure of the water and the reactive force on the curved buckets which tend to change the direction of flow. Except in operating vented at low loads, the water passages are completely filled with water from the intake to the end of the draft tube.

In the guide case (designated the nozzle pipe, needle nozzle, and nozzle tip) of the

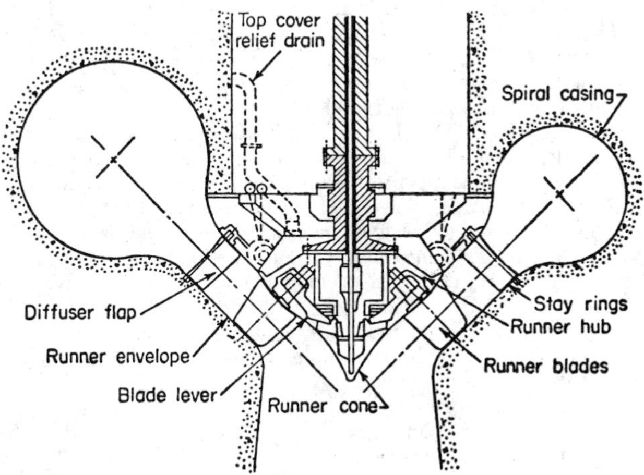

FIG. 8-55. Sectional elevation of diagonal-flow (Dariaz) turbine.

impulse turbine, the pressure head is completely transformed into velocity before striking the runner so that air surrounds both the runner and the full jet issuing from the nozzle tip.

61. Reaction Turbines. Figure 8-51 shows a Francis-type inward-flow reaction turbine. Water enters the spiral case from intake passages or penstocks (Figs. 8-58 and 8-60), passes through the stay ring, guided by the stationary stay-ring vanes, thence through the movable wicket gates through the runner and into the draft tube, through which it flows into the tailrace or tail-water reservoir. The movable wicket gates with axis parallel to the main shaft control the flow of water to the runner and thereby control the power output of the turbine.

Francis turbine runners usually have the upper ends of the buckets attached to a crown and the lower ends attached to a band, thus completely enclosing the water passageway through the runner. Francis turbines are normally used for medium heads ranging from 100 to 1,300 ft.

62. The propeller turbine, which is also of the reaction type, is differentiated from the Francis turbine in that the runner has unshrouded blades (no crown or band). The blades, three to eight in number, are either fixed or adjustable. As the names indicate, in the fixed-blade propeller runner the blades are in a permanent fixed position, while in the adjustable-blade runner the blade angle can be adjusted.

For fixed-blade runners the blade angle is usually set between 16 to 20° where maximum efficiency is obtained. For adjustable-blade runners, the blade angle may vary from 10° minimum to 32° maximum. The blades may be adjusted mechanically by hand or electric motor through a train of gears. However, this method has been largely abandoned in recent years in favor of the oil-pressure-operated blades. This type of turbine is commonly called a Kaplan turbine. Figure 8-53 is a sectional elevation of a Kaplan turbine, and Fig. 8-54 is a sectional view of its rotating element. The blades are adjusted by means of an oil-operated piston located within the main shaft. The operating piston can also be located in the hub of the runner, either above or below the runner blades. The oil is admitted to and discharged from above and below the piston by means of an oil distributor located either on top of the generator shaft above the generator or surrounding the main shaft below the generator. The oil pressure is supplied from the governor oil-pressure system, and the flow of oil is controlled by the governor. The control has a cam so shaped and arranged that blade tilt will vary with the wicket-gate opening so as to produce a maximum-efficiency envelope curve (Fig. 8-63). The greater the wicket gate opening, the greater the angle of tilt and the greater the power output.

Propeller turbines have the same arrangement of water passages as reaction turbines, and the flow of water through these passageways is the same in both types.

63. Axial-flow turbines use the propeller-type runner with either fixed or adjustable blades. Their characteristic feature is that the water approaches the runner parallel to the main shaft and thus avoids having to make a 90° turn after leaving the wicket gates in order to enter the runner, since the axes of the wicket gates extend radially outward from the main shaft, providing a straight or nearly straight through passageway from intake to draft-tube discharge (Fig. 8-59).

The propeller runners described above with either fixed or adjustable blades have the axis of the blades substantially at right angles to the main shaft (Figs. 8-52, 8-53, and 8-54). However, there is a type of propeller runner in which the axis of the blades is at about 45° with the main shaft. These are commonly known as *diagonal-flow turbines.* The blades may be either fixed or adjustable. If adjustable, they are usually known as *Dariaz turbines,* which are sometimes further characterized by having the axis of the wicket gates (either fixed or movable) set at a 45° angle with the main shaft and the spiral case angled accordingly (Fig. 8-55). Some adjustable-blade diagonal-flow runners are so designed that the blades can be closed against one another to shut off the flow of water through the runner, thus eliminating the need for adjustable wicket gates for this purpose.

In general, diagonal-flow runners bridge the gap between the propeller and the Francis runners. Thus their characteristic performance lies somewhere between these two, and they are therefore used for the higher heads in the propeller range and overlap part of the Francis range. Dariaz turbines have been proposed for heads as high as 400 ft.

The regular propeller turbines are used for low heads, ranging from the lowest head that is practical (one installation operates under a 7-ft head) to heads up to 200 ft, thus partly overlapping the range of heads for Francis turbines.

64. Impulse Turbines. The impulse turbine in its modern form consists of one or more free jets of water discharging into an aerated space and impinging on a set of buckets attached around the periphery of a disk (Figs. 8-56 and 8-61). The buckets vary in some details of their construction but in general are bowl-shaped and have a central dividing wall, or splitter, extending radially outward from the shaft. This splitter divides the stream, and the bowl-shaped portions of the bucket turn the water back, imparting the full effect of the jet to the runner. The free jet is formed by the water passing through the nozzle pipe, the needle nozzle, and thence through the nozzle tip.

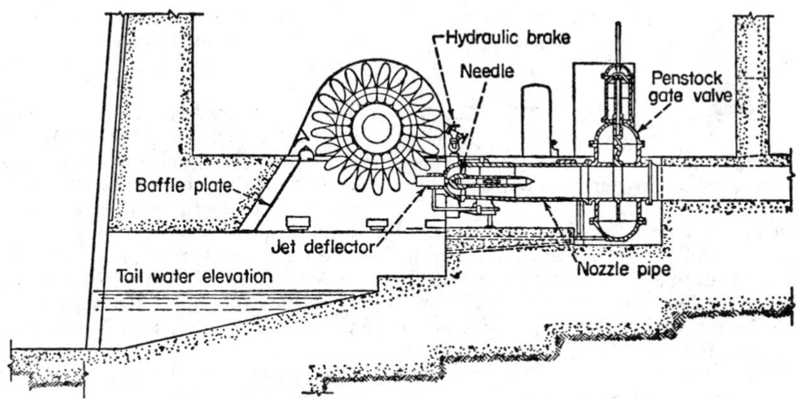

Fig. 8-56. Section through horizontal-impulse turbine.

The size of the jet and thus the power output of the turbine are controlled by a needle in the center of the needle nozzle and needle tip. The movement of the needle is controlled by the governor. A jet deflector is located just outside the nozzle tip to deflect the jet from the buckets to effect sudden load reductions.

Impulse turbines are utilized when the head is too high for practical use of Francis turbines, which is normally any head exceeding 1,000 to 1,300 ft.

POWER-PLANT SETTINGS

65. Plant Arrangement. The setting or arrangement of hydraulic turbines in a power plant varies with the type of turbine, the head, and the type of dam and intake. In the past the most common and most economical setting for heads below 40 ft for either Francis or propeller turbines, where the power output was small, was the *open-flume*, in which the water has a free surface exposed to atmospheric pressure (Fig. 8-57). The turbine is completely submerged in an open chamber, essentially rectangular in form. One disadvantage is the difficulty of lubricating the wicket-gate-operating mechanism.

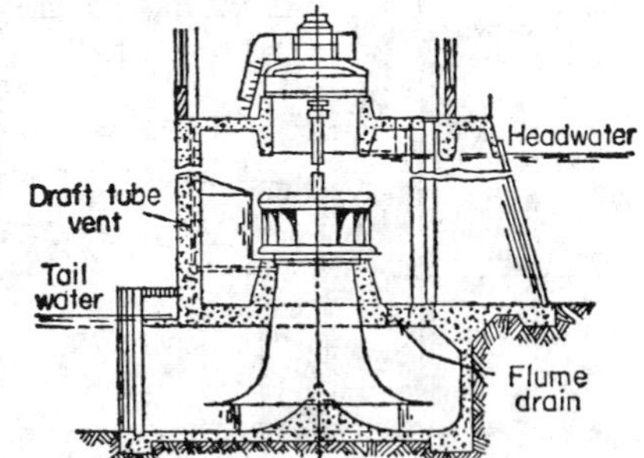

Fig. 8-57. Vertical open-flume setting.

Structural difficulties limit the runner discharge diameters of open-flume turbines to 6 to 8 ft. While open-flume settings ordinarily are used for vertical turbines, they can be used for turbines with horizontal shafts.

The horizontal-shaft turbines can use two or four runners of the Francis type in order to increase the specific speed, resulting in an increased power output for a given speed. With the development of propeller runners with inherently higher specific speeds, the need for multiple-Francis-runner turbines was practically eliminated.

For larger-sized turbines and for heads up to 90 ft, the *vertical* setting with either a complete spiral or a semispiral concrete case can be used (Fig. 8-58). The concrete casing utilizes a stay ring to carry the vertical loads of the generator and superstructure as well as to provide guidance for the water.

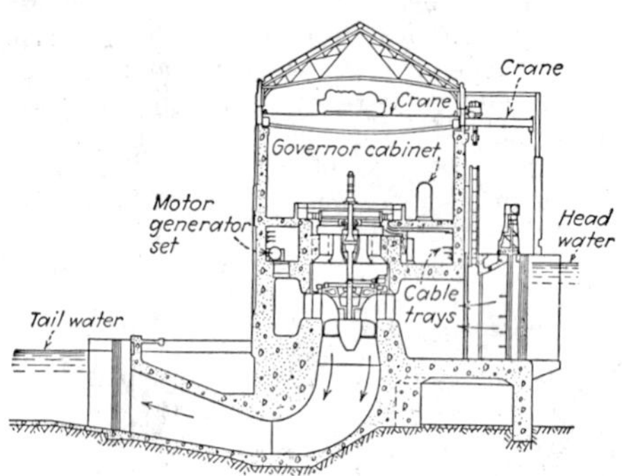

Fig. 8-58. Cross section of typical low-head concrete spiral-case setting, with Kaplan turbine.

The open-flume and concrete-spiral-case settings described above have been largely superseded in recent years by axial-flow turbines, which use either fixed- or adjustable-blade runners. Their principal characteristic feature is the straight-through or nearly straight-through water passageway from intake to draft-tube discharge. The shaft is therefore either horizontal or slightly inclined (Fig. 8-59). The spiral or semispiral case and elbow draft tube, which require substantial widths and depth of excavation, are eliminated. Therefore, with a reduction in height and area of the powerhouse and the turbine's suitability for location directly within the dam, an overall construction-cost savings of up to 35% can be obtained for the power-plant part of the project compared with conventional vertical units. This makes it possible to build or redevelop power plants for low heads or low capacities that have previously been considered uneconomical.

In recent years, axial-flow turbines have been used in tidal plants and designed to operate as pumps or turbines, with water flowing in either direction.

There are four general types of axial-flow turbines. The first, with the generator rotor mounted around the periphery of the turbine runner, has been largely abandoned owing to the difficulty in sealing the large-diameter gap between the rotor and the turbine water passageway. Such installations have had high maintenance costs and excessive outages. The *pit* and *bulb* arrangements locate the generator in series with the turbine runner at a submerged elevation. In the *pit* type, a watertight submerged enclosure is used to house the generator and, when used, the speed increaser. The *bulb*-type unit has the generator enclosed in a streamlined watertight bulb located in the water passageway either on the upstream or on the downstream side of the runner.

The fourth type is the *Tube turbine*, which has the generator located outside the water passages (Fig. 8-59). With this type, a slight bend in the water passageway permits extending the turbine shaft externally. While the unit can be arranged so

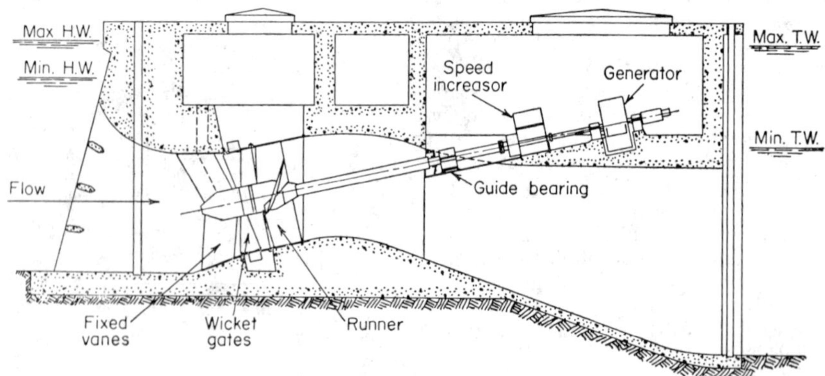

Fig. 8-59. Sectional elevation of axial-flow (tube) turbine.

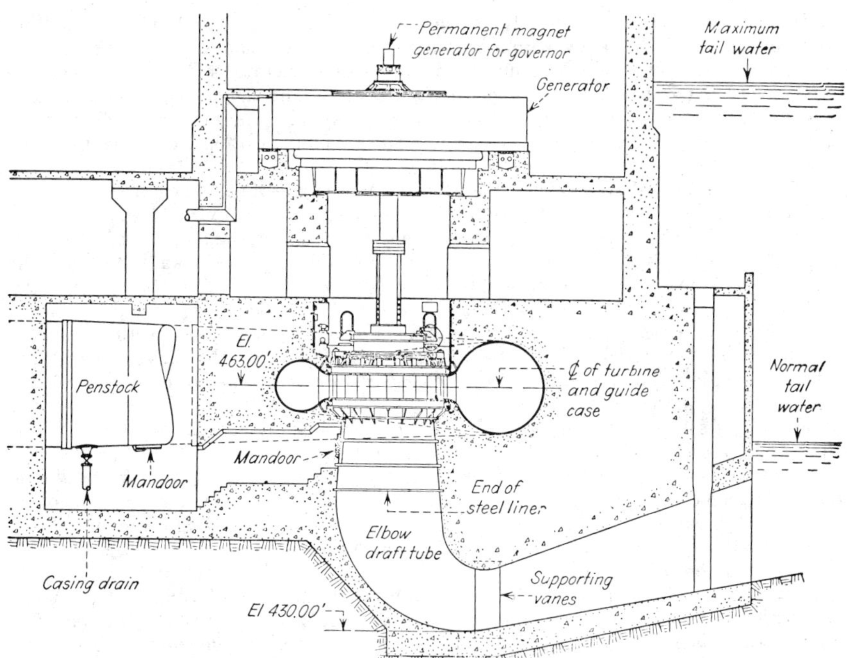

FIG. 8-60. Plate-steel spiral-case setting of vertical Francis turbine, welded casing.

that the generator is either upstream or downstream, the latter is more practical for large low-head units. To reduce excavation, the shaft may be inclined, as shown in Fig. 8-59, thereby raising the generator higher with reference to tail-water elevations.

Any hydraulic losses due to the bend as compared with the pit and bulb straight-through flow arrangements are more than compensated for by the absence of a pit or bulb obstructing the water passageway. Otherwise, the Tube turbine has all the advantages of the bulb turbine, plus the fact that the generator is much more accessible. In some cases, gear-type speed increasers are used to reduce the combined equipment cost and decrease the generator size and weight.

Axial-flow turbines are suitable for heads up to at least 100 ft, with basically the same limitations as apply to the conventional Kaplan or other propeller-type turbines. Maximum unit capacity is limited, however, by maximum practical speed-increaser torques and maximum practical horizontal generator capacities. This appears to be approximately 50 MW at present. Either fixed-position or movable radial wicket gates can be used. Power and efficiency performance are comparable with conventional vertical-shaft propeller turbines.

66. Metal-spiral-case vertical settings (Fig. 8-60) are most frequently used for reaction turbines for heads above 50 ft, up to the upper limit of about 1,300 ft for such turbines. Plate steel, either riveted or welded or a combination of both, can be used for the cases, although in recent years they are practically all entirely field- or shop-welded. Cast-iron cases are restricted to the smaller-sized turbines under low heads. Cast-steel cases are sometimes used for small turbines under high heads, where the forming of heavy plate steel would be difficult.

The metal spiral cases are usually directly connected to a steel penstock or penstock valve. Metal-spiral or -cylindrical-case horizontal-shaft settings can be used for small turbines. The cylindrical case, however, is suitable only for low-head turbines.

67. The horizontal single-jet impulse-wheel arrangement is shown in Fig. 8-56. To minimize the loss of available head, the lower edge of the buckets should be set as close to maximum tail water as possible, but not closer than 3 ft, to ensure that the runner revolves in air at all times.

Impulse turbines have relatively low specific speeds (n_s = 3.5 to 6.0), with resulting low unit speeds. This is overcome on horizontal units by use of two runners (commonly known as the double overhung unit) with the generator mounted between the two runners or by use of two jets per runner, thus doubling the output for a given size of turbine and increasing n_s by $\sqrt{2}$ times, or about 46%. The double overhung unit, or two jets per runner, requires a Y-type of nozzle pipe to bring the water from the penstock to the needle nozzles.

68. The vertical-shaft multijet impulse turbine (Fig. 8-61) has begun to supersede the horizontal-shaft unit because of its numerous advantages over the latter. Since the vertical unit can use four to six jets on one runner, the specific n_s is 2 to 2½ times as great as that of the single-jet single-runner horizontal unit. In addition, with the use of multiple jets, the runner windage and friction loss is less as a percentage of power output. Also, with a properly designed housing for the vertical unit, there is less tendency for the discharged water to interfere with the runner. These features increase the efficiency over horizontal-unit performance 2 to 3%. Another advantage is that with multiple jets the unit can be operated at part load with a reduced number of jets, thus increasing part-load efficiency.

Model tests have indicated that six jets are about the maximum number that can be used on one runner without their interfering with each other.

In vertical units the lower edge of the buckets should be at least 5 ft above maximum tail-water elevation. Any gain in head by setting the unit closer is more than offset by loss in performance. Impulse-turbine discharge entrains a large amount of air, and unless this is completely replaced, a vacuum will be produced if the outlet from the housing is sealed off by the tail water. The vacuum will then draw the tail water up until it drowns out the runner. Thus the roof of the discharge tunnel or passageway for both horizontal and vertical units should be at least 3 ft above the maximum operating tail-water elevation to permit the circulation of free air to the runner (Figs. 8-56 and 8-61). However, in both types of units where extremely high tail water is likely

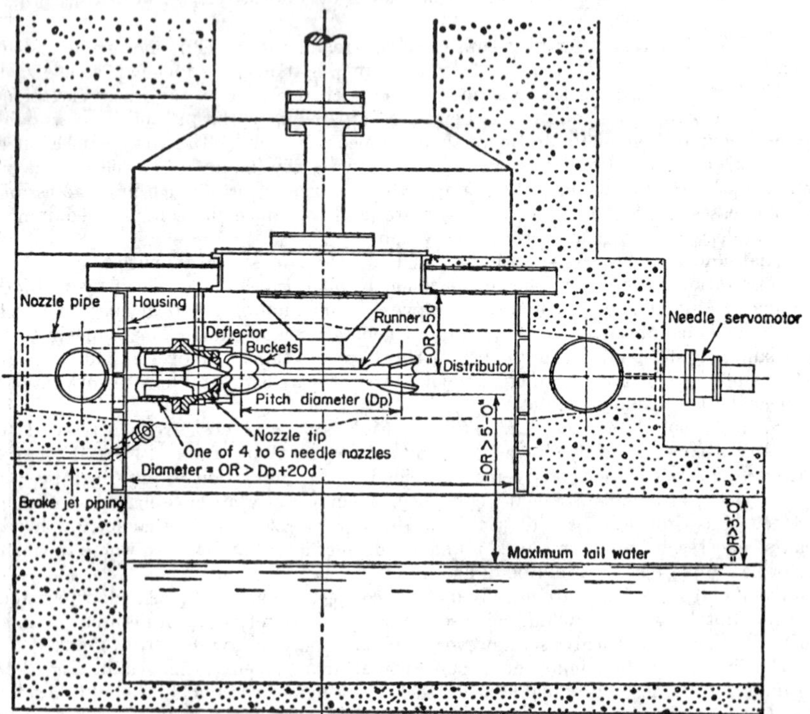

Fig. 8-61. Vertical-shaft multijet impulse turbine.

to occur for a short period of time, compressed air can be used to depress the water elevation, permitting the runner to be set closer to normal tail water.

EFFICIENCY PERFORMANCE

69. Efficiency Performance. In selecting the type of turbine for a given hydroelectric power plant, it is important to consider the efficiency performance of the various types available for the head contemplated. Not only is this true of the maximum efficiency obtainable, but the percent of full load where this maximum occurs and the efficiencies at part loads are also important.

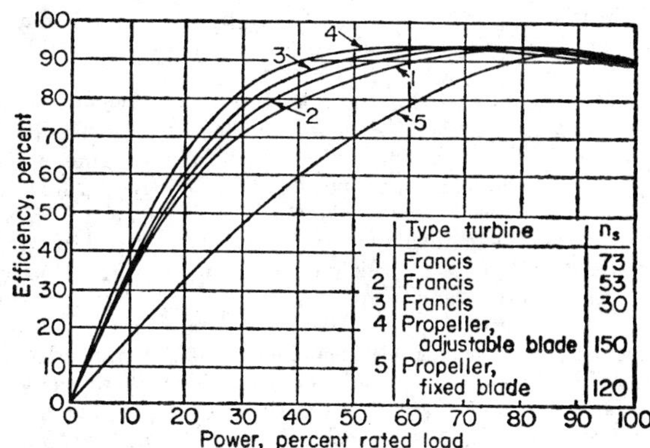

FIG. 8-62. Efficiency-load relations for reaction turbines.

Figure 8-62 shows the efficiency from 0 to 100% rated load of various specific-speed reaction turbines, both Francis and propeller. As the specific speed increases, the percent of full load at which maximum efficiency occurs increases and part-load efficiencies drop. While the maximum efficiency of the adjustable-blade propeller turbine is not so high as the medium specific-speed Francis turbine, its part-load efficiencies are higher.

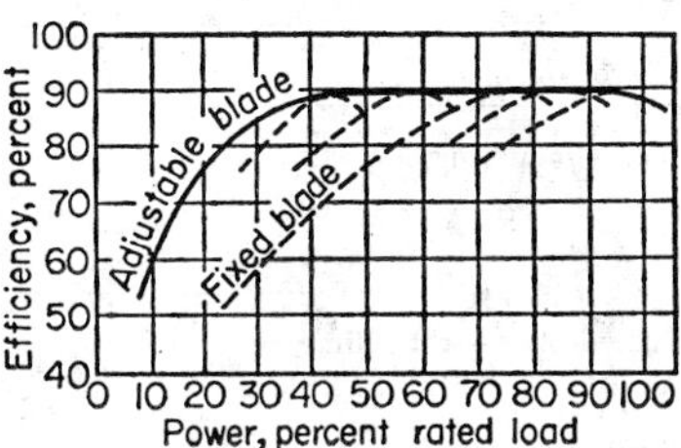

FIG. 8-63. Efficiency-load relations for fixed- and adjustable-blade propeller turbines.

Figure 8-63 shows how the adjustable-blade turbine produces an envelope efficiency curve consisting of the maximum efficiencies of all the blade tilts. Figure 8-63 also shows that the blade-tilt angle producing maximum efficiency on the envelope curve is at a lower tilt angle than the maximum for the normal adjustable-blade runner. Thus, since a fixed-blade runner usually has its blade-tilt angle set for maximum efficiency, its maximum output is about 25% less than that for an adjustable-blade runner of the same size.

Figure 8-64 shows the efficiencies of a horizontal and a vertical impulse turbine. While the maximum efficiency of both is somewhat lower than that for Francis-type turbines, the part-load efficiencies are higher, which is one of the advantages of impulse turbines. The efficiencies shown in Figs. 8-62, 8-63, and 8-64 are all based on field tests of actual installations.

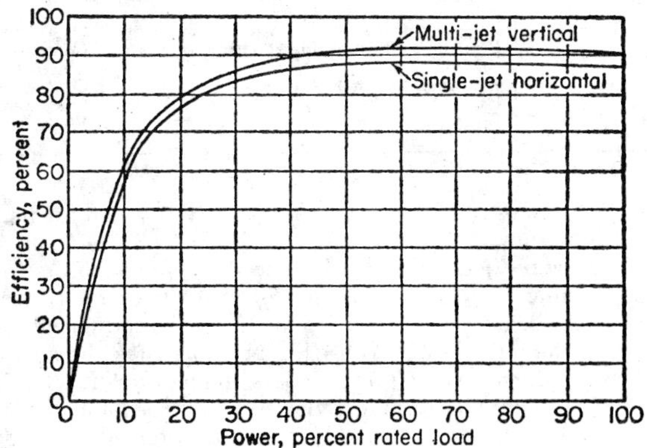

FIG. 8-64. Vertical and horizontal impulse turbine efficiency-load relations.

DESIGN FACTORS

70. Selection of Type of Setting, Turbine, and Case. The type of setting, turbine, and case selected for a hydroelectric power plant, while depending largely upon the available head, also depends upon the size of the turbine, local conditions, type of dam, and economics involved. Thus, no hard-and-fast rule can be established for such selections. Furthermore, there is considerable overlap in head where several types of turbines or cases can be used. Table 8-14 is therefore based upon general practice and should not be taken as limits in head above or below which the respective types cannot be used. The propeller and diagonal turbines listed in the table can use either the fixed-blade or the adjustable-blade type of runner.

71. Selection of Specific Speed and Unit Speed. The speed of the turbine should be as high as practical, as the higher the speed, the smaller the overall size of the turbine and the less costly. Also, since hydraulic turbines are usually connected to electric generators, the higher the speed, the less costly and more efficient the generators.

The speed of the turbine is tied in with the specific speed n_s, which is a characteristic of each design of turbine runner (Par. **55**), which in turn varies with the head under which it will be used. In general, propeller-type runners are designed so that $n_s = 500/H^{1/3}$. For Francis-type runners, $n_s = 700/H^{1/2}$. These two equations immediately indicate the advantage of propeller-type turbines over Francis types for the lower heads. For example, for a 100-ft-head plant, the recommended Francis n_s would be 70, while that for the propeller would be 108. The adoption of the above values for specific speed for propeller and Francis turbines is predicated on setting the elevation of the runner with reference to the tail-water elevation to obtain the proper field cavitation coefficient (Par. **91**).

Table 8-14. Type of Setting, Turbine, and Casing Usually Employed for the Head Available

Head, ft	Type of turbine	Type of casing
	Vertical or horizontal open-flume settings	
Up to 40	Propeller	Rectangular concrete
	Vertical or horizontal axial-flow settings	
Up to 100	Propeller	Welded-plate-steel cylindrical tube
	Vertical encased settings	
15–90	Propeller	Concrete spiral or semispiral
	Vertical or horizontal encased settings	
50–100	Propeller or diagonal flow	Cast-iron or welded-plate-steel spiral
100–200	Propeller, diagonal flow, or Francis	Cast-iron, cast-steel, or welded-plate-steel spiral
200–400	Diagonal flow or Francis	Cast-steel or welded-plate-steel spiral
400–1300	Francis	Cast-steel or welded-plate-steel spiral
1000–5500	Impulse	Welded-plate-steel housing

For impulse turbines an n_s per jet of 5.0 to 5.5 for heads around 1,000 ft gives the best efficiency. For heads around 2,000 to 3,000 ft, n_s for best efficiency should be 4.0 to 5.0 per jet. The specific speed of impulse units can be increased by increasing the number of jets used on a single runner or by increasing the number of runners per unit, since the specific speed of the unit is the specific speed per jet times the square root of the number of jets used. This indicates the advantage of the vertical impulse turbine, which can use up to six jets on one runner.

After having determined the specific speed for the type of turbine selected, the approximate speed of the turbine can be obtained from the equations $n_s = n\sqrt{P}/H^{5/4}$, $n = n_sH^{5/4}/\sqrt{P}$. Since hydraulic turbines are usually directly connected to alternating-current generators, the turbine speed must agree with one of the nearest synchronous speeds as determined from the system frequency. This speed is obtained from the equation $n = 120f/$(number poles in the generator), where f is the system frequency (cycles per second). The number of poles must be an even number.

72. Number of Units. From the standpoint of reducing the number of auxiliaries and the amount of associated equipment and also reducing initial and maintenance costs for the entire plant, the number of units should be a minimum. Also, the larger the unit, the higher the efficiency. However, other considerations such as flexibility of operation, higher-efficiency operation during low-load demands, and minimum loss of capacity during shutdown for repair or maintenance might dictate the use of multiple units, where one unit would be feasible from the standpoint of physical size. For some projects using Francis or fixed-blade propeller runners, the physical size of the unit has been limited to the maximum size of runner that could be shipped in one piece, largely owing to the extra manufacturing costs involved in furnishing split runners. However, since split runners present no serious mechanical difficulties, the tendency in recent years has been to disregard this limitation. Other limitations on size are the availability of machine tools for machining large turbine parts. For a general discussion of large turbines see the paper The Trend towards Larger Turbines by Coulson, Gilcrest, and Haycock presented to the Engineering Institute of Canada, October, 1965.

73. Runaway Speed. If a reaction-type turbine runner is allowed to revolve freely without load and with the wicket gates wide open, it will overspeed to a value called the runaway speed. The runaway speed of a turbine at normal head varies with the specific speed and for Francis turbines ranges from about 170% (normal speed = 100%) at low specific speed ($n_s = 20$) to 195% at high specific speed ($n_s = 100$). For propeller turbines the runaway speed varies with blade angle; the steeper the blade angle, the lower the runaway speed. For fixed-blade propellers with the blades set at 16 to 20°, where maximum efficiency is usually obtained, the runaway speed will be about 255 to 235%, respectively. For adjustable-blade turbines, where the minimum blade angle is sometimes as low as 10 to 12° in order to obtain high efficiency at low load, the maximum possible runaway speed will be about 290 to 270%, respectively. However, with adjustable-blade propeller turbines there is from the standpoint of efficiency an optimum relationship between runner blade angle and wicket-gate opening, usually controlled by a cam in the operating mechanism; the higher the gate opening, the steeper the blade angle. Thus the combination of wide-open gate and minimum blade angle can occur only in the so-called "off-cam" position, which is an extremely rare possibility. In most units this maximum possible "off-cam" runaway speed is reduced by limiting the minimum blade angle to 14 to 16°. Another method is the use of a runaway speed limiter. There are several designs in use. One of the most reliable is a valve designed to open by centrifugal force at speeds of about 130%. The open valve bypasses the runner-blade servomotor piston, thus equalizing the oil pressure on both sides of this piston. If the runner blades are designed to open because of hydraulic unbalance at overspeeds, they will then go to a higher blade angle under off-cam overspeed conditions. The off-cam runaway speed can thus be limited to about 195%, if the runner is designed for a maximum blade angle of 28 to 32°.

The runaway speed for impulse turbines ranges from 180 to 190% of normal speed, depending upon the specific speed of the runner per jet. The higher the specific speed, the higher the runaway speed.

For all turbines, if the maximum head is higher than the normal head, the runaway speed will be increased in proportion to the square root of the head. Therefore, runaway speeds should be based on the maximum operating head rather than on the normal head.

Generators must be designed for the maximum runaway speed. The greater this speed, the greater the cost of the generator.

74. Weight of Runner, Turbine Thrust, and WR^2. The approximate weight of any Francis runner is $0.024D_{th}^3$, where D_{th} is the throat diameter of the runner in inches. For propeller-type runners with fixed blades the constant may be taken as 0.009; for Kaplan-type runners, as 0.018.

The *hydraulic thrust* on Francis runners varies considerably with type, design, specific speed, the pressure between the movable wicket gates and runner, seal design, seal clearance, and the method of venting. It is approximately between 25 and 40%

of the weight of the full head of water acting on the discharge diameter D_d of the runner. The higher the specific speed, the greater the thrust. With propeller-type runners, the thrust is nearly equal to the weight of water on the full area.

Impulse turbines have no hydraulic thrust of any consequence.

The WR^2 of turbine runners, which is their weight times the square of the radius of gyration, varies widely with the type of runner and its design. Thus, to obtain reasonable values, general equations cannot be used, and the WR^2 must be calculated for each specific design, based upon its configuration and distribution of weight.

REACTION-TURBINE ELEMENTS

75. Thrust Bearing. The weight of the rotating parts of the turbine and generator and the hydraulic thrust of a vertical unit are carried by a thrust bearing usually located just above or just below the generator rotor and furnished by the generator manufacturer. Occasionally this thrust bearing is located on the turbine head cover, in which case it is furnished by the turbine manufacturer. On horizontal units the weight of the rotating parts is carried by special heavy horizontal bearings.

76. Runner and Wearing Rings. The number of buckets for Francis runners varies from about 21 for low specific speed to 12 for high specific speed.

The number of blades for propeller-type turbines ranges from eight for low specific speeds to three for high specific speeds.

Most runners are made of cast steel, which can readily be repaired by welding with either mild- or stainless-steel electrodes. Cast iron is sometimes used for low- and medium-head Francis turbines. Built-up runners with pressed-steel or cast-steel buckets welded to the crown and band are sometimes used for Francis turbines. The welded type is suitable for very large runners. Bronze may be used for smaller runners of all types and for medium-high heads. The use of cast stainless-steel runners is increasing, particularly for small and medium-sized runners for high heads and for conditions where pitting may be troublesome.

Propeller runners are practically always made of cast steel, and the surfaces over which pitting may be expected are overlaid with stainless-steel welding, before finishing to the final contour. In place of overlay welding, solid stainless-steel inserts are sometimes welded into the Francis buckets or propeller blades.

The functions of the runner seals for Francis turbines are to prevent excessive leakage loss and thus improve the efficiency; to reduce the hydraulic thrust; to prevent seizure in operation; and to prevent undue wear.

Rolled, forged, and cast steels make excellent wearing rings for low- and medium-head units. To prevent seizure in case of contact, the rotating-ring material should differ from that of the stationary ring, and any material which contains nickel should not be used for either ring. Stainless steel, bronze, or steel with bronze inserts makes excellent wearing rings. For extremely high heads, stainless steel should be used to prevent undue wear and erosion. Wearing rings should be made renewable, or provision should be made for restoration of clearances by welding and remachining.

Seal clearances are made as small as practicable to reduce leakage, particularly on high-head units. Larger clearances are required with water-lubricated main bearings, subject to considerable wear before being readjusted.

77. Main Shaft and Bearing. The main shaft must be rigid and is made of a medium grade of forged steel, with torsional stress ranging from 4,000 to 6,000 lb/in^2.

For water-lubricated bearings a bronze or, preferably, stainless-steel sleeve is installed on the bearing surface. Such a sleeve is also put on the shaft adjacent to the packing box.

Turbines are usually provided with one main bearing located in the head cover as near the runner as practicable. This is usually babbitted and in halves, with an independent low-pressure oiling system. Self-lubricated babbitted bearings, containing an oil reservoir and pumping grooves in the babbitt, are sometimes used.

Water-lubricated bearings of lignum vitae, rubber, or special composition materials are sometimes preferred, particularly for small and medium-sized propeller-type turbines where the bearing is located at the bottom of the head cover cone and where the packing box of an oil-lubricated bearing would be inaccessible and where it would be difficult to avoid water contamination. With the water-lubricated bearing the packing box is placed above the bearing.

78. Spiral Case. The spiral case must be proportioned so as to cause relatively low friction losses, as well as to prevent eddying which would travel into the runner and affect its efficiency. The cases generally used are (1) metal case: cast steel, cast iron, or steel plate, and (2) concrete case. Metal cases are made as complete spirals, customarily with uniform or slightly increasing velocity from the throat to the small end. It is preferable that the water be accelerated as it approaches the case in order to suppress vortices.

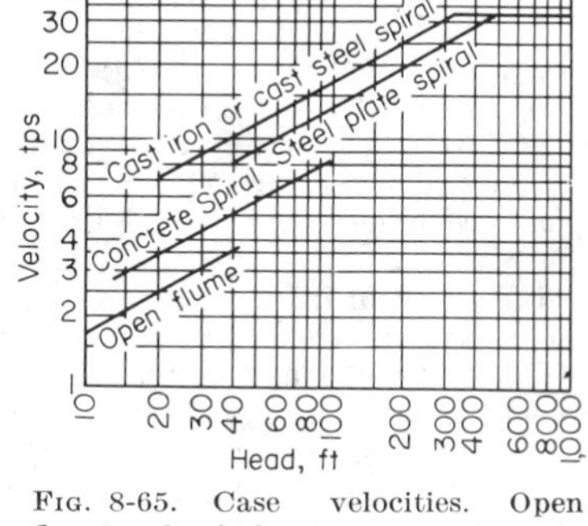

FIG. 8-65. Case velocities. Open flume and spiral.

Concrete cases may be complete spirals or semispirals, rectangular or oval in cross section. There should be no piers close to the turbine.

Rated-load velocities for general practice for cases of various types are shown in Fig. 8-65. Higher velocities are sometimes used with the larger units to reduce size and cost.

78a. Draft Tubes. The draft tube serves the double purpose of (1) allowing the turbine to be set above tail-water level, without loss of head, to facilitate inspection and maintenance, and (2) regaining, by diffuser action, the major portion of the kinetic energy delivered to it from the runner.

At rated load the velocity at the upstream end of the tube for modern units ranges from 24 to 30 ft/s, representing from 9 to 16 ft head. As the specific speed is increased and the head reduced, it becomes increasingly important to have an efficient draft tube. Good practice limits the velocity at the discharge end of the tube to 5 to 7 ft/s, representing less than 1-ft velocity head loss.

Two types of tubes are commonly used: (1) the straight conical or concentric tube and (2) the elbow type. Properly designed, the two types are about equally efficient, over 85%.

The **conical type** is generally used on low-powered units for all specific speeds and, frequently, on large high-head units. The side angle of flare ranges from 4 to 6°, the length from 3 to 4 D_d, and the discharge area from four to five times the throat area (see Fig. 8-57).

The **elbow type** of tube is now used with most turbine installations. With this type the vertical portion begins with a conical section which gradually flattens in the elbow section and then discharges horizontally through substantially rectangular sections to the tailrace. Most of the regain of energy takes place in the vertical portion, very little in the elbow section, which is shaped to deliver the water to the horizontal portion so that the regain may be efficiently completed. Figure 8-66 shows proportions of a good elbow tube, taken as the average proportions from a large number of recent installations. One or two vertical piers are placed in the horizontal portion of the tube, for structural and hydraulic reasons.

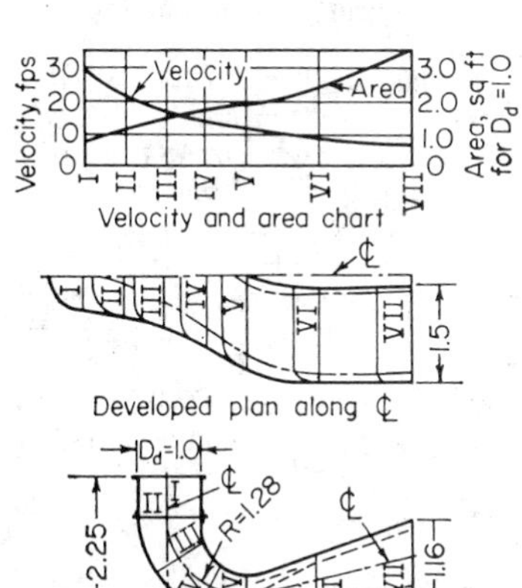

FIG. 8-66. Elbow draft-tube velocity, area and layout.

Small conical tubes are sometimes made entirely of steel plate. Most tubes are made of concrete with a steel-plate lining extending from the upper end to a point where the velocity has been sufficiently reduced (say 15 ft/s) to prevent erosion of the concrete. Sometimes the liner is carried around the elbow. Pier noses are also lined where necessary to prevent erosion and for structural reasons.

79. Stay Ring. That part of the guide apparatus between the spiral case and the wicket gates and containing stationary stay vanes is called the stay ring. The water is accelerated within this space as it approaches the gates. The number of stay vanes employed is usually equal to or one-half the number of gates. They are placed at that angle which will cause the least obstruction to the flow.

The stay ring is cast integral with cast-steel and cast-iron cases and is made separately of cast iron or of welded or cast steel for concrete or steel-plate spiral cases. It should be a continuous ring to facilitate erection, and very rigid, because it serves as a foundation for the rest of the turbine.

80. Wicket Gates and Operating Mechanism. Wicket gates control the power and speed of the turbine. The number of gates ranges from 12 for small units to 28 for large units. The overall dimensions of the turbine decrease as the number of gates increases.

To prevent interference between the gates and the runner buckets, which may cause noise and vibration, the discharge tips of the fully open gates should be kept well away from the inlet edges of the runner buckets of Francis turbines, the radial clearance being large enough to prevent the gate tips from overhanging the curved part of the discharge ring.

The height and angular movement of the gates increase with the specific speed. The angular movement varies from 15° for low specific speed to 50° for high specific speed.

Most wicket gates are made of cast steel; a few, for higher heads, are made of forged steel; some, for lower heads, are weldments built from rolled materials and castings. To reduce wear, the gate tips and ends may be coated with stainless steel welded on before final finishing.

Each gate connection to the operating ring should be provided with a breaking element to protect the gate and other mechanism in case of an obstruction. Each gate should also be provided with stops to prevent it from striking the runner or reversing after the breaking element fails.

One or more **vacuum breakers** or **air valves** are installed in the head cover to admit air to the runner or draft tube, to improve efficiency at low gate openings or to alleviate draft-tube vortex cavitation. An air valve is also necessary on propeller-type units to break the vacuum under the head cover and to help prevent the backslap of a broken water column when the gates are suddenly closed. The air valve is piped to the outside of the powerhouse above the floodwater level.

IMPULSE-TURBINE ELEMENTS

81. Impulse-turbine Elements. The essential elements of an impulse turbine (Figs. 8-56 and 8-61) are a runner, a nozzle pipe, a needle nozzle, a needle tip, a needle, a jet deflector, a housing with a guide bearing, and a shaft.

82. The runner consists of either a plate-steel or forged-steel disk with bolted-on buckets or a cast-steel or stainless-steel disk with integrally cast-on buckets of the same material. The bolted-on buckets can be cast steel, 13% chrome steel, or 18-8 stainless steel. The 13% chrome-steel and 18-8 stainless-steel buckets have greater strength than the cast steel and have greater resistance to pitting caused by cavitation. Thus, they have a considerably longer life. This is essential with multijet turbines.

83. The nozzle pipe forms the water passage which, with the *needle nozzle,* the *needle tip,* and the *needle,* forms the jet and includes that portion of passage extending downstream from the end of the penstock or penstock valve.

The nozzle pipe, needle nozzle, and needle tip are usually made of cast steel or welded plate steel, or a combination of both. For vertical impulse turbines, the nozzle pipe is circular and of decreasing diameter.

The diameter of the upstream portion of the nozzle pipe should be such that the velocity does not exceed $0.10\sqrt{2gH}$.

84. The needle is a moving element inside the needle nozzle and needle tip, actuated by a governor, a servomotor, or a hand mechanism to control the size of jet impinging on the bucket, thus controlling the power output of the turbine.

The needle and its seat in the nozzle tip should be made of a material highly resistant to erosion.

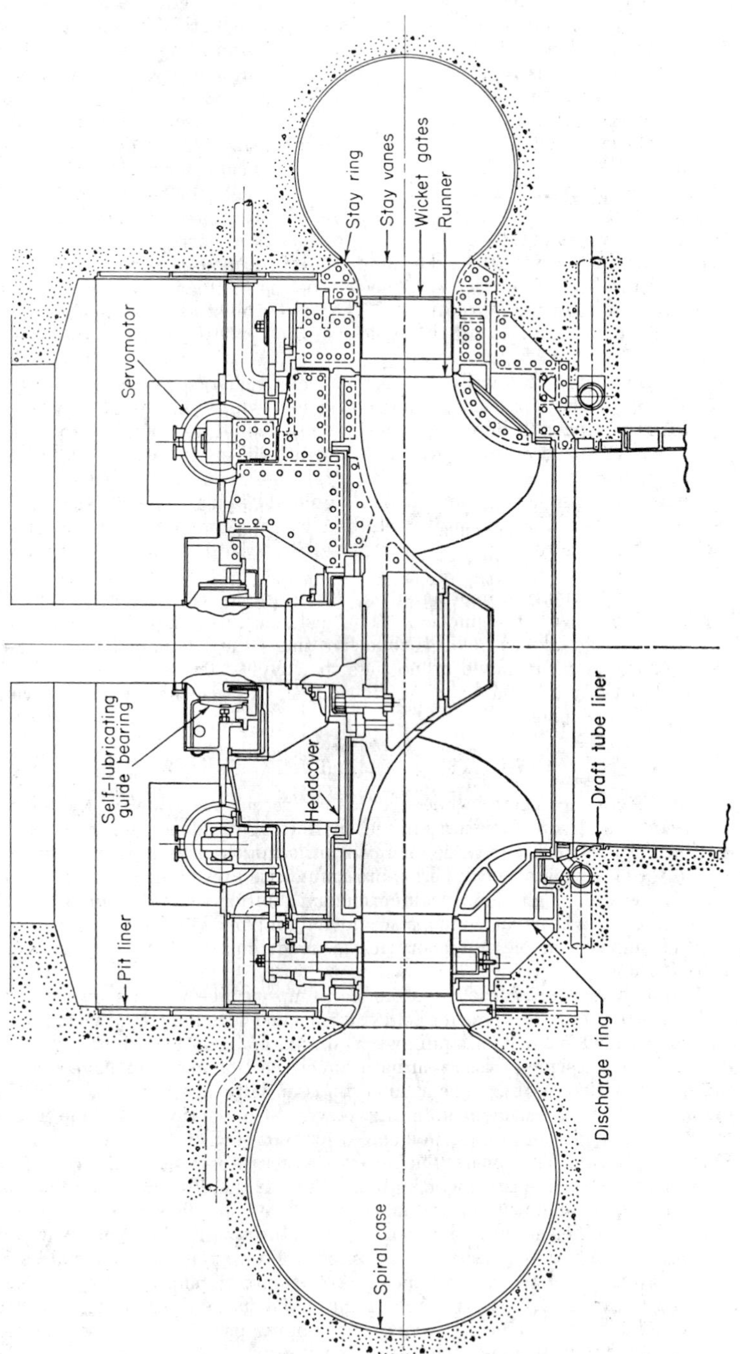

Fig. 8-67. Sectional elevation of reversible pump turbine.

85. Jet Deflector. The inertia of the water flowing through the long penstocks usually employed with impulse turbines prohibits rapid reduction in velocity because of the pressure rise which would occur. Therefore, to minimize the speed rise following a sudden load rejection, it is necessary to reduce the hydraulic power delivered to the runner without changing the flow in the penstock too rapidly. While, in the past, pressure regulators (see Par. **97**) have been used for this purpose, in recent years they have been entirely superseded by placing a governor-controlled jet deflector between the needle tip and the runner. The governor moves this deflector rapidly into the jet, reducing the power output. It is not unusual for the deflector to cut off the entire jet in about 1½s. Since the deflector acts on the jet after it leaves the nozzle, there is no change of flow in the penstock; hence, no pressure rise. The governor then moves the needle at a permissible rate (from the standpoint of pressure rise) with simultaneous automatic withdrawal of the deflector. The jet then is finally reduced the necessary amount to correspond to the reduced load. The needle must also move slowly in the opening direction for oncoming loads to avoid penstock collapse due to large pressure drops.

86. The housing serves primarily to carry off the discharged water to the tail pit below and to support the nozzles. On horizontal-shaft units, at the place where the runner receives the impact from the jet, the housing should be about 10 to 12 times the jet diameter. The more ample width results in higher efficiency. At the place where the runner has been cleared of the discharge, the housing should be as narrow as possible, to decrease windage. Suitable baffling should be used to carry the discharge away from the buckets. The housing should be adequately vented near the center of the runner, to permit the inflow of air to replace that entrained with the discharged water.

For vertical-shaft units, the housing should be of ample size, to prevent discharge water from interfering with the buckets. The distance from the head cover to the center line of the jet should be not less than five times the jet diameter (Fig. 8-61). The diameter of the housing should be not less than $D_p + 20d$, where D_p is the pitch diameter and d is the jet diameter. The housing should also be adequately vented near the center of the runner.

REVERSIBLE PUMP TURBINES

87. Pumped Storage. There has been a trend in recent years toward the increased use of *pumped-storage* hydro facilities for seasonal storage and peaking capacity. In this type of project, surplus low-value energy from either hydro- or thermal plants is used to pump water during off-peak periods to an elevated reservoir, where it becomes available for generating high-value peaking energy. While separate pumps and hydraulic turbines of conventional design can and have been used for this purpose, the development of single reversible pump turbines has made many pumped-storage projects economically feasible.

Figure 8-67 is a cross section of a reversible pump turbine which shows that with the exception of the runner it is essentially a conventional turbine. It has a spiral case, stay ring, movable wicket gates, head cover, discharge ring, and draft tube. The wicket gates must be designed for flow in both directions. A few units have been built without movable wicket gates. The runner is essentially a pump runner modified for optimum performance while generating power. Conventional turbine runners, because of their short blades, are not well suited for pumping.

88. Pump-turbine Performance. The reversible pump turbines have certain fundamental performance characteristics which are inherent in the design. The relationship between pumping and generating performance for a given specific speed is more or less fixed and can be modified only to a minor degree by alterations in the design. For example, if a certain generating capacity is desired, the pumping capacity will be fixed within certain limits. On the other hand, if a certain pumping capacity is desired, the maximum generating capacity is fixed. Figure 8-68 shows the expected generating performance, and Fig. 8-69 shows the expected pumping performance based on model tests of a reversible pump turbine with a specific speed $n_s = 47.3$ generating and 2,600 pumping.

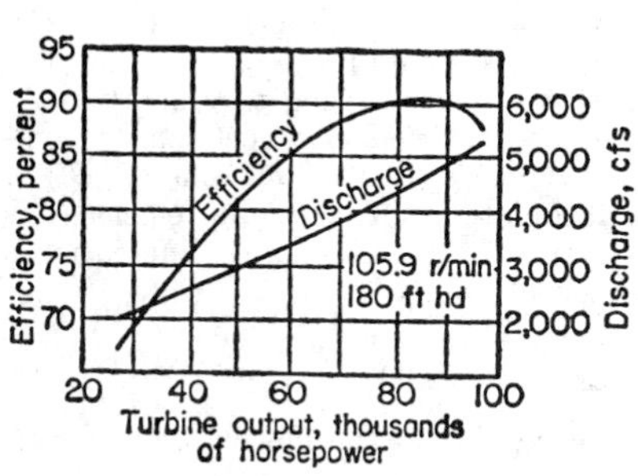

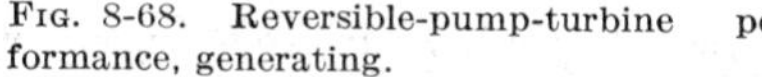
FIG. 8-68. Reversible-pump-turbine performance, generating.

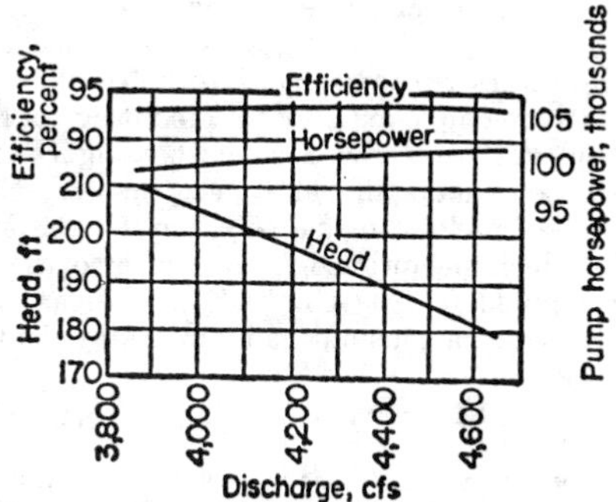

FIG. 8-69. Reversible-pump-turbine performance, pumping.

Because of the normal characteristics of reversible pump turbines the best-efficiency generating occurs at a lower speed than in pumping. This can be compensated for by using a generator-motor capable of operating at two speeds with a constant frequency. Several units of this type have been built, but since two-speed generator-motors cost considerably more than those built for single speed, a careful study should be made of the advantages to be obtained by operating at two speeds.

Single-stage reversible pump turbines can be built for any heads up to 1,200 ft. Beyond this either multiple-stage reversible units or separate pumps and turbines should be used.

The runaway speed for pump turbines is considerably lower than for conventional turbines. It ranges from 150% of normal for low-specific-speed to 175% for high-specific-speed runners.

MODEL TESTS

89. Model tests serve several purposes. They are primarily used to check turbine runner, wicket gate, draft tube, casing, and sometimes inlet works designs for optimum performance. Correctly interpreted, they may also be used as a reliable indication of the performance of the units in the field. In many cases purchasers specify the performance of a homologous model test, which is then often used as an acceptance test of the unit instead of field tests. In such cases the field conditions, particularly for the casing and draft tube, must be reproduced faithfully. Model tests should be run in accordance with the ASME Test Code for Hydraulic Turbines. The International Electrotechnical Commission is in the process of preparing an International Test Code for Hydraulic Turbines Using Laboratory Models for Acceptance Test. When completed this Code will probably be adopted universally by all members, including the United States and Canada.

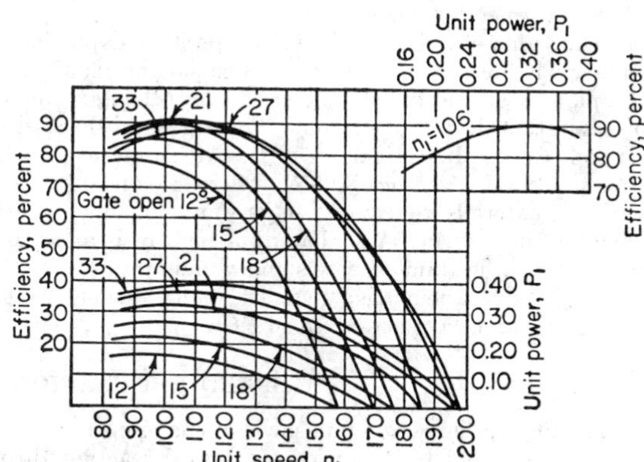

FIG. 8-70. Model test curves for Francis runner. n_s about 65.

Figure 8-70 shows typical model-test results of a reaction-type turbine in which the power P at 1 ft head and the efficiency are each plotted against the speed n at 1 ft head for each of several gate openings.

90. Laws of Proportionality for Homologous Turbines. The laws of proportionality shown in Table 8-13 are used to calculate, from model tests, the power, speed, and discharges of homologous turbines of various sizes for various heads. Actually the laws are employed only in computing the power and speed data for the field unit. The field unit will have a somewhat higher efficiency and power owing to proportionally smaller frictional and bearing losses. Expected field efficiency is customarily computed from the model efficiency by the Moody formula $e' = 1 - (1 - e)(D/D')^{1/5}$, in which the primed letters refer to the field installation. The step-up efficiency is computed for the point of best efficiency only, and the corresponding differential is applied as a constant value from, say, half load to full load.

CAVITATION

91. Cavitation occurs when the pressure at any point in the flowing water drops below the vapor pressure of the water.

The relationship among vapor pressure, barometric pressure, setting of the runner with respect to tail water and net effective head on the turbine which produces cavitation is expressed by the *Thoma cavitation coefficient,*

$$\sigma = (H_b - H_v - H_s)/H \tag{8-33}$$

where H_b is the barometric head, feet of water; H_v is the vapor pressure of water, absolute; H_s is the elevation, feet, of the runner above tail water, measured at the throat of a Francis runner and at the center line of the blades of a propeller runner (if the runner is submerged, H_s becomes negative); and H is the total or net effective head, feet, on the turbine. From the above formula

$$H_s = H_b - H_v - \sigma H \tag{8-34}$$

Thus the setting of the runner depends upon the value of σ, which varies with specific speed n_s of the runner and the individual characteristics of a particular runner design. In practice the model of the proposed runner is first tested with relatively high back pressure (H_s small or negative). Then the back pressure is reduced in increments until the breaking points as indicated by a drop in power, efficiency, and discharge are reached. This breaking point is designated as the critical σ and will vary with gate opening and speed and for propeller turbines with blade angle. Consequently σ must be determined for a range of limiting conditions.

In the absence of cavitation tests the value of σ should be not lower than $\sigma = n_s^{3/2}/2{,}000$ for Francis and propeller runners and $\sigma = n_s^2/25{,}000$ for adjustable-blade propeller runners.

The value of σ at which a plant operates, depending largely upon the setting of the runner with respect to tail water, is called the plant σ. To avoid excessive cavitation, the plant σ should exceed the critical σ. The greater this margin, the less possibility of cavitation during operation. For a general discussion of cavitation phenomena see Knapp, Recent Investigations of the Mechanics of Cavitation and Cavitation Damage, *Trans. ASME,* October, 1955. There is considerable difference in the resistance of various materials to cavitation erosion (pitting). Laboratory tests and experience have shown that materials having a high resistance and suitable for use in hydraulic turbines are the stainless steels and aluminum bronzes, especially when used as welding overlays (see Rheingans, Resistance of Various Materials to Cavitation Damage, *ASME Rept., Cavitation Symp.,* 1956).

SPEED REGULATION

92. Regulation is accomplished by changing the flow of water to the turbine. The flow is controlled by the wicket gates of reaction turbines and by the needle or jet deflector of impulse turbines. The governor, usually supplied with the turbine, moves the gates or needle in response to speed changes resulting from load or head changes.

A schematic diagram of a governor is shown in Fig. 8-71. The parts consist of a speed-responsive device, a power element that changes the gates or needle position, and a follow-up or compensating device that prevents hunting.

The **speed-responsive device** is usually a pair of spring-loaded flyballs driven by an electric motor that receives its power either from the bus line or from an independent generator driven from the main turbine-generator shaft.

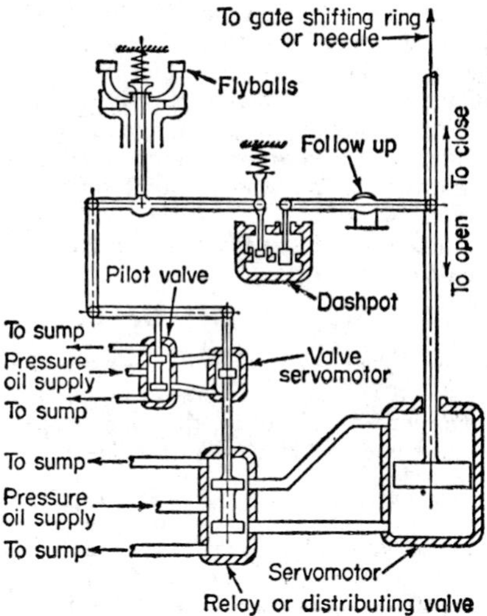

FIG. 8-71. Schematic diagram of governor.

The **power element** consists of one or two oil-operated power cylinders or servomotors which operate the turbine gates or needle. Oil pumps and a pressure tank or accumulator maintain a supply of oil under pressure. A valve operated by the flyballs controls the flow of oil to the servomotors or acts as a pilot valve controlling a larger relay valve, which in turn controls the oil to the servomotors. With a plant consisting of several units, a unit system may be used for each turbine, or there may be a central pumping unit for the turbines. The pump capacity is usually 3⅓ servomotor volumes/min. The capacity of the pressure tank is generally made 20 times the servomotor volume, allowing for 8 volumes of oil and 12 volumes of air. The velocity of oil in the pipelines is kept below 15 ft/s.

The **follow-up** or **compensating device** connects the power piston of the servomotor to the control valve, usually through a dashpot, and causes the motion of gates or needle to stop when they have moved sufficiently to compensate for the load change.

The time for a full stroke is controlled by the rate of flow of oil to the servomotors; most governors have provisions for varying this time. The gate opening changes at a uniform rate over the major portion of the stroke and at a somewhat slower rate at the ends of the stroke. The governor *dead time*, or the elapsed time from the initial speed change to the first movement of the gates, is usually less than 0.2 s.

Electric Governor. In recent years the functional requirements placed upon the hydraulic turbine governor have increased to the point where electrical control of the hydraulic turbine is attractive in view of the simplicity with which electrical signals can be manipulated. The basic elements of an electric hydraulic governing system are: (1) A permanent magnet generator (PMG) or the equivalent, for measurement of turbine speed and for transmittal of such speed signals to the electrical portion of the governor. (2) An electric circuit sensitive to speed variations about some adjustable reference point. (3) Amplifying circuits, to convert speed-reference changes, speed-error signals, and auxiliary signals into a useful electric current. (4) An electrohydraulic transducer, to transform the electric current into a hydraulic-output signal. (5) Hydraulic-amplifying equipment, to deliver suitable power and the desired signal to the gate servomotors as a function of the output of the electrohydraulic transducer. (6) Power supplies for the electrical and the hydraulic portions of the control. (For further particulars of the electric governor see Leum, Electric Governors for Hydro Turbines, *ASME Paper* 62-WA165.)

93. Speed-regulation Requirements. Usually a sufficient measure of the regulation provided is the maximum speed rise resulting from sudden rejection of full load, as from the breaker tripping. A maximum speed rise of 35% of normal speed for this condition is a common limitation.

94. Speed Rise Following Load Reduction. For sudden load reductions the approximate speed rise is

$$(n_x/n) = [1 + 1{,}620{,}000T_xP_x(1 + h/H)^{3/2}/WR^2n^2]^{1/2} \qquad (8\text{-}35)$$

where n_x is the revolutions per minute at the end of time T_x; n is the speed before the load decrease; T_x is the time interval, seconds, for the governor to adjust the flow to the new load; P_x is the reduction in load; h is the head rise caused by the retardation of the flow; H is the net effective head before the load change; WR^2 is the product of the revolving parts, pounds, and the square of their radius of gyration, feet. For values of h, see below. Very rapid gate closure produces a reduction of pressure in the draft tube and the possibility of breaking the water column, with subsequent violent resurge which may damage the turbine.

95. Speed Drop Following Load Increase. For sudden load increases the approximate speed drop

$$(n_x/n) = [1 - 1{,}620{,}000 T_x P_x / WR^2 n^2 (1 - h/H)^{3/2}]^{1/2} \tag{8-36}$$

where P_x is the actual load increase and h is the head drop caused by the increase of the flow. If the speed drop is to be determined for a given increase in gate opening, the governor time T_x for making this increase and the normal change in load for the change in gate opening, under constant head H, can be used in the following formula:

$$(n_x/n) = [1 - 1{,}620{,}000 T_x P_x (1 - h/H)^{3/2} / WR^2 n^2]^{1/2} \tag{8-37}$$

The actual change in load, however, will be $P_x(1 - h/H)^{3/2}$.

For derivation of the above speed-variation formulas and for a more accurate determination, see Strowger and Kerr, Speed Changes of Hydraulic Turbines for Sudden Changes of Load, *Trans. ASME,* 1926, and Rich, "Hydraulic Transients," McGraw-Hill Book Company.

96. Water Hammer in Penstocks. If a gate movement is considered as a series of instantaneous movements with a very small interval between each movement, the pressure variation in the penstock following the gate movement will be the effect of a series of pressure waves, each caused by one of the instantaneous small gate movements. For a steel penstock, the velocity of the pressure wave $a = 4{,}660/\sqrt{1 + (d/100t)}$, where d is the penstock diameter, inches, and t is the penstock wall thickness, inches. The pressure change at any point along the penstock at any time after the start of the gate movement may be calculated by summing up the effect of the individual pressure waves (see *ASME Symp. on Water Hammer,* 1933, and "Water Hammer Analysis" by John Parmakian, Prentice-Hall, Inc.).

Approximate formulas (De Sparre) for the increase in pressure h, feet, following gate closure are given below. They are quite accurate for pressure rises not exceeding 50% of the initial pressure, which includes most practical cases.

$$h = aV/g \qquad \text{(for } K < 1 \text{ and } N < 1) \tag{8-38}$$

$$h = aV/g[N + K(N - 1)] \qquad \text{(for } K < 1 \text{ and } N > 1) \tag{8-39}$$

$$h = aV/g(2N - K) \qquad \text{(for } K > 1 \text{ and } N > 1) \tag{8-40}$$

where $K = aV/2gH$; $N = aT/2L$; V and H are the penstock velocity, feet per second, and head, feet, prior to closure; L is the penstock length, feet; and T is the time of gate closure. For full-load rejection, T may be taken as 85% of the total gate traversing time to allow for nonuniform gate motion.

For pressure drop following a complete gate opening, the following formula (S. Logan Kerr) may be used with T not less than $2L/a$:

$$h = \frac{aV}{g}\,\frac{-K + \sqrt{K^2 + N^2}}{N^2} = \text{pressure drop, ft} \tag{8-41}$$

Pressure variations exceeding 40% rise and about 25% drop should be avoided.

When control directly by the governor causes undesirable pressure variations, a surge tank, a pressure regulator, or a jet deflector may be used.

A **surge tank** is a standpipe with an atmospheric tank, attached to the penstock as close as possible to the casing inlet. The tank provides a reservoir and expansion chamber for the water demand or the water rejection following sudden gate movements, so that sudden accelerations or decelerations of the flow in the penstock are avoided.

97. Pressure regulators may be either of the water-wasting or water-saving type. The **water-wasting type** is a synchronous bypass, generally attached to the turbine casing. It is operated directly from the governor, or the gate mechanism of the turbine, and wastes such an amount as to keep the total water discharge equal at all times to the full-load discharge of the turbine. The bypass is a needle nozzle or a mushroom-shaped disk valve which opens and is partly balanced hydraulically by a piston under pipeline pressure.

The **water-saving type** permits the regulator to open upon rapid closure of the turbine gates, and then close slowly, so that the total water discharge is gradually reduced and finally limited to that through the turbine, adjusted for the new load.

TURBINE TESTS

98. Field testing of hydraulic turbines to determine the absolute efficiency and output involves careful and accurate measurement of the power available in the water supplied to the turbine (water horsepower) and the turbine output (developed horsepower). e = (developed horsepower)/(water horsepower) $= 8.82P/QH$. The tests should be conducted in accordance with the ASME Test Code for Hydraulic Prime Movers.

Because of the difficulties and costs involved in making accurate measurements of horsepower, net head, and discharge in the field, there has been a trend in recent years to dispense with the field test, especially where a laboratory test on a homologous model turbine is available. Instead, an index test is made on the unit in the field, which measures the turbine output and relative discharge under various conditions. Index tests should be conducted in accordance with the ASME Test Code for Hydraulic Prime Movers.

NOTATION

99. Notation.

D = diameter of runner, in
D_d = diameter top of draft tube, in, discharge diameter of runner, in
D_p = pitch diameter of impulse turbine runner, in
d = jet diameter of impulse turbine, in
e = overall efficiency of turbine
g = acceleration of gravity
H = net effective head, ft
h = head change due to load change, ft
n = r/min
n_1 = r/min at 1 ft head $= n/\sqrt{H}$
n_s = specific speed $= n\sqrt{P}/H^{5/4} = n_1\sqrt{P_1}$
P = horsepower
P_1 = horsepower at 1 ft head $= P/H^{3/2}$
Q = discharge, ft³/s
Q_1 = discharge at 1 ft head $= Q/\sqrt{H}$, ft³/s
t = time, s
V = absolute velocity of water, ft/s
w = weight of water/ft³
σ = cavitation coefficient

100. References

Daugherty, R. L. "Hydraulic Turbines"; New York, McGraw-Hill Book Company, 1920.

Creager and Justin "Hydroelectric Handbook"; New York, John Wiley & Sons, Inc.

Barrows, H. K. "Water Power Engineering," 3d ed.; New York, McGraw-Hill Book Company, 1943.

Daugherty, R. L., and Ingersoll "Fluid Mechanics"; New York, McGraw-Hill Book Company.

Rheingans Operating and Maintenance Experience with Pump-turbines in the U.S., Italy, Japan and Brazil, *ASME J. Eng. Power*, July, 1966.

ECONOMICS OF POWER GENERATION

By Franklin J. Leerburger

101. Categories of Generating Facilities. Power-generating facilities fall into a number of categories of ownership, each of which is generally associated with a unique pattern of financing. Industrial or manufacturing corporations usually finance their privately owned generating facilities out of unidentified corporate funds for which there are numerous alternative uses. The conflict in the field of supplying power to the public has given rise to a variety of terms which lead to direct semantic problems. Federal-government-owned facilities have in most cases been financed by the United States Treasury and are true publicly owned facilities. Municipalities and state authorities have generally financed power facilities entirely on funds raised by issuing bonds for the total capital requirements. The bondholders are lenders and not owners of the facilities. The owners are the public, and so this type of ownership is also "public," although different from other kinds of public ownership.

Public-utility corporations are a type of private ownership sometimes designated "investor-owned." While it is true that all the attributes are exercised by the common stockholders through the boards of directors, the bondholders are in somewhat the same position as are those lending money to governmental agencies. The Power Authority of the State of New York has frequently maintained that it is "privately" financed through its bonds even though the ownership rests in the state. The foregoing remarks are intended to explain the basis for the patterns of fixed charges subsequently developed.

The trend of electric power supply in the United States has been away from the independent or isolated generating station, connected to its own privately owned industrial and related loads, to a system of interconnected privately and publicly owned generating stations, supplying a variety of types of loads from large industrial to relatively small domestic ones. The fact that electric energy cannot be stored in that form gives rise to a whole variety of complex technological problems requiring special solutions and to related economic considerations.

This subsection is primarily concerned with costs of electric power from stationary plants and central stations in the United States. Caution must be exercised in comparing costs associated with the generation of electricity by a large, integrated, and pooled system with those from a single generating unit or a single isolated power plant. The reliability of service, accuracy of frequency control, and adherence to a preset voltage level are generally substantially different in the case of the single plant from that of a pooled system. In some industries such as the electrothermal, electrochemical, and metallurgical industries the product cost is significantly influenced by the cost of electricity. However, in most manufacturing and industrial sectors of the United States economy the price of electricity is less important than the availability of an abundant supply at accurately controlled frequency and voltage and with a high index of service reliability.

102. Cost of Power Plants. An industrial power plant designed to meet the requirements of an isolated load involves radically different design considerations with respect to reserve capability, spinning reserve (if any), range of voltage regulation, and frequency control from those of a central station within a system of other generating stations, which in turn are pooled with adjacent systems.

A central station of a utility system, in present practice, is designed to meet not only the existing and prospective loads of the system in which it is to function but also the pooling and integration obligations to adjacent systems. Design will therefore be required to take into account factors such as scheduled maintenance, reserve capability, type of prime mover, fuel, water supplies, geographic conditions, fuel-transportation requirements, transmission limitations, labor costs, and taxes.

Whether the prime mover is a steam turbine, a hydraulic turbine, an internal-combustion engine, or a gas turbine, certain common design considerations, which affect cost, exist. The nuclear steam-electric station, so far as present-day practice is concerned, is a traditional steam-electric layout coupled to a steam generator of novel design. The nuclear heat-generating cycle is substituted for the traditional combustion cycle of fossil fuels. The design considerations which must be taken into account and which are common include subsurface conditions; climatic conditions, which determine the type of equipment housing; quantity, quality, and temperature of water for cooling purposes; and the relative difficulty of leading water to and from the plant. Important in the case of the nuclear plant are its proximity to population centers and the sizes of the exclusion areas.

In the case of fossil-burning thermal plants, the question of location near the fuel source weighed against proximity to the load centers will influence the cost of transportation of fuel on the one hand and the transmission of electricity on the other. Internal-combustion-engine plants are sensitive to loss of capacity as elevations increase above sea level.

Hydroelectric sites are frequently far removed from load centers, involving extensive transmission facilities. It is evident that most of the economic hydroelectric sites have been exploited; future developments will generally be special-purpose plants or those installed as incidental to flood control, navigation, and reforestation. Pumped-storage installations are special and are really analogous to storage batteries in that they make possible the time transfer of kilowatthours of energy of one value to kilowatthours of sufficiently higher value to offset the substantial losses concomitant to the transfer process. The low cost of pumped-storage installations also makes for low-cost peaking and standby capacity.

103. Steam-Electric Plants. The economy of scale has been demonstrated in this type of facility in a most remarkable way in the past few years as manufacturers have become capable of fabricating alloy steels which permit steam-electric units to be built as big as 1 million kW, with even larger sizes projected. In a rural area, with relatively easy transportation problems, low land costs, and easy material-storage conditions, oil-fired steam-electric stations may at present cost less than $110/kW for the first unit, with costs of additional units low enough to bring overall station costs down another $2/kW. Figure 8-72 has been taken from the National Power Survey (NPS),[1] page 70, and clearly suggests the nature of the economy of scale. Under the same assumptions, units of 200,000 kW capacity might cost under $165/kW for the first unit and perhaps $20/kW less for the fully developed station. Coal-fired stations with additional handling facilities for fuel and ash may cost $4 to $5/kW more than oil-fired equivalents, regardless of unit size. Extremely crowded urban conditions may add $6 to $9/kW to the cost of an equivalent rural station. Full-outdoor and semioutdoor stations may save $5 to $10/kW. Special cases in which natural gas is burned instead of oil or coal in outdoor and semioutdoor plants may reduce costs another $3 to $5/kW.

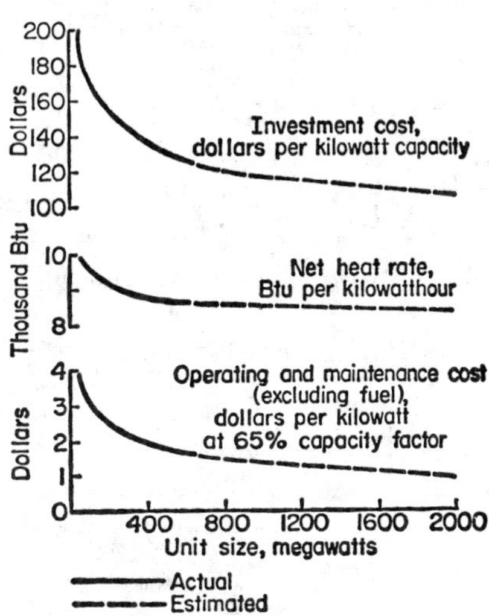

FIG. 8-72. Performance and cost characteristics of coal-fired steam-electric plants (two-unit).

The foregoing costs presume the use of modern high-pressure high-temperature designs such as 2,400 lb/in² and 1000°F superheat and reheat. Extremely high-pressure supercritical stations may have special cost problems which must be considered before adoption. A report to the Joint Congressional Committee on Atomic Energy in July, 1964, introduced estimates of $97/kW for a 1,230,000-kW supercritical plant to be completed in rural Ohio by 1967 at an anticipated heat rate of 8650 Btu/kWh. Higher-pressure and double-reheat plants cost more than conventional plants but effect a reduction in heat rates. Reduction in fuel costs in specific cases must be balanced

[1] Federal Power Commission, National Power Survey, 1964.

against the increased cost of the equipment. An additional plant cost of $5/kW may be expected with pressure increase to 3,500 lb/in².

104. Hydroelectric Plants. Hydroelectric stations encounter a wider range of costs than do steam-electric stations because of widely varying costs of dams, areas flooded for pondage, intake works, discharge arrangements, heads, and rates of water flow. Costs as high as $500/kW of capacity are not unknown, especially for small-sized units. Based on installed capacity rather than on dependable capacity, the St. Lawrence power project cost about $349/kW and the Niagara about $312.

105. Pumped-storage Hydroelectric Plants. This type of station is generally designed for emergencies and for peak-load service and is so special that costs are not comparable with those of high-load-factor stations. The variable factors influencing construction costs cannot be generalized, but they are substantially below the costs of high-load-factor hydroelectric plants. The Cornwall project of Consolidated Edison Company of New York, Inc., would consist of an underground station with reversible pump-turbine and motor-generating units. Eight units, each rated at 225,000 kW but having a capability of 250,000 kW, would cost about $129,400,000 without substations and transmission facilities. This cost corresponds to $64.70/kW of capability,

Table 8-15a. Central-station Nuclear Power Plants

(Operating, under construction, or on order)

Announced	Utility and plant	State	Mfr.	Size (Mwe-net)	Est. cost,* millions of dollars	Year of initial operation
1953	Duquesne (Shippingport)	Pa.	W	135†	69	1957
1955	Commonwealth Ed. (Dresden #1)	Ill.	GE	200	51	1959
	Consolidated Ed. (Indian Pt. #1)	N.Y.	B&W	151‡	107	1962
	Pwr. React. Dev. Co. (Fermi)	Mich.	PRDC	61	66	1963
				412	224	
1956	Yank. At. El. Co. (Yankee)	Mass.	W	175	39	1960
1957	Northern States (Pathfinder)	S.Dak.	AC	58	26	1964
1958	Car's-Va. Nuc. Pwr. Assoc. (CVTR)	S.C.	W	14‡	24	1963
	Phila. El. (Peach Bottom)	Pa.	GA	40	28	1966
	City of Piqua, Ohio (Piqua)	Ohio	AI	11	16	1963
	Puerto Rico Wtr. Res. Auth. (BONUS)	P.R.	CE	16	19	1964
	Rur. Coop. Pwr. Assoc. (Elk River)	Minn.	AC	16‡	14	1962
	Washington Pub. Pwr. Sup. Sys. (NPR)	Wash.	GE	786	315	1966
				883	416	
1959	Consumer Power Co. (Big Rock Pt.)	Mich.	GE	70	28	1962
	Pac. Gas. & Electric (Humboldt Bay)	Calif.	GE	68	24	1963
				138	52	
1960	SCE and SDG&E (San Onofre)	Calif.	W	375	97	1966
1961	Dairyland Power Coop. (LaCrosse)	Wis.	AC	50	19	1966
1963	Conn. Yank. A. P. Co. (Conn. Yankee)	Conn.	W	462	87	1967
	Los Angeles D. W. & P. (Malibu)	Calif.	W	462	83	1970
	Jersey Central (Oyster Creek)	N.J.	GE	515	66	1967
	Niagara Mohawk (Nine Mile Pt.)	N.Y.	GE	500	87	1967
				1,939	323	
1965	Boston Edison (unnamed)	Mass.	GE	540	65	1971
	Commonwealth Ed. (Dresden #2)	Ill.	GE	715	79	1968
	Northeast Utilities (Millstone Pt.)	Conn.	GE	549	81	1969
	Consolidated Ed. (Indian Pt. #2)	N.Y.	W	873	106	1969
	Fla. Power & Lt. (Turkey Pt. #3)	Fla.	W	722	70	1970
	Fla. Power & Lt. (Turkey Pt. #4)	Fla.	W	722	63	1971
	Pub. Ser. Co. of Colo. (Fort St. Vrain)	Colo.	GA	330	69	1971
	Rochester Gas & Elect. (Robt. E. Ginna)	N.Y.	W	420	63	1969
				4,871	596	

Reference: The Nuclear Industry, 1966, USAEC Division of Industrial Participation, Table IV-5.

* Total construction cost. Land, fuel and transmission plant excluded.

† Turbine-generator capacity currently limited to 90 MW.

‡ Does not include capacity added by fossil-fueled superheater. Net electrical output with superheater is: Indian Point 270 MW; CVTR 17 MW; Elk River 22 MW.

or about \$71.70/kW of nameplate capacity. Off-peak pumping has been estimated at 2.6 mills/kWh of incremental cost from system thermal stations, equivalent to 3.9 mills/kWh of hydrogeneration.

106. Internal-combustion Plants. These are generally relatively small in size, and plant capacity is built up by increasing the number of engine-generator sets. The complete "package" diesel-electric unit in standard sizes for immediate full-outdoor installation has brought station costs down to levels of about \$100/kW. Special situations, however, have developed costs that are well above this figure. The forms of internal-combustion prime movers are many and include gas engines, diesel engines, and semidiesel engines, as well as kerosene and gasoline engines. For central-station service, the diesel engine is the more widely used type of internal-combustion prime mover; this is specially true for remote-area protection and for "peak-shaving" use.

107. Jet and Gas-turbine Plants. Jet and gas-turbine prime movers have gone through rapid technological improvements, to which much has been added from the turbopropeller machines and jets and pure gas turbines of the airplane industry. Unit sizes of jet-driven units can be furnished up to 150 MW using multiple units on a single generator shaft, and gas-turbine units have been purchased as large as 200 MW. But such plants burn natural gas or petroleum distillates. These fuels are relatively expensive in costs per million Btu, and the units operate at relatively high heat rates. The combination of fuel conditions makes for high-priced costs per kilowatthour generated. There is some waste heat exhausted to the atmosphere, and this has suggested the use of waste-heat-recovery boilers in special cases. The cost of such plants ranges from \$60 to \$80/kW without waste-heat-recovery equipment.

108. Nuclear Power Plants. Table 8-15 shows costs for some existing and projected plants. The need for containment shells, shielding, instrumentation, and measures to ensure safety and the comparatively low steam pressures and temperatures (and consequent lower thermal efficiency) require larger turbines and piping than do modern fossil-fuel steam-electric stations and in the total view make for higher capital costs for nuclear-powered stations. The predicted competitive advantage of nuclear power over other forms is based on assumed unit sizes in excess of 500,000 kW.

Plants actually constructed have exhibited cost overruns which should have been

Table 8-15b. Central-station Nuclear Power Plants

Utility and plant	State	Mfr.	Size (Mwe-net)	Est. cost* millions of dollars	Year of initial operation
Contracts awarded in 1966:					
Commonwealth Ed. (Dresden #3)	Ill.	GE	715	81	1969
Carolina Power & Lt. Co. (Robinson #2)	S.C.	W	663	76	1970
Consumers Power Co. (Palisades)	Mich.	CE	710	75	1970
Wis. Mich. Power Co. (Point Beach)	Wis.	W	450	60	1970
Common. Ed./Iowa-Ill. (Quad Cities #1)	Ill.	GE	715	90	1970
Common. Ed./Iowa-Ill. (Quad Cities #2)	Ill.	GE	715	80	1971
Northern States (Monticello)	Minn.	GE	472	74	1970
T.V.A. (Browns Ferry #1)	Ala.	GE	1,065	124	1970
T.V.A. (Browns Ferry #2)	Ala.	GE	1,065	123	1971
Duke Power Company (Lake Keowee #1)	S.C.	B&W	822	78	1971
Duke Power Company (Lake Keowee #2)	S.C.	B&W	822	79	1972
Vermont Yankee N.P. Corp. (Vt. Yankee)	Vt.	GE	540	88	1970
PSEG/PE/ACE/DPL† (PSEG area)	N.J.	W	993	125	1971
PSEG/PE/ACE/DPL† (Peach Bottom)	N.J.	GE	1,064	125	1971
PSEG/PE/ACE/DPL† (unnamed)	N.J.	GE	1,064	125	1973
			11,875	1,403	
Announced as planned for nuclear power (contracts not awarded):					
Maine Yankee A. P. Co. (Bailey Pt.)	......	......	700	.....	1972
Long Island Lighting Co. (Shoreham)	......	......	500	.....	1973
PSEG/PE/ACE/DPL†	......	......	993	.....	1974
SCE and SDG&E (Bolsa Chica)	......	......	900	.....	1971
Los Angeles D. W. & P. (Bolsa Chica)	......	......	900	.....	1972
Virginia Electric & Power	......	......	750	.....	1971
Omaha Public Power Dist.	......	......	400	.....	1971
Consumers Publ. Power Dist./Iowa Power & Lt.	......	......	800	.....	1972
Pacific Gas & Electric	......	......	800/1,050	.....	1971

* These are mostly unofficial figures taken from various published sources and are not necessarily on comparable bases.
† Public Service Electric & Gas, Philadelphia Electric, Atlantic City Electric, and Delmarva Power & Light.

expected with a new technology largely adapted from the work of physicists, chemists, and mathematicians and subject to the difficulties of expanding theoretical and laboratory-prototype layouts into workable full-scale facilities. Assuming best available conditions for a fossil-fuel-fired steam plant of 8,000 to 8,500 Btu/kWh and 10,500 Btu/kWh for nuclear-type plants, the heat rejected to the circulating water would be about 50% more for the nuclear fuel than for the fossil-fuel plant, on the assumption of no increase in heat dissipation to the atmosphere.

109. Trends of Costs. The trends of the major elements of public-utility electrical costs as experienced in the four areas of the United States are given in Table 8-16.

Table 8-16. Trends of Construction Costs for Electric Light and Power Construction

Jan. 1	Production plant		Transmission plant
	Steam	Hydraulic	
North Atlantic Division:			
1966	778	899	640
1964	741	844	584
1962	728	811	573
1960	756	796	587
1955	575	609	508
1950	458	490	389
1945	285	308	254
South Atlantic Division:			
1966	798	895	637
1964	764	856	589
1962	753	818	580
1960	783	796	590
1955	592	611	505
1950	464	480	382
1945	291	302	251
North Central Division:			
1966	781	842	602
1964	746	799	557
1962	735	771	550
1960	763	763	560
1955	577	594	484
1950	450	459	365
1945	283	290	241
Pacific Coast Division:			
1966	807	838	622
1964	771	792	569
1962	754	760	561
1960	779	739	567
1955	583	580	480
1950	459	435	366
1945	286	283	243

Base 1911 = 100.
Reference: Handy-Whitman Index of Public Utility Construction Costs, compiled and published by Whitman, Requardt and Associates, Baltimore, Md.

OPERATING EXPENSES

110. General Considerations. In the following discussion of costs of fuel and of other operating expenses, the influence of load variations must be kept constantly in mind. The hour-by-hour loads may be considerably different from any assumed constant-load factor, especially in power plants supplying the needs of manufacturing, which may range from high-load-factor uses as in the making of glass, paper, or aluminum to low-annual-load factors, as in food processing and canning.

111. Steam-Electric Plants. Fossil fuels can be compared most successfully in overall costs per million Btu; these costs will include fuel at sources, fuel-transportation, fuel handling and storage, and ash handling (if any). Because of the increasing interest in air pollution, the costs of minimizing this, which will vary from fuel to fuel, must be considered. In fossil-fuel plants of the northeastern United States, public utilities have been able to buy coal advantageously in large quantities. Rail-transportation costs have been slightly less than mine-mouth coal costs. Total as-fired costs have ranged from 27 to 35 cents/million Btu. New transportation techniques advanced by the railroads include integrated trains, shuttle trains, and high-speed car unloaders. The savings in using these new methods may effect net reductions of 3 to 5 cents/million Btu.

New railroad tariffs on trainload lots are attracting increasing volume of coal shipments for utility generating stations. These rates provide for reductions of up to $1.50/ton on trainloads of 7,000 net tons or more, originating at not more than two mines. The reduction in freight charges on coal was instrumental in 1963 in discontinuing the operation of the only existing coal pipeline, supplying Cleveland Electric Illuminating Company.[1]

Residual fuel oil will have to meet this competition or lose the business, except in small plants which cannot purchase wholesale, mass-delivered coal. Residual fuel oil, which was once a large portion of the total intake of oil refineries in making gasoline, is gradually diminishing as gasoline producers develop better and better methods for producing gasoline. At the beginning of the use of crude petroleum, about one-third was turned into gasoline. With the introduction of "cracking" methods, about one-half the crude was turned into gasoline.

With the commercial success of hydrogenation, more than 70% of crude may end in gasoline, thus substantially "drying up" the source of residuals. This economic factor must not be overlooked in planning new thermal stations. Diesel fuels (distillate) generally run about 70 cents/million Btu in the northeastern region. Plant heat rates[2] at high-load factors and few starts may run as low as 10,300 Btu/kWh. This may also represent the heat rates for gas turbines, but at lower plant factors, heat rates for gas turbines may run as high as 15,000 Btu/kWh. Natural-gas boiler fuel is practical and permissible in the gas fields or near to them. As a boiler fuel in other areas, the FPC has restricted its use significantly as a matter of selective conservation, reserving natural gas to the uses of households and small commercial and industrial enterprises.

The heat rates of steam-turbogenerating stations are functions of pressure, superheat, reheat, number of stages of reheat, condenser inlet temperatures, exhaust pressures at some standard vacuum, and the load or plant factor. Representative but not universally applicable heat rates, based on 2,400 lb/in^2 and 1000/1000°F, seven stages of feed-water heating, and full-load exhaust temperatures at 1.5 in Hg, are: for a nominal 1,000,000-kW unit at 220,000-kW load, 10,150 Btu/kWh; and at 912,000-kW load, 8750 Btu/kWh. Thus, with fuel at $0.23/million Btu, the heat alone would cost about $0.002 (2 mills)/kWh.

Operating labor for 1,000,000-kW units may range from 15 to 25 cents/(kW)(year). Maintenance appears to have both a "fixed" and an "incremental" element; the fixed element may approximate 6% of the cost of fuel, and the variable element at full load may amount to another 5% of the cost of fuel. Thus maintenance may, at full load, approximate 0.2 mills/kWh. At high-load factor, say 80%, the total operating cost might approximate 1.72 to 2.24 mills/kWh. Thus:

Steam-Electric Operating Costs

	Cost, mills/kWh, for a high-pressure conventional plant of 1,000,000 kW capacity in 1 unit (23 cents fuel)	Cost, mills/kWh, for a supercritical plant of 1,230,000 kW in 2 units (17 cents fuel)
Labor	0.02–0.04	0.03
Maintenance	0.2	0.22
Fuel	2.0	1.47
Total	2.22–2.24	1.72

112. Hydroelectric Plants. Operating costs are more particularly related to size

[1] The Peabody Coal Company is reported (January, 1967) to be planning a 275-mi coal slurry pipeline to transport fuel by 1970 from northeastern Arizona to a southern California Edison Company plant near Davis Dam, Nev.

[2] Heat Rates and Thermal Efficiencies; see T. Baumeister and L. S. Marks, "Standard Handbook for Mechanical Engineers," 7th ed.; New York, McGraw-Hill Book Company, 1967.

Table 8-17. Level of Competitiveness, Nuclear vs. Coal Station Installed 1967–1980
(After 5 years of operation)

1. Year plant placed in service	1967		1967		1972		1980	
2. Output, MWe	300		500		500		1,000	
3. Type of plant	Nuclear	Coal	Nuclear	Coal	Nuclear	Coal	Nuclear	Coal
4. Capital cost, $/kWe	180–200	125	160–185	120	145–160	115	125	105
5. Capital cost difference, $/kWe, nuclear vs. coal	50–75		40–65		30–45		20	
6. Cost differential, nuclear over coal station, mills/kWh:								
a. Capital*	1.1–1.5		0.9–1.4		0.7–1.0		0.5	
b. Operation and maintenance	0.2		0.1–0.2		0.1			
c. Total	1.3–1.7		1.0–1.6		0.8–1.1		0.5	
7. Nuclear fuel cost, mills/kWh (5 years after initial operation)	1.8–2.0		1.8–2.0		1.3–1.7		1.0–1.5	
8. Level of competitiveness (cost of coal):								
a. Mills/kWh (items 6+7)	3.1–3.7		2.8–3.6		2.1–2.8		1.5–2.0	
b. Cents/mBtu	35–42		32–41		24–33		18–24	
c. Assumed heat rate, Btu/kWh	8900		8800		8600		8500	

* Based upon capital charge rates adopted by the Federal Power Commission for the National Power Survey—13%/year capital charge rate for the nuclear plant, which includes nuclear liability insurance, and 12%/year for the coal station; plant factor, 80% for both conventional and nuclear.

and capacity and, for small plants under 100,000 kW, might be estimated at $3/(kW) (year). For very large stations, the figure might approximate $2.

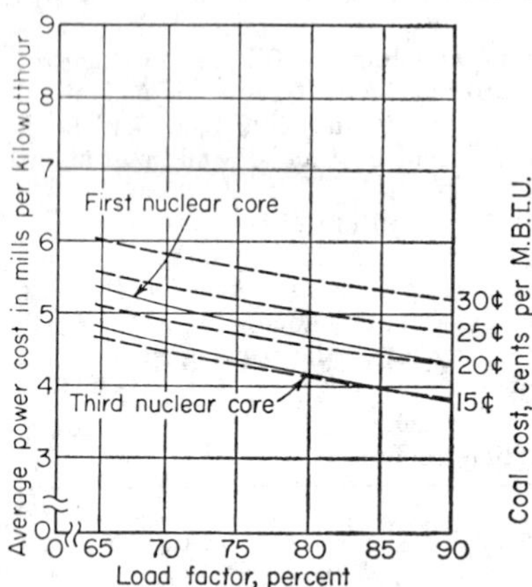

FIG. 8-73. Comparative cost of power from 600-MW nuclear and coal-fired plants (private financing). (*Courtesy F. S. Brown, NRECA Meeting, Feb.* 13, 1966.)

113. Nuclear Plants. The economic promise of nuclear power lies essentially in its potential for achieving fuel costs substantially lower than those of conventional thermal stations, which would make it possible at least to offset the higher capital costs. For detailed costing of fuel fabrication, net burnup, inventory charges, shipping, reprocessing irradiated material, and charges on investment in fuel see Kenneth Davis (Sec. **9**).[1] Totals of these component costs in terms of fuel-cycle costs on the basis of experienced load factors for plants in operation and of projected load factors for plants either under construction or being planned are as follows (in mills per kilowatthour):

Hallam	3.3
Indian Point	5.33*
Yankee	2.8
Big Rock Point	3.75
San Onofre	1.99
Malibu	1.8
Nine-mile Point	2.17
Oyster Creek	1.73–1.65

* Includes fuel oil.

The NPS (Pt. II, page 178) has included estimated comparisons of competitiveness of future nuclear power plants with future coal-fired power plants (see Table 8-17).

The latest estimates at the time of preparing this section were made available by FPC on Feb. 13, 1966, in a paper by F. S. Brown, Trends in the Development of Nuclear Power, from which Fig. 8-73 has been taken.

[1] See Sec. **9**, Economics of Nuclear Power Plants, Par. **47**.

DEVELOPMENT OF OVERALL COST OF PRODUCING ELECTRICITY

114. General Considerations. The cost of money needed to construct a generating station is the amount which must be paid under specific conditions to induce money to be supplied. This depends upon money-market conditions in general and upon the opinions of investors as to the soundness of the project. Thus, if a textile mill or shoe factory is considering investing in a new power plant, alternative uses for the required funds will be considered. The annual range of target returns considered as a criterion by manufacturing or industrial concerns is usually between 8 and 20%. On the other hand, the return on investment earned by public utilities is limited by governmental regulatory agencies, usually falling between 5 and 7% of the difference between original cost and accrued depreciation. Plants built by government institutions or authorities are frequently financed directly from the Treasury and at government long-term borrowing rates or by revenue bonds bearing interest rates of perhaps 4 to 5%.

In considering the sources of construction funds required by privately owned electric utilities, it is of interest to note that NPS at page 22 shows the following:

Source of funds	Percent of total	Percent of security issues
Security issues:		
Common stock	13.8	33.7
Preferred stock	4.4	10.8
Debt	22.7	55.5
Total securities	40.9	100.0
Internal funds:		
Retained earnings	14.0	
Deferred taxes	3.9	
Depreciation and amortization	41.2	
Total funds	59.1	
	100.0	100.0

Depreciation as an accounting concept is an annual allowance which will amortize the original cost, less net salvage realized at time of retirement, over the average estimated service life of the property. Numerous methods are used to compute depreciation; the so-called straight-line method is the most widely adopted, but a significant number of companies use some form of interest method (see Table 8-18).

115. Equal Annual Payments. Use may be made of interest tables which show, for any rate of interest and any number of years, the equal-annual-payment rate which will amortize an amount and also yield an annual return equal to the interest rate on the unamortized remainder.

$$\text{Equal annual payment} = \frac{i(1+i)^n}{(1+i)^n - 1} \tag{8-42}$$

where n = number of years of life and i = interest or rate of return (see Table 8-19; also Grant, "Principles of Engineering Economy").

A frequent error in estimating annual fixed charges on the basis of the first year of service life is caused by using the annual rate of depreciation plus interest or return on the original cost, thus failing to recognize that the interest charged in dollars in years subsequent to the first will diminish because the outstanding principal has been reduced by annual depreciation accruals.

116. Taxes, Insurance, and Other Fixed Charges. In the case of ownership by government or by public authority, no Federal income taxes are paid. Private business, however, has been faced with an ever-mounting tax charge on net income; the tax amounted to 40% in 1945, 42% in 1950, and 52% in 1952 to 1964. Since 1965 the rate has been 48%. Changes in income taxes subsequent to 1967 should be taken into account. The application of the Federal income tax depends upon a number of considerations, but primarily on the sources from which funds are raised, because interest paid on money borrowed by business owners is a deductible expense for tax purposes. A comparison of the rates of return or interest and Federal income taxes for two types

Table 8-18. Range of Rates of Depreciation
(Straight-line method of depreciation accrual)

Account No.	Class of plant	Number of utilities	Depreciation rates: Full range, percent	Range of median 50%, percent	Arithmetic average, percent
	Steam production plant:				
311	Structures and improvements	37	1.35– 3.33	1.70– 2.50	2.15
312	Boiler-plant equipment	37	2.22– 4.00	2.65– 3.30	2.98
313	Engines and engine-driven generators	5	2.00– 5.00	2.50– 3.33	3.16
314	Turbogenerator units	36	1.94– 4.00	2.50– 2.98	2.71
315	Accessory electric equipment	36	2.09– 4.00	2.56– 3.25	2.91
316	Miscellaneous power-plant equipment	37	2.10– 5.00	2.86– 3.50	3.23
	Hydraulic production plant:				
331	Structures and improvements	24	1.23– 2.50	1.56– 1.98	1.74
332	Reservoirs, dams, and waterways	24	1.00– 3.33	1.25– 1.68	1.63
333	Waterwheels, turbines, and generators	23	1.58– 3.74	2.00– 2.50	2.29
334	Accessory electric equipment	21	1.77– 3.76	2.50– 3.00	3.61
335	Miscellaneous power-plant equipment	23	2.00– 8.48	2.50– 3.80	3.42
336	Roads, railroads, and bridges	16	1.00– 3.57	1.33– 2.09	2.02
	Other production plant:				
341	Structures and improvements	15	1.70– 3.33	2.00– 2.50	2.33
342	Fuel holders, producers, and accessories	14	2.00– 6.67	2.50– 4.00	3.46
343	Prime movers	12	2.05– 5.00	3.00– 4.00	3.54
344	Generators	17	2.00–10.00	3.00– 4.00	3.89
345	Accessory electric equipment	15	2.31– 4.25	3.00– 4.00	3.42
346	Miscellaneous power-plant equipment	15	2.00– 6.33	2.50– 4.25	3.73
	Transmission plant:				
351	Clearing land and rights-of-way	11	1.00– 3.87	1.11– 2.22	1.71
352	Structures and improvements	41	1.33– 3.70	2.00– 2.50	2.21
353	Station equipment	44	1.47– 3.93	2.50– 3.00	2.77
354	Towers and fixtures	37	1.30– 4.00	1.83– 2.50	2.22
355	Poles and fixtures	45	2.40– 4.88	2.86– 3.45	3.24
356	Overhead conductors and devices	47	1.17– 4.00	1.74– 2.43	2.15
357	Underground conduit	18	1.20– 4.00	1.67– 2.69	2.28
358	Underground conductors and devices	18	1.50– 4.00	2.27– 3.00	2.58
359	Roads and trails	18	1.00– 5.00	2.00– 2.70	2.40
	Distribution plant:				
361	Structures and improvements	43	0.59–10.00	2.00– 2.50	2.44
362	Station equipment	48	1.34– 4.38	2.72– 3.33	3.00
363	Storage-battery equipment	2	3.93– 4.55	....	4.24
364	Poles, towers, and fixtures	46	2.50– 4.75	3.29– 3.83	3.52
365	Overhead conductors and devices	45	1.50– 4.39	2.13– 3.00	2.54
366	Underground conduit	41	1.33– 4.00	1.67– 2.50	2.15
367	Underground conductors and devices	43	1.50– 4.00	2.29– 3.00	2.56
368	Line transformers	46	2.00– 5.00	2.58– 3.21	2.91
369	Services	46	2.00– 5.00	2.83– 3.81	3.20
370	Meters	46	2.30– 5.00	2.86– 3.46	3.24
371	Installations on customers' premises	30	1.08–10.00	2.88– 4.00	3.79
372	Leased property on customers' premises	14	2.22–10.00	3.10– 4.40	4.28
373	Street lighting and signal systems	44	2.00– 9.50	3.13– 4.00	3.76
	General plant:				
390	Structures and improvements	42	1.33– 5.00	1.79– 2.50	2.19
391	Office furniture and equipment	42	2.37–10.00	3.80– 6.00	4.99
392	Transportation equipment	22	3.17–24.00	6.00–16.00	11.89
393	Stores equipment	40	2.00–10.00	3.80– 5.00	4.41
394	Tools; shop and garage equipment	43	1.25–12.50	3.82– 5.56	4.96
395	Laboratory equipment	40	2.22– 7.50	3.45– 5.00	4.56
396	Power-operated equipment	14	3.08–16.00	4.75– 9.00	7.09
397	Communication equipment	45	1.82–14.29	3.80– 6.67	5.98
398	Miscellaneous equipment	40	2.78– 9.80	3.80– 6.25	5.13

Reference: F.P.C. Electric Utility Depreciation Practices, 1961, Classes A & B Privately Owned Companies.

of private enterprises and for governmental, municipal, and public authority is shown in Table 8-20.

Other taxes which are related to the cost of the plant in one way or another will vary from location to location, but generally there is no legal requirement for government, municipal, or public authorities to pay them. Insurance on the plant is a rela-

Table 8-19. Equal Annual Rate Which Will Amortize Original Investment over Estimated Life and Yield Indicated Return on Difference between Original Investment and Accrued Amortization

Life, years	Annual rate of interest or return, percent							
	3	4	5	6	7	8	9	10
10	0.11723	0.12329	0.12950	0.13587	0.14238	0.14903	0.15582	0.16275
20	0.06722	0.07358	0.08024	0.08718	0.09439	0.10185	0.10959	0.11746
30	0.05102	0.05783	0.06505	0.07265	0.08059	0.08883	0.09734	0.10608
40	0.04326	0.05052	0.05828	0.06646	0.07501	0.08386	0.09296	0.10226
50	0.03887	0.04655	0.05478	0.06344	0.07246	0.08174	0.09123	0.10086
75	0.03367	0.04223	0.05132	0.06077	0.07044	0.08025	0.09014	0.10008
100	0.03165	0.04081	0.05038	0.06018	0.07008	0.08004	0.09002	0.10001

tively small factor. Thus a reasonable allowance for local property taxes and insurance will range from 1 to 4% paid by private enterprise. The rate to be paid for insurance alone on either private or public nuclear power stations is still being developed. Power plants owned by public authorities or agencies obviously have the benefit of lower costs of money and taxes.

117. Total Annual Costs. In estimating fixed charges, it is necessary to decide on a method for accounting for annual depreciation. Assuming that straight-line depreciation is adopted, it then becomes necessary to estimate average service life. The whole subject is too complex to discuss in a paragraph or two, and reference to texts listed in the Bibliography (Par. **125**) should be made (see also Table 8-18). The reciprocal of any straight-line rate is the corresponding imputed average service life

Table 8-20. Comparison of Interest or Return and Federal Income Taxes on Private and Public Projects

	Percentages		
	Industrial and commercial	Public utility	Gov't, public authority, or municipal
Distribution of investment:			
Equity funds (stocks)	100	50	0
Borrowed funds (bonds)	......	50	100
Total	100	100	100
Rate of return or interest:			
On stocks	10*	10†	0.0
On bonds	......	5.0	4.0
Taxable income subject to FIT:			
Average rate	10.0	7.5	4.0
Deduction for interest	......	......	4.0
50% of 5%	......	2.5	0.0
Net taxable income	10.0	5.0	0.0
Return on equity before 48% FIT:			
10% (100%–48%)	19.23	......	0
5% (100%–48%)	......	9.62	0
FIT as percentage of capital:			
19.23%–10%	9.23	......	0
9.62%–5%	......	4.62	0
Interest or return and FIT:			
Return or interest	10.00	7.50	4.00
FIT	9.23	4.62	0
Total	19.23	12.12	4.00

* Many industrial organizations will not consider devoting their own funds to their own power plants unless it is justified on the basis of earning the same return as money devoted to the prime industrial objective. This may run to 20%.

† Many utilities do not earn this much. See public-utility earnings-price ratio studies by leading investment houses. See also Rate of Return Allowed in Public Utility Rate Cases, Arthur Anderson & Co., *Federal Reserve Bull.*, Interest Rates.

of the property. Thus, for overall plant-wide rates and upon assuming the following as applicable to the indicated categories, the corresponding average service lives are readily computed. (The following figures have been used for illustrative purposes.)

Type of station	Annual rate, %	Corresponding average service life, years
Steam-electric	3⅓	30
Hydroelectric	2	50
Diesel-electric	5	20
Nuclear	3⅓*	30

* This rate has no foundation in experience, and it may be considerably more, in the light of the prospective obsolescence associated with the rapid development of the art.

The equal annual payment will, in the selected time or service life, amortize the property and leave an annual amount available to pay interest or return, or a combination of both, as shown in Table 8-21.

Table 8-21. Comparative Fixed Charges for Steam, Nuclear, and Hydroelectric Installations under Private and Public Operations

	Average service life, years	Percentages		
		Industrial and commercial	Public utility	Gov't, public authority, or municipal
Equal annual payments:				
Steam-electric	30	10.608	8.059	5.783
Hydroelectric	50	10.086	7.246	4.655
Diesel-electric	20	11.746	9.439	7.358
Nuclear	30	10.608	8.059	5.783
Average return	..	10	7	
Interest	..	......	......	4
Steam or nuclear:				
1. Equal annual payment	..	10.608	8.059	5.783
2. Less annual depreciation	..	3.333	3.333	
3. Line 1 − line 2	..	7.275	4.726	
4. Less bond interest	..	......	2.500	
5. Income after FIT	..	7.275	2.226	
6. Income before FIT at 48%	..	13.990	4.281	
7. Federal income tax	..	6.715	2.055	
8. Total line 1 + line 7	..	17.323	10.114	
9. Local taxes and insurance	..	4.000	4.000	0.500
10. Total fixed charges	..	21.323	14.114	6.283
Rounded to	..	21.3	14.1	6.3
Hydroelectric:				
1. Equal annual payment	..	10.086	7.246	4.655
2. Less annual depreciation	..	2.000	2.000	
3. Line 1 − line 2	..	8.086	5.246	
4. Less bond interest	..	......	2.500	
5. Income after FIT	..	8.086	2.746	
6. Income before FIT at 48%	..	15.550	5.281	
7. Federal income tax	..	7.464	2.535	
8. Total line 1 + line 7	..	17.550	9.781	
9. Local taxes and insurance	..	4.000	4.000	0.500
10. Total fixed charges	..	21.550	13.781	5.155
Rounded to	..	21.6	13.8	5.2

Analogous diesel-electric costs can be estimated by utilizing the steam or nuclear example, substituting appropriate figures from data given.

The total cost of power may now be estimated by reference to the preceding material, on the basis of private ownership by public utility, related capital structure, and 80% load factor.

118. Load Factor. The load factor of a system is defined as the ratio of the energy consumed to that which might have been consumed in the same period had the peak demand represented a constant load. The load factor as applied to a particular plant is usually referred to as a plant factor inasmuch as the peak load to be carried by a plant within a system and the energy delivered by such a plant are within the control of trained operators. When a thermal plant is new, it is usually allowed to run at a high plant factor, which is gradually diminished over service life to make way for newer and better plants, as illustrated by the following estimates:

Average Age, Years	Load Factor or Plant Factor, %
1–5	70
6–10	65
11–15	55
16–20	40
21–25	24
26–30	16
31–35	9

The assumption that a plant will run at a high plant factor throughout its life whether it be nuclear- or fossil-fuel-fired is based on faulty reasoning, because it assumes that the initial position in the load-duration curve will be maintained even though inevitable obsolescence brings newer and more efficient equipment into service. This decline in plant factor should be kept in mind in computing or estimating future kilowatthours to be generated, the spread of fixed charges, and the incurring of operating expenses.

119. Considerations in Pooling Power Plants. Incremental heat rates from performance data estimated for a single reheat unit rated at 1,000 MW, 2,400 lb/in², 1000/1000°F, and 1.5 in Hg are:

Net load, MW		Incremental heat rate, Btu/kWh
From	To	
220	393	7880; input at 220 MW = $2{,}233 \times 10^6$ Btu/h
393	589	8220
589	785	8580
785	912	8710; input at 912 MW = $7{,}980 \times 10^7$ Btu/h

Incremental maintenance for the same 1,000-MW unit can be estimated as (1) base maintenance = 8% of minimum-load fuel cost and (2) incremental maintenance = 6.4% of incremental fuel cost.

A large generating station is no longer generally designed as an isolated entity but is planned as an addition to existing systems composed of many stations and interconnected with other systems, which are themselves composed of many stations. Diversity of ownership presents some legal, but few engineering or economic, barriers to integration and pooling.

In determining the most economical use of generators to meet interconnected- and combined-system loads, the design problem resolves itself into finding the lowest overall combined cost of production. For the moment, with transmission costs ignored, the algebraic expression for resolving the matter has been designated as the method of lambda dispatching.

F_x = cost of operating generating unit x for 1 h

$$F_t = \sum_1^n F_x = \text{total cost of operating } n \text{ units in the system for 1 h} \tag{8-43}$$

P_x = output of unit x during 1 h

$$P_t = \sum_1^n P_x = \text{total output of } n \text{ units equal to total load during 1 h} \tag{8-44}$$

The search is for a value of F_t, total operating cost for given load

$$P_t = \sum_1^n P_x.$$

Were it possible to express an operating-cost curve for each generating unit by some mathematical equation, it could be demonstrated that the minimum overall cost would be designated as F_t = minimum, when the production or output for each unit contributing to the total output satisfies the following two conditions: (1) the incremental cost at the selected level is equal for all units on the line, and (2) the total output

$$\sum_1^n P_x \text{ equals the total load.}$$

The solution of the problem by the use of the calculus is dependent on finding the incremental cost (condition 1) using a so-called Lagrange multiplier, symbolized λ by the originator of the method (Leon H. Kirchmayer, "Economic Operation of Power Systems," John Wiley & Sons, Inc.).

The mathematically exact solution for

$$F_t = \text{minimum}$$

occurs when

$$(dF_x/dP_x) = \lambda \tag{8-45}$$

for all generating units. Dimensionally λ is the incremental production cost of a generating unit in mills per kilowatthour.

In the practical solution of the problem of reaching an optimum production cost for a given load, it is not possible to find λ by using mathematical production-cost curves and the methods of the calculus. It is therefore essential to resort to trial-and-error approximations of λ. This has been accomplished by assuming an incremental cost or trial value of λ and running each unit at that production level. If the total output fails to equal the load (condition 2 above), a different value of λ is tested and operations are repeated until conditions 1 and 2 are satisfied.

In practice, the problem is quite complex and is further complicated by minimum-load requirements, spinning-reserve requirements, maintenance schedules, and transmission losses. Special computers have been constructed to assist system dispatchers to operate available generating units for maximum system economy.

120. Power Costs and Prices. The very large effect of costs of money and taxes, as illustrated by their influence on fixed charges, makes it essential to determine how a project is to be financed, what kind of ownership is contemplated, and what average service lives are to be expected. With respect to the effect of fixed charges on the overall cost per kilowatthour, it is obvious that high load factor and pushing capability to allowable limits ("stretch-out") will bring about cost reduction. High-load-factor loads have assisted the supplying electrical systems to pass on substantial cost savings; e.g.:

	Average annual price, mills/kWh	Year
ALCOA at Alcoa, Tenn.	4.69	1963
Air Reduction Carbide Plant, Ky.	3.89	1963
Diamond Alkali, Ala.	4.11	1963
Penn-Olin Chemical Co.	4.08	1963
Stauffer Chemical Co.	4.06	1963

These plants are on the TVA system, the total industrial sales of which averaged 4.29 mills/kWh in 1963. The Power Authority of the State of New York supplied ALCOA, Reynolds Metals Company, and General Motors Company with over 3.7 billion kWh in 1963 at an average price of 4.1 mills/kWh.

Table 8-22 shows average-experience price trends from 1930 to 1965.

Table 8-22. Average Revenues per Kilowatthour Sold, Total Electric Utility Industry

Year	Residential, cents	Cents/kWh: Commercial and industrial light and power, Small	Cents/kWh: Commercial and industrial light and power, Large
1965*	2.25	2.13	0.90
1964*	2.31	2.19	0.91
1963*	2.37	2.28	0.93
1962*	2.41	2.37	0.96
1961*	2.45	2.35	0.97
1960*	2.47	2.46	0.97
1955	2.65	2.50	0.94
1950	2.88	2.64	1.01
1940	3.84	3.08	1.06
1930	6.03	4.13	1.41

* Includes Alaska and Hawaii.
Reference: EEI Statistical Yearbook of the Electric Utility Industry for 1965, p. 49, and earlier issues.

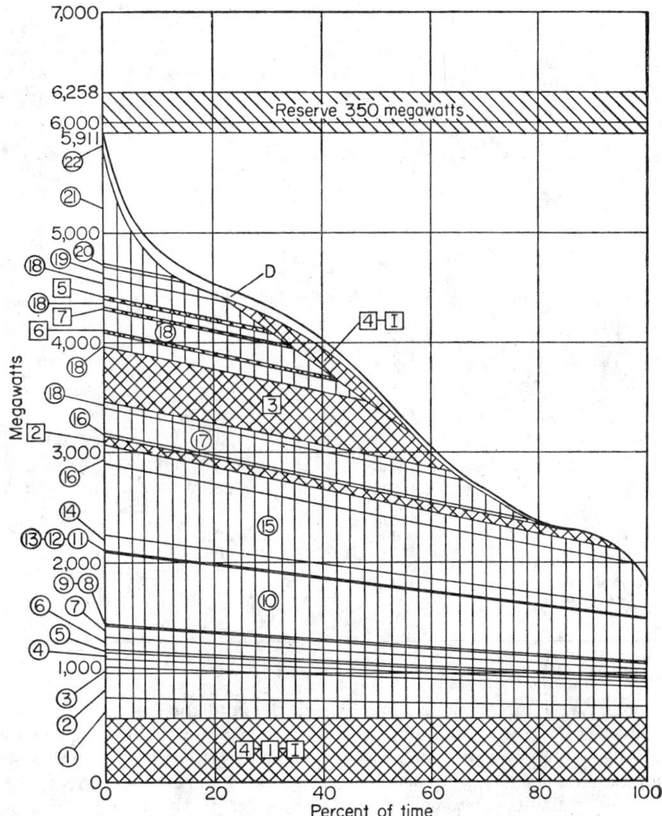

FIG. 8-74. Load-duration curves of 10 major electric power systems. (*A*) Curves (*above*); (*B*) identifications, generating plants "stacked" for load-duration curve, in approximate order of production-system expense (see Fig. 8-74(*B*) on page 8-80).

121. Multipurpose Plants. The multipurpose power plant, in small sizes, can be used as a means of protecting an outlying area of a utility system while providing a low-first-cost plant for peaking operations. The economy of scale is sacrificed for operating expenses which would be excessive at high load factors but which are acceptable for emergency or peak-shaving service. In other situations, both hydro and thermal plants may be used for daytime base-load operation and for providing energy at low incremental costs to pump-storage plants. In the nonutility field, the multipurpose plant frequently can be used for supplying process heat as well as electric power for industrial service. Its usefulness can also be extended to assist a local public utility on an interchange basis in hours when industrial needs are low. The allocation of costs to determine what the costs of heat and of on-peak and off-peak electricity are requires analysis beyond the scope of this book.

Fig. 8-74(*B*). Generating Plants "Stacked" for Load-duration Curve Approximately According to Production-system Expense

Identification number for Fig. 8-74(*A*)	Accumulative capability at time of peak load, MW	Plant factor	Production-system expense, mills/kWh
Imported from other systems	155	100	
4 hydro.	172	100	
1 hydro.	595	100	
1	775	80	3.73
2	1,057	80	3.76
3	1,131	80	4.02
4	1,183	80	4.40
5	1,219	80	4.39
6	1,313	80	4.42
7	1,415	80	4.76
8	1,440	80	5.62
9	1,441	80	20.55
10	2,086	80	5.17
11	2,087	80	8.03
12	2,109	80	5.11
13	2,113	80	7.61
14	2,249	80	6.43
15	2,877	80	6.26
16—1	3,027	80	5.83*
2 hydro.	3,095	91.4	
16—2	3,129	80	5.83*
17	3,401	72.8	6.78
18—1	3,451	65.8	6.88†
3 hydro.	3,951	55.9	
18—2	4,080	43.5	6.88†
6 hydro.	4,103	41.0	
18—3	4,303	36.0	6.88†
7 hydro.	4,315	33.1	
18—4	4,395	31.0	6.88†
5	4,422	29.5	
18—5	4,592	25.0	6.88†
19	4,681	19.0	7.27
20	4,725	16.5	7.17
21	5,561	6.0	7.95
22	5,790	0.5	9.93
Excess hydro. and unidentified steam	5,912		
Total	5,912		

* and † averages for stations.

122. Pooling and Integration. The overall demand for electricity is the composite of class demands by consumers, and these actions are determined by the span of daylight, by living habits, and by industrial and commercial needs. Figure 8-74 represents a so-called load-duration curve which is based on 1 year of experience by an interconnected pool of utility systems. For less than 1 h in the year the combined simultaneous load was slightly in excess of 5,900 MW. For 100% of the time the load never fell below 1,900 MW, which is designated as the "base load." The supply of power to meet the load of each level was composed of imports from surrounding systems [designated I in Fig. 8-74(*A*)] and 7 hydroelectric plants and some 22 steam-electric units or stations or both. In addition, some units and stations aggregating 350 MW were available as reserve at time of peak load. The unlisted small and relatively inefficient stations of all types have been designated D. The stacking of units and plants has been planned in accordance with a pooling program which places the lowest cost producers at the base and the highest ones at peak or in reserve.

123. Pollution Problems and Power Economics. It has already been noted herein that there are few significant hydroelectric sites within the United States remaining for economical exploitation. Except for the possible future development of some direct-conversion process still in the physics-laboratory state of evolution, existing electric loads and all future ones will be supplied by some type of thermal electric power. Two types of objectionable side effects have recently been emphasized in the public press and by a growing public awareness of its environment.

Table 8-23. Energy Consumption and Supply Trends, United States

Line	Description	Trillions of Btu		Projections			
		1964	1965	1970	1975	1980	1985
1	Total energy consumption	51,862	53,962	63,500	74,000	86,500	100,000
2	Bituminous coal and lignite	11,295	12,030	13,300	15,200	18,450	22,100
3	Crude oil	18,742	19,194	21,400	23,850	26,250	28,850
4	Anthracite coal	365	328	350	350	350	350
5	Natural gas (dry)	15,821	16,097	20,100	22,800	25,600	28,000
6	Water power	1,873	2,050	2,400	2,700	3,000	3,400
7	Petroleum products, net	1,958	2,372	2,700	3,000	3,300	3,500
8	Natural-gas liquids	1,774	1,853	2,250	2,600	3,000	3,400
9	Nuclear	34	38	1,000	3,500	6,600	10,400

The calculated consumptions of energy in fuels and nuclear and water power in trillions of Btu are reported by the Bureau of Mines. The energy consumptions include the calculated Btu in the tons of coal, barrels of oil, cubic feet of natural gas, etc., which are made available by production and net imports and consumed for all purposes.

The total energy consumption increased over 4.0% in 1965. During the previous 5 years, the average annual increase was at the rate of about 3.7%.

Bituminous coal and lignite consumption increased about 6.5% in 1965. The average annual gain since 1962 has been about 3.9%.

The increase in crude oil consumption in 1965 was about 2.4% compared with the average annual gain for the past 5 years of about 2.2%.

Anthracite coal decreased in 1965. It has become of minor importance in the total energy.

The volume of natural gas (dry) marketed in 1965 was about 1.7% greater than in 1964. The average annual increase during the past five years has been about 4.7%.

Water-power energy produced and imported by the electric utility industry is calculated in terms of Btu equivalent in fuel required to generate an equal number of kilowatthours. Consumption in 1965 increased about 9.5% over 1964. The average annual increase during the past five years was at the rate of about 4.7%.

Petroleum products, net, represent the excess of imports over exports. Net imports increased in 1965 over 21.0%.

Natural-gas liquids are obtained by processing natural gas. The consumption in 1965 increased some 4.5% compared with an average yearly gain during the past five years of about 5.3%.

Nuclear energy is in terms of the equivalent Btu for the kilowatthours of electricity generated from this source, calculated on the same basis as water power. Nuclear power in 1965 was negligible but is rapidly becoming of great importance.

The projections listed above and as shown in Fig. 8-75 represent an allocation of total energy consumption to each source of energy making up the total.

These side effects are the pollution of the atmosphere and streams by homes, municipalities, and industry. Regardless of the relatively minor contribution to pollutants by power plants compared with all the other sources, power plants are the most conspicuous. Raising river-water temperatures owing to the use thereof for steam-condensing purposes has even been called "thermal pollution."

The reduction of local pollution of all types by ensuring that future plant sites will be at great distances from load centers not only tends to increase the delivered cost of electricity but also merely transfers the objectionable effluents to less articulate neighborhoods. The continuing drive for ever lower costs of electric energy must now take

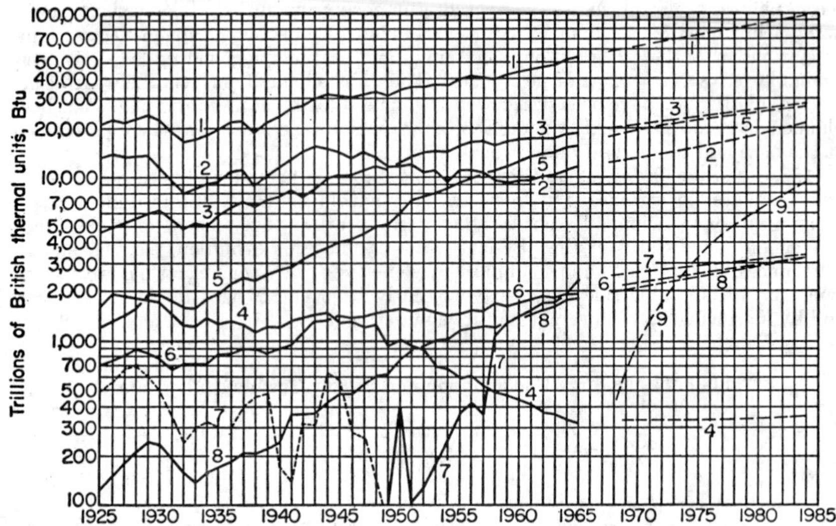

FIG. 8-75. Projection of energy consumption and supply trends to 1985: (1) Total energy; (2) bituminous coal and lignite; (3) crude oil; (4) anthracite coal; (5) natural gas, dry; (6) water power; (7) petroleum products; (8) natural-gas liquids; (9) nuclear. (*Courtesy Ebasco Services, Inc.*)

into account the countervailing socioeconomic demands for greater and greater reliability and continuity of service and for positive efforts to reduce atmospheric pollution. The subject is too complex for full exposition in this limited space. Some interesting comparisons of fuels set forth in the financial section of the *New York Times* for Jan. 22, 1967, follow:

Comparison of Fuels

	% sulfur content	Cost, cents/ million Btu
Bituminous coal	1.8	32
Residual fuel	2.35	33
Natural gas	Minimal	37.5
Nuclear†	0	15.20*

* Future estimates.
† Another probable side effect for future consideration and control is the omission of irradiated materials such as krypton.

124. Future United States Energy-consumption Estimates. The engineering firm EBASCO Services Inc. has produced comparative energy-consumption forecasts to 1985 which have been included here as Table 8-23 and Fig. 8-75.

125. Bibliography

Federal Power Commission, Uniform System of Accounts Prescribed for Public Utilities; Glossary of Important Power Rate Terms, Abbreviations and Units of Measurement; Instructions for Estimating Electric Power Costs and Values; Steam Electric Plant Construction Costs and Annual Production Expenses; Hydro-electric Plant Construction Costs and Annual Production Expenses; Electric Utility Depreciation Practices; National Power Survey, (1) A Report by the FPC—1964, (2) Advisory Reports—Part II—1964.

Agency for International Development, Feasibility Studies—Economic and Technical Soundness Analysis, Capital Project.

Edison Electric Institute, Electric Power Surveys; Statistical Year Books.

Electrical World, Annual Statistical Reports; Steam Station Cost Surveys; Steam Station Design Surveys.

American Society of Mechanical Engineers, Annual Reports on Oil and Gas Engine Power Costs.

Whitman, Requardt, and Associates, Handy-Whitman Index for Public Utility Construction Costs.

U.S. Atomic Energy Commission, The Nuclear Industry—1966; Nuclear Power—Two Years after Geneva, Glenn T. Seaborg.

International Bank for Reconstruction and Development, Economic Choice between Hydro-electric and Thermal Power Development, H. G. Van Dertak.

GRANT, EUGENE L. "Principles of Engineering Economy."

IULO, WILLIAM "Electric Utilities—Cost and Performance"; Washington State University Press.

SECTION 9

NUCLEAR POWER TECHNOLOGY

BY

W. KENNETH DAVIS Vice-President, Bechtel Corporation

HARVEY F. BRUSH Manager of Engineering, Power and Industrial Division, Bechtel Corporation

WESLEY P. SCHIFFRIES Supervising Engineer, Scientific Development Department, Bechtel Corporation; Consultant, San Francisco, California

CONTENTS

Numbers refer to paragraphs

SECTION 9

NUCLEAR POWER TECHNOLOGY

By W. Kenneth Davis, Harvey F. Brush, and
Wesley P. Schiffries

REACTOR TECHNOLOGY

1. Nuclear Power Systems. A nuclear-fueled power-producing system consists essentially of a controlled fission heat source, a coolant system to remove and transfer the heat produced, and equipment to convert the thermal energy contained in the hot coolant to electric power. Regardless of the type of fission heat source used, the basic mechanism is fission of nuclear fuel to produce thermal energy. This thermal energy is removed from the heat source (reactor core) by contacting the fuel with a coolant which can be used directly as the working fluid in the power-conversion cycle or indirectly to heat another fluid to be used as the working fluid.

In some cases an intermediate heat-transfer loop is inserted between the reactor coolant and the working fluid, to increase isolation of the radioactive reactor coolant from the conventional power-producing equipment. The working fluid is then used to drive a turbine-generator set to produce electrical power.

Although a variety of other methods are feasible for the direct conversion of fission energy to electric power (i.e., thermoelectric, thermionic, photoelectric, etc.; see Sec. **12**), these methods are not at present suitable for the production of large quantities of power and are not considered here. Schematic representations of nuclear power systems using the direct, indirect, and indirect with intermediate heat-transfer approaches are shown in Fig. 9-1. A variety of reactor concepts will be discussed.

Nuclear power systems differ in a number of respects from fossil-fuel systems. Some of the more important considerations that differentiate nuclear-fueled plants from fossil-fueled plants are listed below:

1. Nuclear fuel is charged to a power plant infrequently and has a relatively long life, usually measured in months or years, as compared with the continuous fuel-feed requirements for fossil-fueled plants.
2. Burned nuclear fuel is radioactive; it requires remote handling and special processing and disposal.
3. Major portions of a nuclear plant are radioactive during and after operation, requiring special precautions for maintenance of much of the plant.
4. Special system designs are required to prevent radioactivity release during normal operation or due to accidents.
5. Control and instrumentation requirements are strongly influenced by safety requirements and are related to reactor stability, load-following requirements, and the capability of a reactor to increase power output with no additional fuel input.
6. Nuclear fuel is highly processed material generally used in a precise fabricated form, as opposed to fossil fuels, which are essentially raw materials used with only minimal rough processing.
7. The use of nuclear fuel does not require combustion air, thus obviating thermal stack losses and related problems.

These considerations give rise to the general requirements, complexities, and problems of nuclear systems.

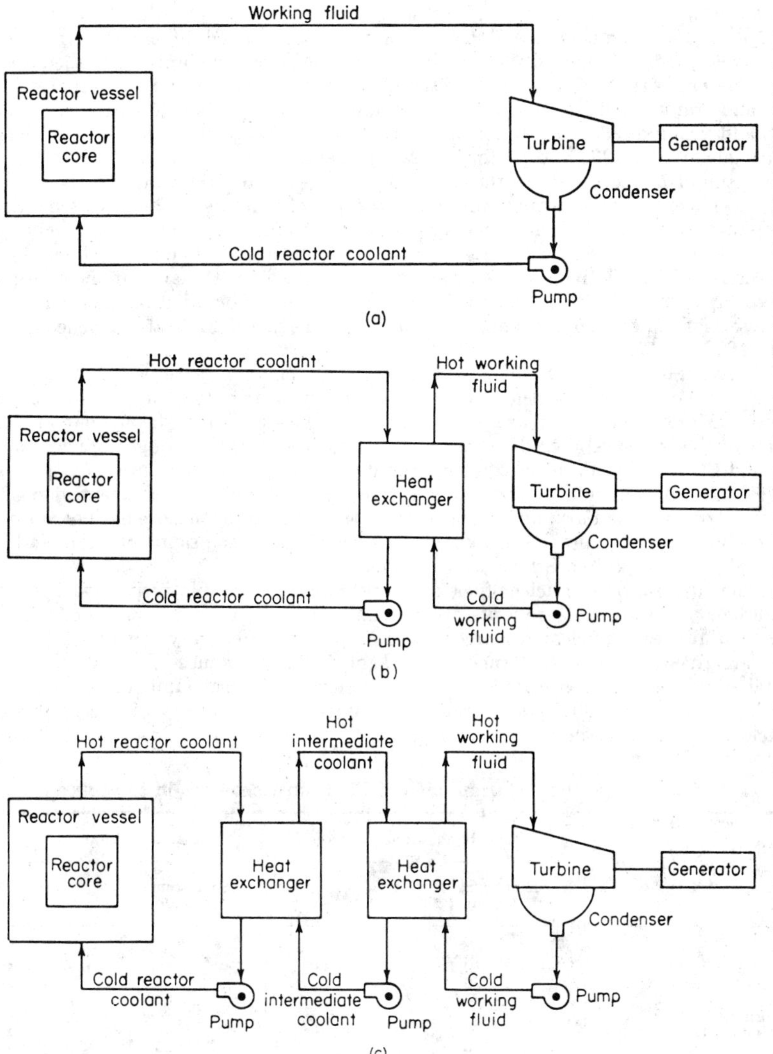

FIG. 9-1. Schematics of nuclear power systems. (*a*) Direct cycle, reactor coolant used as the working fluid; (*b*) indirect cycle, reactor coolant transfers heat to separate working fluid; (*c*) indirect cycle with intermediate loop, reactor coolant transfers heat through intermediate heat-transfer loop to working fluid.

2. Fuel and Fertile Materials. Nuclear-fuel materials are any of the isotopes which undergo fission when bombarded with neutrons and which, under proper circumstances, can be made to sustain an independent fission chain reaction. The principal isotopes (atoms of a specific atomic weight) which meet these conditions are uranium 233, uranium 235, and plutonium 239. Other isotopes such as thorium 232 and uranium 238 also undergo some fission when bombarded with neutrons but are not considered fuels since they cannot be made to sustain an independent chain reaction. Of the three fuel isotopes mentioned above, only uranium 235 occurs naturally, the other two being produced in reactors by the interaction of neutrons with fertile materials.

Fertile materials are isotopes of certain of the heavy elements which absorb neutrons and ultimately produce fissionable isotopes in a process called *conversion*. If more than one fissile atom is produced for each fuel atom consumed, the process is called *breeding*. Two materials that are particularly significant for conversion or breeding are thorium 232 and uranium 238, both of which are readily available in large quantities. Thorium 232 and uranium 238 absorb a neutron each and subsequently yield the nuclear-fuel materials uranium 233 and plutonium 239, respectively.

Uranium 235 is present to the extent of 0.711% in natural uranium and can be used in this concentration as a fuel only in a few types of reactors. The low-fissile-material content in natural uranium imposes significant restraints on reactor design and operation. Therefore, to permit greater design and operating flexibility as well as to extend the useful life of the reactor fuel, fuel enriched in fissile-material content is frequently used. Nuclear fuel can be enriched with any one or combination of the fissile isotopes to essentially any degree desired, but most power-reactor fuels are enriched to about 2 to 4% in U235.

3. Availability of Fuel and Fertile Materials. The raw materials necessary for the production of fuel and fertile materials are widely distributed throughout the earth's crust. Major deposits of uranium ores are widely scattered throughout the world, some of the largest known deposits being in South Africa, the Congo, Canada, and the United States. In addition to direct mining for uranium, some uranium is obtained as a by-product of mining operations for gold and phosphate. Thorium deposits are less well defined than uranium deposits, since the demand for this ore has not yet become large, but sizable quantities are known to exist in the beach sands of India and Brazil as well as in the United States.

Since utilization of nuclear fuels and fertile materials is relatively new, the location and extent of worldwide deposits of uranium and thorium are not well defined. Estimates of domestic nuclear resources based on projected recovery costs have been made and are given in Table 9-1. Included in Table 9-1 are Q values ($Q = 10^{18}$ Btu) for the total fission energy associated with the resources. These numbers when compared with a Q of about 130 for all fossil fuels indicate the immensity of energy content in nuclear fuels and fertile material.

Table 9-1. Magnitude of Uranium and Thorium Resources in the United States

Cost range, dollars/lb of oxide	Reasonably assured resources		Estimated total resources	
	Short tons, thousands	Q units*	Short tons, thousands	Q units*
Uranium:				
Under 10	310	22	660	47
10–15	150	11	350	25
15–30	200	14	640	45
Thorium:				
Under 10	100	7	400	28
10–30	100	7	200	14
30–100	11,000	780	38,000	2,700
100–500	1,000,000	71,000	3,000,000	210,000

Source: AEC, Civilian Nuclear Power, A Report to the President, 1962, for thorium. AEC, The Nuclear Industry—1968, for uranium.
* One Q unit = 10^{18} Btu.

4. Fission of Nuclear Fuel. *Fission* is the process that occurs when a neutron collides with the nucleus of certain of the heavy atoms, causing the original nucleus to split into two or more unequal fragments which carry off most of the energy of fission as kinetic energy. This process is accompanied by the emission of neutrons and gamma rays. Figure 9-2 is a representation of the fission of uranium 235. The energy released as a result of fission is the basis for nuclear-power generation. The release of about 2.5 neutrons/fission makes it possible to produce sustained fissioning.

The *fission fragments* that result from the fission process are radioactive and decay by emission of beta particles, gamma rays, and to a lesser extent alpha particles and

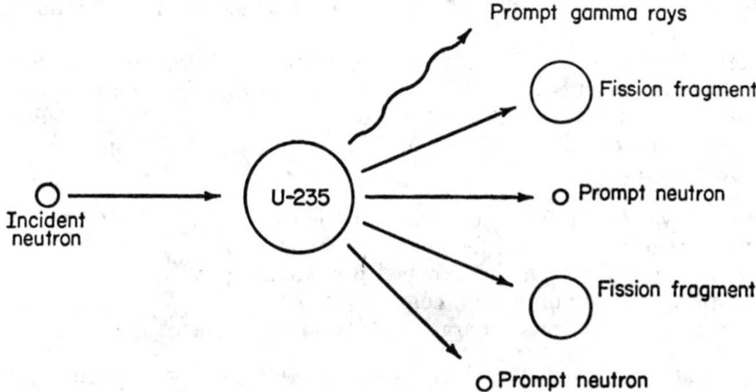

FIG. 9-2. Fission of uranium 235. Incident neutron, upon colliding with U235 nucleus, causes fission to take place, resulting in the production of fission fragments, prompt neutrons, and prompt gamma rays.

neutrons. The neutrons that are emitted after fission, by decay of some of the fission fragments, are called *delayed neutrons*. These are of the utmost importance, since they permit the fission chain reaction to be easily controlled.

The total detectable *energy released* owing to the fission of a single nucleus of uranium 235 is 193 MeV (million electronvolts), distributed as indicated in Table 9-2. As can be seen from Table 9-2, the neutrons emitted as a result of fission of a uranium 235 nucleus carry off 5 MeV of kinetic energy. Since on the average there are about 2.5 neutrons emitted/U235 fission, the average neutron energy is 2 MeV. Actually fission neutrons are emitted with an energy spread of from nearly zero energy to approximately 16 MeV, the bulk of them being in the 1- to 2-MeV energy region.

Table 9-2. Distribution of Fission Energy

	MeV
Instantaneous energy release:	
Kinetic energy of fission fragments	168
Prompt-gamma-ray energy	7
Kinetic energy of prompt neutrons	5
Instantaneous total	180
Delayed energy release:	
Beta-particle decay of fission products	7
Gamma-ray decay of fission products	6
Delayed total	13

NOTE: Although not strictly a result of the fission process, there is an additional 5 to 8 MeV emitted per fission as a result of the capture of neutrons not used in the fission chain reaction. About 1 MeV of this total is emitted over a period of time owing to decay of activation products, and the remainder is emitted immediately upon neutron capture.

The fission neutron energy is important because the probability of neutron capture in fissile material is greater for low-energy neutrons than for high-energy neutrons. It is also true that, for most materials, absorption of neutrons is favored at low neutron energy but relative fission capture is more probable for low-energy neutrons in most combinations of fissile, fertile, coolant, and structural materials.

Most of the reactors in existence today or planned for the near future are called *thermal reactors,* since they depend on neutrons which are in or near thermal equilibrium with their surroundings to cause the bulk of fissions. These reactors make use of the fact that the probability for fission is highest at low energy by slowing down the neutrons emitted as a result of fissioning to enhance fission captures in the fuel. Loss of neutrons to nonfission-capture processes is lessened by minimizing the quantity of nonfissile material in or near the reactor core. The materials used to decelerate fast neutrons to thermal-energy levels are called *moderators*. Effective and efficient moderators must slow the fission neutrons, in the 1- to 2-MeV range, to thermal energy at about

0.025 eV to less than 0.1 eV. This effect must be produced in a small volume and with very little absorption.

5. Nuclear Radiation and Its Effects on Matter. There are only five types of radiation of interest in nuclear power technology, but there are many mechanisms by which these five are produced in reactor systems. The five types of radiation (with electrical charges indicated as + or − for positive or negative, respectively, and mass in atomic mass units) are:

1. Gamma rays (or photons): electromagnetic radiation.
2. Neutrons: uncharged particles, mass approximately 1.
3. Protons: +1 charged particles, mass approximately 1.
4. Alpha particles: helium nuclei, charge +2, mass 4.
5. Beta particles: electrons (charge −1), positrons (charge +1), mass very small.

The principal mechanisms by which each of these types of radiation is produced are described below:

6. Gamma Rays. **Prompt-fission gamma rays** are produced as a result of the fissioning of a U235 (or other fissile-material) nucleus. The gamma rays are emitted within a fraction of a microsecond after fission takes place and are considered to be coincident with the fission process. Prompt-fission gamma rays carry off about 7 MeV/fission, with individual photon energies ranging from less than 0.5 MeV to greater than 1.5 MeV.

Fission-product-decay gamma rays are emitted from the fragments resulting from the fission process and their decay products. These radioactive fission products have half-lives (the time it takes for one-half the atoms originally present to decay) from a fraction of a second to millions of years. In most cases, they emit soft (low-energy) gamma rays and beta particles with energies lower than 1 MeV.

Capture gamma rays are emitted by the nucleus of an atom instantaneously upon the capture of a neutron. The energy of these gamma rays is generally higher than those released by fission or decay. Many elements yield capture gamma rays in the 6- to 8-MeV range.

Activation-decay gamma rays are often emitted from the nucleus after a neutron-capture process, if the new nucleus formed is unstable. Most decays of this type are accomplished by electron emission accompanied by one or more gamma-ray photons. Each unstable (radioactive) isotope has a specific half-life and mode of decay which is an intrinsic property.

Inelastic-scattering gamma rays are emitted from a nucleus that has been excited to a level above its ground state by interaction with an energetic neutron. These are emitted within an extremely short time after the interaction takes place, and the total energy carried off by these photons is less than or equal to the kinetic energy of the incident neutron.

7. Neutrons. **Prompt-fission neutrons** are produced as a result of the fissioning of a fissile material and, as in the case of prompt gamma rays, are considered to be emitted coincidentally with the fission process.

Delayed neutrons are emitted from several of the fission products with apparent half-lives of up to about 2 min. Although half-lives are usually ascribed to the production of delayed neutrons, they are actually emitted within less than a microsecond after the formation of a highly excited nucleus. The half-life actually describes the decay of a fission fragment to the highly excited fission product.

Photoneutrons are produced when a photon with energy greater than the binding energy of a neutron (the energy required to bind a neutron to the nucleus) interacts with a nucleus and ejects a neutron. This process is generally not important, since most isotopes have a high threshold for the reaction and a low probability of occurrence above the threshold. Two isotopes used in some reactor systems which have low thresholds and a fairly significant probability of photoneutron interaction are hydrogen 2 (deuterium) and beryllium 9.

Activation neutrons. Neutron decay of an activated material occurs occasionally in cases other than delayed neutron decay of fission products. The only case which is of some interest in nuclear-reactor technology is the neutron decay of nitrogen 17, formed by the fast-neutron irradiation of oxygen 17.

Reaction neutrons are neutrons ejected from a nucleus by interaction with one of several particles. There are known cases of neutron emission resulting from a nucleus interacting with a neutron, proton, or alpha particle. Important use is made of this process in producing neutron sources for reactor start-up.

8. Alpha particles are produced by the decay of several fission products and a few activated materials as well as by a few neutron-alpha reactions in which an incident neutron interacts with and causes an alpha particle to be emitted from the nucleus.

9. Beta particles (electrons and positrons) are produced by several mechanisms, such as radioactive decay and pair production (in which a photon of high energy is converted into an electron-positron pair).

10. Protons are produced in a few radioactive-decay processes and more frequently by neutron-proton reactions in which an incident neutron causes a proton to be emitted from the nucleus.

11. Nuclear radiation, when it interacts with any material, deposits energy in the material and can have various effects. In chemical compounds the chemical form will be changed, in solids the crystalline structure may be altered, and in any case heat will be generated. For charged particles and gamma rays the mechanism of energy transfer is *ionization* of the material traversed by the radiation. Ionization is the production of electrically charged particles by stripping orbital electrons from the electrically neutral atoms. In the case of neutrons, the primary energy-transfer process is a kinetic-energy exchange caused by collision of the neutrons with nuclei of the matter traversed.

Measurement of radiation intensity is based on the ionization produced either directly or indirectly by the radiation being measured. Charged particles and gamma rays can be measured directly by their ionizing properties, while neutrons are measured by using indirect ionization processes, such as production of an alpha particle in neutron absorption in boron or fission fragments by absorption in uranium 235.

The terms used to describe *energy deposition* and *radiation intensity* are *dose* and *flux*, respectively. The terms used for expressing dose are rad in the general case and rem for biological applications. The rad is defined as the dose which is equivalent to the absorption of 100 ergs/g of any material. The rem is not defined separately, but rather as a normalized index of the damage to living tissue caused by the various types of radiation. The biological effects of the different types of radiation are correlated by a coefficient called the relative biological effectiveness (rbe), which is defined as the ratio of the damage caused by a unit dose of any radiation to that caused by gamma rays. The rem is then defined in terms of the rad and rbe as the dose in rads multiplied by the appropriate rbe. Typical values of the rbe (which are energy-dependent) are:

Type of Radiation	Rbe
Gamma rays	1
Beta particles	1
Fast neutrons	10
Protons	10
Alpha particles	10

The term flux, as opposed to dose, which indicates the energy absorbed by a material in a radiation field, is used to denote the total radiation intensity at a point. *Flux* is defined as the number of particles or photons entering a differential volume sphere of unit cross section in unit time, regardless of direction of the radiation. *Neutron flux* ϕ is equivalent to and frequently represented by the product nv, where n is the neutron density (neutrons per cubic centimeter) at the point of interest and v is the neutron velocity (centimeters per second). Another quite useful and general definition is: Flux is a quantity which when multiplied by the macroscopic cross section in square centimeters for a given reaction gives the number of these reactions taking place in 1 s in 1 cm^3 (or any consistent set of unit time and volume).

The **quantity of radiation** associated with a specific quantity of radioactive material is described by stating the total number of decays that take place in a unit time. A common term used to specify quantity of radioactivity is the *curie*, which is defined as decay at a rate of 3.7×10^{10} disintegrations/s. In addition to quantity of radioactivity, it is necessary to specify the type, number, and energy of each radiation emitted per disintegration to describe completely a radioactive source.

12. Nuclear Cross Sections. Cross sections (or attenuation coefficients) are measures of the probability that a given reaction will take place between a nucleus or nuclei and incident radiation. Cross sections are called either microscopic or macroscopic, depending on whether the reference is to a single nucleus or to the nuclei contained in a unit volume of material.

The **microscopic cross section** is a measure of the probability that a given reaction will take place between a single nucleus and an incident particle. Microscopic cross section is usually denoted by the symbol σ and is expressed in terms of the effective area that a single nucleus presents for the specified reaction. Since these cross sections are usually quite small, in the range of 10^{-22} to 10^{-26} cm^2/nucleus, it is general practice to express them in terms of a unit called the barn, which is 10^{-24} cm^2/nucleus.

Macroscopic cross sections are the products of microscopic cross sections and the atomic density in nuclei per cubic centimeter and are equivalent to the total cross section, for a specific reaction, of all the nuclei in 1 cm^3 of material. Macroscopic cross sections are denoted by the symbols Σ for neutrons and μ for gamma rays and have the units cm^{-1}.

Gamma-ray Cross Sections. Although there are a large number of interaction processes that take place between gamma rays and matter, the most commonly used are the energy-absorption cross section (used to determine gamma heating and dose rates) and the total attenuation cross section (used to determine material gamma-ray attenuation and for shielding design).

Neutron Cross Sections. Neutrons undergo a large number of different interaction processes with matter, and, unlike gamma rays, many of these individual interactions must be evaluated. Neutron cross sections of general use are: fission, gamma-ray production, activation, elastic scattering, inelastic scattering, reaction particle production, total absorption, and total attenuation.

Both neutron and gamma-ray cross sections are energy-dependent properties. Plots of gamma-ray cross sections vs. photon energy for all materials are, over the energy range of interest, smooth curves, whereas for neutron cross sections the curves of many materials show gross variations from a smooth curve. The variations in neutron cross sections show up as peaks and valleys on the cross-section plot; these peaks are called resonances. When a material has a large number of resonance peaks over a portion of the energy range, this portion of the cross-section plot is called a resonance region. The resonance region can have a significant effect on reactor design, since the material U238 which is present in most fuels has a relatively wide resonance region which can cause extensive neutron absorption during the slowing down of neutrons to thermal energy. The importance of resonance region is discussed in the paragraph dealing with Reactor Safeguards (Par. **30**).

The known cross sections for materials potentially useful in reactor systems are used as primary criteria in materials selection. For example, high-neutron-absorption cross-section materials would not normally be used as materials of construction in the vicinity of a reactor core to prevent competition for the neutrons required to sustain the fission process; and high-activation cross-section materials would not be chosen, if they can be avoided, in a region exposed to a high neutron flux during operation, if that region is to be accessible after reactor shutdown.

13. The Chain Reaction. For fission of nuclear-fuel material to occur, neutrons must be absorbed by the fuel nuclei. As a result of the fission process, approximately 2.5 neutrons are released, on the average, for U235 and U233, and almost 3 neutrons are emitted for Pu239 fission. If, on the average, one of these neutrons produced in each fission can be made to cause another fission, the fission process will be sustained in continuous operation. This *self-sustained fission process* is called a *chain reaction.*

Since the chain reaction requires that one neutron from each fission cause another fission, it is worth noting that there are several processes competing for the neutrons produced. These competing processes are nonfission capture in the fuel material, capture in the fuel container (cladding), core structural materials, moderator and coolant, and leakage of neutrons from the core. To permit a chain reaction to take place, it is necessary to design a system in which, after accounting for all neutron losses due to nonfission absorption and leakage, there is still at least one neutron remaining to produce another fission.

The minimum quantity of fuel required for any specific reactor system is called the *critical mass*, and the size associated with this mass is called the critical size. When nuclear fuel is assembled just to the point of a critical mass, the reactor is said to "go critical," i.e., to reach the point of just sustaining a chain reaction. Natural uranium contains only about 0.7% of the fissile isotope U235. Since the U238 which makes up the balance absorbs neutrons, a nuclear reactor which will sustain a chain reaction with natural uranium requires a large critical mass (and size) and the use of moderator and materials of construction which have very low absorption cross sections. To reduce the critical mass required and permit more flexibility in material and design choice, uranium fuel is frequently enriched in U235 content, thereby increasing the fraction of neutron captures that occur in the U235 and cause fission.

14. Nuclear-reactor Cores. A nuclear-reactor core is a collection of fuel and other materials in a configuration which permits a chain reaction to be initiated, sustained, and controlled, and which provides a means for removing the heat generated.

Reactor fuel may be natural uranium, low- (2 to 4% U235), intermediate-, or high-enriched, depending on availability of fuel, cost of enrichment, necessity or desire for minimizing capital cost, breeding requirements, availability of fuel-processing facilities, and many other factors. The choice of fuel type is not usually made on the basis of a single parameter, since an evaluation of the parameters shows that they are, for the most part, highly interrelated.

The common moderators used to slow down fission neutrons in cores utilizing thermal or slow neutrons have a variety of combinations of nuclear properties. Ordinary water is an excellent moderator in the sense that it slows down neutrons better than any other common material and has a relatively low cross section for absorption of neutrons. Despite these qualities, ordinary water cannot be used with natural uranium to produce a chain reaction, since even its relatively low cross section, when combined with the nonfission absorptions in U238, reduces the number of neutrons available for successive fissions to less than the one required for a chain reaction. To produce a chain reaction with natural uranium requires that either heavy water (D_2O) or graphite of extreme purity be used as the moderator. Both these materials are difficult to produce in the purity required and each presents an array of problems associated with their use. The gain obtained with these materials is an improvement in neutron economy, since they have extremely low neutron-absorption cross sections.

Since ordinary water is a good moderator and is widely available, cheap, and a good heat-transfer medium, the approach used in many of today's thermal reactors is to use enriched uranium fuel, thereby allowing for the loss of neutrons by absorption in water.

Another important reason for using enriched fuel is to permit longer *fuel lifetime* and higher *core burnup*. High burnup is desirable because the cost of fabricating fuel for use in a reactor is high, and fewer replacement cores are required over the life of a plant. Also, since the time between refuelings can be longer, the on-line power-producing time is greater if, as in most reactors, the system must be shut down for refueling. Any reactor core, to be useful as a power source, must contain not only enough fuel to achieve criticality, but in addition an *excess of fuel* equal to that which will be consumed to produce the desired power and necessary to overcome the *negative reactivity* effects that build up with the accumulation of fission products, some of which are strong neutron absorbers. Further, since most reactors have a *negative temperature coefficient of reactivity* (i.e., as the temperature increases, the reactivity decreases) some fuel must be supplied to compensate for this effect. In determining the excess initial fuel requirement, account must also be taken of the *positive reactivity* effect of producing fissile Pu239 from U238.

Since enriched uranium is expensive, it is standard practice to reduce the quantity of fuel required by surrounding the core with a material which scatters neutrons readily. This material, called a *reflector*, scatters some of the neutrons leaking out of the core back into the core, thus improving neutron economy. The reflector also serves to increase the fission rate in the outer region of the core and thereby flattens the power-production distribution across the core.

Two terms are generally used to describe the position of a core relative to criticality: *effective multiplication factor* and *reactivity*. The first of these is usually denoted by the symbol k_{eff}, which is defined as the ratio of the number of neutrons produced from fission in each generation to the total number of neutrons lost by all absorption and leakage in the preceding generation. In a system which is just critical the number of neutrons produced in each generation is exactly equal to the number lost by all causes in the preceding generation; therefore the value of k is unity. If only the quantity of fissile material were altered, k would be less than 1 for any fuel less than the critical mass and greater than 1 for more than the critical mass. The effect of k less than 1 would be to cause the fission reactions to die out. For k greater than 1 the fission reactions would increase indefinitely.

For many purposes we are interested, not in the multiplication factor itself, but in the difference between the multiplication factor and unity. This difference is used to define *reactivity* ρ, as follows:

$$\rho = \frac{k_{\text{eff}} - 1}{k_{\text{eff}}} \tag{9-1}$$

Reactivity can then be thought of as a measure of the "distance" a reactor core is from criticality. When a core is exactly at the point of criticality, reactivity is zero; if either some fuel is removed or a neutron-absorbing material is added, the reactivity will be negative and a chain reaction will not be sustained. If some fuel is added or some neutron absorber which was previously present is removed, the reactivity will be positive and the chain reaction will increase steadily. For negative-reactivity values the core is said to be subcritical, and for positive values it is said to be supercritical.

From the preceding statements it can be seen that a reactor core can only contain more fuel than is necessary to make up a critical mass if some material is present which will absorb enough of the extra neutrons produced to prevent the reactivity from becoming positive. The combination of (1) excess fuel over that required for the critical mass and (2) some neutron absorber to prevent the core from going supercritical is the basis for building into the core the fuel that is to be consumed during the power-producing life of the core. The reactivity associated with this excess fuel is frequently referred to as *excess reactivity*. To maintain a net reactivity of zero during operation, a counterbalance, or negative-reactivity effect, must be supplied. The negative reactivity is usually supplied in the form of an effective neutron absorber.

As fuel is consumed and fission-product poisons (neutron absorbers) produced, some of the excess reactivity must be made available to maintain a net reactivity of zero. The most common method of maintaining zero reactivity is to slowly withdraw a neutron absorber from the core. The devices used to control negative reactivity are generally called *control rods*. For long-term reactivity control, negative reactivity is often inserted in the form of removable poison plates or burnable poisons. This point will be discussed in the paragraph dealing with reactor control (Par. **22**).

The insertion of neutron-absorbing control rods reduces net reactivity, and withdrawal increases reactivity. A system which accurately positions control rods can set the reactivity at any desired level within the system limits and start up, shut down, or vary and control the reactor power level. It can be seen that, in addition to making the fuel built into the reactor system available as required to make up for that consumed, the control system can also be used to vary power level as required for load following and to counteract and control the effects of system transients.

In most reactors the heat generated by fission in the fuel must be removed by contacting the fuel pieces with a coolant. The coolant may be gaseous or liquid and may or may not undergo a change of phase. The fuel pieces are generally formed into plate or rod form and are *clad* with a material to prevent loss of fuel and fission products to the coolant and to prevent reactions from taking place which might degrade the fuel quality and shorten core life or even cause the system to be inoperable. The rods or plates are called *fuel elements*, and combinations of them, *fuel assemblies*. The fuel assemblies are put together with passages or channels through which the coolant can flow to remove the heat generated.

The predominant forms of *radiation produced by a nuclear reactor* system are neutrons, electrons, and gamma rays. The role of neutrons in the fission process has already been examined. Neutrons and gamma rays are *penetrating types of radiation* which cause damage to personnel and equipment if the exposure is too great. Since they penetrate matter, some of the neutrons and gamma rays produced in the core leak out and are absorbed in material outside the core. These absorptions outside the core lead to heating, and, in the case of neutrons, they create a secondary source of gamma rays. Fast-neutron interactions with materials can also alter the physical properties of the materials and in some cases reduce their usefulness. The effects of fast neutrons on reactor vessels are discussed in Par. **20** dealing with Reactor Vessels. To prevent damage to personnel and equipment, the radiation leaking out of the core, and any secondary sources of radiation, must be shielded by sufficient material to reduce the radiation intensity to safe levels.

In addition to direct radiation, described above, there are two sources of radiation which are normally brought outside the core. These are *radioactive coolant* (circulated outside the core continuously) and *spent fuel* (removed from core for periodic refueling), both of which are exposed to a neutron flux in the core. To illustrate the extent of the shielding problem created by the leakage of radiation from a reactor, it takes 6 to 9 ft of concrete to reduce radiation levels to safe levels for personnel, depending on the reactor design and power level.

15. Fuel Assemblies. The form of reactor fuel is either homogeneous or heterogeneous. In *homogeneous* form, reactor fuel is dissolved or dispersed in a matrix material, usually a heat-transfer medium which can be circulated to a heat exchanger for heat extraction. Reactors of this type are often called *circulating fuel reactors*. Although this concept of fuel utilization has been proved feasible and some of its advantages are quite desirable, the problems associated with the design and construction of these systems for power-production purposes have not been solved.

Heterogeneous fuel assemblies consist of fuel elements made of fuel encased in a *cladding* material. The basic fuel material, or "meat," can be metallic uranium, UO_2, UC, or some other compound. UO_2 is the material predominantly used owing to its acceptable physical, chemical, and radiation-stability properties. The cladding material is an impervious material which contains the meat and provides a barrier preventing the coolant from contacting the fuel and the fuel and fission products from being released to the coolant. The cladding material should have a low cross section for the absorption of neutrons, high strength, good resistance to corrosion, and good heat-transfer properties (to reduce the temperature rise across the cladding). Materials that have proved to be successful as power-reactor fuel cladding are: stainless steel, zirconium in the form of the alloy Zircalloy, magnesium as the alloy Magnox, and under certain circumstances graphite.

Fuel elements can assume a variety of shapes, the most common one being the thin cylinder or rod. The elements are combined into assemblies to permit handling larger quantities of fuel than a single element and to permit more flexible design of the reactors. Individual elements cannot be made as large as might be desired from a handling point of view: a thick piece of fuel would have a large temperature gradient from its center to the edge and would use neutrons inefficiently, since the outer fuel would shield the inner section from the neutrons required for fission. To enhance the heat-handling ability of a fuel element, extended heat-transfer surfaces are used in the form of fins on the individual fuel elements. Fuel assemblies consist of elements wrapped in bundles welded together with spacers between the elements or set in element grids. In all assemblies it is necessary to ensure adequate flow channels for the coolant.

16. Moderated and Nonmoderated Reactors. Since fissile material absorbs neutrons more readily when the neutrons are slow (low energy) than fast (high energy), it is usually desirable to design reactors to take advantage of this property, although it should be noted that slow, or thermal, reactors are not always more advantageous than fast reactors.

For example, although slow neutrons are absorbed by fuel more readily than fast neutrons, this is also true of the other materials present in the reactor. Conversely, fast neutrons are not absorbed in the fuel so readily as slow neutrons, but neither are they absorbed so readily by the other reactor materials.

Table 9-3. Properties of Common Moderator Materials

Material	Atomic weight	Thermal-absorption cross section, cm^{-1}	No. collisions to thermalize*
Water (H_2O)	18	0.022	19
Heavy water (D_2O)	20	0.000085	35
Beryllium (Be)	9	0.0011	87
Beryllium oxide (BeO)	25	0.0006	103
Carbon (graphite) (C)	12	0.00037	113

* Starting from 2 MeV.

The basic requirement in the design of a slow, or thermal, reactor is that the fission neutrons released, most of which are fast, must be slowed down to or near thermal energy, with little loss by absorption. The best way to do this is to incorporate one of the low-absorption, light elements (i.e., with an atomic mass on the same order as the neutrons). Low-atomic-number nuclei serve best, since the neutrons lose more energy in a single collision with a light nucleus than with a heavy nucleus.

The elements with the best moderating properties are hydrogen, deuterium, beryllium, and carbon. The forms that these elements are usually used in as moderators are ordinary water (H_2O), heavy water (D_2O), beryllium (Be), beryllia (BeO), and graphite (C). Table 9-3 lists some properties of the moderator materials. The element oxygen used in conjunction with the moderator elements, although it is not itself a good moderator, does not substantially degrade the moderating properties of the other elements.

Each of the moderators has advantages and disadvantages. Only the moderators D_2O and graphite in extremely pure form will allow a natural-uranium-fueled reactor to operate. The advantage of this is that natural uranium is widely available and relatively cheap and there is no necessity for the expensive enrichment process. The main disadvantages are that the critical mass is large and the moderators, which are required in large quantity, are expensive. D_2O is costly to produce, and graphite is relatively expensive to purify and fabricate. Ordinary water has none of these disadvantages, but a water-moderated reactor must be fueled with enriched fuel which is expensive, and, owing to the higher absorption cross section of water, the neutron economy of the reactor is lower. Beryllium and beryllia require only slightly enriched fuel. Although they are expensive, the neutron economy is good. The latter materials generally find use in special-purpose reactors such as compact, high-temperature, or mobile reactors.

Nonmoderated reactors make use of *fast, or high-energy, neutrons*. Therefore they do not require and, in fact, must avoid moderator materials in the core. One major advantage of thermal reactors is lost in fast reactors, namely, the ability to control the system reactivity by absorbing neutrons effectively. This is because most materials have very low absorption cross sections at high neutron energy. Control of a fast-neutron reactor can be achieved by one or a combination of methods in which fuel is added or removed by moving a fuel control rod in or out or by moving the core reflector to increase or decrease the neutron leakage from the core. Control of transient effects in fast reactors can be provided by the Doppler effect if the ratio of U238 to U235 atoms is not too low. The negative reactivity effect caused by the Doppler broadening of U238 absorption resonances is very rapid, since there is no time delay required for heat transfer prior to initiation of the effect. In fast reactors, with their shorter effective neutron lifetime (compared with thermal reactors), this rapid action is important, since very-fast-acting negative-reactivity insertion is necessary to prevent transients from causing damage to the core. The fact that most materials do not absorb high-energy neutrons as well as slow neutrons is one of the main reasons a fast reactor is desirable, since this allows a very high burnup of fissile material in the fuel charged to the reactor. Also, greater flexibility is available in the choice of materials of construction and coolants (other than moderators). Thus, more neutrons escape capture in structural materials, allowing them to be captured in fertile material to produce additional fuel.

17. Heat Generation by Reactor Cores. The fission of a single U235 nucleus releases 193 MeV of energy. In addition, there is an average of about 7 MeV of capture gamma-ray energy associated with each fission. This total of 200 MeV is equivalent to 3.2×10^{-11} W/s. Therefore, a rate of 3.1×10^{10} fissions/s produces 1 W of power. Since 1 g of U235 contains 2.56×10^{21} nuclei, the complete fission of all the nuclei in 1 g of U235 would produce 8.2×10^{10} Ws of energy. Fission at the rate of 1 g of U235 per day produces approximately 1 MW of thermal power. Since some of the energy produced in the form of radiation is deposited outside the reactor vessel, fission at a somewhat higher rate is necessary to provide 1 MW of usable thermal power.

In a core loaded uniformly with fuel the *power-production (fission) distribution* is shaped both radially and longitudinally approximately as a cosine curve, with the peak production at the core center. If coolant is supplied uniformly to the fuel throughout the core, a *temperature distribution* similar to this cosine heat-production distribution is set up which would limit maximum reactor power output to a lower level than that if the temperature distribution were nearly flat across the core. Since this situation is undesirable, it is common practice to flatten the power-production distribution by appropriate placement of neutron absorber or fuel, either singly or in combination. Neutron absorbers placed near the core center absorb some of the neutrons in their vicinity, thereby suppressing the fission rate and reducing the central peak power production. Another way to reduce the fission rate in the central region is to use a lower density of fuel, or fuel of lower enrichment, than in the outer region. Conversely, the production of power in the outer regions can be increased relative to the center by using higher enrichment or by placing reflector material outside the core, to increase the neutron population near the outer boundary of the core.

18. Heat Removal from the Core. Heat is removed from reactor cores by thermal conduction across the fuel and cladding materials to a flowing coolant. The heated coolant is used outside the core either as a working fluid or to transfer the heat to another coolant, which is used as the working fluid. The quantity of heat which can be produced in a given volume is not significantly limited by the nuclear processes involved. Rather, in almost all cases, the limit is imposed by the rate at which heat can be removed and transported by the coolant. The major limitations to extremely high heat-removal rates are (1) the relatively low thermal conductivity of most power-reactor fuel materials, which leads to a high temperature drop across the fuel material, and (2) the temperature drop across the film between the fuel cladding and coolant. Under severe conditions, it is possible for the coolant to lose contact with the cladding surface, if substantial film boiling takes place at the surface. In this case local heat removal would become so small that the fuel element would overheat, leading to melting of the fuel and cladding. To reduce the likelihood of overheating the fuel by film boiling, the fuel is fabricated in thin elements to reduce the path length over which heat must travel to the outside. Also, the coolant velocity is maintained well above that required to avoid film boiling. Another potential problem is the buildup of a crud film on the fuel heat-transfer surfaces, which can seriously limit heat-transfer capability. To avoid this problem, the coolant must be maintained at a high level of purity. Coolant cleanup systems (i.e., filters, demineralizers) are used for this purpose.

Since the distribution of heat is not uniform across the reactor core, it is frequently desirable, even after power flattening, to adjust the local rate of heat removal by controlling the rate of coolant flow past the individual fuel assemblies. This is accomplished by *orificing* the fuel assemblies, i.e., controlling the coolant flow in tubes (the coolant and fuel are in a tube, and the moderator is outside). Orifices are placed in some of the assemblies, and the flow openings are set to allow that coolant flow which will maintain an even temperature distribution across the core.

19. Reactor Coolants. Since reactor coolants are of necessity exposed to the core neutron flux, they become radioactive to an extent which depends on the specific properties of the coolant and the extent and type of contamination present. Since the coolant contaminants are predominantly corrosion products, extensive methods are employed to prevent corrosion and remove its products. Coolant radioactivity in itself does not affect the heat-transfer properties of the coolant. But because coolant is normally circulated outside the core, the cost of shielding associated with its radioactivity must be considered in evaluating coolants. Another factor which must be considered in coolant choice is the extent to which radiolytic and thermal decomposition take place in the core-radiation environment. As examples, ordinary water decomposes in the core-radiation field, but it is readily available and cheap to supply, and the decomposition products (hydrogen and oxygen) do not cause any major problems. Organic coolants such as diphenyl and terphenyl decompose, are more costly to make up than water, and under certain conditions may foul the heat-transfer surfaces, reducing heat transfer and increasing the fuel temperature. The only coolants which do not decompose in a radiation field are the elemental ones, such as helium and the liquid metals.

The purpose of a reactor coolant is to extract heat from the core and transfer it to another location for utilization. In terms of this specific purpose, reactor-coolant requirements are essentially the same as for any other coolant application (see Sec. **8**). In addition to the normal requirement (i.e., high thermal conductivity, high specific heat, high heat transfer, low pressure at desired temperature, low pumping-power requirement, and low corrosivity), reactor coolants must be considered in terms of a variety of other properties, such as their moderating value, neutron-absorption cross section, capture- and decay-gamma-ray production, and resistance to radiolytic and thermal decomposition.

For thermal reactors, the only coolant that comes close to having all the desired physical and nuclear properties is heavy water (D_2O). This is an excellent moderator, has extremely low absorption and activation cross sections, and, although it does decompose when irradiated, can be readily recombined to its original form with negligible loss. Unfortunately, D_2O is expensive, and it is difficult and costly to design a loss-free system for it.

In thermal systems that do not require a moderating coolant (i.e., graphite-moderated reactors), a coolant that is essentially ideal for its class is helium. Helium, for a gas, is an extremely good heat-transfer medium, chemically unreactive, radiolytically stable, and has a zero neutron-absorption cross section. The main drawbacks of He are that it is not so good a heat-transfer medium as most liquids, it is expensive, and it is difficult and costly to build a system to maintain acceptably low leakage.

In systems which are based on fast neutrons, liquid metals such as sodium and lithium are excellent coolants, since their heat-transfer properties are outstanding, they do not decompose, and their fast-neutron-absorption cross sections are low. But there are a substantial number of slow neutrons present even in fast reactors and both lithium and sodium absorb these neutrons and become intensely radioactive. In addition, these liquid metals are chemically reactive and must be prevented from being lost from the system and contacting water in order to avoid highly exothermic reactions.

Table 9-4. Properties of Reactor Coolants

Coolant	Melting point, °F	Boiling point,* °F	Specific heat, Btu/(lb)(°F)	Thermal conductivity,† Btu/(h)(ft)(°F)
H_2O	32	212	1.01	0.406
D_2O	39	215	1.01	0.338
Diphenyl	158	490	0.42	0.078
Li	354	2403	1.05	24.000
Na	208	1616	0.31	45.000
NaK (40% K)	65	1518	0.28	15.000
He	...	....	1.24	0.085
Air	...	....	0.24	0.015
CO_2	...	....	0.203	0.009
Steam	...	....	0.48	0.014

* At atmospheric pressure.
† Representative values.

Table 9-4 lists representative values of the more important heat-transfer properties of coolants frequently considered for use in nuclear reactors.

20. Reactor Vessels. The reactor vessel must be an extremely reliable component, since once the plant is built and in operation, the reactor vessel is relatively inaccessible, for two reasons: (1) the vessel becomes highly radioactive during operation because of neutron capture, and (2) a massive biological shield surrounds the vessel.

To assure quality and reliability, the ASME Boiler and Pressure Vessel Code, Sec. III, should be used. This sets forth minimum requirements for the design, construction, inspection, and test of reactor vessels. A major requirement is a complete and detailed stress analysis of all major structural portions of the vessel. The Code requires the owner to prepare a *design specification* which sets forth the functions and design requirements. The vessel manufacturer must prepare a *stress report* which includes a complete set of stress analysis calculations in support of the design and which establishes compliance with the requirements of the design specification. Stresses due to thermal gradients, structural discontinuities, and secondary stresses must be thoroughly analyzed, singly and in combination with other stresses. In addition, consideration must be given to fatigue induced by cyclic operation and thermal-stress ratchet, and, if high temperatures are involved, consideration must also be given to creep. The Code sets forth the manner by which the different stresses are combined and establishes allowable stress values at various temperatures, for a variety of materials of construction.

Since reactor vessels operate in a radiation environment, attention must be given to the effects of radiation on vessel materials. Of special significance is the embrittlement effect fast-neutron bombardment has on steels. This effect is cumulative, being a function of fast flux level, time of exposure, and temperature of exposure. Exposure limits are set on the total number of impinging fast neutrons over the vessel-design lifetime. The term *integrated fast flux* (nvt) is used as the measure of this total exposure. The specified physical effect of design interest is an increase in *nil ductility transition* (ndt) temperature, i.e., the temperature at which the material changes from ductile to brittle behavior.

Steels used in vessel construction have ndt values below the temperatures encountered under normal operating conditions. Materials and operating conditions must therefore be chosen to prevent an increase in ndt to a temperature which will introduce brittle behavior in any portion of the temperature range during operation in the vessel's lifetime. In addition to the use of materials having low initial ndt values, almost all reactor vessels are protected from the direct fast-neutron core leakage by interposition of layers of steel or other form of shielding between the core and vessel. These layers of steel are called *thermal shields* because they have the effect of absorbing gamma rays from the core as well as reducing the intensity of fast neutrons reaching the vessel walls.

21. Steam Generation. The thermal energy produced in the core is transferred by the reactor coolant and ultimately used to produce steam for conventional electric power-producing equipment (see Sec. **8**). Steam can be produced directly in the primary system, or it can be produced externally by transfer of thermal energy through one or more heat exchangers.

In *direct production of steam* in the primary system, water coolant is circulated through the reactor core and raised above the boiling point at the existing pressure. Saturated steam produced in this manner is sent to demisters to reduce entrainment and then to the turbine. Direct in-core boiling of water is an efficient means of removing heat produced in the core and allows more heat to be removed per unit weight of coolant than would be possible if only sensible heat were used. No heat exchangers are required, and coolant recirculation pumps can be smaller for boiling-water systems than for pressurized-water systems.

In the *direct-cycle boiling-water reactor* (BWR) just described, water boils in the reactor vessel, and the steam produced is fed directly to the turbines. This type of reactor does not inherently follow turbine demand. In fact, as more steam is required by the turbine, the reactor pressure decreases, causing increased steam-void formation in the core, which decreases reactivity rather than increasing it, as required. To overcome this deficiency, in the direct-cycle BWR, a system is incorporated to vary the coolant flow with turbine load and thus provide a degree of load following. An early BWR concept called the *dual-cycle system* was developed in which a portion of the steam comes from direct in-core boiling and a portion from an interposed steam generator. This concept can give a wide range of load following, but it has been superseded by the variable-coolant-flow system previously described.

In reactor systems other than the BWR, steam is produced external to the primary system by transferring the primary-coolant thermal energy through one or more heat exchangers which produce steam for the turbines. Such systems, if they use a low operating temperature for the primary coolant, cannot produce high-temperature steam for the turbine. But with primary system coolants such as liquid metals and gases, very high temperatures are possible. Table 9-5 gives typical steam conditions for some of the more common reactor types.

Table 9-5. Primary-coolant and Steam Conditions for Common Reactor Systems

Reactor type	Primary-coolant conditions		Steam conditions	
	Pressure, lb/in² abs	Temperature, °F	Pressure, lb/in² abs	Temperature, °F
Boiling water	1,000–1,500	540–590	900–1,020	540–590
Pressurized water	2,000–2,200	550–600	700–925	500–570
Organic-cooled	120–250	550–650	450–1,000	500–600
Gas-cooled	350–400	1,100–1,400	1,300–1,500	950–1,100
Sodium-cooled	35–60	1,000–1,200	850–2,000	950–1,150

NOTE: Pressure and temperature ranges indicate approximate currently attainable and potential future values.

Since both pressurized- and boiling-water reactors have limited steam temperature, the steam produced is saturated. The relatively low temperatures and steam activation lead to lower thermodynamic efficiency and more expensive turbine equipment than with superheated steam. Consequently, efforts are being made to obtain superheated steam directly from the reactor.

In general, nuclear power systems generate electrical power in the same manner as in fossil-fueled power plants, using conventional turbine-generator equipment beyond the steam generator. Turbines used in nuclear plants are often equipped with integral and/or interstage moisture separators and steam reheaters when the steam to the turbine is saturated. Conventional electric power production is covered in Sec. **8**.

22. Controlling the Reactor. Power-reactor cores are designed to contain sufficient nuclear reactivity to sustain a fission chain reaction at a specified average rate, with controlled variations from zero power (shutdown) to maximum design power, over the design lifetime of the core. To permit efficient and safe operation over this large range of conditions and time, it is necessary to provide a control system which will allow the performance of the basic functions of reactor start-up, maintenance of a desired power level, change in power level to follow electrical load variations, and reactor shutdown. In addition to these basic plant operating functions, it is necessary to provide in the control system means to (1) overcome long-term fuel and poison reactivity changes; (2) limit the maximum reactor power to a level which can be safely handled by the heat-transfer system and electrical generating equipment; (3) limit the rate of change of level to prevent excessive thermal stressing; and (4) provide safeguards to protect plant personnel and the general public from the potential results of equipment failures, malfunctions, or accidents.

Since the power level at which a reactor operates is proportional to fission rate, which is in turn proportional to neutron flux, control of a reactor means controlling the neutron flux. The control function is accomplished by increasing, decreasing, or holding constant the multiplication factor (or reactivity). These operations are performed by addition to, or removal from, the core of neutron absorber, moderator, reflector, or fuel, individually or in combination.

The effect of reactivity changes on neutron flux can be illustrated by a reactor system operating in the steady state. In this case the neutron flux is constant and time-independent, and any increase or decrease in the system reactivity causes a corresponding increase or decrease in flux level. The change will continue indefinitely until the flux falls essentially to zero (decreasing reactivity) or increases to a point where the reactor core is damaged beyond operational use. The flux and power vary as an exponential function of time, with the effective *reactor period* as a parameter. The reactor period T

is defined as the time required for the neutron flux to change by a factor of e (= 2.718). The equation which describes variation in flux (or power) with time for any reactor period is

$$\phi(t) = \phi_0 \exp (t/T) \tag{9-2}$$

where ϕ_0 = the initial flux value, t = time since the reactivity change, and T = reactor period.

The reactor period is defined as

$$T = \frac{l}{k-1} = \frac{l}{k_{ex}} = \frac{l}{\rho} \tag{9-3}$$

where l = neutron lifetime or average time between neutron generations and k_{ex} = the amount by which the multiplication factor was changed (for reactor near criticality, $k_{ex} = \rho$).

For prompt-fission neutrons in a thermal reactor l is on the order of 0.001 s, and it can be seen that even a small change in reactivity would lead to a rapid change in flux or power level if prompt neutrons were controlling. As an example, if ρ = 0.01, T = 0.1 s and the flux would change, according to Eq. (9-2), by a factor exp (t/0.1) or by about 20,000 times per second. A rate of this magnitude would make it extremely difficult, if not impossible, effectively to control the reactor.

The preceding description is based on prompt-fission-neutron lifetime and does not include the effect of *delayed neutrons* which are emitted after fission by several of the fission fragments. This small fraction of delayed neutrons, approximately 0.75% of the total neutrons liberated, has the effect of significantly increasing the effective neutron lifetime, thereby reducing the rate of change of flux for a given reactivity change. The decrease in rate of change of flux caused by the delayed neutrons makes it possible to operate and control reactors effectively and safely.

Since there are six separate delayed neutron groups, a weighted average delay time is found by taking the sum of the mean delay times for the appearance of neutrons from each group. The weighted average delay time for a thermal reactor is on the order of 0.1 s; that is, the effective neutron lifetime is 0.1 s compared with 0.001 s for prompt-fission neutrons. With the reactivity change previously used to illustrate the effect of prompt-neutron lifetime on rate of change of flux (ρ = 0.01), the reactor period for an effective l = 0.1 s is 10 s, and the flux increases according to Eq. (9-2) by a factor exp (t/10), or slightly over 1.10/ s.

To maintain the delayed-neutron effective lifetime as the controlling lifetime, it is necessary to ensure that the maximum insertion of reactivity does not at any time exceed the *delayed-neutron fraction.* Otherwise, the reactor would go critical on prompt neutrons alone, and the effective lifetime would become essentially the prompt-neutron lifetime. Thus are defined the major functions of the control system, i.e., to determine, control, and limit the reactor period.

There are many phenomena which cause changes in reactivity. Some of these are engineered into the system, such as the movement of a neutron absorber into a reactor to decrease reactivity. Others are properties of the system, for example, the reduction in reactivity due to burnup of fuel, or an increase in reactivity caused by production of plutonium from the fertile material U238. Whether the control system initiates the reactivity change or not, reactivity effects must be measured by nuclear detectors and instrumentation, and signals derived from detector outputs must be fed to the control system to check and confirm control settings or institute corrective action.

23. Control-system requirements are derived from the various reactivity changes that occur over the reactor life and from the plant operational and safety requirements. The major reactivity effects are fuel burnup, fission-product poison buildup during operation, Xe135 and Sm149 buildup after shutdown, heavy isotope production, and temperature effects. *Temperature effects* have a special significance in that, either by the nature of the concept or by design, almost any reactor can be made to have an overall *negative temperature coefficient* of reactivity (reduction in reactivity with increase of temperature) or *power coefficient* (reduction in reactivity with increase of power) over all or most of its range of operation. The negative power coefficient creates a requirement for extra fuel to compensate for loss of reactivity during reactor start-up. It acts to prevent major changes from a given operational power level. This property not only increases reactor stability against transient effects but aids the power-reactor control system in following electrical load fluctuations by virtue of their effect on steam and coolant temperatures.

In thermal reactor systems the common control method is insertion or withdrawal of neutron poisons as required to compensate for long-term and short-term operational requirements. In fast reactors, since absorbers are not so effective as in thermal reactors, reflector movement is used to increase or decrease the fraction of neutrons lost from the system, or fuel movement is used to vary the quantity of fuel in the reactor. Neutron absorbers are sometimes used in conjunction with fuel movement to reduce the extent of fuel movement required.

Control systems can be divided into (1) fine and coarse and (2) rapid- and slow-acting for convenience in discussion, although these functions are frequently combined. Table 9-6 lists the common types of control devices used in thermal power reactors. In addition to the control functions indicated, these devices serve to flatten the power distribution in the core to reduce heat-removal problems and thermal stresses. The rod-type control devices are also used to control distortions in flux shape for reactors starting up when xenon isotope Xe135 poisons one part of the core more than the rest.

Table 9-6. Control-device Functions

Device	Functions	Operation
Shim rods	Coarse reactivity control for reactor start-up and approach to power and long-term fuel-depletion reactivity compensation	Restricted to slow- or moderate-speed movements. Contains large negative reactivity
Regulating rods	Brings reactor to full operating power level and maintains power constant. Controls small variations in power output, transients, and short-term fuel depletion	Can move rapidly (with relatively little negative reactivity in rods), or regulating function can be performed by shim rods
Burnable poisons	To hold down large initial excess fuel reactivity and counteract fuel-depletion effect. Long-term shim function	Neutron-absorbing material, which forms a low absorbing isotope on capturing neutron, is dispersed in fuel or moderator
Removable poison curtains	Same as burnable poison	Neutron absorber in form of shroud or curtain placed in core to reduce flux reaching part of fuel. Removed after a portion of initial fuel loading has been consumed
Solution control	Same as burnable poison	Neutron absorber dissolved in coolant-moderator. Concentration varied to control level of reactivity
Safety rods	To shut down reactor rapidly (scram) in the event of emergency	Very rapid insertion. Contains large negative excess reactivity. Sometimes combined with shim controls with interlock to prevent rapid withdrawal
Special safety devices	To scram reactor in extreme emergency where safety rods cannot be used	Some form of quick dumping of neutron poison into core from storage tanks

For rod-type control elements it is necessary that the rod drive and positioning mechanism be extremely reliable and capable of precise position measurement. This permits the degree of reactivity insertion or withdrawal to be ascertained and the control system to be accurately calibrated.

24. Control and Operations. The operation of a power reactor is divided into overlapping but interlocked phases. The operating phases are:

1. *Initial start-up.* Start-up of a core that has never been operated or one that has been operated and shut down for a period of several days before being started up.
2. *Start-up after short shutdown.* Start-up reactor shortly after it has been shut down.

3. *Start-up from power range.* Actually not a start-up, but an increase of power from a reactor which has been reduced in power to an insignificant level relative to power production, but which is still critical and has a high flux level.

4. *Normal operation and load following.* Operation over the normal power range for which the reactor is intended to generate power.

5. *Controlled shutdown.* Intended shutdown of a reactor according to a predetermined plan.

6. *Forced shutdown (reactor scram).* Shutdown of reactor to prevent accidental damage to personnel or system.

In an *initial start-up* of a power reactor which contains a large excess of built-in reactivity, the main problem is associated with the potential of withdrawing negative reactivity too rapidly so that the reactor period becomes excessively small. This problem is compounded by the lack of suitable instruments to detect and measure the very low flux levels existing in a reactor which has never been operated or which has been shut down for a long time. To deal with this problem, the most frequently used approach is (1) to place in the reactor a neutron-emitting source of such strength that, with the reactor multiplication factor, it will create a flux which can be measured and (2) to withdraw control rods at a very low rate while the change in neutron flux and reactor period is observed.

Start-up of a reactor after short shutdown can be handled more rapidly and with more confidence than initial start-up, since there is present readily measurable neutron flux caused by slow decay of delayed neutrons. This neutron flux permits flux and period meters to measure values more accurately, and the probability of getting on a short period is much reduced. Also, xenon poison does not build up in large quantity in a short period after shutdown.

Start-up from the power range is straightforward because the neutron flux is so high that reliable data are available and changes in the period can be readily observed and controlled. In this case the control rods can be withdrawn and set at a position which will yield the desired period.

In *normal operation and load following* a semiautomatic control system is used to compensate for transient reactivity effects, fuel depletion, and demand variations over the entire operating range of the reactor. This system operates on the feedback-comparison principle in which load demand is compared with actual power production. The difference (error or signal) is displayed on instruments for operator action and applied to a servomechanical drive, which moves the control rods in the direction to reduce the error signal to zero. When large power-demand variations are anticipated, the reactor operator can alter rod positions accordingly or the control system can be programmed to follow the load variation without introducing major transients.

In the case of an unexpected large and rapid power-load change, the control system takes up the bulk of the change and the negative power coefficient acts to suppress transient effects. Partial, rather than complete, shutdown has the advantage that start-up from the power range can be performed more rapidly and with greater confidence than from a fully shutdown situation.

Forced shutdown (reactor scram) is to be avoided whenever possible. Reactor scram is accomplished by rapid insertion of control rods, which leads to rapid and complete shutdown. Unless initiated immediately after the scram, subsequent restart is subject to the poison effects of Xe135 buildup and low instrumentation flux. The scramming operation is usually triggered automatically when a predetermined operating level, or limit, is exceeded. Scrams should be reserved for emergencies, i.e., conditions that will damage the system, and should not be considered as a normal shutdown routine.

In some systems several set points are used to trigger a scram. These may involve such operating parameters as a power level or coolant temperature to institute rod reversals and may also employ step setbacks (partial scrams) which come into play below the crucial danger points. In this way, full scrams can often be avoided while the integrity of the system is still protected.

25. Nuclear Instrumentation and Control. Instrumentation and control systems for nuclear power reactors must determine rapidly and accurately the state of the system and report the pertinent descriptive parameters to the system operator and the automatic control system. The detector outputs which actuate the instrumentation and controls are generally reported as meter readings, permanent graphs, and annunciator signals. The overall system, including the detectors, actuators, drive mechanisms, etc., must perform with high reliability and in a fail-safe manner.

These functions must be performed over a very great range, in a complex, overlapped, and interlocked array of detectors, transmitters, meters, control instruments, control actuators, and control devices. Although principal attention relates to nuclear instrumentation and control, the functions must take into account also the thermal characteristics of the entire reactor plant. Generally speaking, thermal characteristics are obtained from conventional process equipment, and these functions will not be treated here (see Sec. **8**).

The reactor instrumentation and control are normally divided into the following functional subsystems:

1. *The reactor-plant control system* provides semiautomatic load-following control in the power range and non-load-following control during start-up and normal shutdown.
2. *The reactor-plant protection system* provides protective response control action to minimize or prevent damage to the system due to equipment failures, malfunctions, or accidents.
3. *The reactor-control instrumentation system* provides the required nuclear data for reactor-plant operating personnel and control and protection systems.
4. *The radiation monitoring system* monitors the reactor-plant and containment system.

26. The reactor-plant control system has the primary function of regulating reactor-plant electrical power output to meet the turbine load demand for all operating conditions over a specified range, typically 5 to 110% of rated plant capacity. This system should be capable of handling step load changes of a predetermined fraction of rated capacity and gradual (ramp) changes up to a specified maximum rate. Specific functions performed by the plant control system are:

Maintaining constant turbine throttle-inlet steam pressure.
Maintaining constant reactor-coolant outlet temperature.
Providing variable flow control in the steam generators to follow load demand while fixed reactor flow is maintained.
Maintaining constant boiler-water level.
Providing steam dump control for large (or complete) loss of load at the turbine.
Providing semiautomatic control for start-up and shutdown operations.

The reactor-plant control system must be designed to meet the requirements of the specific reactor plant. Although not absolutely essential, it is desirable to maintain the reactor-coolant average temperature nearly constant, to maintain turbine efficiency, and to prevent steam inlet temperature and pressure from varying significantly. To achieve these conflicting goals, coolant flow rate and bypass controls are frequently used to reach a compromise.

A reactor control system is made up of a group of units, each of which performs a specific function consistent with overall system requirements. In general, reactor control systems contain units or subsystems, as follows:

The *coolant-flow demand-analyzer* unit receives signals proportional to steam flow and pressure and generates a coolant-flow demand error signal for input to the coolant-flow controller. The flow demand error signal is generated by comparing the measured steam temperature and pressure with set-point values.

The *coolant-flow controller* obtains its input from the coolant-flow demand sensor and provides a signal to the coolant-pump speed controller or bypass valve positioner to maintain or alter coolant flow according to flow demand.

The *coolant-temperature analyzer* provides an error signal to the reactor neutron-flux controller to produce a control-rod movement to maintain constant average or outlet coolant temperature over the power operating range of the reactor. The coolant-temperature error signal produced is obtained by comparing the measured inlet and/or outlet temperatures with the coolant-temperature set points.

The *reactor thermal power computer* develops a nuclear-power demand signal from the measured inlet and outlet coolant temperatures and coolant flow rate. This signal is fed to the neutron-flux controller.

The *neutron-flux controller* provides a control-rod demand signal which operates the control-rod positioner. The signal is produced by combining the power demand signal, coolant outlet-temperature error signal, and the neutron power error signal. The neutron power error signal is produced by comparing the power demand signal with the measured, power-calibrated average core flux.

27. The reactor-plant protection system is required to minimize or prevent damage to the system. This system operates automatically and independently of the reactor control system in the event that the control system fails to perform properly or if there is an accident. The reactor-plant protection system provides protective action by automatically reducing reactor power whenever an unsafe condition is detected in the reactor plant. This system must operate with a high response speed and in a completely fail-safe manner. The plant protection system must rapidly detect either the initial cause of a potential problem or an early effect so that damage is not done before corrective action can be taken. Some of the important abnormal conditions which cause protective action are:

Operator error.
Failure of automatic control system.
Equipment failure.
Loss of power to control system or equipment.
Loss of electrical load to the generator.

Some general requirements for a nuclear-plant-protection system are:

The system should not unnecessarily interfere with normal plant operation and should not cause spurious shutdown.

The annunciator alarm and power set points and scram trips should be fixed (or automatically reset) to prevent plant operation if any part of the protection system is not operating.

The protection system should not normally be subject to operating control by plant personnel.

The protection system should be independent of the operating control system.

The automatic actions of the plant protection system fall into the following categories:

Interlocks are used to prevent electrical or mechanical malperformance by ensuring that a given set of conditions have been met in their proper sequence prior to a subsequent action. These interlocks generally override all operator actions and give visible and audible warnings.

Annunciator signals are provided to alert plant personnel to the existence of or approach to an undesired or unsafe situation. These signals often permit operating personnel to take preventive action before the automatic protection system causes a reactor scram.

Automatic protective actions of the reactor-plant protection system consist of either power setbacks or reactor scrams. These actions occur only if a pre-set danger point has been reached and if prior operator or control-system action has not taken the proper corrective steps. Scrams and setbacks are accompanied by annunciator signals to sound an alarm and indicate the type of action taken and the abnormal condition which caused it.

The *power setback* is intended to correct an abnormal condition without resorting to a scram, and its action is accomplished by removing the reactor plant from automatic load-following control and inserting all control rods at their maximum drive (not scram) rates. The setback action continues as long as the abnormal condition exists, and when the setback is stopped, the reactor is frequently found to be in subcritical condition. If this occurs, operator action will be required to bring the reactor back to criticality before the plant can be restored to automatic control.

Scram action is a rapid, irreversible emergency reduction in power, with shutdown conditions being reached rapidly. The action is indicated only if other means of manual and automatic corrective and protective actions have failed to correct the abnormal condition that is considered potentially dangerous. To prevent excessive thermal shock to the primary system and turbine, the plant protection system must provide trip signals to the turbine protection system and the coolant bypass and throttle valves. To prevent unwarranted loss of on-line time, spurious scrams are avoided by dual instrumentation so that failure of a single instrument would not cause a scram. Some systems use three instrument channels per function and permit a scram (or other action) only if at least two of the three instrument readings agree.

28. The reactor-control instrumentation system consists of the instruments and their associated detectors and transmitters, used for the measurement and transmission of data on neutron flux, temperature, pressure, and flow rate. Since temperature, pressure, and flow measurements are obtained by conventional power-plant process instrumentation, these are not treated here (see Sec. **8**).

The *reactor instrumentation system* provides signals (1) to the automatic control system to regulate reactor power level automatically, (2) to the plant protection system for alarm, setback, and scram purposes, and (3) to reactor operating personnel to facilitate reactor operation from start-up through power operation and shutdown.

The general requirements of a nuclear instrumentation system are:

The instrumentation must cover the entire flux range from start-up to the excess of design operating capacity.

All instrument channels must have a duplicate backup instrument on line. Since no single instrument can cover the entire flux range from start-up to full power, the successive ranges must all have substantial overlaps (usually no less than one decade).

Neutron detectors are the initiating devices in the instrumentation chain. These are located in or near the reactor core in a precisely determined and reproducible position to permit the individual detectors to be replaced and calibrated accurately. The detectors must of necessity operate in a high-flux region, and extensive heating can take place; therefore, these instruments usually must be cooled.

For most reactors the operating range of neutron instrumentation is divided into the following three categories; source or start-up, intermediate, and power ranges. Each range covers a specific portion of the total neutron flux range with positive overlapping between successive ranges. Although there are usually three instrumentation ranges, it is common practice to include four channels (see Fig. 9-3), two of which are identical ranges to cover the power range independently for the automatic control and plant protection systems. For each channel function, no fewer than two instruments are used in duplicate and independent channels.

The *source range and start-up channel* is utilized during start-up operations from shutdown conditions. Source range instruments monitor the neutron flux from the source level to five or six decades above this level. The instruments generally used in this range are high-sensitivity boron trifluoride (BF_3) or U235 proportional counters. The output of these counters is proportional to the ionizing ability of the neutron-induced alpha radiation or of the fission fragments. To protect the detector from loss of sensitivity due to neutron interactions at high flux, the start-up detector is usually withdrawn from the high-flux field during high-power operation. Pulses from the start-up channel are fed to pulse-rate and period computers, which determine the logarithmic count rate and reactor period. These signals are transmitted to meters, recorders, and the control systems in the reactor control room.

The *intermediate-range channels* monitor flux in the intermediate-range portion of the total flux range, starting approximately one decade below the top of the start-up range and extending one decade into the power range. The intermediate-range coverage is generally five to eight decades. A boron- or U235-lined compensated ionization chamber is used for neutron detection in this range. Current from the ionization chamber is fed to a log current and period computer, which converts the input to a

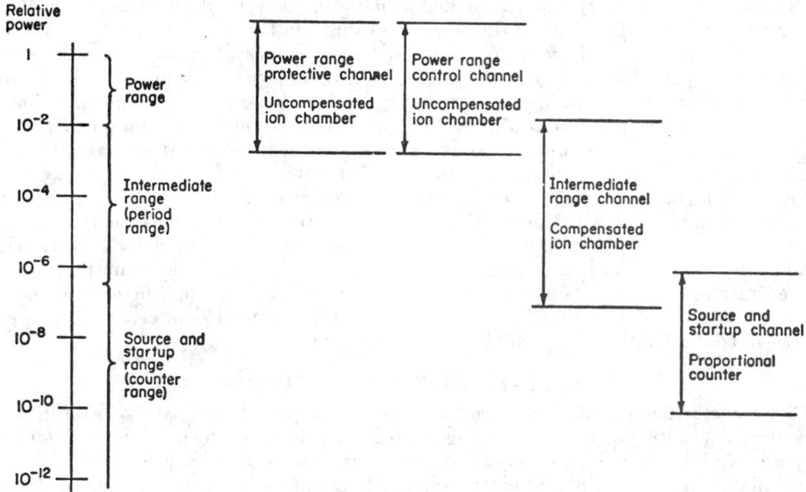

FIG. 9-3. Reactor instrumentation and control ranges.

voltage proportional to the logarithm of the current. This voltage is then differentiated to provide an inverse reactor period signal. The log current and period signals are transmitted to meters, recorders, and the control systems in the reactor control room.

The *power-range control channel* monitors flux from about 1 to 150% of the full reactor power. The instrument commonly used for this function is an uncompensated boron-lined ionization chamber which feeds its current output to a current amplifier. The amplifier output is a signal proportional to the input current which is supplied to the reactor-plant control system.

The *power-range protection channel* monitors flux in the same manner as the power-range control channel, the only difference being that the signal is fed to the reactor-plant protection system to activate the protective functions at predetermined power level, period, temperature, pressure, and flow-rate set points.

29. A radiation-monitoring system is required to monitor the reactor plant, and sometimes surrounding areas, for radiation hazards and to ensure that no legally imposed radiation or activity limits are exceeded either in the plant or in the plant effluents. All accessible areas throughout the reactor plant must be monitored to detect radiation levels in excess of design limits so that alarms may be activated and corrective action taken.

For the *plant monitoring systems,* continuously operating area monitors are used at strategic locations throughout the plant to measure and record gamma-radiation levels in the various areas. *Air-particulate* monitors, which measure the concentration of radioactive particulate matter in the air, are also placed at various potentially contaminated areas throughout the plant.

A *ventilation monitoring system* is provided to measure the level of radioactivity in air being discharged from individual areas and to provide signals to operate isolation dampers and shut down ventilation blowers if a safe radiation level is exceeded. The instruments used to perform these functions are *gamma-ray-sensitive ionization chambers* which are usually located in the exhaust-duct inlets. Signals from ventilation monitors operate remote indicators, recorders, alarms, and the ventilation control system in the control room.

A *waste-gas monitoring system* is provided to measure the concentration of radioactive gases and particulate material suspended in the exhaust from any primary coolant degasification or purification systems and in the discharge line from the decay storage tanks. Signals obtained from the monitors making up this system activate meters, recorders, alarms, and the gas-discharge-line damper valve if the maximum safe concentration or activity set point is exceeded.

The *stack monitoring system* is provided to monitor the total discharge of all ventilation air and gaseous discharge to the stack to ensure that the concentration of airborne activity leaving the stack does not cause specified values of maximum permissible concentration permitted by AEC regulations to be exceeded at ground level in the external environment. Stack monitors usually sample the stack discharge just downstream of the stack particulate filters. If the activity level in the stack exhaust air exceeds the maximum permissible concentration, the monitor signal activates a stack fan cutoff and damper closing operator to seal off the stack.

A *liquid-waste monitoring system* functions to monitor all liquid wastes that are or may be contaminated with radioactivity to ensure that the concentrations of radioactivity in aqueous plant effluent (normally waste streams diluted by mixing with condenser cooling-water discharge) do not exceed AEC limits on maximum permissible concentration. Separate beta-gamma monitors are used on all liquid lines which could become contaminated so that individual loops can be shut off or diverted to prevent excessive contamination of the aqueous plant effluent.

NUCLEAR-SYSTEMS SAFEGUARDS

30. Reactor Safeguards. All reactor plants are designed with safety as a foremost consideration. In addition to designing the plant to minimize the possibility of damage to plant equipment or injury to plant personnel in the event of accidents or malfunctions, major emphasis is placed on the safety of the public in the design and operation of nuclear plants. Because of the legal responsibility of the Atomic Energy Commission for assuring an acceptably low risk to the public from nuclear accidents, it conducts a careful evaluation of all aspects of the plant design, construction, and operation which could affect the safety of the public. Essentially all features of the reactor-plant design are evaluated, including those features provided to permit normal operation of the plant as well as those provided specifically to prevent accidents or to minimize their consequences.

Reactor safeguards may be defined as those features of the plant provided to permit operation over a wide range of normal and abnormal conditions without causing serious damage to the plant. Such safeguards are often referred to as "accident-preventing" engineered safeguards, to distinguish them from the "consequences-limiting" engineered safeguards provided to cope with an accident when it has occurred.

Safety features of reactor systems include: conservative design of the reactor system and components, flexibility designed into the plant to permit it to operate over a wide range of conditions, high-integrity fuel-element design, careful quality control in manufacturing and construction, and in particular the reactor control and instrumentation systems.

Reactors and reactor systems are designed to make maximum use of their *inherent safety characteristics* and thereby reduce the dependence upon external devices to provide the reliability required for safety considerations. In thermal reactors, loss of the moderator under accident conditions and predictable changes in its moderating characteristics over a wide range of operating and accident conditions provide the basis for the control philosophy, stability characteristics, and inherent shutdown capability.

For example, most thermal reactors are designed with a negative moderator temperature coefficient such that as the temperature of the moderator increases, its moderating effectiveness decreases so that power transients are self-limiting. In boiling-water reactors normal reactor control is provided in part by varying the amount of voids, i.e., steam bubbles, within the reactor, thus varying the moderating effectiveness of the water-coolant moderator. In pressurized-water reactors, where voids are prevented during normal operation, boiling or complete loss of coolant moderator provides a rapid and effective shutdown mechanism during accident conditions. Even in some types of solid moderated reactors loss of the moderator may be an effective ultimate shutdown mechanism for extreme accident conditions.

Another important transient limiting characteristic of reactors using natural or low-enrichment uranium, including nearly all reactors of current commercial significance, is the *Doppler effect.* This is an increase in absorption of neutrons in the resonance regions by uranium 238 as its temperature is increased. This produces a decrease in the reactivity, since fewer neutrons are available for fission in the U235. The effect results from a broadening of the resonance peaks (Doppler broadening) which allows more neutrons to be captured in the energy region between the peaks. The Doppler effect is particularly important, since it occurs within the fuel, without the delay associated with the moderator temperature effect.

An ultimate reactor shutdown mechanism for some reactors is disruption and dispersion of the core if all other shutdown mechanisms are ineffective. In most reactors (except solid moderated) loss of moderator would occur before fuel disintegration and would effectively shut down the reactor. Even in the worst case, fuel disintegration would provide a definite upper limit to the magnitude of a reactor excursion which could occur.

The high level of integrity designed and built into fuel assemblies is another feature which enhances the safety of most reactor designs. The low-enrichment UO_2 fuel used in most commercial power reactors in the United States is particularly stable and resistant to damage under extremely severe conditions. Reactivity excursion tests in test reactors with this type of fuel have demonstrated that very little fuel damage occurred even under the most severe reactivity transients that could be produced. These tests have led to a high level of confidence in the safety of such reactor systems and have caused attention to be focused on the hypothetical loss-of-coolant accident as the most severe accident which can be postulated and for which consequences-limiting safeguards are to be designed.

In addition to the inherent reactor characteristics which provide a high degree of reactor control and self-shutdown capability, very reliable *instrumentation and control systems* (Pars. **25** to **29**) are provided to prevent serious accidents and damage to the core. The reactor control system is thus the principal reactor safeguard system provided on all reactors, in addition to providing the means of controlling the reactor to meet its operating requirements of start-up, shutdown, and load following throughout the lifetime of the reactor fuel and under all operating conditions such as loss of all electrical power, tripping of the main turbine-generator, and even gross negligence or willful operator error.

In addition to the normal reactor control system and the shutdown capability it provides, most reactors are provided with *independent secondary shutdown systems* to ensure that the reactor can be shut down even in the event of complete failure of the primary means. Such secondary, or backup, shutdown systems usually are not fast-acting and therefore do not duplicate the scramming function of the primary shutdown system and may not prevent core damage. However, such systems ensure that the reactor can be maintained subcritical and minimize the possibility of further serious damage in the event of a major malfunction or accident.

Water-cooled reactors normally use a borated water solution which can be injected into the core as a secondary shutdown mechanism. Such liquid poisons as boric acid, sodium pentaborate, or potassium tetraborate solutions have been provided for this purpose. Large pressurized-water reactors also use borated water for normal shim control to compensate for fuel burnup and loss of reactivity during core lifetime. In this case, separate boron-injection pumps are provided so that an independent secondary shutdown capability is available.

Some types of reactors, particularly those in which it is not possible to inject liquid poisons, have a completely independent set of control rods provided solely for secondary shutdown capability. Secondary shutdown systems consisting of boronated steel balls which can be dropped into the reactor also have been used.

The design of reactor control systems and the evaluation of reactor-system behavior under normal and a variety of abnormal conditions require a detailed knowledge of the *dynamic behavior of the entire system,* including the neutronic behavior of the core, flow rates and heat-transfer rates in the coolant and in the system components, reactivity feedback effects in the reactor, effects of pressure and temperature transients, and control-system behavior. The complex interrelationships and feedback effects involved require that analog and digital computers be utilized to determine the dynamic behavior of the reactor system. The reactor kinetics equations are well known and can be handled with modern computer technology. Of particular importance is the close coupling of the loop and reactor transient behavior, due primarily to the close relationship between moderator temperature and core reactivity. Furthermore, the very short time scale, on the order of milliseconds, during which core conditions can change radically requires very careful and exact evaluation of reactor-system behavior and control-system design.

Process safety devices normally provided on other types of process systems are also provided on reactor systems. For example, safety valves are provided on the reactor system to prevent reactor-system pressure from exceeding design pressure. Because of the possibility of radioactivity in the reactor coolant, the discharge from the safety valves must be contained to prevent its release to the environment.

Relief valves set to open at pressures below the set points of the safety valves are provided to accommodate large, rapid load changes without overpressurizing the system and, in the case of boiling-water reactors, to minimize the reactivity insertion caused by reduction in the size of steam bubbles under higher system pressure. These relief valves may relieve directly to the atmosphere in the case of closed-cycle systems or to the condenser or containment system in the case of open-cycle systems.

In water-cooled reactor systems consisting of more than one loop and in which it is possible to isolate loops with isolation valves, interlocks are provided on the valves to prevent introduction of a cold slug of water into the reactor system, which could cause a substantial increase in reactivity. Other interlocks are provided to prevent start-up of the reactor under unsafe conditions and to prevent pressurizing of the reactor system before its temperature is sufficiently above its nil-ductility transition temperature. Such interlocks are similar to those used in other types of process systems and are provided wherever necessary to protect against unsafe conditions.

31. Radiological Safeguards. Because of the possible serious effects of exposure to nuclear radiation, many precautions are taken in the design and operation of nuclear plants to protect plant personnel and the public from excessive levels of radiation. Massive shielding is provided around the reactor system to protect plant personnel from the high levels of radiation emitted from the reactor system during both plant operation and shutdown. Systems are also provided to analyze, collect, store, process, and release in a carefully controlled manner the small quantities of radioactive materials released from the reactor system during normal operations.

The levels of radiation exposure and concentrations of radioactive materials in air and water which are permitted for plant personnel and the public as a result of normal operations are defined in Title 10, Part 20, of the Code of Federal Regulations (10 CFR 20), Standards for Protection against Radiation. Occupational exposure limits were set at levels which, over the working life of the individual, are not expected to entail appreciable risk to the individual or to present a hazard more severe than those commonly accepted in other industries. Exposure limits for the public (unrestricted exposure) were established to keep the dose to the whole population at levels comparable with or less than the natural background radiation levels.

Because of the extremely low probability of a major release of radioactivity from a reactor plant, it is recognized that much higher limits may be allowed for a "once in a lifetime" accident situation. Title 10, Part 100, of the Code of Federal Regulations (10 CFR 100), Reactor Site Criteria, provides values of radiation exposure to be used as guides for analyzing the effects of a very severe but unlikely accident. These values are not permissible for upper-limit exposure to the public but serve only as a measure of the relative suitability of a particular site for a particular reactor plant. In conjunction with the site characteristics of population distribution and meteorology and with conservatively assumed accident mechanisms, they provide an indication of the degree of effectiveness required of the various radiological safeguards.

The *major source of radioactivity* in a reactor plant is the fission products produced in the fuel by the fissioning process. After several months of operation, the core of a power reactor contains several hundred pounds of fission products amounting to millions of curies of radioactivity. If an appreciable portion of these fission products were released to the environment, the consequences could be severe. However, the barriers designed to prevent release of fission products are formidable, and the probability of a significant release is negligible. The fuel material in most power reactors (UO_2) and the fuel-element cladding very effectively retain the fission products so that essentially all of them are removed with the spent fuel elements during refueling and ultimately end up in storage at a fuel-reprocessing plant. Very small quantities of fission products may get into the coolant through defects in the cladding or by fissioning of trace amounts of uranium contamination on the cladding, but these can easily be accommodated. The only way significant quantities of fission products can leave the fuel is by overheating and rupturing or melting the cladding and fuel as a result of a severe accident. The extent of the reactor safeguards provided to prevent such accidents (Par. **30**) indicates the extremely low probability of such an event.

The reactor system provides another substantial barrier to the release of fission products. It is designed and built with high integrity and provides reliable means of removing heat to prevent the fuel from overheating. A significant release of fission products from the reactor system would require breaching of the reactor system.

For the purpose of defining requirements for safeguards systems and analyzing their effectiveness, a number of hypothetical accidents to the reactor system are usually postulated to occur. The accident which could lead to the greatest possible release of energy and radioactivity in most water-cooled power reactors is the *loss-of-coolant accident*. This accident is often referred to as the *maximum credible accident*. For this accident, it is postulated, without attempting to define the mechanism by which it could occur, that there is a complete and instantaneous double-ended rupture of the largest reactor-system pipe, allowing rapid flow of reactor coolant out of the reactor system through both ends of the break (or an opening of equal area) and flashing of the coolant to steam as it leaves the system. This is followed by meltdown of some or all of the fuel and cladding and release of a portion of the fission products caused by heating owing to absorption of radiation from the decay of fission products (decay heating). It is postulated that the emergency decay-heat removal systems do not operate. In addition, there is also the theoretical possibility of hydrogen formation from a cladding-water reaction and subsequent reaction of the hydrogen with oxygen, which if it occurred could add substantially to the energy released. A common basis used for analysis of the fission-product release is (as in TID-14844, Calculation of Distance Factors for Power and Test Reactor Sites) that 100% of the noble-gas fission products, 50% of the iodine, and 1% of the solid fission products are released from the melted fuel. Of the iodine released, half of it is assumed to deposit in or near the reactor system so that 25% of the total iodine is assumed to be released from the containment system.

A number of other hypothetical accidents are evaluated during the safety analysis of a reactor plant. In some cases a refueling accident in which a single fuel element is dropped and overheats or is dropped into the core and causes an excursion may be limiting because it may occur outside the reactor containment or while containment integrity is not being maintained. Types of reactors other than water-cooled reactors may have other limiting hypothetical accidents, but the possible consequences are similar.

A number of *design safeguards to minimize release of radioactivity* are provided in all reactor plants. These safeguards control activity release to the environment even if a substantial amount is released from the reactor system. These are the "consequences-limiting" engineered safeguards. The most significant of these is the reactor containment structure, which surrounds the reactor system and completely contains (except for a small amount of leakage) any fission products released from the reactor system. It is designed to withstand the maximum energy release resulting from the loss-of-coolant accident and to maintain its integrity under accident conditions. Various types of reactor containment and their principal characteristics and design bases are discussed in the paragraph dealing with reactor-system containment (Par. **44**).

Other engineered safeguards are provided in many reactor plants to aid the containment system in reducing the possible consequences of the assumed accident. Among these are *containment-pressure suppression systems,* which minimize containment-pressure buildup after a loss-of-coolant accident by quenching the steam. *Spray-system* cooling within the containment is sometimes used to cool the containment atmosphere rapidly after the postulated accident and to reduce containment-pressure buildup. Suppressing containment pressure has the effect of reducing leakage rate from the containment and substantially reducing total integrated leakage of fission products during the course of an accident.

Sprays provide a rapid and effective means of heat transfer, but they rely on pumps, which may possibly fail. Water for the sprays may be taken from inside or outside the containment. After being sprayed, it is collected in the containment sump, pumped through a heat exchanger, cooled, and recirculated to the spray headers. The spray also removes substantial quantities of fission products from the containment atmosphere and thereby reduces the amount available for leakage.

Recirculating cooling and filter systems are provided in some plants to remove fission products from the containment atmosphere and reduce containment pressure. Such systems include coolers, demisters to remove water and water vapor, roughing filters, absolute particulate filters, charcoal absorbers to remove iodine, and blowers to circulate the air. Provisions must be made to ensure that the blowers will operate under accident conditions and that the filters will not plug up or overheat and burn owing to the decay heat of the collected fission products and are sufficiently shielded so that they do not provide a large source of direct radiation external to the containment. Properly designed and tested, such systems can substantially improve plant safety and permit use of sites which might otherwise be considered unsuitable. The need for such systems should be evaluated carefully, since they may impose a significant penalty in initial cost and may require considerable periodic testing to ensure their continued effectiveness.

Core spray and injection systems are provided on water-cooled reactors to remove decay heat and prevent the core from melting in the event of a loss-of-coolant accident. They are designed with redundant components and reliable power supplies.

The first method used to minimize exposure to the public in the event of an accident was remoteness. This has the undesirable effect that the nuclear power plant must transmit power to a load center over an otherwise unnecessarily long transmission line, with its attendant first cost and losses. As more experience is gained in the operation of power reactors, it is expected that safeguards will permit reduction in the distance of power reactors from load centers.

32. Protection against Radiation—Shielding. When nuclear radiation interacts with a material, it may cause chemical and physical changes. The ultimate result of the interaction is the deposition of energy in the material. If the material being irradiated is a chemical compound, its composition will be changed; if it is a solid, its crystalline structure may be changed, and in any case heat will be generated. Excessive quantities of radiation must be avoided to protect personnel and plant.

Safety near a source of radiation requires that the chemical and physical changes due to radiation effects in the human body must be reduced to an insignificant level. To do this, sufficient material is placed between the source and the operators or other personnel to attenuate the energy to a safe level. A shield which reduces the radiation intensity reaching human beings to a safe level is called a *biological shield.*

Radiation intensities near detectors and instruments must be reduced to levels low enough to prevent excessive radiation-induced noise levels and spurious electronic signals from interfering with normal operation of the equipment. Reactor structures, in particular the reactor pressure vessel, must be protected, at least partly, from the fast neutrons that cause most of the radiation damage, by interposing a shield between the core and the vessel. This shield (usually steel) also removes a significant fraction of the total energy leaking from the core before it reaches the biological shield. This reduces the heat-generation rate in the latter shield. The shield between the core and vessel is usually referred to as a *thermal shield.*

Near an intense source of radiation, such as a nuclear reactor, the energy deposited in structures may be sufficiently great to cause significant increases in temperature. Since stresses induced by temperature gradients may cause structural damage, it is necessary to determine the magnitude of heat generated in materials by the various forms of radiation. Even if thermal stress is not a problem, the heat generated may be high enough to require special heat-removal equipment. This is usually true, for example, in a concrete biological shield, where heat generated in the concrete may raise the temperature high enough to drive off a portion of the water. This loss can appreciably reduce the effectiveness of the shield and can lead to structural damage.

To calculate shielding requirements, it is necessary to determine *source terms* for the radiation involved. The usual method used to describe the strength of a radiation source is based on the number of particles or photons emitted from the source. For precise calculations, the source is described in terms of the rate of particles emitted per unit volume of source per unit of time (neutrons per cubic centimeter per second, million electron volts per cubic centimeter per second). The common unit used to describe source strength on a gross scale is the *curie*, 3.7×10^{10} disintegrations/s. To utilize the curie as a source term for calculations, the mode of disintegration (type of radiation), number of particles emitted per disintegration, and the energy of the radiation must be known.

To perform complete shielding analyses for a reactor plant, it is necessary to know the location of all radiation sources throughout the plant. The following paragraphs describe the types and locations of major radioactive sources:

The *reactor core* is the primary source of radiation. Sources which may be unimportant at one time become important at another time. During reactor operation, for example, the radiation from fission and the gamma rays produced by capture of neutrons (capture gamma rays) are most important. After shutdown, however, since the fission neutrons and gamma rays are no longer produced, secondary sources become of prime importance. These secondary sources produce radiation by decay of fission products and neutron-activated materials (and occasionally production of photoneutrons). Fission-product decay in the fuel assemblies is usually considered most important after shutdown, since the radioactive fuel assemblies must be periodically removed to allow refueling. Therefore the handling equipment and storage areas must be shielded.

The *neutron-flux region*, which encompasses a large volume around the core, is important during operation because of the capture gamma rays and inelastic-scatter gamma rays produced in this region. After shutdown the decay of activated materials is, essentially, the only source of radiation. This decay source often does not become important until it is necessary to repair or replace equipment in this region, when access may be limited and material handling complicated. The grid plates, thermal shields, core tank, and biological shield in the neutron-flux region become activated and may emit fairly hard gamma rays. The core coolant is also activated, and since it passes through pipes and equipment throughout the plant, it may be a major source of radiation outside the reactor complex for some time after shutdown, with evident effect on the access time for maintenance and repair. If the gas in the cavity between the core tank and concrete shield is activated, there may be an additional requirement for the holdup of this gas before it can be vented to the atmosphere.

The *core coolant* may be activated in the neutron flux in the core region. In addition to the coolant itself, the impurities in the coolant will be activated. Since the coolant spends a large part of each cycle outside the reactor vessel, it is frequently necessary to supply shielding around the coolant system. In the case of large pieces of equipment such as heat exchangers and tanks the shielding required may be quite thick. If the coolant has to be purified, the portion which is not returned to the coolant system may have to be stored until the activity levels are low enough to permit disposal.

The *fuel elements* in the core after shutdown must be cooled to remove the heat generated by decay of fission products and must be shielded if any work near the core region is necessary. During removal of fuel elements from the core, personnel must be shielded from the decay radiation. The element is lifted out of the core and up into a shielded cask. During this operation, it is necessary to provide sufficient cooling to remove the decay heat and ensure that there will be no path of insufficient shielding to an area in which personnel are working (i.e., through pipes and ducts). The handling cask must be sufficiently shielded to reduce the dose rate at operator and other necessary work positions to acceptable levels. During storage of the irradiated fuel elements, it is required that they be shielded from all accessible areas and that provisions for removing decay heat exist. When fuel elements are shipped to other facilities, it is necessary that the shielding meet the requirements of the AEC, ICC, and state and local governments.

Waste materials which are radioactive and cannot be disposed of with normal plant effluents must be stored in shielded containers which have the means for removing the decay heat generated. The storage capacity must be large enough to handle all the expected waste for a sufficient period to allow the radioactivity to decay to safe levels before being discharged with plant effluents or shipped out for off-site disposal.

Maintenance cells must be provided with enough shielding to reduce dose rates from the hottest (most radioactive) material handled to acceptable levels. When fuel is handled in the cell, it is usually necessary to have a controlled atmosphere and vent system as well as shielded remote-handling equipment. It will also be necessary to have a shielded carrier for the radioactive equipment if it cannot be handled in the shielded fuel-handling cask.

33. Shielding Source Terms. Although there are basic similarities in the methods used to calculate the source terms for shielding for the various types of sources, the differences are great enough to require that each be examined separately.

When a U235 nucleus absorbs a neutron and fissions, there is an almost instantaneous release of gamma rays and neutrons. Each fission, on the average, leads to 2.5 neutrons, and about 7 photons are emitted. The neutrons have energies of 0.01 to 16 MeV, most of them being in the 1- to 2-MeV range. A common assumption made for shielding estimates is that there are two and a half 2-MeV neutrons emitted per fission. The total prompt-photon energy released per fission is about 7 MeV (see also Par. **4**).

For accurate calculations more detailed information is required than given above, but for purposes of shielding estimates it is often adequate to assume that each fission yields seven 1-MeV photons and two and a half 2-MeV neutrons.

In calculating the source of *fission-product decay gamma rays* (*and beta particles*), it is usually assumed that the reactor operates continuously between fuel changes. In this way a conservative estimate of the source strength is obtained by overestimating the buildup of long-lived fission products. K. Way and E. P. Wigner[1] developed equations to describe gamma-ray and beta-particle decay of fission products. Two that are simple and useful are

$$A \simeq 1.4P[t_c^{-0.2} - (t + t_c)^{-0.2}] \tag{9-4}$$

$$P_d \simeq 5.9 \times 10^{-3}P[t_c^{-0.2} - (t + t_c)^{-0.2}] \tag{9-5}$$

where A = total fission-product activity in curies, P_d = total fission-product decay power in watts, P = power of reactor during operation in watts, t = length of time the reactor operated in days, and t_c = cooling (decay) time after shutdown in days.

About 35% of the decays are photon emissions and the remainder are beta particles with mean energies of approximately 0.75 and 0.4 MeV, respectively. With this information and the volume of the source, volumetric source terms for shielding and heat-generation calculations can be approximated with reasonable accuracy.

Capture gamma rays are produced in the core and external to the core. It is common to treat the capture gamma rays produced in the core as part of the core source, lumped together with the fission gamma sources. The complete treatment of capture gammas external to the shield is simple in theory but more complicated in practice. The capture-gamma-ray source is dependent on the thermal-neutron flux at any point in the shield. A complete description of the capture-gamma source in the shield is given by

$$S_v(E, x) = \Sigma_a \phi_{th}(x) f(E) \tag{9-6}$$

where $S_v(E, x)$ = number of capture gamma rays of energy E produced in 1 cm³ of the shield material at distance x from the center (or other base), Σ_a = absorption cross section of the shield material, $\phi_{th}(x)$ = thermal-neutron flux at distance x, and $f(E)$ = number of gamma rays of energy E emitted per neutron capture.

The *decay of activated material* is a function of its past neutron irradiation history and time. Activity in a material builds up in a neutron flux at a rate dependent upon the half-life of the active material, the neutron cross section of the original material, and the thermal-neutron flux in the material. The activity at any time after a material is placed in a neutron flux is

$$A(t) = \Sigma_a \phi_{th}[1 - \exp(-0.693t/t_{1/2})] \tag{9-7}$$

where $A(t)$ = activity of material after being in a neutron flux for time t in disintegrations per second per cubic centimeter, Σ_a = thermal-absorption cross section of the

[1] *Phys. Rev.*, 1948, Vol. 73, p. 1318.

material in cm^{-1}, ϕ_{th} = thermal-neutron flux in neutrons per square centimeter per second, and $t_{1/2}$ = half-life of the material in time units consistent with t.

When the material is removed from the neutron flux, the activity decreases exponentially as a function of time.

$$A(t_d) = A_0 \exp(-0.693t_d/t_{1/2}) \tag{9-8}$$

where $A(t_d)$ = activity remaining at time t_d after removing material from the neutron flux and A_0 = activity at the time the material was removed from the flux.

Equations (9-7) and (9-8) give the decay source term in disintegrations per second per cubic centimeter. To use this value in the case of a decay gamma-ray source, it is necessary to know the mode of decay $N(E)$,

$$S_v(E) = AN(E) \tag{9-9}$$

where $S_v(E)$ = volumetric gamma-ray source term, photons of energy E per second per unit volume, $N(E)$ = number of photons of energy E emitted per disintegration, and A = activity in disintegrations per second per cm^3 [see Eqs. (9-7) and (9-8)].

Inelastic-scatter gamma rays are treated like capture gamma rays except that the flux used includes only the neutrons of energy greater than the threshold (ϕ_{in}) and the cross section is the inelastic-scattering cross section (Σ_{in}), not the thermal-absorption cross section Σ_a. Frequently the flux is considered to include everything above thermal or the fission spectrum of neutron energies is used and the cross section averaged over the chosen spectrum. The equation for the inelastic-scatter gamma-ray source is

$$S_v(E, x) = \Sigma_{in}\phi_{in}(x)f_{in}(E) \tag{9-10}$$

where the terms have been explained above, except for $f_{in}(E)$, which is the number of photons of energy E emitted per inelastic scatter. The term $f_{in}(E)$ is not too well known at present, making accurate estimates quite difficult.

Photoneutrons can be produced in most materials, but for practical purposes only a few materials have threshold energies low enough to make them worth considering in shielding applications. The materials, with their threshold energies, are:

D2	2.2 MeV
Be9	1.67 MeV
C13	4.9 MeV
Li6	5.3 MeV

The equation for a photoneutron source term is

$$S_v = \Gamma_t \mu N_n(E) \tag{9-11}$$

where S_v = volumetric photoneutron source term, neutrons per cubic centimeter per second, Γ_t = gamma-ray flux above the threshold energy, μ = photoneutron cross section (the absorption cross section for photons of energy greater than the threshold energy), and $N_n(E)$ = number of neutrons of energy E emitted per photon absorption above the threshold (this factor is usually taken as equal to 1).

Attenuation of dose and removal of energy from a beam of radiation are accomplished by the same processes but by different computations. In the case of dose attenuation the point of interest is the result not of the process at the point at which it occurs, but of the remaining radiation beam. For energy removal (i.e., calculation of heat generation) we wish to know what happens to the radiation at or near the point of its interaction. The calculation of dose attenuation is a gross calculation which gives the dose outside a shield. The calculation of energy-deposition or heat-generation rates is considered on a more microscopic basis to obtain results that describe the behavior of radiation within a shield.

There are a large number of interaction processes that take place between gamma rays and material, but only three of these processes contribute significantly to the **attenuation of gamma rays** in shields. Although there is overlap in the energy range of operation for the three processes, each is predominant in a different range of photon energies. Above gamma-ray energies of 2 or 3 MeV *pair production* predominates, while below about 0.5 MeV the *photoelectric effect* is strongest. Both the above processes are absorption processes (i.e., the incident photon no longer exists after the interaction). In the energy range from about 0.5 to 2 or 3 MeV the *Compton-scattering* process is the most important mechanism of attenuation. In this case the photons still exist after the interaction.

34. Pair production is the creation of an electron pair when a photon of energy greater than 1.02 MeV interacts with the electric field of a particle. As a result of the pair-production process the incident photon no longer exists in the radiation beam, and the process is characterized as an absorption. The two electrons formed travel a short distance in matter, and their energy is then given up when they recombine to yield two 0.5-MeV photons. Since the distance traveled in material by electrons and low-energy photons produced in the recombination process is quite small, it is usually assumed that all the initial photon energy is given up at the point of interaction.

35. The Compton-scattering process, which predominates in the intermediate energy range from about 0.5 to 2 or 3 MeV, is a scattering interaction of a photon with a free (orbital) electron. As a result of a Compton scattering the incident photon is degraded in energy, and its direction of travel is changed. Since it is difficult to account for the behavior of these scattered photons in attenuation calculations, it is usually assumed that they are absorbed and a semiempirical correction factor called the buildup factor (B) is applied to the results obtained from the absorption calculation.

35a. In the low-energy region up to about $\frac{1}{2}$ MeV the **photoelectric effect** is the dominant attenuation process. The photoelectric effect is the interaction of a photon and a bound electron, with the result that all the energy of the photon is transferred to the electron, ejecting it from the atom. It is usually assumed, for the purpose of calculating energy deposition, that all the incident photon energy is absorbed at the point of interaction.

36. Attenuation of neutrons, as in the case of gamma rays, is primarily based on three energy-dependent processes. Total removal of a neutron (absorption) is significant in shielding mostly at or near thermal energies. The other two processes scatter the neutrons and degrade their energies to low enough values to enable them to be absorbed readily. A secondary effect of the neutron-energy degradation is a reduction in the amount of damage done per neutron (i.e., fast neutrons are more damaging than slow neutrons). Usually the major portion of neutron slowing down is accomplished by elastic scattering, which is similar in effect to billiard-ball collisions. In reactors which have large masses of a heavy material near the core (i.e., thermal shields) a significant portion of the slowing down of fast neutrons to intermediate energies may be accomplished by inelastic scattering.

37. Inelastic scattering is a process in which a neutron interacts to form a compound nucleus which then emits a neutron of lower energy, leaving the original nucleus in an excited state. Part of the kinetic energy of the incident neutron is converted into excitation energy, which is subsequently emitted as one or more *inelastic-scatter gamma rays*. As a result of an inelastic scatter the incident neutron usually loses a large fraction of its incident energy. The process of inelastic scattering can effectively slow down fast neutrons to intermediate, but not to thermal, energies.

38. Elastic scattering is a somewhat simpler process than inelastic scattering. The process appears to be exactly analogous to a billiard-ball type of collision in that the sum of the momenta of the neutron and the atom is the same after an elastic scatter as before. None of the neutron's kinetic energy is carried off as gamma radiation, in contrast to the case of inelastic scattering.

The loss of energy a neutron undergoes as a result of an elastic collision is greatest for light elements and least for heavy elements. Above mass numbers of about 20, the loss of a neutron's energy resulting from an elastic scatter is considered to be negligible. For the light nuclei the average energy loss per collision is high, i.e., a maximum of $\frac{1}{2}$ for hydrogen. It is for this reason that a light material, preferably with a high hydrogen content, is required as part of a neutron shield and is desirable as a moderator material. A material with a high hydrogen content will rapidly slow down the intermediate-energy neutrons to the thermal-energy levels at which the neutrons are readily absorbed by most materials.

39. Capture or absorption of neutrons takes place most readily at low neutron energies. Although there is a finite absorption cross section for neutrons of high energy, it is insignificant with respect to slow-neutron cross sections and also with respect to the scattering cross sections at the high neutron energies. For these reasons it is most efficient in shield design to slow down (moderate) the neutrons to the region of less than a few electronvolts before attempting to capture them.

40. The attenuation of gamma rays interacting with materials is described by a *simple exponential equation.*

$$I = I_0 \exp(-\mu x) \qquad (9\text{-}12)$$

where I = gamma-ray intensity (i.e., dose rate, flux, or current) at position x, I_0 = incident gamma-ray intensity, x = thickness of material traversed, centimeters, and μ = macroscopic gamma-ray cross section, cm^{-1}.

This equation applies accurately only to the case of a collimated beam of photons passing through a material having small dimensions perpendicular to the direction of the incident radiation. This is called *narrow-beam geometry.* If the total cross section (sum of cross sections for photoelectric effect, Compton scattering, and pair production) is used in this equation in the narrow-beam case, the results obtained will be the same as the measured value.

Narrow-beam geometry is used to measure cross sections, since the flux measured is the uncollided flux, and the cross section obtained is not a function of shield thickness. Geometries which more nearly approximate the actual conditions found in a reactor system are called *broad-beam* geometries. In this situation the shield dimensions are large, and scattered radiation must be taken into account. Otherwise the predicted gamma-ray flux will be lower than the actual flux. The scattered portion of the radiation beam is accounted for by inclusion of a *buildup factor* in the exponential equation

$$I = BI_0 \exp(-\mu x) \qquad (9\text{-}13)$$

where the terms are the same as in Eq. (9-12) and B is the buildup factor for the specific material being traversed and particular type of radiation intensity being calculated (i.e., dose, flux, energy absorption).

If the incident radiation is not collimated, the intensity calculated above must be corrected for attenuation due to the effect of distance. For a small (point) source radiating gamma rays in all directions (isotropic emitter) the dose rate will decrease as the square of the distance from the point source. In essence, the radiation is spread uniformly over the area of a sphere of radius equal to the distance from the source to the detector. The *geometric attenuation* for a point isotropic emitter is written as

$$g_p = (4\pi R^2)^{-1} \qquad (9\text{-}14)$$

where g_p = geometrical attenuation for a point isotropic emitter (i.e., the area over which the source radiation is spread = $4\pi R^2$) and R = distance from point source to detector.

The representation of attenuation of gamma rays from a point isotropic source of strength S_p is

$$\phi(R) = \frac{BS_p \exp(-\mu x)}{4\pi R^2} \qquad (9\text{-}15)$$

where $\phi(R)$ = gamma-ray flux at R, photons per square centimeter per second, B = flux buildup factor, S_p = source strength, photons per second, R = distance from source to detector, centimeters, x = thickness of material traversed, centimeters, and μ = total gamma-ray cross section, cm^{-1}.

The use of the buildup factor B in Eq. (9-15) changes the exponential method of calculating attenuation to a semiempirical method, since the buildup factor is not a basic quantity.

The concept of a point isotropic emitter serves a useful purpose in simplifying explanations. The more common geometries actually encountered in reactor shielding can be treated by integrating the equation for the point isotropic emitter over the source geometry. Except in a very few cases the integrations involved turn out to be excessively involved, and a great deal of work has been done to determine reasonable representations of the attenuation functions for various common geometries. A summary of these equations and necessary adjunctive curves is available in the report WAPD-TN-508 (A. Foderaro and F. Obenshain, Fluxes from Regular Geometric Sources, June, 1955) and has been reproduced as Chap. 9, Effect of Geometry of Radiation Source, in TID-7004 (T. Rockwell III (ed.), Reactor Shielding Design Manual). The equations given cover all the common geometries. It is unlikely that a situation will arise in gamma-ray shielding for power reactors which cannot be solved to an acceptable degree of accuracy with the equations tabulated in these references.

In addition to the previously described exponential attenuation methods, a number of more detailed methods have been developed to account for the attenuation processes more accurately than is possible with the exponential method. These are generally used only for special cases where design accuracy is more important than economy, such as for military or space-reactor systems.

41. The behavior of neutrons interacting with material is not readily described by a single simple equation. Although over a large part of the energy spectrum encountered in reactor-shielding work a simple exponential may be used to describe the attenuation of a specified group of neutrons, no such simple solution exists for low-energy or thermal neutrons. In the low-energy cases, special care must be applied to determine the actual phenomenological occurrences and to evaluate the results properly.

Despite the fact that neutrons exhibit complex behavior, it is possible to solve with reasonable accuracy most neutron-shielding problems by using only the fast and slow or thermal groups. This is true because in the intermediate energy range not too many strong neutron interactions take place.

It is not feasible to use an exponential equation with buildup, in the manner of the gamma-ray case, although the situations are very similar. Unfortunately, in the case of neutrons, buildup of scattered neutrons is complicated by nonuniform cross-section variation and by diffusion of neutrons without energy loss.

The *transport equation* (an equation which completely describes spatial distribution, direction of travel, and the number of neutrons of any energy) is the basis for most approaches used to solve neutron-attenuation problems. This complete solution to the problem is unfortunately not available in a readily usable form, but a simplified version of it called the *diffusion equation* is widely used.

Another approach, which sidesteps the problem of calculating thermal and intermediate-energy neutron attenuation, is to set certain specific restrictions on the shield which would lead to overdesign for these neutron groups. If enough good neutron slowing-down material is placed in or after the shield, neutrons removed from the fast beam are rapidly slowed down and absorbed. Therefore, only the uncollided fast-flux attenuation must be calculated, and buildup of lower-energy neutrons can be ignored. It has been found that the fast-neutron dose rate *outside* a shield containing or followed by sufficient hydrogenous material can be obtained by using a simple exponential equation without a buildup factor. This approach, called the *removal method*, requires only a single empirically determined cross section, called the *removal cross section*.

The basic equations used to describe fast-neutron flux are the same ones used for gamma-ray attenuation, with the exception that no buildup factor is used. The *removal-method equation for attenuation of fast neutrons* from a point isotropic emitting source is

$$\phi_R = \frac{S_p}{4\pi R^2} \exp\left(-\Sigma_R X\right) \tag{9-16}$$

where ϕ_R = removal fast flux, neutrons per square centimeter per second, S_p = fast-neutron source term, neutrons per second, R = distance from source to detector, centimeters, Σ_R = removal cross section, cm^{-1}, and X = thickness of shield material, centimeters.

Heat generation in reactor systems takes place owing to the fissioning of fuel and interaction of radiation with materials. If the power or fission rate in the fuel is known, the calculation of heat generation due to fission is straightforward and simple (i.e., the fission rate times the energy released that does not escape from the fuel equals the energy absorbed as heat in the fuel). The essential problem of calculating heat-generation rates thus becomes one of determining where and how much of the radiation that escapes the fuel is absorbed. Significant heat-generation rates may occur in the core, near the core, in the fuel elements, and in general in any high-radiation field.

In almost all cases of interest the major contributor to heat-generation rate is gamma rays. Often the absorption of slow neutrons, which do not by themselves contribute to heat generation, gives rise to the predominant gamma-ray source. The slowing down of fast neutrons frequently leads to a substantial heat-generation rate, but its contribution is usually a small fraction of that produced by gamma rays.

42. Shielding Materials. In choosing a material for duty as a reactor shield it is necessary first to examine the types and intensities of radiation that the material will "see." No useful statements can be made about any material with respect to its shielding efficiency until the radiation is specified.

In the case of *heavy materials* (usually taken as all materials of atomic number equal to or greater than iron) only one general statement pertaining to shielding ability can be made. All heavy materials are relatively efficient gamma-ray shields. In addition most heavy materials are also fair to good *fast*-neutron shields in that they effectively slow down fast and very fast neutrons to intermediate energies by inelastic scattering. Gamma-ray shielding effectiveness is almost directly proportional to material density. But in fast-neutron shielding there are several cases of a heavy element being a worse shield than a somewhat lighter element. A notable example of this is lead, which is less effective as a fast-neutron shield than considerably less dense iron. Lead is one of those unusually stable elements, called the "magic-number nuclei," which have only a few widely spaced energy levels, as opposed to the usual case for heavy elements of many closely spaced levels.

Light materials (lighter than iron) are proportionately poorer gamma-ray shields because of their lower density and also poorer fast-neutron shields owing to their lower cross sections for inelastic scattering. Light materials are better intermediate-neutron shields than heavy materials, but they are seldom used alone as either neutron or gamma-ray shields.

Hydrogen is an exceptional material among the light elements, since inelastic scattering does not take place. Although hydrogen has a low scattering cross section for fast neutrons, the energy loss for each collision is large and the cross section increases rapidly as the neutron energy decreases. For these reasons neutrons are rapidly slowed down after their first collision, and hydrogen is considered a fairly good fast-neutron shield and an excellent intermediate-energy neutron shield.

For neutrons of energy greater than 0.5 MeV inelastic scattering by a heavy material is a considerably more effective means of reducing energies to less than 0.5 MeV, where hydrogen becomes the prime slowing-down medium.

Other very light materials such as lithium, beryllium, and carbon are also good neutron moderators, although to a lesser extent than hydrogen.

A heavy material used alone is adequate for gamma-ray shielding but not for neutron shielding, whereas a hydrogenous material (i.e., water) is useful for neutron shielding but is relatively inefficient for gamma-ray shielding. Neutron shielding with a material such as water is considered only partly sufficient. Although a large thickness will do the job, a much more efficient neutron shield consists of both heavy and hydrogenous material.

Another category of shielding materials, which is not a function of atomic number or density, is that of high *slow-neutron cross section.* Most materials have a sizable cross section for the capture of slow neutrons, but there are some which have unusually high capture cross sections. Two such materials of importance in shielding work are boron, which captures neutrons without emitting hard gamma rays, and cadmium, which is typical of materials that capture neutrons and emit hard gamma rays.

In choosing materials for shielding purposes several properties must be examined. As a basis for comparing fast-neutron shielding effectiveness the removal cross section is the best single parameter. It is important, however, to check all promising materials for neutron "windows," i.e., regions in the neutron energy spectrum for which the material has an unusually low neutron cross section, allowing greater penetration ("streaming") of neutrons in that energy region than in the surrounding energy regions. Iron has definite neutron window, and unless a material with a high neutron cross section in the region of the window (i.e., nickel) is added to the iron, an apparently adequate shield may be insufficient.

Material for gamma-ray shielding can be chosen without danger of gross error by using the ratio of densities as the basis of comparison. The most dense materials are the best gamma-ray shields.

Another important factor may be the production of large numbers of hard gamma rays from slow-neutron capture. If it appears that it is important to suppress capture gamma-ray production, addition of a material such as boron or lithium to shields in high slow-neutron fluxes effectively reduces the production of hard gamma rays.

To ensure that the thermal gradient in high-flux regions is not too steep, it is necessary to check shield materials for good heat conductivity. A good heat conductor not only maintains a low thermal gradient and keeps thermal stresses in reasonable bounds but also prevents excessive temperature rise, which could reduce the shielding effectiveness owing to thermal instability (i.e., pyrolytic decomposition or dehydration).

A material to be used in a high-flux or high-temperature region must be checked to ensure that it is not unstable radiolytically or thermally. For example, polyethylene is an excellent neutron shield because of its very high hydrogen content, but it is not very resistant to either high neutron fluxes or high temperatures. All shield materials must evidently retain their shielding and structural properties in the radiation field to which they are exposed.

Other less direct factors of importance that must be considered are availability, expense, structural properties, and ease of use. These factors, in addition to the properties described above, must be considered in reaching an optimum design consistent with the system requirements. For example, it may be desirable to use a larger amount of a shield material with less than optimum shielding properties if it has high heat conductivity and this is needed to distribute thermal loads more evenly.

43. Nuclear-plant Site Selection. Nuclear power plants must meet all the economic and technical and most of the legal criteria that apply to the siting of conventional fossil-fuel-fired power plants. In addition, the importance of site characteristics in the assessment of public safety results in greater concern in siting nuclear plants than with any other type of industrial facility. Of particular concern are the population distribution with respect to the site and the natural factors which could affect the transport of radioactive material to the public, under normal operating conditions and in the highly unlikely event of an accident which could release radioactive material to the environment.

Proximity to Load Centers. Electrical power can be transmitted over considerable distances by power-transmission lines, but, because of the capital cost of the lines and rights-of-way and transmission losses, an economic penalty is incurred which increases with increasing distance between the generating station and the load center. It is apparent, therefore, that the closer the power-plant site can be located to the load center (while meeting other requirements such as reasonable land cost, adequate cooling water, local zone restrictions, accessibility for fuel shipment, etc.), the lower can be the cost of power delivered to the consumer. Although nuclear plants should be built close to their load centers, regulatory bodies have been very cautious, despite the unparalleled nuclear plant safety record. Most plants built to date have not been in or near heavily populated areas.

Population Distribution. Since power reactors must be located reasonably close to load centers, the population distribution around the site is a necessary consideration in the evaluation of a nuclear power-plant site. Title 10, Part 100, of the Code of Federal Regulations (Reactor Site Criteria) defines the following distances to be considered in reactor siting:

"*Exclusion area.* The area surrounding the reactor, in which the reactor licensee has the authority to determine all activities, including the exclusion or removal of personnel and property from the area.

"*Low-population zone.* The area immediately surrounding the exclusion area which contains residents, the total number and density of which are such that there is reasonable probability that appropriate protective measures could be taken in their behalf in the event of a serious accident.

"*Population-center distance.* The distance from the reactor to the nearest boundary of a densely populated center containing more than about 25,000 residents."

The Reactor Site Criteria provide reference values of radiation exposure to individuals at the boundaries of each of these zones for purposes of evaluating the population-distance characteristics of a site. These distances, the site meteorological conditions, and the amount of radioactive material which could be released from the plant during a major accident are used to evaluate the suitability of the site from the standpoint of safety to the public. In addition to the permanent population surrounding a site, it is also necessary to consider part-time peaks in population, such as during the day or on weekends in recreational areas, and seasonal variations in population, particularly in resort areas. Consideration also should be given to estimates of future increases or changes in population distribution.

To permit placing nuclear plants in desirable locations, it is necessary to provide effective engineered safeguard features to offset, at least in part, the present requirement for large distances from population centers. The trade-off between distance requirements and engineered safeguards is qualitative, and there are no established rules or principles by which such a trade-off can be factored into the evaluation of possible sites for a nuclear power plant. Such a trade-off can, at present, be based only on engineering judgment and on precedents established in the siting of other nuclear plants having different degrees of safeguards and located at various distances from population centers.

Land Use. The use to which the land surrounding a nuclear-plant site is being put, even though it may not be densely populated, may have an effect on the suitability of the site for a nuclear plant. For example, if land is used for agriculture, ingestion of food which has been contaminated by fallout after an accident might conceivably result in a greater radiation dose to the public than might be received from direct exposure to radioactive materials transported downwind from the plants. Of similar concern, but possible as a result of normal operation, is the chance that certain marine life, stationary shellfish in particular, can concentrate the small quantities of radioactivity normally released into the cooling water discharged from the plant. Over a long period of time, the concentration of radioactivity conceivable could build up to levels approaching maximum permissible concentrations.

Meteorology. Because the atmosphere is the principal means by which radioactivity released from a nuclear plant might be transported to the public, site meteorological conditions are considered in selecting a nuclear plant site. Meteorology is of concern both for normal discharges of gaseous radioactive wastes and for the much less likely releases of larger quantities of airborne radioactive material which might result from an accident. A number of meteorological variables are normally evaluated for the site to determine appropriate atmospheric dilution factors. Among these variables are: wind-direction frequencies, in conjunction with the population distribution; wind velocities and the frequencies of each velocity increment; frequency and duration of calms; atmospheric lapse rate; frequency and duration of inversion conditions. Atmospheric dilution is increased, and thus the meteorological conditions are more favorable, the more unstable the atmosphere and the greater the wind velocity.

Other meteorological conditions of concern are the following: precipitation, since it may significantly increase deposition of radioactive materials from the atmosphere, i.e., "rain-out"; possible effects of topography on the local meteorology; seasonal variations in meteorological conditions; and the frequency and severity of storms, particularly tornadoes and hurricanes, which could severely damage the plant. Meteorological information collected at the plant site provides the greatest assurance that it is representative of actual site conditions, provided that sufficiently accurate instrumentation is used and the data are collected over a long enough period of time to be statistically valid. Usually the only meteorological information available during selection and evaluation of a site is data collected at other locations in the same general area, over a long period of time.

In analyzing the consequences of an accident, adverse atmospheric dilution conditions, which might occur only a small percentage of the time, must be taken into account, since an accident could conceivably occur at any time. For considering normal discharges of gaseous radioactive waste, long-term average dilution conditions can be assumed, since it is permissible to average such discharges over an entire year.

Since these normal discharges can be controlled, it is possible, if necessary, to hold up the gaseous waste and discharge it under favorable conditions. There is a tendency, at least for the purposes of accident analysis, to use rather standard unfavorable meteorological conditions, since there is some probability of such conditions occurring at nearly any site. At sites where meteorological conditions are particularly favorable or unfavorable, more careful consideration would be given to meteorological characteristics.

Geology. Investigation of the site geology is necessary to determine the bearing capacity of the soil and the types of foundations which must be used for the major portions of the plant. Test borings are usually made for this purpose, just as for any other large structures. Of particular concern for nuclear plants, because of the implications for public safety, is the possibility of sudden earth movement which could severely damage the plant. Earth slides due to soil instability, subsidence due to consolidation of subsurface materials or to removal of oil or water from subsurface formations, and ground displacements during earthquakes along geologic faults traversing the site—each receives very careful consideration.

If the possibility exists for releasing radioactive liquids from the plant, the ion-exchange and filtering characteristics of the surrounding soil may be important. If the soil could reliably retain any radioactivity released and prevent it from entering water sources or otherwise coming in contact with persons or animals, the site would be that much more favorable.

Seismology is of particular concern in areas of high seismic activity (e.g., California) because of the possibility that the forces which can be produced by earthquakes could be sufficient to damage the reactor system and rupture the containment structure. Careful consideration is given to the general seismic history of the area, including a description of all earthquakes which have been observed at the site, their magnitude or intensity, and the frequency spectrum for which structures should be analyzed. The proximity of the site to known active faults must be determined, and any significant faulting at or near the plant site must be evaluated. Conservative earthquake design factors, usually substantially greater than those required by the Uniform Building Code, are used for critical equipment and structures in areas of high seismic activity.

In coastal areas the possibility of tsunamis may have to be considered. These earthquake-generated sea waves may travel long distances very rapidly and, under certain shore conditions, can build up to substantial heights. Standard seismic design is generally adequate to meet the design criteria based on the factors described above, and with the conservative design factors ordinarily used, reactor systems are adequate for even worse conditions than could be realistically expected to occur.

Hydrology. An important consideration in selecting a site for any power plant is the local hydrology. Present-day types of nuclear plants require substantially greater quantities of cooling water than do modern fossil steam plants because of their higher turbine heat rates (approximately 10,500 Btu/kWh for nuclear vs. about 8000 Btu/kWh for fossil). In areas of limited water supply, cooling towers can be used but at some cost penalty. An additional consideration for nuclear plants is that there be sufficient water flow for the discharge of low-level radioactive liquid wastes. This usually imposes no limitation because of the small quantities of wastes to be discharged and because it is possible to dilute or clean up the wastes to nearly any required concentration. If necessary, it is possible to collect and ship these wastes off site. Radioactive waste discharge is discussed in more detail in Par. **45**, dealing with Reactor-plant Effluents.

Another area of concern is the possibility of **flooding**, which could cause damage to the plant and equipment and cause plant shutdown. Seismic sea waves and hurricanes may increase the possibility of flooding at coastal sites. Seiches (periodic surface oscillations) could result in flooding adjacent to large, enclosed bodies of water. The flooding history of the site must be determined to permit adequate site evaluation and plant design.

The characteristics of the ground water and the level of the water table at the site must be evaluated to ensure that contamination of local water sources by the discharge of liquid radioactive wastes does not occur. If there is any possibility of significant discharge of radioactive contamination to ground water, the absorption characteristics of the soil and the drainage characteristics of the ground water, including its depth and estimated direction of flow, and the characteristics of wells in the area may have to be determined as part of the site evaluation.

44. Reactor-system Containment. Because of the potential hazard represented by fission products contained in the fuel of a nuclear power plant after it has been operated for some period of time, engineered safeguards are built into the plant to prevent serious accidents from occurring and, if they should happen, to minimize the consequences to the public. Of these later "consequence-limiting" safeguards, reactor containment is used for essentially all nuclear power plants in the United States. Aside from acting as a normal building, the primary function of a containment structure is to minimize the consequences of a serious accident which almost surely will not occur in the lifetime of the plant. The reactor system is designed, tested, and operated with such care and has so many redundant and reliable "accident-preventing" safeguards incorporated into the design that an accident of the magnitude for which the containment structure is designed is extremely unlikely.

For water-cooled power reactors, the containment structure is designed for the maximum pressure consistent with the energy released from the design basis accident and for a leakage rate consistent with the amount of fission products released and the site conditions.

A number of types of reactor containment have been used or proposed to meet a wide range of reactor and site requirements. The two principal types of containment used for power reactors are pressure containment ("conventional" or "dry" containment) and pressure-suppression containment.

All pressurized water reactors and the earliest boiling water reactors built used **pressure-containment structures.** These are designed to contain the energy and the fission products released from the reactor system during an accident. Design pressures may range from a few pounds per square inch gage up to as high as 75 lb/in^2 gage. Design leakage rates normally range between 0.1 and 1%/day of the contained air at design pressure. Internal volume may range from several hundred thousand cubic feet up to a few million cubic feet. Most pressure-containment vessels today are constructed of reinforced or prestressed concrete, using a welded steel liner to provide the necessary degree of leak-tightness. Prestressed concrete containment designs are also being considered.

Pressure-suppression containment was first used commercially in a boiling-water reactor plant in California. Since that time all large boiling-water reactor plants have been designed to use this concept. In pressure-suppression containment the reactor system is contained in a relatively small pressure vessel, the dry well, which is connected by large vent pipes to a second vessel partly filled with water, the suppression pool. In the event of a reactor-system piping rupture the steam released from the reactor system passes through the vent pipes and is condensed in the suppression pool. This permits the containment structure to have less volume and/or lower design pressure than a comparable pressure-containment vessel. In addition to condensing the steam, the suppression pool also removes some fraction of any fission products which might be released from the fuel. The dry well and suppression chamber are sometimes enclosed within a low-leakage reactor building which is held at a slight negative pressure by ventilation fans exhausting through filter systems to the stack, thereby collecting and filtering any leakage from the pressure-suppression system.

Multiple-barrier containment is a modification of pressure containment to provide a greater degree of control over leakage than is possible in a single-barrier pressure-containment structure. In this design, leakage past the first barrier is collected within a reduced pressure zone between the first and second barriers and is either exhausted through a filter system and stack or pumped back into the inner vessel. Even if the blowers which maintain the reduced pressure in the intermediate space should fail, there would be a time delay before leakage out of the first barrier would build up sufficient pressure to cause significant leakage out of the second barrier.

Low-leakage buildings such as are used for secondary containment barriers also may serve as primary containment barriers where the possible energy release during an accident is insignificant. For example, low-leakage buildings are used for refueling buildings at power-reactor systems where the reactor system itself is contained in a pressure-containment structure. As with secondary containment buildings, the ventilation system is such that inleakage is assured and the exhaust may be filtered and directed to the stack.

The design pressure of a containment structure is based upon the total amount of energy which could be released from the reactor system during the maximum accident.

For a pressure-containment structure, all possible pipe ruptures are considered and the resulting water and energy releases to the atmosphere are evaluated.

For pressure-suppression containment, the **rate of energy released** from the reactor system is an important design consideration. Three potential sources of energy must be considered:

1. *Stored energy.* When the reactor system is ruptured, as is assumed for the maximum accident, the stored energy in the reactor coolant would be released to the containment structure. Energy stored in the core and primary-system structural materials also may contribute, especially if additional water, added to the core to prevent fuel melting, flashes to steam and transfers the core heat to the containment atmosphere. For gas-cooled, sodium-cooled, and organic-cooled reactors, the stored energy in the coolant is relatively small and does not contribute significantly to containment design pressure.
2. *Nuclear energy* includes the energy produced from reactor excursions (unexpected, large, and rapid increase of power caused by accident or failure) and from fission-product decay heating. The maximum energy from a nuclear excursion in most reactors is usually small compared with the stored energy in the coolant, and so it does not contribute significantly to containment pressure. It may be possible, however, for a nuclear excursion to initiate a metal-water reaction, which in turn could release additional energy. Decay heating is not large by comparison with other sources of energy but is a continuing heat source which must be accommodated by containment heat-removal systems to prevent long-term overpressurizing of the containment structure.
3. *Chemical energy.* In water-cooled reactors, metal-water reactions of the fuel cladding with the coolant and subsequent reaction of the hydrogen produced could add to the total amount of energy released during an accident. In most cases the amount of energy added by this source is smaller than the stored energy release and is released after the peak pressure is reached. Therefore, it should not contribute greatly to the pressure reached in the containment. In gas-cooled or sodium-cooled reactors, where stored energy is small, most of the total energy released may result from reaction of the graphite or sodium with water or air.

Since a containment structure is provided to prevent release of fission products to the environment in the event of an accident, its *leakage* rate at accident pressure is a measure of its effectiveness. Integrated leakage rates as low as 0.1% of the contained air per day have been specified and demonstrated for welded-steel-shell pressure-containment vessels.

An *initial leakage-rate test* is performed on a containment structure after it is completed but before initial plant operation. This test, conducted at full design pressure, is performed with the containment structure in complete operational condition, including all penetrations, isolation valves, and other possible source of leakage. Subsequent integrated leakage-rate tests may be performed at intervals of 1 to 3 or more years, probably at reduced pressures because of the difficulty of performing full-pressure tests. However, the air locks, isolation valves, and other penetrations may be subjected to individual full-pressure leak tests at more frequent intervals, since they are more subject to deterioration and increases in leakage. In some cases, methods of continuously monitoring leak-tightness of the entire containment structure may be used to reduce the frequency of complete integrated tests or local leak tests. Such continuous monitoring can be performed only at a slight pressure and is not so accurate as complete integrated tests, but it will indicate gradual increases in leakage and provide an immediate indication of gross leakage such as might result from a penetration failure or an unclosed opening.

Since *penetrations* in a containment structure are the most likely source for leakage, they require special attention during the design and construction of the containment structure. Openings for access to the inside of the containment vessel during plant operation require air locks with interlocked doors so that a continuous barrier is provided at all times. Such air locks may be required to permit access to equipment during plant operation even though the containment structure is not normally occupied during operation. Similar air locks also may be required to move equipment in and out of the containment structure during operation. Large equipment doors used only for maintenance operations during plant shutdowns may consist of a single cover bolted on and sealed during plant operation.

Electrical penetrations are of particular concern and require special attention because of the large number of wires which penetrate the containment boundary. Attention must be given to the possibility of leakage around the wiring insulation, between the wires and insulation, and between strands of stranded cables. Various methods have been used to prevent this type of leakage, such as employing sealed cable or stripping the insulation and spreading the strands of wires where the wires pass through the sealing compound in potted-type seals. Several different materials may be used to provide the sealing function: hard, strong material to resist pressure and prevent blowout at design pressure, a nonporous material for sealing, and a material which flows at high temperature to fill any cracks that may occur in the primary sealing material. Double-sealed penetrations, with the inner and outer seals able to withstand full containment design pressure and piped to a pressurizing source, are now being used to permit full pressure testing without requiring pressurizing of the containment structure. The contained space between seals can also be sampled to detect leakage, and any leakage may be led to a collection system.

Pipe penetrations through the containment vessel must be suitably designed and anchored so that they do not transmit undue loads to the containment structure. In some cases, bellows-type seals are required for these penetrations, particularly for large, hot pipes subject to thermal deflections. Where the penetration seal is subject to failure, double-sealed penetrations may be necessary. Lines penetrating the containment barrier, which are normally open and which could provide leakage paths in the event of an accident, usually require highly reliable double-isolation valves which can be closed quickly. For example, the main steam line on a direct-cycle boiling-water reactor is essentially an extension of the containment boundary, and thus the line and its isolation valves must have at least as high a degree of leak-tightness and reliability as the containment structure. Other penetrating pipes which are not directly connected to the reactor system or containment atmosphere require at least one isolation valve.

Containment ventilation, heating, and cooling systems often constitute the largest openings in a containment structure and therefore require special attention. In the case of the open, straight-through ventilation system, reliable fast-acting double ventilation valves are provided to permit isolation of the containment system very rapidly. Since these valves may be operated frequently and the sealing surfaces are subject to wear and deterioration, they are tested frequently. Even when closed recirculating cooling systems are provided, a purge system is required to permit ventilating the containment atmosphere at least occasionally. Isolation valves on these systems also must be reliable and leak-tight but may not need to be fast-acting if the system is not to be used during plant operation. Vacuum relief devices, provided on most containment structures to prevent excessive external pressure, also are potential sources of leakage and must be inspected and tested periodically.

Fuel-transfer penetrations are used in some containment structures to permit transfer of fuel from the refueling pool within the containment structure to a spent-fuel storage pit outside containment. On those plants having fuel-transfer penetrations, double-valve locks or water seals are provided to assure the necessary degree of leak-tightness.

Codes for Metal Fabrication. Where containment structures have been fabricated from welded steel plate, they have been designed in accordance with the ASME Boiler and Pressure Vessel Code, Sec. VIII, Unfired Pressure Vessels, and the applicable code case interpretations for containment vessels. More recently, the ASME requirements for steel containment vessels have been incorporated into Subsec. B, Sec. III, Nuclear Vessels, of the Pressure Vessel Code. These codes specify that high-ductility steel plate (i.e., with a low nil-ductility transition temperature) be used and that the maximum plate thickness which can be used (with preheat) without postweld heat treatment be 1.5 in.

Reinforced-concrete containment structures have been used for pressure containment of large pressurized-water reactor plants. Concrete offers the obvious advantage of providing radiation shielding and pressure containment in a single structure so that, if containment shielding is required, a reinforced-concrete pressure-containment structure may be more economical than a steel vessel and separate concrete shield. In addition, there is no limit on the thickness to which a reinforced-concrete structure can be fabricated and thus it may have some advantage over steel. It is somewhat more difficult to provide large openings in a reinforced-concrete structure, however, since the reinforcing bars which are interrupted must be well anchored to reinforcing rings around the openings. After the vessel has been fabricated, it is more difficult to make large construction openings in the vessel wall, such as are frequently used to permit the reactor vessel and other large components to be installed after the containment structure has been completed. In a reinforced-concrete pressure vessel, the concrete is not being used effectively as a structural member since all the tensile forces must be taken by the reinforcing bar. There usually is more steel in the reinforcing bar of a reinforced-concrete pressure vessel than in the steel plate of a comparable steel vessel, since the rebar can be stressed in one direction only, while steel plate is stressed biaxially.

To permit more efficient use of both steel and concrete, many containment structures are now constructed of prestressed concrete. In such structures, steel cables apply a compressive stress to the concrete and maintain it in compression up to design pressure. This technique permits high design pressures to be achieved without great difficulty.

Large concrete caisson structures can also be used for reactor containment. This approach uses a caisson sunk below grade and below groundwater level to house the reactor and the pressure-suppression containment system. Where site conditions are suitable, caissons offer a means of locating a plant belowground with a minimum of excavation.

45. Reactor-plant Effluents. Although there are a large number of individual plant effluents originating in a nuclear power plant, the only ones of interest here are those which are, or might be, radioactive or contaminated with radioactivity. These effluents, which are solids, liquids, and gases, are each handled within the plant and released from the plant in a separate manner.

Radioactive solids are accumulated at the plant and disposed of at infrequent intervals. The materials that make up solid effluents generally consist of contaminated equipment; chemicals, resins, and filters that are used to remove corrosion and fission products from the coolant stream; and filters used to remove particulate matter from gaseous streams. Provision must be made for the monitoring, collection, packaging, and shipment off-site of these accumulated solid materials. The solid materials are packaged to prevent release of activity either in the plant or in transit to the final destination for disposal of solid waste materials (usually a government-owned burial site). Spent nuclear fuel (i.e., fuel that has been burned in the reactor) is not sent to a burial site but is shipped to a fuel-processing plant to recover its useful fuel components and, in some cases, certain of the fission-product isotopes.

Liquid and gaseous radioactive effluents are, under conditions specified by applicable regulations, discharged from the plant to bodies of water and the atmosphere, respectively. The Atomic Energy Commission has established limits for the *maximum permissible concentration* (MPC) of each radioactive isotope which may be released in liquid or gaseous effluents. These limits, given in the Code of Federal Regulations, Title 10, Part 20, are specified at that point where the effluent leaves the plant. The regulations also permit credit to be taken for dilution by the atmosphere or body of water into which the effluent is discharged, such that the limits are not exceeded at the site boundary. Release of liquid and gaseous effluents is permitted only after demonstration that it is not likely that any individual will be exposed to concentrations in excess of the specified limits. For purposes of this restriction, concentrations may be averaged over periods not greater than 1 year. Within the plant, maximum permissible concentrations for a 40-h week are specified at a higher level than for release to unrestricted areas because of the very good control and individual records maintained by plant management.

Liquid effluent from a nuclear plant generally consists of contaminated clean-up and decontamination solutions, laboratory wastes, and small quantities of reactor coolant, usually discharged with and diluted by the condenser cooling water. Liquid wastes are monitored, reused to the greatest possible extent, and then discharged with the condenser cooling water only if the dilute effluent activity concentrations are not greater than the allowable MPC values for water. If the dilute effluent has concentrations greater than permissible for discharge, the liquid waste is stored temporarily to allow for some decay of the activity prior to discharge; or if the activity level is too high, the liquid waste might be accumulated on site for later shipment to a government-owned or -licensed disposal site. In some cases, the liquid wastes are processed to remove certain of the radioactive isotopes to permit discharge of the bulk of the waste without a long holdup period.

Gaseous effluent from a reactor plant may consist of coolant decomposition products, products, minor quantities of fission-product gases, other activated gases that could be dissolved in the coolant, activated air from the region around the reactor vessel, and, in the case of a gas-cooled reactor, the coolant itself. Prior to discharge from the plant stack, the radioactive gas streams are monitored, usually filtered and washed, then diluted with the plant ventilation air and in some cases additional air pumped in for the express purpose of providing additional dilution. Provision is often made for holdup of gases to allow for radioactive decay prior to release.

46. Environmental Dilution of Activity Releases. Evaluations of potential radioactivity releases to the environment are based on standards for exposure at an off-site location as set forth in the Code of Federal Regulations, Title 10, Part 20. Plant effluents must be monitored to determine the level of radioactivity at the discharge from the plant. The extent of dilution of released gases in the atmosphere and aqueous waste into waste bodies must be determined for the maximum, average, and instantaneous release rates from the plant.

At any site, *atmospheric dispersion* conditions may vary over a wide range from day to night, from day to day, and from season to season. Although reasonable estimates of the dilution resulting from atmospheric dispersion can be made for points downwind of a source for a given set of wind and atmospheric stability conditions, the proper weighting of each condition to give an average dilution for a long time period requires more data than are generally available. Atmospheric stability data for a specific site are usually not completely available, since this information is not commonly determined in most areas suitable for reactor plants. The stability information normally employed is the lapse or inversion data used to determine vertical mixing of contaminants introduced into the atmosphere. The *lapse condition* is a decrease in air temperature with increasing altitude (temperature decreases at a rate of about 5.5°F/1,000 ft for an adiabatic atmosphere). With this lapse rate a parcel of air, once displaced upward, tends to continue mixing upward by convection. *Inversions*, which are frequently encountered at night or in the early morning at ground level, are caused by the ground cooling the first few hundred feet of air so that the air temperature increases with altitude in the lower (inverted) layer. Thus, a parcel of air, if displaced upward, tends to return to its original altitude. Information on the frequency of occurrence of lapse and inversion conditions is used for estimating the atmospheric dilutions.

Except where the plant is located immediately upstream from a source of drinking water, *dilution of aqueous effluent* into water bodies is usually of less importance than the atmospheric dilution, since the radioactive concentration in condenser discharge water is usually less than the legal maximum permissible concentrations. However, diffusion equations representing the mixing effects in water bodies have been developed and verified by field investigations. In general, dilution in water is much less rapid than that due to mixing in air. Studies on mixing in water indicate that concentration should decrease from the peak value inversely as the second power of the time allowed for mixing.

ECONOMICS OF NUCLEAR POWER PLANTS

47. General. The economics of nuclear power plants may be developed along the general guidelines that have been previously developed for fossil-fueled power plants. Typically, all costs are broken down into the following categories:

Capital costs, dollars total.
Fuel costs, dollars per year.
Other operating and maintenance costs, dollars per year.

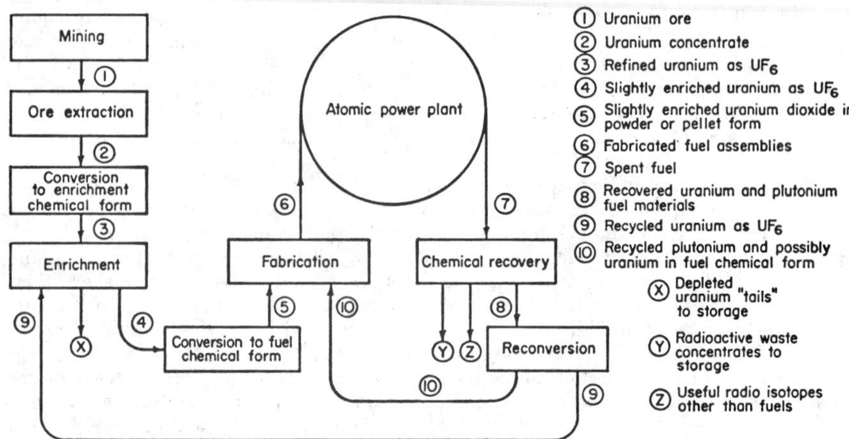

FIG. 9-4. Nuclear fuel cycle (based on water-cooled reactors).

With a knowledge of these total and annual dollar costs, and a knowledge of the pertinent factors relating to production, anticipated plant life, and the costs of invested money, unit costs may be determined for any time period desired. These are most frequently expressed in mills per kilowatthour or cents per million Btu.

Although such unit costs may be determined on a single plant basis, it should be noted that the addition of any new plant will normally provide excess capacity in the system under study. Thus, a more valuable analysis involves the inclusion of the additional costs incurred by the new plant in the total system cost pattern and the redetermination of total system costs. Simplified analysis requires the assumption of an immediate load availability to the new plant at the expense of load availability to the older existing plants. Although more complicated, a system analysis using the lowest system incremental loading cost will provide a more accurate picture of real annual unit costs for the various alternatives considered.

48. Capital Costs. Capital costs are those costs which occur only once and are usually limited to the costs of procurement and construction of the facilities prior to the time of commercial operation. These are normally "capitalized"; i.e., they are treated as an investment which is depreciated over the useful life of the plant rather than being treated as an annual or other shorter-term operating cost.

Determination of those costs which may be allowed within the capital-cost structure often depends on review by the appropriate public-utility commission or other regulatory body and on corporate accounting policies. However, a substantial background of experience and precedence has been accumulated. Additionally, various organizations have prepared standard lists of accounts which may be used as a guide. Table 9-7 lists accounts prepared by the Atomic Energy Commission to facilitate capital-cost estimation (and production-cost analysis) for proposed nuclear power plants.

Allocation of the capital-cost items shown in Table 9-7 over the life of the plant is normally accomplished by applying factors, or percentage rates, which will account for depreciation, return on investment, and taxes applied to income and property value. The product of the total of these factors and the allowable capital costs gives the annual fixed charge in dollars per year for the capitalized investment. Within this structure, several different depreciation methods are used, such as a sinking fund, straight-line depreciation, etc., depending upon the accounting structure already in existence.

49. Fuel Cycle. An understanding of nuclear-fuel costs requires an understanding of the nuclear-fuel cycle, since the fuel-management programs are significantly different from those used for fossil-fueled plants.

The basic fuel cycle is shown in Fig. 9-4. The word "cycle" is intentionally used, for, unlike fossil fuels, a single pass through the reactor does not consume all the nuclear fuel. Further, a valuable new fuel, plutonium, is generated during power production.

Table 9-7. Classification of Construction Accounts, Nuclear Power Plants

Account Number	
	Nuclear Power Production Plant
20	Land and land rights
	201 Land and privilege acquisition
	202 Relocating highways and railroads
	203 Relocating telephone and power lines
21	Structures and Improvements
	211 Ground improvements
	212 Buildings
	218 Stacks
	219 Reactor containment structure
22	Reactor-plant equipment
	221 Reactor equipment
	222 Heat-transfer systems
	223 Fuel-handling and -storage equipment
	224 Fuel-processing and -fabricating equipment
	225 Radioactive-waste treatment and disposal
	226 Instrumentation and control
	227 Feedwater supply and treatment
	228 Steam, condensate, and feedwater piping
	229 Other reactor plant equipment
23	Turbine-Generator Units
	231 Turbine-generators
	232 Circulating water systems
	233 Condensers
	234 Central lubricating system
	235 Turbine-plant boards, instruments, and controls
	236 Turbine-plant piping
	237 Auxiliary equipment for generators
	238 Other turbine-plant equipment
24	Accessory Electric Equipment
	241 Switchgear
	242 Switchboards
	243 Protective equipment
	244 Electrical structures
	245 Conduit
	246 Power and control wiring
	247 Station service equipment
25	Miscellaneous Power-plant Equipment
	251 Cranes and hoisting equipment
	252 Compressed-air and vacuum cleaning systems
	253 Other power-plant equipment
	Transmission Plant
50	Land and Land Rights
51	Clearing Land and Rights-of-way
52	Structures and Improvements
	521 General yard improvements
	522 Substation buildings
	523 Outdoor substation structures
53	Station Equipment
	531 Switchgear
	532 Protective equipment
	533 Main conversion equipment
	534 Conduit
	535 Power and control wiring
	536 Station service equipment
54	Towers and Fixtures
55	Poles and Fixtures
56	Overhead Conductors and Devices
57	Underground Conduit
58	Underground Conductors and Devices
59	Roads and Trails
	General Plant
97	Communication Equipment
	Distributives
98	Indirect Construction Costs
	981 Engineering, design, and inspection
	982 General and administrative
	983 Other indirect costs
	984 Earnings and expenses during construction
	985 Interest during construction
99	Miscellaneous Construction Costs
	991 Construction inventories
	992 Temporary construction facilities
	993 Construction equipment and tools
	994 Construction clearing accounts
	Contingency

Therefore, a recycling of valuable spent-fuel constituents is necessary for the highest economy.

Within the nuclear-fuel cycle, uranium is mined as ore containing 0.1 to 0.3% uranium. This is followed by various extractive metallurgy processes which result in an intermediate product (commonly called yellow cake or U_3O_8), the exact chemical form varying with the process used. This material is then converted to uranium hexafluoride (UF_6) (the only compound of uranium which is gaseous at standard temperature and pressure), in which form it is subsequently enriched in the isotope U235.

The *enrichment process* increases the isotopic ratio of U235 to total U to the desired level by selective partial separation in many gaseous-diffusion stages. The diffusion plant has two principal product streams, one enriched in the isotope U235 and the other depleted. The enriched stream is the principal useful product. The depleted stream ("tails") is at present a waste product. However, it is expected that the "tails" will be used as the blanket fertile material in future fast breeder reactors. The feed-to-product ratio for the enrichment plant is a function of feed enrichment, product enrichment, and tails enrichment. The cost of the enriched uranium product stream is a function of the cost of performing the enrichment service and the cost of the feed material.

The UF_6 product from the enrichment process is then converted to the chemical form in which it will be used as a fuel, normally uranium dioxide, and fabricated into fuel elements for insertion in the nuclear reactor.

The total lead time for all these processes, up to the point of having the fabricated fuel available for insertion into a reactor, is approximately one year. A fuel load typically remains in a reactor 3 to 5 years to achieve the average design fuel burnup. However, during the initial start-up (nonequilibrium) period, one or more portions of the core will normally be removed prior to achieving their average design fuel burnup. Some of these portions may subsequently be reloaded after a condition which is closer to equilibrium is attained.

After the power-production period is completed, fuel is removed from the reactor, stored at the reactor site for approximately ⅓ year to allow radioactive decay of the fission products, and then shipped in specially designed, shielded shipping casks to a chemical reprocessing plant, where the remaining useful fuel components (uranium and plutonium) are recovered. In some cases, selected radioactive nonfuel products are recovered, but the bulk of the fission products is waste which is stored in special permanent waste storage facilities.

The recovered useful fuel components are reconverted to the chemical form required for reuse as fuel. Uranium fuel which has undergone a full power-production period could be made reusable by blending small quantities of highly enriched uranium or plutonium with it, instead of passing the burned fuel back through the enrichment plant. This approach is not generally used because the added cost of the higher enrichment of the makeup fuel overcomes the cost saved by eliminating the UF_6 chemical stages in the fuel recycle. Blending with plutonium has not yet been practiced owing to the lack of plutonium available for recycle and a lack of technological capability to recycle plutonium. In the future, as plutonium becomes more available, it is expected that this approach will be used.

Fuel management, in terms of the feed materials provided to and discharged from the nuclear reactor, is also of basic importance to fuel-cycle costs, since the method used affects both the power-production capability of the nuclear reactor and fuel inventory costs. The fuel-management program and control-rod scheduling program must be interrelated to achieve the maximum power level attainable and to provide a reasonable degree of uniformity of fuel consumption.

Commonly, fuel-management programs have employed *fuel shuffling* at each refueling shutdown. In these programs the reactor is divided into three to five concentric-ring zones of equal size. At each refueling, the spent fuel (fuel that has been in the reactor for its full production period) is removed from the center zone, and each remaining zone is moved toward the center by one zone. New fuel is then placed in the empty outer zone. The approximate annual refueling shutdowns required for this program can be made to coincide with normal utility schedules. This method permits power distribution to be controlled with makeup fuel having a single fuel-enrichment value.

The most recent fuel-management programs provide additional minor variations from the described fuel-shuffling program. These provide for the use of scatter discharge from all but the exterior zone (based on in-core instrumentation monitoring of thermal-power generation distribution), minor sub-ring zones of varying enrichment, since flat-faced fuel elements do not form perfect equally sized concentric rings, and reinsertion of partly spent fuel in subsequent reactor operating periods when the average enrichment has been built back up by intermediate refueling with new fuel.

On-stream refueling, as accomplished in the Canadian heavy-water and United Kingdom gas-cooled reactors, represents an alternative approach to power-distribution flattening and fuel-cycle economics. In this case, fuel-management programs similar to shuffling are also possible but further complicate the plant operations and equipment requirements.

50. Fuel Costs. Fuel costs are affected by the number of functional services which must be performed on the uranium fuel to prepare it for use, the additional services which must be performed to recover the "ash" value of the spent fuel, and the variation in the design data for each batch of fuel employed.

The degree to which these separate services may be procured in package plans is still emerging. The following list indicates the services currently available. It is anticipated that various combinations of these services will be offered in the future:

Natural uranium procurement as U_3O_8 (yellow cake).
Shipping and handling of yellow cake.
Conversion to uranium hexafluoride (UF_6).
Shipping and handling of UF_6.
Toll enrichment (AEC enrichment of customer feed material).
Shipping and handling of enriched UF_6.
Conversion to fuel chemical form (usually UO_2).
Fuel fabrication.
Shipping and handling of fuel.
Spent-fuel storage.
Spent-fuel shipping.
Chemical recovery of valuable spent-fuel constituents.
Conversion of valuable constituents to reenrichment or refabrication form.
Shipping and handling of recovered products.
Disposal of radioactive wastes.

The relative importance of each of these services can be expressed by dividing the resulting dollar cost by the quantity of material involved, thereby developing a unit cost, such as dollars per kilogram of uranium purchased, processed, or handled. However, the unit cost of each of these cost components is likely to vary with time, each being a complex function of the technology applied, volume of business, and market conditions. Thus, the actual costs incurred in supplying a specific batch of fuel to a nuclear reactor will be a complex summation of all the cost components that occur, calculated on the basis of the actual cost at the specific date of their occurrence.

The projected variation of individual nuclear-fuel cost components with time is a complex subject, not too dissimilar from that of projecting the cost of the competing fuels coal, oil, and gas, except that there are more component pieces and that the nuclear industry is still in a much earlier state of development. Thus, technological changes are still occurring rapidly, and cost reductions based solely on expansion of volume are anticipated.

The calculation of nuclear-fuel costs using these data is particularly complex if a high degree of sophistication is desired, such as in the evaluation of competitive fixed-price bids or in the determination of minimum incremental operating costs between two or more plants. However, for project scoping purposes and general familiarization with the principles involved, the following simplified method may be used to determine *equilibrium fuel-cycle costs*:

1. Determine the initial uranium-procurement unit cost in dollars per kilogram of uranium. This cost will vary with the price of uranium yellow cake, the cost of converting it to uranium hexafluoride, and the cost of toll enrichment. This latter item is,

Table 9-8. Typical Nuclear Fuel Design, Performance, and Cost Parameters

	Usual units*	Boiling-water reactor			Pressurized-water reactor				Gas-cooled reactor	
		Core 1	Core 2	Core 3 (equilibrium)	Batch 1	Batch 2	Batch 3	Batch 4 (equilibrium)	Zone 1	Zone 2
A. Design and performance:										
Total reactor power, net	MWe	750	750	750	680	680	680	680	600	600
Total reactor power	MWt	2,344	2,344	2,344	2,020	2,020	2,020	2,020	1,430	1,430
Fuel loading, initial	Mt U/element	0.193	0.193	0.193	0.448	0.448	0.448	0.448	146 MtU total	
Fuel loading, discharged	% of initial	97.63	96.79	96.76	98.0	96.8	96.2	96.2		
No. fuel elements/core	....	796	796	796	156	156	156	156		
No. fuel elements/batch†	....	159	159	159	52	52	52	52		
Fuel exposure	MWD/Mt U	16,700	22,200	22,400	14,400	23,200	27,000	27,000	20,000	20,000
U^{235} enrichment, initial	% U^{235} by wt.	2.01	2.38	2.38	2.21	2.36	2.66	2.90	2.0	2.5
U^{235} enrichment, discharge	% U^{235} by wt.	0.804	0.809	0.796	1.04	0.73	0.70	0.90	0.69	0.69
Plutonium discharge	gPu (fissile)/kgU	4.39	4.87	4.83	4.63	5.69	6.15	6.20	3.4	3.4

* MWe, megawatts electrical.
MWt, megawatts thermal.
Mt U, 1,000 kg (metric tons) of uranium.
MWD, thermal megawatt days.
Pu (fissile), plutonium 239 and 241.
† A loading and unloading schedule must be supplied if all batches do not contain equal numbers of elements and/or fuel-reinsertion fuel management is used.

Table 9-8. Typical Nuclear Fuel Design, Performance, and Cost Parameters.—*Concluded*

B. Typical losses factors:	
Irrecoverable U losses during U_3O_8 to UF_6 conversion	0.5%
Enrichment plant tails	0.25–0.28% U^{235}
Irrecoverable U losses during recovery	1.0–1.5% of U charged
Irrecoverable U losses during conversion	0.3–0.5% of U charged
Irrecoverable Pu losses during recovery	1.3–2.0% of Pu discharged
C. Typical economic factors:	
U_3O_8 cost	6.50–8.00 dollars/lb
U_3O_8 to UF_6 conversion	2.32–1.74 dollars/kg
Separative work	26.00 dollars/kg
Fabrication cost	90.00–50.00 dollars/kg
Pu credit (fissile)	8.00 dollars/g (normally related to value of fully enriched U^{235})
Shipping of spent fuel (including cask rental)	6.00–4.00 dollars/kgU discharged
U conversion to UF_6, sale or reuse form	5.00–4.00 dollars/kgU recovered
Pu conversion to PuO_2, sale or reuse form	0.50–1.0 dollar/gPu (total)
Fuel and working-capital investment charge:	
Material off site	10.5–1.25%
Material on site	12.0–14%
D. Typical time intervals for inventory-period calculation:	
U_3O_8 procurement lead time	60 days
U_3O_8 shipping to UF_6 conversion plant	30 days
UF_6 conversion period/batch	90 days
UF_6 shipment to enrichment plant	10 days
Enrichment period	90 days
Fuel UF_6 to UO_2 conversion and fabrication period	180 days
Shipping to reactor site	30 days
Lead time on reactor site prior to commercial power generation:	
Initial core	240 days
Refueling batches	60 days
Refueling period	14–20 days
Spent-fuel decay cooling period	120 days
Spent-fuel shipping period (full batch)	60 days
Reprocessing period	60 days
U conversion to UF_6	10 days
Pu conversion period	60 days

in turn, dependent upon the enrichment required (which determines the amount of separative work required) and the unit cost of separative work. Tables of enriched uranium values vs. enrichment values are included in Paragraph 2816 Volume 1, of Commerce Clearing House, Inc., Atomic Law Reporter. Specific formulas for the calculation of such values, useful when it is desirable to make alternate cost input assumptions, are developed in the book "Nuclear Chemical Engineering," by Benedict and Pigford, McGraw-Hill Book Company, 1957, Chap. 10.

2. Determine the fabrication unit cost, in terms of dollars per kilogram, of combined uranium.

3. Determine the spent-fuel-element shipping, reprocessing, and reconversion costs in terms of dollars per kilogram of contained uranium.

4. Determine the credit available for the recovered uranium and plutonium. The credit for the uranium recovered is essentially the value of new enriched uranium having the same enrichment as the spent fuel, thereby allowing the use of the tables discussed in item 1 above. The value of the plutonium should also be expressed in terms of dollars per kilogram of uranium.

5. Determine the indirect costs, i.e., return on investment and provision for applicable taxes, by determining the average investment level throughout the procurement period, energy-production period, and spent-fuel recovery period and multiplying by the appropriate interest or fixed-charge rates.

6. Sum items 1 through 5 to give the total unit costs, in terms of dollars per kilogram of uranium.

7. Determine the unit energy production, based upon the average fuel energy production (usually given in terms of megawatt-days thermal per metric ton of uranium) and the thermal efficiency of the plant.

8. Divide item 6 by item 7 to give the unit costs, easily converted to mills per kilowatthour by appropriate conversion factors.

The results of such simplified calculations are likely to be somewhat low, since they do not recognize the various minor material losses. However, they are also not likely accurately to reflect the true investment costs, since the investment value is a complex value varying with energy production.

The complications inherent in a complete fuel-cycle cost, coupled with the need to compute costs over an extended period involving changing cost patterns and also over the several fuel loadings required to reach an equilibrium fuel-flow requirement, have led to the development of computerized calculational methods. Typical data required to evaluate the fuel-cycle costs for fuel-management programs in current commercial use, along with typical values for various reactor types, are given in Table 9-8.

51. Operating and Maintenance Costs. Operating and maintenance costs categories fall in the following groups:

Labor.
Materials, supplies, and services.
Insurance.
Fuel management.
Working capital.

The *plant staff* required is relatively independent of plant size, typically running 60 to 70 men, including all supervision, technical assistance, operations, maintenance, and miscellaneous supporting services. The costs for this staff may vary substantially, being highly dependent upon the local labor market and the labor costs. Utilization of a median value of $13,000/man, including overheads and payroll additives, results in an annual cost of $780,000 to $910,000.

Materials, supplies, and services are also relatively insensitive to plant size, although certain items are directly proportional to the thermal power level and the frequency and extent of power-level changes. Typical costs for a plant of 500 MWe capacity are of the order of $500,000.

Insurance costs may be divided into two component parts, property insurance and liability insurance. Property insurance is normally a direct function of the capital value, typical rates being 0.4% of the capital invested, less land costs. Liability insurance falls into two categories. The first $82 million worth of coverage is procured from private insurance firms for a premium of about $200,000/year. In addition, the federal government supplies indemnity for a fee of $30.00/(year)(MWT). Overall, insurance costs tend to run about 0.8 to 1.0% of the capital investment per year.

Fuel-management services may be provided either by contracting for external services or by adding staff to that already discussed. Typical estimates of the cost of the service are about $100,000/year. This cost is also essentially independent of plant size.

Working capital, other than for the fuel investment which has been discussed separately, is typically computed by using (1) 45 days of average operating and maintenance expenses, (2) the average amount of money invested in materials and supplies, and (3) prepaid expenses such as insurance premiums. From this the average annual accrual of payments for Federal income tax is subtracted to obtain the net working capital.

52. Total Energy Costs. The total unit cost of energy delivered at the bus bar is the sum of all of the preceding cost components which are related to the energy-production period divided by the energy produced during the same period. Typical values of power costs for 800 to 1100 MWe plants starting up in the early 1970s, on the assumption of typical investor-owned utility financing, give an initial unit cost of about 4.5 mills/kWh. If escalation is disregarded, technological improvements and lower fuel costs will tend to reduce this value.

REACTOR DESIGNATIONS

53. Basis of Reactor Concepts. The most frequently used reactor designations are described below. Although a very large number of reactor concepts can be arrived at by various arrangements of the parameters described, not all of them are feasible owing to certain physical, chemical, and nuclear incompatibilities. For example, a fast-spectrum reactor would not be water-cooled or -moderated (nuclear incompatibility), and a sodium-cooled reactor would not normally be water-moderated (chemical incompatibility).

The neutron spectrum, or energy range that is predominant in causing fission, is one of the most important and frequently used of reactor designations. This specific designation is usually omitted only in cases where the other descriptive portions of the designation imply what portion of the spectrum is predominant (i.e., a water-moderated reactor can be assumed to operate predominantly on thermal neutrons). Specific neutron spectrum designations are: thermal, slow, epithermal, intermediate, and fast. A *thermal* reactor operates mostly on neutrons that have reached thermal equilibrium with their surroundings, approximately 0.025 eV at room temperature or, at elevated reactor temperature, up to about 0.1 eV. The term *slow* is sometimes used interchangeably with *thermal* but more often is applied in a broader and more inclusive sense to reactors operating on neutrons which have energies from a fraction of 1 eV (sometimes as low as thermal energy) to a few electronvolts.

Epithermal reactors operate in the energy range above thermal energy but below the energy associated with the resonance-absorption region of fuel materials. *Intermediate* energy reactors generally operate in the range from above the resonance-absorption region of fuel to somewhat less than fast, or from several hundred electronvolts to about 100 keV. *Fast* reactors use neutrons that are predominantly above 100 keV in energy.

The *reactor coolant,* being one of the more important design choices, is usually included in reactor designations. Some materials that are considered for use as reactor coolants are water (ordinary and heavy); gases (helium, carbon dioxide, air, nitrogen, and steam); liquid metals (sodium, lithium, NaK, mercury, and bismuth); and organics (diphenyl and terphenyl). Heavy water is included here as a coolant, but it is more commonly thought of as a moderator, and in fact, heavy-water-moderated systems are often considered in combination with one of the other coolants, such as an organic coolant.

The *moderator* used in a thermal reactor is usually indicated in the reactor designation. Only the elements hydrogen, deuterium, beryllium, and carbon in various physical and chemical forms are usually considered as moderators, and when used, they are usually referred to by name as heavy water, graphite, or organic-moderated reactors. When ordinary water is the moderator, it is often not mentioned in that capacity if the designation mentions water as coolant.

Fuel type expressed in terms of the fuel material is sometimes used as a designation parameter. Fuel-material designations are natural uranium, high-, intermediate-, or low-enriched fuel, U233 or Pu239 spiked fuel, U235, U233, and Pu239 fuels. In another fuel-type category, the physical fuel form is described as oxide (UO_2), carbide (UC), or metal fuels, etc.

Operating fuel cycle is another important designation that is frequently used. The three operating fuel cycles are *burner operation,* in which very little or no additional fuel is produced as the reactor operates; *converter operation,* in which a significant quantity of fuel is produced during operation, but less than the amount consumed; and *breeder operation,* in which more fuel is produced than is consumed. The parameters used to describe the effectiveness of conversion and breeding in reactors are expressed as conversion and breeding ratios, or the ratio of fuel atoms produced per fuel atom consumed. In practice the only reactors that operate as pure burners are those using high-enriched fuel without fertile material in or around the core. Most power reactors operate as converters, since neutrons will always be captured in the fertile U238 present in the low-enriched fuels commonly used. In certain thermal-reactor concepts higher conversion ratios and, in some cases, even breeding can be obtained by using thorium as fertile material with U233 as fuel instead of U238 and U235.

The *fertile-material* designation is used in many reactors that operate in a converter or breeder mode. The important fertile materials are thorium and uranium (U238). When the fertile material is placed as a separate region around the outside of the reactor core, this outer region is called a *blanket.*

Fuel geometry is sometimes described in terms of fuel distribution, such as *heterogeneous* or *homogeneous.* For the more common heterogeneous fuel reactors in which

Table 9-9. Power-reactor Concepts

Designation	Neutron spectrum	Coolant	Moderator	Fuel enrichment	Operating fuel cycle	Fuel geometry
Pressurized water	Thermal	H_2O	H_2O	Low	Converter	Heterogeneous
Pressurized heavy water	Thermal	D_2O	D_2O	Natural to slight	Converter*	Heterogeneous
Boiling water	Thermal	H_2O	H_2O	Low	Converter	Heterogeneous
Boiling heavy water	Thermal	D_2O	D_2O	Natural to slight	Converter*	Heterogeneous
Heavy-water-moderated pressure tube	Thermal	D_2O, H_2O	D_2O	Natural to slight	Converter*	Heterogeneous
Gas-cooled graphite-moderated	Thermal	CO_2, He	Graphite	Natural to low	Converter	Heterogeneous
High-temperature gas-cooled graphite-moderated	Thermal, epithermal	CO_2, He	Graphite	Intermediate to high	Converter*	Heterogeneous
Aqueous homogeneous	Thermal	H_2O	H_2O	Low to intermediate	Converter*	Homogeneous
Liquid-metal-cooled graphite-moderated	Thermal	Na, NaK, Li	Graphite	Low to intermediate	Converter	Heterogeneous
Organic-cooled and moderated	Thermal	Diphenyl, terphenyl	Diphenyl, terphenyl	Low	Converter	Heterogeneous
Organic-cooled heavy-water-moderated	Thermal	Diphenyl, terphenyl	D_2O	Natural to low	Converter	Heterogeneous
Liquid-metal-fueled graphite-moderated	Thermal	Bismuth	Graphite	Low to intermediate	Converter*	Homogeneous
Fast breeder	Fast	Na, NaK, Li	None	Intermediate to high	Breeder	Heterogeneous

* High conversion ratio with U^{235} and U^{238}; potential breeder with thorium and U^{233}.

Table 9-10. Advantages and Disadvantages of Power-reactor Concepts

Reactor designation	Advantages	Disadvantages
Pressurized water	Coolant-moderator cheap. Relatively compact core	Expensive pressure vessel required. Relatively low-pressure and -temperature steam available to turbine. Low to modest conversion ratio
Pressurized heavy water	Fuel can be natural uranium. Has high conversion ratio	Heavy-water inventory expensive. Extremely tight system required to minimize heavy-water losses is expensive. Large, expensive pressure vessel required. Relatively low-pressure and -temperature steam available to turbine
Boiling water	No heat exchanger required. Higher steam pressure and temperature to turbine than for PWR. Coolant-moderator cheap	Void changes in core-effect control. Larger core and vessel required than for PWR due to steam voids in water. Low to modest conversion ratio. Carryover of radioactivity to turbine possible
Boiling heavy water	Same as boiling water, except coolant-moderator not cheap and fuel can be natural uranium; conversion ratio high	Same as for boiling water, except better conversion ratio and heavy-water inventory expensive. Extremely tight system required to minimize heavy-water losses is expensive
Water-cooled heavy-water-moderated (pressure tube)	Fuel can be natural uranium. Has high conversion ratio. High-pressure vessel not required to contain heavy water	Same as for pressurized heavy-water reactor except high-pressure vessel is not required
Gas-cooled graphite-moderated (low-enriched-metal fuel)	Conversion ratio high. Very low corrosion in system. Fuel can be natural uranium	Plants are large and expensive to build. Heat-transfer properties of gases inferior to those of liquids. Nuclear-grade graphite expensive
High-temperature gas-cooled graphite-moderated	Very high temperatures possible. Conversion ratio high (thermal breeding possible with thorium-U233). Very low corrosion in system. Smaller plant than for low-enriched system	Heat-transfer properties of gases inferior to those of liquids. Nuclear-grade graphite expensive. Intermediate- to high-enriched fuel required
Liquid-metal-cooled graphite-moderated	Reactor coolant operates at low pressure. High steam temperature possible. Excellent heat-removal properties	Extreme precautions must be taken to prevent coolant from contacting water. Pumps and valves expensive. Graphite must be canned to prevent contact with sodium. Sodium becomes intensely radioactive. Nuclear-grade graphite expensive
Organic-cooled and -moderated	Fairly high steam pressure and temperature possible with low-pressure primary system. Very low corrosion in system. Pure coolant does not become significantly radioactive	Organic compounds decompose in radiation field. Decomposition products may reduce heat transfer from fuel to coolant. Heat-transfer properties not so good as those of water. Low to modest conversion ratio
Organic-cooled heavy-water-moderated (pressure tube)	Same as organic-cooled and -moderated plus high conversion ratio and less decomposition of organic coolant. Use of low-pressure D_2O system possible, making low-loss system easier to construct. Very low-enriched (possibly natural) fuel can be used	Same as organic-cooled and -moderated except conversion ratio high and heavy-water inventory expensive
Fast breeder	No moderator required. Few nuclear limitations on choice of structural materials. Breeds more fuel than it consumes. Use of liquid-metal coolant allows high temperature at low pressure. Fission-product buildup does not reduce reactivity significantly	Control of fast reactors more difficult than for thermal systems. Large inventory of fuel required. Intermediate- to high-enriched fuel requirements. Extreme precautions must be taken to prevent coolant from contacting water. Sodium becomes highly radioactive

the fuel is of a fixed shape in a specific lattice location, the heterogeneous designation is not ordinarily used. In cases where the fuel is distributed homogeneously throughout the core the method of distribution and the carrier vehicle are often described. Two examples of homogeneous-reactor concepts are: aqueous *homogeneous* reactors, in which the fuel is dissolved in aqueous solution; and liquid-metal-fueled reactors, in which the fuel is dissolved (or dispersed in suspension) in a liquid metal such as bismuth or lead-bismuth eutectic.

Core geometry may be designated for some reactors where the geometry is either unusual or particularly significant to the application. The "usual" power-reactor core geometry consists of fuel assemblies held in grid plates and submerged in the reactor coolant. The moderator in this "usual" geometry, if it is not also the coolant, may be placed around the fuel assemblies in either canned or uncanned form, as dictated by chemical, nuclear, and hydraulic requirements. Another core geometry used in cases where it is desired to separate coolant and moderator is the *pressure-tube reactor* configuration, in which the fuel assemblies are placed in individual pressure tubes. In the pressure-tube reactor, coolant is pumped past the fuel assemblies inside the pressure tubes, and the moderator is outside the tubes. To prevent the liquid moderator, e.g., heavy water, outside the pressure tubes from being heated to high temperature, thus making it necessary to contain it in a pressure vessel, a means of containing the heavy water and separating it from the pressure tubes has been devised. This core geometry consists of a *calandria vessel*, which is a tank pierced by a number of tubes, to contain the heavy water, with another set of tubes placed inside the calandria tubes to contain the fuel assemblies and coolant under pressure. Heat transfer from the inner (pressure) tubes to the moderator is reduced by filling the gap between the two tubes with a gas which has low thermal conductivity. This gas-filled gap also can serve the purpose of detecting leakage from either tube and can permit corrective action to be taken before coolant from the inner tubes can contaminate the heavy water.

The *mode of coolant operation* is frequently designated, in addition to the coolant type. The two modes of operation common to power reactors are: *pressurized*, in which the coolant is prevented from boiling in the reactor vessel by maintaining a sufficiently high pressure; and *boiling*, in which the coolant is allowed to boil in the vessel and core. In addition to these basic modes of coolant operation, considerable effort is being expended to develop *nuclear superheat reactors* for commercial power production. In a nuclear superheat reactor, the saturated, relatively low-temperature steam produced in a boiling-water reactor (BWR) core or in the boiler of a pressurized-water reactor (PWR) is heated to produce dry, high-temperature steam, with a nuclear reactor or a section of one used as the heat source. For boiling-water reactors, the nuclear superheat section may be integral with the BWR core or a completely separate unit. For pressurized-water reactors, the nuclear superheat would be supplied by a completely separate nuclear unit which heats the steam coming from the boiler.

Functional designations are often used to indicate the purpose for which the reactor was designed or the major functional property of interest in the systems application. Some of the more common functional terms are: *test reactor*, *experimental reactor*, *high-flux reactor*, *power reactor*, *isotope-production reactor*, *maritime reactor*, and *mobile reactor*.

Table 9-9 lists various power-reactor concepts with parameters as described in the preceding paragraphs.

The power-reactor concepts listed in Table 9-9 each have advantages and disadvantages which are related to the given parameters. Table 9-10 describes these advantages and disadvantages for each of the power-reactor concepts listed in Table 9-9. These tables do not exhaust the feasible power-reactor concepts but rather list those which either are already well developed or have had substantial development effort expended on them. Many variations can be considered. For example, beryllium or its oxide can replace graphite in graphite-moderated systems, and heavy water can replace ordinary water in an aqueous homogeneous system.

LEGAL AND ADMINISTRATIVE ASPECTS OF NUCLEAR POWER GENERATION

54. Regulation of Nuclear Power. The risk of accidents in nuclear power plants which would result in the release of fission products from the plant is exceedingly small, but not zero. Thus government agencies having a responsibility to protect the health and safety of the public have established extensive controls covering almost every aspect of the utilization of nuclear energy. The control covers the ownership of fuel materials, licensing of nuclear facilities, operation of nuclear facilities, release of radioactive waste, radiation dose to workers, and insurance requirements. These responsibilities are divided among the AEC and various state agencies. The U.S. Atomic Energy Commission Regulations concerning the utilization of nuclear power are contained in Title 10 of the Code of Federal Regulations. The Interstate Commerce Commission, Coast Guard, and Department of Labor also regulate aspects of the use of nuclear materials.

Construction Permit. Although it is possible to do some excavation for a nuclear power plant before approval is received from the AEC, a construction permit is required before plant construction work is undertaken. The following is a summary of the procedure for obtaining a construction permit from the AEC:

1. The applicant submits a document to the AEC, describing the facility, the site, operating procedure, waste disposal, reactor safeguards, emergency procedures, and evaluation of the consequences of limiting accidents. Certain other information on the financial and technical qualifications of the organizations involved is also required. Since this application usually is for a combination license to construct the facility and to receive and use nuclear materials, an estimate of the types and quantities of nuclear materials involved is usually included. This application and all subsequent formal documents, except for any data identified as proprietary, are made available for public inspection in the AEC's public document room.
2. The AEC's Division of Licensing and Regulation makes a detailed review of the application. The AEC staff prepares a document giving their conclusions for presentation to a statutory technical committee called the Advisory Committee on Reactor Safeguards (ACRS). The ACRS reviews the proposal and writes a letter to the AEC, stating its conclusions and recommendations.
3. A public hearing is held near the proposed site to review the application, the AEC staff analysis, and the ACRS recommendations. This hearing, held by a panel selected by the AEC, with participation by all interested parties, is intended to provide the public with a summary of the entire regulatory review.
4. After considering the AEC staff review, the ACRS recommendation, the record of the public hearing, and the findings of the panel, the five commissioners of the AEC make the final decision on the application.

In almost all applications the project schedule is such that final design information is not available for this initial application and the construction permit issued is provisional. The construction permit is only the first step in a regulatory process which covers the entire life of the reactor. Issuance of a provisional construction permit allows the plant to be constructed but implies approval of design features by the AEC only to the extent that definitive design information is included and approval of these design features is requested. A complete or final construction permit is issued only when all design information, site evaluations, and reactor-safeguards evaluation have been submitted and approved as described in the preceding steps.

Operating License. The construction permit must be converted into an operating license before fuel is loaded into the reactor. The AEC must be satisfied that all essential design information has been provided and that the facility has been built as specified. In addition, the applicant must propose technical specifications, which are the key design and operating factors which affect public safety. The purpose of the technical specifications is to separate from the detailed volumes of data, which have been supplied to the AEC, those items which cannot be changed without prior approval from the AEC.

Operator's License. Only operators licensed by the AEC may manipulate the controls of a reactor. These licensed operators must be supervised by a senior operator also licensed by the AEC. The senior operator must be either present at the facility or on call during routine operations and must be present during nonroutine operations such as significant reductions in power, refueling, or start-up. Operator's licenses are issued by the AEC after satisfactory completion of written and oral tests covering reactor theory, reactor design features, operating limits in the technical specifications, emergency procedures, and a physical examination. Licenses are issued for a specific nuclear facility only and must be renewed every 2 years.

Indemnification. For power reactors having a rated capacity of 100 MWe or higher, provision must be made within the United States for financial protection against third-party claims equal to the maximum protection available from the insurance companies (now 74 million dollars). The government provides additional indemnification of 486 million dollars and requires that all power-reactor licensees pay the nominal fee of 30 dollars/(MWt) (year). There is no legal requirement for property-insurance coverage of the facility itself, although this coverage is available from the insurance companies.

55. Ownership of Fuel Materials. Prior to 1964 fissionable material in the United States was subject to absolute government ownership, with material leased as required to operators for use in power reactors. From 1964 until Dec. 31, 1970, the AEC may continue to lease fissionable material for reactor operation, as before, or the operators may purchase fissionable material from the AEC. After Dec. 31, 1970, the AEC may distribute fissionable material to reactor operators only by sale. At July 1, 1973 all existing leases for fuel will be terminated. The only facilities for enrichment of natural uranium in the United States, as used in most power reactors, are operated by the AEC. Starting in 1969 the AEC has authority to enrich privately owned uranium (from United States sources) as a service. Through 1970 the AEC will purchase the isotopes plutonium 239 and 241 for $9.28/g when produced from uranium procured from the AEC. This plutonium price is quite important in determining the total fuel-cycle cost.

Any organization receiving, possessing, or transferring fissionable materials, such as fuel fabricators, must be licensed by the AEC.

NUCLEAR-PLANT SYSTEMS

56. Nuclear Systems and Equipment. Several potentially attractive power-reactor concepts are in various stages of development, ranging from experimental through prototype levels. However, most of the nuclear power generated commercially in the United States is produced in pressurized-water reactors and boiling-water reactors.

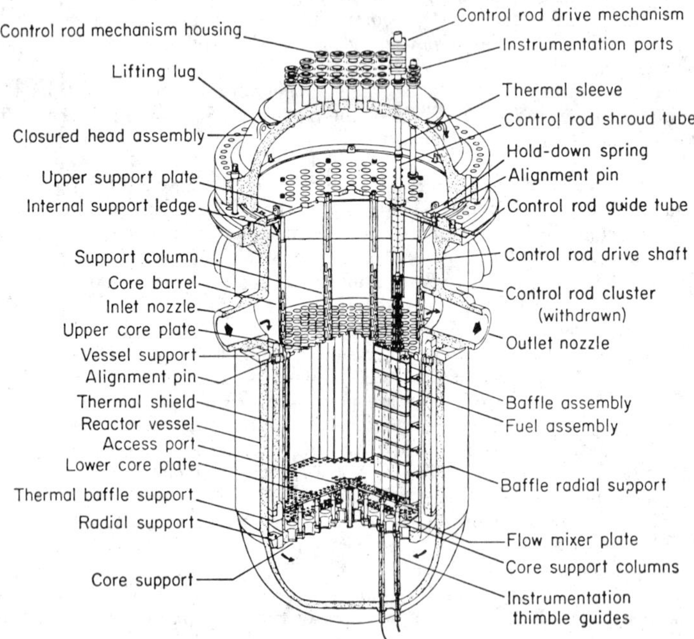

FIG. 9-5. Cutaway of PWR reactor vessel and internals. (*Westinghouse Electric Corporation.*)

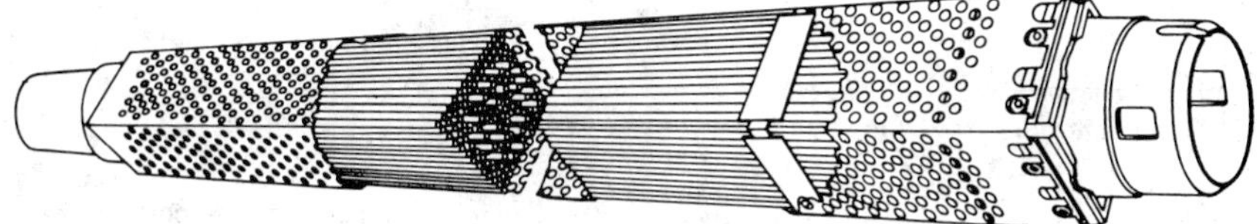

FIG. 9-6. Cutaway view of PWR fuel assembly. (*Babcock and Wilcox.*)

Both make use of light water to cool the core and act as moderator. Current expectations are that these light-water reactors will predominate for a considerable time.

In the PWR water is pumped through the core under pressure of about 2,000 lb/in² abs to remove the nuclear heat. The reactor coolant in this case remains liquid, and steam is produced in a steam generator, where heat is transferred from the reactor coolant in the tubes to water on the shell side. The steam produced in a modern PWR may be dry or saturated and is at a pressure of about 600 to 800 lb/in² abs. After passing through the steam generator, the reactor coolant is pumped back to the reactor vessel and recirculated through the core. Steam generated on the shell side of the steam generator goes to the turbine and condenser, and the condensate is pumped back to the steam generator. Figure 9-5 is a cutaway view of a PWR vessel and its internals. The figure shows the flow path of reactor coolant through the core and the configurations and positions of major components within the vessel.

The *reactor core* in a large, modern PWR may be 10 to 12 ft high and 10 to 14 ft in diameter. The core is composed of *fuel assemblies* such as the one shown in Fig. 9-6. The fuel assembly is made up of UO_2 rods clad in Zircalloy and assembled on a square lattice. These rods are assembled into fuel elements and held in position by spring clips. This method provides mechanical rigidity and proper spacing while minimizing the use of structural material. Spring clips permit axial motion of rods and eliminate thermal bowing caused by a power gradient across the assembly.

Control rods are inserted and withdrawn from above the core, and the control-rod drive mechanisms are located on top of the reactor-vessel head. During refueling the drive mechanisms are disconnected from the control rods. Figure 9-7 shows a cutaway of a control-rod drive mechanism of the magnetic-jack type. All moving parts of this mechanism are contained in a hermetically sealed pressure housing which is welded to the reactor vessel. Electromagnetic forces from externally mounted coils actuate latches within the pressure housing to move the drive shaft up or down in small steps. In case of a reactor scram, the control rod is automatically released and falls into the core by gravity free fall.

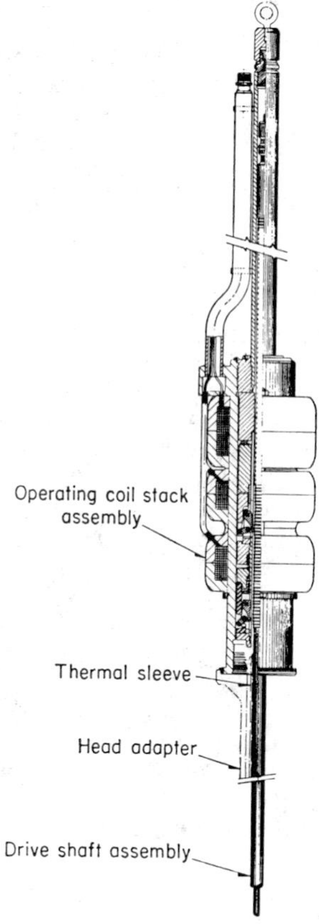

FIG. 9-7. Cutaway view of a control-rod drive mechanism. (*Westinghouse Electric Corporation.*)

Coolant circulation in the primary loop of either a PWR or a BWR is maintained by large *shaft-sealed centrifugal pumps.* In modern plants a widely used type is the *controlled leakage pump,* such as that shown in Fig. 9-8. In this type of pump a three-part seal is used to reduce pressure from the reactor operating pressure to atmospheric pressure in two steps.

Leakage past the first two parts of the seal is controlled and collected and ultimately returned to the

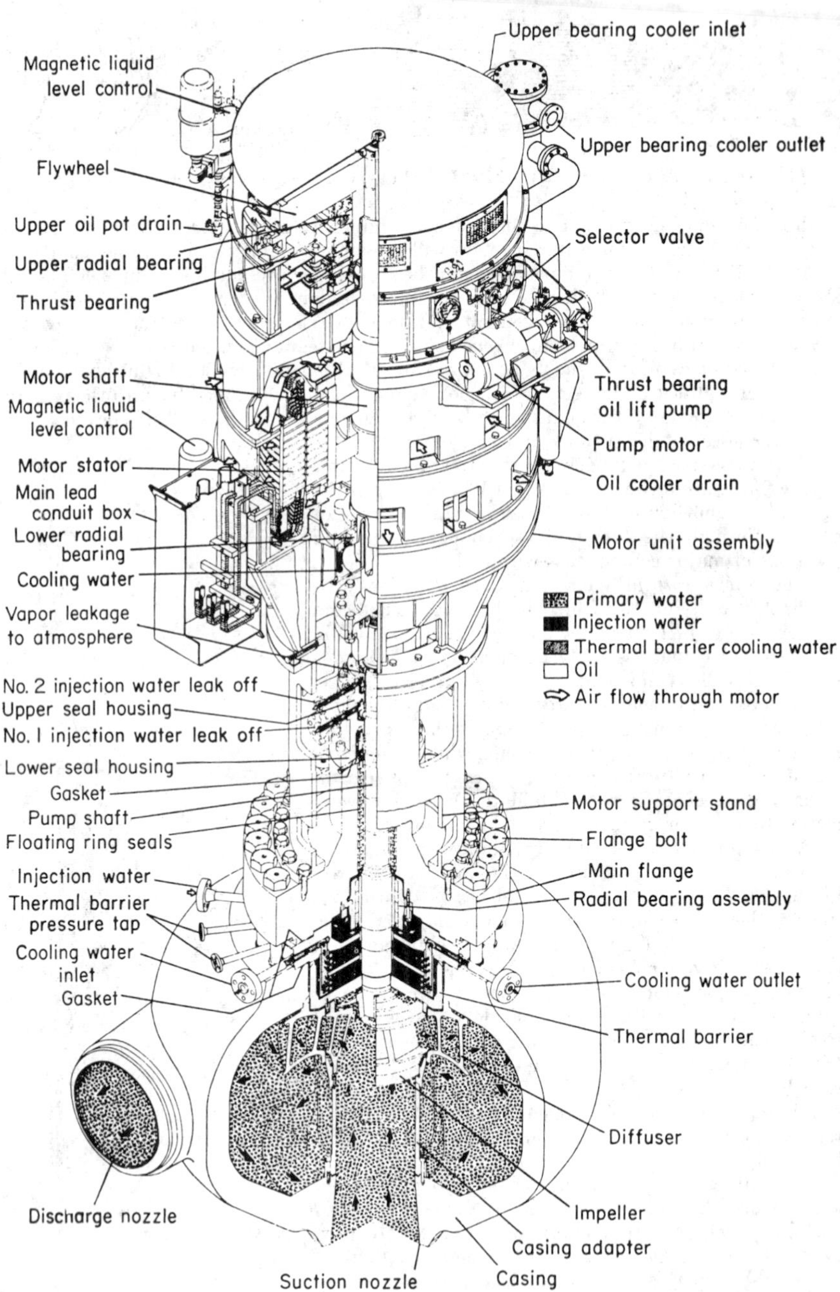

Fig. 9-8. Cutaway of controlled leakage pump. (*Westinghouse Electric Corporation.*)

primary loop. The third part of the seal controls the leakage between the second part and atmosphere, reducing this leakage to a few drops of water per day. The controlled leakage pump has higher efficiency than pumps previously used for reactor-coolant circulation, and, further, a flywheel built on the drive shaft gives this pump better coastdown characteristics in the event of a power failure.

Figure 9-9 is a rendering of a *nuclear steam-supply system.* The figure shows the reactor vessel and primary loop with two steam generators, their associated pumps, and the pressurizer. The pressurizer is used to maintain constant pressure in the primary loop and to provide surge capacity. The pressurizer, shown on the extreme right of the figure, is essentially a small boiler using electrical heaters to maintain constant temperature and therefore a constant vapor pressure above free surface of the water. The reactor service crane shown above the heavily shielded reactor vessel is used to remove the vessel head during refueling operations. A smaller manipulator crane removes, shuffles, and replaces fuel assemblies. The service crane is also used to replace major pieces of equipment as required during the life of the plant.

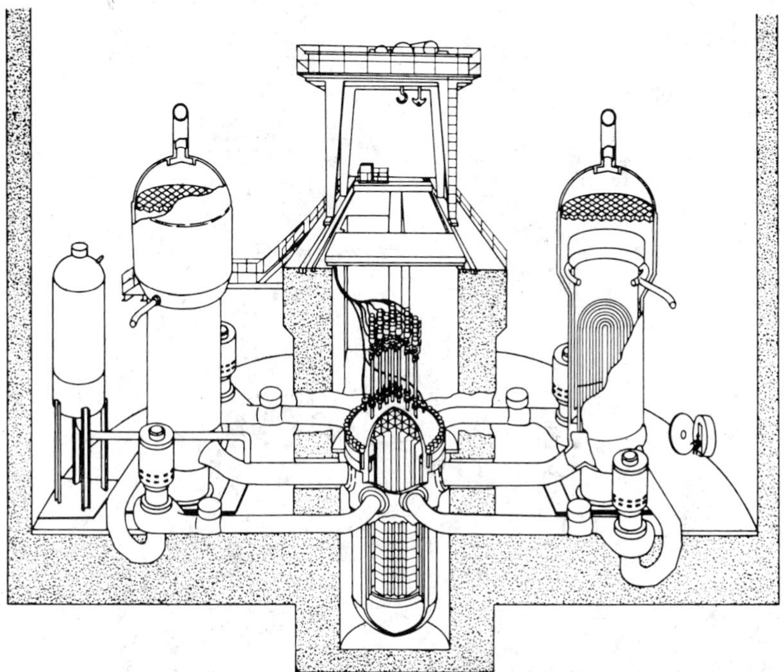

Fig. 9-9. Artist's rendering of PWR nuclear steam-supply system. (*Combustion Engineering.*)

The entire PWR primary circuit, which includes the reactor vessel, pumps, steam generators, and pressurizer, is enclosed within a containment structure. This structure is designed to prevent release of radioactivity to the environment in the unlikely case of an accident. Figure 9-10 is a typical arrangement of a *nuclear steam-supply system and containment vessel.* The containment vessel is designed to withstand the pressure buildup resulting from a design-basis accident, which generally consists of the release of the energy stored in the primary coolant, sensible heat of the primary-loop metal, fission-product decay heat, and energy released as the result of possible metal and water reactions. The containment vessel and all penetrations must be capable of withstanding the maximum pressure which could result from a design-basis accident and not allow more than a specified safe leakage to the external environment.

Auxiliary systems are frequently used with the containment vessel, to limit pressure buildup and to remove radioactivity from the containment atmosphere. Pressure buildup is limited by using water spray to condense steam. Radioactivity is removed by using recirculation blowers and filters. Depending on the stringency of the requirements for a reactor at a specific site, the containment vessel may be a simple unshielded steel structure, or a shielded double-barrier structure, or any of various types between these two extremes. As can be seen in Figs. 9-9 and 9-10, the reactor and vessel are surrounded by a thick concrete shield to prevent activation of the rest of the primary loop and to provide personnel shielding during operation and maintenance.

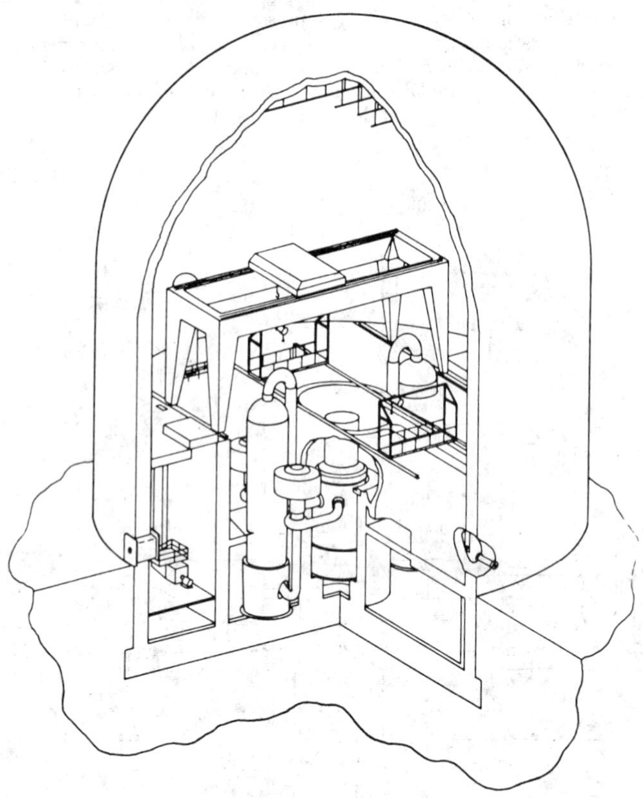

Fig. 9-10. Typical arrangement of PWR nuclear steam-supply system and containment. (*Babcock and Wilcox.*)

In large BWR plants, water is pumped through and allowed to boil in the reactor core to produce saturated steam at about 1,000 lb/in^2 abs and 550°F. The steam produced in a BWR core passes through separators and dryers in the upper part of the reactor vessel before going directly to the turbine. Figure 9-11 is a cutaway view of a BWR reactor vessel and its internals. The figure shows the position of the separators and dryers in the upper part of the vessel, as well as the core location, a temporary control curtain, bottom-entering control-rod drives, and the internal jet pumps.

For large plants the external recirculation pumps in a BWR are used to achieve controlled flow of reactor coolant through the core. Most modern BWR plants use controlled leakage pumps, such as the one shown in Fig. 9-8. Internal jet pumps are used to reduce the required external recirculation flow.

Axial *steam separators* and *steam dryers* have been developed for use in the vessel to preclude the necessity of using a separate steam drum. Steam leaving the reactor vessel has a moisture content of less than 0.1 weight %. Since the steam separators and dryers occupy almost the entire volume above the reactor core, it is necessary to remove these units during refueling operations. To simplify removal of the separators and dryers, they are made as an integral unit which can be removed in a single operation.

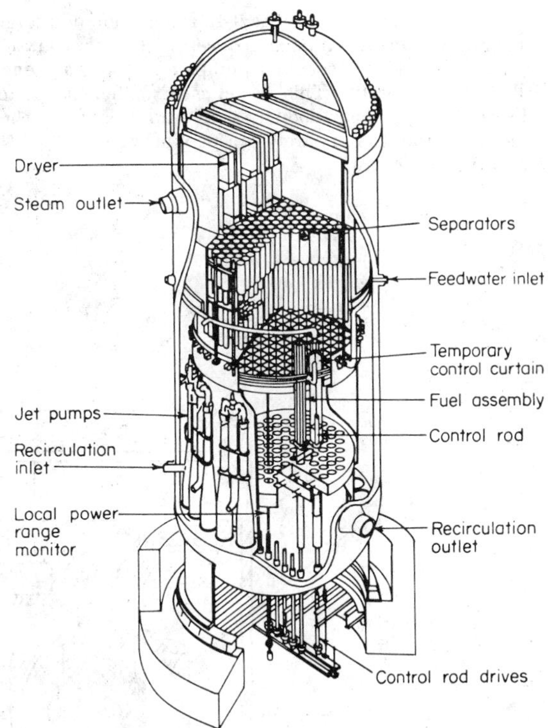

FIG. 9-11. Cutaway of BWR reactor vessel and internals. (*General Electric Company.*)

Control rods enter the BWR core from the underside, and the *control-rod drives* are located below the reactor vessel. Bottom entry of the control rods is used to avoid interference with the steam separators and dryers located above the core. Since the control-rod drives are below the core, they do not have to be removed during refueling operations. Control-rod drives for BWR plants are generally of the locking-piston type, which, since they drive the control rods up, must be highly reliable to ensure that reactor scrams are effectively carried out and that the control rods cannot be released from the core accidentally.

The *temporary control curtain* shown in Fig. 9-11 is necessary to control the excess initial reactivity required to permit high burnup of the BWR fuel. The temporary control curtain is removed from the BWR core after the reactor has operated for sufficient time to allow the control rods to hold down the excess reactivity.

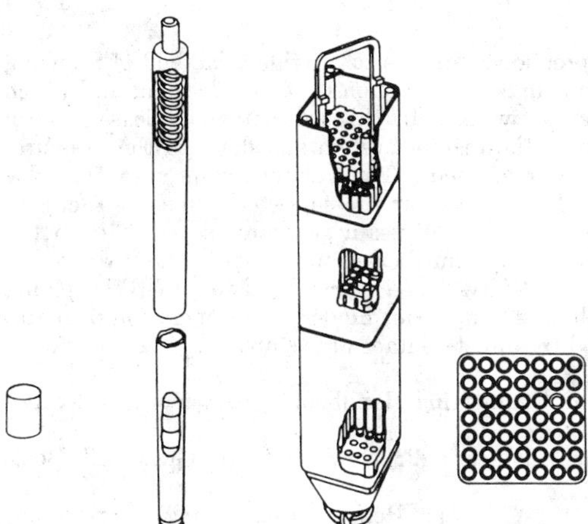

FIG. 9-12. Cutaway view of BWR fuel assembly. (*General Electric Company.*)

Figure 9-12 is a cutaway of a *BWR fuel assembly* and its component parts. The basic component of the fuel assembly is the fuel rod, which is composed of a stack of UO_2 pellets clad with Zircalloy and held down in the rod by a spring. The fuel rods are assembled in a square lattice and held in place by assembly grid plates. The rod assembly is placed in a Zircalloy can to create a coolant-flow channel for the overall fuel assembly.

Modern BWR plants make use of *pressure-suppression containment* to prevent release of radioactivity to the environment in case of an accident. Figure 9-13 is a section of a large BWR plant showing the pressure-suppression containment concept. In this containment system the reactor pressure vessel and its associated pumps and pipes are enclosed in another pressure vessel called the dry well. This vessel is of sufficient capacity to accommodate complete release of the primary-loop coolant and steam for the design-basis accident. Large ducts lead from the dry well to a pool of water, called the pressure-suppression pool, in a separate enclosure.

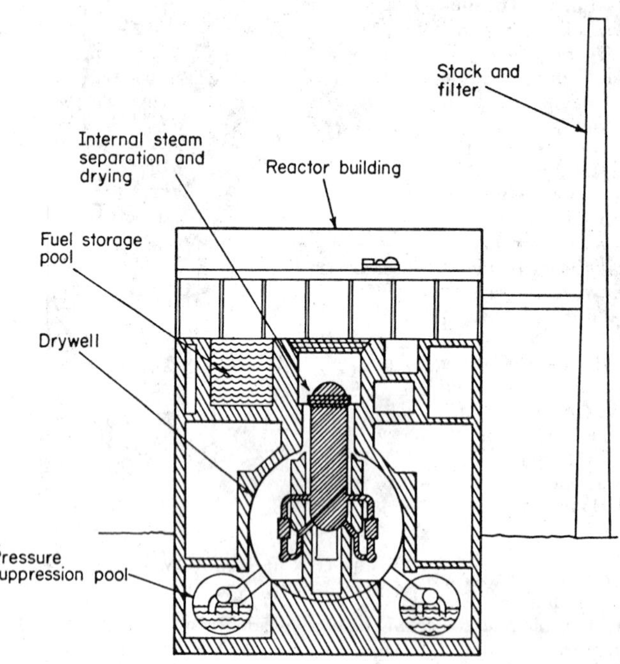

Fig. 9-13. Typical large BWR plant cross section showing pressure-suppression containment. (*General Electric Company.*)

The function of the pressure-suppression system is to provide a method of reducing any potential pressure increase which is in effect at the onset of any accident and which requires no external source of power or water. In this concept any released steam passes through the dry well surrounding the reactor system and flows to the pressure-suppression pool, where the steam is condensed. The volume requirement for this type of containment is considerably lower than for pressure-containment structures.

Unlike the PWR, which produces steam in a steam generator separate from the primary loop, the BWR steam goes directly from the reactor vessel to the turbine and could contain a small quantity of radioactivity. The operating history of BWR plants has shown that this quantity is small, and it does not hinder either operation or maintenance, both of which are performed by direct-contact procedures.

57. Bibliography

Etherington, H. (ed.) "Nuclear Engineering Handbook"; New York, McGraw-Hill Book Company, 1958.

Bonilla, C. F. (ed.) "Nuclear Engineering"; New York, McGraw-Hill Book Company, 1957.

Glasstone, S., and Sesonke, A. "Nuclear Reactor Engineering"; Princeton, N.J., D. Van Nostrand Company, Inc., 1963.

U.S. Atomic Energy Commission, Code of Federal Regulations, Title 10.

PRICE, B. T., HORTON, C. C., and SPINNEY, D. T. "Radiation Shielding"; New York, Pergamon Press, 1957.

SHULTZ, M. A. "Control of Nuclear Reactors and Power Plants," 2d ed.; New York, McGraw-Hill Book Company, 1961.

U.S. Department of Commerce, Meteorology and Atomic Energy; AECU-3066, July, 1955.

DINUNNO, J. J., et al. Calculation of Distance Factors for Power and Test Reactors; *U.S. AEC Rept.* TID-14844, Mar. 23, 1962.

Nuclear Reactors Built, Being Built or Planned in the United States; *U.S. AEC Rept.* TID-8200 (rev.).

Nuclear Power Plant Cost Evaluation Handbook; *U.S. AEC Rept.* TID-7025.

SECTION 10

POWER-SYSTEM ELECTRICAL EQUIPMENT

BY

T. A. GRIFFIN, JR. Assistant Vice President, Consolidated Edison Company of New York, Inc.; Member, Institute of Electrical and Electronics Engineers; Registered Professional Engineer; Member, Power Generation Committee, IEEE; Past Member, Switchgear Committee, IEEE Papers on 345-kV Station Design, Automatic Data Logging, and Nuclear-plant Electrical Design

R. L. WEBB Associate Inside Plant Engineer, Electrical Engineering Department, Consolidated Edison Company of New York, Inc.; Fellow, Institute of Electrical and Electronics Engineers; Registered Professional Engineer; Past Chairman, IEEE Committee on Switchgear; Past Chairman, AEIC Committee on Electric Switching and Switchgear; Chairman, Electric Light and Power Representatives on ASA-C37; Member, U.S. National Committee, CIGRE; Member, CIGRE, U.S. Technical Subcommittee; Advisor, CIGRE Subcommittee on High-voltage Circuit Breakers; U.S. Delegate, IEC-TC-17; Chairman, EEI-AEIC-NEMA Joint Committee on Power Circuit Breakers; Member, NPCC Task Force on System Protection; Developer: AC Generator Loss-of-field Relay; Shaft Vibration Relay; 3-phase Voltage Relay; A-C Machine-winding Temperature Relay; Double Protective System for 345-kV-system Feeder Circuits; Turbine-Boiler-Generator Overall Unit Protective System; author of many AIEE, IEEE, and CIGRE papers

CONTENTS

Numbers refer to paragraphs

SECTION 10

POWER-SYSTEM ELECTRICAL EQUIPMENT

By T. A. Griffin, Jr.
and R. L. Webb

SYSTEM REQUIREMENTS

1. Electrical-system planning should be undertaken for the system as a whole, and each element must be adapted to fit its particular function in the system of which it is a part. Stations or lines of the same capacities may have quite different functions and consequently different design characteristics. For example, a 25,000-kW generating station supplying an electrochemical plant would differ from one supplying general distribution to an urban community and from a run-of-stream hydroelectric plant feeding into a big system. While it is rare that a complete new system is designed and built and most construction is rebuilding or addition to existing systems, the changes should be visualized and planned over the long range for the entire system.

2. The system planning should start from the load and be carried back through the distribution and transmission systems to the generating stations. The first determination must be the size and character of the load, its probable rate of growth and ultimate development, its load factor, the importance of continuity of service and good voltage regulation, the cost of providing service of minimum characteristics, and the value and cost of service of better characteristics. When a system is built to supply a special load, e.g., a steel mill, the whole may be adapted to the special needs, whereas if the system is required to furnish a multiplicity of loads, it must be designed for more average conditions. The system may be required to supply several classes of load, such as residential, commercial, and heavy manufacturing in the same general area, necessitating a somewhat different design from what it would be if all the load were of one class.

3. Existing plant and line facilities are usually an important factor in planning for new load. When important additions to or changes in existing facilities are necessary, the engineer must weigh the costs and advantages of expanding the present system, installing a second system, and curtailing or entirely replacing the old. Rapid load growth or rapid reconstruction of load facilities is more conducive to replacement of an existing supply system than a slow, gradual growth. A rapidly growing load will usually lead to the construction of new generating or substations to supplant or replace existing stations, whereas a slowly growing load can often be met by expansion of existing stations without incurring the large expense of starting a new station.

The capacity, voltage, and number of existing lines may have an important bearing in determining how a new load is to be supplied or a new generating station is to be connected to the system. High-capacity high-voltage lines tend to decrease the number and increase the capacity of supply units, whereas the reverse is true of lower-voltage lower-capacity lines.

4. Selection of voltage and frequency may be determined by existing plant or line facilities; but if a new plant or system is to be built, it should be borne in mind that 3-phase 60-c alternating current is, with a few notable exceptions, almost universal in the United States for new developments.

5. Direct current is limited largely to loads of special requirements for new developments. Even new small industrial generating plants and gas-engine-driven "farm-lighting sets" are now largely alternating current. Direct current is essential for electrochemical purposes and may be generated directly as such but is more commonly obtained by conversion or rectification from an a-c supply. Direct current is practically universal for street-railway, trackless-trolley, and subway-train propulsion and has found application in main-line railway electrification. It is used in high-speed passenger-elevator applications, although individual motor-generator sets of special design are prevalent and may use alternating current as the initial source of supply. Certain other applications requiring adjustable- or variable-speed control, such as some printing-press or steel-mill drives, are common but by no means universal. Individual or group conversion equipment is commonly installed to provide direct current for the parts of the plant for which direct current is essential or highly desirable. The downtown areas of some cities are supplied with direct current or with both direct current and alternating current; but the growth of the d-c load is usually restricted, and in many cities it has been reduced or eliminated. Sections of metropolitan d-c networks have been cut over to the a-c system in the same area, through the use of transformer–silicon rectifier units. Such units having 50- to 250-kW ratings are packaged in oil-filled tanks resembling those for distribution transformers. They permit shutdown of costly d-c converter substations where the peak load has been substantially reduced.

6. The trend away from direct current has been due to the higher first cost of generating equipment and most utilization equipment, higher losses and maintenance costs, and the fact that it is impractical to transmit direct current in large quantities or over great distances. Direct-current generators and motors are generally not practicable in single units in sizes above approximately 5,000 kW.

7. Twenty-five-cycle power is now generally confined to single-phase a-c railway electrification, certain steel-mill applications, and existing generating plants and lines for conversion to direct current. It obtained its large development because of the superior characteristics of 25-c synchronous converters for conversion to direct current as compared with those of higher frequency. Improvements in 60-c converters have made these entirely practicable; and, further, rectifiers are suitable for either frequency. For single-phase railway electrification 25-c power is used because of the greatly superior characteristics of the 25-c railway motor. Some European railway electrifications have been made with still lower frequencies. It has also proved advantageous in some plants requiring large very-slow-speed induction motors because of the better motor characteristics.

8. Sixty-cycle power has become predominant because of the freedom from frequency flicker of incandescent lights and the generally lower cost of 60-c transforming, generating, and motor equipment. It makes possible a greater selection of motor speeds and a top synchronous speed of 3,600 rather than 1,500 r/min.

9. Odd frequencies are obtained for special purposes, e.g., for small very-high-speed motors in some textile mills and for radio. These applications are special, and the odd frequencies are obtained from special frequency-changer sets or through inverter vacuum-tube combinations.

10. Single phase is confined to small lighting and power loads except for special loads, such as railway, electric-welder, and some furnace loads. The service may be either 2- or 3-wire, depending upon the size of load and the service facilities.

11. Three-phase power is almost universal for general usage except for small loads. Generation and transmission lines are usually 3-wire, whereas distribution circuits may be either 3- or 4-wire. Two-phase 3-, 4-, and 5-wire circuits have been installed in the past and are still in operation in some places, but this type of supply is limited to existing installations in restricted areas. Special phase connections, such as 6-phase or 12-phase, are obtained from the standard 3-phase supply by means of special transformer connections. Where single phase is obtained from polyphase circuits, special care must be exercised to ensure approximate balance of the polyphase supply to ensure balanced polyphase generation. This is especially true when the supply is from turbogenerators with solid-type rotors.

12. Voltage Selection. The selection of suitable nominal voltages for different parts of the system has an important bearing on costs and must be made from an analysis of the complete system. It is affected by the size, character, and distribution of load; length, capacity, and type of transmission and distribution circuits; and size, location, and connections of generators. The analysis should start from the load to be served.

13. Preferred Voltages. Most electrical equipment is built in various nominal voltage classes, e.g., 120, 240, and 2,400, and in various steps up to the highest transmission voltages. Trends have been toward fewer voltage classes and wider steps between classifications.

Through the cooperation of Edison Electric Institute and National Electrical Manufacturers Association, agreement was reached on preferred voltage levels, and EEI-NEMA Preferred Voltage Ratings for A-C Systems and Equipment was published by both organizations in 1949. Since then, the work has been recognized and is now available as *USASI Publ.* 84.1-1954, Standard Guide for Preferred Voltage Ratings for A-C Systems and Equipment. Long-range overall economies will be obtained by planning new extensions and systems that will conform to the more recent trends in voltage classifications and will operate well within commercial voltage ranges. It is recognized that extensions to existing lines and stations will generally be made at the existing voltages, but the perpetuation of odd voltages should be restricted in so far as possible.

14. General-purpose lighting is almost universally at nominally 120 V in the United States, although series street lighting and other special applications make use of other voltages. Gas-tube and fluorescent lighting units are frequently supplied through special transformers to give the most suitable voltage for the individual units. The use of 110-V service voltage has been greatly reduced in favor of 115 and particularly 120 V. The lighting voltage may be obtained from a 2-wire 120-V circuit; a 3-wire 120/240-V d-c or single-phase a-c circuit; or a 120/208-V 3-phase 4-wire circuit. Very small motors are usually single-phase alternating current at nominally either 115 or 230 V. Larger motors (above 3 to 5 hp) are usually rated at 230 or 440 V, a-c motors being of the 3-phase type; 3-phase motors rated 230 V generally operate satisfactorily at any voltage between 208 and 250 V, but motors rated at 208 V may be required for special applications when operated at 199 or 208 V, particularly if high start torques or momentary high overloads are encountered; 440 V and, to a lesser extent, 550 V are used in some industrial plants and supplied by some utilities for power purposes; but the trend toward combined light and power circuits is increasing the use of 240 or 208 V for general power applications. In large cities there is a trend toward 265/460-V systems for supply to large buildings and industrial loads.

Large plants using motors of 50- to 100-hp ratings frequently use circuits and motors of 2,300- or 4,000-V rating for the large motors. Only very large motors are economical for voltages above 4,000.

15. Direct-current generators must be at the utilization voltage and must be near the load. **Alternating-current** generators may be at the utilization voltage if near the load but may be at a different voltage if the distance from the load makes the introduction of transformation economical or necessary.

16. Transformers may be obtained for any of the suitable standard voltages, although the higher-voltage transformers are usually limited to units of large capacity, because of cost.

17. A minimum number of transmission and distribution voltages with comparatively large voltage steps will usually be most economical and add to flexibility.

18. Standards of reliability must be coordinated with value of continuity of service. Although no service can be made totally immune from the hazards of failure, greater degrees of reliability can be obtained at greater costs, and these must be balanced against the value of better service. These values will differ with the size and character of load, the probable frequency and duration of outages, the cost of the minimum-type service, and the degree of reliability to which the user is accustomed.

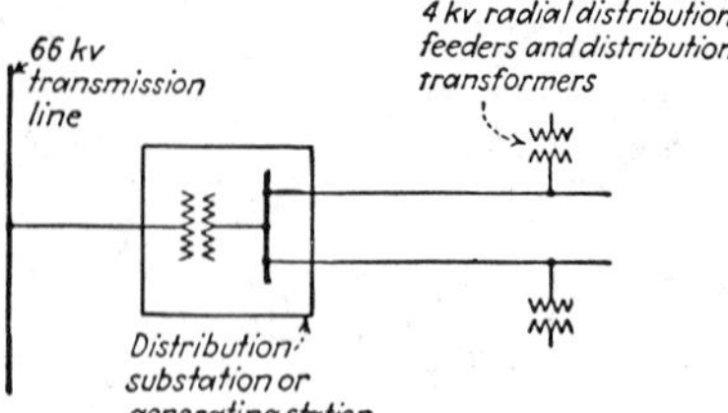

Fig. 10-1. Typical supply to a small community.

19. The small individual customer or **small industrial plant** can usually afford little more than minimum-type service and is usually supplied by a single service connection to a utility's supply lines or a single small generator unit.

20. A small community may be served by diversified distribution lines from a single generating station or from a single transmission line or from a diversity of supply, depending upon such factors as the size of the community and the value of its industries, the probable rate of growth, the cost of one generating unit of large size as compared with several smaller units, and the cost of bringing additional transmission lines to the community (see Fig. 10-1).

21. A large industrial plant is usually supplied by two or more generating units or two or more supply lines, often with its buses sectionalized or segregated. The number and size of supply units may be such that an outage of a single unit at time of peak may result in reduction of plant output; or there may be sufficient reserve, at a higher cost, such that the outage of a single unit will result in no reduction of output.

22. A large city is commonly supplied from one or more generating stations and tie connections such that one or more generating units or tie feeders can be lost at time of peak without the necessity of dropping load. Substations are limited in capacity and supplied from a diversity of supply lines so that the most probable outages are limited to individual distribution feeders on which service can be restored promptly. High-density load areas may be supplied by a low-voltage network which still further limits the size of load affected by a failure (see Fig. 10-2).

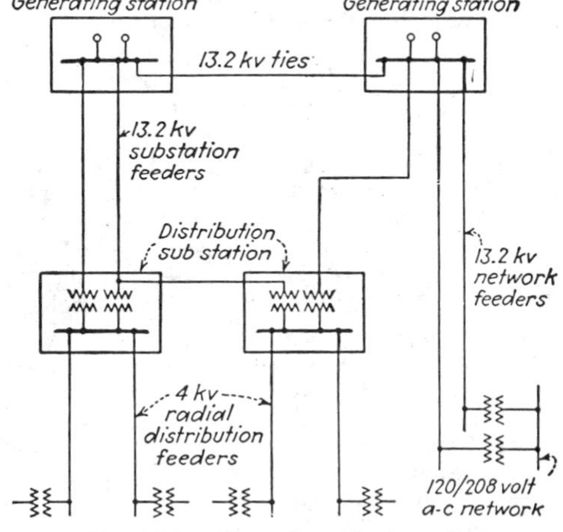

FIG. 10-2. Typical supply to a city.

23. A metropolitan area is usually fed by several generating stations with interconnections between stations and possibly other systems. The general methods of supply and requirements are similar to those of a large city but on a more extensive scale. Units are not only more numerous but generally larger.

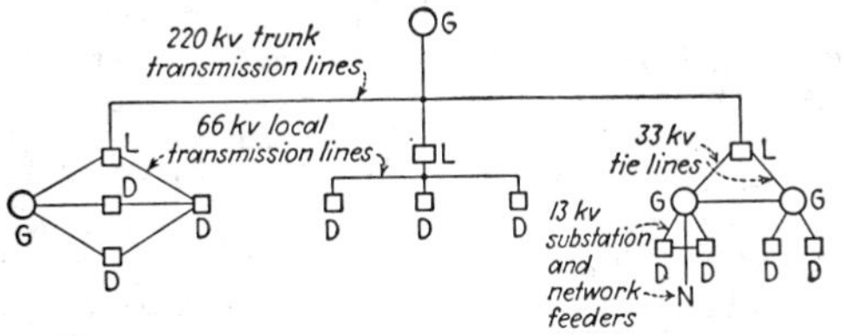

FIG. 10-3. Typical interconnected system. *G*, generating system. *L*, line step-down substation. *D*, distribution substation. *N*, low-voltage network.

24. A large interconnected system consists of several generating stations and loads, interconnected by high-capacity ties. Units of generation or ties can usually be lost without interrupting service (see Fig. 10-3).

25. Increase in system size tends to increase reliability. Within general limits, increase in size of plant or system tends to increase the number of units and degree of diversity, with a decrease in probability of complete or long outage. While mere size of units does not necessarily mean greater reliability, higher grades of construction can usually be provided in large plants and lines without a prohibitive increase in cost. Furthermore, within limits, larger conductors tend to provide a greater margin of mechanical strength and, when insulated, reduce the potential gradients. Very-high-voltage overhead lines tend to become more immune to lightning disturbances and can be made practically lightningproof. On the other hand, outages for maintenance, repair, or replacement of the larger units are generally of longer duration. The multiplicity of units and sectionalizing of buses and lines by automatic breakers, with a pooling of reserves, associated with a large system are a large factor in the net gain in reliability. Greater sectionalization in large systems is necessary in order to limit the magnitude of short-circuit currents to practical limits and is economical, owing to the smaller gain in efficiencies and costs per kilovolt-ampere obtainable with still larger units. In some cases, further increase in size of unit is impractical or results in higher unit costs.

26. The units of loss for ordinary and rare failures must be decided upon in designing a system. These fall into two categories: (*a*) the "units" of load expected to be lost due to failure, (*b*) the "units" of system that can be lost without interrupting service. Furthermore, large systems should determine units of load and system that may be dropped in case of emergency to prevent further spread of trouble and units that may be reenergized in sequence following an extensive outage.

27. The unit of load loss generally consists of all the load connected to a distribution feeder. In urban areas where there are a multiplicity of feeders, large and important loads may be fed by two or more feeders: (*a*) as required to carry the load; (*b*) for the sake of reliability at small increase in cost, so that the loss of a feeder does not interrupt service to that load. In light-load-density areas, such as rural communities and villages, costs may dictate the unit of loss as a moderate-capacity transmission line with a number of small unit substations and local distribution feeders. Automatic reclosing of breakers can be used to limit the outage to a few seconds when the cause of the outage is transitory, such as lightning, as is the case in a great portion of outages of overhead lines.

In very dense areas supplied from low-voltage networks, the unit is generally limited to one section of network mains. Large and important customers may sectionalize their own buses and distribution system and be supplied from a number of network transformer units and connected to different network mains sections.

28. Small substations and generating stations usually form a unit the loss of which, though less frequent than that of a line, results in a loss of load. The damage resulting from failure is usually small, owing to the low voltage and small energy, and can be quickly repaired, or small units can be quickly replaced. In larger stations there should be two or more generators or transformers, as failures of such equipment may require longer outages than buses, breakers, or connections.

29. Large substations and generating stations are usually sectionalized so that a failure, resulting in loss of service, is limited to but a portion of the station or so that service can be immediately restored to all or most of the load. The unit so lost may be comparable to a complete station of smaller size.

30. The unit of system loss without loss of load is the transmission line or portion thereof, the generating or transforming unit, or the bus section that can be lost at any time without loss of load. In generating and transforming units it generally corresponds to the "spinning," or operating, reserve. Large stations and systems are generally operated with enough units in service so that at least one generator, boiler, transformer, or tie line can be lost at any time without loss of load and have in reserve additional units which can be put in operation to meet a second or third contingency. A complete hydro plant tied into an extensive transmission system may be such a unit. When buses are so sectionalized, it is essential that feeders to distributing substations or networks be so distributed among the buses that the feeders left connected to the operating buses after a bus-section tripout can carry the load. With such units of loss established, little additional expenditure can be justified to sectionalize a unit further, but care should be exercised to ensure that failure of one unit does not involve an adjacent unit.

31. Failures on units of loss are ordinarily cleared with a minimum disturbance and loss of service but occasionally are not automatically cleared or involve more than a single unit. It may then become necessary to segregate progressively and manually sections of the system until the trouble can be localized and disconnected. This must be done quickly to limit the extent of the damage and to prevent loss of synchronism between parts of the system. An extensive loss of capacity may result in such serious overload of remaining equipment and lines that sections of load must be dropped to prevent further trouble. This can be done easily in a radial system by opening selected individual feeder breakers or dropping entire substations. In a low-voltage network system it is necessary to drop an entire network section and clear quickly *all* feeders to the network section.

32. Picking up of load in an area that has suffered a complete outage must be done progressively in increments small enough to avoid collapsing voltage on generators or tripping out of tie feeders and to be within the capacity of boilers to respond to sudden changes of load. Time intervals between steps must allow for readjustment of excitation, load division, and boiler-steaming rates. Network sections must be picked up as a unit by simultaneously energizing enough feeders to the area to carry the load temporarily and immediately adding the other feeders. The increased use of domestic refrigeration and oil burners and other automatic motor-driven equipment that may be connected to the lines at the time of restoration of service increases the load and particularly the inrush current to be picked up instantaneously. The load picked up will vary with the type of load in the area, the time of day, the season of the year, and the length of the outage. Data on loads to be expected are very meager, but in a mixed residential and commercial area the load to be expected may be of the order of magnitude of normal load for the season and time of day.

33. Increased reliability can be obtained within limits by a comparatively small increase in costs by the use of good design, construction, and operation; but beyond good standards, costs for improved reliability increase rapidly in proportion to gains. The general trend is toward simplification rather than extreme flexibility and duplication in station construction, generally resulting in increased reliability and savings in costs.

34. The loading of equipment affects its reliability and station costs. The rating of equipment is based upon ordinary loadings, load cycles, ambient temperatures, and life of insulation. Departures from these affect the length of serviceability. Continuous operation under conditions to give the maximum allowable hot-spot temperature, such as 105°C for Class A insulation, would very materially shorten the life of the insulation; but continuous full load is seldom if ever an operating condition, and ambient temperatures of 40°C are at most but a few hours per year. System load factors are generally between 40 and 60%, and plant capacity factors or individual equipment load factors are generally less. A few base-load plants and some industrial plants, such as electrolytic plants, have higher load factors. Conditions are very rare in which the increased reliability and life justify the purchase of equipment with nameplate ratings in excess of expected loads. Generally loads are such that during winter months or under emergency conditions loads in excess of ratings can be safely carried. Some operating companies have established "emergency ratings" based upon station or cable load cycles, which permit temperature rises and hot-spot temperatures in excess of standards for the type of insulation used, the emergency ratings to be limited to a few hours per year, such as 100. Higher ratings are often established for winter than summer conditions, and in some cases hot-spot temperatures are used as a guide for loading rather than nameplates (USASI C57.92 entitled Guide for Loading Oil Immersed Distribution and Power Transformers). Some operators aim to keep loads within nameplate ratings under normal or first-contingency conditions and impose overloads only in case of second contingency, such as the simultaneous loss of a generator and station tie line.

35. The amount and character of insulation affect the cost and reliability of the electric system. Equipment meeting IEEE and USASI standards of test voltages and manufacturers' standards is generally satisfactory, but overinsulation is sometimes used for important locations or where equipment is subject to unusually severe conditions. On important substation or generating-station auxiliary buses, 4.16- or 7.2-kV insulation is often used for 2.3-kV service; and 13.8-kV or 14.4-kV insulation is available for 11-kV to 15-kV switchgear in substations and large generating stations. The impulse strength of high-voltage overhead lines and equipment subject to lightning surges is of most importance. Since overinsulation begins to affect the cost of equipment unfavorably above 69 kV, it is common practice to use reduced insulation levels on equipment above this voltage. The improvements made in lightning-arrester performance and the present-day knowledge of switching impulse voltages make it possible to use one, two, or even three full steps of insulation level below normal for equipment in the extra-high-voltage (345 kV and higher) classes.

Air and porcelain are the most common insulating materials used for high-voltage insulation, except for oil and organic materials in transformers and circuit breakers.

Mica and organic materials, such as impregnated cloth and paper, and organic and inorganic compounds, such as bakelite and Micarta, are used at the lower voltages and intermediate voltages where clearances are close. Oil and most organic materials must be kept dry and free from destructive corona and should not be subjected to high temperatures.

36. Sectionalization and segregation are important factors in ensuring reliability and are designed to limit a failure to a relatively small section of a system and prevent its spreading to another section. Separation by space, barriers, or fire walls are means used to prevent spread of trouble. The degree of segregation is affected by the size of the station; the size, voltage, and type of connections; the importance of continuity; and the costs involved. The greater the energy involved in a possible failure, the greater the need for sectionalizing and segregation. Small and low-voltage plants are usually designed with but one bus section, as the risk of failure and possible damage is small, and maintenance or changes can be done with the bus alive, or the bus may be shut down at a convenient time. Very large stations may have as many as 8 or 10 sections.

37. Multiple-supply systems consist of two or more supplies to a common bus or connection which is operated as a unit but may, if of sufficient importance, be sectionalized automatically or manually. An example is a low- or moderate-pressure steam plant with all boilers and turbines connected to a common steam header and supplied from a common feed water line, an excitation bus for all generators energized by two or more exciters and possibly a battery, and all generators feeding into a common bus. Such a system minimizes the effect of loss of any one auxiliary or generating unit but increases the risk of a total outage.

38. The unit system segregates as far as practicable each unit, its connections, and its auxiliaries. A common example is a high-pressure steam station in which each turbine is connected to its own boiler or boilers alone, each with its own boiler-feed pumps and auxiliaries. The turbine generator has its own direct-connected exciter and is connected to its own bus section, which may be connected to other sections through automatic breakers and reactors, in some cases by way of a synchronizing bus. In bulk-power stations feeding into a transmission system, each generator may be connected through its transformers to a high-tension line. Such a system minimizes the possibility of total outage, but the loss of a boiler or auxiliary results in the reduction or loss of the generator capacity. Costs of large stations are generally lower when built on the unit principle, whereas small stations are cheaper on the multiple principle. A compromise between the two is often applied, with the general trend toward greater use of the unit principle.

39. Protective equipment is an extremely important item in system design, as it is installed to function under abnormal conditions to prevent failure or to isolate trouble and limit its effect. It must function reliably and quickly. It should be selected for greatest reliability, speed of operation, and simplicity and should be consistent with the system design.

40. Breakers and relays are the most common means of energizing or deenergizing a section of an electric system under normal and abnormal conditions. When properly applied they are highly dependable, but the system plan and breaker and relay schemes must be closely coordinated to ensure minimum cost consistent with the system requirements as well as its proper functioning. A complex system of power connections generally requires an involved and expensive relaying scheme and may make complete selectivity impossible. A large system should be designed with a "backup" protection so that, in case of failure of the most immediate breakers or relays, other equipment will function. This is usually accomplished by disconnecting the next larger section of the system, e.g., by clearing a bus section or "stub" bus in case of feeder-breaker failure, rather than by installing duplicate equipment in series on the same circuit. Relay schemes are also installed to reclose a feeder breaker automatically after tripout. These have their principal applications on overhead feeders and have been applied on lines up to 765 kV. Breakers selected for a new station should have ample margin beyond the anticipated short-circuit duties to avoid expensive changes later, as station or tie-line capacities very commonly exceed original plans.

41. Grounding of the system neutral is a common practice as a means of reducing overvoltages from line to ground. Solidly grounding a system through the neutrals of Y-connected generators or transformers reduces the voltage strains to the minimum and permits the lowest setting of protective gaps and voltage ratings of lightning arresters. Grounding through a resistance or reactance limits the fault current and resulting system disturbance in case of line-to-ground failure but increases the voltage to ground of the unfaulted conductors. If the system is grounded through an impedance, the value must be low enough to permit adequate ground current for reliable relaying and to prevent dangerously high voltages due to resonant or arcing conditions. Refer to IEEE Standard 32 for effective grounding requirements. Grounding reactors have had but limited application owing to the danger of high voltages under arcing or resonant conditions. The thermal capacity of grounding resistors or reactors is usually limited to 10 s to 2 min. In American practice high-voltage (69-kV and higher) overhead systems are commonly, but not universally, solidly grounded. Moderate-voltage (13.8-kV) underground systems are frequently grounded through neutral resistors or, when solidly grounded, are grounded through a limited number of generators or transformers to control the ground current under short-circuit conditions. Generators often have a zero-sequence reactance (Z_0) lower than their positive-sequence subtransient reactance (x_d''). In using low-impedance grounding it is customary to use sufficient generator neutral reactance to make these two impedances the same at least. This prevents excessive short-circuit current in the generator windings for line-to-ground faults. It has become common practice in unit generator installations to use high-resistance-generator neutral grounding even though this stresses the insulation for ground faults. The resistance is sufficiently high so that the machine need not be disconnected from the system immediately.

42. Current-limiting reactors are commonly installed on large systems to limit the magnitude of short-circuit currents and reduce the resulting voltage disturbance on the rest of the system. Their cost is often more than offset by the saving in circuit-breaker cost as a result of the lower short-circuit ratings that can be used. They are not generally applied in small stations.

43. Shunt reactors, while used in special instances on some systems for many years, are being found necessary additions to many extra-high-voltage (EHV) systems in more recent years. The higher-voltage systems develop sufficient charging Mvars on a long line that neutralization of it becomes necessary, particularly during light load periods, to avoid severe unexcitation operation of system generators. This condition may appear on cable systems at much lower voltages.

44. Lightning protection by means of lightning arresters and gaps and overhead ground wires is a means of reducing outages and preventing damage to station equipment from lightning disturbances. The amount and kind of protection vary in different applications, depending upon the exposure of the lines, the frequency and severity of lightning storms, the cost of the protection as an insurance value against damage to equipment due to system switching surges, and the value of reduced line outages.

45. The "reserve" available in station and line capacity should be kept at a minimum consistent with the type of service to be rendered. The amount required is affected by the probable frequency and length of outages and, in large systems, the probability of simultaneous outages, such as the loss of a second or a third unit while one is out for overhaul. The reserve may be in plant equipment or tie lines or both.

46. Generating stations are usually provided with at least one reserve generating unit or the equivalent in tie-line capacity, owing to the necessity for outage due to overhaul and the long periods necessary for repairs or replacements. Some industrial stations have no reserve capacity at time of peak load and are planned for overhaul work during light-load periods and to drop nonessential load should a failure occur at time of peak load. Some large generating stations in important load centers have enough reserve to be able to lose the two largest units without dropping load. For example, a station of eight generating units and a tie equivalent to one unit may feed a local load equal to the capacity of seven units and be able to carry its load with two units or one unit and the tie out or, with all units in service, feed the capacity of one unit to the rest of the system. Such a system having several generating stations and adequate interstation ties might establish as its reserve the equivalent of its three largest generating units, on the assumption of not more than two units out simultaneously in any one plant. The cost of providing sufficient system reserve to permit the total loss of a major generating station is generally considered as prohibitive. Systems deriving part or all of their generating capacity from water power must take into consideration periods of low water and the load at such times. Systems that do not have extended periods of light load which may be used for scheduled overhaul need more reserve to guard against forced outages while units are out on schedule than systems with an extended light-load period.

47. Substations can generally be safely operated with less reserve than generating stations owing to (*a*) the fewer and shorter periods of scheduled outages, (*b*) the shorter periods of outage due to repairs or replacement, (*c*) the greater possibility of overloading during emergency, and (*d*) the smaller amount of load affected in case dropping load becomes necessary.

Small distribution substations often have no reserve transformer capacity, as transformer failures are rare and moderate-sized units can be quickly replaced. Large distribution substations usually have one spare 3-phase transformer or 3-phase bank, although a single-phase unit is sometimes provided where banks of single-phase transformers are used. "Unit" substations are sometimes installed in urban areas with excess transforming capacity and low-voltage ties to other "unit" substations which furnish the reserve against loss of any one unit.

48. Station ties are installed as a means of pooling generating-station reserve capacity and to obtain improved economy of operation. Their costs must be balanced against additional plant capacity, as long ties sometimes cost more than additional generation. Hydroelectric developments can seldom be made near a load center; hence, the investment and operating cost of a hydro plant and its ties must be compared with those of a local steam plant. **The outages of ties** are more frequent than those of generating units but of shorter duration, usually for not more than one daily peak period. Outages caused by failure of a transformer forming an essential part of a line, or the failure of large transmission-line towers, may result in a line outage of several days or longer. In planning reserves, simultaneous outages should be anticipated of parallel cables through the same duct bank and manholes or parallel lines on the same towers or, to a lesser extent, on the same right-of-way though on separate towers.

49. A load area or substation may be fed from a single generating station or sometimes from two or more generating stations. Supply from two or more generating stations is advantageous when the connections are such that the load can be fed from the remaining generating stations in case of loss of one or can be sectionalized to lose only *part* of the load with one station out. The connections should be made so that the loss of one station, or the principal ties between the stations, will not result in excessive load or circulating currents. This is particularly important in the case of low-voltage a-c networks where the phase-angle difference between various supply feeders must be kept within a few degrees. If a low-voltage network area is fed from two stations, the supply from each station should be capable of carrying the entire network load in case of the emergency of losing one station; otherwise, the loss of either station may result in the loss of the entire network area and possibly the other station due to the overload thrown on it. Some large systems having extensive network loads divide the networks into areas, each area being fed from one station only, with heavy direct ties between generating stations. As compared with supply from two stations, this reduces the problem of circulating currents in the network and tripping of network switches due to phase-angle differences between network feeders, the need for excess feeder and network transformer capacity to permit supply from *either* generating station, and the damage which might result from loss of synchronism between stations and simplifies the start-up problem in case of total outage of an area.

50. Load control in a single generating station is provided automatically by the governors of the prime movers, with occasional manual or automatic adjustments to (*a*) correct small frequency and time errors (if service is provided for synchronous clocks), (*b*) readjust loadings between units for best economy and loadings, and (*c*) adjust generators in proportion to the bus-section loadings if buses are sectionalized through reactors, to avoid excessive circulating currents in feeders or between substations and in networks. When two or more generating stations are interconnected by ties, one station is usually instructed to hold frequency while the other stations adjust their governors to hold tie-line loadings under the instruction of a load dispatcher.

Where distances between stations are not too great, the loads in each station may be automatically totalized and transmitted to the load dispatcher, where they may be again totalized to indicate the total system load. In a widespread system, loads are periodically telephoned to the dispatcher. Generating stations in an extensive interconnected system carry the load in their load area plus or minus scheduled loads over trunk-line ties. Bulk-power stations with no local load feed into trunk-line ties according to scheduled loadings. Automatic load and frequency control is frequently installed in large systems to aid the operators.

51. Voltage or phase-angle control or both may be required in ties to control reactive kVA flow, voltage, and load distribution, especially if the ties form a loop or network.

52. Stability limits, both steady-state and transient, of stations and ties should be carefully investigated. These limits are of special significance in an extensive system with long overhead lines but are generally beyond ordinary requirements in urban systems with low-impedance underground-cable ties. Stability limits may be increased by (*a*) reducing impedances of ties between generators, (*b*) high-speed generator excitation control, (*c*) high-speed breakers and relays, and (*d*) synchronous condensers with automatic excitation control.

CIRCUIT-OPENING DEVICES

53. Circuit breakers are mechanical switching devices, capable of making, carrying, and breaking currents under normal circuit conditions and also making, carrying for a specified time, and breaking currents under specified abnormal circuit conditions such as those of short circuit. (The medium in which circuit interruption is performed may be designated by a suitable prefix, e.g., air-blast circuit breaker, gas circuit breaker, oil circuit breaker, vacuum circuit breaker, etc.)

54. Fuses are overcurrent protective devices with a circuit-opening fusible part that is heated and severed by the passage of overcurrent through it. (A fuse comprises all the parts that form a unit capable of performing the prescribed functions. It may or may not be the complete device necessary to connect it into an electric circuit.)

55. Knife-blade disconnecting and horn-gap switches are mechanical devices having a movable member adapted to connect or disconnect contact members to which conductors are securely bolted; they are usually operated on dead circuits only but are sometimes operated on energized low-capacity circuits and short lines.

56. Automatic circuit breakers are equipped with a trip coil connected to a relay or other means, designed to open the breaker automatically under abnormal conditions, such as overcurrent.

57. Switches used in the earliest low-voltage electric circuits were generally of the hand-operated, knife-blade type. As currents and voltages increased, it was found that the arc burn while opening the switch damaged or destroyed the contacts. **Circuit breakers** were developed which opened the circuit rapidly by spring or gravity action, thereby reducing the time of arcing and amount of burning. By placing the breaker in a vertical position with a horizontal break, the convection air currents due to the heat of the arc tended to carry the arc upward and away from the breaker, and the magnetic action of the loop through the breaker and arc further tended to increase the length of the arc and extinguish it. Auxiliary or arcing contacts were added which opened after the main or current-carrying contacts parted, reducing the burning on the main contacts. The arcing contacts were frequently made of some material, such as carbon blocks, which was less damaged by arcs than copper. The **brush** type of contact was developed which consisted of curved, laminated, thin copper strips backed by a phosphor bronze spring. The laminated brush, when closing the breaker, slid slightly on solid copper blocks of the stationary contacts with **wiping** action, scouring dirt and oxides from the surfaces, and made good contact over a wide area, thereby reducing the contact resistance and heating. Insulating barriers were sometimes added on the sides of the breaker to prevent the arc from carrying over to adjacent live parts. Later barriers were brought close to the contacts and properly shaped to form **arc chutes** to confine, cool, and **deionize** the arc, aiding in its extinguishment. **Blowout coils** were added, consisting of soft-iron plates on the outside of the arc chutes, connected to cores in conducting coils in series with the arcing contacts. The flow of current produced a strong magnetic field which forced the arc away from the contacts into the chutes to extinguish it. Extensions to the arcing contacts in the form of

horns increase the effective length of the breaker opening, as the arc is carried out on the horns, thereby lengthening it rapidly. Various modifications and refinements of these features have been made and are in use in the present-day air circuit breakers on d-c circuits and low-voltage a-c circuits.

58. Circuit breakers for a-c circuits were immersed in a tank of oil quite early in the development. The high insulating quality of the liquid and the gases formed during the arcing was found effective in quenching the arc and preventing its reestablishment after the current had passed through zero. Breakers for larger interrupting capacities and higher voltages were made by increasing the size of the tank, head of oil, length of stroke and clearances, and improved insulation.

59. Various modifications in contact parts were made from time to time to reduce the burning of the contacts and aid in rapid extinguishing of the arc. The random behavior of the gas bubbles and arc stream in plain-break breakers was eliminated by various arc-controlling means, such as deion grids and oil-blast features, about 1933. These various devices confined the gas pressure developed during the arc to a small volume and used it to force oil or gas through or around the arc and extinguish it consistently with greatly reduced arc length and arcing time. Early breakers often threw some oil in opening abnormal currents. Later, joints were made tight, and a definite vent was provided, fitted with a separating chamber to prevent the escape of oil. Operating mechanisms were changed so that the gas pressure would not tend to reclose the breaker. Developments in breakers and oils reached the point where standard modern transformer oil, such as Westinghouse Wemco C or General Electric 10 C, can be used in modern oil-tight breakers.

These general principles have been further developed, and oil streams have been forced through the arc stream to produce very efficient interrupters.

60. Speed of modern breakers has been increased to 2, 3, 5, or 8 c, measured from energizing of the trip coil to breaking of the arc; even higher speeds can be obtained where conditions warrant a special design. Since 1940 in this country, and somewhat earlier in Europe, designs have been developed eliminating oil and using air or gas as the arc-extinguishing medium. The arc is quenched in arc chutes by a blast of air in the larger ratings, while in the lower ratings blowout coils force the arc into the arc chutes. This development has continued until it is now possible to obtain air circuit breakers at all voltages and at maximum required interrupting capacity. Above 230 kV, oilless breakers are more economical.

61. The proper selection and application of circuit breakers, fuses, and switches are an extremely important element in the design of an electrical system, as these are protective devices which must function properly under normal and abnormal conditions. Fuses and breakers are relied upon to separate a defective portion of the system from the remainder to prevent the spread of damage and to permit the good portion to continue in service.

62. A circuit breaker must carry normal load currents without overheating or damage and must quickly open short-circuit currents without serious damage to itself and with a minimum burning of its contacts. Circuit breakers are rated in maximum voltage, maximum continuous current-carrying capacity, maximum interrupting capacity, and maximum momentary and 3-s current-carrying capacities.

63. The interrupting capacity "in arc kVA" is a product of the voltage of the circuit and the interrupting ability in amperes at stated intervals and a specific number of times. The current taken is the rms value existing during the first half-cycle of arc between contacts during the opening stroke (see USASI Standard C37.03 to C37.011-1964 and later revisions).

64. The generally accepted duty cycle is two openings, with a 15-s interval between them for modern air, air-blast, and oil-tight oil-immersed power circuit breakers. The USASI standards require that a power circuit breaker shall perform at or within its interrupting rating without emitting flame; that, at the end of any performance within its interrupting rating, the circuit breaker shall be in substantially the same mechanical condition as at the beginning; that it shall then withstand rated voltage and its main current-carrying parts shall be in substantially the same condition as at the beginning. It is recognized, however, that, after a breaker performs its duty cycle at or near its interrupting rating, the breaker may have its interrupting ability materially reduced and should be inspected and repaired if necessary (see USASI Standard C37.04-1964).

65. The momentary rating in current should not be exceeded for any period of time, such as during the first half cycle of short circuit, when the total current wave may be asymmetrical.

66. The short-circuit duty is determined by the maximum short-circuit current that the rotating machinery connected to the system at the time of short circuit can pass through the breaker to a point just beyond, at the instant the breaker contacts open. The short-circuit current is determined by the characteristics of synchronous and induction machines connected to the system at the time of the short circuit, the impedance between them and the point of short circuit, and the elapsed time between the starting of the short circuit and the parting of the breaker contacts.

In calculating short-circuit currents on high-voltage a-c circuits it is ordinarily sufficiently accurate to take into account only the reactance of the machines and circuits, whereas in low-voltage circuits resistance as well as reactance may enter into the calculations. In d-c circuits, resistances only are ordinarily used.

For first approximations, the reactance and typical time-decrement curves of the synchronous apparatus may be used. For close calculations, the actual reactances and time characteristics of the equipment should be used and calculations made for single- as well as 3-phase faults. The "per unit" impedance system and the "internal-voltage" method, using "symmetrical components," are often used in more exact calculations. Alternating- or direct-current calculating boards are helpful in calculating short circuits in complicated networks. It will often be necessary to make a series of approximate simplifications of the system. Programs are available for digital-computer studies of system short-circuit currents, both balanced 3-phase and phase to ground.

67. The rate of rise of the recovery voltage affects the duty imposed on the breaker. In opening an a-c circuit, if the recovery voltage at any instant after the current passes through zero exceeds the gap insulation at that instant, the arc will restrike and continue until the next current zero, when interruption will again be attempted. The rate of rise of recovery voltage is a function of the constants of the circuits which supply power through the breaker. The larger the adjacent capacitance to ground before the major inductance limiting the fault current, the slower will be the rise of the recovery voltage. Some breakers radically modify the recovery voltage characteristic by limiting the current, modifying its power factor, etc. (see Bibliography).

68. Closing of power circuit breakers is seldom performed manually, since most breakers are applied on circuits having possible short-circuit currents or voltages which make such practices unsafe. Various means are used for the purpose, such as (*a*) d-c solenoids, (*b*) solenoids operated from an a-c source through a dry-type rectifier, (*c*) compressed air, (*d*) high-pressure oil, (*e*) charged spring, and (*f*) electric motor.

69. Tripping facilities are extremely important. Most breakers are held closed by a latch and are tripped by a tripping magnet acting on a trigger to release the latch. Series overload trip coils which actuate the tripping plunger are common on low-voltage breakers and on some small a-c oil breakers where accurate tripping time is not essential. Direct-current shunt trip coils actuated by relays are practically universally used on large breakers and in important installations. The highest degree of reliability of the wiring and tripping source is essential. Storage batteries at 48, 125, or sometimes 250 V are the most common source.

70. Fuses are used on low- and high-voltage circuits of moderate to high capacity where frequent operation is not expected, e.g., for the protection of distribution transformers, small and medium-size motors, lighting circuits, branch circuits of distribution lines; also industrial plants and commercial buildings up to 600 A (NEC) with time delay at 250 V and 600 V; up to 6,000 A at 600 V current limiting; and in high voltages, up to 250 A at 138,000 V.

71. Knife-blade switches or disconnecting switches are used for opening and closing small-capacity low-voltage circuits and for isolating high-voltage circuit breakers and equipment for maintenance purposes. They are sometimes equipped with horns to break the arc and are used in overhead high-voltage circuits, such as 33,000 V, to break the exciting current of transformers. They are rated in maximum voltage and maximum normal current-carrying capacity. When subject to heavy short-circuit currents, they should be held closed against the magnetic forces tending to open them by latches or other mechanical means. The rms value of the asymmetrical current of the first half cycle of a short circuit should not exceed the momentary rating of the switch, which is usually 30 to 67 times the continuous rating, depending upon its type and size. Supporting insulators must have adequate strength to withstand the magnetic forces as well as the standard voltage tests.

Several forms of a switch known as an *interrupter switch* have been developed. They may be used at any voltage and are suited for use as low-capacity circuit breakers, up to about 4,000 A interrupting capacity. This allows their use for high-voltage line tapping to small loads, etc. Low ampere interruption is obtained in the open air through the aid of a resistance placed in the circuit, plus air blast in some cases. Higher currents are usually interrupted in a separate contact chamber before the disconnecting contacts open in the air.

72. Automatic reclosing of breakers in overhead feeders is frequently used to restore service quickly after a line trips out owing to lightning or other transitory fault. Instantaneous or time-delay reclosing may be provided with a lock-out to prevent more than one to several successive reclosures as desired. If the fault is cleared before the lock-out feature operates, the reclosing device resets itself, permitting a complete cycle of reclosures at a subsequent fault. Such duty cycles should not be more severe than permitted by the standard requiring a derating of the normal interrupting rating. Reclosing or multiple-shot fuses have been used in some applications in place of reclosing breakers. Automatic reclosing should not be applied to station ties where the interruption may result in loss of synchronism (see USASI C37.07-1964, Interrupting Rating Factors for Reclosing Service on Power Circuit Breakers).

LOW-VOLTAGE AIR CIRCUIT BREAKERS

73. Air circuit breakers are used on d-c circuits and low-voltage a-c (up to and including 600 V) circuits for the protection of general lighting, power, and motor circuits. Air circuit breakers are also used on d-c railway circuits for voltages as high as 3,000.

74. The usual construction makes use of two fixed terminals mounted one above the other in a vertical plane, which, when the breaker is closed, are bridged under heavy pressure by a bridging member operated by a system of linkages. Auxiliary and arcing contacts close before and open after the main contacts so as to relieve the main contacts from damage due to arcing. The arcing contacts are easily renewable when this becomes necessary. The breaker is held closed by a latch or the equivalent which may be tripped electrically or mechanically. Modern breakers are "trip-free."

75. Main contacts of older breakers are closed by a "brush-type" bridging member consisting of a group of curved thin copper leaves backed by a bronze spring (see Fig. 10-4). As the breaker is pressed closed, the brush is slightly deformed, giving a sliding or wiping action on the stationary blocks to ensure good contact. Many modern breakers use a solid bridging member with spring-mounted self-aligning contacts. The contact surfaces are of silver, as, with this material, oxidation does not cause excessive resistance and overheating. The contacts are often of the line type (see Fig. 10-5).

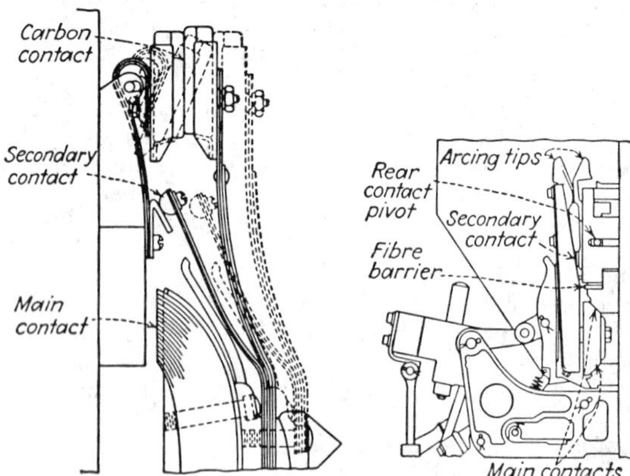

Fig. 10-4. Brush-type bridging member of low-voltage air circuit breaker of old design.

Fig. 10-5. Air circuit breaker with self-aligning, solid bridging member.

76. Arcing contacts of the older breakers are made of a carbon alloy, but modern breakers use a silver-tungsten or copper-tungsten alloy which is arc-resistant. The secondary contacts, where used, are usually of copper or silver alloy.

77. Barriers between poles are generally furnished with breakers on a-c and d-c circuits 250 V and above, and special arc chutes, quenchers, or deionizing chambers are also used throughout the available lines of modern air circuit breakers. These devices are made in different forms by different manufacturers and serve to improve the interrupting performance of the breaker, shortening the arc and arcing time (see Fig. 10-6).

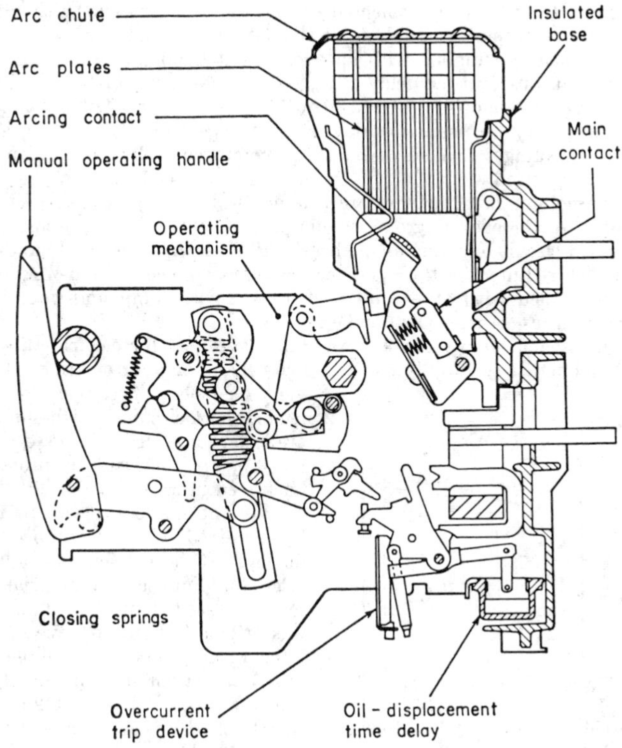

Fig. 10-6. Modern low-voltage air circuit breaker with magnetic air chutes; breaker in open position. (*I-T-E Circuit Breaker Company.*)

78. Ratings. Standard electrically and manually operated breakers are listed in ratings up to and including 6,000 A a-c and 12,000 A d-c. Electrically operated breakers are available in higher current ratings for special applications. Standard breakers are rated on the basis of a temperature rise on the contacts and terminals not to exceed 50°C above an ambient of 40°C. Voltage ratings are 250 to 600 V a-c

Table 10-1. Continuous and Interrupting Current Ratings, 250-V D-C and 600-V A-C Breakers

Frame maximum ampere rating	Range of continuous ampere ratings	Interrupting rating, rms amperes (average 3-phase rms asymmetrical amperes)
225	20–225	15,000
600	70–600	25,000
1,600	225–1,600	50,000
3,000	2,000–3,000	75,000
4,000 and above	4,000 and above	100,000

and 250 to 750 volts d-c, with special breakers available up to 3,000 volts for d-c service. (See Table 10-1 for some ratings and refer to manufacturer's catalogues for more details. Direct-current breakers may be obtained up to 12,000 A continuous rating.)

79. Open-panel or dead-front mounting may be obtained. The breakers are usually mounted on an ebony-asbestos panel, which may be mounted behind a steel panel or enclosed in a cubicle for dead-front or metal-clad draw-out type of construction.

80. Hand operation by means of a lever is common usage, even on large breakers. Electric operation by means of a solenoid or motor mechanisms for either 125 or 250 V direct current is obtainable on all but the smallest sizes of breakers.

Breakers are usually supplied with an overcurrent tripping mechanism which may be of the instantaneous or time-delay type, or a combination of both. Trip devices are adjustable over a range from about 80 to 160% of rating. Other tripping ranges and arrangements may be used, e.g., low-voltage trips (normally 50%), shunt trips connected to overvoltage, reverse current, or overcurrent relays, or to an overspeed device or a control switch.

81. Multiple-pole circuit breakers are commonly used in practically all capacities, one pole being used for each ungrounded line of a circuit, i.e., a two-pole breaker for a 3-wire grounded circuit or a single-pole breaker for a 2-wire grounded circuit.

An extra pole is required for the **equalizer** connection of a compound-wound generator or converter paralleled with other units. Thus a 2-wire compound-wound generator with equalizer requires a three-pole breaker.

82. The high-speed circuit breaker was first developed for electric-railway service and designed to operate so rapidly on a suddenly increasing current that it would prevent flashover of converters in case of feeder short circuit. High-speed breakers are often installed in the d-c circuit of mercury-arc rectifiers to protect against back feed in case of arc-back.

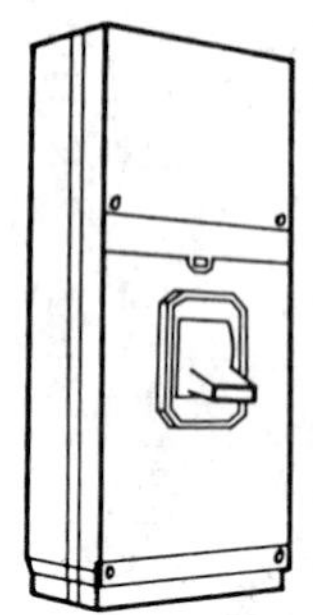

FIG. 10-7. Molded-case air circuit breaker; 3-phase 600-V 1,600-A 42,000 symmetrical A a-c interrupting capacity. (*I-T-E Circuit Breaker Company.*)

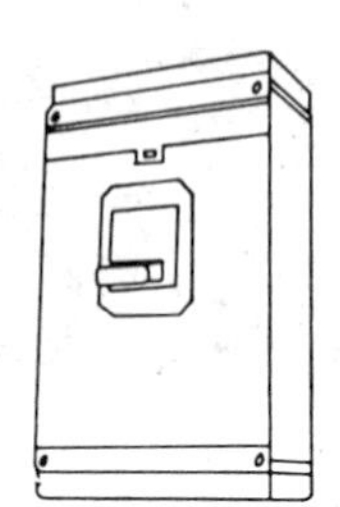

FIG. 10-8. Molded-case air circuit breaker with current-limiting fuses; 3-phase 600-V 1,600-A 100,000-A a-c interrupting capacity. (*I-T-E Circuit Breaker Company.*)

83. The molded-case air circuit breaker has received wide acceptance in the industry. It is particularly adaptable in large buildings and industrial plants to replace fused knife switches. While it was originally built to have only 5,000 A interrupting capacity, the larger units will now open a circuit up to 42,000 A at 600 V a-c or 50,000 A at 250 V d-c. A sketch of this size breaker is shown in Fig. 10-7. The molded-case circuit breaker, in smaller sizes, is adaptable in home lighting circuits where convenience of automatic protection with manual reset of the breaker is desired. The molded-case breaker's interrupting capability and that of a silver-sand type of fuse in series with it have been combined to produce a low-voltage circuit breaker with wide application possibilities. This device, shown in Fig. 10-8, may be used safely in circuits up to 600 V, 3-phase, that are capable of producing 100,000 A (rms asymmetrical) on a copper-bar type of short circuit.

HIGH-VOLTAGE POWER CIRCUIT BREAKERS

84. Power circuit breakers are available today in two main types known as *oil* and *oilless* designs. The oil types are described in Pars. **85** to **106,** and the oilless types are discussed in Pars. **107** to **112.**

High-voltage power circuit breakers are considered to be those applied in circuits from above 1,000 V to the maximum a-c system voltage, which at this writing is 765,000 V in North America.

While the oil-type circuit breaker is still most popular for outdoor service at 34.5 to 230 kV, there is a general trend of preference toward the oilless types (using compressed air and sulfur hexafluoride gas under pressure) for outdoor use at these voltages. At 345, 500, and 735 kV the oil breaker becomes noneconomical, and only oilless types are available.

For indoor service, only oilless circuit breakers are used in new installations. Oil breakers are available for indoor use but are usually applied only when required to match an existing installation. Indoor oilless circuit breakers used in the United States are principally magnetic-air and air-blast types. The vacuum interrupting principle can be expected to make its appearance in circuit breakers in the near future. For 2.5 to 34.5 kV, many indoor-type breakers of the magnetic-air and air-blast designs have been applied outdoors, by placing them in an outdoor metal housing. This practice has been found both economical and practical.

The principal reasons for the swing to oilless circuit breakers are: (*a*) elimination of oil fire hazard; (*b*) elimination of bulk oil handling; (*c*) shorter contact maintenance time and breaker outage time; (*d*) cleanliness; (*e*) higher performance speeds.

Oil Power Breakers

85. Oil circuit breakers are made in two general types, the dead-tank construction and the live-tank construction. The former are available in all classifications of voltage and interrupting ratings for indoor and outdoor application, while the latter have generally been restricted to voltages of 14,400 and below, although applications up to 34,500 V have been made.

86. A dead-tank breaker consists of a steel tank partly filled with oil, through the cover of which are carried porcelain or composition bushings. Contacts at the bottom of the bushings are bridged by a conducting crosshead carried by a wood or composition lift rod, which, in common designs, drops by gravity following contact separation by spring action, thus opening the breaker. Accelerating springs are frequently used to increase the rate of opening over a sizable portion of the total travel. In some designs, such as that shown in Fig. 10-9, the crosshead is opened with a rotary motion by springs. In many designs, two bushings and one crosshead give two breaks per pole, while in some the contacts and crossheads are designed to give four, six, or more breaks per pole with a reduction in the length of the stroke to give adequate arcing distance. An insulating fibrous-composition tank liner aids in preventing the arc from striking the tank walls.

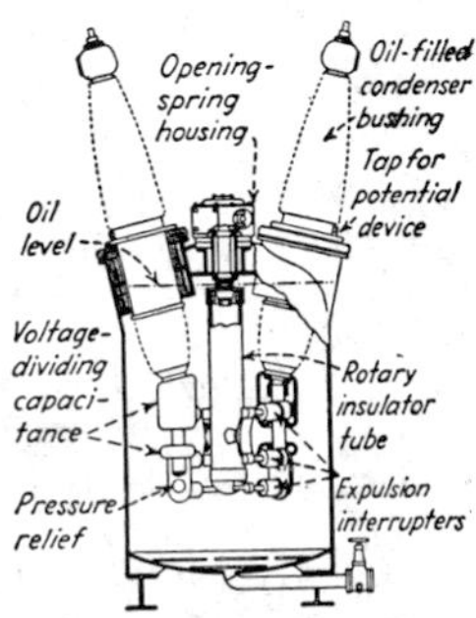

FIG. 10-9. Single-pole unit of Federal-Pacific rotary-break oil circuit breaker, 115 kV and up. Voltage division among the gaps in series per pole is aided by capacitance shielding. The bottom pair of interrupters principally produces pressure to cause oil to flow in the central pair. Oil flow in the upper pair is self-produced.

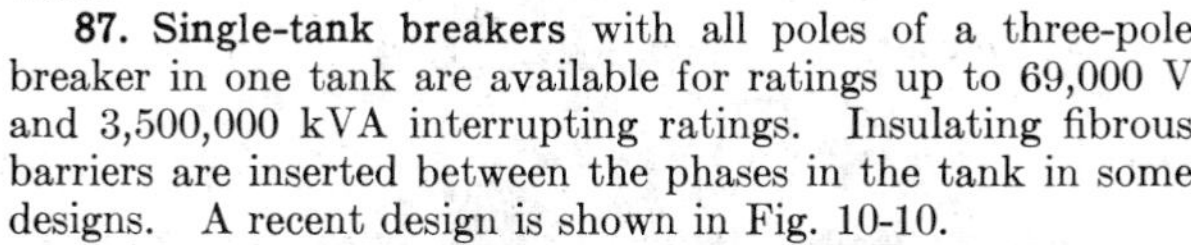

87. Single-tank breakers with all poles of a three-pole breaker in one tank are available for ratings up to 69,000 V and 3,500,000 kVA interrupting ratings. Insulating fibrous barriers are inserted between the phases in the tank in some designs. A recent design is shown in Fig. 10-10.

88. Multitank breakers with each pole in a separate tank are used for higher voltages and interrupting ratings, and generally for the larger outdoor breakers. They are available in many of the smaller ratings, however, and are necessary for the "isolated-phase" type of construction.

89. Oil-tight features are standard in modern oil breakers. A vent with oil-separating features permits the escape of the gases generated by the arc but prevents the escape of the entrained oil.

90. Various types of contacts have been designed to improve the operation of oil circuit breakers and increase the rupturing capacity. Auxiliary contacts or **arcing tips** are provided in nearly all designs to interrupt the circuit after the main current-carrying contacts or contact surfaces have opened, to reduce the pitting on the main current-carrying surfaces. An earlier type of contact consisted of finger-type stationary contacts with a wedge-shaped contact surface on the movable crosshead. Laminated or brush-type contacts consisted of an assembly of thin copper leaves backed by a spring which makes contact with a flat surface. **Inverted-type** brush construction makes use of **brushes** on the stationary members and a rigid movable crosshead so arranged that the magnetic forces of the current tend to press the brushes against the flat surface. This type of contact was often used in breakers of large normal current-carrying capacities, e.g., 2,000 A and above. High-pressure **line** or butt contacts have been used in recent years.

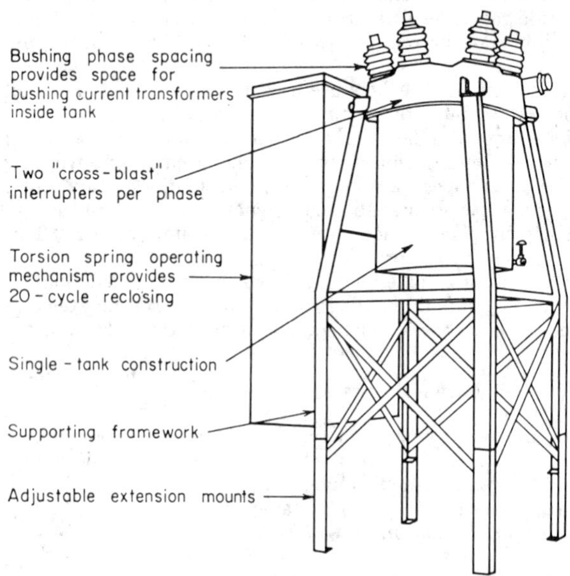

Fig. 10-10. Modern outdoor 3-phase single-tank oil circuit breaker rated at 14.4 kV, 250 MVA, using spring closing mechanism. (*General Electric Company.*)

91. Plain-break breakers have been largely supplanted in recent years by designs having special arc-controlling devices which materially shorten the arc and arcing time, thereby reducing the arc energy. Arc-controlling devices have been added to modernize many of the larger plain-break breakers in service to improve their performance at the original ratings or, in some cases, to obtain higher interrupting ratings.

92. Arc-control Devices. Many types of arc-control devices have been developed since 1935 for oil circuit breakers. Four which are used at present on high-voltage

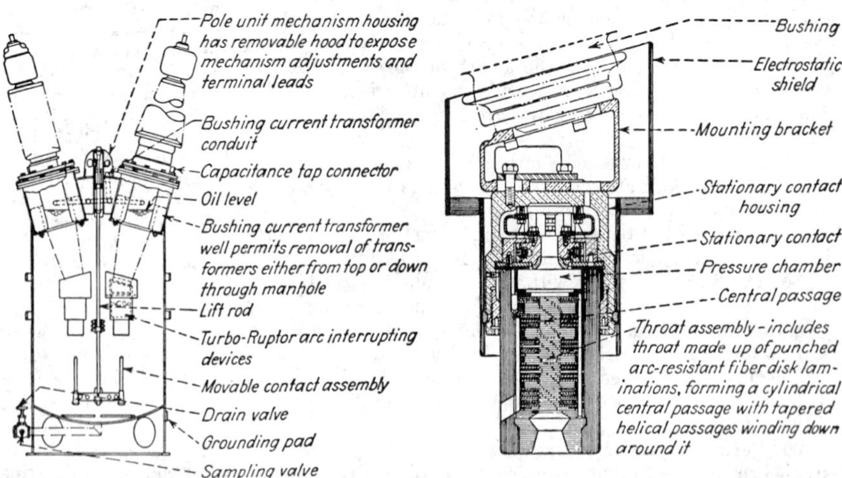

Fig. 10-11. High-voltage outdoor oil circuit breaker, outline and interrupter details. (*Allis-Chalmers Mfg. Company.*)

outdoor circuit breakers are shown in Figs. 10-9, 10-11, 10-12, and 10-13. All make use of oil pressure generated by gas created by the arc, to force fresh oil through the arc path in such quantity as to provide the necessary insulation at current zero to prevent a restrike of the arc and thereby interrupt the circuit. The interrupters contain the high pressures within their chambers and greatly reduce the stress to which the main oil tank would otherwise be subjected. The gases which are generated, after being effectively cooled, pass out through the tank vent pipe to the open air.

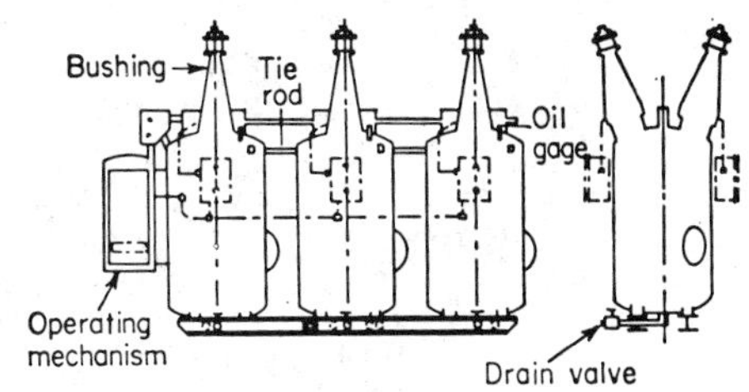

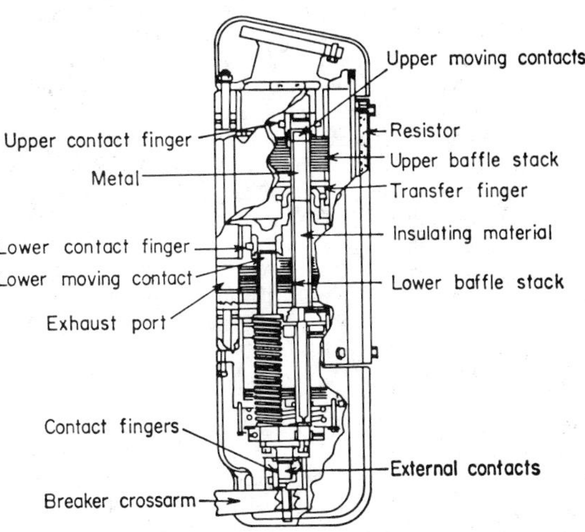

FIG. 10-12. Outline of high-voltage outdoor oil circuit breaker and details of one interrupter. (*General Electric Company.*)

93. Oil-blast features of the General Electric Company are now practically universal on all oil breakers of this manufacturer. The particular designs vary, depending upon the breaker type and rating. In large tank-type high-voltage breakers the interrupter is a special cross-oil-blast design. The construction is shown in Fig. 10-12 for 115 kV, 10,000 MVA, to 161 kV, 15,000 MVA. The gas pressure developed in each interrupting chamber forces a blast of oil horizontally across the arc

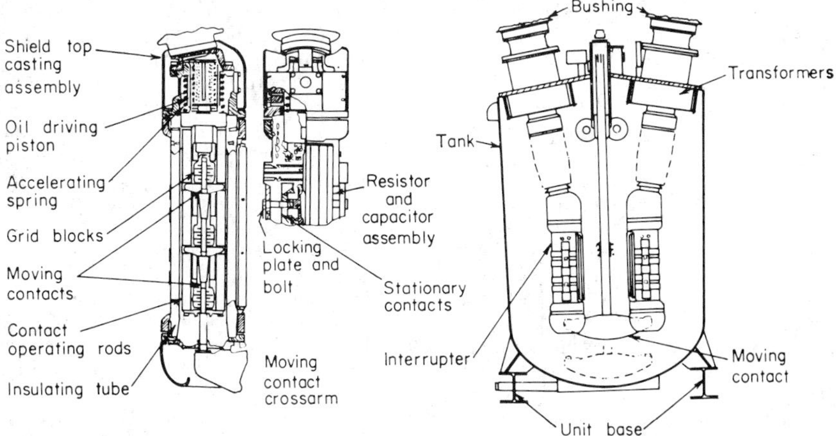

FIG. 10-13. High-voltage outdoor oil circuit breaker outline and interrupter details. (*Westinghouse Electric Corporation.*)

at the interrupting contacts. This blast of oil also prevents the reestablishment of the arc after an early current zero.

In smaller breakers the tank is divided into two portions by insulating barriers with the contacts in one portion. A port is located in the barrier close to one gap of each pole. The gas formed by the arc forces oil across the gap nearest the port and into the other chamber, quickly extinguishing the arc.

94. Expulsion interrupters as used on Federal-Pacific high-voltage breakers (Fig. 10-9) have tulip-type contacts arranged so that the arc is drawn on different surfaces from those which carry load current. Oil circulation is from pressure-producing sources located within or adjacent to the interrupter. Baffles and ducts direct the flow so that fresh oil is forced into the gap as current zero is approached.

95. Deion grids are standard on many Westinghouse oil breakers of 15,000 V and above, and simplified "deion interrupters" are common on lower-voltage breakers. The deion grids surround the arcing contacts and consist of a series of insulating plates having interspersed plates of magnetic material all so disposed and vented that the arc is moved laterally into oil pockets where it vaporizes the oil. The resulting gases of the arc stream are then forced transversely through the conductive gases of arc stream to deionize them and extinguish the arc. The arc is extinguished before the contacts are withdrawn from the grids.

96. Live-tank breakers of General Electric type H designs make use of two small-diameter cylindrical tanks, or "pots," per pole. Each pot is mounted on an insulator and forms part of the circuit. A conducting crosshead above the pots carries two rods, one of which extends through a porcelain sleeve in the top of each pot into the tank and, when the breaker is closed, makes contact with a flexible contact in the bottom of the pot. The breaker is opened by lifting the rods. Insulating baffles in the earlier designs and oil-blast features in the later designs aid in extinguishing the arc and increase the rupturing capacity. The breakers are ordinarily operated by a combination of spring and motor mechanisms (see Fig. 10-14).

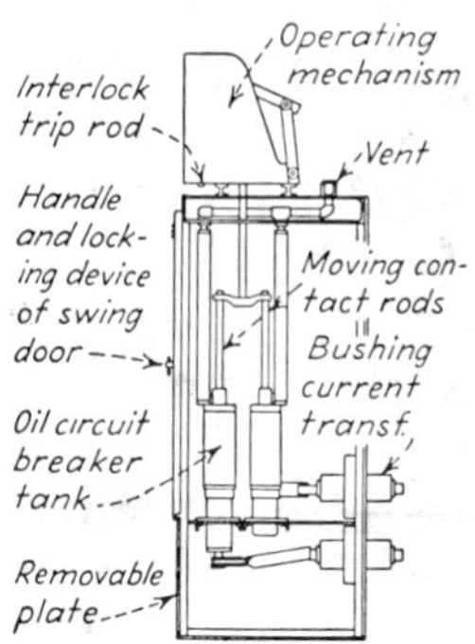

FIG. 10-14. FH-type circuit breaker. (*General Electric Company.*)

97. Bushings for oil circuit breakers are usually of plain porcelain or composition or both on the lower voltages, special construction being used on the higher voltages.

Condenser bushings as furnished by the Westinghouse Electric Corporation are made of alternate layers of untreated kraft paper and metal foil wound around the conductor, the layers of foil being stepped off at both ends to give the desired dielectric gradient. In the high-voltage bushings the insulation is stepped off with the foil and porcelain weather shields placed over the assembly, which is then filled with oil (see Fig. 10-15). Another bushing of similar voltage rating by General Electric Company is shown in Fig. 10-16.

Herkolite bushings of the General Electric Company are made of impregnated-paper insulation wound in sheet form upon the center conductor. They are then cured and machined to size to receive an adapter (see Fig. 10-17). These bushings are light in weight, readily cleaned, of strong insulation, and of relatively high puncture value as compared with the arc-over voltage. They are furnished for breakers of 7.2 and 13.8 kV. **Porcelain bushings** are made both one-piece and two-piece and are universally used for outdoor oil circuit breakers of all voltages.

98. Bushing standardization work has been accomplished on bushings for a number of years. It was started originally to obtain standard dimensions and thereby make bushings, of a given rating and type, interchangeable regardless of manufacturer. The project was expanded to include other items and now is completed for voltage ratings of 15 kV and above. The Sectional Committee in USASI C-76 is working to complete the standardization process. The same bushing may now be used in transformers or circuit breakers.

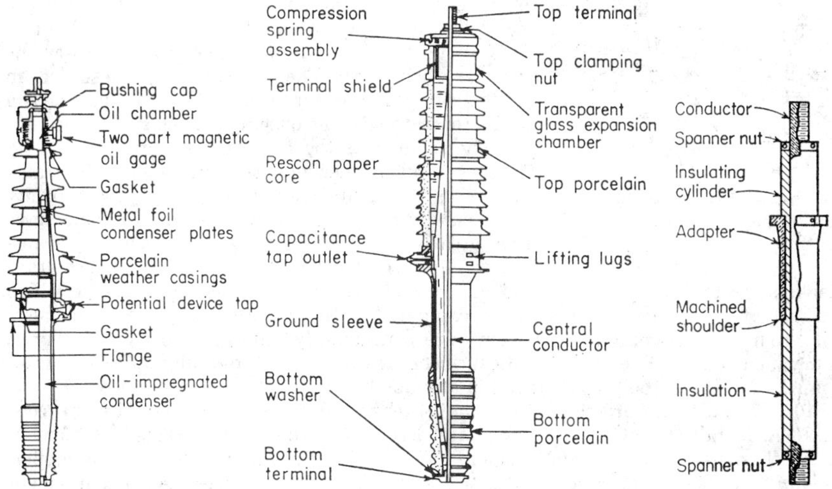

FIG. 10-15. Sectional view of condenser bushing, type O. (*Westinghouse Electric Corporation.*)

FIG. 10-16. Sectional view of type U bushing. (*General Electric Company.*)

FIG. 10-17. Construction of Herkolite bushing, 15,000 V.

99. Bushing potential devices can be applied to power-circuit-breaker bushings of 115 kV and above. The bushing is provided with a capacitance tap connected to a concentric cylindrical metal electrode or capacitance divider inside the bushing. This provides a voltage supply for operating instruments, relays, etc., from high-voltage circuits through a bushing potential device consisting of a high-reactance transformer and a combination protective gap and ground switch, as shown in Fig. 10-18. The output rating is only 35 W from 115 kV. Adjustment is provided to correct any burden to unity power factor. However, the devices are reasonably accurate and much less expensive than potential transformers. High-voltage coupling capacitors and potential devices may be used to 150 W.

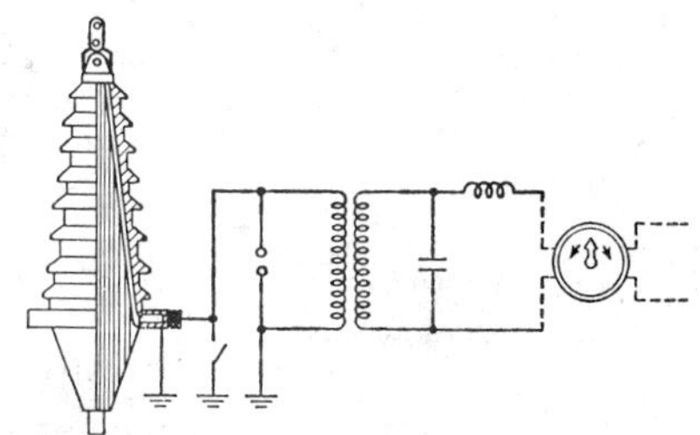

FIG. 10-18. Schematic diagram of condenser-bushing potential device.

100. Current transformers have no primary of their own but are installed around a cable, or conductor in a bushing, which provides its own insulation from the core and secondary winding. The ratios are usually designed to give 5 A secondary current at rating primary current. When subject to heavy short-circuit currents the design must be checked to see that it will withstand the duty imposed. If it is used for close selective relaying, e.g., in differential relaying, the ratios must be carefully selected, and the ratio must hold with sufficient accuracy throughout the range of currents to ensure against improper operation of relays. The secondaries must not be open while the primary is carrying current; otherwise dangerously high voltages may be developed at the secondary terminals. Current transformers for voltages up to 15 kV are of the dry type. Outdoor and high-voltage units are oil-immersed, except for the slip-over or bushing type.

101. Bushing current transformers are used in connection with high-voltage circuit breaker bushings to provide a source of current supply for operating meters, protective

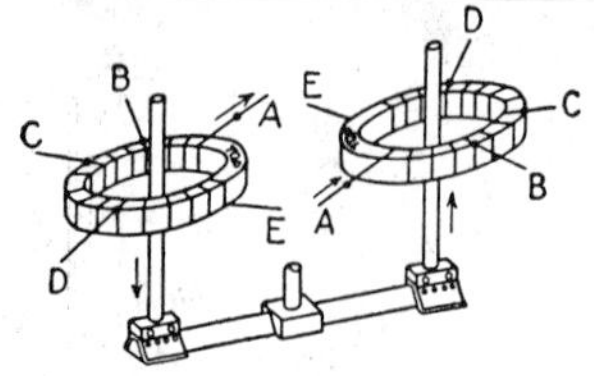

Fig. 10-19. Bushing current transformer.

relays, etc. This type of current transformer consists of a cylindrical ring core, built up of thin iron laminations about which is wound copper wire to form the secondary winding. The transformer is installed in the housing of the bushing, with the high-tension conductor through the bushing, forming the one-turn primary winding of the current transformer (see Fig. 10-19). While former designs of bushing current transformers had comparatively small voltampere capacity and generally not good accuracy, modern designs have been greatly improved, and bushing current transformers are very generally used with high-voltage breakers for indicating instruments, relays and even revenue metering at extra-high voltage, at a considerable saving as compared with separate current transformers. The accuracy and voltampere capacity are considerably reduced in designs for small primary currents. They are usually furnished with taps for obtaining different ratios. The winding should be fully distributed around the core for best performance.

102. The temperature rise of oil-circuit-breaker contacts and oil should not exceed the USASI limit of 30°C, on breakers purchased before March, 1964. Oil circuit breakers purchased since Mar. 6, 1964, should have 30°C or less temperature rise for copper-to-copper contact surfaces and 50°C or less rise for silver, silver-alloy, or equivalent contact surfaces (see USASI Standard C37.04-1964, Par. 04-4.4.2, pp. 8 and 9, or latest revision thereof). Poor contacts at terminals and the use of conductors of insufficient carrying capacity will, in time, cause circuit breakers to overheat and should be avoided.

103. Voltage ratings of oil circuit breakers are determined by the insulation characteristics of the bushings, the distance to the grounded parts, the breaking distance of the contacts, and the interrupter design. The time of interruption of the circuit breaker from the instant of closing of the relay contacts (energizing of the trip coil) is 2 to 8 c for modern designs. The higher speed is advantageous particularly on systems operating near the limits of stability.

104. Circuit-breaker oil should be of the proper quality and carefully maintained. In modern oil-tight breakers, transformer oil such as Westinghouse Wemco C or General Electric 10C or Allis-Chalmers Universal can generally be used. It has the following characteristics:

Flash point	133°C
Burning point	148°C
Freezing point	−40°C
Viscosity point	57 s
Color	Pale amber, clear

The oil level should be carefully maintained, and the oil should be filtered or otherwise reconditioned if its dielectric strength drops to 16,500 V a-c as measured by breakdown between 1-in disks $\frac{1}{10}$ in apart. New oil should have a dielectric strength of at least 30,000 V. Oil should be changed or reconditioned if excessively carbonized. Some operators check oil and contact condition as soon as practicable after each severe operation of large, important breakers. Periodic checks of oil and breaker operation and condition are recommended, the frequency depending upon the service conditions. Generally, annual periodic checks are adequate, but even longer periods have been used successfully. Where operating duties are frequent, a shorter period of time should be used.

105. Indoor oil circuit breakers are generally mounted in steel or masonry compartments. Large dead-tank breakers are usually frame-mounted. Live-tank breakers should always be mounted in cells or metal enclosures.

106. Outdoor-station-type oil circuit breakers are generally frame-mounted up to 69,000-V ratings and floor-mounted above 69,000.

Oilless Power Breakers

107. Oilless power circuit breakers have received wide acceptance in all fields since the 1940s. They have practically displaced the oil circuit breaker from indoor applications and are gaining rapidly in outdoor installations, in the United States. For extra-high-voltage applications, only oilless circuit breakers are available. The oil circuit breaker remains favored for indoor uses only where adverse conditions such as dirty atmospheres, steam, high humidity, and the need to match existing equipment are present. Oilless breakers are manufactured in three main classes, i.e., magnetic-air, compressed-air, and SF_6 (sulfur hexafluoride) types.

108. The vacuum circuit breaker is a new type which has so far been available only as a circuit recloser. It is being developed for medium-voltage circuit-breaker applications and should soon make its appearance. A single vacuum-interrupter bottle is shown in Fig. 10-21.

109. Magnetic air circuit breakers are usually stored-energy-mechanism-operated and interrupt their main circuit in the normal atmosphere under the influence of a strong magnetic field which acts to force the arc deep into a specially designed arc chute. Solenoid mechanisms may be obtained. The chute cools and lengthens the arc to a point where the circuit cannot be maintained by the voltage of the system, and interruption is accomplished. The zone between the main contacts is clear of ionized air by the time interruption is obtained in the arc chute, and so restriking at this point is not a problem. Since the magnetic effect is not great at low currents such as load, transformer magnetizing, cable-charging current, etc., all designs use an air-pump "puffer," actuated by the operating mechanism, which blows a blast of air across the arc and thereby assures its entering the arc chute and giving rapid interruption at the low current values also. When the circuit breaker is opened, the arc transfers from the main arcing contacts to fixed arcing horns which are within the arc chute. The magnetic field is produced by coils in the main current circuit, in some cases wound around a magnetic core which

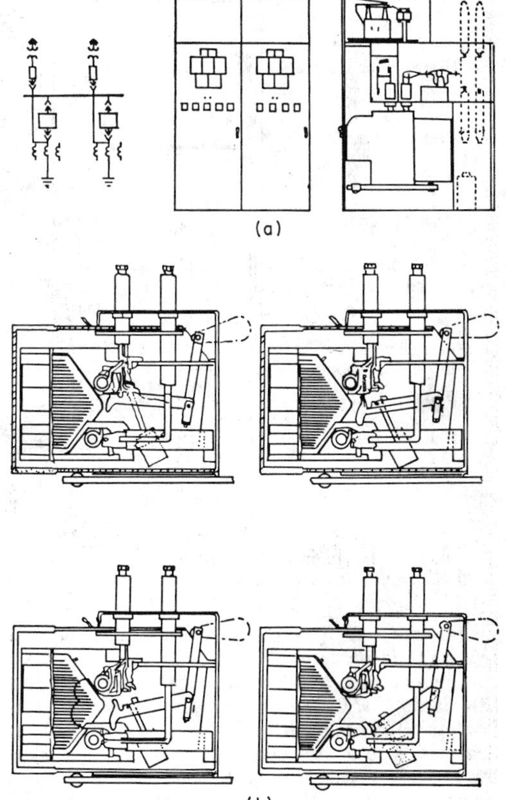

Fig. 10-20. Vertical-lift metal-clad magnetic air-circuit-breaker assembly. (*a*) Two-cubicle assembly; (*b*) circuit breaker interrupting operation. (*General Electric Company.*)

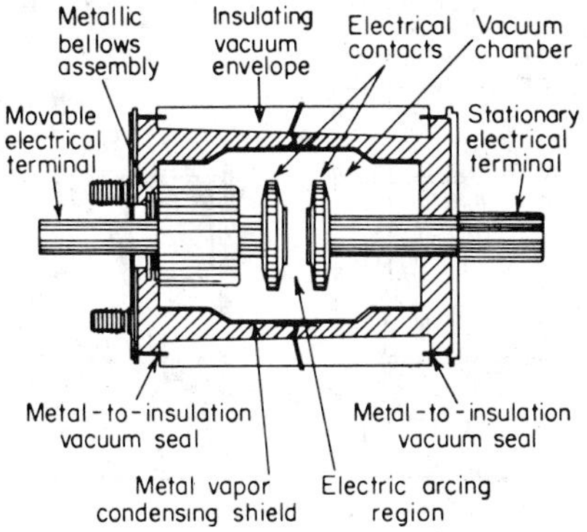

Fig. 10-21. Vacuum-bottle interrupter for a 14.4-kV 500-MVA power circuit breaker. (*General Electric Company.*)

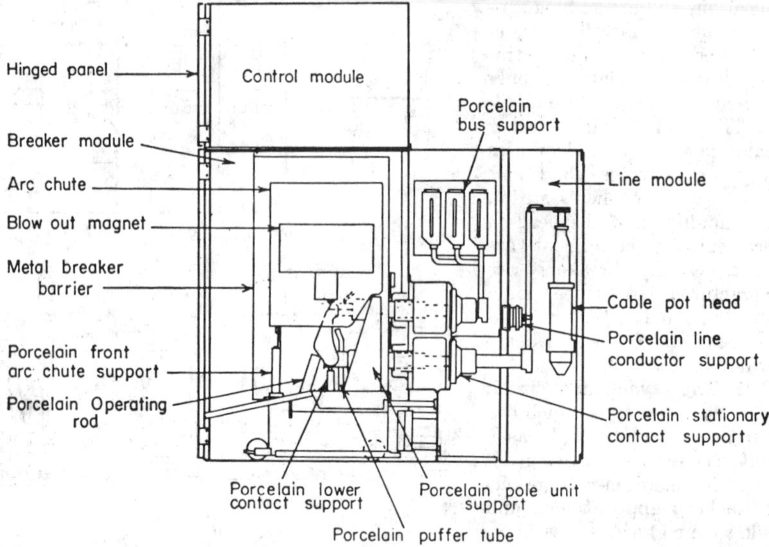

Fig. 10-22. Horizontal drawout metal-clad-type magnetic air circuit breaker in housing. (*I-T-E Circuit Breaker Company.*)

magnetizes soft iron plates in the sides of each arc chute. Some designs do not require the iron core.

Magnetic air breakers may be obtained in any of the ratings of Table 2 of USASI Standards C37.6 and C37.06 through the 13.8-kV ratings. All are designed for use in metal-clad enclosures. Figures 10-20 and 10-22 show the vertical-lift and horizontal-drawout type of breakers, respectively, and also the circuit breakers in place in their metal-clad enclosure. Though the designs shown are for indoor use only, the same circuit breakers are applied in weatherproof housings for outdoor service. When they are so applied, suitable heaters are used in the housings to avoid internal moisture condensation. The gear has proved itself very satisfactory for such service.

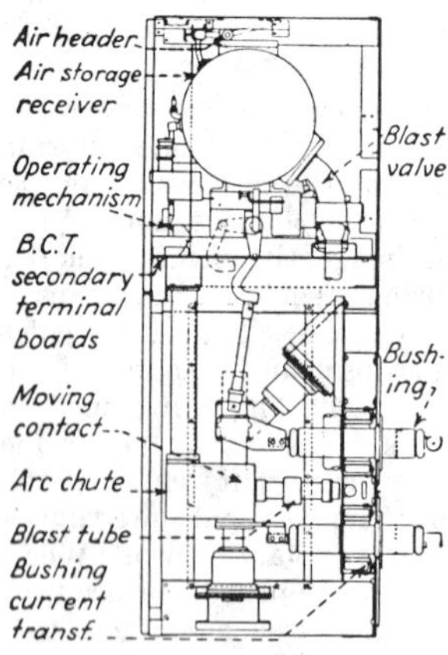

Fig. 10-23. 14.4-kV air-blast power circuit breaker, operating at 250 lb/in² gage. (*General Electric Company.*)

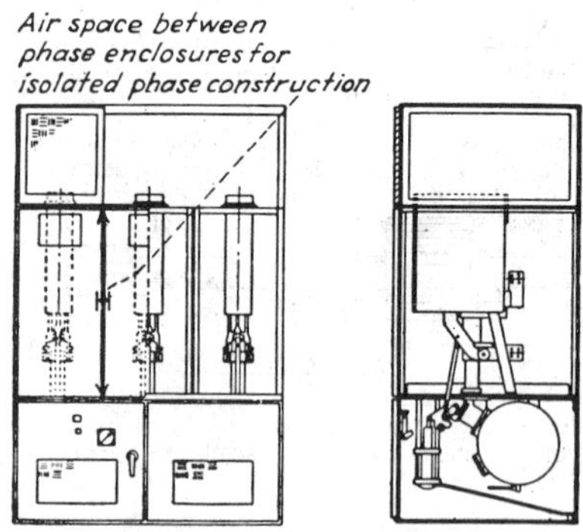

Fig. 10-24. 34.5-kV air-blast breaker in housing, operating at 150 lb/in² gage. See Fig. 10-25 for details. (*Westinghouse Electric Corporation.*)

110. Compressed-air circuit breakers are used to fulfill the heavy-duty requirements of air circuit breakers in high-voltage circuits. They are used mainly to provide the indoor ratings in Table 2 of USAS Standard C37.6 at 14.4 kV and 34.5 kV, though air-blast designs have been used at some lower voltage ratings. They are also used in outdoor applications from 34.5 to 765 kV in ratings given in Table 4 of USAS Standard C37.06.

The first air-blast circuit breaker in the United States was for 13.8-kV generating station service. It was rated 15 kV. The modern design of this circuit breaker and its housing is shown in Fig. 10-23. Compressed air is used to operate the breaker as well as to provide the medium for arc interruption. A blast of air at 150 (Westinghouse) or 250 (General Electric) lb/in² gage is directed across the path of the arc, forcing it into a special arc chute, where the arc products are cooled and deionized to a degree to effect efficient interruption. Contact erosion is small, and maintenance is small and easy to accomplish in these interrupters, which are available at 1,500- and 2,500-MVA duties.

The 34.5-kV air-blast breaker and its housing are shown in Fig. 10-24. The operating principle is shown in Fig. 10-25. This breaker design uses the same general principles as the 14.4-kV breaker, with suitable attention to greater insulation requirements. This design, and the 14.4-kV as well, are for indoor service and for use outdoors, when installed in weatherproof, heated metal housings. They are available at 1,500- and 2,500-MVA duties.

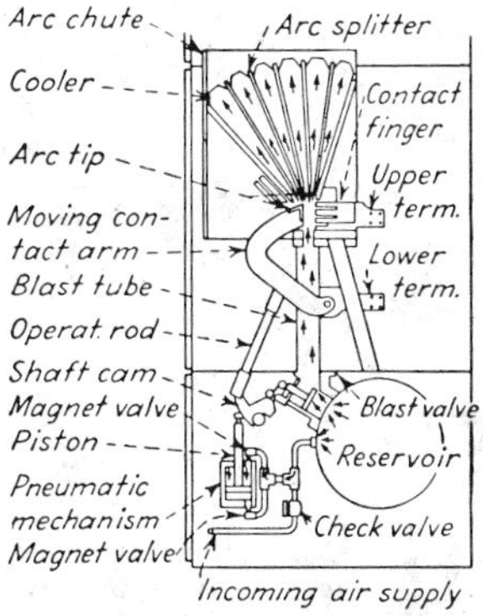

FIG. 10-25. Operating principle of medium-voltage air-blast circuit breaker. Trip coil has opened air valve to admit air to top of piston. Blast valves have released compressed air into arc path. Arc and gases carried upward into arc chute, where arc is quickly extinguished and gases cooled in diffusion chamber. (*Westinghouse Electric Corporation.*)

At 69 kV a different contact and isolating principle was adopted. Compressed air is used for operation and for interruption, but the interrupting contacts do not provide the necessary isolation after interruption. Instead a separate automatic switch, which is not designed for interruption, performs this service while the two interrupting contacts are held open by the rush of air used for arc interruption. This circuit breaker is shown in Figs. 10-26 and 10-27. It has been developed for 3,500- to 5,000-MVA duties.

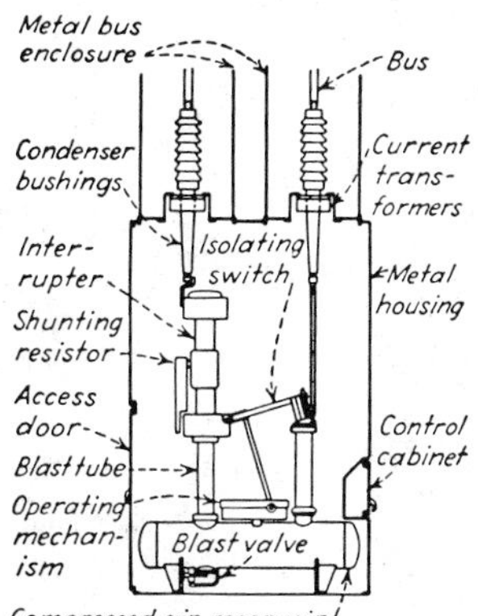

FIG. 10-26. 69-kV air-blast circuit breaker, 2,000 A continuous, 3,500 MVA interrupting capacity. (*Westinghouse Electric Corporation.*)

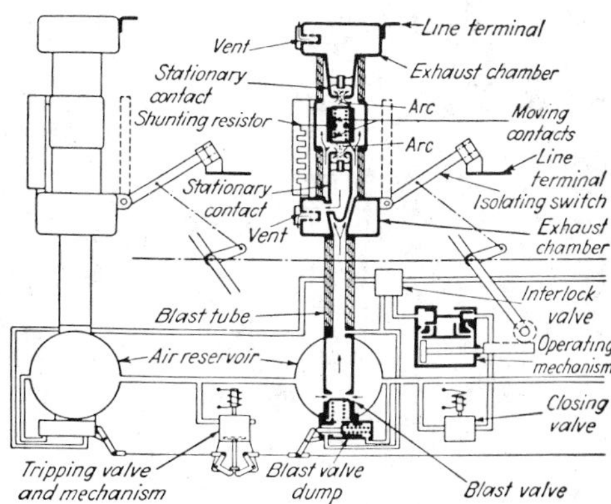

FIG. 10-27. 69-kV air-blast circuit-breaker cross section, showing details of operating and interrupting action. (*Westinghouse Electric Corporation.*)

The 138-kV air-blast circuit breaker was the first fully outdoor development in the United States.

An air-blast breaker by General Electric and an SF_6 unit by Westinghouse are illustrated in Figs. 10-28 and 10-30, respectively. The outdoor circuit breaker was first applied in the United States in 1956. They are now available in voltage and current ratings up to 765 kV and 3,000 A continuous to meet the latest USASI standard requirements. The available ratings are listed in the USASI Standard C37.06, Table 4.

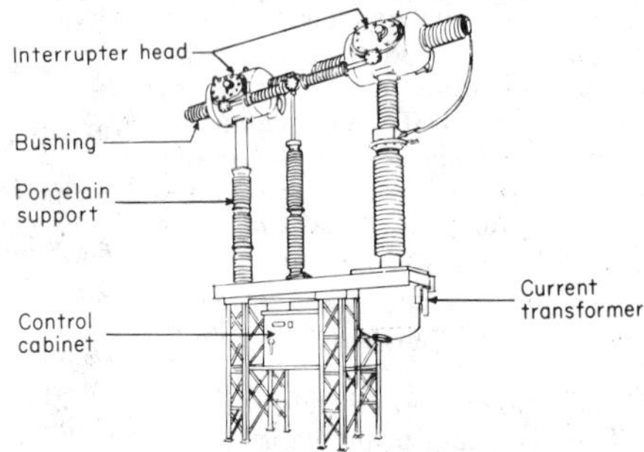

FIG. 10-28. 230-kV air-blast circuit breaker, rated 1,600 A, 20,000 MVA, operating at 500 lb/in² gage. (*General Electric Company.*)

The live-tank construction by the General Electric Company, shown in Fig. 10-28, uses 500 lb/in² gage compressed air in the interrupter heads. The air is blown across the arcs at the two contacts when the breaker is opened, through operation of an air-

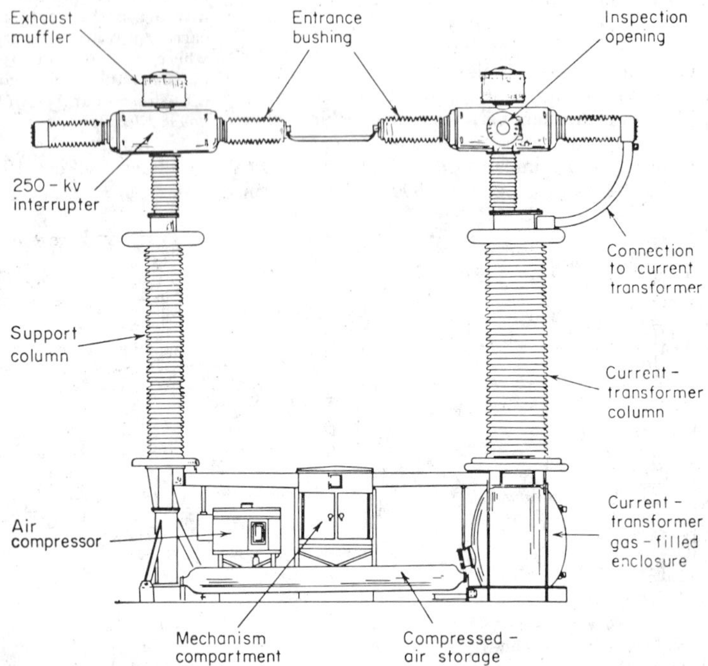

FIG. 10-29. 500-kV 3,000-A 38,000-MVA air-blast circuit breaker. (*General Electric Company.*)

blast valve. The air so used is blown out of the interrupter head to the atmosphere. Current transformers and taps for capacitance-type potential-measuring devices are housed in the tank support columns. The columns are filled with SF_6 gas for insulating purposes. Control rods and compressed-air-supply piping for the interrupter heads are also housed in the porcelain supporting columns. This design has been extended to the extra-high-voltage classes 345 kV, 25,000 MVA: and 500 kV, 38,000 MVA (see Fig. 10-29). The design shown can be extended to 765 kV, 54,000 kVA interrupting rating.

111. SF_6 (sulfur hexafluoride) has been used as an interrupting medium in outdoor circuit breakers for some years by the Westinghouse Electric Corporation. The gas is a very stable compound, has high insulating qualities and good interrupting properties, is inert, nonflammable, and nontoxic, and is odorless. Circuit breakers rated 34.5 to 230 kV, 1,200 to 1,600 A continuous, 500 to 20,000 MVA have been in successful operation for some time. Figure 10-30 shows the 230-kV dead-tank design. It uses

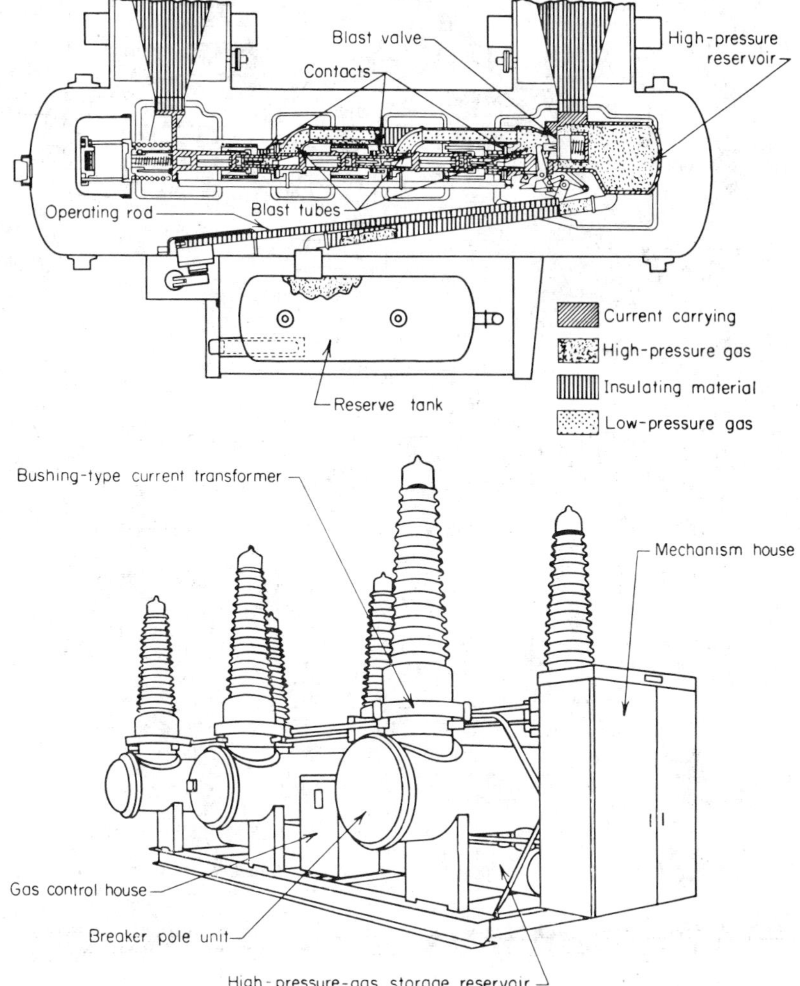

FIG. 10-30. 230-kV SF_6 circuit breaker, rated 1,600 A, 20,000 MVA. (*Westinghouse Electric Corporation.*)

conventional bushing current transformers. Each bushing has a low-capacitance potential device tap.

The SF_6-gas design has also been extended to a live-tank extra-high-voltage circuit breaker. This structure, shown in Fig. 10-31, has been built and operated at 345 kV, 25,000 MVA and 500 kV, 38,000 MVA. It has been designed for 765 kV, 54,000 MVA.

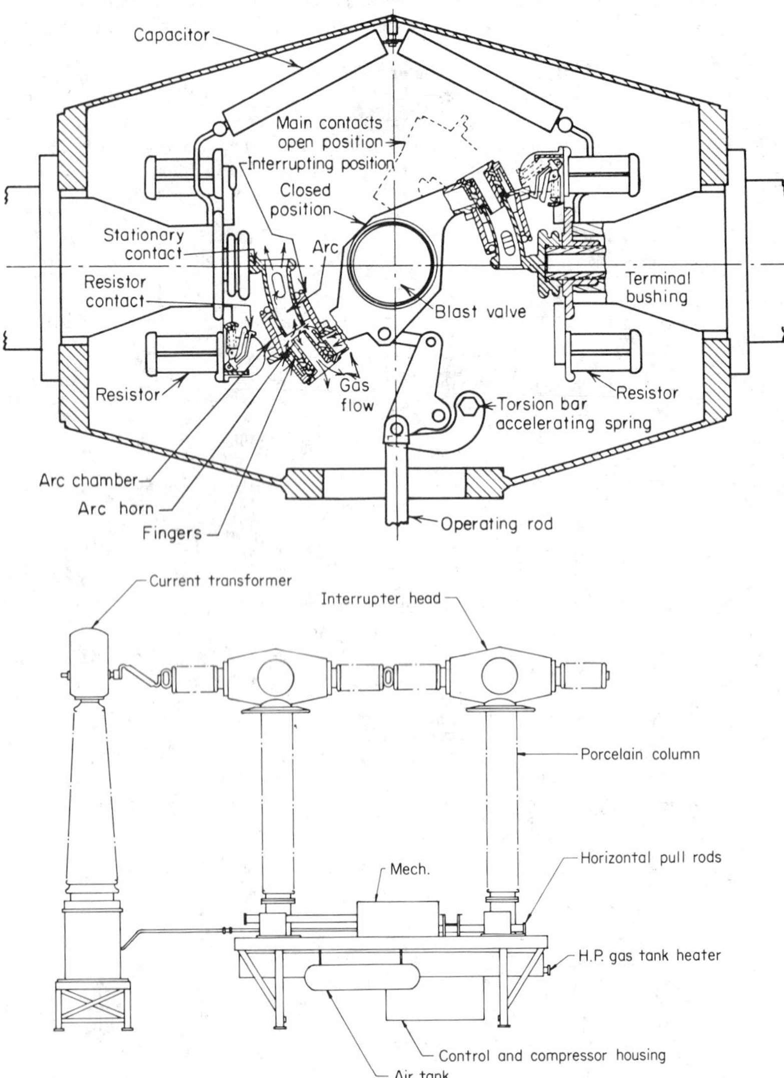

Fig. 10-31. 500-kV 3,000-A 38,000-MVA SF_6 circuit breaker. (*Westinghouse Electric Corporation.*)

112. Preferred ratings of high-voltage circuit breakers at present appear in two sections of the USASI standards. The symmetrical method of rating circuit breakers was adopted by the USASI in 1964. All breakers proved to meet these new standards appear in the USASI C37.06 tables of rating. Those breakers not yet fully proved by all manufacturers in the United States follow the earlier asymmetrical current method of rating, and their rating values appear in USASI C37.6 tables of ratings. These breaker ratings will be transferred to the C37.06 tables as rapidly as the factories can complete the necessary testing to provide proof of capabilities.

FUSES

113. Fuses are the simplest device used for interrupting an electrical circuit under short-circuit, or excessive-overload, current magnitudes. Enclosed fuses are on the market and are approved by the Underwriters' Laboratories Inc., in a wide variety for use up to 250 and 600 V. They may be used in a-c or d-c circuits and have a variation in time-current characteristics that makes them suitable for many special purposes. More important among these is the motor protective fuse, which may be chosen to have a suitable time delay to permit starting inrush current to flow and also the carrying of moderate motor overloads without blowing, and yet be capable of blowing early enough to avoid excessive winding temperature due to overload, and at the same time blow very quickly in case of a short circuit. While a minimum of 10,000 A interrupting capacity is required in low-voltage fuses, some sizes and types today are capable of interrupting up to a 200,000-A a-c circuit or 100,000-A d-c circuit. The standard lines on low-voltage fuses are available in several steps of ampere capacity, each of which is a different physical size, viz., 1 to 30 A, 35 to 60 A, 70 to 100 A, 110 to 400 A, and 450 to 600 A. The current-limiting development in low-voltage fuses extends the above interrupting capacity in sizes from 800 A up to several thousand amperes continuous rating. This is sometimes referred to as the **silversand fuse.**

Low-voltage fuses will carry their rated current continuously, if not heated abnormally at the same time, say by a loose terminal connection, but will blow in 1 to 2 min or less if the current reaches 135% of rating. They must also meet additional performance and other requirements of Underwriters' Laboratories Inc., depending on fuse design.

114. High-voltage fuses are available in a range of capacities of currents up to 720 A, 500 MVA at 14,400 V and 250 A, 2,000 MVA at 138,000 V. They are used for protecting potential transformers, distribution or medium-size power transformers, and, occasionally, branch circuits. In some cases they have been installed in series with protective rod gaps for lightning protection. They are often equipped with contacts so arranged that the fuse and its mounting act as a disconnecting switch. Mechanisms have been used in certain sizes which automatically replace a blown fuse with a new one. When fuses are used to protect transformers or branch circuits in series with other fuses or automatic circuit breakers, care must be exercised to ensure proper time selectivity. Fuse characteristics may be varied somewhat by the choice and disposition of materials, but time and minimum blowing currents are affected by ambients and currents carried immediately previously. Fuse characteristics are usually based on tests starting cold in an 18 to 32°C ambient. Fuse blowing and clearing characteristics should be coordinated throughout the range of short-circuit currents to ensure a proper margin for selec-

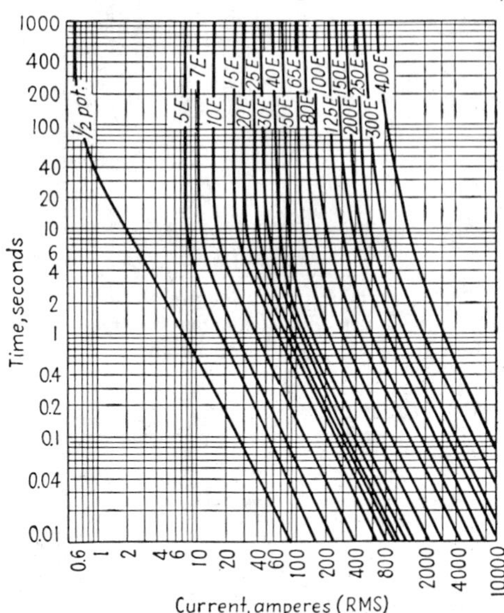

FIG. 10-32. Melting time-current characteristics of type BA deion standard refills for BA-400 fuse.

tivity, as fuse characteristics differ with different types, and the shape of the curves usually differs appreciably from relay characteristics. Figure 10-32 shows typical fuse time-current characteristics.

115. Several types of this fuse are available. (Special attention should be given to 25 c and d-c application.)

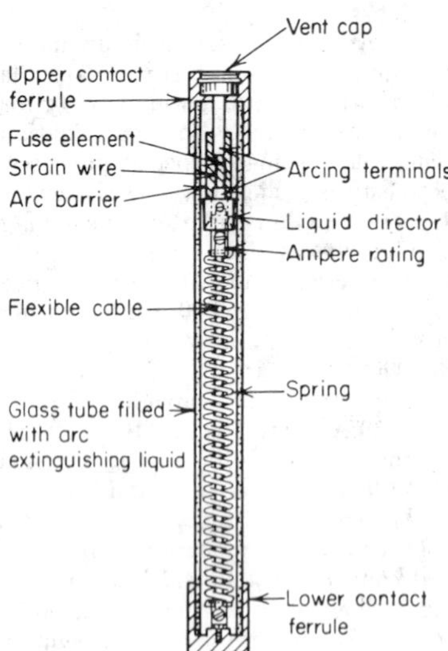

FIG. 10-33. S & C Electric Company type B chemical fuse.

1. Fuse mounted on a porcelain plug set in a weatherproof block or enclosed in a fiber base with a porcelain housing. This type is commonly used on distribution circuits and station apparatus for voltages up to 7,500 and in some cases up to 15,000. They should not be used where the interrupting duty is high.

2. Fuse enclosed in a metal housing filled with Transil oil. This type is used on high current capacities for voltages up to 7,500 for protecting small transformer banks and other station or distribution purposes.

3. Fuse enclosed in a high-strength glass tube filled with special arc-quenching liquid. This type is used on many classes of service. A type of liquid is used which contains carbon tetrachloride and does not freeze or congeal (see Fig. 10-33).

4. An expulsion fuse containing boric acid which generates gas under the arc of the blowing of the fuse wire, aiding in promptly deionizing the arc and opening the circuit (see Figs. 10-34 and 10-35). On the basis of a 3-phase 60-c circuit the interrupting ratings are those of Table 10-2*a*.

Table 10-2a. Interrupting Rating of Boric Acid Fuses*

KV	3-phase Rating, MVA
4.16	250
7.2	325
14.4	500
23	750
34.5	1,000
69	2,000
115	2,000
138	2,000

* Interrupting ratings are based on the circuit capability in symmetrical amperes without effect of the fuse.

5. Current-limiting fuses are those which open with such speed that the fuse arc resistance is introduced during the first half cycle of the short-circuit current and the actual current is reduced to much less than that which would flow were the fuse con-

Table 10-2b. Interrupting Rating of Current-limiting Fuses, Type EJO

Rated volts	Continuous amperes	3-phase interrupting rating, MVA
2,500	1–200	150
5,000	1–200	250
7,500	1–100	500
14,400	1–7	1,500
14,400	10–100	500
23,000	1–7	1,500
23,000	10–100	500
34,500	1	1,500
34,500	10–80	500

ductor intact. In general, these fuses should not be applied to circuits the voltages of which are less than 70% of the fuse-voltage ratings to avoid over-voltages. Typical of these fuses are the General Electric type EJO (see Table 10-2*b*).

116. Limiters are time-delay fusible connectors designed to be installed in low-voltage network mains cables at street-junction points. They are rated in cable sizes and have time-current characteristics to (*a*) allow the cable fault to burn itself clear if it does so promptly, without blowing the limiter; (*b*) blow before the cable insulation away from the fault is roasted and prevent the failure from spreading beyond the junction point; and (*c*) obtain adequate selectivity so that, when installed in a network, only those limiters connected to the faulted cable blow (see Fig. 10-36 for limiter characteristics).

117. The fusible element of a limiter in its simplest form consists of a copper tube flattened in its central portion and reduced in cross section to obtain the proper time-current characteristics. The cable ends are inserted in the tubular portions and connections made by indenting the tubes with a hydraulic press. The fusible element is enclosed in a properly designed asbestos-cement shell, and the whole insulated and made watertight by a rubber sleeve, tape, and cement. Several limiter fusible elements may be joined together to form a junction point or several cables, and the whole enclosed in a shield and insulated to form a limiter "crab joint."

118. The usual applications of limiters are at the street-corner junctions of low-voltage network mains, multiple-

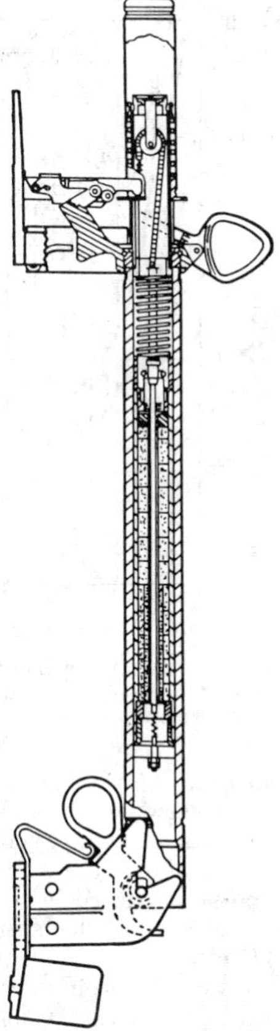

FIG. 10-35. Construction of high-voltage boric acid fuse with target. (*S & C Electric Company.*)

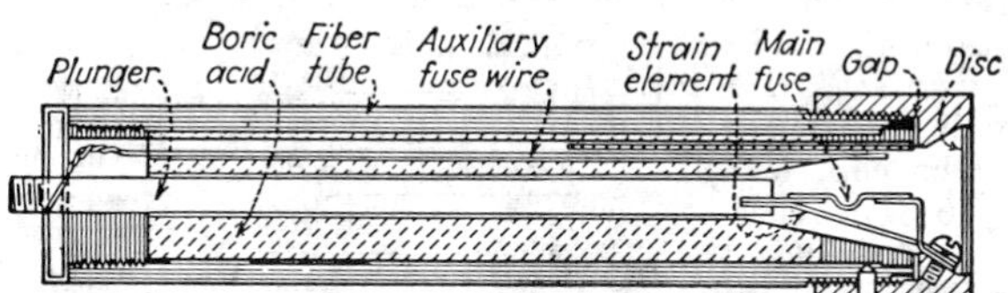

FIG. 10-34. Cross-section view of BA deion power-fuse refill, showing construction details. (*Westinghouse Electric Corporation.*)

transformer secondary cables, multiple-service cables, and intervault ties of building interior networks.

DISCONNECTING SWITCHES

119. Disconnecting switches are used primarily for the purpose of isolating equipment from buses or live apparatus and for sectionalizing buses or circuits, also for transfer, testing, and grounding purposes. They are not intended to break load current except under special conditions and in limited amounts unless designed for this type of duty. High-voltage outdoor disconnects are sometimes equipped with horn gaps but should not be used to break heavy charging currents until the conditions have been carefully analyzed and known to be satisfactory. If mounted vertically, they should be installed so that they open downward. Except for very light duty, disconnects should be equipped with latches or other mechanical mechanisms to prevent their being blown open by the magnetic forces due to short-circuit currents.

120. Designs of disconnecting switches show considerable progress in development of improved types of contact to meet the demands of both indoor and outdoor service at higher voltages and load or fault interruption. Contacts with high-pressure features or provided with secondary motions for producing effective pressures have improved the reliability of operation and increased the serviceability. In most of the modern designs the construction provides for only line or multipoint contact. At high pressures the surfaces are cleaned of dirt, oxides, sleet, etc. Modern designs use silver-plated or silver-inlaid contact surfaces quite generally, as this material is not subject to increased contact resistance and heating due to oxidation.

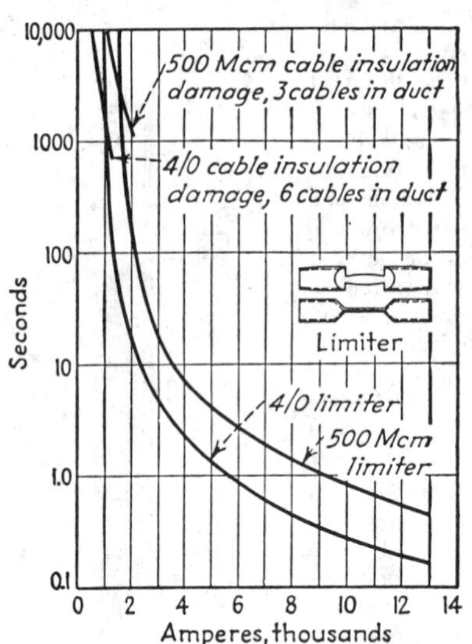

FIG. 10-36. Time-current characteristics of low-voltage limiters for No. 4/0 and 500-Mcmil network mains cables.

Switches are available in many different mechanical designs. Some have been built to provide "snap-open" interrupting contacts which operate after the main current-carrying contacts are open. This provides some measure of small interrupting ability which is useful in some instances. Another design of this general type, but providing for the final interruption of the circuit in a specially suited and enclosed interrupting chamber, has made its appearance. This unit, known as an "interrupter switch," provides a material step in higher interrupting ability (see Par. **128**).

121. A high-pressure contact design for indoor service is illustrated in Fig. 10-37. This type has been adapted for switches rated at 2,500 to 34,500 V.

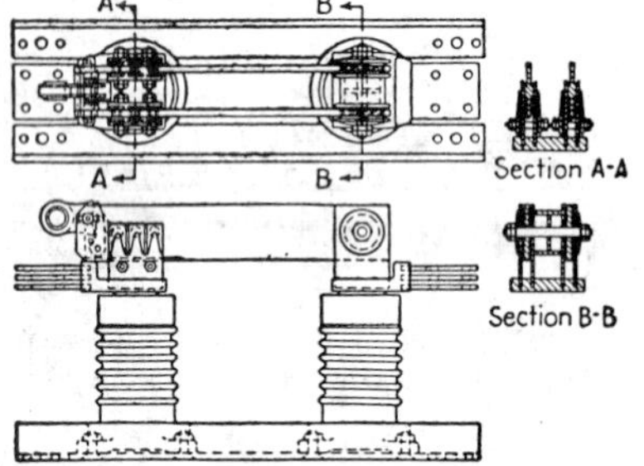

FIG. 10-37. High-pressure contact disconnecting switch, indoor type, 7,500 V, 600 A. (*I-T-E Circuit Breaker Company.*)

The switch illustrated is of 600 A capacity and has 12 contact spots formed in the blades raised above the surrounding surface. The spots are curved in section outline and make contact with the surface of the clip, with effective areas of 0.004 to 0.008 in^2 each. Pressure is supplied by springs independent of the current-carrying parts, with resultant pressures of 3,000 to 9,000 lb/in^2. This pressure is sufficient to give the contacts a self-cleaning action. The current density at the contacts is of the order of 30,000 A/in^2.

The hinge contact surfaces are annular, concentric with the pivot, and ground in with the flat clip surface. The resultant limited area is in constant full contact, regardless of the position of the blade, and is designed to exclude dirt and even atmosphere. This exclusion assures bright hinge contacts with a minimum of deterioration. Opening and closing of the switch are ordinary, except that the engagement stroke is very short; when used in conjunction with a lock provided with a "pry-out," this results in easy operation.

122. In the pressure-contact type several types are available in which the switch is so arranged that, after closing, the blades and stationary tongues, both jaw and hinge end, are automatically bolted together and locked against magnetic expulsion forces. Sufficient pressure is retained in the contacts to wipe them clean during closing.

123. A high-pressure contact design for outdoor service (Figs. 10-38 and 10-39) embraces the feature of the switch blade revolving inside instead of moving parallel with the contact surfaces. The blade enters the contact and is then turned. In opening the circuit the blade is first turned, the pressure thus being relieved before the blade lifts or moves sidewise. The rotation of the blade is accomplished by a yoke connected to a rotating or oscillating insulator which produces a continuous, quick motion with little effort. The total pressure on the contacts varies with the current rating of the switch, the design providing ½ lb of pressure/A of capacity, a 600-A switch having a total pressure of 300 lb and a 2,000-A switch a total contact pressure of 1,000 lb. Switches of this type have been designed for circuits at 7,500 to 765,000 V.

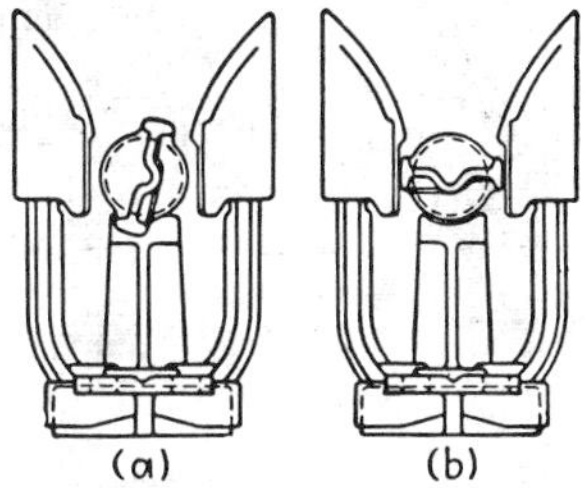

(a) (b)

Fig. 10-38. High-pressure contacts for outdoor disconnects, 34.5 kV, 400 A. (*a*) Switch blade in closed position but not in contact (before rotation); (*b*) switch blade closed and in contact (after rotation.) (*I-T-E Circuit Breaker Company.*)

124. A 550-kV disconnect to fit the highest voltage system in the United States, at this writing, is shown in Fig. 10-40. Good dielectric shielding to reduce radio interference is necessary, as shown. A similar design is planned for 765 kV in the United States in 1969.

125. Manual- and motor-operating mechanisms are available which permit flexibility in the arrangement of disconnecting switches to suit the requirements of the electrical and structural layout. Most low-voltage indoor and outdoor disconnecting switches of moderate current capacity are of the single-pole hook-operated type, while a large portion of the higher-voltage switches are of the gang-operated manual type. Gang-operated switches are often used on moderate-voltage indoor-station circuits, as the mechanism can be readily interlocked mechanically or by key sequence with the oil circuit breakers. Heavy-duty and high-voltage disconnects are usually motor-operated.

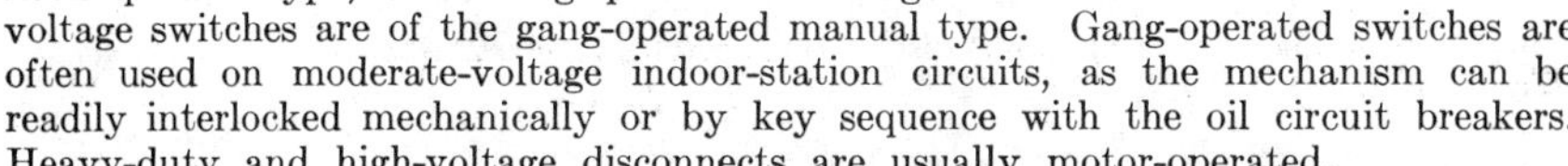

126. The mounting of the disconnects should be rigid, and care must be exercised in erection to ensure proper alignment of blade and clips in operation without undue stresses on the insulators or distortion of contact surfaces.

127. The current rating of a disconnect switch is based upon a temperature rise of not more than 30°C above a 40°C ambient. If the switch is likely to be subject to heavy short-circuit currents, it may be necessary to install a switch of larger capacity than required by normal load, as the rms current of the maximum half cycle of short circuit should not exceed the momentary current rating of the disconnection switch, which is usually from 30 to 67 times its normal rating, depending upon type and size.

128. Interrupter switches combine the functions of air disconnecting switches and a circuit interrupter. Some use interrupters which are limited to the continuous current or less, while others use special interrupting devices which can switch moderate values of short-circuit current.

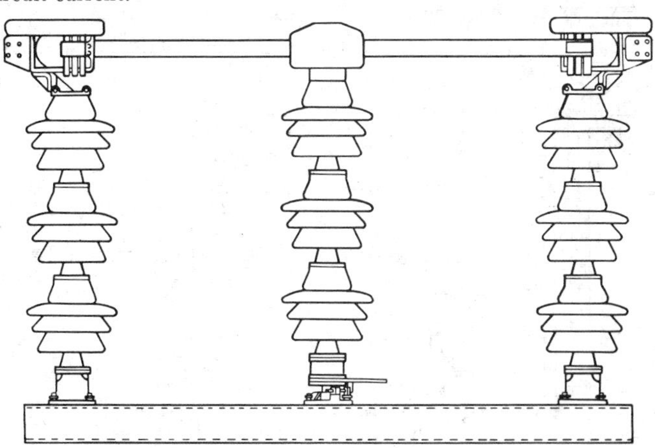

Fig. 10-39. Horizontal double-break high-pressure switch for 1,200-A 115-kV service. (*I-T-E Circuit Breaker Company.*)

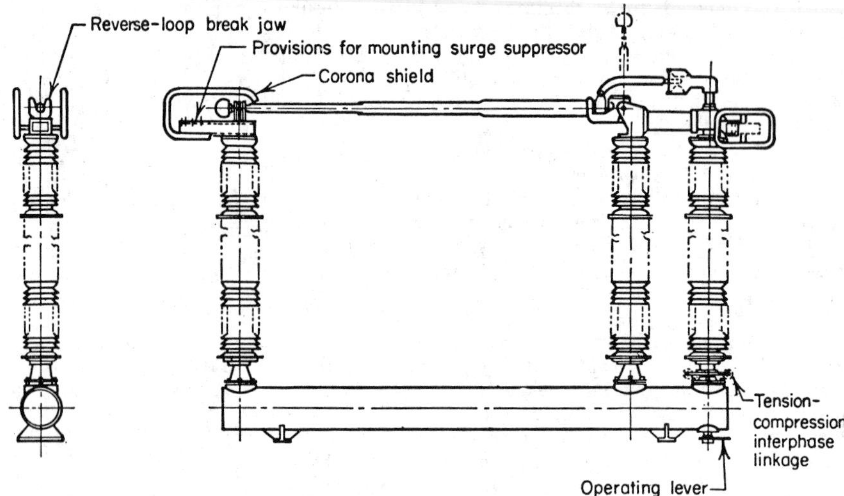

Fig. 10-40. Single vertical-break disconnect switch, 2,000 A, 550 kV. (*Westinghouse Electric Corporation.*)

129. The load-current interrupters may (*a*) insert a resistor in the circuit, following opening of the main switch contact, and interrupt the current drawn between arcing horns in the air, (*b*) use a blast of air or other gas from a compressed-gas cylinder to blow against and to lengthen the open arc after the main switch contacts are open, (*c*) interrupt the circuit by an auxiliary interrupter within a chamber of SF_6 gas after the main switch contacts are open (see Fig. 10-41), or (*d*) first interrupt the circuit in an enclosed SF_6 circuit interrupter and then isolate it by opening of the main switch contacts, after which the circuit switcher recloses. Such interrupters can handle sizable short-circuit currents and can be applied at any voltage by adding interrupters in series as the voltage of the system is increased.

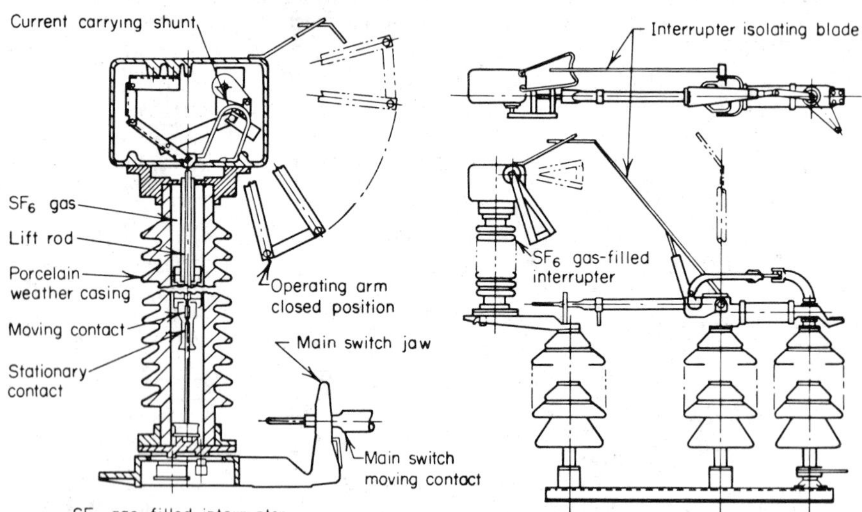

Fig. 10-41. SF_6 interrupter with single vertical-break disconnect switch. (*Westinghouse Electric Corporation.*)

METAL-CLAD AND METAL-ENCLOSED SWITCHGEAR

130. Metal-clad or metal-enclosed switchgear is a type of design in which the electrical equipment is located in an enclosing metal structure, generally factory-assembled.

131. The term "metal-clad switchgear" indicates a type of design in which all the equipment required to control an individual circuit, including bus, circuit breaker, disconnecting devices, current and potential transformers, controls, instruments, and relays, is assembled in one metal cubicle and the circuit breaker is provided with means for ready removal from the cubicle. Circuit breakers can be of the oil or air type, although the trend is strongly to the use of air circuit breakers.

132. Circuit-breaker disconnection is accomplished by "vertical-lift" and "horizontal-drawout" designs as illustrated in Figs. 10-20 and 10-22, respectively. The overall effect is the same in either case.

133. Interlocks are provided in metal-clad assemblies to prevent disconnecting or connecting the circuit breaker if it is closed and to protect personnel from coming in contact with the high-voltage circuits when the circuit breaker is removed from the cubicle.

134. Metal-clad switchgear is used for low- and medium-capacity circuits, for indoor and outdoor installations at 13.8 kV and lower nominal voltages. Ratings range from 4.16 to 13.8 kV and from 75,000 to 1,000,000 kVA interrupting capacity.

135. The term "metal-enclosed switchgear" indicates a type of design in which the major component parts of a circuit, such as buses, circuit breakers, disconnecting switches, and current and potential transformers, are in separate metal housings and the circuit breakers are of the stationary type in the design.

136. Phase segregation in metal-enclosed switchgear is a type of design in which a 3-phase metal housing is divided into three single-phase compartments by means of single metal barriers, as shown in Fig. 10-42.

137. Isolated-phase metal-enclosed switchgear is a type of design in which each phase is enclosed in a separate metal housing and an air space is provided between the housings. This type of gear is illustrated in Figs. 10-26 and 10-43. It is considered to be the most practical and economical way of preventing phase-to-phase short circuits through construction methods, in large and important plants.

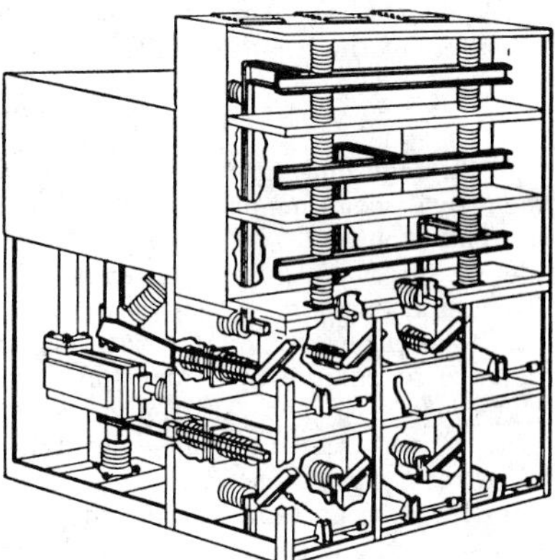

FIG. 10-42. Metal-enclosed segregated-phase construction showing bus, air-blast circuit breaker, and disconnecting switches for 14.4-kV service. (*General Electric Company.*)

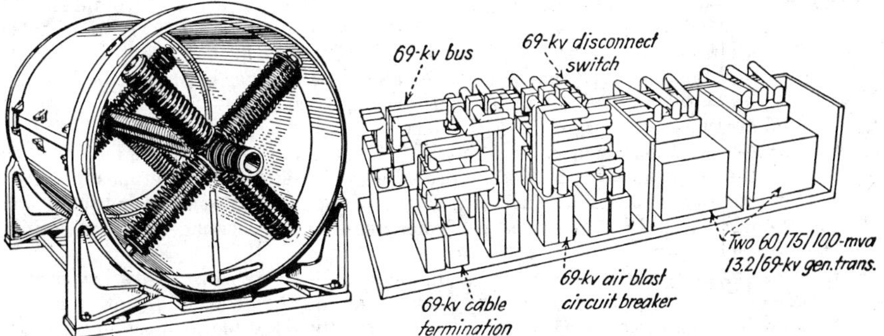

FIG. 10-43. Metal-enclosed, isolated-phase construction showing bus, air-blast circuit breakers, disconnect switches, feeder-cable terminations, and generator transformers for 69-kV service. The cross section of one phase is shown at the left.

138. Metal-enclosed switchgear is used in industrial, commercial, and utility installations, generally on systems that would use switchgear rated from 14.4 to 138 kV, having interrupting duties varying from 1,000,000 to 2,500,000 kVA at 14.4 kV and up to 10,000,000 kVA at 138 kV.

REACTORS

139. Reactors are of two distinct types: (*a*) shunt and (*b*) series, or current-limiting. Additional data on reactors will be found in Secs. **11** and **13** of this handbook.

140. Shunt reactors are connected line-to-line or line-to-ground and are used to neutralize charging current and prevent a voltage rise when the charging current is fed through reactance. A common application is on cables at 22,000 V or above, supplying low-voltage a-c network transformers, the high-voltage windings of which are delta-connected. Under such conditions, if the supply breaker trips and the feeder is left alive by back feed through one or a few network transformers, an objectionable voltage rise will occur on the network as a result of the charging current fed through its high reactance. One or more shunt reactors are installed on the high-voltage cable to reduce the charging current and the voltage rise on the network to acceptable limits. Reactors of this type are usually oil-immersed, built with transformer-type windings on an iron core with air gaps, and installed in transformer-type tanks.

141. Current-limiting series reactors are installed in feeders and ties, in generator leads, and between bus sections to reduce the magnitude of short-circuit currents and the effect of the resulting voltage disturbance.

LIGHTNING ARRESTERS

142. Lightning arresters are protective devices for limiting surge voltages on equipment by discharging or bypassing surge current. They prevent continued flow of follow current to ground and are capable of repeating these functions. They serve as completely automatic surge diverters which function to discharge line overvoltages resulting from lightning strokes, switching surges, or other system disturbances which would otherwise flash over or puncture insulation, perhaps causing line outages and possible damage to equipment. Arresters are designed to chop transient voltages, discharge transient energy, prevent dangerous voltage reflections, and reseal the diversion path of the power-frequency follow (or dynamic) current within a fraction of a cycle.

Arresters have one or more sets of spark gaps which establish the spark-over voltage, aid (by current limiting) in interrupting the follow current, and also prevent any flow of high current under normal conditions. The gaps are connected in series with either nonlinear-resistance valve elements or with an additional arc-extinguishing chamber designed to assist in the interruption of the system frequency current. Arresters have a spark-over response much faster than the breakdown of insulation of protected apparatus, the response voltage being nearly independent of the steepness of the wavefront.

143. Ground connections of arresters should have low resistance and low surge impedance. To improve the effect of the ground connection, the arrester ground should also be tied to the station ground system which connects the frames of the station apparatus. Counterpoises may be required to obtain low resistance in the ground connection.

Where operation discharge counters are installed, the arrester base must be isolated from the normally grounded mounting surface. The counter is connected between the isolated arrester base and ground.

144. Line connections of arresters should be on the incoming side and as near as practical to the principal apparatus to be protected. Arresters are sometimes mounted directly on the tanks of transformers which they protect. They are also mounted inside the metal-enclosed cubicles of switchgear when required for switchgear and rotating-machine protection.

Where arresters are located close to equipment, such as transformers, a low insulation level can be protected. Inherent protection will also be provided for other equipment at greater distances from the arrester. Normally these pieces of equipment require a higher insulation level. Aside from a direct lightning stroke near the remote equipment, protection will be afforded and can be determined (see USASI Standard C62.1, Guide on Arrester Application, in preparation).

145. The voltage rating of an arrester is the designated maximum permissible rms power-frequency voltage which should be applied between its line and ground terminals.

146. Selection of Arrester for Normal Voltage Applications. When an arrester is connected between phase and earth on a 3-phase system, the rated voltage of the arrester must equal or exceed the maximum power-frequency voltage which can be applied to it under normal or abnormal conditions of operation, including fault conditions, if risk of damage to the arrester is to be avoided. While the possible phase-to-earth voltage may vary for a variety of reasons (system faults, regulation, overspeeding of prime movers, resonance, etc.), it has been found generally satisfactory for arrester application to consider only the voltage produced between unfaulted phases and earth as a result of system faults. Experience shows that the majority of systems can be classified into a limited number of groups, depending upon neutral earthing conditions, and that arrester voltage ratings can be determined for these groups of systems without further calculation as follows:

a. Three-phase System with Effectively Earthed Neutral. (See IEEE Standard 32.) If the system neutral is effectively earthed, the voltage between any phase and earth under fault conditions will not exceed 80% of the highest phase-to-phase system voltage. On such systems, the ratio of zero sequence reactance to positive sequence reactance (X_0/X_1) will be between 0 and +3, and the ratio of zero sequence resistance to positive sequence reactance (R_0/X_1) will lie between 0 and +1. The arrester should have a voltage rating equal to at least 80% of the highest phase-to-phase system voltage.

1. On certain low-voltage distribution systems of the 4-wire type where the transformer neutrals and neutral conductor are directly earthed at frequent points along the circuit, the voltage between sound phases and earth under fault conditions may be limited to less than 80% of the highest phase-to-phase system voltage. Under those conditions the arrester voltage rating may be correspondingly less than 80% of system voltage.

2. On extra-high-voltage transmission systems where all transformer banks have directly earthed neutrals, the voltage between sound phases and earth under fault conditions usually will not exceed 75% of the highest phase-to-phase system voltage. Under these conditions the arrester voltage rating may be 75% of the system voltage.

b. Three-phase System with Noneffectively Earthed Neutral. If the system neutral is noneffectively earthed, the voltage between unfaulted phases and earth under fault conditions may exceed 80% and in some instances may even exceed 100% of the highest phase-to-phase system voltage. On such systems the impedance ratios X_0/X_1 and R_0/X_1 will be positive, but with higher values than in (*a*). This category describes systems whose neutrals are earthed through resistance or reactance, including the ground-fault neutralizer suppression coil. The arrester voltage rating should equal or exceed the highest phase-to-phase system voltage.

c. Three-phase System with Isolated Neutral. If the system neutral is isolated from earth, the voltage between unfaulted phase and earth under fault conditions may be 100% of the highest phase-to-phase system voltage or in some instances may exceed 100% of system voltage. On such systems, the zero sequence reactance (X_0) is capacitive, and the reactance ratio (X_0/X_1) is negative. If the reactance ratio X_0/X_1 lies between -40 and $-\infty$, the arrester rating should equal or exceed the highest phase-to-phase system voltage.

If the reactance ratio X_0/X_1 lies between 0 and -40, resonance conditions may exist and no general rule for arrester voltage ratings is practicable.

147. Selection of Arrester for Special Voltage Applications. *a. Abnormal System Voltages.* The selection of the arrester voltage ratings in Par. **146** is based on the assumption that in service the highest system voltage is exceeded only under very abnormal operating conditions and that the probability of an arrester operation coinciding with a voltage exceeding the highest system voltage is very small. If, owing to special circumstances, abnormal system voltages are likely to be a frequent occurrence, and if arrester operations are likely to take place during such conditions, then it may be necessary to use an arrester with a voltage rating slightly higher than recommended in Par. **146,** depending upon the particular circumstances.

b. High Earth Resistance or Excessive Separation. If the arrester earth resistance is high, or if the connections between the arrester and the protected apparatus are of excessive length, the impulse voltages which appear at the protected apparatus may

Table 10-3. Typical Station Arrester Characteristics*

Arrester rating, kV	Line-voltage class		Crest kV sparkover USASI impulse test	Crest *IR* discharge voltage across arrester, 8 × 20 μs current wave, kV	
	Grounded neutral, kV	Ungrounded neutral, kV		5,000 A	20,000 A
144	161	138	437	350	440

* 1966 industry (NEMA) values giving the maximum of the maximums for all makes of arresters.

be substantially higher than those across the terminals of the arrester. In order to obtain the desired degrees of protection for the apparatus, it may be necessary to improve these conditions, to select an arrester with lower protective characteristics, or to use an arrester with a lower voltage rating than indicated in Par. **146.** The latter procedure will, of course, result in a risk of failure of the arrester, if it is required to operate when the power-frequency voltage across its terminals exceeds the arrester voltage rating.

c. Low Apparatus Insulation Strength. If an arrester of lower voltage rating than indicated in Par. **146** is selected in order to obtain satisfactory protection for apparatus with abnormally low insulation strength, a similar risk of arrester failure is incurred (see USASI C62.1, Appendix, Standard Guide on Arrester Application, in preparation).

148. Insulation Coordination. As discussed in Pars. **146** and **147,** the maximum power-frequency voltage which is applied to an arrester should not exceed the voltage rating of the arrester. Furthermore, the maximum dynamic voltage possible at the protective device must be below the rating of the arrester. The surge-voltage sparkover and discharge characteristics of the protective devices, under the worst possible conditions (use maximum arrester characteristic values), determine the protective level provided on lightning surges (see Table 10-3 and Fig. 10-44). At a suitable margin above the protective level is the basic insulation level (BIL) of the apparatus, insulators, etc., which is the $1\frac{1}{2} \times 40$ μs crest impulse voltage that the apparatus or line insulators withstand.

In modern arresters, the impulse protective level is so low that consideration must be given to protection for switching surges which may dictate the margin of protection afforded the equipment. However, the risk involved in the installation of suitable arresters can be kept low by knowledge of switching surge voltages in the circuit, as may be obtained with the aid of system analyzers. At extra-high voltages significant

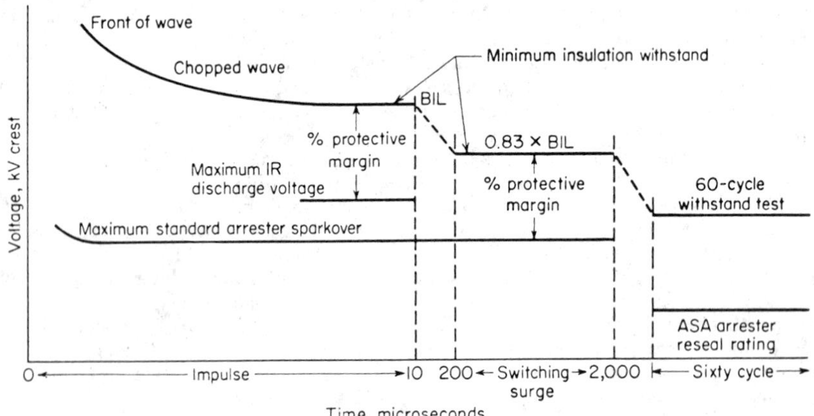

FIG. 10-44. Equipment withstand and arrester voltage characteristics.

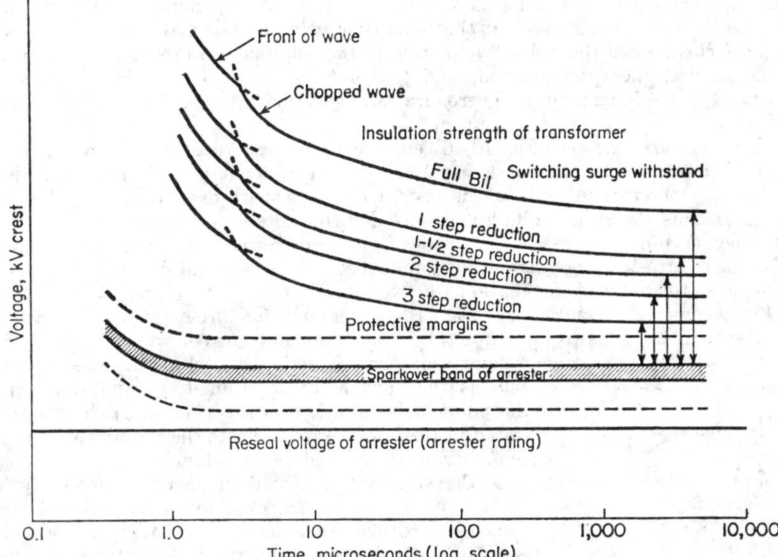

FIG. 10-45. Equipment insulation levels and lightning-arrester protection.

savings in equipment costs may be realized with lowered insulation level and proper lightning-arrester protection (see Fig. 10-45).

As general rules, the arrester spark-over characteristic should be at least 18% below the insulation class rating at 1½ μs (BIL) and at least 15% below the equipment in-

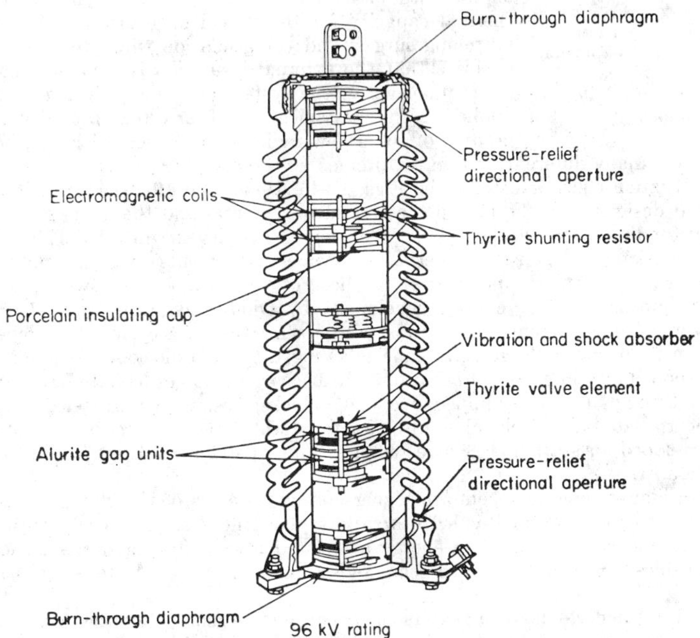

FIG. 10-46. Station-type arrester with nonlinear valve elements. (*General Electric Company.*)

sulation strength for switching surges, using the equipment impulse-withstand capability as the base in each case, at the same time points. The margins selected depend upon the closeness of the protective devices to the equipment protected, the importance of the service, the cost involved, and the probable frequency of dangerous surges.

149. Types of Arresters. There are two general types of high-voltage arrester designs.

Valve-type arresters (see Fig. 10-46) employ a nonlinear resistance known as a valve element in series with the spark gap. This valve element permits practically unrestricted flow of surge currents at the instant the gaps spark over from a voltage surge, but as soon as the surge voltage has subsided, it acutely curtails the flow of power-frequency current and assists the series gap in performing its interrupting action. This type of arrester is not significantly influenced by the amount of short-circuit power available at the point of installation.

Expulsion-type arresters (see Fig. 10-47) employ an arc-extinguishing chamber in series with the gaps to interrupt the power-frequency current which flows after the gaps have been sparked over. Generally, the power-frequency current is that available for a line-to-ground short circuit at the point of installation on the system; so these arresters are rated in rms amperes to indicate their ability to interrupt within stated low and high current limits.

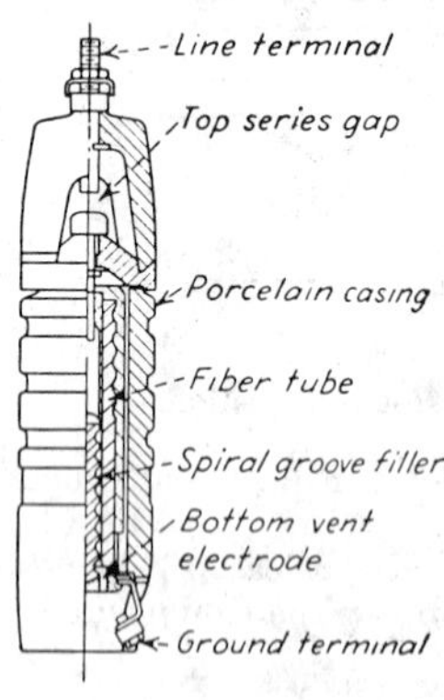

FIG. 10-47. Distribution-type LX expulsion lightning arrester. (*Westinghouse Electric Corporation.*)

150. Classification of Arresters. Arresters are designated as station class (3 to 678 kV), intermediate (line) class (3 to 120 kV), distribution class (3 to 37 kV), and secondary class (175 to 650 V), based on the ability of the arrester to meet the test specifications prescribed in national or international standards. These tests include (but are not restricted to): (*a*) operating-duty-cycle tests to demonstrate the ability of the arrester to interrupt power follow current repeatedly; (*b*) discharge-current withstand tests, both high-current short-duration and low-current long-duration, to demonstrate the adequacy of the electrical, mechanical, and thermal capacity for short- and long-duration lightning and switching surges; and (*c*) insulation withstand tests to demonstrate that the external insulation is equal to that of the minimum level of the system on which the arrester is to be applied. This classification of arresters provides for mass production of a few types of arresters with characteristics which meet many different system requirements.

151. Station-class arresters (see Figs. 10-46, 10-48, and 10-49) represent the most advanced designs with the best protective characteristics and the greatest capabilities for withstanding severe service. The precise spark-over levels which have been achieved have a consistent response to overvoltage waves with fronts ranging from less than 1 μs to crest (lightning-type surges) to those of 200 to 2,000 μs (switching surges). Current-limiting gaps have been developed (by Ohio Brass Company, 1957) which force power follow current interruption within a fraction of a cycle and independently of system frequency voltage zero, thus relieving the valve blocks of a major share of their responsibility in current limitation. If discharge voltages have been substantially reduced through the lower-impedance valve blocks, then much greater durability has been incorporated into their inherent characteristics. Trade names for current-limiting gaps developed for station class arresters are: Alugard (General Electric) and Dynagap (Ohio Brass).

Extra-high-voltage arresters are higher-voltage versions of the station-class design. Specifications for extra-high-voltage arresters place more emphasis on arrester capabilities in respect to discharging energy associated with switching surges on long lines and on cables. Special tests have been devised to confirm the arrester's thermal capacities.

152. Intermediate-class arresters have protective characteristics and operating features which are defined by standards at a level substantially below station grade. Since intermediate arresters are normally unavailable at voltages above 121 kV, they are rarely exposed to damaging switching surges and consequently their protective capabilities are designed principally for the discharge of lightning-type overvoltages. They are often used in protected substations, usually in rural, industrial, or feeder locations, and have recently been applied at underground terminations. The Ohio Brass design has a current-limiting gap. Other intermediate arresters use magnetic

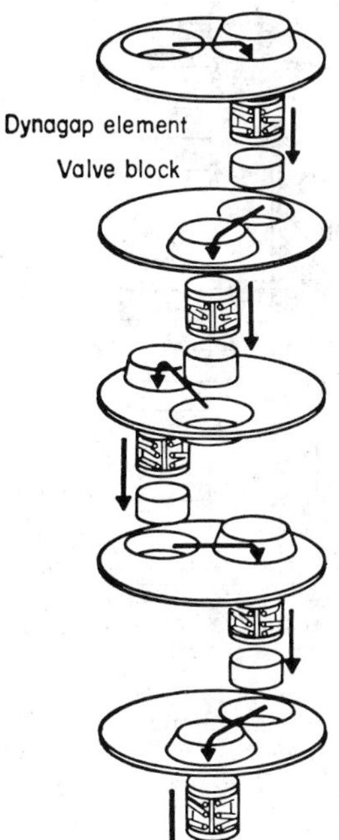

FIG. 10-48. Arrangement of valve blocks in a station-type arrester. (*Ohio Brass Company.*)

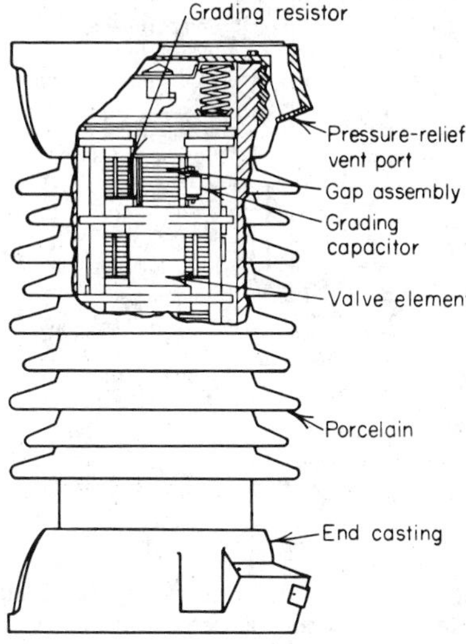

FIG. 10-49. Folded-pole station-type arrester. (*Westinghouse Electric Corporation.*)

or conventional gaps which depend largely on valve block action to cause interruption of power follow current at system frequency.

153. Distribution-class arresters (see Figs. 10-50 and 10-51) are lighter and smaller versions of the intermediate-class arresters. Their voltage range is from 3 to 37 kV, with most installations in the 9- to 12-kV ratings. Their application encompasses the broad field of lightning protection for individual distribution transformers and line protection in place of an overhead ground wire. Their location is generally out on the distribution lines, and they are often mounted directly on the distribution transformers they protect. Characteristics of distribution-class arresters are ample to provide front line protection against lightning surges, but they are usually considered expendable in the case of sustained system overvoltages. For this reason distribution arresters are often equipped with external spark gaps or ground lead isolators, which automatically disconnect the arrester from the line whenever an arrester fault exists.

154. Secondary-class arresters are simple protection devices which are designed to discharge lightning-type overvoltages which may be dangerous to electrical equipment permanently connected to secondary distribution lines. A good example of such an application is in connection with submerged water pumps where the secondary arrester is connected between the service line and ground, ahead of the automatic switching device.

155. Sealed Construction of Arresters. Because deterioration of the operating elements of valve-type arresters results from exposure to the atmosphere, manufacturers employ highly developed sealing systems to enclose the interior circuit permanently. Interiors are evacuated and filled with an inert gas before the final sealing operation.

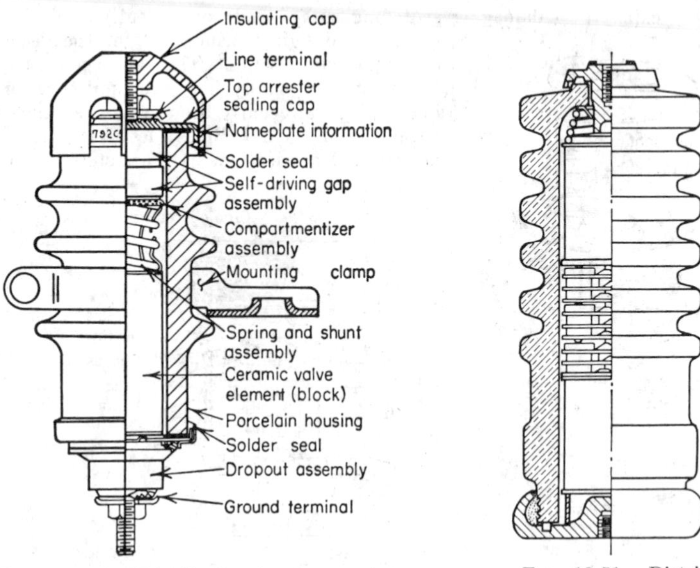

FIG. 10-50. Distribution-type arrester. (*Westinghouse Electric Corporation.*)

FIG. 10-51. Distribution-type arrester. (*Ohio Brass Company.*)

Various systems of pressure relief are incorporated in station- and intermediate-class arresters to provide release of internal gas in the event the arrester is unable to interrupt owing to sustained system overvoltages.

156. Rod and pipe gaps are sometimes used as less expensive protection against lightning surges or as a last line of defense. The breakdown voltage varies with the steepness of the wavefront, and a spark-over is usually followed by a dynamic current resulting in a line outage. Rod gaps are sometimes connected to ground through a fuse which must be replaced after operation.

157. Surge-absorbing capacitors are sometimes installed with or without arresters to modify the shape of the surge voltage wave, reducing the steepness of its front.

158. Shielding in the form of overhead ground wires and towers is an effective protection against lightning and is sometimes installed to prevent direct strokes on the station equipment where the most complete type of protection is required. The ground wires are carried out over the lines to about 2,500 ft from the station where arresters are installed (unless the complete line is shielded by ground wires).

159. Application of lightning-protection equipment is an insurance against outages and equipment failure. The amount and kind of protection needed are influenced by the frequency and severity of storms in the area, the value of reducing line outages due to lightning, and the risk of failure of equipment. Generally the insulation levels of the line in the vicinity of a station and of the station equipment should be so coordinated that the equipment insulation will not be punctured and transformer and circuit breaker bushings will not flashover. Indoor stations connected to underground systems are generally not provided with arresters. However, suitable lightning protection is important at all junction points between underground and overhead systems, to reduce to safe levels lightning or other voltage impulses entering the underground system (see USASI Standard Appendix, C62.1, Arrester Application, in preparation).

160. Rotating apparatus and dry-insulated equipment, because of lower impulse strength as compared with oil-filled transformers, may require special protection, such as low spark-over arresters and surge-absorbing capacitors, when exposed to lightning. Series-reactor coils and current transformers may also require special lightning arresters across the terminals if subject to surges.

TRANSFORMERS

161. Power transformers are essential to the proper coordination and interconnection of the different circuits of any power-system design. Since Sec. **11** covers the features of transformers in detail, no specific discussion of them is included here. They are used for voltage and current transformations; voltage control; phase shifting for load control or to permit correction of system phase relations; phase transformations, e.g., 3-phase to 6-phase; etc.

In loading transformers attention must be given to many things. The transformer type and cooling features and the system load cycles must be carefully coordinated. As a guide in transformer application, the USASI has issued its Guide for Loading Oil-immersed Distribution and Power Transformers, USASI C57.32. It covers the many varied conditions and types of oil-filled and dry-type transformers with which operating groups are concerned.

INDUCTION AND STEP REGULATORS

162. Induction and step voltage regulators have many applications in power systems, particularly in feeder circuits at distribution substations. For details of design and performance, see Sec. **11.**

CONVERSION EQUIPMENT

163. Conversion from alternating current to direct current for power purposes is obtained by three principal means: motor-generators, synchronous converters, and rectifiers. For greater detail than given here, see Sec. **12.**

164. Motor-generator sets are commonly used in small sizes, because of cost and simplicity of operation, and in large sizes where a wide range of voltage control is essential or where service conditions are too severe for successful converter operation.

Motor-generator sets will withstand more severe electrical disturbances without flashover or tripping out than converters, and the d-c voltage is independent of the a-c voltage.

Induction motors are commonly used in smaller motor-generator sizes, while in the larger sizes synchronous motors are more common because of their higher efficiency and ability to improve power factor. Large motor-generator sets with induction-motor drive and equipped with a heavy flywheel have been used in steel mills to reduce the demand on the supply lines during short-time heavy demands, such as reversing heavy motors and loads.

165. Synchronous converters have been commonly used in place of motor-generators wherever practicable, owing to their higher efficiency. Converters have been used almost exclusively for obtaining direct current for 3-wire 250-V Edison service in cities where this is supplied. They were used almost exclusively for d-c railway supply until the time of the commercial development of large-capacity mercury-arc rectifiers.

Edison 3-wire 250-V converters are largely of the "booster" type for voltage control, although earlier machines used an induction regulator between the transformer and the converter. Some shunt-wound (field-controlled) converters with high-reactance transformers have been used in Edison substations where a narrow range of voltage control was adequate. Taps on the transformer are provided and should be set for best average bus voltage.

Converters for industrial application are usually supplied with high-reactance transformers and are compound-wound for flat voltage.

Railway converters are often compound-wound when installed in substations of one or two units with wide spacing between stations. Shunt-wound converters are more frequently used, particularly for supply to city lines with low-circuit resistances, as the drooping voltage characteristic is effective in shifting heavy load swings to other stations and in reducing peak loads on units.

166. Rectifiers have largely replaced converters during recent years for d-c voltages of 600 V and above because of the higher efficiencies, particularly at light loads. They are able to take violent load swings and momentary heavy overloads without damage. While "arc-backs" sometimes occur, quick-acting breakers are used to trip them from the lines, and they may be immediately returned to service, whereas flashover of a converter commutator usually results in serious damage.

The recent development of the single-anode rectifier, such as the **ignitron,** has increased the efficiencies of rectifiers until they are comparable with those of converters at 250 V, although costs are somewhat higher. Three-wire operation must be obtained from balancer sets or some other means of obtaining the neutral. Rectifiers are nonreversible and hence cannot be used for "pump-back" loading and braking. However, the absence of noise and moving parts is advantageous for installations in many applications where noise is objectionable, e.g., in commercial buildings.

FREQUENCY CHANGERS

167. Frequency changers are motor-generator sets consisting of two a-c units mechanically coupled together, each of which has the proper number of poles to run at the same speed as the other end and deliver or receive power to and from another system. They are used to obtain power of a desired frequency when power of another frequency is available, e.g., 25-c power for a-c railway electrification from a 60-c system.

They are also used in a reversible manner as ties between two systems of different frequencies, as between a 60- and a 25-c system, thereby making the reserve capacity of each system available to the other and permitting greater economy of operation by favorable division of load among generating units of both systems. Their advantages as ties are increased when the peak load of one system occurs at a different time from that of the other.

Like station ties of the same frequency, they must also have sufficient capacity, relative to the systems, to hold the two systems together without overloading or pulling out of step. The greater the tendency of one or both systems to change its frequency under normal or fault conditions, the more important this becomes.

168. Five types of frequency changer are generally available, each with its advantages and limitations.

169. Synchronous-synchronous sets are the most common and consist of two synchronous alternators coupled together. The speed and frequency relations between the two units are fixed, and load is controlled by the relative phase positions of the two systems. Hence, a tendency of one system to change in phase relation to the other is immediately reflected in loading of the set and, if too great, will pull it out of step.

Load transfer is controlled by adjustment of the governors of the generators on one or both systems to change the relative phase positions of the two systems and cannot be controlled at the frequency changer.

Voltage and power factor of each unit may be controlled independently by field adjustments of each.

170. When two or more frequency changers are connected between systems, it is essential that the relative field and stator pole positions between the 25- and the 60-c units of all sets be the same in order to parallel. When so operated it is common practice to design one stator of all, or all but one, of the frequency changers so that it can be rotated through a limited angle by means of a frame-shifting device. By this means the division of load between the sets can be controlled, and a set can be added to or dropped from the system at no load. The amount of frame shift required depends upon the design of the sets and the reactance of the connections between them. Because of the different number of poles on the two units of a set, there are only a limited number of field and stator pole positions in which both ends of the set will be in phase with the systems. It is therefore common practice to provide field-reversing switches on one or both ends of the set so that, after the set is up to speed, the field of the driving unit can be reversed (at low field excitation) and poles slipped until the other unit is in phase.

If the field of the driven unit can also be reversed, the number of poles that must be slipped is on the average reduced by one-half.

171. Frequency changers of this type are listed in sizes from 100 to 75,000 kVA. The highest speed possible, if the ratio is 25:60 c, is 300 r/min. Other speeds are obtained at other frequency ratios. The efficiencies at full load and 0.8 power factor vary from about 80% in the smaller sizes to 96% in the larger sizes.

172. The synchronous-induction transformer set consists of a synchronous alternator on the lower-frequency end and a slip-ring induction motor on the other, having the proper number of poles and run at such a speed that the **slip frequency** is the lower frequency of the set. The rings are connected directly or through a transformer to the lower-frequency system.

If the set is a 25- to 60-c combination with a 10-pole 25-c synchronous unit running at 300 r/min and a 14-pole 60-c induction unit at the other end, the synchronous speed of which is 514 r/min, the slip frequency will be $60 \times 214/514 = 25$. The power transmitted mechanically through the shaft to the 25-c unit will be $300/514 = 58.3\%$ of the total. The power transmitted electrically from the slip rings to the 25-c circuit will be 41.7%.

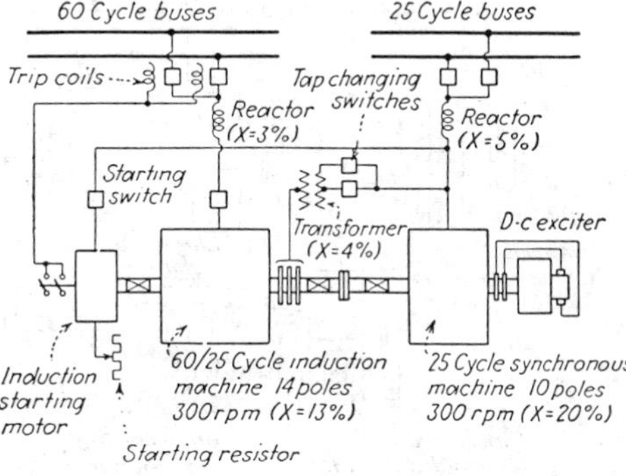

FIG. 10-52. Schematic diagram of synchronous-induction transformer frequency-changer set.

A transformer is usually interposed between the slip rings and the 25-c circuit, as the voltage of the latter is generally too high for good rotor and slip-ring design.

Voltage and power factor of the 60-c unit are controlled by taps on the transformer and on the 25-c unit by field control. Field-reversing switches and frame shifting are provided if the unit is to parallel other frequency changers (see Fig. 10-52).

Owing to the electromagnetic connections between the systems the voltages of the two systems, as well as the frequencies, are tied together so that a voltage disturbance in one system is reflected in the other. This is an advantage when one or more synchronous converters supplied from each system feed a common d-c bus, as it reduces the feedback through the converters and consequent tendency to flashing and damage in case of voltage disturbance on one a-c system.

The pull-out torque is somewhat higher and the phase-angle shift for a given load change is somewhat less than for a synchronous-synchronous set.

The cost and efficiency are about the same. The disadvantages are the somewhat greater complication of equipment and control and difficulties incident to high-capacity high-voltage slip rings. Sets of this type have been built up to 40,000 kW.

173. The synchronous-induction set consists of a synchronous alternator and a squirrel-cage or slip-ring induction unit. As load transfer depends upon slip in the induction unit, such sets are not useful when exact frequency control is maintained on the two systems but can be used when only approximate frequency is required and the relative frequencies of the two systems can be changed. A change from full load in one direction to full load in the other requires about a 4% change in relative frequencies.

As loading is determined by slip and not relative phase positions, frequency changers of this type are practically independent of momentary swings, with the result that systems can be safely interconnected with much smaller capacity frequency changers of this style than with the synchronous-synchronous style.

They have been used as a reserve supply to a particular load with the alternator normally "floating" as a synchronous condenser ready to pick up load if the normal supply is lost and when the two systems are not otherwise substantially tied together.

Frame shifting and field reversing are not required for parallel operation, and there is no control of the division of load between paralleled units with squirrel-cage induction motors. Load division can be obtained by varying the resistance in the rotor of slip-ring induction motors. Parallel operation requires that the load-slip curve of the induction motors be the same; they have generally been limited in capacity to a few thousand kilowatts.

Their efficiencies run about 2% less than those of the corresponding size of synchronous-synchronous sets, and costs are higher except in the smaller sizes designed for speeds higher than 300 r/min with approximate frequency ratios such as 62.5:25 at no load.

174. The synchronous adjustable-speed induction set consists of a synchronous alternator on the higher-frequency end and on the other a slip-ring induction motor with associated regulating equipment to control the voltage and phase angle of the slip frequency of the power received from or supplied to the slip rings. This control may be of the Scherbius or other similar types of control. It is diagrammatically shown in Fig. 10-53.

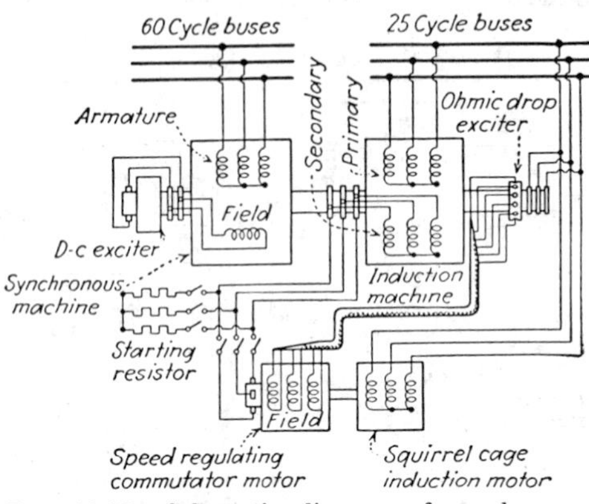

FIG. 10-53. Schematic diagram of synchronous adjustable-speed induction frequency-changer set.

The load-speed characteristics of this style of motor control are similar to those of an adjustable-speed shunt-wound d-c motor in that an increase in load slows down the motor slightly and the speed at any load can be controlled by adjustment of the controlling devices. By this means, local load control is provided, and load can be transferred in either direction at fixed frequency ratio or at ratios differing slightly from the normal. As load transfer is determined by relative speeds and not by phase-angle position, the set like the synchronous-induction set is practically independent of momentary **swings.** It is, therefore, possible to tie together two systems with much less relative frequency-changer set capacity than is possible with synchronous-synchronous sets. Automatic control can be provided to regulate for a desired load transfer in either direction or for fixed frequency ratio with provision to limit the transfer to rating of the set. The power factor of the induction end can be controlled by adjusting the phase angle of the voltage of the regulating set imposed on the rings so that the set can be held at about unity power factor at all loads. While this type of set combines most of the operating advantages of the synchronous-synchronous and synchronous-induction sets, its application has been limited owing to its higher cost, lower efficiency, and greater complication. Its use has been primarily between systems, the frequency control of at least one of which was not close and where the frequency changer capacity was small compared with the system capacities. For additional details on frequency changers, refer to Sec. **6**.

175. Direct-current frequency changers use no main rotating machine. The voltages of the two systems to be interconnected are rectified through high-voltage mercury-arc rectifiers and are connected at high-voltage d-c. Power flow can be controlled much the same as is done in d-c power transmission (see Sec. **14**). The principal advantages of this type of frequency changer are:

1. The system-to-system tie is completely nonsynchronous.
2. The amount of power that may be transferred is not controlled by machine-design limits.

At this writing, the largest such set is used to tie the 50- and 60-c systems in Japan. It is rated at 300 MW.

CONDENSERS

176. Condensers for power purposes are of two general types: "static" and "synchronous."

177. Static condensers, or capacitors, consist of metallic-foil plates separated by paper insulation immersed in oil or a nonflammable insulating fluid such as Askarel and sealed in metallic tanks. They are built in units of standardized capacities and voltages of 230 to 13,800. Banks of units are assembled together to give the desired overall rating. Typical unit sizes are 3 to 15 kvars at 230 volts, either single-phase or 3-phase. At 2,400 to 13,800 V, unit sizes are 15 to 100 kvars and are single-phase only. Discharge resistors or coils are installed to discharge the capacitor when disconnected. The losses are approximately 3½ W/kvar of rating. Their principal field of application is for power-factor correction in industrial plants or on feeders of moderate capacity and voltage where the power factor is objectionably low. While their losses are low and regular attendance is not required, they provide no ready means of adjusting their capacity except by switching on or off as a unit or in blocks. It is sometimes possible to connect the capacitor in parallel with the load, such as a motor, so that it is switched on and off with the load. Their cost and losses must be balanced against reduced transmission losses and the value of the increased kilowatt feeder rating and better voltage obtained (see also Sec. **5**).

178. Synchronous condensers, which are essentially synchronous motors designed to operate without mechanical load and with a wide range of field control, are listed in a wide range of standard rating from 100 kVA at 1,200 r/min air-cooled to 100,000 kVA at 514 r/min hydrogen-cooled. Large hydrogen-cooled units have been successfully built at 1,800 and 3,600 r/min. Hydrogen cooling is particularly applicable to synchronous condensers, as the entire unit can be enclosed without shaft seals, and, owing to the high speeds, the reduction of windage losses is an appreciable factor and conversely makes higher speeds economical. It is generally not economically advantageous in units smaller than 25,000 to 35,000 kVA. Hydrogen-cooled units, being totally enclosed, are readily adaptable for outdoor installation, and several such units are in operation. The full-load losses of a 100-kVA unit are about 10 kW, and those of a 100,000-kVA unit are about 1,375 kW.

179. Excitation may be obtained by the same methods as generators, although the wide range in excitation and the fact that generally not more than one or two units are installed in a station tend to make direct-connected exciters with pilot exciters particularly favorable. Direct-connected exciters are built within the condenser enclosure when the unit is hydrogen-cooled. Electronic excitation has been provided in a few cases, consisting of transformers and a rectifier with suitable firing control to provide the voltage range necessary. Automatic voltage control similar to that used with generators is frequently provided, and high-speed or quick-response excitation is particularly advantageous when the condenser is used to increase line or system stability. A condenser which is required to operate at times underexcited as a shunt reactor should be especially designed for this service, as to both the condenser and its excitation system, in order to provide adequate stability and sufficiently wide range of excitation current.

180. Condensers are used to (*a*) improve the power factor of a line or system, (*b*) regulate the voltage of a line by changing the power factor of the power fed over a line, (*c*) increase the stability of the line by regulating its voltage and power factor. Their cost and capitalized operating costs must be balanced against the capitalized savings in transmission losses and the cost of other expedients, such as increasing line capacity, equipping transformers for tap changing under load, or adding induction voltage regulators. They can occasionally be justified on metropolitan underground-cable systems but are most advantageous on long overhead high-reactance lines.

181. Condensers should be installed at the load or as near as practicable to the load, as at a step-down or distribution substation, when it is intended to regulate one-way lines feeding the load.

182. When connected to two-way lines, as on tie lines between generating stations, or systems, condensers may be advantageous at both ends, particularly if the generators at the receiving end do not have sufficient reactive capacity to supply the wattless current required by the load and that required to regulate the voltage of the line. On the other hand, if substations are tapped to the line between its terminals, it may be found advantageous to install condensers at one or more of the substations in place of or in addition to the terminal condensers. Long, high-capacity lines, in which stability limits the line capacity, may have their limits increased by terminal or intermediate condensers or both, and, similarly, the stable length of a line may be increased by the use of condensers. A resonant line may be obtained approximately by a multiplicity of condensers, installed at different points along it.

The charging current of high-voltage lines may be sufficient to cause an excessive voltage rise at periods of light load requiring the condensers to operate underexcited to neutralize the charging current and reduce the voltage.

ELECTRIC GENERATORS—APPLICATION AND OPERATION

183. Generator Selection. The size, voltage, and characteristics of a generator must be selected with due regard to the load, the relations to the rest of the plant and the system as a whole, and the characteristics and available sizes of the driving medium.

184. The size of generator selected is influenced by overall plant investment, including reserve capacity, annual operating cost, and rate of load growth. Before making a selection one should study alternate combinations of different sizes based upon actual or predicted load-duration curves to obtain minimum overall annual investment and operating costs.

185. The purchase price and installed cost per kilowatt of a generator decrease and the efficiency increases as the size of the unit increases within normal limits of design of the particular type of generator and prime mover, although the relative gains in cost and efficiency decrease with increased size. Operating labor costs are more nearly proportional to the number of units than total plant capacity.

a. **Small plants** having inherently high unit costs and with limited needs for continuity, or where breakdown service can be obtained from another source, often install a single unit with a rating equal to or slightly greater than the maximum integrated demand. An extremely fluctuating load may require a generator of large rating to limit voltage variations, if d-c, to provide adequate commutation limits, and the prime mover must have adequate capacity to drive the generator throughout all fluctuations of load.

b. **In larger and more important plants** with one or more units for reserve, the size of the unit affects the reserve capacity and total cost of the plant. In small plants, the investment costs are usually lowest when the size of the unit is large relative to the load, whereas in large plants the trend is toward a relatively larger number of units.

c. **A plant tied into a large transmission system** can be built economically with larger units than an isolated plant, because of the difference in load and reserve requirements.

186. The daily and annual load factors are important considerations in selecting the size of generator. A high daily and annual load factor can be handled most economically by relatively few large units, whereas a low load factor requires more smaller units. The load characteristics may be such that the greatest economy will be obtained by a few large units for the heavy-load periods, with one or more small units for the light-load periods.

187. The economy of the prime mover throughout its load range has an important bearing on the selection of the unit size. Whereas steam turbines and engines may be designed with a fairly flat water-rate curve from half to full load, internal-combustion engines lose efficiency rapidly at partial loads, tending to favor a greater number of smaller units. Cost of fuel is an important factor in balancing operating economies against investment. The efficiency characteristics of water wheels vary with the type and head but generally tend to decrease rapidly at partial loads. This is an important factor when the plant is built up to the stream capacity and where there is water storage, whereas, if there is excess water or if there is no storage, wheel efficiency, especially at part load, is of less importance. As the rated speed of a water wheel at any given head tends to decrease with increased capacity, the head may dictate smaller units than would be selected by other considerations.

188. The rate of load growth affects the size selected, larger units being more economical in a rapidly growing system than in a static plant. One must include as charges against the larger units in growing plants the capitalized carrying charges of the excess capacity during the periods of excess.

189. The voltage of the generator depends upon the load it is to supply, the distance from the load, and the voltage of the system to which it is connected but may be limited by the size of the unit itself.

Direct-current generators must have the same voltage (except for line drop) as the load.

Alternating-current generators may or may not have the same voltage as the load, depending upon the distance of transmission and the economy of the interposition of transformers. Small a-c generators supplying loads within a few hundred feet usually generate at 120 or 240 V or 120/208 V 4-wire or, in some industrial plants, at 440 V. Larger generators up to capacities of a few thousand kilowatts and feeding loads within a radius of a few miles generate at 2,300- or 4,000-Y V. Large generators and those feeding loads beyond the economical distances of 4,000-Y V are usually designed for 13,800 V, although a number of existing plants have units of intermediate voltages, such as 6,600, 7,800, and 11,000. Very large generators have been built with voltages up to 24,000 in this country, and a 25,000-kW unit has been built in England with a voltage of 33,000. When generators are intended to feed directly into a transmission line or bus, at higher voltage than practical or economical for the generator itself, *savings can sometimes be made by selecting a special generator voltage giving overall minimum generator and transformer costs.*

190. The generator characteristics should meet the load needs and prime-mover capability. Standard-design generators meet ordinary requirements, but special designs may be required. If intended to operate in parallel on a bus with other units, the generator and prime-mover characteristics should be checked with the manufacturers to ensure proper operation. Widely or rapidly fluctuating loads or pulsating driving torque may require special designs. Operation at abnormal power factors, especially leading, or unbalanced or single-phase loads commonly requires special designs. The generator should be designed to withstand the overspeed to which it may be subjected. This is usually 85 to 100% overspeed for water-wheel-driven generators, 25% for engine-driven, and 20% for steam-turbine-driven units. The lower overspeed is permissible in the latter cases, as the prime movers are usually equipped with quick-acting emergency overspeed trips in addition to the governor control. The inertia of the water in hydraulic units is such that sudden cutoff of water supply cannot be provided.

191. Direct-current generators are obtainable from a fraction of a kilowatt up to about 4,000 kW nominally rated and at the various standard voltages up to 1,500 V. Operation at 600 V is confined to units of 25 kW or larger, and voltages of 1,200 or 1,500 are special and not suitable for small machines.

Direct-current generators, primarily for industrial or railway use, are usually rated to permit either a 25 or a 50% overload for 2 h. The "service factor" of 1.15 for a 40° generator permits carrying 115% of rated load continuously when operated at rated voltage under standard service conditions.

192. Direct-current generators may be belt-driven, direct-connected to engine or motor, steam-turbine-driven, or water-wheel-driven. The advent of high-speed steam engines and particularly internal-combustion engines has reduced belt drive with its large space requirements to a minimum. Turbine-driven generators are usually geared, as good turbine design requires a higher speed than is practicable in a d-c generator.

193. Speed. A wide range of speeds is available to meet the needs of the driving medium, although abnormally low speeds increase the size and cost, and abnormally high speeds increase the cost and commutation difficulties. However, 3,600-r/min turboalternators are now being provided with either direct-connected or geared exciters. Small d-c engine-driven generators are listed at 300 to 450 r/min, 1,000-kW units at 150 to 360 r/min, and a 4,000-kW unit at 300 r/min.

194. Excitation. Most d-c generators are self-excited and may be either shunt- or compound-wound, the latter being the common practice for industrial plants where good voltage but not extremely close control is required and constant supervision may not be provided. A voltage spread of 2 to 5% may be expected in "flat"-compounded machines under various conditions of machine load and temperature.

195. Parallel operation of compound-wound generators requires similar shunt and compounding characteristics, similar speed characteristics, and the use of low-resistance series-field equalizer connections.

196. Shunt excitation alone is provided in units supplying loads which do not change rapidly, where constant manual or automatic voltage regulation is provided or where a drooping characteristic is desired, as in some battery-charging applications. No equalizers are required for parallel operation, but the generators should have similar shunt characteristics and the driving media similar speed characteristics. Shunt-wound machines are somewhat less subject to damage from flashing than compound-wound machines and are more stable during load reversals.

197. Separate excitation is provided for generators required to operate over a wide voltage range. As self-excited machines tend to become unstable when operated at low voltage, special construction is required for self-excited exciters designed to be stable on hand regulation down to half voltage. Typical applications of separate excitation are Ward Leonard sets and direct-connected exciters with pilot exciters. Separately excited generators have closer regulation than self-excited shunt machines, are less affected by speed and temperature changes, and have a faster voltage response.

198. Three-wire operation may be obtained by (*a*) two half-voltage generators in series with the neutral tapped to the connection between the units, (*b*) balancer sets, (*c*) "compensators," or "balance coils," connected across slip rings tapped to the armature of the generator. The first method is the most expensive and has the highest losses but is capable of the widest range of unbalance and independent voltage control. Balancer sets are capable of independent voltage control and are cheaper and more efficient than the first method, provided that the unbalance is not great. They are more subject to damage from line-to-ground short circuits, as their capacities are small compared with the main generators. Three-wire generators are the cheapest and most efficient for general application; and while a 10% current unbalance at full load results in a 2% voltage difference, this is usually not too great for practical purposes. If subject to unbalances of more than 10%, special provision may be required, such as dividing the series and commutating fields with half in each polarity.

199. The overload capacity of d-c generators is limited by generator temperatures if sustained and by commutation if momentary. The standard commutation limits are 200% of rating for nominally rated generators and 150% of rating for continuous-rated generators. Momentary overloads beyond this range may require special designs or a larger machine.

200. Fluctuating loads must be studied to determine the equivalent steady-state load for heating purposes, it being recognized that armature copper loss varies as the square of the current. Rapid and wide fluctuations in load or speed affect commutation and may require special designs. Rated load can be thrown on or off a modern nominally rated machine without excessive sparking and without shifting the brushes.

201. Alternating-current generators may be obtained in a wide range of capacities, speeds, and voltages to meet various requirements of load and driving media. The speed must be an even fraction of 7,200 r/min if 60-c and of 3,000 r/min if 25-c.

202. Engine drive, either steam or internal-combustion engine, may be obtained from a fraction of a kilowatt to 7,000 kW, the upper limit being determined by the economic limit of the prime mover. Steam-engine-driven units are rarely manufactured at the present time because of the lower cost and higher efficiencies of turbine-driven units, but gasoline-engine sets in small units and diesel-engine sets in larger units are supplied where conditions are favorable. Engine units, because of the pulsating torque, usually require heavy flywheels and generally require amortisseur windings to reduce electrical surges and provide satisfactory parallel operation. A variety of speeds is obtainable to meet engine requirements, but speeds are generally low, 25-kW units being listed from 257 to 450 r/min and 5,000-kW units from 109 to 200 r/min. Full-load efficiencies vary from 85 to 96% depending upon size and speed.

203. Water-wheel generators are listed in various sizes from 62.5 to 300,000 kVA. The speed is determined by the combined economies of the generator and water wheel and is affected by the head, unit capacity, and type of water wheel. There is a wide range in water-wheel design and characteristics, generally from 100 to 900 r/min for the small sizes, 72 to 514 r/min for the intermediate ratings, and 72 to 180 r/min for the largest ratings. While wheel units are built in both the horizontal and vertical type, overall plant costs and advantages have led to a great predominance in recent years of the vertical type, except for Pelton water wheels. The great inertia of the water, with the necessarily slow governor and gate operation, often requires special flywheel effect. Typical requirements specify sufficient WR^2 to limit the speed change to 20% for full-load change, with a 1-s governor. Units that are a small part of a system or are designed to run at a fixed gate opening do not require so much flywheel effect. Brakes are frequently necessary to stop and hold the generator against leakage through the gates. Full-load efficiencies vary from about 83 to 97% depending upon size and speed.

204. Modern turboalternators are built in the horizontal type in a wide range of capacities at 1,500 r/min for 25-c units and 1,800 or 3,600 r/min for 60-c units. Single-shaft 25-c units have been built up to 160,000 kVA at 1.0 pf, 60-c units at 1,800 r/min up to 690,000 kVA at 0.9 pf, and 3,600 r/min units up to 500,000 kVA at 0.85 pf. Cross-compound multiple-shaft units have been built in ratings exceeding 1,100,000 kVA. Efficiencies vary from 97 to 98.5% for hydrogen-cooled generators. All large units use hydrogen as a cooling medium, and some have also used liquid for cooling the stator conductors.

205. Regulation of an alternator depends not only on the design of the machine but also on the power factor of the load and is large compared with that of a d-c generator. It will vary from 5 to 15% in supplying a 1.0-pf load and from 15 to 30% or even greater in feeding an 0.8-pf lagging load. Constant manual or automatic voltage control is generally provided. Closer regulation can be obtained by designing the generator for lower synchronous reactance, but such a design results in a materially more expensive machine and one which would deliver much greater short-circuit currents and, consequently, is generally undesirable.

206. Various reactances are used in describing the characteristics of alternators, the most commonly used being:

1. **Direct synchronous reactance,** which is used in determining the sustained 3-phase short-circuit currents. A typical value for turboalternators is 100 to 150% and for salient-pole generators from 60 to 125%.

2. **Direct subtransient reactance** is used in determining the symmetrical instantaneous 3-phase short-circuit current. Its value for turboalternators may be from 8 to 20%; and for salient-pole alternators, from 15 to 35% with amortisseur windings and from 20 to 45% without amortisseur windings.

3. **Zero sequence reactance** is used in determining the single-phase line-to-neutral short-circuit current. Its value depends greatly on winding pitch, being a minimum at ⅔ pitch in 3-phase windings, and may vary from 1 to 12% in turboalternators and 2 to 21% in salient-pole alternators.

4. **The negative-sequence reactance** varies from 8 to 20% in turboalternators and from 15 to 40% in salient-pole alternators. The above values are typical for 60-c machines, while those for 25-c machines are generally somewhat lower.

207. Short-circuit currents delivered by alternators rapidly decrease to a value determined by the synchronous impedance and the excitation, but the initial values which determine the mechanical stresses on the windings, station buses, and connections are high. It is common practice, in large stations where power is distributed at generator voltage, to install reactors in the generator leads, not only to reduce the total short-circuit currents but also to reduce the stress on the generator winding. Neutral reactors or resistors of low ohmic value are desirable with generators of low zero sequence reactance when the generator neutral is used for establishing the system ground, so that phase-to-ground fault currents are limited to the 3-phase fault or less in the generator winding.

A method of grounding generator neutrals through a small transformer primary winding, whose secondary winding is connected to a resistor, has become popular in recent years. It is applicable where the machine is connected to the system through a main transformer, as shown in Fig. 10-75. The resistor is usually chosen to limit the maximum generator phase-to-ground fault current to 10 to 20 A. The machine may be tripped immediately by a special voltage relay, or an alarm may be initiated by the relay so that the operator may remove the generator from service slowly or wait until a more convenient system load condition has developed (see Par. **274**).

208. Stability, when used with reference to a power system, is that quality of the system or part of the system which enables it to develop restoring forces between the elements, equal to or greater than the disturbing forces. Hence, in each generator and motor on the system, there is a problem of equilibrium between the mechanical forces acting on the machine shafts and the electrical forces acting on the machine windings. If, in the process of maintaining equilibrium, synchronism is not lost between synchronous machines, and induction machines do not stall or overspeed (in case of induction generators), stability results.

209. The output and inherent stability characteristics of any synchronous generator during a disturbance are determined by its direct-axis flux, by its characteristic reactances, and by the net effect on this flux of the demagnetizing component of armature reaction and of the transitory sustaining effects of induced currents in the rotor faces or damper bars and in the field windings. The lower the machine reactances and the greater the ability of the machine to sustain its flux during disturbances (higher short-circuit ratio), the larger is the synchronizing power and the smaller the drop in its terminal voltage.

Since the rotor-face and damper-bar currents flow for an extremely short time, not usually more than 0.1 s, their effect in sustaining the machine flux is negligible. Therefore, transient reactance instead of subtransient reactance is considered as the characteristic reactance connecting the generator internal characteristic with its external performance and is, therefore, a measure of its inherent stability during the transient period of a disturbance. However, as these currents are effective in damping the oscillations of the rotor, they do add to the stability of the machine.

The induced currents in the generator field circuits are much more effective in sustaining the field flux and, depending upon the time constant of the field and the excitation circuit, may prevent any material decrease in this flux for periods up to 1 or 2 s; this may be sufficient to prevent loss of synchronism in critical cases. As these induced currents die away, the machine flux is reduced by the full effect of the demagnetizing component of the armature reaction. Synchronism or loss of synchronism then depends on whether or not the machine synchronizing power has been reduced below the final load that governor and system-load characteristics place on the unit.

210. On power systems where induction motors constitute a substantial portion of the load, the reduction in the generator terminal voltage accompanying the drop in its flux may be sufficient in severe disturbances to stall some of the induction motors, with increase in current and further drop in generator voltage. This action may become cumulative and result in collapse of voltage to a point where synchronism is lost between some or all of the system generators. This is illustrated in Fig. 10-54, which shows characteristics of an 0.8-pf unity short-circuit ratio turbogenerator when operating alone and carrying a typical load of 70% induction motors and 30% lighting. The maximum load of the same characteristics which the generator could pick up without increase of excitation is represented by the amount AB. On the assumption of no disconnection of motors by undervoltage control equipment, greater loads would collapse the generator voltage below the value indicated at point C.

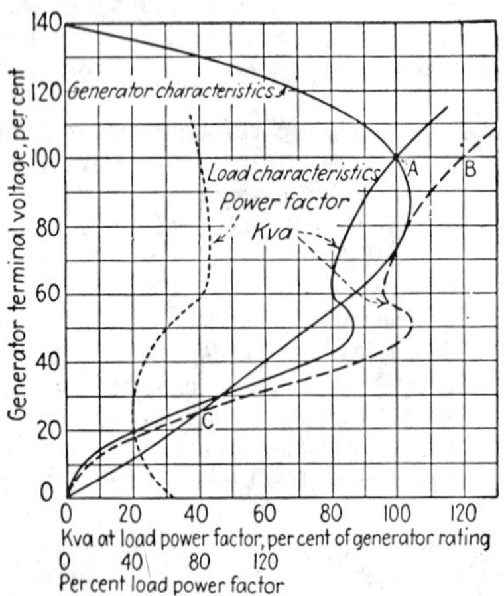

FIG. 10-54. Characteristics of an 0.8-pf 1.0 short-circuit ratio turbogenerator when carrying a typical load of 70% induction motors and 30% lighting.

211. Generator stability is increased by the design of the generator and its excitation system that minimizes the reduction of direct-axis flux during the disturbance. While no single factor evaluates these design features or stability comparisons between generators, the factor of **short-circuit ratio** is a rough approximation, since it is a function of armature reaction and leakage reactance expressed in the form of synchronous reactance and of no-load saturation at rated voltage. The higher the short-circuit ratio, the greater the generator inherent stability during the latter part of the transient period of the disturbance and the subsequent steady-state conditions. Short-circuit ratios of approximately unity or even as low as 0.75 are usually adequate for hand regulation where load swings are comparatively small and gradual. High short-circuit ratios increase the cost of the generator, particularly of high-speed turboalternators, and a lower-short-circuit-ratio generator with fast-acting automatic voltage control is often cheaper.

High-speed voltage regulators are very effective, in working with high-response-rate excitation equipment, in maintaining generator stability even under severe disturbance conditions and low generator short-circuit ratio. Limits of material strength in the field rotor force the use of low short-circuit ratio (0.5 to 0.6 at maximum generator ratings); so high-speed excitation response is needed in the larger generator sizes.

212. When a generator is equipped with a voltage regulator, its stability is better than indicated by the short-circuit ratio, depending upon the characteristics of the regulator and the excitation system and upon the time constant of the field and excitation circuit. If these are such that no appreciable reduction of machine flux occurs during the disturbance, the generator stability is indicated by its transient reactance. Quick response excitation is required to maintain machine flux thus during severe disturbances. Automatic voltage regulators are often necessary when the generating station is connected to large-capacity ties or unduly fluctuating loads or where line stability needs reinforcing. However, owing to the many factors involved, studies of generator performance under specific disturbance conditions are necessary to determine definitely the amount of stability in the machine and its excitation system.

More complete information on generator stability may be found in the following books: "Power System Stability," Vols. I and II, by S. B. Crary, John Wiley & Sons, Inc., and "Electrical Transmission and Distribution Reference Book," by various authors, Westinghouse Electric Corporation. See Bibliography for article references.

213. The generator's rated power factor should be adapted to the system's needs, there being taken into consideration not only peak-load conditions but partial loads when wattless requirements may become a limiting factor in machine operation. Generator outputs may be limited by either armature current or field current. It is generally not advantageous to design generators for less than 80% pf, but rather greater overall economy can be secured by improving the system power factor by condensers or other means or, if the low power factor is of relatively short duration, operating generators at reduced load and at lower than rated power factor. It must be recognized that operation at higher than rated power factor reduces the margin of voltage stability and pull-out torque, a factor requiring special consideration in high-voltage systems where charging current may create a leading power-factor load. Twenty-five-cycle generators feeding primarily d-c loads through converters are commonly rated unity power factor; generators feeding mixed a-c industrial and lighting loads are usually rated 80, 85, or 90% power factor depending upon system needs. The field is frequently the limit in the design of large turboalternators, with the result that a change in the power-factor rating may make a large change in generator cost.

214. Two independent armature circuits, while seldom used, can often be supplied in large generators so that each winding may be connected to a separate bus, the mutual induction between the windings providing a transformer action such that the reactance from the generator to the bus is relatively low, while the reactance from bus to bus through the generator windings is relatively high. When so connected, the loads on the two buses should be closely balanced; otherwise the maximum output of the generator will be reduced, and an appreciable voltage and phase-angle difference will appear between the buses. The maximum output of one winding alone with the other winding carrying no load is somewhat greater than 50% of the generator rating, usually of the order of 60% of the generator rating.

215. Single-phase-operation generators are sometimes required to supply a single-phase a-c electric railway load. The ability of the generators to carry such load should be carefully investigated. Turbogenerators or others with solid-type field structures have practically no ability to carry single-phase load unless specially designed for it, owing to eddy currents induced in the field structure by the pulsating single-phase flux. Short-circuited pole-face windings are installed to damp out this flux and reduce field heating. Salient-pole generators with laminated field structures are less subject to this condition but are frequently built with pole-face windings, and the frames are sometimes mounted on springs to absorb the vibration due to the pulsating torque. If the single-phase load is taken from a 3-phase system, with rotating machinery as load, such as synchronous converters and synchronous and induction motors, the amount of single-phase load imposed on the generator is reduced by the phase-balancing action of rotating load machinery. The resulting phase unbalance may be small enough so that standard 3-phase generators can safely carry it. However, the amount of phase unbalance should be carefully studied with reference to the ability of the generators to carry it in each case.

216. Clean and adequate ventilation is essential for successful operation of generators, as the life of the insulation is directly affected by its temperature. An accumulation of oil and dirt acts as heat insulation, increasing the temperature of the windings, and adds to the fire risk from a static spark or other causes.

Open-frame construction is common for small and slow-speed machines, in which air is taken from and discharged into the surrounding room.

Closed-frame construction is used on the medium-size high-speed machines with inlet and outlet openings separated to reduce or eliminate recirculation. Ducts may be connected to the inlets or the outlets or both, as may be desired. Duct and air-filter designs should be correlated with the generator design to insure that the air-pressure drop is not excessive for the particular machine. Long or restricted ducts may require separate blowers. **Dampers** should be provided to shut off the air in case of fire. Automatic operation is provided in many cases and is necessary with some schemes of fire protection.

217. Dry-type air filters have been used on some small machines but rarely on large ones where large volumes of air are required. Some types use metallic or glass "wool" or fibers coated with oil. Others use perforated plates or screens faced with flannel. The air resistance increases and the dust removal decreases as dirt is accumulated.

218. Closed-circuit ventilation of generators has become standard for medium-size (up to 12,650 kW) units, each generator having its own cooling system. In general, the air circulation is created by impellers on the rotor, but external blowers are used in some cases.

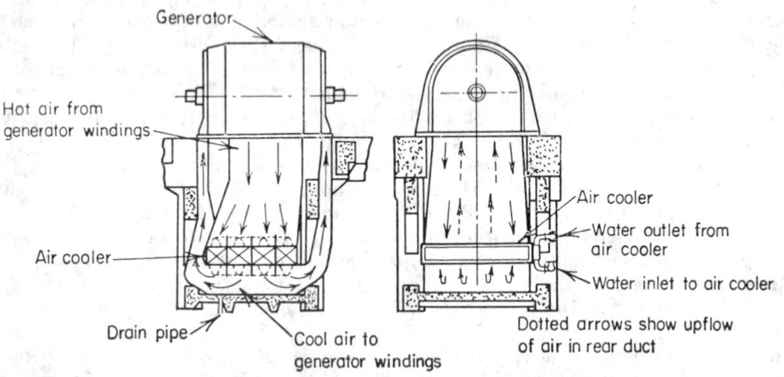

Fig. 10-55. Enclosed ventilating system of generators with air coolers. (*Griscom-Russell.*)

The air is recirculated through the generator and is cooled by surface or tubular air coolers. These are placed in the air duct and cooled by circulating water. See Fig. 10-55. The air leaves the generator at a temperature of 65°C under normal conditions and is cooled to 40°C or less before it reenters the generator.

In closed-circuit systems, dirt is almost completely eliminated, and an excellent opportunity is afforded for fire protection.

219. Air coolers should be mounted below or at the sides of generators rather than above them so that condensation on the coolers, when the unit is shut down, will not drip on the windings. It is desirable to shut off the cooling water as soon as possible after shutdown to reduce the condensation, and to prevent fast temperature cycling of the generator windings.

220. Cooling water for generator air coolers is sometimes taken from the condensate, and in other instances raw circulating water is used. The use of condensate reduces the plant losses somewhat through returning part of the heat losses in the generator to the boilers. Coolers using raw water are usually cheaper because of reduced surface required.

Some systems are designed to use either condensate or raw water through the same cooling coils, but this is not the best practice, as impurities left in the system by the raw water may be carried to the boilers when condensate is again used. Another method employs an auxiliary cooler or heat exchanger, by means of which some of the air is cooled by raw water.

221. Double-bank coolers employ two independent coolers, one using condensate for the higher temperature range and the other raw circulating water. The coolers are superimposed one upon the other, and the air passes through them in series. The air temperature of 150°F is reduced to about 115°F by the condensate cooler and then to 104°F by the raw-water cooler.

Raw-water coolers are generally necessary in addition to condensate coolers where the latter are used, as the amount of condensate is usually not adequate to take care of the generator losses at no load and light load with economically sized coolers.

222. Hydrogen cooling of generators operating at 1,800 and 3,600 r/min has become standard practice for units rated 10,000 kW and above, with attending improvements in generator efficiency, of the order of 1%, based on operation at ½ lb/in² gage hydrogen (see Fig. 10-56).

Since 1957, the use of hydrogen at higher pressures (up to 75 lb/in² gage) has become necessary, in addition to generator design changes, to allow the gas to flow directly over the field and stator winding copper, to permit building the larger-size machines now being demanded by the industry. At the present time orders have been taken for generators as large as 850,000 kW in a single machine, based upon the advantages to be gained by use of higher-pressure hydrogen and direct cooling of the conductors.

The principal advantages resulting from the use of hydrogen cooling for turbogenerators are:

a. Reduced windage and ventilating losses, because of the low density of the hydrogen gas.

b. Increased output per unit volume of active material, because of the high heat-storage capacity, thermal conductivity, and heat-transfer coefficients of hydrogen.

c. Probable increased life of the insulation on the stator winding, because of the absence of oxygen and moisture in the presence of corona.

d. The fire risk of the generator is generally reduced.

223. Direct liquid cooling of stator windings in high-voltage large generators has been used to a limited extent. Oil or water is usually employed and is circulated through the stator winding strands, connections being made at both ends of the machine through insulated tubes. Hydrogen is used for cooling the field winding and other heat-producing parts.

224. The risk of hydrogen explosion is eliminated by maintaining the hydrogen purity well above the explosion limits, which are between 5 and 75% by volume of hydrogen when mixed with air. Hydrogen-cooled generators are constructed with a heavy cylindrical steel shell and domed end bells of sufficient strength to withstand an internal explosion without rupture. Tests indicate that the cooling effect of the large surfaces in a generator reduces the explosion pressures considerably below the theoretical values calculated from the energy available in the explosion mixture. Automatic hydrogen-control devices are furnished to maintain a positive hydrogen pressure above atmospheric of 0.5, 15, 30, 45, or 75 lb/in² gage, and to provide a continuous check on its purity, giving an alarm if the gas pressure becomes abnormally high or low and if the purity becomes less than 90%. The purity of the hydrogen is automatically maintained at about 97% normally.

Oil-sealing glands are furnished around the shafts to prevent the escape of hydrogen or the inflow of air (see Fig. 10-56).

225. In replacing hydrogen by air, or vice versa, an intermediate step is undertaken by scavenging the unit with carbon dioxide to avoid an explosive mixture of air

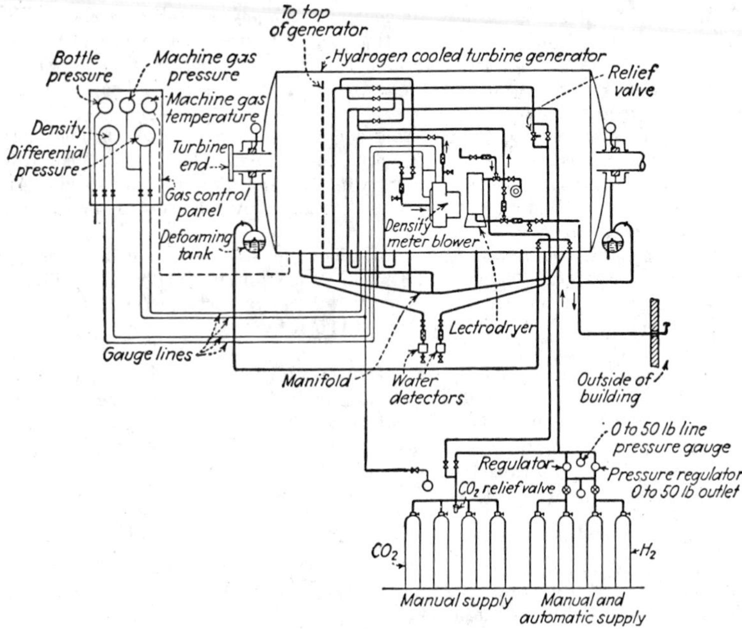

Fig. 10-56. Diagram of gas piping for hydrogen-cooled generator.

and hydrogen at any time. Some operators have concluded that this is an unnecessary precaution and change directly from hydrogen to air or the reverse, taking due care to minimize turbulence and avoid sparks during the process. The vent from the generator shell is piped to the out-of-doors. Most hydrogen-cooled generators operate with about 5.0 to 45 lb/in^2 gage pressure, but pressures are occasionally used up to 75 lb/in^2 gage, which increases the output of the generator but at some sacrifice in hydrogen losses and some lower efficiencies.

The coolers are built in the frame of the generator, and usually distilled water is used in the closed system to reduce the possibility of corrosion. The distilled water is cooled in water-to-water heat exchangers.

The usage of hydrogen in normal operation is from about 250 to 400 ft^3/week for 3,600-r/min generators and from about 450 to 800 ft^3 for 1,800-r/min generators.

226. Synchronizing. Great care must be exercised in connecting an alternator to a live circuit to ensure that the phase rotation, speed, and voltage of the generator and circuit are the same and that the two are in phase. A generator may be synchronized by means of lights, connected as in Fig. 10-57 to be dark when in synchronism

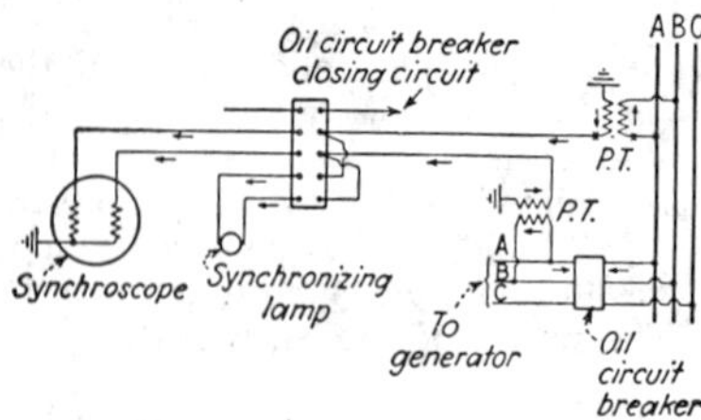

Fig. 10-57. Synchronizing with dark lamp. Arrows indicate instantaneous polarities at the instant when the incoming generator is in phase with the bus.

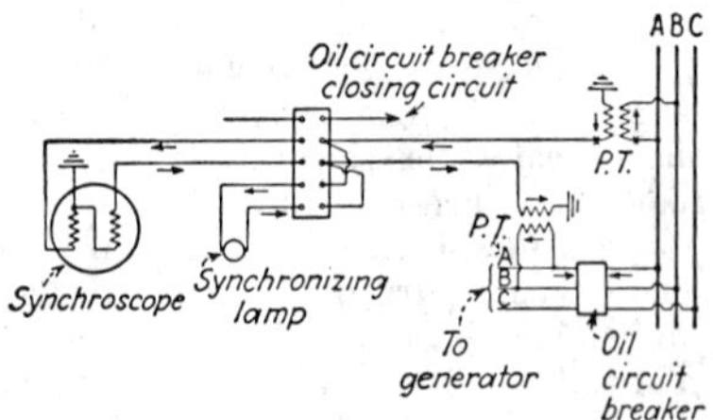

Fig. 10-58. Synchronizing with bright lamp. Arrows indicate instantaneous polarities at the instant when the incoming generator is in phase with the bus.

or bright as in Fig. 10-58, or, better, by means of a synchroscope. The latter device is used practically universally for large machines, as it gives a much more accurate indication of relative phase position than can be obtained with lights. It is good practice to wire the closing circuit of an electrically operated generator circuit breaker through a synchronizing receptacle or switch so that the generator switch cannot be closed unless the synchroscope is also connected in circuit. A synchronizing check relay may be used, with its contacts in series with the circuit breaker, closing control wiring as protection against a personnel error in performing the synchronizing operation. Automatic synchronizing relay equipment is also available, including speed-matching facilities, to perform the complete synchronizing operation.

EXCITATION

227. The importance of the system of excitation cannot be too strongly emphasized. Its availability at all times is of paramount importance. Loss of excitation of a unit on a bus results in a more serious disturbance than that resulting from dropping the generator from the bus, as the remaining units must not only pick up the load dropped but also supply the large reactive current taken by the unexcited generator.

228. Excitation equipment should be designed with a view to the maximum possible continuity of service. Simplicity, ruggedness, "foolproofness," and reserve apparatus are important requirements. The methods in use for securing these results are greatly varied, and much difference of opinion exists on which is best. Each plant must be considered as a separate problem.

229. The excitation systems in general use at the present time are: (1) direct-connected or gear-connected shaft-driven d-c generators; (2) separate prime mover or motor-driven d-c generators; (3) a-c supply through static or mercury-arc rectifiers.

230. Exciters directly connected to the main generators were the earliest form in use and are quite generally favored, being especially suited to the "unit system" of plant operation. The chief arguments in favor of this method are simplicity, high efficiency, the absence of large field rheostats, and the fact that the possibility of loss of excitation on more than one unit at a time is reduced to a minimum. Individual exciters lend themselves most readily to automatic voltage regulators. Each exciter is, in some instances, made large enough to furnish excitation for two generators in emergencies. The most important objections are the possible crippling of a large unit due to trouble with its exciter and the fact that voltage fluctuations in percentage are twice as great as the speed fluctuations.

231. Exciters driven by separate prime movers may be obtained for service with any kind of prime mover. On account of their small size the efficiency of the unit is very poor, this constituting the chief objection to their use. Where exhaust steam is necessary for feedwater heating, the inefficiency is of less importance, but the present trend is not toward this practice. It is sometimes necessary to install prime-mover-driven exciters for starting purposes.

232. Motor-driven exciters are in use in many stations. They are economical, efficient, and reliable but should be supplemented by at least one unit driven by a prime mover, for starting up. Induction motors are almost invariably used and constitute the best type of exciter drive if the a-c supply is reasonably assured. Such motors can be started immediately, which is important. A well-balanced exciter plant consists of units driven both by motors and by prime movers. In some cases motor-driven units are considered as auxiliary to the prime-mover units, and in other cases the motor-driven plant is considered of first importance.

233. Duplex Drive. Another method of securing continuous service is being used with considerable success. The exciter generator is arranged to be driven both by motor and by prime mover. The motor is provided with a relay which opens the power supply to it, if the a-c voltage becomes low. The governor is arranged practically to shut off the steam at normal speed; but after the above relay operates, cutting off the motor, the prime mover continues to drive the unit at but slightly reduced speed, thus preventing any interruption in the excitation supply. This scheme is sometimes employed in connection with heat balance, the prime mover being used to drive the exciter when plant-operating conditions demand.

234. Electronic exciters may be dry-type rectifiers (silicon) or mercury-pool-type rectifiers, although neither type has been used extensively to date. A design using a rotating a-c armature winding and dry-type rectifiers rotating on the same shaft, for direct supply of field current to the generator rotor winding without collector rings or brushes, is receiving considerable attention. Voltage regulation is obtained by field current control in the stationary field pole structure of the a-c exciter generator. Voltage control of mercury-pool rectifiers is obtained by varying the firing time on the a-c voltage wave of the supply power source.

235. The size of the exciter plant depends upon the size of the power plant and the types of generator used. Small low-speed generators require up to 3% of their capacity for excitation. Large high-speed turboalternators may require as little as 0.5%. The exact requirements may be obtained from the manufacturer. The total capacity should be ample to carry the whole excitation load with the spare apparatus out of service. The amount of spare apparatus required is not very definite; practice ranges from providing sufficient spare exciter capacity to replace the exciter of the largest generator, to providing a spare exciter for each generating unit.

236. The exciter voltage in common use is 125 V for the very small plants, and in large plants it is 250, 375, or 500 V. The generators are usually shunt-wound and are excited by flat-compounded pilot exciters. With standard exciters it is possible to run at not more than 15% over standard voltage, which is quite sufficient to take care of the excitation of ordinary alternators at full load. Where overload capacities are to be used, it is frequently desirable to raise the excitation voltage as much as 25% during the peaks. Exciters can usually be arranged for this voltage with very little deviation from standard design.

237. The cost of exciters attached to the shafts of the main generators can best be expressed as a percentage of the cost of these generators; this will vary from 3 to 5%, depending somewhat on the speed.

238. High-speed, or quick-response, excitation has received much attention in recent years because of the importance of generator-voltage behavior in connection with power transmission and system stability. Exciters of relatively high "ceiling" voltage and rapid rate of build-up are employed to increase the transient stability of the system at times of short circuit.

239. Standard exciters with normal self-excitation have a ceiling voltage of approximately 135% of rated voltage and a rate of build-up of the order of 125 V/s. The working range is from 75 to 125% of the rated exciter voltage.

240. In quick-response excitation the rate of voltage rise is of the order of 400 to 600 V/s, the maximum voltage obtainable with standard exciters being about 320 V for a 250-V exciter. The exciter field circuits are usually separately excited.

241. Superexcitation. Effective application of the principle of quick-response excitation to synchronous condensers and generators has led to rates of voltage build-up of the order of 6,000 to 7,000 V/s, and the term "superexcitation" has been applied to such schemes. Exciters designed for such service are of higher rated voltage, 600 V for 250-V excitation, and have a correspondingly higher ceiling voltage, approximately 1,000 V.

The use of high-speed breakers and relaying to remove a-c system short circuits with great rapidity has reduced the trend toward very-high-speed quick-response excitation.

The rate of build-up depends upon the rate of current increase in the exciter field, and it is therefore found advisable to use a pilot exciter to furnish the excitation of the main exciter field. Such pilot exciters are either auxiliary units on the main generator shaft or motor-driven.

Pilot-exciter sizes for quick-response excitation are approximately 1½ to 5% of kilowatt rating of the main exciter. For superexcitation the pilot exciter will approximate 15 to 25% of the main exciter rating.

System stability is improved by high-speed quick-response excitation, as a result of the increase in the system short-circuit kVA during the disturbance. The consequent effect on apparatus and equipment must, however, be carefully considered before such a system is employed.

242. Regulation for voltage control, where the rapid load variation does not exceed

25% of the capacity of the machine and is of relatively high power factor, can be effectively applied to the generator exciter, the exciter being designed for stability down to 30% of its normal voltage.

If the load changes are of the order of 50% of machine capacity or greater, it is generally necessary to excite the field of the exciter, as by a pilot exciter.

243. Automatic voltage regulation is desirable and in many instances essential for plants supplying heavy fluctuating loads or connected to high-tension lines or when the generators are built with low short-circuit ratio.

244. Several types of automatic regulators have been developed to control the voltage of a-c generators. Small machines on moderate-size systems often use slow-acting regulators of the rheostat type. Larger systems and machines, particularly when connected through transformers to high-voltage systems, usually make use of high-speed voltage regulators to provide the needed improvements in generator stability under system disturbance conditions.

245. Rheostatic regulators are popular on small generators. While they have been connected to main-field rheostats, they are more commonly applied to the exciter-field rheostats owing to the higher speeds and lower maintenance associated with the latter. The exciter is usually separately excited if a wide range of control is required. They may be of the direct-acting or indirect-acting type.

246. The direct-acting voltage regulator of the Allis-Chalmers Company type is a device built on the principle of the torque motor and, like the vibrating type, acts upon the exciter field. The mechanical features are such that the regulator has the "overshooting" characteristic, inherent to the vibrating type, and represents a modern application of the principle of exciter-field rheostat control to high-speed voltage regulation. Figure 10-59 shows diagrammatically the construction of the a-c type and its connection to a single generator.

The torque produced by the current in the split-phase stator winding a and the rotor c of the torque motor is counterbalanced by a combination of two springs f (in the "astatic" type), which produce a constant torque whatever the position of the drum may be. In the "static" type, only one spring is used; and for each position of the moving parts, therefore, there is a definite different tension. A sufficiently high resistance is connected in series with the stator winding to prevent variations in temperature and small variations in frequency having any marked effect on the constancy of the voltage which the regulator is set to maintain.

The field rheostat with contact device is an integral part of the regulator. The stationary contacts l, to which the resistance coils g are connected, are arranged concentrically with the rotor spindle in two or four rows, depending upon the size of the regulator. The inner side of these contacts, facing the spindle, is provided with a V-shaped groove which serves as a guide for the moving contacts. The latter have the shape of a sector, with a curved strip of silver or carbon as contact surface and a steel needle as pivot. The latter rests in a jewel cup, which is carried by a leaf spring supported from the rotor spindle. The rotor can be turned through an angle of 60° and in doing so carries the two needle points over the corresponding arc of a circle. This causes the sectors to roll in the groove of the stationary contacts, cutting in or cutting out resistance. The latter, in the case of a generator voltage regulator, is connected in series with the shunt field of the exciter, as shown in Fig. 10-59.

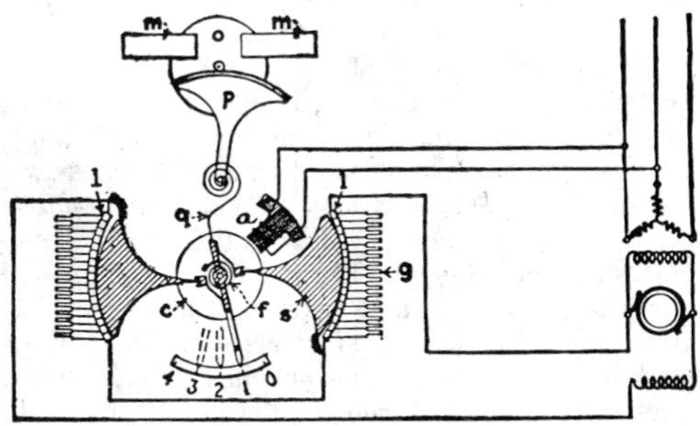

Fig. 10-59. Diagram of Allis-Chalmers direct-acting voltage regulator.

This design of field rheostat eliminates sliding friction and replaces it by rolling friction, which is so small that it can almost be neglected. All moving parts are made of aluminum; therefore their inertia is small; and as only small displacements are required to cover the whole regulating range, the moving system responds very quickly to changes in voltage. Only a few tenths of a second are required to cut in or out all resistance.

The antihunting device consists of an aluminum disk *o* rotating between two permanent magnets *m*. The disk is geared to the rack of an aluminum sector *P*, which can turn concentrically with the rotor spindle and is fastened to the aluminum drum *c* by means of a flexible spiral spring *q*, acting as a recall spring. If a change in voltage occurs, the eddy currents induced in disk *o* tend to resist quick response of the moving system. However, because of the flexible coupling, the drum, which directly controls the contact sectors *s*, will immediately take up a position in which the torque of the various springs and the electrical torque are again balanced. The coupling spring between the drum and the disk is made weak enough to allow powerful overregulation and yet maintain perfect stability.

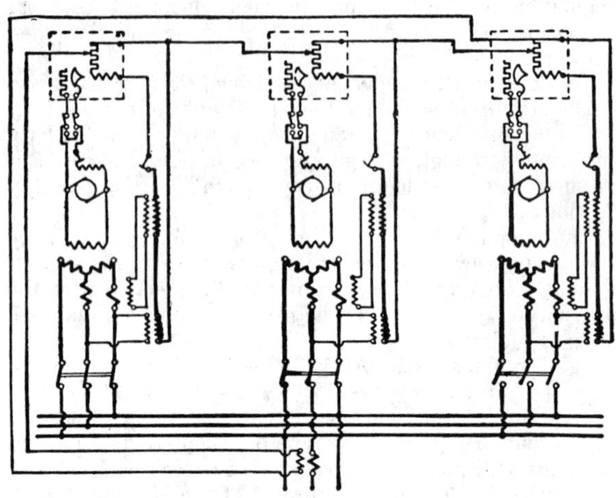

Fig. 10-60. Connections of Allis-Chalmers voltage regulators for three alternators in parallel with line-drop compensation.

Where several alternators are to operate in parallel, each one should be equipped with its own voltage regulator. Stabilizing current transformers must in this case be added, which react on the potential circuit through resistances or auxiliary current transformers. Line-drop compensation may be similarly provided. Figure 10-60 shows the connections for three alternators with current transformers for stabilization and line-drop compensation.

247. Direct-acting regulators of other manufacturers for small units are available in somewhat different designs but operating on the same general principle. The generator voltage is corrected by the direct action of a torque motor on a rheostatic element short-circuiting small steps of resistance in the exciter field as required. The resistance can be varied from practically zero to the maximum required. The voltage-sensitive element is a d-c torque motor. If the voltage being controlled is alternating current, copper oxide rectifiers are installed to provide direct current from the potential transformers. The Westinghouse Silverstat is shown diagrammatically in Fig. 10-61.

248. The indirect-acting-type regulator is used for large generators whose exciters are beyond the range of direct-acting regulators. It makes use of a motor-operated exciter-field rheostat actuated by voltage-sensitive contacts for small changes of voltage and quick-response contacts which short-circuit or introduce a block of resistance in case of large change in voltage until the motor-operated field rheostat moves to take control. The General Electric regulator provides a continuously rotating contact star wheel and raise and lower contacts actuated by a 3-phase torque motor energized from two potential transformers. If the voltage deviates from normal, the raise or lower contact engages the star wheel to complete a circuit and drive the rheostat. The length of contact with the points of the star wheel depends upon the deviation of voltage from normal. Figure 10-62 is a diagrammatic sketch of the General Electric Company's regulator.

The Westinghouse regulator makes use of a "contact-making voltmeter" equipped with a dashpot and energized by direct current obtained through copper oxide rectifiers from two potential transformers. When the floating arm makes contact with the raise or lower contact point to actuate the motor-operated exciter-field rheostat, it

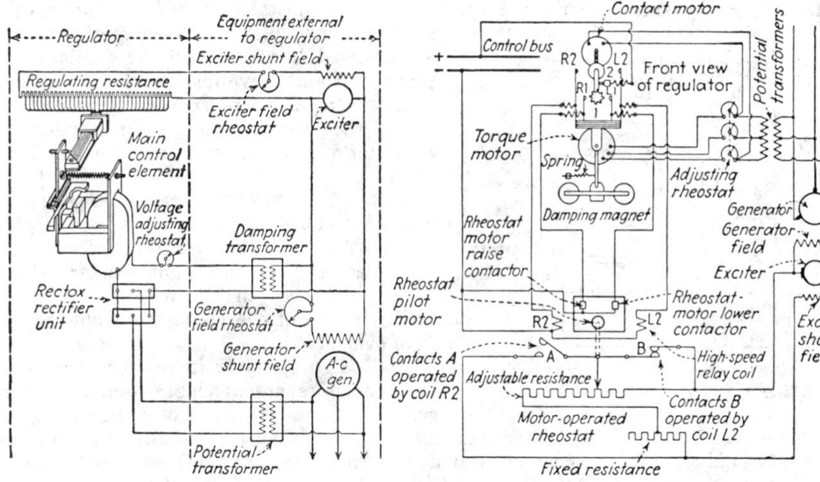

Fig. 10-61. Pictograph schematic wiring diagram of Westinghouse direct-acting Silverstat regulator for a-c application.

Fig. 10-62. Elementary connection diagram of a General Electric type GFA-4 voltage regulator with face-plate-type series rheostat and self-excited exciter.

also energizes coils to separate temporarily and slightly the stationary contacts, serving to widen the setting, thereby providing an antihunting device. On both regulators an additional set of contacts given a wider setting actuate the high-speed contactors for wide changes in voltage. It is usual practice to provide a separate regulator for each generator and its exciter.

249. Parallel operation of generators with regulators is accomplished by means of a suitable type of **cross-current compensation.** The method employs an equalizing reactor or compensator which adds a small voltage, proportional to the reactive current delivered by the generator, to the voltage delivered by the potential transformers. This gives a slight droop to the voltage held by the regulator on reactive loads and serves to divide reactive currents in proportion to load currents. Differential compensation is used when line-drop compensators are installed to increase automatically the voltage as the load increases. With this connection, all the equalizing reactors or compensators are connected in series as in Fig. 10-63. With this connection no current flows in the equalizing reactor under balanced load conditions; but in case of unbalance, the currents flow through the regulators to decrease the excitation of the generator carrying excessive reactive current and increase the excitation of the generators carrying low reactive current.

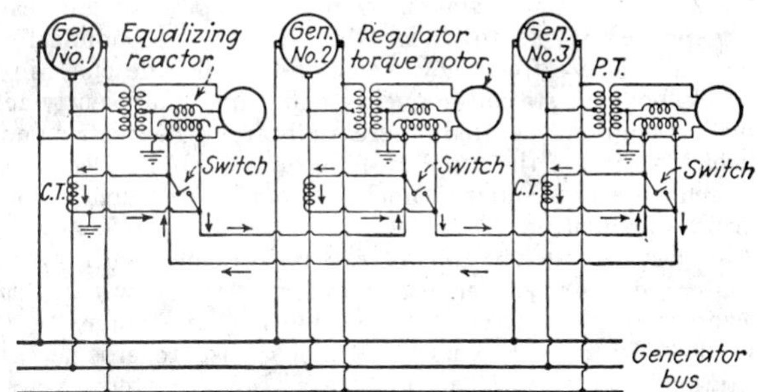

Fig. 10-63. Elementary diagram of application of differential cross-current compensation, showing secondary current flow with balanced load conditions.

250. Flat voltage, or slightly drooping if cross-current compensation is required, is ordinarily held by automatic regulators. The voltage may be changed by hand from time to time, if required, by manipulating an adjustment rheostat.

251. Automatic increase in voltage may be obtained if necessary by means of a line-drop compensator. This is a combination of adjustable reactances and resistances set to simulate line characteristics. It is connected in series with the regulator to subtract from the voltage delivered by the potential transformers a voltage proportional to the line drop, resulting in an increase in generator voltage as load increases (see Fig. 10-64).

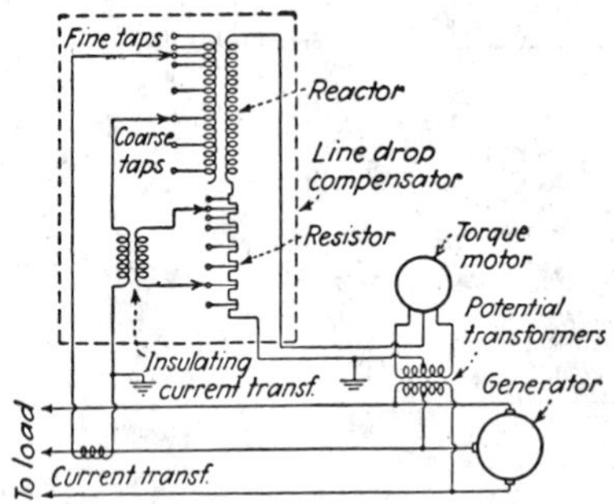

Fig. 10-64. Generator voltage regulator with line-drop compensator.

252. Pilot-exciter voltage-limiting relay may be supplied when the regulator is to be used with a water-wheel generator with direct-connected exciter and pilot exciter. This inserts resistance in the field of the pilot exciter in case of overvoltage, as would result from overspeed.

253. High-speed excitation-response voltage regulators are available from the large generator manufacturers, to provide a higher degree of generator stability under transient conditions than can be obtained from the direct- or indirect-acting types of regulators. Special attention is required in their application; so the manufacturer should be consulted in each case. The more popular arrangements are known as (*a*) Allis-Chalmers Regulex regulator, (*b*) General Electric Buck-Boost Amplidyne regulator, and (*c*) Westinghouse Mag-Amp regulator. In some cases, the proper application of these systems requires special exciter design features. Recently, static systems known as Alterrex by General Electric and "Brushless Excitation" by Westinghouse have made their appearance.

LOAD AND FREQUENCY CONTROL

254. Load and frequency control of interconnected generators introduces problems which are relatively simple in a system having one or two generating stations but which are more difficult in large interconnected systems with many stations scattered over a wide area.

Many systems hold the frequency so close to the standard that electric clocks may be operated from the system with variations of not more than a few seconds.

255. In a single-station system the operator can readily adjust the governors of the prime movers to divide the load most economically between them and, guided by an accurate frequency meter and an electric clock in comparison with a standard clock, hold the station frequency and "time" sufficiently accurate. Clocks are available with a pointer, operated differentially from the electric and standard clock mechanisms, which indicates the number of seconds that the electric clock is fast or slow as compared to the standard time. Automatic frequency control can be provided when conditions justify.

Closer frequency control can be provided with steam turbogenerators than with water-wheel-driven generators owing to the slower governor action and greater speed changes resulting from the great inertia of the water.

256. In large systems a central load dispatcher is necessary to assign loads to various stations and units in accordance with a predetermined schedule, modified from time to time as the actual load differs from the predicted load or as emergencies arise owing to loss of generating units or tie lines. The load dispatching may be by telephone, remote telemetering and signaling, or both.

Load assignment to a particular station varies with the type and function of the station and its relation to the system. Frequency control is sometimes assigned to one of the largest generating stations; stations feeding local load and a tie load carry their local load plus or minus a scheduled tie load, whereas stations with no local load follow a scheduled tie-line load. Automatic load and frequency control has been successfully applied to units in one or more generating stations of interconnected systems. Very large systems are sometimes divided into load districts, each with its own load dispatcher, often with a central load-dispatching agency for general supervision over the districts.

257. Automatic load-frequency control is necessary for maintenance of good overall system operations, proper sharing of load between generating stations, suitable regulation of tie-feeder loading between systems, and maintenance of proper frequency and time control.

The problem of control resolves itself into (*a*) the measurement of a quantity, (*b*) interpretation of the measurement in terms of deviation from a control point, and (*c*) the application of corrections to restore the measured quantity to its normal value. In some cases more than one measurement is required for proper operation of the control equipment. In one of these control systems, developed by Leeds and Northrup Co., measurements of both frequency and load are taken to give various types of combined load-frequency control.

Generator, station, and system loads are measured through the summation of various thermal-converter millivolt outputs. Frequency is measured by a frequency-bridge-type instrument. As the system frequency varies, the bridge circuit is rebalanced by the instrument movement which positions a slide wire used for transmission of a direct voltage.

As all these data are fed into a master controller, it is able to detect the need for more or less generation and to send impulses to the different stations calling for load increase or reduction.

By the use of area requirement, proportional load control, the equipment is able to call for changes at the several generating stations such that they each, in effect, supply the load of their respective areas, thereby causing a minimum of power flow over tie feeders from one station to another.

Within each generating station it is possible to select the units that will be used for regulation and to adjust the percentage of the requested load change that is placed on each machine.

The following three types of **area control** may be used, one at a time, as selected by the system operator:

1. **Flat frequency control** varies the power input to the prime mover so that the system frequency will be corrected to the predetermined value.
2. **Flat tie-line control** varies the input to the prime mover to correct the tie-line load to a predetermined schedule. In this case another system must maintain frequency.
3. **Tie-line bias control** is a modification of (1) and (2) in that the system is allowed to follow its normal regulating characteristic directed toward holding normal frequency. If the frequency deviates from normal, generation is automatically changed to correct it, but only by an amount permitted by the bias setting on a MW/0.1 c of the frequency change. This arrangement has been found to work very satisfactorily on power systems where a large number of stations and generators must be kept under control. If the system frequency change is more than a predetermined amount, say ¼ to ½ c, the control can be made to change from "automatic" to "hand" automatically and sound an alarm, so that the system operator can correct the faulty condition.

ELECTRICAL-SYSTEM PROTECTION

258. The art of electrical-system protection has advanced rapidly since 1950. Available space will not allow a complete treatment of the subject here, but the major recognized principles will be outlined, and a Bibliography is included to permit further study.

259. Protective relay systems are intended to detect abnormal conditions and to isolate or indicate them by initiating the operation of circuit breakers or other devices. The most common electrical hazard against which protection is required is the short circuit. However, there are many other conditions—e.g., undervoltage and overvoltage; open-phase; unbalanced-phase currents; direction of power flow; underfrequency and overfrequency; overtemperature—for which protection is sometimes desired.

260. Overall generating-unit protection has been adopted by some companies recently where protection is afforded to the boiler, the turbine, the generator, and the generator transformer, as a coordinated system. Failure in any part, such as loss of draft air to the boiler, causes relay operations that take the whole unit out of service to the generator electrical circuit breaker (see the Bibliography for references on this item).

261. The quality of protection justifiable may vary widely for different electrical installations. An electrical-supply system for a large community must be designed so that faults in one unit of the system will not affect service to the community after that unit is isolated. It is an accepted practice to design large electrical-supply systems so that service will be maintained to the system as a whole even though several units become defective simultaneously. The service to some types of industrial plant and small stations may receive a minimum of protective relaying, while for service to other types of plant, where continuity of service is highly important, multiple supply circuits are used, and a high quality of protective equipment is desirable.

262. Accurate calculations of short-circuit currents, voltages, phase relations, and system stability characteristics are essential to good protection engineering and for proper selection of circuit breakers. Diagrams giving the impedance data on all major parts of the system are very useful where the system is large and complicated. Accurate knowledge of the behavior of generators under short-circuit conditions and of the characteristics of transformers at the instant of being energized is important in applying protection to the system. It is also necessary to know the utilization load characteristics in many cases, since inrush currents to transformers or to motors being started and the effect on the service voltage as a result of these demands may have a decided effect on the type of protection selected.

263. Complicated networks can be studied more accurately and with less effort by use of a d-c or an a-c calculating board which permits setting up a system in miniature form to study its short-circuit characteristics. The electrical manufacturers and many larger utility companies have such boards available. Some educational institutions are also equipped with them. Digital and analog computers are also available. Programs for digital computers have been prepared to determine system short-circuit currents, settings for the more common relays, load-flow studies, system swings for load changes of different magnitudes, and many other quantities. Analog computers represent a system by miniature polyphase components such as generators, transformers, reactors, capacitors, circuit breakers, lines, cables, loads of varying characteristics, etc. This type of computer, with circuit-switching means, is an excellent tool for determining system switching impulse voltages, ferroresonance behavior, and other transient and dynamic conditions for the proposed system circuits before they are built or designed.

264. Instrument transformer characteristics are exceedingly important in protective relay applications, since good accuracy is frequently necessary under abnormal conditions of current and voltage. It is important that current transformers maintain their ratio reasonably well for overcurrent relay applications, but it is imperative that their ratio error be small where they are used in current-comparison systems with sensitive relays, such as in differential protection schemes.

PROTECTIVE RELAYS FOR A-C SYSTEMS

265. Overlapping of protection where different units of the system come together through circuit breakers is considered excellent practice. Protective relaying which permits sensitivity to faults only within a certain unit part of the system allows a higher quality of system protection by virtue of its high degree of sensitivity and its adaptability to fast speed of operation. Figure 10-65 indicates how protection may be set up to operate for faults only in the protected area with overlapping of the areas around the connecting circuit breakers.

266. Backup protection is important to the proper functioning of a good system of electrical protection. It is the second line of defense which functions to isolate a faulty section of the system in case the primary protection fails to function properly. It may be provided either on the same circuit breakers which would normally be opened by the first line of protection or, still better, so that the second line of protection makes use of different circuit breakers. It is generally satisfactory to isolate more than the faulty section alone in case "backup" protection is called upon to function.

267. Relay selectivity is greatly simplified when protection is used as indicated in Fig. 10-65, since the relays protecting any unit of the system are independent of the other primary relays, and their time of operation may be made very fast, as they are not required to select with other types of protective relay. In other systems where overcurrent relays are used, selectivity may be obtained by proper selection of current and time adjustments in the relays. Selectivity is obtained in some cases by means of "blocking," such as (*a*) use of directional relays where short time of operation is desired for one direction of current flow and long time is essential for the other direction of flow and (*b*) use of overcurrent and undercurrent or voltage relays to open tripping control circuits under special conditions; (*c*) instantaneous plunger-type overcurrent relays may be used as a part of induction overcurrent relays and adjusted to operate instantly on predetermined high values of current, as might occur for faults near a station.

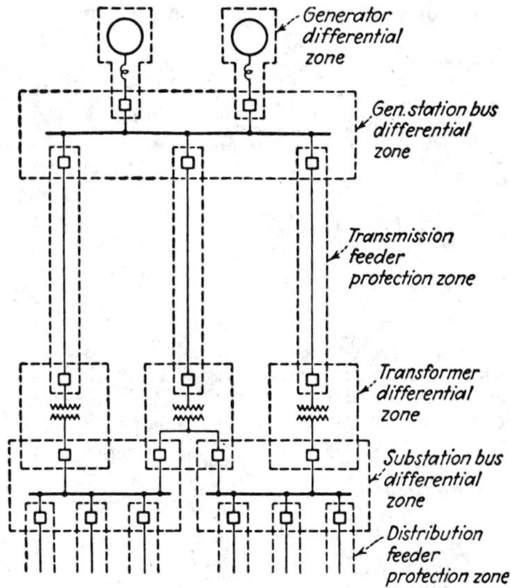

FIG. 10-65. Single-line diagram showing typical protection zones.

There have been much discussion and varied practice as to the time selectivity required between two relays in a circuit to assure that the faster relay operates and its breaker clears the faulted circuit before the second relay-tripping contact closes. It is considered good conservative practice to use 0.5 s between relays where the older 12- to 15-c breakers are in use. This may be reduced to 0.4 s with equal conservatism where modern 8-c breakers are employed. Many users report satisfactory performance where selectivity timing 0.1 s less than the above figures is used, and it is believed that these shorter periods can be employed safely where the relays concerned are kept in a good state of repair and cleanliness. The timing for selectivity between elements of impedance and reactance relays (see Par. **309**) is much the same sort of problem as timing between steps of overcurrent relays; however, the type of timing device is different, and some advantage may be taken of fast resetting of this device after the impedance or reactance element is reset.

268. Relay operation speeds have become extremely important owing to (*a*) the higher standards desired for service continuity, (*b*) the need for reducing damage to electric apparatus during faults, (*c*) applications of rapid reclosing features, and (*d*) the complications of system interconnections plus the resulting stability problems. Figure 10-66 indicates the relation between relay and breaker time and the maximum power that can be transmitted over an interconnecting transmission line.

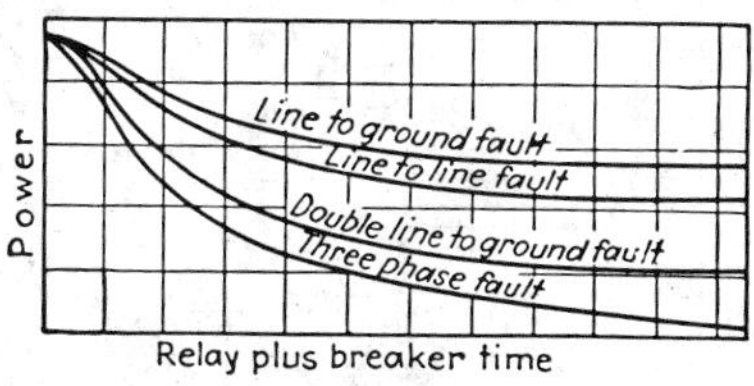

FIG. 10-66. Curves showing allowable power limits vs. time of clearing faults on one transmission system.

269. Generator and large motor stator windings require protection for internal short circuits and occasionally for overheating. In some installations overfrequency, overvoltage, and reverse-power protection may be desirable. Stator-winding short circuits may develop as 3-phase, phase-to-phase, phase-to-phase-to-ground, phase-to-ground, strand-to-strand, or turn-to-turn short circuits. All are reasonably easy to detect with relays except the strand-to-strand and turn-to-turn faults. Even the turn-to-turn type of fault can be detected, however, with certain degrees of unbalance, as indicated below.

270. Protective relaying for generators and other large rotating machinery windings should be (*a*) sensitive to faults in the generator, (*b*) nonresponsive to faults outside the generator, (*c*) fast in operation to prevent serious burning within the machine (0.05 s or less for relay performance), (*d*) equipped to isolate the generator or motor from the system and deenergize the field winding, and in some cases (*e*) equipped so that the fire-extinguishing agent will be automatically released within the machine.

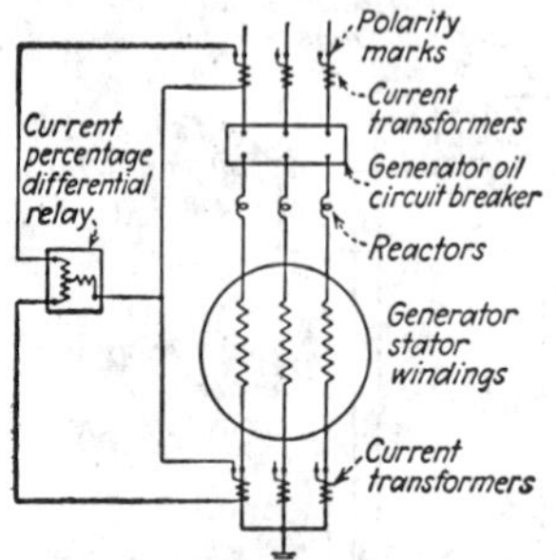

Fig. 10-67. Connections for overall current-differential protection on one phase of a 3-phase Y-connected generator. The other phases are treated similarly.

271. Generator differential protection is the most common and the most effective of all available means for detecting electric failures in stator windings. It acts by comparing the current magnitude at the two ends of a phase winding in its most common form.

Figure 10-67 shows the percentage differential protection around each phase winding. The percentage differential relay has the advantage of high sensitivity at light loads and decreased sensitivity for external short circuits where the current transformers might cause incorrect operation due to ratio errors at high currents. This type of relay permits settings that will detect short circuits nearer the neutral of Y-connected generators than is possible with over-current relays

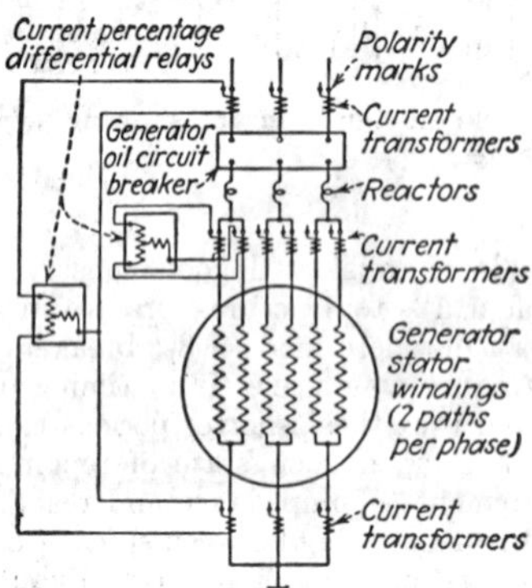

Fig. 10-68. Connections for balanced-winding current differential plus overall differential protection.

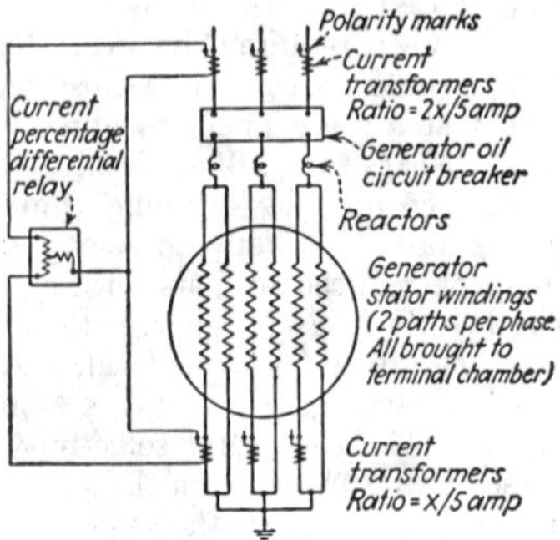

Fig. 10-69. Overall and balanced-winding differential protection for generators.

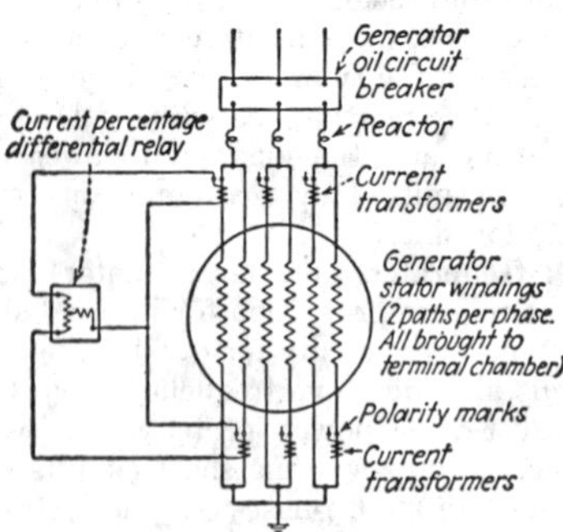

Fig. 10-70. Overall and balanced-winding differential protection for generators.

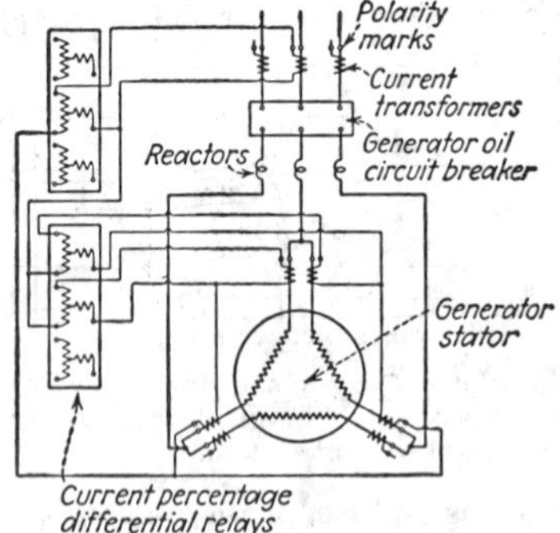

Fig. 10-71. Current differential protection for delta-connected generator, including connections to oil circuit breaker.

used in the differential circuit. The windings may be connected in Y (as in Fig. 10-67) or in delta. In either case the current at the end of each phase is compared with that at the other end of the same phase.

Older schemes not ordinarily used today are (*a*) a double-winding relay where the currents from the two current transformers produce a condition of magnetic balance under normal conditions; (*b*) a self-balancing scheme in which both ends of the stator winding are carried through the same current transformer, thereby producing zero magnetic flux under normal conditions; (*c*) over-current relays connected in the circuit of the percentage differential relay shown in Fig. 10-67.

272. Turn-to-turn short circuits may be detected by comparison means, such as the balanced winding protection indicated in Fig. 10-68, where multiple-path windings are brought out. Current transformers are installed in each circuit of a given phase with their secondaries connected differentially so that current will flow in the relay operating coil only when an unbalanced condition exists. Careful studies are required to determine what settings are required to detect turn-to-turn faults.

A variation of the balanced-winding differential scheme is shown in Fig. 10-69, which reduces the number of relays and current transformers but involves differential connections between current transformers of different ratings whose accuracies may differ under high current conditions. This necessitates less sensitive relay settings.

Where all leads of multiple-path windings are brought out of the machine, a simplified differential arrangement shown in Fig. 10-70 may be used. This method reduces the number of relays and current transformers required to obtain balanced winding and overall differential protection, but it requires less sensitive relay settings and does not give protection to the cables between the generator and its oil circuit breaker.

273. Differential protection on delta windings does not provide protection for faults in the generator cables up to the oil circuit breaker. Figure 10-71 shows an arrangement of connections which will, with one additional relay and one additional set of current transformers, provide full protection to the bus side of the generator oil circuit breaker. This scheme provides greater sensitivity than could be obtained by using one relay and balancing one current transformer against a group of current transformers in the generator leads for each phase.

274. Generator ground-fault protection may be obtained on a machine where only four leads are brought out by using connections shown in Fig. 10-72. Only one relay element is required.

Ground-fault differential protection may be obtained on an ungrounded generator which is connected to a system having a source of ground current as shown in Fig. 10-73.

A grounded generator connected to the system so that it can supply no ground-fault current to system short circuits may be protected for internal ground faults as shown in Fig. 10-74.

Generators connected to the system through two-winding delta-Y transformers may also be protected for internal phase-to-ground faults as shown in Fig. 10-75. The transformer is small (25 to 150 kVA, depending on rating and capacitance to ground of the generator windings), with its high-voltage winding rated about the same as the generator phase-to-phase voltage and its secondary winding rated 120 to 480 V depending on what value provides a suitable resistance to the connected circuit. A special voltage relay (insensitive to third-harmonic voltage) is available to measure the secondary-winding 60-c voltage for ground-fault indication or tripping. This method provides ground-fault protection for the generator leads, transformer secondary winding, and any other direct-connected circuits in addition to the generator winding.

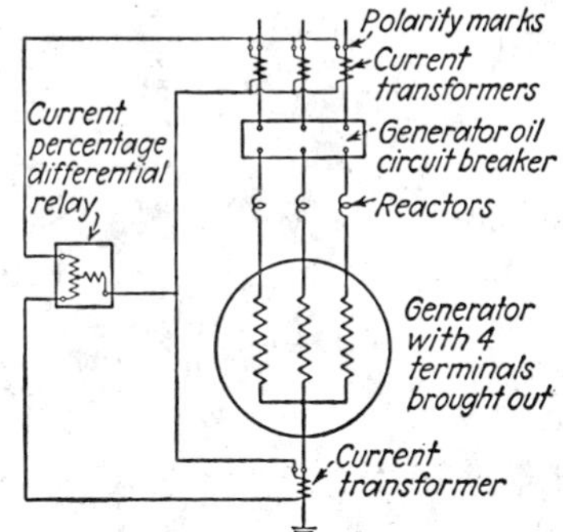

Fig. 10-72. Ground-fault differential protection for four-terminal Y-connected generator.

275. Overcurrent protection of generators is seldom used except in unattended stations or in generators for auxiliary supply. The overcurrent relays must be given high-current and long-time settings to select with bus and feeder protection, which is undesirable; generators may be tripped from the bus in case of sustained system disturbances or failure of relay selectivity; and the protection is inoperative for generator failures unless there is sufficient back feed from the rest of the system. Because of the long-time settings usually required, careful attention must be given to the decrement characteristics of the generator. The generator stator current may be no more than one to two times full-load current after 2 or 3 s following the short circuit. Where over-current protection is necessary for generators, consideration should be given to voltage-restrained or voltage-controlled over-current relays. In some cases these will allow high overcurrent settings for external short circuits and yet will automatically change to low settings for generator circuit failures, when the voltage is low owing to the short-circuit effects.

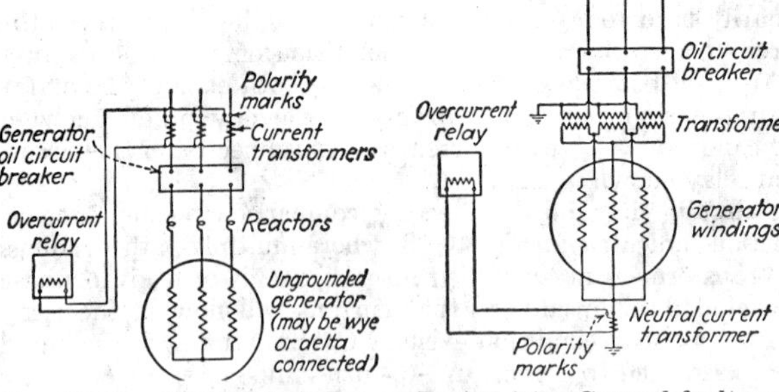

Fig. 10-73. Ground-fault protection for ungrounded generator operating on system with neutral grounded.

Fig. 10-74. Ground-fault protection for grounded generator with infinite zero-sequence impedance in the system.

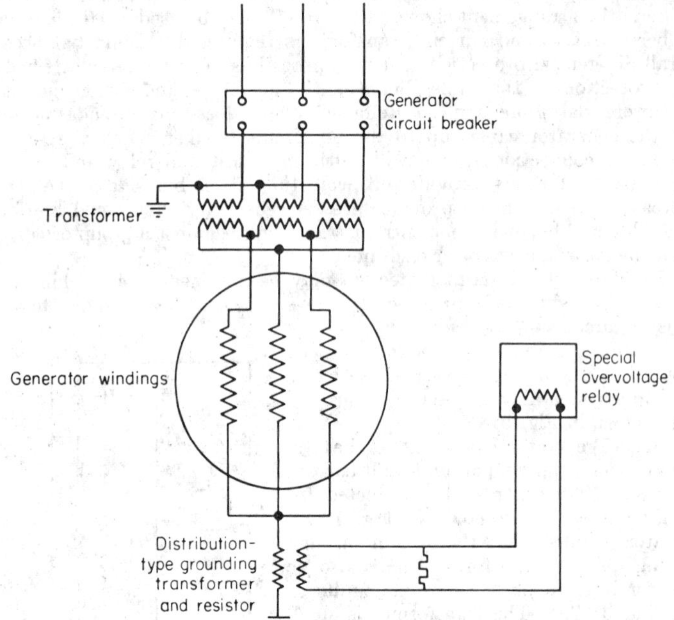

Fig. 10-75. Generator phase-to-ground fault protection with high neutral impedance.

276. Directional protection for generators is seldom used though superior to straight overcurrent protection. The relay must be provided with sufficient time delay to prevent undesired operation on system "swings" where the power may reverse into the generator for a short period of time. Power-directional protection may be required for special purposes, e.g., for generators driven by high-pressure turbines that exhaust into a lower-pressure steam system. Directional relays so applied may be used to **indicate** that the flow of steam has been reduced below that required to provide the no-load losses of the turbogenerator, instead of tripping the machine immediately.

277. Other protection for generators may be required, particularly in unattended stations. Equipment is available for overfrequency or other types of underspeed and overspeed protection, overvoltage protection, overheating, and high vibration protection for the machine windings and bearings. In case of line outages, synchronous condensers should be disconnected by one or more of these means.

278. Out-of-step protection may be required in some generator applications, particularly on small water-wheel machines connected to high-capacity power lines, to remove the generator from the system in case loss of synchronism occurs. Relays which operate on a succession of power reversals and current impulses are available to detect loss of synchronism. Loss-of-field relays of the reverse kvar types may serve the out-of-step requirements, as well as field failure.

279. Field-winding protection for large generators is desirable, since field-winding failures may cause serious effects on the system voltage as well as severe damage to the machine itself. Short-circuited field coils have been known to create sufficient vibration in a rotor to cause mechanical failure of the unit. Loss of field in a generator presents a more serious problem where system stability is involved than the loss of the generator due to opening of its main circuit breaker.

280. Undercurrent and undervoltage protection in the field circuit have been used to detect field-winding failures but cannot be made sensitive owing to the wide range of excitation required for normal operation.

281. Reverse reactive kVA relays may be connected to instrument transformers on the stator winding to detect unusual reductions in field excitation which result in a reverse flow of reactive current as shown in Fig. 10-76. The stabilizer provides greater sensitivity in the relay if the bus voltage is reduced. An undervoltage relay may be used in conjunction with the kvar relay to prevent operations when the bus voltage is not reduced below a predetermined value. Direction-impedance zone-type relays are also available for this purpose. They operate on the basis of one voltage and current phase measurement.

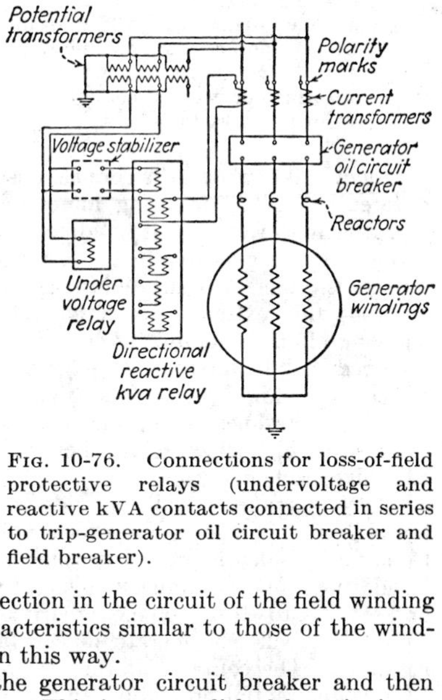

A ground-indicating system is often provided on ungrounded field-supply systems to permit isolation before a second ground develops. While ground-detector lamps are the most common method, greater sensitivity may be obtained by use of sensitive polarized d-c relays.

FIG. 10-76. Connections for loss-of-field protective relays (undervoltage and reactive kVA contacts connected in series to trip-generator oil circuit breaker and field breaker).

282. Overheating of field and amortisseur windings is not generally protected against, but this may be desirable in cases such as isolated installations of synchronous condensers. Temperature relays are available for connection in the circuit of the field winding during starting operation and have heating characteristics similar to those of the windings. Amortisseur windings may be protected in this way.

283. Generator protective relays may trip the generator circuit breaker and then the field circuit breakers or both simultaneously. This is accomplished by tripping a fast multicontact hand-reset control relay with the protective relay. The trip circuit to the field breakers may be made up through this control relay and an auxiliary switch on the generator breaker or may be taken directly from the control relay. The latter method is preferred among protection engineers.

284. Station bus protection deserves the most careful attention, since bus failures are, as a rule, the most serious that can occur to an electrical system. Unless properly isolated, they may develop to involve a complete switch gallery and cause a station shutdown. Much attention has been given to modernizing the bus protection in old stations where complete protection was not originally provided. New station designs should include fast and reliable bus protective equipment.

285. Bus-differential protection is a common method, since it is selective, fast, and sensitive. Current-transformer accuracy must be taken into account, as for generators, but there are various means for using existing current transformers of inferior accuracy where modernization of old installations is considered. Current-transformer secondary windings should be grounded at only one point. This will prevent main-circuit fault currents from entering the control wiring and causing burnouts.

286. Bus-differential protective schemes in use vary widely, depending upon the conditions existing. Several types in common use will be described.

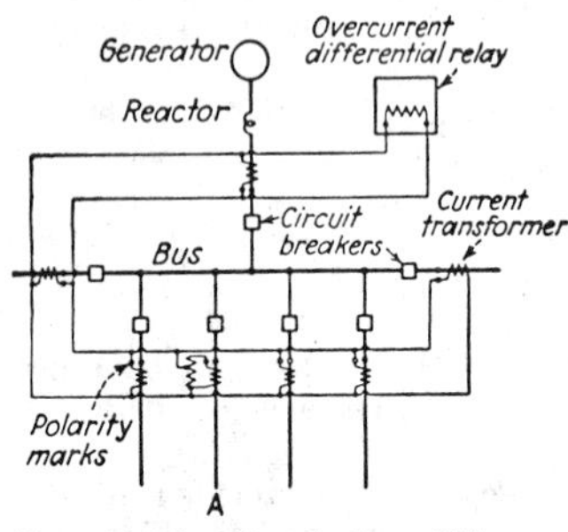

FIG. 10-77. Simple bus differential using overcurrent relays (single-line diagram). All current-transformer ratios are the same except in circuit *A*, where an autotransformer is used to change an unsuited ratio.

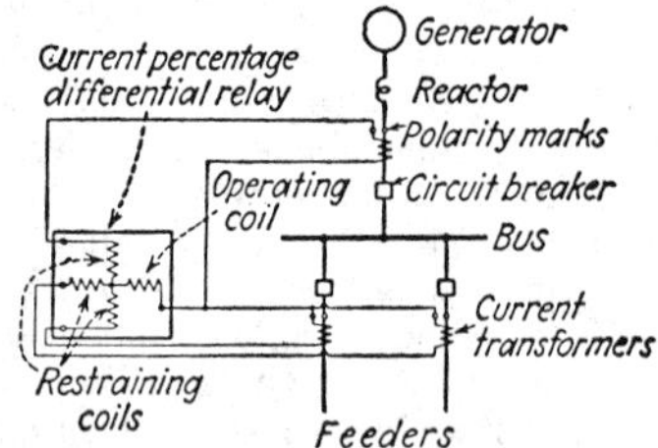

FIG. 10-78. Bus-differential protection using sensitive percentage differential relays.

Figure 10-77 shows the connections for the most common type of overall **bus current differential protection** using an overcurrent relay. In this scheme the ratio of all current transformers must be alike. In some cases it is necessary to alter the effective ratio of existing current transformers when applying bus-differential protection by autotransformers as indicated for circuit *A* (Fig. 10-77).

Generally, extremely sensitive relay settings are not required for bus protection, since bus faults usually involve high currents. However, greater sensitivity can be obtained in bus differential relays by using a **current percentage-differential relay** as indicated in Fig. 10-78. The high fault currents involved and the various current-transformer ratio errors during the first few cycles of fault may cause incorrect operation of sensitive high-speed relays; hence the advantage of the percentage differential relay or a slight time delay.

Since bus faults usually provide an excess of current for relay operation, it has been found desirable in some cases to use percentage differential relays to prevent operation for faults outside the protective zone, due to current-transformer ratio errors, and to use, in addition, an instantaneous overcurrent relay connected in the operating-coil circuit and set at 4 to 10 A. The overcurrent relay will operate instantly in case of bus short circuit but will not operate in case a current-transformer secondary becomes open-circuited. The contacts of the two relays would be connected in series in this case.

Bus-differential protection by use of linear couplers instead of current transformers, connected as shown in Fig. 10-79, is not subject to instrument transformer ratio errors. This is obtained by the use of air cores in the linear couplers, which are constructed much like a bushing-type current transformer with the iron core omitted. As a result, core saturation is not a problem. However, the linear coupler output is measured in millivolts, and it is not suitable for operation of current instruments such as ammeters, wattmeters, etc. Connected as in Fig. 10-79, the total millivolts normally add to zero in the circuit, and no voltage is impressed across the relay. The same condition exists for external system short circuits. For an internal fault condition, current entering the bus is not balanced by the current leaving the bus through the linear couplers, and a net voltage is therefore impressed across the relay, causing it to operate and to trip the circuit breakers. High-speed, sensitive bus protection can be obtained by this method.

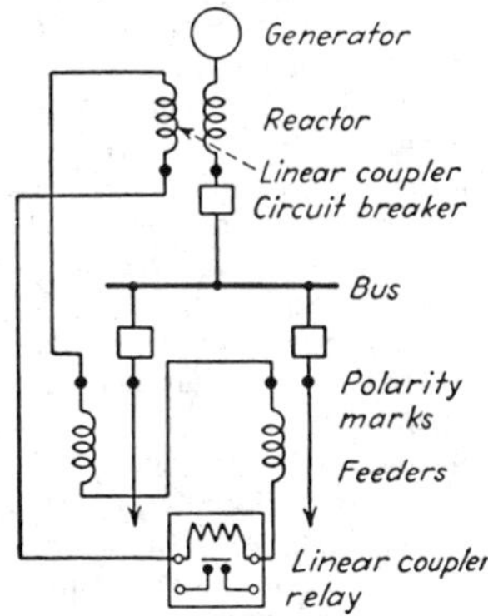

FIG. 10-79. Linear coupler bus-differential protection. (*Westinghouse Electric Corporation.*)

287. In applying bus differential to existing stations, installing large numbers of current transformers on the feeder circuits can be avoided if the feeder circuits are provided with reactors. Current transformers must be in the main circuits to the bus, e.g., in the generator and bus tie connections, and these may be connected as indicated in Fig. 10-80. The relays required include an impedance relay, set so that it will not operate for faults beyond the feeder reactors, an overcurrent relay, which may be connected in the phase or residual circuits of the instrument transformers, and an undervoltage relay, connected to potential transformers on the protected bus section. The overcurrent relays may be used in the residual circuits alone when the bus structure is such that interphase faults cannot occur. They may be set high to assure that a fault condition exists when they operate. The undervoltage relay serves to prevent incorrect operation in case the potential element of the impedance relay is open at the time a fault occurs external to the feeder reactors. A bus fault will cause high current, low impedance, and low voltage, all of which initiate positive action of the relays.

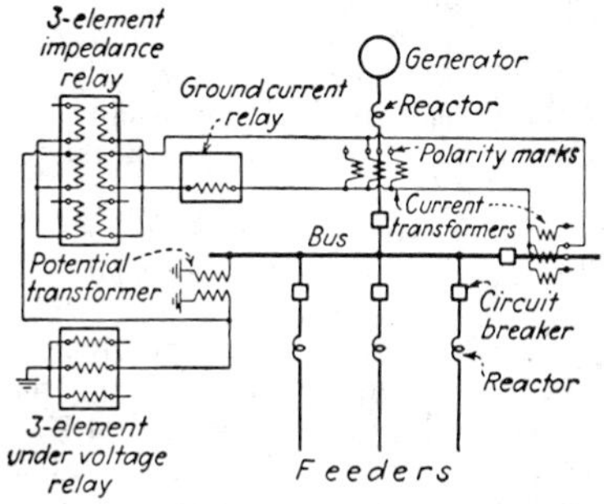

FIG. 10-80. Partial bus differential using impedance, ground-current (could be phase currents), and undervoltage relays.

288. Some station buses permit use of a partial differential protection wherein power-directional and overcurrent relays only are used. This scheme does not require the current transformers to have the same ratio; and where the number of connecting circuits is small and the normal flow of power makes such relaying feasible, it can be applied economically. Figure 10-81 shows bus protection by the directional-comparison method wherein a group of feeders is supplied through a bus section by one generator. In case of bus short circuits, the overcurrent relays operate and both directional relays operate to indicate flow of energy into the bus from all circuits. Time selectivity with the transformer and generator relays is required. The directional-comparison scheme can be

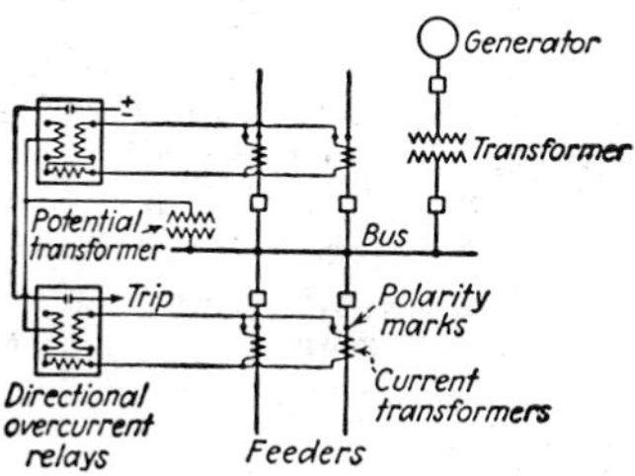

FIG. 10-81. Partial bus-differential protection using directional overcurrent relays.

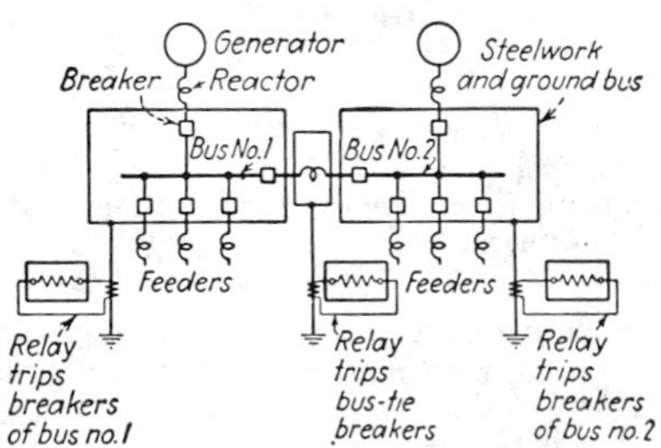

FIG. 10-82. Ground-fault bus protection applicable to grounded systems where faults must involve ground.

used regardless of whether the feeders have reactors, provided that a source of power is available at the ends of all feeders.

289. Ground-fault bus protection provides an excellent means of isolating bus faults where the normally grounded parts surrounding the bus structure can easily be isolated from ground so that a connection can be made from these parts to the ground bus through current transformers. All instrument transformers and secondary wiring of this scheme can be entirely isolated from high-voltage parts. Figure 10-82 shows the general arrangement of the fault bus whereby the flow of ground current is controlled so that the proper relay performance may be obtained in case a fault occurs within a prescribed area. The method may not be applicable to existing installations owing to the difficulty of adequately insulating normally grounded parts. Since the protection is for ground faults only, it is applicable only to a system having a grounded neutral.

The scheme has been found applicable in outdoor as well as indoor installations. Its proper functioning depends upon adequately insulating all grounded parts of the section to be protected except through its grounding current transformer. Sections must be insulated from each other to prevent false operations. Provision should be made for supervising this insulation occasionally.

290. Transformer protection justifiable in different installations will vary widely depending upon the size of the unit, how it is connected in the circuit, its importance as a unit part of the system, and its voltage rating. Small distribution transformers may be protected by fuses, while larger units in feeder circuits may be equipped with various types of protective equipment varying from long-time overcurrent to high-speed differential protection. Transformers such as those connected to feeders supplying an a-c network are often relayed as part of the high-tension feeder.

Transformers may be subjected to short circuits between phase and ground, open circuits, turn-to-turn short circuits, and overheating. Interphase short circuits are rare and seldom develop as such initially, since the phase windings are usually well separated in a 3-phase transformer. Faults usually begin as turn-to-turn failures and frequently develop into faults involving ground.

It is highly desirable to isolate transformers with faulty windings as quickly as possible to reduce the possibility of oil fires, with the attendant destruction and the resulting cost for replacements.

291. Differential protection is the preferred type where breakers are provided on each winding that may be connected to a source of power, due to its simplicity, sensitivity, selectivity, and speed of operation. Figure 10-83 shows a simple arrangement of current percentage differential relaying for a two-winding transformer.

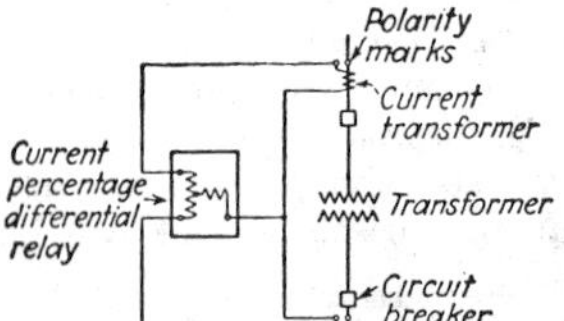

Fig. 10-83. Simplified diagram showing current-differential protection for a two-winding transformer.

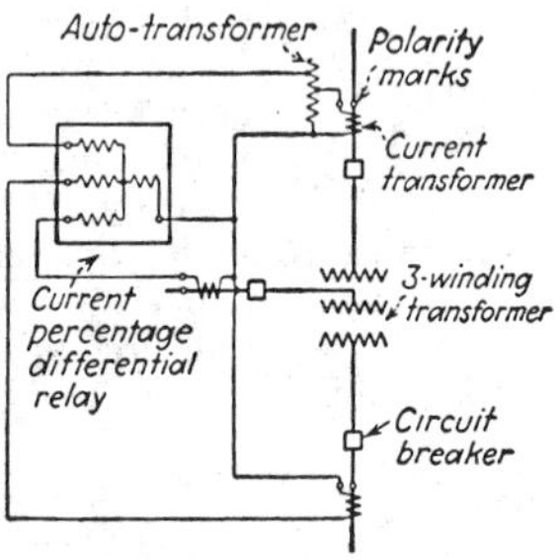

Fig. 10-84. Simplified diagram showing current-differential protection for a three-winding transformer.

Three-winding transformers may be protected in a similar manner as shown in Fig. 10-84. If the current-transformer ratios are not perfectly matched, taking into account the voltage ratios of the transformer, autotransformer or auxiliary current transformers are required in the current-transformer secondary circuits to match the units properly so that no appreciable current will flow in the relay operating coil, except for internal fault conditions.

In applying differential protection to transformers, somewhat less sensitivity in the relays is usually required, as compared with generator relays, since they must remain nonoperative for the maximum transformer tap changes that might be used. It is also necessary to take into account the transformer exciting inrush current which may flow in only one circuit when the transformer is energized by closing one of its circuit breakers. As a rule, incorrect relay operation can be avoided by imposing a slight time delay for this condition.

292. Voltage-load tap-changing (LTC) transformers may be protected by differential relays. The same principles of applying differential protection to other transformers hold here as well. It is important that the differential relay be carefully selected so that the unbalance in the current-transformer secondary circuits will not in any case be sufficient to operate the relay under normal conditions. It is suggested that the current transformers be matched at the midpoint of the tap-changing range. The current-transformer error will then be a minimum for the maximum tap position in either direction.

293. Grounding transformers may be protected as indicated in Fig. 10-85. This scheme does not provide for ground-fault protection of the delta winding but will detect severe turn short circuits on either winding.

294. Current-transformer and relay connections for various types of differential protection are indicated (*a*) in Fig. 10-86 for a Y-delta transformer, (*b*) in Fig. 10-87 for a Y-Y or delta-delta transformer, and (*c*) in Fig. 10-88 for a three-winding Y-delta-Y transformer.

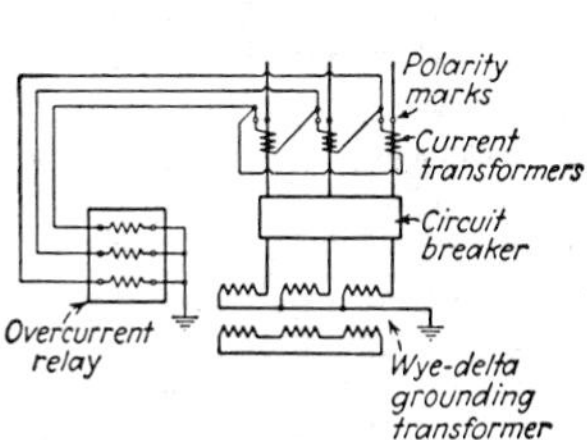

FIG. 10-85. Protection for Y-delta ground transformers.

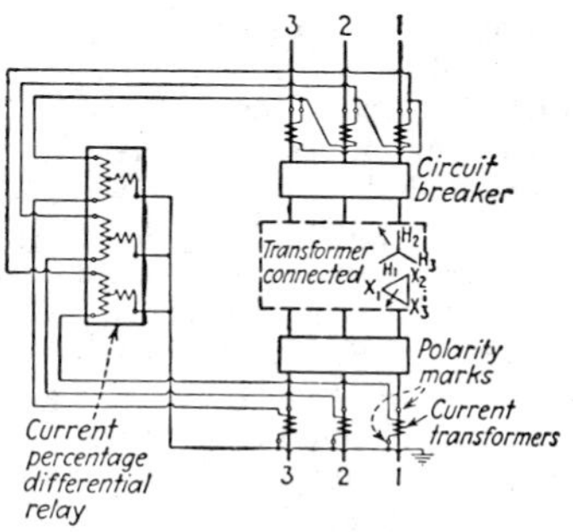

FIG. 10-86. Current-transformer connections for Y-delta transformer differential protection.

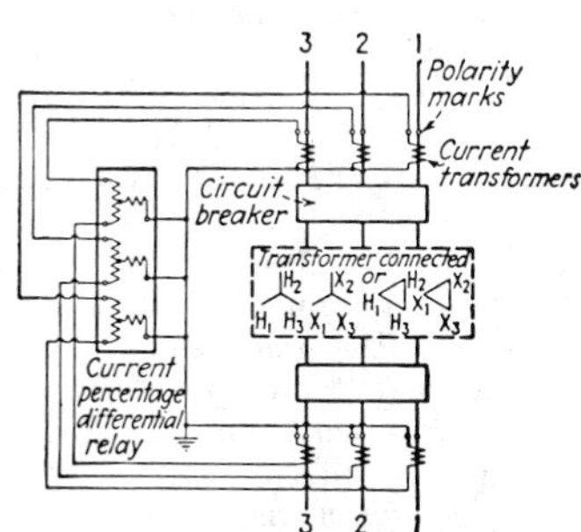

FIG. 10-87. Current-transformer connections for Y-Y or delta-delta transformer differential protection.

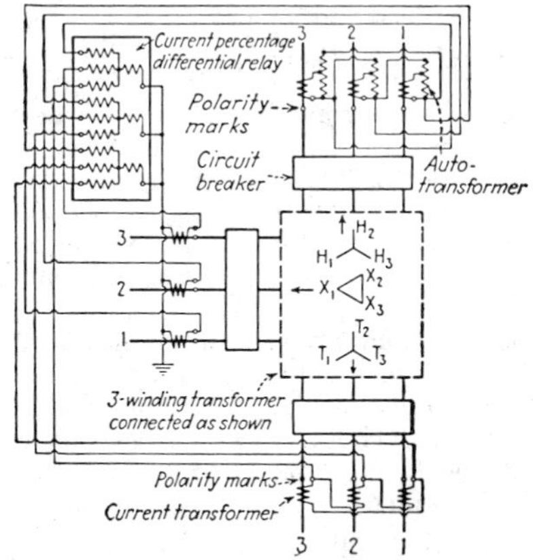

FIG. 10-88. Current-transformer connections for three-winding Y-delta-Y transformer differential protection.

Two rules, frequently used in laying out the wiring for differential protection of transformers whose main windings are connected in Y and delta, are:

1. The current transformers in the leads to the Y-connected winding should be connected in delta; current transformers in the leads to a delta-connected winding should be connected in Y.

2. Make the delta connection of the current transformers a replica of the delta connection of the power transformers; make the Y connection of the current transformers a replica of the Y connection of the power transformers.

Current transformers should be chosen that will give approximately 5 A secondary current at full load on the transformer. This will not be possible in all cases, particularly for transformers having three or more windings, since the kVA ratings may vary widely and may not be proportional to the voltage ratings.

295. Overcurrent protection should be applied to transformers as the primary protection where a differential scheme cannot be justified or as "backup" protection if differential is used. Frequently faster relaying may be obtained for power flow from one direction by the use of power-directional relays.

296. Transformer overheating protection is sometimes provided to give an indication of overtemperature, rarely to trip automatically. Overload relays of the replica type may be connected in the current-transformer circuits to detect overloading of the unit. Others operate on top-oil temperature, while still others operate on top-oil temperature supplemented with heat from an adjacent resistor connected to a current transformer in the circuit. This latter relay is adjusted to operate on a simulated "winding-hot-spot temperature."

297. Gas- or oil-pressure relays are available for attachment to the top or side of transformer tanks to indicate winding faults which produce gas or sudden pressure waves in the oil. Rapid collection of gas or pressure waves in the oil, due to short circuits in the winding, will produce fast operation. This type of protection is slowly gaining in popularity in the United States.

298. Feeder-protection requirements vary more widely than those for any other part of an electrical system. As a general rule, feeders require protection only for short circuits. In some cases there may be need for other types of relays to prevent operation at low voltage, open phase, out-of-step, and overload. Protection for short circuits may be provided in many different ways, the degree of refinement usually depending upon location and relation of the circuits to other circuits of the system.

299. Overcurrent protection is frequently obtained on relatively low-capacity distribution branch circuits and in services from distribution circuits to consumers by the use of **fuses.** While such protection is advantageous from the viewpoint of cost, it does not permit supervision and fuse renewals must be made in the field. Fuse protection has been improved greatly since 1950 by the development of automatic fuse renewals in special disconnect-switch designs. These permit as many as three automatic fuse operations before the circuit is finally interrupted.

300. Overcurrent protection is economically obtained at circuit breakers by the use of series-trip coils in small-capacity circuits where accurate settings are not important. The smaller circuit breakers may be equipped with multiple-trip coils which may be connected directly to current transformers in the main circuit. The trip-coil plungers are adjustable over a moderate range. Built-in series trips with instantaneous or time-delay features are the most common methods of tripping low-voltage air circuit breakers.

301. Further refinement in overcurrent protection may be provided by various types of relays which have a wider latitude of adjustment in minimum operating-current and time adjustments. The induction-disk type of relay is the more common of this class for a-c circuits. It has an inverse definite-time-current characteristic and is applicable to most feeder-protection schemes. Overcurrent relays may be had with a definite-time characteristic wherein an instantaneous overcurrent element is used in conjunction with a definite-time timer, with both operating current and time being adjustable. Relays of these types may be supplemented by instantaneous elements so that currents above certain values will cause instantaneous tripping. Several different inverse time characteristics are available in a-c induction-disk-type relays.

302. Overcurrent relays are commonly used for backup protection where feeders are provided with other, more complicated relay equipment.

Overcurrent relays with current and time adjustments are applicable to radial, loop, and other, more complicated system connections, though they may require a certain type of "blocking" relays as described below. When a short circuit occurs, the relays should trip each breaker necessary to isolate completely the faulted unit from the system, and they should so select with other relays that no other serviceable unit is disconnected. This requires time selectivity on the more simple types of system, while current selectivity with inverse definite-time relays may be used in some cases. Figure 10-89 shows a radial system having feeder sections in series with overcurrent relays only. The relays on each feeder section should be set to protect not only its circuit but the circuit or the feeder section further from the source of power so that backup protection is obtained.

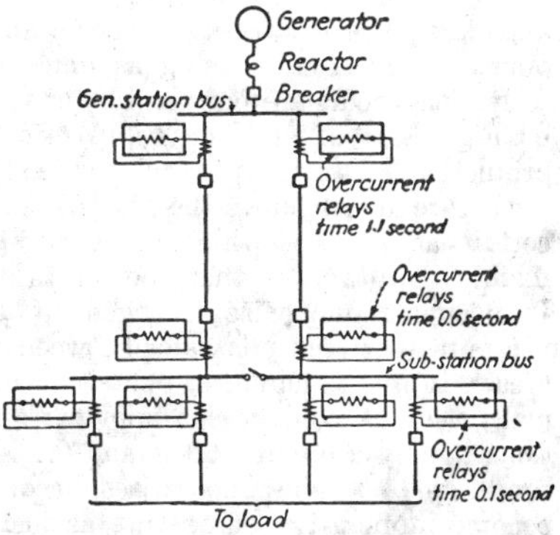

FIG. 10-89. One-line diagram of simple radial system using overcurrent protection on feeders with time selectivity.

303. **On a grounded system** relays may be inserted in the residual circuit of current transformers and set much more sensitively than the phase overcurrent relays to provide protection against feeder ground faults. Blocking relays may be used with ground overcurrent elements, as is done for phase overcurrent relays. Figure 10-90 shows the connections for overcurrent and ground relays on a feeder connected to a system with its neutral grounded.

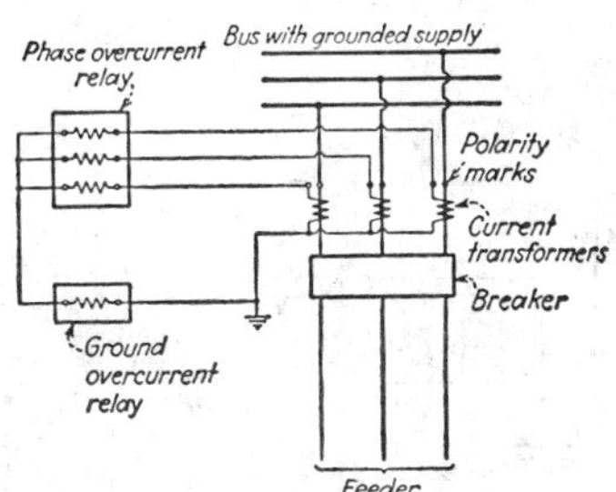

FIG. 10-90. Connections for overcurrent and ground relays on feeder from grounded source of supply.

304. **Blocking devices** are sometimes required to obtain proper selectivity between overcurrent relays. Figure 10-91 shows a loop system provided with overcurrent relays at the source of power and directional overcurrent relays at each substation. The directional elements block operation of the overcurrent relays except for power flow in a given direction. The time given at each circuit breaker is the relay setting in Figs. 10-89 and 10-91. While 0.5 s is indicated between relays for selectivity, this may be reduced to 0.4 s where modern 8-c breakers are used and to 0.35 s for 5-c breakers where each relay is accurately timed by a synchronous timer or other similar device.

Other types of blocking relay may be used as required, e.g., high-set instantaneous overcurrent elements and voltage relays.

Power-directional relays are available as single-phase and polyphase devices. The polyphase relay indicates the true direction of power flow. In using single-phase relays care must be taken in choosing potential connections for each phase to assure proper operation for both balanced and unbalanced faults.

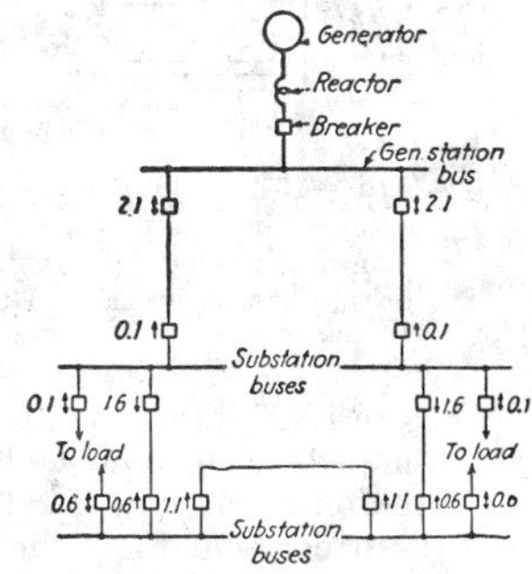

FIG. 10-91. One-line diagram of loop system using overcurrent and directional-overcurrent protection with time selectivity. ↕ 2.1 indicates overcurrent relays with 2.1-s time. ↑ 0.1 indicates direction of 0.1-s overcurrent directional relays.

305. **Differential protection** may be used on short feeders, as already described for buses in preceding paragraphs.

306. **Pilot-wire protection** provides protective features for feeder circuits which are similar to those given by differential protection for buses and equipment. Its application is limited somewhat by cost and circuit resistance and capacitance of pilot wires, but in its present stage of development circuits using as much as 23 mi (one-way distance) of 2-wire No. 19 telephone circuit are permissible for the a-c scheme.

Figures 10-92 and 10-93 illustrate the important parts of two systems of pilot-wire protection. One is an a-c pilot-wire scheme, and the other uses d-c pilot wires.

Figure 10-92 shows the most recently developed method. It is considered to have certain advantages which favor its application in the majority of cases. One of its major advantages is that the relays do not require the use of potential transformers. It is an a-c pilot-wire arrangement using 2 wires only. Normally, it will trip only if a fault, between phases or between phase and ground, develops within the protected area. It may, therefore, operate fast, and relays do function in about $\frac{1}{60}$ s. Normally the voltage between pilot wires is of the order of 5 V; under short-circuit conditions it may approach 60 V and 0.1 A. The relays will not operate on "out-of-step" conditions. If the pilot wires become open-circuited or short-circuited, the relays become inoperative or operate as instantaneous overcurrent relays depending on the condition of the wires and the manufacturer of the relay. An instrument and test switch are provided for manual check of the pilot circuit; however, automatic supervisory instruments are available which will sound an alarm in case of pilot-wire failure. The current transformers used at the ends of the line should have similar accuracy characteristics, to assure correct relay performance.

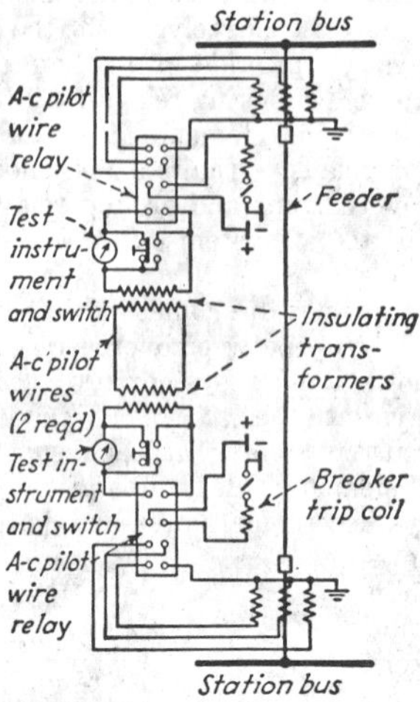

FIG. 10-92. Single-line diagram of a-c pilot-wire protection for feeders, operating on both phase and ground faults.

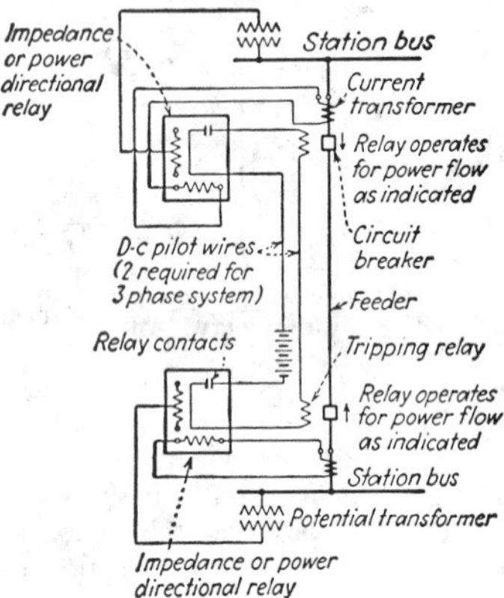

FIG. 10-93. Single-line diagram of directional-comparison pilot wire. No current flows in pilot wire normally.

Figure 10-93 illustrates an older type of scheme which uses 2 d-c pilot wires. Directional-impedance or power relays serve to trip breakers at both ends of the line in case of current or power flow into the line from both ends. This scheme requires that short-circuit current be fed into the faulted feeder at both ends of the circuit. It is generally considered to be limited to a one-way length of 10 mi in the pilot wire. The relays are high-speed types. They may under some conditions operate on out-of-step conditions. An open pilot wire makes the arrangement inoperative. Short-circuited pilot wires make the relays instantaneous directional-impedance or power elements. Means for supervising the condition of the pilot wires are available.

307. Balanced current is an economical and effective method for obtaining fast and selective relaying of multiple feeders operating in parallel with each other between two station buses. This scheme is applicable at the generating-station end of the feeders and also at the substation end if more than two feeders are operated in parallel at all times. Where only two feeders are used, balanced-power protection may be used at the substation. Figure 10-94 shows a system where balanced current is used at the generating station and balanced power is provided at the substation.

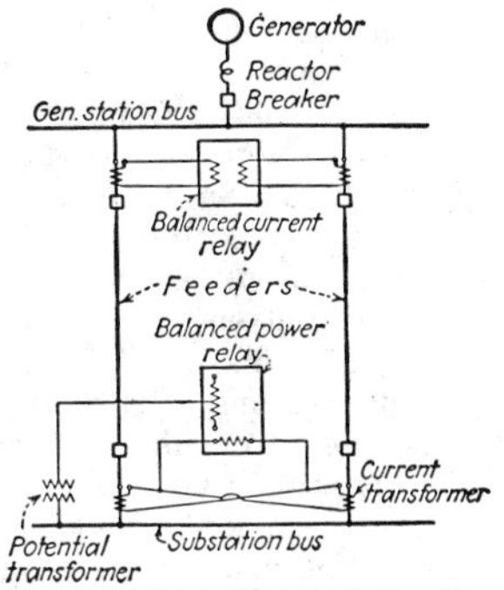

308. Phase-current balance may be used where unbalanced-phase short circuits may cause relatively small currents to flow so that the overcurrent relays are not sufficiently sensitive to detect them. These relays may be set sensitively on systems the phase currents of which are normally well balanced.

309. Impedance and reactance protection may be used on single or parallel feeders and are particularly applicable to lines having intermediate sectionalizing points. The performance of these relays is governed by the impedance or by the reactance of the circuit.

One type of impedance relay has a time characteristic which is directly proportional to the circuit impedance. Its operation is very fast for faults near the station and may be quite long for short circuits at a distance.

FIG. 10-94. Single-line diagram of balanced current and balanced power protection on parallel feeders. Balanced current may be used at substation if another source of current is available there or if several parallel feeders are in service.

Other types of impedance relays and all reactance relays are designed to have step-type time characteristics. They are adjusted to operate instantly for short circuits in the first 80 to 90% of the line impedance and to have sufficient time, in the second step, to select with relays of the adjacent circuit for faults on that feeder. A third timing step is provided so that full backup protection is provided to the adjacent circuit. Figure 10-95 shows how the time of the relays is adjusted to provide full selectivity for faults in any section. Where power may be supplied from both directions it is necessary that the impedance or reactance relays be provided with directional elements for "blocking" purposes.

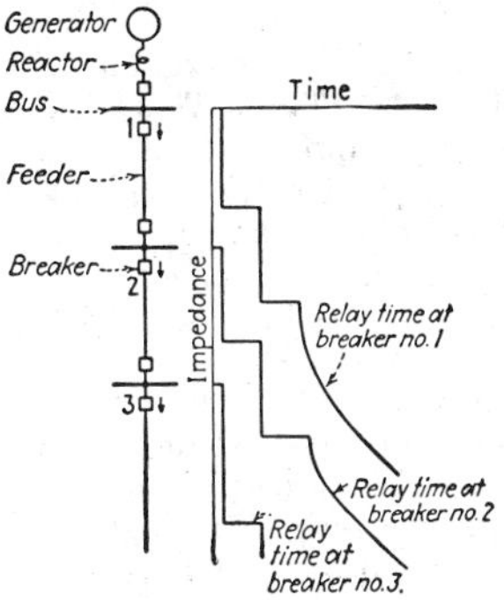

FIG. 10-95. Single-line diagram of impedance relay timing for flow of current from generating station.

Impedance and reactance protection is applicable largely to long feeders. There is a minimum length of line that can be properly protected by this type of relay, and this should be studied and discussed with the relay manufacturer where there is a question.

Impedance and reactance relays may not properly protect a line with a long high-impedance-tapped connection.

Impedance and reactive relays are not generally applicable to ground-fault protection. They have been used with some success, but a careful study should be made in individual cases and discussed with the relay manufacturer.

Impedance and reactive relays provide similar protection. The reactance relay is somewhat more complicated and expensive but is affected less by arc resistance at the point of short circuit. Both types of relay may perform incorrectly on serious system swings and out-of-step conditions.

310. Carrier-current protection uses conventional directional impedance and reactance relays at the line terminals. The high-frequency carrier communication system ordinarily uses the high-voltage circuit conductor of one phase as its transmission circuit. When both terminal relays indicate, by direction, that the protected feeder is faulted, no carrier signal is transmitted and both terminal relays operate quickly. When one terminal relay indicates, by direction, that the fault is outside the protected feeder, a carrier signal is transmitted to the opposite terminal to block tripping by the relays at that station. Carrier-controlled phase and ground relays always "overreach" the impedance to the opposite feeder terminal.

Directional impedance or reactance relays are used together with various auxiliary devices to control the transmission of carrier frequency over the high-voltage line. When the distance relays at both ends of a line section indicate a fault in that section, the relays at both ends of the line operate to trip the breakers instantly. The carrier system consists of (*a*) a reliable source of power such as the station control battery, (*b*) transmitter-receiver unit (frequency ranges from 30 to 250 kc), (*c*) line-tuning unit for tuning the line to the desired frequency, (*d*) high-voltage coupling capacitor for introducing carrier frequency on the line, (*e*) carrier-frequency trap (resonant choke coil) to confine the carrier frequency within the line section to which it applies, and (*f*) protective gaps to prevent damage due to line voltage surges.

Since carrier-current protection does not provide backup for faults in adjacent feeder sections, it is necessary to provide other types of relay for this purpose as well as to back up the carrier-current protection itself. Since carrier-current protection is applied usually for overhead high-voltage transmission-feeder protection, it is customarily provided with impedance or reactance relay-backup protection (see Par. **309**), together with necessary ground relay backup.

311. Out-of-step relays have been developed to operate on a line ohmic basis to disconnect two systems which have lost synchronism with each other. The manufacturers may be consulted for details.

312. Microwave protection is the same as carrier-current protection, except that it uses microwave transmission and receiving equipment for transmitting the carrier frequency signal between the stations, instead of using one phase of the power line. It is usable especially for cable circuits where it may be difficult to obtain sufficient signal strength through the cable conductor. It requires line-of-sight transmission. Hence, intermediate microwave relay stations with suitable cone-type receivers and transmitters may be required for long distances.

313. Transfer-trip relays have become popular in recent years because of the trend toward using a minimum number of high-voltage power circuit breakers. This method permits tapping a high-capacity line by attaching a transformer directly to it, having a size such that the line protective relaying cannot detect a short circuit in the transformer secondary circuit. The transformer protective relays, in this case, may actuate "transfer-trip relays" which will, via carrier, microwave, or telephone-line channels, cause tripping of the high-capacity supply-line circuit breakers. Such transfer-trip relaying sometimes makes use of audio-tone frequency combinations.

314. Mercury-arc rectifiers should be provided overcurrent protection on the a-c side with sufficient time delay to obtain selectivity with relays on the d-c circuit breakers. Other protection may be provided for the transformer, if desired (see Transformer Protection, Par. **290**). The rectifier will usually require protection for high tank pressure and high and low tank temperature. Reverse-current d-c protection may be placed on the cathode breaker to prevent back feed from the d-c bus on an internal failure or "arc-back."

The rectifier auxiliary motors will require the usual protection against overheating, undervoltage, and short circuit. The mercury-condensation pump may also require overheating protection.

315. Station auxiliary equipment may be given essentially the same type of protection as is provided for main station and system circuits. The degree of refinement is sometimes of a lower order. As examples, overcurrent protection for house generators is commonly used, as it would be on a house transformer; and since feeder circuits are usually of the radial type, they require only overcurrent protection. Overcurrent relays are set high or controlled by undervoltage-type relays to prevent false operation on motor-starting current.

316. Squirrel-cage and wound-rotor induction motors require overcurrent protection for short circuits, set sufficiently high to prevent operation on motor-starting current. In addition, large motors which may be subjected to overloads should be equipped with overtemperature relays or induction overload relays having a very long operating time. Locked-rotor protection relays are also desirable in some instances. Small motors are protected against overheating by overload-type fuses or thermal elements in contactors—not fewer than two elements for a 3-phase motor.

Motors for essential auxiliaries have all short-circuit protective relays trip the supply breakers directly. Thermal overload and other relays may give alarm only and allow time for shutdown or other method of correction, by hand control.

Protection against single-phase operation is rarely applied, as overload protection is usually satisfactory for this purpose. Reverse-phase protection may be required at locations where reverse rotation would cause damage, as in elevator motors. Undervoltage tripping is needed in some locations, particularly where across-the-line starting is not permissible and where unexpected starts would be hazardous on restoration of power following a shutdown. Time-delay features are desirable if undervoltage trips are used.

317. Synchronous motors may require any or all of the protective equipment listed for induction motors. Loss-of-field protection is sometimes applied, consisting of undervoltage and undercurrent relays in the field supply circuit or a power-factor relay connected in the armature supply. This relay functions when the power factor reaches a predetermined value at a given load.

Large motors in unattended stations may require bearing-temperature protection.

318. Station lighting circuits may be satisfactorily protected by time-delay overcurrent relays on the bus air circuit breakers and with high-grade low-voltage fuses at distribution panels. A small section of the regular lighting in each important part of a station is often connected to a common supply point and provided with undervoltage selective protection which will transfer this portion of the lighting to a d-c circuit (control or excitation buses) in case the a-c supply fails. The selective relay should return this portion of the lighting to the a-c supply when normal voltage is again available.

319. Relay-operating principles are many and varied for d-c systems. Each type of element may frequently be designed to respond to different system conditions, the fundamental principle of operation always remaining the same.

Alternating-current Relays. The coil and plunger is a simple principle used for both protective and auxiliary relays. It is applicable to both voltage and current and may be made very fast or can be controlled by a time-delay attachment. The plunger principle has been developed in recent years to multiple-plunger designs for use in balance protection through moving arms as in high-speed impedance relays.

The hinged armature design with a coil is more frequently used in a-c auxiliary relays.

Probably the best-known principle used in a-c relays only is the induction disk or cylinder. Shaded-pole or watt-type driving elements are used to control the movement of the rotating element. This principle lends itself readily to a large variety of different applications.

The plunger and induction-disk principles have been combined to obtain a balance or comparison design.

Alternating-current motor designs are used for timers and for high-speed power-directional indication.

Static-element arrangements using transistors, capacitors, resistors, diodes, rectifiers, etc., have recently been developed to take the place of most of the electromechanical arrangements described above. These have the advantages of high-speed performance and low maintenance demands and are not subject to mechanical-movement problems. They are available for transmission-line relaying, timers, overcurrent, overvoltage, and many other purposes. Their principal disadvantage is the tendency of application engineers to question their reliability, since they are of recent origin. Solid-state relays are also being developed for use with low-power transducers that will probably be used instead of current and potential transformers in extra-high-voltage circuits in future years.

Direct-current Relays. The coil-and-plunger design is also used for d-c systems, as for a-c systems, with essentially the same features for control of operating speed. Multiple-plunger designs are not common for direct current. Plungers are used on air circuit breakers to trip the breaker directly when current or voltage reaches predetermined limits.

The hinged armature design with a coil is used in d-c auxiliary relays and in protective relays on circuit breakers, etc.

Polarized relays are widely used where current-directional features or high sensitivity is required. One magnet is of hard steel permanently polarized or of soft iron polarized by a coil supplied with direct current. The other magnet or moving coil is energized by the current or voltage to be supervised. This type of relay is usually very fast. It is made in very light, sensitive designs and with heavy, rugged elements capable of tripping circuit breakers directly.

Motor-driven relays are used for timing and similar functions.

Miscellaneous Relays. Vapor or volatile liquids are used in relays to detect temperature or pressure changes. A gas pressure relay has been designed to detect incipient faults in oil-filled transformers. Also, a tank oil fault pressure relay is used for oil-filled transformers.

Metallic expansion and contraction principles are applied to relays to detect temperature and pressure changes.

Balanced bridge connections are used to operate one or more of the relay types described above, to detect temperature and other changes as selected.

PROTECTIVE RELAYS FOR D-C SYSTEMS

320. Direct-current generators, as a rule, require overcurrent, reverse-current, and overspeed protection. The protection should be fast in operation; and for machines rated 500 V and above, high-speed breakers may be necessary to prevent or reduce the damage due to commutator flashover. Fast reverse-current relays are satisfactory for isolating a faulted machine from others where parallel operation is used. Ground relays are sometimes connected, in a single ground connection used for connecting the machine framework to the station ground bus, to trip the machine in case of internal fault.

321. Excitation generators, if operated on a battery-supplied bus or in parallel with other excitation generators, should be equipped with reverse-current relays to isolate a faulty generator. Reverse-current relays on exciters must be set sufficiently high to prevent incorrect operation on transient-current reversals which may be associated with a-c disturbances, such as short circuits. Where an exciter is operated alone, as in the case of a direct-connected exciter, no protection is required, though loss of field relays may be provided on the generator to initiate opening of the a-c circuit breaker.

Loss of field protection is rarely used on d-c generators but may be obtained by the use of undercurrent relays in the field circuits.

In addition to the above protection, undervoltage, overvoltage, winding overheating, bearing overheating, and reverse-polarity relays may be required in unattended stations.

322. Synchronous converters require the same type of protection on the d-c side as do d-c generators. On the a-c side, overcurrent protection should be provided, having time selectivity with the d-c circuit breakers. If the transformer is air-cooled, air-flow relays should be used to detect loss of ventilation.

Frequently 250-V converters on Edison 3-wire systems are not equipped with reverse-current relays; so they will not be tripped in case of minor current reversals. The d-c breakers may be tripped by the operation of overcurrent relays on the a-c side of the machines. Overspeed protection is extremely important if a source of d-c back feed is available.

323. Direct-current motors should have fast overcurrent protection, together with undervoltage and sometimes overspeed relays. Undervoltage protection is sometimes applied as a part of a starting apparatus.

324. Batteries are usually not provided with protective relays. On control and excitation systems, it is considered better to burn short circuits clear than to risk incorrect relay operation on a battery circuit breaker.

325. Direct-current feeders are usually provided with fast overcurrent protection. Small-capacity branch circuits may be protected by fuses. Special cases may demand directional current protection.

Direct-current feeders for railway service require special attention to protective relays and circuit breakers. High-speed clearing of feeder short circuits may be essential to the proper operation of converting equipment and other apparatus.

STATION DESIGN

326. The general designs of stations differ greatly depending on the conditions which are to be met, and no fixed rule can be established by which a type of station design can be definitely decided.

Factors influencing station design are the size and character of the station and system, the character and size of the load, and the value of the service rendered. See System Requirements, Pars. **1** to **52**.

327. The object of station design is to arrange and house the electrical equipment which is required for the operation of the station from the standpoint of simplicity and reliability of operation.

The purpose of a station is to give reliable service at minimum cost, and a station design must be based on consideration of the causes of disturbances and means for minimizing their effects.

328. Failure of every piece of apparatus must be considered as a possibility, and provisions must be taken for limiting the magnitude and area of such disturbances.

Safety of the men and of major apparatus and equipment from contact with high-voltage circuits, fire, water, etc., at the time of serious trouble should be assured to the maximum extent possible, as it is on such precautions that the restoring of a power station to service, with the least delay, depends.

329. Indoor installations are often used up to 13,800 V. However, in recent years many stations have been built which are essentially of outdoor construction. Climate conditions at the proposed site are a primary factor in determining the extent to which outdoor construction is feasible. Maintainability of various components and cost of maintenance are also important factors in this respect. It has become generally accepted that high-voltage switching equipment associated with the station be placed out-of-doors at 33,000 V and above.

330. Where indoor designs are considered more practical, building design is determined by the arrangement of the apparatus and equipment. Recent practice in large steam plants has been to provide common buildings for housing boilers and turbines. The arrangement of equipment within the buildings should be determined only after a careful study of the economic factors and with full consideration of operation under both normal and emergency conditions.

Simplicity in building design and harmony with the surroundings are desirable. Architectural treatment should be sound, attractive, and modern in its expression of needs, standards, and processes of the time; and the purpose and utility of the buildings should be obvious.

Miscellaneous building requirements such as repair shops, storerooms, offices, toilets, stairways, elevators, protective measures for accidents and fire, ventilation, lighting, and heating must be given careful consideration.

Shipping and erection limitations of the apparatus that must be handled, particularly generators, transformers, large motors, and switching equipment, must be given consideration in the early stages of the building design. The economics of factory-assembled switching and bus structures may modify building designs.

331. Based on the function that the station fulfills in the electrical system of which it is a part, station designs are divided in two major classifications: (1) generating stations, (2) substations.

The design of generating stations deals with the installation of generating units and all mechanical and electrical auxiliary equipment by means of which mechanical energy is produced, changed to electrical energy, and delivered to the system.

Although there are three general types of generating station, viz., **fossil-fuel-fired plants (including gas, oil, coal, etc.), hydroelectric plants, and nuclear-energy plants,** the design of the electrical features receives much the same treatment in all cases. Plants employing nuclear energy plus fossil fuel for superheat have been constructed. The relative economics of the different fuels as predicted over the useful life of the plant will determine the practicability of combination plants. The problem of handling the decay heat of the fission process while a nuclear plant is shut down after an operating period can influence the electrical design of the plant's light and power system (see Secs. **8** and **9**).

The design of substations deals with the installation of switching, transforming, converting, or voltage-modifying equipment and associated structures which may be required at an intermediate station in transmitting the electrical energy from the generating station to the customer.

Substation designs based on their function and type may be segregated into three main groups:

1. **Transmission or primary substations** when they are associated only with the transmission facilities of the power system.
2. **Distributing substations** when they are used to transform, convert, or subdivide the energy for distribution.
3. **Industrial substations** when they are used to transform the energy of the primary distribution system so that it can be used directly by the utilization equipment on the customer's premises.

332. Engineering problems in connection with the design of generating stations and substations are very largely those which have to do with the general design of the apparatus and its capacity to perform a specific duty and with the best possible arrangement of the various circuits, the method of switching, and the use of protective devices.

Elements of station design that are applicable to all electrical installations may be classified, in general, as follows:

1. System of switching connections.
2. Power switchboards; bus and switch structures.
3. Control switchboards.
4. Power and control wiring system.
5. Grounding system.

SYSTEM OF SWITCHING CONNECTIONS

333. The system of switching connections is the key to the entire station design, because upon it depend the selection and arrangement of the protective switching and relaying equipment.

Four major types of connection are included in this system, viz., (1) **main electrical connections,** and connections for (2) **auxiliary power supply,** (3) **excitation supply,** (4) **control power supply.**

The degree of flexibility to be provided in the arrangement of connections will depend upon the type and functions of the plant and upon the reliability of the equipment involved. Recent trends have been toward simplicity rather than extreme flexibility as a means of (1) reducing costs, (2) reducing exposure and likelihood of failure, (3) simplifying operation under normal and emergency conditions and reducing the likelihood of operating errors.

334. The single-line diagram is a schematic diagram prepared to illustrate a system of switching connections in the simplest possible way, using conventional symbols to represent pieces of equipment. This diagram has become the standard form for indicating the system of connections of a generating station or substation and their interconnecting circuits. It is especially valuable to the designing engineer and the operating engineer.

MAIN ELECTRICAL CONNECTIONS

335. Main electrical connections for generating stations reflect the purpose for which the station is built and should be planned after careful consideration of the system as a whole.

System factors which should be considered in determining the type of connections are:

a. Size, number, and voltage of generators or transformers.
b. Size, number, and voltage of feeders.
c. Importance of station and its place in the system.
d. Plan of operation.
e. Plan of maintenance.

Typical single-line diagrams of main electrical connections for generating stations in which the **station output is essentially at generated voltage** follow (see Pars. **336** to **344**).

336. The single-bus scheme is the simplest arrangement. It is confined to small stations where simplicity and economy are of primary importance and where service interruptions can be tolerated. It is commonly used in low-voltage stations and single-unit stations (see Fig. 10-96).

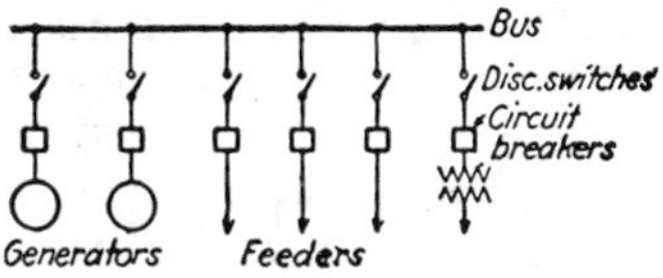

This arrangement has no flexibility; and in case of insulation or equipment failure, a complete station shutdown may result. Maintenance work must be done with the bus alive, or the whole station must shut down.

Fig. 10-96. Single-bus scheme of connections.

337. The spare- or transfer-bus scheme is a variation of the single-bus arrangement (see Fig. 10-97). A spare bus and a spare breaker are provided which can be used on any circuit when, for any reason, the regular circuit breaker must be taken out of service. The transfer may be made through disconnecting switches or nonautomatic oil breakers. If air-break disconnects are used, great care must be taken to ensure that they are not used to open or close circuits except at the lowest voltages. It is a fairly inexpensive way of providing for breaker maintenance without taking the feeder out of service. In case of a bus fault, this arrangement is subjected to the same limitations as the single-bus arrangement, and the scheme is therefore generally confined to small stations.

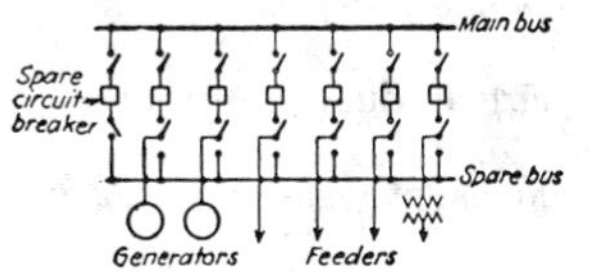

Fig. 10-97. Spare-bus scheme of connections.

338. The double-bus single-breaker scheme is the next step in flexibility at a low cost (see Fig. 10-98). This arrangement facilitates bus maintenance and reduces the duration of outages due to bus failure but does not permit feeder-breaker maintenance without feeder outages. A bus-tie breaker may be provided to permit the transfer of circuits carrying power from one bus to the other without service interruption. This arrangement is confined to small or medium-size stations where the generator capacity is small.

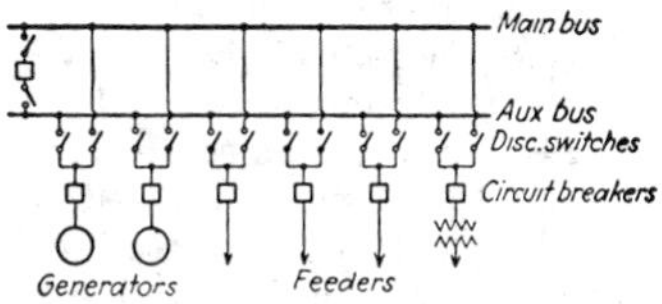

Fig. 10-98. Double-bus, single-breaker scheme of connections.

339. The double-bus double-breaker scheme is a modification of the double-bus single-breaker arrangement. It reduces the likelihood of **extended** outages of any circuit, due to circuit-breaker trouble, and permits breaker maintenances without a feeder outage (see Fig. 10-99).

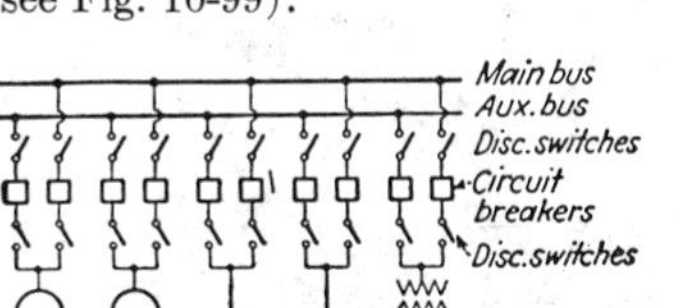

Fig. 10-99. Double-bus, double-breaker scheme of connections.

Although obviously more expensive, this arrangement is often used, as it gives greater flexibility of operation with little complication. In case of a circuit-breaker failure, however, a station shutdown will result until such time as the faulty circuit breaker has been isolated from the rest of the system.

A second, or backup, breaker may be used on each circuit to reduce the likelihood of bus outage due to feeder-breaker failure. In this case the circuits are grouped as shown in Fig. 10-100, requiring two breakers per feeder and three breakers per generator or station tie. This arrangement is generally known as the **H scheme,** and the four breakers are referred to as a **feeder group.** If for any reason the feeder breaker fails to open the circuit at time of a fault, the selector breaker is depended upon to do so. This arrangement may prove to be unsatisfactory where the feeders are few and of large capacity, because of the fact that the opening of a selector breaker causes the loss of two feeders, and overhaul of feeder breakers requires a feeder outage.

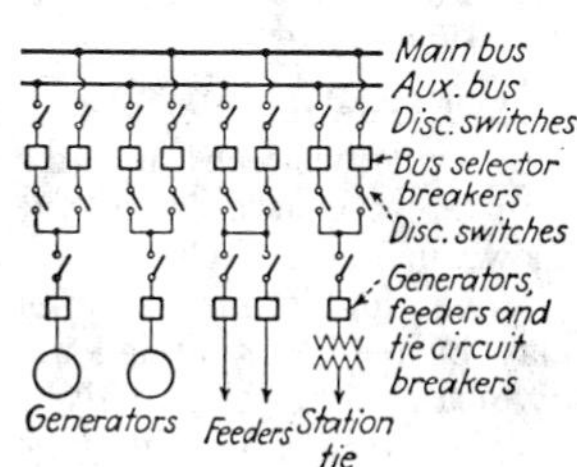

Fig. 10-100. The H scheme of connections.

The arrangements of Figs. 10-99 and 10-100, where applicable, offer maximum flexibility, but, being the most expensive, they may be economically justified only where continuity of service is of prime importance.

340. The group-bus arrangement is a modification of the H scheme in which the two bus-selector circuit breakers are used to supply more than two feeders (see Fig. 10-101). Although this arrangement is not so flexible as the H scheme, it is particularly applicable to large power stations having a large number of feeders, especially if built on the unit-type principle.

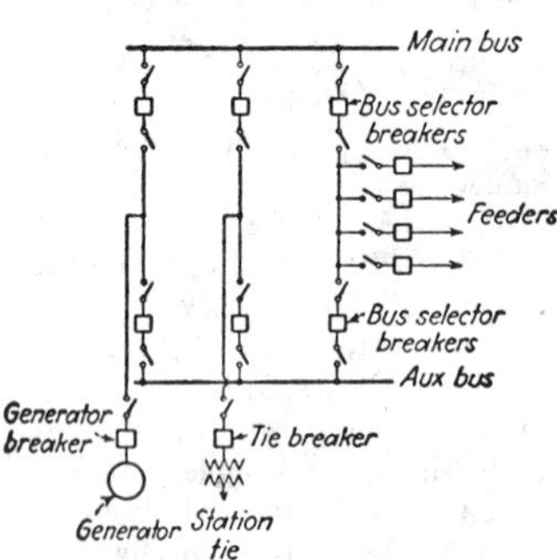

Fig. 10-101. Group-bus scheme of connections.

341. Bus sectionalization is an important feature of the main electrical connections and may be used in conjunction with the above bus and breaker schemes. In stations where the generators and feeders are many, the buses are generally sectionalized by bus-tie breakers and often bus-tie reactors in order to:

1. Limit the short-circuit duty imposed upon the circuit breakers.
2. Reduce the mechanical stresses on buses and equipment.
3. Localize shutdowns due to bus or equipment failure and prevent a complete station shutdown.
4. Provide a form of backup protection against feeder-breaker failure.

The schemes of bus sectionalization generally used may be grouped into four classes:
1. Straight bus.
2. Ring bus.
3. Synchronizing bus.
4. Synchronizing at the load.

342. In the straight-bus sectionalization scheme the buses are divided in sections, and bus-tie circuit breakers are provided between sections (see Fig. 10-102). Bus-tie circuit breakers may be operated normally open or closed depending on the type of the system being supplied. When they are operated closed, a scheme of relay protection should be provided which, in case of bus or equipment failure, will open them, thus isolating the faulted sections.

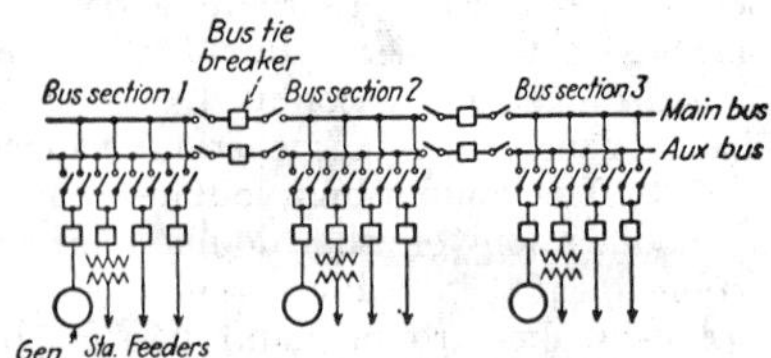

Fig. 10-102. Straight-bus sectionalization.

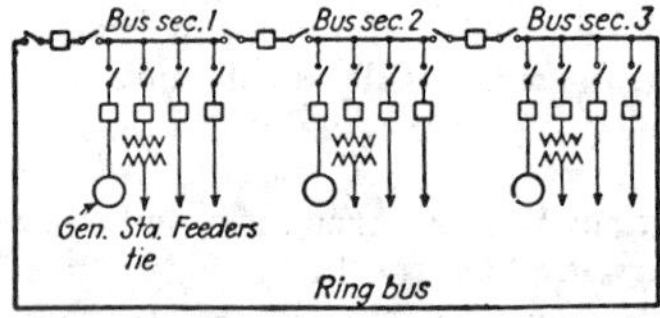

Fig. 10-103. Single-ring bus sectionalization.

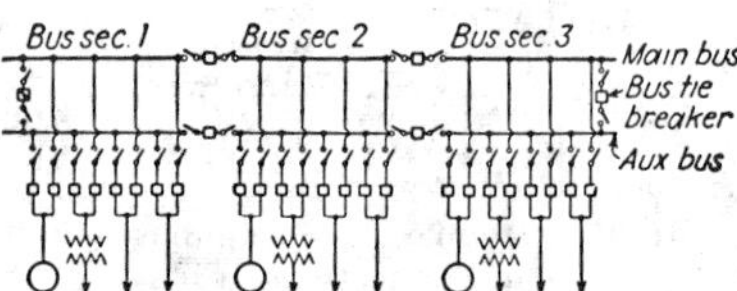

Fig. 10-104. Double-bus scheme of connections arranged for ring-bus operation.

343. The ring-bus sectionalization scheme is used when it is desirable to maintain connection between the bus sections, even in case of loss of one section, and to facilitate transfer of generator capacity between sections. In its simplest form this scheme is shown in Fig. 10-103.

The double-bus single-breaker arrangement and other double-bus arrangements may be readily modified in a ring-bus arrangement by tying together the two buses by means of bus-tie circuit breakers as shown in Fig. 10-104.

In certain cases it may be desirable to maintain continuity of the ring-bus operation in event of trouble in any section, and a **combination ring-bus and transfer-bus** arrangement is used (see Fig. 10-105) or even a double-ring bus.

344. The synchronizing-bus scheme consists essentially of a sectionalized main- and tie-bus arrangement in which both buses are in normal operation (see Fig. 10-106). The main bus is used to supply the load, and the tie or **synchronizing bus** is used for the purpose of tying the buses together and making any of the generators available automatically to any of the main buses. As ordinarily operated, it automatically provides at least two breakers in series between main-bus sections, reducing the likelihood of simultaneous outage of two main sections due to bus-tie breaker failure.

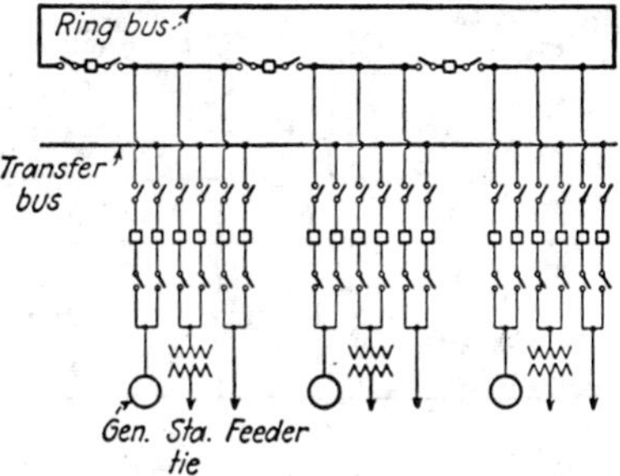

Fig. 10-105. Combination ring-bus and transfer-bus arrangement.

This scheme of connections is simple and flexible and lends itself to the unit type of station design.

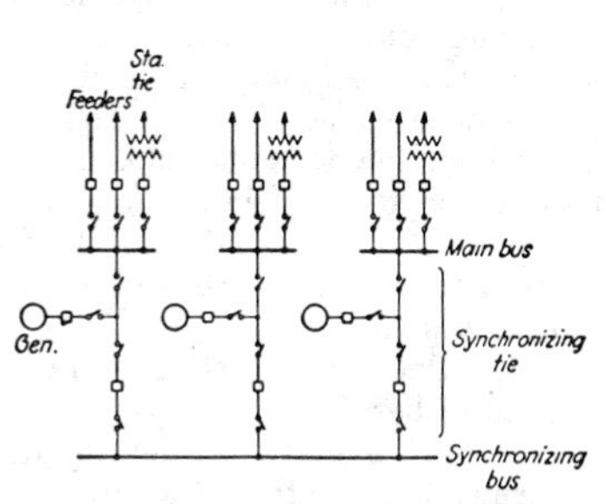

Fig. 10-106. Synchronizing-bus arrangement.

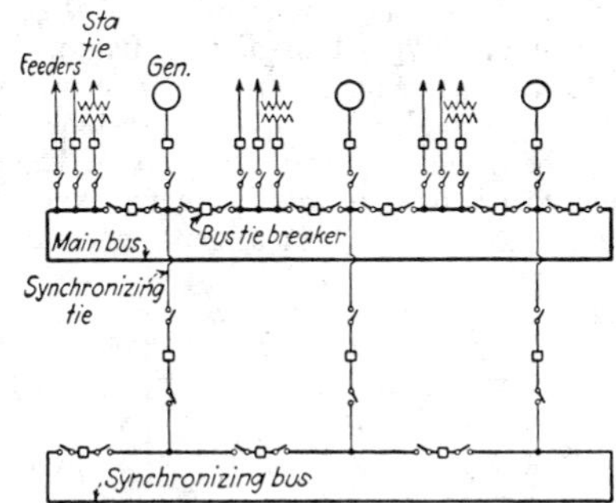

Fig. 10-107. Combination ring-bus and synchronizing-bus arrangement.

Various modifications of this scheme are in use depending upon the degree of flexibility that the size and importance of the station and local conditions seem to warrant. **A typical arrangement** applicable to generating stations of large capacity is shown in Fig. 10-107. In this arrangement each generator is connected normally to one main-bus section and to the synchronizing bus. Flexibility and reliability are enhanced by the selector breaker arrangement between main-bus sections and by the fact that both the main bus and the synchronizing bus are of the ring-bus type.

Another example, typical of synchronizing-bus connections used in generating stations of large capacity, is shown in Fig. 10-108. The major advantages of this arrangement, which is operated with the main bus-tie circuit breakers normally open, are:

1. The loss of a **synchronizing tie circuit** will not affect the station generator capacity. It will not be necessary, therefore, to tie together any of the main-bus sections.

2. The loss of a main-bus section will not cause outages of interstation ties. These ties are available to any of the main-bus sections and provide a direct transfer between stations through the synchronizing bus.

In the bus arrangements of Figs. 10-107 and 10-108, a fault in any one main-bus section will obviously cause an outage of the circuits connected to that section, and these arrangements are, therefore, used only when the station is designed to stand such a loss, which is generally the case in systems of large capacity. Where such a loss cannot be tolerated, a **combination synchronizing-bus and transfer-bus arrangement** may be used (see Fig. 10-109) or, in the extreme, a combination of synchronizing bus and double bus.

345. In most large power systems the generating-station output of units installed in recent times is at higher voltages than that generated or at several voltages, one of

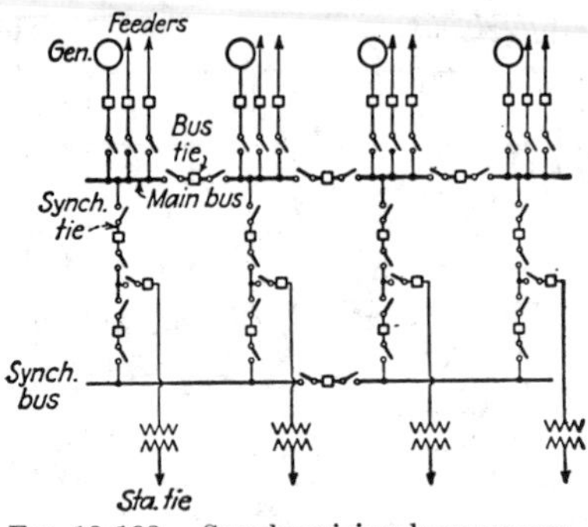

FIG. 10-108. Synchronizing-bus arrangement for large generating station.

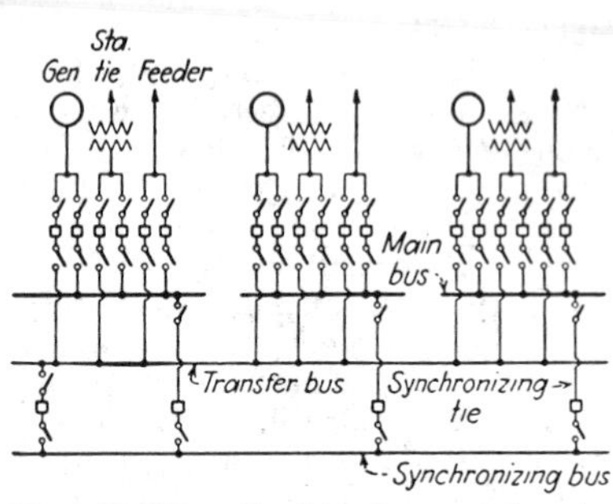

FIG. 10-109. Combination synchronizing-bus and transfer-bus arrangement.

which is the same as that of the generator. To meet these conditions, many switching arrangements have been used varying from a **single low-tension and single high-tension bus arrangement to a double-bus arrangement for both low-tension and high-tension connections.** Typical single-line diagrams of multiple-voltage systems of connections are illustrated in Fig. 10-110. Condensed description and comments follow:

a. Single high-tension and low-tension buses. Lacks flexibility and reliability.

b. The generator and the step-up transformer are treated as a unit. The high-tension bus is used primarily for transfer purposes, and the low-tension bus is used for section auxiliaries. This scheme is commonly used for hydroelectric stations where the load center is at some distance from the station. It is economical but lacks flexibility.

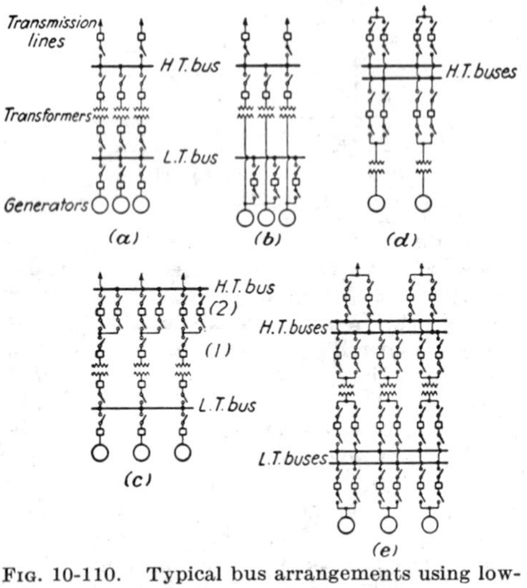

FIG. 10-110. Typical bus arrangements using low- and high-tension buses.

c. This scheme is more flexible than either (*a*) or (*b*). The transmission line and the step-up transformer are treated as a unit. The high-tension bus is used for transfer purposes. Circuit breakers (1) and (2) are sometimes replaced by air-break disconnecting switches.

d. Double high-tension bus. Generator and step-up transformer are treated as a unit. Commonly used in so-called "base-load" stations.

e. Double high-tension and low-tension bus arrangement. Affords maximum amount of flexibility and reliability but is most expensive arrangement.

346. A careful analysis of the abnormal conditions which may take place in a switching and bus arrangement during short circuits is necessary in order that circuit breakers and other protective equipment of adequate capacity may be provided. **A major factor in this analysis because of its effect in planning the system of main connections** is the current-limiting reactor.

347. Primary functions of current-limiting reactors may be classified as follows:

1. To reduce the flow of current into a short circuit so as to protect the apparatus from excessive mechanical forces and from overheating.

2. To balance impedances of two or more circuits to obtain load-current balance in parallel operation.

3. To reduce the magnitude of voltage disturbances caused by short circuits.
4. To localize the effects of short circuits.
5. To reduce the duty imposed on switching equipment during short circuits to be within economical ratings.

The application of reactors in a system of main connections will depend upon the capacity of the station and of the system as a whole and upon operating limitations. **In stations of small capacity,** reactors, as a rule, are not required.

348. Single-line diagrams of typical reactor applications for generating stations are shown in Fig. 10-111. Condensed description and comments follow:

a. **Generator reactors** are used in the main generator leads to protect the generator itself and to limit the current that the generator can supply into a short circuit. In some applications where high-speed 60-c steam turbogenerators having low zero sequence reactance are used on grounded systems, it may be necessary to provide reactors on the generator neutral to protect the generator in case of line-to-ground faults. In hydroelectric systems where slow- or medium-speed multipolar generators are used, generator-neutral reactors are seldom used.

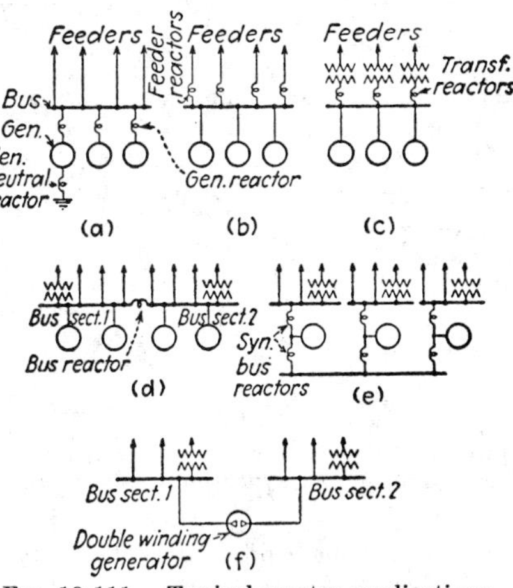

Fig. 10-111. Typical reactor applications.

b. **Feeder reactors** are used to limit and localize the effects of short circuits which occur on the feeders beyond the point at which the reactors are located. They reduce system voltage disturbances due to feeder faults.

c. **Transformer reactors** are similar in their function to feeder reactors and protect the transformer against the effects of short circuits. **They are seldom used** with modern transformers which have a comparatively high inherent reactance but may be required with some autotransformers having low reactance.

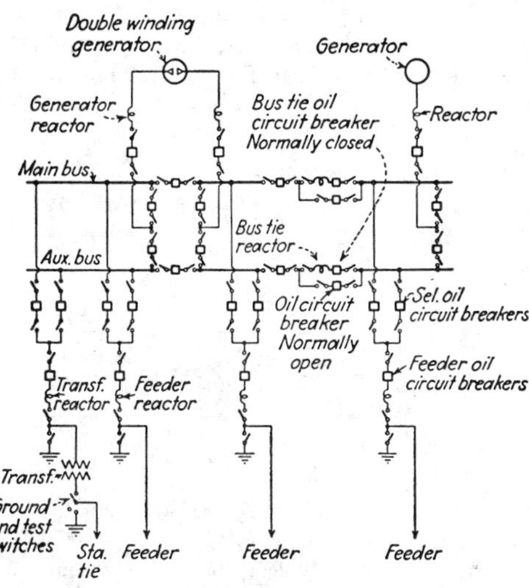

Fig. 10-112. Coordination of reactor and switching equipment for double-bus arrangement.

d. **Bus reactors** are used to tie together separate bus sections. Under normal operating conditions a free exchange of current takes place between sections. The heavy currents and voltage disturbances caused by a short circuit on a bus section are reduced and confined primarily to that section. **Bus reactors, however, do not protect** the generators connected to the faulted section. They facilitate the parallel operation of large systems and are extensively used.

e. **Synchronizing-bus reactors** are a modification of the bus-reactor scheme. The generators are connected to the synchronizing bus through reactors which keep the generators in step and act as bus-tie reactors. In general, with this scheme the voltage regulation between feeder sections is better than when the bus reactors of scheme (*d*) are used. Synchronizing-bus reactors are extensively used on large systems.

f. **Double-winding generators** when connected with each winding to a separate bus section act as bus-tie reactors (see Fig. 10-112).

349. The number of reactors commonly used on various circuits is as follows:

a. For **single-phase circuits,** a single reactor on one side of the line.

b. For **2-phase 4-wire circuits,** two reactors, one for each phase.

c. For **2-phase 3-wire circuits,** two reactors, one on each outside line.

d. For **3-phase circuits,** three reactors, one on each line.

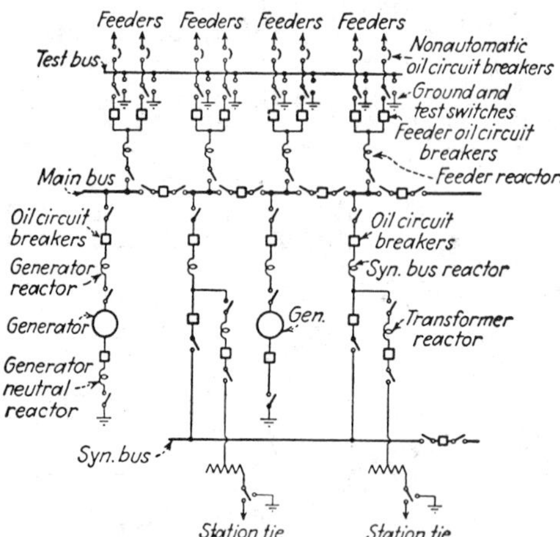

Fig. 10-113. Coordination of reactor and switching equipment for synchronizing-bus arrangement.

350. The preferable location of reactors and the amount of reactance to be provided in a given case involve the following:

a. A careful study of the layout of the system.

b. A determination of the short-circuit currents and the damage that they may cause to generators, transformers, circuit breakers, disconnecting switches, buses, and cables.

c. The time required to clear a short circuit by means of protective relays.

d. Operating conditions as they may be affected by voltage disturbances.

e. System conditions as they may be affected by phase-angle displacement between feeder buses supplying a common network area.

f. Evaluation of reactor and switching costs.

g. Evaluation of Mvar loss imposed on the system by the reactors.

Coordination of switching and reactor equipment representative of general practice for generating stations of large power systems is illustrated in Figs. 10-112 and 10-113.

351. Main electrical connections for substations vary greatly depending upon the purpose and function of the substation, its capacity and voltage, and whether it is to deliver alternating or direct current. Station connection requirements for the three major groups into which substations may be classified, viz., transmission, distribution, and industrial substations (see Par. **331**), will also be influenced by the location of the substation and by the method of connection to the system.

Typical single-line diagrams of bus arrangements for substations are illustrated with condensed descriptions and comments in Figs. 10-114 to 10-128, inclusive.

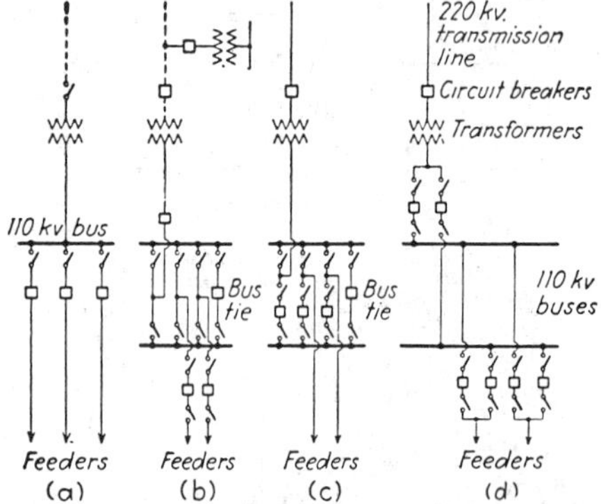

Fig. 10-114. Typical bus arrangements for transmission substations located at the receiving end of a line.

352. Transmission Substations Located at Receiving End of Line (see Fig. 10-114). *a.* The simplest possible arrangement. No switching other than a nonautomatic air-break switch is installed on the high-tension side of the transformer. This arrangement is used only where no feedback is possible over the outgoing feeders. A transformer failure results in a line outage.

b. This arrangement is often used where other substations are tapped off the same transmission line. The automatic high-tension circuit breaker increases service reliability in case of trouble in the other substations. The additional switching may be used depending on the type of service demanded or the severity of storms and trouble.

c. Provides for complete inspection of all feeder circuit breakers. The tie breaker can be used as a spare breaker for any other without interrupting service.

d. Complete double-bus equipment, seldom used owing to its high cost.

353. Transmission substations tapped to transmission lines are by far the most common type. The switching arrangement for such station in its simplest form would be as shown in Fig. 10-114*a* and *b*. Transmission-line trouble in these cases would mean the loss of the substation and its customers unless they were provided with another source of power from some other transmission substation. To improve station reliability if the load and service demand it, the arrangements of Fig. 10-115 may be used.

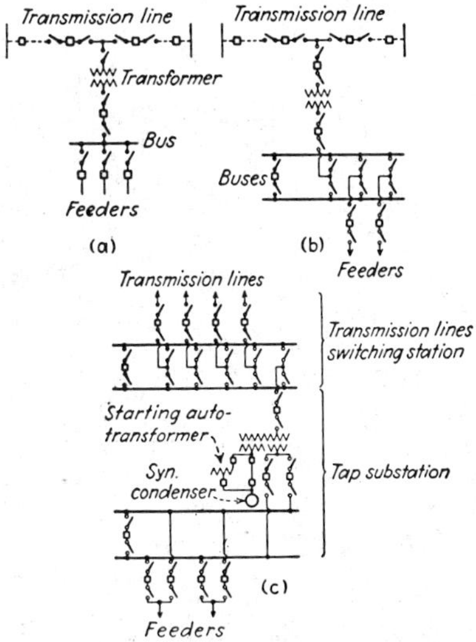

FIG. 10-115. Typical bus arrangements for tap substations and transmission-lines switching station.

a. This arrangement contains the least amount of automatic switching in the transmission line. It improves service continuity to the substation, as the high-tension switching provides the equivalent of two distinct lines.

b. Additional refinements are added which, however, do not increase service continuity of the substations but improve the line reliability in case of a fault in the substation.

c. This arrangement is used in some large substations which are fed from several transmission lines. Automatic switching in the transmission lines is required in such cases regardless of the capacity and function of the substation. The installation of the second breaker in the low-tension bus is seldom used because of the expense. In this diagram, a three-winding transformer and a synchronous condenser used for voltage control are indicated.

354. The breaker-and-a-half design has been used more extensively with the larger capacity transmission feeders made possible by the use of extra-high voltage. As its name implies, it requires three breakers for every two circuits in the scheme. It offers a very high degree of security in that a faulted circuit will in no way affect other operating sections. This design has particular advantages when right-of-way conditions are such that more than one major circuit must share the same right-of-way, thus increasing the possibility of a double circuit outage (see Fig. 10-116).

355. The ring-bus design requires on the average only one breaker per circuit and is therefore less expensive than the breaker-and-a-half design. However, a double circuit outage can remove other circuits from the supply sources. The consequences of such an outage can be minimized by judicious selection of positions for each type of feeder (see Fig. 10-117).

356. This double-synchronizing-bus design of a distribution substation offers very good security, particularly when the feeders are supplying a common network load. In this case the design must be such that the complete loss of one feeder section can be tolerated. The scheme shown here provides a spare transformer, thus permitting the loss of two high-tension feeders without affecting the distribution load (see Fig. 10-118).

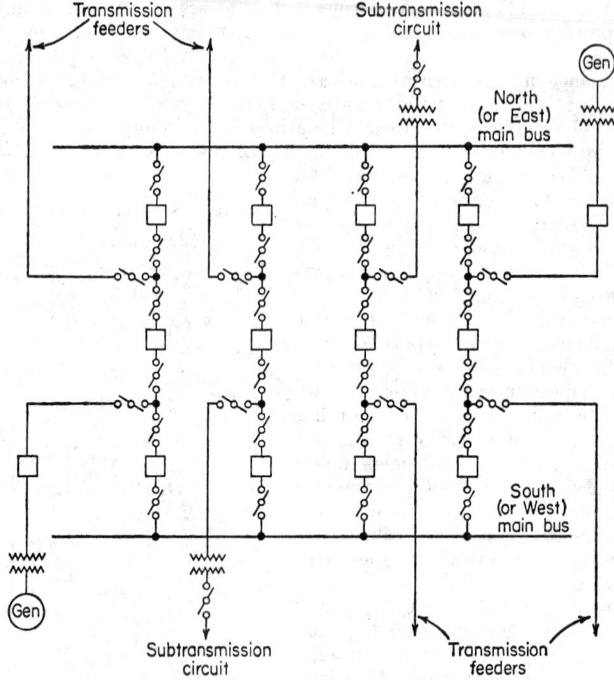

Fig. 10-116. Breaker-and-a-half high-voltage-bus arrangement.

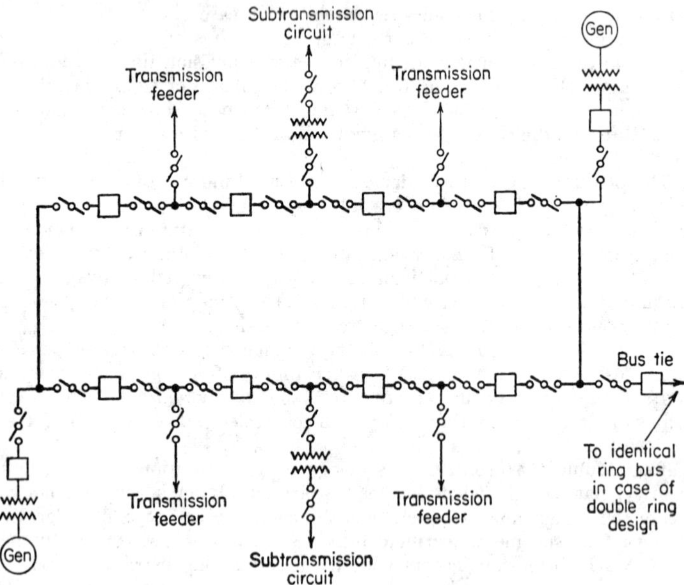

Fig. 10-117. Ring-bus high-voltage-station arrangement.

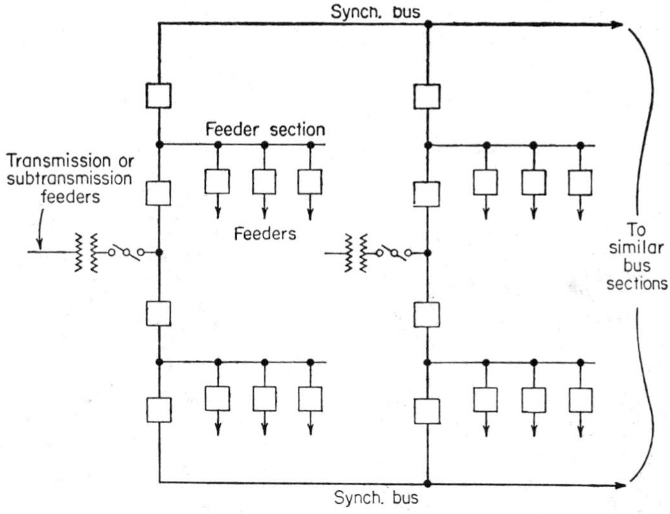

FIG. 10-118. Double-synchronizing bus design for distribution substation.

357. Sectionalization of high-tension buses is common practice on large, important transmission substations. This is done to reduce circuit-breaker duty and to limit the amount of equipment affected by a bus fault. Figure 10-119 illustrates typical single-line diagrams of sectionalized bus installations.

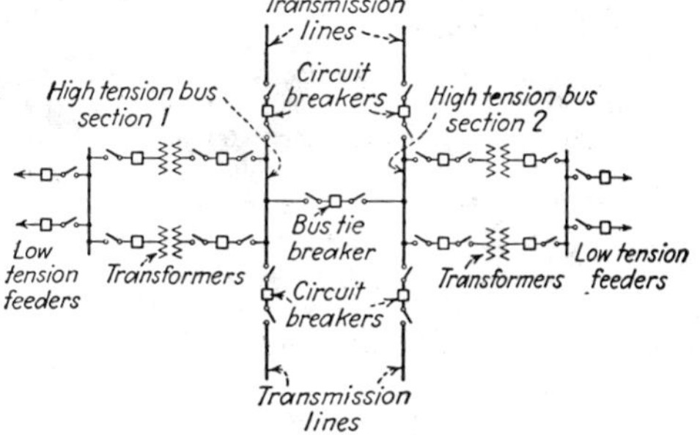

FIG. 10-119. High-tension bus sectionalization for transmission substations.

358. Distribution substations may be of the a-c or d-c type. They are used to supply power from the transmission or subtransmission feeders to local distribution feeders or to feed directly into a low-voltage a-c or d-c city network. In the latter group are included the large metropolitan substations. However, direct current is no longer used in low-voltage distribution systems.

359. Loop-feeder-type substations and **radial-feeder-type substations** are illustrated in Fig. 10-120.

a. The advantage of the loop-feeder arrangement is that, when provided with adequate sectionalizing switching facilities, service reliability to the customer is very high. Feeder loading is low owing to necessity for feeding from either end.

b. The radial-feeder arrangement is more common for ordinary distribution purposes. Important loads may be fed by two or more feeders, giving generally somewhat better service than loop feeders.

360. Metropolitan-type substations, primarily because of their function and the type of load that they serve, are usually provided with a greater flexibility in their switching arrangement and connections than is customary in other types of substation. The arrangement and number of buses, switches, and reactors involve so many variable factors that it is impracticable to specify any rigid principles which will apply to all such substations.

361. Typical single-line diagrams of a-c and d-c distribution substations used on large metropolitan systems are illustrated in Figs. 10-121 to 10-125, inclusive. Figure

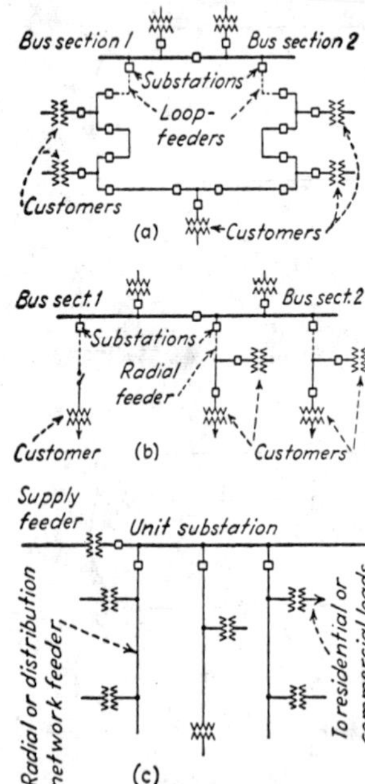

Fig. 10-120. Illustrations of types of substations. (*a*) Loop feeder. (*b*) Radial feeder. (*c*) Unit type with radial or network feeders.

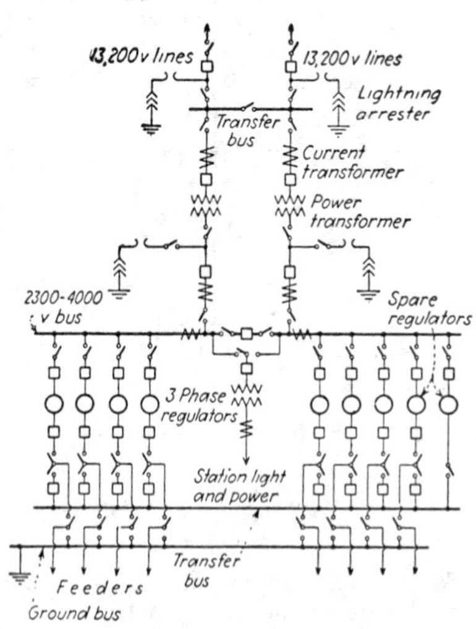

Fig. 10-121. Distribution substation. Typical one-line diagram; single 13,200-volt transfer bus; single 4,000-volt main bus.

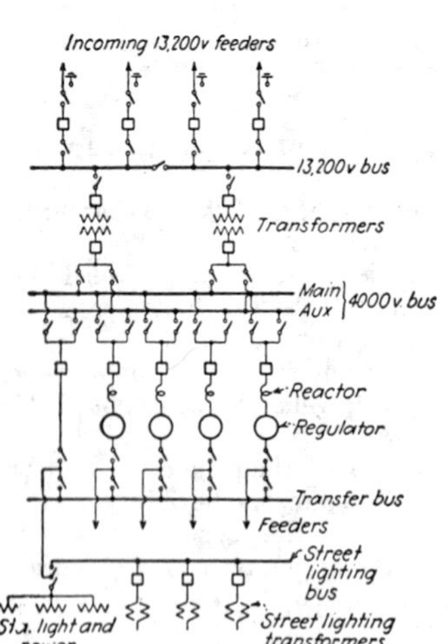

Fig. 10-122. Typical one-line diagram of distribution substation using double 4,000-volt distribution bus.

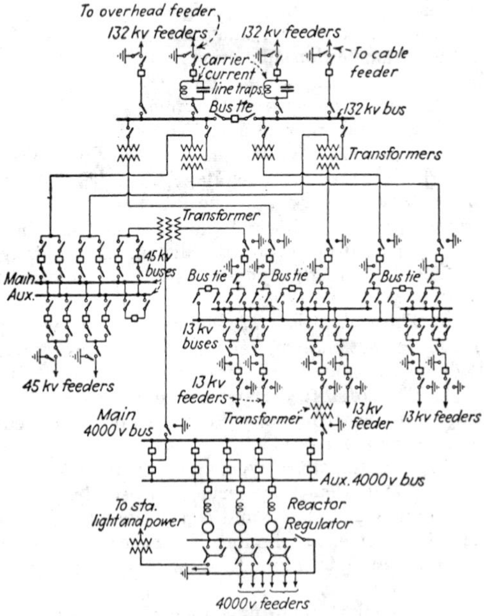

Fig. 10-123. One-line diagram of multiple-voltage distribution substation.

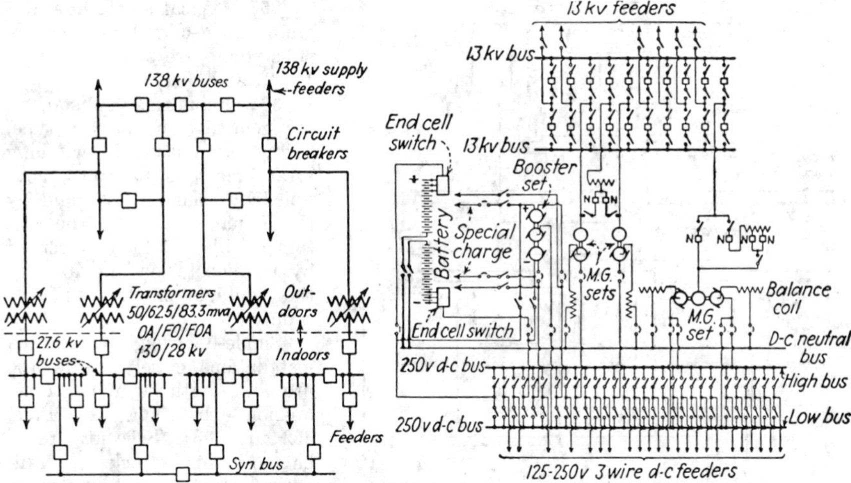

FIG. 10-124. One-line diagram of step-down substation with voltage control and switching for 27.6-kv feeders to low-voltage a-c network and distribution substations.

FIG. 10-125. Typical one-line diagram of d-c substation using double a-c and d-c buses.

10-118 shows one of the more recent types of switching arrangements for service to dense-load areas.

362. Bus connections and switching arrangements for railway substations will depend largely on the function of the substation, viz., whether the substation is to supply urban or interurban systems or trunk-line railway systems, and also on whether the supply to the trolley system is alternating or direct current. In substations of the single-unit type no high-tension automatic switching is generally required on the incoming feeders. In substations where two incoming feeders and two or more converter units are installed, a high-tension bus and automatic switching may be required. For larger substations the high-tension-bus and switching arrangement is more elaborate, as provisions are generally made for bus sectionalization and for the use of current-limiting reactors.

The main-bus and switching arrangement in general is not affected by the type of conversion equipment used, which, in the case of d-c substations, may consist of rotary converters, motor-generator sets, or rectifiers and, in the case of a-c substations, of step-down transformers, frequency changers, or phase-conversion equipment.

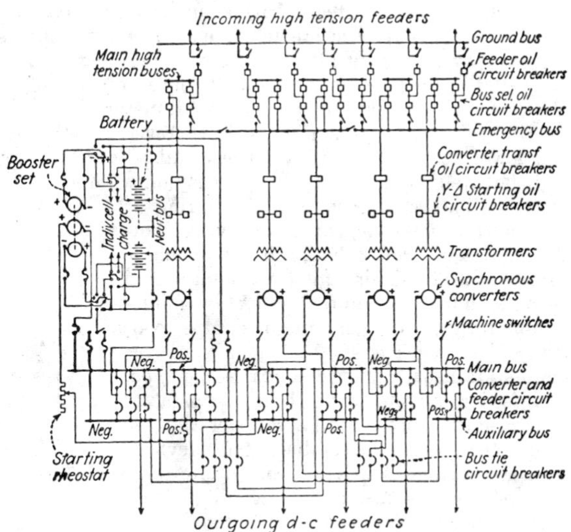

FIG. 10-126. Typical one-line diagram of d-c substation using bus sectionalization throughout.

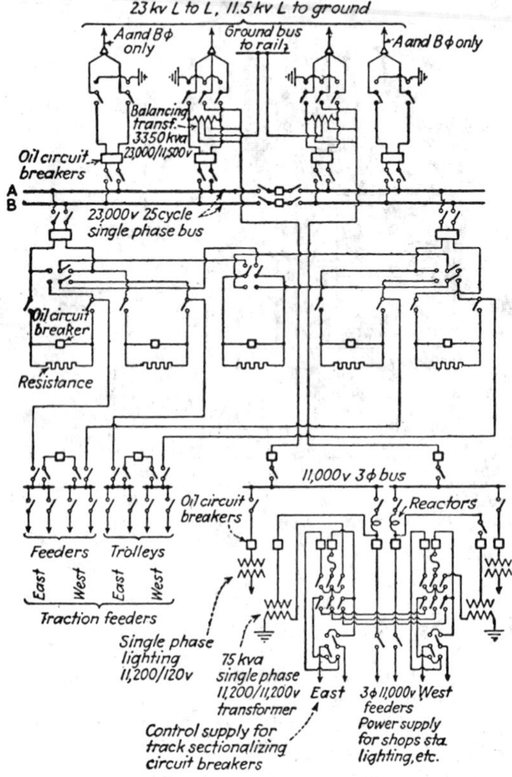

Fig. 10-127. Wiring diagram of an a-c railway substation.

363. Typical single-line diagrams of d-c and a-c railway substations are shown in Figs. 10-127 and 10-128.

364. Industrial substations are generally of the a-c type, and their bus and switching arrangement is naturally governed by their location and by the actual demand of the customer's plant which is served by the substation and also by the voltage of the primary supply.

Single-line diagrams of such installations closely follow the procedure outlined under transmission substations (see Pars. **352** and **353**). Recently plans have been made to electrify railroads, using 60-c power and as high a catenary voltage as feasible.

AUXILIARY POWER CONNECTIONS

365. Electric-driven auxiliaries quite generally use a-c rather than d-c motors because of the greater simplicity and reliability of the motors and the lower cost of the motors and the supply. The increased use of main-turbine bleeding has decreased the use of steam-driven auxiliaries, except for emergency or start-up conditions, such as reserve steam-driven boiler feed pumps and emergency oil pumps. For very small motors 120 V single phase is used, and 240 or 208 V for motors up to 50 or 100 hp, with 2,300 or 4,160 V for larger motors. In some plants 440 or 480 V is used for all sizes except the very small motors.

Squirrel-cage motors are used wherever practicable because of their reliability and lower cost. One- or two-speed squirrel-cage motors are common for circulating and hot-well pumps. Boiler feed pumps have in some cases been of the slip-ring or the squirrel-cage type, sometimes with throttling water control or hydraulic or magnetic variable-speed coupling for controlling pump output. Fans are quite commonly induction-motor-driven, with vane or damper control of air or gases. Mill and conveyer motors are usually of the squirrel-cage type. Coal-feeder motors have been adjustable-speed d-c or, more recently, squirrel-cage a-c with mechanical adjustable-speed devices, such as the Reeves drive, or by intermittent operation of two-speed motors at zero, partial, or full speed. Various other pumps, fans, and miscellaneous motors are usually squirrel-cage, except that large air compressors are frequently of the synchronous type. How-

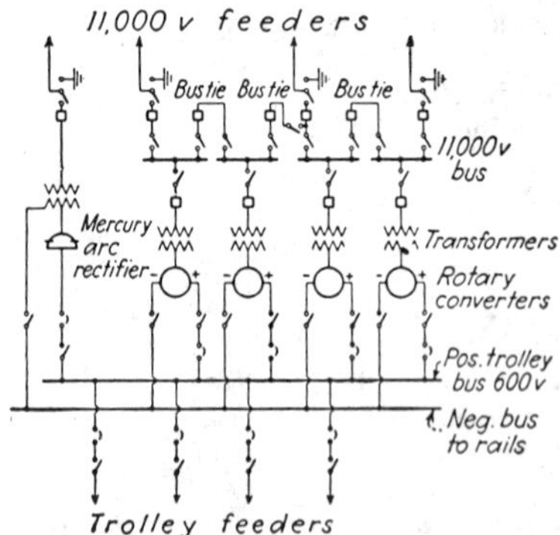

Fig. 10-128. Typical one-line diagram of a d-c railway substation.

ever, synchronous motors are not in common use, because of their somewhat more complicated construction and connections and particularly because the induction motor is better able to pull up to normal speed after a voltage disturbance. Motor-operated valves may be equipped with either d-c or squirrel-cage a-c motors.

366. Switching and bus arrangements for the supply to the station auxiliaries should be designed to provide reliability, simplicity, and low cost. Major considerations in the layout of such a system are the size and nature of the plant and its manner of use, the sources of electric power which may be available, and the amount of duplicate equipment which is provided for the **essential** and **nonessential** auxiliaries.

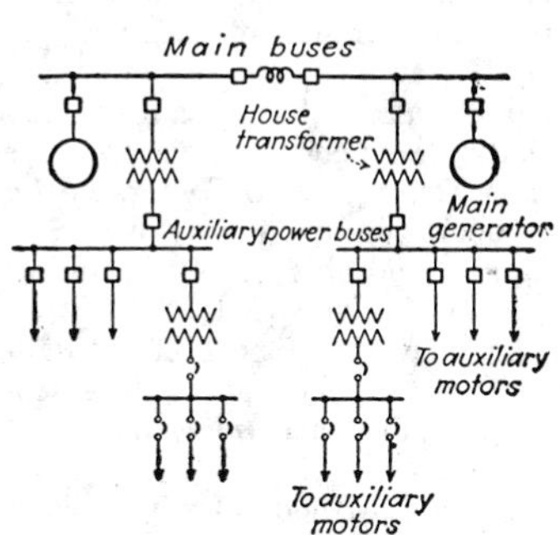

FIG. 10-129. Auxiliary power supplied from main station buses.

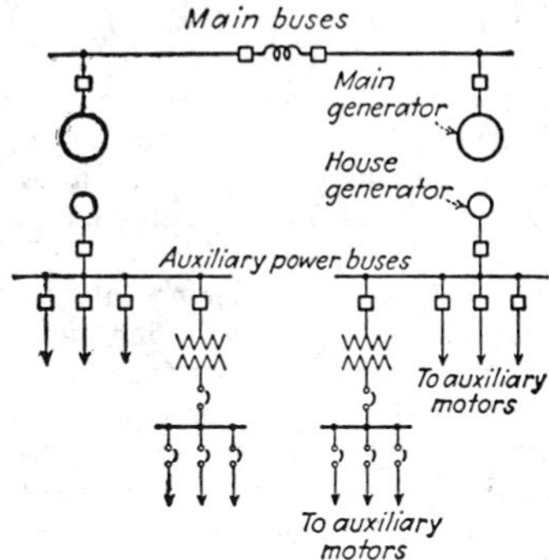

FIG. 10-130. Auxiliary power supplied from house turbogenerators.

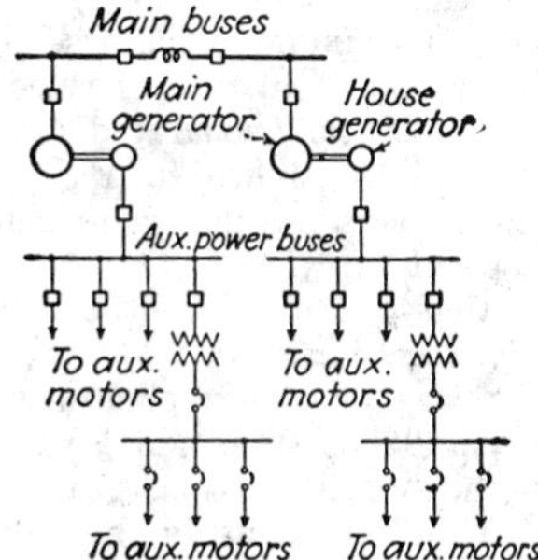

FIG. 10-131. Auxiliary power supplied from house generators directly connected to shafts of main generators.

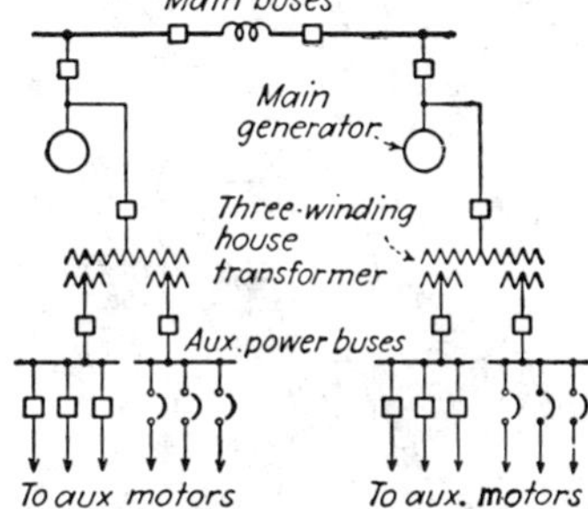

FIG. 10-132. Auxiliary power supplied from house transformers connected to main generator leads.

367. Principal sources of auxiliary power generally used in generating stations are as follows:

1. Main-station buses through house transformers.
2. House turbogenerators.
3. Shaft generators.
4. Transformers connected to generator leads.
5. Motor-generator sets and rotary converters with storage-battery reserve.
6. Direct supply from another station or substation.

368. Single-line diagrams of auxiliary power connections should be prepared as in the case of the main electrical connections to indicate the source of power and the switching arrangement which will meet the operating and service reliability requirements.

Typical single-line diagrams of auxiliary power connections for generating stations and condensed description and comments follow.

369. Auxiliary power supplied from main bus (see Fig. 10-129). First cost, maintenance, and operating costs are low, but service is subject to interruption by system disturbances. This system, however, is satisfactory as supply for nonessential auxiliaries and, when sectionalized and connected to sectionalized main buses, is often used for essential auxiliaries.

370. Auxiliary power supplied from house turbogenerators (see Fig. 10-130). The supply is not affected by system disturbances, but it is not an economical arrangement in either first or operating cost. There are electrical and mechanical complications, and it is not suited for full-voltage starting of large motors.

371. Auxiliary Power Supplied by Auxiliary Generator on Main Unit Shaft (see Fig. 10-131). The supply is not affected by system disturbances, and it is economical and reliable. There are mechanical disadvantages in the design of the main units; automatic voltage regulators may be required on the auxiliary generators, and independent load control is impossible during parallel operation of the main generators. Outage of the main units results in loss of corresponding auxiliary supply.

372. Auxiliary Power Supplied from Transformer Connected to Generator Leads (see Fig. 10-132). First cost, maintenance, and operating costs are low. System disturbances can be reflected into the auxiliary supply, and this supply is subject to interruptions caused by tripping of the main unit throttle on overspeed.

373. The arrangements for auxiliary power supply shown in Figs. 10-129 to 10-132 are considered adequate for small stations where service continuity may not be essential. For the arrangements shown in Figs. 10-130 to 10-132, inclusive, some other source of auxiliary power must be provided for station start-up.

For large stations, however, a combination of various methods of supply is used, and the auxiliary bus and switching arrangement may duplicate any of the arrangements described for the main electrical connections.

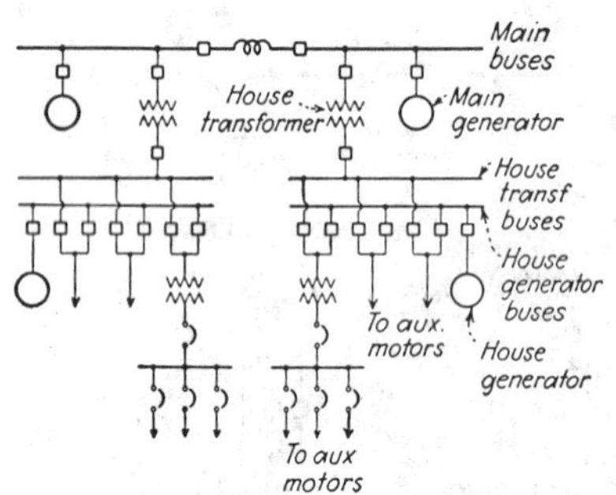

Fig. 10-133. Auxiliary power supplied from two electrically separated sources.

Sectionalized auxiliary buses, connected through transformers to sectionalized main buses, reduce to a minimum the possibility of an extensive outage of auxiliary service.

374. Two electrically separated sources of auxiliary power are shown in Fig. 10-133. In this arrangement the buses are generally operated separately, with the motor load so distributed that the loss of either bus will not seriously affect operation.

375. Auxiliary power supplied from two sources in parallel is shown in Fig. 10-134. In this case the essential auxiliaries are usually connected to the house turbine bus, and the nonessential ones to the transformer bus. The two buses may be tied together by the following methods:

a. *W* and *X* indicate a **straight bus tie** without and with reactor, respectively.

b. *Y* indicates an **induction-motor synchronous-generator-set bus tie.**

Motor-generator sets or reactors may be used between house-generator and house-transformer buses in connection with maintaining proper heat balance by varying the load on the house generator. However, the general use of steam bleeding from one or more stages of the main turbines has resulted in limiting the use of house turbogenerators almost entirely to emergency use. Such an arrangement is shown in Fig. 10-135, where auxiliary power is supplied from **house transformers with spinning house generator for emergency.**

In other cases the house generator is kept solely as a standby set to be used to start the station from a "cold" state.

Duplex units are three-unit sets consisting of a motor supplied from the bus bars, a steam turbine, and the auxiliary to be driven. In some cases the auxiliary is replaced by a d-c generator, which in turn supplies the auxiliaries. In some installations the steam turbine is used as a standby; in others it takes a variable amount of load and is used for heat-balance purposes.

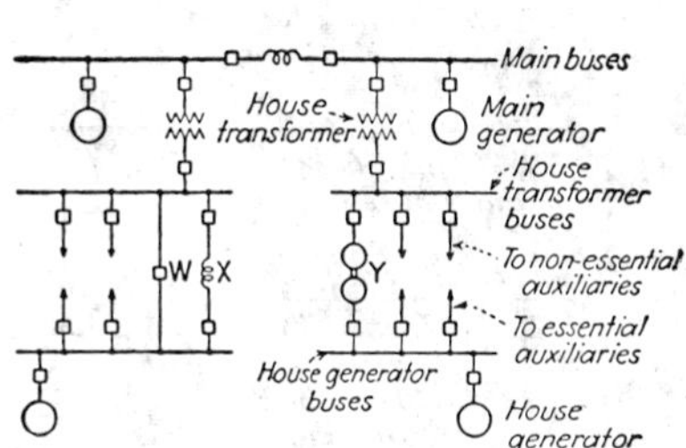

Fig. 10-134. Auxiliary power supplied from two sources in parallel.

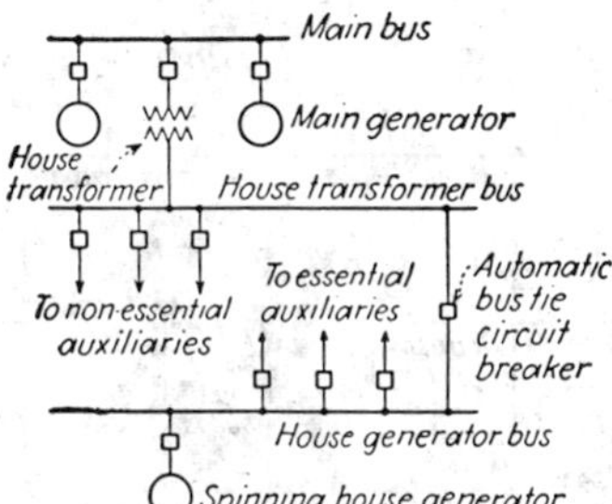

Fig. 10-135. Auxiliary power supplied from house transformers, with spinning house generator for emergency.

376. Typical connections for auxiliary power supply representative of general practice in large generating stations are shown in Figs. 10-136, 10-137, and 10-138. In the arrangement shown in Fig. 10-138 the conventional auxiliary power-supply buses have been eliminated, and high-tension feeders from the main buses supply a number of

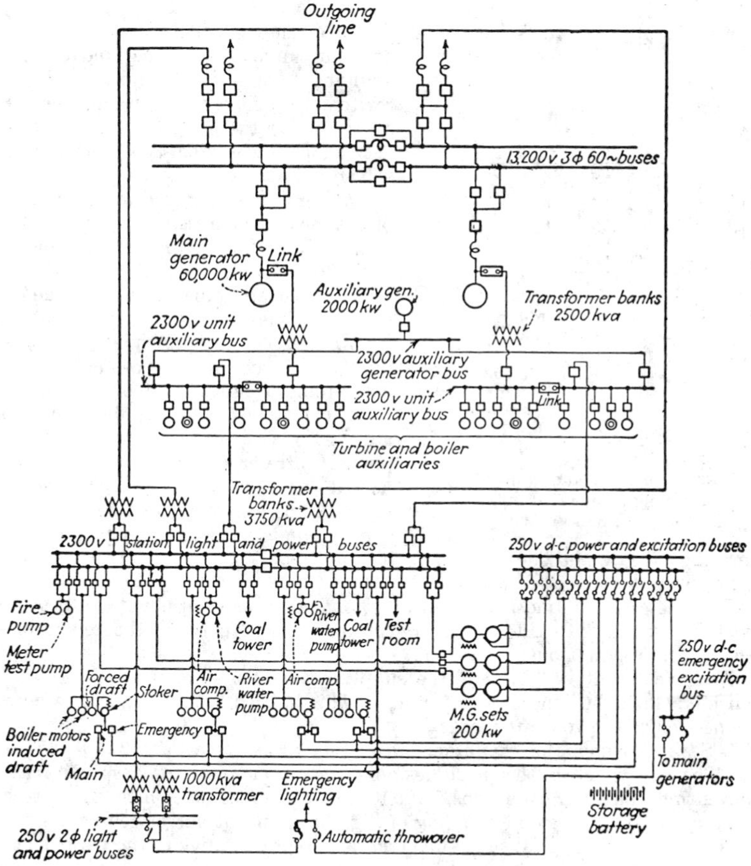

Fig. 10-136. Typical bus arrangement for station auxiliary power supply.

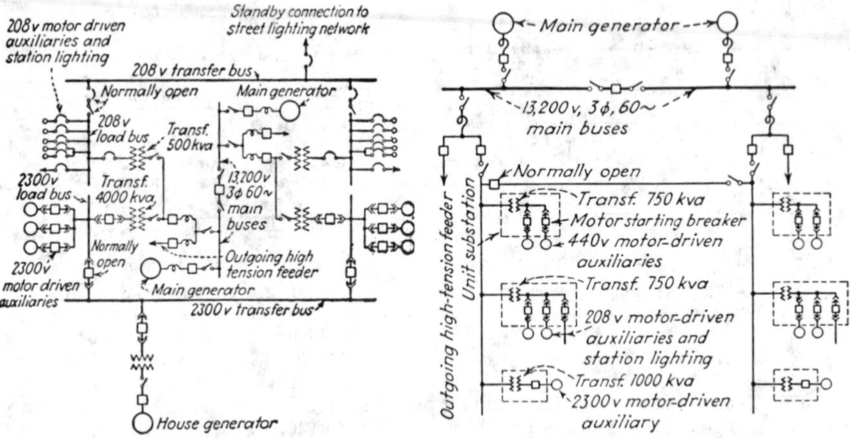

Fig. 10-137. Typical diagram of unit-type bus arrangement for station auxiliary power supply.

Fig. 10-138. Typical diagram of high-tension distribution arrangement for station auxiliary power supply.

transformer unit substations located at the load centers. This arrangement is well adapted to the unit system of design, uses a minimum of switching and transformer equipment, and is economical and reliable. Switching facilities to permit the normal source to be the generator terminals is also used for unit-system-type generators, as shown in Fig. 10-139. The utilization voltage for motor drives is determined on the basis of the most economical combination of transformer, switching, and motor equipment.

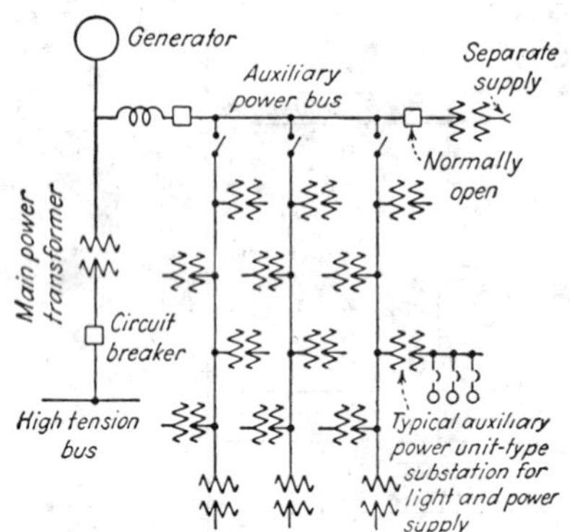

Fig. 10-139. One-line diagram of generating unit auxiliary power supply direct from generator terminals.

Individual motors are often fed over individual feeders from a switch at the switchboard or bus. In other cases, particularly where the "unit" principle is adopted, a number of motors are connected through nonautomatic breakers at or near the motor to a common feeder supplied through an automatic breaker. When so connected, duplicate or reserve units should not be connected to the same feeder, e.g., two boiler feed pumps. Better practice would be to connect units such as one circulating pump, one forced-draft fan, and one induced-draft fan of a unit to a feeder.

Referring to Fig. 10-138:

1. In some cases the most economical voltage of generation is not the best for the auxiliary power bus. In this case a transformer is used between the generator leads and the auxiliary power bus.

2. In some recent installations the supply source for start-up is the same high-tension bus which the generator is feeding. This has led some designers to eliminate the connection to the generator leads.

377. Several systems of excitation are in general use (see Pars. **227** to **253,** inclusive). The one which is the most desirable and economical for a specific case will depend entirely on local conditions. Excitation-system voltages have, until recently, been 125 or 250 V. To avoid handling of high field current, larger generators are now using 375 or 500 V. Typical single-line diagrams representative of the various systems of excitation follow.

1. **Common-bus Arrangement** (see Fig. 10-140). In this arrangement a central-supply source of excitation is provided.

a. The exciters are motor-driven, power for the motors being taken from the main bus. The reserve exciter may be steam- or hydro-driven, and a storage battery is included for emergency excitation. Means are provided for sectionalizing the excitation buses.

b. Direct-connected exciters operating in parallel are used.

2. **Individual-unit Exciter Arrangement** (see Fig. 10-141).

a. This arrangement shows individual-unit exciters, direct-driven and connected directly to the generator field for normal operation. Both reserve and emergency excitation sources are provided.

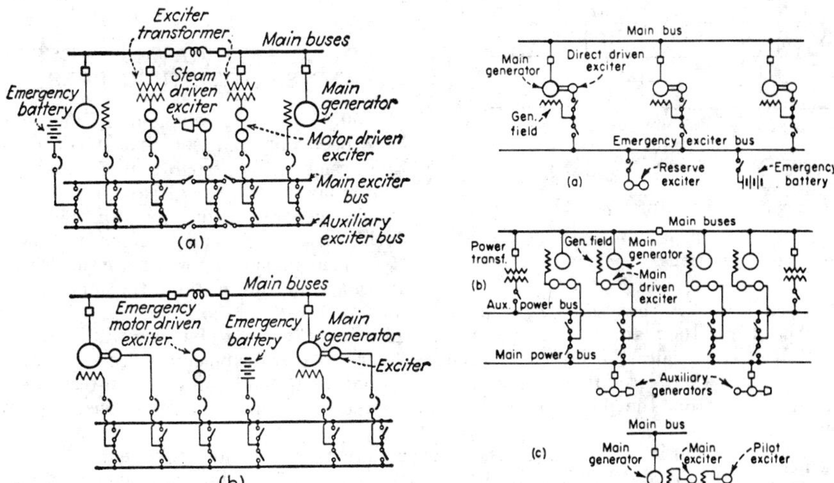

FIG. 10-140. Connections of station exciters using the common-bus arrangement. (*a*) Separately driven exciters operating in parallel. (*b*) Direct-connected exciters operating in parallel.

FIG. 10-141. Typical arrangements of individual-unit exciter supply.

b. This shows individual-unit exciters, motor-driven and connected directly to the generator field. Driving power for the exciter sets is normally obtained from the auxiliary generators which are provided with individual direct-connected exciters.

c. Each generator is equipped with a unit direct-driven main exciter and pilot exciter. No excitation buses are used.

d. The arrangement shown in Fig. 10-141*c* is used for certain high-response rate-regulating systems such as those known as Buck-Boost Amplidyne regulator by General Electric and Mag-Amp regulator by Westinghouse. Allis-Chalmers used a different principle to obtain a similar result in its Regulex regulator system. In all cases the pilot exciter is replaced by a special type of generator whose output is controlled by a voltage-regulating network and fed into the field winding of the main exciter.

378. A source of direct current is supplied for the operation of such equipment as relays, circuit breakers, recording meters, governor motors, field rheostats, and combustion control. Reliability must be of the highest order, and storage batteries in conjunction with motor-generator sets are used for this purpose. Control systems are generally at 125 V except in the larger stations, where 250 V is used. Control systems generally are ungrounded and provided with ground-indicating lights.

Typical single-line diagrams of control systems for generating stations are shown in Fig. 10-142.

a. The double-control bus arrangement is very flexible in operation and is common in large stations.

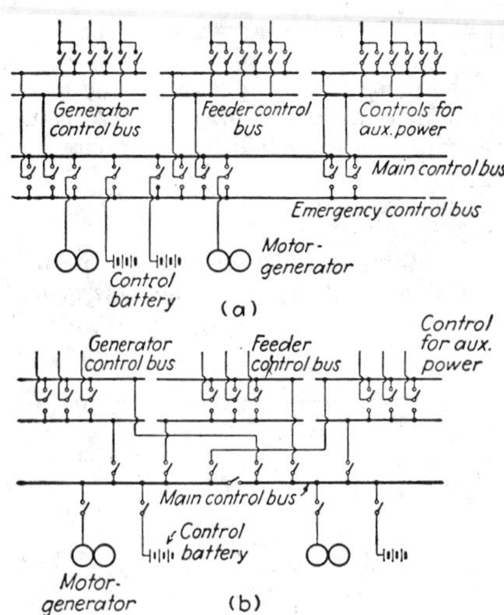

FIG. 10-142. Typical control systems for generating stations.

b. The single sectionalized control bus is slightly less expensive than the double bus but is not so flexible.

379. Because of the recent increase in reliability of rectifying and inverting equipment, brought about by the use of solid-state equipment, many vital a-c plant services are supplied from a **battery-"backed" d-c bus** through the use of inverters. Conversely many d-c services are supplied from a dependable a-c source through rectifiers.

POWER SWITCHBOARDS—BUS AND SWITCH STRUCTURES

380. After the single-line diagrams have been prepared, the next step in station design is the preparation of a **general station layout.**

The general arrangement of the electrical features depends on many conditions, principal among which are the degree of reliability desired and the amount of short-circuit kVA that can be permitted at any point.

The size of the plant and the manner in which its output is to be used will influence the layout, especially the buses. In general, power plants for both utilities and industrial plants, where production operations are continuous and cannot be interrupted, are designed from the point of view of taking reasonable precautions to insure such service.

If the demand is less rigorous, less expensive arrangements are employed.

381. Station layouts should be based, wherever possible, on standard apparatus as obtainable from the manufacturers, and special requirements, unless absolutely necessary, should be avoided. In the selection of the electrical apparatus for any purpose, the following recommendations should be followed:

a. Only that equipment which is necessary for the proper operation and supervision of the station should be installed.

b. Any device that can reasonably be omitted should not be installed, as it introduces unnecessary complications.

c. The equipment selected should be adequate for the service.

382. The logical starting point in the selection of a power switchboard is the **diagram of main connections.** This diagram is essentially an elaboration of the single-line diagram, and in its final form it should indicate the following:

a. Required high-tension switching, including buses and connections for all generators and incoming and outgoing lines.

b. Relation of the various circuits.

c. Required method of operation for both normal and emergency conditions.

d. Kind and capacity of all circuit breakers, disconnecting switches, fuses, transformers, machines, regulators, reactors, etc.

e. Required indications, metering, and protection.

383. Power switchboards of the modern large stations are not switchboards in the original sense of the word. While for small stations or substations the main switching equipment and buses may be mounted directly on or adjacent to the board, for large stations the switching equipment and buses are always mounted remotely in separate buildings or enclosures, or outdoors. In these cases the main switching equipment and buses are generally identified as **bus structures.**

384. Power switchboards are used for the assembly of major electrical equipment for small stations and substations and in industrial plants. They are also used for the supply of station auxiliaries, light and miscellaneous power, in generating stations and substations. They may be grouped into two classes:

1. Panel-type design.
2. Cubicle design, metal-clad.

Depending upon the method of operation they are further identified as:

1. Direct control.
2. Remote mechanical control.
3. Electrically operated.

Station capacity and voltage determine the class of board to be used.

385. Panel materials in common use now are **steel** and **aluminum.**

Marble, because of its high dielectric qualities, was formerly used almost exclusively during the period when instruments were in the primary circuit. It is rarely used for panels at the present time, except where its fine finish and appearance are desired for architectural reasons.

Steel and aluminum panels have been developed of formed construction with turned edges which serve as supports in lieu of stanchions. Such panels are light in weight, being generally ⅛ in thick for steel and $\frac{3}{16}$ in thick for aluminum; are economical in construction, erection, and maintenance; and are foremost in popularity.

386. Panel frames are commonly T iron, angles, and other rolled-steel or aluminum shapes. Manufacturers' catalogues provide very complete lines of fittings for every purpose.

387. Arrangement of panels should be such that the switchboard design provides for future extensions and for the economical distribution of bus-bar material. All bused and other current-carrying parts above 150 V should be adequately protected to guard against accidental contact. Dead-front panels are common and are recommended.

388. Bus-bar materials ordinarily used are copper and aluminum. The shape and dimensions of the buses and their relative position, the current volume, and, in the case of alternating current, the frequency all contribute to fix the effective capacity of the buses. In certain cases it is considered advisable to **insulate all current-carrying parts** against the danger from conducting ionized gases at the time of a fault.

389. Secondary wiring for instruments and controls must be safeguarded against contact with the higher voltages of the power circuit, especially when short circuits create abnormal conditions. The fundamentals of secondary wiring for power boards are the same as those for control boards.

390. Direct-control panel-type switchboards are generally used for small and medium-capacity installations where low cost is of prime importance and a com-

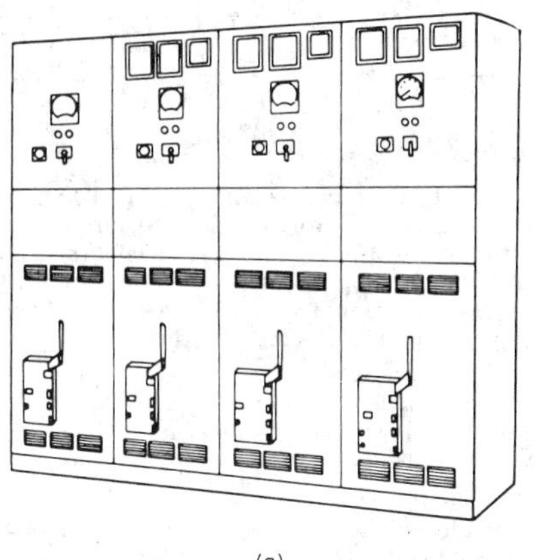

(a)

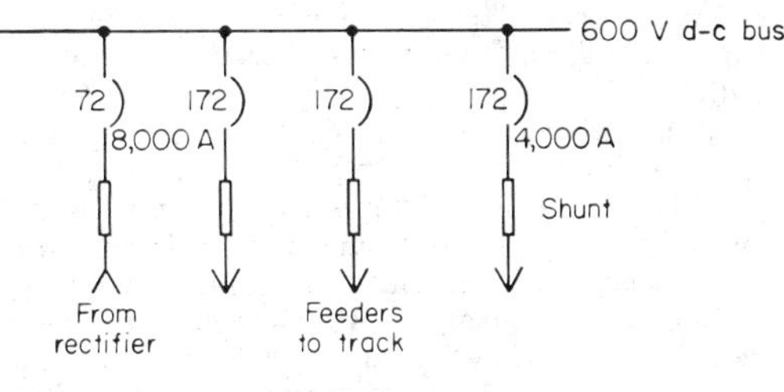

(b)

Fig. 10-143. Direct-current railway switchgear for one source and three loads. (*a*) Four-cubicle switchgear unit with controls and instrumentation; (*b*) single-line diagram of buses and circuit breakers. (*I-T-E Circuit Breaker Company.*)

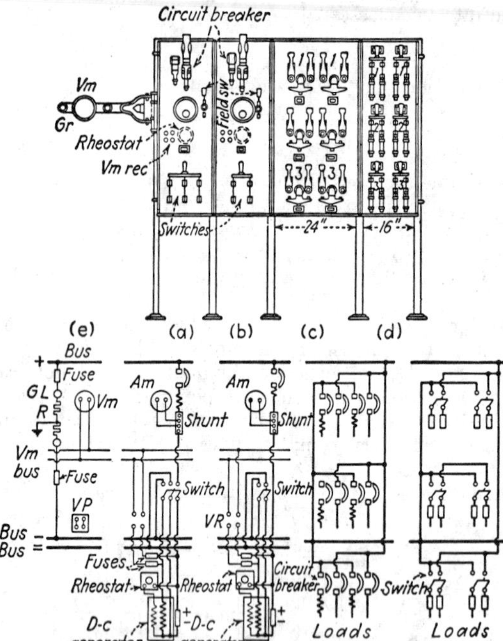

FIG. 10-144. Typical panels and wiring diagrams of small, multipanel d-c switchboards. (*a, b*) Generator panels with 3- and 2-pole main switches. (*c*) Power-feeder panel employing circuit breaker. (*d*) Lighting-feeder panel employing fused switches. (*e*) Bracket voltmeter and ground-detector lamps.

plete installation of only a few panels is required. They are designed for control of incoming lines, generators, motor-generator sets, induction and synchronous motors, feeders, light and power supply, control-power supply, and battery-charging equipment.

391. Live-front switchboards have the circuit breakers and switches mounted directly on the front of the panels and are generally limited to 250 V d-c and 600 V a-c. These are seldom used in modern designs.

392. Dead-front switchboards have the breakers and switches mounted on the rear of the panels and are generally limited to a maximum of 600 V d-c and 2,500 V a-c. Many designs have been used, differing with respect to the degree of enclosure of current-carrying parts. All designs are operated from the front with safety to the operator, and some are equipped with interlocking features which prevent access to live parts. Metal-enclosed designs are favored now.

393. Direct-current switchboards may conveniently be grouped as follows:

a. Single-polarity.
b. Double-polarity with equalizer on pedestal.
c. Double-polarity with equalizer on panel.
d. Three-wire.
e. Multiple-voltage.

394. Single-polarity switchboards find their greatest use in railway work where the negative bus is at approximately ground potential. The negative bus is often located in a basement or pit beneath the machines. In all cases it should be insulated from ground. The circuit breakers and instruments are on the positive side of the board; and the equalizer and negative switches, where negative switches are used, are on small panels or pedestals near the machines. The feeders are frequently without negative switches.

395. Railway switchboards, because of their heavy current capacities and high d-c voltages, are usually of the largest type of panel. The panels, in general, are divided into three sections: the top section for the circuit breakers, the middle section for the instruments and switches, and the lower section for meters when used. Figure 10-143 shows a typical arrangement and single-line diagram for a 600-V railway switchboard.

396. Double-polarity switchboards are generally used where both bus bars must be insulated. They find their greatest use in power work, particularly at 250 V. Figure 10-144 shows a typical arrangement and wiring diagram of a multipanel d-c board for use in small isolated plants.

In some industrial installations, **duplicate buses** are often provided, and the generator and feeder circuits are segregated so that a complete shutdown will not result from trouble on one bus.

Control-power switchboards are used in generating stations and substations to

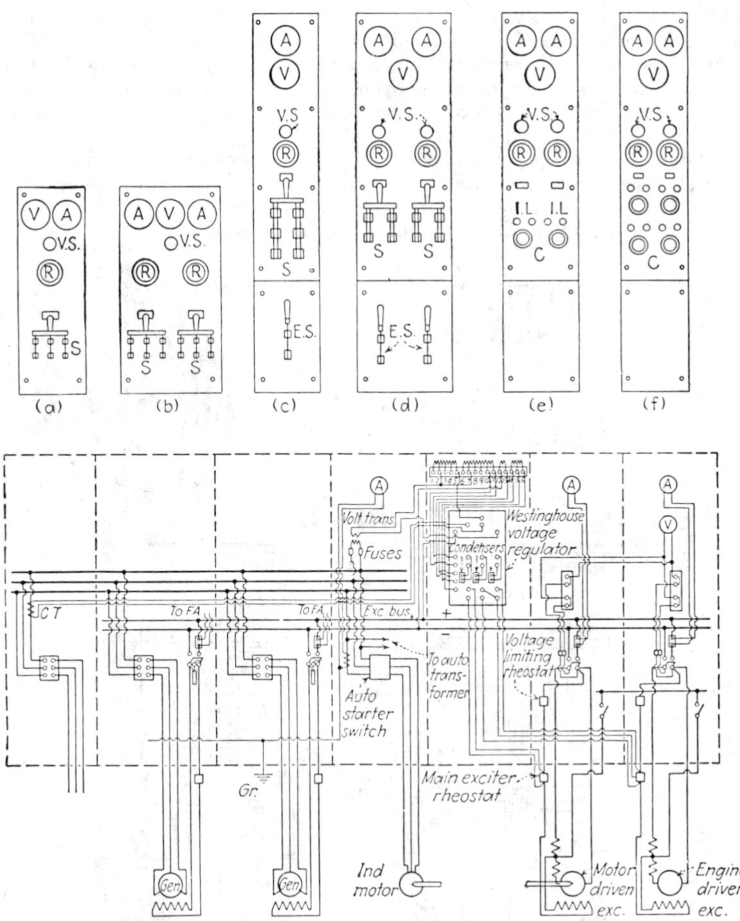

FIG. 10-145. Above, typical exciter panels. (*a*) Forming part of a small board for control of a single exciter paralleling with other exciters. (*b*) Same as (*a*) but controlling two exciters. (*c*) Forming part of a large board and controlling one exciter. (*d*) Same for two exciters. (*e*) Electrically operated exciter panel for control of two machines. (*f*) Same as (*e*) except equipped for double bus. Below, main connections and buses for two exciters operating in parallel and supplying two a-c generators; with voltage-regulator connections.

provide a separate source of energy for the control of electrically operated apparatus. The voltages ordinarily used are 125 and 250 V d-c. The switchboards generally include control for a storage battery, one or two motor-generators, and the required number of feeder panels (see Fig. 10-144). Metal-enclosed construction is popular for such boards.

Exciter switchboards are used where two or more separately driven exciters serve one or more generators (see Fig. 10-145).

Alternating-current battery-charging boards used to charge **electric-vehicle batteries** are usually designed for one or more incoming d-c lines, one or more motor-generator sets or rectifiers, and the required number of charging circuits.

397. Three-wire d-c switchboards are used for combined lighting and power installations. In these boards a third, or neutral, bus is provided for those feeders requiring the neutral lead. Usually, the voltage from positive to negative is 250 V, while from either to neutral is 125 V. Some installations are designed for 500/250 V. The generators require an equalizer on each side. A typical 3-wire d-c switchboard arrangement and wiring diagram for use in a small industrial plant are shown in Fig. 10-146.

Light and power auxiliary switchboards are used in generating stations when some of the station auxiliaries are operated by motors. These boards may control water-driven or steam-driven d-c generators, motor-generator sets, synchronous converters, storage batteries, and the required feeders.

398. Alternating-current switchboards of the **direct-control panel type** for voltages of 500 or less have many characteristics in common with those for direct current. They are used for 120-, 240-, 480-, and 600-V service. The switching is accomplished through knife switches with fuses or through air circuit breakers.

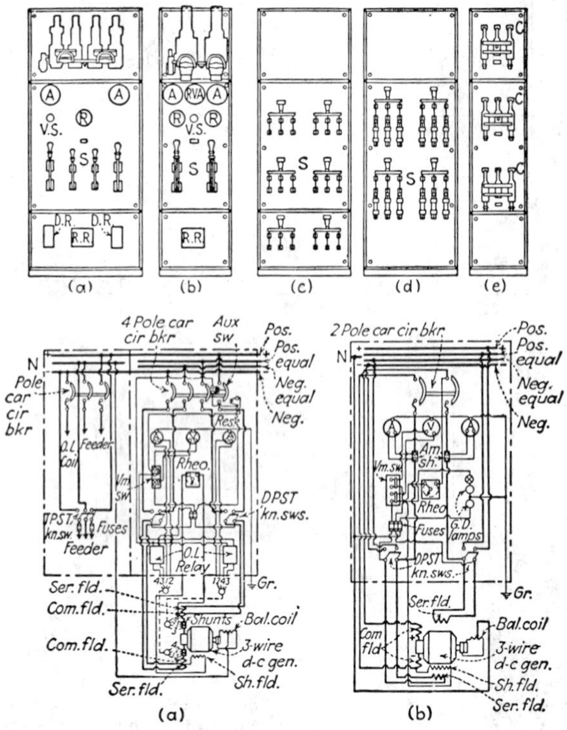

Fig. 10-146. Above, typical large-capacity d-c 3-wire panels. (*a*) Generator panel with four-pole circuit-breaker protection. (*b*) Generator with two-pole protection. (*c*, *d*) Typical light and power panels. Below, typical 3-wire d-c wiring diagrams. (*a*) Four-pole circuit-breaker protection requires but four leads between the machine and switchboard, but the ammeter shunts must be located at the machine, and their leads and those of the special relays must be brought to the board. (*b*) Two-pole protection requires six main leads but no long ammeter leads or special relays.

A typical a-c power switchboard is shown in Figs. 10-147 and 10-148.

Figure 10-148 shows a **dead-front-type** switchboard for generating-station use.

399. Switchboards of 2,500 V or less are similar to those of 500 V or less, except that oil circuit breakers are used in place of the knife switches or low-voltage air circuit breakers. The use of direct-control switchboards of this type is not recommended for stations, for safety reasons.

400. Medium-voltage metal-clad switchboards for 15,000 V or less consist of equipment housed in steel compartments completely assembled by the manufacturer.

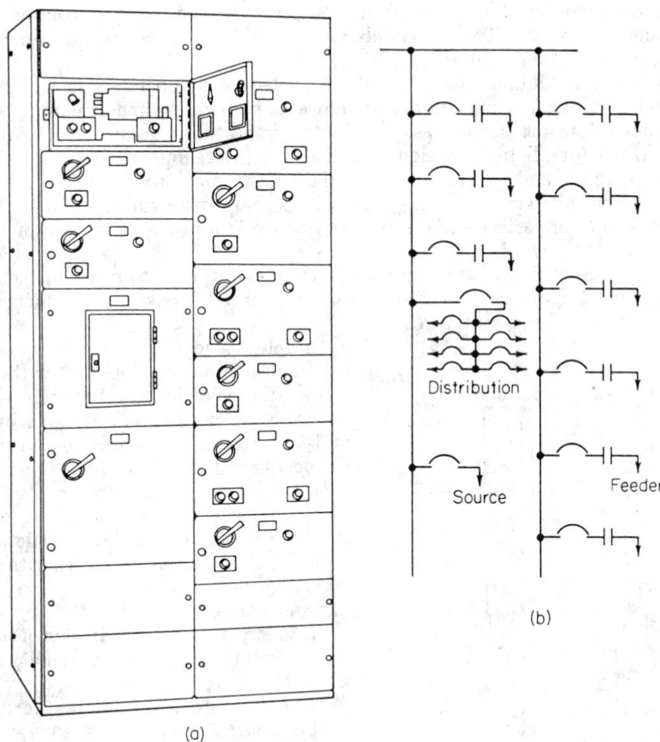

FIG. 10-147. Motor control center. (*a*) Physical arrangement on metal-enclosed switchgear; (*b*) single-line diagram of buses and circuit breakers. (*I-T-E Circuit Breaker Company.*)

This type of switchboard design is generally used for light and power and station auxiliary power supply in large generating stations and industrial plants. It is also used extensively for a-c substation switching.

As all high-voltage parts are enclosed, the equipment is interlocked to prevent mistakes in operation, and all parts of the steel enclosure are grounded, maximum safety to the operator for this class of switchboard is provided. The secondary wiring is shielded, and barriers are provided between phases and between adjacent circuit breakers.

This type of gear is referred to in Pars. **130** to **138**.

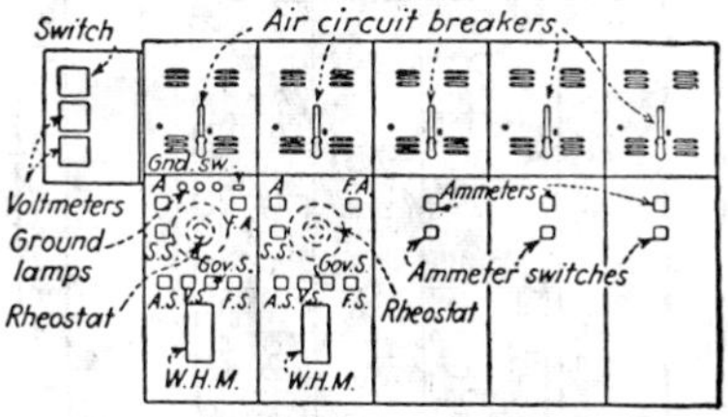

FIG. 10-148. Outline of complete dead-front switchboard for industrial installation, including instrument bracket, two generator panels, and three single-circuit feeder panels (240, 480, and 600 volts), 3-phase, 60 cycles.

401. Electrically operated switchboards may be of the panel type or the metal-clad type. They employ solenoid- or motor-operated mechanisms, some with stored-energy-type designs for circuit-breaker operation, which are, in general, controlled from a central point or control, board. This arrangement has made the location of the control board independent of the location of the power board, and complete isolation of the high-tension equipment has been made possible.

During recent years the employment of switchboards of the metal-enclosed type, which can be remotely or locally operated and are of modular construction, has come into considerable favor. These switchboards are referred to as motor control centers (MCC) or valve control centers (see Fig. 10-147).

In stations of medium or large capacities the electrically operated power switchboard is replaced by an arrangement generally known as **bus-and-switch structure.**

402. Bus-and-switch structures are of great importance because not only do they provide a place for all high-tension equipment, such as buses, circuit breakers, disconnecting switches, and current and potential transformers, but also they are the most important factor in determining the type of switchgear design which will meet the necessary requirements of service reliability, safety to personnel, convenience of operation, and costs.

The size of the plant and the manner in which its output is to be used, the system of connections, and the different available designs of circuit breaker will influence the layout of the switch and bus structures.

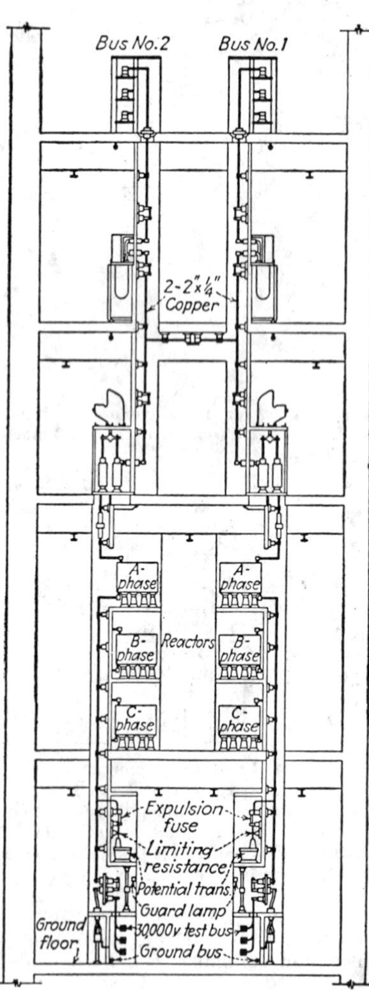

Fig. 10-149. Switch gallery using the group-phase arrangement.

Ample space should be provided for the electrical equipment, together with adequate aisles and passageways between structures, to ensure safety to the operators under normal and emergency conditions and to afford the necessary facility for the making of repairs with the least delay.

Changes in the art should be considered in the design of switching structures, and some freedom permitted in order that advantage may be taken, within reason, of future improvements.

403. Physical sectionalization of the switching structures is a precautionary measure taken to:

a. Reduce to a minimum the possibility of the occurrence of a fault.

b. Keep to a minimum the physical damage produced by a fault.

c. Prevent spreading of the fault.

To what extent this sectionalization can be economically justified depends on the type and size of the plant and on the size and character of the "unit of loss" which the system can tolerate.

In some of the large generating stations this sectionalization is accomplished by means of fire walls and fire doors which divide the switch galleries into sections. Wherever possible, direct passage between sections is eliminated, and the switching equipment is arranged so that failure in one section is not likely to spread to the next.

404. Bus-and-switch structure designs are generally identified as follows:

a. **Based on their phase grouping:**

1. Adjacent-phase or group-phase arrangement (older types of design).

2. Isolated-phase or segregated-phase arrangement.

b. **Based on the type of bus and switch-housing construction:**

1. Masonry-cell type (older types of design).

2. Metal-clad and metal-enclosed type.
3. Metal-enclosed, isolated-phase.

405. Adjacent-phase or group-phase arrangements cover the conventional type of switch-and-bus design and are used for indoor and outdoor installations without limitations as to the type and size of plant or voltage.

Buses, switches, and conductors for all phases are grouped together, the only separation between phases being that offered by the type of housing construction used (see Fig. 10-149).

For indoor installation, the bus-and-switch housings are generally of the cell type or metal-clad with single interphase barriers.

For outdoor installations, metal-clad housings or open-type construction is generally used at low or medium voltages.

406. Isolated-phase or segregated-phase arrangements have been used in many large stations. When masonry-cell-type construction is used, all buses, switches, and conductors of one phase are grouped on one floor or in one section of the switch house and every effort is made to isolate one phase section completely from another. Such isolation reduces the damage and voltage disturbance due to a bus fault and is particularly advantageous when fault ground currents are limited, as by a neutral resistance.

The phases may be isolated horizontally or vertically, but preference has been shown for the vertical-isolation type of design. A typical cell-type isolated-phase arrangement is shown in Fig. 10-150.

With the advent of the metal-clad and metal-enclosed type of construction, there is a definite trend away from the conventional isolated-phase arrangement with cell-type construction.

407. Cell-type bus-and-switch structures have, in past years, been used for medium- and large-capacity plants and for voltages between 2,300 and 34,500. In this type of construction all high-tension equipment and buses are installed in masonry or metal cells or compartments for the purpose of preventing the destructive effects of short circuits spreading and involving the entire bus structure.

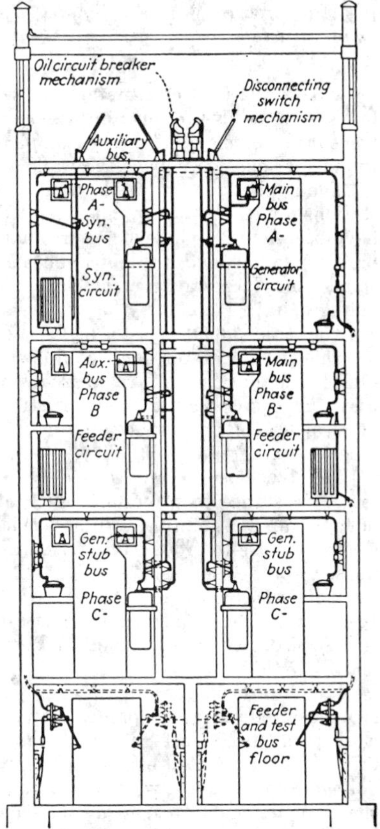

FIG. 10-150. Switch gallery using the vertical isolated-phase arrangement.

For both the group-phase and the isolated-phase arrangements, barriers are installed between phases, and openings between cells are avoided wherever there is danger of permitting the spread of a fault. The general dimensions of the bus-and-switch compartments or cells are usually determined by the type of switching used and by the minimum distances allowable between conductors and ground.

The compartments should be fireproof, and all openings should be closed by lightweight fireproof removable doors to prevent accidental contact with live parts. It is considered good practice to lock all compartment doors as a safety to personnel. In some installations various forms of **key interlock** are used which require opening the doors in sequence, the first key being released only after the main circuit breaker is open. A key is released from each lock in turn, to be used to open the succeeding one.

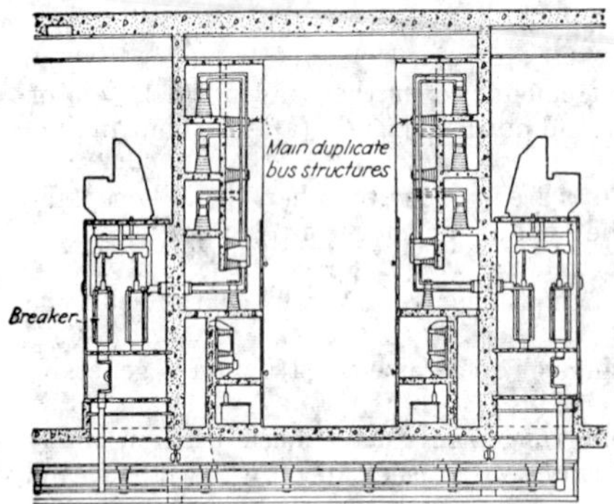

Fig. 10-151. Typical concrete bus-and-switch structure.

In some installations interlocking features are provided to safeguard against the incorrect operation of any equipment or device. In any switching cycle, the circuit breaker must be closed last and opened first.

408. Concrete structures should be as free of conducting material as possible, and reinforcing rods should not be used except where necessary and only in isolated groups. Care should be exercised so that reinforcing rods do not form complete magnetic loops when in the proximity of heavy current-carrying parts. Figure 10-151 shows a typical concrete construction. This type of structure has used many circuit breakers of the type shown in Fig. 10-14.

409. Metal-clad and metal-enclosed bus-and-switch structures are those in which buses and circuit-breaker and disconnecting switches, as well as necessary instrument transformers, relays, etc., are completely enclosed in metal. In this type of design, the metal housings replace the walls and barriers of the old cell constructions.

The applications of metal enclosures have been numerous in central stations, substations, and industrial plants, and many different forms of this type of construction have been built. These structures have been built for indoor and outdoor applications for small- and large-capacity stations, and for voltages up to 138,000.

The major advantages of metal enclosures are safety to operators and equipment, simplified installation and reduced installation costs, extensive factory construction, economical use of floor space, minimum maintenance, and greater reliability.

Metal enclosures are conceded to be of much assistance in providing quick clearing of faults and in limiting troubles which may occur to phase-to-ground failures rather than phase-to-phase failures.

Metal-clad construction is shown in Figs. 10-20 and 10-22. Segregated-phase metal-enclosed construction is given in Figs. 10-42 and 10-152.

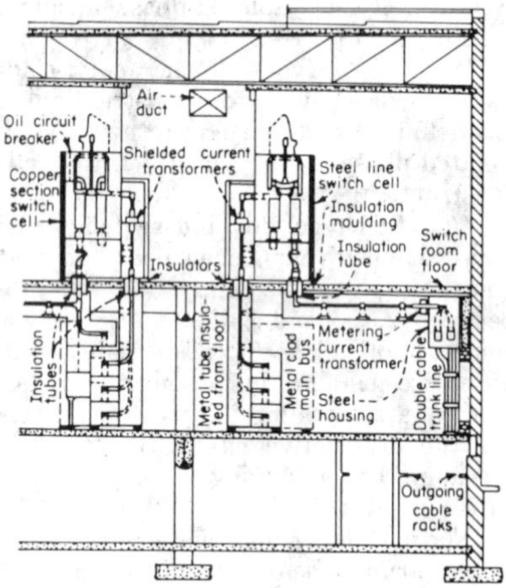

Fig. 10-152. Metal-enclosed bus-and-switch structure.

410. Interlocks are used in metal-enclosed switchgear designs to prevent opening or closing of disconnects in case the circuit breaker is closed, to prevent opening of high-voltage compartments while the circuits are alive, to prevent making circuits alive until the compartments are closed and secured, and other conditions as may be necessary for a particular installation.

411. Isolated-phase metal-enclosed type of bus-and-switch structure, with each phase in a separate metal housing, has largely replaced the conventional isolated-phase cell type of masonry construction because of its higher reliability and lower cost (see Fig. 10-154). Another use of isolated-phase metal-enclosed construction is for generator leads (see Fig. 10-155). A bus of this type has been built and operated up to 138,000 V (see Fig. 10-43 for a 69,000-V installation).

The bus is made in sections which are joined together at the point of installation through silver-surfaced bolted connections. Flexible connectors were used at intervals to allow for mechanical movement of the bus due to thermal expansion and contraction. The insulator-supporting rings are made of high-strength nonmagnetic alloy. The bus covers are of nonmagnetic material such as aluminum or nonmagnetic manganese steel. They are insulated from the support ring at one end to reduce eddy-current loss and to prevent circulating current from support to support. Each ring support is mounted to supporting steel, which is equipped with a ground bus.

Generator sizes have reached the point where it is now necessary to force-cool the metal-enclosed phase-isolated generator leads. A typical bus installation for a total generator current of 30,000 A is shown in Fig. 10-153.

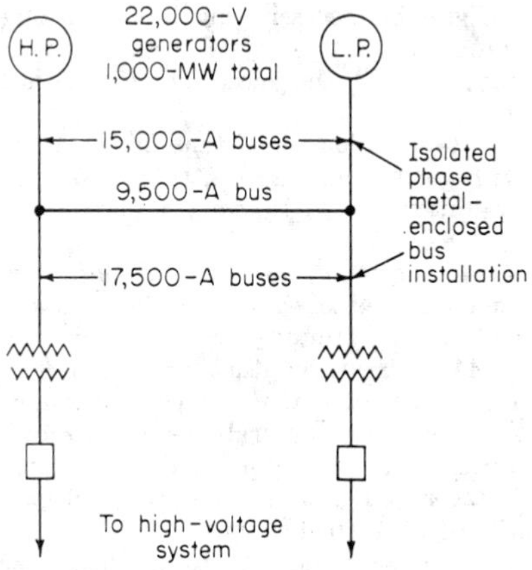

Fig. 10-153. Modern generator and metal-enclosed-bus installation (one transformer slightly exceeds rating of one generator).

412. Materials used for metal enclosures are sheet steel and nonmagnetic materials such as manganese steel, bronze, copper, and aluminum. Nonferrous materials are used, in the case of large-capacity structures, in order to reduce the amount of heating in the enclosures due to hysteresis and eddy currents to within permissible limits (see EEI and NEMA Specification Guides for this type of construction).

413. Open-type bus-and-switch structures for indoor service and for voltages up to 34,500 are considered preferable by some engineers because of the facility with which inspections can be made. Much depends on conditions. The design should be given careful thought in order to have the construction meet operation conditions with reliability and safety.

The present trend for voltages up to 34,500, however, is toward enclosing indoor buses and high-tension or heavy-current-carrying leads, as described in Pars. **402 to 411.**

For higher voltages, the violence of an arc during faults and its destructive effects are much less than for lower voltages, while, on the other hand, the spacings required

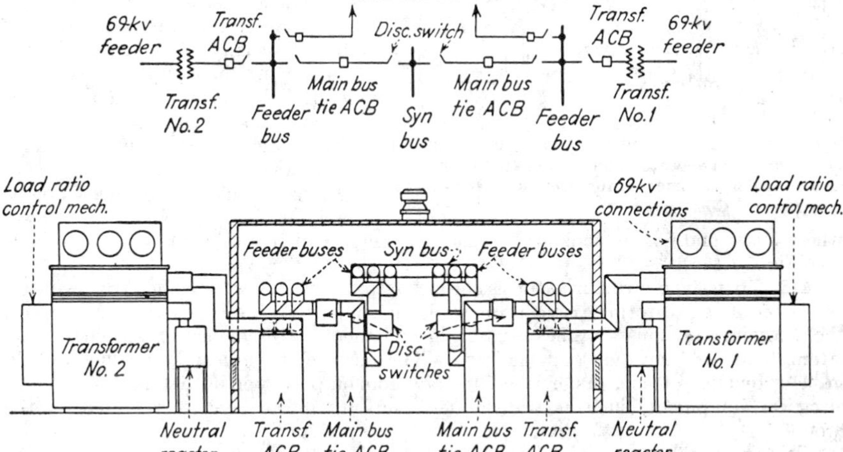

Fig. 10-154. Arrangement of isolated-phase metal-enclosed switchgear for 13.8-kV distribution substation.

are greater, necessitating buildings of excessive proportions. Although there are instances where open-type bus-and-switch structures have been used indoors for voltages above 34,500, the accepted practice is to confine this type of construction to outdoor installations.

414. Outdoor open-type bus-and-switch structures are used generally in connection with generating stations, substations, and industrial plants, and there are no practical limitations as to capacity and voltage. However, enclosed-type structures are used to 138,000 V.

Arrangement and general design characteristics of outdoor switching structures are influenced, of course, by the function and type of the installation and by its capacity, voltage, and ground-area limitations.

415. Structures used for supporting the buses, insulators, and switching equipment are ordinarily made of steel, although sometimes wood and concrete are employed.

Standard structural-steel shapes or truss-type construction are used, and the steel members are either painted or galvanized. For small installations and generally for voltages of 13,200 and below, steel pipe with standard fittings has been used in place of structural-steel shapes.

Likewise, wooden poles and beams have been used for relatively small substations in the lower-voltage classes.

Reinforced concrete has been used occasionally to improve appearance and reduce maintenance costs. The design of the supporting structures is affected by the phase spacings and ground clearances required, by the type of insulator, by the length and weight of buses and other equipment, and by the wind and ice loading. For data on wind and ice loadings, see National Electrical Safety Code (see USASI Standard C2.2-1960 Rev). For recommended clearances and phase spacings, see Par. **457**.

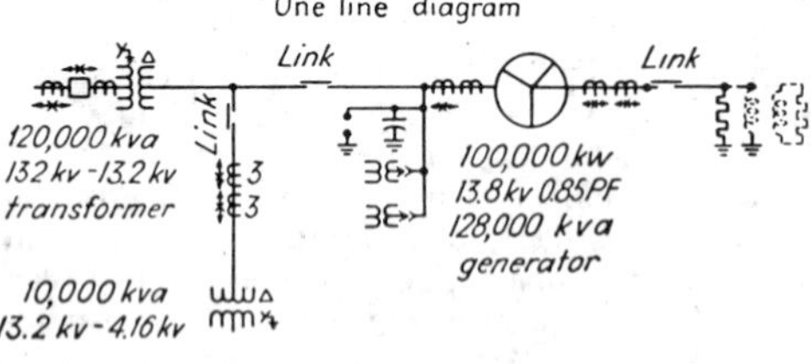

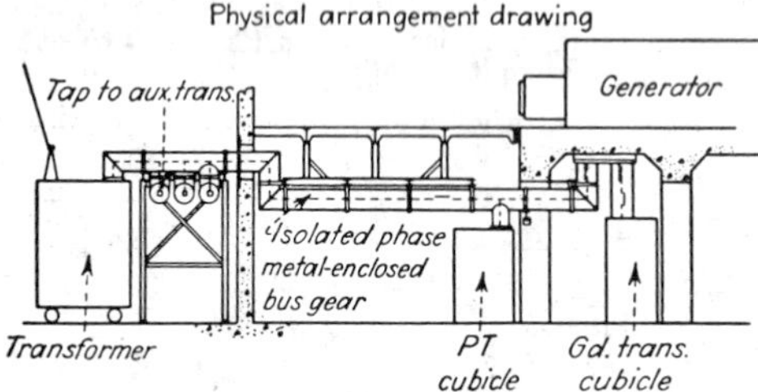

FIG. 10-155. Isolated-phase metal-enclosed switchgear used for generator connection to high-voltage and auxiliary power transformers. (*I-T-E Circuit Breaker Company.*)

416. Outdoor types of switching structure in general use are:

1. Unit-type.
2. Truss-type.
3. Ground-type.

417. Unit-type structures are those which make use of individual units, usually of A-frame design, to support the equipment, and generally they are of the strain-bus type. This construction readily lends itself to the ground configuration and to conditions subject to change. This type is most adaptable for voltages of 44 kV or above where buses and connections are simple and the cost of real estate is not appreciable (see Fig. 10-155).

418. Truss-type structures consist of an assembly of beams or lattice trusses interconnected so as to form a rigid framework for supporting the buses and connections, the incoming transmission lines, and such equipment as disconnecting switches, insulators, lightning arresters, and metering. The buses may be flexible conductors held in place by strain insulators or tubular conductors on post-type insulators. The major electrical equipment, such as transformers and circuit breakers, is mounted on the ground.

This arrangement occupies less ground area than other types of design, and, even though large amounts of steel are required, it is more economical, not only for large transmission transforming and sectionalizing substations but also for small customers where space is valuable.

In this type of design the equipment and connections are generally more congested, presenting some hazards of maintenance. Truss-type structures are most adaptable to the higher voltages (66 kV and above) with complicated systems of connection and to all stations of the lower-voltage class (44 kV and below) (see Fig. 10-157). For aesthetic reasons, some engineers prefer the use of rolled sections instead of lattice work for truss designs.

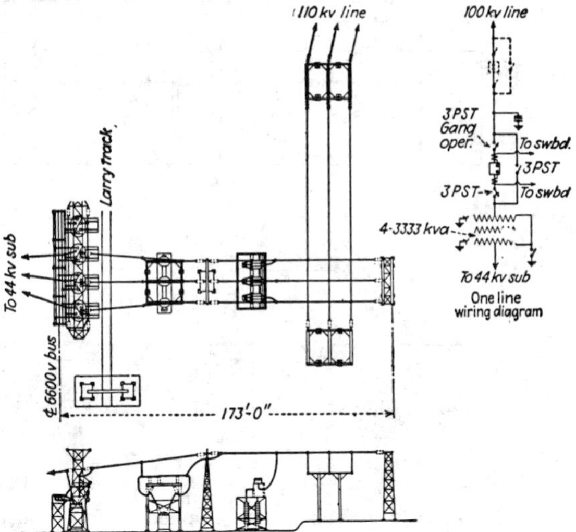

FIG. 10-156. Typical unit-type substation design.

419. Ground-type or flat-type structures are characterized by a general absence of overhead steel structures for supporting the buses and equipment. A-frame structures are generally used for supporting the incoming transmission lines. All other equipment including the buses and connections are kept at a low elevation and are supported on individual lightweight structures or posts close to the ground. Buses and connections are generally of the rigid type. Because of its low arrangement, all equipment is much more easily accessible for inspection and repairs.

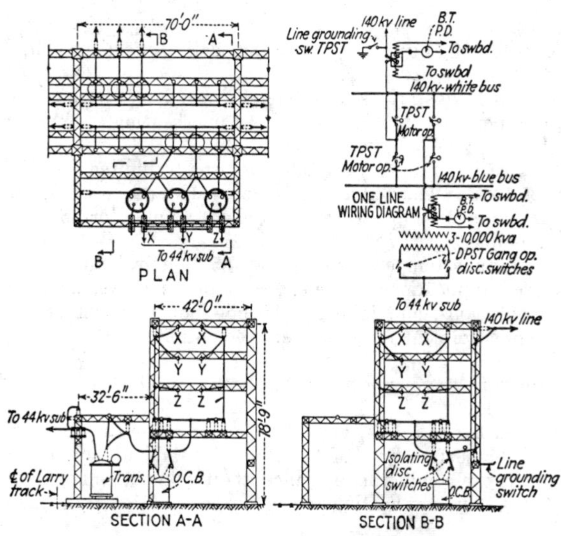

FIG. 10-157. Typical truss-type substation design.

The absence of overhead structures reduces the hazard of accidental electric arcs or fires at one point in the substation being communicated to other circuits.

This design generally requires more ground space than other types of design, and foundation costs are higher owing to their number and scattered locations.

The ground type of structure is considered most reliable, and it has been used on some of the largest and most important substations for voltages of 138 kV and above (see Fig. 10-158).

420. Station buses are a most important part of all power switchgear structures, as they carry the whole energy of the plant in a confined space. They must be carefully designed in order that the construction may provide adequately and economically for the utilization of the electrical energy generated and at the same time have sufficient structural strength to withstand the maximum stresses that may be imposed on the conductors, and in turn on the structure, by heavy currents under short-circuit conditions.

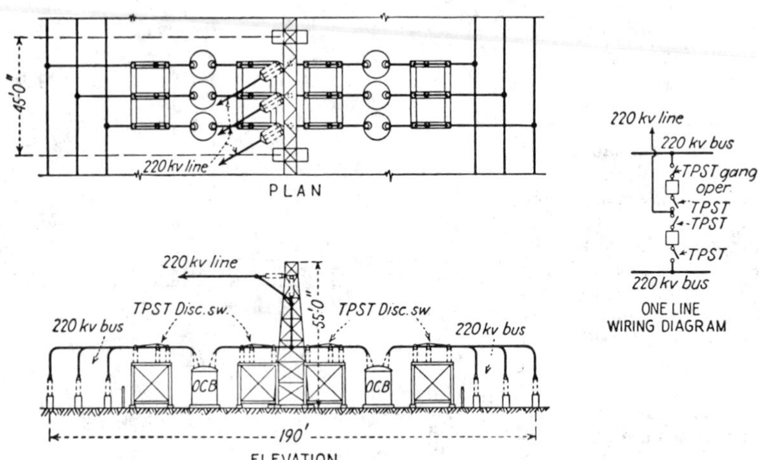

FIG. 10-158. Typical ground-type substation design.

421. Elements of bus design applicable to all installations may, in general, be classified as follows:

a. Current-carrying capacity.

b. Short-circuit stresses.

c. Methods of making connections.

422. The current-carrying capacity of a bus is limited by the heating effects produced by the current.

Buses for generating stations and substations are generally rated on the basis of the temperature rise which can be permitted without danger of overheating equipment terminals, bus connections, and joints.

The permissible temperature rise for plain copper buses is usually limited to 30°C above an ambient temperature of 40°C. This value is the accepted standard of IEEE, NEMA, and USASI. This is an average temperature rise, and a maximum or hot-spot temperature rise of 35°C is permissible.

The standards allow a hot-spot bus temperature rise of 65°C above an outside ambient of 40°C for metal-enclosed applications where silver-contact surfaces are used on equipment terminals, bus connections, and joints. The hot-spot temperature rise is limited to 45°C above the outside ambient for silver-surfaced terminals of outgoing circuits (see USASI C37.20).

For other types of bus installations the engineer should check the latest applicable IEEE, NEMA, and USASI Standards.

Many factors enter into the **heating of a bus,** viz., the type of material used, the size and shape of the conductor, the surface area of the conductor and its condition, the skin effect and proximity effect, the conductor reactance, the thickness and type of insulation, the ventilation, and the inductive heating caused by the proximity of magnetic materials.

423. Bus materials in general use are copper and aluminum. Iron and steel are used in some instances where the amount of current to be carried is relatively small in comparison with the cross section of the bus and the atmospheric conditions do not cause excessive corrosion.

Hard-drawn copper is the most widely used material because of its low resistance with consequent low losses, its low rate of corrosion, and its high thermal conductivity.

Hard-drawn aluminum has not been used so extensively as copper for buses, but certain advantages are claimed on the basis of lightness in weight. The weight of aluminum is about one-third that of copper. Lower relative cost and improved methods of making joints have resulted in increased usage in recent years.

The aluminum ordinarily used for bus work has a conductivity of 63%, as compared with copper at 99%.

For a given current rating and for equal limiting temperatures, the area of the aluminum bus will, in general, be about 133% of the area of the copper bus.

Conductivity of copper alloys is materially reduced by the presence of even small percentages of certain elements, such as phosphorus, aluminum, iron, tin, zinc, or cadmium. The percentage of copper present is no indication of the conductivity, e.g.:

Everdur (95% copper, 4% silicon, 1% manganese)—conductivity 7% of copper.

Monel metal (30% copper, 69% nickel, 1% iron)—conductivity 4% of copper.

Durobronze (97% copper, 2% tin, 1% silicon)—conductivity 13% of copper.

Figure 10-159 shows the effect of alloying elements on the electrical conductivity of copper. The Metals Handbook, Vol. 1, gives a very complete analysis of the effect of various alloys on the conductivity of copper.

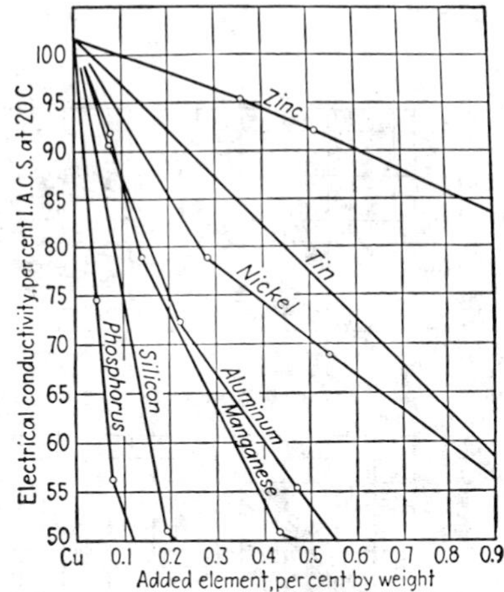

Fig. 10-159. Effect of alloying elements on the conductivity of copper. (*C. S. Smith and E. W. Palmer, Metal Technology, Amer. Inst. Min. & Met. Eng., September,* 1935.)

424. Allowable current density in a bus is the amount of current that the bus can carry per square inch of cross-sectional area without exceeding the permissible temperature rise.

For both a-c and d-c buses, densities may vary from values of 600 and 700 A/in² in heavy-current-carrying copper buses to 1,200 and 1,400 A/in² in light buses under favorable conditions.

For aluminum, densities of only 75% of the above values are usually permitted.

For more specific data on current capacities consult the graphs that follow.

425. Skin effect in a conductor carrying an alternating current is the tendency toward crowding of the current into the outer layer or "skin" of the conductor owing to the self-inductance of the conductor. This results in an increase of the effective resistance of the conductor and in a lower current rating for a given temperature rise. Skin effect is very important in heavy-current buses where a number of conductors are used in parallel, because it affects not only each conductor but each group of conductors as a unit.

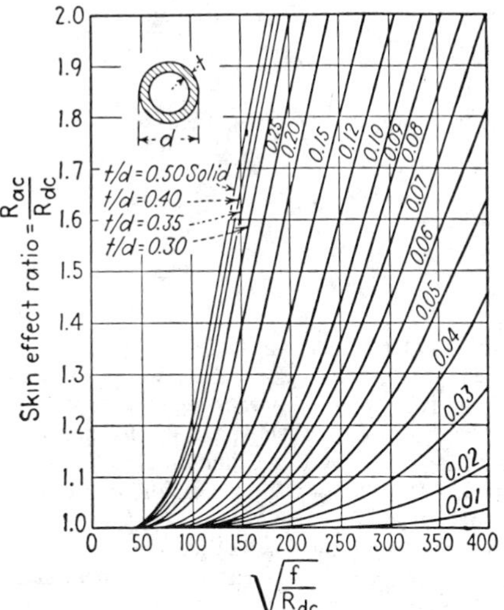

Fig. 10-160. Skin-effect ratio in tubes of varying wall thickness. R_{ac} is a-c resistance, R_{dc} is d-c in ohms per thousand feet, and f is the frequency. (*Dwight, Trans. AIEE, March,* 1922.)

Tubes have less skin-effect resistance than flat conductors of the same cross section, and tubes with thin walls are affected the least by skin effect.

Aluminum conductors are affected less by skin effect than copper conductors of similar cross section because of the greater resistance of aluminum.

The most economical method of reducing skin-effect losses is by proper choice of conductor cross-sectional shape.

Skin-effect ratio is defined as the ratio of a-c to d-c resistance. The curves of Figs. 10-160 and 10-161 show the skin effect in isolated tubes and rectangular conductors, respectively.

426. Proximity effect in a bus is the

distortion of the current distribution caused by the induction between the currents in the go and return conductors. This causes a concentration of current in the parts of the buses nearest each other, thus increasing their effective resistance. In general the proximity effect is directly proportional to the magnitude of the current and inversely proportional to the distance between conductors. The proximity effect must be taken into account for buses carrying alternating current, and on 3-phase buses this effect is less than on single-phase buses.

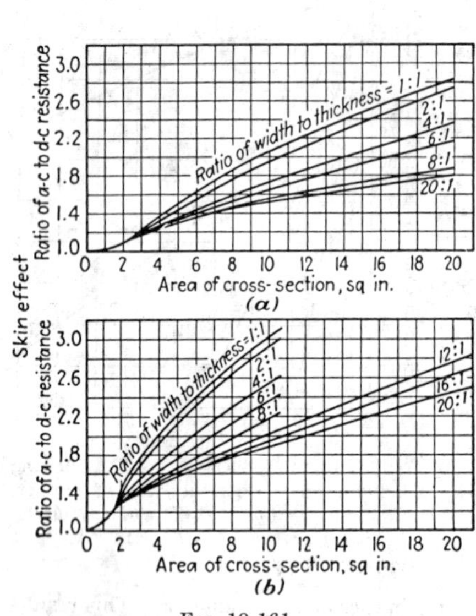

Fig. 10-161.

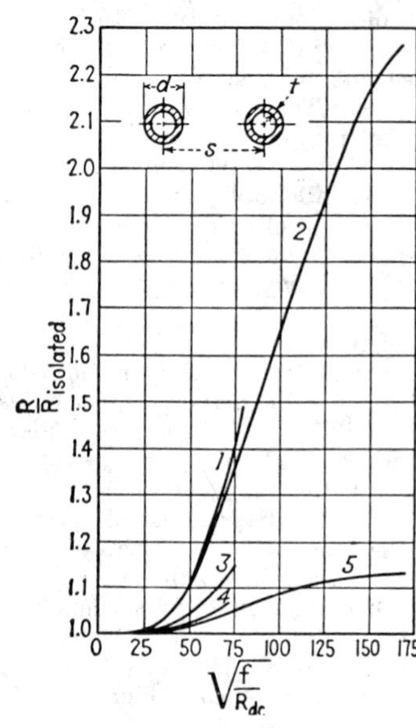

Fig. 10-162.

Fig. 10-161. Skin effect in strap copper conductors. (*a*) At 25 cycles. (*b*) at 60 cycles. (*H. C. Forbes and J. L. Gorman, Elec. Eng., September*, 1938.)

Fig. 10-162. Proximity-effect ratio in tubes: R_{dc} in ohms per 1,000 ft.

1. $s/d = 1.0$; $t/d = 0$ (calculation)
2. $s/d = 1.008$; $t/d = 0.125$ (test)
3. $s/d = 1.5$; $t/d = 0$ (calculation)
4. $s/d = 2.0$; $t/d = 0$ (calculation)
5. $s/d = 2.03$; $t/d = 0.125$ (test)

(*H. B. Dwight, Trans. AIEE, March*, 1922.)

Proximity-effect ratio is taken here as the ratio of a-c resistance of conductors when close together to their normal a-c resistance when isolated. For tubes, when the distance between centers is twice the diameter of the tube, the increase in resistance will be not more than 20% of the a-c resistance when isolated. With spacings in excess of five times the diameter the increase is negligible, being of the order of 2%. Figure 10-162 shows the proximity-effect ratio. Figure 10-163 shows the variation in bus capacity for different bus spacings due to proximity effect.

427. Conductors of various sizes and shapes are available for bus work.

Copper and aluminum are used in bar, tubular, rod, and structural shapes; iron and steel, as a rule, only in tubular form.

Steel structural shapes (angles and T's) have been used in some instances in outdoor construction, the carrying capacity being reinforced, when necessary, with copper bars.

A judicious choice of these conductors is necessary in order to make the most economical use of the material, as the actual current carried by the bus will depend upon its ability to dissipate heat losses. For indoor installations, heat dissipation depends on radiation and convection; for outdoor installations, it is affected in addition by exposure to wind and sun.

428. The maximum bus efficiency is obtained when all parts of the bus are operating at the same temperature. In practice a bus design resolves itself into selecting a conductor or a multiplicity of conductors arranged in such a manner that:

a. The current distribution within the conductors is as uniform as practicable.

b. The surface area of the conductors is sufficient for proper heat dissipation.

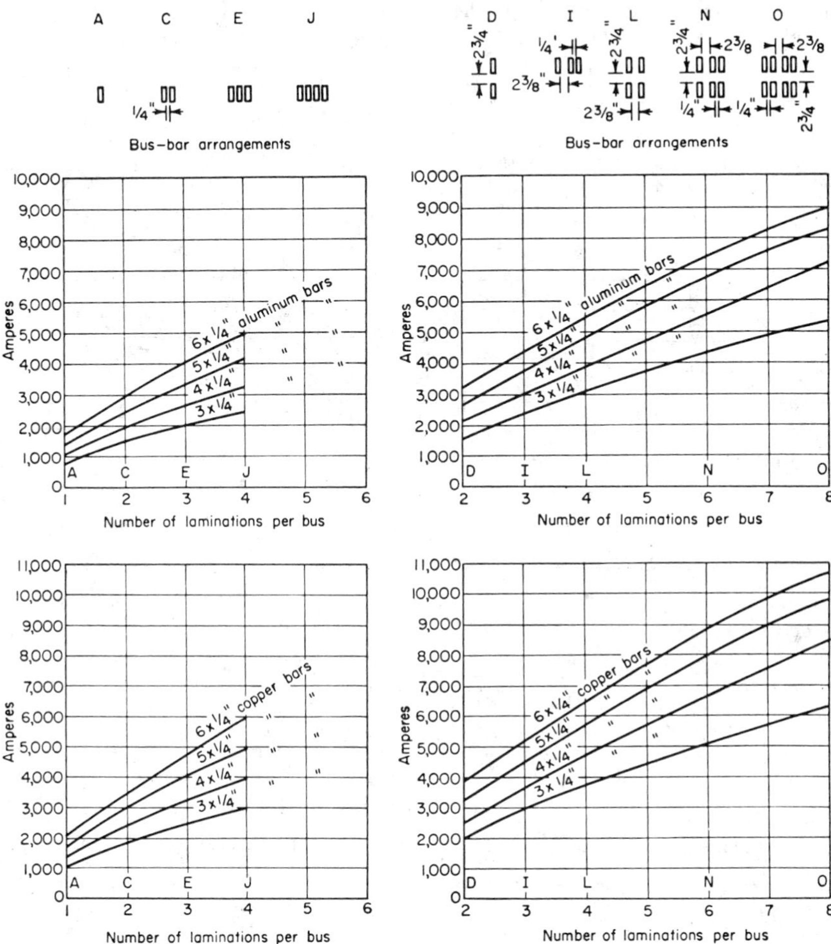

FIG. 10-163. Direct current-carrying capacities of bus bars for 35°C rise in still, unconfined air.

For current-carrying capacities at ambient temperatures other than stated for each graph or table, approximate values can be obtained by using the formula

$$\text{New current rating} = \text{given current rating} \times \left(\frac{\text{new temperature rise}}{\text{given temperature rise}}\right)^{1/2}$$

429. Flat bars are used extensively because of their ease of installation, their good ventilation characteristics, and the fact that their radiating surface in proportion to their cross-sectional area is greater than on any other form of conductor.

For direct current, flat bars are used almost entirely. For multibar buses their current-carrying capacity is somewhat reduced, because of mutual heating between bars. The performance of multibar d-c buses for various arrangements of bars is shown in Fig. 10-163.

For alternating current, the use of flat bars requires careful planning because of skin effect, proximity effect, and reactance. Obviously, for heavy-current buses the current capacity does not increase proportionately to the number of bars in parallel, and consideration must be given to the spacing and position of the bars, it being kept in mind that the bar arrangement must not hinder the dissipation of heat losses.

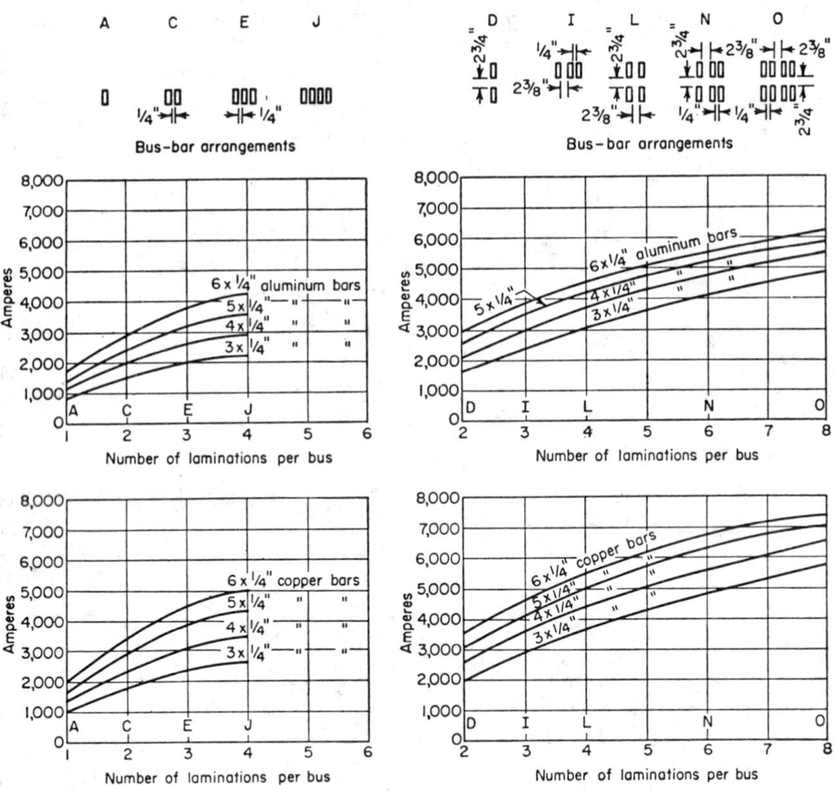

Fig. 10-164. Alternating-current-carrying capacities of bus bars for 35°C rise in still, unconfined air, 25 c/sec. Values apply to single and polyphase circuits.

In long heavy-current low-voltage a-c buses, such as are used in electric-furnace installations, the inductive drop is often a serious consideration, and **interlacing** of the bus bars together with a comparatively short distance between phases has been found to be an **effective means of reducing the bus reactance** and improving the bus rating so that a d-c rating is virtually obtained.

The performance of multibar a-c buses for various arrangements of bars is shown in Figs. 10-165 and 10-166 for 16- to 18-in flat-bar spacings. The current-carrying capacity of a single bar in a horizontal position is about 90% of the carrying capacity of the same bar in a vertical position. This must be taken into consideration when bars in a horizontal position are used for heavy-current buses.

430. Tubular conductors when used on alternating current have a better current distribution than any other shape of conductor of similar cross-sectional area, but they also have a relatively small surface area for dissipating heat losses. These two factors must be properly balanced in the design of a tubular bus.

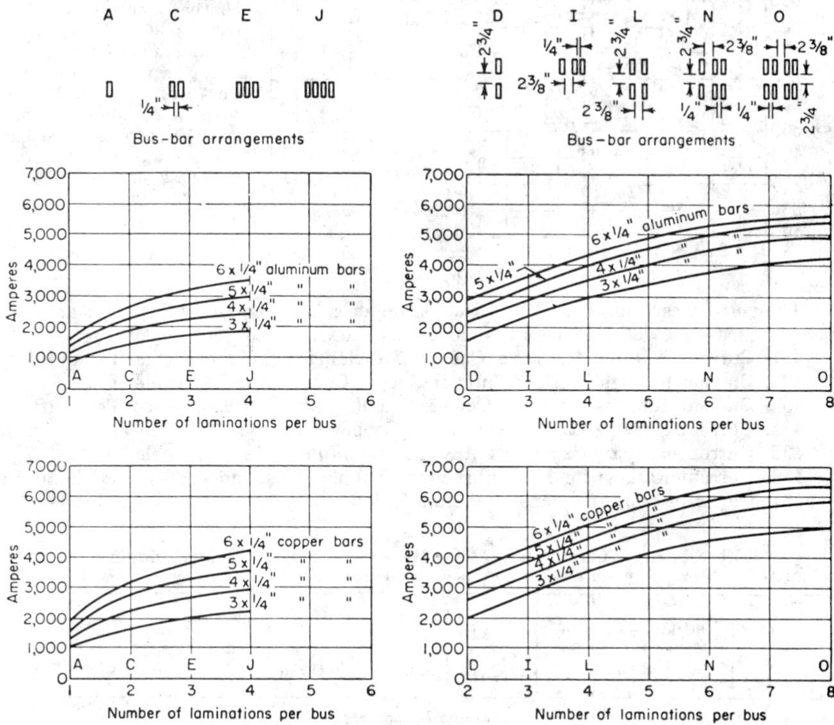

Fig. 10-165. Same as Fig. 10-164, but for 60-c/sec use.

Tubing provides a relatively large cross-sectional area in minimum space and has the maximum structural strength for equivalent cross-sectional area, permitting longer spaces between supports. In outdoor substations spans up to 40 and 50 ft with copper or aluminum tubes up to 6 in diameter are considered practicable, the tendency in design being to use long spans and thus reduce the number of insulator posts to a minimum.

Copper tubes generally used are available in three weights: standard, extra-heavy, and double-extra-heavy pipe sizes.

Aluminum tubes are available in two weights, standard and extra-heavy pipe sizes, or by specified diameters and wall thicknesses.

Taps to and connections between sections of tubing are somewhat more difficult than those of flat bars.

References for **current-carrying capacities** of copper and aluminum tubular buses and solid rods of different dimensions are given in Pars. **431** to **435**. Figure 10-167 shows the capacity of copper and aluminum tubing for various wall thicknesses. The current capacities vary slightly in the different tables.

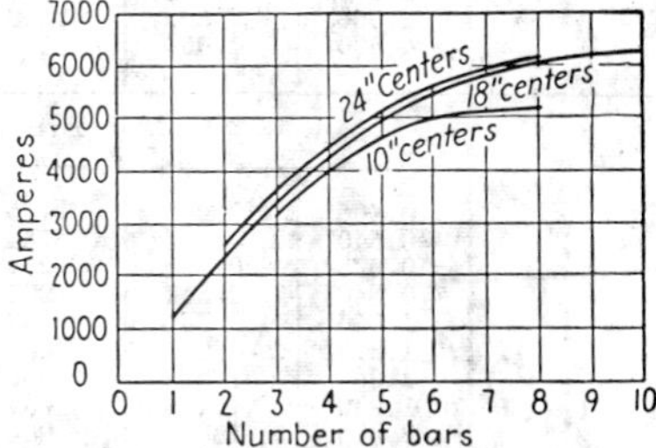

Fig. 10-166. Variation in 60-c/s capacity for different bus spacings, due to proximity effect. (*H. W. Papst, Elec. World, Sept.* 21, 1929.)

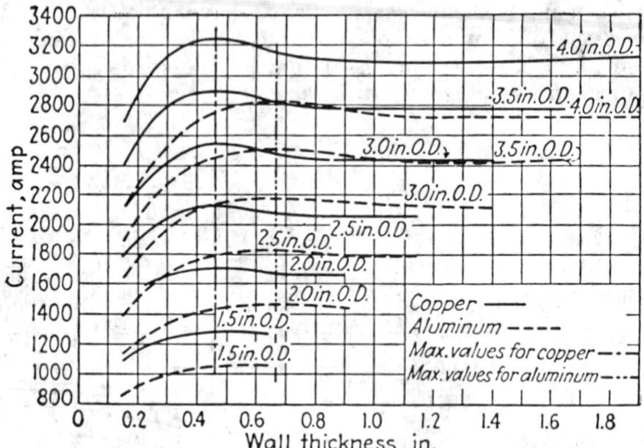

FIG. 10-167. Current-carrying capacity of tubular conductors vs. wall thickness based on a temperature rise of 30°C. Note that maximum current is carried in copper tubes with a wall thickness of slightly less than ½ in, whereas maximum current is carried in aluminum tubes with a wall thickness of about 11/16 in. (*H. W. Papst, Elec. Jour., July*, 1931.)

The current-carrying capacity of round copper tubes increases in terms of amperes per pound as the wall thickness is reduced, as shown for 60-c loading in Fig. 10-168.

431. Current Ratings for Bare Copper Tubular Bus, Indoors (see Table 10-4).

432. Current Ratings for Bare Tubular Copper Conductors (see Table 10-5).

433. Current Ratings for Bare Copper Tubular Bus, Outdoors (see Table 10-6).

434. Current-carrying Capacities of Solid Copper Rods (see Table 10-7).

435. Current-carrying Capacities of Solid Aluminum Rods (see Table 10-8).

436. Current Ratings for Bare Aluminum Tubular Bus, Indoors (see Table 10-9).

437. Current Ratings for Bare Aluminum Tubular Bus, Outdoors (see Table 10-10).

Table 10-4. Current Ratings for Bare Copper Tubular Bus, Indoors*

100% IACS ambient temperature 40 °C, frequency 60 cycles; horizontal position, large rooms without drafts

Size of tube, IPS, in.	Outside diam., in.	Inside diam., in.	Current rating, amp					
			10 C rise	20 C rise	30 C rise	40 C rise	50 C rise	60 C rise
Standard pipe sizes								
1/8	0.405	0.281	97	140	175	203	226	245
1/4	0.540	0.375	140	205	255	298	336	365
3/8	0.675	0.494	178	264	325	384	432	470
1/2	0.840	0.625	232	348	433	508	572	630
3/4	1.050	0.822	290	438	545	645	730	795
1	1.315	1.062	370	562	705	835	945	1030
1 1/4	1.660	1.368	490	748	933	1110	1260	1385
1 1/2	1.900	1.600	558	850	1065	1275	1440	1575
2	2.375	2.062	695	1070	1345	1600	1800	2000
2 1/2	2.875	2.500	885	1370	1760	2050	2350	2550
3	3.500	3.062	1135	1765	2240	2650	3000	3300
3 1/2	4.000	3.500	1360	2120	2650	3200	3650	4000
4	4.500	4.000	1510	2350	2970	3550	4050	4450
Extra heavy pipe sizes								
1/8	0.405	0.205	115	167	206	244	270	295
1/4	0.540	0.294	164	240	296	348	390	425
3/8	0.675	0.421	204	302	375	440	495	540
1/2	0.840	0.542	265	398	490	585	660	715
3/4	1.050	0.736	332	500	615	735	835	910
1	1.315	0.951	435	660	818	980	1105	1210
1 1/4	1.660	1.272	552	845	1050	1260	1425	1560
1 1/2	1.900	1.494	634	970	1210	1440	1650	1800
2	2.375	1.933	804	1235	1560	1850	2100	2300
2 1/2	2.875	2.315	1055	1635	2050	2450	2800	3050
3	3.500	2.892	1305	2020	2580	3050	3500	3850
3 1/2	4.000	3.358	1510	2340	2980	3550	4050	4550
4	4.500	3.818	1715	2670	3400	4050	4650	5100

* From an article entitled Current-carrying Capacity of Bare Cylindrical Conductors for Indoor and Outdoor Service, by C. W. Frick, *Gen. Elec. Rev.*, August, 1931.

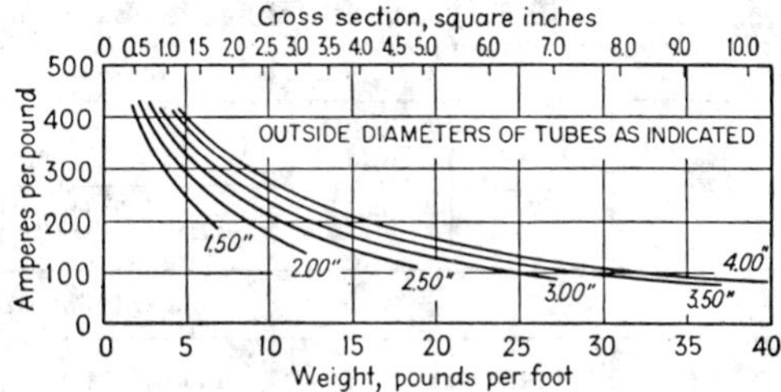

FIG. 10-168. Current ratings of six round copper tubes in terms of tube weight per foot, 60 cycles. (*The American Brass Company.*)

Table 10-5. Current Ratings for Bare Tubular Conductors
[NEMA recommended practice (SG-1, Pt. 3),* 1962]

Indoor Service
The standard 60-c current ratings based on 30°C temperature rise above a 40°C ambient for 98% conductivity IACS copper and 57 to 61% conductivity aluminum conductors are as follows

Size of tube, IPS, in	Capacity, A			
	Aluminum		Copper	
	Standard	Extra-heavy	Standard	Extra-heavy
½	300	350	380	420
¾	400	465	540	590
1	520	610	650	750
1¼	620	735	870	975
1½	735	840	1,020	1,150
2	980	1,125	1,250	1,500
2½	1,215	1,400	1,700	1,975
3	1,470	1,715	2,175	2,475
3½	1,725	2,010	2,575	2,875
4	1,960	2,255	2,850	3,100
4½	2,255	2,645	.....	.....
5	2,645	3,085	3,450	3,850
6	.....	.....	4,000	4,500

Outdoor Service*
When conductors are used for outdoor service, advantage may be taken of favorable conditions, such as air currents, weathering of copper, etc., permitting the following normal values for 98% conductivity IACS copper and 57 to 61% conductivity aluminum

Size of tube, IPS, in	Capacity, A			
	Aluminum		Copper	
	Standard	Extra-heavy	Standard	Extra-heavy
½	400	465	510	580
¾	520	610	710	780
1	675	685	850	1,010
1¼	815	930	1,120	1,250
1½	930	1,080	1,280	1,450
2	1,225	1,470	1,550	1,850
2½	1,470	1,765	2,000	2,400
3	1,790	2,107	2,550	2,950
3½	2,107	2,500	3,050	3,400
4	2,400	2,840	3,400	3,800
4½	2,745	3,235		
5	3,185	3,725	4,100	4,600
6	.....	.....	4,700	5,200

* Since duplication of weather conditions, taken into account in establishing outdoor current ratings, is impractical, check tests, when made, should be made on ratings under SG 1, Pt. 3.

438. Hollow-square conductors are extensively used for bus work, particularly for heavy currents. This type of construction may consist of hollow-square tubing or of a hollow-square arrangement of flat bars (see Par. **429**) or of channel and angle shapes.

Two channels or two angles may be arranged in the form of a hollow square leaving sufficient space between the flanges for ventilation of the inside walls of the bus. When hollow-square tubing is used, ventilating holes are often provided on two opposite sides of the tubing.

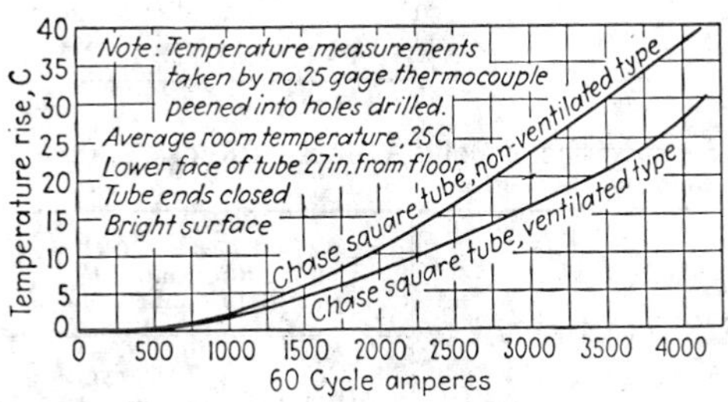

FIG. 10-169. Comparison of temperature rise for ventilated and nonventilated Chase copper tube, hard temper.

These arrangements approach closely the efficiency of a tubular conductor and surpass it appreciably in heat-dissipating surface, resulting in a bus of greater current-carrying capacity for a given weight of material and a given temperature rise, when internally ventilated.

Dimensions and current-carrying capacities of hollow-square tubing and channels

Table 10-6. Current Ratings for Bare Copper Tubular Bus, Outdoors*

100% IACS, ambient temperature 40°C, frequency 60 cycles; partly sheltered locations, wind velocity 2 ft per sec crosswise to conductors

Size of tube, IPS, in.	Outside diam., in.	Inside diam., in.	Current rating, amp					
			10 C rise	20 C rise	30 C rise	40 C rise	50 C rise	60 C rise
Standard pipe sizes								
⅛	0.405	0.281	135	188	228	260	285	310
¼	0.540	0.375	195	270	330	375	415	450
⅜	0.675	0.494	245	345	415	475	525	570
½	0.840	0.625	320	450	550	620	685	745
¾	1.050	0.822	400	560	680	775	860	930
1	1.315	1.062	505	710	860	985	1090	1185
1¼	1.660	1.368	660	930	1130	1290	1430	1550
1½	1.900	1.600	750	1050	1285	1460	1620	1760
2	2.375	2.062	925	1300	1585	1800	2000	2200
2½	2.875	2.500	1175	1640	2010	2300	2550	2800
3	3.500	3.062	1490	2100	2560	2900	3260	3550
3½	4.000	3.500	1780	2520	3040	3500	3900	4250
4	4.500	4.000	1970	2780	3400	3850	4350	4700
Extra heavy pipe sizes								
⅛	0.405	0.205	160	225	270	310	345	370
¼	0.540	0.294	225	320	380	435	485	525
⅜	0.675	0.421	282	395	475	545	605	655
½	0.840	0.542	366	515	620	710	785	850
¾	1.050	0.738	458	655	770	885	980	1060
1	1.315	0.951	590	830	1010	1150	1270	1385
1¼	1.660	1.272	750	1055	1270	1460	1620	1750
1½	1.900	1.494	855	1200	1460	1670	1850	2010
2	2.375	1.933	1075	1510	1850	2100	2350	2550
2½	2.875	2.315	1400	1970	2390	2750	3050	3300
3	3.500	2.892	1715	2420	3000	3400	3750	4100
3½	4.000	3.358	1970	2780	3410	3900	4300	4700
4	4.500	3.818	2230	3150	3880	4400	4900	5400

* From an article entitled Current-carrying Capacity of Bare Cylindrical Conductors for Indoor and Outdoor Service, by C. W. Frick, *Gen. Elec. Rev.*, August, 1931.

Table 10-7. Current-carrying Capacities of Solid Copper Rods

Size, in	Area, sq in	Area, cir mils	Weight, lb per ft	Amp (based on 30°C rise, 20°C ambient) in still unconfined air		
				Direct current	25-cycles a.c.	60-cycles a.c.
⅓	0.196	250,000	0.76	340	330	325
¾	0.442	562,500	1.70	600	580	550
1	0.785	1,000,000	3.05	920	860	800
1¼	1.227	1,560,000	4.72	1250	1170	1100
1½	1.767	2,250,000	6.81	1600	1500	1400

Table 10-8. Current-carrying Capacities of Solid Aluminum Rods

(Aluminum Company of America)

Size, in	Area, sq in	Area, cir mils	Weight, lb per ft	Amperes (based on 30°C rise above 40°C ambient) in still unconfined air	
				Direct current	60-cycles a.c.
½	0.196	250,000	0.230	250	250
¾	0.442	562,500	0.518	445	385
1	0.785	1,000,000	0.920	672	565
1¼	1,227	1,560,000	1.438	925	800
1½	1.767	2,250,000	2.071	1200	1120

and angles in the hollow-square arrangement are given in Pars. **439** to **442** and Fig. 10-169.

439. Physical Characteristics, Weights, and Current-carrying Capacities of Anaconda Square Copper Tubes, Indoors (see Tables 10-11 and 10-12). Anaconda square high-conductivity copper bus-conductor tubes are manufactured with rows of round holes drilled in two opposite sides.

Table 10-9. Current Ratings for Bare Aluminum Tubular Bus, Indoors

61% IACS, ambient temperature 40°C, frequency 60 cycles; horizontal position, large rooms without drafts (From Burndy Engineering Co. Cat. No. 50)

Size of tube, IPS, in.	Outside diam., in.	Inside diam., in.	Current rating, amp					
			10 C rise	20 C rise	30 C rise	40 C rise	50 C rise	60 C rise
			Standard pipe sizes					
⅛	0.405	0.281	76	109	137	158	176	191
¼	0.540	0.364	109	160	199	232	262	285
⅜	0.675	0.494	139	206	254	300	337	367
½	0.840	0.624	181	271	338	396	446	491
¾	1.050	0.824	226	342	425	503	569	620
1	1.315	1.050	289	438	550	651	737	803
1¼	1.660	1.380	382	583	728	866	983	1080
1½	1.900	1.612	435	663	831	995	1123	1229
2	2.375	2.068	542	835	1050	1248	1404	1560
2½	2.875	2.470	690	1070	1375	1600	1835	1990
3	3.500	3.068	885	1375	1745	2065	2340	2575
3½	4.000	3.548	1060	1655	2070	2495	2845	3120
4	4.500	4.026	1180	1835	2320	2770	3160	3470
			Extra heavy pipe sizes					
⅛	0.405	0.205	90	130	161	190	211	230
¼	0.540	0.302	128	187	231	272	304	332
⅜	0.675	0.424	159	236	292	344	387	422
½	0.840	0.546	207	310	383	456	515	558
¾	1.050	0.742	269	390	480	574	651	710
1	1.315	0.958	339	515	638	765	863	945
1¼	1.660	1.278	431	660	820	983	1111	1218
1½	1.900	1.500	494	756	944	1123	1289	1405
2	2.375	1.940	627	963	1218	1442	1639	1795
2½	2.875	2.324	823	1276	1580	1910	2182	2380
3	3.500	2.900	1019	1575	2019	2380	2731	3000
3½	4.000	3.364	1180	1826	2322	2770	3160	3550
4	4.500	3.824	1340	2082	2652	3160	3630	3980

The dimensions of the perforated tubes listed are the same as those with solid walls, but, as would be expected, the weights per linear foot are less and the structural properties are modified. For this reason the separate table 10-12 is presented. The electrical ratings are those for the usual referenced conditions, e.g., the conductor isolated, in still free air, in a large room minimizing reflected heat, continuous loading, maximum temperature rise 30°C in the conductor over maximum 40°C ambient, bus conductor horizontal with the perforated surfaces top and bottom. The increased electrical capacity of the perforated square tube, compared with the same tube with solid walls, is due to the additional cooling of the inside surfaces by a crosswise flow of air in and out of the drilled holes. The heat that escapes in this way more than makes up for the tendency toward higher resistance-heating losses in the reduced conductor sections alongside the drilled holes.

Experience, and judgment based on experience, must be the electrical engineer's guide in selecting perforated square tubes for enclosed bus-conducting structures. In the open, the advantage—cooler operation under maximum loading—is apparent.

Because of the predominant influence of the thickness-to-diameter relationship in determining the skin-effect ratio of square tubes and the minor role played by the d-c resistance, it is customary to use the same skin-effect ratio values for both solid and ventilated designs.

In the ventilated square high-conductivity bus-conductor tube, ventilating holes normally start 1½ tube diameters from the ends and are spaced uniformly along the

Table 10-10. Current Ratings for Bare Aluminum Tubular Bus, Outdoors

61% IACS, ambient temperature 40°C, frequency 60 cycles; partly sheltered locations, wind velocity 2 ft per sec crosswise to conductors
(From Burndy Engineering Co. Cat. No. 50)

Size of tube, IPS, in.	Outside diam., in.	Inside diam., in.	Current rating, amp					
			10 C rise	20 C rise	30 C rise	40 C rise	50 C rise	60 C rise
			Standard pipe sizes					
1/8	0.405	0.281	105	147	178	203	222	242
1/4	0.540	0.364	152	211	257	293	324	351
3/8	0.675	0.494	191	269	324	371	410	445
1/2	0.840	0.624	250	351	429	484	534	581
3/4	1.050	0.824	312	437	530	605	671	725
1	1.315	1.050	394	554	671	768	850	924
1 1/4	1.660	1.380	515	725	881	1006	1115	1209
1 1/2	1.900	1.612	585	819	1002	1139	1264	1373
2	2.375	2.068	722	1014	1236	1404	1560	1716
2 1/2	2.875	2.470	917	1279	1570	1795	1990	2185
3	3.500	3.068	1160	1640	1995	2260	2540	2770
3 1/2	4.000	3.548	1390	1965	2370	2730	3040	3315
4	4.500	4.026	1535	2170	2650	3000	3390	3670
			Extra heavy pipe sizes					
1/8	0.405	0.205	125	176	211	242	269	289
1/4	0.540	0.302	174	250	296	339	379	410
3/8	0.675	0.424	220	308	371	425	473	511
1/2	0.840	0.546	286	402	484	554	612	664
3/4	1.050	0.742	358	511	600	690	765	827
1	1.315	0.958	460	647	788	897	990	1080
1 1/4	1.660	1.278	585	823	991	1140	1265	1365
1 1/2	1.900	1.500	667	937	1140	1302	1445	1570
2	2.375	1.940	840	1179	1441	1639	1832	1990
2 1/2	2.875	2.324	1091	1538	1862	2146	2380	2577
3	3.500	2.900	1338	1890	2340	2652	2930	3200
3 1/2	4.000	3.364	1537	2170	2660	3042	3360	3670
4	4.500	3.824	1740	2460	3030	3438	3822	4210

tube. Holes can be omitted at places where other buses or conductors must be connected.

Ventilated square bus-conductor tubes present a peculiar problem when regarded as beams that must support heavy magnetic loads as the result of accidental short circuits somewhere on the circuits of which they are a part. Perforated square tubes may be expected to respond to *normal* magnetic and gravitational loads in very much the manner of a uniform beam.

The moments of inertia of undrilled square tubes (Table 10-11) may be used for determining the structural characteristics of ventilated square tubes under *normal* working loads by applying the following factors:

With respect to the *X-X* axis.............. 91%
With respect to the *Y-Y* axis.............. 97%

These are average ratios of the moments of inertia of the drilled tubes, calculated for a theoretical average section, to the moments of undrilled tubes.

As the load on a perforated-tube beam is increased, the deflections may be expected to depart from those indicated by the average-area properties and approach those that would be characteristic of beams with moments corresponding to the minimum cross sections, i.e., the cross sections through the centers of the drilled holes. The portions of the tubes between the holes will have some stiffening value. At higher loads, those which impose tensile stresses on the reduced metal sections on the order of the limiting permissible fiber stress, the tube beam will respond in a manner indicated by the moment of inertia of the reduced section. These are the moments, and derived properties, that are presented in Table 10-12. They may be used to determine the probable maximum physical loads that ventilated square tubes can sustain.

Table 10-11. Anaconda High-conductivity Copper Square Bus Tubes, Hard-drawn
(Prepared by American Brass Company)

Nominal size, in.	Outside dimensions, in.			Wall thickness, in.	Nominal weight, lb. per ft	Cross-sectional area copper		Moment of inertia, $in.^4$	Section modulus, $in.^3$	Radius of gyration, in.	D-c resistance, microhms per ft*	Skin-effect ratio†	60-cycle current rating, amp‡
	Square	Diagonal	Corner radius			Sq in.	Mcm						
3 × 1/8	3.00	3.932	0.375	0.125	5.31	1.370	1,745	1.835	1.223	1.157	6.10	1.02	1660
3 × 3/16	3.00	3.932	0.375	0.1875	7.83	2.019	2,570	2.599	1.733	1.135	4.14	1.04	1990
3 × 1/4	3.00	3.932	0.375	0.250	10.24	2.643	3,365	3.272	2.181	1.113	3.15	1.07	2250
3 × 5/16	3.00	3.828	0.500	0.3125	12.31	3.175	4,042	3.723	2.482	1.083	2.62	1.12	2420
3 × 3/8	3.00	3.828	0.500	0.375	14.48	3.736	4,757	4.215	2.810	1.062	2.23	1.19	2540
3 × 1/2	3.00	3.621	0.750	0.500	17.72	4.571	5,820	4.598	3.065	1.003	1.82	1.38	2640
4 × 1/8	4.00	5.243	0.500	0.125	7.15	1.844	2,347	4.475	2.237	1.558	4.54	1.03	2140
4 × 3/16	4.00	5.243	0.500	0.1875	10.58	2.729	3,474	6.431	3.216	1.535	3.06	1.05	2570
4 × 1/4	4.00	5.243	0.500	0.250	13.91	3.589	4,570	8.215	4.108	1.513	2.32	1.09	2900
4 × 5/16	4.00	5.243	0.500	0.3125	17.15	4.425	5,634	9.836	4.918	1.491	1.88	1.15	3140
4 × 3/8	4.00	5.243	0.500	0.375	20.30	5.236	6,667	11.30	5.652	1.469	1.59	1.23	3300
4 × 1/2	4.00	5.036	0.750	0.500	25.47	6.571	8,366	13.06	6.532	1.410	1.27	1.46	3420
5 × 1/8	5.00	6.450	0.750	0.125	8.88	2.290	2,916	8.705	3.482	1.950	3.65	1.03	2610
5 × 3/16	5.00	6.450	0.750	0.1875	13.17	3.398	4,327	12.62	5.048	1.927	2.46	1.07	3130
5 × 1/4	5.00	6.450	0.750	0.250	17.37	4.482	5,706	16.26	6.503	1.905	1.86	1.11	3520
5 × 5/16	5.00	6.450	0.750	0.3125	21.48	5.541	7,055	19.64	7.854	1.882	1.50	1.17	3810
5 × 3/8	5.00	6.450	0.750	0.375	25.49	6.575	8,372	22.76	9.105	1.861	1.27	1.25	4010
5 × 1/2	5.00	6.450	0.750	0.500	33.22	8.571	10,910	28.32	11.33	1.818	0.97	1.51	4180
6 × 1/8	6.00	7.864	0.750	0.125	10.81	2.790	3,552	15.54	5.179	2.360	3.00	1.05	3070
6 × 3/16	6.00	7.864	0.750	0.1875	16.08	4.148	5,282	22.65	7.551	2.337	2.02	1.08	3670
6 × 1/4	6.00	7.864	0.750	0.250	21.25	5.482	6,980	29.36	9.786	2.314	1.52	1.12	4160
6 × 5/16	6.00	7.864	0.750	0.3125	26.32	6.791	8,646	35.66	11.89	2.292	1.23	1.20	4480
6 × 3/8	6.00	7.864	0.750	0.375	31.30	8.075	10,280	41.59	13.86	2.269	1.03	1.28	4720
6 × 1/2	6.00	7.864	0.750	0.500	40.97	10.57	13,460	52.35	17.45	2.225	0.79	1.54	4930
7 × 3/16	7.00	9.278	0.750	0.1875	18.99	4.898	6,236	36.94	10.55	2.746	1.69	1.10	4230
7 × 1/4	7.00	9.278	0.750	0.250	25.12	6.482	8,253	48.07	13.73	2.723	1.28	1.15	4760
7 × 5/16	7.00	9.278	0.750	0.3125	31.17	8.041	10,240	58.64	16.76	2.701	1.03	1.22	5140
7 × 3/8	7.00	9.278	0.750	0.375	37.11	9.575	12,190	68.67	19.62	2.678	0.86	1.31	5420
7 × 1/2	7.00	9.278	0.750	0.500	48.72	12.57	16,010	87.18	24.91	2.633	0.66	1.57	5680
8 × 3/16	8.00	10.69	0.750	0.1875	21.89	5.648	7,191	56.23	14.06	3.155	1.47	1.11	4760
8 × 1/4	8.00	10.69	0.750	0.250	29.00	7.482	9,526	73.40	18.35	3.132	1.11	1.17	5350
8 × 5/16	8.00	10.69	0.750	0.3125	36.01	9.291	11,830	89.82	22.46	3.109	0.89	1.24	5790
8 × 3/8	8.00	10.69	0.750	0.375	42.93	11.08	14,100	105.5	26.38	3.087	0.75	1.33	6110
8 × 1/2	8.00	10.69	0.750	0.500	56.48	14.57	18,550	134.8	33.70	3.041	0.57	1.59	6400

NOTE: Variations from these values must be expected in practice. ASTM Standard Specification B188-54, Seamless Copper Bus Pipe and Tube, controls the tensile properties, conductivity, tests, and inspection. Dimensional tolerances are agreed between manufacturer and purchaser at the time of placing an order.

* D-c resistivity. The resistances are based on the ASTM Standard Specification B188-54. The conductivities of square tubes are, according to size:

Up to 6 in outside diameter and up to 3/16 in wall thickness, conductivity 97.40% IACS minimum.
Up to 6 in outside diameter and over 3/16 in wall thickness, conductivity 97.80% IACS minimum.
All sizes over 6 in. outside diameter, conductivity 98.40% IACS minimum.

† Skin-effect ratio at 60 cps.

‡ 60-cycle current-carrying-capacity ratings have been computed for 30°C temperature rise in still air at 40°C, conductor with new bright mill finish, horizontal and free from all outside magnetic influences.

440. Physical and Electrical Properties of Copper Channels and Angles (see Table 10-13).

441. Physical and Electrical Properties of Aluminum Channels (see Table 10-14).

442. Physical and Electrical Properties of Aluminum Angles (see Table 10-15).

443. Single angles or channels are sometimes used and are easily supported on insulators.

444. The condition of the conductor surface affects the **heat emissivity** or the amount of heat radiated by the conductor for a given temperature rise. Brightly polished bus conductors have a lower carrying capacity than those with dull-black or well-oxidized surfaces because of the reduced heat emissivity. Taping, painting, insulating with various composition materials, all change the heat dissipation of the conductor and its current-carrying capacity for a given temperature rise.

Insulated buses do not have such great capacities as noninsulated buses (see Fig. 10-170).

Table 10-12. Anaconda High-conductivity Ventilated Square Copper Bus Tubes Perforated Walls, Hard-drawn
(Prepared by American Brass Company)

Nominal size, in.	Outside dimensions			Wall thickness, in.	Drillings		Weight, lb per ft	Average cross section*		On X-X axis			On Y-Y axis			D-c resistance, microhms per ft.‡	Skin-effect ratio§	60-cycle current rating, amp‖
	Square, in.	Diagonal, in.	Corner radius, in.		Diameter, in.	Spacing, center to center, in.		Sq in.	Mcm	Moment of inertia,† in.[4]	Section modulus, in.[3]	Radius of gyration, in.	Moment of inertia,† in.[4]	Section modulus, in.[3]	Radius of gyration, in.			
3 × 1/8	3.00	3.934	0.375	0.125	1 1/4	4	5.00	1.294	1,647	1.189	0.7924	1.060	1.697	1.131	1.266	6.46	1.02	2030
3 × 3/16	3.00	3.934	0.375	0.1675	1 1/4	4	7.36	1.904	2,424	1.671	1.114	1.038	2.390	1.593	1.242	4.39	1.04	2430
3 × 1/4	3.00	3.934	0.375	0.250	1 1/4	4	9.62	2.489	3,169	2.087	1.391	1.017	2.993	1.995	1.218	3.35	1.07	2720
3 × 5/16	3.00	3.828	0.500	0.3125	1 1/4	4	11.53	2.983	3,798	2.306	1.537	0.9814	3.545	2.363	1.217	2.79	1.12	2910
3 × 3/8	3.00	3.828	0.500	0.375	1 1/4	4	13.55	3.506	4,464	2.589	1.726	0.9618	4.001	2.668	1.196	2.38	1.19	3040
3 × 1/2	3.00	3.721	0.750	0.500	1 1/4	4	16.48	4.264	5,429	2.619	1.746	0.8880	4.313	2.875	1.140	1.95	1.38	3120
4 × 1/8	4.00	5.243	0.500	0.125	1 1/2	4	6.70	1.733	2,207	3.066	1.533	1.445	4.157	2.078	1.682	4.83	1.03	2620
4 × 3/16	4.00	5.243	0.500	0.1875	1 1/2	4	9.90	2.563	3,263	4.386	2.193	1.423	5.955	2.977	1.658	3.26	1.06	3130
4 × 1/4	4.00	5.243	0.500	0.250	1 1/2	4	13.01	3.368	4,288	5.575	2.787	1.401	7.579	3.790	1.634	2.47	1.09	3520
4 × 5/16	4.00	5.243	0.500	0.3125	1 1/2	4	16.03	4.149	5,282	6.651	3.325	1.381	9.041	4.521	1.610	2.01	1.15	3800
4 × 3/8	4.00	5.243	0.500	0.375	1 1/2	4	18.95	4.905	6,245	7.594	3.797	1.359	10.35	5.175	1.587	1.70	1.23	3990
4 × 1/2	4.00	5.035	0.750	0.500	1 1/2	4	23.68	6.129	7,804	8.439	4.220	1.290	12.07	6.037	1.543	1.36	1.46	4040
5 × 1/8	5.00	6.450	0.750	0.125	1 5/8	4	8.35	2.160	2,751	6.291	2.516	1.827	7.981	3.192	2.058	3.87	1.03	3190
5 × 3/16	5.00	6.450	0.750	0.1875	1 5/8	4	12.38	3.204	4,079	9.089	3.636	1.805	11.53	4.613	2.034	2.61	1.07	3820
5 × 1/4	5.00	6.450	0.750	0.250	1 5/8	4	16.32	4.223	5,376	11.67	4.668	1.783	14.81	5.924	2.009	1.97	1.11	4290
5 × 5/16	5.00	6.450	0.750	0.3125	1 5/8	4	20.16	5.217	6,642	14.05	5.619	1.762	17.82	7.130	1.985	1.60	1.17	4640
5 × 3/8	5.00	6.450	0.750	0.375	1 5/8	4	23.90	6.186	7,877	16.23	6.492	1.725	20.59	8.236	1.943	1.35	1.25	4870
5 × 1/2	5.00	6.450	0.750	0.500	1 5/8	4	31.11	8.052	10,250	20.06	8.022	1.699	25.90	10.36	1.931	1.03	1.51	5050
6 × 1/8	6.00	7.864	0.750	0.125	1 3/4	4	10.20	2.640	3,361	11.76	3.920	2.236	14.18	4.726	2.455	3.17	1.05	3760
6 × 3/16	6.00	7.864	0.750	0.1875	1 3/4	4	15.16	3.923	4,994	17.11	5.703	2.123	20.62	6.872	2.430	2.13	1.08	4500
6 × 1/4	6.00	7.864	0.750	0.250	1 3/4	4	20.02	5.181	6,597	22.12	7.373	2.191	26.64	8.881	2.405	1.61	1.12	5080
6 × 5/16	6.00	7.864	0.750	0.3125	1 3/4	4	24.79	6.415	8,168	26.81	8.937	2.169	32.27	10.76	2.380	1.30	1.20	5480
6 × 3/8	6.00	7.864	0.750	0.375	1 3/4	4	29.46	7.624	9,708	31.19	10.40	2.148	37.52	12.51	2.355	1.09	1.28	5760
6 × 1/2	6.00	7.864	0.750	0.500	1 3/4	4	38.52	9.970	12,690	39.08	13.03	2.105	47.63	15.88	2.324	.84	1.54	5980
7 × 3/16	7.00	9.276	0.750	0.1875	1 7/8	4	17.93	4.639	5,907	28.78	8.223	2.619	33.56	9.588	2.828	1.78	1.10	5150
7 × 1/4	7.00	9.276	0.750	0.250	1 7/8	4	23.71	6.137	7,813	37.39	10.68	2.597	43.56	12.45	2.803	1.35	1.15	5780
7 × 5/16	7.00	9.276	0.750	0.3125	1 7/8	4	29.40	7.609	9,689	45.53	13.01	2.575	53.01	15.15	2.778	1.09	1.22	6240
7 × 3/8	7.00	9.276	0.750	0.375	1 7/8	4	35.00	9.058	11,530	53.23	15.21	2.553	61.91	17.69	2.753	.91	1.31	6570
7 × 1/2	7.00	9.276	0.750	0.500	1 7/8	4	45.91	11.88	15,130	67.33	19.24	2.509	79.13	22.61	2.720	.70	1.57	6860
8 × 3/16	8.00	10.692	0.750	0.1875	2	4	20.69	5.354	6,816	44.79	11.20	3.024	51.29	12.82	3.236	1.55	1.11	5850
8 × 1/4	8.00	10.692	0.750	0.250	2	4	27.39	7.089	9,026	58.38	14.60	3.001	66.82	16.70	3.211	1.17	1.17	6560
8 × 5/16	8.00	10.692	0.750	0.3125	2	4	34.00	8.800	11,200	71.34	17.84	2.979	81.59	20.40	3.185	.94	1.24	7090
8 × 3/8	8.00	10.692	0.750	0.375	2	4	40.52	10.49	13,350	83.69	20.92	2.956	95.64	23.91	3.160	.79	1.33	7470
8 × 1/2	8.00	10.692	0.750	0.500	2	4	53.27	13.79	17,550	106.6	26.65	2.912	122.8	30.71	3.126	.60	1.59	7820

NOTE: Variations from the exact sizes must be expected in practice. ASTM Standard Specification B188-54, Seamless Copper Bus Pipe and Tube, controls the tensile properties, conductivity, tests and inspection. Dimensional tolerances are agreed between manufacturer and purchaser at the time of placing an order.

* The average cross section in square inches and the equivalent in circular mils were determined by dividing the weight per foot by 12 times the weight of 1 cu in. of copper.

† The moments of inertia in this table are those of the minimum transverse sections, those passing through the centers of the ventilating holes. These minimum sections determine the maximum strength of ventilated square tubes as load-supporting members of a structure.

‡ The d-c resistances are based on ASTM Standard Specification B188-54. The conductivities of square tubes both solid and drilled for ventilation are according to size:

Up to 6 in outside diameter and up to 3/16 in wall thickness, conductivity 97.40% IACS minimum.

Up to 6 in outside diameter and over 3/16 in wall thickness, conductivity 97.80% IACS minimum.

All sizes over 6 in outside diameter, conductivity 98.40% IACS minimum.

§ The skin-effect ratios of 60 cps.

‖ 60-cycle current-carrying-capacity ratings have been computed for 30°C temperature rise over still air at 40°C, conductor with new bright mill finish, horizontal and free from all outside magnetic influences.

Ventilated square bus-conductor tube

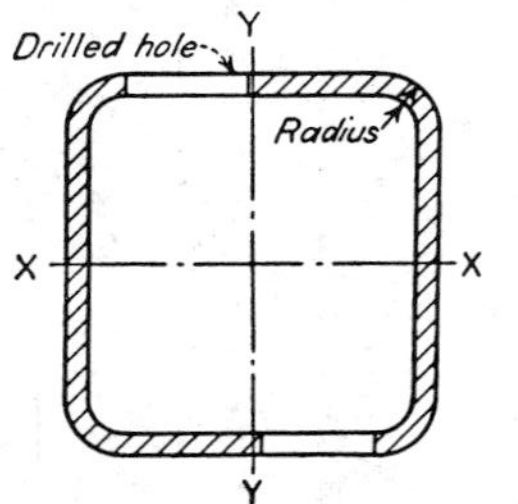

445. Insulation on indoor buses is used extensively for reasons of safety or protection to service. Bus insulation is a safeguard against the spreading of trouble from one section of the bus to another by conducting gases generated by short-circuit currents but, on the other hand, adds inflammable material.

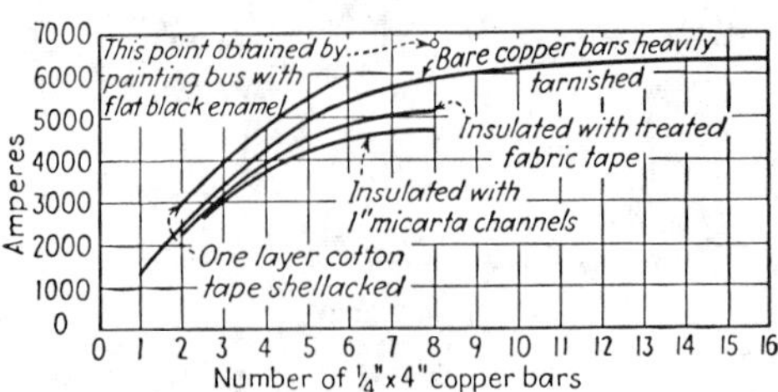

Fig. 10-170. Effect of insulation in limiting current-carrying capacity, 35°C rise, 18-in centers, 60 cycles. (*H. W. Papst, Elec. World, Sept.* 21, 1929.)

No general rule can be given regarding the desirability of insulating buses and connections, as in each case the advantages of insulation and the cost must be properly evaluated. It is quite commonly used on metal-clad buses operating at 2,300 or 4,000 V but less frequently on indoor buses of 13,800 and 27,000 V. Materials generally used for bus insulation are varnished cambric, Micarta, and Herkolite. Insulation can be molded or wrapped around the conductors.

446. Enclosure of buses results in reduced current-carrying capacities. All values given for current-carrying capacities of various bus conductors are based on buses operating in still but unconfined air. Actually, most buses operate enclosed in a compartment or housing, and each housing has its own characteristics which are dependent on the exposed-surface area of the housing and its nature, the enclosed air space, and the type of bus.

In enclosed buses the rate of heat dissipation is reduced, and, in general, the temperature rise above the ambient will depend upon:

a. The temperature rise of the housing due to bus losses and any losses induced in the housing.

b. The temperature gradient between inside air and housing.

c. The temperature gradient between bus and inside air.

Metal enclosures of nonmagnetic material may decrease the current ratings of the buses by as much as 25%, and magnetic housings may decrease these ratings by 35 to 40%.

447. Inductive heating of steel members in the vicinity of buses carrying heavy a-c currents is an important factor in station design and must be given proper consideration, particularly when the steel is in proximity to heavy **single-phase** currents.

The temperature rise of steel members can be greatly reduced by the judicious application of:

a. Short-circuited bands placed around the steel members when the steel members cross the conductors at right angles.

b. Amortisseur grids placed parallel to the **conductors** when the conductors and the steel members are parallel to each other.

448. Bus joints may be made by **welding, brazing, bolting, or clamping** the bus surfaces together. The joints most commonly used are the bolted or clamped type.

The efficiency of a joint is defined as the ratio of the conductivity of the joint to that of an equal length of conductor and should not be less than unity. In the case of flat-bar bolted or clamped joints, two main factors determine this efficiency:

a. Streamline distortion of current flow.

b. Contact resistance of the joint.

Streamline distortion of current flow in a bus-bar joint depends upon the ratio of overlap to the thickness of the bar and, to a certain extent, affects the resistance of the joint. In practice, overlaps of 10 to 20 times the thickness of the bar are used.

The contact resistance of the joint is dependent upon the contact pressure and the condition of the contact surfaces. **In general, the greater the pressure between the contact sufaces, the less will be the contact resistance.**

For a given bus-bar joint, bolts should be chosen of the largest size consistent with other necessary considerations. The selection of bolt sizes and bolt spacings for bus connections is important, and manufacturer's data on the subject should be consulted.

Table 10-13. Physical and Electrical Properties of Copper Channels and Angles

Channel square arrangement

Angle square arrangement

Height, in	Flange width, in	Web thickness, in	Corner radius, in	Weight per ft, lb	X-X axis			Y-Y axis			Resistance,* Ω/ft	Amp† capacity
					I	*S*	*R*	*I*	*S*	*R*		
Single bus channel—physical values											Channel square bus	
3	1⅝16	0.165	0.409	3.19	1.028	0.686	1.116	0.128	0.138	0.394	5.0×10^{-6}	2,200
3	1⅝16	0.216	0.409	4.11	1.284	0.856	1.098	0.160	0.175	0.388	3.9×10^{-6}	2,500
3	1⅝16	0.284	0.409	5.29	1.583	1.055	1.075	0.197	0.222	0.380	3.0×10^{-6}	2,800
4	1¾	0.200	0.463	5.24	3.059	1.529	1.502	0.379	0.303	0.529	3.1×10^{-6}	3,200
4	1¾	0.240	0.463	6.23	3.569	1.785	1.488	0.441	0.357	0.523	2.6×10^{-6}	3,500
4	1¾	0.338	0.463	8.58	4.691	2.346	1.454	0.580	0.482	0.511	1.87×10^{-6}	4,000
5	2³⁄16	0.260	0.464	8.60	7.888	3.155	1.882	0.968	0.619	0.659	1.87×10^{-6}	4,500
5	2³⁄16	0.339	0.464	11.05	9.839	3.936	1.854	1.204	0.783	0.649	1.45×10^{-6}	5,000
6	2¹¹⁄16	0.276	0.615	11.10	14.88	4.960	2.276	1.927	1.001	0.819	1.45×10^{-6}	5,600
6	2¹¹⁄16	0.384	0.615	15.19	19.69	6.562	2.238	2.544	1.347	0.804	1.06×10^{-6}	6,300
7	3³⁄16	0.325	0.645	15.45	28.42	8.120	2.666	3.790	1.661	0.974	1.04×10^{-6}	7,000
8	3¹¹⁄16	0.470	0.595	25.65	60.76	15.19	3.026	8.233	3.154	1.114	0.63×10^{-6}	8,900
9	4½	0.500	0.625	30.77	92.81	20.62	3.414	12.41	4.229	1.248	0.52×10^{-6}	10,000
Single bus angle—physical values											Angle square bus	
3.54	2½ × 2½	³⁄16	¼	3.39	0.8317	0.4705	0.974	0.200	0.229	0.478	4.7×10^{-6}	2,750
4.25	3 × 3	³⁄16	¼	4.11	1.476	0.6958	1.177	0.358	0.341	0.580	3.9×10^{-6}	3,300
4.25	3 × 3	¼	¼	5.41	1.882	0.8874	1.159	0.461	0.429	0.574	3.0×10^{-6}	3,650
4.95	3½ × 3½	¼	⅜	6.33	3.066	1.239	1.368	0.733	0.605	0.669	2.5×10^{-6}	4,200
5.67	4 × 4	¼	⅜	7.30	4.665	1.649	1.572	1.122	0.808	0.771	2.2×10^{-6}	4,800
6.37	4½ × 4½	¼	⅜	8.26	6.741	2.119	1.776	1.629	1.041	0.873	1.94×10^{-6}	5,400
6.37	4½ × 4½	⁵⁄16	½	10.23	8.312	2.612	1.772	1.979	1.270	0.865	1.57×10^{-6}	6,000
7.08	5 × 5	⁵⁄16	½	11.44	11.55	3.268	1.976	2.766	1.595	0.967	1.40×10^{-6}	6,750
8.5	6 × 6	⁵⁄16	½	13.85	20.37	4.801	2.383	4.914	2.354	1.171	1.16×10^{-6}	8,000

I = moment of inertia, in⁴; S = section modulus, in³; R = radius of gyration, in.
Modulus of elasticity = 16,000,000 lb/in². Cross-section area = weight/ft ÷ 3.86 = in².
Tensile strength = 32,000 lb/in². Coefficient of linear expansion = 0.0000167/°C. Coefficient of increase in resistance = 0.00393/°C.
D-c capacity approximately 20 to 25% greater.
* D-c resistance of 20°C.
† Amperes capacity, 60 c, for 30°C rise over 40°C ambient, in still unconfined air, bus not enclosed.

449. Bus-bar surfaces to be joined, both copper and aluminum, should be cleaned with emery cloth, wiped clean, smeared with petrolatum or other special compounds, then rubbed lightly with fine emery cloth and bolted without wiping. This practice excludes air from the freshly abraded surfaces and ensures lower contact resistance; this is important with aluminum, on which oxide film forms instantaneously.

Joints should, where possible, be plated as a means of reducing the effects of oxidation, particularly in high-temperature joints. Silver plating is now standard practice with the principal manufacturers. Tin plating is used to a lesser extent.

450. In jointing bus bars of aluminum to those of copper, care is necessary to avoid galvanic action. Suitable welded copper-aluminum joint inserts are available to be inserted in the bolted joint, so that no bolted connections appear between the two materials. Also means for silver-surfacing the aluminum have been developed to further aid the making of copper-aluminum joints. Where direct-bolted connections must be made between these materials, the instructions given in Pars. **448** and **449** should be followed to provide:

1. Adequate preparation of surfaces.
2. Exclusion of electrolyte.
3. Sufficient pressure.

Table 10-14. Physical and Electrical Properties of Aluminum Channels
(Aluminum Co. of America)

All dimensions, in
Weight, lb/ft
Area, in²
I = moment of inertia, in⁴
S = section modulus, in³
r = radius of gyration, in
Current capacity for 30°C hot spot rise in still but unconfined air. Ratings may be increased 10% for 35°C hot-spot rise. Distance between phases at least 18 in. For a-c currents above 3,000 A, rating should be reduced for closer spacing.

Channel Square Bus

Size: Depth	3			4			5			6		
Size: t	0.170	0.258	0.356	0.180	0.247	0.320	0.190	0.325	0.472	0.225	0.314	0.437
Weight	1.42	1.72	2.06	1.84	2.16	2.51	2.31	3.11	3.97	3.00	3.62	4.48
Area	1.21	1.47	1.76	1.57	1.84	2.13	1.97	2.64	3.38	2.55	3.09	3.82
c	1¾	1¾	1¾	2¾	2¾	2¾	3¾	3¾	3¾	4½	4½	4½
b	1.410	1.498	1.596	1.580	1.647	1.720	1.750	1.885	2.032	1.945	2.034	2.157
Axis X-X:												
I	1.66	1.85	2.07	3.83	4.19	4.58	7.49	8.90	10.43	13.57	15.18	17.39
S	1.10	1.24	1.38	1.92	2.10	2.29	3.00	3.56	4.17	4.52	5.06	5.80
r	1.17	1.12	1.08	1.56	1.51	1.47	1.95	1.83	1.76	2.31	2.22	2.13
Axis Y-Y:												
I	0.20	0.25	0.31	0.32	0.37	0.43	0.48	0.63	0.81	0.73	0.87	1.05
S	0.20	0.23	0.27	0.28	0.31	0.34	0.38	0.45	0.53	0.51	0.56	0.64
r	0.40	0.41	0.42	0.45	0.45	0.45	0.49	0.49	0.49	0.54	0.53	0.52
x	0.44	0.44	0.46	0.46	0.45	0.46	0.48	0.48	0.51	0.51	0.50	0.50
h	3	3	3	4	4	4	5	5	5	6	6	6
w	3 9/16	3 15/16	3 15/16	4 3/16	4 3/16	4⅜	5	5	5	6	6	6
Two channels:												
D-c current capacity	2,400	2,650	2,880	2,900	3,100	3,400	3,500	4,100	4,600	4,300	4,700	5,250
A-c current 60-c capacity	2,350	2,550	2,750	2,800	3,000	3,300	3,400	3,900	4,250	4,150	4,450	4,800

Size: Depth	7			8			10 (car channel)			12		
Size: t	0.314	0.419	0.524	0.303	0.395	0.520	0.375	0.438	0.500	0.387	0.510	0.632
Weight	4.23	5.09	5.95	4.75	5.61	6.79	8.56	9.30	10.02	8.65	10.37	12.10
Area	3.60	4.33	5.07	4.04	4.78	5.78	7.30	7.93	8.55	7.35	8.82	10.29
c	5½	5½	5½	6¼	6¼	6¼	7½	7½	7½	10	10	10
b	2.194	1.299	2.404	2.343	2.435	2.560	3.500	3.563	3.625	3.047	3.170	3.292
Axis X-X:												
I	24.24	27.24	30.25	36.11	40.04	45.37	109.62	114.87	120.03	144.37	162.08	179.65
S	6.93	7.78	8.64	9.03	10.01	11.34	21.92	22.97	24.01	24.06	27.01	29.94
r	2.60	2.51	2.44	2.99	2.90	2.80	3.88	3.81	3.75	4.43	4.29	4.18
Axis Y-Y:												
I	1.17	1.38	1.59	1.53	1.75	2.07	7.10	8.73	8.25	4.47	5.14	5.82
S	0.70	0.78	0.86	0.85	0.93	1.04	2.80	2.93	3.04	1.89	2.06	2.24
r	0.57	0.56	0.56	0.61	0.61	0.60	0.99	0.99	0.98	0.78	0.76	0.75
x	0.52	0.53	0.55	0.55	0.55	0.57	0.93	0.92	0.91	0.67	0.67	0.69
h	7	7	7	8	8	8	10	10	10	12	12	12
w	7	7	7	8	8	8	10	10	10	9	9	9
Two channels:												
D-c current capacity	5,400	6,000	6,500	6,000	6,550	7,200	9,000	9,600	10,000	9,100	10,000	11,000
A-c current 60-c capacity	5,100	5,500	5,800	5,600	5,900	6,400	8,000	8,400	8,600	8,200	8,700	9,000

451. Connections of tubular buses are usually made by clamping, shrink fitting, or flattening of the tubing.

Clamp connectors are in general use, and detailed data for the many available types of connectors will be found in manufacturers' catalogues.

Shrink fitting is the process whereby the end of a tube is heated until it expands several thousandths of an inch and a pluglike connector is inserted in the open ends. When the tube cools, it contracts and shrinks onto the plug, holding it tightly in place. In expanding the tubing, the temperature should be kept below that at which the material would be annealed. "Dry ice" (CO_2) is well suited to the shrinking of inserts. **Splicing connectors** should have a diameter larger than the inside diameter of the tube by 0.004 to 0.008 in, the value varying somewhat with the diameter and thickness of the tubing.

Table 10-15. Physical and Electrical Properties of Aluminum Angles

(Aluminum Co. of America)

All dimensions in inches
Weight in pounds per foot
Area in square inches
I = moment of inertia in in.^{4}
S = section modulus in in.^{3}
r = radius of gyration in inches

Current capacity for 30°C hot-spot rise in still but unconfined air in amperes. Ratings may be increased 10% for 35°C hot-spot rise. Distance between phases at least 18 in. For a-c currents above 3,000 A, rating should be reduced for closer spacing.

Angle Square Bus

Size: Legs	2½×2	2½×2	2½×2½	2½×2½	3 × 3	3 × 3	3 × 3	4 × 3	4 × 3	4 × 3	4 × 3	4 × 4	4 × 4	4 × 4	4 × 4	5×3½	5×3½	5×3½	6 × 4	6 × 4	6 × 4
Size: t	¼	⅜	¼	⅜	¼	⅜	½	¼	⅜	½	⅝	¼	⅜	½	⅝	⅜	½	⅝	⅜	½	⅝
Weight	1.25	1.83	1.41	2.05	1.68	2.48	3.22	1.99	2.92	3.82	4.69	2.28	3.36	4.40	5.42	3.59	4.70	5.78	4.24	5.57	6.88
Area	1.07	1.55	1.19	1.74	1.43	2.10	2.74	1.69	2.49	3.25	3.99	1.94	2.86	3.75	4.61	3.05	4.00	4.92	3.60	4.74	5.85
f_1	¼	¼	¼	¼	5/16	5/16	5/16	⅜	⅜	⅜	⅜	⅜	⅜	⅜	⅜	7/16	7/16	7/16	½	½	½
f_2	⅛	⅛	⅛	⅛	¼	¼	¼	¼	¼	¼	¼	¼	¼	¼	¼	5/16	5/16	5/16	⅜	⅜	⅜
Axis *X-X*:																					
I	0.65	0.91	0.69	0.98	1.18	1.70	2.16	2.68	3.88	4.96	5.95	2.94	4.26	5.46	6.56	7.56	9.77	11.82	13.02	16.95	20.63
s	0.38	0.54	0.39	0.56	0.54	0.80	1.04	0.96	1.42	1.85	2.25	1.00	1.48	1.93	2.36	2.21	2.90	3.56	3.17	4.19	5.17
r	0.78	0.76	0.76	0.75	0.91	0.90	0.89	1.26	1.25	1.24	1.22	1.23	1.22	1.21	1.19	1.58	1.56	1.55	1.90	1.89	1.88
Y	0.78	0.83	0.71	0.76	0.82	0.87	0.92	1.21	1.26	1.31	1.36	1.07	1.12	1.17	1.22	1.58	1.63	1.68	1.90	1.96	2.01
Axis *Y-Y*:																					
I	0.37	0.51	0.69	0.98	1.18	1.70	2.16	1.29	1.86	2.36	2.82	2.94	4.26	5.46	6.56	3.04	3.91	4.70	4.63	6.01	7.27
S	0.25	0.36	0.39	0.56	0.54	0.80	1.04	0.56	0.83	1.08	1.32	1.00	1.48	1.93	2.36	1.15	1.50	1.84	1.50	1.98	2.44
r	0.58	0.57	0.76	0.75	0.91	0.90	0.89	0.87	0.86	0.85	0.84	1.23	1.22	1.21	1.19	1.00	0.99	0.98	1.13	1.13	1.11
x	0.53	0.58	0.71	0.76	0.82	0.87	0.92	0.72	0.77	0.82	0.86	1.07	1.12	1.17	1.22	0.84	0.89	0.94	0.91	0.97	1.02
w	3	3	3½	3½	4	4	4	4	4	4	4	5	5	5¼	5¼	5	5	5	6	6	6
h	2½	2½	2½	2½	3	3	3	4	4	4	4	4	4	4	4	5	5	5	6	6	6
2 angles:																					
D-c current capacity	1,950	2,380	2,150	2,500	2,600	3,200	3,650	3,000	3,700	4,150	4,750	3,300	4,000	4,700	5,300	4,200	4,900	5,500	5,250	6,000	6,600
A-c current 60-c capacity	1,900	2,300	2,100	2,430	2,500	3,000	3,240	2,900	3,500	3,800	4,000	3,200	3,800	4,200	4,400	4,000	4,400	4,700	4,900	5,400	5,600

452. Nonferrous bolts of high strength, high fatigue, and corrosion-resistant properties are gradually replacing steel bolts for bolted bus connections in indoor as well as outdoor installations and are advantageous because their coefficients of thermal expansion are closer to that of copper than those of steel bolts. Among the materials most generally used for such bolts are Everdur or other high-copper-silicon alloys and monel metal.

453. Thermal expansion and contraction of bus conductors is an important factor in bus design, particularly where heavy-current buses or buses of long lengths are involved. A copper bus will expand 0.0112 in/ft of length for a temperature rise of 100°F (38°C). In order to protect insulator supports, disconnecting switches, and equipment terminals from the stresses set up by the thermal expansion of the conductors, provision should be made for expansion by means of expansion joints or, in some cases, by changes of direction in the bus.

Table 10-16. Minimum Clearances for Switchboards

(Westinghouse Switchboard Data Book)

Creeping distance when mounted on same surface of insulating panels		Voltage class (For intermediate voltages, it is advisable to use the distances given for the next higher voltage class)	Striking distances when rigidly supported and clear of surfaces	
Between parts of opposite polarity	**Between live parts and ground**		Between parts of opposite polarity	Between live parts and ground
3/8	1/4	Up to 50	1/4	1/8
1/2	3/8	125	3/8	1/4
5/8	1/2	250	3/8	1/4
1	3/4	600	1/2	3/8
1¼	1¼	750	3/4	3/4
1¾	1¾	1500	1¼	1¼
2½	2½	2500	2	2
3	3	3500	2½	2½

NOTE: For switchboard circuits connected to systems above 150-kva capacity, the distances between parts of opposite polarity should be increased from ⅓ to 1 in.

454. Vibration of long tubular-bus spans has been experienced to some extent, but proper remedies are generally conceded to be possible (see Bibliography, Par. **539**).

455. The spacing of buses is largely a matter of design experience. No definite standardization in this respect has been reached, but practice has indicated average spacings varying with the voltage and application. An AIEE committee report on this subject appears in the June, 1954, *Transactions*, page 636. The minimum electrical clearances outlined in Table 10-17 below have been established for any particular BIL level, predominantly on the basis of lightning-induced surges. In recent years, with the advent of extra-high-voltage circuits, it has been found that switching surges can often produce more severe duty than lightning surges. An IEEE committee report was issued in February, 1965, covering recommendations for minimum electrical clearances for substations, based on switching-surge requirements (see Tables 10-18A and 10-18B).

456. Minimum clearance in inches between live parts or bare conductors for indoor application on switchboard circuits connected to systems up to 150 kVA capacity, or wiring connected to the secondary of instrument transformers, is shown in Table 10-16.

457. Minimum electrical clearances for standard basic insulation levels—outdoor, a-c—have been prepared by the AIEE Committee on Substations, in an effort to correlate data on electrical clearances as now in use. While this material has not yet been standardized, it represents some of the most recent thinking on the subject and is reproduced here in Table 10-17.

458. A station bus must have sufficient mechanical strength to withstand short-circuit stresses. Two factors are involved: (1) the strength of the insulators and their supporting structure, and (2) the strength of the bus conductor acting as a continuous beam from insulator to insulator.

459. Short-circuit stresses in buses and bus supports depend not only upon the

Table 10-17. Minimum Electrical Clearances for Standard BIL Outdoor A-C
(Prepared by AIEE Substations Committee—not yet approved)

Kv class*	BIL level, kv withstand†	Minimum clearance to ground for rigid parts, in.‡	Minimum clearance between phases (or live parts) for rigid parts, in., metal to metal§	Minimum clearance between overhead conductors and grade for personnel safety, inside substation, ft‖	Minimum clearance between wires and roadways, inside substation enclosure, ft
7.5	95	6	7	8	20
15	110	7	12	9	20
23	150	10	15	10	22
34.5	200	13	18	10	22
46	250	17	21	10	22
69	350	25	31	11	23
115	550	42	53	12	25
138	650	50	62	13	25
161	750	58	72	14	26
230	825	65	80	15	27
230	900	71	89	15	27
	1050	83	105	16	28
	1175	94	113	17	29

* Coordinate kv class and BIL when choosing minimum clearances.
† Values are recommended minimums but may be decreased in line with good practice depending on local conditions, procedures, etc.
‡ Values apply to 3,300 ft above sea level. Above this elevation increase above values according to Par. 22-4 of *AIEE Standard No.* 22.
§ These recommended minimum clearances are for rigid conductors. Any structural tolerances, or allowances for conductor movement, or possible reduction in spacing by foreign objects should be added to these minimum values.
‖ These minimum clearances are intended as a guide for the installation of equipment in the field only, and not for the design of electric devices or apparatus such as circuit breakers, transformers, etc. See Tables 10-17A and B.

short-circuit current and the spacing between buses but also upon the vibratory characteristics of the bus structure. The supports respond selectively to different frequencies, and the maximum support stress usually differs from the maximum force set up by the short-circuit current[1] (also see the Bibliography, Par. **539**).

Table 10-18A. Substation Clearances Based on Switching Surges,* Line to Ground
(Prepared by IEEE Substations committee—not yet approved)
TP 65-39

System voltage, kV		Transient voltage crest, kV		Rod-plane withstand distance, in†			Indicated minimum clearance,‡ in			AIEE Guide TP 54–80 clearance	
Nom.	Max.	Times L-N crest	kV	90%	97.7%	99.9%	90%	97.7%	99.9%	BIL	In
230	242	2.8	555	52	54	57	60	62	66	900	71
345	362	2.0	592	57	59	62	66	68	71	1,300	104
		2.3	680	68	71	75	78	82	86		
		2.5	740	78	81	85	90	93	98		
		2.8	830	92	96	102	106	110	117		
500	550	2.0	900	103	108	120	118	124	138	1,800	144
		2.3	1,035	136	144	164	156	166	189		
		2.5	1,125	160	172	200	184	198	230		
		2.8	1,260	214	242	290	246	278	334		

* See 1964 *CIGRE Paper* 415.
† Percentages are probability of withstand values under dry conditions.
‡ 1.15 times withstand probability values.
Test data not available on large air gaps during heavy rainfalls.

[1] For theory and methods of calculation see (1) Schurig, O. R., and Sayre, M. F. Mechanical Stresses in Busbar Supports during Short Circuits; *Trans. AIEE*, 1925, pp. 217–237; (2) Schurig, O. R., Frick, W. C., and Sayre, M. F. Practical Calculation of Short-circuit Stresses for Straight, Parallel Bar Conductors; *Gen. Elec. Rev.*, August, 1926, p. 534 (reproduced in the NEMA Handbook, October, 1927); and (3) Specht, W. Short-cut Methods of Calculating Stresses in Bus Structures; *Gen. Elec. Rev.*, August, 1928, p. 413.

Table 10-18B. Substation Clearances Based on Switching Surges,* Line to Line
(Prepared by IEEE Substations Committee, not yet approved)
TP-65-39

System voltage, kV		Transient voltage, kV, times *L-N* crest		90% rod-rod withstand distance,† in	Indicated minimum clearance,‡ in	AIEE Guide TP 54-80 clearance	
Nom.	Max.	Crest	kV			BIL	In
230	242	4.0	790	58	72	900	89
		4.5	890	68	85		
345	362	3.5	1,035	84	105	1,300	117
		4.0	1,180	96	120		
		4.5	1,330	115	144		
500	550	3.0	1,350	118	148	1,800	176
		3.5	1,580	146	183		
		4.0	1,800	180	226		

* See 1964 *CIGRE Paper* 415.
† Distances listed provide 90% of withstand under dry conditions.
‡ 1.25 times 90% withstand probability values.
Test data not available on large air gaps during heavy rainfalls.

Bus stresses caused by the electromagnetic force acting on the bus conductor bar are of two types:

1. **Lateral stress,** which acts on conductor and supports in the direction of the force.

2. **Longitudinal stress,** which results from the deflection under lateral stress of a flexible bus conductor and which, acting lengthwise in a direction parallel to the conductor, pulls the supports inward.

460. A simplified method for calculating stresses of long, straight, uniform parallel buses is given below. This method is applicable to many types of structure commonly used. For other structures see (2) in footnote to Par. **459.**

The lateral stress P is given by the formula

$$P = \frac{5.4kI_0^2Lp \times 10^{-7}}{d} \qquad \text{(lb/support)} \qquad (10\text{-}1)$$

where I_0 = initial rms value of total asymmetrical short-circuit current, A; L = length of bus-conductor span between centers of supports, ft; d = center spacing between conductors, in; k = shape-correction factor (for tubes, value is 1.0; for rectangular bars, 4 in by ¼ in, at least 6 in between centers, generally between 0.9 for arrangements as shown at 1 and 4 in Par. **465** and 1.1 for arrangements 2 and 3); p = stress factor (see Par. **465**).

The stress factor p is influenced by many variables—the mechanical damping, the frequency of the current, the rate of current decrement— and can be controlled by proper selection of bus span and insulator stiffness.

Values of p **for common types of structure,** and based upon representative rates of mechanical damping and current decrement, are given in Par. **465.**

High stresses due to resonance can be avoided by proper design of the structure (see Par. **462**). The factor p can be determined more accurately if the natural frequencies of the actual structure are checked by test. See curves for 25 and 60 c as given in (2) in footnote to Par. **459.**

The longitudinal stress F on the end insulators of a bus of two equal spans composed of **copper bars ¼ in thick** is given approximately by the formula

$$F\left(F + \frac{26{,}500A}{L^2}\right)^2 = 0.37LsP^2 \qquad (10\text{-}2)$$

where A = total area of bus-bar section, in²; s = stiffness factor of the support (see Par. **463**); and P and L are as defined in Par. **460.** This equation can be solved by trial. For slide-rule solution and for general formula applicable to all sizes of bars and tubes see (2) in footnote to Par. **459.**

For other than two spans multiply F by factor q from Par. **464.**

The total bending load R on the bus support then is:

a. If P is applied perpendicular to axis of insulator, for the end support:

$$R_1 = \sqrt{q^2F^2 + P^2} \qquad (10\text{-}3)$$

For support next to end:
$$R_2 = \sqrt{(q - 0.8)^2F^2 + P^2} \qquad (10\text{-}4)$$

b. If P is applied parallel to axis of insulator, the equivalent bending load for the end support is

$$R_1 = qF + BP/T \qquad (10\text{-}5)$$

For support next to end:
$$R^2 = (q - 0.8)F + PB/T \qquad (10\text{-}6)$$

where B = ultimate cantilever load of insulator applied at center line of bus bar and T = ultimate strength of insulator in tension. For B and T see insulator catalogues.

The structure may be considered safe if the stress R does not exceed the permissible ultimate bending stress B of the insulator, a safety factor of at least 2 being deemed good practice.

461. The maximum stress t in the bus bar due to the lateral load P is in the end span and is given approximately by the formulas

For flat buses:
$$t = 9PL/(a^2b) \quad \text{(lb/sq in)} \qquad (10\text{-}7)$$

For tubular buses:
$$t = 15PLD/(D^4 - d^4) \quad \text{(lb/sq in)} \qquad (10\text{-}8)$$

where L = length of span, ft; a = dimension, in of bar parallel to direction of load P, and b = dimension perpendicular thereto; D = outside diameter, in, and d = inside diameter of tube. The stress t should not exceed the allowable working stress for the material used. The elastic limit for bus-bar copper for this purpose may be considered as being 12,000 to 15,000 lb/in². The working stress will ordinarily be taken at a lower value [see Par. **468** and Eq. (10-15)].

In a laminated bus the average value of t for the group of n laminations is approximately t/n, but the stress in the outer lamination on one side will be higher than the average on account of the forces between laminations and unequal division of current.

The deflection f of the end span of a bus conductor due to the lateral load P may be taken approximately as follows [see Eq. (10-16)]:

For flat bars:
$$f = \frac{112PL^3}{Ea^3b} \quad \text{(in)} \qquad (10\text{-}9)$$

For tubular buses:
$$f = \frac{190PL^3}{E(D^4 - d^4)} \quad \text{(in)} \qquad (10\text{-}10)$$

where E = modulus of elasticity of the bus conductor material, which ranges from 12,000,000 to 17,500,000 lb/in² for copper.

462. Reductions in stresses may be effected by:

1. Reducing the magnitude of the short-circuit current.
2. Increasing the spacing between conductors.
3. Selecting a structure for which the stress factor p is low (see Par. **467**). On the other hand, it may often be desirable to design for a somewhat higher lateral stress (than the minimum) to avoid a high longitudinal stress.

Table 10-19. Stiffness Factor *s* of Typical Bus Insulators*

Duty	Moderate		Heavy	
Voltage	7,500	15,000	7,500	15,000
Factor	7,000	5,100	14,000	10,500

* For general information only. Influenced by design of insulator and flexibility of base and should be determined for specific insulators.

Table 10-20. Values of Factor q for Different Numbers of Bus Spans

Number of equal spans	1	2	3	4	6	8	10
Factor	¾	1	1½	2	2½	2½	2¾

4. Clamping bus bars at supports (as opposed to merely guiding the bars loosely in the supports) to avoid slapping.

5. Avoidance of extra-long spans of buses which have face-to-face arrangement, since the longitudinal stresses increase rapidly with span length for flexible buses.

6. Providing uniform distribution of current by transposition, thus reducing proximity effect.

463. Stiffness Factor of Bus Insulators (see Table 10-19).

464. Values of Factor q for Bus Spans (see Table 10-20).

465. Examples of Values of Stress Factor p for Flat and Tubular Buses Supported by Porcelain Insulators. The values given in Table 10-21 are based upon average rates of current decrement and apply only approximately to systems with widely different decrement characteristics, e.g., systems that employ high-speed excitation regulation. Such systems have to be treated as special problems. The values under arrangements 2, 3, 5, and 6 apply to particular insulators only, mounted on a rigid base, such as solid concrete. For other insulators, different stress-factor values will apply, depending upon the insulator stiffness and mass constants. Flexibility of the structural members upon which the bus support is mounted may also affect the stress factor. In cases of importance, tests may be necessary to determine the natural frequencies or correct stress-factor evaluation.

Table 10-21. Values of Stress Factor p for Flat and Tubular Buses Supported by Porcelain Insulators

Frequency	Bus span, in.	Arrangement 1		Arrangement 2				Arrangement 3	Arrangement 4	Arrangement 5		Arrangement 6
		Any		7,000* 5.35†		14,000* 8.15†		See note	Any	7,000* 5.35†	14,000* 8.15†	See note
		1‡	3‡	1‡	3‡	1‡	3‡	1–3‡	1–3‡	1½–2½-in. pipe		1½–2½
60 cycles	24	4.0–5.3	2.7–3.2	4.8	2.5	3.7	4.2	3.0	5.0	4.2	3.7	3.0
	36	1.8–2.6		3.8	2.3	4.1	2.9	3.3	2.0	2.7–3.4	4.4–4.8	3.0
	48	1.4		2.5	2.1	4.6	2.5	3.3	1.4	2.3–2.7	4.5–4.8	3.0
	54	1.3		2.5	2.0	4.8	2.5	3.6	1.3	2.3–2.7	3.9–4.4	3.5
	60	1.2		2.5	1.7	4.2	2.4	5.6	1.2	2.3–2.8	3.0–3.2	4.0
25 cycles	24	2.8–3.2		2.8–3.4				3.0	3.0	2.5–3.2		3.0
	36	3.5–4.1		2.8–3.4				3.0	3.3	2.5–3.2		3.0
	48	2.5		2.8–3.4				3.0	2.3	2.5–3.2		3.0
	54	2.1		2.8–3.4				3.0	1.8	2.5–3.2		3.0
	60	1.8		2.8–3.4				3.0	1.5	2.5–3.2		3.0

NOTE: The stress factors given for arrangements 3 and 6 apply to three particular insulators as follows:
1. Stiffness 40,000 lb per in equiv. mass 4.0 lb.
2. Stiffness 75,000 lb per in equiv. mass 7.5 lb.
3. Stiffness 120,000 lb per in equiv. mass 13.5 lb.

The stiffness is measured in the direction of the lateral bus-bar deflection at the center of the bus bar.

* Stiffness or pounds per inch of deflection of insulator in direction at right angles to the axis of the insulator and measured at the center of the conductor.

† Equivalent mass of insulator in vibratory motion under bus-bar stresses [see reference (1) Par. **459**].

‡ Number of bus laminations.

466. Short-circuit stresses in outdoor buses of the beam type are calculated in the same manner as for indoor buses under similar conditions [see Eq. (10-3), Par. **460**]. When the bus conductors are clamped at one point only, at or near the middle, and allowed to expand freely in both directions, the longitudinal stress may be neglected.

Structures with long spans held in tension by strain insulators cannot be calculated for stresses by the above procedure, but approximate estimates can be made by following the procedure generally used for calculating mechanical stresses in transmission-line conductors.

467. The total stress in an outdoor bus is the resultant of the stresses due to the short-circuit load, together with the dead, ice, and wind loads.

a. Buses up to 161 kV. The distance between phases and the character of the bus supports and their spacing are such that wind loading may usually be neglected. Ice load of ½ in is usually considered. In calculating the short-circuit stress P, by Eq. (10-1), the stress factor may be taken at value 1, but it is sometimes advisable also to use an impact factor of 2 as a protection of the insulator caps.

b. Buses for 230-kV and Higher Voltages. The spacing between phases is usually so large that the mechanical effects of short-circuit currents may not be the determining factor, and such buses, when properly designed for the mechanical loads only, may be found to satisfy the electrical requirements, such as current-carrying capacity. However, short-circuit duties on modern systems have been increasing rapidly, and the electrical forces should be checked (see Par. **459** for methods of calculation; also see Bibliography, Par. **539**).

c. Buses open at the ends should have a small hole drilled at the bottom, in the center of each span, to provide for drainage of any water that might collect inside the tubing.

468. Calculations of Stresses. Tubing being almost universally used for outdoor buses, the formulas given below are devoted to this conductor shape. Should other shapes be under consideration, standard handbooks on structural design should be referred to for data on the correct section moduli, etc., for the particular case.

If w = total resultant load/in of bus length and w_1, w_2, and w_3, respectively, = loads/in length due to weight of material, ice, and wind, then for case a of Par. **467**

$$w = \sqrt{(w_1 + w_2)^2 + \left(\frac{P}{12L}\right)^2}$$

$$= \sqrt{[0.785(D^2 - d^2)\delta + v_2]^2 + \left(\frac{P}{12L}\right)^2} \quad \text{(lb/in)} \qquad (10\text{-}11)$$

and for case b

$$w = \sqrt{(w_1 + w_2)^2 + (w_3)^2}$$

$$= \sqrt{[0.785(D^2 - d^2)\delta + w_2]^2 + (w_2)^2} \quad \text{(lb/in)} \qquad (10\text{-}12)$$

where δ = weight/in³ of bus material and D and d, respectively, = outside and inside diameters of bus. P, the short-circuit stress per span, and L have the same values as in Par. **460**.

Outside diameter of bus, in.	w_2 = weight in lb of ice per inch length of bus		w_3 = wind load in lb per inch length of bus	
	½ in. ice	1 in. ice	½ in. ice	1 in. ice
1	0.0785	0.211	0.111	0.167
2	0.131	0.302	0.167	0.222
3	0.183	0.418	0.222	0.278
4	0.235	0.522	0.278	0.333
5	0.288	0.627	0.333	0.389
6	0.345	0.732	0.389	0.444
7	0.392	0.836	0.444	0.5
8	0.430	0.940	0.5	0.556

Ice and wind loads may be readily determined from the following tabulation, ice being taken at 57.5 lb/ft³ and wind at 8 lb/ft² of projected area, including a uniform coating of ice.

Large deflections should be avoided for appearance' sake, even if the maximum bending stress may be found to be within safe limits. It is generally satisfactory, in approximation of bus diameter, to allow 1 in of bus outside diameter for every 10 ft of bus span, thus,

$$D = l/120 \qquad \text{(in)} \qquad (10\text{-}13)$$

where l = span length, in.

The inside diameter of the tube may then be assumed at

$$d = D - (0.25 + 0.02D) \qquad \text{(in)} \qquad (10\text{-}14)$$

The maximum fiber stress t can be determined from the following equation, in which w is taken from Eqs. (10-11) or (10-12) according to conditions. Equation (10-15) will be found to be equivalent to Eq. (10-8) but is given again in this form for convenience,

$$t = \frac{wl^2}{8S} = \frac{wl^2}{8[0.098(D^4 - d^4) \div D]} = \frac{1.27wDl^2}{D^4 - d^4} \qquad \text{(lb/in}^2\text{)} \qquad (10\text{-}15)$$

where S = section modulus; for a tubular bus $S = 0.098(D^4 - d^4) \div D$.

Fiber stresses should be less than 24,000 lb/in² for steel, 18,000 lb for hard-drawn copper with an elastic limit of not less than 28,000 lb/in² and less than 10,000 lb for aluminum; otherwise the thickness of the tube must be increased by decreasing the assumed d. If t should exceed the allowable unit stress by a great deal, the assumed D should be increased.

The deflection of the bus f may be determined from the following, all loads assumed uniformly distributed,

$$f = \frac{1}{185} \times \frac{wl^4}{EI} = \text{max. deflection} \qquad \text{(in)} \qquad (10\text{-}16)$$

where E = modulus of elasticity of bus material, which may be taken as 16×10^6 for copper, 10×10^6 for aluminum, and 27×10^6 for steel pipe; and I = moment of inertia for tubular bus $= 0.049/(D^4 - d^4)$.

Equations (10-9) and (10-10), Par. **461,** are derived from the above, the value $\frac{1}{185}$ applying to the condition of the bus assumed fixed at one end and supported at the other (end span). This value becomes $\frac{1}{384}$ for the condition where the assumption may be that the bus is fixed at both ends; if fixed at one end, free but guided at the other, it becomes $\frac{1}{24}$; and for a single beam supported at both ends, but free to move, it is $\frac{5}{384}$.

The deflection of the unloaded bus, by Eq. (10-16), is

$$f = \frac{1}{185} \times \frac{16\delta l^4}{E(D^2 + d^2)} \qquad \text{(in)} \qquad (10\text{-}17)$$

δ being, as above, weight per cubic inch of bus material, and l length in inches.

The dead-load deflection is generally taken care or by giving the bus an equivalent camber, or upward "set," in the middle, so that without wind and ice load the bus will be level.

Table 10-22, gives the maximum allowable spans with different outside diameters and thicknesses of tubular copper, aluminum, and steel buses. From this table the proper-size tubing for any span can at once be determined. To permit putting the table on a single page, the half-inch sizes of tubing from 1 to 7 in outside diameter have been omitted; the corresponding spans for these intermediate sizes can be readily interpolated. The ice coating has been taken as either ½ or 1 in thickness all around, and the wind load as 8 lb/ft² of projected area of tubing including ice. The weights per cubic inch were taken as: copper 0.323 lb, aluminum 0.098 lb, and steel 0.283 lb. The elastic moduli used were: for copper, 16×10^6; aluminum, 10×10^6; steel, 27×10^6.

For any fiber stress t_1, other than as given above, the maximum corresponding

Table 10-22. Limiting Spans of Outdoor Tubular Buses in Feet and Inches*

Thickness, in	Outside diameter, in	8		7		6		5		4		3		2		1	
	Ice thickness, in	½	1	½	1	½	1	½	1	½	1	½	1	½	1	½	1
0.1	Copper	59–11	51–5†	54–9	47–8†	49–4	43–7†	43–7	39–6†	37–3	35–1†	30–8	28–9†	23–2	21–10†	14–1	12–11†
	Aluminum	56–2†	45–2†	52–4†	41–11†	48–0†	38–2†	43–4†	34–5†	38–4†	30–0†	31–8†	24–6†	24–3†	18–5†	14–4†	10–5†
	Steel	71–10†	60–8†	67–2†	56–8†	61–8†	51–9†	55–10†	46–8†	49–8†	41–3†	41–4†	37–7†	31–8	25–9†	18–6	15–7†
0.2	Copper	59–5	59–5	54–3	54–3	48–9	48–9	43–0	43–0	36–9	36–9	30–1	30–1	22–4	22–4	13–3	13–3
	Aluminum	71–2†	57–4†	64–6†	53–4†	59–1†	48–8†	53–2†	43–10†	46–4†	37–10†	38–2	31–1†	28–6	23–0†	16–8†	12–4†
	Steel	80–7†	71–4†	74–0	66–4†	66–7	60–6†	58–9	54–6†	50–4	47–4†	41–0	39–3†	30–7	30–4†	18–1	16–11†
0.3	Copper	58–11	58–11	53–8	53–8	48–2	48–2	42–4	42–4	36–2	36–2	29–5	29–5	21–9	21–9	12–7	12–7
	Aluminum	74–11	65–0†	68–2	59–11†	61–4	53–8†	53–11	49–5†	46–1	41–6†	37–4	34–2†	27–7	25–1†	16–0	12–9†
	Steel	80–6	76–3†	73–4	70–7†	65–11	59–3†	57–11	57–11†	49–5	49–5	40–3	40–3	29–6	29–6	17–2	17–1†
0.4	Copper	58–5	58–5	53–2	53–2	47–8	47–8	41–11	41–11	35–7	35–7	28–8	28–8	21–1	21–1	12–1	12–1
	Aluminum	74–3	69–8†	67–7	64–4†	60–8	54–10†	53–5	52–0†	45–3	44–5†	36–7	36–1†	26–10	25–11†	15–5	12–8†
	Steel	79–10	78–10†	72–8	72–6†	65–2	62–10†	57–1	57–3	48–8	48–8	39–4	39–4	28–10	28–10	16–6	16–6
0.5	Copper	57–11	57–11	52–8	52–8	47–2	47–2	41–4	41–4	35–0	35–0	28–1	28–1	20–5	20–5	11–11	11–11
	Aluminum	73–8	72–5†	66–10	66–10	59–11	59–11	52–6	52–6	44–7	44–7	35–9	35–9	26–0	26–0	15–3	12–9†
	Steel	79–3	79–3	72–0	72–0	64–5	64–5	56–5	56–5	47–11	47–11	38–6	38–6	27–11	27–11	16–1	16–1
Iron-pipe size	Copper	‡61–10	‡61–10	61–7	61–7	51–9	51–9	45–11	45–11	39–9	39–9	33–5	33–5	25–8	25–8	17–2	16–10†
	Aluminum	‡77–1†	‡69–0†	72–4	62–9†	65–9	57–7†	58–5	50–4†	50–7	43–0†	42–6	35–4†	31–1†	24–3†	19–5†	14–6†
	Steel	‡84–5	‡80–4†	77–9	74–4†	70–8	67–11†	62–8	59–10†	54–4	52–11†	45–7	44–8†	35–2	30–8†	23–2	20–4†

NOTE: Maximum spans given in feet and inches for a maximum fiber stress of 18,000 lb per sq in for copper, 10,000 lb for aluminum, 24,000 lb for steel; or for a maximum dead-load deflection of 80th of the span, whichever consideration gives the shortest span. See Par. **468**. (The author acknowledges the assistance rendered by H. J. Glaubitz in the preparation of this table.)

* Aeolian vibrations induced by wind may necessitate smaller spans (see Bibliography).

† Spans marked with daggers are limited maximum allowable fiber stress, while for spans not so marked the maximum allowable dead-load deflection governs.

‡ Inside diameter = 7.981 in.

span L_1 would be

$$L_1 = L\sqrt{\frac{t_1}{t}} \qquad (10\text{-}18)$$

For any ratio q of dead-load deflection to span other than 1:80, the deflections and spans will be in the relation $f_1 = f \times L_1^4/L^4$. Hence,

$$L_1 = L\sqrt[3]{80 \times q} \qquad (10\text{-}19)$$

469. Stresses on disconnecting switches under short-circuit conditions may be sufficient to open them, with disastrous results; therefore modern switch designs embody locks, or overcenter mechanisms, to prevent this from occurring. The force on the switch blade varies as the square of the current and will be increased if the return circuit passes behind the switch, the additional force varying inversely as the distance from the center of the switch blade to the center of the return conductor.

470. Limiting Spans of Outdoor Tubular Buses (see Table 10-22).

471. Bus supports are designed for definite cantilever strength, expressed in inch-pounds and measured 1 in from the base. Ample margin of safety with regard to insulation and structural strength should be provided, and manufacturers' data should be carefully checked and units so selected that allowable values for the particular units are not exceeded. Good practice recommends a safety factor of at least 100% and up to 300%.

CONTROL SWITCHBOARDS

472. The control of large power circuits by electrically operated breakers has caused the "switchboard" to be replaced by the "control board."

Several types of control board are in general use. These are identified in Fig. 10-171. On these are mounted the necessary control switches, lights to indicate the positions of the breakers, necessary indicating and recording instruments, and relays. The arrangements of devices on boards should be simple and distinctive to aid the operator under normal and emergency conditions and to avoid the possibility of confusion and mistakes. The control and indicating devices of each main power circuit should be clearly set off from those of other circuits.

473. The assembly of the panels and the selection of the type of panel depend entirely upon the size and type of station and on local conditions.

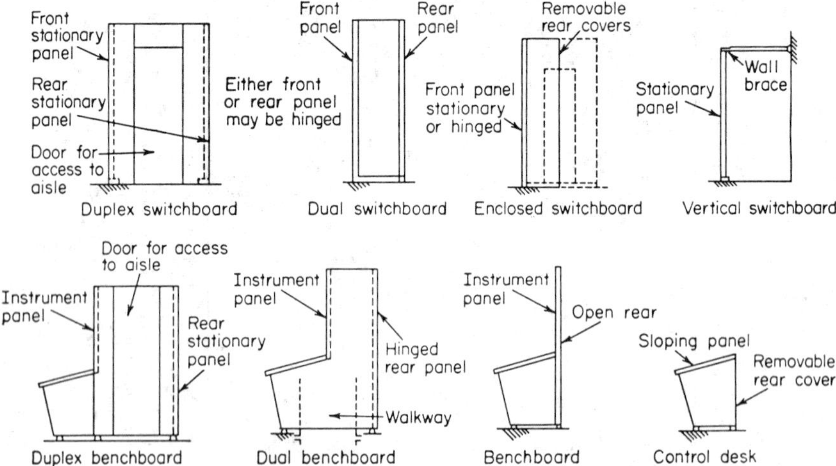

Fig. 10-171. Modern types of power-station control boards.

The panel-type board may prove too long for satisfactory operation in very large stations, although this type is the simplest and least expensive.

The bench board or a combination of control desk and switchboard is generally used, as such designs provide more space for the same length for the mounting of the equipment and can be arranged for closer supervision.

In large switchboards, the panels are generally assembled in the form of an arc or its equivalent. The bench-board type is commonly used for generator controls, while the vertical boards are more commonly used for feeders. Relays are often mounted on separate boards at the rear or near the main control boards in order to concentrate

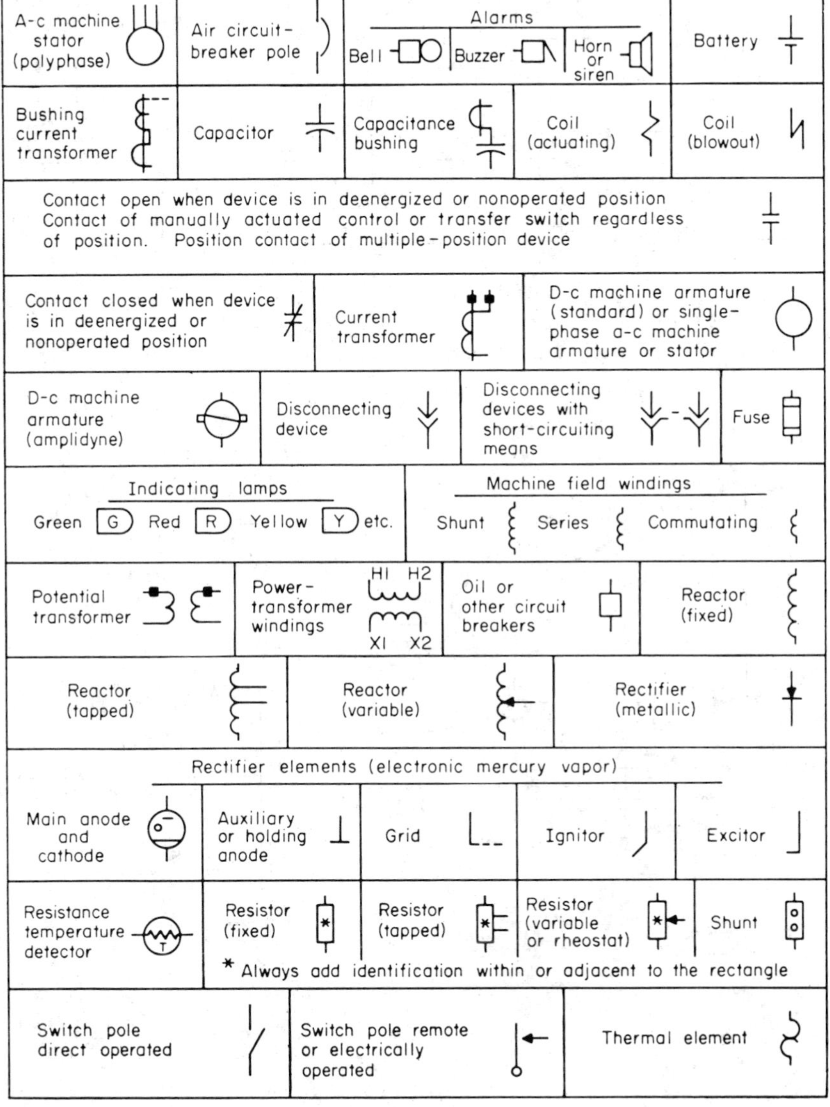

FIG. 10-172. Switchboard-device element symbols.

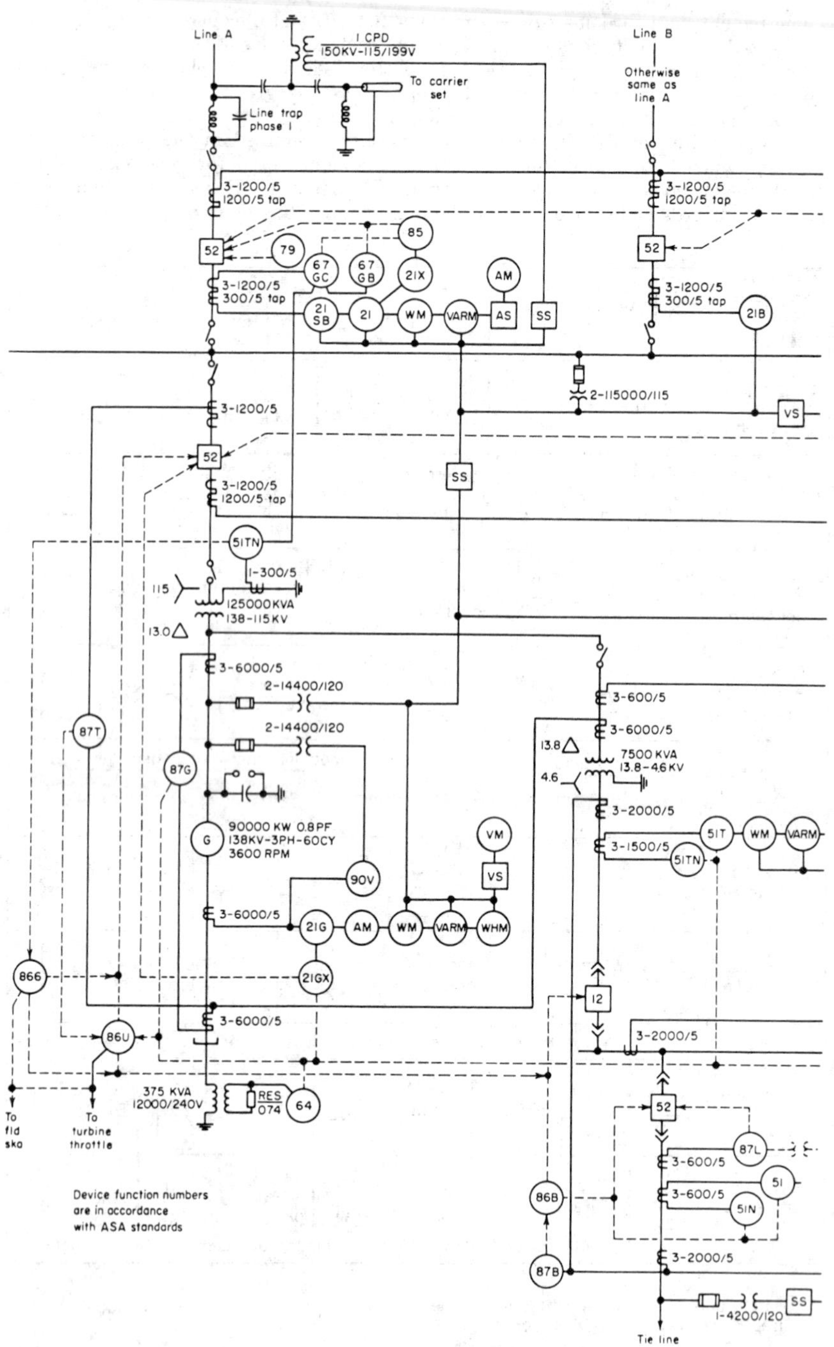

FIG. 10-173. Single-line diagram of main power

the more immediate manual controlling devices and associated indicating lights and instruments within easy range of the operator.

474. Depending on the method of operation, control boards are classified as **manual, automatic,** and **supervisory control.**

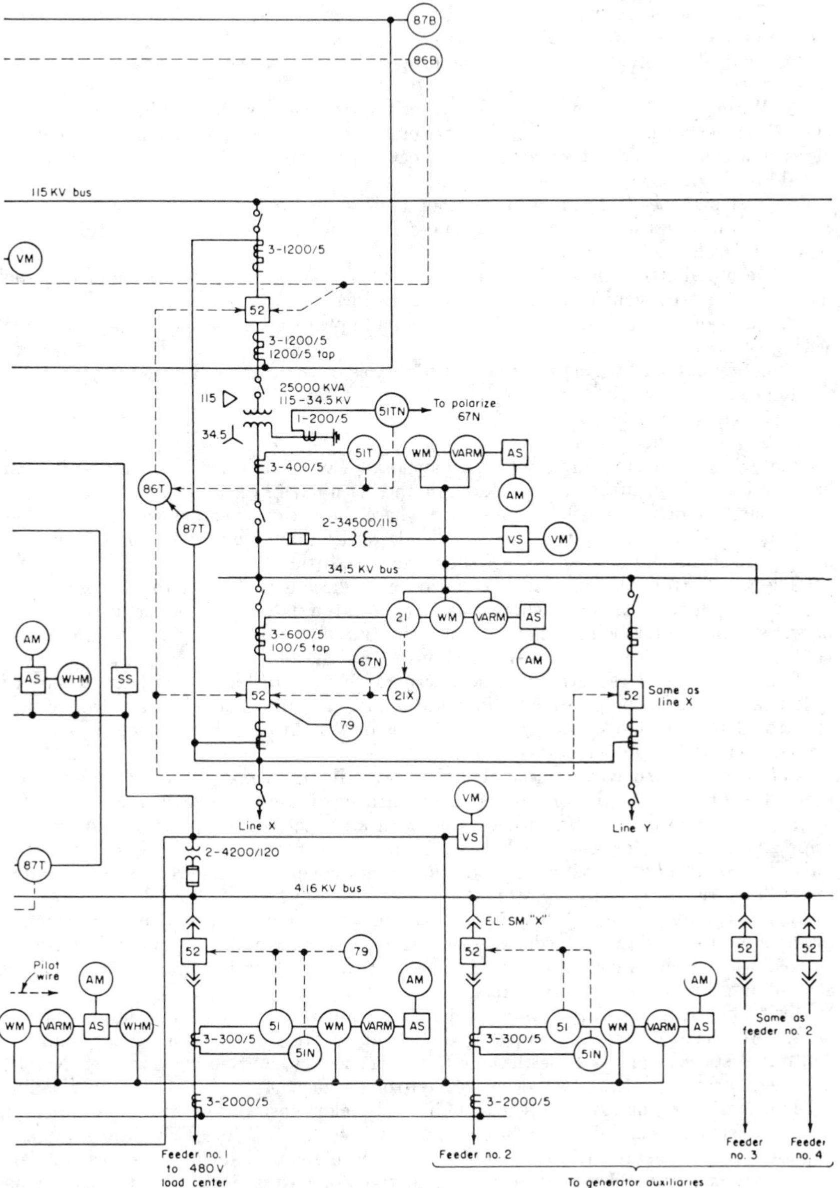

circuits at a generating and switching station.

475. Fundamentals of panel design are as follows:

1. The grouping of the apparatus should be such that the chances of reading the wrong instruments or manipulating the wrong devices, for both normal and emergency operation, are minimized. Control switches should be within easy reach, and indicating instruments within easy sight.
2. Panels for like purposes should be wired alike.
3. Only meters, instruments, and devices of recognized merit in design, construction, and operation should be used.
4. Control-supply buses should be properly protected against accidental ground or short circuits.
5. Wiring and connections to instruments, control switches, and panel devices should be neatly done and in such manner as to facilitate maintenance. Complications in wiring should be avoided. Connections to studs of instruments and devices should be made positive and secure.
6. Test switches, test links, or similar devices should be provided for instruments and relays in cases where, because of operating conditions, the primary circuits cannot be deenergized.
7. Terminal strips or blocks should be provided at bottom, top, or sides of panels as required, for convenient connection of control cables.
8. Fuses of carefully selected capacities should protect the various parts of the small wiring system.

476. Switchboard wiring may be considered as of three general types:

1. Open or flat wiring.
2. Group wiring.
3. Trough wiring.

Flat wiring is cleated directly to the panel and leaves the board at the top or bottom, as may be desired, after passing through the terminal blocks. It is commonly used on the simpler panels but is less practical where the wiring is extensive and congested.

Group wiring is in common use and lends itself to neatness and economy of panel space, with complete accessibility of the wiring at all times.

The wiring from the instruments and other devices is run direct to the right or left of the panel or to both sides, where it is formed into a flat, round, or rectangular group of wires which, after being taken through the terminal blocks, connects to the control cables. It becomes practically essential on congested boards.

Gutter wiring is essentially group wiring used in steel-panel designs, where the wiring is carried up the panel sides in suitably arranged ducts or troughs. In one type, the troughs are formed with perforated steel plates through which the wires can be brought out at the proper points.

477. Switchboard wire for panel wiring is available in many combinations. Generally the primary insulation consisted of varnished cambric, ozone-resistant rubber, or polyvinyl chloride synthetic, protected with an outer covering of fire-retardant unimpregnated cotton or asbestos braid. In more recent times, utilities required that switchboard wire be polyvinyl chloride, insulated, covered with felted asbestos and an outer flame-resistant coated cotton braid. This wire in AWG No. 14 is about ¼ in in diameter, and, in the interest of space saving with small bundles, it is more common practice to use a flame-retardant cross-linked polyethylene or polyvinyl chloride-insulated wire, such as code-designated type TBS wire. This is available from various switchboard and cable manufacturers.

Switchboard wire conductors are solid or stranded and may be had in a variety of sizes. American wire gage Nos. 9, 10, 12, and 14 are commonly used for manually controlled stations and for instrument transformer wiring. Smaller sizes (as No. 16) are used for automatic or supervisory control because of the small current loading.

478. Indicating lamps are used to indicate breaker position, **green** lights indicating **open** breakers and **red** lights indicating **closed** breakers. **White** lights are sometimes included to be energized from potential transformers to indicate **live** circuits, and some stations include orange or other distinctive color to indicate that the circuit has tripped automatically. It is becoming common practice to use low-voltage lamps, such as telephone-circuit lamps, in series with resistors across 120 V rather than 120-V lamps, to reduce size of lamps and reduce the brilliancy and glare and also to reduce the current that is drawn in case the lamp short-circuits. Red and green lights are commonly wired so that they are energized through the tripping coil and circuit of the breaker to supervise the trip circuit. An open in the trip-coil circuit is then indicated by a dark lamp. When so wired it is important that low-voltage low-wattage lamps be used so that the tripping plunger will not be actuated incorrectly, as by short circuiting of the lamp. White lamp bulbs with colored caps are in common usage.

479. Mimic buses are frequently installed on the faces of control boards as an aid to operation, showing in miniature the bus and the circuit connections controlled by each control switch.

480. Auxiliary circuits are provided on switchboards for many different purposes, some of the more commonly used being as follows:

a. Voltmeter bus and switches or plugs which make possible the use of one instrument for several machines or circuits.

b. Current supply for instruments and relays.

c. Control power supply for operation of circuit breakers and switches.

d. Annunciator or alarm circuits.

e. Potential supply for relays.

f. Synchronizing circuits.

g. Test supply.

481. Ground detectors are desirable on ungrounded systems so that, when a ground does occur, immediate steps may be taken to clear it before a second ground occasions a short circuit.

Control circuits especially should be suitably equipped with ground detectors, as a double "ground" may cause the operation of equipment carrying load, with serious consequences.

Lamps, usually a 25-W size, or a **different voltmeter** with zero center scale may be used. Both lamps and coils must be capable of withstanding full-line voltage.

Objections to the use of lamps on control buses for ground detection are:

1. Both lamps may be destroyed by high voltage induced in the control system, and the circuit may be left with a ground on one side and no indication thereof given.

2. Sufficient current can flow through the lamp to energize some secondary relays, and circuit breakers may be opened as a result of double grounds.

Fuses with ground-detector lamps on d-c control circuits should be of the 600-V 1-A type rather than of 250-V rating.

Special relays, with suitable resistors for bridging the circuit, are more often used to give more sensitive indications than lamps can provide, for detecting grounds on control circuits.

482. Instrument and meter-equipment requirements will depend primarily on the size and importance of the installation, its cost, the class of operators, and the indications and records required. The number should be kept at a minimum consistent with the operating conditions.

483. Miniature control switchboards have been developed which permit of greater concentration of control and indicating equipment, such arrangements requiring a minimum of space and making possible the supervision of a large system by one man.

The fundamental principle on which the miniature switchboard is based is that of mounting, close to the apparatus to be controlled, relay switches or contactors which handle the heavier currents necessary for the operation of the ordinary control equipment and devices.

Circuits operated at the switchboard may, therefore, be only of sufficient energy to actuate the relay switches; hence, lower voltages and current values are permissible, making possible smaller control switches and devices.

Miniature meters are used, actuated by telemetering devices, indications being transmitted over a few telephone wires. Only such meters as are essential are included.

The indicating lamps are similar to the type for telephone service and interchangeable with them but designed for longer life.

Control switches of miniature size operating at approximately 60 V are used.

A considerable saving in cost is sometimes possible with this type of board, as a

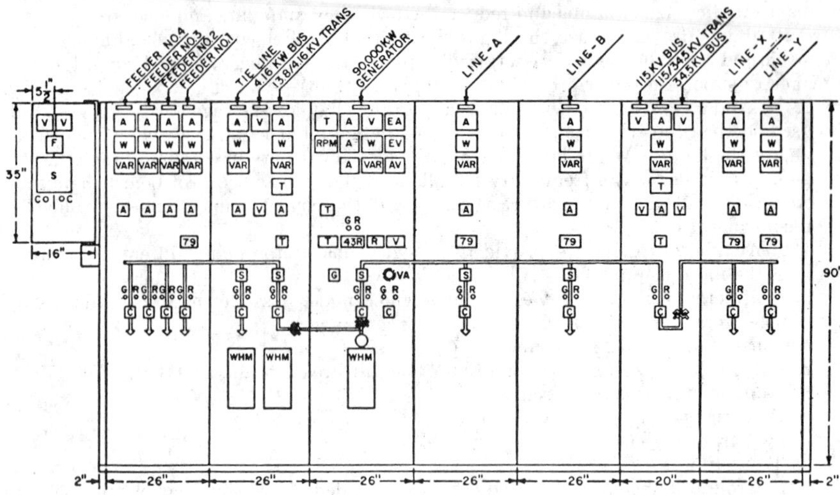

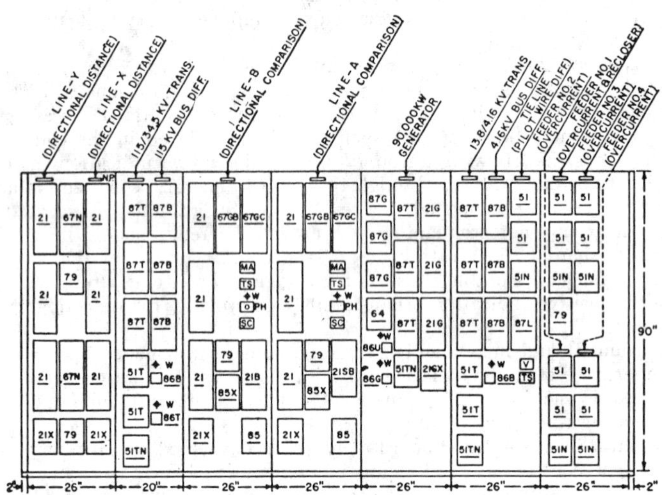

Fig. 10-174. Control and relay sides of duplex switchboard for control and protection of the system shown in Fig. 10-173. In modern practice, placement of nameplate on control side higher than 84 in or lower than 12 in is avoided.

result of the small extent of the board itself; reduction in quantity and size of conduit to a minimum; elimination of long runs of multiconductor control cables of capacity sufficient for the operation of control devices; absence of need for large conduit rooms, etc., now necessary to a proper terminating of control conduits; etc.

484. A typical wiring diagram for manually controlled switchboards is shown in Fig. 10-173.

485. Standard switchboard panels embodying the various devices referred to in the preceding paragraphs and suitable for all classes of work have been developed and catalogued by the various manufacturers, so that it is now possible to buy complete stock panels for many applications.

486. Control. Switchboard devices are identified by USAS device numbers, symbols, or abbreviations. Some of the most frequently used symbols are shown in Fig. 10-172. See USAS Standards Z32.3 for details and additional items.

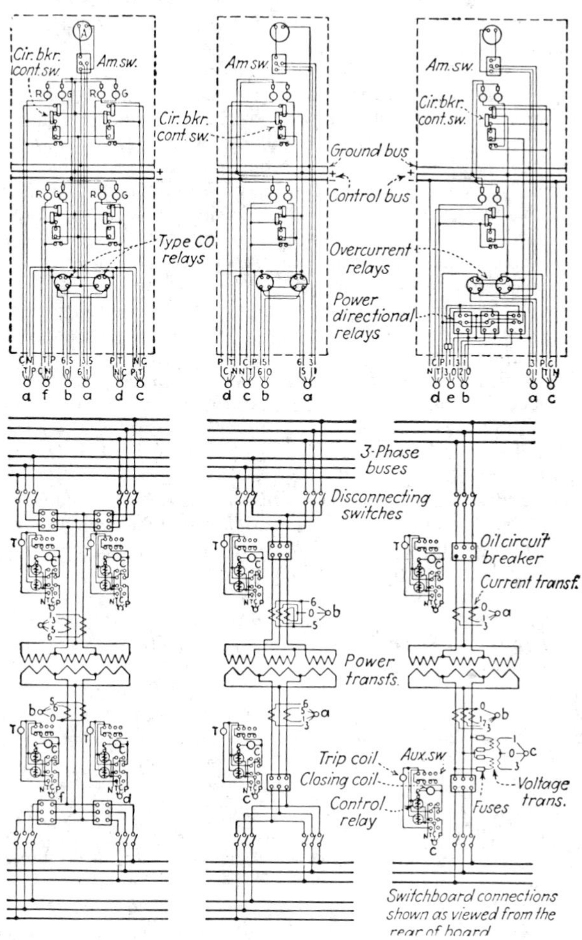

FIG. 10-175. Typical diagrams of main and panel wiring for electrically operated transformer circuits.

487. A single-line diagram of a generator unit connected to a 115-kV transmission system, a 34.5-kV subtransmission system, and a 4.16-kV distribution and auxiliary supply bus is shown in Fig. 10-173. Device numbers identify each element of the system, together with the instrumentation and protection.

Figure 10-174 shows the power switchboard in the duplex design which would be used for control and protection of this installation.

488. Supervisory control equipment is often used to remotely control a number of devices from a distant location, a minimum number of communication channels being employed. The basic approach is to select, at one time, a single operating point and connect the control for that device to the communication channel to permit remote operation. Generally only one device can be controlled at one time.

The functions normally performed by supervisory control include opening and closing circuit breakers and disconnect switches from the master station, operating transformer tap changers and phase-angle regulator tap changers, the control of valves, transmission of set points, etc. In addition to control from the master station the same supervisory equipment is ordinarily used to transmit position indications and metering from the remote to the master station. Another common function is the transmission of alarm conditions from the remote to the master station.

Until recently, most supervisory systems consisted of relay chains at the master and remote stations that were operated by pulse codes to select the proper operating point and then perform the control action. These systems use various pulse codes on either d-c or audio-tone communication channels. Generally the transmission speed is rather slow and is suitable for telegraph-grade channels.

The latest supervisory system uses solid-state components rather then relays to generate and receive selection and operation codes. Generally they operate over audio-tone channels using frequency-shift-type equipment and are capable of very high operating speeds. These systems can be operated over any high-speed communications channel including microwave, telephone, or power-line carrier.

Bit rates of 1,000 bits/s or higher are common. This type of equipment often continually scans the alarm and indication points. Telemetering is usually accomplished by converting from analog to digital form at the remote station and transmitting in digital form, a code similar to that for control being employed. Many modern installations display the telemetered quantities in digital form at the master station.

Various control sequences and code-checking techniques are used to assure operation of the correct device and prevent false operation. By using modern equipment, a very high degree of security is possible.

489. Automatic control is a combination of various devices, including relay operational amplifiers, magnetic amplifiers, and solid-state switches, which are used to automatically operate devices in generating stations and substations without operator intervention. Generally, unattended substations are designed so that all normal operations are performed automatically.

The automatic controls normally provided in unattended stations include automatic reclosing of circuit breakers, control of voltage by transformer tap changers or voltage regulators, switching of capacitors, etc.

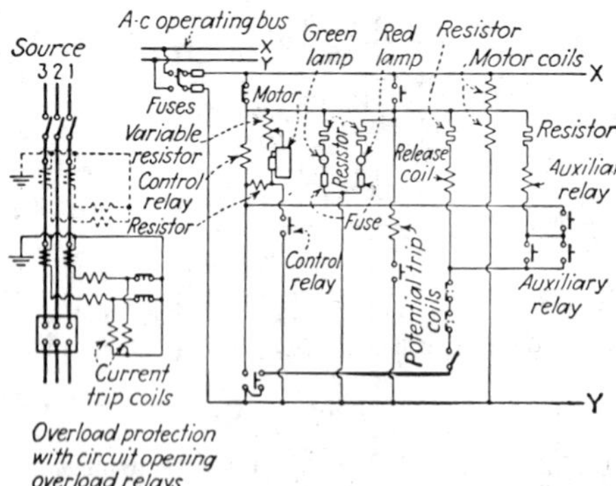

FIG. 10-176. Typical wiring scheme for an automatic reclosing feeder.

Small and moderate-sized hydroelectric stations have been made automatic to start and stop, depending upon water level, time switch, frequency, voltage, or merely the opening or closing of the main transmission line to the station. The generator may be of the induction type but usually is of the synchronous type. In the latter case it is equipped with squirrel-cage winding like a synchronous motor and is connected to the line by closing the main breaker at approximately 95% synchronous speed, after which the field switch is closed. In small sizes the water wheel often operates at fixed gate opening, and the generator at fixed excitation, whereas in larger sizes a governor is installed, and an automatic voltage regulator adjusts the excitation. The usual protective relays of an attended station are included, plus additional protective equipment, such as bearing relays, overvoltage and undervoltage protection, overcurrent or thermal protection, and overspeed protection. Some protective relays shut down the station until reset by hand; others shut it down temporarily until the abnormal condition has disappeared. Interlocks are installed to ensure a proper sequence of steps in starting.

Converter and **synchronous-condenser substations** are other common applications for automatic control. In these, the unit is automatically started, by undervoltage, after a short time delay, and automatically stopped, by underload, after a time delay. The usual protective relays of attendant stations are included plus some other protective devices, such as bearing protection, commutator flashover protection, failure to ground of the windings and overload or overheating, some of which shut down the station until reset by hand, while others shut it down until the abnormal condition has been removed. Interlocks are installed to ensure a proper sequence in starting and, after each step has been completed, that conditions are right for the next step.

Automatic equipment, with standard device numbers for the various relays and devices, has been extensively standardized by the various manufacturers and is available for different equipment and functions (see Fig. 10-176). It has sometimes been applied in whole or in part to equipment in attended stations to relieve the operator for other duties and to reduce the possibilities of operating errors.

POWER AND CONTROL WIRING SYSTEMS

490. The design and construction of the wiring system of a station are of great importance to the rest of the equipment. This system consists of two major parts:

a. Power wiring.

b. Control wiring.

Power wiring includes all main electrical connections throughout a station, other than power switchboards or bus-and-switch structures, and should be arranged so as to minimize and localize possible troubles due to short circuits or grounds.

Control wiring throughout a station is of vital importance and demands careful consideration in every detail. Failure of the control wiring during station disturbances may have disastrous consequences.

The prime requisites of station wiring are simplicity and reliability, and in the design of a station each cable and wire should have a definite place provided for it just as much as any other piece of apparatus.

491. Open wiring is sometimes used for main connections. For voltages up to 27 kV the wires or cables are generally insulated for full potential and are often supported on insulators also good for full potential. For voltages exceeding 27 kV bare wires or cables are employed. Recommended dimensions for spacing of bare conductors are given in Par. **457.**

492. Conduit or trough wiring represents the majority of the station wiring. In conduit practically all the conductors except that which is part of a switchboard or bus structure are enclosed and protected. Conduits in more common use are of the rigid type, although flexible-type conduits are also available. Rigid conduits, when properly installed, are fireproof, moistureproof, reliable, and mechanically strong. In many recent installations, open wiring in troughs has been used successfully in place of conduit wiring. Where applicable, this type of construction is more economical and offers greater flexibility where additional wires must be installed.

493. Materials generally used for rigid-type conduits are iron, nonferrous corrosion-resistant metal alloys, fiber, Transite, tile, and concrete.

494. Rigid-iron conduits are manufactured from steel tubing. They are sold as hot-galvanized, electrogalvanized, sherardized, or black-enameled conduits.

All kinds are enamel-lined. This enamel, to meet the underwriters' requirements, must retain its flexibility under all ordinary conditions of service and changes in temperature and resist the action of acid and alkali. It also prevents rusting to a large degree.

There are many standard conduit fittings made up for use with conduits of these grades. In general, only such fittings should be used as meet the requirements of the Underwriters' Laboratories Inc. Special fittings may or may not need such approval.

Installation of rigid-iron conduit is recommended for exposed and sometimes for concealed work.

Hot-dip galvanized or **sherardized conduit** should be used for outdoor installations, in damp places, or where the conduit is to be concealed in cinder concrete or cinder fill. Conduit for use in ordinary concrete and brickwork in dry locations may be black-enameled.

Black-enameled conduit should not be used where it is required that a good bond be obtained between conduit and concrete.

When conduit is run exposed, it should be suitably supported at intervals along its length, at bends, fittings, dead ends, etc.

In very long runs, attention should be given to the effects of conduit expansion. In general, for 45° or 90° bends in iron conduit 1¼ in and larger, standard factory long-radius bends should be used wherever possible.

Iron conduit should not be employed on alternating currents unless all conductors of the circuit are in the same conduit, because of the heat from magnetic losses in the conduit which would otherwise result.

495. Metallic molding is sometimes used for interior surface wiring, but its field of application is quite limited. It can be used only where the potential difference between conductors does not exceed 300 V; and the power, 1,320 W. Its principal use is for surface wiring of lighting circuits in small industrial or commercial plants, being substituted for open wiring.

496. Fiber conduit consists of tubes made by rolling wet wood pulp or fiber saturated with black insulating compounds around a mandrel.

It is mechanically fairly strong, although somewhat brittle, and may be broken by a sharp blow. It comes in lengths which are joined by screw, socket, sleeve, or drive joints.

Fiber conduits are not fireproof, and, because of the limited strength, they are generally enclosed in a concrete envelope with at least 1 in of concrete free of cracks between adjacent conduits.

Fiber conduit in interior and exterior work finds its field in sizes 2 in and above, where large single-conductor a-c cables are to be installed, and in fact should be considered as a substitute for iron conduit for any service, wherever possible, owing to its low cost.

It is not advisable to use fiber conduit for exposed runs or where the conduit must turn out of a floor or wall, without having adequate support.

497. Transite conduit is a waterproof, though hygroscopic, fireproof, and corrosion-resistant conduit manufactured by building up an asbestos-cement mixture, under pressure, on smooth metal mandrels and is much stronger than fiber. Transite conduit, owing to its relatively high mechanical strength, has sometimes been used for exposed and underground installations without the protection of concrete. It has also been used in place of fiber conduit when embedded in concrete. It usually comes in 5-ft lengths for sizes up to 2½ in and in 10-ft lengths for sizes up to 6 in. A large variety of fittings and couplings is available for use with Transite conduit.

498. Tile conduit is seldom used for indoor installations. It finds its largest field in underground duct-line work. For this work it is made up in single-duct sections approximately 18 in long or multiple-duct sections 36 in long. Bends are not standard. All joints are wrapped with burlap and tar or cement.

499. Corrosion-resistant metal-alloy conduit is usually a copper-alloy pipe, sometimes used in places where a single-conductor cable carrying alternating current must be installed, and fiber or Transite conduit cannot be used. Such conduit should be avoided where possible, however, primarily on account of its high cost.

500. Flexible conduit is built up of spiral metal strips which interlock. It is not moistureproof and hence cannot be used where the action of any considerable amount of moisture is objectionable.

It is sometimes used for wiring to machinery in order to localize machinery vibrations or in congested locations where it may be installed more easily than rigid conduit. A short section is often installed between the end of rigid conduit and a motor-terminal box so that the motor may be more readily disconnected and removed. Flexible conduit is also made up already containing conductors, usually twin-conductor or multiconductor cable, and when so made up has a moisture-repellent gasket placed between the inner and outer strips.

501. Precast concrete conduits are sometimes used, in which case they are installed similarly to tile duct.

502. The allowable number of bends in a conduit run depends upon the length of the run, the size of the conduit, and the number and type of cables to be pulled. In general not more than two or three equivalent 90° bends should be used between pulling points. The number of bends should be kept at a minimum, and long sweeps are preferable to sharp bends where practicable.

503. Wire and cable troughs are available in a wide variety of sizes and designs. Some are made of metal, usually steel, while others are molded of asbestos or other compounds. Some are perforated for ventilation, some tightly enclosed, and others fully open at the top. They make addition of wires or cables, over the route they take, a very simple procedure.

504. Pull boxes are used when a long conduit run contains more than the allowable number of bends. Pull boxes should be sufficiently long to allow for adequate radius of bending of the cable. When individual boxes cannot be provided for each conduit, a common box is used with barriers or other means to separate the cables of each pipe.

505. Vertical conduit runs should have provisions for the adequate support of the cables to prevent undue strain on connections and to avoid the possibility of a heavy conductor slipping through the insulation.

Vertical cables are generally supported by approved clamping devices constructed of or employing insulating wedges inserted in the ends of the conduits or by inserting junction boxes in which supports of approved type are installed.

National Electric Code Recommendations for the support of wires in vertical conduit runs are as follows:

Size of Wire	Max Interval between Supports, Ft
No. 14–No. 0	100
No. 0–No. 0000	80
No. 0000–350,000 cir mils	60
350,001–500,000 cir mils	50
500,001–750,000 cir mils	40
Above 750,000 cir mils	35

506. Moisture in power and control conduits and boxes is often experienced in stations where moisture-laden air from warm rooms travels through the conduits to relatively colder locations.

Adequate ventilation will usually overcome this trouble, but, if it is not possible to remedy it, cable should be lead-covered.

507. Spare conduits from control points and power-distribution boards to various plant sections are relatively inexpensive to install when the plant is constructed. Experience has shown that such facilities can be of great benefit when changes are made in control or additional circuits are needed later.

508. Effective grounding of all control conduits is extremely important, and care should be exercised in making all joints and connections to boxes, etc., so that the continuity of the circuit to the ground connection may not be broken. All conduits should be installed in accordance with the latest rules of the National Board of Fire Underwriters.

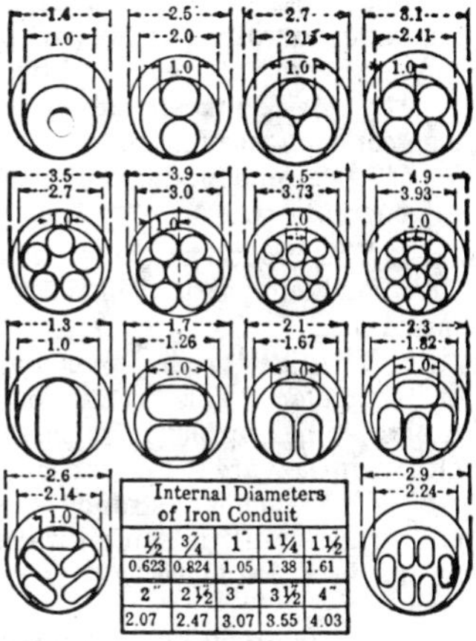

509. Conduits of ample size should be provided for all cables. Figure 10-177 shows the recommended size of conduit necessary for any given number of conductors of any size.

Internal Diameters of Iron Conduit

1/2	3/4	1"	1 1/4	1 1/2
0.623	0.824	1.05	1.38	1.61
2"	2 1/2	3"	3 1/2	4"
2.07	2.47	3.07	3.55	4.03

FIG. 10-177. Conduit sizes for wires. Diagram shows diameter of a group of wires, and diameter of conduit, in terms of diameter of a single wire. Diameter of conduit is for runs of from 50 ft with three 90-deg bends to 150 ft with two 45-deg bends. For more difficult runs, increase to 115%; for less difficult, decrease to 87%.

Common conduits or ducts are not generally used to enclose two circuits, one of which is a reserve to the other, or a power circuit and its protective control wires.

510. Control conduits should be so located in the vicinity of high-voltage circuit breakers and elsewhere that there will be no likelihood of the control being affected by disturbances on the power circuits.

Proper precautions should be taken where iron control conduits pass close to leads carrying heavy alternating currents. Brass sections or short-circuiting copper straps may be used in such cases to avoid overheating due to magnetic losses.

511. Cables for station wiring should be selected to suit requirements. All conductors of cables for conduit or duct service should be stranded to facilitate installation. Bending of insulated cables should be on a radius not less than six times the over-all diameter of the cable.

The essential parts of a cable are:

1. The conductor.
2. The wall of insulating material.
3. The outer protective covering.

Based on the number of conductors, cables may be divided into single-conductor, duplex, concentric, and multiple-conductor cables consisting of 3, 4, or more conductors. Whether single- or multiple-conductor cables should be used for main conductors carrying a-c currents depends on the size, length of run, and whether they are lead-covered or not.

Single-conductor cables are obtainable in sizes up to 3,500,000 cmils, and **3-conductor cables** are obtainable in sizes up to 850,000 cmils.

Cables for indoor use in sizes above 1,000,000 cmils are usually rope core or segmental.

The use of single-conductor cables in the larger sizes (above 1,500,000 cmils), particularly if installed in conduits or ducts, should be avoided if possible. A number of single-conductor cables of smaller sizes connected in parallel will result, generally, in a more efficient and more economical installation.

For circuits of large current-carrying capacities, such as generator connections, where an excessive number of parallel cables would be required, the present trend is toward the use of metal-clad buses in place of cables.

Multiple-conductor cables are generally used for control and instrument wiring. The common sizes of control wiring are No. 12 and No. 10 stranded conductors.

512. Various methods of stranding have been developed, and **the conductor shapes** in general use are:

1. **The "round-shape" conductor,** which has a cross section substantially circular. This conductor is used in both single- and multiple-conductor cables with all types of insulation.

2. **The "sector-shape" conductor,** which has a cross section substantially elliptical or triangular. This conductor is used only in multiple-conductor, paper-insulated cables. These cables have the advantages of lower cost and smaller over-all diameter, although they are stiffer and harder to splice.

3. **The "compact-strand" conductor,** in which the individual wires in all the layers are stranded in the same direction and then rolled to a predetermined shape. This results in a conductor with a very smooth configuration and a slightly smaller diameter.

4. **The "annular" conductor** is generally used for sizes above 1,000,000 cmils. At high currents the "skin effect" becomes appreciable, and annular conductors have a tendency to reduce current losses due to skin effect. In the more common construction, the annular conductor is hemp-cored.

5. **The "segmental" conductor** consists of groups of strands isolated from each other and acting as parallel conductors in carrying the current. This type of construction tends to reduce the losses due to skin effect, and segmental cables are rapidly replacing the rope-cored cables.

513. The principal materials used for cable insulation are:

1. Rubber compound.
2. Saturated paper.
3. Varnished cambric.
4. Synthetic compounds.

514. Rubber insulation is commonly used on power cables for voltages up to 2,500 and on control cables. Rubber insulation has a number of inherent advantages such as flexibility, moisture resistance, and ease of handling. The principal disadvantages which have limited its application are:

1. Rubber tends to age relatively rapidly at temperatures at which other insulations do not deteriorate.

2. Rubber is more quickly damaged at higher voltages by ionization and corona.

3. Rubber is affected by oils and will deteriorate rapidly when in contact with them.

For important work and high ambient temperatures, rubber of better grade than standard "Code" rubber is generally used. Some of the more recent special rubber compounds are designed for higher temperatures or higher voltages than the older standard compounds.

515. Ozone-resisting rubber compounds such as Kerite are available and have been used for voltages up to 13,800. Being moistureproof, they are generally installed without lead, and cables with such insulation have been used for such purposes as generator cables in ducts. Where moisture conditions are very severe, however, it may be desirable to use a protective lead covering. Such insulation is sometimes used for control wiring and auxiliary supply because of its long life. When used on the higher voltages, such as 13,800, it is supplied with a metallic or graphite-ground shield over the insulation to equalize the potential stresses and avoid the effects of corona. The ground shields should be grounded.

516. Saturated paper makes an excellent insulation and is commonly used on higher voltages such as 6,600 and up. Paper cable, because of its hygroscopic nature, is always lead-covered, and its ends should be terminated in sealed potheads to exclude moisture. Figure 10-178 shows a 3-conductor pothead of the compound-filled type with gasketed joints at the top and bottom of the porcelain insulators. Figure 10-179 shows a single-conductor pothead of the "Soldertite" construction, which requires no gaskets, is therefore much less subject to leaks. It is suitable for compound, oil or gas filling. The top and bottom metal fittings are soldered directly to the porcelain by a special procedure.

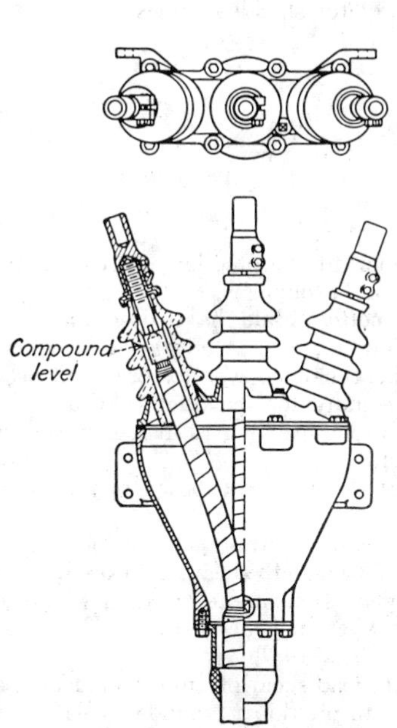

Fig. 10-178. Three-conductor cable pothead. (*G & W Electric Specialty Co.*)

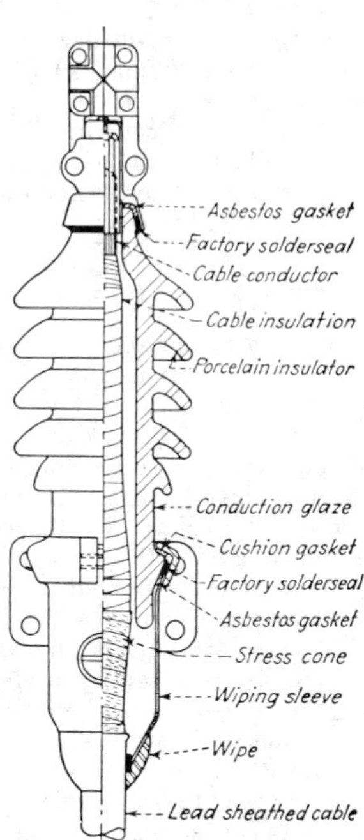

Fig. 10-179. Single-conductor cable pothead of Soldertite design. (*G & W Electric Specialty Co.*)

Large single-conductor lead-covered cables carrying alternating currents should be specially treated to avoid induced sheath currents which would reduce the cable rating. This may be accomplished by grounding the lead sheath of each conductor at one point only. If the runs are long, it may be necessary to install watertight insulating joints in the lead sheath and ground each section of the lead. In very long runs the sheaths of the various conductors of a polyphase circuit may be sectionalized and cross-bonded with each other to form transpositions of lead-sheath circuits so that the induced sheath voltages neutralize each other.

517. Synthetic-compound insulators are being used to some extent for low-voltage work. They appear to have many of the desirable characteristics of rubber with less of its drawbacks. Different compounds have somewhat different characteristics, and choice should be based upon conditions. Polyvinyl chloride as a primary conductor insulation is enjoying considerable popularity.

518. The current-carrying capacity of insulated cables in station work is limited generally by the maximum temperature which the cable insulation will withstand. The actual maximum safe continuous-current load for any given cable is affected by the temperature of the surrounding medium and the rate of radiation. This current value is greater with direct than with alternating current and decreases with increasing frequency.

The **allowable temperature** of insulated cables for the three most important kinds of insulation is as follows:

$T = 90 - E$ for paper (not over 85°C or under 65°C)
$T = 90 - 1\frac{1}{2}E$ for high-temperature varnished cambric—all cable subsequent to 1939 (not over 85°C or under 70°C)
$T = 75 - E$ for standard varnished cambric manufactured prior to 1939
$T =$ for rubber [see NEC Code (not under 60°C)]

where T = temperature in degrees centigrade at surface of conductor and E = rms operating voltage to ground, kV, for single-conductor shielded cables and line to line for belted cables.

519. The protective coverings in general use are:

1. Rubber-filled cotton or synthetic tape.
2. Flame-resisting asbestos and impregnated cotton braid.
3. Weatherproof braid.
4. Lead sheath.
5. Jackets.
6. Armors.

Rubber-filled cotton and **synthetic tapes** are used for protection against mechanical damage and for insulation of metallic sheaths from "ground."

Flame-resisting asbestos and **impregnated cotton braid** have been found to be very good protection of the cable and wire insulation in case of station or apparatus fires. Some users omit the cotton braid and paint the outer asbestos braid with sodium silicate white mineral finish. In such cases the individual conductors of a multiconductor cable are insulated by some users with ozone-resisting rubber, covered with a cotton braid, treated with a flame-resistant finish.

Weatherproof braid is used on wires and cables that are exposed to moisture and where fire hazards are not too important.

Lead sheaths are used on all paper-insulated cables and on other cables where the exposure to moisture is severe or where a metallic "ground" sheath is desired.

Jackets of rubber or synthetic compounds such as **Neoprene** are used where severe exposures to moisture, acids, etc., are present or where it is desired to insulate the lead sheath from the conduit or trough.

Armors are used to protect the cable surface or lead sheath from mechanical damage when exposed or when installed under water and to provide a grounded sheath to nonleaded cables. The armor may also give some protection against fire damage.

High-pressure pipe-type cable systems consist of cables installed in a steel pipe and subjected to a high pressure by the surrounding medium. The surrounding medium can be either oil or inert gas and is kept at a pressure of approximately 200 lb/in^2.

When oil is used, it can normally be fed from one pumping station or, in the case of gas, from one bank of cylinders. Continuity of internal pressure is required.

High-pressure pipe systems are used for transfer of large blocks of power, generally at high voltage. In station design they have been used for generator connections in special cases where their cost can be justified.

520. Station Lighting. It is important to provide sufficient light in all parts of a station for all needs, under normal conditions, and enough light to enable one to find his way about safely and perform the most essential duties, under emergencies. General illumination levels are increasing and are found economic. Light-colored finishes of walls and equipment aid in illumination. Illumination of switchboards and control boards should be arranged to avoid glare from the instrument faces. Totally indirect lighting is a most effective means. Some instruments are now available with a concealed light within the case to overcome the effect of glare, and special nonreflecting glass is available in some instruments to avoid glare from external light sources.

Lighting circuits should be sufficiently sectionalized and diversified so that the loss of a circuit will not leave a room or area in darkness.

521. Emergency-lighting circuits and sources are highly essential. In some cases a portion of the lighting is taken from a source entirely separate from the normal station supply. Selected circuits may be equipped with automatic throw-over switches to connect them to the station battery in case of outage of the normal supply. Such a supply can be used for a limited time only, or the battery will be drawn down dangerously low. In some stations low-voltage batteries with trickle chargers have been installed and feed a few low-voltage lights in the vicinity. Portable batteries and lights are useful to supplement the installed emergency lights but, like all batteries, have a limited life. Perhaps the best emergency source is a small gas-engine plant with capacity to supply the selected emergency-lighting circuits. Where a reliable water pressure is available, water-turbine generator sets are available in small sizes for emergency lighting. A plentiful supply of flashlights should always be at hand, but these are not adequate for operating or repair purposes, and more adequate emergency lighting lighting is essential.

GROUNDING SYSTEM

522. Grounding at stations is highly important and has the following principal functions:

a. Providing the ground connection for grounded neutral systems.

b. Providing the discharge path for lightning arresters, gaps, and similar devices.

c. Ensuring that non-current-carrying parts, such as equipment frames, are always safely at ground potential even though insulation fails.

d. Providing a means of positively discharging and deenergizing feeders or equipment before proceeding with maintenance on them.

A substantial and adequate ground that will not burn off or permit dangerous rise in voltage under abnormal conditions is essential. Extensive damage and dangerous conditions have arisen when inadequate grounds have been provided. Multiple grounds and multiple connections to them are usually desirable to insure ground protection, even though one ground or connection opens owing to burn-off or other condition such as high resistance.

523. The ground connection may be obtained most readily by connection to substantial water pipes where these are available and where regulations do not prohibit. The pipes should be of adequate size to prevent burn-off under fault conditions, and the connection should be preferably on the **street side** of the meter to prevent ground currents from flowing through the meter. While welding the copper ground connection to the pipe may be done under favorable circumstances, the most usual method is by bolted ground clamps. The pipe should be thoroughly cleaned before applying the clamp, and the clamp must be tight. Copper sulfate is sometimes used in preparing the pipe, and the clamp and connection should be thoroughly covered with asphaltum or similar substance to exclude the air and moisture, particularly when the connection is buried in the ground.

524. Ground rods or plates are used where water pipes are not available. The number, size, and arrangement are affected by ground resistance, amount of possible fault current, and potential of the lines. In some cases a counterpoise may be essential. In high-ground-resistance areas, the soil may require treatment, as by salt, to obtain and maintain low-resistance connections. Particular care must be exercised in obtaining the ground for high-voltage stations in high-resistance ground areas to insure that the differences in potential on the surface of the ground may not become dangerous to a man walking on the ground in the neighborhood of the station or ground rods at the time of discharge of a lightning arrester or flashover of an insulator.

525. The frames of electrical equipment should be thoroughly tied in to the ground system, both for safety to personnel and to prevent further damage to equipment in case of failure of insulation of the equipment.

526. Lightning arresters should have the ground connection substantially tied to ground and the frames of the equipment they are protecting. The connection should go as direct to earth as possible, with a minimum of bends or turns.

527. Normally live parts should be grounded before working on them unless proper tools are provided and provisions made for working alive. In some large stations provisions are made for applying the **initial ground** through an adequate circuit breaker, later to be supplemented or replaced by a bolted or clamped working ground. This is particularly important in the case of feeders supplying a low-voltage network where back feed from the network may exist and it is necessary to force out network switches or blow network switch fuses by grounding at the station. In some cases this is accomplished by double-throw disconnects between the feeder breaker and the bus, in which case the disconnects are thrown from the bus to a ground connection while the feeder breaker is open, and the feeder breaker is then closed, thereby connecting the feeder to ground. In other cases a grounding bus and breaker are provided, which may be connected through disconnects to any feeder, and the grounding breaker then closed. In other cases a nonautomatic breaker is provided for each feeder which, through disconnects, is used to isolate, ground, or connect the feeder to a test bus.

When grounding is applied by hand rather than through a breaker, extreme care is essential before applying the ground. Interlocks, such as a mechanical or key sequence, are often provided to ensure that sources of power are disconnected. Test facilities are often provided to test the circuit before applying the ground. These may consist of neon lights or a potential transformer with lights across its secondary or grounding cable with protective resistor and fuse in series. Such checks are negative, and it is recommended that the testing equipment be checked on a known live source before and after testing the circuit in question in order to check the continuity of the testing lamp or stick.

FIRE PROTECTION

528. Fire Protection. Practically all stations contain some combustible material such as organic insulation or oil. Combustible materials should be restricted to those necessary as part of the electrical equipment. Large and important stations should be sectionalized by fire walls and barriers to reduce risk and to avoid spreading of a fire from one section or unit to another within the station area.

529. The amount and kind of fire-extinguishing equipment depend upon the type and extent of the risk and the availability of other fire-fighting means, such as municipal fire departments. Even when the latter are available, equipment should be at hand for immediately fighting small fires without calling on outside help.

530. Sand is useful in forming dikes to prevent the spread of oil and in smothering small fires, as an oil fire on the floor.

531. Carbon tetrachloride is an effective extinguisher for small fires and is a good insulator; it tends to dissolve varnishes. It is available in various portable extinguishers. The fumes, given off rapidly under heat, are poisonous and great care must be exercised, particularly in enclosures, to avoid the danger of breathing them. Carbon tetrachloride is most suitable for small fires.

532. Carbon dioxide is used with rotating machines or enclosed or partially enclosed compartments. When used on open fires the gas must be discharged close to the fire and at its base. It may be obtained in portable extinguishers; or permanently installed units and piping may be made, particularly for totally enclosed generators. While its gases are generally considered nonpoisonous, it is suffocating in nature.

533. Water mixtures with foam-forming ingredients are particularly serviceable for extinguishing a burning-oil surface, particularly when they can be flowed over a stationary or semistationary oil surface. Foam is about as conducting as water. It may be obtained in self-contained portable extinguishers, either of the hand type or larger on wheels or, for large applications, obtained as separate ingredients and mixed with water as used.

534. Dry-powder extinguishers using finely powdered boric acid under pressure of CO_2 have been developed in recent years. They are effective on oil and insulation fires. The powder is nonconducting and may be used directly on live parts. It is also nonpoisonous.

535. Water alone is the universal extinguisher but must be used with care around electrical equipment. It is probably the most useful for large fires. Enclosed or semienclosed air-cooled generators are often built with spray rings, to which a hose may be connected in order to spread water around the end connections and into the air gap. Steam is preferred for such use by some engineers but does not have the cooling effect of water. Water will effectively extinguish oil fires, particularly when applied in a finely divided spray at high pressure. When so used it will extinguish fires in running or falling oil. Permanent systems have been installed over and around oil-filled transformers, regulators, etc., and over and around steam piping and oil lines of high-temperature steam turbines where a leak or failure of an oil line might result in a fire. The piping and installation of a permanently installed system involve a fair amount of expense and may require a special pump to maintain proper pressure at the nozzles for most effective use. Special spray nozzles of various types are available for use with fire hoses, or even the standard fire-hose nozzle is quite effective in directing a stream of water at the base of the fire.

536. Outdoor transformers, switches, and regulators can often be separated sufficiently to prevent the spread of fire, particularly if provided with dikes or pits to prevent the escaped oil from spreading.

Although fires in oil-insulated transformers are rare, the possibility should not be ignored. The amount and kind of fire protection are determined by the location and risk to other equipment and property. When transformers are close together or to other valuable property, fire walls are often installed with pits or retaining walls capable of holding the oil of the transformers. The pits may be drained to a tank and are recommended to be filled preferably with crushed stone to smother the ground fire. Water, especially if applied in a fine spray at high pressure, is an effective extinguisher. Permanent installations of piping and special spray nozzles have been installed in locations where the risk is high. Several types of spray nozzle are available for use on fire hose. The proper water pressure at the nozzle, depending on the design used, is necessary for most effective use. Water mixed with special ingredients to form a tenacious foam is effective, particularly when it can be flowed over a burning-oil surface. Carbon dioxide, if available in sufficient quantities and particularly if used in an enclosed space, as in a vault, is a good extinguishing medium but is not being used extensively for such fires, because of the cost involved and difficulties in supplying sufficient quantities into the fire.

If a fire has burned long enough to raise the temperature of the oil and metallic surroundings to a value above the flash point of the oil, it requires means for cooling the parts, since the oil will otherwise reignite as soon as it comes in contact with air even though it has been extinguished, as by CO_2. Water in some form is most effective in this regard.

537. Extinguishing of fires in air-cooled alternators is accomplished by means of (1) water, (2) steam, (3) carbon dioxide gas (CO_2). Water is, in present practice, in more general use; steam or CO_2, however, is favored by some engineers.

For either water or steam, jets are installed in the end bells of the generator. Water and steam require that the generator be disconnected from the bus and the exciter before they are admitted to the generator housing. Carbon dioxide is favored by many engineers because of its effectiveness, the fact that it is harmless to the insulation, and the ease with which it can be automatically discharged into the housing with no loss of time.

538. The required saturation of CO_2 is not less than 15.6% by volume for extinguishing flame, but percentages of 25 to 40 are actually employed in order to prevent reignition. The 15.6% saturation can be attained in 10 to 15 s, about 1 min being required for the higher percentages. In practice the discharge and suction ducts of air ventilating systems are closed by dampers, and then the CO_2 is admitted.

The required volume of CO_2 for initial discharge may be determined thus,

$$V = v \frac{P}{100 - P} \div k \tag{10-20}$$

where V = volume of CO_2, ft³ required, v = volume of generator air chamber, ft³ P = percentage of CO_2 desired. Losses from leakage require that the installed capacity be in excess of the theoretical, and for ducts up to 2,000-ft³ capacity k should be taken at 0.75; if in excess of 2,000 ft³, at 0.82.

Additional volume required to maintain the desired percentage on the basis of shutdown in ½ h approximates 150% of that for the initial discharge.

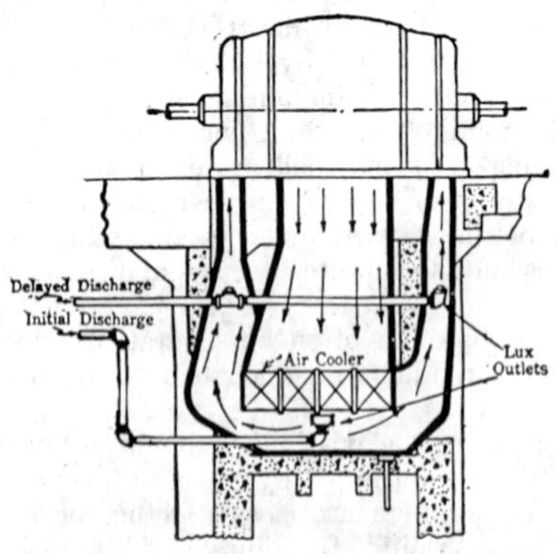

Fig. 10-180. Typical installation of CO_2 generator fire-extinguisher outlets.

One type of system for CO_2 protection as applied to air-cooled generators and other rotating electrical machines is shown in Fig. 10-180. At the instant of trouble, a large volume of liquid CO_2 is automatically discharged into the ventilating ducts. This immediately gasifies and renders the air inert. Smaller quantities are then released automatically at predetermined intervals to compensate for duct leakage and maintain the desired concentration during deceleration. Reignition is thus prevented, and heat radiation permitted.

Operation may be made automatic by means of thermostats in the machine ducts or by differential-relay protection or **manual** from control stations at the switchboard, the machine, or the CO_2 apparatus installation.

BIBLIOGRAPHY

539. References on Power-system Electrical Equipment

Reference Books

Wagner, C. F., and Evans, R. D. "Symmetrical Components"; New York, McGraw-Hill Book Company, 1933.

Dawes, C. L. "A Course in Electrical Engineering—Alternating Currents," 4th ed.; New York, McGraw-Hill Book Company, 1947.

"Electrical Transmission and Distribution Reference Book"; East Pittsburgh, Pa., Westinghouse Electric Corporation, 1950.

Dusenberry, H. S. "Direct Current Motor Manual"; New York, The Macmillan Company, 1950.

Dawes, C. L. "A Course in Electrical Engineering—Direct Currents," 4th ed.; New York, McGraw-Hill Book Company, 1952.

Powel, C. A. "Principles of Electric Utility Engineering"; New York, John Wiley & Sons, Inc., 1955.

Carr, C. C. "American Electricians' Handbook," 8th ed.; New York, McGraw-Hill Book Company, 1961.

Clark, F. M. "Insulating Materials for Design and Engineering Practice"; New York, John Wiley & Sons, Inc., 1962.

Beck, Edward. "Lightning Protection for Electric Systems"; New York, McGraw-Hill Book Company, 1954.

STATION DESIGN

Committee Report, Design of Switchhouses for Generating Stations and Transmission Substations; *Edison Elec. Inst. Serial Rept.* F5, June, 1938.

JONES, R. B. Single-phase Loads from a Three-phase Supply; *Elec. World,* Jan. 2, 1950, Vol. 133, p. 57.

AIEE Committee Report, Basic Structural Design for Transmission Substations Including Light Metals; *Elec. Eng.*, April, 1952, Vol. 71, pp. 344–350.

AIEE Committee Report, Guide to Safety Considerations in Design of Substations; *Trans. AIEE,* June, 1954, Vol. 73, Power Apparatus and Systems, pp. 633–635.

HOOPES, J. E. Modernize Substation Grounding Practice; *Elec. World,* Aug. 25, 1958, pp. 66–67.

BELLASCHI, P. L. Rationalization of Electrical Clearances for Applications at Extra High Voltages, 230 KV to 460/500 KV (*Biblio.*); *Trans. AIEE,* 1959, Vol. 78, Power Apparatus and Systems, pp. 736–743.

CAMBIAS, S. Substation Squeezed to Fit Site, New Orleans; *Elec. World,* Sept. 19, 1960, pp. 95–96.

SWICEGOOD, H. L. Mobile Sub Services Unit Distribution Substations; *Elec. World,* Oct. 24, 1960, pp. 94–95.

SMITH, J. A. Determination of Economical Distribution Substation Size; *Trans. AIEE,* October, 1961, Vol. 79, Power Apparatus and Systems, pp. 663–670.

Committee Report, Design Standardization Methods and Techniques for Substation Facilities (*Biblio.*); *Trans. AIEE,* Oct. 1964, Vol. 83, Power Apparatus and Systems, pp. 1029–1034.

WAGNER, C. L., and OTHERS. Insulation Levels for Virginia Electric and Power Co. 500 KV Substation Equipment; *Trans. AIEE,* Mar. 1964, Vol. 83, Power Apparatus and Systems, pp. 236–241.

HIGBIE, R. M. New Distribution Substation Design Provides High Reliability and High Capacity in Minimum Space; *Power Eng.*, May, 1965, Vol. 69, pp. 59–61.

Committee Report, Minimum Electrical Clearances for Substations Based on Switching Surge Requirements, Second Interim Report; *Trans. IEEE,* June, 1965, Vol. 84, Power Apparatus and Systems, pp. 532–535. (See first report in Vol. 82, December, 1963, pp. 1072–1076.)

WYLIE, C. J. 23 KV Switching Station Designed for Ultimate 2000 MW Plant, Many Lines; *Elec. World,* Sept. 6, 1965, pp. 50–53.

Substation Design Goes into High Gear, Flow Chart; *Elec. World,* Nov. 1, 1965, pp. 87–90.

SYSTEM PLANNING

BANES, A. J. Forced Air Cooled Isolated Phase 20 KV Bus Carries 20,000 Amp. Combined Output at 20 KV; *Elec. World,* Jan. 10, 1966, pp. 60–61.

MILTON, R. M., and OTHERS. TVA's 500 KV System, Stepdown Substation Design; *Trans. AIEE,* January, 1966, Vol. 85, Power Apparatus and Systems, pp. 36–46.

WHITEHEAD, L. E. Building Block Methods Help Design 138 KV Industrial Substations, Quick Estimate Tools for Costs; *Power Eng.*, February, 1966, pp. 38–41.

MCCLOSKA, F. W., and MUSSELMAN, F. L. 330 KV Outdoor Station for Atomic Energy Commission; *Trans. AIEE,* June, 1954, Vol. 73, Power Apparatus and Systems, pp. 628–633.

SEELYE, H. P. What Reserves Should Be Provided in a Modern Power System; *Elec. Light and Power,* April, 1950, Vol. 28, pp. 116–168.

CIRCUIT BREAKERS

HILL, A. W. Circuit Breakers, Oil, Air, Gas; *Westinghouse Eng.*, May, 1960, Vol. 20, pp. 82–85.

General article, High Current Vacuum Interrupter, A New Approach to Power Circuit Breaking; *Elec. Eng.*, January, 1962, Vol. 81, pp. 33–34.

LEE, T. H., and OTHERS. Development of Power Vacuum Interrupters; *Trans. AIEE,* February, 1963, Vol. 82, Power Apparatus and Systems, pp. 629–636.

Specification for A-C Circuit Breakers, *IEC Publ.* 56–7, 1963.

KANE, R. E., and COLCLASER, R. G. 69 KV SF_6 Common Tank Breaker Rated

5000 MVA; *Trans. AIEE,* December, 1963, Power Apparatus and Systems, pp. 1076–1082.

CALVINO, B. J., and FORMICA, A. Behavior of Circuit Breakers in Case of Fault in Circuit with High Natural Frequency, *CIGRE Paper* 131, 1964.

SHAW, A. B., and WHITTAKER, D. Effect of Gas Flow on Post-arc Gap Recovery; *Proc. Inst. Elec. Engrs.* (*London*), January, 1964, Vol. 111, pp. 193–202.

SCHNEIDER, J., and BALTENSPERGER, P. Circuit Breakers for 750 KV; *Brown Boveri Rev.,* January–February, 1964, Vol. 51, pp. 69–75.

BOLTON, E., EHRENBERG, A. C., HAMILTON, F. L., HAWKINS, A. G., MATRAVERS, F. P., and THOMAS, J. A. British Investigations of Short-line Fault Phenomena; *CIGRE,* 109, 1964.

STURGIS, B. K., and QUINN, G. C. How to Insure Right Low-voltage Circuit Protection; *Factory,* Apr. 4, 1964, Vol. 122, pp. 97–102.

COX, V. L. Low Voltage Power Circuit Breakers; *Mag. Standards,* May, 1964, p. 142.

COX, V. L. New Ratings for High-voltage Circuit Breakers; *Mag. Standards,* June, 1964, pp. 163–165.

HUMPAGE, W. D., and STOTT, B. Effect of Autoreclosing Circuit Breakers on Transient Stability in E.H.V. Systems; *Proc. Inst. Elec. Engrs.* (*London*), July, 1964, Vol. 111, pp. 1287–1298.

VEPCO Project to Use 500 KV SF_6: Power Circuit Breakers; *Elec. World,* July 6, 1964, p. 38.

Switching of EHV Circuits; *Trans. AIEE,* December, 1964, Vol. 83, Power Apparatus and Systems, pp. 1187–1223.

HEDMAN, D. E., and OTHERS. Switching of E.H.V. Circuits, Surge Reduction with Circuit Breaker Resistors; *Trans. AIEE,* December, 1964, Vol. 83, Power Apparatus and Systems, pp. 1196–1205.

VAN SICKLE, R. C., and OTHERS. Modular SF_6 Circuit Breaker Design for EHV; *Westinghouse Engr.,* March, 1966, pp. 44–49.

FITZGERALD, J. P. Switching the Magnetizing Current of Large 345 KV Transformers with Double Break Air Switches; *Trans. AIEE,* October, 1965, Vol. 84, Power Apparatus and Systems, pp. 902–906.

BRIGGS, E. E., HAMBRICK, E. D., and UMPHREY, D. M. Hydraulic Operating Mechanisms for High Capacity Circuit Breakers; *Trans. AIEE,* October, 1953, Vol. 72, Power Apparatus and Systems, pp. 874–879.

LINGAL, H. J., and OWENS, J. B. New High-voltage Outdoor Load Interrupter Switch; *Trans. AIEE,* April, 1953, Vol. 72, Power Apparatus and Systems, pp. 324–327.

LEEDS, W. M., and EASLEY, G. J. New Milestone in Circuit Breaker Interrupting Capacity 25 Million KVA at 330 KV; *Trans. AIEE,* April, 1954, Vol. 73, Power Apparatus and Systems, pp. 304–310.

HILL, A. W. Progress in Power Circuit Breaker Development; *Westinghouse Eng.,* May, 1954, Vol. 14, pp. 106–110.

LIGHTNING PROTECTION

GROSS, I. W., LEVESCONTE, L. B., and DILLARD, J. K. Lightning Protection in Extra-high-voltage Stations—Influence of Multiple Circuits; *Trans. AIEE,* October, 1953, Vol. 72, Power Apparatus and Systems, pp. 882–888.

GROSS, I. W., GRISCOM, S. B., CLAYTON, J. M., and PRICE, W. S. High Voltage Impulse Tests in Substations; *Trans. AIEE,* April, 1954, Vol. 73, Power Apparatus and Systems, pp. 210–220.

WAGNER, C. F., MCCANN, G. D., and LEAR, C. M. Shielding of Substations; *Trans. AIEE,* February, 1942, Vol. 61, pp. 96–100.

RUDGE, W. J., JR., COX, V. L., and HUNTER, F. M. Lightning Protection of Switchgear Unit Substations or A-C Rotating Machines Closely Connected to Overhead Lines; *Gen. Elec. Rev.,* October, 1943, Vol. 46, pp. 551–556.

Code for Protection against Lightning, *NBS Handbook* H40, 1945, pp. 1–99.

WITZKE, R. L., and BLISS, T. J. Coordination of Lightning Arrester Location with Transformer Insulation Level; *Trans. AIEE,* 1950, Vol. 69, Pt. II, pp. 964–975.

CARPENTER, T. J., JOHNSON, I. B., and SALINE, L. E. Evaluation of Lightning

Arrester Lead Length and Separation in Coordinated Protection of Apparatus against Lightning; *Trans. AIEE,* 1950, Vol. 69, Pt. II, pp. 933–944.

Gross, I. W., Bliss, T. J., and Dillard, J. K. Lightning Protection in Extra-high Voltage Stations—Analysis, Anacom Study, and Results; *Trans. AIEE,* January, 1952, Vol. 71, Pt. III, pp. 482–492.

Hileman, A. R., Armstrong, H. R., and Dillard, J. K. Lightning Protection in a 120-KV Station—Field and Laboratory Studies; *Trans. AIEE,* 1954, Vol. 73, Pt. III-B, pp. 1143–1152.

Hileman, A. R., Armstrong, H. R., and Ferguson, R. W. Lightning Protection in a 24-KV Station—Field and Laboratory Studies; *Trans. AIEE,* December, 1955, Vol. 74, Pt. III, pp. 1127–1136.

Sporn, Philip, and Gross, I. W. A Quarter Century of Experience in Insulation Coordination: Basic Philosophy, Application and Operating Experience on the AG&E System; *Trans. AIEE,* April, 1957, Vol. 76, Pt. III, pp. 58–73.

AIEE Committee Report, Report on Lightning Arrester Applications for Stations and Substations; *Trans. AIEE,* August, 1957, Vol. 76, Pt. III, pp. 614–627.

Clayton, J. M., and Powell, R. W. Application of Arresters for Complete Lightning Protection of Substations; *Trans. AIEE,* 1958, Vol. 77, Pt. III, pp. 1608–1614.

McElroy, A. J., and Porter, R. M. Digital Computer Calculation of Transients in Electric Networks; *Trans. AIEE,* TP 62–1932.

AIEE Committee Report, A Report on Performance Characteristics of Lightning Arresters; *Trans. AIEE,* April, 1959, Vol. 78, Pt. III-A, pp. 44–46.

Kalb, J. W., and Yost, A. G. New Current Limiting Gap Extends Valve Type Lightning-Arrester Performance; *Trans. AIEE,* 1959, Vol. 78.

Breuer, G. D., Hopkinson, R. H., Johnson, I. B., and Schultz, A. J. Arrester Protection of High-voltage Stations against Lightning; *Trans. AIEE,* 1960, Vol. 79, Pt. III, Power Apparatus and Systems, pp. 414–423. (August, 1960, section.)

AIEE Working Group of Protective Devices Committee, Simplified Method for Determining Permissible Separation between Arresters and Transformers; *Trans. AIEE,* November, 1962, Vol. S82, 63–229, pp. 35–57.

Central Station Engineers, "Electrical Transmission and Distribution Reference Book"; Westinghouse Electric Corporation, East Pittsburgh, Pa.

Koerber, A. R. Transformer Protected Three BIL—Steps Down; *Elec. World,* Oct. 21, 1963, p. 60.

Burton, T. G. Public Service Co. of Indiana Trying Pipe Gap Arresters on 230 KV; *Power Eng.,* April, 1964, pp. 41–42.

Yost, A. G., and Others. Transmission-line Discharge Testing for Station and Intermediate Lightning Arresters; *Trans. AIEE,* January, 1965, Vol. 84, Power Apparatus and Systems, pp. 79–87.

Committee Report, Tests on Pressure Relief Devices for Station and Intermediate Lightning Arresters; *Trans. AIEE,* February, 1965, Vol. 84, Power Apparatus and Systems, pp. 163–172.

EHV Boom Sparks Interest in Energy Absorption; *Elec. World,* Feb. 22, 1965, p. 101.

Griscom, S. B., and Others. Five Year Field Investigation of Lightning Effects on Transmission Lines; *Trans. IEEE,* April, 1965, Vol. 84, pp. 257–280.

Lightning Protection Code, 1965, USASI Standard C5.1 (NFPA 78).

Generators

Edison Electric Institute Committee Report, Field Testing of Generator Insulation; *Trans. AIEE,* December, 1941, Vol. 60, pp. 1003–1011.

Phillips, A. H., Lambert, W. H., and Pattison, D. R. Excitation Improvement—Electronic Excitation and Regulation of Electric Generators as Compared to Conventional Methods; *Elec. Eng.,* June, 1950, Vol. 69, pp. 518–520.

Laffoon, C. M. Hollow Conductor Cooling Boosts Generator Rating 50%; *Elec. Light and Power,* November, 1951, Vol. 29, pp. 100–101; also *Westinghouse Engr.,* November, 1951, Vol. 11, pp. 170–172.

Alke, R. J. D-c Over-potential Testing Experience on High-voltage Generators; *Trans. AIEE,* August, 1952, Vol. 71, Power Apparatus and System, pp. 567–570.

SEELYE, H. P., and BROWN, W. W. Economy of Large Generating Units; *Combustion*, May, 1954, Vol. 25, pp. 45–48.

HOLLEY, C. H., and TAYLOR, H. D. Direct Cooling of Turbine-generator Field Windings; *Trans. AIEE*, June, 1954, Vol. 73, Power Apparatus and System, pp. 542–550.

BECKWITH, S., and ROSENBERG, L. T. New Fully Supercharged Generator; *Trans. AIEE*, June, 1954, Vol. 73, Power Apparatus and System, pp. 477–482.

WALKER, J. H. Generator/Motor Problems in Pumped Storage Installations; *Proc. Inst. Elec. Engrs.* (*London*), April, 1960, Vol. 107, pp. 157–165.

GANTHER, C. E., and MALLET, A. A. Operation of Generating Units during Serious System Disturbances; *Elec. Eng.*, October, 1963, Vol. 82, pp. 626–628.

WIEDEMANN, E. Hydro Electric Generators with Liquid Cooling in Stator and Rotor; *Brown Boveri Rev.*, May 5, 1964, Vol. 51, pp. 267–273.

BAUDRY, R. A., and KING, E. I. Improved Cooling Increases Generator Capabilities; *Elec. World*, Nov. 2, 1964, pp. 70–71.

KING, E. I., and BATCHEHEL, J. W. Effects of Unbalanced Currents on Turbine Generators; *Trans. AIEE*, February, 1965, Vol. 84, Power Apparatus and Systems, pp. 121–125.

BINNS, K. J. Predetermination of the No-load Magnetization Characteristics of Large Turbogenerators; *Proc. Inst. Elec. Engrs.* (*London*), Vol. 112, April, 1965, pp. 720–730.

OLMSTED, L. M. Standby Units Guard Shut Down, Speed Startup; *Elec. World*, Jan. 24, 1966, Vol. 165, pp. 73–74.

GENERATOR EXCITATION

McCLURE, J. B., WHITTLESEY, S. I., and HARTMAN, M. E. Modern Excitation Systems for Large Synchronous Machines; *Trans. AIEE*, 1946, Vol. 65.

PORTER, F. M., and KINGHORN, J. H. The Development of Modern Excitation Systems for Synchronous Condensers and Generator; *Trans. AIEE*, 1946, Vol. 65.

STORM, H. F. Static Magnetic Exciter for Synchronous Alternators; *Elec. Eng.*, December, 1951, Vol. 70, pp. 1084–1088.

GRISCOM, S. B. Operation of Turbine Generators at Leading Power Factors; *Elec. Light and Power*, December, 1951, Vol. 29, pp. 77–80.

MORGAN, W. A. Power System Stability Criteria for Design; *Elec. Eng.*, August, 1952, Vol. 71, pp. 740–744.

ANDERSON, H. C., SIMMONS, H. O., and WOODROW, C. A. System Stability Limitations and Generator Loading; *Trans. AIEE*, June, 1953, Vol. 72, Power Apparatus and Systems, pp. 406–417.

CARLETON, J. T., BOBO, P. O., and HORTON, W. F. New Regulator and Excitation System; *Trans. AIEE*, April, 1953, Vol. 72, Power Apparatus and Systems, pp. 175–181.

HORN, M. E., and CUNNINGHAM, J. C. Transient Voltage and Current Requirements of Main Field Circuit Breakers for Synchronous Machine; *Trans. AIEE*, August, 1954, Vol. 73, Power Apparatus and Systems, pp. 894–900.

HEFFRON, W. G., JR. Simplified Approach to Steady-state Stability Limits; *Trans. AIEE*, February, 1954, Vol. 73, Power Apparatus and Systems, pp. 39–44.

EASTON, V. Excitation of Large Turbogenerators; *Proc. Inst. Elec. Engrs.* (*London*), May, 1964, Vol. 111, pp. 1040–1048.

HOOVER, D. B. Brushless Excitation System for Large A-C Generators; *Westinghouse Engr.*, September, 1964, pp. 141–145.

ALGER, P. L., and BARTON, R. S. Self Excited Synchronous Generator for Isolated Power Supply; *Trans. AIEE*, October, 1964, Power Apparatus and Systems, pp. 1002–1006.

WISZNIEWSKI, A., and SZYMANSKI, A. Automatic Field Suppression in Brushless Excited Turbo Alternators; *Proc. Inst. Elec. Engrs.* (*London*), December, 1965, Vol. 112, pp. 2341–2344.

DRYING ELECTRICAL MACHINERY

Committee Report, Reconditioning Flood Damaged Electrical Equipment, *Edison Elec. Inst. Serial Rept.* F13, December, 1938.

MARCROFT, H. C. Use of Dielectric-absorption Tests in Drying Out Large Generators; *Trans. AIEE*, February, 1945, Vol. 64, pp. 56–60.

SYSTEM PROTECTION

WERRES, C. O. 75,000 Ampere Tests on Current Transformers Used for Bus Differential Protection; *Gen. Elec. Rev.*, August, 1937, Vol. 40, pp. 380–382.

American Institute of Electrical Engineers—Edison Electric Institute Committee Report, Bus Protection; *Elec. Eng.*, May, 1939, Vol. 58, p. 206.

BANY, H. Relay and Circuit Breaker Protection for D. C. Machines; *Gen. Elec. Rev.*, August, 1940, Vol. 43, pp. 312–319.

CROSSMAN, G. C., LINDEMUTH, H. F., and WEBB, R. L. Loss-of-field Protection for Generators; *Trans. AIEE*, May, 1942, Vol. 61, pp. 261–266; discussion, p. 462.

BOSTWICK, M. A., and HARDER, E. L. Protection of Three-terminal Lines; *Westinghouse Eng.*, August, 1943, Vol. 3, pp. 76–79.

McCONNELL, A. J. A New Generator Differential Relay; *Trans. AIEE*, January, 1943, Vol. 62, pp. 11–13; discussion, p. 381.

WEBB, R. L., and MURRAY, C. S. Vibration Protection for Rotating Machinery, *Trans. AIEE*, 1944, Vol. 63, pp. 534–537.

HARDER, E. L., KLEMMER, E. H., and NEIDIG, R. E. Linear Couplers—Field Tests and Experience at York and Middletown, Pa.; *Trans. AIEE*, 1946, Vol. 65.

WARRINGTON, A. R. VAN C. Application of the Ohm and Mho Principles to Protective Relays; *Trans. AIEE*, 1946, Vol. 65.

AIEE Working Group. Protection of Powerhouse Auxiliaries; *Trans. AIEE*, 1946, Vol. 65.

MASON, C. R. New Loss-of-excitation Relay for Synchronous Machines; *Trans. AIEE*, 1949, Vol. 68, pp. 1240–1245.

BAUMAN, H. A., DRISCOLL, J. M., ONDERDONK, P. T., and WEBB, R. L. Protection of Turbine Generators and Boilers by Automatic Tripping; *Trans. AIEE*, December, 1953, Vol. 72, Power Apparatus and Systems, pp. 1248–1255.

Committee Report, Pilot Wire Circuits for Protective Relaying—Experience and Practice 1942–1950; *Trans. AIEE*, April, 1953, Vol. 72, Power Apparatus and Systems, pp. 331–336.

TREMAINE, R. L., and BLACKBURN, J. L. Loss-of-field Protection for Synchronous Machines; *Trans. AIEE*, August, 1954, Vol. 73, Power Apparatus and Systems, pp. 765–772.

BOSTWICK, M. A. Dependable Pilot Wire Relay Operation; *Trans. AIEE*, December, 1953, Vol. 72, Power Apparatus and Systems, pp. 1073–1076.

MATHEWS, C. A. Improved Transformer Differential Relay; *Trans. AIEE*, June, 1954, Vol. 73, Power Apparatus and Systems, pp. 645–649.

GUENZEL, E. L., and MORRIS, W. T. Distribution Circuit Protection; December, 1959, AIEE Power Apparatus and Systems, Vol. 78, pp. 1064–1070.

DALZIEL, C. F., and STEINBACK, E. W. Underfrequency Protection of Power Systems for System Relief; December, 1959, AIEE Power Apparatus and Systems, Vol. 78, pp. 1227–1238.

ADAMSON, C., and TALKHAN, E. A. Selection of Relaying Qualities for Differential Feeder Protection; *Proc. Inst. Elec. Engrs.* (*London*), February, 1960, Vol. 107, pp. 37–47.

Abstracts of AIEE papers, Advances in Relay, Substation Practice Reported; *Elec. World*, Sept. 5, 1960, pp. 39–41.

RUSHTON, J. Fundamental Characteristics of Pilot Wire Differential Protection System; *Proc. Inst. Elec. Engrs.* (*London*), Vol. 108, October, 1961, pp. 409–420.

JOHNSON, I. B. and OTHERS. Phase to Phase Switching Surges on Line Energization; August, 1962, *Trans. AIEE*, Power Apparatus and Systems, Vol. 81, pp. 298–301.

ROCKEFELLER, G. D. Selective Load Shedding; *Westinghouse Engr.*, November, 1963, pp. 187–189.

IEEE Committee Report, Protection of Power Transformers; December, 1963, *Trans. IEEE*, Power Apparatus and Systems, Vol. 82, pp. 1040–1044.

KNABLE, A. H. Grounding and Accompanying Relaying for Medium and Low Voltage Systems; *IEEE Spec. Publ.* T-158, 1963, pp. 67–80.

BLACKBURN, J. L. Protective Relaying for EHV Systems; *Westinghouse Engr.*, May, 1964, pp. 72–77.

IEEE Committee Report, Application of Protective Relays and Devices to Distribution Circuits; October, 1964, *Trans. IEEE*, Power Apparatus and Systems, Vol. 83, pp. 1034–1042.

IEEE Committee Report, Supplement to Recent Practices and Trends in Protective Relaying; October, 1964, *Trans. IEEE,* Power Apparatus and Systems, Vol. 83, pp. 1064–1069.

Sutton, H. J. Resumé of North American Protective Relay Practices and Trends; *CIGRE Paper* 312, 1964.

Penescu, C. Universal Characteristic Transistorized Distance Relay; *CIGRE Paper* 317, 1964.

Brown, R. D., and McClymont, K. R. Power Swing Relay for Predicting Generators Instability; *Trans. AIEE,* March, 1965, Power Apparatus and Systems, pp. 219–224.

Merritt, M. S. Protection of Large Steam Turbine–Generator Units on T.V.A. Systems; *Trans. AIEE,* April, 1965, Vol. 84, pp. 820–826.

Cory, B. J., and Rai, G. B. Simple Phase Detection Relay for Distribution Networks; *Proc. Inst. Elec. Engrs.* (*London*), May, 1965, Vol. 112, pp. 995–999.

Caldwell, J. E. Protection of a Distribution Network; *Proc. Inst. Elec. Engrs.* (*London*), June, 1965, Vol. 112, pp. 1179–1180.

Hunpage, W. D., and Sabberwal, S. P. Developments in Phase-comparison Techniques for Distance Protection; *Proc. Inst. Elec. Engrs.* (*London*), July, 1965, Vol. 112, pp. 1383–1394.

Patton, W. D. Selecting the Right Breaker for a Motor Circuit; *Plant Eng.,* September, 1965, Vol. 19, pp. 120–122.

Petzinger, A. J. Today's Power Systems Require Modern Relay Protection; *Westinghouse Engr.,* January, 1966, Vol. 26, pp. 12–14.

Committee Report, Ground Relaying Practices and Problems; *Trans. AIEE,* Vol. 85, Power Apparatus and Systems, 1965.

Method of Calculation

Kron, G. Tensorial Analysis of Integrated Transmission Systems; *Trans. AIEE,* October, 1952, Vol. 71, Power Apparatus and Systems, pp. 814–821.

Estwick, C. F. Real Power and Imaginary Power in A-C Circuits; *Trans. AIEE,* February, 1953, Vol. 71, Power Apparatus and Systems, pp. 27–35.

Grounding

Committee Report, Principles and Practices in Grounding; *Edison Elec. Inst. Serial Rept.* D9, October, 1936.

Bellashi, P. L. Impulse and 60-cycle Characteristics of Driven Grounds; *Trans. AIEE,* March, 1941, Vol. 60, pp. 123–128.

Johnson, A. A. Grounding Principles and Practice; Generator Neutral Grounding Devices; *Trans. AIEE,* March, 1945, Vol. 64, pp. 92–99.

AIEE Committee Report, Application Guide on Methods of Substation Grounding; *Trans. AIEE,* April, 1954, Vol. 73, Power Apparatus and Systems, pp. 271–275.

Mather, F. Earthing of Low and Medium Voltage Distribution Systems and Equipment; *Proc. Inst. Elec. Engrs.* (*London*), April, 1958, Vol. 105, pp. 97–106.

Kinyon, A. L. Earth Resistivity Measurements for Grounding Grids; *Trans. AIEE,* December, 1961, Vol. 80, Power Apparatus and Systems, pp. 795–800.

Shew, E. B., and Falck, K. J. Grounding Metalclad Switchgear for the Protection of Workmen; *Elec. Eng.,* August, 1963, Vol. 82, pp. 529–531.

Earthing of Consumers' Electrical Installations to Water Pipes; *Eng.,* April, 1964, Vol. 68, p. 148.

Earth Electrode Resistance with Particular Reference to Earth Electrode Systems Covering a Large Area; *Proc. Inst. Elec. Engrs.* (*London*), December, 1964, Vol. 111, pp. 2118–2130.

Jackson, L. A. Grounding of Electric Circuits on Water Pipes; *Eng.,* April, 1965, pp. 57–59.

Abstracts of various papers, Earth Loop Impedance Testing; *Eng.,* May 21, 1965, Vol. 219, p. 890.

Bus Construction

Schurig, O. R., and Sayre, M. F. Mechanical Stresses in Bus Bar Supports during Short Circuits; *Jour. AIEE,* April, 1925, Vol. 44, pp. 365–372.

Forbes, H. C., and Gorman, L. J. Skin Effect in Rectangular Conductors; *Elec. Eng.*, September, 1933.

Dwight, Herbert Bristol. Skin Effect and Proximity Effect in Tubular Conductors; *Trans. AIEE*, February, 1922, pp. 189–198.

Wagner, C. F. Current Distribution in Multi Conductor Single Phase Buses; *Elec. World*, Mar. 18, 1922, Vol. 79, No. 11.

Karman, T. von, and Rubach, T. Resistance of a Body Moving in a Fluid, "Paper in Mathematical Physics"; University of Göttingen, 1911.

Rayleigh, J. W. S. "Aeolian Tones"; New York, Cambridge University Press, 1920.

Thom, A. Experiments on Cylinders Oscillating in a Stream of Water; *Phil. Mag.*, July–December, 1931.

Temple, G., and Brickley, W. G. "Rayleigh's Principle"; New York, Oxford University Press, 1933.

Buchanan, W. B. Vibration Analysis; *Elec. Eng., Trans. AIEE*, 1934.

General Theory of Aerodynamic Instability, *Natl. Advisory Comm. Rept.* 496, 1935.

Milton, R. M., and Chambers, Fred. Behavior of High-voltage Busses and Insulators during Short Circuits; *Trans. AIEE*, August, 1955, Vol. 74, pp. 742–749.

Taylor, D. W., and Stuehler, C. M. Short Circuit Tests on 138 Kv Busses; *Trans. AIEE*, August, 1956, Vol. 75, pp. 739–747.

Power Switching Equipment—NEMA Standards, *Publ.* SG-6.

Con. Ed. Installs First Preassembled Switching Modules; *Elec. World*, Sept. 3, 1962, p. 48.

Curtis, T. E. Switch for Multiple-circuit Load Break Switching with a Common Interrupter; *Trans. IEEE*, December, 1964, Power Apparatus and Systems, Vol. 83, pp. 1161–1167.

Second Interim Committee Report, Minimum Electrical Clearances for Substation Based on Switching Surge Requirements; *Trans. IEEE*, May, 1965, Power Apparatus and Systems, Vol. 84.

Foti, A. Design and Application of EHV Disconnecting Switches; *Trans. AIEE*, October, 1965, Power Apparatus and Systems, Vol. 84, pp. 868–876.

Schurig, O. R., and Kuchni, H. P. Temperature Rise and Losses in Solid Structural Steel Exposed to the Magnetic Fields from A-c Conductors; *Jour. AIEE*, May, 1926, Vol. 45, pp. 446–453.

Higgins, T. J. Formulas for Calculating Short Circuit Stresses for Bus Supports for Rectangular Tubular Conductors; *Trans. AIEE*, August, 1942, Vol. 61, pp. 578–580.

Gerber, J. A. Bus Construction of Welded Wrought Iron Pipe; *Elec. World*, Sept. 18, 1943, Vol. 120, pp. 988–989.

Higgins, T. J. Formulas for Calculating Short Circuit Forces between Conductors of Structural Shape; *Trans. AIEE*, October, 1943, Vol. 62, pp. 659–663.

Station Auxiliaries

Working Group Report, Protection of Powerhouse Auxiliaries; *Trans. AIEE*, 1946, Vol. 65.

Dodds, G. B., and Marter, W. E. Reactance Relays Discriminate between Load Transfer Currents and Fault Currents on 2300 Volt Station Service Generator Bus; *Trans. AIEE*, December, 1955, Vol. 71, Power Apparatus and Systems, pp. 1124–1128.

Nelson, E. Characteristics, Operation and Advantages of Synchronous Motors; *Plant*, December, 1959, pp. 44–47; January, 1960, pp. 49–52.

Larinoff, M. W., and Others. Generator-driven Boiler Feeder Pumps for Large Stations Are Here to Stay; *Power*, April, 1960, pp. 95–97.

Hannakam, L., and Concordia, C. Stability Limits of Synchronous Motors during Power System Disturbances; *Trans. AIEE*, February, 1962, Power Apparatus and Systems, Vol. 81, pp. 1136–1141.

Chidambara, M. R., and Ganapathy, S. Transient Torques in Three-phase Induction Motors during Switching Operations; *Trans. AIEE*, April, 1962, Power Apparatus and Systems, Vol. 81, pp. 47–55.

Beardmore, A. E. Power Station Motor Selection Key to Savings, Reliability; *Elec. World*, Jan. 13, 1964, pp. 56–58.

SIEBER, M. K. What the New NEMA Motor Standards Mean; *Plant Eng.*, July, 1964, pp. 110–140.

WOOD, W. S., and OTHERS. Transient Torques in Induction Motors Due to Switching of the Supply; *Proc. Inst. Elec. Engrs. (London)*, July, 1965, Vol. 112, pp. 1348–1354.

FUSES

HOWELL, M. R. Temperature Coordination of Relays and Fuses; *Elec. World*, July 11, 1942, Vol. 118, pp. 72–74.

WESTRATE, M. C., and STEVENS, G. D. Trace Fuse Blowing to Excitation Surge; D-c Transient of Lightning Surge Magnetizes Transformer Core; *Elec. World*, Apr. 29, 1944, Vol. 120, pp. 1590–1591.

RIEBS, R. E. Effect of Repeated Faults on Fuse Characteristics; *Trans. AIEE*, December, 1952, Vol. 71, Power Apparatus and Systems, pp. 1101–1108.

DEAN, R. H. Recent Developments in Medium-voltage High Breaking Capacity Fuse Links; *Proc. Inst. Elec. Engrs. (London)*, June, 1958, Vol. 105, pp. 263–270.

MATHEWS, W. A. Increasing Interrupting Capacity of Low Voltage Circuit Breaker Systems; *Elec. Eng.*, October, 1963, Vol. 82, pp. 613–617.

CAMERON, F. L. Application of High Voltage Power Fuses; *Westinghouse Engr.*, May, 1963, Vol. 23, pp. 90–93.

PEACH, N. Low Voltage Fuses Can Be Adapted to Varied Jobs; *Power*, September, 1964, Vol. 108, pp. 166–167.

MIKULECKY, H. W. Current Limiting Fuse with Full Range Clearing Ability; *Trans. AIEE*, Vol. 84, Power Apparatus and Systems, pp. 1107–1112.

PEACH, N. Standards for Circuit Protection, Circuit Breakers, Fuses; *Power*, November, 1965, Vol. 109, pp. 177–184.

FIRE PROTECTION

PURCELL, T. E. Carbon Dioxide Fire Protection for Turbine Generators—Colfax Station; *Elec. Jour.*, November, 1935, Vol. 32, pp. 460–464. (Air-cooled.)

BURGOYNE, J. H., KATAN, L. L., and RICHARDSON, J. F. Application of Air Foam to Oils Burning in Bulk; *Inst. Petroleum*, December, 1949, Vol. 35, pp. 795–814.

SLEEPER, H. P. Low Cost Fire Protection for Indoor Unattended Substations; *Elec. Light and Power*, March, 1951, Vol. 29, pp. 90–93.

Potomac Electric to Use Fire Resistant Lube in New 335 MW Turbine Generators; *Power*, May, 1962, p. 64.

BENDER, R. J. What's New in Fire Protection; *Power*, September, 1962, pp. 57–63.

KIRKMAN, H. B. Automatic Fire Protection Cuts Substation Losses, Increases Safety and Public Acceptance; *Elec. Light and Power*, Mar. 3, 1964, pp. 44–46.

Electrical Hazards in Fighting Fires; *Safety Maintenance*, September, 1964, pp. 32–34.

WEAVING, E. J. Methods of Plant Fires Protection; *Safety Maintenance*, May, 1965, Vol. 129, pp. 30–32.

SUPERVISORY CONTROL, LOAD AND FREQUENCY CONTROL

REAGAN, M. E. Self-checking System of Supervisory Control; *Elec. Eng.*, October, 1938, Vol. 57, pp. 600–605.

ALLEN, T. J., and DANIEL, J. Vibracode System of Supervisory Control on Carrier Communication Channel; *Elec. Eng.*, November, 1951, Vol. 70, pp. 100–108.

JOHANNSON, D. E. Telemetering Applications and Operations on a Large Power System; *Elec. Eng.*, May, 1960, pp. 400–404.

ZAMBOTTI, B. Data Logging, Scanning, Alarming, Calculating in Power Plants and Substations; *Elec. World*, Nov. 27, 1961, pp. 35–46.

KOZLOWSKI, R. E. Mozart Substation, the First Supervisory Controlled 12-4 Kv Substation at Hydro-Quebec with Automatic Data Logger; *Trans. AIEE*, April, 1964, Vol. 83, Power Apparatus and Systems, pp. 327–336.

SMITH, H. L. On Line Control Computers Are Increasing; *Elec. World*, Sept. 7, 1964, pp. 56–57.

POPE, H. W., and LEWIS, H. A. Emergency Supervisory Channel Saves; *Elec. World*, Sept. 28, 1964, Vol. 182, pp. 19–21.

ENDER, R. C. System Automation Promises Savings; *Elec. World*, Jan. 24, 1966, pp. 164–165.

SECTION 11

TRANSFORMERS, REGULATORS, AND REACTORS

BY

L. WETHERILL Consulting Engineer (Retired), General Electric Company, Pittsfield, Massachusetts; Fellow, Institute of Electrical and Electronics Engineers; Registered Professional Engineer (Massachusetts)

Help from associates who wrote portions of the material or assisted in its revision is gratefully acknowledged. Included are M. F. Beavers, N. M. Case, R. E. Coates, F. D'Entremont, H. B. Detrick, D. C. Graham, T. Graham, W. M. Johnson, R. B. Kaufman, A. M. Krakower, G. E. Leibinger, G. A. Leiter, J. E. Lenz, B. Mackie, O. P. McCarty, W. G. McMichael, W. J. McNutt, T. E. Rodhouse, S. Sass, G. E. Sauer, H. C. Stanley, D. J. Thomas, A. U. Welch, D. M. West, W. I. Wheelock, and W. A. Williams.

CONTENTS

Numbers refer to paragraphs

SECTION 11

TRANSFORMERS, REGULATORS, AND REACTORS

By L. Wetherill

ELEMENTARY THEORY

1. Elementary theory given in Par. **1** through Par. **10** is developed from the viewpoint of a 3-phase three-leg concentric-cylindrical two-winding transformer, with the primary low-voltage winding next to the core and the secondary high-voltage winding outside the primary winding. This corresponds to a simple generator-step-up transformer of moderate kVA. Much of the information is also applicable to single-phase transformers with windings on two legs, 3-phase transformers with five-leg cores, transformers with the primary winding outside the secondary winding, three-winding transformers, etc.

2. Sinusoidal voltage is induced in windings by sinusoidal variation of flux,

$$E = 4.44 \times 10^{-8} a_C B f N \tag{11-1}$$

where a_C = square inches cross section of core, B = lines per square inch peak flux density, E = rms volts, f = cycles per second frequency, and N = number of turns in winding.

The induced voltage in the primary (excited) winding approximately balances the applied voltage. The induced voltage in the secondary (loaded) winding approximately supplies the terminal voltage for the load.

Voltage ratio is the ratio of number of turns ("turn ratio") in the respective windings. Rated open-circuit terminal voltages are closely proportional to turn ratio. Under load, terminal voltages differ somewhat from turn ratio.

3. Characteristics on Open Circuit. The core loss (no-load loss) and exciting current of power transformers may be obtained from empirical design curves of watts loss per pound of core steel (Fig. 11-1) and of exciting voltamperes per pound of core steel (Fig. 11-2). Such curves are established by plotting data obtained from transformers of similar construction. A separate curve for each size of transformer core may give greater accuracy.

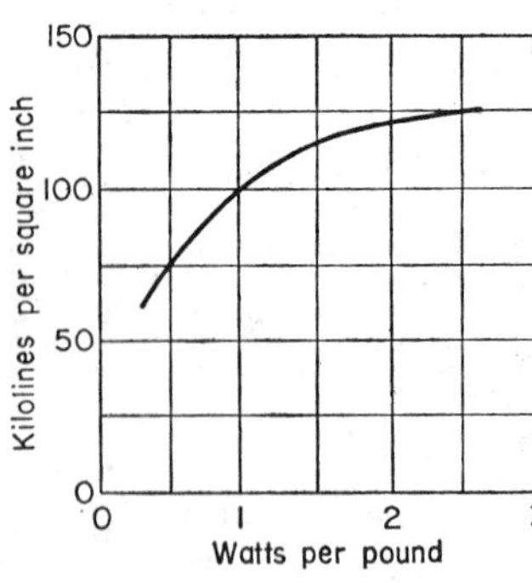

Fig. 11-1. Typical core-loss curve for transformer core steel at 60 c/s.

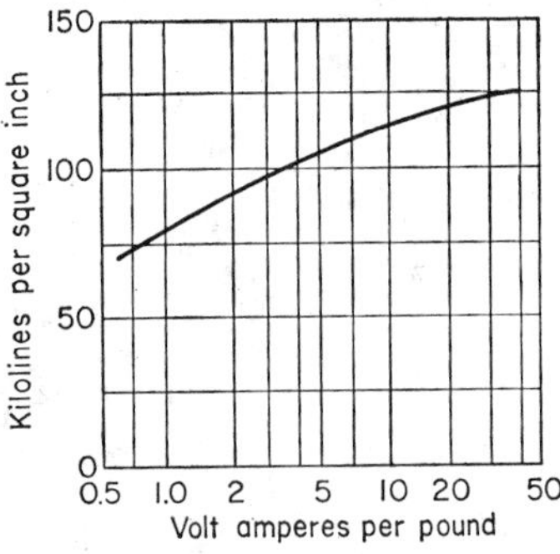

Fig. 11-2. Typical exciting voltampere curve for transformer core steel at 60 c/s.

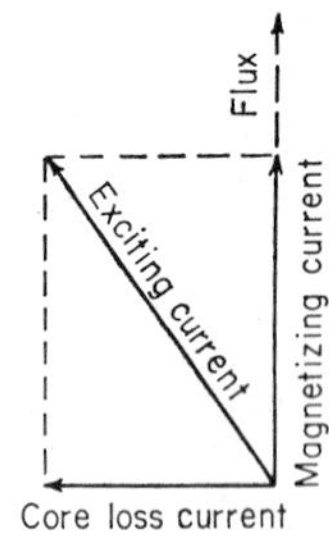

Fig. 11-3. Phasor diagram of equivalent sinusoidal exciting current.

4. Exciting current of a transformer contains many harmonic components because of the greatly varying permeability of the core. For most purposes it is satisfactory to neglect the harmonics and assume a sinusoidal exciting current of the same effective value. This current may be regarded as composed of a core-loss component in phase with the induced voltage (90° ahead of the flux) and a magnetizing component in phase with the flux as shown in Fig. 11-3.

Sometimes it is necessary to consider the harmonics of exciting current to avoid inductive interference with communication circuits. The harmonic content of the exciting current increases as the peak flux density is increased. Performance can be predicted by comparison with test data from previous designs using similar core steel and similar construction.

The largest harmonic component of the exciting current is the third. Higher-order harmonics are progressively smaller. For balanced 3-phase transformer banks the third-harmonic components (or multiples of the third) are displaced by 120 fundamental deg (or multiples of 120 fundamental deg) or 360 harmonic deg and therefore constitute a zero-phase-sequence system. Triple-harmonic currents may flow internally in delta-connected windings and externally in zero-phase-sequence paths in the connected system. The division of third-harmonic exciting current among available paths is not readily calculable.

5. Magnetizing Inrush Current. If an idle transformer is energized at a time in the voltage cycle when the flux in the core would normally be other than the actual residual flux in the core, there will be an exponentially decreasing transient flux component in the core which has the effect of offsetting the initial flux wave. In extreme cases the peak flux may be more than doubled, exceeding saturation of the core, and causing peak magnetizing current several times rated load current. Magnetizing inrush current is important principally because of the possibility of false operation of transformer protective relays.

6. Characteristics on Short Circuit. If the primary winding of a transformer with 1:1 turn ratio is excited with the secondary winding short-circuited, a small exciting current flows in the primary winding, producing mutual flux mostly in the core. In addition a short-circuit current flows forward in the primary and reversed in the secondary, causing leakage flux which passes between the two windings and completes its path through the core. The mutual and leakage flux together make net flux linkages with the secondary to induce voltage to supply the resistance drop in the secondary and make net flux linkages with the primary to induce a countervoltage equal to the applied voltage less the resistance drop in the primary. Figure 11-4 shows the space relationships and the phase relationships neglecting the exciting current. It is apparent that

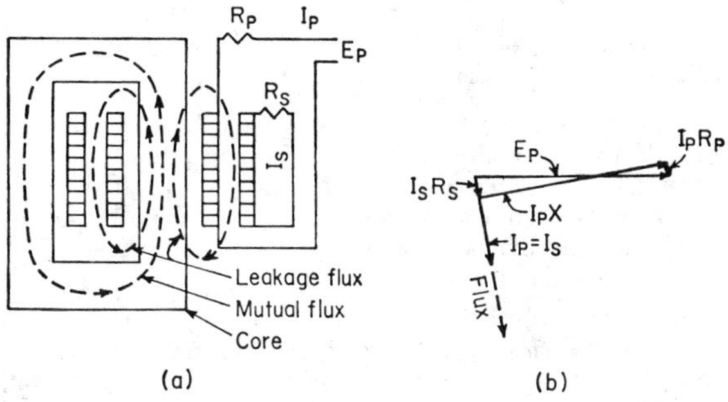

FIG. 11-4. Short-circuited transformer. (*a*) Flux distribution—single phase; (*b*) phasor diagram—1:1 ratio.

$$E_P = I_P(R_P + R_S + jX) = I_P Z \qquad (11\text{-}2)$$

where E_P = rms volts applied to primary (phasor), I_P = rms amperes in primary (phasor), R_P = ohms resistance of primary winding, R_S = ohms resistance of secondary winding, X = ohms reactance (corresponding to the voltage induced in the primary by the mutual and leakage flux), and Z = ohms impedance ($R_P + R_S + jX$).

7. Resistance, Reactance, and Impedance. R_P and R_S are effective a-c resistances. They are greater than d-c resistance calculated from winding-conductor dimensions or measured with d-c, because they include eddy loss in the conductor and stray loss in the core clamps, tank, etc. The reactance of the transformer is X, and the impedance is $Z = R_P + R_S + jX$.

8. Load Loss. The loss on short-circuit test at rated current is the load loss at rated kVA,

$$L_L = I_R^2 R = I_R^2 Z_M \cos \phi \tag{11-3}$$

where I_R = rms amperes rated current, L_L = watts load loss at rated current, R = ohms resistance $(R_P + R_S)$, Z_M = ohms impedance magnitude $[(R^2 + X^2)^{1/2}]$, and ϕ = impedance angle of transformer.

The load loss at another current is

$$L = \frac{L_L I^2}{I_R^2} \tag{11-4}$$

where I = rms amperes and L = watts load loss.

9. Characteristics under Load. Exciting current in the primary winding produces mutual flux mostly in the core. Opposing currents in the primary and secondary windings cause leakage flux which passes between the two windings and completes its path through the core. The magnitude and phase of the mutual flux depend on the voltage. The magnitude and phase of the leakage flux depend on the current. The mutual and leakage flux together generate in the primary a countervoltage equal to the applied voltage less the resistance drop in the primary and generate in the secondary a voltage equal to the terminal voltage plus the resistance drop in the secondary.

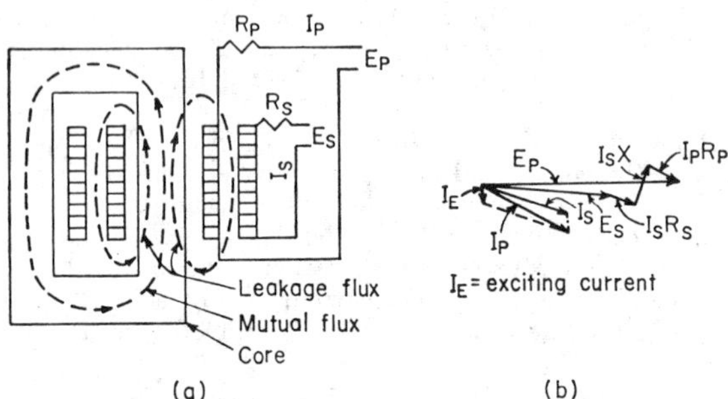

Fig. 11-5. Loaded transformer. (*a*) Flux distribution—single phase; (*b*) phasor diagram—1:1 ratio.

For most purposes the effect of the leakage flux can be represented by the effect of series reactance in the secondary-winding circuit. Figure 11-5 shows the space relationships and the phase relationships in a transformer of 1:1 ratio. It is apparent that

$$E_P = E_S + I_S(R_S + jX) + I_P R_P \tag{11-5}$$

where E_P = rms volts at primary terminal (phasor), E_S = rms volts at secondary terminal (phasor), I_P = rms amperes in primary (phasor), I_S = rms amperes in secondary (phasor), R_P = ohms resistance of primary winding, R_S = ohms resistance of secondary winding, and X = ohms reactance of transformer.

10. Equivalent Circuits. Figure 11-6 shows a circuit which is equivalent to the transformer of Fig. 11-5. The current I_E corresponds to exciting current and flows through a branch with correct impedance to draw the equivalent sine-wave exciting current with rated voltage applied to the primary and the secondary open-circuited. For many purposes the exciting current can be neglected, and this leads to the simpler equivalent circuit of Fig. 11-7, in which a 1:1 ratio transformer is represented by an impedance.

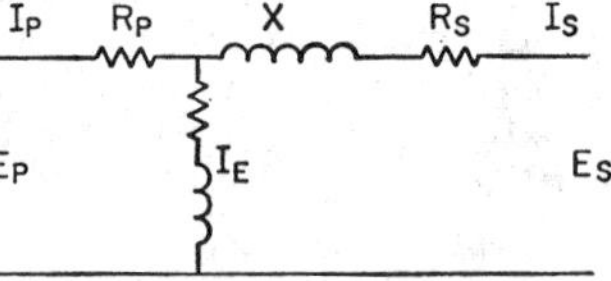

Fig. 11-6. Equivalent circuit of two-winding transformer considering exciting current.

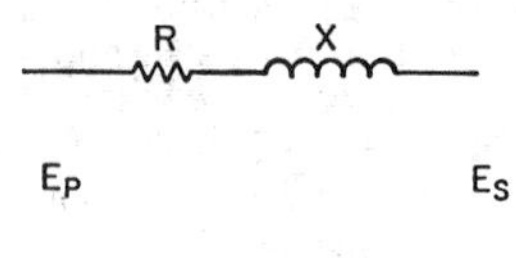

Fig. 11-7. Equivalent circuit of two-winding transformer neglecting exciting current.

11. Effect of Turn Ratio. Equation (11-5) and Fig. 11-7 represent a transformer of 1:1 turn ratio. A transformer of turn ratio T secondary to primary can be transformed into an equivalent 1:1 transformer by imagining the secondary winding replaced by a winding with the same number of turns as the primary winding, but using the same weight of conductor and occupying the same space as the secondary winding. I_S, E_S, and R_S in the real secondary winding become I_ST, E_S/T, and R_S/T^2. The impedance of the load, Z_L, becomes Z_L/T^2. Thus, although Eq. (11-2) to Eq. (11-5) and Fig. 11-4 to Fig. 11-7 were given for 1:1 turn ratio, they can be applied to any turn ratio. The fact that the simple series impedance of Fig. 11-7 may be used as equivalent to a transformer of any turn ratio is very helpful in the analysis of electric power systems. Secondary-winding characteristics corresponding to a fictitious secondary winding of 1:1 turn ratio are called secondary characteristics referred to the primary side. If more convenient, all characteristics can be referred to the secondary side by a reverse process.

12. Per Cent and Per Unit. Current, voltage, and kVA are frequently expressed as per unit or percent of rated value (25% = 0.25 per unit). The procedure is extended to resistance, reactance, and impedance by defining per unit impedance as (ohms impedance) × (rated current in amperes) ÷ (rated voltage in volts). Quantities expressed in percent or per unit are the same regardless of whether they are referred to the primary side or the secondary side.

13. Regulation. From Eq. (11-5) it is apparent that if load current and secondary voltage are at rated value primary voltage must exceed rated value. The excess is called regulation. Regulation for lagging power factor is given exactly by Eq. (11-6) or approximately by Eq. (11-7).

$$G_r = [(R_r + P_r)^2 + (X_r + Q_r)^2]^{1/2} - 1 \tag{11-6}$$

$$G_0 = 100\left[P_rR_r + Q_rX_r + \frac{(P_rX_r - Q_rR_r)^2}{2}\right] \tag{11-7}$$

where G_0 = regulation in percent, G_r = regulation in per unit, P_r = load power factor in per unit, $Q_r = (1 - P_r^2)^{1/2}$, R_r = resistance of transformer in per unit, and X_r = reactance of transformer in per unit.

The calculation of the regulation of a three-winding transformer is considerably more complex. The procedure is given in USAS C57.12.90-1958.

14. Impedance Data. Resistance and reactance of transformers tend to follow normal patterns according to the ratings. Figure 11-8 shows resistance in percent (as determined by measurement of load loss on impedance test). Specific units may vary as much as ±30% depending largely on the evaluation of losses as compared with capital cost. Figure 11-9 shows ranges of reactance in percent. Special designs (transformers with all windings high-voltage, autotransformers, designs with overload ratings, etc.) may have reactances outside the limits shown.

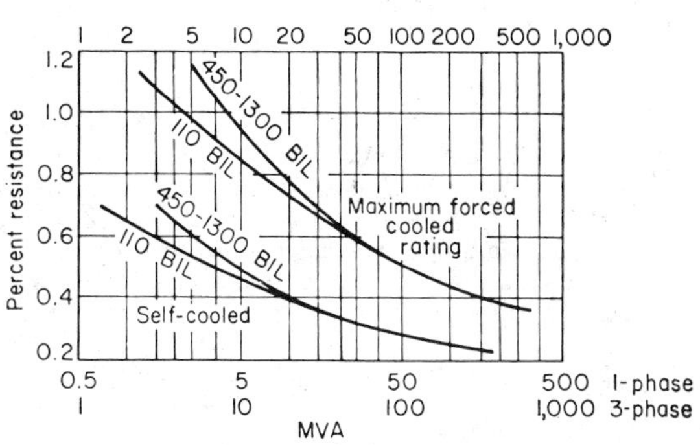

FIG. 11-8. Resistance of typical power transformers.

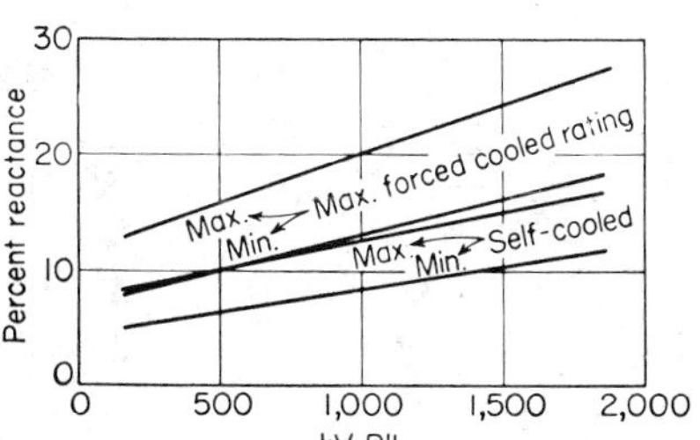

FIG. 11-9. Reactance of typical power transformers.

15. Efficiency is given by Eq. (11-8),

$$F_r = \frac{E_SI_S\cos\theta}{E_SI_S\cos\theta + L_{NS} + L_{LS}} \tag{11-8}$$

where E_S = rms volts at secondary terminals, F_r = per unit efficiency, I_S = rms amperes in secondary, L_{LS} = watts load loss at I_S, L_{NS} = watts no-load loss at $E_S(I_S = 0)$, and θ = impedance angle of load.

16. Three-winding-transformer Load Losses. The load losses of 3-winding transformers, with all three windings carrying loads simultaneously, may be calculated from characteristics obtained by considering each pair of windings as a two-winding transformer,

$$L_T = \left(\frac{I_P}{I_A}\right)^2 \frac{L_{PS} + L_{PT} - L_{ST}}{2} + \left(\frac{I_{SP}}{I_A}\right)^2 \frac{L_{ST} + L_{PS} - L_{PT}}{2} + \left(\frac{I_{TP}}{I_A}\right)^2 \frac{L_{PT} + L_{ST} - L_{PS}}{2} \qquad (11\text{-}9)$$

where I_A = rms amperes reference current referred to winding P; I_P = rms amperes in winding P; I_{SP} = rms amperes in winding S referred to winding P; I_{TP} = rms amperes in winding T referred to winding P; L_{PS} = watts load loss in windings P and S as a two-winding transformer at I_A A; L_{PT}, L_{ST} = similar; and L_T = watts total load loss.

The loss is usually computed at, or corrected to, a temperature of 75°C (85°C for 65°C-rise units).

17. Three-winding-transformer Equivalent Circuit. The equivalent circuit of a three-winding transformer may be determined from the three impedances obtained by considering each pair of windings separately. One form is shown in Fig. 11-10, in which

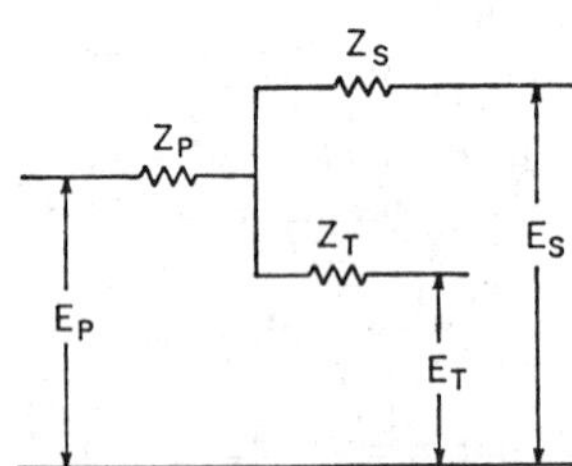

Fig. 11-10. Equivalent circuit of three-winding transformer.

$$Z_P = \frac{Z_{PS} + Z_{PT} - Z_{ST}}{2} \qquad (11\text{-}10)$$

$$Z_S = \frac{Z_{PS} + Z_{ST} - Z_{PT}}{2} \qquad (11\text{-}11)$$

$$Z_T = \frac{Z_{PT} + Z_{ST} - Z_{PS}}{2} \qquad (11\text{-}12)$$

where Z_P, Z_S, Z_T = ohms branch impedances in Fig. 11-10; Z_{PS} = ohms impedance from winding P to winding S in two-winding equivalent circuit of Fig. 11-7; and Z_{PT}, Z_{ST} = similar.

18. Four-winding-transformer Equivalent Circuit. The equivalent circuit of a four-winding transformer may be determined from the six impedances obtained by considering each pair of windings separately. One form is shown in Fig. 11-11, in which

$$Z_P = \frac{Z_{PQ} + Z_{PS} - Z_{SQ}}{2} - \frac{Z_A Z_B}{2(Z_A + Z_B)} \qquad (11\text{-}13)$$

$$Z_S = \frac{Z_{PS} + Z_{ST} - Z_{PT}}{2} - \frac{Z_A Z_B}{2(Z_A + Z_B)} \qquad (11\text{-}14)$$

$$Z_T = \frac{Z_{ST} + Z_{TQ} - Z_{SQ}}{2} - \frac{Z_A Z_B}{2(Z_A + Z_B)} \qquad (11\text{-}15)$$

$$Z_Q = \frac{Z_{TQ} + Z_{PQ} - Z_{PT}}{2} - \frac{Z_A Z_B}{2(Z_A + Z_B)} \qquad (11\text{-}16)$$

$$Z_A = (K_1K_2)^{1/2} + K_1 \qquad (11\text{-}17)$$

$$Z_B = (K_1K_2)^{1/2} + K_2 \qquad (11\text{-}18)$$

$$K_1 = Z_{PT} + Z_{SQ} - Z_{PS} - Z_{TQ}$$

$$K_2 = Z_{PT} + Z_{SQ} - Z_{PQ} - Z_{ST}$$

where Z_A = ohms branch impedance in Fig. 11-11 (complex); Z_B, Z_P, Z_S, Z_T, Z_Q = similar; Z_{PS} = ohms impedance (complex) from winding P to winding S in two-winding equivalent circuit of Fig. 11-7; and Z_{PT}, Z_{PQ}, Z_{ST}, Z_{SQ}, Z_{TQ} = similar.

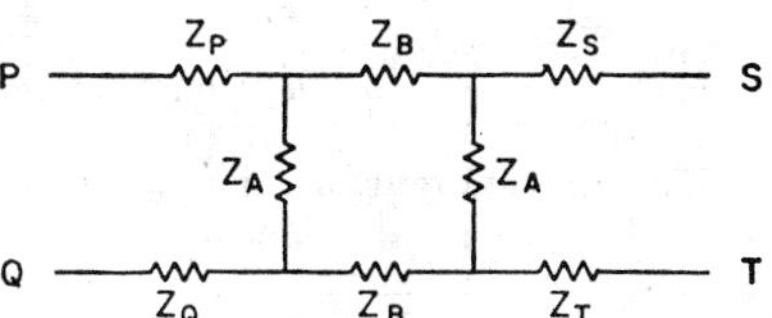

FIG. 11-11. Equivalent circuit of four-winding transformer.

19. Phase - interconnected transformers (i.e., with windings from more than one phase on a single core leg) can be represented by an equivalent circuit only if each winding on a leg is considered as if it were brought out to separate terminals.[1] It is frequently more convenient to use mathematical analysis on circuits containing phase-interconnected transformers, rather than equivalent circuits and a-c calculating boards.

CONNECTIONS

20. Parallel Operation. Two single-phase transformers will operate in parallel if they are connected with the same polarity. Two three-phase transformers will operate in parallel if they have the same winding arrangement (for example, Y-delta), are connected with the same polarity, and have the same phase rotation. If two transformers (or two banks of transformers) have the same voltage ratings, the same turn ratios, the same impedances (in percent), and the same ratios of reactance to resistance, they will divide the load current in proportion to their kVA ratings, with no phase difference between the currents in the two transformers. If any of the above conditions are not met, the load current may not divide between the two transformers in proportion to their kVA ratings and there may be a phase difference between currents in the two transformers.

21. Two unlike transformers connected in parallel will supply current to a load as follows,

$$I_L = \frac{E_P}{\dfrac{1}{(T_1/Z_1) + (T_2/Z_2)} + Z_L \dfrac{(1/Z_1) + (1/Z_2)}{(T_1/Z_1) + (T_2/Z_2)}} \qquad (11\text{-}19)$$

where E_P = rms volts on primary side (phasor), I_L = rms amperes total load current (phasor), T_1 = turn ratio secondary to primary of unit 1, T_2 = turn ratio secondary to primary of unit 2, Z_1 = ohms impedance of unit 1 referred to secondary side (complex), Z_2 = ohms impedance of unit 2 referred to secondary side (complex), and Z_L = ohms impedance of load (complex).

The magnitude of the current in unit 1 is

$$I_{r1} = \frac{\{[T_1R_{r2}I_{rL} + (T_1 - T_2)E_{r1}\cos\theta]^2 + [T_1X_{r2}I_{rL} + (T_1 - T_2)E_{r1}\sin\theta]^2\}^{1/2}}{[(T_1R_{r2} + T_2R_{r1})^2 + (T_1X_{r2} + T_2X_{r1})^2]^{1/2}} \qquad (11\text{-}20)$$

where E_{r1} = rms voltage of secondary terminals in per unit of unit 1, I_{rL} = rms total load current in per unit of unit 1, I_{r1} = rms current in secondary of unit 1 in per unit of unit 1, T_1 = ratio secondary turns to primary turns in unit 1, T_2 = ratio secondary turns to primary turns in unit 2, R_{r1} = equivalent resistance of unit 1 in per unit of unit 1, R_{r2} = equivalent resistance of unit 2 in per unit of unit 1, X_{r1} = equivalent reactance of unit 1 in per unit of unit 1, X_{r2} = equivalent reactance of unit 2 in per unit of unit 1, and θ = impedance angle of load (lagging current positive).

NOTE: Per unit means percent divided by 100, that is, 10% = 0.1 per unit.

The current in the second unit may be determined by using Eq. (11-20) with designation of first and second transformers reversed.

---

[1] COGBILL, B. A. Sequence Impedance of Symmetrical Three Phase Transformer Connections; *Trans. AIEE Paper* 55-671, 1955.

22. Phase-interconnected transformers (i.e., with windings from more than one phase on a single core leg) offer special complication when unlike units are connected in parallel.[1]

23. An autotransformer may be used to force a division of current proportional to the ratings of paralleled transformers or banks when differences of characteristics are so great that parallel operation is otherwise impracticable. The autotransformer must have sufficient winding and core to take up the maximum difference of voltage between the transformers or banks regardless of its cause, and, of course, its windings must be capable of carrying the currents that are involved. The internal tap in the autotransformer must be placed so that the ampere-turns due to the desired currents will be equal on the two sides of the tap. Figure 11-12 shows the arrangement of an autotransformer for two single-phase transformers; the same principles can be applied to 3-phase transformers or banks, making use of either a 3-phase autotransformer or three single-phase units.

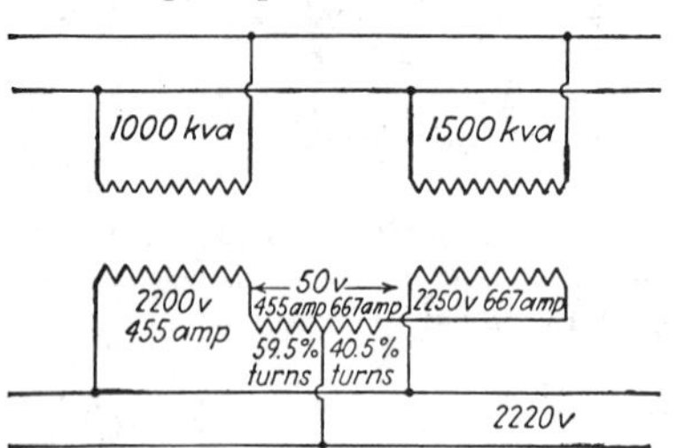

FIG. 11-12. Autotransformer used to force currents proportional to ratings in paralleling two single-phase transformers of unlike characteristics.

24. Three-phase to Three-phase Transformations. The delta-delta, the delta-Y, and the Y-Y-connection are the most generally used; they are illustrated in Fig. 11-13. The Y-delta and delta-delta-connections may be used as step-up transformers for moderate voltages. The former has the advantage of providing a good grounding point in the low-voltage system which does not shift with unbalanced load and has the further advantage of being free from third-harmonic voltages and currents; the latter has the advantage of permitting operation in V in case of damage to one of the units. These connections are not the best for transmission at very high voltage; they may, however, be associated at some point with other connections which provide means for properly grounding the high-voltage system; but it is better, on the whole, to avoid mixed systems of connections. The delta-Y step-up and Y-delta step-down connections are without question the best for high-voltage transmission systems. They are economical in cost, and they provide a stable neutral whereby the high-voltage system may be directly grounded or grounded through resistance of such value as to damp the system critically and prevent the possibility of oscillation.

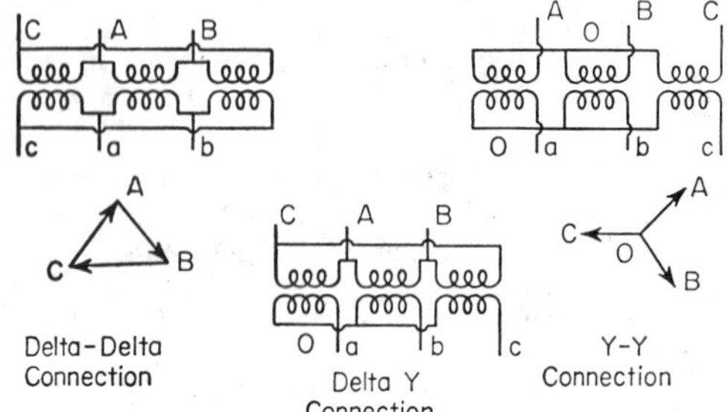

FIG. 11-13. Standard 3- to 3-phase transformation systems.

25. The Y-Y-connection (or Y-connected autotransformer) may be used to interconnect two delta systems and provide suitable neutrals for grounding both of them. A Y-connected autotransformer may be used to interconnect two Y-systems which already have neutral grounds, for reasons of economy. In either case a delta-connected tertiary winding is frequently provided for one or more of the following purposes.

Stabilization of the Neutral. If a Y-connected transformer (or autotransformer) with a delta-connected tertiary is connected to an ungrounded delta system (or poorly grounded Y system), stability of the system neutral is increased. That is, a single-phase short circuit to ground on the transmission line will cause less drop in voltage on

[1] COGBILL, B. A. Sequence Impedance of Symmetrical Three Phase Transformer Connections; *Trans. AIEE Paper* 55-671, 1955.

the short-circuited phase and less rise in voltage on the other 2 phases. A 3-phase three-leg Y-connected transformer without delta tertiary furnishes very little stabilization of the neutral, and the delta tertiary is generally needed. Other Y-connections offer no stabilization of the neutral without a delta tertiary. With increased neutral stabilization the fault current in the neutral on single-phase short circuit is increased, and this may be needed for improved relay protection of the system.

Third-harmonic components of exciting current find a relatively low impedance path in a delta tertiary on a Y-connected transformer, and less of the third-harmonic exciting current appears in the connected transmission lines, where it might cause interference with communication circuits. Failure to provide a path for third-harmonic current in Y-connected 3-phase shell-type transformers or banks of single-phase transformers will result in excessive third-harmonic voltage from line to neutral. Three-phase core-type Y-connected transformers have very little third-harmonic line-to-neutral voltage, and delta tertiaries are not needed to reduce it.

An external load can be supplied from a delta tertiary. This may include synchronous condensers or static capacitors to improve system operating conditions.

26. The open-delta-connection, or V-connection, is an unsymmetrical connection which is used if one transformer of a bank of three single-phase delta-connected units must be cut out because of failure. It is a connection that is sometimes resorted to as an emergency expedient or used as a temporary measure with the intention of completing the delta when conditions of load warrant the addition of a third unit. If one phase of a 3-phase delta-connected transformer of the shell type should fail, operation may be continued at reduced capacity by short-circuiting the damaged phase; if of the core type, operation may be continued by leaving the damaged phase open-circuited, provided that the windings are still capable of withstanding the voltage stresses. Since full-line currents flow in the windings out of phase with the transformer voltages, the normal capacity of the open-delta bank is reduced to 57.7% of its delta rating.

27. The T-connection uses two transformers, the first called the "main" transformer, connected from line to line; and the second, called the "teaser" transformer, connected from the midpoint of the first to the third line. It requires that the midpoint of both primary and secondary windings be available for connections. It has an advantage over the V-connection in being more nearly symmetrical if the proper taps have been provided. As in the case of the V-connection, two transformers of a bank of delta-connected transformers, one of which has failed, may be connected in T, and if 10% taps can be used for the teaser transformer, the transformation will be more nearly symmetrical than if the V-connection were used. Where T-connected transformers are installed, they may later be changed to delta with the addition of one more transformer and an increase in rating of the bank of 73%. In the T-connection (Fig. 11-14) the transformer *AD*, known as the "teaser" transformer, may be a duplicate of the main transformer so as to be interchangeable with it, and it may or may not be provided with an 86.6% tap. Its rated capacity will then be 15.5% more than actually necessary. The main transformer operates at a power factor of 0.866, and therefore, if the two transformers are duplicates, their total rated capacity will be 15.5% greater than the capacity of the load in kVA, or each transformer must have a rating of 0.577 of the kVA delivered. If the transformers are not interchangeable, the teaser may be reduced to a rating of one-half the kVA delivered.

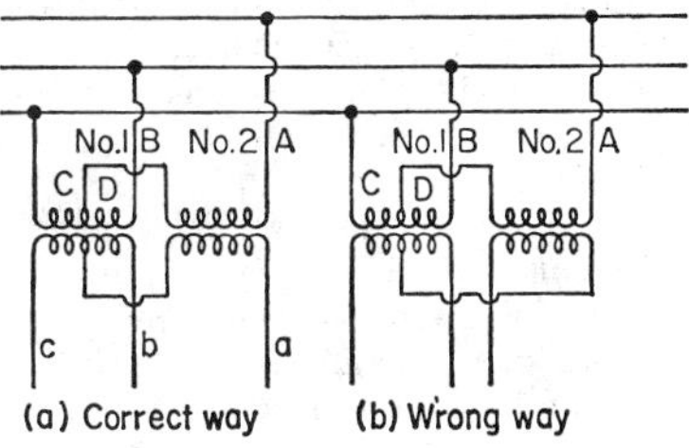

FIG. 11-14. T-connected transformers.

In connecting transformers in T *care should be taken to keep the relative phase sequence of the windings the same;* otherwise the impedance of the main transformer may be excessively high and cause undue unbalance. Figure 11-14 illustrates the right and the wrong way.

28. Transformations between Two- and Three-phase Systems. There are various ways of connecting two or three single-phase transformers to effect the transfer of power

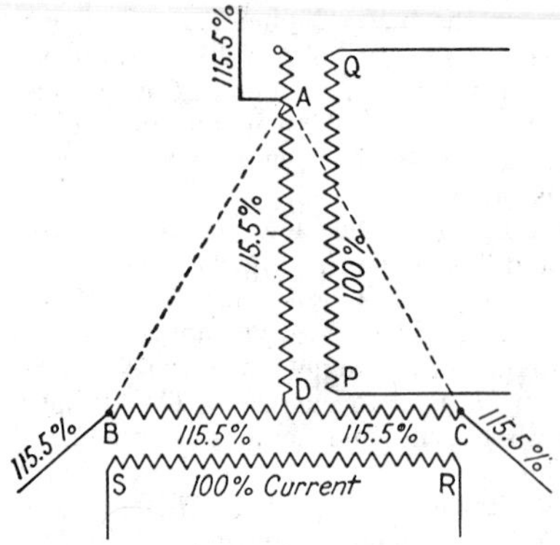

FIG. 11-15. A Scott-connected bank of two single-phase transformers, with winding currents indicated.

between 2- and 3-phase systems. They are described rather extensively in the literature.[1]

29. The Scott connection, which is the best-known and most commonly used arrangement, makes use of two single-phase transformers, one of which, known as the "main" transformer, must have a tap in the middle of the winding on the 3-phase side, while the other, known as the "teaser," must have a tap at 86.6% of the full winding on the 3-phase side. Usually the two units are duplicates, both having full windings and 50 and 86.6% taps. One end of the teaser winding is connected to the 50% tap of the main winding; the 86.6% tap of the teaser winding and the two ends of the main winding become the 3-phase terminals. Figure 11-15, illustrating these connections, shows that a symmetrical 3-phase delta is formed.

Example. **To illustrate the distribution of currents in the windings of a Scott-connected pair of transformers,** assume that the bank is transferring a total of 1,000 kVA from 2,200 V 3-phase to 2,200 V 2-phase, the 1:1 ratio of voltages being selected to afford an easy comparison of currents. The currents in each of the 2-phase windings at rated load will be $500{,}000/2{,}200 = 227$ A. The currents in the 3-phase windings will be $1{,}000{,}000/\sqrt{3} \times 2{,}200 = 262$ A, or 15.5% greater than the rated current flowing normally in a 500-kVA 2,200-V transformer. The 115.5% 3-phase line current flowing in the teaser is in phase with the voltage of the teaser winding but in quadrature with the voltage of the main winding; at the juncture of the teaser winding with the main winding, it divides equally, 57.7% flowing in opposite directions in the two halves of the main winding. The other two line currents which flow in the main winding are 30° out of phase with the voltage of the winding; therefore, a component of 100% is in phase with the voltage of the winding, and this component when added vectorially to the 57.7% quadrature component from the teaser equals the 115.5% line current.

From this illustration we find that the **windings on the 3-phase side of a Scott-connected pair of transformers must be designed to carry 15.5% more current** than the single-phase ratings of the transformers indicate. If two single-phase transformers, having the necessary taps but no extra copper provided for the 3-phase currents, are Scott-connected for a 3-phase 2-phase transformation, the rating of the bank should be reduced to 86.6% of the combined single-phase ratings if no rated currents are to be exceeded.

30. Three-phase to Two-phase Transformation with Three Transformers. These methods of transformation offer some advantages in small installations where the cost of a spare unit is prohibitive and it is desired to provide against the possibility of complete interruption of service due to the loss of one unit of a bank. They are also useful in installations where it is the intention at a later period to change over from 2-phase to 3-phase and it is desired to provide for standard 3-phase connections. The connection shown in Fig. 11-16 is known as the **Taylor connection.** Another of equal practicability is that shown in Fig. 11-17.

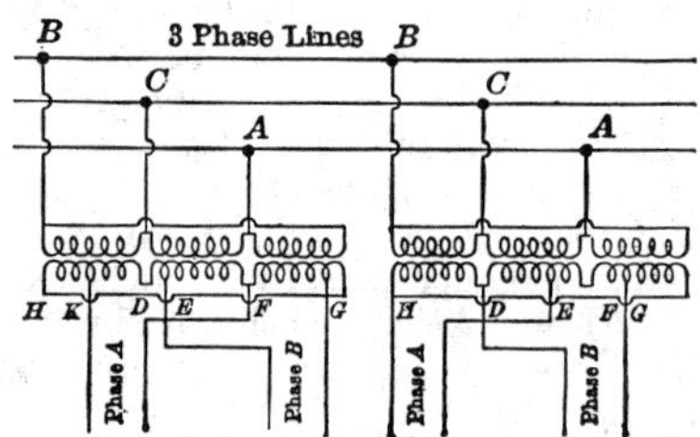

FIG. 11-16. Taylor connections for changing 3-phase to 2-phase.

FIG. 11-17. Another scheme for transforming 3-phase to 2-phase.

[1] BLUME, L. F., BOYAJIAN, A., CAMILLI, G., LENNOX, T. C., MINNECI, S., and MONTSINGER, V. M. "Transformer Engineering," 2d ed.; New York, John Wiley & Sons, Inc., 1951.

31. Three-phase to Six-phase, Double Delta. If two delta-connected banks of transformers are taken, and the polarity of one bank is reversed, the two banks together will furnish 6-phase emf. The relations between the secondaries are indicated in Fig. 11-18. Instead of using two banks of transformers, it is usual to place two equal secondary windings on each transformer and connect one set so as to give opposite polarity to the other set.

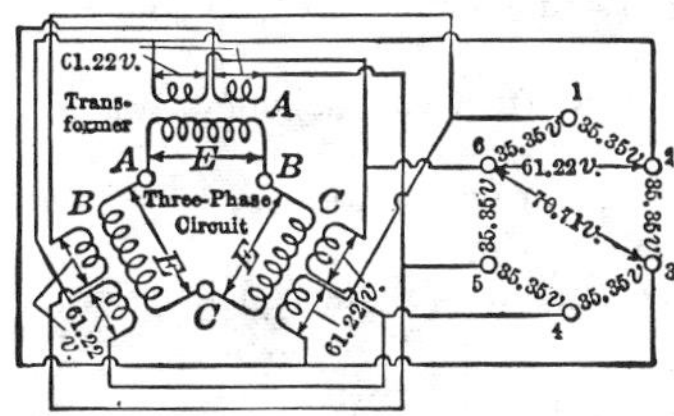

Fig. 11-18. Three-phase to six-phase transformation, double delta.

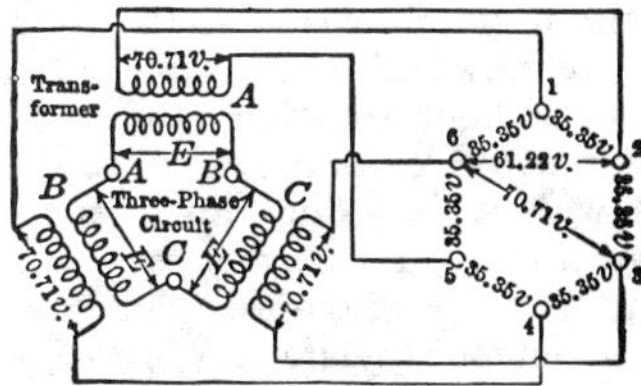

Fig. 11-19. Three-phase to six-phase transformation, double Y.

32. Three-phase to Six-phase, Diametrical. Similarly, if the two sets are connected in Y, one set having its windings reversely connected with respect to the other (Fig. 11-19), we shall have 6-phase emfs. It is not necessary in this transformation, when the neutral connection is required, to have two secondary windings; instead a middle tap may be brought out, all the middle taps of the three transformers being connected together to form the neutral.

33. The interconnected Y-connection shown in Fig. 11-20 is commonly referred to as the **zigzag** connection. It may be used with either a delta-connected winding as shown or a Y-connected winding for step-up or step-down operation. In either case, the zigzag winding produces the same angular displacement as a delta winding and, in addition, provides a neutral for grounding purposes. Owing to the angular relation of voltages of the zig and zag windings, the amount of conductor material required for such a connection is 15% greater than a corresponding Y- or delta-connection. If a transformer consists of zigzag and Y-connections, a third winding, delta-connected, is usually necessary for reasons given under the Y-Y-connection. If the delta-connected winding is included for purposes other than that of providing a third source of power, in some cases it is practical to design it for the same voltage as the zigzag winding and connect it in parallel with the zigzag winding to form the delta-grounded transformer connection.[1]

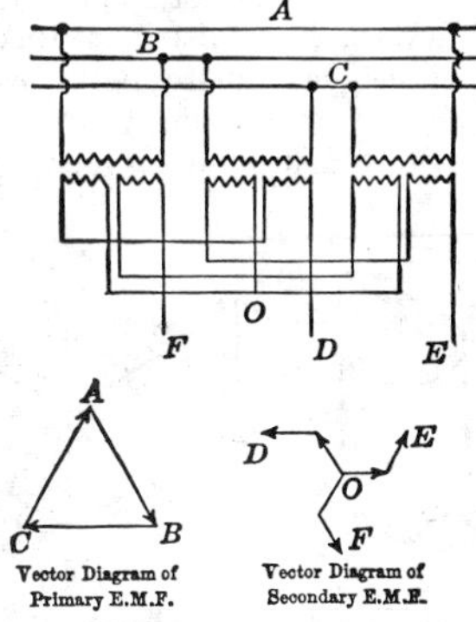

Fig. 11-20. Interconnected Y-connection.

The zigzag connection is used extensively for grounding transformers, the sole purpose of which is to establish a neutral point for grounding purposes; therefore no other windings are required. It is also used in connection with d-c 3-wire distribution systems.

POWER TRANSFORMERS

34. The information given in Pars. **34** to **183** deals particularly with power transformers, which may be defined as transformers used to transmit or distribute power in ratings larger than distribution transformers (usually over 500 kVA or over 67 kV). Some of the following information on power transformers is also applicable to some other transformers which are covered subsequent to Par. **183**.

35. The rating of a power transformer includes the following:

The **rated kVA** is the kilovoltampere output which can be delivered at rated secondary voltage and rated frequency without exceeding the rated temperature rise.

[1] Gross, E. T. B., and Rao, K. J. Analysis of the Delta-grounded Transformer; *Trans. AIEE*, 1953, Vol. 72, pp. 817–826.

The **rated secondary voltage** is the secondary voltage at which the transformer is designed to deliver rated kVA.

The **rated primary voltage** is the rated secondary voltage multiplied by the turn ratio.

The **rated secondary current** is the rated kVA divided by the rated secondary voltage.

The **rated primary current** is the rated kVA divided by the rated primary voltage.

A **tap** is a connection brought out of a winding at some point between its extremities, usually to permit changing the voltage ratio.

A **rated kVA tap** is a tap through which the transformer can deliver its rated kVA output without exceeding rated temperature rise.

A **reduced kVA tap** is a tap through which the transformer can deliver only a reduced kVA without exceeding rated temperature rise.

An **indoor transformer** is a transformer which, because of its construction, must be protected from the weather.

An **outdoor transformer** is a transformer of weatherproof construction.

Design

36. The design of successful commercial transformers requires the selection of a simple form of structure so that the coils may be easy to wind and the magnetic circuit easy to build. At the same time the mean length of the windings and of the magnetic circuit must be as short as possible for a given cross-sectional area, so that the amount of material required and the losses shall be as low as possible. The form of construction should permit of the easy removal of heat by means of ventilating ducts, it should admit of being insulated in a simple and economical manner, and the windings should be of such forms as may be easily reinforced to withstand mechanical stresses.

37. Two Types of Transformer in Common Use. When the magnetic circuit takes the form of a single ring encircled by two or more groups of primary and secondary windings distributed around the periphery of the ring, the transformer is termed a **core-type transformer.** When the primary and secondary windings take the form of a common ring which is encircled by two or more rings of magnetic material distributed around its periphery, the transformer is termed a **shell-type transformer.**

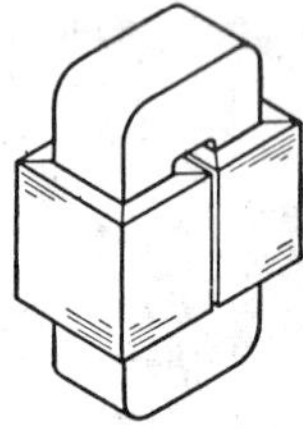

FIG. 11-21. Forms of magnetic circuits for transformers.

38. The characteristic features of the core-type transformer are a long mean length of magnetic circuit and a short mean length of windings; those of the shell type are short mean length of magnetic circuit and long mean length of windings. The result of these features is that for a given output and performance the core type will have a smaller area of core and larger number of turns than the corresponding shell type. Except for certain extremes of current rating the choice between the core- and shell-type construction is largely a matter of manufacturing facilities and of individual preference. In typical design practice the ratio of iron to copper weight is greater in the shell-type transformer. Figure 11-21 illustrates the shell and core forms of magnetic circuits.

39. Design Process. Most power transformers are designed by assuming dimensions, calculating characteristics, comparing calculated characteristics with desired characteristics, and modifying the assumed dimensions better to meet the desired characteristics. Repeating the process leads to close agreement of calculated characteristics with desired characteristics. The repeated calculations, converging on the optimum design, are usually performed by computer. Closeness of agreement of calculated characteristics with tested characteristics depends upon the degree of refinement of the design process, the closeness of agreement of the physical properties of the materials used (particularly the dielectric properties of the insulating materials and the magnetic properties of the core steel) with the properties assumed in the design calculation, and the accuracy of the manufacturing procedures and processes.

Refinement of the design process results from comparison with test data obtained on similar transformers. This applies particularly to core loss, stray loss, noise level, reactance, and dielectric strength. The following calculation methods are mostly approximate.

40. Number of Turns. With an assumed core cross section and flux density the number of turns in each winding is established from Eq. (11-1). The flux density is adjusted to give an integral number of turns in the low-voltage winding, and then an acceptable ratio of open-circuit terminal voltage results from an integral number of turns in the high-voltage winding.

41. Leakage flux density in the main gap (insulation space between windings) for a transformer with one core leg per phase, as shown in Fig. 11-22, is as follows,

$$B_L = \frac{4.52 I_R N}{h_E} \tag{11-21}$$

where B_L = lines per square inch peak leakage flux density, h_E = inches effective length of leakage flux path, I_R = rms amperes rated current of winding, and N = number of turns in winding.

If there is more than one leg per phase, Eq. (11-21) applies to the portion of winding on one leg. The effective length of leakage path is difficult to evaluate accurately. For concentric cylindrical windings it is approximately

$$h_E = \frac{h_P + h_S}{2} + 0.8\left(\frac{b_P + b_S}{3} + b_G\right) \tag{11-22}$$

where b_G = inches radial distance between windings, b_P = inches radial width of winding P, b_S = inches radial width of winding S, h_P = inches length of winding P, and h_S = inches length of winding S.

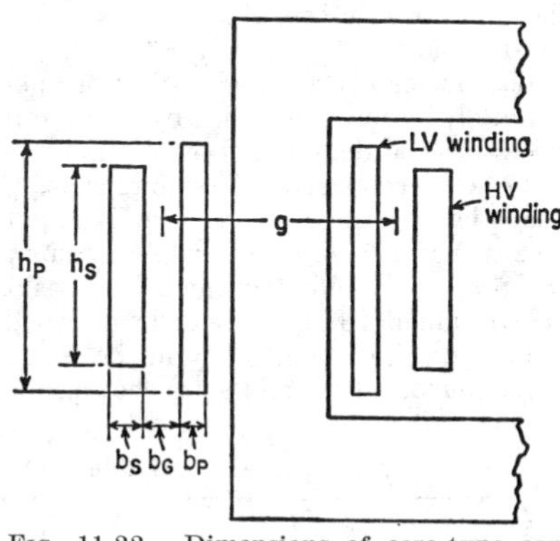

Fig. 11-22. Dimensions of core-type concentric windings for reactance calculations.

42. Leakage reactance may be calculated for a transformer with one set of coils per phase as follows,

$$X_r = \frac{2.01 \times 10^{-7} a_L f I_R N^2}{E_R h_E} \tag{11-23}$$

where a_L = square inches effective cross section of leakage flux path, E_R = rms volts rated voltage of winding, f = cycles per second frequency, h_E = effective length of leakage flux path, in inches, I_R = rms amperes rated current of winding, N = number of turns in winding, and X_r = per unit reactance.

The effective cross section of the leakage flux path is difficult to evaluate accurately. For concentric cylindrical windings it is approximately

$$a_L = \left(b_G + \frac{b_P + b_S}{3}\right)\pi g \tag{11-24}$$

where g = mean diameter of main gap, in.

43. Resistance loss in winding is

$$W_R = 2.57 M^2 \frac{234.5 + C}{309.5} \tag{11-25}$$

$$L_R = H_C W_R \tag{11-26}$$

where L_R = watts resistance loss in winding, M = rms kA per square inch current density, H_C = pounds weight of copper in winding, C = °C temperature, and W_R = watts per pound resistance loss in winding.

44. Eddy loss in the winding may be regarded as caused by circulating current induced in the strand by the magnetic flux passing through the strand. For a two-winding transformer

$$W_E = 2.06 \times 10^{-10} d^2 f^2 B_L^2 \frac{309.5}{234.5 + C} \tag{11-27}$$

$$L_E = H_C W_E \tag{11-28}$$

where B_L = lines per square inch peak leakage flux density, from Eq. (11-21), C = °C temperature, d = inches thickness of strand perpendicular to flux, f = cycles per second frequency, L_E = watts eddy loss in winding, H_C = pounds weight of copper in winding, and W_E = watts per pound average eddy loss in winding.

45. Load loss is the sum of resistance and eddy losses in all windings plus stray loss. The stray loss (in core clamps, tank, etc.) may be predicted from test results on similar transformers.

46. No load loss equals watts per pound determined from Fig. 11-1, multiplied by weight of the core multiplied by a correction factor depending on core configuration and processing and determined by experience.

47. General Design Characteristics. The relationship of power-transformer characteristics to scale factor can be illuminated by considering the effect of increasing all dimensions in the ratio *S*, while retaining the same thickness of core lamination and thickness of conductor strand, but imagining the conductor turns to be reconnected for a terminal voltage proportionate to the insulation thickness. Similarly the effect of increasing the flux density in the ratio *B* and the current density in the ratio *M* can be examined. The results are shown in Table 11-1.

Table 11-1. Scale Effects

Characteristic	At scale factor *S*	At flux density *B**	At current density *M*
Linear dimension	S		
Flux density	....	B	
Current density	....	....	M
Rated current	S^3	B	M
Rated voltage	S		
Rated kVA	S^4	B	M
Weight	S^3		
KVA/lb	S	B	M
Ω reactance	S^{-1}		
% reactance	S	B^{-1}	M
W core loss	S^3	B^2	
% core loss	S^{-1}	B	M^{-1}
W I^2R loss	S^3	....	M^2
% I^2R loss	S^{-1}	B^{-1}	M
W eddy loss	S^5	....	M^2
% eddy loss†	S	B^{-1}	M
W stray loss	S^4	....	M^2
% stray loss†	....	B^{-1}	M

* Applies only to the range in which core loss varies with the square of *B*.
† As a percent of rated current times rated volts.

48. Core dimensions are generally standardized in steps, with only a small number of dimensions varying to meet the requirement of the particular rating. Cold-rolled grain-oriented silicon steel strip in gages of 0.012 to 0.014 in is used with mitered corner joints to take advantage of the good characteristics of this material when carrying flux in the rolling direction.

Insulation

49. Insulation systems used in power transformers comprise liquid systems and gas systems. In both cases some solid insulation is used. Liquid systems include oil, which is most frequently used, and askarel, which is used to avoid combustibility. The gas systems include nitrogen, air, and fluorogases (e.g., sulfur hexafluoride). The fluorogases are used to avoid combustibility and limit secondary effects of internal failure.

Major insulation separates the high-voltage winding from the low-voltage winding. This insulation carries the highest voltage and occupies the most limited space; hence it usually operates at the highest stress. Depending on the construction, layer insulation or coil insulation may be provided between parts of windings. Turn insulation is applied to each strand of conductor or to groups of strands forming a single turn.

50. Oil-insulated Transformers. Low cost, high dielectric strength, and ability to recover after dielectric overstress make mineral oil the most widely used transformer insulating material. The oil is reinforced with solid insulation in various ways. The major insulation usually has solid insulation barriers alternating with oil spaces. The stress on the oil is 50 to 100% higher than the stress on the solid, because of the relatively low dielectric constant of the oil. Therefore, the stress on the oil limits the strength of the structure. Small oil ducts can withstand much higher stress than large oil ducts. Thus suitably spaced solid barriers give the most effective use of the space.

Insulation between adjacent turns is usually solid, to provide mechanical support and to give relatively high dielectric strength against high short-time transient voltages. Solid insulation is sometimes used between layers of a winding or between windings.

Thick solid insulation is used on high-voltage leads in areas of dielectric-stress concentration. The relatively high dielectric constant of the solid material causes the stress on the solid to be only half or two-thirds as much as it would have been if oil occupied the same space.

Most solid-insulation materials used in power transformers are porous, permitting the removal in vacuum treatment of gases and vaporized water and the filling of all cavities and interstices with oil. Any gas pocket inadvertently left in the dielectric field suffers high dielectric stress (twice what oil would have) because of the low dielectric constant of gas. Since entrapped gas not only carries high dielectric stress but also has low dielectric strength, serious loss of dielectric strength results.

Solid materials frequently used include oil-impregnated paper, resin-impregnated paper, pressboard, cloth, vacuum oil-treated wood, and enamel.

51. Askarel-insulated transformers have constructions similar to oil-insulated transformers. The relatively high dielectric constant of the askarel aids in transferring dielectric stress to the solid elements. Askarel has limited ability to recover after dielectric overstress, and thus the strength is limited in nonuniform dielectric fields. Askarels are seldom used over 34.5 kV operating voltage. Askarels are powerful solvents, and material used with them must be carefully selected to avoid damaging the material or contaminating the askarel.

52. Fluorogas-insulated Transformers. Fluorogases have better dielectric strength and heat-transfer capacity than nitrogen or air. Both dielectric strength and heat transfer capacity increase with density and fluorogas transformers operate above atmospheric pressure, in some cases up to 3 atm gage pressure. The gas insulation is reinforced with solid insulation used in the form of barriers, layer insulation, turn insulation, and lead insulation.

It is usually economical to operate fluorogas-insulated transformers at higher temperatures than oil-insulated transformers. Suitable solid insulating materials include glass, asbestos, mica, high-temperature resins, ceramics, etc.

Dielectric stress on the gas is several times as high as the stress on the adjacent solid insulation in series in the dielectric structure. Care in designing is required to avoid overstressing the gas. Sulfur hexafluoride has been used in transformers rated up to 25,000 kVA and up to 138 kV.

53. Nitrogen- and air-insulated transformers are generally limited to 15 kV and lower operating voltages. Air-insulated transformers in clean locations are frequently ventilated to the atmosphere. In contaminated atmospheres a sealed construction is required, and nitrogen is generally used at approximately atmospheric pressure and somewhat elevated operating temperature.

54. Contamination in small traces, particularly in the presence of moisture, may seriously reduce the dielectric strength of insulation materials or structures. Extreme cleanliness is needed in transformer manufacture.

55. Extreme dryness is necessary for the development of full dielectric strength. Appreciable quantities of water may decrease the dielectric strength and cause failure at operating voltage. Transformers are carefully vacuum-dried during manufacture before dielectric test. The initial level of dryness should be preserved through installation and subsequently in service.

56. Effect of Transient Voltages. When an impulse voltage is applied to the terminal of a transformer, the voltage does not divide uniformly throughout the turns of the winding. The initial voltage distribution is such that the line turns take much more than their share of the voltage. The steeper the applied wave, the greater the concentration of voltage on the line end turns. This happens because initially the only current flow is through the capacitance between turns of the winding and from winding to ground. Since the total distributed shunt capacitance to ground is much greater than the series capacitance through the winding to ground, most of the impulse current flows through the shunt capacitance near the line end of the winding, thereby resulting in a large voltage drop across the line-end portions of the winding.

Following the initial period, electrical oscillations occur within the windings. These oscillations impose greater stresses from the middle part of the winding to ground for long waves than for short waves. On the other hand, steep chopped waves impose the greater stresses between turns and coil portions. Switching surges are usually lower in amplitude than lightning surges and of much longer duration. Hence they may cause high stress from winding to ground.

57. Electrostatic shielding may be used to prevent excessive concentration of transient voltages on the line-end turns. The purpose of the electrostatic shielding is to increase the series capacitance of the winding, particularly near the line end, and thereby lower the voltage across the line-end coils and turns. The transient voltage behavior including effect of shielding can be predetermined with reasonable accuracy by calculation or by model tests. Thus the designer can provide adequate insulation safety factor. Most modern high-voltage transformers utilize electrostatic shields in combination with desirable winding arrangements to limit the concentration of impulse voltages near the line end of the winding.

58. Impulse insulation level may be demonstrated by factory impulse-voltage tests using 1.5×40 microseconds (μs) full waves and chopped waves. The full wave demonstrates the basic-impulse insulation level (BIL) for traveling waves coming into the station over the transmission line. The chopped wave demonstrates strength against a wave traveling along the transmission line after flashing over an insulator some distance away from the transformer. These waves do not simulate direct lightning strokes on or near the transformer terminals, which would result in the application of a steep front wave to the transformer winding. Such strokes are usually avoided by ground wires or protecting grounded structures. When it is desired to provide and demonstrate strength against direct lightning strokes, a front-of-wave test using an impulse voltage whose rate of rise is approximately 1,000 kV/μs is used. Such a test is not recognized in the USAS Standards.

59. High-voltage d-c stresses may be imposed on transformers used in terminal equipment for d-c transmission lines. Direct-current voltage applied to a composite insulation structure divides between individual components in proportion to the resistivities of the materials. In general the resistivity of an insulating material is not a constant but varies over a range of 100:1 or more, depending on temperature, dryness, contamination, and stress.

Cooling

60. Removal of heat caused by losses is necessary to prevent excessive internal temperature which would shorten the life of the insulation. Paragraphs **61** to **71** cover the procedure for calculating the internal temperature of oil-insulated self-cooled power transformers of conventional core-type construction using radiators.

61. The average temperature of a winding is the temperature determined by measuring the d-c resistance of the winding and comparing it with the measurement previously obtained at a known temperature. The rise of the average temperature of a winding above ambient temperature is

$$U = B + E + N + T \tag{11-29}$$

where B = degrees centigrade rise of effective oil over ambient, E = degrees centigrade rise of average oil over effective oil, N = degrees centigrade rise of average coil surface over average oil, T = degrees centigrade rise of conductor over coil surface, and U = degrees centigrade rise of average conductor over ambient.

62. Effective oil temperature is the equivalent uniform temperature with equal ability to dissipate heat to the air. The effective oil temperature is approximately the average of the oil entering the top of the radiator and the oil leaving the bottom of the radiator. The oil temperature is approximately the same as the temperature of the adjacent radiator surface exposed to air. A smooth, vertical transformer-tank surface will dissipate heat to the air as follows:

$$D_B = 1.40 \times 10^{-3}B^{1.25} + 1.75 \times 10^{-3}(1 + 0.011A)B^{1.19} \tag{11-30}$$

where A = degrees centigrade ambient temperature, B = degrees centigrade effective oil rise over ambient, and D_B = watts per square inch dissipated to the air.

The first term of Eq. (11-30) covers heat transferred by convection. Usually the radiator consists of parallel flattened tubes with limited accessibility to cooling air, and it is therefore necessary to multiply the first term by an experimentally determined friction factor (less than 1). The second term of Eq. (11-30) covers heat transferred by radiation, on the assumption of low temperature emissivity of 0.95, which applies to most painted surfaces commonly encountered. For any other value of low-temperature emissivity this term should be multiplied by emissivity/0.95 (see Table 11-3). Usually the radiator consists of parallel flattened tubes which radiate heat to each other. The net radiation of heat can be determined by considering the transformer and radiators replaced by a nonreentrant enveloping surface. If the second term of Eq. (11-30) is multiplied by the ratio of the area of the enveloping surface to actual surface (less than 1), the effect of reabsorption of radiation is eliminated. When radiation is small compared with convection, it can be assumed that A = 25°C and that $B^{1.19}$ can be replaced by $0.79B^{1.25}$, and Eq. (11-30) becomes

$$B = \frac{100D_B^{0.8}}{(0.44F + 0.56V)^{0.8}} \quad (°\text{C}) \tag{11-31}$$

where V = ratio of envelope surface area to actual surface area and F = friction factor determined by experiment.

63. The temperature rise of average oil over effective oil, E, is usually negligible for normal transformer designs. It may become important if (1) the center of gravity of the radiators is not elevated sufficiently above the center of gravity of the core and coils, (2) there is unusual loss in the oil space over the core such as might result from high-current leads, (3) a winding has unusually restricted oil ducts, or (4) pumps are used to circulate oil through the radiator without channeling the pumped oil through the oil ducts of the coil. For such cases, E is best evaluated by comparison with performance of previous designs.

64. The temperature rise of average coil surface over average oil, N, carries the loss in the coil through a film of stationary oil into moving oil. For a horizontal pancake coil (vertical axis) most of the heat escapes through the thin oil film on the upper surface and very little heat escapes from the lower surface. On the assumption that all the heat escapes from the upper surface, the temperature rise is

$$N = 13.2D_N^{0.8} \quad (°\text{C}) \tag{11-32}$$

where D_N = watts per square inch dissipated from the coil to the oil.

For a vertical pancake coil (axis horizontal) the heat leaves both sides equally, and

$$N = 14D_N \quad (°\text{C}) \tag{11-33}$$

65. The temperature rise of conductor over coil surface, T, carries the heat from the copper through the solid insulation applied to the conductor and the coil,

$$T = R_T t D_N \qquad (°\text{C}) \tag{11-34}$$

where D_N = watts per square inch dissipated from the coil to the oil, R_T = degrees centigrade per watt per inch thermal resistivity, and t = inch length of path.

66. The components of the winding rise over ambient are determined from Eqs. (11-31), (11-32) or (11-33), and (11-34) by using values of watts per square inch determined from the calculated losses and the design geometry. Then the total rise is determined from Eq. (11-29).

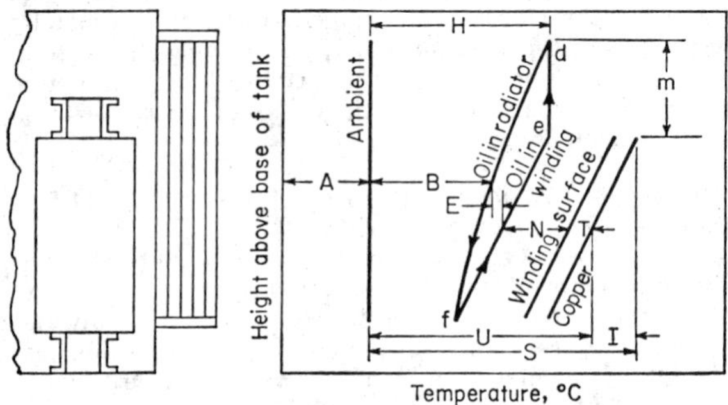

Fig. 11-23. Oil-circulation diagram.

67. Oil Circulation. The oil moves generally upward through ducts in the core and coils, rising in temperature as it goes. It moves generally downward through the radiators, falling in temperature as it goes (see Fig. 11-23). The space above the core and coils is filled with hot oil, so that the height-temperature curve of the circulating oil forms a triangle *def*. The difference in weight of the two columns of oil which furnishes the circulating force is proportional to the area of the triangle,

$$w = 2.50 \times 10^{-5} mI \tag{11-35}$$

where m = inches headroom, I = degrees centigrade top oil rise over average oil, and w = pound-force per square inch circulating force.

I is established by the following relations,

$$L = 222 I G_C \tag{11-36}$$

$$w = G_C R_H \tag{11-37}$$

$$I = 13.5 \left(\frac{L R_H}{m}\right)^{1/2} \tag{11-38}$$

where G_C = gallons per minute rate of circulation of oil, L = watts loss, and R_H = friction opposing oil flow in pound-force per square inch per gallon per minute.

R_H is not easily evaluated except by test, but Eq. (11-38) is useful in evaluating the effect of changing L or m.

The limiting temperature rises are

$$H = B + I \tag{11-39}$$

$$S = B + E + N + T + I = U + I \tag{11-40}$$

where H = degrees centigrade top oil rise over ambient temperature and S = degrees centigrade hot-spot rise over ambient temperature.

Equation (11-40) gives the temperature of the top pancake coil. Values of N and T may need to be separately computed for this coil if it has lower current density or more insulation on the conductor or on the coil. It may be necessary to examine other pancake coils near the top end, appropriately reduced values of I being used to find the hottest coil.

68. Variation of Temperature with Load. If temperature-rise conditions for rated load (or for any load) are known, temperature rises for any other load can be determined,

$$B_2 = B_1 \left(\frac{L_2}{L_1}\right)^{0.8} \tag{11-41}$$

$$N_2 = N_1 \left(\frac{L_2}{L_1}\right)^{0.8} \tag{11-42}$$

$$T_2 = T_1 \frac{L_2}{L_1} \tag{11-43}$$

$$I_2 = I_1 \left(\frac{L_2}{L_1}\right)^{1/2} \tag{11-44}$$

where B_1, N_1, T_1, I_1, and L_1 correspond to the known condition and B_2, N_2, T_2, I_2, and L_2 correspond to the new condition.

The total loss should be used for L_1 and L_2 in Eqs. (11-41) and (11-44). The resistance and eddy loss only should be used for L_1 and L_2 in Eqs. (11-42) and (11-43). The exponent in Eq. (11-42) should be 1.0 for vertical pancake coils. At constant voltage the resistance loss varies with the square of the load kVA and with the resistivity as affected by average temperature of copper according to Eq. (11-25). The eddy loss varies with the square of the load and inversely with the resistivity of the copper according to Eq. (11-27). The stray loss varies with the square of the load and may be assumed to vary inversely with the resistivity of the copper, like the eddy loss. The core loss may be assumed unaffected by load or temperature. For many purposes it is reasonable to assume that the entire load loss varies with the square of the load and with the resistivity.

69. Example. What is the temperature rise in a 30°C ambient at 80% load of a transformer with the following characteristics?

Load loss is two times no-load loss at 75°C.

Load loss is assumed all resistance loss.

Full-load temperature rises with 75°C losses are

$$\begin{aligned} B_1 &= 40 & & \\ N_1 &= 12 & U_1 &= 55 \\ T_1 &= 3 & S_1 &= 65 \\ I_1 &= 10 & H_1 &= 50 \end{aligned}$$

Upon assuming (for a trial) that the average copper temperature will be 74°C the relative load loss is

$$(0.8)^2 \times \frac{234.5 + 74}{309.5} = 0.638$$

and the relative total loss is

$$\frac{0.638 \times 2 + 1}{3} = 0.756$$

Then

$$\begin{aligned} B_2 &= 40(0.756)^{0.8} = 32.0°C \\ N_2 &= 12(0.638)^{0.8} = 8.4°C \\ T_2 &= 3 \times 0.638 = 1.9°C \\ I_2 &= 10(0.756)^{0.5} = 8.7°C \\ U_2 &= 32.0 + 8.4 + 1.9 = 42.3°C \\ S_2 &= 32.0 + 8.4 + 1.9 + 8.7 = 51.0°C \\ H_2 &= 32.0 + 8.7 = 40.7°C \end{aligned}$$

The average copper temperature is 42.3 + 30 = 72.3°C, which is close enough to the assumed 74°C.

70. Example. What is the temperature rise in a 30°C ambient at 140% load for the same transformer? Upon assuming (for a trial) that the average copper temperature will be 125°C the relative load loss is

$$(1.4)^2 \frac{234.5 + 125}{309.5} = 2.28$$

and the relative total loss is

$$\frac{2.28 \times 2 + 1}{3} = 1.85$$

Then

$$\begin{aligned} B_2 &= 40(1.85)^{0.8} = 65.4°\text{C} \\ N_2 &= 12(2.28)^{0.8} = 23.2°\text{C} \\ T_2 &= 3 \times 2.28 = 6.8°\text{C} \\ I_2 &= 10(1.85)^{0.5} = 13.6°\text{C} \\ U_2 &= 65.4 + 23.2 + 6.8 = 95.4°\text{C} \\ S_2 &= 65.4 + 23.2 + 6.8 + 13.6 = 109.0°\text{C} \\ H_2 &= 65.4 + 13.6 = 79.0°\text{C} \end{aligned}$$

The average copper temperature is 95.4 + 30 = 125.4°C, which is close enough to the assumed 125°C. The temperatures are excessive and should have been avoided by reducing the load at a suitable time.

71. Using similar calculations Table 11-2 shows the continuous load at 95°C hot-spot temperature in ambient temperatures from −20°C to 50°C for transformers with 55°C average winding rise and 65°C hot-spot temperature rise at rated load. Loss ratios of 2:1, 3:1, and 5:1 and effective oil rises of 35 and 45°C are covered.

Table 11-2. Loading at 95°C Hot-spot Temperature for a Transformer with 55°C Average Winding Rise and 65°C Hot-spot Rise at Rated Load

Ambient temperature, °C	% of rated kVA					
	Load/no-load loss ratio = 2:1		Load/no-load loss ratio = 3:1		Load/no-load loss ratio = 5:1	
	35°C Eff. oil rise	45°C Eff. oil rise	35°C Eff. oil rise	45°C Eff. oil rise	35°C Eff. oil rise	45°C Eff. oil rise
50	71	67	72	70	74	73
40	85	84	86	85	87	86
30	99	99	99	99	99	99
20	112	113	110	111	110	110
10	124	125	120	123	120	121
0	136	136	132	134	130	132
−10	147	148	143	145	140	142
−20	157	158	152	155	150	152

72. Transient Thermal Conditions. Since transformer temperature takes 15 to 25 h to become constant, it is sometimes desirable to calculate the temperature that will be reached at some earlier time.

$$B_H = B_U - (B_U - B_0)\epsilon^{-h/h_B} \tag{11-45}$$

$$h_B = \frac{K_B(B_U - B_0)}{L_0 - L_D} \tag{11-46}$$

$$K_B = 0.05H_C + 0.06H_S + 0.04H_T + 1.85G_T \tag{11-47}$$

where B_H = degrees centigrade effective oil rise after h h, B_0 = degrees centigrade initial effective oil rise, B_U = degrees centigrade ultimate effective oil rise, G_T = gallons of oil, ϵ = 2.718, h = hours after initial condition, h_B = hours time constant of transformer, H_C = pounds weight of copper in winding, H_S = pounds weight of core steel, H_T = pounds weight of tank, K_B = watthours per degree centigrade thermal capacity of transformer, L_D = watts dissipated at h = 0, and L_0 = watts loss at h = 0.

Equation (11-45) applies whether B_U is greater or less than B_0. However, if B_U and L_0 are zero,

$$B_H = B_0 \epsilon^{-h/h_B} \qquad (11\text{-}48)$$

$$h_B = \frac{K_B B_0}{L_D} \qquad (11\text{-}49)$$

73. The rise of average conductor over average oil ($N + T = Q$) is a single variable for transient calculations. The ultimate value is reached in 10 to 20 min.

$$Q_H = Q_U - (Q_U - Q_0)\epsilon^{-h/h_Q} \qquad (11\text{-}50)$$

$$h_Q = \frac{K_Q(Q_U - Q_0)}{L_0 - L_D} \qquad (11\text{-}51)$$

$$K_Q = 0.05H_C \frac{a/2 + b}{b} \qquad (11\text{-}52)$$

where a = square inches cross section of strand insulation, b = square inches cross section of copper strand, K_Q = watthours per degree centigrade thermal capacity of winding, L_D = watts dissipated at $h = 0$, L_0 = watts loss at $h = 0$, h = hours, h_Q = hours time constant of winding, Q_H = degrees centigrade average conductor rise over average oil after h h, Q_0 = degrees centigrade initial average conductor rise over average oil, and Q_U = degrees centigrade ultimate average conductor rise over average oil.

Equation (11-50) applies whether Q_U is greater or less than Q_0. However, if Q_U and L_0 are zero,

$$Q_H = Q_0 \epsilon^{-h/h_Q} \qquad (11\text{-}53)$$

$$h_Q = \frac{K_Q Q_0}{L_D} \qquad (11\text{-}54)$$

74. Example. If the transformer in Par. **69** operates in a 30°C ambient at 80% load until ultimate temperature conditions are established and then operates at 140% load, what will be the conditions after 4 h at 140% load? It is now necessary to specify additional characteristics of the transformer as follows:

$$\begin{aligned} H_C &= 20{,}000 & a &= 0.018 \\ H_S &= 60{,}000 & b &= 0.090 \\ H_T &= 20{,}000 \\ G_T &= 10{,}000 \\ \text{Total loss} &= 150{,}000 \text{ W at } 75^\circ\text{C} \end{aligned}$$

Then

$$K_B = 0.05 \times 20{,}000 + 0.06 \times 60{,}000 + 0.04 \times 20{,}000 + 1.85 \times 10{,}000 = 23{,}900$$

$$\frac{2 \times (1.4)^2(234.5 + 92.0)/309.5 + 1}{3} = 1.71$$

$$\left(\frac{32}{40}\right)^{1.25} = 0.756$$

$$h_B = \frac{23{,}900(65.4 - 32.0)}{(1.71 - 0.756)150{,}000} = 5.58 \text{ h}$$

$$\begin{aligned} B_H &= 65.4 - (65.4 - 32.0)\epsilon^{-4/5.58} = 49.1^\circ\text{C} \\ N_H &= 23.2 \\ T_H &= 6.8 \\ I_H &= 13.6 \end{aligned} \quad \text{(these reach ultimate value before 4 h)}$$

$$\begin{aligned} U_H &= 49.1 + 23.2 + 6.8 = 79.1^\circ\text{C} \\ S_H &= 49.1 + 23.2 + 6.8 + 13.6 = 92.7^\circ\text{C} \\ H_H &= 49.1 + 13.6 = 62.7^\circ\text{C} \end{aligned}$$

75. Example of Load Cycle. If the transformer examined in Par. **74** operates in a 30°C ambient on a daily cycle of 20 h at 80% load and 4 h at 140% load, what are the temperature rises at the end of the 140% load period? On assuming (for a trial) an effective oil rise at the start of the 140% load period of 35°C, the conditions at the end of the 140% load period are found as follows:

$$30.0 + 35.0 + 23.2 + 6.8 = 95.0\text{°C estimated copper temperature}$$

$$\frac{2 \times (1.4)^2(234.5 + 95.0)/309.5 + 1}{3} = 1.725$$

$$\left(\frac{35}{40}\right)^{1.25} = 0.846$$

$$h_B = \frac{23{,}900(65.4 - 35.0)}{(1.725 - 0.846)150{,}000} = 5.51 \text{ h}$$

$$B_H = 65.4 - (65.4 - 35.0)\epsilon^{-4/5.51} = 50.7\text{°C}$$

$$\left.\begin{aligned} N_H &= 23.2 \\ T_H &= 6.8 \\ I_H &= 13.6 \end{aligned}\right\} \quad \text{(these reach ultimate value before 4 h)}$$

$$U_H = 50.7 + 23.2 + 6.8 = 80.7\text{°C}$$

$$S_H = 50.7 + 23.2 + 6.8 + 13.6 = 94.3\text{°C}$$

$$H_H = 50.7 + 13.6 = 64.3\text{°C}$$

Conditions at the end of the 80% load period are found as follows:

$$30.0 + 50.7 + 8.4 + 1.9 = 91.0\text{°C estimated copper temperature}$$

$$\frac{2 \times (0.8)^2(234.5 + 91.0)/309.5 + 1}{3} = 0.782$$

$$\left(\frac{50.7}{40}\right)^{1.25} = 1.344$$

$$h_B = \frac{23{,}900(32.0 - 50.7)}{(0.782 - 1.344)150{,}000} = 5.30 \text{ h}$$

$$B_H = 32.0 - (32.0 - 50.7)\epsilon^{-20/5.30} = 32.5\text{°C}$$

This is close enough to the assumed 35.0°C. Figure 11-24 shows a complete set of time-temperature curves for the 24-h cycle.

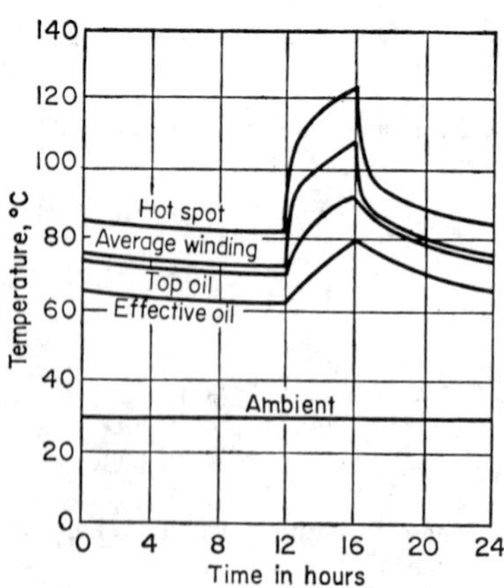

Fig. 11-24. Transformer temperatures during daily load cycle described in Par. **75**.

76. The short-circuit temperature rise is calculated on the assumption that all heat generated in the copper is stored in the copper until the short circuit is over,

$$C_2 = 309.5 \times \left\{\left[\left(\frac{C_1 + 234.5}{309.5}\right)^2 + g\right]\epsilon^{M^2s/10{,}800} - g\right\}^{1/2} - 234.5 \quad (11\text{-}55)$$

where C_1 = degrees centigrade average temperature of winding at start of short circuit, C_2 = degrees centigrade average temperature of winding at end of short circuit, M = rms kiloamperes per square inch current density, s = seconds duration of short circuit, and g = ratio of eddy loss to resistance loss in winding at 75°C.

The temperature resulting from any short circuit may be read directly from Fig. 11-25, which is plotted from Eq. (11-55).

77. Example. If a short circuit resulting in 50 kA/in² current density is held for 3.5 s in a winding with 10% eddy loss ($g = 0.1$) at 75°C, and with a starting temperature of 75°C, what is the average temperature of the winding at the end of the short circuit?

On the curve for $g = 0.1$, at 75°C the value of M^2s is 5,300.

$$5{,}300 + 50^2 \times 3.5 = 14{,}050$$

On the curve for $g = 0.1$, at $M^2s = 14{,}050$, the temperature is 246°C.

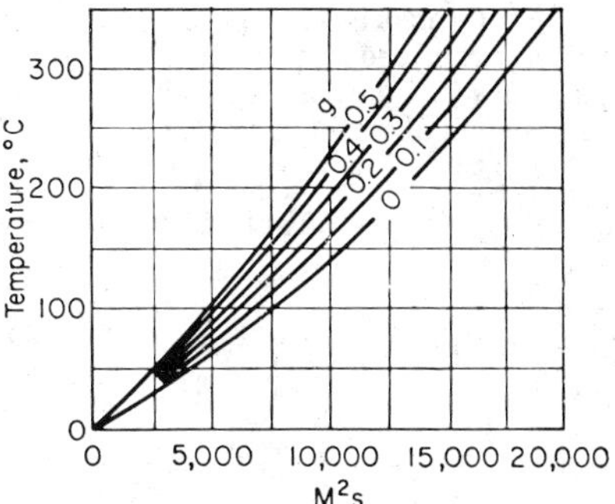

FIG. 11-25. Temperature of windings after short circuit, with all heat stored.

g = ratio of eddy loss to resistance loss at 75°C.

M = rms kiloamperes per square inch current density.

s = seconds duration of short circuit.

Example. For initial temperature of 75°C from the curve of $g = 0.1$ read: M^2s = 5,300

For $M = 50$ and $s = 3.5$:

$(50)^2 \times 3.5$: = 8,750

Total M^2s = 14,050

From curve of $g = 0.1$, read final temperature, 246°C.

78. Fan-cooled transformers use external fans to improve heat dissipation from the radiators, and sometimes internal pumps to circulate the oil through the radiators (and sometimes also through cooling ducts in the core and coils). With fan cooling the effective oil-temperature rise B is determined from test data on the particular arrangement, instead of Eq. (11-31). With oil pumps I is determined from Eq. (11-36). It is usually possible to obtain 67% more capacity with fans and pumps running.

79. Forced-cooled transformers use external oil-to-air heat exchangers requiring both air fans and oil pumps for all operating conditions. The effective oil rise B is calculated from the characteristics of the oil-to-air heat exchanger, and I is determined from Eq. (11-36). Forced-cooled transformers have no continuous load capacity without pump and fans.

80. Water-cooled transformers usually have the oil withdrawn from the transformer at the top of the tank, pumped through an external cooler, and returned to the bottom of the tank. The temperature drop of the oil in passing through the cooler is

$$Y = \frac{L}{G_C \times 111} \tag{11-56}$$

where Y = degrees centigrade drop of oil in cooler, L = watts loss, and G_C = gallons per minute rate of circulation of oil.

The effective oil-temperature rise B may be calculated from the characteristics of the oil-to-water heat exchanger. The top oil rise over average oil, I, is $Y/2$. The other components of temperature rise are calculated as for a self-cooled transformer.

81. Operation at high altitude increases the effective oil rise of air-cooled transformers. USAS C57 provides for a compensating correction of 0.4% of rated kVA of self-cooled transformers or 0.6% of rated kVA of forced-cooled transformers for each 330 ft of additional altitude above 3,300 ft altitude.

82. Effect of Tank Color. Most paint used on transformers has a low-temperature emissivity of about 0.95. Metallic surfaces, particularly polished surfaces, have less low-temperature emissivity and will cause correspondingly higher oil-temperature rise. The same is true of aluminum or bronze paint. For these cases Eq. (11-31) can be used with V' substituted for V, where V' is $V/0.95$ multiplied by the low-temperature emissivity from Table 11-3. On large power transformers with many radiators or heat exchangers the effect is small.

For transformers exposed to intense sunlight the additional temperature rise resulting from the use of aluminum paint is largely offset[1] by the fact that aluminum paint absorbs only about 55% of the impinging solar radiation, while most commonly used paints absorb about 95%.

Table 11-3. Low-temperature Total Emissivity

Aluminum, highly polished	0.08
Copper	0.15
Cast iron	0.25
Aluminum paint	0.55
Oxidized copper	0.60
Oxidized steel	0.70
Bronze paint	0.80
Black gloss paint	0.90
White lacquer	0.95
White vitreous enamel	0.95
Green paint	0.95
Gray paint	0.95
Lampblack	0.95

Load Tap Changing[2,3]

83. Ratio Changes with Shifted Taps. In transformers designed for maintaining a constant voltage on a power system, the ratio of transformation is usually changed by increasing or decreasing the number of active turns in one winding with respect to another winding. Since the turn ratio of the transformer must be changed without interfering with the load, means are provided for shunting the load current from one winding tap to the next. For this purpose an auxiliary **preventive autotransformer** or a bridging reactor is generally used and is designed to limit the resulting circulating current to a safe value during the interval that two adjacent taps are bridged. Because of the circulating and the load current which passes through the current-limiting impedance, arcing always takes place as the power circuit is transferred from tap to tap.

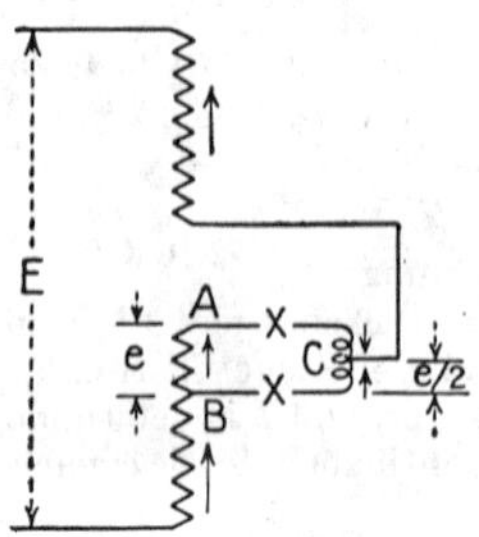

Fig. 11-26. Bridging position for ratio change under load.

Although a variety of switching equipments and transformer connections have been used for the purpose of changing taps under load, the underlying principle remains unchanged and is shown by the transformer connection in Fig. 11-26.

Example. To move from transformer tap A to B, it is first necessary to close the circuit to B, as shown in Fig. 11-26, before opening the circuit at A. During the interval when A and B are both closed on adjacent taps, a circulating current flows through and is limited by the impedance in the loop composed of the tap winding AB and reactor or autotransformer C. With both ends of the reactor connected to A the load current divides equally between the two halves of the reactor. Since the current flows in opposite directions, a negligible amount of reactance is introduced into the circuit and the only loss is the I^2R due to the 50% load current in each half of the reactor winding. With A closed and B open, all the load current flows through one-half of the reactor winding, magnetizing the reactor and thereby introducing into the circuit the induced reactance voltage. It is important, therefore, that the reactance be kept as low as possible to avoid excessive arcing duty on the circuit-interrupting device.

With A and B closed on adjacent taps, the tap voltage e is impressed on the reactor C

[1] Montsinger, V. M., and Wetherill, L. Effect of Color of Tank on the Temperature of Self-cooled Transformers under Service Conditions; *Trans. AIEE*, 1930, Vol. 49, p. 41.

[2] Some of the material under this title is reprinted with certain modifications from "Transformer Engineering," by Blume, L. F., Boyajian, A., Camilli, G., Lennox, T. C., Minneci, S., and Montsinger, V. M., published by John Wiley & Sons, Inc., 2d ed., 1951.

[3] *AIEE Paper* 57-48, Short-circuit Capability Tests of Load Tap Changing Mechanisms, by O. P. McCarty and W. M. Johnson.

and causes a circulating current to flow through the impedance loop. Because of the autotransformer action, a voltage midway between A and B is impressed in the circuit. The load current again divides equally through the reactor windings.

To avoid an excessive voltage drop through the reactor when one side is open and at the same time to keep the circulating current at a low level when in the bridging position, the reactor is usually designed for a magnetizing current of approximately 60% of the normal full-load current.

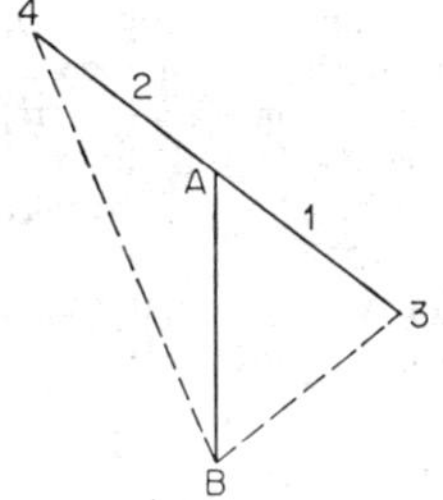

FIG. 11-27. Vector relations for bridging position.

84. Voltage across Reactor. Figure 11-27 shows the voltage relations across the reactor and switching contacts during a tap-changing cycle using a reactor designed for 60% circulating current and with 100% load current at 80% power factor flowing through it. Perfect interlacing between the reactor halves is assumed, and the voltage drop due to resistance of the reactor winding is neglected.

A study of Fig. 11-27 will disclose the fact that increasing the reactance of the reactor to reduce the circulating current will:

1. Increase the voltage across the full reactor winding.
2. Increase the voltage to be ruptured.
3. Introduce undue voltage fluctuations in the line.

Since B-4 and B-3 represent the voltages appearing across the arcing contacts when the bridging position is opened at A and B, the voltage-rupturing duty will increase with:

a. Increase in voltage between adjacent taps.
b. Increase in load.
c. Decrease in power factor of the load.
d. Decrease in the magnetizing current for which the reactor is designed.

85. Load Tap-changer Motor Mechanisms. Tap-changing equipments differ from induction regulators in that they are arranged to give a definite number of voltage steps compared with the relatively unlimited number obtainable with an induction regulator. Motor-operated tap changers must therefore have their controls arranged so that, after a tap change is initiated, operation continues independently of the initiating means until the cycle of one operation is completed.

The mechanical coupling between the operating motor and the tap-changing switches may be through fixed-ratio gears, Geneva gears, cams, springs, or combinations of these. All mechanisms require means for keeping the motor energized until the change of tap is accomplished and for bringing the tap changer to rest on each operating position. The degree of permissible coasting of the motor is determined by the motor mechanism and the switch design.

The need for extremely accurate stopping of the motor is avoided by arranging the parts so that the motor may coast somewhat after the operating position is reached without moving the tap changer around the operating position. This may be accomplished by the inactive sectors of Geneva gears or cams or by the motor travel inherently involved in recharging a spring. Motor mechanisms are provided with limit switches and mechanical stops to prevent operation beyond the limit positions. Operation counters and position indicators are standard auxiliaries on most tap changers. On large station-type units where the control devices are generally mounted on a remote-control panel, remote position indicators, either of the lamp type or of the self-synchronizing type, are generally provided.

86. Automatic Control for Tap Changers. The principle of automatic control for tap changers is, in general, the same as for induction regulators, although several important differences exist owing to the inherent differences in the nature of the two types. The induction regulator is practically instantaneous in its operation, whereas a tap changer generally has a time delay introduced between the voltage-regulating relay and the motor mechanism. The introduction of time delay eliminates unnecessary tap changes and thereby reduces wear on the mechanisms and prolongs the life of the contacts and the oil in which arcing takes place. Voltage regulation may even be improved because over-corrections for voltage changes of only momentary duration are eliminated.

The voltage-regulating relay (or contact-making voltmeter) should be adjusted so that the voltage band width, or spread between voltages at which the raising and lowering contacts close, will be not less than the percentage transformer tap plus an allowance for irregular voltage variations. For example, a tap-changing transformer with 1¼% taps should have a minimum voltage bandwidth of approximately 1¼% + ½% = 1¾%.

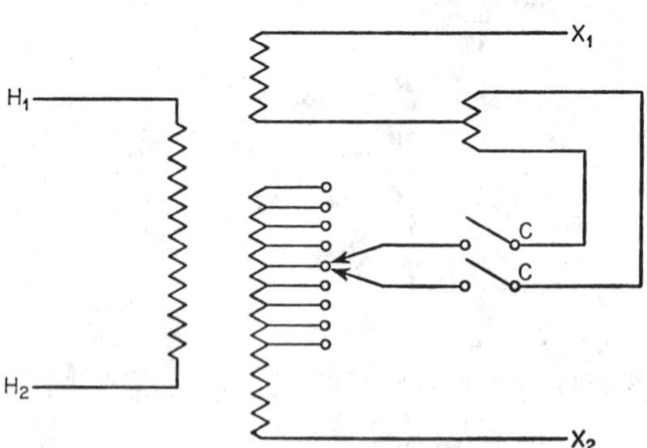

Fig. 11-28. Tap-changing equipment in middle of winding.

87. Voltage Control, a Part of the Power Transformer. The simplest and generally the least expensive connection for voltage control is to provide the necessary taps in the power transformer. For single- or 3-phase delta connection the taps are preferably located on the interior of the winding (Fig. 11-28) so as to avoid the abnormal voltage stresses to which end coils are usually subjected. In the Y-connection the taps may be placed at the neutral end of the winding, and if the neutral is to be solidly grounded, it becomes possible by locating the taps next to ground to use load tap changers designed with greatly reduced insulation; thus, for example, 15-kV apparatus may be placed in the grounded neutral end of a circuit as high as 69 kV.

If the rated current of the transformer exceeds that of the switching equipment, a second core may be added, which constitutes a series transformer (Fig. 11-29). Excitation is derived from taps inserted in the secondary of the power transformers, and by means of the series transformer, the desired voltage is inserted into the circuit. Thus, if a ratio of 3:1 exists in the series transformer, the current handled by the switching equipment becomes one-third of the current in the line.

88. Regulating Transformers (**Single-core**). When the power transformer is not available or it is not desirable to equip the power unit with voltage control, regulating autotransformers are used. In the simplest of these, the necessary taps and switches are placed in the series winding of an autotransformer (Fig. 11-30). For 3-phase circuits, in order that the derived voltage may be in phase with circuit voltage, a Y-connection is commonly used, and hence all the precautions necessary to safeguard the operation of Y-connected autotransformers should be observed. A tertiary winding may or may not be provided,

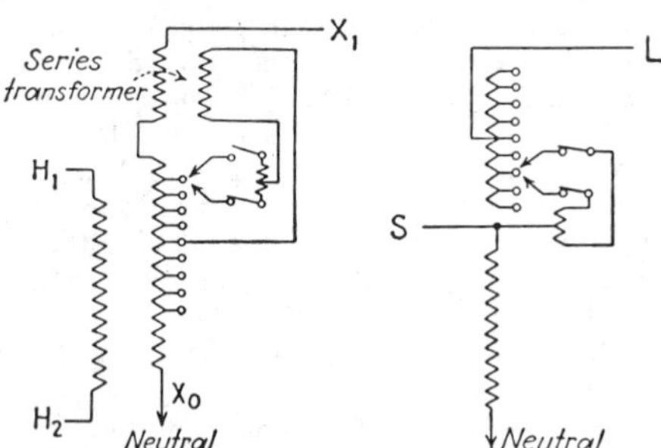

Fig. 11-29. Tap-changing circuit with tap located in interior of transformer winding and an auxiliary series transformer to bring the current and voltage duty on equipment within rating limits.

Fig. 11-30. Regulating single-core autotransformer with taps located in series winding and circuit connected for boost and buck.

depending upon circuit conditions. As the series winding is inserted in the line, adequate insulation must be provided for the tap-changing equipment and taps against the abnormal voltages to which the circuit is subjected.

For very-high-voltage autotransformers the contactor component of the load tap changer may be supported in a compartment on top of the high-voltage common line bushing. The insulation inside the compartment then becomes that required for one tap, and the major insulation to ground is provided by the high-voltage bushing.

Economy in transformer size may be obtained by means of a reversing switch which functions to reverse the connections to the series winding when the regulator is passing through the neutral position. The circuit is so designed and the mechanical sequence is such that the reversing switch operates without rupturing current. The connection diagram (Fig. 11-31) shows the load tap changer provided with nine taps, which gives 17 full-cycle or 33 half-cycle positions. The ratio adjuster is designed with contacts uniformly spaced on the circumference of the circle so as to permit motion through two revolutions.

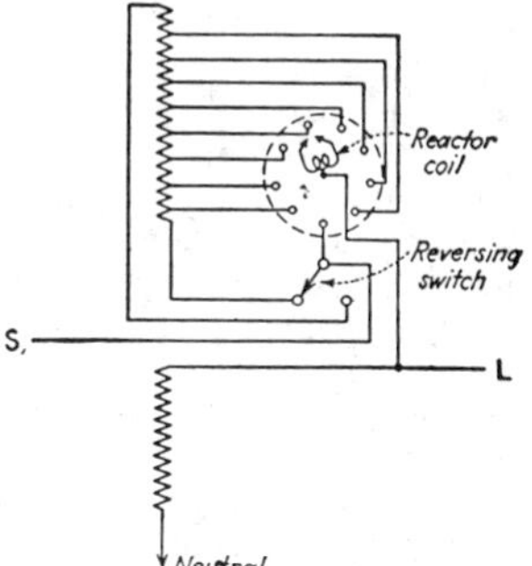

Fig. 11-31. Regulating single-core autotransformer with reversing switch to obtain buck and boost.

89. Two-core Regulating Transformers. In many instances the voltage of the circuit is greater than that for which the switching equipment is designed, and in others the current to be handled exceeds the safe limits of operation. In either case voltage control can be obtained, without the design of special switching equipment, by using two cores (Fig. 11-32), one consisting of an insulating series transformer and the other an exciting transformer, the combination functioning, as far as the circuit is concerned, like an autotransformer. The primary of the exciting transformer is generally connected in Y in order that the derived voltages may be in phase with circuit voltages. The secondary of the exciting transformer provided with the regulating taps is usually connected in delta. The local circuit, consisting of the secondary of the exciting transformer with its taps and the primary of the series transformer being insulated from the main circuit, may be designed for the voltage and current best suited for the available switching equipment. As in this arrangement the size of the core and coils is considerably greater than that normally required in single-core design, it is used only when the voltage or current limitations of the switching equipment prevent the use of a single-core design and when the control cannot be inserted in the grounded neutral of the transformer bank.

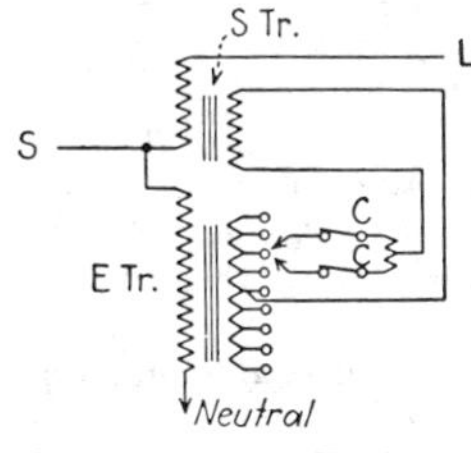

Fig. 11-32. Exciting transformer with taps in secondary and series transformer forming complete isolation for the tap-changing equipment.

90. Tap-changer Designs for Moderate KVA and Current. In the smaller ratings where both the kVA and the current are moderate, the energy to be ruptured in switching from tap to tap becomes relatively so small that light and simple equipments are feasible. A variety of mechanical designs, together with special circuits, has been evolved with the purpose of providing simpler, smaller, and inherently less expensive equipments. The following may be noted:

1. Designing the tap changer so that it is capable of rupturing the current directly on the same switches which select the taps.

2. Designing the circuit so that the tapped winding is reversed in going from maximum to minimum range, thereby securing a substantial reduction in the rating of core and coils for a given output.

3. Using higher switching speed, by means of which the life of the arcing contacts is increased.

91. Tap Changers Designed to Interrupt Current. The contactors C (Fig. 11-28) operate to open the switching circuits so that there is no interrupting duty on the selector contacts which connect to the transformer taps. When the rated current is moderate, it becomes possible to rupture the current directly on the tap-selector switches and thus obtain a major economy in the cost of the mechanical equipment. This method is shown in Fig. 11-31.

92. High-speed Switching. Large units include contactors with high-speed contacts which serve as extinction devices that are specially designed to interrupt repeatedly the high currents and voltages encountered. They may be single- or multiple-break contactors operating in oil or in air with magnetic-arc chutes, oil-blast contactors, or contactors operating in a vacuum. In small units, however, the arcing duty is mild. It is nevertheless necessary to keep in mind that mild arcing duty in the smaller equipments is partly offset by the likelihood of greater frequency of operation. Such units are usually equipped with full automatic control; they are likely to be located on distribution circuits where the voltage is more erratic. Many of them are located out on the lines at considerable distances from substations, and some of them are placed on poles. It is desirable, therefore, to reduce maintenance to a minimum. For these reasons, it is necessary to provide means for high-speed switching on the smaller units where the tap-selector switches are used to rupture current. High-speed action of the tap-changer switches is obtained through Geneva gears or cams which bring the contact fingers to the required high speed at the moment of parting or through a spring drive in which the motor is used to store energy with the release of a spring snapping the contact fingers from one shelf to the next. By these means the duration of the switching arc may be reduced to 1 or 2 c, correspondingly reducing the amount of contact burning and increasing the life of the contacts.

93. Use of Resistors. Another method, used more frequently in load tap changers of European design, makes use of resistors to bridge the tap instead of a reactor or preventive autotransformer. Figure 11-33 shows the tap selectors 1 and 2 connected to alternate taps in the winding. The contactor, or diverter switch, shown connected to 1 and R_1 progressively connects only to R_1, then to R_1 and R_2, then to R_2, and finally to R_2 and tap selector 2. For the next tap change, tap selector 1 moves to an adjacent tap and is then followed by the diverter switch operating in a reverse manner from R_2 and 2 back to R_1 and 1. Since the resistors are designed to carry current for only a very short time, the diverter switch is usually spring-actuated and moves through its sequence in a few cycles of 60-c current. This method has the advantage of relatively small-size resistors but requires a transformer tap for each operating voltage, while the reactor circuit uses the tap bridging position for an operating voltage and thus requires half the number of transformer taps.

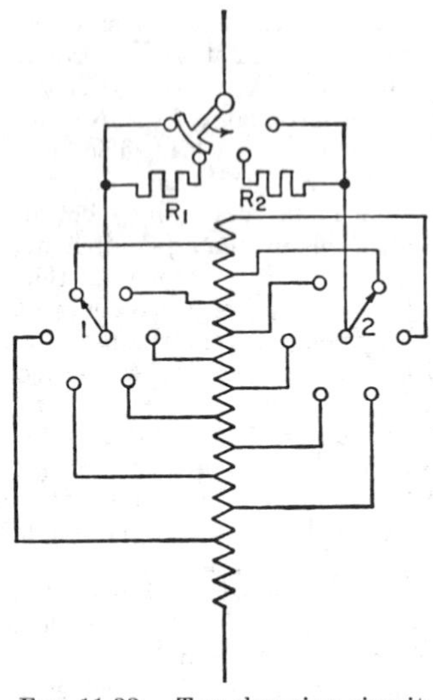

Fig. 11-33. Tap-changing circuit employing tap selectors and contactor, or diverter switch, and resistors to bridge the taps during switchover.

94. Applications for Voltage-control Equipments. The control of transformer ratio under load is a desirable means of regulating the voltage of high-voltage feeders and of primary networks. It may be used for the control of the bus voltage in large distributing substations. It finds a wide field of application in controlling the ratio on step-up transformers operating from power stations whose bus voltage must be varied to suit local distribution.

In industrial work, it is used for the control of current in a variety of furnace operations and electrolytic processes. It also furnishes a convenient means for voltage regulation of concentrated industrial loads.

Many load tap-changer equipments are installed at points of interconnection between systems or between power stations, in order to control the interchange of reactive current, or, in other words, to control the power factor in the tie line. This reactive current may be highly undesirable, especially as it may add to the burden on a fully loaded generating system. It can be increased, eliminated, or reversed by inserting a suitable small ratio of transformation between the systems. It can be varied in amount and in direction of flow to suit varying system conditions if this ratio is variable and under the control of a station operator. Inserting such a ratio of transformation in a tie line by means of a tap-changing equipment is equivalent in its effect on the flow of reactive current to raising or lowering the voltage on one of the systems. Current can be exchanged at any power factor from zero lag to zero lead, without interfering with the voltage maintained on either system.

95. Transformers for Phase-angle Control. Tap-changing equipment is sometimes used in a loop system for phase-angle control for the purpose of obtaining minimum losses in the loop due to unequal impedances in the various portions of the circuit.

Transformers used to derive phase-angle control do not differ materially, either mechanically or electrically, from those used for in-phase control. In general, phase-angle control is obtained by interconnecting the phases, i.e., by deriving a voltage from one phase and inserting it in another.

A simple arrangement given in Fig. 11-34 illustrates a single-core delta-connected autotransformer in which the series windings are so interconnected as to introduce into the line a quadrature voltage. One phase only is printed in solid lines so as to show more clearly how the quadrature voltage is obtained. The terminals of the common winding are connected to the midpoints of the series winding in order that the in-phase voltage ratio between the primary lines *ABC* and secondary lines *XYZ* is unity for all values of phase angle introduced between them.

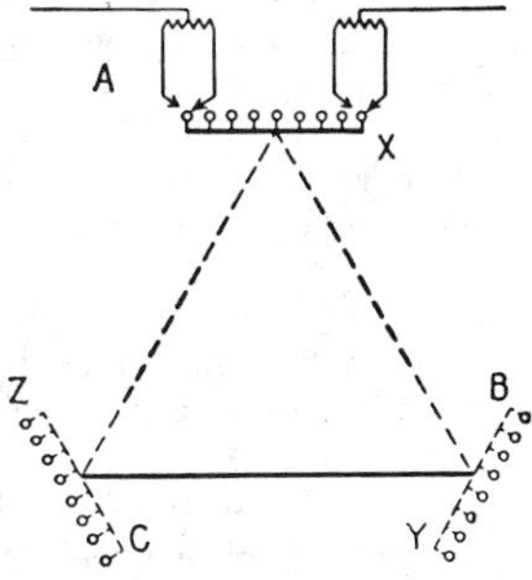

FIG. 11-34. Single-core delta-connected common winding.

AUDIBLE SOUND

96. Source of Sound. Transformers, although they are classed as static apparatus, vibrate and radiate audible sound energy. There are two distinct and different sources. One source is the auxiliary cooling equipment, such as fans, blowers, coolers, and pumps, which are characterized by a broad-band frequency spectrum of approximately equal amplitude, commonly called "white noise." The second and major source of transformer sound is the core. This source is characterized by a range of harmonic tones which are even multiples of the exciting frequency. Thus, for 60 c/s excitation, the core will radiate energy at 120, 240, 360, 480, 600 c/s, and so on.

Alternating flux flowing in the core laminations causes them to change length by the effect known as magnetostriction. While the magnetostrictive effect is very small, being measured in parts per million, it is the principal cause of core noise. It is independent of direction of magnetization; so there are two extensions per cycle of magnetization, which accounts for the fact that the fundamental sound frequency is twice the excitation frequency. Since the magnetostriction characteristic is not a linear function of flux density, many harmonics are generated also. The fundamental or any of the harmonics may be amplified by mechanical resonances in the core or tank structures.

There are also magnetic forces set up between laminations at the joint region where flux passes from one lamination to the next. These forces may likewise produce motion and generate noise, but this is not usually a major source for core constructions which have overlapping laminations in the joint region. It can be a major source for devices which have butt joints in the magnetic circuit, such as gapped iron-core reactors. Of very minor importance are the magnetic forces in the windings produced by load current. In general, the load current has little effect on sound level, but a change in applied voltage has considerable effect.

97. Sound Measurement. Sound waves produce small fluctuations in the atmospheric pressure which are sensed by the human ear. The alternating portion of the total pressure is called "sound pressure." To handle the wide range of sound pressures which are encountered, a quantity called "sound-pressure level," which bears a logarithmic relation to sound pressure, has been introduced. Sound-pressure level is measured in decibels and is defined by[1]

$$P = 20 \log \frac{F}{0.0002} \tag{11-57}$$

where F = rms dynes per square centimeter sound pressure and P = decibels sound-pressure level.

Sound-level-measuring equipment as specified by USASI Standards[2] consists of a microphone, amplifier, frequency weighting network, and indicating meter. Three weighting networks A, B, and C have been standardized. The C network provides an essentially flat response over the frequency range of 20 to 10,000 c/s. The response of the human ear approaches this flat characteristic for very high sound-pressure levels (over 85 dB). At lower levels, the response of the human ear is not linear with frequency, and the A and B networks are weighted to simulate the response of the ear to single frequency tones of 40 and 70 dB, respectively. Weighted values read on the A and B scales are referred to as "sound levels," rather than sound-pressure levels.

The A, or 40-dB, weighting network has been selected for all transformer sound measurements, since it represents the approximate sound level of a transformer at the nearest residence. With this weighting network, frequencies up to 1,000 c/s and above 5,000 c/s are given a negative decibel weighting and from 1,000 to 5,000 c/s slightly positive weighting.

98. Standard Transformer Sound Level. *NEMA Publ.* TR 1 specifies the method for measuring the average sound level of a transformer. The measured sound level is the arithmetic average of a number of readings taken around the periphery of the unit. For transformers with a tank height of less than 8 ft, measurements are taken at one-half tank height. For taller transformers, measurements are taken at one-third and two-thirds tank height. Readings are taken at 3-ft intervals around the string periphery of the transformer, with the microphone located 1 ft from the string periphery and 6 ft from fan-cooled surfaces. The ambient must be at least 7 and preferably 10 dB below that of the unit being measured. There should be no acoustically reflecting surface, other than ground, within 10 ft of the transformer. The A weighting network is used for all standard transformer measurements regardless of sound level.

NEMA Publ. TR 1 contains tables of standard sound levels. For oil-filled transformers from 1,000 to 100,000 kVA self-cooled (400,000 kVA forced-oil-cooled) standard levels are given approximately by Eq. (11-58),

$$L = 10 \log E + K \tag{11-58}$$

where E = equivalent 2-winding, self-cooled kVA (for forced-oil-forced-air-cooled units, use 0.6 × kVA), K = constant, from Table 11-4, and L = decibel sound level.

Example. A transformer rated 50,000 kVA self-cooled, 66,667 kVA forced-air-cooled, 83,333 kVA forced-oil-forced-air-cooled, at 825 kV BIL, would have standard sound levels of 78, 80, and 81 dB on its respective ratings.

Table 11-4. Values of *K* for Eq. (11-58)

High-voltage winding BIL, kV	Self-cooled and water-cooled ratings	Forced-air and forced-oil-forced-air-cooled 25 to 35% above self-cooled rating	Forced-air and forced-oil-forced-air-cooled 67% above self-cooled rating or without self-cooled rating
350 and below	28	30	31
450–650	30	32	33
750–825	31	33	34
900–1,050	32	34	35
1,175	33	35	36
1,300 and above	34	36	37

[1] Handbook of Noise Control, C. M. Harris, McGraw-Hill Book Co., p. 1–14 (1957).
[2] Standard S1.4-1961 published by U.S.A. Standards Institute.

99. Public Response to Transformer Sound. The basic objective of a transformer noise specification is to avoid annoyance. In a particular application, the NEMA Standard level may or may not be suitable, but in order to determine whether it is some criterion must be available. One such criterion is that of audibility in the presence of background noise.[1] A sound which is just barely audible should cause no complaint.

Studies of the human ear indicate that it behaves like a narrow-band analyzer, comparing the energy of a single frequency tone with the total energy of the ambient sound in a critical band of frequencies centered on that of the pure tone. If the energy in the single frequency tone does not exceed the energy in the critical band of the ambient sound, it will not be significantly audible. This requirement should be considered separately for each of the frequencies generated by the transformer core.

The width of the ear-critical band is about 40 c/s for the principal transformer harmonics. The ambient sound energy in this band is 40 times the energy in a 1-c/s-wide band. The sound level for a 1-c/s bandwidth is known as the "spectrum level" and is used as a reference. The sound level of the 40-c/s band is 16 dB (10 log 40) greater than the sound level of the 1-c/s band. Thus, a pure tone must be raised 16 dB above the ambient spectrum level to be barely audible.

The transformer sound should be measured at the standard NEMA positions with a narrow-band analyzer. If only the 120- and 240-c/s components are significant, an octave-band analyzer can be used, since the 75- to 150-c/s and 150- to 300-c/s octave bands each contain only one transformer frequency. The attenuation to the position of the observer can be determined according to Par. **100**.

The ambient sound should be measured at the observer's position. For each transformer frequency component, the ambient spectrum level should be determined. An octave-band reading of ambient sound can be converted to spectrum level by the equation

$$S = B - 10 \log C \tag{11-59}$$

where B = decibels octave-band reading, C = cycles per second octave bandwidth, and S = decibels spectrum level.

Example. Consider the following case:

Transformer sound at 120 c/s by NEMA method = 72 dB
Transformer-sound attenuation to observer (calc. according to Par. **100**) = 35 dB
Ambient sound in the 75- to 150-c/s octave band = 36 dB

$$72 - 35 = 37 \text{ dB at the observer's position}$$
$$36 - 10 \log (150 - 75) = 17.3\text{-dB ambient spectrum level}$$

The 120-c/s transformer sound at the observer's position exceeds the ambient spectrum level by 19.7 dB. This is 3.7 dB greater than the 16-dB differential which would result in bare audibility; so the transformer sound will be audible to the observer.

When transformer sound exceeds the limits of bare audibility, public response is not necessarily strongly negative. Some attempts have been made to categorize public response on a quantitative basis when the sound is clearly audible.[2] For a case where specific knowledge of transformer- and ambient-sound-level frequency composition is not available, some more general guides are useful.[3] Typical average nighttime ambient-sound levels for certain types of communities have been established. These are 30 dB for a "quiet-suburban," 35 dB for a "residential-suburban," and 40 dB for a "residential-urban" community. All sound levels are based on the A scale of weighting. Calculations for typical transformer frequency distributions have been made to determine the nighttime transformer noise which will be audible 50% of the time in these communities. The results are 24 dB for quiet-suburban, 29 dB for residential-suburban, and 34 dB for residential-urban. The NEMA Standard sound level can be

[1] AIEE Committee Report, "Transformer Noise Measurement Methods"; *Trans. AIEE*, 1954, Vol. 73, p. 683.

[2] SCHULZ, M. W., JR., and RINGLEE, R. J. Some Characteristics of Audible Noise of Power Transformers and Their Relationship to Audibility Criteria and Noise Ordinances; *Trans. AIEE*, 1960, Vol. 79, p. 316.

[3] Power Transformer Noise—Prevention and Cure, *Stanford Res. Inst. Rept.*, SRI Project S2410, 1960.

corrected for attenuation with distance to the nearest observer and checked against the above guides for audibility.

100. Sound Attenuation with Distance. A point source in a free field radiates sound in spherical waves. The resultant sound pressure varies inversely with the square of the distance from the source; so the sound level is reduced by 6 dB for each doubling of distance. The sound of auxiliary cooling equipment follows this relation for decrement with distance, since it is the sum of point-source sound contributions.

The transformer tank, which radiates vibrational energy from the core, is a more complex sound source and does not appear as a point source except at substantial distance from the tank. The modes of tank vibration are complicated, and various parts of the tank may act as independent sources, with different amplitudes, phase relations, and frequencies. Studies of scale models[1] and full-size units have uncovered certain useful relationships as follows,

$$A = 20 \log \frac{2.83D}{Q} \tag{11-60}$$

$$Q = 1.7(WH)^{1/2} \tag{11-61}$$

where A = decibels attenuation for distance exceeding Q, D = distance from transformer to observer, H = height of transformer tank, Q = critical distance from transformer beyond which it appears as a point source, and W = width of transformer tank perpendicular to line from transformer to observer.

Equations (11-60) and (11-61) apply in the absence of wind, temperature gradients, and reflecting surfaces other than ground. Each of these factors may significantly influence the observed sound level at a distance from the source, but not always in predictable fashion.

101. Selection of Site. There are a number of methods available for avoiding transformer-noise complaints. Some of the discussion in the previous paragraphs suggests that potential noise problems should be considered when the substation site is selected. It may be possible to take advantage of attenuation with distance to reduce the transformer sound at the nearest observer position to an inaudible level. It may also be possible to choose the site in a location where the normal ambient noise will mask the transformer sound. If these possibilities are kept in mind during the planning stages, more expensive solutions to noise problems may be avoided later.

102. Design Measures. Manufacturers have at their disposal a variety of means of obtaining sound reduction. Most basic is the use of a reduced level of induction in the transformer-core steel. More steel is required, but the magnitude of the magnetostrictive motion is decreased, with a resultant decrease in radiated energy. This means is employed to achieve small to moderate sound reductions. Moderate reductions can also be realized by the use of barriers within the tank. Some of these are "soft" barriers, which operate on the principle of absorbing vibrational energy from the core and reducing its transmission to the tank. Others are "mass" barriers, which operate on the principle of loading the tank to decrease its magnitude of vibration for given energy transmission from the core. To achieve large sound reductions (as much as 25 to 30 dB), some manufacturers employ complete external enclosures of steel. For most substation units, these enclosures can be preassembled and shipped in place over the transformer tank.[2]

When sound originating at the core is lowered by one of the above means, sound originating at the cooling equipment may have to be reduced also. Fans on radiators or in coolers are the principal sources. Their contribution to total sound can be decreased by using a larger quantity of slower-speed fans.

103. Improving Existing Installation. To reduce the sound level of an existing transformer, the most satisfactory method has been found to be the erection of barrier walls on one or more sides of the transformer. The attenuation which can be achieved

[1] JOHNSON, K. A., RINGLEE, R. J., and SCHULZ, M. W., JR. Use of Scale Models for Studying Power Transformer Audible Noise; *Noise Control*, March, 1956, Vol. 2, p. 54.

[2] SCHULZ, M. W., JR., and MCNUTT, W. J. A Way to Get Low Sound Levels in Large Power Transformers—Preassembled Enclosures; *Trans. AIEE*, 1957, Vol. 76, p. 1365.

depends on the transmission loss through the barrier, the diffraction over and around the barrier, and the pressure build-up between the tank and the barrier.

Transmission loss through a barrier wall is a function of the mass of the wall. Structural requirements of most practical masonry barriers ensure sufficient mass to produce 25 to 40 dB attenuation through the wall. The effectiveness is usually limited by diffraction around the edges of the barrier. A theoretical method for calculation of attenuation as limited by diffraction has been formulated[1] as follows,

$$N = \frac{2}{\lambda}\left[(M^2 - U^2)^{1/2} - M + (G^2 - U^2)^{1/2} - G\right] \tag{11-62}$$

where M, U, and G are defined in Fig. 11-35, in any convenient unit (feet or inches), λ = wavelength of harmonic under investigation, in units consistent with M, U, and G, and N = dimensionless parameter given in Fig. 11-35.

The calculation procedure is to determine N from the equation and then find the corresponding attenuation from Fig. 11-35*b*.

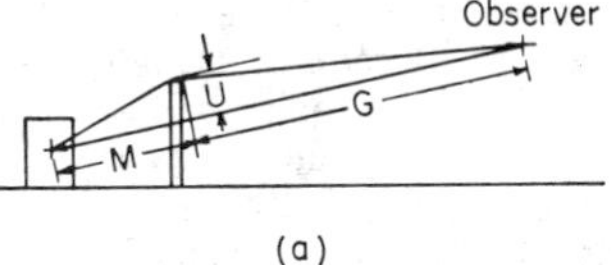

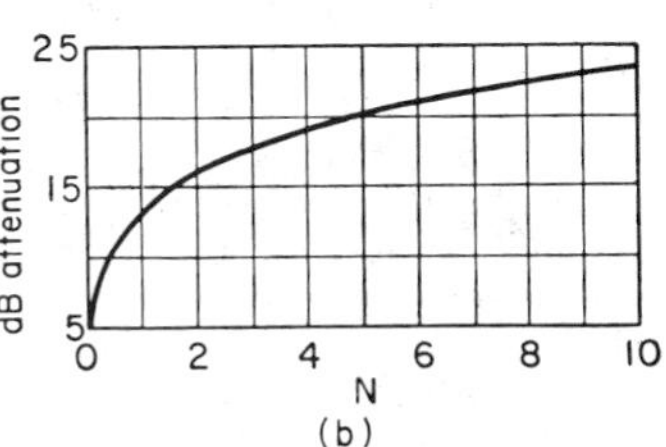

FIG. 11-35. Effectiveness of a barrier in reducing noise level. (*a*) Identification of dimensions for calculation of dimensionless parameter, N, from Eq. (11-62); (*b*) determination of attenuation in decibels.

Test results on models and full-size transformers with two- and three-wall barriers correlate reasonably with Eq. (11-62). Data on four-wall enclosures generally do not correlate. It has been found that approximately 10 dB attenuation can be achieved with a four-wall enclosure having walls 5 ft higher than the transformer.

Enclosures with fewer than four walls should extend at least a distance M beyond the tank, so that attenuation will be limited by diffraction over the top rather than around the ends of the barrier. It should be noted that the sound level on the open side of this type of enclosure will be increased above what it was without the enclosure. Energy is redirected from the critical side of the transformer to the less critical side.

The effective attenuation of an enclosure can be reduced by pressure build-up between the tank and the barrier. The build-up is the result of reflection from hard wall surfaces and reinforcement of direct and reflected waves. Build-up will be most pronounced for spacings between tank and barrier walls which are multiples of the half wavelength of any of the principal sound frequencies. Such spacings should be avoided. Sound-absorbent lining on the interior surface of the barrier walls is helpful in reducing or eliminating build-up.

Masonry enclosures can also be used to hide substation transformers and associated equipment and in that way alleviate complaints which are based on appearance in addition to noise. Some utilities use a three-sided enclosure which resembles the houses in the neighborhood.[2,3] A casual observer on the street may not detect the presence of the substation.

CORONA

104. Corona is localized incomplete dielectric failure, usually in liquid or gas, characterized by abrupt changes in capacitance which cause high-frequency current and voltage. Corona in, or adjacent to, solid insulation causes progressive weakening of the solid insulation. Corona in oil generates gas, which, if trapped or collected in a region of high dielectric stress, may cause trouble. Severe corona in oil carbonizes the oil. Corona may be caused by excessive local stress concentration, gas bubbles,

[1] AIEE Committee Report, Sound Barrier Walls for Transformers; *Trans. AIEE*, 1960, Vol. 79, p. 932.
[2] *Ibid.*
[3] BUCK, F. W. Residential Sub Has Novel Design; *Elec. World*, Dec. 14, 1959, p. 53.

wet fibers, or microscopic contaminants drawn into an area of high stress by electric attraction or oil currents.

105. The continuing trend toward reduced transformer insulation, permitted by improved lightning-arrester characteristics, and more attractive at higher circuit voltages, leads to a progressively smaller margin between test voltage and operating voltage, thus increasing the need to demonstrate absence of significant corona in operation. In addition factory corona measurement may be used to demonstrate that the insulation is uninjured by factory-applied potential and induced-voltage tests.

106. Corona is evaluated by measuring the high-frequency voltage, usually at 1 Mc/s, at the terminal of the transformer. The general test procedure is described in *NEMA Publ.* 107, except that for power transformers the coupling capacitor is replaced by the potential tap in the high-voltage bushing as the means for coupling to the high-voltage circuit and the effect of the capacitive impedance of the bushing at 1 Mc/s is reduced by an adjustable inductive element connected in series with the bushing tap. The voltage-measuring instrument is described in USAS C63.2-1950.

107. Corona in transformers may also be detected by acoustic transducers in the oil or on the tank wall. If a sensitive transducer shows no corona, any corona picked up on the bushing tap originates outside the transformer. If the transducer shows corona, then it can be used to locate the source of corona within the transformer tank by measuring the time interval after the corona voltage appears at the bushing tap until the effect appears at the transducer. Then the distance from the transducer to the source of corona is 1 in for each 15 μs of delay.

RADIO-INFLUENCE VOLTAGE

108. Excessive corona may cause **interference with radio communication.** Suitable maximum limits of corona voltage ensuring compliance with Federal Communication Commission requirements have been established and are shown in *NEMA Publ.* TR 1. For power transformers this limits the high-frequency voltage at 1 Mc/s, measured at about 110% of operating voltage, to 250 μV up to 14.4 kV operating, 650 μV up to 34.5 kV operating, 1,250 μV up to 69 kV operating, and 5,000 μV up to 345 kV operating.

TESTING[1]

109. Standard Tests. The following tests are required by USAS Standards.

1. Resistance measurements.
2. Ratio tests.
3. Phase-relation tests; polarity, angular displacement, and phase sequence.
4. No-load loss and exciting current.
5. Load loss and impedance voltage.
6. Applied-potential dielectric test.
7. Induced-voltage dielectric test.
8. Temperature test. (Test is made on one unit only and is omitted if a unit which is essentially a thermal duplicate has been previously tested.)
9. Regulation and efficiency.
10. Accessories tests.

110. Optional Tests. The following tests are not required by USAS Standards:

1. Impulse-voltage test.
2. Switching-surge test.
3. Radio-influence voltage or corona test.
4. Insulation power-factor test.
5. Audible-noise test.

111. Sequence of tests is immaterial as long as impulse- and switching-surge tests (if required) are made before the applied-potential and induced-voltage tests.

112. Resistance measurements are necessary for the calculation of losses and for that of winding temperatures at the end of the temperature test. The measurements are usually made by the drop-of-potential method or the bridge method, both requiring the use of direct current. The drop-of-potential method is simple, and it is generally found convenient for measurements made in the field. However, the bridge method, while requiring somewhat more cumbersome equipment, is the more accurate, and it

[1] For more information, see USAS C57, Pt. 12.90, and *NEMA Publ.* TR 1 Pt. 9.

is adapted for a wider range of resistance. In making measurements of the resistance of a winding, an accurate measurement of the temperature of the winding should be made at the same time.

113. Ratio Tests. Three methods are commonly used to make ratio tests, depending upon the type of the transformer and the available facilities. A convenient method is application of a known voltage, usually less than normal, to the highest winding and measurement of the voltages on the other windings, suitable voltmeters and potential transformers being used. The ratios of the voltage readings will indicate the ratios of the turns in the various windings. Readings should be taken for all transformer tap positions. A second method, used mostly for factory testing, compares the transformer with a calibrated standard transformer whose ratio is adjustable in small steps. The transformer under test and the standard transformer are connected in parallel with their high-voltage windings energized; the parallel low-voltage windings are connected through a sensitive detector which is made to read zero by adjusting the ratio of the standard transformer. The adjusted ratio of the standard transformer then equals the ratio of the transformer being tested. The third method employs a resistance potentiometer connected across the transformer windings, which are connected in series as an autotransformer. A suitable detector is connected from the junction of the two windings to the adjustable arm of the resistance potentiometer. When the detector shows a zero deflection, the ratio of the resistances gives the turn ratio of the transformer.

114. Polarity tests are of value for paralleling of transformers. There are three general methods of determining the polarity:

1. Comparison with a standard transformer.
2. Inductive kick with direct current.
3. Alternating-voltage test.

115. Polarity Test by Standard Transformer. When a standard transformer of known polarity and of the same ratio as the unit under test is available, connect the high-voltage windings of both transformers in parallel by connecting similarly marked leads together. Connect also similarly marked leads of one end of low-voltage windings of both transformers together, leaving the other ends free. Apply a reduced value of voltage to the high-voltage windings, and measure the voltage between the two free leads. A zero or negligible reading of the voltmeter will indicate that the polarities of both transformers are identical.

116. Polarity Test by Inductive Kick. With direct current passing through the high-voltage winding, connect a high-voltage d-c voltmeter across the outlet terminals of the same winding so as to get a small positive deflection of the pointer when the exciting circuit is closed. Then transfer the two voltmeter leads directly across the transformer to the opposite low-voltage leads. Interrupting the d-c exciting current induces a voltage in the low-voltage winding and causes a deflection of the voltmeter. If the pointer swings in the same direction as before (positive), the polarity is additive. If it swings in the negative direction, the polarity is subtractive.

117. Polarity Test by Alternating-voltage Test. Facing the low-voltage side of the transformer, connect the adjacent left-hand high- and low-voltage outlet leads together. Apply any convenient value of a-c voltage to the full high-voltage winding, and take readings first of the applied voltage and then of the voltage between the right-hand adjacent high- and low-voltage leads. If the latter reading is less than the former, the polarity is subtractive. If it is greater than the former, the polarity is additive.

118. The angular displacement and the phase sequence of the windings of a 3-phase transformer must be known if it is to operate in parallel with other units or if systems are to be interconnected. All manufacturers of transformers follow recognized standard practices in regard to these characteristics, and a diagram showing the angular displacement and the phase sequence is supplied with the transformer. These characteristics may be checked by connecting the H_1 and X_1 leads of the transformer together, exciting the transformer with a low 3-phase voltage, and then measuring the voltages between the various remaining leads. These voltages may then be compared with the vector diagram supplied by the manufacturer.

119. Transformer Losses. The losses of a transformer include the no-load loss (core loss and exciting current loss) and the load loss (resistance loss, eddy-current loss in the windings, and stray loss).

120. The no-load loss consists of hysteresis and eddy-current loss in the core. In addition there is a small resistance loss in the exciting winding due to the exciting current, but it is negligible. For commercial voltage waveforms containing odd harmonics only, the maximum value of the flux is determined by the average value of the voltage. Therefore, the hysteresis loss depends, not upon the rms value of voltage, but upon its average value. The eddy-current loss, on the other hand, depends on the rms value of voltage. For a pure sine wave the rms value is 1.11 times the average value. Thus, if core loss is measured by using a pure sine-wave voltage or voltage with an rms value 1.11 times the average value, the rated rms voltage should be applied to the transformer. If the ratio of rms voltage to average voltage is not 1.11, the applied voltage should be appropriately adjusted by using an empirical correction. Thus core-loss measurement requires the use of an average-voltage voltmeter in addition to an rms-voltage voltmeter.

121. No-load Loss and Exciting-current Tests. Figure 11-36 shows necessary equipment and connections when instrument transformers are used. The potential transformers should be connected nearest to the load and the current transformer nearest to the supply. Since the excitation power factor of transformers may be under 5%, low-power-factor wattmeters and instrument transformers with very small phase error should be used to secure adequate accuracy. Either the high- or low-voltage winding of the transformer may be used for this test by applying the rated voltage of that winding, but it is generally more convenient to use the low-voltage winding. In any case, a full winding should be used if possible. If for some unusual reason only a portion of a winding is used, this portion should be not less than 25% of the full winding. Apply power at normal frequency, and adjust the voltage to the desired value. Record the frequency, average voltage, rms voltage, watts, and amperes, making proper correction for instrument transformers and instruments.

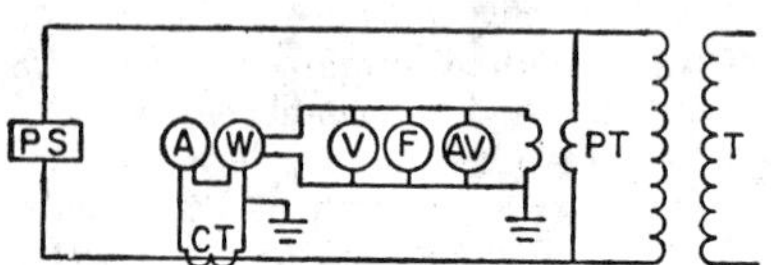

FIG. 11-36. Test connections for no-load loss and exciting current of single-phase transformer. *A*, ammeter. *AV*, average voltmeter. *CT*, current transformer. *F*, frequency meter. *PS*, power supply. *PT*, potential transformer. *T*, transformer under test. *V*, rms voltmeter. *W*, wattmeter.

122. Load loss is loss incident to the carrying of load current. It includes resistance loss and eddy-current loss in the windings and leads due to load current; stray losses in the windings, core clamps, tank, etc., caused by stray flux; and losses due to circulating currents, if any, in parallel windings.

123. The impedance voltage of a transformer is the voltage required to circulate rated current through a winding of the transformer when another winding is short-circuited. It is usually expressed in percentage of the rated voltage of the winding to which the voltage is applied; it comprises an effective resistance component corresponding to the load losses and a reactance component corresponding to the leakage flux linkages of the windings. Measurements of load loss and impedance voltage may be made simultaneously. Figure 11-37 shows the necessary equipment and connections when instrument transformers are used. One of the windings of the transformer is short-circuited, and voltage at rated frequency is applied to the other winding and adjusted to circulate rated currents in the windings. With current and frequency adjusted to the rated values, readings are taken on the ammeter, voltmeter, wattmeter and frequency meter and the proper corrections made for instrument transformers and instruments. Since the power factor is frequently less than 5% on power transformers, low-power-factor wattmeters and instrument transformers with very small phase-angle error should be used to secure adequate accuracy. Immediately following the impedance measurement, the temperature of the windings should be measured.

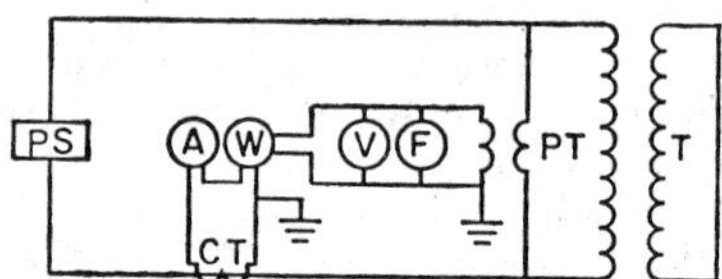

FIG. 11-37. Test connections for load loss and impedance voltage of single-phase transformer. *A*, ammeter. *CT*, current transformer. *F*, frequency meter. *PS*, power supply. *PT*, potential transformer. *T*, transformer under test. *V*, voltmeter. *W*, wattmeter.

124. Separation of Load-loss Components and Correction to Desired Temperature. The resistance loss of the two windings is calculated from the ohms resistance measurements (corrected to the temperature at which the load-loss measurement was made) and the currents which were used in the load-loss measurement. Subtracting the resistance loss from the load loss gives the eddy and stray loss. The resistance component of the load loss increases with the temperature, while the eddy- and stray-loss component decreases with the temperature; therefore, when it is desired to convert the load loss from one temperature to another, e.g., in calculating efficiency at standard temperature, the two components of the impedance loss must be converted separately,

$$L_{R2} = L_{R1} \frac{234.5 + C_2}{234.5 + C_1} \tag{11-63}$$

$$L_{S2} = L_{S1} \frac{234.5 + C_1}{234.5 + C_2} \tag{11-64}$$

where C_1 = degrees centigrade temperature at which loss is known, C_2 = degrees centigrade temperature at which loss is to be found, L_{R1} = watts resistance loss at C_1, L_{R2} = watts resistance loss at C_2, L_{S1} = watts eddy and stray loss at C_1, and L_{S2} = watts eddy and stray loss at C_2.

125. Applied-potential dielectric tests are made between windings and between windings and ground. The winding to be tested is short-circuited and connected to the high-voltage terminal of a suitable step-up testing transformer, and the tank is connected to the grounded return circuit. All terminals not being tested should be grounded to the tank. The voltage should be raised gradually and without interruption from zero to the test value in less than 1 min by an induction-voltage regulator or by generator field control. The voltage is held for 1 min. If the transformer has graded insulation, the value of the voltage applied is limited to the lowest insulation level of the winding. If the neutral bushing is not capable of withstanding the applied potential test, it should be disconnected from the winding.

126. The induced-voltage dielectric test overexcites the transformer and requires a frequency of 120 c/s or more to avoid excessive flux density and exciting current. Required voltage is held for 7,200 c. Ordinarily the test is made at 2 × (rated voltage). On transformers with graded insulation the voltage is adjusted to test the insulation from line terminal to ground. Such a test is made separately on each phase of a Y-connected winding. For a 3-phase core-type transformer it is convenient to ground the neutral or to connect the other two terminals together and ground them. The voltage may be measured by sphere gap or by a capacitance tap in the line bushing.

127. Temperature Test. The temperature test is usually made by the short-circuit method. One winding is short-circuited (usually the low-voltage winding), and suitable voltage is applied to another winding to generate a loss equal to the desired total loss (load loss corrected to standard reference temperature plus no-load loss). This loss is held until the top oil temperature is constant with a constant ambient temperature, thus establishing the top oil rise. The effective oil rise is determined by subtracting from the top oil rise the top oil increment, which is half the difference between the top oil and bottom oil temperatures. The applied voltage is then reduced to give rated current, which is held for 2 h to establish winding rise over effective oil. Top oil and effective oil temperatures are measured, the power is disconnected, and the average winding temperatures are determined by resistance measurements. The total temperature rise over ambient of a winding is the sum of the effective oil rise and the winding rise over effective oil.

128. Measurements of oil temperature are made with thermocouples or thermometers placed in the moving oil above the core and at the bottom of the tank. For transformers with external coolers the top oil increment may be taken as half the difference in external surface temperature of the inlet and outlet connections, eliminating the need to measure the temperature of moving bottom oil. Temperatures of the windings are established by comparing hot resistances measured at the end of the heat run with cold resistances measured before the heat run. The relation between hot and cold resistances is as follows,

$$R_2 = R_1 \frac{234.5 + C_2}{234.5 + C_1} \tag{11-65}$$

where C_1 = degrees centigrade at which resistance is known, C_2 = degrees centigrade at which resistance is desired, R_1 = ohms resistance at C_1, and R_2 = ohms resistance at C_2.

129. Correction Back to Shutdown. Since a drop of temperature occurs in transformer windings between the instant of shutting down a temperature test and the time of measuring the hot resistance, corrections must be added to the temperatures determined by the hot-resistance measurements in order to obtain the temperatures at the instant of shutdown. The correction for a given winding may be determined by plotting a time-temperature curve obtained from a series of resistance measurements made after shutdown and extending the curve back to shutdown. Alternatively, if the loss in a winding does not exceed 30 W/lb of copper, the correction back to shutdown may be determined, for oil-immersed transformers, by interpolation in Table 11-5. For example, at 2 min and 15 W/lb the correction is 4.8°C.

Table 11-5. Correction Back to Shutdown

Time after Shutdown, min	Correction, °C/(W/lb)
1	0.19
1.5	0.26
2	0.32
3.0	0.43
4.0	0.50

130. The regulation of a transformer may be determined by loading it according to the required conditions at rated voltage and measuring the rise in secondary voltage when the load is disconnected. The rise in voltage when expressed as a percentage of the rated voltage is the percentage regulation of the transformer. This test is seldom made, because the regulation is easily calculated from the measured impedance characteristics (see Par. **13**).

131. Efficiency of a transformer is seldom measured directly, because the procedure is inconvenient and the efficiency can be readily calculated (see Par. **15**).

132. Accessories Tests. Appropriate tests are made on auxiliary equipment such as current transformers, winding-temperature indicators, load tap-changing equipment, fans, pumps, and the like, for the purpose of checking calibration, operation, and controls.

133. Impulse-voltage tests are made to determine the adequacy of the transformer to withstand lightning surges. Voltage waves with a nominal front of 1.5 μs and a nominal tail to half value of 40 μs are applied to each terminal to be tested. The first wave is a full wave with a crest value of 50% of the BIL (basic insulation level). It is followed by two chopped waves of 115% of the BIL. Chopping is accomplished by a rod gap in air adjusted to arc after the crest of the voltage wave. The final application is a full wave at 100% of the BIL. Oscillograms are obtained for applied voltage and neutral current. Agreement in voltage and current waveshape between the initial reduced-voltage full-wave test and the final 100% full-wave test indicates that the transformer has passed without damage. The chopped waves stress insulation between turns near the line end. The full wave stresses insulation between the middle of the winding and ground.

134. Switching Surges. For transformers with reduced insulation levels switching surges may be a limiting factor. A test to demonstrate strength against switching surges has been proposed, using an impulse wave with a crest value of 83% of the BIL rising to crest value in not less than 100 μs, with a total duration of not less than 500 μs, and with at least 200 μs at a voltage in excess of 90% of the crest value. Waves are applied with successively higher voltage up to full value, a conventional impulse generator of adequate capacity being used. The voltage waveshape should not change except to the extent that core saturation occurs sooner at higher voltage. Interpretation of neutral current waveshape is more complex.

135. Radio-influence Voltage or Corona Tests. Corona, or local overstress, may interfere with radio communication and may cause deterioration of insulation. The radio-frequency voltage produced by corona is called radio-influence voltage, or RIV, and is measured in microvolts. With reduced transformer insulation levels the margin between operating voltage and the voltage at which corona occurs is reduced; so it may be desirable to demonstrate absence of excessive corona.

A method of measuring radio-influence voltage or corona at any specified voltage is given in *NEMA Publ.* 107. A modified method using the bushing capacitance tap to couple with the transformer winding has been proposed and is in use. NEMA Publ. TR 1 lists limiting RIV values at about 110% operating voltage for reasonable assurance of freedom from interference with radio communication. Assurance of freedom from deterioration of insulation caused by corona in operation or in test requires RIV measurements at higher voltage. On some large high-voltage Y-connected power transformers RIV measurements are made at full induced test voltage. No limits for acceptable RIV values at full induced test voltage have been established.

136. Insulation Power Factor. The development of undesirable insulation conditions may be signalized by an increase in 60-c insulation power factor. For this reason insulation-power-factor measurements are sometimes made on the insulation between windings and other portions of the insulation structure on new transformers and repeated periodically in service. In order to obtain useful comparisons, successive measurements should be made with similar equipment, and readings should be corrected for insulation temperature.

137. Audible Noise Test. Audible noise tests are made according to NEMA Publ. TR 1-9.04, with instrumentation according to USAS S1.4-1961, the 40-dB weighting being used. Usual limits are shown in NEMA Publ. TR 1-0.11. Readings are taken at 3-ft intervals around the transformer, usually at one-third and two-thirds the tank height. The microphone is located 1 ft away from the tight string perimeter of the transformer, except that when necessary the microphone is moved farther away to obtain a distance of 6 ft from the nearest fan-cooled surface. All the microphone readings are averaged to obtain the sound level of the transformer. During measurement the transformer is energized at rated voltage and frequency without load. The ambient noise level should be at least 7 dB below the transformer noise level.

PROTECTION AGAINST LIGHTNING[1]

138. A transformer may be subjected to severe lightning voltages as a result of a direct stroke to the transformer terminal, adjacent bus, or transmission line. Less severe voltages may result from strokes on a distant part of the system or from strokes to ground near the system. Since lightning voltage may exceed the insulation strength of the transformer, protection is necessary.

Volt-time curves are used in evaluating protection, because for short times the insulation strength changes significantly with duration of voltage. Protection is effective if the volt-time curve of the transformer is above the volt-time curve of the protective equipment, so that for any time duration the kilovolts insulation strength of the transformer exceeds the protective level at the same duration. The volt-time curve of transformer insulation has considerable "turn-up," i.e., for durations under 10 μs the kilovolts insulation strength is much greater. Rod gaps in air are unsuitable for protecting transformers, because they have even more turn-up than transformers.

139. Lightning Arresters. The modern lightning arrester has very little turn-up and is an essential adjunct to the transformer whenever there is lightning exposure.

The required lightning-arrester rating depends upon the effectiveness of the neutral grounding. The rating is expressed in percent of rated line to line power-frequency voltage that the arrester will withstand. Effectiveness of system grounding is described by the ratios of the zero-sequence resistance and impedance to the positive-sequence resistance and impedance. An 80% arrester is commonly used when the ratio of zero sequence to positive sequence is between 0.5 and 1.5 for resistance and between 1 and 3 for impedance. Lower ratios may permit 75 or 70% arresters. Higher ratios may

[1] See also Sec. **10**.

require 85 or 90% arresters. Use of the 100% protective level is not economical at high voltages.

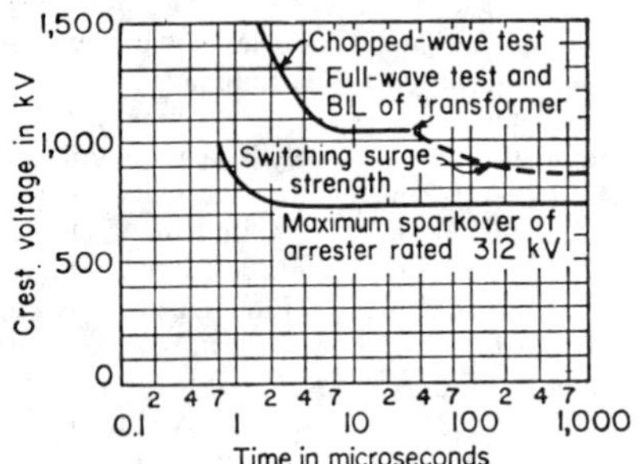

Fig. 11-38. Lightning-arrester protection margin for impulse and switching-surge conditions of a 345-kV transformer with BIL reduced two steps. The arrester curve shown is for 8 × 20-μs current wave of 15,000 A crest.

Figure 11-38 shows the volt-time curve of a 345-kV transformer, with BIL reduced two steps compared with the volt-time curve of a 90% lightning arrester. The volt-time curve of an arrester depends on the amount of lightning current. The volt-time curve shown in Fig. 11-38 corresponds to 15,000 A crest, which is not likely to be exceeded except by direct strokes. For best protection, the arrester should be located as close as possible to the terminals of the transformer, and the arrester ground should be connected by a short, direct conductor to the transformer tank and substation grounding system.

140. The steep rate of rise of lightning current such as may result from a nearby direct stroke will produce discharge voltages higher than shown in Fig. 11-38 and may damage the lightning arrester. Therefore the substation and the first half mile of connected transmission lines should be protected against direct strokes by a suitable combination of grounded masts and ground wires.

INSTALLATION AND MAINTENANCE

141. Careful installation and maintenance of power transformers are needed for long life in service. General requirements from USAS C57.93-1958 and from manufacturer's instructions are summarized below. Many power transformers have in addition special instructions on installation and maintenance which should be carefully observed. The following information applies to the majority of power transformers without special features or unusual ratings.

142. Shipment. Transformers are thoroughly dried, tested, and inspected before shipment. Power transformers are generally shipped as fully assembled as shipping limits permit. If the oil is shipped separately, the tank is filled with dry air or nitrogen. Windings are connected for maximum rated voltage, except that if there are taps above rated voltage, the taps are connected for rated voltage.

143. Inspection on Arrival. Before removal from car, inspect for damage from rough handling. If damage is found, a claim should be filed with the carrier and the manufacturer should be notified.

144. Transformers shipped oil-filled should be inspected for evidence of entrance of moisture during shipment. If the transformer is received in damaged condition, indicating that water or other foreign material has had an opportunity to enter the tank, tests should be made to check the transformer for dryness. Insulation power-factor measurement can be used for this purpose. Power factor of new transformers runs from 0.3 to 0.5% at 25°C. The transformer should be dried if the power factor is over 0.6% at 25°C. In any case a power-factor measurement may be helpful for comparison with subsequent readings.

145. Transformers shipped gas-filled are fitted with a pipe connection on the cover to which a vacuum pressure gage with a range of -10 to $+10$ lbf/in^2 may be connected. A positive or negative pressure indicates that the tank is tight; a continuous zero reading indicates that it leaks.

Inspect the core and coil assembly for signs of damage. Caution: If the transformer is filled with nitrogen, it must be purged before anyone enters the tank in order to avoid asphyxiation. The nitrogen gas can be purged with dry air or in dry weather blown out with a fan. Even in good weather with moderate humidity the exposure of the core and coil assembly to the atmosphere should be limited to a maximum of 24 h. Afterward the transformer should be vacuum-dried at an absolute pressure not to exceed

2 mm Hg for 4 h longer than the period of exposure, to evaporate surface moisture which may have been deposited.

146. Handling. Lifting cables should be held apart by a spreader to avoid damaging bushings or other parts in lifting a transformer. If a transformer cannot be handled by crane, it may be skidded or moved on rollers, provided that care is taken not to tip the transformer or damage the base. A transformer should never be lifted or moved by attaching jacks or tackle to the drain valve, cooler, or radiator connections or other accessories. If a tool or other foreign object is dropped into the transformer, it should be removed before the transformer is energized. Care should be exercised during the entire installation process to protect the transformer against the entrance of moisture.

147. Storage. When a transformer is received from the manufacturer, it should be placed in its permanent location to minimize handling. A transformer shipped gas-filled may be stored gas-filled indefinitely provided that it is equipped with a pressure gage which continues to show a positive pressure in the tank. Otherwise the transformer should be filled with oil and the oil-preservation system made fully effective. Bushings and accessories that are shipped separately should be protected against absorbing moisture until they are installed in the transformer.

148. Oil Sampling. Samples of oil should be taken from the bottom. An oil-sampling valve is provided at the bottom of the transformer tank for this purpose. A metal or glass thief tube, cleaned with gasoline, can be conveniently used to obtain a bottom sample from an oil barrel. Test samples should be taken only after the oil has settled for some time, varying from 8 h for a barrel to several days for a large transformer. Cold oil is much slower in settling. In drawing samples of oil from a sampling valve, some oil should first be discarded so that the sample will come from the bottom of the container and not from the sampling pipe. Examine a sample in a clear glass container for free water, which in any quantity is readily observable. The sample container should be a large-mouthed glass bottle, 1 qt or larger, with cork stopper. The bottle should be cleaned with gasoline and dried before being used. Bottles should be amber color if samples are to be stored and should be tested later for color or sludge-forming characteristics.

149. Testing for Dielectric Strength. The testing fixture should be cleaned thoroughly to remove any particles or cotton fiber and rinsed out with a portion of the oil to be tested. The testing fixture should be filled with oil, both oil and fixture being at room temperature. Allow 3 min for air bubbles to escape before applying voltage. The rate of increase in voltage should be about 3,000 V/s. Five breakdowns should be made on each filling, and then the receptacle should be emptied and refilled with fresh oil from the original sample. The average voltage of 15 tests (5 tests on each of three fillings) is usually taken as the dielectric strength of the oil. It is recommended that the test be continued until the mean of the averages of at least three fillings is consistent.

ASTM Method D877-64 uses 1-in-diameter square-edged electrodes spaced 0.1 in apart. ASTM Method D1816-60T uses special electrodes spaced 0.04 in apart and with continuous oil circulation. The latter test is more sensitive to slight contamination. Strength of new oil should exceed the minimum value for good oil as shown in Table 11-6.

Table 11-6. Dielectric Strength of Oil

KV average dielectric strength by ASTM D877-64	KV average dielectric strength by ASTM D1816-60T	Condition of oil
30 or over	29 or over	Good
26 to 29	15 to 23	Usable
Under 26	Under 15	Poor

150. Filtering to Increase Dielectric Strength. If oil tests below "good," it should be filtered to remove impurities and moisture. It is best to discharge filtered oil into a clean, dry tank and avoid mixing with unfiltered oil. If the filtered oil must be discharged back in the transformer tank, the oil should be withdrawn from the bottom filter-press valve and, after filtering, returned through the top filter-press valve. Oil should not be filtered while the transformer is energized, because the dielectric strength may be temporarily reduced by aeration. If no facilities are available for making dielectric tests, send a sample to the manufacturer marked with the serial number of the transformer.

151. Drying the core and coils may be accomplished by the following means:

1. Heat in oil.
2. Heat in air.
3. Vacuum.
4. Heat and vacuum.

152. Heat in Oil. The oil is heated to a temperature of 75 to 85°C by applying reduced voltage to one winding with another winding short-circuited. It may be desirable to close radiator valves, reduce flow of cooling water, or blanket the tank, to reduce the amount of heat required. If the current in the winding does not exceed one-half or three-quarters of the rating, the winding temperature will not be greatly in excess of the maximum oil temperature and relatively high oil temperature can be obtained without overheating the winding. Ventilation should be obtained by slightly raising manhole covers and protecting the openings from the weather. The cover should be lagged with thermal insulation to prevent recondensation of moisture on the underside. The pumps of forced-oil-cooled transformers should be operated during drying to circulate the oil through the windings. The oil should not be filtered during the drying process. Table 11-7 shows the effect of oil temperature on the maximum short-circuit current that can be held without injury. This should not be exceeded.

153. When to Discontinue Drying. Drying should be continued until oil from both top and bottom of the tank meets the requirements for good oil of Table 11-6. The ventilating openings should then be closed and the oil kept at the same temperature for another 24 h while the oil is tested at 4-h intervals. If the dielectric strength of the oil decreases, moisture is still passing from the insulation into the oil and the ventilators should be opened, the oil filtered, and the drying process continued. If the dielectric strength of the oil remains constant or increases, the transformer should be operated at approximately two-thirds voltage, the same oil temperature being maintained, and the oil should be filtered every 24 h until 24 h elapses without a decrease in oil strength. After a satisfactory two-thirds voltage test, full voltage should be applied until 24 h elapses without a decrease in oil strength. Water-cooled and forced-oil-cooled transformers may require some operation of the cooling system in order to hold the top oil temperature within the 85°C limit during this process.

Table 11-7. Maximum Current for Drying in Oil

Top Oil Temperature, °C	Maximum Percent of Rated Current
75	100
80	85
85	50

154. Heat in Air. The transformer core and coil assembly should be placed in a box with holes near the top and bottom for air circulation. The clearance between the sides of the transformer and the box should be small so that most of the heated air will pass up through the ventilating ducts of the coils and not around the sides. The heat should be applied near the bottom of the box. With some types of transformers it is better to distribute the heat around the lower part of the coils. The best way to obtain heat is from resistance units. Air going into the enclosure should not exceed 90°C. The core and coils should be carefully protected against direct radiation from the heaters. There should be no flammable material near the heaters. The box should be completely lined with asbestos. When forced-air circulation is used, suitable baffles should be placed between the heater and the inlet to the transformer enclosure.

155. Precautions to Be Observed. As the drying temperature approaches the point where fibrous materials deteriorate, great care must be taken to see that there are no points where the temperature exceeds 85°C. Several thermometers should be used. They should be placed in the coils near the top and screened from air currents.

As the temperature rises rapidly at first, the thermometers must be read at intervals of about ½ h. In order to have a reference temperature for insulation-resistance measurements, one thermometer should be placed where it can be read without removing it or changing its position. The other thermometers should be shifted about until the hottest points are found and should then remain at these points throughout the drying period. When possible, the temperature should be checked by the increase-in-resistance method.

156. Vacuum. A vacuum of approximately 50 microns is drawn on the core and coils in the tank without oil. A pump having a capacity of 50 ft³/min and 10 microns capability is recommended. The moisture is evaporated from the insulation and removed as vapor. A cold trap is placed in the vacuum line between the transformer and the vacuum pump, to freeze out the water vapor. The cold trap contains a thimble, which can be filled with liquid nitrogen or a mixture of acetone and dry ice, to lower the temperature and freeze the water vapor. The cold trap can be removed and its contents melted. The condensate will be a mixture of oil and water and should be allowed to separate so that the water can be decanted and measured. A curve showing the rate of water removal vs. time can be plotted to show the progress in drying. A transformer of average size is dry when no more than 24 fluid ounces of water is removed in each of three successive 24-h periods. A typical drying curve is shown in Fig. 11-39.

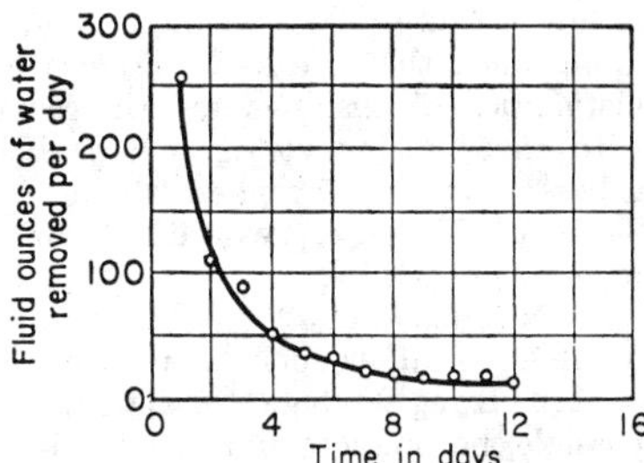

Fig. 11-39. Typical drying curve showing water removed by cold trap at a vacuum of 60 to 65 microns.

157. Heat and vacuum can be applied as in factory processing, with a portable boiler to supply a vaporized petroleum fraction (like kerosene) under vacuum to the transformer tank. The vapor condenses on the core and coils, heating them to the desired temperature, and then drains out of the tank. Simultaneously a vacuum connection to the top of the tank withdraws water vapor and evolved gases, which can be passed through a cold trap.

158. Time required for drying is generally 1 to 3 weeks, depending on the condition and construction of the transformer and the method of drying. In general, the use of high vacuum and a cold trap is faster, safer, and more economical than the use of heat alone.

159. Insulation resistance will indicate the degree of dryness only when the transformer is dried without oil. If the initial insulation resistance is measured at room temperature, it may be high, although the insulation is not dry, but as the transformer is heated up, it will drop rapidly.

As the drying proceeds at a constant temperature, the insulation resistance will generally increase gradually until toward the end of the drying period, when it increases quite rapidly and then levels off at a high value. The drying should continue until the resistance is constant for a period of 12 h.

160. Insulation power-factor readings (at 60 c/s) will indicate the degree of dryness. The power factor will first increase as the temperature increases and then will gradually decrease as drying progresses. Drying should continue until the power factor is constant for a period of 12 h. If power factor is measured on transformers dried in oil by the short-circuit method, the power factor should be used to supplement oil tests as a measure of dryness.

161. Filling without Vacuum. Use extreme care to keep moisture out of the core and coils. The tank should not be opened to the atmosphere until the core and coils are under oil, unless vacuum filling is available. The oil shipping tank or oil drums should not be opened until their temperature is the same as or higher than that of the surrounding air and the transformer is in place and ready to receive the oil. Metal or synthetic rubber hose should be used for filling, because transformer oil is contaminated by natural rubber. Oil should never be added to a transformer without passing through a filter press.

Static charges can be developed when transformer oil flows in pipes, hoses, or tanks. Oil leaving a filter press may be charged to over 50,000 V. To accelerate dissipation of the charge in the oil, ground the filter press, the tank, and all bushings or winding leads during oil flow into any tank. Conduction through oil is slow; therefore it is desirable to maintain these grounds for at least 1 h after the oil flow has ended.

Avoid explosive gas mixtures in any container into which oil is flowing. Arcs can occur along the surface of the charged oil even though all metal is grounded.

162. Filling with Vacuum. The vacuum line should be connected to a tapped opening on a cover-mounted shipping plate or to a valve near the top of the tank. An opening of 2 in minimum is recommended. The oil line can be connected to a suitable opening on a cover-mounted shipping plate or the top filter-press valve. The oil line should always be connected at the top of the tank so that the oil can be deaerated as it enters.

Transformers with operating voltage less than 161 kV and with core and coils not exposed to the atmosphere should be filled under vacuum better than 25 mm Hg absolute pressure. The vacuum should be held 4 h before filling and continued during filling until the core and coils are covered. The vacuum can then be removed for installation of bushings and the remaining oil added without vacuum. Transformers with an operating voltage of 161 kV and above or transformers with core and coils exposed to the atmosphere should be completely filled under a vacuum better than 2 mm Hg absolute pressure. A 2-mm vacuum should be held until the tank is filled to the 25°C level.

The filling rate should be under 1,500 gal/h to facilitate evacuation and complete oil filling of all air pockets and voids.

163. Bring Voltage Up Slowly. When the voltage is first applied, it should, if possible, be brought up slowly to its full value so that any wrong connection or other trouble will be discovered before damage results. After full voltage has been applied successfully, the transformer should preferably be operated for a short period without load. It should be kept under observation until after the first few hours that it delivers load. After 4 or 5 days' service it is advisable to test the oil again.

164. Maintenance schedules vary, depending on the size, complexity, and importance of the unit. A desirable schedule includes hourly readings of ambient temperature, oil temperature, winding-temperature indicator, load current, voltage, and tank pressure and also daily observation of the tank pressure gage, liquid-level indicator, and automatic gas seal equipment and monthly check of control circuits, alarm circuits, and cooling fans. In addition fan-motor bearings should be lubricated every 2 years or after 6,000 h of operation, and oil should be tested after a few days, after 6 months, after 12 months, and then once a year; oil-to-air heat exchangers should be cleaned as necessary; and load-tap-changing equipment should be serviced as called for in the manufacturer's instructions.

165. Internal inspection of transformers in service should be made only after a specific indication of trouble, such as detection of combustible gas, either by a portable gas detector or by a permanent gas-detector relay. The greatest sensitivity is offered by a gas-detector relay on a transformer with the tank completely filled with oil and external provision for expansion of oil. Generation of combustible gas usually indicates internal trouble (not necessarily serious). Analysis of the gas sometimes helps to identify the source. If collection of combustible gas continues without discoverable cause, corona-voltage measurement may establish whether or not there is an internal fault and ultrasonic measurements may locate the fault (see Par. **107**).

166. Idle Water Coolers. When a transformer is idle and exposed to freezing temperatures, water should be completely drained from the coolers.

167. Operation without Cooling. A liquid-cooled transformer should not be run continuously, even at no load, without the cooling liquid. In an emergency, forced-oil air-cooled transformers may be operated without fans and pumps (1) at rated load for approximately 1 h, starting at full-load temperature rise, (2) at rated load for approximately 2 h, starting cold (at ambient temperature), (3) at rated voltage and no load for approximately 6 h, starting at full-load temperature rise, and (4) at rated voltage and no load for approximately 12 h, starting cold.

When only a portion of the cooling equipment is operating, the transformer may be operated at reduced load approximately as indicated in Table 11-8.

Table 11-8. Operation with Limited Cooling Equipment

Percent of Cooling Equipment in Operation	Percent of Rated Load That May Be Carried
33	50
40	60
50	77
80	90

LOADING PRACTICE

168. Temperature Limitation of Loading. Ordinarily the kVA that a transformer should carry is limited by the effect of reactance on regulation or by the effect of load loss on system economy. At times it is desirable to ignore these factors and increase the kVA load until the effect of temperature on insulation life is the limiting factor. High temperature decreases the mechanical strength and increases the brittleness of fibrous insulation, making transformer failure increasingly likely, even though the dielectric strength of the insulation material may not be seriously decreased. Overloading of transformers should be limited by reasonable consideration of the effect on insulation life and the probable effect on transformer life.

169. The insulation life of a transformer is defined as the time required for the mechanical strength of the insulation material to lose a specified fraction of its initial value. Loss of 90% of the tensile strength is the usual basis for evaluating paper.

170. The aging of insulation is a chemical process which occurs more rapidly at higher temperatures according to the Arrhenius reaction-rate theory, as expressed in Eq. (11-66),

$$h = \epsilon^{[K_1 + K_2/(C+273)]} \tag{11-66}$$

where C = °C temperature of insulation, K_1, K_2 = constants determined by test, and h = hours of life.

Use of this equation permits results of relatively short-duration tests at relatively high temperatures to be extrapolated to indicate probable insulation life at moderate temperatures. USAS C57.92-1962 contains loading recommendations for power transformers with 55°C average winding-rise insulation systems based upon extrapolated life tests. *NEMA Publ.* TR 98-1964 contains corresponding recommendations for power transformers, with 65°C average winding-rise insulation systems. Figure 11-40 shows the corresponding curves of rate of loss of life as a function of temperature.

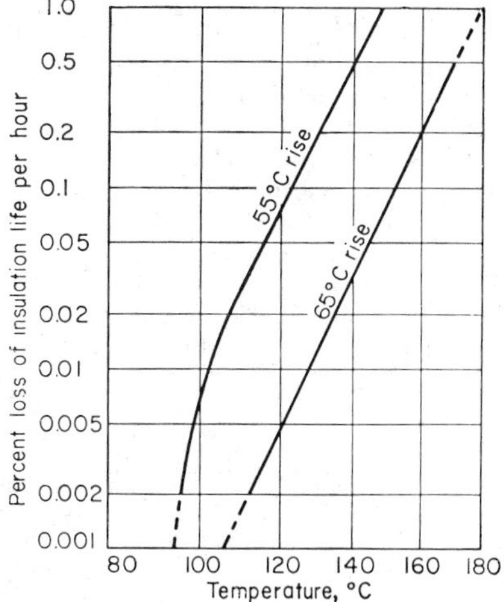

Fig. 11-40. Loss of insulation life as affected by temperature. 55°C curve from USAS C-57.92-1962. 65°C curve from *NEMA Publ.* TR 98-1964.

The concept of insulation life has been extended to cover life tests on distribution transformers operated at high temperature and subjected to periodic overvoltage and overcurrent tests until failure occurs. This approach has not yet been developed to the point of affecting overloading practice on power transformers.

171. To determine the aging of the insulation resulting from a specific daily load cycle, (1) establish an approximately equivalent

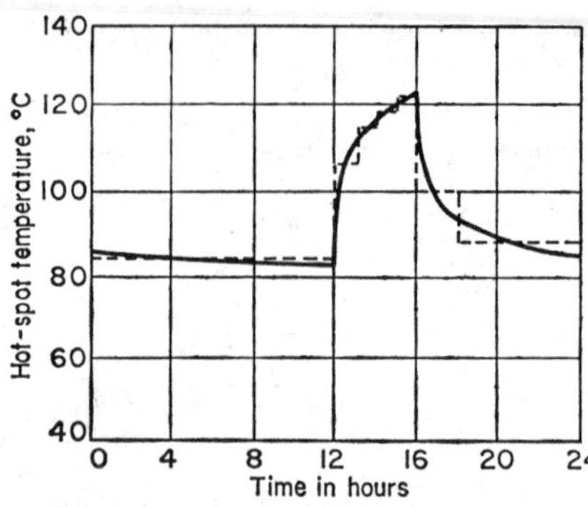

FIG. 11-41. Equivalent stepped curve of hot-spot temperature for loss-of-life calculation of daily load cycle.

stepped load cycle; (2) calculate the resulting curve of hot-spot temperature by the methods of Par. **72,** (3) replace the hot-spot temperature curve by an approximately equivalent stepped curve, (4) calculate the percent aging for each step from the applicable curve of Fig. 11-40, and (5) add the aging for all the steps in the daily cycle. The result is the fraction of insulation life used up each day. The reciprocal is the number of days of total insulation life if the same load cycle repeats every day.

172. Example. Consider the transformer used in the example of Par. **75,** with a daily load cycle of 4 h at 140% load and 20 h at 80% load in a 30°C ambient. The hot-spot temperature curve shown in Fig. 11-24 is reproduced in Fig. 11-41, together with an equivalent stepped curve. The calculation of loss of life per day is shown in Table 11-9, with steps below 95°C neglected.

Table 11-9. Calculation of Loss of Life per Day on Daily Load Cycle

Duration of step, h	°C temp. of hot spot	%/h loss of life	% loss of life on step
12	83.5	.....	0
1	106	0.015	0.015
1	116	0.046	0.046
1	119	0.064	0.064
1	123	0.095	0.095
2	100	0.006	0.012
6	87.5	.....	0
Total % loss of life per day..........			0.232

A loss of 0.232% of the insulation life each day gives an insulation life of 431 days, or 1.2 years. For comparison, a transformer with a 55°C average winding-rise insulation system (65°C hot-spot rise) operating in a 30°C ambient would have a hot-spot temperature of 95°C and would use 0.0010% of the insulation life each hour. This gives an insulation life of 11.4 years. The shortening of the insulation life from 11.4 to 1.2 years is a measure of the severity of the load cycle. The actual transformer life may, of course, be shorter or longer, depending on exposure to overvoltage, overcurrent, shock, contamination, etc.

Table 11-10. Percent Daily Peak Load for Normal Life Expectancy with 30°C Cooling Air or 25°C Cooling Water

Duration of peak load, h	Self-cooled or water-cooled, with % load before peak of			Forced-air-cooled up to 133% of self-cooled rating, with % load before peak of			Forced-air-cooled over 133% of self-cooled rating or forced-oil-cooled, with % load before peak of		
	50%	70%	90%	50%	70%	90%	50%	70%	90%
0.5	189	178	164	182	174	161	165	158	150
1	158	149	139	150	143	135	138	133	128
2	137	132	124	129	126	121	122	119	117
4	119	117	113	115	113	111	111	110	109
8	108	107	106	107	107	106	106	106	105

173. Daily overload cycles consistent with normal life expectancy for air-cooled power transformers in 30°C ambient and for water-cooled power transformers with 25°C ingoing water are given in Table 11-10, which was taken from USAS C57.92-1962.

174. Ambient temperature affects load capacity by an amount dependent on the type of cooling, as shown in Table 11-11.

Table 11-11. Effect of Ambient Temperature on kVA Capacity

Type of cooling	% of rated kVA decrease in capacity for each °C increase over 30°C air or 25°C water	% of rated kVA increase in capacity for each °C decrease under 30°C air or 25°C water
Self-cooled*..............	1.5	1.0
Water-cooled†...........	1.5	1.0
Forced-air-cooled*........	1.0	0.75
Forced-oil-cooled*........	1.0	0.75

* From 0 to 50°C air temperature.
† Up to 35°C water temperature.

For ambient temperature of air-cooled transformers use the average value over a 24-h period or 10°C under the maximum temperature during the 24-h period, whichever is higher. For ingoing water temperature use the average value over a 24-h period or 5°C under the maximum temperature during the 24-h period, whichever is higher.

175. Limitations. At abnormally high temperatures it may be necessary to remove some oil to avoid overflow or excessive pressure. The temperature of the top oil should never exceed 100°C (90°C if there is contact with the atmosphere). The hot-spot winding temperature should not exceed 150°C. Short-time peak loading should not exceed 200% rating for 30 min.

LOSS EVALUATION[1]

176. Loss evaluation is a procedure by which the buyer and seller achieve an economic balance in adding material to the transformer design to get lower losses. It is achieved by establishing a value in dollars per kilowatt for load loss and a similar value for no-load loss.

An incremental investment in capacity is required to generate power to supply loss and bring it to the transformer. In addition there is a continuing expense for fuel to supply the lost power. The continuing expense is converted to present worth and added to the incremental investment to give the total present worth of the loss.

Loss evaluations for generator step-up transformers usually come between $150 and $400 per kilowatt for load loss and $200 and $500 per kilowatt for no-load loss. Loss evaluations for substation transformers usually come between $65 and $550 per kilowatt for load loss (based on self-cooled rating of multiple-rated transformers) and $300 to $1,000 per kilowatt for no-load loss.

177. The following equations are commonly used to establish loss evaluations,

$$V_L = S + 8{,}760EF_L/R \tag{11-67}$$

$$V_N = S + 8{,}760EF_N/R \tag{11-68}$$

where E = dollars/kWh cost of energy (0.001 to 0.008), F_L = ratio of average load loss to rated load loss, F_N = ratio of average no-load loss to rated no-load loss (1.00 for continuous operation), R = per unit (%/100) annual carrying charge on system investment (covers insurance, taxes, depreciation, and return on investment), S = dollars/kW system investment (100 to 300), V_L = dollars/kW evaluation of rated load loss, and V_N = dollars/kW evaluation of rated no-load loss.

178. Loss evaluation is an important factor in purchasing new transformers, as in many cases the loss evaluation of the total loss equals or exceeds the price of the transformer.

[1] MASON, H. J. *Distribution Mag.*, April, 1962.

Small Power Transformers

179. Small Power Transformers. Standard kV and kVA ratings have been established in USAS C57.12.10-1965 for small power transformers up to 5,000 kVA single-phase and up to 10,000 kVA 3-phase for voltages up to 67 kV, as shown in Table 11-12. Insulation levels, taps, impedance, accessories, and many mechanical features are specified. Not all combinations of kVA ratings and kV are included.

Table 11-12. Standard Ratings of Small Power Transformers

Low-voltage ratings, V		High-voltage ratings, V	kVA ratings	
			1-phase	3-phase
480	7,200	2,400	833	750
480Y/277	7,560	4,160	1,250	1,000
2,400	12,000	4,800	1,667	1,500
2,520	12,470Y/7,200	6,900	2,500	2,000
4,160Y/2,400	12,600	7,200	3,333	2,500
4,360Y/2,520	13,090Y/7,560	12,000	5,000	3,750
4,800	13,200	13,200	.....	5,000
5,040	13,200Y/7,620	13,800	.....	7,500
8,320Y/4,800	13,800Y/7,980	22,900	.....	10,000
8,720Y/5,040	14,400	26,400		
6,900		34,400		
		43,800		
		67,000		

Autotransformers

180. Part of an autotransformer winding is common to both primary and secondary circuits. The common portion is called the common winding, and the remainder is called the series winding. The high-voltage terminal is called the series terminal, and the low-voltage terminal is called the common terminal. Part of the power passes from one winding to the other by transformation, and the rest passes directly through without transformation. Figure 11-42 shows an autotransformer compared with an equivalent two-winding transformer. Both have the same ratio of secondary voltage to primary voltage, T, and both have the same power output. The fraction $1 - T$ of the power is transformed, and the fraction T passes through without transformation. The fraction $1 - T$, called the "co-ratio," is a measure of the required size of the core and coils as compared with a two-winding transformer. In addition the losses and reactance are reduced in approximately the same ratio. For a low value of $1 - T$ the economy of an autotransformer compared with a transformer is attractive.

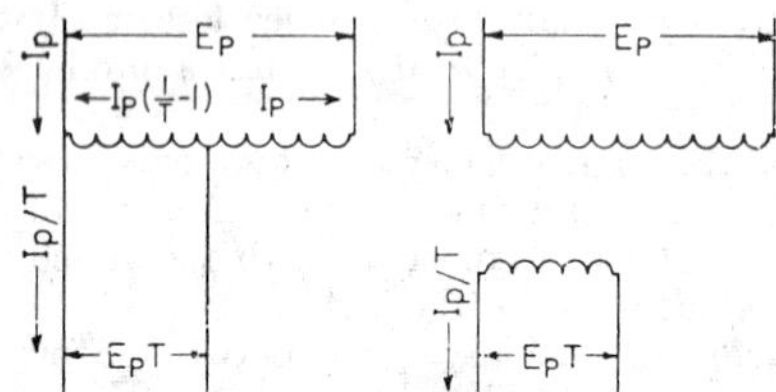

FIG. 11-42. Comparison of autotransformer with two-winding transformer.

181. The following special characteristics of autotransformers may need consideration:

A metallic connection exists between the primary and secondary circuits; this is generally of little consequence with low-voltage circuits, but with high-voltage systems the neutral point must be grounded for safe operation.

The impedance of an autotransformer is normally lower than that of the equivalent two-winding transformer, and the short-circuit current is higher.

Impulse voltage on the high-voltage side of an autotransformer may produce sizable impulse voltage on the low-voltage side. Voltage ratios of 3 or more should normally be avoided for this reason.

Taps near the neutral are relatively ineffective, because turns tapped out at the neutral come out of both circuits and increase core flux density without greatly changing the voltage ratio. The high-voltage rating is more effectively adjusted by taps in the series winding. The low-voltage rating is more effectively adjusted by tapping turns out of the common winding while turns are added to the series winding.

Inversion of the neutral may occur under abnormal conditions in Y-connected autotransformers with ungrounded neutrals. If the voltage on a series terminal is lower than the voltage on the corresponding common terminal, high voltage tends to appear on the neutral. Inversion of the neutral can occur on power-frequency voltage or on transient voltage. Grounding the autotransformer neutral, use of a delta tertiary, and use of three-leg 3-phase cores all help to prevent inversion of the neutral.

182. Autotransformers are commonly used to connect two transmission systems at different voltages, frequently with a delta tertiary winding. Similarly autotransformers are attractive for generator step-up transformers when it is desired to feed two different transmission systems. In this case the delta tertiary winding is a full-capacity winding connected to the generator, and the two transmission systems are connected to the autotransformer winding. The autotransformer not only has lower losses than the straight transformer, but the smaller size and weight permit larger ratings to be shipped.

Bibliography

183. Books

Massachusetts Institute of Technology, Department of Electrical Engineering, "Magnetic Circuits and Transformers"; New York, John Wiley & Sons, Inc., 1943.

"Transformer Reference Book"; Milwaukee, Allis-Chalmers Manufacturing Company, 1951.

Blume, L. F., Boyajian, A., Camilli, G., Lennox, T. C., Minneci, S., and Montsinger, V. M. "Transformer Engineering," 2d ed.; New York, John Wiley & Sons, Inc., 1951.

Bean, R. L., Chackan, N., Moore, H. R., and Wentz, E. C. "Transformers for the Electric Power Industry"; New York, McGraw-Hill Book Company, 1959.

Stigant, S. A., Lacey, H. M., and Franklin, A. C. "J & P Transformer Book," 9th ed.; London, Johnson & Phillips, Ltd., 1961.

Standards

NEMA Publ. TR-1.

NEMA Publ. 107.

USAS C57.

CONSTANT-CURRENT TRANSFORMERS

184. A constant-current transformer is a transformer that automatically maintains an approximately constant current in its secondary circuit under varying conditions of load impedance when its primary is supplied from an approximately constant-potential source. The most usual type, the "moving-coil" design, has separate primary and secondary coils which are free to move with respect to each other, thereby varying the magnetic leakage reactance of the transformer.

Fundamentally, the equivalent circuit may be considered as a conventional distribution transformer with a self-adjusting variable-series reactance. This reactance automatically adjusts itself to a value which, when added to the load impedance, permits a constant current to flow. The amount of reactance is determined by the moving-coil position, which in turn is maintained by the force of magnetic repulsion between the coils opposing the force of gravity on the moving coil. The desired output current sets up a force of repulsion which floats the moving coil in the position which produces this current. A state of mechanical equilibrium is attained whereby the force of repulsion exactly balances the weight of the moving coil. Any change in load or line voltage is immediately counteracted by a movement of the floating coil to restore the mechanical-electrical balance.

Constant-current transformers are used primarily for series lighting circuits. Figures 11-43 and 11-44 illustrate typical examples of commercial products.

185. The design of a moving-coil constant-current transformer requires an inherently high reactance, as contrasted with the low leakage reactance of a power or distribution transformer. Another peculiarity of this transformer is that variation of load means variation of secondary voltage, rather than current, as with a constant-potential transformer. The reactance at full load (coils together) should be as low as possible consistent with cost, power factor, and satisfactory lamp operation. The minimum (coils together) reactance value is usually about 50%, dictated largely by the requirement to limit the starting current in the cold filament of an incandescent lamp to no more than twice normal. The maximum (coils apart) reactance is generally slightly over 100%, so that at the no-load (short-circuit) condition, no more than rated current will flow.

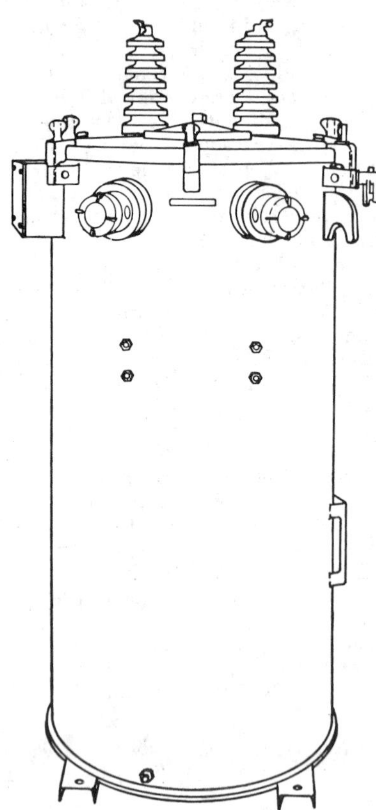

FIG. 11-43. Moloney constant-current transformer rated 20 kW.

In some designs, the moving coil is made of aluminum wire, so that it is light enough to "float" with the normal repulsion between coils; other designs provide counterweights to assist repulsion and overcome the added weight of copper-wound coils. In either case, small adjusting weights are provided to compensate for slight changes in coil weight.

Owing to its inherently high reactance, the constant-current transformer presents a low power-factor load to the power line, varying from approximately 20% lagging at one-quarter load to 75% lagging at rated load. Power-factor correction capacitors are an optional accessory, providing a power factor of approximately 65% lagging at one-quarter load to unity at full load.

Substation types are available which provide an integral package, including the accessory components necessary for the control and protection of the transformer. The usual accessories include a primary solenoid switch, open-circuit protector, primary fuses or fuse cutouts, and primary and secondary lightning arresters.

Pole-type transformers are arranged so that they can be mounted outdoors on poles or platforms. The subway, or vault-type, transformer is of such construction that it may be installed safely in a subway vault or manhole. Both types are generally oil-immersed.

186. The rating of a constant-current transformer is expressed in kilowatts output at the secondary terminals with rated primary voltage and frequency, and with rated secondary current and unity secondary power factor. These transformers are usually designed to deliver rated kilowatt output at 95% of their rated primary voltage and at 99.5% power-factor load.

Standard output ratings for constant-current transformers are 10, 15, 20, 25, and 30 kW. Sizes both smaller and larger than these have been built. Standard primary-voltage ratings are 2,400, 4,800, 7,200, and 12,000 V. Other voltage ratings can also be furnished. The standard rated output current of these transformers is 6.6 or 20 A.

The maximum winding-temperature rise for these oil-immersed transformers is 55°C, based on an 8-h operating period for pole types and 15 h for vault types, at any load between 50 and 100% of its rating. The method recognizes the fact that a moving-coil constant-current transformer usually reaches its maximum temperature at minimum load and that 50% load is the minimum at which such a transformer may be expected to operate under usual service conditions.

187. Regulation of a constant-current transformer is the maximum departure of the secondary current from its rated value, expressed in percent of the rated secondary current. For purposes of demonstration, rated primary voltage is applied at rated

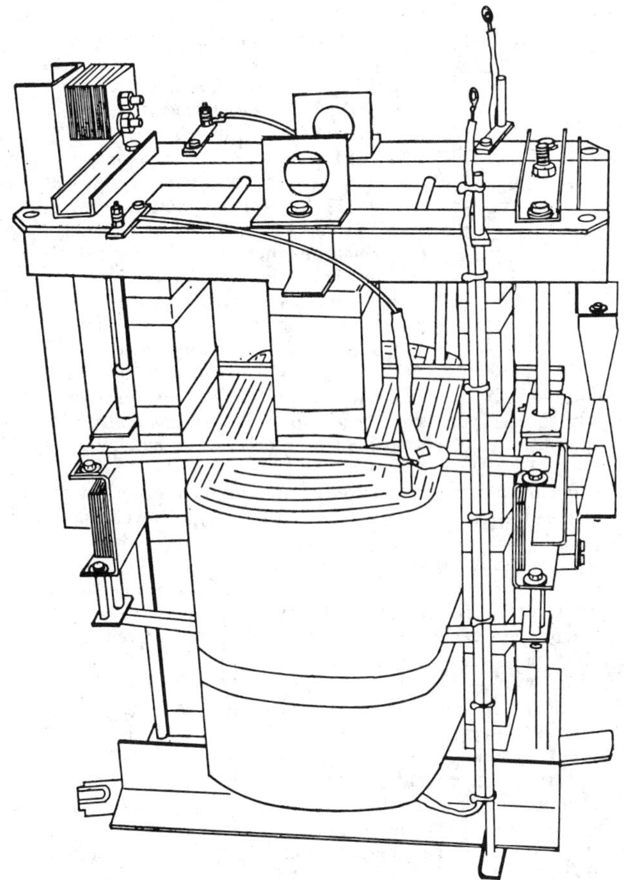

Fig. 11-44. Interior view of General Electric constant-current transformer.

frequency, with a connected load at rated power factor, and the secondary current is then measured between the limits of rated load and short circuit. A regulation of 1% or less is generally attained on moving-coil designs.

188. Efficiency of a constant-current transformer is the ratio of the useful power output to the total power input. The efficiency may be obtained by direct measurement of input and output or by separate determination of all the component losses. Since either of these measurements is difficult to obtain with reasonable accuracy, it is customary to use the "conventional" efficiency as the basis for comparison of transformer performances, even though it neglects stray losses. The conventional efficiency is determined from the core loss when excited at rated voltage and frequency with the secondary circuit open and from the resistance loss (by using resistance measured with direct current) corrected to 75°C. However, the stray losses are generally significant in a moving-coil-type design and may be estimated at about 30% of the "conventional" losses.

189. The loading capability of a constant-current transformer of the moving-coil type is determined to a large extent by the power factor of the secondary load circuit. Early lighting circuits consisted entirely of incandescent filament-type lamps, and the resulting power factor of 99.5% or higher made it possible to use the actual load kilowatts as a basis for estimating the constant-current transformer kilowatts capacity required. However, the widespread use of mercury-vapor and fluorescent lamps with individual insulating transformers on series circuits has resulted in lighting-load power factors on the order of 80 or 90%. The effect of load power factor on the output capability of a constant-current transformer is illustrated by the curve of Fig. 11-45, which is generally typical for all automatically operated constant-current transformers.

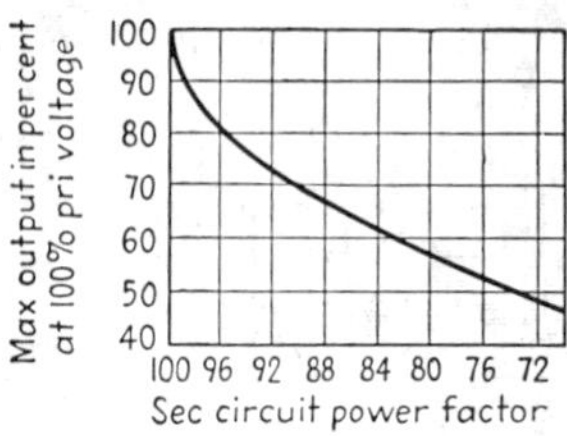

FIG. 11-45. Effect of secondary-circuit power factor on output of constant-current transformer.

It will be noted from the slope of the curve that the greatest effect on output is at circuit power factors near unity; e.g., a load power factor of 90% results in only 70% of the kilowatts load capability of the transformer, compared with supplying a load at unity power factor. Manufacturers of products to be used on series lighting circuits generally specify the kilowatts capacity of constant-current transformers needed to supply these products. These allowances are generally conservative, and a more precise computation of actual load power factor may sometimes result in more efficient utilization of transformer capacity.

190. Constant-current transformers of the static type have no moving parts and operate on the principle of a resonant network, such as the monocyclic-square circuit, shown in Fig. 11-46. This network usually consists of two inductive and two condensive reactors, each of equal reactance at the power frequency. With such a network, the secondary current is independent of the impedance of the connected load but is directly proportional to the primary voltage. This circuit has the advantages over the moving-coil type of no moving parts and an inherently high power factor; however, since the secondary current varies directly with primary voltage, it does not compensate for variations in line voltage. Regulation is also adversely affected by harmonics set up by certain types of load, such as open-circuited series insulating transformers and gaseous-vapor lamps, which cause high harmonic content due to the arc characteristics of the lamp. In the case of the moving-coil transformer, these harmonics are suppressed by the inherently high reactance of the transformer, and regulation is not affected.

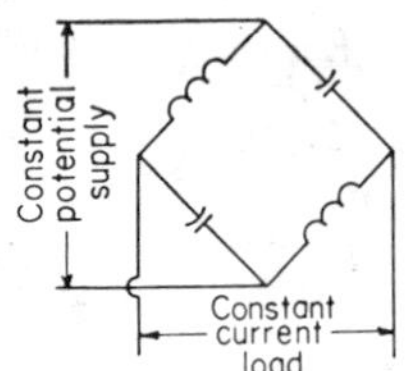

FIG. 11-46. Monocyclic-square circuit.

DISTRIBUTION TRANSFORMERS

191. Distribution transformers are generally considered as transformers 500 kVA and smaller, 67,000 V and below, both single-phase and 3-phase. Although the majority of the units are designed for pole mounting, some of the larger kVA sizes above the 18-kV class are built for station or platform mounting. Typical applications are for supplying power to farms, residences, public buildings or stores, workshops, and shopping centers.

Distribution transformers have been standardized as to high- and low-voltage ratings, taps, type of bushings, size and type of terminals, mounting arrangements, name plates, accessories, and a number of mechanical features, so that a good degree of interchangeability results for transformers in a certain kVA range of a given voltage rating. They are now normally designed for 65°C rise.[1]

The most popular primary voltages are 12,470Y/7,200, 13,200Y/7,620 and 12,000 V delta. Many of the 2,400- and 4,800-V primary systems are being converted to 7,200 and 7,620 V. In order to facilitate change-over, special series-multiple transformers are available. There is also increasing interest in higher-voltage distribution systems such as 24,900Y/14,400 and 34,500Y/19,900 V. Secondary voltage for pole-type units is usually 120/240 or 240/480.

192. Magnetic cores, in general, are composed of cold-rolled silicon steel strip. They take various forms, all designed so that the magnetic flux will pass through the sheet in the direction of rolling in order to secure the maximum benefit of the superior magnetic quality of this material. For an appreciable portion of the 24-h day, the average distribution transformer (particularly the pole-mounted 5- to 167-kVA range) is lightly loaded. Because of this, the loss in the core is a significant portion of the total daily loss. Cores for these units are therefore designed for low exciting current and for relatively low core loss, to minimize the operating cost. Low-loss cold-rolled silicon strip has contributed materially to reduced losses, weights, and dimensions.

193. Coils are usually wound or assembled in a concentric arrangement, with cooling ducts to maintain reasonable differentials between oil temperature and the average coil and hot-spot temperatures. In general some form of coil bracing is used.

As a matter of practical operating procedure in practice, distribution transformers are subjected to considerable overloads for short time peaks. This requires limited temperature on overload for long insulation life. It also has led in some instances to thermal uprating of the insulation system—usually by chemical means—to improve aging. There may be a varnish or other treatment to promote bonding and increase the mechanical strength. Vacuum oil treatment of the complete transformer is usually provided to impart high impulse strength to the windings. Aluminum windings are in some cases replacing copper. This is particularly so for the secondary windings, where full-width aluminum strip is employed in instances where it is deemed to be economical.

194. To cool the unit, the radiating surface of the tank itself suffices in the smaller ratings. In the larger ratings, auxiliary cooling is provided by the addition of fins or tubes. By these means, the height, size, and weight are held to desirable minimums. Special attention is given to sealing the transformers from the atmosphere. Likewise, careful attention is given to the external finish and fittings to assure reliable service for many years of exposure to the elements. External connectors are good for either aluminum or copper conductors.

195. The conventional pole type consists of core and coils securely mounted in an oil-filled tank, with the necessary terminals brought out through their appropriate bushings. The high-voltage bushings may be two in number for urban and suburban installations or one bushing plus a ground terminal on the tank wall connected to the ground end of the high-voltage winding for use on multiple-grounded circuits. The conventional type includes just the basic transformer structure without any protective equipment. The desired overvoltage, overload, and short-circuit protection is obtained by using lightning arresters and primary fuse cutouts separately mounted on the pole or crossarm closely adjacent to the transformer. The primary fuse cutout provides a means of visually detecting blown fuses on the system primary and also serves to remove the transformer from the high-voltage line, either manually when desired or automatically in the event of an internal coil failure.

196. The self-protected transformer (Figs. 11-47 and 11-48) has an internally mounted, thermally controlled secondary circuit breaker for overload and short-circuit protection; an internally mounted protective link in series with the high-voltage winding to disconnect the transformer from the line in the event of an internal coil failure;

---

[1] Dakin, T. W. Electrical Insulation Deterioration Treated as a Chemical Rate Phenomenon; *Trans. AIEE*, 1948, Vol. 67, pp. 113–122.

Beavers, M. F., and Lipsey, G. F. Voltage Stress as a Factor in Thermal Evaluation Program for Insulation Systems in Distribution Transformers; *Trans. IEEE*, 1964, Vol. 83, pp. 902–909.

and a lightning arrester or arresters integrally mounted on the outside of the tank for overvoltage protection. On most of these transformers, except some 5-kVA ratings, the circuit breaker operates a signal light when a predetermined winding temperature has been reached, as a warning before tripping. If the signal is unheeded and the breaker trips, the breaker may be reset and the load restored by an external handle.

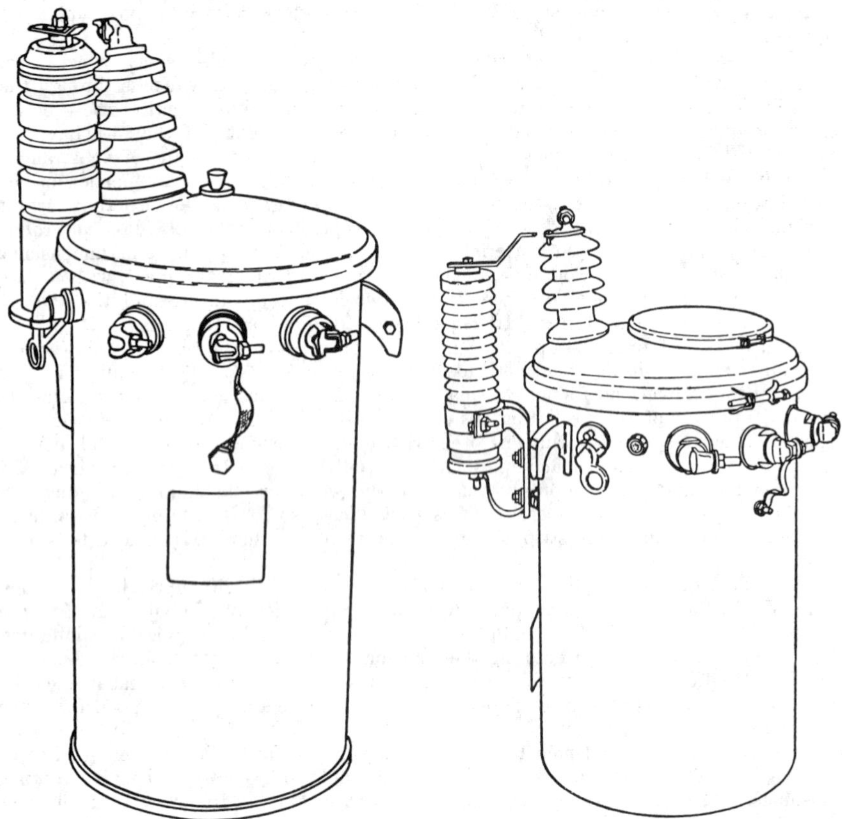

FIG. 11-47. Westinghouse completely self-protected 10-kVA distribution transformer.

FIG. 11-48. General Electric self-protected pole-type distribution transformer rated 25 kVA, 12,470 Grd Y/7,200-120/240 V, 65°C rise.

Usually this can be accomplished with the normal breaker setting. If, however, the load has been a long-sustained one which has allowed the oil to reach a high temperature, the breaker may soon trip again; or it may be impossible to reset so that it will remain closed. In such cases, the trip temperature may be set up by an auxiliary external control handle to allow reclosing of the breaker for the emergency until a larger transformer may be installed.

197. Three-phase self-protected transformers are similar to the single-phase units except that a three-pole circuit breaker is used. The breaker is arranged to open all three poles in case of a serious overload or fault on one of the phases.

198. The self-protected transformer for secondary banking is another variation. Such transformers are provided with two secondary breakers to sectionalize the low-voltage circuits, confining the outage to just the faulted or overloaded section, leaving the entire transformer capacity available for supplying the remaining sections. These are also made for single- and 3-phase.

199. "Station-type" distribution transformers are normally rated 250, 333, or 500 kVA. A "pole/station type" distribution transformer is shown in Fig. 11-49. For distribution to low-voltage a-c networks in areas of high-load-density **network transformers** are available in even higher ratings.

200. Losses and Characteristics.[1] For the pole-type ratings 100 kVA and smaller full-load efficiencies range from 97 to 99%, and impedance is generally less than 2%.

201. Recent trends indicate considerable interest in hermetically sealed construction and in resin or epoxy encapsulation of core and coils. There is widespread interest in pad-mounted and residential underground units. Underground service is particularly applicable to new residential developments, where the elimination of poles and overhead wiring is most desirable, and where installation can be carried out in cooperation with the construction contractor.

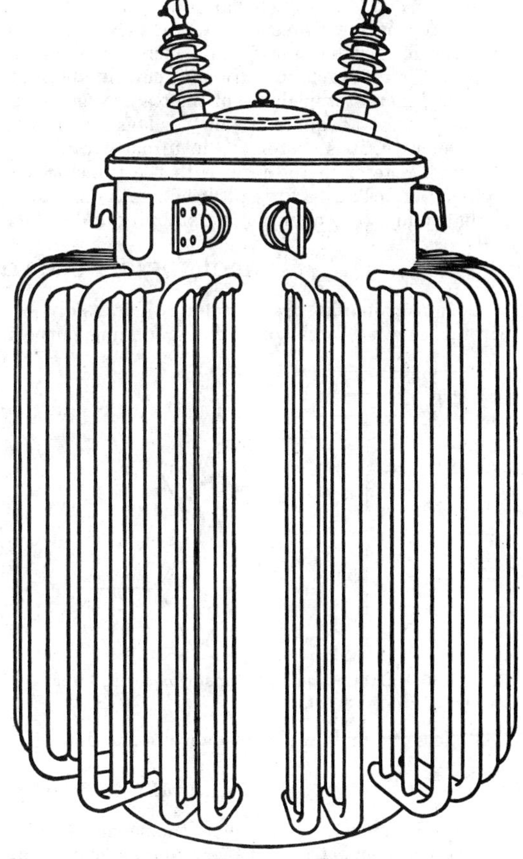

Fig. 11-49. Pennsylvania pole/station-type distribution transformer rated 250 kVA, 7,620/13,200Y-277/480Y volts, 65°C rise.

FURNACE TRANSFORMERS

202. Furnace transformers supply power to electric furnaces of the induction, resistance, open-arc, and submerged-arc types. The secondary voltages are low, occasionally less than 100 V, but generally several hundred volts. Sizes range from a few kVA to over 50 mVA, with secondary currents over 60,000 A. High currents are obtained by parallel connection of many winding sections. Current is collected by internal bus bars and brought through the transformer cover by the bus bars or by high-current bushings.

203. The power input to the furnace is controlled by adjusting the output voltage of the furnace transformer. Optimum performance of the furnace may require adjustment of the secondary voltage over a range of 3:1 or more. This may be accomplished by a regulating transformer between the high-voltage power source and a fixed-ratio furnace transformer. More frequently, regulation is obtained by taps in the high-voltage winding. In addition to taps in the high-voltage winding, a delta-Y-switch in the high-voltage winding is often used to extend the range of voltage by an additional ratio of 1.73.

[1] Beavers, M. F. Effect of Overload and Resulting Temperature on Load Losses in Distribution Transformers; *Trans. IEEE*, 1963, Vol. 82, Supplement, pp. 599–609.

204. Motor-operated off-load tap changers are usual, but occasionally on-load tap-changing equipment is justified by the saving in melt time and reduced breaker maintenance. The load-tap-changing duty is more severe than on the usual power transformer, not only with respect to frequency of operation, but also because of the extreme range, which results in large kVA increments per tap.

205. Circuit reactance furnishes current stability for a-c arc furnaces. In the larger sizes the inherent impedance of the transformer and its associated secondary conductors is sufficient for adequate stability. This is not generally true for smaller arc furnaces. Consequently, it is customary in furnace transformers rated 7,500 kVA and below to include a reactor in the tank with the transformer. This reactor is connected in the high-voltage circuit and is furnished with taps to permit adjusting the total reactance to that required to maintain arc stability under the existing service conditions.

GROUNDING TRANSFORMERS

206. A grounding transformer is a transformer intended primarily for the purpose of providing a neutral point for grounding purposes. It may be a two-winding unit with a delta-connected secondary winding and a Y-connected primary winding which provides the neutral for grounding purposes, or it may be a single-winding 3-phase autotransformer with windings in interconnected Y or zigzag. With the latter, the windings consist of six equal parts, each designed for one-third the line-to-line voltage; two of these parts are placed on each leg and connected as in Fig. 11-50. In the case of a ground fault on any line, the ground current flows equally in the three legs of the autotransformer, and the interconnection offers the minimum impedance to the flow of the single-phase fault current.

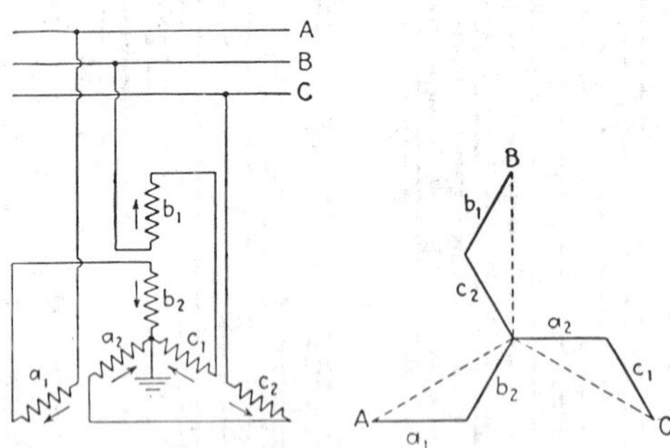

Fig. 11-50. Grounding autotransformer with interconnected Y or "zigzag" windings.

INSTRUMENT TRANSFORMERS

207. Functions. Instrument transformers are used to insulate measuring and control devices connected in the secondary circuit from the primary-circuit operating voltages. To provide complete protection, the secondary circuit should be grounded at one point. Metal cases should also be grounded. They are likewise used to transform the primary current or voltage to values suitable for standard ratings for instruments, meters, relays, and other measuring or control devices. The normal secondary ratings are 5 A for current transformers and 120 V for potential transformers.

The primary winding of a current transformer is connected in series with the load for which the current is to be measured or controlled; the primary winding of a potential transformer is connected in parallel with the load for which the voltage is to be measured or controlled (see Fig. 11-51). The secondary windings provide a current or voltage that is substantially proportional to the primary values for the operation of measuring instruments and control devices.

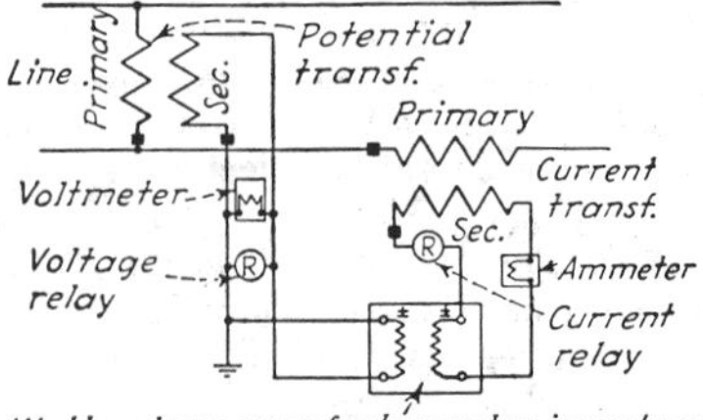

Fig. 11-51. Potential and current transformers as commonly connected to insulate meters and to transform current and voltage to convenient values.

208. Polarity. When instrument transformers are used with measuring or control devices that respond only to the magnitude of the current or voltage, the direction of the current flow does not affect the response and the connections to the secondary terminals can be reversed without affecting the operation of the devices. When instrument transformers are used with measuring or control devices that respond to the interaction of two or more currents, the correct operation of the devices depends on the relative phase positions of the currents, in addition to the magnitudes. To show the relative instantaneous directions of current flow, one primary and one secondary terminal are identified with a distinctive polarity marker; these indicate that at the instant when the primary current is flowing into the marked primary terminal the secondary current is flowing out of the marked secondary terminal (see Fig. 11-52).

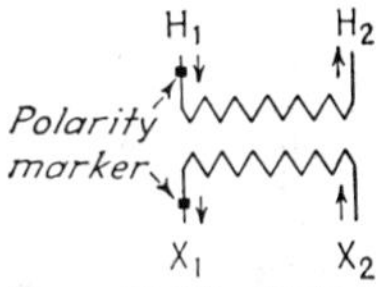

FIG. 11-52. Polarity definition. Arrows indicate instantaneous relative direction of currents in the windings.

209. Errors in Current Transformers. There are two types of errors that affect the accuracy of the measurements made with current transformers. The ratio error is the true ratio of the primary to the secondary current, divided by the name-plate ratio,

$$F_{CR} = R_{CT}/R_{CN} \tag{11-69}$$

where F_{CR} = ratio correction factor of current transformer, R_{CT} = true ratio [(primary current)/(secondary current)], and R_{CN} = name-plate ratio [(primary current)/(secondary current)] of current transformer.

The phase-angle error is the angle of lead of the current leaving the marked secondary terminal over the current entering the marked primary terminal.

Relative Importance of Ratio and Phase-angle Errors. A ratio-correction factor of 1.010 indicates that the secondary current is lower than the correct value by 1% and that all measuring or control devices connected in the secondary circuit will have 1% less current than the primary current divided by the marked ratio. The phase-angle error does not affect current-actuated devices, such as ammeters or over-current relays, but the operation of devices that respond to the products, the sums, or the differences of currents is affected by the phase-angle error.

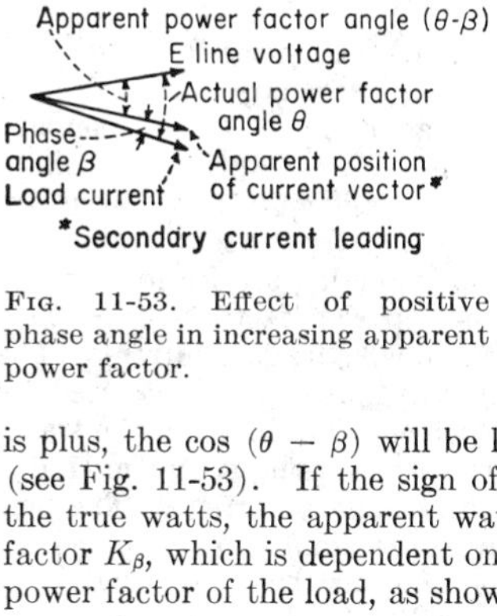

FIG. 11-53. Effect of positive phase angle in increasing apparent power factor.

A wattmeter is a device that responds to the product of the voltage applied to the potential terminals, the current through the current coils and the power factor, which is the cosine of the angle between the voltage and current. If the current is supplied from the secondary of a current transformer with unity name-plate ratio, unity ratio-correction factor, a phase-angle error of β, and primary current lagging the voltage, then the wattmeter will not indicate the true watts, $EI \cos \theta$, but will indicate $EI \cos (\theta - \beta)$. If the sign of β is plus, the $\cos (\theta - \beta)$ will be larger than the $\cos \theta$ and the wattmeter will read high (see Fig. 11-53). If the sign of β is minus, the wattmeter will read low. To obtain the true watts, the apparent watts should be multiplied by the phase-angle correction factor K_β, which is dependent on both the phase-angle error of the transformer and the power factor of the load, as shown in Eq. (11-70),

$$K_\beta = \frac{\cos \theta}{\cos (\theta - \beta)} = \frac{1}{\cos \beta + \sin \beta \tan \theta} \tag{11-70}$$

where K_β = phase-angle correction factor of current transformer, β = angle of lead of secondary current over primary current, and θ = angle of lag of load current behind load voltage.

Summarizing the Effect of Current-transformer Errors. Ratio correction factors. Above 1.000 the secondary current is low, and below 1.000 the secondary current is high. Phase-angle errors with lagging load current. Positive phase-angle errors cause wattmeter readings to be high if the load-current power-factor angle is greater than the

Table 11-13. Constants for Phase-angle Correction Factor

% power factor of line	C_1	C_2
50	0.000503	0.0005
60	0.000390	0.0004
70	0.000296	0.0003
80	0.000218	0.00022
90	0.000141	0.00015
95	0.000096	0.0001

phase-angle error, and negative phase-angle errors cause the wattmeter readings to be low.

The value of K_β can be determined with sufficient accuracy for most metering applications with lagging power-factor loads by one of the following equations,

$$K_\beta = 1 - C_2\beta \tag{11-71}$$

$$K_\beta = 1 - C_1\beta + (C_1\beta)^2 \tag{11-72}$$

where C_1 and C_2 = constants, from Table 11-13.

Equation (11-71) has an error less than ±0.0005/unit for β within the range of ±30′. Equation (11-72) has an error less than ±0.0006/unit for β within the range of ±90′, or less than ±0.0003 per unit for β within the range of ±60′.

210. Transformer Correction Factor. The transformer correction factor to be applied to the reading of the wattmeter for both the ratio and phase-angle errors is given by Eq. (11-73),

$$F_T = F_{CR}K_\beta \tag{11-73}$$

where F_T = transformer correction factor.

If F_{CR} and K_β are both between 0.985 and 1.015, Eq. (11-74) may be used with an error under ±0.0003,

$$F_T = F_{CR} + K_\beta - 1.000 \tag{11-74}$$

211. Classification of Errors. The errors in a current transformer change with the voltage required by the burden connected across the secondary terminals and the magnitude of the secondary current. The USASI has established classifications of accuracy performance of current transformers for metering service for a range of line power factor from 100% down to 60% lagging current. The accuracy classes are designated by the percent error limits of the transformer correction factors at 100% rated secondary current. The percent error limits are doubled at 10% current and, although not specified in the 1954 edition of USAS C57.13, the 100% current limits also apply at the secondary current corresponding to the maximum continuous thermal-current rating of the current transformer. The USAS accuracy classes and limits are shown in Fig. 11-54 and in Table 11-14.

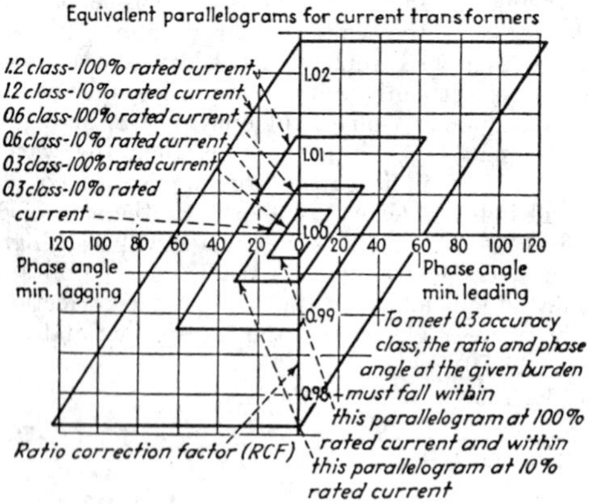

Fig. 11-54. Accuracy-class parallelograms for current transformers. The ratio correction factor and phase angle shall be limited by the indicated parallelograms for the correction-factor limits listed in Table 11-14.

The complete USAS accuracy specification comprises the accuracy class from Table 11-14 followed by the designation

Table 11-14. Limits of Current-transformer Correction Factor, F_T, for Metering Service

Accuracy class	Limits of transformer correction factor				Limits of power factor (lagging) of metered line
	100% rated current*		10% rated current		
	Minimum	Maximum	Minimum	Maximum	
1.2	0.988	1.012	0.976	1.024	0.6–1.0
0.6	0.994	1.006	0.988	1.012	0.6–1.0
0.3	0.997	1.003	0.994	1.006	0.6–1.0

* These limits also apply at the maximum continuous thermal-current rating factor.

of the burden from Table 11-15. For example, "0.3B-0.2" describes a transformer of 0.3 accuracy class when loaded with a B-0.2 burden on the secondary terminals.

212. Standard Burdens. The standard USAS burdens are listed in Table 11-15. The burden characteristics are specified in terms of resistance (ohms) and inductance (millihenrys) and do not change with frequency. The reactance (ohms), voltamperes, and burden power factor all change with frequency. The values of resistance and inductance listed are for current transformers with 5-A secondary. For other ratings corresponding burdens may be derived from Table 11-15. The resistance and inductance change inversely with the square of the change in rated current. For example, for a 1-A secondary-current transformer the burden would have 25 times as much resistance and inductance as shown in Table 11-15.

213. Effect of Phase-angle Errors on Metering 3-phase 3-wire Power with a Two-stator Watthour Meter. With the meter connected as indicated in Fig. 11-55 with balanced load, if the marked ratio and ratio correction factor are both 1.000 and the phase-angle error β is the same for both current transformers,

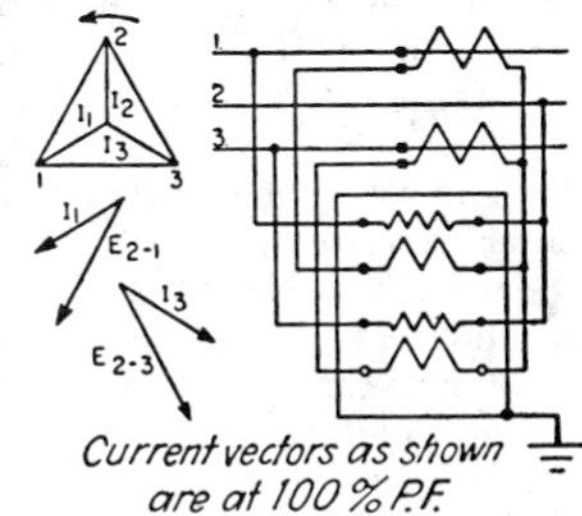

Current vectors as shown are at 100% P.F.

FIG. 11-55. Metering connections for 3-phase 3-wire with two watthour-meter elements.

$$K_\beta = \frac{E_1 \cos(\theta + 30) + E_1 \cos(\theta - 30)}{E_1[\cos(\theta + 30 - \beta) + E_1 \cos(\theta - 30 - \beta)]} = \frac{1}{\cos\beta + \sin\beta \tan\theta} \tag{11-75}$$

This is identical with Eq. (11-70) for a single-phase measurement.

Table 11-15. Standard Burdens for Standard 5-A Secondary-current Transformers
(Table 13-21.110 from USAS C-57.13)

Designation of burden	Standard burden characteristics		Standard secondary-burden impedance ohms and power factor and standard secondary voltampere burdens					
			For 60-c and 5-A secondary current			For 25-c and 5-A secondary current		
	Resistance, Ω	Inductance, mH	Impedance, Ω	VA*	Power factor	Impedance, Ω	VA*	Power factor
B-0.1	0.09	0.116	0.1	2.5	0.9	0.0918	2.3	0.98
B-0.2	0.18	0.232	0.2	5.0	0.9	0.1836	4.6	0.98
B-0.5	0.45	0.580	0.5	12.5	0.9	0.4590	11.5	0.98
B-1	0.5	2.3	1.0	25	0.5	0.617	15.4	0.81
B-2	1.0	4.6	2.0	50	0.5	1.234	30.8	0.81
B-4	2.0	9.2	4.0	100	0.5	2.468	61.6	0.81
B-8	4.0	18.4	8.0	200	0.5	4.936	123.2	0.81

* The burden may also be designated by means of the voltampere characteristic: e.g., 25 VA at 5 A or 50 VA at 5 A.

214. Standard Accuracy Classes for Relaying. The USASI has established accuracy classifications of current transformers for relaying service, consisting of three factor parts: percent ratio error limit, performance class, and secondary-terminal-voltage rating.

The percent ratio error limit specified is either 2.5% or 10%.

The established secondary-terminal-voltage values are: 10, 20, 50, 100, 200, 400, and 800, corresponding to the USAS standard burdens carrying 100 A. The percent ratio error limit should not be exceeded at the secondary-terminal-voltage rating. When the rated secondary current is not the standard 5-A value, then the secondary-terminal-voltage ratings in the relay accuracy classification are the values specified times 5 divided by rated secondary current. The performance class of the transformer is denoted by either the letter L or the letter H. The L-class is a current transformer that is capable of operating with any secondary burden up to and including a burden that will produce the accuracy-class secondary terminal voltage at 20 times rated secondary current over a current range of from rated to 20 times rated secondary current without exceeding the accuracy-class percent ratio error limit.

The H-class is a current transformer that is capable of producing any secondary terminal voltage up to and including the accuracy-class voltage with any secondary current over the range of 5 to 20 times rated secondary current without exceeding the accuracy-class percent ratio error limit. The H-class is also capable of operating with any secondary burden up to and including the burden that will produce the accuracy-class terminal voltage at 5 times rated secondary current over the range of from rated to 5 times rated secondary current without exceeding the percent ratio error.

Examples of relay accuracy classifications would be 2.5 L 800 and 2.5 H 800.

The relaying accuracy performance of a current transformer with no leakage flux entering the core can be calculated from the open-circuit excitation characteristics and the total impedance of the secondary circuit, including the resistance of secondary-winding carrying current. The leakage flux can be considered to be negligible if the secondary turns are uniformly distributed and the turns of a wound primary transformer are also uniformly distributed around the magnetic circuit. The leakage flux can likewise be considered negligible for a bar, window, or bushing type of current transformer if the secondary turns are uniformly distributed around the magnetic circuit and the distance from the magnetic circuit to the return-current circuit is adequate to keep the leakage flux at a low value relative to the normal flux in the core. The relaying accuracy performance of a current transformer that has appreciable leakage flux entering the core cannot be determined by calculation unless all design details and dimensions are known as well as the effect of adjacent current-carrying conductors. Relaying performance characteristics for this type of transformer are furnished by the manufacturer and are based on tests or are calculated from the design and dimensions of the particular type.

215. Short-time Current Ratings. The mechanical short-time current rating is the rms value of the a-c component of a completely offset primary-current wave which the transformer is capable of withstanding with the secondary short-circuited.

The thermal short-time current rating is the rms symmetrical primary current that can be carried for 1 s with the secondary winding short-circuited without exceeding the maximum temperature specified for the class of insulation used in the transformer.

The mechanical forces are proportional to the square of the peak current, which is 2.82 times the rms value of the a-c component. If the maximum peak short-circuit current that can be obtained due to a fault divided by 2.82 is less than the mechanical rating of the transformer, then the transformer is satisfactory for the application.

The 1-s thermal rating is based on the assumption that all heat is stored. On the same basis the 3-s and 4-s ratings are 58% and 50%, respectively, of the 1-s rating.

216. Causes of Errors. The errors in a current transformer are due to the energy required to produce the flux in the core that induces the voltage in the secondary winding that supplies the current through the secondary circuit. The total ampere-turns available to provide secondary current are equal to the primary ampere-turns minus the ampere-turns required to produce the core flux.

A change in secondary burden alters the flux required in the core and changes the core-exciting ampere-turns; leakage flux entering the core changes the magnetic characteristics of the core and affects the core-exciting ampere-turns.

217. Types of Construction. The types of current transformers are: the wound primary type, which consists of primary and secondary windings completely insulated and permanently assembled on the magnetic circuit; the bar type, which is similar to the wound primary type except that the primary is a single straight conductor of the bar type; the window type, which has a secondary winding completely insulated and permanently assembled on the magnetic circuit and a window through which a conductor can be passed to provide a primary winding, and the bushing type, which is a special window type designed to fit over apparatus bushings, with the conductor through the bushing as the primary winding.

Current transformers are classified in accordance with the major insulation used, as dry-type, compound-filled, molded, or liquid-immersed.

218. Safety Precautions. The secondary winding should always be short-circuited before disconnecting the burden. If the secondary circuit is opened, with primary current flowing, all the primary ampere-turns are magnetizing ampere-turns and usually will produce an excessively high secondary voltage across the open circuit. All instrument-transformer secondary circuits should be connected to ground; when instrument-transformer secondaries are interconnected, only one point should be grounded. If the secondary circuit is not grounded, the secondary becomes, in effect, the middle plate of a capacitor, with the high-voltage winding and ground acting as the other two plates.

219. Errors in Potential Transformers. There are two types of errors that affect the accuracy of the measurements made with potential transformers. The ratio error is the difference between the true ratio of the primary to secondary voltage and the ratio that is marked on the name plate. The phase-angle error is the difference in the phase position of the voltage applied to the secondary burden and the voltage applied to the primary winding.

The ratio error is expressed as a ratio correction factor by which the secondary-voltage value should be multiplied to obtain a secondary voltage that is directly proportional to the primary voltage,

$$F_{PR} = R_{PT}/R_{PH} \tag{11-76}$$

where F_{PR} = ratio correction factor of potential transformer, R_{PT} = true ratio (primary/secondary) of potential transformer, and R_{PN} = name-plate ratio (primary/secondary) of potential transformer.

The phase-angle error is designated by the symbol γ, is expressed in minutes, and is defined as positive when the voltage applied to the burden from the marked to the unmarked secondary terminal leads the voltage applied to the primary from the marked to the unmarked terminal.

220. Relative Importance of Ratio and Phase-angle Errors. The effect of the ratio and phase-angle errors of potential transformers is the same as described for current transformers in Par. **208**, except that with a lagging power-factor load a positive-potential transformer phase-angle error will cause the wattmeter to read low. To obtain the true watts, the apparent watts should be multiplied by the phase-angle correction factor K_γ,

$$K_\gamma = \frac{\cos\theta}{\cos(\theta+\gamma)} = \frac{1}{\cos\gamma - \sin\gamma\tan\theta} \tag{11-77}$$

where K_γ = phase-angle correction factor of potential transformer, θ = angle of lag of load current behind load voltage, and γ = angle of lead of secondary voltage over primary voltage.

221. Classification of Errors. The errors in a potential transformer change with the current required by the burden connected across the secondary terminals and the magnitude of the secondary voltage. The USASI has classified the accuracy performance of potential transformers for metering service. The method involves standard burdens, which are listed in Table 11-16, and accuracy performance limits, which are listed in Table 11-17. The limits of transformer correction factor apply over a range of 90 to 110% rated voltage and from zero burden to the specified burden. The classes and limits are graphically represented as parallelograms in Fig. 11-56. The complete USAS accuracy classification of a potential transformer must include the secondary burden such as 0.3X or 0.6Z.

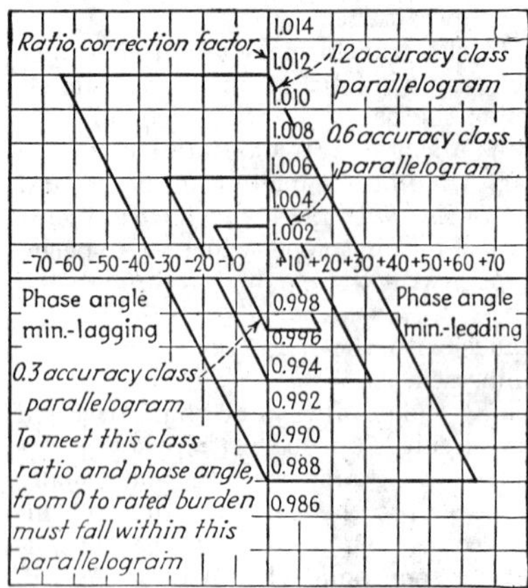

FIG. 11-56. Accuracy-class parallelograms for potential transformers. The ratio correction factor and phase angle shall be limited as indicated for the limits listed in Table 11-17. These limits shall apply from 10% above rated voltage at rated frequency and from zero burden on the potential transformers to the rated burden on a 120-V basis.

222. Potential transformers are made for all the standard rated circuit voltages. They are usually dry-type or molded for voltages below 23 kV and liquid-filled for the higher voltages.

223. The measurement of power by using a wattmeter and both current and potential transformers requires a correction for the ratio and phase-angle errors of both of the instrument transformers and a correction for the phase angle α in minutes of the potential circuit of the wattmeter. Figure 11-57 shows the relative phase positions of the primary and secondary voltages and currents of the instrument transformers if both β and γ are positive. If the potential circuit of the wattmeter is inductive by a small amount, then the current in this circuit will lag the voltage and the phase angle will have the same effect as a negative potential-transformer phase angle. The total phase-angle correction factor is

$$K_S = \frac{\cos(\theta + \beta - \gamma + \alpha)}{\cos\theta} \tag{11-78}$$

where K_S = total phase-angle correction factor, α = angle of lag of potential circuit of wattmeter, β = angle of lead of secondary current over primary current, γ = angle of lead of secondary voltage over primary voltage, and θ = angle of lag of secondary current behind secondary voltage.

The true watts, with all corrections included, is

$$P = WR_{CN}R_{PN}F_{CR}F_{PR}K_S \tag{11-79}$$

where F_{CR} = ratio correction factor of current transformer, F_{PR} = ratio correction factor of potential transformer, P = watts drawn by load, R_{CN} = name-plate ratio of current transformer, R_{PN} = name-plate ratio of potential transformer, and W = watts reading of wattmeter.

Table 11-16. Standard Burdens for Potential Transformers
(From Table 13-11.110, USAS C57.13)

Burden designation	Secondary voltamperes*	Burden power factor
W	12.5	0.10
X	25	0.70
Y	75	0.85
Z	200	0.85
ZZ	400	0.85

* At 120 V for secondary winding; at 69.3 V for tertiary winding.

Table 11-17. Limits of Potential-transformer Correction Factor

Accuracy class	Limits of transformer correction factor		Limits of power factor of metered load (lagging)	
	Minimum	Maximum	Minimum	Maximum
1.2	0.988	1.012	0.6	1.0
0.6	0.994	1.006	0.6	1.0
0.3	0.997	1.003	0.6	1.0

In watthour meters the phase angle corresponding to α is corrected for in the meter by the lag adjustment or by making an overall calibration and adjustment for all errors, including the ratio and phase-angle errors of the instrument transformers.

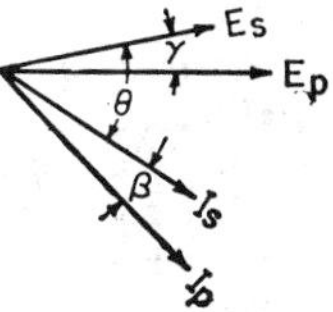

Fig. 11-57. Vector relations in current and potential transformers.
E_P = primary voltage I_P = primary current
E_S = secondary voltage I_S = secondary current

Figures 11-58 and 11-59 show typical ratio-correction-factor and phase-angle curves for a current and potential transformer.

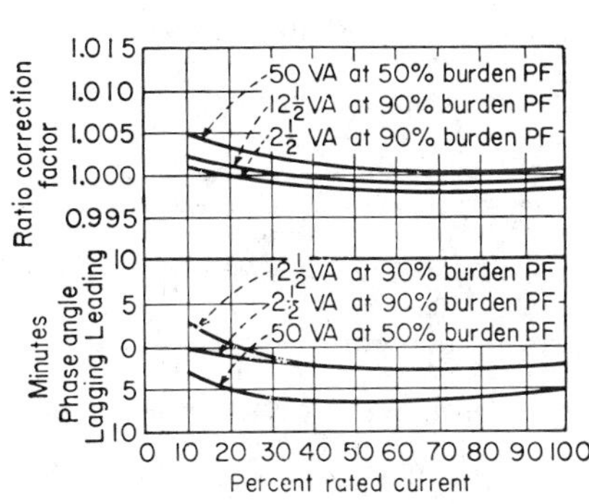

Fig. 11-58. Typical curves of ratio correction of phase angle for a current transformer.

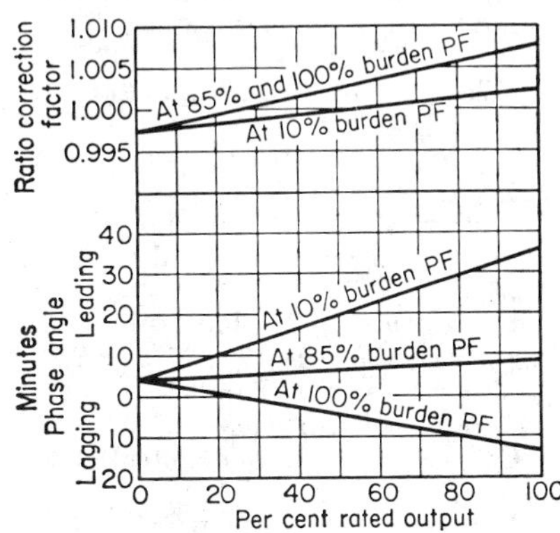

Fig. 11-59. Typical curves of ratio correction factors and of phase angle for a potential transformer.

224. Example. Consider a load measured with 50/5-A current transformer, with a 12.5-VA 90%-power-factor burden characteristic curve according to Fig. 11-53, with 2,300:115-V potential transformer rated 200 VA with a 75-VA 85%-power-factor burden characteristic curve according to Fig. 11-54, and with a lag angle of 6 minutes in the wattmeter potential circuit. What is the load true watts when the instruments read 500 W, 115 V, and 5A, calibration errors of instruments being neglected?

From Fig. 11-58:

$$F_{CR} = 0.9997 \qquad \beta = -2'$$

From Fig. 11-59:

$$F_{PR} = 1.0015 \qquad \gamma = +5'$$

Stated above,

$$\alpha = +6'$$

$$\theta = \cos^{-1} \frac{500}{5 \times 115} = \cos^{-1} 0.86957 = 29°35.47'$$

From Eq. (11-78),

$$K_S = \frac{\cos\,(29°35.47' - 2' - 5' + 6')}{\cos\,(29°35.47')} = \frac{0.86974}{0.86957} = 1.0002$$

From Eq. (11-79),

$$P = 500 \times 10 \times 20 \times 0.9997 \times 1.0015 \times 1.0002 = 100{,}140 \text{ W}$$

225. Bibliography on Instrument Transformers

Silsbee, F. B. A Method of Testing Current Transformers, *NBS Sci. Paper* 309, 1918.

Camilli, G. Current Transformer Design; *Trans. AIEE*, 1940, Vol. 59, p. 835.

Camilli, G., and Ten Broeck, R. L. A Proposed Method of Determination of Current Transformer Errors; *Trans. AIEE*, 1940, Vol. 59, p. 547.

Wentz, E. C., and Sonnemann, W. K. Current Transformers and Relays for High Speed Differential Protection with Particular Reference to Offset Transient Currents; *Trans. AIEE*, 1940, Vol. 59, p. 481.

Park, J. H. Accuracy of High-range Current Transformers; *NBS Jour. Res.*, April, 1935, Vol. 14, pp. 367–392.

Kaufmann, R. H., and Camilli, G. Overvoltage Protection of Current Transformer Secondary Windings and Associated Circuits; *Trans. AIEE*, 1943, Vol. 62, p. 467.

Boyajian, A., and Camilli, G. Orthomagnetic Bushing Current Transformer; *Trans. AIEE*, 1945, Vol. 64, p. 137.

"Electrical Metermen's Handbook"; Meter and Service Committee, Edison Electric Institute, 1940.

Specht, T. R. Biased-core Current Transformer Design Method; *Trans. AIEE*, 1945, Vol. 64, p. 635.

Specht, T. R., and Wentz, E. C. Peak Voltage Induced by Accelerated Flux Reversal in Cores Operated above Saturation Density; *Trans. AIEE*, 1946, Vol. 65, p. 254.

Wentz, E. C. Evolution of Standard Lines of Current Transformers for High Overcurrent Capacity; *Trans. AIEE*, 1944, Vol. 63, p. 658.

Wiggins, A. M. Parallel Operation of Current Transformers for Totalizing Two or More Circuits; *Elec. Jour.*, 1929, p. 379.

Camilli, G. Cascade Potential Transformers; *Gen. Elec. Rev.*, February, 1936, p. 95.

Agnew, P. G. Accuracy of the Formulas for the Ratio, Regulation and Phase Angle of Transformers, *NBS Sci. Paper* 211, 1914.

Brooks, H. B. Testing Potential Transformers, *NBS Sci. Paper* 217, 1914.

Price, L. D. Potential Transformer Connections for Three-phase, Four-wire Metering; *Elec. Jour.*, August, 1927, p. 377.

Camilli, G. Potential Transformer Design; *Trans. AIEE*, 1943, Vol. 62, p. 483.

MOBILE TRANSFORMERS

226. Mobile Transformers and Mobile Substations. A mobile transformer or autotransformer is usually mounted on a semitrailer with lightning arresters and disconnecting switches. A mobile substation has in addition switchgear and measuring and protective relaying equipment. The unit is moved over the highway by a tractor truck. State and Federal highway regulations limit maximum weight and size. Mobile units are used to restore electric service in emergencies, to permit maintenance without service disruption, to provide service during major construction, and to reduce system investment.

The mobile unit is designed to be a multiple-purpose package delivering maximum kVA for allowable weight. Performance and design characteristics vary considerably from those of a conventional transformer. The margin between the voltage level of the insulation structure (BIL) and the operating voltage is generally smaller, the average winding temperature rise over ambient is generally higher (75°C), the overload capability is less, and losses and impedance tend to be higher. The circuitry of the mobile unit is generally more complicated, in order to meet a variety of operating situations in a particular utility system.

RADIO TRANSFORMERS

227. Power Transformers. The purpose of the power transformer in radio-receiver applications is to change the residential lighting-circuit voltage to a value which, when applied to a vacuum-tube or a solid-state rectifier (either half wave or full wave) and filtered properly, may be used to supply the appropriate bias voltages and currents for the active devices (tubes, transistors, etc.) in the radio. The power transformer may also be used to change the lighting-circuit voltage to a value suitable for the filaments of tubes or any lamps which may be in the radio.

228. Filter Reactors. One or more filter reactors may be employed in a rectified a-c supply, in either T- or Π-filter systems, to reduce the ripple voltage to a level which is not audible at the loudspeaker.

229. Audio-frequency Transformers. Three types of audio-frequency transformers may be employed in radio receivers—input, interstage, and output. In the average receiver, only the output transformer is used. Interstage coupling between amplifying stages is accomplished by impedances common to the input and output circuits of the amplifying stages.

Input Transformers. These operate between the source of a-c potential (most commonly the last intermediate-frequency amplifier in a radio) and the first amplifying vacuum tube or transistor of the audio amplifier. The turns ratio for this transformer is determined by the normal applied voltage on the primary and the desired value of voltage to be applied to the grid of the first tube or the base of the first transistor.

Interstage Transformers. Everything mentioned above for the input transformer applies to interstage transformers, with the exception that interstage transformers are used between amplifying stages of audio amplifiers.

Output Transformers. These operate between the last vacuum-tube or transistor stage of the audio amplifier and the load circuit, which in radios is the coil of the loudspeaker. Ordinarily, the output transformer for a power-output stage has a step-down ratio, because the loudspeaker impedance is relatively low compared with the output impedance of either a tube or a transistor amplifier.

230. Radio-frequency Transformers. This term is used to describe a class of transformers which operate at a frequency much higher than the audio range. This is the frequency of the carrier of the incoming radio signal, or in superheterodyne radios it is the difference between the frequency of the incoming carrier and the frequency of the radio's oscillator. This difference frequency is called the intermediate frequency, and transformers through which it passes are called intermediate-frequency transformers.

The radio-frequency transformers perform essentially the same functions as audio-frequency transformers (turns ratio determined by desired voltages), but they have three important differences. As already mentioned, they encounter much higher frequencies. In addition, they handle considerably less power than audio transformers. Finally, one or both windings of a radio-frequency transformer are frequently shunted by a capacitor so that a tuned circuit which attenuates all but the desired frequency is formed.

RECTIFIER TRANSFORMERS

231. Rectifier transformers furnish power to rectifiers at the required a-c input voltage for the desired d-c output voltage. They are built in sizes ranging up to 15,000 kVA, and occasionally higher. The secondary voltage is generally low, varying from less than 50 V for some electrolytic processes to as high as 1,000 V for other applications.

The secondary current is generally high and may reach many thousands of amperes. High currents are obtained by parallel connection of many winding sections. Current is collected by internal bus bars and brought through the transformer cover by high-current bushings.

232. Phase-shifting transformer connections may be used to produce 12 phases, 24 phases, or even more, in order to reduce the current harmonics in the a-c input. Auxiliary phase-shifting transformers or phase interconnections of windings on the rectifier transformers themselves may be used. When two secondary windings are used (as in the double-Y-circuit), the same impedance must be provided between the primary and each secondary winding to get equal commutation angles and d-c voltage on the two secondary circuits.

233. The small number of turns in the low-voltage winding restricts the core size, particularly with phase shifting. For example, the use of delta- and Y-secondaries to get 12-phase output requires turns in the ratio of 1.73. Whole-number ratios meeting this requirement adequately are limited, as, for example, 4:7, 7:12, etc. Phase-shift requirements of 15° or smaller values are even more restrictive. When delta- and Y-primaries are used to obtain 12-phase output, it is customary to place all windings on a single core by mounting one set of primaries with associated secondaries on the upper part of the core and mounting the other set of primaries with associated secondaries on the lower part of the core. It is then necessary to provide an extra core yoke at half height to transfer flux between core legs, as made necessary by the phase difference of 30° between the flux in the upper half of the core and the flux in the lower half of the core.

234. Rectifier transformers supplying mechanical rectifiers or mercury-arc rectifiers must withstand repeated severe overcurrent and overvoltage conditions. Lightning arresters or nonlinear resistors are sometimes used to minimize the overinsulation required.

SPECIALTY TRANSFORMERS

235. General-purpose specialty transformers are dry-type distribution transformers which are generally used with primaries connected to secondary distribution circuits to supply lighting and small power loads at still lower voltages. They are available for primary voltages of 120, 240, 480, and 600 V, with ratings ranging from 25 VA to 500 kVA, 60 c/s.

236. Control transformers are dry-type constant-potential isolation transformers. They are generally used with their primary windings connected to secondary distribution circuits of 600 V or less. Proper selection of a control transformer will supply the correct power at a reduced voltage for lighting and control loads up to 250 VA.

237. Machine-tool transformers are similar to control transformers with capacities up to 1,500 VA for localized lighting and machine control devices, such as solenoids, contactors, and relays on both portable and stationary tools. They are primarily used to provide 120-V output from 240- and 480-V 60-c/s lines. They are also available for 25- and 50-c/s operation at various voltages.

238. Class 2 transformers are dry-type isolation transformers suitable for use in National Electrical Code Class 2 circuits. These transformers are generally used in remote-control, low-energy power, and signal circuits for the operation of bells, chimes, furnace controls, valves, relays, solenoids, and the like. They are 120-V primary units of either the energy-limiting or the non-energy-limiting type. The energy-limiting type has sufficient inherent impedance to limit the output current to a thermally safe value even under short-circuit conditions. The energy-limiting units with maximum open-circuit voltages of 10 or 16 V are used as bell and chime transformers. Energy-limiting transformers are also available with an open-circuit voltage of 30 V and ratings up to 45 VA. Non-energy-limiting Class 2 transformers must be provided with a temperature-sensitive device to limit maximum temperature. Non-energy-limiting units have a rated load voltage of 24 V and are available in ratings up to 75 VA.

239. Signaling transformers are dry-type constant-potential step-down isolation transformers, which are generally used with their primary windings connected to secondary distribution circuits to supply signaling systems not subject to Class 2 circuit limitations. They are available for 120- or 240-V circuits. An output-voltage selection of 4, 8, 12, 16, 20 or 24 V is available by properly connecting the four output leads. Units up to 1,000 VA are available.

240. Luminous-tube transformers, for supplying power to neon or other gaseous signs, are manufactured in sizes ranging from about 50 to 1,650 VA. The secondary voltage ranges from 2,000 to 15,000 V. The voltage requirements depend upon the length of gas tubing in the circuit; i.e., the longer the tubing, the higher the required voltage. The current furnished by the transformers ranges from about 18 to 120 mA.

The essential requirements for luminous-tube transformers are (1) constant current and (2) the fact that the short circuit should exceed full-load current by not more than 25%. These requirements necessitate a transformer design having a high leakage reactance. This is generally accomplished by the use of a magnetic shunt between the primary and secondary windings. Figure 11-60 shows a cross section of a typical design. It will be noted that the secondary is divided into two sections which are connected in series. The midpoint of the secondary is usually grounded, which reduces the voltage stresses on the windings.

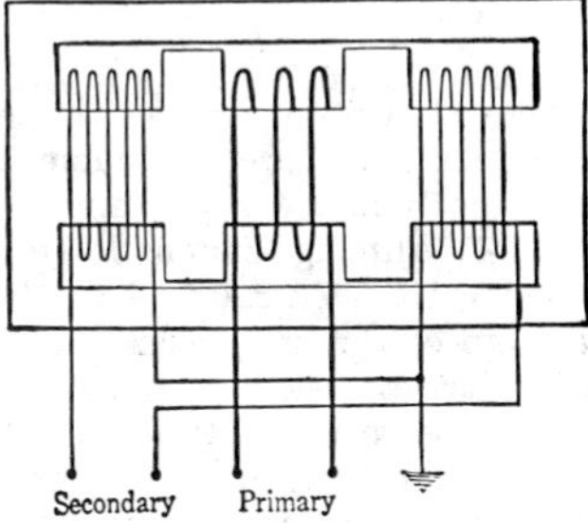

FIG. 11-60. Core and coil arrangement of a typical neon-sign transformer showing windings and leakage paths in magnetic circuit.

241. Ignition transformers are dry-type step-up transformers of the high reactance type used to provide ignition for gas or domestic oil burners. Such transformers are limited to primary voltages of 120 or 240 V. Secondary voltages are limited to 15,400 V, and they usually range from 6,000 to 14,000 V. Secondary-current ratings range from 20 to 28 mA and capacities from 140 to 430 VA.

242. Toy transformers are step-down transformers of the low-voltage secondary type which are intended primarily to supply current to electrically operated toys. They are usually portable, and, because of their intended use, special attention is paid in their design to safety and the elimination of fire hazard; input to the primary winding must be limited by design to 660 W even when the secondary winding is short-circuited, a condition that must be withstood without creating a fire hazard. Such transformers are not authorized for primary voltages higher than 150 V, and secondary voltages may not exceed 30 V between any two output terminals.

TESTING TRANSFORMERS

243. Testing transformers, used for making low-frequency high-potential tests, have been developed for higher voltages to keep ahead of the application of higher transmission voltages. Voltages of 1,500,000 or more are often required. Single units for 1,000,000 V above ground have been built, but it is usually more economical to obtain such potentials by connecting two or more units in "cascade," or "chain." In the most common form of cascade connection, the high-voltage windings of all units are connected in series, and each winding is connected to its own tank. The tank of the first unit in the series is grounded, and that of each remaining unit is insulated from ground. All units except the one on the line end are provided with a tap in the high-voltage winding near the high-voltage end for supplying the primary of the succeeding unit. It is practical to extend this scheme of connection to three or even four units. Testing sets embodying this connection have been built to provide 2,000,000 V between terminals with the midpoint of the circuit grounded or with one terminal grounded. Testing transformers are usually "short-time"-rated. However, for special purposes, a rating of several thousand kVA may be required, and the time rating may be continuous.

244. The testing voltage is usually adjusted by varying the supply voltage to the primary of the testing transformer. The most common methods of varying the supply voltage are by means of an induction regulator, a regulating transformer, or a generator with adjustable field.

245. The recognized standard for measuring high voltage is the sphere gap. Since it is necessary to flash over the gap to obtain a measurement, some kind of indicating meter is desirable. Testing transformers are usually provided with a bushing capacitance tap for connection to a crest voltmeter or a voltmeter winding for connection to an rms voltmeter. In either case a sphere gap is used for calibration. An indicating instrument greatly expedites the work of testing, and under normal conditions it will give reasonable accuracy.

REACTORS

246. A reactor, sometimes called a choke coil, is a device which stores energy in its magnetic circuit. The stored energy determines the rating as follows:

$$Q = 2 \times 10^{-3}\pi fLI^2 \qquad (11\text{-}80)$$

$$Q = 9.84 \times 10^{-12} \times \frac{VB^2f}{\mu} \qquad (11\text{-}81)$$

where Q = kVA rating, f = cycles per second frequency, L = henrys inductance, I = rms amperes, V = cubic inches volume of magnetic field, B = lines per square inch peak flux density, and μ = permeability (for constant-permeability material).

247. Although magnetic steel is often used for a part of the magnetic circuit of a reactor, the kVA rating depends on the high-density energy storage in the nonmagnetic portion of the flux path. This may be one gap the length of the coil or many small gaps between steel sections distributed inside the coil. The function of steel in the reactor is to concentrate the energy storage into small volumes at high flux density, with consequent reduction of coil size and electrical losses. Overloading iron-core reactors results in saturation of the steel and a substantial reduction of inductance. Where constant inductance is necessary over a range of load, such as in current-limiting service, in tuned circuits, or for precision absolute measurements, coreless ("air-core") reactors are used.

248. The equivalent of adjustable reactance can be obtained from a d-c-controlled saturable core reactor. Such a reactor is made with uninterrupted steel magnetic circuits, at least two a-c coils per phase and coils for adjustable d-c excitation. During part of each cycle of alternating current the high reactance of the unsaturated core is encountered, and during the remaining part of each cycle the low reactance of the saturated core is met with. The amount of d-c excitation determines the relative proportions of the cycle in which these two levels of reactance occur. In large reactors of this type the d-c control power in kW may be less than 1% of the controlled a-c kVA.

FIG. 11-61. General Electric dry-type shunt reactor rated 8,000 kVA and 13,200 V.

249. Applications of reactors in electric power systems include (1) series reactors for limiting fault current, (2) neutral grounding reactors for limiting fault current, (3) shunt reactors for supplying line charging current, and (4) switching reactors used in load-tap-changing transformers. The dry-type reactor in Fig.

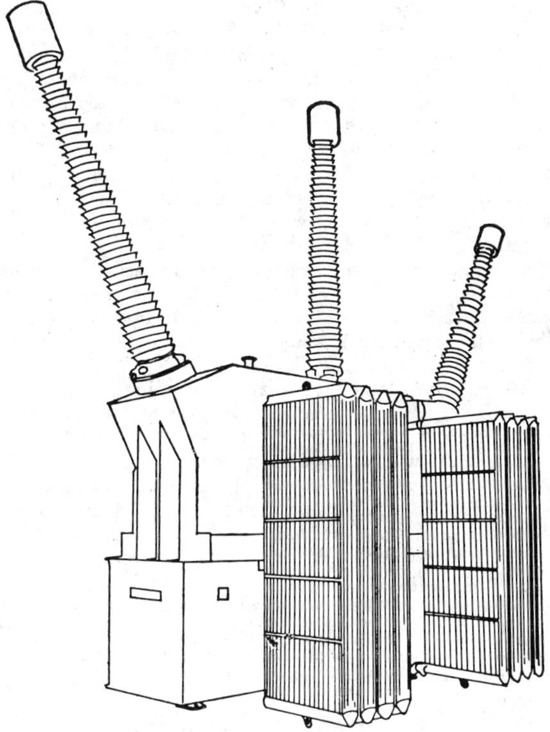

FIG. 11-62. Allis-Chalmers oil-immersed shunt reactor rated 70 mVA, 500 kV, 1,425 kV BIL, 3-phase.

11-61 is suitable for a series application, a neutral grounding application, or a shunt-reactor application. Figure 11-62 shows an oil-immersed shunt reactor.

250. Ballast reactors, either fixed- or adjustable-reactance, are used for stabilizing current in nonlinear resistance loads such as molten glass or in arc loads such as arc furnaces, welding arcs, fluorescent lamps, or vapor lamps.

251. For controlling d-c power output from a rectifier, a variant of the saturable reactor, the amplistat reactor, or magnetic amplifier,[1] is often used. Connected in series with the line between the rectifier transformer and the rectifier, the amplistat carries a d-c component of load current. By using core steel with a sharp knee in the magnetization curve for the amplistat a very small additional d-c input to a control winding suffices to modify the impedance characteristics so as accurately to control the transmitted power with a fast response.

252. Miscellaneous applications include (1) resonant reactor-capacitor combinations for excitation of nuclear-accelerator magnets, for filtering harmonics from the a-c or d-c side of rectifiers, or for constant-voltage to constant-current transformation (monocyclic network); (2) energy storage for high-power arcs; (3) balancing reactors for equalizing current in parallel circuits; (4) smoothing reactors for rectified d-c circuits; (5) adjustable reactors (usually d-c-controlled) for fine control of heating or lamp loads; and (6) current-limiting reactors for motor starting.

253. Communications applications include (1) loading (Pupin) coils, (2) smoothing reactors, (3) filtering reactors, and (4) reactors for oscillator circuits. Many of these reactors use magnetic ferrite materials in place of steel because the absence of eddy currents gives faster response.

[1]STORM, H. F. "Magnetic Amplifiers"; New York, John Wiley & Sons, 1955.

254. Dry-type current-limiting or shunt reactors are widely used in electrical power systems up to 15 kV, occasionally to 34.5 kV. The trend is toward outdoor installation, especially of the larger units. Small- and moderate-sized current-limiting reactors are often stacked into 3-phase assemblies with center phase reversed to avoid net repulsive force. Large shunt reactors have the phases mounted separately. Dry-type reactors utilize inorganic materials such as Fiberglas, concrete, and porcelain for resistance to weathering and reduced fire hazard.

255. Design Considerations. Reactors subject to severe short-time overloading, such as current-limiting reactors, must withstand large mechanical forces. There is a net radial outward force, which in a circular coil is resisted by hoop stress in the conductor. There is also an axial compressive force requiring the use of strong conductor spacers and short conductor spans. Bracing is often needed to withstand forces between adjacent reactors. Because of the high flux density in reactors containing steel in the magnetic circuit, they have, at normal current, heavy pulsating forces on the steel at all gaps in the circuit. Structural design must be extremely stiff to avoid excessive vibration.

In reactors of large size, whether coreless or containing steel, there are strong magnetic fields in the coils. To avoid excessive eddy-current losses, the conductor must be subdivided into small insulated strands, transposed so that they have the same flux linkages. Metal parts near coreless reactors may be heated by the magnetic field of the reactor, especially long, magnetic steel parts which parallel the reactor field, or closed conducting loops which link part of the field. When such reactors are oil-immersed, the steel tank is usually shielded by an internal concentric conducting cylinder. Similar shielding is often used for small reactors in high-frequency circuits.

256. Transient overvoltage protection for reactors in power circuits is most effectively obtained by lightning arresters, to provide a safe margin below the basic impulse level of the reactor insulation. Such arresters will also limit the inductive voltages created in the reactor by current chopping in switches. For current-limiting reactors where continuous reactor voltage is low, it is economical to use nonlinear resistors shunting the reactor.

STARTING COMPENSATORS

257. Autotransformer starters are used for starting induction motors to reduce the current drawn from the supply lines. The prime advantage of the autotransformer starter is that the autotransformer, through transformer action, allows more current to flow to the motor than is drawn from the lines. The ratio of line current to motor current is dependent upon the excitation current of the autotransformer and the tap

Table 11-18. Duty Cycles for Motor-starting Autotransformers

Sequence	Heavy-duty cycle	Medium-duty cycles		
		Manual motor controllers, 250 hp and below	Magnetic motor controllers, 200 hp and below	Magnetic motor controllers, above 200 hp to 3,000 hp
On	1 min	15 s	15 s	30 s
Off	1 min	3 min and 45 s	3 min and 45 s	30 s
Repeat	4 times (5 c total)	3 times (4 c total)	14 times (15 c total)	2 times (3 c total)
Rest	2 h	2 h	2 h	1 h
On	1 min	15 s	15 s	30 s
Off	1 min	3 min and 45 s	3 min and 45 s	30 s
Repeat	4 times (5 c total)	3 times (4 c total)	14 times (15 c total)	2 times (3 c total)
Tap	Lowest tap	65%	65%	65%
Load	Motor with rotor blocked or connected to equivalent inductive load			
Current	....	300% of motor full-load current	300% of motor full-load current	300% motor full-load current rent
Power factor	....	50% or less	50% or less	50% or less

of the autotransformer being used. A typical example is as follows: Assume a 100-hp 3-phase 440-V 123-A full-load-current motor with a locked rotor current of six times full-load current. Assume excitation current of the autotransformer to be 25% of motor full-load current. When connected to the 50% tap of the autotransformer, the current to the motor will be half of motor locked rotor current, or 369 A. The current drawn from the lines, on the other hand, will be 25% of motor locked rotor current plus 25% motor full-load current, or 215 A.

Manual starters require manual operation to switch the autotransformer out of the power circuit after the motor has accelerated to full speed. They are normally an open-transition-type starter. The motor is momentarily disconnected from its power source during the time it takes to cut the autotransformer out of the circuit. Magnetic-type starters can be either an open-transition type, like the manual-type starter, or a closed-transition type, whereby the motor is not momentarily disconnected from the lines during the time the autotransformer is switched out of the power circuit.

Since the autotransformer is in the power circuit only during the acceleration period of the motor, the autotransformers are designed to specific duty cycles for economy reasons. Table 11-18 shows several typical duty cycles. Special-design autotransformers are required for such applications as frequent starting and extremely long motor accelerating times due to large inertia loads.

VOLTAGE REGULATORS

258. Voltage Regulation. It is the function of a transmission and distribution system to deliver power to the user at a voltage within limits which are acceptable for the operation of his equipment. It is impractical to regulate large systems solely by means of generator regulators; so other means for correcting the voltage are commonly used throughout the system. Among these are transformer-load tap changers, synchronous condensers, switched capacitors, and step-voltage regulators. The induction-voltage regulator was commonly used but has been generally superseded by other, less expensive schemes in this application.

259. Voltage regulators find their principal use for bus or individual feeder regulation at substations or on the feeders between the substation and the load. They are installed on foundations, on platforms, or directly on poles, depending on their size and the practice of the user. A voltage range of 10% above to 10% below the circuit voltage is commonly provided, and it is usual to have adjustable limit switches which increase the current capacity to as much as 1.6 times the rated value as the available range of voltage adjustment is reduced to 5% above and 5% below normal.

Regulators are also used within specific load areas where voltage problems exist. Air- or askarel-insulated regulators are used to regulate lighting circuits in factories where heavy intermittent loads cause deviations in the general power-supply voltage which are too great for successful operation of lights. Similar regulators are sometimes built into industrial or military equipment, such as furnaces, rectifiers, or radar sets, to provide control of voltage, current, or power. Small regulators, either automatic or hand-operated, are available for testing sets, laboratories, motor speed control, and other applications requiring controlled or steady voltage. These may be induction regulators or step types in which very small steps are provided by running a movable contact directly on the exposed turns of a toroidal winding. Other types of control voltage are by d-c saturation of iron-core reactors or by variable impedance with solid static devices. Electronic controls, including a variety of detecting and amplifying circuits, are commonly used for automatic control of small regulators in this field.

260. Step-voltage regulators (Figs. 11-63 and 11-64) are single- or 3-phase autotransformers with voltage-regulating taps in one of the windings, usually the series winding. The tap-changing circuits and motor-driven mechanism are similar to those used in power transformers. With the commonly used connections, the output voltage is in phase with the input voltage.

Thirty-two ⅝% voltage steps are provided in standard regulators for a range from 10% below to 10% above normal voltage. During switching, the taps are bridged by a current-limiting reactor which also serves as an autotransformer on alternate positions to give a voltage intermediate between taps. Thus, a winding with eight taps is used with a nine-point dial switch and a reversing switch to give 32 voltage steps. Refinements of these mechanisms include self-positioning, spring-operated contacts loosely coupled to a driving motor or, alternatively, electrically positioned contacts directly geared to the motor. Long contact life is obtained by using a sintered mixture of tungsten and copper for the arcing tips.

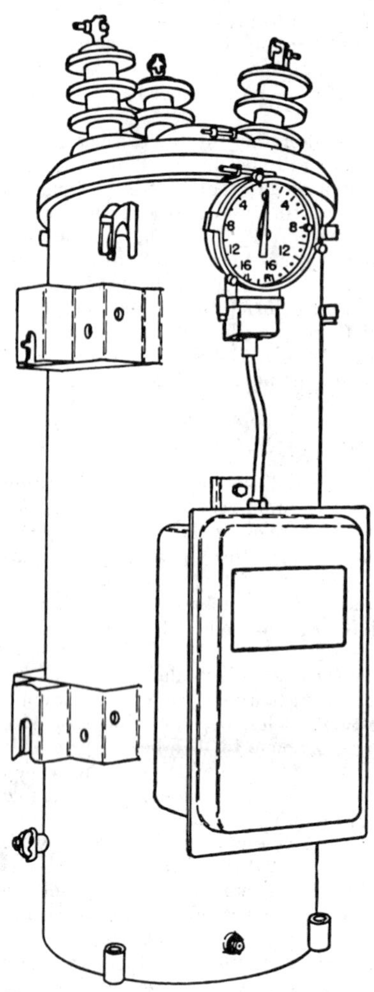

FIG. 11-63. Typical General Electric single-phase step-voltage regulator.

Each switching operation involves an arc which carbonizes a small amount of the insulating oil. To prevent this material from causing dielectric breakdown under operating voltage, dependence is placed on vertical insulating surfaces upon which the carbon cannot settle by gravity and on shields and other configurations which prevent electrostatic deposit of carbon on insulations subject to intense electric fields. In regulators below 34.5 kV operating voltage, windings may be immersed in the same oil as the switch, in which case similar precautions are used to protect them.

261. The induction regulator resembles a wound-rotor induction motor with the rotor limited to movement through one pole pitch and driven to required positions by a motor. The exciting winding, which is connected to the supply circuit, is assembled in slots on the rotor, and the series winding for raising or lowering the output voltage is placed in slots in the stator. In single-phase regulators, a short-circuited winding on the rotor in quadrature relation to the exciting winding limits the internal impedance of the regulator when operated near its neutral position. The axis of the regulator rotor is usually vertical.

Present applications are largely in electronic or industrial systems requiring the unique advantages of stepless voltage changes with relatively fast response time, free of harmonic distortions.

262. The kVA rating of a regulator is the product of the load amperes and the voltage of the series winding in kilovolts. Standard ratings of regulators are available up to 250 kVA single-phase and 2,500 kVA 3-phase for line voltages of 2,500, 4,330, 5,000, 7,500, 12,000, 13,800, 23,000, 34,500, 46,000 and 69,000 for a voltage range of 10% raise or lower. The kVA of the regulated circuit for these regulators is 10 times the regulator kVA, so that circuits up to 25,000 kVA can be regulated. Single-phase regulators for 8,700 V and less are commonly insulated for use, line to neutral, in circuits of voltage 1.73 times their nominal rating.

Standard regulators are oil-immersed and self-cooled. Units above 500 kVA are usually suitable for the addition of fans for forced cooling. The additional output so obtained is usually 25% of the self-cooled rating.

263. Wide-range regulators are required for many industrial applications, particularly in the electrochemical field. Commonly, these regulators are used to control the output of a solid-state rectifier installation by varying the a-c voltage applied to

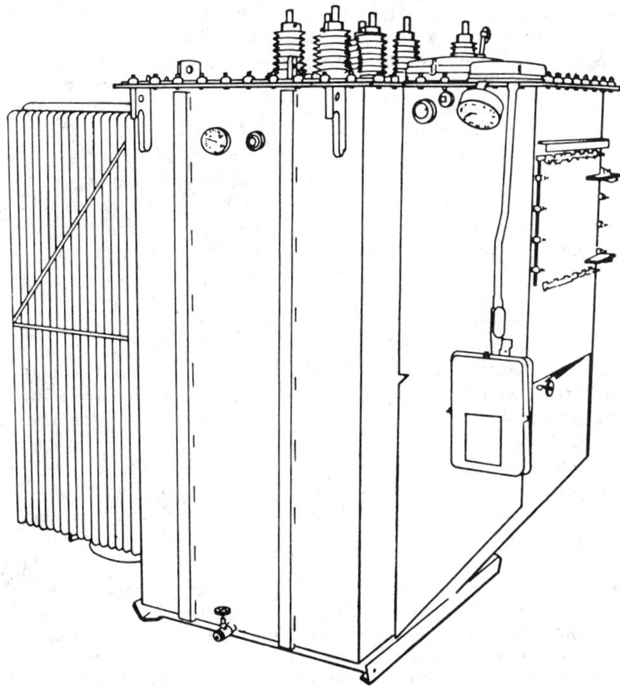

FIG. 11-64. Typical Allis-Chalmers 3-phase step-voltage regulator.

the rectifier-transformer primary windings. These regulators cover the range with a combination of a main transformer winding with a number of coarse taps with load tap-changing equipment and a vernier transformer with load tap changing, induction regulator, or saturable reactor.

The circuits and sequence of operation are arranged to permit the vernier unit to traverse each coarse tap in fine increments. At the end of the range of the vernier unit the coarse tap changer makes a tap change, and the vernier unit is connected to traverse the next coarse tap. The circuits and switching sequence are arranged so that load current is not broken by the coarse mechanism and little or no rupture duty is seen by this mechanism.

The required sequencing of operation of the coarse- and fine-operating mechanisms is accomplished by relays actuated from drum or dial switches coupled to the mechanism-operating shafts.

264. Automatic control of voltage in transmission, distribution, and industrial circuits is performed in regulators by a servo control having three principal elements: (1) a voltage-sensing device which monitors the output voltage and sends a signal to the control circuits; (2) an amplifying section, with or without time delay, which transmits the signal; and (3) a motor drive which responds to the signal by moving the tap changer or induction-regulator rotor in the direction to correct the voltage.

Voltage regulators for industrial circuits are designed to correct for relatively short-term voltage fluctuations or changes in load characteristics so that correction is initiated as soon as a need is detected. Correction starts as soon as the motor starts to run on induction regulators and takes 1 to 5 s on step regulators, depending on the time required for the tap changer to complete an operation.

Voltage regulators for transmission and distribution circuits are designed to correct for source voltage or load trends so that a time delay is provided. The time delay withholds the signal until accumulated time outside the voltage limit less accumulated time inside the voltage limit exceeds the time-delay setting. The time delay is usually adjustable and is seldom set for less than 10 or more than 60 s. Time delay is accomplished by means of motor-driven cams operating switches, resistance-capacitance networks, or thermal delay relays. On modern regulators the latter two means have generally replaced the mechanical time delay in order to reduce the number of moving parts and thus reduce maintenance.

The voltage-sensing device is connected to the output circuit of the regulator through a potential transformer. The controls for feeder-voltage regulators usually include a line-drop compensator consisting of variable-resistance and -reactance elements connected in series with the voltage-sensing device. A current proportional to line current is circulated through the compensator by means of a current transformer in the load circuit. Adjustment of the compensator impedances to correspond to those of the line permits the regulator to hold constant voltage at a predetermined point on the circuit, regardless of the magnitude or power factor of the load.

265. The voltage-sensing device consists of a voltage-regulating relay or a static voltage sensor. The voltage-regulating relay has a movable member held in a balanced position at a predetermined voltage. Contacts supported by the movable member close at voltages near the balanced point for raising or lowering operation of the regulator. These contacts energize relays in the amplifying section of the control. On modern regulators, the voltage-regulating relay is commonly replaced with a static sensing circuit which eliminates the need for contacts in the sensing circuit. The static-sensor output is applied to a solid-state or magnetic amplifier and discriminator circuit, which in turn provides the signals for raising or lowering voltage. For either sensing device, the zone of voltage between the initiation of a raising and a lowering control action is known as the voltage band. The voltage band must be more than the minimum correction obtainable through the regulator, or hunting of the regulator will result. A minimum band of 1½ V on the basis of a 120-V secondary in the potential transformer is now common with new regulators. As the maximum voltage applied to loads near the regulator is determined by the top of the band and the minimum voltage reaching the most distant loads by the bottom of the band, the voltage band becomes a lost voltage which cannot be corrected by the regulator. Thus a reduction in the voltage band will permit regulating larger loads or longer feeders without exceeding the permitted voltage limits at any point on the circuit.

The voltages at the upper and lower limits of the voltage band may vary, owing to temperature effects or other causes. Such variation is held within ±1% on standard regulators with Class I control.

266. Bibliography on Voltage Regulators

OKERBERG, R. D. Regulators vs. Switched Capacitors; *Allis-Chalmers Elec. Rev.*, 1954, Vol. 19, No. 1, pp. 10–14.

BUTLER, R. M., and SAMSON, D. R. Regulators and Capacitors Used Jointly Prove Effective and Economical; *Elec. Light & Power*, Dec. 15, 1954, Vol. 32, No. 14, pp. 94–98.

WILLIAMS, W. A. Considerations Affecting the Use of Time Delay in Feeder Voltage Regulators; *Distribution Mag.* (General Electric), July, 1955, Vol. 17, No. 3, pp. 9–11.

BEST, I. W., and BREM, H. C. What about ⅝% Feeder Voltage Regulation?; *Allis-Chalmers Elec. Rev.*, Third Quarter, 1958, Vol. 23, pp. 4–6.

EATON, E. F. Universal Compensator Settings for Voltage Regulators; *Distribution Mag.* (General Electric), July, 1961, Vol. 23, No. 3, pp. 18–20.

MASSARA, J. M. Wide Range Step Voltage Regulators; *Power*, February, 1963, Vol. 107, No. 2, pp. 68–70.

USAS Requirements, Terminology and Test Code for Step Voltage and Induction Voltage Regulators, C57.15-1965, published by the United States of America Standards Institute.

SECTION 12

CONVERSION OF ELECTRIC POWER

BY

HAROLD WINOGRAD Consulting Engineer (Retired), Allis-Chalmers

JOHN B. RICE Chief Development Engineer, Control-Switchgear, General Products Division, Allis-Chalmers

J. F. SELLERS Chief Engineer, D-C Machines, Allis-Chalmers

R. E. APPLEYARD Manager, Product Engineering, Hydrogenerator Department, Allis-Chalmers

R. L. ROBERTSON Senior Staff Engineer, Systems Control Products, General Products Division, Allis-Chalmers

ROLAND W. URE, JR. Fellow Physicist, Westinghouse Research Laboratories

H. P. MEISSNER Professor of Chemical Engineering, Massachusetts Institute of Technology

M. C. DEIBERT Department of Chemical Engineering, Massachusetts Institute of Technology

CLIFFORD MANNAL Consultant, Engineering Services, General Electric Company

NORMAN W. MATHER Professor of Electrical Engineering, Princeton University

V. C. WILSON Research and Development Center, General Electric Company

CONTENTS

Numbers refer to paragraphs

SECTION 12

CONVERSION OF ELECTRIC POWER

RECTIFIERS AND OTHER STATIC CONVERTERS

By Harold Winograd and John B. Rice*

1. Rectifying Devices and Symbols. Static power converters covered in this section utilize rectifying devices to transform electric power. They have been applied for converting alternating-current power to direct current and the reverse, for frequency conversion, and for control or regulating functions.

A rectifying device conducts current effectively in only one direction. A variety of such devices have been used. The newest type is the semiconductor silicon rectifier, which has largely superseded the other types, including mercury-arc rectifiers, for practically all applications. One notable exception is the use of mercury-arc rectifiers for high-voltage d-c power transmission (see Sec. **14**). The operating principles and characteristics of rectifying devices are discussed in Pars. **43** to **76**.

Rectifying devices can be considered as unidirectional switches which connect two parts of a circuit. There are two classes:

a. **Diode,** which has two main electrodes, **anode** and **cathode.** The current flows in the direction from anode to cathode. A diode is always in a closed position to permit current flow in one direction. The current values are determined by the external circuit.

b. **Controlled rectifier,** which has one or more control electrodes in addition to the anode and cathode. The *start* of conduction can be controlled by applying a suitable voltage or current to the control electrodes. After current flow has started, the electrodes lose control, and the device behaves like a diode. The control electrodes can regain control after the current is brought to zero by the action of the circuit.

There are some devices with control electrodes that can also interrupt the current. Their application for power conversion is of limited interest at this time, and they will not be considered.

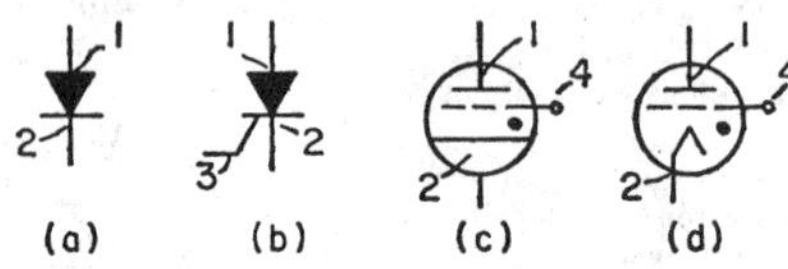

Fig. 12-1. Graphical symbols for rectifier elements. (*a*) Semiconductor diode; (*b*) thyristor (SCR); (*c*) pool-cathode mercury-arc tube; (*d*) thyratron tube. 1, Anode; 2, cathode; 3, gate; 4, grid.

2. Graphical Circuit Symbols. In Fig. 12-1 are shown the graphical symbols of several types of rectifying devices. The symbols of Fig. 12-1*a* and *b* are used in most of the circuit diagrams of this section. The symbols represent a **rectifier circuit element.** In all the devices of Fig. 12-1, the direction of current is from anode to cathode.

RECTIFIER CIRCUITS

3. Simple Rectifier Circuits. The operation of rectifier circuits for conversion of a-c power to d-c is illustrated in Fig. 12-2.

In Fig. 12-2*a*, a load resistance R_d is connected to an a-c circuit through a diode. At the right is shown the sine wave of the a-c voltage *e*. During the positive half-cycle, current flows through the diode and the load. The voltage e_d across the resistor, indicated by the heavy line, follows the a-c voltage wave (the voltage drop across the diode

* J. B. Rice contributed the subsections on Self-commutated Inverters, Semiconductor Rectifiers, and Par. **27**.

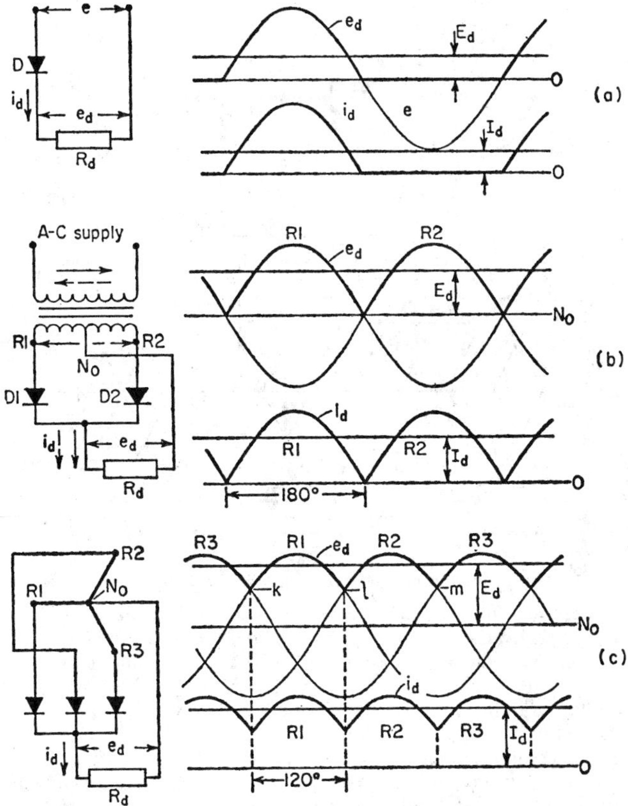

FIG. 12-2. Simple rectifier circuits.

and a-c circuit impedance being disregarded). During the negative half-cycle, the flow of current is blocked by the diode, and the load voltage is zero. The current i_d follows the voltage wave. The average values of the direct current and voltage are I_d and E_d.

In Fig. 12-2*b* a transformer is interposed between the a-c and d-c circuits. The load resistance is connected between the midtap of the secondary winding and the common connection of the diodes. During one half-cycle of the a-c voltage, terminal $R1$ is positive with respect to N_0, and current flows into the load resistance through diode $D1$, as indicated by the solid arrow. During the second half-cycle, $R2$ is positive to N_0, and current flows into the load circuit through diode $D2$, as indicated by the broken arrow. Direct current flows into the load circuit during both half-cycles, doubling the average value in relation to the peak.

In Fig. 12-2*c* the voltage phasors $R1$-$R2$-$R3$ represent a 3-phase winding of a transformer or generator with a neutral terminal N_0. The load resistance is again connected between the neutral and the common connection of the diodes. As indicated by the traces at the right, each of the phases supplies current to the load circuit during one-third of a cycle while its voltage is more positive than that of the other phases. Thus phase $R1$ conducts between k and l, phase $R2$ conducts from l to m, etc. The d-c voltage wave e_d is indicated by the heavy lines.

As seen in Fig. 12-2, the ripple magnitude in the d-c voltage and current waves decreases as the number of phases is increased, while the ripple frequency increases. For most power-conversion applications, 6-phase rectifier circuits, or multiples of 6 phases, are used. They are usually made up of 3-phase groups which are shifted in phase from each other to obtain a larger total number of phases.

4. Phase Control. In Fig. 12-3 is shown a 6-phase rectifier circuit, which inter-

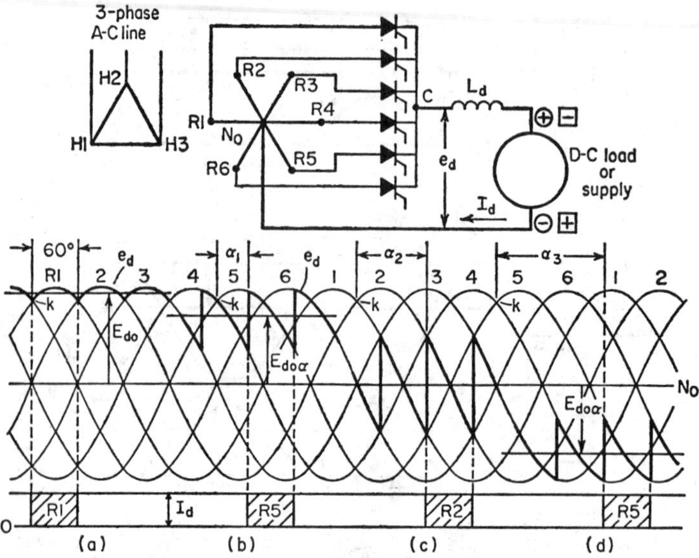

FIG. 12-3. Operation of phase control for rectification and inversion.

connects a 3-phase a-c line and a d-c circuit. The delta/6-phase star rectifier transformer is represented by its voltage phasors. The 6-phase winding is connected to the d-c circuit through controlled rectifier elements, the conduction start of which can be controlled. An inductance L_d irons out the ripples in the direct current I_d, which is assumed to be flat as indicated at the bottom of the figure. The phase voltages $R1$ to $R6$ are represented by sine waves.

In Fig. 12-3*a* conduction of the rectifier elements is started at the intersection points of the phase voltages, so that each conducts the direct current while its phase voltage has the highest value, similar to diodes. The resulting d-c voltage is indicated in heavy outline, and its average value E_{do} is a maximum. As in Par. **3**, the effects of voltage drops and impedances are disregarded.

In Fig. 12-3*b* the conduction start of each rectifier element is delayed by an angle α_1 from the intersection point of the phase voltages. During this delay, the preceding phase continues to carry the current. The result is that the average value of the d-c voltage is reduced below that of Fig. 12-3*a*.

If the conduction start is delayed by an angle α_2 of 90°, the average d-c voltage is zero, the positive and negative areas of the voltage wave being equal. For this condition, the d-c load is effectively short-circuited and the voltage ripple is absorbed by the inductance L_d.

In Fig. 12-3*d* the start of conduction is delayed by a still larger angle α_3. The resulting average d-c voltage is negative. Since the direction of current in the rectifier circuit cannot change, a reversal of the d-c voltage represents a reversal of power. To make a current flow in the circuit while the rectifier d-c voltage is negative requires an external voltage source in the same direction as the current and of sufficient magnitude to overcome the negative rectifier voltage, which can be considered as a back-emf.

When the rectifier voltage is positive and power flows from the a-c supply to a d-c load, the d-c circuit of Fig. 12-3 will have the plus and minus polarities enclosed in circles. The load might be a d-c motor or some other power-consuming device. When the voltage and power flow are reversed, the circuit will have the polarities enclosed by squares. The load would be replaced by a d-c supply, which might be a d-c generator driven by a prime mover or an a-c motor; it could be another rectifier unit which is connected to another a-c supply.

The method of varying the d-c voltage by varying the starting point of rectifier elements in their voltage cycle is called **phase control.** A rectifier unit operates as a **power rectifier** when it transmits power from an a-c to a d-c circuit. It operates as a

power inverter when the power flow is from a d-c to an a-c circuit. The functions are called **rectification** and **inversion.** In some applications a rectifier unit performs both functions alternately. The average value of the d-c voltage with phase control is given by Eq. (12-2).

At the bottom of Fig. 12-3 the transformer phase currents, shown by crosshatched blocks, are shifted from the peak point of the phase voltage by the angle α. This is reflected on the a-c line side as an increase in the power-factor angle. For the theoretical conditions of Fig. 12-3, in which the effect of circuit impedances is neglected, the power factor is equal to cos α.

5. Rectifier Transformer. A rectifier unit is usually provided with a rectifier transformer which has one or more of the following functions: (*a*) To transform the available a-c supply voltage to a value needed for obtaining the desired d-c voltage. (*b*) To provide the number of phases required to obtain the desired waveshapes of the d-c voltage and current and of the a-c line current. (*c*) To isolate the d-c circuit from the a-c supply circuit. (*d*) The transformer reactance serves to limit fault currents which might otherwise cause damage. Under certain favorable conditions a transformer can be dispensed with (see Par. **12**).

Because of the reversible power flow, the terms "primary" and "secondary" cannot be used conveniently for the two windings of a rectifier transformer. The winding connected to the a-c circuit is called the **a-c winding.** The winding connected to the d-c circuit through the rectifier elements is called the **d-c winding.**

RECTIFIER CIRCUIT THEORY

6. Letter Symbols. In Table 12-1 is a list of letter symbols used in the paragraphs on rectifier circuit theory and other parts of this section. Other symbols are explained where used.

Table 12-1. Letter Symbols Used in Power Conversion

α = phase-control angle of retard
β = phase-control angle of advance (for inverter)
β_0 = phase-control angle of advance for $\gamma = 0$
γ = inverter margin angle
$\cos\delta$ = distortion component of power factor
$\cos\phi'$ = displacement component of power factor
D_x = commutating-reactance transformation constant
e_c = commutating voltage (instantaneous values)
e_d = instantaneous values of d-c voltage
VR = total voltage regulation in volts
E_d = average d-c voltage under load, with all voltage drops considered
E_d' = average d-c voltage under load, with only reactance voltage drop E_x considered
E_{do} = theoretical no-load d-c voltage; $E_{do\alpha} = E_{do}$ with phase control
E_f = forward voltage drop of rectifier element
E_{ii} = initial reverse voltage
E_L = line-to-line voltage (rms) of a-c system
E_m = rms of harmonic in d-c voltage
E_n = line-to-neutral voltage (rms) of a-c system
E_{pf} = peak forward voltage
E_{pi} = peak reverse voltage
E_r = d-c voltage drop caused by resistance losses
E_s = line-to-neutral voltage (rms) of transformer d-c winding
E_x = d-c voltage drop caused by commutating reactance
F_x = I_cX_c/E_s = commutating-reactance factor
i_c = commutating current (instantaneous values)
I_c = direct current commutated in a set of commutating groups
I_d = rectifier d-c load current (average)
I_H = total rms value of harmonic components of I_L
I_L = a-c line current in rms amperes
I_m = harmonic component (rms) of I_L
I_p = transformer a-c winding coil current (rms)
I_s = transformer d-c winding coil current (rms)
I_1 = fundamental component of I_L
I_{10} = fundamental component of I_L, u and α being assumed zero
I_{1P} = watt component of I_1
I_{1Q} = reactive component of I_1
m = order of harmonic
P_{do} = $E_{do}I_d$ = theoretical d-c power in watts
p = number of phases in a commutating group
q = total number of rectifier phases
s = circuit factor (1 for single-way, 2 for double-way)
u = commutating angle (angle of overlap)
u_0 = commutating angle when α is zero
u_α = commutating angle when α is not zero
X_c = line-to-neutral commutating reactance in ohms, referred to d-c winding of rectifier transformer
X_{cn} = line-to-neutral commutating reactance in ohms, referred to transformer a-c winding
X_L = line-to-neutral reactance in ohms of a-c system

7. Commutation. In the rectifier circuit of Fig. 12-4*a* a winding with *p* symmetrical phase voltages e_1, e_2, etc., which have an rms value E_s, is supplying a direct current I_c to a load circuit through rectifier elements *D*1, *D*2, etc. This might be a d-c winding of a rectifier transformer, and the reactance X_c in series with each phase might be the transformer leakage reactance. Although the phasor diagram and sinusoidal voltage waves in Fig. 12-4 represent a 3-phase circuit, the derived equations are applicable to any *p*-phase rectifier, *p* being an integer of 2 or higher. A large inductance L_d irons out the d-c ripples, and the direct-current wave is assumed to be flat. The effect of the losses in the circuit and rectifier elements will be considered later. For the present, it will be assumed that they are included in the d-c output.

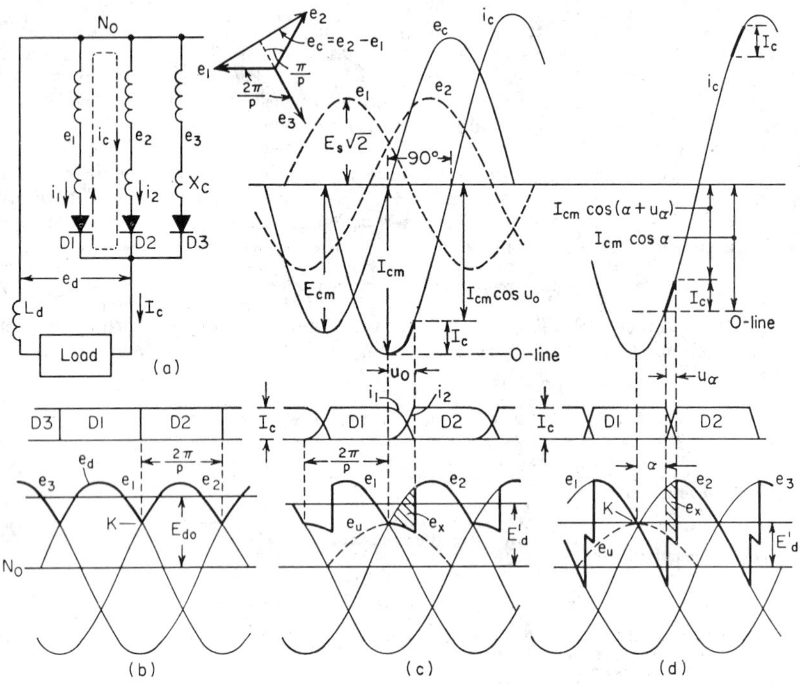

FIG. 12-4. Commutation in a rectifier circuit.

The current and voltage waves of Fig. 12-4*b* would be obtained if the reactance X_c were disregarded. Each phase conducts the direct current during an angle $2\pi/p$, when it has the highest voltage, and the current is transferred instantly to the next phase at the intersection point of the voltage waves. This condition is approached near no load. The average value of the d-c voltage, which is called the "theoretical no-load d-c voltage," is

$$E_{do} = s\sqrt{2}E_s(p/\pi)\sin(\pi/p) \tag{12-1}$$

The symbol *s* is a circuit factor which is 1 for single-way circuits and 2 for double-way circuits, which have two commutating groups in series (see Pars. **11** and **12**).

If the d-c voltage of Eq. (12-1) is reduced by phase control, as explained in Par. **4**, its value is

$$E_{do\alpha} = E_{do}\cos\alpha \tag{12-2}$$

The rms value of the phase current

$$I_s = I_c/\sqrt{p} \tag{12-3}$$

The reactance prevents instantaneous transfer of current between phases. As shown in Fig. 12-4*c*, the current I_c is transferred or commutated from phase 1 to phase 2 during an angle u_0. During this interval, the 2 phases are short-circuited through their rectifier elements. The voltage difference e_c between the phases produces a circulating current i_c in a direction to reduce the current i_1 and increase the current i_2. When i_1 is brought down to zero, rectifier element $D1$ prevents current reversal, thus opening the circuit of phase 1.

At the top of Fig. 12-4c are shown the sine waves of the commutating voltage e_c and the commutating current i_c (if the short circuit were maintained), which is shown displaced from the voltage by 90° because the circuit resistance was disregarded. The error from this assumption is relatively small, because the reactance is usually in the range of 5 to 10 times the resistance.

During commutation, the current i_2 of the incoming phase follows the wave of i_c, as indicated by the heavy line at the lower part of i_c. The following relations can be obtained from Fig. 12-4c by inspection and substitution.

$$E_{cm} = \sqrt{2}E_s \cdot 2 \sin (\pi/p) \tag{12-4}$$

$$I_{cm} = E_{cm}/2X_c = \sqrt{2}E_s \sin (\pi/p)/X_c \tag{12-5}$$

$$I_c = I_{cm} - I_{cm} \cos u_0 \tag{12-6}$$

$$\cos u_0 = 1 - I_c/I_{cm} = 1 - \frac{I_c X_c/E_s}{\sqrt{2} \sin (\pi/p)} \tag{12-7}$$

The rms value of the phase current, including the effect of commutation,

$$I_s = (I_c/\sqrt{p}) \sqrt{1 - p \cdot \psi(u)} \tag{12-8}$$

$$\psi(u) = \frac{(2 + \cos u) \sin u - (1 + 2 \cos u)u}{2\pi(1 - \cos u)^2} \tag{12-9}$$

Function $\psi(u)$ and factors involving it are given in Fig. 12-5.

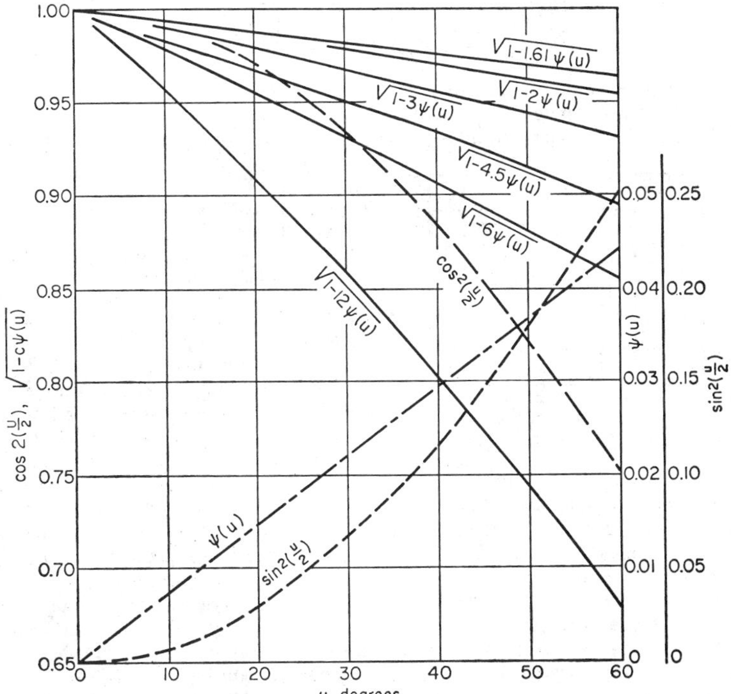

FIG. 12-5. Commutation correction factors (CCF) as functions of the commutating angle u (without phase control).

During commutation, when 2 phases are short-circuited, the terminal voltage is the average of the two phase voltages, represented in Fig. 12-4*c* by the wave e_u. The result is that the average d-c voltage E_d' is reduced below E_{do} by the average value E_x of the shaded area e_x.

$$E_x = spI_cX_c/2\pi \qquad (12\text{-}10)$$

$$E_d' = E_{do} - E_x \qquad (12\text{-}11)$$

Dividing Eq. (12-10) by Eq. (12-1),

$$E_x/E_{do} = \frac{I_cX_c/E_s}{2\sqrt{2}\sin(\pi/p)} \qquad (12\text{-}12)$$

If controlled rectifier elements are used and the start of conduction of each phase is delayed by an angle α from the intersection point K of successive phase voltages, the operating condition will be as shown in Fig. 12-4*d*. The commutation is shifted to a higher point on the i_c wave, and the commutating angle is smaller. The equation of the commutating angle can be derived in the same way as in Fig. 12-4*c*.

$$I_c = I_{cm}\cos\alpha - I_{cm}\cos(\alpha + u_\alpha) \qquad (12\text{-}13)$$

By substitution from Eq. (12-5),

$$\cos(\alpha + u_\alpha) = \cos\alpha - \frac{I_cX_c/E_s}{\sqrt{2}\sin(\pi/p)} \qquad (12\text{-}14)$$

By substitution from Eq. (12-12),

$$\cos(\alpha + u_\alpha) = \cos\alpha - 2E_x/E_{do} \qquad (12\text{-}14a)$$

The rms value of the phase current

$$I_s = (I_c/\sqrt{p})\sqrt{1 - p\psi(u,\alpha)} \qquad (12\text{-}15)$$

The approximate ratio of $\psi(u,\alpha)$ to $\psi(u)$ of Eq. (12-9) is given in Fig. 12-6 as a function of the phase-control angle α.

As in Fig. 12-4*c*, the terminal d-c voltage during commutation lies on e_u. The shaded area e_x lost during commutation is the same as in Fig. 12-4*c*. The average value of the d-c voltage, including the effect of phase control and commutation,

$$E_d' = E_{do}\cos\alpha - E_x \qquad (12\text{-}16)$$

Equations (12-13) to (12-16) are applicable over the complete range of phase control, including inversion. As α is increased, the commutation period will move up the i_c wave. The heavy line near the top of i_c in Fig. 12-4*d* is in a position close to maximum inversion.

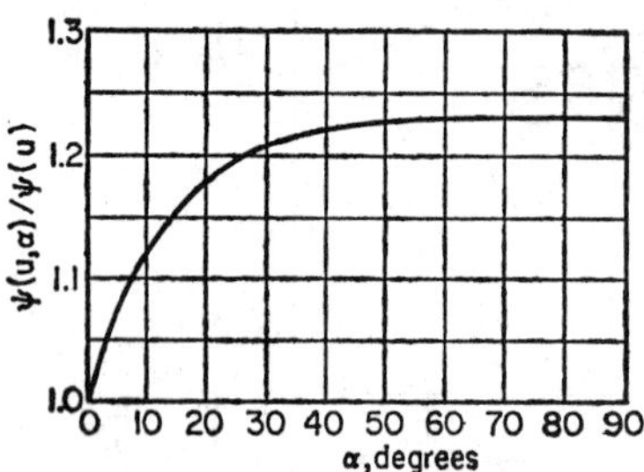

Fig. 12-6. Effect of phase control on factor $\psi(u)$.

The circuit of Fig. 12-4*a*, in which the direct current is commutated between successive phases, is called a **commutating group,** and X_c is the **commutating reactance.** The ratio I_cX_c/E_s appears in several of the preceding equations. This ratio is called the **commutating reactance factor** F_x; it is a per unit quantity and is used as a parameter for many of the curves.

The curves of Fig. 12-7 can be used for obtaining the reactance voltage-drop ratio E_x/E_{do} and the commutating angles for several values of p. The commutating angle u_0, with zero phase control, can be obtained on the right side of the graph. The corresponding commutating angle with phase control, u_α, can be obtained on the left side of the graph, for values of α up to 90°, by projecting across from the right side.

8. Commutating Reactance. In Eq. (12-10) and others and in many of the figures, the commutating reactance X_c in ohms is used. The methods for measuring this react-

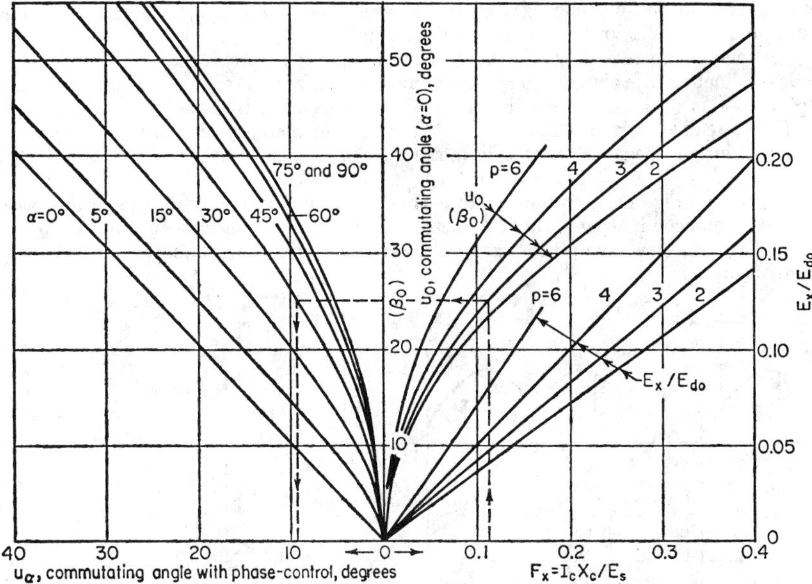

FIG. 12-7. Commutating angles in function of commutating reactance factor and phase-control angle; also reactance voltage-drop ratio E_x/E_{do}. (Dotted lines indicate sequence for obtaining commutating angle when operating with phase control.)

ance are specified in Ref. 11 (Par. **78**). The value in ohms is usually given on the transformer nameplate, in addition to the short-circuit impedance in percent. For some circuits, such as that in Fig. 12-10, the commutating reactance is the same as the short-circuit reactance. For others, there could be a significant difference between them.

If the commutating reactance is given in percent or per unit, its ohmic value referred to the transformer a-c winding is

$$X_{cn} = X_{c(pu)}E_n/I_L \tag{12-17}$$

in which $X_{c(pu)}$ is the per unit reactance for the a-c line current I_L. The value of X_{cn} referred to the d-c winding is

$$X_c = (E_s/E_n)^2X_{cn} \tag{12-18}$$

The product of per unit reactance and per unit load current, $X_{c(pu)}I_{d(pu)}$, is sometimes used as a parameter instead of the commutating reactance factor I_cX_c/E_s. For a specific circuit, these factors differ by a fixed ratio.

Reactance X_L in the a-c circuit ahead of the rectifier transformer can be transferred to equivalent commutating reactance of the d-c winding by Eq. (12-19).

$$X_c = X_L(E_s/E_n)^2/D_x \tag{12-19}$$

Values of transformation constant D_x are given in Tables 12-2*a* and *b*, Par. **14**.

9. Circuit Duty. A comparison of the operating conditions of a power rectifier and a power inverter is shown in Fig. 12-8. The waveshapes of the voltages and currents were drawn for a polyphase circuit with controlled rectifier elements. The anode-to-cathode voltage during the nonconducting period, shown for phase 2, is the difference between the phase voltage (anode potential) and the d-c voltage e_d (cathode potential) shown in heavy outline.

During conduction, the anode-to-cathode voltage is zero (disregarding the voltage drop in the rectifier element). At the conclusion of conduction, there is a steep reversal of the voltage. This is a critical part of the operating cycle, as the rectifying device changes from a good conductor to a good insulator, and the charge carriers remaining from the conducting period are swept out. The steep change in voltage could produce oscillations and overshoots of voltage.

For rectifier operation, the initial reverse voltage E_{ii} increases as the phase-control angle α is increased; this also increases the rate of current change (di/dt) and the number of residual charge carriers at the end of the conducting period. The product $E_{ii}\, di/dt$ is called **commutation factor.** Arc-backs in mercury-arc tubes usually occur at this point of the cycle, and this factor has been considered an indicator of their probability.

As seen in Fig. 12-8, the anode-to-cathode voltage is negative during the greater part of the nonconducting period for rectifier operation and positive for inverter operation. The peak reverse and forward voltages E_{pi} and E_{pf} are important circuit-duty quantities.

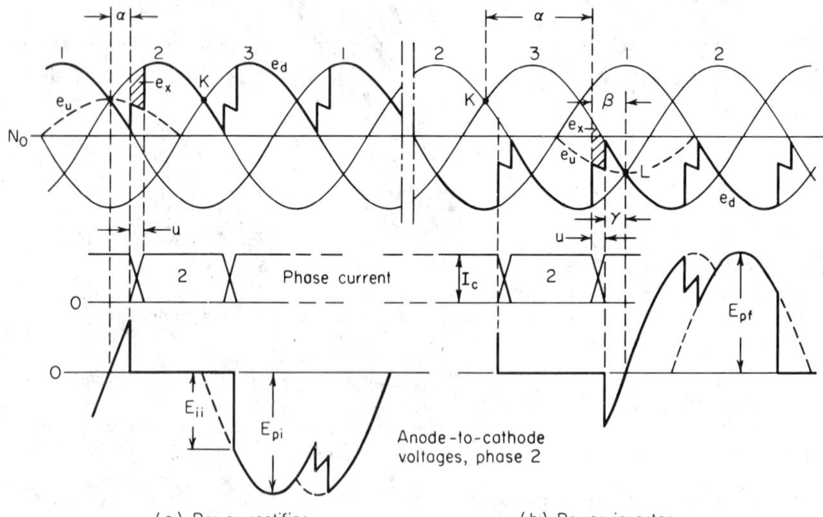

FIG. 12-8. Direct-current and anode-to-cathode voltages for power rectifier and inverter.

10. Inverter Operation. The commutation of current between two successive phases, such as phases 2 and 3 in Fig. 12-8*b*, can occur only between points K and L, while the voltage of phase 3 is higher than that of phase 2. This imposes a basic requirement for inverter operation. The maximum phase-control angle α for the start of conduction must allow time for commutation and a margin angle γ before point L. The margin angle is needed by the outgoing rectifier element for regaining its control properties, so that it could block forward voltage. It should include a safety margin for possible variations of the commutating angle caused by current or voltage variations. Failure to complete commutation and regain control before point L would cause phase 2 to continue conducting into the positive half-cycle of its voltage. This would cause reversal of the d-c voltage and would result in a fault.

Points K and L are separated by 180°. For inverter operation, it is sometimes more convenient to designate the start of conduction by the angle of advance β from point L instead of angle of retard α from point K. Substituting $(180° - \beta)$ for α in Eq. (12-16),

$$E_d' = -(E_{do} \cos \beta + E_x) \qquad (12\text{-}20)$$

It can be seen from Eq. (12-20) and the effect of the shaded area e_x in Fig. 12-8*b* that the reactance voltage drop E_x increases the negative d-c voltage of an inverter.

There is a reverse symmetry between points K and L of Fig. 12-8. When Fig. 12-8*b* is viewed upside down, it resembles Fig. 12-8*a*. If the angle of advance β_0 were made just equal to the commutating angle $(\gamma = 0)$, it would have the same value as the com-

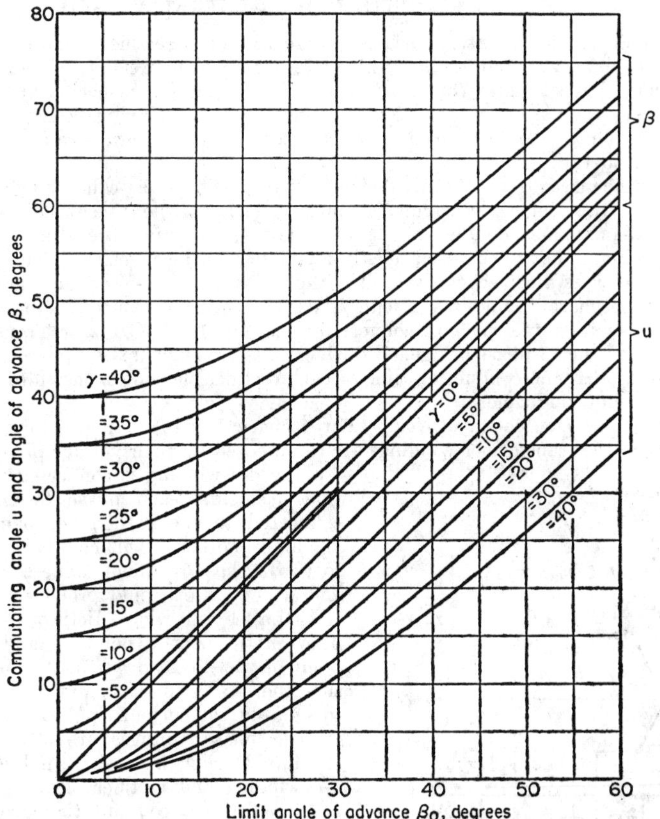

FIG. 12-9. Relations of angle of advance, margin angle, and commutating angle for inverter operation. Angle β_0 is obtained from Fig. 12-7.

mutating angle u_0 for rectifier operation starting at point $K(\alpha = 0)$. It is therefore possible to obtain β_0 by means of Eq. (12-7) or from the curves on the right side of Fig. 12-7. When β_0 is known, the values of β and u can be obtained from Fig. 12-9 for a number of values of the margin angle γ.

Another useful relation can be obtained from the symmetry in Fig. 12-8. Insofar as the voltage waves are concerned, angle γ in Fig. 12-8*b* corresponds to α in Fig. 12-8*a*, and Eq. (12-16) can be used to obtain E_d' for inverter operation by substituting γ for α, keeping in mind that the resulting voltage is negative.

$$E_d' = -(E_{do} \cos \gamma - E_x) \tag{12-21}$$

By equating Eqs. (12-20) and (12-21)

$$E_x/E_{do} = (\cos \gamma - \cos \beta)/2 \tag{12-22}$$

In the design of an inverter, it is usually necessary to start with a value for the margin angle, which is determined by the deionizing or turn-off characteristic of the rectifying device and considerations of the specific application. From the economic viewpoint, it is desirable to make γ as small as feasible in order to obtain the highest possible power factor.

The correspondence between inverter γ and rectifier α makes it possible to apply to inverter operation certain data and curves for which α is a parameter, such as Figs. 12-21, 12-26, and 12-31.

RECTIFIER CIRCUITS MOST GENERALLY USED

11. Circuit Classifications. There is a large variety of rectifier circuits. Some are shown in Tables 12-2*a* and 12-2*b*. Still more are given in Refs. 1, 5, and 11 of the Bibliography, Pars. **77** and **78**. The choice of a circuit for a specific application is determined by the number of phases required to obtain a desired waveshape of voltage and current, the type and arrangement of rectifier devices, performance characteristics, and cost considerations.

The rectifier circuits can be divided into two classifications called **single-way** and **double-way**. In a single-way circuit, the phase currents of the transformer d-c winding are unidirectional; in a double-way circuit, the phase currents are alternating. The single-way circuits have been used mostly for mercury-arc converters. The polyphase circuit of this class most generally used is the 6-phase double-Y circuit shown in Fig. 12-12. The 6-phase double-way circuit is used mostly for semiconductor converters and for high-voltage mercury-arc converters (see Sec. **14**, D-C Power Transmission). It is also called the "3-phase bridge." Both circuits are made up of 3-phase commutating groups, so that the conducting time of each rectifier element is one-third of a cycle plus the commutating time.

12. Six-phase Double-way Circuit (Three-phase Bridge). In this circuit, shown in Fig. 12-10, the transformer windings are represented by their voltage phasors. The 3-phase d-c winding is connected to six rectifier elements, each phase terminal being connected to the cathode of one and the anode of another. This circuit is equivalent to two 3-phase rectifiers connected in series, and its operation could probably be understood best by drawing a fictitious connection between the neutral point N_0 and the midpoint of the d-c load circuit. It is assumed that the d-c circuit has sufficient inductance to iron out the current ripples.

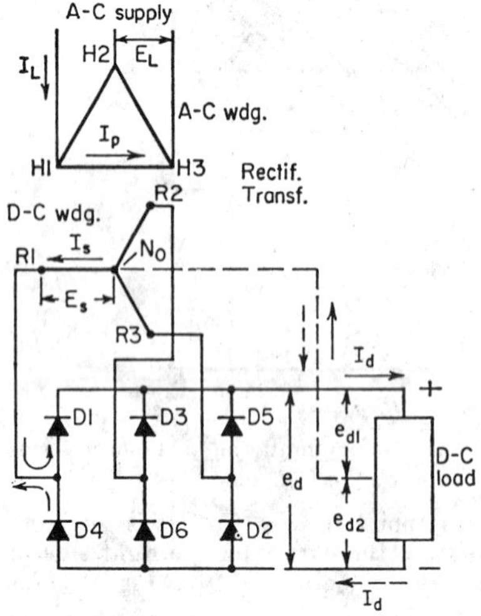

Fig. 12-10. Six-phase double-way (3-phase bridge) rectifier circuit (circuit 6 of Table 12-2*b*).

The phase voltages are shown in Fig. 12-11*a*. During the positive half-cycles, the d-c winding and rectifier elements *D*1, *D*3, *D*5 constitute a commutating group which produces the d-c voltage wave e_{d1} and causes the flow of a direct current I_d through one-half of the d-c circuit, as indicated by the solid arrows. As seen in Fig. 12-11*c*, this current is conducted in sequence by each of the three elements and transferred to the next element during the commutating angle u.

During the negative half-cycles of the phase voltages, the d-c winding and rectifier elements *D*2, *D*4, *D*6 constitute another commutating group, which produces the d-c voltage wave e_{d2} and causes the flow of a direct current I_d through the other half of the d-c circuit, as indicated by the broken arrows. Since the direct currents in the neutral connection are equal and opposite, they cancel each other and this connection can be eliminated. The path of the direct current in Fig. 12-10 at any instant selected at random, such as Q in Fig. 12-11, is as follows: $R3 \rightarrow D5 \rightarrow$ load circuit $\rightarrow D6 \rightarrow R2 \rightarrow N_0 \rightarrow R3$. The current flows through 2 phases and two rectifier elements in series.

The total d-c voltage e_d is the sum of the voltages e_{d1} and e_{d2} and has a 6-phase waveshape as shown in Fig. 12-11*b*. The average values of the d-c voltages can be obtained by means of Eqs. (12-1) and (12-11); $p = 3$ for each commutating group; $s = 2$.

$$E_d' = E_{d1}' + E_{d2}' = 2.34E_s - E_x \tag{12-23}$$

E_x can be obtained from Eq. (12-10), in which $I_c = I_d$ and $s = 2$.

The transformer d-c winding has alternating currents, as seen in Fig. 12-11*c*. The currents of the a-c winding (Fig. 12-11*d*) have the same shape as in the d-c winding but changed in magnitude by the transformation ratio *h*. The a-c line current at terminal *H*1 (Fig. 12-11*e*) was obtained by taking the difference between the currents in windings *H*1-*H*3 and *H*2-*H*1. Formulas for calculating the rms values of the currents are given in Table 12-2*b*.

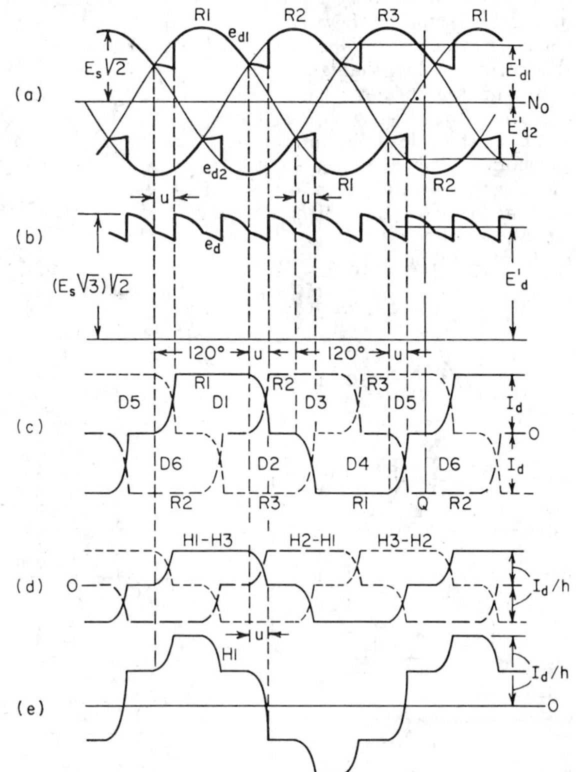

FIG. 12-11. Voltage and current waves for circuit of Fig. 12-10.

The a-c and d-c windings can be connected in wye or delta without affecting the mode of operation or their voltampere rating. If the d-c winding has a delta connection, the phase-to-neutral voltage E_s in Eq. (12-23) is equal to $1/\sqrt{3}$ times the delta voltage.

A 6-phase double-way converter can be connected directly to a 3-phase supply circuit, without a transformer, if the voltage of the available a-c circuit will give the desired d-c voltage and if it is not necessary to isolate the d-c circuit. Such arrangements have been used for supplying d-c motors and for other applications. Because of the absence of transformer reactance, reactors may have to be used in series with the a-c terminals to limit fault currents. If the mismatch between the a-c and d-c voltages is not too large, an autotransformer could be used to provide the required a-c voltage.

13. Six-phase Double-Y Circuit. This single-way circuit, shown in Fig. 12-12, consists of two 3-phase commutating groups ($p = 3$) operating in parallel. Each group consists of a 3-phase d-c winding with three rectifier elements and conducts one-half of the direct current I_d. The d-c voltage waves of the two groups are shown in heavy outline in Fig. 12-13*a* and *b*.

The voltage waves e_{d1} and e_{d2} are shown superposed in Fig. 12-13*c* to make up the terminal d-c voltage wave e_d, which is the average of these waves and has a 6-phase waveshape. The average value of the voltages, obtained from Eqs. (12-1) and (12-11),

$$E_d' = E_{d1}' = E_{d2}' = 1.17E_s - E_x \tag{12-24}$$

The instantaneous difference between voltages e_{d1} and e_{d2} is shown shaded in Fig. 12-13*c* and is reproduced in Fig. 12-13*d*. It is an alternating voltage with a basic frequency of three times that of the phase voltages. The voltage is absorbed by the midtapped iron-core reactor, called **interphase transformer** (**IT**), which is connected between the neutral points N_1 and N_2. The midpoint N_0 is the negative d-c terminal.

The triple-frequency magnetizing current of the interphase transformer flows between the 3-phase groups and is superposed on the phase currents. Since negative current cannot flow through the rectifier elements, a certain minimum d-c load current is required to permit the flow of the full magnetizing current. For the usual interphase transformer design, this is about 1 percent or less of the converter rated current. At this current, called the **light transition load,** the theoretical d-c voltage, from Eq. (12-1) for $p = 3$, is $1.17E_s$. At no load the interphase transformer is ineffective, so that the circuit operates as a 6-phase star circuit ($p = 6$), and the theoretical d-c voltage is $1.35E_s$. There is, therefore, about a 15% rise in the d-c voltage between the light transition load and no load.

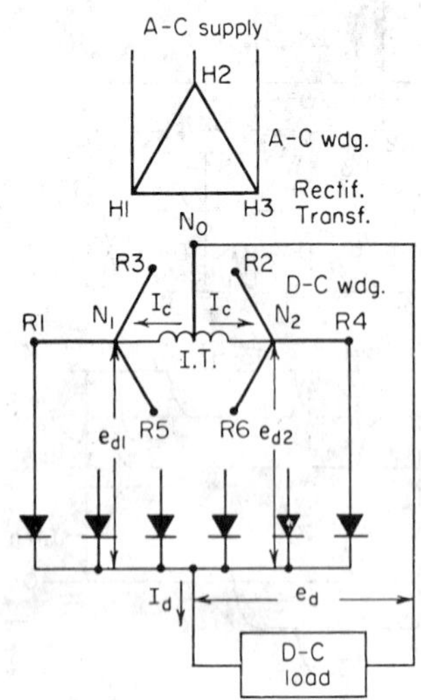

FIG. 12-12. Six-phase double-Y rectifier circuit (circuit 4 of Table 12-2*a*).

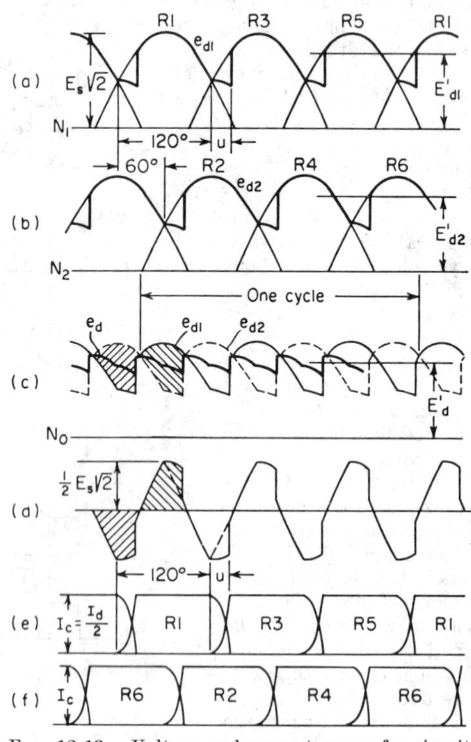

FIG. 12-13. Voltage and current waves for circuit of Fig. 12-12.

The ripple in the d-c voltage waves of the 3-phase commutating groups increases when phase control is used, which causes an increase in the interphase transformer voltage. Figure 18 in Ref. 28, Par. **79**, has curves giving the rms and average values of the interphase transformer voltage for phase-control angles up to 90°.

In Fig. 12-13*e* and *f* are shown the phase currents of the transformer d-c winding. The a-c winding and line currents are the same as in Fig. 12-11*d* and *e*. The rms values of the currents and other circuit quantities are given in Table 12-2*a*.

There are other rectifier circuits employing interphase transformers for paralleling out-of-phase commutating groups or circuits. One of these, the 12-phase circuit No. 8 in Table 12-2*b*, consists of two parallel 6-phase circuits. The basic frequency of the interphase transformer is six times that of the a-c supply.

14. Tables of Circuits. In Tables 12-2*a* and 12-2*b* are given the essential data for the more generally used rectifier circuits. Similar information for these and other circuits is given in Refs. 1, 5, and 11 of the Bibliography, Pars. **77** and **78**. The names of the circuits are the same as in Ref. 11. The double-way circuits of Table 12-2*b* are also called "bridge" circuits. The letter symbols used in the tables are defined in Table 12-1.

The circuit relations are given for zero phase control and are based on a flat-top direct-current wave (inductive d-c circuit). When there is a considerable departure from this assumption, the values should be used with caution; this applies particularly to circuits 1 and 5. The waveforms of the voltages and currents are shown without the effects of commutation. The rms values are given for the rectangular currents, followed by the commutation correction factors (CCF). According to Ref. 11, the rating of rectifier transformers should be based on rectangular current shape, disregarding the effect of commutation. The voltampere (VA) rating in the tables is in accordance with this; they are given in terms of the theoretical d-c power $P_{do} = E_{do}I_d$. These references also specify the methods for measuring the commutating reactance and load losses of rectifier transformers.

For inverter operation, the peak forward voltage is the same as the peak reverse voltage given in the tables.

Delta- or wye-connected a-c windings can be used for circuits 2, 4, 6, 7, and 8. The following notes apply to specific circuits:

Circuit 2. The transformer d-c winding is connected in zigzag instead of straight Y in order to avoid d-c magnetization of the core.

Circuit 3. The voltamperes of the a-c winding is higher than the a-c line voltamperes because a triple-frequency component of the winding current circulates in the Δ. A Y-connected a-c winding can be used with a Δ-connected tertiary, which has a voltampere rating of $0.74P_{do}$. The per unit commutating reactance $X_{c(pu)}$ is based on the rated a-c line voltamperes.

Circuit 4. The peak reverse voltage is given for zero load current. For currents above the light transition load, the peak reverse voltage is $E_s\sqrt{6}$.

Circuit 5. The voltage of the d-c winding is designated by $2E_s$ to fit Eq. (12-1) for double-way circuits ($s = 2$). The commutated current I_c is equal to $2I_d$ because the d-c winding current changes from $+I_d$ to $-I_d$ during each commutation.

Circuit 7. The voltage from the middle terminal N to the other d-c terminals has a 3-phase waveform. A zigzag d-c winding is used to avoid d-c magnetization of the core during unbalanced operation, when the middle terminal is carrying current. The waveforms of the currents and the value of the reactive d-c drop are for balanced operation.

Circuit 8. In some transformer designs, a separate a-c winding is provided for each d-c winding. The total voltamperes rating of the a-c windings is then $1.05P_{do}$. The commutating reactance X_c is for each d-c winding. If the two rectifier sections were connected in series instead of parallel, the circuit would be called "cascade" instead of "multiple." These terms are defined in Ref. 11*a* and *c*, Par. **78**.

D-C VOLTAGE REGULATION

15. D-C Voltage and Voltage Drops. Equation (12-16) and others gave the d-c voltage E_d', considering only the voltage drop E_x caused by the commutating reactance. Values of E_x and E_{do}, the theoretical no-load d-c voltage, are given in Tables 12-2*a* and 12-2*b* for the circuits shown. The d-c voltage is reduced also by the drops E_f in the rectifier elements and E_r caused by resistance or load losses. The terminal d-c voltage, considering all the drops and voltage reduction by phase control (when used), is

$$E_d = E_{do}\cos\alpha - E_x - E_r - sE_f \tag{12-25}$$

The drop E_f is a characteristic of the rectifying devices used and ordinarily varies linearly with the current. If a rectifier element represents a number of series-connected devices, the sum of their drops should be used. In double-way circuits, such as that of Fig. 12-10, s is 2 because current flows through two elements in series. In Eqs. (12-1) and (12-10) s is 2 for double-way circuits.

The value of E_r is obtained by dividing the load losses P_r by the direct current.

$$E_r = P_r/I_d \tag{12-26}$$

The load loss of the transformer is usually the major component of P_r. However, losses in reactors, conductors, and rectifier assembly (except rectifying devices) should be included. Losses which do not vary with the load current, such as the transformer

Table 12-2a. Data on Single-way Rectifier Circuits

	1	2	3	4
CIRCUIT DIAGRAM	$h = \frac{E_L}{E_S}$; H1, H2, E_L, R1, R2, E_S, I_d	$h = \frac{E_L}{E_S}$; H1, H2, H3, E_L, R1, R2, R3, E_S, I_d	$h = \frac{E_L}{E_S}$; H1, H2, H3, E_L, R1–R6, E_S, I_d	$h = \frac{E_L}{E_S}$; H1, H2, H3, E_L, R1–R6, E_S, I_C, I_d
CIRCUIT NAME	DIAMETRIC (FULL WAVE)	DELTA, 3-PHASE, ZIG-ZAG	DELTA, 6-PHASE STAR	DELTA, 6-PHASE, DOUBLE-WYE
CIRCUIT CHARACTERISTICS	$I_c = I_d$; $D_x = 1$; $p = 2$; $q = 1$; $S = 1$	$I_c = I_d$; $D_x = 1$; $p = 3$; $q = 3$; $S = 1$	$I_c = I_d$; $D_x = 3$; $p = 6$; $q = 6$; $S = 1$	$I_c = 0.5\,I_d$; $D_x = 1$; $p = 3$; $q = 6$; $S = 1$
D-C VOLTAGE WAVE FORM	$E_S\sqrt{2}$; 0, π, 2π	$E_S\sqrt{2}$; 0, π, 2π	$E_S\sqrt{2}$; 0, π, 2π	$1.22\,E_S$; 0, π, 2π
N.L. AVE. E_{do}	$0.9\,E_S$	$1.17\,E_S$	$1.35\,E_S$	$1.17\,E_S$
REACT. D-C DROP E_X (VOLTS)	$0.318\,I_d X_c$	$0.477\,I_d X_c$	$0.955\,I_d X_c$	$0.239\,I_d X_c$
E_X/E_{do}	$0.353\,F_x$; $0.707\,X_c(pu)$	$0.408\,F_x$; $0.866\,X_c(pu)$	$0.707\,F_x$; $1.5\,X_c(pu)$	$0.408\,F_x$; $0.5\,X_c(pu)$
RECTIF. ELEMENT CURRENT WAVE FORM	I_d; 0, π, 2π	I_d; 0, π, 2π	I_d; 0, π, 2π	I_c; 0, π, 2π
(AVE.); RMS	$(I_d/2)$; $0.707\,I_d$	$(I_d/3)$; $0.577\,I_d$	$(I_d/6)$; $0.408\,I_d$	$(I_d/6)$; $0.289\,I_d$
C.C.F. FOR RMS	$\sqrt{1-2\psi(u)}$	$\sqrt{1-3\psi(u)}$	$\sqrt{1-6\psi(u)}$	$\sqrt{1-3\psi(u)}$
PK. REVERSE VOLT.	$2\sqrt{2}\,E_S$	$\sqrt{6}\,E_S$	$2\sqrt{2}\,E_S$	$2\sqrt{2}\,E_S$ (WHEN $I_d = 0$)
TRANSFORMER D-C WINDING CURRENT WAVE FORM RMS C.C.F.	SAME AS FOR RECTIF. ELEMENT	SAME AS FOR RECTIF. ELEMENT	SAME AS FOR RECTIF. ELEMENT	SAME AS FOR RECTIF. ELEMENT
VA RATING	$1.57\,P_{do}$	$1.71\,P_{do}$	$1.81\,P_{do}$	$1.48\,P_{do}$
TRANSFORMER A-C WINDING CURRENT WAVE FORM	I_d/h; 0, π, 2π	$0.577(I_d/h)$; 0, π, 2π	I_d/h; 0, π, 2π	$0.5(I_d/h)$; 0, π, 2π
RMS	I_d/h	$0.471(I_d/h)$	$0.577(I_d/h)$	$0.408(I_d/h)$
C.C.F.	$\sqrt{1-4\psi(u)}$	$\sqrt{1-4.5\psi(u)}$	$\sqrt{1-6\psi(u)}$	$\sqrt{1-3\psi(u)}$
VA RATING	$1.11\,P_{do}$	$1.21\,P_{do}$	$1.28\,P_{do}$	$1.05\,P_{do}$
A-C LINE CURRENT WAVE FORM	SAME AS TRANSFORMER A-C WINDING	$1.154(I_d/h)$; $0.577(I_d/h)$; 0, π, 2π	I_d/h; 0, π, 2π	I_d/h; $0.5(I_d/h)$; 0, π, 2π
RMS		$0.816(I_d/h)$	$0.816(I_d/h)$	$0.707(I_d/h)$
C.C.F.		$\sqrt{1-4.5\psi(u)}$	$\sqrt{1-3\psi(u)}$	$\sqrt{1-3\psi(u)}$
FUNDAMENTAL RMS-I_{10}	$0.9(I_d/h)$	$0.675(I_d/h)$	$0.778(I_d/h)$	$0.675(I_d/h)$
VA RATING	$1.11\,P_{do}$	$1.21\,P_{do}$	$1.05\,P_{do}$	$1.05\,P_{do}$

Table 12-2b. Data on Double-way Rectifier Circuits

	5	6	7	8
CIRCUIT DIAGRAM	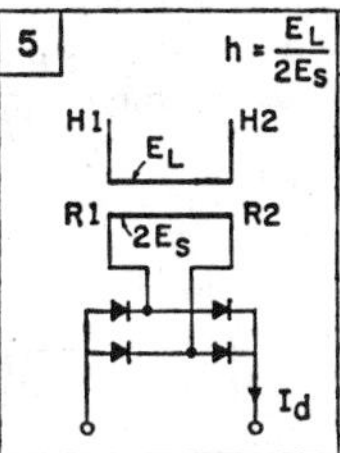 $h = \frac{E_L}{2E_s}$; H1, H2, E_L, R1, R2, $2E_s$, I_d	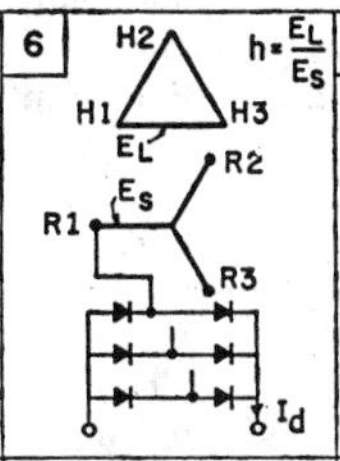 $h = \frac{E_L}{E_s}$; H1, H2, H3, E_L, E_s, R1, R2, R3, I_d	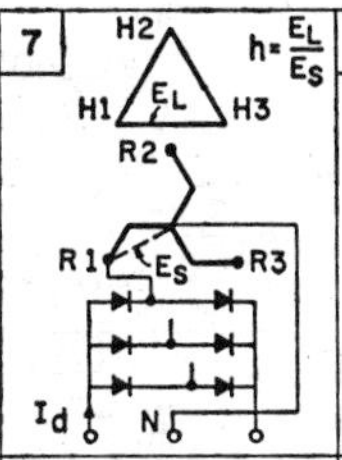 $h = \frac{E_L}{E_s}$; H1, H2, H3, E_L, R1, R2, R3, E_s, I_d, N	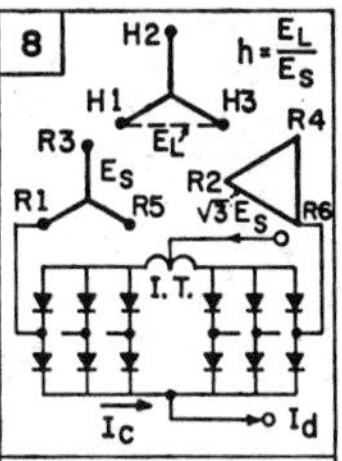 $h = \frac{E_L}{E_s}$; H1, H2, H3, E_L, R1, R2, R3, R4, R5, R6, E_s, $\sqrt{3}E_s$, I.T., I_c, I_d
CIRCUIT NAME	DIAMETRIC, DOUBLE-WAY (1-PHASE BRIDGE)	DELTA, 6-PHASE, WYE, DOUBLE-WAY (3-PHASE BRIDGE)	DELTA, 6-PHASE, ZIG-ZAG, DOUBLE-WAY, 3-WIRE	WYE, 12-PHASE, MULTIPLE Y-Δ, DOUBLE-WAY
CIRCUIT CHARACTERISTICS	$I_c = 2I_d$; $D_x = 1$; $p = 2$; $q = 2$; $s = 2$	$I_c = I_d$; $D_x = 1$; $p = 3$; $q = 6$; $s = 2$	$I_c = I_d$; $D_x = 1$; $p = 3$; $q = 6$; $s = 2$	$I_c = I_d/2$; $D_x = 1$; $p = 3$; $q = 12$; $s = 2$
D-C VOLTAGE WAVE FORM	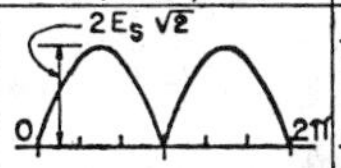 $2E_s\sqrt{2}$; 0, 2π	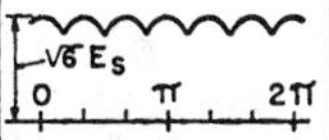 $\sqrt{6}E_s$; 0, π, 2π	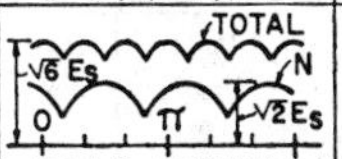 TOTAL, N, $\sqrt{6}E_s$, $\sqrt{2}E_s$; 0, π, 2π	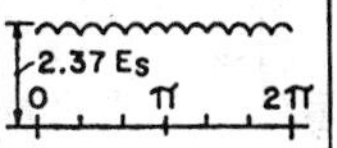 $2.37 E_s$; 0, π, 2π
N.L. AVE. E_{do}	$1.8 E_s$	$2.34 E_s$	$2.34 E_s$ (TOTAL) $1.17 E_s$ (TO N)	$2.34 E_s$
REACT. D-C DROP E_x (VOLTS)	$1.274 I_d X_c$	$0.955 I_d X_c$	$0.955 I_d X_c$	$0.955 I_c X_c$
E_x/E_{do}	$0.353 F_x$; $0.707 X_c$(pu)	$0.408 F_x$; $0.5 X_c$(pu)	$0.408 F_x$; $0.5 X_c$(pu)	$0.408 F_x$; $0.5 X_c$(pu)
RECTIF. ELEMENT CURRENT WAVE FORM	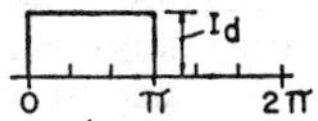 I_d; 0, π, 2π	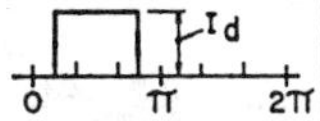 I_d; 0, π, 2π	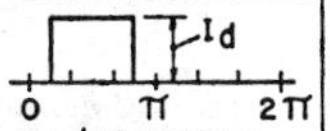 I_d; 0, π, 2π	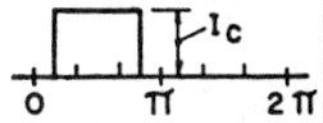 I_c; 0, π, 2π
(AVE.); RMS	$(I_d/2)$; $0.707 I_d$	$(I_d/3)$; $0.577 I_d$	$(I_d/3)$; $0.577 I_d$	$(I_c/3)$; $0.577 I_c$
C.C.F. FOR RMS	$\sqrt{1-2\psi(u)}$	$\sqrt{1-3\psi(u)}$	$\sqrt{1-3\psi(u)}$	$\sqrt{1-3\psi(u)}$
PK. REVERSE VOLT.	$2E_s\sqrt{2}$	$\sqrt{6}E_s$	$\sqrt{6}E_s$	$\sqrt{6}E_s$
TRANSFORMER D-C WINDING CURRENT WAVE FORM	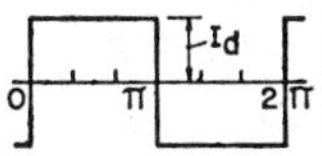 I_d; 0, π, 2π	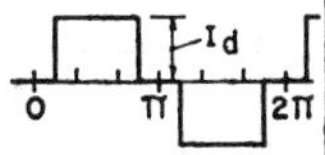 I_d; 0, π, 2π	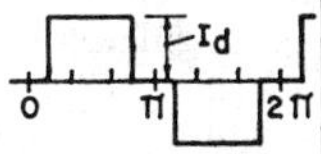 I_d; 0, π, 2π	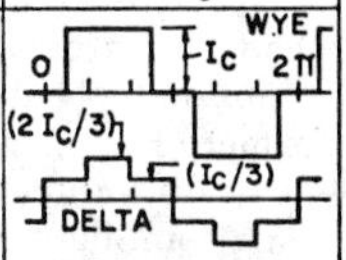 WYE, I_c, 0, 2π; $(2I_c/3)$, $(I_c/3)$, DELTA
RMS	I_d	$0.816 I_d$	$0.816 I_d$	WYE — $0.816 I_c$; DELTA — $0.471 I_c$
C.C.F.	$\sqrt{1-4\psi(u)}$	$\sqrt{1-3\psi(u)}$	$\sqrt{1-3\psi(u)}$	$\sqrt{1-3\psi(u)}$
VA RATING	1.11 P_{do}	1.05 P_{do}	1.21 P_{do}	1.05 P_{do} (2 WDGS.)
TRANSFORMER A-C WINDING CURRENT WAVE FORM	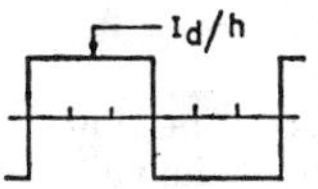 I_d/h	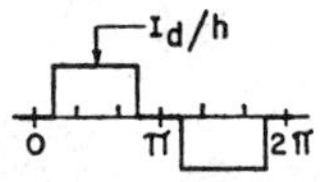 I_d/h; 0, π, 2π	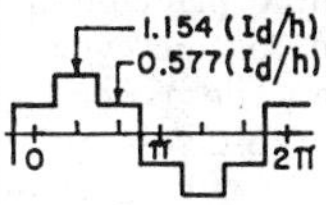 $1.154(I_d/h)$, $0.577(I_d/h)$; 0, π, 2π	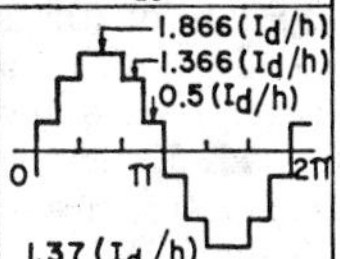 $1.866(I_d/h)$, $1.366(I_d/h)$, $0.5(I_d/h)$; 0, π, 2π
RMS	I_d/h	$0.816(I_d/h)$	$0.816(I_d/h)$	$1.37(I_d/h)$
C.C.F.	$\sqrt{1-4\psi(u)}$	$\sqrt{1-3\psi(u)}$	$\sqrt{1-3\psi(u)}$	$\sqrt{1-1.61\psi(u)}$
VA RATING	1.11 P_{do}	1.05 P_{do}	1.05 P_{do}	1.01 P_{do}
A-C LINE CURRENT WAVE FORM	SAME AS TRANSFORMER A-C WINDING	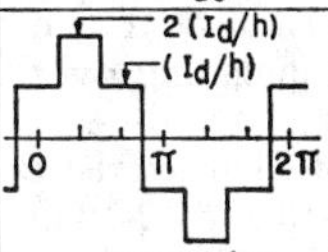 $2(I_d/h)$, (I_d/h); 0, π, 2π	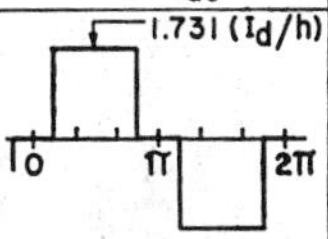 $1.731(I_d/h)$; 0, π, 2π	SAME AS TRANSFORMER A-C WINDING
RMS		$1.41(I_d/h)$	$1.41(I_d/h)$	
C.C.F.		$\sqrt{1-3\psi(u)}$	$\sqrt{1-3\psi(u)}$	
FUNDAMENTAL RMS-I_{10}	$0.9(I_d/h)$	$1.35(I_d/h)$	$1.35(I_d/h)$	$1.35(I_d/h)$
VA RATING	1.11 Pdo	1.05 Pdo	1.05 Pdo	1.01 Pdo

core loss, should not be included. Both E_r and E_x are directly proportional to the load current in the normal operating range. In Fig. 12-14 are shown voltage-regulation lines E_d of a static converter and their makeup. The voltage regulation VR is the change in the terminal voltage when the direct current is changed from its rated value to no load or light transition load (for circuits with interphase transformers). The phase-control angles α_1 and α_2 are arbitrary, the latter being in the inversion range. For inverter operation, the voltage drops increase the negative terminal voltage. The applied d-c voltage, which has to be equal and opposite to the negative inverter voltage, is

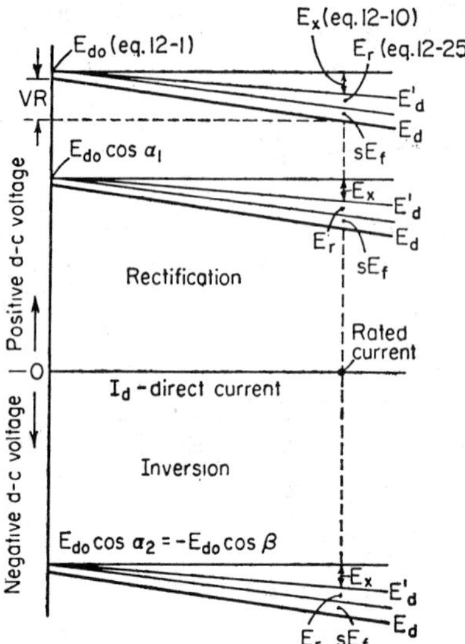

FIG. 12-14. Voltage-regulation lines and their makeup.

$$E_d = E_{do} \cos \beta + E_x + E_r + sE_f \qquad (12\text{-}27)$$

16. Voltage Control. The d-c voltage of a converter with controlled rectifier elements can be varied over the full range by phase control. One objection to this method is that it lowers the power factor. For converters with diodes, or where phase control cannot be used for the required range, the d-c voltage can be varied by changing the voltage of the transformer d-c winding with one of the following methods.

a. Taps on the rectifier transformer or an autotransformer ahead of it. The power has to be interrupted to change taps.

b. A step voltage regulator (which can be operated under load) ahead of the transformer or as part of it.

c. An induction voltage regulator.

A combination of phase control and one of these methods is sometimes used.

For converters with diodes, saturable reactors with d-c control windings can be used to regulate the d-c voltage. The reactors can be connected in series with the a-c circuit or in series with individual rectifier elements. With the latter arrangement, the reactors operate on the same principle as in magnetic amplifiers and can control the start of conduction in the voltage cycle, as is done with phase control. The use of saturable reactors also lowers the power factor.

17. Extended Voltage-regulation Characteristics. The linear voltage-regulation characteristics in Fig. 12-14 are based on the linearity of the voltage drops, particularly E_x, which is generally the largest drop. Equation (12-10) for E_x is based on normal commutation between two successive phases of a commutating group. For the single-phase circuits 1 and 5 in Tables 12-2*a* and 12-2*b* ($p = 2$), this condition prevails over the complete load range to short circuit. For other circuits, the linearity continues over the normal operating range, including overloads, encountered in industrial and railway service.

The complete voltage-regulation characteristics, to short circuit, for the commonly used polyphase circuits of Figs. 12-10 and 12-12 are shown in Fig. 12-15. The per-unit values of the d-c voltage E_d'/E_{do}, including the effects of the commutating reactance, are given as a function of the commutating reactance factor F_x. Curve 1 is for uncontrolled rectifier elements and inductive d-c circuit. Curve 2 is for resistive d-c circuit (Refs. 31 and 32, Par. **79**).

Curve 1 consists of three sections, which represent three operating modes. Section *a-b* is linear; point *b* corresponds to a value of 0.612 for F_x and a commutating angle of 60°. Because of the common or closely coupled transformer windings in the two commutating groups of these circuits, commutation in one group depresses the a-c voltage and delays the start of commutation in the other group after a commutating

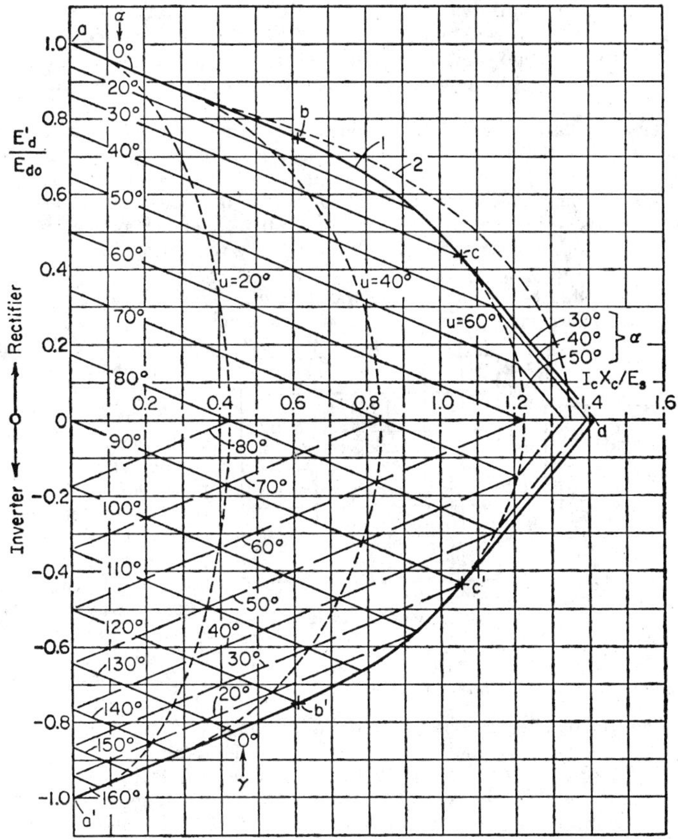

FIG. 12-15. Complete voltage-regulation characteristics, to short circuit, for the circuits of Figs. 12-10 and 12-12.

angle of 60° has been reached. This introduces an inherent delay angle which has the same effect as a phase-control angle. Over section *b-c* of curve 1, the commutating angle remains at 60° while the delay angle increases from 0 to 30°. At point *c*, F_x is 1.06. Between points *c* and *d*, the commutating angle increases from 60 to 120°, while the inherent delay angle remains at 30°. Section *c-d* is a straight line; it intersects the abscissa at $F_x = \sqrt{2}$.

Below curve 1 are shown voltage-regulation curves for an inductive d-c circuit and a number of phase-control angles (α). The straight lines are parallel to line *a-b*. They start at a point on the ordinate equal to cos α and terminate at a curve corresponding to a commutating angle of 60°. The dotted portion of this curve is a continuation of section *b-c* of curve 1. If α is 30° or less, the regulation curve coincides with curve 1 beyond the intersection point; for values of α from 30 to 60°, it continues on a line parallel to line *c-d* beyond the intersection point. The other dotted curves in Fig. 12-15 are regulation curves for constant commutating angles of 40 and 20°.

A mirror image of curve 1 is drawn below the abscissa for a theoretically maximum inverter voltage, with a margin angle $\gamma = 0$ (see Par. **10**). The lines sloping upward parallel to $a' - b'$ are for different margin angles. They are symmetrical with the α lines above the abscissa. This reverse symmetry between γ and α are discussed in Par. **10**. The α lines below the abscissa can be marked with a corresponding inverter angle of advance $\beta = 180° - \alpha$.

For these circuits, the reactance factor F_x is equal to 1.225 times the product of per-unit current and per-unit reactance ($I_{d(pu)} \cdot X_{c(pu)}$). For a reactance of 0.08 (8%), for example, F_x is 0.098 at rated current and 0.294 at three times rated current. For a specific operating condition, E_d' can be obtained from Fig. 12-15 if E_{do} is known. The terminal d-c voltage E_d can be obtained by subtracting from E_d' the voltage drops E_r and E_f.

18. Voltage-regulation Curves for 12-phase Converter with A-C System Reactance. In a 12-phase converter made up of two 6-phase sections, with a common reactance in the a-c circuit, commutation in one section interferes with commutation in the other section after a commutating angle of 30° is reached. This introduces an inherent delay angle, which causes an increased drop in the d-c voltage. The effect is a function of the reactance ratio. An analysis of this effect and voltage-regulation curves are contained in Ref. 33 of the Bibliography, Par. **79**.

19. Semicontrolled Six-phase Double-way Circuit. The d-c voltage of a converter with an inductive d-c circuit and a full complement of controlled rectifier elements can be regulated down to zero with a phase-control angle α of 90° and to maximum negative with an α of 180° (voltage drops being disregarded).

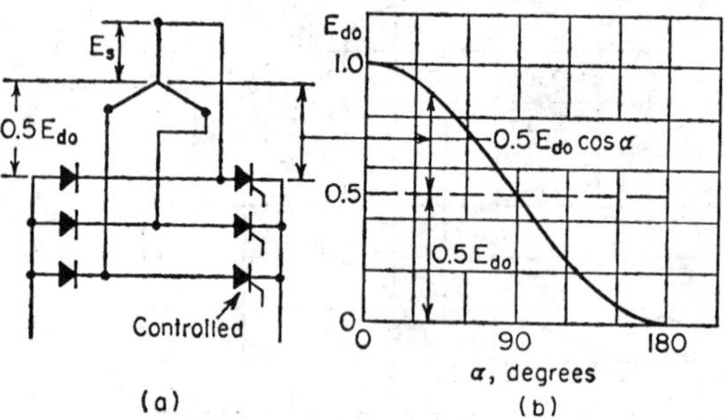

FIG. 12-16. Semicontrolled 6-phase double-way circuit and its phase-control curve at no load.

The d-c voltage of a 6-phase double-way circuit with three controlled and three uncontrolled elements, as shown in Fig. 12-16*a*, can be regulated down from maximum to practically zero. The voltage as a function of the control angle is shown in Fig. 12-16*b*. This circuit consists of two 3-phase commutating groups in series (Par. **12**). If the maximum d-c voltage is E_{do}, the voltage of the uncontrolled group is $0.5E_{do}$. The voltage of the controlled group is $0.5E_{do} \cos \alpha$; with a range of 0 to 180° for α, the voltage of this group can be varied from maximum positive to maximum negative. Actually, the maximum negative voltage cannot be obtained because of the margin angle required for inverter operation. The d-c terminal voltage, the voltage drops being taken into account, is

$$E_d = 1.17E_s(1 + \cos \alpha) - E_x - E_r - 2E_f \qquad (12\text{-}27a)$$

Because the waveshapes of the d-c voltages for the two groups are not the same when phase control is used, the total d-c voltage contains triple-harmonic components (see Par. **27**). The total d-c voltage cannot be made negative. For applications which require inversion, six controlled rectifier elements have to be used (see Par. **79**, Ref. 42, p. 217).

POWER FACTOR

20. A-C Line Current. The a-c line current of a static converter (see Fig. 12-11) is nonsinusoidal. It can be resolved into a sinusoidal fundamental (of the same frequency as the a-c voltage) and harmonic components. The line current I_L and its components, in relation to the line-to-neutral voltage E_n, are shown in the phasor diagram of Fig. 12-17; I_1 and I_H are the rms fundamental and harmonic components of I_L; I_{1P} and I_{1Q} are the power and reactive components of I_1.

$$I_L = \sqrt{I_1^2 + I_H^2} \qquad I_H^2 = \Sigma I_m^2$$

in which I_m represents the individual harmonic components.

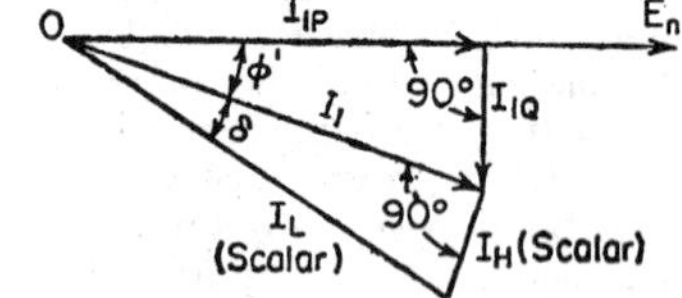

FIG. 12-17. Phasor diagram showing components of converter a-c line current.

The total power factor is the ratio of the power component to the total current:

$$\text{pf} = I_{1P}/I_L = (I_{1P}/I_1)(I_1/I_L)$$

$$I_{1P}/I_1 = \cos\phi' \qquad I_1/I_L = \cos\delta$$

$$\text{pf} = \cos\phi' \cos\delta \tag{12-28}$$

The total power factor is the product of two factors. The **displacement factor** $\cos\phi'$ is the power factor of the fundamental component. The **distortion factor** $\cos\delta$ indicates the deviation of the line current from a sine wave; it is unity when there are no harmonics. For most applications the displacement factor is more significant than the total power factor in evaluating the effects of the converter load on the power factor of the a-c system. The international (IEC) and American (USASI) standards for static power converters designate the displacement factor for power-factor specification.

21. Displacement Power Factor. The power factor of the fundamental current component is less than unity, for two reasons. As seen in Fig. 12-4, commutation shifts the center of gravity of the phase currents in the lagging direction. Phase control shifts the conducting periods and currents with respect to the voltages.

A phasor diagram of the fundamental current for rectifier and inverter operation is shown in Fig. 12-18. For inversion, the current phasor I_1 is swung by phase control to a position lagging behind the voltage E_n by an angle ϕ_2'. The angle is less than 180° because of the margin angle requirement for inversion (Par. **10**). The power component I_{1P} is reversed, but the reactive component I_{1Q} remains in the same direction as for rectification. This is an important aspect of inverter operation. The a-c system has to supply the reactive kVA required by the inverter, as if it were a lagging-power-factor load. The kvars can be supplied by synchronous machines or capacitors, or a combination of both. For the type of inverter considered so far, generators are required on the system to provide the a-c voltages for commutation; they can supply the needed kvars for most industrial inversion applications. Converter installations of a d-c power-transmission system generally include capacitors for this purpose (see Sec. **14**).

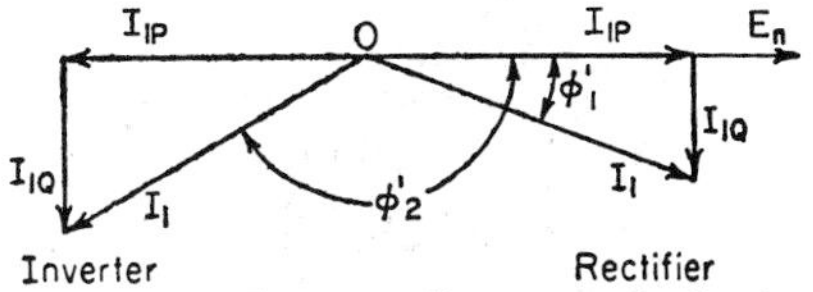

FIG. 12-18. Phasor diagram of fundamental current for rectifier and inverter operation.

22. Determination of Displacement Power Factor. The curves of Figs. 12-20 and 12-21 are applicable for the circuits of Figs. 12-10 and 12-12 and for others based on them, such as circuit 8 of Table 12-2*b*. They cover an operating range of 0 to 0.4 for the commutating reactance factor F_x, which is on the linear part of the voltage-regulation characteristic in Fig. 12-15. The curves were obtained from a Fourier analysis of the a-c line-current wave, with the assumption that the d-c circuits are inductive (flat-top currents).

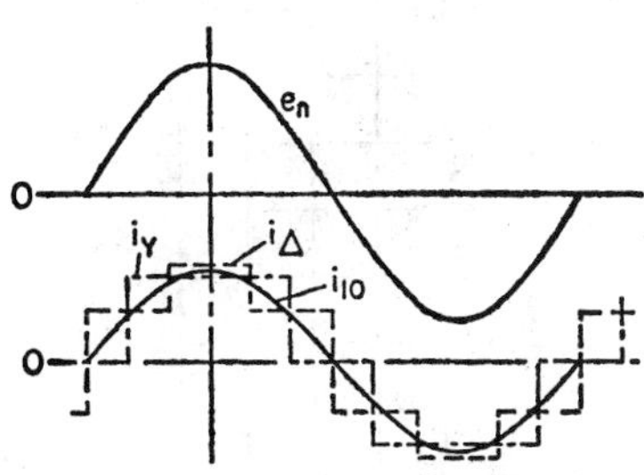

FIG. 12-19. Derivation of theoretical fundamental current I_{10}.

The per-unit base of the ordinate scales in Figs. 12-20 and 12-21 is I_{10}, the theoretical fundamental current component when the commutating angle u, phase-control angle α, and losses are assumed to be zero. The a-c line current then has a step-type wave centered with respect to the line-to-neutral voltage E_n as shown in Fig. 12-19. The dotted current wave is for the circuits of Figs. 12-10 and 12-12 with a delta-connected a-c winding. The dot-and-dash wave is for a wye-connected a-c winding. The two waves have the same rms value and harmonic content. They differ only in the phase position of the harmonics.

From the symmetry in Fig. 12-19, the fundamental component of the current I_{10} is in phase with the line-to-neutral voltage. For a sinusoidal voltage, the power is transmitted by the fundamental component of the current and is equal to $3E_nI_{10}$. For the assumed conditions, the d-c voltage is E_{do} and the power $E_{do}I_d$, which is equal to the a-c power. From this, the equation for I_{10} is

$$I_{10} = (E_{do}I_d)/3E_n \tag{12-29}$$

For operation with a commutating angle and phase control, the equations for the per-unit values of the in-phase and quadrature components of the fundamental current, obtained by a Fourier analysis, are

$$\frac{I_{1P}}{I_{10}} = \frac{\cos \alpha + \cos (u + \alpha)}{2} \tag{12-30}$$

$$\frac{I_{1Q}}{I_{10}} = \frac{\sin 2(u + \alpha) - \sin 2\alpha - 2u}{4[\cos \alpha - \cos (u + \alpha)]} \tag{12-31}$$

The curves of Fig. 12-20 give the values of these components and the corresponding power-factor angles as a function of the phase-control angle α for several values of the commutating reactance factor. The change from rectification to inversion occurs at

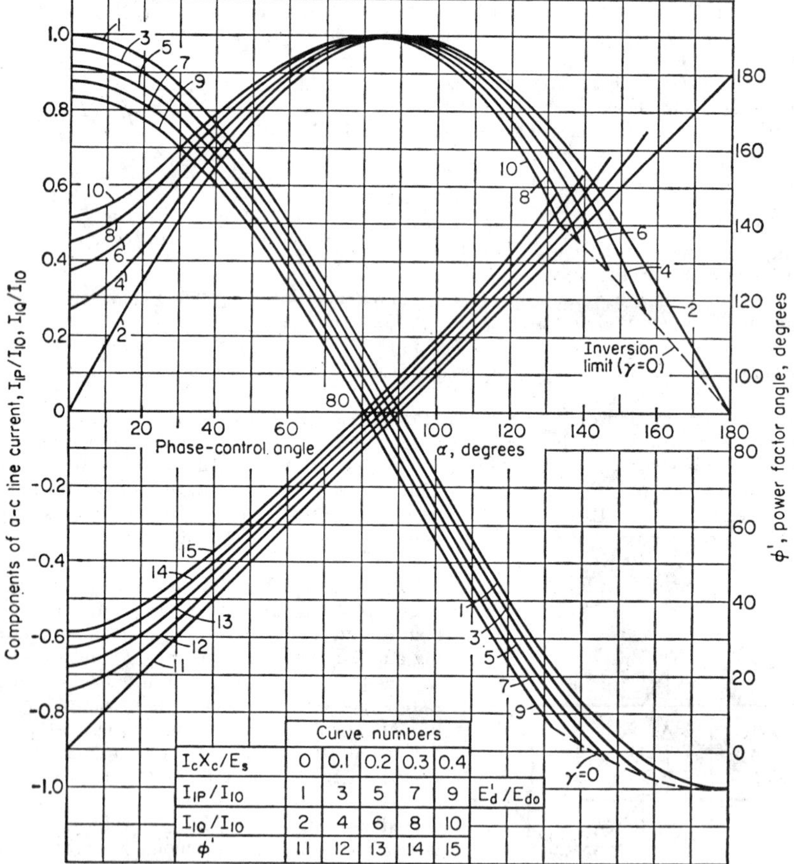

FIG. 12-20. Fundamental in-phase and reactive a-c line currents, power-factor angle ϕ', and d-c voltage E_d'/E_{do}, in function of phase-control angle α and commutating reactance factor I_cX_c/E_s, for circuits of Figs. 12-10 and 12-12.

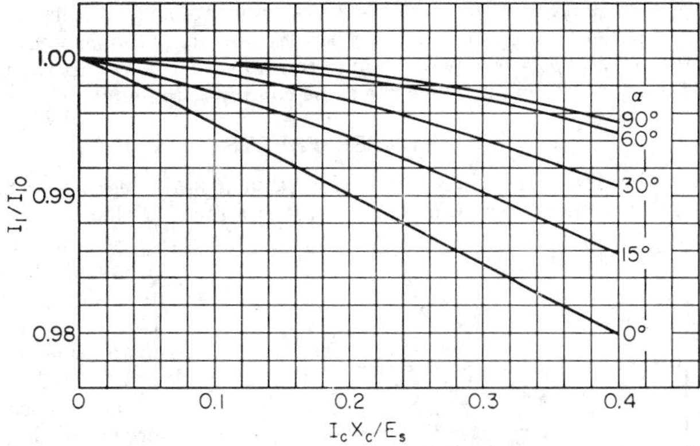

FIG. 12-21. Fundamental a-c line current, in function of commutating reactance factor and phase-control angle, for circuits of Figs. 12-10 and 12-12.

the intersection points of curves 1, 3, 5, 7, 9 with the abscissa. All the curves terminate at the commutation-limit points for inversion, i.e., the points for which the angle of advance β is equal to the commutating angle and the margin angle γ is zero.

The curves of Fig. 12-21 give the per-unit values of the total fundamental current as a function of the commutating reactance factor for several values of α. These curves can be applied for inversion by substituting the margin angle γ for α. An equation for the displacement power factor can be obtained from Eq. (12-30), with the aid of Fig. 12-21.

$$\cos\phi' = \frac{I_{1P}}{I_1} = \frac{\cos\alpha + \cos(u+\alpha)}{2}\frac{1}{I_1/I_{10}} \tag{12-32}$$

The displacement power factor can be determined also from a ratio of d-c voltages. From the discussion preceding Eq. (12-29), the a-c voltamperes are equal to $E_{do}I_d$ when u and α are zero. For other operating conditions, this has to be multiplied by the ratio I_1/I_{10} from Fig. 12-21. The a-c line watts are equal to the d-c watts adjusted for losses, which are represented by the voltage drops E_r and sE_f in Eqs. (12-25) and (12-27). For rectifier operation, the losses have to be added to the d-c watts to obtain the a-c watts. Adding E_r and sE_f to Eq. (12-25) results in a voltage E_d' of Eq. (12-16), so that $E_d'I_d$ gives the watts on the a-c side. For inversion, the losses have to be subtracted from the d-c watts input to obtain the a-c watts. Subtracting E_r and sE_f in Eq. (12-27) results in a voltage E_d' of Eq. (12-20), and $E_d'I_d$ is the a-c watts. The minus sign in Eq. (12-20) does not affect the voltage magnitude. For either direction of power flow, E_d' represents the d-c voltage including the effect of the reactance voltage drop. The displacement power factor, which is the ratio of a-c watts to fundamental voltamperes, is

$$\cos\phi' = \frac{E_d'I_d}{E_{do}I_d(I_1/I_{10})} = \frac{E_d'}{E_{do}}\frac{1}{I_1/I_{10}} \tag{12-33}$$

For approximate calculations, the correction factor I_1/I_{10} can be disregarded.

The effect of the transformer exciting current on the power factor is usually slight. It can be included in the power-factor calculation, when necessary, by adding its power and reactive components to the corresponding components of the converter line current. The displacement pf is then designated by $\cos\phi$.

Since the a-c power is equal to $E_d'I_d$, the power component of the line current

$$I_{1P} = E_d'I_d/3E_n \tag{12-34}$$

Dividing Eq. (12-34) by Eq. (12-29),

$$E_d'/E_{do} = I_{1P}/I_{10} \tag{12-35}$$

Therefore curves 1, 3, 5, 7, and 9 of Fig. 12-20 also give the ratio E_d'/E_{do}.

While the curves of Figs. 12-20 and 12-21 are applicable specifically to the circuits of Figs. 12-10 and 12-12 ($p = 3$), they can be used for other 6-phase circuits, such as circuit 3 of Table 12-2*a* ($p = 6$), by obtaining from Fig. 12-7 the equivalent commutating reactance factor which will produce the same commutating angle.

CONVERTER FAULTS

23. Types of Faults. The three major faults in a static converter which cause overcurrents are a d-c short circuit, a short circuit of a rectifier element during rectifier operation, and a short circuit of an element during inverter operation. These are indicated by dotted lines in Fig. 12-22.

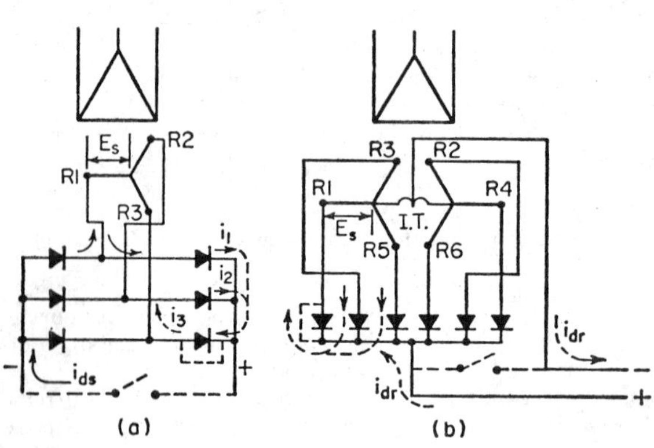

FIG. 12-22. Currents in 6-phase double-way and double-Y rectifier circuits during faults.

24. D-C Short Circuit. In the 6-phase double-way circuit of Fig. 12-22*a*, a d-c short circuit connects the two rectifier elements of each phase in reverse parallel, so that it is equivalent to a short circuit of the transformer d-c winding. The positive part of each phase current flows through one element, the negative part through the other element, as indicated by the solid-shaft arrows for phase *R*1. The current i_{ds} in the d-c circuit is equal to the sum of either the positive or negative parts.

The waveshapes of the currents resulting from a d-c short circuit are shown in Fig. 12-23. The dotted sine waves are the sustained phase currents after the initial transient has passed. The short circuit was assumed to start at an instant when the sustained

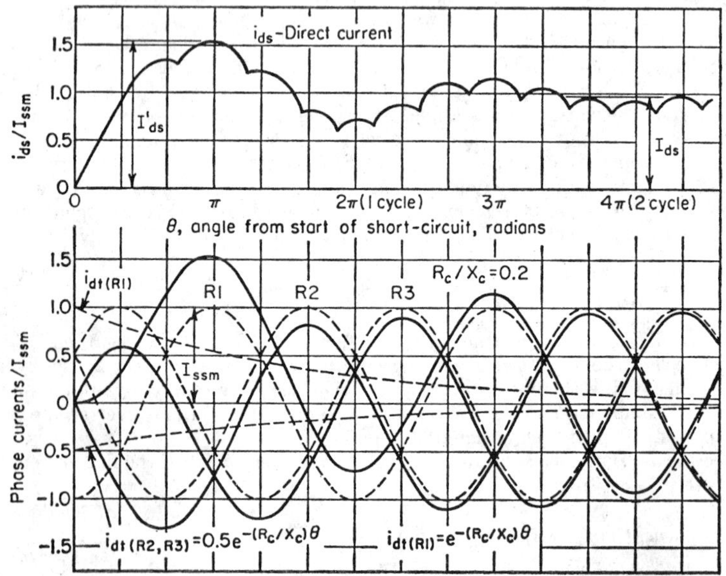

FIG. 12-23. Phase-current and d-c waves during d-c short circuit in Fig. 12-22*a*.

current of phase $R1$ would have been at a negative maximum, so that the decremental offset transient i_{dt} starts with an equal positive value. The offset transient for phases $R2$ and $R3$ was obtained similarly. The solid waves are the actual phase currents, which are equal to the sum of the sustained and transient components. The direct-current wave was obtained by adding the positive parts of the phase currents. The currents in the transformer a-c winding follow the waveshapes of the currents in the d-c winding.

The instantaneous currents in Fig. 12-23 are given as a ratio to I_{ssm}, the crest of the sustained phase currents.

$$I_{ssm} = \sqrt{2}E_s/Z_c = \sqrt{2}E_s/\sqrt{X_c^2 + R_c^2} \tag{12-36}$$

in which X_c and R_c are reactance and resistance referred to the phase-to-neutral voltage E_s. They include the impedances of the a-c supply circuit, the rectifier transformer, the connections to the rectifier (if significant), and the equivalent resistance of the loss in the rectifier elements. The average value of each half-cycle of sustained current is $0.318I_{ssm}$. The sustained direct current for this circuit consists of three half-cycles, and its average value is

$$I_{ds} = 0.955I_{ssm} \tag{12-37}$$

The average sustained direct current can be obtained also from Eq. (12-38), in which I_{dn} is the rated direct current and $Z_{c(pu)}$ the per-unit impedance (at rated current).

$$I_{ds} = 1.1I_{dn}/Z_{c(pu)} \tag{12-38}$$

The transient crests of the phase and direct currents occur at about one-half cycle after the start of the short circuit. Their magnitudes are determined by the transient current i_{dt} and are therefore a function of the R/X ratio.

Phase-current crest: $$I_{st} = K_sI_{ssm} \tag{12-39}$$

Direct-current crest: $$I_{ds}' = 1.05K_sI_{ds} \tag{12-40}$$

The values of K_s as a function of the R/X ratio are given in Fig. 12-25. This ratio is usually in the range of 0.1 to 0.3.

As can be seen in Fig. 12-23, the first positive part of the current in phase $R1$, which is carried by one rectifier element, has a higher amplitude and is of longer duration than a half-cycle of sustained short-circuit current. This has to be considered in applications of semiconductor rectifying devices.

The short-circuit conditions in the single-way circuit of Fig. 12-22*b* are similar to those of Fig. 12-22*a*, except that the alternate parts of the current in each phase of the a-c winding are contributed by currents in two opposite phases of the d-c winding. For example, $R1$ would carry the positive part of a phase current shown in Fig. 12-23 and $R4$ the negative part. The sustained direct current consists of six half-cycles, and its average value is

$$I_{ds} = 1.91I_{ssm} \tag{12-41}$$

It can be obtained also from Eq. (12-38). Equations (12-36), (12-39), and (12-40) and the curve for K_s are also applicable.

A d-c short circuit can be cleared by interrupting the d-c circuit or the a-c supply by means of circuit breakers. If controlled rectifier elements are used, the fault could be interrupted also by current suppression. This is effected by removing the firing voltages of the control electrodes. The nonconducting elements are prevented from starting. The others are prevented from restarting after their individual currents drop to zero. If there is inductance in the short-circuited part of the d-c circuit, the firing control may have to be shifted to inverter operation in order to regenerate the stored energy and bring the current to zero.

25. Short-circuited Rectifier Element (Rectifier Operation). A rectifier element can become short-circuited by an arc-back in a mercury-arc tube or by a breakdown

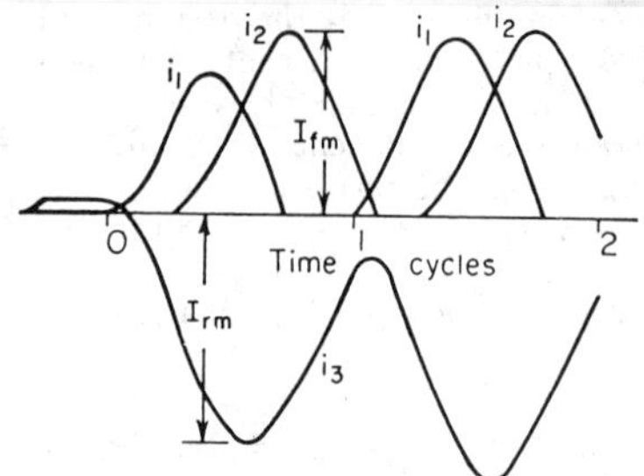

FIG. 12-24. Phase-current waves when a rectifier element in Fig. 12-22 is short-circuited.

of a semiconductor device. In Fig. 12-22 this fault is represented by a dotted-line circuit across a rectifier element, and the currents are indicated by dotted-shaft arrows.

In rectifier operation, the highest fault current is obtained if the fault is initiated at the end of the conducting period (see Par. **9**). This is also the most likely point for the occurrence of such a fault, because of the voltage reversal during the transition from a conducting to a nonconducting state. As indicated in Fig. 12-22*a*, reverse current flows through the faulted element into its phase winding from the other phases.

The typical shapes of the currents are shown in Fig. 12-24 for two cycles. This type of fault is usually cleared in one cycle or less, so that the peak values of the currents during the first cycle are of interest. They are given by the following equations (Ref. 14, Par. **78**):

Peak reverse current of faulted rectifier element,

$$I_{rm} = K_r I_{ssm} \quad (12\text{-}42)$$

Peak forward current of unfaulted rectifier elements,

$$I_{fm} = K_f I_{ssm} \quad (12\text{-}43)$$

The value of I_{ssm} is given by Eq. (12-36). The values of K_r and K_f can be obtained from the curves of Fig. 12-25.

For this type of fault in the circuit of Fig. 12-22*b*, conditions during the first cycle are substantially the same as for Fig. 12-22*a*. Practically all the reverse current of the faulted element is contributed by the other two phases of the same wye, as indicated by the arrows. Until the interphase transformer (IT) becomes saturated, its reactance limits the contribution from the other wye to a small value, so that Eqs. (12-42) and (12-43) can be used to obtain the peak currents in the wye of the faulted element.

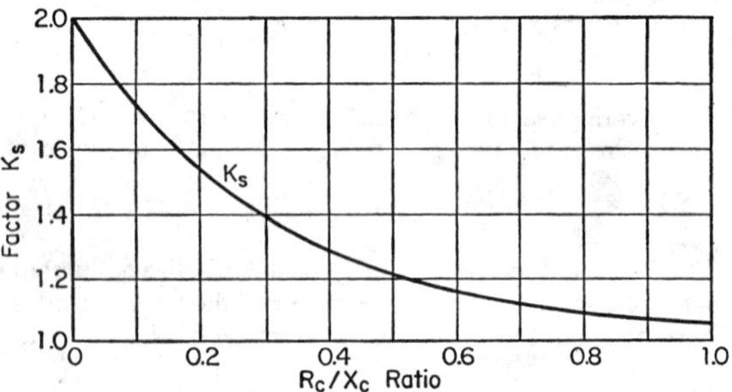

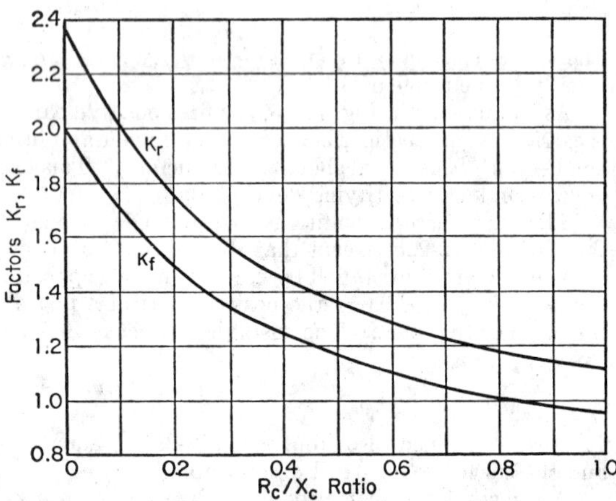

FIG. 12-25. Factors used for calculating fault currents, as functions of R_c/X_c ratio.

If another source of direct current is connected to the d-c circuit of Fig. 12-22*b*, it will contribute a reverse current i_{dr} to the faulted rectifier element, as indicated by the arrows. Methods for evaluating this current are outlined in Ref. 14 (Par. **78**). Reverse current from the d-c side cannot flow in the circuit of Fig. 12-22*a*, for it is blocked by the rectifier elements on the other side of the circuit.

This type of fault with a-c feed only (i.e., without feed from the d-c side) can be cleared by interrupting the a-c supply with a circuit breaker on the supply side or rectifier side of the transformer; also by circuit breakers or fuses in series with the rectifier elements. Fuses are generally used for semiconductor rectifiers (see Par. **51**). Anode breakers are used in many mercury-arc rectifier installations. Backfeed from the d-c side can be interrupted by a circuit breaker in the d-c circuit. If fuses or circuit breakers are used in series with the rectifier elements, the interrupter of the faulted element will clear both the a-c and the d-c feeds. If controlled rectifier elements are used, the a-c feed can be interrupted by current suppression (see Par. **24**).

26. Inverter Fault. This fault occurs when a rectifier element conducts during the positive nonconducting period (see Fig. 12-8*b*), which causes a transient reversal of the inverter d-c voltage. In effect, the d-c power supply to the inverter is short-circuited across a phase winding of the inverter transformer. The resulting current consists of d-c and a-c components and could attain a high value if not interrupted or limited. A similar fault would result if the a-c voltage of the inverter were interrupted. The current obtained for either condition can be readily calculated if the inductance and resistance of the circuit and the voltage-regulation characteristics of the d-c supply are known. The current through the faulted rectifier element is in the forward direction.

A sustained fault of this type can be cleared by interrupting the d-c supply. If the d-c supply is obtained from another static converter with controlled rectifier elements, the supply can be interrupted by current suppression without tripping a circuit breaker and normal operation can be resumed immediately if the cause of the fault has disappeared. If the fault is caused by a momentary transient, the inverter can recover at the next commutation of the faulted element if the current did not exceed the commutation limit. Usually a reactor is connected in the d-c circuit of an inverter to limit the rate of rise of current, so that transient faults can clear themselves by recommutation.

HARMONICS

27. D-C Voltage Wave. The d-c voltage wave of a converter contains a ripple component superimposed on the average d-c value. As shown in Fig. 12-4, the ripple is made up of portions of sine waves; its amplitude and shape depend upon the number of rectifier phases, the phase-control angle α, and the commutating angle u (see Par. **7**).

The voltage ripple can be resolved by a Fourier series into harmonic components having frequencies equal to $q, 2q, 3q, \cdots, mq$ times the frequency of the a-c circuit, where q is the number of rectifier phases and m is the order of the ripple harmonic. The d-c voltage of a 6-phase rectifier, for example, has sixth, twelfth, eighteenth, $\cdots$, $(q \cdot m)$th harmonics of the a-c line frequency. Equations (12-44*a*) and (12-44*b*) give the harmonics as a function of α and u.

$$\frac{A_m}{E_{do}} = \frac{1}{(qm)^2 - 1}\left\{\begin{array}{l} -\cos\alpha \sin qm\alpha - \cos(\alpha + u)\sin qm(\alpha + u) \\ +qm[\sin\alpha\cos qm\alpha + \sin(\alpha + u)\cos qm(\alpha + u)] \end{array}\right\} \qquad (12\text{-}44a)$$

$$\frac{B_m}{E_{do}} = \frac{-1}{(qm)^2 - 1}\left\{\begin{array}{l} \cos\alpha\cos qm\alpha + \cos(\alpha + u)\cos qm(\alpha + u) \\ +qm[\sin\alpha\sin qm\alpha + \sin(\alpha + u)\sin qm(\alpha + u)] \end{array}\right\} \qquad (12\text{-}44b)$$

$$C_m/E_{do} = \sqrt{(A_m/E_{do})^2 + (B_m/E_{do})^2} \qquad (12\text{-}44c)$$

where A_m = amplitude of sine component of mth harmonic, B_m = amplitude of cosine component of mth harmonic, and C_m = amplitude of total harmonic.

Figure 12-26 shows the *rms* values of the sixth and twelfth harmonics of the a-c line frequency for a 6-phase rectifier as functions of α and reactance factor I_cX_c/E_s.

The harmonic components of the d-c voltage can cause a flow of harmonic currents in the d-c circuit, the current of any frequency being equal to the harmonic voltage divided by the circuit impedance at that frequency. In the analysis of rectifier circuits in the preceding articles, the d-c current wave is assumed to be a straight line, on the assumption that there is sufficient inductance in the d-c circuit to reduce the harmonic currents to negligible values. Insofar as the current and voltage relations in the rectifier

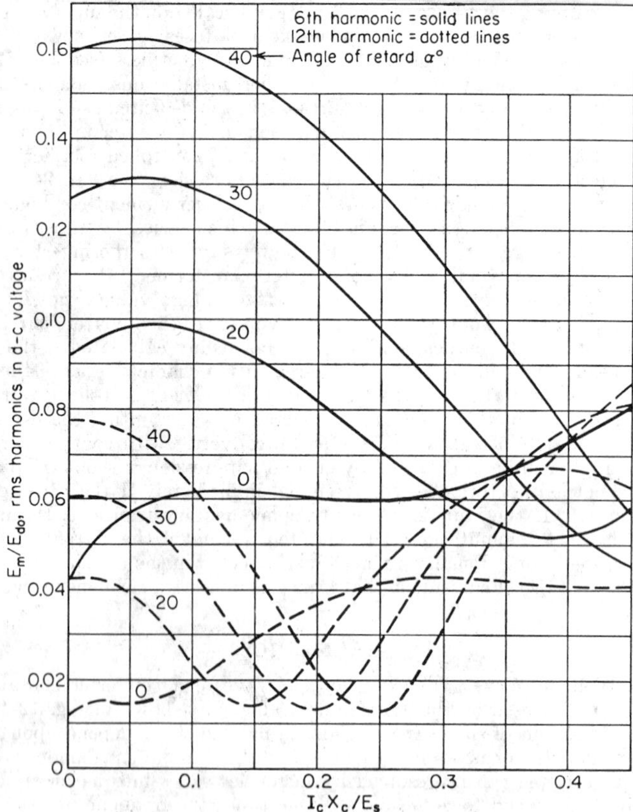

FIG. 12-26. Sixth and twelfth harmonics in d-c voltage of 6-phase rectifier circuits with a $p = 3$.

circuits are concerned, this assumption leads to satisfactory results in most cases. The presence of harmonic currents and voltages in the d-c circuits, however, may cause additional losses in the load, such as electrolytic cells, supplied by rectifiers. If the harmonic currents are large, they may have a detrimental effect on the commutation of rotating d-c machines.

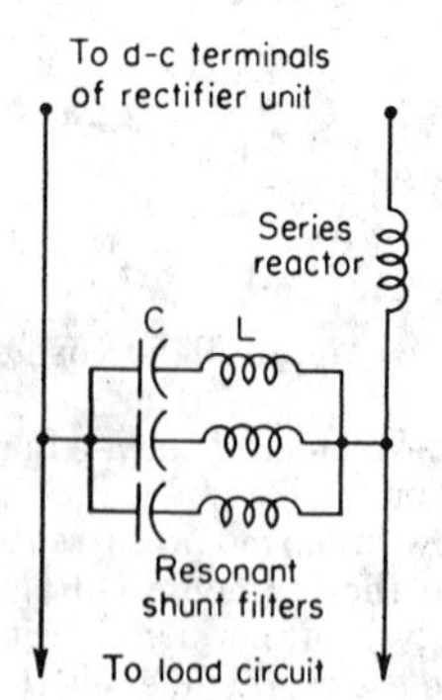

FIG. 12-27. Filter for rectifier output voltage.

The harmonics in the current can be reduced to harmless values by means of a reactor in series with the d-c circuit. Sometimes a series reactor in combination with resonant shunt filters, as shown in Fig. 12-27, is used. Each branch of the shunt filter consists of a reactor in series with a capacitor, tuned for resonance at the frequency of the harmonic which it is to suppress.

The total rms current ripple for inductive loads can be calculated with sufficient accuracy for most purposes by dividing the total rms voltage ripple by the load impedance at the lowest harmonic frequency. The result is pessimistic since it ignores the higher impedance of the load to the higher harmonics. Figures 12-28 and 12-29 show the total rms voltage ripple for various values of phase retard, α, and reactance factor for 6-phase and 12-phase rectifiers.

Figure 12-30 shows the first-order harmonic and the total

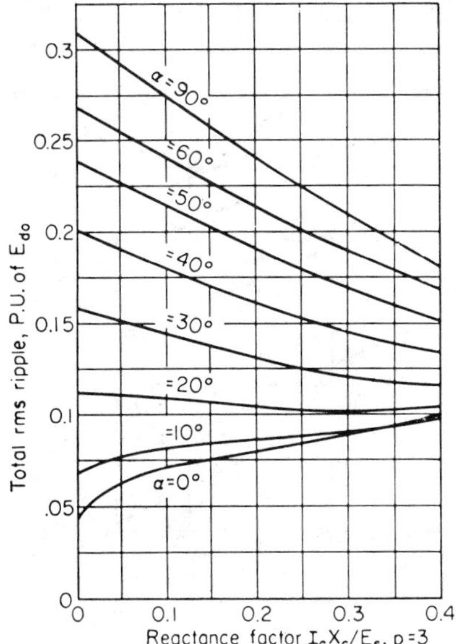

FIG. 12-28. Total rms ripple in d-c voltage of 6-phase rectifier circuits with a $p = 3$.

rms ripple voltage at *no load* for a 6-phase controlled rectifier and for a "semicontrolled" 6-phase double-way rectifier (see Par. **19**). The ripple is plotted against voltage reduction by phase control. The markedly larger ripple for the full-controlled rectifier with inductive load, at low values of output voltage, is due to the fact that the d-c voltage wave reverses during part of each cycle. This reversal does not occur in the semicontrolled circuit.

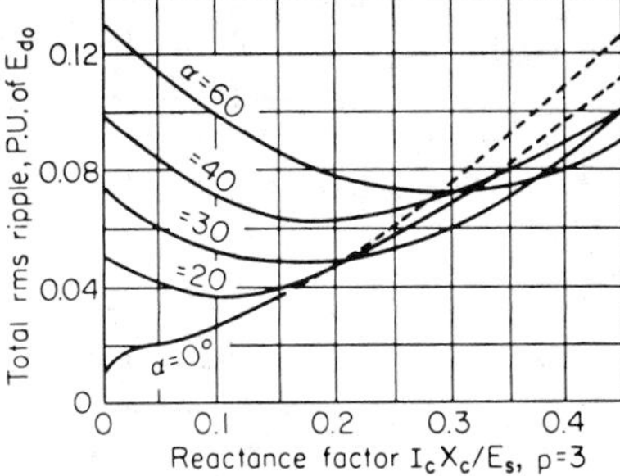

FIG. 12-29. Total rms ripple in d-c voltage of 12-phase rectifier circuits with a $p = 3$. NOTE: Dotted portions of curves are valid only if two 6-phase groups share negligible primary impedance.

28. Harmonics in A-C Line Current. The a-c line current of a static converter (see Fig. 12-11) is nonsinusoidal and can be resolved into a fundamental and harmonic components. The fundamental components were discussed in Par. **22**.

The frequencies of the harmonic components are equal to $mq \pm 1$ times the fre-

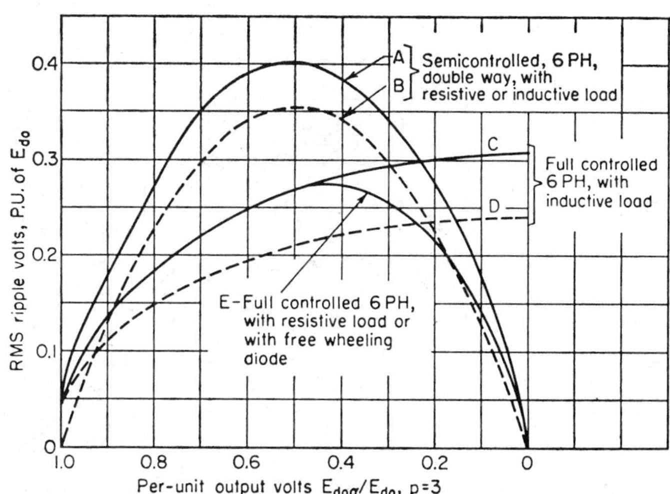

FIG. 12-30. Total rms ripple and first-order harmonics, at no load, in voltage of 6-phase controlled and semicontrolled rectifiers. Curves *A*, *C*, and *E* total rms ripple. Curve *B* rms value of third harmonic. Curve *D* rms value of sixth harmonic.

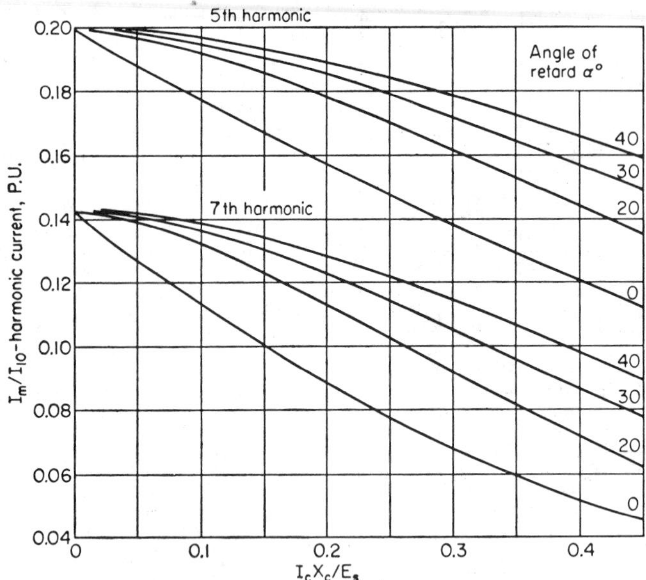

Fig. 12-31. Per unit rms fifth and seventh harmonics in a-c line current of 6-phase rectifier circuit with a $p = 3$.

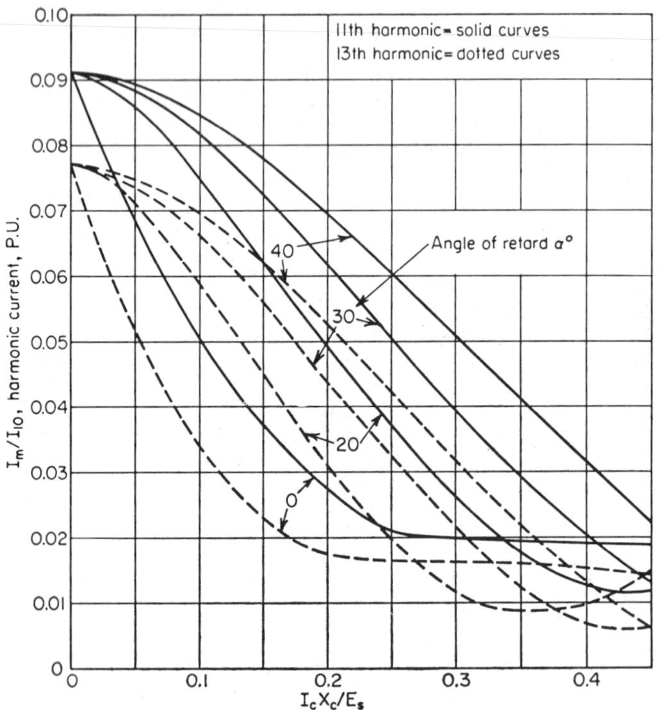

Fig. 12-32. Per unit rms eleventh and thirteenth harmonics in a-c line current of 6-phase or 12-phase rectifier circuit with a $p = 3$.

quency of the fundamental, m being an integer 1, 2, 3, etc. For a 6-phase converter ($q = 6$), for example, the line current has a fifth, seventh, eleventh, thirteenth, etc., harmonics. For each harmonic in the d-c voltage there are two harmonics in the a-c line current. Phase control increases their magnitude.

The first four current harmonics for the circuits of Figs. 12-10 and 12-12 are given in Figs. 12-31 and 12-32, as functions of the phase-control angle α and the commutating reactance factor (Ref. 28, Par. **79**). The rms values of the harmonics are expressed in per unit of I_{10}, which can be obtained from Eq. (12-29) or from Tables 12-2*a* and 12-2*b*. The curves in these figures are applicable also to other 6-phase circuits, such as circuit 3 of Table 12-2*a*, if the equivalent commutating reactance factor for equal u or E_x/E_{do} is obtained from Fig. 12-7.

Figure 12-32 is applicable also for a 12-phase circuit with $p = 3$, such as circuit 8 of Table 12-2*b*. In general, harmonics present in a rectifier circuit with a number of phases which is a multiple of 6 have the same values as the corresponding harmonics in the 6-phase circuit with the same p. This applies also to the harmonics in the d-c voltage.

The curves of Figs. 12-26, 12-31, and 12-32 can be used for inverter operation if margin angle γ is substituted for α.

29. Phase Multiplication and A-C Filters. Under some adverse conditions, harmonics in the a-c line current of static converters can cause interference in telephone lines which are exposed to transmission or distribution lines of the power system. From the viewpoint of the a-c system, a converter installation can be considered as a generator of harmonic currents which flow into various parts of the system in accordance with their relative impedances at the harmonic frequencies. Parallel resonances in the system for one or more of the harmonic frequencies could amplify their effect.

Serious telephone interference occurred when the first large mercury-arc rectifier installation for production of aluminum was started in 1938. The problem was solved by multiphasing (Ref. 23, Par. **79**). A converter installation for electrochemical service usually consists of a number of parallel 6-phase units. Multiphasing is effected by phase-shifting the rectifier transformers from each other to produce a new symmetrical polyphase rectifier system. This can be done by means of phase-shifting autotransformers, shown in Fig. 12-33, which are connected ahead of the rectifier transformers.

Another method is to provide the required phase shifts in the a-c windings of the rectifier transformers. A typical multiphased arrangement is shown in Fig. 12-42. A 48-phase rectifier system was obtained with eight 6-phase units. Theoretically, the lowest harmonics are the forty-seventh and forty-ninth, and the line current is practically a sine wave. Residuals of lower harmonics, produced by unbalances, are generally too small to cause trouble.

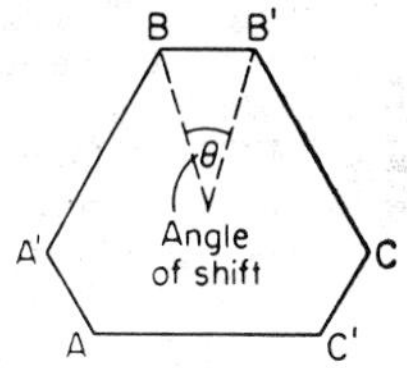

FIG. 12-33. Diagram of phase-shifting autotransformer used for multiphasing.

In installations where multiphasing is not practical, problems caused by harmonic currents can be solved usually by connecting to the a-c line terminals shunt filters with low impedance at the frequencies of the offending harmonics. The filters reduce to harmless values the proportion of these harmonics going into the a-c system.

30. Telephone Influence Factor (TIF). Different frequencies of noise induced into a telephone circuit by a voltage or current in a power circuit have different interfering effects, depending on the characteristics of the telephone circuits, receivers, and other factors. The relative interfering effect is given by a TIF weighting curve vs. frequency, which was developed from extensive tests and has been revised from time to time to reflect changes in the telephone circuits and instruments.

To obtain the overall influence of a current in a power circuit, the current value of each frequency is multiplied by the corresponding weighting factor, and the square root of the sum of the squares of these products is obtained. The result is called the **I·T product.** The same procedure is used to obtain a **kV·T product** for the interfering effect of a voltage, expressed in kilovolts. The **voltage TIF** is frequently used instead of kV·T; it is obtained by dividing the kV·T number by the voltage in kilovolts (Refs. 12 and 15, Par. **78**).

LINE-COMMUTATED FREQUENCY CONVERTERS

31. Frequency Converter with D-C Link. The basic circuit diagram of a static frequency converter used for interconnecting two power systems is shown in Fig. 12-34. This type has been used in place of a rotating frequency changer when the frequency ratio of the a-c systems is not fixed. The d-c link provides a flexible tie, and the power transfer in either direction can be controlled independently of frequency variations.

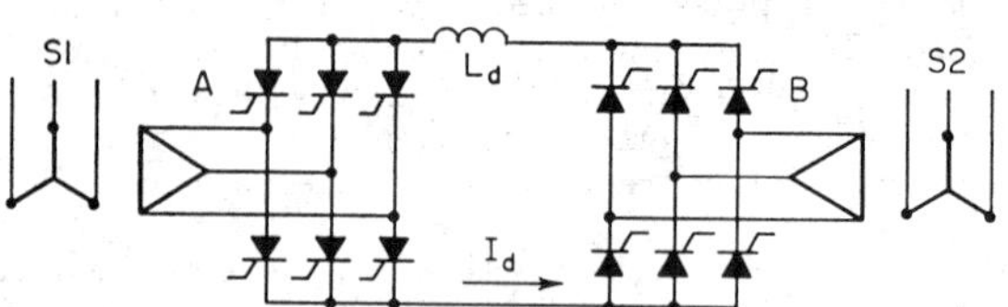

Fig. 12-34. Power-circuit diagram of static frequency converter with d-c link.

The frequency converter consists of two static converters A and B which are connected to the power systems $S1$ and $S2$ on the a-c side, and their d-c circuits are connected in series so that the current I_d is common to both. The current cannot reverse, and reversal in the direction of power flow is effected by reversal of the d-c voltages. For power flow from $S1$ to $S2$, converter A is operated as a power rectifier, and its d-c voltage is in the same direction as the current; converter B operates as an inverter, so that its voltage is opposite to the current and bucks the rectifier voltage. The operation is interchanged for power flow in the opposite direction.

The rectifier voltage, obtained from Eq. (12-25), has to be equal to the inverter voltage, Eq. (12-27), plus the resistive drop in the d-c circuit. In normal operation, the direct current and power are usually regulated by phase control of the rectifier. It is desirable to operate the inverter with the minimum angle of advance β needed to obtain the required margin angle γ. The d-c reactor L_d is usually made sufficiently large to limit the rate of current rise during a transient inverter fault so that the inverter will recover within one cycle by recommutation (Refs. 27 and 30, Par. **79**).

Figure 12-34 is the basic circuit of a d-c power transmission system.

32. Cycloconverter. A cycloconverter is used to convert a-c power of one frequency to a-c of a lower frequency, which is variable. In Fig. 12-35 is shown a power-circuit diagram of a 3-phase to 3-phase cycloconverter supplying power to a load, such as a motor. The supply phases $R1$-$R2$-$R3$ represent the winding of a generator or the d-c winding of a rectifier transformer.

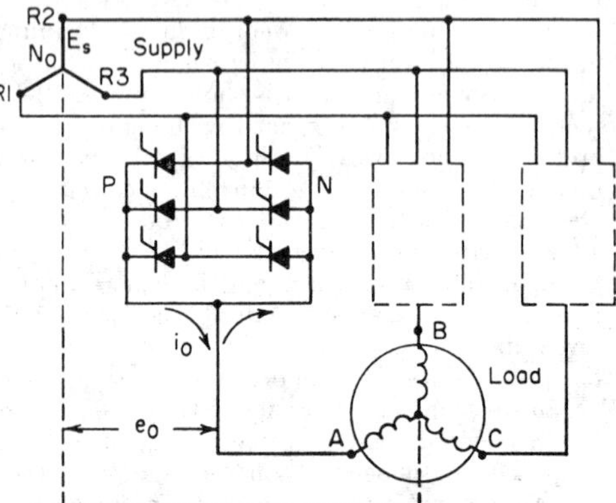

Fig. 12-35. Power-circuit diagram of 3-phase cycloconverter.

Each phase of the load circuit is supplied from all the phases of the supply through two groups of controlled rectifier elements, P and N. When one group is conducting, the other is blocked. Group P delivers the positive half-cycle of the load current, group N the negative half-cycle. Each group together with phases $R1$-$R2$-$R3$ can be considered simply as a 3-phase controlled converter, the output average voltage of which can be varied by phase control from a maximum in the rectifier range, through zero, and into the inverter range (see Par. **4**).

The operation is illustrated in Fig. 12-36 for one of the load phases. At the top are shown the sine waves of phase voltages $R1$-$R2$-$R3$. The conducting parts of these voltages, shown in heavy outline, constitute the output voltage wave e_o (with the effect of commutation neglected). Voltage e_o' is the *average* of the output voltage wave. The phase control of groups P and N is so modulated that the average output voltage approaches a sine wave of the desired frequency.

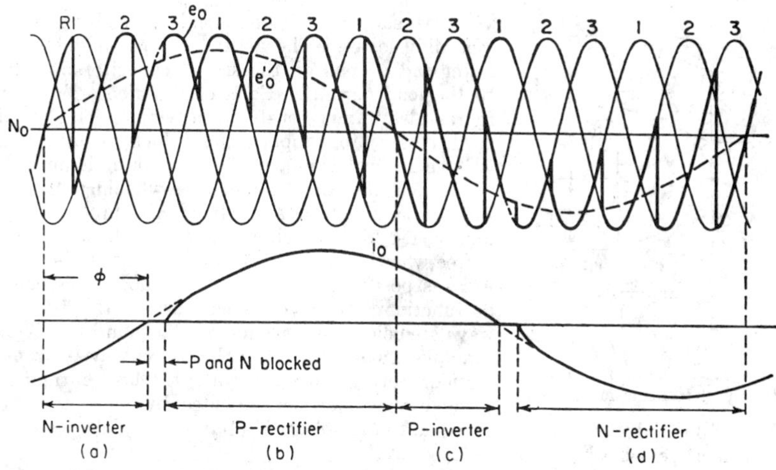

FIG. 12-36. Output phase voltage and fundamental current of cycloconverter.

Because of the ripples in the voltage wave, the current wave also has ripples. In Fig. 12-36, i_o is the fundamental component of the current wave; it is shown lagging behind the voltage e_o' by the power-factor angle ϕ. The operation of the P and N groups during one cycle of the output voltage is indicated below the current wave. During interval a, the voltage is positive while the current is negative; this involves power reversal, so that group N, which conducts the negative current, operates as inverter. During interval b, the voltage and current are positive, so that group P operates as rectifier. The operation during intervals c and d is similar to intervals a and b. An interval of zero current precedes current reversal, for the purpose of blocking one group before the other starts conducting, in order to prevent a short circuit of the power supply.

It is obvious that the magnitude as well as the frequency of the output voltage can be varied by phase control. The highest crest voltage of e_o' is equal to the uncontrolled average voltage, which is $1.17E_s$ for the 3-phase groups of Fig. 12-35, minus the voltage drops.

A cycloconverter can produce outputs of one or more phases. For a balanced 3-phase load, the connection between the neutral points of the supply and load can be omitted. Six-phase groups can be used for P and N, instead of the 3-phase groups shown, thereby reducing the magnitude and increasing the frequency of the ripple in the output voltage. A cycloconverter can be operated without sine modulation of the output voltage wave; a voltage of approximately rectangular shape is obtained then (see Par. **79**, Ref. 37 and Ref. 42, p. 69).

SPECIAL RECTIFIER CIRCUITS

33. Voltage-multiplier Circuits. With this type of circuit, a d-c voltage is obtained which is two or more times the crest of the a-c voltage from which it is derived. There are several variations of such circuits (Ref. 29, Par. **79**). In all of them, capacitors are charged through rectifier elements, and the voltage multiplication is effected by the series connection of the capacitors or of a capacitor and the a-c charging voltage.

The circuit of Fig. 12-37, known as the Cockcroft-Walton circuit, consists of a number of multiplier stages. During the half-cycles when terminal $R2$ of the transformer d-c winding is positive, capacitor $C1$ is charged to the crest of the voltage E_s through diode $D1$. When $R1$ is positive, capacitor $C2$ is charged to twice the crest of E_s by the transformer voltage in series with $C1$, through diode $D2$. When $R2$ is positive and $C1$ is charged, $C3$ is charged to twice the crest voltage by $C2$ through $D3$. Similarly, the capacitors of the remaining stages are charged to twice the crest of the a-c voltage. In Fig. 12-37, the voltage of terminal P is six times the crest of E_s. The

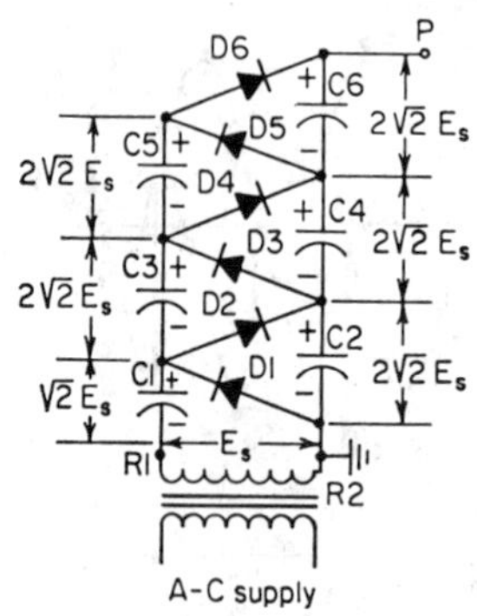

FIG. 12-37. Cockcroft-Walton voltage-multiplier circuit.

voltages shown are theoretical no-load values, disregarding losses. The actual voltages under load are lower and depend on the size of capacitors in relation to the load current. Each of the diodes shown may represent a number of diodes in series. This circuit is used mostly for applications which require a high voltage at low current, in the range of milliamperes.

34. Rectifier Circuit with Free-wheeling Diode. In Fig. 12-38 is shown a rectifier unit supplying an inductive d-c load, which is shunted by diode D_{fw}, called a free-wheeling diode. When the entire d-c voltage wave is positive, the diode carries no current and has no function. However, when parts of the d-c voltage wave are below the zero line, as shown in Fig. 12-4*d*, for example, diode D_{fw} takes the current away from the rectifier during those intervals; the load current flows through the diode, as indicated by the dotted arrow in Fig. 12-38, and the d-c voltage is equal to the voltage drop of the diode.

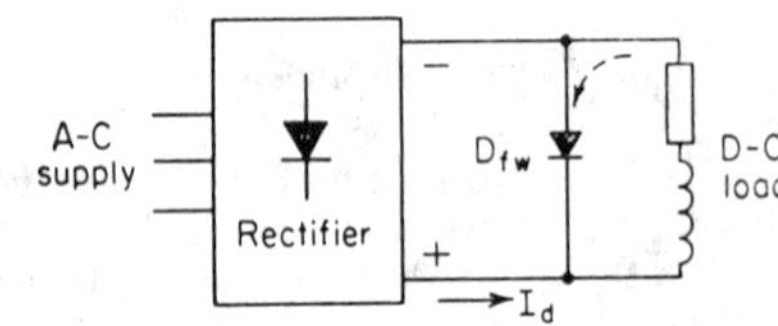

FIG. 12-38. Rectifier with free-wheeling diode.

When a free-wheeling diode is used, Eq. (12-2) for the rectifier d-c voltage with phase control is applicable only to the angle α when the voltage wave touches the zero line. For 6-phase circuits, such as those of Figs. 12-10 and 12-12, this occurs when α is 60°. To reduce the average voltage to zero, α has to be larger than 90°. The suppression of the negative portions of the voltage wave reduces the ripple of the d-c voltage. The free-wheeling diode also improves the power factor on the a-c side because the conducting angle of the phase currents is reduced, so that the center of the current is closer to the center line of the voltage. These improvements apply only for operation with phase control beyond the point when the voltage wave touches the zero line.

A detailed analysis of rectifier operation with a free-wheeling diode is given in Chaps. 13 and 20 of Ref. 5, Par. **77**. In some of the literature on mercury-arc rectifiers, the free-wheeling diode is called the "zero anode."

SELF-COMMUTATED INVERTERS

35. General. In inverters, power flows from the d-c to the a-c side. Therefore all inverters regardless of type must operate so that during the major part of each SCR's conducting period the a-c voltage is negative. Yet for current to commutate from one SCR to the next, it is necessary that the voltage of the incoming SCR be instantaneously more positive than that of the outgoing SCR. The different ways in which this positive commutating voltage is obtained can be used to classify inverter circuits. Line-commutated inverters and frequency converters, which have been covered in Pars. **31** and **32**, get their commutating voltage from the a-c system. Self-commutated inverters furnish their own commutating voltage and can therefore supply passive loads. Several self-commutated inverter circuits are described below. These and other inverter circuits are analyzed in detail in Ref. 4, Par. **77**.

36. Parallel Capacitor-commutated Inverter. A phase-controlled rectifier can be operated as an inverter into a passive load if sufficient capacitance is connected across the output to cause the total a-c current to lead the voltage. The commutating voltage is then supplied by the capacitor. Figure 12-39 is a simple single-phase parallel-inverter circuit. The load-voltage waveshape depends upon the load-circuit constants and the value of the d-c inductor L_d. It will approach a sinusoid and the current will be flat-topped if the load is highly inductive and if L_d is large. Figure 12-39 shows the waveshapes of a-c voltage and total current for these conditions. The half-cycle average value of the output at no load is $E_{o,\text{avg}} = (E_d/\cos\beta)(N_2/N_1)$, where N_2/N_1 is the transformer turns ratio and β is the angle of advance.

The operation of the circuit is as follows: Assume that SCR1 is conducting. At t_1 when SCR2 fires, capacitor C has a charge on it of the polarity shown in Fig. 12-39. Firing SCR2 establishes a short circuit across the capacitor, and the discharge through the two SCRs will commutate the current from SCR1 to SCR2. For C to be charged in the correct polarity requires that the zero crossing of the total current, including the capacitor current, lead the zero crossing of the voltage as shown in Fig. 12-39*b*, although the load current may be lagging. Frequency may be varied by varying the gate pulse-repetition rate provided that there is enough time for SCR1 to turn off before its anode becomes positive again at t_2 in Fig. 12-39. Too high a frequency will not allow enough time. With a highly inductive load the circuit will be sharply resonant, C being effectively in parallel with the load inductance, and it is necessary to operate near the resonant frequency. Moreover the operating frequency must be slightly higher than the natural frequency so that total current will lead the voltage, as explained above. The circuit may be self-excited by feeding the output voltage back to the gates through a phase-shift network that provides the proper angle of advance, β, of the gate pulse.

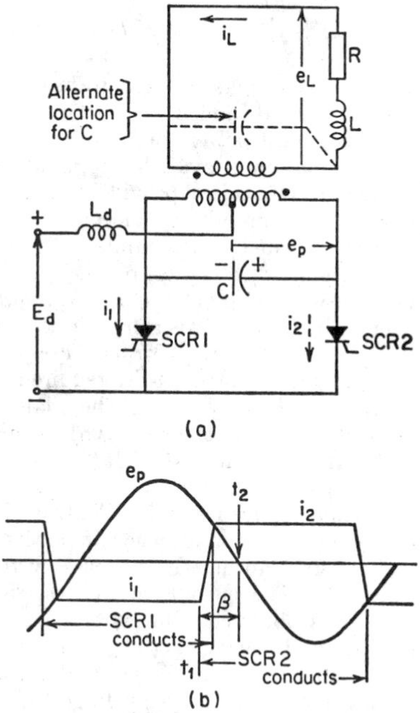

FIG. 12-39. Parallel capacitor-commutated inverter.

A 3-phase circuit can be made by connecting in Δ the secondary windings of three such circuits as Fig. 12-39, whose gating signals are 120 elec deg apart.

37. Impulse-commutated Inverter. Inverters in which the commutating current is a pulse of short duration compared with one a-c period are called "impulse-commutated" inverters. Figure 12-40 shows a basic single-phase inverter circuit with center-tapped transformer. It differs from the parallel inverter in having an inductor in series with the commutating capacitor and in the addition of the "pump-back" diodes $D1$ and $D2$. If SCR1 is conducting, the capacitor C will become charged in the polarity shown. If SCR2 is now fired, the capacitor will discharge through L, SCR2, and SCR1. The current in SCR1 is the difference between its initial current and the commutating pulse current. When the current in SCR1 reaches zero, it will block current reversal and the remainder of the pulse will flow through $D1$ until the pulse current reaches zero, which completes the commutating transient.

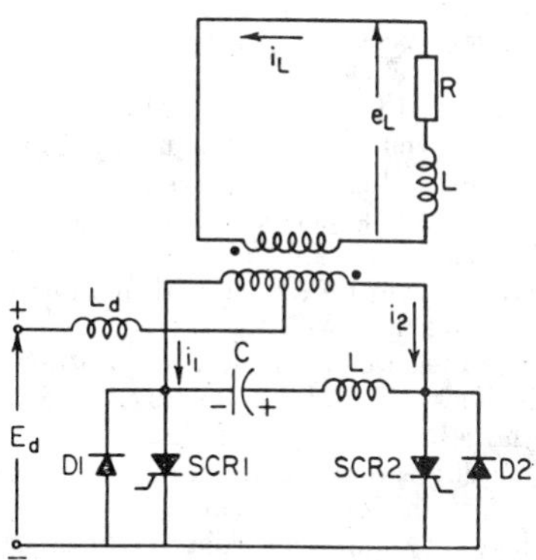

FIG. 12-40. Impulse-commutated inverter.

Prior to commutation, capacitor C has been charged to a potential $2E_d$ and during commutation this charge must furnish the load current through the transformer long enough for SCR1 to turn off. If the commutating circuit has a high q at its natural frequency, the commutating current pulse will be approximately a half sinusoid. The values of L and C are chosen so that the instantaneous value of this current is greater than the load current just long enough to allow SCR1 to turn off under worst conditions (highest load current and longest predicted turnoff time of an SCR). Various combinations of L and C will satisfy these conditions, but there is an optimum combination for minimum energy dissipation during the commutating transient (Ref. 4, Par. **77**). The following equations will give near optimum values:

$$C = I_{LM}t_{om}/2E_d \qquad \text{and} \qquad L = E_d t_{om}/I_{LM}$$

where I_{LM} = max. current to be commutated, t_{om} = max. turn-off time of any SCR at that current, and E_d = lowest d-c supply voltage.

A separate power source for charging the commutating capacitor may be used to avoid variation of available commutating energy with varying d-c voltage.

Diodes *D*1 and *D*2 provide a path for lagging-power-factor load current after commutation. The impulse-commutated inverter can operate into an inductive load without adding as much capacitance as a parallel inverter. The amount of inductive-load kVA is limited, however, by the size of the commutating circuit constants. The load-voltage waveshape of impulse-commutated inverters is, typically, rectangular. However, in polyphase inverters addition of voltages from different phases can be used to obtained stepped waveforms. For loads which cannot tolerate substantial harmonic content, filters are generally required. The frequency can be varied over a wide range by varying the frequency of the pilot oscillator which drives the SCR gates. The upper limit of frequency is generally determined by the turnoff time of the SCRs and/or the switching losses in the SCRs. Special fast-turnoff SCRs are commercially available for inverter applications.

Output-voltage Control. Control of the output voltage of an inverter is often necessary. For example, when inverters are used to drive a-c motors at variable speed and constant torque, it is necessary to maintain output volts per cycle approximately constant. The output voltage can be varied by varying the d-c input voltage. If the d-c is obtained from a rectifier, this can be done either by phase control of the rectifier or by varying the a-c input voltage to the rectifier. Sometimes it may not be possible to vary the source d-c voltage, e.g., in the case of a battery supply. In that case the average d-c voltage to the inverter may be varied by using a d-c chopper with "time-ratio control," where the ratio of on time to off time is varied. Alternatively "pulse-width modulation" (a form of time-ratio control) can be used in the inverter. This varies the width of each rectangular block of output voltage. Another method is to add the voltages of two inverters which can be shifted in phase displacement from each other (Ref. 4, Par. **77**). Induction regulators, variable transformers, or saturable reactors in the a-c output can also be used.

APPLICATION OF STATIC CONVERTERS

38. Application Fields. Static power converters are used for converting alternating current to direct current and direct current to alternating current, for frequency conversion, and for control and regulating functions. Their biggest application is for converting alternating to direct current, and they have largely superseded rotating-type converters because of the higher efficiency, lower installed cost, and other advantages of static converters. In Fig. 12-41 are shown efficiency curves for silicon and mercury-arc rectifiers and for a motor-generator set, each rated at 1,000 kW.

The application of metal-tank mercury-arc rectifiers was started in the United States about 1924 and subsequently spread to various industries where d-c power is used, including the electrochemical industry, which is its largest user. The silicon rectifier, introduced about 1955, has now superseded the mercury-arc type in practically all applications because of its higher efficiency, lower cost, and other advantages. New applications have also been found because of the flexibility and wide range of ratings of silicon diodes and thyristors. A large number of motor-generator sets and mercury-arc converters are in operation, and many of them will remain in service for a long time. In some electrochemical installations, mercury-arc tubes have been replaced by silicon diodes to increase the conversion efficiency. The diodes are connected to the existing rectifier transformers. Silicon rectifiers have also superseded the copper-oxide, selenium, and mechanical rectifiers. Brief discussions of several static-converter applications follow.

39. Electrochemical Service. Direct-current power is used for the electrolytic production of aluminum, magnesium, chlorine, and other products. The load circuit usually consists of many electrolytic cells connected in series. The rated current per circuit ranges from several thousand to 200,000 A or more, depending on the process and production volume. There has been a trend to higher currents. Operation is continuous with approximately rated current. Installations with mercury-arc rectifiers usually have circuit voltages in the range of 500 to 1,000 V. In newer installations, with silicon-diode rectifiers, there has been a downward trend in the voltages, and the economical range has been extended below 250 V, which was considered the economical limit for mercury-arc rectifiers.

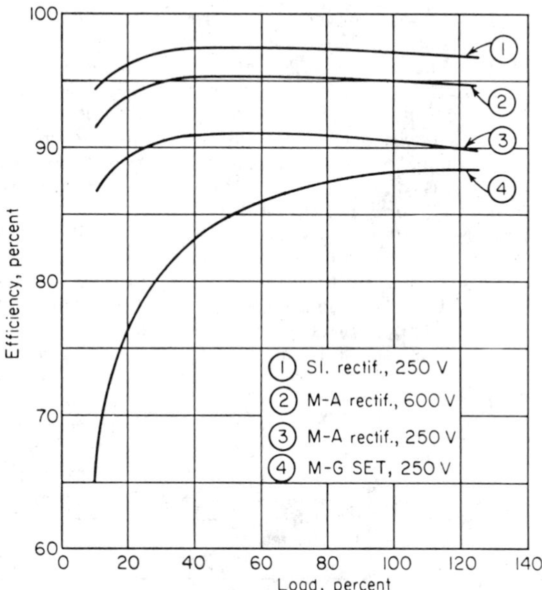

FIG. 12-41. Overall efficiency curves of silicon and mercury-arc rectifiers and motor-generator set.

A single-line diagram of a typical power supply with silicon-diode converters for a high-current electrochemical load is shown in Fig. 12-42. Eight units are connected in parallel, each consisting of two rectifier sections. The a-c power is supplied through a circuit breaker and a step regulator, which is used for adjusting the d-c voltage and to control the maximum power demand. The a-c circuit breaker is the only one used. Fuses interrupt the circuits of any faulted diodes. Overcurrent relays, connected to the d-c transductors and current transformers of individual units, provide protection for overloads or transformer faults by tripping the a-c circuit breaker.

Each rectifier unit has a 6-phase connection. The installation is multiphased to obtain a 48-phase system by phase-shifting the voltages of individual units in accordance with the table in Fig. 12-42. This can be effected by either a phase-shifting autotransformer or a zigzag connection of the transformer a-c windings (see Par. **29**).

A harmonic component in the a-c supply voltage may cause an increase or decrease in the average value of the d-c voltage, depending on the phase position of the harmonic in relation to the fundamental voltage of the transformer d-c winding (Ref. 24, Par. **79**). Such voltage harmonics could be generated, for example, by a parallel resonance or near resonance in the a-c system at the frequency of one or more of the harmonics in the a-c line current of the converters. The lower harmonics, such as the fifth, seventh, eleventh, and thirteenth, are likely to be the most influential, because the effect on the d-c voltage decreases as the harmonic order increases. When a number of parallel 6-phase units are displaced in phase from each other, the phase position of a particular harmonic in relation to the fundamental voltage will be different for each of the units. Consequently, there will be a difference in the d-c voltage characteristics of the parallel units, causing a current unbalance. In installations where this is a problem, the individual units can be provided with voltage-adjusting means, of limited range, for balancing the currents. Another method is to install resonant shunt filters on the a-c side to trap the harmonic currents which cause the generation of harmonic voltages.

40. Land Transportation. Static converters are used to supply d-c power to electric railway systems for subway, surface, and interurban lines. Series d-c motors are used, because of their favorable torque characteristics. This was the first major application for mercury-arc converters. More recent installations have silicon diodes. The d-c voltage is usually 600 to 700 V and unit ratings 1,000 to 3,000 kW. For heavy traction service, the units have a 2-h rating of 150% and a 1-min rating of 300%, following 100% load (Ref. 11, Par. **78**). Sometimes other short-time overload ratings are required.

Static converters have been applied also on electric locomotives and multiple-unit cars which operate on railroad lines electrified with single-phase 25-c power. Instead of the commutator-type a-c motors, d-c motors are used with a single-phase rectifier. A reactor is connected in series with the motors because of the large ripple in the voltage wave (Ref. 40, Par. **79**). In the past, 25-c power was used on account of the commutator a-c motors. Very likely, 60-c power will be used for any future electrification.

On a diesel-electric locomotive, the d-c motors are supplied by a d-c generator, which is driven by the engine. On more recent designs, the d-c generator has been replaced by a 3-phase a-c generator and a silicon-diode rectifier. The elimination of the d-c generator with its commutator reduces maintenance and effects other economies. It also eliminates design problems of large d-c generators for high-power locomotives.

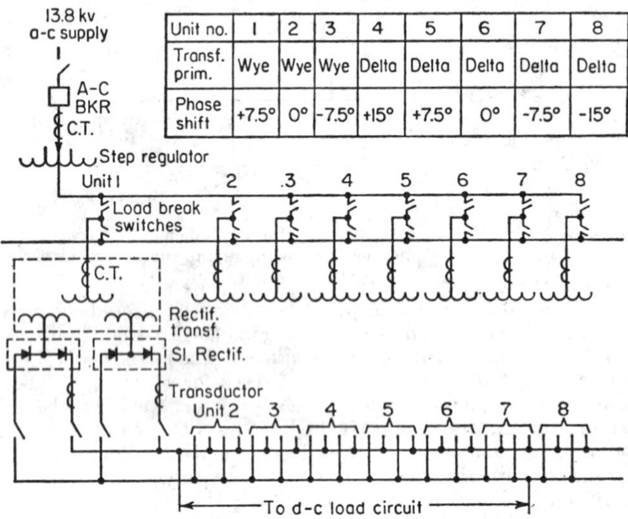

Unit no.	1	2	3	4	5	6	7	8
Transf. prim.	Wye	Wye	Wye	Delta	Delta	Delta	Delta	Delta
Phase shift	+7.5°	0°	-7.5°	+15°	+7.5°	0°	-7.5°	-15°

FIG. 12-42. Single-line diagram of a silicon-rectifier power supply for a high-current electrochemical load.

Static converters are used also to supply d-c power, usually at 275 V, for mine haulage. Low-headroom portable units mounted on mine cars are generally employed. Higher voltages are used for some surface mines.

41. Adjustable-speed and Reversing D-C Motor Drives. There is a large and increasing application of these drives, with static converters, for machine tools and many industrial processes. The introduction of thyristors (SCRs) broadened this application field. Probably the most demanding and largest application is for the main and auxiliary drives in metal-rolling mills. The largest of these is for hot-strip mills.

Most of the steel hot-strip finishing mills built in the United States since about 1945 are supplied by static converters. A modern mill of this type has six or seven rolling stands with a total horsepower of about 60,000 or more. The converter kilowatts rating is about 0.8 times the horsepower. They are designed to carry frequently repeated loads of 175% and occasional 200% loads. The d-c voltage is usually about 700 V and is regulated to 175% load current. Because of the high speed of modern mills, they have to be slowed down periodically for threading the strip into the coiler; this is effected by regenerative braking of the mill motors with an inverter or by dynamic braking with a resistor (Ref. 41, Par. **79**).

A more recent application of static converters is for hot reversing mills such as are used for rolling plates, blooms, or slabs. The work is passed back and forth through the rolls, and fast reversal of the motors is required. Typical load-swing ratings of the converters are 225% of rated current frequently repeated and occasional 275%. Reversing drives are used also for some cold mills and for mill auxiliaries such as screw-down and roll-out-table drives. For the heavy-duty auxiliaries, load swings of three to four times rated current may be applied.

The basic circuit for a reversing drive is shown in Fig. 12-43*a*. It consists of two controlled rectifier circuits *A* and *B*, one for each direction of motor current. Each circuit with the transformer can be operated by phase control over the complete voltage range of rectification and inversion. The controls are so designed that the rectifier elements of one circuit are blocked before the other starts conducting. For the large drives, two or more converters are usually connected in parallel to obtain a 12-phase waveshape. Circuit breakers and reactors are connected in the d-c circuit. The motor voltage, current, and speed for a reversing sequence are indicated in Fig. 12-43*b*. (Par. **79**, Ref. 42, pp. 40 and 59.)

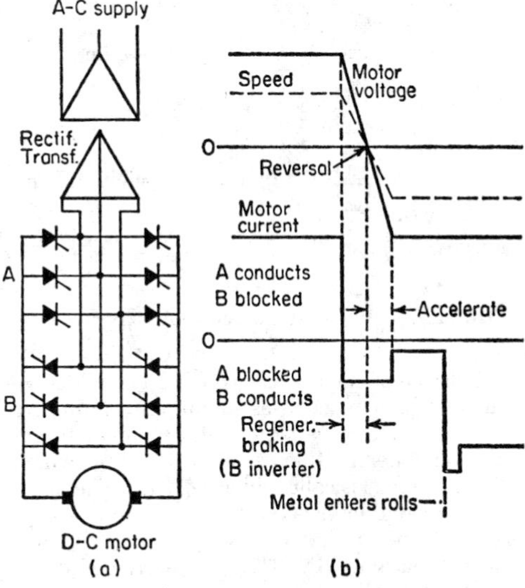

FIG. 12-43. (*a*) Power-circuit diagram of a reversing motor drive; (*b*) diagram of reversing sequence.

42. Other Applications. Static converters are used to supply d-c power at constant voltage (usually 250 V) in industrial plants for cranes, conveyors, machine tools, and many other needs. Unit ratings range up to 3,000 kW. They have a 2-h rating of 125% and a 10-s rating of 200%. Diodes are generally used. The d-c voltage can be regulated with saturable reactors, where necessary.

Other well-known applications of static converters are:

a. High-voltage d-c power transmission (see Sec. **14**).

b. High-voltage power supplies for r-f transmitters, and high-voltage d-c experimental installations.

c. Field excitation for d-c motors and synchronous machines.

d. D-c supply for electrolytic plating, cleaning, anodizing.

e. Power supplies for research laboratories. For some high-energy particle accelerators, a power supply with controlled rectifiers is used for pulsing the guide magnet. The converter is operated alternately as power rectifier and inverter to bring the current up and then bring it down by regenerating the stored energy (Ref. 44, Par. **79**).

SEMICONDUCTOR RECTIFIERS

43. Diodes. A diode is a device consisting of one or more rectifying junctions mounted in an encapsulation with two terminals. A cross section of a typical silicon diode is shown in Fig. 12-44. The majority of silicon medium- and high-current diodes are manufactured with a threaded stud base which serves both as a terminal and as a means of attachment to a cooler or heat sink. The other terminal is commonly a cable lug on the end of a flexible lead. Diodes are usually made with cathode base, although some manufacturers offer both polarities. Other, later designs of large diodes and thyristors are provided with flat-surface terminals instead of the stud and flexible lead. They are designed for clamping to flat surfaces of heat sinks. The increased thermal-contact area of the flat terminals increases the heat-dissipation capability.

The silicon wafer is sliced from a single crystal of extremely pure silicon. The wafer

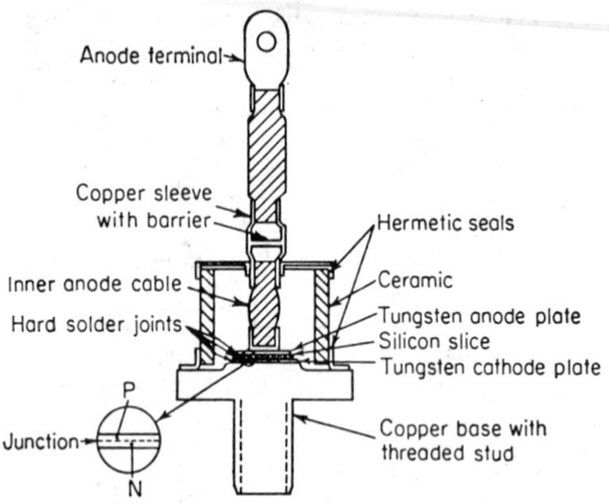

Fig. 12-44. Typical high-current silicon diode.

with the anode and cathode plates is called a cell. An enlarged section is shown at the left of Fig. 12-44. The rectifying action takes place at the junction, as explained in the following paragraphs.

The atoms of pure germanium and silicon have four valence electrons which are used in the bond between the atoms in the crystal lattice. If a minute quantity of an element with five valence electrons such as arsenic, phosphorus, or antimony is added and its atoms enter into the crystal structure, the fifth valence electron (not required for the bond) becomes available for conduction. This makes the semiconductor n-type, because of the negative charge of the free electrons. The added element is called a donor. When a three-valence element, such as aluminum, indium, or gallium, is added to a semiconductor crystal, one valence bond of its atoms is unfilled. These electron vacancies are called holes and behave like positive charges, which makes the semiconductor p-type; the added element is called an acceptor. The addition of these elements is called "doping." If a semiconductor crystal is doped to make one side n-type and the other side p-type, the p-n junction has rectifying properties, as illustrated in Fig. 12-46. Reference 2 of the Bibliography, Par. **77**, is an early definitive treatise on this subject.

When a cell is connected in a circuit so that the anode is positive with respect to the cathode (forward-biased), the voltage gradient causes the flow of electrons from the n region across the junction toward the anode, while the holes from the p region flow in the opposite direction. The flow of these charge carriers constitutes a current from anode to cathode, which by convention is opposite to the direction of electron flow.

When the polarity on the cell is reversed, the mobile charge carriers are drawn away from the junction and conduction stops. A schematic representation of a reverse-biased junction is shown in Fig. 12-45.

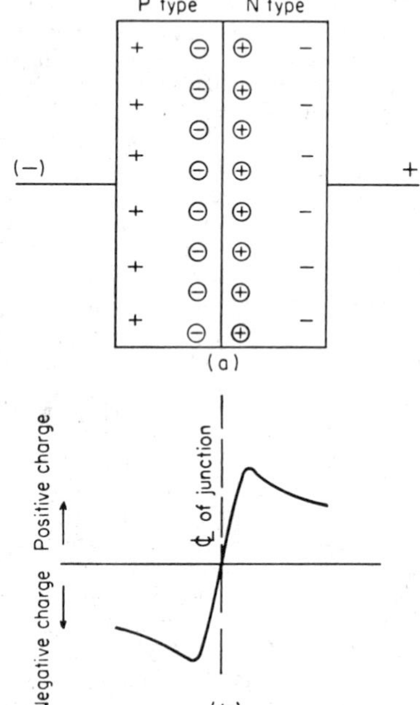

Fig. 12-45. Reverse-biased p-n junction.

The unencircled minus and plus signs represent the free electrons in the n region and holes in the p region. The impurity atoms are bound in the crystal lattice and cannot move. Therefore, the crystal lattice is left with an excess positive charge in the vicinity of each donor atom and an excess negative charge in the vicinity of each acceptor atom. These bound charges, shown circled, set up a space charge with a voltage distribution as shown in Fig. 12-45b. It is this space charge which blocks the flow of electrons and holes. Germanium diodes were perfected before silicon but have now been largely superseded by silicon diodes because of the higher temperature and voltage capability of the latter. Germanium is still used extensively for transistors, however.

44. *V-I* Characteristics. Figure 12-46

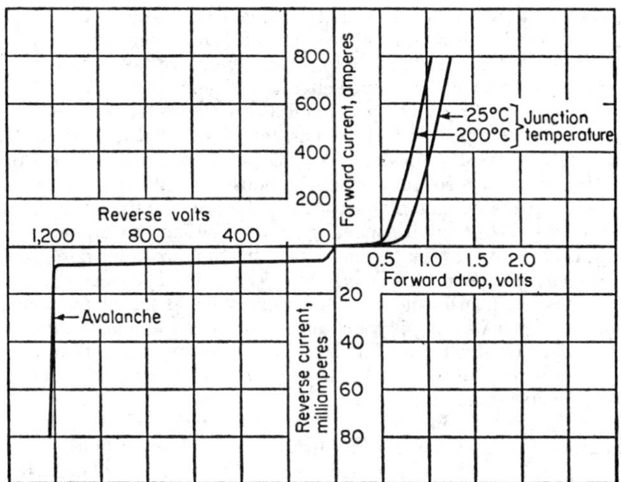

Fig. 12-46. Reverse and forward *V-I* characteristics of a typical 250-A silicon diode.

shows the forward and reverse characteristics of a typical silicon diode. Different scales are used for the forward and reverse portions. In the forward direction practically no forward current flows until approximately 0.5 to 0.75 V is applied. This is called the "threshold" voltage. From then on the forward drop is roughly linear. In the reverse direction very little current flows until a voltage exceeding rated voltage is applied, when there is a steep increase of current. This is called "avalanche breakdown" if the current is distributed over the junction area, and no damage will be done if the generated losses do not cause excessive heating. However, if the current is concentrated in a small area owing to a crystal defect or an edge breakdown, the diode may be permanently damaged. So-called "controlled-avalanche" diodes are designed to be self-protecting against overvoltages, provided that the energy to be dissipated is within specified limits.

45. Transient Voltage Protection. Ordinary semiconductor diodes must be protected against reverse overvoltage transients because even overvoltages of extremely short duration (e.g., less than 1 μs) can destroy them.

The necessary protection can be provided by proper application of nonlinear resistors such as selenium surge limiters across the diodes or, in double-way rectifier circuits, across the d-c terminals. Capacitors may also be used to limit transient voltages. For example, in the case of a double-way rectifier a capacitor bank of proper size across the d-c output will perform the same function as the voltage limiter. Enough resistance must be connected in series with the capacitance to provide critical or near-critical damping. Neither of these protective schemes can be depended upon in practice to limit the transient peaks to much less than two times the normal circuit-imposed reverse voltage. Therefore, it is common practice to select diodes having a transient peak reverse voltage rating of two to three times the normal reverse voltage imposed by the circuit.

A ubiquitous type of voltage transient encountered by the power-supply designer is the "commutation spike" due to the "hole-storage" effect. In simple terms, before a diode can block, the many free charges accumulated during forward conduction must be swept out. During this sweeping out, substantial reverse current flows for a very short time. When sweeping out is complete, the diode begins to block, thereby rapidly reducing the reverse current to zero. This current-chopping action results in a high voltage spike appearing across the diode owing to the energy stored in the circuit inductance by the reverse current. This is a very fast, relatively low-energy transient, and therefore suppression means must be located physically very close to the diode. Usually small capacitors with series resistors are located adjacent to the diodes.

46. Series Operation. In high-voltage rectifiers it is sometimes necessary to connect two or more diodes in series. Since reverse leakage currents vary considerably even among diodes of the same type and rating, reverse voltage will not divide equally unless the diodes are matched for reverse characteristics or other steps are taken. Voltage division may be forced by shunting each diode with a resistor which draws a reverse current several times the highest leakage current of any of the diodes. This will balance the voltage during the normal blocking period. However, the recovery time following conduction will also vary among diodes so that the first diode to recover will tend to block all the voltage momentarily. To prevent this, it is common practice to shunt each diode with a small capacitor also. These same capacitors then protect against the hole-storage transients mentioned in the previous paragraph. Another scheme for forcing voltage division that has seen limited use is the voltage-divider transformer. This is a transformer having as many equal secondary windings as there are series diodes. Each winding is then connected across one of the diodes (or groups of parallel diodes). Controlled-avalanche diodes need no protection when connected in series. It is necessary only that the sum of all the avalanche voltages exceed the circuit-imposed reverse voltage.

47. Parallel Operation. Frequently the current rating of a power supply is too high to be handled by one diode per circuit element. Two or more diodes must then be connected in parallel. Unbalance between parallel paths can be caused by unequal forward voltage characteristics of the diodes and by unequal resistance and reactance of the bus bars or cables in the parallel paths. Current sharing can be forced by placing a relatively high equal resistance in each parallel path. Some power supplies are built in a "coaxial" arrangement where the diodes and heat sinks are arranged in a circle and return cables or buses are arranged to be coaxial with the center conductor. This makes the resistances and reactances of all parallel paths equal.

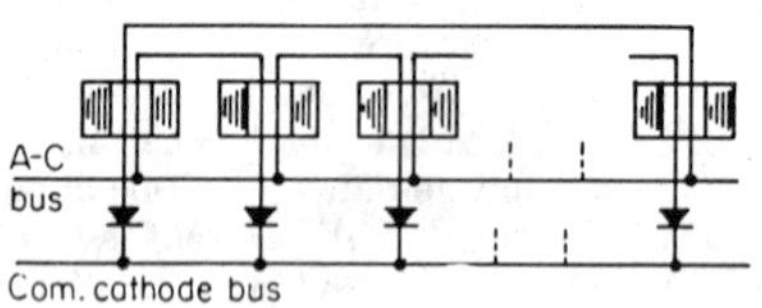

Fig. 12-47. Paralleling reactor arrangement.

Another method of forcing current balance is the use of "paralleling reactors," sometimes called current-balancing transformers, as shown in Fig. 12-47. Each reactor is a laminated magnetic core linked in opposing polarity by the anode currents of two diodes. If the two diode currents become unequal, the current difference excites a magnetic flux which induces a voltage in such a direction as to try to equalize the currents. It should be pointed out that the cores must be designed not to saturate at the highest expected current.

If reactors are inserted in the parallel paths with reactance large compared with the impedance variance between paths, they will force good current sharing, especially since their inherent resistance will also help substantially in forcing balance.

48. Cooling. The maximum junction operating temperature of most silicon power diodes is in the neighborhood of 200°C or slightly less. This limit is imposed not by the silicon but by other materials such as the solder used in the bonds or by internal stresses caused by thermal expansion. (The limitation for germanium devices is much lower—less than 100°C.) In order to dissipate the forward losses in a high-current

Table 12-3. Typical Thermal-resistance Values of High-current Silicon Diodes

Avg. current rating, A*	θ_{SC}, °C/W†	
	Typical	Max.
100	0.45	0.55
160	0.25	0.30
250	0.15–0.20	0.17–0.24

* For 120° conduction.

† 120° or 180° conduction.

diode without exceeding the allowable junction temperature, it is necessary to mount the diode on some kind of heat sink. A flat plate of aluminum or copper of sufficient area can dissipate the heat by radiation and natural convection. However, the same current or more can be handled with a saving in space by mounting the diode on a multiple-finned heat sink such as shown in Fig. 12-48, especially if forced-air cooling is used. A wide range of heat sinks, designed for either forced-air or convection cooling of standard semiconductor devices, are commercially available. Even more current can be carried by a given diode if it is mounted carefully on a copper or aluminum heat sink that is cooled by water. However, water cooling of semiconductor rectifiers has not received wide acceptance because of the operating and maintenance problems associated with water-cooling systems, especially where electrically insulating water connections must be made, as is almost always the case in rectifier cooling systems.

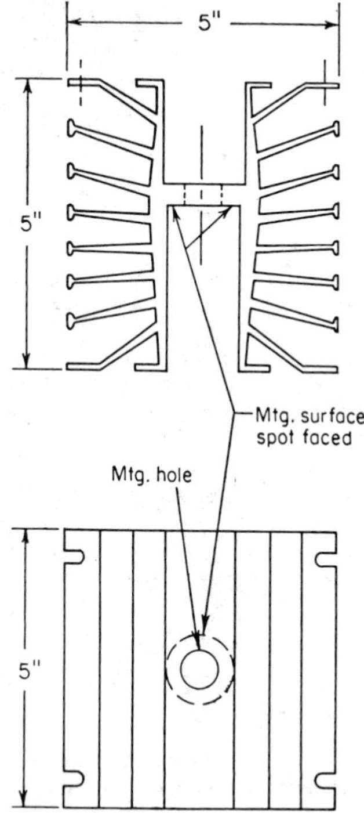

FIG. 12-48. Typical air-cooled heat sink.

49. Thermal Calculations. The ultimate temperature rise of an air-cooled silicon diode above ambient air when operating under constant load is given by

$$T_{JA} = P(\theta_{JC} + \theta_{CS} + \theta_{SA}) \qquad (12\text{-}45)$$

where T_{JA} = average ultimate temperature rise in degrees centigrade of the junction above ambient air, P = average forward power dissipation in watts at the given current and conduction angle (this can be taken from the diode data sheet), θ_{JC} = thermal resistance in degrees centigrade per watt between junction and case as given in the diode data sheet (Table 12-3 shows typical values), θ_{CS} = thermal resistance in degrees centigrade per watt between the diode case and heat sink (see Table 12-4 for typical values), and θ_{SA} = thermal resistance from heat sink to ambient for the given air flow (this can be obtained from the heat-sink data sheet; Table 12-5 shows data for heat sinks similar to Fig. 12-48).

Following is an example illustrating the use of Eq. (12-45).

Example. A three-phase double-way rectifier having four 250-A silicon diodes paralleled in each leg is to deliver a constant d-c current of 2,000 A to an inductive load. The forward losses are shown in Fig. 12-49. Ambient temperature is 50°C. The diode manufacturer specifies a maximum junction operating temperature of 190°C. Determine heat-sink size and cooling-air velocity required.

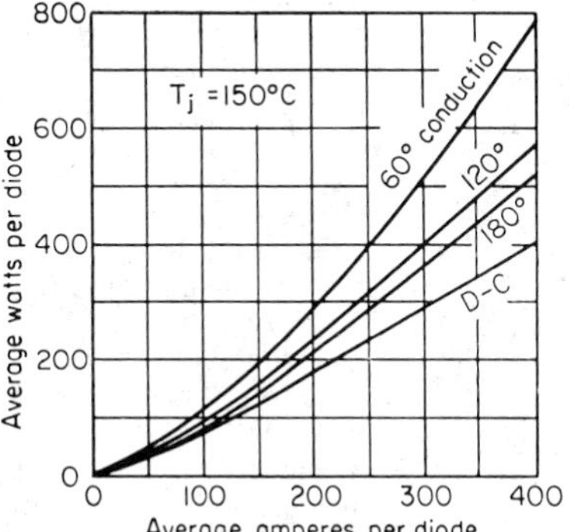

FIG. 12-49. Forward power dissipation of a typical 250-A silicon diode.

There are 3 times 4, or 12, diodes in each side of the circuit. Therefore, average current per diode is 2,000 ÷ 12 = 167 A. The estimated maximum unbalance is 10%. Therefore, the maximum average diode current is 184 A. From Fig. 12-49, the average loss per diode at 184 A read from the 120° conduction curve is approximately 215 W. From the device data sheet, θ_{JC} is 0.18°C/W. The device has a 1⅝-in hexagonal base. From Table 12-4, θ_{CS} is 0.08°C/W. It has been determined from previous calculations

Table 12-4. Thermal Resistance of Diode Case to Heat Sink for High-current Stud-mounted Silicon Rectifiers

Hex size across flats, in	θ_{CS}, °C/W	
	Dry	With silicone grease
$\frac{11}{16}$	0.30	0.25
$1\frac{1}{16}$	0.3	0.15
$1\frac{1}{4}$	0.16	0.10
$1\frac{5}{8}$	....	0.08

that, for the maximum expected d-c fault, the junction temperature will increase 30°C before the circuit breaker interrupts. Therefore, the maximum allowable junction operating temperature under load is 190 − 30, or 160°C. From Eq. (12-45),

$$160 - 50 = 215(0.18 + 0.08 + \theta_{s-A})$$

Then $\theta_{s-A} = {}^{110}\!/_{215} - 0.26 = 0.25$°C/W.

Table 12-5 shows that there is a choice of a 235 in³ heat sink, similar to Fig. 12-48, with an air velocity of 200 ft/min, a 150 in³ heat sink with 500 ft/min, or a 64 in³ heat sink with 1,000 ft/min.

50. Transient Thermal Impedance. Loads of short duration cause transient temperature rises in a semiconductor rectifier. The maximum diode transient junction temperature for any load, I_1, following a continuous load I_0 is given by

$$T_{J(t_p)} = (P_1 - P_0)\theta_{JA(t_p)} + T_{J0} \qquad (12\text{-}46)$$

where P_0, P_1 = diode forward loss in watts corresponding to I_0 and I_1, $\theta_{JA(t_p)}$ = junction to ambient "transient thermal impedance" corresponding to t_p, t_p = duration of pulse, and T_{J0} = junction temperature just prior to application of transient load.

Reference 39 (Par. **79**) shows how to use the superposition principle to calculate peak junction temperature rise for any duty cycle. The "transient thermal impedance" of a body is defined by $\theta_{(t)} = T(t)/P$, where $T(t)$ is the temperature rise in degrees centigrade above ambient at time t and P is a constant heat input to the body in watts applied at $t = 0$ and with the body at room temperature. The transient thermal impedance $\theta_{JC(t)}$ from junction to case is given on diode data sheets. For high-current diodes with loads of very short duration (typically less than 5 s) this value may be used to calculate junction rise above ambient, because the diode case and heat sink do not have time to heat up appreciably. For longer times, however, $\theta_{JA(t)}$ from junction to ambient must be used. The value of $\theta_{JA(t)}$ is a characteristic of a complete rectifier

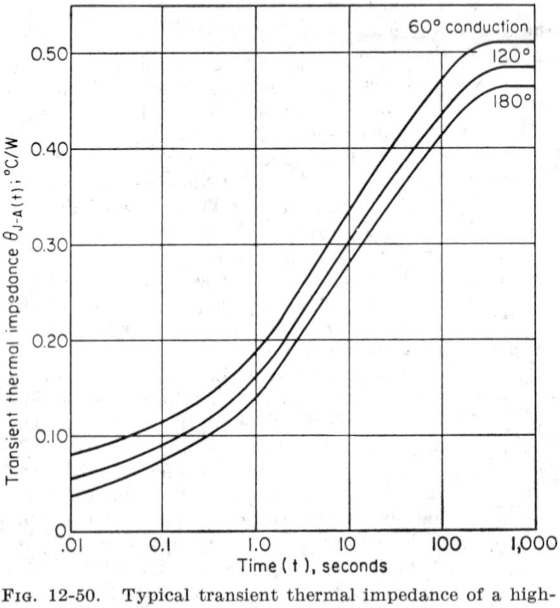

Fig. 12-50. Typical transient thermal impedance of a high-current silicon diode on a forced-air-cooled heat sink.

Table 12-5. Typical Thermal Resistance of Heat Sinks Similar to Fig. 12-48

Heat sink Vol. $= D \times W \times H$, in³	Thermal resistance, °C/W			
	Natural convection at 150 W	Air velocity		
		200 ft/min	500 ft/min	1,000 ft/min
64	0.63	0.44	0.28	0.20
150	0.46	0.29	0.20	0.16
235	0.38	0.21	0.15	0.13
350	0.33	0.18	0.115	0.10

assembly consisting of diode, heat sink, and cooling air flow. The equivalent thermal circuit is too complicated for ready theoretical solution. It must ordinarily be obtained by test. This is done by passing any constant forward current I_F through a diode mounted on the specified heat sink and with specified air flow. Diode case temperature is read at intervals during the heating cycle, and temperature rise is plotted as a function of time. The junction temperature is obtained by adding the calculated junction-to-case rise for the given load to the measured case temperature. Then for any time t, $\theta_{JA(t)} = (T_{J(t)} - T_A)/P$, where P is the forward loss in watts corresponding to I_F. (See Fig. 12-50.)

51. Overload Protection. Because of their very small thermal capacity, proper protection of semiconductor rectifiers against overloads and fault currents requires high-speed protective equipment and careful coordination of protective devices with the rectifier thermal characteristics. Two problems must be considered: first, protection of the rectifier against excessive overloads or short circuits in the load circuit; second, protection against failure of a semiconductor device.

Figure 12-51*a* is a schematic diagram of a typical large diode rectifier supplied from a

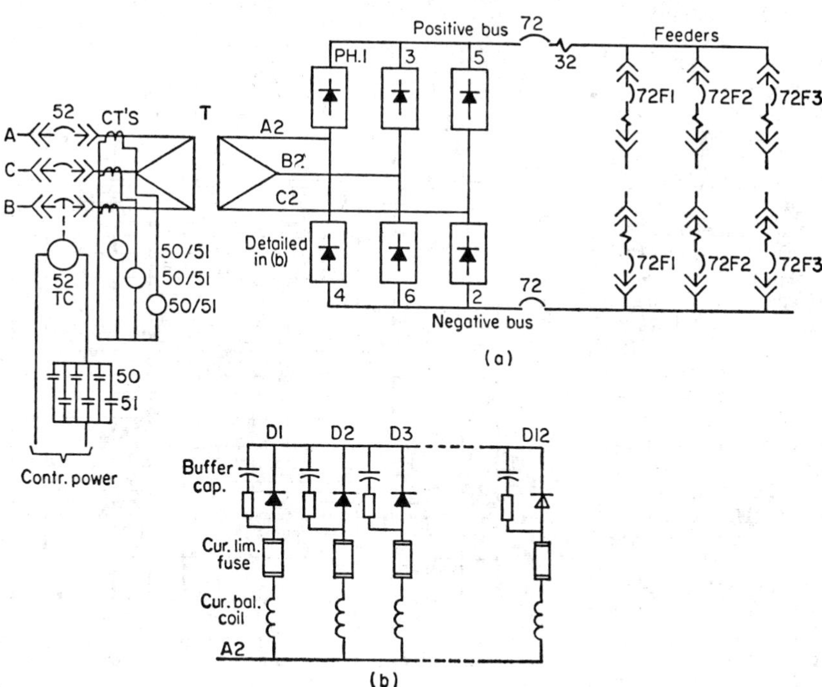

Fig. 12-51. Schematic diagram of silicon rectifier unit.

medium-voltage 3-phase 60-c system. Device 52 is typically a medium-voltage air-magnetic circuit breaker; 52TC is the circuit-breaker trip coil; 51*a*-*b*-*c* are induction overcurrent relays with instantaneous trip attachments 50. The rectifier is a 6-phase double-way circuit. Each leg has 12 diodes in parallel as detailed in Fig. 12-51*b*. A current-balance coil and a current-limiting fuse are connected in series with each diode. These are high-speed fuses developed especially for protection of semiconductor rectifiers. The waveshape of the "let-through" current of these fuses when subjected to a heavy fault is shown in Fig. 12-52. Also shown in Fig. 12-51 are the buffering capacitors and resistors for suppression of voltage surges. Device 72 is a main d-c circuit breaker with a reverse current trip device 32. Devices 72 F-1, 2, and 3 are d-c feeder circuit breakers either manually or electrically operated, with self-contained overcurrent trips.

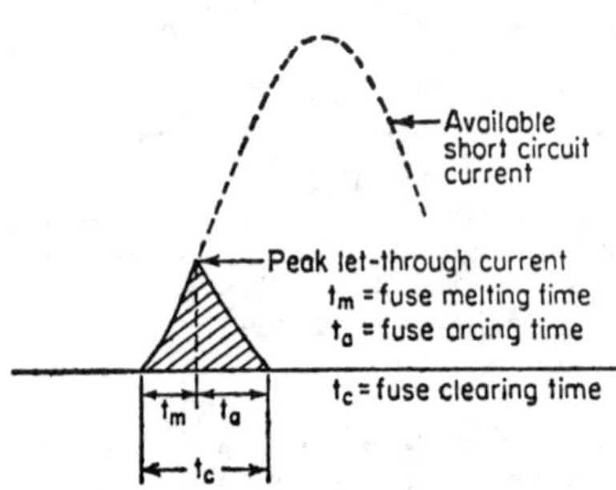

Fig. 12-52. Current-limiting-fuse let-through current.

Many variations of this arrangement are used. For smaller rectifiers, up to about 300 kW, the supply is often 480 V, in which case 52 is usually a manually operated air circuit breaker with built-in overcurrent trip. If there are no other power sources or regenerative loads which can feed current into a rectifier internal short circuit, the main d-c circuit breaker 72 may be omitted.

52. D-C Overloads and Faults. Figure 12-51*a* is used here to explain the coordination problem. Main d-c circuit breaker 72 is usually made to trip on reverse current only. The feeder breakers 72F are ordinarily set lower than devices 50/51 so that they will isolate an overload feeder without tripping 52. However, a heavy fault, whether on the load side of a feeder circuit breaker or on the rectifier bus, must be cleared by 52 before any diodes or diode fuses are damaged. To ensure this, a coordination chart such as shown in Fig. 12-53 must be plotted which contains the time-current characteristics of the diodes, diode fuses, and circuit breaker 52. The curve for 52 is the time-current characteristic of 50/51 plus the interrupting time of the breaker. For standard low- and medium-voltage circuit breakers having interrupting times of 3 to 8 c at 60 Hz, it is quite common

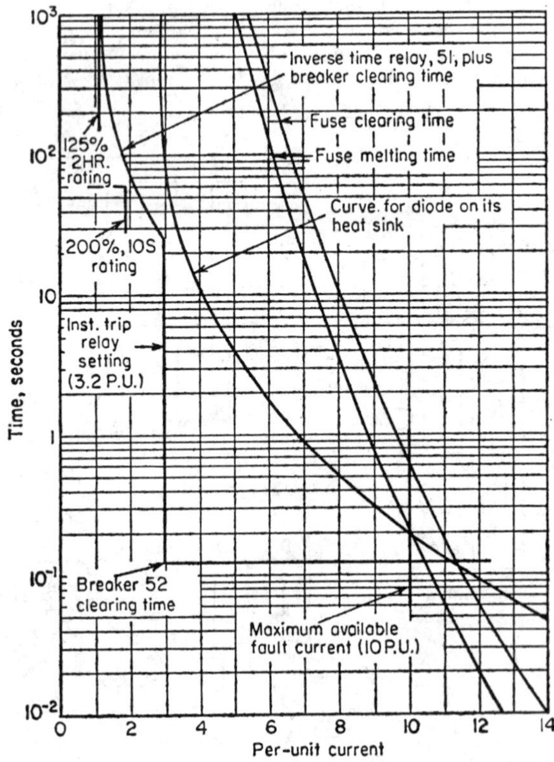

Fig. 12-53. Overload coordination curves for a silicon rectifier unit.

that more diodes are required in parallel to withstand a through fault than are required to carry the rated continuous and overload currents.

The curve for each circuit component is plotted as a function of its per unit current corresponding to rated full load. Thus, for the diode curve, one per unit is the average current in amperes that flows through one parallel path at rated load, with allowance for unequal current sharing. The procedure for plotting a point on the diode curve is as follows:

1. Assume a value of diode per unit current I_{dpu}.
2. Convert this to maximum average amperes per diode, I_{d1}, by multiplying by rated amperes per parallel path and by the estimated unbalance factor.
3. From the diode data sheet find the forward loss P_1 in watts corresponding to I_{d1}.
4. The curves are plotted assuming that all overloads start with the diodes at junction temperature t_{Jo} corresponding to one per unit load at maximum ambient. Maximum allowable junction temperature is 190°C. Therefore maximum allowable transient thermal impedance is

$$\theta_{JA(t_1)} = \frac{190 - t_{Jo}}{P_1 - P_0},$$

where P_0 is the forward loss at one per unit current.

5. A transient thermal impedance curve has been previously constructed for the specified diode and heat sink with specified cooling. From this curve determine t_1 corresponding to $\theta_{JA(t_1)}$.
6. Plot the point I_{dpu}, t_1.

In plotting the fuse melting curve from published curves the equivalent rms value of the nonsinusoidal anode current is used.

A faster protective scheme, not much used in this country, uses a high-speed "bypass" switch between the transformer and rectifier assembly. When a fault occurs on the d-c side, this switch applies a 3-phase short circuit on the secondary side of the transformer, thus removing all fault current from the rectifier. The short circuit is then cleared by the primary circuit breaker. Short-circuiting switches are available which operate in 1 to 4 ms, depending on the type.

53. Diode Failure (Short Circuit). After the rating and number of parallel paths of diodes and fuses have been determined from the d-c overload and fault coordination curve, it is necessary to check coordination for a diode failure in the rectifier.

In Fig. 12-51*a*, if a device in one leg becomes short-circuited, a line-to-line short circuit is established across the transformer secondary during the normal reverse blocking period of that leg. It is usually desired that the diode fuse of the faulted diode shall clear before circuit breaker 52 trips, so that service will not be interrupted. In addition it is necessary to ensure that the faulted diode fuse clears the fault before any of the good diodes or fuses are damaged. If the melting time of the fuse is less than about 0.01 s, it can be assumed that all the heat is stored in the diode junction and the I^2t rating of the diode may be used. This rating is usually given on the application data sheet. Coordination can then be checked from the inequality

$$3H_{fm} \leq [n/(1+k)^2]H_d \tag{12-47}$$

where H_{fm} is the fuse-melting I^2t constant in (amperes)^2 seconds. This can be calculated from the fuse-melting time curve by using any point for which t is 0.01 s or less. The factor 3 before H_{fm} is based on the pessimistic assumption that the fuse total clearing time is three times the melting time. This is a conservative but realistic assumption based on experimental data. H_d is the diode rated (amperes)^2 seconds, n is the number of parallel paths per phase, and k is the maximum deviation between highest diode current and average diode current in per unit of average diode current.

When there are as many parallel paths as in Fig. 12-51, the required selectivity is no problem. Three or more parallel paths when properly fused will result in the faulted device fuse blowing alone. It is quite commonly specified that large rectifiers be able to operate under specified conditions with one fuse blown in each leg. In that case the above coordination checks must be made for $n - 1$ parallel paths. Smaller power supplies with only one parallel path usually have only three fuses placed in the a-c line.

54. Diode Failure Monitoring. Large power supplies with many semiconductor devices in parallel are usually provided with a visual indication of a device failure. Since devices almost always fail by short circuit, which will blow the device fuse, it is convenient to monitor the fuse condition. This can be done by sensing the voltage across the fuse or the voltage across the device itself. The simplest system consists of a small indicating light connected, through a suitable voltage-dropping resistor, across the fuse or the rectifying device.

For essential service sometimes enough diodes are provided so that the power supply can furnish full rated power with one fuse blown in each leg of the rectifier. Monitoring systems are then sometimes specified which will perform some or all of the following functions:

1. Give an audible alarm each time the first fuse in any leg blows even if the alarm has been previously silenced manually after indicating the blowing of a fuse in another leg.
2. Indicate visually which legs have blown fuses.
3. Provide a separate signal to trip a circuit breaker and drop load if a second fuse blows in any leg.

55. Thyristors (Silicon-controlled Rectifiers). The construction of a thyristor, or SCR, is similar to that of the diode except that there are four layers (*p-n-p-n*) instead of two (*p-n*) and three junctions instead of one. Figure 12-54 is a cross section of the active parts of an SCR. The anode (usually the case in SCRs) is connected to the bottom *p* layer and the cathode to the top *n* layer. The control, or gate, lead is connected to the *p* layer, second from the top. The bottom junction (see Fig. 12-54) is called the anode junction. The middle junction is the control junction, and the top junction is the cathode junction. Figure 12-55 is a cross section of a typical completed SCR showing one way in which the gate lead may be brought out of the sealed package.

If reverse voltage (anode negative with respect to cathode) is applied to the *p-n-p-n* structure, it is at once apparent that the cathode and anode junctions will be reverse-biased and block. If a forward voltage (anode positive) is applied, the anode and cathode junctions will be forward-biased but the control junction will be reverse-biased. Therefore, if the gate lead is left floating, the device will block in both directions. However, if enough current is injected at the gate, the control junction potential barrier will be neutralized and the SCR will turn on, or "fire." The turn-on process starts at the gate connection and propagates across the entire junction. There is a time associated with this propagation that is called "turn-on time." This is typically 2 to 10 μs. Excellent detailed explanations of the gate turn-on mechanism may be found in Ref. 3, Par. **77**, or in the handbooks of SCR manufacturers.

The voltampere characteristic of a typical SCR is shown in Fig. 12-56. Note that the device can also be fired by exceeding the forward breakover voltage V_{BO}. This is not harmful provided that breakover occurs at less than the maximum rated forward blocking voltage. After the SCR is fired, its forward characteristic is similar to that of a diode. Its reverse characteristic is also like that of a diode. Once an SCR has

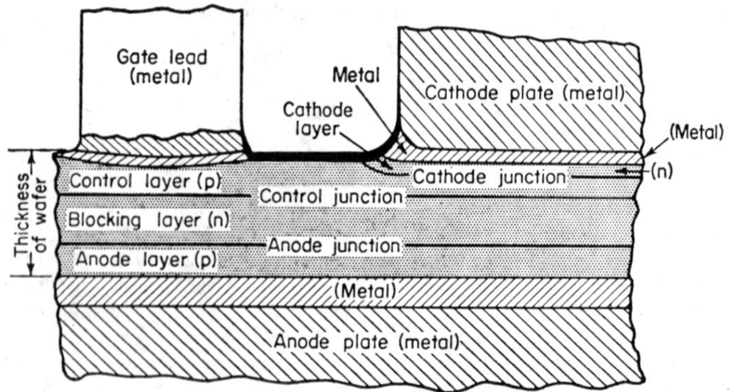

Fig. 12-54. Cross section of active part of a silicon-controlled rectifier.

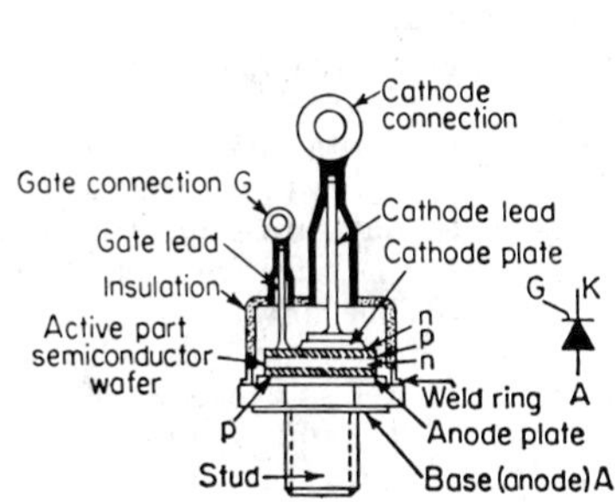

FIG. 12-55. Cross section of a silicon-controlled rectifier.

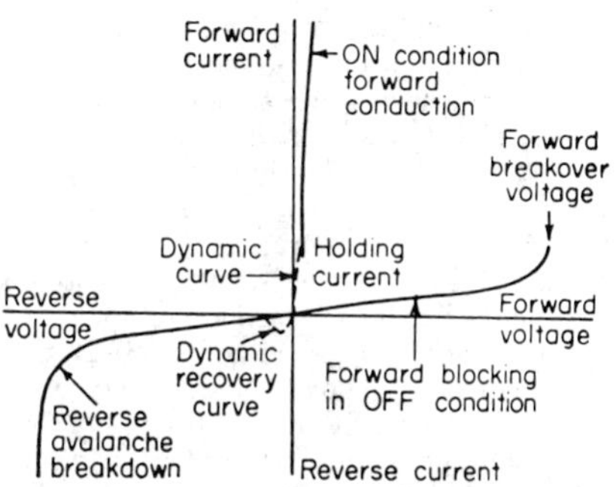

FIG. 12-56. Reverse and forward *V-I* characteristics of a typical silicon-controlled rectifier.

been fired, it will remain on, even though the gate current is removed, until the anode current is reduced by circuit action to less than a certain minimum value called the "holding current." This ability to control the firing of an SCR is used to control the output voltage of a rectifier (Par. **4**).

It should be pointed out that it takes a finite time (on the order of 15 to 100 μs) after anode current reaches zero for an SCR to recover its forward-blocking ability. This is called the "turnoff time."

56. Firing Characteristics. Figure 12-57 shows the gate firing requirements of a typical design of high-current SCR. Curve *A* is the voltage-current relationship of a gate with the highest resistance that will be encountered. Curve *B* is for the lowest gate resistance. The firing points of all SCRs of this type will fall within the small shaded area at any temperature between −65 and +125°C. The firing-circuit load line must be above and to the right of this area to ensure positive firing of all SCRs.

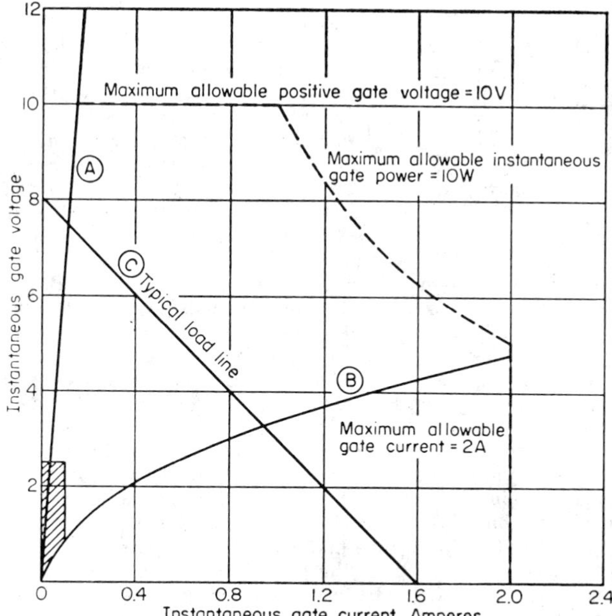

FIG. 12-57. Typical gate characteristic of a 235-A rms silicon-controlled rectifier.

Line *C* is a satisfactory load line which has a peak power of 3.2 W. This is well below the maximum allowable instantaneous power of 10 W. In designing the gating circuit the duty cycle of the gate pulse must be kept low enough so that the maximum allowable average gate power of the SCR is not exceeded.

57. Operating Limitations. (1) The maximum junction operating temperature of an SCR is 125 to 150°C depending upon the design. The SCR will withstand 190 to 200°C without permanent damage just as a silicon diode, but gate control is not guaranteed above the 125 to 150°C temperature. (2) An SCR is sensitive to voltage transients in the forward direction as well as the reverse. While forward breakover will not necessarily damage the SCR, the resultant circuit malfunction can cause damaging currents to flow. SCRs are given a maximum forward as well as a reverse voltage rating. (3) An SCR may be falsely fired by the rate of change, *dv/dt,* of applied forward voltage even though the peak voltage remains within rating. Such *dv/dt* firing will not damage the device, but the resultant circuit malfunction may produce damaging fault currents. SCRs are given a maximum *dv/dt* rating. (4) Since forward current flow starts at the gate connection when the SCR is gate-fired, an excessive rate of rise, *di/dt,* of anode current can cause local overheating around the gate connection. High-current SCRs are given a maximum *di/dt* rating.

58. Transient Voltage Protection. The same measures described previously to guard diodes against transient overvoltages are also used for SCRs. However, it is often necessary to place reactors in series with each SCR as well as capacitors shunting them. This arrangement can be designed to satisfactorily limit the applied *dv/dt* and *di/dt* during commutating transients and switching transients.

59. Series Operation. Series operation of SCRs requires the same shunting resistors and capacitors as series operation of diodes. In addition it is necessary to gate simultaneously two or more SCRs whose gates are at different potentials. In large power supplies, it is common to use separate pulse transformers, or separate secondaries on one transformer, to obtain the necessary isolation. Another method is so-called "slave firing," as shown in Fig. 12-58. Here when SCR1 fires, SCR2 is gated on by the charging current of the capacitor C_1.

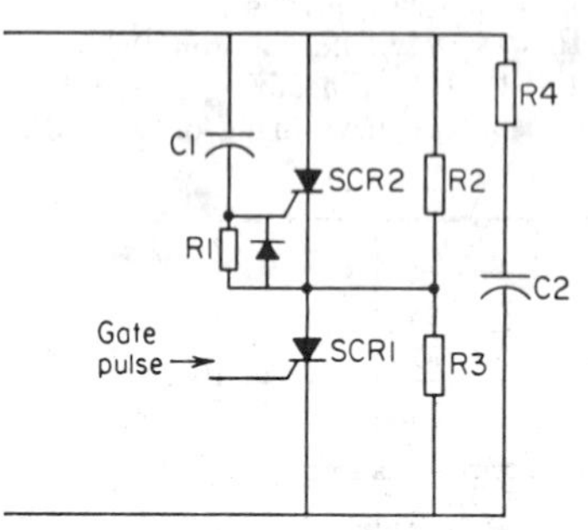

FIG. 12-58. Slave firing circuit for silicon-controlled rectifier.

60. Parallel Operation. The same methods as enumerated for paralleling diodes can also be used for SCRs. However, the series-reactor method is preferred for SCRs for two reasons. First, the same reactors which aid current sharing can also serve for controlling *dv/dt.* Second, the difference in turn-on and turnoff time is an added problem in paralleling SCRs, and the series linear reactors are especially effective in mitigating this.

It is, of course, necessary to gate simultaneously all parallel SCRs. This can be done from a common gate pulse supply. However, it is advisable to insert equal resistance in series with each gate to ensure good current sharing by all the gates.

61. Cooling and Temperature Rise. The problems in cooling SCRs and calculating their temperature rise are exactly the same as described for diodes except that the junction maximum operating temperature is lower (125 to 150°C), and in some inverter applications switching losses may have to be considered.

62. Overcurrent Protection. The overcurrent protection of SCRs is similar to that for diodes. The lower temperature limit does not hold for through faults since the SCRs are not expected to maintain control. In case of an SCR failure, the good SCRs are expected to maintain control, but this is usually no problem in large power supplies, because the fault current is shared by a number of parallel paths.

One problem, peculiar to SCRs, is the possibility of abnormally high currents in case of spurious firing of an SCR. In the case of regenerative loads on a double-way rectifier such a failure when operating in the inverter mode may result in a dead short circuit across the load through the misfired device. It is desirable, therefore, to coordinate the I^2t ratings of the fuse and the SCR in series with it. However, this may require such a low fuse rating that it is not feasible because of additional parallel paths required to carry normal loads.

POOL-CATHODE MERCURY-ARC RECTIFIERS

63. Elementary Tube. An elementary mercury-arc tube is shown in Fig. 12-59. It has two main electrodes, an anode and a mercury-pool cathode, enclosed in a vacuum envelope. The cathode is caused to emit electrons. If the tube is connected into an electric circuit, electrons will flow from the cathode to the anode and into the external circuit when the anode is at a positive potential to the cathode. This constitutes a flow of current, and by convention the direction is from the anode to the cathode, opposite to the direction of electron flow. When the anode is made negative to the cathode, the electrons are repelled by the anode and the current flow ceases. The rectifying property is due to the fact that the anode does not emit electrons and can receive them only when it is positive to the cathode.

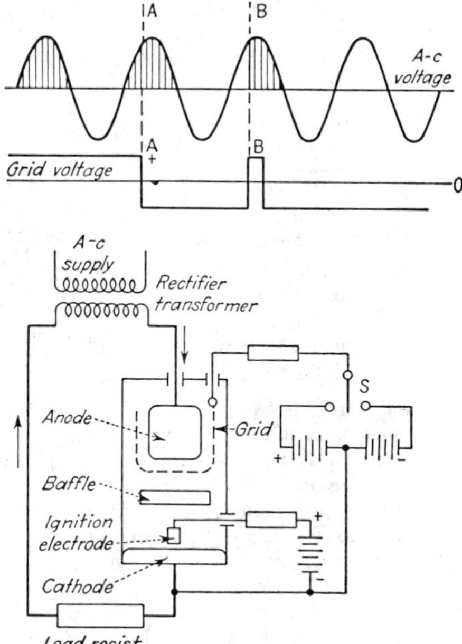

FIG. 12-59. Elementary mercury-arc rectifier tube and circuit.

Electrons are emitted at cathode spots on the mercury surface. These small, luminous spots have a current density of several thousand amperes per square centimeter and move irregularly over the surface. The emission is believed to be caused by a high voltage gradient, of the order of 1 million V/cm, produced by a cathode voltage drop of about 10 V applied over a very small space. A cathode spot can carry 15 to 30 A. At higher currents additional spots are created automatically.

A cathode spot can be initiated by breaking contact between a starting electrode and the mercury when current is flowing in the direction from the electrode to the cathode. It can be initiated also by passing a current pulse through an ignitor of high-resistivity material immersed in the mercury (Ref. 21, Par. **79**).

64. Ionization. When a tube is conducting, the movement of electrons from the cathode to the anode causes ionization of mercury vapor. In this process, electrons are removed from the mercury atoms by excitation or collision with other electrons, leaving the atoms with a positive charge. These are called **positive ions,** and they are an essential element in the conduction of current, because they neutralize the negative space charge of the electrons and permit their flow to the anode with a relatively low arc-drop voltage. The ionized conducting medium is called **plasma.**

In the space between the anode and cathode during conduction, there is a continuous process of ionization and recombination of ions and electrons to form neutral atoms. Mercury atoms are also condensed on the cooled tube surfaces and returned to the cathode pool. When conduction is terminated, the positive ions are neutralized within a short time interval by recombination with electrons. This process is called **deionization.**

A baffle is generally provided between the anode and cathode to control ionization and to shield the anode from the mercury vapor emitted at the cathode.

65. Control Electrode. A control grid, with openings for the passage of current, is placed in front of the anode. If the grid is made negative with respect to the cathode, it will repel electrons and prevent the start of conduction, even when there is cathode emission and the anode is positive to the cathode. When the grid voltage is changed from negative to positive, conduction can start, provided that the other conditions are satisfied. After conduction has started, the grid loses control and can regain it only after the anode current is brought to zero by the action of the external circuit and after the space in the vicinity of the grid has become deionized.

The grid action is illustrated by the diagram at the top of Fig. 12-59; the shaded areas indicate conduction. The tube thus operates as a unidirectional switch which can be closed by grid control. The same result can be obtained by controlling the initiation of a cathode spot with an ignitor, because conduction can start only when there is emission at the cathode.

66. Arc-drop Voltage. When a tube is carrying current, there is a voltage drop between the anode and cathode terminals. The product of this voltage and the current is the power loss in the tube. The arc drop consists of the anode drop (3 to 6 V), the drop in the arc path (0.2 to 0.5 V/in), and the cathode drop (9 to 10 V). A typical arc-drop curve of a 500-A single-anode rectifier tube is shown in Fig. 12-60. High-voltage tubes with an extensive grid or baffle system have a higher arc-drop voltage.

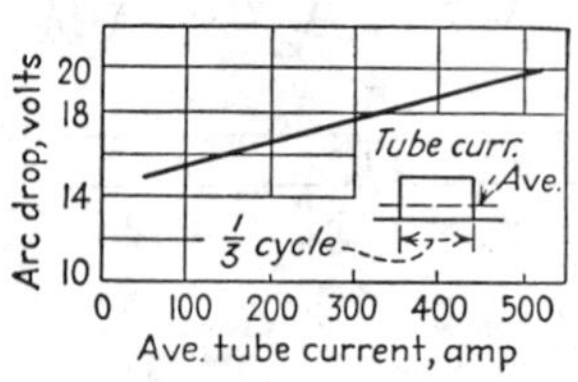

FIG. 12-60. Typical arc-drop curve of single-anode rectifier tube rated at 500 A, 800 V.

67. Arc-backs and Their Causes. If a cathode spot is accidentally formed on an anode when it is negative to the cathode, it will lose its rectifying action and reverse current will flow. This type of fault is called an *arc-back*. In normal operation arc-backs are infrequent and, with adequate protection, cause no damage. Their occurrence is random and unpredictable, but the probability of occurrence increases as the current and voltage are increased. This is one limitation on the rating of a mercury-arc tube, particularly for d-c voltages above 300 V.

Arc-backs are believed to be caused by the residual ionization in the vicinity of an anode at the conclusion of its conducting period. As shown in Fig. 12-8, the anode-to-cathode voltage reverses at that point, and the positive ions remaining from the conducting period are pulled in by the anode, resulting in a very small negative current. Under some adverse conditions, this ion current can initiate a cathode spot. The ion current increases at higher load currents and voltages. Phase control, which increases the initial reverse voltage, also increases the ion current and the arc-back probability. Arc-backs may be caused also by abnormal operating conditions, such as poor vacuum or overtemperature (Refs. 22 and 26, Par. **79**).

68. Tube Types. From the design point of view, there are two principal classifications of mercury-arc rectifiers, the multianode and single-anode types. In a multianode type a number of main anodes, usually 6, 12, or more, are assembled in a single vacuum enclosure with a common cathode. The anodes are connected to a multiphase rectifier transformer, so that there is a continuous main arc to the cathode. A single-anode tube consists of one main anode and a cathode in an individual vacuum enclosure, and 6, 12, or more tubes are used for a rectifier unit.

There is another type of multianode rectifier tube, which is used for high-voltage d-c power transmission. The anodes are connected in parallel to one phase of the transformer, so that the main arc is intermittent, as in a single-anode tube. Six tubes are required for a 6-phase converter (see Sec. **14**).*

There are two types of single-anode mercury-arc tubes, the ignitron and excitron. In an ignitron, the cathode spot is initiated before the conducting period of each cycle by a pulse of current passed through an ignitor which is partly immersed in the cathode mercury. The timing of the pulse is used for controlling the anode firing point in the cycle. In an excitron, a cathode spot is maintained continuously on the mercury, when in operation, by a small pilot arc from an excitation anode, and the anode firing is controlled by a grid. Both types have been made in pumped and sealed construction. The pumped tubes have a demountable, gasketed top cover and a vacuum valve for connection to an evacuating system.

* The term "valve" has been used for tubes in European technical literature published in English and in the English version of IEC (International Electrotechnical Commission) publications. It is a term of long standing in Great Britain for electron tubes. This term has crept into usage in the United States for mercury-arc tubes applied for d-c power transmission, because of the European source of this technology.

69. Materials and Processing. The anodes, grids, and baffles of industrial tubes are usually made of graphite. The enclosure and other parts are made of steel. Glass and ceramics have been used for insulating bushings. Rectifier tubes with glass envelopes have also been made, mostly in Europe. The water-cooled parts of sealed tubes are made of stainless steel to prevent corrosion, which would cause diffusion of hydrogen ions through the steel and impair the vacuum in the tube.

In operation, mercury-arc tubes have to be free of gases or vapors other than mercury vapor, and the gases have to be removed from the anodes, grids, and other parts which operate at high temperatures. Pumped tubes are degassed with current, which is increased gradually to or above the rated current. Sealed tubes are usually degassed by subjecting them to a high temperature. The internal parts are heated by radiation from the tube enclosure. For some tube designs, a treatment at an elevated temperature is followed by current degassing. Tubes which are opened for reconditioning have to go through a degassing process.

70. Ignitron. In Figs. 12-61 and 12-62 are shown two designs of sealed ignitron tubes. The ignitor, used for cyclic ignition of the cathode spot, is a pencillike rod of a refractory material, such as boron carbide, partly immersed in the mercury. The auxiliary, or holding, anode is used for maintaining an auxiliary arc at the cathode during part of a cycle, after the ignitor is pulsed. This stabilizes operation when the load current is too small for a stable cathode spot. It is needed also to start conduction in a 6-phase double-way circuit and for operation of a 6-phase double-Y-circuit below the light transition load.

The various tube components are marked in the figures. The anode heater in Fig. 12-62 prevents condensation of mercury on the anode assembly when operating at light or no load. The walls of an ignitron tube, which are connected solidly to the cathode, are not shielded from the arc. They carry part of the current by ion collection and intermittent cathode spots, which disappear at the end of the conducting period.

In the lower part of Fig. 12-64 is shown a typical ignitor firing circuit for a pair of tubes, which are connected to opposite phases of the rectifier transformer. A capacitor is connected to one phase of the excitation transformer in series with a phase-shifting reactor. When the capacitor voltage reaches a certain value, the iron-core reactor in series with the ignitors saturates and a pulse of current is discharged from the capaci-

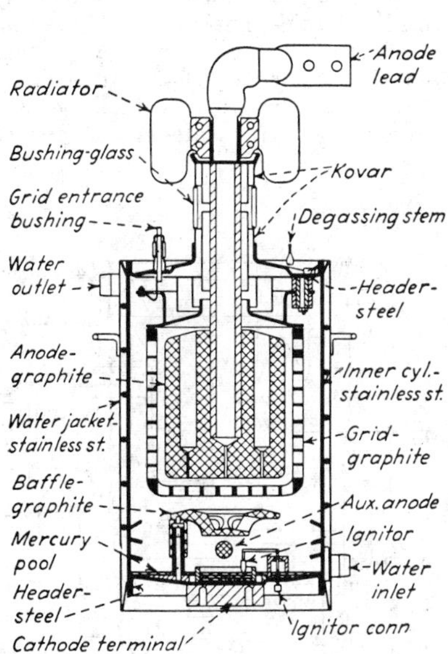

Fig. 12-61. Cross section of a large sealed ignitron tube with a glass-Kovar anode bushing. (*Westinghouse Electric Corporation.*)

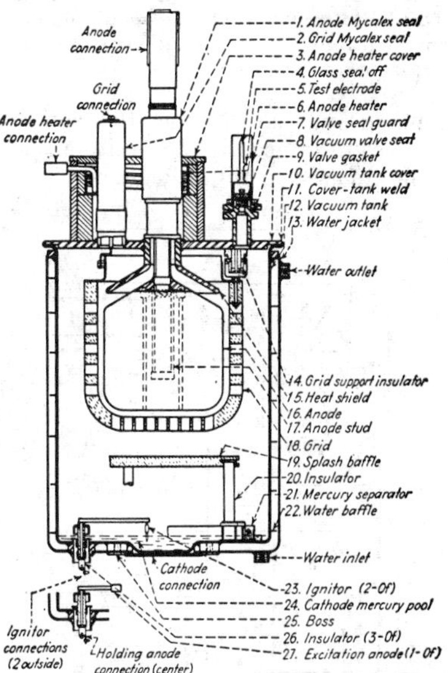

Fig. 12-62. Cross section of a large sealed (pumpless) ignitron tube, with a Mycalex anode bushing. (*General Electric Company.*)

tor through an ignitor to initiate the cathode spot. The two ignitors receive a current pulse during alternate half-cycles. The diodes direct the pulse of each polarity to the corresponding ignitor and prevent current reversal. The d-c winding of the phase-shifting reactor is used to vary the phase position of the current pulses for phase control of the rectifier d-c voltage. The grids are connected to a grid transformer. When auxiliary anodes are used, they are supplied from a similar transformer.

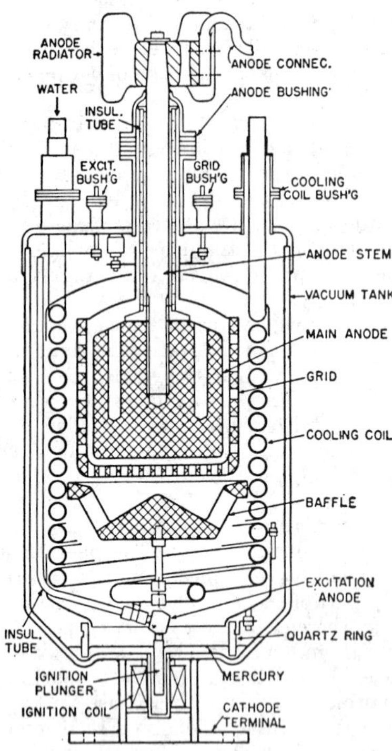

Fig. 12-63. Cross section of a large sealed excitron tube with vitreous-enamel anode bushing. (*Allis-Chalmers Manufacturing Company.*)

71. Excitron. In Fig. 12-63 is shown a cross section of a sealed excitron tube. The internal cooling coil is insulated from the tank and has two functions: (1) it provides passages for the cooling water; (2) it shields the tank walls from the arc, to prevent initiation of cathode spots on the walls. The vitreous-enamel bushings used for the insulated terminations are made by coating steel disks with vitreous enamel and fusing them together.

The excitation anode is supplied with direct current in series with the ignition coil, as shown in Fig. 12-65. Before the circuit is energized, the anode is connected to the cathode by the ignition plunger. When energized, the coil pulls down the plunger, drawing an arc from the anode. When the plunger submerges, the arc is transferred to the mercury and is maintained as long as the circuit is energized.

In Fig. 12-65, the control grids are shown connected to a grid transformer. Phase control is effected by a variable bias voltage between the center tap of the transformer winding and the cathode; this shifts the point at which the grid-to-cathode voltage changes from negative to positive. For many applications, grid voltages with a steep front are used to obtain more precise control.

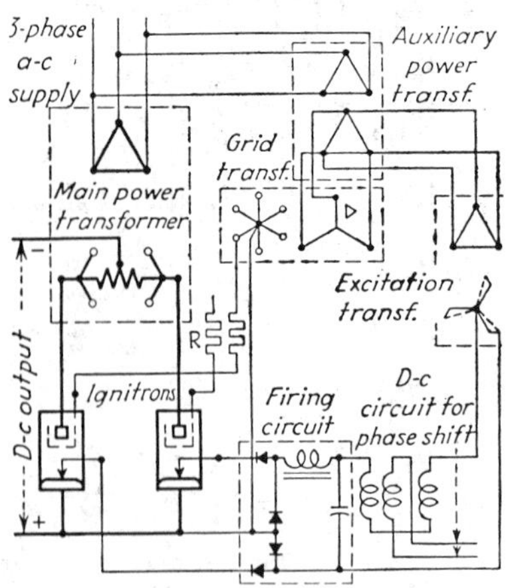

Fig. 12-64. Typical circuit diagram of ignitron rectifier.

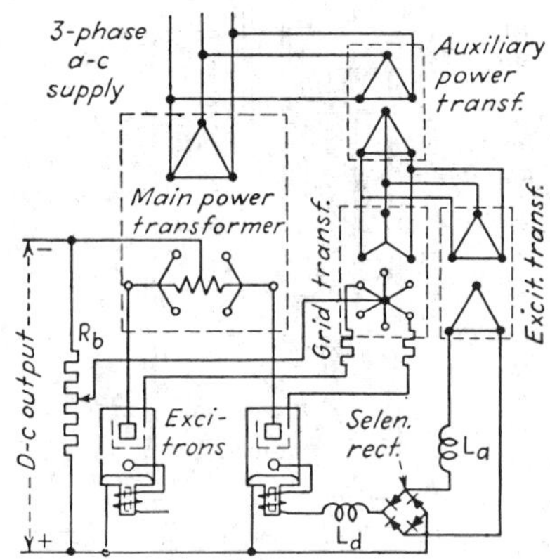

Fig. 12-65. Typical circuit diagram of excitron rectifier.

In some excitron-tube designs, the cathode is insulated from the tank (instead of an insulated cooling coil) to prevent transfer of cathode spots to the tube wall.

72. Cooling and Evacuating Auxiliaries. Besides the excitation equipment, a mercury-arc rectifier is provided with a cooling system and evacuating equipment (when pumped tubes are used). These are shown schematically in Fig. 12-66.

The cooling water is circulated between the tubes and a heat exchanger; there the heat is transferred to raw water from a water supply. The flow of raw water is regulated by a thermostatically controlled valve. The cooling equipment is usually kept at the rectifier potential and is connected to the water supply through an insulating pipe. Water-to-air heat exchangers are used in many installations. Sodium chromate is added to the circulated water to prevent corrosion.

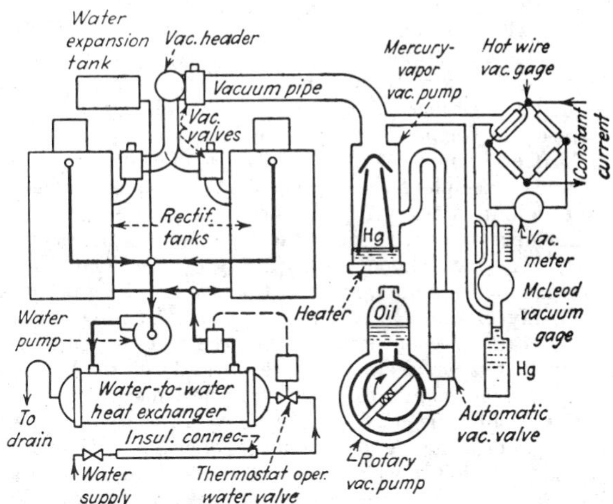

FIG. 12-66. Vacuum and cooling systems of a pumped mercury-arc rectifier.

The evacuating equipment of a pumped rectifier consists of two vacuum pumps connected in tandem and vacuum measuring devices. In the mercury-vapor pump, mercury vapor rises through a nozzle and is deflected downward to compress gases received from the rectifier. The mercury vapor is condensed. The gases pass to a rotary vacuum pump, where they are compressed and discharged to the atmosphere through oil.

In the McLeod vacuum gage, a sample of gas in a calibrated bulb is compressed by mercury into a small glass measuring tube, and the pressure is read on a scale. The hot-wire vacuum gage operates on the principle that heat transfer from a heated wire at low pressure is proportional to the pressure. A wire exposed to the vacuum system is supplied with constant heating current. The change in the wire resistance, caused by a change of its temperature, is a measure of the pressure and is indicated on a calibrated meter through a Wheatstone bridge. The normal operating pressure is usually below 3 microns (1 micron = 0.001 mm Hg pressure).

OTHER TYPES OF RECTIFIERS

In the following are brief descriptions of other types of rectifying devices that have been used for static power converters. They have been largely superseded by silicon diodes or thyristors, but many continue to operate in existing equipment.

73. Thermionic Gas Tubes. There are two types—diodes, also called "phanotrons," and thyratrons, which have a control grid. In both types, electron emission is obtained from a heated cathode. Mercury vapor, derived from a drop of mercury, or an inert gas is used as the source of positive ions during conduction. The operating principle is similar to that of a mercury-arc tube.

Alternating- to direct-current converters with high-voltage phanotron tubes have been used to supply plate power for r-f oscillators. Thyratron tubes have been used mostly in converters for variable-speed motor drives.

74. Copper-Oxide and Selenium Diodes. These have been called "metallic" rectifiers. They are polycrystalline semiconductor devices as contrasted with the monocrystalline silicon and germanium diodes. In the metallic rectifiers, the rectifying junction (also called barrier layer) is between the semiconductor and a metal. The direction of current is from the semiconductor to the metal.

In a copper-oxide rectifier cell, the metal is copper and the semiconductor is cuprous oxide, which is produced on the copper by oxidation at a high temperature. The rated reverse voltage per cell is 4 to 8 V rms. The operating temperature limit is about 60°C.

In a selenium cell, selenium is deposited on a metal plate and heat-treated. A eutectic-alloy counterelectrode is deposited over the selenium. The rectifying junction is between the selenium (semiconductor) and the counterelectrode. The rated reverse voltage per cell is about 20 to 40 V rms; some higher-voltage diodes have also been made. The maximum operating temperature is 85 to 100°C.

Metallic rectifiers are thermally rated devices. They are subject to aging, which increases the forward voltage drop.

75. Germanium Diodes. This is a monocrystalline semiconductor diode, and its operating principle is the same as that of a silicon diode (Par. **43**). The junction temperature limit is below 100°C, as compared with about 200°C for silicon diodes.

The development of commercial germanium diodes preceded silicon diodes by several years. They were superseded quickly by the silicon diodes because of the higher permissible junction temperature of the latter and higher peak reverse voltage ratings.

76. Mechanical Rectifier. In a mechanical rectifier, a set of synchronously operated contacts is connected between the rectifier transformer and the d-c load circuit. Although it is not a static device, it is placed in that category because the operation of the contacts simulates static rectifier elements.

Each contact is closed at the start of a conducting period and is opened at the end of a commutating period. The contact must be opened when the current reaches zero and before it reverses, to avoid a sustained short circuit between phases. A saturable reactor is connected in series with each contact. When the current drops to near zero, the reactor is desaturated and absorbs the voltage difference between the phases, with a very small exciting current. This provides an interval or step of near zero current, which allows time for the contact to open the circuit.

The 6-phase double-way circuit has been used for the mechanical rectifier. It was made for current ratings of about 5,000 A at 400 V d-c. It has a higher efficiency than mercury-arc rectifiers. Practically all the applications have been for electrochemical service.

BIBLIOGRAPHY

77. Books

1. MARTI, O. K., and WINOGRAD, H. "Mercury-arc Power Rectifiers"; New York, McGraw-Hill Book Company, 1930.

2. SHOCKLEY, W. "Electrons and Holes in Semiconductors"; Princeton, N.J., D. Van Nostrand Company, Inc., 1950.

3. GENTRY, F. E., GUTZWILLER, F. W., HOLONYAK, N., and VON ZASTROW, E. E. "Semiconductor Controlled Rectifiers"; Englewood Cliffs, N.J., Prentice-Hall, Inc., 1964.

4. BEDFORD, B. D., and HOFT, R. G. "Principles of Inverter Circuits"; New York, John Wiley & Sons, Inc., 1964.

5. SCHAEFER, J. "Rectifier Circuits"; New York, John Wiley & Sons, Inc., 1965.

78. Standards and Committee Reports

11. American Standards Institute. Standards for:

a. Pool-cathode Mercury-arc Power Converters, USAS C34.1-1958.

b. Pool-cathode Mercury-arc Rectifier Transformers, USAS C57.18-1964.

c. Semiconductor Power Rectifiers, USAS C34.2.

12. AIEE Committee Report, Inductive Coordination Aspects of Rectifier Installations; *Trans. AIEE*, 1946, p. 417.

13. AIEE Committee Report, Mercury-arc Power Converters in North America; *Trans. AIEE*, 1948, p. 1031.

14. AIEE Committee Report, Protection of Electronic Power Converters; *Trans. AIEE*, 1950, p. 813.

15. AIEE Committee Report, Inductive Coordination Aspects of D-C Systems Supplied by Rectifiers; *Trans. AIEE*, 1951.

79. Miscellaneous Literature

21. Slepian, J., and Ludwig, L. R. A New Method for Initiating the Cathode of an Arc; *Trans. AIEE,* 1933, p. 693.

22. Kingdon, K. H., and Lawton, E. J. The Relation of Residual Ionization to Arc-back in Thyratrons; *Gen. Elec. Rev.,* November, 1939, p. 474.

23. Marti, O. K., and Taylor, T. A. Wave Shape of 30- and 60-phase Rectifier Group; *Trans. AIEE,* 1940, p. 218.

24. Evans, R. D. Harmonics and Load Balance of Multiphase Rectifiers; *Trans. AIEE,* 1943, p. 182.

25. Steiner, H. C., Zehner, J. L., and Zuvers, H. E. Pentode Ignitrons for Electronic Power Converters; *Trans. AIEE,* 1944, p. 693.

26. Winograd, H. Development of Excitron Type Rectifier; *Trans. AIEE,* 1944, p. 969.

27. Willis, C. H., Kuenning, R. W., Christensen, E. F., and Bedford, B. D. Design of an Electronic Frequency Changer; *Trans. AIEE,* 1944, p. 1070.

28. Christensen, E. F., Willis, C. H., and Herskind, C. C. Analysis of Rectifier Circuits; *Trans. AIEE,* 1944, p. 1048.

29. DeBliux, E. V. High Voltage Rectifier Circuits; *Gen. Elec. Rev.,* April, 1948, p. 42.

30. Winograd, H. Electronic Frequency Changer Used as Nonsynchronous Tie between A-C Power Systems; *Trans. AIEE,* 1953, Pt. I, p. 263.

31. Dortort, I. K. Extended Regulation Curves for Six-phase Double-way and Double-wye Rectifiers; *Trans. AIEE,* 1953, Pt. I, p. 192.

32. Witzke, R. L., Kresser, J. V., and Dillard, J. K. Influence of A-C Reactance on Voltage Regulation of Six-phase Rectifiers; *Trans. AIEE,* 1953, Pt. I, p. 47.

33. Witzke, R. L., Kresser, J. V., and Dillard, J. K. Voltage Regulation of 12-phase Double-way Rectifiers; *Trans. AIEE,* 1953, Pt. I, p. 689.

34. Dortort, I. K. Current Balancing Reactors for Semiconductor Rectifiers; *Trans. AIEE,* 1958, Pt. I, p. 452.

35. Gutzwiller, F. W. Overcurrent Protection of Semiconductor Rectifiers; *Elec. Mfg.,* April, 1959, p. 106.

36. Gutzwiller, F. W. Rectifier Voltage Transients: Cause, Detection, Reduction; *Elec. Mfg.,* December, 1959, p. 167.

37. Chirgwin, K. M., Stratton, L. J., and Toth, J. R. Precise Frequency Power Generation from an Unregulated Shaft; *Trans. AIEE,* 1960, Pt. II, p. 442.

38. McMurray, W., and Shattuck, D. P. A Silicon Controlled Rectifier Inverter with Improved Commutation; *Trans. AIEE,* 1961, Pt. I, p. 531.

39. Gutzwiller, F. W., and Sylvan, T. D. Power Semiconductor Ratings under Transient and Intermittent Loads; *Trans. AIEE,* 1961, Pt. I, p. 699.

40. Ogden, H. S. A Unique Propulsion System for Electric Multiple-unit Cars for Philadelphia-area Commuter Service; *IEEE Trans. on Appl. Ind.,* 1964, p. 329.

41. Roumanis, P. J. Silicon Controlled Rectifiers on Steel Mill Drives; *Iron Steel Engr. Year Book,* 1964, p. 909.

42. Record of the IEEE Industrial Static Power Conversion Conference, 1965; *IEEE Publ.* 34 C 20. (Contains the papers presented at the conference.)

43. Duff, D. L., and Ludbrook, A. Reversing Thyristor Armature Dual Converter with Logic Crossover Control; *IEEE Trans. on Ind. Gen. Appl.,* 1965, p. 216.

44. Winograd, H., Millikin, A. D., and Hedin, R. A. Power Supply for Ring Magnet of Zero-gradient Synchrotron; *IEEE Trans. on Nucl. Sci.,* April, 1966, p. 46.

MOTOR-GENERATORS

By J. F. Sellers

80. Application Conditions. The use of motor-generator sets is limited to those cases where one or more special requirements inhibit the use of more economical static rectifiers. In some instances of larger kilowatts output, the a-c motor can be connected directly to the power line without a transformer, which is always required for a static converter. In this case the cost of static and rotating conversion becomes more nearly equal. In other cases, d-c voltage range, effect of peak loads on power-line stability, power factor, efficiency, installation, or maintenance cost will have major influence on the choice.

81. Type of Drive Motor. In converting alternating to direct current, three different types of a-c motor may be used, the choice depending on the specific load or power-line criteria:

Synchronous motor.

Induction motor, with or without flywheel.

Wound-rotor induction motor with flywheel.

The synchronous type is used when power-factor correction is needed and where sudden heavy loads will not cause a serious system disturbance.

Simple squirrel-cage induction motors are ordinarily used for motor-generator sets of 500 kW and smaller. In special cases such as power supply for excitation of a large a-c generator, a flywheel is added to the rotating system, so that, in case of a power outage of not more than 2.0 s, the motor can recover speed with 70% of rated voltage applied. In some cases the flywheel is made large enough to provide 250% of rated exciter output for 0.25 s, with a-c power off, and the induction motor is specified capable of recovering speed with 70% voltage applied.

The inertia value of the flywheel, plus the inertia of the a-c and d-c machines, makes up the inertia constant H of the total motor-generator set. Usually the flywheel, when used, is more than 90% of the total inertia, for all types of applications. The inertia constant is

$$H = (WK^2)(0.231)(\text{r/min})^2/(\text{kW} \times 10^6) \qquad (12\text{-}48)$$

where WK^2 = inertia of the entire motor-generator set, in lb·ft², r/min = full load speed, and kW = exciter continuous rating.

The required H factor, which is usually specified as 5 for exciter motor-generator sets, can be determined as follows:

The energy given up by the total inertia when speed decreases from r/min_{fl} to r/min_t occurs in t s, where

$$t = H[1 - (\text{r/min}_t \div \text{r/min}_{fl})^2] \quad \text{(seconds)} \qquad (12\text{-}49)$$

when constant rated kilowatts load is on the exciter. Figure 12-67 shows the variation of time t s relative to $\text{r/min}_t \div \text{r/min}_{fl}$, for different values of inertia factor H, when constant full load is maintained.

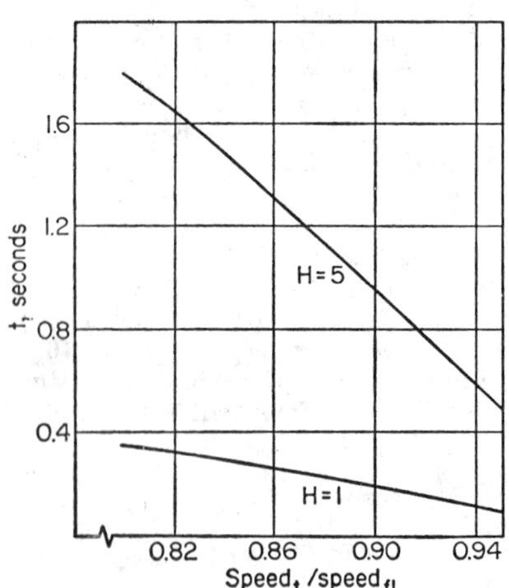

Fig. 12-67. Time vs. speed ratio for two values of M.

A wound-rotor induction motor with flywheel is used when loads of 200% of full load or higher are applied to the d-c generator for no more than 5 s and the power system would have a serious problem with this repeated demand. In this case the rotor winding is connected to a variable resistance, such as a liquid slip regulator. The flywheel is provided to keep the motor speed from falling below 80% of rated continuous load speed. The control of the variable resistance in the rotor circuit keeps the current drawn from the a-c line to not more than 100 to 125% of rated motor current. The integrated motor input will be the horsepower-seconds of the complete load cycle, plus the motor and generator load losses and idle loss of the motor-generator set.

Such sets are widely used for ore hoists and reversing mills in metal rolling. An analysis of the work-cycle load must be made, and an rms and average load value determined. The ratio of peak load to rms load determines the design criteria of the d-c generator, and the average load value plus losses determines the induction-motor rating. On special short-pulse load applications it has been possible to build flywheel motor-generator sets with d-c generators having peak loads of 400% of continuous current rating, driven by an induction motor having one-third the continuous generator rating.

The flywheel design is based on the practical limit of 20% maximum speed reduction at any point of the load cycle. The rim speed of a solid steel-plate wheel is approximately 25,000 ft/min at no-load speed, and the thickness is made to suit the required WK^2. This gives the lightest flywheel and most economical cost.

82. The Ward Leonard drive is the most commonly used d-c variable speed drive and can give a combination of constant torque and constant power output, when constant d-c current is used in the armature loop circuit. As shown in Fig. 12-68, the a-c motor *A* drives d-c generator *B*. The armature of this generator is connected in a closed loop with the armature of d-c motor *C*. With forced ventilation on *C*, the system can be operated continuously, at rated current, from 100% V down to 10% of rated volts. An adjustable speed regulator is usually provided with this d-c system in place of generator field rheostat *E*, which works in the generator field circuit to provide a constant preset speed. The d-c motor field is set at its maximum rated value, to produce the maximum possible torque per ampere of armature current. The speed regulator is stable without hunting, over a generator voltage range of 10 to 100% of rated volts.

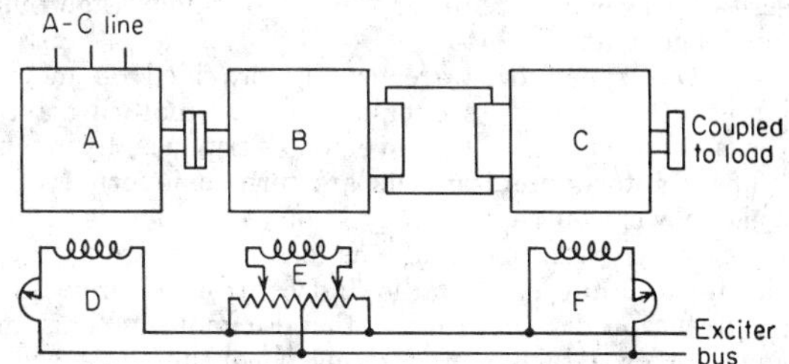

FIG. 12-68. Ward Leonard drive. (*A*) Drive motor; (*B*) d-c generator; (*C*) d-c motor; (*D*) a-c motor field rheostat; (*E*) d-c generator field rheostat; (*F*) d-c motor field rheostat.

The d-c motor may also be operated up to 200 to 600% of base speed by shunt field weakening with rheostat *F*. The speed regulator would continue slightly to adjust the generator voltage depending on load and field temperature to maintain constant preset motor speed in the field-weakening range.

Generator field rheostat *E* is usually a potentiometer type, or the equivalent, so that the voltage applied to the motor armature is reversible. In case the generator voltage becomes less than the counter volts generated in the motor, the armature current reverses and the system is regenerative, returning power to the a-c line. This regeneration is repetitive in a hoisting load cycle and in many applications for metal rolling. When very fast reversal of d-c motor speed is needed, a separate static excitation source is usually provided for each d-c motor and generator. This may provide field forcing of 7 to 10 times the maximum value of the field *IR* drop, and with relatively low-inertia motor armature, reversing time from forward base speed to reversed base speed is obtained in 1.5 s, without exceeding approximately 50% of continuously rated armature current. Shorter motor reversal time may be obtained by use of static rectifier power to the d-c motor armature, or a d-c generator with a laminated magnet frame. A solid generator magnet frame has induced eddy currents which will allow initially only a limited rate of voltage change, regardless of the amount of field forcing voltage applied.

83. Certain accessories and methods of construction have been found necessary on medium and large synchronous induction and flywheel motor-generator sets:

a. **All bearings of the motor-generator set should be insulated,** with a grounding strap at one bearing. This permits easy checking of the bearing insulation system.

b. **High-pressure oil lift** should be used on flywheel and drive motor bearings in starting. Because of high bearing loading and relatively long starting time, oil must be forced into the bottom of the bearing before starting; otherwise pulling of babbitt and overheating occur.

c. **Forced ventilation** is usually provided to each of the rotating elements. The flywheel has a close-fitting cover to reduce the windage loss. The cover has air openings at floor level, with an outlet at the top, to prevent overheating of the enclosed air.

d. **Temperature indicators** are usually embedded in the winding on the commutating coil of each d-c generator. These can be used for monitoring machine temperature or for sounding an alarm if the rms current loading is more than the load cycle for which the d-c generator was specified.

e. **Synchronous or induction motors are started** either across the a-c line or through a starting reactor, depending on the a-c system characteristics.

f. **Voltage regulators** are used to maintain the output voltage of the d-c generators, independent of overload or change in drive motor speed.

g. **Load division** between two or more d-c generators of a motor-generator set, connected to a common load, is best obtained by load-balancing windings in a controlled rectifier amplifier which monitors the shunt field supply to each generator. The input to each load-balance winding is from the voltage drop across the generator commutating field. Previously this was accomplished by cross-connected cumulative and differential series fields, which requires the use of much large bus bar and heavy series field coils on the generators.

h. **Overspeed trips** are usually provided on motor-generator sets of 300 kW and larger. This prevents overspeeding by motoring action of the d-c generator, should the a-c power fail. The overspeed trip usually opens the generator circuit breaker.

84. Motor-generator sets are termed inverted when they produce a-c power from a d-c power source. This type of conversion is used for specialized a-c requirements in the range of 20 kW and smaller. Many are used on shipboard and supply frequencies up to 400 c/s, for radio and control instruments. It is necessary to have a very accurate speed regulator on the d-c motor. This may be either a centrifugal-type governor or a tachometer giving a continuous speed signal into a controlled rectifier amplifier which regulates the shunt field strength of the d-c motor. A voltage regulator is also used to correct the rather poor inherent voltage regulation of small a-c generators.

85. Dynamotors. Another type of motor-generator set converts d-c voltage of one value to d-c voltage of a higher or lower value. The two functions of motor and generator may be combined into one armature core, in which case it is called a dynamotor (Fig. 12-69). The core has two separate windings, one connected to a commutator on one end of the core, and the other to a commutator at the other end of the core. A common magnetic field maintains a constant ratio of voltage between the input and output windings. The voltage regulation of the output winding is quite poor, but armature reaction is almost zero since ampere-turns of the motor winding current are practically equal and opposite to that of the generator winding current. The example shown has twice as many turns in one armature winding as the other, giving a 2:1 ratio between input and output voltage.

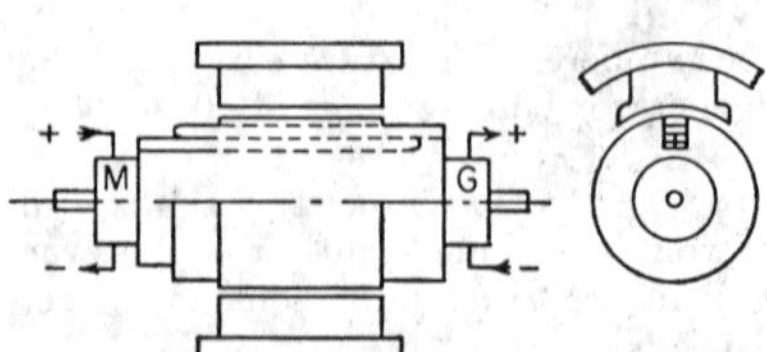

Fig. 12-69. Dynamotor with 2:1 ratio between input and output d-c voltages.

86. Balancer Sets. A d-c to d-c motor-generator set, which is used to provide fractional voltage to a 3-wire d-c system, is called a balancer set (see Fig. 12-70). The set is made up of two duplicate d-c machines, each capable of operating as a motor or a generator. The common point between the two machines is called the neutral line and carries the difference in current load between the two half-voltage systems. The machine which operates across the heavier loaded half-voltage system becomes a generator, and the function changes inherently from generator to motor with any change in unbalance load location. Each machine has one-half the unbalanced load in its armature, modified by the loss current required for the two balancer machines. The one with motor action will have one-half unbalanced amperes plus the loss amperes, and the generator will have one-half unbalance amperes minus the loss amperes, giving a probable actual difference of 25 to 30% in armature amperes. Likewise the neutral line voltage will not be exactly one-half the incoming

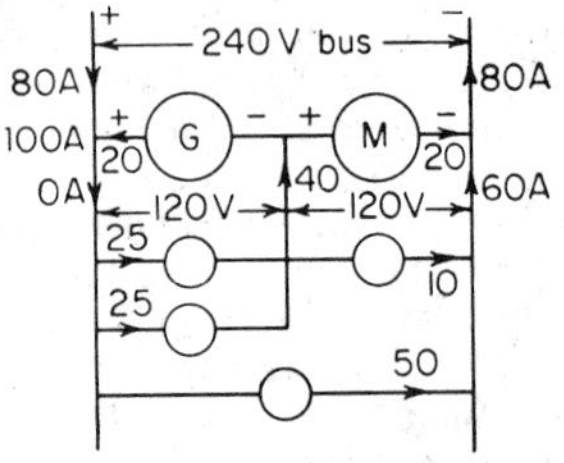

Fig. 12-70. Balancer motor-generator set with assumed unbalanced load.

d-c line voltage, owing to the direction of the internal *IR* drops in the armatures; so a small series field is added to the main poles of each machine, which is cumulative when generating. This arrangement is sometimes modified to have the armature current of one machine excite the series field of the other machine, which gives a more stable system with respect to speed and equality of terminal voltage.

SYNCHRONOUS CONVERTERS

87. A synchronous (or rotary) converter transforms energy from alternating to direct current, or vice versa (when it is termed an inverted converter), through the rotation of a single core and armature winding in a special field frame which has windings similar to that of a d-c generator. The a-c system is connected to armature winding through slip rings, and the d-c system is connected to the winding through a commutator, at the end opposite the slip rings. The continuing gains in per unit ampere and back voltage rating of static rectifiers have made this rotary type of conversion almost obsolete economically for new applications, especially with respect to maintenance cost.

The synchronous converter is a combination of synchronous motor driving a d-c generator, the net armature current being the difference between the a-c motor amperes and the continuous d-c amperes. Since only one armature winding is used, there is a definite and practically constant ratio between the emfs at the a-c and d-c terminals. With no d-c load the synchronous converter operates purely as a synchronous motor. The speed is determined by the frequency and number of poles. Most converters of 100 kW and larger have commutating poles, with excitation about 40% of that on a conventional d-c machine.

88. Excitation. In the synchronous converter, as in the synchronous motor, the magnetic flux and the corresponding net exciting ampere-turns are determined solely by the impressed voltage. If it is attempted to vary the exciting ampere-turns and flux by variation of the main field excitation, the latter variation is neutralized by an equivalent change in excitation brought about by a change in phase and value of the armature current, so that the flux and excitation remain constant. Increased excitation in the main field winding produces a leading current in the armature (leading with respect to the line voltage) which, in the majority of transmission lines serving synchronous converters, is beneficial to line power factor and voltage. Conversely, underexcitation produces a lagging current which is detrimental to the power factor and voltage of such lines.

88a. Ratio of A-C Voltage to D-C Voltage. Upon assuming the synchronous converter to be operating without d-c load, it will be clear that the d-c voltage between brush arms will, through the action of the commutator, have a value equal to the maximum instantaneous value of the alternating voltage, proper consideration being given to the relative points in the armature winding to which, at any instant, the brushes and collector rings are connected. In the single-phase, 2-phase, and 6-phase diametrically connected converters, the d-c brushes and collector rings are connected to equivalent points on the armature winding, so that the ratio between the a-c and d-c voltages is simply the ratio between the effective and the maximum alternating voltages. In the 3-phase converter, the collector rings are connected to points 120 elec deg apart, while the d-c brushes are connected to points 180 elec deg apart, so that the voltage ratio is affected by this difference. The theoretical ratios are shown in Table 12-6.

Table 12-6. Theoretical Voltage Ratios

No. Converter Phases	Ratio of Alternating Voltage
Single-phase	0.71 of direct voltage
2-phase	0.71 of direct voltage
3-phase	0.61 of direct voltage
6-phase double delta	0.61 of direct voltage
6-phase diametrical	0.71 of direct voltage

These theoretical ratios are based on the assumptions that the impressed a-c waveform and the counter-emf waveform of the converter are both sine waves, that there is no loss in the converter, and that the d-c brushes are at the no-load neutral position. Variations in waveform are small in commercial circuits and apparatus. The effect

Table 12-7. Approximate Current Ratios

No. Phases	Alternating Current/Terminal
Single-phase	1.50 times direct current
2-phase	0.75 times direct current
3-phase	1.00 times direct current
6-phase	0.50 times direct current

of the resistance of the windings, brushes, and brush contact is appreciable and may vary the ratios from 2 to 4%. Changes in brush position, even within the small limits permitted by commutating conditions, may affect the ratio 1%. A lagging current in the converter winding has the same effect as narrowing the pole face and will thus increase the ratio of a-c voltage to d-c voltage. In general, therefore, actual ratios are slightly higher than the theoretical ratios, assuming conversion from alternating to direct current, by 2 to 4%.

The current ratio, neglecting the losses and assuming 100% power factor, is the inverse ratio of the a-c and d-c voltage. For all practical purposes an average efficiency of 94% may be assumed, in which case the current ratios will be as shown in Table 12-7.

88b. Converter Effective Armature Current and Losses. The ratio of the effective or resultant armature current to the external direct current varies with the number of phases or, more properly, with the number of connections per pole to the armature winding. The larger the number of such connections, the smaller will be the effective current. The distribution of the current among the different conductors at 100% power factor is such that the maximum loss occurs in the tap coils and the minimum loss in the coils midway between taps. For the relative converter losses on the basis of d-c generator losses, see Table 12-8.

Table 12-8. Converter Losses Relative to D-C Generator Losses Taken as Unity

Calculations based on 3% rotational lo se
(Calculated by C. E. Wilson)

No. of collector rings	Relative armature loss in complete winding					Relative maximum loss in one conductor				
	Power Factor 100%	Power Factor 98.5%	Power Factor 94%	Power Factor 86.6%	Power Factor 76.6%	Power Factor 100%	Power Factor 98.5%	Power Factor 94%	Power Factor 86.6%	Power Factor 76.6%
2	1.451	1.522	1.734	2.160	2.940	3.121	3.653	4.358	5.342	6.808
3	0.587	0.627	0.753	1.005	1.468	1.249	1.594	2.048	2.673	3.596
4	0.391	0.426	0.532	0.746	1.137	0.751	1.017	1.367	1.861	2.596
6	0.274	0.304	0.400	0.589	0.935	0.430	0.614	0.874	1.249	1.825
12	0.209	0.236	0.326	0.500	0.824	0.249	0.354	0.525	0.792	1.232

89. Armature reaction is small compared with that of a conventional d-c generator since the resultant armature current is small, as shown in Fig. 12-71. It varies between 7 and 20%, and the converter operates very nearly like a compensated d-c generator with respect to distortion of flux distribution across the face of the main pole. Because of the cyclical variation of resultant armature ampere-turns, the commutating poles are built with an internal gap which uses two to three times the mmf at the face of the commutating pole, thus reducing the inequality between the pulsating armature mmf and the constant mmf of the commutating pole.

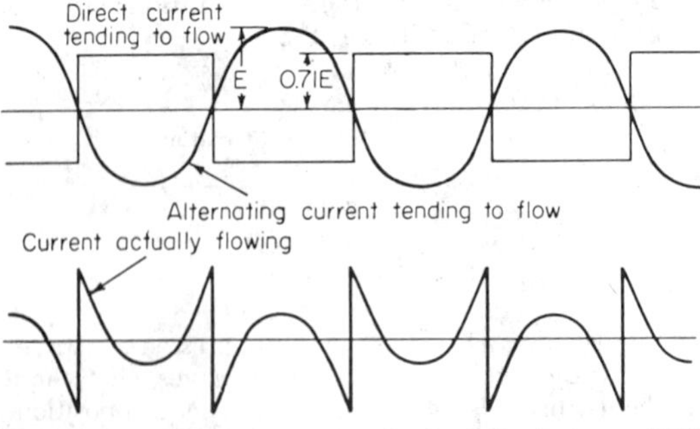

FIG. 12-71. Armature current in conductor, 100% power factor.

The d-c voltage of the converter can be changed by:

a. Changing the applied a-c voltage by transformer taps or an induction regulator.

b. Over a small range by shunt field control. This is limited by commutation at rated current load to about 96% lagging to 96% leading power factor.

c. By use of a compounding series field, to give flat compounding at rated load, with 100% power factor. At light loads the power factor will be considerably lagging. It is necessary to have approximately 15% reactance in the transformer to make the series field produce flat compounding (Table 12-9).

Table 12-9. Power-factor Variation with Load in Compound Converters

Load	1/8	1/4	2/4	3/4	4/4	5/4	6/4
Power factor, %	40	65	92	98.5	100	99	98.5
Lag or lead	lag	lag	lag	lag	...	lead	lead

90. General design methods for a synchronous converter follow closely those for a d-c generator. However, the main pole face contains a damper winding which has about 25% of the total area of the armature conductors. This provides the starting torque when energized from the a-c line. It is customary for starting to use either:

a. 50% transformer primary taps, or

b. Switch transformer from Y to Δ on the primary.

On converters of 300 kW and larger the commutator brushes must be lifted in starting, leaving only two pilot brushes down for voltage indication. The shunt field is also separately excited on larger converters, so that the converter comes into step with the right polarity.

Three-wire d-c systems use the center point of Y-connected transformer secondaries as the midvoltage point for distribution of the unbalanced d-c current.

The a-c collector rings usually are fitted with metal graphite brushes, and this usually causes the most trouble in operating a converter. The metal dust accumulates on exposed insulating tubes, and unless the construction is made completely accessible for maintenance, dangerous creepage paths build up between a-c phases. As a rule, collector rings have spiral grooves to secure cool operation and reduce selective current collection in the metal graphite brushes, which have very low voltage contact drop.

SYNCHRONOUS CONDENSERS

By R. E. Appleyard

91. Definition. Synchronous condensers are devices for adjusting circuit power factor to unity or near unity. They increase circuit capacity through both minimizing kVA for given kilowatts and improving system stability.

92. Synchronous condensers are essentially synchronous motors designed to operate at zero power factor, without connected mechanical load. Typical characteristic curves are shown in Fig. 12-72. By adjusting the excitation, a synchronous condenser can supply either inductive or capacitive kVA. When connected to a circuit in which the current lags the voltage, one in which the average load is inductive, the synchronous condenser is operated overexcited so that it supplies some of the inductive reactive kVA and adjusts the power factor toward unity. If, on the other hand, the average circuit load is capacitive and the current leads the voltage, the synchronous condenser is operated underexcited so that it supplies some of the capacitive reactive kVA to adjust the power factor toward unity. Since it is more commonly required that synchronous condensers supply inductive reactive kVA, they are usually designed for overexcited capability about twice the underexcited capability.

The family of V-curves in Fig. 12-73 illustrates the way in which a synchronous condenser improves system stability. With constant excitation a drop in system voltage will increase reactive kVA output of the condenser if it is operating overexcited and decrease the reactive kVA output if it is operating underexcited. Both these effects tend to maintain constant system voltage. The amount of the change in reactive kVA output for a given change in system voltage depends on the degree of saturation in the magnetic circuit of the condenser; the effect is greatest in designs with high saturation and, expressed in percent of rated voltage, is about the same as the change

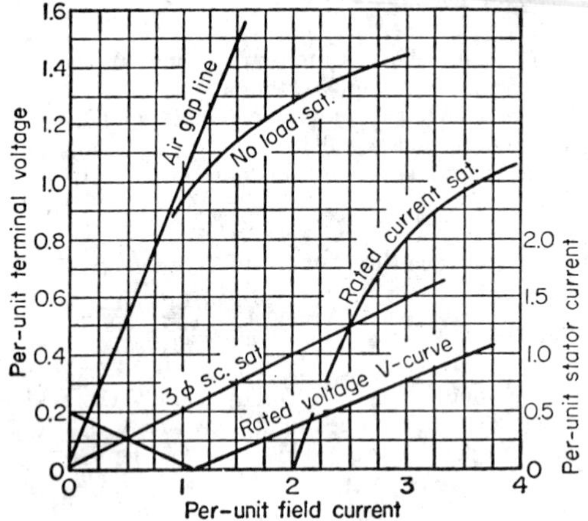

FIG. 12-72. Typical characteristic curves of air-cooled synchronous condenser.

in system voltage. That is, a 5% drop in system voltage will produce approximately 5% increase in reactive kVA output of an overexcited synchronous condenser.

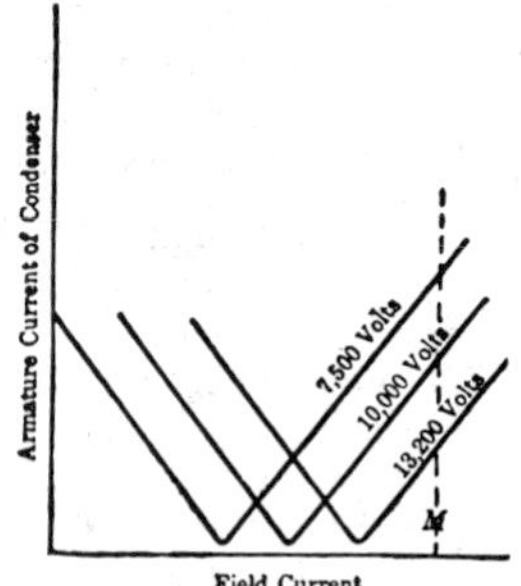

FIG. 12-73. Typical V-curves of synchronous condenser.

93. Performance under Transient Conditions. The foregoing has considered steady-state operation. Under transient conditions, with high-speed regulating systems, the effect of a voltage change will be increased to an initial value limited by the subtransient reactance of the condenser and will vary with time similar to a short-circuit current decrement curve. Figure 12-74 illustrates the change in condenser reactive kVA output with time for a sudden 5% decrease in system voltage and fixed condenser excitation. With a high-speed voltage-regulating and excitation system, the minimum change in reactive kVA could be held to about the "knee" of this decrement curve, i.e., to the value at about 0.20 s.

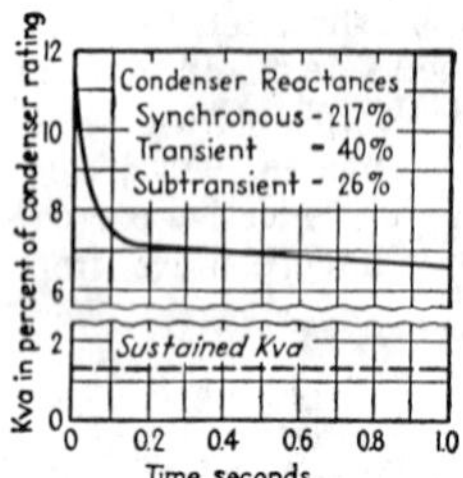

FIG. 12-74. Transient performance of a synchronous condenser following a sudden change in terminal voltage.

94. Data on Speeds, Voltages, and Losses. Table 12-10 gives data on speeds, voltages, and losses for standard ratings of large 3-phase 60-c synchronous condensers. These data are typical for United States manufacturers. They are subject to variation with specific applications; in particular, the speed may be higher than that shown in the table.

95. Construction. The construction of synchronous condensers is similar to that of large high-speed synchronous motors except that the field windings are heavier to accommodate the greater excitation current required for zero-power-factor overexcited operation, and the shaft and bearings may be smaller, since there is no mechanical load connected to it. Almost without exception, they are salient-pole machines with eight poles or more. The limited thermal capacity of the higher-speed machines generally precludes their use as synchronous condensers.

96. Ventilation. In the smaller sizes synchronous condensers are usually open, self-ventilated machines. For larger sizes, particularly with voltage ratings of 6,900 V and above where cleanliness of the windings is particularly important, recirculating ventilation systems with air coolers and, possibly, filters are provided. In the largest sizes hydrogen cooling is employed as a means both of improving efficiency through reduced windage losses and of reducing machine size through improved heat transfer.

The ventilation systems employed with air cooling are generally similar to those for large synchronous motors. Air enters the machine axially at both ends, is carried between the salient poles, passes through the stator core radially along ducts provided at intervals along the length of the core, and is carried around the stator frame to suitable discharge openings. If air coolers are provided, they are mounted on the frame or

Table 12-10. Date on Speeds, Voltages, and Losses of Large 3-phase, 60-cycle Synchronous Condensers

Rating, kva	Speed, rpm*	Voltage air cooled	Full-load losses, air cooled	Voltage hydrogen cooled	Full-load losses in hydrogen at ½ lb. pressure
2,500	900	4150/6900/13,800	80/85/90		
3,000	900	4150/6900/13,800	89/95/100		
4,000	900	4150/6900/13,800	107/114/120		
5,000	900	4150/6900/13,800	125/133/140		
7,500	900	4150/6900/13,800	170/178/185		
10,000	900	4150/6900/13,800	215/223/230	6900/13,800	191/197
15,000	900	4150/6900/13,800	300/310/320	6900/13,800	267/275
20,000	900	6900/13,800	....	6900/13,800	340/347
20,000	720	6900/13,800	395/405	6900/13,800	
25,000	720	13,800	485	13,800	415
30,000	720	13,800	560	13,800	482
40,000	600	13,800	710	13,800	615
50,000	600	13,800	860	13,800	741
60,000	600	13,800	1010	13,800	867
75,000	600	13,800	....	13,800	1055
75,000	514	13,800	1235	13,800	
100,000	514	13,800	1610	13,800	1370

* Hydrogen-cooled machines may be built for speeds higher than listed at the manufacturer's option.

in ductwork attached to the frame and the discharge from the coolers is returned to the ends of the machine.

97. Hydrogen-cooled Synchronous Condensers. The advantages of hydrogen over air as a cooling medium for synchronous condensers are the same as for other large electric machines: its lower density reduces windage losses; its improved surface heat-transfer characteristic reduces temperature drop at the coil surfaces; it virtually eliminates the insulation-deterioration effects of corona discharge; it does not sustain combustion and therefore eliminates the fire hazard. In addition, the enclosure design is simplified since there are no shaft extensions through the enclosure requiring running seals. Therefore hydrogen cooling is frequently specified on synchronous condensers, particularly for the larger, higher-speed ratings.

All hydrogen cooling systems are designed to operate above atmospheric pressure so that any leakage will be outward and the gas inside the machine will remain at high purity. The minimum nominal operating pressure is 0.5 lb/in^2 gage. Most hydrogen-cooled synchronous condensers have ratings also at 15 lb/in^2 gage hydrogen pressure, and some at 30 or 45 lb/in^2 gage. As the hydrogen pressure is increased, its heat-transfer capability increases and the output of the machine for the same temperature rise may be increased.

Figure 12-75 shows a typical relationship between hydrogen pressure and output for a conventionally hydrogen-cooled condenser. By "conventionally" is meant a machine in which the hydrogen is circulated over the outside of the windings. Some other types of large rotating electric machines, notably steam-turbine-driven generators, are cooled by hydrogen circulated through ducts within the windings in direct contact with the copper. This system has not yet found commercial application to synchronous condensers.

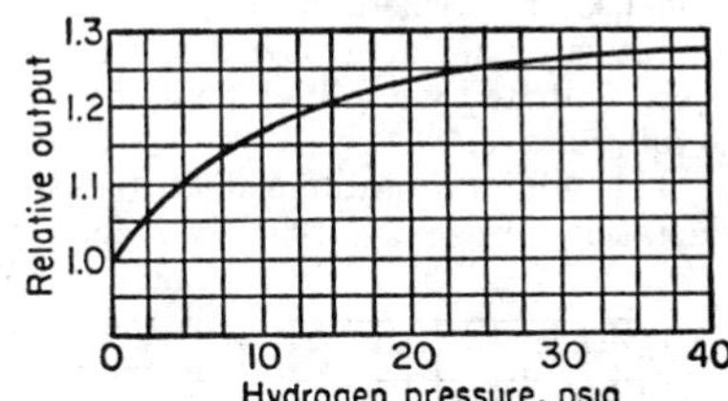

Fig. 12-75. Effect of hydrogen pressure on output for a conventionally hydrogen-cooled machine.

As the hydrogen pressure is increased, the windage loss increases proportionally. Since this is a relatively small part of the total loss, however, the total loss as a percent of the increased reactive kVA output ordinarily does not increase with hydrogen pressure.

The flattening off of the curve in Fig. 12-75 as pressure is increased is due to the fact that the increased hydrogen pressure affects only the surface heat transfer and the specific heat of the gas itself. A substantial part of the copper temperature rise is the drop through the insulation wall and the temperature differences within the stranded conductor due to eddy-current effects. There is probably little economical gain in designing conventional hydrogen-cooled synchronous condensers for pressures above 30 lb/in² gage.

Winding temperature detectors for hydrogen-cooled synchronous condensers are usually resistance elements placed between the stator coil sides toward the axial center of the core. These cannot sense the different distribution of temperature between surface drops, insulation drops, eddy-current effects, etc., corresponding to different hydrogen pressures. It is not, therefore, a safe practice to load a synchronous condenser by observation of the winding temperature; this could result in internal coil temperatures damaging to the insulation system. Ordinarily these machines should not be loaded above their nameplate ratings.

In addition to the controls and protective devices usually provided with large rotating electric machines, hydrogen-cooled synchronous condensers require systems for filling with gas and maintaining suitable purity, pressure, and temperature.

Exciters for synchronous condensers are frequently direct-connected. For hydrogen-cooled machines, the exciters are placed in the hydrogen enclosures. This requires some special consideration of commutator brush operation and maintenance. The film on the commutator surface necessary to proper brush operation is mainly an oxide. Special brush grades are necessary to maintain it, and preliminary operation in air may be required to establish a suitable film. Manufacturers' operating instructions should be observed. In order to change brushes, it would be undesirable to require emptying hydrogen from the entire machine enclosure. The machine may be designed with a separate compartment for the exciter within the machine enclosure, arranged so that it may be sealed at standstill from the rest of the enclosure. Then only the smaller compartment need be emptied of hydrogen before opening to maintain the brushes. The condenser field-collector assembly is located in this same enclosure so that its brushes may be maintained in the same way.

The enclosure for a hydrogen-cooled synchronous condenser is designed to be explosion-resistant when operating a rated gas pressure.

98. Starting Methods. Most synchronous condensers which are applied as voltage regulators or to improve system stability are started and placed on the line under the control of a voltage regulator. A principal factor in selecting the method of starting is the amount of disturbance that the system can tolerate during the starting period. The need for a condenser to regulate voltage or improve system stability implies a relatively "soft" system at the point where the condenser is connected. Therefore, reduced-voltage starting methods, which minimize system disturbance, are most commonly employed. Of the various reduced-voltage methods, a tap on the winding of the power transformer to which the condenser is connected and a separate autotransformer are two frequently used. In either case, the starting voltage may be quite low (on the order of 20%) so that the current inrush may be 100% current or less. The condenser is accelerated as a conventional synchronous motor, by torques produced in a starting winding placed in the pole faces. The condenser is synchronized on reduced voltage by applying field at near synchronous speed, and a transfer is then made to full voltage.

The minimum system disturbance in transferring to full voltage will be with the excitation adjusted so that the reactive kVA supplied to the system is the same before and after the transfer. A method for determining this is indicated in Fig. 12-76. Curve A is the V-curve corresponding to full voltage. Curve B is the V-curve corresponding to the starting voltage, multiplied at each ordinate by the percent starting voltage so that it corresponds to kVA at system voltage. Where these two curves intersect in the overexcited range, the kVA after the transfer will be the same as that before the transfer.

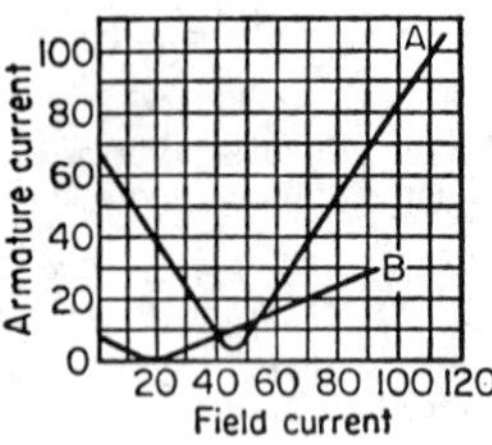

Fig. 12-76. Determining optimum excitation for transfer from starting voltage to full voltage.

It is important also that the transfer be made rapidly. In most reduced-voltage starting methods the condenser is momentarily disconnected from the system during the transfer to full voltage. The rotor decelerates during this period so that there is a phase displacement between the condenser voltage and the system voltage. In the typical switching arrangement where this is accomplished in 5 to 10 c this phase displacement is not significant.

In the extreme application where the system is very large the synchronous condenser may be started across the line. Inrush kVA may be on the order of 500%. Special attention must be given to the starting winding design to prevent thermal unbalances or overheating. In the largest high-speed ratings this starting method may be impracticable because of limitations in starting winding design.

At the other extreme, where the starting kVA must be kept to the very minimum, a direct-connected wound-rotor induction motor may be provided for starting. The motor has two fewer poles than the synchronous condenser so that the unit may be accelerated and synchronized to the system much as a generator would be. With this method the starting inrush kVA can be limited to a fraction of the condenser rating.

Other starting methods such as neutral reactor starting, part-winding starting, and series-parallel starting are possible but less commonly used.

A limitation to minimizing the inrush kVA with reduced-voltage starting methods is the obtaining of sufficient torque to break loose the shaft bearings. A high-pressure lubrication system may be applied to float the shaft journals on an oil film where very low starting voltages are desired. Another limitation is the thermal capacity of the starting winding. At the lower starting voltages more total energy is transferred to the rotor during the starting period because of the various losses in the machine, and these result in added temperature rise in the starting winding.

99. Excitation Systems. Direct-connected exciters are the most usual source of field current for synchronous condensers. Because of the extremely wide range of excitation required, a pilot exciter or other means of separate excitation is usually provided for the main exciter. This pilot excitation may be a part of the voltage-regulating system, as in the various rotating-amplifier and magnetic-amplifier types of high-speed voltage regulators.

Static exciters employing the various solid-state semiconductors have been used for synchronous condensers and may find more widespread application in the future because of their inherently quick response.

100. Other Devices for Correcting Power Factor. Static capacitor banks have replaced synchronous condensers as sources of reactive kVA in many applications where the principal need is to supply leading reactive kVA. They can be connected to transmission and distribution systems wherever the load conditions require them. Although many installations are of fixed capacity, some are arranged for switching in increments. Switching may be manual or automatic, using voltage, power factor, load, or a combination of these for control signal. Capacitor banks are relatively less flexible than synchronous condensers, but their cost may be substantially less. In recent years capacitor installations have grown at a considerably faster rate than condenser installations.

FREQUENCY CHANGERS

101. Definition. Frequency changers receive alternating-current electrical energy at one frequency and deliver it at a different frequency. Their applications are to interconnect power systems of different frequency and to provide power to specific loads that require a frequency different from that of the available power system.

102. Application. Most parts of the world have standardized on either 50 or 60 c/s as a generating frequency for alternating-current power. Traces remain, however, of earlier generating systems from which the standard frequency evolved. In the United States, for example, minor amounts of power are generated at 25, 40, and 50 c/s. When it becomes necessary to interconnect one of these nonstandard generating systems with the standard one, a frequency changer is employed.

Some power applications require frequencies other than the usual 50 or 60 c/s. If the power requirement is very large and the frequency requirement is below about 500 c/s, a special generator capable of producing power at that frequency may be used. For smaller amounts of power, or for higher frequencies than are practicable for direct electromechanical energy conversion, or where the economy of the application precludes a separate generator, frequency changers may be employed. Table 12-11 shows some of the applications for the various ranges of frequency.

103. Types. Frequency changers may be generally classified as static and rotating.

Table 12-11. Frequency-changer Applications

Frequency, c/s	Application
15–25	Single-phase commutator motors for electric traction and special industrial applications
50–60	Lighting, heating, and general industrial applications
120–180	Machine tools for woodworking, textile, spinning, and other special applications
500–10,000	Inductive heating, melting, and electrolytic applications
20,000–300,000,000	Special inductive heating and melting, and dielectric heating
20,000,000 and above	Radio transmission and special inductive heating processes

Mercury-arc rectifiers and the various controlled solid-state devices are examples of static frequency changers. They are commonly employed when the desired output is direct current and are used also to synthesize current of a required frequency from the available power supply. The rotating frequency changers include motor-generator sets in which the generator is a synchronous machine and the motor is synchronous, induction, or direct-current; and the induction frequency changer, which is essentially a wound-rotor induction machine with provision for taking the slip frequency power off the rotor winding.

104. Synchronous-synchronous sets are the most common and consist of two synchronous alternators coupled together. The speed and frequency relations between the two units are fixed, and load is controlled by the relative phase positions of the two systems. Hence, a tendency of one system to change in phase relation to the other is immediately reflected in loading of the set and, if too great, will pull it out of step.

Load transfer is controlled by adjustment of the governors of the generators on one or both systems to change the relative phase positions of the two systems and cannot be controlled at the frequency changer.

Voltage and power factor of each unit may be controlled independently by field adjustments of each.

105. When two or more frequency changers are connected between systems, it is essential that the relative field and stator pole positions between the 25- and the 60-c units of all sets be the same in order to parallel. When so operated, it is common practice to design one stator of all, or all but one, of the frequency changers, so that it can be rotated through a limited angle by means of a frame-shifting device. By this means the division of load between the sets can be controlled, and a set can be added to or dropped from the system at no load. The amount of frame shift required depends upon the design of the sets and the reactance of the connections between them. Because of the different number of poles on the two units of a set, there are only a limited number of field and stator pole positions in which both ends of the set will be in phase with the systems. It is therefore common practice to provide field-reversing switches on one or both ends of the set so that, after the set is up to speed, the field of the driving unit can be reversed (at low field excitation), and poles slipped until the other unit is in phase.

If the field of the driven unit can also be reversed, the number of poles that must be slipped is on the average reduced by one-half.

Frequency changers of this type are listed in sizes of 100 to 75,000 kVA. The highest speed possible, if the ratio is 25 to 60 c, is 300 r/min. Other speeds are obtained at other frequency ratios. The efficiencies at full load and 0.8 power factor vary from about 80% in the smaller sizes to 96% in the larger sizes.

106. The synchronous-induction transformer set consists of a synchronous alternator on the lower frequency end and a slip-ring induction motor on the other, having the proper number of poles and run at such a speed that the **slip frequency** is the lower frequency of the set. The rings are connected directly or through a transformer to the lower frequency system.

If the set is a 25- to 60-c combination with a 10-pole 25-c synchronous unit running at 300 r/min and a 14-pole 60-c induction unit at the other end, the synchronous speed of which is 514 r/min, the slip frequency will be $60 \times 214/514 = 25$. The power transmitted mechanically through the shaft to the 25-c unit will be $300/514 = 58.3\%$ of the total. The power transmitted electrically from the slip rings to the 25-c circuit will be 41.7%.

A transformer is usually interposed between the slip rings and the 25-c circuit, as the voltage of the latter is generally too high for good rotor and slip-ring design.

Voltage and power factor of the 60-c unit are controlled by taps on the transformer and on the 25-c unit by field control. Field-reversing switches and frame shifting are provided if the unit is to parallel other frequency changers (see Fig. 12-77).

Owing to the electromagnetic connections between the systems the voltages of the two systems, as well as the frequencies, are tied together so that a voltage disturbance in one system is reflected in the other. This is an advantage when one or more synchronous converters supplied from each system feed a common d-c bus, as it reduces the feedback through the converters and consequent tendency to flashing and damage in case of voltage disturbance on one a-c system.

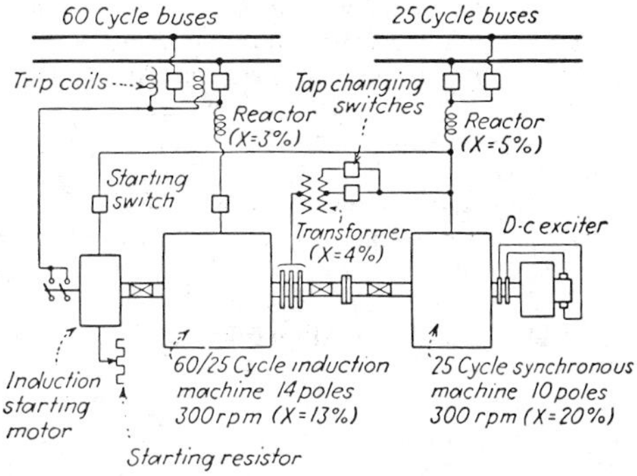

FIG. 12-77. Schematic diagram of synchronous-induction transformer frequency-changer set.

The pull-out torque is somewhat higher and the phase-angle shift for a given load change is somewhat less than for a synchronous-synchronous set.

The cost and efficiency are about the same. The disadvantages are the somewhat greater complication of equipment and control and difficulties incident to high-capacity high-voltage slip rings. Sets of this type have been built up to 40,000 kW.

107. The synchronous-induction set consists of a synchronous alternator and a squirrel-cage or slip-ring induction unit. As load transfer depends upon slip in the induction unit, such sets are not useful when exact frequency control is maintained on the two systems but can be used when only approximate frequency is required and the relative frequencies of the two systems can be changed. A change from full load in one direction to full load in the other requires about a 4% change in relative frequencies.

As loading is determined by slip and not relative phase positions, frequency changers of this type are practically independent of momentary swings, with the result that systems can be safely interconnected with much smaller capacity frequency changers of this style than with the synchronous-synchronous style.

They have been used as a reserve supply to a particular load with the alternator normally "floating" as a synchronous condenser ready to pick up load if the normal supply is lost and when the two systems are not otherwise substantially tied together.

Frame shifting and field reversing are not required for parallel operation, and there is no control of the division of load between paralleled units with squirrel-cage induction motors. Load division can be obtained by varying the resistance in the rotor of slip-ring induction motors. Parallel operation requires that the load-slip curve of the induction motors be the same; they have generally been limited in capacity to a few thousand kilowatts.

Their efficiencies run about 2% less than those of the corresponding size of synchronous-synchronous sets, and costs are higher except in the smaller sizes designed for speeds higher than 300 r/min with approximate frequency ratios such as 62.5 to 25 at no load.

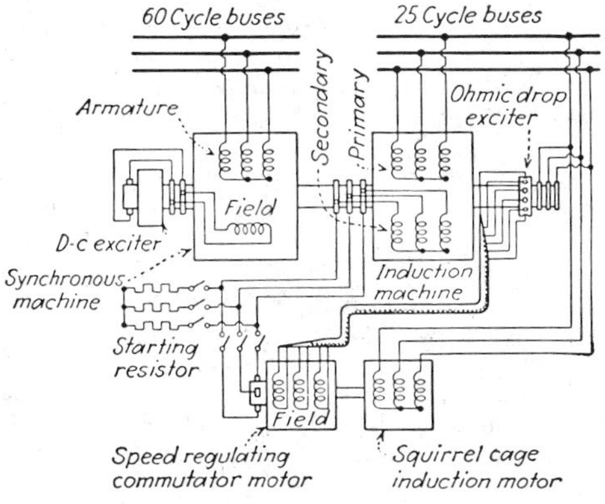

FIG. 12-78. Schematic diagram of synchronous adjustable-speed induction frequency-changer set.

108. The synchronous adjustable-speed induction set consists of a synchronous alternator on the higher-frequency end and on the other a slip-ring induction motor with associated regulating equipment to control the voltage and phase angle of the slip frequency of the power received from or supplied to the slip rings. This control may be of the Scherbius or other similar types of control. It is diagrammatically shown in Fig. 12-78.

The load-speed characteristics of this style of motor control are similar to those of an adjustable-speed shunt-wound d-c motor in that an increase in load slows down the motor slightly, and the speed at any load can be controlled by adjustment of the controlling devices. By this means, local load control is provided, and load can be transferred in either direction at fixed frequency ratio or at ratios differing slightly from the normal. As load transfer is determined by relative speeds and not by phase-angle position, the set like the synchronous-induction set is practically independent of momentary swings. It is therefore possible to tie together two systems with much less relative frequency-changer set capacity than is possible with synchronous-synchronous sets. Automatic control can be provided to regulate for a desired load transfer in either direction or for fixed frequency ratio with provision to limit the transfer to the rating of the set. The power factor of the induction end can be controlled by adjusting the phase angle of the voltage of the regulating set imposed on the rings so that the set can be held at about unity power factor at all loads. While this type of set combines most of the operating advantages of the synchronous-synchronous and synchronous-induction sets, its application has been limited owing to its higher cost, lower efficiency, and greater complication. Its use has been primarily between systems, the frequency control of at least one of which was not close and where the frequency-changer capacity was small compared with the system capacities.

MAGNETIC AMPLIFIERS

By R. L. Robertson

109. A magnetic amplifier is a device which utilizes saturable core reactors, either alone or in combination with other circuit elements to obtain amplification or control. The magnetic amplifier is one of the earliest known types of electrical amplifying devices (rotating, electronic, magnetic, and solid-state). It is sometimes referred to as a transductor, direct-current transformer, or saturable reactor. The development in the late 1940s and early 1950s of square-loop core material and improved rectifiers provided the basis for widespread application of both high-gain and large-power-handling-capability magnetic amplifiers. The magnetic amplifier is now being made obsolete in certain applications by thyristors and other multiple-junction solid-state semiconductor devices.

The major advantage of magnetic-amplifier devices is their ability to sum signals and provide complete multiple signal and load-circuit isolation for d-c signals.

A *saturable reactor* is a magnetic-core device in which magnetomotive forces applied to a control winding alter the saturation of the core, changing the a-c impedance of the anode, or output, winding and modulating the output from an a-c source.

Output, or anode, windings are those windings associated with the load through which power is delivered to the load.

Control windings are those windings by which control magnetomotive forces are applied to the core. Control windings may be further classified as to function as signal, reference, bias and feedback windings.

110. The elementary saturable reactor in Fig. 12-79*a* is composed of a core assembly with output winding of turns N_L, control winding with turns N_c, an a-c power source, a load R_L, and a d-c source with adjustment R_c and filter choke L.

The operation of this circuit is based upon the reduction of impedance or effective inductance (flux linkages per ampere) of the core as the d-c control component of magnetization is increased. Full saturation of the core will reduce the impedance of the reactor to a minimum and give a maximum load current I_L. Progressive increases in level of d-c magnetization in the hysteresis loops of the core assembly are shown in Fig. 12-79*b*, *c*, and *d*.

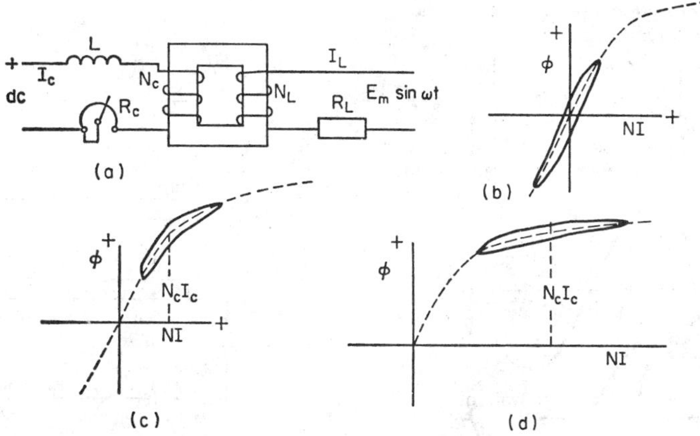

FIG. 12-79. (*a*) Elementary saturable reactor circuit; (*b*, *c*, *d*) progressively increasing d-c control.

This elementary reactor of Fig. 12-79*a* is not practical because of the a-c voltage induced into the control winding. This reactor will behave like a transformer with a short-circuited secondary unless a high impedance such as the filter choke is in series with the control winding. Because of this, saturable reactors are built so that a-c-induced voltages are eliminated from the control circuit.

Two basic two-core saturable-reactor connections are available. Each reactor assembly has an output winding and a control winding. The control windings are always connected in series. The output windings may be connected in series as in Fig. 12-80*a* or in parallel (Fig. 12-80*b*).

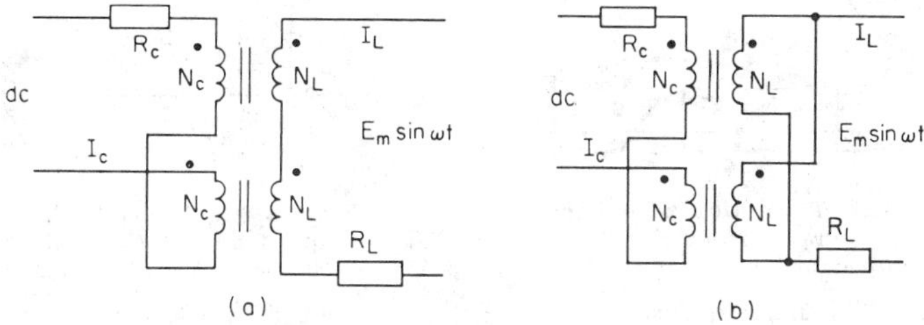

FIG. 12-80. (*a*) Series saturable reactors; (*b*) parallel saturable reactors.

111. Series saturable-reactor operation waveshapes may assume different forms depending upon the relative impedance in the control circuit. With the assumptions of negligible winding resistance and small magnetizing currents, performance of the reactors is governed by Faraday's law $e = N d\phi/dt$. This law states that voltage can exist across a winding only if the core is unsaturated and flux linkages can change.

The series reactor with low control impedance operates with instantaneous waveforms as shown in Fig. 12-81. When neither core is saturated, the applied voltage divides evenly across them. When one core reaches the saturation level, voltage across it drops to zero. The voltage across the other core also goes to zero, since with zero control impedance this unsaturated core is effectively short-circuited by the control winding of the saturated core, through the d-c source.

During the interval when the reactor voltages are zero, the applied voltage appears across the load, and current flows through the load limited only by the load resistance. Owing to current transformer action in the unsaturated core, the control current flows only when load current flows. They are related by the turns ratio; thus $N_cI_c = N_LI_L$.

The series-connected saturable reactor with high control impedance

$$\left[R_c\left(\frac{N_L}{N_c}\right)^2 \geqq 2R_L\right]$$

or a filter choke in the control source operates with a set of waveshapes as shown in

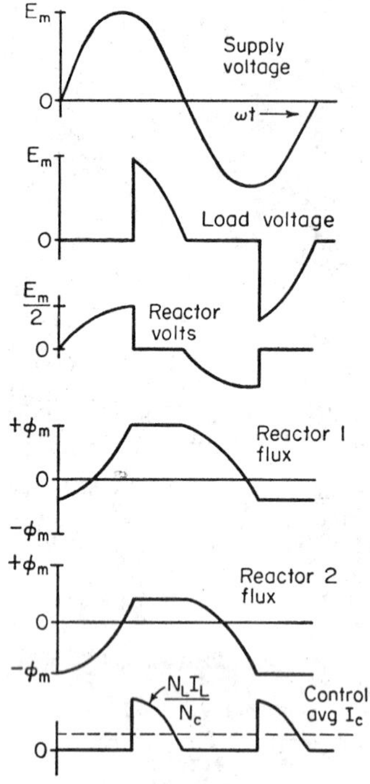

FIG. 12-81. Theoretical waveshapes for series saturable reactor with low control impedance.

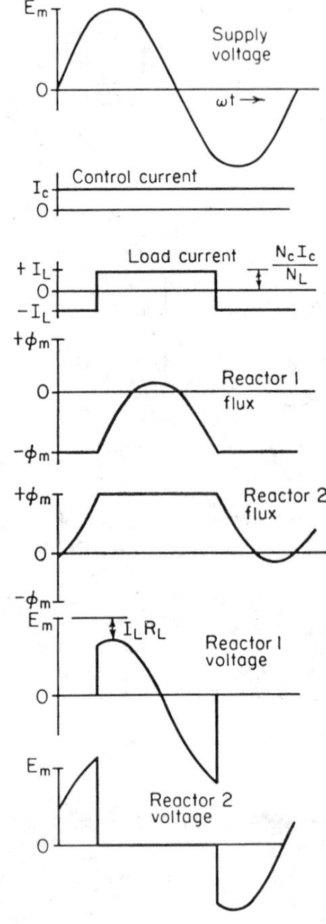

FIG. 12-82. Theoretical waveshapes for series saturable reactor with high control impedance.

Fig. 12-82. This mode of series-reactor operation is called the d-c current transformer or transductor operation; it is characterized by rectangular wave current switching. The rectangular switching of the a-c load current results from the fact that voltage cannot exist across a saturated core.

For analysis it is assumed that a core is driven far into positive saturation by a control current I_c (magnetizing force N_cI_c). At some particular instant negative load current can increase until equal magnetizing force N_LI_L occurs so that net NI is nearly zero, when voltage appears across the core, preventing further load current change, until the core once again reaches positive saturation. During this time interval, the line voltage is reversed, and since both cores are saturated, an immediate load-current reversal occurs and load current increases until the equal ampere-turn relationship is satisfied in the other core.

112. The parallel-connected reactor circuit is shown in Fig. 12-80*b*. When either core is saturated, the total supply voltage is applied to the load. During the intervals prior to core saturation, the ampere-turn balance relationship is maintained by circulating current in the output windings of the two reactors. Waveshapes are shown in Fig. 12-83.

From observation of the ampere-turn relationship, load current flows as a result of the control NI contribution in each core; thus the parallel-reactor formula becomes $N_L I_L = 2N_c I_c$. Hence for either series- or parallel-reactor connections, we obtain a constant kVA for a particular core assembly; since load current is doubled for parallel, required reactor voltage ratings are full line voltage instead of half line for the series connection.

113. General Reactor Characteristics. The extremes of operation of a saturable reactor are first the zero control, or exciting current. Second is the region where the cores are completely saturated and maximum load current is determined by average supply voltage and load resistance. The region between these extremes is defined as the proportional region where the saturable reactor obeys the equal ampere-turns relationship $N_c I_c = N_L I_L$. Reactor performance is independent of control polarity; thus the control characteristics assume the V-curves of Fig. 12-84.

Basic formulas and characteristics for proportional region operation of saturable reactors are tabulated in Fig. 12-85 for both series and parallel connections. The load resistance R_L is adjusted to include the output winding resistances as appropriate. All quantities are expressed as average values; rms to average ratios or form factors are not constants in these nonlinear and nonsinusoidal operating devices. Depending upon reactor type and operating point, form factors may range from 1.0 up to approximately 1.5.

Saturable-reactor usage has mainly been for theater lamp dimming, heating-element control, a-c motor speed, and torque control. The series saturable reactor is most often utilized as a transductor or d-c current transformer for signal isolation and measurement of large direct currents. The d-c bus itself forms the single-turn control winding.

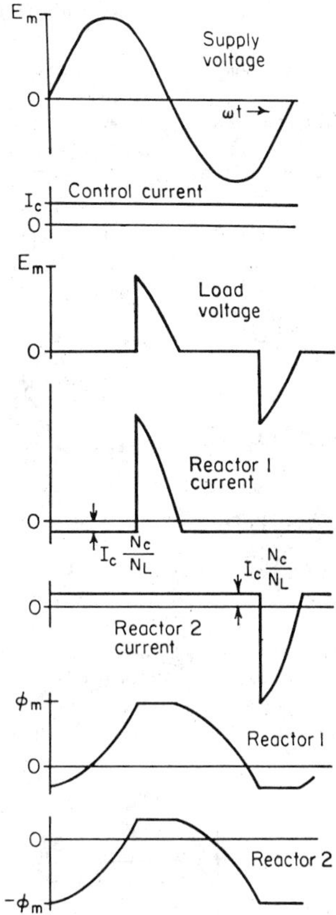

FIG. 12-83. Theoretical waveshapes for parallel reactor with high control impedance.

114. Reactor construction as previously described uses multiple-core assemblies with series control-coil connections to cancel the fundamental induced voltages. Alternatively, the mutual flux linkages between control and output windings can be eliminated by variations in core configuration and common coils.

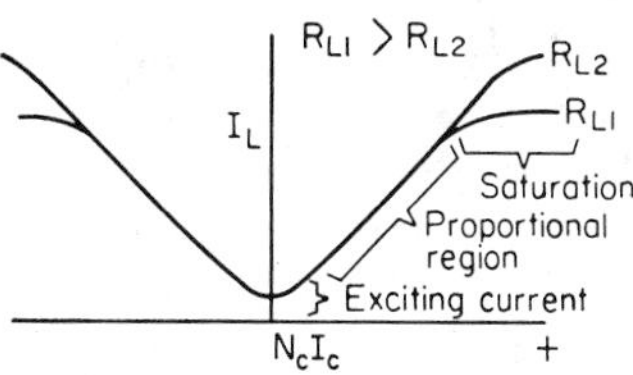

FIG. 12-84. Saturable-reactor control characteristic.

The magnetic circuit may be two or more individual cores or three- or four-legged cores. Output windings may be series, parallel, or common coils linking multiple cores or legs. Control windings may be series-connected or common-coil linking multiple cores or legs.

Physical limitations often control the reactor configuration. Winding insulation levels for control windings limit the two separate core assemblies to fractional kVA ratings. Multiple-leg or common-control coil designs are used for higher kVA. Winding geometry favors parallel-connected reactors for high-power units, and since a discharge path exists for the stored energy, parallel reactors are often selected. Leakage flux is minimized by the four-legged design. The series reactor has superior transient-response characteristics. The parallel reactor has higher gain.

115. Saturable-reactor external-feedback windings can be used to increase gain. Reactor output current is rectified and applied to an additional control winding to aid core saturation. The necessary control current can be reduced by the amount of feedback, and current and voltage gain are improved at the expense of increased response time. Waveforms and operating modes are basically unchanged. The use of external feedback windings on saturable reactors has been outdated by the self-saturating principle. Equal improvement in performance is more easily achieved with better material utilization by adding rectifiers in series with the reactor windings and thus making a self-saturating magnetic amplifier.

Configuration	series (Fig. 12-80a)	Parallel (Fig. 12-80b)
Control current I_c	$\frac{I_L N_L}{N_c}$	$\frac{I_L N_L}{2N_c}$
Control voltage E_c	$\frac{I_L N_L R_c}{N_c}$	$\frac{I_L N_L R_c}{2N_c}$
Time constant T	$\frac{1}{4f} \frac{R_L N_c^2}{R_L N_L^2}$	$\frac{1}{f} \frac{R_L}{N_L} \left(\frac{N_c^2}{R_c} + \frac{N_L^2}{2R_o}\right)$*
Voltage gain K_E	$\frac{I_L R_L}{I_c R_c}$	$\frac{I_L R_L}{I_c R_c}$
Current gain K_I	I_L/I_c	I_L/I_c
Power gain (avg) K_p	$K_E K_I$	$K_E K_I$
Output voltage (max) E_{max}	$\frac{2E_m}{\pi}$	$\frac{2E_m}{\pi}$

* R_o = individual reactor output winding resistance

Fig. 12-85. Reactor characteristics in proportional region.

116. The self-saturating magnetic amplifier is formed by connecting half-wave rectifying elements in series with each winding of a saturable reactor. Trade names such as Amplistat, Magamp, and Magnestat have been used to describe these self-saturating reactor and rectifier circuits. The basic self-saturating-type magnetic-amplifier device is shown in Fig. 12-86 with rectifier, reactor core assembly, load, and high-impedance control source. The significant control curves and instantaneous waveshapes as a function of time are shown. An operating point *a* can be established at any point on the saturation curve by control d-c ampere-turns. As the a-c supply voltage increases on its positive half-cycle, an opposing

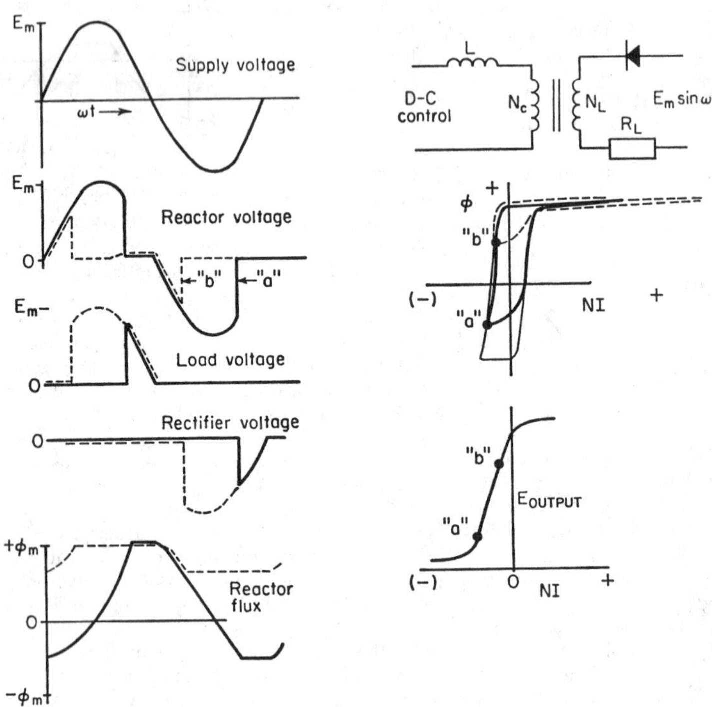

Fig. 12-86. Half-wave self-saturating magnetic amplifier.

and almost equal voltage is induced in the output winding as flux is changing from *a* to positive saturation. Load or magnetizing current is low over this portion of the cycle. When the core reaches saturation, no further flux change can occur and therefore the supply voltage appears across the load. With a higher initial flux *b* the flux reaches saturation earlier in the cycle, and greater output voltage is applied to the load. During the reversal of supply voltage, the control-current source resets the core flux to its original value. For the balance of the negative-supply half-cycle the supply voltage appears as reverse voltage across the rectifier.

117. Full-wave and polyphase magnetic-amplifier circuits are built up from this basic half-wave unit. A magnetic-amplifier circuit can be made from every possible d-c rectifier configuration for single and polyphase, full and half wave by simply connecting a reactor in each rectifier anode circuit. Similar circuits can be arranged for a-c output. Some of the most common magnetic-amplifier configurations are shown in Fig. 12-87*a* through *e*. Commutating diodes are required across inductive loads to provide a path for load discharge current and allow the cores to reset.

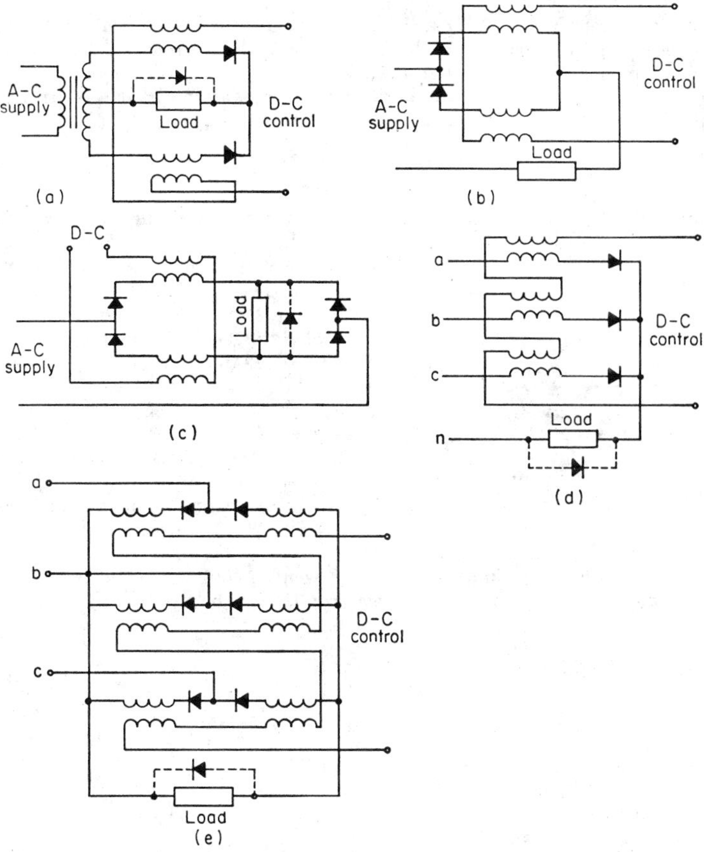

FIG. 12-87. Magnetic-amplifier circuits. (*a*) Center tap; (*b*) doubler a-c output; (*c*) bridge; (*d*) 3-phase half wave; (*e*) 3-phase bridge.

Output waveshapes and form factors of magnetic amplifiers very nearly resemble the equivalent rectifier circuits with phase control or firing delay. The major deviations result from the saturated reactance drop and *IR* losses in the reactors. The more nearly ideal the core characteristics, the better representation by the rectifier phase-control conversion formulas in Table 12-2.

118. The basic design formula for magnetic-amplifier reactor a-c supply voltage is the familiar transformer equation

$$E_{\text{rms}} = 4.44 f N_L B_m A_c \times 10^{-8}$$

where f = cycles per second, maximum core flux density B_m = maxwells per square inch, and core area A_c = square inches.

Control-range ampere-turns (maximum output to cutoff) is established by the slope and width of the core dynamic flux-current (hysteresis) loop. It can be increased by rectifier leakage. Typical values for NI/in and B max. for some of the more common core materials (assumed gapless) are tabulated assuming 60-c operation.

	Thickness, in	B_m, Mx/in²	NI/in
Grain-oriented steel	0.014	110	0.4
Grain-oriented 50% Ni	0.002	100	0.25
50% Ni	0.006	65	0.16
80% Ni	0.006	35	0.07

Control ampere-turns $N_c I_c$ = NI/in $\times L_c$ where NI/in is material magnetizing force and L_c is mean length of core. If the core is other than toroidal, further allowance must be made for air gaps or cross-grain flux.

The control time constant on the basis of linear operation about a control point is expressed by

$$T_c = \frac{1}{2fN_L}\left(\frac{\Delta E_0}{\Delta NI}\right) \Sigma \frac{N_c^2}{R_c} + \cdots + C_2 \qquad (12\text{-}50)$$

Time constant (63% response) is expressed in cycles of supply frequency where $\Delta E_0/\Delta NI$ is the slope of the control output characteristic and C_2 is a residual delay of 0.5 to 1.0 c. All connected control windings and their equivalent resistance must be included in the control turns squared per ohm summation.

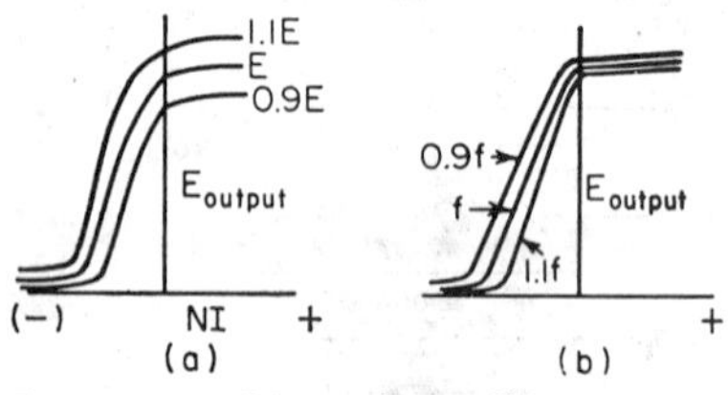

Fig. 12-88. Magnetic-amplifier output variation. (a) Supply-voltage effect; (b) supply-frequency effect.

119. Ratings, Performance, and Limitations. All magnetic amplifiers react to supply voltage and frequency variations with a change in output. Figure 12-88 illustrates these effects.

Reactor weights, power gains per cycle, and power ratings are shown in Fig. 12-89 for some typical designs, configurations, and core materials. Most reactor designs and ratings are limited by winding temperature.

To overcome partly the limitations of size, weight, cost, and transient response of magnetic amplifiers, special high-frequency supplies can be used. Special high-speed (½-c response) configurations of magnetic amplifiers were invented by R. A. Ramey. These configurations utilize an auxiliary a-c supply voltage in series with the control source to provide for core reset. These reactor designs were extremely limited in application owing to the complexity of control windings and auxiliary supplies and the introduction of alternating current into the control-signal source.

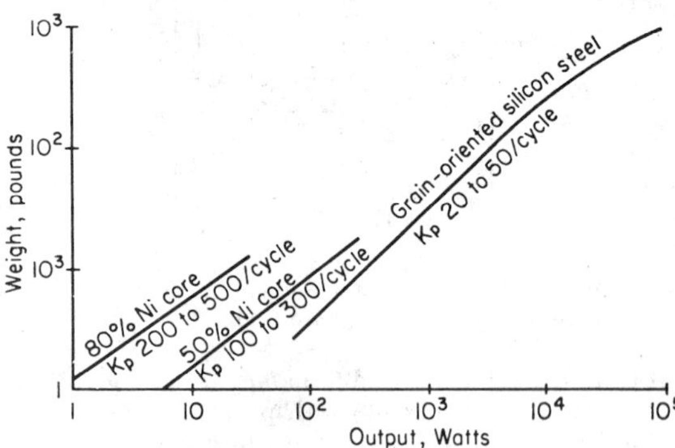

Fig. 12-89. Typical design magnetic-amplifier ratings, weight and response.

The major advantages of magnetic amplifiers are their static, rugged, reliable construction materials and their ability to isolate, impedance-match, and sum signals. The major magnetic limitations are limited gain response and a residual output which cannot be reduced to zero. Other limitations are the inability to absorb regenerative power from a load and complexity and inefficiency of reversible output circuits. The phase-controlled output with its ripple and harmonic distortion may cause problems.

120. Applications. Some representative magnetic-amplifier applications are servo systems, motor and generator field supplies, d-c motor armature supplies, primary and secondary control of induction motors, and logic and switching systems. One of the most recent applications is as a gating source for thyristor amplifiers.

121. References

1. GEYGER, W. A. "Magnetic Amplifier Circuits"; New York, McGraw-Hill Book Company, 1954.

2. STORM, H. F. "Magnetic Amplifiers"; New York, John Wiley & Sons, Inc., 1955.

3. DORNHOEFER, W. J., and KRUMMENACHER, V. J. Applying Magnetic Amplifiers; *Elec. Mfg.*, March, April, August, and September, 1951.

4. RAMEY, R. A. On the Mechanics of Magnetic Amplifier Operation; *Trans. AIEE*, Vol. 70, Pt. II, pp. 2124–2128.

THERMOELECTRIC CONVERSION

BY ROLAND W. URE, JR.

INTRODUCTION

122. Joule Heat. When an electric current flows in a conductor, heat is given off. The "Joule heat" is the part of this heat which is given by

$$q_j = \rho J^2 \qquad (12\text{-}51)$$

where q_j is the rate of Joule-heat generation per unit volume of conductor, ρ is the electrical resistivity, and J is the current density.

123. Thermal Conduction. Thermal conduction is the flow of heat in a temperature gradient in a material by processes which do not involve a net flow of material in any part of the system. The rate at which heat is conducted across a surface is given by

$$\mathbf{Q}_k = -\kappa \nabla T \qquad (12\text{-}52)$$

where $\mathbf{Q}_k$ is the heat current density, κ is the thermal conductivity, and T is the temperature.

124. Peltier Heat. When an electric current flows across a junction between two conductors with different properties, heat is given off or absorbed. Part of this heat is Joule heat (always given off), but there is an additional heat called the "Peltier heat." This is given by

$$q_p = -\Pi_{12} J_{12} \qquad (12\text{-}53)$$

where q_p is the rate of Peltier heat absorption per unit area of the junction, Π_{12} is the Peltier coefficient of the couple, and J_{12} is the current density taken as positive for current flow from material 1 to material 2 at the junction in question. The quantity q is taken as positive if heat is absorbed by the system.

125. Seebeck Voltage. Consider a circuit composed of two different materials. If the two junctions between the materials are held at different temperatures, then a voltage V_{12} will be produced in the circuit. For small temperature differences this voltage is given by

$$V_{12} = \alpha_{12}\,\Delta T \qquad (12\text{-}54)$$

where ΔT is the temperature difference between the junctions and α_{12} is called the Seebeck coefficient of the couple. (α_{12} has often been called the thermoelectric power. However, it is not a "power" but a materials coefficient.) The quantity α_{12} is positive if the Seebeck voltage is in a direction to produce current flow from material 1 to material 2 at the cold junction.

The absolute Seebeck coefficient α_1 is a property of one material only. The relation between the absolute Seebeck coefficients of two materials α_1 and α_2 and the Seebeck coefficient of a couple composed of these two materials α_{12} is

$$\alpha_{12} = \alpha_1 - \alpha_2$$

An α with either no subscripts or with two subscripts will be used for the Seebeck coefficient of a couple. An α with a single subscript will denote the absolute Seebeck coefficient of a single material.

126. Thomson Heat. In a homogeneous conductor in which an electric current is flowing and a temperature gradient is present, heat will be given off or absorbed in addition to the Joule heat. The difference between the total heat given off and the Joule heat is called the "Thomson heat." The rate of Thomson heat absorption per unit volume of material is given by

$$q_t = \tau \mathbf{J} \cdot \nabla T \tag{12-55}$$

where τ is the Thomson coefficient of the material. Thus positive τ means that heat is absorbed by the material when current flows toward the high-temperature region.

127. Nernst Voltage. Consider a conductor in which a temperature gradient and an external magnetic field exist perpendicular to each other. The effect of these two fields is to induce an electric field, called the Nernst field, in a direction perpendicular to both the magnetic field and the temperature gradient. This electric field is given by

$$E_x = -N_{xy}{}^i B_z(\partial T/\partial y) \qquad \text{with} \qquad J_x = J_y = \partial T/\partial x = 0 \tag{12-56}$$

where $N_{xy}{}^i$ is the isothermal Nernst coefficient.

128. Ettingshausen Heat. Consider a conductor with an electric current flowing in a direction perpendicular to an external magnetic field. Then in addition to the Joule heat it is found that a heat flow is produced in the sample in the direction perpendicular to both the electric current and the magnetic field. This is the Ettingshausen heat. The Ettingshausen coefficient is usually defined in terms of the temperature gradient which is produced if a sample is thermally insulated so that the Ettingshausen heat flux is zero. For currents small enough so that the Joule heat can be neglected, this temperature gradient is given by

$$\partial T/\partial y = \epsilon J_x B_z \qquad \text{with} \qquad J_y = Q_y = \partial T/\partial x = 0 \tag{12-57}$$

where ϵ is the Ettingshausen coefficient and Q is the heat flux.

129. Kelvin Relations. The thermoelectric coefficients are related as follows:

$$\begin{aligned} \Pi &= T\alpha \\ \tau_1 &= T(d\alpha_1/dT) \\ N_{xy}{}^i T &= \kappa\epsilon \end{aligned} \tag{12-58}$$

130. Figure of Merit. An important parameter in thermoelectric devices is the figure of merit for the materials. The figure of merit for a couple is defined as

$$Z = \alpha_{12}^2/[(\rho_1\kappa_1)^{1/2} + (\rho_2\kappa_2)^{1/2}]^2 \tag{12-59}$$

where the subscripts refer to the two materials making up the couple. In discussions of particular materials, it is convenient to define a figure of merit for a single material as

$$Z_i = \alpha_i^2/\rho_i\kappa_i \tag{12-60}$$

where α_i is the absolute Seebeck coefficient. The relation between the figure of merit for a couple and the figure of merit of the two individual materials is

$$Z_{\text{couple}} = (Z_1^{1/2} \pm \beta Z_2^{1/2})^2/(1 + \beta)^2 \tag{12-61}$$

where $\beta = (\rho_2\kappa_2/\rho_1\kappa_1)^{1/2}$. The positive sign is used if the Seebeck coefficients of the two materials have opposite signs. If the two materials are similar except that their Seebeck coefficients have opposite signs, the figure of merit of the couple is approximately equal to the average of the figures of merit of the two individual materials.

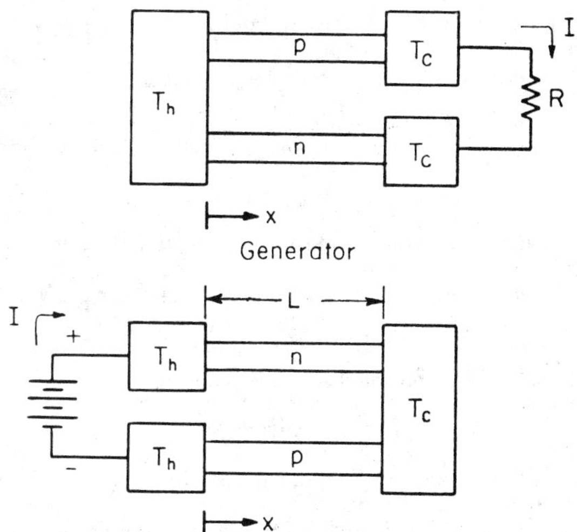

FIG. 12-90. Schematic drawing of a Seebeck-effect thermoelectric generator and a Peltier-effect refrigerator. The elements marked T_h and T_c are heat reservoirs and are assumed to have zero electrical resistance. The elements marked p and n are called arms and usually have positive and negative absolute Seebeck coefficients, respectively.

131. Thermal Energy Transfer. Figure 12-90 shows a single thermoelectric couple whose operation is based on the Seebeck-Peltier effects. The performance of a single couple of this type will be discussed in Pars. **132** through **140**. There is no heat transfer between the reservoirs except through the thermoelectric elements. There must be two active arms, labeled p and n in Fig. 12-90. The lengths of the arms are L_p and L_n, and the cross-sectional areas are A_p and A_n.

The rate of heat removal from a thermal reservoir q is given by the sum of two terms—q_P, the rate of Peltier heat absorption, and q_k, the rate of heat conduction down the arms n and p. The Peltier heat is given by

$$q_P = -\Pi_{pn} I_{pn} = -T\alpha_{pn} I_{pn} \tag{12-62}$$

where I_{pn} is the total current flow from the p material to the n material at the reservoir in question. The rate of heat conduction from the hot reservoir is given by

$$q_h = -A_n \kappa_n \left.\frac{dT_n}{dx}\right|_{x=0} - A_p \kappa_p \left.\frac{dT_p}{dx}\right|_{x=0} \tag{12-63}$$

and the rate of heat conduction into the cold reservoir is given by

$$q_c = -A_n \kappa_n \left.\frac{dT_n}{dx}\right|_{x=L_n} - A_p \kappa_p \left.\frac{dT_p}{dx}\right|_{x=L_p} \tag{12-64}$$

The temperature gradients are evaluated by determining the temperature distribution from the differential equation

$$\frac{d}{dx}\left(\kappa \frac{dT}{dx}\right) - \tau J \frac{dT}{dx} + \rho J^2 = 0 \tag{12-65}$$

For the case that the thermal conductivity, electrical resistivity, and Seebeck coefficient are independent of temperature [$\tau = 0$ by Eq. (12-58)], the thermal energy transfer to the hot and cold reservoirs is the sum of three separate components as follows:

1. The Joule heat delivered to each reservoir per unit time is $I^2R/2$, where

$$R = (\rho_n/\gamma_n) + (\rho_p/\gamma_p) \tag{12-66}$$

is the resistance of the two arms in series, ρ_n and ρ_p are the resistivities of the n and p arms, and γ_n and γ_p are

$$\gamma_n = A_n/L_n \qquad \gamma_p = A_p/L_p \tag{12-67}$$

The two arms may have different lengths.

2. The rate of transport of zero-current heat between the two reservoirs is $K\,\Delta T$, where

$$K = \kappa_n\gamma_n + \kappa_p\gamma_p \tag{12-68}$$

is the thermal conductance of the two arms in parallel and κ_n and κ_p are the thermal conductivities of the n and p arms.

3. The rate of Peltier heat absorption from each reservoir is given by

$$-T\alpha_{pn}I_{pn} \tag{12-69}$$

where α is the Seebeck coefficient of the couple p and n.

POWER GENERATION

132. General. The important design parameters for a power generator are the efficiency, power output, and heat input. The efficiency η is defined as the ratio of the electrical power dissipated in the load P_o to the thermal power input q_h to the hot junction,

$$\eta = P_o/q_h \tag{12-70}$$

The reduced efficiency η_r is defined as the ratio of the efficiency to the Carnot efficiency,

$$\eta_r = \eta T_h/\Delta T \tag{12-71}$$

where $\Delta T = T_h - T_c$. The thermal power input to the hot junction is given by the sum of the three effects discussed in Par. **131**,

$$q_h = \alpha T_h I - \tfrac{1}{2}I^2R + K\,\Delta T \tag{12-72}$$

where R and K are given by Eqs. (12-66) and (12-68). In a power generator, the positive direction for current is from the p arm to the n arm at the cold junction, and the current is given by

$$I = \alpha\,\Delta T/(R + R_L) = \alpha\,\Delta T/[R(s + 1)] \tag{12-73}$$

where R_L is the load resistance and $s = R_L/R$. The electrical power output is

$$P_o = I^2R_L = (\alpha\,\Delta T)^2 s/[R(s + 1)^2] \tag{12-74}$$

The output voltage is

$$V = IR_L = \alpha\,\Delta T s/(s + 1) \tag{12-75}$$

and the efficiency is

$$\eta = \frac{2s\,\Delta T}{T_h(2s + 1) + T_c + 2(1 + s)^2RK\alpha^{-2}} \tag{12-76}$$

There are five parameters which must be selected to optimize the generator design: T_h, T_c, γ_n, γ_p, and s. [Once γ_n and γ_p are determined, R is given by Eq. (12-66). Thus adjustment of s is done by varying the load resistance.] Three design criteria will be considered in the next three sections. In some applications other criteria may be important, and designs different from those considered here may be most suitable.

133. Maximum Efficiency between Fixed Temperatures. The parameters of a couple which has been designed to maximize the efficiency when operating with fixed hot- and cold-junction temperatures will be given in this section. This type of operation assumes that T_c and T_h do not change as R_L, γ_n, and γ_p are varied. The shape ratio which maximizes the efficiency is

$$\gamma_n/\gamma_p = (\kappa_p\rho_n/\kappa_n\rho_p)^{1/2} \tag{12-77}$$

With this shape ratio, $RK\alpha^{-2} = Z^{-1}$. The load resistance which gives maximum efficiency is

$$R_L = Rs_e \tag{12-78}$$

where $s_e = (1 + Z\bar{T})^{1/2}$ and $\bar{T} = (T_h + T_c)/2$. Values for s_e are given in Fig. 12-91.

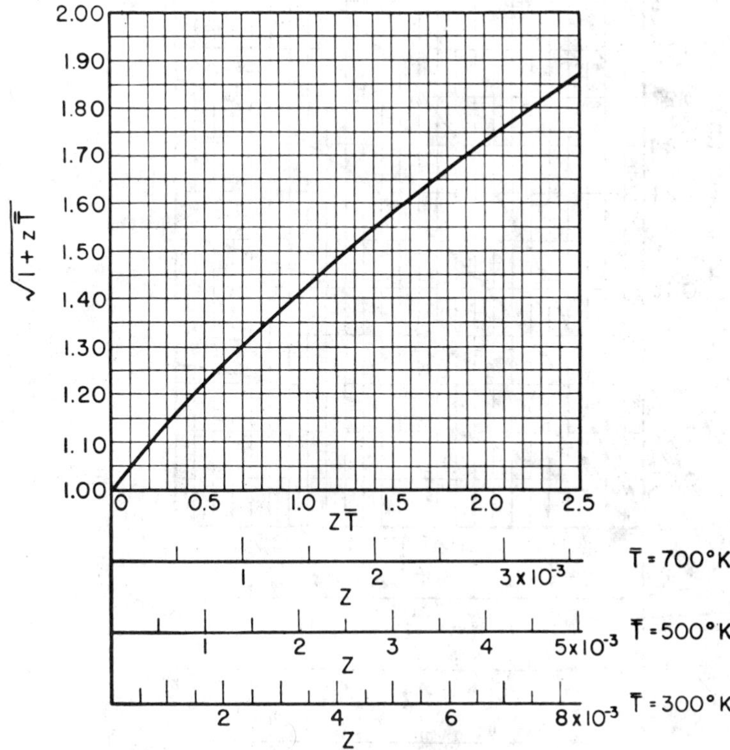

FIG. 12-91. Values of $(1 + Z\bar{T})^{1/2}$ as a function of $Z\bar{T}$, or of Z for three values of $\bar{T}$.

The efficiency with both the geometry and the load resistance optimized is

$$\eta = \frac{\Delta T}{T_h} \frac{s_e - 1}{s_e + (T_c/T_h)} \tag{12-79}$$

The optimum efficiency is shown in Fig. 12-92. With optimum load and geometry, the shape factor required to give a power output P_o is

$$\gamma_n = \frac{A_n}{L_n} = \frac{P_o(1 + s_e)^2}{(\Delta T)^2 s_e \alpha} \left(\frac{\rho_n}{Z\kappa_n}\right)^{1/2} \tag{12-80}$$

and the internal resistance is

$$R = \frac{\alpha}{\gamma_n} \left(\frac{\rho_n}{Z\kappa_n}\right)^{1/2} = \frac{(\alpha\,\Delta T)^2 s_e}{P_o(1 + s_e)^2} \tag{12-81}$$

The output voltage is given by Eq. (12-75), with s_e substituted for s.

For quick calculations using materials in which ρ_n and ρ_p are not too different and

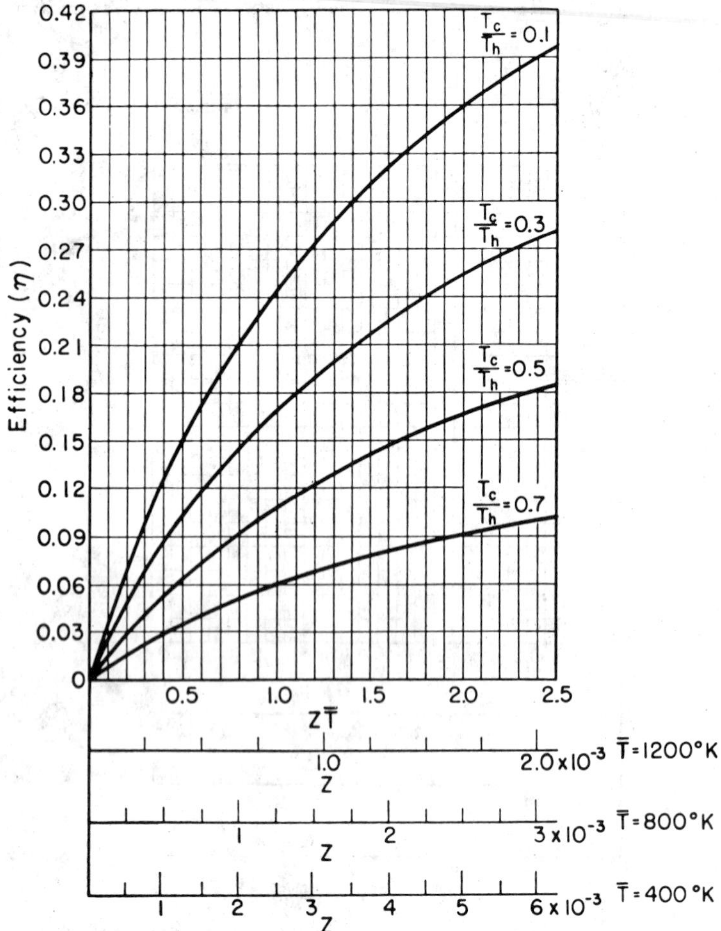

FIG. 12-92. Efficiency of a thermoelectric couple as a function of $Z\bar{T}$, or of Z for three values of $\bar{T}$. Designed for maximum efficiency.

κ_n and κ_p are similar, the following expressions can be used,

$$\gamma_n \approx \gamma_p \approx 2P_o(1 + s_e)^2\bar{\rho}/[s_e(\alpha\,\Delta T)^2] \tag{12-82}$$

$$R \approx 2\bar{\rho}/\gamma_n \tag{12-83}$$

$$Z \approx \alpha^2/4\bar{\rho}\bar{\kappa} \tag{12-84}$$

where

$$\bar{\rho} = (\rho_n + \rho_p)/2 \qquad \text{and} \qquad \bar{\kappa} = (\kappa_n + \kappa_p)/2 \tag{12-85}$$

134. Maximum Power Output between Fixed Temperatures. A generator operating with fixed hot- and cold-junction temperatures can also be designed to maximize the power output instead of the efficiency. As is well known, the load resistance which gives this maximum power output is $R_L = R$ or $s = 1$. With this load resistance, the current, power output, and load voltage are given by Eqs. (12-73), (12-74), and (12-75) with $s = 1$. The size factors which maximize the efficiency are again given by Eq. (12-77), and the efficiency is

$$\eta = \frac{2\,\Delta T}{3T_h + T_c + 8Z^{-1}} \tag{12-86}$$

For $Z\bar{T} = 3$, the efficiencies given by Eqs. (12-79) and (12-86) differ by about 10%, and this difference becomes smaller for smaller $Z\bar{T}$.

However, the usual objective in designing for maximum power output is to minimize the size or weight of a generator or to minimize the thermoelectric material required. Thus the factor which should be maximized is, not the efficiency, but the power output divided by the sum of the size factors, P_o/γ_t, where $\gamma_t = \gamma_n + \gamma_p$. (The length of the elements should also be minimized. This will be discussed in Par. **144.**) The size-factor ratio which maximizes P_o/γ_t is

$$\gamma_n/\gamma_p = (\rho_n/\rho_p)^{1/2} \tag{12-87}$$

With this size-factor ratio, the internal resistance is

$$R = [\rho_n + 2(\rho_n\rho_p)^{1/2} + \rho_p]/\gamma_t$$

$$= [\rho_n + (\rho_n\rho_p)^{1/2}]/\gamma_n \tag{12-88}$$

The size factor required to produce a power output P_o is

$$\gamma_n = 4P_o[\rho_n + (\rho_n\rho_p)^{1/2}]/(\alpha\,\Delta T)^2 \tag{12-89}$$

The efficiency is given by Eq. (12-76), with $s = 1$ and RK replaced by

$$RK = \rho_n\kappa_n + (\rho_n\rho_p)^{1/2}(\kappa_n + \kappa_p) + \rho_p\kappa_p \tag{12-90}$$

The performance of devices designed by the procedures of Pars. **133** and **134** are similar unless κ_n and κ_p are extremely different.

135. Constant Heat Input with Fixed Cold-junction Temperature. In applications such as generators utilizing solar energy, the hot-junction temperature is not fixed but varies as the load resistance and the size of the thermoelectric elements are changed. It is thus of some interest to determine the optimum design under conditions of fixed heat flux. However, in the approximation being used here (α, ρ, and κ independent of temperature, and no heat losses except through the thermoelectric arms), there is no optimum design. As the load resistance is made larger and γ is made smaller, the hot-junction temperature rises and the efficiency and power output increase. In the limit, the load resistance, the internal resistance, and the hot-junction temperature all approach infinity, the efficiency approaches unity, and the power output approaches the fixed input heat flux.

There are two approaches which can be taken to the design. (1) If the effect of the heat loss through the supporting structure of the device is taken into account, then the efficiency and power output will go through a maximum as the hot-junction temperature is increased by increasing the load resistance and decreasing γ. Thus there will be an optimum design when the inevitable heat losses in the device are taken into account. (2) There is an upper limit to the temperature at which available thermoelectric materials can be used. Thus the device can be designed by choosing a fixed hot-junction temperature on the basis of performance of available materials and then designing the device by the procedures of Par. **133** or **134**.

REFRIGERATION

136. General. The important performance parameters for a refrigerator are the coefficient of performance φ, the power input P, and the rate of heat removal from the cooled body, q_c. The coefficient of performance is defined as $\varphi = q_c/P$, while the reduced coefficient of performance φ_r is the ratio of the actual coefficient of performance to the theoretical coefficient of performance of a Carnot refrigerator, $\varphi_r = \varphi\,\Delta T/T_c$. The relations describing the performance are written in terms of a parameter t which gives the current in the thermoelectric couple as

$$I = \alpha\,\Delta T/[R(t-1)] \tag{12-91}$$

The applied voltage is

$$V = t\alpha\,\Delta T/(t-1) \tag{12-92}$$

and the power input is

$$P = t(\alpha \, \Delta T)^2/[R(t-1)^2] \tag{12-93}$$

The rate of heat removal from the cold junction is given by the sum of the three terms discussed in Par. **131**,

$$q_c = \frac{\alpha^2 \, \Delta T}{2R(t-1)^2}[T_c(2t-1) - T_h - 2(t-1)^2 KR\alpha^{-2}] \tag{12-94}$$

The coefficient of performance is

$$\varphi = \frac{T_c(2t-1) - T_h - 2(t-1)^2 KR\alpha^{-2}}{2t \, \Delta T} \tag{12-95}$$

137. Maximum Coefficient of Performance between Fixed Temperatures. The shape ratio which maximizes φ is given by Eq. (12-77). The value of the parameter t

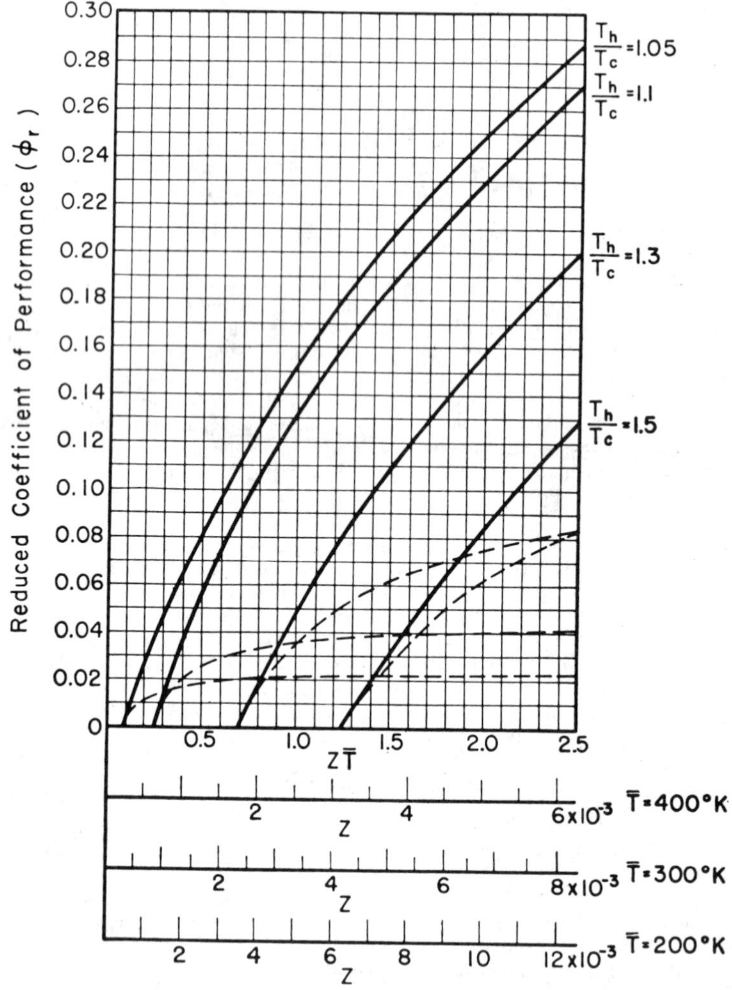

FIG. 12-93. Reduced coefficient of performance of a thermoelectric couple as a function of $Z\overline{T}$, or of Z for three values of $\overline{T}$. The solid curves are for a device designed for maximum coefficient of performance [Eq. (12-96)]. The dashed curves are for a device designed for maximum heat-pumping rate [Eq. (12-97)].

which maximizes φ is $t_e = (1 + Z\bar{T})^{1/2} = s_e$. Under these conditions, the coefficient of performance is

$$\varphi = \frac{T_c}{\Delta T} \frac{t_e - (T_h/T_c)}{t_e + 1} \tag{12-96}$$

The reduced coefficient of performance for this case is shown in Fig. 12-93.

138. Maximum Heat Pumping between Fixed Temperatures. The value of the parameter t which maximizes q_c for fixed T_c and T_h is $t_q = T_h/T_c$. With this value for t and the shape ratio which maximizes φ, the coefficient of performance is

$$\varphi = \frac{T_c^2 - 2Z^{-1}\,\Delta T}{2T_hT_c} \tag{12-97}$$

This parameter is shown by the dotted lines in Fig. 12-93. Note that, in contrast to the generator case, there is a significant difference between the coefficients of performance given by Pars. **137** and **138**. This difference becomes greater as the temperature difference is made smaller. In fact, when optimized for maximum φ, the heat-pumping rate goes to zero as the temperature difference goes to zero. When the temperature difference is small, it is often advantageous to use a design which lies somewhere between the extremes of maximum φ and maximum q_c.

139. Maximum Temperature Difference. The maximum temperature difference which a thermoelectric couple will produce is

$$\Delta T_{\max} = \tfrac{1}{2}ZT_c^2 \tag{12-98}$$

At this temperature difference, the coefficient of performance is zero, and the designs discussed in Pars. **137** and **138** become identical.

140. Heat Pump. In some applications, the major interest is in the rate at which heat is delivered to the hot junction q_h. The coefficient of performance φ_h is defined as $\varphi_h = q_h/P$. From the first law of thermodynamics it is easily seen that $\varphi_h = \varphi + 1$. The maximizing t and γ_n/γ_p given in Par. **137** and **138** maximize the corresponding heat-pump designs.

MULTICOUPLE DEVICES

141. Load Matching—Series-couple Devices. The single-couple devices which have been discussed so far are essentially very-low-voltage high-current devices. The optimizing load resistances of Pars. **133** and **134** are much smaller than the normal resistance of most devices which would be powered by a thermoelectric generator. To work into higher-resistance loads, the couple can be split into a number of couples of smaller size arranged electrically in series and thermally in parallel as shown in Fig. 12-94. In the following discussion the symbols without subscript m refer to the single-couple device, while the symbols with the subscript m refer to a device having m couples, each with a size factor $\gamma_m = \gamma/m$. The output voltage is given by $V_m = mV$ and the output current by $I_m = I/m$; so the power output is unchanged. The internal resistance is given by $R_m = m^2R$, and the matching load resistance is $R_{Lm} = m^2R_L$.

The procedure is to design a

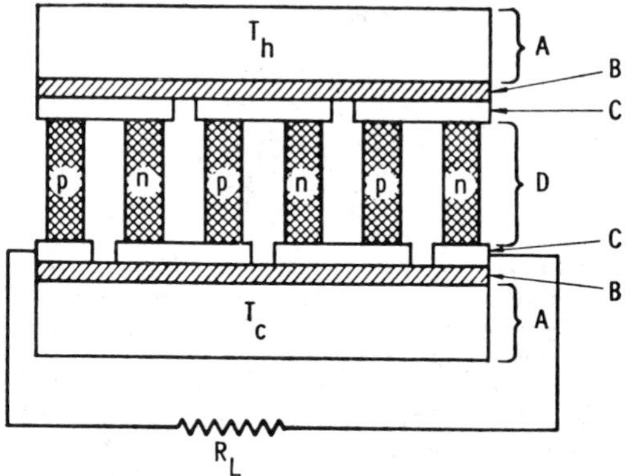

Fig. 12-94. A single-stage thermoelectric device having three couples thermally in parallel and electrically in series. Material B is an electrically insulating material of high thermal conductivity so that the straps C are at the same temperature as the adjacent heat reservoir.

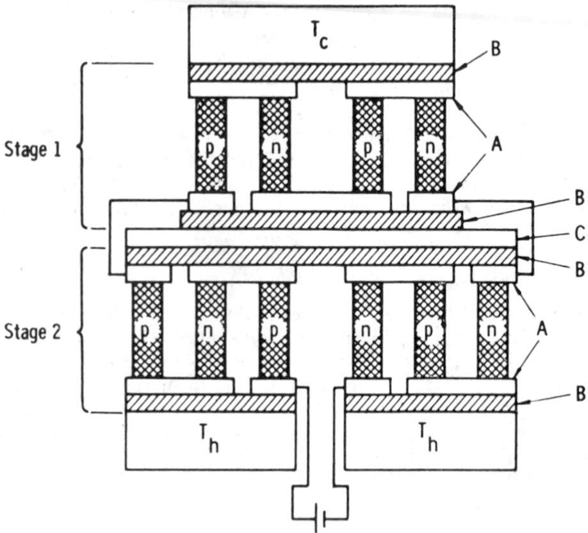

FIG. 12-95. A two-stage thermoelectric device having two couples in stage 1 and three couples in stage 2. All the couples are electrically in series. The material B is an electrically insulating material of high thermal conductivity, so that the temperature drop across it is a minimum. C is a material of high thermal conductivity, inserted to distribute the heat across the device so that all couples in the same stage operate between the same two temperatures.

single-couple device which will supply the power desired P_o by using the procedure discussed in the previous sections. The relations above give the number of couples m which will supply this power into a specified load resistance R_{Lm} or at a specified voltage or current V_m or I_m.

142. Multistage Devices. A number of couples can also be used by placing couples thermally in series as shown in Fig. 12-95. There are several reasons for using this type of structure. Thermoelectric generators normally operate with a large temperature difference across them. Most materials have a high figure of merit over only a limited temperature range. In principle, the multistage structure is the most efficient way of using different materials in different temperature ranges. On assuming that there is zero temperature drop across the thermal insulation between stages (material B in Fig. 12-95), it can be proved that a multistage device is more efficient (or has higher coefficient of performance) than a single-stage device and that the efficiency increases as the number of stages increases. The efficiency of an N-stage generator is

$$\eta_N = 1 - \prod_{j=1}^{N}(1 - \eta_j) \tag{12-99}$$

where η_j is the efficiency of the jth stage. The intermediate temperatures can be chosen to optimize the overall efficiency. The geometry of the stages must be adjusted to produce the desired intermediate temperatures. The overall efficiency is also maximized by maximizing the efficiency of each stage. Harman[1] has considered the design of a two-stage device.

Multistage devices are important for refrigeration applications, since temperature differences larger than the maximum ΔT for a single stage can be produced in this way. The coefficient of performance of an N-stage refrigerator is

$$\varphi_N = \left[-1 + \prod_{j=1}^{N} (1 + \varphi_j^{-1})\right]^{-1} \tag{12-100}$$

[1] HARMAN, T. *Jour. Appl. Phys.*, October, 1958, Vol. 29, No. 10, p. 1471.

where φ_j is the coefficient of performance of the jth stage. The overall coefficient of performance is close to a maximum if the intermediate temperatures are adjusted so that $\varphi_1 = \varphi_2 = \varphi_3 = \cdots$.

If the number of stages N in Eq. (12-99) or (12-100) is allowed to approach infinity, an expression can be given for the efficiency of a theoretical infinite-stage device. This result is an upper limit on the efficiency or coefficient of performance obtainable with a given set of materials operating over a given temperature range. The efficiency of such an infinite-stage generator is

$$\eta_\infty = 1 - \exp\left(-\int_{T_c}^{T_h} \epsilon\, dT/T\right) \tag{12-101}$$

where $$\epsilon = [(1 + TZ)^{1/2} - 1]/[(1 + TZ)^{1/2} + 1]$$

and the coefficient of performance of an infinite-stage refrigerator is

$$\varphi_\infty = \left[\exp\left(\int_{T_c}^{T_h} \frac{dT}{T\epsilon}\right) - 1\right]^{-1} \tag{12-102}$$

Even on the assumption that there is no temperature drop across the thermal insulation between the stages, the increase in efficiency on going from one to many stages is small for a generator. For a refrigerator, the situation is similar if the temperature difference across the device is smaller than two-thirds the maximum temperature difference. However, when the temperature difference is close to or larger than the maximum temperature difference, the increase in coefficient of performance on increasing the number of stages may be significant.

In practice it is difficult to keep the temperature difference across the electrical insulation between stages small. For this reason multistage generators are seldom used. Multistage refrigerators have been built, but the number of stages is usually made just large enough to produce the temperature difference required.

143. Segmented Arms. Different materials can be used at different temperatures in a generator by segmenting the arms as shown in Fig. 12-96. If the materials making up the arm (for example, n' and n'') are similar in their electrical properties, the efficiency of this device is close to that for a multistage device using the same materials. However, if the materials are badly mismatched, there can be a large difference between the efficiency of a segmented device and the equivalent multistage device, the segmented device always having the lower efficiency.

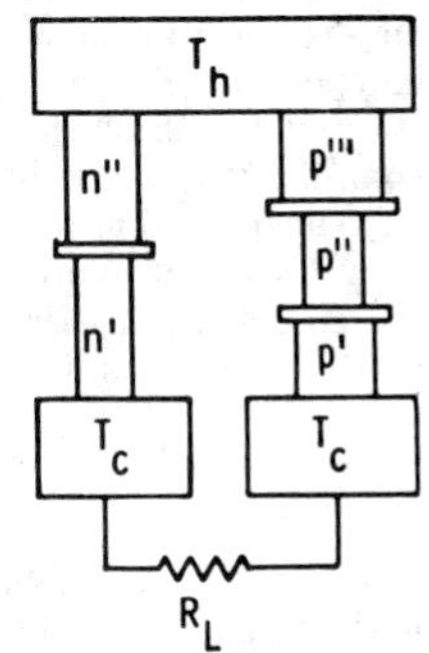

FIG. 12-96. A segmented thermoelectric generator in which the two arms are made of a number of different materials.

The lengths of the individual segments are calculated from the desired interface temperatures and the condition that the total heat flowing into any interface must be zero. The optimization of device performance is difficult, since the relative lengths of the individual segments are a function of the current through the device and hence of the load resistance. It is not possible, in general, to determine analytically the optimum values of R_L and the γ's. The design problem has been solved exactly[1] by the use of iterative procedures on a high-speed computer. An approximate procedure which is accurate enough for most applications is to calculate the length of the individual segments by considering the heat-flow problem under the condition of zero current. With these lengths determined, the average values of electrical resistivity, thermal conductivity, and Seebeck coefficient for the entire arm can be calculated. Then the design procedures of Par. **133** or **134** can be used.

[1] SWANSON, B., SOMERS, E., and HEIKES, R. *Jour. Heat Transfer*, February, 1961, Vol. 83, p. 77.

CONTACTS, TEMPERATURE-DEPENDENT MATERIALS, AND MAGNETIC FIELD DEVICES

144. Element Length and Contact Resistance. The only way in which the size of the arms enters the performance relations is through the size factor $\gamma = A/L$. The performance is independent of the particular A or L provided that their ratio remains constant. However, the volume of thermoelectric material, V (and therefore the weight of the device), is given by $V_n = \gamma_n L_n^2$, and thus the amount and weight of the thermoelectric material decrease rapidly as the length of the arms is reduced. There are two factors which place a lower limit on the length of the arms. (1) The heat flux density increases as L (and therefore A) is reduced. It may require heavy and expensive structures at the hot and cold junctions to concentrate the heat into the smaller cross-sectional area of a shorter element. Also, there may be larger heat losses through the shorter supporting structure involved with shorter elements. Ingenious design of a device may obviate these possible disadvantages. (2) The contact resistance has not been taken into account in the previous calculations. The contact resistance at the two cold junctions R_c written in terms of the contact resistance per unit area r is

$$R_c = (r_n/A_n) + (r_p/A_p) \tag{12-103}$$

It will be assumed that the contact resistances at the hot and cold junctions are equal. Then the internal resistance is the sum of Eq. (12-66) plus twice Eq. (12-103). Thus all the formulas derived for the zero-contact-resistance case can be used, provided that an apparent resistivity ρ' defined as

$$\rho' \equiv \rho + 2(r/L) \tag{12-104}$$

is used in place of the resistivity. Hence, as the length of the elements is decreased, the apparent resistivity increases and thus gives a decrease in the performance of the device. The length of the element must be determined by balancing the decrease in performance against the reduction in weight and volume of thermoelectric material.

145. Materials Parameters Varying with Temperature. The exact calculation of device performance for arbitrary temperature dependence of ρ, κ, and α must be done by numerical methods. Expressions for the temperature distribution within an element have been derived for some special temperature dependences of ρ, κ, and α. For example, the case of ρ having a linear variation with temperature and τ, rather than α, independent of temperature has been discussed.[1] However, even with these very restricted forms of the temperature dependence, the expressions for φ, η, and q_c are so complicated that it is impossible to maximize them with respect to R_L, γ_n/γ_p, etc.

An approximate procedure which is often used is to employ average values of ρ, κ, and α in the expressions derived for the case of the materials parameters independent of temperature. In general, simple averages are used. The simple average of the quantity $x(T)$ is defined as

$$\bar{x} \equiv (\Delta T)^{-1} \int_{T_c}^{T_h} x(T)\, dT \tag{12-105}$$

However, Eqs. (12-66) and (12-68) give the correct value for the internal resistance and the thermal conductance at zero current if α, κ, and ρ are replaced by $\bar{\alpha}$, $\bar{\kappa}$, and ρ_{avg} defined by

$$\rho_{\text{avg}} = (\bar{\kappa}\, \Delta T)^{-1} \int_{T_c}^{T_h} \rho(T)\, \kappa(T)\, dT \tag{12-106}$$

The average figure of merit ordinarily used is

$$Z_{\text{avg}} = \bar{\alpha}^2 [(\overline{\rho_n \kappa_n})^{1/2} + (\overline{\rho_p \kappa_p})^{1/2}]^{-2} \tag{12-107}$$

rather than the average Z. Note that $\bar{\kappa}\rho_{\text{avg}} = \overline{\rho\kappa}$.

[1] Burshtein, A. I. *Soviet Phys. Tech. Phys.*, July, 1957, Vol. 2, No. 7, p. 1397.

A comparison between the exact machine calculation and approximate calculations using the average parameters discussed above has shown that the approximate calculation for a generator agrees to better than 5 to 10% with the exact calculation.[1] However, for a refrigerator, the exact and approximate calculations sometimes differ by 20 to 25%. Thus, the approximate calculation is usually sufficiently accurate for generator-design purposes but may not be for refrigerator design.

146. Nernst-Ettingshausen Devices. Power generators and refrigerators can be made by utilizing Nernst and Ettingshausen effects, respectively. The basic device configuration is shown in Fig. 12-97. Note that these types of devices require a magnetic field for their operation. Therefore consideration must be given to the thermal-boundary conditions at the surfaces of the active elements. The two cases usually considered are the isothermal and the adiabatic cases. The isothermal electrical resistivity $\rho_{xx}{}^i$ is defined by

$$E_x = \rho_{xx}{}^i J_x \qquad \text{with} \qquad \nabla T = B_x = B_y = 0$$

where B is the magnetic field. The adiabatic electrical resistivity $\rho_{xx}{}^a$ is defined by

$$E_x = \rho_{xx}{}^a J_x \qquad \text{with} \qquad Q_x = B_x = B_y = 0$$

Here Q is the heat current density. Similarly, the isothermal thermal conductivity κ^i is defined by

$$Q_y = -\kappa_{yy}{}^i(\partial T/\partial y)$$

with

$$\partial T/\partial x = \partial T/\partial z = \mathbf{J} = B_x = B_y = 0$$

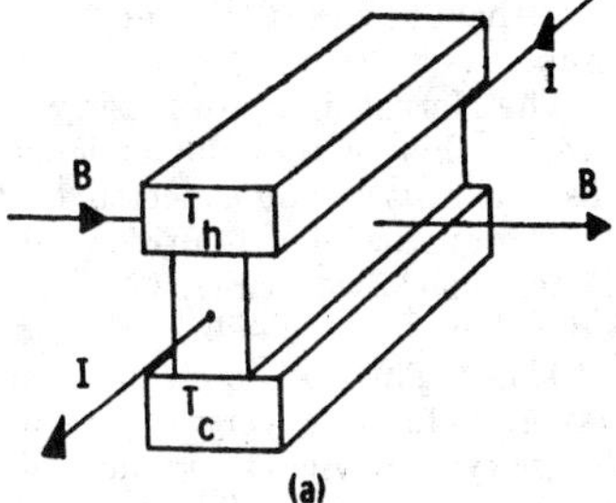

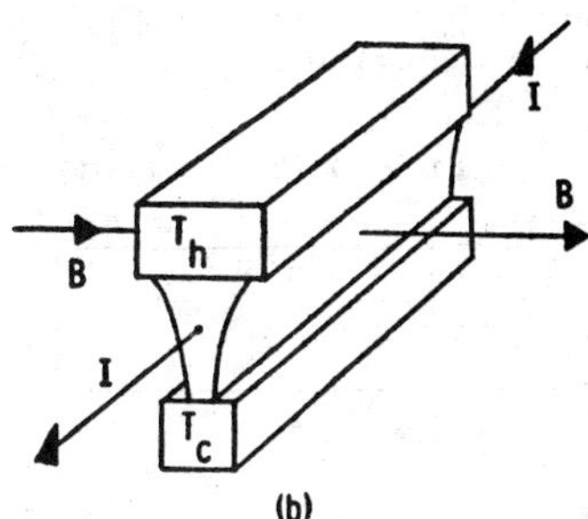

FIG. 12-97. Nernst-Ettingshausen devices. The elements marked T_h and T_c are heat reservoirs which are assumed to be electrically insulating and at constant temperature along their length. The active element runs between the two heat reservoirs and carries a current I along its length and has a magnetic field B perpendicular to it. (*a*) Rectangular device having a performance similar to a single-stage Seebeck-Peltier device; (*b*) tapered device, having performance similar to multistage Peltier-Seebeck device.

The internal resistance of a Nernst-Ettingshausen device can be matched to customary power supplies or power loads by making the element long in the direction parallel to the current and thin in the directions perpendicular to it. With this sort of shape, the device will have a very small temperature gradient in the direction parallel to the current. This configuration is "isothermal." The isothermal Nernst figure of merit is given by

$$Z_{xy}{}^i = (B_z N_{xy}{}^i)^2/\rho_{xx}{}^i \kappa_{yy}{}^i \tag{12-108}$$

Kooi, Horst, Cuff, and Hawkins[2] have shown that the performance of an isothermal Nernst or Ettingshausen device having a rectangular cross section as shown in Fig. 12-97*a* is given by Eqs. (12-79), (12-96), and (12-98), provided that the *adiabatic* Nernst figure of merit $Z_{xy}{}^a$ is used in place of the Peltier figure of merit. This adiabatic figure of merit is given by

$$Z_{xy}{}^a = (B_z N_{xy}{}^i)^2/\rho_{xx}{}^a \kappa_{yy}{}^i \tag{12-109}$$

The relation between these two figures of merit is

$$Z^i T = Z^a T/(1 + Z^a T) \tag{12-110}$$

[1] SHERMAN, B., HEIKES, R., and URE, R. *Jour. Appl. Phys.*, January, 1960, Vol. 31, No. 1, p. 1.
[2] KOOI, C. F., HORST, R. B., CUFF, K. F., and HAWKINS, S. R. *Jour. Appl. Phys.*, June, 1963, Vol. 34, No. 6, p. 1735.

Note that Z^i varies through the range from 0 to 1 only, while Z^a varies from 0 to infinity.

The Nernst figure of merit will be large only at temperatures below room temperature; so this type of device is expected to be useful only at low temperatures. For this reason it is not expected that the Nernst generator will find many applications.

The analog of a multistage device for the Ettingshausen-type device is a shaped element as shown in Fig. 12-97*b*. In this case the hot side of the device is made much wider than the cold side so that there is a much higher heat-pumping capacity near the hot side. There is no electrical insulation between stages as in the multistage Peltier device. This is a very big advantage, since it is extremely difficult in a multistage Peltier-type device to reduce the temperature drop across the electrical insulation between the stages. The Ettingshausen refrigerator appears promising in applications in which it is desired to produce a large temperature difference at temperatures below room temperature.

Table 12-12. Properties of *p*-type Materials for Thermoelectric Devices as Reported in the Literature

	Temp., °C	Seebeck coeff., μV/°C	Electrical resistivity, Ω·cm	Thermal conductivity, W/(cm)(°C)	Figure of merit, (°C)$^{-1}$
1. $Bi_{0.5}Sb_{1.5}Te_3$[a]					
	25	210	0.98×10^{-3}	1.27×10^{-2}	3.5×10^{-3}
	100	215	1.3	1.24	2.8
	200	200	2.0	1.18	1.8
	300	115	2.1	1.23	0.5
2. PbTe + 0.1 atomic % Na[b]					
	25	205	1.30×10^{-3}	2.0×10^{-2}	1.7×10^{-3}
	100	265	2.8	1.75	1.4
	200	345	6.8	1.51	1.2
	300	360	13.8	1.37	0.8
3. $AgSbTe_2$[c]					
	100	228	4.8×10^{-3}	7.5×10^{-3}	1.4×10^{-3}
	300	240	4.3	8.3	1.6
	400	238	4.0	8.8	1.6
	500	228	3.8	9.3	1.5
4. PbTe + 1.0 atomic % Na[b]					
	100	108	0.60×10^{-3}	2.5×10^{-2}	0.7×10^{-3}
	300	238	3.0	1.63	1.2
	500	277	5.8	1.38	0.9
	600	230	5.4	1.39	0.7
5. $Ge_{0.3}Si_{0.7} + 1.8 \times 10^{20}$ cm^{-3} B[d]					
	300	175	1.35×10^{-3}	4.8×10^{-2}	4.7×10^{-4}
	500	213	1.74	4.5	5.7
	700	244	2.2	4.2	6.3
	900	275	2.8	4.6	5.9
6. $(GeTe)_{0.95}(Bi_2Te_3)_{0.05}$[e]					
	300	167	1.22×10^{-3}	2.0×10^{-2}	1.1×10^{-3}
	400	177	1.13	2.0	1.4
	500	203	1.30	1.85	1.7
	600	203	1.48	2.0	1.4
7. ZnSb[f]					
	100	197	1.8×10^{-3}	2.2×10^{-2}	1.0×10^{-3}
	200	223	2.3	1.70	1.2
	300	232	3.7	1.63	0.9
	400	224	5.4	1.58	0.6

[a] $Bi_{0.5}Sb_{1.5}Te_3$ + 1.75% Se, ROSI, F. D., HOCKINGS, E. F., and LINDENBLAD, N. E. *RCA Rev.*, March, 1961, Vol. 22, No. 1, p. 99, Fig. 10.
Thermal conductivity calculated from electrical resistivity measured at temperature and thermal conductivity measured at room temperature, on assuming lattice thermal conductivity independent of temperature.

[b] PbTe, FRITTS, R. W. *Trans. AIEE*, January, 1960, Vol. 78, Pt. I, p. 817.
Thermal conductivity and figure of merit, RICHARDS, J. D. Private communication, 1960.

[c] $AgSbTe_2$, ROSI, F. D., DISMUKES, J. P., and HOCKINGS, E. F. *Elec. Eng.*, June, 1960, Vol. 79, No. 6, p. 450.
Thermal conductivity calculated from electrical resistivity measured at temperature and thermal conductivity measured at room temperature, on assuming lattice thermal conductivity independent of temperature.

[d] $Ge_{0.3}Si_{0.7}$, DISMUKES, J. P., EKSTROM, L., STEIGMEIER, E. F., KUDMAN, I., and BEERS, D. S. *Jour. Appl. Phys.*, October, 1964, Vol. 35, No. 10, p. 2899. Additional data courtesy Dr. F. D. Rosi, RCA Laboratories, Princeton, N.J.

[e] $(GeTe)_{0.95}(Bi_2Te_3)_{0.05}$, MILLER, R. C., and URE, R. In Snyder, N. W. (ed.), "Energy Conversion for Space Power"; New York, Academic Press, Inc., 1961, p. 27.

[f] ZnSb, MILLER, R. C. In Heikes, R. R., and Ure, R. W. (eds.), "Thermoelectricity: Science and Engineering"; New York, Interscience Publishers, Inc., 1961.

MATERIALS

147. Materials for Seebeck-Peltier Devices. The figure of merit published in the literature for the most important materials for Seebeck-Peltier devices is shown in Figs. 12-98 and 12-99, and the properties of the materials are given in Tables 12-12 and 12-13.

148. Materials for Ettingshausen-Nernst Devices. The most important materials for Nernst-Ettingshausen devices are bismuth and the bismuth-antimony alloys.[1]

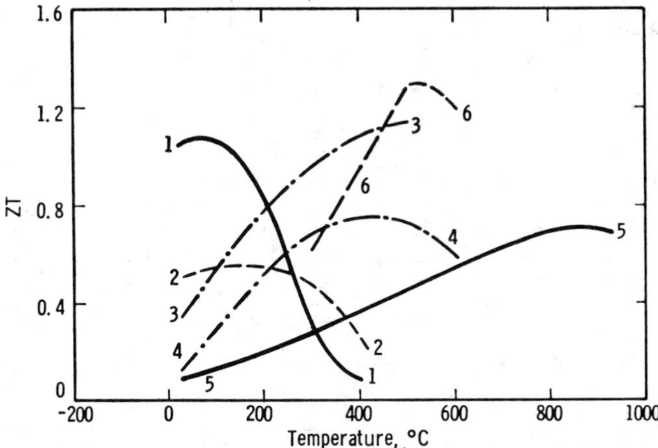

FIG. 12-98. Figure of merit times temperature, ZT, for p-type thermoelectric materials. Numbers identifying materials are the same as in Table 12-12:

1. $Bi_{0.5}Sb_{1.5}Te_3$.
2. PbTe + 0.1% Na.
3. $AgSbTe_2$.
4. PbTe + 1.0% Na.
5. $Ge_{0.3}Si_{0.7}$.
6. $(GeTe)_{0.95}(Bi_2Te_3)_{0.05}$.

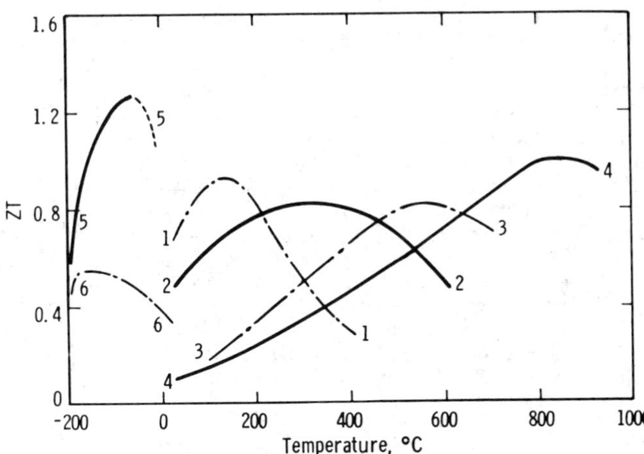

FIG. 12-99. Figure of merit times temperature, ZT, for n-type thermoelectric materials. Numbers identifying materials are the same as in Table 12-13:

1. $Bi_2Te_{2.4}Se_{0.6}$ + 0.5% CuBr. 2. PbTe + 0.03% PbI_2. 3. PbTe + 0.25% Bi. 4. $Ge_{0.3}Si_{0.7}$. 5. $Bi_{88}Sb_{12}$ with magnetic field. 6. $Bi_{88}Sb_{12}$ in zero magnetic field.

[1] CUFF, K. F., HORST, R. B., WEAVER, J. L., HAWKINS, S. R., KOOI, C. F., and ENSLOW, G. M. *Appl. Phys. Letters*, Apr. 15, 1963, Vol. 2, No. 8, p. 145; HARMAN, T. C., HONIG, J. M., FISCHLER, S., and PALADINO, A. E. *Solid State Electronics*, July, 1964, Vol. 7, No. 7, p. 505.

Table 12-13. Properties of *n*-type Materials for Thermoelectric Devices as Reported in the Literature

	Temp., °C	Seebeck coeff., μV/°C	Electrical resistivity, Ω·cm	Thermal conductivity, W/(cm)(°C)	Figure of merit, $(°C)^{-1}$
1. $Bi_2Te_{2.4}Se_{0.6}$ + 0.05 wt. % CuBr[a]					
	25	−192	1.07×10^{-3}	1.53×10^{-2}	2.3×10^{-3}
	100	−216	1.34	1.43	2.4
	200	−218	1.76	1.56	1.7
	300	−185	1.90	2.0	0.9
2. PbTe + 0.03 mole % PbI_2[b]					
	25	−138	0.42×10^{-3}	2.7×10^{-2}	1.6×10^{-3}
	200	−206	1.38	1.95	1.6
	400	−260	4.0	1.47	1.2
	500	−255	5.2	1.38	0.9
3. PbTe + 0.25% Bi[c]					
	100	− 85	0.57×10^{-3}	2.8×10^{-2}	0.5×10^{-3}
	300	−154	1.57	1.78	0.9
	500	−218	3.5	1.30	1.0
	700	−209	5.0	1.22	0.7
4. $Ge_{0.3}Si_{0.7}$ + 1.5×10^{20} cm^{-3} P[d]					
	300	−176	1.17×10^{-3}	4.3×10^{-2}	6.2×10^{-4}
	500	−227	1.66	4.1	7.6
	700	−279	2.2	4.0	8.8
	900	−257	1.84	4.3	8.4
PbTe + 0.1 mole % PbI_2[b]					
	200	−120	0.52×10^{-3}	3.2×10^{-2}	0.8×10^{-3}
	300	−150	0.87	2.6	1.0
	400	−174	1.43	2.2	1.0
	500	−194	2.2	1.8	0.9

5. $Bi_{88}Sb_{12}$ in magnetic field which maximizes the figure of merit, as follows:[e]

Temperature, °K	75	100	125	150	200	220
Magnetic field, kG	0.4	1.0	2.2	4.2	12	17

6. $Bi_{88}Sb_{12}$ in zero magnetic field[e].

[a] $Bi_2Te_{2.4}Se_{0.6}$ + 0.05 wt. % CuBr, Ure, R. W., McHugh, J. P., Bauerle, J. W., Milnes, M., and Gallo, C. P. In Heikes, R. R., and Ure, R. W. (eds.), "Thermoelectricity: Science and Engineering"; New York, Interscience Publishers, Inc., 1961, Fig. 13.11, p. 426.

[b] PbTe. Same as footnote *b* of Table 12-12.

[c] PbTe + 0.25% Bi, Lubell, M. S., Snyder, P. E., Metz, E. P., and Petrolo, J. *Thermoelectricity Quart. Rept.* 4, November, 1960, to U.S. Navy, Bureau of Ships, Contract NObs-78365.

[d] $Ge_{0.3}Si_{0.7}$ Same as footnote *d* of Table 12-12.

[e] $Bi_{88}Sb_{12}$, Wolfe, R., and Smith, G. *Rept. Intern. Conf. Phys. Semiconductors*, Exeter, England, 1962, Institute of Physics and Physical Society, London, p. 774.

149. References

Heikes, R. R., and Ure, R. W., Jr. "Thermoelectricity: Science and Engineering"; New York, Interscience Publishers, Inc., 1961.

Goldsmid, H. J. "Thermoelectric Refrigeration"; New York, Plenum Press, 1964.

Proc. IEEE, Special Issue, May, 1963, Vol. 51, No. 5, pp. 699–724.

Elec. Eng., Special Issues, May and June, 1960, Vol. 79, Nos. 5 and 6, pp. 353–493.

Proc. Symp. Thermoelec. Energy Conversion; in Klein, P. H. (ed.), *Advanced Energy Conversion*; 1961, Vol. 1, Nos. 1–4, pp. 1–365; January–June, 1962, Vol. 2, pp. 1–312.

Cadoff, I. B., and Miller, E. (eds.) "Thermoelectric Materials and Devices"; New York, Reinhold Publishing Corporation, 1960.

Egli, P. H. (ed.) "Thermoelectricity"; New York, John Wiley & Sons, Inc., 1960.

Harman, T. C., and Honig, J. M. "Thermoelectric and Thermomagnetic Effects and Applications"; New York, McGraw-Hill Book Company, 1967.

FUEL CELLS

By H. P. Meissner and M. C. Deibert

150. Fundamentals. A fuel cell is a device in which, as in a battery, a fuel and an oxidant are caused to combine isothermally, converting the chemical energy released directly to electrical energy. Unlike conventional batteries (see Sec. **24**), in which fuel, oxidant, and reaction products are self-contained, a fuel cell has no storage capacity, and so proper provision must be made for the supply of reactants to the cell and for the withdrawal of reaction products during operation.

To be suitable for fuel-cell use, the overall power-producing reaction must be separable into two so-called *half-cell reactions*, one for each of the two electrodes. Half-cell reactions are electrochemical, in that they involve electrons and ions as well as the cell reagents. Electrons are produced in the half-cell reactions at the electrode to which fuel is charged; consequently this electrode is the anode. These electrons travel through the external circuit to the other electrode, the cathode, where they enter into an electron-consuming half-cell reaction with the oxidant. The electric circuit is completed by travel of ions through an electrolyte contained between the two electrodes. Depending on the system, either positively charged ions (cations) travel from the anode to the cathode, or negatively charged ions (anions) travel in the opposite direction. Occasionally both types of ion transfer occur simultaneously. The net reaction product of the fuel-oxidant combination is formed at that electrode at which ions are discharged. A schematic representation of the components and operation of a fuel cell is shown in Fig. 12-100.

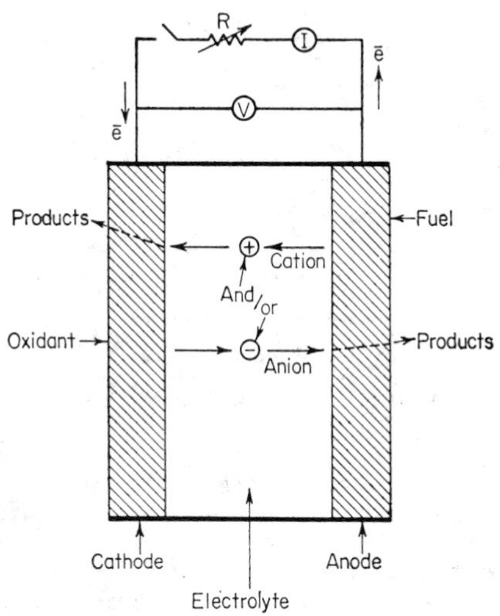

FIG. 12-100. Components and operation of a fuel cell.

151. Hydrogen-Oxygen Fuel Cells. In cells consuming gaseous hydrogen as fuel and gaseous oxygen as oxidant, operating with an aqueous acid solution as electrolyte, the half-cell reactions are, respectively,

Anode: $2H_2 \rightarrow 4H^+ + 4e^-$

Cathode: $O_2 + 4H^+ + 4e^- \rightarrow 2H_2O$

Overall: $2H_2 + O_2 \rightarrow 2H_2O$

The net reaction in this fuel cell is the production of water, which in this case is formed at the cathode.

The theoretical reversible potential difference between the anode and cathode of a hydrogen-oxygen fuel cell, 1.23 V, is calculated from the difference of the free energy of the reaction product and the fuel and oxidant. Theoretical potentials are determined in this way for the different possible fuel-cell reactions from published data on the free energies of the reaction partners or from tabulations of voltages of the half-cell reactions involved. These theoretical potentials are at best only approximated in cells and then only when measured in the absence of current flow. The cell voltage drops below this open-circuit potential when current is drawn from the cell, giving rise to the concept of "voltage efficiency," namely, the ratio of the actual to the theoretical reversible voltage.

Decreased potentials result from various irreversibilities in the cell, which are a consequence of the electrical resistance of the electrolyte and electrodes (resistance polarization), the accumulation of produced ions and reaction products and the depletion of consumed ions and reactants in the electrolyte near the electrode surfaces (concentration polarization), and the reluctance of the fuel and oxidant to undergo reaction at each electrode (activation polarization). Undesirable ions and products produced by intermediate irreversible reactions also decrease cell potentials, even at open circuit.

152. Output and Efficiency. Faraday's law, which states that 26.8 Ah of current is produced per gram equivalent of oxidant consumed, applies to the fuel cell as to all other electrochemical operations. Moreover, if the cell is well designed, so that no fuel or oxidant escapes from it without reacting, then the current efficiency (actual ampere-hours vs. theoretical ampere-hours from Faraday's law) is close to 100%. Voltage efficiencies, however, are poorer, in that the cell potential always falls with increasing current densities (amperes per unit area of anode or cathode) as shown in Fig. 12-101. Since cell power is the product of voltage and current, power efficiency (actual power vs. theoretical power at perfect voltage and current efficiencies) must parallel voltage efficiency very closely.

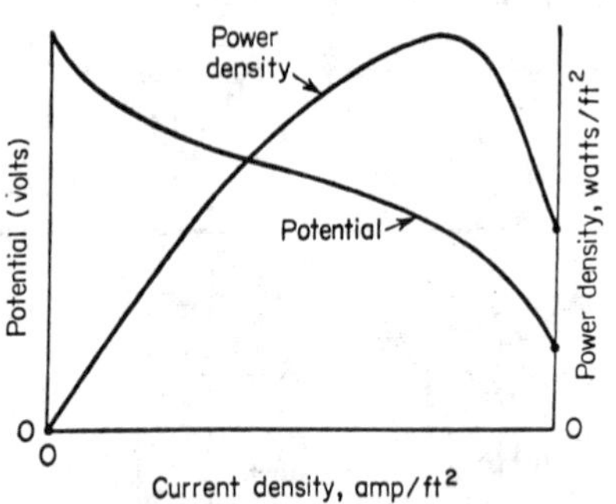

Fig. 12-101. Operational characteristics of a fuel cell.

The typical variation of actual electrical power output (*IE*) with current density is also shown in Fig. 12-101. Inspection shows this power curve to go through a maximum at some intermediate current called the peak power current. Fuel cells are usually operated at a fraction of the peak power current to balance the size of the cell required for a given power requirement and the efficiency of the energy conversion. Design for operation in this current region also provides capacity for temporary increases of the total power output in times of high demand of the order of 50 to 100% above normal design output, with commensurate temporary efficiency loss and decreased load voltages.

Most fuel cells are operated at current densities resulting in energy-conversion efficiencies of 60 to 80%. That part of the chemical energy not converted to electrical energy appears in the cell as heat, which must be removed by appropriate cooling.

153. Fuel-cell Types. Most fuel cells may be classified into three groups, namely, low-temperature cells (up to about 250°C), intermediate-temperature cells (about 400 to 700°C), and high-temperature cells (above 900°C). The electrolytes used in these cells are, respectively, aqueous solutions, molten salts (carbonates), and solid ceramic materials. Electrolyte characteristics determine the operating temperature ranges, in that aqueous solutions in general develop superatmospheric pressures in excess of about 140°C, and pressure-cell construction introduces serious problems of design and safety. Pressure operation is nevertheless sometimes deliberately undertaken with aqeuous electrolytes, as in the Bacon hydrogen-oxygen cell, to gain the advantage of the faster half-cell reaction rates with the resultant reduction in polarization which higher temperatures always bring.

The operating temperature ranges of the molten carbonate cells is set at the low end by carbonate melting points, on the high end by salt vaporization and corrosion. The electrolytes in high-temperature cells are thin sheets of ceramic materials, through which oxygen will travel in ionic form without excessive resistance at temperatures in excess of 900°C.

The ceramics used are generally some form of calcium zirconate, and the high temperature is required to reach a condition of low electrical resistance in these materials. Work on development of ceramics capable of operating at temperatures below 900°C is being intensively pushed. Low-temperature cells have the advantages of easy start-up, relatively simple construction, and (usually) minimized corrosion, with relatively simple heat-balance equipment needed. Their disadvantages are that the electrode reactions are sluggish and inefficient, especially with hydrocarbon fuels. Costly precious-metal catalysts are required for all currently available low-temperature cells.

Basic electrolytes (strong aqueous KOH at the concentration of maximum conductivity) are preferred over acid electrolyte with fuels like hydrogen and hydrazine, not only because of reduced corrosion, but because the oxygen half-cell functions more efficiently in base than in acid. Unfortunately, basic electrolytes cannot conveniently be used with hydrocarbon or methanol fuels, since the CO_2 formed as combustion product reacts with the hydroxide to form carbonate, an unsatisfactory electrolyte.

Strong phosphoric acid has been used as electrolyte in hydrocarbon cells with interesting results, but the corrosion problems encountered are formidable.

Several types of cells exist which cannot be conveniently classified as above. *Ion-exchange membrane cells* use a solid as electrolyte and yet operate at under 100°C. The membranes currently used are polyelectrolytes such as a polystyrene sulfonate. *The biochemical cells*, while always operated at low temperatures, are unique in involving enzymes as catalysts. In the so-called indirect biochemical cells, microorganisms convert a fuel such as a waste carbohydrate to a form more usable in a fuel cell, such as hydrogen. In the direct biochemical cell, the enzymes themselves participate in the half-cell reactions.

154. Fuel-cell Feed Materials. In theory, any spontaneous chemical reaction can be made the basis of fuel-cell operation. In low-temperature cells, two types of feed materials require consideration: (*a*) those which will readily participate in fuel-cell reactions, and (*b*) those which cannot yet be made to react efficiently in a low-temperature cell but which would yield cheap power because of their low cost. Few fuels have been found which will enter effectively into electrochemical reaction at (low-temperature) anodes, with hydrogen, hydrazine, and the alkali metals heading the list and methanol showing promise. The hydrocarbons, which, of course, are the most attractive economically, unfortunately cannot as yet be made to participate efficiently in anode reactions. For oxidants, only hydrogen peroxide, chlorine, and relatively pure oxygen have been used with any degree of success. For cheap terrestrial power, it is obvious that air must be used as the oxidant, but the nitrogen which dilutes the oxygen in air disadvantageously affects cathode performance.

Electrode reactions proceed rapidly in high-temperature cells, and the rate difficulties which limit the materials which may be used as fuels and oxidants in low-temperature operations disappear in consequence. All fuels appear to be suitable, with difficulties arising for some fuels because of their thermal decomposition on surfaces other than the electrodes. Thus hydrocarbons may crack to carbon, which deposits in the wrong places, and admixture with steam to prevent such decomposition introduces further complications. Air can be used successfully as oxidant in these cells.

The intermediate-temperature cells, like the high-temperature cells, can operate successfully on various hydrocarbon fuels and air. For a successful complementary pair of electrode reactions to occur in a molten carbonate electrolyte, carbon dioxide must be mixed in proper proportions with air fed to the cathode. In using a hydrocarbon fuel, this carbon dioxide is most readily obtained from the product combustion gases, which are evolved at the anode; however, the recycle of this CO_2 introduces complications in design.

The only reliably operable fuel cells currently available are low-temperature cells utilizing hydrogen or hydrazine as the fuel and pure oxygen or air as the oxidant. They utilize expensive noble-metal catalysts.

155. Fuel-cell Electrodes. Fuel-cell electrode construction depends upon whether the feed materials are gaseous or liquid and on the temperature of operation, i.e., on the cell type. The half-cell reactions at the electrodes always occur at the electrolyte-electrode interface, to which the reactant must somehow find its way. At this interface, either the oxidant accepts or the fuel surrenders electrons to the electrode, forming or consuming ions which can exist only in the electrolyte. When the rate of these half-cell reactions per unit surface area of electrode-electrolyte interface is low, a great deal of electrode surface in contact with the electrolyte must be provided. To maximize these half-cell reaction rates, a suitable catalyst coating on the electrode surface must be provided.

For low-temperature cells operating on gaseous fuels and oxidants, porous gas-diffusion-type electrodes have been developed which generally contain a precious-metal catalyst (usually platinum). These electrode structures serve as barriers or retaining walls between electrolyte and the gaseous-reactant supply chamber, so designed that the electrolyte penetrates only a small distance on that side of the porous electrode which is catalyst-coated. The reacting gas enters from the other side, meeting the electrolyte somewhere within the electrode structure. The desired interface position between gas and electrolyte is maintained within the electrode, by pressure balance, contact-angle (wettability) adjustment, or the use of a fine capillary reservoir for the electrolyte.

The limitations on intermediate- and high-temperature fuel-cell electrodes are largely those of short life, since corrosion is severe at elevated temperatures in the presence of oxygen, especially in contact with molten salts.

156. Cell Engineering Problems. Depending on the reagents used, the open-circuit potential developed in a single fuel cell ranges from less than 1 to perhaps 2.5 V. As pointed out earlier, choice of operating conditions depends upon an economic balance, with typical current densities in low-temperature hydrogen-oxygen cells ranging from 100 to 250 A/ft^2. Large currents require correspondingly large electrode areas, which should be spaced as closely as possible for minimized *IR* drop due to electrolyte resistance. Gas-diffusion-type electrode membranes are usually held parallel to each other in a cell at perhaps 0.010 to 0.100 in spacing. The problems of manufacturing large areas of electrodes uniform in porosity and thickness, strong enough to withstand distortions due to thermal expansion, gravity, and the like, are great. To economize on space, the reactant supply chambers behind each electrode are made small, so that a typical complete cell is perhaps 1 in or less in thickness. Variations in fuel and oxidant feed rates and concentration across the face of these electrodes are difficult to avoid, but undesirable, since they may give rise to corrosion, stray currents, localized heating, and the like. Cooling and removal of products of combustion from a cell of this general shape are again not easy, making scale-up a difficult problem.

Completeness of fuel utilization may be reduced by the presence of unconsumed reagents in a purge stream or in the products of combustion. Unless any such losses can be minimized by recovery and recycle, they may result in a serious loss in efficiency.

To satisfy the voltage requirements of most applications, a battery of fuel cells must be operated in series. In addition to two electrical connections, each cell must be provided with several piping connections, one to introduce fuel, another to introduce the oxidant, and a third to remove the products of reaction. If a coolant is to be circulated, then two more connections are involved at appropriate intervals in a battery of cells. These many connections complicate the typical fuel-cell battery, making reliable and well-controlled operation more difficult.

An integrated fuel-cell battery requires a considerable amount of auxiliary equipment for voltage control, for metering the reagent feed streams, for processing the product streams, and for cooling. While these auxiliary devices do not appear to represent an insuperable challenge in pioneering development, they nevertheless constitute an important part of the weight, volume, and cost of a battery system.

Current military specifications illustrate the types of performance expected of a fuel-cell battery. An integrated 300-W hydrazine-air fuel-cell unit should be operable in two modes, either at a regulated 28 V or at a variable voltage up to 28 V to charge conventional batteries. Physical limitations on the unit are that it weigh under 32 lb when containing sufficient hydrazine fuel for a 12-h operating period, that its total volume be under 1,500 in^3 with no dimension over 15 in, and that it be essentially silent in operation. It must contain protective devices to prevent damages on the circuit by overloads, short circuit, or polarity reversal. Total cell capacity must be realized in operating in ambient temperatures between $-25°$ and $+125°$F.

The above-described fuel cell, as with most others, requires silent and efficient integrated d-c motor-blower and motor-pump combinations. In delivering full load this 300-W fuel cell will require approximately 60 W of parasitic loads with currently available relay and solenoid devices and motor-blower and pump units.

157. Bibliography

Basic Electrochemistry

1. Conway, B. E. "Theory and Principles of Electrode Processes"; New York, The Ronald Press Company, 1965.

2. Delahay, Paul "Double Layer and Electrode Kinetics"; New York, Interscience Publishers, Inc., 1965.

3. Glasstone, Samuel "Introduction to Electrochemistry"; Princeton, N.J., D. Van Nostrand Company, Inc., 1942.

Reviews

1. Bockris, J. O. M. (ed.) "Modern Aspects of Electrochemistry"; London, Butterworth & Co. (Publishers), Ltd., 1954, Vol. I, 1959; Bockris, J. O. M., and Conway, B. E. (eds.) 1964, Vol. III.

2. "Advances in Electrochemistry and Electrochemical Engineering"; electrochemistry volumes edited by P. Delahay, 1961, Vol. 1, 1963, Vol. 3, 1965, Vol. 5; electrochemical engineering volumes edited by C. W. Tobias, 1962, Vol. 2, 1964, Vol. 4.

3. *Trans. Symp. Electrode Processes*, E. Yaeger (ed.), New York, John Wiley & Sons, Inc., 1961.

Technology

1. BAKER, B. S. (ed.) "Hydrocarbon Fuel Cell Technology"; New York, Academic Press, Inc., 1965.

2. MITCHELL, WILL (ed.) "Fuel Cells"; New York, Academic Press, Inc., 1963.

3. YOUNG, G. J., JR. "Fuel Cells"; New York, Reinhold Publishing Corporation, 1960, Vol. I, 1963, Vol. II.

4. Fuel Cell Systems; *ACS Adv. Chem. Ser.* 47, 1965.

5. BAGOTSKII, V. S., and VASIL'EV, YU B. (eds.) "Fuel Cells, Their Electrochemical Kinetics"; Moscow, Nauka Press, 1964.

6. WILLIAMS, K. R. (ed.) "Introduction to Fuel Cells"; Amsterdam, Elsevier Publishing Company, 1966.

7. BERGER, C. (ed.) "Handbook of Fuel Cell Technology"; Englewood Cliffs, N.J., Prentice-Hall, Inc., in press.

MAGNETOHYDRODYNAMICS

BY CLIFFORD MANNAL AND NORMAN W. MATHER

158. Introduction. Recognition that the magnetic force on a conductor can produce desirable engineering results when the conductor is a gas or liquid is not a new concept in electrical engineering. Designs for arc-expulsion gaps, which use the self-induced magnetic forces to extinguish the arc, and for liquid-metal pumps are old in the art. These, however, are power-consuming rather than power-*generating* applications.

Since 1959, substantial effort has been devoted to exploring the conditions under which a conducting fluid in motion through a magnetic field might generate useful electrical power (Refs. 15 and 16, Par. **173**) or, inversely, might convert electric energy into thrust for rocket propulsion in space (Ref. 14). Only the former is discussed here. The terms plasmadynamics, magnetofluiddynamics, electrogasdynamics, and magnetohydrodynamics (abbreviated MHD) as well as other similar combinations have been used to describe the combination of disciplines required to treat the phenomena. Common usage appears now to favor magnetohydrodynamics as the generic term.

The motivation for the development and use of MHD generators in central-station application is the promise of a significant improvement in plant thermal efficiency because much higher temperatures are allowable in MHD generators, which employ no rigid moving parts or close tolerances. As presently envisaged, the MHD generator would best be suited as a "topping" unit on an otherwise conventional turbine-generator station. The exhaust temperature of the MHD unit, which would be comparable with present central-station maximums, would be used to generate steam for the turbine generators. The limiting Carnot efficiency for a station might by this means be raised from the present maximum of about 65% ($T_1 = 1540°R$, $T_2 = 540°R$) upward toward 88% ($T_1 = 4500°R$ or $2500°K$, $T_2 = 540°R$), although the fraction of this limit that may actually be achieved will be somewhat reduced in the latter case. If, for example, the net thermal efficiency η_1 of an MHD generator exhausting into a "bottom" steam plant is 0.25 and the bottom plant efficiency η_2 is 0.40, the overall plant efficiency is $\eta_1 + \eta_2(1 - \eta_1) = 0.55$.

The only practical design configurations that have appeared have d-c outputs taken from electrodes at the sides of fluid-flow channels. Ionized gas (plasma) is most frequently the conducting fluid proposed. Recent studies of 2-phase systems, in which both liquid and gas phases of a metal such as mercury or potassium are used, indicate possible advantages for the latter (Refs. 25, 37 to 39, and 51).

Table 12-14 lists the major lines of engineering endeavor. There is concurrently a large supporting scientific effort, both theoretical and experimental, devoted to a basic understanding of the phenomena and the mathematical relations describing idealized fluids in ideal geometries, as well as the design, construction, and operation of

Table 12-14. MHD Engineering Endeavor

Primary goal	Primary energy source	Working fluid	Cycle	Field	Form of power
Central-station topping cycle	Fossil combustor	Seeded air	Open	Magnetic	Direct current
Central-station topping cycle......	Nuclear reactor	Seeded He or A	Closed	Magnetic	Direct current
Space-vehicle primary power	Nuclear reactor	Liquid metal (Na, K, Hg)	Closed	Magnetic	Direct or alternating current
Primary power generation	Combustor or reactor	Seeded air or He	Open or closed	Electrostatic	High-voltage direct current (e.g., Ref. 48)
Thrust for space vehicles	Electric energy	Gas	Open	Magnetic	—(e.g., Refs. 6 and 14)

engineering test rigs (in the United States, England, Poland, Japan, Russia, and elsewhere). Estimates place the total world expenditure for this work in the vicinity of 10 million dollars annually. A complete bibliography in this field would list perhaps 1,000 original papers. Representative references are given in Par. **173**.

159. Fundamental Equations. Figure 12-102 shows the basic scheme for an MHD generator having quasi-one-dimensional flow of ionized gas (i.e., plasma) channeled through a perpendicular, static magnetic field. The charged particles experience a transverse force which is equivalent to an electric field $\mathbf{u} \times \mathbf{B}$ (see Table 12-15, Glossary) due to their velocity **u** through the magnetic field. This generates, between electrodes spaced a distance d apart, an open-circuit voltage of

$$V_{oc} = \int_0^d (\mathbf{u} \times \mathbf{B}) \cdot d\mathbf{l} \qquad (12\text{-}111)$$

An electric field $\mathcal{E}$ is therefore present between the electrodes such that $\mathcal{E} + \mathbf{u} \times \mathbf{B} = 0$,

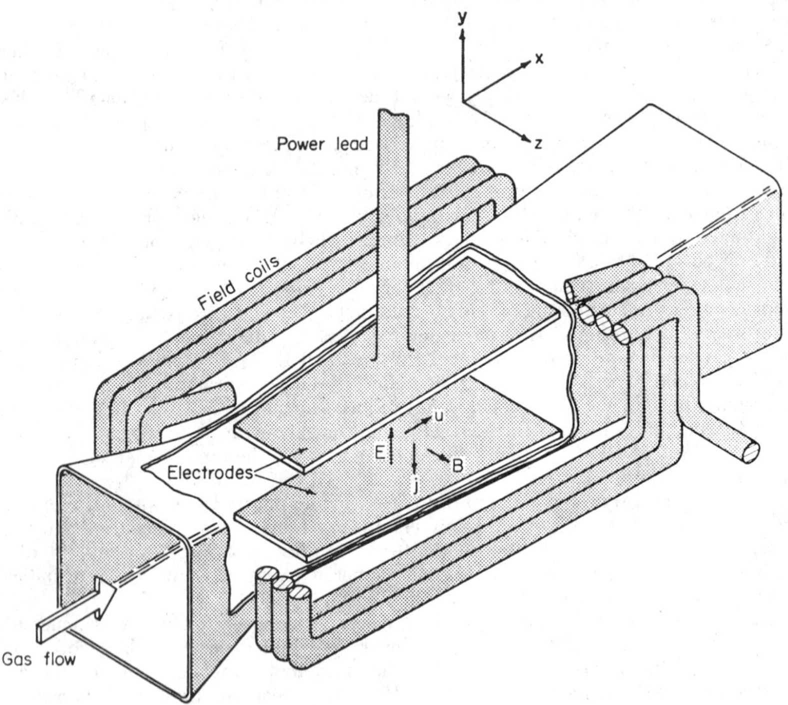

Fig. 12-102. Basic elements of a magnetohydrodynamic converter.

Table 12-15. Glossary

NOTE: Except where explicitly noted otherwise, MKSA units used throughout.

A_c = cross-sectional area of flow channel, square meters
A_e = electrode area, square meters
a = speed of sound in fluid, meters per second
= $(\gamma p/\rho)^{1/2} = (\gamma RT/W_m)^{1/2}$
B = magnetic induction field, webers per square meter = teslas
C = electron rms thermal velocity, meters per second
= $(3kT_e/m_e)^{1/2}$ for Maxwellian distribution
c_p = specific heat at constant pressure, $J \cdot m^3/kg \cdot °K$
c_v = specific heat at constant volume, $J \cdot m^3/kg \cdot °K$
d = channel width, meters
$\mathcal{E}$ = electric field, volts per meter
$\mathcal{E}_H$ = Hall-effect electric field, volts per meter
$\mathcal{E}^*$ = electric field in moving frame, volts per meter
ϵ = internal energy of fluid, $J \cdot m^3/kg$
e = electronic charge, 1.60×10^{-19}, coulombs
H = total enthalpy of fluid, $J \cdot m^3/kg$
h = static enthalpy, $J \cdot m^3/kg$
= $H - u^2/2$
j = current density, amperes per square meter
K = generator coefficient or electrical efficiency
= $\mathcal{E}/uB$
k = Boltzmann's constant, 1.38×10^{-23} joules per degree Kelvin
l = length, meters
L = length of MHD channel, meters
L_i = interaction length, meters
M = Mach number
= u/a
m_a = atom mass, kilograms
m_e = electron mass, 9.11×10^{-31} kilograms
m_i = ion mass, kilograms
n_e = electron particle density, cubic meters
P = power density, watts per cubic meter
p = pressure, newtons per square meter; also, atmospheres, where 1 atm = 1.01×10^5 N/m²
$\hat{p}$ = pressure tensor, newtons per square meter
Q_{en} = electron-neutral collision cross section, square meters
R = universal gas constant, 8.31×10^3 J/(kg mole)(°K)
R_i = internal resistance, ohms
R_L = external load resistance, ohms
s = surface or area variable, square meters
T = temperature, degrees Kelvin; also, degrees Fahrenheit and degrees Rankine = °F + 459 are used
T_a = temperature of background gas, degrees Kelvin
T_e = electron temperature, degrees Kelvin
u = fluid velocity, meters per second
V = electrode voltage
V_i = ionization voltage = ionization energy, electronvolts (since e = 1 in electronvolt units)
V_{oc} = open-circuit electrode voltage
v = electron-drift velocity, meters per second
W_m = molecular weight of gas, atomic units
$\mathbf{x}_1$ = unit vector in x direction
$\mathbf{z}_1$ = unit vector in z direction
β = $1/en_e$ Hall constant, cubic meters per coulomb
γ = ratio of specific heats
= c_p/c_v
λ = electron mean-free path, meters
μ_e = electron mobility, m²/V·s
= $v/\mathcal{E}$
$\approx 3.6 \times 10^{-16} T_e^{1/2}/(pQ_{en})$
ρ = mass density, kilograms per cubic meter
σ, σ_0 = fluid conductivity in absence of magnetic field, mhos per meter
$\hat{\sigma}$ = tensor conductivity, including effect of magnetic field, mhos per meter
τ = electron mean free time, seconds
= λ/C
τ_i = ion mean free time, seconds
ω = electron cyclotron frequency, s⁻¹
= eB/m_e
= $\mu_e B/\tau$
ω_i = ion cyclotron frequency, s⁻¹
= eB/m_i (if singly ionized)

corresponding to the zero-current condition. When current $\mathbf{j}$ flows, owing to connecting an external load R_L, the electrode voltage V is reduced because of the electrical resistance of the fluid. In general,

$$V = -\int_0^d \boldsymbol{\mathcal{E}} \cdot d\mathbf{l} = R_L \int_{A_e} \mathbf{j} \cdot d\mathbf{s} \tag{12-112}$$

Both electrons and ions are affected by the field, but the electrons, having much greater mobility, conduct almost all the current. Since the current carriers are predominantly electrons, another electric field, the *Hall field*, is created,

$$\boldsymbol{\mathcal{E}}_H = \beta \mathbf{j} \times \mathbf{B} \tag{12-113}$$

Here $\beta = 1/n_e e$ is the Hall constant. The current component which is created by this field, the *Hall current*, is given by $-\mu_e \mathbf{j} \times \mathbf{B}$, in which $\mu_e = \omega\tau/B$ is the electron mobility.

Segmented electrode structures are used to minimize the Hall current. In the ideal limit of zero Hall current, the effect of the Hall field cancels the perpendicular force on the electrons as they conduct current in the transverse magnetic field. In that case the current flow is described by *Ohm's law*,

$$\mathbf{j} = \sigma_0(\boldsymbol{\mathcal{E}} + \mathbf{u} \times \mathbf{B}) \tag{12-114}$$

in which σ_0 is the conductivity. When there is Hall current, it is added into the right-hand side of this expression to give a form of *generalized Ohm's law*,

$$\mathbf{j} = \sigma_0(\boldsymbol{\mathcal{E}} + \mathbf{u} \times \mathbf{B}) - \mu_e \mathbf{j} \times \mathbf{B} \tag{12-115}$$

A vector diagram of this relation is shown in Fig. 12-103. A theoretical discussion of the generalized Ohm's law is given in Refs. 1, 2, 4, and 5.

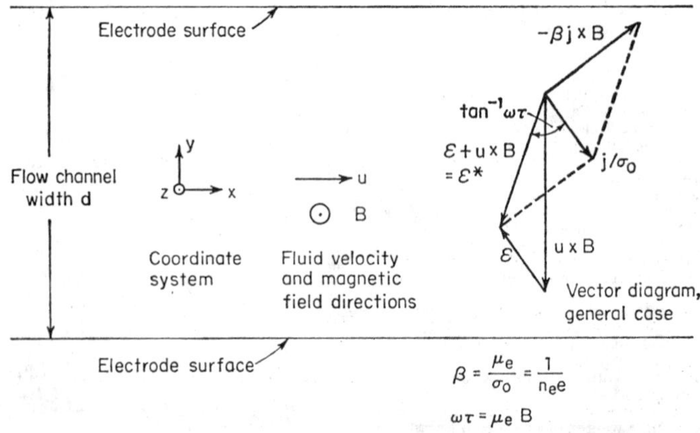

FIG. 12-103. Vector quantities involved in magnetohydrodynamic conversion.

Various forms of generalized Ohm's law are obtained depending upon the approximations made in deriving the current-field relationship from the equations of motion [frequently referred to as the "transport" or "momentum" equations, which are derived from the Boltzmann equation of the kinetic theory of gases (see, e.g., Ref. 1)] of the constituent parts of the fluid and their interactions. The form given here, Eq. (12-115), is appropriate for weakly ionized gases in thermal equilibrium at moderate temperatures. It also has the equivalent form, neglecting ion current, given by

$$\mathbf{j} = \hat{\boldsymbol{\sigma}} \cdot (\boldsymbol{\mathcal{E}} + \mathbf{u} \times \mathbf{B}) = \hat{\boldsymbol{\sigma}} \cdot \boldsymbol{\mathcal{E}}^* \tag{12-116}$$

or

$$j_\nu = \sum_k \sigma_{\nu k} \mathcal{E}_k^* \qquad (\nu, k = x, y, z) \tag{12-117}$$

where, when $\mathbf{B} = \mathbf{z}_1 B_z$,

$$\hat{\boldsymbol{\sigma}} = \| \sigma_{\nu k} \| = \sigma_0 \begin{Vmatrix} \dfrac{1}{1 + \omega^2\tau^2} & \dfrac{-\omega\tau}{1 + \omega^2\tau^2} & 0 \\ \dfrac{\omega\tau}{1 + \omega^2\tau^2} & \dfrac{1}{1 + \omega^2\tau^2} & 0 \\ 0 & 0 & 1 \end{Vmatrix}$$

where $\omega = \dfrac{eB}{m_e}$ (electron cyclotron angular frequency)

$\tau = \dfrac{\lambda}{C}$ (electron mean free time)

Also,

$$\omega\tau = \mu_e B$$

Here the conductivity is anisotropic unless $\omega\tau \ll 1$, as is indicated by the conductivity tensor $\hat{\boldsymbol{\sigma}}$.

The dimensionless product $\omega\tau$ is an important characteristic number in MHD design. On the microscopic scale this number indicates the average angular travel of electrons between collisions as they tend to move helically about the magnetic field lines due to the perpendicular forces that act on moving charges in a transverse magnetic field. Typically, $\lambda \sim 10^{-7}$ m, $C \sim 10^5$ m/s, so that $\tau \sim 10^{-12}$ s. Also, $\omega = 1.76 \times 10^{11}B \sim 10^{12}$ for $B \approx 6T$. These values give $\omega\tau \sim 1$. The mean free path λ is inversely proportional to pressure. Low pressures and high values of B give large values of $\omega\tau$.

When $\omega\tau = 1$, the current vector is directed 45° to the left of the $\boldsymbol{\mathcal{E}}^*$ vector and for large values of $\omega\tau$ the current vector is nearly perpendicular to $\boldsymbol{\mathcal{E}}^*$ (predominantly Hall current). In weakly ionized gases, if both the equivalent characteristic number for the ions, $\omega_i\tau_i$, and $\omega\tau$ are large simultaneously, the angle is reduced. The conductivity then becomes low owing to a phenomenon called "ion slip" (see Ref. 5). Since ω_i is much smaller than ω (because of the much larger ion mass) even though τ_i is increased because of the smaller ion thermal velocity, the $\omega_i\tau_i$ product is ordinarily negligible.

160. Generator Configurations. If the geometry of Fig. 12-102 is used so that $\mathbf{u} = \mathbf{x}_1 u_x$ and $\mathbf{B} = \mathbf{z}_1 B_z$, and if segmented instead of continuous electrodes are used so that the condition $j_x = 0$ can be established by electrically isolating each pair of electrodes along the flow channel, the generalized Ohm's-law relation, Eq. (12-115), has the following components:

$$j_x = \sigma_0 \mathcal{E}_x - \mu_e j_y B_z = 0 \qquad \text{or} \qquad \mathcal{E}_x = \frac{\mu_e B_z j_y}{\sigma_0} = \omega\tau \frac{j_y}{\sigma_0} \qquad \text{(Hall field)}$$

$$j_y = -\sigma_0 (u_x B_z - \mathcal{E}_y)$$

$$j_z = 0$$

If the circuit for j_y is completed through an external load, the arrangement is known as the *Faraday generator configuration* (Fig. 12-104*b*). The open-circuit voltage in this case, with uniform conditions over the cross section of the channel assumed, is

$$V_{oc}\,|_F = -\int_0^d \mathcal{E}_y\, dy = -\int_0^d u_x B_z\, dy = -u_x B_z d \tag{12-118}$$

Alternatively, if the condition $\mathcal{E}_y = 0$ is achieved by short-circuiting opposite segmented electrode pairs, the current components are

$$j_x = \sigma_0 \mathcal{E}_x - \mu_e j_y B_z$$

$$j_y = -\sigma_0 u_x B_z$$

$$j_z = 0$$

from which

$$j_x = \sigma_0(\mathcal{E}_x + \mu_e u_x B_z^2) = \sigma_0(\omega\tau u_x B_z + \mathcal{E}_x) \tag{12-119}$$

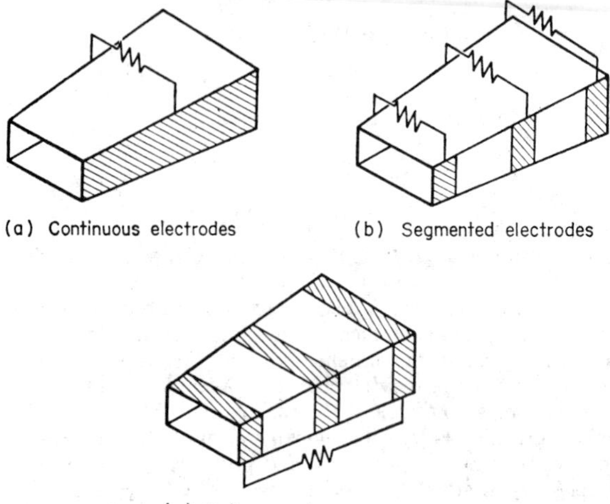

Fig. 12-104. Alternative electrode arrangements.

If an electrical load is connected between the first and the last segmented electrodes, it is called the *Hall generator configuration* (Fig. 12-104*c*). The field $\mathcal{E}_x$ is negative in this case, which means that the voltage developed is positive in the downstream direction. Its load current j_x has larger magnitude than the corresponding $-j_y$ in the Faraday configuration when $\omega\tau > 1$. This condition is enhanced by designing for relatively low pressure and using large magnetic fields. The open-circuit terminal voltage, with uniform conditions assumed throughout the length L of the channel between load-circuit connections (somewhat of an oversimplification), is

$$V_{oc}\,|_H = -\int_0^L \mathcal{E}_x\,dx = \int_0^L \omega\tau u_x B_z\,dx = \omega\tau u_x B_z L \qquad (12\text{-}120)$$

A third configuration, in which output current is obtained from continuous instead of segmented electrodes along the sides of the channel (Fig. 12-104*a*), is a variation of the Faraday generator configuration. In this case $\mathcal{E}_x = 0$, and both j_x and j_y components contribute to the load current. The open-circuit voltage is the same as for the Faraday generator with segmented and isolated electrode pairs, namely, $-u_x B_z d$.

In any of these configurations the power available at a pair of electrodes per unit electrode area is jV, that is, electrode current per unit area times electrode voltage. However, if uniform conditions can be assumed, it is more convenient to use the power density in the flow channel $\mathbf{j}\cdot\boldsymbol{\mathcal{E}}$ (watts per cubic meter). Detailed comparisons of various configurations have been summarized in Refs. 3 and 5 (see also Refs. 20 and 32).

161. Power Generated. The fundamental power relations for the Faraday generator configuration with segmented electrodes are briefly presented here. Since in this configuration the current is just j_y, the velocity u_x and the magnetic field B_z, it is the usual practice to drop the subscripts.

The electrical power density generated internally in the fluid, juB, owing to conversion of fluid energy into electrical energy is proportional to the velocity times the rate of change of momentum (i.e., velocity $\times$ force) in the fluid,

$$\rho\mathbf{u}\cdot\frac{d\mathbf{u}}{dt} = -\mathbf{u}\cdot\text{grad}\,p + \mathbf{u}\cdot(\mathbf{j}\times\mathbf{B}) \qquad (12\text{-}121)$$

in which ρ is the mass density and p is the pressure, here taken as a scalar quantity. (More generally, the pressure term is $-\mathbf{u}\cdot\mathbf{div}\,\hat{\mathbf{p}}$ in which the viscous forces in the fluid are included as well as the hydrostatic pressure. Also in the more general form a term $\rho_e\mathbf{u}\cdot\boldsymbol{\mathcal{E}}$, usually negligible, is added to the right-hand side.) The second term on the right is the internally developed electrical power juB.

The density of electrical power actually delivered is, as noted above, $j\mathcal{E}$, and is less than juB owing to dissipation of energy in the electrical resistance of the fluid. The difference remains in the fluid, although the result is a degradation of energy since kinetic energy is converted into thermal energy. The fluid enthalpy H (sum of internal energy ϵ, kinetic energy per unit mass $u^2/2$, and the pressure–specific volume product p/ρ) is therefore affected only by the power actually delivered,

$$\rho \frac{dH}{dt} = \mathbf{j} \cdot \boldsymbol{\mathcal{E}} \tag{12-122}$$

The ratio of electrical power delivered to that generated internally, a dimensionless quantity, is widely used,

$$K = \frac{\mathcal{E}}{uB} = \frac{V}{V_{oc}} = \frac{R_L}{R_i + R_L} \tag{12-123}$$

The last two forms are based on the existence of uniform conditions throughout the channel. Here R_i is the internal resistance, and R_L is the load resistance. The condition for maximum power is $K = 0.5$ (that is, $R_L = R_i$). In this case $\mathcal{E} = uB/2$, and the power density in the fluid is

$$P_{\max} = \frac{\sigma_0 u^2 B^2}{4} \qquad (\text{W/m}^3) \tag{12-124}$$

If K is different from 0.5, the power density is

$$P = \sigma_0 u^2 B^2 K(1 - K) \qquad (\text{W/m}^3) \tag{12-125}$$

Target values for the various quantities in these equations might be $\sigma_0 = 10$ mhos/m, $u = 950$ m/s, $B = 6$T, giving a maximum available power density of 8×10^7 W/m³ or 80 W/cm³. In actual practice these values may not be achievable. Furthermore, considerations of efficiency, secondary effects, etc., may make a value of K of about 0.75 optimum (Ref. 40). (See Table 12-17 for values as determined in typical design studies.)

162. Flow Relations. In general the equations governing the power generation and flow cannot be solved in closed form; so numerical computation of individual cases is necessary. However, a few simple cases are soluble and, although highly approximate, are useful for the insight they give. The simplest is obtained for quasi-one-dimensional constant-velocity flow with ideal gas flow assumed. Here, the Faraday generator configuration with segmented electrodes is also assumed. In this case the current component due to the Hall field is zero, and the *generalized Ohm's law* reduces to

$$j = \sigma(\mathcal{E} - uB) = -\sigma uB(1 - K) \tag{12-126}$$

where σ is the scalar conductivity, previously denoted as σ_0.

The *momentum equation* for steady-state one-dimensional flow is obtained by setting equal to zero the $\partial u/\partial t$ part of the expansion of du/dt in Eq. (12-121), giving

$$\rho u \frac{du}{dx} + \frac{dp}{dx} = jB \tag{12-127}$$

With constant velocity flow as assumed here, the du/dx term vanishes.

The *energy equation*, Eq. (12-122), becomes

$$\rho u \frac{dh}{dx} = j\mathcal{E} \tag{12-128}$$

since $H = u^2/2 + h$ and u is constant. For a perfect gas $h = c_p T$, giving

$$\rho u c_p \frac{dT}{dx} = j\mathcal{E} \tag{12-129}$$

A *continuity*, or *mass-conservation*, law also applies to the fluid flow. In this case the product of density and channel area must be constant, i.e.,

$$\frac{d}{dx}(\rho A_c) = 0 \tag{12-130}$$

where A_c is the cross-sectional area of the channel.

From these equations the electrical efficiency $K = \mathcal{E}/uB$ becomes

$$K = \rho c_p \frac{dT}{dp} \tag{12-131}$$

For a perfect gas

$$c_p T = \frac{\gamma}{\gamma - 1} \frac{p}{\rho} \tag{12-132}$$

which can be combined with Eq. (12-131) to give

$$K = \frac{\gamma}{\gamma - 1} \frac{d \ln T}{d \ln p} \tag{12-133}$$

This is integrated to give the relation between p and T in the constant-velocity channel,

$$\frac{T}{T_0} = \left(\frac{p}{p_0}\right)^{K(\gamma-1)/\gamma} \tag{12-134}$$

where T_0 and p_0 are the temperature and pressure at the entrance to the generator channel.

The momentum equation and Ohm's law when combined and integrated under the same conditions yield

$$p = p_0 - \sigma u B^2 (1 - K)x \tag{12-135}$$

in which $x = 0$ at the entrance, where $p = p_0$. This may be written in the dimensionless form

$$\frac{p}{p_0} = 1 - \frac{x}{L_i} \qquad L_i = \frac{1}{1 - K} \frac{p_0}{\sigma u B^2} \tag{12-136}$$

where L_i may be termed the "interaction length" (Ref. 24) and is an approximate measure of the channel length required in an actual design to extract an appreciable part of the gas energy (cf. G. W. Sutton in Ref. 8, pp. 123–126; also Ref. 40). Only 0.6 to 0.8 of p_0 would be utilized in an actual design, but the average value of σ is likely to be about the same fraction of the inlet value so that taking $L \approx L_i$ is approximately correct.

The pressure relation and other quantities found in a similar manner are shown in Fig. 12-105. Also shown is the Mach number variation. The Mach number is

$$M \equiv \frac{u}{a} = \frac{u}{\sqrt{\gamma p/\rho}} = \frac{u}{\sqrt{\gamma RT/W_m}} \tag{12-137}$$

where a is the local sound speed, R is the universal gas constant, and W_m is the molecular weight of the gas. For constant velocity u,

$$\frac{M}{M_0} = \left(\frac{T}{T_0}\right)^{-\frac{1}{2}} = \left(\frac{p}{p_0}\right)^{-\frac{1}{2}K(\gamma-1)/\gamma} \tag{12-138}$$

More detailed channel-flow solutions are given in Refs. 3 and 5. The quasi-one-dimensional equations arranged for computer programming, and the results obtained in a particular set of design studies, are given in Refs. 40 and 56.

163. Duct Design. Design of an MHD duct depends upon detailed solution of the fundamental magnetohydrodynamic equations, as outlined above, and the thermodynamical relations for the fluid. Since general closed-form solutions do not exist, numerical methods are necessary and many particular solutions are required for optimization. These must take account of material limitations at high temperature, wall friction, boundary layers, possible eddy-current losses at the entrance and exit of the magnetic field region, electrode drops, etc. Although economic considerations greatly limit the choice of possibilities (and are used together with approximate general solutions and estimates to determine the appropriate conditions and assumptions to be used at the beginning of a design calculation), the number of variables is large and their ranges are often ill defined. The problem is to find a system having overall high efficiency consistent with the capital investment required. The various desiderata are often conflicting so that arbitrary limitations are necessary in optimization, and the scientific and engineering data needed for detailed calculations are currently fragmentary in many areas and of unknown reliability (see Ref. 3).

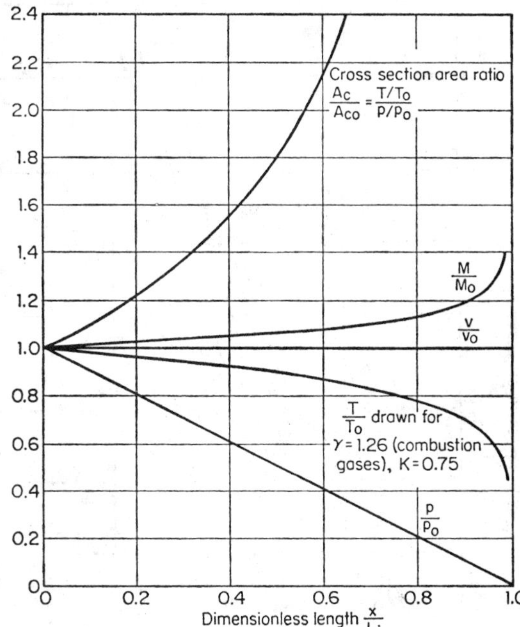

FIG. 12-105. Values of ratios for normalized length x/L_i.

The choice of maximum cycle pressure and temperature is basic to everything else in the design. This will be strongly influenced by components outside the channel itself, namely, the combustor, air preheat equipment, the various heat exchangers, and the temperature and pressure limitations of gas-cooled reactors if used in closed-cycle loops. Figure 12-106 shows in a general way the relation between pressure and temperature in the channel itself for various constant generator lengths and constant total power output (Ref. 24). High pressure reduces the size of heat exchangers, but it also reduces the gas conductivity and must be compensated by increased temperature or increased generator length.

Then a choice of input-flow velocity is necessary. Here a clear optimum in terms of performance exists (Ref. 17). Both σu^2 and σu have maximums for a given total enthalpy at the channel entrance since increased u means decreased temperature and, on assuming the electrons are in thermal equilibrium with the gas, almost exponentially decreased σ. Also, the length of the channel is inversely proportional to u, as indicated in the expression for L_i given above [Eq. (12-136)]. In the open-cycle optimization study reported in Ref. 56 the optimum velocity including the effect of wall friction was found to be 750 m/s ($M \approx 0.8$). In one case the maximum cycle temperature was 2645°K (4301°F), and 2640°K (4292°F) in another. Maximum cycle pressures were 5.175 and 5.750 atm, respectively. For closed-cycle systems using seeded inert gases, the optimum Mach numbers are expected to range from 0.4 to 0.6 depending upon the optimization criteria used (Ref. 17).

Proportions. From the standpoint of magnet design and minimization of end-loss effects, a thin, rectangular duct of moderate length is desirable. However, such proportions are undesirable from the standpoint of erosion (i.e., duct-wall and electrode life), pressure drop, and heat loss to the walls. Conversely, a short, wide-aperture duct may be expected to make end-loss problems and magnet design more difficult. Problems with lack of homogeneity of current flow over the larger distances may be severe.

Size. If power density in the main stream can be made independent of size, then large sizes with their more favorable volume-to-surface ratio will be more efficient. Both the heat loss to the walls and the weight (or cost) of the magnet are proportionately

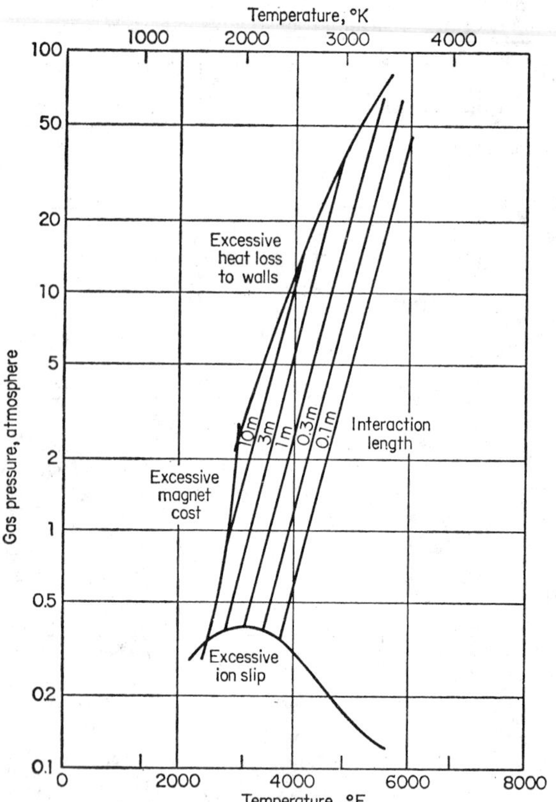

FIG. 12-106. Gas pressure vs. temperature for various interaction lengths. Curves are for a 500-MW closed-cycle MHD channel using cesium-seeded helium and a magnetic field of 6 T (Ref. 24). Limits imposed by heat loss, magnet cost and ion slip are shown. (*Adapted from Ref. 24.*)

smaller in large size units. Sizes below about 500 MW are not expected to be economically justifiable for central-station installations, but in special applications, for example, for power generation in space craft, the overall system balance may be more favorable to small-sized units (see Ref. 24).

Wall Materials. Moderate experience with ceramic and other high-temperature materials available at present (see Table 12-16 and Ref. 21) indicates that durabilities of between 10^4 and 10^5 h cannot easily be obtained. Furthermore, the physical bulk of the duct eliminates on economic grounds the very expensive materials. Suggested alternatives are cooling by laminar gas flow along the inner wall surface or cooling from the back together with the use of more conventional materials (see Refs. 24, 46, and 58).

164. Properties of High-temperature Air. Since a proposed application of MHD power generation is as a topping unit in a fossil-fueled plant, designs which use the combustor gas directly as the working fluid have obvious advantages. Temperatures for fossil-fuel combustion (with preheated or oxygen-enriched air) are in the range of 4200 to 5000°F. The conductivity of high-temperature air is therefore of interest and is given by the lowest of the curves in Fig. 12-107.

Over a limited range, the conductivity is proportional to $1/p^x$, where p is the pressure and x is ~0.5 at 5000°F and ~0.3 at 10,000°F. At atmospheric pressure ($p = 1$) the conductivity of normal air is too low to serve as the working fluid, and the conductivities of various combustion-product gases from the burning of fossil fuels are all similarly low. Increasing the conductivity by decreasing the density by, say, 10^{-4} would yield almost adequate conductivity at the cost, however, of a prohibitive demand on the size of the duct and the heat-exchanging surfaces.

165. Seeded Air and Inert Gases. Conductivity is given by summing $en\mu$ (charge, density, and mobility) for all charged species present. But electrons, because of their high mobility, contribute most of the conductivity in an ionized gas even though, to a high degree of approximation, there are an equal number of ions in the gas. At low degrees of ionization, the mobility of electrons is inversely related to the collision frequency for momentum transfer with neutral particles. As the degree of ionization increases, electron-ion and electron-electron collisions cause the electron mobility to be reduced. This effect is so strong that at less than 1% ionization the conductivity is no longer increasing linearly with electron density, as at first, but instead

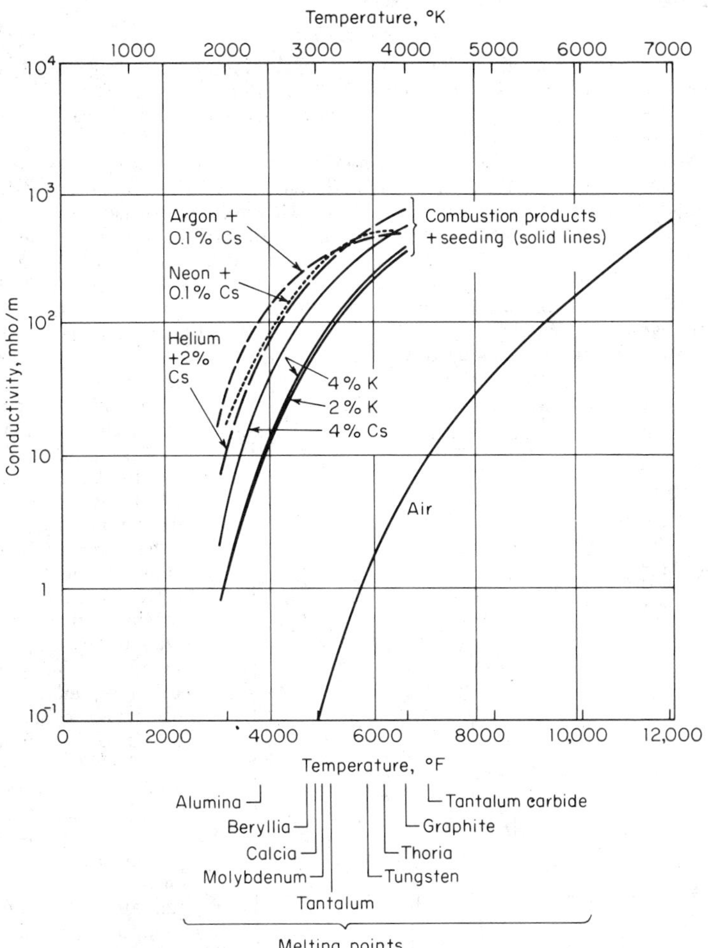

Fig. 12-107. Conductivity of air, seeded combustion gases and seeded argon, helium and neon at atmospheric pressure (all curves, except air, from Ref. 31).

increases more slowly toward the fully ionized value owing to the progressive reduction of electron mobility. A high degree of ionization is therefore not necessary for the achieving of an appreciable fraction of the fully ionized conductivity limit. However, at temperatures that are practicable in MHD applications the thermal ionization of air, combustion-product gases, and the inert gases is so low that the electron density is orders of magnitude below that necessary to obtain suitable conductivities.

A large increase in conductivity is obtained by seeding the gas with a small percentage of materials which have much lower ionization potentials than air, combustion products, and the inert gases. First ionization potentials of the air components are ~14 V, and those of the inert gases are even higher, while those of the alkali metals range from 3.89 V for cesium and 4.34 V for potassium to 5.4 V for lithium. Some calculated conductivities of seeded gases are shown as functions of temperature in Fig. 12-107 (Ref. 31) and as a function of temperature and pressure in Fig. 12-108 (Ref. 40). In general, these curves are in close agreement with measured values (see, e.g., Ref. 23).

Over the temperature range of interest, the conductivity of a seeded gas can be approximated by the semiempirical relation

$$\sigma = AT^a e^{-eV_i/2kT} \tag{12-139}$$

where A and a are empirical constants and V_i is the ionization voltage of the alkali-metal seed. For the curves shown in Fig. 12-107, a is approximately 1 for combustion products seeded with potassium, 0.6 for combustion products seeded with cesium, and 3 for argon seeded with cesium.

The exponential factor in Eq. (12-139) stems from the Saha equation used to calculate equilibrium density of electrons for given densities of alkali seed and neutral atoms as a function of temperature (see, e.g., Ref. 31). However, this density is significantly reduced by the presence of electron negative components in the gas such as, for example, OH in the combustion products and Cl present as an impurity. (The seeding compound used should be chosen to avoid such components.) The remainder of the expression is empirical. Since the electron density is not exactly proportional to an exponential, and since the mobility of electrons is analytically very complex, no simple formula can be written for conductivity that will apply over any very great range of the variables.

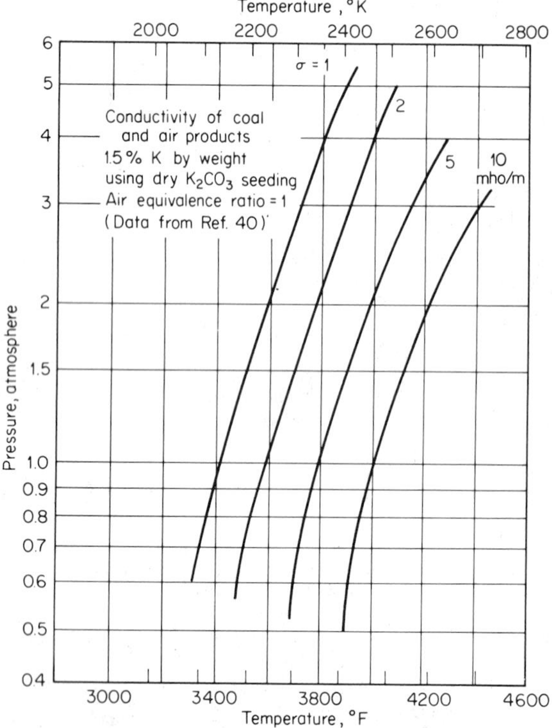

FIG. 12-108. Pressure vs. temperature for various conductivities of coal and air products plus 1.5% K by weight, using dry K_2CO_3. Air equivalence ratio = 1. (*Adapted from Ref.* 40.)

Somewhat unfortunately, the collision cross sections of the alkali metals are large compared with those of the other gases of interest so that, as the percentage of seed atoms is increased, the electron density increases as desired but the mobility begins to be markedly decreased, and an optimum seeding percentage exists. This is about 0.1% for Cs and K in argon and about 0.3% for these same alkali metals in neon (Ref. 31). At high temperatures, the electron-ion and electron-electron collisions, discussed above, dominate the collisions and limit the mobility and conductivity still further.

166. Nonequilibrium Ionization. If the electron density in an ionized gas is large enough for the frequency of electron-electron collisions to be comparable with that of electron-atom and electron-ion collisions, one may have an electron equilibrium temperature different from that of the atom-ion gas. This is possible because electrons exchange energy with each other very readily and hence rapidly set up an equilibrium distribution among themselves, at the same time exchanging energy with atoms and ions very poorly because of the large mass ratio in these cases.

In fact, the theory of the process assumes that the average energy loss by an electron in such a collision is $\delta(m_e/m_a)\frac{3}{2}k(T_e - T_a)$ where δ is a constant depending upon the kind of atom involved (about 2 for monatomic gases but much larger for molecular gases), m_e is the electron mass, m_a is the atom mass, T_e is the electron temperature, and T_a is the background gas temperature (Refs. 29 and 41).

The maintenance of such a temperature difference requires: (1) a source of electrons to maintain a sufficient electron density; (2) a way of supplying energy to them independent of the background gas; and (3) disposal of the energy transferred to the background gas by the electrons so that they do not come into thermal equilibrium with each other. The last condition is realized in an MHD channel by conversion of gas thermal energy into kinetic energy.

The achievement of sufficient electron density may be accomplished by using electron guns to supply electrons to the gas or by some scheme of enhancement, i.e., nonequilibrium ionization. The latter has produced great interest and activity since it was pointed out that the ionization equilibrium level should be determined entirely by the electron temperature, and if that were elevated, as it might be due to Joule heating as current flows, the degree of ionization would also be enhanced (Ref. 29).

The energy-balance relation for electrons in the Faraday generator configuration, with no Hall current allowed, has been found to be equivalent (Refs. 26 and 36) to

$$\frac{T_e}{T_a} = 1 + \frac{2\gamma}{3\delta}(\omega\tau)^2M^2(1 - K)^2 \tag{12-140}$$

where γ is the ratio of specific heats, δ is the collision constant noted above, $\omega\tau$ is the dimensionless parameter previously discussed, M is the Mach number of the flow, and K is the electrical efficiency, also previously discussed. From this expression it is clear that $\omega\tau$ and M should both be large to compensate for a small value of $1 - K$ if nonequilibrium ionization and conductivity enhancement is to occur. Figure 12-107 shows that doubling the conductivity of argon seeded with Cs requires a temperature increase of about 200°K in the region of 2000°K.

Assuming that this is the electron temperature increase required and using $\gamma = 1.67$, $\delta = 2$, $K = 0.75$, and $M = 0.5$, the value of $\omega\tau$ must be 3.4. Obviously, still larger values would seem desirable, but evidently nonuniformities are likely to occur when $\omega\tau$ is made large, and these may be expected greatly to limit the gains predicted from this simple theory (Ref. 26).

167. Electrodes. Electrical contact with the working fluid occurs at the electrodes. When ionized gas is the working fluid, the requirements are rigorous: (1) Ability to withstand the corrosive and erosive actions, at high temperature, of the fluid, which usually has an alkali-metal content. (2) Ability to emit copious quantities of electrons at the cathode side (cathode terminals are positive when generating power). (3) Good thermal conductivity so that liquid cooling is practicable.

Electrode materials used experimentally have included carbon (graphite), tantalum, tungsten, tungsten alloys, copper, stainless steel, silicon carbide, ZrO_2, and ZrB_2. Carbon and tungsten emit electrons thermionically in accordance with the Richardson-Dushman emission equation. In the presence of an alkali-seeded gas stream, however, the electrode surface (if not too hot) acquires a thin layer of the Cs or K metal, and this results in greatly enhanced emission (e.g., see Ref. 5). Since the latter effect can easily be poisoned by impurities, control of certain types of impurities is important for proper electrode operation.

When the electrodes have appreciable emission, the cathode voltage drop, and perhaps the anode drop also, tends to be very low ($<$10 V). In other cases, electrode voltage drops of approximately 40 to over 100 V have been measured. When large currents are drawn from relatively cool electrodes, arcing and point emission are frequently observed.

In a normal flow situation, the main body of the gas is highly turbulent, but a laminar boundary layer is present at the wall surface, with the fluid velocity dropping to zero at the surface itself. This layer provides thermal insulation for walls and electrodes, but because of the lower temperature it also has much less electrical conductivity, and appreciable electrode voltage drops may occur. When current flows, however, the high resistivity of the boundary layer causes appreciable Joule heating immediately adjacent to the electrodes, with the result that the heat flux into the electrode surface may be many times larger than that with no current flow. This effect, which compli-

Table 12-16. Properties of Structural Refractory Materials for MHD Applications

Material	Melting point, °C	Density, gm/cm³	Oxidation resistance	Thermal shock resistance	Thermal conductivity, cal·cm/(cm²)(sec)(°C)	Thermal expansion, cm/(cm)(°C)	Electrical resistivity, Ω·cm	Temp. of appreciable volatility (max. service temp.)
Oxides:								
SiO_2	1710	2.20	Excellent	Excellent	0.002(1200°C)	5×10^6(20–1280°C)	5×10^4(1000°C)	1400–1500°C
Al_2O_3	2030	3.96	Excellent	Fair	0.0173(1800°C)	8×10^6(20–1580°C)	1×10^6(1100°C)	1950°C
ZrO_2*	2650	5.56	Excellent	Good	0.0058(1400°C)	5.5×10^6(20–1200°C)	3×10^2(1200°C)	2300
ThO_2	3330	9.64	Excellent	Poor	0.006(1400°C)	9.5×10^6(20–1400°C)	1.5×10^4(1200°C)	2400
HfO_2	2780	9.68	Excellent	Good	n.d.	n.d.	n.d.	2300
Mixed oxides:								
$CaZrO_3$	2320	1.54	Excellent	?	n.d.	n.d.	n.d.	n.d.
$SrZrO_3$	2700	n.d.	Excellent	?	n.d.	n.d.	n.d.	n.d.
$BaZrO_3$	2700	2.25	Excellent	?	n.d.	n.d.	n.d.	n.d.
Carbides:								
ZrC	3510	6.7	Poor	Good	0.049(20°C)	6.73×10^6(24–5000°C)	6.34×10^{-5} (rm. temp.)	nonoxidizing† 2240°C oxidizing ?
TaC	3860	14.5	Poor	Good	n.d.	8.2×10^6(25–800°C)	2.0×10^{-5} (rm. temp.)	nonoxidizing 3780°C oxidizing—pyropharic at room temp.
HfC	3910	12.2	Poor	Good	n.d.	n.d.	1.09×10^{-4} (rm. temp.)	nonoxidizing 2240°C oxidizing ?
Carbon:								
Graphite	3700s	1.80	Poor	Excellent	0.131(1200°C)	1.0×10^6(20–1200°C)	$1.2 - 8 \times 10^{-1}$ (rm. temp.)	nonoxidizing 2700°C oxidizing 500°C
Pyrolytic graphite	3700s	2.20	Poor	Good (delamination)	⊥ = 0.02(2100°C) ‖ = 0.2(2100°C)	⊥ = 200×10^6(20–1200°C) ‖ = 0.5×10^6(20–1200°C)	⊥ = 2×10^{-1} ‖ = 6.10^{-5} (rm. temp.)	nonoxidizing 2800°C oxidizing 800°C
Nitrides:								
BN	3000s	2.25	Good	Excellent	⊥ = 0.364 ‖ = 0.760 } 1000	⊥ = 4.17×10^6 ‖ = 0.43×10^6 } 25–1000°C	1.2×10^4 (1000°C)	nonoxidizing 2980°C oxidizing 1100°C

* Stabilized (cubic phase).
† Oxidizing and nonoxidizing refer to environment.
s = sublimes. n.d. = no data. ⊥ = perpendicular crystal orientation. ‖ = parallel crystal orientation.

cates the electrode cooling problem, has been considered in some detail by Kerrebrock (Ref. 30).

Much difficulty with electrode isolation has been experienced because of leakage currents between segmented electrodes. Tantalum (frequently used as support for alumina or other refractory duct linings) and graphite have caused particular trouble by depositing on insulator surfaces to create low-resistance leakage paths. This has been reportedly overcome in one case by making the electrodes free-standing on posts away from the main wall and using elaborate ceramic shields behind the electrodes to prevent the deposit of conducting materials (Ref. 55).

168. Structural Materials. Table 12-16 lists structural refractory materials that may be used in MHD applications (see also Ref. 21). With the exception of graphite, the materials are difficult to form, brittle, low in tensile strength, in some cases very expensive, and, in others, at present available only in laboratory quantities. In experimental work, tantalum has been frequently used as a structural base for refractory liners and, especially in closed systems, to provide gastight integrity.

In a power generator of some hundreds of megawatts output, wall design may be easier than in small-sized units because of improved surface-to-volume ratio. The radiative component from the hot gas is proportionally less, and the walls are protected by a stagnant gas layer of poor thermal conductivity. Liquid cooling of duct walls in that case can be accomplished without great complication or serious degradation of overall thermal efficiency. In order to avoid large thermal gradients in the walls, a metal-ceramic matrix wall design has been proposed (Refs. 24 and 46).

169. Magnets. Were it not for the recent rapid development of new superconducting materials, the cost of operating and cooling the large magnet required might well be prohibitive. However, superconducting magnets promise to reduce operating costs by more than an order of magnitude and make feasible higher field strengths than can be achieved economically in conventional designs.

Many problems remain before a practical superconducting MHD magnet is achieved. However, it is anticipated that such magnets can be built to give fields as large as 6 T (Ref. 56). The largest reported superconducting magnet (as of early 1967), built at the Avco Everett Research Laboratory, produced a field of 4 T in a 12-in-diameter bore, uniform for a length of 45 in. Total weight of the magnet was 16,000 lb. The winding consisted of Nb-Zr wires embedded in 0.040- by 0.500-in copper strips. Cooling to 4.2°K was accomplished by means of liquid helium, at which temperature the magnet was fully superconducting (Ref. 50).

170. Experimental Status. Experiments with high-temperature seeded gases in thermal equilibrium have been in general agreement with theoretical predictions, provided that the experimental conditions were carefully controlled. The kinds of difficulties encountered include nonuniform conditions in the gas, poor contact between electrodes and the flowing gas, and leakage currents along the insulator walls (Refs. 33, 34, and 55). In addition, materials failures due to high temperature, erosion, chemical incompatibility, and thermal shock have plagued most experiments. As a consequence, long-duration tests have been limited to relatively small ($\sim$10 kW) experiments, while large power-generation tests have been limited to short runs (10 to 20 s). In the duration tests, electrode and wall lives of 100 h or more have been achieved, representing an order-of-magnitude improvement in 3 years' time, while in the power-generation tests outputs up to 33 MW have been achieved (Refs. 23, 47, and 57).

Great difficulty has been reported in eliminating leakage currents between segmented electrodes (Refs. 47 and 52), and in one case special electrode construction having shielded-post supports has been used (Ref. 55). This construction was designed to increase the leakage path and reduce the tendency of the alkali seed metal to form conducting layers on the insulator surfaces. Electrode voltage drops have caused difficulty in some experiments (e.g., Ref. 45) and not in others (e.g., Ref. 47), and whether electrodes have to be in the gas stream rather than flush with the wall is uncertain.

Among the electrode materials, tantalum and stainless steel have both been found satisfactory insofar as electrode voltage drop is concerned (Ref. 47). Screening tests in a test duct have been reported on graphite, tungsten, coated tungsten, tungsten

Table 12-17. Conceptual Design Studies

Item	T. R. Brogan, J. F. Louis, R. J. Rosa, and Z. J. J. Stekly (Avco), Ref. 8, pp. 147–165, and Ref. 9, pp. 243–258	J. J. W. Brown (General Electric), Ref. 28 (in Ref. 9, pp. 223–241). See also Tables 18, 19, 20.	W. E. Gunson, E. E. Smith, T. C. Tsu, and J. H. Wright (Westinghouse), Ref. 20	L. L. Prem and W. E. Perkins (Atomic International), Ref. 39. Also see Ref. 25.
Fuel and oxidant:				
Date of report	1961–1962	1962†	1963	1964
Fuel	Coal	Coal	Atomic GCR	Fossil
Air preheat T/p	1600/16	2034/7.69	....	None
Oxygen enrichment	20.5	None	....	None
MHD generator:				
Type of cycle	Open	Open	Closed	Closed
Working fluid	CG + alkali	CG + 1% K	He + alkali	Potassium 2-phase
Magnetic field	No data	10	10	1–2
Fluid temperature (static):				
°F	~5000ϕ	4440–3633	2450–1750θ	1602ϕ
°K	~3033ϕ	2721–2273	1616–1228θ	1145ϕ
Pressure (static)	16–1.05	7.5–1.45	4.42–1.89	~2.7
Velocity	No data	735	1,775	~100
Channel length	No data	6.0	27.5	**
Generator coefficient K	No data	0.8	~0.8	>0.7
MHD electrical power	352	361	315.5	No data
Power consumed by d-c/a-c inverter	14	18.2[a]	16.5[a]	No data
Net MHD output	338	342.8	299.0	~190[c]
Steam turbines:				
Gross turbine power	275.5[b]	280[b]	400.0[b]	~810
Net electrical output	152	154.1	201.0	~810
Plant output:				
Net plant output	490	496.9	500.0	1,000[c]
Plant heat rate	6380	6830	7240	~7600
Net plant efficiency	53.5	50	47.1	~45[d]
Costs:				
Plant capital cost	No data	138.50[e]	No data	No data
Fuel cost	1.63	1.04–2.07[f]	1.09–2.17	1.14–2.28
At fuel price	23	15–30	15–30	15–30

NOTE: CG = combustion gases.
GCR = gas-cooled reactor.
† With cost revisions to 1964.
‡ Preheat temperature reduced to 1668°F when a carbonizer is used to produce char for the first stage and volatile fuel for the second stage of the two-stage cyclone combustor.
ϕ = Maximum cycle temperature.
θ = Assumes that nonthermal ionization is feasible.
** The actual power-generating section uses a liquid-metal working fluid and is very short, as is the magnet; however, a relatively long drift tube is required ahead of the generator.

alloys, zirconium, zirconium boride, titanium boride, and zirconia cermet (Ref. 21). Of these, the zirconium boride (ZrB_2) was found to be the best electrode material at present available.

Tests on refractory insulators for wall materials led to the conclusion that no single available material could withstand the required high temperatures (Ref. 21). More recent work has led to the development of the "peg-wall" concept, in which closely spaced, water-cooled metallic segments separated by thin refractory sections for electrical insulation are used (Refs. 24 and 46). Endurance times in excess of 1 week have been achieved with this type of construction.

When $\omega\tau$ is greater than unity, the Hall field should be larger than the $\mathbf{u} \times \mathbf{B}$ field (unless it is short-circuited out by wall leakage currents, either in the stagnant gas layer at the surface or in the surface conduction). Attempts to achieve high Hall fields have usually given less than 5% of the expected values (Refs. 26, 45, 47, 52, and 55). Furthermore, tests at high values of $\omega\tau$ have had noticeably large random fluctuations of electrode voltage and current (Ref. 45), indicating possible instabilities with this mode of operation (Ref. 26). If so, the nonequilibrium enhancement of conductivity by this means may be unachievable, although it can perhaps be achieved by other means such as electron-beam ionization (Ref. 20).

Experimental studies with liquid metals in MHD power applications have been undertaken in several laboratories, but only preliminary reports have been made (Refs. 37 and 51). The state of these studies up to 1965 was reported in a survey (Ref. 25) which indicated that, although the prospects of liquid-metal MHD generators appear

of Central-station MHD-Steam Plants

S. Way and W. E. Young (Westinghouse), Ref. 40		M. Rosner (Brown Boveri), Ref. 53	T. C. Tsu, W. E. Young & S. Way* (Westinghouse and *Univ. Calif. Santa Barbara), Ref. 56		Units
1964		1966	1966		
Coal	Coal	Fuel oil	Coal	Char	
2197/4.16	2482/4.16	1537/∼2.8	1769‡/4.70	1606/5.20	°F/atm
None	None	None	None	None	%
Open	Open	Open	Open	Open	
CG + 1.5% K	CG + 1%K	CG + 0.4% K	CG + 0.5% Cs	CG + 0.7% Cs	Gas + seed
5.8	6.5	5	6	6	T
4356–3691	4467–3783	4172–3681	4302ϕ	4293ϕ	°F
2675–2306	2737–2357	2573–2300	2645ϕ	2640ϕ	°K
4.00–0.908	5.00–1.039	2.56–1.19	4.5–1.1	5.0–1.1	atm
780	750	870–670	750–567	750–561	m/s
11.27	10.4	10	17.45	12.98	m
0.80	0.786	0.82–0.60	0.78	0.75	
376.5	311.7	135	400	400	MW
7.5	6.2	7	6	6	MW
369.0	305.5	128	394	394	MW
328.9[b]	386.35[b]	378[b]	No data	No data	MW
223.5	294.5	331	398.77	384.16	MW
592.5	600.0	459	792.77	778.16	MW
8189	6837	7440	6771	6567	Btu/kWh
41.7	49.9	45.9	50.40	51.97	%
99.65[g]	138.50[h]	100	No data	No data	$/kW
1.23–2.46	1.024–2.05	1.11–2.23	1.01–2.03	0.985–1.97	Mills/kWh
15–30	15–30	15–30	15–30	15–30	Cents/10^6 Btu

[a] Includes cryogenic power for magnet.
[b] Includes shaft-connected compressor power.
[c] Data given in Ref. 39 (125 MW of MHD power and 1,000 MW for whole plant) are inconsistent with other data given; these values estimated from the other data.
[d] 14% higher than steam cycle alone.
[e] Estimated minimum cost.
[f] Includes seed makeup cost.
[g] Assumes $100/lb superconducting magnet wire. For $200/lb wire cost is 6.73% greater.
[h] Assumes $100/lb magnet wire. For $200/lb wire, cost is 7.34% greater.

favorable and perhaps even more so than for plasma MHD generators, there are major areas of uncertainty yet to be resolved through experimentation.

Combustors for use in fossil-fueled systems have had some experimental development, based on ramjet or turbojet afterburner designs, with seeming success although the long-term endurances are still unknown (Ref. 23). The development of a fuel-oil-fired combustor using steel construction has been reported (Ref. 44). In recent optimization studies of a coal-fired MHD-steam plant, a two-stage cyclone combustor is proposed which would operate at temperatures close to those which are usual for such equipment (Ref. 56). In this case a carbonizer is used to produce "char" for the first stage (the cyclone section) and volatile fuel for the second stage. Excess air can then be used in the first stage to keep temperatures there at moderate levels (see also Ref. 40).

171. Conceptual Designs. *Open-cycle—Fossil-fueled.* Since the standard steam plant extracts 85% of the available heat energy, efficiency can be appreciably increased only by raising the input temperature. However, increases in steam temperature and pressure beyond 3,500 lb/in^2 and 1050°F produce very small net increases in plant economy, as measured in terms of final cost of delivered power (mills per kilowatthour), because of substantially increased capital costs for equipment to operate at the higher temperatures and pressures.

The temperature in a fossil combustor can be as high as 4500°F. Carnot efficiencies approaching 90% could result if this upper temperature were used in the thermal cycle, and it is precisely this regime that the gas MHD unit is designed to exploit. With such

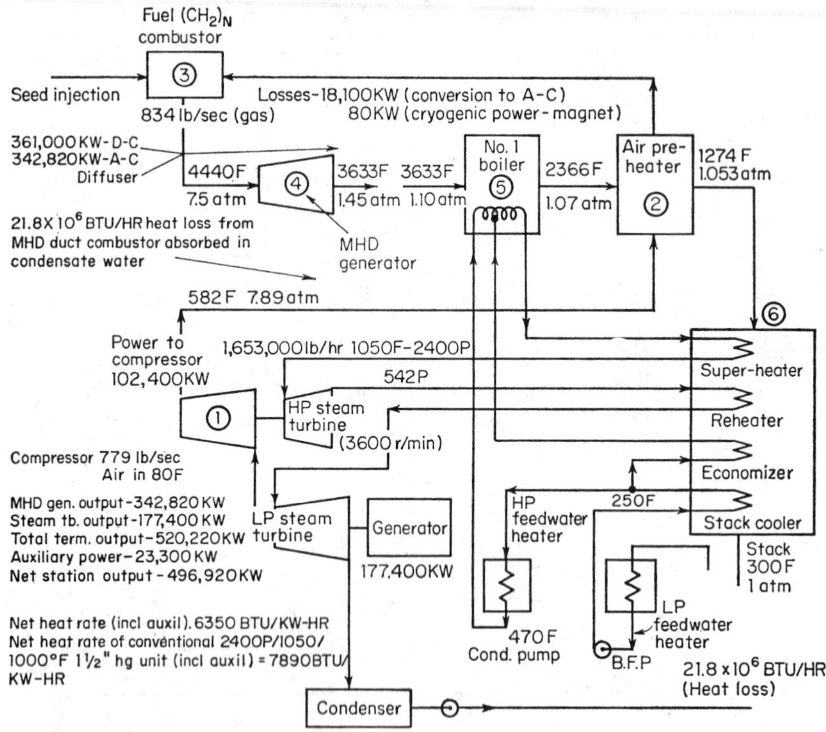

Fig. 12-109. MHD/steam-turbine combined cycle (estimated data).

a topping unit, energy is first removed from the combustor gas by MHD, with an accompanying temperature drop from 4500 to 3500°F. The gas is then used to generate high-quality steam in a conventional Rankine-cycle steam plant and is finally vented at a flue temperature of 300°F.

The cycle is illustrated in greater detail in Fig. 12-109. Starting at (1), intake air is compressed, preheated (2) to 2000°F in a regenerator, and fed, along with seed material and fuel, into the combustor (3). Seed material may be any low-cost Na or K compound such as K_2CO_3. The gas reaches a temperature of 4500°F in the combustor and passes through the MHD duct (4), where both temperature and pressure drop. At the exit of the duct the seed material is recovered, and the gas then provides heat to a steam boiler (5) and to the MHD cycle air preheater (2). The remaining gas energy is extracted in the steam cycle by superheat, reheat, and economizer cycles (6).

Liquid-metal Topping Unit. The very high temperatures required to maintain a plasma in a conducting state can be avoided if liquid metals are used as the working fluid. Conceptual studies (Refs. 25, 37 to 39, and 51) have been made with mercury, mercury-potassium alloy, potassium, cesium, and sodium. All such systems generate a high-velocity jet of liquid by mixing liquid with vapor from the boiler (nuclear- or fossil-fueled). Expansion of the vapor accelerates the liquid and tends to condense the vapor. The liquid or liquid-vapor mixture passes first through an MHD duct, then a condensing unit which generates standard quality steam for a conventional steam plant, and then through the boiler.

The overall efficiency of such a system is calculated to be approximately 55% for temperatures as low as 2200°F, in contrast with plasma MHD topping cycles, which require initial temperatures of 4000°F for the same efficiency.

The primary unsolved questions relate to the efficiency of the condenser-ejector for generating the high-speed liquid flow (Ref. 38) or, in other conceptual designs, the efficiency of the drift-tube momentum exchange and condensation (Ref. 39) on the

Table 12-18. Summary—Estimated Plant Generation Costs

(Adjusted to July, 1964, basis)

	Combined MHD-steam nat.-gas-fired, mills/kWh	Combined MHD-steam coal-fired, mills/kWh	Conventional natural-gas-fired plant, mills/kWh	Conventional coal-fired plant, mills/kWh	Combined gas-steam nat.-gas-fired turbine, mills/kWh	Nuclear BWR, mills/kWh
15 cents/10⁶ Btu fuel:						
Capital costs*	2.36	2.77	1.58	2.08	1.58	2.75
Fuel costs	1.08	1.02	1.34	1.29	1.24	1.40†
Oper. and maint. costs	0.55	0.60	0.25	0.30	0.35	0.44‡
Seed costs	0.02	0.02				
Total cost on feeders	4.01	4.41	3.17	3.67	3.17	4.59
30 cents/10⁶ Btu fuel:						
Capital costs*	2.36	2.77	1.58	2.08	1.58	2.75
Fuel costs	2.16	2.05	2.67	2.58	2.47	1.40†
Oper. and maint.	0.55	0.60	0.25	0.30	0.35	0.44‡
Seed costs	0.02	0.02				
Total cost on feeders	5.09	5.44	4.50	4.96	4.40	4.59

* 14% capital charge and 0.8 load factor.
† Fuel cost as projected to 1973.
‡ Includes insurance.

stability of the fluid film required in the proposed single-component cycle (Refs. 25 and 37). Experimental and theoretical studies of this device are reported as giving high efficiency and high stagnation pressures.

172. Economics. The primary target of the MHD topping unit is the production of power at lower cost, through the exchange of added capital investment and operating costs to obtain lower fuel costs per kilowatthour of power produced. Because of the great complexity and extensive structural interrelation of the modern central-station plant, evaluation of such incremental savings is difficult. It requires the determination of the savings (due to increased efficiency resulting from the MHD addition) from which the increased equipment and operating costs must be subtracted. To be significant the calculation must be done in considerable detail, and few such studies have been made.

Table 12-17 represents an assemblage of the best data available for common reference. Of the studies listed, that by J. J. W. Brown presents the most detailed analysis of power costs in terms of several system options (Tables 18 to 20; see also Ref. 40).

Table 12-19. Preliminary MHD Power-plant Cost Breakdown, Natural-gas-Fired

(Adjusted to July, 1964, basis)

Item No.	Estimated costs (000 omitted) Min.	Max.
1. MHD generator, including combustor and seeding system	$ 7,035	$10,630
2. Air compressor, preheater, start-up fan, etc	5,902	9,160
3. Electrical, including inverter, high yard, generator kvar adder, controls, instrumentation, etc	10,935	13,883
4. Misc. equipment, piping, etc	500	500
5. Site work, structural, contingency, installation, engineering and management, start-up cost, and interest during construction	7,749	10,698
Plant cost, MHD portion	$32,121	$44,871
6. Gas-fired conventional steam-turbine generator portion, including boiler, etc	$26,525	$26,525
Total plant cost	$58,646	$71,396
$/kW, total plant (497-MW base)	$118	$143.65
Range, total plant, %	100	120
Range, MHD portion only, %	100	140

Table 12-20. Summary—Estimated Plant Costs and Heat Rates

(Adjusted to July, 1964, basis)

500-MW plants	Plant costs, $/kW	Plant heat rates, Btu/kWh
Combined cycle, MHD-steam, natural-gas-fired	118	7,200
Combined cycle, MHD-steam, coal-fired	138.5	6,830
Conventional natural-gas-fired	79	8,900
Conventional coal-fired	104	8,600
Combined cycle, gas-steam turbine, natural-gas-fired	79	8,250
Nuclear, BWR	137.5	

He concludes that coal plus MHD has no economic advantage over either the conventional gas- or the coal-fired station in the range of fuel costs up to 30 cents/10^6 Btu and that MHD, when proved, would be a less favorable choice than a nuclear reactor in regions where fuel costs exceeded 30 cents/10^6 Btu.

In the period 1963 to 1966 nuclear-power generation costs have decreased. The principal reduction (in excess of 0.5 mill/kWh) has been due to both technical and economic improvements in the fuel cycle. Currently, larger plants (about 1,000 MW) are being constructed with capital costs as much as $25/kW below those for the 500-MW size. Nuclear plants are now being constructed to deliver power for under 4 mills/kWh, and, over the next decade, it is not unreasonable to expect that technologic advances will produce further reductions.

There are still, however, many areas where low-cost coal is plentiful and where mine-mouth power, using the advances made possible by modern power-transmission techniques, will continue to be the system of choice. A balanced view of the future would indicate that a very appreciable portion of all electric power will continue to be generated by fossil fuel. In such future plants the trade-offs between plant capital cost and over-all efficiency will undoubtedly tend toward increased sophistication and complexity.

Later studies both in the United States and abroad arrive at more optimistic evaluations of the economic advantage of a MHD topping unit. The economic balance of many complex factors may shift in the next decade under the influence both of a developing competitive technology and of a changing cost structure.

173. References.

Books

1. Spitzer, L., Jr. "Physics of Fully Ionized Gases"; New York, Interscience Publishers, Inc., 1st ed., 1956, 2d ed., 1962.

2. Pai, S. I. "Magnetogasdynamics"; Englewood Cliffs, N.J., Prentice-Hall, Inc., 1962.

3. Coombe, R. A. (ed.) "Magnetohydrodynamic Generation of Electrical Power"; London, Chapman & Hall, 1964.

4. Kulikovskiy, A. G., and Lyubimov, G. A. "Magnetohydrodynamics"; Reading, Mass., Addison-Wesley Publishing Company, Inc., 1965.

5. Sutton, G. W., and Sherman, A. "Engineering Magnetohydrodynamics"; New York, McGraw-Hill Book Company, 1965.

6. Jahn, R. G. "Physics of Electric Propulsion"; New York, McGraw-Hill Book Company, 1968.

Conference Proceedings

7. Symposia on Engineering Aspects of Magnetohydrodynamics, held annually since 1960. (Although no preprints or proceedings were issued for the First Symposium, held in Philadelphia, Feb. 18–19, 1960, the program is printed as Appendix I of Ref. 8, below. Preprints for all other symposia were issued and may be purchased, insofar as supply remains, through Prof. J. A. Fox, Department of Mechanical Engineering, University of Mississippi, University, Miss. 38677. Proceedings of the Second and Third Symposia were published in book form, Refs. 8 and 9 below.)

8. Mannal, C., and Mather, N. W. (eds.) Engineering Aspects of Magnetohydrodynamics; *Proc. Second Symp.*, Philadelphia, Mar. 9–10, 1961, New York, Columbia University Press, 1962.

9. Mather, N. W., and Sutton, G. W. (eds.) Engineering Aspects of Magnetohydrodynamics; *Proc. Third Symp.*, Rochester, Mar. 28–29, 1962, New York, Gordon and Breach Science Publishers, 1964.

10. Magnetoplasmadynamic Electrical Power Generation; *Rept. Symp. IEE (London)*, Newcastle-upon-Tyne, Sept. 6–8, 1962. Inst. Elec. Engrs., London.

11. McGrath, I. A., Siddall, M. W., and Thring, M. W. (eds.) Advances in Magnetohydrodynamics; *Proc. Colloq. Sheffield Univ.*, October, 1961, New York, The Macmillan Company, 1963.

12. *Proc. Intern. Symp. Magnetohydrodynamic Elec. Power Generation*, Paris, July 6–11, 1964, 4 vols., published by ENEA (European Nuclear Energy Agency).

13. *Intern. Symp. Magnetohydrodynamic Elec. Power Generation*, Salzburg, Austria, July 4–8, 1966. [Preprinted papers; publication by ENEA (or IAEA) is expected.]

Review Articles

14. Resler, E. L., Jr., and Sears, W. R. The Prospects for Magneto-aerodynamics; *Jour. Aeron. Sci.*, April, 1958, Vol. 25, pp. 235–245, 258.

15. Sporn, P., and Kantrowitz, A. Magnetohydrodynamics: Future Power Process? *Power*, November, 1959, Vol. 103, No. 11, pp. 62–65.

16. Steg, L., and Sutton, G. W. Prospects of MHD Power Generation; *Astronautics*, August, 1960, Vol. 5, pp. 22–25.

17. Rosa, R. J. Physical Principles of Magnetohydrodynamic Power Generation; *Phys. Fluids*, February, 1961, Vol. 4, pp. 182–194.

18. Mannal, C. Recent Progress in Magnetohydrodynamics; *IEEE Trans. Nuclear Sci.*, September, 1963, Vol. NS-10, pp. 8–17.

19. Somers, E. V. Coal-fired MHD Steam Plants—Present and Future; *Proc. Am. Power Conf.*, 1963, Vol. 25, pp. 83–92.

20. Gunson, W. E., Smith, E. E., Tsu, T. C., and Wright, J. H. MHD Power Conversion; *Nucleonics*, July, 1963, Vol. 21, No. 7, pp. 43–47.

21. Goldberg, D. C., Young, W. E., and Hunstad, R. L. Materials Requirements for Magnetohydrodynamics, *Metals Eng. Quart.*, November, 1963, Vol. 3, No. 4, pp. 47–54.

22. Lindley, B. C. Magnetoplasmadynamic Generation—Future Power Plant; *Proc. Inst. Mech. Engrs.*, 1963–1964, Vol. 178, Pt. 3H, pp. 48–59.

23. Brogan, T. R. MHD Power Generation; *IEEE Spectrum*, February, 1964, Vol. 1, pp. 58–65.

24. Rosa, R., and Kantrowitz, A. MHD Power, *Intern. Sci. Technol.*, September, 1964, No. 33, pp. 80–86, 89, 90, 92.

25. Petrick, M. Liquid Metal Magnetohydrodynamics, *IEEE Spectrum*, March, 1965, Vol. 2, pp. 137–151.

26. Kerrebrock, J. Magnetohydrodynamic Generators with Nonequilibrium Ionization; *Jour. AIAA*, April, 1965, Vol. 3, pp. 591–601.

Journal and Conference Papers

27. Bernstein, I. B., Fanucci, J. B., Fischbeck, K. H., Jarem, J., Korman, N. I., Kulsrud, R. M., Lessen, M., and Ness, N. An Electrodeless MHD Generator; in Ref. 8, pp. 255–276.

28. Hurwitz, H., Jr., Kilb, R. W., and Sutton, G. W. Influence of Tensor Conductivity on Current Distribution in an MHD Generator, *Jour. Appl. Phys.*, 1961, Vol. 32, pp. 205–216.

29. Kerrebrock, J. L. Conduction in Gases with Elevated Electron Temperature; in Ref. 8, pp. 325–346.

30. Kerrebrock, J. L. Electrode Boundary Layers in Direct-current Plasma Accelerators; *Jour. Aerospace Sci.*, August, 1961, Vol. 28, pp. 631–644.

31. Frost, L. S. Conductivity of Seeded Atmospheric Pressure Plasmas; *Jour. Appl. Phys.*, October, 1961, Vol. 32, pp. 2029–2036.

32. Harris, L. P., and Cobine, J. D. The Significance of the Hall Effect for Three MHD Generator Configurations; *Trans. ASME*, 1961, Vol. 83, Ser. A (*Jour. Eng. Power*), pp. 392–396.

33. Way, S., Corso, S. M., Hundstad, R. L., Kemeny, G. A., Stewart, W., and Young, W. E. Experiments with MHD Power Generation; *Trans. ASME*, 1961, Vol. 83, Ser. A (*Jour. Eng. Power*), pp. 397–408.

34. Sutton, G. W., Hurwitz, H., Jr., and Poritsky, H. Electrical and Pressure Losses in a Magnetohydrodynamic Channel Due to End Current Loops; *Trans. AIEE*, 1961, Vol. 80, Pt. I (*Commun. Electronics*, January, 1962, No. 58), pp. 687–694.

35. Brown, J. J. W. Some Aspects of MHD Power Plant Economics; in Ref. 9, pp. 223–241.

36. Hurwitz, H., Jr., Sutton, G. W., and Tamor, S. Electron Heating in Magnetohydrodynamic Generators; *Jour. Am. Rocket Soc.*, 1962, Vol. 32, pp. 1237–1243.

37. Petrick, M., and Lee, K-Y Performance Characteristics of a Liquid Metal MHD Generator; *Argonne Natl. Lab. Rept.* AN-6870 (TID-4500, 34th ed.), July, 1964.

38. Brown, G. A., and Lee, K. S. A Liquid Metal MHD Power Generation Cycle Using a Condensing Ejector; paper 60 in Ref. 12, 1964, Vol. 2, pp. 929–938.

39. Prem, L. L., and Perkins, W. E. A New Method of MHD Power Conversion Employing a Fluid Metal; paper 63 in Ref. 12, 1964, Vol. 2, pp. 971–984.

40. Way, S., and Young, W. E. The Feasibility of Large-scale MHD Power Generation; paper 98 in Ref. 12, 1964, Vol. 3, pp. 1483–1495.

41. Kerrebrock, J. L. Nonequilibrium Ionization Due to Electron Heating: I. Theory; *Jour. AIAA*, 1964, Vol. 2, pp. 1072–1080.

42. Kerrebrock, J. L., and Hoffman, M. A. Nonequilibrium Heating: II. Experiments; *Jour. AIAA*, 1964, Vol. 2, pp. 1080–1087.

43. Fishman, F. J. Effect of Electrode Nonuniformities along the Magnetic Field in MHD Generators; *Advanced Energy Conversion*, December, 1964, Vol. 4, pp. 223–236.

44. Gallant, H. Development of a Combustion Chamber for MHD Generators; *Brown Boveri Rev.*, December, 1964, Vol. 51, pp. 817–820.

45. Klepeis, J., and Rosa, R. J. Experimental Studies of Strong Hall Effects and $U \times B$ Induced Ionization; *Jour. AIAA*, September, 1965, Vol. 3, pp. 1659–1666.

46. Novack, M. E., and Brogan, T. R. Water Cooled Insulating Walls; *Advanced Energy Conversion*, July, 1965, Vol. 5, No. 2, pp. 95–102.

47. Brown, R., Lindley, B. C., and McNab, I. R. Further Development, Operation and Experiments with the IRD Closed-cycle MPD Loop; paper SM-74/36 of Ref. 13.

48. Damon, E. L., and Gourdine, M. C. Electrogasdynamic Power Generation; paper SM-74/197 of Ref. 13.

49. Hals, F. A High Temperature Regenerative Air Preheater for MHD Power Plants; paper SM-74/68 of Ref. 13.

50. Kantrowitz, A., and Stekly, Z. J. J. A Model MHD Type Superconducting Magnet; paper SM-74/256 of Ref. 13.

51. Petrick, M. MHD Generators Operating with Two-phase Liquid Metal Flows; paper SM-74/196 of Ref. 13.

52. Ralph, J. C., and Baniton, K. F. MHD Power Generation Experiments Using Seeded Inert Gases; paper SM-74/12 of Ref. 13.

53. Rosner, M. The Oil Fired MHD Power Plant; paper SM-74/28 of Ref. 13.

54. Rosner, M. Efficiency of Closed-loop MHD Generators Utilizing Thermal Nonequilibrium Ionization; paper SM-74/29 of Ref. 13.

55. Schneider, R. T., and Wilhelm, H. E. Experimental Investigation of Closed Loop MHD Power Generation; paper SM-74/160 of Ref. 13.

56. Tsu, T. C., Young, W. E., and Way, S. Optimization Studies of MHD-Steam Plants; paper SM-74/179 of Ref. 13.

57. Dragoumis, P., and Brogan, T. R. The Development of MHD Power Generators; AVCO Corporation, September, 1966.

58. Yeh, H. Status of Magnetohydrodynamic Power Generation for Terrestrial Applications; *Paper* 66–1013 presented at AIAA Third Annual Meeting, Boston, Mass., Nov. 29–Dec. 2, 1966.

THERMIONIC CONVERSION

V. C. Wilson

174. Definition of Thermionic Converter. A thermionic converter is a vacuum or vapor-filled device with a hot electrode to emit electrons and a cold electrode to collect the electrons. It is a high-temperature static heat engine that obeys the Carnot cycle.

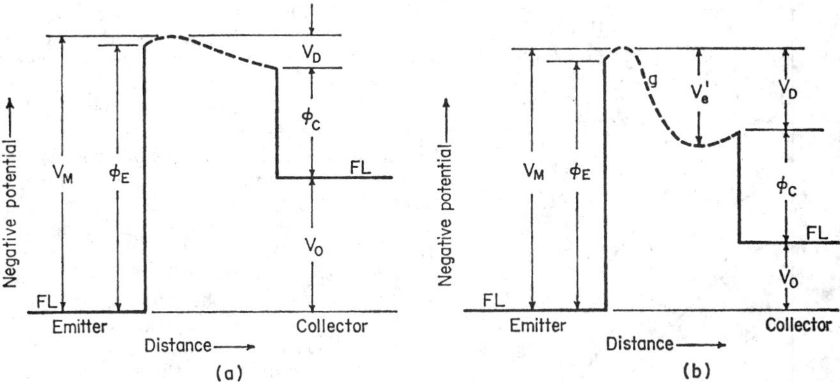

FIG. 12-110. Thermionic-converter potential diagrams. (*a*) Extinguished mode; (*b*) obstructed mode.

The electrons emitted from the emitter are the source of current. The heat put into the emitter is the driving force that provides the potential. This may be understood by reference to Fig. 12-110*a*. Heat lifts the electrons over the work-function barrier of the surface of the metal, ϕ_E. An electron just outside the metal has kinetic energy of motion, and it has potential energy relative to the electrons in the metal equal to ϕ_E or in some cases V_m, which is the greatest negative potential in the system. As the electrons are collected, they may fall through a potential V_D to the surface of the collector. They then fall through an additional potential ϕ_C to the Fermi level of the collector. Any additional potential V_o is available to drive the electrons through an external circuit. Thus, V_o is the output voltage of this generator. For an ideal converter with no heat loss and with V_D equal to 0, the converter would obey the Carnot efficiency

$$\eta = \frac{T_E - T_C}{T_E} = \frac{V_E - \phi_C}{V_E} \tag{12-141}$$

175. Types of Thermionic Converters. *a. Vacuum Converters.* From Fig. 12-110*a* it is apparent that the work function of the emitter should be greater than the work function of the collector. The standard barium strontium oxides, widely used for emitter materials in thermionic vacuum tubes, have high temperature coefficients of the work function, so that with the material evaporated from the emitter to the collector there is still a difference $\phi_E - \phi_C$. However, space charge is a serious problem in the vacuum converter.

H. F. Webster[1*] has worked out the theory of the vacuum converter. Figure 12-111 shows the output one can obtain at T_E equal to 1100°K for various work-function differences and as a function of electrode spacing. J. E. Beggs[2] constructed a number of vacuum converters which in their performance agreed extremely well with Webster's calculations. As may be seen in (*b*), below, the output power of the vacuum converter is much less than for a cesium converter; so little effort is being made to develop vacuum thermionic converters.

b. Cesium-vapor Converters. If cesium vapor is introduced between the electrodes of the thermionic converter and if the temperature of the emitter is sufficiently high, some Cs ions will be generated as the Cs atoms strike the hot emitter. These ions overcome the electron space charge, enable more current to be conducted, and permit increasing the spacing between the electrodes. The Cs may be used for other purposes. A coating of Cs on the collector develops a very low-work-function collector, which is desirable. By adjusting the Cs pressure and the temperature of the emitter a partial coating of Cs can be obtained on the emitter, and in this way the work function of the emitter may be adjusted.

If the work function is too high, insufficient current flows; if the work function is too low, insufficient voltage is generated. The optimum work function is between 2.5

* Superscript numbers correspond with those of the References, Par. **180**.

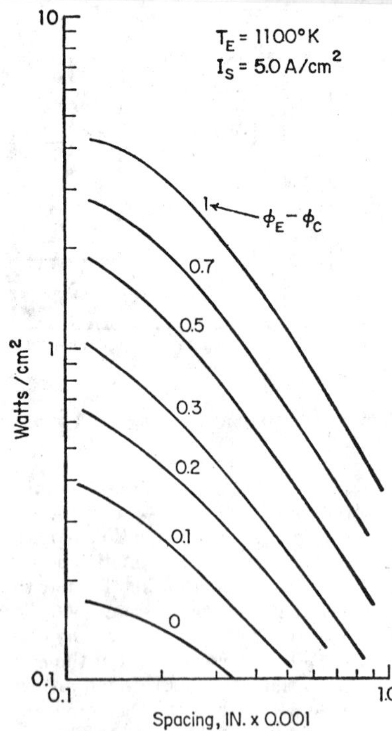

FIG. 12-111. Maximum-output power density from a vacuum thermionic converter vs. emitter-collector spacing for seven different values of work function difference $\phi_E - \phi_C$, in electronvolts.

and 3 eV. The Cs pressure necessary to obtain the optimum emitter work function is much greater than the pressure necessary to overcome space charge and to obtain a low-work-function collector. Because of this high Cs pressure, 1 to 10 Torr, electrons going from the emitter to the collector scatter on the Cs ions and atoms; so it is advantageous to keep the electrodes fairly close. In present-day converters, the electrodes are spaced about 0.01 in. apart.

c. Special Converters. In the Cs diode converter, to generate the ions requires operating the emitter at 1600°K or greater. Many problems, both in heat sources and in materials, could be simplified if the operating temperature could be lowered. Several writers have proposed that this could be accomplished by creating the ions in a small auxiliary gas discharge adjacent to the emitter. This approach has the additional advantage that the discharge could be obtained in rare gases rather than in Cs. The extra electrode to create the auxiliary discharge might also be used to turn on and off the converter and in this way generate alternating current. Hernqvist[3] developed a scheme for alternately directing the emission current to one or the other of two electrodes and in this way generating alternating current.

Attempts have also been made to generate ions by intense radiation or by fission-particle recoils.[4] Another possibility is to utilize the Cs to overcome space charge and to reduce the work function of the collector, but to adjust the work function of the emitter by coating thorium or barium on the surface of the emitter. To date, all these schemes have produced less output than has been obtained from the direct high-pressure Cs diode converter.

176. Cesium-vapor Diode Converters. Although a diode thermionic converter is an extremely simple device, the physics involved is surprisingly complicated. The electric potential distribution between the electrodes and the generation of ions depends upon the way in which the thermionic converter is operated. Figure 12-112 is a plot of the voltage generated vs. the current density. At open circuit, point K, the electrons reaching the collector just equal the leakage currents from the collector. If one starts to draw current from the converter, the voltage drops rapidly along the line K to A. In the region from H to A the current is limited by the number of ions that are generated at the hot emitter.

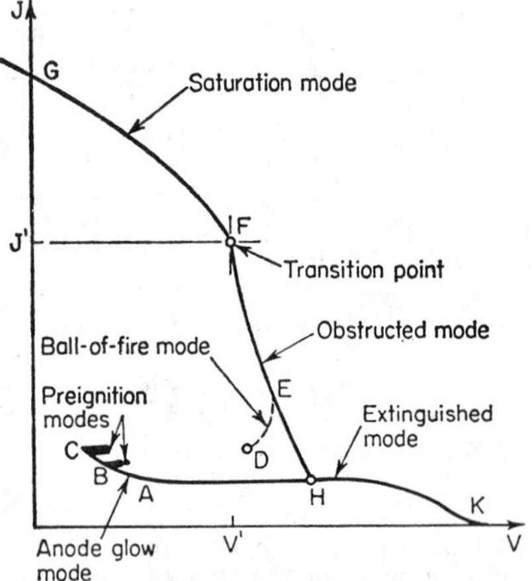

FIG. 12-112. Identification of the various modes of operation for a high-cesium-pressure diode thermionic converter.

At point B, V_D of Fig. 12-110*a* has become so large that some electrons arriving at the collector have obtained sufficient energy to generate Cs ions adjacent to the collector. This creates a glow adjacent to the collector. As one tries to draw more current, the glow pulls away from the collector and develops adjacent to the emitter. This transition is sometimes referred to as a ball-of-fire mode and usually occurs very rapidly.

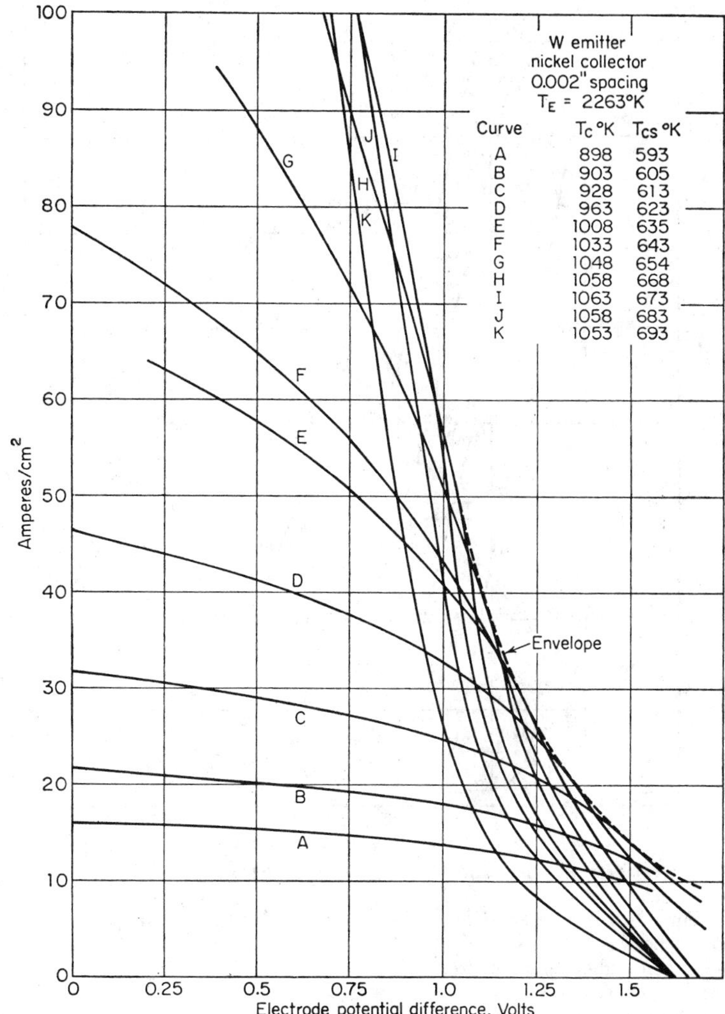

Fig. 12-113. Characteristic *J-V* curves, output current density vs. potential for a high-performance thermionic converter.

It is believed that the potential distribution at this point shifts from that of Fig. 12-110*a* to that of Fig. 12-110*b*. In this obstructed mode there is a double sheath adjacent to the emitter. Closest to the emitter is an excess of electrons such that the negative potential increases as one moves away from the emitter. Next, there is a high density of positive ions creating a potential well of magnitude V_e'. Electrons are accelerated into this well and then rethermalize to a temperature much higher

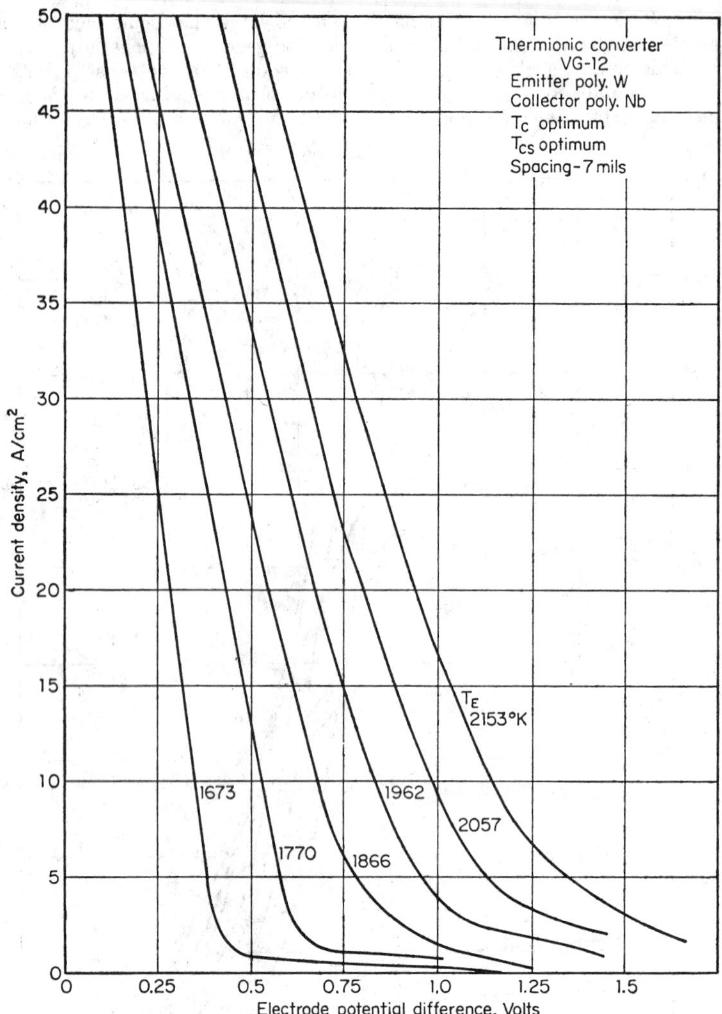

Fig. 12-114. Envelopes of *J-V* curves for various emitter temperatures T_E.

than the temperature of the emitter. The Cs atoms and Cs ions are at a temperature intermediate between the temperature of the two electrodes. However, the electron gas is at a much higher temperature.

The high-temperature electrons ionize some of the Cs by a two-step process. Cesium has an excitation level at about 1.5 V. A three-body encounter between two excited Cs atoms and a high-energy electron furnishes enough energy to ionize one of the atoms. This requires 3.86 eV. The high ion density throughout the plasma gives the thermionic converter an extremely low internal impedance in the obstructed mode. At the transition point *F*, all the electrons emitted from the emitter enter the plasma, and the negative space charge next to the emitter disappears. One might think that this would produce a sharp limit to the current density. However, as the output voltage decreases, V_D increases and forces the positive space charge closer to the emitter, causing increased emission by the Schottky effect.

177. Output Characteristics. Figure 12-113 shows a family of *J-V* curves for a

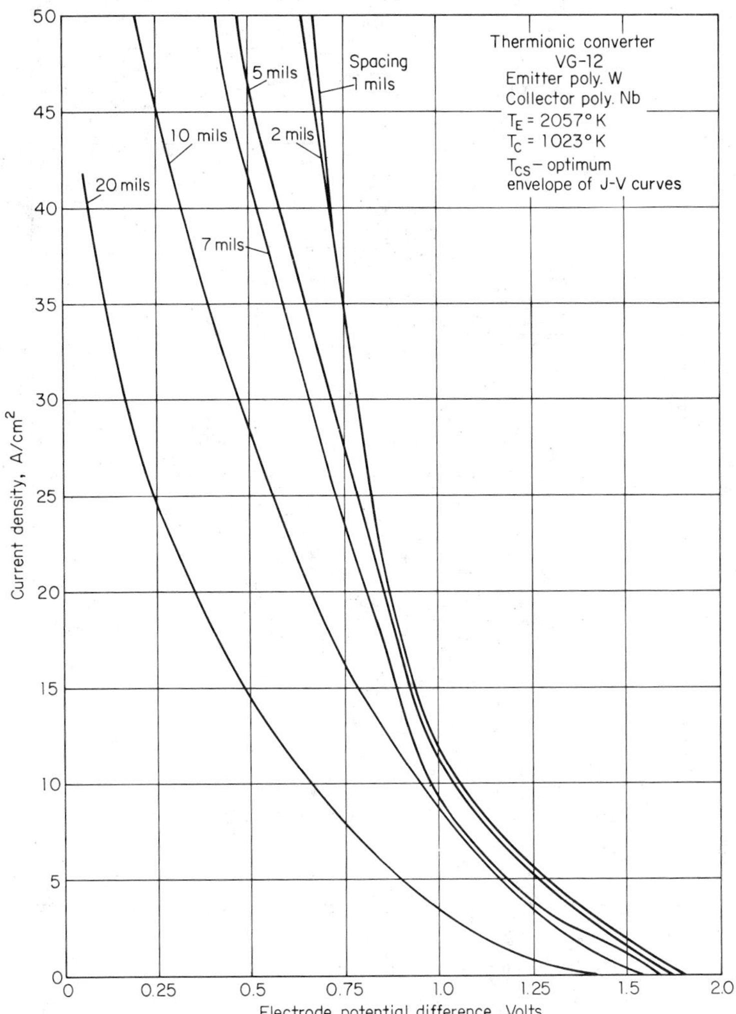

FIG. 12-115. Effect of varying the interlectrode spacing.

thermionic converter. The abscissa gives the electrode potential difference and the ordinate is the amperes per square centimeter of emitter surface. These data apply to a particular converter which had a polycrystalline tungsten emitter and a nickel collector. The spacing was 0.002 in., and the temperature of the emitter was 2263°K. At this high emitter temperature, the thermionic converter goes into the discharge mode at very low current densities. From the table given in Fig. 12-113 it is apparent that the difference between the different curves is related to a difference in the Cs bath temperature, T_{Cs}.

From curve A to curve K the Cs pressure increases from 3 to 40 Torr. This increasing pressure increases the Cs coverage on the emitter surface and reduces the emitter work function. Curve A has too high a work function so that the current density is limited to about 16 A/cm². Curve K on the other hand has a much lower work function due to the high Cs coverage, and the saturation must be several hundred amperes per square centimeter. This high Cs coverage and low emitter work function

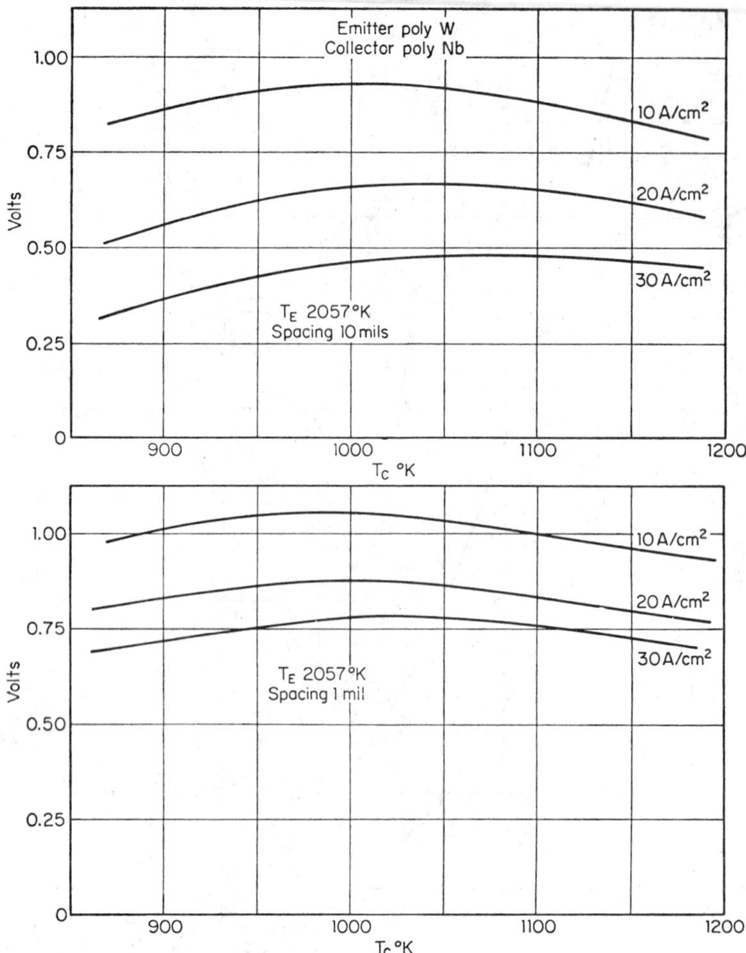

FIG. 12-116. Effect of varying the collector temperature T_C.

produce a lower output voltage. See, for example, curve K at 10 A/cm^2. Also, the high electron scattering at the high Cs vapor density reduces the output voltage.

As will be shown subsequently, the optimum efficiency is obtained at about 30 A/cm^2 current density. At this current density curve E gives the maximum output voltage. This is at a Cs pressure of about 8 Torr. From these curves it is apparent that thermionic converters are high-current-density and low-voltage electrical generators.

Figures 12-114 through 12-116 illustrate the effect of changing the various parameters such as emitter temperature, spacing, and collector temperature for a converter with a polycrystalline tungsten emitter and a niobium collector. Notice that in Fig. 12-113 an "envelope" of the J-V curves was drawn. Figure 12-114 shows how such envelopes vary with variations in emitter temperature T_E. At 10 A/cm^2 each 100°K increase in T_E increases the output voltage about 0.1 V. Figure 12-115 illustrates various spacings. Notice that for moderate current densities one does not need spacings less than 2 mils. Figure 12-116 shows that the output power is comparatively insensitive to the collector temperature.

The efficiency of a thermionic converter is defined as the ratio of the electric power out to the heat power in. In most experimental arrangements it is difficult to measure the heat into the emitter because of heat losses inherent in the heat source. Therefore, the efficiency is usually calculated from the sum of all the heat losses from the emitter. The heat input must equal this sum. The calculated efficiency is then

$$\eta_c = \frac{(V_E - \Delta V_S)J}{q_t} \tag{12-142}$$

where V_E is the measured output voltage at the electrodes, J is the measured current per unit of emitter area, ΔV_S is the calculated voltage drop in the electrical leads, and q_t is the sum of the thermal losses,

$$q_t = q_e + q_r + q_g + q_s + q_k \tag{12-143}$$

where q_e is the electron cooling or heat of evaporation of the electrons leaving the emitter, q_r is the energy flux density associated with interelectrode thermal radiation per

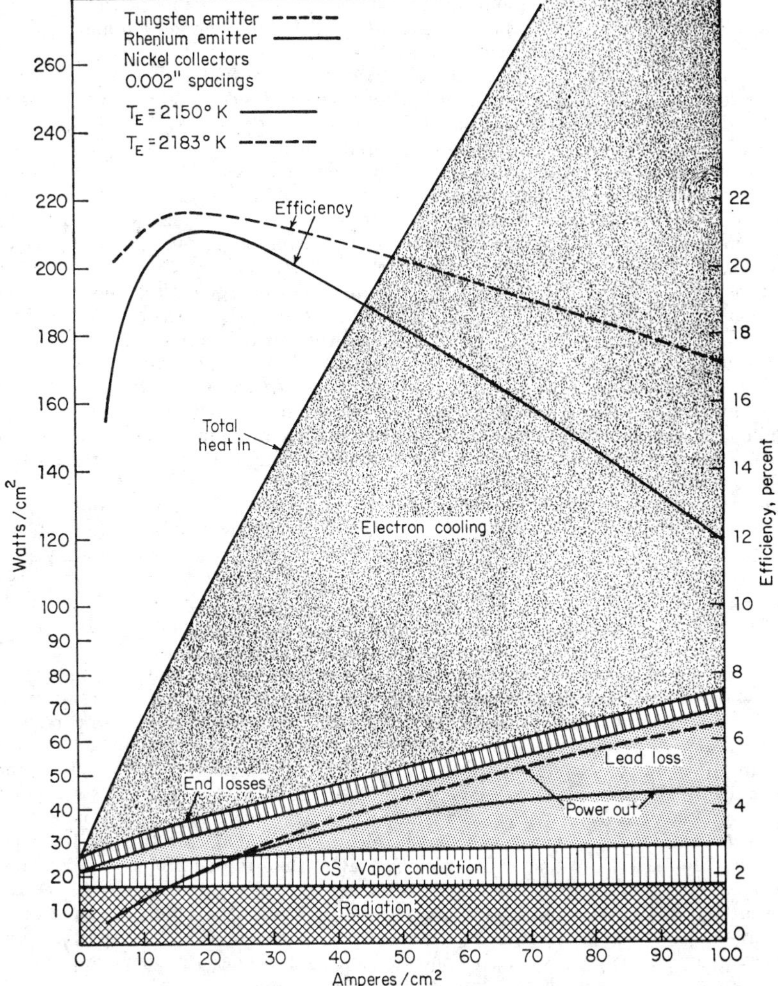

FIG. 12-117. Total power leaving the emitter vs. current density for two converters at T_E 2150 and 2183°K. Also shown are the converter-output power densities and corresponding calculated efficiencies.

unit emitter area, q_g is the heat conduction of the Cs vapor per unit of emitter area, q_s is the heat conduction through the leads per unit of emitter area, and q_k is additional heat conduction through structural members such as spacers or emitter supports, per unit emitter area. This latter term usually contains all miscellaneous heat losses such as radiation from hot surfaces that are not part of the emitter working area. For details of methods of calculating these q's see Ref. 6, Par. **180**.

The various terms and their importance in determining efficiency are illustrated by Fig. 12-117, which is based on cylindrical geometry. The radiation q_r is independent of the current density. It is of course important to have the electrode surfaces as shiny as possible so that their emissivities will be low and the radiation term small. As shown in Fig. 12-113, operating a thermionic converter at high current densities requires a high Cs vapor pressure. So the second term, the Cs-vapor heat conduction q_g, increases slightly with increasing current density.

In Fig. 12-113 the voltages shown are electrode potential difference between the two electrodes. These electrodes must have electrical leads attached to them; in particular the emitter lead must be carefully designed for the application and the current density anticipated. If the electrical resistance is too high, then too much voltage drop is lost in the lead. If the electrical resistance is unusually low, i.e., if it is a very heavy lead, then too much heat will be lost down the lead.[7] Figure 12-117 assumes that a different-sized lead would be used at each current density. This explains why the lead loss q_s increases with increasing current density.

Figure 12-117 assumes that the thermionic converter is a cylindrical converter with the thermionic action on the sides of the cylinder and radiation shields on the ends of the cylinder. As may be seen in the figure these end losses stay constant with increasing current density. If there are no mechanical emitter supports other than the emitter electrical lead, the end radiation losses are the only terms for q_k.

It is apparent from this figure that, at high current densities, about 30 A/cm^2, the biggest term is the electron cooling term. If the work function of the emitter is, say, 3 V, for every ampere of current lifted over this work-function barrier, 3 W of heat is required. One cannot avoid this heat loss. In fact, this is just where the electric energy is generated from heat. As soon as the electrons are outside the emitter surface, they have potential energy relative to the electrons in the metal and the object in designing the converter is to collect these electrons with a minimum of potential loss. The output power for two converters is also shown in Fig. 12-117. Dividing the output electric power by the total heat in, one may obtain the two efficiency curves. A converter with the tungsten emitter has slightly higher efficiency, and, more important, its efficiency remains high at higher current densities. This result is due to the fact that the tungsten emitter adsorbs cesium slightly better than the rhenium emitter does, and hence the tungsten emitter operates at somewhat lower Cs vapor pressures and the scattering of the electrons in the Cs is less for this converter.

178. Future Improvements. The present trend in thermionic-converter development is to develop electrode materials that will more strongly adsorb Cs vapor, which would permit the converters to operate at lower Cs pressures and therefore at lower voltage drop within the converter plasma. Lower Cs pressures also permit opening up the spacing and therefore make the engineering problems much less severe.

Kitrilakis[8] has discovered that, if an oxygen pressure of between 10^{-6} and 10^{-8} Torr is maintained in the converter, this is sufficient to keep a partial coating of oxygen on the emitter surface. This oxidized surface is very efficient in adsorbing Cs vapor, and as a result the converter operates extremely well. Currently the problem is to dispense the correct amount of oxygen between the electrodes to maintain this oxygen pressure for long life. H. F. Webster[9] has demonstrated that if the proper crystal faces of tungsten and rhenium are exposed, particularly if the (110) tungsten surface is exposed, the surface adsorbs Cs quite strongly. Recently the author built a converter with predominantly a (110) tungsten emitter surface. This converter appeared to operate as well as a polycrystalline-tungsten emitter converter at one-fifth the Cs vapor pressure of the polycrystalline converter and at five times the spacing.

Figure 12-110 shows that to improve the output voltage V_o one should reduce V_D and ϕ_C. The search for improved emitter surfaces is really to reduce the vapor pressure, which in turn will help reduce V_D. The other obvious opportunity is to reduce ϕ_C. To date very little effort has been put on finding the ideal collecting material. However, it has been found that those materials which tend to form an oxide surface work best for collectors. For example, nickel seems to keep a monolayer of oxygen on the surface, and this in turn has a high work function so that when Cs is adsorbed on the surface the Cs atoms are ionized. The ionized Cs monolayer has a lower work function than bulk liquid Cs. This is probably because the ionized layer forms an electric dipole on the surface.

179. Applications. *a. Space-vehicle Power Supplies.* 1. *Solar-heated.* As mentioned in Par. **174**, the thermionic converter is a high-temperature heat engine. The input temperature is about 2000°K; the output temperature, about 1000°K. Because of this high heat-rejection temperature, the thermionic converter is an ideal heat engine for use in outer space. To operate a heat engine, it is just as important to remove the rejected heat as to put heat in at a high temperature. Since the only way to remove heat in outer space is by radiation, and since radiation varies as the fourth power of the temperature, it is apparent that this high reject temperature results in very lightweight and comparatively small radiator areas.

For a satellite orbiting the earth, which may go into the shadow of the earth periodically, some power storage must be provided. It turns out that electrical batteries are very heavy and do not store the energy as well as it could be stored as thermal energy at high temperature. One proposal is to use mixed oxides for the energy storage. The concept is to concentrate solar energy by a large mirror onto a tungsten container holding the oxides. Thermionic converters would be placed on the other end of the container. The system would operate at the melting and freezing temperature of the oxide. During the time the vehicle is in the sun, the mirror would concentrate more energy than is needed to operate the converters, and some of the oxide would melt, thus absorbing the extra energy. During the time the vehicle is in the shadow of the earth, the energy would be taken from the oxide, and in this way some of the oxide would refreeze. It has been demonstrated that rhenium and also possibly tungsten can be used to contain the oxides.

E. Batutis[10] has made studies of this type of system. Figure 12-118 is a summary of his findings. The three oxide compositions ranging in melting points from 1890 to 2010°K are the most promising. The use of beryllia-magnesia at a melting point of 2150°K places very severe requirements on the accuracy of the mirror. The use of the lower-melting-point oxides results in a poor efficiency for the thermionic

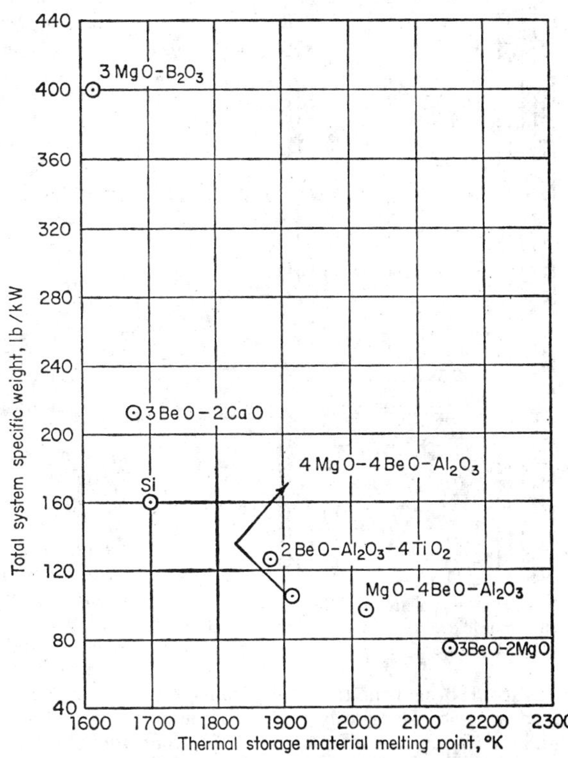

FIG. 12-118. Solar-heated thermionic converter with thermal storage. System specific weight vs. thermal storage-material melting point.

converter, and therefore many mirrors and many converters are required, and this raises the specific weight of the system. Currently very little effort is being expended to develop the solar-heated thermionic converters for space power supplies, primarily because it is feared that in outer space the mirrors will not maintain their high polish and high degree of solar-energy-collecting ability. If successfully developed, solar thermionic converters could probably be used for power supplies of 100 W to perhaps 10 kW.

2. *Nuclear-heated thermionic converters.* Several studies have been made to see how thermionic converters may best be heated by nuclear energy for use in space power supplies. If the converters are put in a heat exchanger, between the nuclear reactor and the radiator, or if the converters are built as an integral part of the radiators, the heat from the nuclear reactor must be carried to the converter at about 2000°K.

To date, no materials are available which will hold a liquid metal to transmit heat at this high temperature without serious corrosion problems. If the thermionic converters are built inside the reactor, then the heat is generated where it is to be used and it is not necessary to transport the heat from the nuclear source to the emitter.

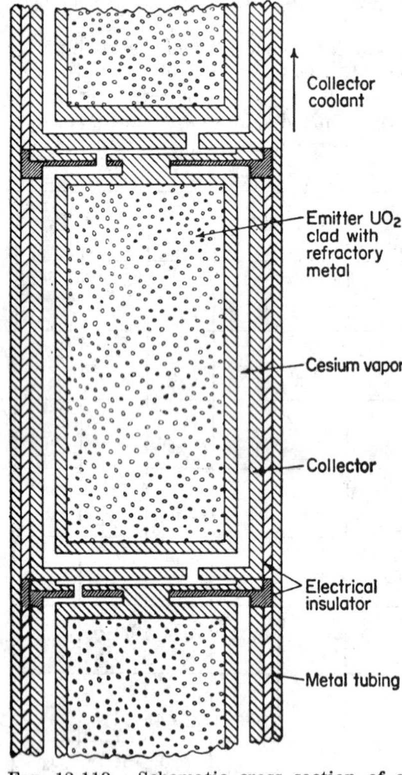

Fig. 12-119. Schematic cross section of a thermionic converter assembly inside a nuclear-reactor fuel rod.

One method of building thermionic converters inside a nuclear reactor is illustrated by Fig. 12-119. The nuclear reactor would be built in a conventional fashion, with the nuclear fuel in fuel rods and with a coolant removing the unconverted heat from the reactor. The thermionic converters would be built inside the nuclear fuel rods. A nuclear fuel would be divided into small, cylindrical pellets perhaps 1 in long and ½ in in diameter. These pellets would be clad in a refractory metal and supported from the electrical lead to the emitter.

The heat would be generated by neutrons entering the nuclear fuel inside the refractory-metal emitter. Thus, the heat would flow to the emitter surface by thermal conduction. The collectors would be metallic cups closely spaced about the emitter, with Cs vapor between the collector and emitter. The collectors in turn would be insulated but bonded to an outer metal tubing against which the coolant would flow to remove the unconverted heat. All the thermionic converters for a single fuel-rod assembly would operate with a common Cs vapor. Cs vapor passages would be provided between converters.

In this design the thermionic converters appear like small flashlight batteries assembled in series electrically. If each converter had 10 cm^2 of operating emitter surface and were operating at 25 A/cm^2, the current would be 250 A. If the converters generate 1 V/cell, for 30 such cells in one fuel rod the output voltage would be 30 V. Each fuel rod would develop 30 × 250, or 7500, W. Nuclear thermionic power supplies should be useful for space applications from 10 to 1,000 or perhaps 10,000 kW. Studies indicate that these supplies of 100 to 1,000 kW should weigh 15 to 10 lb/kW.

b. Thermionic Converters for Central-station Power Plants. Because thermionic converters are high-temperature heat engines, they could be used as topping devices for steam turbines in central-station power plants. The converters would be built inside the reactors as explained above. In this case the reject heat would be used to generate steam for the steam turbine. If, for example, the thermionic converters were 25% efficient, then 25% of the heat generated in the nuclear fuel would be brought out as electrical power on bus bars. The other 75% of the heat would be passed on to steam turbines. If the turbines and associated generators were 40% efficient in converting the heat to electricity, then 40% of the 75%, or 30%, of the nuclear heat would be converted in the steam-turbine generators, and the entire system would be 55% efficient in converting the nuclear heat to electricity.

Thus, the efficiency for the central-station plant would be raised from 40 to 55% by the introduction of thermionic converters. The question remains whether the extra complexity would permit a system to be built at a sufficiently low price to take advantage of this increased efficiency and result in a lower cost in mills per kilowatthour. One study[11] indicates that, at 30 W/cm^2 of electricity generated at 23% efficiency, the cost of electricity would be 4.6 mills/kWh as compared with a reference plant without thermionic converters of 4.9 mills/kWh.

To install thermionic converters in present-day low-energy neutron-flux reactors would require the use of molybdenum and niobium for the refractory materials. This limits the maximum temperature for the emitter and makes it questionable that 23% efficiency can be achieved. The more desirable refractory metals such as tungsten and rhenium for the emitters have very high, slow neutron-resonance-adsorption cross sections, and probably the economy in neutrons would not be sufficiently good. On the other hand, for future breeder reactors, which will operate at a high neutron energy to improve the breeding ratio, these materials are usable, and it is possible that the second generation of breeder reactors will find thermionic converters of value.

c. Undersea and Remote-area Nuclear Power Supplies. If one is not fighting an economic battle to get minimum cost per kilowatthour of electricity, thermionic converters utilizing molybdenum for the emitters and niobium for the collectors would be about 15% efficient. For undersea operation the waste heat might be removed merely by thermal convection. The system would be extremely simple, with essentially no moving mechanical parts. Except for the convection cooling, it would be a static system.

d. Gas-fired Converters. It has been proposed to make gas-fired or oil-fired thermionic converters for small, remote power supplies. To obtain the high emitter temperatures requires special burners. Usually the air is preheated before mixing with the fuel. Such burners have been developed and are quite efficient. Most flames contain hydrogen, which diffuses through refractory metals. Therefore the problem to date in this application has been to obtain some kind of protective coating for the emitter to prevent hydrogen diffusion.

A second gas-fired thermionic-converter application is for home heating. In this case it is visualized that a small thermionic converter might operate on a pilot light. When heat was required in the house and the large flame came on, this flame would heat larger thermionic converters. When the converters reached temperature and started to develop electricity, the electricity would be used to blow the hot air through the house. To date thermionic converters have not been sufficiently reliable or economical to justify this application.

e. Solar-heated Earthbound Converters. No doubt solar-heated thermionic converters could be used in remote desert areas, say, along the Nile River, to pump water. However, the economics of this situation do not warrant serious development at this time.

180. References

1. WEBSTER, H. F. *J. Appl. Phys.*, 1959, Vol. 30, p. 488.

2. BEGGS, J. E. *Advanced Energy Conversion,* 1963, Vol. 3, p. 447.

3. HERNQVIST, K. G. *RCA Rev.*, March, 1961, pp. 7–26.

4. JAMERSON, F. E., LEFFERT, C. B., and REES, D. B. *Thermionic Converter Specialists Conf.*, October, 1964, p. 219.

5. HOUSTON, J. M. *Rept. Twentieth Ann. M.I.T. Conf. Phys. Electronics,* 1960, p. 72. PSAROUTHAKIS, J. *Therm. Conv. Spec. Conf.*, 1960, p. 100.

6. Block, F. G., Hatsopoulos, G. N., and Wilson, V. C. *Therm. Conv. Spec. Conf.*, 1965, p. 379.

7. Houston, J. M. *J. Appl. Phys.*, 1959, Vol. 30, p. 481.

8. Kitrilakis, S. S., and Lieb, D. *Therm. Conv. Spec. Conf.*, 1966, p. 348.

9. Webster, H. F. *J. Appl. Phys.*, 1959, Vol. 30, p. 488; *Advan. Electronics*, 1962, Vol. 17, p. 200.

10. Batutis, E. *Final Tech. Rept.*—Thermal Energy Storage R&D Program, NASA Contract NAS 5-826, December, 1961.

11. Wilson, V. C. *IEEE Spectrum*, 1964, Vol. 1, p. 75.

SECTION 13

A-C POWER TRANSMISSION

REVISED BY

V. J. CISSNA Systems Studies Engineer, Power System Planning Branch, Tennessee Valley Authority (Retired); Fellow, Institute of Electrical and Electronics Engineers

CONTENTS

Numbers refer to paragraphs

SECTION 13

A-C POWER TRANSMISSION

REVISED BY

V. J. CISSNA

CLASSIFICATION OF SYSTEMS

1. Transmission Systems. Modern a-c power systems usually consist of the following elements: (1) generating stations; (2) step-up transformer stations; (3) transmission lines; (4) switching stations; (5) step-down transformer stations; (6) primary distribution lines or networks; (7) service transformer banks; (8) secondary lines or networks.

Essentially, elements 2 to 5 are the transmission system, and elements 6 to 8 are the distribution system. The difference between transmission systems and distribution systems depends upon the function. The function of the transmission system is the transmission of bulk power to load centers and large industrial users beyond the economical service range of the regular primary distribution lines. The function of the distribution system is to deliver power from generating stations or transmission substations to the various consumers.

2. General Features of Design. Fundamentally, transmission-system design is the selection of the necessary lines and equipment which will deliver the required amounts of power with the quality of service demanded for the lowest overall average annual cost over the period of time for which service may be required or for the life of the equipment. At the same time, the system should be capable of expansion with a minimum of changes to existing facilities.

Electrical design of a-c systems involves the following features: (1) selection of voltage; (2) conductor size; (3) line regulation; (4) losses; (5) corona; (6) voltage control; (7) system stability; (8) system protection, under which may be grouped (*a*) circuit-breaker duties, (*b*) circuit-breaker arrangement, (*c*) relaying, (*d*) insulation coordination, (*e*) lightning arresters, (*f*) neutral grounding, (*g*) station grounds, (*h*) overhead ground wires, (*i*) counterpoises.

Mechanical design includes (1) sag and stress calculations; (2) conductor composition; (3) conductor spacing (minimum spacing to be determined under electrical design); (4) kind and types of insulators; (5) selection of conductor hardware.

Structural design includes (1) selection of the type of structures to be used; (2) stress calculations; (3) foundations; (4) guys and anchors.

Miscellaneous features of transmission-line design are (1) line location; (2) acquisition of right-of-way; (3) profiling; (4) locating structures; (5) inductive coordination which is involved in locating the line and also in electrical calculations; (6) means of communication.

ECONOMICS

3. Choice of Voltage. Standard transmission voltages are established in this country by the United States of America Standards Institute (USASI). Table 13-1 shows the standard voltages listed in USA Standard C-84, all of which are in use at present.

Some power systems rate their transmission voltages by the nominal voltages; others may use the maximum voltage.

Table 13-1. Standard System Voltages, kV

Rating		Rating	
Nominal	Maximum	Nominal	Maximum
34.5	36.5	161	169
46	48.3	230	242
69	72.5	345	362
115	121	500	550
138	145	700	765

In the design of a system, the voltage selected should be the one best suited for the particular service on the basis of economic considerations. The voltages 345 kV and up are known as extra-high voltages (EHV). Extra-high voltage is used for the transmission of large blocks of power for longer distances than would be economically feasible at the lower voltages. It may be used also for interconnections between systems, or it may be superimposed upon large power-system networks to transfer large blocks of power from one part to another. Extra-high-voltage systems require an entirely new concept of system design. Whereas voltages 230 kV and below are relatively simple and well standardized in design and construction practices, EHV is far from standardized at this time and requires complete and thorough reconsideration of all normally standardized design features, such as necessity for bundled conductors, switching surges which control the insulation, corona, radio interference, lightning protection, line-charging current, clearances, and construction practices.

4. Conductor Selection. There may be instances in the selection of conductors in which the classical methods embodied in **Kelvin's law**[1] can be applied. Kelvin's law states: "The most economical area of conductor is that for which the annual cost of the energy wasted is equal to the interest on that portion of the capital outlay which may be considered as proportional to the weight of the conductor." It is developed from Eq. (13-1),

$$\text{Yearly cost} = \frac{3CI^2r}{1{,}000} + \frac{pwa}{100} \tag{13-1}$$

where C is the cost per kilowatt-year of the energy wasted, I is the current per wire, r is the resistance per mile of one conductor, w is the weight per mile of all conductors, a is the percent annual cost of money, and p is the cost per pound of conductor, which must include erection variations and structure variations due to the different conductors studied. The minimum cost occurs when $(3CI^2r)/1{,}000$ equals $(pwa)/100$.

It is seldom that a line can be set up for a simple I^2r loss solution by Kelvin's law; rather, the I^2r must be an integrated total for the period under study.

Nearly all power systems today are complex networks, and the addition of one line causes load changes in many if not all of the other lines of the system. It then becomes necessary to find the specification for the line by means of an a-c network analyzer or a digital computer. From 1930 until the late 1950s the a-c network analyzer was the instrument for solving problems in large power networks. During the 1950s the digital computer was perfected and power-system programs developed which now permit the solution of any power-system network problem. The digital computer is much faster and more accurate and has practically displaced the a-c network analyzer in power-system analysis (see Sec. **27** for information on computers).

In setting up a transmission-system analysis on the a-c network analyzer or digital computer, a system impedance diagram is drawn in which equivalent π or T representations of the transmission lines are used, as given in Pars. **17** to **21**. With the digital computer, as with many system arrangements, line voltages and future load conditions as desired can be tried out in detail. One of the computed quantities can be the total losses of the entire system; this permits analysis of the effect of the proposed changes on the system as a whole. The loss computation may be broken up into areas, if desired. Such detailed analysis is not practicable with the a-c network analyzer.

5. Type of Construction. It is very difficult to assign dollar values to the advantages and disadvantages of various types of structures such that a yearly cost comparison will represent the truly comparative value of each. Such cost comparisons are better considered as incomplete and as only one of the important factors to be con-

[1] Still, Alfred "Electric Power Transmission"; New York, McGraw-Hill Book Company, 1927.

Table 13-2. Transmission-line Data of Typical Lines in the United States

Company and line name	Normal rating		Steel or wood structures	Cir-cuits	Length, miles	Structures per mile	Conductor, Mcm or AWG*	Insulators, disk	Conductor spacing, ft†	Overhead ground wires		Counter-poise§
	Mva per cir	Kv								No.	Mtl‡	
Central Maine Power Co :												
Windsor-Rockland	.....	115	Wood	1	26.71	10.9	266.8 A	8	14 H	2	Cw	None
Boston Edison Co.:												
Braintree-Whitman	130	110	Wood	1	7.4	11.8	350 C	8	14 H	2	St	Li
Connecticut Light and Power Co.:												
Devon-Norwalk	66.7	115	Steel	2	23.3	8	4/0 C	9	10 & 12 V	2	Cw	Li
Niagara Mohawk Power Corp.:												
Utica-Schenectady	120	115	Wood	1	70.5	9	795 A	8	12.5 H	2	Cw	None
Metropolitan Edison Co.:												
Hosensach-Montebello	250	230	Steel	1	105	4.1	795 A & 1033.5 A	16	26 H	2	St	Li-Cr
Philadelphia Elec. Co.:												
Conowingo Ply. Mtg	426	220	Steel	1	57.7	4.5	795 A	16	25.5 H	2	St	Li-Cr
Pennsylvania Power and Light Co.:												
Sunbury-Siegfried	175	220	Steel	1	.....	4.6	1.14″ A	16	28 H	2	St	Li
Sunbury-Hummelstown	60	132	St & Wd	1	48.8	8	C-Cw .55″	8	12 H	2	St	Li
American Gas and Elec. Co.:												
Tanners Cr.-Madison	100	138	Steel	2	82	5.5	636 A	10	13 V	1	A	None
Sporn-Kanawah	500	330	Steel	..	63	4.1	1275 A	18	21.5 V	1	A	None
Virginia Electric and Power Co.:												
Petersburg-Suffolk	.....	110	Steel	2	67.9	8	336.4 A	9	10 V	2	St	
South Carolina Elec. & Gas Co.:												
Fairfax-Graniteville	30	115	Wood	1	59.1	6.4	336.4 A	7	16 H	2	St	Li
Georgia Power Co.:												
Yates-Americus	50	115	Wood	1	111.27	6.94	477 A	7	14 H	2	St	Cr
Tennessee Valley Authority:												
Johnsonville-Cordova	600	500	Steel	1	118.36	4.6	3–971.6 A,B	24	4 OH	2	Aw	None
Typical Wood	.....	161	Wood	1	50	7.7	636 A	11	17.5 H	2	St	None
Kentucky Utilities Co.:												
Dix–Green River	100	138	Steel	1	139	5.8	556.5 A	10	16.5 H	2	St	
North Indiana Public Service Co.:												
Mich. City–Aetna	96	138	Steel	2	23	5.5	397.5 A	9	13 V	1	St	None
Union Elec. Co. of Missouri:												
Osage-Moberly	100	161	Wood	1	85.8	7	556.5 A	10	15.5 H	2	St	None
Kansas Gas and Elec. Co.:												
Wichita-Topeka	600	345	Wood	1	133	8	2–795 A,B	18	27 H	2	St	None
Neosho-Marmaton	60	138	Wood	1	38	8	336.4 A	10	14.5 H	2	St	None

Table 13-2. Transmission-line Data of Typical Lines in the United States.—*Concluded*

Company and line name	Normal rating		Steel or wood structures	Cir-cuits	Length, miles	Structures per mile	Conductor, Mcm or AWG*	Insulators, disk	Conductor spacing, ft†	Overhead ground wires		Counter-poise§
	Mva per cir	Kv								No.	Mtl‡	
Kansas City Power & Light Co.: Kansas City–Moberly	75	154	Wood	1	103	7	556.5 A	10	15.5 H	2	St	None
Minnesota Power & Light Co.: Aurora SS–Va. SS	80	115	Wood	1	25.4	9	336.4 A	7	11 H	2	Cw	
Miss. Power Company: Hattiesburg-Meridian	60	110	Wood	1	104.65	6.6	226.8 A	7	16 H	2	St	Cr
Gulf States Utility Co.: Neches-Riverside	60	138	Wood	1	51.37	9	4/0 C	9	14.5 H	1	St & Cw	None
Houston Light and Power Co.: West Junction–Peters	100	132	Wood	1	49.7	9.5	4/0 C	9	14.5 H	2	St	None
Oklahoma Gas and Elec. Co.: Osage-Enid	25/47.5	138	Wood	1	49	8	336.4 A	9	15.5 H	2	St	None
Texas Power and Light Co.: Lake Cr–Temple	100	138	Wood	1	32.9	10	636. A	9	14.5 H	2	St	None
Lower Colorado River Authority: HiCross-Comal	60	138	Wood	1	41.4	8	336.4 A	9	14.6 H	2	St	None
Public Service of Colorado: Leadville-Poncha	20	115	Wood	1	57.04	7.14	4/0 A	7	12.5 H	no	....	None
Montana Power Co.: Anaconda-Grace	50	161	Wood	1	273.9	7.5	250 CH	8	13.5 H	no	....	None
Washington Water Power Co.: Spokane-Colfax	80	110	Wood	1	52.8	7.35	250 C	6	10.5 H	no	....	None
Utah Power and Light Co.: Goshen–Reac. Test Sta.	20	132	Wood	1	43.17	7.05	397.5 A	8	13.5 H	2	St	
Idaho Power Co.: Strike-Caldwell	120	138	St & W	1 & 2	73.3	St 4.4 W 8.25	715.5 A	10	13.5 V & H	1 & 2	St	None
Bonneville Power Administration: Olympia-Covington	....	230	Steel	1	59.84	4.9	795 A	14	27 H	end 2 mi	St	Cr
Seattle Dept. of Lighting: Diable-Bothell	180	230	Steel	2	81.5	4.5	795 A	14	17.35 V	no	....	None
Pacific Gas and Elec. Co.: Sunol-Moss	175	220	Steel	2	61.6	4.35	795 A	13 fog	16.5 V	no	....	None
City of Los Angeles: Hoover Dam–Los Angeles	160	275	Steel	1	263.7	5.5	1.4" CH 500	24	32.5 H	2	St	Li cont.
Southern California Edison Co.: Megunden-Mesa	200	220	Steel	1	119.5	3.9	605 A	15 or 14 fog	23 H	2	St	None

* A = ACSR, Aw-Alumoweld, C = copper, H = hollow. † = H = horizontal, V = vertical plane. ‡ Cw = Copperweld, B = bundled, St = steel, A = ACSR. § Li = linear, Cr = crowfoot.

sidered in making the selection. For instance, flexible construction, including wood-pole H-frame and narrow longitudinal-base steel towers, which are not designed to carry a broken conductor, will almost always prove cheaper than rigid structures, which are designed to carry a broken conductor; however, while giving excellent service in many cases, flexible construction has not been generally accepted for the most important lines.

6. Length of span is likewise a matter of other considerations from those which can be readily evaluated in dollars. For straight-line construction, the lowest cost is generally obtained by longer spans than has been considered good practice. In recent years, spans have been gradually increased as the problems of conductor swing, sleet jump, vibration, and unbalanced tensions have been analyzed and compared with operating experience. For very long spans the savings are small, and considerable refinement is required in the design, layout, and construction.

7. Cost data, stated as the total cost per mile, for various types of construction are seldom a reliable basis for estimates or comparisons because of the wide possible variation in practically every item of the total; this includes, not only a very wide range in local conditions such as right-of-way, climate, and service requirements, but a considerable choice in standards of design and refinement of details as well as some variation in the unit cost of materials. It is usually found that a line which is inherently expensive because of large conductor, high voltage, or very valuable right-of-way will and should be planned and built throughout on a more conservative and therefore more costly basis than less important lines. Local conditions and special requirements can be taken into account only by adding together all the various items of cost. The cost of structures can be closely estimated from manufacturers' estimates on preliminary designs. The manufacturers also will furnish accurate estimates on the cost of conductors and insulators. In figuring right-of-way costs in some sections of the country the value of marketable lumber should be taken into account. A relatively large amount of general expense is connected with transmission-line construction, including equipment, roads and trails, camps, offices, maintenance of motor equipment, and other items which are best estimated as a percentage of the total labor and material. Overhead costs such as engineering, management, tests, inspection, and legal costs may usually be estimated from other construction.

8. Transmission-line Data. Table 13-2 gives some characteristic data of lines in the United States of 110 kV and higher voltages. These data were taken mainly from a 25-page table containing 70 items of characteristic data for each line. This table appeared in the Aug. 24, 1953, *Electrical World* and covers mainly lines which have been built since the end of World War II. All parts of the country are included.

ELECTRICAL PROPERTIES OF CONDUCTORS

9. The conductors most commonly used for transmission lines are ACSR (aluminum cable steel-reinforced), copper, Copperweld, and high-strength aluminum alloy. Tables of the electrical characteristics of the most commonly used conductors will be found in Sec. **4**. Characteristics of other conductors can be obtained from the manufacturers.

The tables in Sec. **4** contain all necessary information for the determination of resistance, reactance, and capacitance per mile when combined with the spacing factors X_d and X_d'. Tables 13-5 and 13-6 cover spacings up to 40 ft in 0.1-ft steps. Wider spacing factors can be found, as explained in the footnotes under the tables. Capacitance is given in the tables in terms of the equivalent shunt reactance in megohms. Susceptance in micromhos is the reciprocal of the shunt reactance in megohms.

The system employed in the tables[1] was devised by W. A. Lewis, Jr., and is made possible by the GMR method of calculating inductance. The Lewis system makes use of the fact that the expressions for inductive reactance and shunt capacitive reactance each can be broken up into two parts. The expression for inductive reactance per mile is

$$X = 0.004657f \log \frac{D}{\text{GMR}} \qquad (13\text{-}2)$$

[1] Wagner, C. F., and Evans, R. D. "Symmetrical Components"; New York, McGraw-Hill Book Company, 1933, App. VII.

where D = equivalent spacing in feet, GMR = geometric mean radius in feet as given in the conductor tables of Sec. **4,** and f = frequency in cycles per second. Geometric mean radius for ACSR is given at 60 c. However, 60-c values of GMR can be used at other commercial power-system frequencies with small error. X also can be expressed as

$$X = 0.004657f \log \frac{1}{\text{GMR}} + 0.004657f \log D \tag{13-3}$$

The first term on the right is designated as X_a, and the second term as X_d. When the spacing is 1 ft, X_d becomes zero. If the spacing is greater than 1 ft, X_d has a positive value which is added to X_a; and if the spacing is less than 1 ft, X_d has a negative value which is subtracted from X_a. The expression for capacitive shunt reactance per mile,

$$X_c = \frac{4.099 \times 10^6}{f} \log \frac{D}{r_c} \tag{13-4}$$

where r_c = conductor radius in feet, likewise can be expressed in two parts (see footnote, Table 13-6),

$$X_a' = \frac{4.099 \times 10^6}{f} \log \frac{1}{r_c} \qquad \text{and} \qquad X_d' = \frac{4.099 \times 10^6}{f} \log D$$

X_d' is added to or subtracted from X_a', depending upon the magnitude of D. Values of X_d and X_d' are given in Tables 13-5 and 13-6, respectively. The internal inductive reactance K is found from Eq. (13-11), $K = 0.004657f \log (r_c/\text{GMR})$. Series reactance ($+jX$) and shunt reactance ($-jX$) values per mile at 60 c are given, but values at other frequencies can be determined readily by ratios of the frequencies. $+jX$ **reactance** at other frequency f is found by multiplying the 60-c values by **f/60.** $-jX$ **reactance** at other frequencies is found by multiplying the 60-c values by **60/f.**

Bundle conductors, or multiple conductors as they have been called, consist of two or more conductors per phase supported by one insulator assembly. Edith Clarke in the 1932 *Transactions of the AIEE* developed general formulas for the inductance and capacitance of bundle conductors.

$$L_\phi = \frac{1}{n}\left[0.08047 + 0.74113 \log \frac{24(S_{gm})^n}{d(M_{gm})^{n-1}}\right] \quad \text{(mH/mi)} \tag{13-5}$$

From Eq. (13-6) inductive reactance is found to be

$$X = \frac{1}{n}\left[K + 0.2794 \log \frac{24(S_{gm})^n}{d(M_{gm})^{n-1}}\right] \quad (\Omega\text{/mi at 60 c}) \tag{13-6}$$

$$C_\phi = \frac{0.03883n}{\log \dfrac{24(S_{gm})^n}{d(M_{gm})^{n-1}}} \quad (\mu\text{F/mi}) \tag{13-7}$$

In the above, n = number of conductors per phase (bundle); d = diameter of conductor in inches; S_{gm} = geometric mean distance between conductors of different phases in feet, found by taking the mean distance from all conductors of one phase to all conductors of the other phases; M_{gm} = geometric mean distance in feet between the n conductors of one phase; K = internal conductor reactance defined in Eq. (13-11).

The reactive and capacitive shunt reactances for bundled conductors can be found

Table 13-3

Bundle	X_{aeq}	X'_{aeq}
2 conductors	$\frac{1}{2}(X_a - X_s)$	$\frac{1}{2}(X'_a - X'_s)$
3 conductors	$\frac{1}{3}(X_a - 2X_s)$	$\frac{1}{3}(X'_a - 2X'_s)$
4 conductors	$\frac{1}{4}(X_a - 3X_s)$	$\frac{1}{4}(X'_a - 3X'_s)$

Table 13-4. Effective Resistance and Internal Reactance of Steel-strand and Copperweld Conductors[1] (Overhead Ground Wire Types)

Current,[2] amp	Diameter, in.	No. strands	Resistance, ohms per conductor per mile		Reactance, ohms per conductor per mile			
					K, internal (inductive)		X_a, series (inductive)	
			60 cycles	25 cycles	60 cycles	25 cycles	60 cycles	25 cycles
			Siemens Martin Steel Strand					
5	1/4	7	12.25	12.20	0.551	0.231	1.021	0.426
15	1/4	7	12.54	12.51	0.621	0.258	1.091	0.454
25	1/4	7	13.04	13.02	0.698	0.295	1.168	0.491
5	3/8	7	5.44	5.42	0.528	0.220	0.948	0.395
15	3/8	7	5.49	5.47	0.594	0.249	1.014	0.424
25	3/8	7	5.62	5.57	0.651	0.272	1.071	0.447
5	1/2	7	3.40	3.37	0.535	0.231	0.920	0.391
15	1/2	7	3.43	3.39	0.574	0.244	0.960	0.405
25	1/2	7	3.47	3.41	0.614	0.261	1.000	0.422
			High-strength Steel Strand					
5	3/8	7	5.87	5.83	0.511	0.214	0.912	0.389
15	3/8	7	5.94	5.86	0.557	0.235	0.959	0.410
25	3/8	7	6.03	5.99	0.581	0.246	0.983	0.421
			Copperweld Conductor—40% Conductivity					
10	3/8	7	1.19	1.16	0.17	0.08	0.67	0.29
50	3/8	7	1.20	1.16	0.18	0.09	0.68	0.30
100	3/8	7	1.24	1.19	0.21	0.10	0.71	0.31
200	3/8	7	1.32	1.27	0.17	0.09	0.67	0.30
10	1/2	7	0.765	0.730	0.15	0.07	0.62	0.27
50	1/2	7	0.792	0.740	0.15	0.07	0.62	0.27
100	1/2	7	0.792	0.750	0.16	0.08	0.63	0.28
200	1/2	7	0.820	0.776	0.16	0.07	0.63	0.27
			Copperweld Conductor—30% Conductivity					
5–120	5/16	3	2.92	......	0.18	.....	0.71	
5–120	5/16	7	2.45	......	0.20	.....	0.73	
0.1–160	3/8	7	1.53	1.50	0.19	0.09	0.69	0.30
5–200	1/2	7	1.03	0.99	0.18	0.09	0.65	0.29
5–215	21/32	19	0.62	......	0.19	.....	0.63	

Alumoweld at 25°C, 60 cycles (from Copperweld Steel Co.)

Current	Diameter, in.	No. strands	Resistance, ohms per conductor per mile	Reactance, ohms per conductor per mile	GMR, feet
Small	0.546	7 nr 5	1.240	0.707	0.002958
Small	0.496	7 nr 6	1.536	0.721	0.002633
Small	0.433	7 nr 7	1.937	0.735	0.002345
Small	0.385	7 nr 8	2.440	0.749	0.002085
Small	0.343	7 nr 9	3.080	0.763	0.001858
Small	0.306	7 nr 10	3.880	0.777	0.001658

[1] The resistance and K values for the steel-strand conductors were obtained from tables compiled by the Indiana Steel and Wire Company. Copperweld values from *Eng. Rept.* 37, Computation of Zero Sequence Impedances of Power Lines and Cables, by Joint Subcommittee on Development and Research, Edison Elec. Inst. and Bell Telephone System, July 22, 1936, Table 8, p. 330. Resistance is at 20°C.

[2] Current at which resistances and reactances are correct.

Table 13-5. Values of X_d

Inductive reactance,[1] ohms per conductor per mile, at 60 cycles

D, equivalent[2] spacing, ft	0	0.1	0.2	0.3	0.4	0.5	0.6	0.7	0.8	0.9
1	.0000	.0116	.0221	.0318	.0408	.0492	.0570	.0644	.0713	.0779
2	.0841	.0900	.0957	.1011	.1062	.1112	.1159	.1205	.1249	.1292
3	.1333	.1373	.1411	.1449	.1485	.1520	.1554	.1588	.1620	.1651
4	.1682	.1712	.1741	.1770	.1798	.1825	.1852	.1878	.1903	.1928
5	.1953	.1977	.2001	.2024	.2046	.2069	.2090	.2112	.2133	.2154
6	.2174	.2194	.2214	.2233	.2252	.2271	.2290	.2308	.2326	.2344
7	.2361	.2378	.2395	.2412	.2429	.2445	.2461	.2477	.2493	.2508
8	.2523	.2538	.2553	.2568	.2582	.2597	.2611	.2625	.2639	.2653
9	.2666	.2680	.2693	.2706	.2719	.2732	.2744	.2757	.2769	.2782
10	.2794	.2806	.2818	.2830	.2842	.2853	.2865	.2876	.2887	.2899
11	.2910	.2921	.2932	.2942	.2953	.2964	.2974	.2985	.2995	.3005
12	.3015	.3025	.3035	.3045	.3055	.3065	.3074	.3084	.3094	.3103
13	.3112	.3122	.3131	.3140	.3149	.3158	.3167	.3176	.3185	.3194
14	.3202	.3211	.3219	.3228	.3236	.3245	.3253	.3261	.3270	.3278
15	.3286	.3294	.3302	.3310	.3318	.3326	.3334	.3341	.3349	.3357
16	.3364	.3372	.3379	.3387	.3394	.3402	.3409	.3416	.3424	.3431
17	.3438	.3445	.3452	.3459	.3466	.3473	.3480	.3487	.3494	.3500
18	.3507	.3514	.3521	.3527	.3534	.3540	.3547	.3554	.3560	.3566
19	.3573	.3579	.3586	.3592	.3598	.3604	.3611	.3617	.3623	.3629
20	.3635	.3641	.3647	.3653	.3659	.3665	.3671	.3677	.3683	.3688
21	.3694	.3700	.3706	.3711	.3717	.3723	.3728	.3734	.3740	.3745
22	.3751	.3756	.3762	.3767	.3773	.3778	.3783	.3789	.3794	.3799
23	.3805	.3810	.3815	.3820	.3826	.3831	.3836	.3841	.3846	.3851
24	.3856	.3861	.3866	.3871	.3876	.3881	.3886	.3891	.3896	.3901
25	.3906	.3911	.3915	.3920	.3925	.3930	.3935	.3939	.3944	.3949
26	.3953	.3958	.3963	.3967	.3972	.3977	.3981	.3986	.3990	.3995
27	.3999	.4004	.4008	.4013	.4017	.4021	.4026	.4030	.4035	.4039
28	.4043	.4048	.4052	.4056	.4061	.4065	.4069	.4073	.4078	.4082
29	.4086	.4090	.4094	.4098	.4103	.4107	.4111	.4115	.4119	.4123
30	.4127	.4131	.4135	.4139	.4143	.4147	.4151	.4155	.4159	.4163
31	.4167	.4171	.4175	.4179	.4182	.4186	.4190	.4194	.4198	.4202
32	.4205	.4209	.4213	.4217	.4220	.4224	.4228	.4232	.4235	.4239
33	.4243	.4246	.4250	.4254	.4257	.4261	.4265	.4268	.4272	.4275
34	.4279	.4283	.4286	.4290	.4293	.4297	.4300	.4304	.4307	.4311
35	.4314	.4318	.4321	.4324	.4328	.4331	.4335	.4338	.4342	.4345
36	.4348	.4352	.4355	.4358	.4362	.4365	.4368	.4372	.4375	.4378
37	.4382	.4385	.4388	.4391	.4395	.4398	.4401	.4404	.4407	.4411
38	.4414	.4417	.4420	.4423	.4427	.4430	.4433	.4436	.4439	.4442
39	.4445	.4449	.4452	.4455	.4458	.4461	.4464	.4467	.4470	.4473
40[3]	.4476									

Values of X_d computed from the formula $X_d = 0.004657f \log_{10} D$. See Par. **9**.

[1] For all conductors of all materials. For values at other frequencies f, multiply 60-cycle values by $f/60$.

[2] D is center-to-center distance between conductors.

[3] For spacings greater than 40 feet, factor the spacing into two or more values within the range of the above tabulation. The required value of X_d will be the sum of the values of X_d for the factors. For example, find X_d for a spacing of 175.7 feet. One pair of suitable factors is:

10 × 17.57	X_d for 10′ = .2794
by interpolation	X_d for 17.57′ = .3478
	X_d for 175.7′ = .6272

also by means of the $X_a + X_d$ method, by determining the equivalent X_a and X_a' of the conductor bundle. The expressions for the equivalents are given in Table 13-3. These expressions are for three conductor bundles on equilateral spacing and for four conductor bundles on square spacing. The subscript s indicates the spacing of the conductors within the bundle in feet. Values for X_a and X_a' will be found in the conductor tables in Sec. **4**. Values for X_s and X_s' can be obtained from Table 13-5 for X_d and Table 13-6 for X_d' if s is greater than 1 ft. If s is less than 1 ft, the values are negative and must be calculated from Eq. (13-8).

$$X_d = 0.004657f \log s \tag{13-8}$$

$$X_d' = \frac{4.099 \times 10^8}{f} \log s \tag{13-9}$$

Table 13-6. Values of X_d'

Shunt capacitive reactance,[1] ohms per conductor per mile, at 60 cycles

D, equivalent[2] spacing, ft	0	0.1	0.2	0.3	0.4	0.5	0.6	0.7	0.8	0.9
1	0,000	2,828	5,409	7,784	9,983	12,030	13,945	15,743	17,439	19,043
2	20,565	22,013	23,393	24,712	25,974	27,185	28,349	29,469	30,548	31,589
3	32,595	33,568	34,510	35,423	36,308	37,168	38,004	38,817	39,608	40,379
4	41,130	41,863	42,578	43,276	43,958	44,625	45,277	45,915	46,539	47,151
5	47,751	48,338	48,914	49,479	50,034	50,578	51,113	51,638	52,154	52,661
6	53,160	53,650	54,133	54,607	55,075	55,535	55,988	56,434	56,873	57,306
7	57,733	58,154	58,569	58,978	59,382	59,780	60,173	60,561	60,944	61,322
8	61,695	62,064	62,428	62,787	63,143	63,494	63,841	64,184	64,523	64,858
9	65,190	65,517	65,842	66,162	66,480	66,794	67,104	67,412	67,716	68,017
10	68,316	68,611	68,903	69,193	69,479	69,763	70,044	70,323	70,599	70,872
11	71,143	71,412	71,678	71,942	72,203	72,462	72,719	72,974	73,226	73,477
12	73,725	73,971	74,215	74,457	74,698	74,936	75,172	75,407	75,640	75,871
13	76,100	76,327	76,553	76,777	76,999	77,219	77,438	77,656	77,871	78,086
14	78,298	78,510	78,719	78,927	79,134	79,339	79,543	79,746	79,947	80,147
15	80,345	80,542	80,738	80,933	81,126	81,318	81,509	81.699	81,887	82,074
16	82,260	82,445	82,629	82,811	82,993	83,173	83,352	83,531	83,708	83,884
17	84,059	84,233	84,406	84,578	84,749	84,919	85,088	85,256	85,423	85,589
18	85,755	85,919	86,082	86,245	86,407	86,568	86,727	86,887	87,045	87,202
19	87,359	87,514	87,669	87,824	87,977	88,129	88,281	88,432	88,582	88,732
20	88,881	89,029	89,176	89,322	89,468	89,613	89,758	89,901	90,044	90,186
21	90,328	90,469	90,609	90,749	90,888	91,026	91,164	91,301	91,437	91,573
22	91,708	91,843	91,977	92,110	92,243	92,375	92,507	92,638	92,768	92,898
23	93,027	93,156	93,284	93,412	93,539	93,665	93,791	93,917	94,042	94,166
24	94,290	94,413	94,536	94,658	94,780	94,902	95,022	95,143	95,263	95,382
25	95,501	95,619	95,737	95,855	95,972	96,089	96,205	96,320	96,436	96,550
26	96,665	96,779	96,892	97,005	97,118	97,230	97,342	97,453	97,564	97,674
27	97,784	97,894	98,003	98,112	98,221	98,329	98,436	98,544	98,651	98,757
28	98,863	98,969	99,075	99,180	99,284	99,389	99,492	99,596	99,699	99,802
29	99,904	100,007	100,108	100,210	100,311	100,412	100,512	100,612	100,712	100,811
30	100,910	101,009	101,107	101,206	101,303	101,401	101,498	101,595	101,691	101,787
31	101,883	101,979	102,074	102,169	102,264	102,358	102,452	102,646	102,639	102,732
32	102,825	102,918	103,010	103,102	103,194	103,285	103,376	103,467	103,558	103,648
33	103,738	103,828	103,917	104,007	104,096	104,184	104,273	104,361	104,449	104,536
34	104,624	104,711	104,798	104,884	104,971	105,057	105,143	105,228	105,314	105,399
35	105,484	105,568	105,653	105,737	105,821	105,905	105,988	106,071	106,154	106,237
36	106,320	106,402	106,484	106,566	106,647	106,729	106,810	106,891	106,972	107,052
37	107,133	107,213	107,292	107,372	107,452	107,531	107,610	107,689	107,767	107,846
38	107,924	108,002	108,079	108,157	108,234	108,312	108,389	108,465	108,542	108,618
39	108,694	108,770	108,846	108,922	108,997	109,072	109,147	109,222	109,297	109,371
40[3]	109,446									

[1] For all conductors of all materials. For values at other frequencies f, multiply 60-cycle values by $60/f$.

[2] D is center-to-center distance between conductors.

Values of X_d' computed from the formula $X_d' = [(4.099 \times 10^6)/f] \log_{10} D$. See Eq. (13-4). This is an approximate formula with negligible error if D/r_c ratios are greater than 5:1. For closer spacing, $x_a' + x_d'$ should not be used to obtain susceptance. Calculate from capacitance exact equations which are given in Sec. 2.

[3] For spacings greater than 40 feet, factor the spacing into values within the range of the above tabulation. The required value of X_d' will be the sum of values of X_d' for the factors. For example, find X_d' for a spacing of 175.7 feet. One pair of the suitable factors is:

10×17.57	X_d' for 10′ = 68,316
by interpolation	X_d' " 17.57′ = 85,037
	X_d' " 175.7′ = 153,353

where s is in feet and f is frequency in cycles per second. Equation (13-9) is correct as long as the ratio of spacing s to conductor radius r is 5 or more.

The value of X_{aeq} is added to X_d (the spacing factor which is determined for the mean spacing between the conductors of the different phases). X_{aeq}' and X_d' are handled in a like manner.

Example of Use of the $X_a + X_d$ Method. To find impedance per mile $Z = R + jX$ and capacitive susceptance b per mile for a 60-c 250,000-cmil stranded copper line on 15.1-ft equivalent spacing D. From Sec. **4,** R per mile $= R_a$, at 50°C $= 0.257\ \Omega$. Re-

Table 13-7. Approximate X_o/X_p Ratios for Transmission Lines

Number of circuits	Number of overhead ground wires	Circuit voltage		
		22 kv	115 kv	230 kv
1	0	5	3.5	3
1	1	3.5	2.75	2.5
1	2	2.75	2.4	2.2

sistances at 50°C are given, since this temperature comes close to being an average operating temperature and gives more conservative results than lower temperatures.

$$\begin{aligned} X_a &= 0.487\ \Omega \text{ from Sec. } \mathbf{4} \\ X_d &= 0.3294\ \Omega \text{ from Table 13-5} \\ \overline{X} &= 0.8164\ \Omega = \text{sum of } X_a \text{ and } X_d \\ Z \text{ per mile} &= R + jX = 0.257 + j0.816\ \Omega/\text{mi} \end{aligned}$$

X_d is given to four places in Table 13-5 in order to show the actual effect of spacing differences. Three places to the nearest number may be used in calculations.

b is determined from the capacitive shunt reactance $X_a' + X_d' = X_c$.

$$\begin{aligned} X_a' &= 110{,}756\ \Omega \text{ from Sec. } \mathbf{4} \\ X_d' &= 80{,}542\ \Omega \text{ from Table 13-6} \\ \overline{X_c} &= 191{,}298\ \Omega = \text{sum of } X_a' \text{ and } X_d' \\ b &= \frac{1}{X_c} = 0.00000523 \text{ mho/mi} \end{aligned}$$

Line-charging current in amperes per mile of conductor can also be found by the use of the capacitive reactance tables (Table 13-6). $I_c = E_{l-n}/(X_a' + X_d')$ amperes per mile per phase. E_{l-n} = volts to neutral.

Three-phase line-charging kVA per mile is found from

$$\text{kVA} = [(E_{l-l})^2/(X_a' + X_d')] \times 1{,}000$$

wherein E_{l-l} = kilovolts line to line.

10. Zero-phase-sequence Reactances and Capacitances. The inductive and capacitive reactances explained in the foregoing are also known as positive-phase-sequence reactances and are used in the conventional balanced-load flow problems of 3-phase circuits. When earth-return currents due to faults or other causes must be calculated, negative-phase-sequence and zero-phase-sequence impedances must be determined. Negative-phase-sequence quantities are the same as the positive sequence for transmission lines. It must be understood that precise determination of the zero-phase-sequence quantities is impossible because of the variability of the earth-return path. Methods[1] of calculating zero-phase-sequence impedances and capacitances have been developed which give results sufficiently accurate for all practical purposes. The reader may consult the reference below[1] for detailed development of the methods. They apply to bundled conductors if distances are mean distances among all conductors.

The zero-sequence impedance is a function of conductor size, spacing, relative position of conductors with respect to overhead ground wires, electrical characteristics of overhead ground wires, and the resistivity of the earth-return circuit. Table 13-8 lists some of the more commonly used formulas for calculating the zero-phase-sequence impedance and capacitance per mile of overhead transmission lines as taken from the

[1] Wagner, C. F., and Evans, R. D. "Symmetrical Components"; New York, McGraw-Hill Book Company, 1933. Also Joint Subcommittee on Development and Research, Edison Electric Institute and Bell Telphone System, *Eng. Rept.* 37, Computation of Zero-sequence Impedances of Power Lines and Cables, July, 1936.

joint subcommittee report.[1] The capacitance equations make simplifying assumptions that all conductors have equal capacitance to ground and to each other. This simplification results in small error. The impedance formulas for twin-circuit lines assume conductors of equal impedance symmetrically arranged with respect to ground wires of equal impedance. The expressions for the individual impedance terms of the formulas are obtained from Wagner and Evans.[2]

Zero-sequence mutual impedance for use with twin circuits not having common terminals can be found from the impedance formulas of Table 13-8 by subtracting the single-circuit impedance from twice the twin-circuit impedance. For example, the zero-sequence mutual impedance of a twin-circuit line with one overhead ground wire is $3\{Z_{14} - [(Z_{17})^2/Z_{77}]\}$. This expression also holds true if the conductors of the two circuits are different as long as the spacings are identical.

In estimating short-circuit currents for circuit-breaker duties, zero-sequence reactance is often obtained by means of X_o/X_p ratio multipliers. X_p is the positive-phase-sequence reactance of the circuit. Some representative multipliers are listed in Table 13-7. An X_o/X_1 multiplier of 2.5 can be used for most EHV lines.

For twin-circuit lines on the same structures, the zero-sequence reactance of the two circuits in parallel will be approximately 85% of that of one circuit.

11. Zero-phase-sequence Impedance and Capacitance Formulas. The formulas for zero-sequence impedance involve only self- and mutual impedances of ground-return circuits. Conductors and ground wires are designated by numbers as shown in Fig. 13-1. Self-impedances are indicated by subscript notations Z_{11} for conductors, since all conductors are assumed to be of equal self-impedance. Self-impedances of ground wires also are assumed to be equal, and Z_{77} therefore represents the self-impedance of either ground wire. Mutual impedances between conductors are assumed to be equal, as are the mutual impedances between conductors and ground wires. Z_{12} represents the mean mutual impedance between wires within groups; and since group 2 is assumed to be symmetrical with group 1, the same notation (Z_{12}) applies to the mean mutual impedance between wires within group 2. Mean mutual impedance between any wire of group 1 and any wire of group 2 is indicated by Z_{14}. The numbers within the conductor groups have no significance as regards phases, since only zero-phase-sequence voltages and currents are considered. Mean mutual impedance between any conductor of either group 1 or group 2 and either ground wire is indicated by Z_{17} owing to the assumed symmetry and transpositions. Mutual impedance between ground wires is indicated by Z_{78}. The same scheme of notation applies to the self- and mutual potential coefficients in the capacitance formulas.

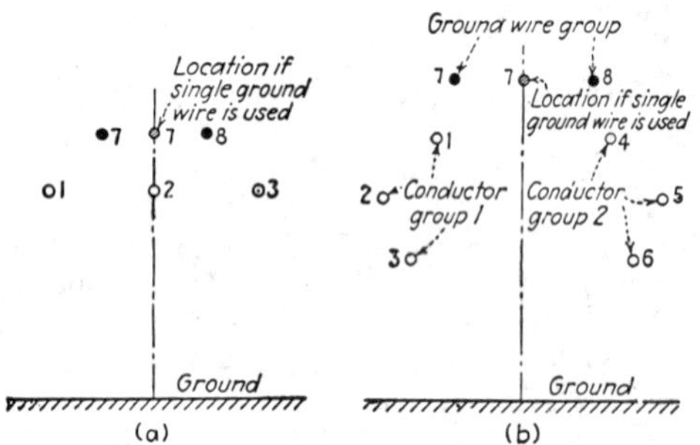

FIG. 13-1. (*a*) Single-circuit configuration. (*b*) Twin-circuit configuration.

Explanation of Terms in Formulas. Self-impedance of one conductor of group with earth return,

$$Z_{11} = 0.00159f + R_a + j(X_{ec} + K) = \Omega/\text{mi} \qquad (13\text{-}10)$$

where f = frequency in cycles per second and R_a = resistance of one conductor in ohms per mile. Table 13-4 gives characteristics of the most commonly used ground wires.

Internal reactance of one conductor,

$$K = 0.004657f \log \frac{r_c}{\text{GMR}} = \Omega/\text{mi} \qquad (13\text{-}11)$$

[1] Joint Subcommittee on Development and Research, *op. cit.*
[2] *Op. cit.*

Table 13-8. Zero-phase-sequence Impedance and Capacitance Formulas

Reference, Fig. No.	No. of ground wires	Zero sequence impedance, ohms per mile	Zero sequence capacitance, microfarads per mile (susceptance, $b_0 = 2\pi f C_0 \times 10^{-6}$)	Total fault current I_{GW} in ground wire or wires
Single three-phase circuit				
13- 1*a*	0	$Z_0 = Z_{11} + 2Z_{12}$	$C_0 = \frac{0.0777}{A_{11} + 2A_{12}}$	
13- 1*a*	1	$Z_0 = Z_{11} + 2Z_{12} - 3\frac{(Z_{17})^2}{Z_{77}}$	$C_0 = \frac{0.0777}{A_{11} + 2A_{12} - \frac{3(A_{17})^2}{A_{77}}}$	$I_{GW} = -\frac{Z_{17}}{Z_{77}} I_F$
13- 1*a*	2	$Z_0 = Z_{11} + 2Z_{12} - 3\frac{2(Z_{17})^2}{Z_{77} + Z_{78}}$	$C_0 = \frac{0.0777}{A_{11} + 2A_{12} - 3\left[\frac{2(A_{17})^2}{A_{77} + A_{78}}\right]}$	$I_{GW} = -\frac{2Z_{17}}{Z_{77} + Z_{78}} I_F$
Twin three-phase circuits				
13- 1*b*	0	$Z_0 = \frac{1}{2}[Z_{11} + 2Z_{12} + 3Z_{14}]$	$C_0 = \frac{0.1554}{A_{11} + 2A_{12} + 3A_{14}}$	
13- 1*b*	1	$Z_0 = \frac{1}{2}\left[Z_{11} + 2Z_{12} + 3Z_{14} - 6\frac{(Z_{17})^2}{Z_{77}}\right]$	$C_0 = \frac{0.1554}{A_{11} + 2A_{12} + 3A_{14} - 6\left[\frac{(A_{17})^2}{A_{77}}\right]}$	$I_{GW} = -\frac{Z_{17}}{Z_{77}} I_F$
13- 1*b*	2	$Z_0 = \frac{1}{2}\left[Z_{11} + 2Z_{12} + 3Z_{14} - 6\frac{2(Z_{17})^2}{Z_{77} + Z_{78}}\right]$	$C_0 = \frac{0.1554}{A_{11} + 2A_{12} + 3A_{14} - 6\left[\frac{2(A_{17})^2}{A_{77} + A_{78}}\right]}$	$I_{GW} = -\frac{2Z_{17}}{Z_{77} + Z_{78}} I_F$

I_F = total ground-fault current flowing in the circuit in question.

The geometric mean radius is also known as the "geometric mean distance" (GMD) (see Sec. **4,** Par. 139).

External reactance of one completely transposed conductor is

$$X_{ec} = 0.004657f \log \frac{D_e}{r_c} = \Omega/\text{mi} \tag{13-12}$$

where D_e = equivalent depth of earth return in feet and r_c = actual radius of the conductor in feet. Equation (13-12) is derived from the expression for Z_{11} in terms of D_e and the GMR of the conductor.

$$D_e = 2{,}160 \sqrt{\frac{\rho}{f}} \tag{13-13}$$

where ρ = earth resistivity in ohms per meter cube.

The reactance term $j(X_{ec} + K)$ in Eq. (13-10) is taken from the joint subcommittee report. This term can be replaced by the term $j(X_a + X_d)$, in which X_a = reactance of the conductor at 1-ft spacing (see Par. **9**). Tables of X_a are given in Sec. **4,** 4-29 to 4-35 X_d is the reactance beyond 1 ft and is expressed as

$$X_d = 0.004657f \log D \tag{13-14}$$

in which D = distance in feet to the return conductor. In calculations for Z_{11} in the zero sequence, D becomes D_e, the equivalent depth of earth return. It will be seen that $X_a = K + X_{ec}$ when X_{ec} is calculated for D_e = 1 ft by Eq. (13-12).

Mean **mutual impedance** between two conductors of a group:

$$Z_{12} = 0.00159f + jX_{12} = \Omega/\text{mi} \tag{13-15}$$

in which

$$X_{12} = 0.004657f \log \frac{D_e}{\sqrt[3]{D_{12}D_{13}D_{23}}} \tag{13-16}$$

D_{12}, D_{13}, and D_{23} (Fig. 13-1) are the center-to-center distances between conductors in feet.

Mean mutual impedance between one ground wire and one conductor,

$$Z_{17} = 0.00159f + jX_{17} = \Omega/\text{mi} \tag{13-17}$$

$$X_{17} = 0.004657f \log \frac{D_e}{\sqrt[3]{D_{17}D_{27}D_{37}}} \tag{13-18}$$

D_{17}, D_{27}, D_{37} are center-to-center distances between one ground wire and the conductors of group 1. Equation (13-18) applies to the mutual impedances for both one- and two-circuit lines with one and two ground wires if the two-circuit lines are completely symmetrical about the vertical axis. If one offset ground wire only is used on a two-circuit line,

$$X_{17} = 0.004657f \log \frac{D_e}{\sqrt[6]{D_{17}D_{27}D_{37}D_{47}D_{57}D_{67}}} \tag{13-18a}$$

D_{47}, D_{57}, D_{67} are distances in feet from conductors 4, 5, 6 to ground the wire.

Mean mutual impedance between one conductor of group 1 and one conductor of group 2 (two-circuit line),

$$Z_{14} = 0.00159f + jX_{14} = \Omega/\text{mi} \tag{13-19}$$

$$X_{14} = 0.004657f \log \frac{D_e}{\sqrt[9]{D_{14}D_{25}D_{36}(D_{15}D_{26}D_{34})^2}} \tag{13-20}$$

D_{14}, D_{25}, D_{36}, D_{15}, D_{26}, D_{34} are distances in feet between conductors of the two groups of conductors.

Self-impedance of one ground wire is Z_{77}. The self-impedance equation for Z_{11} [Eq. (13-10)] applies to any self-impedance. Table 13-5 gives electrical characteristics of some common ground wires.

Mutual impedance between ground wires,

$$Z_{78} = 0.00159f + jX_{78} = \Omega/\text{mi} \tag{13-21}$$

$$X_{78} = 0.004657f \log \frac{D_e}{D_{78}} \tag{13-22}$$

in which D_{78} is the distance between ground wires in feet.

The foregoing zero-phase-sequence reactance formulas can be simplified by breaking up the logarithmic expressions into two parts, as in the $X_a + X_d$ method. By taking Eq. (13-16), for example, the mutual reactance

$$X_{12} = 0.004657f \log \frac{D_e}{\sqrt[3]{D_{12}D_{13}D_{23}}} \qquad (\Omega/\text{mi})$$

This breaks up into

$$0.004657f \log D_e - 0.004657f \log \frac{1}{\sqrt[3]{D_{12}D_{13}D_{23}}}$$

Let the first half of the expression be called X_ρ, since D_e is a function of ρ (the earth resistivity). Then

$$X_{12} = X_\rho - X_{D12} \qquad (\Omega/\text{mi}) \tag{13-23}$$

$$X_{11} = X_\rho - X_a \qquad (\Omega/\text{mi}) \tag{13-24}$$

X_{aeq} is the equivalent X_a (Table 13-3).

$$X_{14} = X_\rho - X_{D14} \qquad (\Omega/\text{mi}) \tag{13-25}$$

$$X_{17} = X_\rho - X_{D17} \qquad (\Omega/\text{mi}) \tag{13-26}$$

$$X_{73} = X_\rho - X_{D78} \qquad (\Omega/\text{mi}) \tag{13-27}$$

The subscripts D_{12}, D_{14}, D_{17}, and D_{78} above are the mean distances among the wires of the phases. If three conductor bundles are involved, the mean of 18 distances is found. The X_d for the mean distance is the mean of the X_d values for the distances. These values can be obtained from Table 13-5 and averaged. However, owing to the indefiniteness of D_e, the bundle may be considered to be a single conductor at the center of the bundle.

Formulas for calculating self- and mean mutual potential coefficients in capacitance equations are

$$A_{11} = 2 \log \frac{4h_m}{d_c} \tag{13-28}$$

$$A_{12} = \log \left[1 + \frac{4h_m^2}{(\sqrt[3]{D_{12}D_{23}D_{13}})^2}\right] \tag{13-29}$$

$$A_{14} = \log \left\{1 + \frac{4h_m^2}{[\sqrt[9]{D_{14}D_{25}D_{36}(D_{15}D_{26}D_{34})^2}]^2}\right\} \tag{13-30}$$

$$A_{17} = \log \left[1 + \frac{4h_m h_g}{(\sqrt[3]{D_{17}D_{27}D_{37}})^2}\right] \tag{13-31}$$

Equation (13-31) applies for both one- and two-circuit lines with one and two ground wires if the two circuit lines are completely symmetrical about the vertical axis.

If one offset ground wire only is used, on a two-circuit line,

$$A_{17} = \log \left[1 + \frac{4h_m h_g}{(\sqrt[6]{D_{17}D_{27}D_{37}D_{47}D_{57}D_{67}})^2} \right] \tag{13-32}$$

$$A_{77} = 2 \log \frac{4h_g}{d_g} \tag{13-33}$$

$$A_{78} = \log \left[1 + \frac{4(h_g)^2}{(D_{78})^2} \right] \tag{13-34}$$

where d_c = diameter of conductors in feet, d_g = diameter of ground wires in feet, h_m = mean height of conductors above ground in feet, h_g = mean height of ground wires above ground in feet.

LINE-PERFORMANCE CALCULATIONS

12. The electrical performance of any single-phase or symmetrical polyphase system may be calculated by using one conductor of the system and voltage to neutral. In single-phase calculations, voltage to neutral is one-half the line-to-line voltage; current is the line current; impedance is that of one conductor; and power per conductor is one-half the total, as are losses. Voltage regulation is obtained directly. In 3-phase calculations, voltage to neutral is line-to-line voltage divided by the square root of 3; current is the line current; impedance is that of one conductor; and power is one-third the total 3-phase power, as are losses. Voltage regulation is obtained directly. Three-phase calculations may also be made by using equivalent single-phase quantities representing the total 3-phase power of the circuit. In such calculations the impedance is that of one conductor; voltage is line to line; current is the square root of three times the 3-phase line current (3-phase kVA/E in kilovolts line to line); total losses are the equivalent current squared times the resistance of one conductor; voltage regulation is obtained directly. This method is often preferred owing to the fact that results are in total 3-phase power and line-to-line voltage. Line-performance calculations fall into two general classifications: (1) lines having negligible capacitance effects; (2) lines having capacitance effects which cannot be neglected. In the following calculations, the $R + jX$ convention and counterclockwise rotation of vectors are used.

Circuits with negligible capacitance may be calculated with a minimum of labor by the use of complex quantities. Voltage and currents have the following relationships:

$$\dot{E}_s = \dot{E}_r + \dot{I}_r\dot{Z} \tag{13-35}$$

$$\dot{I}_s = \dot{I}_r \tag{13-36}$$

$$\dot{E}_r = \dot{E}_s - \dot{I}_s\dot{Z} \tag{13-37}$$

In calculations using Eq. (13-35), it is convenient to use E_r as the reference (positive x axis), which eliminates one of the complex terms.

Equation (13-35) can then be written $\dot{E}_s = E_r + (I_r \cos \theta_r \pm jI_r \sin \theta_r)(R + jX)$, in which θ_r = power-factor angle of the load which determines whether the plus or minus sign shall be used. If the power factor is lagging, the minus sign is used; and if it is leading, the plus sign is used. In using Eq. (13-37), it is convenient to employ E_s as the reference. Equation (13-37) can then be expressed as

$$\dot{E}_r = E_s - (I_s \cos \theta_s \pm jI_s \sin \theta_s)(R + jX)$$

in which θ_s = sending-end power-factor angle determining which of the signs ($\pm$) shall be used. As above, the minus sign is used with lagging power factors and the plus sign with leading power factors.

A sample calculation follows, using Eq. (13-35) and equivalent single-phase quantities:

Given the 3-phase system of Fig. 13-2, to find E_s, the voltage at point A, the total line losses, and the power and power factor at the sending end.

$R + jX$ of section 2 of the line (10 mi 2/0 7-strand copper, on 6 ft equivalent spacing at 50°C and 60 c) $= 4.81 + j7.49$. $R + jX$ of section 1 (10 mi 4/0 19-strand copper on 6-ft equivalent spacing at 50°C and 60 c) $= 3.03 + j7.14$. The 5,000-kV 85% lagging power-factor load is expressed by the IEEE standard as $5{,}000 + j3{,}104$ (kW and kvar).

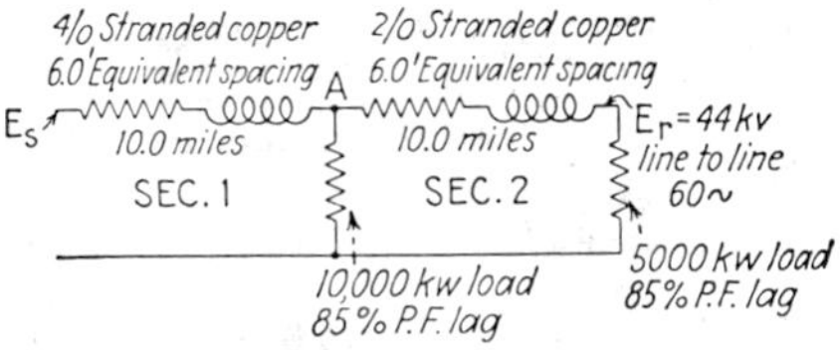

Fig. 13-2. Transmission line with intermediate load.

$$\text{Equivalent current at receiving end} = \frac{5{,}000 - j3{,}104^*}{44} = 113.64 - j70.54 \text{ A}$$

$$\text{Voltage at } A = \dot{E}_a = 44{,}000 + (113.64 - j70.54)(4.81 + j7.49)$$

$$\begin{array}{rrl}
 & 113.64 - j70.54 & = \text{equiv. A} \\
 & 4.81 + j\,7.49 & = R + jX \\
\hline
 & +546.6 \quad - j\,339. & \\
 & +528.3 \quad + j\,851. & \\
\hline
 & 1{,}075.0 \quad + j\,512. & = \dot{Z}\dot{I} \text{ of sec. 2} \\
 & 44{,}000 \qquad\qquad & = \dot{E}_r \\
\hline
\dot{E}_a = & 45{,}075 \quad + j\,512. & = 45{,}075 \text{ V scalar at } A
\end{array}$$

From Eq. (13-36), $\dot{I}_{a2} = \dot{I}_r$, wherein $\dot{I}_{a2}$ = input current to line section 2. In order to obtain the power input into line section 2, the multiplication $\dot{E}_a \underset{\cdot}{I}_r$† is made. Conjugate of current is chosen in order that lagging reactive kVA will be plus and conform to the new IEEE convention. The results of this multiplication are the same as would be obtained by the multiplication $I_{a2} \times (E_a \cos \theta_a + jE_a \sin \theta_a)$, in which E_a = scalar voltage at A, I_{a2} = scalar current in line section 2 at A, and θ_a = angle between them (power-factor angle). The effect of the multiplication $\dot{E}_a \underset{\cdot}{I}_{a2}$, therefore, is the same as if I_{a2} is the reference for E_a. Note that by the present American standard, lagging current has a minus sign, while lagging vars have a plus sign.

$$\begin{array}{rl}
45.075 + j.512 & = \dot{E}_a \quad \text{(kV)} \\
113.64 \;\; + j70.54 & = \underset{\cdot}{I}_{a2} = \underset{\cdot}{I}_r \text{ (conj.)} \\
\hline
5{,}122. \quad + j \quad 58 & \\
-\;36. \quad + j3{,}180 & \\
\hline
5{,}086 \quad + j3{,}238 & = \text{input to line sec. 2} \\
10{,}000 \quad + j6{,}200 & = \text{load at } A \\
\hline
15{,}086 \quad + j9{,}438 & = \text{output line sec. 1 to } A
\end{array}$$

Shunt load at A, 10,000 kW at 85% lagging pf = 10,000 kW + j6,200 kvars. Add algebraically to input to line sec. 2.

$$\begin{array}{rrl}
(15{,}086 - j9{,}438)^*/45.075 = & 334.7 - j209.4 & = \text{equiv. A in line sec. } 1 = \dot{I}_{a1} \\
 & 3.03 + j7.14 & = R + jX \text{ of line sec. 1} \\
\hline
 & 1{,}014 + j634.5 & \\
 & 1{,}495 + j2{,}389.8 & \\
\hline
 & 2{,}509 + j1{,}755.3 & = \dot{Z}\dot{I} \text{ of line sec. 1} \\
 & 45{,}075 \qquad\qquad & = E_a \\
\hline
 & 47{,}584 + j1{,}755.3 & = \dot{E}_s = 47{,}616 \text{ V scalar} \\
\text{Power at sending end by } \dot{E}_s \underset{\cdot}{I}_s & 47.584 + j1.755 & = \dot{E}_s \quad \text{(kV)} \\
 & 334.7 + j209.4 & = \underset{\cdot}{I}_s = \underset{\cdot}{I}_{a1} \text{ (conj.)} \\
\hline
 & 15{,}926 + j\;587 & \\
 & -367 + j9{,}964 & \\
\hline
 & 15{,}559 + j10{,}551 & = \text{input to line sec. 1}
\end{array}$$

* Change sign to obtain vector current.

† Dr. C. L. Fortescue (*Trans. AIEE*, 1923, Vol. 42) showed that true power could be obtained by the product of the vectorial voltage and conjugate of current ($\dot{E}\underset{\cdot}{I}$) or conjugate of voltage and vectorial current ($\underset{\cdot}{E}\dot{I}$). The only difference between the two multiplications is that the first gives lagging reactive kVA a plus sign, whereas the second gives it a minus sign.

The power factor at the sending end is

$$\text{Cosine of the angle } \tan^{-1}\frac{10{,}551}{15{,}559} = 82.8\% \text{ lagging}$$

Total transmission losses are

$$15{,}559 - 15{,}000 = 559 \text{ kW}, \quad \text{or} \quad \frac{559}{15{,}000} \times 100 = 3.73\%$$

Voltage regulation in line section 1 is $[(47{,}616 - 45{,}075)/45{,}075] \times 100 = 5.64\%$. Voltage regulation in line section 2 is $[(45{,}075 - 44{,}000)/44{,}000] \times 100 = 2.44\%$. Overall voltage regulation between the two ends of the line is

$$\frac{47{,}616 - 44{,}000}{44{,}000} \times 100 = 8.21\%$$

and is not the sum of the two section percentages. When accurate losses are required, slide-rule calculations by the conjugate-multiplication method cannot be used, and straight I^2R calculations must be made. As a check on the above calculations, I^2R loss calculations follow. Line section 1, $[(334.7)^2 + (209.4)^2] \times 3.03 \times 10^{-3} = 472$ kW compared with $15{,}559 - 15{,}086 = 473$ kW. I^2R losses in line section 2 are

$$[(113.64)^2 + (70.54)^2] \times 4.81 \times 10^{-3} = 86 \text{ kW}$$

which is the same as $15{,}086 - 15{,}000 = 86$ kW. The above conjugate multiplications and ZI calculations have been made with a calculating machine.

13. Exact solutions of transmission circuits making use of the so-called "auxiliary-circuit constants" provide the most satisfactory method for obtaining the performance of circuits when capacitance must be considered. In the following discussion, complex quantities are used throughout. The relationships between voltages and currents in a long transmission are given by the following equations:

$$\dot{E}_s = \dot{E}_r\dot{A} + \dot{I}_r\dot{B} \tag{13-38}$$
$$\dot{E}_r = \dot{E}_s\dot{A} - \dot{I}_s\dot{B} \tag{13-39}$$
$$\dot{I}_s = \dot{I}_r\dot{A} + \dot{E}_r\dot{C} \tag{13-40}$$
$$\dot{I}_r = \dot{I}_s\dot{A} - \dot{E}_s\dot{C} \tag{13-41}$$

Vector diagrams for Eqs. (13-38) and (13-39) are shown by Fig. 13-3*a* and *b*, respectively.

Fig. 13-3. Vector diagram for $\dot{E}_s = \dot{E}_r\dot{A} + \dot{I}_r\dot{B}$. (*b*) Vector diagram for $\dot{E}_r = \dot{E}_s\dot{A} - \dot{I}_s\dot{B}$.

In Eqs. (13-38), (13-39), (13-40), and (13-41), $\dot{A}$, $\dot{B}$, and $\dot{C}$ are the auxiliary or derived circuit constants. $\dot{A} = a_1 + ja_2$ is a vector numeric term which might be called the "distributed admittance modifier" of a line, since it gives the impedance volts effect upon line voltage due to the current in the distributed admittance when the line is not short-circuited and also gives the effect upon load current of the impedance volts of the load current acting upon the distributed admittance. Equation (13-38), $\dot{E}_s = \dot{E}_r\dot{A} + \dot{I}_r\dot{B}$, becomes $\dot{E}_s = \dot{E}_r\dot{A}$ when a line is open-circuited at the receiving end and $\dot{I}_r\dot{B}$ is zero. $\dot{E}_s = \dot{E}_r\dot{A}$ can be written $\dot{E}_s/\dot{E}_r = \dot{A}$, from which the value of E_s required to produce a desired value of E_r can be found. Also, E_r can be calculated for given values of E_s. The magnitude of the a_1 constant is an approximate measure of the maximum error introduced into voltage calculations of circuits if capacitance is neglected. Average values of this constant at 60 c are 0.999 for a 25-mi line, 0.995 for a 50-mi line, and 0.979 for a 100-mi line. This means that the maximum charging-current effects on voltage are approximately 0.1, 0.5, and 2.1%, respectively, indicating that small errors will result if capacitance is neglected in calculations for 60-c overhead lines less than 50 mi in length. Equation (13-40), $\dot{I}_s = \dot{I}_r\dot{A} + \dot{E}_r\dot{C}$, becomes $\dot{I}_s = \dot{I}_r\dot{A}$ when a line is short-circuited at the receiving end and $\dot{E}_r\dot{C}$ is zero. $\dot{I}_s = \dot{I}_r\dot{A}$ can be written $\dot{I}_s/\dot{I}_r = \dot{A}$, from which the difference in short-circuit current at the sending end and at the receiving end of a line can be determined. Since $\dot{E}_s/\dot{E}_r = \dot{A}$ on open circuit and $\dot{I}_s/\dot{I}_r = \dot{A}$ on short circuit, $\dot{A}$ is the **open-circuit voltage ratio** of the line and also the **short-circuit current ratio.** $\dot{B} = b_1 + jb_2$ is the **impedance characteristic** of the circuit and is the $R + jX$ of the circuit in ohms as modified by the distributed admittance. $\dot{C} = c_1 + jc_2$ is the **leakage-current characteristic** of the circuit, which, it will be seen from Eq. (13-40), gives the open-circuit leakage current in amperes when the line is open at one end and I_r is zero. C_1 is conductance in mhos, and C_2 is susceptance in mhos and may be called the **charging-current characteristic.**

A, *B*, *C* constants, as shown in Eqs. (13-38), (13-39), (13-40), and (13-41), apply to lines which have uniform electrical characteristics throughout their length. That is, $\dot{Z}$ per mile and $\dot{Y}$ per mile are uniform. If construction changes, thereby changing the $\dot{Z}$ and $\dot{Y}$ per mile, it is necessary to determine special (generalized) constants which will be discussed later. The generalized constants call for the addition of a fourth auxiliary constant $\dot{D} = d_1 + jd_2$, which is obtained by combining two or more series circuits having different $\dot{Z}$ and $\dot{Y}$ per mile. The *D* constant is simply another *A* constant, with the line viewed from the opposite end. The *A*, *B*, *C*, *D* constants are all functions of the frequency and the physical characteristics of the circuit and are not functions of power flow. Hence, the values of the *A* and *D* constants when determined must be used with reference to the end of the line for which they were determined, regardless of direction of power flow. The *D* constant substitutes for the *A* constant in Eqs. (13-39) and (13-40), which means that, if the nomenclature of Eqs. (13-38), (13-39), (13-40), and (13-41) is maintained for power-flow calculations in the reverse direction, the values of the *A* and *D* constants must be interchanged in order to keep them in proper relationship to the ends of the line. Another circuit characteristic which can be expressed in terms of the *A*, *B*, and *D* constants is the **equivalent or short-circuit impedance** required in short-circuit calculations for long lines and in impedance-relay settings. This may be obtained from Eq. (13-39), $\dot{E}_r = \dot{E}_s\dot{A} - \dot{I}_s\dot{B}$. With the line short-circuited, $\dot{E}_r = 0$, and $\dot{E}_s\dot{A} = \dot{I}_s\dot{B}$, or $\dot{I}_s = \dot{E}_s\dot{A}/\dot{B}$. $\dot{E}_s\dot{A}/\dot{B}$ may be expressed as $\dot{E}_s/(\dot{B}/\dot{A})$. $\dot{B}/\dot{A}$ is, therefore, the short-circuit impedance of the line. A nonuniform line will have different short-circuit impedances depending upon which end of the line is short-circuited. When viewed from one end, it will be $\dot{B}/\dot{A}$, and when viewed from the other, it will be $\dot{B}/\dot{D}$.

14. Methods of Calculating the Exact Circuit Constants. The *A*, *B*, C constants for lines of uniform electrical characteristics may be readily calculated, either by infinite series or by hyperbolic functions. The expressions for the *A*, *B*, *C* constants in terms of infinite series and hyperbolic functions are as follows:

$$\dot{A} = a_1 + ja_2 = 1 + \frac{\dot{Z}\dot{Y}}{2} + \frac{\dot{Z}^2\dot{Y}^2}{24} + \frac{\dot{Z}^3\dot{Y}^3}{720} + \frac{\dot{Z}^4\dot{Y}^4}{40{,}320} + \cdots \tag{13-42}$$

$$= \cosh \sqrt{\dot{Z}\dot{Y}} \tag{13-43}$$

$$\dot{B} = b_1 + jb_2 = \dot{Z}\left(1 + \frac{\dot{Z}\dot{Y}}{6} + \frac{\dot{Z}^2\dot{Y}^2}{120} + \frac{\dot{Z}^3\dot{Y}^3}{5{,}040} + \frac{\dot{Z}^4\dot{Y}^4}{362{,}880} + \cdots\right) \tag{13-44}$$

$$= \sqrt{\frac{\dot{Z}}{\dot{Y}}} \sinh \sqrt{\dot{Z}\dot{Y}} \tag{13-45}$$

$$\dot{C} = c_1 + jc_2 = \dot{Y}\left(1 + \frac{\dot{Z}\dot{Y}}{6} + \frac{\dot{Z}^2\dot{Y}^2}{120} + \frac{\dot{Z}^3\dot{Y}^3}{5{,}040} + \frac{\dot{Z}^4\dot{Y}^4}{362{,}880} + \cdots\right) \tag{13-46}$$

$$= \sqrt{\frac{\dot{Y}}{\dot{Z}}} \sinh \sqrt{\dot{Z}\dot{Y}} \tag{13-47}$$

In the above expressions, $\dot{Z} = L(R + jX)$ Ω/mi, and $\dot{Y} = L(g + jb)$ mhos/mi of a circuit of length L. The term g in the expression $\dot{Y} = L(g + jb)$ is usually neglected in overhead-line calculations. Also, the fourth term of the series is seldom required, and usually the third can be omitted.

Sample Calculations of Auxiliary Circuit Constants by Infinite Series. Line length 200 mi, conductor 636,000 cmils — 26/7 strand ACSR, 20-ft equivalent spacing, frequency 60 c. From tables of Sec. **4,** R per mile at 50°C = 0.162 Ω, X per mile = $j0.776$ Ω, Y per mile = 5.44×10^{-6} mho.

$$\dot{Z} \text{ for 200 mi} = 32.4 + j155.2$$

$$Y \text{ for 200 mi} = 1{,}088 \times 10^{-6} \text{ mho} = +j0.001088 \text{ mho}$$

$$\begin{array}{rl} 32.4 + j155.2 &= \dot{Z} \\ +j0.001088 &= \dot{Y} \\ \hline -0.1688 + j0.0352 &= \dot{Z}\dot{Y} \end{array} \qquad \begin{array}{rl} -0.1688 + j0.0352 &= \dot{Z}\dot{Y} \\ -0.1688 + j0.0352 &= \dot{Z}\dot{Y} \\ \hline 0.028493 - j0.00594 & \\ -0.001239 - j0.00594 & \\ \hline 0.027254 - j0.01188 &= \dot{Z}^2\dot{Y}^2 \end{array}$$

$$\begin{array}{rl} 0.02725 - j0.01188 &= \dot{Z}^2\dot{Y}^2 \\ -0.1688 \quad + j0.0352 &= \dot{Z}\dot{Y} \\ \hline -0.00459 + j0.00200 & \\ +0.00042 + j0.00096 & \\ \hline -0.00417 + j0.00296 &= \dot{Z}^3\dot{Y}^3 \end{array}$$

It is now seen that $\dot{Z}^3\dot{Y}^3$ when divided by 720 and 5,040 will be negligible and can be omitted.

$$\begin{array}{rl} \dot{A} = 1.00 & \\ -0.0844 + j0.0176 &= \dot{Z}\dot{Y}/2 \\ +0.001135 - j0.000495 &= \dot{Z}^2\dot{Y}^2/24 \\ \text{Neglect } \dot{Z}^3\dot{Y}^3/720 & \\ \hline = \quad 0.91673 + j0.017105 &= \text{sum of terms} \end{array}$$

$$\begin{array}{rl} \dot{B} = 1.00 & \\ -0.02813 + j0.005866 &= \dot{Z}\dot{Y}/6 \\ +0.00227 - j0.000099 &= \dot{Z}^2\dot{Y}^2/120 \\ \text{Neglect } \dot{Z}^3\dot{Y}^3/5040 & \\ \hline 0.97209 + j0.005767 &= \text{sum of terms} \\ 32.4 + j155.2 &= \dot{Z} = \text{multiplier} \\ \hline 31.496 + j0.1868 & \\ -0.895 + j150.868 & \\ \hline +30.601 + j151.0548 & \\ \text{Use } 30.60 + j151.05 & \end{array}$$

$$\begin{array}{rl} \dot{C} = \text{from } B \text{ constant series, sum of terms,} & \\ 0.97209 + j0.005767 & \\ + j0.001088 &= \dot{Y} = \text{multiplier} \\ \hline = -0.0000063 + j0.0010576 & \end{array}$$

The constants are carried out to more places than usual in order to obtain a comparison with calculations by the use of hyperbolic functions, which follow.

Calculation of the auxiliary circuit constants by the use of hyperbolic functions is more complicated than by the use of infinite series when the usual line lengths are involved. However, the use of hyperbolic functions provides more information about the circuit and may be desirable for this reason. Hyperbolic functions are simpler and more satisfactory in working with line lengths of one-quarter wavelength or more. A sample calculation form using hyperbolic functions is shown in Fig. 13-4. The circuit assumed is the same as in the example for infinite series.

In the calculation sheet (Fig. 13-4) the vector mathematics is based upon the principle that, when two vectors of length l_1 and l_2 at the angles θ_1 and θ_2, respectively, are multiplied, the product is $l_1 \times l_2$ at the angle $\theta_1 + \theta_2$. When the square root of such a vector product is taken, the result is $\sqrt{l_1 \times l_2}$ at the angle $(\theta_1 + \theta_2)/2$. When l_1 is divided by l_2, the result is l_1/l_2 at the angle $\theta_1 - \theta_2$.

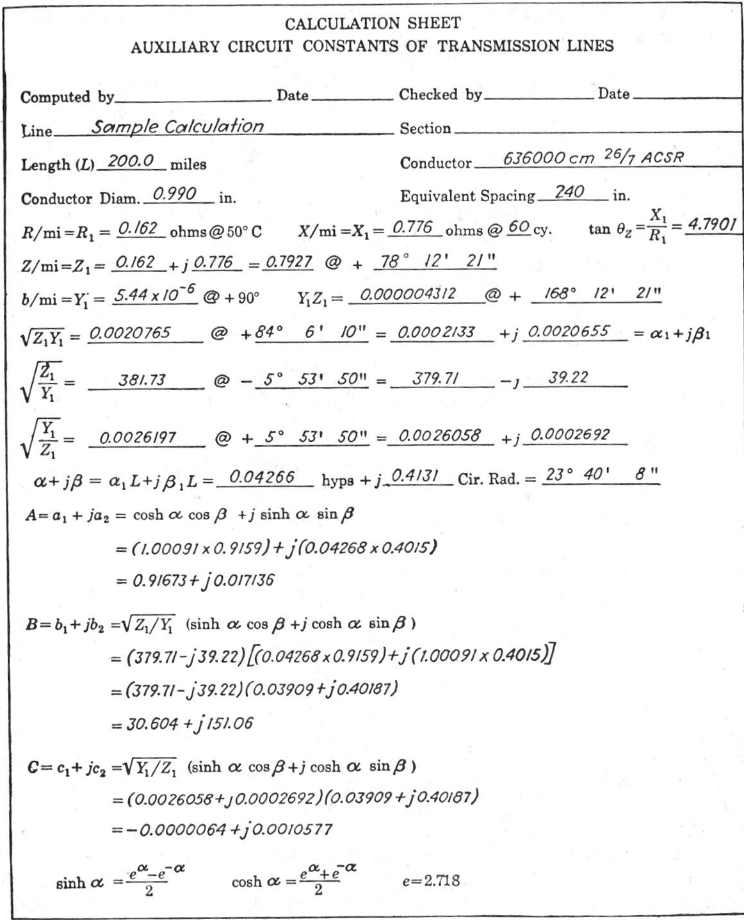

CALCULATION SHEET
AUXILIARY CIRCUIT CONSTANTS OF TRANSMISSION LINES

Computed by ______ Date ______ Checked by ______ Date ______

Line *Sample Calculation* Section ______

Length (L) *200.0* miles Conductor *636000 cm 26/7 ACSR*

Conductor Diam. *0.990* in. Equivalent Spacing *240* in.

$R/\text{mi} = R_1 = 0.162$ ohms @ 50° C $X/\text{mi} = X_1 = 0.776$ ohms @ 60 cy. $\tan\theta_Z = \frac{X_1}{R_1} = 4.7901$

$Z/\text{mi} = Z_1 = 0.162 + j\,0.776 = 0.7927$ @ + 78° 12′ 21″

$b/\text{mi} = Y_1 = 5.44 \times 10^{-6}$ @ + 90° $Y_1Z_1 = 0.000004312$ @ + 168° 12′ 21″

$\sqrt{Z_1Y_1} = 0.0020765$ @ + 84° 6′ 10″ $= 0.0002133 + j\,0.0020655 = \alpha_1 + j\beta_1$

$\sqrt{\frac{Z_1}{Y_1}} = 381.73$ @ − 5° 53′ 50″ $= 379.71 - j\,39.22$

$\sqrt{\frac{Y_1}{Z_1}} = 0.0026197$ @ + 5° 53′ 50″ $= 0.0026058 + j\,0.0002692$

$\alpha + j\beta = \alpha_1 L + j\beta_1 L = 0.04266$ hyps $+ j\,0.4131$ Cir. Rad. = 23° 40′ 8″

$A = a_1 + ja_2 = \cosh\alpha\cos\beta + j\sinh\alpha\sin\beta$

$= (1.00091 \times 0.9159) + j(0.04268 \times 0.4015)$

$= 0.91673 + j\,0.017136$

$B = b_1 + jb_2 = \sqrt{Z_1/Y_1}\,(\sinh\alpha\cos\beta + j\cosh\alpha\sin\beta)$

$= (379.71 - j\,39.22)[(0.04268 \times 0.9159) + j(1.00091 \times 0.4015)]$

$= (379.71 - j\,39.22)(0.03909 + j\,0.40187)$

$= 30.604 + j\,151.06$

$C = c_1 + jc_2 = \sqrt{Y_1/Z_1}\,(\sinh\alpha\cos\beta + j\cosh\alpha\sin\beta)$

$= (0.0026058 + j\,0.0002692)(0.03909 + j\,0.40187)$

$= -0.0000064 + j\,0.0010577$

$\sinh\alpha = \frac{e^{\alpha} - e^{-\alpha}}{2}$ $\cosh\alpha = \frac{e^{\alpha} + e^{-\alpha}}{2}$ $e = 2.718$

FIG. 13-4. Calculation sheet for auxiliary circuit constants of transmission lines.

It will be noted that in Eqs. (13-43), (13-45), and (13-47) $\dot{Z}$ and $\dot{Y}$ are the values for the whole line. Therefore, the expressions $\sqrt{\dot{Z}\dot{Y}} = \sqrt{\dot{Z}_1L \times \dot{Y}_1L} = L\sqrt{\dot{Z}_1\dot{Y}_1}$. Also, in the expression $\sqrt{\dot{Z}/\dot{Y}} = \sqrt{\dot{Z}_1L/\dot{Y}_1L}$ the L's cancel out, and $\sqrt{\dot{Z}_1/\dot{Y}_1}$ has the same value. In Fig. 13-4 the term $\sqrt{\dot{Z}_1/\dot{Y}_1}$ is known as the **surge impedance** of the line at the frequency for which it is calculated and is independent of the line length. Its reciprocal $\sqrt{\dot{Y}_1/\dot{Z}_1}$ is known as the **surge admittance.** $\sqrt{\dot{Z}_1\dot{Y}_1} = \alpha_1 + j\beta_1$ is the **propagation constant** at the frequency used, in which α_1 = **attenuation constant** measuring the decrement in voltage and current per mile in the direction of travel and β_1 = **wavelength constant** per mile in circular radians. $2\pi/\beta_1$ is therefore the wavelength of the line or the distance in miles required for the current and voltage vectors to rotate 360°.

15. Sample Calculation of Line Performance by the Exact Circuit Constants. This calculation is made in the same manner as illustrated under lines with negligible capacitance, using complex quantities. The 200-mile line constants of Fig. 13-4 will be used.

Problem. If 75,000 kW is delivered at 154 kV, 3-phase, 60-c, 99% leading pf (obtained by use of synchronous condensers), find E_s, input to the line, losses, and angle between E_s and E_r. Phase-to-phase voltages and equivalent single-phase current will be used in the calculations.

$$75{,}000 \text{ kW at } 99\% \text{ leading pf} = (75{,}000 + j10{,}685)^*/154 \text{ kV} = (487.0 + j69.4) \text{ equiv. A}$$

From Eq. (13-38), Par. **13**,

$$\dot{E}_s = \dot{E}_r\dot{A} + \dot{I}_r\dot{B}$$

$$\dot{E}_s = 154{,}000(0.9167 + j0.0171) + (487 + j69.4)(30.6 + j151.06)$$

$$\begin{aligned} 487 + j69.4 &= \dot{I}_r \text{ equiv.} \\ 30.6 + j151.06 &= \dot{B} \\ \hline 14{,}902 + j\ 2{,}123 & \\ -10{,}483 + j73{,}566 & \\ \hline 4{,}419 + j75{,}689 &= \dot{I}_r \text{ equiv.} \times \dot{B} \end{aligned}$$

$$\begin{aligned} & 154{,}000(0.9167 + 0.0171) \\ &= 141{,}172 + j\ 2{,}633 = \dot{E}_r\dot{A} \\ & \quad 4{,}419 + j75{,}689 = \dot{I}_r\dot{B} \\ \hline & 145{,}591 + j78{,}322 = \dot{E}_s \end{aligned}$$

Angle between E_s and E_r = $\tan^{-1}(78{,}322/145{,}591) = 28°17' = \delta$

$\cos 28°17' = 0.88066$ $\quad E_s$ scalar $= 145{,}591/0.88066 = 165{,}320$ V

From Eq. (13-40)

$$\begin{aligned} \dot{I}_s &= \dot{I}_r\dot{A} + \dot{E}_r\dot{C} \\ &= (487 + j69.4)(0.916 + j0.0171) + 154{,}000(-0.0000064 + j0.0010577) \\ &= 444.3 + j234.8 \text{ equiv. A} \end{aligned}$$

Sending-end power and reactive power are obtained by the conjugate multiplication $\dot{E}_s\underset{\cdot}{I}_s$ (Par. **12**).

$$\begin{aligned} 145.591 + j78.322 &= \dot{E}_s \quad \text{(kV)} \\ 444.3 - j\ 234.8 &= \underset{\cdot}{I}_s \text{ (conj.)} \\ \hline 64{,}686 - j34{,}185 & \\ 18{,}390 + j34{,}798 & \\ \hline 83{,}076 + j\quad 613 &= \text{sending-end kW and kVAr} \end{aligned}$$

Sending-end power factor is practically unity (slightly lagging). Line losses are $(83{,}076 + j613) - (75{,}000 - j10{,}680) = 8{,}076$ kW and $+j11{,}293$ kVAr. If the line should be open-circuited at the receiving end with 165 kV at the sending end, E_r would rise to $\dot{E}_s/\dot{A} = 165/(0.9167 + j0.0171) = 180$ kV.

16. Generalized Constants.[1] The auxiliary-circuit constants discussed heretofore have been constants of circuits with uniform electrical characteristics throughout their length. It is frequently desirable to combine various circuits in series and in parallel and to include step-up and step-down transformers. Table 13-9 lists the most useful generalized constants with the methods of calculation. Generalized constants are those which will apply to any general circuit. It will be noted that no shunt admittances are shown for the transformers. Such shunt admittances are variable and for normal loads are lost in the assumptions as to loads and power factors. When transformers are connected to long transmission lines with the low-tension breakers open, the shunt admittance will have an appreciable effect in reducing the open-circuit voltage rise, and it is frequently desirable to know this effect. The manufacturers can supply curves of "no-load loss in kilowatts" against "voltage" and "magnetizing kVA" against "voltage," with voltage covering the probable range expected (see Fig. 13-16, Par. **29**). From these curves it is possible by trial-and-error calculations to arrive at the open-circuit voltage which will be experienced. Transformer impedances, when used in the generalized constants, must be reduced to an equivalent $R + jX$ to neutral. This is accomplished by the following equation,

$$Z_{tr}\ (\Omega) = (R + jX) = (\%\,IR + j\%\,IX)\,\frac{\text{kV}^2 \times 10}{\text{kVA}} \tag{13-48}$$

* Change sign of vars to obtain vector amperes.

[1] Evans, R. D., and Sels, H. K. Power Limitations of Transmission Systems; *Trans. AIEE*, 1924, Vol. 43, pp. 26–38.

Table 13-9. Formulas for Generalized Circuit Constants

No.	Type of network		Equivalent constants A_E	B_E	C_E	D_E
1	Series impedance	E_S — Z — E_r	1	Z	0	1
2	Shunt admittance	E_{sn} — Y — E_{rn}	1	0	Y	1
3	Uniform line	E_S —[A B C]— E_r	A	B	C	A
4	Two uniform lines	E_S [$A_2B_2C_2$]—[$A_1B_1C_1$] E_r	$A_1A_2 + C_1B_2$	$B_1A_2 + A_1B_2$	$A_1C_2 + A_2C_1$	$A_1A_2 + B_1C_2$
5	Two nonuniform lines or networks	E_S [$A_2B_2C_2D_2$]—[$A_1B_1C_1D_1$] E_r	$A_1A_2 + C_1B_2$	$B_1A_2 + D_1B_2$	$A_1C_2 + D_2C_1$	$D_1D_2 + B_1C_2$
6	General network and sending transformer impedance	E_S — Z_{TS} —[A B C D] E_r	$A + CZ_{TS}$	$B + DZ_{TS}$	C	D
7	General network and receiving transformer impedance	E_S [A B C D]— Z_{TR} — E_r	A	$B + AZ_{TR}$	C	$D + CZ_{TR}$
8	Two networks in parallel	E_S [$A_1B_1C_1D_1$] ∥ [$A_2B_2C_2D_2$] E_r	$\frac{A_1B_2 + A_2B_1}{B_1 + B_2}$	$\frac{B_1B_2}{B_1 + B_2}$	$C_1 + C_2 + \frac{(A_1 - A_2)(D_2 - D_1)}{B_1 + B_2}$	$\frac{D_1B_2 + D_2B_1}{B_1 + B_2}$

NOTE: All constants in this table are complex quantities. $A = a_1 + ja_2$ and $D = d_1 + jd_2$ are numerical values. $B = b_1 + jb_2$ = ohms. $C = c_1 + jc_2$ = mhos. As a check on calculations of $ABCD$ constants, note that $AD - BC = 1$.

in which kV is the line-to-line kV corresponding to the voltage for which percent IR and percent IX are determined, and kVA = total 3-phase kVA of the transformer bank regardless of Δ- or Y-connections.

Formulas for generalized constants for eight networks are given in Table 13-9, with which it is possible to calculate practically all the combinations met with in practice. Combinations 6 and 7 are forms of combination 5 in which the transformers are general networks with $A = 1$, $B = Z_{tr}$, $C = 0$, and $D = 1$.

17. Nominal π representation. Transmission lines having appreciable capacitance can be represented by the nominal π as in Fig. 13-5, in which half the capacitive susceptance in mhos of one conductor is connected at each end of the line. Line calculations are similar to those shown in Par. **12** after the load current and the current in Y_r are combined. The current at the sending end is found by adding the resultant current at the receiving end to the current in Y_s. Since there may be an appreciable angle between E_r and E_s, the current in Y_s and Y_r may not be in phase, and the current in Y_s is found by multiplying $\dot{E}_s = (E_{s1} + jE_{s2})$ by $jb/2$. The term jE_{s2} may lead or lag E_{s1}, depending upon the X/R ratio of the line and the power factor of the load. The nominal-π-method of calculation in effect is the same as if an A constant equal to $1 + \dot{Z}\dot{Y}/2$ of Eq. (13-42) were used, with no modification of the impedance. The use of the auxiliary-circuit constants has practically superseded the nominal-π-method for long-line calculations. The nominal-π-representation, however, still has use in "a-c calculating-board" and digital-computer studies involving lines of moderate length (usually under 100 mi).

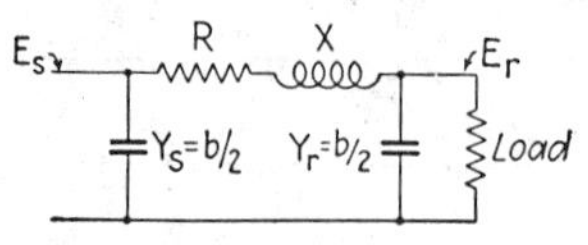

Fig. 13-5. Nominal π line.

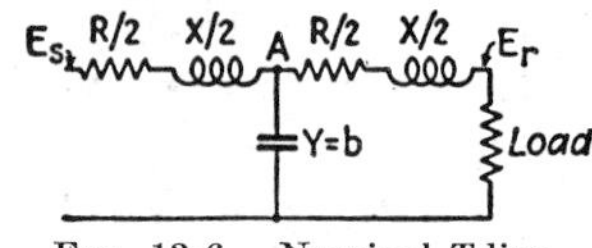

Fig. 13-6. Nominal T line.

18. The nominal T representation of a transmission line is shown in Fig. 13-6. The total line susceptance b in mhos of one conductor is concentrated at A, the midpoint of the line. Calculations are similar in form to those shown in Par. **12**, except that Y in mhos, substituting for the load at point A (Fig. 13-2, Par. **12**), is multiplied by

$$\dot{E}_A = E_{A1} + jE_{A2} = E_r + \dot{I}_r\left(\frac{R}{2} + \frac{jX}{2}\right)$$

and the resulting current $\dot{I}_Y$ is added to $\dot{I}_r$ algebraically to obtain the current $(\dot{I}_A)$ delivered to point A. $\dot{E}_s$ is

$$\dot{E}_s = \dot{E}_A + \dot{I}_A\left(\frac{R}{2} + \frac{jX}{2}\right)$$

$$\dot{I}_s = \dot{I}_A$$

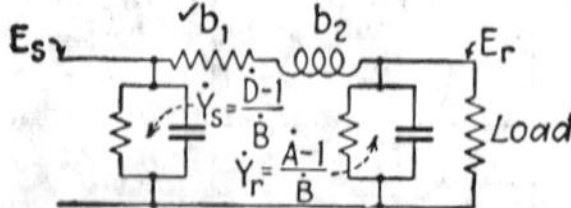

Fig. 13-7. Equivalent π line.

19. Equivalent π. A general (nonuniform) transmission line may be represented by an equivalent π which will give the same terminal conditions as the exact circuit constants. An equivalent π is shown in Fig. 13-7. For a uniform line $\dot{Y}_s$ takes the same form as $\dot{Y}_r$, with the A constant replacing the D constant. For all except the longest lines, the real or power term of $\dot{Y}_s$ and $\dot{Y}_r$ can be neglected. The equivalent-π and T representations are the foundations for digital-computer and a-c calculating-board system setups involving long lines. The equivalent π is used most, since it requires fewer a-c board units; but the equivalent T has the advantage that a-c board readings of line reactive power and reactive components of line current require no corrections in working up results. With the equivalent π the capacitances representing $\dot{Y}_s$ and $\dot{Y}_r$ when π line units are not available are connected to buses representing the stations at the ends of the lines, and consequently measured line reactive quantities must be corrected for end-shunt currents. In a-c calculating-board work $\dot{Y}_s$ and $\dot{Y}_r$, which are in mhos, are converted to shunt impedances. For most a-c board work the capacitive reactance in ohms of $1/\dot{Y}_s$ can be used as $-jX_s = b_2/(1 - d_1)$, and of $1/\dot{Y}_r$ as $-jX_r = b_2/(1 - a_1)$. The exact expressions are $-jX_s = B^2/(b_2 - d_1b_2 + d_2b_1)$ and $-jX_r = B^2/(b_2 - a_1b_2 + a_2b_1)$. If the equivalent π is used for line-performance calculations, it is handled in the same manner as the nominal π.

20. Equivalent T. A long transmission line may be represented by an equivalent T (Fig. 13-8) which will give the same terminal conditions as will be obtained by the use of the exact circuit solution of Par. **15**. The equivalent T shown is for a nonuniform line and is obtained from the equivalent π, the end shunts of which

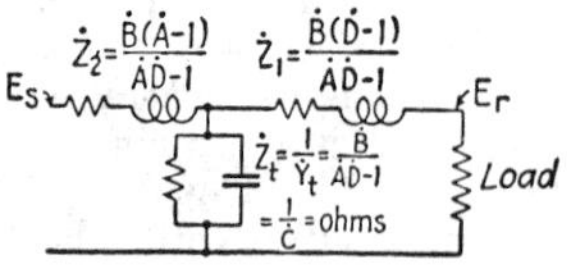

Fig. 13-8. Equivalent T line.

are expressed in ohms, by means of a Δ to y mesh conversion. The staff of the T shown in Fig. 13-8 represents impedance in ohms. If the line is uniform and $\dot{A} = \dot{D}$, then

$$\dot{Z}_1 = \dot{Z}_2 = \frac{\dot{B}}{\dot{A}+1}$$

and

$$\dot{Z}_t = \frac{\dot{B}}{\dot{A}^2 - 1} = \frac{\dot{B}}{\dot{A}+1} \times \frac{1}{\dot{A}-1} = \frac{1}{\dot{C}}$$

Since

$$\frac{\dot{B}}{\dot{A}\dot{D}-1} = \frac{1}{\dot{C}} \quad \text{and} \quad \frac{\dot{B}}{\dot{A}^2-1} = \frac{1}{\dot{C}},$$

then

$$\dot{Z}_1 = \frac{\dot{D}-1}{\dot{C}} \quad \text{and} \quad \dot{Z}_2 = \frac{\dot{A}-1}{\dot{C}}$$

for a nonuniform line, and for a uniform line

$$\dot{Z}_1 = \dot{Z}_2 = \frac{\dot{A}-1}{\dot{C}}$$

The expressions for $\dot{Z}_1$ and $\dot{Z}_2$, in terms of $\dot{A}$, $\dot{D}$, and $\dot{C}$ constants, are simpler for a nonuniform line but are somewhat more difficult to handle with a uniform line. c_1 should not be neglected in calculating $\dot{Z}_1$ and $\dot{Z}_2$ for long lines but usually can be neglected in $\dot{Y}_t$.

21. Use of Equivalents. When a study is to be made of an unbalanced condition to ground, impedance diagrams making use of the foregoing equivalent representations are prepared. Figure 13-9 shows such a diagram for a single phase-to-ground fault at bus C. The diagram includes loads and capacitances. Capacitances probably need be considered only when long lines are involved. The diagram makes use of the equivalent π of Fig. 13-7. Whether a fault is phase to ground, 2-phase to ground, or phase to phase, the individual sequence diagrams are as shown but their interconnections are changed. Important points are: (1) Only 4-wire 3-phase loads appear in the zero sequence. This includes grounded neutral capacitor banks which are a path for zero-sequence currents. (2) Negative- and zero-sequence loads are subjected only to negative- and zero-sequence potentials, accomplished by the network connections shown.

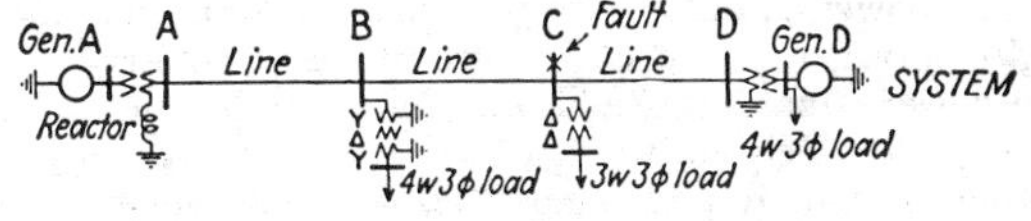

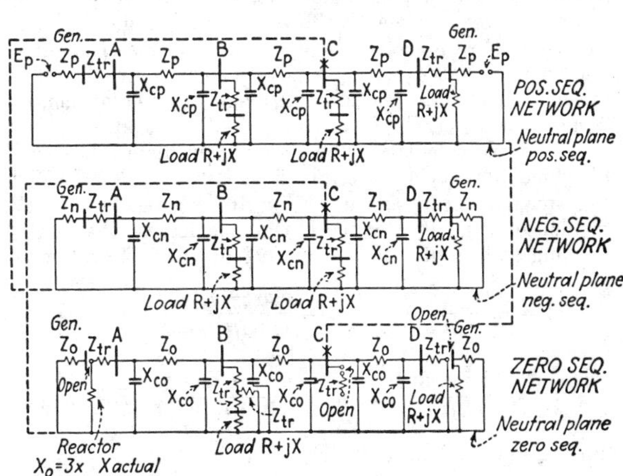

FIG. 13-9. Connections of sequence networks.

22. Power Equations. In Par. 12, it was stated that true power is obtained by the product of conjugate of voltage and vectorial current $\underset{\bullet}{E}\dot{I}$ or vectorial voltage and conjugate of current $\dot{E}I$. This method has been expanded[1] to include the characteristics of the circuit and the angle between voltages. By utilizing the product $\dot{E}I$ which follows the convention used in Par. **12,** the following power equations have been developed:

For lines having negligible capacitance,

$$P_s + jQ_s = \frac{E_s^2 \cos \phi}{Z} - \frac{E_rE_s}{Z} \cos (\phi + \delta) - j\left[\frac{E_rE_s}{Z} \sin (\phi + \delta) - \frac{E_s^2 \sin \phi}{Z}\right] \quad (13\text{-}49)$$

$$P_r + jQ_r = \frac{E_rE_s}{Z} \cos (\phi - \delta) - \frac{E_r^2 \cos \phi}{Z} - j\left[\frac{E_r^2 \sin \phi}{Z} - \frac{E_rE_s}{Z} \sin (\phi - \delta)\right] \quad (13\text{-}50)$$

For lines having appreciable capacitance the exact circuit constants are used, and

$$P_s + jQ_s = E_s^2 \frac{d_1b_1 + d_2b_2}{B^2} - \frac{E_rE_s}{B} \cos (\phi' + \delta)$$
$$- j\left[\frac{E_rE_s}{B} \sin (\phi' + \delta) - E_s^2 \frac{d_1b_2 - d_2b_1}{B^2}\right] \quad (13\text{-}51)$$

$$P_r + jQ_r = \frac{E_rE_s}{B} \cos (\phi' - \delta) - E_r^2 \frac{a_1b_1 + a_2b_2}{B^2}$$
$$- j\left[E_r^2 \frac{a_1b_2 - a_2b_1}{B^2} - \frac{E_rE_s}{B} \sin (\phi' - \delta)\right] \quad (13\text{-}52)$$

where $Z = \sqrt{R^2 + X^2}$; $\phi = \tan^{-1} (X/R)$; δ = angle between E_s and E_r; a_1, a_2, b_1, b_2, d_1, d_2 = numerical values of line constants; $B = \sqrt{b_1^2 + b_2^2}$; and $\phi' = \tan^{-1} (b_2/b_1)$.

If E_s and E_r are expressed in kilovolts line to line, $P + jQ$ quantities will be total 3-phase power in megawatts and megavoltamperes. If line-to-neutral voltages are used, values of $P + jQ$ are one-third of the values for line-to-line voltages.

The power equations are particularly useful in studying the power flows for various angles between E_s and E_r and are the basic equations for system stability calculations. The power equations are also useful in studies involving lines in network transmission systems in which voltages at the various points on the network must be kept within certain limits and angles must be balanced over the various parallel paths of the network. Power and reactive flows are readily determined. It is not necessary to make the whole $P + jQ$ calculations in every case, as frequently power is all that is required, and in such cases only the power terms of the equations need be considered. For example, if P_r only is wanted in long-line calculations, $P_r = (E_rE_s)/B \cos (\phi' - \delta) - E_r^2[(a_1b_1 + a_2b_2)/B^2]$ from Eq. (13-52) is used.

A sample calculation for a simple problem follows: By using the line the constants of which have been determined in Par. **14,** 75,000-kW load at 85% lagging pf at the high-tension receiving bus, E_s = 169 kV, E_r = 154 kV, to find the net synchronous condenser capacity required, the line losses, the angle between E_s and E_r, and the sending-end input. Equations (13-51) and (13-52) will be used. Let

$$l = \frac{a_1b_1 + a_2b_2}{B^2} \qquad m = \frac{a_1b_2 - a_2b_1}{B^2} \qquad A = a_1 + ja_2 = 0.91673 + j0.017136$$

$$l' = \frac{d_1b_1 + d_2b_2}{B^2} \qquad m' = \frac{d_1b_2 - d_2b_1}{B^2} \qquad B = b_1 + jb_2 = 30.604 + j151.06$$

Since the line is uniform, $l = l'$ and $m = m'$.

$$l = (0.91673 \times 30.604 + 0.017136 \times 151.06)/[(30.604)^2 + (151.06)^2] = 0.00129$$
$$m = (0.91673 \times 151.06 - 0.017136 \times 30.604)/[(30.604)^2 + (151.06)^2] = 0.00581$$
$$B = \sqrt{(30.604)^2 + (151.06)^2} = 154.13$$

[1] FORTESCUE, C. L., and WAGNER, C. F. Some Theoretical Considerations of Power Transmission; *Trans. AIEE,* 1924, p. 16.

First solve for δ. From $P_r = (E_r E_s/B)\cos(\phi' - \delta) - (E_r)^2 l$, it is seen that

$$\cos(\phi' - \delta) = \frac{P_r B}{E_r E_s} + \frac{E_r}{E_s} lB$$

Using E_r and E_s in kilovolts line to line, P_r is in megawatts, and

$$\cos(\phi' - \delta) = \frac{75 \times 154.13}{154 \times 169} + \frac{154}{169} \times 0.00129 \times 154.13 = 0.62534$$

$$(\phi' - \delta) = 51°17'37''$$

$$\phi' = \tan^{-1}\frac{151.06}{30.604} = 78°32'50''$$

$$\delta = \phi' - (\phi' - \delta) = 78°32'50'' - 51°17'37'' = 27°15'13''$$

$$jQ_r = -j\left[(E_r)^2 m - \frac{E_r E_s}{B}\sin(\phi' - \delta)\right]$$

$$= -j\left[(154)^2 \times 0.00581 - \frac{154 \times 169}{154.13} \times 0.78036\right]$$

$$= -j(137.790 - 131.770) = -j6.020 \text{ MVA (leading)}$$

75,000 kW at 85% lagging pf $= 75.000 + j46.500$ (MW and MVA)

Total net leading condenser capacity $= 46.500 + 6.020 = 52.520$ MVA

Sending-end conditions are

$$P_s = (E_s)^2 l' - \frac{E_r E_s}{B}\cos(\phi' + \delta) \qquad \begin{aligned} \phi' &= 78°32'50'' \\ \delta &= 27°15'13'' \\ \phi' + \delta &= 105°48'3'' \end{aligned}$$

$$\cos(\phi' + \delta) = -\cos[180° - (105°48'3'')] = -\cos 74°11'57'' = -0.2723$$

$$P_s = (169)^2 \times 0.00129 - \left(\frac{154 \times 169}{154.13} \times -0.2723\right)$$

$$= 36.844 + 45.980 = 82.824 \text{ MW}$$

$$+jQ_s = -j\left[\frac{E_r E_s}{B}\sin(\phi + \delta) - (E_s)^2 m'\right]$$

$$= -j\left[\frac{154 \times 169}{154.13} \times 0.9622 - (169)^2 \times 0.00581\right]$$

$$= -j(162.474 - 165.939) = +j3.465 \text{ MVA (lagging)}$$

Line I^2R loss $= P_s - P_r = 82.824 - 75.000 = 7.824 \text{ MW} = 10.4\%$

Losses may be obtained as above, or a **loss equation** may be set up from the power terms of Eqs. (13-51) and (13-52) as follows:

$$P_s - P_r = (E_s)^2 l' + (E_r)^2 l - \frac{E_r E_s}{B}[\cos(\phi' + \delta) - \cos(\phi' - \delta)]$$

$$\cos(\phi' + \delta) - \cos(\phi' - \delta) = 2\cos\phi'\cos\delta$$

and

$$P_s - P_r = \text{loss} = (E_s)^2 l' + (E_r)^2 l - \frac{2E_r E_s}{B}\cos\phi'\cos\delta \qquad (13\text{-}53)$$

$$= \text{megawatts when } E_s \text{ and } E \text{ are in kilovolts line to line}$$

The foregoing loss equation is the most satisfactory one for long-line calculations. However, it is not satisfactory for slide-rule use, since four to six significant figures are usually required. Excellent results are obtained with seven-place trigonometry tables and a calculating machine. For short lines in which capacitance can be neglected, losses are best found by straight I^2R calculations.

23. Charts and diagrams are very useful in determining transmission-line performance, and their accuracy when they are carefully constructed is usually sufficient for all practical purposes. Their use can be supplemented by calculations where necessary; and while losses can be obtained, approximately, from some of the diagrams, they are best obtained by calculations.

24. The Perrine-Baum[1] regulation diagram originally presented by F. A. C. Perrine and F. G. Baum and based upon the relationships shown in Fig. 13-10*a* and 13-10*b* is very useful in studying transmission-line performance. This diagram presents the "power–power-factor grid" which is the foundation for many of the later diagrams using the exact equations in their construction. The diagram of Fig. 13-10*a* is drawn

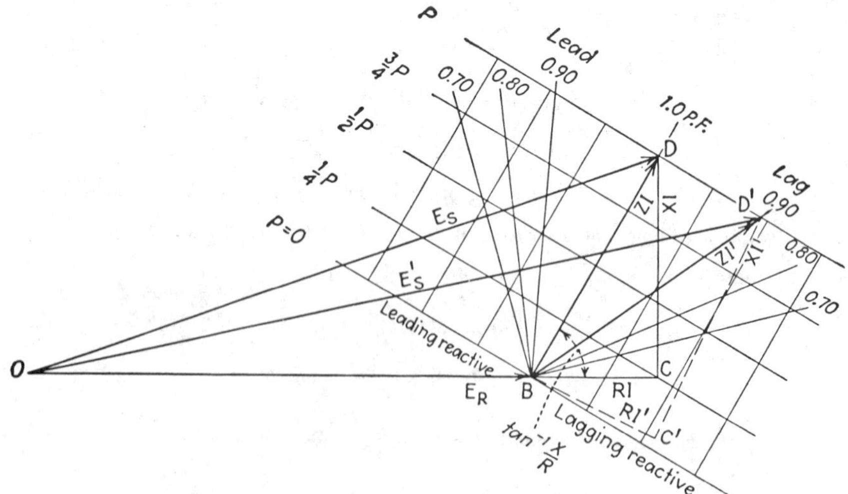

Fig. 13-10*a*. Perrine-Baum diagram.

with voltage as the scale and is simply a plot of $\dot{E}_s = \dot{E}_r + \dot{I}_r\dot{Z}$, for E_r constant when capacitance is negligible. E_r is used as reference. Line *OB* represents E_r to scale, and the line *BD* represents the *ZI* drop at the same scale as for E_r for power *P* at unity power factor. If the power remains the same and the load power factor is changed to 0.9 lagging, voltage conditions are as shown by triangle *BC'D'*, E_s' being the sending-end voltage required to maintain E_r constant. The line *DD'* is perpendicular to line *BD*, and it is this relationship which makes possible the construction of the "power–power-factor" grid. The power-factor lines are drawn at angles $\cos^{-1} 0.9$, $\cos^{-1} 0.8$, $\cos^{-1} 0.7$, etc., with line *BD*. Voltage regulation is found from measurements of E_s, E_s', etc., when

$$\text{Regulation } (\%) = \frac{E_s - E_r}{E_r} \times 100 \qquad \frac{E_s' - E_r}{E_r} \times 100 \qquad \cdots$$

The line *DD'* represents the reactive component of the load in kVA at the same scale as that used for power. E_s and E_r can be line-to-neutral voltage or line-to-line voltage. If line-to-neutral voltage is used, line *BD* is the voltage drop I_rZ in which I_r = actual line current of the total 3-phase power. Line *BD* can be scaled in total 3-phase power

[1] Perrine, F. A. C. "Electrical Conductors"; Princeton, N.J., D. Van Nostrand Company, Inc., 1903.

or in one-third of the 3-phase power as desired. If line-to-line voltage is used in the construction of the chart, line BD is the voltage drop $\sqrt{3}I_rZ$, in which I_r = actual line current for the total 3-phase power. The scale of BD is in total 3-phase power. This diagram is very useful for the study of lines for which E_s can be varied to hold E_r constant or for lines over which E_r is kept within the maximum allowable variation from E_s by means of condensers. For example, if $E_s = OD$ is the maximum value of E_s allowable, it would be necessary to supply condenser capacity equal to DD' in order to carry the same load at 0.9 lagging pf. The diagram can be extended to include a line-capacitance correction as illustrated in Fig. 13-10*b*. The principle applied is that of the nominal π. Line BD of the grid (Fig. 13-10*a*) is shifted to position B_1D_1 of Fig. 13-10*b*. The line $BE = RI_c/2$, and $EB_1 = XI_c/2$. R = line resistance, X = line reactance, and $I_s/2$ = one-half the line-charging current, equal to the current in the condenser of the norminal π at the receiving end of the line at the value of E_r for which the grid is constructed.

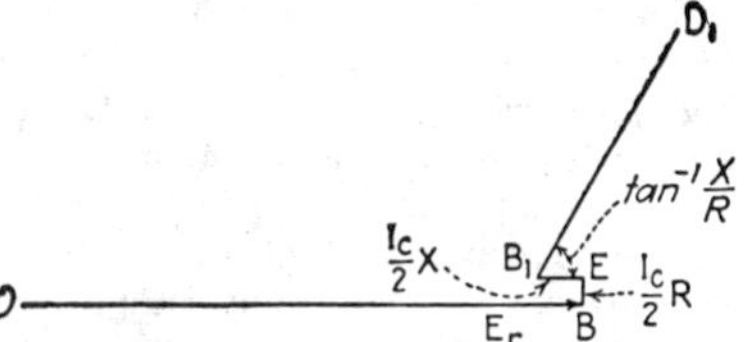

FIG. 13-10*b*. Capacitance correction for Perrine-Baum diagram.

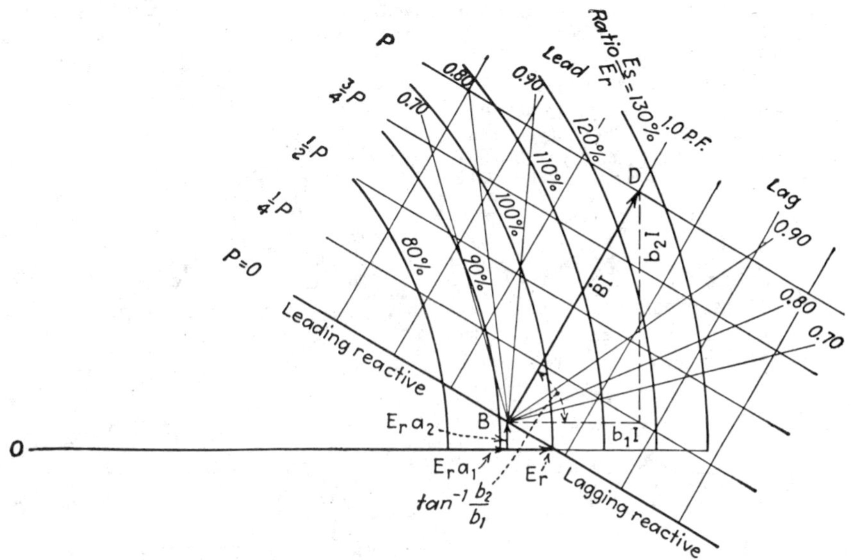

FIG. 13-11. Regulation diagram for E_r constant and E_s variable.

25. Regulation Diagram for E_r Constant and E_s Variable. Regulation diagrams based upon Eq. (13-38) ($\dot{E}_s = \dot{E}_r\dot{A} + \dot{I}_r\dot{B}$) are modeled after the Perrine-Baum diagram (Par. **24**). The auxiliary-circuit constants are used in the calculations for the construction of the chart, and voltage is the scale as in the Perrine-Baum diagram. Concentric circles of E_s marked in percentage of E_r are drawn, from which percent regulation can be found by subtracting 100 from E_s in percent. Figure 13-11 shows such a chart. E_r is used as reference; origin B of the power–power-factor grid is located as shown. The slope of BD is found from the angle $\tan^{-1}(b_2/b_1)$. The grid is constructed and scaled in the same manner as in the Perrine-Baum diagram. This diagram, by the use of conversion factors, can be used for other values of E_r than that for which it was constructed. For example, if a chart has been constructed for E_{r1} and it is desired to know the performance at another receiver voltage E_{r2}, kilowatt and kvar values entering the chart are multiplied by $(E_{r1})^2/(E_{r2})^2$ to obtain new values of kilowatts and kvar, which must then be located on the chart. The voltage ratio for the point located is read; and, having E_{r2}, E_s is found from the marked ratio. If the amount of power at a given power factor which can be delivered for a given ratio of E_s/E_{rn} is wanted, the kilowatts and kvars at the point of intersection of the ratio circle and the power-factor line are read and multiplied by $(E_{rn})^2/(E_{r1})^2$ to obtain the actual power and kvar for the new receiver voltage E_{rn}. This diagram is often rotated to fit the axes of cross-section paper or cloth and the power markings made to fit the divisions of the paper. This is accomplished by the use of a scale-conversion factor which is used to multiply the lengths of the E_s/E_r radii and the line BD to make them fit the paper.

26. Regulation Diagram for E_s Constant and E_r Variable.[1] This diagram is also based upon Eq. (13-38) ($\dot{E}_s = \dot{E}_r\dot{A} + \dot{I}_r\dot{B}$) and is particularly useful in determining voltages at the ends of radial lines served from load centers, where E_s must be maintained practically constant, and gives values of E_r direct. Figure 13-12 shows such a diagram. The grid of the diagram is constructed first, and the procedure is the same as for the Perrine-Baum diagram and the diagram of Par. **25.** The voltage used to determine the current for calculating the length of BD can be any one of the values of E_r which may be expected, but it is usually more convenient to use the lowest. When the values for the grid have been determined, line BD is drawn at an angle $\tan^{-1}(b_2/b_1)$ with the reference (X) axis and the grid is constructed. The line of centers is next located as indicated. This is a line through B at an angle $\tan^{-1}(a_2/a_1)$ with the reference axis. If the circuit has negligible capacitance, the A constant will be $1 + j0$, and the line of centers will coincide with the reference axis.

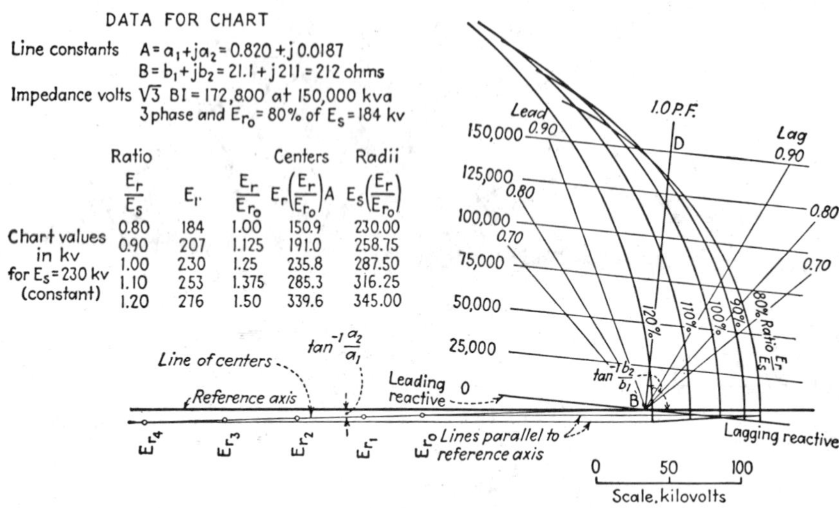

Fig. 13-12. Regulation diagram for E_s constant and E_r variable.

With the line of centers located, the first circle is drawn. The center is at a distance $-E_{r0}\dot{A} = -E_{r0}(a_1 + ja_2)$ from B. E_{r0} is the voltage for which the grid was constructed. The radius of the first circle is E_s. As many other circles as desired can be drawn, but the distance to the centers of the circles and the radii of the circles must be modified to fit the grid. This is accomplished as follows: Let $E_{r1}, E_{r2}, \cdots, E_{rn}$ be other values of E_r. The distances of the centers of circles from B will be $E_{r1}(E_{r1}/E_{r0})\dot{A}$, $E_{r2}(E_{r2}/E_{r0})\dot{A}$, $\cdots$, $E_{rn}(E_{rn}/E_{r0})\dot{A}$. The circles are for constant E_s, but for each value of E_r the length of E_s (the radius) must be modified as follows: $E_s(E_{r1}/E_{r0})$, $E_s(E_{r2}/E_{r0})$, $\cdots$, $E_s(E_{rn}/E_{r0})$. The points of zero angle between E_s and E_r are the intersections of the circles with lines drawn through the centers, parallel to the reference axis. The circle passing through

[1] Cissna, V. J. Voltage Regulation Diagram for Constant Sending Voltage; *Elec. Eng.*, May, 1936, p. 562.

point B, the origin of the grid, represents $E_r = E_s/A$, the open-circuit voltage of the circuit. The circles are marked in percent ratios of E_r/E_s. The diagram can be used for other values of E_s than that for which it was drawn by determining conversion factors as explained in Par. **25.** In this case, if the diagram was drawn for sending voltage E_{s1}, the conversion factors are $(E_{s1})^2/(E_{s2})^2$ to enter the diagram and $(E_{s2})^2/(E_{s1})^2$ to convert diagram values of power to actual values of power at the new value E_{s2}. Values of E_r are found from the marked ratios.

27. Power-circle diagrams, giving both sending- and receiving-end voltage, power and reactive-power conditions, can be constructed from the power equations (Par. **22**) and are probably the most useful diagrams available for the study of transmission-line performance. These diagrams take a variety of forms, but only one of the simpler forms will be described. Equations (13-51) and (13-52) will be used. Equation (13-51) can be written

$$P_s + jQ_s = E_s^2 \frac{d_1b_1 + d_2b_2}{B^2} + jE_s^2 \frac{d_1b_2 - d_2b_1}{B^2} - \frac{E_rE_s}{B}[\cos(\phi' + \delta) + j\sin(\phi' + \delta)]$$

This is the equation for a circle with centers at

$$E_s^2 \frac{d_1b_1 + d_2b_2}{B^2} + jE_s^2 \frac{d_1b_2 - d_2b_1}{B^2}$$

and radius E_rE_s/B.

Equation (13-52), it will be seen, represents a circle with center at

$$-E_r^2 \frac{a_1b_1 + a_2b_2}{B^2} + jE_r^2 \frac{a_1b_2 - a_2b_1}{B^2}$$

and the same radius E_rE_s/B. Both circles can be drawn on one common grid, thereby greatly facilitating comparison of sending- and receiving-end conditions. Either E_s or E_r can be varied and nests of circles drawn, giving the same results that are obtained from the regulation diagrams of Par. **25** or **26,** respectively, plus sending-end conditions. Figure 13-13 shows a power-circle diagram for E_r constant, in which case the receiving-end circles are concentric and the lengths of the radii vary with E_s. The sending-end circles are not concentric, since the centers are located from $(E_s)^2l'$, $+j(E_s)^2m'$ with E_s varying.

$$l' = \frac{d_1b_1 + d_2b_2}{B^2} \quad \text{and} \quad m' = \frac{d_1b_2 - d_2b_1}{B^2}$$

The centers for the sending-end circles are located on a straight line through 0, with coordinates as shown. The scale of the diagram is total 3-phase power and reactive power in megawatts and megavoltamperes when E_s and E_r are line-to-line kilovolts. It is usually desirable to mark angles on the concentric circles in order that δ, the angle between E_s and E_r, can be read directly. The angles start from line RB, which is at angle $\phi' = \tan^{-1}(b_2/b_1)$ with the power axis. Line RB is the line of zero angle between E_s and E_r. There is no straight line for the sending-end circles corresponding to RB. The zero-angle points should be marked on each sending circle by drawing parallel lines through the points of centers $S_1, S_2, \cdots, S_n$ at an angle ϕ' with the power axis. It is necessary to have these zero-angle points accurately marked in order that angle δ can be transferred from receiving end to sending end, or vice versa. This is done by measuring, for example, the arc of the 110% receiving-end circle (Fig. 13-13) between line RB and radius E_rE_s/B with a compass and transcribing it to the 110% sending-end circle to obtain the location of the radius E_rE_s/B as shown. The intersections of the radii with the 110% circles shown give the power, reactive power, and power factors at the sending and receiving ends of the line. The difference in lengths of P_s and P_r (Fig. 13-13) represents the line loss.

Figure 13-13, while it represents an actual circuit, is an unusual example of a power-circle diagram, as will be seen from the circuit constants. When the line has negligible capacitance, both 100% voltage circles pass through the origin O of the power grid and, as capacitance increases, move away from the origin in the directions as shown in Fig.

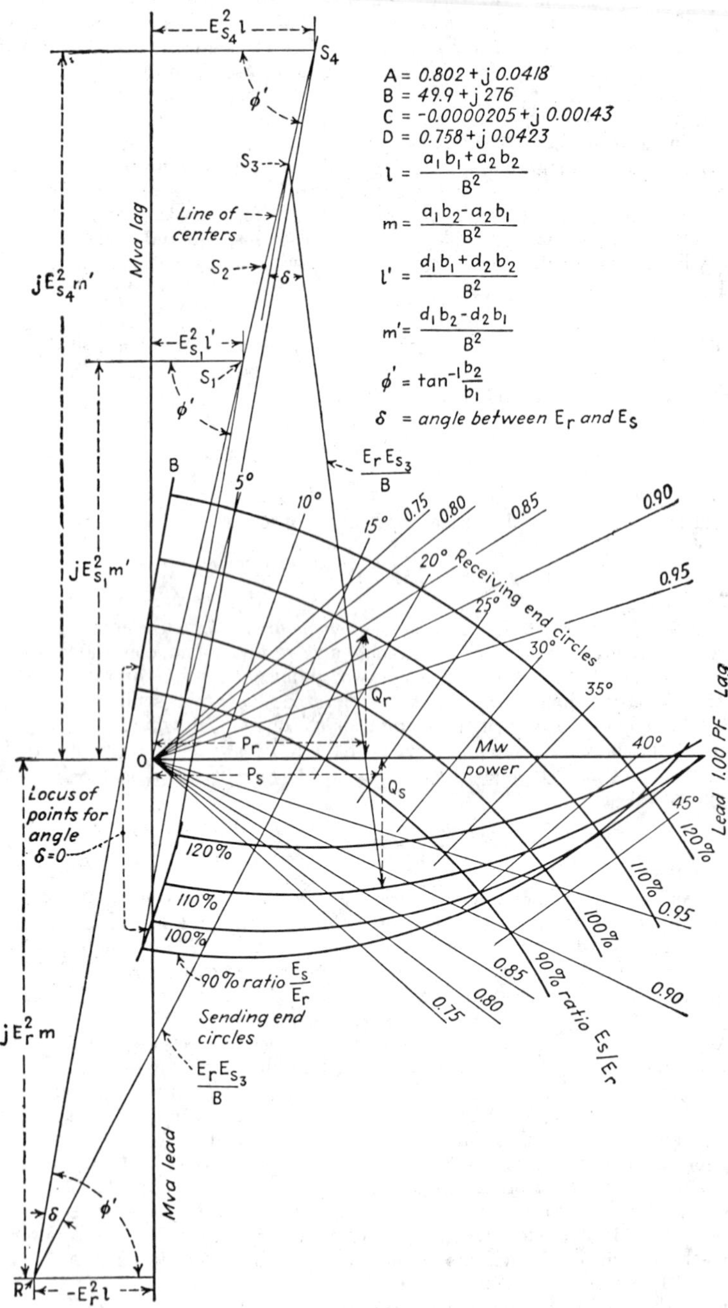

FIG. 13-13. Power-circle diagram for E_r constant.

13-13. Only the arcs of circles covering the normal load range of the circuit are shown, but each arc can be drawn as a complete circle or extended to the left ($-\delta$ angles). When the circles are extended into $-\delta$ angles, power flow in the opposite direction is represented, with E_s constant at the same value as that of E_r for positive angle δ. The sending-end circles then become receiving circles, the receiving circles become sending circles, and the ratio markings of the circles are E_r/E_s instead of E_s/E_r. The marked values, however, are unchanged. The continuation of the circles to the left may be useful if there is power flow in both directions at flat voltage ($E_s = E_r$) and the A and D constants of the line differ appreciably. Power-circle diagrams are used in the same manner as the Perrine-Baum diagram and the regulation diagrams of Pars. **25** and **26.** The diagram for E_r constant can be used for other values of E_r by the use of the conversion factors described in Par. **25,** with the same conversion factor applied to *both* sending- and receiving-end quantities. If a power-circle diagram is constructed for E_s constant, the sending-end circles are concentric and the receiving-end circles have different centers. The diagram for E_s constant can be used for other values of E_s by the use of the conversion factors described in Par. **26,** with the same conversion factor applied to *both* sending- and receiving-end quantities. Since it is often desirable to use a diagram for other values of the voltage assumed to be fixed, it is preferable to mark the circles in ratios instead of in values of the variable voltage. If E_s is variable, marked ratios should be E_s/E_r; and if E_r is variable, marked ratios should be E_r/E_s. When a power-circle diagram is used as a regulation diagram, the radius E_rE_s/B represents the position of E_s in the receiving-end diagram and E_r in the sending-end diagram.

VOLTAGE CONTROL

28. Step-type Voltage Regulators. The method of controlling the voltage on transmission circuits is largely determined by the X/R ratio of, and the duty imposed upon, the circuit. The development of step-type voltage regulators and transformers with built-in step-type regulators (load-ratio control or tap changing), all in sizes up to the limits imposed by railroad facilities, has provided a means of voltage control which, in a large measure, replaces or supplements the use of synchronous or static condensers and frequently with considerable savings in cost. Step-type voltage regulators installed in lines having an X/R ratio of less than 1 will often solve the voltage difficulties for a fraction of the cost of condensers. The voltage-control range of these regulators is ordinarily from 10% buck to 10% boost, or a total range of 20%. The usual application of a regulator is to boost the voltage. Consequently, care must be taken in establishing its location, since its action is to boost the voltage received at the point where it is installed. Another application is to regulate the output voltage of step-down transformer banks. Large substations may require the use of both regulators and condensers to provide the most economical installation. In this case, since automatic voltage control cannot be applied to both simultaneously, it should be applied in the manner which will effect the greatest saving in line losses.

Step-type voltage regulators are occasionally installed in the high-voltage transmission lines of a network system. In such installations the steps of voltage control in percent buck or boost have little meaning in relation to change in voltage obtained. One result of tap change is to shift the flow of reactive power over the circuits connected to the regulator. The net result is that voltage change is the percent step change less the change in voltage drop between the regulator and its source of supply. The calculation of the effects of such regulators in a network system is practically impossible, and if such an installation is being considered, it should be made the subject for analysis by means of the a-c calculating board or digital computer.

29. Synchronous condensers provide one of the principal means for the control of voltage on transmission systems. The application of synchronous condensers on a transmission system is seldom for the sole purpose of maintaining a fixed power factor. The effectiveness of synchronous condensers in controlling transmission-line voltage is dependent upon the X/R ratio of the circuit. The higher the ratio, the greater the effectiveness. Condenser standards usually call for a leading to lagging capacity ratio of 2:1 for air-cooled machines and a ratio of 2.5:1 for hydrogen-cooled machines (see

Fig. 13:15). The most economical application of a condenser for voltage control will be one which makes full use of the leading and lagging capacity over the normal load cycle. However, it is difficult to obtain this ideal application, owing to the fact that one of the main functions of synchronous condensers is to maintain voltage during emergency conditions such as one transmission circuit out of service. If the condensers are selected to be of sufficient capacity to maintain voltage during the station peak load with the worst emergency condition existing, there will be a surplus of condenser capacity during all normal loading conditions.

The study of synchronous condenser requirements is greatly facilitated by the use of regulation diagrams for E_s constant (see Pars. **26** and **27**). The power equations (Par. **22**) are also very useful in calculations of condenser requirements. Figure 13-14 is a regulation diagram for E_s constant, with transformer taps taken into account. If a no-load overvoltage of approximately $6\frac{1}{2}\%$ is allowable, with the condenser on full buck (lag), a 30,000-kVA/15,000-kVA rating is suitable.

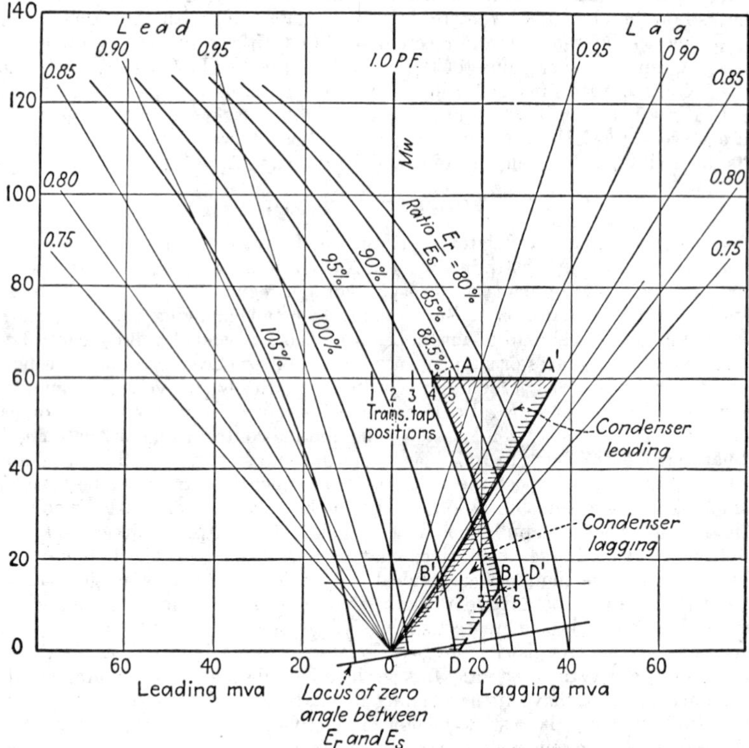

FIG. 13-14. Regulation diagram for E_s = 161 kV constant.

Condenser calculations for correction of high-tension bus voltage at a point where several circuits radiate to loads are much more complicated. The load and power factor served from the bus must be accurately determined. If low-tension load power factors only are available, they must be corrected to the high-tension side. A simple method of correction is to calculate the I^2X loss of the transformer bank and to add it to the reactive component of the load. For two winding transformers the I^2X loss in percent of the transformer bank capacity can be calculated from

$$\% \, I^2X = \left(\frac{E_n}{E_r}\right)^2 \times \left(\frac{\text{kVA load}}{\text{kVA}_{tr}}\right)^2 \times \% \, IX \qquad (13\text{-}54)$$

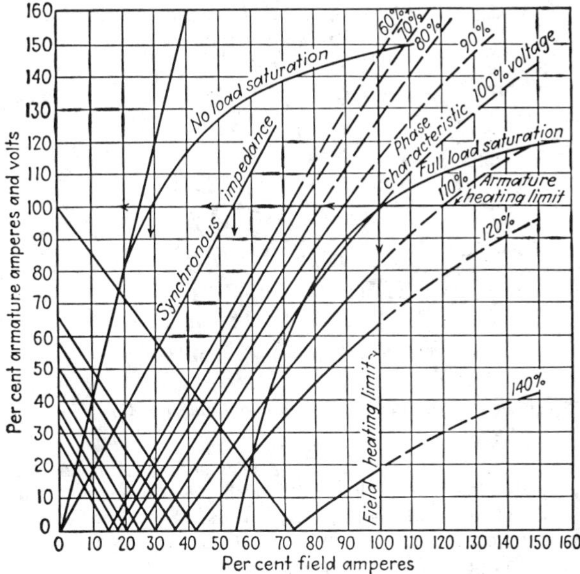

Fig. 13-15. Calculated characteristic curves on typical 3-phase 60-c, air-cooled synchronous condenser. (*General Electric Company.*)

in which E_r = actual output voltage, E_n = nameplate rated output voltage, kVA_{tr} = rating of the bank for which the percent reactive drop is given. Percent IZ as given on the transformer nameplate can be used for percent IX if the percent IR is not high. With an IR of 20% of the percent IZ, the percent IZ, when used for percent IX, will be approximately 2% in error. To obtain the percent I^2X loss in three-winding transformers, the three transformer impedances can be converted into the equivalent Y (Sec. **6**), and calculations can be made by the same methods as used for short-line performance. The I^2X loss in high-voltage transformers can be quite large, especially for transformers having a forced-air rating. For example, a high-voltage transformer having a percent IX of 14%, when operating on a (kVA load)/kVA_{tr} ratio of 1.333:1 at rated voltage, will have an I^2X loss of 25% of the self-cooled rating. The magnetizing current of the bank, which at normal voltage is in the order of $1\frac{1}{2}$ to 2% of the bank rating for high-voltage transformers, can usually be neglected, since errors in estimating of load power factors will usually cover it up. At overvoltages, the magnetizing kilovoltamperes become appreciable, as will be seen from Fig. 13-16.

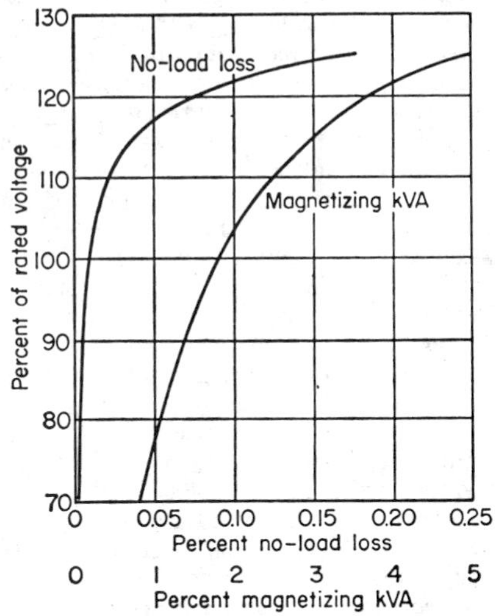

Fig. 13-16. Magnetizing kVA and no-load loss for a 175-MVA 240-kV transformer, in percent rating. (*Allis-Chalmers Company*).

30. Shunt capacitors have been developed until they are now very reliable and in general will provide condenser capacity at a lower cost than for synchronous condensers. Their action, however, is not the same as that of synchronous condensers. Their load

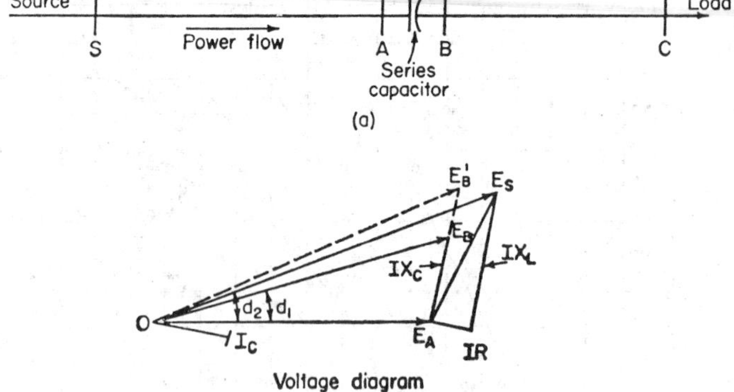

FIG. 13-17. Direction of power flow (*a*) and voltage diagram (*b*) for a series capacitor used to neutralize circuit reactance.

in kVA varies as the square of the voltage, whereas synchronous machines maintain approximately constant kVA for sudden voltage changes and respond quickly to voltage changes when controlled by fast voltage regulators. Static capacitors, when used in place of synchronous condensers for voltage control, must have a total capacity equal to the sum of the leading and lagging ratings of the synchronous machine, or else they must be supplemented by shunt reactors to take the place of the underexcited capacity of the synchronous machine. They also must be switched in blocks small enough so that excessive voltage changes will not result. The size of the voltage steps will depend upon the system, but in general 2½% steps will not be objectionable.

Since circuit breakers must be used for switching the capacitors, the response to voltage changes must be delayed as it is with step-type voltage regulators so that momentary voltage dips will not cause the capacitors to switch unnecessarily. Capacitor switching in large blocks such as 5,000 kVA or more at 13.8 kV when several blocks are used creates a severe duty on circuit breakers. The switching in of a block, when several other blocks are energized already, can cause a heavy momentary current inrush. This inrush current must be determined and switching facilities selected accordingly. The oilless breakers are most suitable for capacitor switching. Special oil circuit breakers with resistor-bridged interruptors also are used.

31. Shunt reactors[1] provide a means of reducing overvoltages during light load periods for stations served from long lines. They perform the same function as underexcited synchronous condensers. They also can be used at generating stations troubled by leading power factor which results in reduced d-c field current and low internal voltage. This in turn reduces the transient and steady-state stability limits of the plant. In these instances, reactors can be installed on either the high-tension or the low-tension side of the generator step-up transformers.

Extra-high-voltage systems may encounter a similar problem due to the line charging kVA, which is a function of the square of the transmission voltage. For a generating station to charge a 200-mi 500-kV open-circuit line on which the charging current will be in the order of 400,000 kVA, the station generation must have a leading reactive capability of 400,000 kVA under positive field excitation. The generator voltage also must not exceed permissible limits. If the generation does not have the required capability, shunt reactors should be installed on the 500-kV bus. In this location, the voltage rise through the transformers will be held down, which in turn reduces the line charging kVA.

32. Series capacitors also may be used to improve voltage conditions by neutralizing part of the circuit reactance.[2] The reduction in reactance also increases the power limit of the circuit and improves stability. They may be used in circuits of any volt-

[1] CLARKE, EDITH Steady State Stability in Transmission Systems; *Trans. AIEE*, 1926, pp. 22–41.

[2] JOHNSON, A. A., BARKLE, J. A., and POVEJSIL, D. J. Fundamental Effects of Series Capacitors in High-voltage Transmission Lines; *Trans. AIEE*, 1951, Vol. 70, Pt. 1, p. 526.

age. Figure 13-17 shows the voltage diagram of a series capacitor. Lagging current through a capacitor causes a voltage rise, whereas a leading current causes a voltage drop. Figure 13-17*b* is simplified by neglecting line charging current. If the capacitor is sized to compensate for all the reactance of line section *SA*, IX_C will be equal to IX_L and E_B will move to E_B'. All but the *IR* drop of the line will be compensated.

Heavy short-circuit currents through the capacitors cause an excessive voltage rise and heating and can damage the capacitors. Therefore, provision must be made for the capacitors to be short-circuited automatically, by special calibrated spark gaps and short-circuiting switches, within a time short enough so that damage cannot result. The short-circuiting switches must be opened very quickly so as not to impair power flow more than momentarily. Detailed study is required to determine their best location in the line and whether or not the capacitors must be distributed to avoid excessive voltage rise at the capacitor output terminals under heavy load conditions.

SYSTEM STABILITY[1]

33. Power Limits of Transmission Lines. There is a definite limit to the power which may be transmitted over any transmission line at any given values of the sending and receiving voltages. This limit is readily determined by means of the power equations (Par. **22**). The real power terms of Eq. (13-51) (Par. **22**) are

$$P_r = \frac{E_r E_s}{B} \cos (\phi' - \delta) - E_r^2 \frac{a_1 b_1 + a_2 b_2}{B^2}$$

It is obvious that this expression becomes a maximum when $\delta = \phi'$ and

$$\cos (\phi' - \delta) = 1$$

and

$$P_r(\max) = \frac{E_r E_s}{B} - E_r^2 \frac{a_1 b_1 + a_2 b_2}{B^2} \tag{13-55}$$

Another expression which is frequently used in determining power flows for circuits with negligible resistance and capacitance may be derived from the real power terms of Eq. (13-49) (Par. **22**),

$$P_r = \frac{E_r E_s}{Z} \cos (\phi - \delta) - \frac{E_r^2 \cos \phi}{Z}$$

With negligible resistance, ϕ becomes 90°, $\cos (\phi - \delta) = \sin \delta$, and $Z = X$. The second part of the expression drops out, since $\cos \phi$ becomes zero. The expression for power flow at any angle δ, then, is

$$P_r = \frac{E_r E_s}{X} \sin \delta \tag{13-56}$$

Equation (13-56) is useful in calculations involving synchronous machine power, short lines of high X/R ratio, and also approximate calculations for longer lines. Maximum power by the use of this equation is obtained when $\delta = 90°$ and

$$P_r(\max) = \frac{E_r E_s}{X} \tag{13-57}$$

In the above expressions, E_s and E_r are used as line-to-line kilovolts, giving results in total 3-phase power in megawatts. If E_s and E_r are in volts, power is in watts.

Regulation diagrams are also very useful in the study of power limits. The diagrams of Par. **26** for E_s constant and Par. **27** when drawn for E_s constant are probably the most useful, since they show the peculiar reactive conditions which exist when angle δ approaches angle ϕ' in magnitude. Figure 13-18 shows such a power-circle diagram for the receiver end only, with the circles extended to show the maximum power for the various values of E_r. The points of maximum power are the points where tangents to the various circles are perpendicular to the real power axis and angle δ = angle ϕ'. Line PQ' shows such a tangent. The distance *OP* represents the maximum power and

[1] See bibliography (Par. **38**).

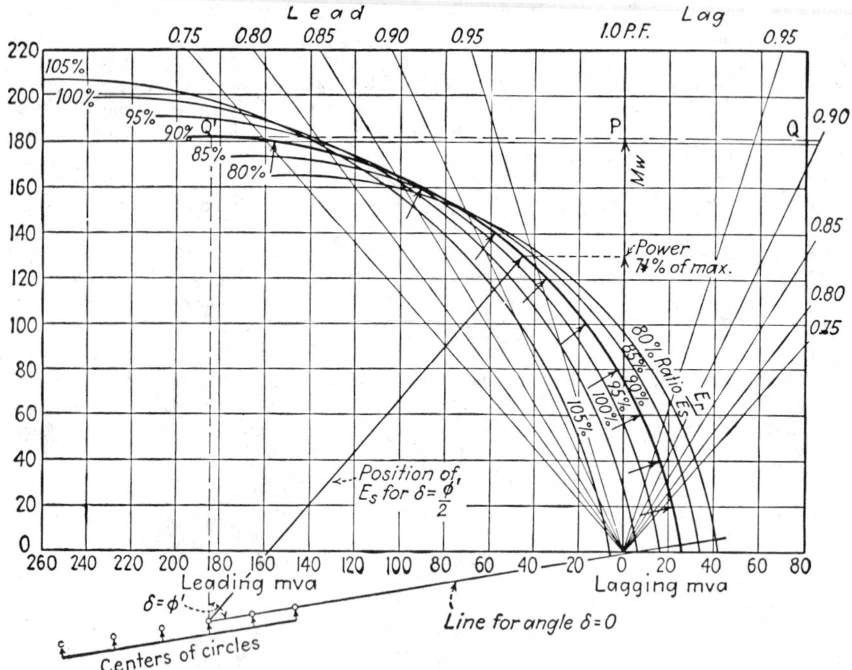

FIG. 13-18. Receiver-end power-circle diagram showing maximum power conditions.

QQ' the condenser capacity required if load power factor is 90%. It will be noted that, as angle δ approaches ϕ', a drop in receiver voltage requires more condenser capacity for the same power delivered, instead of less capacity, as is the case at low values of δ.

The study of power limits plays an important part in system-stability problems.

34. Steady-state stability, as it is generally considered in the study of power-system stability, may be defined as the ability of a system to maintain synchronism among its various generating stations and loads during the desired range of system loading when there is no aperiodic disturbance on the system. This condition is perhaps better called **operating steady-state stability.** The operating steady-state stability limit of any circuit is always less than the **theoretical steady-state (static) stability limit,** which is determined by the power limits (Par. **33**) of circuits involved, with the circuit loads increased by infinitesimal increments until the power limits are reached. In the power-limit equations of Par. **33**, E_s and E_r can be terminal voltages of lines, internal voltages of synchronous equipment, or an internal voltage and a bus voltage. In steady-state stability, the synchronous reactance of generators and synchronous motors must be used. This is very difficult to obtain[1] owing to the fact that it is a variable, dependent upon the external system characteristics and satura-

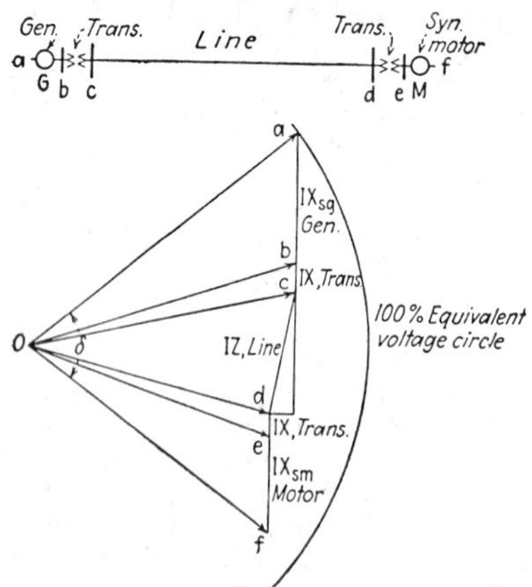

FIG. 13-19. Steady-state stability with fixed internal voltages.

[1] CRARY, S. B., SHILDNECK, L. O., and MARCH, L. A. Equivalent Reactance of Synchronous Machines; *Trans. AIEE*, January, 1934, Vol. 53, p. 124.
KINGSLEY, CHARLES, JR. Saturated Synchronous Reactance; *Elec. Eng.*, March, 1935, p. 300.

tion. In general, synchronous reactance is important only as regards stability, in systems in which the plants may experience low leading power factors.

Figure 13-19 illustrates the principle of steady-state stability. *Oa* and *Of* represent internal voltages of a generator and a synchronous motor, respectively. With these voltages maintained constant (fixed excitations), the steady-state stability limit is reached when angle δ is equal to $\tan^{-1}(X/R)$ of the total circuit (*a* to *f*). X_{sg} and X_{sm} are synchronous reactances. This illustration does not correspond to any usual power-system condition, since no case will be met with in which all load is synchronous motor load unless it should be a remote plant delivering all its load through frequency changers. In actual operation, the bus voltages will be maintained approximately constant by means of synchronous condensers or other equivalent means, the line angle δ of Fig. 13-19 can be exceeded, and greater power can be transmitted.

Transmission lines are never operated at their static-stability limits. The maximum steady-state operating limit will depend upon the magnitude of instantaneous load changes, the voltage control used, and its ability to follow quickly any load changes. The action taking place on a line as load is increased, on the assumption of sufficient condenser capacity to restore voltage, is: (1) An increment of load is applied. (2) Both sending and receiving voltages drop, the amount depending upon the magnitude of the load increment. (3) The angle between E_r and E_s and the internal angle of the generator spread, attempting to take new positions as required by the expression

$$P_r = \frac{E_r E_s}{B} \cos(\phi' - \delta) - E_r^2 \frac{a_1 b_1 + a_2 b_2}{B^2}$$

as it is applied to each section of the circuit for the new lower voltages. The internal voltages alone are constant. (4) The voltage regulators on the condensers and the generator bring the voltage back to normal, and the angles swing back to values between those for the previous load and those of the initial swing as load was added. This action is repeated as additional load is picked up, the swings becoming greater and greater for the same load increments, as will be seen from power-circle diagrams, until finally the angles on the application of load overswing the maximum angle, and instability results. While this action is, strictly speaking, a series of light transient conditions, it is treated as operating steady-state stability, since it is the result of normal load increases and is not connected with disturbances. In the expression above for P_r, the safe or operating steady-state stability limit probably will obtain for values of δ equal to approximately $\frac{1}{2}\phi'$, corresponding to 70 to 75% of the power limit of the circuit, although with good load and frequency control it can be exceeded. Figure 13-18 shows how increments of angle δ increase for equal increments of power received.

The operating steady-state stability limit is in reality the safe overload limit of the system or any part thereof, and it will be found that operation at such loads is accompanied by high losses. Therefore, when system load grows to the point where extra precautions must be taken to maintain steady-state stability, it is time to take measures to reinforce the system.

The series capacitor (Par. **32**) provides a means of increasing the power limit of a highly reactive circuit by neutralizing part of the reactance of the circuit. Too much neutralization will impair the flow of synchronizing power, and this must be guarded against. Under favorable conditions, it may be possible to double the normal circuit capability, i.e., the load that can be carried without excessive losses (see Breuer *et al.* reference, Par. **38**).

35. Transient Stability. A power system may be said to have transient stability if the various generating stations will regain equilibrium following aperiodic system disturbances. In some instances synchronous motor loads are included if they are an appreciable part of the system load. System disturbances which cause the greatest trouble are those due to line faults. A line fault, in addition to the changes in generator loadings which it causes before being cleared, will trip out a radial line, causing a loss of load, or else it will trip one of the interconnecting lines of a network, thereby requiring a readjustment of phase angles all around and at the same time increasing the impedance between plants and impairing the flow of synchronizing power. In general, transient instability results when the flow of synchronizing power between generating stations during and following a fault is insufficient to overcome the speed changes acquired by the generators during and following the disturbance quickly enough to prevent one or more of the stations from falling out of step. Frequently, the flow of synchronizing power, while sufficient to keep plants from falling out of step, is so great that it will cause the relays on interconnecting lines to trip out. There are relay systems in use today designed to prevent such tripping, but it is still one of the major problems of transient stability. System design usually requires that transient stability shall be maintained if the most important interconnecting line of a network shall be tripped out owing to one or more of the four types of fault which may be experienced. These faults in the order of increasing severity are (1) line to ground, (2) line to line, (3) two line to ground, (4) 3-phase. Transient stability for 3-phase faults in the main lines of a transmission system is very difficult to obtain under all conditions, and not all systems are able to afford the expense involved. Even the Hoover Dam lines of the City of Los Angeles[1] are not designed for transient stability during 3-phase faults. A 2-phase to ground fault under the most unfavorable condition was set up as the criterion for stability, and to obtain this required the highest-speed high-voltage oil circuit breakers yet built, special carrier-type pilot relays, intermediate line sectionalizing, and generators of high-inertia and low-reactance characteristics.

Transient-stability studies are essentially a study of the momentary speed changes of rotating equipment, which are functions of their inertias. A fault may be cleared in less than 0.1 s, but during this time an impulse may be imparted to a generator which causes its inertia to carry it out of step. The inertia characteristics of rotating machines may be compared by means of their so-called "inertia constants" H or by means of acceleration factors C. The inertia constant of a machine is defined as

$$H = \frac{E_{\mathrm{kWs}}}{\mathrm{kVA}} \tag{13-58}$$

in which E_{kWs} = stored energy in kilowattseconds at *rated speed* and kVA = kilovolt-ampere rating of the machine.

$$E_{\mathrm{kWs}}{}^{*} = 2.31 \times WR^2 \times (\mathrm{r/min})^2 \times 10^{-7} \tag{13-59}$$

in which WR^2 = moment of inertia of the rotating parts of the machine in pounds times feet squared. The inertia constant H, which is the value as furnished by the manufacturers, it will be noted from the definition [Eq. (13-58)], applies to the **rated speed** of the machine. H is sometimes used in calculations for machine acceleration; and if these calculations are carried into appreciable speed changes, H should be varied directly as the square of the speed $(\mathrm{r/min})^2$.

The acceleration factor C is an acceleration multiplier used in transient-stability calculations.

$$C = \frac{180f}{E_{\mathrm{kWs}}} \tag{13-60}$$

in which f = frequency corresponding to the revolutions per minute used in calculating E_{kWs}. From Eqs. (13-58) and (13-60), it will be seen that the acceleration factor can also be expressed as

$$C = \frac{180f}{\mathrm{kVA} \times H} \tag{13-61}$$

$C \times \Delta P$ gives the rate of acceleration in electrical degrees per second per second, from which it is possible to calculate the angular change in rotor position for a given short-time interval (usually from 0.02 to 0.1 s). ΔP is the average change in power

---

[1] Scattergood, E. F. Engineering Features of Boulder Dam—Los Angeles Line; *Elec. Eng.*, May, 1935, p. 494.

* Wagner, C. F., and Evans, R. D. Static Stability Limits and the Intermediate Condenser Station; *Trans. AIEE*, 1928, Vol. 47, pp. 94–121.

output of the generator during the time interval. It will be noted from Eqs. (13-59) and (13-60) that C is variable with frequency in the numerator and $(\text{r/min})^2$ in the denominator. Therefore, the acceleration factor varies inversely as the speed (revolutions per minute). Curves for C can be plotted and average values for the time interval used; but in most transient-stability studies this is unnecessary, since the speed change while a machine is falling out of step is usually negligible. The acceleration factor for a generating station having several generating units with different speeds can be found by adding the E_{kW_s} of all machines and applying Eq. (13-60). Or the acceleration factor for each machine can be calculated, and the station-acceleration factor C_T can be determined from

$$\frac{1}{C_T} = \frac{1}{C_1} + \frac{1}{C_2} + \frac{1}{C_3} + \cdots + \frac{1}{C_n} \tag{13-62}$$

36. The major factors affecting transient stability may be listed as follows: (1) **Generator** $WR^2 \times (\text{r/min})^2$. The greater this quantity, the lower the acceleration factor. (2) **System impedance,** which must include the transient reactances of all generating units. This affects phase angles and the flow of synchronizing power. (3) **Duration of the fault** chosen as the criterion for stability. Duration will be dependent upon the circuit-breaker speeds and the relay schemes used. (4) **Generator loadings** prior to the fault which will determine the internal voltages and ΔP (the change in output). (5) **System loading,** which will determine the phase angles among the various internal voltages of the generators.

Transient stability as treated in system design involves mainly an investigation of the foregoing five items in order to determine where best results may be obtained, and the final result is a balancing of the effects obtainable from the various items to obtain the minimum overall cost. Item 1 can be considered only in new plants or units. Item 2 is very important and is best studied by means of a system-impedance diagram in which all quantities are reduced to a common voltage or kVA base. The proportions of transient reactances of generators as parts of the total impedances of the paths of the flow of synchronizing power will be seen, and the effects of possible changes can be gaged. The effects of circuit changes also can be studied. In item 3, circuit breakers and relays have been the principal factors to date in bringing system stability to its present stage of perfection. High-voltage a-c circuit breakers are available with interrupting speeds of 2, 3, 5, and 8 c (60-c basis), and it is possible that still faster speeds will be developed. This contrasts with speeds of 15 to 20 c in the older types of breaker. Relay speeds as fast as ½ c (60-c basis) have been developed for some applications, and 2- to 3-c operation is common. Fast clearing of faults is considered to be the most important factor in maintaining transient stability. Item 4, generator loadings, ordinarily is fixed by the generation schedules, and therefore the most advantageous loadings from a stability standpoint cannot be specified by the design engineer. It may be possible on systems subject to lightning storms to shift the generator loadings in order to favor the weaker stations, and the design engineer should call attention to such possible load shifts in order that they may be taken advantage of if operating conditions permit. Leading power-factor loads on generators should be avoided because they reduce the internal voltages, increase the initial angle within the generators, and reduce transient-stability limits. Item 5, system loads, ordinarily cannot be changed. Together with the generator loadings, they fix the phase angles over the system and determine the margins in angles available for swings. Practically all the other factors are concerned with limiting the swings to low values in order that these margins may be as small as possible.

The installation of a new plant on a large system or the addition of a large generator to an existing plant requires a transient-stability study on a digital computer. Alternating-current network analyzers are not suitable for such studies when large network systems are involved. Transient-stability studies show unacceptable circuit arrangements, and studies can be made rapidly enough so that many conditions can be covered. Transient-stability studies can be used to show apparent impedances seen by relays on main trunk lines and interconnections during system disturbances and thereby to determine the necessity for special relaying schemes.

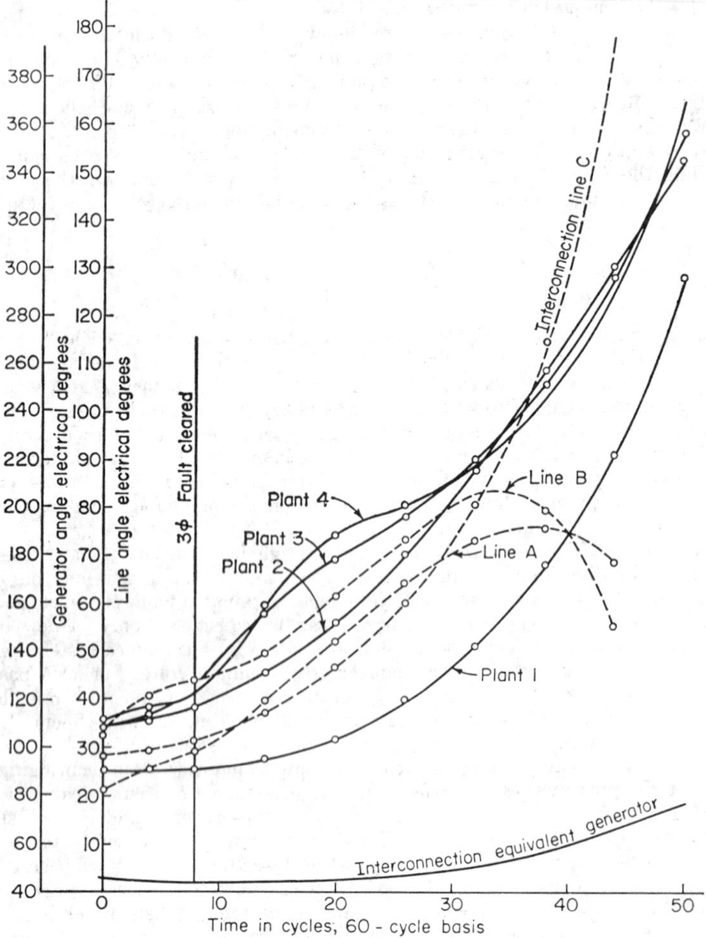

FIG. 13-20*a*. Transient-stability swing curves for a 3-phase fault. (*Tennessee Valley Authority.*)

Figure 13-20*a* illustrates, in a set of swing curves, a condition which caused an interconnection between two large systems to trip out. The curves for plants 1 to 4 show all to be speeding up and staying in synchronism but leaving the interconnection generation behind. The fault-clearing time is 8 c. With the new EHV circuit breakers, this time probably would be not more than 4 c (60-c basis), resulting in a more stable system. As the swing curves are run, voltage, watts, and vars at each end of heavily loaded lines can be read for each time point of the swing curves and the impedance seen by the relays at each end of the line calculated and plotted on a vector-impedance diagram. Figure 13-20*b* shows such a diagram for line *B* of Fig. 13-20*a*. The relays at both ends of the line are assumed to have the same impedance circles, and the receiving-end apparent-impedance curve is as it appears to the receiving-end relays.

It is seen that the apparent impedance at the sending end enters the second-zone tripping reach and stays for 6 c, which is close to tripping time. The apparent-impedance curves may perform as shown, or they may approach the relay tripping zones and retrace their paths, or they may turn back on small close hooks. If they go right on through the relay zones, an out-of-step condition is shown. This is what happened with the apparent-impedance plotting for line *C* of Fig. 13-20*a*. As the swing curves

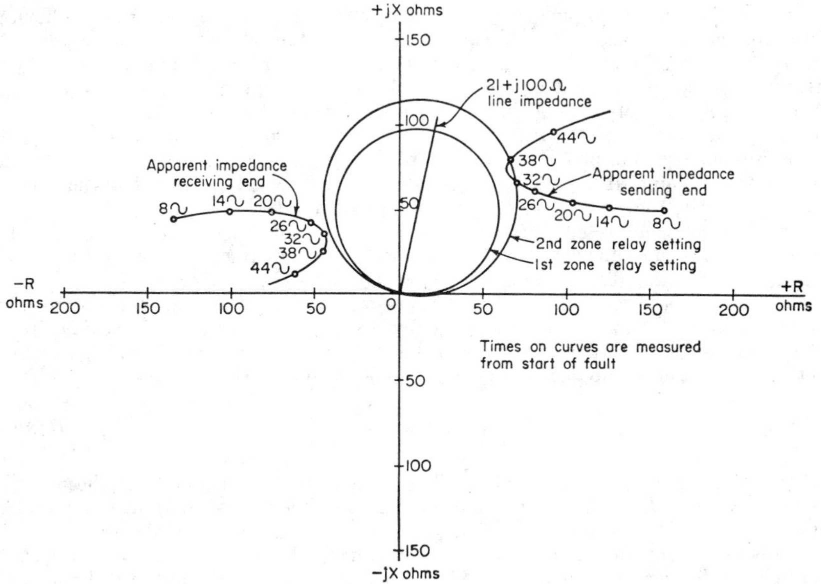

FIG. 13-20*b*. Transient-stability apparent-impedance curves for line *B* of Fig. 13-20*a*. (*Tennessee Valley Authority*.)

on Fig. 13-20*a* were being run, the angles across lines *A*, *B*, and *C* were recorded and plotted. The curve for line *B* shows the angles for the various points of the curves of Fig. 13-20*b*. Lines *A* and *B* are 78° impedance-angle lines. This angle was exceeded, but, with the help of other lines, the system remained intact after the interconnection was lost.

It should be remembered that, at an angle of 60° between equal sending and receiving voltages, the relays see an impedance drop equal to that for a 3-phase short circuit at the end of the line. Whether or not they trip will depend on the type of relay.

37. Transmission-system Relaying. The relaying of the lower-voltage transmission lines is relatively simple and is covered briefly in Sec. **10**. The relaying of large high-voltage network systems is more complex, and there are many points to be considered. Carrier-current pilot wire is generally used. The two most popular types are:

1. Directional comparison.
2. Phase comparison.

There may be times when a combination of the two types is required.

The **directional-comparison** scheme consists in general of phase and ground directional fault detectors which are designed for three functions as follows:

a. First line of defense. This provides for carrier-controlled simultaneous tripping of the line circuit breakers at the terminals of the line on internal faults and the blocking of tripping for normal through faults. Simultaneous fast reclosing of the line breakers is monitored by the carrier auxiliary relays.

b. Carrier off. If the carrier equipment becomes inoperative, the fault detectors, which usually consist of distance and ground relays, provide protection. This means sequential tripping for faults near the ends of the lines and the blocking of instantaneous reclosing.

c. Backup relaying. This function acts to trip a circuit breaker, the relays of which are directional toward a line or bus which has not been cleared owing to failure of a circuit breaker or failure of relays. This function is always ready, whether the carrier is on or off, and is the most difficult of the functions to apply. It will be discussed later.

The **phase-comparison** scheme of relaying compares the currents and phase angles at each end of the line by means of composite voltages created by the currents flowing. The voltages are transmitted by carrier current to the opposite ends of the line for comparison, and if they are not in agreement, the line is tripped. This type of relaying requires the use of distance and ground or overcurrent and ground relays for backup purposes and for protection of the line when the carrier is inoperative.

The combination of the two schemes consists of directional-comparison relays for phase-to-phase and 3-phase faults and phase-comparison relays for ground faults. This scheme is applied when a closely paralleling circuit can induce currents sufficiently large to operate the conventional ground relays incorrectly.

Backup relaying presents probably the greatest problem in the relaying field. The distance relays normally used for transmission-system relaying contain three zones of distance elements. These elements may all be directional toward the line, or two elements may be directional toward the line with the third element directional away from the line. The third zones are usually assigned the function of backing up the relays on the other lines connected into the station. The determination of third zone settings is that of the apparent impedance, which is found from

$$Z_r = \frac{Z_l I_l}{I_r} \tag{13-63}$$

where Z_r is the relay setting in ohms, I_r is the vector current flowing in the relay, Z_l is the impedance $R + jX$ of the faulted line, and I_l is the vector current in the faulted line.

It is seen from Eq. (13-63) that the value of Z_r, the apparent impedance, may be quite large if several circuits are feeding current into the faulted line with the inoperative breaker and may be a value representing many times the impedance of the lines on which the backup relays are installed. It is these sensitive backup settings which limit the ability of the circuits to carry emergency loads and to withstand surges; see Par. **38**. These settings may be reduced in some cases by allowing sequential tripping, in which case the line making the heaviest contribution to the fault is made to trip first, which then causes the relay current in the remaining relays to increase, thereby permitting the use of less sensitive settings. A more satisfactory but more expensive scheme is to make the third-zone elements separate elements and to connect them and a ground relay through a timer. These elements are directional toward the line and connected to different current transformers from those to which the regular relays are connected. If this breaker fails to trip, the timed relays trip the bus-tripping relay, which clears all other circuit breakers connected to the bus.

38. References to Literature on System Stability

SHAND, E. B. The Limitations of Output of a Power System Involving Long Transmission Lines; *Trans. AIEE*, 1924, p. 59.

DOHERTY, R. E., and DEWEY, H. H. Fundamental Considerations of Power Limits of Transmission Systems; *Trans. AIEE*, 1925, p. 972.

FORTESCUE, C. L. Transmission Stability; *Trans. AIEE*, 1925, p. 984.

"Westinghouse Electrical Transmission and Distribution Reference Book"; E. Pittsburgh, Pa., Westinghouse Electric Corporation, 1950.

CRARY, S. B. "Power System Stability"; New York, John Wiley & Sons, Inc., Vol. I, 1945, and Vol. II, 1947.

BUTLER, J. W., SCHROEDER, T. W., and RIDGEWAY, W. Capacitors, Condensers, and System Stability; *Trans. AIEE*, 1944, Vol. 63, p. 1130.

KIMBARK, E. W. "Power System Stability"; New York, John Wiley & Sons, Inc., Vol. 1, 1948, Vol. 2, 1950, and Vol. 3, 1956.

BREUER, G. D., RUSTEBAKKE, H. M., GIBLEY, R. A., and SIMMONS, H. O., JR. The Use of Series Capacitors to Obtain Maximum EHV Capability; *Trans. IEEE Power Group*, November, 1964, p. 1090.

CORONA[1]

39. Electric corona occurs when the potential of a conductor in air is raised to such a value that the dielectric strength of the surrounding air is exceeded. Corona mani-

[1] See References (Par. **41**).

fests itself by bluish tufts or streamers appearing around the conductor, being more or less concentrated at irregularities on the conductor surface. This discharge is accompanied by a hissing sound and by the odor of ozone. In the presence of moisture, nitrous acid is produced. Corona is due to ionization of the air. The ions are repelled from and attracted to the conductor at high velocity, producing other ions by collision. The ionized air is a conductor (however, of high resistance) and increases the effective diameter of the metallic conductor.

Corona on transmission lines causes power loss and radio and television interference.[1] Corona loss is probably the most difficult item which the transmission engineer is called upon to determine. It is affected by many things: conductor diameter, number of conductors per phase, phase spacing, conductor surface condition, weather, altitude, temperature, and voltage. A long, heavily loaded line may have corona at the sending end for approximately half its length because of voltage differences. Another, crossing a mountain ridge, will have corona only at the higher elevations. Weather may not be uniform along the line. Precise loss determination, therefore, is impossible.

The present-day knowledge of corona is still incomplete and is the result of study and experiments by many investigators. One of the earliest investigations was made about 1910 by the General Electric Company under the direction of F. W. Peek, Jr. Peek's studies became the accepted standard for selecting conductor sizes to avoid corona troubles. Previously, Prof. H. J. Ryan had proposed a formula for safe corona voltage which he called "critical pressure at which the sudden increase in the brush discharge takes place." He recommended that lines be operated below this voltage. The voltages which he recommended were of the same general order as those found from Peek's studies. Peek's studies were made before bundle conductors were considered in high-voltage transmission and before the days of radio. His studies dictated that lines should be operated below the fair-weather critical disruptive voltage which is the corona-loss starting voltage. Lines so built have been remarkably free of radio interference. These lines have been the single-conductor lines up to 220 kV built previous to the construction of the Boulder Dam–Los Angeles 275-kV lines in the early 1930's. This being the case, experience apparently dictates that Peek's operating-voltage recommendations should continue to be used on new single-conductor lines of the voltages under EHV. Peek's conclusions[2] follow:

The **disruptive gradient** of the air g_o is constant for conductors of all materials and all frequencies at 25°C and 76 cm barometer and is 21.1 kV/cm effective or 29.8 kV/cm crest. $g_o = e_o/[r \ln (S/r)]$, wherein e_o = disruptive **critical voltage** in kilovolts to neutral and is the voltage at which corona loss starts; r = conductor radius in centimeters. S is the actual distance between conductors in **centimeters.** e_o in effective kilovolts to neutral, **fair weather,** is found from

$$e_o = 21.1 m_o r \delta \ln (S/r) \tag{13-64}$$

δ = air-density factor, equal to $3.92b/(273 + t)$ in which b = barometric pressure in centimeters and $273 + t$ = absolute ambient temperature in degrees centigrade. ln = 2.3026 log. m_o = surface condition or **irregularity factor** and is unity for smooth, polished, round conductors, solid or tubular. m_o = 0.93 to 0.98 for roughened and weathered solid conductors and 0.80 to 0.87 for stranded cables up to 1 in in diameter. The value of 0.8 corresponds approximately to new cable and increases with age to

Table 13-10. Barometric Correction Factors $b_h/29.92$

Altitude, ft	$b_h/29.92$	Altitude, ft	$b_h/29.92$	Altitude, ft	$b_h/29.92$	Altitude, ft	$b_h/29.92$
0	1.00	2,000	0.92	5,000	0.82	9,000	0.71
500	0.98	2,500	0.91	6,000	0.79	10,000	0.68
1,000	0.96	3,000	0.89	7,000	0.77	12,000	0.63
1,500	0.94	4,000	0.86	8,000	0.74	14,000	0.58

[1] Transmission System Radio Influence; *IEEE Comm. Rept., Trans. IEEE Power Group,* August, 1965, p. 714.

[2] PEEK, F. W., JR. "Dielectric Phenomena in High Voltage Engineering"; New York, McGraw-Hill Book Company, 1929.

approximately the higher value. It is approximately the same for the standard strandings but with rope lay cables may be as low as 0.7. The values of e_o when used for 3-phase circuits apply to equilateral triangular spacing. For lines with flat configuration, whether horizontal or vertical, Peek states that e_o should be decreased 4% for the center conductor and increased 6% for the two outer conductors, the spacing used being $S_{1-2} = S_{2-3} = S_{1-3}/2$. e_o for **wet weather** is approximately **80%** of the fair-weather calculated values. Calculated values of $\sqrt{3}e_o$ (kilovolts line to line) at 25°C and 76 cm barometer are given in Table 13-11. Altitude corrections are given in Table 13-10. These and other tables based on the same formula have been used for many years as guides in avoiding excessive corona on transmission voltages up to 230 kV.

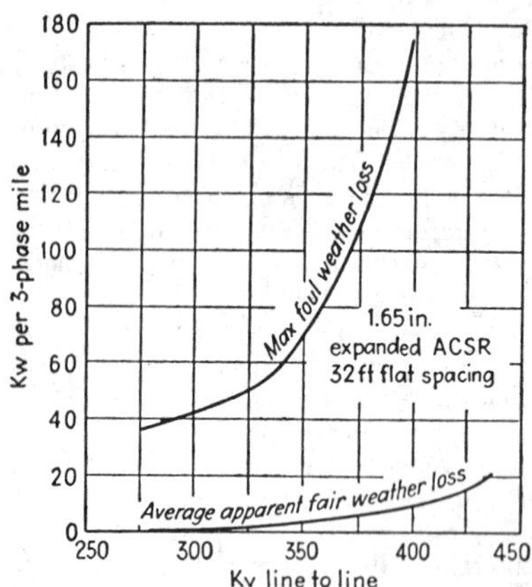

Fig. 13-21. Comparison of fair- and foul-weather corona loss. (*Gross, Wagner, Naef, and Tremaine, Trans. AIEE, 1951, Vol. 70, p. 75.*)

Peek determined empirical fair-weather loss formulas for a 2-mi test line. However, no foul-weather formulas were determined, and as fair-weather losses have little significance except at high altitudes, the loss formulas are omitted here.

Peek found that corona loss is proportional to the frequency. This holds true for the range of frequencies used in the tests (47 to 120 c/s). The law departs from the linear relation at low frequencies. At zero frequency, i.e., direct current, the loss is from one-fourth to one-half the 60-c loss for the maximum voltage. Humidity has no effect on the critical voltage or on the loss; smoke lowers the critical voltage and increases the loss; heavy winds have no effect on the critical voltage or on the loss; fog, sleet, rainstorms, and snowstorms all lower the critical voltage and increase the loss. Wet weather causes a very marked increase in the loss. This is shown by Fig. 13-21, which is plotted from data of a much later study.

To find the voltage limit at any other barometer reading b_h, in inches, with temperature remaining constant, multiply the voltage values by $b_h/29.92$. Table 13-10 gives approximate values of $b_h/29.92$ for various altitudes.

For temperatures other than 25°C, the voltage value, as modified by the barometric correction, must be corrected for the new temperature t_1, in degrees centigrade, by multiplying by the temperature-correction factor $298/(273 + t_1)$, where 298 = absolute temperature at 25°C, and $273 + t_1$ = new absolute temperature.

For 3-phase configurations with all conductors in the same plane, use 96% of the corrected values above for the center conductor and 106% for the two outer conductors. For **wet-weather** values, use 80% of the fair-weather values.

40. Later Corona Research. Peek's findings were accepted without question until the Boulder Dam–Los Angeles line came up for study in the early 1930's. It was then decided to conduct new corona investigations to check Peek's formulas. These tests were conducted at Stanford University, and it was determined that the formulas were not of sufficient accuracy for use on the large conductors under consideration. From the test results, W. S. Peterson[1] developed empirical formulas for fair-weather critical starting voltage and also for fair-weather corona losses. No foul-weather loss formulas were developed, and bundle conductors were not investigated.

Since then every step-up in transmission voltage has required extensive corona research, which has been extended to include radio interference (*RI*). However, no new corona loss formulas have been established to date. When the American Gas and Electric Company[2] decided to superimpose a higher voltage network upon their 138-kV

[1] Peterson, W. S. Discussion; *Trans. AIEE*, 1933, Vol. 52, p. 62.

[2] Sporn, Philip, Gross, I. W., Peterson, E. L., and St. Clair, H. P. The 300/315 KV Extra-high-voltage Transmission System of the American Gas and Electric Company; *Trans. AIEE*, 1951, Vol. 70, Pt. 1, p. 64.

Table 13-11. Fair-weather Corona Limits of Voltage, in Kilovolts, between Conductors (3-phase) at Average Sea Level, 76 Cm (29.92 In.) Barometer and 25°C (77°F) Temperature. Equilateral Spacing

A.W.G. and cir mils	No. of wires	O.D., in.	Spacing, ft												
			3	4	5	6	8	10	12	14	16	20	24	28	32
Stranded Conductors (From formula $e_0 = \sqrt{3}\, g_0 m_0 r \delta \log_e \frac{S}{r}$, where $m_0 = 0.87$)															
4	7	0.232	..	56	59	60	63	65	67	68	69	72			
3	7	0.260	..	62	64	66	69	72	74	75	77	79			
2	7	0.292	..	..	71	73	77	79	81	83	85	87			
1	7	0.328	..	..	78	81	84	87	90	92	94	97			
0	19	0.373	..	..	..	90	94	97	100	102	104	108			
00	19	0.419	..	..	..	99	104	107	111	113	115	119			
000	19	0.470	..	..	..	..	114	118	122	125	127	132			
0,000	19	0.528	..	..	..	..	126	130	134	138	141	145	149		
250,000	19	0.574	..	..	..	..	135	140	144	148	151	156	160	164	
300,000	19	0.629	..	..	..	..	...	151	156	160	163	169	173	177	
350,000	19	0.679	..	..	..	..	...	161	166	170	174	180	185	189	
400,000	19	0.726	..	..	..	..	...	170	175	180	184	190	196	200	204
450,000	19	0.770	..	..	..	..	...	179	184	189	193	200	206	211	215
500,000	37	0.813	..	..	..	..	...	187	193	198	202	209	215	221	225
800,000	37	1.029	..	..	..	..	...	...	234	241	246	255	263	269	275
1,000,000	61	1.152	..	..	..	..	...	...	257	264	270	281	289	296	302
Solid wires, $m_0 = 0.93$															
4	..	0.204	51	54	56	58	60	62	64	65	66	68			
3	..	0.229	..	59	62	64	66	68	70	72	74	76			
2	..	0.258	..	..	69	70	74	76	78	80	82	84			
1	..	0.289	..	..	75	77	81	83	86	88	90	92			
0	..	0.325	..	..	..	85	89	92	95	97	99	102			
00	..	0.365	..	..	..	94	98	102	105	107	110	113			
000	..	0.410	..	..	..	..	109	113	116	119	121	124			
0000	..	0.460	..	..	..	..	120	125	128	131	134	138			

network system, they set up the 500-kV Tidd test line to conduct research leading to the selection of the most suitable voltage. This research continued for a period in excess of 3 years, and continuous corona observations were made over this period. Also, for the first time, extensive investigations were made of radio interference and radio-interference voltages (*RIV*). From the overall test results, annual corona kilowatthour losses were estimated which, together with kilowatt demands at times of heaviest foul-weather losses, were used to determine the most economical voltage and conductor. Other considerations entered into the selection, such as ice melting.

Final result was the selection of 315 kV (now rated 345 kV) and an expanded ACSR conductor of 1,269,300 cmils and 1.6 in in diameter. Since the Tidd line research, all systems from 460 to 700 kV have had the benefit of corona research before construction. In this research, *RI* and *RIV* have had an increasingly important part, and in some cases this has been the factor determining the conductor characteristics.

Radio-influence voltage is a radio-frequency emanation set up by the transmission line which is of appreciable magnitude at voltages below which corona becomes measurable. It is greatly increased by heavy corona. It has been found that *RI* is more readily minimized by the use of bundle conductors on voltages 500 kV and above. The Bonneville Power Administration has chosen a single 2.5-in-diameter expanded ACSR conductor for its 500-kV system.[1]

Many other research projects have been carried out since the Tidd project and have been reported in the *Transactions of the AIEE* and *IEEE*. Project engineers of new EHV systems may find information to fit their conditions in these published results.

41. References to Literature on Corona

Cozzens, Bradley, and Peterson, Wm. S. Symposium on Operation of the

[1] Osipovich, A. A., and Poland, M. G. First 500 kV Transmission Line Designs for the Bonneville Power Administration's Grid; *Trans. IEEE Power Group*, January, 1964, p. 28.

Boulder Dam Transmission Line—Corona Experience on Transmission Line; *Trans. AIEE,* 1939, Vol. 58, p. 137.

LIPPERT, G. D., PAKALA, W. E., BARTLETT, S. C., and FAHRNKOPF, C. D. Radio Influence Tests in Field and Laboratory—500 KV Test Project of the American Gas and Electric Company; *Trans. AIEE,* 1951, Vol. 70, Pt. I, p. 251.

REICHMAN, J. Radio Interference Studies on Extra-high-voltage Lines; *Trans. AIEE,* June, 1961, p. 261.

NIGOL, O., and CASSAN, J. G. Corona Loss Research at Ontario Hydro Coldwater Project; *Trans. AIEE,* June, 1961, p. 304.

HARMON, R. W. Effect of Bundle-conductor Field Influence on EHV Transmission Line Design; *Trans. AIEE,* June, 1961, p. 316.

IEEE Committee Rept., Transmission System Radio Influence, August, 1965, p. 714.

LINE INSULATION

42. Requirements. The operating performance of a transmission line depends largely on the insulation. Good practice requires a dry flashover of the assembled insulator of three to five times the nominal operating voltage and a leakage path approximately twice the shortest air-gap distance. Modern practice, especially for the higher voltages, tends toward the higher limit. Special cases of salt fog, dust, or chemical-laden air will require separate consideration.

The insulator not only must have sufficient mechanical strength to support the greatest loads of ice and wind that may be reasonably expected, with an ample margin, but must be so designed as to withstand severe mechanical abuse, lightning, and power arcs without dropping the conductor. It must prevent an arc-over for practically any operating condition except lightning, under any conditions of humidity, temperature, rain, or snow, and with such accumulations of dirt as are not periodically washed off by rains.[1]

43. Insulator Materials. Porcelain is a ceramic product obtained by the high-temperature vitrification of clay, finely ground feldspar, and silica. High-grade electrical porcelain of the proper chemical composition free from laminations, holes, and cooling stresses is the recognized dielectric for insulating high-voltage lines. The perfection of methods of factory control and testing which has been achieved since 1930 has brought the product to a standard of uniformity that was impossible in the early days of the art. The trouble with porcelain insulators of early manufacture was due largely to lack of standardized factory methods. This difficulty has been overcome.

Good electrical porcelain has a coefficient of linear expansion of approximately $0.03 \times 10^{-4}/°C$. It has a tensile strength of 7,000 to 9,000 lb/in^2 and a compressive strength of 40,000 to 60,000 lb/in^2. A smooth surface is obtained by coating the insulators with a very fine film of colored glass known to the trade as "glaze." It is very important that this glaze should fit the body, i.e., have the same coefficient of expansion as the porcelain; otherwise, after a period of years fine surface cracks may appear which under some conditions will progress through the porcelain, causing failure of the dielectric. The crystalline structure of the porcelain has a decided effect upon its characteristics. Also see Sec. **4**.

A very tight, finely ground, or high-fired body, while good electrically, is likely to be brittle and to show weakness under thermal changes. A coarse, low-fired body may be good mechanically but poor from a puncture-strength standpoint. Thus the porcelain best suited for electrical insulation must be a compromise between puncture strength, mechanical strength, and ability to withstand thermal shocks.

Glass insulators are available in pin type (annealed glass) for voltages up to and including 63 kV. Toughened- (tempered) glass suspension insulators are available for all transmission voltages. Toughened-glass insulators are the equal of porcelain in all respects except that they are somewhat more susceptible to breakage from small-arms fire. They have one important advantage in that they do not require periodic testing for electrical deterioration (see Par. **188**). A toughened-glass insulator is a good insulator if inspection shows it to be unbroken.

---

[1] KAMINSKI, J., JR. Long Time Mechanical and Electrical Strength in Suspended Insulators; *Trans. AIEE,* August, 1963, p. 446.

44. Insulator Design. The design of insulators must be carefully considered by the manufacturer. It is to be observed that a multipart unit is in reality a number of condensers in series, since the cement is conducting. In order that the voltage across each shell shall be a proportionate share of the total voltage, the capacitances of the individual shells are adjusted by varying the area of the cement joints. Uniform voltage distribution prevents puncturing and cascading in case of flashover. In accordance with a fundamental law of electrostatics, the porcelain surfaces of an ideal insulator should be either parallel or normal to the lines of electrostatic force for maximum flashover. Porcelain has a permittivity of 4 to 5; and if the insulator shape does not conform to the above law, portions of the air will be overstressed, resulting in corona.

This ideal form is, however, dependent upon the method of attaching to the structure, and this must be modified to develop the necessary length of leakage path, the best wet flashover characters, and rugged mechanical connections. The porcelain surfaces must be readily washed by the action of rain and must be easily accessible for manual cleaning if necessary. The fundamental electrical requirement of insulator design is that the insulator shall flash over without the formation of a conducting path over the surface and with an ample margin of safety against puncture. This requirement must be obtained, insofar as possible, against any form of lightning-voltage wave, surge frequency, or climatic conditions and should be maintained even after the destruction of the more fragile parts of the insulator.

In damp, dirty atmospheres, especially when long, dry, dusty periods are commonly succeeded by heavy fogs, it is necessary to provide insulators with a leakage surface out of proportion to the flashover value. This is obtained in the design shown in Fig. 13-22 without the expense of higher flashover characteristics and corresponding increase in puncture strength. Post insulators are used for voltages up to 230 kV.

It is now conceded that the cause of nearly all line flashovers can be traced to abnormal conditions on the line, due generally to lightning discharges. This phenomenon is studied in the laboratory by subjecting insulators to voltage impulses by means of a "lightning generator." It has been found that the voltage at which an insulator string or gap will flash over is considerably higher for lightning surges than for 60-c voltages.

The effective 60-c voltage and the minimum crest voltage of two forms of lightning waves required to flash over insulator strings of various numbers of units are shown in Fig. 13-23. In the same diagram, similar values are given for various spacings of the standard rod gap. The scale of rod-gap spacing is made to correspond

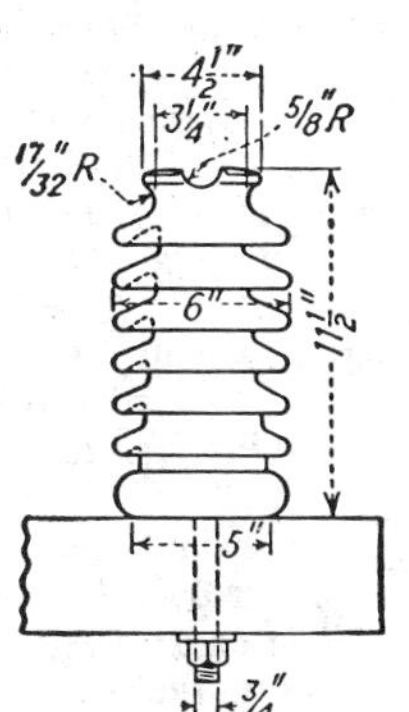

FIG. 13-22. Post-type insulator, 35 kV.

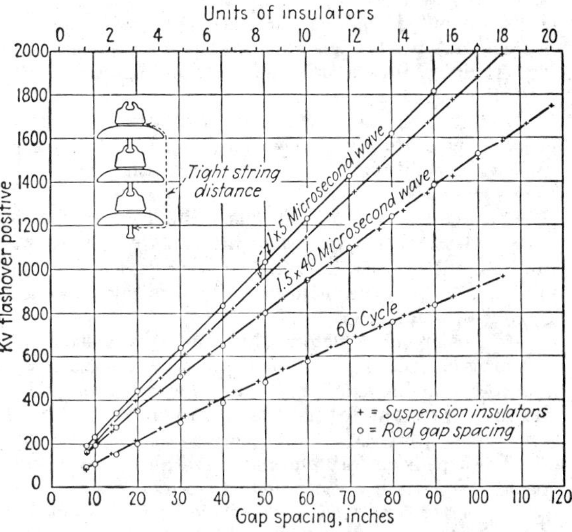

FIG. 13-23. Sixty-cycle and crest-voltage flashover values for insulator strings and rod gaps.

to the tight-string distance from the cap of the top unit to the stud of the bottom, thus giving a direct comparison between the flashover value of the rod gap and of an equal distance over the insulator. These values are plotted from tabulations made by the EEI and NEMA Subcommittee[1] as the averages of a very large number of carefully controlled tests.

45. Pin-type insulators (Figs. 13-24 and 13-25) are designed for 60-c dry flashovers up to about 200 kV, although the use of pin insulators over 69 kV is rare. The latter is equivalent to somewhat less than three 5¾-in suspension units in flashover value. The relatively low margin of insulation and the hazard of carrying so high a voltage on a single comparatively fragile unit make the pin-type unit unsuited for voltages above 69 kV.

The hazard of mechanical shock or heavy power arcs causing a deep fracture, with possible puncture of the unit, is reduced by forming pin insulators, for 20 kV and above, in two or more separate shells, cemented together with neat portland cement. Thus, a crack in any one shell is prevented, by the cement joint, from passing through the entire insulator and grounding the line. In the two-piece unit shown in Fig. 13-24, *ab* is the leakage path and *ABC* the wet flashover path.

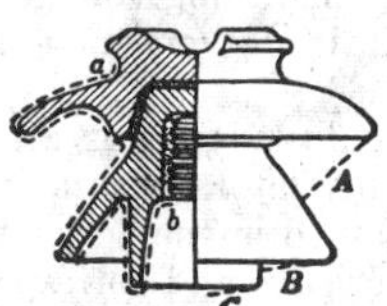

FIG. 13-24. Leakage and wet flashover paths of pin-type insulators.

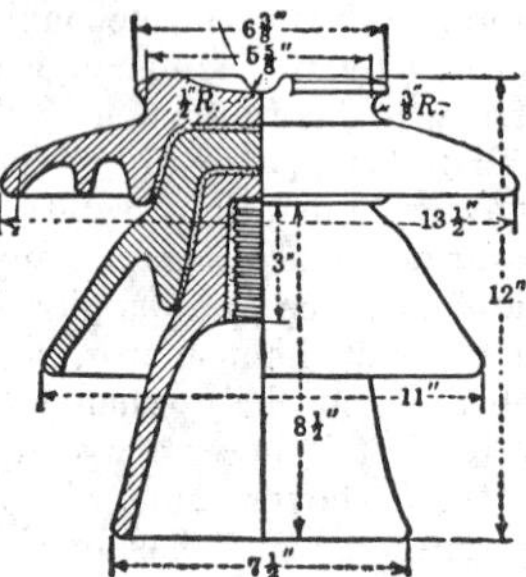

FIG. 13-25. Typical pin-type insulators for 69 kv.

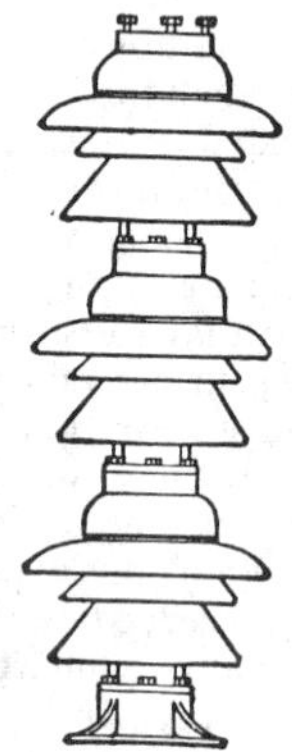
FIG. 13-26. Insulator stack for 115 kv.

Insulators are furnished either with a thread formed in the porcelain or with a zinc or galvanized-iron thimble cemented into the pinhole. The metal thimble is generally used with metal pins, and the porcelain thread with wood pins. Modern construction favors the metal pin.

For switch and bus supports, insulators are furnished with metal caps and pins cemented to the porcelain and fitted with holes to which the proper bus connections can be made. These units are so designed that a sufficient number may be assembled in stacks to give the required voltage rating. A stack similar to that shown in Fig. 13-26 is used for 115 kV.

46. Post-type insulators (Fig. 13-22) are used in place of pin-type insulators on crossarms and as side-post insulators with special saddle-type bases in vertical configurations on the sides of poles. They are more substantial than pin-type insulators and more resistant to rifle fire and to thrown stones. They also are practically free from radio interference. They are available in voltages up to and including 230 kV.

47. Suspension insulators are used almost exclusively on lines over 69 kV, on long spans, and with heavy conductors. Modern cap-and-pin units have had a very satisfactory performance record and have been developed to meet the requirements of the highest voltages and heaviest construction with simplicity and economy.

The flashover of the suspension insulator string is very closely proportioned to the air distance to ground and approximately equal to the rod-to-rod flashover for the

[1] Flashover Characteristics of Rod Gaps and Insulators, Joint EEI and NEMA Subcommittee; *Elec. Eng.*, June, 1937.

same distance for 60-c and operating surges (see Fig. 13-23). Individual test results vary widely from the averages shown in Fig. 13-23, owing to humidity and air pressure and also to undefined atmospheric conditions. In average practice the

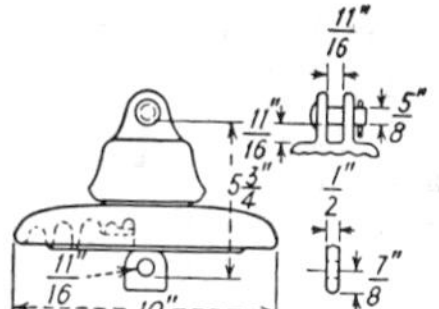

FIG. 13-27. Clevis-type insulator.

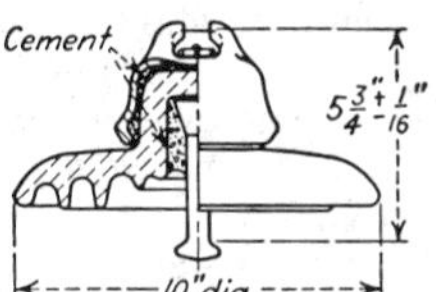

FIG. 13-28. Ball-and-socket type. (*General Electric Company.*)

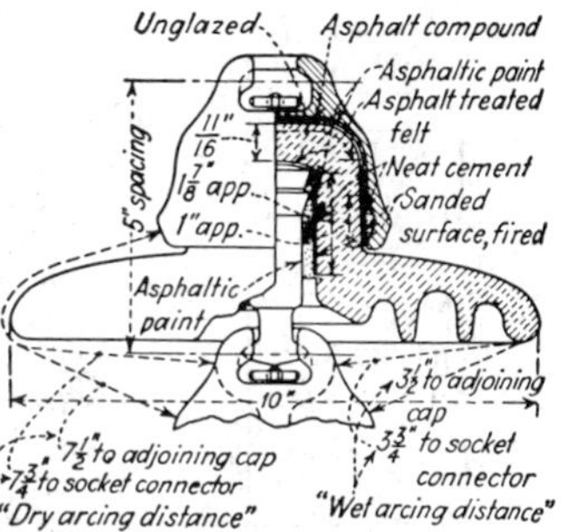

FIG. 13-29. R. Thomas & Sons 5- × 10-in. "chip-proof" insulator. Leakage distance 11¾ in.

number of units used in a string is approximately proportional to the line voltage with a slight increase for the highest voltages and with some allowance for the length of the unit. Four to five units have generally been used at 69 kV, 7 to 8 at 115 kV, 8 to 10 at 138 kV, 9 to 11 at 161 kV, 14 to 20 at 230 kV. Twenty-four 5-in units are used on the Hoover Dam 287-kV line (see Table 13-2 for additional line data).

Figures 13-27 and 13-28 illustrate two standard cap-and-pin units of the 15,000-lb class. The choice is a matter of individual preference, although the ball-and-socket connection is more popular on account of the advantage in hot-line work.

Figure 13-29 shows a 15,000-lb unit for 5-in spacing. It is designed particularly to resist breakage of the skirts during handling and installation.

Fog- or smog-type insulators are used where contamination from dust, chemicals, or salt spray would cause standard types of suspension insulators to flash over. Figure 13-30 shows the Ohio Brass Company's smog-type insulator.

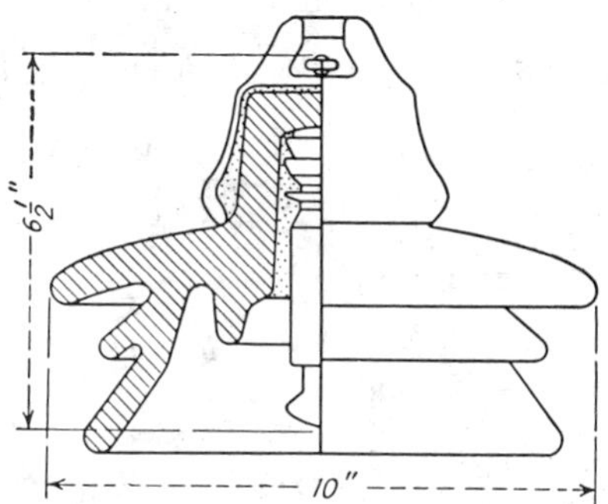

FIG. 13-30. Smog-type insulator. (*Ohio Brass Company.*)

The foregoing applies to transmission voltages below EHV. The American Electric Power Company 345-kV lines use 18 units. Five-hundred-kilovolt lines vary from 24 to 35 units, with the 35-unit strings used at high altitudes. Hydro-Quebec uses 35 units on its 700-kV system. Most EHV lines use units rated at 25,000 and 36,000 lb.

Present-day suspension insulators conform to USAS C29.2, and standards have been established for 15,000- and 25,000-lb ratings. The dimensions are the same for both, 5¾ in length and 10 in diameter. The standard should be consulted for complete information. Units are now color-coded to indicate their strength ratings; 15,000-lb units are brown, and 25,000-lb units are gray. To date there are no USAS color standards for the higher-strength units. There are also sizes other than 5¾ by 10 in. For example, 5 by 10 in and 7 by 12 in are available. Only insulators of recent manufacture follow the USAS color code. It is common practice to use a factor of safety of 2 for the maximum stress applied to insulators.

48. Strain Insulators. The assembly of suspension units arranged to dead-end the conductor at a structure is termed a "dead-end," or "strain," insulator. Such assemblies must carry the full conductor tension and should be designed with an ample factor of safety for the maximum expected ice and wind loads; and the ultimate strength of the insulators and hardware should be compared with the ultimate of the conductor in order to consider the effect of possible extreme loads greater than assumed in the general design. It is general practice to protect the dead-end string especially against damage from arcs by using one to three additional units and installing arcing horns or rings.

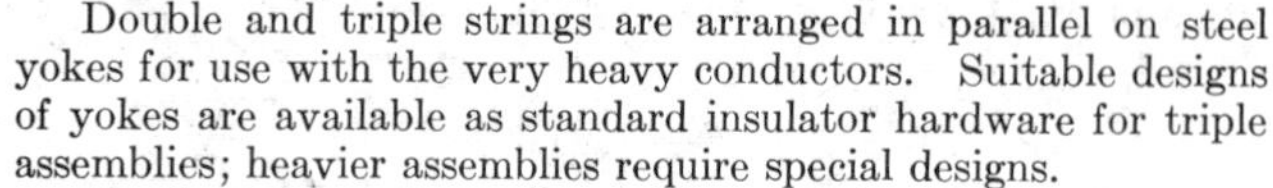

Fig. 13-31. Voltage gradient in percent of applied voltage for 4-, 10-, and 16-unit strings of 5¾- × 10-in. standard suspension insulators.

Double and triple strings are arranged in parallel on steel yokes for use with the very heavy conductors. Suitable designs of yokes are available as standard insulator hardware for triple assemblies; heavier assemblies require special designs.

49. Voltage gradient over a string of insulators is not the same for each unit. Owing to the capacitance of the hardware of each unit to ground, the charging currents are not equal from unit to unit, and the voltage gradient over the string is not uniform. The voltage drop over the line unit is greatest. This effect was the cause of concern in the early use of long insulator strings and is of special importance on insulators for radio frequencies (see Fig. 13-31). These curves are for voltages below EHV.

50. Protective Devices. The damage to insulators from heavy arcs is a serious maintenance problem, and several devices have been developed in attempting to ensure that an arc will hold free of the insulator string. Such devices are useful, but improvements in the design of overhead ground-wire protection and fast relaying have resulted not only in reducing the insulator damage but in improving the whole performance of the line.

The first protective measure consisted in attaching small horns to the clamp. It was found, however, that horns with a large spread both at the top of the insulator and at the clamp were required to be effective (see Fig. 13-32). Especially under lightning impulse, the arc tends to cascade the string, and tests show that the gap between horns should be considerably less than the length of the insulator string. Protection by arcing horns thus results in either a reduced flashover voltage or an increase in the number of units and length of the string.

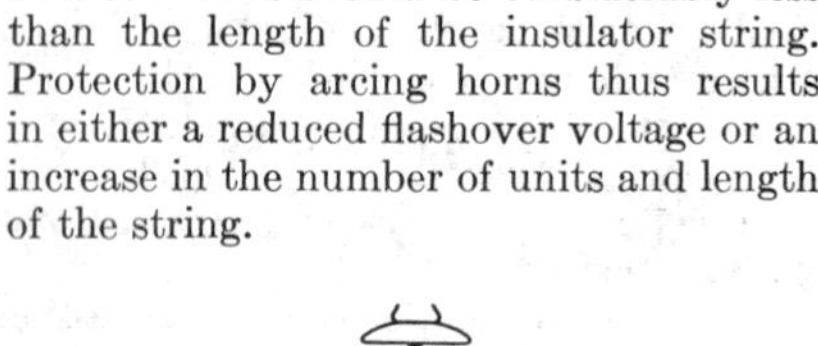

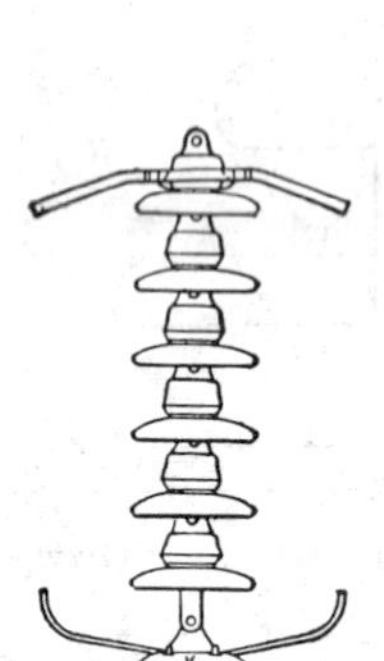

Fig. 13-32. Suspension string with arcing horns.

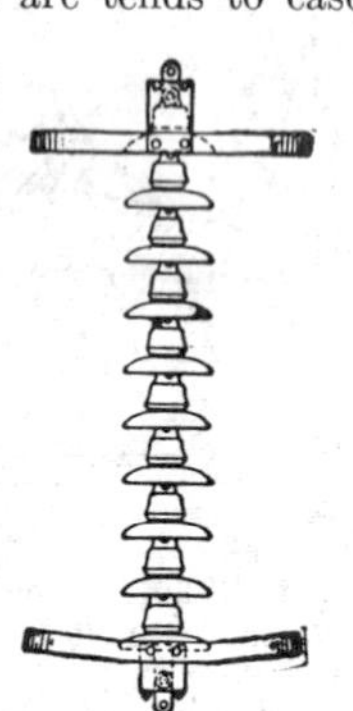

Fig. 13-33. Grading shields on suspension string.

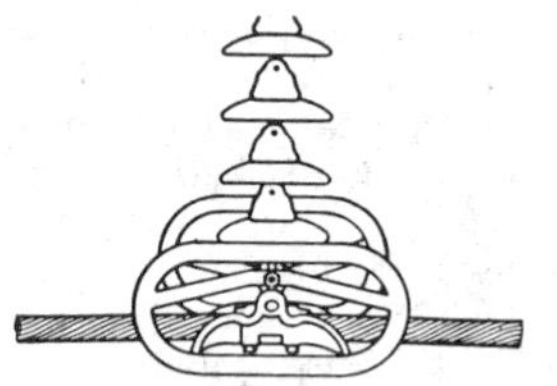

Fig. 13-34. Ohio Brass Company control ring.

The arcing ring, or grading shield (Fig. 13-33), is a more efficient protection. Lightning-impulse test shows that, if the proportions of ring diameter to length of insulator string are correct, cascading of the insulator can be prevented even for very steep waves. The effectiveness of the grading ring is due to the tendency to equalize the gradient over the insulator and to produce a more uniform field. Protection of the insulator is not therefore dependent on simply providing a shorter arcing path, as is the case with horns. Efficient rings are rather large in diameter, and, on suspension strings, a check should be made to ensure clearances to the structure at least as great as from ring to ring.

Recent experience with single-conductor high-voltage transmission above 250 kV has shown that the potential stress on insulator strings can be controlled better by a ring design such as that shown in Fig. 13-34, the corona formation on the line hard-

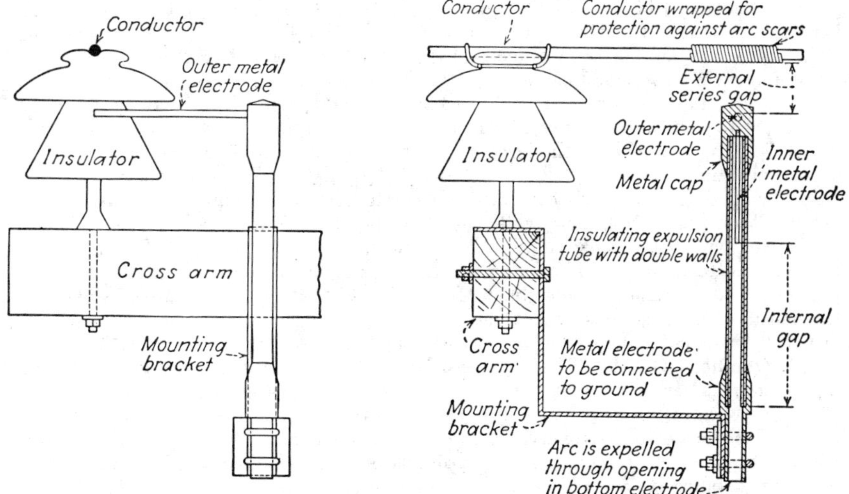

FIG. 13-35. Expulsion-tube gap for 46 kV.

ware being reduced at the same time. The shield (control) rings shown in Fig. 13-34 can be varied in shape and size as required. Bundle-conductor lines need no control rings or arcing rings, unless the bundle is two conductors one above the other.

51. The expulsion tube (Fig. 13-35) is probably the most successful form of insulation protection. The tube is arranged to provide a path to ground of sufficient length to prevent a flashover due to usual operating surges but considerably less than over the insulator. The tube is so designed as to extinguish a flashover so quickly, by the explosive force of the arc over the gap within the tube, that an operation of the circuit breaker does not follow. No fuse is used. As a matter of insulator protection only, the saving in cost of insulator replacements seldom warrants the expense of tube installation; but as a means of eliminating line outages, the expulsion tube offers promise of being second only to well-designed overhead ground-wire protection.

52. Insulator Standards. NEMA *Publication*, High Voltage Insulator Standards, and AIEE Standard 41 have been combined in USAS C29.1 through C29.9. Standard C29.1 covers all electrical and mechanical tests for all types of insulators, pin, post, spool, strain apparatus, and suspension. The standards for the various insulators covering flashover voltages, wet, dry, and impulse, radio influence, leakage distance, standard dimensions, and mechanical-strength characteristics are as follows: C29.2, suspension; C29.3, spool; C29.4, strain; C29.5, low- and medium-voltage pin; C29.6, high-voltage pin; C29.7, high-voltage line post; C29.8 apparatus pin; C29.9, apparatus post. These standards should be consulted when specifying or purchasing insulators.

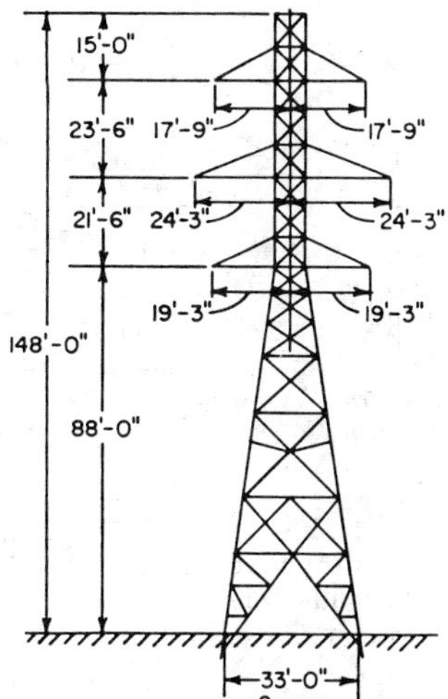

FIG. 13-36. American Electric Power System 330-kV tower. (*Trans. AIEE*, 1951, *Vol.* 70, *p.* 67.)

LIGHTNING PROTECTION

53. Protection Methods. The design of transmission lines for protection against lightning is too broad a subject to be treated here. For lines below the EHV levels, the methods given in the report of the AIEE Lightning and Insulation Subcommittee[1] have proved very satisfactory. Lightning protection for EHV lines is still in the development stage. One method has been proposed by J. M. Clayton and F. S. Young.[2] The first EHV lines, which were 330 kV, applied the methods of the AIEE subcommittee, using only one overhead ground wire (Fig. 13-36), and the results were not satisfactory.[3] It appears to be generally accepted that one overhead ground wire should not be used on EHV lines. Because of the large size of the towers, a single overhead ground wire apparently cannot intercept all strokes, particularly side flashes from the main stroke. Also, owing to the large spacing between conductors and ground wire, the coupling between them is reduced, thereby increasing the potential difference between conductors and tower on strokes to ground wire or tower. Even two ground-wire towers are not immune to lightning trouble. This may be due to steeper wavefronts than have previously been considered possible.[4] It is believed that, with very steep wavefronts, the neutralizing reflections do not get back up the taller towers in time to reduce the flashover voltages. The following discussion applies to voltages below EHV.

54. Overhead Ground Wires. Transmission lines of 115 kV and higher voltage in lightning areas nearly always are protected by means of overhead ground wires. Lines of lower voltages frequently are protected by other means, but the more important lines will be protected by overhead ground wires. In many instances the structures will be provided with supplementary grounds or counterpoises to hold the structure footing resistances to values such that the required line performance will be obtained. A prime consideration of the overhead ground wire is that it shall be installed so as to intercept lightning strokes and prevent them from striking the conductors in the spans away from the towers, and the towers must be so designed that they will intercept strokes and keep them off the conductors. It is now recognized that the poor performance of lines built before 1930 was due to insufficient height of the ground wires above the conductors, coupled with high structure-ground resistance.

55. Protector tubes (Par. **51**) are especially applicable to the lightning protection of lines with pin-type insulators. Such lines cannot be protected with much success by overhead ground wires. The use of protector tubes on the top conductor of a 26-kV pole-top-pin line[5] has given good protection. In this case the pole-top insulator is higher than usual above the crossarm in order that the top conductor may intercept lightning strokes.

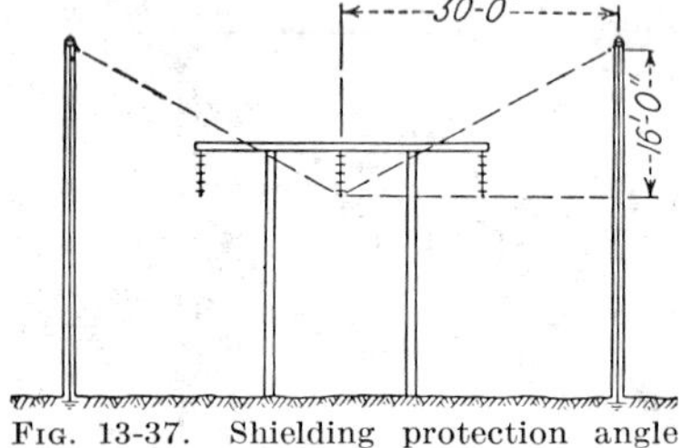

Fig. 13-37. Shielding protection angle for conductors between two ground wires.

56. Wood Insulation.[6] The impulse insulation of wood can be utilized frequently to supplement that of insulator strings on lines having overhead ground wires. The use of wood insulation obtained by means of horn gaps in pole ground wires and wood strain insulators in the guys has been found unsatisfactory on lines without overhead ground wires. No insulation has been found which will withstand the direct lightning strokes which would be experienced without properly installed overhead ground wires.

57. Shielding. Shielding, in this discussion, means protecting a line or station by means of overhead ground wires supported on structures other than those supporting the conductors. Stations frequently are protected by means of overhead ground wires supported by tall grounded structures of sufficient height so that the wires and

[1] *Trans. AIEE*, 1950, Vol. 69, p. 1187.

[2] Clayton, J. M., and Young, F. S. Estimating Lightning Performance of Transmission Lines; *Trans. IEEE*, November, 1964, p. 1102.

[3] Miller, Dr. Charles J., Jr. Why EHV Lines Show Excess Outage; *Elec. World*, Sept. 5, 1955, p. 74.

[4] Miller, C. J., Jr. Anomalous Flashovers on Transmission Lines; *Trans. AIEE*, 1956, Vol. 75, Pt. III, Power Apparatus and Systems, pp. 897–907.

[5] Sels, H. K., and Gothberg, A. W. Lightning Protection of Wood Pole Lines; *Elec. Eng.*, June, 1940.

[6] Melvin, H. L. Experience with Wood as Lightning Insulation; *Trans. AIEE*, June, 1933.

structures will intercept any lightning strokes which might otherwise strike the station structures or conductors. The shielding of transmission lines in this manner has not been used extensively but where used has been very satisfactory. Experience with the type of construction shown in Fig. 13-37 indicates complete shielding within a line through the ground wire on a slope of one vertical to two horizontal. This construction is used on the 96-mi wood-pole portion of the Tennessee Valley Authority's 161-kV Wheeler-Norris line.[1] This line, built for 220 kV and operated at 161 kV, went into service in 1936 and to date has had no known outages due to lightning and no structures or insulators damaged. It is located in territory having approximately 60 thunderstorm days per year and is probably the only strictly "lightningproof" overhead high-voltage transmission line so far built in this country. However, age is taking its toll of the structures, and, with decreased importance of the line and increased maintenance costs, it is gradually being rebuilt with conventional H frames. With this type of construction, no consideration need be given to low ground resistance or to counterpoising. Only the shield-wire poles are grounded. This type of shielding has been used to protect short high-power 13-kV lines with complete satisfaction. Single-pole lines can be protected by a shielding wire on one side only. Important lines of the lower transmission voltages can be made practically lightningproof by shielding.

58. Lightning Arresters. Lightning arresters are used at transmission substations to coordinate the impulse insulation of the line with the impulse insulation of the equipment in the substation. Standard arresters listed by the manufacturers for use on solidly grounded neutral systems are given a maximum rms voltage rating (cutoff voltage) of 75 to 80% of the maximum line-to-line system rms voltage. This voltage should not be exceeded during abnormal system voltages caused by loss of load or overspeeding of generators. Ground faults under certain conditions may also cause excessive voltages to be impressed upon lightning arresters. So-called "solidly grounded neutral systems" of the lower transmission voltages may experience overvoltage due to ground faults if neutral grounds are of fairly high resistance. Also, lightning arresters at the ends of long lines where the receiving-end transformers do not have grounded neutrals on the high-tension side may be subjected to overvoltages due to ground faults. This overvoltage is a function of the positive-, negative-, and zero-phase-sequence impedances. Paragraphs **79** to **82** explain some of the features that must be considered.

For grounded-neutral arresters to be used satisfactorily on grounded-neutral systems, whether solidly grounded, grounded through reactors or resistors, or grounded by limited transformer capacity, the relationship of zero-phase-sequence impedance to positive-phase-sequence impedance *at the station where the lightning arresters are located* should not be such that the maximum phase-to-neutral voltage during ground faults will exceed the maximum permissible line-to-neutral voltage of the arresters. The fault voltage can be determined by the curves of Fig. 13-47, if the sequence impedances have approximately the same X/R ratios. When system neutral-ground connections are of appreciable resistance and the lower transmission voltages are involved, calculations, by the method of symmetrical components, should be made to determine the maximum voltage which may be experienced on the unfaulted phases. The **maximum system line-to-neutral voltage** should be used with the curves of Fig. 13-47 and in calculations. For example, if the maximum system voltage is 110% of normal, it is found from Fig. 13-47 that the Z_o/Z_p ratio should not exceed 3.8:1. If the maximum system line-to-neutral voltage is as high as 116.5%, the Z_o/Z_p ratio permissible reduces to 2:1 (see also Sec. **10**).

59. References to Literature on Lightning Protection

Fortescue, C. L., and Conwell, R. N. Lightning Discharges and Line Protection Measures; *Trans. AIEE,* 1931, p. 1090.

Subcommittee on Lightning and Insulators, Lightning Performance of 220 Kv Lines; *Rept.* II, *Trans. AIEE,* 1946, Vol. 65, p. 10.

Gross, I. W., LeVesconte, L. B., and Dillard, J. K. Lightning Protection in Extra-high-voltage Stations. Influence of Multiple Circuits; *Trans. AIEE,* 1953, Vol. 72, Pt. III, p. 882.

[1] Evans, Llewellyn, and Daniels, H. C. *Elec. World,* 1939, Vol. 111, p. 1434.

AIEE Committee Report, Application and Performance of 13–138-kv Line Expulsion Lightning Arresters (Line Protector Tubes); *Trans. AIEE*, 1953, Vol. 72, Part III, p. 151.

HOUSLEY, J. ELMER, and HARPER, JOHN D. Protection of Transmission Lines over Mountainous Region Where Lightning Incidence Is High; *Trans. AIEE*, 1951, Vol. 70, p. 124.

WAGNER, C. F., and HILEMAN, A. R. A New Approach to Calculation of Lightning Performance of Transmission Lines, II; *Trans. AIEE*, Pt. III, Power Apparatus and Systems, December, 1959, Vol. 78, p. 996.

GRISCOM, S. B. The Prestrike Theory and Other Effects in the Lightning Stroke; *Trans. AIEE*, Pt. III, Power Apparatus and Systems, December, 1958, Vol. 77, p. 919.

FISHER, F. A., ANDERSON, J. G., and HAGENGUTH, J. H. Determination of Lightning Response of Transmission Lines by Means of Geometrical Models; *Trans. AIEE*, Pt. III, Power Apparatus and Systems, February, 1960, p. 1725.

SWITCHING SURGES

60. Switching surges are sudden changes in the electrostatic and electromagnetic fields of a circuit caused by switching. If the load is switched out, the resulting surge may have an oscillatory characteristic depending upon the circuit constants. The oscillations also are subject to harmonics which increase the transient voltage.

In an oscillation, the energy passes from one form to the other, so that, on the assumption of no loss,

$$\tfrac{1}{2}Li^2 = \tfrac{1}{2}Ce^2$$

or

$$e/i = \sqrt{L/C} \tag{13-65}$$

where e = voltage produced by a sudden suppression of the current i; L and C = line inductance and capacitance, respectively. If the line resistance and leakage are negligible, the term $\sqrt{L/C}$ is usually called the *natural*, or *surge, impedance* of the line.

From Eq. (13-65), the maximum voltage possible to occur on suppression of the current I is

$$e = I\sqrt{L/C} \tag{13-66}$$

Any change in load will alter either e or i; and as the energy of the electrostatic and electromagnetic fields cannot change in zero time, e and i must pass through some transient values before the steady state is again reached.

If the resistance is equal to or greater than $\sqrt{L/C}$, the transient is nonoscillatory and dies out. If it is less than $\sqrt{L/C}$, the transient is oscillatory.

An oil circuit breaker almost invariably opens the circuit at the zero point of the current wave. When opening an unloaded line or an unloaded transformer, the circuit voltage will be a maximum when the circuit is opened. The line or network will then be left with a charge equal to the crest value E_m of the line-voltage wave. This becomes a traveling wave whose value cannot exceed $2E_m$ on reflection. Ordinarily, it attenuates rapidly so that this value is not approached. However, if the arc in the circuit breaker is reestablished during the next half cycle of voltage, the surge may theoretically reach a value of $3E_m$. With a current transient, however, the maximum voltage is given by Eq. (13-66).

Since the current on short circuit may be large, the voltage rise theoretically may be very high. The fact that circuit breakers tend to interrupt the current near the zero point of the wave prevents the high voltages that otherwise might occur. Investigations by W. W. Lewis and others show that voltages of two to five times normal may occur during switching.

For many years switching surges were not a cause of concern, since the insulation of transformers was designed with high margins in the basic impulse insulation levels (BIL). In recent years, owing to the improved capability and reliability of lightning arresters, together with close tolerances in spark-over voltages, it has been possible to reduce the BIL of EHV transformers as much as two steps. This brings some switching surges into the protection zone of lightning arresters and makes such surges a factor in their design. Manufacturers are working on the design of circuit breakers to make them restrike-free and to hold switching surges to not more than 2.5 times the crest line-to-neutral voltage or the rated voltage of the circuit. Circuit breakers have already been produced which limit 500-kV switching surges to 900 and 950 kV to ground.

61. References to Literature on Switching Surges

AIEE Committee Rept. Switching Surges, One-phase to Ground Voltages; *Trans. AIEE*, Power Apparatus and Systems, June, 1961, p. 240. (Discusses survey reports on many types of switching surges, but includes no theory or mathematical analyses; contains an extensive list of reference articles which cover all phases of switching-surge problems.)

GEHRIG, E. H., GENS, R. S., and TUPPER, G. A. Application of New Concepts to 500 kV System Insulation Co-ordination; *Trans. AIEE*, January, 1964, p. 41.

Symposium, Four Papers on Switching of Extra-high-voltage Circuits; *Trans. IEEE*, December, 1964, pp. 1187–1222.

UNDERGROUND CABLES

62. Underground cables are used for transmission of power where it becomes impracticable to use overhead construction. Such location may be congested urban areas, where right-of-way cost would be excessive or local ordinances prohibit overhead lines for reasons of safety, in or around plants and substations, or crossings of wide bodies of water, which for various reasons would not permit of overhead crossings. The kind of insulation and type of cable used will depend upon the voltage and service requirements.

63. Insulating materials for cables used in the transmission of power are (1) rubber and rubberlike compounds and (2) oil-impregnated paper. Rubber and rubberlike compounds are not used for voltages in excess of 35,000. The paper-insulated cables are suitable for use on all voltages, and installations up to 345 kV are in use at this time. There appears to be no reason to believe that paper-insulated cables cannot keep pace with system voltages as they increase. For voltages of 115 kV and above, the paper-insulated oil-impregnated cables with or without sheaths may be installed in pipes, or they may be provided with strong sheaths and installed in ducts or tunnels. In all cases the paper insulation is subjected to pressure by means of either oil or nitrogen gas. Cables for 69 kV and below are more frequently (1) low-pressure, not over 15 lb/in^2, or (2) medium-pressure, not over 45 lb/in^2. High-pressure cables, up to 200 lb/in^2, installed in pipes so far have not been found generally to be economical for voltages of 69 kV and below.

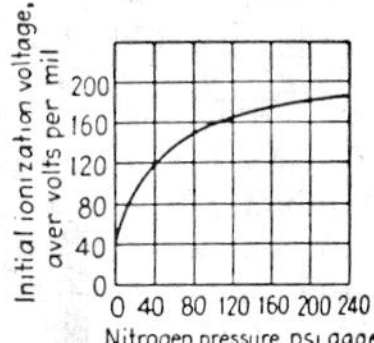

FIG. 13-38. Ionization voltage vs. gas pressure.

The effect of pressure on the dielectric strength of oil-impregnated-paper insulation was discovered by Emanueli of Italy in the middle 1920s and resulted in the production of Pirelli cable, in which the pressure is applied from the center out. Hochstadter of England in the late 1920s developed the principle of pressure applied from the outside radially inward, and it is this principle that is used in the high-pressure cables now made in the United States.

The effect of pressure on the cable insulation is to eliminate voids and thereby prevent ionization of the gas which would appear in the voids. Voids cannot be prevented without the pressure, regardless of the care taken in winding the paper tapes. Bending the cable in handling and on installation—also expansion and contraction due to temperature changes—will cause voids in insulation of cable not under pressure. Continued presence of ionization will damage the insulation and eventually cause failure. The presence of ionization is detected by means of the power-factor change as a test voltage is applied. High-pressure cables are so designed that ionization is completely absent. Figure 13-38 shows a typical curve of ionization voltage vs. pressure for gas-filled solid cable.[1] Standard thickness of paper insulation for oil- and gas-filled pressure-type cables is given in Table 13-12.

64. Conductors may be copper or aluminum and may have round, oval, or sector-shaped cross sections. They may have regular or compact stranding. The compact

[1] SHANKLIN, G. B. Low-, Medium-, and High-pressure Gas-filled Cable; *Trans. AIEE*, 1942, Vol. 61, p. 719.

Table 13-12. Thickness of Paper Insulation for Pressure Cables

System Voltage, Kv	Insulation Thickness, Mils
46	205
69	285
115	435
138	505
161	590
230	835

stranding differs from the regular stranding in that the spiraling of all layers is in the same direction. The cable is then compressed until the outside is practically smooth and internal spaces are essentially eliminated. The round type may be hollow, with the hollow center used as an oil duct. When conventional stranding is used for the hollow conductors, it is wound over a spirally wound steel or copper tube. Segmental stranding also is used. When segmental stranding is used for hollow conductors, no spiral center tubing is necessary.

65. Cable current-carrying capacity is dependent upon the loading cycle and the dissipation of the I^2R and dielectric losses into the surrounding media. The loading cycle will fix the watts conductor loss for any given cable, and the potential stress and temperature will determine the dielectric loss. The dielectric loss normally can be neglected with pressure-type cables below 230 kV. This heat must be conducted away from the cable at a rate fast enough to prevent damage to the insulation. This means that for present-day installations the design must be such that the maximum normal copper temperature for high-voltage cable should not exceed 70°C and the emergency maximum temperature should not exceed 90°C. The path over which the heat must be conducted away has a thermal resistance frequently represented as R_{th} which is expressed in degrees centigrade difference per watt transferred per foot of cable [°C/(W)(ft)]. The thermal resistance can be determined if the thermal resistivity of the various parts of the path is known. Insulation resistivity and the resistivity of other cubic materials are given in degrees centigrade difference required to transmit 1 W through 1 in³ of the material [°C/(W)(in³)]. Surface resistivity is expressed in degrees centigrade difference to transmit 1 W across 1 in² of surface such as from conductor to insulation, pipe to earth, etc., and is expressed as degrees centigrade per watt per square inch. Figure 13-39[1] shows typical temperature gradients from the cable to the thermal sink (remote earth). The calculation of current-carrying capacity is too involved to be covered here, and the reader is referred to the various AIEE and IEEE papers listed at the end of this part covering current-carrying capacity, temperature rise, conductor temperatures, etc.

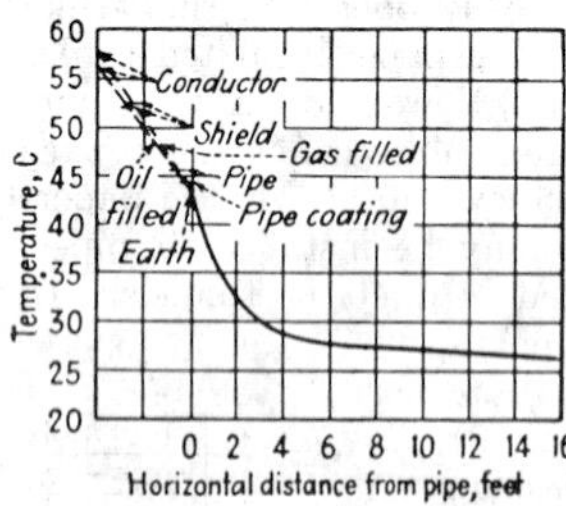

FIG. 13-39. Comparison of conductor-to-earth temperature gradient for oil- and gas-cable samples. Current 455 A; conductor 600 M cmils; paper 480 mils; oil 700 SSU at 100°F; gas, nitrogen at 200 lb/in².

Carrying capacity and dissipation of losses are really one problem, especially as more and more cables are installed in close proximity to each other and the earth does not have time to cool down during the load cycle. Therefore it may be necessary to install the cables other than by direct burial, so that forced cooling can be applied.

66. Electrical characteristics of cables, resistance, reactance, and capacitance can be determined for cables other than pipe types with reasonable accuracy. Cables for transmission use will be sufficiently large so that all skin and proximity effects must be taken into account along with temperature coefficients in determining the resistance. Skin effect is discussed in Sec. **4.** Reactance and capacitance for cables other than pipe type can be calculated with reasonable accuracy by means of the formulas given in Sec. **2.**

There is no precise method of calculating the electrical characteristics of pipe-type cables, and tests on the final installation provide the only accurate means of determining these quantities. It is necessary to have the quantities in usable form for design

[1] BULLARD, W. R., PETTEE, A. D., and RHODES, G. L. 115-kv High-pressure Oil-filled Pipe Cable Installations at New Orleans; *Trans. AIEE*, 1948, Vol. 67, p. 475.

purposes, and at this time there is no such general information available in the technical press. It is therefore necessary to consult the various manufacturers for this information. The Insulated Power Cable Engineers Association sponsored tests of resistance and reactance of large conductors in steel pipe or conduit,[1] but the spacings were for low-voltage cables and no capacitance tests were made. Empirical formulas for resistance and reactance were developed, and these formulas will give roughly approximate results for high-voltage cables. J. H. Neher (*Trans. IEEE Power Group,* August, 1965, p. 795) presents zero-phase-sequence impedance formulas for pipe-type cables.

C. T. Hatcher reports[2] test values for the Harbirshaw (Phelps-Dodge) 350,000-cm 138-kV Jamaica–Valley Stream (N.Y.) cable per mile as follows:

a. 0.21 Ω resistance a-c (temperature not given).

b. 0.33 Ω reactance.

c. 0.386 μF capacitance.

The nominal carrying capacity of this cable at a 70% loss factor is 415 A.

The cable shown in Fig. 13-43, which also is of Phelps-Dodge design, has calculated electrical characteristics as follows:

a. 0.045 Ω/mi d-c resistance at 70°C.

b. 0.063 Ω/mi a-c resistance at 70°C.

c. 0.550 Ω/mi reactance.

d. 0.550 μF/mi capacitance.

The a-c resistance includes the effect of losses due to induced currents, etc., in other metallic parts of the system.

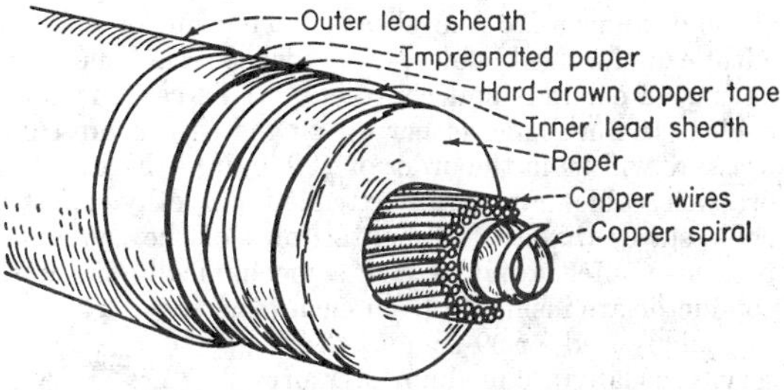

Fig. 13-40. Construction of 132-kv Pirelli hollow-conductor cable.

66a. Present trend in the United States for voltages 110 kV and up is toward the use of pipe-type cables with pressures up to 200 lb/in² by means of gas or oil as the pressure medium. Until the end of World War II the Pirelli-type cable was the most popular, and many hundreds of miles are in satisfactory service. Figure 13-40 shows the construction of Pirelli 132-kV cable. Pirelli cables are classed as medium-pressure cables and are installed one cable per duct in underground installations.

High-pressure cables are of two general types:

1. Those without a sheath over the insulation.

2. Those with a plastic or lead sheath over the insulation.

The first type is usually provided with a temporary lead sheath for protection during shipment and handling prior to pulling in. The sheath is ripped off

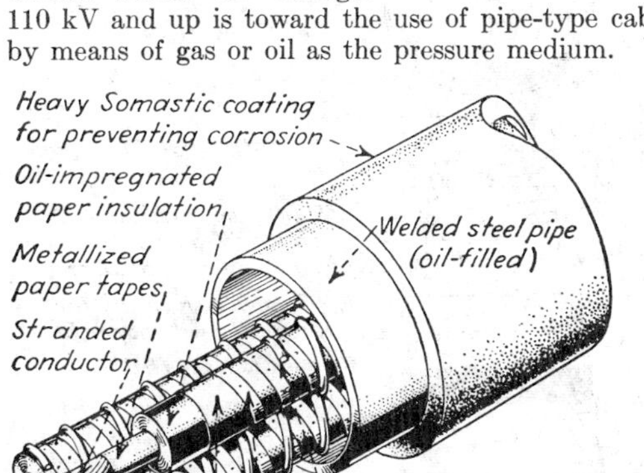

Fig. 13-41. Cutaway view of Oilostatic cable installation.

[1] Salter, E. H. Problems in the Measurement of A-c Resistance and Reactance of Large Conductors; *Trans. AIEE,* 1948, Vol. 67, p. 1390.

Del Mar, William A. Reactance of Large Cables in Steel Pipe or Conduit; *Trans. AIEE,* 1948, Vol. 67, p. 1409.

Wiseman, R. J. A-c Resistance of Large Size Conductors in Steel Pipe or Conduit; *Trans. AIEE,* 1948, Vol. 67, p. 1745.

[2] McCormack, J. E., Hatcher, C. T., Wyatt, K. B., Del Mar, W. A., Merrell, E. J., Palmer, J. H., and DeTurk, E. F. A 138,000-volt Polyethylene-sheathed Compression Cable—Pipe Line Type; *Trans. AIEE,* 1948, Vol. 67, p. 447. Discussion, p. 474.

during the pulling in. Figure 13-41 shows a cross section of an Oilostatic sheathless-cable installation made by the Okonite Company. Figure 13-42 shows a schematic diagram of an Oilostatic installation. With both high-pressure types, three single-conductor cables are pulled simultaneously into the pipe. Cables provided with lead or plastic sheaths over the insulation normally must have oval cross sections, although the large cable shown in Fig. 13-43 has a rounded triangular cross section. The three flat sides permit expansion without damage to the sheath. This out-of-round feature is particularly necessary with lead-sheathed cables; otherwise the lead sheath will fold and crack as the impregnating oil expands and contracts. This cable is commonly referred to as *compression* cable. It may have either oil or gas as the pressure medium, and in either case the pressure will be in the order of 200 lb/in². Figure 13-43 shows a section of a large compression cable having an extruded polyethylene sheath. The manufacturer of this cable states that the segmental form of the conductor is designed to reduce skin and proximity effects and to give a mechanically stable form. The three segments of the conductor are insulated from each other by paper tapes.

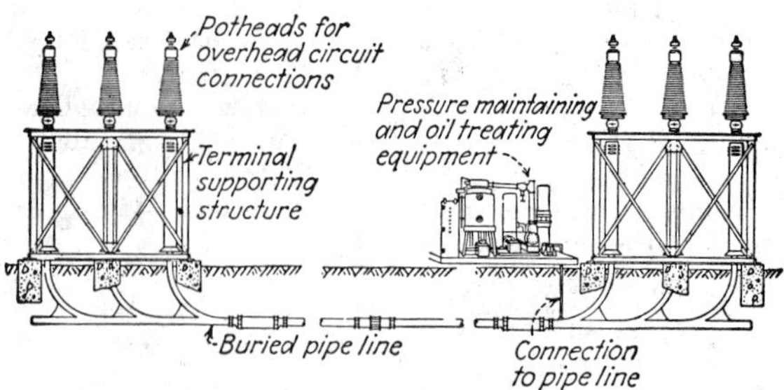

Fig. 13-42. Schematic layout of a typical Oilostatic installation.

Cables used for 69-kV and 46-kV service usually are medium-pressure types, and a 3-phase installation may consist of three single-conductor cables pulled into separate ducts or it may consist of a 3-conductor cable pulled into a duct. Such cables normally are not buried directly in the ground. Cables for voltages under 46 kV may be low- or medium-pressure impregnated-paper cables or rubber-insulated cables. Varnished-cambric-insulated cables are no longer used for transmission circuits.

67. Cable Installation. Installation of high-voltage Pirelli-type cables is little different from that of the lower-voltage cables. The 3-phase circuits consist of three single-conductor cables pulled into regular nonferrous ducts and spliced in conventional manholes.

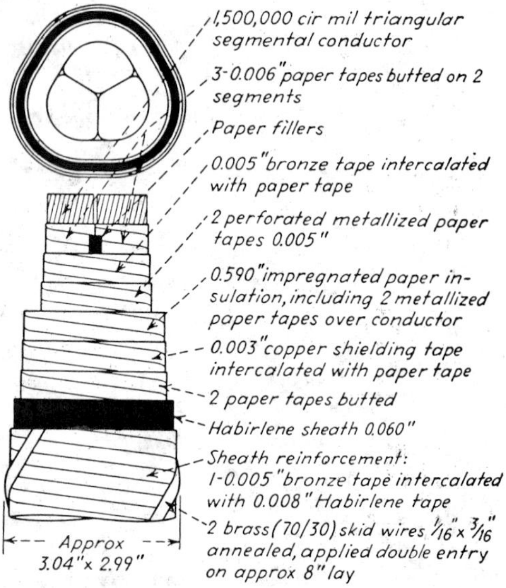

The installation[1] of pipe-type cables is considerably more complex than for the medium-pressure and rubber-insulated cables. The pipe-size requirements outlined in the specifications of the AEIC (Par. **70**) suggest that the area of the circle circumscribing the three cables over the skid wires when all three cables are touching each other should not exceed two-thirds the cross-sectional area of the inside of the pipe.

Fig. 13-43. Oil compression cable for 161 kV. Single-conductor, 1,500,000 cmils, triangular segmental Habirlene sheathed. (*Phelps-Dodge Copper Products Corp.*)

[1] Gillette, R. W., and Hull, F. M. Pipe-type Cable Installation Techniques; *Trans. AIEE*, 1954, Vol. 73, Pt. III, p. 587.

In no case should the pipe be large enough to permit the cables to become crossed and thereby susceptible to damage during pulling. The pipes must be of high-grade steel with carefully welded joints. Any weld roughness on the inside of the pipe must be avoided in order to prevent damage to the cable. The welds must be tested under pressure for leaks. This can be done by the use of double (tandem) plugs, rubber-faced, which can be inserted in the pipe bridging the joint and expanded. Nitrogen gas under pressure is then admitted and leaks detected by a water bath. The pipe is covered with a suitable coating to prevent corrosion and to provide insulation against possible electrolysis voltages. The pipe is laid directly in the ground, and present practice calls for not more than two pipes to a single trench. Before the cables are pulled into the pipe, it should be thoroughly cleaned and dried out. Manholes are installed as required for pulling purposes and need be only large enough for making splices. Temporary steel manholes[1] have been used for the purpose of making splices and then removed. Pulling lengths up to 3,000 ft offer no difficulties on straight runs, and in one case a 6,750-ft pull was made in crossing the Narrows from Long Island to Staten Island, N.Y. In this case pulling was facilitated by partially filling the pipe with oil before pulling to provide buoyancy and lubrication. In making pulls, attachment is made to the conductor and care taken not to exceed 10,000 lb/in^2 on copper cable. To date aluminum cable has not been used for pipe-type cable.

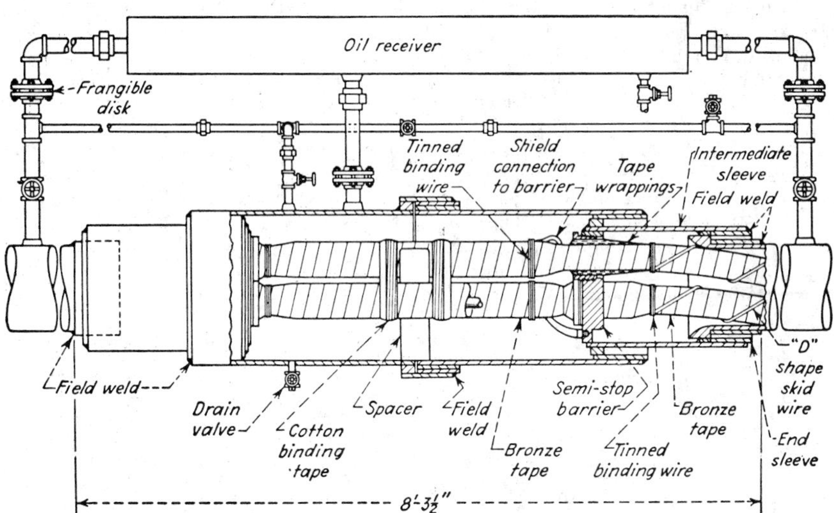

Fig. 13-44. Oil-filled semistop as used with a compression cable. (*McCormack, Hatcher, Wyatt, Del Mar, Merrell, Palmer, and DeTurk, Trans. AIEE*, 1948, *Vol.* 67, *p.* 447. *Discussion, p.* 474.)

68. Operation. Pressure is maintained on oil-filled pipes by means of automatic pumping plants which keep constant pressure on the cables and draw upon storage tanks which have capacity to handle moderate leaks if they develop and to take care of the volume changes which occur with changes in temperature. Semistops which place a barrier to the flow of the pressure medium but do not interfere with the movement of impregnating compound are built into splices at points between pumping stations so that pressure can be cut off from a pipe section in case of trouble. These stops also are provided with bypasses so that a pump can take over an adjoining section in case of pump trouble. Safety diaphragms also are provided with oil-filled pipes for the relief of excessive pressures which may occur. Figure 13-44 shows an oil-filled semistop.

[1] Ref. 1, Par. **65**.

Gas pressure in gas-filled pipes is maintained by means of nitrogen flasks located as needed, and no pumping equipment is required. Gas leaks are an ever-present problem,[1] and while they are seldom large, they must be detected and located.

The operation of pressure-type cables is much simpler than with the old-type solid and rubber cables with which constant watch and tests were necessary for signs of excessive ionization. The proper maintenance of pressure is assurance against ionization, so that cable operation requires proper supervision of facilities for maintaining pressure and temperature or overcurrent alarms to warn of overloading.

69. Long-distance transmission of power by means of underground cables poses some problems not met with in overhead transmission. From the per mile capacitance of the Jamaica–Valley Stream cable, in Par. **66,** it is determined that the charging current is about 12 A/mi. Approximately 35 mi of cable energized from one end would have a charging current of 415 A. This means that a 35-mi cable could not deliver power at unity power factor and that the receiving-end power factor must always be lagging at full load. Long-distance transmission thus is a study of conductor sizes, shunt reactors, and the frequency of their application, together with the magnitude of the lagging component of the load current. For example, if the cable is loaded to 415 A output at 86.6% power factor, the lagging component of the load is 207.5 A. The first shunt reactor, then, must be located where the line-charging current component becomes 415 A and the net leading reactive component becomes 207.5 A, less the current corresponding to the I^2X loss due to the 415-A load. The magnitude of the shunt reactor current at this point, then, is 415 less the above-mentioned I^2X component. The other shunt reactors are determined in a similar manner. The inphase component of current at the receiving end is approximately 359 A, and this current could be obtained with a smaller conductor and smaller, closer-spaced shunt reactors or a larger conductor and larger, farther-spaced reactors. In each case the receiving-end power factor is adjusted to make the full-load amperes of the load equal to the cable rating. It is obvious that synchronous condensers cannot be used for voltage control of long cable circuits and that regulating transformers must be used for this purpose.

70. Cable Specifications. The Association of Edison Illuminating Companies (AEIC)[2] publishes complete specifications for paper-insulated cables of the pressure types as well as for solid types up to 69 kV. The manufacturers comply with these specifications throughout. Main items covered for pressure cables are:

1. Voltages and operating limits.
2. Relation of cable size to pipe size.
3. Maximum operating temperatures.
4. Conductors.
5. Insulation.
6. Sheath.
7. Factory tests.
8. Tests after installation.
9. Reels and shipment.

71. References to Literature on Underground Cables

Publications of the Insulated Power Cable Engineers Association, Montclair, N.J., covering Current Carrying Capacity of Impregnated Paper, Rubber and Varnished Cambric Insulated Cables, and General Specifications for the same type of cables.

Morris, M., and Burrell, R. W. Current-carrying Capacity of Pipe Cable under Steady-state and Transient Cyclic Loading Conditions; *Trans. AIEE,* 1954, Vol. 73, Pt. III, p. 650.

AIEE Committee Report, The Effect of Loss Factor on the Temperature Rise of Pipe Cables and Buried Cables; *Trans. AIEE,* 1953, Vol. 72, Pt. III, p. 530.

Neher, J. H. Procedures for Calculating the Temperature Rise of Pipe Cables and Buried Cables for Sinusoidal and Rectangular Loss Cycles; *Trans. AIEE,* 1953, Vol. 72, p. 541.

[1] Piper, John D. Location of Gas Leaks in Pipe-encased Gas Pressure Cable Lines; *Trans. AIEE,* 1948, Vol. 67, p. 10.

[2] 40 West 40th Street, New York, N.Y.

SHANKLIN, G. B., and BULLER, F. H. Cyclical Loading of Buried Cable and Pipe Cables; *Trans. AIEE*, 1953, Vol. 72, p. 535.

WISEMAN, R. J. An Empirical Method for Determining Transient Temperatures of Buried Cable Systems; *Trans. AIEE*, 1953, Vol. 72, Pt. III, p. 545.

MEYERHOFF, L., and EAGER, G. S., JR. A-c Resistance of Segmental Cables in Steel Pipe; *Trans. AIEE*, 1949, Vol. 68, p. 816.

CLARK, H. W. Medium-pressure Gas-filled Cable, Washington, D.C.; *Trans. AIEE*, 1951, Vol. 70, p. 418.

CUNHA, S. H., GUNNING, M. P., and FARNHAM, D. M. 120-kv Self-contained Compression Cable Installation at Montreal; *Trans. AIEE*, 1954, Vol. 73, Pt. III, p. 491.

MOLLERHOJ, J. S. Flat-type Oil-filled Cable; *Trans. AIEE*, Power Apparatus and Systems, June, 1961, p. 861.

OUDIN, J. M., and TELLER, R. A. Some Considerations Concerning Extra-high-voltage A-C Cable Power Transmission; *Trans. AIEE*, Power Apparatus and Systems, December, 1961, p. 861.

BURRELL, R. W. Applications of Oil-cooling in High-pressure Oil-filled Pipe-cable Circuits; *Trans. IEEE Power Group*, September, 1965, p. 795.

THOMAS, E. R., and KERSHAW, R. H. Impedance of Pipe-type Cable; *Trans. IEEE Power Group*, October, 1965, p. 953.

SYSTEM CONNECTIONS

72. Types of Transmission System. There are very few simple transmission systems, and these are mainly small isolated systems. An interconnected system may include all possible types of individual system. Three convenient classifications of transmission systems are (1) radial, (2) ring, and (3) network.

In all types of system the object is to provide the quality and continuity of service required by the various loads or load areas. The proper determination of the system connections, therefore, becomes of first importance in obtaining the desired results at the minimum cost.

73. Radial systems were the simplest early form of transmission system. A community would have its generating plant and distribution system, from which radial transmission lines would be built to neighboring communities. The radial service was unreliable, and some of the lines would be reinforced by one or more parallel circuits. Then some of the communities on the ends of radial lines would be interconnected directly, if not too far apart, to obtain better continuity of service. While such a system, as expanded, has characteristics of a network system, it may still be classed as a radial system as long as the power comes from one central source. Service by means of a single radial circuit is the least reliable but at the same time the least expensive and, in sparsely settled parts of the country, is the only means whereby service can be economically justified.

74. Ring systems are in general confined to large centers of population and consist of a transmission ring encircling the load area into which one or more generating stations are connected and a series of step-down substations where power is fed into sections of the load area. The ring may be single-circuit or multicircuit or a combination of single-circuit sections and multicircuit sections. This type of system provides very high quality and continuity of service, and substation interruptions are practically limited to those due to trouble within a substation itself. The ring substations frequently feed into primary distribution systems which interconnect the ring substations on the low-tension sides.

Establishment of an open-wire high-tension ring system around an existing load area to provide additional feed-in points to a primary distribution network may meet with difficulties. If the primary network is 11 kV or higher and is served by 3-phase cables direct from the generator buses and the high-tension ring is to be served from the same buses, the problem is not simple. Cables inherently have low power angles (phase angle between E_s and E_r) due to the low X/R ratios of the cables, whereas high-tension systems inherently have high power angles which are aggravated by the step-up and step-down transformer impedances. The result is that for the new ring substations to carry the desired loads the (1) ring voltage must be higher than desired, or (2) reactors must be installed in the feeder cables from the plants, or (3) phase-shifting transformers must be used at the ring substations, or (4) the primary network must be broken up. The problem is one of balancing power angles from the plant or plants by way of the high-tension ring against those by way of the feeders from the plants and is one that requires careful analysis.

75. Network systems in general are the result of the expansion of the foregoing types of system. A representative example is the expansion of two radial systems until they meet and are interconnected in two or more places. The majority of transmission systems in this country are network systems. The important substations on a network will have two or more sources of supply from different directions, providing a very reliable and high quality of service. A network system is the most complicated to operate and is the type that requires careful analysis for transient stability.

76. Transformer substations are usually located at automatic sectionalizing points on rings or networks. In some instances a transformer substation may be connected to two parallel lines by means of automatic oil circuit breakers, without sectionalizing the lines. This complicates the relaying somewhat but in general is a satisfactory scheme when economy is paramount. It is more economical if the transformer substations step down to the primary distribution voltage of the load area or to the utilization voltage of a large industrial customer. However, if the main transmission voltage is high (161 kV or higher), this is not always possible owing to right-of-way difficulties, and it is necessary to provide step-down substations which transform to lower or secondary transmission voltages. Frequently as many as three secondary transmission voltages may be employed, e.g., 161-kV main transmission, 115-, 69-, and 34.5-kV secondary transmission voltages stepping down from 161 kV to either of the lower voltages or in the full series of transformations. The various transformer substations are the preferred locations for condensers and step-type regulators for the control of transmission voltages.

77. Sectionalizing switches are of two general types: (1) **Sectionalizing air-break switches** are frequently installed in long radial lines in order to facilitate the locating of line breakdowns. Accessibility is the first consideration in selecting the location of such switches. The use of sectionalizing air-break switches is questionable in lines the loss of which does not cause service interruptions. (2) **Sectionalizing oil circuit breakers** are installed in long double-circuit lines for the purpose of increasing the transient-stability limit. An installation of this type was the original two-circuit Hoover Dam–Los Angeles 287-kV line,[1] which was divided into three sections by means of two intermediate switching stations. These switching stations utilized the ring-bus scheme of connection as shown in Fig. 13-45. This scheme permits one breaker to be taken out of service for inspection or maintenance without disturbing service or relay protection. This might be called an ideal application for a ring bus. In addition to increasing the transient-stability limit, the sectionalizing stations reduce the synchronous-condenser capacity required and improve relaying by permitting a higher ratio of relay pickup current to load current.

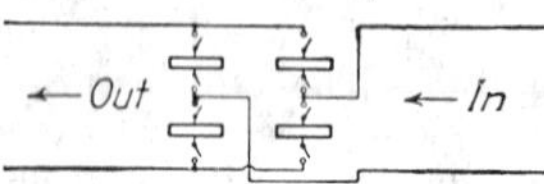

Fig. 13-45. Intermediate sectionalizing station; twin-circuit line.

78. Ungrounded-neutral vs. Grounded-neutral Systems. In the early days of power transmission there were advocates for each system, and both systems were used with satisfactory results. The transmission systems in the United States today are predominantly grounded neutral. The isolated-neutral systems were free from interruptions due to grounds on one conductor but were frequently subject to equipment failures when swinging grounds occurred (arcing-ground phenomena[2]). Occasionally a conductor of an isolated-neutral system would break, both ends would fall to the ground, and service would be maintained. This could happen only for small currents;

[1] Scattergood, E. F. Engineering Features of Boulder Dam–Los Angeles Lines; *Elec. Eng.*, May, 1935, p. 494.

[2] Peters, J. F., and Slepian, J. Voltages Induced by Arcing Grounds; *Trans. AIEE*, 1923, Vol. 42, p. 478.

Clem, J. E. Arcing Grounds and Effect of Neutral Grounding Impedance; *Trans. AIEE*, 1930, Vol. 49, p. 970.

and with load growth and increased danger to the public, this feature became one of the major hazards of the isolated-neutral system. Transformers were not so reliable as at present, and failures were more common. The isolated neutral systems could quickly resume service with open-Δ-connected transformers. With radial lines, the isolated-neutral systems, on the whole, were subject to fewer and shorter outages than the grounded-neutral systems, which suffered many outages due to the poor quality of insulators available. As transmission systems increased in extent and voltages became higher, the advantages of the isolated neutral largely disappeared. The cost of isolated-neutral transformers for voltages above 73 kV is appreciably higher, since they must have insulation for full line-to-line voltage throughout. The charging current with one conductor grounded became a problem, especially from telephone interference, making it necessary to disconnect a grounded circuit, and left only the advantage of freedom from interruptions due to momentary single-phase-to-ground faults and lightning flashovers, which did not cause an insulator failure, and the advantage of open-Δ-operation. Today few new systems in this country are isolated-neutral systems. There are, however, other systems which from the design standpoint are classed as ungrounded-neutral systems, although they have their neutrals grounded. These are systems grounded through tuned reactances which are designed to neutralize the charging current to ground when one conductor becomes grounded (Petersen coil, Par. **83**) and systems grounded by means of grounding banks of such limited capacity that the neutral shift approaches full line-to-neutral voltage with one phase grounded. The last-named system is generally an isolated-neutral system to which there have been added small grounding transformers to furnish the minimum current required for satisfactory operation of ground relays.

79. Solidly grounded neutral systems, according to the usually accepted meaning, are those in which the neutrals of all transformer banks supplying power to the transmission system and all high-tension neutrals of all receiving transformer banks connected to the transmission system are connected directly to low-resistance station grounds. This permits the use of transformers with graded insulation and one bushing on the high-tension side of single-phase transformers, which results in reduced transformer costs. A possible disadvantage is that ground-fault currents may be excessive, requiring the use of circuit breakers of higher capacity than are required for 3-phase faults. Inductive-coordination problems also may be more difficult. The design considerations connected with neutral grounding are becoming better known, and it is now understood that the solidly grounded neutral is not always best and that neutral grounding must be subjected to a study of balance among the positive-, negative-, and zero-sequence-system impedances in order to obtain the most desirable results. Neutrals can be ungrounded if the windings have full insulation throughout. Transformers with reduced insulation at the neutrals ordinarily should not be ungrounded. While zero-sequence voltages might not exceed the neutral insulation, reflections at the neutral, due to traveling waves, may be sufficient to cause breakdown.

80. Resistance grounding of high-tension transmission neutrals is not used today. It is used on 13.8-kV systems now classed as distribution but at one time classed as transmission. Resistance grounding of high-tension neutrals causes badly distorted voltages to ground during ground faults unless the resistance is such that the system has an essentially isolated neutral. The least distortion in voltage during faults for a given ratio of Z_o/Z_p occurs when the X/R ratios of the three sequence impedances are the

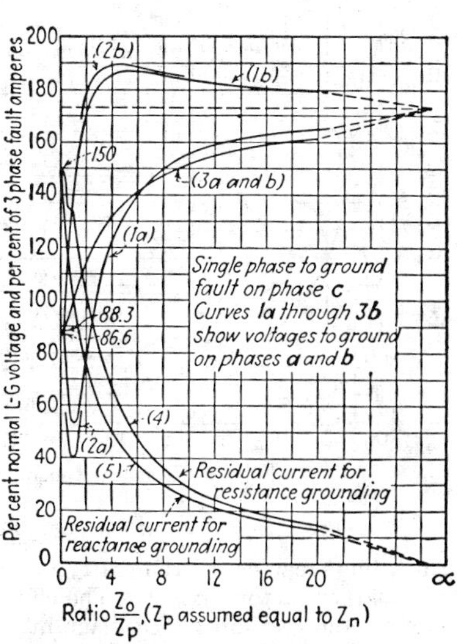

FIG. 13-46. Fault currents and voltages with resistance and reactance grounding.

same. Since resistance grounding tends to unbalance the X/R ratio of the zero-sequence impedance, it can cause distorted voltage conditions during faults. Figure 13-46 shows sample curves of fault voltages to neutral on the unfaulted conductors in percent against ratios of Z_o/Z_p for resistance grounding and for reactance grounding in a system, with reactance predominating. Curve 3*a* and *b* is for reactance grounding. Curves 1*a*, 1*b*, 2*a*, and 2*b* are for resistance grounding.

81. Reactance grounding by the use of neutral reactors or by limited transformer capacity provides an effective means of limiting ground-fault currents in systems in which reactance predominates. A large measure of ground-fault-current reduction can be obtained with reactance grounding, and insofar as lightning arresters and other protective devices are concerned the system will comply with all the requirements of AIEE Standard 32 for an effectively grounded system. On a solidly grounded neutral system having a Z_o/Z_p ratio of 0.5:1 at the generating high-tension bus, the ratio may increase to 4:1 or more at a distant substation depending upon the line length and configuration, owing to the inherent difference in the positive- and zero-sequence reactances X_p and X_o of the transmission line. The Z_o/Z_p ratio or X_o/X_p ratio, if the X/R ratios of the sequence impedances are approximately balanced, at any point on the system determines the neutral-voltage shift at that point and the rise of voltage on the unfaulted conductors during ground faults. The curves of Fig. 13-47 show ratios of sequence voltages and currents to 3-phase line-to-neutral voltage and 3-phase short-circuit current during ground faults for various ratios of Z_o/Z_p and are useful in determining neutral-reactor values, grounding-transformer reactances, and lightning-arrester voltage ratings. The ratio to be selected depends upon the system and will be determined to a large extent by the expected dynamic overvoltages for which the system is designed. In general, the voltage rise in good conductors during faults should be held within the same limits as regards line-to-neutral voltages, which usually will mean that the Z_o/Z_p ratio should not exceed 4:1 at points remote from neutral grounds and will be 1:1 to 2:1 at grounding points.

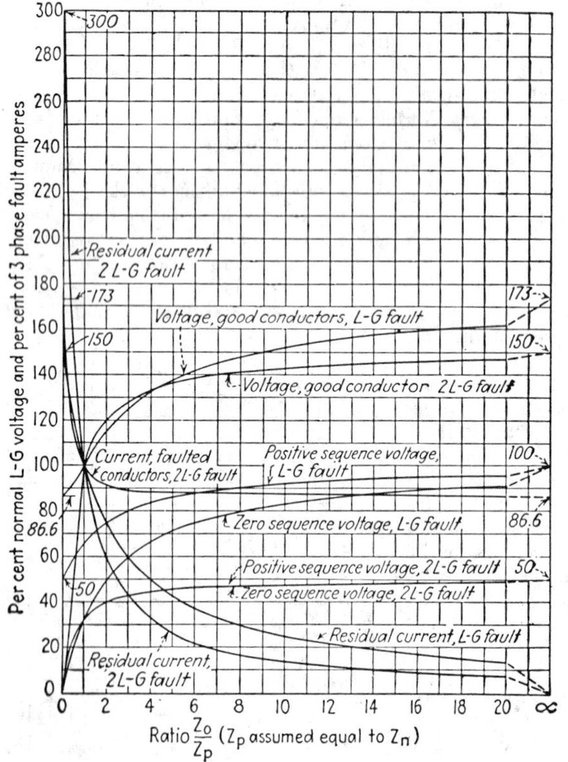

Fig. 13-47. Fault currents and voltages.

On extended high-voltage transmission systems, the zero-sequence capacitance should be considered in determining the zero-sequence impedance at various points and also in determining neutral-grounding reactor current-carrying capacities. The zero-sequence capacitance increases the reactor current. The effect of capacitance can be determined by analyzing a zero-sequence impedance diagram as illustrated in Fig. 13-9 (Par. **21**). The zero-sequence impedance at any point is the resultant of all parallel and series-parallel paths to that point. Neutral-grounding reactors of this type usually are given short-circuit time ratings of 10 to 60 s, depending upon the duty.[1] The maximum initial symmetrical rms current which they may be called upon to carry is used as the current rating, and the voltage rating is equal to this maximum current times the ohms of the reactor. They are usually a nonsaturating type and may be air-cooled or oil-filled self-cooled. The oil-filled type is preferable for outdoor use and for all applications on the higher voltages.

82. Grounding transformers in transmission systems are transformers the purposes of which are the supply of ground-fault current and maintenance of the neutral shift within the desired limits. Grounding-transformer impedances can be determined by the method used for the determination of neutral-grounding reactor values, i.e., by determining the Z_o/Z_p ratios required to obtain the desired results. Z_o/Z_p ratios of less than 1:1 are, in general, undesirable. Figure 13-47 shows a set of curves which are useful for the determination of Z_o/Z_p ratios required. They are applicable to all systems regardless of size as long as the X/R ratios of the positive- and zero-sequence impedances are close together, which is usually the case. The curves are calculated by the method of symmetrical components and give conditions at the point of fault. After the impedance of the grounding transformer has been determined, its kVA fault rating is commonly obtained from the maximum rms fault current which can appear in the neutral of the transformer. This current ($3 \times I_{o\,max}$) of the bank times normal system line-to-neutral voltage, usually for 10 to 60 s, then becomes the rating. Grounding transformers are sometimes installed on the low-tension sides of step-down transformers having Δ-connected low-tension windings, in order to establish 4-wire 3-phase distribution. In such cases, the grounding transformer must have a continuous rating in addition to the fault-current rating if the continuous load is more than 10% of the fault-current rating. The continuous rating is determined by the maximum emergency unbalance in load on the 4-wire 3-phase system and is the total unbalance-neutral current in amperes times the line-to-neutral voltage. Grounding transformers are of two general types: zigzag and Y-Δ (see Fig. 13-48).

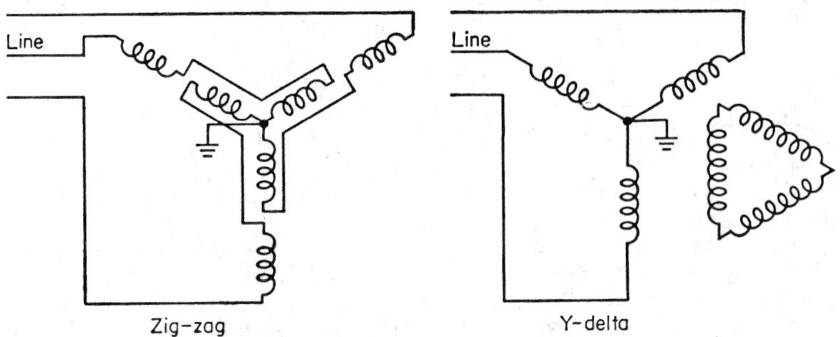

FIG. 13-48. Grounding transformers.

83. The Petersen coil[2] and other arc-extinction coils are neutral-grounding reactors having values of reactance such that the reactive current passed by the neutral reactor when a single line-to-ground fault occurs approximately neutralizes the capacitive current of the system to ground, thereby reducing the fault current to a value that will not sustain an arc. The result is practical elimination of circuit interruptions for momentary single line-to-ground faults due to lightning, birds, tree limbs, etc. Systems utilizing the Petersen coil are essentially isolated-neutral systems and require the same system and equipment insulation. The Petersen coil has wide use in Germany, where it is applied on all system voltages including 230 kV. There are Petersen-coil installations in the United States, ranging in voltage from 23 to 230 kV, all of which are reported to give satisfactory results.[3] However, their use is declining. The balancing of the charging current and the neutral-reactor current is not critical, and quadrature

[1] AIEE Standard 32, Neutral Grounding Devices, effective 1947.

[2] AIEE Committee Report, Guide for Application of Ground Fault Neutralizers; *Trans. AIEE*, 1953, Vol. 72, Pt. III, p. 183.

[3] BROWN, H. H., and GROSS, E. T. B., Experience with Resonant Grounding in a Large 34.5 kV System: *Trans. IEEE Power Apparatus and Systems*, 1966, Vol. PAS 85, No. 5, p. 541.

unbalances of as much as 25% in some systems do not appear to affect the arc-extinction ability. This unbalance has been taken advantage of by one European manufacturer to avoid patent difficulties. It also permits system switching within limits without retuning. There is, however, a limit to the arc-unbalance current which can be broken whether from quadrature or inphase unbalance, and this should be given study in making Petersen-coil applications.

The field of application of Petersen coils appears to be limited in this country. New systems for which the transformers can be bought with full-line insulation can make use of Petersen coils, if they have no interconnections with other systems or if the interconnections provide a break in the zero-phase-sequence circuit. They could, of course, be interconnected with other Petersen-coil systems. Existing ungrounded-neutral systems can be equipped with Petersen coils with the practical elimination of overvoltage troubles due to arcing grounds. Isolated sections of transmission systems, the grounded-neutral transformers of which have full insulation, can be equipped with Petersen coils, usually with great improvement to service.

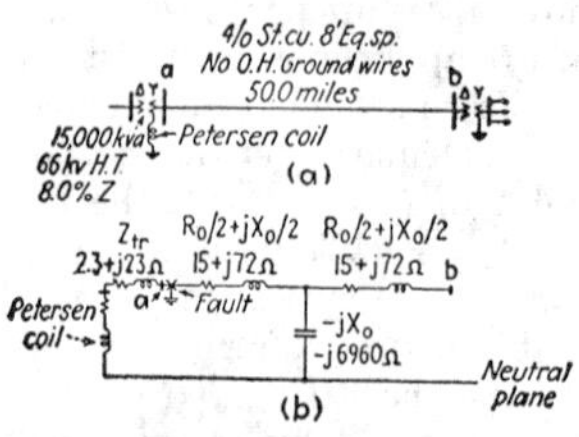

FIG. 13-49. Petersen coil application.

The determination of the reactive ohms of the Petersen coil is not difficult. They are made equal to the system zero-sequence capacitive ohms, and this requires consideration of only the zero-sequence network of the system. The zero-sequence capacitance should be found by tests if at all possible, although experience indicates[1] that calculated values of zero-sequence capacitance if increased 10 to 15% will provide sufficiently accurate results. The coils as built are provided with a tap range sufficiently wide to take care of unforeseen variations in the balancing required. The zero-phase-sequence impedance and capacitance may be calculated by the methods of Par. **11**. Figure 13-49*a* shows a simple system the zero-sequence-impedance diagram of which is given in Fig. 13-49*b*. The transmission line is represented by the nominal *T*, since it will be of sufficient accuracy for this length of line.

SUBSTATION GROUNDING

84. Station Grounding. Grounding systems are provided at substations for two main purposes: (1) safety to operating personnel and to the public and (2) provision of connections to earth for transformer and other power-equipment neutrals. The requirements for each purpose differ, and it is possible for the station grounding system to be satisfactory for one purpose and not for the other.

Substation safety requirements call for the grounding of all exposed metal parts of switches, structures, transformer tanks, metal walkways, fences, steelwork of buildings, switchboards, instrument-transformer secondaries, etc., so that a person touching or near any of this equipment cannot receive a dangerous shock if a high-tension conductor flashes to or comes in contact with any of the equipment listed. This function in general is satisfied if all metalwork between which a person can complete contact, or which a person can touch when standing on the ground, is so bonded and grounded that dangerous potentials cannot exist. This means that each individual piece of equipment, each structural column, etc., shall have its own connection to the station grounding system (mat). These connections should be of heavy copper and should be protected against mechanical damage. In order that all ground potentials around a station shall be equalized, the various ground cables or buses in the yard and in the substation building, especially if the building is at a distance from the main switchyard, should be bonded together by heavy multiple connections and tied into the main station ground. This is necessary in order that appreciable voltage differences to ground may not exist between the ends of signal wires, control cables, or other conductors which may run from the switchyard to the substation building. Ground cables should never be run through conduits of magnetic material. Heavy ground currents such as those which may flow in a transformer neutral during ground faults should not be localized

[1] CHAMPE, W. C., and VAN VOIGTLANDER, F. System Analysis for Petersen Coil Application; *Trans. AIEE*, 1938, Vol. 57, p. 663.

in ground connections (mats or groups of rod) of small area, since the potential gradients in the earth around the ground connections may be dangerous. Ground mats composed of heavy bonded cables and covering a large area are the safest and most satisfactory means of reducing potential gradients in the earth's surface at large substations where heavy currents to ground can be obtained. Another point to be considered, in substations which can have very heavy ground-fault currents, is the tendency of the fault current in the ground mat to follow the path closest to the buses or circuits carrying fault current. This usually causes difficulty only on the low-voltage side of a substation, and the ground-grid conductors buried immediately below these buses and lines must

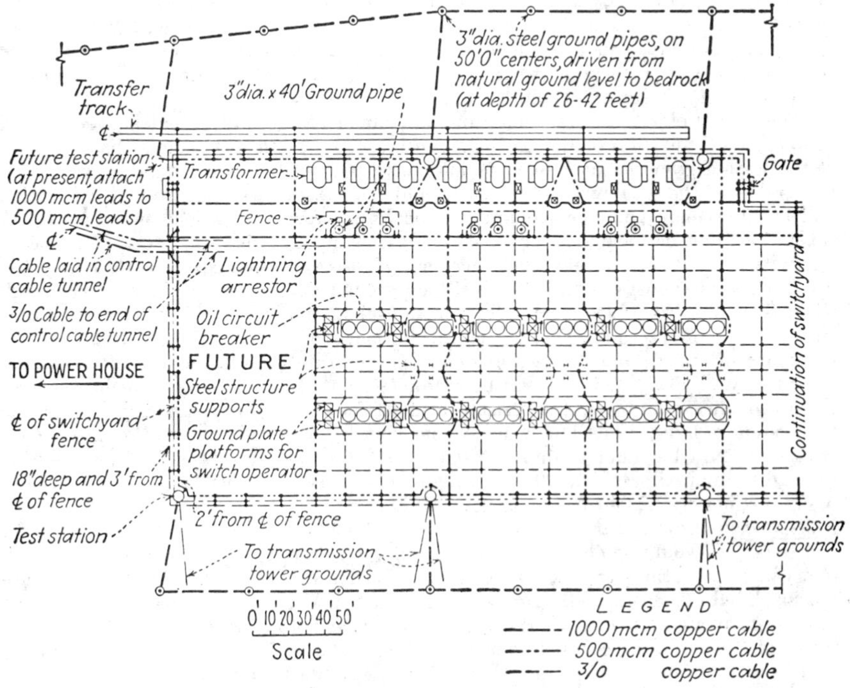

FIG. 13-50. Grounding plan for section of switchyard.

1. All cables within the yard to be buried deep enough for protection against mechanical damage.
2. 1,000-M-cmil cables connected to test stations except where noted otherwise.
3. Distance between parallel conductors not to exceed 20 ft where possible.
4. All ground connections from transformers, lightning arresters, and oil circuit breakers to be 500-M-cmil cable.
5. Cable intersections bound with copper wire and brazed.

be sufficiently heavy to carry the currents. Welded, brazed, or bolted connections are preferable to soldered connections where heavy currents may be carried. The ground mats should be connected to water mains if permission can be obtained and to any other large buried metalwork in the vicinity and in addition should be connected to driven grounds distributed over the ground mat and in the vicinity in sufficient number to keep voltage gradients within safe values during maximum faults to ground. If the switchyard is on soil of high resistivity so that it is impossible to obtain suitably low resistance from rods driven within the yard, the main ground may be located outside the enclosure and connected to the safety ground mat of the yard. Figure 13-50 illustrates the safety grounding scheme of a high-voltage substation and portions of the main driven ground system. The effective resistance of the complete grounding system measured 0.197 Ω as tested by the last method described in Par. **86**. In general, safety grounding is adequate if potentials which may exist between metalwork and ground and on the surface of the ground are equalized and therefore bear no relationship to the resistance value of the ground connection to absolute earth. For small substations, the same principle is advisable, since it provides adequate equalization of potentials in the surface of the ground and eliminates a hazard to the personnel. The conductor sizes in the mat may be reduced but for mechanical reasons should not be smaller than 1/0 and should preferably be larger. The mat should extend not less than 3 ft outside the substation fence, if a metal fence is used, and the fence posts should be connected to the mat in order to avoid the possibility of dangerous potential differences between the fence and the surrounding ground. If the fenced enclosure is of too large an area to be covered by the ground mat, a buried cable encircling the fence and connected to the posts and to the ground mat by multiple connections will provide adequate safety to personnel and the public.

The second objective in station grounding applies to grounded-neutral systems and concerns the limitation of the rise of potential of the earth in the immediate vicinity of the station ground mat above absolute earth potential during ground faults. This rise is due to the resistivity of the earth and the resistance of the ground connection and at times may endanger communication lines entering the field of influence. A substation may be located on ground which is underlaid with solid rock of low conductivity. Individual ground rods may show normal resistance when tested by the usual probe methods, but the overall station ground resistance may be relatively high when measured from a point remote from the substation. It is this ground resistance which enters into zero-phase-sequence calculations. It is therefore essential that the resistance of the ground connection should be accurately measured in order that the effects may be evaluated. It will be readily seen that, from the standpoint of zero-phase-sequence quantities, the resistance of the station ground can vary directly as the system voltage. The lower the system voltage, the lower the station ground resistance should be. For example, 500,000 kVA ground fault at 132 kV would correspond to approximately 2,200 A in the ground connection, whereas the current would be approximately 22,000 A at 13.2 kV. If the ground connection should have 3.1 Ω resistance to absolute earth, the *IR* drop would be 220 V, caused by the 132-kV fault, and the effect would be negligible. The 2,200 V *IR* drop due to the 13.2-kV fault current can cause serious trouble to communication lines entering the station if they are not insulated or neutralized. In addition, the 2,200-V *IR*-drop component of voltage in the zero-phase sequence would cause a rise in voltage to neutral on the order of 25% in one of the unfaulted conductors owing to the effects explained in Par. **80**.

From the foregoing it is seen that no definite values for the resistance of ground connections to absolute earth can be given and that the adequacy of a station ground is determined by the maximum allowable impedance drop of the ground connection and by the distribution of this drop over sufficient area of ground so that dangerous gradients cannot be experienced.

85. Low-resistance station grounds are frequently difficult to obtain, but, as stated before, the lowest-resistance grounds are necessary only for the lower transmission voltages and for distribution voltages, and the values will be limited by the maximum *IR* drop which can be permitted in the ground connection. High-voltage substations are frequently located in places where connections cannot be made to underground piping systems or other buried metalwork which will give low values of ground resistance. In such cases, the use of driven grounds will provide the most convenient means of obtaining a suitable ground connection. The arrangement and number of driven grounds will depend upon the station size and the nature of the soil. Figure 13-51 shows one of the several main-station ground mats for an isolated heavy-duty substation. This mat is outside the substation yard and has a resistance of approximately 0.5 Ω.

The best soils for ground mats are wet and marshy, followed by clay or clay loam. Sand and sandy soils are of higher resistance, making it difficult to obtain low-resistance ground connections.

The size of the rods used is determined mainly by the depth to which they must be driven, although small rods can be driven to considerable depths by the use of driving

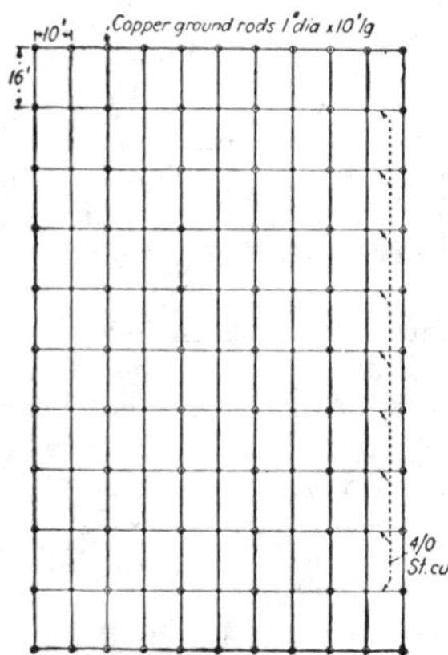

FIG. 13-51. Ground mat placed in ground at depth of 2 ft; all conductors 500,000 cmils stranded copper except as noted.

collars. Figure 13-52 shows the relationship between rod size and resistance obtained. Driving more rods in a given space will help reduce resistance, but the reduced resistance is not a function of the number of rods. Figure 13-53 shows the effect of spacing and number of rods in square areas of the resistance. These curves apply to ¾-in by 10-ft rods. The rods or pipes for permanent stations should be of noncorroding materials. Galvanized rods or pipes may be used for temporary installations. Figure 13-54 shows the effect of increased length of rods in uniform soil. Usually the improvement is much greater than indicated owing to the fact that the rods penetrate into better-conducting earth as they are driven deeper.

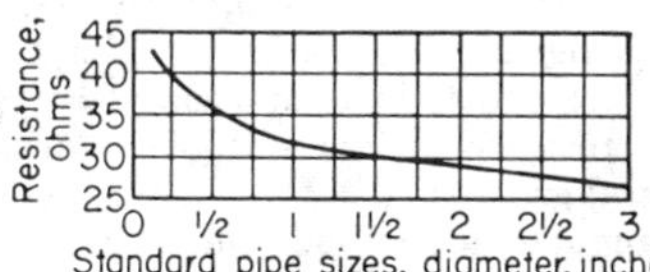

FIG. 13-52. Relation between pipe diameter and ground resistance. (*NBS Tech. Paper* 108, *June*, 1918.)

In general, it is advisable to obtain reduced ground resistance by the use of a more extensive mat and more ground rods rather than by treating the earth around the rods with salt, because of the impermanence of the treatment. However, treatment of the soil is sometimes the only means whereby suitable resistance can be obtained.

It is not possible to describe all methods of obtaining ground connections of suitably low resistance. The problem sometimes presents great difficulties and calls for considerable extra expense. Substations should not be located on solid rock with little or no topsoil, since the cost of obtaining a low-resistance ground would be excessive. Such a ground would require the use of an extensive counterpoise system with many drilled "wells," in which electrodes would be inserted in treated filling, with provision made for renewing the treatment.

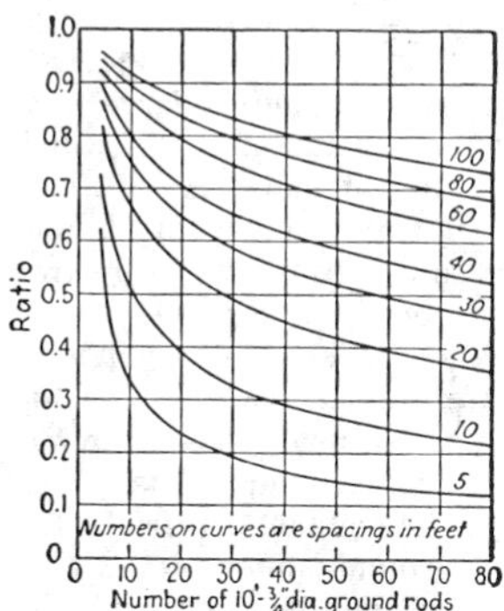

FIG. 13-53. Ratio of conductivity of ground rods in parallel on an area to that of isolated rods. (*H. B. Dwight, Trans. AIEE*, 1936, *Vol.* 55, *p.* 1936.)

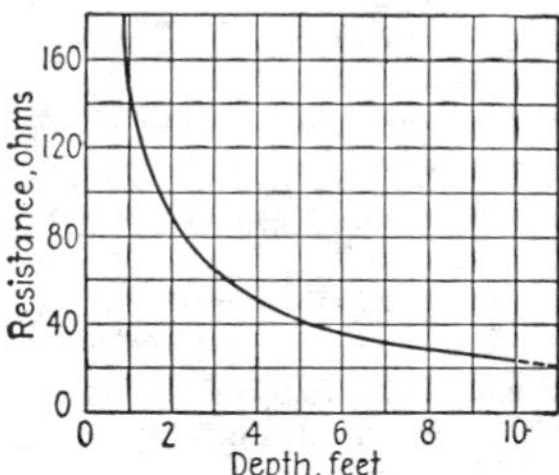

FIG. 13-54. Variation of resistance of driven pipes with depth. Soil fairly wet. External diameter of pipe 1.02 in. (*NBS Technologic Paper No.* 108.)

86. Ground-resistance Test Methods. The measurement of ground resistance is necessary both at the time of initial service and at periodic intervals thereafter to determine the adequacy and permanence of the ground connection. The measurement of the resistance of a ground connection with respect to absolute earth is quite difficult, and all results are approximations of varying degrees of accuracy. The simpler methods of testing the resistance of small-area grounds are tenable owing to the fact that practically all the resistance of a ground connection is concentrated in the earth immediately surrounding the connection. For a single $\frac{3}{4}$-in by 3-ft rod ground, approximately 90% is within a 6-ft radius. For large isolated mats, the 90% zone may extend beyond a 1,000-ft radius. Extensive grounding systems consisting of mats connected to pipelines, overhead ground wires, etc., will have still greater radii for the 90% zone, which makes it quite difficult to determine the resistance accurately by any test methods.

There are many methods of testing ground resistance, all of which have limited fields of use. All the best methods circulate some form of alternating current through the ground under test in amounts from a few milliamperes, as in bridge methods and with some of the patented ground testers, up to 100 A or more. The amount of current used depends upon the method, and methods using very small currents will give results as accurate as methods using heavy currents if the ground under test is one for which the test method is suitable. Methods of testing ground resistance fall into three general groups:

a. **Triangulation methods,** in which two auxiliary test grounds are used and the series resistance of each pair of grounds is measured. Measurements may be made by the voltmeter-ammeter method or by means of a suitable bridge. For accurate results the resistance of the auxiliary grounds and the ground under test should be in the same order of magnitude. Results may be meaningless if the test grounds have more than 10 times the resistance of the ground under test. This method is suitable for measuring the resistance of tower footings, isolated ground rods, or small grounding installations. It is not suitable for measurement of low-resistance grounds.

b. **Ratio methods,** in which the series resistance R of the ground under test and a test probe is measured by means of a bridge. A slide-wire potentiometer is then connected to the two ground connections, with the slider of the potentiometer connected to a second test probe. The potential of the slider to ground is adjusted to zero. If D is the total slide-wire resistance and d_1 is the resistance from the slider to the ground under test, the resistance of the ground under test is $(d_1/D) \times R$. The Groundometer, a self-contained test instrument, makes use of this principle. This method is much more satisfactory than triangulation methods, since ratios of test-probe resistance to the resistance of the ground under test may run as high as 300:1 with test instruments such as the Groundometer. This method has its limitations in testing low-resistance grounds of large area, as do all methods which do not make use of reference grounds or probes miles from the large-area ground under test.

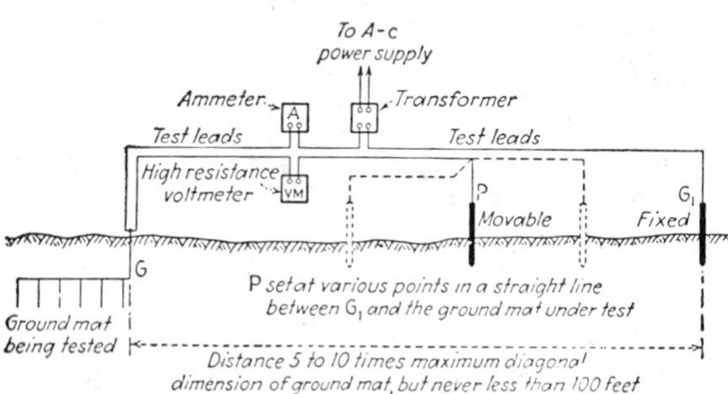

FIG. 13-55. Field setup for making ground-resistance tests by means of fall-of-potential method.

c. **Fall - of - potential methods,** which include those using close-in reference grounds, usually less than 1,000 ft from the ground under test, and a special method using remote reference grounds, usually many miles away from the ground under test. The principle of the fall-of-potential method using close-in reference grounds is illustrated in Fig. 13-55. Alternating current is circulated through ground G and a fixed test ground G_1, which may consist of one or more driven grounds. A high-resistance voltmeter is connected to ground G and to a movable test probe P. P is moved along a line from G to G_1, and voltmeter readings E are taken simultaneously with ammeter readings I. $E/I = \Omega$ is plotted as shown in Fig. 13-56. The resistance shown on the flat part of the curve or at the point of inflection is taken as the resistance of the ground. This method may be subject to considerable error if stray currents are present. It is frequently applied by using only one test-probe reading, with the test probe located midway between G and G_1. Results will be satisfactory if there are no buried pipes or other conductors near the test probe. When such conditions are suspected, the curve should be plotted as in Fig. 13-56. Self-contained test instruments which make use of this method are available; among them are the Ground Ohmer and Megger ground tester. These instruments give better results than the voltmeter-ammeter method, since they are designed to eliminate the effects of stray currents. The above methods are subject to limitations in testing low-resistance large-area grounding systems. They are also difficult to use in congested urban districts.

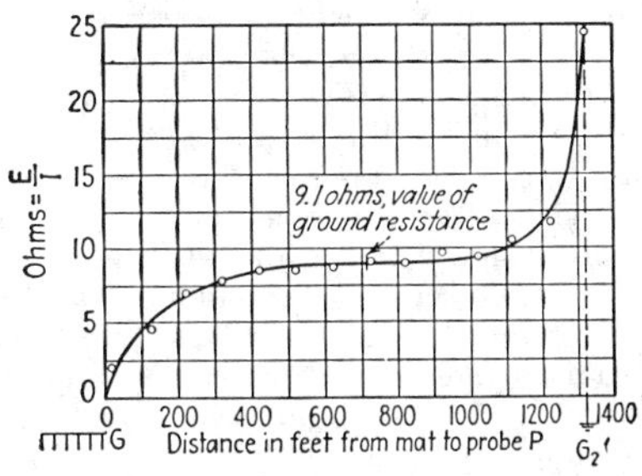

FIG. 13-56. Ground-resistance curve for a small substation ground mat.

A **special fall-of-potential method**[1] has been developed in an effort to overcome the limitations inherent in all previously described methods of measuring the resistance of extensive low-resistance grounding systems usually located at generating plants and large substations. This method is based upon the difference in potential of the grounding system to be tested with respect to a remote reference ground when the grounding system is energized by a ground-return current. This method requires an adequate a-c supply to energize the grounding system under test, with the energizing source located at a considerable distance beyond the influence of the grounding system, and a conductor to establish a remote reference ground. Transmission lines entering a generating plant or substation are particularly suitable for use as power-supply circuits for this type of test. Telephone lines serving the generating plant or substation are very useful in establishing connections to remote reference grounds. A vacuum-tube voltmeter having a resistance of 100,000 Ω or more must be used in order that the total voltage drop produced by the test current flowing into the grounding system may be impressed across the voltmeter. The vacuum-tube voltmeter must be of the type not affected by d-c potentials which may be present, or a blocking condenser must be used in the meter circuit.

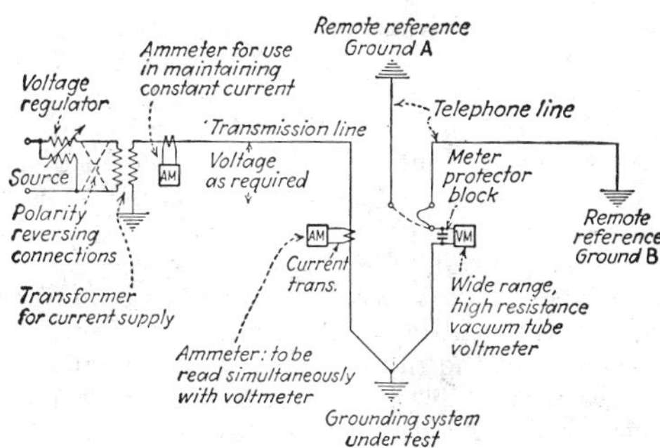

FIG. 13-57. Setup for measuring resistance of grounding system by fall-of-potential method, using high-resistance voltmeter and remote current supply.

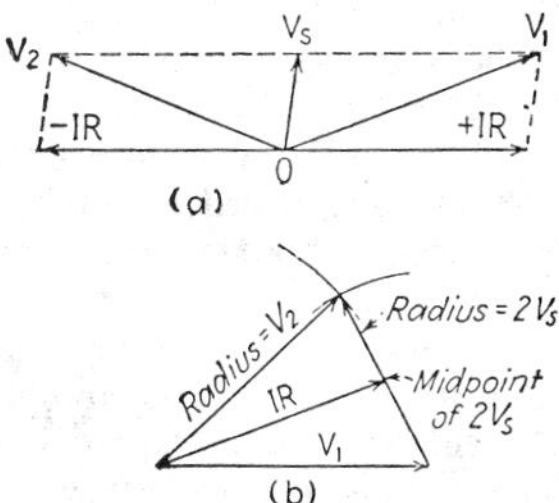

FIG. 13-58. (*a*) Voltages in ground-resistance tests. (*b*) Graphical determination of IR. V_s = standing ground potential between grounding system and remote reference ground. V_1 = voltage reading with $+I$ A to ground. V_2 = voltage reading with $-I$ A to ground (polarity reversed). IR = voltage produced by I A flowing in station ground of R Ω.

Figure 13-57 shows the schematic circuit for determining the resistance of a grounding system by this modified fall-of-potential method. The energizing transformer is shown at the remote end of the transmission line but can be located at the station under test if desired. The choice of location will depend upon the availability of an adequate power supply.

[1] Joint Subcommittee on Development and Research, Edison Electric Institute and Bell Telephone System, Low Resistance Grounds; *Eng. Rept.* 31, June, 1935.

The farther the reference ground is established from the grounding system under test and from all overhead and underground metallic conductors connected to the grounding system under test, the more accurate are the results to be expected. The precaution must be taken, also, to select reference-ground circuits widely separated from the current-supply circuit in order to avoid induced potentials in the remote reference-ground circuit from the power-supply circuit energizing the grounding system. With the cooperation of commercial telephone companies, as many remote reference grounds as are found desirable can be established, and the average effective value of the grounding-system resistance can be determined by the several tests.

The voltages involved in determining the effective resistance R of a grounding system are shown in Fig. 13-58*a*. V_s is a standing ground potential usually found between two widely separated grounds. This potential must be measured prior to application of the test current. The test current is then applied, and voltage V_1 and current I readings taken. If the voltage reading V_1 is more than 10 times V_s, V_s can be neglected, since its effect on results will be not more than 1%. If V_s cannot be neglected, it must be subtracted vectorially as follows: The polarity of the current energizing the grounding system is reversed, and the test repeated. The voltage reading on reversed polarity is designated as V_2. CAUTION: The current energizing the grounding system must be held at the same value during both sets of readings; or if the current cannot be held the same, V_2 as read should be changed directly as the ratio of the two current readings for use in calculating R. Having V_1, V_2, and V_s, IR can be found graphically (Fig. 13-58*b*) or from

$$(IR)^2 = \frac{V_1^2}{2} + \frac{V_2^2}{2} - V_s^2 \qquad (13\text{-}67)$$

Having IR, R is readily determined. The values of R determined by means of the several reference grounds will usually be of negligible difference in magnitude. If differences are relatively large, the test setup should be checked, and the tests rerun.

TRANSFORMER CONNECTIONS

87. Delta-delta-connected (Δ-Δ) transformers are used mainly on the lower transmission voltages. This is due to the fact that they must be insulated for full line-to-line voltage; and for voltages above 73 kV the cost increase, as stated in Par. **78**, is appreciable over Y-connected transformers with graded insulation. The Δ-Δ-connected transformers have one advantage in that the bank can be operated open Δ at 86.6% of the capacity of the two remaining transformers.

88. The delta-star (Δ-Y) **connection** is in common use for both step-up and step-down purposes. When used as a step-up transformer, the high-tension winding is Y-connected; and when used for step-down purposes, the low-tension winding is usually Y-connected in order to provide a grounded neutral for secondary transmission or for primary distribution. Δ-connected high-tension windings, however, are seldom used for transmission voltages of 138 kV and above. The Δ-Y-connection almost completely suppresses the triple harmonics even with the neutral solidly grounded. Triple harmonics which can appear on power systems are the third and its odd multiples. Y-connected windings on the higher voltages are usually provided with graded insulation, the neutral-end turns of which may have very little insulation if the neutral is solidly grounded. If neutral impedance (reactor or resistor) is used, the neutral insulation must be equal to or greater than the maximum ZI drop of the neutral impedance. If the neutral is to be left ungrounded on either grounded neutral systems or ungrounded neutral systems, the neutral insulation should be the same as it is on the line side to avoid traveling-wave troubles.

89. Star-star-connected (Y-Y) transformers are used infrequently on high-voltage transmission systems. When used with both neutrals grounded, if single-phase or 3-phase shell type, they must be used with Y-connected generators, and a solid neutral connection must be provided between the generator, or generators, and the low-tension transformer neutral in order to minimize triple-harmonic troubles. All three types of Y-Y-connected transformer can be used with both neutrals ungrounded with satisfactory results or with neutrals grounded if of the 3-phase core type. The triple harmonics are nearly suppressed in 3-phase core-type transformers.

90. Star-star-connected transformers with a Δ-connected third winding (tertiary) (Y-Δ-Y) overcome the difficulties of the simple Y-Y-connection. The tertiary winding may be for the suppression of harmonics only, in which case no connections are brought out with 3-phase transformers. Y-Δ-Y-transformers are frequently used to supply two distribution voltages or a distribution voltage and a secondary-transmission voltage. If the service supplied from the Δ-connected winding is 4-wire 3-phase, the neutral must be obtained from a grounding transformer (see Par. **82**). A common use for the tertiary winding is for the operation of synchronous condensers when the low-tension Y-voltage is above the range of available condenser voltages or if the main purpose is the regulation of the high-tension voltage. Three-winding transformers all windings of which are used are frequently rated with two outputs: (1) the individual output of each secondary winding alone with the other secondary winding carrying no load and (2) a simultaneous loading rating in which each secondary winding is given a rated loading with the primary winding loading the resultant of the two secondary loadings.

91. Autotransformers are generally used for transforming from one transmission voltage to another when the ratio is 2:1 or less. Such transformers are best connected in Y with the neutral solidly grounded and when so connected should be provided with a closed-Δ tertiary winding of adequate capacity for the suppression of harmonics and for ground-fault duty. Autotransformers are superior to separate-winding transformers owing to lower cost, greater efficiency, smaller size and weight, and better regulation. Autotransformers may also be obtained with zigzag-connected windings or with Δ-connected windings. Both these types are free from triple-harmonic troubles but in general are more expensive. Δ-connected autotransformers have a possible disadvantage in that they insert a phase shift into the transformation, which means that the system being served must be radial or else it must be served by similar transformations at other points.

92. Transformer Fusing. Transformers are frequently protected by high-tension fuses when economic consideration will not permit the installation of oil circuit breakers. If the low-tension feeders are equipped with oil circuit breakers and the time-current characteristics of the fuses are properly coordinated with the feeder relays, satisfactory operation will result. It is essential, however, that fuses of known and accurate time-current characteristics be used and that they have *adequate short-circuit interrupting capacity*. Fuses are available for voltages up to 161 kV with limited short-circuit capacities. Fusing in general is not so satisfactory as the use of circuit breakers.

LINE AND STRUCTURE LOCATION

93. Preparation for Construction. The cost of preparing for transmission-line construction is a considerable part of the total—under some conditions as much as 25%. Right-of-way and clearing are more or less fixed by local conditions, but the cost of surveys, accompanying maps, profiles, and engineering layout is to some extent governed by judgment. Many times in the past the overall costs have been increased by right-of-way difficulties and by delays in receiving proper materials because of inadequate preparations. This engineering work, properly carried out, makes it possible to purchase the right-of-way and complete the clearing well in advance of construction and to purchase every item of material and deliver it to the correct location.

The work of locating and laying out a line does not require great refinement, but every check must be made to prevent errors. With inexperienced surveyors or draftsmen it must be assumed that errors will be made, and every possible device must be used to discover these errors before construction is started.

94. Location. The general character of the country in which the line is to be located should be determined, for it has a definite bearing on the type of design. In extreme cases, such as difficult mountainous sections or in highly developed areas near cities, this may be a determining factor in the selection of the conductor and type of structures.

On heavy trunk lines, minor repairs and replacements are not an important item, and accessibility may often be rightly sacrificed to obtain the economy of a more direct route. Light wood lines must, however, be readily accessible for inspection and repairs. Line location is a matter of judgment and requires a man of wide general experience capable of correctly weighing the divergent requirements for cheap right-of-way, low construction costs, and convenience in maintenance. A man trained in operation inclines toward too devious a route; a surveyor, toward too direct a line. In mountainous country or in thickly populated areas, it is generally not advisable to attempt a direct route or try to locate on long tangents. Small angles of a few degrees cost little more and add little to the length of line. Most designs provide suspension structures for line angles of 5 to 15° which are not excessively costly. High, exposed ridges should be avoided, to afford protection against both wind and lightning.

No effort should be spared to secure the best available maps from every source, including U.S. Geological Survey maps, both published and preliminary, state geological survey maps, and aerial topographic maps.

Following a general reconnaissance by ground and air, for which 10 to 20 days/100 mi should be allowed, and the assembling of all available maps and information, control points can be established for a general route or areas selected for more detailed study which may prove to be determining factors in the location of the line.

With this preliminary work completed, the major difficulties should have been determined, and any advice or suggestions and any general rules or future plans must be thoroughly discussed. The policy as to such matters as right-of-way condemnation, telephone coordination, navigable-stream crossings, air routes, airports, and crossings with other utilities must be decided as definitely as possible.

Preliminary specifications, which definitely state all such matters of general policy and give a general outline of the line design, should be issued before the final survey is started. These should include (1) outline drawings of the various structures with the important dimensions, (2) conductor sag curves and a sag template, (3) the maximum spans and angles for each type of structure, and (4) the requirements for right-of-way and clearing. Estimated costs are valuable, especially comparative costs of the various types of structure. With this information the field engineer can often, in a difficult section, choose the location best suited to the design.

95. Aerial maps can often be secured at much less cost than preliminary surveys and in highly developed areas may be used to advantage for completely laying out the line without sending surveyors into the area until after the right-of-way has been secured.

Photographs taken at approximately ½ mi to the inch give sufficient detail for most work. Such maps can be photographically enlarged about four times for special detail. With a ½-mi-to-the-inch scale, the route of the line can be determined within a width of about 3 mi and sufficient landmarks located on a fairly accurate map to serve as a guide for flying the line.

96. Location Survey. If the preliminary work has been thoroughly done, the actual survey party can be divided into four divisions, each of which, if well organized with experienced men, can complete at least a mile a day in average weather and country. Their operations may be carried out separately or, to better advantage, nearly concurrently by allowing a full week's separation between successive operations and using good judgment in transferring reinforcements as needed.

The work falls naturally into the following: (1) an alignment party, picking the exact location and cutting out the line; (2) a staking party, driving stakes at 100-ft stations and locating all obstructions; (3) a level party, taking elevations and side slopes; (4) a property and topography party, locating property lines.

A field drafting force located at a convenient point for receiving field notes can complete the final plan and profile drawings as fast as the survey can be made.

The method of procedure and size of survey organization depend upon the character of the country, the length and type of line, the experienced men available, and the schedule which must be maintained. In level, sparsely populated country, satisfactory, but incomplete, property surveys and profiles have been made during an open dry winter for a wood H-frame line 50 mi in length in approximately 4 months' time, with a crew hired for the job averaging eight men and an engineer.

On a development of considerable size involving the construction of several hundred miles of line, the survey for a 65-mi steel-tower line in rather difficult country, including 25 mi of inaccessible mountainous country, was completed with property maps and profiles in the form for permanent records in 2 months' time, with an experienced and thoroughly organized crew averaging 20 men and a locating engineer.

The factors which make an organization suitable for one job and entirely impossible for the other are evident.

97. Right-of-way purchase requires careful preliminary work, checking the ownership reported by the field men against the records to discover faulty titles, transfers, joint owners, foreclosed mortgages, etc. This work should be started with the survey, as progress is usually slow and is often the determining factor in a schedule.

Generally, right-of-way is not purchased in fee, but a perpetual easement is secured in which the owner grants the necessary rights to construct and operate the line but retains ownership and use of the land. The width of the right-of-way may be stated as a definite width or in general terms, but the easement must provide for (1) a means of access to each structure; (2) permission to erect all structures and guys; (3) all trees and brush to be cleared over a width at least 10 ft greater than the spread of the conductors, so that a free working space is allowed for erection; (4) the removal of all trees which would not safely clear the conductor if the conductor were to swing out under maximum wind and all trees which would not safely clear the conductor if they were to fall; (5) the removal of all buildings, lumber piles, haystacks, etc., which constitute a fire hazard. One of the major causes of serious line outages is the neglect to adhere strictly to conservative rules for clearing.

98. The plan and profile should be permanent drawings, as they are the key construction drawings, right-of-way record, and permanent property record.

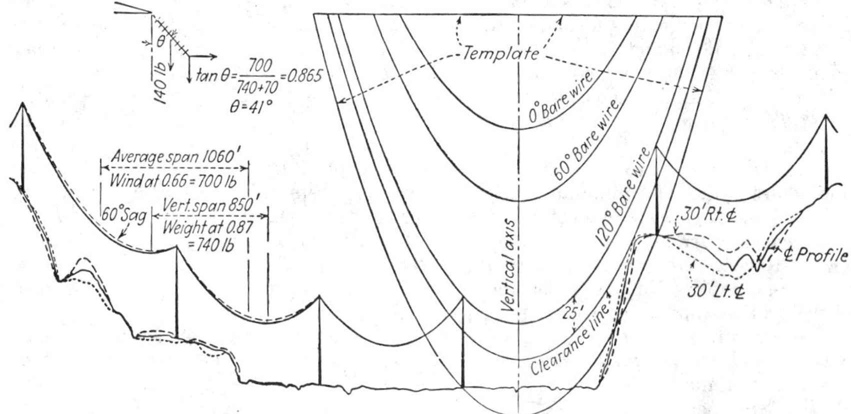

Fig. 13-59. Sag template determines clearance of suspended conductor from ground.

The vertical scale should be considerably exaggerated, with a horizontal scale not greater than is required for scaling distances with necessary accuracy. Horizontal distances seldom are required closer than 10 ft, while vertical distances should be shown to the closest foot. A scale of 400 ft to the inch horizontally and 40 ft to the inch vertically is often used and gives a most compact drawing with sufficient accuracy for most conditions. Larger scales may be preferred for short spans; a smaller scale is not recommended.

99. The location of structures on the profile with a template is essential for both correct design and economy. With care, reasonable competence, and some experience, this method can be relied upon to give (1) ample clearances, (2) average spans and structure heights within a few percent of the design requirements, (3) structures carrying very nearly their design loads, and (4) exactly the correct quantity of material purchased and delivered to the proper site.

A celluloid template, shaped to the form of the suspended conductor, is used to scale the distance from the conductor to the ground and to adjust structure locations and heights to (1)provide proper clearance to the ground, (2) equalize spans, and (3) grade the line (see Fig. 13-59). The vertical axis of the template must be held vertical.

The template is cut as a parabola on the maximum sag (usually at 120°F) of the ruling span (see Par. **116**) and should be extended by computing the sag as proportional to the square of the span for spans both shorter and longer than the ruling span. By extending the template to a span of several thousand feet, clearances may be scaled on steep hillsides. The form of the template is based on the fact that, at the time when the wire is erected, the horizontal tensions must be equal in all spans of every length, both level and inclined, if the insulators hang plumb. This is still very nearly true at the maximum temperature. The template, therefore, must be cut to a catenary or, what is approximately the same thing, a parabola. The parabola is accurate to within about one-half of 1% for sags up to 5% of the span, which is well within the necessary refinement.

Since vertical ground clearances are being established, the 120° no-wind curve is used in the template. Special conditions may call for clearance checks. For example, if it is known that a line will have high temperature rise due to the current to be carried, conductor clearance should be checked for the estimated maximum conductor temperature. One crossing over a navigable stream was designed for 190°F at high water. Ice and wet snow many times cause weights several times that of the ½-in radial ice loading, and conductors have been known to sag to within reach of the ground. Such occurrences are not considered in line design, and when they happen, the line is taken out of service until the ice or snow drops. Checks made afterward have nearly always shown no permanent deformation.

The template must be used subject to a "creep" correction when aluminum conductors are involved. "Creep" is a perpetual nonelastic conductor stretch which continues for the life of the line. All conductors of all materials are subject to creep, but to date only aluminum conductors have had intensive study. Creep is not substantial in other conductors, but the conductor manufacturers should be consulted. The IEEE Committee Report, Limitations on Stringing and Sagging Conductors, in the December, 1964, *Transactions of the IEEE Power Group* discusses creep, and the reader should examine that report.

Creep causes a continuous slow increase in the sag of the line which must be estimated and allowed for. The aluminum-conductor manufacturers will furnish creep-estimating curves. These curves are at approximately constant temperatures, around 60 to 70°F, and plot stress against elongation, one curve for each period of time, 1 h, 1 day, 1 month, 1 year, 10 years, etc. The values are integrated values for the period and are considered to be reasonable estimates. The temperature used is taken as a reasonable average of the year's temperature across the center of the United States.

Precise values for creep are impossible to determine, since they vary with both temperature and tension, which are continuously varying during the life of the line. From Fig. 3 of the above committee report, it is found that a 1,000-ft span of 954,000-cmil 48/7 ACSR when subjected to a constant tension of approximately 18% of its ultimate strength at a temperature of 60°F will have a sag increase of approximately 5.5 in in 1 day, 13 in in 10 days, 27 in in 1 year, 44 in in 10 years, and 52 in in 30 years.

Unless it is known that the line will have a life of less than 10 years, not less than 10 years creep should be allowed for. Creep has come into consideration in transmission-line design only during the past 15 years, and to date no standards have been established for handling it. Probably the simplest approach is to check all close clearance points on the profile with a template made with no creep allowance and to specify higher structures at these points if the addition of liberal creep sag infringes on the required clearances. It is possible to prestress the creep out of small conductors, but for large conductors this requires time and special tensioning facilities not normally available. Also the time lost in constructing an EHV line will more than pay for the extra structure height required to compensate for the creep. Prestressing changes the modulus of elasticity, and this new modulus should be used in the design.

The vertical weight supported at any structure is the weight of the length of wire between low points of the sag in the two adjacent spans. For bare-wire weights, this distance between low points can be scaled by using a template of the sag at any desired temperature. The maximum weight under loaded conditions should be scaled from a template made for the loaded sags. For most problems, the horizontal distance may be taken as equal to the conductor length. Distances to the low point of the sag may be computed by Eq. (13-88) (Par. **122**).

Uplift. On steep, inclined spans the low point may fall beyond the lower support; this indicates that the conductor in the uphill span exerts a negative or upward pull on the lower tower. The amount of this upward pull is equal to the weight of the conductor from the lower tower to the low point in the sag. Should the upward pull of the uphill span be greater than the downward load of the next adjacent span, actual uplift would be caused, and the conductor would tend to swing clear of the tower.

It is important that abrupt changes in elevation of the structures should not occur, so that the conductor will not tend to swing clear of any structure even at low temperatures. This condition would be indicated if the 0° curve of the template can be adjusted to hang free of the center support and just touch the adjacent supports on either side. In northern states it would be well to add a curve to the template for the below-zero temperatures experienced.

Insulator Swing. The uplift condition should not even be approached in laying out suspension insulator construction; i.e., each tower should carry a considerable weight of conductor. The minimum weight that should be allowed on any structure may be logically determined by finding the transverse angle to which the insulator string may swing without reducing the clearance from the conductor to the structure too greatly and by requiring that the ratio of vertical weight to horizontal wind load be kept such as not to allow the insulator to swing beyond this angle. The maximum wind is usually assumed at a temperature of 60°F. The wind pressure in pounds per square foot to be used in swing calculations is a matter of judgment and depends upon local conditions. The National Electrical Safety Code, 6th edition, gives 30° as the minimum angle of conductor swing to be used in calculations where proximity to other circuits is involved. Under high-wind conditions it is reasonable to require somewhat less than normal clearances. Generally a clearance corresponding to about 75% of the flashover value of the insulator is adequate. The insulator will swing in the direction of the resultant of the vertical and horizontal forces acting on the insulator string as shown in Fig. 13-59.

100. Long spans, necessitated by rough country, may be considerably longer than contemplated in the design and may involve a number of factors including (1) proper clearance between conductors, (2) excessive tensions under maximum load, and (3) structures adequate to carry the additional loads.

Safe clearance between conductors is often based on the Safety Code formula, in which the spacing a in inches is given as proportional to the square root of sag. d is in inches.

$$a = 0.3 \text{ in/kV} + 8\sqrt{d/12} \tag{13-68}$$

This relation was developed for and is useful on comparatively short-span lines of the smaller conductors and for voltages up to 69 kV; but for very long spans and heavy conductors, the formula results in spacings considerably larger than have operated without difficulty. It also results in spacings that are questionably small for very light conductors on long spans. Percy H. Thomas proposed an empirical formula which takes into account the weight of the wire and its diameter, requiring less spacing for heavy wires and a greater spacing for small wires by the ratio of diameter in inches to weight in pounds per foot D/w (discussed in Par. **125**) as a means of determining the required conductor spacing for the average span of the line. The factor C [Eq. (13-92)] includes an allowance to permit the standard spacing to be used on somewhat longer spans than average construction. The same formula, however, may be used to examine the spacings which have been successfully used on maximum spans and a value for C selected from experience for determining the safe spacing required for an occasional unusually long span. An example is given in Par. **155**.

Excessive tensions on very long spans may be avoided by dead-ending at both ends and computing such a stringing sag as will result in the same maximum tension as elsewhere in the line. Such a span will be found to have considerably greater stringing sag and lower stringing tension than the normal span. Sag curves or charts are often prepared giving the sag for dead-end spans of various lengths such that the maximum tension under loaded conditions will be the same.

Dead-end construction is costly, and consideration should be given to avoiding this additional expense. It is common practice to permit spans up to double the average span without dead ends, although spans of this length may require additional spacing between wires. A careful examination of some trial figures on the sags and tensions developed in a long span will often indicate how great a span may be carried on suspension structures. The maximum loaded tension which would occur in a long span, if this span were dead-ended and sagged to the same stringing tension as the rest of the line, compared with the maximum tension for normal span lengths, is a good indication of the necessity for dead-end construction.

In case a number of long spans are encountered in a line or section of line, it may prove more economical to reduce the tension in the entire section to the long-span values and accept an increase in sag and corresponding reduction in span length in order to avoid dead ends.

MECHANICAL DESIGN OF OVERHEAD SPANS

101. Conductor Loads. The span design consists in determining the sag at which the conductor shall be erected so that heavy winds, accumulations of ice or snow, and low temperatures, even if sustained for several days, will not stress the conductor beyond the elastic limit, cause a serious permanent stretch, or result in fatigue failures from continued vibrations.

The dead weight of the conductor and the weight of accumulated ice or snow act vertically; the wind load is assumed to act horizontally and at right angles to the span; the resultant is the vectorial sum. Under combined vertical and horizontal loading the conductor swings out into an inclined plane whose angle with the vertical is the angle between the direction of the vertical force and the resultant force. The resulting deflection is measured in this inclined plane.

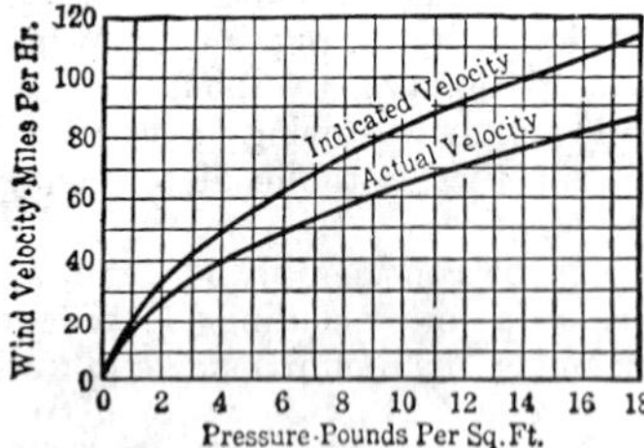

FIG. 13-60. Wind velocity and pressure.

Wind pressure in pounds per square foot, p, as a function of the actual wind velocity in miles per hour V, is given by Buck's formula[1] for cylindrical surfaces,

$$p = 0.0025V^2 \qquad (13\text{-}69)$$

which is generally accepted in span computations. The pressure on flat surfaces is generally taken as

$$p = 0.004V^2 \qquad (13\text{-}70)$$

The relation between actual wind velocity and indicated wind velocity is shown in Fig. 13-60. However, this relation is not entirely definite, and with any wind-velocity data, correction factors should be obtained from the U.S. Weather Bureau.

102. Assumed Simultaneous Weather Conditions in the United States. See Table 13-13.

103. Safety Code loadings have generally been accepted as a guide in determining the thickness of ice, wind velocity, and temperature which may be expected in any section of the country (see Fig. 13-61). These loading assumptions are convenient as a basis of design, in that the loads caused by ice, wind, and low temperatures are assumed to occur simultaneously; however, consideration should be given to past experience and local conditions.

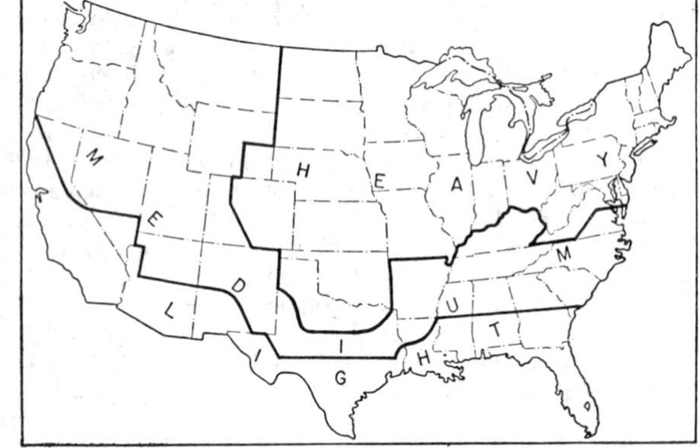

FIG. 13-61. NESC district loading map of United States for mechanical loading of overhead lines. (*6th ed.*)

[1] FOWLE, FRANK F. A Study of Sleet Loads and Wind Velocities; *Elec. World*, 1910, Vol. 56, p. 995. LAMB, E. H. Behavior of Overhead Transmission Lines in High Winds; *Jour. IEE*, 1928, Vol. 66, p. 1079.

Table 13-13. Assumed Simultaneous Weather Conditions in the United States*

Loading district	Radial thickness ice, in.	Horizontal wind pressure, lb/sq ft on projected area of cable	Temp, deg F	Constant K, lb per ft†	
				For bare conductors of copper, steel, copper alloy, copper-covered steel, and combinations thereof	For bare conductors of aluminum (with or without steel) and weatherproof and similar covered conductors of all materials
Heavy loading (H).......	0.5	4	0	0.29	0.31
Medium loading (M).....	0.25	4	+15	0.19	0.22
Light loading (L)........	0	9	+30	0.05	0.05

* Nat. Elec. Safety Code, 6th ed., Bur. Stds. Handbook H81, 1961.
† To be added to the resultant of the horizontal and vertical forces in the direction of the resultant.

For instance, accumulation of ice and snow on the conductors is rare in Minnesota, but extremely low temperatures are common; ice loads considerably greater than heavy loading but without extreme winds have occurred on several occasions from Maryland to New England, as well as in many other locations.

Unit wind and ice loadings for conductors are found by the following formulas,

$$\text{Wind load (lb/ft)} = \frac{p}{12} D \tag{13-71}$$

$$\text{Ice load (lb/ft)} = 0.311[(D + 2r)^2 - D^2] \tag{13-72}$$

in which p = wind pressure in pounds per square foot, D = diameter of conductor in inches, r = radial thickness of ice. Ice is taken at 57 lb/ft³.

104. Stresses in a Span. The high-tension stress in the conductor is the result of attempting to support a vertical load by a member, i.e., the conductor, extending in a very nearly horizontal direction or nearly at right angles to the direction of the load. The slope of the conductor at the support is generally only a few degrees below the horizontal, which causes a stress in the conductor many times the weight supported. From mechanics, the horizontal tension in the wire t is equal to the weight supported V (which is the length of wire $l/2$ times the weight per foot w) divided by the tangent of the angle of slope θ (see Fig. 13-62).

105. The Parabola and Catenary Curves. The slope of the conductor at a support, which is the factor determining the conductor tension, indicates the form of the curve assumed by the conductor. In Fig. 13-62 it will be seen that the cosine of the angle of slope θ at the support is the ratio of the horizontal tension t to the resultant tension T in the conductor, or $\cos \theta = t/T$. The mathematical form of the slope of the tangent to a catenary is $\cos \theta = c/y$, in which c = ordinate of the low point of the curve and y = ordinate of the point of tangency. It is thus evident that the span must form a catenary and that the ordinate y should be considered as a dimension of a span corresponding to the tension at the point of support; or, more definitely, this imaginary dimension y is a length of wire whose weight is equal to the tension T. Similarly, the ordinate c corresponds to the horizontal tension existing at the low point of the sag and should also be considered a dimension of the span.

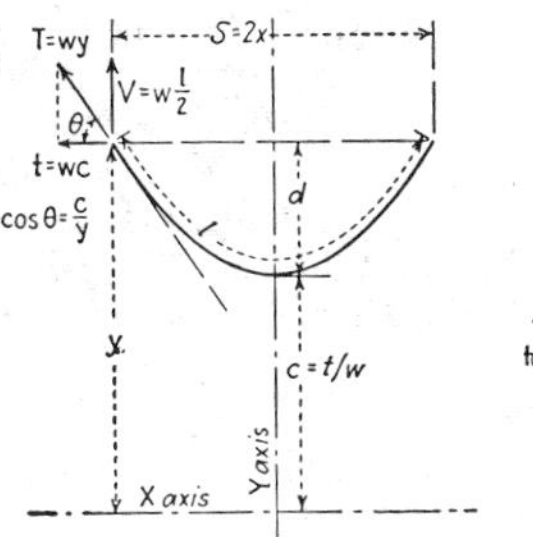

FIG. 13-62. The catenary.

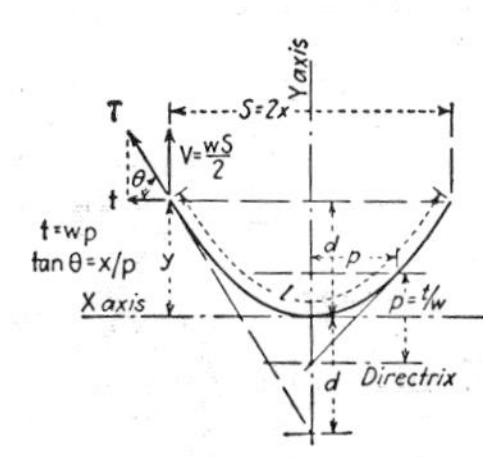

FIG. 13-63. The parabola.

In Fig. 13-63 the weight of wire in the half span is approximately equal to wx, since $2x$ approximately equals l, if the sag is not too great. The tangent of the slope $\theta = wx/t$, which is the mathematical form of the tangent to a parabola; i.e.,

$$\tan \theta = x/p$$

in which p = distance from the directrix to the focus of the parabola. This demonstrates that the wire in a span assumes approximately the form of a parabola.

The principal formulas useful in computations are given below (as illustrated in Figs. 13-62 and 13-63), where w = weight in pounds per foot of wire, T = tension in pounds at the support, and t = horizontal component of the tension; S = span, and d = sag, both in feet. Formulas based on the catenary are accurate; those on the parabola are approximate for very large sags. In most cases the parabola formulas are more simple and sufficiently accurate for almost any practical problem; the error, for sags up to 6% of the span, is about ½%; and for a sag of 10% of the span, the parabola formula results in a sag about 2% too small. The catenary formulas are, however, used in the unit-span dimensions.

Catenary equations:

$$y = c \cosh \frac{x}{c} \tag{13-73}$$

$$\frac{l}{2} = c \sinh \frac{x}{c} = \sqrt{y^2 - c^2} \tag{13-74}$$

$$d = y - c = c\left(\cosh \frac{x}{c} - 1\right) \tag{13-75}$$

Parabola equations:

$$x^2 = 2py \tag{13-76}$$

$$d = \frac{wS^2}{8t} \tag{13-77}$$

$$l = S\left(1 + \frac{8}{3}\frac{d^2}{S^2}\right) \text{ approx.} \tag{13-78}$$

$$T = t + wd \tag{13-79}$$

The dimensions y and c of the catenary give actual dimensions of the curve corresponding to tension in the wire; that is, $y = T/w$, $c = t/w$. The relation $T = t + wd$ is very useful.

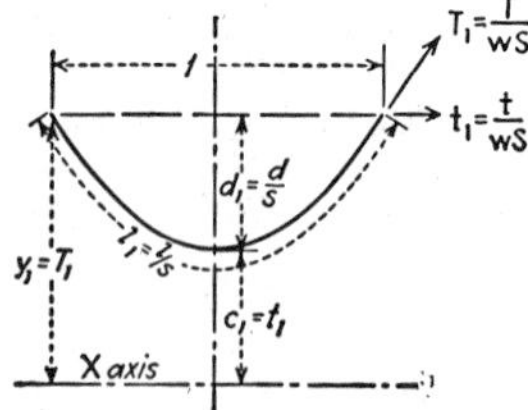

FIG. 13-64. Unit span.

106. The Unit Span. A considerable saving in tedious arithmetic and a gain in accuracy are made by using the unit-span dimensions as devised by Percy H. Thomas.[1] On the unit-span basis the catenary corresponding to the actual span of wire is not considered to have the large physical dimensions of the span itself but to be reduced, in proportion, to a curve of unit length.

Each dimension of the unit curve is therefore in the ratio of $1/S$ to the dimensions of the actual span, including the dimensions y and c corresponding to tension. In the following discussion the unit dimension y_1 will be referred to as the tension factor T_1. In Figs. 13-62 and 13-64, the unit dimensions are as follows:

Unit sag:
$$d_1 = d/S = y_1 - c_1 = c_1[\cosh (1/2c_1) - 1] \tag{13-80}$$

Unit length:
$$l_1 = l/S = 2c_1 \sinh (1/2c_1) \tag{13-81}$$

Tension factor:
$$T_1 = y_1 = y/S = T/wS = c_1 \cosh (1/2c_1) \tag{13-82}$$

$$t_1 = c_1 = c/S = t/wS \tag{13-83}$$

[1] THOMAS, P. H. Sag Calculations for Suspended Wires; *Trans. AIEE*, 1911, Vol. 30, p. 2229.

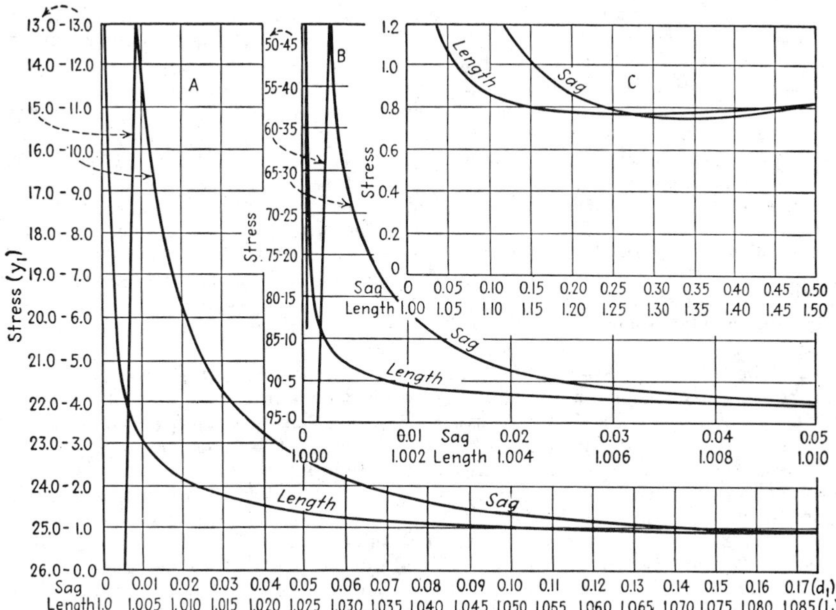

FIG. 13-65. Thomas chart. Curve *A* is good for sags from 2 to 15%; curve *B* for sags less than 2%; curve *C* for very large sags, being especially useful for spans on steeply inclined slopes. Since the chart is too small for accurate work, the data from which the curves were plotted are given in Table 13-14. Figure 13-66 is a portion of curve *B*, plotted to a larger scale.

All spans having the same ratio of sag to span or the same ratio of length to span, or the same tension factor, are represented by the same unit curve; and thus a comparatively brief tabulation listing the unit dimensions may be prepared by assuming values of c_1 and calculating d_1, l_1, and y_1 by Eqs. (13-80), (13-81), and (13-82). If, for any span, the sag, the length, or the tension is known, the corresponding unit dimension is readily figured, and the unknown unit dimensions are read from the tabulation. Interpolations are best made by plotting these values in a curve as shown in Fig. 13-65. Three sets of curves are shown, each plotted to a different scale. Working charts should be plotted to convenient scale from data given in Par. **111**. A very elaborate tabulation of unit-span dimensions was prepared by Martin[1] and published by the Copperweld Steel Company. In this tabulation, interpolations may be made directly from the table (see Par. **115**).

107. Span Calculations. The problems of span design may be divided into three classes: (*a*) The sag in a span resulting from a conductor of given weight at a given tension (or the reverse). These problems are direct solutions of Eqs. (13-76) to (13-79) or may be read from the Thomas chart as described in Par. **106**. (*b*) The sags or tensions resulting from unequal spans or differences in elevation of supports. These effects are generally of minor importance, and it is usually necessary only to investigate a few limiting conditions by graphic methods. Special cases may be solved by the methods outlined in Par. **123**. (*c*) The sags resulting from changes in loading or temperature. These problems are complicated, but the two simultaneous equations involved are so readily solved as two intersecting curves that, in common with most design problems, nothing is gained by attempts at algebraic solutions. The following paragraphs describe this solution.

[1] MARTIN, JAMES S. "Sag Calculations by the Use of Martin's Tables"; Copperweld Steel Company, 1931, rev. 1961.

108. The Thomas and the Martin methods are the simplest methods of making sag calculations that apply to any type of conductor. No approximations are introduced, and the accuracy is limited only by the scale to which the charts are drawn (see Fig. 13-65 and Table 13-14).

In the following paragraphs, the unit sag is often referred to as the sag, the unit length as the length, and the unit tension factor as the tension. In discussing the catenary, the unit dimensions give a clearer picture, while consideration of the stretch and temperature expansion seems to apply more naturally to actual dimensions. There is some advantage in not attempting to distinguish too closely between the actual and unit dimensions, as the two are, for many purposes, interchangeable. Likewise, in the following paragraphs, it will be noted that the "length" is considered alternatively as the length of the catenary arc and as the length of the physical wire. Some confusion will be avoided if it is realized that, in the span, these two must be the same, and it is only for the purpose of visualizing the process of computation that the two lengths are separated.

109. Length of the Wire in the Span. The process of erecting the wire and adjusting the sag, usually considered as adjusting the tension, is actually a matter of adjusting the length of wire in the span. A little wire is taken out if the sag is too great or a little added in the span if the sag is too small. Changes in the length of wire due to elastic stretch and to expansion and contraction from temperature, with corresponding changes in sag, produce similar adjustments. In this adjustment two sets of simultaneous conditions must be satisfied: (1) the length of catenary arc determined entirely by the form of the curve dependent on the sag and reflected in the tension must be equal to (2) the length of the wire determined by the stress and elastic stretch.

110. The Length Curve. All values of the first of these two simultaneous conditions, i.e., the length of the wire as determined by the length of the catenary arc and the y dimension of many possible forms of the catenary, are plotted as the length curve of the Thomas chart (Fig. 13-65). Any particular point on this length curve represents the length and tension for a particular unit catenary.

The length of the unit catenary, corresponding to the actual span under consideration, is the basis from which the sags and tensions are determined for any temperature or loading condition. This length of the catenary is not used directly; instead, the "unstressed length" of the wire is more convenient. The unstressed length is found by subtracting the elastic stretch of the wire from the catenary length.

$$\text{Stretch/ft} = P/E \tag{13-84}$$

in which P = stress in pounds per square inch and E = modulus of elasticity.

111. The effect of change in temperature is considered to take place after the wire is relieved of all stress, thus eliminating the complicated adjustments between change in length due to temperature and change in stretch from the resulting change in tension. The process is as if the wire were removed from the span and laid on the ground before the rise in temperature takes place.

$$\text{(Change in length)/ft} = \alpha t \tag{13-85}$$

where α = linear coefficient of expansion and t = change in temperature. This gives a new unstressed length P_1 at the new temperature. This is one point on the curve of the second of the two simultaneous conditions described in Par. **107**, viz., the curve of the length of the wire as determined by the elastic stretch. Points on this curve may be computed from the modulus of elasticity [Eq. (13-84)] or read from a stress-strain curve of the conductor. If the modulus is constant, this will be a straight line. The intersection of the elastic-stretch curve with the catenary-length curve is the solution of these two simultaneous conditions and the tension at the new temperature. It is as if the same wire lying on the ground at the new temperature were lifted back into the span and stretched until the ends just reached the supports. The intersection is the point at which the stretch is just sufficient to allow the wire to hang freely.

Points on the chart (Fig. 13-65) are unit dimensions; the stretch is the stretch per unit length; and the tension is expressed as the unit tension factor T/wS.

Table 13-14. Stress and Length in Terms of Ratio of Sag to Span
Unit span dimensions*

c_1	y_1	d_1	l_1	c_1	y_1	d_1	l_1
Horizontal stress factor	Stress factor	Unit sag	Unit length	Horizontal stress factor	Stress factor	Unit sag	Unit length
100.0000	100.001 3	.001 250	1.000 004 2	6.2500	6.270 0	.020 01	1.001 066
90.9091	90.910 5	.001 375	1.000 005 1	5.8824	5.903 6	.021 26	1.001 205
83.3333	83.334 8	.001 500	1.000 006 1	5.5555	5.578.1	.022 52	1.001 351
76.9231	76.924 7	.001 625	1.000 007 1	5.2632	5.286 9	.023 77	1.001 503
71.4286	71.430 3	.001 750	1.000 008 2	5.0000	5.025 0	.025 02	1.001 668
66.6667	66.668 5	.001 875	1.000 009 4	4.7619	4.788 2	.026 27	1.001 839
62.5000	62.502 0	.002 000	1.000 010 7	4.5455	4.573 0	.027 53	1.002 017
58.8235	58.825 7	.002 125	1.000 012 0	4.3478	4.376 6	.028 78	1.002 205
55.5555	55.557 8	.002 250	1.000 013 5	4.1667	4.196 7	.030 04	1.002 402
52.6316	52.633 9	.002 375	1.000 015 0	4.0000	4.031 3	.031 29	1.002 606
50.0000	50.002 5	.002 50	1.000 017	3.8462	3.878 7	.032 55	1.002 819
45.4545	45.457 3	.002 75	1.000 020	3.7037	3.734 2	.033 80	1.003 040
41.6667	41.669 7	.003 00	1.000 025	3.5714	3.606 5	.035 06	1.003 270
40.0000	40.003 1	.003 13	1.000 026	3.4483	3.484 6	.036 31	1.003 508
38.4615	38.464 8	.003 25	1.000 028	3.3333	3.370 9	.037 57	1.003 754
35.7143	35.717 8	.003 50	1.000 033	2.9412	2.983 8	.042 60	1.004 825
33.3333	33.337 1	.003 75	1.000 037	2.5000	2.550 2	.050 17	1.006 680
31.2500	31.254 0	.004 00	1.000 043	2.2727	2.328 0	.055 22	1.008 086
29.4118	29.416 0	.004 25	1.000 048	2.0000	2.062 8	.062 83	1.010 444
28.5714	28.575 8	.004 38	1.000 051	1.8519	1.919 8	.067 91	1.012 194
27.7777	27.782 3	.004 50	1.000 054	1.6667	1.742 2	.075 56	1.015 068
26.3158	26.320 5	.004 75	1.000 060	1.5625	1.643 2	.080 68	1.017 154
25.0000	25.005 0	.005 00	1.000 067	1.4286	1.517 0	.088 40	1.020 542
22.7273	22.732 8	.005 50	1.000 081	1.3514	1.444 9	.093 56	1.022 973
20.8333	20.839 3	.006 00	1.000 096	1.2500	1.351 3	.101 34	1.026 881
20.0000	20.006 3	.006 25	1.000 104	1.1905	1.297 0	.106 55	1.029 660
19.2308	19.237 3	.006 50	1.000 113	1.1111	1.225 5	.114 41	1.034 093
17.8571	17.864 1	.007 00	1.000 131	1.0638	1.183 5	.119 68	1.037 224
16.6667	16.674 2	.007 50	1.000 150	1.0000	1.127 6	.127 63	1.042 19
15.6250	15.633 0	.008 00	1.000 171	0.9091	1.050 1	.141 00	1.051 19
14.7059	14.714 4	.008 50	1.000 193	0.8333	0.987 9	.154 55	1.061 09
13.8889	13.897 9	.009 00	1.000 216	0.7143	0.896 5	.182 26	1.083 69
13.1579	13.167 4	.009 50	1.000 241	0.6250	0.835 8	.210 83	1.110 13
12.5000	12.510 0	.010 00	1.000 267	0.5555	0.796 2	.240 61	1.140 57
11.6279	11.638 7	.010 75	1.000 308	0.5000	0.771 54	.271 54	1.175 20
10.6383	10.650 1	.011 75	1.000 368	0.4545	0.758 42	.303 87	1.214 23
10.0000	10.012 5	.012 50	1.000 417	0.4167	0.754 44	.337 77	1.257 88
9.0909	9.104 7	.013 75	1.000 504	0.3846	0.758 04	.373 43	1.306 45
8.3333	8.348 3	.015 00	1.000 600	0.3571	0.768 18	.411 04	1.360 21
7.6923	7.708 4	.016 26	1.000 704	0.3333	0.784 14	.450 80	1.419 52
7.1428	7.160 4	.017 51	1.000 817	0.3125	0.805 46	.492 96	1.484 73
6.6667	6.685 4	.018 76	1.000 938				

* THOMAS, PERCY H. Sag Calculations for a Suspended Wire; *Trans. AIEE*, 1911, Vol. 30, p. 2229. Refer to Fig. 13-64.

112. Change in loading changes the value of w in the equation $T_1 = T/wS$. Thus, in plotting the curve of the stretch of the wire on the Thomas chart, the slope of the stretch curve is quite different if the ice and wind are removed and the value of w correspondingly changed.

113. Example Calculation (see Fig. 13-66). Required: the sag at which a 600-ft span of 4/0 19-strand hard-drawn bare-copper conductor shall be erected at 60°F so that, under heavy loading, ½-in. ice, 4-lb wind at 0°F, the tension shall not exceed one-half the ultimate strength. The following data are taken from Table 13-15:

Ultimate strength	9,617 lb
Maximum tension T	4,808 lb
Conductor weight bare w	0.6533 lb per ft
Resultant weight w_r	1.679 lb per ft
Area	0.1662 sq in.
Modulus of elasticity E	14,500,000 (initial)
Coefficient of expansion X	0.0000094

Dimension y = 4,808/1.679 = 2,864 ft, and the unit-span dimensions corresponding to a catenary of y = 2,864 ft for a 600-ft span are:

Tension factor T_1 = 2,864/600 = 4.77 (point M, Fig. 13-66).

Unit length stressed l_1 = 1.00186 (read from chart, point M').

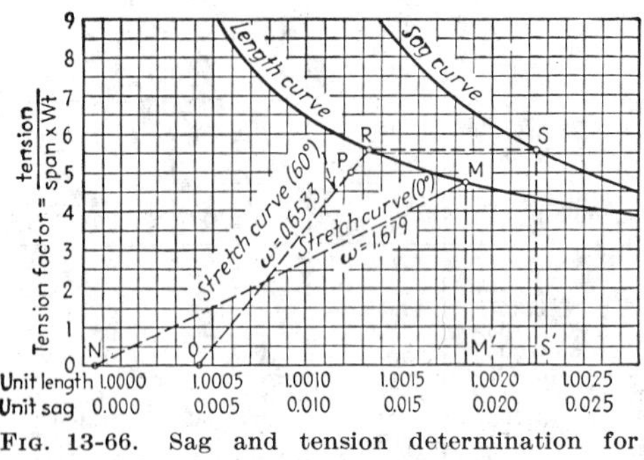

FIG. 13-66. Sag and tension determination for copper conductor.

Owing to the tension of 4,808 lb, or a stress of 4,808/0.1662 = 28,929 lb/in², the conductor is stretched 28,929/14,500,000 = 0.001995 ft/ft, or, for the unit length of 1.00186 ft, a total of 1.00186 × 0.001995 = 0.001999 ft.

The unstressed length at 0°F (l_0 at 0°F) = 1.00186 − 0.001999 = 0.99986 (point N).

A change in temperature from 0 to 60°F results in an increase in length of 60 × 0.0000094 = 0.000564 ft/ft or a total increase of 0.000564 < 0.99986 = 0.000564 ft.

The unstressed length at 60°F (l_0 at 60°F) = 0.99986 + 0.000564 = 1.000424 (point O).

It is now assumed that with the load removed and starting with the unstressed length computed above, the conductor is stretched until the stretch curve intersects the length curve. Since the modulus is constant, the stretch curve is a straight line determined by any two points, e.g., the unstressed length and the length corresponding to any tension factor such as 5.00, or a tension, for the bare cable, of

$$5.00 \times 600 \times 0.6533 = 1{,}960 \text{ lb}$$

Table 13-15. Physical Properties of Conductor Materials

This table lists the physical properties of the more common conductor and overhead ground-wire materials. National Electrical Safety Code recommended initial and final moduli of elasticity are given where available and are denoted by an asterisk (*). (See also Sec. **4**.)

Material	Ultimate strength, lb/in²	Modulus of elasticity, initial	Modulus of elasticity, final	Coefficient of expansion/°F
Copper, stranded, hard-drawn	57,000– 59,000			
3–12 strand		14,000,000*	17,000,000*	.000,009,4*
7–19 strand		14,500,000*	17,000,000*	.000,009,4*
37 strand		14,500,000	17,000,000	.000,009,4
Solid, hard drawn	49,000– 63,000	14,500,000*	17,000,000*	.000,009,4*
All-aluminum cable, stranded, hard-drawn	23,000– 30,000		10,000,000	.000,012,8
ACSR all sizes	40,000– 56,000	See Par. **119**	$E_aH_a + E_sH_s$ See Par. **119** for symbols	$(E_aH_a/E_{as})\theta_a + (E_sH_s/E_{as})\theta_s$
Steel	190,000		29,000,000	.000,006,4
Aluminum	23,000– 30,000		10,000,000	.000,012,8
No. 2 and No. 4 7/1	54,000– 55,000	11,600,000*	12,600,000*	.000,010,1*
No. 6 to No. 4/0 6/1	43,000– 49,000	10,250,000*	11,500,000*	.000,010,5*
Copperweld, solid	80,000–170,000	22,000,000*	24,000,000*	.000,007,2*
Stranded	100,000–170,000	20,500,000*	23,000,000*	.000,007,2*
Copperweld—copper:				
No. 4 to 350,000 cmil Cu eq	65,000– 96,000			
No. 6A to No. 2A	86,000– 96,000	16,500,000*	19,000,000*	.000,008,5*
No. 2F to No. 1/0F	71,000– 73,000	15,500,000*	18,000,000*	.000,009,0*
Galvanized-steel strand, all strengths	45,000–190,000		27,000,000	.000,006,7
Stainless-steel strand, Page type 301	235,000		26,000,000	.000,009,2
Bronze, stranded 8.5 to 85 cond. ASTM B105 alloys	67,000–135,000		16,000,000	.000,009,4
Alumoweld 7-strand	180,000		23,000,000	.000,007,2

This will give a stretch of

$$1.000424[1{,}960/(0.1662 \times 14{,}500{,}000)] = 0.000813$$

The unit length for tension factor 5.00 at 60°F = l_0 at 60°F = 1.000424 + 0.000813 = 1.001237 (point P).

The line OP extended intersects the length curve at R or at a tension factor of 5.60, which, for the 600-ft span of 4/0 copper, corresponds to

$$5.60 \times 600 \times 0.6533 = 2{,}195 \text{ lb}$$

The unit sag for a tension factor of 5.60 is read from the sag curve (point S') as 0.02242, or an actual sag for the 600-ft span of 0.02242 × 600 = 13.45 ft, which is the sag at which the conductor should be erected at 60°F.

Practically all computations can be made by slide rule with sufficient accuracy. Multiplications such as 1.00186 × 0.001995 can be made on the slide rule as

$$(1.00 \times 0.001995) + (0.00186 \times 0.001995) = 0.001995 + 0.00000371 = 0.001999$$

The intersection of the stretch curve with the length curve may be made very accurately by trial from the values given in Martin's tables, as described in the introduction to those tables.

The computations are repeated in condensed form below and include the sags for 120°F.

			Point
T_1 =	4,808/(1.679 × 600)	= 4.77	M
l_1 =		= 1.00186	M'
Stretch =	1.00186[4,808/(0.1662 × 14,500,000)]	= 0.001999	
l_0 at 0 F =		= 0.99986	N
Change 60 F =	0.99986(60 × 0.0000094)	= 0.000564	
l_0 at 60 F =		= 1.00042	O
l_0 at 120 F =		= 1.000988	
Stretch at T_1 =	5.00 = 1.000424[5.00 × 600 × 0.6533/(0.1662 × 14,500,000)]	= 0.000813	
l_1 at 60 F when T_1 =	5.00	= 1.001237	P
l_1 at 120 F when T_1 =	5.00	= 1.00180	
T_1 at 60 F =	5.60		
Tension =	5.60 × 600 × 0.06533	= 2,195 lb	
d_1 at 60 F =	0.02242		
Sag =	0.02242 × 600	= 13.45 ft	
T_1 at 120 F =	4.88		
Tension =	4.88 × 600 × 0.6533	= 1,913 lb	
d_1 at 120 F =	0.0257		
Sag =	0.0257 × 600	= 15.42 ft	

114. Conductor Materials. See Table 13-15.

115. Martin's Tables. A very complete tabulation of unit-span values has been prepared by James S. Martin and published by the Copperweld Steel Company. The unit values are given in such small steps that these tables may be used in exactly the same manner as trigonometric functions. The stress factor as given in Martin's tables is the reciprocal of the stress factor in the Thomas chart, that is, WS/T. Martin's tables also include an additional dimension, which may be described as the average stress factor between the stresses at the support and at the low point. In computing the stretch it is evident that this average should be used rather than the maximum at the support. This may have an appreciable effect on extraordinarily large sags.

116. Stresses Due to Unequal Spans. It is impractical to dead-end and erect each span separately; therefore, each span in a level line must, as erected, have approximately the same tension. Any change in temperature has a much greater effect on short spans than on long, but changes in loading have a greater effect on long spans than on short. However, high tensions on short spans are equalized by a very slight movement of supports, while this is not the case on long spans. Computations for sagging are usually made on a somewhat longer span than the average, often designated as the **ruling span.** A formula commonly used to compute the ruling span is as follows: average span $+\frac{2}{3}$(longest span − average span). The movement of suspension insulators and deflection of poles may generally be relied upon to equalize the tensions in occasional long or short spans. Sag charts for erecting conductor should usually not cover a range greater than one-half to double the average span, as extreme cases should be investigated.

The effect of insulators swinging longitudinally and increasing or reducing the tension may be computed with the Thomas chart method by taking the horizontal movement of the insulator as an increase or decrease in the length of wire in the span, exactly as in the case of changes in length due to temperature (see Par. **110**). If the insulator swings toward the span, the effect is to increase the length of wire by the amount of the swing. The small change in span length resulting from the swing is negligible.

117. Sagging Charts for Erecting Conductors. From the computations outlined in preceding paragraphs, sag and tension curves or tables are prepared for erecting the conductors giving the tension for a range of temperatures in 10° intervals and the sag for a range of spans in 20-ft intervals. Actually, even though the work of preparing these tables has been carried out with great accuracy and the work of erection is done with great care, the irregular profile and unequal span lengths usually encountered in a line result in conductor tensions, especially under maximum loading, slightly different from the computed result (also see Par. **174**). It is impossible to avoid these irregularities, and the differences are of little importance on the average line if (1) the calculations are based on a span length representative of the particular line, (2) some allowance is made in the maximum tension in case of particularly rough country, and (3) correction measures are applied for very long spans or great differences in elevation.

118. Initial and Final Sags. The modulus of elasticity of a wire for usual computations is assumed to be constant and of a definite value for any given material. Actually, however, individual tests to determine the modulus of elasticity show a considerable variation. The first loading of a wire gives a slightly curved stress-strain diagram; i.e., the modulus is not strictly constant. If the test is continued to a load approaching the elastic limit and then backed down to no load, the wire will return along a straight stress-strain curve. Subsequent loading of the wire to the same maximum value will follow this return stress-strain curve on a constant modulus. From a large number of careful tests the average initial stress-strain curve and the average final modulus may be determined.

A conductor erected in a span and not previously stressed to the maximum design tension will stretch, under maximum load, along the initial stress-strain curve; upon release of the load, the wire will contract along the final modulus and will not return to the initial length by the amount of the permanent set. This results in a slightly greater sag than that at which the conductor was originally installed. Also, the conductor will never quite reach the same maximum tension if the same maximum load is applied a second time. Often sag computations do not warrant consideration of these refinements. However, for conductors stressed beyond the theoretical elastic limit and for composite conductors such as ACSR at high maximum tensions, the final sags may be considerably larger than the initial, and these conditions must be considered.

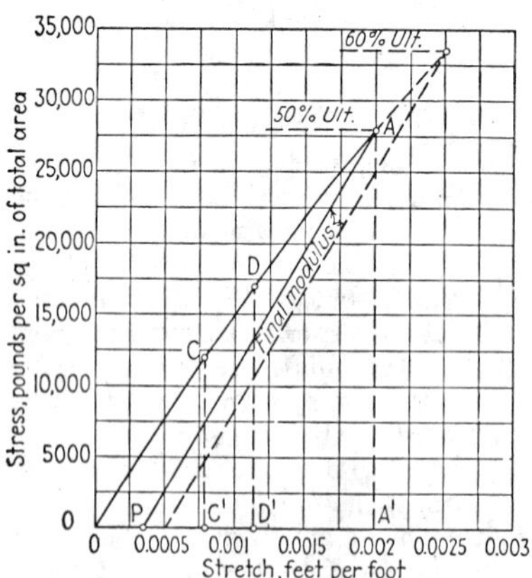

FIG. 13-67. Stress-strain curve for hard-drawn seven- and nine-strand copper cable.

Such computations are made on the Thomas chart by considering that the maximum tension is the result of the conductor stretching on the initial stress-strain curve from the tension at which it is erected to the maximum loaded tension. By reversing this process on the chart, the initial unstressed length of the conductor is computed by subtracting the initial stretch from the loaded length, while the final unstressed length is found by subtracting the stretch as determined by the final modulus from the loaded length. The initial and final bare-wire sags at stringing temperatures are computed as usual from these initial and final no-load lengths, respectively, the initial curve being used for computing the initial sag and the final curve for computing the final sag. A typical stress-strain curve for copper is shown in Fig. 13-67. The stretch under various conditions is best read directly from the chart rather than attempting to compute from initial and final values of the modulus of elasticity. Thus, from Fig. 13-67, OA' is the initial stretch for a load of half the ultimate, which is subtracted from the length of the conductor under maximum load to give the initial unstressed length; and PA' is the final stretch, which subtracted from the length under maximum load gives the final unstressed length.

In drawing the stretch line to intersect the length curve on the Thomas chart it is not necessary to draw the entire stretch curve OCD but is more convenient to plot any two points on the Thomas chart corresponding, for instance, to the points C and D, which have a stretch of OC' and OD', respectively. A line drawn through these two points to intersect the length curve will give the initial bare-wire tension. In making such computations, various short cuts will suggest themselves.

119. Sag computations for ACSR are made exactly as described in Pars. **113** and **118** except for the complications introduced by the composite stress-strain diagram. Most of these difficulties may be eliminated by reading the stretch for any condition or assumed tension directly from the diagram, as described for copper conductors in Par. **118**, and not attempting to compute the stretch from the modulus. The stress-strain diagram must be obtained from tests on actual conductors.

A typical stress-strain diagram for ACSR is shown in Fig. 13-68. It will be noted that the final modulus curve shows a distinct break at P. This is due to the permanent set in the aluminum strands, which throws all the load on the steel strands at low tensions. For loads from M to P the aluminum strands are slightly loose on the steel, and the conductor stretches as if it consisted of the steel core alone; for loads greater than P, the aluminum and steel strands work together as a composite conductor.

Changes in temperature change the relative length of aluminum and steel strands and thus change the point at which the steel and aluminum begin to work together. It is only occasionally that conductor tensions fall below the point P; often, therefore, computations need not take this effect into account. This is the case in the computation given below.

Changes in temperature actually occur while the conductor is under tension and not, as assumed in computations, in the unstressed condition. Such temperature changes change the relative amount of stress in the steel and aluminum even though the total tension is kept constant. Such a temperature change therefore produces a different expansion or contraction in the composite conductor, depending on the conductor tension existing at the time the temperature change occurs. The coefficient of expansion is therefore not the same at all tensions as it is in a conductor of only one material. A very accurate method of computing sags, taking into account all changes in modulus and the coefficient of expansion, is described in a pamphlet published by

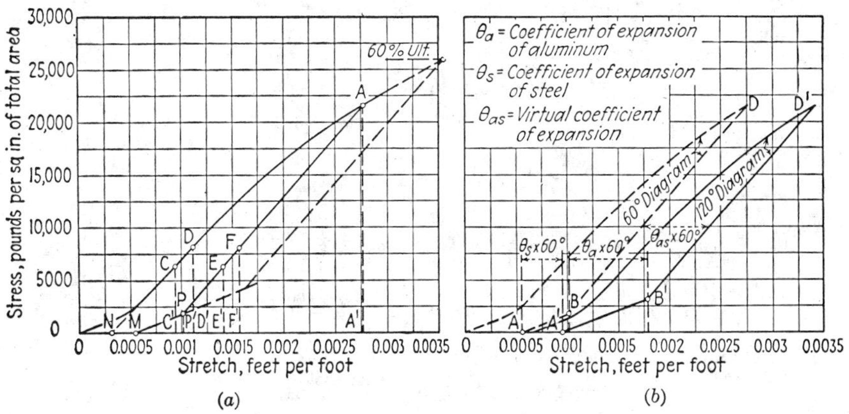

FIG. 13-68. Stress-strain diagram, No. 4/0 ACSR.

the Rome Cable Division of the Aluminum Company of America.[1] This pamphlet also contains catenary tables and creep curves for some conductors. Creep is discussed in considerable detail.

However, fairly accurate computations may be made by use of the "virtual coefficient of expansion," which assumes the temperature change to occur under no load. The coefficient of expansion thus obtained is slightly higher than the actual. The virtual coefficient of expansion is used in the following example and is obtained thus: Where E_a and E_s are the modulus of the aluminum and steel strands, respectively, and E_{as} is the modulus of the composite cable, H_a and H_s are the percent area of aluminum and steel, respectively, and θ_a, θ_s, and θ_{as} are the coefficients of expansion of the aluminum, steel, and composite cable. Average values of E_a, E_s, θ_a, and θ_s are given in Par. **114**.

$$\theta_{as} = \theta_a \frac{E_a H_a}{E_{as}} + \theta_s \frac{E_s H_s}{E_{as}} \tag{13-86}$$

$$E_{as} = E_a H_a + E_s H_s \tag{13-87}$$

The coefficient of expansion for 4/0 ACSR, in which $H_a = 0.857$ and $H_s = 0.143$, is 0.0000107.

Example Calculation. Required the sag at which a 600-ft span of 4/0 ACSR shall be erected at 60°F so that, under heavy loading, ½-in ice, 4-lb wind at 0°F, the tension shall not exceed one-half the ultimate strength; also the final sag at 60°F which will occur after the conductors have been loaded to the maximum tension. Symbols are as given in Par. **113**. The following data are taken from Table 13-15.

			Point
Ultimate strength		8,420 lb	
Maximum tension		4,200 lb	
Conductor weight bare		0.292 lb per ft	
Resultant weight		1,396 lb per ft	
Area		0.1939 in²	
Modulus of elasticity		Read stretch from Fig. 13-68	
Coefficient of expansion		0.0000107	
$T_1 = 4{,}200/(1{,}396 \times 600)$		5,026 (Fig. 13-69) (point *M*)	
l_0 at 0°F		1.00166 (point *M′*)	
Initial			
Stress	$= 4{,}200/0.1939$	= 21,712 psi (Fig. 13-68*a*)	(*A*)
Stretch		= 0.00280 (Fig. 13-68*a*)	*O-A′*
l_0 at 0°F	$= 1.00166 - 0.00280$	= 0.99886	
Change 60°F	60×0.0000107	= 0.00064	
l_0 at 60°F	$= 0.99886 + 0.00064$	= 0.99950	

It is not necessary to plot the foregoing points on the Thomas chart. As described in Par. **118**, it is more convenient to plot two points on the stretch curve such as *C* and *D* (Fig. 13-69), corresponding to points *C* and *D* (Fig. 13-68*a*), than to draw in the entire stretch curve *OCD* (Fig. 13-68*a*). The intersection of a line through these two points with the length curve gives the tension factor at 60°F bare wire.

			Point
Stress when $T_1 = 9.00$	$= 9.00 \times 0.292 \times 600/0.1939$	= 8,130 psi	
Stress when $T_1 = 7.00$	$= 7.00 \times 0.292 \times 600/0.1939$	= 6,320 psi	
Initial stretch at 8,130 psi		= 0.00112 (Fig. 13-68*a*)	*O-D′*
Initial stretch at 6,320 psi		= 0.00094	*O-C′*
Length at $T_1 = 9.00$	$= 0.99950 + 0.00112$	= 1.00062 (Fig. 13-69)	
Length at $T_1 = 7.00$	$= 0.99950 + 0.00094$	= 1.00044	
T_1 at 60°F	=	= 8.44	R_i
Initial tension	$= 8.50 \times 600 \times 0.292$	= 1,489 lb	
d_1 at 60°F	=	= 0.01475	S_i
Initial sag	$= 0.01475 \times 600$	= 8.85	
Final			
Stress	$= 4{,}200/0.1939$	= 21,650 psi (Fig. 13-68*a*)	*A*
Stretch	=	= 0.00222	*M-A′*
l_0 at 0°F	$= 1.00166 - 0.00222$	= 0.99944	
l_0 at 60°F	$= 0.99947 + 0.00064$	= 1.00011 (Fig. 13-69)	*V*
Stress when $T_1 = 9.00$	$= 9.00 \times 0.292 \times 600/0.1939$	= 8,130 psi	
Stress when $T_1 = 7.00$	$= 7.00 \times 0.292 \times 600/0.1939$	= 6,320 psi	
Final stretch at 8,130 psi		= 0.00102 (Fig. 13-68*a*)	*M-F′*
Final stretch at 6,320 psi		= 0.00086	*M-E′*
Length at $T_1 = 9.00$	$= 1.00011 + 0.00102$	= 1.00113 (Fig. 13-69)	
Length at $T_1 = 7.00$	$= 1.00011 + 0.00086$	= 1.00097	
T_1 at 60°F	=	= 6.67	R_f
Final tension	$= 6.67 \times 600 \times 0.292$	= 1.168 lb	
d_1 at 60°F	=	= 0.0188	S_f
Final sag	$= 0.0188 \times 600$	= 11.28 ft	

[1] Rome Cable Division, Aluminum Company of America, ACSR Graphic Method for Sag-Tension Calculations, 1961 ed.

It will be noted in Fig. 13-68*a* that in the above example the point *P*, where the steel starts to take the entire stress, is at 1,800 lb/in², or a tension of 350 lb. This corresponds to a tension factor for this conductor of $350/(600 \times 0.292)$, or 1.99. This stress of 1,800 lb/in² causes a stretch of 0.00046 (distance *MP* as measured on Fig. 13-68*a*) and would result in a length of 1.00013 + 0.00046 = 1.00059.

This point of unit tension factor 1.99 and length 1.00059 is plotted on the Thomas chart (Fig. 13-69), point *U*, to illustrate that this break in the final stress-strain curve does not in this case have any effect on the solution of the problem.

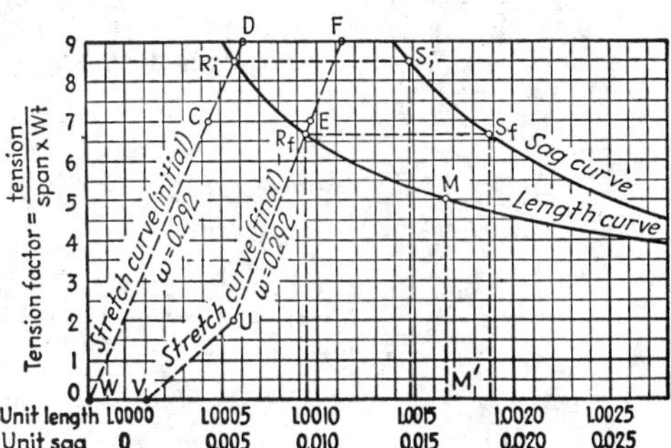

FIG. 13-69. Sag and tension determination for ACSR.

As a matter of interest, the point *U* is plotted on the Thomas chart, which corresponds to the point *P* on the stress-strain chart (Fig. 13-68*a*). The point *W* on the Thomas chart corresponds to the point *N* on the stress-strain diagram. This point *N* is actually a more convenient reference point than point *O*.

The effect of change in temperature is accurately accounted for in the method described in the Aluminum Company's pamphlet. However, fairly accurate results can be obtained by assuming that temperature expansion and contraction take place at no load. It is not necessary to redraw the stress-strain diagram for each temperature, as by this assumption the stress-strain diagrams for all temperatures are identical except for the location of the point at which the aluminum begins to pick up load, i.e., point *B*. The method of computing this change is illustrated in Fig. 13-68*b*. The 120° stress-strain diagram is indicated as being offset from the 60° diagram by the virtual coefficient of expansion, or 0.00064 ft, showing that the composite conductor increases this amount in length. However, between points *A* and *B*, where the steel alone is acting, the conductor changed in length only by the expansion of the steel, or $60° \times \theta_s = 0.0004$ ft, shown as AA_1. The aluminum strands expanded 0.00077 ft ($60° \times \theta_a$) at the same time; and while this change in the length of the aluminum strands did not change the length of the conductor, it did move the point at which the aluminum begins to take up stress from *B* to *B′*.

120. Unbalanced Ice Loads.[1] The jump of the conductor resulting from ice dropping off one span of an ice-covered line has been the cause of many serious outages on long-span lines where conductors are arranged in the same vertical plane. The vertical spacing required to prevent "sleet-jump" trouble does not seem to be a matter which can be readily calculated. However, the trouble has been practically cleared up by horizontally offsetting the conductors from 18 in to 3 ft on medium-voltage lines. Apparently the conductor jumps in practically a vertical plane, and this should be true if no wind is blowing, since then all forces and reactions are in a vertical direction.

Fairly accurate computations can be made of the static condition resulting from one span only being free of ice, although the process is tedious. The insulators on each side of the bare span will swing longitudinally away from this span; and the insulators at each next successive pair of towers, on each side of the bare span, will swing in the same direction, but each a lesser amount, since each insulator picks up a certain amount of horizontal tension, equal to $V \tan \theta$, in which θ is the angle of swing of the insulator and *V* is the vertical load supported (see Fig. 13-70). Also the sag in loaded spans adjacent to the unloaded span will have more than normal sag by the reduction in tension caused by adding a length of wire equal to the difference in swing of the insulators at the two ends of this span; or the length of wire in such a span $= l + d_1 - d_2$, where l = length of wire with the insulators vertical and d_1 and d_2 = horizontal displacement of the clamp on the two ends of the span. (The changes in span lengths are negligible.)

[1] GREISSER, V. H. Effects of Ice Loading on Transmission Lines; *Trans. AIEE*, 1913, Vol. 32, p. 1829. HEALY, E. S., and WRIGHT, J. A. Unbalanced Conductor Tensions; *Trans. AIEE*, 1926, p. 1064.

If an assumed value is taken for θ_1, the corresponding tension in the bare span may be computed from the length $l - d_1 - d_1$ and the horizontal tension h_1 taken by insulator 1 computed as the vertical weight on the insulator V times $\tan \theta$. The horizontal tension in the first loaded span must on this assumption be $T - h_1$, and the length of wire in this span may be computed from this tension The length of wire in this first loaded span must be $l + d_1 - d_2$, from which the deflection d_2 can be figured.

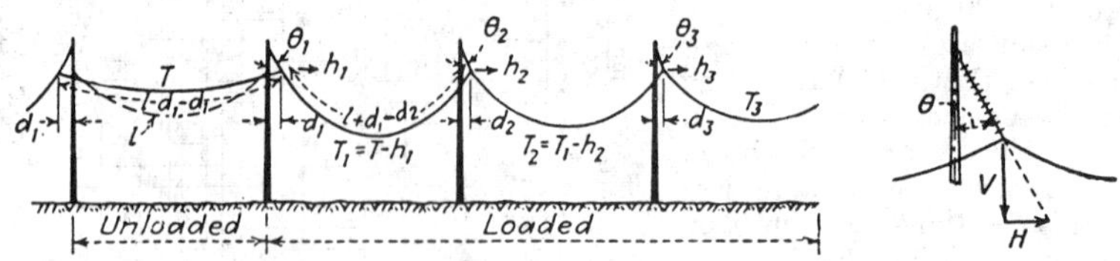

FIG. 13-70. Ice loading on all but one span.

If the assumption as to the swing of the first insulator was too great, successive computations of each following insulator will not become smaller but will after the first few insulators remain the same or increase, which is obviously impossible. If the assumption as to the swing of the first insulator is too small, the third or fourth insulator will apparently swing to the left, which cannot be true. Very small differences in assumptions will result in one or the other of these incorrect results. The correct assumption must be between the two.

121. Vibration.[1] The failure of conductors under tensions much below maximum design stresses has been caused by fatigue, resulting from rapid vertical vibrations of the conductor (from 15 to possibly 100 c/s) caused by steady nonturbulent winds blowing across the line. Most steel ground wires and telephone wires are subject to a slight but apparently harmless vibration of this kind. The vibration is caused by wind eddies on the leeward side of the conductor which swing from the upper to the lower side at regular intervals, the rate depending on the diameter of the conductor and the wind velocity. For this reason large-diameter lightweight conductors are especially subject to vibration. The conductor has a natural vibration frequency, depending on the ratio of tension to weight; and if the frequency of the swing of the eddies matches with that of the conductor, vibration results. Thus conductors of a high natural frequency or ratio of tension to weight tend toward dangerous vibration. In addition to the fundamental natural frequency, a conductor will vibrate in many "overtones," multiples of the fundamental, by dividing itself into a series of loops and nodes. Long spans, as they have a greater range of "overtones," vibrate more than short spans. A steady wind of not over about 10 mi/h is required to sustain the eddies at leeward side of the conductor at a constant frequency. Vibration seems to be much more severe in the prairie country of the Middle West and areas of similar topography, where steady winds are common.

When ACSR first came into use on transmission lines, it was strung for a maximum design tension of 50% of ultimate or more, to permit use of fewer structures. It was not long before vibration began to cause strand breakage. This breakage occurred at pin insulators, suspension clamps, and even at the early two-piece compression joints. The steel core seldom broke, but frequently all aluminum strands broke with square breaks and pulled apart as much as an inch or more, depending on the span length, leaving the steel core to carry the load. Frequently this caused no trouble as the aluminum cooled the steel until a short circuit occurred, when the steel overheated, softened, and pulled apart with a necking break. The development of dampers (Par. **168**) and armor rods practically eliminated vibration breakage.

On account of vibration, recent practice tends toward limiting the tension without ice or wind to 25% of the ultimate or less. Experience in different areas varies greatly,

[1] MONROE, R. A., and TEMPLIN, R. L. Vibration of Overhead Transmission Lines; *Trans. AIEE*, 1932, Vol. 51, pp. 1059–1073.
Aluminum Company of America, Overhead Conductor Vibration, Engineering Data, Sec. 4, 1961 ed.

but conductors under low normal tensions seem to be comparatively free from such violent or continuous vibration as to cause damage. Extra-high-voltage lines with bundle conductors in general are not built with tensions as high as have been used in the past. It has been found that, with the lower tensions, bundle conductors with spacers to keep the conductors from wrapping together during short circuits have very little vibration. The spacers have a pronounced damping effect, but it has not been determined yet that damaging vibration can be completely eliminated in bundle conductors by reduced tensions and spacers. It appears that dampers can be omitted in such cases.

122. Supports at Different Elevations. For the usual cases encountered in a line, the slight local variations from the calculated sags and tensions, due to the difference in elevation of supports at the ends of a span, are of no importance. The differences in tension due to this and to the method of supporting the conductor in sagging are discussed under Conductor Installation (Pars. **173** to **175**).

The sag d (Fig. 13-71), measured vertically to a tangent to the conductor which is parallel to a line through the supports, will be very nearly equal to the sag in a level span of a length equal to the slope distance S_1. For very large sags on inclined spans, Martin suggests a slight correction to the slope distance S_1 which gives very accurate results. The sag in the inclined span is more accurately equal to the sag in a level span of a length equal to $S_1 + (S_1 - S)$. Generally, however, the difference between the horizontal and slope distance is negligible, and the sag is taken as the same as the sag in a level span of the same horizontal length. This latter statement would be theoretically correct if the conductor hung in the form of a parabola, i.e., if the weight of wire were measured in feet of horizontal distance instead of along the length of the wire.

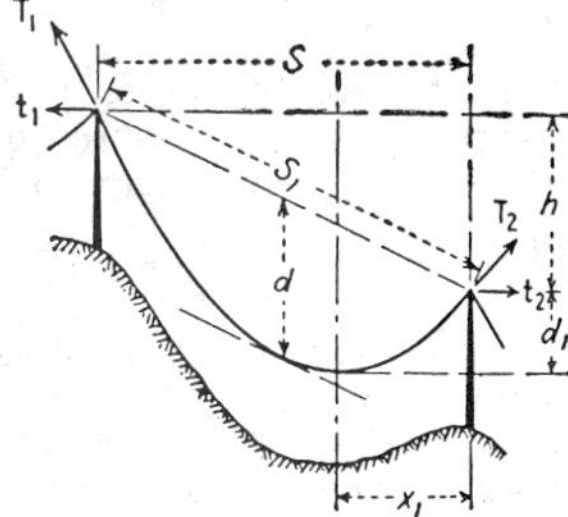

FIG. 13-71. Span with support at different elevations.

The dimensions to the low point of the sag d_1 and x_1 shown in Fig. 13-71 are obtained from the parabola and are as follows:

$$d_1 = d\left(1 - \frac{h}{4d}\right)^2 \qquad (13\text{-}88)$$

$$x_1 = \frac{S}{2}\left(1 - \frac{h}{4d}\right) \qquad (13\text{-}89)$$

On steep hillsides the low point of the sag may fall outside the low support; in such a case $h/4d$ would be greater than 1.

The horizontal components of the conductor tensions t_1 and t_2 (Fig. 13-71) must be equal, and as the vertical component is greater on the uphill side, the resultant tension T_1 must be greater than T_2. The catenary formula applied to this case is

$$T_1 = T_2 + wh \qquad (13\text{-}90)$$

in which w = weight per foot of the wire and T_1 and T_2 = resultant tension in the uphill and downhill support, respectively (also see Par. **99** for a discussion of the vertical loads).

123. Inclined-span Calculations. The mathematics of the inclined catenary is quite complicated and will not be analyzed here. However, workable methods[1] have been devised and may be used when necessary. On the other hand, the equations of the inclined parabola are very simple, and calculations for any usual span may be made on the parabola formulas with sufficient accuracy for almost any practical purpose.

The equation of the inclined parabola is exactly the same form as for the parabola

[1] NASH, JOHN F., and NASH, JOHN F., JR. Calculations for Cable and Wire Spans; *Trans. AIEE*, 1945, Vol. 65, p. 685, and discussion, p. 984.
HUSTON, JOHN M. Design of Hillside Spans Simplified; *Elec. World*, June 2, 1952, p. 86.

with level supports. This is stated in mathematical language as follows: The equation of a parabola, referred to a tangent and the diameter through the point of contact, is

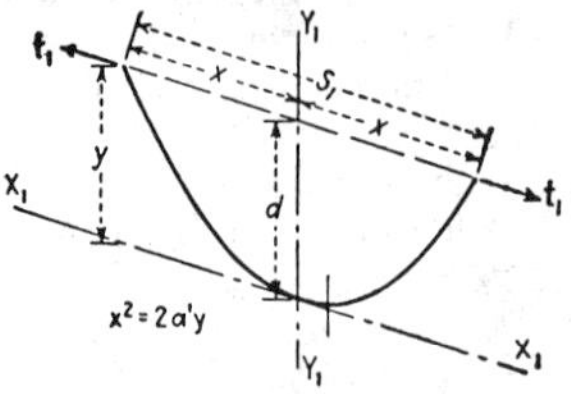

$$x^2 = 2py$$

Thus, if in Fig. 13-72 the y axis is vertical and the x axis inclined, y is measured vertically and x is measured parallel to the x axis. The sag is

$$d = \frac{wS_1^2}{8t_1} \tag{13-91}$$

FIG. 13-72. Inclined parabola referred to nonrectangular coordinates.

The sag d is measured vertically, and S_1 = slope distance between supports, t_1 = tension in the direction of the x axis, and w = weight per foot, also as measured parallel with the x axis. In other words, the total weight of wire in the span is approximately wS_1.

SUPPORTING STRUCTURES

124. Types of Supporting Structure. Numerous types of structure are used for supporting transmission-line conductors, e.g., self-supporting steel towers, guyed steel towers, self-supporting aluminum towers, guyed aluminum towers, self-supporting steel poles, flexible and semiflexible steel towers and poles, wood poles, wood H-frames, and concrete poles. The type of supporting structure to use depends upon such factors as the location of the line, importance of the line, desired life of the line, money available for initial investment, cost of maintenance, and availability of material. Because of the wide conductor spacing required for electrical clearances and insulation, the high tensile stresses used in conductors and ground cables to pull these cables up to a sag which will keep the heights of the structures within reason, the long spans necessary for crossing ravines in mountainous country, and the reliance to be placed on a major trunk line, high-voltage lines exceeding 138,000 V are generally built of self-supporting steel towers. A line built with self-supporting steel towers is the most satisfactory in all respects, as it requires less inspection and has a maximum life with minimum maintenance costs. However, high-strength aluminum-alloy towers are now available, and while at this time their first cost is more, they have the advantage of better resistance to corrosive atmospheres than steel.[1] The structural configurations and design details are the same as with steel, with the added problem of greater deflections when stresses are applied owing to the lower modulus of elasticity of aluminum. The effect of long-time creep of aluminum is yet to be determined. To date, aluminum towers have had little use comparatively, but the use is increasing. Self-supporting steel poles are frequently used in congested districts where right-of-way is limited and short spans are necessary. The advent of EHV has brought a great variety of new structural configurations. Details of some of these have been published. *Electrical World*, Nov. 15, 1965, pp. 95–118, contains outline drawings of 35 towers and six wood-pole H-frame structures as applied to EHV, as well as a tabulation of specification items of EHV lines in the United States and Canada.

Wood poles are used extensively where they are readily available. Medium- and lower-voltage lines can be built economically with such poles fitted with either steel or wood crossarms. **Wood H-frames** composed of two poles tied together at the top with wood or steel crossarms have been successfully used for the higher-voltage lines up to 345 kV. To take full advantage of the transverse strength, such poles can be braced internally for at least a portion of their height with wood "X"-bracing.

Concrete poles have been used in some parts of the world where timber is scarce and where the ingredients for making concrete are readily obtainable. They are generally cast in units, by using standard forms, and transported to the site, although they may be manufactured where used. Concrete poles should always have sufficient steel

[1] SELLERS, A. H., and WILLIAMS, J. E. All-aluminum Transmission Tower Line; *Trans. AIEE*, June 1961, p. 169.

reinforcement to take care of the bending stresses due to wind loads, pulls from cables, etc., in addition to being designed as columns under vertical loads. In all structures conductor configuration and the effect of various forces which may act upon them must be taken into account.

125. Conductor Spacing and Clearances. *a. Horizontal Configuration.* The minimum spacing of conductors on structures where pin insulators are used on medium-length spans will generally depend upon the least separation that can be used at midspan without the cables approaching too closely under adverse wind or ice-loading conditions.

With suspension insulators a different problem exists, as the swing of the insulator string has to be considered and clearances to the structure determined. This will generally give conductors a spacing at the supports which will be greater than the required midspan separation. A good rule to follow is to calculate the swing of the insulator string, both with the wind on the bare conductor and the wind on the ice-coated conductor with the corresponding vertical loads acting at the point of conductor suspension, to determine which condition gives the maximum deflection. The vertical loads should be taken on a length of span which is two-thirds the span for the horizontal loads. This will allow a certain amount of leeway in using a standard height structure at a location where the ground is lower than at the two adjacent structures. After the length of the insulator string has been determined electrically and the angle of insulator swing calculated, a normal electrical clearance is established to the structure from the deflected position of the conductor, which, when applied to the 3 conductors in their relative positions, will determine the necessary horizontal separation of the phases at the supports. This separation should then be checked to see whether or not it is sufficient for the midspan separation required. Midspan separations that will not be subject to arc-over if the conductors begin to swing out of step are usually inherent on high-voltage lines owing to the clearances required at the structures. On very long spans and on the longer spans of low-voltage lines the usual spacings may be insufficient. Thomas[1] proposed a horizontal-spacing formula for the determination of safe midspan spacings in windy territory where gusts and strong eddies might cause wires to start swinging at different periods,

$$\delta = CdD/w + A + L/2 \qquad (13\text{-}92)$$

in which δ = horizontal spacing in feet; C = an experience factor discussed later; d = percent sag of the condition to be studied; D = overall diameter of the conductor; w = conductor weight, in pounds per foot, used in calculating d; A = arcing distance of the line voltage (1 ft/110 kV); L = length, in feet, of the swinging portion of the insulator string. Thomas proposed an experience factor of 4 for copper and 3.5 for ACSR. It has since been found that, in country not subject to frequent violent winds, values of C as low as 1 will provide safe midspan spacings. Thomas was doubtful whether or not the added $L/2$ distance is necessary, since insulators seldom swing out of step. This doubt seems to have been justified. Spacing is further discussed in Par. **155**.

b. Vertical Configuration. Where the conductors are arranged in vertical configuration, the same electrical clearances will apply for the same voltage as for horizontal configuration; but it may be necessary to increase the vertical separation somewhat to prevent the conductors from coming together or approaching too closely at the center of the span when unequal ice-loading conditions occur or the ice falls off a lower conductor first (see Par. **120**).

In Fig. 13-73, θ = angle of insulator swing from vertical, H = horizontal span, V = vertical span, w = weight of conductor with or without ice load per lineal foot, w_e = resultant wind on ice-coated or bare cable (corresponding to loading condition assumed for w) per lineal foot, and w_i = weight of insulator string including hardware. Then

$$\tan \theta = Hw_e/(Vw + w_i/2) \qquad (13\text{-}93)$$

Ground-wire cables, if used, are located above the conductors for lightning-protec-

[1] THOMAS, PERCY H. Formula for Minimum Horizontal Spacing; *Trans. AIEE*, 1928, Vol. 47, p. 1323.

tion purposes and in such a position that there is no danger of contact with the conductors at midspan. As ground-wire cables are generally strung with less sag than the conductor cables, ample clearance at midspan is readily obtainable.

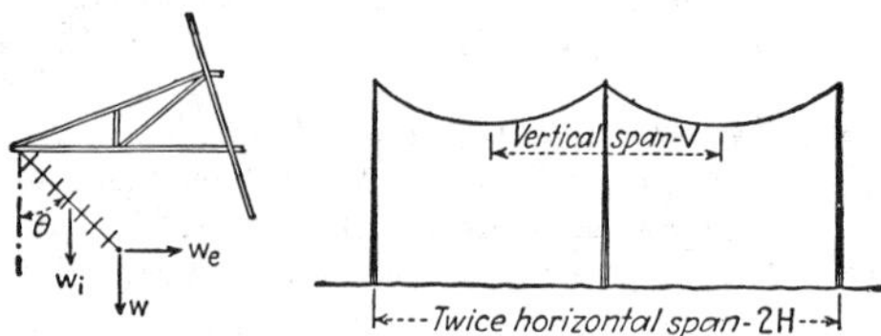

FIG. 13-73. Determination of suspension insulator swing.

The above considerations taken together with the maximum vertical sag to be used and the height required for the conductors above the ground level will determine the height and width of the supporting structure. Extensions can be used where the terrain requires a higher structure than normal.

126. Transverse forces acting on towers or poles are due to wind on the conductors and ground cables (and ice coating if in ice districts), wind on the structures, and horizontal components of the tensions in the cables at angle turns in the line (see Fig. 13-74).

The stress due to an angle in the line is computed by finding the resultant force produced by the wires in the two adjacent spans. For example, in Fig. 13-74, if the change in the direction of the line is the angle a and the stresses t in the adjacent spans are equal to each other, the resultant force

$$F = 2t \sin (a/2) \qquad (13\text{-}94)$$

Table 13-16 gives the resultant force F due to a tension t of 1,000 lb in each wire of two adjacent spans. The resultant force due to each wire may be thus computed and the moments about the ground line determined. These moments may be added to those produced by the wind pressure to find the maximum stress.

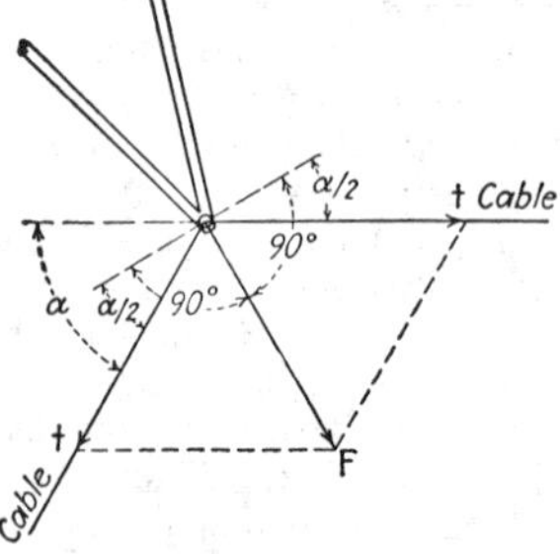

FIG. 13-74. Determination of transverse forces.

In applying wind loads to the structure, it is common practice to consider that the effective area on the leeward side is one-half that on the windward side, owing to shielding. Wind pressure on the structure is usually taken at 13 to 20 lb/ft² on 1½ times the exposed area of one face, with the wind acting in a horizontal direction.

Table 13-16. Resultant Force Due to Equal Tensions of 1,000 Lb in Adjacent Spans

Angle		Resultant F, lb	Angle		Resultant F, lb
a deg	$a/2$ deg		a deg	$a/2$ deg	
10	5	174.4	70	35	1147.2
20	10	347.2	80	40	1285.6
30	15	517.6	90	45	1414.2
40	20	684.0	100	50	1532.0
50	25	845.2	110	55	1638.4
60	30	1000.0	120	60	1732.0

127. Longitudinal forces acting on towers or poles are due mainly to the maximum tension which is assumed to exist in the conductor and ground-wire cables if broken. Ordinarily, especially with suspension insulator strings, these tensions are normally balanced in the adjacent spans; but if a cable breaks, either from overstress or from burning off, a distinct force is produced along the line due to unbalanced tension. If the break occurs on a cable at the end of a crossarm, there is in addition to the longitudinal force a torsional force introduced which must be resisted by the structure. Wind acting in the direction of the line is not ordinarily a factor, as the maximum tension in the cables is produced when the wind is blowing transverse to the line. As to the reduced stress which occurs in a span from the breaking of a conductor with the suspension insulator string deflecting in the direction of the line, the best practice is to ignore this reduction in tension, as the force due to breaking may cause an impact which more than offsets the reduction in tension. Special release clamps were devised for use on the Plymouth Meeting–Siegfried line of the Philadelphia Electric Company so that, if an insulator string deflected to an angle of 20° in the direction of the line, the clamping mechanism would release the pressure with only the friction in the saddle holding the conductor. This reduced the tension in the conductor considerably; and by assuming a lower value for the tension in the conductor due to a break, a more economical structure was obtained.

128. Vertical forces acting on towers or poles are those caused by the weight of that portion of the cables, plus ice loading if any, which is supported by the structure in question. In addition, there are the weights due to insulators and accessories and the weight of the structure itself. If a structure is located in a valley, there may actually be uplift on it, if the vertical components of the tensions in the cables exceed the downward loads.

129. Combined Forces. In determining the maximum forces acting on towers or poles, it is necessary to combine the transverse forces, longitudinal forces (including torsion), and vertical forces so that they act simultaneously. Several different combinations of loading conditions may be desirable, as follows:

a. A condition with all cables intact and the full transverse and vertical forces acting.

b. A condition with all cables intact, except the number it is desired to assume broken, with the transverse and vertical forces computed for each particular cable, according to whether or not it is assumed broken. The longitudinal forces due to broken cables must be combined with the transverse and vertical forces at all points of support where the cables are assumed broken. It is customary, when more than one cable is assumed broken, to consider all breaks in the same span and at the supports which will produce the maximum overturning moment, the maximum torque, or a combination of both.

c. A condition in some localities where extra-heavy vertical loads, caused by an unusually large formation of ice on the cables, may occur. These loads are combined with the weight of the structure.

d. A condition with vertical loads acting upward at the cable supports.

NOTE: It is not customary to combine transverse and longitudinal loads with the loads specified under (*c*) and (*d*).

Other factors may enter into the determination of the maximum forces acting on supporting structures in special cases, such as the horizontal and vertical components of tensions in guy cables and the addition of pole-top transformers, switches, and working platforms.[1]

The proper number of cables to assume broken is a debatable question and depends upon what margin of safety is desired and the amount of money it is desired to invest for this security. Generally speaking, the minimum number of cables to assume broken for tangent suspension single-circuit towers should be either one ground cable or any one conductor, and for double-circuit towers either one ground cable and one conductor or any two conductors on the same side of the tower and in the same span, by using the different cable supports for application of the forces to determine the maximum stress in each member of the tower. Anchor or dead-end towers should be able to withstand all or any number of cables broken. Generally, the condition of broken cables on one side of the tower will produce greater stresses in the web members than if all the cables are considered broken, owing to the unbalanced torsional forces existing when only the cables on one side of the tower are broken.

130. Types of Tower. Towers may support single, double, or multiple circuits.

[1] FARR, F. W., FERGUSON, C. M., MCMURTRIE, N. J., STEINER, J. R., WHITE, H. B., and ZOBEL, E. S. A Guide to Transmission Structure Design Loadings; *Trans. IEEE Power Group*, November, 1964, p. 1073.

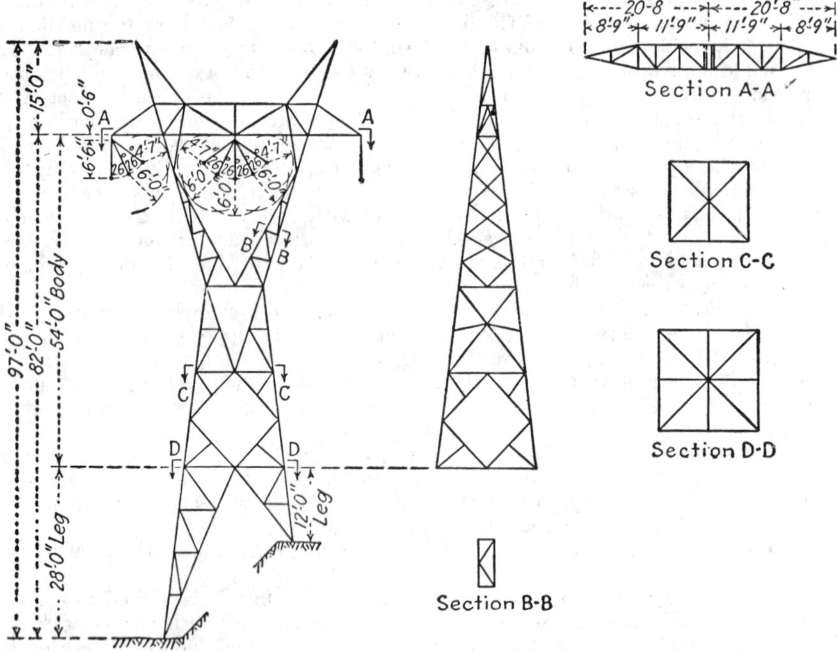

FIG. 13-75. Tennessee Valley Authority 161-kV single-circuit tangent suspension corset-type tower. (*Designed by Blaw-Knox Co.*)

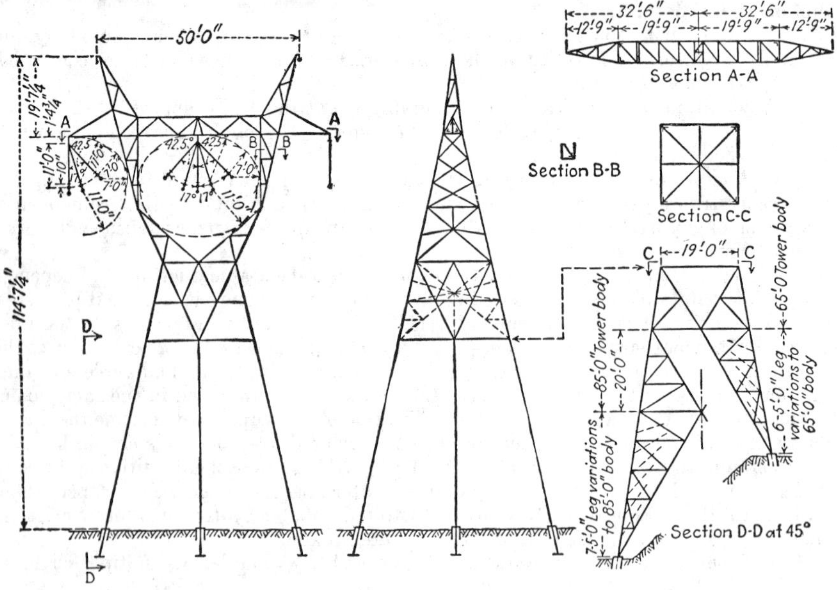

FIG. 13-76. City of Los Angeles 287-kV tangent suspension rotated-type tower. (*Designed by American Bridge Company.*)

The first two types are generally used for transmission-line work except in congested areas where right-of-way is very expensive and it is desired to transmit large blocks of power over one tower line. In such a case three or more circuits may be supported by the towers.

a. Self-supporting or Rigid Towers. On both single- and double-circuit tower lines of any considerable length, at least three kinds of tower are required for economic reasons:

1. A tangent suspension tower which can be used for normal spans where no angles in the line occur (see Figs. 13-75 and 13-76).

2. An angle suspension tower which can be used for normal spans with a small angle turn in the line or with longer spans on tangents.

3. An angle tower which can be used for normal spans with a large angle turn in the line, with extra-long spans on tangent, or as a full dead-end tower for anchoring. Insulators may be either suspended or in the strain position.

Very often it is desirable to introduce a fourth kind of tower with insulators always in the strain position to take care of exceptionally large angle turns in the line; in extremely long spans on tangent; and also, where required, as a full dead-end tower. When this type of tower is provided, the tower listed under (3) may be of lighter construction and not used for dead-end purposes.

Double-circuit towers with the vertical configuration of conductors, as used on different lines, are very much alike in appearance, generally being square in cross section. It is customary to locate the middle conductors outside the upper and lower conductors, for reasons explained in Par. **120**.

With single-circuit towers and the conductors arranged in horizontal configuration, a different problem arises which has resulted in the design of special patented structures for the higher-voltage lines with wide conductor spacing. The shape of these towers has been developed with a view to minimizing the weight of steel required in the superstructure and also reducing the size of footings by minimizing the effect of torsion. The more common types are the Blaw-Knox tower (see Fig. 13-75), or corseted type, as originally used on the Plymouth Meeting–Conowingo line of the Philadelphia Electric Company; and the American Bridge Company's rotated tower (see Fig. 13-76), used on the first Hoover Dam–Los Angeles line and also by the Bonneville system and on lines of the Tennessee Valley Authority. Either of these types serves the purpose for which intended. The theory behind the rotated tower is that the greatest overturning moment is caused by a combination of the transverse forces and longitudinal forces, due to broken cables, acting simultaneously, which produces a resultant force acting at an angle of approximately 45° with the direction of the line. In this case the whole four tower legs are resisting the overturning moment, thereby reducing foundation loads and consequently costs. Under normal conditions of loading, with only the transverse forces acting, two legs on the diagonal separation will take care of the overturning moment. Obviously the greatest advantage of the rotated type over the nonrotated type is on tangent towers and towers used for small-angle turns in the line when the transverse and longitudinal forces are approximately equal.

Figure 13-77*a* shows a TVA 500-kV conventional-design tangent self-supporting tower for a bundle-conductor line having three 971,600-cmil ACSR conductors/phase. The overhead ground-wire clamps are suspended and insulated from the tower by means of distribution-type guy strain insulators. The overhead ground wires are composed of seven strands of No. 9 Alumoweld and are used for carrier-current communication channels.[1] Each ground-wire insulator is provided with a spill-over gap to protect it during lightning discharges.

It is interesting to compare Fig. 13-77*a* with Fig. 13-77*b*. Both show 500-kV towers, but Fig. 13-77*b* is designed for a narrower right-of-way. The wind side swing of the conductors in a span is half the sag at 30° side swing, and this is common to both towers. Therefore the saving in right-of-way for Fig. 13-77*b* is 40 ft plus 7 ft 7 in less 30 ft 3 in, or 17 ft 4 in on each side, or a total of approximately 35 ft.

b. Semiflexible Towers. Such towers have been used to some extent for the voltages

[1] Farmer, G. E. Ths Use of Insulated Ground Wires on a Transmission Line for Communication Purposes; *Trans. IEEE*, Power Apparatus and Systems, December, 1963, p. 884.

under EHV. This type of tower has a narrow base in the direction of the line. The ground cables are strung tightly to take up unbalanced loads due to broken conductors and form part of the structural system. In case a conductor breaks, the unbalanced load will be taken up by the ground cables and transmitted by them to the next anchor tower.

Fig. 13-77*a*. 500-kV tangent self-supporting tower. (*Tennessee Valley Authority.*)

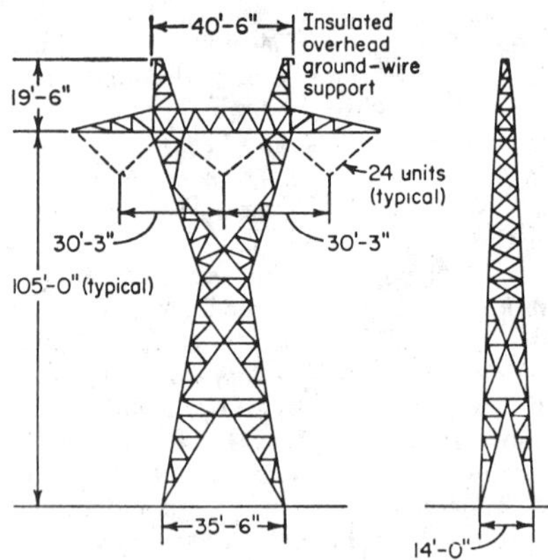

Fig. 13-77*b*. 500-kV semiflexible steel tangent tower of Arkansas Power and Light Company.

With the advent of EHV and bundle conductors, semiflexible self-supporting towers are receiving more consideration, and some are being used. With the heavy bundle conductors, the breaking of one conductor is not serious, and the breaking of all conductors of a phase is practically nonexistent. Possible causes are airplanes in trouble and tornadoes, which no practical tower could withstand. Figure 13-77*b* shows such a tower as used on the 500-kV system of the Arkansas Power and Light Company. The overhead ground wires are insulated from the towers as they are in Fig. 13-77*a*, for communication purposes. Figure 13-78 shows a steel-saving semiflexible tower used by the Pacific Gas and Electric Company. Note the X-guying used between tower legs to obtain the required lateral strength.

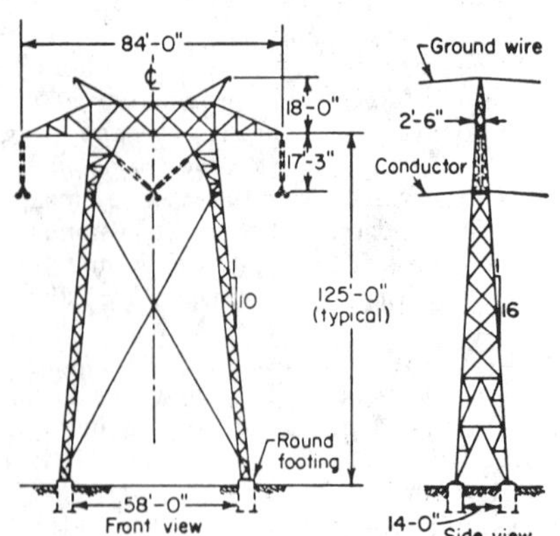

Fig. 13-78. 500-kV steel tower used in valley areas by Pacific Gas and Electric Company.

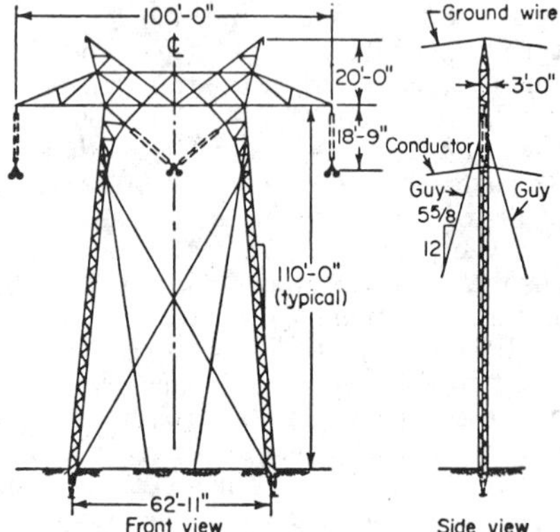

Fig. 13-79. 500-kV steel tower used in mountainous areas by Pacific Gas and Electric Company.

Guyed towers overcome the weakness of semiflexible towers in line with the line. They can be used for single-conductor lines or for any other service. Guyed towers can be damaged by the guys being struck by heavy farm equipment or other heavy vehicles. Right-of-way therefore should be carefully chosen. Trouble from livestock rubbing against the guys should not cause damage, for high-strength strand is used and the guys are tightly strung. Figure 13-79 shows a guyed steel tower used by the Pacific Gas and Electric Company in mountainous country. A feature of this tower is that the legs do not have to be of equal length. This tower has the same internal X-guying as the tower of Fig. 13-78. The self-supporting feature of the tower of Fig. 13-78 is replaced by four guys in the direction of the line and with an increase in strength.

Figure 13-80 shows a Kaiser aluminum guyed "V"-tower used on the Greene County–Meridian 230-kV line of the Mississippi Power Company. Figure 13-81 shows the guying diagram of this tower. V-towers are a comparatively recent development (see line survey appearing in *Electrical World*, Nov. 15, 1965).

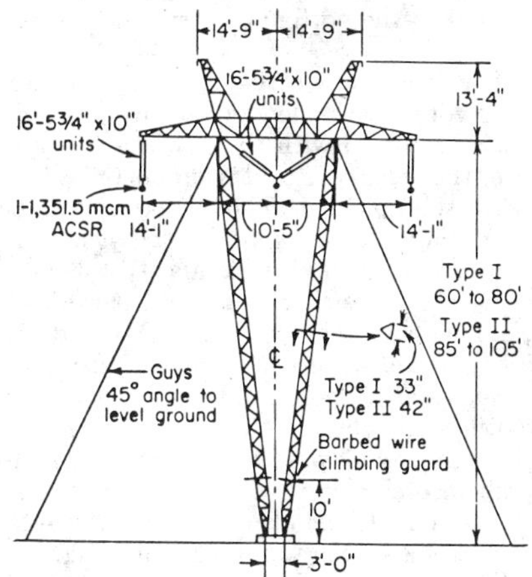

Fig. 13-80. Aluminum V tower of Mississippi Power Company (230 kV).

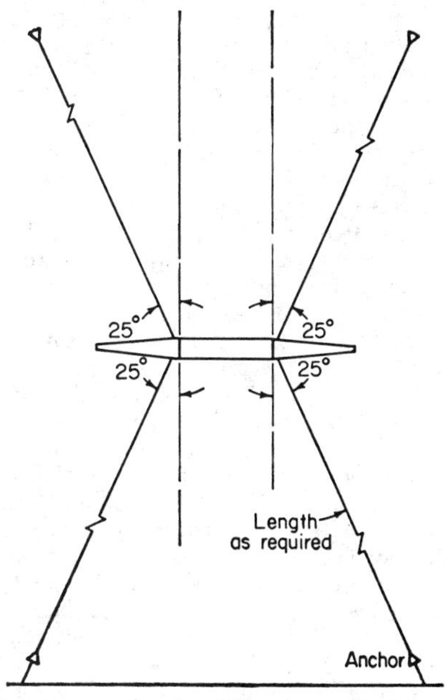

Fig. 13-81. Guying diagram of tower shown in Fig. 13-80.

131. Stresses in Towers—Design. Most stresses in towers can be computed analytically if the structure is statically determinate, but graphic methods are quicker and just as satisfactory, and there is less likelihood of error. This is true if the foundations are assumed to be rigid. In actual practice, with towers set in the ground, an uneven settlement of the foundations may produce a statically indeterminate structure, and excess stresses may develop which must be considered and taken care of in the overload factor provided against failure.

Tension members may be designed for their full net area of section with bolt holes deducted, but a compression member will fail soon after the elastic limit of the material is reached, and factors of safety based on the ultimate strength of the material are unsatisfactory and erroneous.

Torsional stresses present a special problem, and they must be combined with either transverse or longitudinal shearing stresses in all cases where they exist. The combination that is greatest determines the size of the web members where the four faces of the tower are made alike. With a tower of square cross section, the torsional stresses may be considered as evenly distributed on the four faces of the tower; but if the cross section of the tower is rectangular in shape, the distribution of the stresses due to torsion is very uncertain; and where there is much departure from a square section, all the torsion should be considered as taken on the two faces which form the short sides of the rectangle, which is of course uneconomical for the larger structures. With the new types of corseted and rotated single-circuit tower, the torsional stresses are considered as equally distributed on the four faces of the tower below the waist and at the point where all faces are alike. Torsional stresses may exist in the corner posts of towers having a square cross section if a tension system of diagonal bracing is used for web members.

For short panels in narrow-faced towers, it is economical to use a tension and compression system of web bracing without horizontal struts, as the stresses in the corner posts will be reduced considerably, with torsional stresses eliminated entirely. The bracing should make an angle of 30 to 45° with the horizontal, with all web members in any one panel of the same size to divide the loads equally. Where a tension system of diagonal bracing is used with horizontal struts taking the compression, the angle of the bracing may be increased to 50° or more from the horizontal. Long, unsupported lengths of main members can be broken up by the use of redundant members. The lower panels of towers should be so designed that variable-length leg extensions can be employed, which are interchangeable, to take care of sloping ground. Square extensions of any desired length may be used where towers higher than standard are necessary, with variable-length leg extensions fitted to the bottom of the square extensions if required.

132. Stresses in Tower Foundations. The calculation of the stresses in tower foundations is quite involved, as it is necessary to consider the downward thrust, uplift, and shear. On one side of the tower the shear is combined with the downward thrust (which also includes the weight of the structure, the weight of cables supported, etc.), and on the other side the shear is combined with the uplift. All these forces may be considered as being applied at the top of the footing, as this is where the system of web bracing usually terminates. Although the center of earth resistance to shear is some distance below the top of the footing, the type of steel anchor now used with spread-angle members will take care of the bending moment and shearing forces, thus doing away with the necessity of reinforcing the stub angles and the use of special shear plates. A horizontal strut, at the bottom of the tower just above the ground, will redistribute the applied external shear equally to the two footings and eliminate the "spreading-out" and "pinching-in" shears, but this strut is usually omitted for practical reasons. Therefore, the distribution of external shear is unequal. Torsional forces which are assumed to be equally distributed in the four faces of a square tower also cause a horizontal shearing force at the top of each footing, and the distribution to the two footings in any one face is affected by the system of web bracing used, in a similar manner to the direct shear. Any horizontal load applied to a tower with sloping faces, where a horizontal strut at the ground is omitted, will also produce an outward thrust, or tendency to spread the tower legs which are in compression, and an inward thrust, or tendency to "pinch in" the tower legs which are in tension. Symmetrical vertical loads applied to a tower with sloping faces cause equal horizontal shearing forces at the top of the footings in a diagonal direction at angles of 45° with the axes of the tower, which may be resolved into components in the four faces. Unsymmetrical vertical loads, such as are produced by installing a single circuit on one side of a double-circuit tower, unequal loadings of cables on the two sides of a tower, etc., require a special investigation. By considering the effect of the transverse, longitudinal, and vertical loads, together with the torsional forces, each acting independently, and deriving their respective horizontal shear-

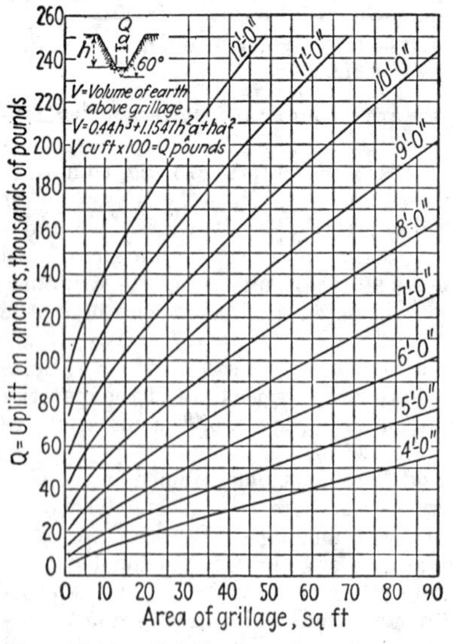

FIG. 13-82. Depth of anchors in earth for various uplifts.

ing forces at the top of the footing, a tabulation can be made from which the maximum resultant shearing force acting in any direction on any one footing may be obtained.

While the point of application of the resultant of earth resistance against shear is unknown, it is safe in good soil to assume that the earth will offer some lateral resistance to displacement, depending on the amount of bearing surface in the footing, so that it is not absolutely necessary to resist the combined shear and downward thrusts on the bottom of the footing alone. In the case of concrete foundations in good soil, the lateral resistance of the earth may safely be assumed to be as much as 25% of the maximum bearing pressure on the bottom of the footing.

The general dimensions of the footing must be such as to provide suitable resistance to uplift when considered with the depth to which it is set in the ground. The minimum depth of setting should be not less than 6 ft, but excessive depths will increase the cost of excavation considerably. It is customary in ordinary good soil to consider that earth offering resistance to uplift will be that due to the volume of an inverted frustum of a cone or pyramid of earth whose sides make an angle of 30° with the vertical. This can be modified for wet soils. Ordinary earth may be considered as weighing 100 lb/ft³ and concrete 150 lb/ft.³

The factor of safety against overturning of the tower should be at least 50%, and preferably 75%, above the uplift produced by the working loads. Figure 13-82 shows a chart giving depths of footings required for different uplift values.

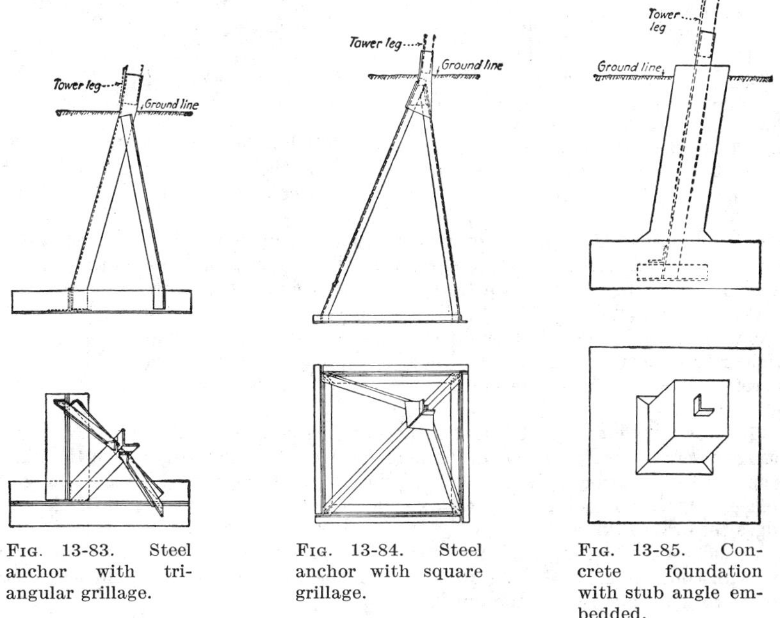

FIG. 13-83. Steel anchor with triangular grillage.

FIG. 13-84. Steel anchor with square grillage.

FIG. 13-85. Concrete foundation with stub angle embedded.

133. Grillage Footings. Steel anchors set in the ground with protective coatings are economical and have been successfully used for tangent and small-angle towers. The most satisfactory type of steel anchor is of the pyramidal type, either triangular (see Fig. 13-83) or square (see Fig. 13-84), built of steel angles and with a grillage bolted to the lower end. This type of anchor will resist the shearing and bending stresses satisfactorily. The grillage itself should present a flat surface on the bottom, with a bearing surface great enough for the soil with which it is used. Average good, dry, sandy, or gravelly soil will withstand without undue settlement a load of 3 tons/ft² for working loads. This should be modified for the poorer soils such as clay and alluvial silt. Where the soil is subjected to flood conditions or is likely to be permanently wet, the grillages may be encased in concrete to increase the bearing surface and also provide additional resistance against uplift.

Galvanized-steel anchors buried in the ground will last nearly as long as the superstructure under ordinary soil conditions, but they are not satisfactory, without additional protection, in soil having a sulfur content such as cinder fill.

134. Concrete foundations are generally used on large-angle and dead-end towers and for all special structures requiring great strength such as river-crossing towers and towers at the ends of extremely long spans. There are several types which can be used (see Figs. 13-85, 13-86, and 13-88). The type whereby the stub angle is embedded in concrete and takes the natural slope of the tower legs is the most economical as far as the concrete is concerned (see Fig. 13-85) but requires more accuracy in setting, as there is not the chance for vertical adjustment that exists where anchor bolts are used. Figure 13-87 shows a typical concrete anchor for fairly heavy tower construction which is both economical and easy to set. The maximum bearing pressure on the bottom of a concrete footing should not exceed 3 tons/ft² for the working loads. If appreciable tension stresses exist in a concrete base, it should be reinforced with steel rods. Also, the shaft of the footing should be reinforced for tension stresses due to flexure. Anchor bolts should be of sufficient diameter to take the full uplift load and long enough to develop the bond value of the concrete. If anchor bolts are bent, the necessity for using washers at the lower end is eliminated.

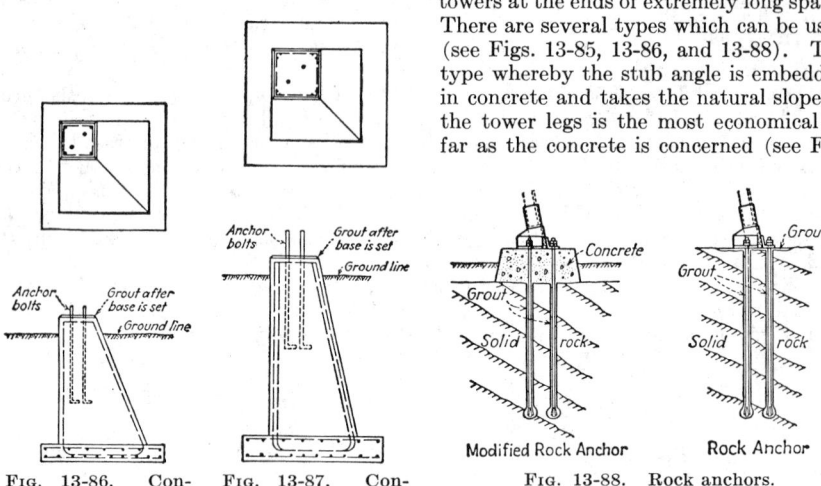

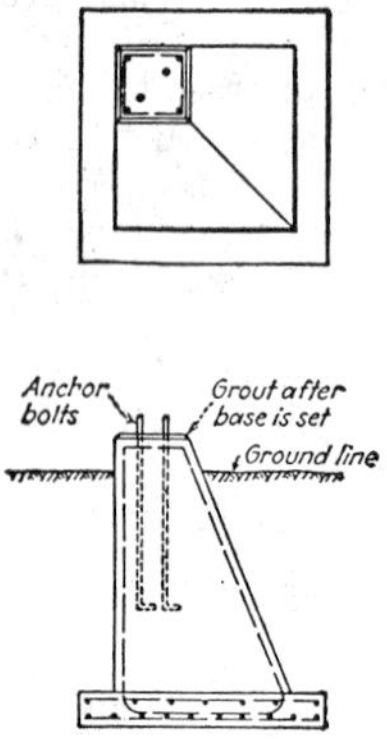

FIG. 13-86. Concrete footing with anchor bolts; for single-circuit tower with sloping legs.

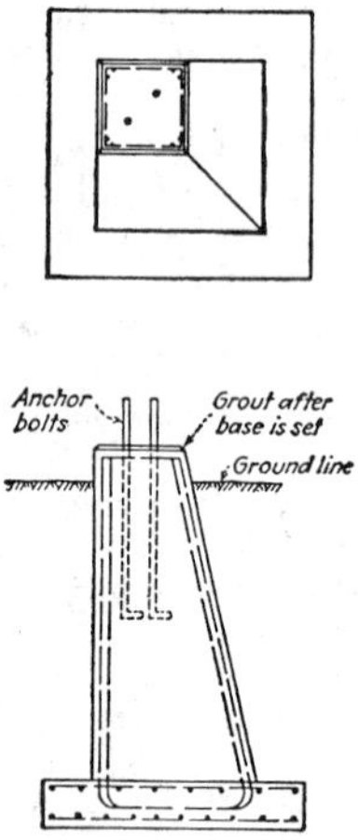

FIG. 13-87. Concrete footing with anchor bolts; for double-circuit square tower.

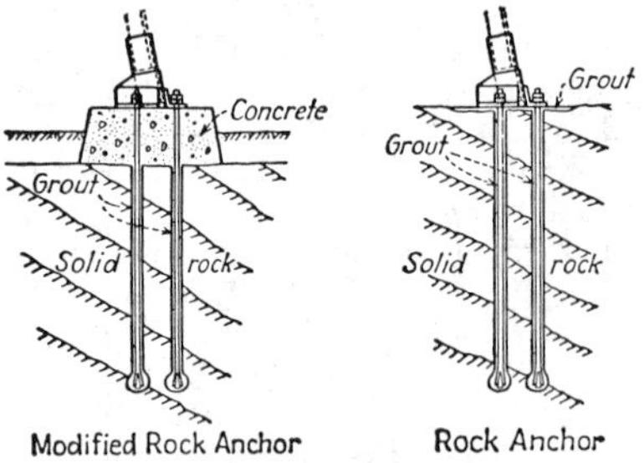

FIG. 13-88. Rock anchors.

135. Rock anchors may be used in lieu of steel anchors with grillages or concrete footings where the rock is solid throughout. Holes somewhat larger than the diameter of the bolt can be drilled in the rock to receive the anchor bolts, which are then grouted in place. Anchor bolts for use in rock should be split at the bottom and furnished with wedges (see Fig. 13-88).

If solid rock is encountered below the surface, a modified form of footing may be used with concrete on top of the rock and anchor bolts extending a sufficient distance into the rock to take care of the uplift forces.

136. Special structures are required for transposition towers where it is not expedient to make the transpositions on a standard tower by the use of special crossarms. Long spans over rivers and bays and crossings over important highways and trunk-line railroads frequently require towers which either are much higher than normal or must have a larger factor of safety against collapse. Anchor towers near substations, towers for mounting switches, and towers for turning 90° angles also may come under this classification. Such special structures are designed to suit local requirements and are subject to regulations of the U.S. Army Engineer Corps in the case of navigable-river crossings, to state public-utility commissions or other bodies for highway crossings, and to the particular regulations of railroads which are concerned.

137. Insulator and Ground-cable Attachments (see Fig. 13-89). Pin insulators are generally mounted on pins bolted directly to the crossarms of the tower. The method of attaching suspension insulator strings varies. One form of attachment is the U-bolt fastened on the underside of the crossarm to which the insulator hardware is attached. This device will give flexibility both longitudinally and transversely. Another attachment is in the form of a bent plate or angle fastened to the underside of the crossarm with a fairly large hole to receive a hook or shackle at the top of the insulator string.

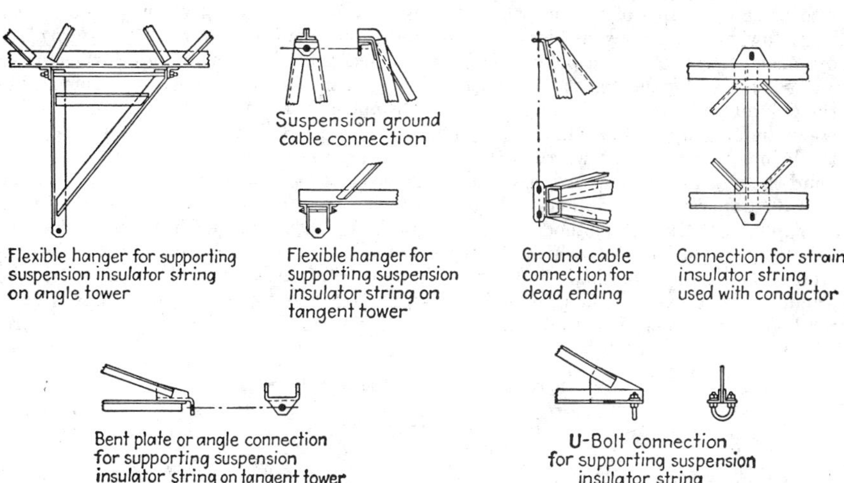

Fig. 13-89. Insulator and ground-cable attachments.

In order to keep the conductor spacing to a minimum on high-voltage lines and take advantage of clearances to steelwork, the point of attachment of the insulator string may be dropped several inches below the crossarm, in which case flexible hangers are required, i.e., hangers which are hinged at the crossarm and free to pivot in the direction of the line. These hangers may be made of plates, shapes, or bent round rods, with suitable connection at the bottom for receiving the insulator string. With suspension-type insulators in the strain position, horizontal pull-off plates are required.

In order to minimize failure of ground cables due to vibration, they are preferably suspended, and the attachment at the tower consists of an angle or bent plate with a hole to receive the suspension clamp. Patented rigid ground-cable clamps may be obtained if desired, which can be bolted directly to the steel structure. These are generally of the V-groove type.

Ladders and Step Bolts. Ordinary steel towers are provided with step bolts on one corner post for climbing the tower. Special high structures, such as river-crossing towers, should have ladders extending up to the level of the top crossarm. Such ladders should be provided with guard cages supported on the sides of the ladder.

138. Tower Tests. When an important transmission line is built using towers of new design, at least one type of tower (generally the tangent suspension tower or the type which is to be used most frequently) should be tested, first with the working loads as specified for the design of the tower applied, and finally with the ultimate loads which the tower is expected to withstand, applied. Equivalent concentrated loads may be used in some cases to avoid applying a multiplicity of small loads at different points, which would cause delay in shifting loads, but care should be taken to see that all combinations of loads or individual loads which will produce the maximum stress in each member are applied. After the tower has successfully withstood all the specified loads, a destruction test is desirable to determine the overload factor. This can generally be made with the test loads which cause the maximum stresses in the greatest number of members on the tower in place, by increasing the transverse loads indefinitely until failure occurs. After a test is completed, members should be examined for elongation of bolt holes, straightness, etc.

Towers should be tested with the protective coating which is to be used in service on the steel, and the foundations should be the same as those for which the towers are designed. If it is impossible to test towers on earth foundations, they may be tested on rigid foundations but a test on rigid foundations will undoubtedly show a greater overload factor than may be expected in service.

139. Erection of Towers. Towers may be assembled on the ground and then lifted into place by means of self-propelled derrick cranes or latticed-steel gin poles.[1] Very large towers are usually assembled in sections, and the sections are lifted into place by means of cranes or gin poles. Towers in inaccessible locations may be transported and assembled by the use of helicopters. If necessary, the towers can be erected from the ground up by the use of gin poles moved from corner to corner and with the erection crew climbing up on the partly completed structure. This method may be used for small jobs which do not warrant the use of heavy cranes or for very tall towers beyond the reach of cranes, such as those for river crossings.

140. Concrete Foundations. Two types of concrete footing may be used with steel poles. The first method consists in building a latticed-steel framework for use below the ground level to which the superstructure is bolted and encasing this in concrete. The second method employs a plain or reinforced-concrete block set in the ground with anchor bolts provided for supporting the superstructure. The second method offers better opportunity for vertical adjustment in erecting the superstructure.

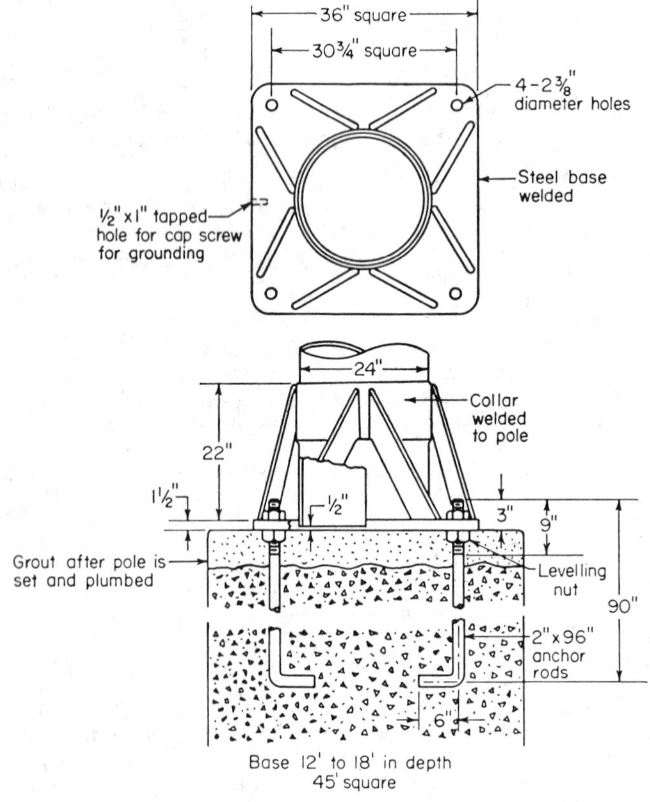

Fig. 13-90. Foundation for steel pole.

[1] Richardson, W. B. New Techniques Speed Construction of 500 kV Lines; *Elec. World*, Jan. 15, 1965, p. 27.

The concrete footing must be able to resist both the vertical forces acting and the overturning moment on the superstructure due to wind loads, broken cables, etc. The maximum toe pressure on the base, which will occur at the side toward which the pole deflects under load, must be kept within the limits suitable for the soil in question. Not much reliability can be placed on the lateral resistance of the earth for some time after poles are erected, and there must be sufficient weight in the concrete footing to keep the pole from overturning or leaning perceptibly.

Figure 13-90 shows a concrete foundation with special leveling provisions for the steel pole of Fig. 13-91. The foundation is precast, 45 in square and 12 to 18 ft in depth as required. The precast blocks are placed in holes and leveled, and the hole around the block is filled with low-cement-content fluid concrete.

Tubular steel poles can be set in holes on a concrete base and the hole then filled with concrete. Uncoated carbon-steel poles rust rapidly on the inside at the ground line and to a lesser extent on the outside. Therefore, they should be heavily galvanized inside and out, at least from the bottom end to 18 to 24 in above the ground line. Dipping the bases in asphalt also gives a good measure of protection.

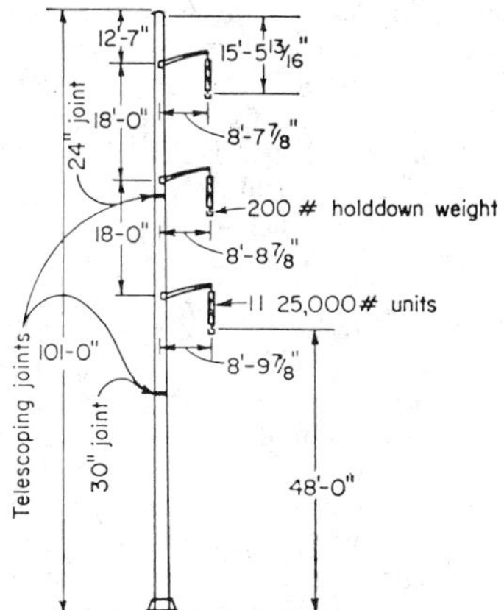

Fig. 13-91. Steel pole for 230 kV. (*Arizona Public Service Company.*)

141. Tubular steel poles are being used on city streets and in congested areas where a wide right-of-way cannot be gained. They have been used for voltages up to and including 230 kV. Vertical configuration of conductors is used for all high-voltage lines. Insulators may be side post or suspension[1] on cantilever arms or a combination[2] of the two. Figure 13-91 shows a 230-kV pole used on a line of the Arizona Public Service Company in Phoenix. These poles are of tubular steel in three sections with telescoping joints. The poles are tapering, with a diameter of 24 in at the base and 10.8 in at the top. Their foundations are described in Par. **140**. The mast arms are 8 ft long, of tubular steel, with brackets bolted to the poles with two ¾-in through bolts. The poles are spaced approximately 300 ft. Insulator side swing is reduced by a 200-lb combined hold-down weight and corona shield. The poles present a pleasing appearance and have elicited no objections even with a line installed on each side of a 60-ft street. The poles, side arms, and accessories were furnished by the Union Metal Manufacturing Company of Canton, Ohio.

The New Orleans Public Service Company 230-kV line[2] is of similar construction but is designed for hurricane-force winds. The poles are of 12-sided, elliptical, high-strength steel, with the short diameter, which is in line with the line, 75% of the long diameter. The insulators are a combination of 12 suspension insulators and a swivel-ended strut (side-post) insulator equal to 12 suspension insulators, to prevent side swing of the suspension insulators. Some poles have side-post 230-kV insulators only. The poles have no base for bolting to a foundation but do have baseplates and are set in concrete in holes 25 ft deep. The holes are made by driving steel casings 32 in in diameter to a depth somewhat deeper than 25 ft and cleaning them out.

142. Pole Tests. A new type of steel pole should always be tested on rigid foundations, with the transverse, longitudinal, and vertical loads applied simultaneously, to introduce all direct bending and torsional stresses. Crossarms should be tested for the additional torsional stresses introduced, where pin insulators are used, and combined with the longitudinal loads and the heaviest vertical loads specified.

---

[1] Ramthun, M. K., Pitzel, B. H., and Campbell, D. W. Stream Lined 230 kV Transmision Passes Overhead in City's Streets; *Elec. World*, June 29, 1964, p. 94.

[2] Stumpf, M. W., and Mouton, R. A. 12 Sided Single Poles Carry 760 MVA Capacity Line (New Orleans, La.); *Elec. World*, Nov. 16, 1964, p. 94.

PROTECTIVE COATINGS FOR STEEL STRUCTURES

143. Galvanizing. For important transmission lines, where a long life is desired, it is almost universal practice to galvanize fabricated-steel towers and poles which are field-bolted, after the other shopwork has been completed, for under such conditions galvanizing is more economical than painting. The method of "hot-dip" galvanizing is also used for bolts and nuts, with the threads rerun for nuts after galvanizing. This has practically superseded the sherardizing, or "dry galvanizing," of bolts and nuts which was in use for a number of years. All galvanizing should be in accordance with the Standard Specifications for Zinc (Hot Galvanized) Coatings on Structural Steel Shapes, Plates and Bars and Their Products, as given by ASTM Designation A123, which calls for an average coating of 2 oz of zinc/ft^2 of surface.

To test the uniformity in thickness of the galvanized coating, the so-called **Preece test** is used. This is described in the Standard Methods of Determining Weight and Uniformity of Coating in Zinc-coated (Galvanized) Iron or Steel Articles, as given by ASTM Designation A90.

Structures located near industrial plants, if subjected to sulfuric acid and fumes, should not be galvanized.

144. Painting. Painting is sometimes resorted to for fabricated-steel towers and poles, and generally shop-riveted, welded, or special steel poles are painted. Towers located near industrial plants in a smoky atmosphere should preferably be painted. The base coat should be of a mixture of red lead and raw linseed oil in something like the proportion of 33 lb of dry red lead to a gallon of oil with a little dryer added. The outer coats may be of any good all-weather paint. To keep structures in good condition, painting is necessary every 2 or 3 years. With structures having a larger number of small members, this may make the cost of maintenance very high. On structures having few pieces and large, flat surfaces, painting may be economical.

Where steel structures are buried in the ground, a special problem sometimes arises at the ground level where moisture occurs. At this point, galvanizing may deteriorate after a short interval of time, especially if the soil has a sulfur or acid content. A paint made from asphaltum compounds will often prove useful for protection at these points.

Newly galvanized steel should not be painted until it has a chance to weather for a period of 6 months, and the galvanized surfaces should then be clean. Aluminum paint is ordinarily used to paint galvanized towers when the galvanizing has deteriorated.

145. Weathering, or self-painting, steels have been developed by the major steel companies. These are steels which are not treated with any kind of protective coating but the chemical composition of which is such that they may be said to "paint" themselves. They are installed completely uncoated but thoroughly cleaned of all mill scale and foreign matter. In a few years, a dense dark-brown oxide with a purplish cast has formed which becomes a permanent protective coating to all surfaces exposed to the weather. A slight loss of thickness occurs, which eventually stops, as the corrosion rate is nonprogressive. These steels have been available for many years but only recently have come into extensive use for transmission towers. Two of the steels are U.S. Steel's Cor-Ten and Bethlehem's Mayari-R.

WOOD AND MISCELLANEOUS STRUCTURES

146. Wood poles are considerably cheaper than steel for most types of construction. The lower cost is due, in part, to the more conservative basis of design normally adopted for steel. Generally, steel structures are designed to support safely one or more broken conductors, while wood structures are often not so designed. It is logical that the reasons for choosing the more expensive steel construction should require conservative design throughout and that conditions justifying the cheaper and shorter-lived wood structures would warrant accepting some of the more theoretical hazards.

For voltages of 69 kV and lower, wood is quite generally used; only the very heavy lines of this class justify the expense of steel.

Wood-pole construction for many years has been used for all voltages up to and including 230 kV. With the advent of EHV wood-pole construction has been extended to 345 kV. H-frames and various modifications have been designed, the most popular using the main crossarm as the bottom member of a truss (see Par. **160**).

Butt-treated cedar and full-treated pine are used almost exclusively in transmission-line construction; the use of untreated poles has been practically abandoned as uneconomical since the supply of chestnut and northern cedar poles has been exhausted. Treated fir has in late years been supplied in some quantity from the Northwest and, with the establishment of adequate treating plants, promises to be a valuable source of transmission-line poles.

Cedar poles resist decay, but satisfactory life is not secured unless the butt is treated. The pole is usually treated from the butt to about 2 ft above the ground line. The balance of the pole is not treated. Pine and fir require complete treatment of practically all the sapwood. This treatment is applied under pressure.

No effective protection has been devised against woodpecker damage. Some localities are often subject to serious epidemics of woodpecker trouble.

147. Preservative Treatment. Pole decay is due to a fungus which requires air, moisture, warmth, and food for its subsistence; the wood of the pole constitutes its food. The conditions most favorable to the growth of the fungus are found at the ground line. The preservative has toxic or antiseptic properties which make the wood either poisonous or unfit food for the fungus.

Preservatives and preserving methods conforming to the standards of the American Wood Preservers Association (AWPA)[1] should be used in the treatment of poles. There are many wood preservatives, including those using poisonous salts such as copper, mercury, zinc, and arsenic compounds. However, there are only two included in AWPA recommendations for poles, Standard C4, and they are:

1. Coal-tar creosote, AWPA Standard P1.
2. A 5% solution of pentachlorophenol in a petroleum distillate, AWPA Standard P8 (commonly called "penta").

By AWPA Standard M1, pentachlorophenol is not recommended for use in coastal waters. Coastal waters are defined as salty waters. One other preservative is increasing in popularity. This is AWPA Standard P11, a creosote-pentachlorophenol mixture in which pentachlorophenol is not less than 2% of the mixture. All of these preservatives are applied by the following methods.

The open-tank method, applied to cedar poles, consists in boiling the butts of the poles in a tank of creosote oil, after which the oil is allowed to cool or the poles are transferred to a cold tank of oil. The duration of the hot and cold treatment, usually 8 h or more, depends on several factors, the most important of which is the degree of seasoning. The treatment is based on the fact that the wood cells expand with heat and on cooling draw the creosote into the wood under atmospheric pressure. The sapwood of unseasoned poles has annular rings of a nearly impervious fiber which prevent penetration of the oil. In seasoning, this fiber dries and breaks open. To ensure penetration of the greater part of the sapwood, which is usually less than 1 in in depth, an incision process has been developed and is almost universally used. Narrow cuts, parallel with the wood fibers, are made to a depth of about ½ in at frequent intervals around the circumference of the pole for a distance above and below the ground line. Complete penetration is obtained to a depth somewhat greater than the depth of the incisions even on unseasoned poles.

Pressure treatment is applied to pine and fir. The poles, on a truck, are run into a steel cylinder and subjected to a steam treatment for a period of several hours at a temperature which will not damage the wood cells, usually specified at not more than 259°F. The pressure is then removed and a vacuum applied. The steam treatment opens up the wood cells and allows the preservative to penetrate. The length of time required for the steam and vacuum treatment depends on the condition of the wood, the amount of oil that is to be injected, and the depth of penetration desired. From this point in the process, one of two methods may be followed. The **full-cell,** or **Bethel, process** allows all the preservative injected to remain in the wood. This process is generally used for piling and underwater work when it is desired to exclude water from

[1] Standard of Recommended Practices, American Wood Preservers Association, 839 17th Street N.W., Washington, D.C.

the wood and to resist the attack of marine borers. The **empty-cell process** draws off excess oil and secures protection from decay by the coating of oil left on the walls of the wood cells. The empty-cell process is adequate and preferable for usual structures and is almost exclusively used for poles and arms.

The empty-cell treatment is obtained by either the Rueping or the Lowry process. The Rueping process seems to be in more general use, although the Lowry process is equally successful.

In the Rueping process, following the steam treatment, an air pressure is applied. While still under pressure, hot oil is forced into the cylinder. The oil is held under this pressure and maintained at a temperature of about 200°F, by steam coils within the cylinder, for a period of several hours. Upon removing the oil and reducing the pressure, the compressed air within the wood cells forces out the surplus oil. The amount of oil retained depends on the pressures applied and the time of treatment, although it is possible to remove only a part of the oil that has been injected.

The **Lowry process** is similar to the Rueping process except that no compressed air is used. After the preservative has been forced into the wood under pressure, a high vacuum is quickly created, causing a sudden expansion of the air within the wood cells and thus driving out surplus preservative (see also Sec. **4**).

148. Strength Calculations. As used in a line, the pole is a cantilever beam, fixed in the earth at the butt and supporting the transverse wind load from the conductors of a length equal to half the sum of adjacent spans. Computation of the safe load that may be carried is a matter of simple mechanics outlined in Fig. 13-92. Some slight approximations have been introduced for simplicity.

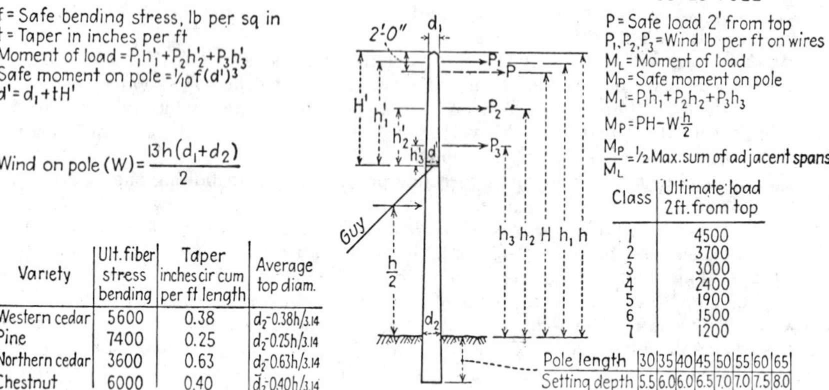

Variety	Ult. fiber stress bending	Taper inches circum per ft length	Average top diam.
Western cedar	5600	0.38	d_2-0.38h/3.14
Pine	7400	0.25	d_2-0.25h/3.14
Northern cedar	3600	0.63	d_2-0.63h/3.14
Chestnut	6000	0.40	d_2-0.40h/3.14

Class	Ultimate load 2 ft. from top
1	4500
2	3700
3	3000
4	2400
5	1900
6	1500
7	1200

Pole length	30	35	40	45	50	55	60	65
Setting depth	5.5	6.0	6.0	6.5	7.0	7.0	7.5	8.0

FIG. 13-92. General pole-strength calculations.

The fact that, if the pole were a part of a perfect cone, the maximum fiber stress might occur at a point above the ground line is of more theoretical interest than practical use. The difference between the load carried at the critical section and at the ground line is less than may readily be caused by irregularities in the pole.

Poles are almost universally classified according to the ASA dimensions (see Sec. **4**), which have been arranged so that the nominal ultimate strength is the same for all lengths and species of the same class. Poles are classified as Class 1, 2, 3, etc., up to Class 10, and the minimum circumference 6 ft from the butt is specified for each class and each species to give the desired nominal strength. The nominal ultimate was computed from conservative average ultimate fiber stresses from a very large number of tests.

The top diameters are specified but are given only as a minimum and are the same for all species. Actually, the taper of various kinds of timber while fairly uniform is quite different for different species, and the average top diameter of ASA-class poles will be considerably larger than this minimum.

149. Pole tests give very erratic results, and tests on a few poles should never be given great consideration. Designs should if possible be based on accepted average unit fiber stresses rather than test results unless a considerable number of duplicate tests can be made and averaged.

150. Factor of Safety. It has been found from experience with heavy transmission-line construction that a factor of safety of 2 on the accepted average ultimate is conservative. On light construction, this is sometimes slightly reduced, but a material reduction is usually not justified in view of the deterioration of wood with age. On sustained loads, such as heavy angles, a liberal additional factor of safety is desirable to prevent the pole's warping and giving the appearance of being overloaded. When possible, guys should be attached close to the load to eliminate heavy continued bending.

151. Setting Depth. The strength of the pole foundation is difficult to reduce to figures and is not of such primary importance to the safety of the structure as in the design of a tower. Failure of the foundation, in the sense that failure is used in the design of steel towers, i.e., a considerable movement of the pole in the ground, is of little consequence except for the inconvenience and expense of straightening up the line and retamping the poles. The setting depth for poles of various lengths has been pretty well established by general practice and is almost universally used (see Fig. 13-92). These depths seem somewhat illogical in that no account is taken of the strength of the pole or of the quality of the soil; however, this appears more reasonable when it is considered that the desired result is not to obtain a rigid foundation but to prevent the pole from "kicking" out of the ground.

152. Wood crossarms are now generally manufactured of creosoted yellow pine or untreated Douglas fir. Untreated pine arms of the timber commercially available are not satisfactory. Untreated fir arms are widely used and are apparently giving a life comparable with that of the poles. Arms should be of the highest-quality timber. The smaller arms, up to 5 by 6 in and 10 or 12 ft in length, can generally be supplied on standard crossarm specifications, although structural specifications give very satisfactory arms. Heavy arms for H-frames, that is, 6- by 8- and 6- by 10-in timbers and 3- by 8- and 3- by 10-in plank, 20 to 35 ft long, are best purchased as high-grade structural timbers. Structural timbers are furnished under the rigid specifications and inspection of the large timber manufacturers' associations.

Plank Arms. The eccentric connection of large arms to the pole, especially when carrying heavy conductors, is not desirable, and a number of designs make use of two planks, one on each side of the pole attached together at the ends. Generally two 3- by 8-in planks are used in place of a 6- by 8-in timber arm. The plank-arm construction has several advantages in addition to the better connection to the pole, although the end hardware is somewhat complicated, and in many designs the strength of the crossarm against longitudinal loads is somewhat reduced.

153. Wood vs. Steel Arms. Wood crossarms are lower in cost than steel arms of the same strength and, aside from the shorter life and possibility of being shattered by lightning, are satisfactory. On wood-pole construction the longer life of steel arms is of little value, and the possibility of lightning damage is the price paid for the lightning insulation of the wood. Lightning damage to arms is usually not a major operating problem; and on lines thoroughly shielded with overhead ground wires, crossarm and pole damage is practically eliminated. With pin-type construction, the insulation is at a somewhat lower level than with the usual suspension-insulator construction for the same voltage line, and 2 or 3 ft of ungrounded wood arm is generally considered preferable in such designs to the solidly grounded steel-arm design.

154. Design of arms must provide for carrying the vertical load with an ample margin of safety, but often neither the arm nor the connecting hardware is well suited for carrying the full load of a broken conductor as is required of steel towers. Crossarms on single-pole construction have practically no resistance to longitudinal loads. If a heavy conductor breaks, the arm will swing around to very nearly a longitudinal position, restrained only by the attachment of the unbroken conductors. This would be likely to result in badly bent hardware and probably a split and disfigured arm but little serious damage; the major damage would be the broken conductor and not the effects of the break. H-frame construction (Fig. 13-95*c*) is better adapted to such loads, but the effect of a break is much the same, in that the deflection of the poles and movement of the poles in the ground relieve the greater part of the load and usually result only in some minor damage to the arms and hardware, which is easily repaired.

Double-arm construction can be considered very little, if any, stronger than twice the strength of one arm. To obtain the additional strength of any truss action requires sufficient bolts and keys to develop the shear. The shear is several times the applied load and makes a very heavy joint necessary.

The common sizes used, 5 by 6 in for lighter single-pole construction and 6 by 8 in for H-frame, allow ample vertical strength for ordinary spans. Conservative unit stresses should be used in vertical load on the arms.

The connecting hardware, as generally used, is not designed as would be necessary in a framed structure, such as a truss, where movement in a joint would cause serious secondary stresses in the main members. Only one ¾-in bolt is ordinarily employed, even in heavy H-frame construction. In types of construction carrying wire heavier than has been general practice and in very long spans, the use of these connections, based entirely on experience, should not be followed without a careful check. The same applies to designs carrying heavy angles on crossarm construction where the entire angle load must be transmitted through the bolts to the pole. For such angles, the three-pole structure is a more positive arrangement.

155. Conductor arrangement and spacing in wood construction with short spans over comparatively level terrain are determined largely by the line voltage. A wide variety of conductor arrangements will be found in past practice. However, with the use of larger conductors and longer spans, the conductor configuration and separation are often a matter of providing the safest arrangement with ample spacing, especially for the occasional longer than normal spans encountered in rolling country. The conductor arrangement should provide spacing for these occasional long spans, as it is generally more economical to design a standard structure with spacings suitable for a span about 50% longer than normal rather than to use too many special structures. The dimension *B* (Fig. 13-93) is often determined by the span length rather than by the line voltage, especially for the lower-voltage pin-type construction.

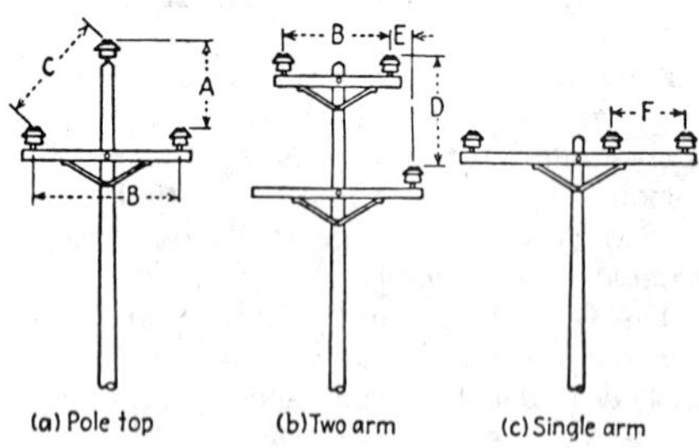

Voltage	Spacing			
	A	B	D	F
23 kv	3'-0"	5'-0"	2'-6"	3'-0"
34.5 kv	3'-4"	5'-0"	3'-0"	3'-4"
46 kv	4'-0"	7'-0"	5'-0"	5'-0"
69 kv	6'-0"	8'-0"	6'-0"	

Fig. 13-93. Typical spacings for pin-type construction.

On the **pole-top-pin construction** (Fig. 13-93*a*) the dimension *A* should raise the center conductor well above the plane of the other two, so that the horizontal spacing may be considered as the dimension *B*. Irregular movement of the conductor would be expected to be either horizontal owing to wind or vertical owing to ice, and the diagonal separation *C* should be very effective in keeping conductors from getting together. The separation *C* is often somewhat less than *B*. Dimension *D* is often made somewhat less than *B* to keep down the length and cost of the pole and to give a symmetrical appearance. However, this vertical arrangement is not a desirable one for long spans. Sleet jump has been the cause of real service failures; and if spans greater than 300 to 400 ft are probable, an offset *E* of at least 1 ft should be provided in areas subject to ice storms.

The H-frame design gives the best conductor arrangement and mechanical strength for long-span construction and may be used as a special structure for especially long spans in almost any type of line. Pin-type insulators may be used on H-frames, but suspension insulators are generally preferred.

It should be noted that the conductor spacing is primarily a function of the sag and that a conductor arrangement entirely satisfactory for a large or high-strength conductor would be hazardous for a small conductor, with correspondingly greater sags, in the same length of span. Also, a conductor spacing safe for a light loading district should not be used with the heavy sags required for heavy loading conditions. It seems logical to assume that a small or lightweight conductor would require more spacing than a larger or heavier conductor in the same span and with the same sag, as is indicated by the Thomas spacing formula (see Par. **125**).

The Thomas formula may be used as follows to apply the experience obtained with especially long spans that have given a satisfactory performance to the spacing of occasional long spans on new construction.

A 1,860-ft span of 1/0 copper, with 192-ft sag at 60°, has operated successfully at 44 kV for some 25 years on 15-ft horizontal spacing. The constant C of the Thomas formula for this span is 1.9, determined as follows: Sag under 12-lb wind at 30° is 192 ft, or 10.3%. Ratio D/w, that is, ratio of diameter to resultant weight per foot under a 12-lb wind, is 0.368/0.4914 = 0.75. Arcing distance for 44 kV at 1 ft/110 kV is about 4 in. Length of insulator is 0, as the span was dead-ended. From Eq. (13-92),

$$\text{15-ft spacing} = C(10.3 \times 0.75) + 0.33 + 0$$
$$C = 1.9$$

Applying this constant of 1.90 to other conductors and other spans results in Table 13-17 for 46-kV dead-ended spans sagged for heavy-loading conditions.

Table 13-17. Smallest Allowable Spacing, Ft

Copper	Span		
	800 ft	1200 ft	1600 ft
1/0	7.7	12.3	16.3
2/0	6.3	9.9	13.1
4/0	...	6.7	8.8

156. On suspension construction the spacings are usually determined by the clearance required for swing of the suspension insulators as discussed under steel tower design (see Par. **125**). A detailed layout is required for each conductor, as the size and material have a marked effect on the swing characteristics (see Fig. 13-94*c*). A fairly conservative assumption, which results in reasonable design, requires that the clearance from the conductor to a grounded structure shall be at least 0.75 the dry flashover distance of the insulator or the "tight-string" distance (see Fig. 13-23, Par. **44**) under an 8-lb wind on the bare conductor at a temperature of 60°F. This may be modified in details, and it is common practice to allow somewhat reduced clearances to wood members. Typical layouts are shown in Fig. 13-94.

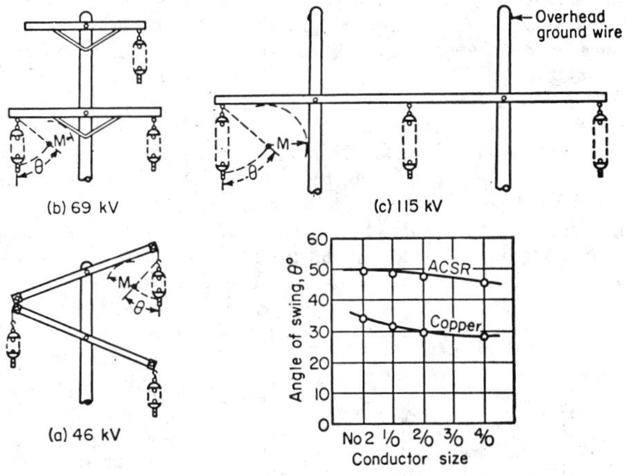

Fig. 13-94. Typical suspension-insulator arrangements on wood construction.

The swing should be taken for a somewhat more unfavorable case than level spans.

The usual range of conditions encountered would be fairly well covered if a vertical span of three-fourths to two-thirds the horizontal span is assumed; i.e., the clearances provide for cases where it is necessary to locate a structure somewhat below the elevation of the adjacent supports. The insulator-swing calculations are discussed in Par. **125**.

157. Angle structures in general use are shown in Figs. 13-95 and 13-96. The design is a matter of providing clearance from the conductor to the structure and to the guys under all conditions and at the same time of attaching guys as close to the load as possible to keep bending stresses down to a conservative value. On small angles where the loads are small, the angle pull may be carried as a bending in the pole and arm; but on larger angles, the loads should be carried directly by the guys, insofar as possible.

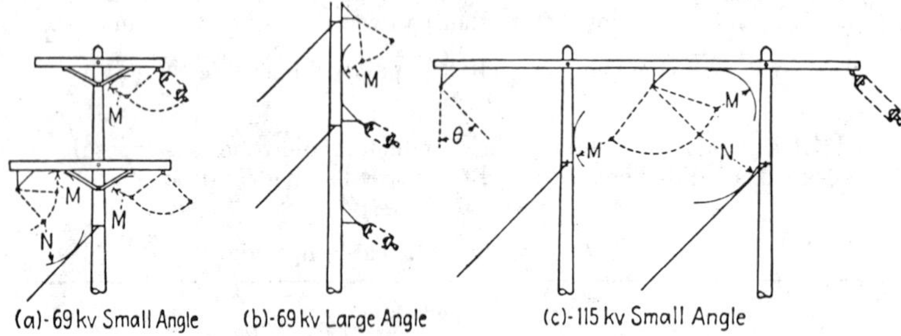

Fig. 13-95. Suspension-angle structures.

Figure 13-95*a* shows the usual small-angle construction, illustrating how it may be necessary to offset the arms to give clearance to the inside conductor. Figure 13-95*c* illustrates a similar design or small angles on heavy H-frame construction where the angle is so small that, if the maximum wind should blow from right to left in the illustration, it would cause the insulator to swing somewhat to the left of vertical. Therefore, clearance M must be provided, not only to the pole on the right, in the illustration, but also to the pole on the left. In the above designs the guy is attached some distance below the crossarm in order to give a clearance N from the conductor to the guy, which is somewhat greater than the flashover distance across the insulator. The clearances N and M (Figs. 13-94 and 13-95) indicate the "normal clearance" and "minimum clearance," respectively. The normal clearance should be at least equal to the procelain insulator.

The bracket on the pole as shown in Fig. 13-94*b* is used for larger angles where the mechanical stresses are too great for crossarm construction but for which the angle pull is not sufficient to swing the insulator string away from the pole under a wind from the right or at locations where a heavy vertical load is encountered on an angle structure. A similar three-pole design is used for H-frame construction. The position of the insulator may be computed for various combinations of loading as shown in Par. **158**.

The simplest angle structure is illustrated in Fig. 13-95. The fewest pieces of hardware and the most direct transfer of stress to the guy are obtained. However, this design can be used only where the angle load is sufficient to hold the insulator string away from the pole under all conditions.

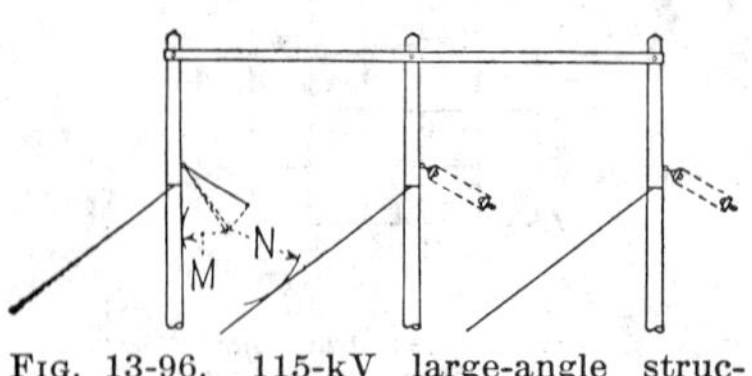

Fig. 13-96. 115-kV large-angle structure.

Angles greater than about 50° are usually dead-ended in a structure similar to Fig. 13-96, as it is not advisable to carry too large an angle on the usual suspension clamp. Erection is difficult on large conductors, and guying becomes complicated for very large angles on suspension construction.

If grounded guys are used, a ground wire should be carried up the pole; and when the guy is attached close to the insulator, contact should be made with the insulator hardware to avoid the possibility of burning the pole from leakage or splintering from

Table 13-18. Clearances for Various Lengths of Insulator String

Insulator, $5\frac{3}{4}''$ units	Normal clearance, in.	Min. clearance 0.75 normal, in.
4	$25\frac{1}{4}$	19
6	$36\frac{3}{4}$	27
8	$48\frac{1}{4}$	36
12	$71\frac{1}{4}$	50
16	$94\frac{1}{4}$	70

lightning. It is common practice to use one or two additional insulators on such angle structures.

158. Clearances on angles should be the same as on tangent construction, i.e., under normal conditions must be somewhat greater than the flashover distance over the insulator but under maximum wind conditions may be reduced to 0.75 of normal, with some slight further reduction if this clearance is to ungrounded wood (see Table 13-18).

The greatest swing, i.e., angle load and wind in same direction, may occur with wind on the bare wire at 0°F, but usually the combined ice and wind load is limiting because of the larger conductor tension. In the case of the wind blowing against the angle, clearances must be computed for a high temperature and resulting low conductor tension. Under normal conditions, for example, 60°F, full clearance should be maintained, equal at least to the dry-flashover distance of the insulator.

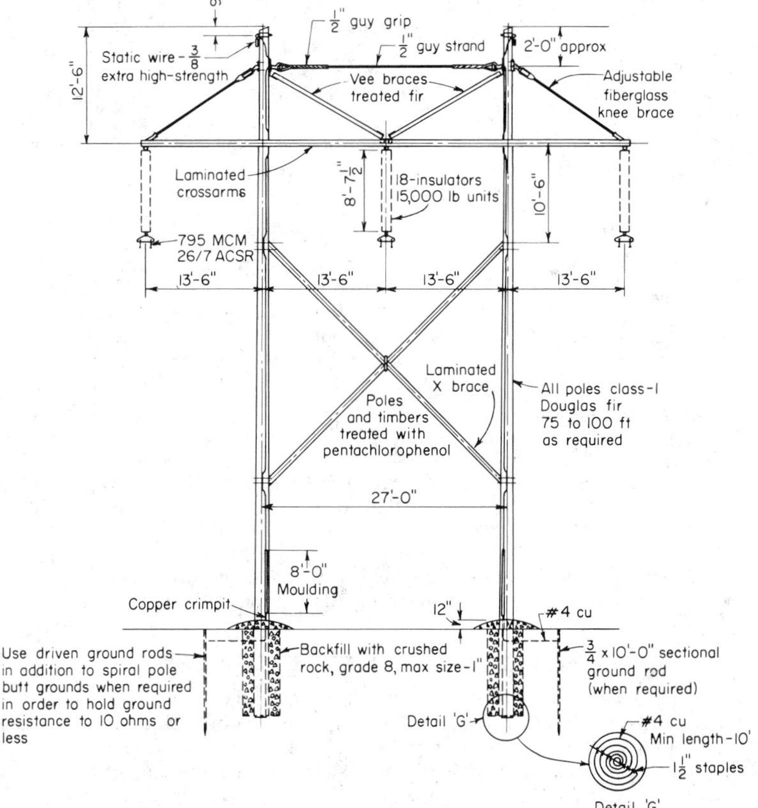

FIG. 13-97. 345-kV wood H-frame structure of Kansas Gas and Electric Company.

The angle load, i.e., the transverse component of the conductor tension t at an angle a, is found as follows:

$$\text{Angle load} = 2t \sin (a/2) \tag{13-95}$$

$$\tan \theta = \frac{(\text{angle load}) \pm (\text{wind load})}{(\text{vertical load}) + (\frac{1}{2} \text{ weight of insulator})} \tag{13-96}$$

in which θ = swing of the insulator from the vertical; the vertical load is the weight of the conductor supported by the insulator, or the weight per foot times the distance between the low points of the sag in the adjacent spans; and the horizontal load is the wind load on the spans supported by the insulator.

159. Pole ground wires should be installed on all poles, at all voltages, in lightning areas:

1. To prevent splitting of poles by lightning.
2. To provide a direct connection to ground and prevent pole burning if an insulator breaks down. This is especially important on the pin-type lines of Fig. 13-93. Since the ground wire on these lines has relatively high resistance to ground, the wire can be as small as number 6 galvanized iron and the ground connection can be several wraps of the wire around the butt of the pole. Low ground resistance is not so necessary on medium-voltage wood-pole lines as it is on steel tower lines and high-voltage H frames.

160. Special wood-pole H-frame structures are used on EHV lines at an appreciable saving over metal towers. Figure 13-97 shows an H-frame structure with trussed crossarm as used on the Kansas Gas and Electric Company 345-kV lines (see Table 13-2). This line uses two 795-M-cmil ACSR conductors/phase on 18-in bundle spacing and 27-ft phase spacing. The insulator suspension hardware is not grounded, and full advantage is taken of the impulse insulation of the crossarm. Lightning flashovers would be expected to take place between the conductors and the ground wires on the poles and not to follow the insulator string and crossarm. All poles and timbers are penta-treated fir.

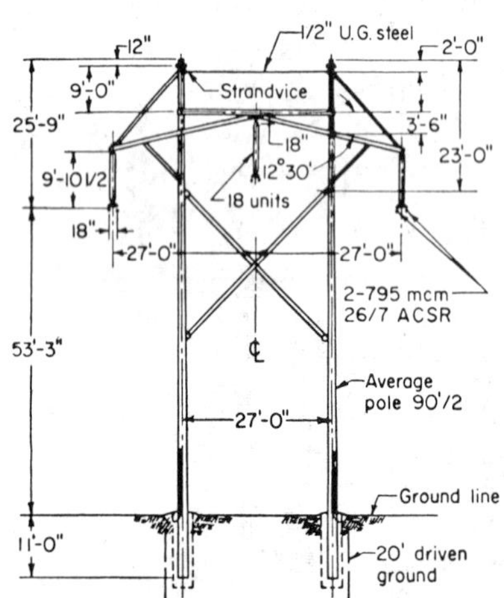

Fig. 13-98. 345-kV wood K-frame structure of Northern States Power Company.

Figure 13-98 shows a modified H-frame wood-pole structure, designated a K-frame, as used by the Northern States Power Company on its 345-kV system.[1] This structure also is designed to carry two 795-M-cmil ACSR conductors/phase on 18-in bundle spacing and 27-ft horizontal phase spacing, but the center conductor is approximately 6 ft higher than the outside phases. This structure also takes full advantage of the impulse insulation of the crossarm.

161. Reference to Literature on Wood and Miscellaneous Structures

"Wood Handbook"; U.S. Department of Agriculture, September, 1935. (Fully rev. 1955.)

LINE ACCESSORIES (LINES UNDER EHV)

162. Suspension-clamp (see Fig. 13-99) designs are fairly well standardized for the usual conductors. Simple, light, well-designed clamps in both malleable iron and forged steel are available for almost any conductor. The seat and clamping surfaces should be smooth, without any projections or sharp bends, and should be formed to support the conductor on long, easy curves at the comparatively sharp bends formed

[1] Weber, L. C., Glass, E. C., and Alexander, G. W. Application of Statistical Methods in the Design and Uprating of Wood-pole Transmission Lines; *Trans. IEEE Power Group*, August, 1965, p. 725.

at horizontal and vertical angles. Heavy, complicated clamps, unless very carefully designed, are generally avoided, to allow as much freedom as possible at the support. For the same reason care is exercised to avoid rigid connections of any kind.

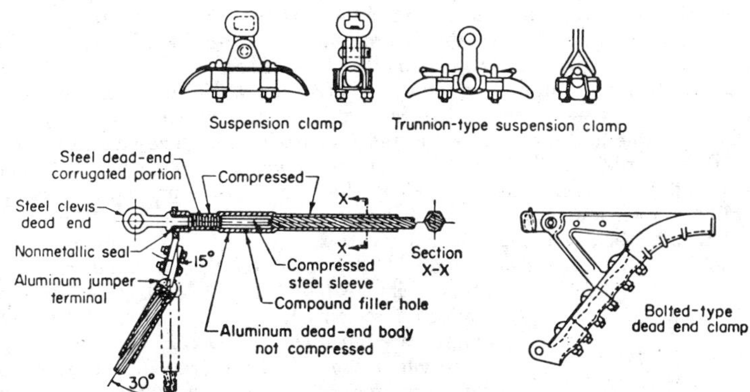

FIG. 13-99. Conductor clamps.

163. Trunnion-type clamps are designed to give an almost completely flexible connection by supporting the clamp on a pivot, approximately on the axis of the conductor (see Fig. 13-99). Thus any vibration of the conductor tends to be transmitted through the clamp, eliminating much of the heavy binding stresses caused by a fixed support.

The suspension clamp is intended primarily to support the weight of the conductor and to prevent any longitudinal movement from accidental unequal tensions in adjacent spans. It is generally considered desirable but not always essential that the suspension clamp hold the conductor in case of a break. For large conductors under heavy tensions it is difficult to design a light, flexible connection that will not slip under such a contingency.

164. Slip, or Releasing, Clamps. Several especially heavy lines have been designed on the proposition that, since suspension clamps could not reasonably be secured that would positively hold the conductor, a clamp would be used that would hold under all ordinary conditions but would slip at something like one-half the maximum conductor tension in case of a break. This arrangement justified a considerable reduction in the exceedingly large longitudinal design loads on the towers and resulted in a considerable saving in tower and foundation costs. Several designs of slip clamp and releasing clamp have been used.

165. Dead-end clamps of the bolted type are available for practically all copper and aluminum conductors. However, for the larger ACSR conductors the compression-type dead-end clamp is generally used (see Fig. 13-99). This is very similar to the compression splice used on ACSR.

The dead end for the steel core, which may have a clevis or an eye-type end, is pressed on after the aluminum sleeve has been slipped out over the conductor. The aluminum sleeve is then slipped back over the steel sleeve until the aluminum body makes contact with the shoulder of the steel sleeve. The electrical connection tongue on the aluminum body is aligned with the clevis or eye of the steel-core dead end as required, after which the aluminum body is filled with the nonoxidizing compound furnished with the body and the body compressed. Similar pressed-on dead ends are available for copper, Copperweld, and other conductors. Both the Aluminum Company of America and the Thomas and Betts Company furnish ACSR dead ends in all sizes required. The Alcoa body has a smooth, hexagonal surface after compression, and the Thomas and Betts body (also hexagonal after compression) has narrow, uncompressed ribs between the compressions. These compressions are the same as on the sleeves (Fig. 13-105).

166. Copper Liners. Both suspension and bolted dead-end clamps are often, although not universally, supplied with a copper liner which acts as a protection to the conductor and prevents possible electrolytic action between the copper conductor and the zinc galvanizing on the clamp. Such protection is quite generally used with Copperweld conductors because of the comparatively thin copper layer on the steel. Electrolytic action, while rare, has occurred and has caused damage to the conductor.

167. Armor rods are quite generally used on ACSR lines as a protection against fatigue of the aluminum strands from vibration. Armor rods consist of a bundle of aluminum rods, somewhat larger in diameter than the strands of the conductor, laid parallel to the length of the conductor and arranged to form a complete covering. These are spirally twisted by a tool to lie approximately parallel with the lay of the strands in the cable and are clamped in place at each end. The suspension clamp is attached at the center, with the armor extending 2 or 3 ft on each side. The bending stresses caused by vibration are reduced by the increased diameter and area of metal and distributed over a longer section of conductor.

168. The Stockbridge damper is a device for damping vibration out of the entire span. Such dampers have been used on ACSR copper and steel conductors and ground wires as illustrated in Fig. 13-100. The cause of conductor vibration and the action of the Stockbridge damper are outlined in Par. **121**.

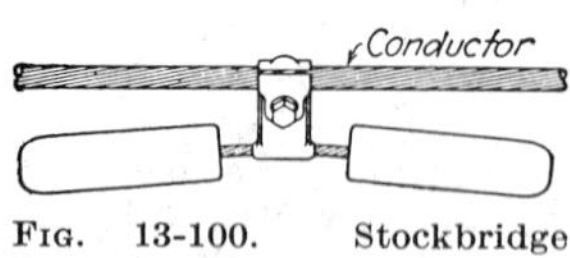

FIG. 13-100. Stockbridge damper.

Overhead Ground-wire Vibration. Overhead ground wires are especially subject to vibration; in fact, most steel ground wires will often be found in rather irregular vibration of small amplitude which generally does not appear to have any ill effects. Ground-wire attachments should be made with at least as great care as given the conductor clamps. Rigid clamps have been almost entirely abandoned in favor of a suspension clamp similar to that used on the conductor and attached by links or shackles so as to give a perfectly flexible connection.

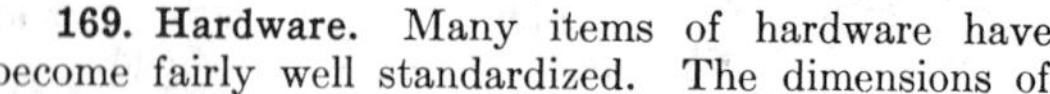

169. Hardware. Many items of hardware have become fairly well standardized. The dimensions of the eye of eye bolts, the length of thread on various-length bolts, end links, and hardware for suspension insulators are quite uniform. It is usually possible to obtain about identical stock material from a number of manufacturers. Many other items such as shackles, guy clamps, and crossarm braces are furnished in such a wide variety that considerable care is required to choose the most commonly used but suitable stock items. Much expense and confusion in both construction and maintenance are saved by limiting the number of hardware items.

170. Wood Braces and Guys. With the use of wood to increase the impulse insulation as a protection against lightning, steel crossarm braces have been replaced with wood on a number of lines. Connections are made by pressed-steel fittings. The use of a 48-in wood brace in place of steel, for additional wood insulation, is roughly the equivalent in lightning-flashover strength of adding one suspension unit to the insulation. The effect on 60-c flashover is, however, negligible.

To obtain equal wood insulation at guyed structures to that which may be obtained on unguyed construction requires long wood insulators in the guys. These guy insulators are quite efficient because of the high tensile strength of clear wood; an ultimate strength of 6,000 lb/in^2 on the net section is conservative. A 2- by 2-in fir insulator will develop the full strength of a $\frac{3}{8}$-in Siemens-Martin guy strand. The design of the connection to the pressed-steel fitting requires only that several bolts of insufficient diameter be used to give the necessary bearing area between the wood and the shank of the bolt. The bolts should be placed alternately through the face and side of the stick to prevent splitting (see footnote 6, Par. **56**).

171. Guys. The various grades of guy strand are almost universally furnished in accordance with ASTM specifications. The ultimate strength for each size and grade is given in Sec. **4**. The so-called "double-galvanized" is generally used. Common guy strand is not ordinarily employed in transmission construction, as the best-quality galvanizing is not furnished in this grade. Siemens-Martin strand is most commonly used for the lighter lines and high-strength for heavy construction.

More than one size of guy strand is not economical for a line, and often the same size may be used for several designs. The ⅜-in size, in either Siemens-Martin or high-strength grade, is most generally used both for guys and for overhead ground wires.

In the usual wood-pole construction great refinement is not required in designing guys. Usually it is sufficient to determine the number of guys, of the size and quality to be used on the line, required to support the load, an additional guy being employed for any fractional part. In transmission construction a factor of safety of 2 is general for guys, although this may be somewhat reduced.

A common problem in guy design is illustrated in Fig. 13-101. The ratio of the guy load L to the conductor load T is the same as the ratio of the length of the guy B to the distance A. The length B is readily scaled from a sketch drawn to scale.

$$L = T\frac{B}{A} \quad \text{or} \quad L = \frac{T}{\sin\theta} \qquad (13\text{-}97)$$

If the conductor load T_c is above the point of attachment of the guy,

$$T = T_c h_c / h \qquad (13\text{-}98)$$

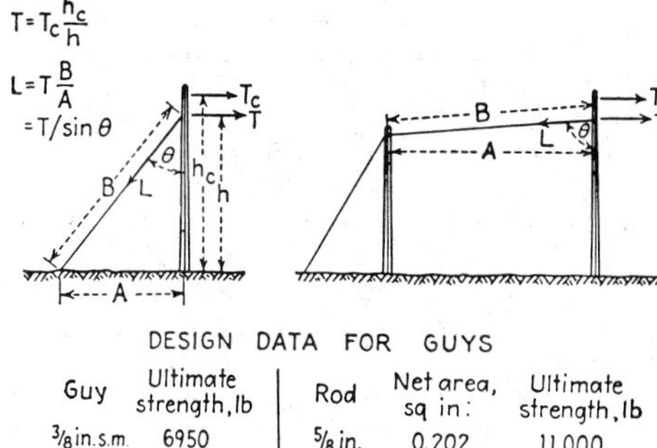

DESIGN DATA FOR GUYS

Guy	Ultimate strength, lb	Rod	Net area, sq in.	Ultimate strength, lb
⅜ in. s.m.	6950	⅝ in.	0.202	11,000
⅜ in. h.s.	10,800	¾ in.	0.302	16,500
7/16 in. s.m.	9350	1 in.	0.551	30,500
7/16 in. h.s.	14,500			

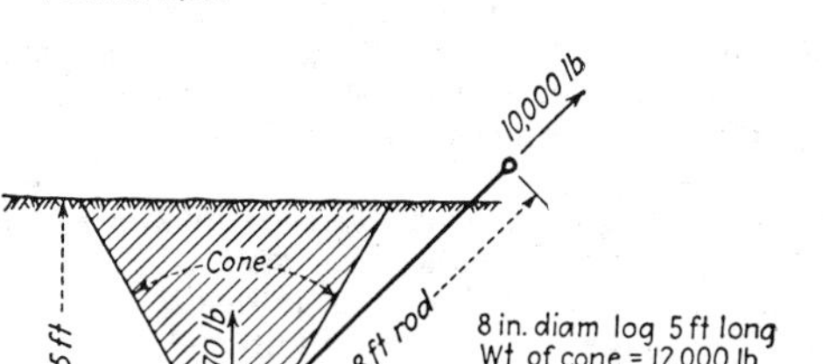

Fig. 13-101. General guy and log anchor-strength calculations.

172. Guy Anchors. The same problems apply to the design of guy anchors as to tower footings (see Par. **132**). (1) The area of the anchor must be sufficient to prevent "cutting" through the earth; i.e., the earth immediately above and in contact with the anchor must not fail in bearing. (2) The anchor must be buried deep enough to prevent the vertical component of the pull from lifting the cone of earth bodily. It is of course impossible to give data applying to all the innumerable conditions that will be encountered. However, for average conditions a log 5 ft long and 8 in in diameter will usually safely hold about 10,000 lb or approximately the safe strength of two ⅜-in high-strength guys (see Fig. 13-101). Such an anchor must be set to bear directly on undisturbed earth by placing the log in a slight undercut at the bottom of the hole, and the anchor rod must be laid in a trench so that the rod is directly in the line of the pull. Many metal anchors, usually of patented design, develop excellent holding power by bearing against undisturbed earth.

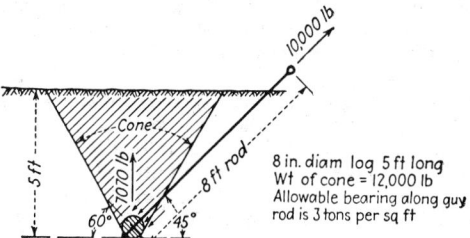

CONDUCTOR AND OVERHEAD GROUND-WIRE INSTALLATION

173. Wire stringing should be handled by an experienced crew, not only to prevent damage to the conductor and overhead ground wire but also to maintain the sags and tensions specified in the design. Correct sags are essential to give the required mechanical safety, but it is equally important that the actual sag in the line correspond to that used in the design, to ensure proper clearances to the ground.

174. Wire-stringing Equipment. All transmission conductors and overhead ground wires should be strung over free-running snatch blocks or rollers made for this purpose. Both conductors and ground wires of any material are easily damaged, and with the long spans and heavy conductors used in modern construction, satisfactory sags cannot be obtained at reasonable cost except by eliminating all possible friction at the supports. **Dynamometers** for measuring the tension are of value as a means of knowing the tension at all times but cannot be relied upon to set the sag. The final sag should be adjusted by sighting. Walkie-talkie radios are widely used as standard erection equipment; better and more efficient work is obtained by having direct communication between the reel crew, the pulling crew, and the men sagging.

Sags are measured by setting sights on the structures at each end of the span at a vertical distance below the conductor support equal to the sag. This method is convenient and well within the necessary accuracy, even for inclined spans. For average inclined spans the sag is taken as the sag for a level span of the same horizontal length, although the sag for a level span equal to the slope distance is more nearly correct. Except in extreme cases the horizontal and slope distances are practically the same. On long, inclined spans, when the low point of the sag falls below the ground level of the lower tower, it is more convenient to measure the vertical sag below the lower support, as given by Eq. (13-88) (Par. **122**).

175. Accuracy of Sagging. Friction in sagging blocks prevents the wire from reaching exactly the same tension at all points. As the wire is pulled up, the sag tends to be greater in spans farthest from the pulling point; and when slacked back somewhat, more sag is thrown into the nearer spans. These effects are usually fairly well eliminated by allowing some time for the tension to equalize and by skillful handling.

Curves of "span vs. sag" and "span vs. tension" for possible stringing temperatures are used in sagging. The actual conductor temperature in sagging is very important. The IEEE Committee Report on Stringing and Sagging Conductors, *Trans. IEEE Power Group*, December, 1964, p. 1235, recommends that the temperatures be obtained by direct measurement at the time of sagging by means of a thermometer placed inside a short length of the conductor and suspended at least 15 ft above the ground. Accurate temperature also is important in making allowance for creep during stringing. Creep elongation starts as soon as the conductor is pulled into the air, and it is important that this elongation be allowed to take place and not to pull the conductor back up to the calculated sag. One way to do this is to use a temperature curve which will indicate the calculated sag plus the additional sag due to creep up to the time of clipping in. The creep sag must be estimated from the manufacturer's curves.

In spans of varying length a greater sag tends to form in the long spans; and on steep grades the sags at the higher elevations tend to be less than at the bottom of the hill. These effects are not of importance except in extreme cases and are due to the fact that the wire is, and must be, supported on rollers in such a way as to be entirely free to travel. Thus the tension in the wire on each side of the roller must be equal irrespective of the slope of the wire away from this support, and the resultant on the support is not vertical but in the direction of the bisector of the angle between the slopes of the wires as they leave the roller on each side. At a support between a short and long span (Fig. 13-102*a*) the wire on the short-span side is more nearly horizontal, OA, while on the long-span side the wire may have a considerable slope OB. The tension on each side must be equal, $OA = OB$, but the horizontal component of the tension is therefore less in the long span BD than in the short span AE. The resultant OC is inclined.

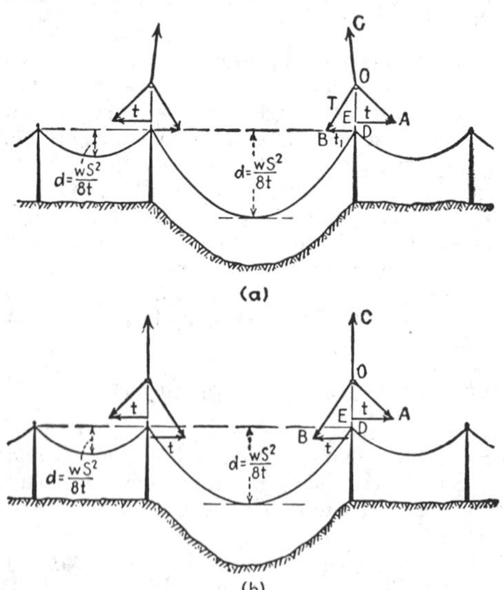

FIG. 13-102. Illustrating the change in tension in long spans. (*a*) On rollers. (*b*) Clamped in theoretically correct position.

It is theoretically possible, although very difficult practically, to clamp the conductor at the correct position so that the resultant will be vertical and the horizontal tensions equal as in Fig. 13-102*b*. This is the condition assumed in the computations and office location and, for all except extreme cases, is the reasonable assumption.

Similarly, the different slopes of the wire leaving the roller on hillsides with spans of equal length but at different elevations cause the horizontal component of the tension to be less in the wire with the greatest slope (see Fig. 13-103). With a series of spans on a slope this effect tends to accumulate, for the horizontal tension t_2 at the upper support must be the same as the horizontal tension t_2 at the lower support, while the resultant tension T_2 is less than the tension T_1 because of its smaller slope. The relation $T_1 = t_2 + wD$ and $T_2 = t_2 + w\delta$ (see Fig. 13-103) is discussed in Par. **122.**

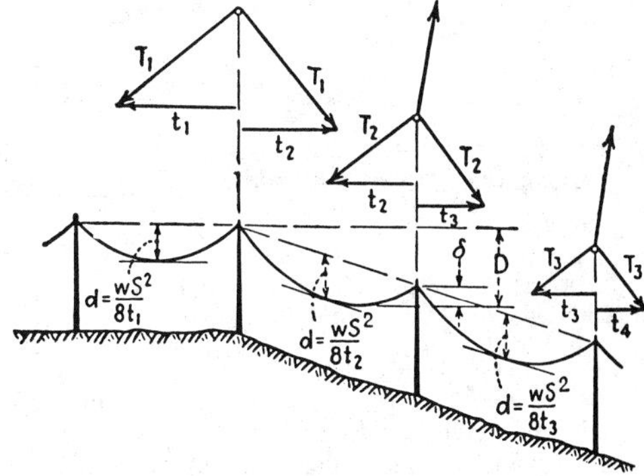

FIG. 13-103. Illustrating the change in tension on hillsides. (This diagram is much exaggerated.)

The differences in sag are not usually carried from one conductor pull to the next, but each pull is sagged to approximately the correct tension independent of the other; thus when the snubs, between pulling sections, are removed, differences in tension tend to equalize. For this reason it is best not to clamp in the conductors too close behind the sagging crew. Often skillful sagging reduces these effects by using the friction in the blocks to prevent the conductor from "collecting in the low spots."

These irregularities are of little consequence, generally, especially when it is realized that the important consideration is to have equal tensions under maximum load rather than under bare-wire conditions.

In extreme cases, provision for the above conditions may be made by special sags, allowing somewhat higher tensions in spans above the normal level of the line and providing extra clearance in low sections so that slightly larger than normal sags may be used. Occasionally special methods must be devised.

176. The McIntyre joint (Fig. 13-104) is used chiefly on small sizes of cable. It consists essentially of seamless copper or aluminum tubing, oval in section, into which each conductor is pushed from opposite ends, until the conductors project about 2 in beyond the ends of the sleeve. The tube is then twisted the required number of turns by special tools. The joint shown in Fig. 13-104 is used for 1/0 to 4/0 ACSR and steel ground wires. A similar joint requiring only one sleeve with 3½ turns is used on copper up to 500,000 cm.

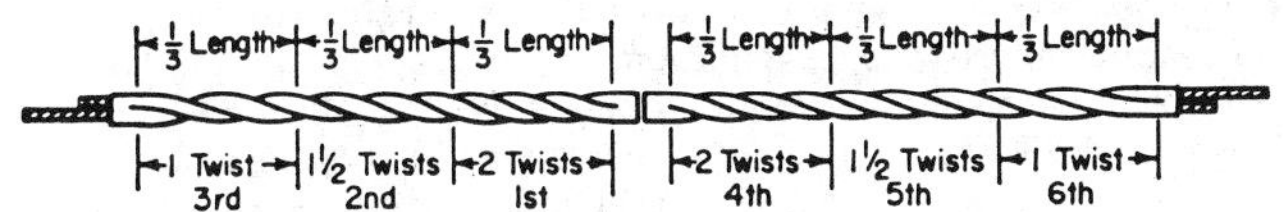

FIG. 13-104. McIntyre or twisting joint.

For steel-reinforced aluminum (ACSR), the simple twisted-sleeve joint can be used in sizes up to 4/0 and develops about 80% of the strength of the cable.

177. Compression joints are used with the large ACSR conductors and the large "all-aluminum" conductors including the high-strength types. As with the compression dead ends, the most widely used joints are those made by Thomas and Betts and the Aluminum Company of America (Alcoa). Cutaway drawings of these joints are shown in Fig. 13-105. They consist of aluminum sleeves and steel sleeves for ACSR. Installation procedures call for the aluminum cable and the insides of the aluminum sleeves to be thoroughly cleaned.

If the conductor is weathered, the strands should be unlayed and all scale removed. The aluminum sleeve is then slipped on the cable and backed out of the way.

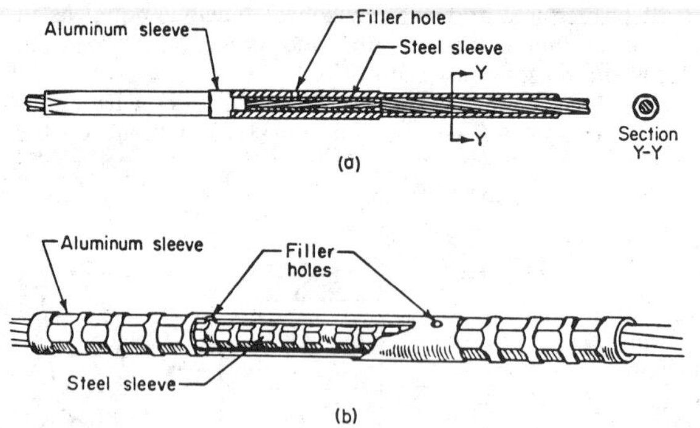

FIG. 13-105. Compression joints. (*a*) Aluminum Company of America; (*b*) Thomas and Betts.

The aluminum on each cable end is next carefully cut back, care being taken not to nick the steel core, for a distance equal to one-half the length of the steel sleeve plus a distance of ½ in or more, depending upon the size of the conductor, so that the elongation of the steel sleeve on compression will not interfere with the free lay of the aluminum strands. The conductor ends are then marked by tape or other suitable means to center the sleeve. The steel sleeve is put in place and compressed, working from the center out. The aluminum sleeve is next slipped into place and filled with the nonoxidizing compound furnished with it, the filler holes are plugged, and the joint is ready for compression. The sleeves are compressed by working from the center out. The center section of the aluminum sleeve over the steel sleeve is not compressed. When the compression is completed, the Alcoa sleeve is hexagonal and smooth from overlapping compressions, while the Thomas and Betts sleeve, also hexagonal, has uncompressed ribs between the compressions as shown in Fig. 13-105*b*.

178. The Anaconda seamless connector (Fig. 13-106) consists of a short piece of copper tube with its center reduced for a short distance to a diameter slightly less than that of the die to be used in making the joint. The cross section is greater, however, than that of the cable being connected. The wire or cable is inserted to the full depth of the opening in each end of the connector (see Fig. 13-106*a*). The connector is then forced down over the cable by the use of a split die (Fig. 13-106*b*), the die being applied to the reduced portion and drawn over each end by means of a portable drawbench. This forces the metal down on the cable and causes the strands to be firmly embedded in the sleeve. This joint develops a strength greater than that of the cable itself.

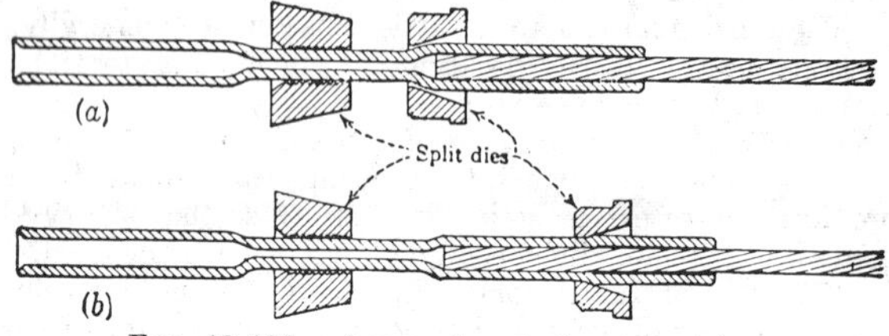

FIG. 13-106. Anaconda seamless connector.

179. Overhead Ground-wire Installation. Overhead ground wires should receive no less care in erecting than the conductors, for the usual zinc or copper protective coating is very easily destroyed. Ground wires should be sagged in the same way as the conductors except that the important factor in ground-wire sags is to maintain ample clearance to the conductors. Generally, ground wires are sagged to about 80% of the conductor sags, thus ensuring proper clearance even under ice loads. McIntyre joints are generally used for splicing, but a much more efficient joint can be made with the pressed-steel joint similar to the ACSR joint illustrated in Fig. 13-105.

TRANSPOSITIONS[1]

180. Transpositions are made for the purpose of reducing the electrostatic and electromagnetic unbalance among the phases which can result in unequal phase voltages for long lines, such as one connecting a remote generating station to a load center. Untransposed lines also can cause inductive interference with paralleling wire com-

[1] FOWLE, FRANK F. The Transposition of Electrical Conductors; *Trans. AIEE*, 1904, Vol. 23, p. 659. VON VOIGTLANDER, F. Transposition Practices; *Elec. Eng.*, January, 1943.

munication lines. However, communication interference in the past has been largely with overhead long-distance telephone and telegraph lines. Many of these lines are now going underground, and other overhead lines are being replaced by microwave radio.

For some time, transpositions have been little used. With the large power-system networks comprising most of the country's transmission lines the unbalance of an untransposed line is largely smoothed out by the phase-balancing effect of the rotating equipment scattered over the system. However, the advent of long extra-high-voltage lines opens the subject to further study, and it is possible that transpositions may be found necessary in some cases.[1]

OPERATION AND MAINTENANCE

181. Operation. Effective operation of a system is as essential to good service as is excellent engineering design. In fact, a well-designed system may fall short of its service requirements owing to faulty operation. Aside from switching lines and power units to meet the load conditions of the system, operation consists not only in restoring service promptly after an interruption but also in detecting and removing faulty apparatus, thus actually preventing the development of faults.

182. A chief system operator should be in absolute control of the system, and if it is a small system, he should have direct communication with and direct control of every part of it. If it is a large system, it will not be possible for one man to supervise all switching operations, and area dispatchers must be located at convenient points. These dispatchers will have the same authority over their areas as the chief system operators for small systems. The area dispatchers will call upon the chief system operator only for approval of unusual switching operations, particularly those involving interruptions to important loads. They will, however, make reports each shift as convenient on routine switching operations and will report major interruptions as soon as possible, with cause if known. Dummy boards are useful at dispatching centers. These boards should show the one-line diagrams of the circuits at all stations under the dispatcher's supervision, and provision should be made to show whether switches are open or closed. These boards must be kept correct up to the minute, even if it is necessary to do so by temporary means. It is best to anticipate system changes so that the dummy board can show the changes as soon as they are made. During normal operating conditions no switching should be done, including that of generators, without the dispatcher's permission. All dispatching orders should be reported back in order to prevent misunderstandings and should be recorded in log books both by the dispatcher and by the operator who will do the switching. The logs should show a record of all transactions, with particular care about times of receiving orders, of opening and closing switches, and of cases of trouble.

Emergency routines should be set up for all stations and should be followed at times of catastrophic storms when all means of communication with the dispatcher are interrupted. These routines will list the sequence for doing emergency switching on the operator's responsibility in an effort to restore service.

183. Supervisory control systems make it possible for operators at one transmission substation to operate several nearby substations as well as their own and thereby reduce operating personnel. Supervisory control also makes it possible for one central dispatching office to operate all the transmission substations serving a large metropolitan area. The supervisory may utilize carrier-current, microwave radio, or telephone channels for transmitting information and operating switches.

184. Sleet or glaze formation on lines is highly undesirable, and many companies prevent it by raising the temperature of the wires with current. Ice will form on conductors over a small temperature range which is on the order of -3 to $+2°C$. The current required to prevent ice formation may be found according to Clem[2] from

$$I^2 = \frac{\theta\sqrt{dv}}{8.18 \times R} \times 10^4 \qquad (13\text{-}99)$$

[1] Holley, Henry, Coleman, Dorothy, and Shipley, R. Bruce Untransposed EHV Line Computations; *Trans. IEEE Power Group*, March, 1964, p. 291.

[2] Clem, J. E. Currents Required to Remove Conductor "Sleet"; *Elec. World*, Dec. 6, 1930, p. 1053, and Jan. 31, 1931, p. 245.

in which I = current, θ = temperature rise in degrees centigrade above surrounding air, R = conductor resistance in ohms per mile at 20°C, d = diameter of conductor, v = wind velocity in miles per hour.

With the lines in service, it is usually difficult to obtain sufficient current. However, the necessary current sometimes may be obtained by transferring load from other lines to the line in trouble. Dead lines may be heated by short-circuiting them at one end and sending the necessary current from the other end. The approximate voltage to neutral is $E = I\sqrt{R^2 + X^2}$, where R = resistance per wire and X = reactance per wire.

Melting the ice after it has formed is considerably more difficult and requires more current than is required to prevent formation. Clem's article gives the various formulas required to calculate the melting current.

185. Emergency crews are stationed at locations always available by telephone or radio so that every important section of the transmission system can be reached by a crew within a reasonably short time. A light truck, provided with two-way radio and with the necessary tools and materials for making immediate repairs, is used by many companies. In addition to spotlights on the truck, a spotlight operated from a portable storage battery is frequently very useful. Small houses containing spare parts, such as insulators, lengths of cables, and clamps, should be located at intermediate and accessible points along the line in sparsely settled country. Such houses should be kept locked. Some companies employ concrete construction with iron doors. A routine inspection and checking of materials in such houses is advisable.

186. Line repairs and replacements are often made only after the line is "dead." The line crew should notify the dispatching office when a particular line or section is desired. The line should then be not merely cleared through the oil circuit breakers but opened by disconnecting switches as well. If it is to be out of service for an extended period and there is danger of lightning, the line should be grounded out at its terminals to prevent the possibility of double-voltage reflections flashing over switches or insulators at the terminals. If the line is not equipped with grounding switches, it may be grounded out by equipment such as is used by line crews. Before the line crew is allowed to work, the line should be short-circuited and well grounded on each side of the location where the men will work, with the grounding equipment in full sight of the workmen. The grounding equipment should consist of heavy extra-flexible copper cables, which should be attached by means of "hot-line" tools and clamps, the line being considered to be "hot" until the grounding equipment is applied. Ground chains are not safe and should not be used. Reliance should not be placed in grounding switches or grounding cables at the ends of the line.

In order to make repairs and replacements without interrupting the service, special "live-line" tools have been devised whereby insulators may be replaced, conductors spliced, etc., on lines of all voltages while the line is "hot."

187. Periodic inspection should be maintained over all lines, with the frequency of inspection depending on the country traversed and the importance of the lines. In some densely settled areas, patrols of once a week may be considered necessary, whereas important lines in areas not subject to heavy storms or other hazards may not require inspections oftener than once in two months. The patrolman may cover the line on foot, on horseback, by automobile, or by helicopter, depending upon the characteristics of the right-of-way. Close and accurate patrolling is not obtained by one man in an automobile, in general, even when the line follows a highway. Helicopter patrol is by far the best over mountainous and sparsely settled country, and 200 mi a day can be covered readily. The helicopter can fly as slowly and as close to the line as is necessary, and it has the great advantage that the patrolman is looking down on the line instead of up against the bright sky as a background. Tower and wood-pole structure numbers should be fastened to the tops of the towers or structures in such a manner that they can be read without trouble by the helicopter patrolman. Helicopter patrol cannot be used over urban areas or congested industrial areas because of governmental restrictions on height of flight. Patrols on foot are best in such areas. Horseback patrol is best in cattle-range country, if aerial patrol is not available.

The patrolman should report landslides, washouts, danger timber, or anything else that is a potential danger to the line, such as piles of brush or straw which if burned could cause hot gases to short-circuit the line. Of course he must also be on a close lookout for damaged conductors, insulators, and structures. A pair of field glasses is usually considered indispensable to a patrolman.

Patrolmen on ground patrols should keep the dispatcher informed as to their whereabouts and should call in from all patrol telephone stations and from substations as they reach them. They should call in not less often than morning, noon, and night. If a storm comes up while a patrolman is out on a line, he should call the dispatcher as soon as possible, telling where he can be reached. Patrol cars and helicopters are radio-equipped.

Tree gangs whose sole duty is to remove brush, trim trees, and remove danger timber have been found to be advantageous by large companies. The use of chemical sprays to kill brush along rights-of-way is satisfactory from the standpoint of killing the brush but leaves a potential fire hazard. Care must be taken in spraying that wind does not blow the chemicals over growing crops. This danger has been found to be a disadvantage in helicopter spraying.

188. Faulty insulators and faulty insulator sections may be detected while they are in service. In general, the methods employed for faulty-insulator detection are based upon the measurement of the voltage gradient (Fig. 13-31) across the individual units of a string of suspension insulators or across the parts of multipart pin-type insulators. For safety reasons, none of the test methods should be used in wet weather.

189. In the "buzz-stick" method[1] the faulty unit is detected by the appearance and sound of the sparks obtained, first by a feeling-out process followed by a shorting-out process on the individual parts of the insulator or insulator string. The feeling-out process consists in touching the caps of cap and pin-type insulators or the cement of multipart pin-type insulators with a single prong on the end of an insulating stick. If the static spark obtained during the feeling-out process shows no gradient across several of the units, the short-circuiting process should not be tried. The shorting-out process consists of short-circuiting out the individual insulator units or parts by means of a metallic U-shaped fork on the end of an insulating stick. No spark or a weak spark on short circuiting indicates a defective unit.

190. The Doble method[2] is, in effect, a spark-gap voltmeter which is safe to use and which gives high accuracy in measuring potentials in the field on live transmission-line insulators.

There are two general types of Doble safety tester: the type A single-prong tester for multipart pin-type insulators on either wood or steel construction, and the type B two-prong tester for multiunit suspension-type insulators on either wood or steel construction. The type B tester may also be used on small, one-piece, pin-type insulators on wood construction.

In both types of testers, the equipment consists of a micrometer spark gap in series with a capacitor and a special telephone-type headset with which to listen to gap sparkover. The telephone receiver is heard through a rubber hose connected to a highly insulated hollow tube which is long enough so that there is no danger to the operator. This also serves as the handle for operating the set. The spark gap of the type A tester (Fig. 13-107, left) is set so as not to spark over on the potential across a good shell of a particular pin-type insulator. Therefore a spark heard indicates a bad shell. Moisture on the insulators, particularly early morning dew, throws off the calibration and causes a bad shell indication. Consequently testing should not be done unless the insulators are perfectly dry. Each shell of a pin-type insulator has a different potential across it and requires a different gap setting so that it is not practicable in testing in the field to test other than the top shell, which usually is the only one to fail.

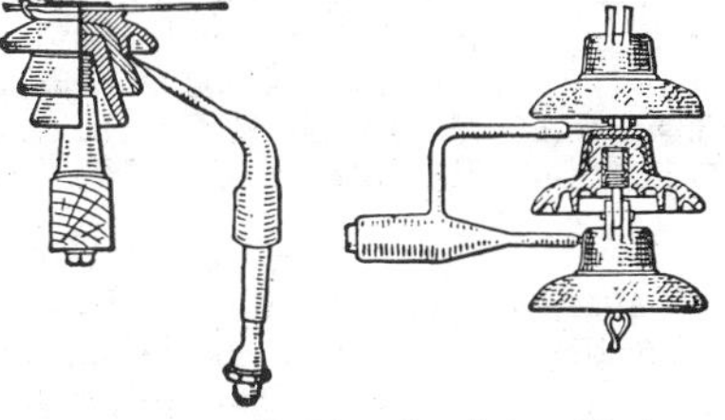

Fig. 13-107. Doble insulator testers using principle of spark-gap voltmeter. Left, for pin-type insulators. Right, for suspension insulators.

In the type B tester (Fig. 13-107, right) both sides of the electrical circuit of the tester are terminated by exposed metal tips arranged so as to bridge readily a single

[1] Johnson, T. F. *Elec. World*, 1919, Vol. 74, p. 568.

[2] Doble, F. C. Progress in Field Testing of Insulators; *Elec. World*, 1923, Vol. 81, p. 1397.

insulator disk or section. In the circuit between the two contact tips, a protective insulating condenser is built into the tester in such a manner as to make the impedance between its terminals greater than the impedance of a single good disk. Thus, in operation, the tester *does not short-circuit* the disk under test. The tester may be considered as a voltmeter which indicates the voltage between the points on the insulator touched by the tips of the tester.

The degree of defect in the disk under test is indicated by the size of the maximum gap at which a noise is heard in the tester, as compared with the size of a maximum gap for a good disk in the same position in a string; a totally dead disk gives no sound, irrespective of the gap setting. In practice one gap length is fixed in advance and used for all units of the string. The operator judges if a unit is defective from the noise he hears in the headset.

This apparatus is now in general use for testing insulators on lines at all voltages from 11 kV up.

The Doble field power-factor testing equipment measures the power factor of the insulators and bushings at a potential of 10,000 V or greater, when they are out of service. It has been found that high power factors and anomalous values of the power factor of oil-circuit-breaker bushings frequently show that operating hazards exist, such as deterioration of the main insulation of the bushing, moisture or carbon particles in the oil, formation of ice, carbon deposits on the bushing. Power-factor measurements with this equipment have been adopted by many companies as a criterion for servicing.

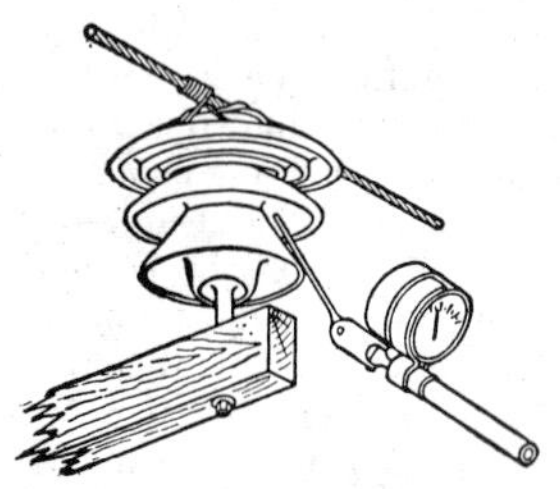

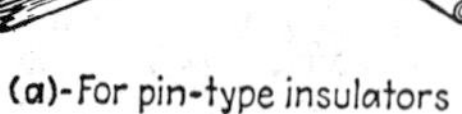

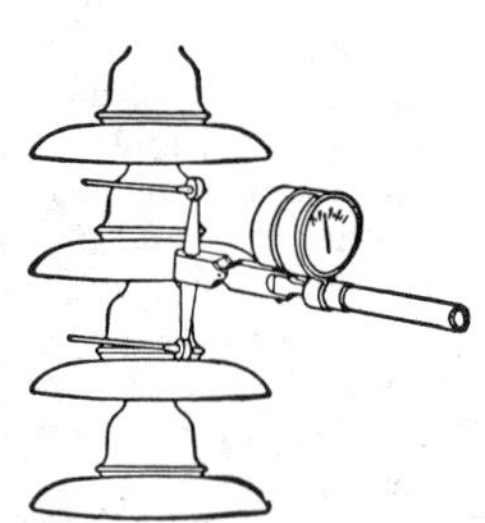

(a)-For pin-type insulators (b)-For suspension insulators

FIG. 13-108. The R. and I. E. live-line insulator tester.

191. The R. and I. E. live-line insulator tester detects defective insulators, either multipart pin-type or suspension, by comparing the measured voltage distribution over insulators or insulator strings while in service with characteristic curves plotted for good insulators of the same type. It is suitable for use on all transmission voltages. The tester employs a single-prong head (see Fig. 13-108*a*) for multipart pin-type insulators, a small two-prong head (see Fig. 13-108*b*) for suspension strings or small one-piece insulators, and a large two-prong head for multipart pedestal insulators. Visual indication is given by means of a meter which shows a deflection in proportion to the voltage gradient. Since relative indications only are required, the meter is calibrated simply in units of deflection. Tests may be made of all shells of multipart pin-type insulators and all units of suspension strings with equal facility. As with the Doble tester, tests should be made only on perfectly dry insulators.

ADDITIONAL BIBLIOGRAPHY

192. General. Aside from the footnotes and other references throughout this section, the references listed below should be helpful.

DWIGHT, H. B. "Transmission Line Formulas"; Princeton, N.J., D. Van Nostrand Company, Inc., 1913.

DWIGHT, H. B. "Electrical Elements of Power Transmission Lines"; New York, The Macmillan Company, 1954.

WOODRUFF, L. F. "Principles of Electric Power Transmission and Distribution," 2d ed.; New York, John Wiley & Sons, Inc., 1938.

LOEW, E. A. "Electrical Power Transmission"; New York, McGraw-Hill Book Company, 1928.

CLARKE, EDITH "Circuit Analysis of A. C. Power Systems"; New York, John Wiley & Sons, Inc., Vol. I, 1943; Vol. II, 1950.

DEWEESE, FRED C. "Transmission Lines"; New York, McGraw-Hill Book Company, 1946.

SECTION 14

DIRECT-CURRENT POWER TRANSMISSION

BY

P. G. ENGSTRÖM Director, Semiconductor and Electronics Division, Allmänna Svenska Elektriska Aktiebolaget, Sweden

ASSISTED BY

H. MÅRTENSSON Chief Engineer, High Voltage Direct Current Department, ASEA

E. UHLMANN Consulting Engineer, H.V.D.C. Department, ASEA

S. ANNESTRAND Head, High Voltage Group, U.S. Department of the Interior, Bonneville Power Administration

S. BERNERYD Manager of H.V.D.C. Valve Design, H.V.D.C. Department, ASEA

B. FUNKE Manager of H.V.D.C. Valve Development, H.V.D.C. Department, ASEA

I. LIDÉN Electrical Engineer, H.V.D.C. Department, ASEA

H. STACKEGÅRD Manager of Project Planning Department, ASEA

REVIEWED BY

E. C. STARR Consulting Engineer, U.S. Department of the Interior, Bonneville Power Administration

CONTENTS

Numbers refer to paragraphs

SECTION 14

DIRECT-CURRENT POWER TRANSMISSION

1. General. From the beginning of electric power history it has been recognized that d-c lines and cables are less expensive than those for 3-phase a-c transmission. It was soon made obvious, however, that alternating current is more advantageous than direct current for generation, low-voltage distribution, and consumption. To utilize the savings of a d-c line, the generated a-c power has to be converted to d-c power and transmitted over the d-c line to another converter station where the power is converted back to alternating current. The lack of reliable high-voltage power-conversion equipment had, however, made application of d-c systems impractical up to the middle of the 1950s, when the development of the high-voltage mercury-arc valve resulted in a commercially competitive position for d-c transmission.

2. History. Over the years, many attempts have been made to develop converters for H-V d-c transmission.[1] Converters based on mechanical switches were tested in England and Sweden in the 1920s and 1930s. The best-known system was developed by the Swiss engineer Thury (in 1889). It consisted of d-c generators and motors connected in series on the d-c side. It was used in Europe from 1890 to 1937. In the United States, the General Electric Company built converters for d-c lines during the 1930s. These converters used mercury-arc valves with relatively low ratings and were in operation from 1937 to 1945. The first commercial d-c installation, which remains in operation, was the Gotland transmission in Sweden, which was commissioned in 1954. This scheme has been followed by many others in different countries, as shown in Table 14-10.

OVERHEAD D-C LINES

3. General. In present power transmission technique there is no question about the superiority of the a-c 3-phase system for power generation, distribution, and short- to medium-distance transmission. However, on the assumption that the direct voltage is equal to the crest value of the alternating voltage, the conductor efficiency in d-c transmission is double that of a single-phase a-c system and 50% higher than that of a 3-phase system.

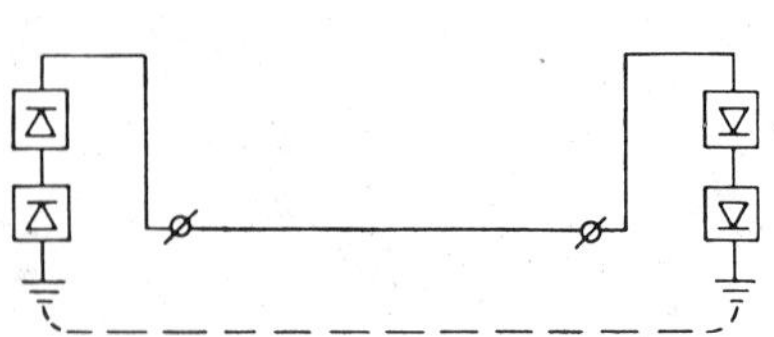

FIG. 14-1. Direct-current monopolar transmission arrangement.

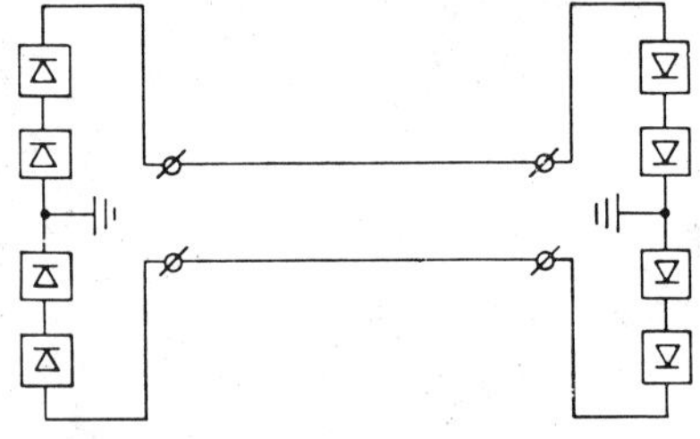

FIG. 14-2. Direct-current bipolar transmission arrangement.

In d-c transmission ground return can be used as one conductor. This means that each separately insulated transmission conductor, together with the ground-return path, forms a separate electric circuit. Based upon this fundamental principle, the following basic circuit arrangements can be considered.

[1] The History of H.V.D.C. Power Transmission; *Direct Current,* 1963.

4. Monopolar Arrangement. Here, only one transmission pole is installed, and ground return is permanently used, as shown in Fig. 14-1. Monopolar transmission is used in systems of comparatively low power rating, mainly with cable transmission.

5. Bipolar Arrangement. It is mechanically most suitable to design an overhead-line tower for two insulated conductors, one suspended on each side of the center post. These can be arranged as plus and minus poles in bipolar transmission, as shown in Fig. 14-2. In bipolar transmission ground return is not necessarily used but is normally provided for to increase the transmission availability in case of pole failure.

6. Homopolar Arrangement. A two-pole tower design can also be used in a homopolar transmission where both poles have the same polarity. This is, of course, applicable for overhead lines feeding a monopolar cable transmission, but owing to reduced corona the use of the homopolar arrangement, as shown in Fig. 14-3, can also be considered for very large overhead transmission systems where two bipolar circuits are used and the insulator chains of each tower are carrying two separate conductors with the same polarity.

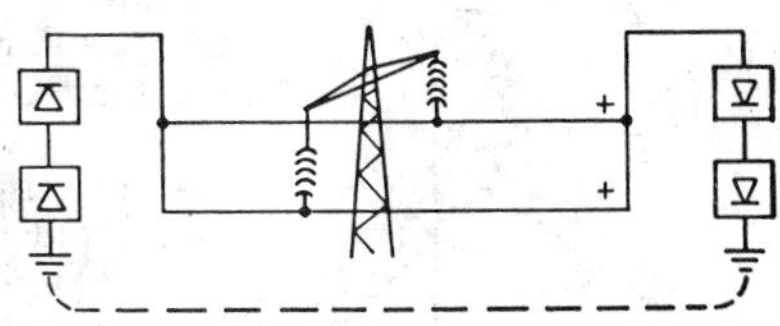

Fig. 14-3. Direct-current homopolar transmission arrangement.

7. Mechanical Tower Design. The principle for mechanical design of the transmission towers and calculation of conductor forces is similar to that for a-c transmission, as described in Sec. **13**. As the d-c tower normally carries only two insulated conductors, the natural arrangement is a center mast and a crossarm which holds the two-pole insulator chains. When an overhead ground wire is used, this is most conveniently supported by an extension of the center mast.

8. Self-supporting Tower. As an example of self-supporting or rigid two-pole towers, Fig. 14-4 shows the design for the Danish overhead line in the Konti-Skan 250-kV transmission.[1] This is a steel tower which can be equipped with two fully insulated poles.

9. Guyed Tower. On the Swedish side of the Konti-Skan transmission, guyed aluminum towers are used.[1] The design is shown in Fig. 14-5. The tower is equipped with a center-torque crossbeam which prevents the tower from turning in case of conductor breakage.

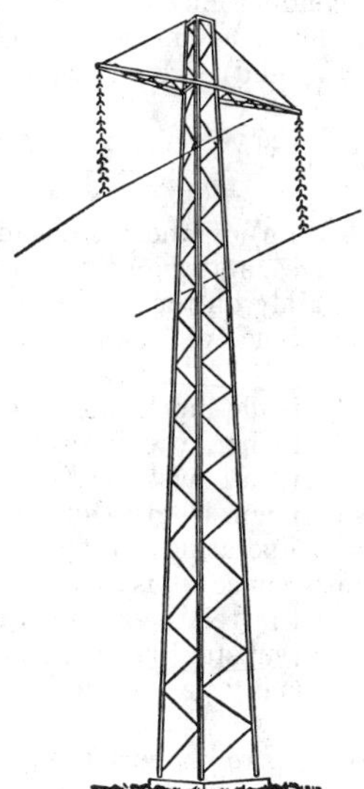
Fig. 14-4. Self-supporting tangent tower, Danish side, Konti-Skan transmission.

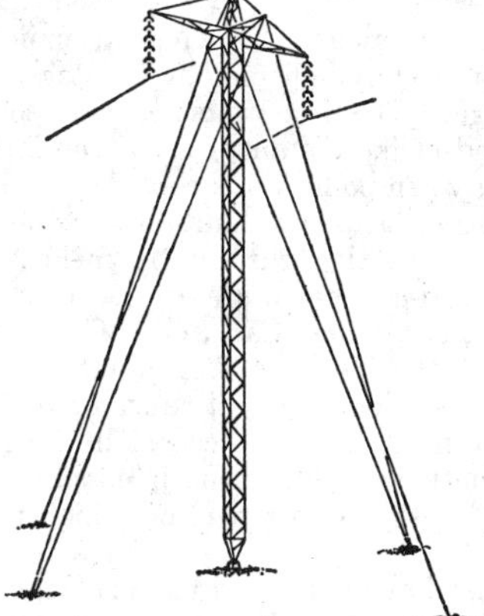
Fig. 14-5. Guyed tangent tower, Swedish side, Konti-Skan transmission.

[1] The Konti-Skan H.V.D.C. Project; *CIGRE Rept.* 408, June, 1964.

The guyed aluminum tower for d-c transmission is especially suitable for helicopter erection, as its weight is low. The weight of the Konti-Skan tower design for a transmission rating of ±250 kV at 500 MW is around 1,500 lb.

10. Insulator Arrangement. In addition to normal single-string arrangements, a V-string insulator arrangement, shown in Fig. 14-6, should be considered. For direct current, this arrangement is especially attractive, owing to the fact that here the ratio between required creepage distance and air distance to ground is higher than for the normal one-string insulator support.

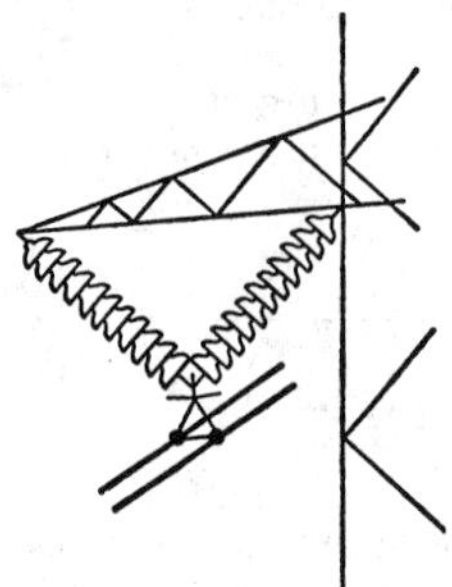

Fig. 14-6. V-string insulator arrangement.

11. Ground Wire. As in the case of a-c overhead lines, a ground wire can be erected on top of the tower to limit lightning effects. Because of the limited consequences of a line fault (see Pars. **13** and **37-39**) the demand on ground wire is more limited. For d-c lines often the main purpose of the ground wire is to decrease the resulting footing resistance per tower to ensure proper operation of the ground-protection system, and therefore a larger protection angle can often be accepted, whereby the height of the tower is limited. It is recommended that ground wires be used on the line close to the station, following the same practice as in some a-c systems having mild lightning exposures.

12. Electrode Line. As stated in Par. **3**, almost without exception H-V d-c transmission is equipped with ground-return facilities. It is normally not possible, and never essential, to locate the electrode in the converter station (see Pars. **49** and **50**). These are interconnected via an electrode line. This line may follow the main d-c line and may even serve as a ground wire. The insulation of this line is somewhat dependent on the electrode-line length, but normally around 100 kV BIL is sufficient, as transient grounding via electrode capacitors is made in the converter station. Wooden-pole lines are quite suitable, and often the conductors can be erected on existing a-c line towers for part of the distance. It is advisable to use two separately insulated conductors to prevent transmission trip-out due to an electrode-line fault. As an electrode line is cheap, it is reasonable to use a particularly suitable electrode site (sea or land), even if the distance from the converter station is comparatively great.

LINE INSULATION

13. Requirements. The insulation in H-V d-c (as in H-V a-c) overhead transmission lines is stressed by lightning overvoltages, internal overvoltages, and the normal operating voltage. The first two stresses impose requirements mainly on the insulator chain length and strike distance, while the latter influences the choice of leakage distance, especially when pollution is considered.

For d-c systems, the fault-clearing characteristics can be made much more effective than for alternating current by means of grid control. Further, the internal overvoltages in d-c systems are appreciably lower than in present a-c systems (see Overvoltage on a D-c Line, Par. **37**). The lower requirements on chain length thus imposed mean that the insulators' capability to withstand normal operating voltage will be the factor considered in all zones of pollution. As a consequence, it is advantageous to use d-c insulators with especially long leakage paths even in relatively clean areas. This reduction of chain length possible in clean areas is of substantial economic importance, as the largest part of a high-voltage power line usually passes through such areas.

The insulator must also have sufficient mechanical strength to support the greatest loads of ice and wind that may be expected (see also Sec. **13**).

14. Insulator Materials. Insulators of porcelain or toughened glass can be used in d-c transmission lines. Both types can be made to have a sufficiently high mechanical strength. In the case of direct current, it is important to choose the material in such a way that the internal resistance is high. The sodium content especially must be kept low so that the internal leakage current at normal temperature and normal service voltage does not exceed a few microamperes.

Insulators of organic materials have been developed, but so far they have not been found suitable for outdoor usage.

15. Atmospheric Pollution. The severity of the insulator pollution depends both on the atmospheric pollution and on the amount of dissociable matter, for example, salt. Also, the frequency and intensity of rain are of vital importance, since rain washes the insulators periodically—in effect, it causes an equilibrium state of insulator pollution. For districts exposed to salt storms, a different mechanism applies.

Table 14-1 gives characteristic data for atmospheric pollution in equilibrium state for various districts.

The actual pollution on insulators is usually referred to as the amount of dissociable matter present—generally it is given in terms of equivalent NaCl-deposit density. This is defined as the amount of NaCl which gives the same electrical conductivity as the dirt when dissolved in a given quantity of distilled water.[1]

The values for insulator pollution given apply mainly to alternating current. In the case of direct current, the electrostatic forces may play a more dominating role and consequently affect the pollution.

Because of the different mechanisms of dirt-particle attraction to the insulator and differences in washing properties of various parts of the insulator, the dirt becomes very unevely distributed over the insulator surface. The equivalent NaCl-deposit density is, in most cases, greatest on the undersurface, while the amount of nondissociable matter is greatest on the upper surface. An example is shown in Fig. 14-7. Furthermore, the dirt is unevenly distributed along an insulator chain, which can be seen in Fig. 14-8.

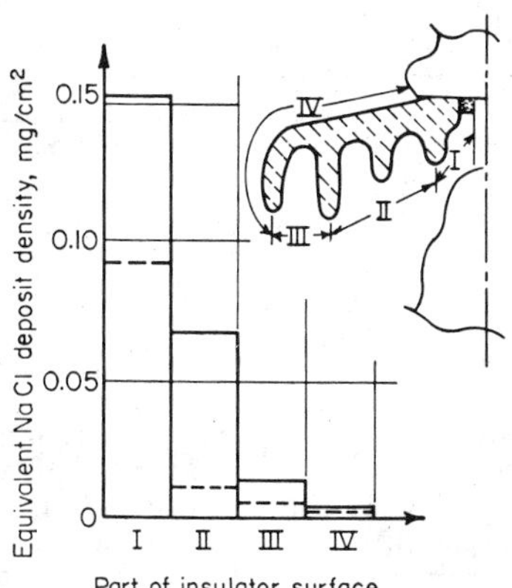

FIG. 14-7. Equivalent NaCl deposits on different parts of a d-c insulator at live end of +250-kV chain. Solid line (——) = after 32 months operation in a ±250-kV bipolar long-term test. Broken line (- - -) = after 9 months operation on the ±250-kV bipolar Konti-Skan line.

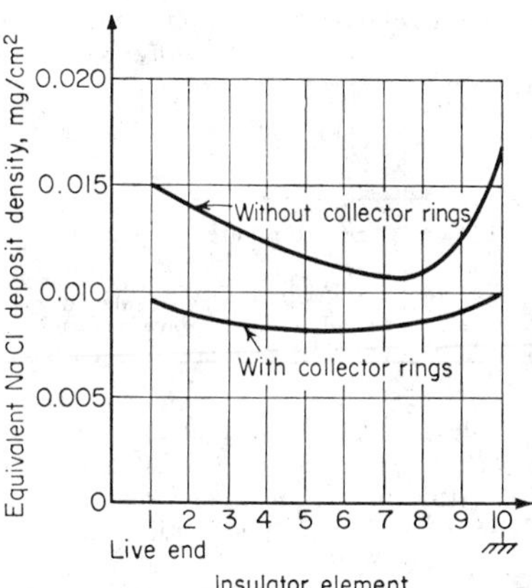

FIG. 14-8. Measured deposit density on undersurfaces of two insulator chains, showing the influence of grading rings at both ends, after 4 months' operation at +250 kV.

There is a pronounced tendency for the elements at the ends of the chain to be more polluted than others.

16. Necessary Leakage Paths. Usually, the leakage properties of insulators are characterized by the length of their leakage paths. Sometimes it has been found convenient to distinguish between protected and unprotected leakage paths or to introduce some form factor, etc. In the following, the necessary leakage path for a d-c line will first be discussed under the simple assumption that insulators of "good design" are used. Later, the factors which contribute to this good design will be discussed.

[1] FORREST, J. S., et al. Research on the Performance of High Voltage Line Insulators in Polluted Atmospheres; *Jour. IEE*, 1960, Vol. 107/A.

Table 14-1. Characteristic Data for Atmospheric Pollution in Equilibrium State for Various Districts

District	Atmospheric pollution, mg/(cm^2)(month)	Equivalent NaCl deposit density on line insulators, mg/cm^2
Clean areas	0.01–0.2	0.02–0.05
Industrial areas	0.2–0.8	0.05–0.10
Heavily polluted and coastal areas	0.8 –4	0.10–0.40

Table 14-2 gives the recommended specific creepage distance in different zones of pollution. The figures in Table 14-2 are based on a-c experience, in which case the a-c rms voltage to ground is comparable with the d-c voltage to ground. However, for zone 1 the figure 2.3 has been recommended instead of the 2.8 cm/kV normally used. The reason is that the requirement of withstanding internal overvoltages in a-c systems automatically leads to leakage paths of 2.8 cm/kV phase to ground or more. But there is little doubt that smaller leakage paths are sufficient to sustain the continuous voltage stress, and cases have been reported where leakage paths as low as 2.0 cm/kV have given satisfactory service. When data for existing a-c lines passing through a certain district are available, such data should be utilized as a basis for determination of necessary creepage distance for passage of the d-c line through that district.

Table 14-2. Recommended D-C Specific Creepage Distance in Different Zones of Pollution

Zone No.	Description	D-c specific creepage distance, cm/kV
1	Agricultural area, woodlands	2.3
2	Outskirts of industrial areas. Some distance from the sea	4.0
3	Industrial area. Direct vicinity of the sea	5.2
4	Direct vicinity of extremely dirty industries such as certain chemical industries, power stations	7.0

Table 14-2 is too simplified to allow immediate application to practical design projects. The table makes no reference to the frequency and intensity of rain, which may be of vital importance. An extreme case is very dry districts with extremely long periods between rainfalls, which may constitute special problems both for a-c and d-c lines. Disregarding the effect of rain, it may be difficult to decide to which zone a certain district should be referred. Finally, the performance of an insulator is not adequately described by the leakage path. In spite of these inadequacies, the table may give useful guidance.

17. Insulator Design. For H-V d-c transmission lines suspension insulators are recommended. Antifog cap and pin insulators having a ratio between leakage path and height of 2.8 to 3.2 have been found suitable. A specific creepage distance of 2.3 cm/kV and a figure of 3.0 for the above-mentioned ratio lead to an impulse level of about 1,400 kV for a 400-kV-to-ground transmission. This value would in most cases seem acceptable, considering the good fault-clearing properties of grid-controlled rectifiers. This example is illustrative only as higher or smaller values are obtained when other leakage paths are employed, or if the insulators have a different relation of leakage path to height.

The mechanical requirements do not differ from those of a-c insulators (see Sec. **13**). The electrical requirements, on the other hand, are different from those for alternating current. In direct current, the voltage distribution is resistive, instead of capacitive and resistive as in the case of alternating current. Even under dry conditions the air humidity will wet the dirt on the insulator surfaces so that the external leakage currents exceed the internal currents. Consequently, the voltage distribution will be affected by the distribution of dirt.

Drying processes must also be taken into account. The leakage current density, for instance, will be highest near the pin of a cap and pin suspension insulator and cause an effective drying. As a consequence, the surface voltage gradient may become very high there, although the equivalent NaCl-deposit density is high. The voltage distribution is also affected by the shape of the petticoats.

The self-cleaning properties of a d-c insulator are very important, and therefore the surface of a good d-c insulator must be made very smooth. The protected leakage path should be about 60 to 70% of the total path. This means that the sheds of an insulator will be relatively large. The petticoats on the undersurface can be made deep, but if so, the distance between them must be large enough. Figure 14-9 shows (broken lines) a good a-c insulator which did not behave well under d-c conditions. By reduction of the height of the inner petticoat, as shown by the solid line, the performance was improved considerably. This latter insulator is used in the New Zealand link in the 300-mi-long overhead transmission line, operating at ±250 kV. In clean districts 12 elements are used in each insulator chain, giving a specific creepage distance of 2.3 cm/kV.

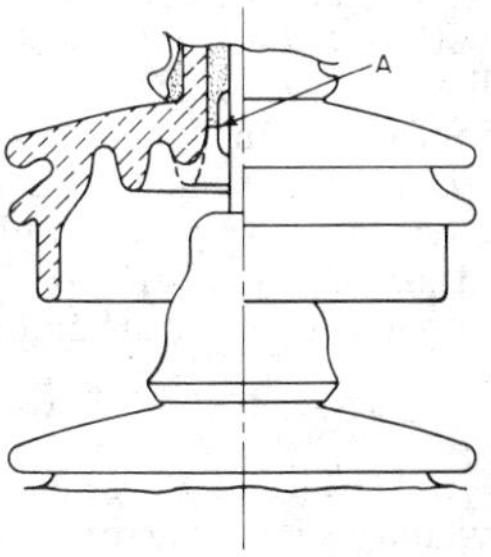

FIG. 14-9. Antifog cap and pin suspension insulator, showing good design for direct current (solid lines) and alternating current (broken lines). *A* indicates deformable material.

18. Electrolytic Phenomena. On the whole, the effects of internal leakage currents are of no consequence to d-c insulators, at least if precautions regarding the material composition are taken. The effects of external electrolysis are more easily detected.

Surface currents are transported by anions and cations. Anions are neutralized at the cathode, where metals are deposited as a visible cathode growth, while hydrogen is liberated. Cations, which are neutralized at the anode, may escape as a gas or chemically attack the anode.

At the anode, secondary processes lead to the formation of hydroxides and carbonates. As these compounds require more space than the pure metal, a swelling of the anode will result. The swelling is most pronounced if the anode has small geometric dimensions, for example, if the pin in a cap and pin insulator is the anode.

To avoid insulator cracking due to pin growth, a sleeve of a weak material which deforms can be placed around the pin as shown in Fig. 14-9 by *A*. In principle the shape of the insulator shall be such that mechanical-tension forces are avoided.

19. Collector Rings to Reduce Pollution. The nonlinear deposit distribution along the chain is caused by the forces acting on the dirt particles in the air surrounding the chain. These are forces due to wind, gravitation, electric forces on charged particles, and electric forces on uncharged particles in a nonuniform field. Therefore, under the influence of a unidirectional electric field, both charged and uncharged dirt particles in the air surrounding a d-c chain are set into motion and attracted especially to the energized and grounded end of the chain where the electric field strength is highest. The nonlinear deposit distribution causes a nonlinear voltage distribution along the chain, which, to a certain extent, reduces the breakdown voltage.

To reduce the insulator pollution and to improve the deposit distribution along the chain, collector rings can be mounted at the top and bottom of the insulator chain. The electrostatic field around the insulators will then change in that the gradient concentration now occurs on the surface of the ring itself. As a consequence, the dirt particles will move to the collector ring instead of the chain.

The reduction in insulator pollution by the use of grading rings is greater at positive than at negative polarity. Also, the influence of the collector rings on the pollution is greater on the undersurface of the insulator elements than on the upper. Figure 14-8 shows the influence of collector rings on the dirt distribution along a d-c insulator chain.

20. Greasing to Reduce Influence of Pollution. Another means for reducing the effect of dirt accumulation on flashover voltage is to apply coatings on the insulator. Silicone grease in a 0.5- to 2-mm-thick surface layer has been used with good results. Also, different kinds of petroleum grease are available. These coatings reduce the external leakage currents considerably by encapsulating the dirt particles in the grease and isolating them from each other. When the grease, however, has absorbed a certain quantity of dirt, its effectiveness is reduced suddenly and flashover may occur without warning.

If the grease is replaced regularly, coatings of this kind can be helpful in high-pollution districts. The time between preparations of the insulators can be between a couple of months in an extremely dirty area to a couple of years in cleaner areas.

21. Insulator Tests. As a consequence of the requirements for adequate leakage path in a d-c insulator there is also a need for pollution tests, in addition to those outlined in Sec. **13** for a-c insulators. Investigations have shown that the results of d-c rain tests can differ very much from those of d-c pollution tests. Pollution tests can be carried out indoors or outdoors. No standards exist for indoor tests, but test methods have been proposed[1] which give reproducible results and give the same ranking to insulators as do the long-term outdoor tests. In polluting insulators artificially it is important to do this under voltage. Otherwise, misleading results can be obtained.

CORONA

22. Corona Phenomena. Corona is associated with ionization phenomena in the vicinity of conductors and occurs when the electric field strength is high enough to cause a breakdown of the surrounding air (see Sec. **13**). The discharges give rise to current pulses which cause power losses and radio interference. At negative polarity the corona is characterized by frequently repeated pulses of a few picocoulombs, the so-called Trichel pulses. At positive polarity the pulses are much less frequent but contain, instead, charges up to thousands of picocoulombs. The current pulses have a duration of only a fraction of a microsecond (shorter for negative polarity than for positive).

Although the physics behind corona discharges is well known, it is almost impossible to express the losses and the interference in an exact mathematical form, because the corona starts at local points on the conductor where the electric field strength can be considerably higher than the nominal average field strength. Therefore, not only the conductor roughness, but also the accumulation of dirt particles, bird droppings, water droplets, etc., is of vital importance.

Since the conductor surface conditions change from time to time, owing to washing by rain, etc., the losses, as well as the interference, can be treated only in a statistical manner.

23. Corona Losses. In principle, the corona-loss problem is not so serious for d-c as for a-c lines. For an overhead transmission line the corona losses are dominated by the discharges on the conductors, while the corona on the metal elements and insulators will be a small part of the total losses. The intensity of the discharges on conductors depends very much on their surface conditions. In many cases, the losses are high for new lines because of intensive corona from scratches and foreign material on the cables. After a certain time, the conductors will, however, age, and as a consequence the average corona losses decrease.

It is characteristic of d-c corona that the charge released must be carried to ground or to a conductor of opposite polarity (this is not the case for alternating current). Thus a space charge, which reduces the conductor surface voltage gradient, will be formed around the conductor. The effect is so pronounced that it is usually more reasonable to characterize the corona performance by the line voltage than by the surface gradient.

Also because of the large d-c space charges, the losses from a monopolar and bipolar transmission (see Pars. **4** and **5**) are quite different. In the bipolar case, charge mixing takes place as negative ions reach the direct vicinity of the positive conductor, and vice versa, thus increasing the gradient on both conductors. For this reason, losses from bipolar lines are larger in comparison with monopolar lines than would be expected from the nominal surface gradients.

24. Monopolar and Homopolar Corona Losses. Although the corona-discharge mechanism is very different for positive and negative polarity, the average corona

[1] Annestrand and Schei *Direct Current*, 1967, Vol. 12, No. 1, pp. 1-8.

losses are about the same for both polarities. This is illustrated in Fig. 14-10, where results from d-c corona-loss measurements on monopolar lines are presented. Compared with a-c, the d-c corona losses increase more slowly with voltage.

For a given conductor, the d-c losses decrease approximately with the square of the height of the conductor. The influence of bundling is not pronounced, and, in fact, very little can be gained as regards corona losses by using multiple conductors, unlike the case with alternating current.

The conclusions above are also applicable to homopolar d-c lines. However, the average corona losses for a homopolar H-V d-c line are slightly less than the sum of the losses of the corresponding monopolar lines.

If earth conductors are erected on the towers, corona on these may occur, especially if the size of the conductor is small. Near the converter station, the electrode line is often erected on the towers. In this case the line should, from the corona point of view, be characterized as a bipolar one.

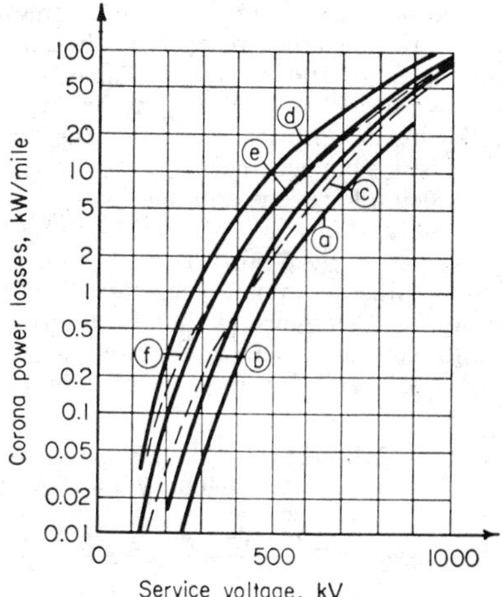

FIG. 14-10. Measured d-c corona losses on monopolar test lines. (*a*) Fair weather, smooth Al tube, d = 100 mm, H = 10 m, positive polarity; (*b*) fair weather, smooth Al tube, d = 37 mm, H = 10 m, positive polarity; (*c*) fair weather, smooth Al tube, d = 37 mm, H = 10 m, negative polarity; (*d*) rain, smooth Al tube, d = 37 mm, H = 10 m, positive polarity; (*e*) fair weather, ACSR conductor, d = 27.7 mm, H = 10 m, positive polarity; (*f*) fair weather, ACSR conductor, d = 27.7 mm, H = 10 m, negative polarity.

25. Bipolar Corona Losses. The total average corona losses for a bipolar line are greater than the sum of the losses of two monopolar lines with the same conductors and operating at the same voltages. This is due to the recombination of positive and negative charges in the space between the conductors.

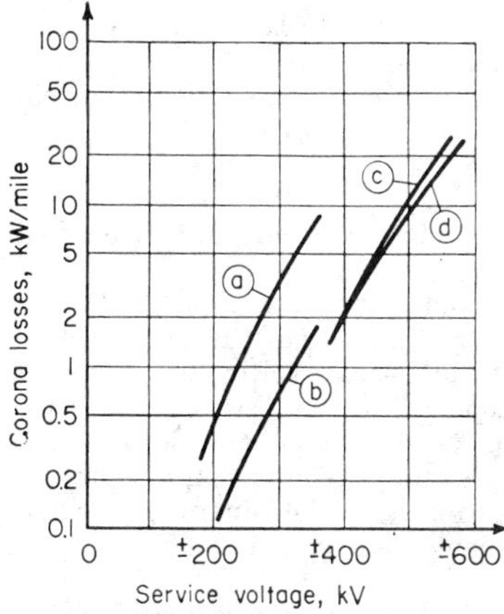

FIG. 14-11. Measured fair-weather bipolar d-c corona losses. (*a*) Simplex, ACSR, d = 36.2 mm, D = 6.0 m, H = 13.0 m; (*b*) simplex, ACSR, d = 36.2 mm, D = 12.0 m, H = 13.0 m; (*c*) simplex, ACSR, d = 61.0 mm, D = 10.5 m, H = 15.0 m; (*d*) duplex, ACSR, d = 46.2 mm, D = 10.5 m, H = 15.0 m.

Figure 14-11 shows the corona-loss characteristics. The losses are approximately equal for both conductors. As in the monopolar case, the corona losses are relatively independent of conductor bundling.

If earth conductors are used, precautions as in the case of monopolar lines, such as ensuring that the distance from the earth wire to conductor is sufficiently great, must be taken to limit the normal surface gradient to ensure that corona losses do not occur.

26. Influence of Weather. Direct-current corona losses increase during bad weather, for example, in rain, in snow, and with hoarfrost. Compared with alternating current, the increase, however, is small, and the ratio between bad- and fair-weather losses very seldom exceeds 10. This constitutes one of the advantages of H-V d-c systems compared with a-c. The increase during bad weather is less if the conductors have a high surface gradient. For instance, it can be noted that at 20 kV/cm the power-loss ratio due to corona between good and bad weather is 1:10, whereas at 28 kV/cm this ratio is reduced to 1:2.

27. Annual Mean Losses. As in the case of

a-c corona, the average annual corona losses can be calculated if the corona-loss characteristic and the weather conditions during an average year are known. This is illustrated by the example in Table 14-3. A 500-kV 3-phase a-c and a ±400-kV bipolar d-c overhead transmission line having the same capability have been compared. The following assumptions regarding the weather conditions have been considered:

5,880 h fair weather.

2,000 h fog, dew, and hoarfrost.

880 h rain.

As shown, the fair-weather corona losses are roughly the same for the two systems, while both the annual mean losses and especially the maximum losses are considerably lower for direct current than for alternating current. While for an a-c system the maximum corona losses may amount to the same magnitude as the load losses, they will be considerably less for a d-c system.

Table 14-3. Comparison for Corona Power Losses between a Bipolar ±400-kV D-C and a 3-phase 500-kV A-C Line

(D-c line conductor 2 by 1.8 in. Pole separation 34.5 ft. Average height 71 ft
A-c line conductor 3 by 1.4 in. Phase spacing 36.6 ft. Average height 61 ft)

	Corona losses, kW/mi	
	±400 kV d-c	500 kV a-c
Average losses in fair weather	2.0	2.0
Minimum losses in fair weather	1.0	0.2
Maximum losses in a short line section under the worst weather conditions	15	200
Maximum losses for the whole line under the worst weather conditions	10	30
Annual mean losses for the whole line	3.7	9.0

In effect, these quantitative differences will cause the d-c corona losses to be of minor economic importance for many practical cases. Another consequence is that the choice of conductor arrangement and conductor type in d-c systems generally can be made without special regard to the corona-loss problem.

28. Radio Interference from H-V D-C Overhead Lines. An H-V d-c power line generates radio interference principally in three different ways:

1. By pulses, occurring at the ignition of the valves, being transmitted via the switchyard to the line.
2. By corona pulses on the line.
3. By partial discharges on the insulators.

The first source can be reduced or eliminated by taking precautions in the terminal station. The latter two are dealt with in the following.

29. Characteristics of Measuring Instruments. Radio-interference measurements are generally carried out with an instrument specified by NEMA, CISPR, or variations thereof. In principle, such apparatus consists of a quasi-peak voltmeter connected to a bandpass filter, via a weighting circuit. The response characteristic of the instrument varies with the pulse-repetition frequency, pulse duration, bandwidth of the apparatus, etc. In the case of random pulses, the reading becomes roughly proportional to the rms values of the entrance voltage multiplied by the square root of the apparatus bandwidth. In other words, the instrument reading is a function of the input power within the apparatus passband.

Since the characteristics of occurrence of d-c and a-c corona pulses are different, the same instrument reading will not give the same subjective disturbance (on radio receivers tuned for acceptable listening conditions). Up to three times lower signal-to-noise ratio can be allowed for direct than for alternating current for achieving the same quality of reception.

30. Line-radiation Factors. The radio-interference level due to radiation from a horizontal line decreases, in a first approximation, with the square of the distance from the line.

For a vertical conductor, on the other hand, the radio-interference level is inversely proportional to the distance from the line. The line insulators and the towers constitute such vertical conductors and, because of the insulator capacitance, may transmit a considerable current to earth, at least for frequencies above a few megacycles per second.

Corona on multiconductor systems may be regarded as being composed of two sets of pulses. One set has equal polarity on all conductors; the corresponding radiation follows the inverse-square-distance law. The other set has one polarity on one conductor and the opposite polarity on the other conductors. The radiation from the latter will be much less and will roughly be inversely proportional to the cube of the distance to the line. Therefore, it can usually be neglected. Thus, the radiation from a multiconductor system is roughly the same as for a system with one single conductor. A typical frequency spectrum of the interference due to corona on a line is shown in Fig. 14-12.

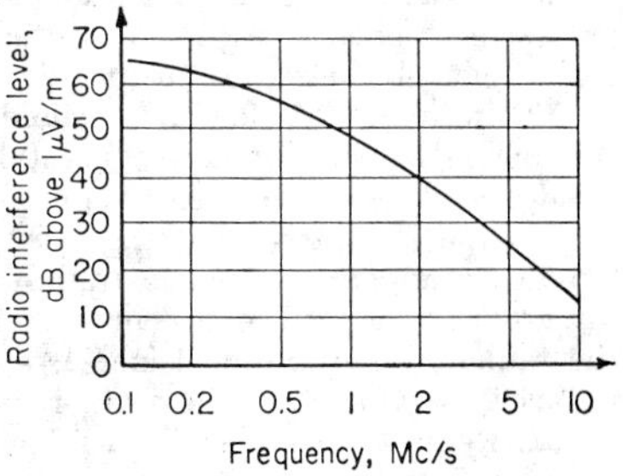

Fig. 14-12. Measured frequency spectrum of radio interference at 30 m distance from a bipolar d-c line operating at ±400 kV. Line parameters: cable 1 by 2.4 in; pole separation 34.5 ft; average height 63 ft; ±400 kV d-c.

31. Radio Interference from Monopolar and Homopolar D-C Lines. Radio interference from corona on the conductors is considerably less with negative voltage than with positive. The polarity dependence can be explained as an effect of the rms characteristic of the measuring instrument. It is possible to demonstrate that the small, frequent negative pulses will give a considerably lower pulse-current rms value than the larger, infrequent positive pulses, even if the total negative corona losses are somewhat higher than the positive. In Fig. 14-13

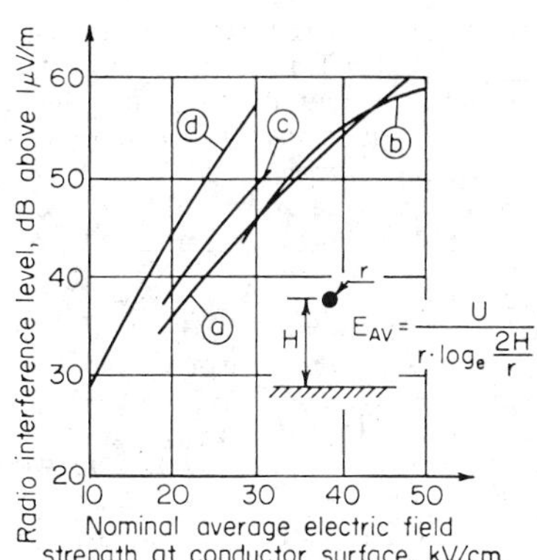

Fig. 14-13. Measured fair-weather radio interference level at 1 Mc/s and at 30 m distance from a monopolar d-c line. E_{Av} = nominal field strength; H = height of conductor over earth; r = individual cable radius; U = service voltage. (*a*) Simplex, smooth Al tube, d = 37.0 mm, H = 12.5 m; (*b*) simplex, ACSR, d = 27.7 mm, H = 12.5 m; (*c*) simplex, ACSR, d = 36.2 mm, H = 13.0 m; (*d*) simplex, smooth Al tube, d = 100.0 mm, H = 10.0 m.

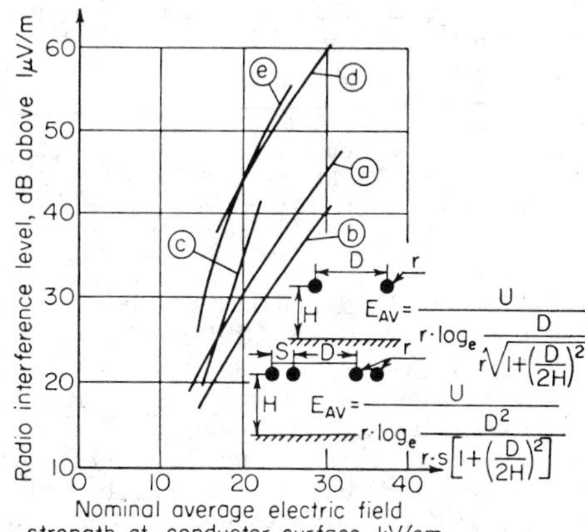

Fig. 14-14. Measured fair-weather radio-interference level at 1 Mc/s and at 30 m distance from a bipolar d-c line (d = $2r$). Symbols as in Fig. 14-13. S = cable spacing; D = pole spacing. (*a*) Simplex, ACSR, d = 36.2 mm, D = 6.0 m, H = 13.0 m; (*b*) simplex, ACSR, d = 36.2 mm, D = 12.0 m, H = 13.0 m; (*c*) duplex, ACSR, d = 31.7 mm, D = 10.0 m, H = 10.0 m; (*d*) simplex, ACSR, d = 61.0 mm, D = 10.5 m, H = 15.0 m; (*e*) duplex, ACSR, d = 46.2 mm, D = 10.5 m, H = 15.0 m.

characteristics of the radio-interference level, measured according to NEMA, from a monopolar transmission line operating at positive polarity are given. The method of computing voltage gradients is also shown.

If the line had been working at negative polarity, the radio-interference level would have been about 10 to 20 dB lower. The radio interference from a homopolar line appears to be the same as, or slightly less than, that from a monopolar line with the same voltage.

32. Radio Interference from Bipolar D-C Lines. The fair-weather radio-interference values for bipolar lines are presented in Fig. 14-14. In this case the interference is almost entirely caused by the positive conductor.

33. Influence of Conductor Bundling on Radio Interference. By using bundled conductors with the same cross-sectional area as the single conductor, the radio-interference level can be reduced. This is due to the fact that the interference is very sensitive to the electric field strength at the conductor surface, and this is reduced on a bundled conductor, as demonstrated in Fig. 14-14, with computation methods for voltage gradients of such conductors. For a single conductor with a constant field strength the radio-interference level increases approximately quadratically with the conductor radius r. A voltage gradient of 24 to 26 kV/cm can generally be recommended.

34. Radio Interference during Bad Weather Conditions. With alternating current, the radio interference during rain is higher than for the fair-weather value, the increase being dependent on both the rain intensity and the conductor surface voltage gradient. With direct current, just the opposite occurs, as illustrated by Fig. 14-15. This reduction is also present during snow. Wind, in most cases, increases the interference level, but in a very irregular manner.

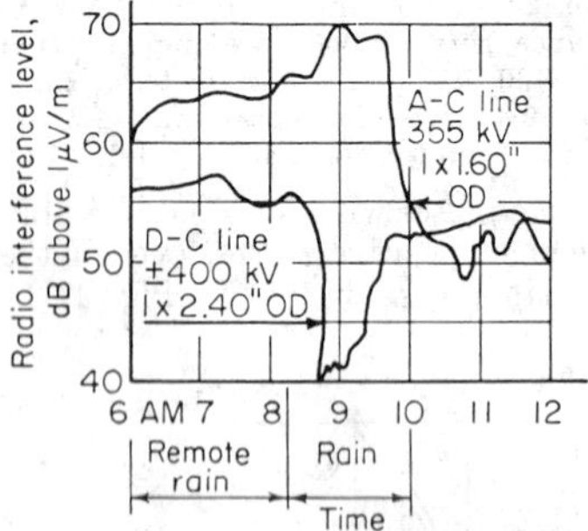

Fig. 14-15. Measured radio-interference level at 0.83 Mc/s and 30 m distance from a bipolar d-c and a 3-phase a-c line, before, during, and after rain.

35. Radio Interference in the Extreme-high-frequency Range. The noise in the television-frequency range, i.e., above 50 Mc/s, is dominated by the insulator corona. In this case, the interference during fair-weather conditions will be very small but will increase during bad-weather conditions. Generally, the noise level in this frequency range is very low and can be ignored in most practical cases.

36. References to Literature on Insulation, Corona Losses, and Radio Interference

Witt, H. Insulation Levels and Corona Phenomena on H-V D-C Transmission Lines; Techn. Dr. thesis, Chalmers University, Gothenburg, Sweden, 1961.

Hyltén-Cavallius, N., et al. Insulation Requirements, Corona Losses, and Radio Interference for High Voltage Direct Current Lines; *Trans. IEEE Paper* 63–998.

Morris, R. M., and Rakoshdas, B. An Investigation of Corona Loss and Radio Interference from Transmission Line Conductors at High Direct Voltages; *Trans. IEEE*, Power Apparatus and Systems, January, 1964.

Hyltén-Cavallius, N., et al. Corona Losses, Radio Interference and Insulator Requirements for HVDC Lines. Studies Regarding Insulator Interference for Frequencies between 30 and 1500 MHz; *CIGRE Paper* 407, 1964.

Kovalskaja, O. T., et al. Investigation of Corona Losses on Experimental Section DC Electrical Transmission Line; *Direct Current Sci. Res. Inst. Bull.*, 1960, Vol. 5.

Gehrig, E. H., et al. BPA's 1100 KV DC Test Project. Part II. Radio Interference and Corona Loss; *Trans. IEEE*, Power Apparatus and Systems, Vol. 86, No. 3, pp. 278–290, March, 1967.

Methods of Measuring Radio Noise; *NEMA Publ.* 107, 1960.

Specification for CISPR Radio Interference Measuring Apparatus for the Frequency Range 0.15 Mc/s to 30 Mc/s, *IEC Publ.* 1, 1961.

Clark, F. M. "Insulating Materials for Design and Engineering Practice"; New York, John Wiley & Sons, Inc., 1962.

OVERVOLTAGES AND LINE PROTECTION

37. Overvoltages on a D-C Line. Direct-current lines as well as a-c lines are exposed to lightning strokes leading to overvoltages, which are dealt with in Sec. **13.**

Switching surges induced by the terminals on a d-c line are limited, and such overvoltages normally do not exceed 1.7 times the rated operational line voltage (E_d). This factor is considerably lower than that for a-c systems.

As described under Line Insulation (Pars. **13** to **21**) the comparatively low overvoltage level leads to a requirement for insulators with a high ratio between effective creepage and flashover distances.

There are remote possibilities for certain malfunctions in the converter control or in the valves giving higher overvoltages, theoretically reaching 2.2 times E_d. Proper means for protection, such as lightning arresters or rod gaps, should be installed to absorb and limit such overvoltages to a lower level.

38. Ground Wires. The demand for ground wires on d-c lines is less than on a-c lines, because the control system in the d-c terminals will limit the fault current and thereby reduce the risk of damage to insulators.

The use of overhead ground wire on a d-c line naturally decreases the lightning strokes to pole conductors, but it also adds to the corona. It reduces the influence between poles during transients and also reduces the resulting footing resistance (resistance from tower to ground), which is important for the function of the line protection, as shown in Par. **39**. An economic evaluation has to be made to determine whether complete overhead ground wires are justified. In all circumstances, overhead d-c lines should be equipped with station protection (line-end protection); i.e., the switchyard should be protected from transient overvoltages from lightning very close to the station through the use of overhead ground wire, preferably as far as 2 to 3 mi from the station.

In cases where the d-c transmission includes both overhead line and cable sections, it is important to equip the overhead line with line-end protection close to the cable end as well as the converter stations.

39. Line-protection System. The grid control of the converters in an H-V d-c transmission system, as described in Par. **80**, provides a means for the clearing of line faults and the subsequent reestablishment of normal operation. This is done by blocking the converters in case of line faults and re-deblocking after interruption of the fault current. Relays can, of course, control these functions.

The current-control system, however, makes it possible to stop the fault current without relay function and to reduce the fault time. Moreover, the current after the line discharge will be only a small fraction of the rated current. The reduction of fault current and fault time prevents damage to insulators, limits the ionization in the neighborhood of the insulators, and facilitates rapid re-deblocking.

The fact that the fault current is small causes certain difficulties in indicating the fault. However, the rapid reduction of the voltage, or a persistent decrease to a low value for some time (2 to 3 c), is used as a criterion of a flashover. If, as a consequence, the current order of the rectifier is brought to zero or even a negative value, the converter will rapidly bring the voltage down and the arc will be quenched. By suitable circuitry it is then possible to restart the transmission after an appropriate time (about 0.2 to 0.5 s). It is also possible to make two or more attempts if the first one is not successful and then block permanently if the flashover should be persistent.

To trace a line fault it is necessary that the footing resistance at any point of the line be limited to a maximum of about 100 Ω. Where tower grounding is poor, this might mean that crowfoot counterpoise arrangements should be used or that several towers should be interconnected. These interconnections can then most preferably be achieved through the use of an overhead ground wire for a limited length of the line.

CABLES FOR H-V D-C TRANSMISSION

40. Types of H-V D-C Cables. Power-transmission cables used for a-c transmission can also be used for d-c. Owing to better utilization of insulation and absence of capacitive currents, they are normally capable of transmitting more d-c power than a-c.

A cable especially designed for H-V d-c transmission will, however, have somewhat different design from that of an a-c cable.

There are in principle four different types of cable to consider: solid-type insulation, oil-filled, gas-pressure, and plastic-insulated. The solid-type insulation cable is probably the most economical for limited voltages—at present around 300 kV. The conductor temperature has to be limited to about 50°C at present to prevent migration of impregnating oil in the insulating material.

The oil-filled cable always operates with oil pressure and can be used for the highest voltages. Oil supply has to be arranged, and this requires high pressure on long submarine cables. A variety of the oil-filled cable—the flat cable—for which no external oil supply is required, has also been used for d-c transmission.

The gas-pressure cable has, as its name implies, gas instead of oil as the pressure medium. This type of cable has been used for direct voltage up to 250 kV to ground.

Development work on plastic-insulated cables for H-V a-c systems is going on. When a reliable result has been reached for alternating current, they will probably prove suitable for H-V d-c systems.

41. D-C Cable (Performance). A power cable used for d-c transmission has no capacitive leakage currents, and the power transmission is limited by the conductor losses only.

How much the cable insulation can withstand in a-c operation is normally limited by the maximum voltage stress in service and at impulse overvoltages. For d-c cables a stress of three to seven times that for an a-c cable may be used. In the Konti-Skan cable (solid insulation) 25 kV/mm is used.

42. Cable Insulation at D-C Stresses. In a-c cables the stress created by the electrical field is distributed in inverse proportion to the capacitance of the cable dielectric. This always gives the highest stresses close to the conductor.

In direct current the voltage distribution is determined by insulation resistance and space charges and is dependent on temperature. In a cold cable (with uniform dielectric temperature) the voltage distribution is the same as for alternating current, but at high conductor-to-sheath temperature gradient the stress may become highest on the surface, as shown in Fig. 14-16.

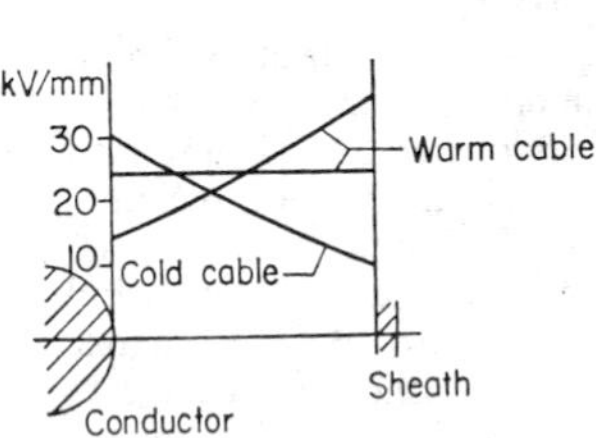

FIG. 14-16. Stress distribution in an H-V d-c cable during cold and warm conditions.

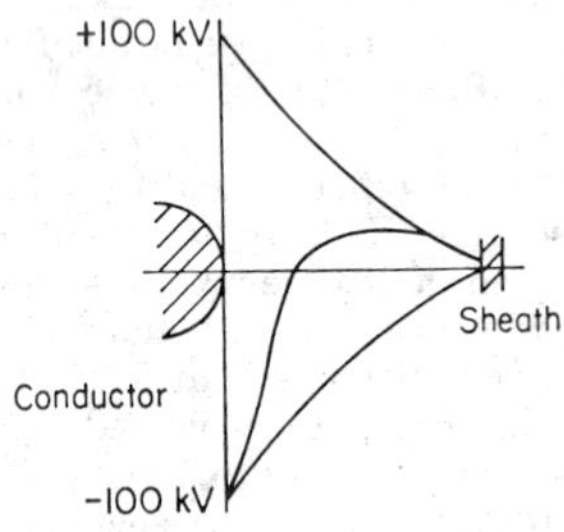

FIG. 14-17. Stress distribution in changing the polarity of a warm d-c cable.

In d-c transmission there is normally a demand for fast regulation, including reversals. In this circumstance the cable stresses are increased (see Fig. 14-17). As the stress is proportional to the time derivative of the potential, it can be seen that the unit stress becomes high. Overvoltages from connected d-c overhead lines must be borne in mind, even though overvoltage protection is used. The cable can also be exposed to voltages of power frequency in case of valve malfunctions.

43. Design of D-C Cables. As no dielectric losses exist, the insulation paper can be chosen to give the best dielectric strength. Highly dense paper is therefore used. To refine insulation performance even more, this paper can be used close both to the conductor and to the outside surface.

For a solid-type cable a high-viscosity compound is needed to prevent oil migration, but the viscosity must still be low enough so that good flexibility is obtained.

In a d-c cable there is a leakage current through the insulation between conductor and sheath. In underground cables this leakage current should be given a special path to the armoring, designed to ground at the joints to prevent corrosion. For sea cable without corrosion protection the current leakage is evenly distributed, and the corrosion risk is very minor. Where corrosion protection between the lead sheath and armoring is used this should be interconnected at intervals to prevent high voltages between the two. By locating the earth electrode at sufficient distance from the cable, the current leaking in is kept low, and corrosion is prevented as well.

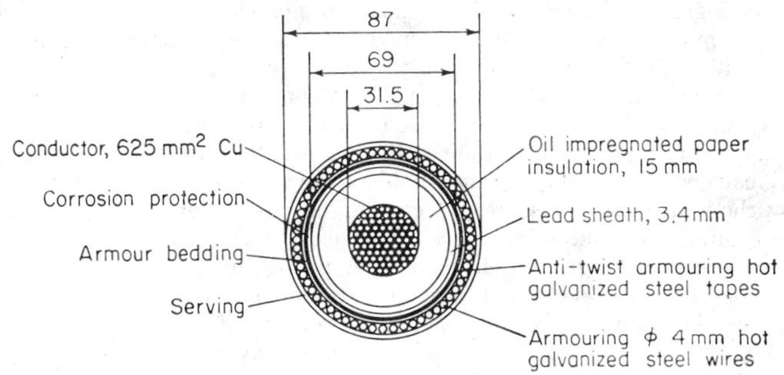

FIG. 14-18. Konti-Skan 250-kV d-c transmission from Gothenburg, Sweden, to Ålborg, Denmark, submarine cable cross section. Weight in air = 24.5 kg/m; in water = 19.6 kg/m.

The armoring can be made as for a-c cables. Due to the absence of eddy-current losses, steel wire can be used. The armoring is designed to stand the stresses of laying and repairing the cable (around 3×10^4 N for Konti-Skan at a 225-ft depth). Figure 14-18 shows one of the cable designs used for the Sweden-Laeso part of the Konti-Skan transmission.

The auxiliaries for d-c cables are the same as for a-c cables. It must be kept in mind, however, that a long creepage distance is required for the pothead at the terminals.

44. Cable Erection. Sea cables are normally transported and laid from a ship with special laying facilities. In favorable cases cable may be transported in one separate ship or barge and laid with the assistance of a separate cable-laying vessel. The latter method was used for Konti-Skan.

Underground cables are basically erected in the same way as a-c cables. As the d-c cable design is normally simpler and the losses are lower, the erection can be accomplished without special cooling facilities. In many cases simple sand bed can be used instead of concrete ductwork.

GROUND RETURN

45. General Aspects. It has been proved possible to use the ground as return path for direct current in H-V d-c transmissions, thus decreasing losses as well as installation costs. Ground return may be used in single-pole transmission or as a spare "conductor" in case of a fault on one pole of a bipolar transmission (see Pars. **3** to **12**).

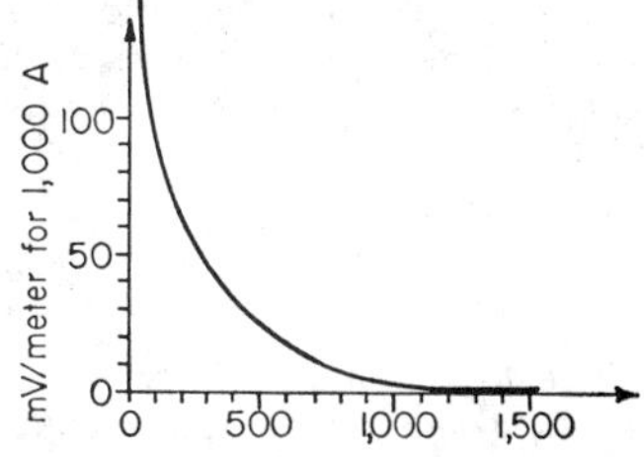

FIG. 14-19. Voltage drop in the ground return path per length unit.

The voltage drop in the ground-return path is concentrated at the ground or sea electrodes because the direct current will follow good conducting layers several miles under the earth surface between the electrodes. The voltage drop per length unit in an actual case may be seen in Fig. 14-19. The electrode may be placed in the ground, in the sea, or on the seashore.

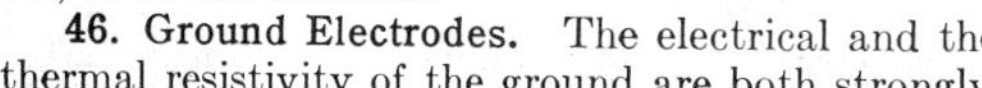

46. Ground Electrodes. The electrical and the thermal resistivity of the ground are both strongly dependent on its moisture content and increase greatly when the water content is below about 15%. The electrode should therefore always be buried below the ground water level. The temperature increase around the electrode must be kept below about 75°C. According to Rusck,[1] the following correlation exists,

$$U \approx \sqrt{2\delta_0 \cdot \lambda \cdot \rho}$$

where U = electrode voltage drop to a remote reference point, δ_0 = temperature rise on electrode surface in degrees centigrade, λ = ground thermal conductivity in watts per meter per degree centigrade, and ρ = ground electrical resistivity in ohm-meters.

The ground must be thoroughly examined and constants λ and ρ measured before the electrode site is fixed. ρ will normally increase with particle size; typical values are $\rho \approx 10$ for clays and $\rho = 10^4$ for certain kinds of sand. The thermal conductivity does not vary that much; typical values are 0.5 to 2.5 W/(m) (°C).

The moisture around the electrode will decrease because of influence from the electrical (electroosmosis) and thermal fields (thermoosmosis), but this is compensated for by the hydrostatic pressure. A ground electrode should not be placed in very fine-grained ground, to avoid its drying out because of electroosmosis.

A suitable area for the ground electrode is found from thermal and electrical properties and groundwater level, and possibly also experiments with small electrodes on site to check the computed results. The electrode itself may be arranged according to Fig. 14-20. The coke, with a resistivity of approximately 0.3 Ω·m, is the actual electrode, and the graphite electrode (standard cathodic protection electrode) will just transmit the direct current from the electrode line to the coke. This arrangement will protect the graphite electrode from anode corrosion.

The maximum allowable voltage drop is given by the foregoing correlation, and the electrode resistance is a function of the electrode dimensions. For a straight electrode of a depth of 7 ft, $\lambda = 2$ W/(m)(°C), $\delta_0 = 75$°C, $r = 0.25$ m, the length per ampere and the voltage drop are given by Fig. 14-21. It is often convenient to arrange the coke electrode in a six-arm star. In this case, the lengths in Fig. 14-21 should be multiplied by 1.6. The resistance for such a star electrode could be calculated according to the formula

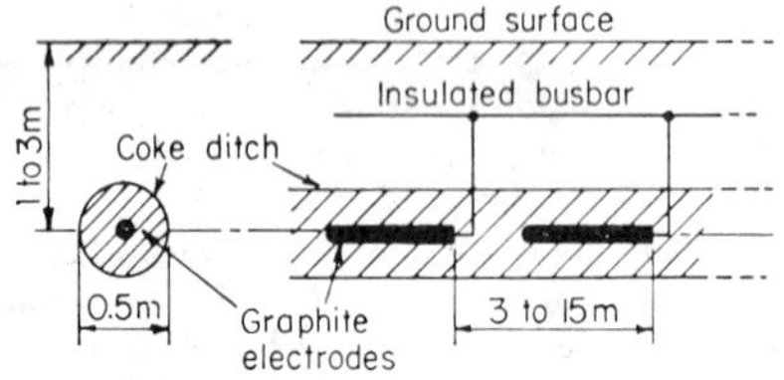

FIG. 14-20. Typical arrangement of a ground electrode.

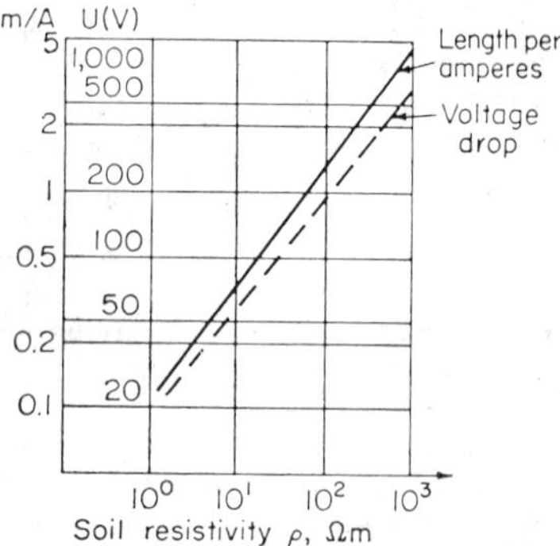

FIG. 14-21. Length per ampere and voltage drop of a ground electrode.

$$R = \frac{\rho}{L \cdot \pi} \cdot \left(\ln \frac{2L}{a} + 1.43 \right) \qquad \text{(ohms)}$$

where ρ = soil resistivity in ohm meters, L = total length of conductor in meters, $a = \sqrt{2 \cdot d \cdot r}$ in meters, r = equivalent radius of coke bed in meters, and d = depth of conductor in meters.

47. Sea Electrodes. If possible, the electrodes should be arranged in the sea. A negative electrode may consist of simple copper conductors laid on the sea bottom.

[1] RUSCK, S. H.V.D.C. Power Transmission: Problems Relating to Earth Return; *Direct Current*, 1962, Vol. 7, No. 11, pp. 290–298, 300.

A positive electrode should be protected so that fish do not get within around 30 ft of the electrode. Graphite or magnetite electrodes should preferably be used. For the Gotland scheme (200 A), 12 magnetite electrodes were used, mounted inside an area protected from the open sea by a stone wall.

It is also possible to lay an electrode on the sea bottom in the manner shown in Fig. 14-22. The electrode must be protected by a nonconducting cage which prevents fish or persons from coming into contact with it.

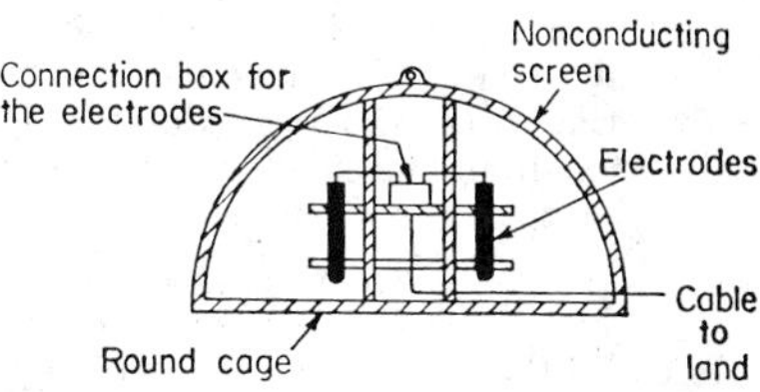

FIG. 14-22. Arrangement for a sea electrode featuring a nonconducting protective cage.

48. Shore Electrodes. It is sometimes convenient to place the electrode on the shore. The seawater should pass by the graphite or magnetite electrodes. This could be arranged via pumps, or the natural tide may be used. The Konti-Skan anode (1,000 A) electrode consists of 25 graphite electrodes as shown in Fig. 14-23.

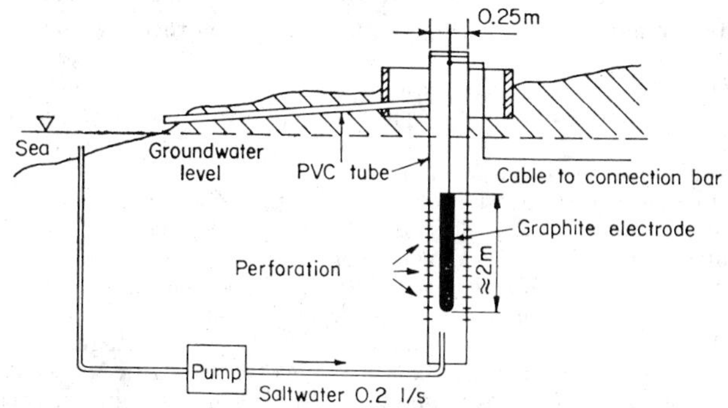

FIG. 14-23. Arrangement of the shore anode electrode for the Konti-Skan transmission (1,000 A).

49. Electrode Line. In order to avoid corrosion (see Par. **50**), the electrode should be placed at least 2 to 3 mi away from the converter station. Either a cable or an overhead line can be used. In the station, a transient ground via a capacitor (>10 μF) and a spark gap as overvoltage protection should be arranged.

50. Corrosion. To avoid corrosion on metallic structures, cables, or pipelines, the electrode should be well away from such structures, i.e., several hundred yards from small structures, or several miles from cables or pipelines. Protection could also be achieved by giving the metallic structure a small (about 1 V) negative potential (so-called cathodic protection).

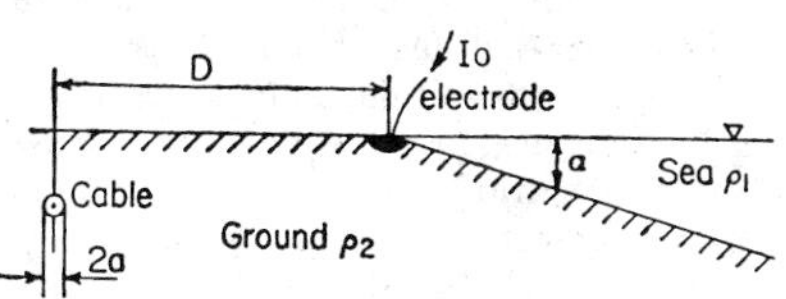

FIG. 14-24. Arrangement of a shore electrode showing mathematical functions involved in figuring current leakages.

For cables and pipelines, the leaking d-c density must be kept below specified values, normally 0.1 to 1 μA/cm². A very approximate formula for such currents occurring in cables or pipelines at a distance D m from a shore or sea electrode is given below (see Fig. 14-24).

$$i_{c,\max} = \frac{1}{2\pi a \cdot R} \cdot \frac{I_0}{2[\alpha/\rho_1 + (\pi - \alpha)/\rho_2]} \cdot \frac{1}{D^3} \qquad (\text{A/m}^2)$$

where I_0 = direct current in amperes, ρ_1 = sea resistivity in ohm-meters, ρ_2 = earth resistivity in ohm-meters, R = resistance of cable armoring or pipeline in ohms per meter, a = cable (pipeline) radius in meters, and α = angle between ground and sea bottom in radians.

At low values of D the formula above gives far too high values for the current leaking into the cable, owing to the influence of the cable on the potential field. The expression in the above equation must then be multiplied by a correction factor. However, for submarine cables the approximation gives reasonable values.

For a ground electrode the corresponding formula would be

$$i_{c,\max} = \frac{I_0 \cdot \rho_2}{4\pi^2 \cdot a \cdot R \cdot D^3} \qquad (\text{A/m}^2)$$

Here also the expression should be multiplied by a correction factor at low values of D.

51. References to Literature on Ground Return

Lundholm, R. D.C. Transmission with Return Current through the Earth for H.V.D.C. Transmission; *Direct Current,* 1953, Vol. 1, No. 4, pp. 79–86.

Rusck, S. H.V.D.C. Power Transmission: Problems Relating to Earth Return; *Direct Current,* 1962, Vol. 7, No. 11, pp. 290–298, 300.

Köhler, A. H.V.D.C. Transmission: Earth Electrode Arrangements; *Direct Current,* 1965, Vol. 10, No. 1, pp. 18–24.

HIGH-VOLTAGE CONTROLLED RECTIFIERS (VALVES)

52. The main requirements of a high-voltage controlled rectifier (valve) for H-V d-c transmission are:

Low forward voltage drop during conduction.

The valve must withstand high negative and positive voltages without breaking down.

The firing instant must be controllable.

A reasonably short commutation margin during inverter operation.

The easiest present solution to this is the mercury-arc valve. High-vacuum switching tubes and thyristors are under development.

53. General Design of Mercury-arc Valves. In low-voltage mercury-arc valves (Sec. **12**) the requirements of low voltage drop while conducting and controllability by grid control are fulfilled. The ability to withstand high voltages is, however, normally very unsatisfactory, because the whole reverse voltage is concentrated into a narrow region. Increasing the grid-anode distance is not a solution, because of the formation of a space-charge sheath at the anode surface during deionization over which the whole voltage is concentrated. By careful design and processing it may be possible to increase the voltage rating in such a valve to about 100 kV peak inverse voltage at small currents (10 A). To arrive at higher voltages at high currents, it is necessary to distribute the voltage stresses over a considerable region by inserting voltage-grading electrodes between the cathode and the anode (i.e., by dividing the voltage over several gaps).

The high-voltage valve is usually of the excitron type; cooling medium may be air, water, or oil; the valve may be pumpless or continuously pumped. Single-anode valves as well as multianode valves with the anodes in parallel have been constructed.

Tank and cathode design does not differ much from low-voltage valves described in Sec. **12**, Pars. **63** to **72**. The cooled condensation surfaces are, however, relatively large compared with low-voltage valves, to assure an accurately controlled mercury-vapor pressure. The main difference is found in the design of the anode assembly. Two different lines of approach are observed: the few-gap valve and the multigap valve.

In the former the voltage stresses per gap are very high. Such a valve will therefore be very sensitive to vacuum conditions and surface contaminations. The current-carrying capability may be high.

The multigap valve has much lower voltage stresses per gap and is not very sensitive to vacuum conditions and surface cleanliness. The tendency is therefore to use more parallel anodes on the same tank in a multigap valve.

To compensate for inferiority in voltage rating, two to three valves with few gaps have to be connected in series in each bridge arm.

54. The Few-gap Valve. A typical design of a valve with relatively few grading electrodes is shown in Fig. 14-25. The anode assembly, consisting of two control grids, four grading electrodes, and an anode cooled by vaporizing mercury, is mounted on an oil-cooled steel tank. The valve is of pumped construction, and coated aluminum rings are used as gaskets. As electrode material graphite, steel, or molybdenum may be used.

The high interelectrode capacitances in this valve may be sufficient to give the grading electrodes an acceptable voltage division.

The arc drop is about 35 V.

A six-pulse bridge of such valves may have the following rating:

Direct voltage, 50 kV.

Direct current, 300 to 900 A (depending on size).

So far the few-gap valve has not been used in commercial transmissions outside the U.S.S.R.

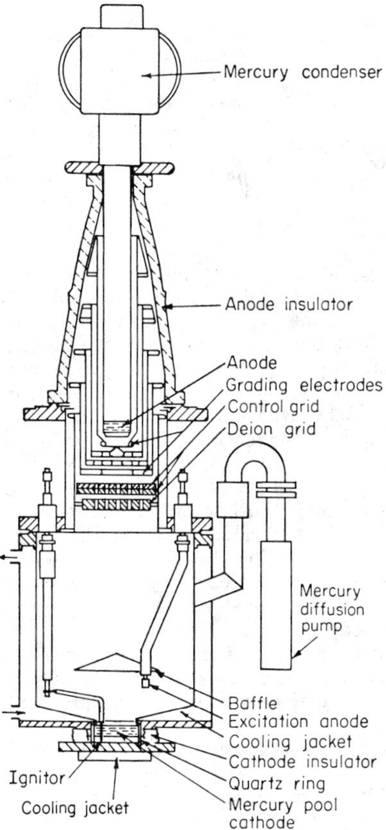

FIG. 14-25. Typical design of a few-gap mercury-arc valve for moderate voltages.

55. The Multigap Valve. The design of a valve with a high number of grading electrodes is shown in Fig. 14-26. The valve consists of a water-cooled stainless-steel tank on which several anode assemblies are mounted. The anode, grading electrodes, and control grid are mounted in a porcelain cylinder. The grading electrodes (about 20 in a valve for 125 kV) are connected to an external capacitive-resistive voltage divider, which gives the most favorable voltage control. The anode porcelain is sealed to the tank by a rubber gasket secured by a mercury seal. All other seals are of permanent type (kovar-glass-porcelain).

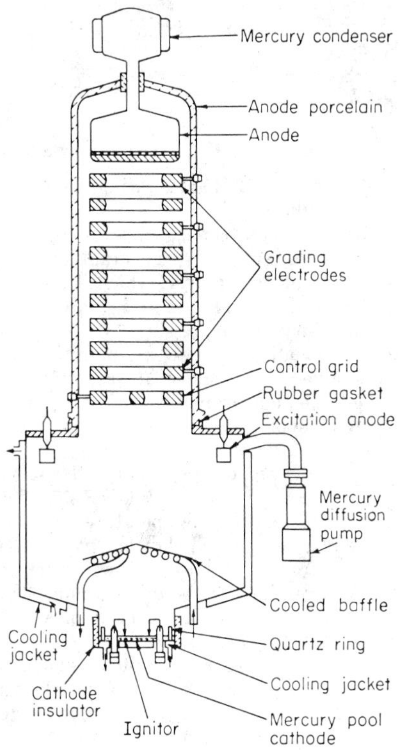

FIG. 14-26. Principal design of the multigap mercury-arc valve for voltages above 100 kV.

Arc voltage drop is about 50 V.

Typical rating for a six-pulse bridge of four-anode valves of this type is:

Direct voltage, 130 kV. Direct current, 1,200 A.

Higher ratings in current can be achieved by using more anodes on a common tank. There is an inherent possibility of developing the direct voltage further, and a rating of 200 kV is within reach. The multigap valve is practically the only type of high-

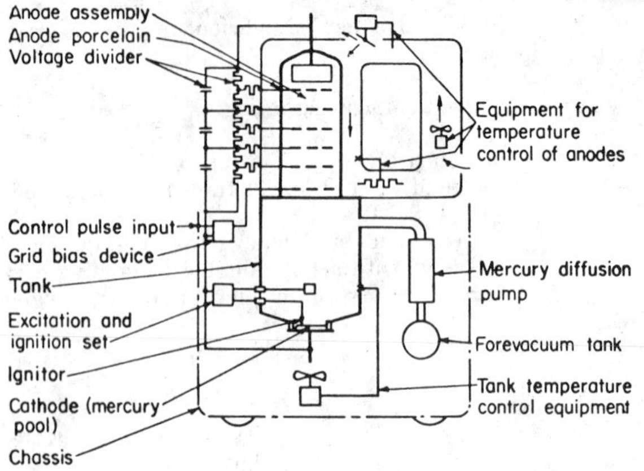

FIG. 14-27. High-voltage mercury-arc valve assembly.

voltage valve now used commercially.

56. Auxiliaries. The auxiliaries necessary for proper functioning are usually placed in a chassis together with the tank. Figure 14-27 shows a sketch of an H-V valve including principal auxiliaries.

57. Vacuum Equipment. Even though present-day sealing techniques permit construction of pumpless valves, it may be advantageous to work with pumped valves. The evacuating system described in Sec. **12**, Par. **72**, could be used. In Fig. 14-27 a somewhat different system, which could be called "semi-pumpless," is shown. In this design a mercury-diffusion pump with high backing pressure compresses the gas into a forevacuum tank acting as a "vacuum reservoir." This tank has to be pumped with a normal rotating forevacuum pump about once a year, which is done in conjunction with routine maintenance work.

The residual pressure in the valve should be at least one order of magnitude less than the mercury-vapor pressure 3×10^{-3} mm Hg. A hot-wire (Pirani) gage is normally used for monitoring the vacuum in the tank, and a simple mercury U-gage for monitoring the forevacuum (see Sec. **12**, Par. **72**).

58. Cooling Systems. It is essential that the temperature of different parts of the valve be kept within certain limits. The anode assembly must be considerably warmer than the tank, to avoid mercury condensation. In the valve shown in Fig. 14-27 this is achieved by a circulating hot-air system. At standby the air must be heated, while at load, when valve losses are high, the air has to be cooled by taking up air from outside. The temperature is controlled by a thermostat for the heater and a thermostat-controlled shutter. Normal air temperature is 80 to 100°C.

The tank cooling system which determines the mercury-vapor pressure can use air, water, or oil as the cooling medium. At moderate current ratings air cooling may be used. The valve is then usually provided with tank temperature-regulating equipment, as indicated in Fig. 14-27.

At higher ratings an indirect system with demineralized water in the closed loop is advantageous (Sec. **12**, Par. **72**). The temperature-regulating equipment is then placed at ground potential, and water at the correct temperature is supplied to the valves through insulating pipes.

The operating tank temperature for a few-gap valve is usually about 15 to 25°C and for the multigap valve in the range 30 to 40°C. The accuracy should be 1 to 2°C.

Table 14-4. Losses and Auxiliary Power for a Four-anode Valve, 125 kV, 1,200 A

	Power consumption at full load, watts	Power consumption at standby, watts
Arc drop	20,000	
Other circuit losses*	9,000	
Excitation equipment	1,500	
Vacuum equipment and misc	800	800
Anode air fan	2,500	2,500
Anode air heater	......	8,000
Total	33,800	11,300

* Loss in anode reactor, current divider, and voltage divider.

59. Ignition and Excitation Equipment. The valve in Fig. 14-27 is equipped with circuits for:

a. Ignition at start of the valve (new ignition).

b. Feeding excitation current.

c. Rapid reignition (if the cathode spot should extinguish suddenly).

The circuits for (*b*) and (*c*) are shown in Fig. 14-28.

The excitation current is supplied by a transformer connected to the two excitation anodes in a two-pulse single-way connection. The transformers are designed as leakage-flux transformers, thereby securing a considerable overlap between the anode currents (conducting interval about 280 elec deg).

Two possibilities for rapid reignition may be arranged:

Should the arc spot transfer to the tank, the high-voltage drop in the nonlinear resistor caused by the main current will produce a reignition by igniter 1.

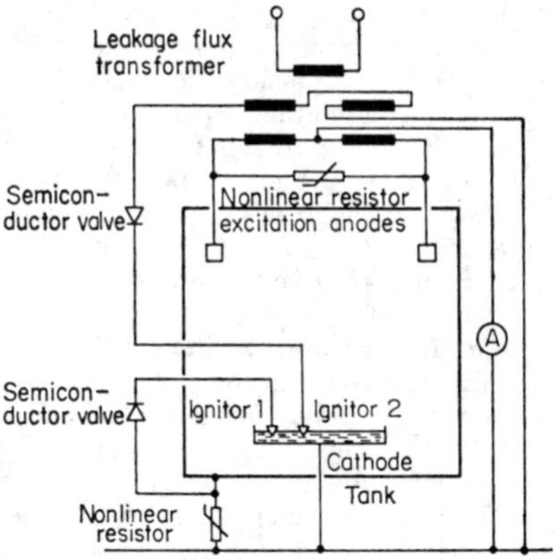

FIG. 14-28. Excitation and reignition circuits in a high-voltage mercury-arc valve.

Should the excitation current suddenly become extinguished, the leakage flux in the transformer will drive a current through the igniter 2 and start a new cathode spot.

60. Grid-biasing Device. This produces a negative direct voltage which is impressed on the control grids. Upon this the positive control-grid pulses are then superimposed. These firing pulses are transmitted to the valve from ground potential by an isolating transformer or by a light-beam-transmission system.

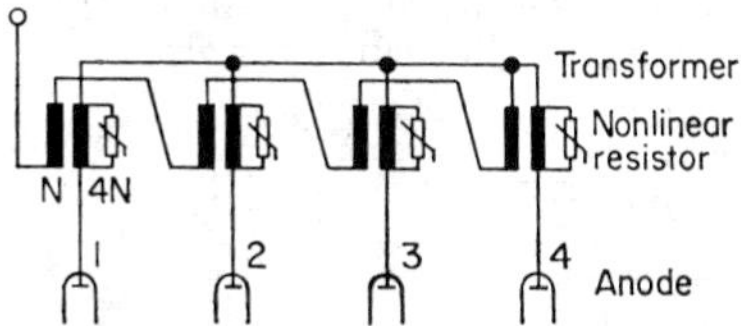

FIG. 14-29. Current-divider circuits for a four-anode high-voltage mercury-arc valve.

61. Current Divider. For a multianode valve it is necessary to use a current divider, which is used to ensure that all parallel anodes pick up at the same instant and that the current is equally shared between them. Figure 14-29 shows the circuit for a valve with four parallel anodes.

62. Valve Losses and Auxiliary Power Consumption. Typical figures for losses and auxiliary power demand (for a four-anode valve rated 125 kV, 1,200 A in six-pulse connection) are given in Table 14-4. Of these losses about one-fourth to one-third goes to the tank cooling medium, while the rest is dissipated into the ambient air in the valve hall.

63. Valve Disturbances. As in low-voltage valves, arc-backs may occur randomly. An arc-back rate of 2 to 4/(year)(valve) seems to be reachable. For a given design the arc-back rate is influenced by the following operating factors: (*a*) rate of change of current at the end of the commutation interval; (*b*) rate of rise of and (*c*) magnitude of reverse voltage; (*d*) mercury vapor pressure. Other disturbances that can occur are: arc-through, in which the grid fails to block the valve at positive anode voltage, and misfire, when it fails to fire the anode in spite of a correct firing signal. These disturbances are largely avoidable by careful design of the valve and of the grid control circuits.

An arc-back means that a short-circuit current appears not only in the faulty valve. The current is increased to short-circuit value also in the valve carrying forward current at the arc-back instant, which is usually an order of magnitude greater than the normal current. This might under certain conditions occasion an arc-back in this valve when reverse voltage is applied. This phenomenon, which entails a permanent phase-to-phase a-c short circuit and must be cleared by circuit-breaker operation, is called a consequential arc-back (CAB).

Arc-through can appear in inverter operation if the margin of commutation is too small. The minimum commutation angle ("commutation margin") is determined by the deionizing properties of the anode assembly and accuracy of grid control. A typical value for the deionization of a multigrid valve is about 5 elec deg. A normal operating value of the margin of commutation is about 15 elec deg.

64. Valve degassing of high-voltage valves is performed in essentially the same way as for low-voltage valves described in Sec. **12**, Par. **69**.

65. Valve Maintenance. Routine checking of auxiliary voltage dividers and tightness of the valve, etc., should be carried out at least once every year.

Owing to the intense ion bombardment during the deionizing interval the electrodes suffer from wear. Therefore some of the electrodes must be replaced at certain intervals. For a multigap valve this interval is 5 to 10 years, while it is expected to be considerably shorter for a few-gap valve.

66. High-vacuum Switching Tubes. Standard high-vacuum tubes (triodes) have too high anode-to-cathode voltage to be considered for H-V d-c transmission. There are switching tubes using the magnetron principle that have considerably reduced, but still unacceptably high, losses. Tubes with sufficiently high ratings for H-V d-c application have not yet been developed.

67. Semiconductors. Series connection of thyristors (SCR, or semiconductor-controlled rectifiers) to achieve a high enough withstand voltage may be a possible solution in the future. Thyristors for a few kilovolts inverse voltage are available. The complexity in control systems and in voltage division in series-connecting a great number of thyristors is a serious problem, but acceptable solutions are within reach. The total valve losses will be higher for a thyristor arrangement than for mercury-arc valves. The thyristor valves of today are not economically competitive with the commercial mercury-arc valves for higher ratings of d-c terminals. Further development of the thyristor itself to achieve higher voltage and current per thyristor is necessary before it will replace the valves now in use. It seems possible, however, that thyristors may become practical in the near future for lower power and voltage applications, where the cost per kilowatt is comparatively high for mercury-arc valves.

68. Reference to Literature on High-voltage Valves

Lamm, U. Mercury-arc Valves for High Voltage D.C. Transmission; *IEE Paper* 4538 P, *Proc. IEE (London)*, 1964, Vol. 111, No. 10, pp. 1747–1753.

TERMINAL DESIGN

69. Main Circuit Arrangements. A d-c line terminal consists of converters acting as rectifiers at the sending end and as inverters at the receiving end of the transmission. The converters, in principle, are series-connected on the d-c side of the valve groups and parallel-connected on the a-c side of the converter transformers (see Figs. 14-30 and 14-31). Thus, the d-c line potential is built up by a number of converter voltages and is therefore a multiple of the direct voltage rating per valve group.

When the terminal is feeding a bipolar line with ground return as spare conductor (Fig. 14-30), the circuits should be arranged in such a way that the two poles can be operated independently of each other. An outage of one pole, or equipment allocated to that pole, should not restrict the use of the other. As far as possible, this design philosophy should apply to each converter unit.

70. Connection to Generators. If the d-c line terminal is located at the point of power generation, the generators can be connected directly to the converter transformers, as shown in Fig. 14-30. A common generator bus bar, to which converter transformers are connected, can also be arranged, provided that the fault current can be handled. For economical reasons, double transformation between generators and valve groups should be avoided.

71. Connection to an A-C System. The converter transformers can also be connected to a high-voltage a-c system, in which case a-c harmonic filters are installed in order to absorb harmonic currents generated by the converters (see Fig. 14-31).

A minimum of three to four times transmitted power is required for proper regulation of the converters. For a-c systems not meeting this requirement, additional short-circuit power can be obtained through the use of synchronous condensers instead

Table 14-5. Typical Approximate Auxiliary Power Requirements

Supply to	Power for converter ratings 20–200 MW		Accuracy in voltage	Reliability
	KVA	KW		
Grid timing.	....	....	Ratio ±3%	1st grade
Per converter.	1	1	Angular ±3°	
Grid control.	....	....	Normal ±5%	1st grade
Converter control per pole.	2	0.2	Transient −15%	
Regulation and measuring per pole	10	0.5	Normal ±5% Transient −15%	
Valve auxiliary.	....	....	Normal ±5%	1st grade
Per valve.	8–20	5–16		
Cooling, ventilation.	....	....	....	2d grade, partially 1st grade
Per converter.	20–200			
Degassing.	100–1,000	....	....	2d grade

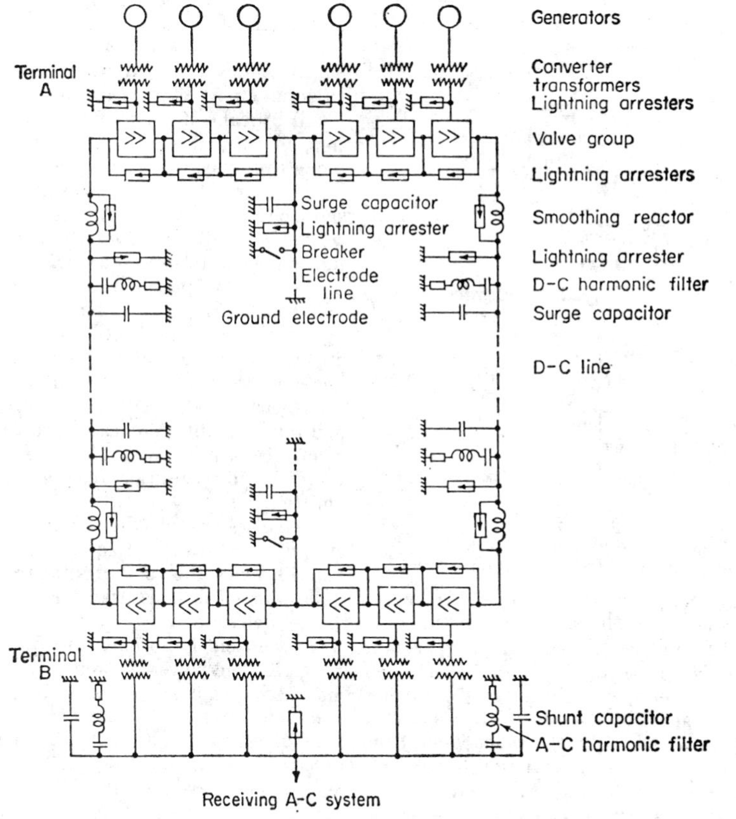

Fig. 14-30. A transmission arrangement with the sending end at point of generation, the receiving end connected to an a-c system via a bipolar d-c line, and ground return used as a spare conductor.

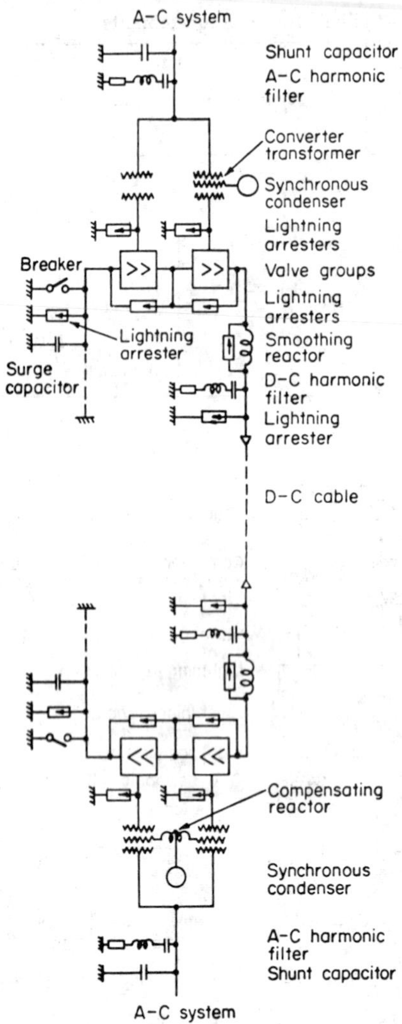

FIG. 14-31. A d-c link between two a-c systems, with one cable and ground return.

of static capacitors for phase compensation.

72. Reactive-power Compensation. The reactive-power demand of an H-V d-c converter station is 50 to 60% of the active power during normal operation. Synchronous condensers, static capacitors, or a combination of both may be used for power-factor compensation. The synchronous condensers may be needed when the converter station is working on an a-c system having a comparatively low short-circuit power level. The condensers can be connected to a tertiary winding of the converter transformers. At the inverter end, however, the converters can mutually have an adverse influence on their operational condition owing to a common commutating reactance, for instance, in a synchronous machine. It would therefore be necessary to connect the machine through a compensating reactor to the tertiaries of the converters as shown in Fig. 14-31.

73. A-C Filters. It is often convenient to supply part of the reactive-power demand by capacitors, combined with reactors, to form filters for the current harmonics produced by the converters. The largest current harmonics will then be short-circuited by these filters, instead of traveling into the a-c network, where interference with telecommunication systems or resonance phenomena might arise. With such filters, the alternating voltage at the filter connection point can be regarded as sinusoidal, and only the reactances between this point and the valves take part in the commutation process. With filters, interference between series-connected valve groups is avoided, and therefore compensation reactors may be omitted.

Separate tuned filters for the lower current harmonics at frequencies 5, 7, 11, and 13 times the fundamental frequency are normally installed. For higher harmonics, a damped filter is used. Figure 14-32 indicates a normal arrangement of filters on the a-c network.

The total reactive-power generation from these filters is normally 10 to 20% of the active power of the d-c link. The remaining reactive-power demand may be produced by shunt capacitors or synchronous condensers or a combination of both. The quality factor of the tuned filter branches is adjusted by series resistances to values between 50 and 150, the exact value being determined by the a-c network frequency variation, component variation because of temperature variation, tuning facilities, etc.

In addition to these basic converter harmonics, other harmonics, for instance, three, four, eight, and nine, can arise under certain network conditions. These can be caused by nonlinear transformer characteristics, asymmetries in the d-c control system, etc., and can be noticeable through resonances arising in the a-c network. By careful symmetrizing of the a-c system and accurate adjustment of the valve firing intervals, these harmonics can normally be held within acceptable limits.

The a-c filter can be connected directly to the a-c network or on tertiaries to the converter transformers. The capacitor units must be designed to withstand the current harmonics produced by the converter, plus their own capacitive current.

So that the tuning of these filters during normal operation can be checked, special equipment is installed by which the phase angle between harmonic current and harmonic voltage for each tuned filter may be measured.

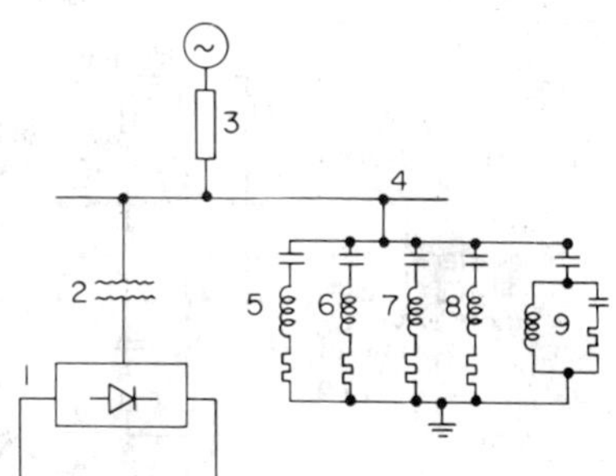

Fig. 14-32. Diagram showing connections of a-c filters. (1) Valve groups; (2) converter transformers; (3) network impedance; (4) a-c filter block; (5) fifth-harmonic filter; (6) seventh-harmonic filter; (7) eleventh-harmonic filter; (8) thirteenth-harmonic filter; (9) high-pass filter.

74. Special Requirements for Generators and Synchronous Condensers. Machines feeding converters should be designed with due consideration to the current harmonics in the armature windings, which in turn produce additional currents in the dampers of the rotors. The machine reactances should preferably not vary in different axes, so that constant commutation reactance at varying control angles can be maintained.

75. Converter transformers are basically designed as power transformers for high voltages. The valve winding of the converter transformer, in operation, has a d-c potential to ground, which must be considered in building up the insulation of the winding. Current harmonics, giving eddy-current losses, may result in windings with more parallel parts than normal. Arc-backs in the valves give short-circuit duty with a frequency higher than on normal power transformers, and the converter transformers must be designed to withstand many such faults without damage.

76. Valve Groups. The valve group itself consists of six high-voltage valves connected in two-way, six-pulse connection (Fig. 14-33). The use of two or more series-connected valves instead of a single valve in each leg has been contemplated. Such an arrangement has many disadvantages and complications and is considered to be undesirable.

A bypass valve is connected between the d-c terminals of each valve group. In the event of disturbance in the valve group, all six main valves are blocked, and the direct current is transferred to the bypass valve. After a period of time of about 0.5 to 1.0 s, during which the main valve has had time to recover, the direct current is automatically transferred back to the main valves. When the main valves are out of service for a longer period, for instance because of maintenance work on the valves or a persistent fault in the valve group, the bypass switch is closed and the isolators can be opened.

77. Damping Circuits Connected to the Valve Groups. An anode damping circuit (see Fig. 14-33), consisting of an anode reactor in parallel with a damping resistor, is connected on the anode side in series with each valve. This circuit will suppress the oscillations in the anode current at the instant of current pickup in the valve. If the current oscillations are not sufficiently damped, they may cause an extinction of the cathode spot, thus necessitating the restarting of the excitation current.

The main purpose of these damping circuits is to prevent such a disturbance, but they also reduce the radio noise transmitted from the conductors connected to the valves. If a low radio-noise level is demanded, it may be advisable to put additional high-frequency reactors in the conductors in the cathode side of the valve. Instead of air-cooled reactors without iron cores, the cathode conductor can pass through rings of magnetic material, such as ferrite, or ring cores of thin transformer plate.

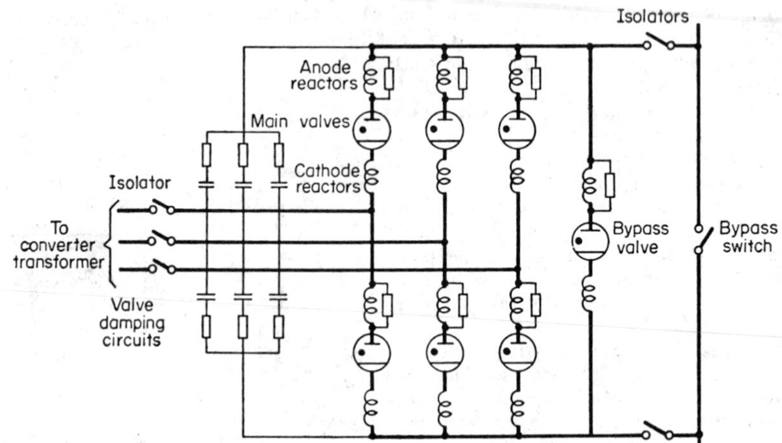

FIG. 14-33. A mercury-arc valve group.

In order to reduce the overswing in the recovery voltage across the valve, valve damping circuits are installed. These circuits reduce the steepness of the recovery voltage, which also gives a more favorable working condition for the valve.

78. Smoothing and Filtering Arrangements on the D-C Side. In the case of both an overhead d-c line (Fig. 14-30) and a d-c cable (Fig. 14-31), a reactor is needed for smoothing the outgoing direct current and for damping oscillations in the direct voltage and direct current in connection with disturbances. In addition, special filtering circuits may be installed in order to reduce telephone interference from the d-c line. These are arranged as damping circuits between the d-c line and earth where a capacitor is the main component. The capacitor units must have internal or external resistive voltage dividers to distribute direct voltage equally between units.

79. Overvoltage Protection. In present HVDC transmission, conventional lightning arresters are connected to the a-c bus bar and the valve side of the converter transformer, normally both between phases and to ground. Another lightning arrester protects the smoothing reactor. Special arresters, self-extinguishing against direct voltages, are connected across the d-c side of the valve group and between the d-c line and earth. Overhead lines may also need a surge-protection capacitor per pole (Fig. 14-30) and some kind of line-end protection (overhead ground wires).

The electrode line with a comparatively low voltage level needs special protection equipment. In addition to a spark gap, a surge-protection capacitor is connected between the electrode line and earth at the station. When a spark-over in this gap occurs, for instance, due to damage of the electrode line, a short-circuit device is automatically closed. Normal operation can therefore continue for a short period, in spite of a failure on the electrode line.

80. Grid Control Equipment. The function of the grid control equipment is described in Pars. **114** to **128**. The semiconductor technique is fully employed in the grid control equipment. Considerations due to the requirements set up by the valves on the pulse firing equipment have led to the use of rectangular grid pulses with the same length as the ideal load-current interval, that is, 120° during steady-state conditions.

The control pulse is added as a positive voltage to a negative bias voltage produced in each valve. Thus, when no pulses are generated, the valve is effectively blocked for new ignitions. Owing to the design of the control pulse generator, the pulse generation is instantaneously interrupted or started for all main valves of a group when a blocking or deblocking order is provided. This action is a condition necessary for the successful control of certain disturbances, especially arc-backs.

As the grid pulses have to be supplied at the high potential of the valve but the generation of the pulses is performed at ground potential, some kind of isolating and transmitting device must be used. This device may be an isolating transformer or a light beam with a photocell and amplifier at valve potential.

81. Protection. The converter transformer is protected against overcurrent by use of relays in the three a-c phases. This protection is also a backup for the arc-back protection of the different rectifier groups.

A harmonic voltage protection for the d-c side will prevent the equipment in the d-c side, e.g., the smoothing reactor, from being overloaded as a result of alternating currents of different frequencies arising in the event of a failure of one or more main valves or because of asymmetry in the control system or the network.

The differential protection acts in the event of an earth fault occurring in the d-c switchgear, and its object is to give an alarm or to trip the faulty pole, depending on the magnitude of the differential current. The differential protection is based on the comparison of the direct current in the outgoing line and in the midpoint of the station.

The arc-back protection equipment compares the alternating current to the valve group with the outgoing direct current. From the a-c side of the groups an arc-back can be looked upon as a short circuit. This means that, at arc-back, a large alternating current occurs without any increase in direct current. The increase of the alternating current in relation to the direct current causes the protection to block the group and deblock it again within 0.5 to 1.0 s.

The commutation-failure protection also compares the alternating current and the direct current. At commutation failure the direct current will, during certain intervals, pass through the bridge without flowing through the converter transformer. Because of the decreasing alternating current, while the direct current remains constant, a persistent commutation failure blocks the group and deblocks it again within 0.5 to 1.0 s. Furthermore, some of the individual control units are protected against overcurrent, voltage drop, etc.

82. Auxiliary Power. A d-c terminal must be supplied with auxiliary power of high reliability. The requirement for different kinds of auxiliary power is described and summarized in Table 14-5. Operation is normally blocked for supply voltages below 60% of normal. Deblocking is delayed about 1 s after the restoring of the voltages, to allow for proper valve ignition.

83. The grid timing supply voltage, which is the basis for the correct timing of the valve ignition, must give an exact image of the commutating voltage in the a-c system, undisturbed by converter operation (see Par. **60**). The voltage is supplied from the a-c side via magnetic or capacitive voltage transformers to the grid control equipment.

84. Auxiliary power for the grid control equipment must be free from switching surges and therefore should be supplied from the valve supply, or motor-generator sets in case the auxiliary power for the valves is fed from the a-c net.

85. Auxiliary Power for the Valves. Power for excitation, grid biasing, and temperature regulation is fed to the valves on high potential via isolating transformers or rotating transmissions. The power is preferably supplied from auxiliary power generators directly connected to synchronous compensators or to main power generators, the inertia of which provides undisturbed voltage for continuous excitation of the valves even during fault conditions.

86. Auxiliary power for cooling of transformers, reactors, and resistors, as well as for ventilation, control equipment, and the degassing process, must be supplied separately to prevent disturbances in this system from extending to the grid control and valve supply systems.

For equipment with short time constants, standby units for the cooling are normally required to prevent cooling-equipment failures from interrupting the operation.

87. Ventilation of Terminal Building. Typical loss-distribution and ventilation requirements are shown in Table 14-6.

The high direct voltage supplied to the equipment in the valve hall causes airborne dust to pollute insulating surfaces of the equipment, the walls, roof, etc. Surface treatment must therefore be suitable for easy cleaning. In addition, ventilation air has to be fairly free from contamination, and the tabulated values for cleanliness are estimated to give reasonable time intervals between each maintenance and cleaning.

Table 14-6. Typical Loss Distribution and Ventilation Requirements for Terminal Building

	Valve hall per 125-MW group		Degassing room, 3 valves	Control room	Clean workshop	Assembly room
	Stand-by	Full load at 35°C hall temperature				
Losses to Room, kW Anode system, kW Tank system, kW	 60 0 15	 110 45 70	 25 25 40	 ∼5		
Air temperature, °C: Maximum, 12 h/year Maximum, months Maximum, cont.	 45 35 25		 35	As convenient to the staff Supply of room ventilation air to grid-control cubicles	 22	 22
Air: Dust and salt content average, mg/m³	<0.02		<0.5		<0.02	<0.5
Type of filtering	Oil, dry, or electrostatic, as convenient		As convenient		Dry	As convenient
Air: Relative humidity, %	<100		<100		<50	<100
Average mercury content, mg/m³	<0.07		<0.07		<0.07	<0.07
Pressure	Overpressure		Overpressure	Overpressure	Highest overpressure	Overpressure

The clean workshop, where the interior parts of the valve are exposed to the air during maintenance and assembly, must be dust-free and the room designed for easy everyday cleaning.

88. Cooling of Valves. See Pars. **62** and **87** regarding valve losses and Sec. **12**, Par. **72**, regarding arrangement of valve-cooling equipment.

As the seven valves in a converter are combined as a common operating unit, part of the coolers, heaters, temperature-control circulation pumps, and fans can become common to one converter and installed at ground potential.

If air is used as the temperature-regulation medium, the inner and outer surfaces of the insulating ducts from ground to the valve platforms should have a creepage distance of 1.4 to 1.8 cm/kV. If water is the medium, the resistivity of the water can be the decisive factor. For the insulating plastic tubes a leakage current of 1 mA/tube is acceptable, which, at 250 kV d-c and water resistivity of 10^6 Ω·cm, may result in a 40-ft tube length, depending on the tube area required. The water system has to be built of noncorrosive material such as stainless steel or plastic and the water continuously demineralized during circulation.

89. Insulation Levels. The series connection of the converters on the d-c side between d-c line and ground requires a graded phase-to-ground insulation of the converter equipment. Only the converter next to the line must have an insulation corresponding to the full d-c line voltage above ground. The remaining converters require gradually decreased insulation to ground. The insulation between phases is related to the converter voltage and is equal in all converters. This gives different requirements for phase-to-phase and phase-to-ground insulation in the converters.

90. Clearances. The selection of clearances is based primarily on the short-wave (for example, 1.5/40 μs) impulse-withstand value, and the same clearance values as in a-c installations are applied.

The difference between impulse test voltage and the related rated voltage normally decreases with increasing voltage. For test levels above 1,000 kV the withstand values of the clearances selected must be controlled with respect to the long wave (for example,

100/2,000 μs) and peak direct, or peak of mixed alternating- and direct-voltage values appearing.

In indoor installations, where space is expensive and clearances are minimized, improvement of air-withstand capability may be obtained by the increase of surface radius of the equipment.

91. Creepage Distances on Insulators. Creepage distances for the outdoor switchyard support and suspension insulators are selected according to the same rules as for the line insulators (see Pars. **13** to **21**).

Indoors, in the valve hall, the electrostatic field leads to an accumulation of dust on the insulators, as fully purified ventilation air is not continuously supplied. However, losses in the equipment normally create temperatures which prevent the development of dew on the insulators. Creepage distances can therefore be selected between 1.4 and 1.8 cm/kV operating voltage.

92. Arrangements for Radio-interference Suppression. The valve ignition gives rise to high-frequency electromagnetic radiation from the valves. The corresponding change of charge in the stray capacitances of the equipment and connections introduces high-frequency oscillations in the system. The frequency spectrum is 0.1 to 10 Mc/s.

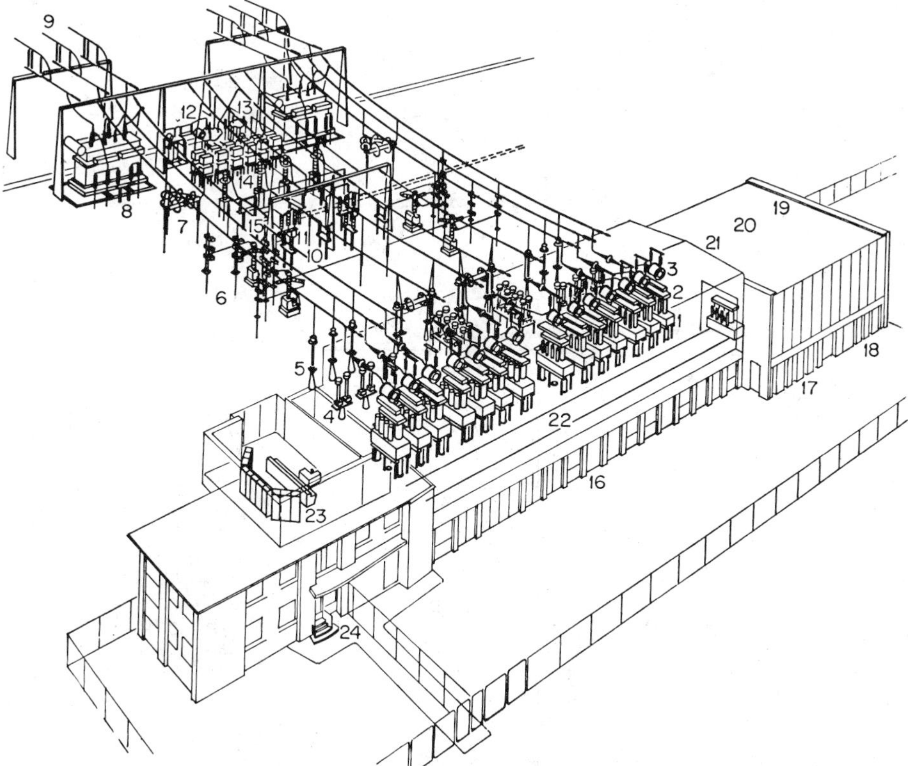

FIG. 14-34. General arrangement of a converter station comprised of two converter groups. (1) Valve; (2) current divider; (3) anode reactor; (4) isolating transformers for grid pulse and valve auxiliary power; (5) voltage divider (omitted in modern equipment except for grid control); (6) damping equipment for the valve group; (7) lightning arrester; (8) main transformer; (9) a-c switchgear apparatus; (10) isolator for converter group; (11) bypass switch. Additionally, the converter station contains the following d-c pole equipment: (12) smoothing reactor; (13) impedance for d-c damping circuit; (14) capacitor bank for d-c damping circuit; (15) isolator for d-c line.

Possible interference with radio communication can be reduced by:

a. Selection of a valley as site for the plant.

b. Designing of the valve hall as a screen for electromagnetic radiation.

c. Limiting the length and height of conductors in the converter switchyard and installing ground wires over the switchyard.

d. The selection of a switchyard insulator type which gives limited contribution to the radio-interference level.

These precautions should be considered in designing the plant, as they will be unnecessarily expensive to introduce at a later date.

The radio-interference level can easily be brought to a level of about 50 dB above 1 μV/m at a distance of 1,000 ft from the station boundary.

If extremely low radio-communication signals are received in the neighborhood of the station, further reduction of the radio-interference level emanating from the station can be obtained by cathode reactors, skin-effect dampers, *RC* circuits between conductors and ground, filters in outgoing conductors, and screening of the converter switchyard, whichever proves to be the most efficient solution in the specific case.

93. Practical Arrangements for Valve-hall Screening. Sufficient screening effect can be obtained if the valve-hall walls and roof are constructed of sheet steel or sheet

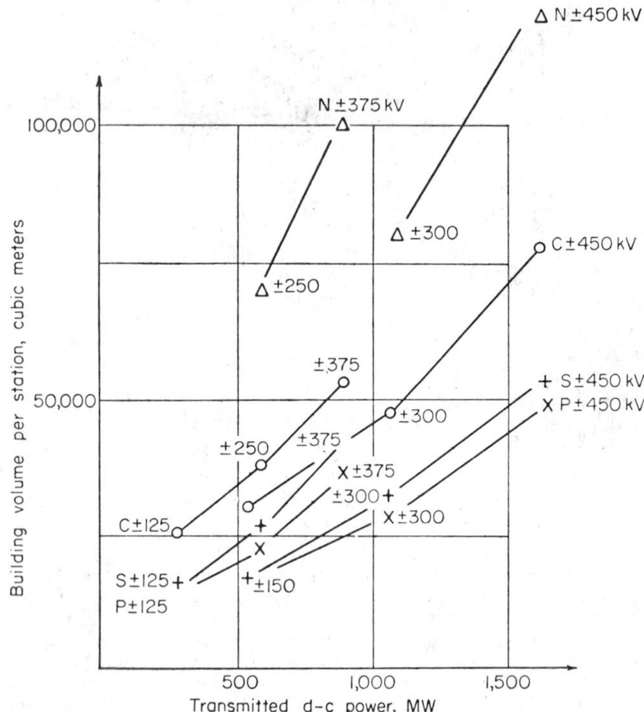

Fig. 14-35. Approximate building volume requirements for converter stations.

aluminum, with a minimum thickness of 0.5 mm, and combined with a floor concrete reinforcement of 5 × 5 cm mesh. As an alternative to the sheet metal, a 2 × 2 cm mesh wire with a minimum 0.1 cm wire diameter can be used.

Perfect electrical contact between the sheets and mesh parts is secured by overlapping 5 cm, and welding every 30 cm, along the edges. The mesh should be of welded or galvanized type to allow good contact in every cross point. The screen should be well grounded in many points around the building and connected to the station grounding system. No unconnected metal parts should be allowed to penetrate the screen. The damping factor obtained is 40 to 50 dB.

94. Converter-station Layout. Figure 14-34 shows the general arrangement of a converter station of two converter groups, and numbers given in the text below refer to this figure.

Means for reactive-power compensation, such as synchronous condensers, or capacitor banks with harmonic filters, are often installed in the same plant.

The valves are designed for erection indoors in a valve hall (16). Requirements on steel structures for suspension of anode reactors, current dividers, and bus bars, plus shielding requirements for radio harmonic suppression, along with necessary improvement of valve temperature control and protection against pollution for outdoor valve design, make the outdoor alternative less attractive. Accommodation should also be arranged for valve cooling (17) and ventilation equipment (18).

Large multianode mercury-arc valves of semipumpless type are preferably transported to the site in sections and assembled and processed in accommodations there. The anodes are mounted on the valve tank in a "clean" workshop (19), and exterior parts are arranged on the valve in an assembly room (20).

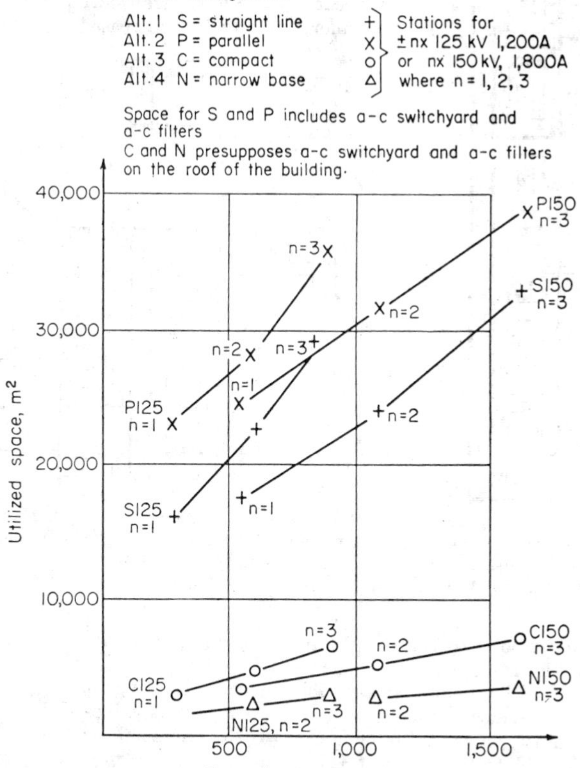

FIG. 14-36. Approximate space requirements for converter stations.

The valves are degassed in a special degassing room (21). Equipment is installed for independent degassing of three to six valves simultaneously. Temporary degassing facilities can be arranged in the valve hall.

Transport between the maintenance area and the insulated valve platforms in the valve hall is secured via the transport passage (22) and temporary bridges to the valve platforms.

Even if the plant can be operated unattended, a local control room is arranged (23).

Where staff is required for the operation, offices and living quarters can be arranged in the station (24).

95. Space and Building Volume Requirements. The arrangement of the equipment can meet different demands on site. The following four alternatives are the basis for space and building volume requirements given in Figs. 14-35 and 14-36.

1. If sufficient premises are available at low cost, an arrangement with the converter groups side by side, as in Fig. 14-37, the transformers forming one line and the valves another, gives the most accessible layout, with the equipment of the same type placed together simplifying the installation and maintenance arrangements. The size of the valve itself is not critical since, normally, more than 95% of the volume in the valve hall is required for insulation, conductors, and apparatus other than valves and current dividers.

For instance, a 75% reduction of the valve volume would result in less than 30% reduction of the volume required for valves and related insulation.

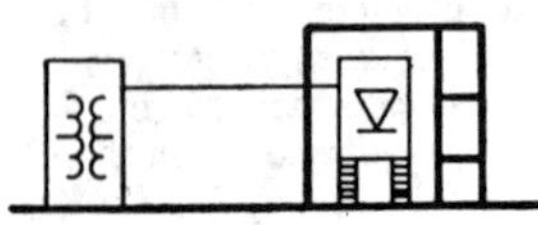

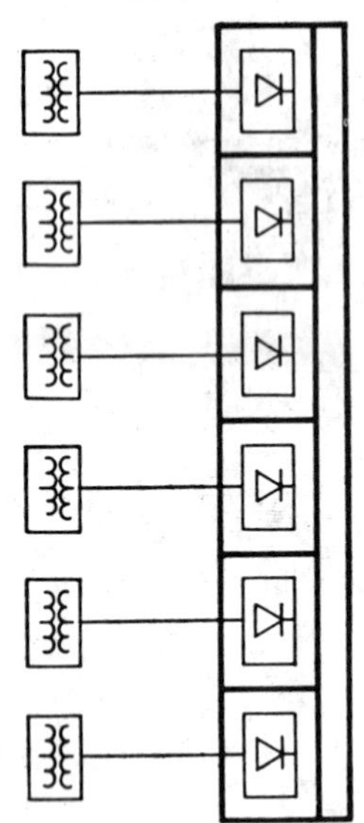

Fig. 14-37. In-line arrangement of converter terminal, with transformers outside in one line and the valves inside in one line.

2. If it is difficult to find land for a long building on an even level, the valve groups can be arranged on both sides of the valve transport passage. This gives a smaller building but increased switchyard area, as this has to extend to both sides of the buildings and allow for interconnection of the two sides (see Fig. 14-38).

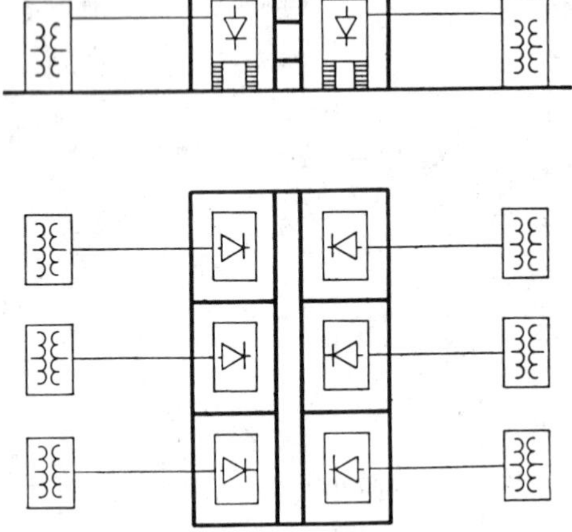

Fig. 14-38. Arrangement of converter terminal with transformers outside on two sides of the building, with the valves inside.

3. High cost of land, limited available space, and stringent requirements for low radio harmonic levels lead to a compact arrangement with valves and converter switchgear indoors, transformers still outside close to the valve-hall wall, and a-c switchgear and possibly a-c filters at different levels on the roof. Converter transformers and valves remain on one level (see Fig. 14-39).

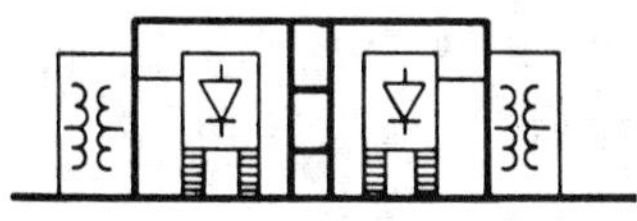

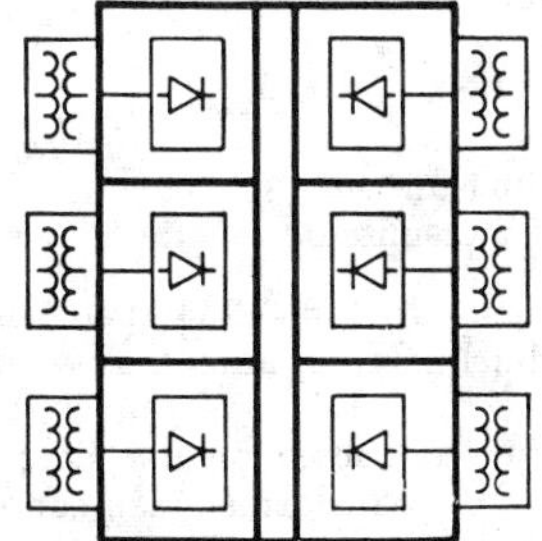

Fig. 14-39. Compact arrangement of converter terminal, with transformers close to the walls and valves and d-c switchgear inside.

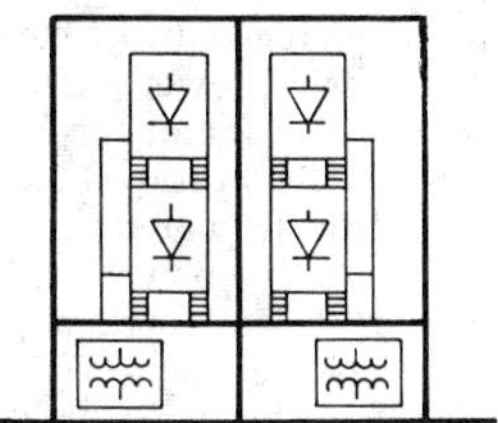

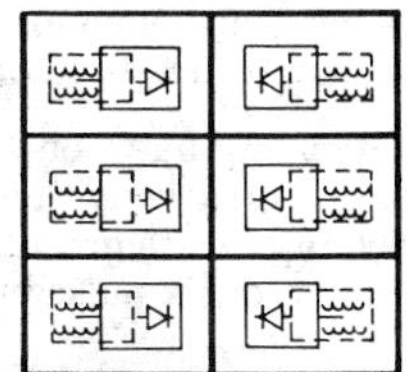

Fig. 14-40. Arrangement of converter terminal with both transformers and valves inside on different levels.

4. If space limitations are excessive or land costs are extremely high, the plant can be built multistory, with transformers and d-c reactors on the ground level, valves and converter switchgear on two higher levels, and a-c switchgear and filters on the roof. A disadvantage is that access for maintenance and exchange of equipment will be more complicated (see Fig. 14-40).

The sound basis for alternatives 1 and 2 has been proved in practice. Alternatives 3 and 4 are fully feasible, according to present experience.

96. References to Literature on Terminal Design

Lidén, I. The Design of the D.C. Connection across the English Channel; *Jour. ASEA*, 1958, Vol. 31, No. 6, pp. 70–74.

Engström, G. D.C. Power Transmission: Operation and Control; *Elec. Jour.*, 1960, Vol. 164, pp. 1048–1055.

Engström, G. Operation and Control of H.V.D.C. Transmission; *Trans. IEEE Paper* 1963-83.

Engström, G., and Stackegård, H. High Voltage Direct Current Design Considerations; *Transmission and Distribution*, 1965, Vol. 17, No. 8, pp. 30–34.

Lidén, I., Svidén, S., and Uhlmann, E. The Gotland D.C. Link: The Layout of the Plant; Pt. I, *Direct Current*, 1954, Vol. 2, No. 1, pp. 2–7; Pt. II, *Direct Current*, 1954, Vol. 2, No. 2, pp. 34–39; also published in *Jour. ASEA*, 1954, Vol. 27, No. 10, pp. 141–154, and *ASEA Pam.* 7434E.

CONVERTER LINKS BETWEEN A-C AND D-C NETWORKS

The following symbols and definitions are used in this discussion:

N = number of six-pulse bridges, all of them series-connected on d-c side; primary windings of N transformers are connected to 3-phase a-c network

E = rms value of phase-to-phase fundamental wave voltage in 3-phase a-c network

I = rms value of fundamental wave current in a-c network

E_d, I_d = mean value of direct voltage and direct current, respectively, in d-c line

E_0 = a-c network voltage behind transient network reactance

E_{00} = reference alternating voltage

E_{d0} = ideal no-load direct voltage

I_{d00} = reference direct current
P = active power passing through converter station = E_dI_d
$\cos\varphi$ = power factor for fundamental wave
k = current ratio = $1.35Nm$
m = transformer ratio
Q = reactive power
α = rectifier control angle
γ = inverter margin of commutation
d = transformer's direct voltage drop per unit (see Sec. **9**)
Q_k = short-circuit capacity of a-c network
Q_c = capacity of capacitors
R = resistance of total d-c loop
e_{dr} = per unit value for direct voltage, including line losses
p_r = power measured on inverter side of the transmission

97. Relationship between A-C and D-C Quantities. In Fig. 14-41 is shown a schematic diagram for a converter station feeding a d-c line from an a-c network. The following definitions are given:

N = number of six-pulse bridges, all of them series-connected on d-c side; primary windings of N transformers are connected to 3-phase a-c network

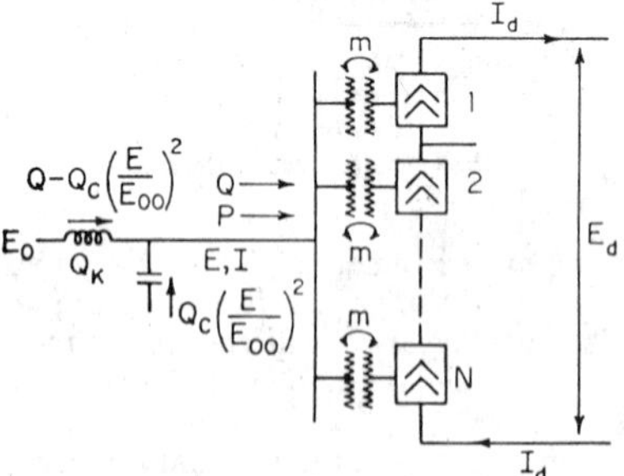

Fig. 14-41. Schematic diagram for converter station feeding a d-c line from an a-c network.

E = rms value of phase-to-phase fundamental wave voltage in volts in 3-phase a-c network

I = rms value of fundamental wave current in amperes in a-c network

E_d, I_d = mean value of direct voltage and direct current, in volts and amperes respectively, in d-c line.

Assuming that there are no losses in the terminal, the power P passing through the converter station is

$$P = \sqrt{3}EI\cos\varphi = E_dI_d \qquad \text{(watts)} \qquad (14\text{-}1)$$

$\cos\varphi$ being the power factor for the fundamental wave. It can be shown that a very good approximation is obtained by putting

$$\frac{I}{I_d} = \frac{k}{\sqrt{3}} \qquad (14\text{-}2)$$

where $k = 1.35Nm$, and where m is the transformer ratio, which may be adjusted by tap changers. The factor 1.35 is valid for six-pulse bridges (see Sec. **12**).

From the above equations is derived

$$\cos\varphi = \frac{E_d}{kE} = \frac{E_d}{E_{d0}} \qquad (14\text{-}3)$$

kE being the ideal no-load direct voltage E_{d0} (see Sec. **12**). The power factor is dependent on the control angle of the converters. In this way, the converter links between the a-c and d-c systems are represented by a current ratio k defined by the number of valve bridges N and the transformer ratio m and by a voltage ratio adjustable via grid control. The power factor depends only on the ratio between direct voltage and direct no-load voltage. The difference between these two voltages may be caused by inherent reactive voltage drop, by control-angle delay, or by both.

98. Active and Reactive Power in the A-C Network. A d-c terminal with the direct voltage E_d and the direct current I_d loads the a-c network by

$$\text{Active power } P = E_dI_d \qquad \text{(watts)} \qquad (14\text{-}4)$$

$$\text{Reactive power } Q = P\tan\varphi = E_dI_d\sqrt{\left(\frac{kE}{E_d}\right)^2 - 1} \qquad \text{(vars)} \qquad (14\text{-}5)$$

99. Power-factor Limits. According to Eq. (14-3) and Sec. **12** we have

$$\cos \varphi = \frac{E_d}{kE} = \cos \alpha - d \cdot \frac{I_d}{I_{d00}} \cdot \frac{E_{00}}{E} \tag{14-6}$$

in which α is the rectifier control angle. For inverter operation, α is replaced by γ, which is the inverter commutating margin. d is the transformer's direct voltage drop per unit (see Sec. **12**), which for the converter connection used here means half the short-circuit reactance, defined for a reference alternating voltage E_{00} and a reference direct current I_{d00}. For rectifiers $\alpha_{\min} = 0$, and for inverters $\gamma_{\min} = \gamma_0$ (normally about 15°). Thus, the following maximum values apply:

Rectifier:

$$\cos \varphi_{\max} = 1 - d \frac{I_d}{I_{d00}} \cdot \frac{E_{00}}{E} \tag{14-7}$$

Inverter:

$$\cos \varphi_{\max} = \cos \gamma_0 - d \frac{I_d}{I_{d00}} \cdot \frac{E_{00}}{E} \tag{14-8}$$

With constant voltage E on the a-c side of the converter, the maximum power factor drops from 1 and $\cos \gamma_0$, respectively, linearly with the direct load current.

100. A-C Network Loaded by a Converter. The short-circuit capacity Q_k of the a-c network at the converter connection point and the capacity Q_c of the capacitors are defined by the reference voltage E_{00} (see Par. **99**). The capacitor represents the resulting capacitance at the power frequency of the a-c filters, including separate capacitor banks for reactive-power compensation. E_0 is the network voltage behind the transient network reactance and E the alternating voltage on the condenser bank.

The following relative quantities per unit are defined:

Direct voltage:

$$e_d = \frac{E_d}{kE_0} \cdot \frac{Q_k - Q_c}{Q_k} \tag{14-9}$$

Alternating voltage:

$$e = \frac{E}{E_0} \cdot \frac{Q_k - Q_c}{Q_k} \tag{14-10}$$

Active power:

$$p = \frac{P}{Q_k} \cdot \frac{kE_{00}I_d}{Q_k} \cdot \left(\frac{E_{00}}{E_0}\right)^2 \tag{14-11}$$

Direct current:

$$\frac{p}{e_d} = \frac{kE_{00}I_d}{Q_k} \cdot \frac{E_{00}}{E_0} \tag{14-12}$$

Power factor (see Par. **97**) is $\cos \varphi = e_d/e$. A constant for the converter station that does not vary with load is $a = d(Q_k - Q_c)/kE_{00}I_{d00}$.

The equation for the a-c network with active and reactive load (Par. **98**) is

$$e = \sqrt{1 + (p/e_d)^2 - 2(p/e_d)\sqrt{1 - e_d^2}} \tag{14-13}$$

Figures 14-42 and 14-43 show the alternating voltage at the converter station and the reactive power to the converter, respectively, as a function of the direct voltage for some d-c values.

101. D-C Load Characteristics. Equation (14-6) in Par. **99** can be expressed in per unit values,

$$e_d = e \cos \alpha - a \frac{p}{e_d} \tag{14-14}$$

This gives, together with Eq. (14-13), the current p/e_d as a function of the voltage e_d:

$$\frac{p}{e_d} = \frac{\cos^2 \alpha - e_d^2}{\cos^2 \alpha \sqrt{1 - e_d^2} + ae_d + \cos \alpha \sqrt{a^2 + 2ae_d\sqrt{1 - e_d^2} + e_d^2 \sin^2 \alpha}} \tag{14-15}$$

This relation is illustrated in Fig. 14-44 for $\cos \alpha = 1$ and for some values of the terminal constant a.

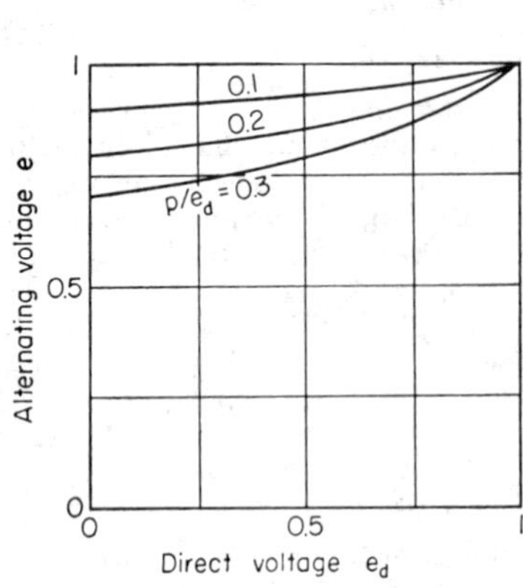

Fig. 14-42. The alternating voltage (e) at the converter station as a function of the direct voltage (e_d) for d-c values (p/e_d).

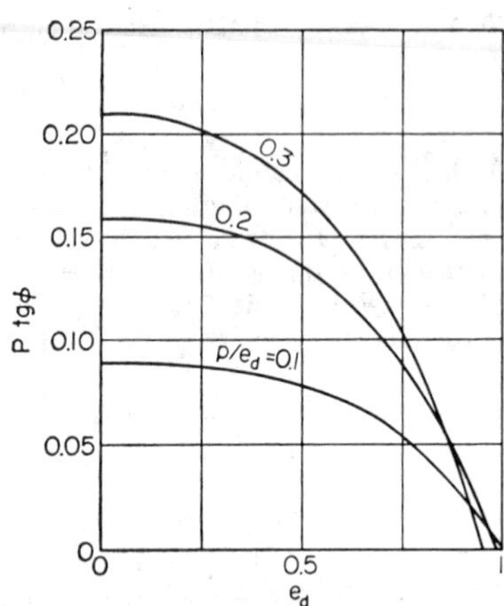

Fig. 14-43. Reactive power (p tan φ) as a function of the direct voltage (e_d) for d-c values (p/e_d).

102. Active Power Maxima with Constant Source Voltage E_0 behind the Transient Network Reactances. The curves in Fig. 14-44 plotted against $a \cdot p$ instead of p/e_d give the curves shown in Fig. 14-45, from which it is obvious that there is a maximum power which can be transferred through the converter. If the converter is power- or frequency-controlled and the control system demands more power than this maximum, the converter has a tendency to give more current but less power. Thus there is a risk for instability.

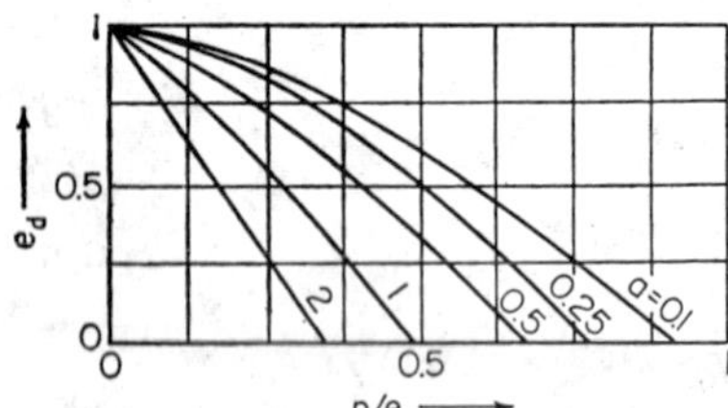

Fig. 14-44. Direct current (p/e_d) as a function of the direct voltage (e_d) giving values of the terminal constant (a).

The power maxima, calculated from Par. **101**, are shown in Fig. 14-46 for different values of cos α and the plant constant a.

Note that according to Par. **100**

$$ap = \frac{I}{kE_{00}I_{d00}} \cdot d \cdot \left(\frac{Q_k - Q_c}{Q_k}\right)^2 \cdot \left(\frac{E_{00}}{E_0}\right)^2 \tag{14-16}$$

For $Q_k \gg Q_c$ ap is nearly independent of Q_k. Therefore, the different curves a indicate the influence of the short-circuit capacity Q_k on the power maxima, with all other data unchanged.

103. Power Limits with Constant Voltage (E) on the Bus Bar at the Converter Terminal. If the voltage E on the a-c side of the converter is kept constant, which can be achieved for slow changes of the converter load by means of normal voltage control via on-load tap changers or excitation of synchronous machines, Par. **102** gives the following expression for the power maximum:

$$P_{\max} = \frac{kE_{00}I_{00}}{4k} \cdot \left(\frac{E}{E_{00}}\right)^2 \cdot \cos^2 \alpha \tag{14-17}$$

104. Examples of Calculation of Power Maximum. To demonstrate the described calculations, two examples are given, both on the following assumptions: short-circuit capacity on the a-c side of the converter transformers Q_k = 2,700 MVA; inherent reactive direct voltage drop in converter transformers 8% or d = 0.08; produced reactive power in shunt capacitor at rated alternating voltage Q_c = 200 Mvars; rated direct current $I_d = I_{d00}$ = 2,000 A.

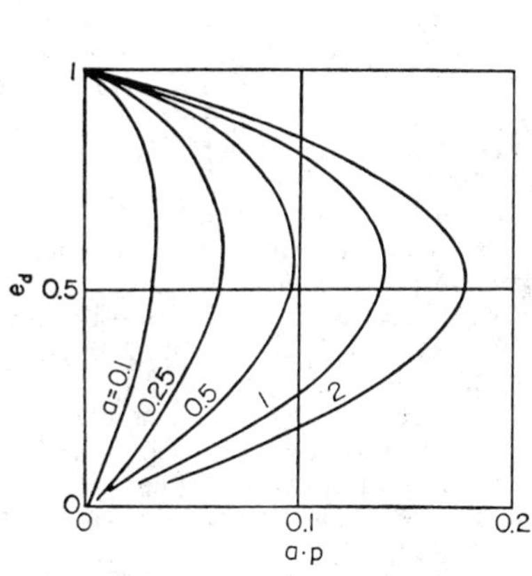

Fig. 14-45. The curves in Fig. 14-44 plotted against $a \cdot p$ instead of p/e_d.

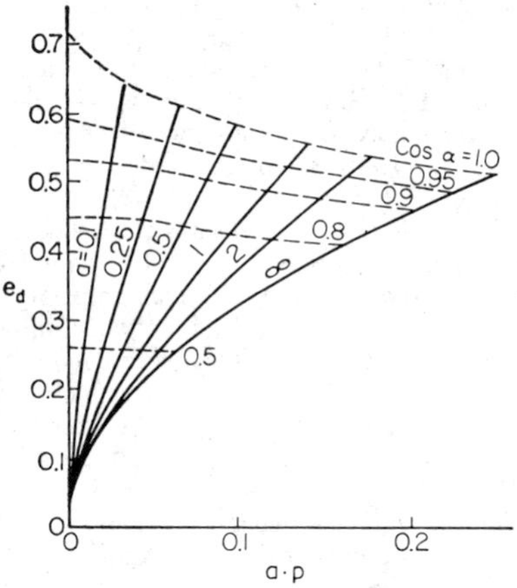

Fig. 14-46. Power maxima for different values of cos α and terminal constant (a).

Case a. Assume that the voltage behind the reactances, the source voltage E_0, is constant and further that $kE_{00} = kE_0 = 400$ kV. According to Par. **100**,

$$a = \frac{0.08 \cdot 2{,}500}{400 \cdot 2.0} = 0.25$$

and Par. **101** gives

$$ap = \frac{P}{400 \cdot 2{,}000} \cdot 0.8 \left(\frac{2{,}700 - 200}{2{,}700}\right)^2$$

or

$$P = 11{,}600ap \qquad \text{(MW)}$$

From the curve for $a = 0.25$ in Fig. 14-46, we find the corresponding ap and P_{max} values as shown in Table 14-7.

Table 14-7. Power Limits for Data from Par. 104 and for Constant Source Voltage and Constant Voltage at the Converter Terminals

cos α	Constant source voltage E_0		Constant voltage E on a-c side of converter transformers, P_{max}
	ap	P_{max}	
1.0	0.0643	746 MW	2,500 MW
0.95	0.0575	667 MW	2,250 MW
0.9	0.0521	604 MW	2,020 MW
0.8	0.0436	506 MW	1,600 MW
0.5	0.0214	248 MW	625 MW

Case b. Assume constant bus-bar voltage E at the a-c side of the converter transformers, and

$$kE_{00} = kE = 400 \text{ kV}$$

According to Par. **103**,

$$P_{max} = \frac{400 \cdot 2{,}000}{4 \cdot 0.08} \cdot \cos^2 \alpha = 2{,}500 \cos^2 \alpha \qquad \text{(MW)}$$

105. Reasons for Limitation of Power. Under Par. **104** calculations have been made for a terminal which could be rated 800 MW. From Table 14-7, it is obvious that rated power could not be reached with constant source voltage E_0 (Case *a*). With constant a-c bus-bar voltage E (Case *b*) there are ample margins, even in case of a very low value for cos α corresponding to a large control angle. Thus, the importance of controlling a-c bus-bar voltage is well demonstrated. As such a control does not act instantaneously, only the amount of power corresponding to Case *a* can be switched on suddenly. In order to reach rated power or even higher values, the bus-bar voltage E has to be restored.

The basic reason for this is the relatively large inherent voltage drop in the terminal and in the a-c system. This drop is amplified by the shunt capacitor. In increasing power on the d-c side, the consumption of reactive power is also increased, giving a higher voltage drop in the network impedance. This results in the reactive-power generation in the capacitor being decreased, thus producing a higher voltage reduction than that corresponding only to the changes in direct power.

106. Influence on Power Maximum from D-C Line Losses. In a point-to-point transmission the lowest direct voltage will be at the inverter d-c terminals. The calculations above can be used for the inverter, or receiving, terminal. If the transmitted power through the rectifier, or sending terminal, and over the line is to be calculated, the line losses must be considered.

If R is the resistance of the total d-c loop in ohms, we define a per unit value as

$$r = \frac{Q_k - Q_c}{(kE_{00})^2/R} \tag{14-18}$$

Let e_{dr} be the per unit value for the direct voltage including line losses and in the same way p_r the power measured at the inverter side of the transmission.

$$e_{dr} = e_d - r\frac{p}{e_d} \tag{14-19}$$

$$p_r = p - r\left(\frac{p}{e_d}\right)^2 \tag{14-20}$$

In all curves, for instance in Fig. 14-45, changes according to the above equation have to be made, and reduced power limits for the rectifier will be obtained. For long d-c lines this reduction may be important.

107. Calculation Example Including D-C Line Losses. The most economical voltage drop in a d-c line is about 40 V/mi. Assuming the length of the line = 600 mi, and with rated current 2,000 A, the total resistance will be

$$R = \frac{40 \cdot 600}{2{,}000} = 12\ \Omega$$

Other assumptions, the same as in Par. **105**, give

$$r = \frac{2{,}700 \cdot 200}{(400)^2} \cdot 12 = 0.488/\text{unit}$$

In a point of the characteristics where $e_d = 0.8$ and for $\alpha = 0$ and $a = 0.25$, the power will be $p = 0.213$ according to Par. **103**, d-c line losses being disregarded. With the above line losses the power maximum at the receiving end of the d-c line will be, according to Par. **106**,

$$p_r = 0.213 - 0.188\left(\frac{0.213}{0.8}\right)^2 = 0.178$$

108. Current Limitations. It is uneconomical to operate a transmission under steady-state conditions near the power maximum, where a small increase of power requires a big increase of current and losses. But, during transient conditions, the control device can ask for power exceeding the power limit. To avoid instabilities, the control equipment should include a current limitation.

109. Sudden Power Change. For the conditions before the change (index 1), per unit values according to the following are used:

$$e_{d1} = \frac{E_{d1}}{kE_1} \tag{14-21}$$

$$e_1 = \frac{E_1}{E_0} \cdot \frac{Q_k - Q_c}{Q_k} \tag{14-22}$$

If the prechange voltage E_1 at the inverter terminal is known, the source voltage E_0, which is not changed during the transient, can be calculated from

$$e_1 = \frac{1}{\sqrt{1 + 2(p_1/e_{d1})\sqrt{1 - e_{d1}^2} + (p/e_{d1})^2}} \tag{14-23}$$

The per unit values after the change (index 2) are:

$$e_{d2} = \frac{E_{d2}}{E_{d1}} \cdot e_{d1} \cdot e_1 \tag{14-24}$$

$$e_2 = \frac{E_2}{E_1} \cdot e_1 \tag{14-25}$$

$$P_2 = \frac{P_2}{P_1} \cdot \frac{P_1}{e_{d1}} e_1^2 \tag{14-26}$$

All these quantities with index 2 correspond to the definitions in Par. **100**, and all the formulas shown previously are valid and based on the source voltage E_0 established by the prechange conditions.

110. Sudden Break of Transmitted Power. According to Par. **100**, the value of e_1 is 1 when the direct current is zero. In Par. **109** the voltage rise at the converter terminals will be

$$\frac{E_2}{E_1} = \frac{1}{e_1} = \sqrt{1 + 2(P_1/e_{a1})\sqrt{1 - e_{d1}^2} + (p_1/e_{d1})^2} \tag{14-27}$$

In Fig. 14-47 the voltage rise is demonstrated, $\cos \varphi_1$ being the power factor of the converter before the break.

111. Power Transfer over a D-C Line. Index 1 refers to the sending end and index 2 to the receiving end of the transmission line. The power in a line according to Fig. 14-48 is

$$P = \frac{E_{d1}^2 - E_d^2}{2R} \pm \frac{(E_{d1} - E_{d2})^2}{2R} \tag{14-28}$$

The first term gives the power in the middle of the line, and the second term gives half the line losses. The + sign gives the power in the sending end and the − sign in the receiving end.

112. Power Transfer by a D-C Link. A d-c link between two points in a common a-c system or between two separated a-c systems, as illustrated in Fig. 14-49, gives the following loading on the connected a-c systems:

Active power in sending end:

$$P_1 = k_1E_1 \cos \varphi_1 \cdot \frac{k_1E_1 \cos \varphi_1 - k_2E_2 \cos \varphi_2}{R} \tag{14-29}$$

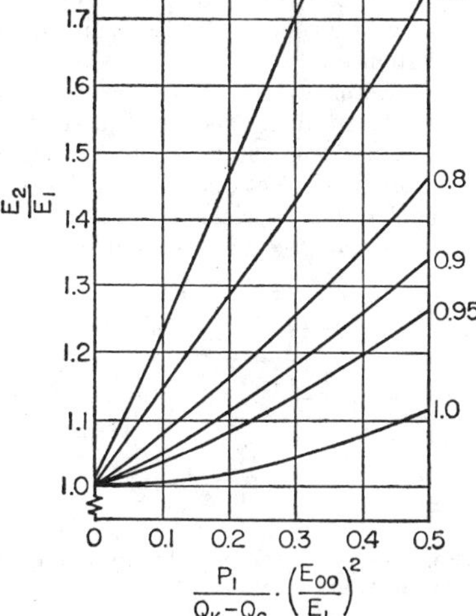

FIG. 14-47. Voltage rise at the converter terminals with sudden break of the transmitted power.

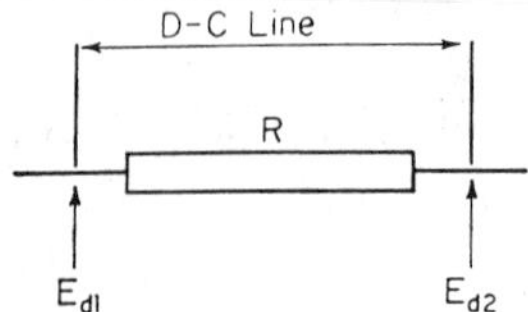

FIG. 14-48. A d-c line with load.

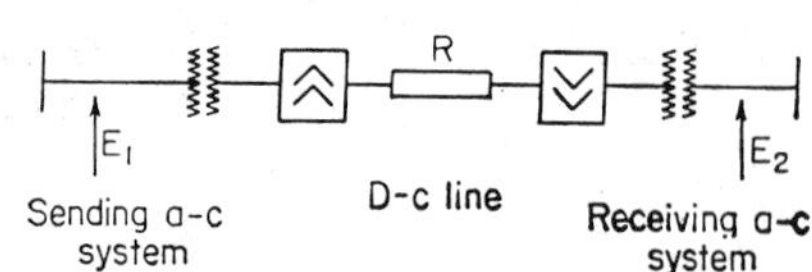

FIG. 14-49. A d-c link between two points in a common a-c system or between two separated a-c systems.

Reactive power in sending end:

$$Q_1 = k_1E_1 \sin \varphi_1 \cdot \frac{k_1E_1 \cos \varphi_1 - k_2E_2 \cos \varphi_2}{R} \qquad (14\text{-}30)$$

Active power in receiving end:

$$P_2 = k_2E_2 \cos \varphi_2 \cdot \frac{k_1E_1 \cos \varphi_1 - k_2E_2 \cos \varphi_2}{R} \qquad (14\text{-}31)$$

Reactive power in receiving end:

$$Q_2 = k_2E_2 \sin \varphi_2 \cdot \frac{k_1E_1 \cos \varphi_1 - k_2E_2 \cos \varphi_2}{R} \qquad (14\text{-}32)$$

The factor k is defined in Par. **97**.

113. References to Literature on Converters as Links between A-C and D-C Networks

UHLMANN, E. Alternating Voltage, Direct Voltage Regulation and Power Factor of Converter Stations Operating on A.C. Systems of Finite Short-circuit Capacity; *Proc. IEE (London)*, 1955, Vol. 102C, pp. 284–289.

UHLMANN, E. Stabilisation of an A.C. Link by a Parallel D.C. Link; *Direct Current,* Vol. 9, No. 3, August, 1964.

CONTROL AND OPERATION OF HIGH-VOLTAGE D-C LINKS

114. General Requirements and Control Principles. The fundamental objects of an H-V d-c control system are:

a. To control a system quantity such as d-c line current, transmitted power or frequency of either of the two connected a-c networks with sufficient accuracy and speed of response.

b. To ensure safe inverter operation also in presence of system disturbances.

c. To fulfill the above objectives at minimum reactive power consumption.

Besides these items valid for normal operation the control system shall, if possible, also ensure correct operation at large disturbances or at least minimize the consequences when the fault is cleared.

The principle of direct-current control is illustrated by Fig. 14-50 from which it is clear that the transmitted current is

$$I_d = \frac{E_{d1} - E_{d2}}{R} \qquad (14\text{-}33)$$

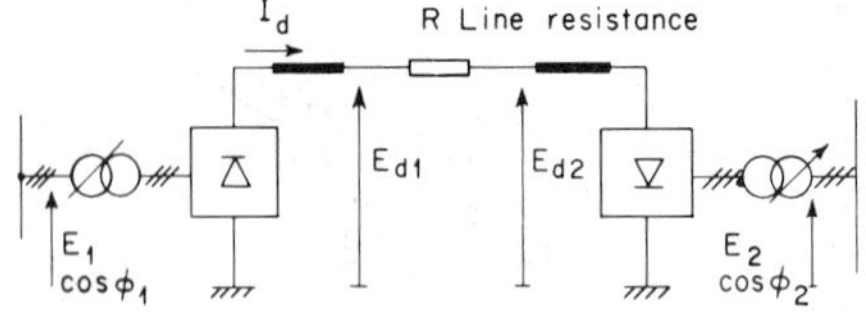

FIG. 14-50. Principle diagram for H-V d-c transmission.

Because of the rather small value of R, the current will vary rapidly with changes in the difference between terminal direct voltages, $E_{d1} - E_{d2}$.

Two methods are available for the control of E_{d1} and E_{d2}. One is to change the converter transformer ratio by tap-changer control. The other is to change the ratio between direct voltage and alternating voltage by grid control, i.e., change of the delay angle α. See Par. **97**.

The former is a slow method because of the highly limited speed of the transformer tap changer. The delay angle α, on the other hand, may be changed very rapidly by the grid control system. However, this is accomplished at the cost of an increased amount of reactive power being consumed by the converter. See Par. **98.**

To minimize the amount of reactive power consumed it is therefore usual to operate either the rectifier on minimum delay angle or the inverter on minimum margin of commutation γ. The latter operation mode is the normal one for the inverter and the value of γ is determined by the requirement for secure inverter operation without commutation failures even at reasonable disturbances in the a-c system. A value between 15° and 18° is usually chosen. A larger value would give a decreased risk for commutation failures but at the same time increased stresses on the valves and increased demand for reactive power.

The d-c line voltage may then be adjusted by tap-changer control in the inverter.

The current and thus also the transmitted power may then be rapidly controlled by grid control in the rectifier; i.e., control of the rectifier direct voltage E_{d1} in Fig. 14-50 is accomplished by means of delay angle control.

At steady-state operation it is suitable to operate with a delay angle in the range 12–18° el. A smaller delay angle would give less demand for reactive power but also would limit the possibility to rapidly increase the rectifier voltage by decreased delay angle.

As the direct voltage is determined by the inverter the suitable delay-angle range for the rectifier may be obtained by tap-changer control in the rectifier.

115. Static Characteristics of a Converter. Assume a converter with a basic constant-current-control system with a current order I_d. The converter is also provided with a constant γ control system, which prevents the margin of commutation angle from decreasing below the set value.

In Fig. 14-51 are shown the characteristics for such a converter when the load is increased from zero. From point A to point B (to the left in Fig. 14-51) the load could be assumed to be a resistance decreasing from infinity to point A. The load resistance is too high and the converter voltage is too low for the converter to be able to deliver the ordered current. The converter operates at the minimum limit value of the delay angle, which usually is set at 5°.

At point B, the resistance of the load has decreased to such a value that the desired current is achieved if the converter voltage has its maximum value. At a further decrease of the resistance the converter must increase its delay angle to keep the direct current at the value I_d. This occurs in the region B-F in which the converter operates in current-control mode.

At point C, the converter changes polarity and, if the overlap angle is neglected, has a delay angle equal to 90° el. From this point the converter works as an inverter and the load must be active, e.g., a direct voltage source in series with a resistor.

At point F, the delay angle α has increased to a value where the margin of commutation has reached its minimum value γ_0 and the γ control system takes over. F represents the point at which the inverter has its maximum voltage.

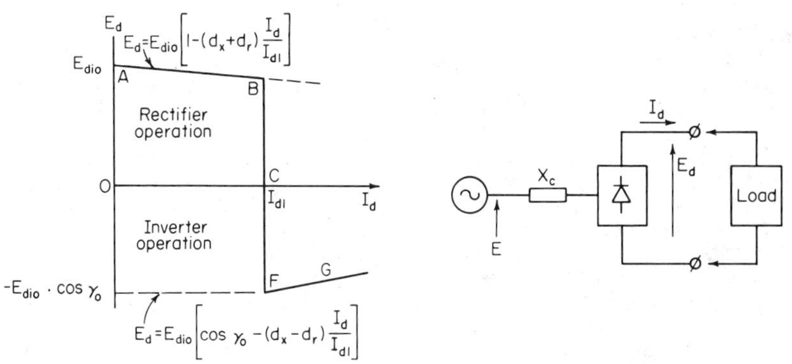

FIG. 14-51. Voltage-current characteristics for a current-controlled converter d_r = per unit resistive voltage drop $[P_{cu}/(E_{d0} \cdot I_d)]$.

From F through G the margin of commutation is kept constant. Thus when the voltage of the load, and by this the direct current and the overlap angle u of the converter, is increased, the converter direct voltage is decreased. This gives the converter a characteristic in this region corresponding to a negative resistance.

From this discussion we can sum up the following modes of operation for a converter:

a. Rectifier operation against the minimum limit in the delay angle (from A to B in figure). This occurs when the converter voltage is not high enough to generate the desired current.

b. Constant current control (between B and F). There are no principle differences between rectifier and inverter operation.

c. Control on constant margin of commutation (from F and further). This is the normal inverter mode of operation.

116. Cooperation between Sending and Receiving Terminals. Consider an H-V d-c transmission with the principal configuration shown in Fig. 14-50. Each station may consist of a series connection of a number of six-pulse converters. The terminal with index 1 is operating as rectifier and the other as inverter.

The rectifier has the characteristic *A-B-C* of Fig. 14-51 and the inverter is principally represented by the curve *C-F-G*. The inverter is, however, given a current reference I_{r2} which is slightly less than the reference of the rectifier I_{r1}. The difference

$$\Delta I = I_{r1} - I_{r2}$$

is called the current margin and is normally in the order of 10–15% of rated current.

The cooperation between the rectifier and inverter terminals with characteristics as above is illustrated by Figs. 14-52 and 14-53. The d-c line voltage drop can either be regarded as neglected or included in any of the characteristics for the terminals.

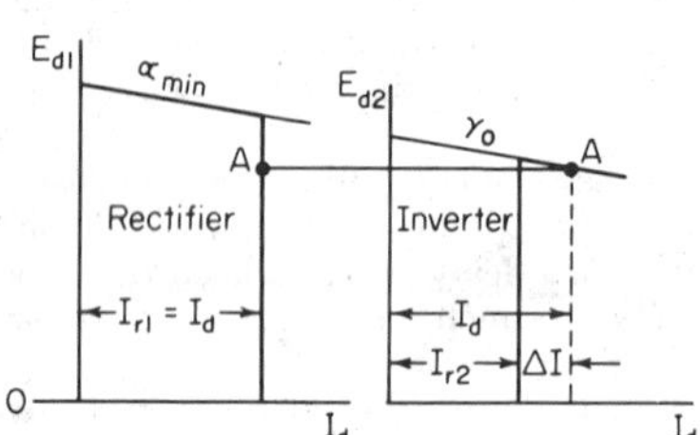

FIG. 14-52. Direct voltages plotted against the direct current with higher maximum voltages on the rectifier side.

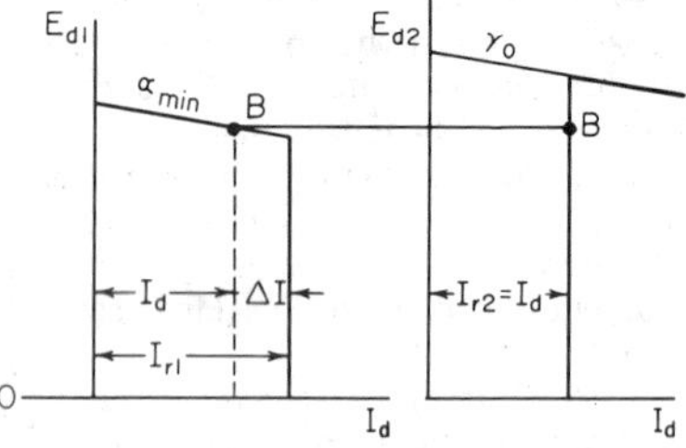

FIG. 14-53. Direct voltages plotted against the direct current with higher maximum voltages on the inverter side.

Contrary to Fig. 14-51, where the line voltage has been defined as positive for rectifier operation, the inverter characteristic has been defined as positive for inverter operation in Figs. 14-52 and 14-53.

Figure 14-52 depicts the usual operation mode with the minimum delay angle in the rectifier giving a higher direct voltage than the minimum extinction angle in the inverter. The interconnection point A between the two characteristics gives a stable operation point. The direct voltage is determined by the inverter and the current by the rectifier. If the alternating voltage is decreased in the rectifier terminal or increased in the inverter terminal, the condition may change in such a way that the inverter takes over the current control as depicted in Fig. 14-53 operation point B. With more than one converter connected in series such a transition will also occur when one rectifier is disconnected (blocked) or when another inverter is connected in series (deblocked).

The reason for introducing a margin between the current orders in rectifier and inverter is obvious from the two figures, as zero margin or a negative margin would not have given any stable operation point.

A change of power-flow direction is usually performed in an H-V d-c transmission by a change of the polarity of the line voltage. This is illustrated in Fig. 14-54 where the complete converter characteristics have been drawn for both stations. The initial operation point is A and the current margin is ΔI. If the latter is removed in station 2 (initially inverter) and the current order in station 1 is reduced by the amount ΔI, the broken line characteristics will be valid and the new operation point B is obtained. Thus the d-c line has changed polarity and the two converter stations have changed mode of operation.

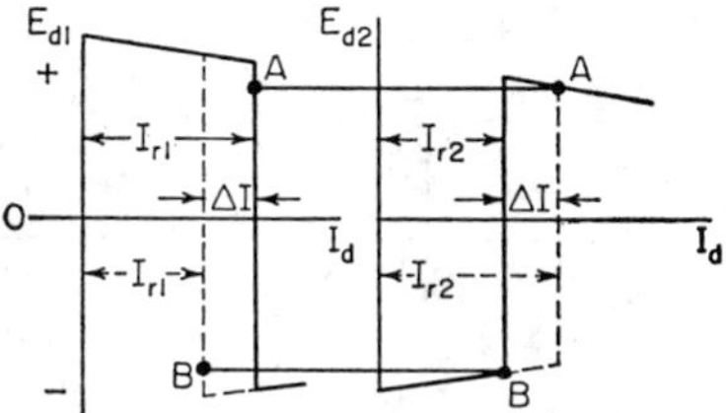

FIG. 14-54. Change of power-flow direction.

In practice, the reversal of power-flow direction also requires other operations, such as switching the minimum limit of α from about 100° el for the inverter to about 5° el for the rectifier and vice versa.

When a ground fault occurs on the d-c line, the steady-state current through the fault equals the current margin. This is easily explained by the fact that the rectifier continues to feed the d-c line with a current equal to the reference value, and the inverter demands a current which is ΔI lower. The difference ΔI must pass the fault. In practice the d-c line earth fault is detected by a special protection unit, which forces the rectifier to full inverter operation, and by this the current through the fault is reduced to zero.

117. Closed-loop Current Control. The rectifier normally is provided with a basic feedback current control system, simply illustrated by Fig. 14-55.

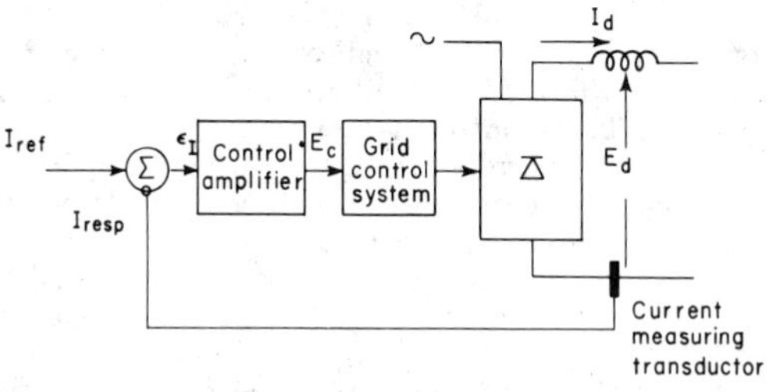

The control amplifier is used to give suitable static gain and dynamic compensation for stabilization of the control loop. Control pulses are generated and transmitted to valve potential by the grid control system.

Grid control system designs may differ considerably but they belong mostly to one of two types, here called the individual phase control system and the equidistant firing control system, respectively.

FIG. 14-55. Simplified block diagram for a constant current control system.

In the individual phase control system the valves connected to the same phase are controlled separately from other valves. Such a system may be so designed that the output voltage from the control amplifier (in this case approximately integrating) is compared to a monotonously increasing voltage, one for each valve, which starts at that zero crossing of the commutation voltage which corresponds to $\alpha = 0$. At equality between the two voltages a control pulse is generated.

The increasing voltage may be a constant ramp function giving a linear relationship between the delay angle and the output of the control amplifier.

A drawback, however, is that this gives a nonlinear loop gain as there is a cosine relationship between the delay angle and the converter direct voltage. See Eq. 14-6.

If we use a cosine control function, instead of the ramp, as shown in Fig. 14-56, we obtain an arccosine relationship between the output of the control amplifier E_c and α and thus linearity between E_c and the converter voltage E_d. The cosine control function is obtained by 90° el phase shift of the commutation voltage.

Figure 14-56 indicates that there is no principal difference between current control in rectifier and inverter regions; the control voltage E_c determines the delay angle between $\alpha = 0°$ and $\alpha = 180°$ if no restrictions are made.

The grid control system may also have the property of a voltage-controlled pulse oscillator with a steady-state frequency of $6f_0$ for a six-pulse converter where f_0 is the a-c network frequency. In this case the control amplifier controls the frequency of the oscillator. A change of delay angle for the converter is performed by a transient

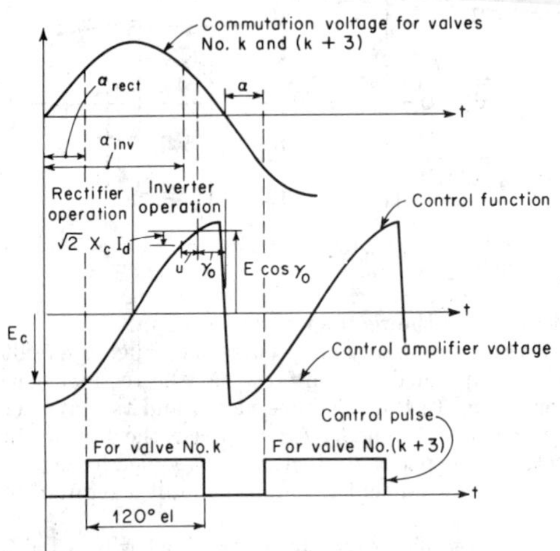

Fig. 14-56. Generation of control pulses on individual phase control system with a cosine control function.

increase or decrease of the oscillator frequency and thus a phase shift of the pulse sequence from the device is obtained. This type of control system gives equidistant firing since the distances between all consecutive firings in steady-state operation are equal to the pulse period time of the oscillator.

The voltage-controlled oscillator may be designed in different ways but measures must always be taken to prevent it from falling out of synchronism. The control pulses must always be generated within the region corresponding to α_{min} and γ_{min}.

In addition to providing equidistant firing, the system also has the property of being inherently nonlinear. However, some types of such systems may be linearized if special measures are taken.

118. Control on Minimum Margin of Commutation. The control on minimum margin of commutation may, analogously to current control, be of the individual phase or equidistant firing type. Two further alternatives exist for both types, namely, a predictive system or a feedback control system.

A predictive system is a feedforward-type control system which, based upon instantaneous measurements of direct current and commutation voltage, calculates the correct time instant for firing. This type of system is, of course, inherently individual for each valve or phase. However, if the most critical valve is chosen and allowed to determine the firing for all valves in an equidistant way, an equidistant control system is obtained.

A typical feedback control system measures the resulting γ, compares it to a reference value γ_r and uses the difference $E_\gamma = \gamma_r - \gamma$ to control a voltage-controlled pulse oscillator as described for the current control system. In this case an equidistant firing control system is obtained. However, an individual phase control system using the feedback principle is also possible. Both types of system, predictive and feedback, have their advantages. The predictive system provides better possibilities for prevention of commutation failures at disturbances, whereas the feedback principle gives better precision of control and may be simpler, especially when combined with a voltage-controlled oscillator system for current control.

The operation of a predictive γ control system of the individual phase type will now be described.

Consider the two equations:

$$\cos \alpha - \cos (\alpha + u) = \sqrt{2} \cdot \frac{I_d \cdot X_c}{\hat{E}} \tag{14-34}$$

$$\gamma = \pi - (\alpha + u) \tag{14-35}$$

Equation 14-34 gives the relationship between delay angle α, overlap angle u, direct current I_d, commutation reactance X_c and the commutation voltage E.

Substitution of $\pi - \gamma$ for $\alpha + u$ gives

$$-\hat{E} \cos \alpha + \sqrt{2} \cdot I_d \cdot X_c - \hat{E} \cos \gamma = 0 \tag{14-36}$$

If a voltage proportional to $\sqrt{2} \cdot I_d \cdot X_c - \hat{E} \cos \gamma$ is compared to a cosine voltage $\hat{E} \cdot \cos \omega t$ with $t = 0$ at the zero crossing of the commutation voltage, and the valve is fired at equality, a delay angle is obtained which is determined by Eq. 14-36. This is illustrated in Fig. 14-57.

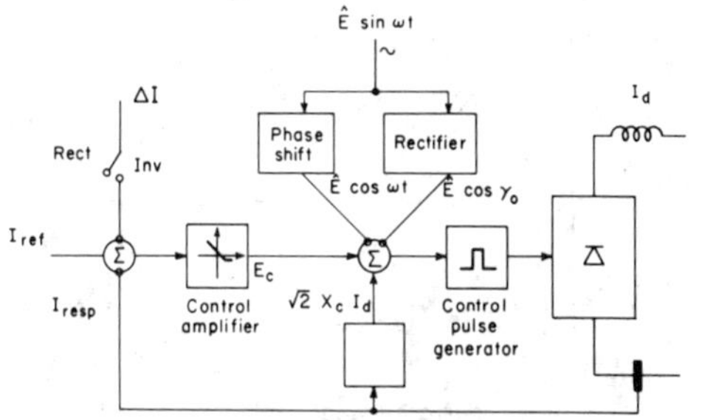

The voltage $\hat{E} \cdot \cos \omega t$ is exactly the control function defined in Fig. 14-56 and in fact this voltage may be used common to rectifier and inverter operation if the system is built as shown in the block diagram Fig. 14-57. As indicated in the figure the control amplifier voltage is limited to positive values.

FIG. 14-57. A combined rectifier-inverter control system of individual-phase predictive type.

When the control amplifier is not limited, the current feedback is closed and the direct current is controlled to the desired value independent of other input quantities. For inverter operation ΔI is connected and when the current is controlled by the rectifier, i.e., $I_{ref} = I_{resp}$ also in the inverter, the current feedback loop of the inverter is broken up because the control amplifier is forced against its zero limit. This means that E_c equals 0 and the γ control system just described is obtained. The delay angle is determined according to Eq. 14-36 giving the desired commutation margin γ.

119. System Control. The basic current control system has been described in the previous paragraph. Since the line voltage is almost constant as long as the number of converters connected in series in each terminal is not changed, the direct-current control is almost equivalent to power control. However, occasionally converters may be connected or disconnected, which results in a considerable change of direct voltage. It might then be favorable to control the current in such a way that the transmitted power is kept constant. One way of doing this is to provide each terminal with a current order calculator which supplies the reference to the current control system. This reference is calculated from the desired transmitted power by dividing it by the measured d-c line voltage.

As identical current orders have to be supplied to both terminals a telecommunication link must be used for system control. This may either transmit a current order or a power order.

The common order is either set manually in any of the terminals or set from a control system of higher order, which may also consider the situation in the sending or receiving a-c networks as described in the next paragraph.

120. Interconnection of Nonsynchronous and Synchronous A-C Networks. The interconnection between the sending and receiving a-c networks is either nonsynchronous or synchronous; i.e., there may or may not be a parallel a-c connection. In the first case the frequency might also be different, e.g., 50 Hz and 60 Hz. From the point of view of control this is, however, of minor interest.

In the nonsynchronous case the transmitted power on the d-c link is sometimes made dependent upon the frequency in any of the networks either to completely control this frequency or just to improve the stability in any of the networks during certain transient conditions.

In the synchronous case the transmitted power may be partly controlled from the phase difference or a time derivative of the phase difference (frequency difference) which might have a stabilizing effect at disturbances as described by the example in the next paragraph.

121. Breaking and Reclosing of Parallel A-C Ties. The following alternatives for control of the d-c link are assumed:

a. Constant power.

b. Constant phase-angle between the a-c sides.

c. As (*a*), with the addition of a term proportional to the frequency difference.

d. As (*b*), with the addition of a term proportional to the frequency difference.

It is also assumed that during the dead time no controls in the a-c network act and that the d-c link can carry the necessary overload.

Case a. After the opening of the a-c line the frequency difference Δf between the

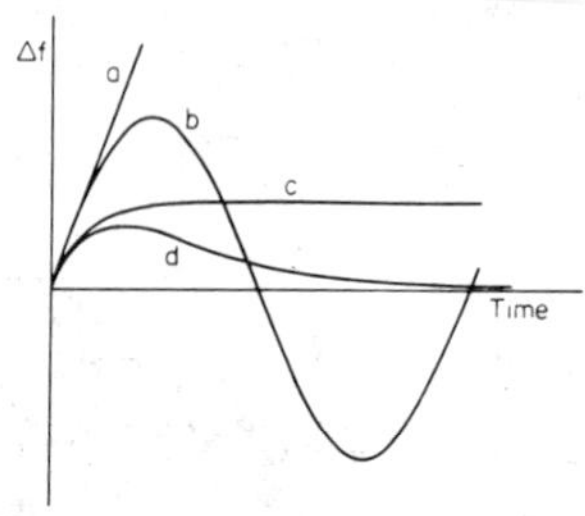

Fig 14-58a

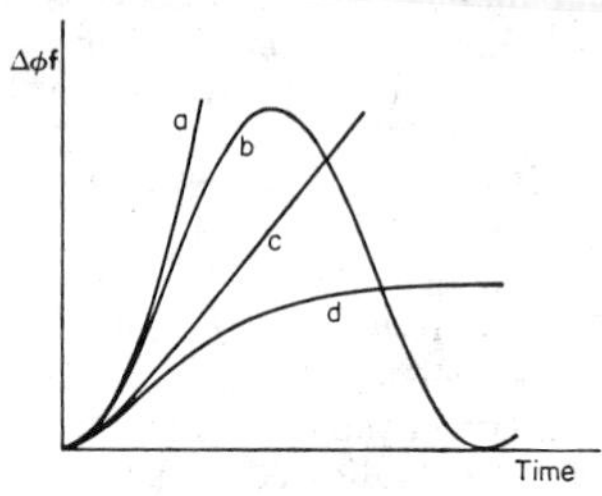

Fig. 14-58b.

Fig. 14-58. The functions of frequency difference (Δf, Fig. 14-58a) and the phase angle ($\Delta \phi f$, Fig. 14-58b) in a parallel d-c link stabilizing an ac link.

two a-c networks increases continuously (curve *a*, Fig. 14-57), along with the phase angle $\Delta\varphi$ (curve *a*, Fig. 14-58). The a-c line must be reclosed before Δf and $\Delta\varphi$ have reached allowable values.

Case b. The d-c link works as an a-c link. The oscillation of Δf and $\Delta\varphi$ is shown as curve *b* in the above-mentioned figures.

Case c. This is shown in curves *c*. The frequency deviation is limited, and the phase-angle difference increases more slowly than in case *a*, giving more time for reclosing. It will even be possible to wait one or more revolutions before the a-c line is reclosed.

Case d. This is shown by curve *d*. The phase-angle deviation is limited and the frequency difference brought down to zero after some time. The a-c line can be reclosed at any time.

122. Reference to Literature on Control and Operation of H-V D-C Links.

Engström, G. Operation and Control of H.V.D.C. Transmission; *Trans. IEEE Paper* 1963-83.

Forssell, H. The Gotland D.C. Link: The Grid Control and Regulation Equipment. *Direct Current*, vol. 2, no. 5, June, 1955 and no. 7, December, 1955.

MULTITERMINAL ARRANGEMENTS

123. Two-pole Transmission with Earth Return. The two rectifier poles, rectifier 1 and 2, and/or the two inverter poles, inverter 1 and 2, of a two-pole transmission can be located at different places, with ground used as a balance conductor (see Fig. 14-59). The power can be taken from different generator stations or a-c systems and may be sent to different load centers.

124. Series-connected Tappings. Some of the series-connected converters may be located at different places. Figure 14-60 shows this arrangement for two inverters.

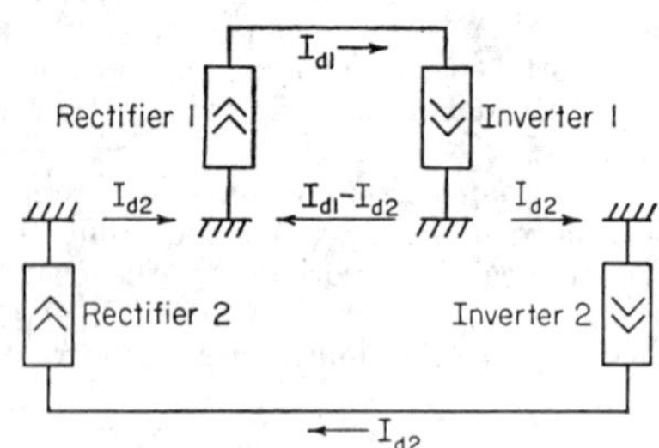

Fig. 14-59. Two-pole transmission with converters located at different places.

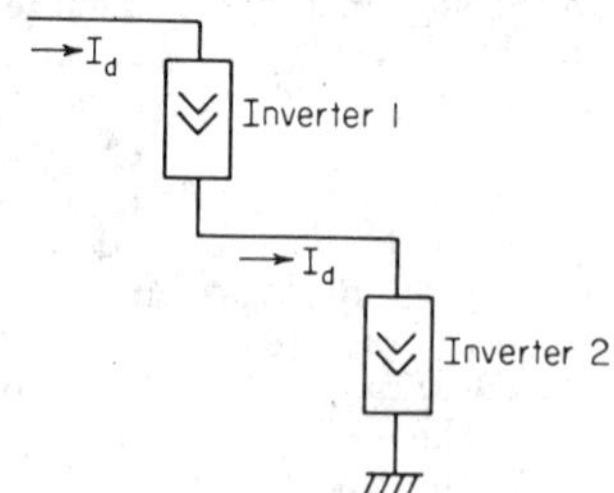

Fig. 14-60. Series-connected inverters.

The current in both inverter groups is the same, and for this reason the flexibility of operation is limited. To avoid a big demand for reactive power, a large tap-changer range is necessary. Series-connected terminals are most suitable, if both converter

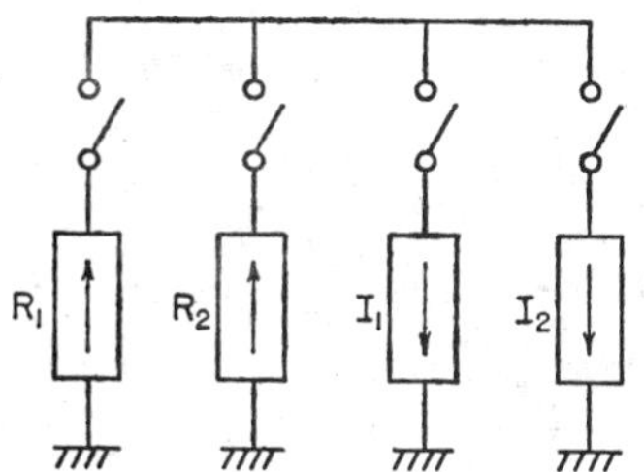

FIG. 14-61. Two rectifier and two inverter terminals on the same d-c line.

stations belong to the same a-c network, with nearly the same power-demand fluctuation in both points, but with neither point capable of receiving the total power.

125. Shunt-connected Tappings. Two kinds of terminals may be connected to the same d-c line (see Fig. 14-61), the only difference between them being the current direction with respect to the line. The power-flow direction, i.e., which terminal will be acting as rectifier or inverter, is determined by the current settings in the same way as in the point-to-point transmission (see Par. **120**). If $I_{rx}^{(1)}$ is the current setting to the rectifier x, and $I_{rx}^{(2)}$ the current setting to the inverter x, the rule for these settings is

$$\sum_x \cdot I_{rx}^{(2)} = \sum_x I_{rx}^{(1)} - \Delta I$$

All terminals will have the current according to the setting by respective controls, but the terminal with the lowest maximum voltage will have this current changed the amount ΔI (see Fig. 14-62, and compare it with Figs. 14-54 and 14-55). The control-program regulator, acting on the settings of the current regulators, must assure that the condition mentioned above is met, especially when the power distribution is changed.

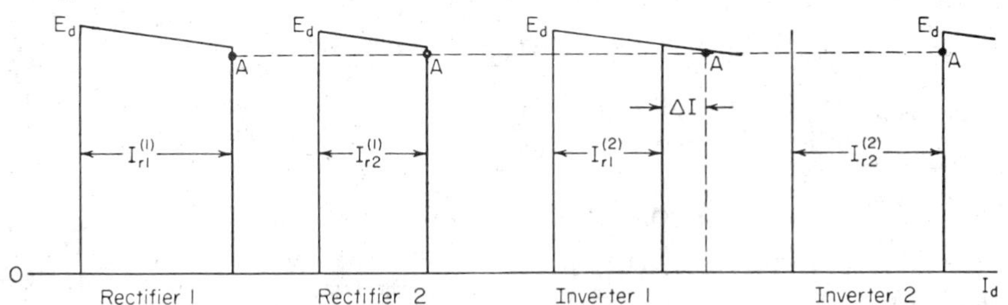

FIG. 14-62. Two rectifiers and two inverters connected to the same d-c line. Inverter 1 has the lower maximum direct voltage.

126. Operation of a Sending Terminal. It is always possible to connect to the line a rectifier terminal having a zero current setting. In loading this terminal, the central program regulator has to change the current settings of all the other terminals according to a scheduled program, in order to meet the condition mentioned in Par. **125.** In the same way, it is possible to take a rectifier terminal out of service.

127. Operation of a Receiving Terminal. An inverter terminal can be handled in the same way as a rectifier terminal, providing it does not have the lowest maximum voltage. The disconnectors to the d-c line have to be opened before the inverter is blocked.

If the inverter terminal, which has to be disconnected, has the lowest maximum voltage, then it will not be possible to reduce its current to zero. The current ΔI will remain in this station. If there is time enough to maneuver the tap changers, the lowest maximum voltage can be changed over to another terminal. If the time available is not sufficient for such a maneuver, the d-c line voltage can be reduced by grid control or the system can be made dead for the time necessary to disconnect the inverter terminal by a quick-acting isolator.

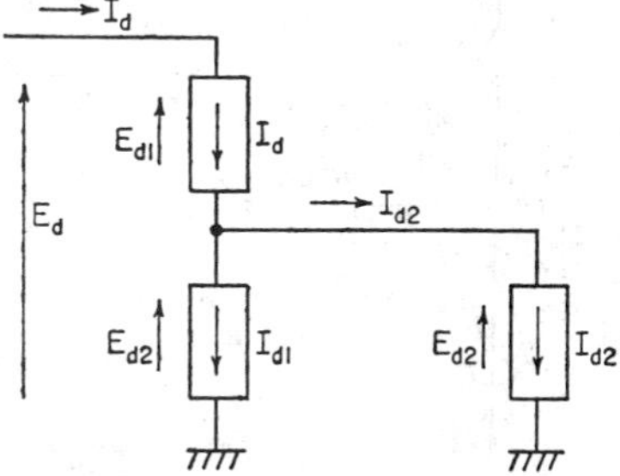

FIG. 14-63. Combination of series- and shunt-connected terminals.

128. Combination of Series- and Shunt-connected Tappings. As shown in Fig. 14-63, the series-connected valve groups are common for the two parallel terminals and are placed in the same station. This arrangement gives great flexibility for the power flow and a decreased transmission voltage to the terminal corresponding to the reduced power to this point.

129. Change of Power-flow Direction. If the change of the power-flow direction is made in the same way as in a point-to-point transmission (see Par. **121**), all rectifiers will be inverters and all inverters will be rectifiers. This means that the voltage polarity will be changed. If the power-flow direction must be changed only in one terminal, it is advantageous not to change the voltage polarity on the d-c line but instead to change the connection of the respective terminal to the d-c line. As there are different insulation levels to ground for the valve groups, the connections to all the valve groups in series must be changed. If there are two poles with different polarity in the same terminal, the simplest way to establish the power-flow direction change is to change the converters used in the plus and minus pole, as shown in Fig. 14-64. In the position of the disconnections shown, the terminal operates as an inverter, and in the other position as a rectifier, without changing the polarity of the line.

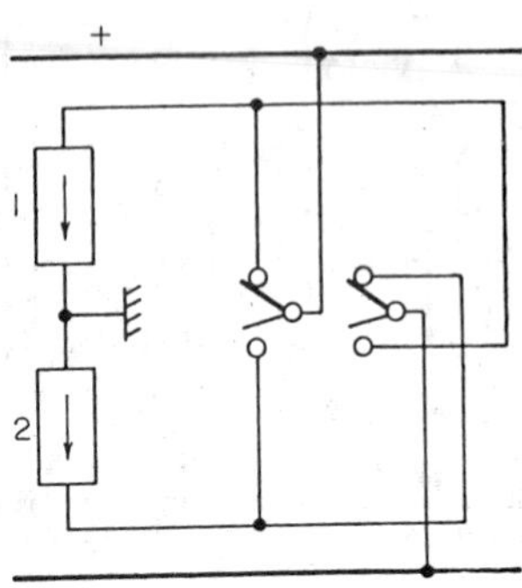

FIG. 14-64. Power-flow direction change in a tapping terminal of a bipolar transmission.

130. Reference to Literature on Multiterminal Arrangements

LAMM, U., UHLMANN, E., and DANFORS, P. Some Aspects on Tapping of H-V D-C Transmission; *Proc. Am. Power Conf.*, 1963, Vol. 25, pp. 736–744; also published in *Direct Current*, 1963, Vol. 8, No. 5, pp. 124–129.

ECONOMICS AND EFFICIENCY

131. Choice of D-C Line Voltage. For a d-c link the transmission voltage can be chosen freely to meet the actual demand owing to the fact that it is not directly connected to the a-c system.

The optimization of voltage to give lowest possible line costs for a certain transmission capacity and distance follows the same rules as for a-c lines, but the limitations set by corona interference are not so pronounced as in alternating current. As in a-c transmission, theoretically the loss cost and current-dependent cost should be equal, and this sum should be equal to the voltage-dependent costs. This would give too high a voltage, considering corona and terminal equipment cost. It can be shown that, for a system voltage which is a given fraction of the theoretically most economical, the d-c line cost per kilowatt is two-thirds of that of alternating current in the single-circuit arrangement. Practical experience verifies that the overhead line cost for d-c transmission is around two-thirds of that for a-c single circuit at the same transmitted power.

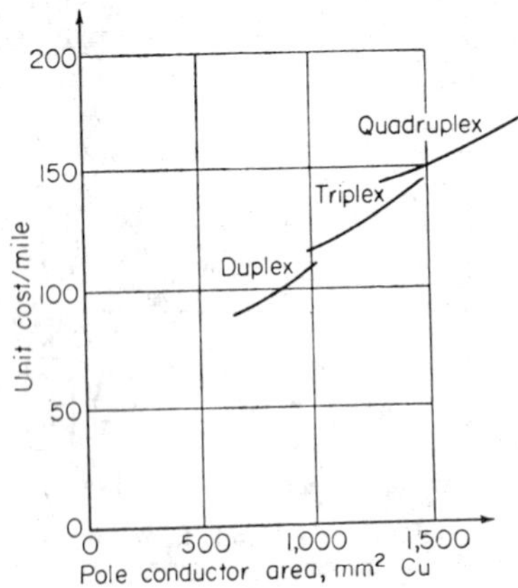

FIG. 14-65. Relative overhead d-c line cost as a function of the conductor area for a ±400-kV bipolar line.

132. D-C Line Cost. The d-c line cost varies according to Fig. 14-65. The d-c line voltage has a small influence on the total line cost for a given conductor size. In this figure an increase of voltage from 400 to 500 kV means a cost increase of only 3 to 5% throughout the conductor area range. This

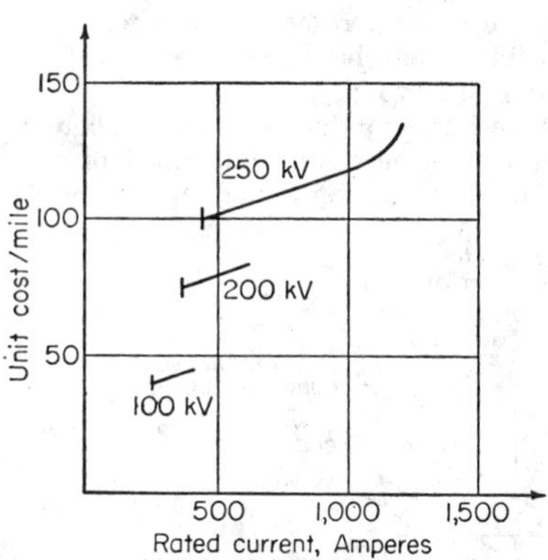

FIG. 14-66. Relative cost for solid-insulation d-c cable.

means that an increase in line voltage has little influence on the optimization of the total line cost. However, the voltage dependence on station cost has a significant influence on the resulting cost. Optimization of line voltage alone is of little value. The absolute line cost is, to a large extent, dependent on local regulations and conditions, such as transport and right-of-way costs.

133. Cable Cost. As indicated in Par. **145**, d-c cables are considerably less costly than a-c cables for the same rating. A resulting cable-cost ratio of 1:3 in comparing direct and alternating current for a given project is not abnormal. The relative cost of solid-insulation cables as function of power rating has a characteristic similar to what is shown in Fig. 14-66. Owing to the fact that the conductor must not exceed a certain temperature, current density has to be decreased rapidly above a certain current rating, as is illustrated in the curves.

134. D-C Converter Terminal Cost. The converter terminal is built up of a number of series-connected six-pulse converters. The series is either grounded in one end, forming a monopolar terminal, or in the middle, forming the two converter poles in a bipolar transmission. As the six-pulse converters make up by far the largest portion of the terminal equipment, the specific station cost is mainly dependent on the size of the individual converter. In Fig. 14-67 the influence of the main data on relative converter-station cost is shown. As can be seen, the specific cost is close to equal for ±300 kV at 1,080 MW, ±450 kV at 1,620 MW and ±600 kV at 2,160 MW, all of which refer to a current rating of 1,800 A.

The curves refer to turn-key cost for fully phase-compensated stations, with an a-c voltage of 230 kV, at 60 c/s. The cost increase for 50 c/s is around 2 to 3%. A monopolar terminal costs about 2 to 3%/kW more compared with a bipolar terminal with the same current rating. The influence of a-c network voltage on d-c terminal cost is shown in Table 14-8.

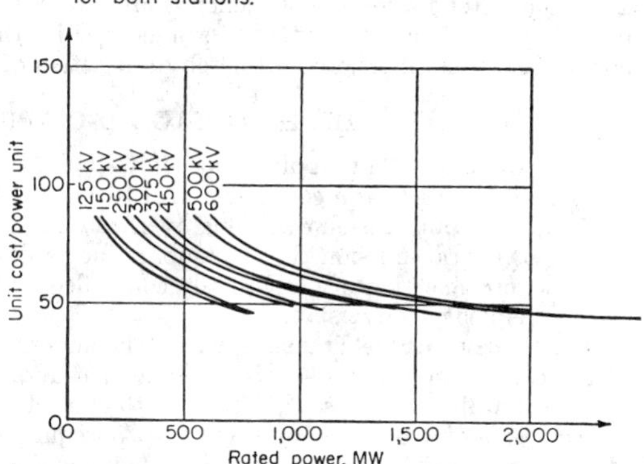

FIG. 14-67. Specific converter-station cost, a-c 230 kV.

Table 14-8. Influence of Voltage on H-V D-C Terminal Costs

132 kV	550 kV BIL	98%
230 kV	900 kV BIL	100%
800 kV	1,925 kV BIL	105%

When a d-c terminal is located at a generating station, direct connection of the separate converters to individual generators should be considered, as no tap changers on converter transformers are required. As the generator regulators can give voltage control, no phase-compensating shunt-capacitor batteries are needed, as the generators can supply reactive power. Furthermore, with the converters as the sole generator load, no harmonic filters are required, as the generators can be designed to withstand the harmonics. The resulting cost reduction would be 20 to 25%.

135. Line Losses. The losses on a d-c line are resistive conductor losses and corona losses. The latter are, in most cases, relatively small. There is no skin effect to consider, and a somewhat higher current density compared with alternating current should be justified. A normal value is 1.1 A/mm² for copper. It is not critical to reach the exact optimum, as a reduction in installed cost is counteracted by increased line losses.

136. Station Losses. The valves introduce a certain forward voltage drop, but in H-V direct current this has small influence on the total station losses. The major part of the total losses comes from the converter transformers. In Fig. 14-68 the itemized losses are given as a function of relative converter load.

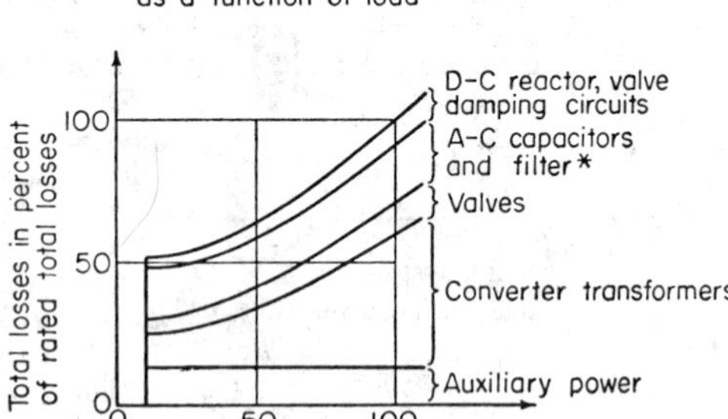

*No switching of shunt capacitors has been assumed at reduced load. In practice the capacitor losses will decrease in steps as the load is reduced.

FIG. 14-68. Total converter-station losses as a function of load. Transmission data ±375 kV, 1,000 MW, a-c voltage 230 kV, 50 Hz. Rated total losses = 1.25% rated load.

For a fully compensated converter station the efficiency is around 98.7%. It must be understood that the efficiency can, to a large extent, be influenced by the design of the equipment, such as converter transformers, and is thereby a function of the loss evaluation.

HIGH-VOLTAGE D-C APPLICATIONS

137. General. The applications of H-V d-c transmission may be classed in four different categories (or a combination of two or more of these):

1. Power transmission over long overhead lines.
2. Power transmission through submarine cables or underground cables.
3. Interconnection of individually controlled systems.
4. Frequency conversion.

138. Overhead Line Transmission. The advantage of direct current is that the cost of the transmission line itself is lower than if alternating current were used. Against this, the terminal stations required at both ends of the d-c lines are more expensive than the corresponding ones for an a-c line. Consequently, the H-V d-c system can be more economical than the a-c system only when the line is long. It is not possible to give an exact figure for the break-even distance, since the economical comparison between

Table 14-9. Staff Required for Maintenance of a Bipolar Converter Station
(Three valve groups per pole)

Kind of work	Category of staff *A*		Category of staff *B*	
	No.	Time/year, weeks	No.	Time/year, weeks
1. Renovation of valves	2	45	1	37
	2	5		
2. Detailed check of valves	2	6		
3. Smaller check of valves	2	3		
4. Check of control, protections	2	2		

d-c and a-c alternatives depends so much on local conditions, such as requirements placed on line, performance, properties of connecting a-c systems, etc. Studies show, however, that under normal conditions it is advantageous to consider the d-c system for overhead lines when the transmission distance is 300 mi or more. In areas with high cost for right-of-way, H-V d-c becomes feasible at a shorter distance.

139. Underground and Submarine Cable Transmission. As in the case of overhead lines, the extra cost of converter stations should be paid for by the saving in cable and associated costs. The difference between cable costs for d-c and a-c power is more pronounced than for overhead lines. Thus the cable cost does not have to be such a large portion of the total transmission cost as for overhead lines to make direct current preferable, based on cost comparison for the cable and terminals. Distances of 20 to 30 mi or more are feasible from the point of view of H-V d-c application.

There is a practical limit to the possible noninterrupted length of an a-c cable due to the capacitive charging currents. For underground cables, these can be compensated for by intermediate shunt reactors. However, this cannot be practically done for sea cables. The charge currents are especially pronounced in extra-H-V cables, and necessary shunt reactors give a substantial increase in transmission cost and also in the required land and intermediate terminals. Direct-current cables have no steady-state capacitive charging currents influencing the design.

140. Interties between Separate A-C Systems. Interconnection between power systems is justified where sufficient production or load diversity exists and can be partly justified in limiting required spinning reserve and to enable increase of the maximum power-production unit size.

In spite of the fact that diversity is normally small compared with the power-network size, it often represents a large-size intertie. It is both costly and technically difficult, and sometimes practically impossible, to hold a weak a-c intertie stable between two independently controlled power systems. In these cases, the asynchronous d-c intertie provides a very good solution. In this application, the d-c terminal cost might be somewhat lower than in other applications, as sufficient short-circuit power normally is available without installation of synchronous condensers, and, in many cases, the demanded reactive power is available from the a-c network.

Direct-current transmissions can also be used to supply small, remote systems, for instance, on islands, where the power flow through the link likewise controls the local-system frequency.

The asynchronous nature of a d-c link gives the further possibility of running pump storage stations at different speeds for pumping and generating operations. Such a mode of operation gives maximum efficiency and can be obtained without any switching in the main circuits.

141. Frequency Conversion. When it is desirable to interconnect two power systems having different nominal frequencies, the H-V d-c system has such technical advantages that it can be used even where the length of the d-c line is nil. The specific cost of a static-frequency converter station will be a little lower than for two d-c terminals with the same rating, thanks to simplifications which can be introduced when sending and receiving terminals are located in the same station.

OPERATION AND MAINTENANCE

142. Operation. The operation of an H-V d-c converter station is similar to that of an a-c generator station, which means that remote control is also possible. Normal disturbances are automatically taken care of within the converter station itself, and automatic reclosure after such a disturbance is used. This applies to both valve disturbances and ground faults along the d-c line.

One of the converter stations may have an operator, and from this station the power, frequency, current, etc., are manually set. The operator receives the essential information about the other station via some kind of communication channel. This information includes fault signals, position of circuit breakers, etc. The operator can stop and start up the transmission. The degree of automization of the starting-up procedure will be determined from case to case. A normal procedure provides for most of the station isolators to be operated from the station itself.

Table 14-10. Particulars

System name	Gotland	English Channel	Sardinia	New Zealand
Commissioning year	1954	1961	1966	1965
Power transmitted, MW	20	160	200	600
Direct voltage, kV	100	±100	200	±250
Valve groups per station	2	2	2	4
Direct voltage per valve group, kV	50	100	100	125
Direct current, A	200	800	1,000	1,200
Parallel anodes per valve	2	4	4	4
Rms value of current per valve, A	115	460	575	690
Reactive power supply	Synchr. condenser Static capacitors	Static capacitors	Synchr. condenser Static capacitors	Synchr. condensers Static capacitors
Converter station location	Västervik Visby	Lydd, England Echinghen, France	Codrongianos San Dalmazio	Benmore Haywards
A-c grid voltage	Västervik 130 kV Visby 30 kV	Lydd 275 kV Echinghen 225 kV	Codrongianos 230 kV S. Dalmazio 220 kV	Benmore 16 kV Haywards 110 kV
Overhead d-c line distance	....	....	290 km (180 mi)	575 km (354 mi)
Cable arrangement	1 cable, earth return	1 cable per pole	2 parallel cables, earth return	1 cable per pole
Cable distance	96 km (60 mi)	64 km (40 mi)	116 km (72 mi)	42 km (25 mi)
Earthing of the d-c circuit	For full current in two sea electrode stations	Midpoint earthed in one station	For full current in two sea electrode stations	For full current in one earth and one sea electrode station
Control	Constant frequency on Gotland	Constant power in either direction	Constant power or constant frequency on Sardinia or a mixture of both	Constant power from Benmore to Haywards
Reversal of power flow	Effected manually	Controlled by power-setting device	Controlled by power-setting device and frequency-regulator equipment respectively	Possible, not normal
Emergency change of power flow	....	On manual or automatic order to preset value	....	....
Number of circuits	1	1	1	1
Tower structure, main	....	....	Steel	Steel
Tower structure, electrode	Wood, nonstayed	....	Steel, nonstayed	Steel, wood, nonstayed
Number of insulator disks, main, inland	....	....	16	13
Number of overhead ground wires	....	....	1	1, 2
Type of cable insulation	Solid	Solid	Solid	Gas
Cable insulation thickness, mm	7	English 9.3 French 7.5	11.8	14.4
Cable armoring, ϕ, mm	4	6	6.4	5.9
Conductor material, overland	....	....	ACSR	ACSR
cable	Copper	Copper	Copper	Copper
Copper equivalent conductor area, mm², overland	....	....	....	502
cable	90	French 344 English 339	420	516
Thickness of lead sheath, mm	3.0	2.5	2.55	3.55
Conductor cross section, overland	....	....	683.2	974
cable, mm²	90	French 344 English 339	420	516
Approximate overall diameter, cable, mm	50	English 83.5 French 70.4	.... 83.4 × 79.7	111
Main reason for choosing H-V d-c system	Long sea crossing, frequency control	Sea crossing, asynchronous link	Long sea crossing, earth return	Sea crossing plus overhead line
Power company	Statens Vattenfallsverk, Vällingby 1, Sweden	Central Electricity Generating Board, Guildford, England Electricité de France, D.E.R.T. Paris 9e, France	Ente Nazionale per l'Energia Elettrica, Rome, Italy	New Zealand Electricity Department, Wellington, New Zealand

of H-V D-C Systems

Japan 50–60 c/s	Konti-Skan	Vancouver	Pacific intertie 1	Pacific intertie 2
1965	1965	1968–1969	1968	1971
300	250	78–312	1,440	1,440
2 × 125	250	130–260	±400	±400
2 + 2	2	1–2	6	
125	125	130	133	200
1,200	1,000	600–1,200	1,800	1,800
4	4	4	6	2
690	575	347–694	1,040	1,040
Static capacitors A-c grids	Synchr. condenser Static capacitors	Static capacitors	Static capacitors	Static capacitors
Sakuma	Gothenburg	Arnott	Dalles 1	Dalles 2
	Aalborg	Stratford	Sylmar	Mead
275 kV, 50 c/s	Gothenburg 130 kV	Arnott 138 kV	Dalles, 1, 230 kV	Dalles 2, 230 kV
275 kV, 60 c/s	Aalborg 150 kV	Stratford 138 kV	Sylmar 230 kV	Mead, 230 kV
...	86 km (53 mi)	41 km (25.5 mi)	1,330 km (825 mi)	1,350 km (840 mi)
...	1 cable, earth return	1–3 cables, earth return	...	...
...	87 km (54 mi)	28 km (17.5 mi)	...	...
One point earthed direct	For full current in two sea electrode stations	For full current in two sea electrode stations	For full current in two earth electrode stations (intermittent)	For full current in two earth electrode stations (intermittent)
Constant power in either direction	Constant power in either direction	Constant power in either direction	Constant power in either direction	Constant power in either direction
Controlled by power-setting device	Controlled by power-setting device	Controlled by power-setting device	Controlled by power-setting device	Controlled by power-setting device
On manual or automatic order to preset value	On manual or automatic order to preset value	On manual or automatic order to preset value	On manual or automatic order to preset value	On manual or automatic order to preset value
...	1	1	2	2
...	Steel, nonstayed and aluminum, stayed	Steel, nonstayed	Steel, nonstayed	Steel, nonstayed
...	Al. stayed, steel, nonstayed	Steel, nonstayed	Aluminum, stayed Celilo, suspended from d-c line Sylmar, suspended from existing a-c line	Aluminum, stayed
...	13		24	24
...	1		1, partly 2	1
	Sweden *Denmark*			
...	Solid Oil		...	...
...	15 12.4		...	...
...	4 4		...	...
	ACSR	ACSR	ACSR	ACSR
...	Copper Copper	Copper	...	...
...	527		748	748
...	625 620		...	...
...	3.4 3.2		...	...
...	910		1165	1165
...	625 620		...	...
...	87 129 × 92		...	...
Rapid control, low losses, asynchronous link	Sea crossing, building in stages	Sea crossing, building in stages	Long distance, fast control	Long distance, fast control
Electric Power Development Company, Tokyo, Japan	Statens Vattenfallsverk, Vällingby 1, Sweden Elsam, Skaerbaek Pr. Fredericia, Denmark	British Columbia Hydro and Power Authority, Vancouver 1, B.C., Canada	U.S. Department of the Interior, Bonneville Power Administration, Portland 8, Ore. The Department of Water and Power of the City of Los Angeles, Los Angeles, Calif.	U.S. Department of the Interior, Bureau of Reclamation, Denver, Colo.

The communication channel is also used for changes in the transmitted power. In case of an interruption of this communication link, the power on the transmission line remains at the value existing before the interruption.

143. Maintenance. At certain intervals, say once a year, all converter equipment including the valves should be subjected to a checking procedure. For the control equipment special measuring points are available when the doors for the control cubicles are opened and voltages, currents, and oscillographic traces are compared with reference values and oscillograms. If certain differences are found, the corresponding control unit of plug-in type is replaced by a spare unit.

The valves should have a detailed check every year, covering evacuation system, leakage paths, and resistance values in the voltage divider, insulation and grid contacts, and functioning of the auxiliary equipment. Between the major checks, a smaller check of all valves is also made once a year.

All valves should also have an internal renovation every 5 to 10 years, which requires disassembly of the anodes and electrodes, inspection and cleaning of these, followed by reassembling, degassing, and testing.

For a bipolar converter station with three valve groups per pole a summary of the number of staff required for the above maintenance is given in Table 14-9.

In summary, the total staff required for each terminal is:

4 men, category *A*. Those in this category have a knowledge of basic electrical and mechanical functions, along with experience from electrical testing (measuring with dial instruments, oscilloscope, relay testing) and mechanical assembly. At least one man in this category could be trained as reserve for *B*.

1 man, category *B*, who could to some extent be used as reserve for *A*. This man must have a basic knowledge of mechanics, along with experience from a mechanical workshop, and special training in H-V d-c valve assembly.

1 man, category *C*, supervision engineer. This man must have a technical engineering knowledge of electrical theory and some mechanical theory, along with experience from electrical testing and organizing and leading such work. He also must have special training for work on H-V d-c valve and control equipment.

144. H-V D-C Systems. Table 14-10 shows particulars of H-V d-c systems now operating and planned.

145. General Reference List

LANE, F. G., RATHSMAN, B. G., LAMM, U., and SMEDSFELT, K. S. Comparison of Transmission Cost for High Voltage A-C and D-C Systems; *CIGRE*, 1956, Vol. 16, No. 4, *Rept.* 417; also *Direct Current*, 1956, Vol. 3, No. 1, pp. 27–36.

ENGSTRÖM, G. Operation and Control of D-C Power Transmission; *Trans. IEEE Paper* 1963-83.

LAMM, U. H-V D-C Transmission, General Background and Present Technical Status; *Trans. IEEE Paper* 1963-82.

LAMM, U. What Is the Place of H-V D-C Transmission in Today's Power Systems? *Elec. World*, 1963, Vol. 159, No. 20, pp. 98–99, 129–130, 132.

LAMM, U. High Voltage Direct Current—New Applications and New Systems; *Jour. ASEA*, 1963, Vol. 36, No. 7, pp. 91–96.

ENGSTRÖM, P. G., and STACKEGÅRD, H. The Role of H-V D-C Today; *Transmission and Distribution*, 1964, Vol. 16, No. 9, pp. 28–32.

SECTION 15

POWER-SYSTEM INTERCONNECTIONS

BY

NATHAN COHN Senior Vice President, Technical Affairs, Leeds and Northrup Company; Fellow, Institute of Electrical and Electronics Engineers.

W. E. PHILLIPS Head, Electric Power Section, Research and Development Department, Leeds and Northrup Company; Senior Member, Institute of Electrical and Electronics Engineers.

DOUGLAS W. TURRELL Senior Scientist, Development Division, Leeds and Northrup Company; Senior Member, Institute of Electrical and Electronics Engineers.

WILLIAM RUSSELL CLARK President, Eddystone Machinery Company; formerly Staff Assistant to the Senior Vice President, Technical Affairs, Leeds and Northrup Company; Fellow, Institute of Electrical and Electronics Engineers.

JOHN BERDY Applications Engineer, Electric Utility Engineering Operation, General Electric Company; Member, Institute of Electrical and Electronics Engineers.

ARTHUR N. ROBERTSON Electronics and Communications Engineer, American Electric Power Service Corporation; Member, Institute of Electrical and Electronics Engineers.

FREDERICK R. NELSON Assistant Head, Meter and Communications Section, American Electric Power Service Corporation; Member, Institute of Electrical and Electronics Engineers.

CONTENTS

Numbers refer to paragraphs

SECTION 15

POWER-SYSTEM INTERCONNECTIONS

CONTROL OF GENERATION AND POWER FLOW

BY NATHAN COHN

1. Growth of Power Production. The world's production of electric energy has grown rapidly in recent years. Comparative production in major nations, from both thermal and hydroelectric sources, is shown in Fig. 15-1. The United States, with about 6% of the world's population, accounts for more than 35% of the world's energy production. Growth of electric power in the United States[1,2,*] has been at an annual compound rate of about 7%, generating capability doubling about every 10 years. Growth of installed capacity from 1920 to 1965 is shown in Fig. 15-2.

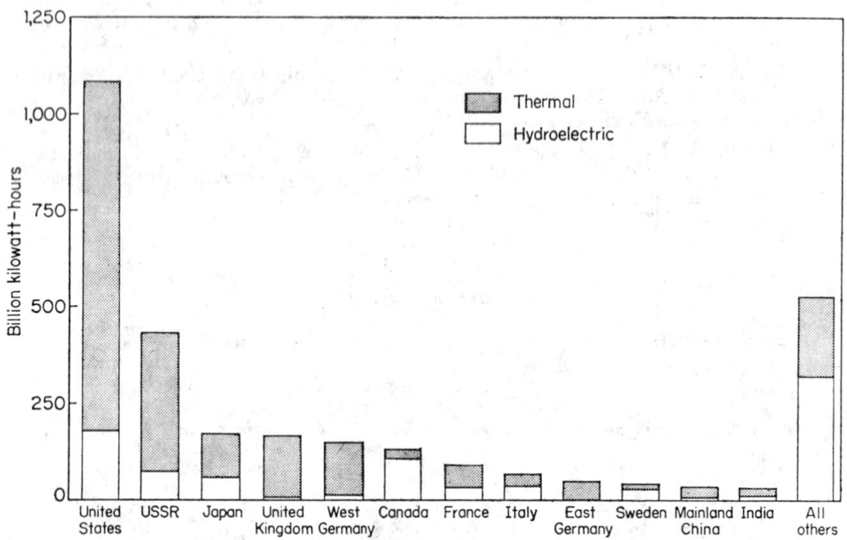

FIG. 15-1. Comparative electric power production in major nations.

2. Interconnected Systems. Adjacent electric power systems have found it advantageous, for reasons of reliability and economy, to interconnect and to operate in parallel.

Advantages of Interconnection. In periods of need, interconnected companies can draw on one another's rotating reserves, increasing reliability of system operation and

* Superior numbers correspond with those of the References, Par. **28**.

contributing to continuity of service to customers. In normal operating periods, adjacent companies can schedule bulk power transfers over intercompany ties to take advantage of energy-cost differentials in the respective areas. Load diversity, seasonal conditions, time-zone differences, or shared investment in larger, and hence more efficient, generating units may make excess generation available in one area at a cost lower than energy could at that time be generated in adjacent areas. Bulk transfers are correspondingly scheduled. Resultant savings are equally shared by participating companies.

Interconnections in the United States. Interconnections between adjacent operating utilities have been steadily expanded in the United States.[3] Major interties in the nation are shown in Fig. 15-3. The extent to which these ties establish interconnected operating areas is illustrated in Fig. 15-4. There are, as shown by the segmented map of Fig. 15-4, five major interconnected areas in the United States and eastern Canada.

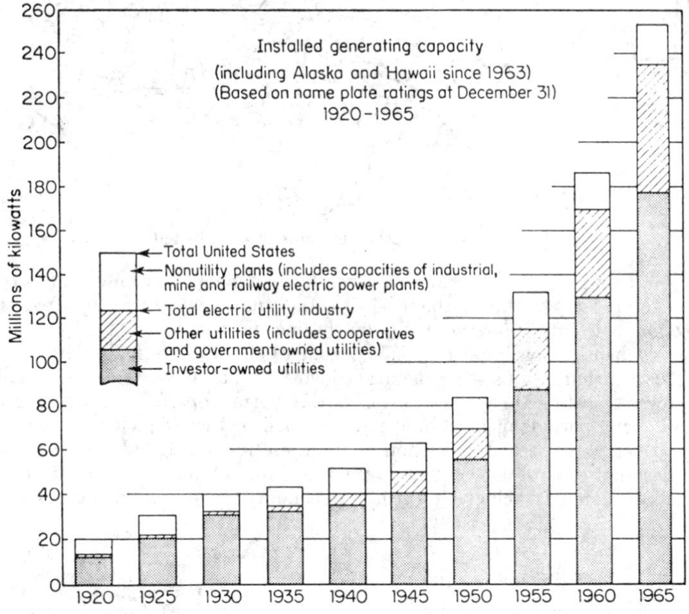

FIG. 15-2. Growth of U.S. installed capacity, 1920 to 1965.

The largest of these interconnections extends to the East from the Rocky Mountains and includes the Middle West, the Gulf Coast, the eastern seaboard, and eastern Canada. Constituent groups within this interconnection include the 115 operating utilities of the Interconnected Systems Group (ISG), the 12 operating companies of the Pennsylvania–New Jersey–Maryland pool (PJM), and the 31 operating utilities of the Canadian-Eastern United States group (CANUSE). The many utilities of this interconnection, some publicly owned, some investor-owned, having a peak load greater than 130 million kW, operate continuously in parallel.

Similar parallel operation is achieved in each of the other four interconnections.

Intertie closures between the interconnected systems, some in actual operation and some projected for the future, extending still further the areas of parallel operation, are shown by the broken lines in Fig. 15-4. At the time of printing this Handbook, the ISG-PJM-CANUSE interconnected system, the Northwest–Rocky Mountain interconnected system, and the Pacific Southwest interconnected system have all been tied

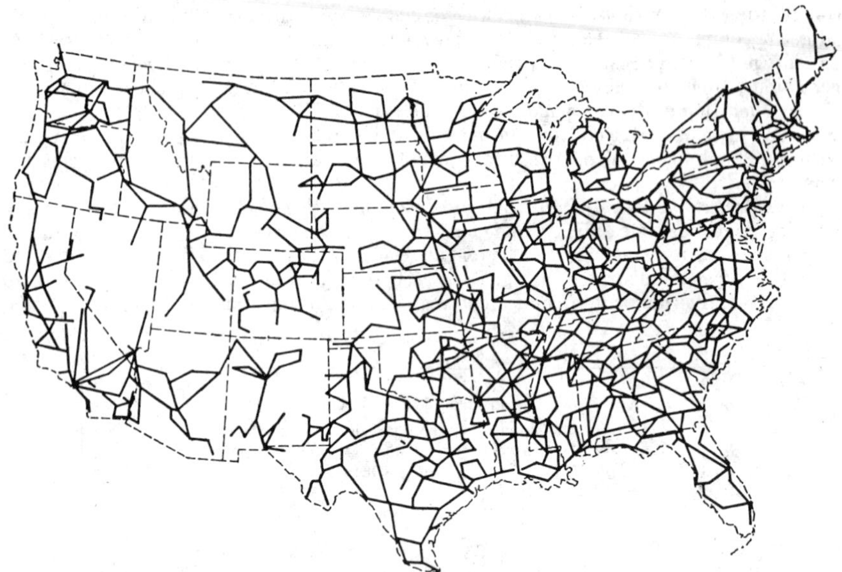

FIG. 15-3. Major interconnections in the United States.

together for test periods, operating successfully as a single nationwide interconnected system, encompassing within it about 94% of the country's generating capacity.

Obligations of Interconnection. While sharing in the benefits of interconnected operation, each participating utility is expected to share correspondingly in its obligations. These include adjusting generation levels to match load changes, maintaining intertie power transfers at scheduled levels during normal operating conditions, assisting neighbors during periods of need, and participation in the frequency regulation of the system. Later portions of this section define these regulating requirements more specifically, describe control concepts and executions for fulfilling them, and discuss techniques for concurrently achieving optimum economy within the utility's own system.

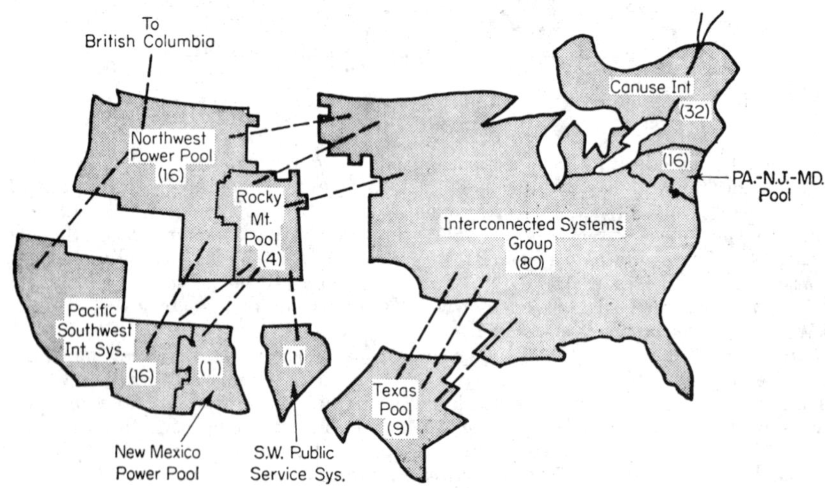

FIG. 15-4. The five major U.S. interconnected systems.

3. Control Areas. Each interconnected system is made up of one or more control areas, each of which is defined[4] as that portion of an interconnected system to which a common generation control scheme is applied. It may also be regarded as that portion of the interconnected system which is expected to regulate its own generation to follow its own load changes. It may consist of a single utility, or a part of one, or a whole group of pooled utilities. In each case a control area would include all the generating units, loads, and lines that fall within its prescribed boundaries. A simplified schematic of a control area is shown in Fig. 15-5. All the control areas of an interconnection, taken together, should account for all the generation, load, and ties of the interconnected system.

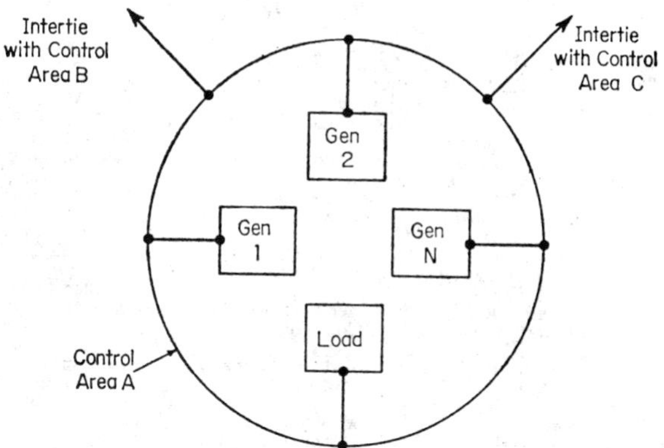

Fig. 15-5. Simplified schematic of a control area.

A *single-area system* is one in which the entire interconnected system is encompassed within one control area. One control system would regulate the entire interconnection and would not distinguish between the locations of load changes within the interconnection.

A *multiple-area system* is one in which there are many control areas, each with its own control system, each normally adjusting its own generation in response to load changes within its own area. All the major interconnected systems in the United States and eastern Canada operate on a multiple-area basis.

ISG-PJM-CANUSE Control Areas. The 158 utilities of the ISG-PJM-CANUSE interconnection of Fig. 15-4 are grouped into 88 control areas, shown schematically as connected circles in Fig. 15-6. Interties between adjacent companies are shown as single lines, though in most cases there are multiple ties between areas. The numbered circles and the additional companies which some of them contain are identified in Table 15-1.

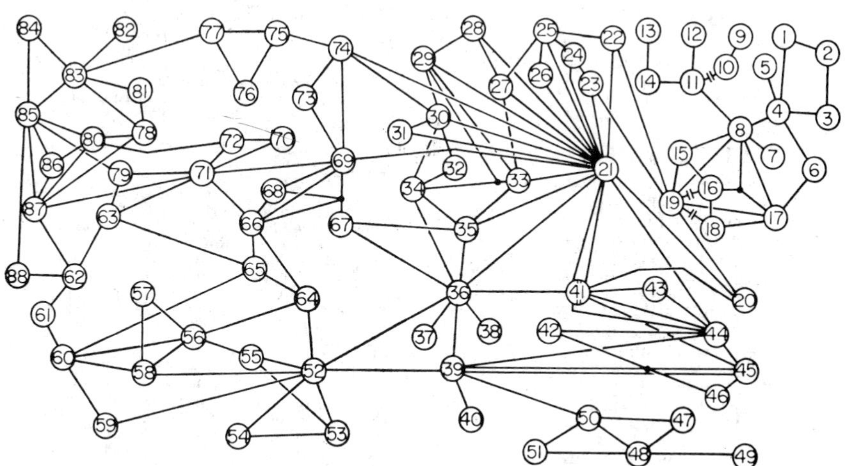

Fig. 15-6. Interconnected control areas of midwestern and eastern United States and Canada.

Table 15-1. Utilities in the ISG-PJM-CANUSE Interconnection Shown in Fig. 15-6

1. Boston Edison Company
   Cambridge Electric Company
2. New Bedford Gas & Electric Company
3. Montauk Electric Company
4. New England Electric System
   Central Vermont Public Service Corporation
   Citizens Utilities Company
   Green Mountain Power Corporation
   Public Service of New Hampshire Company
   Fitchburg Gas & Electric Company
   New England Power Company
5. Central Maine Power Company
   Bangor Hydro Electric Company
6. Connecticut Valley Electric Exchange (CONVEX)
   Connecticut Light & Power Company
   Hartford Electric Light Company
   United Illuminating Co.
   Western Massachusetts Electric Company
7. Power Authority State of New York (included in area 8)
8. Niagara Mohawk Power Corporation
   Rochester Gas & Electric Company
9. Saguenay Power Company
10. Hydro Electric Power Commission of Quebec
11. Hydro Electric Power Commission of Ontario
12. Great Lakes Power Company
13. Consumers Power Company
14. Detroit Edison Company
15. New York State Gas & Electric Corporation
16. Central Hudson Gas & Electric Corporation
17. Consolidated Edison Company of New York
18. Orange & Rockland Utilities, Inc.
19. Pennsylvania-New Jersey-Maryland Pool (PJM)
    Atlantic City Electric Company
    Baltimore Gas & Electric Company
    Delaware Power & Light Company
    Jersey Central Power & Light Company
    Luzerne Electric & Gas Division
    Metropolitan Edison Company
    New Jersey Power & Light Company
    Pennsylvania Electric Company
    Pennsylvania Power & Light Company
    Philadelphia Electric Company
    Potomac Electric Power Company
    Public Service Electric & Gas Company
20. Virginia Electric & Power Company
21. American Electric Power System
    Indiana Michigan Electric Company
    Ohio Power Company
    Appalachian Power Company
    Kentucky Power Company
22. Cleveland Electric Illuminating Company
23. Allegheny Power System
    Monongahela Power Company
    Potomac Edison Company
    West Penn Power Company
24. Duquesne Light Company
25. Ohio Edison Company
    Pennsylvania Power Company
26. Toledo Edison Company
27. Columbus & Southern Ohio Electric Company
28. Dayton Power & Light Company
29. Cincinnati Gas & Electric Company
30. Public Service Co. of Indiana
31. Indianapolis Power & Light Company
32. Southern Indiana Gas & Electric Company
33. Ohio Valley Electric Corporation
34. Louisville Gas & Electric Corporation
35. Kentucky Utilities Company
    East Kentucky R.E.C.C.
36. Tennessee Valley Authority
37. Memphis Light, Gas & Water Division (included in area 36)
38. Tapoco, Inc. (included in area 36)
39. The Southern Company
    Alabama Power Company
    Georgia Power Company
    Gulf Power Company
    Mississippi Power Company
40. Savannah Electric & Power Co. (included in area 39)
41. Carolina Power & Light Company
42. Greenwood County Electric Power Commission
43. Yadkan, Inc.
44. Duke Power Company
45. South Carolina Electric & Gas Company
46. South Carolina Public Service Authority
47. Orlando Utilities Commission
48. Florida Power & Light Company
49. Jacksonville Department of Electricity & Water
50. Florida Power Corporation
51. Tampa Electric Company
52. Middle South System
    Arkansas-Missouri Power Company
53. Central Louisiana Electric Company
54. Gulf States Utilities Company
55. Southwestern Electric Power Company
56. Public Service of Oklahoma Company
    Southwestern Power Administration
57. Western Farmers Electric Cooperative
58. Oklahoma Gas & Electric Company
59. Empire District Electric Company
60. Kansas Gas & Electric Company
    Western Light & Telephone Company
    Central Kansas Power Company
61. Kansas Power & Light Company
62. Omaha Public Power Company
63. Iowa Power & Light Company
64. Southwestern Power Administration—Arkansas, Missouri
    Associated Electric Cooperative, Inc.
65. Kansas City Power & Light Company
    St. Joseph Light & Power Company
    City of Independence Missouri
    Public Utility, Kansas City, Kansas
    Missouri Public Service Company
    Missouri Power & Light—Western
66. Union Electric Company
    Missouri Edison Company
    Missouri Power & Light Company
    Missouri Utilities Company
67. Electric Energy, Inc.
68. Central Illinois Public Service Company
69. Illinois Power Company
70. City of Muskotine, Iowa
71. Iowa-Illinois Gas & Electric Company
72. Eastern Iowa Light & Power Cooperative
73. Central Illinois Light Company
    City, Water, Light & Power, Springfield
74. Commonwealth Edison Company
    Commonwealth Edison of Indiana
    Northern Indiana Public Service Company
    Central Illinois Electric & Gas Company
75. Wisconsin Electric Power Company
    Wisconsin-Michigan Power Company
    Madison Gas & Electric Company
76. Wisconsin Power & Light Company
77. Wisconsin Public Service Company
    Consolidated Water Power Company
    Manitowoc Public Utilities
    Marshfield Electric & Water Utilities
78. Interstate Power Company
79. Iowa-Southern Utilities Company
80. Iowa Electric Light & Power Company
81. Dairyland Power Cooperative
82. Minnesota Power Company
83. Northern States Power Company
84. Rural Power Cooperative Association
85. U.S. Bureau of Reclamation, region 6
86. Corn Belt Power Cooperative
87. Iowa Public Service Company
88. Nebraska Public Power System
    Consumers Public Power District—East

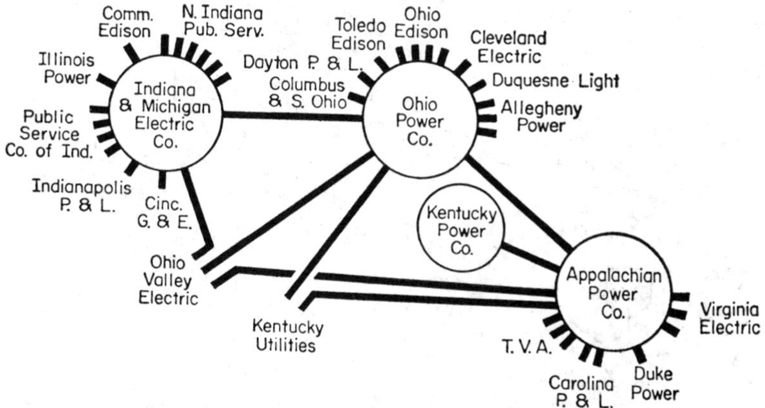

FIG. 15-7. Major operating companies and tie lines of the American Electric Power Company system.

Each of the control areas of Fig. 15-6 has generation, load, and ties after the manner of Fig. 15-5 and with varying degrees of size, complexity, and geographical extent. Illustrative of the extent of a single control area is the seven-state American Electric Power System, circle 21 in Fig. 15-6. Its four operating companies, which have common ownership, the ties between them, and its 40 interties with 19 other control areas are shown schematically in Fig. 15-7.

A large group of independently owned utilities that operate as a single control area is the PJM pool, shown as circle 19 in Fig. 15-6. Its 12 participating companies, the ties between them, and its 17 major interties with six other control areas are illustrated schematically in Fig. 15-8. Prior to its interconnection with ISG and CANUSE, the PJM pool operated as a single-area interconnection.

Western Control Areas. Control areas for both the Northwest–Rocky Mountain interconnected system and the Pacific Southwest–New Mexico interconnected system are shown in Fig. 15-9.

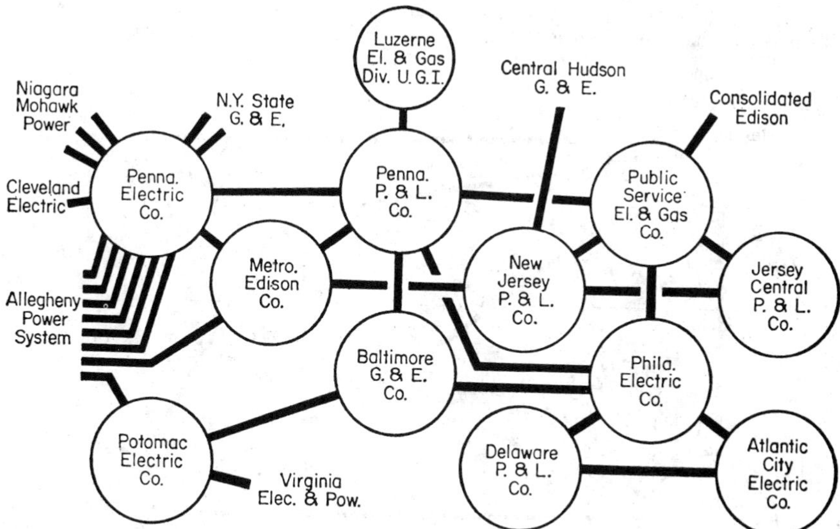

FIG. 15-8. Operating companies and major tie lines of the Pennsylvania–New Jersey–Maryland pool.

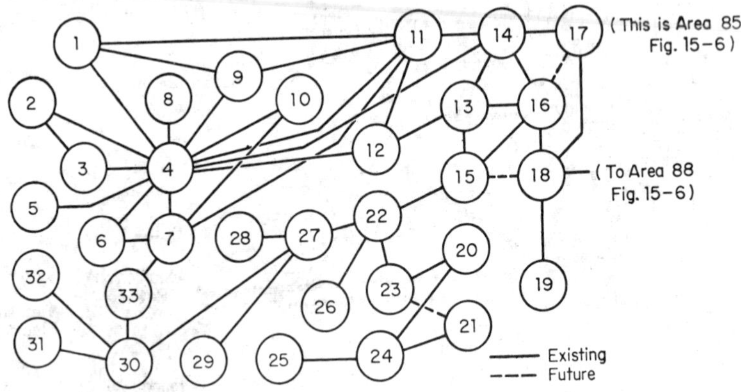

Fig. 15-9. Interconnected control areas of the western United States and Canada. For northwest-southwest parallel operation, ties between areas 7 and 33 and between areas 15 and 22 are closed. For east-west parallel operation, ties between areas 14 and 17, areas 17 and 18, and areas 18 and 88 (Fig. 15-6) are closed. Note that area 17 is area 85 of Fig. 15-6.

For parallel operation of the Northwest and Pacific Southwest, the ties between area 7 and area 33 and between area 15 and area 22 in Fig. 15-9 are closed. For East-West parallel operation, ties from areas 14 and 18 to area 17 (which is area 85 of Fig. 15-6), and between area 18 of Fig. 15-9 and area 88 of Fig. 15-6 are closed.

Texas Control Areas. The control areas of the Texas interconnection and their interties are shown schematically in Fig. 15-10.

4. Operating Objectives of Generation and Power-flow Control. Automatic control of generation and power flow is an essential need for the smooth, neighborly and effective operation of a widespread interconnected system. On a multiple-area interconnection the regulating or control objectives are threefold, as follows:

Objective 1. Total generation of the interconnection as a whole must be matched, moment to moment, to the total prevailing customer demand.

Objective 2. Total generation of the interconnected system is to be allocated among the participating control areas so that each area follows its own load changes and main-

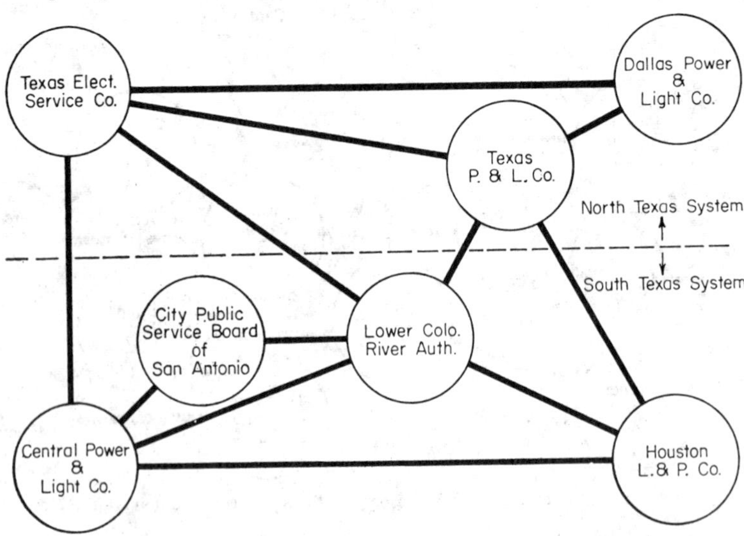

Fig. 15-10. Interconnected control areas of Texas systems.

tains scheduled power flows over its interties with neighboring areas. This objective is achieved by *area regulation.*

Objective 3. Within each control area, its share of total system generation is to be allocated among available area generating sources for optimum area economy. This objective is achieved by *economic dispatch.*

On a single-area system, objective 2 does not apply.

Relative Priority of Control Objectives. For all practical purposes, and except for short-term stored-energy effects discussed later, electric energy as produced and distributed on interconnected power systems cannot be stored. It must be made as it is used. Matching total generation of the interconnection to total load of the interconnection is therefore its first and paramount objective if continuity of service to customers is to be maintained. Within the limits of intertie transfer capabilities and any other applicable factors of area safety or security, objective 1 as listed above will therefore take precedence over the other two objectives.

As between objectives 2 and 3, it is generally felt that each area has an obligation to the interconnection to place higher priority on good execution of area regulation than on optimum area economic dispatch. In other words, if an operating conflict develops in an area, making it difficult for the control simultaneously to achieve objectives 2 and 3, each area will generally be expected to subordinate its desire for optimum economic dispatch to its responsibility to provide area regulation that coordinates effectively with the needs of the interconnection as a whole. Each interconnected system will itself establish criteria on the degree of departure from good area regulation that it considers acceptable for its constituent control areas.

Evaluation of Control Performance. Relative control performance[5] of an area control system in achieving objectives 2 and 3 is illustrated in the curves of Fig. 15-11. The figure includes four sets of hypothetical curves. Each pair plots, on a common time axis, departures from fully effective area regulation (on the left) and departures from optimum area economic dispatch (on the right). Departures from fully effective area regulation are identified as an *area control error* above or below zero.

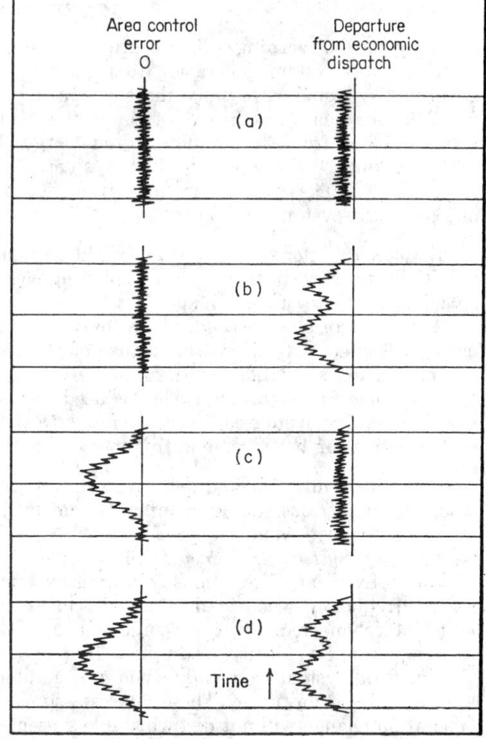

Fig. 15-11. Relative control performance of an area control system. (*a*) Excellent area regulation and dispatch; (*b*) excellent regulation, but at the expense of internal area economy; (*c*) area regulation sacrificed to achieve internal area economy; (*d*) poor regulation and internal area dispatch.

The curves at (*a*) show excellent control performance for both area regulation and economic dispatch. At (*b*) area regulation is again excellent, but this time it is achieved at the expense of internal area economy. At (*c*) area regulation has been sacrificed to achieve internal area economy. At (*d*) control fails to achieve either good area regulation or internal economic dispatch.

From the viewpoint of the interconnection as a whole, control as reflected in the curves at (*a*) and (*b*) would be regarded as excellent, the control area in both instances completely fulfilling its obligation to the

Table 15-2. Utilities in Interconnections Shown in Fig. 15-9

1. Puget Sound Power & Light Company
2. Seattle City Light
3. Tacoma City Light
4. Bonneville Power Administration
5. West Kootenay Power & Light Company
6. Portland General Electric Company
7. Pacific Power & Light Company
8. British Columbia Hydro & Power
9. Chelan PUD
10. Glant County PUD
11. Washington Water Power Company
12. Idaho Power Company
13. Utah Power & Light Company
14. Montana Power Company
15. U.S. Bureau of Reclamation, region 4
16. Pacific Power & Light Company, Wyoming Division
17. U.S. Bureau of Reclamation, region 6 (area 85 of Fig. 15-6)
18. U.S. Bureau of Reclamation, region 7
19. Public Service Company of Colorado
20. Plaines Electric Gas & Telephone Cooperative
21. El Paso Electric Company
22. Arizona Public Service Company and Salt Water Power District
23. Public Service Company of New Mexico
24. U.S. Bureau of Reclamation, region 5
25. Community Public Service Company
26. Tucson Electric Light & Power Company
27. U.S. Bureau of Reclamation, region 3
28. Imperial Irrigation District
29. Arizona Electric Power Cooperative
30. Southern California Edison Company
31. San Diego Gas & Electric Company
32. Los Angeles Department of Water & Power
33. Pacific Gas & Electric Company

interconnection. The area regulation at (*c*) and (*d*) might be regarded by the interconnection as less than satisfactory, depending upon the magnitude and duration of the departures of area control error from zero, as compared with performance criteria that the interconnection would regard as reasonable.

GOVERNING

5. System Governing. There are two components in the governing action that matches total system generation to total system load. One derives from the speed governing systems[6] with which most generating units are equipped. The other results from the frequency coefficient of connected load. In addition, there is a short-term transient effect from the spinning stored energy of the system when load changes occur.

The sequence of steps by which system load changes are accommodated by corresponding changes in system generation is, on assuming, for example an *addition* of load to the system, as follows:

1. *Change in stored energy.* The additional increment of load applied to the system is initially satisfied from the stored spinning energy of the system, resulting in a corresponding decrease of system speed or frequency.
2. *Change in effective load.* The total system load, because of its frequency coefficient, will effectively decrease because of the frequency decrease, correspondingly releasing already available generation to serve the newly added load increment.
3. *Change in generation.* The reduced frequency will also cause generating unit-speed governing systems to increase input to their prime movers, increasing generation and arresting further change in frequency.

For a *drop* in connected system load, system stored energy and system frequency would both *increase,* the remaining system load would effectively *increase,* and total system generation would be *decreased.*

Effect of System Frequency. The above steps of system governing will result in sustained frequency deviations following system load changes, causing a decrease in system frequency when load is added and an increase when load is dropped. On a large interconnected system, a sudden load change of about 2% of system spinning capacity would typically result in a frequency change of about 0.1 Hz. Similarly, the effect on system frequency when a large block of generation is lost is of this same approximate magnitude. These figures assume that the impact of the sudden load or generation change will not disturb stable system operation and that the complete interconnection remains in synchronism. Unless specifically stated to the contrary, this same assumption applies throughout this section.

Index to Mismatch of Generation and Load. A *changing* system frequency is the index of a mismatch between total system generation and total system load. Governing action is in process while such acceleration or deceleration of system speed is taking place. It has been effectively completed and the match between generation and load restored when frequency is steady, even though it is at a value other than the normal frequency schedule, 60 Hz in the United States.

Importance of Constant Frequency. A constant frequency at its normal scheduled value is not necessarily, of itself, a major system operating objective.[7] The deviations from normal value that result from governing action might, therefore, not be regarded as objectionable. Synchronous time would be affected but could readily be corrected periodically. Much more significant, however, is the fact that the automatic control equipment which carries out the functions of area regulation and economic dispatch utilizes prevailing frequency as a parameter in the control execution. As is discussed later, this is the factor that permits the automatic controls in the many control areas of an interconnection to operate simultaneously without interaction or hunting between them. More specifically, proper allocation of total system generation to each of the control areas, under this type of area regulation, occurs only when system frequency is at its normal value. For this reason, maintaining system frequency constant, which is to say restoring it to its normal value after departures owing to system governing action, becomes a system control objective.

On the large ISG-PJM-CANUSE interconnection, frequency typically has a moment-to-moment bandwidth of about 0.01 to 0.02 Hz, with superimposed deviations of the band average, on a time scale of minutes, usually well within ±0.01 Hz of nominal 60 Hz. A section of frequency chart from this interconnection, made on an exceptionally high-resolution measuring instrument, is shown in Fig. 15-12.

Natural Regulation and Supplementary Control. System governing actions are sometimes referred to as *natural regulation.* Subsequent regulating steps, executed manually or automatically through the speed changers of one or more system generator governors to correct frequency deviations, area generation allocations, tie-line loadings, economic dispatch, or combinations of these conditions, are defined as *supplementary control.*

6. Stored-energy Considerations. Restoration of system frequency to its nominal value, for a given connected load, is a matter of restoring system stored spinning energy to the value that corresponds to normal frequency. Such restoration is achieved by supplementary control creating overgeneration if frequency is to be raised or undergeneration if frequency is to be lowered. This in effect creates a new mismatch between total generation and total load, system governing action follows, and a new frequency level is established corresponding to the new level of stored spinning energy in the system. Such supplementary control is continued until normal frequency is restored.

It is significant to note that supplementary control action, because of the mismatch between generation and load it creates, is always followed by governing action. Within the limits of governor sensitivities and dead band, there will correspondingly be reallocation or shifting of generation between units which have remained on governing control only and those to which supplementary control has been applied.

Stored-energy Equations. Stored spinning energy may be written as a function of the square of system frequency, as follows,

$$S_1 = \left(\frac{F_1}{F_0}\right)^2 S_0 \tag{15-1}$$

where S_1 is spinning stored energy at prevailing system frequency F_1 and S_0 is spinning stored energy at scheduled frequency F_0.

For a small change in frequency, ΔF, a change in stored energy, ΔS, will be

$$\Delta S = \left(\frac{F_0 + \Delta F}{F_0}\right)^2 S_0 - S_0 \tag{15-2}$$

When ΔF is small compared with F_0, this becomes

$$\Delta S = \frac{2\,\Delta F}{F_0} S_0 \tag{15-3}$$

Numerical Example. At scheduled frequency F_0, system stored energy S_0 is the product of the system spinning capacity and the average inertia constant for the system at that frequency. One reference[8] suggests inertia constants of 2 to 6 kWs/kVA for hydro units and 5 to 9 kWs/kVA for steam units. Upon assuming that a reasonable constant for all rotating equipment of the system is 6 kWs/kVA, the stored spinning

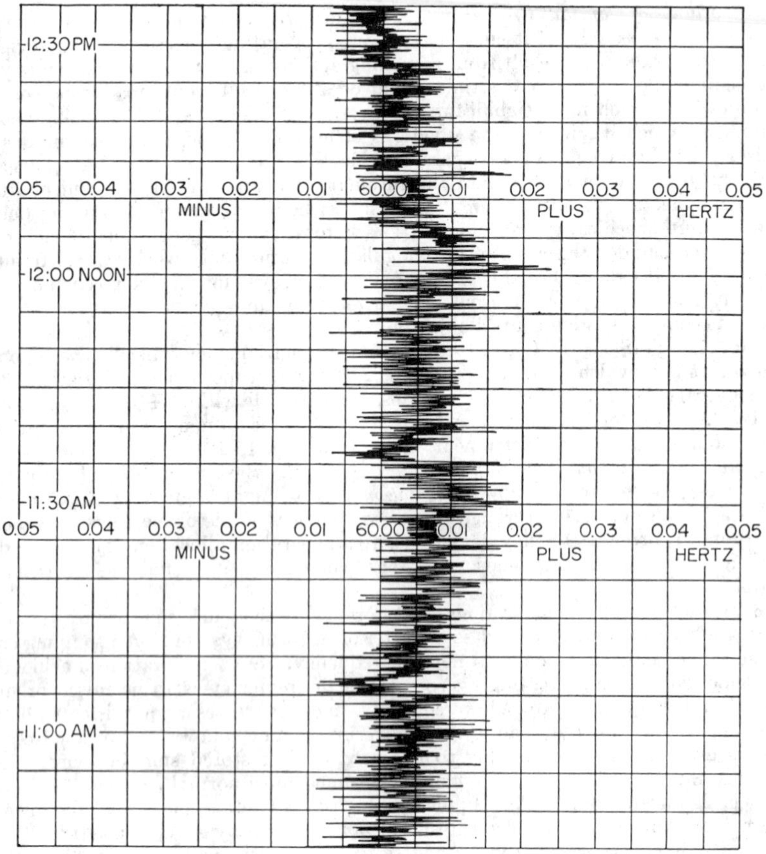

Fig. 15-12. Frequency-deviation chart of ISG-PJM-CANUSE interconnection.

energy at normal frequency for a 150,000 MVA system would be on the order of 900,000 MWs.

On such a system a sudden load increase of 25 MW, for example, could be served from system stored energy for a period of 6 s with a frequency drop of 0.005 Hz. Similarly, stored spinning energy would have to be increased by 150 MWs to raise system frequency 0.005 Hz.

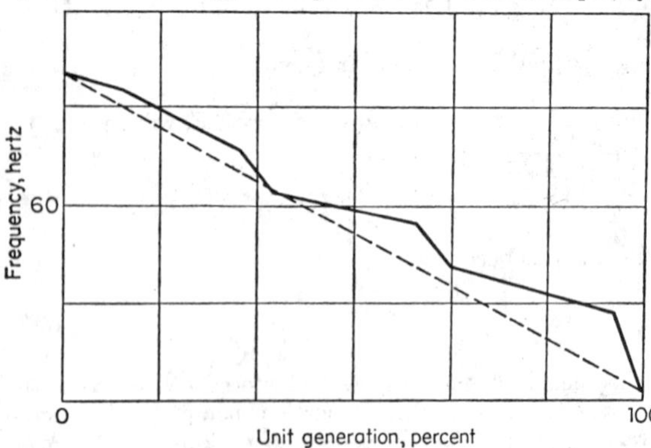

Fig. 15-13. Typical generating-unit governing characteristic.

7. Unit Governors. A typical generating unit governing characteristic is illustrated by the solid-line curve of Fig. 15-13, which is a plot of unit output vs. frequency. Such curves are not linear over the full range of unit output. They have inflection or break points, for example, at inlet valve-opening points on multivalve turbines.

Steady-state Speed Regulation. For a single unit *steady-state speed regulation* is the change in steady-state speed expressed as a percent of rated speed when the output of the unit is gradually reduced from rated to zero power. It is sometimes referred to as *percent droop.* It is represented by the slope of the broken line in Fig. 15-13.

Steady-state Incremental Speed Regulation. For a single unit, the *steady-state incremental speed regulation* is defined, at a given steady-state speed and power output, as the rate of change of steady-state speed with respect to power output expressed in percent of rated speed. It is sometimes referred to as *percent incremental droop.* In Fig. 15-13 it is represented by the slope of each of the segments that make up the overall characteristic.

Dead Band. The *dead band* of a speed-governing system is the measure of its insensitivity to changes in system speed and is expressed in percent of rated speed.

Hypothetical Governing Characteristic. For many aspects of power systems' operations and control analysis it is convenient and acceptable to consider that unit governing characteristics are linear over the small range of system speed being considered. Such a characteristic, solid line *GG,* is shown in Fig. 15-14.

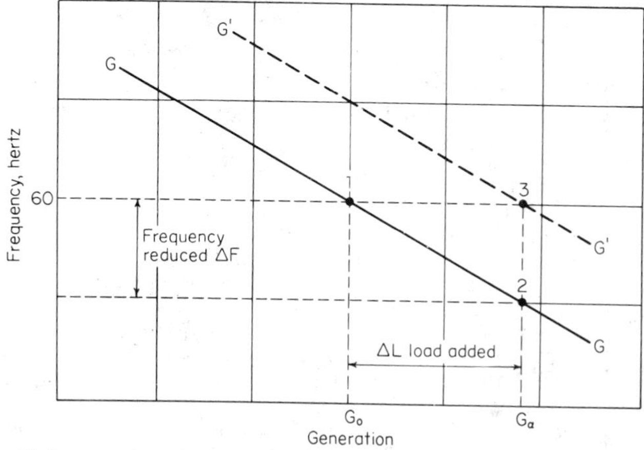

FIG. 15-14. Unit governing characteristics (considered linear over a small range of system speed).

Shifting the Governing Characteristic. Applying supplementary control to the speed changer of a speed-governing system has the effect of shifting the unit characteristic parallel to itself, illustrated by the broken line *G'G'* in Fig. 15-14. A complete cycle of governing action and supplementary control for an isolated unit with a governing characteristic as shown in Fig. 15-14 and with a connected load having zero frequency· coefficient is as follows: Point 1 on *GG* defines initial conditions of generation and load G_0 and 60 Hz. Load is increased by ΔL, and governing action achieves balance at point 2. Supplementary control shifts the generating characteristic to *G'G'*, on which frequency is restored and the new load accommodated at point 3.

Alternative Expressions for Governing Characteristic. In addition to stating speed regulation or governing characteristics in terms of percent of rated speed, it is sometimes desirable to express these parameters either in percent capacity per 0.1 Hz or megawatts per 0.1 Hz. Conversions may be made in accordance with the following relations,

$$N = \frac{100}{6D} \tag{15-4}$$

$$N' = \frac{M}{6D} \tag{15-5}$$

where N is the governing characteristic in percent of unit capacity per 0.1 Hz, D is the steady-state incremental speed regulation in percent, N' is the governing characteristic expressed in megawatts per 0.1 Hz, and M is the unit capacity in megawatts.

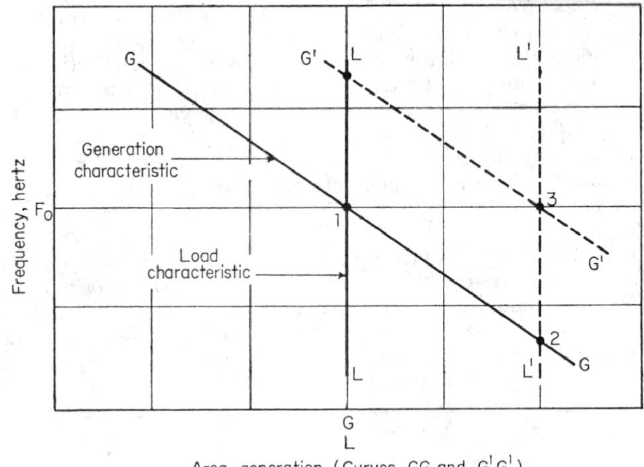

Fig. 15-15. Composite-area governing characteristics, with zero frequency coefficient.

It is convenient to note from Eq. (15-4) that the product of N and D is always $16\frac{2}{3}$.

Equations (15-4) and (15-5) can also be applied to comparable area and system parameters.

8. Area Governing. A typical control area will have many generating units of varying types, sizes, and ages, with varying speed-regulating characteristics and dead bands. Taken in the aggregate, operating units of an area may be regarded as having a composite governing characteristic similar to curve *GG* in Figs. 15-15 and 15-16. Typical area generation governing characteristics expressed as percent of rated speed fall in the range of about 16 to 5%, corresponding according to Eq. (15-4) to a range of about 1 to 3.5% of capacity per 0.1 Hz.

Area-load Frequency Characteristic. The frequency coefficient of connected load in a control area defines its *area-load frequency characteristic.* The latter is the change

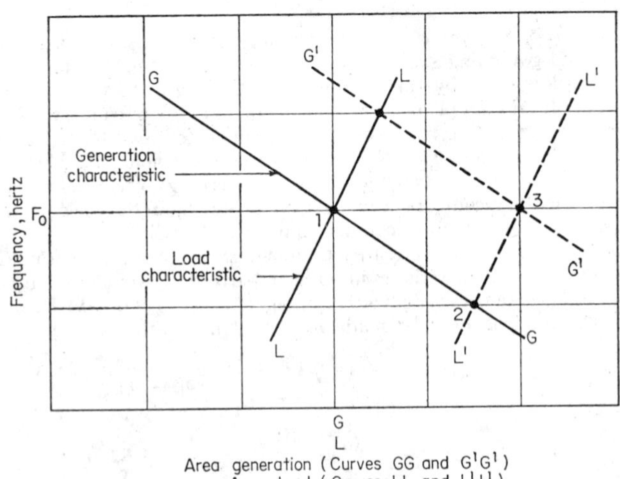

Fig. 15-16. Composite-area governing characteristics, with load-frequency characteristic at one-third the generation governing characteristic.

in total area load that results from a change in system frequency. It is usually expressed in percent of connected load per 0.1 Hz. This characteristic is reported[9] to cover a fairly broad span in various areas. Typical averages are in the range of 0.3 to 0.5%/0.1 Hz. In Fig. 15-15 the load-frequency characteristic *LL* is drawn to illustrate zero frequency coefficient. In Fig. 15-16 the characteristic *LL* is shown with a typical value of about one-third the generation governing characteristic.

Matching Area Generation and Load—Isolated Operation. Consider an area with governing characteristics first as in Fig. 15-15 and then as in Fig. 15-16, operating isolated. In both graphical representations, balance between area generation and area load exists where the load frequency characteristic *LL* intersects the generation governing characteristic *GG*. In both figures, initial conditions are as at point 1. Load is then increased to the level defined by *L′L′*. A new point of balance is achieved at point 2, by generation governing only in Fig. 15-15, by both generation governing and the effect of area-load frequency characteristic in Fig. 15-16. In each case supplementary control is then applied, shifting *GG* to a new position *G′G′*, and as shown in each of the figures balanced conditions with frequency returned to normal are achieved at point 3.

Matching within Dead Band—Isolated Operation. Upon assuming an appreciable overall dead band for the aggregate of its unit governors, the area generation governing characteristic for small load changes that fall within the limits of the dead band may be considered to be a vertical line, such as *GG* in Fig. 15-17. The load frequency characteristic is as shown by *LL*. On isolated operation and starting from balanced conditions as point 1, a change in load ΔL too small to initiate generator speed governor action can be accommodated on a sustained basis after the initial response from stored spinning energy by the frequency coefficient of area load. Area generation remains as it was before the load increase. This condition is illustrated by point 2 in the figure, at the intersection of *GG* and *L′L′*.

Area Frequency-response Characteristic. The combined effect of area generation governing and change in area load with frequency is defined as the *area frequency-response characteristic.* It is usually expressed in percent of area spinning capacity per 0.1 Hz and is the arithmetic sum (algebraic difference) of its two components, expressed to the same base. It will typically fall in the range of 1 to 4%/0.1 Hz. Its magnitude when expressed in megawatts per 0.1 Hz will vary with the level of area load and with the magnitude and nature of area spinning capacity. Its magnitude can be determined for a given set of operating conditions by tripping a significant block of load or generation when an area is operating isolated and observing the resulting change

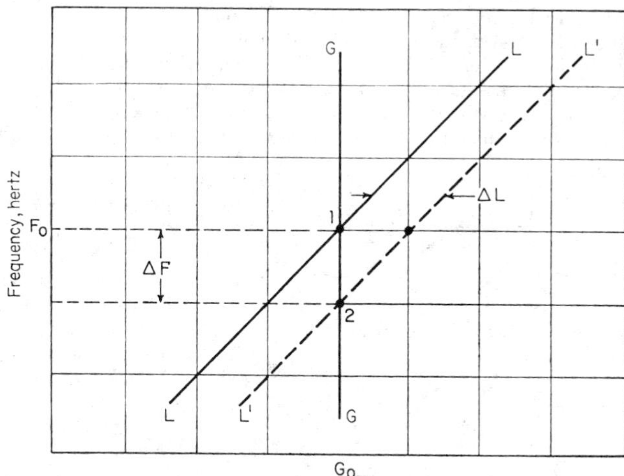

FIG. 15-17. Area generation governing characteristic for small load changes (*GG* vertical line).

in frequency. When operating interconnected, noting the power-flow change on area interties when a disturbance outside the area causes a significant change in system frequency provides data for a computation of the area frequency-response characteristic.

Area Responses when Interconnected. When an area operates as part of an interconnection, it responds not only to its own load changes but also to the load changes that occur in other areas. The governing characteristic curves of Figs. 15-15, 15-16, and 15-17, though referred to earlier as applicable to an isolated area, may be regarded as applicable to a complete interconnected system, which in itself is really a large isolated area. The *GG* curves of these figures represent the composite characteristic of all the generation governors in operation within the system. The *LL* curves are the aggregate of all area load-frequency characteristics. The arithmetic sum of the *GG* and *LL* characteristics represents the *system frequency-response characteristic* and is typically on the order of 2% of spinning capacity per 0.1 Hz for large interconnections.

Because governors cannot recognize the origin or location of a load change, all areas of an interconnection share in system load changes in proportion to their respective frequency-response characteristics. Supplementary control is required to allocate generation changes to the control areas in which the load changes occurred.

BIAS CONTROL

9. Area Regulation. Various control techniques have evolved over the years for achieving effective area regulation in the constituent areas of an interconnection. In one technique, an area was assigned the task of controlling frequency while other areas endeavored to hold power flow on interarea tie lines at fixed levels. In another technique, one area still had the frequency-control assignment, while other areas sought to maintain interarea ties at levels that varied with system frequency. Both these methods have important limitations, as analyzed in Ref. 10, and result in inequitable distribution of regulation requirements and in interarea hunting. The preferred technique, and the one which is standard for the large interconnections in the United States and Canada, does not assign frequency control to any one area. Instead, all participating areas regulate their interarea ties in a manner that permits predetermined departures from scheduled flows for variations in system frequency. When this regulation is properly executed, the following desirable operating conditions are automatically achieved:

1. Under normal conditions, with each area able to carry out its control obligations, load changes are assigned to and absorbed by the areas in which they originate, interarea power transfers are held at scheduled levels, all areas share in the control of system frequency, and normal system frequency is obtained.

2. Under abnormal or emergency conditions, when one or more areas are unable to carry out their control assignments, and assuming the interconnection remains in synchronism, all other areas will automatically assist the areas that are in need by permitting interarea power transfers to depart from normal schedules in the direction that will help the trouble areas, and frequency will be permitted to depart from normal, to the extent required to provide the necessary assistance over the interarea ties to the trouble areas.

This type of automatic control is identified as *net interchange tie-line bias control.*

Interchange Flow Paths. When an area has bulk transfer ties with more than one additional area, scheduled interchange transfers between them, though maintained at

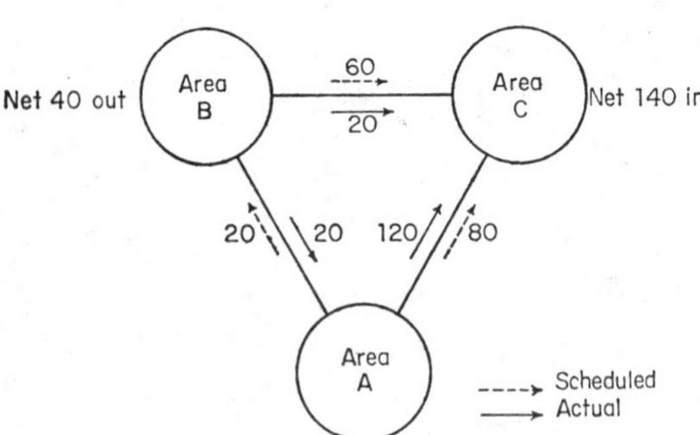

Fig. 15-18. Power flow in individual tie lines at different levels and in direction opposite to schedules.

scheduled levels, will not necessarily flow directly between them. Flows are likely to split over parallel paths through other areas. In some cases, flow on an individual tie line may be in the direction opposite to its schedule, though in the aggregate each area is achieving the correct net of all of its prevailing schedules.[11] This can be seen by considering area A of Fig. 15-5 tied to areas B and C as shown in Fig. 15-18. Scheduled interarea transfers are as shown by the broken-line arrows. Actual flows may be as shown by the solid-line arrows. Scheduled and actual flows on individual ties are summarized in Table 15-3.

Table 15-3. Scheduled and Actual Tie-line Power Flows as Shown in Fig. 15-18

Interarea tie	Scheduled flow	Actual flow
AB	20 to *B*	20 to *A*
AC	80 to *C*	120 to *C*
BC	60 to *C*	20 to *C*

It will be noted in Fig. 15-18 that the actual *net* interchange of each area with the interconnection as a whole does match the algebraic sum of its schedules with adjacent areas. This is summarized in Table 15-4.

Table 15-4. Scheduled and Actual Net Interchange Flows as Shown in Fig. 15-18

Area	Net of area interchange schedules	Actual area net interchange
A	100 out	100 out
B	40 out	40 out
C	140 in	140 in

In applying automatic control, and as suggested by Tables 15-3 and 15-4, an area having ties with more than one other area cannot regulate just one interarea schedule. It must regulate to the net of all of its interarea schedules.

Net Interchange as a Measure of Area Balance. An objective of area regulation is to match area generation changes to area load changes while interarea transfers are maintained on schedule. The aggregate of area load changes or the extent to which they have been matched by area generation changes cannot be measured directly. However, deviations of net interchange from schedule are an index of the extent to which area load changes have not been matched by area generation changes and become a direct measure of the generation required in the area. The following relations apply,

$$G = L + T_0 \tag{15-6}$$

$$G_1 = L + T_1 \tag{15-7}$$

where G is the area generation required to match prevailing area load L and maintain area net interchange schedule T_0, interchange "out" is $+$, interchange "in" is $-$, and G_1 is the prevailing area generation, with prevailing area load L and actual net interchange T_1. Subtracting Eq. (15-6) from Eq. (15-7),

$$G_1 - G = T_1 - T_0 \tag{15-8}$$

Equation (15-8) states that the change required in area generation to follow its own load changes and at the same time restore net interchange to schedule is given by the deviation of area net interchange from schedule.

10. Net Interchange Tie-line Bias Control. Area regulation based on Eq. (15-8) would represent *constant net interchange control,* sometimes referred to as *flat tie-line control,* and would have the limitations[10] already referred to. Expanding Eq. (15-8) to include a frequency bias factor provides the effective and cooperative type of area regulation identified as *net interchange tie-line bias control.*[10,12]

Area Control Error. In operating on net interchange tie-line bias control, deviation from desired area generation is defined as the *area control error* and is given by

$$E = (T_1 - T_0) - 10B(F_1 - F_0) \tag{15-9}$$

where E is the area control error, in megawatts, T_1 is the area net interchange in megawatts, power "out" being considered as positive, T_0 is the area net interchange schedule in megawatts, F_1 is system frequency, in hertz, F_0 is system scheduled frequency, and B is area frequency bias, in megawatts per 0.1 Hz, and is considered to have a negative sign.

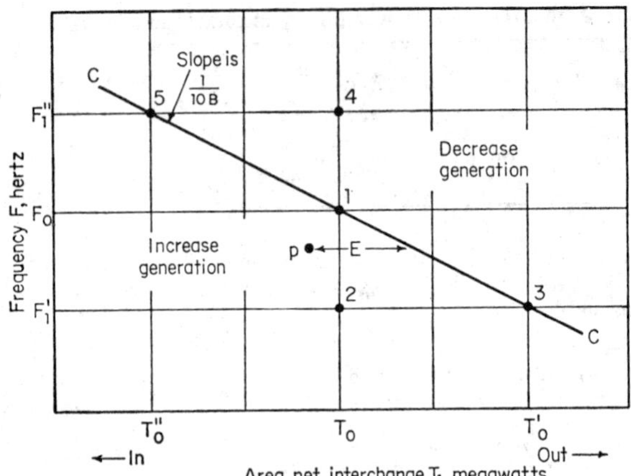

FIG. 15-19. Area tie-line bias zero-control-error curve.

The objective of area regulation is to compute E continuously and automatically and to adjust area generation as required to reduce E to zero.

While no longer preferred terminology, much of the literature on tie-line bias control utilizes the term *area requirement* in defining deviation of area generation from the desired value. For ready cross reference, area requirement is arithmetically equal to area control error but is of opposite algebraic sign.

Bias Regulating Characteristic. The regulating characteristic of a net interchange tie-line bias control which operates to reduce E of Eq. (15-9) to zero is shown as curve CC in Fig. 15-19. Point 1 on CC defines the scheduled net interchange T_0 at normal frequency F_0. The slope of curve CC is the reciprocal of $10B$.

At each point in time, a plot of prevailing system frequency at prevailing net interchange will define a point on the coordinates of Fig. 15-19. When this point falls on CC, the area control error will be zero. The controller will be in balance, and it will not change area generation. When a point such as p does not fall on CC, its horizontal

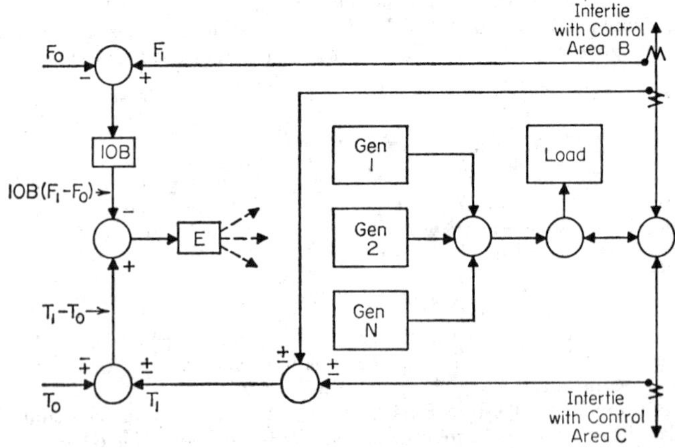

FIG. 15-20. Computation of an area control error.

displacement from curve CC is the control error E. When it falls above the curve, E is plus, indicating a need to decrease area generation. When it falls below the curve, E is minus, reflecting a need to increase area generation. In each of the latter cases, the automatic control would adjust area generation until E is reduced to zero, and the point defining the resultant net interchange and frequency falls on curve CC.

Rotating characteristic CC clockwise to a vertical position would yield the control curve for *constant net interchange control.* Rotating the characteristic CC counterclockwise to a horizontal position would provide the control curve for *constant frequency control.*

Computation of Area Control Error. A schematic diagram of the computation of area control error in accordance with Eq. (15-9) for the control area of Fig. 15-5 is shown in Fig. 15-20. While this area has only two interties that need be summated to obtain area net interchange T_1, most areas will have many ties. All must be included in the net interchange summation.

Power flow on tie lines normally swings or oscillates at a period of a few seconds. To obtain an accurate measurement of net interchange, and to ensure that any power that is being wheeled through the area be excluded from the measurement, tie-line telemetering equipment should provide simultaneity of measurement of all ties that enter into the computation.

The broken lines leaving block E in Fig. 15-20 represent the application of control to area generators to reduce E to zero (see Par. **15** for a summary of pertinent techniques and considerations applicable to this facet of generation control).

Functions of the Bias Factor. The inclusion of the bias factor $10B(F_1 - F_0)$ in Eq. (15-9) fulfills two functions:

1. It coordinates area regulation with area governing responses to load changes that occur in other control areas. Depending on the bias setting, the bias factor will permit such area governing responses to persist, will add to them, or will diminish them.

2. It assigns to each control area a share of the frequency control burden. For example, if net interchange is on schedule for all control areas, but frequency is too low, then for each area the factor $T_1 - T_0$ is zero but the bias factors are not. All areas, therefore, have an area control error E proportionate to the frequency deviation $F_1 - F_0$ and to the respective bias settings. All areas will therefore participate in overgenerating to restore system stored energy and system frequency to their normal values.

Bias as a Schedule Shift. The bias factor of Eq. (15-9) may be regarded as causing a shift in area net interchange schedule with frequency. Equation (15-9) may be rewritten,

$$E = T_1 - T_0' \qquad (15\text{-}10)$$

where T_0' is the biased scheduled interchange for the area at frequency F_1 and is equal to $T_0 - 10B(F_1 - F_0)$.

In Fig. 15-19, for example, point 2 defines normal interchange schedule T_0 and frequency F_1'. The point does not fall on CC, however, and the new biased schedule for this frequency is T_0', defined by point 3 on CC. Similarly, point 4 is not on CC, and for frequency F_1'' the new biased schedule is T_1'', defined by point 5 on CC.

An Illustration of Bias Action. The sketches in Fig. 15-21 illustrate how the bias factor coordinates area regulation with area governing responses to remote load changes, and also illustrate the concept of schedule shift with frequency. In this example,[13] area A, in which a load change of five units in magnitude occurs, is assumed to have one-fifth the interconnection capacity. The remaining four-fifths of the interconnection capacity is shown as another single area B. Governing characteristics and bias settings are assumed to be of the same percentage magnitude in each area and, hence, when expressed in megawatts per 0.1 Hz are proportional to respective area capacity. Both areas are on net interchange bias control, and the interchange schedule is zero.

At sketch (a) balanced conditions prevail prior to addition of five units of load at area A. At (b), the effect of governing responses to these load changes is shown. Four units of the new load at A are picked up by B as the result of governing action and appear as departure from normal on the B-A tie. Upon assuming that the bias

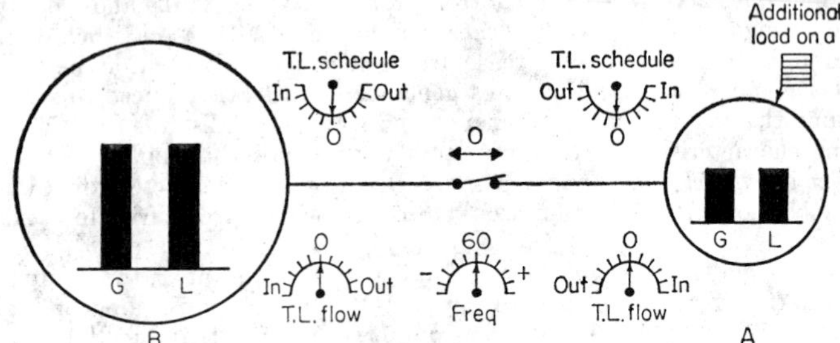

(a) Balanced conditions prior to load change at A

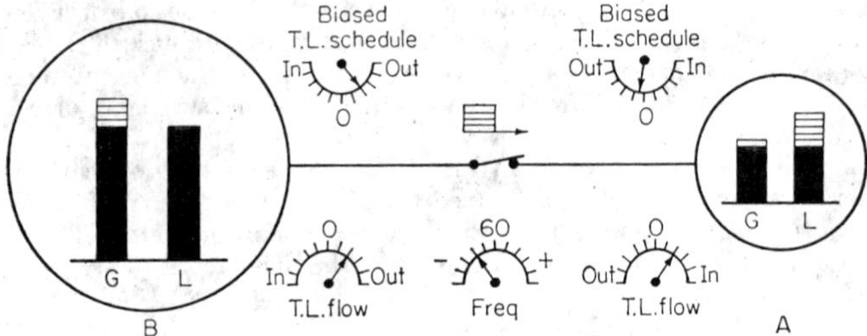

(b) Conditions following governing action in response to load increase at A

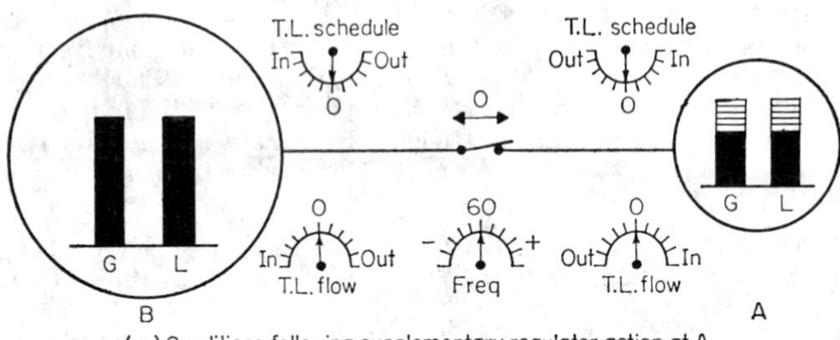

(c) Conditions following supplementary regulator action at A

Fig. 15-21. Governing and supplementary regulation responses, illustrating schedule shifts with frequency.

factor at B is set to match its governing characteristic, the new biased schedule for area B will call for four units of flow from B to A. Since this flow already exists, E for area B will remain at zero, permitting the governing contribution of area A to persist on the B-A tie.

At area A, however, the plot of F_1 and T_1 will fall considerably below and to the left of its CC curve, like point p of Fig. 15-19. There will be a minus control error E, and control will act to increase generation correspondingly. As generation at A is increased, frequency will be raised and governing action at B will reduce its generation contribution correspondingly. In this hypothetical example, these steps will continue until fully balanced conditions are restored as at sketch (c), with all five units of added load in area A being accommodated by five units of added generation in that area.

BIAS SETTINGS

11. Influence of Bias Settings on Control. The magnitude of the bias setting, B in Eq. (15-9), has relatively little effect on the ability of a tie-line bias controller to respond to load changes that occur within its own area. Even with zero bias the controller could fulfill this function. The bias setting has considerable influence, however, on the response of the controller to load changes in other areas and on the coordination of the controller with area governing and with controller actions in other areas.[14,15]

Response to Remote Load Change. Load changes that occur outside a given control area are considered from the viewpoint of that area as *remote* changes. To demonstrate bias control responses to such changes and their effect on significant system and area parameters, consider two areas connected as in Fig. 15-21. Assume that both areas have the same percentage governing and bias characteristics and that a step-function load change sufficient to cause a discernible change in system frequency occurs at area A. Assume also that the bias control at area A is inoperative, while the bias controls at area B, which represents all the rest of the interconnection, do respond.

The nature of the changes in system frequency F, tie-line power flow T, effective load L at area B and generation G at area B is shown in Fig. 15-22.

Initial steady-state conditions apply from t_0 to t_1. The step-function load change, an increase in this example, occurs at area A at t_1. From t_1 to t_2, bias control acts at area B, resulting in the new steady-state conditions shown from t_2 to t_3.

Effect of Bias Ratio. The action from t_1 to t_2 (Fig. 15-22) depends on the ratio R of the bias setting in area B to its combined frequency-response characteristic. When $R = 1$, the several parameters do not change from t_1 to t_2. There is full coordination of the bias control with governing responses at area B. This was the condition in sketch (*b*) (Fig. 15-21).

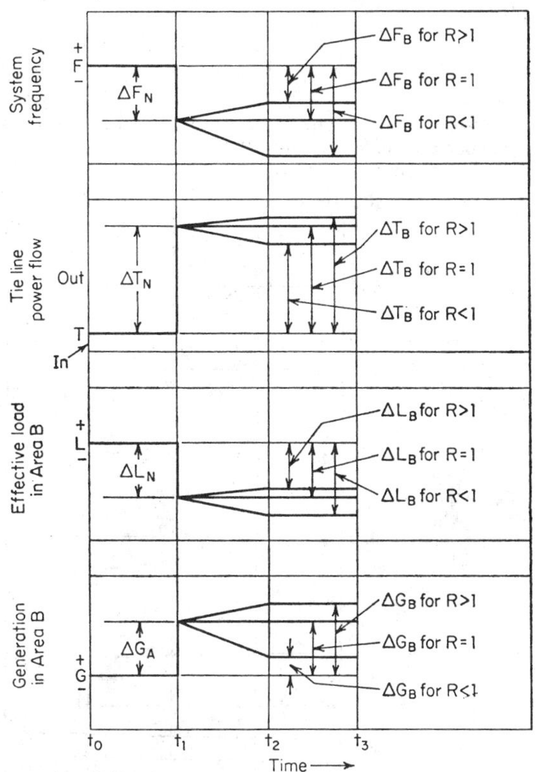

FIG. 15-22. Changes in system frequency, tie-line power flow, effective load, and generation resulting from a remote load change.

When $R<1$, bias control imposes additional changes on all four parameters in the direction that causes frequency, tie-line flow, and effective load to move farther away from their initial steady-state values, while generation is moved in the direction of its initial value.

When $R>1$, bias control imposes additional changes on frequency, tie line, and load in the direction toward their respective initial steady-state values, while generation is moved farther away from its initial value.

The curves drawn in Fig. 15-22 are for the condition that R is the same percentage below unity when $R<1$ as it is above unity when $R>1$. It will be noted that the imposed effects are smaller where $R>1$ than they are for $R<1$.

Other Factors Affecting Bias Control Responses. In addition to the bias ratio, other factors which influence the magnitude of one or more of the imposed effects of bias control on the parameters of Fig. 15-22 when R is not unity are:

Size ratio. This is the ratio of the size of the *disturbance* area to the size of the total system, based on initial generation magnitudes, and is designated Y.

Governing ratio. This is the ratio in the *nondisturbance* area of its generation governing characteristic to its frequency-response characteristics and is designated P.

Regulating ratio. This is the ratio in the *nondisturbance* area of the amount of its generation subject to control by the bias controller to the total generation within the area and is designated Q.

Magnitude of disturbance. This is the magnitude of the disturbance, expressed in percent of total system generation, and is designated d.

Imposed Effects on Frequency Deviation. The imposed effect on frequency deviation expressed as a percentage of the initial frequency deviation is designated $\%\Delta F$ and is given by

$$\%\Delta F = \frac{(Y-1)(R-1)}{R - Y(R-1)}(100) \tag{15-11}$$

Curves based on Eq. (15-11) are shown in Fig. 15-23. It will be noted that $\%\Delta F$ is independent of P, Q, and d.

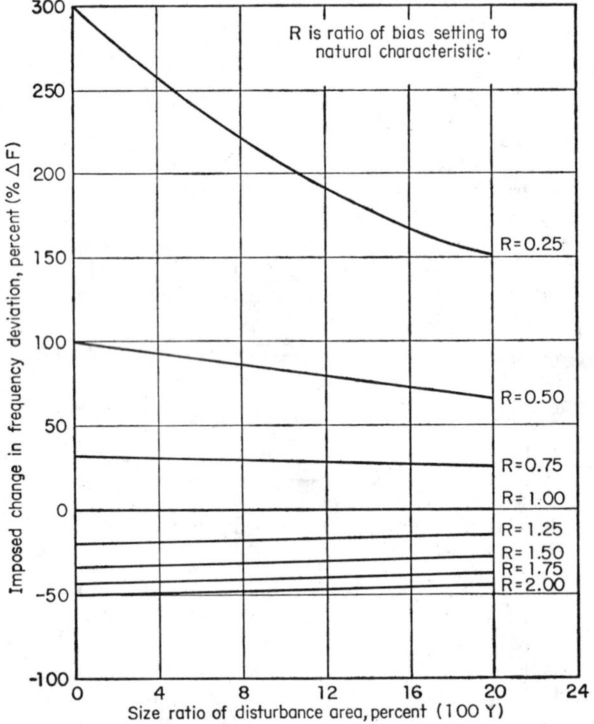

FIG. 15-23. Imposed effects on frequency deviation.

Imposed Effects on Tie-line Deviation. The imposed effect on tie-line deviation expressed as a percentage of the initial tie-line deviation is designated $\%\Delta T$ and is given by

$$\%\Delta T = \frac{Y(R-1)}{R - Y(R-1)}(100) \tag{15-12}$$

Curves of Eq. (15-12) are shown in Fig. 15-24. It will be noted that $\%\Delta T$, like $\%\Delta F$, is independent of P, Q, and d.

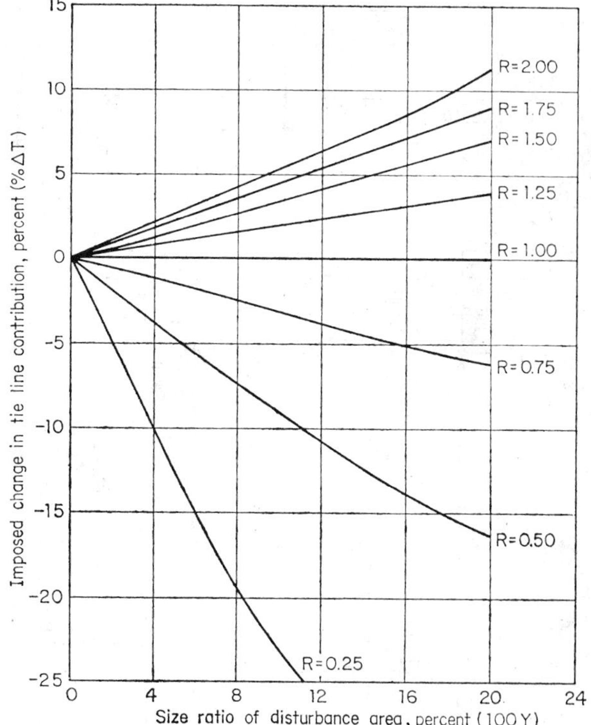

Fig. 15-24. Imposed effects on tie-line deviation.

Imposed Effects on Generation Change. Imposed effects on generation changes in the nondisturbance area may be expressed in alternative ways. As a percentage of the change in initial *regulated* generation in the area, it is designated $\%\Delta G$, and is given by

$$\%\Delta G = \frac{d}{Q} \frac{[1 + PQ(Y - 1)](R - 1)}{R - Y(R - 1)} \tag{15-13}$$

As a percentage of the initial change in *total* generation in the area it is designated $\%\Delta G'$ and is given by

$$\%\Delta G' = \frac{[1 + PQ(Y - 1)](R - 1)}{R - Y(R - 1)} \tag{15-14}$$

Curves based on Eqs. (15-13) and (15-14) are shown in Figs. 15-25 and 15-26, respectively.

Summary of Bias Effects. The foregoing equations for bias control performance in response to remote load changes show that:

1. Bias settings which match the area frequency-response characteristic impose no further changes on frequency, tie-line power flow, or local area generation.
2. Bias settings which are a given percentage lower than the area frequency-response characteristic impose a relatively greater change on frequency, tie-line power flow, or local area generation than bias settings which are the same percentage higher than the frequency-response characteristic.
3. Bias settings that differ from the frequency-response characteristic impose relatively large changes on system frequency.
4. Bias settings that differ from the frequency-response characteristic in the range of 50% below to 150% above the characteristic cause little change in tie-line flow.

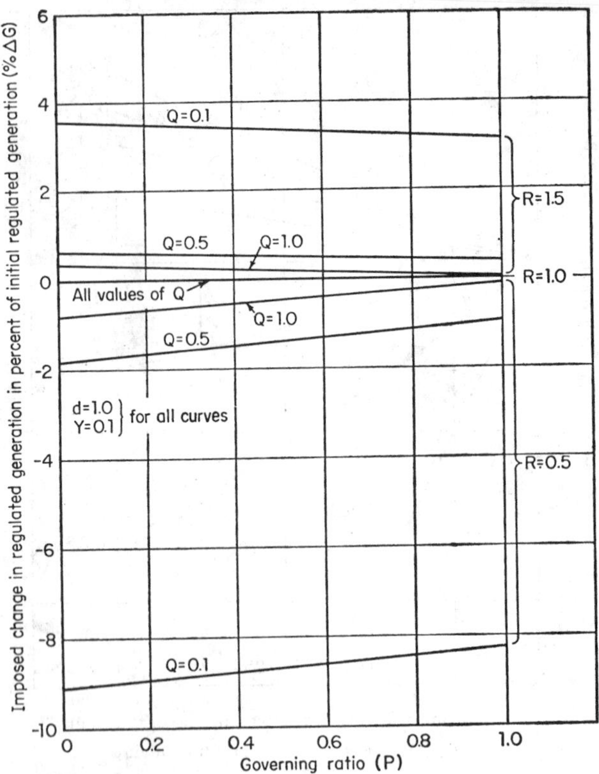

FIG. 15-25. Imposed effects on initial regulated generation.

5. The greater the percentage the area-load frequency characteristic is of the area frequency-response characteristic, the greater the changes in area generation caused by bias settings that differ from the frequency-response characteristic.

6. The larger the amount of local spinning generation not subject to bias control, the greater the changes in area generation caused by bias settings that differ from the frequency-response characteristic.

12. Bias Control on an Isolated Area. An important question related to tie-line bias control is what the nature of its performance will be if the area becomes disconnected from the interconnected system. Under such conditions, if the bias controller remains in operation and can reduce area control error to zero, it will perform like a *constant frequency control.* The frequency F_1 it will seek to maintain is a function of the normal frequency schedule F_0, the bias setting B, and the tie-line schedule T_0 for which the control was set. It is given as follows,

$$F_1 = F_0 - \frac{T_0}{10B} \tag{15-15}$$

where F_1 is the frequency to which the area will be forced by the tie-line bias controller action. Standard algebraic convention is for B to be negative, while T_0 is positive for outgoing, negative for incoming, power schedules.

This relation may also be written

$$F_1 = F_0 - \frac{W}{10B'} \tag{15-16}$$

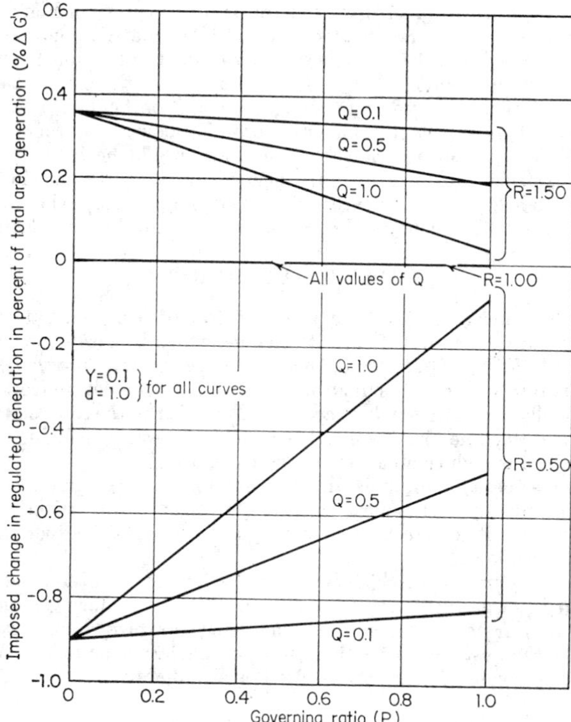

FIG. 15-26. Imposed effects on total area generation.

where W is the tie-line schedule T_0 expressed as a percentage of the area spinning capacity. It is positive for outgoing, negative for incoming, power schedules. B' is the bias B expressed in percent of area spinning capacity per 0.1 Hz and is negative.

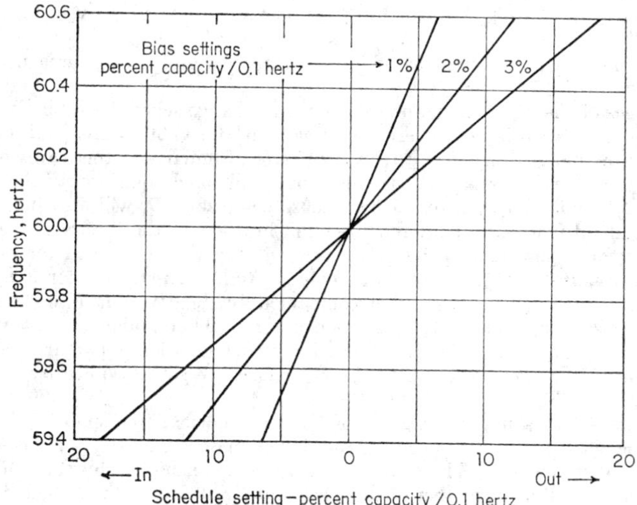

FIG. 15-27. Isolated bias controller as a frequency regulator.

The relationships of Eq. (15-16) are illustrated in the curves of Fig. 15-27. With a net interchange schedule of zero, frequency would be regulated to normal 60 Hz. For other interchange schedules, the frequency that is held on isolation is in the direction to minimize the regulating burden on the isolated area. If there were incoming interchange before isolation occurred, the frequency would be held at a level lower than normal, minimizing the effect of the loss of incoming power. If there were outgoing interchange before isolation occurred, the frequency would be held at a level higher than normal, utilizing the local load frequency characteristic and increased stored spinning energy to help absorb the excess generation in the area. The controller can be shifted to regulation of frequency at its normal value as local conditions permit.

SCHEDULE DEVIATIONS

13. Inadvertent Interchange. The time integral of area net interchange minus the time integral of the scheduled net interchange, which is to say the time integral of the term $T_1 - T_0$ in Eq. (15-9), is defined as the *inadvertent interchange* of the area. It develops in two ways. One results from the *intentional* deviation from normal interchange schedule due to frequency bias action. The other is the *unintentional* deviation from interchange schedule that results from metering errors, schedule-setting errors, or failure of control to reduce area control error to zero.

Operating areas seek to minimize inadvertent interchange, and when it develops, they endeavor, utilizing techniques consistent with operating criteria of the interconnection, to reduce it to zero by creating equal amounts of inadvertent interchange in the opposite direction.

The Intentional Component. The intentional inadvertent interchange, represented by D expressed in megawatthours, accumulated by an area during periods of system frequency deviation from normal is a function of the time error t, expressed in seconds, accumulated by the system during the frequency deviation periods. Where an area has operated to zero control error, the relation[5] is as follows,

$$D = \frac{B(t)}{6} \tag{15-17}$$

where B is the area bias, in megawatts per 0.1 Hz.

Balanced-system Inadvertent Interchange. The summation of accumulated inadvertent interchange for all the areas of an interconnected system, represented by D_s, may be written as follows,

$$D_s = \Sigma(T_1 - T_0)h \tag{15-18}$$

where h is the time, in hours, for which net interchange for each area has averaged $T_1 - T_0$.

If all areas of the interconnection set their interchange schedules properly, each "in" schedule for one area will have a corresponding "out" for another area and the algebraic sum of all interchange schedules, ΣT_0, will be zero. Similarly, if there are no unilateral tie-line metering errors, area "in" tie-line flows will have equal "out" flows in other areas and ΣT_1 will also be zero. Under these conditions D_s will also be zero, which is to say that all inadvertent interchange "in" in areas of the interconnection will be balanced by an equal summation of inadvertent interchange "out" for other areas.

Unbalanced-system Inadvertent Interchange. All inadvertent interchange in one direction is not always necessarily accompanied by inadvertent interchange in the opposite direction elsewhere on the interconnection. Such unbalance occurs if schedules are not properly set, so that ΣT_0 is not equal to zero. Consider a two-area interconnection as in Fig. 15-21, where the bias characteristics are set as shown by AA and BB in Fig. 15-28.

Area B is set to a net interchange of T_{0b} for a normal frequency F_0. Instead of area A being set for the same schedule, the $A'A'$ line of Fig. 15-28, it is set to schedule T_{0a} for normal frequency. The actual net interchange achieved by the control action of the two areas will be at neither of these schedules but will be that which corresponds to the intersection of the two bias control curves AA and BB, namely, T_1 at frequency F_1.

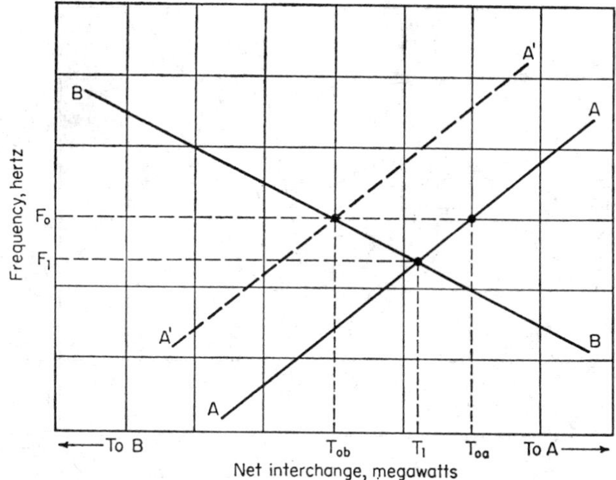

FIG. 15-28. Inadvertent interchange in a two-area interconnection.

For area A, inadvertent interchange is the time integral of $T_1 - T_{0a}$. Since, with respect to area A, T_1 is algebraically larger than T_{0a}, it represents inadvertent energy "out." For area B, the inadvertent interchange is the time integral of $T_1 - T_{0b}$. With respect to area B, T_1 is algebraically larger than T_{0b}; thus inadvertent interchange for this area is also energy "out." With both areas having inadvertent interchange in the "out" direction, there would be no opportunity to balance accounts by rescheduling interchange between them. Comparable conditions can result from unilateral tie-line metering errors. Operating areas exercise care to avoid such unbalances.

Correcting Inadvertent Interchange. One procedure for correcting accumulated inadvertent interchange is for an area with inadvertent interchange in one direction to make a bilateral exchange agreement with another area having inadvertent interchange in the opposite direction. Each then offsets its net interchange schedule appropriately, and by regulating to the new interchange schedule levels, each corrects its respective inadvertent interchange accumulations without interfering with or upsetting schedules of other control areas.

On a widespread interconnection it is sometimes difficult to establish such bilateral agreements. An alternative practice is for any area with an inadvertent interchange accumulation to make a unilateral schedule adjustment when, in correcting for inadvertent interchange, it will also correct for system accumulated time error.

Such unilateral action will be beneficial to other areas that have inadvertent interchange accumulations in the opposite direction. On the other hand, any area which has no inadvertent interchange, or has inadvertent interchange in the same direction, would have its inadvertent interchange accumulations increased by such a unilateral procedure. This adverse effect, distributed over many areas, may be small. Continued actions of this type can, however, undesirably increase the inadvertent interchange accumulations of some of the areas of the interconnection.

Effects of Unilateral Schedule Shifts on Frequency and Time Error. When an area A with a bias control characteristic such as AA in Fig. 15-29 undertakes a unilateral correction of its inadvertent interchange, the conditions with respect to the rest of the interconnection designated area C, which has a composite bias control characteristic CC, are shown in Fig. 15-29. Area A offsets its interchange schedule by shifting AA in the amount of ΔT_{0a} to $A'A'$. By assuming that the bias settings in area A and area C are each proportional by the same factor to their respective sizes and that area controls operate to reduce area errors to zero, the following relation defines the resultant change in actual tie-line flow,

$$\Delta T_0 = \Delta T_{0a}(1 - R_a) \tag{15-19}$$

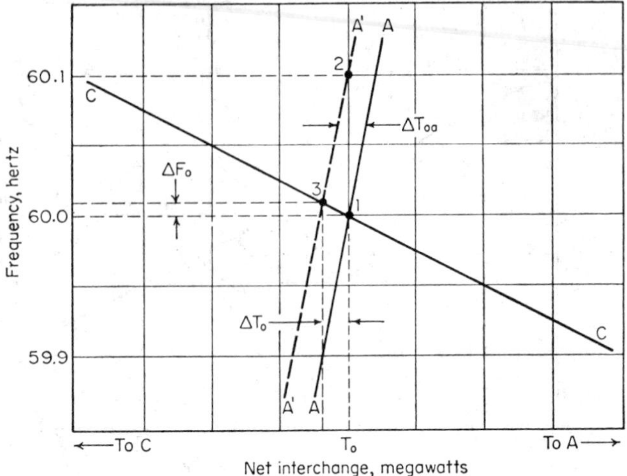

Fig. 15-29. Effects of unilateral schedule shifts on frequency.

where ΔT_0 is the actual change in net interchange achieved as the result of shifting area A schedule by ΔT_{0a} and R_a is the ratio of the spinning generating capacity in area A to the spinning capacity of the complete interconnection.

The corresponding change in system frequency is given by

$$\Delta F_0 = 0.1(q)(R_a) \tag{15-20}$$

where ΔF_0 is the frequency deviation resulting from the shift in area A interchange schedule by ΔT_{0a} and q is the shift in net interchange schedule expressed as a fraction of area bias; that is, $q = -T_{0a}/B_a$.

Customary shifts in net interchange schedules for unilateral inadvertent interchange correction are frequently in the range of 20 to 30% of the area bias, corresponding, respectively, to $q = 0.2$ and $q = 0.3$.

Curves showing the effect of unilateral net interchange schedule changes of this magnitude on system frequency and on the corresponding time required to correct for a 1-s time error as a function of R_a are drawn in Fig. 15-30. It will be noted that relatively small changes in system frequency and correspondingly large periods for a

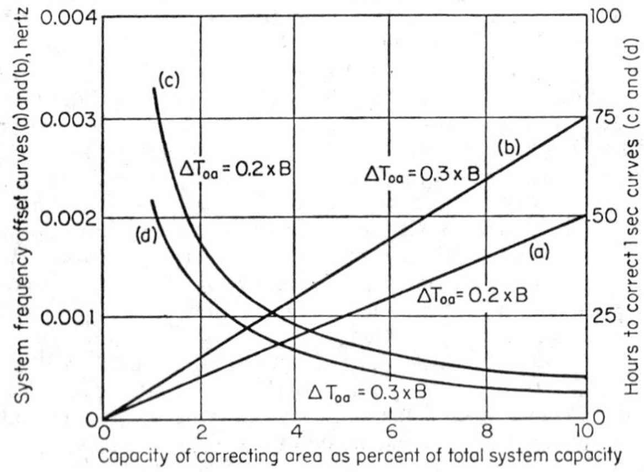

Fig. 15-30. Effect of unilateral net interchange schedule changes on frequency, and time required for correction of a 1-second time error.

1-s time-error correction result from shifts of this magnitude in net interchange schedule. For example, the frequency change is less than 0.001 Hz, and the time for 1-s time-error correction is about 17 h when $q = 0.3$ and $R_a = 0.03$.

TIME ERROR

14. System Time-error Correction. Normal system frequency is an automatic by-product of effective application of net interchange tie-line bias control in all control areas of an interconnection. When all net interchange schedules add algebraically to zero, when there are no tie-line metering errors, and when all bias controllers reduce their respective area control errors to zero, normal system frequency, 60 Hz, is automatically obtained. Improper schedule settings, metering errors, ineffective control and system disturbances—all contribute to frequency deviations from normal and to corresponding accumulations of system synchronous time error.

Error Accumulation. The time error accumulated by a deviation of frequency from normal is given by

$$t = \frac{60(f)(m)}{F} \tag{15-21}$$

where t is the time error in seconds, f is the average frequency deviation in hertz, m is the time during which frequency deviation f has persisted in minutes, and F is normal system frequency in hertz. For a 60-Hz system, Eq. (15-21) becomes

$$t = f(m) \tag{15-22}$$

Thus, an 0.02-Hz average deviation that persists for 50 min will cause accumulation of a 1-s time error.

Error Correction. Conventional practice for correcting system time errors is for all participating areas, on relayed signal from a designated control center where comparisons of system time to standard time are made, to offset the system frequency schedule in their respective controllers. The objective is for *all* areas to make this offset simultaneously, of the same magnitude, and for the same time duration. Executed in this way, the time-error correction can be carried out without upsetting interchange schedules in any of the areas.

The time required to correct the prevailing system time error is defined by the same equations (15-21) and (15-22) which define the time-error accumulations.

Thus, we may rewrite Eq. (15-21) as follows,

$$m = \frac{F(t)}{60f} \tag{15-23}$$

where m is the time in minutes to achieve the time-error correction, t is the prevailing time error in seconds, and f is the frequency offset in hertz. On a 60-Hz system, we may write

$$m = \frac{t}{f} \tag{15-24}$$

Thus, on a 60-Hz system, an 0.05-Hz shift in frequency schedule will correct a 1-s time error in 20 min. An 0.02-Hz shift will correct a 1-s error in 50 min.

ECONOMIC DISPATCH

15. Objectives of Economic Dispatch. The objective of area economic dispatch is to allocate the total generation required of the area to alternative available sources in order to achieve best possible area economy consistent with safe, effective operation. The nature of this problem, for the simplified area of Fig. 15-5, is shown in schematic form in Fig. 15-31. Area load can be satisfied, alternatively and at different costs, by adjusting outputs at generators 1, 2, ···, N or by scheduling new levels of bulk power transfer over interties with area B and area C.

Factors Influencing Economic Dispatch. Generators will be of different sizes and efficiencies, fuel costs will vary, and there will be variations in transmission losses from

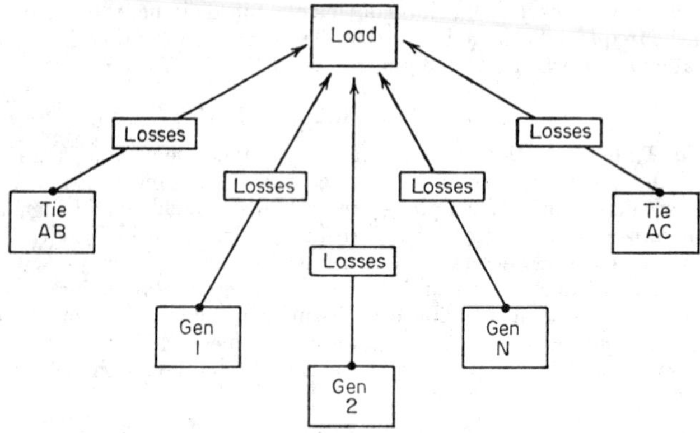

Fig. 15-31. Alternative sources for satisfying area load.

the various generating sources to load centers. Each of these factors influences the cost at which power can be generated and delivered to users within the area. In addition, each intertie represents opportunity for possible purchase of power at prices that may be more attractive than local area delivered costs.

Factors Overriding Economy. There are, at the same time, factors that limit achievement of highest possible area economy. For example, the operating range of generators may be restricted to certain high or low limits, system security may require location and amounts of spinning reserve that differ from the most economic allocation, there may be transmission-line limitations, or there may be operating requirements related to stream flow or storage at hydrounit locations.

Finally, and of particular importance, the demand for generation change from the area controller may exceed the permissible rate of generation change of the sources next in line to change generation for optimum economy. In this case, the control will bypass the economic allocation schedules and assign generation changes, in what is termed *area assist action,* to faster-responding units in order more effectively to reduce area control error to zero. Subsequently, as system conditions permit, a reallocation will automatically be made to achieve optimum internal area economy.

Economic Dispatch Control. Factors which are used for computing area economic

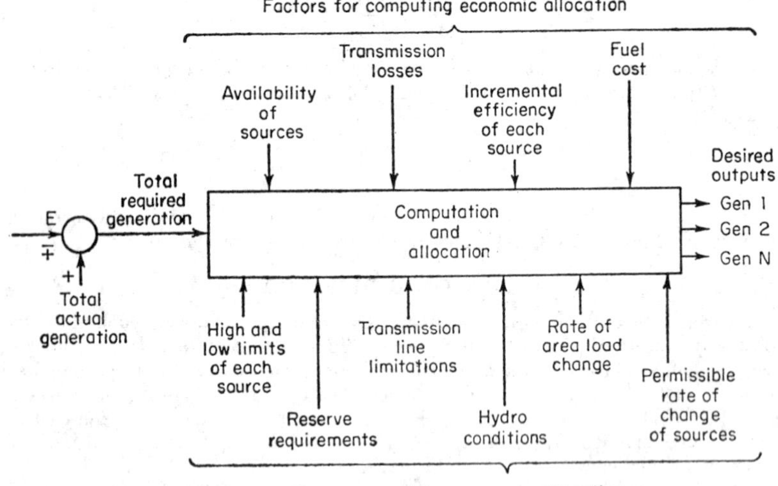

Fig. 15-32. Allocating total required generation to individual area sources.

dispatch and those which sometimes make it necessary to override economic allocations are shown in Fig. 15-32. This schematic also illustrates how a single control system combines the objectives of economic dispatch with the requirements of area regulation.

Area control error E, computed in accordance with Eq. (15-9) and Fig. 15-20, combined as a feed-forward parameter with total actual generation as a feedback defines the total generation required of the area. By appropriate computation and allocation, this total requirement is divided among available area sources. When control causes these individual assignments to be achieved, area control error will have been reduced to zero and optimum available area economic dispatch will simultaneously have been established.

16. Coordination Equation. For a steam turbogenerator unit the equation that coordinates delivered cost considerations with parameters related to unit heat rate, fuel cost, and transmission losses is as follows,

$$\lambda_n = \frac{(dH_n/dP_n)f_n}{1 - (\partial P_L/\partial P_n)} \tag{15-25}$$

where λ_n is the *incremental cost* of *delivered power* for source n, dH_n/dP_n is the *incremental heat rate* for source n, f_n is the cost of incremental fuel for source n, adjusted to include other varying costs such as maintenance cost at source n, and $\partial P/\partial P_n$ is the *incremental transmission loss* for source n.

Equation (15-25) may be rewritten,

$$\lambda_n = \frac{dF_n/dP_n}{1 - \partial P_L/\partial P_n} \tag{15-26}$$

where dF_n/dP_n is the *incremental generating cost* at source n and is equal to the product of dH_n/dP_n and f_n.

The individual terms of Eqs. (15-25) and (15-26) are more fully defined in the paragraphs that follow. For detailed derivation and discussion of these equations and their parameters see Refs. 16 to 23.

Incremental Heat Rate. The incremental heat rate of a steam turbogenerator at any particular output is the ratio of a small change in heat input per unit time to the corresponding change in power output. At each unit output it is given by the first derivative of the unit input-output curve. It is usually expressed in Btu per kilowatt-hour. It appears as dH_n/dP_n in the numerator of Eq. (15-25).

A hypothetical curve of incremental heat rate vs. unit output is shown in Fig. 15-33. In practice, curves are not this smooth. Discontinuities and slope reversals are fre-

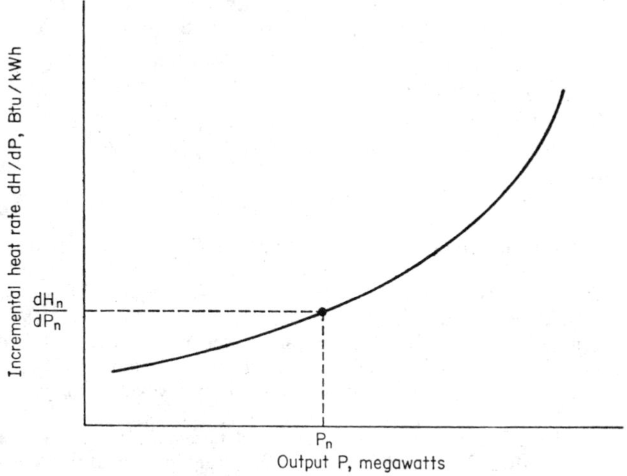

FIG. 15-33. Hypothetical curve of incremental heat rate vs. unit output.

quently encountered. For such curves, approximations of the actual curves are used, drawn so that the curves are continuous and have no slope reversals. Such approximations are in general well within the basic accuracy of the method.

Cost of Incremental Fuel. For a given generating source, the cost of incremental fuel is defined as the ultimate replacement cost of the fuel that would be consumed to supply an additional increment of power. In Eq. (15-25) it is combined with other incremental costs, such as maintenance, and the combination is designated f_n.

Incremental Generating Cost. The incremental cost of generated power is the product of incremental heat rate and the cost of incremental fuel adjusted to include all variable costs, including maintenance. At any particular value of generation it is the ratio of the additional cost incurred in producing an increment of generation to the magnitude of that increment of generation. It is usually expressed in mills per kilowatthour. It appears as dF_n/dP_n in the numerator of Eq. (15-26).

Total Transmission Losses. Total transmission losses do not enter directly in the coordination equation, but they provide the basis for deriving the incremental loss factors that do appear in the equation.

The preferred equation for total transmission loss, derived in Ref. 21, is

$$P_L = \sum_m \sum_n P_m B_{mn} P_n + \sum_n B_{n0} P_n + K_{L0} \qquad (15\text{-}27)$$

where P_L is the total transmission loss, P_m is the power output of source m, P_n is the power output of source n, B_{mn} are the constants related to the nature and characteristics of the area, it being noted that B_{mn} is not necessarily equal to B_{nm}, B_{n0} is a constant related to source n, and K_{L0} is a constant that may be regarded as representing total system losses under the imaginary condition of zero system power supply.

Incremental Transmission Losses. The incremental transmission loss for a source is the fraction of power loss incurred by transmitting a small increment of power from that source to another point, in this case the hypothetical load center of the area. It is defined for a given source as the partial derivative of Eq. (15-27) with respect to that source. For source n this may be written

$$\frac{\partial P_L}{\partial P_n} = \sum_m 2B_{nm} P_m + B_{n0} \qquad (15\text{-}28)$$

where $\partial P_L/\partial P_n$ is the incremental transmission loss for source n. It appears in the denominator of Eqs. (15-25) and (15-26).

Incremental transmission losses for a given area may rise fairly rapidly with increased power output. Total transmission losses for a given output may be only 5 or 6%, but incremental losses at that same output may be many times these values.

There are three types of B constants in Eq. (15-28). For source n, these are:

B_{nm}, identified as *self-constants*, which are always positive. B_{mn}, identified as *mutual constants*, which may be positive or negative. B_{n0}, identified as the added constants, which may be positive or negative.

Methods for determining B constants for a wide variety of system conditions are discussed in Ref. 21.

Incremental Delivered Power. The *incremental fraction of delivered power* from source n is represented by the denominator of Eq. (15-25), namely, $1 - \partial P_L/\partial P_n$. When multiplied by 100, it may be expressed as *percent incremental delivered power*.

Penalty Factor. The *penalty factor* for source n is the reciprocal of the incremental fraction of delivered power and hence is given by $(1 - \partial P_L/\partial P_n)^{-1}$.

17. Other Sources. The coordination equations (15-25) and (15-26) apply to steam turbogenerator units. It is frequently necessary to consider other types of sources, including hydrounits, intertie points, and nonconforming loads.

Hydro Plants. Hydro plants may be handled by assigning an equivalent incremental generation cost to each plant (see Ref. 24).

Tie Points. Interties with adjacent areas may be considered as generating sources with the numerator of Eq. (15-26) representing the incremental cost of purchased power at the intertie points. This is a particularly significant computation in determining the possible economic advantages of bulk transfer over interties between adjacent areas.

Large Nonconforming Loads. A *nonconforming load* is one that does not vary linearly with total area load. When the magnitude of such a load is large compared with the generation at or near the substation that serves this load, it should receive special consideration in the computation of incremental transmission losses. A nonconforming load is telemetered to the computing network and is treated as a negative power source. In this way, appropriate factors for such a load and its B constants can be introduced into the computation (see Ref. 21 for additional details).

18. Equal Incremental Costs. Optimum operating economy for a group of sources is achieved when they are loaded to equal incremental costs. The incremental costs that are to be compared and equalized to achieve such economic dispatch depend on whether there are significant differences in incremental transmission losses for the several sources.

General Case. For the general case, applicable to control areas whose sources have significantly differing incremental transmission losses, economic dispatch is achieved when sources are loaded to equal incremental costs of delivered power. The corresponding relationship for n sources of an area is

$$\lambda = \lambda_1 = \lambda_2 = \cdots = \lambda_n \tag{15-29}$$

where λ is the incremental cost of delivered power for area as a whole, and $\lambda_1, \lambda_2, \cdots, \lambda_n$ are the incremental costs of delivered power for sources $1, 2, \cdots, n$, respectively.

Cases Where Incremental Transmission Losses Can Be Ignored. Where the incremental transmission losses for the several sources of an area, or the differences between them, are small, the denominators of Eq. (15-26) as applied to the several sources become essentially equal. For such cases typically encountered in small, closely knit areas, economic dispatch can be achieved by loading the sources to equal incremental generating costs. Economic dispatch is thus achieved when

$$\frac{dF_1}{dP_1} = \frac{dF_2}{dP_2} = \cdots = \frac{dF_n}{dP_n} \tag{15-30}$$

Within Stations. For a given station having n sources tied to a common bus, any applicable incremental transmission losses are the same for all n units. Thus, *within* such a station optimum economy is achieved when its units are loaded to equal incremental generating costs in accordance with Eq. (15-30).

When the f_n factor of Eq. (15-25), the cost of incremental fuel and other varying costs, is the same for all units within such a station, optimum economy within the station is achieved by loading its units to equal incremental heat rates, as follows:

$$\frac{dH_1}{dP_1} = \frac{dH_2}{dP_2} = \cdots = \frac{dH_n}{dP_n} \tag{15-31}$$

Sources Out of Range. In applying the principle of loading to equal incremental costs to achieve economic dispatch it should be noted that some of the stations or units of the area then in operation may be *out of range* of the incremental cost being used at that particular time as the loading reference. Units with *lower* incremental costs over their full range will have been fully loaded by earlier incremental cost allocations. Units with *higher* incremental costs over their full range may be in operation for reserve purposes or in anticipation of imminent increases in demand.

CONTROL APPLICATION

19. General Practice. Modern practice is to apply automatic control to most or all of the generating units of a control area, thereby achieving economic dispatch while simultaneously fulfilling the obligations of area regulation. Figure 15-34 shows in schematic form how the desired output for each of the sources of an area, computed in accordance with the schematic of Fig. 15-32, is achieved by a closed-loop execution. The computed desired output is compared with a feedback from actual output from each source, and control action is applied to the source until the two are matched.

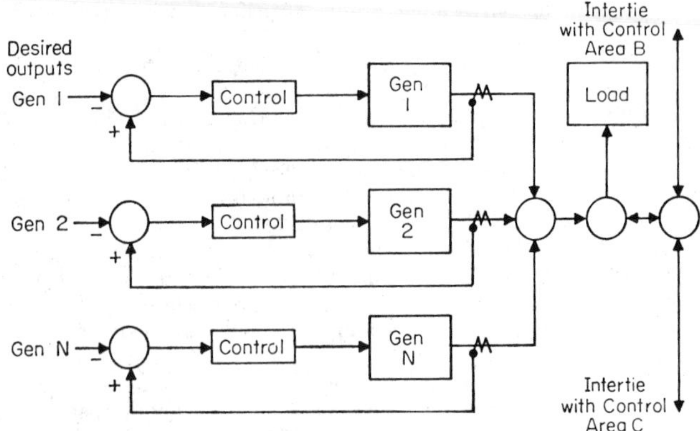

FIG. 15-34. Method of achieving computed allocations on area sources.

The combination of Figs. 15-20, 15-32, and 15-34 provides a complete schematic representation of how all sources of the area of Fig. 15-5 are automatically controlled to maintain area control error at zero while simultaneously achieving area economic dispatch, tempered as required by prevailing area conditions.

20. Control Classification. Control systems have been classified[25] by the portions of the area control problem which they seek to solve and by the nature of the programming techniques used for achieving economic dispatch. Classifications are as follows:

Class I. In this category, control is applied to area regulation only. Modern controls seldom have this limited objective.

Class II. In this category, control is applied to both area regulation and economic dispatch. Generation allocation programs are manually preset.

Class III. In this category, control is applied to both area regulation and economic dispatch. Generation allocation programs are based on stored incremental generating cost data and are automatically computed and applied in real time.

21. Manually Set Programming. Class II controls are sometimes referred to as being of the *flexible-programming* type. Allocation programs are manually set and periodically reset as required to meet changed system conditions. For example, programs must be reset when units which are subject to control are brought on or taken off the line.

Example. A control execution[26] of this class which utilizes as the programming reference a feed-forward from area control error algebraically combined with a feedback from actual area generation is shown schematically in Fig. 15-35. For simplicity, this schematic applies to an area having two generating sources G_A and G_A' and two interties AB and AC. For each source, desired generation as a function of total-area required generation is programmed with manually set dials to approximate the loading curves shown at the upper right of the figure. The allocation curve for each source is made up of straight-line segments by applying suitable settings to *base point* and *slope* (or *participation*) dials. At all points, the curves are set so that the sum of source generation allocations is equal to the total generation required of the area. Thus,

$$(G + G') - E = P + P' \tag{15-32}$$

where G is the output of source G_A, G' is the output of source G_A', E is the area control error, and hence $-E$ is the generation change required to reduce the area control error to zero, P is the programmed allocation for source G_A, and P' is the programmed allocation for source G_A'.

Equation (15-32) may be rewritten as follows:

$$E = (G - P) + (G' - P') \tag{15-33}$$

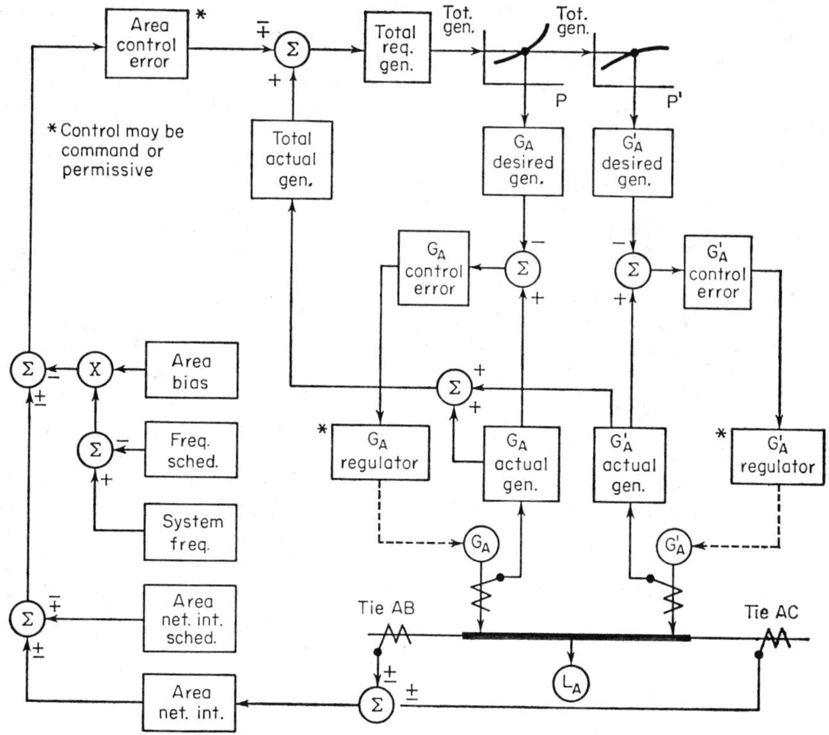

FIG. 15-35. Flexible programmed analog control system (Class II).

Let the *source control errors* for sources G_A and G_A' be represented by S and S', respectively; then

$$S = G - P \tag{15-34}$$

and

$$S' = G' - P' \tag{15-35}$$

Substituting Eqs. (15-34) and (15-35) in Eq. (15-33) yields

$$E = S + S' \tag{15-36}$$

The generalization of this analysis is that, in an execution of this type applied to an area having n sources, the area control error is equal to the algebraic sum of all of the source errors,

$$E = \Sigma S_n \tag{15-37}$$

An important characteristic of such a control execution is that it predictably computes the generation assignment to each source. That is to say that the programmed allocation for each source is independent of the rate at which other controlled sources respond to their allocations. Hunting between units is thereby minimized.

Many control systems of this type are in service[27] in the United States and elsewhere in the world. Areas that are predominantly hydro find the flexibly set schedules useful. This type of control also coordinates well with digital computers, as noted later in this section.

22. Stored Cost Programming. Class III controls utilize a *lambda reference* for allocating total required area generation to individual sources. Generation at each source is automatically adjusted until its incremental cost of delivered power, source lambda, matches a common area lambda, thereby achieving economic dispatch in accordance with Eq. (15-29). This execution is sometimes referred to as being of the *fixed-programming* type. It has the advantage that program allocations automatically adapt to the number of sources in service and require no reprogramming as do the Class II executions when the number of units on the line is changed.

Computing Lambda Reference. The common lambda to which each source lambda is matched can be established by automatic adjustment until the resultant control action reduces area control error to zero; or it can be automatically computed to satisfy the criterion that total source allocations shall equal the algebraic sum of area control error and existing total area generation. This latter criterion is the same as expressed for one type of Class II control in Eq. (15-32).

Another way of applying the criterion for area lambda computation that derives from Eq. (15-32) is automatically to adjust the lambda reference until the area control error is equal to the algebraic sum of source control errors, in accordance with Eq. (15-37).

For computation techniques based on either Eq. (15-32) or Eq. (15-37), the area reference lambda is established predictively, without waiting for area control error to be reduced to zero. In these executions, the allocation to each source is again independent of the rate at which other control sources respond to their respective allocations.

Example. A control execution of this type for a simplified two-source area having

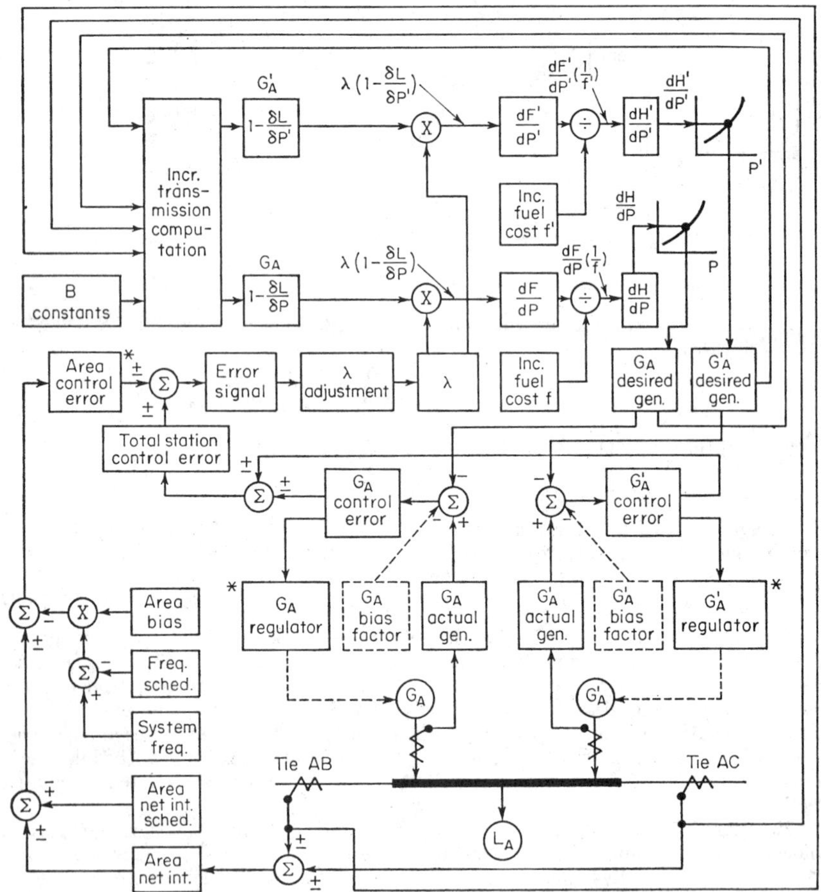

Fig. 15-36. Predictive lambda reference computer-control system (Class III).

two interties is shown in Fig. 15-36. The area lambda reference is adjusted until area control error equals the summation of source control errors in accordance with Eq. (15-37). Incremental transmission loss for each source is computed in accordance with Eq. (15-28) from stored B constants and from real-time telemetered measurements of source outputs and intertie power flows. By using stored data for cost of incremental fuel, the incremental heat rate for each source is computed on the basis of Eqs. (15-26) and (15-25). From stored relationships of incremental heat rate vs. source output, represented by the curves in the upper right of the figure, the desired outputs for each source are obtained. Control is applied to each source until desired output is matched by actual output.

23. Control Executions. Two general techniques are available for applying control signals to individual sources. One is termed *permissive control,* the other *command control.* The former puts a priority, when a choice must be made, on area regulation as against source regulation. The latter does not.

Permissive Control. In the permissive control execution, source generation is permitted to change only when, in correcting a source control error, it will also help reduce area control error.

Command Control. In the command control execution, source generation is changed whenever a source control error exists, regardless of whether the action will reduce or augment area control error. This type of control has also been referred to as *mandatory control.*

24. Unit Bias. In applying command control to the individual generator it is helpful to include a unit frequency bias[25] in the computation of unit control error. This will coordinate the control with governor responses to remote load changes and avoid opposition to such responses by the command control. A unit bias may also be used to apply a specific governing characteristic to a generating unit, overriding nonlinearities or dead band in the unit governor.

Such unit bias factors are shown as broken-line blocks in Fig. 15-36.

25. Analog and Digital Techniques. Computer control of the general Class III type is widely used.[27] The execution is sometimes analog, sometimes digital, and sometimes a combination of the two. The scope of an installation, the probabilities of extensive change during its life, and the availability of standby control equipment are among the factors that influence the choice of technique best suited for a given project.

Analog Execution. Many existing installations are of the analog type. All the computation and control steps are carried out with analog-type equipment. In addition to area regulation and economic dispatch operation, the assembly is usually arranged to permit computation of advantageous bulk transfers with neighboring areas. Provision is usually made to readily expand a centralized installation as additional units are added in the area. Malfunctions in one section of such a computer control do not necessarily interfere with operation of other parts of the system.

Digitally Directed Analog Execution. In digitally directed analog control assemblies, a flexibly programmed analog console of the Class II type shown in Fig. 15-35 is supplemented by a digital computer. The latter is programmed to compute an area economic dispatch every few minutes and is arranged to automatically set the allocation dials of the analog console to correspond to these economic dispatch computations. The digital computer is available for other services, such as determining when units should be put on or taken off the line, computing advantageous interchanges with neighboring areas, collecting data for bulk transfer billings, and monitoring system security. When the digital computer is in use for such other services, or should it be down for repair or maintenance, the analog console continues in service for area regulation and source allocation control.

Direct Digital Control. The most recent approach, particularly in large control areas, is to execute the complete assignment of area regulation and economic dispatch with a properly programmed digital computer, without an intermediate analog control console. As in the digitally directed analog execution, the digital computer would perform other significant computational and data-gathering functions. It would be supplemented with a suitable standby analog system for use when the digital computer is not available for area control.

26. Protection. Components of a typical computer control system may be spread out over the thousands of square miles of a control area and would be linked together by long telemetering and control channels. Protective features should be incorporated into the computer control system so that it will not apply improper control action to generating sources when components fail or do not function properly. There should also be comparable protective action when system, area, or station conditions are beyond the corrective scope or capability of the control system.

Typical conditions which may be utilized to interrupt or suspend control automatically, partly or completely, are:

1. System frequency swings to abnormally high or low value.
2. Voltages of power supplies are outside their normal ranges.
3. A telemetering or control channel is lost.
4. Normal telemetering or control signals are not received.
5. Abnormal telemetering or control signals are received.
6. A high or low limit is reached on an individual generator.
7. A high limit is reached on a tie line.

Suitable equipment is incorporated into the computer control system and takes programmed protective action, with suitable alarm warnings to operators, when one or more of these abnormal conditions are detected.

27. Installation Data. Reference 27 is a survey of computing and control practices for the control of generation and power flow on the interconnected systems of the United States. Its bibliography includes references to papers that describe typical analog, digitally directed analog, and direct digital control systems. For additional data on concepts and practices see the additional bibliography listed at the end of this subsection.

REFERENCES

28. Bibliography on Control of Generation and Power Flow

1. Statistical Year Book of the Electric Utility Industry for 1965; *Publ.* 66-49, New York, Edison Electric Institute.

2. National Power Survey; Federal Power Commission, 1964.

3. Status of Interconnections and Pooling of Electric Utility Systems in the United States; *Publ.* 63–39A, New York, Edison Electric Institute, 1963.

4. Definitions of Terminology for Automatic Generation Control on Electric Power Systems; *IEEE Publ.* 94, November, 1965.

5. Cohn, Nathan Considerations in the Regulation of Interconnected Areas; IEEE Winter Power Meeting, New York, *Paper* 66-187, February, 1966.

6. Recommended Specification for Speed-governing of Steam Turbines Intended to Drive Electric Generators Rated 500 kW and Larger; *IEEE Publ.* 122 (*AIEE* 600), December, 1959

7. Cohn, Nathan Common Denominators in the Control of Generation on Interconnected Power Systems; presented before Systems Operation Committee, Pennsylvania Electric Association, May, 1957. Reprint 461-5(8), Leeds & Northrup Company.

8. "Westinghouse Electrical Transmission Reference Handbook," 4th ed.; Westinghouse Electric Corporation, Chap. 13, Pt. X, p. 486.

9. The Effect of Frequency and Voltage on Power System Load; IEEE Winter Power Meeting, 1966, *Paper* 31-CP 66-64, Jan. 30–Feb. 4.

10. Cohn, Nathan Power Flow Control—Basic Concepts for Interconnected Systems; *Proc. Midwest Power Conf.*, Chicago, Ill., 1950, Vol. 12, pp. 159–175. Also *Elec. Light and Power*, 1950, Vol. 28, No. 8, pp. 82–94, and No. 9, pp. 100–107.

11. Cohn, Nathan Control of Power Flow in System Interconnections; presented before Electric Section, Wisconsin Utilities Association, November, 1954. *Reprint* ND4-56-461(9), Leeds & Northrup Company.

12. Cohn, Nathan Automatic Control of Power Systems, Symposium on Reliability of Bulk Power Supply in Large Interconnected Power Systems; *Proc. IEEE Intern. Conv.*, New York, March, 1966, Pt. 12, *Paper* 50.1.

13. Cohn, Nathan Principles and Applications of Tie Line Bias Control and Economic Loading; *Tenth Ann. Am. Public Power Assoc. Eng. Operations Workshop*, New Orleans, La., January, 1966. *Reprint* E7-3111 RP, Leeds & Northrup Company.

14. Cohn, Nathan Some Aspects of Tie-line Bias Control on Interconnected Power Systems; *Trans. AIEE,* 1956, Vol. 75, Pt. III, pp. 1415–1428.

15. Cohn, Nathan A Step-by-step Analysis of Load-frequency Control Showing the System Regulating Responses Associated with Frequency Bias; presented before the 1956 Meeting of the Interconnected Systems Committee, Des Moines, Iowa, March, 1956. *Reprint* 461-5(6), Leeds & Northrup Company.

16. Kaufmann, P. G. Load Distribution between Interconnected Power Stations; *Jour. IEE,* 1943, Vol. 90, Pt. II, No. 14, pp. 119–130.

17. George, E. E. Intrasystem Transmission Losses; *Trans. AIEE,* 1943, Vol. 62, pp. 153–158.

18. Ward, J. B., Eaton, J. R., and Hale, H. W. Total and Incremental Losses in Power Transmission Networks; *Trans. AIEE,* 1950, Vol. 69, pp. 626–632.

19. Kirchmayer, L. K., and Stagg, G. W. Analysis of Total and Incremental Losses in Transmission Systems; *Trans. AIEE,* 1951, Vol. 70, pp. 1197–1204.

20. Harder, E. L., Ferguson, R. W., Jacobs, W. E., and Harker, D. C. Loss Evaluation, Part II. Current-power-form Loss Formulas; *Trans. AIEE,* 1954, Vol. 73, pp. 716–731.

21. Early, E. D., Watson, R. E., and Smith, G. L. A General Transmission Loss Equation; *Trans. AIEE,* 1955, Vol. 74, Pt. III, pp. 510–520.

22. Brownlee, W. R. Coordination of Incremental Fuel Costs and Incremental Transmission Losses by Functions of Voltage Phase Angles; *Trans. AIEE,* 1954, Vol. 73, Pt. III, pp. 529–541.

23. Kirchmayer, L. K., and Stagg, G. W. Evaluation of Methods of Coordinating Incremental Fuel Costs and Incremental Transmission Losses; *Trans. AIEE,* 1952, Vol. 71, Pt. III, pp. 513–521.

24. Fereshetian, H., Liechty, M. D., and Brown, N. E. Coordination of Desired Generation Computer with Area Control; *Proc. Am. Power Conf.*, Chicago, Ill., 1959, Vol. 21, pp. 554–563.

25. Cohn, Nathan Methods of Controlling Generation on Interconnected Power Systems; *Trans. AIEE,* 1962, Vol. 80, Pt. III, pp. 270–282.

26. Cohn, Nathan Area-wide Generation Control—A New Method for Interconnected Systems; *Proc. Am. Power Conf.*, Chicago, Ill., 1953, Vol. 15, pp. 316–344. Also *Elec. Light and Power,* June, 1953, Vol. 31, No. 7, pp. 167–175; July, 1953, No. 8, pp. 97–108; August, 1953, No. 9, pp. 77–83.

27. Cohn, Nathan State of the Automatic Control Art in the Electric Power Industry of the United States; *Proc. Sixth Joint Automatic Control Conf.*, Troy, N.Y., June, 1965. Also *IEEE Spectrum,* 1965, Vol. 2, No. 11, pp. 67–77.

ADDITIONAL BIBLIOGRAPHY

Morehouse, S. B. Automatic Economic Loading Practices of Interconnected Power Systems in the U.S.A.; CIGRE, Paris, *Paper* 315, 1966.

Kirchmayer, L. K. "Economic Operation of Power Systems"; New York, John Wiley & Sons, Inc., 1958.

Kirchmayer, L. K. "Economic Control of Interconnected Systems"; New York, John Wiley & Sons, Inc., 1959.

Cohn, Nathan "Control of Generation and Power Flow on Interconnected Systems"; New York, John Wiley & Sons, Inc., 1966.

TRANSDUCERS FOR POWER-SYSTEM MEASUREMENTS

By W. E. Phillips

29. Transducers. Numerous transducers are used for the measurement of a-c watts, voltage, current, and frequency. The principles employed in typical transducers and the methods of connection for the measurement of watts and vars will be described.

PRINCIPLES USED IN A-C WATT TRANSDUCERS

30. Thermal Wattmeter. The thermal wattmeter produces a d-c millivoltage proportional to input watts. It is in all respects a true wattmeter—the polarity of its output reverses with reversal of watts; it measures the product of voltage, current, and the cosine of the angle between them; it is independent of waveshape to a high degree because of the fact that its output depends on heating (rms) values.

A single-element thermal converter, complete with its internal potential transformer, is shown in Fig. 15-37. Two such elements are used for conventional 3-phase 3-wire metering.

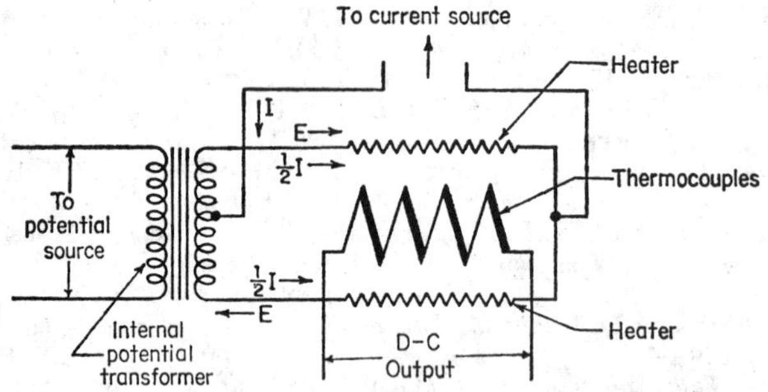

Fig. 15-37. Thermal wattmeter.

The current and potential circuits are connected to the thermal converter. The current I divides so that one-half flows through each heater. The potential causes a current designated as E to flow through both heaters in series. These two currents are in the same direction in the top heater and in opposite directions in the lower heater. The vector sum of the currents in the upper heater will be greater than that in the lower heater when the phase angle between the two currents I and E is less than 90°. Thus more heat will be produced in the upper heater than in the lower heater, since the heaters have equal and constant resistances. Small thermocouples are located adjacent to the heaters, and the thermocouple junction at the upper heater is raised to a higher temperature than at the lower heater. The output emf of the thermocouples is proportional to this difference in heating, which in turn is proportional to watts input to the thermal converter.

Upon using E as a reference vector, E and I may be expressed as follows when the current is leading the voltage by $\theta°$:

$$\text{Vectorial } E = \dot{E} = E + j0$$

$$\text{Vectorial } I = \dot{I} = I \cos \theta + jI \sin \theta$$

The current in the upper element is

$$\dot{E} + \tfrac{1}{2}\dot{I} = \sqrt{(E + \tfrac{1}{2}I \cos \theta)^2 + (j\tfrac{1}{2}I \sin \theta)^2}$$

The heating in the upper element is proportional to the current squared. Thus

$$\text{Heating in upper element} = (E + \tfrac{1}{2}I \cos \theta)^2 + (j\tfrac{1}{2}I \sin \theta)^2 \tag{15-38}$$

The current in the lower element is

$$\dot{E} - \tfrac{1}{2}\dot{I} = \sqrt{(E - \tfrac{1}{2}I \cos \theta)^2 + (-j\tfrac{1}{2}I \sin \theta)^2}$$

The heating in the lower element is proportional to the current squared.

$$\text{Heating in lower element} = (E - \tfrac{1}{2}I \cos \theta)^2 + (-j\tfrac{1}{2}I \sin \theta)^2 \tag{15-39}$$

The difference in heating is proportional to

$$\text{Eq. (15-38)} - \text{Eq. (15-39)} = (E^2 + EI\cos\theta + \tfrac{1}{4}I^2\cos^2\theta - \tfrac{1}{4}I^2\sin^2\theta)$$
$$- (E^2 - EI\cos\theta + \tfrac{1}{4}I^2\cos^2\theta - \tfrac{1}{4}I^2\sin^2\theta) \quad (15\text{-}40)$$
$$= 2EI\cos\theta$$

Since 2 is a constant of calibration, the difference in heating is proportional to the product of E, I, and the cosine of the angle between them, which is power.

31. Magnetic-core Multiplier Wattmeter. The basic principle of the magnetic-core multiplier watt transducer is the use of a magnetic-type oscillator, controlled by the voltage input, to switch the current input in a manner to produce a d-c voltage output proportional to a-c watts. A simplified diagram of the watt transducer is shown in Fig. 15-38.

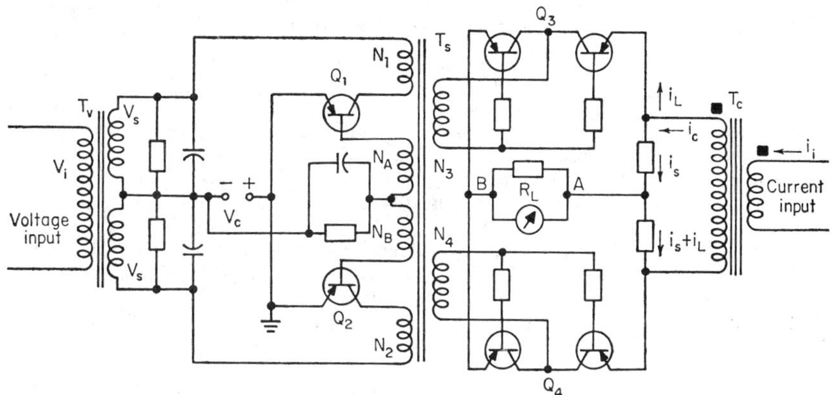

FIG. 15-38. Magnetic-core multiplier wattmeter.

The saturable transformer T_s in combination with transistors Q_1 and Q_2 and voltages v_s and V_c constitutes a magnetic oscillator the frequency of which is a function of the voltages V_c and V_s. The underlying principle of the magnetic oscillator is that it requires a constant volt-second input to drive the core from saturation in one direction to saturation in the opposite direction. The iron used in T_s is of the square-loop type so that the maximum flux density is constant.

Note that V_s is polarized with V_c, and this in combination with the polarity of the windings on T_s is such that, when the two voltages are subtractive, transistor Q_1 is conducting and, when the two voltages are additive, Q_2 is conducting. Transistors Q_1 and Q_2 conduct alternately, with one or the other conducting at all times.

The impressed voltage, time, and flux-linkage change during the magnetic-oscillator switching are related as follows:

$$T_1(V_c - V_s) = 2N_1\Phi \quad (15\text{-}41)$$

where T_1 is the time required to go from saturation in one direction to saturation in the opposite direction when Q_1 is conducting.

$$T_2(V_c + V_s) = 2N_2\Phi \quad (15\text{-}42)$$

where T_2 is the time required to go from saturation in one direction to saturation in the opposite direction when Q_2 is conducting.

Since we have made

$$N_1 = N_2 \quad (15\text{-}43)$$

Eqs. (15-41) and (15-42) yield the following:

$$V_s = V_c \frac{T_1 - T_2}{T_1 + T_2} \tag{15-44}$$

Since the magnetic-core switching frequency is on the order of 5 kHz, the period T_1 plus T_2 for one complete switching cycle is small compared with the period of 60 Hz. Thus, Eq. (15-44) is the relation between the instantaneous input voltage and the times T_1 and T_2 since the bias voltage V_c has a constant magnitude.

During the period T_1 when the core flux is changing due to the conduction of Q_1, the input-current transistor switch Q_3 is conducting a current i_L, which is proportional to the current input i_i, through resistor R_L across which is developed the d-c output voltage. The voltage drop developed across R_L with Q_3 conducting will be considered positive. When Q_2 and Q_4 are conducting, the voltage developed across R_L will again be proportional to the current input i_i, but of the opposite polarity.

The product of the instantaneous voltage and current is a measure of a-c power. Therefore, if Eq. (15-44) is multiplied on both sides by i_i, the result is

$$i_i v_s = V_c \left(\frac{i_i T_i}{T_1 + T_2} - \frac{i_i T_2}{T_1 + T_2} \right) \tag{15-45}$$

As noted earlier, V_c is a constant and can be taken care of in the calibration of the watt transducer. The expression in parentheses of Eq. (15-45) describes exactly the operation of the circuit. As was indicated, the time T_1 plus T_2 is small compared with the period for 60 Hz so that the multiplication for practical purposes is made on an instantaneous basis and accordingly results in a measure of instantaneous power.

The magnetic-core multiplier watt transducer operates in all four quadrants so that the polarity of the output reverses with a reversal in the direction of power flow and has an output of 5 V d-c for nominal full-load input.

32. Watt Transducer Employing Zener-diode Exponential Circuit. The circuit for a watt transducer in which the unique feature is a zener-diode exponential circuit is shown in Fig. 15-39. The zener-diode exponential section consists of three zener diodes connected in series with appropriate resistors connected across each zener diode. This circuit is a squaring device so that the current flowing through it is a squared function of the voltage applied across the circuit.

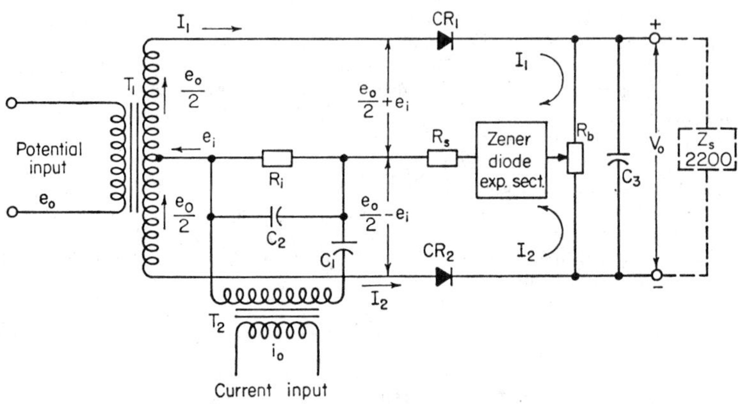

FIG. 15-39. Watt transducer employing zener-diode exponential circuit.

The voltage and current inputs to the watt transducer are applied, respectively, to transformers T_1 and T_2.

The operation of the watt transducer will be described with a unity power-factor load, i.e., with the current and voltage inputs in phase. The voltage e_i which is developed across R_i is proportional to the current input, and on the positive half cycle this voltage and voltage $\frac{1}{2}e_o$, which is proportional to the applied voltage, are summed in an additive sense. The vector sum of these two voltages will produce a current I_1 in the upper branch of the circuit. The current I_1 is proportional to the square of the summed voltages due to the action of the zener-diode exponential circuit. This may be expressed as follows:

$$I_1 = k_1(e_i + \tfrac{1}{2}e_o)^2 \tag{15-46}$$

Expanding Eq. (15-46),

$$I_1 = k_1(e_i^2 + e_ie_o + \tfrac{1}{4}e_o^2) \tag{15-47}$$

The voltage developed across the lower branch of the circuit on the negative half cycle is the vector difference of $\frac{1}{2}e_o$ and e_i. The current I_2 in the lower branch may be expressed as

$$I_2 = k_2(e_i - \tfrac{1}{2}e_o)^2 \tag{15-48}$$

Expanding Eq. (15-48),

$$I_2 = k_2(e_i^2 - e_ie_o + \tfrac{1}{4}e_o^2) \tag{15-49}$$

Current I_1 develops a voltage $\frac{1}{2}(I_1R_B)$ across the upper half of resistor R_B, and the current I_2 develops a voltage $\frac{1}{2}(I_2R_B)$ of opposite polarity across the lower half of R_B, and the output voltage V_o is the difference of these two voltages.

The output voltage V_o may be expressed as

$$V_o = \tfrac{1}{2}(I_1R_B) - \tfrac{1}{2}(I_2R_B) \tag{15-50}$$

Substituting the expressions for I_1 and I_2 from Eq. (15-47) and (15-49) and since the resistances in the two branches are equal and k_2 is equal to k_1, the reduced expression becomes

$$V_o = k_1(2e_ie_o) \tag{15-51}$$

Since the multiplication is made on an instantaneous basis, V_o is proportional to the product of current and voltage, which is power. The values 2 and k_1 in Eq. (15-51) are proportionality constants.

Currents I_1 and I_2 are developed on alternate half cycles so that for most accurate results the waveshape of the inputs should be symmetrical.

33. Hall Watt Transducer. The Hall watt transducer utilizes the Hall effect to produce a d-c millivolt output proportional to the product of a-c current and voltage and the cosine of the phase angle between current and voltage, i.e., proportional to watts. The Hall crystal is located in a magnetic field so that the flux, which is proportional to the current, is perpendicular to the axis of the crystal. The voltage is applied to two opposite faces of the flat crystal along an axis perpendicular to that of the flux, and the d-c millivolt output is generated across the two remaining sides of the crystal, along an axis mutually perpendicular to the flux and applied voltage axis.

PRINCIPLES USED IN A-C VOLTAGE AND CURRENT TRANSDUCERS

34. Thermal Voltage and Current Transducers. A voltage transducer operating on the thermal principle—similar to that shown for the thermal wattmeter in Fig. 15-37—produces a millivolt output proportional to the square of the input voltage. In the voltage transducer, only the upper heater is used, and the current input is deleted. Since the transducer operates on the heating principle, it has the advantage of being substantially independent of the waveshape of the voltage and the output is an accurate measurement of the rms value of the voltage.

A current transducer operating on the same principle as the voltage transducer described in the preceding paragraph has the same characteristics with respect to output, etc. In the current transducer, the voltage input is deleted.

35. Rectifier-type Voltage and Current Transducers. The circuit for a typical rectifier-type voltage transducer is shown in Fig. 15-40. This transducer provides full output for an input voltage of either 120 or 69.3 V. A device of this type is characterized by a relatively high output—5 to 7.5 V d-c—and high speed of response. The transducer is normally calibrated in rms values. Since the unit measures the average voltage, a change in the waveshape of the applied voltage will cause an error.

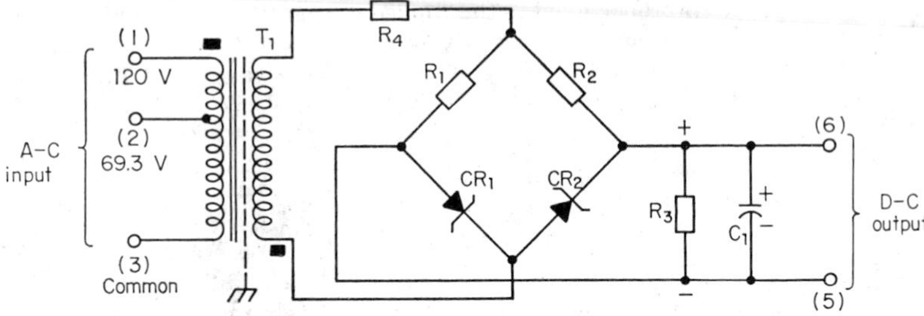

FIG. 15-40. Rectifier-type voltage transducer.

A current transducer of the rectifier type may typically employ the same circuit except that the input transformer T_1 is replaced by a current transformer. The current transducer has the same characteristics as the voltage transducer with respect to the magnitude of the output, speed of response, etc.

PRINCIPLES USED IN FREQUENCY TRANSDUCERS

36. Precision-frequency Deviation Transducer. A precision-frequency deviation transducer operating on the principle of measuring the rate of change of phase angle between the power-system frequency and a crystal-controlled oscillator frequency produces a d-c millivolt output having a high order of accuracy, stability, and resolution. The accuracy and stability at zero frequency difference between the two sources is equal to that of the crystal oscillator. Typical full-scale ranges are ± 0.05 and ± 0.25 Hz.

Figure 15-41 is a block diagram of the precision-frequency transducer circuitry. A detailed explanation of its operation is contained in Ref. 2.

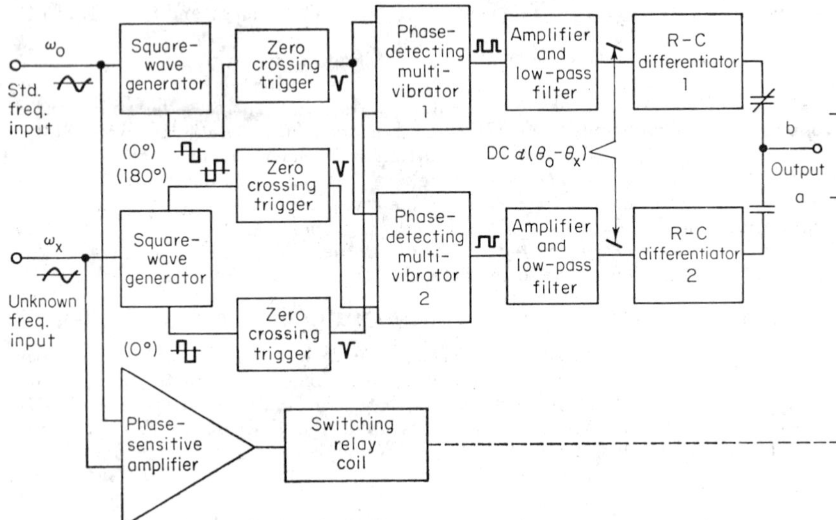

FIG. 15-41. Block diagram of precision frequency transducer.

37. Magnetic-frequency Deviation Transducer. The magnetic-frequency transducer produces a d-c millivolt output proportional to the deviation of power-system frequency from a reference which is determined by a preset bias potential. At zero frequency deviation the output is zero, and as the power-system frequency deviates above and below the preset value of frequency, the polarity of the output reverses.

The conversion from frequency to direct current is based on generating a constant volt-second area pulse for each half cycle of the input frequency. The saturable transformer T_1 of Fig. 15-42 has a square-loop hysteresis core and is driven alternately from saturation in one direction to saturation in the other direction by each half cycle of the input frequency. The area of the volt-second pulses is independent of the time taken to switch the core; however, in a practical case the core is switched by a square-wave voltage.

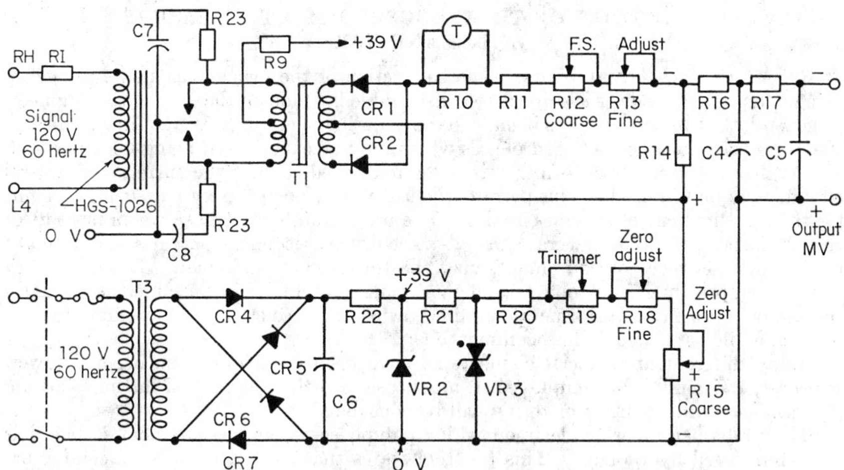

FIG. 15-42. Magnetic-frequency transducer.

The magnetic-frequency transducer is used with typical ranges of ±1 to ±10 Hz deviation from a nominal 60 Hz.

38. Resonance-type Frequency Deviation Transducer. The operation of a frequency transducer of this type depends on two series-resonant circuits tuned to different frequencies. In Fig. 15-43 these circuits are represented by L_1C_1 and L_2C_2. One circuit is tuned above, and one below, the nominal frequency (for example, 60 Hz).

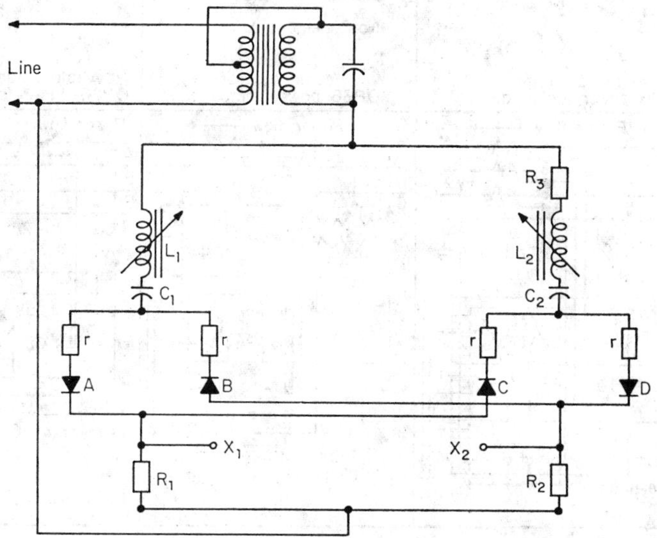

FIG. 15-43. Resonance-type frequency transducer.

Since the resonant circuits are tuned to different frequencies, they present a different impedance to the voltage across them and a different a-c current flows in each.

The four rectifiers are connected to provide a d-c potential across the output $X_1 = X_2$, which is dependent only on the frequency input to the transducer. A voltage-regulating transformer in the input circuit maintains a constant voltage across the transducer circuitry for a ±15% input voltage variation.

CONNECTION OF TRANSDUCERS USED TO MEASURE WATTS AND VARS

39. Alternating-current Watts. The connection of the primary elements for measuring polyphase a-c power is subject to considerable latitude on the part of the engineer. In general, the problem resolves itself into the equating of cost against required accuracy, the degree of unbalance between phases which may be experienced being kept in mind.

Totalizing current transformers are sometimes used so that the currents of several circuits may be totalized. This permits the use of only one primary measuring element for the measurement of several circuits. The use of totalizing transformers introduces certain hazards in that several current-transformer secondary circuits are brought into a common piece of equipment, viz., the totalizing transformer. The possibility of trouble spreading from one secondary circuit to another is thus increased. Also the use of totalizing transformers is limited owing to the fact that all circuits must be in parallel, i.e., supplied from a common bus.

Many different and special connections have been used for measuring a-c power. However, the connections employed in most cases are those described in the following paragraphs. Only 3-phase metering will be considered.

The fundamental principle upon which polyphase-power measurement is based is that enunciated by Blondel. This theorem states that for a system of N conductors the total power can be measured by $N - 1$ wattmeter elements.

If a 3-phase system having a grounded neutral return is examined, it is in reality a 4-wire system and consequently three elements are required for correctly measuring

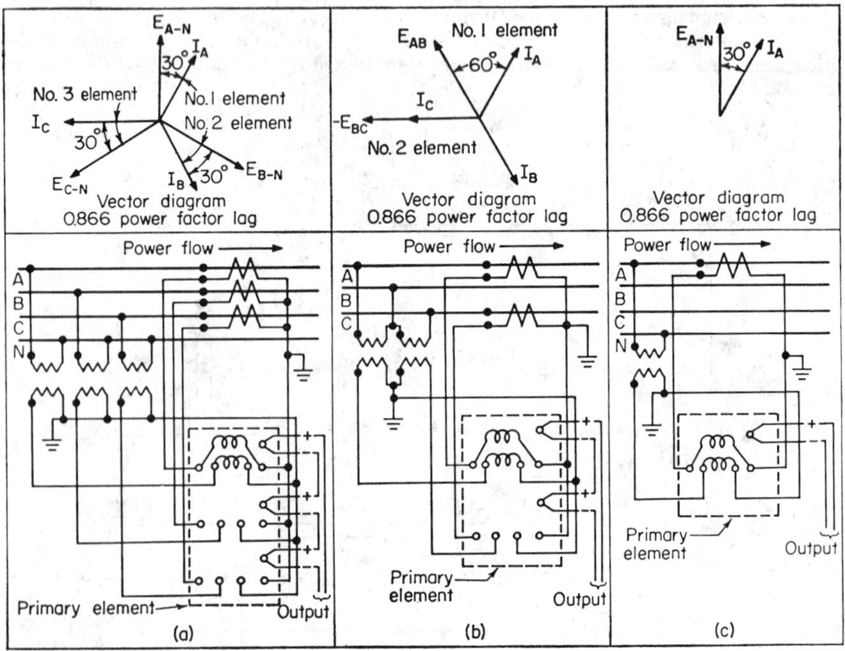

Fig. 15-44. Connections for the measurement of 3-phase power.

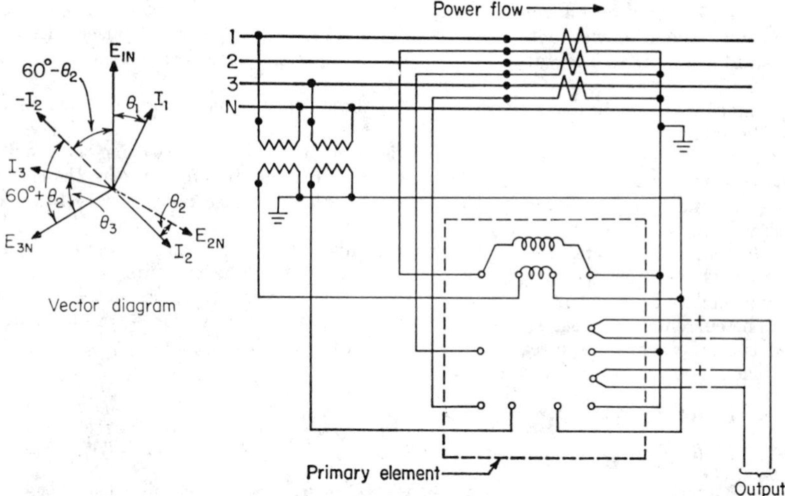

FIG. 15-45. Two-and-one-half-element metering.

the power under any condition of load unbalance. If we consider the above system further and state as a premise that no current flows through the ground-return circuit, it can be considered to be a 3-wire system and the power can be measured correctly by two wattmeter elements. As a further step in considering such a system, if we

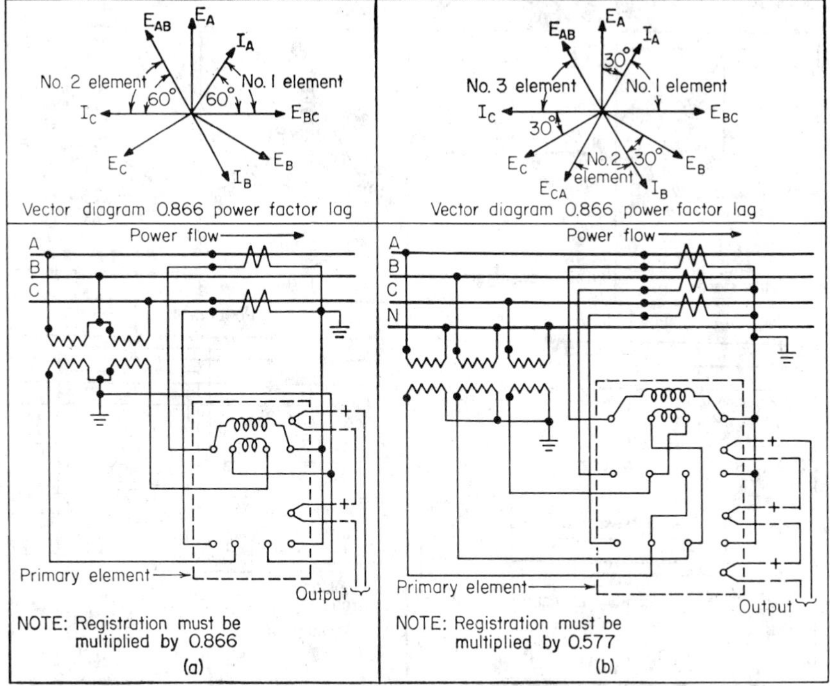

FIG. 15-46. (*a*) Var measurement of 3-phase three-wire circuit using cross phasing; (*b*) var measurement of 3-phase four-wire circuit using cross phasing.

state that the load on all 3 phases is identical, i.e., balanced voltages and currents and the same value of power factor, one element measuring the power in 1 phase only has a constant relation to the total 3-phase power and can be used as a measure of total 3-phase power.

The connections and vector diagrams for these types of metering are shown in Fig. 15-44*a* to *c*.

A modification of a two-element meter is often employed in measuring the power in a 3-phase circuit with a grounded neutral. This is commonly referred to as 2½-element metering and has the advantage that it measures the 3-phase power correctly for unbalanced currents.

The connections and vector diagram for this type of metering are shown in Fig. 15-45. The vector diagram shows balanced voltages. The currents are unequal in magnitude and displaced by different phase angles with their respective voltages. The currents in phases 1 and 3 react with their respective voltages, and the current in phase 2 is reversed and reacts with the potentials of phases 1 and 3.

The metering can be expressed as

$$\text{Metered phase 1 power} = E_{1n}I_1 \cos \theta_1 \tag{15-52}$$

$$\text{Metered phase 3 power} = E_{3n}I_3 \cos \theta_3 \tag{15-53}$$

$$\text{Metered phase 2 power} = E_{1n}I_2 \cos (60° - \theta_2) + E_{3n}I_2 \cos (60° + \theta_2) \tag{15-54}$$

If the voltages are equal in magnitude, then as far as magnitude is concerned we can write $E_{1n} = E_{2n} = E_{3n}$ and Eq. (15-54) reduces to $E_{2n}I_2 \cos \theta_2$. The total for the 3 phases is the sum of the power measured on the individual phases which is correct for total 3-phase power.

40. Var Measurement. As in the measurement of a-c watts, there is considerable latitude in the connections which may be employed in the measurement of vars, but

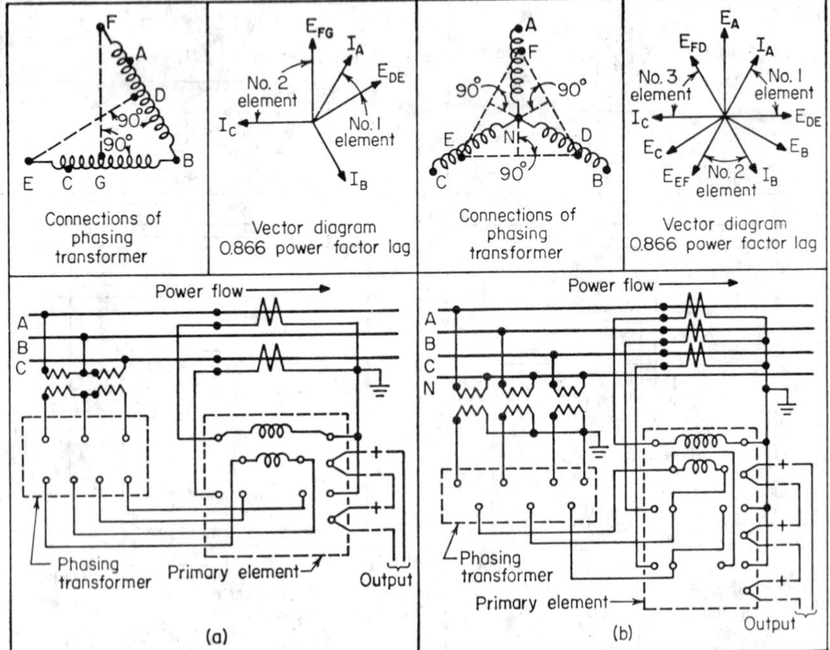

FIG. 15-47. (*a*) Var measurement of 3-phase three-wire circuit using phasing transformers; (*b*) var measurement of 3-phase four-wire circuit using phasing transformers.

the method which is used universally for the measurement of reactive voltamperes on the 3-phase circuits is that of cross phasing the potential circuits with respect to the current circuits. This cross phasing may be accomplished merely by connecting the voltage coils across the phases opposite to that in which the current coil is connected, as shown by Fig. 15-46*a* and *b*, or by using phasing transformers, as shown by Fig. 15-47*a* and *b*. The latter method, which is most generally employed, has the advantage of greater accuracy but is more complicated.

Unfortunately, there is no method available for accurately metering vars of a 3-phase circuit when both currents and voltages are unbalanced. A detailed analysis of the errors is beyond the scope of this section. The errors arising from unbalanced load conditions are most readily determined by the use of symmetrical components. This is covered in a comprehensive manner by A. E. Knowlton.[1]

BIBLIOGRAPHY

41. References on Transducers

1. Knowlton, A. E. "Electric Power Metering"; New York, McGraw-Hill Book Company.

2. *Trans. IEEE,* Power Apparatus and Systems, Spec. Suppl., 1963, pp. 295–303.

POWER-SYSTEM TELEMETERING

By Douglas W. Turrell

42. Introduction. The electric power industry uses many channels of various types of stationary or point-to-point telemetering for remote indication, recording, and control. This telemetering can be divided into the broad catagories of analog and digital according to the method of processing and transmitting the input information.

ANALOG TELEMETERING

43. Transmission. Analog telemetering is extensively used, since most transducers produce a d-c electrical analog of a primary quantity. This signal can be transmitted directly to a remote location by a cable or open pair of wires. The wire pair can be classified as a d-c telemetering link. With more involved processing of the electrical analog at the transmitting and receiving locations, the information can be transmitted via an a-c carrier link such as a telephone system, power-line carrier, or radio. The most common form of radio link is microwave.

44. Voltage Telemeter. The d-c voltage from a low-resistance transducer or other source can be transmitted directly over a wire pair. The receiver must have a high input resistance and is typically a potentiometer recorder or an amplifier with high input resistance. The receiver must also have good common and differential mode rejection of interference such as 60-Hz signals (60 dB or more). Such interference restricts the usable line length for millivolt transducer signals.

45. Current Telemeter. A d-c signal from a high-resistance source can also be transmitted over a wire pair. Thus changes in line and load resistance do not alter calibration. Some form of amplification is used at the transmitting end to obtain the current source. The receiver may be an indicating or recording meter. A resistor can also be inserted in the line and the voltage drop across it applied to potential input-type devices.

46. Pulse and Frequency Telemeters. Variable-duration pulses from a pulse-duration transmitter (impulse telemetering) or a variable-frequency signal from a variable-frequency transmitter are also transmitted over wire pairs. These devices make the telemetering system less sensitive to the changes in the characteristics of the wire line—i.e., line length and loading are not critical. Some degradation in static and dynamic accuracy is experienced owing to the pulse duration or variable frequency-modulation and -demodulation devices. Power-system operators usually require that static errors due to such telemetering be less than 1% of full scale and the response of the output to step changes at the input be less than 1 s to 90% of the final value of the output. These requirements have restricted most analog telemetering to frequency-

Table 15-5. Electric-power-system Analog Telemetering Classifications

Primary elements	Telemeter transmitter	Carrier system			Telemeter receiver	End device
Power Voltage Current Level Temperature Etc.	Voltage ampl. Current ampl. Pulse duration Variable frequency		Wire line	→	Voltage ampl. → → Pulse duration Variable frequency	Meter Recorder Analog control system Analog-to-digital converter
	Freq. or PDM transmitters	Narrow-band carrier-current transmitters	Power line	Narrow-band carrier-current receivers	Freq. or PDM receivers	
	Freq. or PDM transmitters	Narrow-band audio-tone transmitters	Cable or open wire Telephone system Power-line carrier Microwave	Narrow-band audio-tone receivers	Freq. or PDM receivers	

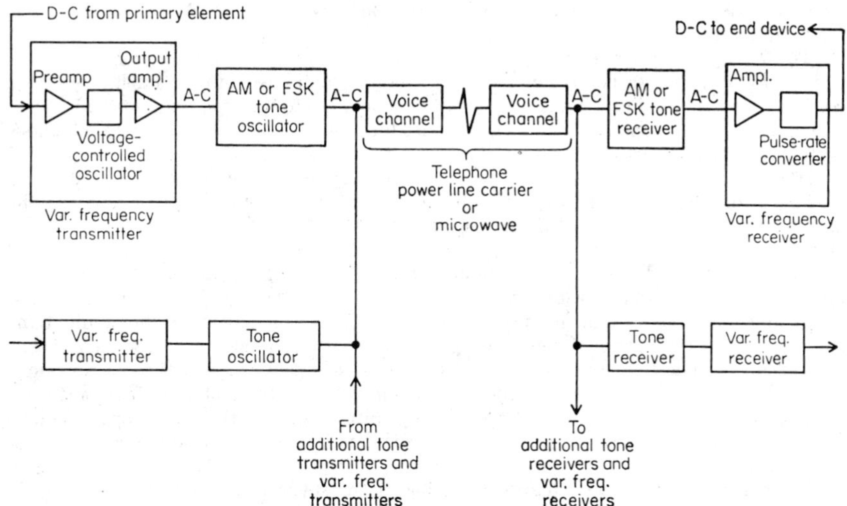

Fig. 15-48. Typical audio-tone multiplexed analog telemetering system.

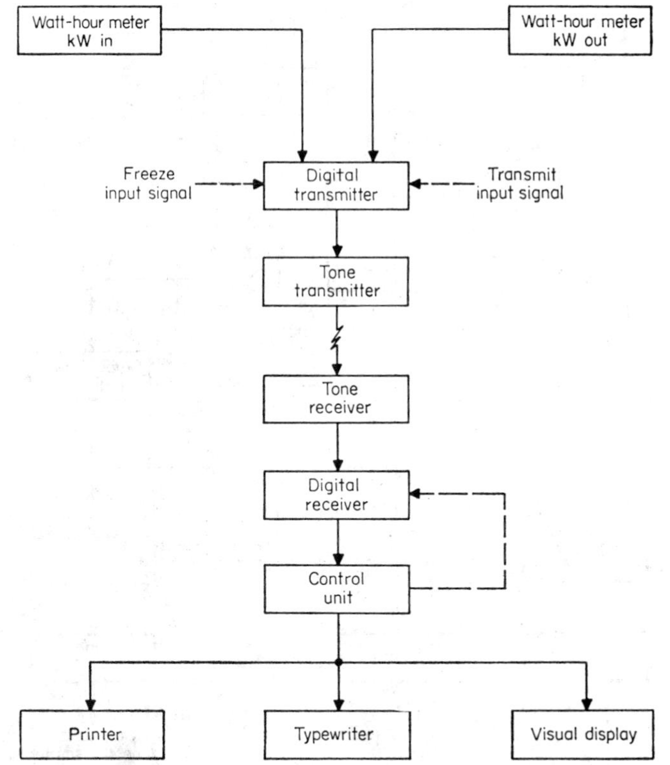

Fig. 15-49. Basic block diagram of digital telemetering system.

type systems. Available information indicates that over 90% of power-system analog telemetering is of the variable-frequency type.

47. Multiplexed Systems. Several analog signals are usually telemetered from generating stations, transmission-system tie points, and distribution substations to a common receiving point. Thus it is usually economical to multiplex the several signals and transmit the composite over an a-c carrier link such as a leased or private telephone system, power-line carrier, or microwave.

Frequency multiplexing techniques are universally used in preference to time multiplexing. Advantages in economy, flexibility, ease of implementation, and interchangeability of components are obtained because of a standardized frequency multiplexing approach.

Narrow-band audio-frequency signals (tones) are used as subcarriers of the information. These tones are generated by transmitting oscillators and passed through narrow-band filters to eliminate harmonics. A group of tones are spaced in the audio band and applied to the carrier (telephone, power-line carrier, microwave) as a composite audio signal. The tone receivers at the output of the carrier system each contain a narrow-band filter that responds only to a specific tone and the modulation on this tone.

The tones are keyed on and off or frequency-shifted at a variable rated for frequency-type telemetering and for a variable time duration when pulse-duration-type telemetering is used.

Table 15-5 provides a comparative listing of the several analog telemetering tech-

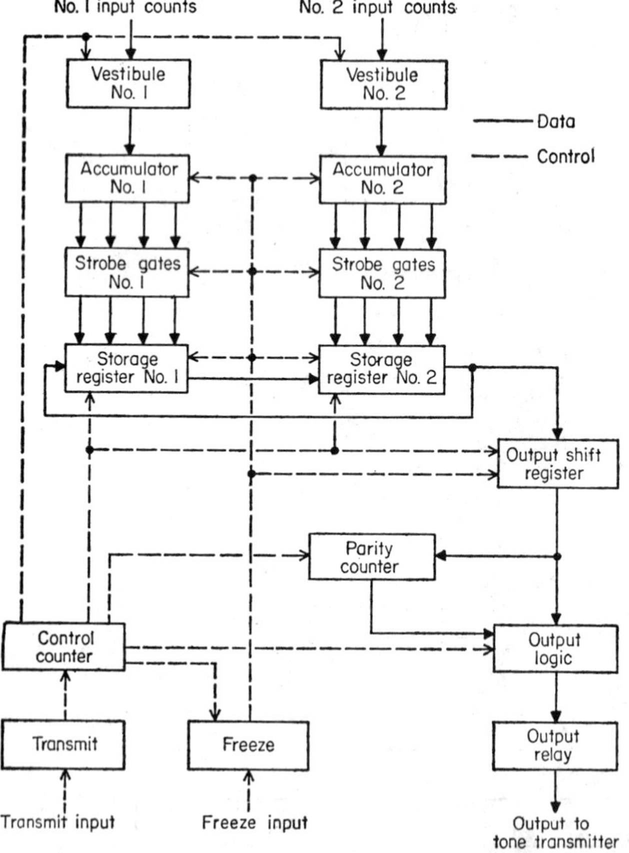

FIG. 15-50. Functional block diagram of transmitter.

niques, including narrow-band frequency-shift-keyed (FSK) power-line carrier. Figure 15-48 shows the essential circuit elements of a frequency-type telemeter combined with a frequency-multiplexed narrow-band tone system and an audio carrier.

DIGITAL TELEMETERING

48. Conversion. Digital telemetering refers to the generation and transmission of a pulse code for each element of a set of discrete levels of a primary quantity or signal. For example, an analog signal is sampled periodically and converted to a group of binary digits (bits) uniquely representing the value of the sample. This conversion (quantization) is accomplished by an analog-to-digital converter. More simply, a counter may be connected to the pulse output of an integrating device.

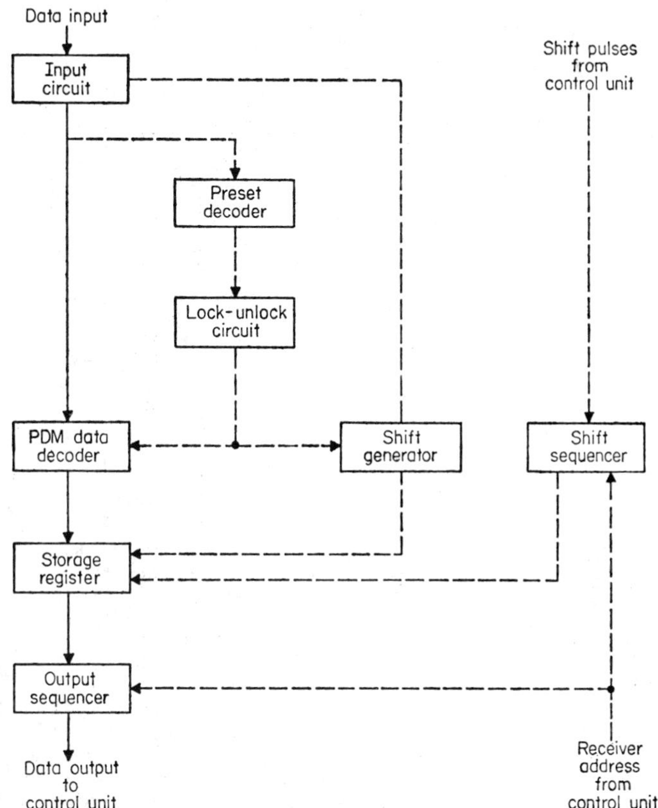

FIG. 15-51. Functional block diagram of receiver.

Once the sample and conversion or pulse counting is completed, the resultant digital word can be stored, telemetered, and processed in a variety of ways with no degradation in accuracy. However, during transmission, noise impulses may change the code word. Hence coding at the transmitter and decoding at the receiver for error detection and means for error correction are important considerations in digital telemetering and digital data transmission in general.

49. Pulse-duration Modulated Digital Telemetering. Digital telemetering is widely used in the electric power industry for the periodic transmission of the accumulated pulse count from kilowatthour meters. These data are used for billing, accounting, and control. A basic telemetering system is shown in Fig. 15-49 and functional block

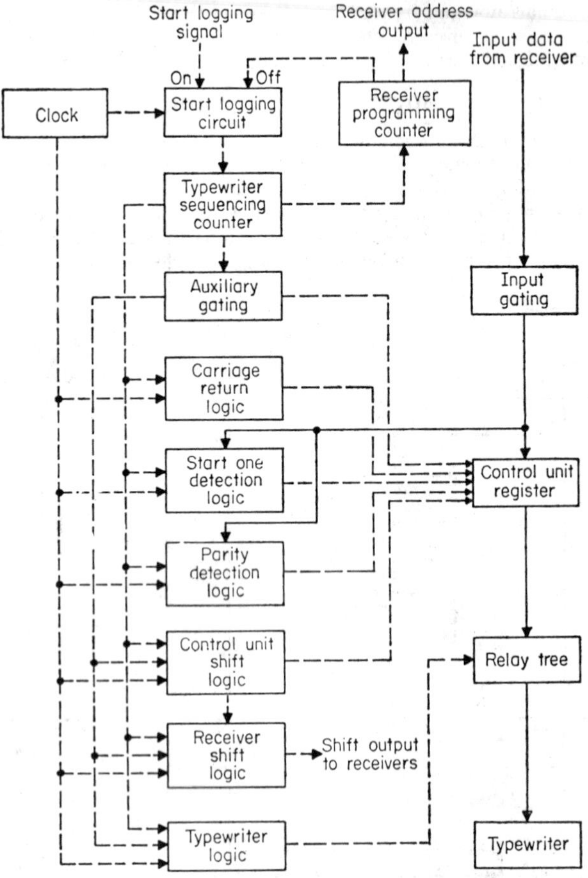

Fig. 15-52. Functional block diagram of control unit.

diagrams in Figs. 15-50, 15-51, and 15-52. A typical rack-mounted solid-state system, consisting of a transmitter, receiver, and control unit, and output typewriter, is shown in Fig. 15-53.

Referring to Figs. 15-49 and 15-50, the digital transmitter receives a FREEZE command from the receiver station as a pulse of specified duration. This causes the transfer of the pulse count in each accumulating counter to a storage register, as indicated in Fig. 15-50. Upon receipt of a TRANSMIT command of another specified duration, the transmitter transfers the contents of each storage register in sequence to a serial output register (shift register). The ONE and ZERO bits in the shift register are then serially transmitted over the telemetering channel. In pulse-duration modulation (PDM) telemetering the ONE and ZERO bits are changed to long pulses and short pulses to represent the ONES and the ZEROS of the data, and parity pulses are added according to a coding algorithm to permit error detection at the receiver. The long and short pulses can be transmitted directly over a d-c link or used to control an amplitude-keyed (AM) or frequency-shift-keyed (FSK) audio tone in the same manner as a pulse-duration analog transmitter. With digital telemetering, however, the pulses are only two discrete durations, which represent the binary levels.

The receiver shown in Fig. 15-51 detects the presence of the telemetered pulses and decodes the long and short pulses into the binary levels. The binary levels are shifted into a storage register at the pulse rate (bit rate) of the telemetering. When the storage

FIG. 15-53. Digital telemetering transmitter, receiver, and output typewriter.

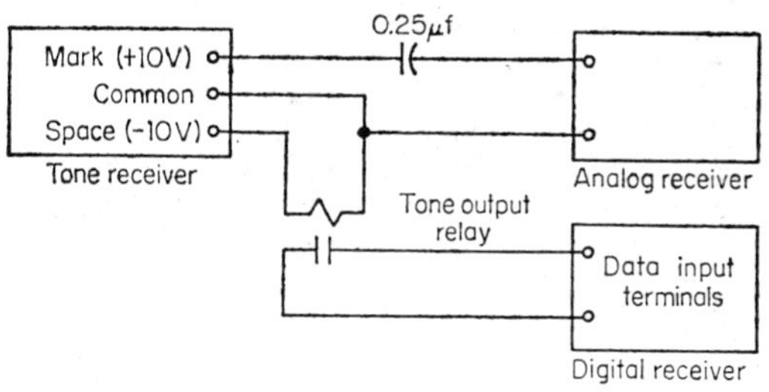

(a) Multiplex connections

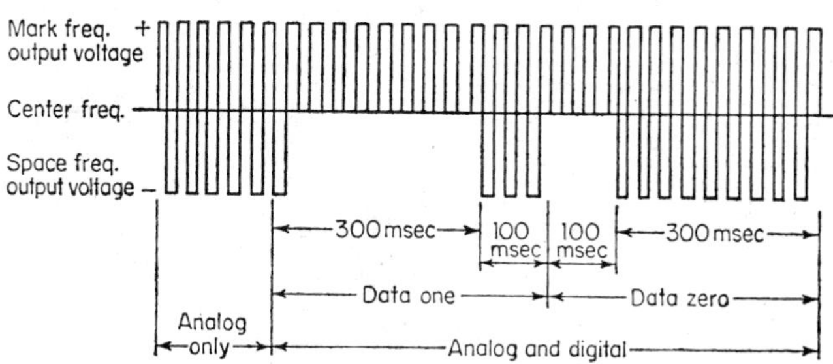

(b) Output of tone receiver

FIG. 15-54. Connections and tone output for multiplex operation at receiver.

register is filled, control is transferred to a digital control unit as shown in Fig. 15-52. The control unit processes the telemetered bits at high speed according to conventional digital principles for typewriter display, visual display, digital-analog conversion, or digital-computer use.

50. Digital-analog Multiplexing. For infrequent transmission of digital data such as meter readings, the digital telemetering may time-share a tone channel with a slow-changing analog signal such as a generator output. The analog telemetering is interrupted periodically, and the digital readings are transmitted over the tone channel for several seconds. Alternatively, a three-frequency FSK tone set may transmit a variable-frequency analog signal and a pulse-type digital signal simultaneously. The analog information is contained in the rate of keying of the tone transmitter, and the digital pulses are used to switch the keying from the mark-space frequencies to the mark-center frequencies, and vice versa. The connections and output waveform of the tone receiver are shown in Fig. 15-54.

51. Time-multiplexed Digital Readings. The equipment for telemetering of the accumulated count of kilowatthour meters can be modified to transmit the quantized value of any number of primary quantities. This is done by time-sequencing the additional readings into the output shift register of the transmitter shown in Fig. 15-50. Additional control logic, switching, and storage registers are required for the transmitter and receiver.

Quantities that may be quantized and telemetered in this manner are typically:

Tie-line kilowatts, kvars, and kVA. Tie-line kilowatthours. Bus voltage. Station load (kilowatts, kvars, kVA). Station kilowatthours. Generated kilowatts, kvars, and kVA. Generated kilowatthours. Temperatures (ambient and equipment). Frequency. Time error. Water levels.

The real-time requirements of electric power-system control may dictate that many such readings be telemetered every second from a single station. Thus the cost of many channels of slow-speed (7.5 to 60 bits/s) telemetering using a multiplicity of tone sets and digital transmitters and receivers must be balanced against the cost of time-multiplexing the readings using a single digital transmitter-receiver at each end with high-speed data sets and a full duplex audio bandwidth channel (600 to 2,400 bits/s). The probability of loss of all information from a single station in time multiplexing is also an important consideration.

52. Telemetering and Supervisory Control. Time-multiplexed digital telemetering to obtain numerous readings in rapid sequence over a high-speed system has not been extensively applied to electric power systems. Rather, the more complex systems available offer the added capabilities for transmission of alarms, state of circuit breakers and other equipment, plus off-line data from punched cards, paper tape, and operations

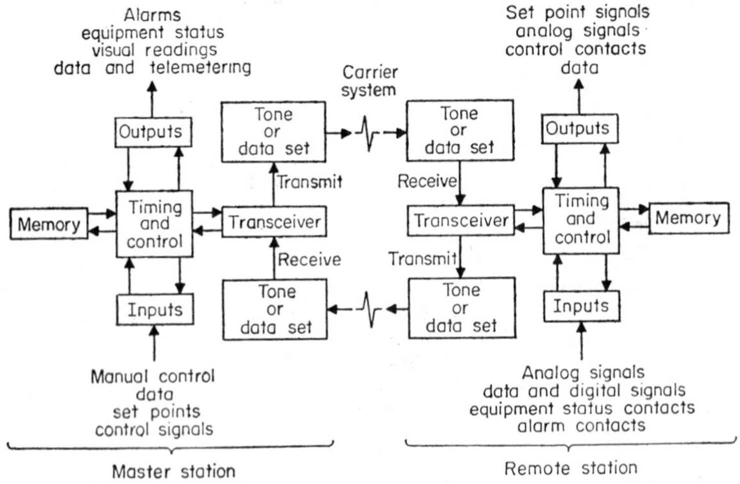

Fig. 15-55. Generalized block-diagram digital supervisory control system.

recorders at the remote locations. Remote stations also have the capabilities to operate control switches, control motors, printers, etc., by decoding the digital commands from a master station. Such a broad spectrum of capabilities places the more comprehensive systems in the category of all-digital supervisory control. Thus telemetering of quantized signals from primary elements is only one of the many functions of such systems. Control of a digital supervisory system is accomplished by prewired control programs at the master and remote stations, with operator intervention permitted for remote and local operation of the end equipment. Figure 15-55 is a generalized block diagram of such a system.

53. Bibliography on Power System Telemetering

GENERAL

Telemetering, Supervisory Systems and Associated Channels; Committee on Telemetering, *AIEE Publ.* S-111, June, 1959, Vol. I. (Includes 1,648 references through 1958.)

Telemetering, Bibliography; Committee on Telemetering, *AIEE Publ.* S-111-A, April, 1961. (648 references through 1960.)

FOSTER, L. E. "Telemetry Systems"; New York, John Wiley & Sons, Inc., 1965. (Radio-frequency telemetry—theory and practice, analog and digital.)

POWER-LINE CARRIER

"Electrical Transmission and Distribution Reference Book," 4th ed., Westinghouse Electric Corporation, 1950, Chap. 12.

DATA TRANSMISSION

BENNETT, W. R., and DAVEY, J. R. "Data Transmission"; McGraw-Hill Book Company, 1965.

PETERSON, W. W. "Error-correcting Codes"; Cambridge, Mass., The M.I.T. Press, 1961.

SUPERVISORY CONTROL

RYAN, F. M. Supervisory Control Systems; *Control Eng.*, January, 1963, pp. 77–86.

RECORDERS FOR POWER SYSTEMS

BY WILLIAM RUSSELL CLARK

54. Introduction. An important instrument in the electric power field is the graphic recorder. It provides concurrent and historic records of various functions and important operating conditions affecting the performance of a power plant or a complete interconnected power system.

55. Direct-writing Type. The earliest recorders were of the direct-writing type, both mechanically and electrically activated. Many of these recorders are still used. A schematic diagram of the electrically activated type is shown in Fig. 15-56.

Operation. In this recorder an electrical signal energizes the moving coil of a deflecting galvanometer, the recording means being attached to or controlled by the coil. The recorder is calibrated against a standard external spring or against the restoring force of the suspension supporting the moving coil of the galvanometer.

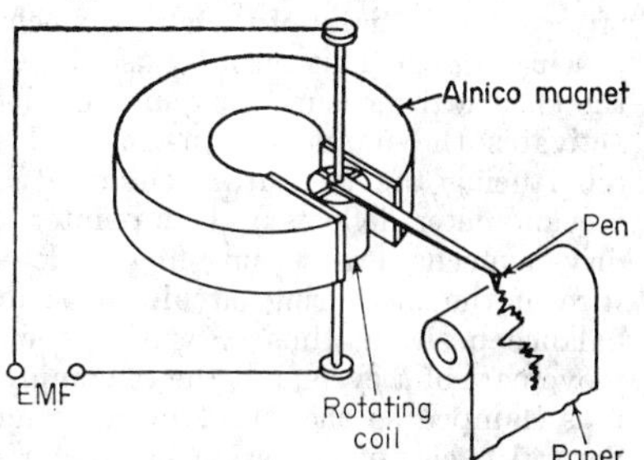

FIG. 15-56. Schematic of direct-writing electrical recorder.

Records can be made by various methods; the one most commonly used employs a pen with ink at the end of the recording stylus, which makes a record of its movements on a motor-driven paper chart. Pressure-sensitive paper can be used to record the position of the stylus on the recording element, and, in some cases, an electrical spark is used between the recording stylus and a spark-sensitive recording paper. To overcome friction between the recording stylus and the paper, the recording stylus, in some electrical recorders, is clamped against a typewriter ribbon, which is pressed against the recording chart, thus making an intermittent dot record of the location of the recording stylus. The stylus is left free to move to a new position for approximately half the recording cycle time.

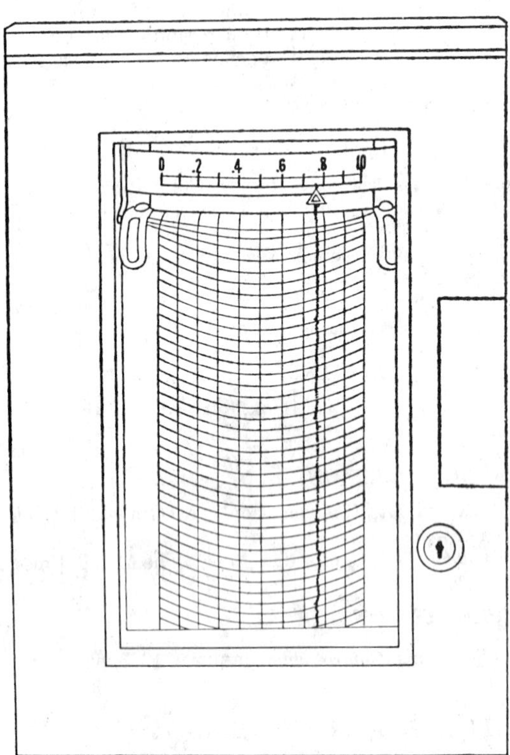

FIG. 15-57. Typical direct-writing electrical recorder (15¾ × 10½ × 10 in). (*Courtesy of Esterline Angus Instrument Company.*)

In some direct-writing electrical recorders a mirror is fastened to the moving coil of the deflecting system, and a light beam is reflected from that mirror onto a light-sensitive paper, thus providing a record in which friction forces are reduced to a minimum. This type of transmission of record permits the fastest response time in recorders.

Applications. The direct-writing electrical recorder records a d-c electrical current or voltage. In power-plant applications they are used for recording volts, amperes, watts, voltamperes, or temperature. They can be used whenever an emf of sufficient magnitude can be obtained from a transducer which has a known relation to and is a direct indication of the magnitude of the measured variable. Electronic amplifiers can be used between the unknown signal and the recorder to permit recording of electrical signals too small to activate the recorder directly. Figure 15-57 shows a typical direct-writing recorder.

NULL-BALANCE RECORDERS

56. Electromechanical Recorders. The desire to achieve better accuracy, better response to small signals, and wider charts led to development of the null-balance type of recorder, initially of the electromechanical type and currently of the electronic type.

Operation. This recorder uses a galvanometer to detect the degree of unbalance of the electrical measuring circuit but does not itself drive the pen. The galvanometer activates the input of a mechanical amplifier, which provides sufficient torque for rebalancing the measuring circuit. When the measuring circuit is unbalanced, the galvanometer deflects and its pointer is clamped in this off-balance position. Motor-driven mechanical sensing fingers detect the off-balance position and cause the slide wire of the measuring circuit to be motor-driven toward, and eventually reach, the balance position, thus measuring the unknown input. The galvanometer is free to move part of a cycle, but the remaining time, when it is activating the torque amplifier, it is clamped so that no damage or undue strain is put on the galvanometer system. Several cycles of operation are required to balance the system, the number of cycles being dependent upon the degree of unbalance. In most cases a single, continuously running motor supplies all the power for operating the recorder, including driving the paper-drive assembly. The balancing mechanism also operates a pointer which indicates the value of the unknown being measured on a scale and in addition drives a pen which continuously records on the chart paper the value of the indicated measurement.

The galvanometer is sometimes more delicate and sensitive than that used in the direct-writing electrical recorder, but it is used in such a way that no strain is put on the suspensions and hence no changes are caused in the galvanometer's operating characteristics. Sufficient torque is available in this recorder to operate alarm signals, control contacts, control slide wires, retransmitting slide wires, and selsyn motors.

Applications. These recorders were used to measure all the quantities measured by direct-writing recorders and also provided most of the features available with electronic recorders, but they are being replaced by the electronic recorder.

The type of measurement made by this recorder is called a null-balance measurement. This means that, for an emf measurement, the current to the detector (galvanometer) is essentially zero for all balance positions of the instrument; hence, the effect of lead resistance and the loading of the unknown source by the recorder are negligible. This highly desirable condition, which results in greater accuracy of measurement, is not possible when direct-writing electrical recorders are used. Other measuring circuits similar to those used with electronic recorders were also used with these recorders.

Although this electromechanical recorder was a reliable instrument, it had the following limitations:

1. It required a minimum time of about 12 s for full-scale rebalance.
2. The voltage range owing to the limitation in sensitivity of the galvanometer was limited to about 8 mV, with a corresponding limitation for current or resistance measurements.
3. Severe vibration or shocks caused faulty or irregular operation of the recorder.

57. Electronic Recorders. In this recorder the galvanometer-mechanical amplifier system is superseded by an electronic amplifier motor system. Figure 15-59 shows schematically the wiring diagram of a typical electronic recorder employing solid-state components throughout.

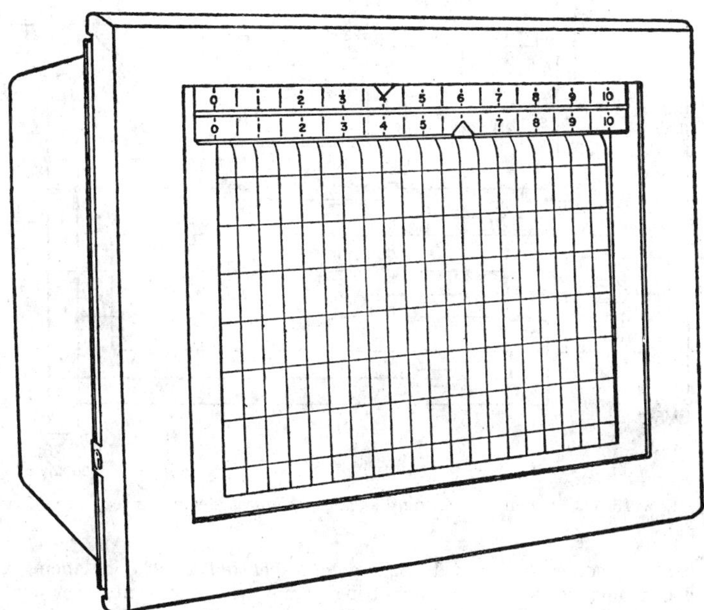

Fig. 15-58. Typical electronic recorder (12 × 15 × 13 in). (*Courtesy of Leeds & Northrup Company.*)

Operation. The essential recorder parts are:

1. Measuring circuit (in this case a d-c potentiometer).
2. Detector.
3. Amplifier.
4. Balancing motor and slide-wire drive system.
5. Display system, which includes a method for indicating and/or recording the value of the measurement being made.

The operation of the recorder, when used for voltage measurements, is as follows: The unknown signal being measured is opposed by an adjustable calibrated emf; the exact value of this calibrated emf depends on the position of the slide-wire contact in the potentiometric measuring circuit. When these two voltages are not equal, an unbalanced current flows in the detector circuit. The detector, through its converter, then changes the input d-c current from the measuring circuit to an output a-c voltage. The amplifier amplifies this voltage until it is of sufficient magnitude to operate the balancing motor, which is of the reversible type. The motor, through a mechanical linkage, adjusts the recorder slide wire in the potentiometric circuit until the calibrated emf it produces is so nearly equal to the unknown emf that the resulting signal to the amplifier will no longer produce sufficient power to drive the motor. Sufficient amplification is supplied in the amplifier so that the calibrated emf is equal to the unknown signal to within the stated accuracy of the recorder. When the error current becomes substantially zero, then a null, or balanced, condition exists. Figure 15-58 shows a typical electronic recorder. This instrument has two complete systems and will draw two independent records.

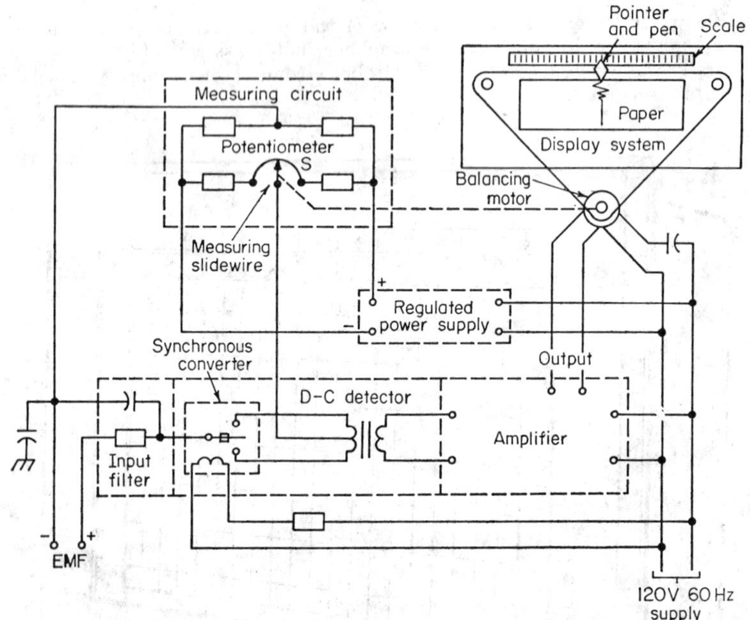

FIG. 15-59. Schematic wiring diagram of typical electronic recorder.

Damping. In some recorders a tachometer, operated by the balancing motor, produces a damping voltage in the measuring circuit which prevents overshooting of the balance point. Most present-day recorders obtain this damping voltage by resistors and capacitors in the input circuits in which the changing voltage caused by the motion of the slide wire itself generates a voltage of correct polarity for damping the recorder.

Interference Reduction. Most recorders provide filtering in the low-level input circuit to reduce the effect of pickup in the unknown signal circuit (interference). This filtering usually consists of resistors and capacitors which are the same as those which produce the damping voltage. The pickup is produced by electrostatic and electromagnetic coupling of the low-level leads to the recorder, with power lines and power equipment. If this pickup voltage should get into the low-level measuring circuit, it would be chopped by the converter and amplified by the amplifier. While this unwanted signal may not drive the balancing motor of the recorder, it could completely paralyze and block the amplifier and prevent its operation. By supplying sufficient filtering it is relatively easy to reduce the effects of this pickup to a negligible amount. This unwanted interference is called "transverse interference" or "normal mode interference."

Another type of interference called "longitudinal interference" or "common-mode interference" can develop between the input leads of the recorder and ground. A capacitor placed between one input lead and ground is usually sufficient to minimize the amount of such interference which can enter the recorder circuit. In some applications the voltage between the input leads and ground is large, and a double-shielded input transformer, plus a completely isolated input measuring circuit, may be required to reduce its effect to negligible proportions.

Measuring Circuit Supply. In earlier recorders of the electronic type as well as the electromechanical type, current for the measuring circuit had to be obtained from batteries, and a standardizer had to be supplied in the recorder which would periodically standardize the potentiometer current. In more recent instruments a d-c regulated power supply powered from the a-c line and using, in most cases, two stages of zener-diode regulation supplies the measuring circuit with the required stabilized current. This regulated supply replaces both the batteries and the automatic standardizer formerly used. Bridge measuring circuits require only one stage of stabilization.

58. Measuring Circuits. There are several types of measuring circuits which can be used with the null-balance recorders. The most common is used for measuring emf and employs a potentiometer circuit. A Wheatstone bridge can be used as a measuring circuit for measuring resistance directly. Currents can be measured directly by using a null-current circuit. Measuring circuits of present-day recorders are mounted on interchangeable cards, thus making it relatively easy to change ranges.

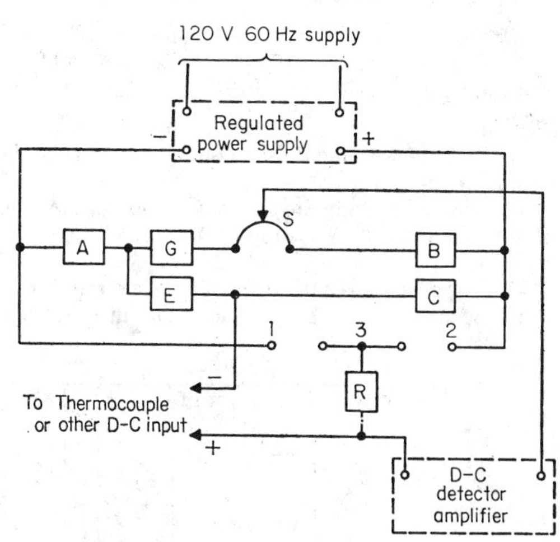

Fig. 15-60. Potentiometer circuit.

Potentiometer Circuit. Figure 15-60 shows a typical potentiometer circuit which by adjustment of the circuit constants can be used for measuring d-c voltages or measuring temperatures directly by thermocouples. The circuit has two branches of equal resistance supplied by a regulated power supply. The balancing motor of the recorder drives the contact of the slide wire S until the emf being measured is balanced by the potential difference existing between the slide-wire contact and the junction of resistors E and C. The value of resistances S, E, and G determines the desired span of the recorder and the suppression or elevation of the zero of the recorder. A range of 20 to 30 mV is said to have 20-mV zero suppression, while a range −50 to +50 mV has a zero elevation or negative suppression of 50 mV.

Fail-safe operation is provided by resistors A and R. They introduce in the measuring circuit a voltage of sufficient value to move the recorder to a predetermined down-scale or up-scale position if an open circuit occurs in the unknown being measured. When terminal 3 is connected to 2, up-scale fail-safe is provided and when connected to 1, down-scale fail-safe is provided.

A potentiometric type of measurement has the following characteristics:

1. The measurement is made by indirect comparison of the unknown emf with a precision standard.
2. At the balance condition practically no current flows in the source of measured emf; so there is negligible voltage drop in the source. As a result the instrument measures the true emf or open-circuit voltage of the source.
3. Since the resistance between the contact and the measuring slide wire is in the detector input circuit, this resistance has no effect on the measuring accuracy.

This circuit is used for measuring temperature by thermocouples. The variation in thermocouple output caused by the two cold junctions of the temperature-measuring thermocouple can be compensated for by using the proper combination of materials for the E coil in the measuring circuit. As the recorder temperature varies, the value of the E coil varies to compensate the measuring circuit for these cold-junction voltage changes. Thus the recorder can be made to always read the true temperature of the thermocouple hot junction.

Current Circuit. Current can be measured by passing it through a known-value resistance and measuring the voltage drop across this resistance by a potentiometer circuit recorder. This measurement is useful for many applications, but it is not a null-current measurement. A true null-current type of measurement can be made by the circuit shown in Fig. 15-61. The range of this recorder is the value of the slide-wire voltage S divided by the value of the series resistor R. The current I_e produced through R by the voltage E obtained from the slide wire opposes the unknown current I_x. Any difference between these currents produces a voltage drop across the detector, which in turn causes the recorder slide wire to be adjusted until the currents are made equal.

The null-current type of circuit has the advantage that it presents essentially zero resistance to the source of unknown current, which is highly desirable in current measurements.

Bridge Circuit. Figure 15-62 shows a Wheatstone-bridge circuit used in a recorder for measuring resistance by direct current. This circuit uses two moving slide wires, and the values of the resistance in each arm are the same. It has the following advantages:

1. The measurement is made by comparing the unknown resistance RX with precision resistances R and S.
2. The measuring accuracy of the bridge does not depend upon the constancy of the regulated power-supply voltage. It is only necessary that the voltage be high enough.
3. The resistances between the slide wires and their contacts do not affect the measuring accuracy due to their location in the circuit.

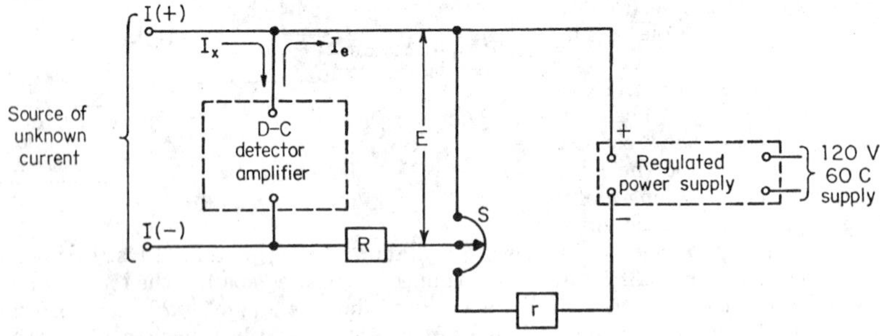

FIG. 15-61. Null-current circuit.

4. By using a three-lead type of measurement in which the a and b leads are of the same resistance, the measuring accuracy of the circuit is not affected by the length of lead. The resistance of the c lead is in the power supply.

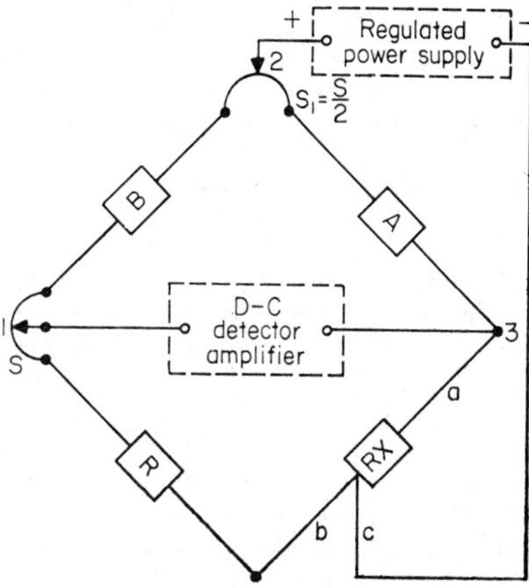

FIG. 15-62. Wheatstone-bridge circuit.

Frequency Circuit. For measuring frequency, a bridge measurement is used as indicated in Fig. 15-63. Its operation depends on the impedance difference between two arms of the bridge resulting from the applied frequency. This difference occurs because resistor A of the bridge is in parallel with the capacitor C and resistor B of the bridge has capacitor $C1$ in series with it. Slide wire S is connected between the two arms A and B and permits a variable amount of resistance to be included in each. As the frequency varies, the position of the two slide wires S and $S1$ varies accordingly. Both these slide wires are mounted on the same molding and move in unison. The bridge is supplied by alternating current of the frequency being measured. The amplifier accepts an a-c input rather than the d-c input previously described. A double-pole double-throw mechanical converter with the proper arrangements of resistances and capacitances provides the required damping and elimination of unwanted harmonics in the input circuit. The amplifier drives the motor M, which in turn drives the slide wire so that the circuit is always balanced for the frequency being measured. The recorder can have a span of ± 10 Hz down to ± 0.5 Hz and can be supplied with double ranges within the values of the spans given.

Frequency recorders can be supplied with a device for recording the time error on

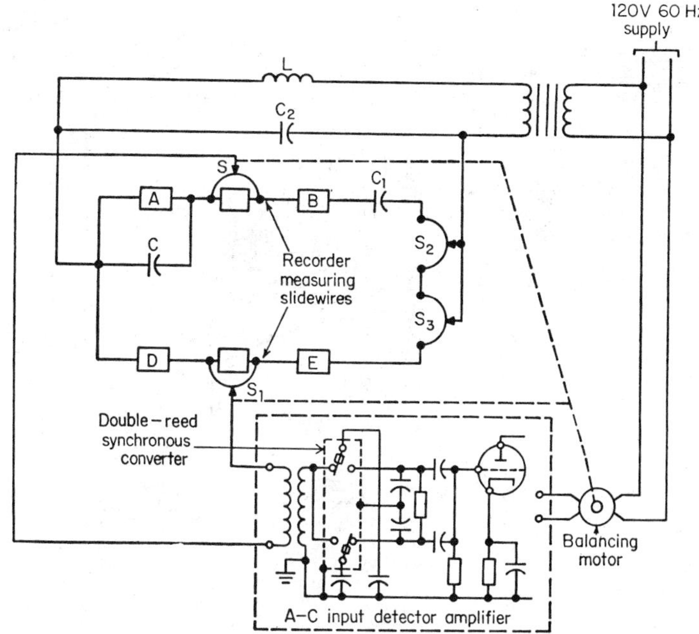

FIG. 15-63. Frequency-bridge circuit.

the same chart. This instrument continuously indicates the difference in seconds between system frequency and a master frequency standard.

59. Applications. The recorders are available as strip-chart recorders or round-chart recorders, as well as indicators. The recorder has sufficient torque to operate alarm signals, controls, retransmitting slide wires, and selsyn motors. A selector switch driven by the same motor that supplies the motivation for the rest of the recorder operations is used in the input circuit to permit recordings of multiple inputs by the same recorder. In these cases a multiple-point printing wheel is used to print the measurements on the chart paper. Each input is represented by a specific printed number, and, in some cases, each input point has its own specific color. Multipen recorders are also available with separate and complete measuring systems.

The electronic null-balance type of recorders finds wide use with suitable transducers for measuring power-system parameters such as watts, voltamperes, reactive voltamperes, volts, amperes, frequency and temperature.

Wheatstone-bridge recorders, particularly of the multipoint type, are used with resistance thermometers for precise measurement of the temperature of electric generators, generator bearings, and transformers and their associated cooling media.

The major use of electronic recorders for interconnected system operation is in central-area dispatching and control offices. The customary practice is to telemeter into such central locations and display on self-balancing recorders important operating variables such as power generation at each station, power flow on each intertie, and significant loads, reactive power, and voltage in important area locations.

Figure 15-64 is an example of such centralized recording used at the Canton Operating Center of the American Electric Power Service Corporation, where the electronic recorders provide the operating personnel moment-to-moment records of power generation and power-flow conditions throughout the seven-state system shown in Fig. 15-7.

Fig. 15-64. Typical centralized control station. (*Courtesy of American Electric Power Service Corporation.*)

60. Bibliography on Recorders

REFERENCES

1. Dickey, P. S., and Hornfeck, A. J. Electronic Type Instruments for Industrial Processes; *Trans. ASME*, 1945, Vol. 67, pp. 393–398.

2. Williams, A. J., Jr., Clark, W. R., and Tarpley, R. E. Electronically Balanced Recorder for Flight Testing and Spectroscopy; *Trans. AIEE*, April, 1946, Vol. 65, pp. 205–208.

3. Clark, W. R., Wills, W. P., Hornfeck, A. J., Williams, A. J., Jr., Parsegian, V. L., Keinath, G., and Hellman, R. K. Electronic Recording Instruments; *Trans. AIEE*, 1947, Vol. 66, pp. 36–43.

4. Williams, A. J., Jr. AC Null-type Recorder with Balancing Amplifier Which Provides Damping and Suppresses the Quadrature Component; *AIEE Paper* 53–244, June, 1953.

5. Maddock, A. J. Servo-operated Recording Instruments; *Proc. IEE* (*London*), 1956, Vol. 103, pp. 617–632.

6. Keinath, G. Recorder Survey: Recording Surfaces and Marking Methods; *NBS* (*U.S.*) *Circ.* 601, 1959.

7. Specifications for Automatic Null-balancing Electrical Measuring Instruments; USAS C39.4/624.

PROTECTION OF POWER-SYSTEM INTERCONNECTIONS

By John Berdy

61. Introduction. Interconnections between power systems provide increased service reliability and economy in system design and operation. The majority of these interconnections are high-capacity transmission lines which tie neighboring utility systems together at one or more points. These tie lines not only are capable of providing support during emergencies but also provide for economical energy interchange and sharing of reserves and permit utilization of larger, more efficient generator-unit sizes.

Realization of the advantages afforded by interconnections depends to a large extent on the establishment of economical as well as highly reliable transmission tie lines. While the achievement of such transmission lines requires consideration of numerous system and electrical-design problems, particular attention must be given to the design of adequate protective relaying systems. The protective relaying not only must be capable of high-speed clearing of all faults but also must have a high degree of reliability and must provide coverage for more contingencies than might be justified in other system areas.

This section describes the protective relays, schemes, and practices used to meet the speed and reliability requirements of high-capacity transmission interconnections. Since it is beyond the scope of this section to describe all the techniques used to provide protection on a power system, discussion will be limited to the basic concepts of transmission-line protection and the practices most commonly used.

62. Protective Schemes. The protective relays and schemes used for transmission-line protection are basically computing systems using a combination of analog computing units and digital logic functions. Until a few years ago, these computing and logic functions were accomplished solely with electromechanical devices. At present, there is a rapid evolution toward the use of semiconductor components in static control and logic circuitry to accomplish these same functions.

While the operating principles of electromechanical devices differ appreciably from those of static devices (devices which use semiconductor components), electromechanical relay and static relay units have the same basic operating characteristics and the schemes using these devices perform, functionally, in the same manner. Therefore, the comments apply to electromechanical as well as to static relaying schemes.

63. Relay Units. The most common method of detecting faults on transmission lines is by impedance measurement. Relay units, having inputs of current and voltage, respond to the ratio of voltage to current and therefore to impedance or a component of impedance, depending on the relay design. Since impedance is a measure of distance along a transmission line, these relay units are commonly called distance relays.

This impedance-measurement approach is used because it provides an excellent way of discriminating between faults and normal system conditions and between faults in a specific area and faults elsewhere in the system. Discrimination is obtained by limiting relay operation to a certain range of impedance. The operating limits of distance units are usually expressed in terms of impedance or in terms of impedance components—resistance and reactance. In most instances, these impedance expressions are equations of simple geometric figures, like a straight line or a circle, which can be plotted on a rectangular-coordinate system using resistance as the abscissa and reactance as the ordinate. Figure 15-65 illustrates some of the obtainable distance relay characteristics plotted on this coordinate system. The relay location or point of measurement is at the origin, and the relay will operate for impedances which fall within the shaded area. Since the reactance component of system impedance is usually inductive, the distance relay units are designed to measure inductive impedances. Therefore, the relay characteristics are always shown in the first quadrant, where resistance and reactance are positive.

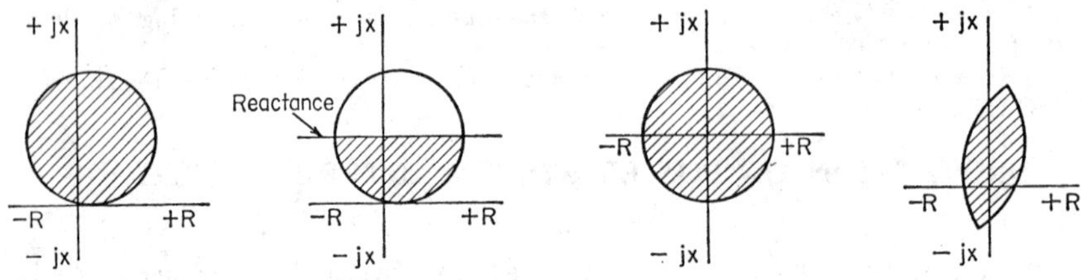

FIG. 15-65. Distance-relay characteristics.

This coordinate system, called the $R - X$ diagram, is a widely used relaying tool. With this diagram, the operating characteristic of a distance relay can be superimposed on the same graph with almost any system characteristic, thus making the response of the relay immediately apparent.

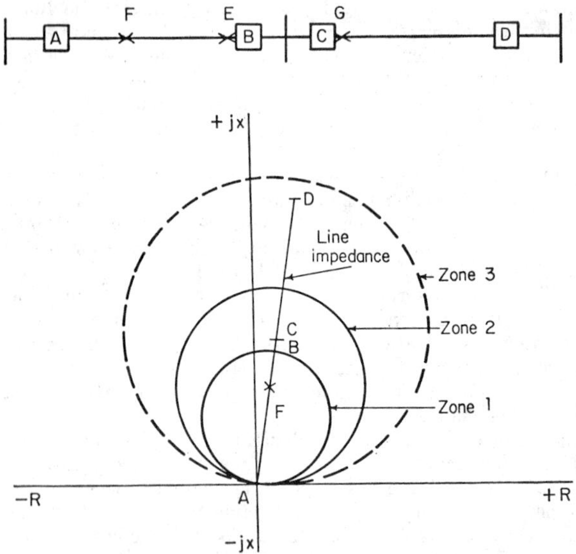

FIG. 15-66. Distance-relay protection of a transmission line.

The protection of a transmission line with distance relays is usually accomplished with two units. Using the relay characteristic of Fig. 15-65*a*, this type of protection is shown in Fig. 15-66. The relay characteristics are superimposed on the characteristic of line $A - B$. A fault on line $A - B$ is represented by the impedance measured from A to the point of fault. For instance, the impedance $A - F$ represents the impedance to a fault at F. The zone 1 relay is set to protect about 90% of the line and will trip circuit breaker A in high speed for faults within its zone. The reach of this unit is restricted to 90% of the line length because it is not possible to determine line impedances with

100% accuracy and because the relay unit cannot readily discriminate between faults at E near circuit breaker B and faults at G near circuit breaker C in the next line. The impedance between points E and G is essentially zero. If the zone 1 relay at A were permitted to reach to circuit breaker B, it would probably trip circuit breaker A incorrectly for faults beyond C. See Fig. 15-66, top.

The remaining 10% of the line is protected by the zone 2 relay. To assure that this zone will "see" all faults in this 10% area, the reach of this unit is set greater than the impedance of line $A - B$. With this setting, this unit will see faults in the next line section, and therefore a predetermined time delay is associated with this zone, giving the relays at C time to trip for faults in line $C - D$. The second zone time is normally 0.2 to 0.5 s.

A third relay unit may be added at A to provide backup protection for line $C - D$. This unit, shown as a broken-line circle, zone 3, would be set to reach beyond D and would have considerable time delay associated with it to permit the relays at C to trip for faults in line $C - D$. There would be a similar set of relays at B looking toward A, a set at C looking toward D, and a set at D looking at C, and so on.

An additional set of relays is used to provide protection for faults between phase conductors and ground. Where reliable, high-speed ground-fault protection is required, it is the practice to use a form of distance relaying for this function.

Aside from their ability to discriminate between faults and other conditions, etc., distance relays are less susceptible to incorrect operation due to false information supplied to the relays. Therefore, distance-type relaying is almost always used where dependable and secure performance is mandatory.

The above type of protection is normally allocated to less important lines on a power system or for backup protection on an interconnection, since it provides high-speed protection for only about 90% of the line. The requirements for primary protection for transmission tie lines are that the protective scheme trip both line circuit breakers (A and B) simultaneously and in high speed for faults anywhere on the line. This is accomplished through the use of pilot relaying systems.

64. Pilot Relaying Systems. Pilot relaying protection for transmission lines utilizes protective relays at the line terminals and a communication channel between relays to achieve simultaneous high-speed tripping of all terminals. The relays utilize the channel to compare conditions at the line terminals to determine whether a fault is within the protected line section or external to it. If it is determined that the fault is internal to a protected line section, all terminals are tripped in high speed. If it is determined that a fault is external to that line section, tripping is blocked.

The location of the fault is indicated by either the presence or the absence of a pilot signal. If the presence of a signal blocks tripping, it is called a blocking pilot system. If the presence of a signal causes tripping, it is called a tripping pilot system.

Two basic relaying schemes are used in the blocking system, directional comparison and phase comparison. In the tripping pilot system, there are three basic schemes: direct underreaching, permissive underreaching, and permissive overreaching. Of all these methods, directional-comparison blocking is the most widely used in the United States, and therefore this section will be limited to a discussion of this scheme.

65. Directional-comparison Blocking Scheme. The directional-comparison scheme uses distance relays to detect line faults. Distance relays at the line terminals are set to see faults on the entire length of the transmission line. When these relays detect a fault, they will trip the circuit breaker unless they receive a blocking signal from another terminal. Blocking signals are initiated only when other distance relays at each terminal determine that a fault is external to the protected line. A typical directional-comparison arrangement is shown in Fig. 15-67 for line $C - D$. Figure 15-67a shows the relay units at terminal C, and Fig. 15-67b shows the units at terminal D. Normally, these diagrams would be superimposed. However, for this discussion and for added clarity, they are shown separately. At each terminal, there is a trip relay MT which is set to see faults on the entire length of line $C - D$ and blocking relays MB which are set to see faults external to the line section. To assure that a tripping relay will see faults on the entire length of the line $C - D$, the reach of these relays must be set greater than the impedance of line $C - D$. With this setting, the tripping relays will see faults in the adjacent line section. Since the tripping units MT cannot

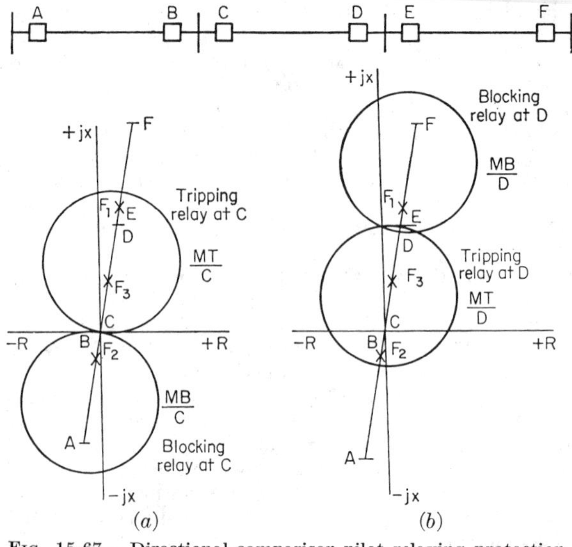

FIG. 15-67. Directional-comparison pilot relaying protection of a transmission line.

be permitted to trip for these external faults, relays *MB* are used at each line terminal to detect these external faults and to send a signal back to the *MT* terminal to block tripping. As an illustration of how this scheme operates, consider the faults F_1, F_2, and F_3 in Fig. 15-67. For fault F_1, the tripping relay at *C*, *MT/C*, will detect this fault, and so will the blocking relay at *D*, *MB/D*. The scheme is designed so that the blocking relay *MB/D* will operate quickly to send a blocking signal to *C* and prevent tripping by *MT/C*. Similarly, for the fault at F_2, the blocking relay at *C*, *MB/C*, operates to send a signal to *D* to block tripping by *MT/D*. For fault F_3, which is a fault on the protected line, both tripping relays *MT/C* and *MT/D* will see this fault. The blocking relays cannot see this fault, and therefore no blocking signals are transmitted. With the absence of any blocking signal, both *MT/C* and *MT/D* will trip their respective circuit breakers in high speed. While this arrangement may seem complicated and to require considerable coordination, it can provide high-speed tripping of both line terminals. A scheme using static relays and a high-speed communications chan-

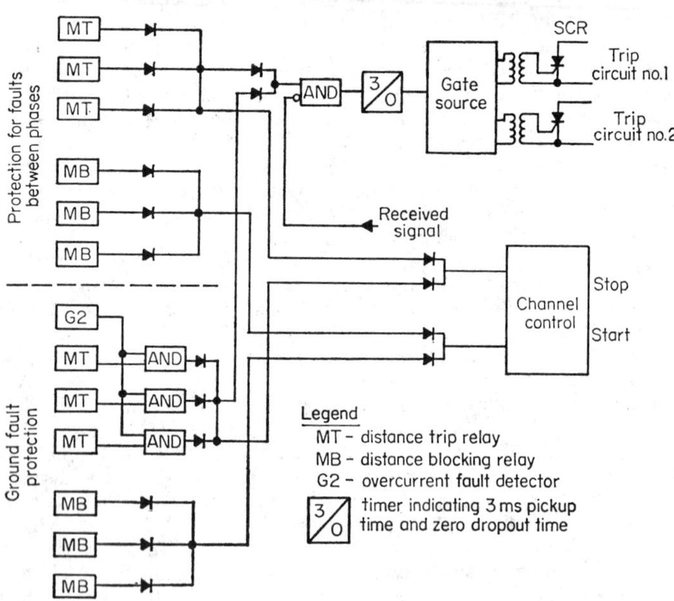

FIG. 15-68. Simplified logic diagram of a static directional-comparison pilot relaying scheme using phase and ground distance relays.

nel is capable of initiating a trip signal to a circuit breaker in 12.5 to 21 ms (¾ to 1¼ c on a 60-c basis), depending on the type of fault. A simplified logic diagram of a complete static directional-comparison relay terminal is shown in Fig. 15-68. Similar types of distance units are used for detecting faults between phases and between phase and ground. The communications channel can be power-line carrier, microwave link, or audio-tone channel operating over a wire line. The power-line carrier channel is the most widely used in the United States.

66. Protection Practices. The protective relaying systems and practices employed in high-voltage bulk power systems were evolved for one purpose: maintenance of maximum power-system reliability. These protective systems and practices are designed to provide the following:

1. High-speed clearing of all faults, with both primary and backup protection.
2. Removal of a minimum of transmission facilities during the isolation of a fault.
3. Maximum protective-system reliability.

High-speed clearing of faults is achieved through the use of static relay units and static pilot relaying schemes. The static relay units have an operating time between ¼ and ¾ c, while the pilot schemes will generally provide operating times between ¾ and 1¼ c. High-speed relaying and modern air-blast circuit breakers are capable of providing total fault-clearing times between 2 and 3 c on a 60-c basis.

The removal of a minimum of transmission facilities during the isolation of a system fault is accomplished through the use of local backup schemes. In these the detection and clearing of faults, with either primary or backup protection, are all performed at the stations closest to the fault. This arrangement will be described in more detail in Par. **67**.

Finally, maximum protective-system reliability is attained through the use of static relay units, through distance-type relaying for all types of faults, and through redundancy. Experience to date with static relays indicates that less maintenance is required with these types and that they are highly dependable. Distance-type relays are least susceptible to incorrect operation due to false information supplied to the relay and therefore will provide dependable and secure performance. It is common practice to duplicate the protective relaying function, the current and voltage transformers which provide information to the relays, the control circuitry, and in some instances the station battery which is the d-c power source for the control circuits. In addition, it is the practice to separate electrically the primary and backup protection control circuitry so that no single failure in these circuits can nullify all protection.

67. Typical Transmission-line Protection. Consider the system area shown in Fig. 15-69, and the protection of line section $C - D$.

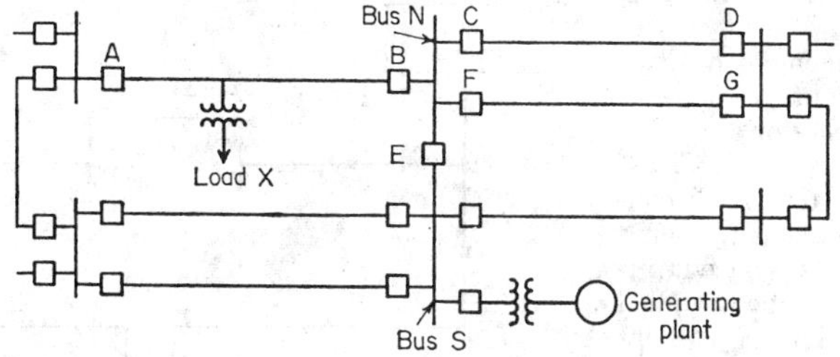

FIG. 15-69. One-line diagram of a portion of a power system.

The primary protection of line $C - D$ would be a directional-comparison pilot relaying scheme as described in Par. **62**. Some form of distance relays would be used to detect faults between phases and between phases and ground. Again, this scheme would provide simultaneous high-speed tripping of both terminals (C and D) for all faults.

To take care of a possible failure in the primary protection, a second complement of relays is provided at each line terminal. These backup relays are usually entirely separate from the primary protection, having separate sources of information and control circuits. If system stability is a problem and high-speed tripping is mandatory at all times, the backup protection may be identical to the primary protection, constituting a duplicate directional-comparison pilot scheme. Where stability is not a consideration and there are no other severe problems, relay backup may be provided by two zones of distance relaying. Figure 15-70 shows this complement of relays on $R - X$ diagrams for terminal C.

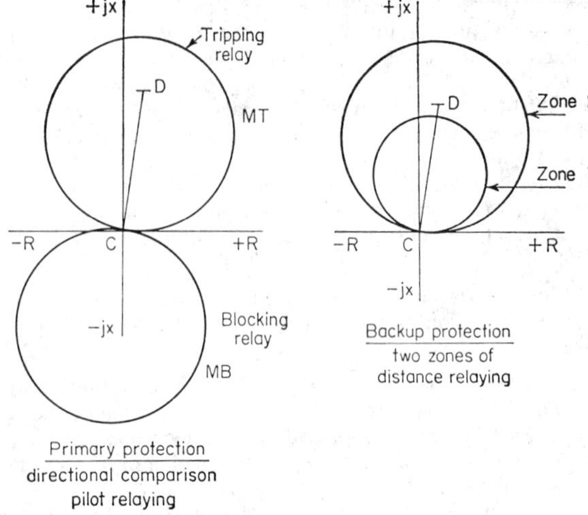

Fig. 15-70. Primary and backup protection of line C-D shown in the system of Fig. 15-69.

In addition to relay backup, steps are taken to protect against the possible failure of a circuit breaker to clear a fault. Breaker-failure backup is provided by a timer which is started by the line relays. If a fault is not cleared within a specified interval of time, this timer will operate to trip the circuit breakers necessary to isolate a fault.

This combination of relay and circuit-breaker-failure backup at a line terminal is called local backup and is shown in Fig. 15-71 for terminal C. When a fault occurs on the protected line, both the primary and the backup protection will operate to trip line circuit breaker C and will also energize the timer to start the circuit breaker backup function. If circuit breaker C fails to clear the fault, the line relays will remain picked up, permitting the timer to time out and trip the necessary other circuit breakers, B, E, and F, to isolate the fault. See Fig. 15-69.

This arrangement has two decided advantages. First, for either a relay or a circuit-breaker failure, the least amount of system is disconnected in the clearing of the fault. Bus N is isolated, but load X can still be supplied from A and bus S is kept intact, thus keeping a generating plant and other interconnecting lines in operation. Second, this approach provides faster backup clearing times. Backup times can be as much as 0.5 s faster than those provided by other backup methods.

Since most transmission-

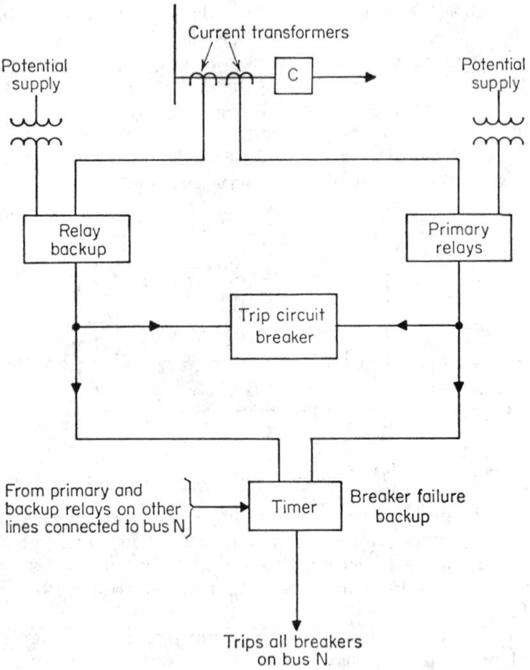

Fig. 15-71. Functional diagram of a local backup scheme.

line faults are of a transitory nature, it is the practice to provide control circuitry which will automatically reclose a line after it has been tripped out because of a fault. In most instances, a single high-speed reclosure will be attempted only after the simultaneous, high-speed tripping of all line circuit breakers by the primary pilot relaying system. If the fault persists, the line relays will trip the line out again and usually no further attempts will be made to reclose the line automatically. However, some utilities attempt to automatically reclose the line two or three times, with time delay between each attempt. If all attempts at automatic reclosure fail, a system operator may attempt to reclose the line manually after some time interval. Of course, if this fails, it indicates a permanent fault and the line is taken out of service for repairs.

This practice of high-speed automatic reclosure of a line:

1. Returns a line quickly to service, thus maintaining the integrity of the power system.
2. Minimizes the effect of a line outage on critical loads.
3. Permits loading of the transmission system more nearly to its stability limit.

The utility industry has used this practice for years with excellent results. Experience has shown that successful reclosure can be achieved 80 to 90% of the time.

High-speed reclosing is not used in all instances. In some cases, a reclosure into a persistent fault might cause system instability and therefore would not be desirable.

68. Other System Arrangements. The protection of lines in a system area such as shown in Fig. 15-69 usually presents no difficult problems. It is generally simple to provide high-speed reliable protection.

On the other hand, there are system designs which create relaying problems and which make it difficult to meet the protective requirements. For example, some systems utilize capacitors in series with line sections to compensate for some of the system inductive reactance, as shown in Fig. 15-72. The use of this series-capacitor compensation provides several benefits from a system design and operating point of view: improved stability limits, improved voltage regulation, a desired load division, or the maximum utilization of the load-carrying capability of the interconnection.

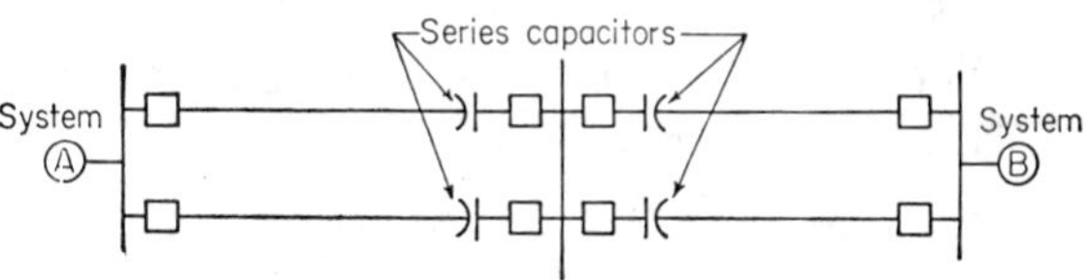

FIG. 15-72. One-line diagram of a series-capacitor-compensated system.

Because of economic considerations, most of the series-capacitor installations are at or near line terminals. From a line-relaying point of view, this location creates the most problems, since the capacitor can adversely affect relay operation. The adverse effects of series capacitors on protective relays have been described in the literature[1] and therefore will not be discussed in detail here. The negative reactance of series capacitors affects the phase position of system voltages and currents and thereby introduces incorrect information as to the fault location to distance-type relays. Therefore, distance relays and

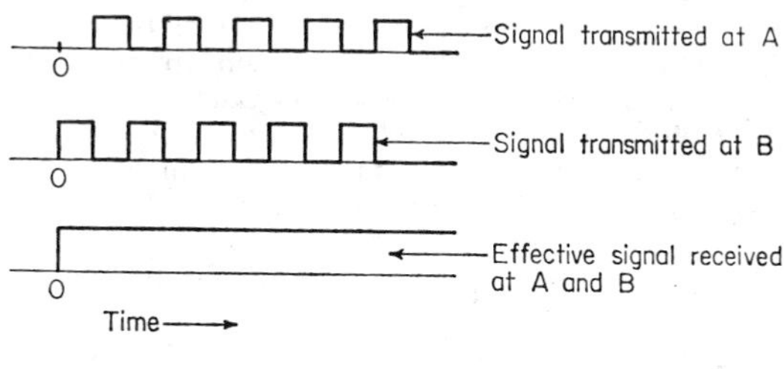

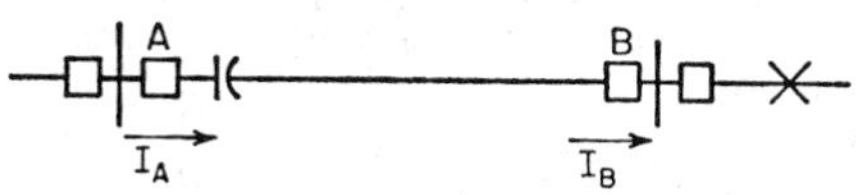

FIG. 15-73. Phase-comparison relaying; diagram shows relationship between currents and signals at line terminals for external faults.

the directional-comparison pilot relaying scheme will not generally be applicable on series-capacitor compensated systems.

The protective arrangement used for the protection of series-capacitor compensated lines is the phase-comparison pilot blocking scheme. A special phase-comparison relay and a communication channel detect line faults by comparing the phase position of currents entering and leaving the transmission line. This comparison is accomplished by the transmission of a signal from each terminal every half cycle (60-c basis). When a fault occurs external to a protected line section, the currents entering and leaving a line section are essentially in phase, as shown in Fig. 15-73. Under this condition, terminal *A* will transmit a signal during one half cycle, and terminal *B* will transmit a signal on the alternate half cycle. In effect, a continuous signal is received at all terminals, and tripping is blocked. When a fault occurs on the protected line, the currents at the line terminals will be 180° out of phase, as shown in Fig. 15-74. Under this condition, both terminals transmit a signal during the same half cycle. During the other half cycle, no signal is transmitted from either terminal, and therefore tripping is permitted.

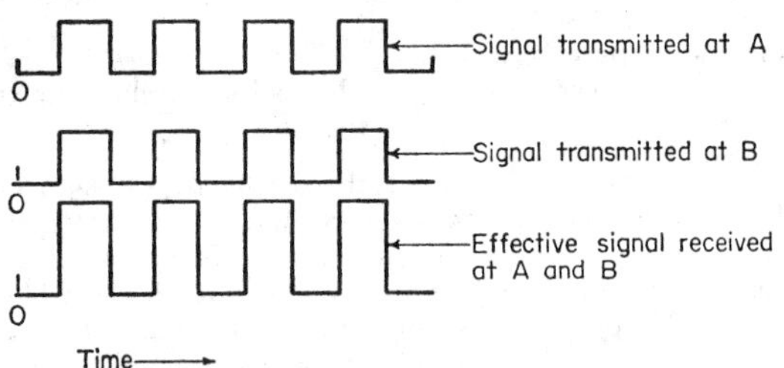

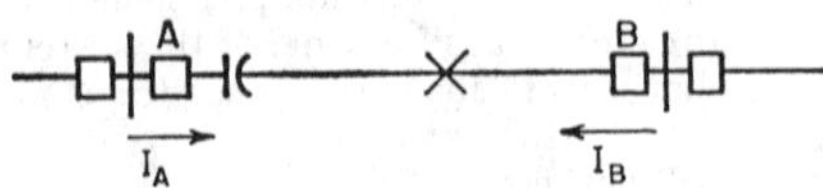

FIG. 15-74. Phase-comparison relaying; diagram shows relationship between currents and signals at line terminals for internal faults.

The complete scheme for the protection of series-capacitor compensated lines uses a combination of a phase-comparison relay and distance-type units. The distance units are used primarily as fault detectors and supervising relays, while the phase-comparison relay provides protection for all types of faults. The usual complement of protection for a series-capacitor compensated line includes two complete phase-comparison pilot relaying systems, one system used for primary protection, and the duplicate system used for backup.

In general, this type of protection is more complex than the directional-comparison arrangement, it requires considerably more application effort, and it results in slightly slower operating times. Because of the inherent characteristics of the phase-comparison approach, maximum relay operating times will be around 1¾ c.

69. Protection of Other System Elements. The protection of other system elements, such as power transformers, buses, etc., follows the same pattern as the protection of transmission lines. In all instances, two levels of protection (primary and backup) are provided, and consideration is given to circuit-breaker failure protection.

In most instances, the primary protection is high-speed, and the backup may be slightly slower. However, if stability or fault damage is a problem, it is common practice to use high-speed relaying for both primary and backup protection.

Automatic reclosing is not generally used after clearing a transformer fault. It is assumed that the fault is permanent, and therefore reenergization of the transformer would only cause further damage to the transformer.

Automatic reclosing may or may not be used in bus protection. Some utilities automatically reclose a single energized line to a bus which has been tripped out because of a fault. If this line holds in, indicating that the fault was transitory in nature, the remainder of the lines will be automatically reclosed to the bus.

70. Reference on Interconnection Protection

1. BERDY, J. Protection of Circuits with Series Capacitors; *Trans. IEEE,* Power Apparatus and Systems, February, 1963.

POWER SYSTEMS COMMUNICATIONS—POWER LINE CARRIER

BY A. N. ROBERTSON

CARRIER CURRENT

71. Carrier current provides a means of conveying speech, metering indications, control impulses, etc., from one station to another by existing transmission lines without interfering with their normal function of transmitting power.

Fundamental justification lies in utilizing existing power lines of sturdy construction to gain reliability, economy, or special characteristics. High-voltage lines are so constructed that they are less affected by the elements than any other circuits except underground cable.

72. Elements of a carrier channel comprise the sending terminal assembly including line matching and tuning, a coupling means, receiving-station coupling and terminal assembly, and the high-voltage transmission path for transmitting high-frequency energy between the terminals.

73. Coupling capacitors are the most widely used and effective coupling means, the paper capacitor being the standard carrier coupling device. Made up of a large number of sections connected in series to obtain the proper voltage, they are stacked mechanically to provide units in the higher-voltage insulation classes. The capacitor is mounted on a metal base to provide convenient installation and to provide space for connection to its lower terminal. The base contains a 60-Hz drain coil and may also contain protective gaps, grounding switches, and part or all of the coupling network.

Coupling capacitors may be provided with a tap or an auxiliary capacitance in their base, which is used as a voltage source for a potential device. This tap is so arranged that it will be maintained at a fixed nominal voltage above ground regardless of the voltage class of the unit.

74. Line tuning is required to tune out the reactance of a capacitor with a suitable inductance. The simplest application involves the coupling of a single frequency between a single line conductor and ground using a single coupling capacitor. The capacitor is series-resonated with a variable inductor at the operating frequency of the carrier terminal equipment, thus providing an efficient path for coupling the carrier signal to the line conductor.

75. The carrier terminal with its coaxial cable is matched to the impedance of the power line by an adjustable impedance-matching transformer. The coupling capacitor may be used for more than one frequency if more tuning elements are added.

76. Transmitters generally require a resistive load to be capable of modulation without excessive distortion. Therefore, where there is more than one transmitter coupled to the line at a common point, there should be a resonant path for each frequency. To permit independent tuning, antiresonant traps must be inserted in each path to reject all except the resonant frequency.

The practical upper limit for resonant tuning is two frequencies. Above this, broadband tuning provides a most satisfactory solution. Where future expansion is expected, it may be preferable initially to design the circuit with broadband tuning, since a reasonable number of additional carrier circuits may then be introduced at any time without disturbing the existing circuit or circuits. The line tuner may also include safety devices such as a protective gap, 60-Hz drain coil, and grounding switch.

77. Line traps are used to make the transmission line appear as a simple two-terminal line used in telephone circuits. They direct the carrier wave over a given circuit, increase efficiency, smooth out frequency characteristics, minimize interference, prevent interruption of the communication channels when the ground switches are closed, block off spur lines, and allow transmission during a nearby external fault.

Commercial traps are available with inductances of approximately 180 μH to 1.8 mH, but there is no theoretical limit to the inductance which can be provided. Size, weight, and cost limit production to an approximate 2-mH range. They can be tuned to one or two frequencies or may be broadbanded.

Since a wave trap is connected directly to the power line, it must be rated for the full operating current and should be able to withstand the same dynamic and thermal-limit current as the other station equipment. Operating currents are within the range of 100 to 3,000 A. In some instances, coils rated for above 2 mH have been installed to eliminate the need for tuning capacitors. This eliminates trouble caused by the possibility of the tuning capacitors being ruined by overvoltages or lightning. The present cost of these extremely large inductances precludes their use at higher operating currents.

TRANSMISSION-LINE CHARACTERISTICS

78. Attenuation is a measure of the loss of energy between the transmitting and receiving terminal and depends on many factors: frequency, conductor size and spacing, line configuration, presence of ground wires or parallel circuits, transpositions, ground resistivity, and weather conditions. The type of coupling used and the phase to which it is applied control the total attenuation from terminal to terminal and must be considered in planning the carrier channel.

Recently a method for the theoretical treatment of carrier channels has been developed and has proved its value in studying power-line-carrier transmission conditions. Natural modes of propagation are applied to the phase currents and voltages of power-line carrier signals between two line terminals.

For each mode, there is a set of phase voltages and currents which bear a constant relationship to each other along the length of the line. Each mode is represented by its own characteristic impedance, a constant specific attenuation, and its own phase velocity. Practically no energy is transferred between individual modes.

79. Modal analysis can become complicated on multicircuit lines, but its best application appears to be for the longer lines of extra-high-voltage systems. These systems usually consist of a single-circuit 3-phase horizontal line. This type of line may operate in three modes:

Mode 1. The current in the middle phase is roughly twice that in the outer phases and flows in the opposite direction. The attenuation is least with this mode.

Mode 2. The currents in the outer phases are equal and opposed, while the middle phase does not contribute anything to the transmission. Attenuation is greater than for mode 1.

Mode 3. The currents flow in the same direction in all 3 phases and are approximately equal in magnitude. Owing to earth return, attenuation is highest with this mode.

Various forms adopted in practice for transmission over this type of line generally represent a combination of two or three modes. To achieve the lowest possible losses, the available transmission power should be fed to the line by mode 1. This becomes more important the longer the line and the higher the carrier frequency. At a certain distance from the transmitter, mode 2 and 3 components are attenuated to a low level. Therefore, with long lines and high carrier frequencies, only proper coupling using mode 1 will cause the signal to arrive at the receiving point.

80. Line reflection occurs at the end of a line, at the junction of two or more lines, and at substations. Spur lines having attenuation of 10 dB or more do not generally produce either a serious loss or a serious distortion of the frequency characteristics. Short spur lines of low total attenuation produce severe losses over a wide frequency range.

81. Transmission-line loop circuits may cause a partial cancellation of carrier energy arriving over two sides of the loop or, at least, distortion of a carrier telephone channel. Corrective measures consist in the judicious use of line traps or line terminations, changing the coupling mode, or shifting the carrier frequences.

82. Carrier-frequency noise on the lower-voltage transmission lines is caused chiefly by defective insulators and loose hardware and on the higher-voltage lines by corona discharge. All circuits are subject to the noise effects of atmospherics, line faults, switching surges, and faulty apparatus. Noise increases substantially in bad weather as the combined result of lightning, corona from drops of water or particles of snow or ice on the conductor, and leakage over insulators. Background noise is generally greater at the lower frequencies, but disturbances and lightning storms may produce high noise levels throughout the entire spectrum. Noise during thunderstorms can produce a value 10 times or more the fair-weather noise. All these factors must be considered for best utilization of the transmission line.

The proper characteristic of the noise must be considered for each application, and receiver bandwidth as well, since the noise response is usually a function of bandwidth.

It is not ordinarily possible to reduce appreciably the noise level present at a given receiving point in a carrier system. Therefore, the only practical way to improve signal-to-noise ratio is to raise the signal level at the receiving point. It is not usually feasible to raise signal levels by increasing the transmitting power, because appreciable gains in terms of decibels require inordinately large increases in power. While it is practical to raise the signal level from a 1-W transmitter to 100 W or from a 1/10-W transmitter to 10 W, raising the level of a 10-W transmitter by 20 dB requires an increase to 1,000 W, or 100 times the original power. A much more practical solution is to reduce the channel attenuation by using every means available.

CARRIER COMMUNICATION

83. Two-frequency duplex telephony uses different frequencies for the two directions. This is a simple scheme in which transmitters and receivers at both ends operate continuously during a conversation. This system is readily adaptable to extension over wire line telephone lines through PBX's, but it is not fundamentally party-line, nor does it conserve frequencies. Hence, each length requires an additional two frequencies within the carrier spectrum unless widely separated by many line sections.

84. Single-frequency duplex telephony uses one frequency for both directions, conserves frequencies, provides party-line service, and is accomplished with equipment similar to two-frequency duplex, except that a relay is added to transfer from send to receive. A pushbutton in the handset controls this relay. The requirement for manual control prevents operation through a PBX.

85. In single-frequency automatic simplex, the carrier transmitter is started and stopped automatically by the presence or absence of sound at the microphone. Voice-operated relays (either electronic or mechanical) start and stop the transmitter as well as block and unblock the receiver in correct sequence and with the proper rate of change to accomplish a smooth transition and also to hold the transmitter inoperative on local noise while signals are being received.

86. Amplitude modulation is used for the majority of carrier telephone installations at the present time, utilizing both double-sideband and single-sideband transmission. With double-sideband transmission, both the carrier and two sidebands are transmitted. Therefore, it will generally occupy a frequency band of 10 kHz.

87. With single-sideband transmission, the frequency band assigned to a channel will be only half as wide. By using both upper and lower sidebands in the single-sideband system, two telephone-quality voice circuits can be obtained in the same bandwidth as one double-sideband channel. The carrier itself need not be transmitted but may be reinserted at the receiving end. The reduction in bandwidth and improvement in signal-to-noise ratio are important advantages.

88. Frequency modulation is also applicable to power-line carrier. Because of the limited frequency bands available, some of the advantages of this method must be sacrificed by restricting the modulation index. To avoid distortion, the frequency characteristic of the power system must be relatively flat throughout the width of both sidebands.

89. Carrier transmitters are generally arranged for mounting in relay racks or rack-type cabinets. Attention is given to harmonic suppression by using high-capacitance tank circuits; to frequency stability by using crystal-controlled or stabilized oscillators; and to high modulation percentages by using automatic gain control circuits. Output power requirements range from about 2 W for short and medium hauls to 25 W for most applications. In two-channel single-sideband systems, powers as high as 100 to 200 W are common. Most sets today are designed to operate from a standard 48-V d-c source but have an optional alternative power pack, enabling them to be operated from either 115-V 60-c a-c or 125-V d-c supplies. The complexity of the more sophisticated devices precludes their use with a station battery owing to the lower insulation characteristics of solid-state components.

90. Receivers may be simple for short hauls where maximum frequency utilization is not required. For general use, selectivity should be as high as obtainable, consistent with commercial fidelity. Automatic gain control circuits which hold constant output over at least a 40-dB range are usually incorporated. Many receivers for long-range work are equipped with special squelch circuits to block the receiver during periods of no received carrier.

91. Calling methods include the use alone, or in combination, of loudspeakers, code ringing, and selective ringing. Code ringing normally utilizes an audible tone with audio filter and time-delay relay at the receiver to prevent false operation from voice or from line noise. Selective ringing is obtained similarly by the addition of standard selectors, in which case a distinctive "ring back," or revertive ring, indicates that selective equipment at the call station functions properly to actuate the ringer.

CARRIER RELAYING

92. Directional-comparison Relaying. Carrier-current channels used in place of pilot wires in directional-comparison relay schemes have proved effective, fast, and economical. The resulting system is selective as to line sections protected, and the function delegated to the carrier is only to close a relay. A typical arrangement consists of directional relays at the two ends of a line section to determine whether power flows into or out of the section. If fault current flows into one end and out of the other, the line section is sound and the fault is external. If fault current flows into the line section at both ends, an internal fault must be present. Carrier transmitters at the two ends of the line section are each arranged to transmit only when overload power flows out of the section, in which case the line is sound and carrier receiving relays operate to prevent tripping of the circuit breakers. If fault current flows inward at each end, neither set transmits and the circuit breakers at both ends are allowed to trip from fault detector relays. Each line section so protected is independent of all other line sections. The carrier system is required to transmit and receive over its own line section during an overload or fault outside its section. Therefore, line traps at the ends of the section are necessary to prevent an external fault occurring a short distance outside the section from short-circuiting out the carrier path. Receivers must not respond to the high-frequency arc component of the fault current. Consequently, they are made relatively insensitive. Transmission over a line section is not required in both directions at the same time, which permits a single-frequency two-way channel. Total elapsed time from the occurrence of the fault to energizing the trip coil is usually 1 to 3 Hz, but even greater speeds are required as the transmission system becomes more sophisticated.

93. This requirement for ever greater speeds has fostered the development of static relaying, in which all relays are replaced by their equivalent electronic circuits. Overall operating time is $\frac{3}{8}$ to $\frac{3}{4}$ Hz.

94. Phase-comparison relaying, by balancing the current phase angles at the near and far ends of the transmission line, is the counterpart of directional-comparison relaying. It has the advantages of simplicity and high speed, although not so fast as directional-comparison relaying, and is immune to the effects of hunting and instability.

95. Transfer trip relaying provides a means for high-speed control of a remote circuit breaker. It is especially applicable for limiting transformer damage, in those cases where a high side circuit breaker is used, by fast operation of the far-end circuit breaker. Although the carrier equipment is basically simple, special precautions must be taken in the design of the entire system to assure positive action when needed with maximum protection against false tripping. A continuously transmitted guard channel is usually employed with line traps at both ends and equipment designed for maximum rejection of noise and spurious signals. For tripping the remote circuit breaker, the channel is shifted a small amount, which causes the remote trip relay to operate.

96. Frequency-shift blocking may be used where power-line sections have a high attenuation or other irregularities. Where the attenuation is between 40 and 60 dB, frequency-shift equipment often provides the only satisfactory channel. Because of the narrow band used, the response time of such channels is somewhat longer than for the assemblies previously described. This additional delay is usually secondary to reliability in importance, but the delay does limit the application of frequency-shift blocking in the relay system. A continuous carrier-frequency signal is transmitted over the high-voltage line. Upon occurrence of a fault, frequency is changed slightly under the control of the protective relays. This slight change in frequency actuates a relay at the receiving end, which allows blocking or tripping as required. This system has the advantage of being relatively insusceptible to interference. Also, continuous channel supervision is readily provided.

The receiver limiter and discriminator circuits which are employed ensure that the receiver output will be actuated only by a carrier signal of exactly the right frequency and also having more power than any other signal having only slightly different frequencies. Therefore, this equipment can be used on power circuits where other types of carrier would be entirely inadequate.

OTHER USES FOR POWER-LINE CARRIER

97. Carrier may be used for telemetering and supervisory control. These channels cover distances ranging from one line section of a few miles in length up to many sections totaling several hundred miles as required for power-interchange readings between large interconnected systems. Choice of the carrier method depends on channel length, available frequency band, speed of response, accuracy required, and whether or not full-time operation is essential. Carrier channels for remote control, supervisory control, automatic load control, or telemetering generally use a narrow bandwidth, compared with telephone channels. For short-haul work, simple on-off keying is adequate and may sometimes be accomplished as a secondary function of telephone or relaying equipment. For long-haul work, where both attenuation and noise are important considerations, better performance results from using some form of continuous carrier.

98. The straight continuous-wave system consists of an unmodulated transmitter keyed by supervisory transmitting contacts or by the sending meter.

99. In the frequency-shift system, transmission on one frequency or the other is continuous, which minimizes the effect of interference from noise or other carrier channels and permits continuous compensation by automatic-volume-control circuits for a change in signal level. A typical system using a 25-W transmitter is capable of producing reliable indications through attenuations up to about 60 dB, representing several hundred miles of average transmission line with intermediate stations. When several quantities are to be transmitted between two points, economy in both equipment and spectrum results from multiplexing with audio tones, one of which is used for each quantity. The tones are shifted or frequency-modulated as required for each individual telemetering indication, and the entire group is used to modulate the carrier.

100. Supervisory control of unattended stations requires transmission of signals of the same general type as for telemetering. Since it is generally desired to control many operations, some form of multiplexing is indicated. Tones, impulses, or combinations of both have been successfully employed. In general, a similar type of system is used for conveying the indications back to the control point.

BIBLIOGRAPHY

101. Selected References on Power-line Carrier

Podszeck, H. K. "Carrier Communication over Power Lines"; Berlin, Springer-Verlag OHG, 1963.

Application Guide for Power Line Carrier; *IEEE Tech. Paper* 54–12.

Natural Modes of PLC on Horizontal 3-phase Lines; *IEEE Paper* 63–936.

"Communication System Engineering Handbook"; McGraw-Hill Book Company, 1967.

For complete bibliography see *IEEE Paper* 31-TP-65-25.

POWER SYSTEMS COMMUNICATIONS—MICROWAVE RADIO, MOBILE RADIO, AND TONE MULTIPLEX SYSTEMS

By F. R. Nelson

MICROWAVE RADIO

101a. Microwave radio is used extensively in the power industry for point-to-point communications. Broad bandwidths are available in the microwave region of the radio spectrum, permitting the simultaneous transmission of many types of information over a single radio link. Propagation characteristics at microwave frequencies (i.e., those above 890 MHz) limit direct and reliable transmission to line-of-sight distances. In practical radio systems links are connected in tandem through intermediate repeater stations to permit long-range high-quality communications service over distances on the order of hundreds or thousands of miles.

102. Frequencies available to microwave users in the electric power industry and similar utility enterprises fall into four general groups: (1) 952 to 960 MHz; (2) 1,850 to 2,200 MHz; (3) 6,575 to 6,875 MHz; (4) 12,200 to 12,700 MHz. Higher frequencies are available to users on terms of developmental operation. All transmitters must be licensed by the Federal Communications Commission (FCC) under rules governing the Power Radio Service, a subpart of the Industrial Radio Services.

Operation on frequencies below 1,850 MHz is generally restricted to low densities of multiplexed channels because of limitations on bandwidth occupancy permitted under FCC rules. These frequencies find much application for local operations or as tributary links to multihop microwave systems operating in one of the higher groups of authorized frequencies. Occupancy of bandwidth up to 20 MHz is permitted in that band of frequencies around 12,500 MHz. The susceptance of signals in this latter range of frequencies to absorption and scattering by rainfall sometimes limits their application to short path distances. This factor merits close consideration in those regions noted for heavy rainfall and, in general, when a high level of reliability is necessary for transmission-line protective relaying, remote supervisory control, and similar critical services.

103. Radio equipment for operation in each of the several groups of regularly licensed frequencies is available in various configurations and with options as to transmitter power-output rating, multichannel bandwidth capability, and receiver sensitivity. Equipment employing frequency-modulation techniques has become standard for most applications of point-to-point microwave radio, particularly in the higher ranges of frequencies. Amplitude-modulated equipment is seldom practical owing to the extremely critical requirements for amplifier linearity to achieve acceptable levels of distortion with wideband AM signals. Forms of pulse modulation are not widely used in the power industry because of the relative complexity of channel combining and synchronizing devices.

Typical characteristics of radio apparatus used on frequencies around 2,000 MHz are: transmitter power output in the range of +30 to +37 dBm (1 to 5 W expressed in decibels referred to 1 mW), receiver FM improvement threshold on the order of −83 to −87 dBm, and usable modulation base band of 300 Hz up to around 250 to 750 kHz. Microwave radio (rf) equipment available for use in the frequency band around 6,700 MHz has the following general characteristics: transmitter power output from +20 to +35 dBm, receiver FM threshold ranging from −76 to −84 dBm, and information-handling base-band capability up to approximately 1.5 MHz.

Microwave radio equipment available prior to about 1960 typically used standard low-power-level vacuum tubes in basic circuitry. The lighthouse tube was commonly used in final transmission-frequency stages at the lower microwave frequencies, and the reflex klystron tube was used in the upper portions of the microwave spectrum.

Advancements in ultrahigh-frequency and superhigh-frequency components and techniques have resulted in the development and availability of radio transmitting and receiving equipment employing solid-state signal generating, amplifying, and multiplying states for those frequencies up to on the order of 7,000 MHz. Recently developed transmitters employ transistorized amplifiers and frequency multipliers up to intermediate frequencies on the order of 70 MHz. Varactor multiplying chains are used to produce final transmission frequencies at acceptable power levels. The same techniques and types of components are used for generation of local oscillator signals in superheterodyne microwave receivers.

Microwave transmitter and receiver apparatus used in most applications by electric power systems includes auxiliary features to ensure maximum reliability of service and to facilitate maintenance procedures. On a typical link of a system handling highly important channels, radio equipment at all terminal and repeater stations is installed in duplicate and is under control of local sensing and switching devices to minimize interruptions due to failure of primary radio apparatus. Duplicate radio equipment is generally operated in the "hot-standby" mode, i.e., energized and available for service with minimum delay. Since near-absolute frequency and phase locking of transmitters is not practical at higher microwave frequencies, the energized standby transmitter must be effectively disconnected from the common antenna so as not to cause interference to the working system. Ferrite devices or diode switches are commonly used for waveguide switching on Power Radio Service equipment operating above 6,575 MHz. Switching devices provide isolation on the order of 70 dB to the unwanted transmitter output while attenuating the desired signal by as little as 0.5 dB. Switching times range from 100 μs to on the order of 10 ms, depending upon the devices used and the type of fault triggering the switch.

104. Security against communications interruption due to power failure is normally provided by on-site emergency generator, station battery, or both. Especially stringent

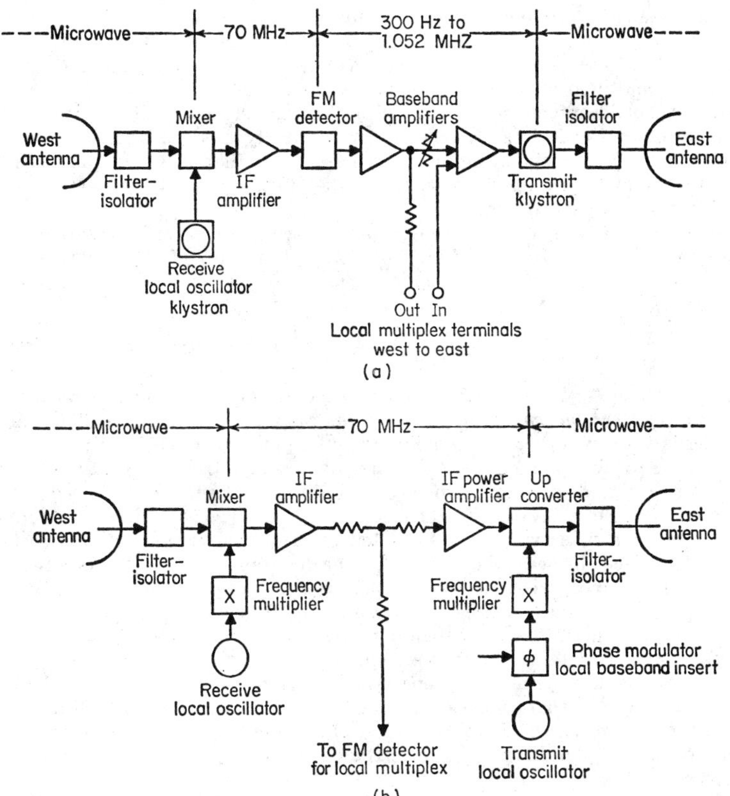

FIG. 15-75. Radio equipment interconnection at microwave repeater stations: (a) base-band repeat with detection and modulation of through channels; (b) if heterodyne repeat, amplification of main-line FM base band at 70 MHz and conversion to transmit microwave frequency.

requirements for power continuity are faced when various forms of direct-acting line and station protective relaying channels are handled through the microwave system. Rotary or static inverters are frequently used for service backup to a-c-powered radio apparatus. Direct battery input at 24 or 48 V is standard option with the majority of microwave radio r-f units now available, thus assuring excellent security against power failure at reasonable cost.

Additional auxiliary features normally supplied as part of the radio apparatus are fault-alarming and service-channel facilities. Alarm-encoding and -transmitting devices are used almost universally at "standby" equipped stations to alert personnel at maintenance centers to failures or marginal operations. Service-channel voice facilities are normally separate from through-traffic channelizing equipment. This voice channel is used for coordination of maintenance and testing procedures between all microwave terminal and repeater stations.

Two alternative techniques are commonly used in the arrangement of radio equipment at intermediate repeater stations. The more usual arrangement is with back-to-back connection of basic terminal station apparatus. Through-traffic (base-band) signals are demodulated at receiver output, amplified, and remodulated for transmission to the next station. This type of repeater station is shown in simplified form in Fig. 15-75*a*.

105. The i-f heterodyne-repeater technique, long employed in common carrier microwave systems, has recently been introduced for use on industrial microwave networks (Fig. 15-75*b*). In a repeater of this type, the through-traffic FM signal is amplified at the i-f level and combined in an up-converter with a signal from the transmit local oscillator to obtain proper microwave frequency. Noise products normally associated with FM detection and modulation are avoided and problems of channel-level variations in multihop microwave systems are minimized.

106. Paraboloid antennas consisting of a dish-shaped surface illuminated by a feed horn mounted at the focus of a reflector are normally used on electric-power-system microwave installations operating on frequencies higher than 960 MHz. It is the function of the paraboloid antenna to increase the effective radiated power by focusing radio energy into a narrow beam.

Gain of an antenna is expressed in decibels relative to that of an isotropic antenna, i.e., one radiating equally well in all directions. Efficiency of the typical paraboloid antenna is on the order of 55%. Gain in decibels of a 55% efficient dish antenna can be calculated from

$$G = 20 \log D + 20 \log f - 52.5 \qquad \text{(dB)}$$

where D = diameter in feet of paraboloid reflector and f = frequency in megahertz.

An arrangement known as the periscope antenna is frequently used in the Power Radio Service and by other industrial users on frequencies above 6,575 MHz. In such a system, the paraboloid antenna is mounted as near as possible to the associated radio (r-f) equipment and is oriented to direct the radio beam upward toward a reflector at or near the top of a tower. The reflector is positioned so as to direct the radio beam toward the next microwave station. Reflectors are available in both rectangular and elliptical face surfaces, the latter allowing high aperture efficiency with some reduction in mechanical wind-loading characteristics. Reflectors of both face types are available with either flat surfaces or with provision for some degree of concave curving so as to increase the total effective antenna-reflector gain. Overall gain of a periscope antenna system is determined by the types and sizes of the paraboloid antenna and reflector and by the distance of separation between them. Figure 15-76 illustrates the gain at various separations for a system employing a paraboloid antenna 4 ft in diameter with a variety of reflectors. It will be noted that the gain of the combined system exceeds the 36-dB gain of the paraboloid antenna alone over wide sections of the curves.

Advantages of the periscope antenna system are: (1) an overall antenna system gain offsetting loss of a waveguide run to a tower-mounted paraboloid antenna; (2) a reduction in echo distortion caused by impedance mismatches and waveguide imperfections; (3) cost reduction by eliminating long waveguide runs and the need for waveguide pressurizing and dehydrating devices. A disadvantage of the periscope antenna system is the more stringent requirement for tower rigidity. Permissible tower sway under all anticipated conditions of wind and ice loading must be restricted to approximately one-half that permitted with an equivalent tower-mounted paraboloid antenna, to keep variations in beam direction in the vertical plane within proper limits.

107. Propagation characteristics at frequencies above 890 MHz are such that careful attention must be given to such effects as obstruction, reflection, refraction, and diffraction. A primary requirement for reliable microwave communications with conventional transmitters and receivers and practical sizes of antennas is that proper beam clearance must be maintained over earth, trees, and man-made obstructions. In practice, clearance in excess of optical line of sight is necessary.

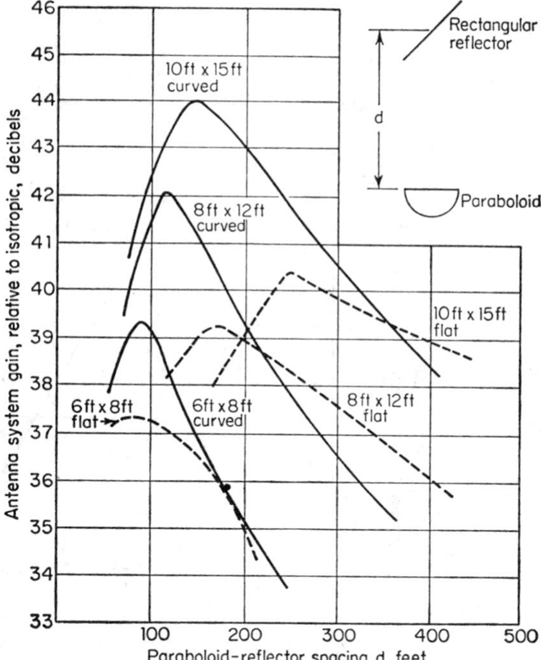

FIG. 15-76. Periscope antenna system gains at 6,700 MHz. Four-foot paraboloid with 36-dB gain used with typical sizes of rectangular reflectors.

Free-space loss (equivalent to unobstructed propagation via space wave alone) is assumed to apply with proper beam-to-obstruction clearance and when atmospheric conditions are normal. Path free-space loss in decibels is readily calculated as

$$L = 20 \log d + 20 \log f + 36.6$$

where d = path distance in miles and f = frequency in megahertz.

108. Fades, or variations with time, in path loss are encountered during abnormal propagation conditions. The most common type of fading is that due to multipath transmission. Combinations of irregularities and fluctuations in atmospheric temperature, humidity, and pressure cause more than one and often many propagation paths to exist between the transmitting antenna and the receiving antenna. As the atmospheric conditions vary, the routes and distances of paths also vary, causing signals of differing phases and amplitudes to arrive at the receiving antenna at the same instant. Multipath, or interference, fading is characterized by rapid fluctuations of received carrier power to levels of degradation on the order of 40 dB or more below normal signal. The characteristics of multipath fading with time approach the Rayleigh distribution shown in the graph in Fig. 15-77.

Security against multipath fading is obtained by providing adequate margin of received signal power in excess of that required for normal propagation conditions. Fade margins ranging from 30 to 40 dB are typical in industrial microwave systems. The technique of space diversity takes advantage of the low order of correlation of fading of signals on the same frequency at two antennas properly spaced from one another in the vertical plane. This arrangement is sometimes used in paths and systems requiring extreme reliability. Frequency-diversity operation requires simultaneous use of two different microwave frequencies and is generally not permitted in the industrial bands.

Attenuation fading is characterized by slow fluctuations in received signal level.

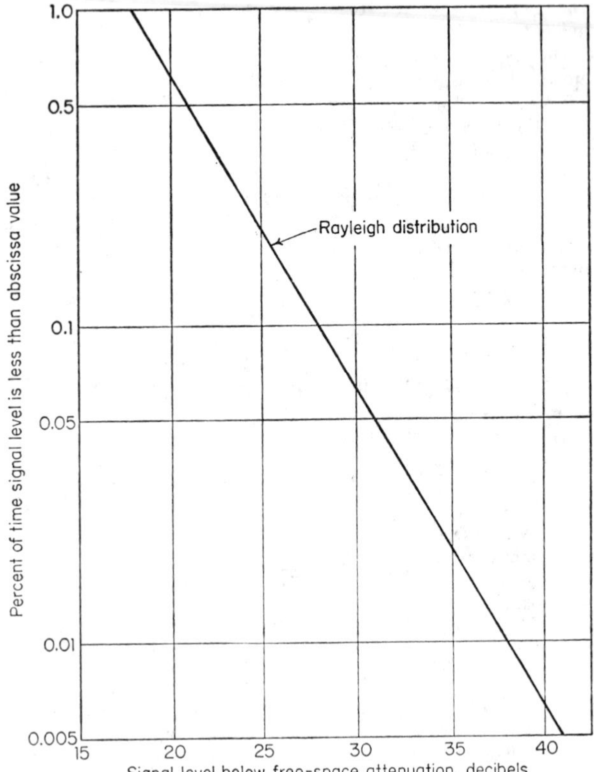

Fig. 15-77. Statistical distribution of interference fading for single microwave path, non-diversity.

Three types of attenuation fading are generally recognized: (1) inverse bending; (2) ducting; (3) rain attenuation. The first two of these are usually caused by changes in the refractive index or stratification in levels of atmosphere through which the beam passes. Severity and frequency of occurrence of inverse bending (sometimes called earth bulge) and ducting types of attenuation fading vary between regions of the country. Effects may be minimized or eliminated by providing additional signal margin or designing the system with shorter paths. Effects of rain attenuation are considered negligible on frequencies below about 7,000 MHz.

109. Multiplex equipment is used to derive separate channels within the microwave radio base band. Multiplex system characteristics and arrangements are based upon speech-transmission requirements as standard. Overall signal-transmission bandwidth on each of the derived channels is thus approximately 300 to 3,400 Hz.

Maximum numbers of multiplex channels which can be accommodated in a radio link are dependent upon a number of factors: (1) modulation and noise characteristics of r-f equipment; (2) maximum bandwidth occupancy permitted by the FCC in the various frequency bands; (3) channel derivation and modulation methods in the multiplex equipment.

Multiplex techniques and apparatus are available to permit sharing of microwave base-band capability on time-division and frequency-division bases. Electric-power-system requirements generally dictate use of multiplex equipment for a wide variety of communications functions and a complex combination of long-distance and short-range channels. These needs, coupled with the need of reasonable cost, have led to almost universal adoption of frequency-division types of multiplex.

Various types of frequency-division multiplex are in use. Modulation techniques of amplitude modulation, frequency modulation, and phase modulation are employed. Frequency-modulation and phase-modulation multiplexing systems are generally superior to amplitude-modulation equipment as regards performance in noise background; this advantage, however, is at the cost of additional per channel bandwidth requirements. Applications of phase-modulated multiplex are generally limited to digital information transmission and special applications. Among AM multiplexing methods, single-sideband suppressed carrier exhibits especially attractive features in terms of both minimum-base-band spectrum occupancy and, for voice signal transmission, minimization of loading on the radio system. With SSB-SC (single-sideband suppressed-carrier) multiplexing techniques, the carrier portion of the signal and one of the two symmetrical information-bearing sidebands are eliminated from the transmitted signal, thus achieving high spectrum efficiency. Restoration of signal components at receiving terminals employing SSB-SC techniques requires local reinsertion of carrier tone, duplicating, or nearly so, the frequency and level of carrier at the point of channel origin.

Needs for standardization of multiplex operating specifications have, in the case of SSB-SC applications, led to general adoption by manufacturers of recommendations by the CCITT (International Telegraph and Telephone Consultative Committee). Rec-

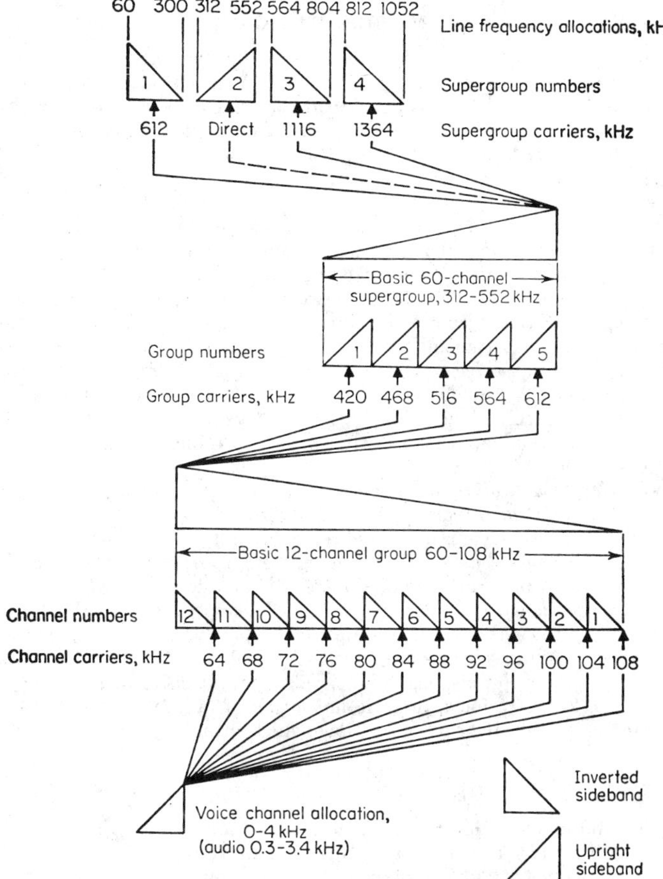

FIG. 15-78. Multiplex modulation and allocation plan for up to 240 channels, four supergroups, single-sideband, suppressed-carrier.

ommendations of the CCITT plan for use of three successive modulation steps, as necessary, to insert channels and blocks of channels into proper position in the base-band frequency spectrum. In the first step, individual voice-capability portions of the audio spectrum are converted into sideband products each occupying a maximum allotment of 4 kHz within the limits of 60 to 108 kHz. This array of 12 channels, comprising a basic group, is next beat against an appropriate group carrier signal, and the desired composite sideband is accepted for insertion within one of the five available group positions within the basic supergroup. A final modulation step is required for transmission of more than 60 channels. This converts the basic supergroup block of frequencies to other adjacent portions of the spectrum. A simplified diagram of the CCITT modulation and allocation plan is shown in Fig. 15-78. This diagram is carried to four supergroups (240 channels), a typical upper limit in industrial microwave-channel usage. Main-route common carrier microwave systems commonly employ a further step of modulation which multiplexes blocks of 10 supergroups into master groups for transmission of up to 2,400 channels.

Special-purpose channeling devices are available for operation in other sectors of the base band or for direct compatibility and insertion with voice-grade channels in the CCITT plan. Examples of the latter are group or supergroup bandwidth terminals used for high-speed binary-data transmission and high-resolution facsimile.

MOBILE RADIO

110. Mobile radio is employed in the power industry for maintenance-crew dispatching, construction coordination, and related functions vital to the day-to-day operation of electric systems. Frequencies are available on a shared basis and are selected through coordination with other users in the Power Radio Service. All transmitters must be licensed by the Federal Communications Commission under rules governing this service.

111. Frequencies available for mobile radio service fall into four general groups, those around: (1) 37 MHz; (2) 48 MHz; (3) 153 and 158 MHz; (4) 456 MHz. Generally only one frequency is granted to each applicant for operation on a single-frequency simplex basis. Communication is permitted between mobile stations and land stations or between mobile stations. Operation between base stations is permitted only in emergency or in connection with a call of immediate importance to mobile units.

FCC rules require that transmitter output powers be limited to the minimum required for satisfactory service. Land-base stations generally operate with output ratings on the order of 25 to 300 W, and mobile units are rated at 10 to about 75 W. Portable transmitter-receiver sets with output ratings of 1 W or less are commonly used for short-range communications with other portable sets, mobile stations, or associated base stations.

Quarter-wavelength vertically oriented antennas are commonly used for mobile stations. Folded coaxial and unipole antennas are normally used for base stations for lower angle of radiation and effective gain in omnidirectional operation. Directional antennas are sometimes used with base-station transmitters to reduce interference between adjacent stations or to provide coverage in specific directions.

Tone or impulse signaling is permitted in mobile operations in the Power Radio Service as necessary to establish or maintain the primary voice service. Tone-operated squelch or selective calling schemes are frequently used to minimize reception of unnecessary traffic. Tone or impulse signaling is also permitted on a secondary basis or frequencies above 25 MHz for automatic indication of failure of power-system equipment or for indication of abnormal conditions which, if not promptly corrected, would result in failure of equipment.

TONE MULTIPLEX EQUIPMENT

112. Tone multiplex equipment finds use in a number of basic communications media for those types of information transmission requiring narrower bandwidth than a standard voice channel. Typical of such transmission requirements are those for telemetry, load control, supervisory control and acknowledgment, substation monitoring and alarm, and teletypewriter and slower forms of digital-data transmission.

Techniques of data encoding, tone channel modulation, demodulation, and decoding permit the simultaneous transmission over a voice-quality communications circuit of up to approximately 24 tone multiplex channels. Methods of tone channel separation are generally the same as those employed in the division of radio base bands into individual voice multiplex channels.

113. Frequency-division separation techniques are most usually applied because of the relative simplicity of apparatus and lower equipment costs. The three main classes of tone-channel-deriving equipment as defined by their modulation methods are: amplitude modulation (AM); frequency-shift keying (FSK); phase-shift keying (PSK).

With AM *tone-transmission methods,* the presence or absence of carrier indicates to receiver detection and logic circuits the instantaneous existence of mark or space keying conditions, respectively. Since AM detectors are sensitive to gain variations in the transmission path, thresholds must be set above which a mark condition is indicated and below which a space condition is indicated.

In FSK *systems* a constant-amplitude tone is shifted between discrete and separate frequencies dependent upon the instantaneous condition of transmitter keying circuits. In most FSK systems two-frequency keying is employed; i.e., transmission is instantaneously on one or the other of two frequencies (mark or space). Keying signals are presented to the transmitter in such forms as (1) switch on, switch off; (2) voltage present, voltage absent; (3) voltage positive, voltage negative. Frequency-shift-keying tone systems are also available for three-frequency operation, permitting transmission of mark frequency, center frequency, and space frequency. Use of this form of channel permits transmission of control or selector functions based on raise-hold-lower types of operations.

The PSK *method of modulation* is one in which the mark-space information is indicated by a shift in phase of a constant-amplitude constant-frequency carrier. Unambiguous detection of mark and space conditions in PSK systems requires some means of deriving a reference phase-locked waveform at the receiver. Alternative techniques for phase determination include reconstruction of reference from preceding signal elements or the transmission of a separate, unkeyed reference tone.

114. Applications. Frequency-shift-keying tone equipment generally finds widest application in the electric power industry owing to its high signaling reliability and relative simplicity. Frequency-shift-keying systems of various bandwidths are employed, depending upon keying speeds associated with the different functions. Equipment is widely available with standard spacings of 120 and 170 Hz between individual channels. These assignments allow simultaneous accommodation of 24 or 18 channels, respectively, within circuits providing 300 to 3,400 Hz bandwidth and permit maximum keying rates of approximately 30 and 42.5 Hz, respectively.

Frequency-shift-keying tone devices employing wider spacings between channel allocations and wider frequency shifts between mark and space frequencies are used for functions requiring extreme reliability and security, high-speed response, and minimum sensitivity to noise. Equipment of this type finds its application in circuits for transmission-line protective relaying and transferred tripping. Bandwidths of channels in these categories permit simultaneous operation of only four or sometimes two such channels over a standard voice-grade circuit.

Proper planning of tone-channel levels and frequencies and recognition of the harmful effects of amplitude distortion, phase distortion, and noise influences permit highly satisfactory implementation of tone multiplex equipment over wire lines, power-line carrier and microwave, or various tandem combinations of these media.

BIBLIOGRAPHY

115. Selected References on Microwave Radio

Bray, W. J. The Standardization of International Microwave Radio-relay Systems; *IEE (London), Paper* 3412E, March, 1961.

Hathaway, S. D., and Evans, H. W. Radio Attenuation at 11 KMc/S and Some Implications Affecting Relay System Engineering, *Bell System Tech. Jour.*, January, 1959, Vol. 38, No. 1, pp. 73–97.

JASIK, HENRY (ED.) "Antenna Engineering Handbook"; New York, McGraw-Hill Book Company, 1961, Chaps. 12, 13, and 33.

PARRY, C. A. CCITT Recommendations for Multichannel Radio Relays and White Noise; *Trans. AIEE,* May, 1959, Vol. 78, Pt. 1 (Communication and Electronics), No. 42, pp. 107–117.

YEH, L. P. Noise-loading Test of Complete Frequency-division Multiplex Voice Point-to-point Communication System, *Trans. AIEE,* March, 1962, Vol. 81, Pt. 1 (Communication and Electronics), No. 59, pp. 40–43.

For complete bibliography on power-system use of microwave, see *IEEE Paper* 31-TP-65-25.

116. Selected References on Mobile Radio

JASIK, HENRY (ED.) "Antenna Engineering Handbook"; New York, McGraw-Hill Book Company, 1961, Chap. 22.

LYTEL, ALLAN H. "Two-way Radio"; New York, McGraw-Hill Book Company, 1959.

SHEFFER, A. System Alarm Signals Transmitted on Conventional Radio Frequency; *Elec. World,* July 2, 1962, Vol. 158, No. 1, pp. 32–33.

SECTION 16

POWER DISTRIBUTION

BY

LEONARD M. OLMSTED Senior Editor, *Electrical World*; Senior Member, Institute of Electrical and Electronics Engineers; Former Chairman, Pennsylvania Electric Association Transmission & Distribution Committee; Formerly Division Engineer, Duquesne Light Company; Formerly Special Consultant, United States Air Materiel Command; Registered Professional Engineer, Pennsylvania

JULIUS BLEIWEIS Administrative Manager, Northeast Power Coordinating Council; Formerly Distribution Editor, *Electrical World*; Member, Institute of Electrical and Electronics Engineers; Past Chairman of several IEEE Power Group committees; Formerly Engineer, Consolidated Edison Company of New York, Inc.

CONTENTS

Numbers refer to paragraphs

SECTION 16

POWER DISTRIBUTION

By Leonard M. Olmsted and Julius Bleiweis

ELEMENTS OF DISTRIBUTION PRACTICE

1. Distribution Defined. "Distribution" includes all parts of an electric utility system between the bulk power source or sources and consumers' service switches. Bulk power sources may be generating stations or major substations supplied over transmission lines. A typical distribution system consists of (1) subtransmission circuits operating between 13 and 66 kV which deliver energy to the distribution substations, (2) distribution substations which convert the energy to a lower voltage for local distribution and regulate the voltage delivered to load centers, (3) primary circuits, or "feeders," which operate between 2.4 and 13.5 kV and supply load to a well-defined geographic area, (4) distribution transformers on poles, on pads, or in vaults near the consumers which convert the energy to utilization voltages, (5) secondary mains, or "secondaries," which deliver the energy along the street or alley to within a short distance of the users, and (6) service drops which deliver the energy from the secondary mains to consumers' service switches. Figure 16-1 shows the component parts of a typical distribution system.

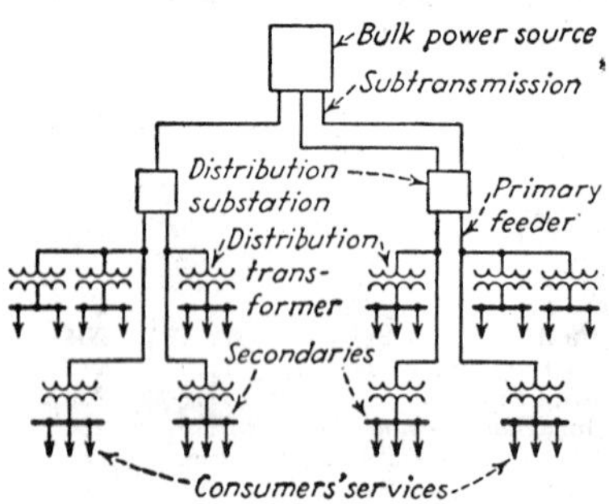

Fig. 16-1. Typical distribution system showing component parts.

2. Distribution investment constitutes approximately 50% of the capital electric utility systems in the United States. Where underground distribution is extensive, as in large cities, distribution investment is nearly 60%. In a typical 100-million-dollar system with distribution chiefly overhead, distribution is 35 to 40% of the total investment.

3. The function of distribution is to receive electric power from bulk power sources and to distribute it to consumers at voltage levels and with degrees of continuity that are acceptable to the various types of consumer.

Utilization voltage between 110 and 125 has been cited as "favorable" by Edison Electric Institute and National Electric Manufacturers Association. Drop in customers' wiring boosts the minimum for service-switch voltage to about 113. Numerous regulatory commissions allow variations of about ±5% from nominal voltage for lighting service; if nominal is 120 V, these limits become 114 to 126 V, or very nearly the "favorable" range. High-grade service in commercial areas usually meets closer limits, on the order of ±3%, while rural service sometimes must be allowed a few volts more variation. Instantaneous dips in voltage which cause lamp "flicker" must be limited to 6 or 8% when they occur infrequently and to 3 or 4% when several times per hour. Frequent dips, such as those caused by elevators and industrial equipment, must be limited to 1½% or less, depending on their frequency.

Uninterrupted service is increasingly essential. Short and infrequent interruptions are tolerated by residential and small commercial customers, but large commercial and industrial customers, hospitals, theaters, and public buildings often require continuous service.

4. Certain basic economic principles should be followed in designing a distribution system to supply adequate and continuous service to load under consideration now and in the future for lowest possible cost. Among them are: (1) The entire distribution system—subtransmission, distribution substations, primary feeders, transformers, secondaries, and services—should be considered as a whole in order that money saved in one part will not be lost in increases elsewhere. (2) Additions to distribution capacity should follow load growth closely as to time and location; inherent short-time overload capacity should be used to keep excess for emergencies to a minimum. (3) Distribution plant should be designed to permit additions with minimum change in existing construction; uncluttered pole lines help to alleviate agitation for undergrounding. (4) Maximum use should be made of existing plant; serviceable plant should not be replaced by new, more efficient plant unless savings in operation and maintenance are greater than amortization charges on plant replaced.

A distribution system designed in accordance with these principles can be expanded in small increments to meet load growth with minimum modification and expense. This flexibility allows system capacity to be kept close to actual load requirements and realizes minimum yearly cost per peak kilovoltampere of actual load. It minimizes need for predicting location and magnitude of future loads which are only scientific guesses.

5. Overhead and underground distribution are both used on large metropolitan systems. In smaller cities and less congested districts of larger cities the entire system may be overhead. Cost of an overhead system per kilovoltampere of load decreases as load density increases. But overhead construction becomes unwieldy at load densities in the neighborhood of 100 to 150 kVA per 1,000 ft because of the heavy transformers and conductors and the tall poles it requires. Hence densely built-up sections are customarily served by underground distribution despite its cost penalty of three to seven times the cost of conventional overhead construction.

Recent emphasis on improving the appearance of residential areas and suburban shopping centers has spurred development of less costly light-duty underground and semiunderground distribution systems. Costs have been cut drastically by resorting to direct-buried cable or plastic conduits, by eliminating manholes and transformer vaults, and by adopting switching practices comparable with those of conventional overhead distribution. In many locations, utilities have been able to offer such distribution as an inducement for total-electric homes or as an alternative to conventional overhead where the property owner contributes the cost differential.

6. Rural service has been extended to most farmers and rural dwellers by the vigorous efforts of utilities, cooperatives, and government agencies. Rural construction must be of the least expensive type consistent with durability and reliability because there are often only three to five customers per mile of line. Despite the added cost, however, a few sections are being placed underground to avoid unsightly exposures.

7. Higher primary distribution voltages are used by many utilities in their rapid growth areas. A 1954 survey of 90 companies by *Electrical World* showed 36 using 12.5 and 13.2 kV and 15 more proposing these voltages for future use. The trend has continued, some utilities using primary voltages as high as 20/34.5 kV in sparse areas. Four-wire Y-systems have become almost universal, many older Δ-systems having been uprated 73% by establishing a neutral conductor and reconnecting existing equipment phase to neutral. Thus 2.4-kV systems were converted to 4-kV, and 7.2-kV to 12.5-kV.

When the voltage boost is more than converting from Δ to the corresponding Y, extensive changes may be required. Existing distribution transformers must be replaced, and frequently the line insulation level must be raised. Progressive utilities minimize this conversion cost by placing the higher voltage in new growth areas and changing existing distribution only as load growth in their service areas demands.

CLASSIFICATION AND APPLICATION OF DISTRIBUTION SYSTEMS

8. Distribution systems may be classified in various ways:

a. As to current—alternating or direct.

b. As to voltage—120-V, 2,400-V, 13,800-V, etc.

c. As to scheme of connection—radial, network, multiple, and series.

d. As to loads—general light and power, industrial, railway, street lighting, etc.
e. As to number of conductors—2-wire, 3-wire, 4-wire, etc.
f. As to type of construction—overhead or underground.

Alternating-current circuits may be further classified as to phases: single-phase, 2-phase, or 3-phase; and as to frequency: 25-c, 60-c, etc.

9. Application of Systems. The various kinds of distribution system are applied in American practice in the following ways.

Alternating-current systems are almost universally used. They are the most economical method yet devised, largely because of the ease of converting to higher voltage for the bulk power system, then dropping to distribution voltage, all with economical static transformers instead of rotating machinery. By proper design and use of appropriate protective equipment, voltage stability and service reliability can be matched to almost any consumer requirement. The ultimate in service is provided by the low-voltage secondary networks now serving dense business-district loads of over 150 cities. Where this type of service is deemed insufficiently reliable because it depends on the utility system, self-contained emergency generators are sometimes provided, often with switching to pick up essential services, such as exit lights, hospital operating rooms, etc.

10. Single-phase lines are used for lighting, appliances (ranges, water, heaters, etc.), and small power. They are the most extensive and least expensive per kilovoltampere of load supplied of all systems.

11. The 3-phase, 4-wire systems, operated at 2,400/4,160, 4,800/8,300, and 7,200/12,470 V, are in very general use because of better efficiency and line capacity resulting from the higher voltage. The fourth wire of these Y-connected systems is a neutral operating at ground potential. Single-phase branches supply lighting, and the "phase" wires supply polyphase power. Separate phases may be regulated independently without regard to load balance. The 7,200/12,470-V system is probably the system most widely used today, but higher voltages are also employed extensively.

12. The 3-phase, 3-wire systems are most applicable to balanced loads, such as power use. Satisfactory regulation on each phase can be effected with two single-phase regulators, provided that single-phase load is reasonably balanced among the phases.

13. The 2-phase systems are rarely used today except in older sections of a few cities where 2-phase was selected in preference to 3-phase because of greater ease of balancing loads.

14. Direct-current systems today are usually vestiges of old Edison systems established in the formative days of electric utilities. Some continue to operate as 120/240-V 3-wire networks and provide highly reliable service, particularly where reserve storage batteries remain in use. But these systems are more costly and less flexible than a-c networks which have almost entirely replaced them.

15. Series systems have been used for street lighting since the earliest electric lighting. They are inherently high-voltage systems, well suited to most street lighting because simultaneous control is usually desired for all lamps on a circuit. Originally they were d-c systems, but a-c systems were developed later for incandescent street lighting. Many series systems remain in use today, but there is a marked trend toward multiple lighting from general-service secondary mains with photoelectric cells or other devices to control lighting.

CALCULATION OF VOLTAGE REGULATION AND I^2R LOSS

16. General. When a circuit supplies current to a load, it experiences a drop in voltage and a dissipation of energy in the form of heat. In d-c circuits voltage drop is equal to current in amperes multiplied by ohmic resistance of the conductors, $V = IR$. In a-c circuits voltage drop is a function of load current and power factor and the resistance and reactance of the conductors. Heating is caused by conductor losses, traditionally called "copper losses"; for both d-c and a-c circuits they are computed as the square of current multiplied by conductor resistance in ohms. Watts $= I^2R$, or kW $= I^2R/1{,}000$. Capacitance can be neglected in distribution circuits because its effect on voltage drop is negligible for the circuit lengths and operating voltages used. The chief problem in circuit design is to select a conductor that will carry the load current without exceeding a specified voltage drop and a safe operating temperature.

Limiting factor for overhead conductors is usually voltage drop; for underground cables it may be either voltage drop or current-carrying capacity.

17. Percent voltage drop or percent regulation is the ratio of voltage drop in a circuit to voltage delivered by the circuit, multiplied by 100 to convert to percent. For example, if the drop between a transformer and the last customer is 10 V and the voltage delivered to the customer is 230, the percent voltage drop is $10/230 \times 100 = 4.35\%$. Often the nominal or rated voltage is used as the denominator because the exact value of delivered voltage is seldom known.

18. Percent I^2R or percent conductor loss of a circuit is the ratio of the circuit I^2R or conductor loss, in kilowatts, to the kilowatts delivered by the circuit (multiplied by 100 to convert to percent). For example, assume a 240-V single-phase circuit consisting of 1,000 ft of two No. 4/0 copper cables supplies a load of 100 A at unity power factor.

$$I^2R = 100^2 \times 2 \times 0.0512 = 1{,}024 \text{ W} = 1.024 \text{ kW}$$

$$\text{Load delivered} = 240 \times 100 = 24{,}000 \text{ W} = 24 \text{ kW}$$

$$\text{Percent } I^2R \text{ loss} = 1.024/24 \times 100 = 4.26\%$$

19. Direct-current voltage drop is easily calculated by multiplying load amperes I by ohmic resistance R of the conductors through which the current flows (see Table 16-2 for ohmic resistance per 1,000 ft of various conductors). *Example.* A 500-ft d-c circuit of two 4/0 copper cables carries 200 A. What is the voltage drop? Resistance of 1,000 ft of 4/0 copper cable is 0.0512 Ω.

$$\text{Drop} = IR = 200 \times 0.0512 = 10.24 \text{ V}$$

If 240 is the delivered voltage,

$$\text{Percent regulation} = 10.24/240 \times 100 = 4.26\%$$

20. Voltage-drop calculations for 3-wire d-c circuits require separate computations for each conductor if the load is appreciably unbalanced. Drop in the neutral that carries the unbalanced current must be *added* to drop in the heavier loaded leg and *subtracted* from drop in the lighter loaded leg to obtain voltage drops from leg to neutral.

21. I^2R or conductor loss in d-c or a-c circuits is calculated by multiplying the square of the current in amperes by ohmic resistance of the conductors through which the current flows. The result is in watts.

22. In d-c circuits percent voltage drop and percent conductor loss are identical, and their ratio is unity.

$$\text{Percent voltage drop} = IR/V \times 100$$

$$\text{Percent } I^2R = I^2R/VI \times 100 = IR/V \times 100$$

In a-c circuits the ratio of percent conductor loss to percent voltage regulation is given approximately by the following formula:

$$\frac{\%\ I^2R \text{ loss}}{\%\ \text{voltage drop}} = \frac{\cos\phi}{\cos\theta\cos(\phi-\theta)} \tag{16-1}$$

where θ = power-factor angle and ϕ = impedance angle, i.e., $\tan\phi = X/R$.

23. Table 16-1 gives kilowatt-feet per 1% voltage drop for d-c circuits of three different voltages for several sizes of conductor. Kilowatt-feet is the product of load in kilowatts and circuit distance in feet over which load is carried. Kilowatt-feet calculated for any d-c circuit when divided by the appropriate value in Table 16-1 gives percent voltage drop. For voltages different from those in the table, multiply kilowatt-feet in the table by the square of the ratio of the new voltage to the voltage in the table.

24. Use of Table 16-1. *Example* 1. What is the voltage drop when 100 d-c A flows 1,500 ft over a 2-wire 120-V, 556-M cmil aluminum circuit? First determine kilowatt load and circuit distance. kW·ft = 100 A × 0.12 kV × 1,500 ft = 18,000 kW·ft. From Table 16-1 the value 2,420 kW·ft per 1% voltage drop for 556-M cmil aluminum is obtained. Voltage drop = 18,000/2,420 = 7.43%, or 8.9 V. The circuit has 7.43% I^2R loss, which is equal to 0.0743 × 12 kW = 0.89 kW. *Example* 2. A mine 1 mi from

Table 16-1. Kilowatt-feet per 1% Voltage Drop, D-C Circuits

Conductor size, Awg or M cmil		120-V, 2-wire	120/240-V, 3-wire balanced loads	600 V, 2-wire
Annealed copper	Equiv. aluminum			
6	4	182	728	4,550
4	2	290	1,160	7,250
2	0	460	1,840	11,500
0	000	732	2,928	18,300
00	0000	923	3,692	23,050
0000	336	1,465	5,860	36,650
350	556	2,420	9,680	60,500
500	795	3,460	13,840	86,500
1000	....	6,910	27,650	173,000
1500	....	10,400	41,600	260,000
2000	....	13,800	55,200	345,000

a motor-generator station must have 100 kW direct current at not less than 575 V. Maximum voltage of motor-generator is 600. What conductor should be used? kW·ft = 100 kW × 5,280 ft = 528,000. Permissible percent drop = $^{25}\!/_{575}$ × 100 = 4.35%.

$$\text{Min. kW·ft per 1\%} = 528{,}000/4.35 = 121{,}000$$

From Table 16-1, by using the 600-V column and applying the factor $(575/600)^2 = 0.92$ to make the values applicable to 575 V delivered, it is found that 0.92 × 86,500 = 79,600 kW·ft per 1% drop corresponds to 500 M cmil copper conductor; and 0.92 × 173,000 = 159,200 kW·ft, to 1,000 M cmil. *Answer*. Use 750 M cmil copper for the circuit.

25. Calculating Voltage Drop in A-C Circuits. The voltage drop per mile in each round wire of 3-phase 60-c/s line with equilateral spacing D in between centers or in each wire of a single-phase line D in between centers is

$$\dot{V} \text{ drop} = \dot{I}R + j\dot{I}\left(0.2794 \log \frac{D}{r} + 0.03034\,\mu\right) \quad \text{vector volts} \tag{16-2}$$

where I is in vector amperes, R = 60-c/s resistance of the wire per mile in ohms, r = radius of round wire in inches, and μ = permeability of the wire (unity for nonmagnetic materials such as copper or aluminum). j in Eq. (16-2) denotes an angle of 90°; $+j$ means 90° leading; $-j$, 90° lagging. Thus the expression for vector current lagging the reference voltage is $\dot{I} = I_x - jI_y = I/\overline{\theta°}$ with reference to a conveniently chosen horizontal axis of reference—usually sending- or receiving-end voltage. The dot beneath $\dot{I}$ or $\dot{V}$ indicates vector values. Voltage drops determined in this manner are also vectors and are with respect to the reference axis.

When wire is stranded, an equivalent radius must be used for r in Eq. (16-2). $r = 0.528\sqrt{A}$ for 7 strands, $r = 0.5585\sqrt{A}$ for 19 strands, $r = 0.5675\sqrt{A}$ for 37 strands, where r = equivalent radius in inches and A = area of metal in square inches.

Frequency is 60 c/s for the constants in parentheses in Eq. (16-2), which gives reactance X in ohms per mile. For 25 c/s, multiply by 25/60. The equation is sometimes written

$$V \text{ drop per mile} = I(R + jX) = \dot{I}\dot{Z} \quad \text{vector volts} \tag{16-3}$$

where I is in vector amperes and $\dot{Z} = Z/\underline{\phi°}$ Ω/mi at 60 c/s.

26. Three unsymmetrically spaced wires, a, b, and c, of a 3-phase circuit with correct transpositions can have voltage drop in each wire calculated by Eq. (16-2) by substituting for D the geometric mean of the three interaxial distances:

$$D = \sqrt[3]{D_{ab}D_{bc}D_{ca}}$$

27. The Vector Method. In Eq. (16-3), $\dot{I}$ is in vector amperes,

$$\dot{I} = I_x - jI_y = I/\overline{\theta°}$$

where θ = angle that the current lags (or leads) the voltage. The sending-end voltage is usually chosen as the axis of reference in drawing the vector diagram. For example, consider Fig. 16-2, where sending voltage $\dot{V}_S = V_S\underline{/0°}$, load current $\dot{I} = I\overline{/\theta°}$, circuit impedance $\dot{Z} = R + jX = Z\underline{/\phi°}$, and $\dot{V}_L = \dot{V}_S - \dot{I}\dot{Z}$ (all vectors).

Let $\dot{V}_S = 230\underline{/0°}$, $\dot{I} = 50\overline{/37°}$, $\dot{Z} = R + jX = 0.2\underline{/26°}$

$$\dot{V}_L = 230\underline{/0°} - 50\overline{/37°} \times 0.2\underline{/26°} = 230\underline{/0°} - 10\overline{/11°}$$

$$= 230 - 10 \cos 11° + j10 \sin 11°$$

$$= 230 - 9.9 + j1.92 = 220.1 + j1.92 = 220.1 \text{ (very nearly)}$$

Neglecting the last term $+j1.92$ simplifies final calculation and gives the load voltage accurate to within 0.1% in the usual case. This method is sufficiently accurate for all engineering calculations. Expressed in simplest form,

$$V \text{ drop} = IZ \cos(\phi - \theta) = IR \cos\theta + IX \sin\theta \tag{16-4}$$

where I and Z = absolute values, not vectors; ϕ = impedance angle; and θ = power-factor angle by which current lags (or leads) voltage. Calculating drop in the above example by this method,

$$V_{\text{drop}} = 50 \times 0.2 \cos(26° - 37°) = 10 \cos(-11°) = 9.9 \text{ V}$$

28. The impedance Z is the hypotenuse of a right triangle in which the base is resistance R and the leg is reactance X. $Z = R + jX = Z\underline{/\phi}$. Inductive reactance is positive, and capacitive reactance is negative. Impedance angle ϕ is the angle that hypotenuse Z makes with horizontal leg R in the right triangle of Fig. 16-3. The angle is an absolute value; i.e., it has no relation to the axis of reference in a vector diagram, as do current and voltage. Alternating current causes voltage drop in resistance in phase with current and a drop in inductive reactance that is a quarter cycle behind current.

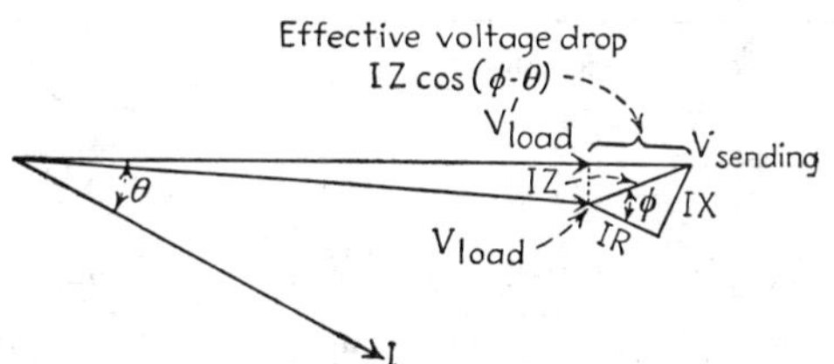

FIG. 16-2. Vector diagram showing effective voltage drop.

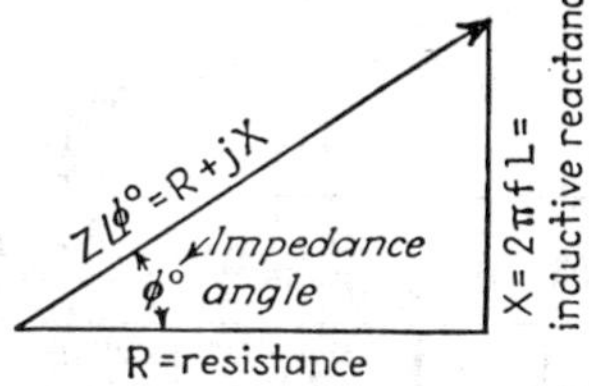

FIG. 16-3. Impedance diagram.

29. Impedance Values. Table 16-2 gives 60-c/s impedance per conductor per 1,000 ft for common sizes of wire and cable. Impedances are given in the form $R + jX = Z\underline{/\phi°}$. The $Z\underline{/\phi°}$ form is the most convenient to use when current is in the form $I\overline{/\theta°}$. Values of $I\overline{/\theta°}$ for 100 or 10 kW of load at various lagging power factors for several voltage classes are given in Table 16-3.

30. Example of Calculation of Single-phase Line Drop. Given 250-kW load, 80% power factor, 7,200 V, 60 c/s, single phase supplied by a No. 0 aluminum circuit, 15,000 ft in length, conductors spaced 28 in on centers. What is the line drop? The load voltage will be the axis of reference = $7{,}200\underline{/0°}$. From Table 16-3, 250 kW at 80% power factor = $2.5 \times 17.4\overline{/36.8°} = 43.4\overline{/36.8°}$ A. From Table 16-2,

$$Z = 2 \times 15 \times 0.208\underline{/36.2°} = 6.24\underline{/36.2°}$$

$$\text{Voltage drop} = IZ \cos(\phi - \theta) = 43.5 \times 6.24 \cos(36.2° - 36.8°)$$

$$= 271.44 \cos(-0.6°) = 271.44 - j0.48 = 271 \text{ V (very nearly)}$$

The last term $-j0.48$ can usually be neglected in such calculations. In this case, sending voltage is 7,471, and voltage drop = 271/7,200 × 100 = 3.76%.

Table 16-2. 60-c/s Impedance per 1,000 Ft of Single Conductor;

A. Overhead Wire—Copper

(Stranded hard-drawn copper, 97.3% conductivity, Nos. 6 to 4/0; stranded annealed copper, 100% conductivity, 350 to 1,000 Mcm)

Wire size	Distance between centers of conductors, in.					
	8 in.	14 in.	28 in.	44 in.	58 in.	72 in.
No. 6	.4221 + j.110 .436/14.6°	.4221 + j.1227 .4420/16.10°	.4221 + j.1389 .4445/18.20°	.4221 + j.1490 .4472/19.45°	.4221 + j.1553 .4500/20.20°	.4221 + j.1605 .4515/20.82°
No. 4	.2655 + j.1048 .286/21.5°	.2655 + j.1177 .2902/23.90°	.2655 + j.1332 .2968/26.62°	.2655 + j.1438 .3020/28.42°	.2655 + j.1500 .3042/29.48°	.2655 + j.1550 .3080/30.25°
No. 2	.1670 + j.0995 .1945/30.8°	.1670 + j.1122 .2014/33.95°	.1670 + j.1282 .2102/37.54°	.1670 + j.1384 .2170/39.63°	.1670 + j.1448 .2213/40.90°	.1670 + j.1499 .2243/41.90°
No. 1/0	.1051 + j.0944 .1412/41.9°	.1051 + j.1068 .1499/45.40°	.1051 + j.1227 .1617/49.40°	.1051 + j.1330 .1695/51.62°	.1051 + j.1394 .1750/53.00°	.1051 + j.1442 .1784/53.90°
No. 2/0	.0834 + j.0913 .1238/47.6°	.0834 + j.1042 .1332/51.35°	.0834 + j.1200 .1460/55.20°	.0834 + j.1304 .1545/57.42°	.0834 + j.1369 .1602/58.62°	.0834 + j.1418 .1645/59.55°
No. 4/0	.0526 + j.0862 .101/58.55°	.0526 + j.099 .1120/62.00°	.0526 + j.1149 .1263/65.40°	.0526 + j.1253 .1360/67.22°	.0526 + j.1315 .1418/68.20°	.0526 + j.1367 .1455/68.88°
350 Mcm	.0312 + j.078 .084/68.2°	.0312 + j.0913 .0965/71.15°	.0312 + j.1082 .113/73.7°	.0312 + j.119 .123/75.3°	.0312 + j.127 .131/76.2°	.0312 + j.132 .1358/76.7°
500 Mcm	.0221 + j.076 .0792/73.6°	.0221 + j.0872 .09/75.8°	.0221 + j.1058 .108/78.3°	.0221 + j.115 .117/79.1°	.0221 + j.118 .1202/79.4°	.0221 + j.127 .1288/80.1°
1000 Mcm	.0118 + j.0692 .070/81.7°	.0118 + j.079 .08/81.55°	.0118 + j.098 .0986/83.13°	.0118 + j.109 .1096/83.82°	.0118 + j.115 .116/84.15°	.0118 + j.120 .121/85.64°

B. Overhead Wires—ACSR and Aluminum

(Typical ACSR Nos. 4 to 3/0; stranded EC-grade hard-drawn aluminum, 60.97% conductivity, 1/0 to 795 Mcm)

Wire size	Distance between centers of conductors, in.					
	8 in.	14 in.	28 in.	44 in.	58 in.	72 in.
No. 4 ACSR 6/1	.424 + j.116 .439/15.3°	.424 + j.129 .443/17.0°	.424 + j.145 .448/18.9°	.424 + j.155 .451/20.0°	.424 + j.161 .453/20.9°	.424 + j.166 .455/21.4°
No. 2 ACSR 6/1	.267 + j.117 .292/23.5°	.267 + j.130 .297/26.0°	.267 + j.146 .304/28.7°	.267 + j.156 .309/30.3°	.267 + j.162 .312/31.1°	.267 + j.167 .316/31.9°
No. 1/0 ACSR 6/1	.168 + j.115 .204/34.4°	.168 + j.128 .211/37.4°	.168 + j.144 .222/40.3°	.168 + j.154 .228/42.5°	.168 + j.160 .232/43.5°	.168 + j.165 .235/44.8°
No. 3/0 ACSR 6/1	.106 + j.109 .152/45.8°	.106 + j.122 .162/48.9°	.106 + j.138 .174/52.5°	.106 + j.148 .182/54.4°	.106 + j.154 .187/55.4°	.106 + j.159 .191/56.3°
No. 1/0 Alum. 7 str.	.168 + j.0940 .192/29.2°	.168 + j.107 .199/32.5°	.168 + j.123 .208/36.2°	.168 + j.133 .214/38.3°	.168 + j.140 .218/39.8°	.168 + j.144 .222/40.6°
No. 3/0 Alum. 7 str.	.106 + j.0887 .138/39.9°	.106 + j.102 .147/43.9°	.106 + j.118 .158/48.0°	.106 + j.128 .166/50.6°	.106 + j.134 .171/51.7°	.106 + j.139 .175/52.6°
No. 4/0 Alum. 19 str.	.0837 + j.0848 .119/45.3°	.0837 + j.0976 .129/49.4°	.0837 + j.114 .142/53.7	.0837 + j.124 .150/56.0	.0837 + j.130 .155/57.2°	.0837 + j.135 .159/58.2°
336.4 Mcm Alum. 19 str.	.0528 + j.0795 .0955/56.4°	.0528 + j.0923 .106/60.3°	.0528 + j.108 .120/63.9°	.0528 + j.119 .130/66.1°	.0528 + j.125 .136/67.1°	.0528 + j.130 .140/67.9°
556.5 Mcm Alum. 37 str.	.0322 + j.0735 .0803/66.3°	.0322 + j.0861 .0920/69.5°	.0322 + j.102 .107/72.5°	.0322 + j.112 .116/74.0°	.0322 + j.119 .123/74.9°	.0322 + j.124 .128/75.5°
795 Mcm Alum. 37 str.	.0228 + j.0692 .0728/71.8°	.0228 + j.0820 .0851/74.5°	.0228 + j.0980 .101/76.9°	.0228 + j.108 .110/78.1°	.0228 + j.115 .117/78.8°	.0228 + j.120 .122/79.2°

NOTE: To obtain impedance in the form of $R + jX$ for 25- or 50-c/s, multiply the $+jX$ values in above table by ratio of new frequency to 60. Resistances are sufficiently accurate for 25 or 50 c/s.

Vector Ohms, $R + jX = Z/\underline{\phi°}$ at 25°C

C. Cables—Copper

(Stranded annealed copper conductors, 100% conductivity)

Single-conductor cables lying against each other in same duct				3-conductor cables	
Cable size	0 to 1 kv	5 kv	15 kv	15 kv	25 kv
No. 6	.411 + j.0503 = .414/6.97°	.411 + j.0557 = .415/7.73°			
No. 4	.2585 + j.0473 = 2630/10.37°	.2585 + j.0545 = .2641/11.9°	.2585 + j.0780 = .271/16.76°	.2585 + j.045 = .262/9.86°	
No. 2	.1626 + j.0450 = .1686/15.48°	.1626 + j.0515 = .1708/17.6°	.1626 + j.0723 = .179/23.79°	.1626 + j.0426 = .1682/14.68°	
No. 1/0	.1023 + .0405 = .1101/21.59°	.1023 + j.0475 = .113/24.9°	.1023 + j.0655 = .123/32.2°	.1023 + j.0396 = .110/21.15°	.1023 + j.0428 = .1110/22.65°
No. 2/0	.0812 + j.0392 = .0904/25.75°	.0812 + j.0460 = .0935/29.53°	.0812 + j.0631 = .105/37.1°	.0812 + j.0383 = .897/25.25°	.0812 + j.0413 = .0911/26.96°
No. 4/0	.0512 + j.0368 = .0630/35.70°	.0512 + j.0424 = .0665/39.6°	.0512 + j.0588 = .0794/47.75°	.0512 + j.0356 = .0624/34.8°	.0512 + j.0383 = .0640/36.8°
350 Mcm	.0312 + j.0358 = .0475/48.93°	.0312 + j.0412 = .0519/52.7°	.0312 + j.0540 = .0634/58.34°	.0312 + j.0331 = .0455/43.2°	.0312 + j.0354 = .0473/48.6°
500 Mcm	.0221 + j.0329 = .0396/56.10°	.0221 + j.0401 = .0458/61.1°	.0221 + j.0514 = .0569/64.6°	.0221 + j.0296 = .0370/53.2°	.0221 + .0316 = .0386/55.0°
1000 Mcm	.0118 + j.0306 = .0328/68.90°				

D. Cables—Aluminum

(Stranded EC-grade aluminum conductors)

Single-conductor cables lying against each other in same duct				Single-conductor cables, buried, 6-in. spacing
Cable size	0 to 1 kv	5 kv	15 kv	0 to 1 kv
No. 6	.674 + j.0503 = .676/4.27°	.674 + j.0557 = .676/4.72°		
No. 4	.423 + j.0473 = .426/6.38°	.423 + j.0545 = .426/7.34°	.423 + j.078 = .430/10.45°	
No. 2	.266 + j.0450 = .269/9.6°	.266 + j.0515 = .271/10.95°	.266 + j.0723 = .276/15.2°	.266 + j.0980 = .283/20.1°
No. 1/0	.168 + j.0405 = .173/13.55°	.168 + j.0475 = .175/15.79°	.168 + j.0655 = .180/21.3°	.168 + j.0915 = .191/28.6°
No. 2/0	.134 + j.0392 = .140/16.3°	.134 + j.046 = .142/18.95°	.134 + j.0631 = .148/25.22°	.134 + j.0890 = .161/33.5°
No. 4/0	.0845 + j.0368 = .0921/23.53°	.0845 + j.0424 = .0946/26.65°	.0845 + j.0588 = .103/34.83°	.0845 + j.0840 = .119/44.8°
350 M cmil	.0521 + j.0358 = .0632/34.5°	.0521 + j.0412 = .0664/38.35°	.0521 + j.054 = .075/46.0°	.0521 + j.0780 = .0938/56.2°
500 M cmil	.0375 + j.0329 = .0499/41.26°	.0375 + j.0401 = .0549/46.92°	.0375 + j.0514 = .0636/53.88°	.0375 + j.0740 = .0820/63.1°
1,000-M cmil	.0209 + j.0306 = .0371/55.67°			

NOTE: To obtain impedance in the form of $R + jX$ for 25- or 50-c/s, multiply the $+jX$ values in above table by ratio of new frequency to 60. Resistances are sufficiently accurate for 25 or 50 c/s.

Numerous such calculations for secondary systems can be done with Table 16-1 for unity-power-factor a-c loads. For other power factors, use Table 16-4.

31. Calculation of 3-phase Line Drops with Balanced Loads. In a 3-wire 3-phase circuit with balanced and equal loads on each phase, the drop in phase-to-phase voltage is 1.73 times the drop in each conductor when each conductor is treated as a single-phase circuit with no return wire.

$$V_{\text{drop}} \text{ (phase to phase)} = 1.73IZ \cos(\phi - \theta) \qquad (16\text{-}5)$$

where θ = power-factor angle, ϕ = impedance angle, I = amperes in each conductor, and Z = ohmic impedance of one conductor. For example, given a 5,000-ft 3-phase 12,470-V circuit of three No. 1/0 ACSR conductors on 28-in centers carrying 100 A/phase at 80% power factor, what is the line-to-line voltage drop?

$$Z = 5 \times 0.222\underline{/40.3^\circ} \qquad \text{and} \qquad I = 100\overline{/36.8^\circ}$$

$$\text{Line-to-line voltage drop} = 1.73 \times 5 \times 0.222 \times 100 \cos(40.3^\circ - 36.8^\circ)$$

$$= 192 \cos(3.5^\circ) = 191.6 + j11.7 = 191.6 \text{ V (very nearly)}$$

32. Calculation of Voltage Drop in Unbalanced Unsymmetrical Circuits. If there are n different wires, $a, b, c, d, \cdots, n$ carrying currents $I_a, I_b, I_c, \cdots, I_n$, respectively, whether 2-, 3-, or 6-phase, the voltage drop in wire a per mile at 60 c/s is

$$I_aR_a + j\left[0.2794\left(I_a \log\frac{1}{r} + I_b \log\frac{1}{D_{ab}} + I_c \log\frac{1}{D_{ac}} + \cdots + I_n \log\frac{1}{D_{an}}\right) + 0.03034\mu I_a\right] \quad \text{vector volts} \qquad (16\text{-}6)$$

where currents are in vector amperes; R_a = 60-c/s ohmic resistance of conductor a per mile; r = equivalent radius, in inches, of conductor a; D_{ab}, D_{ac}, D_{an} = distances,

Table 16-3. Conversion of 100 or 10 kW into Amperes and Power-factor Angle

Load voltage		Lagging power factor and power-factor angle							
		Unity	95%	90%	85%	80%	70%	60%	50%
Kv	Phases	$\overline{/0^\circ}$	$\overline{/18.2^\circ}$	$\overline{/25.8^\circ}$	$\overline{/31.8^\circ}$	$\overline{/36.8^\circ}$	$\overline{/45.6^\circ}$	$\overline{/53.2^\circ}$	$\overline{/60^\circ}$
		A per 100 kW of load							
24.9	3	2.31	2.44	2.57	2.72	2.88	3.30	3 85	4.62
12.5	3	4.62	4.87	5.13	5.43	5.76	6.59	7.70	9.23
4.8	3	12.1	12.7	13.4	14.2	15.1	17.2	20.1	24.1
4.16	3	13.9	14.6	15.5	16.4	17.4	19.9	23.2	27.8
2.4	3	24.1	25.4	26.7	28.3	30.1	34.4	40.2	48.2
14.4	1	7.0	7.1	7.8	8.2	8.7	9.9	11.6	13.9
7.2	1	13.9	14.2	15.5	16.4	17.4	19.8	23.2	27.8
4.8	1	20.8	21.9	23.1	24.5	26.0	29.8	34.7	41.7
2.4	1	41.6	43.8	46.2	49.0	52.0	59.6	69.4	83.4
Volts	Phases	A per 10 kW of load							
480	3	12.1	12.7	13.4	14.2	15.1	17.2	20.1	24.1
240	3	24.1	25.4	26.7	28.3	30.1	34.4	40.2	48
208	3	27.8	29.2	30.9	32.7	34.8	39.7	46.3	55.6
277	1	36.0	38.0	40.0	42.5	45.0	51.7	60.0	72.1
240	1	41.6	43.8	46.2	49.0	52.0	59.6	69.4	83.4
208	1	48.1	50.6	53.5	56.6	60.1	68.7	80.2	96.3
120	1	83.3	87.6	92.4	98.0	104	120	139	167

in inches, between centers of conductors a and b, a and c, and a and n; and μ is the permeability of conductor a (unity for nonmagnetic material). To get the drop in b, replace all a's by b's and all b's by a's in Eq. (16-6); similarly, to get the drop in c, interchange a's and c's; likewise for n. For 25 c/s, multiply that part of Eq. (16-6) that is in brackets by $^{25}/_{60}$. Equation (16-6) gives voltage drop for any degree of load unbalance, power factor, or conductor arrangements. In using this formula calculations are made easier by choosing voltage to neutral as the reference axis.

33. Approximate Method of Calculating Voltage Drop in Unbalanced, Unsymmetrical Circuits. Equation (16-6) requires laborious calculations and is used only when exact results are necessary. Voltage drops sufficiently accurate for engineering purposes can be made by using an equivalent impedance for each conductor. The reactance component of the equivalent impedance is computed from a spacing D equal to the geometric means of the interaxial distances of the other conductors to the conductor being considered. For instance, if there are four conductors a, b, c, and n for conductor a, $D = \sqrt[3]{D_{ab}, D_{ac}, D_{an}}$; for conductor b, $D = \sqrt[3]{D_{ab}, D_{bc}, D_{bn}}$.

Table 16-4. Kilowatt-feet per 1% Voltage Drop, 60-c A-C Circuits

Conductor size 8-in flat spacing, AWG or M cmils		0.80 power factor		0.50 power factor	
Copper	Aluminum	120-V, 2-wire	120/240-V, 3-wire, balanced	120-V, 2-wire	120/240-V, 3-wire, balanced
No. 6	No. 4	143	557	111	420
No. 4	No. 2	210	809	149	557
No. 2	0	298	1140	193	709
0	000	410	1533	239	862
00	0000	474	1760	263	937
0000	336	614	2220	310	1083
350	556	803	2820	371	1262
500	795	910	3210	399	1370

34. Vector and connection diagrams are drawn in computing voltage drops in unbalanced circuits. Figure 16-4 shows an unbalanced 4-wire 3-phase 4,160/2,400-V circuit with assumed loads, power factors, and equivalent line impedances. Phase-to-neutral drops between source and load are given by the following:

$$\begin{aligned} V_{na} - V_{n'a'} &= I_aZ_a + I_nZ_n \\ V_{nb} - V_{n'b'} &= I_bZ_b + I_nZ_n \\ V_{nc} - V_{n'c'} &= I_cZ_c + I_nZ_n \end{aligned} \qquad (16\text{-}7)$$

Phase-to-phase drops between source and load are given by the following:

$$\begin{aligned} V_{ba} - V_{b'a'} &= I_aZ_a - I_bZ_b \\ V_{ac} - V_{a'c'} &= I_cZ_c - I_aZ_a \\ V_{cb} - V_{c'b'} &= I_bZ_b - I_cZ_c \end{aligned} \qquad (16\text{-}8)$$

In computing line-to-neutral drop in phase a, it is convenient to choose V_{na} as the axis of reference.

$$\begin{aligned} V_{na} - V_{n'a'} &= I_aZ_a + I_nZ_n = (100\underline{/20^\circ})(1.2\underline{/49^\circ}) + (43.2\overline{/32.2^\circ})(0.5\underline{/40^\circ}) \\ &= 120\underline{/29^\circ} + 21.6\underline{/7.8^\circ} = 126.4 + j61.9 \end{aligned}$$

Load voltage $V_{n'a'} = 2{,}400 - 126.4 - j61.9 = 2{,}273.6$ V (very nearly).

Likewise, in computing line-to-neutral drop in phase b, it is convenient to choose

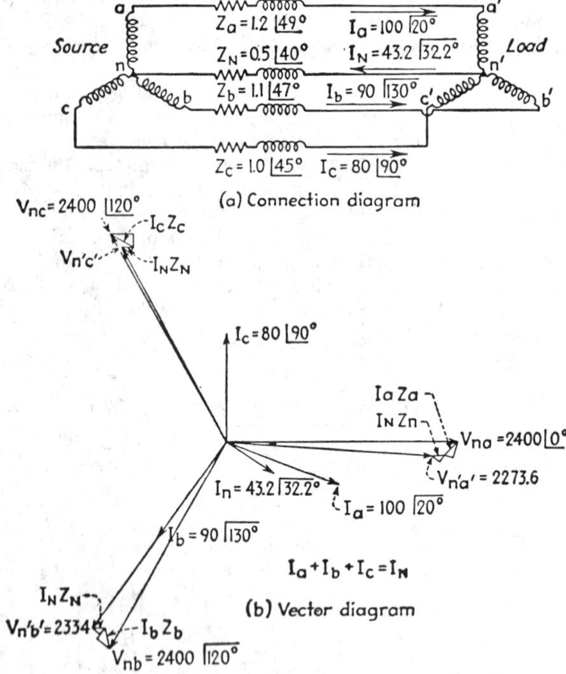

FIG. 16-4. Connections and vector diagrams for unbalanced loads and unsymmetrical circuit.

V_{nb} as the axis of reference. The vector diagram of Fig. 16-4 must be rotated in a counterclockwise direction 120°; then $I_b = 90/\overline{10°}$, and $I_n = 43.2\underline{/87.8°}$.

$$V_{nb} - V_{n'b'} = I_b Z_b + I_n Z_n = (90/\overline{10°})(1.1\underline{/47°}) + (43.2\underline{/87.8°})(0.5\underline{/40°})$$
$$= 65.8 + j76.6$$

Load voltage $V_{n'b'} = 2{,}400 - 65.8 - j76.6 = 2{,}334.2$ V (very nearly).

35. The 2-phase Drop. In a 2-phase 4-wire circuit, drop is the same as in a single-phase circuit carrying the same current and is calculated as in Par. **30**. In a 2-phase 5-wire circuit carrying unbalanced load, drop in any one of the four phases is the vector sum of drops in the phase wire and the neutral. Drop is calculated in the same manner as in Par. **33** for an unbalanced 3-phase, 4-wire system.

36. Drop in the neutral conductor of a 4-wire 3-phase circuit or a 5-wire 2-phase circuit makes resultant drop on the more heavily loaded phases greater than it would be for the same current under balanced conditions. Likewise, net drop is less on lighter-loaded phases than for the same current when balanced.

37. Evenly Distributed Loads; Voltage Drop and Conductor Loss. When loads of equal amounts and the same power factor are supplied by a circuit at equal intervals of distance, the total voltage drop from beginning to end of circuit can be calculated by assuming all load concentrated at the *middle* of the circuit; total I^2R or conductor loss can be calculated by assuming all load concentrated at a point one-third the total distance from the source. Total voltage drop $= (D/2)IZ \cos(\phi - \theta)$; total conductor loss $= (D/3)I^2R$, where D = total distance, I = total load current, and R and Z = resistance and impedance per unit distance, θ = power-factor angle, and ϕ = impedance angle. This method expedites calculations of distribution secondaries.

38. Annual conductor-loss factor is the ratio of average yearly conductor loss to conductor loss at peak load and is used to determine the total annual energy dissipated in conductor losses. It is related to, but not a direct function of, annual load factor, which is the ratio of average yearly load to annual peak load. Annual kilowatthours of conductor losses can be calculated by multiplying annual loss factor by hours in the year by kilowatt loss at peak load.

THE SUBTRANSMISSION SYSTEM

39. Definition. Subtransmission is that part of the distribution system between bulk power source or sources (generating stations or power substations) and the distribution substation. In metropolitan areas subtransmission circuits are usually 3-conductor lead-covered cables located in ducts underground. In outlying districts of large cities and in most small cities subtransmission circuits are overhead on poles.

40. Voltages of subtransmission circuits range from 13 to 138 kV, but 34 kV is common. Three-phase 3-conductor circuits are universally used, and the lead sheath of cable used to avoid congestion in built-up areas is usually grounded.

41. Conductor sizes most commonly used for overhead subtransmission vary from No. 2 to 556-M cmil ACSR or stranded aluminum alloy. The larger sizes are sometimes stranded EC-grade aluminum. Such conductors have to a large degree replaced copper.

42. Underground subtransmission ordinarily uses 3-conductor copper cable varying from 1/0 to 500 M cmils.

43. KVA-miles per 1% Regulation, Subtransmission. The size of subtransmission conductors is determined by (*a*) magnitude and power factor of the load, (*b*) distance the load must be carried, (*c*) operating voltage, (*d*) permissible voltage regulation, and (*e*) conductor loss. Table 16-5 gives kVA-miles per 1% voltage drop for common sizes of 15-kV shielded cables and overhead conductors, at 13.8 and 34.5 kV subtransmission voltage, for representative power factors. The table gives also the ratio of percent conductor loss to percent regulation; hence, having determined percent regulation, multiply by the conductor-loss factor to obtain percent conductor loss. Percent conductor loss multiplied by kilowatts delivered gives actual conductor loss in kilowatts.

Table 16-5. Voltage Regulation and Conductor-loss Ratios, 3-phase 60-c/s, 13.8- and 34.5-kV Subtransmission

Conductor size	Load voltage, 13,800				Load voltage, 34,500			
	95% pf		80% pf		95% pf		80% pf	
	kVA-miles per 1% reg.	% cond. loss per % reg.	kVA-miles per 1% reg.	% cond. loss per % reg.	kVA-miles per 1% reg.	% cond. loss per % reg.	kVA-miles per 1% reg.	% cond. loss per % reg.
	Underground* subtransmission							
Copper:								
1/0	2,020	1.00	2,070	1.21	19,850	0.956	20,050	1,163
4/0	5,420	0.932	5,400	1.095	35,000	0.881	32,800	0.980
350	8,220	0.864	7,610	0.952	51,750	0.806	45,350	0.844
500	10,870	0.800	9,520	0.834	65,650	0.750	54,950	0.749
750	14,350	0.722	11,780	0.703	84,100	0.688	66,900	0.651
	Overhead† subtransmission							
Aluminum:								
No. 2ACSR	1,200	0.937	1,190	1.101	7,510	0.937	7,440	1.101
1/0ACSR	1,760	0.865	1,640	0.950	11,000	0.865	10,270	0.950
2/0ACSR	2,100	0.821	1,880	0.870	13,160	0.821	11,730	0.870
4/0ACSR	2,940	0.724	2,420	0.706	18,390	0.724	15,110	0.706
336 Al.	4,030	0.622	3,060	0.561	25,200	0.622	19,150	0.561
556 Al.	5,300	0.498	3,690	0.412	33,150	0.498	23,070	0.412

* Underground conductor is annealed copper in 3-conductor cable.
† Overhead wire is ACSR at 60-in equivalent spacing; stranded aluminum in M-cmil sizes.
To obtain percent loss, multiply conductor-loss ratio from table by percent regulation.
To obtain conductor loss in kilowatts, multiply percent conductor loss (divided by 100) by kilowatts delivered.
For voltages slightly different from values in table, multiply kVA-miles in table by square of ratio of new voltage to value in table. *Example*: For 14,000 V, multiply kVA-miles by $(14{,}000/13{,}800)^2 = 1.028$.

Values in Table 16-5 are based on the approximate formulas

$$\text{kVA-miles/1\% drop} = \frac{\text{line kV}^2 \times 10}{Z \cos (\phi - \theta)} = \frac{\text{line kV}^2 \times 10}{R \cos \theta + X \sin \theta} \qquad (16\text{-}9)$$

$$\text{Conductor-loss ratio} = \frac{\%\ \text{conductor loss}}{\%\ \text{voltage drop}} = \frac{\cos \phi}{\cos \theta \cos (\phi - \theta)} \qquad [\text{see Eq. (16-1)}]$$

where R, X, and Z = 60-c/s resistance, reactance, and impedance, respectively, in ohms per mile of a single conductor; θ = power-factor angle; and ϕ = impedance angle. Equations (16-9) and (16-1) are sufficiently accurate for engineering purposes.

44. Regulation and Conductor-loss Ratios for Subtransmission. See Table 16-5.

45. Examples of How to Use Table 16-5. Determine the percent voltage drop and conductor loss when a 3,000-kVA load at 95% power factor is carried 3 mi over No. 4/0 3-conductor 15-kV type H cable. Assume receiving-end voltage to be 13,800: kVA-miles = 3,000 kVA × 3 mi = 9,000. From Table 16-5 the value 5,420 kVA-miles per 1% voltage drop is obtained for No. 4/0 cable at 95% power factor. Voltage drop = 9,000/5,420 = 1.66%. From Table 16-5 conductor-loss ratio is 0.932; conductor loss = 0.932 × 1.66 = 1.55%. Actual conductor loss = 0.0155 × 3,000 × 0.95 = 44.2 kW.

What size cable should be used to supply 5,000 kVA of 95% power-factor load at 5 mi, receiving-end voltage 13,800 with a permissible drop of 5%? What is conductor loss? kVA-miles = 5,000 kVA × 5 mi = 25,000. kVA-miles per 1% drop to satisfy 5% limitation = 25,000/5 = 5,000 kVA·mi. From Table 16-5, it is found that No. 4/0 3-conductor cable corresponds to 5,420 kVA·mi per 1% regulation; therefore No. 4/0 is the size to use. Voltage drop with No. 4/0 = 25,000/5,420 = 4.6%. From Table 16-5 conductor-loss factor is 0.932. Conductor loss = 4.6 × 0.932 = 4.3%. Actual conductor loss = 4.3/100 × 5,000 × 0.95 = 204 kW.

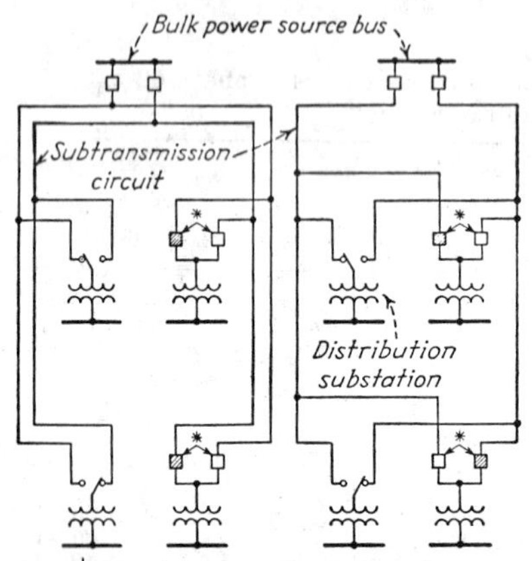

FIG. 16-5. A form of radial subtransmission acceptable to many customers.

46. Subtransmission[1] may be simple radial circuits, parallel or loop circuits, or interconnected circuits forming a grid or network. Cost and reliability of power supply are the most important factors that determine the type to use. Load distribution, topography, and number and location of bulk power sources are determining factors also.

47. The radial form is the simplest and lowest in first cost; however, it is not generally used, because a large block of load would be interrupted and customers over a wide area would be without service in the event of a fault on the circuit.

48. A form of radial subtransmission acceptable to many customers is shown in Fig. 16-5. Each subtransmission circuit serves as normal feed to certain distribution stations and emergency feed to others. In each substation, manual or automatic switches properly interlocked transfer the transformers to the auxiliary circuit if the normal circuit fails. This arrangement requires considerable spare capacity in each circuit and does not prevent an extensive service interruption for a short time. But it is sufficiently reliable for many small substations and industrial customers.

49. Parallel-, or loop-circuit, subtransmission is shown in Fig. 16-6. Uninterrupted service to all distribution substations will be maintained during any single fault on a circuit. Circuits must have sufficient reserve capacity so that no circuit will be overloaded when one is out of service. In this arrangement two "parallel" circuits are, in reality, a sectionalized loop supplying several substations, each located between two automatic sectionalizing breakers. The two circuits should follow different routes.

[1] Paragraphs **46** to **50** digest part of Chap. 20 on Distribution Systems by John S. Parsons and H. G. Barnett, "Electrical Transmission and Distribution Handbook," Westinghouse Electric Corporation.

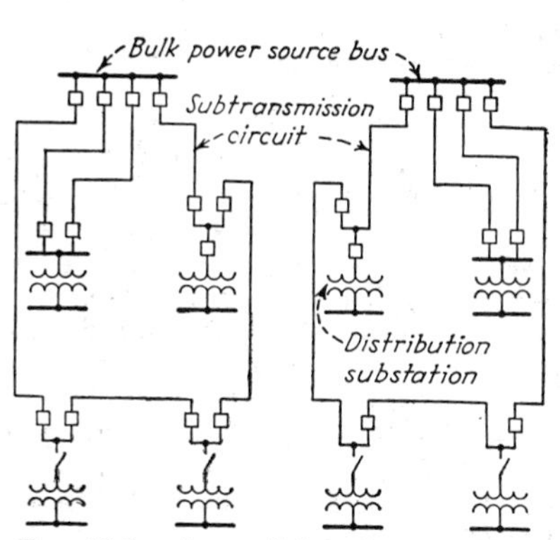

FIG. 16-6. A parallel or loop arrangement of subtransmission.

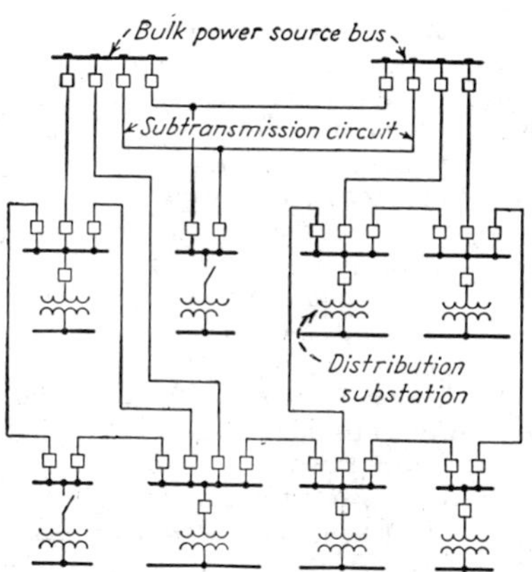

FIG. 16-7. The network or grid form of subtransmission.

50. Network, or grid, subtransmission, shown in Fig. 16-7, provides the best service reliability to distribution substations, particularly when supplied from two or more bulk power sources. Power can flow from any source to any substation. Network subtransmission can be extended to supply additional substations with little new construction. But it requires many circuit breakers and is difficult to relay.

SUBSTATIONS

51. A distribution substation is an assembly of equipment designed to receive energy from a higher voltage system, convert it to a form suitable for local distribution, and distribute it to feeders through switching equipment designed to protect service from faults. In many cases provision is included for regulating voltage.

Energy may be received (*a*) at higher voltage and distributed at lower voltage, (*b*) as alternating current and distributed as direct current, or (*c*) as alternating current at one frequency and distributed at another frequency.

52. Substation Operation. Substations may be attended by operators or designed for automatic or remote control. If attended by operators, control equipment, and sometimes converting equipment, are placed within a building. If unattended, the transformers and other equipment may be placed outdoors. Some substations in the more important parts of cities are attended because of their large size, varied equipment, and importance of service dependent upon them. But many large new substations are either automatic or remotely controlled.

53. Automatic substations are those in which switching operations are so controlled by relays that transformers or converting equipment is brought into or taken out of service as variations in load may require and feeder circuit breakers are closed and reclosed after being opened by overload relays.

54. Remote-control substations are often used where they may be placed within a suitable distance from attended stations or control centers. In such cases pilot-wire cables give the operator indications of circuit-breaker operations in the unattended substation and enable him to open or close breakers as desired. Microwave radio and carrier current are often used for remote-control links at distances beyond the economic reach of pilot wire.

55. A transformer substation transforms energy received from the subtransmission to a voltage suitable for local distribution, regulates voltage delivered to feeders when necessary, and serves as emergency supply for feeders in adjacent areas. Transformer capacity varies from a few hundred kVA in small rural stations to 250,000 kVA in large metropolitan substations. Large substations of 30 to 40 years ago required an excessive number and mileage of 2,400- and 4,160-V feeders. More recent practice has been to use one to four or five feeders of higher primary voltage, say 12.5 kV, and limit stations to 10,000 to 25,000 kVA. Such moderate-sized substations have given more flexibility to match load requirements, and simplification of the primary feeder system. The need to reach more load centers increases subtransmission mileage, but its layout has been simplified and relatively fewer high-voltage circuit breakers are provided. Subtransmission and its associated substation transformers are often treated as a unit with automatic switching confined to the low-voltage side. Basic designs are illustrated and discussed in Pars. **56** to **69**.

Distribution substations may be supplied at 66 kV or even higher. The combination of increased feeder capacity and higher voltage supply has reestablished a trend to larger distribution substations, but their design shows many of the simplifications which will be discussed for lower-voltage substations.

56. The simplest form of transformer substation[1] is shown in Fig. 16-8*a*. Equipment consists of a high-voltage disconnecting switch capable of interrupting transformer exciting current, a 3-phase transformer or three single-phase transformers, a primary feeder breaker equipped with overcurrent relays and automatic reclosing, disconnecting switches on both sides of the breaker, and a bypass switch for use in servicing the breaker. Where transformers are single-phase, a spare unit facilitates emergency restoration; otherwise a mobile transformer may be used.

57. Lift-up or draw-out-type breakers are commonly used in modern distribution substations. These breakers provide the disconnecting-switch features shown in Fig. 16-8, interlocked in such a way that disconnects cannot be opened until the breaker is tripped.

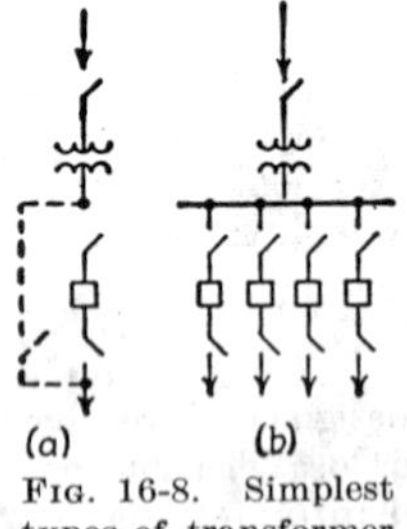

Fig. 16-8. Simplest types of transformer substations.

58. The substation of Fig. 16-8*b* is similar to that of Fig. 16-8*a*, except that it has a low-voltage bus and several primary feeders and associated breakers instead of one. Usually the primary breakers are interlocked with the high-voltage disconnecting switch in order that the switch will never interrupt load current.

59. Fuses, or protective links, are sometimes installed in high-voltage leads of the transformer, either inside or outside the tank, to prevent damage to the transformer in case short-circuit current is too small to trip the subtransmission circuit breaker.

60. Voltage-regulation equipment in the form of automatic tap changing under load built into the transformer, or one or more separate voltage regulators, may be required in the transformer substations of Fig. 16-8. Tap changing under load built into a 3-phase transformer provides voltage regulation in the least space and usually at lowest cost.

61. A transformer substation with auxiliary subtransmission supply is shown in Fig. 16-9. A double-throw disconnecting switch or two interlocked disconnecting switches capable of interrupting transformer exciting current are used on the high-voltage side of the transformer. In event of failure of the normal circuit, the transformer can be transferred to the auxiliary supply by manual operation of the double-throw switch, thereby restoring service quickly to loads fed through the substation. In the multiple-feeder substation of Fig. 16-9*b* a breaker in the secondary leads of the transformer is usually installed to simplify switching procedure and interlock circuits. The double-throw disconnecting switch of Fig. 16-9*a* and *b* can be replaced by two manually operated high-voltage breakers, one in each supply circuit, interlocked so that only one can be closed at a time. The breaker in the secondary leads of the transformer is, in general, omitted when high-voltage breakers are used. Continuity of service is satisfactory for many industrial customers and is often the first step in building a larger station.

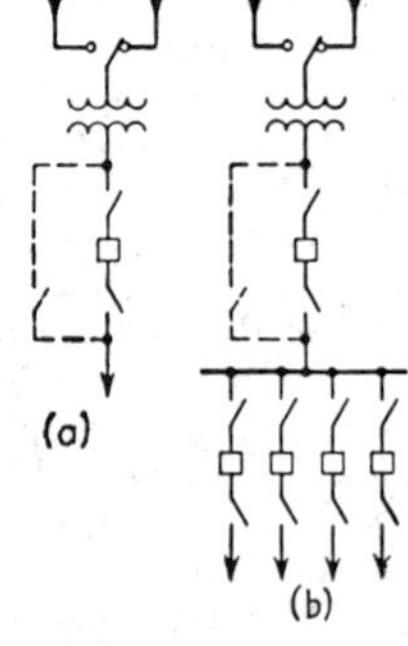

Fig. 16-9. Transformer substations with auxiliary supply to shorten service interruptions.

62. Substations with duplicate transformer and subtransmission supply are shown in Fig. 16-10. Transformer capacity is divided into two banks, and one bank is normally connected to each supply circuit. Load-break disconnects between the two primary bus sections

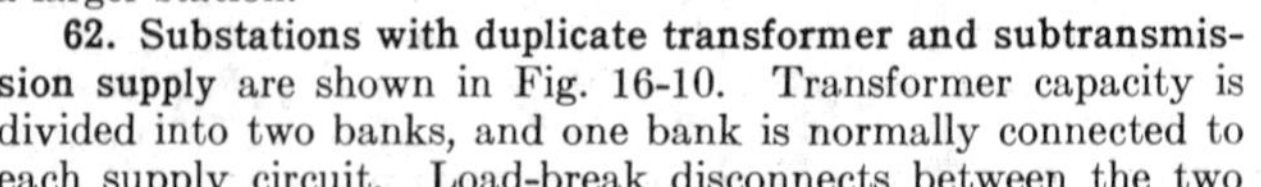

[1] Paragraphs **56** to **68** digest part of Chap. 3 on Subtransmission and Distribution Substations by D. N. Reps, "Electric Utility Engineering Reference Book," Westinghouse Electric Corporation, 1959, Vol. 3.

operate open normally and are interlocked with the transformers' secondary switches or breakers to prevent the transformers from operating in parallel. Each transformer must be capable of carrying the entire load of the substation. Both are operated normally with the low-voltage bus split to reduce the duty on primary feeder breakers. The substation in Fig. 16-10*b* differs from that of Fig. 16-10*a* in that the only time both transformers cannot be used to supply substation load is when one transformer fails. If one subtransmission circuit fails, the associated transformer can be switched to the remaining supply circuit, after its low-voltage breaker has been opened, thereby making both transformers available for carrying substation load. The two substations give a continuity of service that is adequate for many classes of customer.

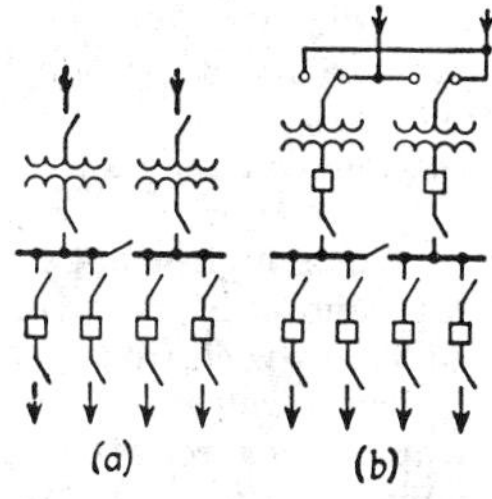

FIG. 16-10. Distribution substations with duplicate transformers and subtransmission circuits.

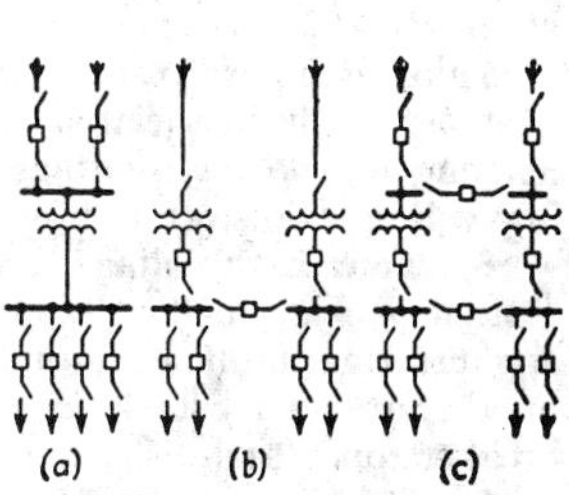

(a) (b) (c)

FIG. 16-11. Three automatic throw-over substations.

63. Three types of automatic throw-over substation are shown in Fig. 16-11*a*, *b*, and *c*. In Fig. 16-11*a* two high-voltage breakers are used to obtain automatic throw-over. When the preferred or normal circuit fails, its breaker opens and the auxiliary circuit breaker closes automatically. Return to the preferred circuit can be manual or automatic. In case of failure of one transformer, the entire load is interrupted until repairs can be made or a new transformer installed.

64. Automatic throw-over by means of a bus-sectionalizing breaker and breakers in the low-voltage leads of two transformers is shown in Fig. 16-11*b*. No high-voltage breakers are used. The bus-sectionalizing breaker is normally open. It closes automatically when either of the two transformer breakers opens and reopens automatically when both transformer breakers are closed. Failure of a supply circuit or a transformer results in loss of half the transformer capacity until repairs are made; therefore, sufficient reserve capacity must be kept in each transformer to supply the entire load of the substation.

High-voltage circuit breakers in supply lines (Fig. 16-11*a* and *c*) are normally controlled by directional-overcurrent relays which trip when fault current flows from the substation high-voltage bus toward the power source. When a fault occurs on a supply circuit, the associated high-voltage breaker at the substation and at the source end of the circuit trips and disconnects the faulty circuit from the system. The two high-voltage breakers at the station are also equipped with overcurrent relays, which have a longer time setting than the directional-overcurrent relays previously mentioned, to trip both breakers if the transformer fails. The opening of these two breakers may be undesirable, because the two supply circuits may be part of a loop or a ring. The substation can be modified by addition of high-voltage transformer breakers. With these additional breakers the supply-circuit breakers do not open when a transformer fault occurs.

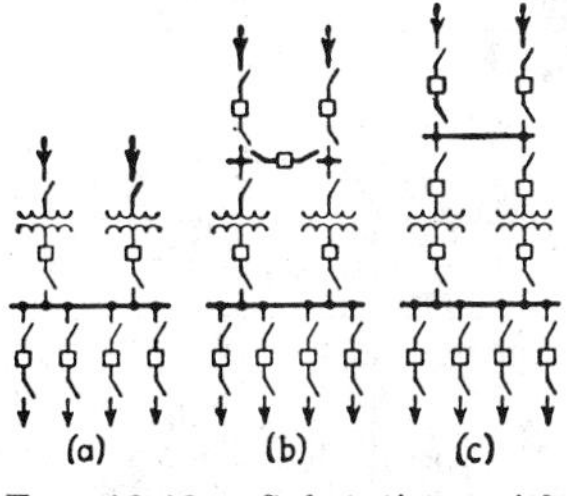

FIG. 16-12. Substations with duplicate supply circuits and transformers.

65. Three types of substation with duplicate supply circuits and transformers are shown in Fig. 16-12. No loads supplied by the substations are interrupted when a transformer or supply-circuit fault occurs. If one transformer fails, the other carries the entire load; therefore each transformer should have a self-cooled rating of 75% of the substation rating and be equipped with automatic air blast.

66. The substation of Fig. 16-12*a* gives high-quality service with a minimum amount of equipment. Breakers in the low-voltage leads of the transformers are controlled by directional overcurrent relays with a short time setting to protect against a transformer or supply-circuit fault and by overcurrent relays with a longer time setting to protect against low-voltage bus faults. This substation requires no high-voltage breakers.

67. Three high-voltage breakers and a high-voltage bus are required in the substation of Fig. 16-12*b* to allow both transformers to remain in service in event of a supply-circuit fault. All three high-voltage breakers are normally closed. The two supply circuits are tied together through the high-voltage sectionalizing breaker and may be part of subtransmission loops or rings. The various breakers are equipped with time-delay overcurrent and directional relays to protect the substation in the following manner: A faulted supply circuit is disconnected from the substation by opening of the associated high-voltage breaker. A transformer fault is cleared by opening of the high-voltage bus-sectionalizing breaker and its associated high- and low-voltage transformer breakers.

68. Four high-voltage breakers and a high-voltage bus are required in the substation of Fig. 16-12*c* to prevent opening either subtransmission breaker in event of a transformer fault. A supply-circuit fault is cleared by the associated subtransmission breaker at the substation. When a transformer fault occurs, its high- and low-voltage breakers open and disconnect the transformer from the system.

69. Modern unit substations minimize the possibility of a low- or a high-voltage bus fault. The likelihood of a bus fault in a modern factory-built-and-tested unit substation is remote. A modern unit substation consists of one or more 3-phase transformers equipped with automatic load tap changing, the necessary high-voltage disconnecting switches or breakers, and low-voltage breakers and bus work. The switchgear section or sections, containing all necessary disconnecting switches, buses, circuit breakers, relays, and auxiliaries, are completely metal-enclosed and mounted on or bolted directly to the associated transformer or transformers.

A single-bank unit substation usually embodies the layout of Fig. 16-8*a* or *b*. Multiunit stations resemble Fig. 16-10*a*, 16-11*b*, or 16-12*a*. Subtransmission and primary feeders are usually brought to the substation underground to avoid the congestion of overhead circuits near the station.

70. Bus Construction. High-voltage buses are made of copper or aluminum structural shapes or bars supported by insulators of suitable type. More important buses are surrounded by fireproof barriers with generous clearances; smaller ones are often mounted on a pipe or structural-steel framework. Manufacturers have developed standardized types of bus insulators and mechanical supports which simplify construction and assembly.

71. Substation transformers are of various types, sizes, and classes. Three-phase transformers are finding increasing use in new substations because of their almost negligible failure rate and their lesser cost compared with three single-phase transformers. Transformers are oil-and-air-cooled, oil-and-water-cooled, or air-cooled. Where water is not expensive, it is sometimes used to cool large transformers. Oil circulation, with an external radiator, is used extensively. Forced-air blast is applied to the external radiators of oil-filled transformers to give them higher ratings, particularly for emergencies. Indoor-type air-cooled transformers have coils cooled directly by air, and the hazard of an oil fire is eliminated.

Table 16-6. Power-transformer Regulation; 3-phase Transformers at Full Load, 60 C/s: Percent Voltage Drop through Transformer with Full-load Current

3-phase kVA	Voltage class, high-voltage side								
	15 kV			34.5 kV			69 kV		
	Unity pf	80% pf	50% pf	Unity pf	80% pf	50% pf	Unity pf	80% pf	50% pf
750	1.2	4.4	5.6						
1,000	1.1	4.3	5.5	1.5	4.2	5.4			
2,500	0.9	4.0	5.2	1.0	4.3	5.7	1.2	5.1	6.7
5,000	0.8	3.9	5.2	0.9	4.3	5.7	1.0	4.9	6.5
10,000	...	...	...	...	...	...	0.8	4.8	6.5

72. Regulation of 3-phase banks at full load is given in Table 16-6 for typical designs. The percent regulation at full load and unity power factor is also the percent copper loss at full load and the percent resistance.

73. Circuit breakers are rated by voltage, current-carrying capacity, and interrupting ability. The breakers are provided with operating mechanisms which are, in turn, actuated by power applied through suitable relays. Lift-up- or draw-out-type breakers which incorporate a disconnect feature on each side are used extensively, particularly in factory-assembled unit substations.

Air circuit breakers are often used instead of oil up to 15 kV in these units. And oil as well as vacuum reclosers have gained wide acceptance for circuit protection in small and medium-sized substations (see Sec. **10**).

74. Relays used in distribution, particularly in substations, are as follows: **Overcurrent** relays operate to open the circuit when two to three times rated current flows. They operate independently of voltage and require current coils only. A very inverse time-current relationship over a wide current range facilitates coordination with other relays and protective devices.

Directional, or **reverse-energy,** relays operate when the flow of energy, usually less than full load, is in a reversed direction. They operate on the wattmeter principle and require both current and voltage coils.

The **definite-time-delay** relay is an overcurrent relay, with mechanical means for limiting the travel of the relay contact to a definite and predetermined time interval, e.g., 2 s. This feature may be applied to other types of relay.

The **differential,** or balanced, relay operates on the principle that the current entering and the current leaving a section are equal normally but unequal when a fault develops within the section. Current transformers are placed at each terminal of the protected section and joined through a pilot-wire circuit. The necessity for pilot wire is the chief limitation of differential protection.

75. Application of Relays. Radial circuits are protected from short-circuit currents by overcurrent relays, set to operate "instantaneously" (a small fraction of a second) for heavy faults. Currents in the remote fault or overload range cause slower operation governed by the inverse characteristics.

Ring, or loop, circuits are protected by definite-time or differential relays, or both. If more than four substations or industrial vaults are served on one ring, time settings at the point of supply become too great and differential protection must be used for additional sections or for the entire ring if uniformity is desired.

Where several lines are connected in parallel, as at the incoming bus of a large substation, the overcurrent relay is used at the supply end of each line and the directional type at the receiving end. The directional relays are set to operate only on reversal of the normal direction of energy flow.

Differential relays are effectively used for protection of a substation from failure in a transformer on the substation bus. Such relays are arranged to cut out the faulty unit on both sides without interfering with operation of other units on the same bus. They are similarly effective in protecting various sections of a ring circuit by cutting out a faulted section without interrupting service to users of energy from the ring. The chief limitation of differential protection is the necessity for pilot wires.

All relay systems derive their low-voltage circuits through current and potential transformers, provision for which must usually be independent of similar circuits provided for ammeters, voltmeters, and wattmeters. Choice of relays is thus, to some extent, determined by the amount of auxiliary equipment required to produce the desired result.

76. Control of Reclosers. Oil reclosers won a place in substation design, now shared by vacuum reclosers, because they offered the required interrupting capability at a cost substantially below that of circuit breakers. Part of the savings came from use of built-in direct-trip and timed-reclosing functions which provide acceptable protection for most applications. Many larger reclosers have provision for adding control relays where circumstances require.

77. Automatic Substations. Most utilities have found that fully automatic control for the majority of distribution substations is more economical than manual operation. Feeder breakers reclose automatically after tripping on a fault or overload but lock open after a predetermined number of unsuccessful reclosures. Supply lines also can be arranged for automatic reclosing or left open until a traveling operator can check conditions.

78. Remote indication of breaker and switch positions at automatic substations, augmented by some additional information, is often brought to the system control center or to a regional attended substation. Usually this is accomplished by means of a multiconductor signal cable connected to contacts on breakers and other equipment at the automatic substation and to position indicators and alarm devices at the control center. In many cases, the information is transmitted by carrier-current or microwave-radio channels. A recent development is feeding these signals to a computer-actuated monitoring system which scans conditions continuously and prints out all abnormalities.

79. Remote control can often be superposed on remote-indication channels to permit a control-center operator to open or close breakers and switches at an unattended substation. Often the added investment in remote-control facilities is well justified by savings in operation expense. And time saved by direct control can be helpful to the system operator when trouble requires prompt corrective action.

PRIMARY RADIAL DISTRIBUTION

80. Primary Radial. The primary distribution system takes energy from the low-voltage buses of distribution substations and delivers it to primaries of distribution transformers. Most primary circuits are radial. They consist of primary feeders that take the energy to a load area, and subfeeders and laterals that distribute it to individual transformers. Manually operated sectionalizing and tie switches are used to isolate faulted sections of circuits and transfer unfaulted sections to adjacent circuits.

81. The tree type of radial circuit is shown in Fig. 16-13. Primary laterals are taken off at points along the lines as service is required. The current tapers off, and conductors are reduced in size as one proceeds from the "trunk" toward the top of the "tree." It is an economical and simple circuit but limited to situations where voltage drop between the first tap and the end of the circuit does not exceed 3 to 5%.

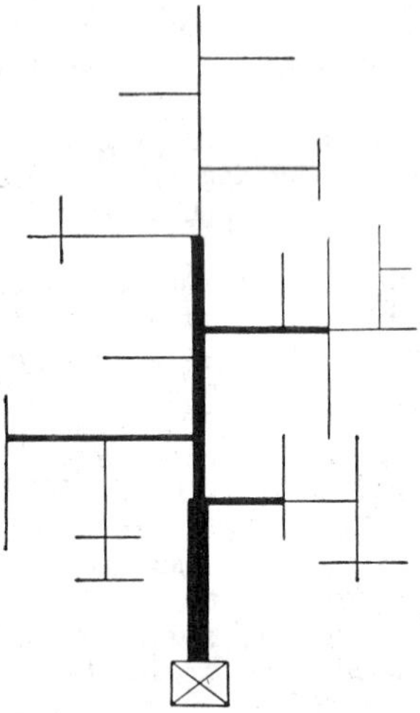

FIG. 16-13. Tree type of radial primary circuit.

82. A loop within a primary radial circuit is often established to good advantage. The loop improves voltage by providing minimum impedance for load current and realizes the advantage of diversity in demand among a large group of customers. It improves service continuity because a break in a loop does not interrupt supply to any transformers. Sometimes loops are compounded to form grids. The disadvantage of both, however, is that a broken wire may remain in hazardous condition without notice because no customer is out of service.

83. The voltage most commonly used in primary distribution is 7,200/12,470 3-phase 4-wire with the neutral grounded. Single-phase 7,200-V laterals are extended over a wide area to supply 7,200-120/240-V distribution transformers. This system has largely superseded 2,400/4,160-V 4-wire distribution, once used extensively, because it more easily satisfies the voltage and capacity requirements of today's distribution customers. There are some systems operating at 14,400/24,940 V or 20,000/34,500 V in sparsely developed areas, as well as a few 3-wire Δ-systems at 2,400, 4,800, and 7,200 V.

84. The conductor sizes most commonly used in overhead primaries range from No. 4 ACSR to 336.4-M cmil stranded aluminum. Laterals are often No. 4 or No. 2 ACSR, but feeders run to the larger sizes. Aluminum, ACSR, and alloys of aluminum have largely displaced copper from primary-circuit construction, although they may be specified on a "copper-equivalent" basis.

Aerial cable is often used for primary conductors where clearances are too close for open wire or tree trimming is not practicable. Cable in some cases comprises three rubber-neoprene- or polyethylene-insulated conductors lashed to a bare messenger which serves as the neutral for 3-phase circuits. In other cases, the phase conductors are supported from the messenger by insulating spacers; this construction is commonly called "spacer cable." Single-phase taps in either case are usually one insulated conductor and messenger or an insulated conductor with neutral strands spiralled around it.

85. The conductor sizes most commonly used in underground primary distribution vary from No. 6 to 500-M cmil copper. Feeders are usually 3- or 4-conductor paper-insulated lead-covered cable, while laterals may be single-conductor because of numerous taps to transformers. On some systems, interconnected cable sheaths are used for the neutral conductor; others use a separate bare neutral or a fourth conductor in the cable. Stranded aluminum conductors and aluminum sheaths have been used in some cases, but the increased outside diameter usually requires excessive duct space.

Table 16-7. Voltage Regulation of 5- and 15-kV Cables and Overhead Lines for Primary Distribution, kVA-miles per 1% Voltage Drop, Balanced Load

Underground 5- and 15-kV 1-conductor copper cables		Circuit operating voltage and load power factor							
		2,400 V, 1-phase		4,160 V, 3-phase		7,200 V, 1-phase		12,470 V, 3-phase	
Cable size	Voltage rating, kV	95% pf kVA-mi	80% pf kVA-mi	95% pf kVA-mi	80% pf kVA-mi	95% pf kVA-mi	80% pf kVA-mi	95% pf kVA-mi	80% pf kVA-mi
No. 6	5	13.3	15.1	80	90				
No. 4	5	20.8	22.8	125	137				
No. 4	15	20.2	21.5	121	129	182	194	1,088	1,161
No. 2	5	32.0	33.9	192	204				
No. 2	15	30.8	31.4	185	187	277	283	1,664	1,698
No. 0	5	48.7	49.4	293	297				
No. 0	15	46.4	45.0	278	270	417	405	2,504	2,431
No. 2/0	5	59.6	58.9	358	354				
No. 2/0	15	56.3	53.0	338	318	507	477	3,042	2,864
No. 4/0	5	89.6	82.1	538	494				
No. 4/0	15	81.4	71.5	488	429	733	644	4,396	3,868
350-M cmil	5	128.4	109.8	771	660				
500-M cmil	5	162.8	130.6	978	785				
Overhead line, aluminum									
Wire size	Equiv. spacing, in								
No. 4 ACSR	28	12.2	12.8	73	77	110	115	660	690
No. 2 ACSR	28	18.2	18.1	109	109	165	165	985	980
No. 0 ACSR	28	26.7	24.7	160	148	240	220	1,445	1,335
No. 3/0 ACSR	28	38.0	32.5	228	196	340	290	2,050	1,760
336-M cmil Al	28	65.0	51.0	390	305	585	460	3,520	2,755
556-M cmil Al	28	87.4	62.7	525	375	785	565	4,730	3,390

NOTES: For slightly different receiving voltages, multiply kVA-miles from the table by the square of the ratio of the new voltage to the voltage in the table. *Example.* For 4,000 V multiply by $(4{,}000/4{,}160)^2 = 0.924$.

Regulation of aluminum cable can be estimated with reasonable accuracy as that of copper cable two sizes smaller.

Regulation of copper overhead conductors can be estimated with reasonable accuracy as that of aluminum conductors (or ACSR) two sizes larger.

Recent improvements in polyethylene insulation have extended the use of aluminum-conductor cable for direct-buried systems where overall size is not detrimental.

Adoption of higher voltage for primary distribution avoids extensive use of cable because it provides more circuit capacity with fewer circuits and facilitates locating substations farther from congested areas.

86. Voltage Regulation of Primary Distribution. Table 16-7 of kVA-miles per 1% regulation can be used to determine the voltage drop of an existing circuit when load data are known or to determine conductor size to limit drop to a set value, or to guide selection of voltage for a new line. *Example.* What is the voltage drop in a 4,160-V

3-phase circuit that supplies 100 A balanced load at 80% power factor over No. 3/0 ACSR overhead wires for 2 mi?

$$\text{kVA-mi} = 2 \times 100 \times 4.16 \times 1.732 = 1{,}443 \text{ kVA-mi}$$

Table 16-7 shows 196 kVA-mi per 1% drop for No. 3/0 ACSR. Drop = 1,443/196 = 7.36%. *Example.* What is the proper underground cable to supply 1,500 kW at 95% power factor with 4% voltage drop in 2 mi to deliver 7,200 V 3-phase? kVA-mi = 1,500/0.95 × 2 = 3,158 kVA-mi. Table 16-7 shows 2.504 kVA-mi per 1% drop for No. 0 cable at 12.470 V, 3-phase. This would be $(7{,}200/12{,}470)^2 \times 2{,}504 = 813$ kVA-mi/1% drop at 7,200 V. Regulation with No. 0 cable = 3,158/813 = 3.9%. Thus 3 No. 0 is the proper cable to install for 4% voltage drop.

87. Ratings of primary feeders vary from 500 to 2,500 kVA at 4.8 kV and below. Cost considerations dictate a large feeder, but high-quality service calls for small blocks of load. Improved line sectionalizing with reclosers and fuses is making 2,000-kVA feeders feasible. These devices assume greater importance at 12.5 kV where economy demands feeder ratings of 5,000 to 10,000 kVA.

88. Permissible voltage drop in a primary feeder is an important factor in design. In general, drop between the supply substation and the first transformer of the longest feeder at peak load should not exceed 10%, and then only with voltage regulators for that particular feeder. Feeder drops ordinarily should not exceed 5%, and present practice of bus regulation sets an even lower limit. Permissible voltage drop on any primary feeder between the first and last distribution transformer cannot exceed 2 to 3% at peak load for urban and suburban distribution. Where secondary mains are omitted, as for farms served by individual transformers, the 2 to 3% drop usually allowed for secondary drop can be added to give 5 to 6% primary drop between the first and last transformer. The design objective is, of course, to hold delivered voltage to all customers within an acceptable range.

Pole-mounted voltage regulators and shunt capacitors have gained wide acceptance for supplementary control of voltage on long feeders. Such devices permit loading conductors to several times the limit set by their own voltage drop, sometimes even to their thermal limits.

THE COMMON NEUTRAL

89. Multigrounded common-neutral distribution has been used extensively with large economies. It finds application on 4-wire Y-primary circuits where the station transformer's Y-point is grounded and secondary neutrals are more or less continuous along most of the primary routes. The secondary neutral must be made continuous

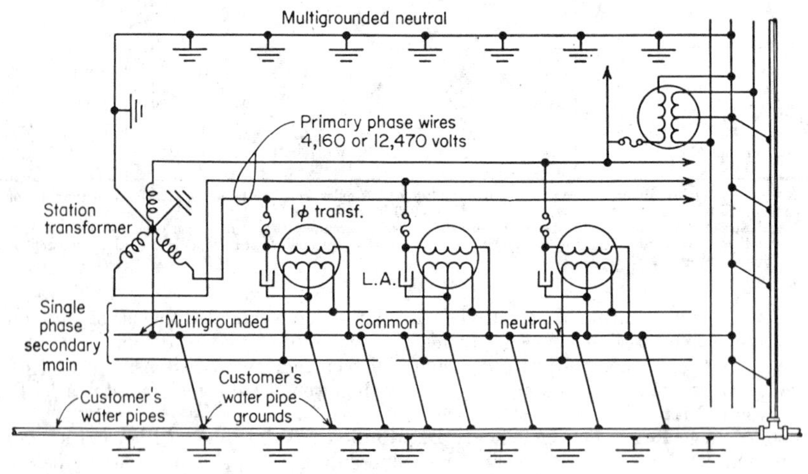

Fig. 16-14. Common-neutral method of distribution.

and grounded adequately, usually to customers' water pipes. This multigrounded secondary neutral is used as the primary neutral also and is called the common neutral because it carries the unbalanced or residual currents of both systems. The 2,400/4,160-V and 7,200/12,470-V 3-phase 4-wire circuits use the common neutral extensively. Figure 16-14 illustrates common-neutral circuits.

90. Adequate Grounding of Neutral. Certain rules must be observed in forming the common neutral. The most important is that it must be grounded adequately. A common neutral is considered adequately grounded when it has a minimum of one ground to a public or private water-pipe system at intervals not in excess of 1,000 ft or one driven ground at intervals not in excess of 500 ft.

91. Two continuous-neutral paths from any point in the common neutral to the supply station are *desirable* to provide a low-impedance path to the station for primary-neutral current to flow in case the common neutral should break. Without two paths, all primary-neutral current beyond the break would be shunted into water pipes. The more extensive the secondary grid, the easier it is to establish the second path. On many side streets supplied by primary laterals it is impossible to establish a second neutral.

92. The second neutral return path may consist wholly or in part of secondary or primary overhead neutral wires and sheaths or neutral conductors of primary or secondary cables. Water pipes should not be considered as constituting the neutral-return path.

93. Single-phase 2,400-V primary common-neutral laterals should be limited to 100 kVA of connected transformer capacity in order to limit current flowing in the earth–water-pipe ground to approximately 20 A. Roughly half the primary-neutral current flows in the earth or water pipes in the common-neutral system.

94. Minimum Size of Common Neutral. The overhead common neutral should have a minimum cross section equal to half of the largest primary-phase conductor.

95. In underground distribution the secondary neutral in multiple with all cable sheaths serves as the common neutral and is equivalent in all cases to more than one half the area of the largest phase conductor.

96. A single common-neutral conductor may serve for two primary circuits when they are on the same pole line but may not serve for more than two circuits.

97. The maximum allowable unbalanced primary current in the common-neutral system is approximately 40 A. Approximately half this current flows in the earth–water-pipe ground. Loads should be kept well balanced among the three phases to keep to a minimum the current flowing in the earth–water-pipe ground.

98. No disconnecting devices should be installed in the common neutral. Secondary neutrals should form a solid grounded grid whose numbers of conductors and total cross sections are increased as the station of supply is approached.

99. The common neutral has many advantages compared with 3-wire Δ-systems or 4-wire two-neutral systems. It requires only one lightning arrester and fuse cutout per transformer compared with two for the other systems. It requires only one primary wire for single-phase extensions instead of two. It requires only two primary wires to supply open-Y-Δ-transformers compared with three for the Δ-system. It requires only three primary wires for 3-phase compared with four wires in the two-neutral 4-wire system.

100. The common-neutral single-phase voltage drop for the same load is about 25% less than for the ungrounded neutral system. Line-to-neutral short circuits are 20 to 30% greater, assuring faster and more positive fuse blowings.

101. Conversion of 2,400-V Δ to 2,400/4,160-V common neutral can be done easily and at minimum expense. The work is to (1) establish a grounded-neutral bus at the station; (2) make the secondary neutral continuous over the 2,400-V primary route, and connect this neutral grid to the station neutral; (3) reconnect single-phase transformers to the common neutral solidly, removing one lightning arrester and fuse cutout; (4) reconnect three single-phase transformers or 3-phase transformers from Δ-Δ to Y-Δ; (5) reconnect the two single-phase regulators and potential transformers in each circuit at the station to the grounded-neutral bus and install an additional regulator, current transformer, and potential transformer in each circuit; (6) reconnect station transformer from Δ to Y and ground its Y-point to the neutral bus. An article giving in further detail the steps to take is that in *Elec. World,* Feb. 16, 1946, by Fred T. Bear, entitled Methodical Distribution Change-over from 2.4 to 4.16 kV.

102. Common-neutral and Telephone Circuits. Common-neutral distribution does not affect telephone plant, particularly when telephone circuits are in cable on the same pole line with the common neutral. It has been found that noise level in telephone circuits can be lowered appreciably by connecting telephone cable sheath to the multi-grounded common neutral at appropriate locations.

VOLTAGE REGULATION

103. Voltage regulation in a-c substations is usually accomplished by means of voltage regulators or switched capacitors placed on each feeder or on a section of bus serving several feeders. The latter may take the form of load tap changing on the transformer which supplies voltage to the bus. Group regulation is used only where feeders are relatively short and serve loads of similar characteristics. Automatic control is necessary for best service; this is secured through line-drop compensators which actuate, through relays, the regulator or tap-changer motors or the capacitor switches.

104. Circuit Regulators. In a-c radial primary systems it is often necessary to regulate the voltage of each feeder separately. Regulators are commonly arranged to raise or lower the voltage delivered to the feeder in amounts up to 5 or 10% of delivered voltage. Single-phase regulators are commonly 12 to 72 kVA and regulate one phase of a 3-phase circuit. Three-phase regulators are correspondingly larger; they may be used to regulate a 3-phase circuit or a group of circuits having well-balanced loads.

The regulator consists of a transformer supplied at bus voltage on the primary side and delivering the amount to be added to bus voltage on its secondary side. The line loops through the secondary winding to receive the effect of its regulating voltage.

105. Regulators are of two types: step and induction. In the step type, voltage is changed in definite amounts, for example, 1¼%, as the position of the sliding contacts is changed by the control circuit. The induction type has no moving contacts but changes voltage by rotating the core which carries the secondary winding, thus changing the amount of magnetic flux linking the secondary coils. The induction type was for years the only regulator available. But the step type has proved popular and has supplanted the induction type on many systems; in addition, its light weight and low magnetizing current in the smaller sizes have made it particularly attractive for pole mounting to supplement voltage on long circuits.

Regulators are commonly controlled by contact-making voltmeters, usually supplemented by line-drop compensators to factor line current into the voltage delivered.

106. The line-drop compensator consists of resistance and inductive reactance, which function as shown in Fig. 16-15. The two sections are tapped through dial switches or potentiometers, and the voltage drops therein are led into the voltmeter and regulator control circuits in such proportions that the counterpressure in the voltmeter circuit is equivalent to drop in the line. Dials of the compensator are marked in volts at full load on a 120-V base. The resistance section is set to give a full-load resistance drop of the same percentage as the resistance drop in the feeder conductor when carrying full load. The reactance section is set correspondingly. With proper settings, the voltmeter indicates delivered voltage.

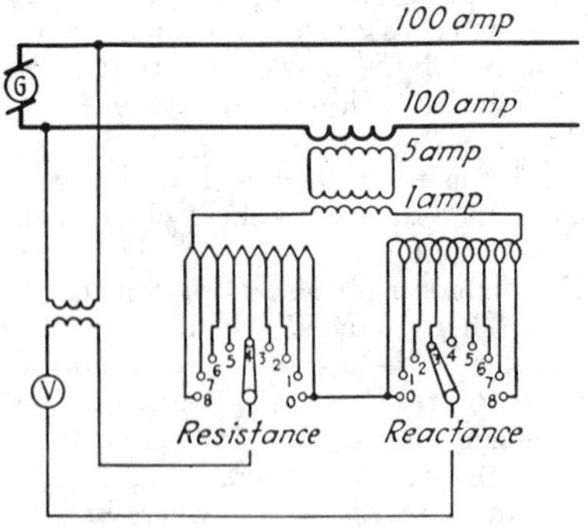

Fig. 16-15. Circuits of typical line-drop compensator.

When used in conjunction with automatic control of feeder regulators, the voltmeter is supplemented by a relay. This relay is normally in balance, but as load changes or bus voltage varies, the relay closes the circuit to the motor driving the regulator which it repositions sufficiently to restore a balance.

107. Calculation of Compensator Setting for a Single-phase Feeder. Given a 60-c/s feeder of No. 0 copper wire, 5,000 ft long, single-phase, wires 14 in apart, current-transformer ratio 100:5 A, potential transformer ratio 2,400 to 120 V, how should the compensator be set?

The resistance drop on No. 0 wire, carrying 100 A for 5,000 ft, will be

$$2 \times 100 \times 5 \times 0.1023 = 102.3 \text{ V} = 4.3\%$$

The inductive drop of No. 0 wire, at 14-in spacing, carrying 100 A for 5,000 ft, will be $2 \times 100 \times 5 \times 0.1068 = 106.8$ V $= 4.45\%$. The resistance and the reactance should each be set at 5 V ($= 0.042 \times 120$) to give approximately constant voltage at the end of the feeder at all loads.

A refinement of compensator setting permits boosting the delivered voltage automatically as load comes on, much like the overcompensation of some d-c generators. If, in the above example, it was desired to add 3% to delivered voltage during its 100-A peak and that peak were at 90% power factor, the 5-V settings based on line drop would be boosted by $100/(90 - j44) \times 0.03 \times 120 = 3.2 + j1.6$. Adding 3.2 to the 5-V resistance setting gives 8 V as the closest dial setting. The reactance dial would go to 7 V as the closest setting to the calculated 6.9 V.

108. Compensation of Polyphase Circuits. The calculation of compensator settings for polyphase circuits carrying unbalanced loads is made similarly for each of the conductors of the circuit, including the neutral if there is one.

Compensators are needed in only one phase if load on the different phases is balanced. In such case the one-way drop is taken as a percentage of the phase-to-neutral voltage of the circuit.

109. Bus regulation of unit-type substation transformers is usually provided by automatic load tap changing built into the transformer. This method provides bus regulation in the least space and usually at the lowest cost.

110. Switched shunt capacitors are often applied at distribution substations or on primary circuits in the load area, to regulate voltage. They accomplish this by drawing leading reactive current through the inductive reactance of the supply system to the point where they are connected. In such applications, the switch is often actuated by a voltage relay resembling that described above for regulator control. Some installations, however, are controlled by time clocks, air temperature, or even by radio signals.

PROTECTIVE EQUIPMENT

111. Protection from Excess Currents. Distribution circuits are protected from ground and short-circuit currents by fuses or circuit breakers, so arranged as to disconnect the faulted equipment promptly from its source of supply. Fuses are used almost exclusively for the protection of cables in low-tension light and power systems and for transformers in primary distribution systems, where the amount of energy controlled does not exceed about 250 kVA per phase. Circuit breakers or reclosers are used for larger amounts of energy and in cases where the operation of the overload device is so frequent as to make the use of fuses impractical.

Reclosers came into use initially for sectionalizing primary circuits. But subsequently they were built for increasingly heavy currents and for voltages as high as 46 kV. Control devices on these larger reclosers have operating characteristics closely resembling those of relays, and the reclosers have gained wide acceptance as replacements for circuit breakers for feeder protection in many substations.

112. Types of Distribution-circuit Fuses. In cable-junction boxes, operating at voltages up to 250, copper strips, having a narrow section in the middle, are found effective and economical as fuses. They are little subject to corrosion and not likely to operate unnecessarily because of deterioration. They are usually designed to fuse at about twice their rated carrying capacity. Many cable systems are fused with "limiters"—fuselike reduced sections inserted into cable runs. Enclosed fuses of the cartridge type are best adapted to use in situations where the percentage of moisture in the air is not continuously high, as the filler materials are somewhat hygroscopic and moisture impairs their ability to extinguish the arc.

Transformer fuses for sizes up to 200 A are of silver, copper, tin, or special alloys. They are mounted within a fiber tube, proportioned to utilize the explosive force of a confined arc to expel conductive gases. The fiber tube may be exposed on a porcelain insulator or enclosed in a porcelain box; both types operate in air. Other types of "primary cutouts" immerse the fusible link in oil or in chemical compounds like compressed boric acid.

113. Sizes of Wire Fused by Electric Current. See Sec. 4.

114. Application of Fuses. On outdoor installations, where there is ample clearance, open or enclosed fused cutouts are used for the protection of transformers or other equipment at distribution voltages. Oil-immersed fuses are sometimes employed for distribution transformers in vaults or in enclosures, and occasionally overhead. The boric acid fuse, which has a fusible link threaded through a block of boric acid, has been very successful for protecting potential transformers at high voltages and for line transformers up to about 500 kVA.

115. Distribution Transformer Fusing. Most utilities fuse distribution transformers to protect the system against faults in the transformer. For safety to linemen, enclosed cutouts are widely used to fuse distribution transformers at 7,200 V and below where lines are maintained while energized by use of rubber gloves. Open-type cutouts are common for higher voltages and where live-line work is done with special tools or from insulated lift trucks. Fused cutouts should be selected with interrupting ratings equal to the short-circuit duties expected, because lack of capacity can lead to destruction of cutouts. Cutouts available today offer a wide range of interrupting capabilities.

116. The rating of transformer fuses depends largely on the kVA rating of the transformer. For 2,400-V single-phase and 4,160-V Y-systems the old rule of 1 A/kVA has been followed more or less for years. But many utilities have found that setting 10 or 15 A as the minimum-size fuse eliminates the frequent blowings formerly experienced with smaller links during lightning storms without appreciably increasing the failure rate of small transformers. Fusing practices are approaching a degree of uniformity today because of new standard fuse characteristics developed by EEI and NEMA in joint study and adopted in 1962 as USA Standard C37.43-1962.

The **EEI-NEMA study group** determined that the entire range of primary fuse applications from 6 to 200 A could be covered by a series of nine "preferred" sizes, each about 50% higher in rating than the next smaller one. It further determined characteristics such that fuses made with tin or other low-temperature fusible section, designated as USAS type T, would coordinate with the next smaller size for current up to 24 times the rating of the smaller link. A similar series for copper or silver fusible sections, designated USAS type K, coordinates to 13 times the rating of the smaller link. All such fuses are specified to carry 100% rated current continuously without any part of the fuse holder exceeding 30°C rise above 40°C ambient temperature. **For ratings smaller than 6 A,** where it seemed unlikely that conventional fuse elements would provide the desired longevity and close tolerances, the study group suggested dual-element links which would not coordinate with each other but would protect the smallest single-element link of the type T preferred series. Subsequently, to aid users whose coordination problems could not be met by either preferred series, about six intermediate sizes were interspersed between characteristic curves of each of the preferred series; these coordinate with adjacent ratings of intermediate links but not with adjacent preferred ratings.

117. A schedule for fusing distribution transformers with USAS type T preferred links is shown in Table 16-8. The T characteristic provides protection over a wider range of current conditions than the K, and more nearly approximates the overload capabilities of a transformer. If type K links were used for this purpose, slightly larger links would be required for some transformer ratings, to take full advantage of the transformers' overload capabilities.

118. Fused cutouts are used extensively for line sectionalizing in radial primary distribution. They are employed on main feeders to a limited extent, on 3-phase branches more often, and quite frequently on single-phase laterals. When properly coordinated, they protect the balance of the circuit against trouble in one section and often minimize damage on the section they isolate. But most overhead line faults are of a temporary nature, and blowing the line fuse prolongs the interruption. In some cases, the feeder circuit breaker can be set for a first trip fast enough to protect line fuses, followed by automatic reclosing for a delayed second trip which permits line fuses to blow on a permanent fault. In other cases, oil reclosers are installed at certain sectionalizing points to provide the fast initial trip followed by delayed subsequent trips to blow line fuses.

119. Location of fuses and reclosers on a primary circuit is determined by circuit arrangement, operating record, and the judgment of operating personnel. Single-phase laterals with extensive exposure to trees are usually fused where the lateral leaves the 3-phase main. Single-phase primary extensions on private property are often fused where the circuit leaves the public street. Overhead single- or 3-phase branches supplied by underground circuits are usually fused at the standpipe pole to protect the cable from faults on the overhead circuit.

120. Proper coordination of fuses is necessary where fuses are used in series. The maximum number of line-sectionalizing fuses permissibly employed in series varies with different companies from one to six. When fuses are properly coordinated, a fault on any part of a radial circuit will cause the fuse nearest the fault on the substation side to blow, thereby isolating the faulted section. No other fuses nearer the substation should blow or begin to melt if fuses are chosen correctly.

In determining sizes of fuses that will coordinate properly it is necessary to calculate maximum short-circuit current at the several points on the circuit where fuses are to be installed. Short-circuit current can be obtained by use of charts or approximated by addition of the impedances of various sections of circuit arithmetically and assuming supply voltage maintained constant at the substation bus. Values obtained by this approximate method are 10 to 20% greater than actual values. Maximum short-circuit current at the various points where fuses are to be installed having been determined, the next step is to choose the correct sizes of fuses that will coordinate with each other when a fault occurs.

Coordination was a principal objective in setting up EEI-NEMA Standard TDJ-110, later designated as USAS C37.43-1962. The T series comprises nine preferred ratings from 6 to 200 A, specified to coordinate between adjacent sizes to currents 24 times the smaller fuse rating. The K series is similar except that faster clearing limits coordination to 13 times the rating of the smaller fuse. Intermediate ratings were provided in both series for users with special problems; these were designated "nonpreferred."

121. Time-current fuse curves or fuse-coordination tables must be used to determine the correct sizes of fuses for coordination. A time-current curve shows the length of time required for a certain fuse to melt or clear a circuit for various currents passed through the fuse. Figure 16-16 shows typical time-current characteristics for the type-T (slow) series fuse links (General Electric). The curves are for maximum total clearing time.

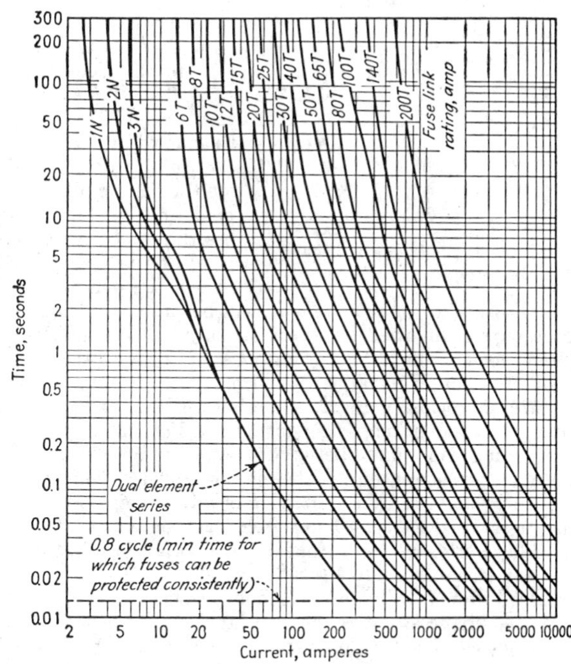

FIG. 16-16. Total clearing-time circuit characteristic curves, USA Standard type-T fuse links.

Some utilities obtain from the manufacturer two sets of curves, one the maximum total clearing time and the other the minimum melting time. From the second they plot a damaging-time curve, assuming 75% of melting time to be the threshold of damage. When these two sets of curves are superposed and fuses selected such that the maximum clearing of the smaller does not overlap the damage

Table 16-8. Fusing Schedule for Distribution Transformers, Using USA Standard Type-T Fuse Links, Ratings in Amperes

Transformer size, kVA*	2,400 V, 3-phase Δ	2,400 V, 1-phase; or 4,160 V, 3-phase Y	4,800 V, 3-phase Δ	4,800 V, 1-phase; or 8,330 V, 3-phase Y	7,200 V, 3-phase Δ	7,200 V, 1-phase; or 12,470 V, 3-phase Y
5	10	6	6	6 or 3†	6	6 or 2†
10	15	10	10	6	6	6
15	25	15	15	10	10	6
25	40	25	25	15	15	10
37½	65	40	25	25	25	15
50	65	40	40	25	25	15
75	100	65	65	40	40	25
100	140	100	65	40	65	40
167	200	140	100	65	100	65
250	...	200	200	100	140	100

* kVA per phase for 3-phase units or banks.
† Dual-element fuses desirable for 1, 2, and 3 A.

limit of the larger, the two fuses will coordinate properly; i.e., the smaller fuse will clear a fault before the larger fuse begins to melt.

122. Fuse-coordination tables that are easier to use than time-current curves have been prepared by various manufacturers. Table 16-9 is a coordination chart published by General Electric Company for its type-T fuse links when used in 50-, 100-, or 200-A expulsion-type cutouts. Dual-element links for small transformers are designated 1N, 2N, and 3N; such special links are also mentioned in the standards.

123. The coordination chart is used in the following manner: In Fig. 16-17 fuses are to be located at points 1, 2, 3, and 4 in series. Maximum short-circuit currents at these points have been calculated to be 110, 255, 1,900, and 2,750 A, respectively. Start with a 6T fuse at point 1, determined by the size of the transformer. The maximum short circuit is 110 A at point 1. Enter the coordination chart under column 1 for the 6T fuse, and go to the right horizontally to the 10T column. The 10T fuse will be protected adequately for the 110 A. Note that the 8T fuse will be protected for only 33 A and hence will not coordinate with the 6T transformer fuse. Having determined the 10T fuse for point 2, where the maximum short circuit is 255 A, the chart is entered under column 1 for the 10T fuse; and going across the table horizontally, the 15T fuse is selected, as it will be protected for 255 A. Thus the 15T fuse is chosen for point 3. In a similar manner the 40T fuse is selected for point 4. It will be protected adequately for 1,900 A. This same procedure is followed for every fusing point on the system. In some cases, the fuse size at a particular location must be made larger than the size necessary to coordinate properly in order to carry the normal load current.

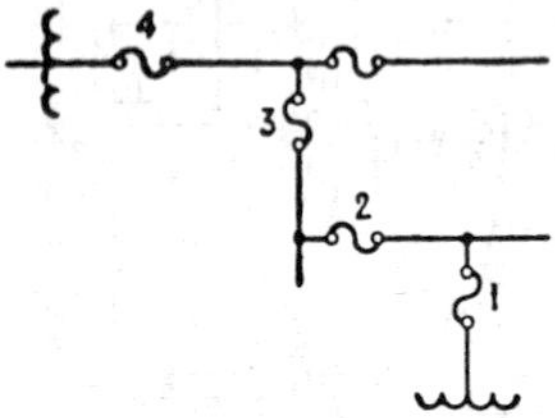

Fig. 16-17. Fuse coordination diagram.

124. If reclosing fuses, sometimes called "two- or three-shot repeater fuses," are used, the same principles as outlined in Par. **123** apply, except that a greater spread in fuse ratings is required between points 2 and 3 of Fig. 16-17 if 2 is a reclosing fuse and 3 is a single-shot cutout. The fuse at 3 must be sufficiently large to prevent its blowing or being damaged by the blowing of all fuses at 2 in quick succession. Repeater fuses are rarely used today. Their place has been taken by the smaller ratings of oil reclosers.

125. Reclosers have considerable ability to distinguish between temporary and permanent faults, unlike fuse links, which interrupt either type indiscriminately. Reclosers automatically test the line by successive operations, giving temporary faults repeated chances to clear or be cleared by subordinate protective devices. Should the fault not clear, the recloser recognizes it as a permanent fault and locks open. Single-phase reclosers usually have hydraulic control, but 3-phase reclosers may be either hydraulically or electronically controlled. Single-phase reclosers of the lower ratings are closed by stored energy in the springs of toggle mechanisms, but the larger ratings, as well as many of the 3-phase designs, use a coil for closing power. Most reclosers interrupt fault current with contacts in oil-filled chambers, but recent designs have been built around power-rated vacuum switches.

Reclosers operate according to curves published by the manufacturer. Figure 16-18 shows typical time-current characteristics for a single-phase unit which operates initially on the fast curve A, then transfers in response to its set sequence to delayed curve B or C. Each curve represents total time required to clear a fault, the fast curve A showing maximum values with all deviations negative, while both delayed curves show average values subject to +10%.

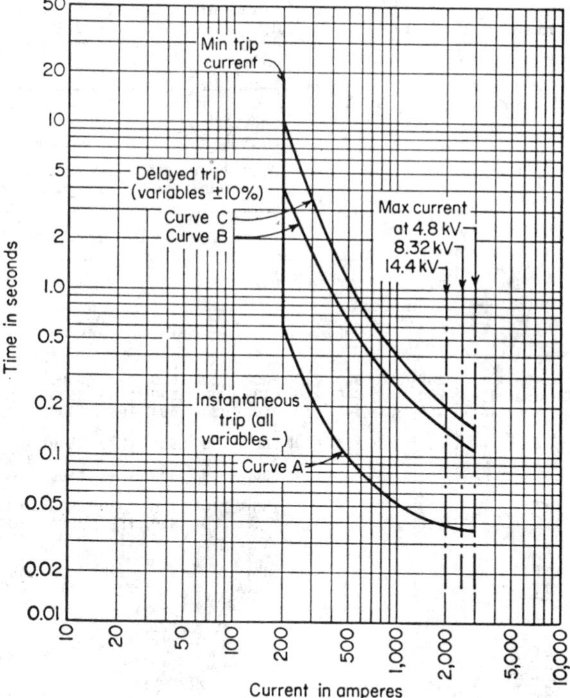

FIG. 16-18. Time-current curves for 100-A hydraulically controlled single-phase recloser (Line Material Industries).

A recloser can be set for up to four operations in sequence. It can follow one curve for all its operations, or it can be set to follow the fast curve for initial operations, then shift to a delayed curve for subsequent operations. The latter timing permits fast clearing of temporary faults before branch fuses blow but holds on long enough to blow branch fuses when permanent faults occur. Such a recloser also overrides current inrush when a circuit is reenergized after a long interruption.

126. Application of Reclosers. Recloser ratings include seven items: nominal voltage, maximum design voltage, impulse withstand voltage, frequency, continuous current, minimum trip current, and interrupting current. These values must be considered in applying reclosers, and time-current curves, operating sequences, and certain basic rules must be taken into account.

The type and voltage of the system, as well as the load current, available fault current, and X/R ratio at the application point, are factors in determining the required rating of the recloser. Inrush current must be considered in selecting the minimum trip level, normally 200% of the coil's continuous rating for hydraulically operated reclosers but determined by control components in electronic types.

An electronically controlled unit would be preferred where system conditions are changing, because control components can be adjusted during a short down period. Hydraulically controlled units must be untanked to change the current capability or operating characteristics.

Use of a 3-phase recloser instead of three single-phase units on a grounded Y-system guards against transformer burnouts and current back feeds. And accessories provide for special requirements, such as different time-current curves for a ground fault from those for a phase fault.

Four basic requirements are mandatory for a successful recloser application. Its interrupting rating must equal or exceed the available fault current. Its continuous current rating must equal or exceed the maximum line loading. The minimum trip

Table 16-9. Coordination Chart (Fuse Links in Series)

(Protective characteristics of General Electric USA Standard Type T and dual-element small fuse links when used in 50-, 100-, or 200-A expulsion-fuse cutouts and connected in series)

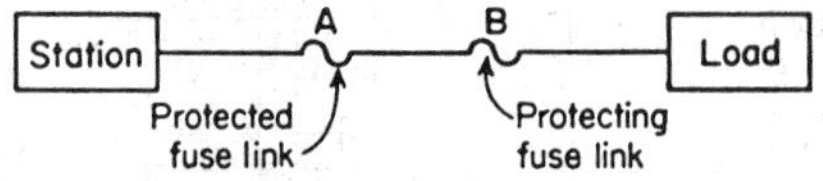

When the protecting fuse link is used in single-element fuse cutout

(1) Rating, A, of protecting link	Rating, A, of protected fuse link (A in diagram)														
	6T	8T	**10T**	12T	**15T**	20T	**25T**	30T	**40T**	50T	**65T**	80T	**100T**	**140T**	**200T**
	Maximum short-circuit rms A to which fuse link will be protected														
1N dual el.	**250**	395	**540**	710	**950**	1220	**1550**	1930	**2500**	3100	**3950**	4950	**6300**	**9600**	**15,000**
2N dual el.	**250**	395	**540**	710	**950**	1220	**1500**	1930	**2500**	3100	**3950**	4950	**6300**	**9600**	**15,000**
3N dual el.	**250**	395	**540**	710	**950**	1220	**1500**	1930	**2500**	3100	**3950**	4950	**6300**	**9600**	**15,000**
6T	...	33	**365**	650	**950**	1220	**1500**	1930	**2500**	3100	**3950**	4950	**6300**	**9600**	**15,000**
8T	...	...	125	480	850	1220	1500	1930	2500	3100	3950	4950	6300	9600	15,000
10T	...	...	...	74	**620**	1130	**1500**	1930	**2500**	3100	**3950**	4950	**6300**	**9600**	**15,000**
12T	...	...	...	...	135	770	1400	1930	2500	3100	3950	4950	6300	9600	15,000
15T	...	...	...	...	...	100	**880**	1750	**2500**	3100	**3950**	4950	**6300**	**9600**	**15,000**
20T	...	...	...	...	...	....	105	1150	2300	3100	3950	4950	6300	9600	15,000
25T	...	...	...	...	...	....	....	190	**1500**	3100	**3950**	4950	**6300**	**9600**	**15,000**
30T	...	...	...	...	...	....	....	....	115	1900	3950	4950	6300	9600	15,000
40T	...	...	...	...	...	....	....	....	....	310	**2350**	4950	**6300**	**9600**	**15,000**
50T	...	...	...	...	...	....	....	....	....	....	150	3400	6300	9600	15,000
65T	...	...	...	...	...	....	....	....	....	....	....	270	**4300**	**9600**	**15,000**
80T	...	...	...	...	...	....	....	....	....	....	....	....	660	9200	15,000
100T	...	...	...	...	...	....	....	....	....	....	....	....	....	**6000**	**15,000**
140T	...	...	...	...	...	....	....	....	....	....	....	....	....	....	**6,600**

NOTE: Boldface marks C37.43—1962 "preferred" ratings.

setting must sense fault current throughout the desired zone of protection. And the time-current curves must coordinate with those of protective devices on both sides of the recloser.

127. Coordination with Load-side Fuses. For good coordination with load-side fuses, a line recloser should be set for two operations on the *A* curve, followed by operations on one of the delayed curves. The first recloser operation should clear about 80% of temporary faults. The second operation should clear an additional 10%. The recloser then shifts to delayed tripping and, if the fault persists, permits current to flow long enough to melt the appropriate fuse link before the third operation occurs. Two rules apply:

1. For all possible values of fault current, the minimum melting time of the fuse must exceed the recloser's fast clearing time by a factor which reflects the number of fast operations and the interval between fast operations.

2. For all possible values of fault current, the maximum clearing time of the fuse must not exceed the maximum delayed clearing time of the recloser.

The coordination range is fixed by the above rules. The maximum current value is the intersection of the fuse's minimum melting curve with a reference curve determined as the product of the recloser's fast clearing curve and the factor reflecting the sequence of fast operations. The minimum current value is the intersection of the fuse's maximum clearing curve with the recloser's delayed clearing curve, or the minimum trip current if the curves do not intersect. A typical case is shown as Fig. 16-19.

128. Typical Example of Recloser-fuse Coordination. Figure 16-20 represents a 3-phase circuit to be protected by a 3-phase recloser OCR_1 at the substation. Available phase and ground-fault currents at application points and 3-phase ends are plotted, as are load currents at application points. Fuses at sectionalizing point *ABC*/29 and the recloser must be so coordinated that any fault beyond the fuses to *ABC*/30 and *ABC*/31 will be cleared by the recloser on its fast curve, if temporary, or by the fuses, if persistent. In addition, the recloser must interrupt any faults between *ABC*/27 and *ABC*/29.

The recloser selected for this application has an interrupting capability of 4,000 A at 14.4 kV, more than adequate for the 1,500 A available at the substation. Its 140-A continuous rating is adequate for the 135-A load current, and its 280-A minimum trip value is low enough to sense the lowest expected fault current, 340 A, in the protected zone.

Logical choice for fuse F_1 would be 40T, and a plot of its minimum melting and maximum clearing curves against the recloser's tripping characteristics (similar to Fig. 16-19) shows a maximum coordination limit of 1,250 A, proving that the recloser would protect the fuse with its fast operations. All delayed curves of the recloser are slower than the 40T fuse's maximum clearing curve; hence the minimum coordination limit becomes the 280-A minimum trip current of the recloser, and the recloser will hold in long enough on its first delayed operation to permit the 40T fuse to isolate the fault.

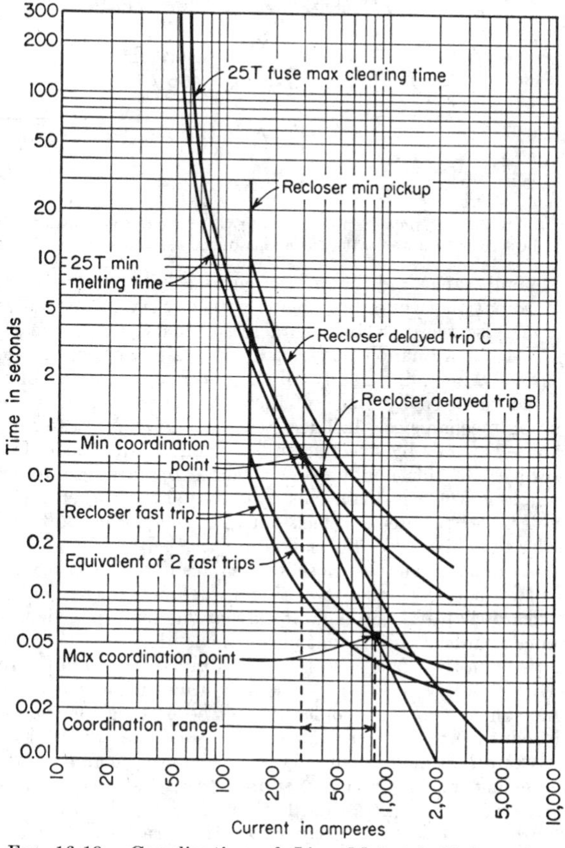

FIG. 16-19. Coordination of Line Material 70-A recloser with 25-T USA Standard fuse link.

Fuse group F_2, however, has 75 A of load current and must not be smaller that 65T, which, with inherent overload capability, should handle the load. The total clearing curve of this link fails to coordinate with the delayed *B* curve of the recloser. But the *C* curve gives coordination down to the 280-A minimum pickup of the recloser. The upper limit of coordination between the fast curve and the 65T link is 2,000 A; so the recloser, with two fast trips on the *A* curve followed by slow trips on the *C* curve, meets requirements for this case.

129. Coordination with source-side fuses is required between the recloser of Fig. 16-20 and fuses on the high side of the substation transformer. Time-current curves can be used to select fuse links for F_s, provided that the recloser's delayed-tripping curve *C* has its time increased by a factor reflecting the cumulative effect of its two fast and two delayed operations. In addition, the high-side fuse curves should be moved to the right to reflect the 3.7 turn ratio of the substation transformer. Such a plot for a 60ES high-side fuse shows coordination with the low-side recloser up to 2,000 A, or well above the 1,500-A maximum fault on the recloser.

130. Coordination of Reclosers with Circuit Breakers. When the substation has a circuit breaker for backup protection, its operation must be coordinated with that of all reclosers on the load side. The circuit breaker's control relays tend to integrate the recloser's total clearing time. Accordingly, the recloser's cumulative curve is superposed upon the circuit breaker's relay curve. If they do not intersect within the avail-

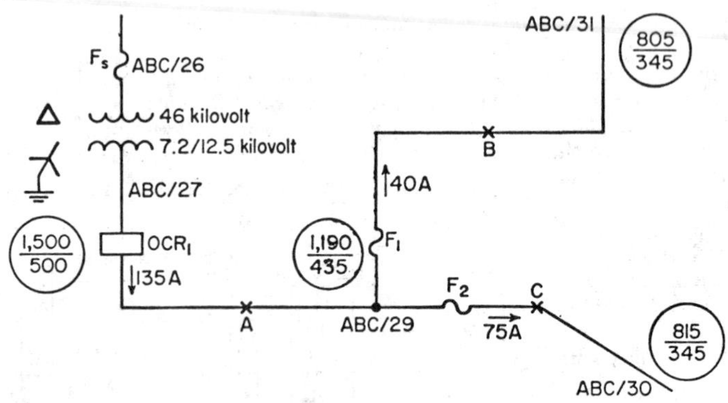

FIG. 16-20. Typical 3-phase circuit, illustrating coordinated protection with an oil recloser at the substation and with fuses on source side as well as load side.

able fault range and the recloser curve is below the relay curve, coordination is assured. Coordination is usually difficult with extremely inverse relays but is readily attained for inverse relays whose characteristic curves more nearly parallel those of reclosers.

131. Coordination between Reclosers in Series. Reclosers operated only by series trip coils, including nearly all single-phase and many of the smaller 3-phase reclosers, are coordinated by study of their time-current curves. As long as they have similar timing mechanisms, their time-current curves are essentially parallel and coordination can be attained by selecting suitable coils. Margins of less than 2 c should be avoided, however, as this might cause them to operate simultaneously. Limited cascading—completion of fast operations by the smaller recloser and then by the larger one before the smaller recloser can lock out and isolate the fault—sometimes can be tolerated in order to achieve coordination.

Example. Three reclosers of adjacent coil sizes (50, 70, and 100 A) are adjusted for two fast (A) and two extra-retarded (C) operations. Time-current characteristics indicate that at 1,000 A fault current the fast operations would be less than 2 c apart; hence a fault of this magnitude on the load side of the 50-A recloser could cause all three to perform their fast operations simultaneously. On the retarded cycles, the separation at 1,000 A is only 3 to 7 c; so simultaneous retarded operations might occur as well. One solution to this problem would be moving all reclosers farther out on the circuit to reduce the available fault current and take advantage of the greater time difference between the time-current curves at the lower current. But the preferred method would be setting the 50-A recloser set on the two-fast–two-retarded sequence, keeping the 70-A on two-fast–two-extra-retarded, and setting the 100-A for one fast and three extra-retarded. With these settings, all three reclosers would make their initial operations simultaneously but would achieve coordination on subsequent operations.

Reclosers with closing coils actuated by line voltage, and particularly those with electronic control, offer more alternatives in planning coordination. In addition, they can be provided with settings for ground faults different from those for phase-to-phase faults. Thus, coordination may be more complex to plan but is usually possible.

132. Sectionalizers. A sectionalizer resembles an oil circuit recloser but lacks the interrupting capability. Its operating controls are set to count current surges and open its contacts after the set number of surges while the recloser or reclosing circuit with which it coordinates is open. If the fault is temporary, both the sectionalizer and the recloser reset, ready for the next fault. If the fault is persistent, however, the recloser operates on its sequence but the sectionalizer isolates the fault before the recloser starts its final operation; thus lockout is avoided, and only that portion of the circuit beyond the sectionalizer is interrupted.

Application of sectionalizers requires that the backup recloser or reclosing circuit breaker be capable of sensing fault current to the end of the sectionalizer's zone of protection, that the minimum fault current be sufficient to actuate the sectionalizer, that the momentary and short-time ratings of the sectionalizer not be exceeded, and that the sectionalizer not be placed between two reclosers.

133. Lightning Protection. A major cause of outages on overhead distribution lines is the effect of lightning. The flow of power across an arc established by lightning causes fuses to melt and in some cases burns out transformers or other equipment.

The lightning arrester's function is to provide a path by which lightning readily finds its way to ground without flashing over insulation of line equipment and without permitting power current to follow the lightning discharge. The number of service interruptions resulting from lightning may be kept within reasonable limits if overhead portions of distribution lines are adequately protected by lightning arresters.

134. Arrester Density. Arresters should be placed as close as possible to apparatus they protect. In districts subject to frequent lightning storms, arresters should be located at each transformer pole and at each cable standpipe pole. Experience has shown that, in built-up districts of a city, overhead lines are effectively shielded from lightning by nearby buildings and arresters may be less frequently needed.

135. Arrester Grounds. The effectiveness of an arrester is dependent upon its having a ground connection with a resistance not greatly in excess of 50 Ω. In soils which permit it, this may be had by driving a galvanized-iron pipe or copper-clad steel rod 8 to 10 ft into the earth. In sandy soils it may be necessary to drive two or more grounds a few feet apart and to drive them deeper to secure the desired resistance.

Where neutral conductors of service drops are grounded to metallic water pipes, the ground terminal of the arrester may advantageously be interconnected with the neutral conductor of the secondary main. The dependability of grounds is further assured by interconnecting the neutral conductors of secondary mains of all adjacent transformers, thus forming a neutral bus grounded at many points.

136. Three-point protection of a transformer can be obtained by connecting the case of the transformer directly (or through an insulating gap) to the ground terminal of the arrester that is interconnected to the secondary neutral. Assuming the secondary neutral to have several water-pipe grounds, there is no need for a driven ground at the transformer. The three-point method limits voltage between high- and low-voltage windings and between high-voltage windings and core or tank to the discharge voltage of the lightning arrester and limits voltage between low-voltage winding and the core or tank to a few hundred volts. Figure 16-21 illustrates the three-point protection method.

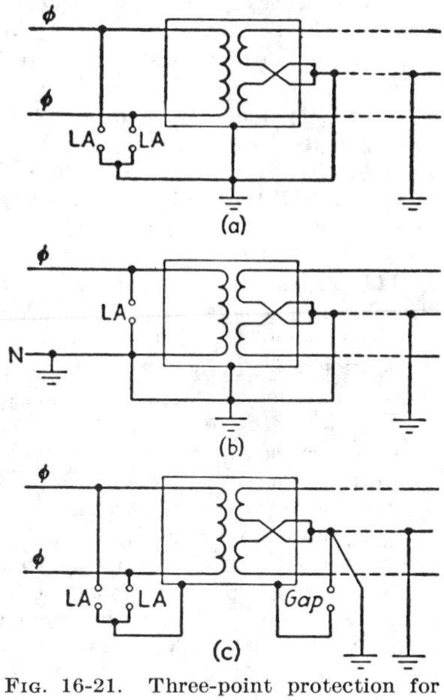

Fig. 16-21. Three-point protection for single-phase transformers. (*a*) Simplified. (*b*) With multigrounded primary neutral. (*c*) With insulated tank.

137. Lightning-arrester protection is not necessary on the primary neutral conductor of a transformer connected to multigrounded common neutral. A lightning arrester on the primary neutral of a single-phase transformer is necessary if the primary neutral is grounded at the station only.

138. Lightning Protection of Cables. Cables that operate at 2.4 kV and higher and are connected to exposed overhead lines usually require lightning arresters on cable-terminal poles or at junctions of cable and overhead line. It is important that the arrester ground be connected to the lead-cable sheath as close as possible to the pothead in addition to a driven- or secondary-neutral ground at the pole.

139. The valve-type lightning arrester is the device most generally used for protection of conventional distribution apparatus. Operating experience of several years has demonstrated its ability to provide a high degree of protection. Latest designs have the ability to discharge surge currents of high crest magnitude. Recent data obtained with the fulchronograph have shown that the valve-type arrester should also be capable of discharging surge currents of long duration.

DISTRIBUTION TRANSFORMERS

140. Distribution transformers convert electrical energy from primary voltages (2.4 to 34.5 kV) to utilization voltages (120 to 600). Momentary drops in lighting voltage caused by the starting current of motors often necessitate use of separate transformers where 3-phase motors 5 hp and larger must be served from radial circuits.

141. Standard Ratings of Single-phase Distribution Transformers. By agreement between users and manufacturers, certain features of line-transformer design have been standardized for sizes up to 500 kVA and for pressures up to 15,000 V. Capacities are 5, 10, 15, 25, 37½, 50, 75, 100, 167, 250, 333, and 500 kVA.

Voltage rating on primary windings are 2,400, 4,800, 7,200, 7,620, 12,000, 13,200, and sometimes as high as 34,500. On the secondary side windings are built for 3-wire operation at voltages of 120/240 or for 240/480. For certain uses a 600-V secondary winding is available. Bushings for both primary and secondary terminals are preferably of the stud type with provision for mechanical connection to line wires without soldering. They are located on the side of the case, except that primary bushings for 7,200 V and higher are cover-mounted. Supporting lugs are arranged to permit mounting either by bolting to the pole or by hanging on crossarms. Where necessary, provision is made for a grounding connection to the case or from the secondary neutral terminal to the case.

Similar standards have been promulgated by EEI and NEMA for 3-phase pole-type transformers up to 500 kVA.

142. Electrical characteristics typical of single- and 3-phase transformers of the 7,200/12,470-V class are given in Table 16-10. All values are in percent and are approximately correct for transformers rated 2,400 and 4,800 V. The column that gives percent voltage drop through the transformer with a 50% power-factor load is useful in calculating momentary drop caused by starting a motor. For example, a 5-hp 230-V single-phase motor draws 100 A locked-rotor current with a power factor close to 50%. What voltage drop does this cause in a 25-kVA single-phase transformer? From Table 16-10 the value of 11.5% drop is obtained for full load at 50% power factor. Motor load = 0.23 × 100 = 23 kVA. Momentary voltage drop = $(^{23}/_{25})$ 1.5% = 1.4%.

143. Transformers are installed on poles in the following ways: Transformers 167 kVA and smaller are bolted directly to the pole or hung on crossarms by means of steel hangers attached securely to the transformer. Banks of three single-phase transformers are hung side by side on heavy double arms, usually located low on the pole, or on a "cluster" bracket which spaces them around the pole.

Transformers 250 to 500 kVA in size are usually installed on a wooden- or steel-pole step and are secured to crossarms by means of steel hangers.

Three or more transformers 167 kVA and larger are installed on a platform supported by two poles set 10 to 15 ft apart. The transformer-platform structure is often placed on the customer's premises to reduce the distance that secondaries must be run and to avoid pole congestion on public thoroughfares.

144. Transformers are installed in street vaults, in manholes, on pads at ground level, or within buildings when underground construction is employed. They must be of the "subway" type when subject to occasional submersion. A recent trend for underground residential distribution is to put them in prefabricated enclosures ventilated via top gratings.

When installed within buildings where the possibility of submersion is remote, the overhead or inside types of transformer and cutout are used. Transformer vaults within a building are of fireproof construction, except when transformers are askarel-filled or dry-type.

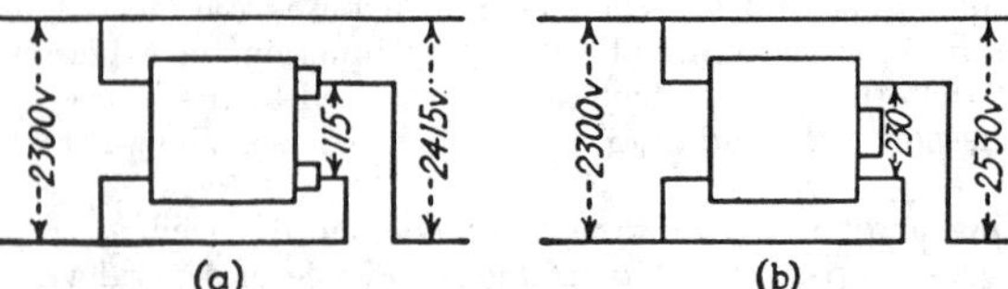

Fig. 16-22. Booster-transformer connections.

145. Booster Transformers. Where it is desired to raise the voltage by a fixed percentage, as when line drop is excessive, this may be accomplished by a transformer used as a booster. This is a transformer so connected that the secondary is in series and in phase with

Table 16-10. Electrical Characteristics of Single-phase and 3-phase 60-c/s Distribution Transformers*

Single-phase transformers—voltage rating 7,200 to 120/240

Size, kVA	Impedance, % $Z/\underline{\phi°}$	% voltage drop through transformer with full-load current at			% copper loss	% no-load loss
		Unity pf	80% pf	50% pf		
5	2.2/17.2°	2.1	2.1	1.6	2.10	0.80
10	1.6/28.3°	1.5	1.6	1.4	1.41	0.67
15	1.6/31.2°	1.4	1.6	1.4	1.37	0.49
25	1.6/38.2°	1.3	1.6	1.5	1.26	0.42
37½	1.5/42.2°	1.2	1.5	1.4	1.15	0.43
50	1.5/42.2°	1.1	1.5	1.4	1.11	0.41
75	1.8/47.7°	1.1	1.5	1.5	1.03	0.37
100	1.5/57.3°	1.0	1.5	1.8	0.98	0.35
167	1.9/59.0°	1.0	1.8	2.4	0.98	0.28
250	2.7/71.0°	0.9	2.2	2.7	0.88	0.28
333	3.0/74.6°	0.9	2.4	2.9	0.80	0.25
500	2.8/74.3°	0.8	2.3	2.6	0.73	0.22

Three-phase transformer—voltage rating: 12,470 to 208Y/120

Size, kVA	Impedance, % $Z/\underline{\phi°}$	% voltage drop through transformer with full-load current at			% copper loss	% no-load loss
		Unity pf	80% pf	50% pf		
15	2.3/13.5°	2.2	2.1	1.6	2.24	0.82
30	1.7/17.7°	1.6	1.6	1.3	1.62	0.60
45	1.6/20.3°	1.5	1.6	1.2	1.48	0.49
75	1.4/17.2°	1.3	1.4	1.1	1.35	0.42
112½	1.4/31.0°	1.2	1.4	1.2	1.22	0.44
150	1.8/49.7°	1.2	1.8	1.8	1.18	0.41
225	1.6/46.5°	1.1	1.6	1.6	1.10	0.37
300	1.4/44.4°	1.0	1.4	1.4	1.04	0.34
500	2.2/60°	1.1	2.0	2.2	1.05	0.29

Three-phase transformer—voltage rating: 12,470 to 208T/120†

Size, kVA	Impedance, % $Z/\underline{\phi°}$	Unity pf	80% pf	50% pf	% copper loss	% no-load loss
15	2.4/33.4°	2.0	2.4	2.1	2.04	0.82
30	2.3/49.4°	1.6	2.3	2.3	1.48	0.67
45	1.9/42.6°	1.5	1.9	1.8	1.38	0.55
75	1.9/46.4°	1.3	1.9	1.8	1.26	0.49
112½	2.0/53.3°	1.2	2.0	2.0	1.15	0.43
150	1.7/49.2°	1.2	1.7	1.7	1.11	0.41
225	1.8/56.3°	1.1	1.8	1.8	1.04	0.37
300	1.8/56.3°	1.0	1.8	1.8	0.98	0.35
500	3.0/70.1°	1.0	2.5	3.0	0.99	0.29

* Courtesy of Westinghouse Electric Corporation.

† Many 3-phase transformers have two cores and coils connected T–T instead of the conventional three cores and coils.

the main line, and thus the primary voltage is raised by the amount of the secondary voltage, 5% in (*a*) and 10% in (*b*) of Fig. 16-22, where the ratios of transformation are 20:1 and 10:1, respectively.

In installing boosters care must be taken not to open the primary of the booster while the secondary is carrying the line current. Dangerous voltages would be induced in the primary coils in such a case. The safest way of connecting or disconnecting a booster is to have the main line open.

146. The use of boosters in a Δ-connected 3-phase system is not so simple as in single-phase circuits. The booster secondary is looped into the line, and voltage is taken for the primary from an adjoining phase, as in Fig. 16-23. Insertion of a booster on one phase affects the voltage on two phases, as shown diagrammatically in Fig. 16-24. The effect of a booster in each phase is shown in Fig. 16-25. Three boosters are required, therefore, to balance a 3-phase 3-wire circuit.

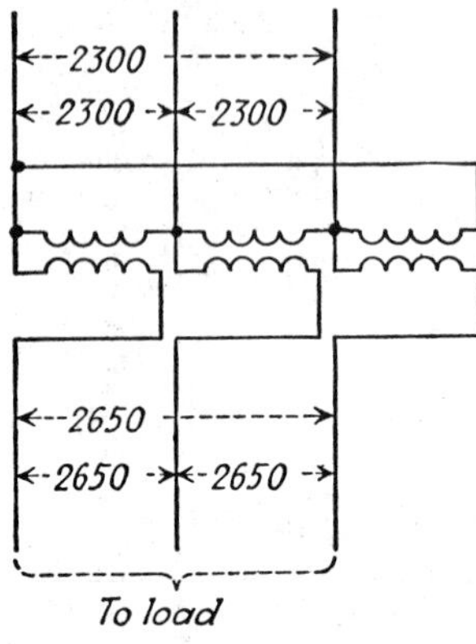

Fig. 16-23. The connection of boosters in 3-phase circuit.

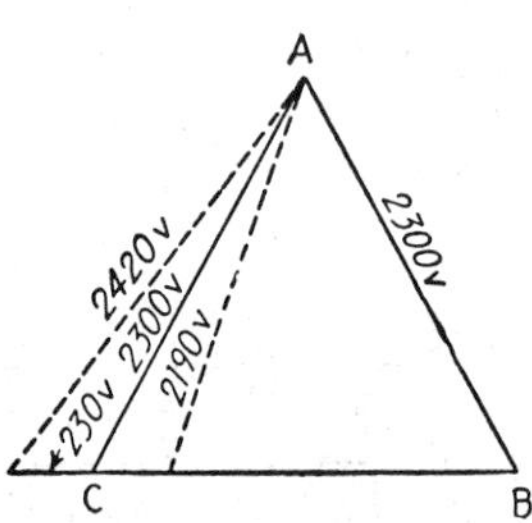

Fig. 16-24. The effect of a booster in one phase of a delta 3-phase circuit.

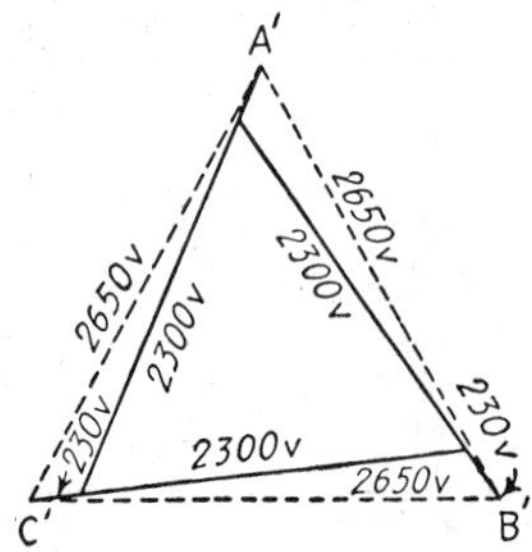

Fig. 16-25. The effect of a booster in each phase of a delta 3-phase circuit.

147. Automatic Boosters or Pole-mounted Regulators. Fixed boosters are seldom used today, except for occasional 120/208-V customers whose motors demand a full 220 V to handle their loads. For voltage boost on primary circuits, their place was first taken by automatic boosters which contactors cut in and out of service under control of a voltage relay. Later improvements provided two steps, then four steps, until today 16- or 32-step regulators built for pole mounting cover the customary ±10% range. Contact-making voltmeters and line-drop compensators give them essentially the same characteristics as the larger station regulators.

148. Capacitors are extensively used for voltage regulation. The usual method is to install shunt capacitors, either fixed or switched, well out in the load area. But where voltage fluctuates, series capacitors may be used to neutralize part or all of the inductive reactance of the distribution system, thus reducing the voltage drop.

149. Types of 3-phase Transformation. In transforming from generating voltage to higher voltage, or vice versa, it is desirable that use of Y-connection on both sides of the transformer be avoided. The Y-Y-connection is favorable to development of higher-harmonic currents, which sometimes cause inductive interference with communication circuits unless

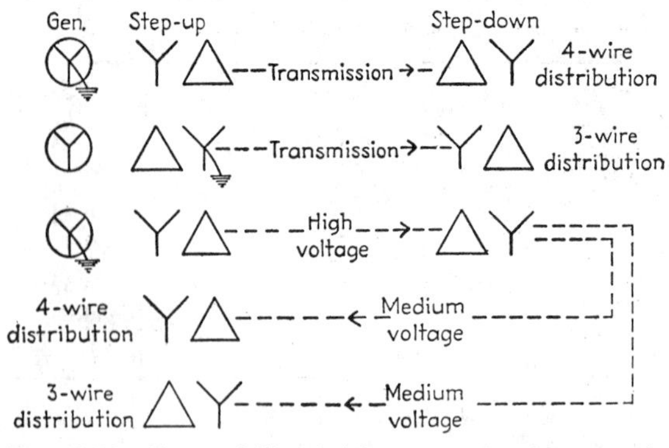

Fig. 16-26. Types of Y and delta connection for transmission lines.

a Δ-connected tertiary winding is provided on the same core. It is general practice, therefore, to use Y-Δ-connection for the step-up transformers, if the generator neutral is grounded, and Δ-Y in stepping down from transmission to a 4-wire 3-phase distribution system. When the generator neutral is not grounded, step-up transformers may be Δ-Y, giving the benefits of a grounded neutral on the transmission system (see Fig. 16-26).

The grounded neutral must be derived through separate Y-connected grounding transformers when transmission is Δ-connected on the high side of the step-up transformers, if the system is of such extent as to make grounding of the neutral advisable.

Interconnections between two large systems sometimes require Y-Y connection. In such cases, a Δ-connected tertiary winding is added to the transformer to damp out harmonic currents. Sometimes the tertiary winding can be used to supply local distribution (see also Sec. **11**).

Open-Δ-connection enables small power customers to receive 3-phase service from two transformers connected to a 3-phase circuit, thus reducing the investment in transformers.

Open-Δ from a 3-wire system is the usual Δ-connection with one transformer omitted. The connection from a 4-wire system is shown in Fig. 16-27, 2-phase wires and neutral being used on the primary side of the transformers. Current in each of two single-phase transformers connected in open-Δ is 73% greater than in each of three transformers connected in closed Δ.

150. The 2-phase 240-V service from a 3-phase 4-wire system may be secured with three transformers, connected as shown in Fig. 16-28. The unit at the left has a ratio of 10:1 and is connected from phase to neutral. The other two have ratios of 9:1, with their

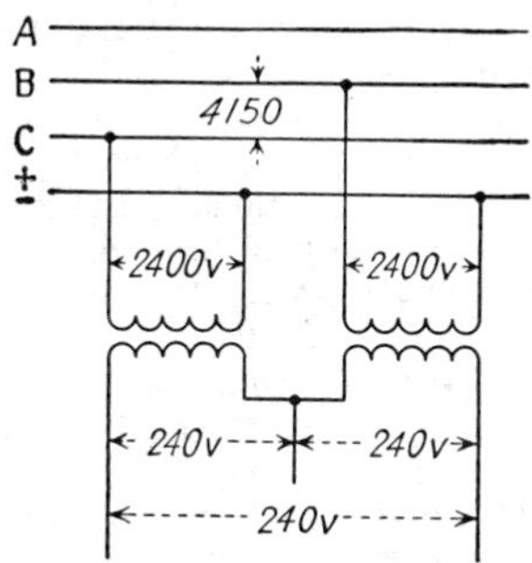

FIG. 16-27. Open-delta connection from 4-wire, 3-phase system.

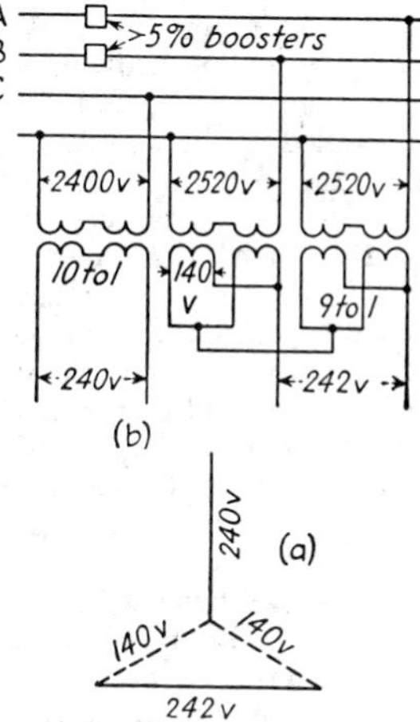

FIG. 16-28. Two-phase service from 4-wire, 3-phase with three transformers.

secondary coils in multiple and arranged as two limbs of a Y-connection to give 240 V across the outer wires. The 3-phase system is therefore unbalanced by this arrangement, since half the energy is taken from one phase. Capacities of the transformers should be selected accordingly.

The **Scott connection** gives an accurate transformation but requires one of the transformers to have an 86.6% tap.

SECONDARY RADIAL DISTRIBUTION

151. Secondary mains operate at utilization voltage and serve as the local distributing main. In radial systems secondary mains that supply general lighting and small power are usually separate from mains that supply 3-phase power because of the dip in voltage caused by starting motors. This dip in voltage, if sufficiently large, causes an objectionable lamp flicker. Single-phase and 3-phase systems are usually supplied by separate transformer installations, except in commercial and industrial areas.

152. Secondary mains supplying general lighting and small power are usually single-phase 3-wire mains operating at 120 V line to neutral and 240 V line to line.

Incandescent lamps, fans, heating devices, small fractional-horsepower motors, and other appliances rated 115 V are supplied from the 120-V line and neutral. Electric ranges, larger single-phase motors up to 5 hp, and large appliances rated 220 or 230 V are supplied 240 V.

153. Secondary mains supplying 3-phase power are commonly operated 3-wire 240-V. The 3-phase mains are on the same poles or in the same duct line (but in separate ducts) with single-phase lighting mains. Separate single-phase and 3-phase services are extended to customers who require both types of service.

In large commercial and industrial installations power is often delivered at 480 V to effect an economy in conductor investment.

154. European practice is to supply 220 V to lighting and appliances. This effects a saving in distribution and interior wiring but results in less efficient incandescent lamps and other small appliances. Recent large commercial buildings and factories in America have been served at 265/460 or 277/480 V because most permanent lighting is fluorescent, which operates efficiently at 265 V, and 460 V is well suited for the numerous 3-phase motors. Such installations have small dry-type autotransformers to supply 120 V for portable lights and tools and for business machines; these autotransformers are located near the 120-V loads and supplied from the 265-V system.

155. Fractional-horsepower motors up to about ¾ hp are regularly supplied by single-phase 120-V mains. Industry committees, sparked by sudden acceptance of home air conditioning, several years ago agreed to permit starting currents not to exceed 46 A for manually started 120-V motors. Special design enabled motors up to ¾ hp to meet this limitation. Frequent starting is more serious in human reaction to its light flicker, and the same agreement allowed only 23 A to start automatically controlled devices. Larger motors up to 5 hp are usually served at 240 V, although 3- and 5-hp motors may require extra care in distribution design to avoid troublesome flicker. Motors larger than 7½ hp are usually connected 3-phase.

156. Light and Power from One Secondary Main. In a **radial** system, 3-phase service is usually supplied from a separate secondary main if voltage is affected by elevator motors or other intermittently used load. If separation of light and power service is not necessary, the nature of the connection may depend upon the relative size of light and power loads. When power load is predominant, lighting load may be served by providing additional capacity in one of the transformers and bringing in a neutral from it for the lighting service. The neutral for lighting service is sometimes derived from a transformer connected to one phase of 240- or 480-V power circuits giving 120/240 V for lighting. This is the usual procedure where power is served at 480 V. When the lighting load is predominant, service is often provided at 120/208 Y, 4-wire, which requires a three-element meter.

157. Transformer and Secondary-main Economy, Overhead Distribution. Several independent studies have been made to determine the proper combination of transformer and secondary main that provides satisfactory voltage regulation and costs a minimum per kVA of load served. All these studies indicate that, for 120/240-V single-phase distribution, overhead secondary mains should be three No. 4 or three No. 2 copper or the equivalent in aluminum (No. 2 or No. 0), the latter being preferred when large appliances are to be served.

158. Permissible length of the three No. 0 aluminum secondary mains depends on the load density. On the assumption of evenly distributed loads and 2½% drop in the mains, for 7½ kW/1,000 ft the permissible length is 750 ft; for 15 kW/1,000 ft, 520 ft; for 22½ kW/1,000 ft, 420 ft; and for 30 kW/1,000 ft, 350 ft. Widespread use of ranges and motor-driven appliances establishes an additional limit for flicker at 300 to 400 ft.

159. Transformer size should be such that the initial peak load is between 75 and 100% of rated capacity. In medium-load densities 25- and 50-kVA transformers will fulfill this requirement. Transformers should be allowed to remain in service until their winter peak load reaches at least 150% of rated capacity. When this occurs the "hot-spot" winding temperature is approaching 110°C—the maximum safe temperature.

160. Load growth should be taken care of by *installing additional transformers* or by *increasing the size* of the existing transformers where secondary-main regulation permits. In general overhead distribution, the 37½- or 50-kVA is the largest size that should be installed except where a heavy "spot" load requires a larger transformer or where the mains feed three or four ways from the transformer.

161. The three No. 0 aluminum single-phase secondary mains should not be replaced by larger conductors to improve secondary-main regulation. Additional transformers should be installed and the ends of existing mains transferred to the new transformers.

162. Underground systems, although more costly than overhead systems, should be designed initially with capacity for growth. The design engineer where possible should use computers and a variety of computer programs now available to optimize an underground design.

163. Sizes of Transformers and Mains. Results of an economic study to determine the proper combination of transformers and mains for 120/240-V single-phase underground distribution for load densities from 12½ to 200 kVA/1,000 ft are shown in Table 16-11. Values are based on three-way distribution from the transformer, evenly distributed loads, and regulation (transformers plus secondary mains) of 4%.

164. Local load conditions often require that the transformer vault be located near a large spot load. In order to supply load in the area and maintain satisfactory voltage the mains must be larger and longer than the values in Table 16-11.

Table 16-11. Economical Initial Design, $^{120}\!/_{240}$, Single-phase, Underground Distribution

	Load density, kva per 1,000 ft (97% power factor)					
	12½	25	50	75	100	200
Transformer size, kva	25	50	75	75	150	2–100
Initial load, %	85	88	73	81	73	66
Conductor size	No. 4	No. 1/0	No. 1/0	No. 1/0	No. 4/0	No. 4/0
Permissible circuit length, ft	530	550	350	275	350	200

165. Load densities in underground distribution systems should, if possible, be at least 50 kVA/1,000 ft.

166. Load growth in underground secondary distribution should be handled by increasing transformer capacity in its present location, provided that drop in secondary mains does not exceed permissible limits. If the permissible drop has been exceeded, mains should be increased in size or a transformer installed in a new location and the ends of existing secondary mains transferred to it. Both methods are expensive.

167. Subway-type transformers should not be replaced or relieved of load until the calculated hot-spot winding temperature exceeds 110°C, provided, of course, that voltage at the ends of mains is satisfactory. To calculate hot-spot winding temperature, the maximum load and top-oil (or case) temperatures must be measured. Maximum case temperature has been found to be within 3°C of top-oil temperature. It is assumed in making the calculation that the difference between hot-spot-winding and top-oil temperature is 20°C at full load and that this difference varies as the square of load. This is a conservative assumption. For example, assume maximum case temperature 67°C when 130% load is on the transformer.

$$\text{Calculated winding hot spot} = 67°\text{C} + 3°\text{C} + 20°\text{C}(1.30)^2 = 114°\text{C}$$

Fans to supplement natural air movement have been used to boost safe capability of vault transformers.

168. Any 3-phase 220-V motors up to 10 or 15 hp located 300 to 500 ft from the single-phase transformer installation can be economically served underground by installing a second single-phase transformer in open-Δ with the existing single-phase lighting transformer and installing one additional conductor, thereby changing the 3-wire lighting main into a 4-wire light-and-power main. The new transformer and the lighting transformer each carry approximately 60% of the 3-phase kVA. The transformer on the "leading" phase should be the lighting transformer, because motor current lags the voltage of this transformer by the motor power-factor angle plus 30°. Resultant current in the lighting transformer is minimum when it is connected to the "leading" phase. A 15-hp motor, 350 ft from the lighting transformer, would cause a dip in lighting voltage of 6% when it is started, assuming the lighting mains to be No. 4/0 copper and the lighting transformer to be 50 kVA. This 6% dip would not be objectionable if it happened only a few times per day.

Table 16-12. Voltage Regulation of 600-V Aluminum Cables in Underground Secondary Distribution, kVA-Ft* per 1% Voltage Drop

No. and size of aluminum cables	120/240 V, 1-phase, 3-wire						No. and size of aluminum cables	120/208 V, 3-phase, 4-wire					
	Balanced loads, pf			10% unbalanced loads,† pf				Balanced loads, pf			10% unbalanced loads,‡ pf		
	97%	80%	50%	97%	80%	50%		97%	80%	50%	97%	80%	50%
3 No. 4	680	780	1,140	600	690	1,000	4 No. 4	1,020	1,180	1,710	880	1,000	1,460
3 No. 2	1,070	1,200	1,670	940	1,050	1,460	4 No. 2	1,600	1,800	2,510	1,380	1,520	2,140
3 No. 0	1,660	1,810	2,420	1,460	1,580	2,110	4 No. 0	2,500	2,720	3,630	2,150	2,300	3,090
2 2/0 & No. 2 ±	2,060	2,200	2,850	1,720	1,830	2,370	3 2/0 & No. 2 ±	3,100	3,300	4,280	2,520	2,640	3,450
2 4/0 & No. 0 ±	3,170	3,210	3,890	2,640	2,670	3,240	3 4/0 & No. 0 ±	4,750	4,820	5,830	3,860	3,850	4,700
2 350-M cmil, & 2/0 ±	4,860	4,560	5,050	4,050	3,800	4,210	3 350-M cmil & 2/0 ±	7,290	6,830	7,580	5,930	5,740	6,110
2 500-M cmil & 4/0 ±	6,490	5,800	6,100	5,410	4,830	5,080	3 500-M cmil & 4/0 ±	9,740	8,790	9,150	7,910	6,950	7,380
2 1,000-M cmil & 500-M cmil ±	10,400	8,200	7,780	8,660	6,840	6,490	3 1,000-M cmil & 500-M cmil ±	15,590	12,310	11,670	12,680	9,840	9,410

No. and size of aluminum cables	240 V, 3-phase, balanced loads, pf			480 V, 3-phase, balanced loads, pf			600 V, 3-phase, balanced loads, pf		
	97%	80%	50%	97%	80%	50%	97%	80%	50%
3 No. 4	1,360	1,570	2,280	5,450	6,270	9,120	8,550	9,840	14,290
3 No. 2	2,140	2,400	3,350	8,560	9,600	13,400	13,410	15,040	20,980
3 No. 0	3,330	3,620	4,830	13,310	14,490	19,340	20,880	22,730	30,290
3 2/0	4,130	4,390	5,700	16,510	17,480	22,810	25,860	27,600	35,720
3 4/0	6,330	6,420	7,770	25,340	25,690	30,090	39,690	40,230	48,680
3 350-M cmil	9,720	9,110	10,110	38,890	36,450	40,430	60,910	57,120	63,240
3 500-M cmil	12,980	11,590	12,200	51,930	46,380	48,810	81,320	72,540	76,380
3 1,000-M cmil	20,790	16,410	15,570	83,180	65,630	62,270	130,210	102,850	97,650

* kVA-ft = load in kVA × one-way distance in feet (never go-and-return distance).

† 110% in leg 1, 90% in leg 2, half of neutral current assumed to flow in earth and water pipes; regulation calculated between leg 1 and neutral.

‡ 110% in phase 1, 100% in phase 2, 90% in phase 3, half of neutral current assumed to flow in earth and water pipes; regulation calculated between phase 1 and neutral.

169. Voltage-regulation Tables. Table 16-12 gives the kVA-feet per 1% voltage drop for the several sizes of 600-V single-conductor cable used in underground secondary distribution for power factors of 97, 80, and 50%. Table 16-13 gives the same information for several sizes of overhead aluminum conductor on racks. The tables can be used to determine voltage drop quickly on any secondary circuit if load, circuit length, and size are known. If the voltage of the circuit under consideration is different from voltages in the tables, kVA-feet per 1% in the table should be multiplied by the square of the ratio of circuit voltage to the voltage in the table. If the power factor is significantly different from 97, 80, or 50%, values can be interpolated that are reasonably accurate. kVA-feet per 1% under the 50% power-factor column can be used in calculating drop caused by starting current of motors that have power factors in the neighborhood of 50%.

Aluminum and ACSR have gained extensive use in secondaries and services. Hence

Tables 16-12 and 16-13 are shown for aluminum conductors. Values for copper can still be determined with satisfactory accuracy by using Tables 16-12 and 16-13 for a conductor of equivalent resistance. *Example.* Use No. 2 aluminum values for No. 4 copper.

Table 16-13. Voltage Regulation of Aluminum Conductors in Overhead Secondary Distribution, kVA-Ft* per 1% Voltage Drop

No. and size of aluminum conductors	Spacing, in	120 V, 1-phase, 2-wire† pf			No. and size of aluminum conductors	Spacing, 240 V, in	120/240 V, 1-phase, 3-wire: Balanced loads, pf			120/240 V, 1-phase, 3-wire: 10% unbal. loads,‡ pf		
		97%	80%	50%			97%	80%	50%	97%	80%	50%
2 No. 4 ACSR	8	270	260	340	3 No. 4 ACSR	8	720	710	920	630	620	800
2 No. 2 ACSR	8	370	380	460	3 No. 2 ACSR	8	1,000	1,010	1,230	870	890	1,080
2 No. 0 ACSR	8	560	530	590	3 No. 0 ACSR	8	1,500	1,420	1,570	1,320	1,240	1,380
					2 3/0 Al & 1/0 ACSR ±	8	2,320	2,080	2,230	1,940	1,730	1,890
					2 4/0 Al & 1/0 ACSR ±	8	2,840	2,430	2,500	2,270	1,940	2,000
					2 336-M cmil Al & 3/0 Al ±	8	4,090	3,200	3,020	3,270	2,560	2,410
					2 556-M cmil Al & 297-M cmil ±	8	5,880	3,590	3,620	4,700	2,870	2,890
					2 795-M cmil Al & 397-M cmil ±	8	7,440	4,800	4,040	5,950	3,840	3,230

No. and size of aluminum conductors	Equiv. spacing, in	240 V, 3-phase, balanced loads, pf			480 V, 3-phase, balanced loads, pf			600 V, 3-phase, balanced loads, pf		
		97%	80%	50%	97%	80%	50%	97%	80%	50%
3 No. 4 ACSR	10	1,310	1,400	1,570	5,250	5,600	7,300	8,200	8,720	11,350
3 No. 2 ACSR	10	1,980	2,010	2,310	7,940	8,040	9,240	12,400	12,550	14,430
3 No. 0 ACSR	10	2,990	2,790	3,060	11,950	11,180	12,250	18,700	17,400	19,180
3 3/0 ACSR	10	4,400	3,770	3,800	17,600	15,100	15,200	27,500	23,600	23,700
3 3/0 Al	10	4,590	4,100	4,300	18,350	16,400	17,200	28,550	25,650	26,800
3 4/0 Al	10	5,600	4,760	4,850	22,400	19,050	19,400	35,000	29,800	30,200
3 336-M cmil Al	10	8,050	6,200	5,800	32,200	24,800	23,200	50,300	38,800	36,250
3 556-M cmil Al	10	11,420	7,870	6,810	45,700	31,500	27,250	71,600	49,400	42,600
3 795-M cmil Al	10	14,250	8,750	7,620	57,000	35,000	30,500	89,800	54,700	47,600

* kVA-ft = load in kVA × one-way distance in feet (never go-and-return distance).
† Half of neutral current assumed to flow in earth and water pipes.
‡ 110% in leg 1, 90% in leg 2, half of neutral current assumed to flow in earth and water pipes; regulation calculated between leg 1 and neutral.

BANKING OF DISTRIBUTION TRANSFORMERS

170. Banking. Tying together the secondary mains of adjacent transformers supplied by the same primary feeder is known as **banking**. The practice of banking is usually applied to the secondaries of single-phase transformers, and all transformers in a bank must be supplied from the same phase of the primary circuit. Banked distribution transformers differ from the low-voltage a-c network in that one circuit supplies all transformers where secondaries are banked together, whereas different circuits supply adjacent transformers in an a-c low-voltage network. A survey by *Electrical World* in 1954 revealed that 53% of 90 utilities queried either used banking or included it in future plans. However, only a few companies operated a large percentage of their transformers banked.

171. Advantages claimed for banking compared with secondary radial distribution are (1) reduction in lamp flicker caused by starting motors, (2) less transformer capacity required because of greater load diversity among a larger group of customers, (3) better average voltage along the secondary, and (4) greater flexibility for load growth.

172. Secondaries may be banked in three ways: straight line, loop, and grid. The straight-line arrangement is illustrated in Fig. 16-29. It is the easiest to establish and requires the shortest length of secondary copper, but it does not realize the full advantages of banking because the secondaries at the two ends are radial. Voltage flicker is not reduced on these radial ends as on the loop or grid layout, and more transformer capacity may be required to prevent overloading when an end transformer is out of service.

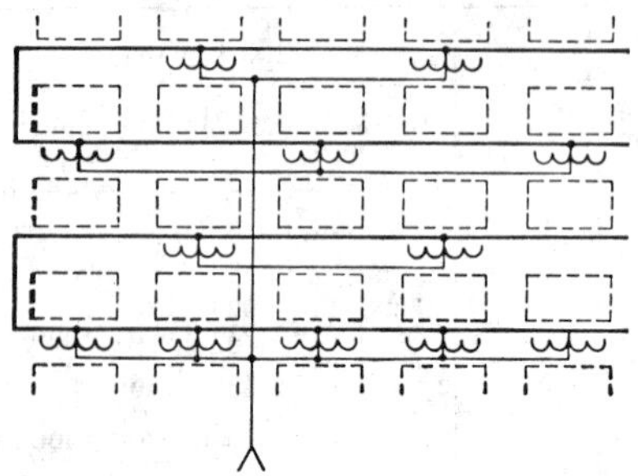

Fig. 16-29. Straight-line type of secondary bank.

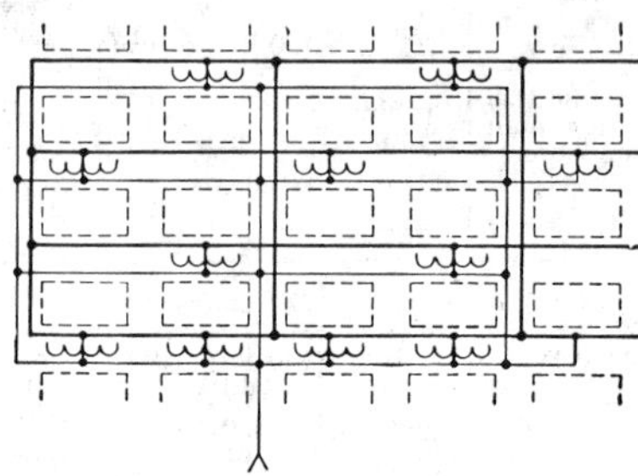

Fig. 16-30. Grid type of secondary bank.

173. The loop arrangement, accomplished by tying the open ends of Fig. 16-29, is a most desirable form for general use. It requires 10 to 15% more secondary copper than the line type but provides the maximum service reliability obtainable with banking.

174. The grid arrangement, shown in Fig. 16-30, requires 15 to 20% greater length of secondary conductor than the loop but permits use of smaller conductors and less transformer capacity. The grid produces heavy fault currents and burns clear most secondary faults quickly.

175. The number of transformers in a bank should be not less than 3, and the usual number banked together is 6 to 10. Various factors, including the necessity of transferring banks from one circuit to another in emergency and considerations of local topography, make it desirable to limit transformers in a line bank to about 10 and on a grid system to about 20.

176. All transformers in a bank are usually about the same size, and satisfactory loading is obtained by locating them properly. The largest transformer rating should exceed the smallest by no more than 67%, in order to minimize transformer capacity and the danger of burning up a small transformer when a large one is out of service.

177. Primary supply to banked distribution transformers usually consists of radial single-phase laterals. A looped single-phase lateral, which permits all transformers to remain energized when a fault burns the primary loop in two, has been used to good advantage to supply banks of 3 to 12 transformers.

178. Methods of Protecting Banked Distribution Transformers. Most companies that bank transformers consider the possibility of cascading and use protective devices to prevent it. By cascading is meant disconnection of some or all of the good transformers in a bank following a transformer or secondary fault. In a well-designed system, cascading occurs *infrequently* for any method of protection used. But the possibility of cascading varies widely with the type of protection chosen.

179. The most popular form of protection is shown in Fig. 16-31. The chance of cascading is practically nil with this arrangement. Sectionalizing fuses are chosen small enough to blow in event of a nonclearing secondary or transformer fault. This small sectionalizing fuse limits the amount of load that can be transmitted over the secondary tie and also increases the chance of having fuses open unnecessarily. To effect proper coordination of fuse blowings when faults occur, the sectionalizing fuse must be larger than the largest service-entrance fuse

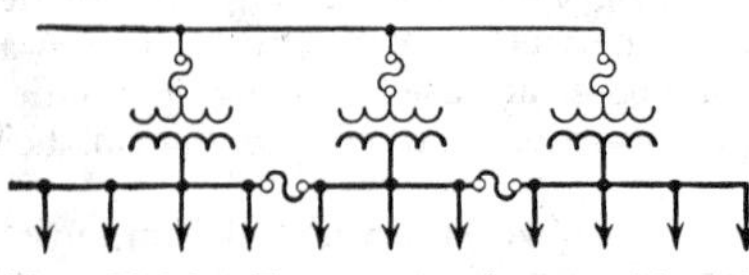

Fig. 16-31. Fuses on primary side of transformers; fuses for sectionalizing secondaries.

on either side of the connected secondary sections. This calls for a 100T (overload capacity provides coordination) sectionalizing fuse when the service-entrance fuse is 100 A.

180. A commonly used rule is to fuse each section point at 75 to 100% of normal full-load current of the smaller transformer. This rule does not ensure short-circuit coordination between service entrance, sectionalizing, and primary fuses.

Another method is to select the sectionalizing fuse on the basis of minimum fault current that will flow when a line-to-neutral (120-V) short circuit occurs near the sectionalizing point. The magnitude of fault current in the second stage of a line-to-neutral short circuit is dependent chiefly upon spacing of transformers and size of conductors. Transformer size has little effect on fault current in the second stage.

181. The transformer primary fuse for the banking shown in Fig. 16-31 must be chosen to coordinate with the secondary sectionalizing fuse. Table 16-14 shows the minimum-size primary fuse to coordinate properly with a 100T sectionalizing fuse and a 100-A service-entrance fuse.

Table 16-14. Minimum Size of Transformer Primary Fuse to Coordinate with 100T Secondary Sectionalizing Fuse (All Fuses T-rated)

Transformer, kVA	2,400 V	4,800 V	7,200 V
5	15T	10T	6T
10	15T	10T	6T
15	15T	10T	6T
25	25T	15T	10T

182. Annual load measurements on each transformer and accurate methods of estimating the load per transformer are considered essential to proper operation of banks. The single-phase primary supply to each bank is usually tapped off the main circuit through fuses. These fuses can be used to isolate the bank if cascading occurs.

183. Internal secondary breakers and primary protective links can be used to protect banked distribution transformers. The CSP transformer has these two features installed within the transformer bank. A transformer failure causes primary protective links to blow and the secondary breaker associated with the faulty transformer to trip. If more than one secondary breaker trips, which sometimes occurs, the remaining transformers must carry the entire load. If any of them are seriously overloaded, their breakers trip and prevent any transformer from burning out. Most secondary line faults are burned clear, but a sustained fault may cause two or more secondary breakers to open and resultant overload on remaining transformers may cause additional breakers to open. Usually reports of extreme low voltage bring trouble men to the area, and the fault is cleared before all the transformers' breakers open. The grid or loop arrangement of secondaries is preferable for this type of banking.

184. Use of CSPB transformers, developed specifically for banked operation, is illustrated in Fig. 16-32. Within the transformer case are protective primary links and two secondary sectionalizing breakers, one in each of the two circuits supplied by the transformer. The two 3-wire circuits are brought out of the transformer by means of 5 wires, the neutral being common to both circuits, and extend in two directions from the transformer.

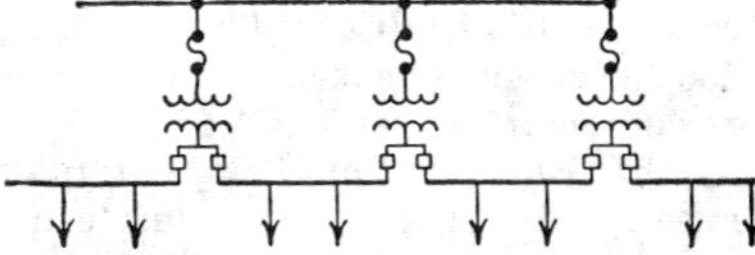

FIG. 16-32. Protective links on primary and internal secondary breakers in each transformer facilitate banking.

One sectionalizing breaker in each transformer is tripped by current through its respective circuit. The other sectionalizing breaker is tripped by the entire secondary current of the transformer and is designed to protect the transformer against excessive winding temperatures. When a transformer fault occurs, the transformer is isolated from the bank by blowing of protective links and tripping of two sectionalizing breakers. In event of a sustained secondary fault, sectionalizing breakers on each side of the fault (one on each transformer) open and isolate the faulty section. No transformer capacity is disconnected from the bank. Cascading is impossible with the CSPB transformers. The loop arrangement is highly desirable for the bank, though straight-line secondaries can be used.

APPLICATION OF CAPACITORS

185. Correction of Power Factor. When load on distribution circuits has a power factor below 85% during heavier load periods, it is desirable to add shunt capacitors to supply part of the lagging component of current. The cost is frequently justified by the value of circuit and substation capacity released. In general the installed cost of static capacitors is least on 2,400- to 12,500-V distribution and in distribution substations.

Corrective equipment should be located as far out on the distribution system as possible to get savings on the greatest amount of system investment. This is usually on feeder branches. But there are cases, particularly in underground distribution, where secondary capacitors are justified despite higher cost per kvar.

Development of low-cost 15-kV oil switches has promoted use of shunt capacitors switched for supplementary voltage control. Time clocks are the most common actuator, but contact-making voltmeters also are used.

186. Capacitor Installations. Capacitors are available in 25- to 100-kvar units suitable for pole mounting in banks of 3 to 12. They should be connected through fuses. Fuse and cutout ratings should be as listed in Table 16-15.

Table 16-15. Fusing for Shunt Capacitors, USA Standard Links

3-phase kvar	2,400-V Δ, fuse links	4,160-V Y, fuse links	4,800-V Δ, fuse links	7,200-V Δ, fuse links	12,470-V Y, fuse links
75	....	12K or T	10K or T	8K or T	
150	50K or T*	25K or T	25K or T	12K or T	10K or T
300	100K or T†	50K or T*	40K or T	25K or T	15K or T
450	....	80K	80K or T	50K or T*	25K or T
600	....	100K†	100K†	65K or T	40K or T

* In 100-A cutout.
† In 200-A cutout.

Normal procedure is to install capacitors on a feeder when load is approaching full current rating of the conductor, with low power factor, and installing a capacitor will permit the feeder to carry the load for which it was designed.

For example, a 4-kV feeder cable is rated 1,000 kW at 90% power factor, or 1,110 kVA. It happens to supply an area where power factor is 70% and the feeder has reached its capacity of 1,110 kVA, but the load is only 777 kW instead of 1,000 kW. This circuit may be brought up to rated load by a capacitor installation equal to the difference between the reactive component of a load of 1,000 kW at 70% and the same component of 1,000 kW at 90% power factor.

At 70% power factor 1,000 kW is $1{,}000/0.7 = 1{,}428$ kVA. Reactive component at this power factor is $0.714 \times 1{,}428 = 1{,}020$ kvar. At 90% power factor reactive component is $1{,}110 \times 0.436 = 484$ kvar. Capacitor rating required is $1{,}020 - 484 = 536$ kvar. This reduces kVA for 1,000-kW load from 1,428 at 70% to 1,110 at 90% power factor and releases $1{,}428 - 1{,}110 = 318$ kVA of feeder capacity.

187. Effect of Shunt Capacitors on Voltage. Capacitor charging current tends to raise feeder voltage. The effect can be calculated as the product of capacitor current and circuit reactance from the last point where voltage is controlled, usually the substation. For example, 300 kvar of capacitors is connected to a 1/0 ACSR 4,160-V feeder 10,000 ft from the substation. Reactance of 1/0 ACSR on crossarms at 14-in flat spacing (Table 16-2) is $j0.128\ \Omega/1{,}000$ ft; charging current per phase of 300-kvar capacitors at 4,160 V $= 300/4.16 \times 1.73 = 41.7$ A. Then the voltage boost in 10,000 ft of 1/0 ACSR $= 41.7 \times 10 \times 0.128 = 53.376$ V/phase. This boost is 2.22% of 2,400 V to neutral; hence 120-V customers get 2.66 V increase.

Voltage correction by use of shunt capacitors is a valuable tool. Often capacitors achieve required correction at lower cost than any other method and pay a double dividend in power-factor improvement. But their effect on voltage remains the same during light load periods and may cause too high customer voltage unless care is exercised in applying them. Switched installations become especially valuable for this condition because kvar in excess of what can be tolerated at light load can still be installed for its contribution at peak; it is switched off during light loads.

POLES AND STRUCTURES

188. Overhead construction is only 15 to 60% as costly as underground and is therefore an economic necessity in most locations. In some cities, pole lines along streets have been avoided by locating them in alleys or along rear-lot lines. In addition, much can be done to improve the appearance of pole lines by simplifying the devices which must be mounted on them.

189. Wood poles have been used almost universally for overhead distribution lines because of the abundance of the material, ease of handling, and cost. The life of wood poles is materially extended by impregnation with wood preservatives. Cedar, pine, and fir are best suited by their proportions and properties for use as distribution poles.

190. Concrete poles reinforced with steel have been employed chiefly for street-lighting standards, where a neat appearance is demanded. But recently some concrete poles have been used for general distribution as well, usually with a minimum of attached wires and apparatus.

191. Steel poles, ordinarily set in concrete, were long used to support trolley overhead and street lights. More recently, in a more ornamental form and bolted to concrete foundations, they have been used extensively for parkway lighting and, to a limited extent, for distribution where appearance demands.

192. Aluminum poles also are employed for parkway lighting standards and for certain other locations. They are bolted to concrete foundations to avoid the attack of fresh cement on aluminum.

193. Types of Loading. Poles carrying overhead distribution lines are subject to vertical and horizontal forces, of which some are continuous and others are applied only under abnormal or occasional conditions. Normal vertical forces are the weight of wires, transformers, and other equipment, and these are less than normal horizontal forces in many cases. Abnormal vertical load is imposed when wires are coated with ice, which may increase their normal weight 200 to 400%. For example, the weight of six covered No. 6 copper wires 100 ft long is normally 67 lb; but ice to a radial thickness of 0.5 in increases their weight to about 370 lb.

Normal horizontal forces acting upon a pole are the unbalanced component of wire tension at turns and corners, the side pull of service drops, and the horizontal component of weight when the pole is not vertical. Abnormal horizontal stresses are imposed by wind pressure, by breakage of conductors, or by failure of supporting guys.

194. Application of Loading. Vertical loading of wires and equipment is applied through crossarms and other attachments to the pole. These forces are amply sustained by poles chosen to meet requirements of transverse forces, except that, for line transformers, poles having 1 in greater diameter than line poles may be chosen. Transverse forces from unbalanced conductor tension at corners and bends are normally the greatest forces acting on the line. These are usually carried by guy cables secured to suitable anchorages, which relieve the pole itself of the stress. In some cases, the pole is underbraced and carries the entire load.

195. Ice Loading. When wires are loaded with ice, conductor tension is increased in direct proportion to the added weight of ice and may become two to four times as great as normal. This stress is borne by the conductors and, through them, communicated to the pole and the guying system. Where ice loading occurs, the guying system must have a suitable factor of reserve to meet abnormal loadings. The tension of conductors being increased with ice loading, elasticity in the wire permits a slight increase in length which makes tension less than the calculated amount for nonelastic conductors and supports.

196. Wind Loading. Loading due to wind pressure becomes appreciable in the design of poles and structures when wind velocities of over 40 mi/h are prevalent. Such forces are most noticeable on overhead lines when the direction of wind is at right angles to the direction of wires, both because the area exposed is greatest at that angle and because the force exerted is sustained by the pole without the aid of guying.

The area of conductors exposed to wind is much increased by a coating of ice, and the combination of ice with high wind is the most severe loading condition to which a line is subjected. In many parts of the United States such a condition is never experienced, and it is very rare even where ice coatings occur almost every winter.

197. Strength of Wood Poles. The strength of a wood pole must be sufficient to withstand transverse forces, such as wind pressure, on the pole and conductors; unbalanced pull on conductors when they are broken; and side pull on curves and corners where guys cannot be provided. These forces place the fiber of the wood under tension, and the load which a pole will carry is determined by the inherent strength of its wood fiber under tension and the moment of forces. The moment is

$$M = PL + P_1L_1 + P_2L_2 + \cdots \qquad (16\text{-}10)$$

in which P = force, in pounds, acting at one crossarm; L = height, in feet, at which the arm is attached; P_1, P_2, etc. = forces acting on other arms; and L_1, L_2, etc. = respective heights. If s = fiber stress in pounds per square inch and c = circumference at the ground in inches, the allowable moment of a pole of given size is

$$M = 0.0002638sc^3 \qquad (16\text{-}11)$$

Thus for a pole having a ground-line circumference of 40 in and an allowable fiber stress in an emergency of 2,500 lb/in², the maximum allowable moment is

$$M = 0.0002638 \times 2{,}500 \times (40)^3 = 42{,}200\ \text{lb}\cdot\text{ft}$$

If the average height of attachments is 30 ft, total force is 42,200/30 = 1,406 lb.

For normal unbalanced forces such as those of an unguyed corner, fiber stress should not exceed about 15% of that at the breaking point; but for wind pressures and unbalanced force of broken wires which are abnormal and do not persist, the stress is usually taken at about 50% of that at the breaking point.

Equation (16-11) for allowable moment is based on the assumption that the ground line is the weakest point. This is not always the case with northern cedar or other woods having a taper of 1 in in 5 to 6 ft of length. The point of greatest fiber stress is that at which diameter is 1.5 times diameter where the load is attached; i.e., with a pole having a diameter at the crossarm of 8 in, the greatest stress is where the diameter is 12 in. With a northern cedar pole this would be 5 × (12 − 8) = 20 ft below the crossarms and with the usual sizes of pole about 10 ft aboveground.

When the point of maximum stress is above ground line, the allowable moment is determined from the formula

$$M = 0.001781sc^2(c_1 - c) \qquad (16\text{-}12)$$

in which c = circumference at the point of attachment and c_1 that at the point of maximum stress. Decay usually occurs at the ground line, and as the pole ages, its ground-line circumference is reduced to make it the point of greatest stress.

For resisting moments, see Table 16-16.

198. Example of Calculation of Pole Size. Assume that a western cedar pole carries two crossarms, one of which carries a normal transverse pull of 400 lb at a height of 33 ft and the other 200 lb at 31 ft. What should be the ground-line circumference of the pole? $M = PL + P_1L_1 = 400 \times 33 + 200 \times 31 = 19{,}400$ lb·ft, and

$$19{,}400 = 0.0002638 \times 800c^3$$

whence $c^3 = 19{,}400/0.211$ and $c = \sqrt[3]{92{,}000} = 45.2$ in. With a decrease of 3 in in circumference for each 8 ft in length the circumference at a point 34 ft above ground would be (34/8) × 3 = 12.75 in less than at the ground, or 45.2 − 12.75 = 32.5 in. This would require a pole having a top diameter of about 10.3 in.

If the allowable fiber stress were increased to 2,500 instead of 800 lb, as permissible

Table 16-16. Resisting Moments of Wood Poles

Pole-circumference at ground, in	Lb-ft at one-half ultimate fiber stress ratings of						
	8,400 lb/in²	8,000 lb/in²	7,400 lb/in²	6,600 lb/in²	6,000 lb/in²	5,600 lb/in²	4,000 lb/in²
28	24,300	23,150	21,400	19,150	17,350	16,250	11,550
30	29,900	28,500	26,350	23,500	21,350	19,950	14,250
32	36,300	34,600	32,000	28,500	25,950	24,200	17,300
34	43,600	41,500	38,400	34,200	31,150	29,000	20,750
36	51,750	49,300	45,600	40,700	36,950	34,450	24,650
38	60,750	57,850	53,500	47,800	43,400	40,550	28,900
40	70,950	67,550	62,500	55,700	50,650	47,250	33,800
42	82,200	78,250	72,400	64,400	58,700	54,700	39,150
44	94,450	89,950	83,200	74,100	67,450	62,900	44,950
46	107,950	102,800	95,100	84,600	77,100	71,900	51,400
48	122,600	116,750	108,000	96,200	87,550	81,700	58,350
50	138,600	132,000	122,100	108,700	99,000	92,300	66,000
52	155,850	148,400	137,300	122,400	111,300	103,900	74,200

under storm conditions, the moment could be increased to

$$M = 0.0002638 \times 2{,}500c^3 = 60{,}700 \text{ lb}\cdot\text{ft}$$

If this were a northern cedar pole, the ground-line circumference would be

$$c^3 = \frac{19{,}400}{550 \times 0.0002638} = 133{,}600$$

$$c = \sqrt[3]{133{,}600} = 51.2 \text{ in}$$

With a decrease of 3 in in circumference for each 5 ft in length, the circumference at a point 34 ft above ground would be $(34/5) \times 3 = 20.4$ in less than at the ground.

$$c \text{ at top} = 51.2 - 20.4 = 30.8 \text{ in}$$

199. Wind Pressure. Design of pole lines to withstand wind pressure must be considered when lines are in exposed positions. The pressure on flat surfaces normal to the direction of wind may be calculated from the formula

$$P = 0.004V^2 \tag{16-13}$$

in which P is in pounds per square foot and V = velocity in miles per hour. For cylindrical surfaces such as wires and wood poles

$$P = 0.0025V^2 \tag{16-14}$$

In using values of V taken from Weather Bureau reports it must be noted that the observed values are taken at elevations considerably above those at which wires are usually carried. The following correction[1] should be made before using Bureau values of velocity recorded prior to Jan. 1, 1928, but velocities recorded since that date are substantially the actual or correct ones.

Indicated velocity, mph	30	40	50	60	70	80	90	100
Actual velocity, mph	26	33	41	48	55	62	69	76

In closely built-up districts in cities these "actual" values are still somewhat high. Where lines are in exceptionally exposed locations, as in passing over hills where there are no trees, the full indicated values may exist.

Wind pressure on a 40-ft pole, of which 34 ft is aboveground, with 7-in top and 14-in butt, would be calculated as follows:

$$\text{Projected area} = [(7 + 14)/2] \times 34 \times 12 = 4{,}284 \text{ in}^2 = 29.75 \text{ ft}^2$$

[1] Fowle, F. F. A Study of Sleet Loads and Wind Velocities; *Elec. World*, 1910, Vol. 56, p. 995.

Upon assuming a wind velocity of 60 mi/h, $P = 0.0025 \times (60)^2 = 9.0$ lb/ft². Total pressure on pole = $29.75 \times 9.0 = 268$ lb. This force acts at the center of gravity of the projected area, which is a long, narrow trapezoid, or 15.1 ft from the ground line. The moment of this force about the ground line is therefore

$$268 \times 15.1 = 4{,}047 \text{ lb}\cdot\text{ft}$$

With the diameter of No. 4 polyethylene weatherproof ACSR wire taken as 0.32 in, total wind pressure at 60 mi/h on a 120-ft span would be 28.8 lb; on 20 wires it would be 576 lb (see Table 16-17); this resultant force would act about 3 ft below the pole top, or 31 ft from the ground line. The resulting moment would be 17,856 lb·ft. The sum of the moments from wind pressures on pole and wires would be

$$4{,}047 + 17{,}856 = 21{,}903 \text{ lb}\cdot\text{ft}$$

200. Tension in Other Conductors. Transverse force on triple-braid weatherproof copper and neoprene weatherproof aluminum conductors runs very similar to that on ACSR and aluminum of equivalent conductivity because the covering used on copper is enough thicker to compensate for the slightly smaller conductor diameter. The reduced weight of aluminum tends to lower the resultant force somewhat. But distribution is turning increasingly to bare conductors; tables showing forces on these are determined in the same manner as for power transmission (Sec. **13**).

Table 16-17. Ice and Wind Loading of Polyethylene Weatherproof Aluminum Wires*

Size, AWG or M cmils	Diameter, in	Weight			Transverse force			Resultant force, weight and wind + constant		
					4-lb wind		9-lb wind, no ice			
		0.5-in ice	0.25-in ice	No ice	0.5-in ice	0.25-in ice		0.5-in ice	0.25-in ice	No ice
ACSR:										
No. 4	0.320	0.592	0.259	0.0817	0.440	0.273	0.240	1.047	0.596	0.304
No. 2	0.418	0.705	0.342	0.134	0.472	0.306	0.314	1.159	0.679	0.391
No. 0	0.523	0.827	0.431	0.191	0.507	0.341	0.392	1.280	0.770	0.486
No. 3/0	0.627	0.991	0.563	0.290	0.542	0.376	0.470	1.440	0.897	0.603
Stranded aluminum:										
No. 0	0.493	0.759	0.372	0.141	0.497	0.331	0.370	1.217	0.718	0.446
No. 3/0	0.589	0.888	0.472	0.211	0.530	0.363	0.442	1.344	0.815	0.540
No. 4/0	0.647	0.974	0.540	0.261	0.549	0.382	0.485	1.428	0.882	0.601
336-M cmil	0.822	1.226	0.737	0.404	0.607	0.440	0.617	1.678	1.079	0.787
556-M cmil	1.014	1.574	1.025	0.632	0.671	0.505	0.761	2.021	1.363	1.039
795-M cmil	1.215	1.964	1.353	0.897	0.738	0.572	0.911	2.408	1.689	1.329

Selected data from National Electrical Safety Code, 5th ed.

* Transverse force for triple-braid weatherproof copper is approximately equal to that listed for ACSR or stranded aluminum of equivalent conductivity but weights and resultant loading are 10 to 20% more.

201. Transformer Loading. Poles are subjected to normally heavy loads when used to support transformers or when changes in direction of the line impose transverse stresses due to the unbalanced component of wire tension. Transformer loads are chiefly vertical but have a transverse component when the pole is bent or drawn away from a vertical position. It is therefore usual to select poles having a top diameter about 1 in greater than would otherwise be needed when a transformer installation of more than 25 kVA is to be placed on the pole selected. For three-unit installations of 300 kVA or larger, it may be desirable to support the transformers on a platform carried by two poles set 10 to 15 ft apart, although hardware designed to mount three transformers closely spaced around the pole sometimes supports somewhat larger transformers. Recent reductions in transformer sizes and weights permit larger units to be pole-mounted than was formerly considered good practice.

202. Unbalanced Loads. Transverse forces imposed upon a pole where wires form an angle are offset by guys, if it is practicable to install them. If it is not practicable to guy, the pole must be underbraced and must have sufficient stiffness to withstand the stresses imposed upon it. The force applied to the pole at the point of attachment by wires carried on the crossarm may be calculated as in Fig. 16-33. Divergence from a straight line in 200 ft of line is determined by joining two points, each 100 ft from the corner pole, by a straight line. Distance A, from the corner pole to this line, is then a measure of transverse force applied by line wires on the pole.

$$\text{Transverse force} = 2A \times \frac{T}{100} = 0.02AT \qquad (16\text{-}15)$$

where T = sum of the tensions of all the wires carried by the pole.

For example, if A is 10 ft and there are nine wires having a total tension of

$$9 \times 200 = 1{,}800 \text{ lb}$$

the transverse force to be normally sustained by the corner pole is

$$0.02 \times 10 \times 1{,}800 = 360 \text{ lb}$$

When lines are carried around a curve, each pole is, in effect, a corner pole and the force is calculated accordingly. For right-angle turns the value of A is 70.7, and this is the most severe condition ordinarily found. In such cases tensions should be reduced at the corner pole as much as possible by shortening spans and by transferring part of the stress to lower points on the corner pole through head guys. This practice is desirable with heavy conductors, even though guying may be practicable, as it serves to reduce stresses on guy cables and anchors which sustain the corner pole. Stresses at dead-end poles are distributed in a similar manner.

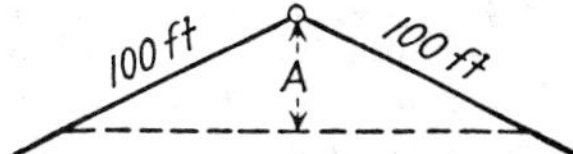

FIG. 16-33. Determining divergence to an angle in the line.

203. Safety-code Requirements. The National Electrical Safety Code, in Part 2, sets up minimum safety requirements for loading, strength, and clearance for those parts of supply lines which are involved in crossings of railroad or communication circuits or which come into such proximity to these as to create a "conflict." It also includes joint use of lines and separate use on any public way. The code recognizes differences in degree of estimated hazard, which are assumed to depend upon voltage of the supply line and relative importance of the railroad or communication circuits. The rules are thus less rigorous for distribution voltages below 8,700 phase-to-ground (15,000 phase-to-phase) than for circuits at higher voltages and are more rigorous for crossings over main lines than for minor lines.

STRUCTURAL DESIGN OF POLE LINES

204. Selection of Poles. The height of poles is determined by clearances over obstructions and number of crossarms or racks to be carried. The 35-ft pole is the most popular. In a large metropolitan system, more than half the company-owned poles are 35 ft, and most of the others are either 30 ft or 40 ft.

205. Poles for Joint-line Construction. The 35-ft pole is the minimum generally used on public streets with joint-line construction. The 30-ft pole may be used in alleys and on rear-lot lines. Joint use by two or more public-service companies is encouraged by municipalities and state regulatory bodies because it makes maximum use of investment and avoids having pole lines on both sides of the street.

206. Pole Spacing. Poles should be spaced 100 to 150 ft apart in suburban and urban districts in order to provide convenient points for service attachments and to keep the service lengths to a minimum.

207. Pole Setting. Experience has proved that the following practice is conservative, as regards the depth of setting. Corner poles should be set about 6 in deeper.

Size pole, ft.	30	35	40	45	50	55	60	70
Depth, ft.	5	5.5	6	6.5	6.5	7	7	7.5

208. Underbracing. Where poles supporting unbalanced stresses, from dead ending or change of direction, are so situated that it is not practical to support them by guying, they must be underbraced to withstand the force imposed with as little deviation from original position as possible. This normally requires that the pole have more than usual diameter and be no taller than absolutely necessary for clearance.

For very long spans or heavy conductors, steel poles are often required. If wood poles are used, top diameters of 8 to 10 in are required for such positions to avoid bending. In addition, the pole is underbraced by timbers bolted to it below ground line and at the butt, as shown in Fig. 16-34. An alternative method, using blocks of concrete, is also good. Small boulders may be used in backfill to advantage.

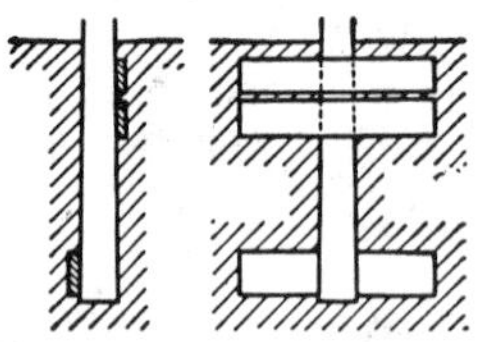

Fig. 16-34. Underbracing for pole with unbalanced stress.

In the use of plank or concrete, the pole is set at a slight angle in a direction opposite that from which stress is to be applied, to allow for compacting of soil when wires are pulled up. The area of plank or concrete should be about 4 ft^2, both top and bottom. Where steel poles are used to secure strength, they are usually set in a concrete base of such dimensions as to bear the stresses imposed.

209. Guying and Bracing. Where the direction of a line changes, tension of the wire should be supported by guying. Guys are secured in various ways, depending upon space and clearance required. Where nothing prevents, the guy cable may be secured to an **anchor** of timber or a patent anchor. On public thoroughfares guys cannot be run directly to anchors without interfering with traffic and must be run to **stubs** at such height as to permit free passage of traffic beneath. Guys over roadways should clear by 18 ft and those over pathways by at least 8 ft (see Fig. 16-35).

Where side-arm construction is used, it is necessary at corners to guy crossarms as well as poles. At heavy corners a "head" guy is run from the base of the corner pole to the upper part of the next pole in line. Tension in corner spans may thus be reduced, relieving strain on the corner pole. In straightaway lines the head guy is used to limit damage in case several poles go down; this is sometimes called "storm guying."

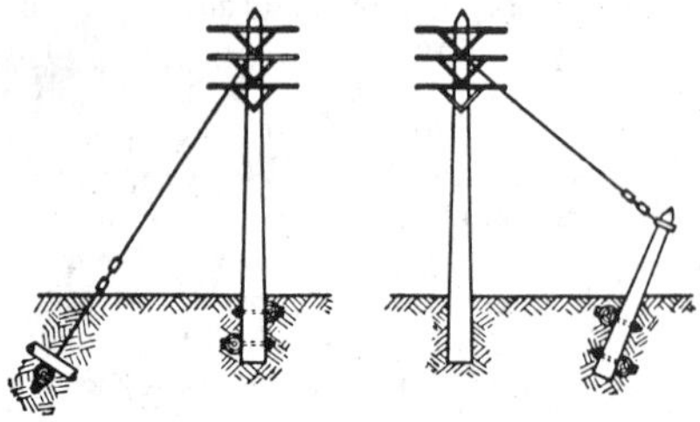

Fig. 16-35. Anchor and stub guy.

210. Galvanized-steel cable is generally employed for guying because of its high tensile strength. Such cable is made in sizes varying in diameter by steps of $\frac{1}{16}$, from $\frac{1}{4}$ in up. Ultimate breaking strength of $\frac{1}{4}$-in cable is about 3,000 lb; $\frac{5}{16}$-in, 5,300 lb; $\frac{3}{8}$-in, 6,900 lb; and $\frac{1}{2}$-in, 12,000 lb. Some utilities use special guy strands of higher ultimate strength or protected by a layer of copper or aluminum.

The tension having been calculated, guy cables should be selected such that strain will be one-fourth to one-fifth the ultimate breaking strength. Anchors should be placed at a distance from the pole not less than one-quarter the height of guy attachment. In general, $\frac{1}{4}$-in cable is used for smaller loads, whereas the $\frac{3}{8}$-in size is standard for corner poles where ordinary two- to three-arm distribution line is supported.

211. Pole Clearance, Horizontal. The location of poles should be chosen to provide sufficient clearance from driveways, fire hydrants, street traffic, railroad tracks, buildings, fire escapes, and other projections. Clearance from main tracks of railroads should be not less than 12 ft and from side tracks, 7 ft. Except in alley lines, poles should be set not closer than 6 ft from buildings, in order to minimize interference with firemen's ladders.

212. Vertical clearances given in Table 16-18 are taken from the National Electrical Safety Code, sixth edition. They apply to crossings where span lengths do not exceed 175 ft in heavy-loading districts, 250 ft in medium-loading districts, or 350 ft in light-loading districts. These clearances are based on a temperature of 60°F, with no wind and voltages not over 50,000 to ground. For longer spans and higher voltages, greater clearances are required, depending upon sag and tension in the span.

Table 16-18. Minimum Vertical Clearances above Ground or Rails, Ft

Type of location	Guys, messengers, etc.	Open-supply wires, voltage to ground			Trolley conductor, voltage to ground	
		0 to 750	750 to 15,000	15,000 to 50,000	0 to 750	Over 750
When crossing above:						
Railroads	27	27	28	30	22	22
Streets, alleys, and roadways	18	18	20	22	18	20
Private driveways	10	10	20	22	18	20
Walks for pedestrians only	8	15*	15	17	16	18
When wires are along:						
Streets or alleys	18	18	20	22	18	20
Roads in rural districts	14	15	18	20	18	20

* 12 ft for max. of 300 V to ground.

213. Clearances at Wire Crossings. Where wires of intersecting lines cross in open span without making contact on a joint pole, clearances between bottom wires of the higher line and top wires of the lower line must be sufficient to guard against possibility of accidental contact under varying wind, temperature, and ice loading. Clearances fixed for such crossing spans by the National Electrical Safety Code (Table 16-19) apply at 60°F with no wind and for spans not exceeding 175, 250, or 350 ft, in heavy-, medium-, and light-loading districts, respectively. For longer spans, greater clearances are required, depending upon sag and tension in the span.

Table 16-19. Wire-crossing Clearances, Ft

Nature of wires crossed over	Communication guys	Services, guys, 0- to 750-V cable	Open-supply wires, voltage to ground		
			0 to 750	750 to 8,700	8,700 to 50,000
Communication circuits	2	2	4	4	6
Aerial supply cables	4	2	2	2	4
Open supply wires, 0 to 750 V	4	2	2	2	4
Open supply wires, 750 to 8,700 V	4	4	2	2	4
Opensupply wires, 8,700 to 50,000 V	6	6	4	4	4
Trolley conductors	4	4	4	6	6
Services, guys, lightning-protection wires	2	2	2	4	4

214. Clearances of Conductors from Poles and Buildings. Conductors of one line should, in general, be not less than 4 ft from those of another and conflicting line. Where wires pass near the pole of another line but are not attached, they should not interfere with climbing space. Wires should clear sides of buildings 3 ft if below 8,700 V to ground and 8 ft if at voltages between 8,700 and 15,000. Wires of all voltages up to 15,000 to ground should have a vertical clearance of not less than 8 ft over buildings. Crossing over roofs should, however, be avoided if practicable.

215. Horizontal Separation of Conductors. For supply conductors, at voltages up to 8,700 between conductors, the minimum horizontal clearance provided by the National Electrical Safety Code is 12 in. For higher voltages it is required that 0.4 in be added for each 1,000 V above 8,700. For sags of more than 24 in, separations greater than 12 in are required and are determined, for wires of No. 2 AWG or larger, as

$$\text{Separation} = 0.3\text{ kV} + 8\sqrt{S/12} \qquad \text{(in)} \tag{16-16}$$

where S = sag, in inches. For wires smaller than No. 2 AWG the rule is

$$\text{Separation} = 0.3\text{ kV} + 7\sqrt{(S/3) - 8} \qquad \text{(in)} \tag{16-17}$$

Multiconductor, spacer, and other cabled types of supply-circuit construction are exempt from the above phase-spacing requirements.

216. Climbing Space. Safety to linemen requires suitable provision for climbing through lower wires on a pole to gain access to wires on upper arms. This is accomplished by making distances between pole pins on a crossarm not less than 30 in for voltages up to 8,700 to ground and 36 in up to 15,000. Climbing space through supply wires under 300 V or communication circuits should be not less than 24 in. Apparatus such as transformers, arresters, and cutouts should be on one side of the pole only and, where possible, on crossarms outside the climbing space.

217. Vertical Separation of Crossarms. Usually, 2 ft from center to center is provided for arms carrying conductors operating up to 8,700 V to ground. When wires of 8,700 to 50,000 V to ground are carried above circuits less than 8,700 V, minimum vertical clearance should be 4 ft. On joint poles, 4 ft should be maintained between communication circuits and supply wires up to 8,700 V to ground and 6 ft at voltages of 8,700 to 50,000 (see the 6th edition of National Electrical Safety Code for additional information).

LINE CONDUCTORS

218. Conductor Factors. Copper and aluminum are the metals most used as conductors in distribution systems. Proportions are fixed by the combined effect of conductivity, weight, strength, and cost. Recent years have seen such a shift in availability and cost that aluminum has gained almost universal use in distribution, supplanting copper, which was preferred for many years. Sodium also is thought to have a place in distribution, but methods to permit its use have been devised so recently that its future cannot now be evaluated.

219. Conductor Materials. Aluminum has the advantage of about 70% less weight for a given size, but its conductivity is only about 61% that of annealed copper. For distribution, it is commonly rated as equivalent to a copper conductor two AWG sizes smaller, which has almost identical resistance. Its tensile strength is less than copper, and it is commonly used, particularly in the smaller sizes, by stranding aluminum around a steel core of proper size to give the desired tensile strength. In larger sizes, the tensile-strength requirements of distribution are satisfied by stranded aluminum without the reinforcing steel. (For further data about properties of conductors, see Sec. **4.**)

Both copper and aluminum are suitable for use as substation buses, being available in flat bars, tubes, and rods. For very heavy currents, channel shapes are used to make up box-type buses, which are the most economical for such applications.

220. Use of Copper. Where copper is used for overhead circuits with span lengths of 200 ft or more, it is commonly used in the "hard-drawn" form because of its greater tensile strength. For common types of local distribution circuit where spans are shorter and flexibility is desirable, "medium-hard-drawn," or annealed, copper is used. Mechanical connectors are extensively used for joints and taps on overhead copper.

Underground cables are usually made of stranded soft copper because of its greater flexibility. The smaller size of copper conductors helps to offset unfavorable price levels because of savings in insulating and sheathing material as well as the ability to put maximum carrying capability in a given size of duct. But the development of such synthetic insulations as polyethylene has made aluminum more competitive for underground use.

221. Use of Aluminum. In rural line work, where long spans and conductors of high tensile strength are an economic necessity, the combined requirements of conductivity and strength have been met with aluminum stranded around a steel core sized to give the required strength. Such a cable is known as "aluminum cable steel-reinforced" and is commonly designated as ACSR. Development of high-strength aluminum alloys has led to such alternative cables as "aluminum conductor alloy-reinforced" (ACAR) and "all-aluminum-alloy conductor" (AAAC), which also combine conductivity with tensile strength.

Urban distribution uses similar conductors for the portions where light loads are served but swings to stranded aluminum where large conductors are required.

Jointing of aluminum requires special care to secure good contact and to guard against corrosion. Jointing is often done with compression devices, although mechanical connectors packed with corrosion-inhibiting compound can be used.

222. Use of Steel. Steel conductors are rarely used for distribution circuits because of their high resistance. But steel with a heavy covering of copper, known as Copperweld, or with a heavy covering of aluminum, known as Alumoweld, has conductivity approaching 40% that of copper and can be used in some applications. Such coated conductors are also very attractive as high-strength strands or reinforcements for composite cables, which get improved conductivity from strands of hard-drawn copper over the Copperweld or hard-drawn aluminum over the Alumoweld.

Conductors reinforced with steel have impedances which increase somewhat as current density increases. Voltage drops are correspondingly higher than those of copper or aluminum conductors of equal conductivity.

Copperweld and Alumoweld are generally more durable than galvanized-steel cables. They have therefore been used to some extent for guy cables.

OPEN-WIRE LINES

223. Crossarms. Southern pine and Douglas fir are the best woods for crossarms because of thin, straight grain, high tensile strength, and durability. Experience indicates that a cross section 3½ in wide by 4½ in high is ample for the average distribution line. Main lines are commonly built with six-pin arms, and smaller lines use four-pin arms. Minimum spacing of pins is 12 in, and spacings of 14 to 16 in are commonly used. Minimum spacing of pole pins is 30 in to provide climbing space.

224. Double crossarms are installed on poles at corners, at terminals, and at other points where unusual loads are to be supported.

225. Vertical racks are installed on poles to support secondary and multiple street-lighting wires. They are available with two, three, or four spool insulators. Rack construction is less expensive than crossarms and has supplanted them to a large extent. When services run to houses on both sides of the street, two racks are required, one on each side of the pole. In addition, several pole-top designs mount insulators directly on the pole, eliminating the use of crossarms.

226. Wire Stringing. In erecting wire, the **tension** should be sufficient to prevent too much sag in the spans and yet not so great as to stress the wire unduly. For practical purposes the approximate formula given by Rankine may be used,

$$t = \frac{S^2 w}{8d} \qquad \text{(lb)} \qquad (16\text{-}18)$$

in which t = tension in pounds, S = span length in feet, w = weight per foot of conductor, and d = sag in feet at the center of a horizontal span. If span length is doubled, tension must be quadrupled in order to keep sag the same. If tension is the same on several spans of different lengths, sag is different in each span. The sag of any span when tension is known is found by changing Eq. (16-18) to the form

$$d = \frac{S^2 w}{8t}$$

227. Sag Tables. Maximum tension in a span is limited by strength of wire and supports. It is customary to design distribution so that the tension will not exceed about 50% of the tensile strength of the wire or 2,000 lb stress on supports, whichever is less. Tensions under normal conditions are, of course, considerably less. The amount of sag in a typical span is a convenient measure of wire tension and is commonly used. In addition, the same sag values are often employed for a range of wire sizes, thus achieving a more uniform appearance (see Sec. **13** for a detailed treatment of sag and tension problems).

Sags given in Table 16-20 are selected from standard sheets of a large utility.

228. Expansion and Contraction of Spans with Temperature Changes.[1] Changes in sag due to expansion and contraction under varying temperatures are of much importance in the erection of conductors. Lines erected during winter months are likely to be too slack during the summer, and allowance should be made accordingly. The length of wire in a span, elastic stretching due to load being disregarded, varies in pro-

[1] THOMAS, P. H. Sag Calculations for Suspended Wires; *Trans. AIEE*, 1911, p. 2229.

Table 16-20. Sags for Typical Distribution Conductors (Heavy Loading District—60°F)

Size, AWG or M cmils	Conductor material	Sags (in) for span lengths (ft) of							
		80	100	125	150	175	200	250	300
Open wire:		. .							
No. 0	Al. alloy, bare	. .	10	16	23	31	40	27*	38*
No. 3/0	Al. alloy, bare	. .	10	16	23	31	40	32*	46*
336.4	Aluminum, bare	. .	10	16	23	31	40	75	108
No. 0	Al. alloy, polyeth.	. .	10	16	23	31	40	59*	90*
No. 3/0	Al. alloy, polyeth.	. .	10	16	23	31	40	70	101
336.4	Aluminum, polyeth.	. .	18	27	38	51	66		
Cabled secondaries:		. .							
3 No. 0	Al. alloy, insul.	. .	10	16	23	31	40		
3 No. 3/0	Al. alloy, insul.	. .	18	27	38	51	66		
4 No. 3/0	Al. alloy, insul.	. .	18	29	43	60	80		
Cabled service drops:		. .							
3 No. 4	Aluminum, insul.	32	52	73†					
3 No. 0	Aluminum, insul.	51	79	116†					
4 No. 3/0	Aluminum, insul.	68	108	171†					

* Taken up to 2,000-lb tension limit for spans over 200 ft.
† For 120-ft service drops, 450-lb limit.

portion to the coefficient of expansion and range of temperature,

$$L_t = L_o(1 + \alpha_o t) \tag{16-19}$$

in which α = coefficient of expansion, t = temperature in degrees Fahrenheit, and L_o = length of wire at zero temperature. The linear coefficient of expansion of aluminum is 0.000024/°C or 0.0000133/°F. The coefficient for ACSR is very nearly that of the steel, 0.0000112/°C or 0.0000062/°F. Copper's coefficient is 0.000017/°C or 0.0000094/°F.

In practice, pole supports have a certain degree of flexibility, which tends to take up part of the slack caused by expansion and to prevent excessive strains being placed on wires by contraction during cold weather.

JOINT-LINE CONSTRUCTION

229. In alley-line distribution, where both power and communication systems are maintained, **and on highways,** where two or more utilities would otherwise maintain pole lines, there are advantages in the use of joint poles, which are often utilized.

230. Basis of Joint Use. Poles are used jointly under a **joint-ownership agreement** or under a **lease agreement.** Under joint ownership, the cost of providing the pole is borne jointly by the companies which share in its ownership. Division of expense is, in general, made in proportion to the space allotted to respective users. Clearance space, required between power and communication circuits and between the lowest attachment and ground, is disregarded in determining percentage of ownership. Clearance between higher-voltage and lower-voltage power circuits is chargeable to the higher-voltage circuits.

In case of leased space, the lessee acquires only the right to occupy a specified space. The owning company installs and maintains the pole and includes all charges in the rental price. The lessee usually installs his own attachments and maintains them, though pin space is sometimes leased where space for only a few wires is required.

231. Construction Specifications. The type of construction, clearances, and relative levels of different classes of circuit should be provided for by a suitable specification, forming part of the agreement under which joint use is entered into. The general purpose is that construction of all parties be such as not to jeopardize the service or equipment of any of the other parties to the agreement. Some more important parts of such a specification are outlined in the paragraphs following.

232. Relative Levels. Conductors should be assigned to attachments placed, in general, so that the higher voltages are at the higher levels. This places the highest voltage near the pole top and communication circuits at a lower level. Where 600-V railway trolley and feeders for it are carried on joint poles, they are necessarily placed at the height required for the trolley-contact arm. Private communication circuits should be carried in the zone assigned to this class of circuits. The foregoing plan tends to place the stronger conductors above the weaker and provides access to lower voltage and communication circuits with minimum hazard to linemen or service.

233. Vertical Clearances. Spacing of attachments is adjusted in accordance with requirements of safety in operation and maintenance. This usually results in spacing of 4 ft between the top arm or cable carrying communication circuits and the bottom arm or attachments carrying power circuits. The private communication circuit of a supply company is sometimes placed 2 ft below the lowest electric-supply arm or attachments. If the lowest power circuit is operated at above 8,700 to ground, clearance to top arm or cable of the communication company is made 6 ft. Electric-supply wires at voltages below 8,700 to ground are carried 2 ft below the primary circuit, except that, if the higher-voltage circuit is operated at above 8,700 to ground, the vertical clearance is 4 ft.

234. Grades of Construction. Strength of poles must be such as to withstand wind and ice loading normally experienced in the locality where the line is built, for all the conductors which are to be carried. These conditions vary materially in different parts of the United States, there being no ice in some parts and a greater prevalence of excessive wind in some places than in others. Weather Bureau records are usually available for exact information as to wind velocities and ice formation.

235. The size of conductors should be such that they will not be subjected to more than 60% of their ultimate strength under the maximum loading of the line or 25% of ultimate strength for final unloaded tension at 60° Fahrenheit. Tensions are usually kept lower to minimize fatigue from vibration. The minimum sizes that may be safely used are No. 8 for medium- or hard-drawn copper, No. 1 for stranded aluminum, and No. 6 for ACSR in spans up to 150 ft with heavy loading, 175 ft with medium or light loading.

236. Inductive Coordination. Where the voltage of supply circuits does not exceed 7,500, effects of induced currents in communication circuits are usually sufficiently compensated by transpositions in communication circuits to prevent crosstalk. Where circuits are operated at higher voltages, or where the joint-line portion of the circuit exceeds 5 mi in length, it is sometimes necessary to use transpositions in both power and communication circuits. Electrical unbalances should be avoided as far as practicable in both systems.

237. Aerial-cable Construction. Lead-covered cables have been carried on steel messenger cables, supported by poles, in some cases where conditions made underground impracticable and the circuits were of such importance that it was desirable to avoid exposing them to the hazards of open-wire lines. The cable is carried by hangers on a galvanized-steel cable, commonly called the "messenger," in a manner similar to that used for communication cables. The point of attachment to poles is made as low as clearance from buildings and roadways will permit. It is important that provision be made to permit expansion and contraction of the cable under variations of temperature.

Increasingly aerial cable is being applied to distribution. A common application is for a second feeder on the same poles with an open-wire feeder where it is desired to avoid simultaneous trouble on both circuits. But some utilities are using aerial cable to stormproof important feeders or to maintain circuits through trees where trimming is prohibited or excessively costly. This cable is usually one, two, or three insulated conductors spiraled around or lashed to a bare high-strength neutral which supports the assembly. The lashed assembly is better suited to primary circuits because it provides more convenient access for tapping on transformers and laterals. Special hardware items and techniques have been developed to facilitate installation of the cable.

238. Spacer-cable construction provides many of the advantages of aerial cable at lower cost. It consists of conductors with less insulation than is customary for the circuit voltage, supported from a messenger by insulating spacers which supplement the insulation on the conductors.

UNDERGROUND RESIDENTIAL DISTRIBUTION

239. Residential underground systems are not a new concept, but until about 1960 cost usually ruled out this design method. Public demand and improved technology subsequently have made underground residential distribution (URD) a common supply method.

The majority of residential underground primary systems are single-phase. Where loads are heavy, however, some utilities use 3-phase primaries. System voltages generally are 2.4/4.16-kV Y, with an increasing trend to 7.2/12.5-kV Y or 7.6/13.2-kV grounded neutral.

240. Types of Cable. Nonleaded cable is used extensively for URD primaries and secondaries. Polyethylene rubber or other rubberlike insulation is practically the universal choice. The primary cable usually has a concentric neutral of tinned copper wires and a semiconducting jacket over rubber, rubberlike, or polyethylene insulation. Secondary and service cable is also rubber-, rubberlike-, or polyethylene-insulated but generally without the concentric neutral or semiconducting layer used on higher-voltage cables. For both primaries and secondaries, the conductors may be either copper or aluminum.

241. Cable Installation. Direct burial is the least expensive of installation practices now being used in underground residential distribution systems. Care must be taken in trenching to the proper depth, installing of the cable, and, when necessary, protecting the cable from damage by digging or other means with a concrete slab or corrugated nonmetallic sheathing. Primary, secondary, and telephone cable are usually buried in that order and separated by partially backfilling over each circuit. Where state codes permit, random-lay configuration is used, the primary, secondary, and telephone cable being laid at the same trench depth and not separated by earth fill.

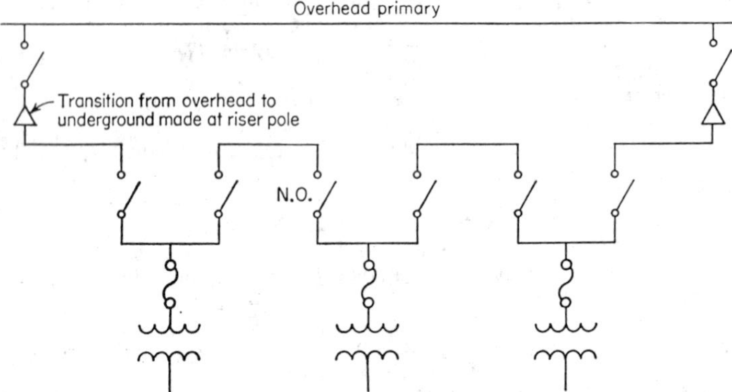

FIG. 16-36. Load feed providing two sources of power.

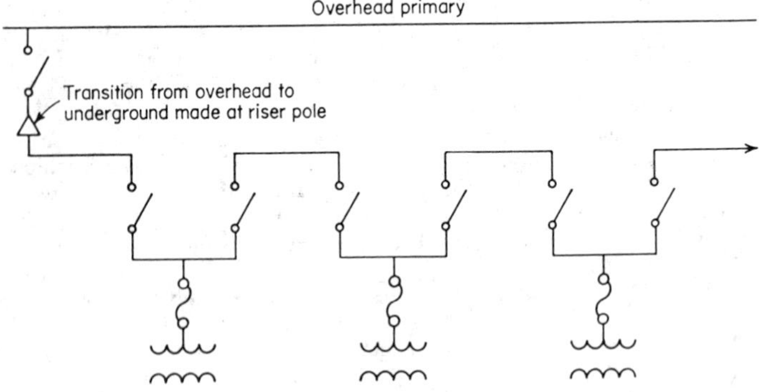

FIG. 16-37. Radial system supplied from one source.

Table 16-21. Characteristics of Polyethylene-insulated 15-kV Cables

(Courtesy of Reynolds Metals Company)

High-molecular-weight polyethylene (Insulation thickness 0.220 in—conductor jacket 0.030 in)						Cross-linked polyethylene (insulation thickness 0.175 in—conductor jacket 0.030 in)					
Size, AWG or M cmil	No. of strands	Neutral nom. OD, in		Ampacity		Size, AWG or M cmil	No. of strands	Neutral nom. OD, in		Ampacity	
		Under	Over	In air*	Direct burial†			Under	Over	In air‡	Direct burial§
		Solid aluminum						Solid aluminum			
4	Solid	0.78	0.91	80	115	4	Solid	0.68	0.81	95	130
2	Solid	0.83	0.96	105	150	2	Solid	0.73	0.86	125	170
1	Solid	0.86	0.99	125	175	1	Solid	0.75	0.88	150	190
1/0	Solid	0.90	1.03	140	200	1/0	Solid	0.80	0.93	165	220
2/0	Solid	0.93	1.06	165	225	2/0	Solid	0.88	1.01	190	250
3/0	Solid	0.97	1.13	190	260	3/0	Solid	0.89	1.05	220	285
4/0	Solid	1.02	1.18	220	295	4/0	Solid	0.93	1.09	255	325
250-M cmil	Solid	1.07	1.27	245	325	250-M cmil	Solid	0.98	1.18	285	360
300-M cmil	Solid	1.12	1.32	280	365	300-M cmil	Solid	1.02	1.23	325	405
		Stranded aluminum						Stranded aluminum			
4	7	0.81	0.94	80	115	4	7	0.70	0.83	95	130
2	7	0.86	0.99	105	150	2	7	0.76	0.89	125	170
1	19	0.90	1.03	125	175	1	19	0.80	0.93	150	190
1/0	19	0.95	1.08	140	200	1/0	19	0.85	0.98	165	220
2/0	19	0.98	1.11	165	225	2/0	19	0.93	1.06	190	250
3/0	19	1.03	1.19	190	260	3/0	19	0.95	1.11	220	285
4/0	19	1.09	1.25	220	295	4/0	19	1.00	1.16	255	325
250-M cmil	19	1.15	1.35	245	325	250-M cmil	37	1.06	1.26	285	360
300-M cmil	19	1.20	1.40	280	365	300-M cmil	37	1.11	1.31	325	405
		Stranded copper						Stranded copper			
4	7	0.81	0.94	100	120	4	7	0.70	0.83	120	160
2	7	0.86	0.99	135	160	2	7	0.76	0.89	160	210
1	19	0.90	1.03	155	180	1	19	0.80	0.93	180	240
1/0	19	0.95	1.08	180	210	1/0	19	0.85	1.01	210	275
2/0	19	0.98	1.11	210	240	2/0	19	0.93	1.09	240	310
3/0	19	1.03	1.19	240	280	3/0	19	0.95	1.15	280	360
4/0	19	1.09	1.25	280	325	4/0	19	1.00	1.20	325	410

* In air based on 75°C conductor temperature, 40°C ambient.
† Direct burial based on 75°C conductor temperature, 20°C ambient—100% load factor.
‡ In air based on 90°C conductor temperature, 40°C ambient.
§ Direct burial based on 90°C conductor temperature, 20°C ambient—100% load factor.

Table 16-22. Lightning-arrester Characteristics*
(Courtesy of Westinghouse Electric Corporation)

Arrester rating max. line to ground, kV rms	Minimum 60-c spark-over, kV rms	Maximum impulse spark-over ASA front of wave, kV crest		Discharge voltage, kV crest, for discharge currents of 8 × 20 μs waveshape with following maximum crest amplitudes						Withstand test voltage of arrester insulation†		
				1,500 A		10,000 A		65,000 A		Impulse test 1½ × 40‡ μs full-wave kV crest	Alternating-current 60-c test voltages, kV rms	
		Avg.	Max.	Avg.	Max.	Avg.	Max.	Avg.	Max.		1 min dry	10 s wet
3	6	17	19	9.5	10	13.3	13.8	18	19	48	34	24
6	11	32	35	18	19	25	26	33	35	71.8	45	38
9–10	18.0	42	50	29	30	40	41	50	56	77.7	51	50
12	23.5	52	60	36	38	51	52	67	71	85.6/85.6	50/55	40/45
15	27	63	75	43	45	61	62	76	85	114/125	75/75	80/85
18	33	75	90	52	54	73	74	93	103	128/128	90/90	85/93
21	37.5	94	100	60	63	83	87	114	120	157/179	110/115	95/115

* All values apply to both the positive and negative polarity surges.

†Due to the fact that porcelain arresters rated 12 kV and above have two positions for the mounting clamps; the first value was measured with the clamp in the upper position; the second value was measured with the clamp in the lower position.

‡ Using 7,000 Ω/in^3 water.

242. Rigid duct systems are sometimes used. Several types of plastic, fiber, and concrete duct suitable for burial are available. They provide some degree of mechanical protection and are resistant to corrosion. The additional protection provided by the duct simplifies the backfilling operation and minimizes mechanical damage. Maintenance of the system during operation is aided by the ease of removing and installing cables.

243. Cable-in-duct (sometimes referred to as preassembled cable) consists of a flexible duct with insulated cable preinstalled at the factory. This system combines the installation ease of a direct-buried cable with the advantage of being able to pull in new conductors when necessary. It also affords some degree of mechanical protection.

244. Terminations. The shielding of a primary cable must be terminated so that no damage can occur from concentration of electrical stress at the terminating point. Terminations of the deenergized variety can be disconnected or connected with hot-line tools and require no taping. This type of connection is not suitable for disconnection when energized at normal voltage and carrying no load.

245. Lightning protection may be obtained from arresters at the riser pole, at the end of a radial feeder, or at the normally open tie point of a loop. If an arrester with low enough spark-over is installed at the riser pole, protection is not necessary at the normally open tie point. Caution should be followed in relating the size or spark-over value of the arrester (Table 16-22) to the basic insulation level of the system.

246. Types of System. The loop-feed system usually has two sources of power. The primary system would normally have an opening near the midpoint, so that half the system would be supplied from each source. After a cable fault the normally open point is closed to pick up as much of the deenergized section as possible from the remaining live source.

247. Radial service is supplied from one source. Failure of the primary may interrupt all load fed by the radial primary.

248. The secondary pedestal may be used as a connection box for service extensions to the loads. This avoids making service connections directly to the transformer.

Secondary connection box or pedestal

Service to load

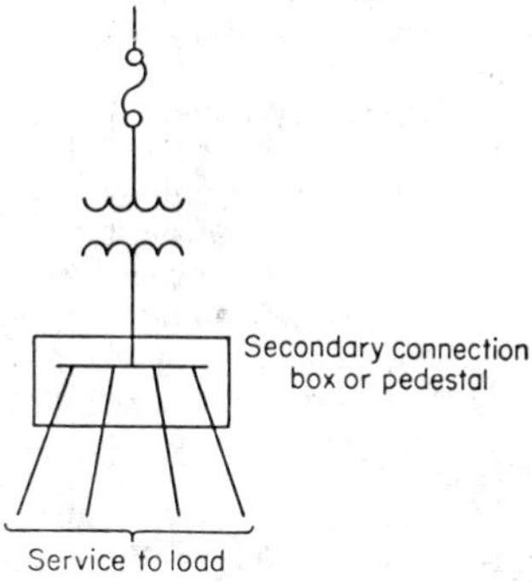

FIG. 16-38. Connections to load from secondary enclosure.

249. Transformers are generally installed as pad-mounted units, set on concrete slightly above grade. However, the trend appears to be toward a totally underground design in which the transformer is set in a prefabricated enclosure, usually of concrete or bituminous material.

Transformer size varies with the amount of load and number of homes fed from each bank. In addition, the rating of transformers installed below grade depends on the free air area of the protective grating. In certain cases it may be advisable to install a baffle in the enclosure to produce a chimney effect and prevent mixing of hot and cool air in the enclosure.

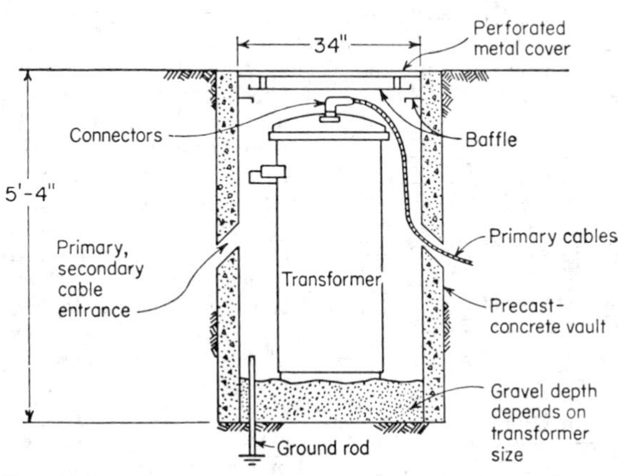

FIG. 16-39. Transformer vault flush with surface, adjacent to curb or sidewalk. Gravel is used as foundation.

UNDERGROUND URBAN SYSTEMS

250. Underground construction is necessitated in more built-up portions of cities by heavy transformers and lines and by

Table 16-23. Insulation Thickness for Multiple-conductor Cables *
(Varnished-cloth-insulated)

Rated circuit voltage, phase-to-phase V	Conductor size, AWG or M cmils	Insulation thickness, mils	
		Grounded neutral	Ungrounded neutral
0–600	14–8	45	45
	7–2	60	60
	1–4/0	80	80
	213–500	95	95
	501–1,000	95	95
	Over 1,000	110	110
1,001–2,000	12–2	80	80
	1–500	95	95
	501–1,000	95	95
	Over 1,000	110	110
3,001–5,000	8–4/0	95	95
	213–1,000	110	110
	Over 1,000	110	110
8,001–10,000	6 and larger	140	140

*Copyright 1962 by Insulated Power Cable Engineers Association. Adapted by permission.

the multiplicity of service connections to buildings. Feeders are often laid in conduits along main routes, even though local distribution mains may not be sufficiently heavy to necessitate putting them underground. Where alley and back-lot routes are available, local distribution is preferably overhead, except in business districts with heavy load densities.

251. Solid systems have not been used as generally in America as in Europe. In such systems, cable is laid in the ground, with a covering of steel armor or of vitreous duct filled with bituminous compound to exclude moisture. It is accessible only by excavation. The Edison-tube system is the only solid type of underground construction which has been extensively used in the United States.

252. Draw-in conduit systems have the advantage of minimum disturbance of expensive pavements and interference with traffic. They consist of ducts as many as local and through-line requirements demand (with allowance for future growth), and manholes, through which access is had to the ducts for drawing in cables or withdrawing them for repairs or alterations.

Manholes are placed at all junction points and corners and in straight runs at intervals of 500 to 700 ft, the length being fixed by allowable cable-drawing tension and utility practice. Smaller holes are required at points where services are taken off from a distribution main.

253. Duct Materials and Arrangement. Ducts are of fireproof material or of a material which does not support combustion and is not depreciated by continuously moist surroundings. Vitrified clay tile, terra-cotta, and concrete pipe are the fireproof types most often used. Bituminized fiber, laid in concrete, is also extensively employed. The tile is mostly in 2-ft lengths, while the concrete, or "stone," pipe and fiber are commonly in 5-ft lengths. The principal advantage of fiber is its reduced weight and consequent saving in freight and labor. Ducts are also made of an asbestos and cement composition which is fire-resisting and known as Transite. Some types of single duct are shown in Fig. 16-40.

For electricity supply cables, the duct line is preferably built up of sin-

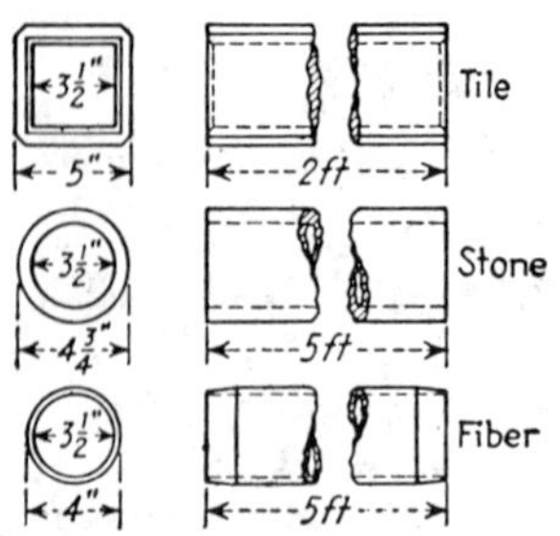

Fig. 16-40. Types of single duct for underground conduits.

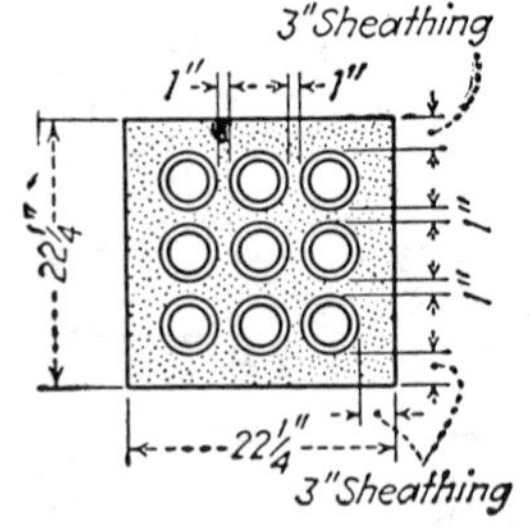

Fig. 16-41. Arrangement of conduit with concrete sheath.

Table 16-24. Insulation Thickness for Rubber-insulated Cables *

Rated circuit voltage, phase-to-phase, V	Conductor size, AWG or M cmils	Insulation thickness, mils	
		Grounded neutral	Ungrounded neutral
0-600	14-9	47	47
	7-2	62	62
	1-4/0	78	78
	225-500	94	94
	525-1,000	109	109
	Over 1,000	125	125
1,001-2,000	14-8	78	78
	7-2	94	94
	1-4/0	109	109
	225-500	125	125
	525-1,000	141	141
	Over 1,000	141	141
3,001-5,000	8-4/0	156	156
	225-1,000	172	172
	Over 1,000	188	188
8,001-10,000	6-1,000	219	281
	Over 1,000	234	297
12,000-15,000	2-1,000	297	422
	Over 1,000	312	438
23,001-28,000	1-1,000	500	
	Over 1,000	516	

* Copyright 1962 by Insulated Power Cable Engineers Association. Adapted by permission.

gle duct in order to have two walls of duct material between adjacent cables. There is also often about 1 in of concrete between walls of adjacent ducts, as shown in Fig. 16-41. The bank of ducts is surrounded by concrete 3 in thick, which acts as protection against later excavations and as support when a ditch is opened across the line.

254. Laying Out a Conduit Line. The number of ducts must be sufficient for local distribution, for feeders, for transmission, and for 5 to 10 years' growth.

The **maximum number of ducts** which it is advisable to put into a line is governed chiefly by temperature. Space available for training cables is limited; and if more than 20 are carried through a manhole, a large part of the load is endangered by a failure of any of the cables. Where conditions are such that a very large line must be used, protection may be had by building double manholes. More than four ducts in each layer is to be avoided where possible on account of the difficulty of training cables.

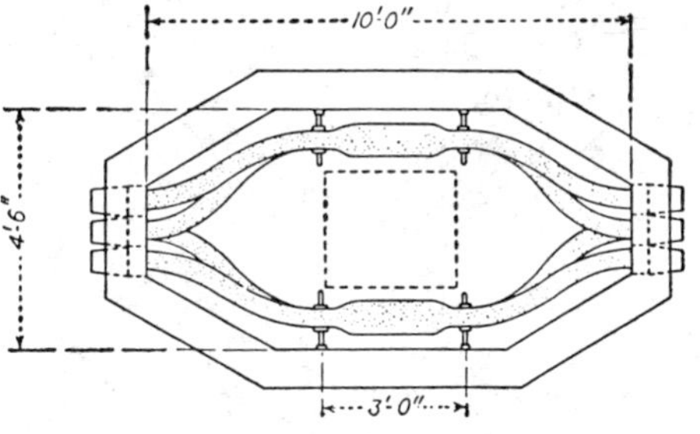

255. Location of Manholes. Manholes must be provided in sufficient number to permit drawing in cable without overstraining. Thus **manhole spacing** should be not over 600 ft with large cables. Where distribution is overhead in alleys and underground on the street, manholes should be at alley intersections as far as possible. Where numerous underground service connections are needed, manholes must be at intervals of about 100 ft. Large manholes are required for transformers to get suffi-

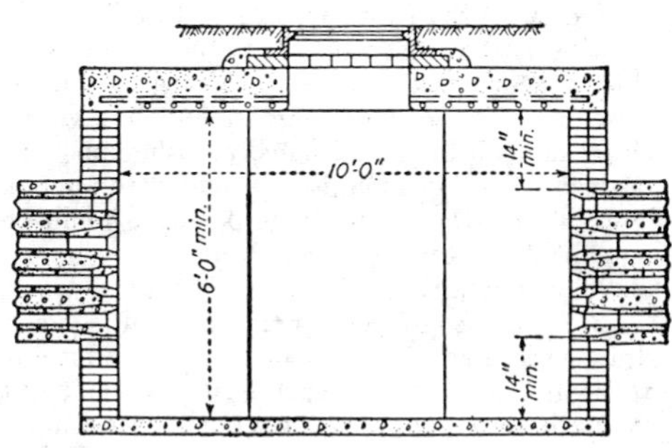

FIG. 16-42. Straight-type manhole.

Table 16-25. Table of Skin-effect Coefficients

Cir mils × frequency	Coefficient		Cir mils × frequency	Coefficient	
	Copper	Aluminum		Copper	Aluminum
10,000,000	1.000	1.000	80,000,000	1.158	1.069
20,000,000	1.008	1.000	90,000,000	1.195	1.085
30,000,000	1.025	1.006	100,000,000	1.23	1.104
40,000,000	1.045	1.015	125,000,000	1.332	1.151
50,000,000	1.07	1.026	150,000,000	1.433	1.206
60,000,000	1.096	1.04	175,000,000	1.53	1.266
70,000,000	1.126	1.053	200,000,000	1.622	1.33

cient room and proper ventilation. Steel gratings are sometimes used on transformer manholes to improve ventilation.

At intersections a square design is preferable. It is usual to provide manholes 6 by 6 ft at junctions where two six-duct lines cross—and larger as needs may require.

The size and shape of manholes are often governed by local obstructions such as gas or water pipes or conduit lines of other companies. Depth must be sufficient to give headroom and yet not be so great as to carry the floor of the manhole below sewer level. Service manholes may be 5 ft high inside, but junction manholes should be 6 or 7 ft. In some cases a shallow form known as a **handhole** is used for distribution laterals. These are placed above the conduit line so that only the top row of ducts enters the handhole. Distributing mains are thus accessible for service taps, and through lines in lower ducts are not in the way.

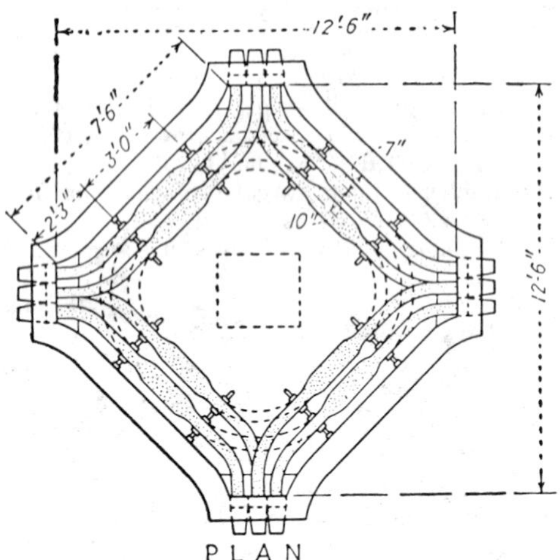

Fig. 16-43. Nine-duct X-type manhole.

256. Installation of Conduit System. In streets where conduit and pipe utility systems already occupy space, a survey should be made to select a position in which obstruction of the new conduit system will be as little as possible. This is done by noting the position of manholes of existing systems and by consulting map records of other utility systems.

Exact depth of the ditch cannot be determined until the depth of pipes and conduits crossing it have been disclosed by excavation. Alignment and grades should be fixed with surveying instruments to ensure drainage and avoid pockets. Dips in which water may stand and freeze should be avoided. The line may be curved slightly to clear manholes, catch basins, or other obstructions. Where no sewer is available for drainage, a sump should be provided to facilitate pumping.

Width of the ditch at the bottom should be about the width of the concrete jacket outside the ducts. The earth may then be used as the form for the bottom and sides of the concrete jacket.

257. Manhole Construction. Vaults or manholes provided to facilitate cable drawing and splicing are constructed of brick or concrete. Brick walls should be 8 in thick and laid up with good-quality cement mortar. The roof must have sufficient strength to support the heaviest street traffic passing over it, which necessitates use of steel reinforcement in most cases. The entrance is covered by a cast-iron cover having a roughened surface and strength to bear the weight of trucks passing over it. Both circular and square covers are used, the circular cover having the advantage that it cannot fall into the vault in handling. Square or rectangular covers are preferred where low-tension junction boxes and transformers are to be installed.

See Fig. 16-52 for details on transformer vaults.

258. Cable supports should be provided on the walls of the vault to prevent undue strain on the sheath and permit an orderly arrangement. It is usual practice to provide racks of such a design as to permit two cables to be carried on each.

UNDERGROUND CABLES

259. Types of Cable. Cables placed in underground duct systems are continuously exposed to moisture and are usually of the lead-sheathed type. Nonleaded moisture-resistant cables, rated 600 V, are used extensively for secondary mains. Cables are classed as single-conductor, 2-conductor, 3-conductor, etc., according to the number of separately insulated conductors enclosed by a single sheath (see Fig. 16-44).

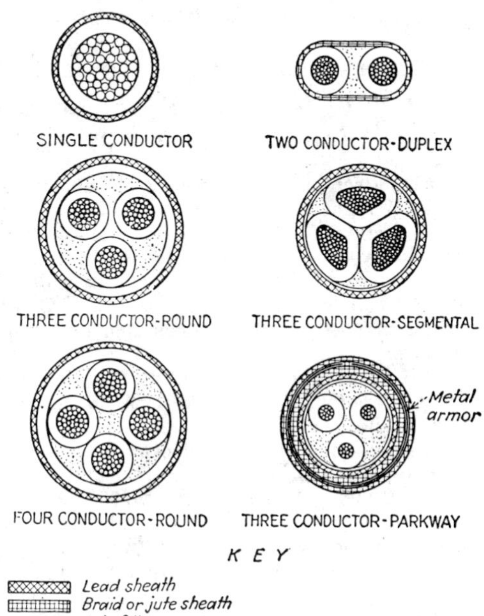

Fig. 16-44. Cross sections of typical cables.

260. Cable Insulation. Electric supply cables are insulated with rubber, varnished cloth, or impregnated paper. *Rubber* is used extensively in cables rated 600 V to 35 kV; *polyethylene,* 600 V to 138 kV; *varnished cloth,* 600 V to 28 kV; *impregnated paper*—solid type to 69 kV; pipe type, pressure, or compression to 345 kV. Impregnated paper is used for higher voltages because of its low-dielectric-loss characteristic and comparative low cost. Joints are insulated with material matching the cable insulation and covered with material replacing the cable sheath; lead-covered cable takes lead sleeves. Terminals are equipped with suitable potheads to exclude moisture. (Tables 16-23 and 16-24 are taken by permission from IPCEA-NEMA Standards and publications on varnished-cloth- and rubber-insulated wire and cable for transmission and distribution of electrical energy.)

Single-conductor cables are used where taps are made frequently as on distributing mains and service connections and for street-lighting circuits. Where the diameter of a multiple-conductor cable would exceed the maximum diameter which can readily be drawn into a duct, it is necessary to use single-conductor cables in separate ducts. This condition is found with 3½-in ducts when conductors of a 3-conductor cable exceed 600,000 cmils.

Multiple-conductor cables have the advantage of occupying but a single duct and of being lower in cost than three single-conductor cables. They are used for primary feeders and subtransmission lines quite generally. Where 3-conductor cables have conductors of over 250,000 cmils, they must usually be the **sector** type to avoid an outside diameter too great to be safely drawn into the duct.

261. Thickness of Cable Insulation. Thickness of insulation of a cable is determined in part by mechanical stresses incident to handling and, at higher voltages, by the dielectric strength necessary to withstand voltages imposed under service conditions. In a system with grounded neutral electric stress is somewhat lower than under certain conditions in a system without a grounded neutral. Thickness of impregnated-paper and varnished-cambric insulation under various conditions for multiple-conductor cable of various sizes and voltage ratings is shown in Tables 16-23 and 16-24.

262. Cable Diameters. Overall diameter D of a cable may be computed from the diameter of its conductors d, the thickness of its conductor insulation T, its belt insulation t, and its lead sheath S, as follows:

Single-conductor: $$D = d + 2T + 2S \tag{16-20}$$

2-conductor: $$D = 2(d + 2T + t + S) \tag{16-21}$$

3-conductor: $$D = 2.154(d + 2T) + 2(t + S) \tag{16-22}$$

4-conductor: $$D = 2.414(d + 2T) + 2(t + S) \tag{16-23}$$

These formulas apply to conductors of circular cross section. For sector-type 3-conductor cables, the overall diameter

$$D_3 = D - 0.35d \text{ (approx.)} \tag{15-24}$$

263. Electrical Characteristics of Cables. **Skin effect** is an a-c phenomenon (Secs. **2** and **4**) which materially affects cables of large cross section, because currents passing through strands around the outer surface of the cable encounter less inductance and impedance than near the center. This causes outer strands to carry more current, proportionately, than inner strands. It is desirable, therefore, to build up large cables about a core of nonconducting material. Cables of over 500,000 cmils are often made in this manner for 60-c systems, and cables of more than 1,000,000 mils for 25-cycle systems.

264. Skin-effect Coefficients. The increase in effective resistance due to skin effect is approximately proportional to the product of the frequency and the circular mils, as shown in Table 16-25.

To determine the skin effect of a copper cable having an area of 1,000,000 cmils, carrying current at 60 c/s, refer to the table, opposite the product 60,000,000. The coefficient is 1.096. Resistance of 1,000,000-cmil cable/1,000 ft being 0.01035, the effective resistance at 60 c/s is $0.01035 \times 1.096 = 0.01134$, or 9.6% more than with continuous current. The resistance of a 1,500,000-cmil cable is increased 19.5% at 60 c/s. Current-carrying capacity of large cables is reduced in proportion to the reciprocal of the skin-effect coefficient; i.e., if the coefficient is 1.096, the capacity is only $1/1.096 = 91.2\%$ of that with continuous currents.

265. Electrostatic Capacity. With overhead circuits, charging current of their electrostatic capacity is negligibly small at usual power frequencies and at voltages below 35,000. The effect of capacitance upon voltage drop and line losses at higher voltages is fully discussed in Sec. **13.** With underground cables, insulated with impregnated paper, capacitance becomes appreciable at voltages above 20,000 because separations are small and the specific inductive capacity of the insulation is about 3.3 (air being 1).

266. Charging current of a single-conductor cable is

$$I = \frac{0.106EfK}{1{,}000 \log_e(D/d)} \text{ A/1,000 ft} = \frac{0.0463EfK}{1{,}000 \log(D/d)} \text{ A/1,000 ft} \tag{16-25}$$

in which E = kilovolts to neutral, f = frequency in cycles per second, K = dielectric constant, D = diameter over the insulation, and d = diameter of the conductor.

Charging current of 3-phase three-core cable is affected by arrangement of conductors (round or sector) and by relative thicknesses of conductor insulation T and belt insulation t. Simmons has put these relations into usable form by working out logarithmic denominators of the equation for various ratios of thickness of insulation to diameter of conductor. This has been termed the **geometric factor.** Charging current of a three-core 3-phase cable is

$$I = \frac{3 \times 0.106EfK}{1{,}000G_2}\ \text{A/1,000 ft} \qquad (16\text{-}26)$$

For impregnated-paper cable, K is 3.3, and the equation for 60-c/s circuits becomes

$$I = \frac{3 \times 3.3 \times 0.106 \times 60E}{1{,}000G_2} = 0.063\frac{E}{G_2} \quad \text{(amperes)} \qquad (16\text{-}27)$$

Values of G for single-conductor and G_2 for 3-conductor cable may be taken from Table 16-26.

267. Geometric Factors of Cables. See Table 16-26.

Intermediate values may be found by interpolation.

268. *Example.* Find 60-c/s charging kVA for 33-kV cable having three 350,000-cmil sector-type conductors each with 10/32 in of paper and a 5/32-in belt.

$$T = 0.313\ \text{in} \qquad t = 0.156\ \text{in} \qquad d = 0.681\ \text{in} \qquad t/T = 0.5$$

$$(T + t)/d = (0.313 + 0.156)/0.681 = 0.69 \qquad E = 33/1.73 = 19\ \text{kV}$$

Interpolating in Table 16-26, we find $G_2 = 2.78$.

269. For sector-type cable, G_2 must be multiplied by the sector factor for 0.69, which is seen to be 0.86 in the sector-factor column, Table 16-26. For such a cable,

$$G_2 = 0.86 \times 2.78 = 2.39 \qquad \text{and} \qquad I = (0.063 \times 19)/2.39 = 0.5\ \text{A/1,000 ft}$$

Charging kVA $= 3IE = 3 \times 0.5 \times 19 = 28.5$ kVA/1,000 ft, and for a cable having a length of 20 mi it would be $20 \times 5.28 \times 28.5 = 3{,}010$ kVA. For single-conductor cables, $t = 0$ and $(T + t)/d = T/d$, which is used to get the value of G from the values for single-conductor cable in Table 16-26.

270. Selection of Duct Position for Cables. Cables used in local distribution should be in the top row so that manholes can be built for service laterals without sinking them below the top row of ducts. Ducts should be selected for through lines so that they may be trained with the least interlacing with other cables.

Table 16-26. Table of Geometric Factors of Cables

Ratio $\frac{T+t}{d}$	G Single conductor	Sector factor	Three-conductor cables: G_1 at ratio t/T = 0	0.5	1.0	G_2 at ratio t/T = 0	0.5	1.0
0.2	0.34	.....	0.85	0.85	0.85	1.2	1.28	1.4
0.3	0.47	0.690	1.07	1.075	1.08	1.5	1.65	1.85
0.4	0.59	0.770	1.24	1.27	1.29	1.85	2.00	2.25
0.5	0.69	0.815	1.39	1.43	1.46	2.10	2.30	2.60
0.6	0.79	0.845	1.51	1.57	1.61	2.32	2.55	2.95
0.7	0.88	0.865	1.62	1.69	1.74	2.55	2.80	3.20
0.8	0.96	0.880	1.72	1.80	1.86	2.75	3.05	3.45
0.9	1.03	0.895	1.80	1.89	1.97	2.96	3.25	3.70
1.0	1.10	0.905	1.88	1.98	2.07	3.13	3.44	3.87
1.1	1.16	0.915	1.95	2.06	2.15	3.30	3.60	4.05
1.2	1.22	0.921	2.02	2.13	2.23	3.45	3.80	4.25
1.3	1.28	0.928	2.08	2.19	2.29	3.60	3.95	4.40
1.4	1.33	0.935	2.14	2.26	2.36	3.75	4.10	4.60
1.5	1.39	0.938	2.20	2.32	2.43	3.90	4.25	4.75
1.6	1.44	0.941	2.26	2.38	2.49	4.05	4.40	4.90
1.7	1.48	0.944	2.30	2.43	2.55	4.17	4.52	5.05
1.8	1.52	0.946	2.35	2.49	2.61	4.29	4.65	5.17
1.9	1.57	0.949	2.40	2.54	2.67	4.40	4.76	5.30
2.0	1.61	0.952	2.45	2.59	2.72	4.53	4.88	5.42

271. Installation of Cable. Cables are drawn into ducts by a line attached to a source of power. This line is put through the duct by use of detachable rods which are pushed into the duct as they are joined together. Cables are secured to the pulling line by exposing the copper and making a secure mechanical connection or by means of cable grips, which are more quickly attached and removed.

The maximum outside diameter that can be drawn into 3½-in duct without undue strain is 2⅞ to 3⅛ in. Where there is a curve in the duct line or where the length of the pull exceeds 400 ft, the maximum size should be a little less than 3 in.

Where several single-conductor cables are to be drawn into the same duct, the reels are set up in tandem and all cables are drawn in simultaneously.

272. Cable Training. Sufficient length must be left in manholes to permit training cables on racks around the manhole walls, as shown in Figs. 16-42 and 16-43, and for jointing. Radius of bends should be not less than 10 times the diameter of cable.

In large cables there is periodic bending with rising and falling loads, and duct-mouth protection is often necessary to prevent cracking of lead sheaths. This may consist of a piece of galvanized metal inserted below the cable and arranged to prevent the sheath from being pressed against sharp edges of the duct mouth.

Cables are protected from damage due to a burnout of an adjoining cable by covering them with asbestos tape applied with cement.

273. Bonding Sheaths. Large cable systems are subject to stray return currents, particularly near substations supplying power for electric-railway service. Stray currents cause corrosion of lead sheaths by electrolysis where they leave the lead. Such corrosion may be avoided by bonding sheaths together with heavy copper conductors and by providing suitable cables bonding cable sheaths to the return circuit of the railway system.

274. Jointing, or Splicing. In jointing single-conductor cables, the lead sheath is removed about 6 in back from the end, and enough insulation is cut away to permit a soldered connection to be made. When the connection is complete, the bare parts are wrapped with tape until the equivalent of cable insulation has been applied. A lead sleeve which has previously been slipped over one of the cables is now wiped on the two cable sheaths so as to enclose the joint. Air space around the joint is then filled with hot insulating compound poured into a small hole in one end of the sleeve; a similar hole is left in the other end to allow air to escape. These holes are then closed by soldering. The joint should be allowed to cool before it is moved, so that the compound will hold the parts rigidly in place.

In jointing 3-conductor cables, the lead must be removed about 10 in to facilitate taping the conductors (see Fig. 16-45). In making joints for 6,600 V and higher, it is important that as little air remain in taping as possible. If paper tape is used, each layer should have compound poured over it before the next is applied.

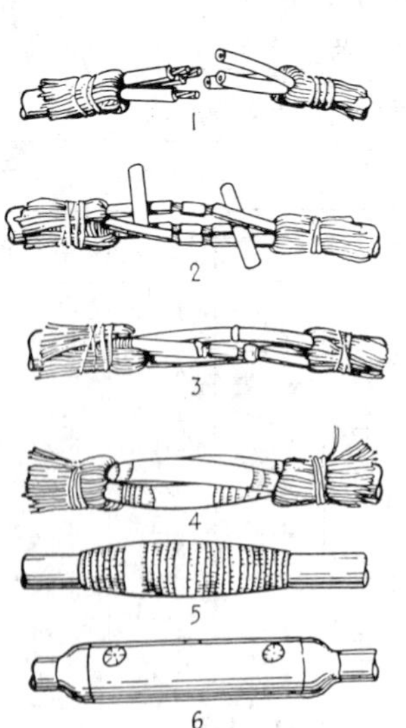

Fig. 16-45. Successive stages in cable splicing.

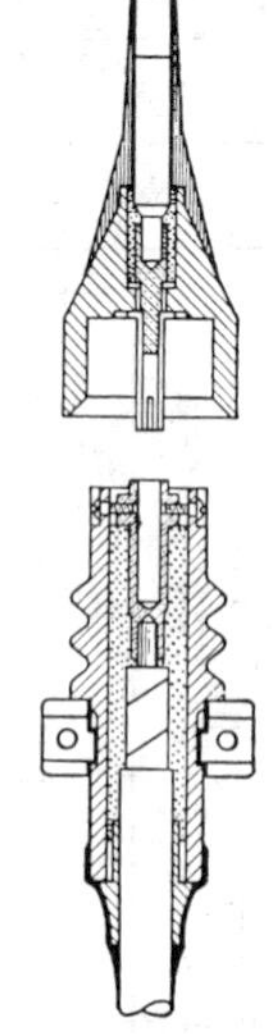

Fig. 16-46. Single-conductor disconnecting pothead.

275. Cable Terminals. Where cables are connected to overhead portions of a system, they are equipped with **potheads** to protect cable insulation from the weather and provide means of disposing of static stresses set up at cable ends. On lines operating at voltages at which line equipment is handled alive, it is usual to provide potheads of a disconnecting type, such as that shown in Fig. 16-46, for single-conductor cable, or Fig. 16-47, for multiple-conductor cable. These consist of a pot for insulating compound, with porcelain tubes for outlets to the overhead line. The tube is covered by a weatherproof cap, with a terminal so arranged that removing the cap disconnects the circuit. Thus, the cable terminal acts as a disconnective, useful in locating trouble, and makes unnecessary other testing equipment at pole terminals. Where mains of one circuit are near those of another, potheads form a useful means of restoring service by temporarily transferring supply to an adjoining circuit.

276. Subway Junction Boxes. Junction boxes are used in cable systems to interconnect distribution mains at points where it is desired that the connection be opened, at times, for construction or operating purposes and where no overhead disconnectives are available in the circuit arrangement. In low-tension networks, such boxes may include sectionalizing fuses or copper links. They are four-way at main junctions. At feeder ends, junction boxes are usually five- or six-way, thus providing for disconnecting any or all of the branch mains. Tap boxes are used where a branch from a main requires disconnection.

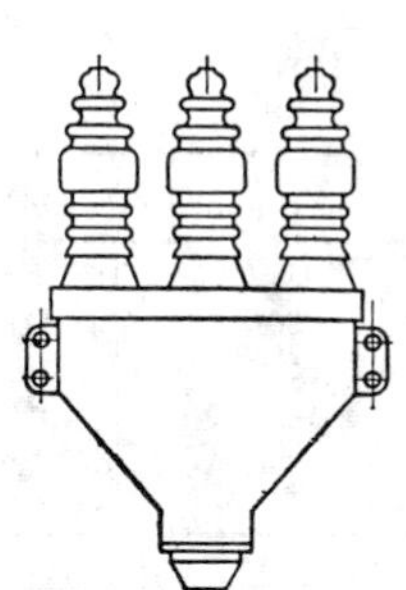

FIG. 16-47. Three-conductor pothead of the disconnecting type.

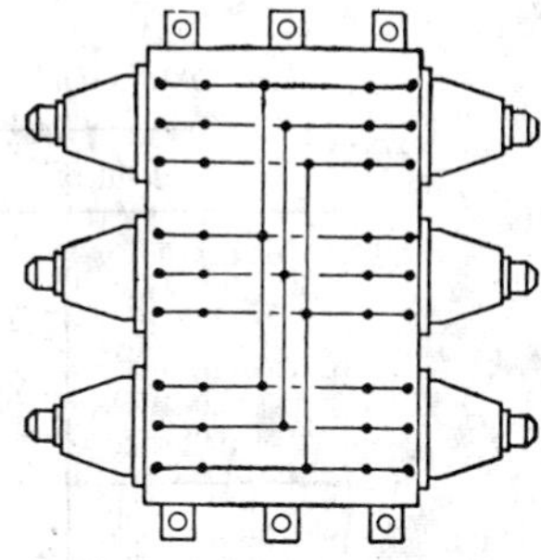

FIG. 16-48. Six-way disconnective cable junction box.

277. Subway Switches. Primary disconnective blades have an arrangement for opening by an insulated handle. Where the connection is opened at periodic intervals, it is often supplemented by an oil switch, the disconnective being used only when repair or alteration work requires that the branch be cut dead (see Fig. 16-48). Such boxes are installed in manholes or transformer vaults and should be placed as conveniently accessible as possible. The manhole should provide ample working space for the box and entering cables.

278. Ambient earth temperatures vary with location, season, and depth. Usually the earth temperature increases directly with the depth in winter and decreases with the depth in summer. At a depth of 3½ ft, typical temperatures are as follows:

	Temperature, °C	
	Summer	Winter
Northern US	20 to 25	2 to 15
Southern US	25 to 30	10 to 20

279. Duct temperature is dependent on the amount and duration of loading of cables drawn into it and on the ability of the duct structure to conduct heat to soil around it. Various studies of the thermal resistivity of concrete structures have given a basis for the following formula for resistivity of duct structures,

$$R = 0.012 \times 100 \log \frac{4d}{L} \text{ per ft length} \tag{16-28}$$

where d = depth from surface to center line of ducts, L = square root of cross-sectional area of duct structures, and 100 = resistivity of average clay-loam soil.

For 16 ducts having cross section 27 × 27 in and average depth of 42 in,

$$R = 1.2 \log {}^{168}\!/_{27} = 0.95°\text{C/W}\cdot\text{(ft)}$$

Table 16-27. Maximum Conductor Temperatures for Impregnated-paper-insulated Cable*

Rated voltage, kV	Conductor temperature, °C		
	Normal operation	Emergency operation	
Solid-type multiple conductor belted			
1	85	105	
2–9	80	100	
10–15	75	95	
Solid-type multiple conductor shielded and single conductor			
1–9	85	105	
10–17	80	100	
18–29	75	95	
30–39	70	90	
40–49	65	85	
50–59	60	75	
60–69	55	70	
Low-pressure gas-filled			
8–17	80	100	
18–29	75	95	
30–39	70	90	
40–46	65	85	
Low-pressure oil-filled and high-pressure pipe type			
		100 h	300 h
15–17	85	105	100
18–39	80	100	95
40–162	75	95	90
163–230	70	90	85

Rise of temperature in the duct system is $WR = 0.95W$, where W = average watts per foot during a load cycle of 24 h. This is the sum of watts loss per foot (I^2r) at maximum load, multiplied by the average loss factor of load on all cables in the duct line. On the assumption of 12 cables with total loss at maximum load of 60 W and average loss factor of 60%, rise of temperature in this 16-duct line is

$$WR \times \text{loss factor} = 60 \times 0.60 \times 0.95 = 34°\text{C}$$

280. Temperature rise in cables from conductor to the outside of the lead sheath is W_1R_1, in which W_1 = watts loss per foot in the conductor and R_1 = thermal resistance of insulation and sheath.

$$R_1 = \frac{3.65G_1}{n} + \frac{4.93}{D} \quad \left[°\text{C/W}\cdot(\text{ft})\right] \tag{16-29}$$

In this formula G_1 is taken from Table 16-26, n = number of conductors in a multiple-conductor cable, and D = outside diameter of the sheath in inches.

For example, a cable having 3 sector conductors of 500,000 cmils may have the thickness of insulation on each conductor, t, 0.203 in, and the outer belt T = 0.109 in. Diameter of conductors d is 0.814 in, and D is 2.875 in. Then $(T + t)/d = 0.312/0.814 = 0.383$, and $t/T = 0.109/0.203 = 0.538$.

In Table 16-26, G_1 when $(T + t)/d = 0.4$ and t/T is 0.5 is 1.27. By prorating this for 0.383 and 0.538, G_1 is approximately 1.24. For a sector-type cable this value is multiplied by sector factor 0.76, making $G_1 = 1.24 \times 0.76 = 0.942$.

Thermal resistivity of the cable is

$$R_1 = \frac{3.65 \times 0.942}{3} + \frac{4.93}{2.875} = 2.86°\text{C/W}\cdot(\text{ft})$$

281. Maximum Allowable Copper Temperature. The rise of copper temperature in cables varies as the square of current carried and inversely as the thermal resistance of insulation, sheath, and duct system. The maximum allowable copper temperatures for paper- and rubber-insulated cables as adopted by the Insulated Power Cable Engineers Association are given in Table 16-27.

The maximum allowable copper temperature for *varnished-cambric-insulated cables* as adopted by the Insulated Power Cable Engineers Association is $(90 - 1\frac{1}{2}E)$°C, with a maximum of 85°C and a minimum of 70°C, where E = line-to-line kilovolts.

282. Maximum Current Capacity of Cables. Tables 16-28 to 16-33 give maximum current-carrying capacity of paper- and rubber-insulated single- and 3-conductor shielded and nonshielded cables.

The tables were obtained from a publication of the Insulated Power Cable Engineers Association entitled "Power Cable Ampacities," Vol. 1, Copper Conductors; Vol. 2, Aluminum Conductors (1962), published by the IEEE (at that time the AIEE). They represent conservative views based on operating experience and laboratory work and are intended as a guide in selecting cables for safety and reliability.

LOW-VOLTAGE A-C NETWORKS

283. Applicability. The a-c network is economically practicable only in the higher-load-density areas. Load densities which are favorable are those running from about 125 kW per city block upward, or a minimum of 15,000 kW/mi².

The load is spread along the mains with approximate uniformity, except that there are occasional large individual blocks of load in department stores, theaters, and tall office buildings which are best served by local transformer installations in or near the premises. Such installations can often be connected into the low-voltage network and serve as points of supply for small users nearby.

The a-c network, if it is to be a dependable source of supply, must include protective equipment on both primary and secondary sides of the transformers. This equipment is expensive and can be justified only when loads are large.

284. Basis of Network Design. A network is made up of primary feeders with branches supplying transformers, which in turn supply energy to the low-voltage network. The most prevalent primary-feeder voltage is in the range of 14 kV.

The arrangement of supply from transformers to the network must be such that when a feeder is out of service the load normally carried by its own transformers will be carried by the transformers of other feeders. This can be accomplished by a transfer of energy through the low-voltage network or by placing two or more transformers supplied from different feeders in the same transformer vault. In the case of very large loads in tall

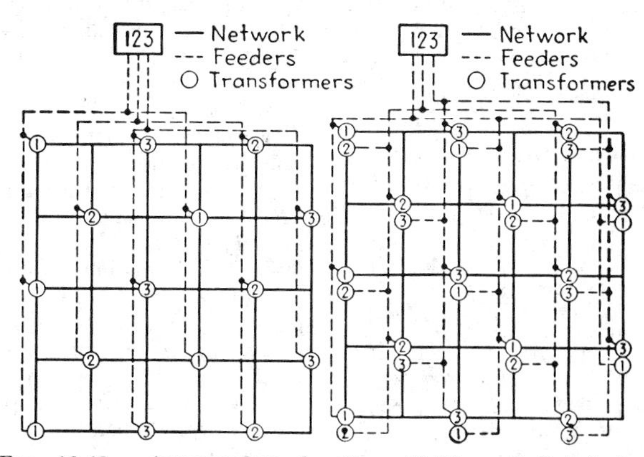

FIG. 16-49. Arrangement of radial feeders and transformers for an a-c network.

FIG. 16-50. Radial feeders supplying duplicate transformers for a-c network.

buildings, it may be desirable to place more than one transformer room in the building, with one on an upper floor.

285. Primary Feeders. The arrangement of primary feeders and transformer vaults shown in Fig. 16-49 is used in areas where load densities are near the lower limit but secondary cables have been provided for future growth with capacity to permit transfer of energy from neighboring vaults when a feeder is out.

In this stage, where large blocks of load are to be supplied, there should be two or more transformers per vault and a like number of primary feeders brought into the vault to supply them. Figure 16-50 shows an installation where loads are too great to be supplied through the secondary network when a feeder is out.

The feeders shown in these two diagrams are radial circuits with as many branches as are needed to reach transformers normally carried by the feeder.

286. Secondary Mains. The secondary network, serving as the tie between transformers on different feeders and as a voltage-steadying medium, should be made up of conductors of ample size to meet requirements.

Where loads are not so great as to require transformers near each intersection, mains must be large enough to carry the combined load of two feeders when either of them is out of service. Under this condition it may be permissible to allow somewhat more drop in voltage than the 2 or 3 V considered permissible under normal conditions.

Starting currents of motors and particularly of elevator and other intermittently operated equipment may affect lighting service. The conductor should be proportioned so that motors do not produce an objectionable amount of flicker in lighting service from the same main. Initially, conductors should provide capacity for as much future load as is practicable, as the cost of replacement is high after many service connections have been made.

With transformer sizes of 150 kVA (3-phase) feeding into four-way junctions, conductors of 4/0 AWG are usually adequate. For units of 300 kVA and larger, the use of 500,000 cmils or two sets of 4/0 in parallel is usual. The impedance drop of two sets of 4/0 is less than that of a 500,000-cmil circuit.

287. Limitation of Size of Conductors. It has been found advisable from experience and from many tests to limit a-c network mains to a maximum of 500,000 cmils. Such tests have shown that 4/0 cables burn clear in about 3 s if grounded on one phase, with a flow of 3,000 A. When two phases are firmly crossed, it requires 60 s to burn clear and takes 6,000 A.

A 4/0 cable is fused throughout its length when 5,500 A has flowed in it for about 2 min, and a 500,000-cmil cable is fused by 9,000 A in the same time. If larger cables are used, the arc tends to persist long enough to destroy the cable before it burns itself clear.

288. Operation of Networks. Three-phase 4-wire distribution is commonly used for networks to provide both light and power from the same service and meter. Lighting is connected from phase wires to neutral at 120 V, with 208 V for 3-phase power service. Motors being commonly designed for 220 V do not give satisfactory service at less than 90% of rated voltage.

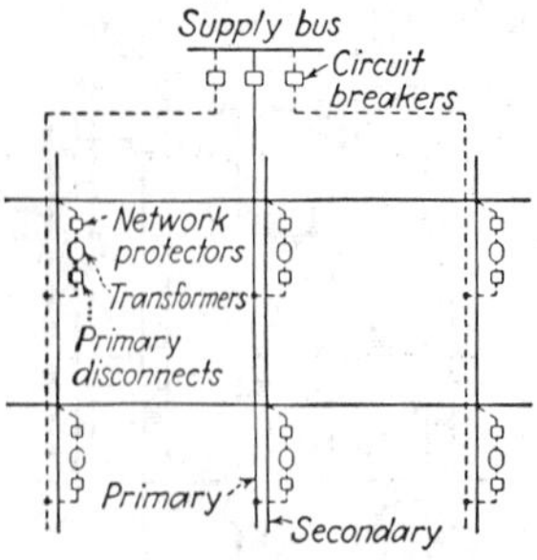

Service from networks is protected from interruption by isolation of transformers or primary cables when either has failed. Figure 16-51 shows the connections of feeders and protective breakers. The primary connection of the transformer is provided with oil-immersed disconnects by which supply from the feeder may be isolated. This disconnect also has a position in which the feeder is grounded to prevent its being accidentally energized while work is in progress.

Fig. 16-51. Protective equipment, network feeders.

The network protector is arranged to open automatically when primary supply is cut off and to reclose automatically when it is reenergized. This breaker is backed up by fuses in the secondary circuit of the transformer which are not designed to operate except in case the breaker should fail to function as intended.

Table 16-28a. Maximum-current Capacity of Solid-type Impregnated-paper-insulated 3-conductor Cables, Shielded Copper Conductor *

(One loaded cable in a duct. See correction factors below for three, six, and nine loaded cables in a duct bank. Earth temperature = 20°C.)

Conductor size, AWG or M cmils	Rated line voltage, grounded neutral					
	15,000		25,000		37,000	
	Max. copper temp. 80°C (RHO 90)		Max. copper temp. 75°C (RHO 120)		Max. copper temp. 70°C (RHO 90)	
	% load factor		% load factor		% load factor	
	50	75	50	75	50	75
2	151	145				
1	173	166	167	160		
1/0	199	190	191	182		
2/0	224	214	219	208		
3/0	256	244	249	237	236	224
4/0	294	279	284	270	269	255
250	324	307	313	296	299	283
350	394	372	379	358	360	340
500	481	453	461	434	437	411
750	598	560	570	534	538	503
1,000	690	644	657	613	617	574

Table 16-28b. Correction Factors for Various Ambient Earth Temperatures *

°C	Rated line voltage, grounded neutral		
	15,000	25,000	37,000
10	1.08	1.09	1.10
20	1.00	1.00	1.00
30	0.91	0.90	0.89
40	0.82	0.80	0.76
50	0.71	0.67	0.61

Table 16-28c. Correction Factors for Three, Six, and Nine Loaded Cables in Duct Bank *

Conductor size	Number loaded cables in duct bank					
	3		6		9	
	% load factor		% load factor		% load factor	
	50	75	50	75	50	75
2–4/0	0.94	0.91	0.87	0.80	0.82	0.74
250–400	0.93	0.89	0.85	0.78	0.79	0.71
500–750	0.91	0.87	0.83	0.75	0.76	0.67

*Tables **16-28a**, b, and c copyright 1962 by Insulated Power Cable Engineers Association. Adapted by permission.

Table 16-29a. Maximum Current Capacity of Solid-type Impregnated-paper-insulated 3-conductor Belted-type Cables, Copper Conductor *

(One loaded cable in a duct. See correction factors in Table 16–29c for three, six, and nine loaded cables in a duct bank. Earth temperature = 20°C.)

Conductor size, AWG or M cmils	Rated line voltage, grounded neutral			
	8,000		15,000	
	Max. copper temp. =80°C (RHO 90)		Max. copper temp. =75°C (RHO 90)	
	% load factor		% load factor	
	50	75	50	75
6	79	77		
4	103	100	101	97
2	135	131	131	127
1	156	150	150	145
1/0	179	172	172	166
2/0	204	196	197	189
3/0	234	225	225	216
4/0	270	259	258	247
250	299	286	285	273
350	368	351	349	333
500	456	432	429	408
750	578	544	540	510
1,000	677	634	630	592

Table 16-29b. Correction Factors for Various Ambient Earth Temperatures *

°C	Rated line voltage, grounded neutral	
	8,000	15,000
10	1.08	1.09
20	1.00	1.00
30	0.92	0.90
40	0.83	0.79
50	0.72	0.67

Table 16-29c. Correction Factors for Three, Six, and Nine Loaded Cables in Duct Bank *

Conductor size	Number loaded cables in duct bank					
	3		6		9	
	% load factor		% load factor		% load factor	
	50	75	50	75	50	75
6–2/0	0.95	0.92	0.90	0.85	0.86	0.80
3/0–400	0.94	0.90	0.83	0.82	0.84	0.76
500–750	0.92	0.89	0.85	0.79	0.79	0.72

*Tables **16-29a, b,** and **c** copyright 1962 by Insulated Power Cable Engineers Association. Adapted by permission.

Table 16-30a. Maximum Current Capacity of Rubber- or Thermoplastic-insulated 3-conductor Concentric Cables, Copper Conductor *

(One loaded cable in a duct. See correction factors in Table 16-30c for three, six, and nine loaded cables in duct bank. Earth temperature = 20°C.)

Conductor size, AWG or M cmils	Rated line voltage, grounded neutral					
	8,000		15,000		25,000	
	Max. copper temp. 60°C		Max. copper temp. 60°C		Max. copper temp. 60°C	
	% load factor		% load factor		% load factor	
	50	75	50	75	50	75
6	76	73				
4	99	95				
2	128	123	131	125		
1	147	140	149	142		
1/0	168	160	170	162	151	144
2/0	192	183	194	184	172	163
3/0	219	209	221	210	195	185
4/0	250	237	252	239	221	210
250	275	261	276	262	251	238
350	333	314	333	314	275	260
500	403	379	402	378	330	311
750	491	460	489	458	398	373
1,000	552	515	551	514	484	452
					546	509

Table 16-30b. Correction Factors for Various Ambient Earth Temperatures *

°C	Rated line voltage, grounded neutral		
	8,000	15,000	25,000
15	1.06	1.06	1.06
20	1.00	1.00	1.00
25	0.93	0.93	0.93
30	0.87	0.86	0.86
35	0.79	0.78	0.78
40	0.71	0.70	0.69

Table 16-30c. Correction Factors for Three, Six, and Nine Loaded Cables in Duct Bank *

Conductor size	Number loaded cables in duct bank					
	3		6		9	
	% load factor		% load factor		% load factor	
	50	75	50	75	50	75
6–2/0	0.94	0.89	0.86	0.78	0.82	0.72
3/0–350	0.92	0.87	0.83	0.75	0.78	0.68
500–1,000	0.90	0.84	0.77	0.70	0.73	0.64

When a secondary cable fails in a system having ample transformer capacity to supply short-circuit current, the cable main burns off very quickly and clears itself. There is a violent fluctuation of voltage while the cable is burning, and it may be followed by voltage below normal on one side of the point where the main is open. A failure is sometimes cleared by fuses at junction points of the network, but the trend in design is toward the omission of junction boxes.

289. Limiters are used by many companies to protect the insulation of cables from heat generated by a short circuit. The limiter is a restricted copper section, usually enclosed in a porcelain envelope and installed at junction points in the secondary main.

290. Transformer Vaults. When transformers are placed in a street or alley, a vault is constructed especially to meet requirements of network equipment. If placed

Table 16-31a. Maximum Current Capacity of Single-conductor Paper-insulated Solid-type Concentric-strand Copper *

(Three similar loaded cables in a duct bank, one cable per duct. For more than three cables in a duct bank see correction factors in Table 16–31c)

Conductor size, AWG or M cmils	Rated line voltages, grounded neutral							
	8,000		15,000		25,000		37,000	
	Max. copper temp. 85°C		Max. copper temp. 80°C		Max. copper temp. 75°C		Max. copper temp. 70°C	
	% load factor		% load factor		% load factor		% load factor	
	50	75	50	75	50	75	50	75
1/0	258	242	246	231	232	219		
2/0	296	277	282	265	266	250		
4/0	391	364	373	348	352	329	329	308
350	530	490	505	467	474	440	443	412
500	657	603	623	574	586	541	545	505
750	835	761	791	723	742	680	689	633
1,000	983	890	931	846	872	795	807	739
1,500	1,221	1,096	1,154	1,039	1,096	973	995	903
2,000	1,400	1,249	1,320	1,181	1,230	1,105	1,131	1,020

Table 16-31b. Correction Factors for Various Ambient Earth Temperatures *

°C	Rated line voltage, grounded neutral			
	8,000	15,000	25,000	37,000
10	1.07	1.08	1.09	1.10
20	1.00	1.00	1.00	1.00
30	0.92	0.92	0.90	0.89
40	0.83	0.82	0.80	0.76
50	0.73	0.71	0.67	0.60

Table 16-31c. Correction Factors for More than Three Similar Loaded Cables in Bank *

Conductor size	Number single-conductor cables in duct bank		
	3	6	9
1/0–4/0	1.0	0.91	0.86
350–60	1.0	0.88	0.82
750–1,000	1.0	0.86	0.80
1,500–2,000	1.0	0.85	0.78

*Tables **16-31a, b,** and **c** copyright 1962 by Insulated Power Cable Engineers Association. Adapted by permission.

Table 16-32a. Maximum Current Capacity of Single-conductor Rubber- or Thermoplastic-insulated Concentric-strand Cables, Copper Conductor *

(Three similar loaded cables in a duct bank. For more than three cables in a bank see correction factors in Table 16-32c. Load factor 75%. Earth temperature = 20°C.)

Conductor size, AWG or M cmils	Rated line voltage, grounded neutral						
	8,000			15,000		25,000	
	Max. copper temp.			Max. copper temp.		Max. copper temp.	
	60°C	70°C	80°C	60°C	75°C	60°C	70°C
6	89	98	105				
4	116	127	137				
2	151	166	179	151	173		
1	172	190	205	172	197	170	188
1/0	197	217	234	197	226	195	215
2/0	225	248	267	225	258	222	244
3/0	258	284	306	257	294	253	279
4/0	295	325	350	293	337	288	318
350	395	435	469	391	449	384	424
500	485	534	576	479	550	468	517
750	608	670	724	599	689	584	647
1,000	709	782	846	698	804	678	751
1,500	865	957	1,036	848	981	822	914
2,000	976	1,082	1,175	959	1,112	927	1,031

Table 16-32b. Correction Factors for Various Earth Temperatures *

°C	Rated line voltage, grounded neutral						
	8,000			15,000		25,000	
	Max. copper temp.			Max. copper temp.		Max. copper temp.	
	60°C	70°C	80°C	60°C	75°C	60°C	70°C
15	1.08	1.06	1.04	1.06	1.04	1.06	1.05
20	1.00	1.00	1.00	1.00	1.00	1.00	1.00
25	0.91	0.94	0.95	0.93	0.95	0.93	0.95
30	0.82	0.87	0.90	0.86	0.90	0.86	0.89
35	0.71	0.79	0.85	0.79	0.85	0.78	0.83
40	0.58	0.71	0.80	0.70	0.79	0.70	0.77

Table 16-32c. Correction Factors for More than Three Similar Loaded Cables (60°C Conductor Temp.) in Duct Bank *

Conductor size	Number single-conductor cables in duct banks		
	3	6	9
6–1	1.0	0.94	0.90
1/0–4/0	1.0	0.94	0.84
350–750	1.0	0.92	0.83
1,000–2,000	1.0	0.90	0.82

Table 16-33a. Maximum Current Capacity of Single-conductor Rubber- or Thermoplastic-insulated Concentric-strand Cables, Aluminum Conductor *

(Three similar loaded cables in a duct bank 75% LF. For more than three cables in a bank, see correction factors in Table 16-33c. Earth temperature 20°)

Conductor size, AWG or M cmils	Rated line voltage						
	8,000			15,000		25,000	
	Max. alum. temp.			Max. alum. temp.		Max. alum. temp.	
	60°C	70°C	80°C	60°C	75°C	60°C	70°C
6	69	76	82				
4	90	99	107				
2	117	129	139	118	135		
1	134	148	159	134	154	133	147
1/0	154	169	182	154	176	152	167
2/0	176	193	208	175	201	173	191
3/0	201	221	238	200	229	197	218
4/0	230	254	273	229	263	225	248
350	309	340	367	306	351	300	331
500	380	418	451	376	431	367	406
750	481	529	571	473	544	462	510
1,000	565	623	672	556	640	541	598
1,500	705	778	840	691	796	670	743
2,000	815	900	973	799	922	772	858

Table 16-33b. Correction Factors for Various Earth Temperatures *

°C	Rated line voltage, grounded neutral (60°C conductor temp.)		
	8,000	15,000	25,000
15	1.08	1.06	1.06
20	1.00	1.00	1.00
25	0.91	0.93	0.93
30	0.82	0.86	0.86
35	0.71	0.79	0.78
40	0.58	0.70	0.70

Table 16-33c. Correction Factors for More than Three Similar Loaded Cables in Duct Bank *

Conductor size	Number single-conductor cables in duct bank		
	3	6	9
6–1	1.0	0.90	0.87
1/0–4/0	1.0	0.89	0.84
350–750	1.0	0.86	0.82
1,000–2,000	1.0	0.84	0.78

*Tables **16-33a, b,** and **c** copyright 1962 by Insulated Power Cable Engineers Association. Adapted by permission.

within a building, the room is provided at a location which has adequate means of entrance and exit by which equipment can be got in or out. In a very tall building, more than one room is sometimes needed, in which case one is placed on an upper floor, thereby reducing the cost of riser mains in the building.

The general arrangement of a vault placed beneath the sidewalk is shown in Fig. 16-52. This vault is arranged for two transformers with network protectors and oil

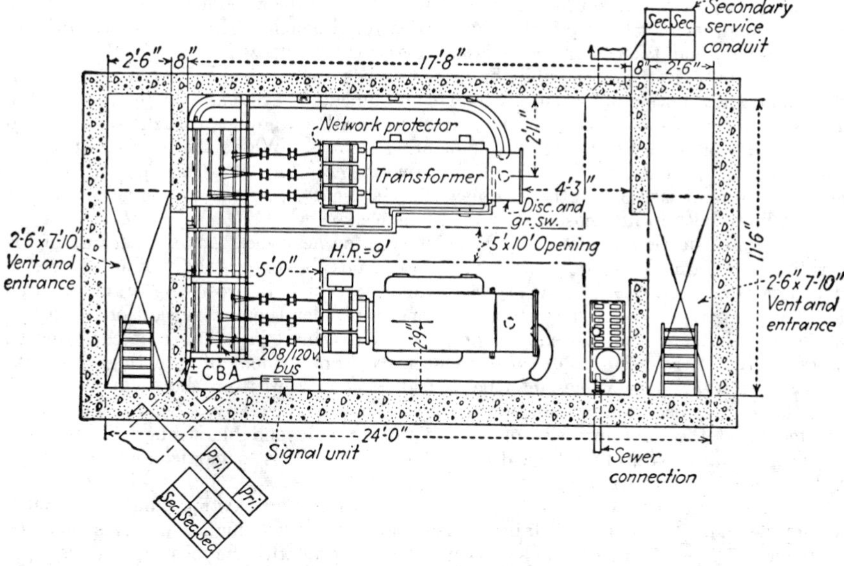

FIG. 16-52. General arrangement of a network vault under a sidewalk (plan view).

disconnects. The roof consists of removable slabs of sidewalk (not shown), and access is had at either end by iron steps going into space provided for circulation of air. The entrance is covered by a grid at sidewalk level which permits air circulation.

Drainage is provided by a sewer connection. Secondary terminals pass from the network protector to the 208-V bus, thence to street mains.

LINES FOR RURAL SERVICE

291. Basic Conditions. Rural distribution differs from urban in that consumers are scattered over a much wider area and load units are generally small. Distances being great, the voltage must often be higher than 4,000, and, the load in kilowatts per mile being low, the cost of line construction must be kept as low as is consistent with a reasonable degree of permanence and reliability. Individual transformers are required in many cases. Rural construction since the late 1930's has made electric service available to practically every farm. Efforts are directed now to bolstering capacity to serve the growing loads.

292. Poles and Spans. Design of lines for rural service differs from urban lines in several respects. Costs are reduced by using longer spans and as few accessories as possible. Longer spans mean greater sag and higher poles to get proper clearance at the low point of the span. The increase in sag may, however, be reduced by use of higher tensile stresses in conductors. This is possible when steel is employed in conjunction with copper or aluminum wires. Steel is combined with copper in a high-strength strand known as Copperweld, which has 30 or 40% of the conductivity of a copper conductor of equal size, or in a similar aluminum and steel conductor known as Alumoweld. Where greater conductivity is needed, one or more strands of hard copper are stranded with or around the Copperweld, such conductor being known as Copperweld copper, or hard aluminum with Alumoweld to make AWAC. Steel is also stranded with aluminum wires into ACSR cable. Such types of conductor have ample conductivity for the general average of rural lines, and No. 4 has been used widely.

In level country spans of 400 to 600 ft are practical, while in hilly country spans of 800 to 900 ft are occasionally possible.

293. Location of Lines. Rural-service lines are carried along main highways, where the largest number of users may be reached. Branches along intersecting roads are extended as may be warranted by service requirements. In some cases, private rights-of-way, maintained for transmission lines, may be utilized.

Lines should, as far as possible, be placed on the opposite side of the roadway from communication lines, unless supply and communication systems occupy joint poles. Lines should not be carried from one side of the road to another in such manner as to obstruct use of the route for telephone or other service. Where the number of utility services using a highway route is such that two or more lines must occupy one or both sides of the road, use of jointly occupied poles is usually preferable to erection of conflicting lines, both in first cost and in maintenance.

294. Pole Locations. Poles should not be set too near the roadway, because of the possibility of damage by traffic and avoidance of possible expense of moving in case of widening the roadway. The most practicable location is between the drainage ditch and the property line. Where such locations are under the supervision of county or state officials their approval should, of course, be secured in advance of construction work. Often permission must also be secured from owners of abutting property.

295. Voltage. Rural lines may be extended 5 to 50 mi from the point of supply, and the voltage used for distribution must be chosen accordingly. Loadings are often so small that the minimum size of wire required for dependable strength is sufficient to meet requirements of voltage drop and line loss. This is particularly true when voltages above 4,000 are used.

Within a radius of 3 to 5 mi, it is possible to extend an existing 2,300- or 4,000-V 4-wire system. When this distance is materially exceeded, higher voltages must be employed; 7,200 V from a 12,500-V 4-wire source is probably the most common supply for transformers at distances over 5 mi. Some systems use 14,400/24,900-V 4-wire in sparse areas, and a few are 34,500 or even 46,000 V.

Single-phase circuits are most economical where loads are small and power units do not exceed 10 hp. Three-phase circuits are desirable when loads require their use.

296. Limitations of voltage and distance are illustrated by the following figures showing kilowatt-miles corresponding to a 5% line drop at 80% power factor for a circuit of No. 6 copper, or its equivalent in other metals:

Kilowatts × Miles for 5% Voltage Drop, Power Factor 80%

System and voltage	2,400	4,160	7,200	8,330	12,500	13,800	20,000	34,500
Single-phase	48	...	432	....	1,290	1,735	3,700	
Three-phase	96	290	864	1,160	2,590	3,460	.....	21,600

Values for other sizes are approximately in proportion to relative cross section.

297. Conductors and Spans. Because of the economy of using long spans, the choice of span lengths and conductor strength is of much importance in planning rural lines. Single-phase lines are commonly taken from a 3-phase system with neutral grounded. The grounded conductor is carried on a metal bracket without an insulator about 2 ft below the phase wire, which rests on an insulator carried on the top of the pole. No crossarm is required, except on a main line of more than one phase.

Conductivity of No. 4 ACSR or Copperweld is adequate for the greater part of a rural-line system. The strength of such conductors is ample for spans of 350 ft under a coating of ice ½ in thick with an 8-lb wind at a sag of 8 ft at 60°F. Where no ice is encountered, the span may be as much as 600 ft with a sag of 8 ft at 60°F. These span lengths are for level ground.

Where the elevations permit greater amounts of sag, the span of two No. 4 ACSR conductors may be increased up to 1,300 ft under the ice and wind loading abovementioned or to 2,000 ft where no ice is likely to be experienced. The sag would, of course, be much greater than the 8 ft allowed on level ground.

298. Poles. The strength of poles should be determined for the height required by the methods described in Pars. **188** to **203.**

The length of poles required in any situation must be such as to allow for depth of setting and height of wire supports needed to give proper clearance above ground at the low point of the span. Such clearances should be not less than 17 ft over intervening land, 18 to 20 ft over roadways, and 28 ft over railroads. In the case of road and railroad crossings, the necessary clearance may sometimes be more readily had by placing one end of the span near the crossing, thus avoiding having the low part of the span over the crossing. In rolling or hilly areas, it is desirable to locate poles on higher elevations to permit use of longer spans and greater sags.

Where no ice loading is likely, 30-ft poles can be used for two conductors of a single-phase branch on level ground or on long even slopes with span lengths to 400 ft. Where ice and wind loading is expected with some regularity, it is necessary to use 35-ft poles for spans exceeding 300 ft. At corners or angles, poles should be supported by guying or bracing to support unbalanced longitudinal stresses.

299. Crossarms are required for the main 3-phase circuits and for lines supplying any user taking 3-phase service for power. A two-pin arm is often used with the third phase on the pole top. The grounded neutral is carried on the side of the pole about 2 ft below the arm.

300. Transformer Installations. Transformers usually supply not more than one or two customers; and sizes, therefore, are small as compared with the average used in urban work, 5 to 10 kVA being average for single-phase installations. Where points of use are more than 1,200 ft apart, it is usually most economical to provide separate transformers. When two users are within 1,200 ft, it is usually economical to place a transformer midway between and construct a secondary main of No. 4 ACSR to supply both. For loads in excess of 3 kVA per farm, larger sizes of wire and shorter spacings are necessary. Rural loads on some systems have grown to the point where 10- and 15-kVA transformers are required.

Transformer capacity may usually be selected on a basis of overloads up to 50% for 1 to 2 h. Many power applications on farms are intermittent, or not over 2 h duration. Pumping for drainage or irrigation is likely to require rated capacity more nearly equal to load. Secondary circuits should be properly grounded to water pipes, if available, or to a pipe driven about 8 ft into the soil.

DEMAND AND DIVERSITY FACTORS

301. Demand Factor. The ratio of maximum use of energy to total load connected, expressed as a percentage, is termed the demand factor of an installation. For example, if a residence having equipment connected with a total rating of 6,000 W has a maximum demand of 3,300 W, it has a demand factor of 55%.

Demand factors of various types of users are important in designing distribution systems and in determining service rates, when demand factors for various classes of users have been selected by averaging a considerable number of each class. They may be applied to the choice of meter and transformer sizes for new installations and to the design of secondary mains and services. Where rates are based upon a demand charge, the demand factor is of great assistance in predicting cost of service.

302. Typical Demand Factors. For three to five rooms, individual demand is usually 60 to 80% of connected wattage, while in larger residences it is 45 to 65%.

Table 16-34. Diversity Factors

Elements of system between which diversity factors are stated:	Diversity factors for			
	Residence lighting	Commercial lighting	General power	Large users
Between individual users	2.0	1.46	1.45	
Between transformers	1.3	1.3	1.35	1.05
Between feeders	1.15	1.15	1.15	1.05
Between substations	1.1	1.10	1.1	1.1
From users to transformer	2.0	1.46	1.44	
From users to feeder	2.6	1.90	1.95	1.15
From users to substation	3.0	2.18	2.24	1.32
From users to generating station	3.29	2.40	2.46	1.45

Stores, offices, shops, and factory buildings may have demands of 50 to 90% of connected wattage. Summer air conditioning and winter electric heating usually set the maximum demands and make for high demand factors. Loft spaces, warehouses, and the like, usually have considerably lower demands.

303. Diversity factor is defined as the ratio of the sum of demands of a group of users to their coincident maximum demand. Stated in this form, it is a figure greater than unity. The ratio may obviously be stated in the reciprocal relation and expressed in percent, as for demand, load, and power factors. The effect of diversity may be observed at the local transformer, at the substation, and at the power source.

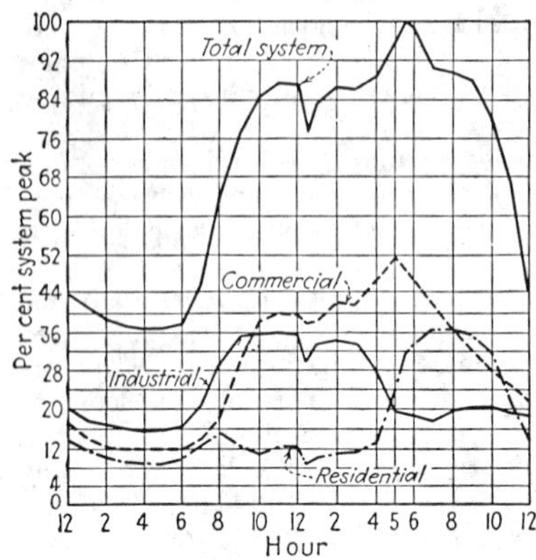

FIG. 16-53. Characteristic metropolitan load pattern.

304. Diversity between Classes of Users. The daily-load diagram of a utility is a composite of demands made by various classes of user. Industrial users make their heaviest demands in the morning, and a considerable part of the load has disappeared before demand for lighting in the afternoon nears its peak. Commercial users make their heavier demands in the afternoon and early evening. The highest demand for residential lighting occurs from 7 to 8 P.M., when commercial demand has receded from its peak and is rapidly falling away (see Fig. 16-53).

Air conditioning is shifting these curves for some systems to cause daytime peaks during hot weather. Electric house heating builds heavy evening and morning loads during cold weather.

305. Diversity in the Feeder System. The diversity of demands by transformers on a radial feeder makes the maximum load on the feeder less than the sum of the transformer loads. The diversity factor of lighting feeders ranges from 1.1 to 1.5, while that of mixed light-and-power feeders is likely to be 1.5 to 2 or more. At the substation there is also a diversity factor of 1.05 to 1.25 between the sum of feeder maxima and the substation maximum. A large system has a further diversity factor between substations of 1.05 to 1.25.

306. Total diversity factors in a large system are somewhat as in Table 16-34.

DISTRIBUTION ECONOMICS

307. Kelvin's Law. Lord Kelvin, in 1881, established the general principle that the most economical size of conductor was that for which investment charges were equal to cost of energy losses, this being the point at which their sum is a minimum. Practical evaluation of investment cost and energy loss should, however, also take into account fixed charges on supply-system capacity absorbed by line loss. Refinements to reflect annual revenue requirements of investment, energy losses, future replacement, taxes, etc., have largely supplanted the Kelvin approach for large power systems. However, they are much too complex to include here; hence the Kelvin method is offered as a guide.

308. General Formula. The variable in choice of conductor size which controls the other factors is R, resistance per 1,000 ft. Investment charges on the conductor are inversely proportional to R; investment charges on generating capacity absorbed in supplying losses and cost of losses in the conductor are directly proportional to R. Total annual cost is

$$Y = \frac{a}{R} + bI^2R + cI^2R \qquad (16\text{-}30)$$

in which a, b, and c = constants to be determined for any particular case from total cost of conductor installed, of power-station capacity, and of producing energy dissipated by flow of current in conductor. Formula (16-30) implies the possibility of an approximate determination of annual loss in kilowatthours in the conductor. It does not take into account voltage regulation, which is often the factor that determines conductor size. Kelvin's law should be applied with due consideration to voltage drop.

309. Application of Overhead Lines. For bare-copper wire there is a practically constant relation between weight per 1,000 ft and resistance per 1,000 ft, which, for solid wire at 20°C, is $WR = 31.4$ and for stranded copper is $WR = 32$. For triple-braid weather-resisting solid wire at 20°C, $WR = 40.5$ from No. 4 to No. 0; and for stranded conductors, WR has an average value of 42.4 from No. 00 to 300,000 cmils. With copper at \$0.36 per pound and labor of erection at \$0.04 per pound, the cost per 1,000 ft of conductor is $0.4W$. For bare wire, stranded, $W = 32/R$, and with fixed charges on the conductor at 12%, the annual cost for the circuit conductors is $a/R = (0.12 \times 0.4 \times 32)/R = 1.536/R$. For covered stranded conductors

$$\frac{a}{R} = \frac{0.12 \times 0.4 \times 42.4}{R} = \frac{2.035}{R}$$

Similar relationships can be derived for aluminum, ACSR, and other line conductors.

310. Application to Cables. Cost of lead-sheathed cables does not vary in direct proportion to cross-sectional area of conductor, as it does in the case of overhead conductors, because relative thickness of insulation is greater and the insulation may be covered by a sheathing of lead. However, costs per 1,000 ft/1,000 cmils of adjacent sizes of cables are such as to permit their use in deriving values in the formula for annual cost.

If M is area of conductor in thousands of circular mils and P is price per 1,000 ft/1,000 cmils, cost per 1,000 ft of conductor, with its proportionate share of insulation and sheath, is MP. $M = 10.56/R$, and the cost is $10.56P/R$ at 25°C.

With fixed charges at 12%, the annual cost is

$$\frac{a}{R} = \frac{(10.56 \times 0.12)P}{R} = \frac{1.27P}{R} \quad \text{(dollars)} \tag{16-31}$$

311. Fixed Charges on Supply-system Capacity. When the circuit is supplied at generated voltage from a power-station bus, fixed charges on the capacity required to supply circuit loss must be included as part of cost. Where the line is supplied through transformers and over a transmission line, cost of "generating capacity" must include cost per kilowatt of all transformations and intermediate lines. Because distribution losses are too small to be responsible for installing additional supply-system capacity, it is customary to reduce supply-system charges by 25 to 50% in studies of distribution economics.

Even so, the fixed charges on generating capacity are higher for distribution lines in remoter parts of the system than for circuits supplied at a point of production. If generating capacity is valued at K dollars per kilowatt and the fixed charges are 12%, annual cost is

$$bI^2R = 0.12KI^2R/1{,}000 = 0.00012KI^2R \tag{16-32}$$

312. Calculation of Energy Loss. Where load characteristics are established with such stability that annual loss factor may be computed, annual loss in kilowatthours = loss in kilowatts at peak × 8,760 × loss factor. The loss factor of typical loads may be found with approximate accuracy by computing the loss in kilowatthours for one week in March, June, September, and December and multiplying the total of these 4 weeks by 13 to get total annual loss. The loss factor is then the ratio of annual loss to loss at peak hour of the year multiplied by 8,760 h.

313. Cost of Conductor Loss. Loss factor for the type of load carried by a circuit having been determined, annual loss for maximum load current I at loss factor F is

$$\text{Annual loss} = 8{,}760FI^2R/1{,}000 \text{ kWh} \tag{16-33}$$

Cost of energy loss at p cents per kilowatthour is

$$cI^2R = 8.76pFI^2R \tag{16-34}$$

314. Summary of Annual Cost. The sum of the three elements entering into the annual cost is, according to Eq. (16-30),

$$Y = \frac{a}{R} + (b + c)I^2R$$

By Kelvin's law Y is least when $a/R = (b + c)I^2R$. This is so when

$$I^2R^2 = \frac{a}{b + c} \qquad \text{and} \qquad IR = \sqrt{\frac{a}{b + c}} \tag{16-35}$$

For bare conductors, $a = 1.536$, $b = 0.00012K$, $c = 8.76pF$;

$$IR = \sqrt{\frac{1.536}{0.00012K + 8.76pF}} \tag{16-36}$$

For any assumed value of I, maximum amperes carried by the circuit, the most economical value of R and from this the best size of conductor are readily found. Likewise, for any conductor, the maximum load in amperes that is most economical in annual cost may be found. Values of K, cost of generating capacity; p, cost of energy per kilowatthour; and F, loss factor, must be chosen to fit each case.

Where load varies from year to year, either the value of I must be selected to represent the range of maxima or the present worth of losses at future loads must be considered.

315. Example. Assume generating-station capacity to cost $150 per kilowatt, transmission-line capacity $30 per kilowatt, and step-up transformer capacity $20 per kilowatt. All three costs are applied at 50% value to reflect the incremental cost of marginal capacity. With energy cost p at 2 mills per kilowatthour and loss factor F at 15%, the value of IR for *bare overhead* conductors is

$$\sqrt{\frac{1.536}{0.00012 \times 100 + 8.76 \times 0.002 \times 0.15}} = 10.22$$

Upon assuming current of 200 A, the most economical value of R is 10.22/200 = 0.0511. Area of the conductor then is 10,400/0.0511 = 203,523 cmils copper. If 250,000 cmils copper is assumed, its most economical maximum load current would be $I =$ 10.22/0.0440 = 232 A.

STREET-LIGHTING SYSTEMS

316. Characteristics. The lighting of streets, parkways, and other public thoroughfares is about the only service in which the electric supply utility is often responsible also for the utilization equipment. This responsibility, moreover, involves turning the lights on when they are needed and off when natural light has sufficient intensity to meet public needs. For many years this service was powered by special circuits which could be energized only during periods when lighting was required. These special circuits usually had many lights connected in series, thus operating at a fairly high voltage and low current, and were turned on and off manually or by time-clock-controlled switches. But extensive improvements have culminated in photoelectric controls having a high order of reliability at reasonable cost; with these, it has become feasible and economical to connect street-light luminaires directly to the area distribution and eliminate the separate circuits and control systems.

317. Multiple lighting is extensively used today and is usually connected directly to local 120/240-V distribution. The lights ordinarily are controlled by photocells, but for some special applications they may be left on continuously or may be switched by time clocks, pilot wires, carrier-current signals, or other means.

318. Lamp units used for multiple street lighting originally were incandescent bulbs resembling those in general use but ranging in size from 100 to 1,000 W. Today, however, there is a much broader range, including fluorescent or color-corrected mercury units for commercial white-way lighting, conventional mercury for most efficient lighting of important highways and interchanges, and occasionally sodium units where the distinctive yellow hue is deemed valuable to warn motorists of potential danger. All types come in a wide range of mountings designed to direct the light output in a variety of patterns to match requirements, as well as in a choice of designs to suit the environment.

319. Series street lighting is still extensively employed, however, and, because it involves many features not found in other aspects of distribution, must be considered here. Unlike general distribution, it is powered by a variable-voltage constant-current system.

320. Constant-current transformers are commonly used to supply current, usually 6.6 or 20 A to the many lamps connected in series. The moving-coil constant-current transformer is a single-phase device with inherently high leakage reactance. Movement of the coil varies this reactance automatically to maintain constant-current output. The transformer is capable of regulating current within 1% for all loads within its rating, at any supply voltage within 5% of normal, and with any ordinary variation in frequency and temperature. Full-load power factor is approximately 75%; three-fourths load, about 56%. Input kVA is substantially constant over the entire range of loads. In certain designs, capacitors improve the low power factor to a value nearly unity for normal loads.

The **indoor,** or **station, type** of constant-current transformer is supplied from the substation high-tension bus. Indoor ratings vary from 5 to 70 kW, the 30- and the 60-kW being most popular. The secondary winding of larger ratings is usually designed with two coils, and leads from each coil are cross-connected to supply two series circuits. The **pole-** and **submersible-type** constant-current transformers are built in sizes from 2 to 30 kW in oil-filled tanks, usually with capacitors. They are mounted in the immediate vicinity of the lamps they supply, thus realizing economy in circuit construction. Distribution primaries supply the constant-current transformers with single-phase power.

321. Series circuits are controlled (1) manually, (2) by time clocks, and (3) by relays operating on the cascade principle. Time switches with astronomical dials are available which automatically vary lighting hours as seasons change. Time clocks are electrically driven from the 120-V circuit; hence no winding is required. After an extended outage to the circuit supplying electric drive, the clock must be reset.

322. Cascade control is commonly used to turn "on" and "off" a series circuit that is contiguous to a circuit that is manually controlled. Current in the controlling circuit operates a coil in the controller switch that closes the primary supply to the constant-current transformer of the controlled circuit. Figure 16-54 illustrates this method.

323. Series Systems. Current in a series circuit passes directly, without transformation, through the lamps and circuit. All parts of an ungrounded series circuit are subject to high voltages and must be insulated accordingly. Most circuits operate at 6.6 A. Circuits serving large lamps in the heart of a city often operate at 20 A. Series circuits are usually carried on the same crossarms with primary wires in overhead distribution and in separate ducts or pipes in underground distribution. In parkways

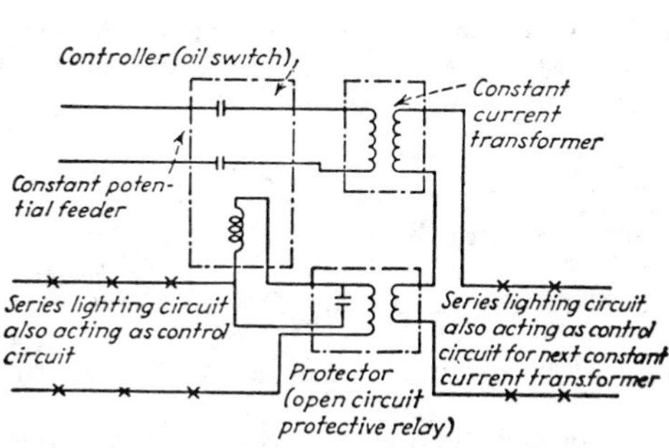

Fig. 16-54. Cascade control of constant-current transformers.

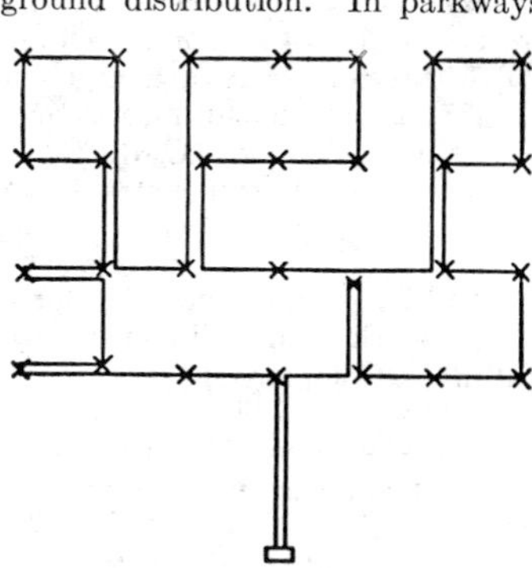

Fig. 16-55. Series circuit.

series-circuit cables are often buried. Individual lamps are provided with series cutouts which automatically close the circuit when lamps are removed. A *film cutout* designed to puncture at about 1,000 V is connected across terminals of each lamp; if the lamp filament breaks or burns out, the film cutout is punctured, the lamp short-circuited, and the circuit closed.

324. Types of Series Circuit. Series circuits may be laid out on the *parallel-* or the *open-loop* plan. In the parallel loop, going and returning wires of the circuit are carried along the same route; in the open-loop plan, the circuit goes out along one street and returns by another. The parallel loop requires greater mileage of wire to supply lamps in a given area than the open-loop method. The disadvantage of the open loop is that it is more difficult to arrange test points by which an open-circuited loop can be located. The circuit of Fig. 16-55 is arranged in part on each system and illustrates a fair compromise between test points and conductor economy.

Three disadvantages of the series circuit are (1) the necessity for insulating lamps, leads, fixtures, and circuit for full operating voltage to ground; (2) the large exposure of circuit wire in which a break causes interruption to all lamps; and (3) the economic necessity of installing street-lighting wires on the same cross-arm with primaries, causing wire congestion and exposing other distribution to possible faulting if a street-lighting wire breaks.

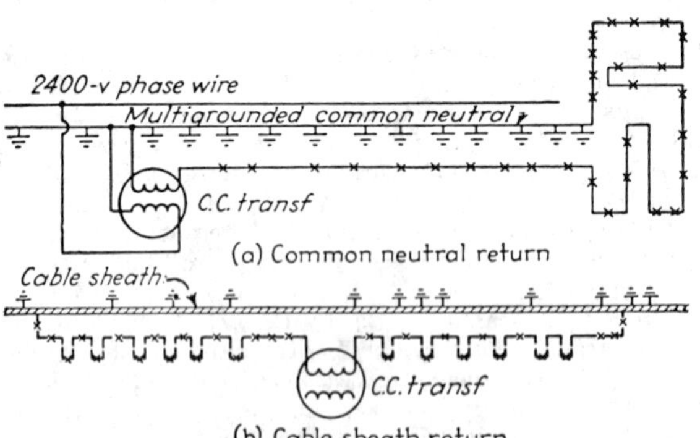

FIG. 16-56. Grounded operation of series circuit.

325. Grounded Operation of Series Circuits. A series circuit supplied by a single-circuit constant-current transformer can be designed with the multigrounded common neutral as the return leg of the circuit. In underground distribution cable sheaths can be used as the return leg. Figure 16-56 illustrates common-neutral and "sheath"-neutral return for series circuits. The method effects a substantial saving in wire and cable.

326. The series insulating transformer (type IL) is installed in, on, or near a lighting standard or pole to provide low voltage for one or two lamps. An IL transformer is commonly used to transform 6.6 to 20 A for a single lamp rated 5,000 lm or larger. The 20-A lamps have longer life, better efficiency, and cost less than 6.6-A lamps.

327. The 20-A series circuit is used extensively to supply lamps 5,000 lm and larger in the downtown sections of cities. It is more economical than the 6.6-A circuit with an IL transformer at each large lamp.

328. The conductor most commonly used for overhead series circuits is No. 6 copper or the equivalent in ACSR. It is the smallest size with strength to withstand wind and ice loads. For underground, No. 6 cable rated 5 or 10 kV is commonly used.

BIBLIOGRAPHY

329. Books for General Reference

BECK, EDWARD "Lightning Protection for Electric Systems"; New York, McGraw-Hill Book Company, 1954.

KURTZ, E. "The Lineman's and Cableman's Handbook," 4th ed.; New York, McGraw-Hill Book Company, 1964.

SANFORD, FRANK "Electric Distribution Fundamentals"; New York, McGraw-Hill Book Company, 1940.

SKROTZKI, B. G. A. "Electric Transmission and Distribution"; New York, McGraw-Hill Book Company, 1954.

"Electrical Transmission and Distribution Reference Book"; East Pittsburgh, Westinghouse Electric Corporation, 1944, particularly Chaps. 20 and 21.

"Electric Utility Engineering Reference Book"; East Pittsburgh, Pa., Westinghouse Electric Corporation, 1959, Vol. 3, Distribution Systems.

References to Transactions of AIEE (now IEEE)

Slepian, J., and Strom, A. P. Arcs in Low-voltage A-c Networks; 1931, p. 847.

Searing, H. R., and Thomas, E. R. Voltage Regulation of Cables Used for Low-voltage A. C. Distribution; 1933, p. 114.

Xenis, C. P. Short-circuit Protection of Networks by the Use of Limiters; 1937, p. 1191.

Coleman, J. O'R., and Davis, R. F. Inductive Co-ordination of Common-neutral Power Distribution Systems and Telephone Circuits; 1937, p. 17.

Olmsted, L. M. Capacitors and Automatic Boosters for Economical Correction of Voltage on Distribution Circuits; 1939, p. 49.

Parsons, John S., and Wallace, J. M. Fusing on Distribution Systems; 1944, p. 89.

Linder, F. W. Graphical Method of Calculating Fault Currents on Rural Distribution Systems; 1945, p. 16.

Hendrickson, P. E., Johnson, I. B., and Schulz, N. R. Abnormal Voltage Conditions Produced by Open Conductors on 3-phase Circuits Using Shunt Capacitors; 1953, Vol. 72, Pt. III, p. 1183.

Bodicky, Andrew Submarine Cables; 1953, Vol. 72, Pt. III, p. 1227.

Hawley, D. C. Power Cable Crossings on Bridges and Viaducts; 1953, Vol. 72, Pt. III, p. 1312.

AIEE Group on Substation Grounding Practices Application Guide on Methods of Substation Grounding; 1954, Vol. 73, Pt. III, p. 271.

Bankus, H. M., and Gerngross, J. E. Unbalanced Loading and Voltage Unbalance on 3-phase Distribution Transformer Banks; 1954, Vol. 73, Pt. III, p. 367.

AIEE Committee on Substations, A Guide to Safety Considerations in the Design of Substations; 1954, Vol. 73, Pt. III, p. 633.

Anderson, A. S., and Ruete, R. C. Voltage Unbalance in Delta Secondaries Serving Single-phase and 3-phase Loads; 1954, Vol. 73, Pt. III, p. 928.

Clayton, J. M., and Hileman, A. R. A Method of Estimating Lightning Performance of Distribution Lines; 1954, Vol. 73, Pt. III, p. 953.

deBellis, A. M., and Griscom, S. B. A New Approach to the Problem of Higher Distribution Voltages; 1954, Vol. 73, Pt. III, p. 1508.

Duncan, T. C., Neubauer, J. P., Comly, J. M., Lawrence, R. F., and Maxwell, Miles Economics of Various Secondary Voltages for Commercial Areas; 1954, Vol. 73, Pt. III, p. 1512.

Brieger, L., Xenis, C. P., Bisson, A. J., and deLellis, J. Distribution Equipment Used on 265/460-volt Networks and Its Operating Features; 1954, Vol. 73, Pt. III, p. 1525.

Schwab, R. L., and Stohr, E. W. Secondary Network Equipment for 250- to 600-volt Systems; 1954, Vol. 73, Pt. III, p. 1531.

Barnett, H. G., and Lawrence, R. F. Service Voltage Spread and Its Effect on Utilization Equipment; 1954, Vol. 73, Pt. II.

References to Periodicals

Blake, D. K. Low-voltage A-c Networks; *Gen. Elec. Rev.*, February, March, April, 1928, pp. 82, 140, 186.

Olmsted, L. M. Load Division in Networks; *Elec. Jour.*, June, 1934, p. 226.

Gamble, L. R., and Starr, F. M. Shunt Capacitors on Distribution Circuits; *Gen. Elec. Rev.*, October, 1936, pp. 466–474.

Arvidson, C. E. Calculating Overhead Primaries; *Elec. World*, Oct. 7, 1939. Residential Loads, Diversities; *Elec. World*, Oct. 21, 1939. Design of Radial Primaries; *Elec. World*, Nov. 4, 1939. Fixing Regulator Settings; *Elec. World*, Nov. 18, 1939. Transformer-secondary Design for Service; *Elec. World*, Dec. 2, 1939. Checking Transformer and Secondary Conditions; *Elec. World*, Dec. 16, 1939.

Benner, P. E. Banking Secondaries; *Elec. Light and Power*, April, 1939.

PARSONS, JOHN S. Vaults for A.C. Secondary Networks; *Elec. World,* Mar. 23, 1940, p. 886; Apr. 20, 1940, p. 1201; May 18, 1940, p. 1512.

BULLARD, W. R. Voltage Drops in Radial Overhead Distribution Circuits; *EEI Bull.,* November, 1940.

BARNETT, H. G. Secondary Network Planning; *Elec. World,* Aug. 9, 1941, p. 422; Aug. 23, 1941, p. 575; Sept. 6, 1941, p. 718.

JENSEN, CLAUDE Grounding Principles and Practice; Establishing Grounds; *Elec. Eng.,* February, 1945, p. 68.

FORBES, A. D. The Banked Secondary Transformer; *Westinghouse Eng.,* March, 1945, p. 43.

BEAR, FRED T. Methodical Distribution System Changeover from 2.4 to 4.16 Kv; *Elec. World,* Feb. 16, 1946, p. 86.

PARSONS, JOHN S. Banking Distribution Transformers; *Westinghouse Engr.,* March, 1946, p. 39.

PARTRIDGE, K. L. Lessons from Secondary Banking Practice; *Elec. World,* July 20, 1946, p. 62.

BROOKES, A. S. Primary Aerial Cable Designs; *Elec. World,* Nov. 9, 1946, p. 96.

SAYLES, E. V. Planned Substations Can Cost Less; *Elec. World,* Nov. 22, 1947, p. 82.

CAMPBELL, H. E. How to Apply Secondary Capacitors; *Elec. World,* July 16, 1949, p. 88.

FURANNA, A. L. Underground Secondary Cable Rings Serve Homes for $130 Each; *Elec. World,* Dec. 17, 1949, p. 90.

NELSON, E. E., and VAN ANTWERP, G. S. Utility and Developer Build Low-cost Underground System; *Elec. World,* Jan. 2, 1950, p. 50.

TYNES, REX A. Southwest Co-ops Go to 14.4/25-kv Distribution System; *Elec. World,* Oct. 23, 1950, p. 102.

STEIN, J. J. New Transformer Enclosures for Buried Cable Systems; *Elec. World,* Nov. 20, 1950, p. 107.

TOPPING, C. E. The "How to" of Aluminum in Distribution; *Elec. World,* Feb. 11, 1952, p. 97.

GREEN, ARTHUR T. Choice of Right Wire Size Simplified by Charts and Tables; *Elec. World,* May 19, 1952, p. 154.

TREMAINE, R. L., and CUTTINO, W. H. What Control Is Best for Switched Capacitors? *Elec. World,* June 16, 1952, p. 120.

SCIOTTI, FRED Aerial Cable Storm-proofs 4-kv Distribution System; *Elec. World,* Oct. 19, 1952, p. 114.

RIEBS, R. E., and AMUNDSEN, R. H. How Do Reclosers Affect Fuse Link Performance? *Elec. World,* Feb. 8, 1954, p. 88.

CAMPBELL, H. E. Problems Posed by Home Motor Loads; *Elec. World,* May 17, 1954, p. 144.

BARNETT, H. G., and MANNING, L. W. 120/240-v Can Serve Bigger Loads; *Elec. World,* May 17, 1954, p. 147.

LYMAN, W. J., and HILL, V. E. Service Reliability Standards Aid Planning and Operation; *Elec. World,* May 17, 1954, p. 150.

MINDER, P. M. Effective Capacitor Fusing Today; *Elec. World,* July 12, 1954, p. 103.

EEI-NEMA Joint Committee, Distribution Transformers Can Be Selected without Detailed Calculations; *Elec. World,* July 12, 1954, p. 122; July 26, 1954, p. 116; Aug. 9, 1954, p. 116.

BAUGH, CLIVE E. 12 Kv Has Cost Edge over 4 Kv; *Elec. World,* July 26, 1954, p. 87.

SKAFF, PAUL A. Pushing UG Conduit Saves $6/Ft on Single-duct Runs; *Elec. World,* Oct. 31, 1955, p. 84.

ANDREWS, F. E. EEI Underground Handbook Survey Shows Residential Distribution Adaptable; *Elec. World,* Dec. 12, 1955, p. 107.

ANDERSON, C. B. Sectionalize 4-kv Underground System Safely; *Elec. World,* Aug. 12, 1957, p. 96.

Porter, L. F. Aluminum Cable Used for Secondary Mains; *Elec. World,* Aug. 12, 1957, p. 101.

Fryer, L. T. Underground System Costs Pared for Residential Service; *Elec. World,* June 1, 1959, p. 46.

Lawrence, R. F., and Smith, L. G. First 240/480-V Underground Serves New Home Area Economically; *Elec. World,* July 4, 1960, p. 42.

Olmsted, L. M. Distribution Design Swings to Cable; *Elec. World,* Feb. 6, 1961, p. 50.

Ewald, G. A. Aerial Cable Is Key to Storm-proofing; *Elec. World,* Feb. 6, 1961, p. 57.

Himel, Newton Suburb Puts Distribution Underground; *Elec. World,* Feb. 6, 1961, p. 62.

Fleming, B. D. Cheap Residential Underground May Prove Costly in Long Run; *Elec. World,* May 29, 1961, p. 58.

Hally, H. G. 7.62 Kv to Utilization Transformer Cuts Residential Underground Cost; *Elec. World,* June 12, 1961, p. 56.

Johnson, W. H. 240/480-V Design Cuts Cost of Underground Residential Distribution; *Elec. World,* Aug. 14, 1961, p. 76.

Kemnitz, L. A., and Smith, J. C. Power and Telephone in Common Trench Cut Underground Cost; *Elec. World,* Jan. 15, 1962, p. 48.

Kaye, Albert, and Mackinnon, W. F. Direct-buried Distribution Serves 3,500 Total-electric Homes; *Elec. World,* Nov. 5, 1962, p. 54.

Danforth, C. F. Huge Development in Virginia Served Underground; *Elec. World,* Aug. 19, 1963, p. 24.

Gilton, R. F., and Massara, J. M. Underground Turnkey Package Offered; *Elec. World,* Jan. 13, 1964, p. 54.

Kaupie, R. T., and Stoelting, H. O. Reduce Underground Equipment BIL; *Elec. World,* Mar. 2, 1964, p. 28.

Chun, C. Y. K., Kresser, J. V., Patton, A. D., and Richardson, J. F. Optimize Hawaiian Underground Distribution; *Elec. World,* Aug. 10, 1964, p. 64.

Blankenburg, R. C., and Lauder, J. W. URD Achieves Cost Break-through; *Elec. World,* Nov. 9, 1964, p. 23.

Bleiweis, Julius Trends in Distribution; *Elec. World,* Nov. 16, 1964, p. 101.

Thomas, J. F. Soil Heat Tests Yield Useful Thermal Data; *Elec. World,* Feb. 22, 1965, p. 90.

Bleiweis, Julius Tomorrow's URD Takes Shape; *Elec. World,* Mar. 22, 1965, p. 103.

Chiles, R. E., and Smith, B. E. 34.5-kv Cable Used for URD Primary; *Elec. World,* Mar. 22, 1965, p. 111.

Duffy, E. K. Pre-package Service Facilities to Lower URD Costs; *Elec. World,* Mar. 22, 1965, p. 137.

Strickland, C. B. Town within Town Served by 13.2-kv Underground; *Elec. World,* Oct. 4, 1965, p. 62.

Medek, J. D., and Steeve, E. J. Devices Help Find URD Cable Faults; *Elec. World,* Oct. 4, 1965, p. 102.

Bettersworth, T. A. Partial Underground Proves Attractive Alternative; *Elec. World,* Dec. 6, 1965, p. 24.

Cook, R. F. Economics Spurs Use of URD; *Elec. World,* Nov. 22, 1965, p. 26.

Moore, D. Utility Speeds Underground Program; *Elec. World,* Dec. 13, 1965, p. 59.

Lund, V. E. Combining Systems Brings Beautility; *Elec. World,* Jan. 3, 1966, p. 34.

Xenis, C. P. Aluminum Conductor Use for Networks Widens; *Elec. World,* Jan. 10, 1966, p. 50.

Bythewood, R. C. Georgia Power Now after Mobile Home Business; Jan. 17, 1966, p. 34.

Smith, B. E. Medium Load Areas Get Underground Distribution; *Elec. World,* May 23, 1966, p. 32.

ANDREWS, D. L., and HAASCH, D. E. Dry-type Transformer on Each House a URD Feature; *Elec. World*, June 20, 1966, p. 101.

References to Miscellaneous Publications

Insulated Power Cable Engineers Association Tables of Current Carrying Capacity of Impregnated Paper, Rubber and Varnished Cambric Insulated Cables; AIEE (now IEEE), 1962, Vol. 1, Copper Conductors; Vol. 2, Aluminum Conductors.

Sag and Tension Charts of the Copper Wire Engineering Association.

Anaconda Tables of Current Ratings for Overhead Copper Conductors, Anaconda Wire and Cable Company, Hastings-on-Hudson, N.Y.

"Kaiser Aluminum Electrical Conductor Technical Manual"; Kaiser Aluminum and Chemical Sales, Inc., Chicago, Ill.

"Application Manual of Overcurrent Protection for Distribution Systems"; General Electric Company, Schenectady, N.Y.

National Electrical Safety Code, Pt. 2, Safety Rules for the Installation and Maintenance of Electric Supply and Communication Lines; *NBS Handbook* 81, Washington, D.C. 20402, Government Printing Office.

SECTION 17

WIRING DESIGN—COMMERCIAL AND INDUSTRIAL BUILDINGS

BY

W. T. STUART Editor, *Electrical Construction and Maintenance*

CONTENTS

Numbers refer to paragraphs

SECTION 17

WIRING DESIGN—COMMERCIAL AND INDUSTRIAL BUILDINGS

By W. T. Stuart

BASIC INSTALLATION RULES AND INSPECTION

1. The National Electrical Code (NEC) establishes the minimum standards of wiring design and installation practice in the United States. Its rules are written to protect the public from fire and life hazards. It is revised periodically by a committee drawn from industry associations, insurance groups, organized labor, and representatives of municipalities. It is sponsored by the National Fire Protection Association, adopted by the National Board of Fire Underwriters as its Regulations, and approved by the American Standards Association as an American Standard. It forms the basis of more than 600 municipal electrical wiring ordinances, most of which adopt successive editions of the Code as issued.

Semiannually a List of Inspected Electrical Appliances is issued by the **Underwriters' Laboratories, Inc.,** which has been established by the National Board of Fire Underwriters. One function of the Laboratories is to examine and pass upon electrical materials, fittings, and appliances in order to determine if they comply with the standard-test specifications set up by these laboratories. It is to be noted that there are numerous types of electrical appliance which are not submitted by the manufacturers to the Underwriters' Laboratories but, it is believed, are in every way just as safe and serviceable as those submitted and approved by these laboratories. Both the National Electrical Code and the List of Inspected Electrical Appliances may be obtained on application to any of the fire underwriters' offices or by making such request directly to the National Board of Fire Underwriters, 85 John Street, New York City, or to the Underwriters' Laboratories, Inc., 207 East Ohio Street, Chicago, Ill.

2. Legal Status of Code. The rules in the Code as issued by the National Board of Fire Underwriters are enforced by being incorporated in the ordinances passed by various cities and towns, covering the installation of electric wiring.

When installing any electrical equipment, first ascertain whether local installation rules in the form of ordinances are enforced in the community. If so, follow such rules; if none exists, follow the requirements of NEC.

3. 1965 Edition of NEC. Where reference is made in this section to installation rules, the 1965 edition of NEC is used as a basis.

4. Code Not a Design Manual. Design of an installation in accordance with the Code minimizes fire and accident hazards but does not guarantee satisfactory or efficient operation of the system. Other design standards are necessary to accomplish the latter purposes. These features are covered in Pars. **71** to **85** of this section.

5. License. In some areas the installation of electric wiring is controlled by city, county, or state license often combined with installation rules.

6. Rules of Electric Service Companies. Electric lighting and power companies generally issue certain rules of their own, based to a large extent on peculiar requirements which are necessary in order to give the best possible service to the greatest number of customers.

These rules are concerned mostly with matters of distribution engineering. They relate to locations and details of service entrance, provision for meters, the kind of electricity furnished by the company, its frequency and voltage, the types and sizes of motors, rules in connection with starting characteristics of such motors, and similar matters.

The electric-service company usually supplies copies of its rules at no charge.

7. Inspection. Every electrical installation should be inspected wherever an experienced inspector is available to ensure that it complies with local and NEC rules. Such inspection is usually mandatory in cities having electrical ordinances. In some areas the fire underwriters maintain inspectors who check electrical wiring, while in others the municipality makes a check through its electrical inspectors. Where inspection is not mandatory, it is always advisable to request the most convenient fire underwriters' bureau to make the necessary inspection.

Federal and state buildings usually require inspection by authorized Federal or state inspectors. In these instances inspection includes not only safety considerations but the requirements of the particular job specifications. Other inspection may be required but it is often waived.

METHODS OF WIRING

8. Wiring Methods Classified. The discussion of wiring methods in this section relates to **interior circuits for light, heat, and power** and does not cover signaling or communication systems.

Numerous methods of wiring are authorized by NEC, most of them used to a greater or lesser extent in commercial and industrial buildings. Those of interest can be grouped as follows:

a. Raceways for general use.
  1. Rigid metal conduit.
  2. Electric metallic tubing (EMT).
  3. Nonmetallic conduit.

b. Cable-assembly systems for general use.
  1. Nonmetallic sheathed cable.
  2. Underground feeder and branch-circuit cable.
  3. Metal-clad cable (armored cable).
  4. Mineral-insulated metal-sheathed cable (MI).
  5. Aluminum-sheathed cable (ALS).

c. Conductor systems for general use.
  1. Open wiring on insulators.
  2. Concealed knob-and-tube wiring.

d. Cable-assembly systems for limited use.
  1. Service-entrance cable.
  2. Nonmetallic extensions.
  3. Underplaster extensions.

e. Raceway systems for limited use.
  1. Flexible-metal conduit.
  2. Liquid-tight flexible-metal conduit.
  3. Underfloor raceway.
  4. Cellular-metal-floor raceway.
  5. Wireways.
  6. Continuous rigid cable supports.

f. Special systems.
  1. Busways.
  2. Multioutlet assemblies.

9. Installation Methods. Requirements to be met in installing each of the foregoing systems are found in the current edition of the NEC. The requirements are specific and detailed and change somewhat as the art progresses; hence reference should be made to the Code for the exact circumstances under which each system is permitted or prohibited, together with the precise rules to be followed in installation.

The discussion in the following paragraphs compares the systems generally and indicates the major limitations on use of each.

Table 17-1. Conductor Insulations

Trade name	Type letter	Insulation	Thickness of insulation	Outer covering
Heat-resistant	RH RHH	Heat-resistant rubber	14–12 . . . 2/64 in 10 . . . 3/64 in 8–2 . . . 4/64 in 1–4/0 . . . 5/64 in 213–500 . . . 6/64 in 501–1,000 . . . 7/64 in 1,001–2,000 . . . 8/64 in	Moisture-resistant flame-retardant nonmetallic covering
Moisture-resistant	RW	Moisture-resistant rubber	14–10 . . . 3/64 in 8–2 . . . 4/64 in 1–4/0 . . . 5/64 in 213–500 . . . 6/64 in 501–1,000 . . . 7/64 in 1,001–2,000 . . . 8/64 in	Moisture-resistant flame-retardant nonmetallic covering
Moisture- and heat-resistant	RH-RW	Moisture- and heat-resistant rubber	14–10 . . . 3/64 in 8–2 . . . 4/64 in 1–4/0 . . . 5/64 in 213–500 . . . 6/64 in 501–1,000 . . . 7/64 in 1,001–2,000 . . . 8/64 in	Moisture-resistant flame-retardant nonmetallic covering
Moisture- and heat-resistant	RHW	Moisture- and heat-resistant rubber	14–10 . . . 3/64 in 8–2 . . . 4/64 in 1–4/0 . . . 5/64 in 213–500 . . . 6/64 in 501–1,000 . . . 7/64 in 1,001–2,000 . . . 8/64 in	Moisture-resistant flame-retardant nonmetallic covering
Heat-resistant latex rubber	RUH	90% unmilled grainless rubber	14–10 . . . 18 mils 8–2 . . . 25 mils	Moisture-resistant flame-retardant nonmetallic covering
Moisture-resistant latex rubber	RUW	90% unmilled grainless rubber	14–10 . . . 18 mils 8–2 . . . 25 mils	Moisture-resistant flame-retardant nonmetallic covering
Thermo-plastic	T	Flame-retardant thermoplastic compound	14–10 . . . 2/64 in 8 . . . 3/64 in 6–2 . . . 4/64 in 1–4/0 . . . 5/64 in 213–500 . . . 6/64 in 501–1,000 . . . 7/64 in 1,001–2,000 . . . 8/64 in	None
Moisture-resistant thermo-plastic	TW	Flame-retardant moisture-resistant thermoplastic	14–10 . . . 2/64 in 8 . . . 3/64 in 6–2 . . . 4/64 in 1–4/0 . . . 5/64 in 213–500 . . . 6/64 in 501–1,000 . . . 7/64 in 1,001–2,000 . . . 8/64 in	None
Heat-resistant thermo-plastic	THHN	Flame-retardant heat-resistant thermoplastic	14–12 . . . 15 mils 10 . . . 20 mils 8–6 . . . 30 mils 4–2 . . . 40 mils 1–4/0 . . . 50 mils 250–500 M cmils . . . 60 mils	Nylon jacket
Moisture- and heat-resistant thermo-plastic	THW	Flame-retardant moisture- and heat-resistant thermoplastic	14–10 . . . 3/64 in 8–2 . . . 4/64 in 1–4/0 . . . 5/64 in 213–500 . . . 6/64 in 501–1,000 . . . 7/64 in 1,001–2,000 . . . 8/64 in	None
Moisture- and heat-resistant thermo-plastic	THWN	Flame-retardant moisture- and heat-resistant thermoplastic	14–12 . . . 15 mils 10 . . . 20 mils 8–6 . . . 30 mils 4–2 . . . 40 mils 1–4/0 . . . 50 mils 250–500 M cmils . . . 60 mils	Nylon jacket
Thermo-plastic and asbestos	TA	Thermoplastic and asbestos	Th'pl'. Asb. 14–8 . . . 20 mils 20 mils 6–2 . . . 30 mils 25 mils 1–4/0 . . . 40 mils 30 mils	Flame-retardant nonmetallic covering
Thermo-plastic and fibrous braid	TBS	Thermoplastic	14–10 . . . 2/64 in 8 . . . 3/64 in 6–2 . . . 4/64 in 1–4/0 . . . 5/64 in	Flame-retardant nonmetallic covering

Table 17-1. Conductor Insulations.—*Concluded*

Trade name	Type letter	Insulation	Thickness of insulation	Outer covering
Synthetic heat-resistant	SIS	Heat-resistant rubber	14–10 . . . . $\frac{2}{64}$ in 8 . . . . $\frac{3}{64}$ in 6–2 . . . . $\frac{4}{64}$ in 1–4/0 . . . . $\frac{5}{64}$ in	None
Mineral-insulated metal-sheathed	MI	Magnesium oxide	16–4 . . . . 50 mils 3–250 M cmils . . . . 55 mils	Copper
Silicone-asbestos	SA	Silicone rubber	14–10 . . . . $\frac{3}{64}$ in 8–2 . . . . $\frac{4}{64}$ in 1–4/0 . . . . $\frac{5}{64}$ in 213–500 . . . . $\frac{6}{64}$ in 501–1,000 . . . . $\frac{7}{64}$ in 1,001–2,000 . . . . $\frac{8}{64}$ in	Asbestos or glass
Fluorinated ethylene propylene	FEP	Fluorinated ethylene propylene	14–10 . . . . 20 mils 8–2 . . . . 30 mils	None
	FEPB	Fluorinated ethylene propylene	14–8 . . . . 14 mils	Glass braid
			6–2 . . . . 14 mils	Asbestos braid
Varnished cambric	V	Varnished cambric	14–8 . . . . $\frac{3}{64}$ in 6–2 . . . . $\frac{4}{64}$ in 1–4/0 . . . . $\frac{5}{64}$ in 213–500 . . . . $\frac{6}{64}$ in 501–1,000 . . . . $\frac{7}{64}$ in 1,001–2,000 . . . . $\frac{8}{64}$ in	Nonmetallic covering or lead sheath
Asbestos and varnished cambric	AVA and AVL	Impregnated asbestos and varnished cambric	(Dimen. in mils) 1st Asb. / VC / AVA 2d Asb. / AVL 2d Asb. 14–8 (solid only) . . . — 30 20 25 14–8 . . . 10 30 15 25 6–2 . . . 15 30 20 25 1–4/0 . . . 20 30 30 30 213–500 . . . 25 40 40 40 501–1,000 . . . 30 40 40 40 1,000–2,000 . . . 30 50 50 50	AVA-asbestos braid or glass AVL-lead sheath
Asbestos and varnished cambric	AVB	Impregnated asbestos and varnished cambric	VC / Asb. 18–8 . . . 30 20 6–2 . . . 40 30 1–4/0 . . . 40 40	Flame-retardant cotton braid (switchboard wiring)
			Asb. / VC / 2d Asb. 14–8 . . . 10 30 15 6–2 . . . 15 30 20 1–4/0 . . . 20 30 30 213–500 . . . 25 40 40 501–1,000 . . . 30 40 40 1,001–2,000 . . . 30 50 50	Flame-retardant cotton braid
Asbestos	A	Asbestos	14 . . . . 30 mils 12–8 . . . . 40 mils	Without asbestos braid
Asbestos	AA	Asbestos	14 . . . . 30 mils 12–8 . . . . 30 mils 6–2 . . . . 40 mils 1–4/0 . . . . 60 mils	With asbestos braid or glass
Asbestos	AI	Impregnated asbestos	14 . . . . 30 mils 12–8 . . . . 40 mils	Without asbestos braid
Asbestos	AIA	Impregnated asbestos	Sol. / Str. 14 . . . 30 mils / 30 mils 12–8 . . . 30 mils / 40 mils 6–2 . . . 40 mils / 60 mils 1–4/0 . . . 60 mils / 75 mils 213–500 . . . . . . 90 mils 501–1,000 . . . . . . 105 mils	With asbestos braid or glass
Paper	. . . .	Paper	. . . .	Lead sheath

10. General Provisions Applying to All Wiring Systems. The types of wiring discussed may be used for voltages up to 600, unless otherwise indicated. Each type of insulated conductor is approved for certain uses and has a maximum operating temperature. If this is exceeded, the insulation is subject to deterioration (see Table 17-3). Each conductor size has a maximum current-carrying capacity, depending on type of insulation and conditions of use. These ratings should not be exceeded (see Table 17-1). Conductors may be placed in multiple usually in large sizes only.

Conductors of more than 600 V should not occupy the same enclosure as conductors carrying less than 600 V, but conductors of different light and power systems of less than 600 V may be grouped together in one enclosure if all are insulated for the maximum voltage encountered. In general, communication circuits should not occupy the same enclosure with light and power wiring.

Boxes should be installed at all outlets, at switch or junction points of raceway or cable systems, and at each outlet and switch point of concealed knob-and-tube work.

11. Provisions Applying to All Raceway Systems. The number of conductors permitted in each size and type of raceway is definitely limited to provide ready installation and withdrawal. For conduit and electrical metallic tubing see Tables 17-8, 17-9, and 17-10. Raceways, except surface-metal molding, must be installed as complete empty systems, the conductors being drawn in later. Conductors must be continuous from outlet to outlet without splice, except in auxiliary gutters and wireways.

Conductors of No. 6 AWG and larger must be stranded. Raceways must be continuous from outlet to outlet and from fitting to fitting and shall be securely fastened in place.

All conductors of a circuit operating on alternating current, if in metallic raceway, should be run in one enclosure to avoid inductive overheating. If, owing to capacity, not all conductors can be installed in one enclosure, each raceway used should contain a complete circuit (one conductor from each phase).

12. Rigid-metal conduit and **electrical metallic tubing** are the two systems generally employed where wires are to be installed in raceways. Both conduit and tubing may be buried in concrete fills or may be installed exposed. Wiring installed in conduit is approved for practically all classes of buildings and for voltages both above and below 600. Certain restrictions are placed on the use of tubing.

Conduit consists of standard-weight steel pipe (preferably either galvanized or cadmium-plated, although it may be black-enameled for use indoors and where not subject to severe corrosive influences) or of aluminum. Electrical metallic tubing (EMT) has the same internal diameter as conduit but a materially thinner wall of higher-quality steel.

Fittings and connectors used with conduit may be threaded or threadless. EMT fittings are usually threadless; however, threaded fittings of an approved type may be used.

Sizes of EMT above 2 in have the same external diameter as the equivalent rigid-conduit size.

Nonmetallic rigid conduits, in approximately the same dimensions as rigid-metal conduits, are also a general-use raceway. Some restrictions are imposed, affecting particularly installations exposed to possible mechanical injury. Grounding continuity is provided by an additional grounding conductor pulled into the raceway with the circuit conductors or as part of a cable assembly.

Nonmetallic PVC rigid conduits are commonly assembled with matching fittings by adhesives. Field bends are made by softening the plastic in a hot air stream of several hundred degrees from an electric heater-blower.

Nonmetallic PVC raceways of relatively flexible construction and with conductor already drawn in are used for direct burial in airport, highway, parkway, and similar installations.

Asbestos-cement and fiber conduits are extensively used in underground distribution. They may be installed directly in earth or encased in concrete envelopes.

13. Cable-assembly systems are used extensively for concealed wiring not embedded in masonry or concrete. They may also be installed exposed in dry locations and, depending upon the particular construction and ratings, in wet locations. Branch-circuit sizes are conventionally 600-V-rated. Cables rated for 5 through 15 kV are frequently used for primary distribution feeders in large commercial and industrial electrical systems.

In industrial plants and commercial utility areas cable assemblies are often installed in expanded metal trays, ladder racks, or other approved cable-support systems.

Armored cable and nonmetallic-sheathed cables are almost universally used in single-family house wiring in the United States and in many multifamily occupancies. Armored cable is used in extending branch circuits from outlet boxes on rigid conduit or EMT systems to lighting fixtures in suspended ceiling work.

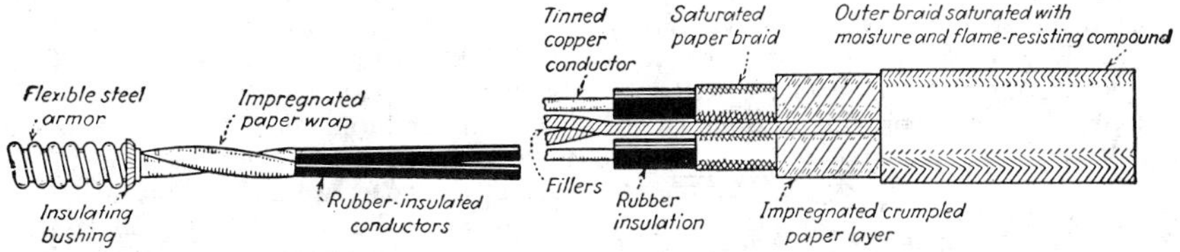

FIG. 17-1. Armored cable (BX). FIG. 17-2. Nonmetallic-sheathed cable.

Metal-clad type MC cable applies to constructions using interlocked armor, close fittings, or flexible corrugated tube over No. 4 copper, No. 2 aluminum, or larger conductors.

14. Two other metal-sheathed cables of special construction are recognized by the Code. Mineral-insulated metal-sheathed cable (MI) is copper-clad, containing one or more conductors and insulated with highly compressed refractory mineral insulation. It is widely used in industrial power and control wiring and in either wet or dry locations.

Aluminum-sheathed cable (ALS) consists of one or more insulated conductors in an impervious, continuous, closely fitting tube of aluminum and may also be used in wet or dry locations.

Both MI and ALS must be terminated and connected by means of fittings designed and approved for the purpose.

15. Open wiring on knobs and cleats is rarely encountered in current work. Open feeders are still used in some industrial construction where low cost is a consideration, no safety hazard is involved, and appearance is unimportant.

16. Cable-assembly Systems for Limited Use. Several cable assemblies have been developed for particular uses, rather than for complete wiring systems for a building. The NEC should be consulted for specific requirements in each case. These assemblies are discussed in Pars. **17** to **20**.

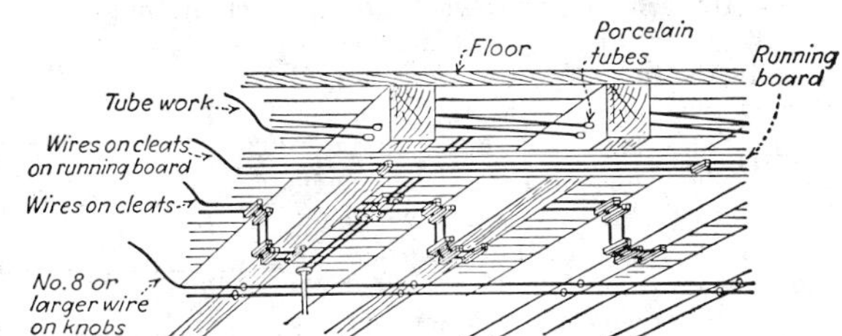

FIG. 17-3. Methods of supporting open wiring.

17. Service-entrance cable is a form of armored or nonmetallic-sheathed cable specifically approved for service-entrance use. It is available in four types: ASE, with interlocked metallic-armor protection; SE, without armor; SD, for service drops, construction similar to SE; USE, underground service-entrance cable suitable for direct burial in the ground.

18. Nonmetallic surface extensions are 2-wire assemblies limited to exposed work in office (or residence) occupancies, where additional outlets are to be installed in the same room with the outlet from which the extension originates. The location must be dry and not subject to corrosive vapors. The voltage should not exceed 150 V between conductors.

19. Underplaster extensions may be used as a concealed-wiring method to install additional outlets on an existing branch circuit. They may be buried in the plaster finish of walls or ceilings in buildings of fire-resistive construction. A flat, oval form of armored cable is usually employed, although rigid or flexible conduit or electrical metallic tubing may also be used.

20. Raceway Systems for Limited Use. In general, the raceway systems developed for special purposes and discussed in Pars. **22** to **27** are of more commercial importance and find a more varied use than do the special cable-assembly systems previously discussed. This is particularly true of underfloor and cellular raceways for concealed work and of wireways and busways for exposed work. In cases where great flexibility in the use of electric power is of importance, the application of one of these special systems should be considered. In each case, NEC should be consulted for specific installation rules.

21. Flexible-metal conduit, consisting of a flexible metallic tube roughly similar to the armor of armored cable, is used generally with rigid-conduit or electrical-metallic-tubing systems, to provide flexible connections at motor terminals, for instance, or in place of the rigid product where installation of the latter would be difficult owing to numerous bends, close working quarters, etc. The conductors are installed after the flexible conduit is in place.

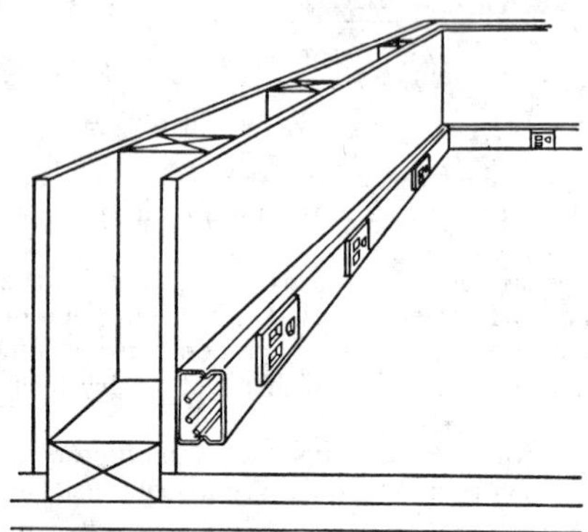

Fig. 17-4. Typical surface raceway with plug receptacles.

22. Surface metal raceways are flat, rectangular wireways used for exposed work in dry locations. They are frequently used to install additional outlets in a building already wired, where concealment of conductors is difficult, and are also used for special purposes, e.g., installation of cove lighting and for show-window reflectors. Unless made of a metal at least 0.040 in thick, they are limited to use on circuits not exceeding 300 V.

23. Liquid-tight flexible-metal conduit is, as the name suggests, a type of flexible-metal conduit having an outer jacket impervious to liquids and terminated in liquid-tight fitting. It is most widely used for connecting motors to rigid-conduit systems or fixed-equipment enclosures.

24. Underfloor raceways are employed in buildings of fire-resistant construction to provide readily accessible raceways in the floor slab for light and power, telephone, and signal circuits. One, two, or three ducts are installed, depending on the desired uses. Junction boxes which mark each end of a run of raceway, and the tops of which are flush with the floor covering, make it possible to locate accurately the run of duct and, hence, to install additional outlets with the special tools provided by the manufacturer. Owing to its flexibility, this type of construction is particularly suitable for large office areas or where outlet locations are subject to change.

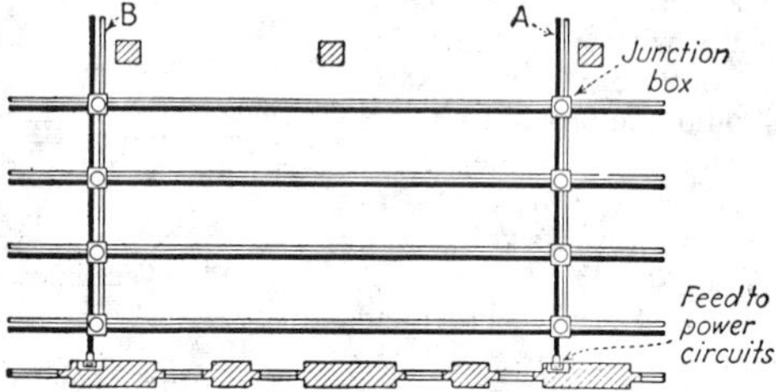

Fig. 17-5. Layout of double underfloor duct system. (*A*) For power circuits. (*B*) For signal and telephone circuits. (*General Electric Co.*)

25. The cellular-metal-floor raceway involves a cellular-steel floor, which is a structural load-carrying element whose hollow cells form the wire raceway and a system of transverse headers, together with the necessary fittings and adapters. The headers are also wire raceways, providing electrical access from distribution points to any predetermined number of cells. The system can be designed to provide overall floor and ceiling electrical service for conductors not larger than No. 0 AWG, not only for light and power but also for telephone and signal circuits. The large internal-cell areas (normally on 6-in centers) afford adequate conductor space, while the complete floor and ceiling coverage provides for great flexibility in use during the building life, since access can be had to headers and cells at any time for additional outlets, new or rerouted circuits, etc.

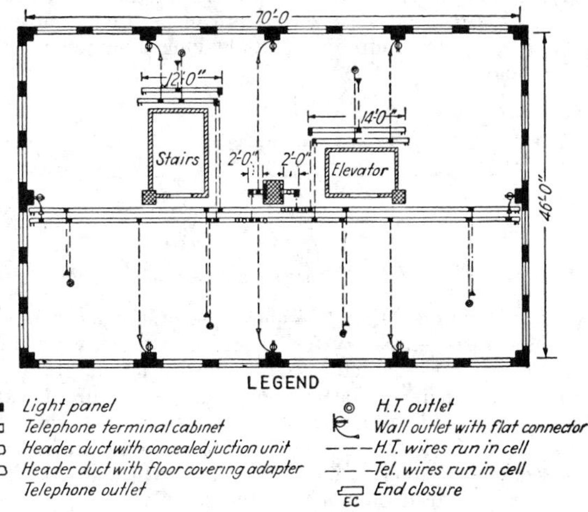

Fig. 17-6*a*. Cellular-floor wiring layout.

Cellular-concrete-floor raceways are precast slabs with tubular "cells" designed to line up in a continuous raceway. Cells terminate in metallic header ducts and other special fittings for connection to other parts of the electrical systems. Fittings approved for the purpose are inserted into the cell to provide for outlets.

Structural raceways are formed-steel members which may be assembled to provide for the installation of electrical wires and cables. Such assemblies also provide for the installation of wiring devices in vertical members which may be concealed.

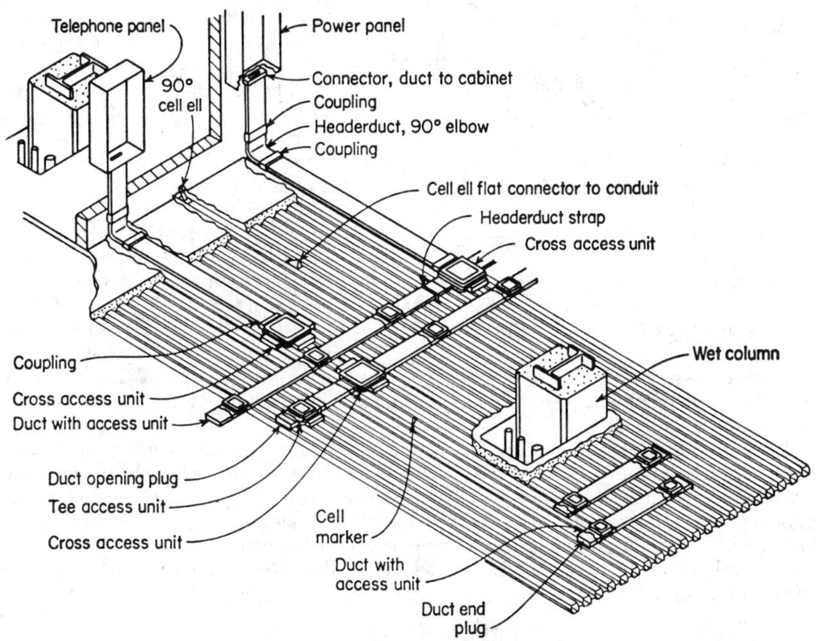

Fig. 17-6*b*. Floor ducts and access units.

26. Wireways provide a convenient, exposed, rectangular metal raceway or trough (usually 4 by 4 in in section) for electric conductors of sizes to 500,000 cmils. The product is available in several standard lengths, which are bolted together for continuous runs. Access at any point is through hinged covers and conduit knockouts. A complete array of fittings assures flexibility for various installation conditions.

Owing to their size, wireways can be used to advantage for large numbers of conductors, e.g., a group of circuits leaving a branch circuit panelboard or feeder distribution board.

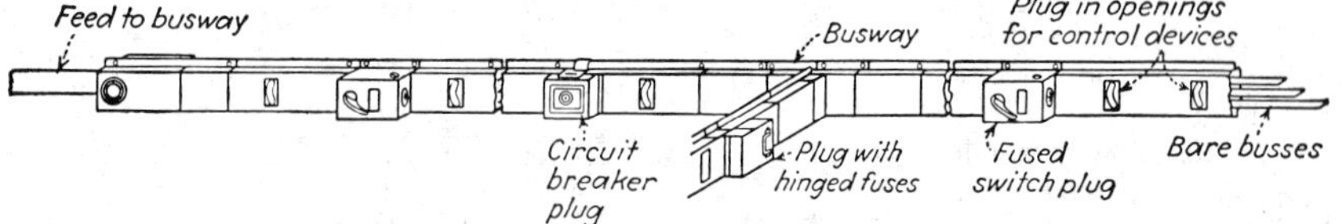

FIG. 17-7. Units of busway distribution system. (*Bull Dog Electric Products Co.*)

27. Busways (Fig. 17-7) are one of the more important recent developments for exposed heavy-capacity feeder and circuit wiring in industrial plants because of their flexibility in use, which makes them readily adaptable to future needs and to changing conditions such as relocation or revamping of production lines. The initial investment can be confined to immediate requirements and additions made at any time as requirements increase. The system consists essentially of interconnected prefabricated lengths or sections of steel or aluminum duct which enclose bus bars mounted on insulators. Regularly spaced openings in the sides of the duct permit plugging in branch-circuit control devices of the circuit-breaker, fuse, or fused-switch type, for convenient control of individual or group motor drives, lighting or heating circuits, etc. The ease of relocating both the duct and control devices makes its use advantageous for supplying power to machines on assembly lines, mass production manufacturing, and other applications where flexibility of electric supply is essential. Busways are available in capacities ranging from about 125 to about 3,000 A, for 3-phase 3- or 4-wire systems.

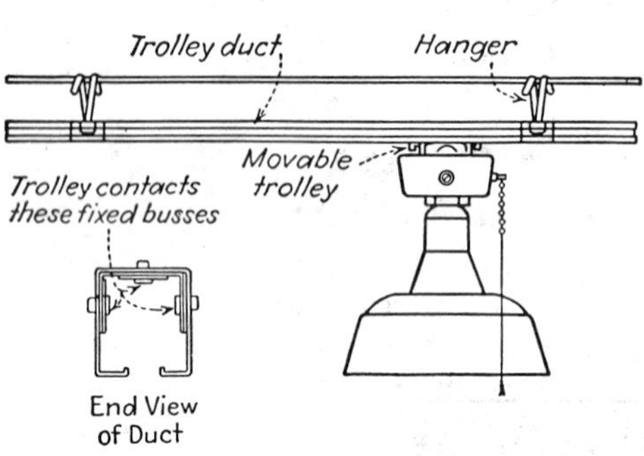

FIG. 17-8. Trolley duct used for movable lighting fixture. (*Bull Dog Electric Products Co.*)

The so-called trolley duct (Fig. 17-8) is a variation of the busway in which the metal duct and electrical buses (either single-phase or 3-phase) are so arranged that access is had to the buses at any point in the run. Current is collected from the buses by movable trolleys to which are wired portable or movable electrical devices. In industrial plants the system is used to supply power to cranes and hoists, to portable tools on assembly lines and benches, etc. It has found some application in drafting rooms, stock departments, and similar locations, where ability to move lighting units quickly is of advantage.

28. Multioutlet assemblies are surface-mounted raceways of metal or plastic with plug receptacle outlets at spaced intervals or provisions for the insertion of receptacles as desired. Multioutlet assemblies are widely used where a number of cord-connected appliances must be served (as along the back of a workbench or laboratory table). They are also used to provide greater convenience for the attachment of portable cords. In this application they are usually installed along the top of the baseboard (as around the perimeter of a private office).

TYPES OF CONDUCTORS

29. Conductors for Building Wiring. The various types of conductor available for interior wiring, together with their sizes, insulations, and uses, are indicated in Table 17-1. Data on type R, for many years the most common building wire, is omitted, as this type of insulation has been largely replaced by TW and other types. Rubber and thermoplastic insulations are available in a number of compounds and constructions for resistance to heat, moisture, or other environmental conditions.

Other insulations used in building wiring include magnesium oxide, fluorinated ethylene propylene, silicone rubber, and the long-familiar varnished-cambric and asbestos constructions.

30. Dimensions of rubber-covered and thermoplastic-covered conductors are given in Table 17-2.

31. Current-carrying Capacity (Ampacity). As the conductors of an electrical wiring system offer some resistance, a current-carrying conductor dissipates heat. Under practical conditions of installation and operation the temperatures reached must not result in the destruction of the insulation or risk to surrounding materials.

Tables of maximum allowable current-carrying capacity are given in the NEC. The following tables are based upon the 1965 Code tables but are, for brevity, not necessarily complete in detail.

Allowable ampacities for insulated conductors are based upon an allowable temperature rise above an ambient of 30°C, 86°F. A partial list of temperature ratings for types of insulated conductors is given in Table 17-3.

32. Allowable ampacities for copper conductors and aluminum conductors in accord-

Table 17-2. Dimensions of Rubber-covered and Thermoplastic-covered Conductors

Size, AWG Mcmils	Types RF-2, RFH-2, R, RH, RHH, RHW, RH-HW, RW		Types TF, T, THW, TW, RU, RUH, RUW		Types THHN, THWN		Types FEP, FEPB	
	Approx. diam., in	Approx. area, in^2	Approx. diam., in	Approx. area, in^2	Approx. diam., in	Approx. area, in^2	Approx. diam., in	Approx. area, in^2
(1)	(2)	(3)	(4)	(5)	(6)	(7)	(8)	(9)
18	0.146	0.0167	0.106	0.0088				
16	0.158	0.0196	0.118	0.0109				
14	2/64 in 0.171	0.0230	0.131	0.0135	0.105	0.0087	0.105 0.105	0.0087 0.0087
14	3/64 in 0.204	0.0327						
14	....	......	0.162	0.0206				
12	2/64 in 0.188	0.0278	0.148	0.0172	0.122	0.0117	0.121 0.121	0.0115 0.0115
12	3/64 in 0.221	0.0384						
12	....	......	0.179	0.0251				
10	.... 0.242	0.0460	0.168	0.0224	0.153	0.0184	0.142 0.142	0.0159 0.0159
10	....	......	0.199	0.0311				
8	.... 0.311	0.0760	0.228	0.0408	0.201	0.0317	0.189 0.169	0.0280 0.0225
8	....	......	0.259	0.0526				
6	0.397	0.1238	0.323	0.0819	0.257	0.0519	0.244 0.302	0.0467 0.0716
4	0.452	0.1605	0.372	0.1087	0.328	0.0845	0.292 0.350	0.0669 0.0962
3	0.481	0.1817	0.401	0.1263	0.356	0.0995	0.320 0.378	0.0803 0.1122
2	0.513	0.2067	0.433	0.1473	0.388	0.1182	0.352 0.410	0.0973 0.1316
1	0.588	0.2715	0.508	0.2027	0.450	0.1590		
0	0.629	0.3107	0.549	0.2367	0.491	0.1893		
00	0.675	0.3578	0.595	0.2781	0.537	0.2265		
000	0.727	0.4151	0.647	0.3288	0.588	0.2715		
0000	0.785	0.4840	0.705	0.3904	0.646	0.3278		
250	0.868	0.5917	0.788	0.4877	0.716	0.4026		
300	0.933	0.6837	0.843	0.5581	0.771	0.4669		
350	0.985	0.7620	0.895	0.6291	0.822	0.5307		
400	1.032	0.8365	0.942	0.6969	0.869	0.5931		
500	1.119	0.9834	1.029	0.8316	0.955	0.7163		
600	1.233	1.1940	1.143	1.0261				
700	1.304	1.3355	1.214	1.1575				
750	1.339	1.4082	1.249	1.2252				
800	1.372	1.4784	1.282	1.2908				
900	1.435	1.6173	1.345	1.4208				
1,000	1.494	1.7531	1.404	1.5482				
1,250	1.676	2.2062	1.577	1.9532				
1,500	1.801	2.5475	1.702	2.2748				
1,750	1.916	2.8895	1.817	2.5930				
2,000	2.021	3.2079	1.922	2.9013				

Table 17-3. Conductor Applications

Trade name	Type letter	Max. operating temp.	Application provisions
Table 17-3*a*			
Heat-resistant rubber	RH	75°C 167°F	Dry locations
Heat-resistant rubber	RHH	90°C 194°F	Dry locations
Moisture-resistant rubber	RW	60°C 140°F	Dry and wet locations For over 2,000 V, insulation shall be ozone-resistant
Moisture- and heat-resistant rubber	RH-RW	60°C 140°F 75°C 167°F	Dry and wet locations For over 2,000 V, insulation shall be ozone-resistant Dry locations For over 2,000 V, insulation shall be ozone-resistant
Moisture- and heat-resistant rubber	RHW	75°C 167°F	Dry and wet locations For over 2,000 V, insulation shall be ozone-resistant
Latex rubber	RU	60°C 140°F	Dry locations
Heat-resistant latex rubber	RUH	75°C	Dry locations
Moisture-resistant latex rubber	RUW	60°C 140°F	Dry and wet locations
Thermoplastic	T	60°C 140°F	Dry locations
Moisture-resistant thermoplastic	TW	60°C 140°F	Dry and wet locations
Heat-resistant thermoplastic	THHN	90°C 194°F	Dry locations
Moisture- and heat-resistant thermoplastic	THW	75°C 167°F	Dry and wet locations
Moisture- and heat-resistant thermoplastic	THWN	75°C 167°F	Dry and wet locations
Thermoplastic and asbestos	TA	90°C 194°F	Switchboard wiring only
Thermoplastic and fibrous outer braid	TBS	90°C 194°F	Switchboard wiring only
Synthetic heat-resistant	SIS	90°C 194°F	Switchboard wiring only
Table 17-3*b*			
Mineral insulation (metal-sheathed)	MI	85°C 185°C 250°C 482°F	Dry and wet locations with type O termination fittings For special application
Silicone-asbestos	SA	90°C 194°F 125°C 257°F	Dry locations For special application
Fluorinated ethylene propylene	FEP or FEPB	90°C 194°F 200°C 392°F	Dry locations Dry locations—special applications
Varnished cambric	V	85°C 185°F	Dry locations only. Smaller than No. 6 by special permission
Asbestos and varnished cambric	AVA	110°C 230°F	Dry locations only
Asbestos and varnished cambric	AVL	110°C 230°F	Dry and wet locations
Asbestos and varnished cambric	AVB	90°C 194°F	Dry locations only

ance with the temperature rating of the insulation are given for installation in conduit and for installation in free air in Tables 17-4 to 17-7.

33. Conductor and conduit diameters and areas are frequently necessary to calculate fill. Nominal values for conductors are given in Table 17-2, for conduit and tubing in Table 17-10. Table 17-12 gives resistances of conductors in ohms per 1,000 ft.

34. Permissible Percent Raceway Fill for Conductor Combinations. See Table 17-10.

35. Number of Conductors in One Conduit or Tubing. The number of conductors of a certain size that may be installed in a given-sized conduit or electrical metallic tubing is limited to provide for ready installation and withdrawal without injury to conductor or insulating covering. Tables 17-8 and 17-9 give these values for commonly used conductors.

Table 17-4. Allowable Ampacities of Insulated Copper Conductors

(Not more than three conductors in raceway or cable or direct burial, based on room temperature of 30°C, 86°F)

Size AWG M cmils	Temperature rating of conductor (see Table 17-3)					
	60°C (140°F)	75°C (167°F)	85°–90°C (185°F)	110°C (230°F)	125°C (257°F)	200°C (392°F)
14	15	15	25*	30	30	30
12	20	20	30*	35	40	40
10	30	30	40*	45	50	55
8	40	45	50	60	65	70
6	55	65	70	80	85	95
4	70	85	90	105	115	120
3	80	100	105	120	130	145
2	95	115	120	135	145	165
1	110	130	140	160	170	190
0	125	150	155	190	200	225
00	145	175	185	215	230	250
000	165	200	210	245	265	285
0000	195	230	235	275	310	340
250	215	255	270	315	335	
300	240	285	300	345	380	
350	260	310	325	390	420	
400	280	335	360	420	450	
500	320	380	405	470	500	
600	355	420	455	525	545	
700	385	460	490	560	600	
750	400	475	500	580	620	
800	410	490	515	600	640	
900	435	520	555			
1,000	455	545	585	680	730	
1,250	495	590	645			
1,500	520	625	700	785		
1,750	545	650	735			
2,000	560	665	775	840		

Correction factors, room temp. over 30°C, 86°F

°C	°F						
40	104	0.82	0.88	0.90	0.94	0.95	
45	113	0.71	0.82	0.85	0.90	0.92	
50	122	0.58	0.75	0.80	0.87	0.89	
55	131	0.41	0.67	0.74	0.83	0.86	
60	140	....	0.58	0.67	0.79	0.83	0.91
70	158	....	0.35	0.52	0.71	0.76	0.87
75	167	....	....	0.43	0.66	0.72	0.86
80	176	....	....	0.30	0.61	0.69	0.84
90	194	....	....	....	0.50	0.61	0.80
100	212	....	....	....	....	0.51	0.77
120	248	....	....	....	....	....	0.69
140	284	....	....	....	....	....	0.59

* The ampacities for types FEP, FEPB, RHH and THHN conductors for sizes AWG 14, 12, and 10 shall be the same as designated for 75°C conductors in this table.

36. In considering the ampacity of conductors and conduit fill it is important to note that derating of ampacity applies for increased ambient temperatures and for more than 3 conductors, excluding neutrals, in a conduit.

37. Flexible cords and fixture wire in some cases may be as small as No. 18 AWG; hence these are exceptions to the general rule that no conductor smaller than No. 14 should be used in light and power wiring. Owing to the heat generated in the lamp heat-resistant wiring is required in fixtures.

TYPES OF CIRCUIT

38. Services and Feeders. No limit is placed on the electrical capacity of service conductors and service protection employed in bringing the electric supply into a

Table 17-5. Allowable Ampacities of Insulated Copper Conductors
(Single conductor in free air, based on room temperature of 30°C, 86°F)

Size AWG M cmils	Temperature rating of conductor (see Table 17-3)						
	60°C (140°F)	75°C (167°F)	85°–90°C (185°F)	110°C (230°F)	125°C (257°F)	200°C (392°F)	Bare and covered conductor
14	20	20	30*	40	40	45	30
12	25	25	40*	50	50	55	40
10	40	40	55*	65	70	75	55
8	55	65	70	85	90	100	70
6	80	95	100	120	125	135	100
4	105	125	135	160	170	180	130
3	120	145	155	180	195	210	150
2	140	170	180	210	225	240	175
1	165	195	210	245	265	280	205
0	195	230	245	285	305	325	235
00	225	265	285	330	355	370	275
000	260	310	330	385	410	430	320
0000	300	360	385	445	475	510	370
250	340	405	425	495	530	...	410
300	375	445	480	555	590	...	460
350	420	505	530	610	655	...	510
400	455	545	575	665	710	...	555
500	515	620	660	765	815	...	630
600	575	690	740	855	910	...	710
700	630	755	815	940	1005	...	780
750	655	785	845	980	1045	...	810
800	680	815	880	1020	1085	...	845
900	730	870	940	....	....	...	905
1,000	780	935	1000	1165	1240	...	965
1,250	890	1065	1130				
1,500	980	1175	1260	1450	....	...	1215
1,750	1070	1280	1370				
2,000	1155	1385	1470	1715	....	...	1405

Correction factors, room temp. over 30°C, 86°F

°C	°F	60°C	75°C	85°–90°C	110°C	125°C	200°C	Bare and covered
40	104	0.82	0.88	0.90	0.94	0.95		
45	113	0.71	0.82	0.85	0.90	0.92		
50	122	0.58	0.75	0.80	0.87	0.89		
55	131	0.41	0.67	0.74	0.83	0.86		
60	140	....	0.58	0.67	0.79	0.83	0.91	
70	158	....	0.35	0.52	0.71	0.76	0.87	
75	167	....	....	0.43	0.66	0.72	0.86	
80	176	....	....	0.30	0.61	0.69	0.84	
90	194	....	....	....	0.50	0.61	0.80	
100	212	....	....	....	....	0.51	0.77	
120	248	....	....	....	....	....	0.69	
140	284	....	....	....	....	....	0.59	

* The ampacities for types FEP, FEPB, RHH, and THHN conductors for sizes AWG 14, 12, and 10 shall be the same as designated for 75°C conductors in this table.

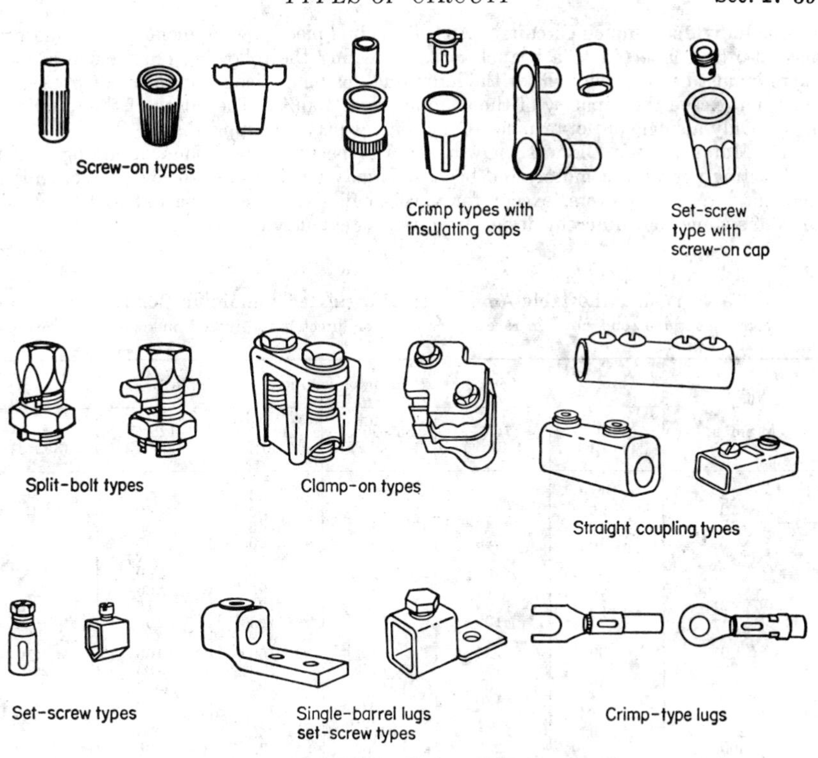

FIG. 17-9. Types of wire connectors.

building, since one supply only should be introduced whenever possible. Near the point of entrance of the supply, the heavy-service conductors are tapped by feeders which conduct the electricity to panelboards at various load centers in the building, where the final branch circuits which supply individual lighting, heating, and power outlets originate (see Fig. 17-10). No limits are placed on the electrical capacity of feeders, but for practical purposes they are limited in size by the difficulty of handling large conductors and raceways in restricted building spaces, by voltage drop, and by economic considerations.

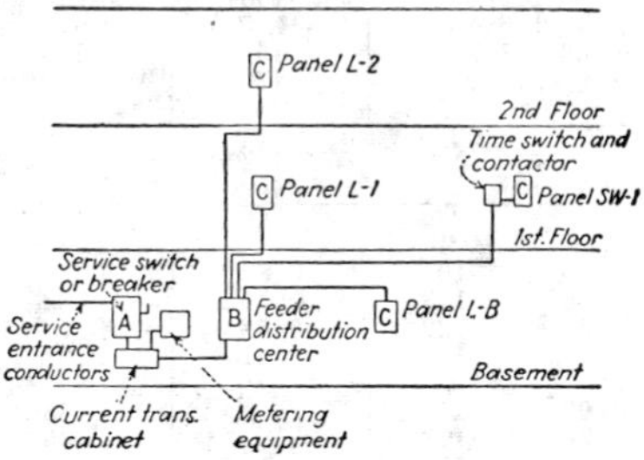

FIG. 17-10. Riser diagram showing location of (*A*) service, (*B*) feeder, (*C*) branch circuit overcurrent protective devices.

Each lighting fixture, motor, heating device, or other item of utilization equipment must be supplied by one of the types of branch circuit of Pars. **39** to **41**.

39. Branch Circuits for Grouped Loads. The uses and limitations of the common types of branch circuit are outlined in Table 17-13. It will be noted that lighting branch circuits may carry loads as high as 50 A. Such heavy-duty circuits are extensively employed in commercial and industrial occupancies. Branch circuits supplying convenience outlets for general use in other than manufacturing areas are usually limited to a maximum of 20 A, as the type of outlet required for heavier-capacity circuits usually will not accommodate the connection plug found on portable cords or lamps, motor-driven office machinery, etc.

40. Individual Branch Circuits. Any individual piece of equipment (except motors) may also be connected to a branch circuit meeting the following requirements: Conductors must be large enough for the individual load supplied. Overcurrent protection must not exceed the capacity of the conductors or 150% of the rating of the individual load. Only a single outlet or piece of equipment may be supplied.

41. Motor Branch Circuits. Owing to the peculiar conditions obtaining during the starting period of a motor, and because it may be subjected to severe overloads at frequent intervals, motors, except for very small sizes, are connected to branch circuits of a somewhat different design from that previously discussed.

Table 17-6. Allowable Ampacities of Insulated Aluminum Conductors

(Not more than three conductors in raceway or cable or direct burial, based on room temperature of 30°C, 86°F)

Size AWG M cmils	Temperature rating of conductor (see Table 17-3)					
	60°C (140°F)	75°C (167°F)	85°-90°C (185°F)	110°C (230°F)	125°C (257°F)	200°C (392°F)
12	15	15	25†	25	30	30
10	25	25	30†	35	40	45
8	30	40	40†	45	50	55
6	40	50	55	60	65	75
4	55	65	70	80	90	95
3	65	75	80	95	200	225
2*	75	90	95	105	115	130
1*	85	100	110	125	135	150
0*	100	120	125	150	160	180
00*	115	135	145	170	180	200
000*	130	155	165	195	210	225
0000*	155	180	185	215	245	270
250	170	205	215	250	270	
300	190	230	240	275	305	
350	210	250	260	310	335	
400	225	270	290	335	360	
500	260	310	330	380	405	
600	285	340	370	425	440	
700	310	375	395	455	485	
750	320	385	405	470	500	
800	330	395	415	485	520	
900	355	425	455			
1,000	375	445	480	560	600	
1,250	405	485	530			
1,500	435	520	580	650		
1,750	455	545	615			
2,000	470	560	650	705		

Correction factors, room temp. over 30°C, 86°F

°C	°F	60°C	75°C	85°-90°C	110°C	125°C	200°C
40	104	0.82	0.88	0.90	0.94	0.95	
45	113	0.71	0.82	0.85	0.90	0.92	
50	122	0.58	0.75	0.80	0.87	0.89	
55	131	0.41	0.67	0.74	0.83	0.86	
60	140	....	0.58	0.67	0.79	0.83	0.91
70	158	....	0.35	0.52	0.71	0.76	0.87
75	167	....	....	0.43	0.66	0.72	0.86
80	176	....	....	0.30	0.61	0.69	0.84
90	194	....	....	....	0.50	0.61	0.80
100	212	....	....	....	....	0.51	0.77
120	248	....	....	....	....	....	0.69
140	284	....	....	....	....	....	0.59

* For 3-wire single-phase service and subservice circuits, the allowable ampacity of RH, RH-RW, RHH, RHW, and THW aluminum conductors shall be for sizes 2-100 A, 1-110 A, 1/0-125 A, 2/0-150 A, 3/0-170 A, and 4/0-200 A.

† The ampacities for types RHH and THHN conductors for sizes AWG 12, 10, and 8 shall be the same as designated for 75°C conductors in this table.

PROTECTION AND CONTROL

42. Overcurrent protection for circuits and equipment is generally required in all ungrounded conductors of wiring systems. Overcurrent protection is provided by circuit breakers or fuses, sometimes by a combination of both.

A desirable characteristic in system design is the coordination of overcurrent protection to limit response to a fault or overload to the nearest overcurrent protective device.

43. Overcurrent protective devices are designed to provide in each type a characteristic time-current response. These time-current characteristics may be found in manufacturers' technical data literature and are important to coordination design. In general a circuit breaker or fuse will carry its rated current indefinitely, open in-

Table 17-7. Allowable Ampacities of Insulated Aluminum Conductors
(Single conductor in free air, based on room temperature of 30°C, 86°F)

Size AWG M cmils	Temperature rating of conductor (see Table 17-3)						
	60°C (140°F)	75°C (167°F)	85°–90°C (185°F)	110°C (230°F)	125°C (257°F)	200°C (392°F)	Bare and covered conductor
12	20	20	30*	40	40	45	30
10	30	30	45*	50	55	60	45
8	45	55	55*	65	70	80	55
6	60	75	80	95	100	105	80
4	80	100	105	125	135	140	100
3	95	115	120	140	150	165	115
2	110	135	140	165	175	185	135
1	130	155	165	190	205	220	160
0	150	180	190	220	240	255	185
00	175	210	220	255	275	290	215
000	200	240	255	300	320	335	250
0000	230	280	300	345	370	400	290
250	265	315	330	385	415	...	320
300	290	350	375	435	460	...	360
350	330	395	415	475	510	...	400
400	355	425	450	520	555	...	435
500	405	485	515	595	635	...	490
600	455	545	585	675	720	...	560
700	500	595	645	745	795	...	615
750	515	620	670	775	825	...	640
800	535	645	695	805	855	...	670
900	580	700	750	....	...	...	725
1,000	625	750	800	930	990	...	770
1,250	710	855	905				
1,500	795	950	1020	1175	...	...	985
1,750	875	1050	1125				
2,000	960	1150	1220	1425	...	...	1165

Correction factors, room temp. over 30°C, 86°F

°C	°F	60°C	75°C	85°–90°C	110°C	125°C	200°C	Bare and covered conductor
40	104	0.82	0.88	0.90	0.94	0.95		
45	113	0.71	0.82	0.85	0.90	0.92		
50	122	0.58	0.75	0.80	0.87	0.89		
55	131	0.41	0.67	0.74	0.83	0.86		
60	140	....	0.58	0.67	0.79	0.83	0.91	
70	158	....	0.35	0.52	0.71	0.76	0.87	
75	167	....	....	0.43	0.66	0.72	0.86	
80	176	....	....	0.30	0.61	0.69	0.84	
90	194	....	....	....	0.50	0.61	0.80	
100	212	....	....	....	....	0.51	0.77	
120	248	....	....	....	....	....	0.69	
140	284	....	....	....	....	....	0.59	

* The ampacities for types RHH and THHN conductors for sizes AWG 12, 10, and 8 shall be the same as designated for 75°C conductors in this table.

stantly on short circuits, and hold on overloads for various intervals depending upon its design characteristics and percent of overload.

44. Molded-case circuit breakers are made in sizes of 10 to 2,000 A for circuits up to 600 V. Interrupting capacities, depending upon frame size, go up to 75,000 rms A. Combined with current-limiting fuses, IC ratings up to 200,000 rms A are available.

Power circuit breakers, also called large air breakers, are heavy-duty switching and protective devices with high interrupting capacity (up to 150,000 rms A).

Circuit breakers are frequently used for routine branch-circuit switching (as in lighting branch-circuit panelboards) or in the panelboard mains. For larger circuit breakers motor drives may be applied, providing for remote control.

Table 17-8. Maximum Number of Conductors in Trade Sizes of Conduit or Tubing

(New work or rewiring—types RF-2, RFH-2, R, RH, RW, RHH, RHW, RH-RW; new work—FEP, FEPB, RUH, RUW, T, TF, THHN, THW, THWN, TW)

Size AWG or M cmils	Maximum number of conductors in conduit or tubing (based upon % conductor fill, Table 17-10, for new work)											
	½ in	¾ in	1 in	1¼ in	1½ in	2 in	2½ in	3 in	3½ in	4 in	5 in	6 in
18	7	12	20	35	49	80	115	176				
16	6	10	17	30	41	68	98	150				
14	4	6	10	18	25	41	58	90	121	155		
12	3	5	8	15	21	34	50	76	103	132	208	
10	1	4	7	13	17	29	41	64	86	110	173	
8	1	3	4	7	10	17	25	38	52	67	105	152
6	1	1	3	4	6	10	15	23	32	41	64	93
4	1	1	1	3*	5	8	12	18	24	31	49	72
3	.	1	1	3	4	7	10	16	21	28	44	63
2	.	1	1	3	3	6	9	14	19	24	38	55
1	.	1	1	1	3	4	7	10	14	18	29	42
0	.	..	1	1	2	4	6	9	12	16	25	37
00	.	..	1	1	1	3	5	8	11	14	22	32
000	.	..	1	1	1	3	4	7	9	12	19	27
0000	.	..	..	1	1	2	3	6	8	10	16	23
250	.	..	..	1	1	1	3	5	6	8	13	19
300	.	..	..	1	1	1	3	4	5	7	11	16
350	.	..	..	1	1	1	1	3	5	6	10	15
400	.	..	..	..	1	1	1	3	4	6	9	13
500	.	..	..	..	1	1	1	3	4	5	8	11
600	.	..	..	..	..	1	1	1	3	4	6	9
700	.	..	..	..	..	1	1	1	3	3	6	8
750	.	..	..	..	..	1	1	1	3	3	5	8
800	.	..	..	..	..	1	1	1	2	3	5	7
900	.	..	..	..	..	1	1	1	1	3	4	7
1,000	.	..	..	..	..	1	1	1	1	3	4	6
1,250	.	..	..	..	..	..	1	1	1	1	3	5
1,500	.	..	..	..	..	..	...	1	1	1	3	4
1,750	.	..	..	..	..	..	...	1	1	1	2	4
2,000	.	..	..	..	..	..	...	1	1	1	1	3

* Where an existing service run of conduit or electrical metallic tubing does not exceed 50 ft in length and does not contain more than the equivalent of two quarter bends from end to end, two No. 4 insulated and one No. 4 bare conductors may be installed in 1-in conduit or tubing.

In operation the circuit-breaker switch in the "on" position is normally latched by an armature. A thermal-magnetic or fully magnetic element, on sensing overload, trips the armature opening the switch.

Fuses contain a metal link which responds critically to I^2R and melts on overload to open the circuit. Various constructions are employed to speed or delay response.

Plug fuses are limited to circuits of not more than 30 A and 150 V to ground. Type S fuses are a noninterchangeable version.

Cartridge fuses contain the current-responsive link inside an insulating tube with terminals at each end. Many types are available to meet various system design requirements.

Standard single-element fuses are available in 250- and 600-V sizes up to 600 A. Renewable fuses permit the replacement of links without discarding the enclosure and terminals.

Terminals of fuses up to 60 A are the end ferrules; over 60 A, a knife-blade extension of the ferrules. Large current-limiting and high-capacity fuses have terminals for stud mounting.

Table 17-9. Maximum Number of Conductors in Trade Sizes of Conduit or Tubing

Rewiring—types TF, T, THW, TW, RUH, RUW

Size AWG or M cmils	Maximum number of conductors in conduit or tubing											
	½ in	¾ in	1 in	1¼ in	1½ in	2 in	2½ in	3 in	3½ in	4 in	5 in	6 in
18	13	24	38	68	93	152						
16	11	19	31	55	75	123	176	270				
14	5	10	16	29	40	65	91	143	192			
12	4	8	13	24	32	53	76	117	157	202		
10	4	6	11	19	26	43	61	95	127	163	257	
8	1	4	6	11	15	25	36	56	75	96	152	219
6	1	2	4	7	10	16	23	36	48	62	97	141
4	1	1	3	5	7	12	17	27	36	46	73	106
3	1	1	2	4	6	10	15	23	31	40	63	91
2	1	1	1	4	5	9	13	20	27	34	54	78
1	..	1	1	2	4	6	9	14	19	25	39	57
0	..	1	1	2	3	5	8	12	16	21	33	48
00	..	1	1	1	3	4	7	10	14	18	28	41
000	..	..	1	1	2	4	5	9	12	15	24	35
0000	..	..	1	1	1	3	5	7	10	13	20	29
250	..	..	..	1	1	2	4	6	8	10	16	23
300	..	..	..	1	1	2	3	5	7	9	14	20
350	..	..	..	1	1	1	3	4	6	8	12	18
400	..	..	..	1	1	1	2	4	5	7	11	16
500	..	..	..	..	1	1	1	3	4	6	9	14
600	..	..	..	..	1	1	1	3	4	5	7	11
700	..	..	..	..	..	1	1	2	3	4	7	10
750	..	..	..	..	..	1	1	2	3	4	6	9
800	..	..	..	..	..	1	1	1	3	4	6	9
900	..	..	..	..	..	1	1	1	2	3	5	8
1,000	..	..	..	..	..	1	1	1	2	3	5	7
1,250	..	..	..	..	..	...	1	1	1	2	4	6
1,500	..	..	..	..	..	...	1	1	1	1	3	5
1,750	..	..	..	..	..	...	1	1	1	1	3	4
2,000	..	..	..	..	..	...	...	1	1	1	2	4

Rewiring—types FEP, FEPB, THHN, THWN

Size AWG or MCM	Maximum number of conductors in conduit or tubing											
	½ in	¾ in	1 in	1¼ in	1½ in	2 in	2½ in	3 in	3½ in	4 in	5 in	6 in
14	13	24	39	69	94	154	220					
12	10	18	29	51	70	114	164	252				
10	6	11	18	32	44	72	104	160	215	276		
8	3	6	10	19	26	42	60	93	125	160	252	
6	1	4	6	11	15	25	37	56	76	98	154	222
4	1	2	4	7	9	16	22	35	47	60	94	136
3	1	1	3	6	8	13	19	29	39	51	80	116
2	1	1	3	5	7	11	16	25	33	43	67	97
1	1	1	1	3	5	8	12	18	25	32	50	72
0	..	1	1	3	4	7	10	15	21	27	42	61
00	..	1	1	2	3	6	8	13	17	22	35	51
000	..	1	1	1	3	5	7	11	14	18	29	42
0000	..	..	1	1	2	4	6	9	12	15	24	35
250	..	..	1	1	1	3	4	7	9	12	20	28
300	..	..	1	1	1	3	4	6	8	11	17	24
350	..	..	..	1	1	2	3	5	7	9	15	21
400	..	..	..	1	1	1	3	5	6	8	13	19
500	..	..	..	1	1	1	2	4	5	7	11	16

Table 17-10. Dimensions and Percent Area of Conduit and of Tubing
(Areas of conduit or tubing for combinations of wires)

Trade size	Internal diameter, in	Area, in								
		Total 100%	Not lead-covered			Lead-covered				
			1 cond. 31%	2 cond. 31%	3 cond. and over 40%	1 cond. 55%	2 cond. 30%	3 cond. 40%	4 cond. 38%	Over 4 cond. 35%
½	0.622	0.30	0.16	0.09	0.12	0.17	0.09	0.12	0.11	0.11
¾	0.824	0.53	0.28	0.16	0.21	0.29	0.16	0.21	0.20	0.19
1	1.049	0.86	0.46	0.27	0.34	0.47	0.26	0.34	0.33	0.30
1¼	1.380	1.50	0.80	0.47	0.60	0.83	0.45	0.60	0.57	0.53
1½	1.610	2.04	1.08	0.63	0.82	1.12	0.61	0.82	0.78	0.71
2	2.067	3.36	1.78	1.04	1.34	1.85	1.01	1.34	1.28	1.18
2½	2.469	4.79	2.54	1.48	1.92	2.63	1.44	1.92	1.82	1.68
3	3.068	7.38	3.91	2.29	2.95	4.06	2.21	2.95	2.80	2.58
3½	3.548	9.90	5.25	3.07	3.96	5.44	2.97	3.96	3.76	3.47
4	4.026	12.72	6.74	3.94	5.09	7.00	3.82	5.09	4.83	4.45
5	5.047	20.00	10.60	6.20	8.00	11.00	6.00	8.00	7.60	7.00
6	6.065	28.89	15.31	8.96	11.56	15.89	8.67	11.56	10.98	10.11

Dual-element fuses contain a series of current-responsive elements of different characteristics (such as long delay for motor starting plus high speed and high IC for short circuits).

High-capacity and large current-limiting silver-sand fuses are designed for use on low-impedance heavy-duty mains and feeders with IC ratings to 200,000 V rms.

45. Service Equipment and Distribution Centers. The NEC provides that the incoming service to a building or plant must be controlled near the point of entrance by not more than six sets of switches and fuses or six circuit breakers. In large installations, this "service equipment" is frequently combined with a "distribution center"

Table 17-11. Multiplying Factors for Converting D-C Resistance to 60-c A-C Resistance

Size	Multiplying factor			
	For nonmetallic sheathed cables in air or nonmetallic conduit		For metallic sheathed cables or all cables in metallic raceways	
	Copper	Aluminum	Copper	Aluminum
Up to 3 AWG	1	1	1	1
2	1	1	1.01	1.00
1	1	1	1.01	1.00
0	1.001	1.000	1.02	1.00
00	1.001	1.001	1.03	1.00
000	1.002	1.001	1.04	1.01
0000	1.004	1.002	1.05	1.01
250 M cmils	1.005	1.002	1.06	1.02
300 M cmils	1.006	1.003	1.07	1.02
350 M cmils	1.009	1.004	1.08	1.03
400 M cmils	1.011	1.005	1.10	1.04
500 M cmils	1.018	1.007	1.13	1.06
600 M cmils	1.025	1.010	1.16	1.08
700 M cmils	1.034	1.013	1.19	1.11
750 M cmils	1.039	1.015	1.21	1.12
800 M cmils	1.044	1.017	1.22	1.14
1,000 M cmils	1.067	1.026	1.30	1.19
1,250 M cmils	1.102	1.040	1.41	1.27
1,500 M cmils	1.142	1.058	1.53	1.36
1,750 M cmils	1.185	1.079	1.67	1.46
2,000 M cmils	1.233	1.100	1.82	1.56

Table 17-12. Properties of Conductors

Size AWG M cmils	Area, cmils	Concentric lay stranded conductors		Bare conductors		D-c resistance, Ω/1,000 ft at 25°C, 77°F		
						Copper		Aluminum
		No. wires	Diam. each wire, in	Diam., in	Area,* in²	Bare cond.	Tin'd. cond.	
18	1,624	Solid	0.0403	0.0403	0.0013	6.51	6.79	10.7
16	2,583	Solid	0.0508	0.0508	0.0020	4.10	4.26	6.72
14	4,107	Solid	0.0641	0.0641	0.0032	2.57	2.68	4.22
12	6,530	Solid	0.0808	0.0808	0.0051	1.62	1.68	2.66
10	10,380	Solid	0.1019	0.1019	0.0081	1.018	1.06	1.67
8	16,510	Solid	0.1285	0.1285	0.0130	0.6404	0.659	1.05
6	26,250	7	0.0612	0.184	0.027	0.410	0.427	0.674
4	41,740	7	0.0772	0.232	0.042	0.259	0.269	0.424
3	52,640	7	0.0867	0.260	0.053	0.205	0.213	0.336
2	66,370	7	0.0974	0.292	0.067	0.162	0.169	0.266
1	83,690	19	0.0664	0.332	0.087	0.129	0.134	0.211
0	105,500	19	0.0745	0.373	0.109	0.102	0.106	0.168
00	133,100	19	0.0837	0.418	0.137	0.0811	0.0843	0.133
000	167,800	19	0.0940	0.470	0.173	0.0642	0.0668	0.105
0000	211,600	19	0.1055	0.528	0.219	0.0509	0.0525	0.0836
250	250,000	37	0.0822	0.575	0.260	0.0431	0.0449	0.0708
300	300,000	37	0.0900	0.630	0.312	0.0360	0.0374	0.0590
350	350,000	37	0.0973	0.681	0.364	0.0308	0.0320	0.0505
400	400,000	37	0.1040	0.728	0.416	0.0270	0.0278	0.0442
500	500,000	37	0.1162	0.814	0.520	0.0216	0.0222	0.0354
600	600,000	61	0.0992	0.893	0.626	0.0180	0.0187	0.0295
700	700,000	61	0.1071	0.964	0.730	0.0154	0.0159	0.0253
750	750,000	61	0.1109	0.998	0.782	0.0144	0.0148	0.0236
800	800,000	61	0.1145	1.031	0.835	0.0135	0.0139	0.0221
900	900,000	61	0.1215	1.093	0.938	0.0120	0.0123	0.0197
1,000	1,000,000	61	0.1280	1.152	1.042	0.0108	0.0111	0.0177
1,250	1,250,000	91	0.1172	1.289	1.305	0.00863	0.00888	0.0142
1,500	1,500,000	91	0.1284	1.412	1.566	0.00719	0.00740	0.0118
1,750	1,750,000	127	0.1174	1.526	1.829	0.00616	0.00634	0.0101
2,000	2,000,000	127	0.1255	1.631	2.089	0.00539	0.00555	0.00885

* Area given is that of a circle having a diameter equal to the overall diameter of a stranded conductor.

The values given in the table are those given in *NBS Circ.* 31 except that those shown in the eighth column are those given in Specification B33 of the American Society for Testing and Materials.

The resistance values given in the last three columns are applicable only to direct current. When conductors larger than No. 4/0 are used with alternating current, the multiplying factors in Table 17-11 should be used to compensate for skin effect.

Table 17-13. Branch-circuit Requirements

(Type FEP, FEPB, R, RW, RU, RUW, RH-RW, SA, T, TW, RH, RUH, RHW, RHH, THHN, THW, and THWN conductors in raceway or cable)

Circuit rating	15 A	20 A	30 A	40 A	50 A
Conductors (min. size):					
Circuit wires*	14	12	10	8	6
Taps	14	14	14	12	12
Fixture wires and cords					
Overcurrent protection	15 A	20 A	30 A	40 A	50 A
Outlet devices:					
Lampholders permitted	Any type	Any type	Heavy duty	Heavy duty	Heavy duty
Receptacle rating	15 max. A	15 or 20 A	30 A	40 and 50 A	50 A
Maximum load	15 A	20 A	30 A	40 A	50 A

* These ampacities are for copper conductors where derating is not required.

at which the feeders to the various power and lighting panelboards are protected. Modern practice is toward completely dead-front construction in which no live parts are exposed on the front of the board. Large installations are usually switchboards of the

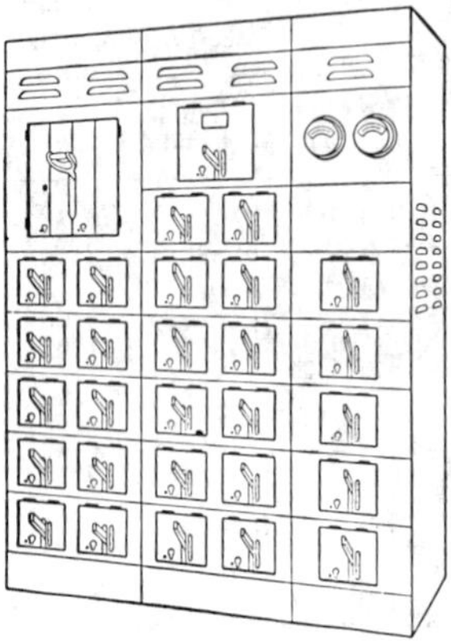

Fig. 17-11. Dead-front distribution center of the free-standing switchboard type. (*Bull Dog Electric Products Co.*)

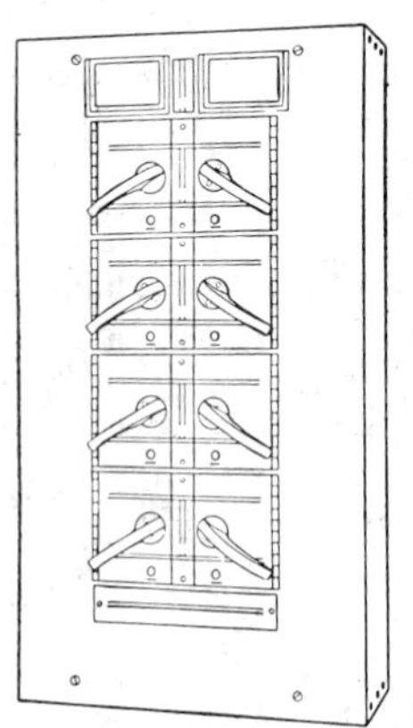

Fig. 17-12. Dead-front subdistribution center or panelboard for heavy circuits, wall-mounting type. (*Bull Dog Electric Products Co.*)

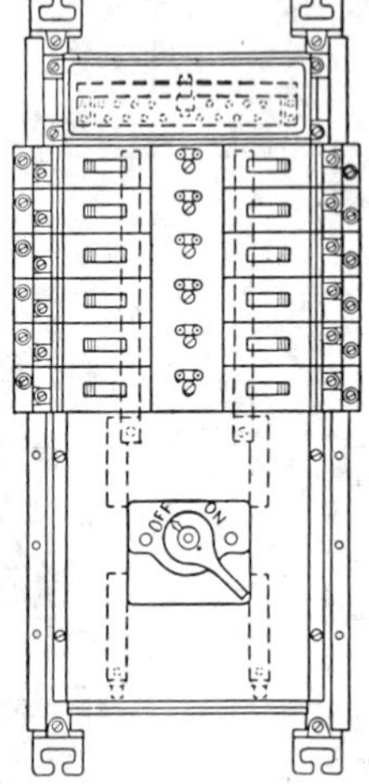

Fig. 17-13. Conventional circuit-breaker panelboard; cabinet not shown. (*Westinghouse Electric Corp.*)

"free-standing" type (Fig. 17-11), accessible from the front and the rear, while smaller-capacity equipment may be wall-mounted (see Fig. 17-12). Provision for metering equipment of the utility company supplying the service is usually made at the service equipment.

46. Panelboards are used as control and protection points for groups of branch circuits serving lighting, heating, and power outlets in a given building area, usually a floor or section of a floor. They may be of circuit-breaker, switch-and-fuse, or fuse type and are set in a sheet-steel enclosure with a wiring gutter to accommodate the large accumulation of branch-circuit conductors terminating at the board.

47. Local Control Equipment. For switching lighting in a small area involving, typically, a few hundred watts, conventional manually operated tumbler switches are usually employed. A variety of different operating mechanisms and styles are available for the same general purpose. Relay switches employing a small relay in the lighting circuit controlled by a low-voltage circuit and switch at the point of control are also used. Both methods are adaptable to controlling a light or lights from a number of points of control. For large loads, magnetic contactors operated by conveniently located wall switches are used. The contactors may operate a branch circuit, a portion of the panelboard bus, or the panelboard mains.

Panelboard circuit breakers or circuit switches are also commonly used to switch general-area lighting where the panelboard is conveniently located for the purpose.

In large buildings, central control is often used by extending control circuits from local contactors to a console in the custodian's control center.

Automatic control by time switches, program systems, photocell controllers, or similar unattended devices may also be applied by conventional remote-control measures.

Motor control is broadly similar in commercial buildings, with the important difference that control devices are usually of special types designed and rated for the purpose and include motor-overcurrent protective features. Otherwise the methods of relaying and remote control are conventional. At control centers, however, motor circuits are usually monitored by appropriate pilot lights.

PROTECTIVE GROUNDING

48. Purpose of Grounding. Secondary a-c distribution systems should be grounded at the neutral conductor if the maximum voltage to ground does not exceed 150 V and may be grounded if this voltage is above 150 but does not exceed 300 V. This is to guard against imposition of a dangerous high voltage in case a breakdown in the transformer or crossing of primary- and secondary-circuit wires occurs.

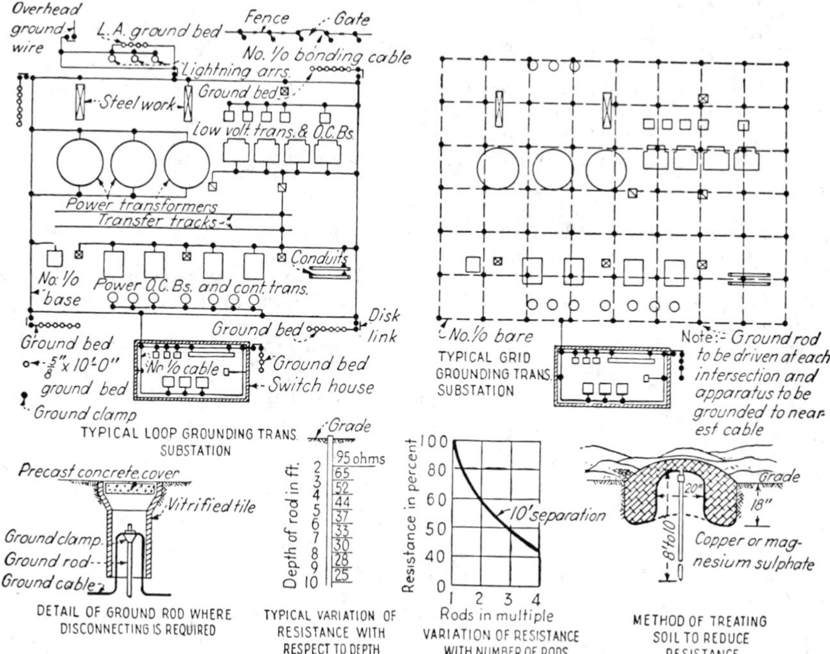

FIG. 17-14. Complex equipment-grounding details showing connections, wire sizes, and grounds.

Conduits, metal raceways, cable armors, and metal cases or frames of equipment must be grounded (or isolated, as an alternative) so that, if this metal enclosure should come into contact with any of the circuit wires within it, no dangerous current would be passed to a person who touched the enclosure, since it is kept at ground potential.

The path to ground must be as low in resistance as possible, as otherwise sufficient current may not flow to open an overcurrent protective device and clear the fault. In that case, a voltage buildup will occur on metallic parts, with consequent hazards.

49. Size and Location of Ground Connection. The service neutral should be grounded at the point of entrance ahead of any disconnecting equipment with a copper wire or bus not smaller than indicated in the following table:

Size of Largest Service Conductor	AWG Size of Copper Grounding Conductor
2 or smaller	8
1 or 0	6
00 or 000	4
Over 000–350,000 cir mils	2
Over 350,000–600,000 cir mils	0
Over 600,000–1,100,000 cir mils	00
Over 1,100,000 cir mils	000

Metallic enclosures are commonly grounded through the grounding conductor used for the service neutral, although conduit or electrical metallic tubing may be used.

The grounding connection should, wherever possible, be made to a continuous underground water-piping system. Resistance of such a ground will usually be less than 0.1 Ω, thereby ensuring effectiveness.

50. Polarization of Wiring. The conductor to be grounded must be continuously identified throughout the system to avoid errors in connections. For No. 6 AWG or smaller, this is accomplished by finishing the insulating covering with a white or natural-gray finish. Ends of conductors larger than No. 6 AWG are painted white or gray where exposed in outlets or panelboards.

SYSTEMS OF INTERIOR DISTRIBUTION

51. Standard Secondary-voltage Types. Common interior distribution systems for building or plants having appreciable loads are:

a. 3-phase 4-wire 120/208-V serving power and lighting.

b. Single-phase 3-wire 115/230-V serving lighting with 3-phase 3-wire 240- or 480-V for power.

c. 3-phase 4-wire 277/400-V serving power and fluorescent or mercury lighting with single-phase 115/230-V circuits for other utilization provided from the power system by means of air-cooled transformers.

For very large buildings and industrial plants distribution is often provided at higher voltage, notably 13.2, 4.1, or 2.3 kV stepped down to utilization voltage at strategically load-centered substations.

52. Three-wire Single-phase Systems. The 3-wire 115/230-V single-phase system (Fig. 17-15) is very commonly used for interior wiring for lighting and miscellaneous purposes but not for motor loads much in excess of 5 hp. The neutral wire is grounded, hence should not be fused at any point. The branch circuits may be 2-wire 115- or 230-V or 3-wire 115/230-V. The neutral conductor carries only the unbalance in load between the two ungrounded conductors. For circuits up to 200 A, it should have the same capacity as the ungrounded conductors. A factor of 0.7 may be applied to unbalanced loads above 200 A in determining its size.

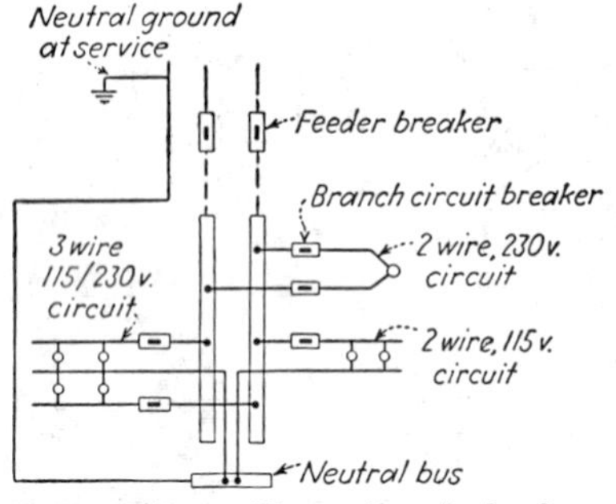

FIG. 17-15. Three-wire, single-phase system showing types of circuits.

53. The **3-phase 3-wire system** is usually employed where motors form a substantial load and where lighting is supplied from a separate single-phase system or by transformer. The usual voltage is 240 or 480. Branch circuits may be either 2-wire single-phase or 3-wire 3-phase.

54. Three-phase 4-wire Systems. The 3-phase 4-wire system (Fig. 17-16) is widely used. Branch lighting circuits are connected between any one of the phase wires and the neutral wire. Power is taken from the 3-phase wires. The neutral wire is grounded. The voltage between phase wires is usually 208, and between any phase wire and neutral it is 120 V.

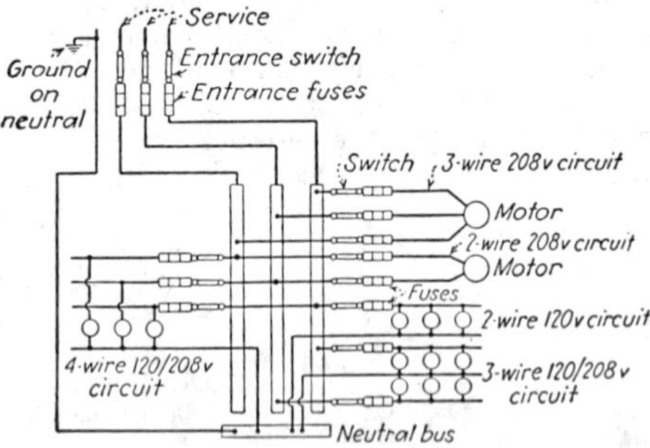

FIG. 17-16. Four-wire 3-phase system showing types of circuits.

Circuits that may be supplied include 2-wire 120- or 208-V; 3-wire 120/208- or 208-V; and 4-wire 120/208-V. In the case of a 3-wire circuit consisting of 2-phase wires and the neutral, the "neutral" differs from that of a 3-wire single-phase circuit in that with both sides of the circuit evenly loaded it will carry current equal to the current in the phase wires.

55. Two-wire Systems. The 2-wire system is used where current is supplied at 115 V, as when obtained from a 115-V generator, and on installations connected to utility companies' lines when the installation is of small capacity. It is limited in capacity; requires excessively large copper for heavy loads; and, in general, is not considered a modern complete distribution method.

56. Two- and 3-wire d-c systems are similar in connections (except grounding on the premises) and use to the 2- and 3-wire single-phase a-c systems.

57. Two-phase distribution may be effected with 4 or with 3 wires. In the former case there is a pair of wires for each phase, while in the latter there is 1 wire for each phase and a common wire for both phases. The circuits must be balanced on either side, just as in the case of a 3-wire single-phase system. Where 3 wires are used, the common wire should be 1.4 times as large as either of the other two, since it must carry 1.4 times as much current. Motors are connected to both phases and employ all 3 or all 4 wires, as the case may be. With 4 wires, the lamps are connected to each phase as though the supply were single-phase, and care should be exercised to balance the phases as nearly as possible.

DESIGN CONSIDERATIONS

58. Energy Sources. In general, there are two possible sources of electrical energy. The one most generally employed is the distribution system of the electric utility which serves the territory; the other is a private generating plant. Under modern economic conditions the cost of purchased energy is normally lower than the total cost of production in an isolated plant, unless an industrial plant has heavy uses for process steam, when it may be feasible to generate energy to a point of favorable heat balance and purchase the excess requirements.

Self-contained automatic gas-, gasoline-, or diesel-engine generators are, however, widely used for an emergency source in the event of power failure. The voltage and frequency of such sources are usually similar to the normal supply. The emergency source is connected to the system or a portion of the system by means of an automatic or manual throw-over switch.

59. Power-plant Wiring. Such plants may be operated independently of the purchased-power supply or, in some cases, in parallel with it. The former method requires an interior-distribution layout so designed that the total load can be divided into two portions, one for the local generating plant, and the remainder connected to the source of purchased power. If the two sources are operated in parallel, special problems arise, e.g., maintaining satisfactory synchronous operation and the proper protection of both sources against interruptions caused by overloads or circuit troubles. No plan of parallel operation should be considered without the approval of the utility supplying the purchased power. Under either plan careful attention should be given to reliability and quality of service as well as its cost.

60. Choice of Interior Distribution System—General. If the energy supply is to be purchased, the utility company supplying the service should always be consulted regarding the types of service available. Where the supply is taken directly from utility circuits at utilization voltage, the choice is limited to the types available. Where the load is such as to warrant primary service with step-down transformers on or adjacent to the property, or if energy is generated on the premises, greater leeway is possible in the choice of secondary voltages and distribution system.

61. For commercial buildings, the type of service will usually be determined by the required load. For small buildings a single-phase 115/230-V will occasionally serve. In some areas the single-phase service is augmented by a 230-V 3-wire 3-phase separate power service. "Network" power is conventionally 120/208, 3-phase, 4-wire, but for larger buildings 277/480-V 3-phase 4-wire service is increasingly favored. For large buildings and "high-rise" structures primary service is often supplied, typically at 4,160 V or 13.2 kV, and transformed at load centers within the building, a practice now common in industrial plants.

62. Industrial plants vary in size from establishments occupying a portion of a single building to plants covering hundreds of acres. The demand for electrical energy varies from a few kilowatts in small plants to thousands of kilowatts in large plants. The methods of interior wiring described in this section are applicable to individual buildings; but where the industry occupies several buildings not connected with each other, or often many buildings scattered over a large ground space, the complete system of wiring may be comparable with a distribution system serving a local area in which the load density is ordinarily of high order.

Many industrial plants can be served adequately and economically at the standard secondary or service voltages, depending on the demand and the distances from the service entrance to the individual loads. However, when the demands exceed the capacity economically available from the secondary distribution system, it is standard practice to install a substation or transformer station on the premises. This is supplied from the primary distribution system at any of various voltages ranging from 2,300 V upward, depending on the total demand and the availability of primary supply.

As in commercial buildings, the most satisfactory interior systems are separate light and power at standard voltages (for power 230, 460, or 550 V) or a 4-wire 3-phase 120/208-V system. Another popular system employs 460-V 3-phase power distribution with 120/240-V 3-wire lighting tapped off through air-cooled transformers. Where plants are large, it may be practicable to install a primary-network system similar to a utility network. In this case, transformers and network protectors are located at various load centers, and secondaries are interconnected so that failure of one transformer station will not interrupt service to the section normally supplied through it. With multiple transformer locations it may also be practicable to loop the primary supply cables or circuits so that each transformer bank can be fed from two directions.

Primary-voltage supply (usually 2,300 or 4,000 V) may also be used advantageously to supply large individual motors or special apparatus.

63. Special Design Considerations. Factors affecting the **wiring method** selected arise owing to the type of occupancy, requirements for flexibility or accessibility of the system, or the nature of equipment to be served. For example, stage and auditorium lighting of theaters introduces problems in handling large numbers of circuits. In hazardous locations (as defined in NEC), precautions must be taken against arcs and sparks igniting flammable liquids, vapors, dusts, etc. Requirements of flexibility and accessibility may dictate an underfloor raceway system or cellular-metal-floor construction in fire-resistive commercial buildings or the use of wireways and busways for a mass-production industry. Circuits to serve large resistance welders must be carefully designed to avoid excessive reactance drop.

64. Outside Wiring. In large industrial plants the outside wiring or feeder system is frequently of an extensive character and requires careful planning, especially where the total number of individual loads is large and the power demands are considerable. If the distribution system is aerial, in open wire, consideration should be given to protection from lightning and the possibility of foreign contacts. Underground systems, although higher in first cost, afford greater security from interruptions.

65. Standard Equipment. Standard voltages are desirable because motors, transformers, and other equipment for such voltages are of lower cost than special equipment for nonstandard voltages, and replacement parts are much more readily available.

66. Voltage Drop. Each lamp, motor, heating device, and other piece of electrical utilization equipment is designed to operate at a certain design voltage. Its performance and efficiency are adversely affected if connected to a circuit operating at a voltage somewhat lower. Incandescent lamps and heating devices, particularly, are extremely sensitive to this condition. In the former, 1% loss (or drop) in voltage produces about 3% loss in light output; 5% voltage drop reduces the light output 16%; a 10% voltage drop reduces the light output about 30%. The starting and pull-out torque of motors is seriously affected by reduced voltage, and current drawn is increased, so that the heat rise will be above normal.

The drop in voltage through the conductor is caused by its resistance to the flow of current and, in a-c circuits, by the reactance of the circuit. The power required to overcome the resistance is dissipated in heat, which in severe cases may be sufficient to cause deterioration of the conductor insulation. In addition, whether the electricity is generated or purchased, this power loss appears as a component of the monthly energy cost.

In view of the above, it is essential that every installation be designed with voltage drop and resistance losses fully in mind. Except for short and lightly loaded runs, voltage drop will be the determining factor in design. It has been taken fully into account in the material which follows.

DESIGN STANDARDS—GENERAL

67. Necessity for Modern Design Standards. Before beginning the actual layout of the wiring system, it is essential that a modern standard of wiring design be consulted and followed, in order to avoid early obsolescence of the electrical investment due to inadequacy, with its inability to keep pace with electrical developments. Growth in electrical usage resulting from development of new equipment, better appreciation of the value of high levels of lighting, and the practically universal use of motor drives is so rapid that a wiring system designed merely for the initially contemplated load will be hopelessly inadequate within a few years.

The NEC requires that certain minimum loadings for lighting be assumed in calculating branch circuits and feeders. These are useful as irreducible minimums but, in general, are too low to be used as a design standard (see Pars. **68** to **70**).

68. National Electrical Code Demand Factors for Feeders. The demand factors indicated in Table 17-14 may be applied to the feeders for general lighting when computed as outlined in Par. **70**. The unit values are based on minimum loads; hence feeders based on these values may not provide sufficient capacity and should be increased.

69. National Electrical Code Branch-circuit Requirements. For general illumination in the occupancies listed, a load of not less than the "watts per square foot" shown in Table 17-14 must be included for each square foot of floor area, computed from the outside dimensions of the building or space.

For occupancies not listed in Table 17-14, and for special lighting and appliance loads, capacity shall be provided for the specific load but not less per outlet than indicated below:

Outlets	Amp
Supplying fixed or special appliances	Rating of Appliance
Supplying heavy-duty lampholders	5
Other	1½

70. National Electrical Code Feeder Requirements. The computed feeder load shall be the sum of the branch-circuit loadings computed from Table 17-14.

Table 17-14. General Lighting Loads by Occupancies

Type of Occupancy	Unit Load/Ft², W
Armories and auditoriums	1
Banks	2
Barbershops and beauty parlors	3
Churches	1
Clubs	2
Courtrooms	2
Dwellings (other than hotels)	3
Garages—commercial (storage)	½
Hospitals	2
Hotels and motels, including apartment houses without provisions for cooking by tenants	2
Industrial commercial (loft) buildings	2
Lodge rooms	1½
Office buildings	5
Restaurants	2
Schools	3
Stores	3
Warehouses (storage)	¼
In any of the above occupancies except single-family dwellings and individual apartments of multifamily dwellings:	
Assembly halls and auditoriums	1
Halls, corridors, closets	½
Storage spaces	¼

Table 17-15. Calculation of Feeder Loads by Occupancies

Type of occupancy	Portion of lighting load to which demand factor applies (wattage)	Feeder demand factor, %
Dwellings—other than hotels	First 3,000 or less at	100
	Next 3,001 to 120,000 at	35
	Remainder over 120,000 at	25
Hospitals*	First 50,000 or less at	40
	Remainder over 50,000 at	40
Hotels and motels—including apartment houses without provision for cooking by tenants*	First 20,000 or less at	50
	Next 20,001 to 100,000 at	40
	Remainder over 100,000 at	30
Warehouse (storage)	First 12,500 or less at	100
	Remainder over 12,500 at	50
All others	Total wattage	100

* The demand factors of this table shall not apply to the computed load of subfeeders to areas in hospitals, hotels, and motels where entire lighting is likely to be used at one time as in operating rooms, ballrooms, or dining rooms.

ADEQUACY STANDARDS FOR COMMERCIAL, PUBLIC, AND INDUSTRIAL STRUCTURES*

71. Outlet Location—Commercial and Public Occupancies. *a. Ceiling Outlets for General Illumination.* The lighting layout should be made according to the type of lighting equipment to be installed. The ceiling-outlet location can then be determined by the spacing needed for the desired lighting result.

b. Stores—Convenience Outlets. There shall be installed at least one convenience outlet for each 400 ft^2 of floor area or major part thereof, these outlets to be uniformly distributed over the entire area. No part of the floor area shall be more than 15 ft from an outlet.

c. Stores—Outlets for Window Illumination. Provision for show-window illumination by (1) a junction box on wall or column above transom bar or false ceiling to which circuits of show-window reflectors or spotlights can be connected or (2) individual outlets spaced as recommended below.

For medium-screw lamp holders, outlets for reflectors located 12 to 18 in apart. For mogul-screw lamp holders, outlets for reflectors located 15 to 24 in apart. For fluorescent lamps, as required by the lighting layout. The specific layout is to be governed by the standard loads of Table 17-14. When more than one circuit per window is necessary, adjacent outlets should be on different circuits.

For spot- or floodlights, at least two outlets per window, symmetrically placed. Additional outlets for every 8 ft (or major fraction) in excess of 16 ft.

d. Stores—Convenience Outlets in Show Windows. In or near the floor of each window, at least one convenience outlet for each 50 ft^2 of platform or floor-window-display area, but not less than two outlets per window.

e. Stores—Outlets for Case Lighting. Outlets in the floor or wall for termination of the circuits for showcase and wall-case lighting, suitably located for connection to the lighting equipment and governed by the standard loads of Table 17-14.

f. Offices and Schools—Convenience Outlets. In each separate office with 400 ft^2 or less of floor areas, at least one convenience outlet for each 20 lin ft of wall space. In each separate office larger than 400 ft^2, at least four convenience outlets for the first 400 ft^2 and at least two outlets for each additional 400 ft^2 or major fraction thereof. Outlets should serve all parts of the office space. Certain offices, e.g., professional and display offices, require many more outlets than specified above and must be individually studied where such probable occupancy is known.

For typical schoolrooms, at least one convenience outlet along the front wall and at least one along the rear wall.

* Abstracted from Handbook of Interior Wiring Design, Ref. 4, Par. **118**.

g. Offices and Stores—Fans. Unless made unnecessary by provision for complete air conditioning, outlets for fans on the basis of at least two (approximately 7 ft above the floor) for each 400 ft² or major portion thereof.

h. Stores—Signs. Where no other equivalent provision is made for sign lighting, a rigid raceway, not smaller than 1 in from a cabinet to a suitable point on the front face of the building for each intended or probable individual-store occupancy. Feeder and service capacity based upon a sign load of not less than 50 W/lin ft of store frontage on the principal street. Space provision for time switch.

72. Outlet Location—Industrial Occupancies. *a. Ceiling Outlets for General Illumination.* The many special considerations of industrial lighting make it necessary to develop the outlet locations around the lighting layout. Local lighting units frequently supplement general illumination; special directional overhead sources are often necessary. In planning outlet locations, these must be considered as part of the complete installation.

In halls and corridors lighted from a single row of outlets, the spacing between outlets shall not exceed 20 ft.

b. Convenience Outlets. At least one convenience outlet in each bay in both manufacturing and storage spaces.

73. Standard Loads for General Illumination. The number of branch circuits and the capacities of the feeders and service shall be based upon the loads and outlets per circuit specified in Tables 17-14 and 17-15 as minimum. Where a known load is to be supplied that will exceed the loading here specified, the circuits, feeders, and service shall be based upon such known load.

The watts per square foot for each occupancy were determined by considering average conditions found in such occupancies, together with the accepted methods and equipment used in providing such illumination levels. In certain cases the standard loads include allowance for convenience outlets as well as for general illumination.

The calculated wattage standards were based on the usual types of fluorescent and incandescent lamp.

74. Circuit Control. Suitable provision shall be made for the control of all circuits except those supplying convenience outlets only. Control of the latter is recommended.

In a retail store, individual control at the panelboards is usually preferable. In most other occupancies, control should be provided locally.

Provision shall be made for control of all circuits to a single sign by one switch or circuit breaker. A time switch shall also be provided.

In a retail store, each show-window circuit shall be individually controlled. A time switch for group control shall be provided.

Group control of circuits for general illumination or for decorative effect, by remote-control switches, is desirable in certain spaces, e.g., large offices, large reading rooms, museums, art galleries, and ballrooms.

75. Branch Circuits. The minimum number of branch circuits shall be based as follows on the standard loads as given in Tables 17-14 and 17-15.

a. For 2-wire 15-A circuits, the load per circuit shall not exceed 1,000 W.

b. For multiwire 15-A circuits, the load shall not exceed 1,000 W between each outside wire and the neutral.

c. For heavy-duty lamp circuits, the load per circuit shall not exceed 1,500 W for No. 10 wire, 2,500 W for No. 8, and 3,000 W for No. 6.

The following considerations of wire size and circuit runs shall apply:

d. No wire smaller than No. 12 shall be used for any circuit.

e. Where the run from a panelboard to the first outlet of a lighting-branch circuit exceeds 50 ft, the size of wire used shall be at least one size larger than that determined by any of the above considerations. The calculated size may be used between outlets.

f. Where the run from a panelboard to the first outlet of a convenience outlet circuit exceeds 100 ft, No. 10 wire shall be used for that run.

g. No runs longer than 100 ft between panelboard and the first outlet of a lighting-branch circuit shall be made, unless the intended load is so small that the voltage drop can be restricted to 2% between panelboard and any outlet on that circuit. To avoid this condition, panelboards should be relocated, or additional panelboards installed.

Table 17-16. Standard Loads for Illumination in Commercial and Public Interiors

Occupancy	Watts per sq ft
1. Armories:	
Drill sheds and exhibition halls. This does not include lighting circuits for demonstration booths, special exhibit spaces, etc	5
2. Art galleries:	
a. General	3
b. On paintings—50 watts per running ft of usable wall area	
3. Auditoriums	4
4. Automobile show rooms	6
5. Banks:	
a. Lobby	4
b. Counters—50 watts per running ft including service for signs and small motor applications, etc.	
c. Offices and cages	5
6. Barber shop and beauty parlors. This does not include circuits for special equipment	5
7. Billiards:	
a. General	5
b. Tables—450 watts per table	
8. Bowling:	
a. Alley runway and seats	5
b. Pins—300 watts per set of pins	
9. Churches:	
a. Auditoriums	2
b. Sunday-school rooms	5
c. Pulpit or rostrum	5
10. Club rooms:	
a. Lounge	2
b. Reading rooms	5
The above two uses are so often combined that the higher figure is advisable. It includes provision for convenience outlets	
11. Court rooms	5
12. Dance halls. No allowance has been included for spectacular lighting, spots, etc	2
13. Drafting rooms	7
14. Fire-engine houses	2
15. Gymnasiums:	
a. Main floor	5
b. Shower rooms	2
c. Locker rooms	2
d. Fencing, boxing, etc	5
e. Handball, squash, etc	5
16. Halls and interior passageways—15 watts per running ft	
17. Hospitals:	
a. Lobby, reception room	3
b. Corridors—10 watts per running ft	
c. Wards. Including allowance for convenience outlets for local illumination	5
d. Private rooms. Including allowance for convenience outlets for local illumination	5
e. Operating room	5
f. Operating tables or chairs:	
Major surgeries—3000 watts per area	
Minor surgeries—1500 watts per area	
This and the above figure include allowance for directional control. Special wiring for emergency systems must also be considered	
g. Laboratories	5
18. Hotels:	
a. Lobby. Not including provision for conventions, exhibits	5
b. Dining room	4
c. Kitchen	5
d. Bedrooms. Including allowance for convenience outlets	3
e. Corridors—10 watts per running ft	
f. Writing room. Including allowance for convenience outlets	5
19. Library:	
a. Reading rooms. This includes allowance for convenience outlets	6
b. Stack room—12 watts per running ft of facing stacks	
20. Motion-picture houses and theaters:	
a. Auditoriums	2
b. Foyer	3
c. Lobby	5
21. Museums:	
a. General	3
b. Special exhibits—supplementary lighting	5
22. Office buildings:	
a. Private offices, no close work	4
b. Private offices, with close work	5
c. General offices, no close work	4
d. General offices, with close work	5
e. File room, vault, etc	3
f. Reception room	2
23. Post office:	
a. Lobby	3
b. Sorting, mailing, etc	5
c. Storage, file room, etc	3
24. Professional offices:	
a. Waiting rooms	3
b. Consultation rooms	5
c. Operating offices	7
d. Dental chairs—600 watts per chair	
25. Railway:	
a. Depot—waiting room	3
b. Ticket offices—general. On counters 50 watts per running ft	5
c. Rest room, smoking room	3
d. Baggage-checking office	3
e. Baggage storage	2
f. Concourse	2
g. Train platform	2
26. Restaurants, lunchrooms, and cafeterias:	
a. Dining area	3
b. Food displays—50 watts per running ft of counter (including service aisle)	
27. Schools:	
a. Auditoriums	3
If to be used as a study hall—5 watts per sq ft	
b. Class- and study rooms	5
c. Drawing room	7
d. Laboratories	4
e. Manual training	5
f. Sewing room	7
g. Sight-saving classes	7
28. Showcases—25 watts per running ft	
29. Show windows:	
a. Large cities:	
Brightly lighted district—700 watts per running ft of glass	
Secondary business locations—500 watts per running ft of glass	
Neighborhood stores—250 watts per running ft of glass	
b. Medium cities:	
Brightly lighted district—500 watts per running ft of glass	
Neighborhood stores—250 watts per running ft of glass frontage	
c. Small cities and towns:	
300 watts per running ft of glass frontage	
d. Lighting to reduce daylight window reflections—1000 watts per running ft of glass	
30. Stores, large department and specialty:	
a. Main floor	6
b. Other floors	6
31. Stores in outlying districts	5
32. Wall cases—25 watts per running ft	

Table 17-17. Loads for General Illumination from Overhead Sources in Industrial Occupancies

Occupancy	Watts per sq ft	Occupancy	Watts per sq ft
1. Aisles, stairways, passageways: 10 watts per running ft		*b.* Dark goods:	
2. Assembly:		(1) Cutting, pressing, etc	*4.5
a. Rough	3.0	(2) Stitching, trimming, etc	*4.5
b. Medium	4.5	20. Hangars—airplane:	
c. Fine	*4.5	*a* Storage—live	2.0
d. Extra fine	*4.5	*b.* Repair department	*3.0
3. Automobile manufacturing:		21. Hat manufacturing:	
a. Assembly line	*4.5	*a.* Dyeing, stiffening, braiding, cleaning, refining:	
b. Frame assembly	3.0	(1) Light	2.0
c. Body assembly	4.5	(2) Dark	4.5
d. Body finishing and inspecting	*4.5	*b.* Forming, sizing, pouncing, flanging, finishing, ironing:	
4. Bakeries	4.0	(1) Light	3.0
5. Bookbinding:		(2) Dark	6.0
a. Folding, assembling, pasting	3.0	*c.* Sewing:	
b. Cutting, punching, stitching, embossing	4.0	(1) Light	4.5
6. Breweries:		(2) Dark	*4.5
a. Brew house	3.0	22. Icemaking:	
b. Boiling, keg washing, etc	3.0	*a.* Engine and compressor room	2.0
c. Bottling	4.0	23. Inspection:	
7. Candymaking	4.0	*a.* Rough	3.0
8. Canning and preserving	4.0	*b.* Medium	4.5
9. Chemical works:		*c.* Fine	*4.5
a. Hand furnaces, stationary driers, and crystallizers	2.0	*d.* Extra fine	*4.5
b. Mechanical driers and crystallizers, filtrations, evaporators, bleaching	2.0	24. Jewelry and watch manufacturing	*4.5
c. Tanks for cooking, extractors, percolators, nitrators, electrolytic cells	3.0	25. Laundries and dry cleaning	4.5
10. Clay products and cements:		26. Leather manufacturing:	
a. Grinding, filter presses, kiln rooms	2.0	*a.* Vats	2.0
b. Moldings, pressing, cleaning, trimming	2.0	*b.* Cleaning, tanning, stretching	2.0
c. Enameling	3.0	*c.* Cutting, fleshing, stuffing	3.0
d. Glazing	4.0	*d.* Finishing, scarfing	4.5
11. Cloth products:		27. Leatherworking:	
a. Cutting, inspecting, sewing:		*a.* Pressing, winding, glazing:	
(1) Light goods	4.5	(1) Light	2.0
(2) Dark goods	*4.5	(2) Dark	4.5
b. Pressing, cloth treating (oil cloth, etc.):		*b.* Grading, matching, cutting, scarfing, sewing:	
(1) Light goods	3.0	(1) Light	4.5
(2) Dark goods	6.0	(2) Dark	*4.5
12. Coal breaking, washing, screening	2.0	28. Locker rooms	2.0
13. Dairy products	4.0	29. Machine shops:	
14. Engraving	*4.5	*a.* Rough bench- and machinework	3.0
15. Forge shops:		*b.* Medium bench- and machinework, ordinary automatic machines, rough grinding, medium buffing, polishing	4.5
a. Welding	2.0	*c.* Fine bench- and machinework, fine automatic machines, medium grinding, fine buffing, polishing	*4.5
16. Foundries:		*d.* Extra-fine bench- and machinework, grinding:	
a. Charging floor, tumbling, cleaning, pouring, shaking out	2.0	(1) Fine work	*4.5
b. Rough molding and coremaking	2.0	30. Meat packing:	
c. Fine molding and coremaking	4.0	*a.* Slaughtering	2.0
17. Garages:		*b.* Cleaning, cutting, cooking, grinding, canning, packing	4.5
a. Storage	2.0	31. Milling—grain foods:	
b. Repair, washing	*3.0	*a.* Cleaning, grinding, rolling	2.0
18. Glassworks:		*b.* Baking or roasting	4.5
a. Mixing and furnace rooms, pressing and Lehr glass-blowing machines	3.0	*c.* Flour grading	4.5
b. Grinding, cutting glass to size, silvering	4.5	32. Offices:	
c. Fine grinding, polishing, beveling, etching, inspecting, etc	*4.5	*a.* Private and general:	
19. Glove manufacturing:		(1) No close work	3.0
a. Light goods:		(2) Close work	4.5
(1) Cutting, pressing, knitting, sorting	4.5	*b.* Drafting rooms	7.0
(2) Stitching, trimming, inspecting	4.5	33. Packing and boxing	3.0
		34. Paint manufacturing	3.0

Table 17-17. Loads for General Illumination from Overhead Sources in Industrial Occupancies.—*Concluded*

Occupancy	Watts per sq ft
35. Paint shops:	
a. Dipping, spraying, firing, rubbing, ordinary hand painting and finishing	3.0
b. Fine hand painting and finishing	*3.0
c. Extra-fine hand painting and finishing (automobile bodies, piano cases, etc.)	*3.0
36. Paper-box manufacturing:	
a. Light	3.0
b. Dark	4.0
c. Storage of stock	2.0
37. Paper manufacturing:	
a. Beaters, grinding, calendering	2.0
b. Finishing, cutting, trimming	4.5
38. Plating	2.0
39. Polishing and burnishing	3.0
40. Power plants, engine rooms, boilers:	
a. Boilers, coal and ash handling, storage-battery rooms	2.0
b. Auxiliary equipment, oil switches and transformers	2.0
c. Switchboards, engines, generators, blowers, compressors	3.0
41. Printing industries:	
a. Matrixing and casting	2.0
b. Miscellaneous machines	3.0
c. Presses and electrotyping	4.5
d. Lithographing	*4.5
e. Linotype, monotype, typesetting, imposing stone, engraving	*4.5
f. Proofreading	*4.5
42. Receiving and shipping	2.0
43. Rubber manufacturing and products:	
a. Calendars, compounding mills, fabric preparation, stock cutting, tubing machines, solid-tire operations, mechanical goods building, vulcanizing	3.0
b. Bead building, pneumatic-tire building and finishing, inner-tube operation, mechanical-goods trimming, treading	4.5
44. Sheet-metal works:	
a. Miscellaneous machines, ordinary benchwork	3.0
b. Punches, presses, shears, stamps, welders, spinning, medium benchwork	4.5
c. Tin-plate inspection	*4.5
45. Shoe manufacturing:	
a. Hand turning, miscellaneous bench- and machinework	2.0
b. Inspecting and sorting raw material, cutting and stitching:	
(1) Light	4.5
(2) Dark	*4.5
c. Lasting and welting	4.5
46. Soap manufacturing:	
a. Kettle houses, cutting, soap chips and powder	3.0
b. Stamping, wrapping and packing, filling and packing soap powder	4.5
47. Steel and iron mills, bar, sheet, and wire products:	
a. Soaking pits and reheating furnaces	2.0
b. Charging and casting floors	2.0
c. Muck and heavy rolling, shearing (rough by gage), pickling and cleaning	2.0
d. Plate inspection, chipping	*4.5
e. Automatic machines, light and cold rolling, wire drawing, shearing (line by line)	4.5
18. Stone crushing and screening:	
a. Belt-conveyor tubes, main-line shafting spaces, chute rooms, inside bins	2.0
b. Primary breaker room, auxiliary breakers under bins	2.0
c. Screens	3.0
49. Storage-battery manufacturing:	
Molding of grids	3.0
50. Store and stock rooms:	
a. Rough, bulky material	2.0
b. Medium or fine material requiring care	3.0
51. Structural-steel fabrication	3.0
52. Sugar grading	5.0
53. Testing:	
a. Rough	3.0
b. Fine	4.5
c. Extra-fine instruments, scales, etc.	*4.5
54. Textile mills:	
a. Cotton:	
(1) Opening and lapping, carding, drawing, roving, dyeing	3.0
(2) Spooling, spinning, drawing, warping, weaving, quilling, inspecting, knitting, slashing (over beam end)	4.5
b. Silk:	
(1) Winding, throwing, dyeing	4.5
(2) Quilling, warping, weaving, finishing:	
Light goods	4.5
Dark goods	6.0
c. Woolen:	
(1) Carding, picking, washing, combing	3.0
(2) Twisting, dyeing	3.0
(3) Drawing in, warping:	
Light goods	4.5
Dark goods	6.0
(4) Weaving:	
Light goods	4.5
Dark goods	6.0
(5) Knitting machines	4.5
55. Tobacco products:	
a. Drying, stripping, general	3.0
b. Grading and sorting	*4.5
56. Toilets and wash rooms	2.0
57. Upholstering:	
Automobile, coach, furniture	4.5
58. Warehouse	2.0
59. Woodworking:	
a. Rough sawing and benchwork	2.0
b. Sizing, planing, rough sanding, medium machine- and benchwork, gluing, veneering, cooperage	4.5
c. Fine bench and machinework, fine sanding and finishing	6.0

NOTE: Figures given are for average design loads for general lighting. In those cases marked with an asterisk (*) the values provide only for large-area lighting applications. Consideration should be given to high-intensity supplementary lighting needs. Values assume the use of conventional incandescent, metallic-vapor, and fluorescent-lamp sources customary in the application or type of establishment.

h. No convenience outlet shall be supplied by the same branch circuit that supplies ceiling or show-window lighting outlets.

i. Outlets for show-window spotlights shall be on separate circuits from general show-window outlets.

j. The number of convenience outlets included on one circuit should be as indicated in the following tabulation:

Location	Maximum
Barber shops, beauty parlors, etc.	2
Medical, dental, and similar offices	2
Store show windows (for spotlights)	3
Display areas in retail stores	6
School classrooms	6
Manufacturing spaces	6
Office spaces	8
Storage spaces	10

76. Panelboards. The number and location of panelboards shall be based on the number of branch circuits and the distance or runs, as specified in Par. **75**.

On each panelboard, provide one spare circuit for each five active circuits. Where flush-type cabinets are used, provision shall be made for bringing a corresponding number of circuit conductors to the ceiling of the story served or to the ceiling of the story immediately below or both. This may consist of circuit conductors or empty raceways terminating in boxes suitably located for future extensions.

77. Feeders. *a. Carrying Capacity.* The carrying capacity of each feeder shall be based on the number of circuits it supplies, computed as follows:

1. Overhead lighting circuits—1,000 W for each 15-A circuit.
2. Convenience outlet circuits—1,000 W/circuit.
3. Spare panelboard circuits—500 W/circuit.
4. Nonitemized and heavy-duty additional circuits—specific load for which designed.

To the total of these four, the demand factors permitted by NEC may be applied.

b. Provision for Future. In all feeder calculations, make provision to increase the capacity of the initial system by 50% to provide for future growth at minimum expense. If the ultimate feeder size does not exceed No. 4 wire, the excess capacity should be installed immediately. Otherwise, one of the following methods should be adopted:

1. The installation of oversize raceways to permit replacing conductors.
2. Arrangement of the installed equipment so that additional feeders may be added at a minimum of expense.
3. The installation of feeders of excess size.

c. Voltage Drop. Feeders shall be of such size that the total voltage drop from any panelboard to the point where connection is made to the lines of the utility does not exceed 2%. To compute such drop, the ultimate demand as calculated above shall be used if the wire size based on such demand is installed immediately. Otherwise the voltage drop shall be computed on the basis of the carrying capacity of the wire used.

78. Feeder-distribution Center. A feeder-distribution center (panelboard, switchboard, or group of enclosed switches or circuit breakers) shall be provided for the control and protection of each feeder.

Unless oversize feeders are originally provided, provision shall be made for connection and protection of feeders of increased size or of supplemental feeders by providing additional protective devices initially or by designing the original equipment so that space, bus capacity, and facilities for connections will be available.

Table 17-18. Service-entrance Adequacy Standards

Initial load, amp	Service switch, amp	Conductor size, gage No. or cir mils
1– 23	60	8
24– 33	60	6
34– 47	100	4
48– 60	100	2
61– 67	100	1
68– 83	200	0
84–100	200	00
101–117	200	000
118–133	200	0000
134–150	400	0000
151–167	400	250,000
168–183	400	300,000
184–200	400	350,000
201–217	400	400,000
218–267	400	500,000

79. Service Conductors and Equipment. The minimum capacity of the service required for the initial load depends on the total number of feeders supplied by it. Where applicable, demand factors as permitted by NEC may be used. To determine the probable ultimate capacity of the service, 50% should be added to the calculated initial load.

Where the calculated initial load does not exceed 267 A, service conductors and equipment having the capacity needed for the ultimate load should be included as part of the original installation. The recommended equipment for various initial loads is given in Table 17-18.

Where the initial load exceeds 267 A, a study should be made of each individual case to determine what provision should be made for a future increase.

CONDITIONS OF ADEQUACY FOR POWER WIRING

Note: Electrical-power applications are so numerous and so diversified that it is impossible to compile specific standards covering all cases. The following, therefore, is a general recommendation to be observed as closely as conditions will permit.

80. Adequacy Factors. A power-wiring installation may be considered adequate when due weight has been given to each of the following factors:

a. Safety and reliability.
b. Avoidance of excessive voltage drop.
c. Avoidance of excessive copper loss.
d. Flexibility in changing locations of equipment.
e. Provision for supplying increased loads.
f. Provision for economical maintenance.

All these are of importance in an industrial plant and must be studied for each plant. Power applications in other buildings are of a more nearly permanent character; hence flexibility and provision for increase in the load may not be so important as in an industrial plant.

81. Safety and Reliability. The NEC contains specifications for conductor sizes as follows:

a. For a motor-branch circuit, conductor carrying capacity not less than 125% of the full-load current of the motor.

b. For services, feeders, or subfeeders supplying more than one motor, conductor carrying capacity not less than 125% of the full-load current of the largest motor plus the sum of the full-load currents of all other motors. By special permission of the inspection authority, a demand factor may be applied.

The NEC also contains requirements for the overcurrent protection of motor circuits and motors.

82. Voltage Drop and Copper Loss. To ensure satisfactory operation, the total voltage drop from the service entrance to any motor should not exceed 5%.

The copper loss depends on the resistance of the wire and the square of the current which is carried. For motors operating at or near full load any considerable part of the time, it may be economical to increase the size of conductors to lessen the copper loss.

For a-c motors, voltage-drop calculations should consider both resistance and reactance of the circuit on an 80% power factor except where the power factor is definitely known.

For electric heating equipment, the total voltage drop in the conductors should not exceed 2% as a general rule. In the case of a high-wattage heater for a special application, the drop to be allowed should be based upon the recommendations of the manufacturer of the heater.

83. Flexibility in Changing Equipment Locations. The incidental power applications in a commercial or public building may usually be considered as practically permanent and as requiring no special provisions for flexibility. In the majority of industrial plants, changing the location of the motors is a more or less common occurrence, and suitable provisions should be made to meet the probable future conditions. Some degree of flexibility may be secured by means of one or more of the following methods:

a. Busways arranged for the use of plugging-in devices.

b. Wireways carrying feeders and motor-branch circuits.

c. An underfloor raceway system, where a large number of small motors is to be supplied and the building is suited to this type of construction.

d. Oversize raceways, where changes in motor sizes are anticipated.

84. Provision for Supplying Increased Loads. Since the changes in layout and equipment that occur in industrial plants and professional buildings often involve increased load, it is advisable to make provisions in the original installation for increasing the capacity of the wiring system. For example, conduit embedded in concrete implies a permanent job, and future demands must be considered in the early stages of the layout. The following methods are suggested:

a. The service equipment, feeders, and subfeeders may be made large enough to provide some excess capacity over the present needs; or raceways and other equipment may be installed of sufficient size to permit the installation of conductors that will provide for a definite increase in capacity. Demand factors less than 100% should seldom be applied.

b. Panelboards should be of the sectional type with interchangeable units, permitting the substitution of control units of larger ratings. Distribution centers should be so designed that the original control units can be replaced by units of larger size and provision made for the control of additional circuits.

85. Distribution Centers. Every feeder, subfeeder, or branch-circuit distribution center should be provided with a switch and fuses or a circuit breaker for each circuit originating at the distribution center. At such centers, working clearance must be allowed for maintenance and future expansion.

PROCEDURE IN DESIGN—WIRING FOR LIGHTING

86. The following outline should be used for the wiring layout of lighting loads:

a. Location and wattage of lighting outlets.

1. Ceiling and other outlets for general lighting.
2. Show-window, showcase, and wall-display-case outlets.
3. Miscellaneous lighting outlets.

b. Location of convenience outlets.

c. Branch circuits and switch control.

d. Panelboards.

e. Feeders.

f. Feeder-distribution center.

g. Service conductors and equipment.

h. Miscellaneous applications (communication systems, emergency systems, etc.).

It is assumed that the type of service has been determined (voltage, frequency, single or polyphase) and that building plans are available or can be prepared. Usually an elevation and one plan for each floor are sufficient; but on floors of large areas or with complex electrical demands, several may be needed.

87. Location and Wattage of Lighting Outlets. *a. Ceiling Outlets for General Lighting.* Locate the outlets, and indicate these on each floor plan (Fig. 17-20) with standard symbols (Par. **88**); the wattage to be allowed for each outlet may be calculated directly by illuminating-engineering methods.

b. Show-window, Showcase, and Wall-display-case Outlets. Usually one junction box per show window is installed to handle the necessary circuits. Where individual outlets for each reflector are to be installed, the maximum spacings (Par. **71c**) should be used as a guide. With either method, outlets for spot- or floodlights must be installed, or provision allowed in the circuits.

Showcase outlets are usually placed in the floor, adjacent to the location of the cases. In stores having unfinished basements, circuits may be carried from the panelboards to the basement and the run completed after the case locations have been determined (see Fig. 17-20).

Outlets for wall-display cases can usually be located in the wall or a column so as to be just above the cases. Wiring can then be extended along the tops of the cases to the lighting equipment.

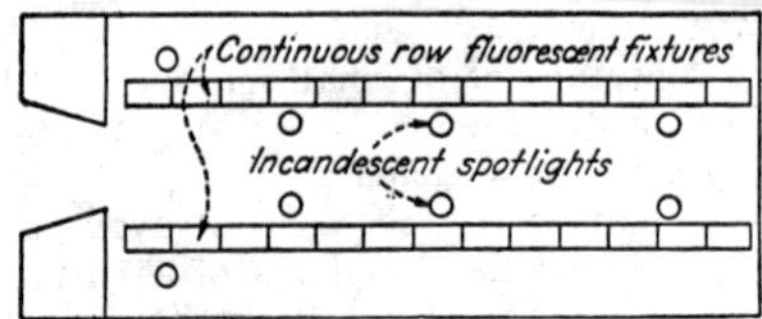

Fig. 17-17. Locating outlets for general lighting.

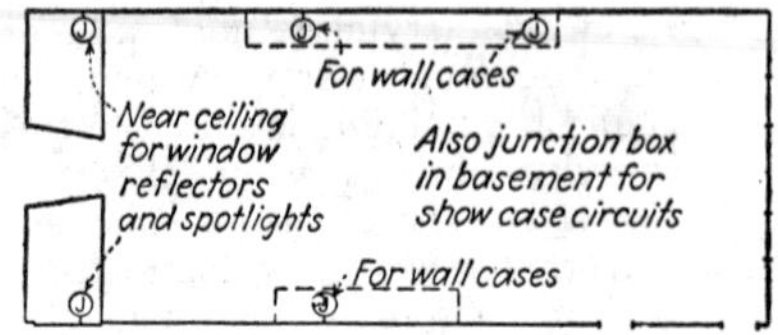

Fig. 17-18. Locating outlets for window and showcase lighting.

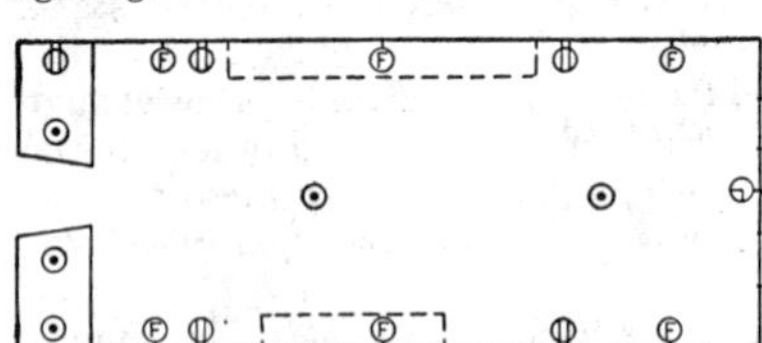

Fig. 17-19. Locating the convenience outlets.

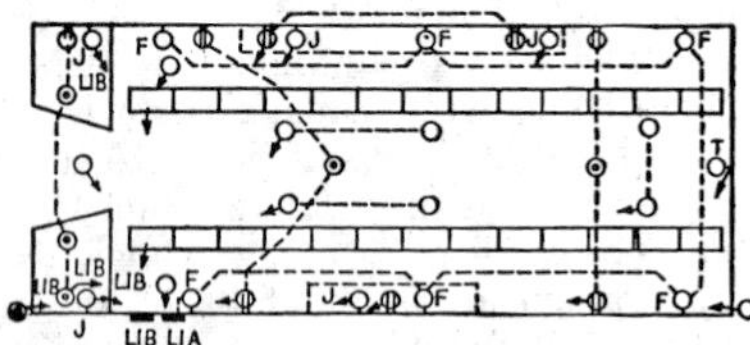

Fig. 17-20. Finished small-store layout. Circulating may be on plan or described in specifications.

c. Miscellaneous Lighting Outlets. Lighting demands of a miscellaneous nature must always be considered. The following brief list may aid:

1. Exit lights—emergency and regular exits.
2. Aisle lights—for auditoriums and similar rooms.
3. Yard lights—for loading platforms, delivery entrances, etc.
4. Entrance lights—brackets, lanterns, etc.
5. Floodlighting—protective lighting or advertising.
6. Signs: exterior—advertising; interior—directional, advertising.

These and others are important lighting applications in large buildings and may be of consequence in small structures. A complete wiring layout must give them full consideration.

88. Location of Convenience Outlets. Paragraphs **71** and **72** specify the number of convenience outlets to be installed per unit area, per bay, or per linear foot of wall.

For large floor areas an excellent way to achieve uniform spacing is by the use of underfloor raceway or cellular-metal-floor systems. Where columns are relatively close, outlets may be located in them and in the walls. It is important to study the building plans for column locations, floor construction, and other pertinent items before the final location is made. Those intended for a specific purpose (fans, clocks, etc.) must receive individual consideration (see Fig. 17-19).

89. Branch Circuits and Switch Control. After indicating the outlets on a floor plan (Fig. 17-20), the number of circuits may be estimated by noting that Par. **75** specifies that the initial wattage per 15-A branch circuit should not exceed 1,000 W. This is based on consideration of voltage drop, copper loss, and provision for future increase. By adding heavy-duty convenience-outlet circuits, etc., an approximation of the number required is obtained. Other factors such as switch control and length of run will confirm or alter this preliminary estimate.

Control of circuits may be provided by local switching or by switching at the panelboard. Distances from the source of natural light, division between private and public areas in a commercial occupancy, three-way control, cleaning routine, etc., all have some bearing and may increase the number of circuits.

There are times when the initial load per circuit should be restricted to less than 1,000 W, for example, when the distance from the panelboard to the first outlet is more than 100 ft. Voltage drop in the branch circuit should not exceed 2%. In such cases the allowable load must be cut down to 800 or even 600 W/circuit, the exact figure depending on individual circumstances.

If the lamp holders used are of the mogul or other type rated at more than 300 W, heavy-duty circuits protected at 20 to 50 A may be used. It is suggested that at least one voltage-drop calculation be made for such circuits (see Par. **107**).

Switches are almost as important as outlets in many branch circuits and should be carefully located. Special methods of control, e.g., relay switching, time switch, or photoelectric cell for signs, schoolroom lighting, and floodlighting, are phases of complete switching layouts.

90. Panelboards. Branch circuits originate at panelboards where the overcurrent protection for each circuit is installed. Switch control is not mandatory but is frequently desirable, as in retail stores or in large industrial areas where centralization is desirable even though local control is provided. The panelboard itself may include a main switch which may be remotely operated so that all branch circuits may be controlled simultaneously.

A summary of the general considerations which determine the number and location of panelboards is given below.

a. Panelboards should be placed as near as possible to the center of the load they supply.

b. The number of branch circuits from one panel board should not exceed 42.

c. No run from a panelboard to the first outlet of a circuit should exceed 100 ft.

d. Panelboards should always be accessible.

e. Convenience from the standpoint of panelboard switching must be considered.

f. Locations should be such that feeders are as short as possible and have a minimum of bends and offsets.

g. At least one lighting panelboard per active floor is necessary.

If the wiring layout has been completed for the other parts of the building, there will be available the circuit information for all panelboards. This should be compiled as shown in Table 17-19 and included in the specifications.

91. Feeders. The design procedure for feeders is indicated in Par. **77**. In calculating the size, a load of 1,000 W is assumed for each 15-A circuit, 500 W for each spare circuit, and the specific initial load for all other circuits. The exact wire size depends on the voltage regulation and on current, which in turn depend on the type of distribution system employed. The amperage computed is the initial load. To obtain the probable load, 50% must be added to provide for normal increases, for changes in occupancies, for expansion of facilities, etc. After checking for voltage drop, the minimum size of wire to provide carrying capacity is determined.

If the ultimate size of feeder is not installed immediately, provide for future growth by one of the following methods:

a. The installation of raceways larger than those needed for the initial feeder size. Where this necessitates only one size larger raceway, for example, 3- instead of 2½-in conduit, the extra cost is not great in view of the benefits gained.

b. The installation of spare raceways so that extra capacity may be provided when required later by installing additional conductors. This requires a careful original layout. It is not good practice to parallel two conductors of unequal size; hence the installation should be planned to utilize the additional feeder capacity by sectionalizing panelboards or by changing the connections so as to supply only certain panelboards by the new feeders.

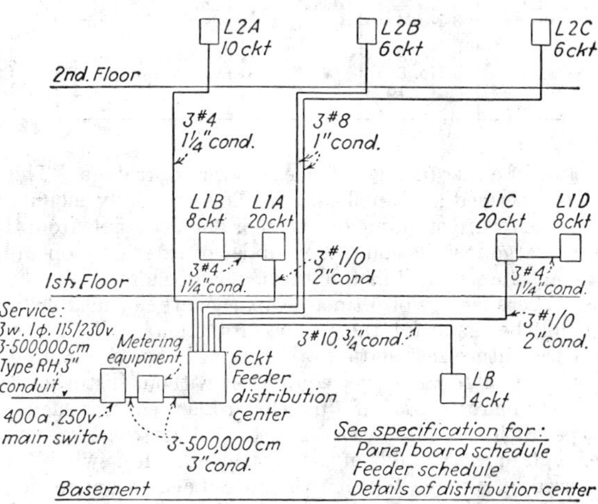

FIG. 17-21. Lighting riser diagram.

Table 17-19. Typical Panelboard Schedule

Panelboard designation	Location	No. 15-amp branch circuits	Mains—capacity, amp	Branch-circuit equipment
L1A	Store 1	20	Lugs only, 100	Fuse and 30-amp switch
L1B	Store 1	8	Switch and fuse, 60	Fuse
L1C	Store 2	20	Lugs only, 100	Fuse and 30-amp switch
L1D	Store 2	8	Switch and fuse, 60	Fuse
L2A	Suite 1	10	Lugs only, 60	15-amp circuit breaker
L2B	Suite 2	6	Lugs only, 60	15-amp circuit breaker
L2C	Suite 3	6	Lugs only, 60	15-amp circuit breaker
LB	Basement	4	Lugs only, 30	15-amp circuit breaker

When more than three feeders are installed, a separate riser- or feeder-layout diagram should be prepared to clarify the location and general connections of these electrical "main arteries" (see Figs. 17-8 and 17-21). Complete information on all feeders should be tabulated in a schedule (Table 17-20) and included in the specifications.

92. Feeder-distribution centers provide for the protection and control of individual feeders. For 3-phase 4-wire service, both lighting and power feeders are controlled at one center. Where power and lighting are supplied from separate systems, a distribution center is provided for each.

The modern types of feeder-distribution center include dead-front panelboards, dead-front switchboards, and assemblies of enclosed, externally operable units, consisting of either fusible switches or circuit breakers. The equipment for a small installation is termed a "panelboard." This is usually mounted in a cabinet on or in a wall. A switchboard standing on the floor and accessible from the front or rear is more suitable for a large installation. Assemblies of externally operable switches or circuit breakers are adaptable to all installations, small or large.

Table 17-20. Typical Feeder Schedule

Feeder No.	Conductors	Size of raceway, in.	From	To panel	Panel location
FL1	3 No. 1/0	2	Distribution panel	L1A	Store 1, front
FL1	3 No. 4	1¼	Panel L1A	L1B	Store 1, front
FL2	3 No. 1/0	2	Distribution panel	L1C	Store 2, front
FL2	3 No. 4	1¼	Panel L1C	L1D	Store 2, front
FL3	3 No. 4	1¼	Distribution panel	L2A	North wall, head of stairs, 2nd floor
FL4	3 No. 8	1	Distribution panel	L2B	West wall, center 2nd floor
FL5	3 No. 8	1	Distribution panel	L2C	East wall, center 2nd floor
FL6	3 No. 10	¾	Distribution panel	LB	Basement, foot of stairs

Where feeders of ultimate capacity are not installed immediately, suitable provision should be made for their later protection. All that is necessary is to provide space for the future installation of larger switches or circuit breakers and means of making connections to the larger equipment without disturbing such of the original equipment as may be retained. If a panelboard is used, it is suggested that it be of the sectional type, with space in the cabinet to contain the larger equipment and with buses large enough to carry 150% of the initial load. If a switchboard or assembly of unit devices is used, it is suggested that the buses be as recommended by panelboards and that the switchboard or assembly be specially designed to accommodate the larger equipment.

93. Service Conductors and Equipment. The loads computed for individual feeders give an immediate indication as to the size of service conductors. At this point, power feeders are often joined with the lighting load. In this case, the sum of all feeders (modified by demand factors if applicable) will determine the service requirements.

Paragraph **79** provides a 50% margin for load growth with ultimate size installed originally if the initial load is no more than 267 A.

Where the calculated future load exceeds 400 A, that is, 267 plus 50%, an individual study should be made. Due weight should be given to the following:

a. In any building having an expected life of 10 years or more, it is highly probable that some additional service capacity will be needed.

b. In many cases, additional capacity can be provided only by tearing out and completely replacing the original service conductors and service equipment. The larger the service the greater the loss involved in this procedure.

c. Considerable additional expense is involved in providing 50% excess capacity in the case of a heavy service. This is a nonproductive investment until some part of the excess capacity is utilized.

94. Miscellaneous Applications. In addition to general lighting loads, it is frequently necessary to provide for certain special systems which are often considered a part of the "lighting installation" rather than the "power." Among these are communication systems (bells and buzzers, telephones), emergency lighting systems, burglar alarms, watchman systems, clock systems, television reception, etc. Most of these operate at low voltages (under 30) from transformers connected to the main wiring system, although certain of them require special standby battery service.

The layout for these systems is too dependent on the individual application and the specific manufacturer of the material used to consider in detail. The following should, however, be of value as a guide:

a. In every case, investigate the possibility of special electrical requirements, from either a legal or a special-convenience standpoint.

b. Secure from the manufacturers of the necessary equipment full details regarding the installation, with special reference to its demands on branch circuits, feeders, or service-entrance equipment.

c. Indicate the details of the installation in the specifications and on the drawings, using separate floor plans if any confusion of the main-wiring plans might result from the presence of too many symbols and circuit lines.

PROCEDURE IN DESIGN—WIRING FOR MOTOR LOADS

95. National Electrical Code Requirements for Motors. Motor circuits are subject to numerous and detailed requirements in NEC, so that only the basic general rules for **motors used on continuous duty** are abstracted here. Figure 17-22 diagrammatically represents a motor circuit, with the component parts of feeder and branch-circuit equipment lettered for identification. The following paragraphs correspond to the diagram identification.

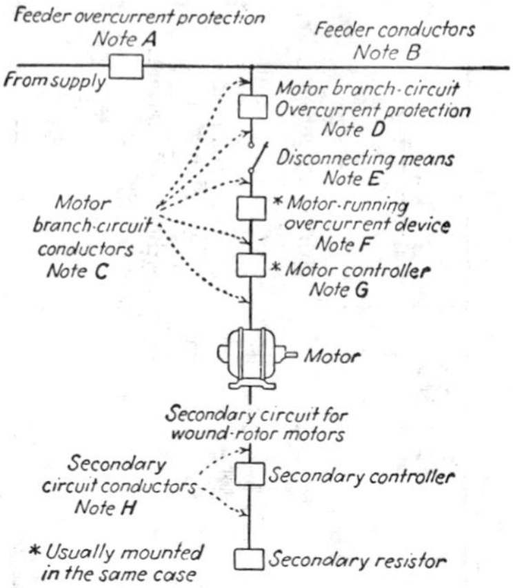

a. Feeder Overcurrent Protection. This should not be greater than the rating or setting of the branch-circuit protective device for the largest motor [see (*d*)] plus the sum of the full-load current of the remaining motors supplied by the feeder.

b. Feeder conductors shall have a current-carrying capacity not less than 125% of the full-load current of the highest-rated motor plus the sum of the full-load currents of the remaining motors supplied by the feeder. (Demand factors may be permitted.)

Fig. 17-22. Motor feeder and branch circuit.

c. Motor branch-circuit conductors supplying one motor shall have a capacity not less than 125% of the motor full-load current. If the circuit supplies two or more motors, determine the size in the manner used for feeder conductors.

d. Motor branch-circuit overcurrent protection must be capable of carrying the starting current of the motor. The code indicates the maximum permissible size as a percentage of full-load current of the motor, depending on its type, starting method, and locked-

Table 17-21. Overcurrent Protection for Motors

Full-load current rating of motor, A	For running protection of motors		Maximum allowable rating or setting of branch-circuit protective devices							
			With code letters: Single-phase, squirrel-cage, and synchronous. Full voltage, resistor or reactor starting, code letters F to V inclusive Without code letters: Same as above (4)		With code letters: Single-phase, squirrel-cage, and synchronous. Full voltage, resistor or reactor start, code letters B to E inclusive. Autotransformer, start, code letters F to V inclusive Without code letters: (Not more than 30 A) Squirrel-cage and synchronous, autotransformer start, high-reactance squirrel-cage* (5)		With code letters: Squirrel-cage and synchronous autotransformer start, code letters B to E inclusive Without code letters: (More than 30 A) Squirrel-cage and synchronous autotransformer start, high-reactance squirrel-cage* (6)		With code letters: All motors code letter A Without code letters: D-c and wound-rotor motors (7)	
(1)	Maximum rating of nonadjustable protective devices, A (2)	Maximum setting of adjustable protective devices, A (3)	Fuses	Circuit breakers (nonadjustable overload trip)	Fuses	Circuit breakers (nonadjustable overload trip)	Fuses	Circuit breakers (nonadjustable overload trip)	Fuses	Circuit breakers (nonadjustable overload trip)
Table 17-21*a*										
1	2	1.25	15	15	15	15	15	15	15	15
2	3	2.50	15	15	15	15	15	15	15	15
3	4	3.75	15	15	15	15	15	15	15	15
4	6	5.0	15	15	15	15	15	15	15	15
5	8	6.25	15	15	15	15	15	15	15	15
6	8	7.50	20	15	15	15	15	15	15	15
7	10	8.75	25	20	20	15	15	15	15	15
8	10	10.0	25	20	20	20	20	20	15	15
9	12	11.25	30	30	25	20	20	20	15	15
10	15	12.50	30	30	25	20	20	20	15	15
11	15	13.75	35	30	30	30	25	30	20	20
12	15	15.00	40	30	30	30	25	30	20	20
13	20	16.25	40	40	35	30	30	30	20	20
14	20	17.50	45	40	35	30	30	30	25	30
15	20	18.75	45	40	40	30	30	30	25	30
16	20	20.00	50	40	40	40	35	40	25	30

17	25	21.25	60	50	45	40	35	40	30	30
18	25	22.50	60	50	45	40	40	40	30	30
19	25	23.75	60	50	50	40	40	40	30	30
20	25	25.00	60	50	50	40	40	40	30	30
22	30	27.50	70	70	60	50	45	50	35	40
24	30	30.00	80	70	60	50	50	50	40	40
26	35	32.50	80	70	70	70	60	70	40	40
28	35	35.00	90	70	70	70	60	70	45	50
30	40	37.50	90	100	80	70	60	70	45	50
32	40	40.00	100	100	80	70	70	70	50	50
34	45	42.50	110	100	90	70	70	70	60	70
36	45	45.00	110	100	90	100	80	100	60	70
38	50	47.50	125	100	100	100	80	100	60	70
40	50	50.00	125	100	100	100	80	100	60	70
42	50	52.50	125	125	110	100	90	100	70	70
44	60	55.00	125	125	110	100	90	100	70	70
46	60	57.50	150	125	125	100	100	100	70	70
48	60	60.00	150	125	125	100	100	100	80	100
50	60	62.50	150	125	125	100	100	100	80	100
52	70	65.00	175	150	150	125	110	125	80	100
54	70	67.50	175	150	150	125	110	125	90	100
56	70	70.00	175	150	150	125	125	125	90	100
58	70	72.50	175	150	150	125	125	125	90	100
60	80	75.00	200	150	150	125	125	125	90	100
62	80	77.50	200	175	175	125	125	125	100	100
64	80	80.00	200	175	175	150	150	150	100	100
66	80	82.50	200	175	175	150	150	150	100	100
68	90	85.00	225	175	175	150	150	150	110	125
70	90	87.50	225	175	175	150	150	150	110	125
72	90	90.00	225	200	200	150	150	150	110	125
74	90	92.50	225	200	200	150	150	150	125	125
76	100	95.00	250	200	200	175	175	175	125	125

Table 17-21*b*

78	100	97.50	250	200	200	175	175	175	125	125
80	100	100.00	250	200	200	175	175	175	125	125
82	110	102.50	250	225	225	175	175	175	125	125
84	110	105.00	250	225	225	175	175	175	150	150
86	110	107.50	300	225	225	175	175	175	150	150
88	110	110.00	300	225	225	200	200	200	150	150
90	110	112.50	300	225	225	200	200	200	150	150
92	125	115.00	300	250	250	200	200	200	150	150

Table 17-21. Overcurrent Protection for Motors.—*Concluded*

Full-load current rating of motor, A	For running protection of motors		Maximum allowable rating or setting of branch circuit protective devices							
			With code letters: Single-phase, squirrel-cage, and synchronous. Full voltage, resistor or reactor starting, code letters F to V inclusive Without code letters: Same as above (4)		With code letters: Single-phase, squirrel-cage, and synchronous. Full voltage, resistor or reactor start, code letters B to E inclusive. Autotransformer, start, code letters F to V inclusive Without code letters: (Not more than 30 A) Squirrel-cage and synchronous, autotransformer start, high-reactance squirrel-cage* (5)		With code letters: Squirrel-cage and synchronous autotransformer start, code letters B to E inclusive Without code letters: (More than 30 A) Squirrel-cage and synchronous autotransformer start, high-reactance squirrel-cage* (6)		With code letters: All motors code letter A Without code letters: D-c and wound-rotor motors (7)	
(1)	Maximum rating of nonadjustable protective devices, A (2)	Maximum setting of adjustable protective devices, A (3)	Fuses	Circuit breakers (nonadjustable overload trip)	Fuses	Circuit breakers (nonadjustable overload trip)	Fuses	Circuit breakers (nonadjustable overload trip)	Fuses	Circuit breakers (nonadjustable overload trip)
94	125	117.50	300	250	250	200	200	200	150	150
96	125	120.00	300	250	250	200	200	200	150	150
98	125	122.50	300	250	250	200	200	200	150	150
100	125	125.00	300	250	250	200	200	200	150	150
105	150	131.50	350	300	300	225	225	225	175	175
110	150	137.50	350	300	300	225	225	225	175	175
115	150	144.00	350	300	300	250	250	250	175	175
120	150	150.00	400	300	300	250	250	250	200	200
125	175	156.50	400	350	350	250	250	250	200	200
130	175	162.50	400	350	350	300	300	300	200	200
135	175	169.00	450	350	350	300	300	300	225	225
140	175	175.00	450	350	350	300	300	300	225	225
145	200	181.50	450	400	400	300	300	300	225	225
150	200	187.50	450	400	400	300	300	300	225	225
155	200	194.00	500	400	400	350	350	350	250	250
160	200	200.00	500	400	400	350	350	350	250	250

165	225	206.00	500	500	450	350	350	350	250	250
170	225	213.00	500	500	450	350	350	350	300	300
175	225	219.00	600	500	450	350	350	350	300	300
180	225	225.00	600	500	450	400	400	400	300	300
185	250	231.00	600	500	500	400	400	400	300	300
190	250	238.00	600	500	500	400	400	400	300	300
195	250	244.00	600	500	500	400	400	400	300	300
200	250	250.00	600	500	500	400	400	400	300	300
210	250	263.00	800	600	600	500	450	500	350	350
220	300	275.00	800	600	600	500	450	500	350	350
230	300	288.00	800	600	600	500	500	500	350	350
240	300	300.00	800	600	600	500	500	500	400	400
250	300	313.00	800	700	800	500	500	500	400	400
260	350	325.00	800	700	800	600	600	600	400	400
270	350	338.00	1000	700	800	600	600	600	450	500
280	350	350.00	1000	700	800	600	600	600	450	500
290	350	363.00	1000	800	800	600	600	600	450	500
300	400	375.00	1000	800	800	600	600	600	450	500
320	400	400.00	1000	800	800	700	800	700	500	500
340	450	425.00	1200	...	1000	700	800	700	600	600
360	450	450.00	1200	...	1000	800	800	800	600	600
380	500	475.00	1200	...	1000	800	800	800	600	600
400	500	500.00	1200	...	1000	800	800	800	600	600
420	600	525.00	1600	...	1200	...	1000	...	800	700
440	600	550.00	1600	...	1200	...	1000	...	800	700
460	600	575.00	1600	...	1200	...	1000	...	800	700
480	600	600.00	1600	...	1200	...	1000	...	800	800
500	...	625.00	1600	...	1600	...	1000	...	800	800

* High-reactance squirrel-cage motors are those designed to limit the starting current by means of deep-slot secondaries or double-wound secondaries and are generally started on full voltage.

Table 17-22. Full-load Current,* Three-phase A-C Motors

Hp	Induction-type, squirrel-cage, and wound-rotor, A					Synchronous type, unity power factor, A†			
	110 V	220 V	440 V	550 V	2,300 V	220 V	440 V	550 V	2,300 V
½	4	2	1	0.8					
¾	5.6	2.8	1.4	1.1					
1	7	3.5	1.8	1.4					
1½	10	5	2.5	2.0					
2	13	6.5	3.3	2.6					
3	....	9	4.5	4					
5	....	15	7.5	6					
7½	....	22	11	9					
10	....	27	14	11					
15	....	40	20	16					
20	....	52	26	21					
25	....	64	32	26	7	54	27	22	5.4
30	....	78	39	31	8.5	65	33	26	6.5
40	....	104	52	41	10.5	86	43	35	8
50	....	125	63	50	13	108	55	44	10
60	....	150	75	60	16	128	64	51	12
75	....	185	93	74	19	161	81	65	15
100	....	246	123	98	25	211	106	85	20
125	....	310	155	124	31	264	132	106	25
150	....	360	180	144	37	...	158	127	30
200	....	480	240	192	48	...	210	168	40

For full-load currents of 208- and 200-V motors, increase the corresponding 220-V motor full-load current by 6 and 10%, respectively.

* These values of full-load current are for motors running at speeds usual for belted motors and motors with normal torque characteristics. Motors built for especially low speeds or high torques may require more running current, and multispeed motors will have full-load current varying with speed, in which case the nameplate current rating shall be used.

† For 90 and 80% pf the above figures shall be multiplied by 1.1 and 1.25, respectively.

The voltages listed are rated motor voltages. Corresponding nominal system voltages are 110 to 120, 220 to 240, 440 to 480, and 550 to 600 V.

Table 17-23. Full-load Currents in Amperes, Direct-current Motors

(The following values of full-load currents are for motors running at base speed)

Hp	120 V	240 V
¼	2.9	1.5
⅓	3.6	1.8
½	5.2	2.6
¾	7.4	3.7
1	9.4	4.7
1½	13.2	6.6
2	17	8.5
3	25	12.2
5	40	20
7½	58	29
10	76	38
15	....	55
20	....	72
25	....	89
30	....	106
40	....	140
50	....	173
60	....	206
75	....	255
100	....	341
125	....	425
150	....	506
200	....	675

Table 17-24. Full-load Currents in Amperes, Single-phase Alternating-current Motors

The following values of full-load currents are for motors running at usual speeds and motors with normal torque characteristics. Motors built for especially low speeds or high torques may have higher full-load currents, and multispeed motors will have full-load current varying with speed, in which case the nameplate current ratings shall be used.

To obtain full-load currents of 208- and 200-V motors, increase corresponding 230-V-motor full-load currents by 10 and 15%, respectively.

The voltages listed are rated motor voltages. Corresponding nominal system voltages are 110 to 120, 220 to 240, 440 to 480.

Hp	115 V	230 V	440 V
1/6	4.4	2.2	
1/4	5.8	2.9	
1/3	7.2	3.6	
1/2	9.8	4.9	
3/4	13.8	6.9	
1	16	8	
1 1/2	20	10	
2	24	12	
3	34	17	
5	56	28	
7 1/2	80	40	21
10	100	50	26

rotor current. These maximum values for various full-load currents and motor types are found in Table 17-19.

e. Disconnecting Means. Each motor and controller shall have an indicating-type disconnecting switch or circuit breaker, with a current-carrying capacity not less than 115% of the motor nameplate current rating, so arranged as to disconnect all ungrounded conductors. In general, up to 50 hp the switch is actually rated in horsepower. Groups of motors driving a single machine or piece of apparatus or protected by one set of overcurrent devices or in a single room within sight of the disconnecting means may be observed by a single disconnecting switch.

f. Motor-running Overcurrent Device. Integral-horsepower motors shall have running overcurrent protection not greater than 125% of the motor full-load current (see Table 17-19). The device may be shunted out for starting if during this period fuses or a time-limit circuit breaker rated or set at not more than 400% of the motor full-load current is in the circuit. Variations are permitted for fractional-horsepower motors and certain special cases.

g. A motor controller (device normally used to start and stop the motor) shall be provided for each motor (or group of motors as in Par. **96e**). It normally should be in sight of the motor and, for a-c use, shall be capable of interrupting the locked-rotor current. In general, the controller is rated in horsepower.

h. Secondary Circuit Conductors. The secondary conductors of a wound-rotor a-c motor between slip rings and controller shall have a capacity not less than 125% of the full-load secondary current. Between controller and resistor, the capacity of conductors should be at least 110% of full-load secondary current for continuous duty.

96. Types of Circuit Layout. Where wiring is to be installed for two or more motors, it may be possible to use any one of five different types of wiring layout, as follows:

a. A separate circuit run to each motor from a power panelboard or a distribution center (see Fig. 17-23).

b. A feeder or subfeeder may be carried around the building with branch circuits tapped to the feeder at various points, no branch-circuit distribution center being used (see Fig. 17-24).

c. A feeder or subfeeder may be carried

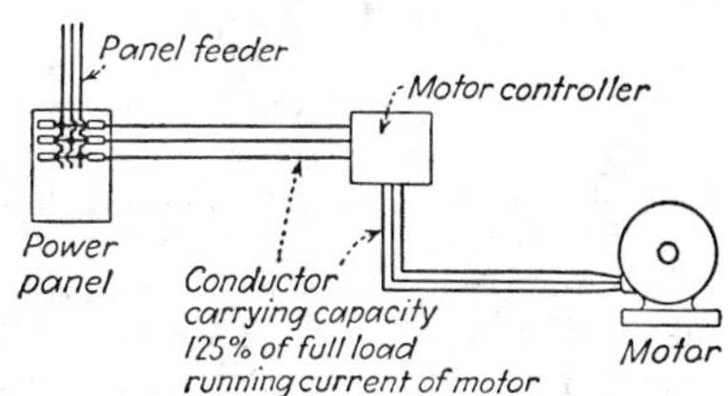

FIG. 17-23. Individual circuit to each motor.

around the building with subfeeder taps having no individual overcurrent protection, carried direct to the disconnecting means or controller for each motor. In this case, the branch-circuit overcurrent device is usually omitted, and the motor branch circuit originates at the controller (see Fig. 17-25).

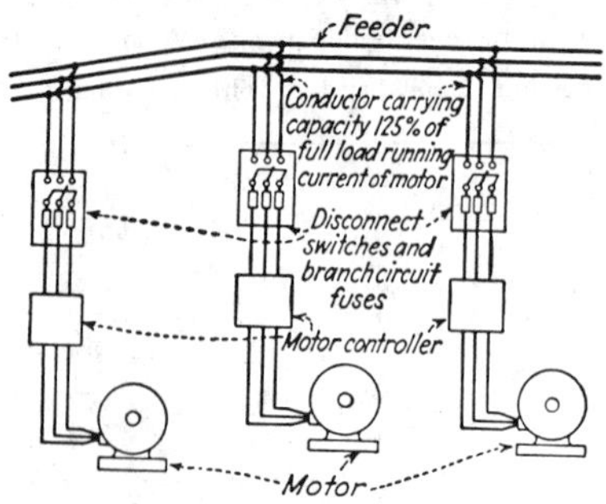

FIG. 17-24. Common feeder to several motors.

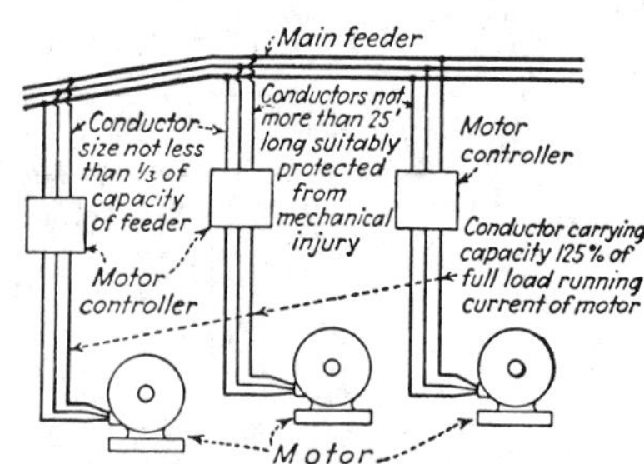

FIG. 17-25. A common feeder with small wire taps to motors.

d. A feeder or subfeeder may be carried direct to the disconnecting means or controller for each one of the group of motors. Otherwise the layout is the same as in (*c*) (see Fig. 17-26).

e. A group of small motors, each having a full-load current rating not exceeding 6 amp, may be supplied by a 15-, 20-, or 25-A branch circuit (see Fig. 17-27).

The NEC or local ordinance covers the conditions under which each of the foregoing types can be used and the installation requirements applying in each case.

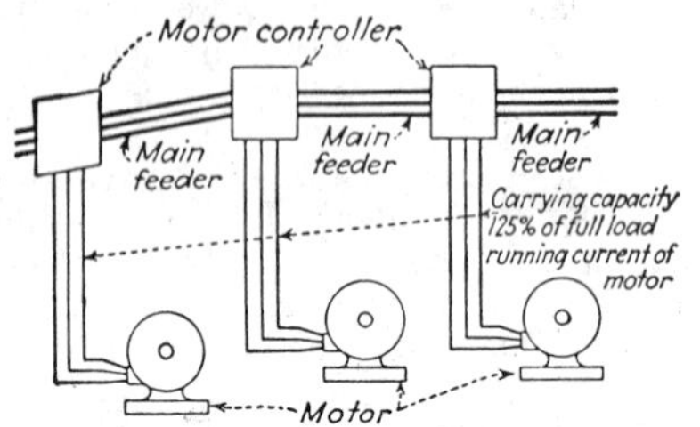

FIG. 17-26. Common feeder looped to motor controllers.

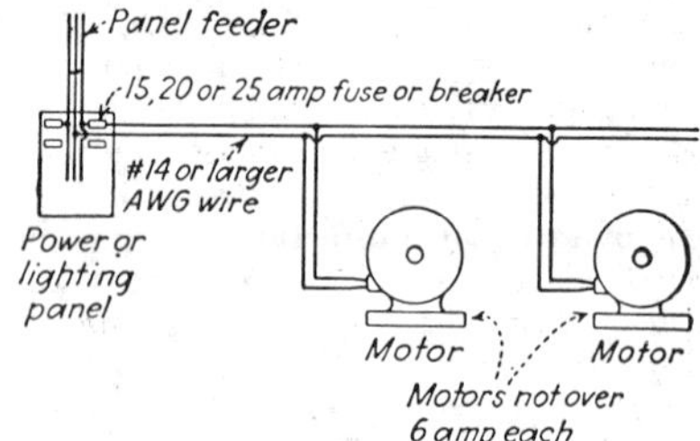

FIG. 17-27. Connecting small motors to one circuit.

97. Application of Various Types of Layout. Type *a* can be used under any condition. It is usually the most satisfactory for supplying the miscellaneous power loads in a commercial or public building and is also common in industrial plants.

The use of types *b*, *c*, and *d* is limited chiefly to industrial plants where a large number of motors is used to drive individual machines. Type *b* requires a branch-circuit overcurrent device for each motor. In type *c*, no branch-circuit overcurrent device is required, but the conductors from the subfeeder to the controller must be larger than in type *b*. Type *d* will show a saving in cost over either type *b* or type *c* if the subfeeder can be economically brought direct to each controller. Types *b* and *c* lend themselves well to busway layouts.

Type *e* is a means of serving small motors in offices and similar locations, permitting them to be connected to lighting branch circuits. Usually it has little application in a factory.

For power applications in industrial plants, the first four types may be considered on a par as regards serviceability. The choice of type should be made on the basis of economy in cost of installation and flexibility, i.e., adaptability to changes in sizes and locations of motors.

98. Busway Distribution. The rapidly changing requirements of the mass-production industries, particularly of those utilizing assembly lines where equipment may be replaced or shifted physically at regular or frequent intervals, have outmoded the usual conduit-wiring system for such applications owing to lack of flexibility and because of the cost of changes. Busway systems should be considered for such layouts. Since each problem requires individual study, no detailed design data can be presented. Manufacturers of this type of equipment should be consulted for information. Figure 17-28 shows a combination busway and trolley duct system serving approximately 50 kW in lighting and 250 motors totaling 750 hp. Forty percent excess capacity has been provided for load growth.

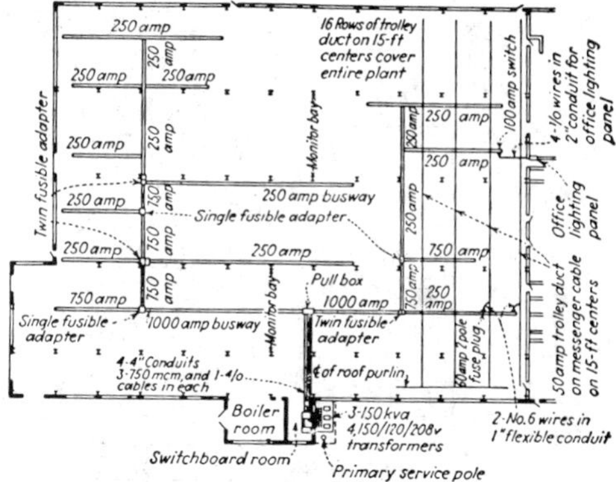

Fig. 17-28. Combination busway and trolley-duct system for industrial plant.

99. Voltage Drop and Carrying Capacity of Conductors. In addition to having sufficient carrying capacity according to NEC, conductors should be of such size that the total voltage drop to any motor will not exceed 5% under full load.

On any system operating at 208 volts or higher, it is recommended that the drop in motor branch circuits should not exceed 1%, allowing a drop of 4% in the feeders. With the minimum conductor sizes of NEC, the feeder drop will exceed 4% only where a feeder is unusually long. Where the drop exceeds 3%, the annual cost of the kilowatthours consumed in copper loss should be computed and consideration should be given to the installation of larger conductors in order to reduce this loss (See Par. **117**).

In an industrial plant it is almost always desirable to install service and feeder conductors of larger sizes than are required for the initial load. Besides providing for load increases, the excess size will also have the advantage of reducing the copper loss.

100. Distribution Centers. In a typical building with numerous motors, there will be required a main feeder-distribution center and, depending upon the type of layout employed, one or more subfeeder-distribution centers or one or more branch-circuit distribution centers; or both subfeeder and branch-circuit centers may be necessary.

A branch-circuit distribution center may consist of an assembly of cubicles, each cubicle being a complete unit containing equipment required for one motor, including the branch-circuit overcurrent device, disconnecting means, motor controller, and the running overcurrent device. Such an assembly constitutes one type of dead-front switchboard.

Suitable provision should be made for changes in motor sizes and for additional loads. Panelboards should preferably be of the sectional type with interchangeable control units permitting the substitution of units of different ratings. Switchboards and switch or circuit-breaker assemblies should be so designed that the control units can be replaced by units of larger size and additional control units can readily be installed and should be provided with buses having considerable excess carrying capacity.

Table 17-25. Typical Motor Schedule for 208-volt, 3-phase, 60-cycle Operation

Motor No.	Location	Hp	Speed, rpm	Type[1]	Drive	Controller[1]
1	1st floor	5	1800	Squirrel cage	Pump	Type C
2	Basement	7½	1800	Squirrel cage	Blower	Type C
3	Basement	10	1200	Wound-rotor	Grinder	Type A
4	Basement	25	600	Wound-rotor	Compressor	Type B

[1] Unless standard makes and types are used, these must also be described.

Drawings and Specifications. Where wiring is to be installed to any considerable number of motors, floor plans of the building or buildings should be secured. On these the location of each motor should be shown, together with the horsepower rating, the kind of machine driven, and the location of the controller. Assign a number to each motor, and prepare specification sheets giving for each motor its number, location, horsepower, speed, description of machine driven, and type of controller to be used, as in Table 17-25.

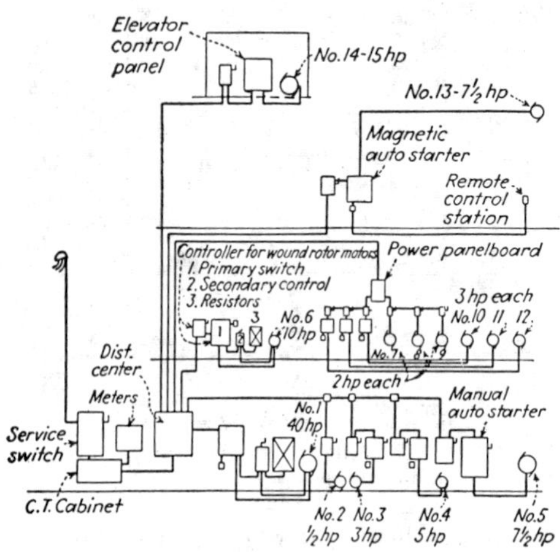

Fig. 17-29. Typical riser diagram for power. Wire sizes may be noted on plan or in specifications (see ref. 7, Par. 118).

After details of the wiring have been determined, the floor plans should be completed by showing the location of service entrance, switchboards, panelboards or other distribution centers, and similar apparatus. Data should be provided, either on the floor plans or in the form of diagrams, covering conductor sizes and raceway sizes for the service, feeders, subfeeders, and branch circuits (see Fig. 17-29 for typical riser diagram).

PROCEDURE IN DESIGN—WIRING FOR SPACE HEATING

101. Circuits serving space heating are generally analogous to those designed for heavy-duty lighting loads. Fixed resistance-type heating units vary in load from several hundred watts to tens of kilowatts.

Two general classes of electric space-heating systems are in common use, supplementary and fully electric.

Supplementary electric space-heating equipment is installed to aid or supply a readily controllable margin to the heat input from a fuel-fired heating system.

Wiring design follows essentially the same requirements as those for heavy-duty fixed-appliance circuits. Branch circuits, feeders, and associated distribution assemblies should exceed the rated load by 25%. Voltage-drop considerations may require larger conductors.

Fully electric space-heating systems may consist of many combinations of pieces of heating equipment; however, for wiring design they may be classified according to a few general types.

Heat pumps are motor loads, and wiring-design requirements follow generally the rules for motors and controllers. However, many heat pumps are supplemented by additional resistance-heating elements; so the nameplate full-load current plus the additional capacity required for continuous loads guides the selection of conductor sizes.

Central resistance heaters combine heating elements, blowers, filters, and controls in a common housing, with connections for incoming fresh- and return-air ducts and outgoing warm-air ducts. Central hydronic heaters employ immersion heaters and are equipped with a circulating pump and fittings for connection to the piping system. Such equipment is conventionally provided with junction boxes for the connection of the incoming power-supply circuit and external control wiring.

Central heating equipment is served by an individual branch circuit with ampacity of 125% or more of the nameplate full-load current rating. Larger conductors may be required by voltage-drop considerations. A disconnect switch should be provided near the installation; however, if conveniently accessible, the branch circuit switch or circuit breaker may serve as the disconnecting means.

Duct heaters are installed in the air-duct system. Wiring requirements are similar to those for central heaters.

Individual electric resistance heaters, including fan types, panels, baseboard, and embedded-cable units, may be wired on individual branch circuits or one of the approved multioutlet branch circuits.

Wiring located in a heated ceiling must be installed at least 2 in above the heated ceiling and considered as operating at an ambient of 50°C and ducted accordingly. If wiring is run above at least 2 in of insulation, however, no derating is required.

If readily accessible, the branch-circuit switch or circuit breaker may serve as the disconnecting means. Thermostats with an "off" position opening all ungrounded conductors may also serve as the disconnecting means.

Electric space-heating units are a continuous load and must not exceed 80% of the rating of branch circuits and feeders serving the load.

Voltage drop is of special importance in wiring design for space-heating systems. Such heating systems are designed, typically, to a close margin with the maximum heat loss. As the heat output of an electric resistance heater varies as the square of the voltage, it is apparent that the wiring system and supply must provide the full nameplate voltage.

PROCEDURE IN DESIGN—PRIMARY-SUPPLY CONSIDERATIONS

102. Type of Primary Equipment. Although standard practice until recently has dictated that transformers, primary switchgear, and control equipment be installed in fireproof transformer vaults, fire-resistive substation rooms, or outdoors, recent developments in such apparatus provide an alternative choice whereby this isolation is not required unless the voltage exceeds 15,000.

Transformers not above 15,000 V between terminals, if air-cooled or if filled with **a liquid that will not burn,** may be installed in buildings without a vault or fire-resistive room if mechanical protection, ventilation, and inaccessibility to unauthorized persons are provided. Otherwise, in general, they must be installed in a fire-resistive enclosure.

Switches, circuit breakers, and control apparatus connected to a circuit not exceeding 15,000 V between conductors need not be in a vault or fire-resistive room if they are of the so-called "metal-clad," "truck," or "cubicle" type. Otherwise, in general, they should be installed in a fire-resistive enclosure.

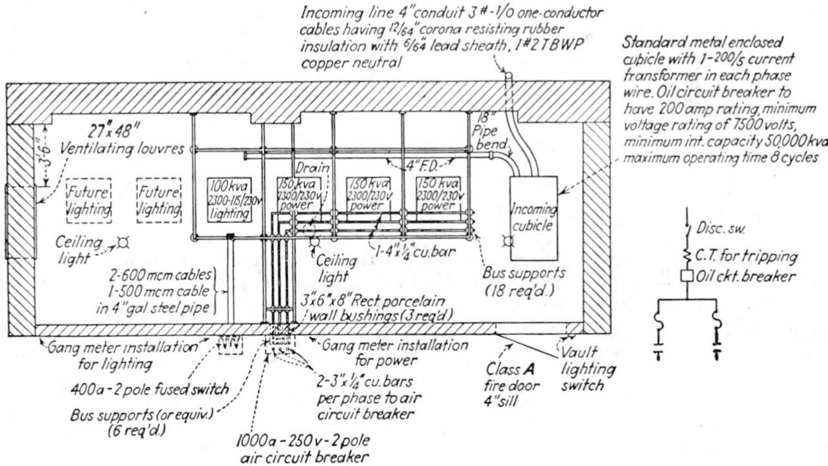

Fig. 17-30. Typical transformer-vault layout.

The choice of equipment depends on a study of the individual case. Where building electrical requirements can be served by one transformer location, which can be a basement vault built at little additional expense while the building is under construction, the extra cost of the newer type of transformers may not be warranted. Where numerous transformer locations are required, e.g., in a large industrial plant, the ability to locate these at load centers with equipment directly on the floor or suspended under the roof or ceiling without fire-resistive enclosure may warrant the use of metal-clad switchgear and transformers approved for such use.

CIRCUIT AND CONDUCTOR CALCULATIONS

103. The current in the individual line conductors of the distribution systems described in Pars. **51** to **57** may be determined from the following formulas, in which I = conductor current, W = power in watts, pf = power factor, E_p = voltage between wires, E_g = voltage between phase wire and neutral.

Single-phase, 2-wire: $$I = \frac{W}{E_p \times \text{pf}} \qquad (17\text{-}1)$$

Single-phase, 3-wire: $$I = \frac{W}{2E_g \times \text{pf}} \qquad (17\text{-}2)$$

2-phase 3-wire: $$I \text{ (outside wires)} = \frac{W}{2E_p \times \text{pf}} \qquad (17\text{-}3)$$

2-phase 3-wire: $$I \text{ (common wire)} = \frac{W}{\sqrt{2} \times E_p \times \text{pf}} \qquad (17\text{-}4)$$

2-phase 4-wire: $$I = \frac{W}{2E_p \times \text{pf}} \qquad (17\text{-}5)$$

3-phase 3-wire: $$I = \frac{W}{\sqrt{3} \times E_p \times \text{pf}} \qquad (17\text{-}6)$$

3-phase 4-wire: $$I = \frac{W}{\sqrt{3} \times E_p \times \text{pf}} \qquad (17\text{-}7)$$

or $$I = \frac{W}{3E_g \times \text{pf}} \qquad (17\text{-}8)$$

D-c 2-wire: $$I = \frac{W}{E_p} \qquad (17\text{-}9)$$

D-c 3-wire: $$I = \frac{W}{2E_g} \qquad (17\text{-}10)$$

NOTE: $\sqrt{2} = 1.41 \qquad \sqrt{3} = 1.73$

104. Conductor Resistance and Voltage Drop. The resistance of a circular mil-foot of commercial copper wire (a wire 1 ft long and having a cross-sectional area of 1 cmil) is usually quoted at 10.6 to 10.8 Ω at 24°C (75°F). For most wiring calculations 10.7 Ω/mil·ft is a sufficiently accurate assumption (see Table 17-2 for circular-mil areas of conductors). On this basis, the resistance of any commercial copper conductor is

$$R = \frac{10.7 \times l}{\text{cmils}} \quad \text{(ohms)} \qquad (17\text{-}11)$$

where l = length of conductor in feet and cir mils = area of conductor in circular mils.

From Ohm's law of $I = E/R$, the voltage drop in a conductor is $E = IR$. Upon substituting for R, the voltage drop E becomes

$$E = \frac{10.7 \times l \times 2 \times I}{\text{cmils}} \qquad (17\text{-}12)$$

Also
$$I = \frac{E \times \text{cmils}}{10.7 \times l \times 2} \qquad (17\text{-}13)$$

and
$$\text{cmils} = \frac{10.7 \times l \times 2 \times I}{E} \qquad (17\text{-}14)$$

In the above, l = length, in feet, of the circuit and is multiplied by 2 for the total length of wire.

Formula (17-12) gives volts lost for a given-sized wire and specific current flowing.

Formula (17-13) indicates the current producing a given voltage drop in a given-sized wire.

Formula (17-14) indicates the correct wire size for a certain voltage drop and specific current flowing.

105. Power Loss. By assuming that 10.7 Ω is the resistance of a circular mil-foot of copper conductor, the power loss in watts in any conductor may be found thus,

$$P = \frac{10.7 \times I^2 \times l}{\text{cmils}} \qquad (17\text{-}15)$$

where P = power lost in the conductor in watts, I = current in amperes in the conductor, l = length of the conductor in feet, and cmils = area of the conductor in circular mils.

For a 2-wire circuit (direct-current or single-phase):

$$P = \frac{2 \times 10.7 \times I^2 \times l}{\text{cmils}} \qquad (17\text{-}16)$$

For a 4-wire 2-phase circuit (on assuming balanced currents):

$$P = \frac{4 \times 10.7 \times I^2 \times l}{\text{cmils}} \qquad (17\text{-}17)$$

For a 3-wire 3-phase circuit (on assuming balanced currents):

$$P = \frac{3 \times 10.7 \times I^2 \times l}{\text{cmils}} \qquad (17\text{-}18)$$

where P = power, in watts, lost in the circuit; I = current, in amperes, which flows in each of the wires of the circuit; l = length (one way) of the circuit; and cmils = cross-sectional area in circular mils of each of the wires. The above formulas can be used only when all the wires of the line are of the same size.

Where resistance of the conductor is determined from tables, power loss in watts = I^2R.

106. Calculation of Simple Circuits. Calculation of d-c circuits may be made directly from the foregoing equations. The method to be used in calculating an a-c circuit is determined by the characteristics of the circuit under consideration. Where the load is small, the circuit is reasonably short, and the conductors lie close together, the approximate methods of the foregoing equations can be used; these disregard the effect of line reactance. The results obtained may be subject to less error than other factors entering into ordinary wiring calculations.

In practice, voltage-drop tables and curves are resorted to in designing circuits, or mechanical computers of the slide-rule type are employed. The voltage drops of Table 17-26 are useful where power factor and reactance drop can be neglected.

107. Use of D-C or 100%-pf A-C Circuit, Voltage-drop Table (17-26). *a. To Find Wire Size for Given Drop:*

1. Find "kiloampere-feet" by multiplying current in amperes of length of one wire in feet and dividing by 1,000.

2. Starting with the given voltage drop, follow column down to the number of kilo-ampere-feet nearest to the number calculated. Find the correct size of wire at the extreme left column.

3. With short runs, table may indicate a size of wire smaller than permitted by Code regulations. In such cases, the wire size must be increased to meet Code requirements.

b. To Find the Drop in Volts That Will Be Produced by a Given Size of Wire

1. Find the kiloampere-feet as above.

2. Starting with the given size of wire, follow the horizontal line to the right to the number of kiloampere-feet nearest the actual number calculated. Follow this column up, and find the drop in volts.

c. Example. A 23-kW balanced lighting load is to be supplied from 3-wire 120/240-V mains. The length of the run between service switch and distribution panel is 250 ft. The voltage drop is not to exceed 2%. What size of conductor should be used?

Table 17-26. Voltage Drop, D-c or 100% Pf A-c Circuits*

Wire size AWG or cir mils	Voltage drop at 49 C, per kiloampere-feet								
	1	2	3	4	5	6	7	8	10
14	0.177	0.354	0.531	0.705	0.886	1.06	1.24	1.42	1.77
12	0.282	0.563	0.845	1.13	1.41	1.69	1.97	2.25	2.82
10	0.448	0.895	1.34	1.79	2.24	2.69	3.13	3.58	4.48
8	0.712	1.42	2.14	2.85	3.56	4.27	4.98	5.69	7.12
6	1.11	2.22	3.33	4.44	5.55	6.66	7.77	8.88	11.1
4	1.76	3.53	5.29	7.06	8.82	10.6	12.3	14.1	17.6
2	2.88	5.75	8.63	11.5	14.4	17.3	20.1	23.0	28.8
1	3.54	7.07	10.6	14.1	17.7	21.2	24.8	28.3	35.4
1/0	4.46	8.91	13.4	17.8	22.3	26.7	31.2	35.6	44.6
2/0	5.62	11.2	16.9	22.5	28.1	33.7	39.3	44.9	56.2
3/0	7.08	14.2	21.2	28.3	35.4	42.5	49.6	56.6	70.8
4/0	8.92	17.8	26.7	35.7	44.6	53.5	62.4	71.3	89.2
250,000	10.5	21.0	31.6	42.1	52.6	63.1	73.7	84.2	105
300,000	12.6	25.2	37.8	50.4	63.0	75.6	88.2	101	126
350,000	14.7	29.3	44.0	58.7	73.3	88.0	103	117	147
400,000	16.7	33.4	50.2	66.9	83.6	100	117	134	167
500,000	20.8	41.5	62.3	83.1	104	125	145	166	208
750,000	30.5	60.9	91.4	122	152	183	213	244	305
1,000,000	39.5	79.0	118	158	198	237	277	316	395

NOTE: See Par. **107** for use of table.

* From Ref. 4 in Bibliography (Par. **118**).

Solution. On a balanced 3-wire system, the current in each of the outside wires would be 95.8 A. The kiloampere-feet would equal

$$\frac{95.8 \text{ A} \times 250 \text{ ft}}{1{,}000} = 22.0 \text{ kA}\cdot\text{ft}$$

Since the permitted percentage voltage drop is 2, the actual drop permitted is

$$0.02 \times 240 = 4.8 \text{ V}$$

To determine the wire size required, start at the top of the column marked 5 V (which is nearest to 4.8). Follow down until the figure 22.3 is reached (which is nearest 22.0). This would indicate the use of 1/0 conductors. The actual drop would then be

$$\frac{22.0}{22.3} \times 5 \text{ V} = 4.93 \text{ V}$$

108. Reactance Voltage Drop. Reactance in an a-c circuit depends on conductor size, spacing between conductors, their arrangement, frequency of the supply, and the presence or absence of magnetic materials, e.g., steel conduit. It may be decreased for a given-sized conductor by reducing the spacing between conductors or by placing

Table 17-27. Reactance to Neutral of 3-phase, 60-cycle Circuits*

(Copper cables in nonmagnetic conduit, three-stranded 0-600 V; rubber insulated, double braided)

Size of copper conductor AWG or cir mils	D-c resistance per 1,000 ft at 60°C (140°F†), 12.3 ohms/mil ft	Size of conduit, in	60 cycles reactance to neutral‡	
			X_{min} per 1,000 ft, ohms§	X_{max} per 1,000 ft, ohms§
14	3.00	½		
12	1.87	½		
10	1.18	¾		
8	0.740	¾	0.041	0.045
6	0.465	1¼	0.040	0.055
4	0.292	1¼	0.037	0.050
2	0.185	1½	0.034	0.048
1	0.146	1½	0.035	0.044
1/0	0.116	2	0.034	0.048
2/0	0.092	2	0.032	0.044
3/0	0.073	2	0.032	0.041
4/0	0.058	2½	0.031	0.045
250,000	0.049	2½	0.030	0.044
300,000	0.0408	3	0.030	0.045
350,000	0.0350	3	0.029	0.044
400,000	0.0306	3	0.029	0.043
450,000	0.0272	3	0.029	0.040
500,000	0.0245	3	0.028	0.038
550,000	0.0222	3½	0.028	0.040
600,000	0.0204	3½	0.028	0.038
750,000	0.0163	3½	0.028	0.035
1,000,000	0.0123	4	0.028	0.038
1,250,000	0.0098	5	0.028	0.041
1,500,000	0.00816	5	0.028	0.038
1,750,000	0.00700	5	0.027	0.037
2,000,000	0.00612	6	0.027	0.039

† Temperature correction is 0.34% per deg C.
‡ In steel conduit, reactances will be larger, possibly by 25%.
§ X_{min} = cables touching in conduit.
X_{max} = cables widely separated in conduit.
* From Ref. 8 in Bibliography (Par. **118**).

Table 17-28. A-c/D-c Resistance Ratio*

Conductor size, AWG or MCM	A-c/d-c resistance ratio	
	Single conductors in air or separate nonmetallic conduit	Multiconductor cable or 2 or 3 single-conductor cables in same steel conduit
Up to 3	1.00	1.00
2 and 1	1.00	1.01
1/0	1.00	1.02
2/0	1.00	1.03
3/0	1.00	1.04
4/0	1.00	1.05
250	1.005	1.06
300	1.006	1.07
350	1.009	1.08
400	1.011	1.10
500	1.018	1.13
600	1.025	1.16
700	1.034	1.19
750	1.039	1.21
800	1.044	
1000	1.067	
1250	1.102	
1500	1.142	
1750	1.185	
2000	1.233	

* From Ref. 14 in Bibliography (Par. **118**).

Table 17-29. Three-phase Circuit Length for 1-volt Drop—0.8 Power Factor

(Anaconda Wire and Cable Co.)

Single-conductor cables in close proximity in conduit in air—60°C operating temperature

Size, AWG or cir mils	Resistance max. a-c at 60 C per 1,000 ft, ohms	Reactance at 60 cycles per 1,000 ft, ohms	Current in amperes																		
			3	5	10	15	20	25	30	40	50	60	80	100	120	150	200	250	300	400	500
			.Permissible length of circuit (one way), feet																		
Installation in nonmetallic conduit or iron conduit in air																					
14	3.024	.05090	79	47	24	16	12	9													
12	1.901	.04742	125	75	37	25	19	15	12	9											
10	1.196	.04414	196	118	59	39	29	24	20	15											
8	0.7451	.04704	309	186	93	62	46	37	31	23	19	15									
6	0.4824	.04353	468	281	141	94	70	56	47	35	28	23	18								
4	0.3034	.04080	722	433	217	144	108	87	72	54	43	36	27	22							
3	0.2406	.03956	892	535	268	178	134	107	89	67	54	45	33	27	22						
2	0.1908	.03837	1098	659	324	220	165	132	109	82	66	55	41	34	27	22					
1	0.1513	.03738	1344	806	403	269	201	161	134	101	81	67	50	46	34	27					
Installation in nonmetallic conduit only in air																					
1/0	0.1200	.03636	1577	946	473	315	237	189	158	118	95	79	59	47	39	32	24	*	*	*	*
2/0	0.09518	.03522	1899	1139	569	380	285	228	190	142	114	95	71	57	47	38	28				
3/0	0.07546	.03434	2265	1359	679	453	340	272	226	170	136	113	85	68	57	45	34	27			
4/0	0.05931	.03387	2682	1609	805	536	402	322	268	201	161	134	101	80	67	54	40	32	27		
350,000	0.03651	.03249	3664	2198	1099	733	550	440	366	275	220	183	137	110	92	73	55	44	37	27	
500,000	0.02555	.03140	4479	2687	1343	896	672	537	448	336	269	224	168	134	112	90	67	54	45	34	27
600,000	0.02164	.03125	4844	2906	1453	969	727	581	484	363	291	242	182	145	121	97	73	58	48	36	26
750,000	0.01739	.03063	5364	2872	1600	1073	805	644	536	402	322	268	201	161	134	107	80	64	54	40	32
1,000,000	0.01339	.02982	5998	3599	1799	1199	899	720	600	450	360	300	225	180	150	120	90	72	60	45	36
Installation in iron conduit only in air																					
1/0	0.1200	.03636	1502	898	449	299	225	180	150	112	90	75	56	45	37	30	22	†	†	†	†
2/0	0.09518	.03522	1787	1072	536	357	268	214	179	134	107	89	67	54	45	36	27				
3/0	0.07546	.03434	2111	1266	634	422	317	253	211	158	127	106	79	63	53	42	32	25			
4/0	0.05931	.03387	2473	1484	742	495	371	297	247	185	148	124	93	75	62	49	37	30	25		
350,000	0.03651	.03249	3298	1979	990	660	495	396	330	247	198	165	124	99	82	66	49	40	33	25	†
500,000	0.02555	.03140	3961	2376	1188	792	594	475	396	297	238	198	148	119	99	79	59	48	40	30	24
600,000	0.02164	.03125	4246	2547	1274	849	637	510	425	318	255	212	159	127	106	85	64	51	42	32	25
750,000	0.01739	.03063	4654	2792	1396	931	698	558	465	349	279	233	175	140	116	93	70	56	47	35	28
1,000,000	0.01339	.02982	5143	3086	1543	1029	772	617	514	386	309	257	193	154	129	103	77	62	51	39	31

* These values calculated with the reactance component increased 20% based on experimental data.

† These values calculated with the reactance component increased 50% based on experimental data.

NOTE: 1. For other voltage drops, multiply the lengths in the table by the desired drop in volts. For example: No. 4 wire will carry 15 amp 144 ft for a drop of 1 volt. If a drop of 3 volts is permissible, the permissible length = 144 × 3 = 432 ft.

2. For other currents, multiply the lengths in the table by the ratio of the current in the table to the new current. For example: the table shows a permissible circuit length of 536 ft for a No. 4/0 cable in nonmetallic conduit carrying 15 amp at a drop of 1 volt. If it should be desired to carry 150 amp at a drop of 1 volt, the permissible length = 536 × 15/150 = 54 ft.

3. In using these tables check to make sure the NEC permissible current is not exceeded.

4. A-c/d-c resistance ratio in accordance with Table II, NEMA report of Determination of Maximum Permissible Current Carrying Capacity of Code Insulated Wires and Cables for Building Purposes, June 27, 1938.

5. Sizes No. 14 to 8 AWG are solid wires; No. 6 AWG and larger are stranded, Class B stranding in accordance with ASTM Designation, B8-36.

6. Reactance computed for equivalent spacing equal to insulated cable over-all diameter plus 10%.

7. For single phase reduce the above circuit lengths by 14%.

them in nonmagnetic duct rather than in steel conduit. Its effect on voltage regulation in heavy or long a-c circuits is frequently greater than the effect of resistance drop and hence must be considered in the design of such circuits. Table 17-27 gives a-c reactance of 3-phase feeders per 1,000 ft of length when installed in nonmagnetic conduit.

109. Skin effect in interior-wiring calculations is ordinarily of little consequence unless conductors are larger than 500,000 cmils. Its effect is to increase the apparent resistance of a conductor to a-c flow. Both it and proximity effect have been considered in compiling the a-c/d-c resistance ratios of Table 17-28.

110. Calculation of Circuits Where Reactance and Power Factor Must Be Con-

sidered. The voltage-drop formula is

$$E = I\sqrt{R^2 + X^2} \qquad (17\text{-}19)$$

where E = voltage drop, I = current in amperes, R = a-c resistance in ohms (determined from ratios of Table 17-28), X = reactance in ohms (given per 1,000 ft in Tables 17-27 and 17-29).

111. Mershon Diagram. Where the percent resistance drop and the percent reactance drop of a circuit have separately been determined, the total drop can be calculated graphically by means of the Mershon diagram. The diagram can also be used to check circuit calculations arrived at by other means.

112. Conductors in Multiple. Owing to the factors discussed above, it is advisable to use two or more conductors in multiple on heavy a-c circuits where otherwise a single conductor in excess of 500,000 cmils would be required. Multiplied conductors should be of the same size; and if they are divided between two or more raceways, each raceway must contain a complete circuit.

113. Voltage Drop at Elevated Temperatures. If a circuit is designed to operate normally at an elevated temperature, say 75°C (permitted by NEC for certain building-wire insulations), it is essential that voltage drop be checked at the operating temperature, because of the increase in resistance. If tables or curves based on this temperature are not available, the resistance at the elevated temperature can be determined if the resistance at, say, 15°C is known by applying the correction factor of 0.393%/°C in the following formula:

$$r = r_{15} \times [1 + 0.00393(t - 15)] \qquad (17\text{-}20)$$

where r = resistance/1,000 ft at desired temperature, r_{15} = resistance/1,000 ft at 15°C t = new temperature at which resistance is desired.

With resistance per 1,000 ft known, the volts drop can readily be determined.

ECONOMICS OF WIRING DESIGN

114. Economy in interior wiring installations is influenced by many considerations. Obviously any conductor must fulfill the requirements of mechanical strength, carrying capacity, and reasonable voltage drop. However, annual losses may dictate a higher initial investment in larger copper. Different types of distribution system require different amounts of copper to transmit equal amounts of power. Not all types of wiring can be installed for the same unit cost. Primary supply to large individual motors may be warranted by elimination of transformer investment and reduction in conductor losses. Spare capacity for future load growth can be installed initially at less expense than if provided after construction is completed.

In many cases, there is little choice to be had; but frequently the above considerations, plus items of reliability of service, low maintenance expense, ease of accessibility, etc., will dictate one of a number of possibilities.

115. Copper Required for Different Systems. If it is assumed that the weight of copper for a single-phase 2-wire 120-volt system is 100%; then to transmit 50 kW of unity power-factor load a distance of 100 ft, with not more than 2% drop, using type RH conductors, will require the following amounts of copper for other common systems:

System	Per Cent
Single-phase, 2-wire, 120-volt	100
Single-phase, 3-wire, 120/240-volt	53
3-phase, 3-wire, 240-volt	20.9
3-phase, 4-wire, 120/208-volt	26.4

116. The relative cost of common types of wiring method will not be uniform throughout the country and depends to a certain extent on relative quantities of labor and material required, on the contractor's familiarity or unfamiliarity with the material in question, etc.

117. Annual charges consisting of fixed charges on the investment and cost of copper losses may be evaluated in arriving at economical wire sizes where conductor and raceway investment will be heavy. Power loss may be determined from the formulas of Par. **105**. By reducing the information to a table, total annual costs of various-sized conductors may readily be determined.

BIBLIOGRAPHY

118. General References on Wiring of Buildings

1. Stetka, Frank, and Brandon, Merwin "NFPA Handbook of the National Electrical Code"; New York, McGraw-Hill Book Company, 1966.

2. Beeman, D. L. "Industrial Power Systems Handbook"; New York, McGraw-Hill Book Company, 1955.

3. Croft, Terrell "American Electrician's Handbook," 8th ed., rev. by Clifford C. Carr; New York, McGraw-Hill Book Company, 1961.

4. Handbook of Interior Wiring Design; New York, Industry Committee on Interior Wiring Design, 1946, and Residential Wiring Handbook, 1954.

5. *IEEE No.* 241, Electrical Systems for Commercial Building, Institute of Electrical and Electronic Engineers, Inc., 1964.

6. Industrial Terminal Innovations; *Elec. World,* Feb. 11, 1933, p. 198.

7. Kehoe, A. H., and Jones, Bassett Vertical A-c Distribution Networks; *Elec. Eng.,* April, 1931, p. 292.

8. Kun, Emil Circuits in Conduit—Simplified Wiring Calculations; *Elec. Jour.,* July, 1935, Vol. 32, No. 7.

9. McPartland, J. F. "Electrical Systems for Power and Light"; New York, McGraw-Hill Book Company, 1964.

10. McPartland, J. F., and Novak, W.J. "Electrical Design Details"; New York, McGraw-Hill Book Company, 1960.

11. McPartland, J. F., and Novak, W. J. "Electrical Equipment Manual," 3d ed.; New York, McGraw-Hill Book Company, 1965.

12. McPartland, J. F., and Novak, W. J. (eds.) "Practical Electricity"; New York, McGraw-Hill Book Company, 1964.

13. National Electrical Code, National Board of Fire Underwriters, Boston, Mass, 1965.

14. Rosch, S. J. The Current Carrying Capacity of Rubber Insulated Conductors; *Elec. Eng.,* March, 1938, Vol. 57, p. 155.

The following monthly publications should be consulted for current discussions of wiring problems and practice:

15. *Electrical Construction and Maintenance,* New York, McGraw-Hill Publishing Company.

16. *Qualified Contractor,* Washington, D.C., National Electrical Contractors Association.

SECTION 18

MOTORS

REVISED BY

JOSEPH TENO Principal Research Engineer, Avco Everett Research Laboratory; Formerly Professor of Electrical Engineering, Lehigh University.

CONTENTS

Numbers refer to paragraphs

SECTION 18

MOTORS

DIRECT-CURRENT MOTORS

1. Direct-current motors are practically identical in construction to direct-current generators; with minor adjustments, the same direct-current machine may be operated either as a motor or as a generator. The main difference between the motor and generator is the way in which they are used or in the direction of energy conversion, i.e., electrical to mechanical for the motor and mechanical to electrical for the generator (see Sec. **7** for the principles of operation, design aspects, constructional details, and general operational procedures for direct-current machines). This subsection will deal exclusively with the operational characteristics and performance of direct-current motors.

2. Counter EMF. It was pointed out in Sec. **7,** Par. 11, that, since a motor armature revolves in a magnetic field, an emf is generated in the conductors which is opposed to the direction of the current and is called the counter emf. The applied emf must be large enough to overcome the counter emf and also to send the armature current I_a through R_m, the resistance of the armature winding, the brushes, and the series field coils; or

$$E_a = E_b + I_a R_m \qquad \text{(volts)} \tag{18-1}$$

where E_a = applied emf and E_b = counter emf. Since the counter emf at zero speed, i.e., at starting, is identically zero and since normally the armature resistance is small, it is obvious in view of Eq. (18-1) that, unless measures are taken to reduce the applied voltage, excessive current will circulate in the motor during starting. Normally, starting devices consisting of variable series resistors are used to limit the starting current of motors (see Pars. **158** through **166** for a complete discussion of direct-current motor starters).

3. Shifting of Brushes. It may be seen from Fig. 7-7 that, while the brushes of a generator are shifted forward in the direction of motion to help commutation, those of a motor have to be shifted backward. In each case, however, the armature reaction reduces the flux per pole. Brush shift will also affect the speed regulation of the motor. A shift of the brushes with rotation makes the speed drop more with an increase in load.

4. Torque Equation. The torque of a motor is proportional to the number of conductors on the armature, the current per conductor, and the total flux in the machine. The formula for torque is

$$\text{Torque} = 0.1175 Z \phi I_a \frac{\text{poles}}{\text{paths}} \times 10^{-8} \qquad \text{(lb at 1-ft radius)} \tag{18-2}$$

where Z = total number of armature conductors, ϕ = total flux per pole, and I_a = armature current taken from the line.

5. The speed equation is

$$E_b = E_a - I_a R_m = Z\phi \frac{\text{r/min}}{60} \frac{\text{poles}}{\text{paths}} \times 10^{-8} \qquad \text{(volts)} \tag{18-3}$$

or

$$\text{r/min} = 60 \frac{E_a - I_a R_m}{Z\phi} \frac{\text{path}}{\text{poles}} \times 10^{8} \tag{18-4}$$

For a given motor the number of armature conductors Z, the number of poles, and the number of armature paths are constant. The torque can therefore be expressed as

$$\text{Torque} = \text{const} \times \phi I_a \tag{18-5}$$

and the speed, likewise, is expressed as

$$\text{r/min} = \text{const} \times (E_a - I_a R_m)/\phi \tag{18-6}$$

6. Shunt-motor Speed and Torque. In this case E_a, R_m, and ϕ are constant, and the speed and torque curves are shown as curves 1 (Fig. 18-1); the effective torque is less than that generated by the torque required for the windage and the bearing and brush friction. The drop in speed from no load to full load seldom exceeds 5%; indeed, since ϕ, the flux per pole, decreases with increase of load, owing to armature reaction, the speed may remain approximately constant up to full load.

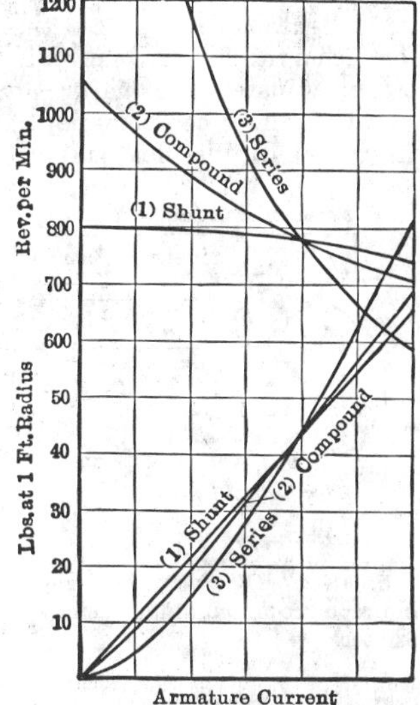

FIG. 18-1. Motor characteristics.

7. Speed regulation of a d-c motor is defined as the percent change in speed from full load to no load based on full-load speed. Figure 18-1 shows typical speed-regulation curves of the three types of d-c motors.

8. Speed of a d-c motor can be varied by changing the field strength of the motor, by varying the armature voltage, or by inserting a resistance in the armature circuit. Equation (18-6) illustrates how these factors are related to speed. Variation of the field strength gives constant horsepower but variable torque over the speed range on assuming constant armature current. Variation of speed by armature resistance is inefficient and also gives poor speed regulation.

9. Speed Changes of Shunt Motors under Rapidly Fluctuating Loads. When the load on a shunt motor increases slowly, the flux per pole decreases as the result of armature reaction and the speed [Eq. (18-6)] remains approximately constant. If, however, the load changes rapidly, the flux per pole cannot change rapidly, owing to the self-induction of the field coils; the machine then operates for the instant as a constant-flux machine, and the speed drops rapidly to allow the counter emf to decrease and the necessary current to flow.

The instantaneous speed drop is affected by the armature circuit resistance, the armature circuit inductance, the terminal voltage, the speed of the motor, and the moment of inertia of the motor. The armature circuit resistance is the most important factor, since the speed drop is directly proportional to the voltage drop in the armature circuit. Accurate calculation of the speed drop requires lengthy, precise calculation of all the factors involved.

A 1,000-hp 60-V 200-r/min motor designed with a normal volume of core iron and with proportions chosen to give a minimum impact speed drop would have approximately 5.5% impact speed drop. This can be reduced to approximately one-half the above value by doubling the size of the motor and proportioning the machine parts properly. Machines of smaller sizes will approach 8% impact speed drop.

10. Speed and Torque of Series Motors. Equations (18-6) and (18-5) apply to motors of all continuous-current types. In the case of series motors the flux ϕ increases with the armature current I_a; the torque would be proportional to I^2_a were it not that the magnetic circuit becomes saturated with increase of current. Since ϕ increases with load, the speed drops as the load increases. The speed and torque characteristics are shown in curves 3 (Fig. 18-1).

11. Excessive Speeds of Series Motors on Small Loads. If the load on a series motor becomes small, the speed becomes very high, so that a series motor should always be geared or direct-connected to the load. If it were belted and the belt were to break, the motor would run away and would probably burst.

12. Speed Adjustment of Series Motors. For a given load, and therefore for a given current, the speed of a series motor can be increased by shunting the series winding or by short-circuiting some of the series turns so as to reduce the flux. The speed can be decreased by inserting resistance in series with the armature.

13. The compound motor is a compromise between the shunt and the series motors. Because of the series winding, which assists the shunt winding, the flux per pole increases with the load, so that the torque increases more rapidly and the speed decreases more rapidly than if the series winding were not connected; but the motor cannot run away under light loads, because of the shunt excitation. The speed and torque characteristics for such a machine are shown in curves 2 (Fig. 18-1). The speed of a compound motor can be adjusted by armature and field rheostats, just as in the shunt machine (Par. **8**).

14. Indirect compounding is used on some d-c motors. In this case, the heavy strap-wound series field is replaced by a wire-wound field similar to a small shunt field. This field is excited by an unsaturated d-c exciter, usually separately driven at constant speed. This exciter is excited by the line current of the motor for which it supplies the series excitation (see Fig. 18-2). The output voltage and the current from the exciter are proportional to the main motor current; so a given proportionality exists between the load current of the motor and its wire-wound series-field strength. The use of a reversing switch and rheostat in the armature circuit of the series exciter permits variations in strength and even polarity of the series field. This furnishes an easy method of changing the compounding of the motor if desired for various speeds, to maintain constant-speed regulation over a speed range. If desired, the series exciter rheostat can be mechanically connected to the shunt-field rheostat to accomplish this automatically.

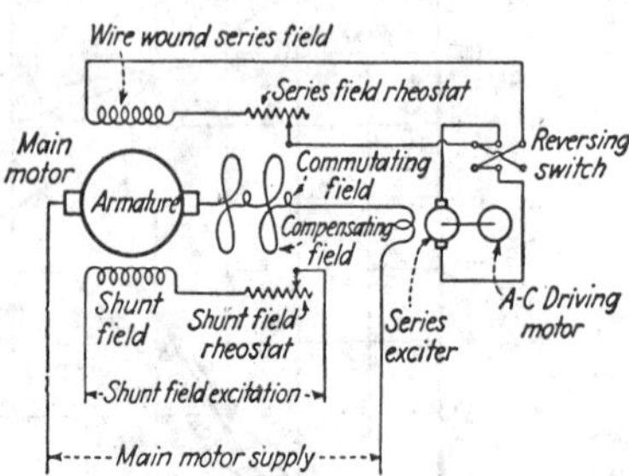

FIG. 18-2. Direct-current motor with indirect compounding, using a series exciter.

15. Automatic Speed Regulation. Numerous types of speed regulators may be used to control the speed of a d-c motor. These regulators may be of the electronic type, the magnetic-amplifier type, or the rotating-regulator type. Intelligence for the regulator is usually obtained from a small speed-indicating generator driven from the motor. The regulator controls either the armature terminal voltage or the strength of the shunt field of the motor, depending on the application.

16. Unstable Operation with Rising-speed Characteristic. Upon suddenly increasing the load torque on a differential motor designed for a rising-speed characteristic, the speed for a brief instant decreases. This results in a momentary drop of counter emf, which admits a larger armature current, in turn weakening the resultant field strength and further reducing the counter emf. The attendant increase in armature current increases the armature torque, and the reactions are such that the latter continues to increase until it exceeds the load torque and starts to accelerate the armature. The increase in speed continues until the rising counter emf finally limits the armature current to a value at which the armature torque equals the load torque, and the speed becomes constant. These reactions, with change of load, occur quite rapidly; if, however, the field cores are large and massive, changes in flux attendant upon sudden changes in mmf will lag by an appreciable time interval, on account of eddy currents in the cores. The presence of an appreciable flux lag, with very rapidly changing loads, results in unstable operation; e.g., when the load is suddenly increased, the speed will drop appreciably before it begins to accelerate; and when the load is suddenly removed, the speed will rise appreciably before it begins to decrease. The effect of the armature inertia will accentuate these defects in speed regulation. Such defects are not found in motors whose speed decreases with increasing load, i.e., that have a drooping speed characteristic.

17. Reversal of Direction of Rotation. In order to reverse a motor, it is necessary to reverse the current in the field coils or in the armature but not in both. In a commutating-pole machine, the commutating-pole winding must be considered as part of the armature and not as part of the field system. If the armature current is reversed, the current in the series field must not be reversed; otherwise, the series-field effect will be in the opposite direction to that of the shunt field.

18. Short-time Ratings. The effect of time and enclosure on motor rating may be seen from the following: A given frame will have a rating of 12 hp at 500 r/min as an enclosed machine on continuous duty, or 19 hp at 500 r/min as an open machine on continuous duty, or 31 hp at 500 r/min with a 1-h rating, or 40 hp at 500 r/min with a ½-h rating. The temperature rise on full load is 40°C as an open machine and 50°C as an enclosed machine. The horsepower is proportional to the speed over a range of 30% above or below the rated speed.

19. Effect of Voltage on Weight, Cost, and Efficiency. If two machines are built on like frames for the same output and speed but for different voltages, the number of commutator segments will be proportional to the voltage and the commutator length will be proportional to the current. The low-voltage machine will be the heavier because of the long commutator; but for moderate outputs, there will not be much difference in cost between a 120- and a 600-V machine; the cost of the extra copper on the low-voltage commutator is compensated for by the cost of the extra labor on the high-voltage commutator.

The losses will be affected in the following way: The windage, bearing friction, excitation, and iron losses will be unchanged; the brush friction loss will be the smaller in the machine with the higher voltage; the contact resistance loss will be the smaller in the machine with the higher voltage, because the contact drop is the same in each case, and the loss is therefore proportional to the current. The armature copper loss will be unchanged if the same amount of armature copper is used in each machine, because, since the number of conductors is directly proportional to the voltage, the section of each conductor will be proportional to the current, and the current density will be the same in each machine. If, however, because of the space taken by the insulation on the large number of conductors, the total amount of copper is the smaller in the machine with the higher voltage, then the copper loss will increase with the voltage, and, for the same copper loss in each machine, the output of the high-voltage machine will be less than that of the low-voltage machine.

20. Effect of Speed on Weight, Cost, and Efficiency. The output of a machine equals the product of (volts per conductor) × (current per conductor) × (number of conductors). For a given frame, the volts per conductor is directly proportional to the speed, and the product of (current per conductor) × (number of conductors) is constant for a constant-current density and a constant weight of armature copper. The output is therefore directly proportional to the speed. As a matter of fact, the flux density must decrease as the speed and the frequency increase, in order to keep down the iron loss; but the current density may increase with speed owing to better ventilation. However, the output is directly proportional to the speed, over a considerable range. The higher the speed, the larger the output, and therefore the longer the commutator and the heavier the machine.

At very low speeds the number of conductors becomes large, and, on account of the amount of insulation required, the total amount of copper is less than normal; the rating of such machines must therefore be reduced faster than the speed is reduced.

The efficiency generally increases with speed until a point is reached where the bearing friction and windage losses become an appreciable portion of the total losses.

21. Effects of Voltage Variation. Large-sized d-c motors are designed for operation on rated voltage. If the shunt excitation is held constant, then the speed of the motor will vary directly with the change in voltage from normal. If an attempt is made to hold the speed constant with a variation of applied armature voltage, the field strength will have to be increased to maintain constant speed, if the voltage is reduced below normal, and vice versa for an increased armature voltage. Within a very few percent this is permissible; but if such operation is attempted with appreciable changes in armature voltages, trouble will result. In the case of increased voltages, the field coils will overheat, and the core loss will increase very rapidly, with change in voltage resulting in an overheated armature. In the case of a reduced voltage, weak field operation will result in instability and overheating of armature coils owing to increased armature current for a given output at reduced voltage, and poor commutation will be encountered owing to larger armature current and weakened main field. Flashover between commutator bars may result with weak-field and heavy-armature-current operation. Such operation as this on large-sized motors is not recommended by manufacturers (see Table 18-1).

Table 18-1. General Effect of Voltage Variation on D-c Motor Characteristics

Voltage variation	Starting and max. run torque	Full-load speed	Efficiency			Full-load current	Temp. rise full load	Maximum overload capacity	Magnetic noise
			Full load	¾ load	½ load				
Shunt wound									
120% voltage	Increased 30%	110%	Slight increase	No change	Slight decrease	Decrease 17%	Main field increase. Commutating field and armature decrease	Increased 30%	Slight increase
110% voltage	Increased 15%	105%	Slight increase	No change	Slight decrease	Decrease 8.5%	Main field increase. Commutating field and armature decrease	Increased 15%	Slight increase
90% voltage	Decreased 16%	95%	Slight decrease	No change	Slight increase	Increase 11.5%	Main field decrease. Commutating field and armature increase	Decreased 16%	Slight decrease
Compound wound									
120% voltage	Increased 30%	112%	Slight increase	No change	Slight decrease	Decrease 17%	Main field increase. Commutating field and armature decrease	Increased 30%	Increased slightly
110% voltage	Increased 15%	106%	Slight increase	No change	Slight decrease	Decrease 8.5%	Main field increase. Commutating field and armature decrease	Increased 15%	Increased slightly
90% voltage	Decreased 16%	94%	Slight decrease	No change	Slight increase	Increase 11.5%	Main field decrease. Commutating field and armature increase	Decreased 16%	Decreased slightly

NOTE: This table shows general effects, which will vary somewhat for specific ratings.

On smaller-sized motors excited from the same voltage supply as furnishes power to the armature, reduction in supply voltage will result in a reduction in speed but not so great as in the case of separately excited motors, since the field flux is not reduced in proportion to voltage reduction owing to saturation in the magnetic circuit. Conversely, an increase in voltage will result in an increase in speed, but not so great as in the case of the separately excited motor.

Thus the desirability of operating a motor at the rated voltage is seen.

22. Development of commutatorless d-c motors has been and is currently being pursued by many groups. Various solid-state switching devices such as transistors, thyristors, and silicon-controlled rectifiers are being substituted in an effort to eliminate the commutator, which has placed limitations on speed ratings, voltage ratings, and winding-inductance values, to say nothing of maintenance problems encountered because of brush and commutator wear. Operationally, the solid-state switching devices are triggered by control circuits to switch the current in the armature coils at exactly the same time as the commutator would switch the current or (for that matter) at other more appropriate times if the application requires it. The reader is referred to the Bibliography, Par. **23,** for papers dealing with this subject matter in detail.

BIBLIOGRAPHY

23. Bibliography on D-C Motors

HANCOCK, S. Speed Regulation and Stability of D-C Motors; *Elec. Jour.*, February, 1922, p. 46.

HATHAWAY, C. B. Parallel Operation of Commutating-pole Motors and Generators; *Elec. Jour.*, June, 1922, p. 263.

Frazier, R. H., et al. Precise Speed Control for D-C Machines; *Bull. Mass. Inst. Technol.*, June, 1935.

Burt, R. F. Adjustable Voltage Skip Hoist Control; *Iron Steel Engr.*, August, 1943, pp. 38–40.

Caldwell, G. A., and Formhals, W. H. Electrical Drives for Wide Speed Ranges; *Trans. AIEE*, February, 1942.

Langsdorf, A. S. "Principles of D-C Machines," 5th ed.; New York, McGraw-Hill Book Company, 1940.

Shoults, Rife, and Johnson "Electric Motors in Industry"; New York, John Wiley & Sons, Inc., 1942.

Siskind, C. S. "Direct Current Machinery"; New York, McGraw-Hill Book Company, 1952.

Liwschitz-Barik, M., and Whipple, C. C. "Direct-current Machines," 2d ed.; Princeton, N.J., D. Van Nostrand Company, Inc., 1956.

Puchstein, A. F. "The Design of Small Direct Current Motors"; New York, John Wiley & Sons, Inc., 1961.

Kincer, R. D., and DeWitt, B. F. Modifications of Basic Drive Motors and Applications; *Wescon Tech. Papers*, 1965, Vol. 9, Pt. 3, No. 5.2.

Newell, A. F., and Williamson, K. H. Regenerative Braking on D-C Motors; *Mullard Tech. Communs.*, January, 1966, Vol. 8, No. 79, pp. 283–290.

Kawano, S. Speed Characteristics of Separately Excited D-C Motor Taking Commutation into Account; *Elec. Engr. Japan*, June, 1964, Vol. 84, No. 6, pp. 32–40.

Yates, W. W. Brushless D-C Motors; *Machine Design*, March, 1966, Vol. 38, No. 5, pp. 136–142.

Bhagwat, P. G., and Wolfendale, E. Silicon Controlled Rectifiers Replace Commutator in D-C Motor; *Control*, January, 1964.

Tustin, A. Electrical Motors in Which Commutation Is by Switching Devices Such as Controlled Silicon Rectifiers; *Proc. IEEE*, January, 1964.

Bauerlein, G. Brushless D-C Motor with Solid-state Commutation, *IRE Intern. Conv. Record*, 1962, Vol. 10, Pt. 6, pp. 184–190.

Hagevat, P. G., and Wolfendale, E. Silicon Controlled Rectifiers Replace Commutator in D-C Motor; *Control*, June, 1964, Vol. 8, No. 67, pp. 24–27.

Kincer, R. D., and Rakes, R. G. Brushless D-C Motor, *Wescon Tech. Papers*, 1964, Vol. 8, Pt. 9.

Miyairi, S., and Tsunehiro, Y. Analysis of Characteristics of Commutatorless Motor as D-C Motor, *Elec. Engr. Japan*, September, 1965, Vol. 85, No. 9, pp. 51–62.

SYNCHRONOUS MOTORS

24. Structure. Synchronous motors are similar to a-c generators in mechanical construction, the principal difference being that they are provided with amortisseur windings in the pole faces for starting, while a-c generators may or may not have such windings. Like a-c generators they are usually of the revolving-field type, except for some small machines. Also, some very small synchronous motors are built with fields having polar projections but without windings. These machines are excited by the reactive component of the armature current and depend upon reluctance torque to hold them in synchronism, i.e., the torque resulting from the difference in reluctance between the direct and quadrature axes of the machine. The following discussion does not apply to these very small machines.

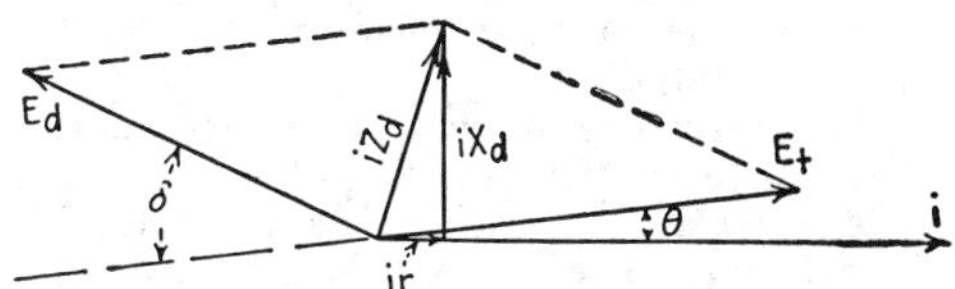

Fig. 18-3. Round-rotor diagram.

25. Diagrammatic Representation. Owing to its simplicity and common usage the "round-rotor" method of representation, in which the impedances are considered constant for all positions of the rotor, will be given first in spite of its considerable inaccuracy. A single phase of a polyphase machine will be considered for the sake of simplicity.

26. Theory of Operation. Consider a synchronous machine driven at such a speed and at such excitation that its voltage E_d is exactly the same in frequency and magnitude as the voltage E_t of the supply circuit, for which the voltage and frequency may be assumed constant. Assume also the phase of the machine voltage in exact opposition to that of the supply voltage. As these two exactly neutralize, there will be no current when the machine is connected to the supply circuit. If the driving power is removed, the machine will begin to slow down; and E_d falls back, yielding a resultant emf and current such that the resultant emf equals iZ_d (Fig. 18-3), where Z_d = synchronous impedance of the motor. As Z_d is almost all synchronous reactance, the current will lag almost 90° behind the resultant emf and for small values of δ will be nearly in phase with E_t.

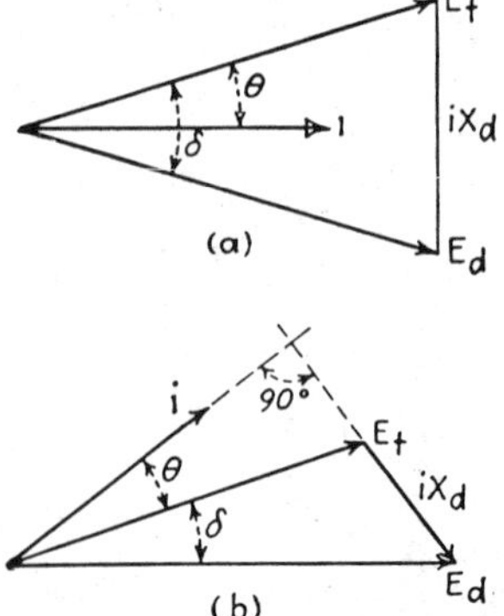

Fig. 18-4. Different form of round-rotor diagram (resistance neglected). (a) Lagging current; (b) leading current.

A form of this diagram more convenient for the calculation of synchronous-motor performance is shown in Fig. 18-4*a*. In this diagram, resistance is neglected, and, as before, iX_d lags behind i by 90°. If E_t, i, θ, and X_d are known, E_d and δ can easily be calculated. Figure 18-4*b* is the same as Fig. 18-4*a* except that the current leads the voltage instead of lagging. This results in greater excitation voltage E_d being required than before.

27. Two-reaction Diagram. In the above, no account was taken of the nonuniformity of the field structure resulting from the definite pole construction that is ordinarily used. This leads to considerable errors, particularly in calculations involving displacement angle. More accurate results in such cases are obtained by the use of the two-reaction diagram as follows:

In accordance with the Blondel two-reaction theory the armature current is resolved into two components: one, the direct-axis component producing magnetization whose axis coincides with the center lines of the poles; and the other, the quadrature-axis component producing magnetization whose axis is midway between the field poles. The direct-axis component tends either to magnetize or to demagnetize the field poles; i.e., it adds to or subtracts from the magnetization produced by the direct current in the field winding. The quadrature-axis component, on the other hand, produces magnetization midway between poles, or what is frequently called "cross magnetization." The total voltage drop in the machine is the vector sum of the drops due to the direct- and quadrature-axis components of current separately, the direct-axis drop being the product of the direct-axis current by the direct-axis impedance, and the quadrature-axis drop being the product of the quadrature-axis current and the quadrature-axis impedance.

Figure 18-5 is a vector diagram of a synchronous motor drawn in accordance with the above. It is for leading current, and resistance is neglected as in Fig. 18-4. The direct-axis reactance drop lags the direct-axis current by 90°, and the quadrature-reactance drop lags the quadrature-axis current by 90° also.

E_d is the excitation voltage corresponding to the d-c excitation, and E_t is the terminal voltage, making δ the displacement angle of the machine. The current i leads E_t by the angle θ. The direct- and quadrature-axis currents i_d and i_q are at right angles and in phase, respectively, with E_d. The direct and quadrature drops are calculated as mentioned above, and E_d is the vector sum of E_t and these drops.

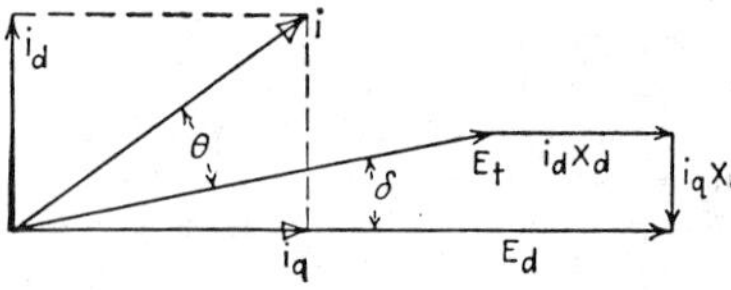

Fig. 18-5. Two-reaction diagram.

28. Power-factor Correction. Synchronous motors were first used on account of their ability to raise the power factor of systems having large induction-motor loads, since the power factor of such systems is frequently quite low. In many cases, however, they also cost less and have higher efficiencies than the corresponding induction motors. This is especially true for low-speed motors.

Many synchronous motors are designed to operate at unity power factor; i.e., sufficient excitation is supplied to meet the requirements of the motor only, and no reactive current is either drawn from or supplied to the power system. Sometimes it is desirable to have the motors furnish a part of the reactive current required by the system. In such cases they are usually designed to operate at 80% leading power factor and draw current which leads the voltage rather than lags, as is the case with induction motors and transformers.[1] For a particular motor this means greater excitation than that necessary for unity power factor.

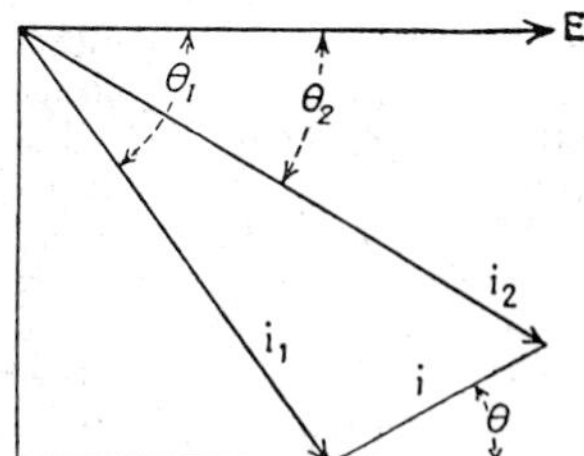

FIG. 18-6. Power-factor correction.

Power-factor correction problems can be readily solved by resolving all currents into energy and reactive components and taking the algebraic sums of each separately or by means of a vector diagram as shown in Fig. 18-6.

In Fig. 18-6 E, i_1, and cos θ_1 are the voltage, current, and power factor, respectively, of the power supplied to the plant in question before adding the synchronous motor; i and cos θ are the current and power factor, respectively, of the synchronous motor that is added. Note that θ is drawn upward from the horizontal instead of downward as in the case of θ_1, because the synchronous motor is overexcited and draws leading current from the power system instead of lagging. The total current drawn from the system is i_2, the vector sum of i_1 and i; and the resultant power factor is cos θ_2. If the power factor of the synchronous motor were unity, the current vector i would be parallel to E, and the resulting power factor cos θ_2 would be less.

29. Synchronizing Power. The following definitions are from the USAS C50, Par. 3.161:

"**Synchronizing power** is the power at synchronous speed corresponding to the torque developed at the air gap between the armature and field. This torque tends to restore the rotor to the no-load position relative to line voltage.

"P_r, the synchronizing coefficient, is determined by dividing the shaft power by the corresponding angular displacement of the rotor. It is expressed in kilowatts per electrical radian. Unless otherwise stated, the values given will be for rated voltage, load, power factor, and frequency.

"**The displacement angle** is the angle of the rotor relative to the terminal voltage under load referred to the no-load position; i.e., zero displacement angle is at no load."

From the vector diagram (Fig. 18-5) it can easily be shown that the displacement angle is

$$\delta = \tan^{-1} \frac{ix_q \cos \theta}{ix_q \sin \theta + E_t} \tag{18-7}$$

Then a close approximation of P_r will be 57.3 times the ratio of the full-load kilowatts output to the full-load displacement angle expressed in electrical degrees; i.e.,

$$P_r = \frac{57.3 \text{ kW}}{\tan^{-1} \dfrac{x_q \cos \theta}{ix_q \sin \theta + E_t}} \tag{18-8}$$

In working with per unit quantities, a convenient expression is the ratio of P_r to the kilovoltamperes corresponding to the rating of the motor. Upon neglecting the armature losses, this becomes

$$\frac{P_r}{\text{kVA}} = \frac{57.3 \cos \theta}{\tan^{-1} \dfrac{x_q \cos \theta}{x_q \sin \theta + 1}} \tag{18-9}$$

[1] Note that a synchronous condenser or an overexcited synchronous motor has a leading power factor while an overexcited synchronous generator has a lagging power factor. Similarly an induction motor draws a lagging current but an induction generator supplies a leading current.

If we expand the USASI definition of P_r to cover any load, then

$$\frac{P_r}{\text{kVA}} = \frac{1}{x_q} \qquad \text{(at no load)} \tag{18-10}$$

By interpolation P_r/kVA can be determined for any load and any power factor if x_q is known. x_q can be calculated from the design of the motor.

30. Application to Pulsating Loads. If the load driven by a synchronous motor is a periodically pulsating one such as a reciprocating pump or compressor, the inertia of the rotating parts must be carefully selected, as otherwise severe pulsations in armature current and power drawn from the system may result. The USASI rule covering this is as follows (USAS C50, Par. 3.160):

"Pulsating Armature Current. When the driven load such as that of reciprocating pumps, compressors, etc., requires a variable torque during each revolution, the combined installation shall have sufficient inertia in its rotating parts to limit the variations in motor armature current to a value not exceeding 66% of full-load current.

"The basis of determining the variation shall be by oscillograph measurement and not be ammeter readings. A line shall be drawn on the oscillogram through consecutive peaks of the current wave. The variation is the difference between the maximum and minimum ordinates of this envelope. This variation shall not exceed 66% of the maximum value of the rated full-load current of the motor. (The maximum value of the motor armature current to be assumed as 1.41 times the rated full-load current.)"

Figure 18-7 is an oscillogram of current input to a 200-hp 3,810-V 100-r/min 3-phase

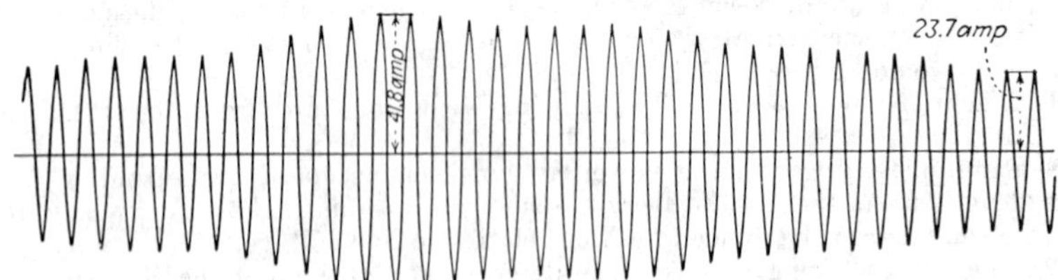

Fig. 18-7. Current-pulsation oscillogram of a 200-hp 100%-pf 3,810-V 100-r/min synchronous motor driving a double-acting duplex ammonia compressor. Current in one phase is shown with line voltage = 3,750, line amperes = 21.5, field amperes = 32, and power factor = 100% approximately. Both cylinders in operation, head pressure = 152 lb/in², suction pressure = 27 lb/in².

60-c synchronous motor driving a duplex double-acting ammonia compressor. The current variation in accordance with the above rule is 18.1 A, which is 51% of 35.4 A, which in turn is 1.41 times 25.1 A, the rated motor current.

31. Natural-frequency Oscillations. The percentage variation in power and current drawn from the line will ordinarily be less than the percent variation in torque causing it, but this is not always the case, because there may be partial resonance between the frequency of certain impulses of the compressor and the natural frequency of the unit on the power system.

The USASI formula for natural frequency of a synchronous machine on a large power system is

$$F = \frac{35{,}200}{\text{r/min}} \sqrt{\frac{P_r f}{WR^2}} \tag{18-11}$$

A quantity called "compressor factor" and designated by the letter C has been incorporated in the USASI standards. It is defined as

$$C = \frac{0.746WR^2\,(\text{r/min})^4}{10^8 P_r f} \tag{18-12}$$

It is a criterion for the comparison of flywheel effects of various synchronous motor-compressor units. For a tangential effort curve of a *given shape*, the power pulsation expressed as a percentage of the average is practically a function of C only. Thus it is possible to determine values of C necessary to limit the power pulsation to any desired percentage if the tangential effort curve is known.

A large amount of work has been done on this subject, and values of C corresponding to 66% power pulsation have been adopted for various types of compressors having various methods of unloading. The NEMA table of compressor factors is given in NEMA Motor and Generator Standards, 1941, MG 10-45.

The definition of C given in Eq. (18-12) above does not indicate its character so very well because most of the quantities in Eq. (18-12) do not depend upon the compressor at all. If Eq. (18-11) is solved for WR^2 and this value is substituted in Eq. (18-12), it becomes much clearer. We then obtain

$$C = 9.25\left(\frac{\text{r/min}}{f}\right)^2 \tag{18-13}$$

Even this is not self-evident, since the compressor alone does not have a natural frequency. However, by specifying limiting values of C, we limit the ratios of speed to natural frequency which may be used and so limit the ratios of electrical power variation to torque variation for each harmonic in the torque curve of the compressor. This is true because this last ratio depends principally upon the ratio of speed to natural frequency, the damping of the amortisseur winding of the motor not being so very important in operating well away from resonance, as is ordinarily the case. Thus we limit the magnitude of each harmonic in the resulting power- or current-input curve and so limit the resulting pulsation.

Solving Eq. (18-12) for WR^2 gives the following convenient expression:

$$WR^2 = \frac{1.34P_rfC}{\left(\dfrac{\text{r/min}}{100}\right)^4} \tag{18-14}$$

From this it is seen that the flywheel effect required for a given type of compressor varies inversely as the fourth power of the speed and directly as the compressor factor, the frequency of the system, and P_r. Frequently a judicious modification of the design of the compressor will make a material difference in the permissible values of C.

32. Starting. It is usually desirable to reduce the load as much as possible during the starting period, because it may result in a smaller and cheaper motor which will in turn result in a smaller starting kVA. Considerably larger currents at much lower power factors are required to start most synchronous motors than to operate them at full load in synchronism.

Frequently the load can be materially reduced quite easily, as is the case with most reciprocating compressors, positive-pressure blowers, and centrifugal pumps or blowers. There are, however, some machines, such as cement mills, that cannot be unloaded for starting, and others, such as mine fans, that have large and heavy rotating parts and can sometimes be partially unloaded. Still others, such as induced-draft fans, have heavy rotating parts and cannot be unloaded very much. Each of these cases requires careful consideration, particularly those where both large starting loads and heavy rotating parts are involved.

33. "**Locked rotor torque** (**static torque**) of a motor is the minimum torque which it will develop at rest for all angular positions of the rotor, with rated voltage applied at rated frequency.

34. "**Pull-in torque** of a synchronous motor is the maximum constant torque under which the motor will pull into synchronism, at rated voltage and frequency, when its field excitation is applied.

"The maximum induction-motor speed to which a motor will bring its load depends on the power required to drive it. Whether the motor can pull the load into step from this speed depends on the characteristics of the load and on the inertia of the revolving parts. Consequently the pull-in torque cannot be determined without having the WR^2 as well as the torque of the load.

35. "**Nominal pull-in torque** of a synchronous motor is the torque it develops as an induction motor when operating at 95% of synchronous speed with rated voltage applied at rated frequency.

"Note: This quantity is useful for comparative purposes when the inertia of the load is not known."

It should not be used to compare motors of different frequencies, however, as a motor that will pull into step from 95% speed (3 c slip) at 60 c will pull into step from about 88% speed (about 3 c slip) at 25 c.

36. "Pull-out torque of a synchronous motor is the maximum sustained torque which the motor will develop at synchronous speed for 1 min, with rated voltage applied at rated frequency and with normal excitation."

37. Speed-Torque Curves. The usual type of revolving-field synchronous motor is started by applying voltage to the stator winding with the rotor circuit closed through a suitable resistance. The currents induced in the amortisseur winding with which the rotor is equipped and in the closed-field circuit react upon the revolving field produced by the stator currents in nearly the same manner as in an induction motor, producing the torque necessary to bring the motor and its load to within a few per cent of synchronous speed. The phenomenon is complicated somewhat by the definite pole construction with its unequal air gap and exciting winding, which introduce a second field revolving at a speed corresponding to twice the slip.

This is called the "negative-sequence" or "backward" field. It produces a component of torque which adds to the above torque below half speed and subtracts above this point. If the negative-sequence field is excessive, a half-speed cusp in the speed-torque curve will result. This effect is usually unimportant in a motor having a well-designed amortisseur winding.

38. Pull into Step. If the speed attained under starting conditions is high enough, d-c excitation can be applied and the machine will go to full speed and operate in synchronism; i.e., it will "pull into step." When d-c excitation is applied, an additional torque is produced which alternates at slip frequency. This torque is superimposed upon the torque produced by the amortisseur winding and produces pulsations in speed which will depend upon the inertia of the rotating parts. These pulsations must be sufficient to bridge the gap between induction and synchronous operation if the machine is to pull into synchronism; and if, in the case of a particular motor, the inertia of the load is increased, the motor must be allowed to come closer to synchronism before applying excitation; i.e., the pull-in torque will be reduced.

The maximum slip from which it is possible to pull in depends upon the design of the motor, the inertia of the rotating parts, the characteristics of the system from which power is drawn, and the phase angle at which voltage is

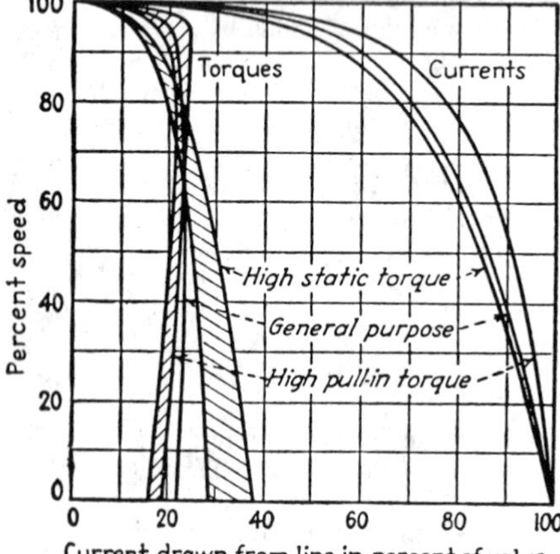

Fig. 18-8. Approximate starting performance of synchronous motors.

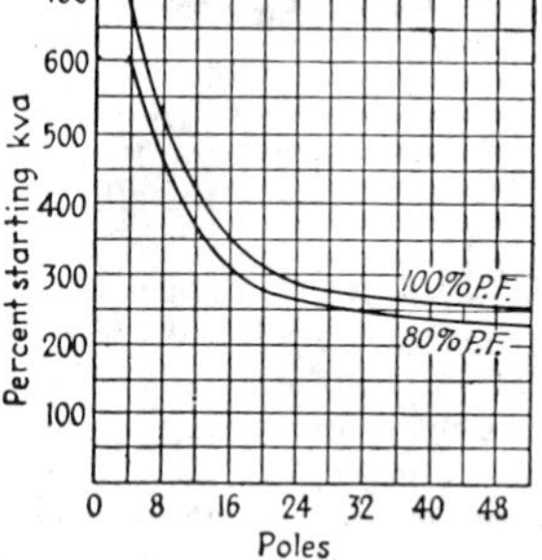

Fig. 18-9. Approximate blocked-rotor kVA of synchronous motors.

applied. Shoults, Crary, and Lauder, Par. **50,** give the following expression for average slip for the most unfavorable phase angle:

$$s < k_2 \sqrt{\frac{E_d e}{\pi f H (x_d + x_1)}} \tag{18-15}$$

As shown by the above equation, the maximum permissible slip at the time excitation is applied varies inversely as the square root of the inertia of the revolving parts. This means that the pull-in torque of a particular motor depends upon the inertia of the load and the characteristics of the power system to which it is connected as well as upon the design of the motor itself.

39. Typical torques and currents of synchronous motors during starting are shown in Figs. 18-8, 18-9, 18-10, which are self-explanatory. The torques shown in Fig. 18-8 are in such form that they indicate the useful in-phase component of stator current and are thus often called "torque factors."

40. Methods of Starting. Synchronous motors are usually started with the field circuit closed through a suitable resistance. This is done both on account of holding down the high induced voltage which would otherwise appear at the field terminals and also on account of the torque contributed by the induced field current, especially at the higher speeds. The pull-in torque developed by a particular motor depends to a considerable extent upon the resistance across the field circuit while it is being started.

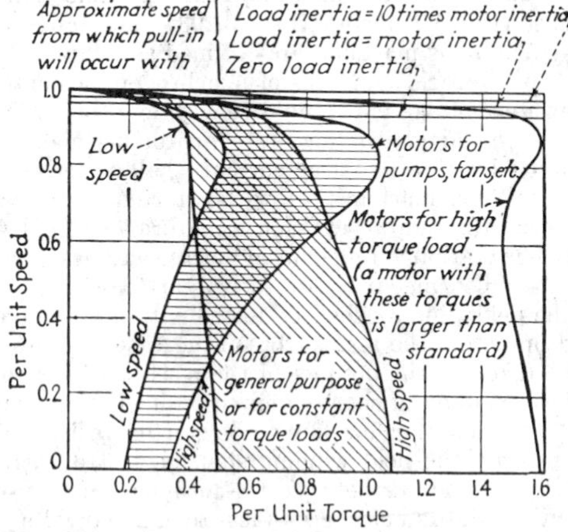

Fig. 18-10. Typical synchronous-motor starting torques at full voltage.

The following methods of starting are extensively used:

1. Across-the-line starting consists simply in closing the main-line switch with the field closed as mentioned above and, after the speed has stopped increasing, applying d-c excitation to, and disconnecting the external resistance from, the field circuit.

2. Reduced-voltage starting is the same as the above, except that reduced voltage is first applied by means of a transformer (usually an autotransformer), full voltage not being applied until after the motor has gained as much speed as possible on the reduced voltage.

3. Reactance starting is the same as reduced-voltage starting, except that the first step is obtained by reactance in series with the armature instead of by means of a transformer. It requires more current from the line for the same motor torque on the first step than if a transformer is used but has the advantage of not opening the circuit when the transfer to full voltage is made. This transfer is made by simply short-circuiting the reactance.

4. A modification of item 2 is sometimes used in which the change from one transformer tap to the next is made by means of reactances and resistances in order not to open the circuit during the transfer.

5. The Korndorfer method is somewhat similar to item 4. By this method the motor is first connected to suitable taps of an autotransformer and then started by connecting the autotransformer to the line. The transfer to full voltage is made by first opening the neutral of the autotransformer, leaving the motor running with sections of the autotransformer in series, and then short-circuiting these sections. The objection to this is that the reactance of the sections of the autotransformer left in the circuit during the transfer may be so high that there is little gained over straight reduced-voltage starting.

6. Part-winding starting requires a motor with armature windings arranged for

the particular scheme being used. The simplest arrangement is to make a two-circuit armature winding such that one circuit can be used to start the motor with the other being left disconnected. The second circuit is connected after the motor has gained as much speed as possible on one circuit. Several circuits and various combinations with reactance to provide additional steps are sometimes used.

41. Large low-speed synchronous motors are well suited for almost any constant-speed drive provided that suitable provision can be made for starting. As compared with induction motors they usually have somewhat higher efficiency and lower cost and need less floor space but require an exciter and have higher starting currents. Provision should be made for reducing the load as much as practicable when the motor starts, and the starting requirements after this has been done should be considered carefully in the design of the motor. Care should be exercised to see that adequate line and transformer capacity feeding the motor is installed, as otherwise objectionable variations in voltage may result.

42. Standard Dielectric Tests of Field Windings. In the USAS C50 (Rotating Electrical Machinery, Par. 3.203) the standard dielectric tests for field windings of synchronous machines are outlined as follows:

"(*a*) Field windings of synchronous generators shall be tested with ten times the exciter voltage but in no case with less than 1,500 volts.

"(*b*) Field windings of synchronous machines, including motors which are to be started with alternating current, shall be tested as follows:

"When machines are to be started with the field short-circuited or with the field closed through an exciting armature, the field winding shall be tested with ten times the excitation voltage but in no case with less than 2,500 volts nor more than 5,000 volts.

"When machines are to be started with a resistor in series with the field winding, the field winding shall be tested with a voltage equal to twice the rms value of the *IR* drop across the resistor but in no case with less than 2,500 volts. The *IR* drop shall be taken as the product of the resistance and the current which would circulate in the field winding, if short-circuited on itself at the specified starting voltage.

"When machines are to be started with field open-circuited and sectionalized while starting, the field winding shall be tested with 1½ times the maximum rms voltage which can occur between the terminals of any section under the specified starting conditions but in no case with less than 2,500 volts or ten times the rated excitation voltage per section, whichever is the larger.

"When machines are to be started with the field open-circuited and connected in series while starting, the winding shall be tested with 1½ times the maximum rms voltage which can occur between the field terminals under the specified starting conditions but in no case with less than 2,500 volts or ten times the rated excitation voltage, whichever is the larger."

43. Multispeed Synchronous Motors. It is possible to build two-speed synchronous motors for less than the cost of two single-speed motors. This is usually done by means of two stator windings (reconnecting one stator winding is possible but usually less desirable) and by reconnecting the field winding. Five slip rings are required for this rotor reconnection.

For large machines for pumped storage plants where a slightly higher speed is desired for pump operation of a water wheel than for normal operation of the same wheel, it is possible to build speed ratios such as 4:5, and some such large machines are now being built.

44. Starting Synchronous Motors by External Means. Synchronous motors driving frequency-changer sets, d-c generators, or for other applications where the load may be greatly reduced during the starting period are sometimes started from a small wound-rotor induction or d-c motor on the end of the shaft. This is done only when the relatively large starting current inherent to the synchronous motor is objectionable because of the voltage dip it produces. When oil pressure is supplied and autotransformer starting is used, the starting kVA taken by an unloaded synchronous motor may be as low as 20% or less of the kVA rating depending mainly on the time to reach synchronism.

45. Use of Clutches. Clutches of various kinds have been used between synchronous motors and their loads, the motor being started light with the clutch open and the clutch engaged after the motor is in synchronism and fully excited. This results in very little disturbance to the system, especially if other than across-the-line starting is used. Sometimes the clutch is made an integral part of the motor. Combining the synchronous motor with clutches permitting speed variations opens the field of variable-speed drive to synchronous motor applications. Clutch types include (1) mechanically operated friction clutches, (2) electrically operated friction clutches, (3) revolving stator with brake, (4) centrifugally operated friction clutches, (5) squirrel-cage or eddy-current electrical clutches, (6) planetary clutches, and (7) hydraulic clutches.

46. Phase-wound damper windings are sometimes used to obtain a motor which, during starting, is substantially a wound-rotor induction motor. External resistors are inserted in the rotor circuit to permit starting with high-torque and low-stator current. When up to speed, the rotor windings are supplied with d-c excitation, and the motor is then a synchronous motor and carries its load as such.

47. Effect of Differing Impressed and Induced Waveshapes on Power Factor. When the impressed and induced emf waves are of different shape, the unbalanced emf harmonics produce wattless harmonic currents, which though small compared with the full-load current may be considerable with respect to the no-load fundamental current. Thus the power factor at no load may be less than unity even with the fundamentals of the current and emf in phase.

48. Appropriate Field of Application for Synchronous Motors. Since the synchronous motor may be operated with unity power factor (and with leading power factor when desired), it is obviously especially appropriate for ratings for which an induction motor inherently has a very poor power factor. This is the case to a greater degree the lower the speed and the higher the frequency. Thus for drives requiring a motor speed of 200 r/min, the induction motor would have a very low power factor at 60 c. Even at 25 c the power factor for such a speed is rather low. At small loads, say, half load and quarter load, the power factor of 60-c slow-speed motors is exceedingly low (often less than 0.6 at half load). Consequently this point is of even greater importance for motors which do not run most of the time at or near full load.

Synchronous motors are more suitable for high-voltage winding than induction motors because their inherently longer air gap permits the use of wider slots without a sacrifice in characteristics. This permits the use of fewer stator slots per pole.

49. Nomenclature

C = compressor factor as defined in USASI Standards
e = per unit system voltage at point where constant
E_t = per unit voltage at terminals of motors
E_d = per unit voltage due to current in field windings
f = system frequency in cycles per second
F = natural frequency of oscillation in cycles per minute of motor with its load on an infinite system
H = inertia factor in kilowattseconds per kVA rating

$$= \frac{2.31 \times WR^2 \times (\text{r/min})^2}{\text{kVA} \times 10^7}$$

i = per unit motor current
i_d = per unit direct-axis component of motor current
i_q = per unit quadrature-axis component of motor current
i_1 = per unit line current before adding synchronous motor
i_2 = per unit line current after adding synchronous motor
kW = kilowatts output of motor at full load
kVA = kilovoltamperes corresponding to motor rating
r/min = rated speed of motor in revolutions per minute
s = per unit slip of motor when excitation is applied
WR^2 = moment of inertia of rotating parts expressed in pounds weight times feet radius of gyration squared
x_1 = per unit line reactance between motor and point in system where voltage is constant and has a value e
x_d = per unit direct-axis synchronous reactance of motor
δ = displacement angle of motor

θ = angle between voltage and current at motor terminals
θ_1 = angle between voltage and current of load before adding motor
θ_2 = angle between voltage and current of load after adding motor

BIBLIOGRAPHY

50. References for Synchronous Motors

Part-winding Starting of Synchronous Motors; *Elec. World,* Jan. 16, 1932, p. 142.

EDGERTON, H. E., ET AL. Pulling into Step of Synchronous Induction and Salient Pole Synchronous Motors; *Bull.* 70, *Mass. Inst. Technol.,* April, 1931.

EDGERTON and LYON Transient Torque-angle Characteristics of Synchronous Machines; *Trans. AIEE,* 1930, Vol. 49, pp. 686–698.

SHOULTS, CRARY, and LAUDER Pull-in Characteristics of Synchronous Motors; *Trans. AIEE,* 1935, Vol. 54, pp. 1385–1395.

HYDE, M. A. Synchronous Motor with Phase-connected Damper Winding for High-torque Loads; *Trans. AIEE,* 1931, Vol. 50, pp. 600–606.

DOHERTY, R. E., and NICKLE, F. A. Synchronous Machines; *Trans. AIEE,* 1926, Vol. 45, pp. 913, 927; *Trans. AIEE,* 1927, Vol. 46, p. 1.

PARK, R. H. Two-reaction Theory of Synchronous Machines—Generalized Method of Analysis, I; *Trans. AIEE,* July, 1929, Vol. 48, p. 716. Two-reaction Theory of Synchronous Machines, II; *Trans. AIEE,* June, 1933, Vol. 52, p. 352.

TITTEL, J. Variable-speed Synchronous Machines for Hydro-electric or Pumped Storage Power Stations; *CIGRE Paper* 109, May, 1954.

WOLL, R. F. A-C Motor Rerate—New Industry Standard, *Westinghouse Engr.,* November, 1965, Vol. 25, No. 6, pp. 175–181.

EDWARDS, J. D., GILBERT, A. J., and HARRISON, E. H. Application of Thyristors to Excitation Circuits of Synchronous Motors, *IEE Conf. Publ.* 17, 1965, Pt. 1, pp. 158–167.

TANAKA, Y. Synchronous Motor with Transistorized Phase Controller, *Elec. Engr. Japan,* November, 1964, Vol. 85, No. 11, pp. 41–50.

MOORE, R. C. Synchronous Motors Help Power Factor, *Power,* December, 1965, Vol. 109, No. 12, pp. 73–75.

CAHILL, D. P. M., and ADKINS, B. Permanent-magnet Synchronous Motor, *Proc. IEE,* December, 1962, Vol. 190, Pt. A, No. 38, pp. 483–491.

INDUCTION MACHINES

Theory of the Polyphase Induction Motor[1]

51. Principle of Operation. An induction motor is simply an electric transformer whose magnetic circuit is separated by an air gap into two relatively movable portions, one carrying the primary and the other the secondary winding. Alternating current supplied to the primary winding from an electric power system induces an opposing current in the secondary winding, when the latter is short-circuited or closed through an external impedance. Relative motion between the primary and secondary structures is produced by the electromagnetic forces corresponding to the power thus transferred across the air gap by induction. The essential feature which distinguishes the induction machine from other types of electric motors is that the secondary currents are created solely by induction, as in a transformer, instead of being supplied by a d-c exciter or other external power source, as in synchronous and d-c machines.

52. Construction Features. The normal structure of an induction motor consists of a cylindrical rotor carrying the secondary winding in slots on its outer periphery and an encircling annular core of laminated steel carrying the primary winding in slots on its inner periphery. The primary winding is commonly arranged for 3-phase power supply, with three sets of exactly similar multipolar coil groups spaced one-third of a pole pitch apart. The superposition of the three stationary, but alternating, magnetic

[1] ALGER, P. L. "The Nature of Polyphase Induction Machines"; New York, John Wiley & Sons, Inc., 1951.

fields produced by the 3-phase windings produces a sinusoidally distributed magnetic field revolving in synchronism with the power-supply frequency, the time of travel of the field crest from one phase winding to the next being fixed by the time interval between the reaching of their crest values by the corresponding phase currents. The direction of rotation is fixed by the time sequence of the currents in successive phase belts and so may be reversed by reversing the connections of one phase of a 2- or 3-phase motor.

Figure 18-11 shows the cross section of a typical polyphase induction motor, having in this case a 3-phase four-pole primary winding with 36 stator and 28 rotor slots. The primary winding is composed of 36 identical coils, each spanning 8 teeth, one less than the 9 teeth in one pole pitch. The winding is therefore said to have 8/9 pitch. As there are three primary slots per pole per phase, phase A comprises four equally spaced "phase belts," each consisting of three consecutive coils connected in series. Owing to the short pitch, the top and bottom coil sides of each phase overlap the next phase on either side. The rotor, or secondary, winding consists merely of 28 identical copper or cast-aluminum bars solidly connected to conducting end rings on each end, thus forming a "squirrel-cage" structure. Both rotor and stator cores are usually built of silicon-steel laminations, with partly closed slots, to obtain the greatest possible peripheral area for carrying magnetic flux across the air gap.

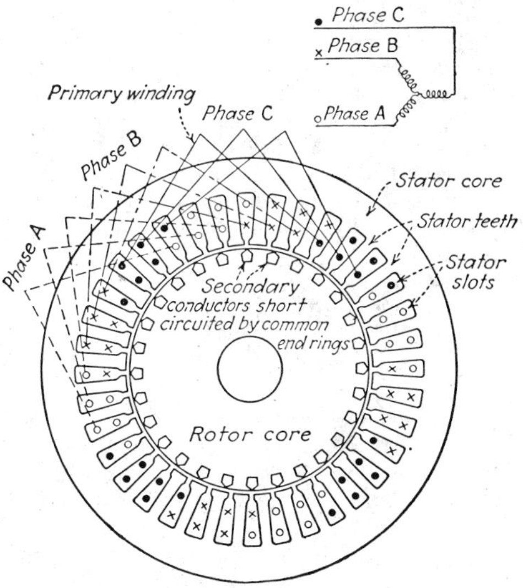

FIG. 18-11. Section of squirrel-cage induction motor, 3-phase, four-pole, 8/9-pitch stator windings.

Inverted constructions, with the primary winding on the rotor, are sometimes used, or single- or 2-phase instead of 3-phase windings; and numerous variations in arrangement are possible.

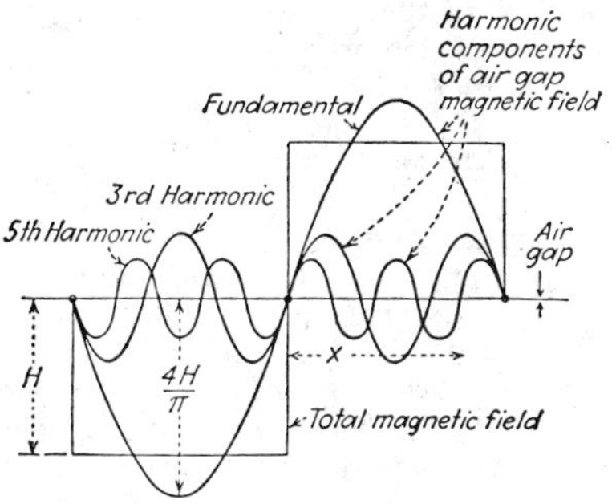

53. The Revolving Field. The key to understanding the induction motor is a thorough comprehension of the revolving magnetic field.

The rectangular wave in Fig. 18-12 represents the mmf, or field distribution, produced by a single full-pitch coil, carrying H At. The air gap between stator and rotor is assumed to be uniform, and the effects of slot openings are neglected. To calculate the resultant field produced by the entire winding, it is most convenient to analyze the field of each single coil into its harmonic components, as indicated in Fig. 18-12 or expressed by the following equation:

FIG. 18-12. Magnetic field produced by a single coil.

$$\mathrm{H} = \frac{4\mathrm{H}}{\pi}\left(\sin x + \frac{1}{3}\sin 3x + \frac{1}{5}\sin 5x + \frac{1}{7}\sin 7x + \cdots\right) \tag{18-16}$$

When two such fields produced by coils in adjacent slots are superposed, the two fundamental sine-wave components will be displaced by the slot angle θ, the third-harmonic components by the angle 3θ, the fifth harmonics by 5θ, etc. Thus, the higher-harmonic components in the resultant field are relatively much reduced as compared with the fundamental. By this effect of distributing the winding in several slots for each phase belt, and because of the further reductions due to fractional pitch and to phase connections, the harmonic fields in a normal motor are reduced to negligible values, leaving only the fundamental sine-wave components to be considered in determining the operating characteristics.

The alternating current flowing in the winding of each phase therefore produces a sine-wave distribution of magnetic flux around the periphery, stationary in space but varying sinusoidally in time in synchronism with the supply frequencies. Referring to Fig. 18-13a, the field of phase A at an angular distance x from the phase axis may be represented as an alternating phasor $I \cos x \cos \omega t$ but may equally well be considered as the resultant of two phasors constant in magnitude but revolving in opposite directions at synchronous speed:

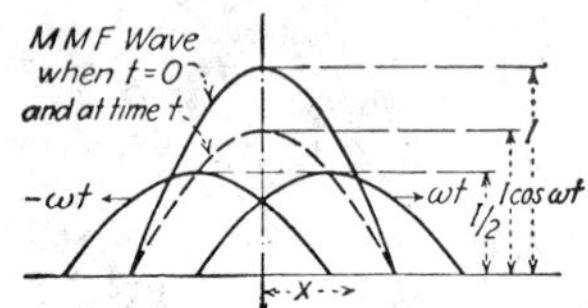

FIG. 18-13a. Resolution of alternating wave into two constant-magnitude waves revolving in opposite directions.

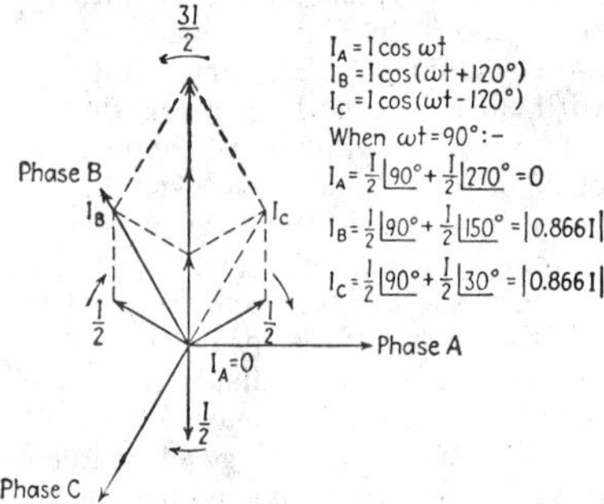

FIG. 18-13b. Resolution of alternating emf of each phase into oppositely revolving constant-magnitude components, shown at instant when phase A current is zero ($\omega\tau = 90°$).

$$I \cos x \cos \omega t = \frac{I}{2} [\cos (x - \omega t) + \cos (x + \omega t)] \tag{18-17}$$

Each of the right-hand terms in this equation represents a sine-wave field revolving at the uniform rate of one pole pitch, or 180 elec deg, in the time of each half cycle of the supply frequency. The synchronous speed N_s of a motor is therefore given by

$$N_s = \frac{120f}{P} \qquad \text{(r/min)} \tag{18-18}$$

where f = line frequency in cycles per second and P = number of poles of the winding.

Considering phase A alone (Fig. 18-13b), two revolving fields will coincide along the phase center line at the instant its current is a maximum. One-third of a cycle later, each will have traveled 120 elec deg, one forward and the other backward, the former lining up with the axis of phase B and the latter with the axis of phase C. But at this moment, the current in phase B is a maximum, so that the forward-revolving B field coincides with the forward A field, and these two continue to revolve together. The backward B field is 240° behind the backward A field, and these two remain at this angle, as they continue to revolve. After another third of a cycle, the forward A and B fields will reach the phase C axis, at the same moment that phase C current becomes a maximum. Hence, the forward fields of all 3-phases are directly additive, and together they create a constant-magnitude sine-wave-shaped synchronously revolving field with a crest value $\frac{3}{2}$ the maximum instantaneous value of the alternating field due to one phase alone. The backward-revolving fields of the 3 phases are separated by 120°, and their resultant is therefore zero so long as the 3-phase currents are balanced in both magnitude and phase.

If a 2-phase motor is considered, it will have two 90° phase belts per pole instead of three 6° phase belts, and a similar analysis shows that it will have a forward-revolving constant-magnitude field with a crest value equal to the peak value of 1 phase alone and will have zero backward-revolving fundamental field. A single-phase motor will have equal forward and backward fields and so will have no tendency to start unless one of the fields is suppressed or modified in some way.

While the harmonic-field components are usually negligible in standard motors, it is important to the designer to recognize that there will always be residual harmonic-field values which may cause torque irregularities and extra losses if they are not minimized by an adequate number of slots and correct winding distribution. An analysis similar to that given for the fundamental field shows that in all cases the harmonic fields corresponding to the number of primary slots (seventh and nineteenth in a nine-slot-per-pole motor) are important and that the fifth and seventh harmonics on 3-phase, or third and fifth on 2-phase, may also be important.

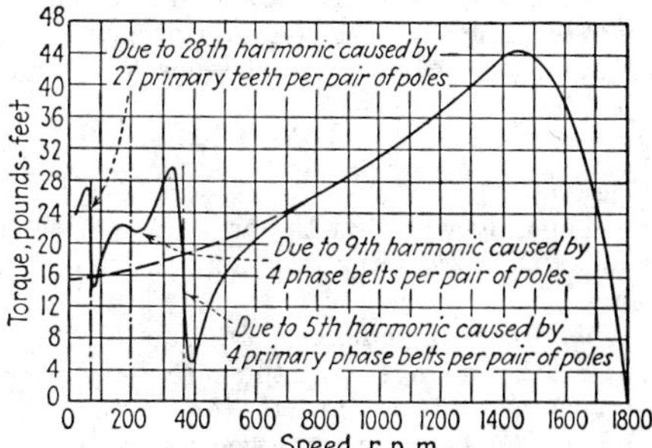

FIG. 18-14. Speed-torque curve of 2-phase motor showing harmonic torques.

The third-harmonic fields and all multiples of the third are zero in a 3-phase motor, since the mmfs of the 3-phases are 120° apart for both backward and forward components of all of them. Finally, therefore, a 3-phase motor has the following distinct fields:

a. The fundamental field with P poles revolving forward at speed N_s.

b. A fifth-harmonic field with 5 P poles revolving backward at speed $N_s/5$.

c. A seventh-harmonic field with 7 P poles revolving forward at speed $N_s/7$.

d. Similar thirteenth, nineteenth, twenty-fifth, etc., forward-revolving and eleventh, seventeenth, twenty-third, etc., backward revolving harmonic fields.

Figure 18-14 shows a test speed-torque curve obtained on a 2-phase squirrel-cage induction motor with straight (unspiraled) slots. The torque dips due to three of the forward-revolving fields are clearly indicated. These harmonic phenomena have been treated in numerous technical articles.[1]

54. Torque and Slip. When the rotor is stationary, the revolving magnetic field cuts the short-circuited secondary conductors at synchronous speed and induces in them line-frequency currents. To supply the secondary IR voltage drop, there must be a component of voltage in time phase with the secondary current, and the secondary current, therefore, must lag in space position behind the revolving air-gap field. A torque is then produced corresponding to the product of the air-gap field by the secondary current times the sine of the angle of their space-phase displacement.

At standstill, the secondary current is equal to the air-gap voltage divided by the secondary impedance at line frequency, or

$$I_2 = \frac{E_2}{Z_2} = \frac{E_2}{R_2 + jX_2} \tag{18-19}$$

where R_2 = effective secondary resistance and X_2 = secondary leakage reactance at primary frequency.

The speed at which the magnetic field cuts the secondary conductors is equal to the difference between the synchronous speed and the actual rotor speed. The ratio

[1] DREESE, E. E. Synchronous Motor Effects in Induction Machines; *Trans. AIEE*, July, 1930, Vol. 49, p. 1033.
KRON, G. Slot Combinations of Induction Motors; *Elec. Eng.*, December, 1931, Vol. 50, p. 937.
LIWSCHITZ, M. M. Harmonics in Induction Motors; *Trans. AIEE*, November, 1942, Vol. 61, p. 797.
MORRILL, W. J. Harmonic Theory of Noise in Induction Motors; *Trans. AIEE*, August, 1940, Vol. 59, p. 474.
ALGER, P. L., KU, Y. H., and PAN, C. H. T. Speed-torque Calculations for Induction Motors with Part Windings; *Trans. AIEE*, Power Apparatus and Systems, April, 1954, No. 11, p. 151.

of the speed of the field relative to the rotor, to synchronous speed, is called the slip s,

$$s = \frac{N_s - N}{N_s}$$

or

$$N = (1 - s)N_s \tag{18-20}$$

where N = actual and N_s = synchronous rotor speed.

As the rotor speeds up, with a given air-gap field, the secondary induced voltage and frequency both decrease in proportion to s. Thus, the secondary voltage becomes sE_2, and the secondary impedance $R_2 + jsX_2$, or

$$I_2 = \frac{sE_2}{R_2 + jsX_2} = \frac{E_2}{(R_2/s) + jX_2} \tag{18-21}$$

The only way that the primary is affected by a change in the rotor speed, therefore, is that the secondary resistance as viewed from the primary varies inversely with the slip.

55. Rotor Impedance. In practice, the effective secondary resistance and reactance, R_2 and X_2, change with the secondary frequency, owing to the varying "skin effect," or current shifting into the outer portion of the conductors, when the frequency is high. This effect is employed to make the resistance, and therefore the torque, higher at low motor speeds, by providing a double cage, or deep-bar construction, as shown in Fig. 18-15. The leakage flux between the outer and inner bars makes the inner-bar reactance high, so that most of the current must flow in the outer bars at standstill, when the frequency is high. At full speed, the secondary frequency is very low, and most of the current flows in the inner bars, owing to their lower resistance. Or the secondary impedance may be controlled by employing a wound rotor and connecting an adjustable resistance across the rotor slip rings.

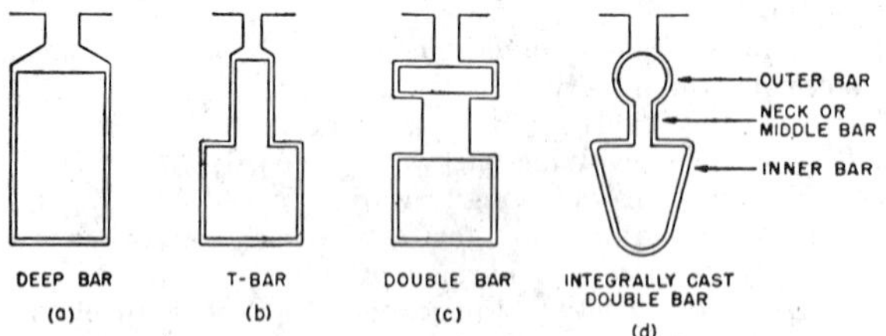

FIG. 18-15. Alternative forms of squirrel-cage rotor bars.

56. Circle Diagram.[1] The voltage-current relations of the polyphase induction machine are roughly indicated by the circuit of Fig. 18-16. The magnetizing current I_M proportional to the voltage and lagging 90° in phase is nearly constant over the operating range, while the load current varies inversely with the sum of primary and secondary impedances. As the slip s increases, the load current and its angle of lag behind the voltage both increase, following a nearly circular locus. Thus, the circle (Heyland) diagram (Fig. 18-17) provides a clear picture of the motor behavior.

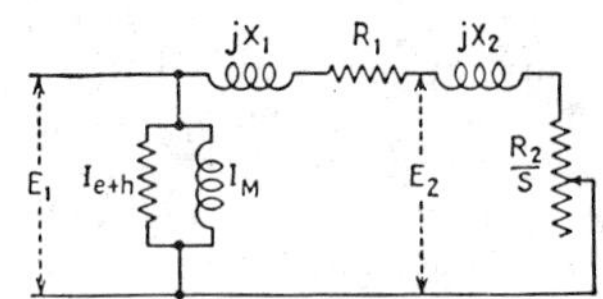

FIG. 18-16. Equivalent circuit for circle diagram.

The data needed to construct the diagram are the magnitude of the no-load current ON and of the blocked-rotor current OS and their phase angles with reference to the line voltage OE. A circle with its center on the line NU at right angles to OE is drawn to pass through N and S. Each line on the diagram can be measured directly in amperes, but it also represents voltamperes or power, when multiplied by the phase voltage times number of phases. The line VS drawn parallel to OE represents the total motor input with blocked rotor, and on the same scale VT represents the corresponding primary I^2R loss. Then ST represents the power input to the rotor at standstill, which, divided by the synchronous speed, gives the starting torque.

[1] JEFFREY, F. Circle Diagram and the Induction Motor; *Allis-Chalmers Elec. Rev.*, September, 1939, Vol. 4, p. 5, Serial.

BEHREND, B. A. "The Induction Motor and Other Alternating Current Motors: Their Theory and Principles of Design"; New York, McGraw-Hill Book Company, 1921.

At any load point A, OA is the primary current, NA the secondary current, and AF the motor input. The motor output is AB, the torque $\times$ (synchronous speed) is AC, the secondary I^2R loss is BC, primary I^2R loss CD, and no-load copper loss plus core loss DF. The maximum power-factor point is P, located by drawing a tangent to the circle from O. The maximum output and maximum torque points are similarly located at Q and R by tangent lines parallel to NS and NT, respectively.

The diameter of the circle is equal to the voltage divided by the standstill reactance or to the blocked-rotor current value on the assumption of zero resistance in both windings. The maximum torque of the motor, measured in kilowatts at synchronous speed RY, is equal to a little less than the radius of the circle multiplied by the voltage OE.

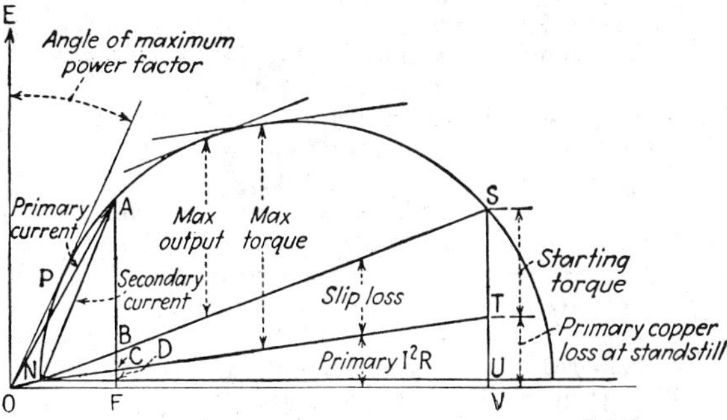

FIG. 18-17. Circle diagram of polyphase induction motor.

While convenient for visualizing overall performance, the circle diagram is too inaccurate for most purposes. The magnetizing current is not constant, but decreases with load, owing to the primary impedance drop, and all the circuit "constants" vary over the operating range, owing to magnetic saturation and skin effect. Hence, designers generally employ the (Steinmetz) equivalent circuit for performance calculations, using impedances adjusted to fit the conditions at each point.

57. Equivalent Circuit.[1] Figure 18-18 shows the polyphase motor circuit usually employed for accurate work. The advantages of this circuit theory over the vector- and circle-diagram methods are that it facilitates the derivation of simple formulas or charts for calculating torque, power factor, and other motor characteristics and that it enables impedance changes due to saturation or multiple squirrel cages to be readily taken into account. Also, to those familiar with such circuit calculations, it provides any necessary degree of accuracy with a minimum expenditure of time for calculations.

Impedances

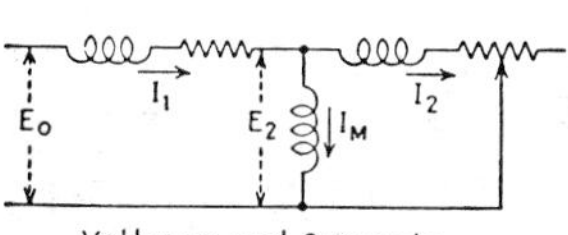

Voltages and Currents

FIG. 18-18. Equivalent circuit of polyphase induction motor.

Formulas for calculating the constants from test data are given in Form B, Par. **75,** and their definitions are given in Form A below.

Inspection of the circuit reveals several simple relationships which are useful for estimating purposes. The maximum current occurs at standstill and is somewhat less than E/X. Maximum torque occurs when $s = R_2/X$, approximately, at which point the current is roughly 70% of the standstill current. Hence, the maximum torque is approximately equal to $E^2/2X$. This gives the basic rule that the percent maximum torque of a low-slip polyphase motor at a constant impressed voltage is about half the percent starting current.

By choosing the value of R_2, the slip at which maximum torque occurs can be fixed at any desired value. The maximum-torque value itself is affected, not by changes in R_2, but only by changes in X and to a slight degree by changes in X_M.

[1] ALGER, P. L. Induction Motor Performance Calculations; *Trans. AIEE*, July, 1930, Vol. 49, p. 1055. "The Nature of Polyphase Induction Machines"; New York, John Wiley & Sons, Inc., 1951, Chap. 5.

VEINOTT, C. G. Performance Calculations on Induction Motors; *Trans. AIEE*, September, 1932, Vol. 51, p. 743.

Form A

Definitions of Equivalent-circuit Constants

Unless otherwise noted, all quantities except watts, torque, and power output are per phase for 2-phase motors and per phase Y for 3-phase motors.

E_0 = impressed voltage (volts) = line voltage ÷ $\sqrt{3}$ for 3-phase motors
I_1 = primary current (amperes)
I_2 = secondary current in primary terms (amperes)
I_M = magnetizing current (amperes)
R_1 = primary resistance (ohms)
R_2 = secondary resistance in primary terms (ohms)
R_0 = resistance at primary terminals (ohms)
X_1 = primary leakage reactance (ohms)
X_2 = secondary leakage reactance (ohms)
$X = X_1 + X_2$
X_0 = reactance at primary terminals (ohms)
X_M = magnetizing reactance (ohms)
Z_1 = primary impedance (ohms)
Z_2 = secondary impedance in primary terms (ohms)
Z_0 = impedance at primary terminals (ohms)
Z = combined secondary and magnetizing impedance (ohms)
s = slip (expressed as a fraction of synchronous speed)
N = synchronous speed (revolutions per minute)
m = number of phases
f = rated frequency (cycles per second)
f_t = frequency used in locked-rotor test
T = torque (foot-pounds)
W_0 = watts input
W_H = core loss (watts)
W_F = friction and windage (watts)
W_{RL} = running light watts input
W_s = stray-load loss (watts)

$$Y = G - jB = \frac{1}{Z} = \frac{1}{R + jX} = \frac{R}{Z^2} - \frac{jX}{Z^2}$$

$$|Z| = \sqrt{R^2 + X^2}$$

The magnetizing reactance X_M is usually eight or more times as great as X, while R_1 and R_2 are usually much smaller than X, except in the case of special motors designed for frequent-starting service. These relationships enable the solution of the equivalent circuit to be expressed in the form of an infinite series, of which only the first few terms normally need be considered.

To facilitate the derivation of formulas and charts from which any desired characteristics of a polyphase motor can be determined, it is convenient to make use of the following symbols:

$$a = \frac{I_M}{I_1} = \text{ratio of no-load current to primary current at assumed load}$$

$$b = \frac{I_1 X}{E_0} = \text{ratio of apparent leakage reactance drop at assumed load to impressed primary voltage}$$

$$ab = \frac{I_M X}{E_0} = \frac{X}{X_M + X_1} = \text{"leakage factor"}$$

58. Power Factor. The value of ab, the so-called "leakage factor" of the motor, is a very important constant which determines the maximum motor power factor. The power-factor chart (Fig. 18-20) is derived by equivalent-circuit calculations for a series of motors having different leakage-factor values. This particular chart assumes a value of primary resistance drop in each case equal to 0.04 at the load point at which $a = 2b$, which is fairly representative of general-purpose polyphase motors in the size range 10 to 100 hp. For larger motors the chart gives power-factor values up to 2% high, while for smaller motors the chart values may be low by 1 to 3%, unless correction is made for the particular value of primary resistance used.

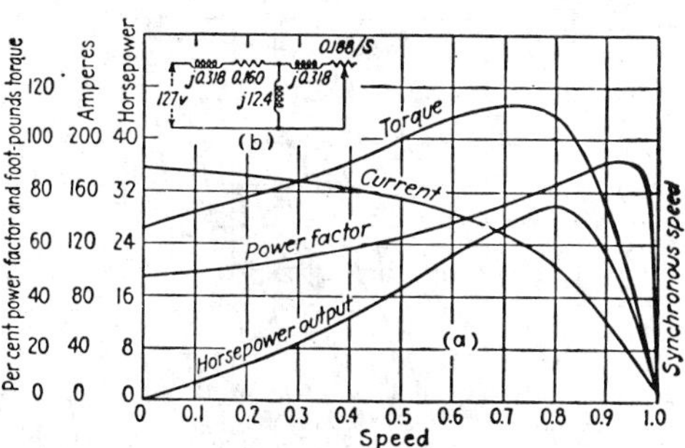

Fig. 18-19. (*a*) Performance curves of polyphase induction motor; (*b*) equivalent circuit.

For example, considering the motor of Fig. 18–19,

$$ab = \frac{0.636}{12.4 + 0.32} = 0.050$$

The maximum power factor occurs when the reactive voltamperes consumed in the leakage reactance are equal to the magnetizing voltamperes.

At maximum power factor,

$$b = a\left(1 + \frac{ab}{4}\right) \qquad \text{(approx.)} \qquad (18\text{-}22)$$

or

$$I_1 = \sqrt{\frac{I_M E_0}{X}}\left(1 + \frac{ab}{8}\right) \qquad \text{(approx.)} \qquad (18\text{-}23)$$

Figure 18-20 gives a maximum power factor for $ab = 0.050$ equal to 0.916. At this point, $a = b = \sqrt{0.050} = 0.223$, and the current is

$$I_1 = \frac{bE_0}{X} = \frac{(0.223)(127)}{0.636} = 44.5\ \text{A}$$

These values check closely with the curves of Fig. 18-19.

A rough formula for the power factor as a function of primary current is

$$\text{Power factor} = 1 - \frac{(a + b)^2}{2} + 3a^2b^2 \qquad \text{(very approx.)} \qquad (18\text{-}24)$$

59. Torque and Output Formulas. The maximum torque of a polyphase induction motor occurs when the slip is equal to

$$\text{Slip at max. } T = \frac{R_2}{\sqrt{R_1^2 + X^2}}\left(1 + \frac{ab}{4}\right) \qquad \text{(approx.)} \qquad (18\text{-}25)$$

The maximum torque itself is equal to

$$T_{\max} = \frac{3.52mE_0^2}{N_s(R_1 + \sqrt{X^2 + R_1^2})}\left(1 - \frac{3ab}{4}\right) \qquad \text{(ft·lb, approx.)} \qquad (18\text{-}26)$$

The maximum output occurs when the slip is

$$\text{Slip at max. output} = \frac{R_2}{R_2 + \sqrt{X^2 + (R_1 + R_2)^2}}\left(1 + \frac{ab}{4}\right) \qquad \text{(approx.)} \qquad (18\text{-}27)$$

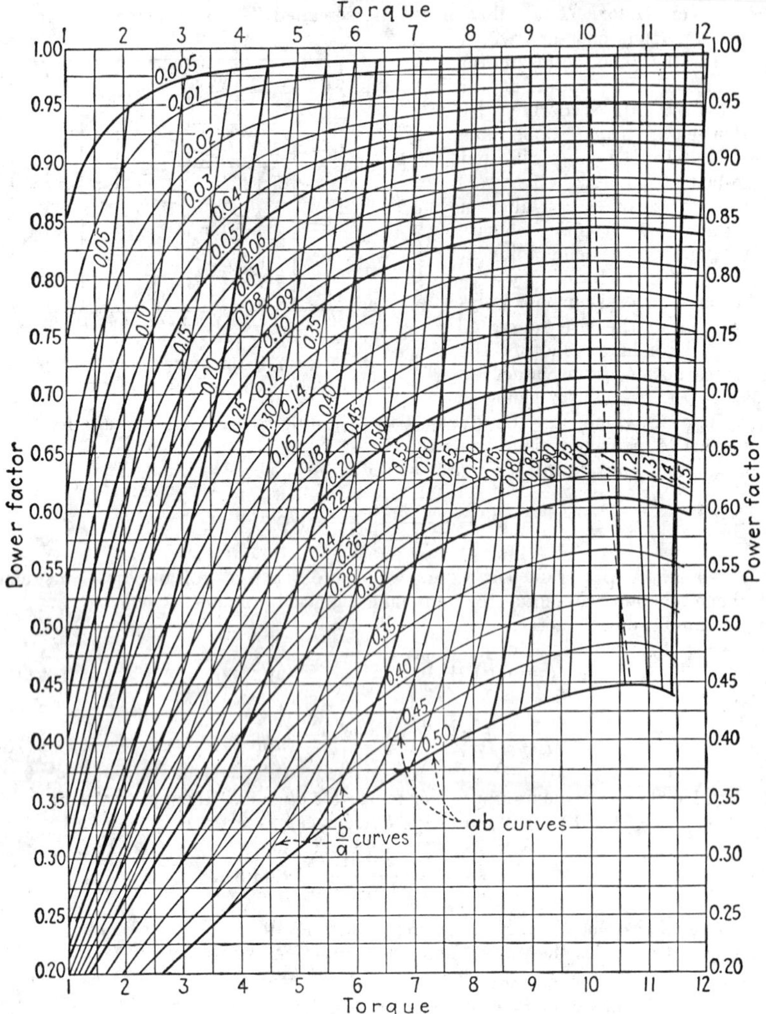

FIG. 18-20. Power-factor chart for polyphase induction motor.

and the maximum output is equal to

$$\text{Max. output} = \frac{mE_0^2}{1{,}500[(R_1 + R_2) + \sqrt{X^2 + (R_1 + R_2)^2}]}\left(1 - \frac{3ab}{4}\right) \quad \text{(hp, approx.)} \tag{18-28}$$

At standstill, the slip is 1, and the torque is

$$\text{Starting torque} = \frac{7.04mE_0^2R_2}{N_s[X^2 + (R_1 + R_2)^2]}\left(1 - \frac{ab}{2}\right) \quad \text{(ft·lb)} \tag{18-29}$$

At any speed $(1 - s)$ times synchronous speed, a fraction s of the secondary power input is consumed in secondary copper loss, and the remainder $1 - s$ is shaft output. When the motor is driven backward against its torque, the value of s exceeds unity and both the electrical input to the rotor and the shaft input are consumed in secondary copper loss. If the motor is driven above its synchronous speed, its slip and torque become negative and it behaves as a generator supplying electric power to the system to which it is connected.

The energy dissipated in the secondary copper loss of the rotor during its starting period is equal to the total energy stored in the flywheel effect of the motor rotor and its shaft-connected load, plus whatever amount is incurred in overcoming the friction or load torque during the starting period. Thus, excessive motor heating occurs if a motor is required to start a high-inertia load or to start or reverse frequently.

60. Secondary Resistance and Slip.[1] For loads less than half the maximum output, the slip is closely proportional to the load torque. A relation between current and slip, which is accurate within a few percent over the range from ½ to 1½ times normal load, is expressed by the following equation:

$$\frac{IR_2}{E} = s\left(1 + \frac{a^2 - b^2}{2} - \frac{IR_1}{E} + 2a^2b^2\right) \tag{18-30}$$

This equation may be used to determine R_2 when one or more points on a current-slip curve are established.

All the equations (18-24) to (18-30) inclusive are simply approximations useful for estimating or quick-calculation purposes. When exact results are required, equivalent-circuit calculations should be made.

DESIGN OF INDUCTION MACHINES

61. Armature windings[2] of induction machines are designed to give low-leakage reactance and as nearly sinusoidal a current distribution around the periphery as possible. Otherwise, with the relatively small length of air gap between stator and rotor, the extra leakage fluxes produced would cause reduced output and much increased core losses. In consequence, two-layer fractional pitch windings are normally used, with a relatively large integral number of slots per pole, usually 9 or 12. It is also usual to connect the windings of large motors with two or more circuits in parallel in each phase. The same voltage being impressed on each circuit ensures equality of the magnetic fluxes linking them, preserving a balanced magnetic pull, despite any equalities of air gap. For 3-phase power supply, windings may be connected either Y or Δ.

The "winding factor," or ratio of effective to total number of turns, is equal to the product of the pitch factor K_p by the distribution factor K_d. K_p is the ratio of the voltage produced in the actual coil spanning $100p$ % of one pole pitch to the voltage in a full-pitch coil and is given by the equation

$$K_p = \sin\frac{p\pi}{2} \tag{18-31}$$

K_d is the ratio of the voltage produced in the actual series-connected coils of one circuit to the voltage that would be produced if all the coil voltages were in time phase. For a normal winding, it is given by the equation

$$K_d = \frac{\sin(\pi/2m)}{n\sin(\pi/2mn)} \qquad \text{or approx.}\ \frac{2m}{\pi}\sin\frac{\pi}{2m} \tag{18-32}$$

where m = number of phase belts per pole and n = number of coils per phase belt. The usual values of K_pK_d are given in Table 18-2.

The number of turns in the primary winding is fixed by the required voltage and

[1] AGER, R. W. Transient Overspeeding of Induction Motors; *Trans. AIEE*, December, 1941, Vol. 60, p. 1030.

MORGAN, T. H., BROWN, W. E., and SCHUMER, A. J. Induction-motor Characteristics at High Slip; *Trans. AIEE*, August, 1940, Vol. 59, p. 474.

CHAPMAN, F. T. "Study of the Induction Motor"; New York, John Wiley & Sons, Inc., 1930.

LYON, W. V. "Applications of the Method of Symmetrical Components"; New York, McGraw-Hill Book Company, 1937.

CONCORDIA, C. Induction Motor Damping and Synchronizing Torques; *Trans. AIEE*, 1952, Vol. 71, Pt. III, p. 364.

[2] ADAMS, C. A., CABOT, W. K., and IRVING, G., JR. Fractional Pitch Windings for Induction Motors; *Trans. AIEE*, 1907, Vol. 26, p. 1485.

ADAMS, C. A. Design of Induction Motors; *Trans. AIEE*, 1905, Vol. 24, p. 649.

the permissible magnetic flux,

$$N = \frac{100E}{4.44fK_pK_d\phi} \qquad \text{(turns in series/phase)} \qquad (18\text{-}33)$$

where E = rms volts between terminals of one phase, f = frequency in cycles per second, and ϕ = flux per pole in megalines.

The flux density in core and teeth cannot exceed 90,000 to 100,000 lines/in² without a material increase in magnetizing current due to magnetic saturation, and the value of ϕ is limited accordingly, for a given core section.

Table 18-2. Winding Factor = K_pK_d

Pitch p	3-phase 60-deg phase belts	2-phase 90-deg phase belts	3-phase 120-deg phase belts
1.00	.956	.902	.827
.95	.953	.899	.824
.90	.944	.890	.816
.85	.930	.877	.804
.80	.909	.858	.786
.75	.883	.834	.764
.70	.852	.804	.737
.65	.815	.769	.705
.60	.774	.730	.669
.55	.727	.686	.629
.50	.676	.638	.585

62. Winding Connections.[1] From the foregoing, the following simple rules to be followed in connecting induction-motor windings for different voltages, numbers of poles, or phases are derived:

Assume that a winding has S slots and is to be connected for 3 phases, P poles, with an impressed voltage V and a frequency f. The coil pitch p is preferably chosen to be as nearly equal to ⅚ as possible, giving a slot span of $5S/6P$, approximately; and the corresponding value of winding factor is found from Table 18-2. A Y-connection will normally be chosen, giving $E = V/\sqrt{3}$. A suitable value of flux per pole ϕ is determined from the core section and the desired flux density or by calculation from Eq. (18-33), if the number of turns in a previous winding on the same core is known. Equation (18-33) then gives the value of N for the new conditions, and the number of turns in each coil is made equal to $3cN/S$, where c = number of parallel circuits per phase. The values of c and p are adjusted to make the turns per coil a convenient whole number, or a Δ-connection may be used to attain this purpose.

(A) Three-phase, two speed, one winding constant horsepower

SPEED	L_1	L_2	L_3	OPEN	TOGETHER
1 LOW	T_1	T_2	T_3		T_4, T_5, T_6
2 HIGH	T_6	T_4	T_5	T_1, T_2, T_3	

(B) Three-phase, two-speed, one winding constant torque

SPEED	L_1	L_2	L_3	OPEN	TOGETHER
1 LOW	T_1	T_2	T_3	ALL OTHERS	
2 HIGH	T_6	T_4	T_5		T_1, T_2, T_3

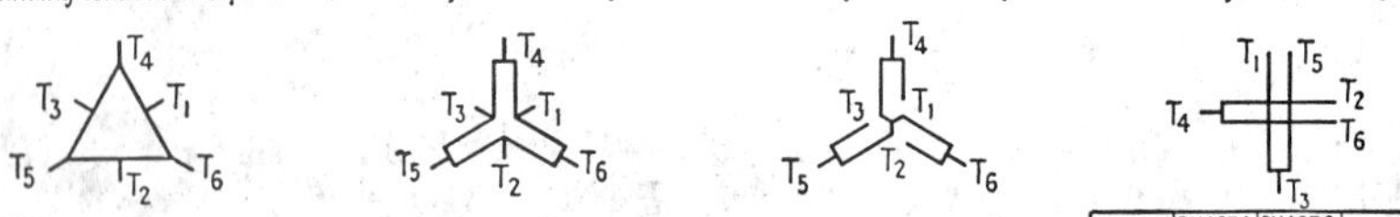

(C) Three-phase, two-speed, one-winding variable torque

SPEED	L_1	L_2	L_3	OPEN	TOGETHER
1 LOW	T_1	T_2	T_3	ALL OTHERS	
2 HIGH	T_6	T_4	T_5		T_1, T_2, T_3

(D) Two-phase, two-speed one winding variable torque

	PHASE 1		PHASE 2		
SPEED	L_1	L_3	L_2	L_4	OPEN
1 LOW	T_1	T_5	T_2	T_6	T_3, T_4
2 HIGH	T_1, T_5	T_3	T_2, T_6	T_4	

FIG. 18-21. Stator connections of 2 : 1-speed pole-changing single-winding induction motors Incoming power lines are denoted by L_1, L_2, L_3, etc. Motor terminals are denoted by T_1, T_2, T_3, etc. When the motor is wired to the controller according to the diagram that should be furnished by the controller manufacturer, the operation of the controller will make connections according to one of the figures shown, thus giving the desired speed.

[1] DUDLEY, A. M. "Connecting Induction Motors," 3d ed.; New York, McGraw-Hill Book Company, 1936.

Circuit balance requires that P/c and $S/3c$ each be an integer. The number of slots per phase belt $S/3P$ is normally an integer also, but fractional values are occasionally used, in which case the irregularities between phase belts must be spaced as uniformly as possible around the periphery to minimize extra reactance and losses from harmonic fluxes.[1]

It is possible to arrange a normal 3-phase winding for reconnection with twice as many poles, if six leads are brought out, as indicated in Fig. 18-21. On the high-speed connection, there are two, or a multiple of two, circuits in each phase, alternate poles being connected in series in each circuit, and each phase belt spans 60 elec deg. By reversing half of the circuits in each phase and connecting them in series with the other half of the circuits, a consequent pole winding is produced, with twice as many poles and 120° phase belts.

It is customary to provide general-purpose (1 to 200 hp) polyphase motors with dual voltage windings, adapted for connecting for either 220- or 440-V power supply. Figure 18-22 shows the connections used, with nine leads brought out.

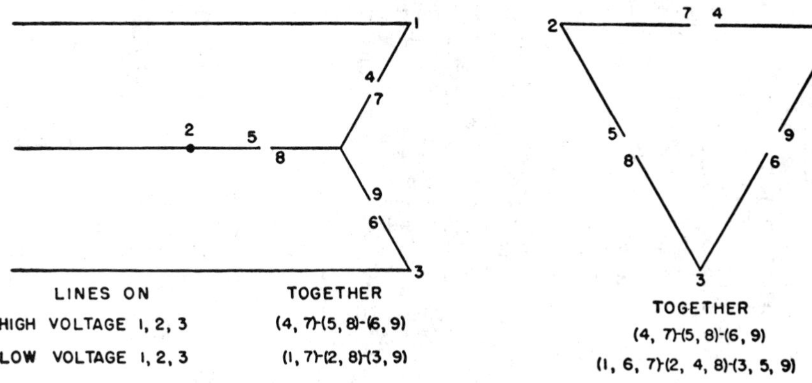

FIG. 18-22. Dual-voltage-winding connections.

63. Insulation.[2] For motors smaller than about 100 hp at 1,800 r/min, it is usual to employ "random-wound" stators, with semiclosed slots, and coils formed of many turns of enameled (synthetic-resin-coated) round copper wire. Additional asbestos or glass-fiber coverings may be used for high-temperature service. Composite paper-mylar or similarly formed slot tubes are used to protect against failures to ground, and varnished cloth, paper, or Mylar separators are placed between coils to prevent phase-to-phase failures. The whole winding is given one or more varnish dips and bakes to provide additional moisture resistance and end-winding protection.

In the larger motors, preformed coil windings are usual, with open slots and rectangular conductors. The insulated wires are wound on a form, the slot portion is hot-molded to minimum dimensions, and the coils are then pulled out to shape, wrapped with tape or sheet insulation, and finally varnish-dipped and baked before assembly.

[1] CALVERT, F. J. Amplitudes of Magnetomotive Forces for Fractional Slot Windings; *Trans. AIEE*, 1938, Vol. 57, p. 777.

[2] HERMAN, C. J. Motor Insulation Life as Measured by Accelerated Tests and Dielectric Fatigue; *Trans. AIEE*, 1953, Vol. 72, Pt. III, p. 986.

LEAPE, C. G., MCDONALD, J., and GIBSON, G. P. A Method of Evaluating Insulation Systems in Motors; *Trans. AIEE*, 1953, Vol. 72, Pt. III, p. 793.

BUSH, W. J., and DEXTER, J. F. Thermal Endurance of Silicone Magnet Wire Evaluated by Motor Test; *Trans. AIEE*, Power Apparatus and Systems, August, 1954, No. 13, p. 1005.

GAIR, T. J. Insulation Systems for Random-wound Motors Evaluated by Motorette Tests; *Trans. AIEE*, Power Apparatus and Systems, February, 1955, p. 1702.

AIEE Working Group on Evaluation of Insulation for Rotating Machines: New Performance Standards for Electrical Insulation of Rotating Machines; *Trans. AIEE*, Power Apparatus and Systems, December, 1954, No. 15, p. 1542.

DEKIEP, J., HILL, L. R., and MOSES, G. L. The Application of Silicone Resins to Insulation for Electric Machinery; *Trans. AIEE*, March, 1945, Vol. 64, p. 94.

Up to 6,600 V, varnished cloth or mylar sheet wrappings in the slot portion are normal; for higher voltages, high temperatures, and large machines, mica tape is usually employed. Insulation to ground is designed to withstand the standard high-potential test of two times normal plus 1,000 V. Insulation between turns is designed with a higher factor of safety to withstand the nonuniform voltage distribution that occurs during switching transients. Mechanical bracing is provided for large motor end windings, to withstand the high magnetic forces due to starting currents.

It is now becoming customary to evaluate the performance of the complete motor insulation system in accordance with AIEE test codes,[1] especially to prove the suitability of new synthetic materials that are coming into use in place of the cotton, paper, and other natural materials previously used. Small and low-voltage machines normally employ Class A insulation (limiting temperature 105°C), while large and high-voltage machines employ Class B insulation, with a limiting hottest-spot temperature of 130°C. Silicone varnishes, new wire enamels, and other synthetic materials that withstand higher temperatures are available, but these are somewhat limited in use for reasons of cost or novelty and various practical objections to high-temperature machines.

64. Reactance.[2] The two most important constants of the induction-motor equivalent circuit are the magnetizing reactance X_M and the total leakage reactance X. The no-load current, which is practically the same as the magnetizing current, is

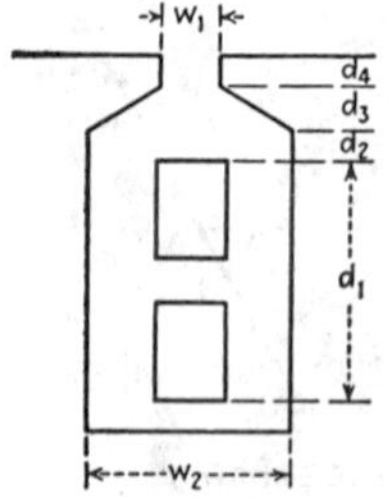

Fig. 18-23. Slot dimensions.

$$I_0 = \frac{E}{X_M + X_1} \qquad \text{(amperes)} \qquad (18\text{-}34)$$

If the length of radial air gap is g in inches, the air-gap diameter D in inches, and the axial core length L in inches, the relation between flux per pole and magnetizing current, with magnetic saturation neglected, is

$$\phi = \frac{5.75 m I_M D L N K_p K_d}{P^2 g_e \times 10^6} \qquad \text{(megalines)} \qquad (18\text{-}35)$$

The effective gap length g_e is equal to the actual gap length g multiplied by the fringing coefficient (>1) derived from Fig. 18-17.

Combining Eqs. (18-33) and (18-35) to eliminate ϕ,

$$X_M = \frac{E}{I_M} = \frac{25.5 f m D L N^2 K_p^2 K_d^2}{P^2 g_e \times 10^8} \qquad (\Omega/\text{phase}) \qquad (18\text{-}36)$$

Equations (18-35) and (18-36) neglect the ampere-turns consumed in the iron, which in practice amount to 10 to 30% of the air-gap ampere-turns for usual values of flux density, at rated voltage. The actual no-load current to be expected will therefore be 1.10 to 1.30 times that indicated by Eq. (18-36), or even higher at overvoltages or for highly saturated magnetic designs.

The leakage reactance X is a measure of all the nonuseful magnetic flux produced by a current in either winding, i.e., all flux producing no fundamental frequency voltage in the other winding. It is usually considered as composed of four parts, the primary and secondary slot leakages, the coil-end leakage, and the differential leakage.

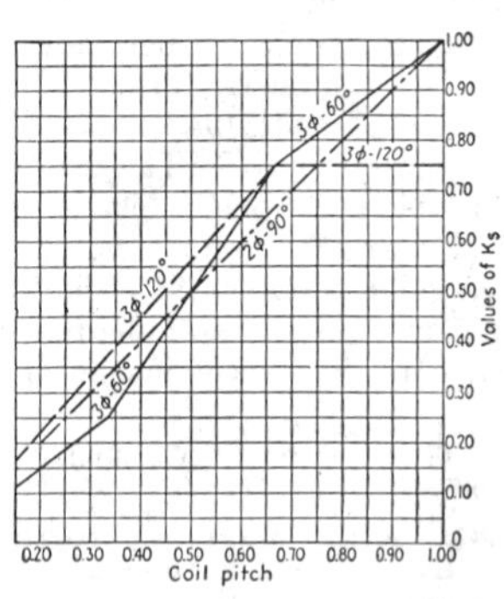

Fig. 18-24. Slot-leakage pitch factor.

[1] AIEE No. 1C Test Code for Evaluation of Systems of Insulating Materials for Random-wound Electric Machinery, January, 1954.

[2] Alger, P. L. Calculation of the Armature Reactance of Synchronous Machines: *Trans. AIEE*, April, 1928, Vol. 47, p. 493.

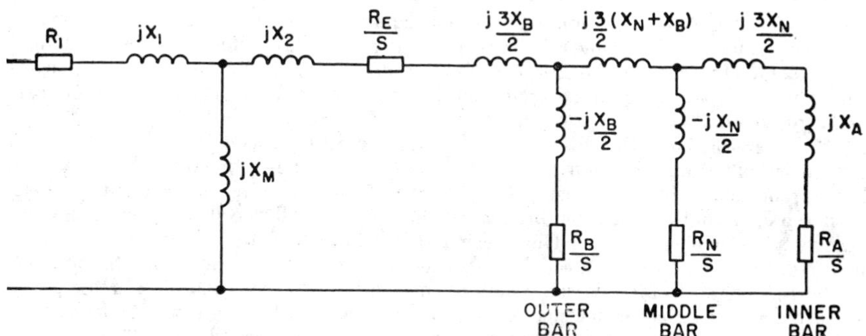

FIG. 18-25. Equivalent circuit of polyphase motor with triple squirrel cage (Fig. 18-15*d*).

In Fig. 18-23, the slot-leakage coefficient is equal to

$$L_s = K_s\left(\frac{d_4}{w_1} + \frac{2d_3}{w_1 + w_2} + \frac{d_2}{w_2} + \frac{d_1}{3w_2}\right) + (1 - K_s)\frac{d_1}{12w_2} \qquad (18\text{-}37)$$

where K_s = a factor dependent on coil pitch, as indicated in Fig. 18-24.

To refer the secondary slot reactance to the primary, it is necessary to multiply its leakage coefficient by the square of the ratio of the secondary to the primary effective current per slot for equal air-gap ampere-turns, so that the combined reactance of primary and secondary slots is given by

$$X_s = \frac{8fmLN^2}{10^7S_1}\left(L_{s1} + \frac{S_1K_{p1}^2K_{d1}^2}{S_2K_{p2}^2K_{d2}^2}L_{s2}\right) \qquad (\Omega/\text{phase}) \qquad (18\text{-}38)$$

For a squirrel-cage secondary winding, the values of K_s, K_{p2}, and K_{d2} are all equal to unity. If a deep-bar or double squirrel-cage secondary winding is used, the secondary slot reactance referred to primary will be much larger at full speed than at starting, owing to the concentration of the current in the upper portion of the slot, or low-reactance paths, when the secondary frequency is high. The extra reactance at full speed, ΔX, is usually about equal to the increase of the secondary resistance at starting above its value at speed as determined by the full-load slip.[1]

The end leakage reactance is less subject to exact calculation, as it depends on the length and configuration of both primary and secondary coil ends, but it is approximately represented by

$$X_s = \frac{4fmDN^2}{10^7P^2}K_{s1} \qquad (\Omega/\text{phase}) \qquad (18\text{-}39)$$

The differential leakage reactance is most simply calculated as the difference between the total air-gap reactance and the useful, or fundamental sine-wave, component.[2] Assuming normally well-distributed windings in both primary and secondary, it is approximately given by Eq. (18-40):

$$X_d = 0.83X_M\left(\frac{6K_1 - 1 + 5\sigma^2}{5S_1^2} + \frac{\sigma K_2 - 1}{5S_2^2}\right) \qquad (\Omega/\text{phase}) \qquad (18\text{-}40)$$

where K_1 and K_2 are the ratios of actual to effective air gap, g/g_e, due to the primary and secondary slot openings, respectively; S_1 and S_2 are the primary and secondary slots per pole; and σ is the angle of slot skew, as a fraction of one stator slot pitch.

The sum of X_s, X_e, and X_d equals X. Usually, it is assumed that

$$X_1 = X_2 = 0.5X$$

[1] ALGER, P. L. "The Nature of Polyphase Induction Machines"; New York, John Wiley & Sons, 1951, Chap. 8.

[2] ALGER, P. L., and WEST, H. R. The Air-gap Reactance of Polyphase Machines; *Trans. AIEE*, 1947, *Tech. Paper* 47-209.

for circuit-calculation purposes. For single squirrel-cage or wound-rotor motors, X may be assumed constant over the entire speed range, except for a reduction at high currents due to saturation. For deep-bar or double squirrel-cage motors, however, the value of X will increase materially as the speed increases, owing to decreasing secondary frequency.

Figure 18-25 shows the extended equivalent circuit[1] used to calculate the combined impedance of a multiple squirrel cage (Fig. 18-15). A deep, or oddly-shaped, single bar is represented by considering it as formed of two or more independent sections, each represented by a distinct branch in the circuit. All the flux crossing the slot outside a given bar is included in the horizontal branch of the circuit, as mutual to it and all the inner bars. All the flux crossing below the next to the inmost bar appears in the right-hand branch as self-reactance of the inmost bar. The flux crossing any given bar creates a self-reactance of that bar represented by one-third its depth over width ratio, $d/3w$, and a mutual reactance with all inner bars, $d/2w$. This flux also creates an additional self-reactance of all the inner bars equal to $d/2w$. In the circuit, therefore, the flux crossing any bar appears as a $3X_s$ term in its mutual reactance with all inner bars, an $X_s/2$ term in the self-reactance of the bar itself, and another $3X_s$ term in the reactance of all the inner bars, where X_s is the self-reactance or $d/3w$ term for the particular bar.

65. Resistance. The primary-winding resistance may be simply calculated by the formula

$$R_1 = \frac{3.31mN^2}{10^6} \frac{\text{MLC}}{\text{copper section}} \qquad (\Omega/\text{phase at } 75°\text{C}) \qquad (18\text{-}41)$$

where MLC = mean length of one conductor in inches and copper section = total cross-sectional area of copper, or section per slot times number of slots, in square inches.

The secondary resistance referred to primary is given by the same formula, using rotor values, except for a multiplier equal to the square of the effective turn ratio, and a resistivity factor K_R (ratio of resistivity of secondary conductors to that of copper).

$$R_2 = \frac{3.31mN^2}{10^6} K_R \frac{K_{p1}^2 K_{d1}^2}{K_{p2}^2 K_{d2}^2} \frac{\text{MLC}}{\text{copper section}} \qquad (\Omega/\text{phase at } 75°\text{C}) \qquad (18\text{-}42)$$

In practice, the value of R_2 at starting may be increased materially above that given by Eq. (18-42), by the effects of rotor iron and skin-effect losses.

66. Losses. The primary and secondary currents and corresponding I^2R losses at any load, or slip, value are found by equivalent-circuit calculations. The friction and windage losses are usually 1 to 2% of the rated output, the percent value decreasing with increasing motor size. The no-load core losses are markedly dependent on many features of design, but a rough average figure is 2 to 3% of the rated output. In general, the percent core loss decreases if the slot openings or the ratio of secondary to primary slots is made smaller or if the air-gap length is increased. A large proportion of the core loss and the stray-load losses is normally due to slot-frequency pulsation losses, so that high-quality silicon steel is usually employed for core laminations. Typical values of stray-load loss are 1 to 2% of the output, but values as low as 0.5% and as high as 4% are encountered.

It is interesting to note that all the tooth-frequency core and stray losses are produced by a magnetic drag effect opposing the motion of the rotor teeth through the magnetic field, just as if they were true friction losses. Hence, sufficient slip-frequency secondary current must flow at no load to provide a torque equal to the sum of the friction and windage and the high-frequency core losses, not merely the friction-windage losses alone.[2]

67. Temperature Rise. USASI Standards require that the full-load temperature

[1] Symposium on Design of Double-cage Induction Motors; *Trans. AIEE*, 1953, Vol. 72, Pt. III, p. 621.
ALGER, P. L., and WRAY, J. H. Double and Triple Squirrel Cages for Polyphase Induction Motors; *Trans. AIEE*, 1953, Vol. 72, Pt. III, p. 637.
BABB, D. S., and WILLIAMS, J. E. Circuit Analysis Method for Determination of A-C Impedances of Machine Conductors; *Trans. AIEE*, 1951, Vol. 70, Pt. I, p. 661.

[2] SPOONER, T., and KINCAID, C. W. No Load Induction Motor Core Losses; *Trans. AIEE*, April, 1929, Vol. 48, p. 645.
SPOONER, T. Squirrel Cage Induction Motor Core Losses; *Trans. AIEE*, 1925, Vol. 44, p. 155.

rise of general-purpose motor windings with Class A insulation shall not exceed 40°C if measured by mercury thermometers or 50°C if measured by the resistance method. For special-purpose motors, including splashproof and totally enclosed designs, a rise of 50°C (55°C for enclosed machines) by thermometer, or alternatively 60°C (65°C for nonventilated machines) by the resistance method, is allowed. The latter method gives a more realistic measure of the limiting temperatures, especially for enclosed machines, whose windings are inaccessible to thermometers. With usual values of efficiency and torque, self-ventilation with fans not exceeding the rotor diameter is adequate to meet these requirements for open motors. For totally enclosed motors, it is usual to provide external fans of a diameter nearly as large as the outside of the stator laminations, which blow air over the end shields and the back of the core.

Since the apparent efficiency of an induction machine normally reaches its maximum value in the neighborhood of full load and decreases materially at high overloads, the I^2R losses on overloads increase more rapidly than as the square of the power output. Hence, a motor which has 50°C rise at rated load may have 100°C or even greater rise at 150% load continuously. This rapid increase in temperature on overloads means that little gain in permissible output can be made by use of Class B or other more temperature-resistant insulation on a given motor, unless the electrical design is changed to a higher maximum output. Low temperatures at high loads are secured by reducing the number of turns in the motor winding and so increasing its maximum output, at the expense of higher no-load losses, higher starting current, and lower light-load power factor.

68. Frame Construction. The mechanical structures, or frame and end shields, which support and enclose the core and windings, serve three distinct purposes. First, they transmit the torque to the motor supports and so are designed to withstand twisting forces and shocks. Second, they serve as a ventilating housing, or means of guiding the cooling medium into effective channels. Third, they shield live and moving motor parts from human contact and from injury caused by falling objects or weather exposure.

A great variety of designs are employed to meet these requirements and to adapt machines to particular service conditions. For fractional-horsepower motors, rolled-steel shells directly enclosing the motor laminations are usual, and cooling is chiefly by air entering and leaving each end shield separately. For general-purpose motors, cast-iron frames are normally used, with cooling air admitted at each end and discharged from the midlength of the frame after passing over the back of the core. Large machines most often have steel frames fabricated by welding.

Formerly, distinct types of frame were used for open, dripproof, splashproof, and totally enclosed motors, but the recent trend has been to concentrate on the "protected" and totally enclosed types only. A modern protected motor frame is proof against falling objects and dripping liquids and still provides adequate openings for a free flow of ventilating air. It is also designed with smooth contours for good appearance. These advantages, together with the economic benefits of consolidating open and dripproof types in a single design, have caused the new protected types almost entirely to supersede the wide-open frames formerly in general use.

Totally enclosed motors have found a much wider field of use in recent years, wherever dust, corrosion, wetness, or explosion risks are encountered. Improved fan designs and internal ventilation, together with better styling and higher production with attendant lower costs, have further increased their popularity. Nonventilated designs are generally employed in the fractional-horsepower sizes and external fan cooling in the integral-horsepower sizes (see Par. **97**).

69. Rotor Construction. Rotors are built up of laminated steel with overhung slots, mounted on relatively stiff shafts to prevent noise and vibration due to tooth-frequency magnetic forces. The number of rotor slots is usually equal to $2P$ more or less than the number of stator slots; and in the smaller sizes, they are generally skewed about one slot pitch, to minimize magnetic noise and smooth out torque variations. Form windings, used only for large machines or where rheostatic speed control is required, are held in place against centrifugal forces by nonmagnetic binding wire on the ends. By far the greater number of American motors have squirrel-cage windings, formed of brazed or welded copper in the larger sizes and usually of cast aluminum in sizes below about 100 hp at 1,800 r/min.

Motors designed for frequent starting duty, which require a high-heat storage capacity to minimize squirrel-cage winding temperatures, frequently use brass or other high-resistance alloys, which have a larger volume of metal for a given resistance (see Par. **78**).

TESTING OF POLYPHASE INDUCTION MACHINES

70. General. Proof of guaranteed performance, the determination of torque or efficiency of driven machines, and the evaluation of design changes are some of the purposes that require accurate tests of induction machines. Normally, running-light, locked-rotor, resistance, and dielectric tests only are made on standard motors. Input-output tests or segregated-loss tests are made when accurate efficiency determination is required. The inconvenience of making input-output tests and the inaccuracies inherent in any method which determines the losses as a small difference between two large quantities make the segregated-loss methods of test preferable in many cases. Such tests are especially necessary when actual performance under the varying conditions of service is to be determined from a limited number of factory or laboratory test runs. Experience has shown that the equivalent-circuit method of calculation enables accurate predictions of efficiency and other performance data to be made, provided the circuit "constants" are determined in advance by careful tests.

The AIEE Test Code for Induction Machines[1] gives authoritative procedures for conducting all usual tests, and many of the data contained in the following sections are derived from this source.

71. Running-light Test. The motor is run at no load with normal frequency and voltage applied, until the watts input becomes constant. On slip-ring motors, the brushes are short-circuited. Readings of amperes and watts are taken at one or more values of impressed voltage, with rated frequency maintained. Accurately balanced phase voltages and a sine-wave form of voltage are necessary for good results, requiring operation of the test alternator and transformers well below magnetic saturation. The watts input at rated voltage will be the sum of the friction and windage, core loss, and no-load primary I^2R loss. Subtracting the primary I^2R loss at the temperature of test from the input gives the sum of the friction and windage and core loss. Segregation of the core loss from the windage and friction is not necessary for normal efficiency or other rated-voltage performance calculations. However, the segregation can be made, if desired, by taking amperes and watts input readings, at rated frequency, at different voltages varying from 125% of normal down to about 15% voltage, or the point of minimum current. Plotting the input watts, less primary I^2R, against the square of the voltage and extrapolating the lower part of the curve in a straight line to intercept the zero-voltage axis determines the friction and windage. Typical data of such a test are shown in Fig. 18-26.

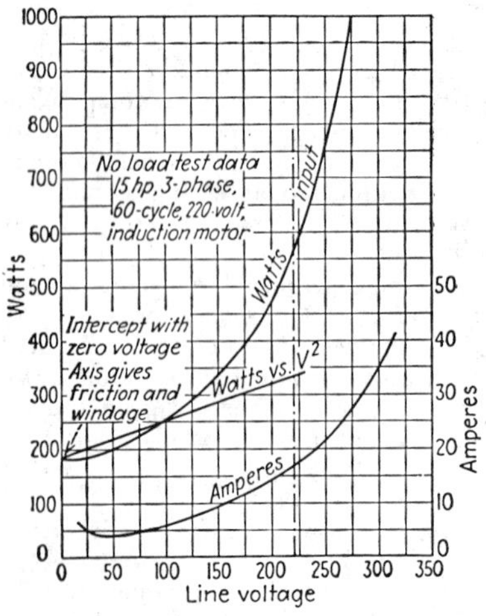

FIG. 18-26. No-load excitation curves.

The value of the magnetizing reactance X_M in Fig. 18-18 is determined from the no-load current at rated voltage I_0 by Eq. (18-34), by using the value of primary leakage X_1 determined from locked-rotor test data.

72. Locked-rotor Test. The motor is blocked so it cannot rotate; a reduced voltage of rated frequency is applied to the terminals; and readings of volts, watts, and amperes are taken. Readings should be taken quickly, and the temperature of the windings

[1] American Standard Test Code for Polyphase Induction Motors and Generators; USASC50.20-1954.

should be observed before and after the test to minimize errors due to changing resistance values. In the case of machines with closed-slot rotors or very small air gaps, magnetic saturation of the leakage paths will occur, and it may be desirable to take readings at half or full voltage to establish the actual value of starting current. Equivalent-circuit performance calculations, however, should be based on data taken at approximately rated current.

When only low-voltage test data are available, the locked-rotor current at higher voltages can be estimated by the formula

$$I = \frac{V - V_0}{V_t - V_0} I_t \qquad (18\text{-}43)$$

where V_t, I_t = test values of voltage and locked-rotor current, V, I = corresponding values at a different voltage, and V_0 = intercept of test current-voltage curve with zero-voltage axis, obtained by extrapolating the test curve as a straight line through points in the approximate range of 50 to 200% current. V_0 represents the voltage due to flux of saturation density crossing closed slot bridges and similar leakage flux paths.

The motor impedance per phase is determined from the volts, amperes, and watts readings. The total resistance component for a 3-phase motor is

$$R = \frac{W}{3I^2} \qquad (\Omega/\text{phase Y}) \qquad (18\text{-}44)$$

and the reactance component is

$$X = \sqrt{\frac{V^2}{3I^2} - R^2} \qquad (18\text{-}45)$$

where W = watts input, I = line current, and V = voltage between lines.

Normally, the primary and secondary leakage-reactance values X_1 and X_2 are assumed equal, each having the value $X/2$.

The primary resistance is measured with direct current, a current about one-quarter of full-load value being preferably used, and readings being taken quickly to avoid errors due to temperature changes during the test. The primary resistance per phase Y is equal to one-half the resistance between any two terminals.

Subtracting the primary resistance at the temperature of test from the resistance component of the total impedance gives the effective secondary resistance at standstill. The starting torque may be calculated from this value by the equation

$$\text{Starting torque} = \frac{7.04KmI^2R_{2e}}{N_s} \qquad (\text{ft}\cdot\text{lb}) \qquad (18\text{-}46)$$

where I = amperes starting current per phase at specified voltage; m = number of phases; N_s = synchronous speed in r/min; R_{2e} = resistance component of motor impedance, less primary resistance at temperature of test, in ohms per phase; K = an empirical constant, usually approximately 0.9, which allows for nonfundamental secondary losses.

In practice, it is usual to measure the torque produced, by means of a lever arm and scale, in which case Eq. (18-46) provides a useful check on the accuracy of the measurements.

In the case of deep-bar or double squirrel-cage motors, the effective secondary reactance at speed is materially higher than at standstill, owing to the progressive shifting of the secondary current from the low-reactance high-resistance paths into the low-resistance high-reactance paths as the secondary, or slip, frequency decreases. Hence, for accurate performance calculations, it is necessary to determine the motor reactance at low secondary frequency. If a low-frequency supply is available, the locked-rotor test may be repeated at 15 c, or at most 25 c, for a 60-c motor. Calculation of the low-frequency reactance by Eq. (18-45) and multiplying this by the ratio of the rated to the test frequency will give the proper value to use in operating performance calculations.

Alternatively, the reactance value at speed may be obtained by adding an amount

ΔX to the reactance determined by full-frequency locked-rotor test. The value of ΔX is approximately

$$\Delta X = R_{2e} - R_2 \tag{18-47}$$

where R_2 = secondary resistance at full-load slip, determined by the slip test of Par. **73.**

73. Slip Test. Whenever feasible, a current-slip curve should be taken under actual load conditions, with rated voltage and frequency maintained at the motor terminals. Measurements at a few points in the neighborhood of full-load current are usually sufficient; but for slip-ring motors a wider range should be covered, owing to the variable resistance of the brushes. The slip is normally too small to be determined by tachometer readings and should therefore be measured with a slip meter or stroboscopically. The slip-meter method makes use of a revolution counter differentially geared to the motor under test and to a small synchronous motor driven from the same power supply at the same synchronous speed. Care must be taken to correct the observed values of slip for the difference between the test temperature and the standard value of 75°C or the temperature attained in a full-load heat run with an ambient temperature of 25°C.

In practice, the value of current corresponding to an assumed value of R_2/s is calculated exactly by the equivalent circuit; the corresponding value of s is read off the slip-current curve; and the true value of R_2 is obtained by multiplying R_2/s by this value of s. However, R_2 may be approximately determined as follows:

Very roughly, the secondary resistance is equal to

$$R_2 = 1.1 \frac{E \cdot s}{I_1} \qquad (\Omega/\text{phase})\ (\text{approx.}) \tag{18-48}$$

where E = terminal voltage per phase, s = ratio of revolutions per minute of slip to synchronous speed, and I_1 = observed phase current.

The coefficient 1.1 varies over a range of about 1 to 1.2, depending on the motor characteristics and the value of the test load.

In case direct slip measurements are not practicable, the value of R_2 determined by Eq. (18-44) in a low-frequency locked-rotor test may be used. Or, in the case of a wound rotor, the actual resistance between slip rings may be measured and multiplied by the square of the ratio of primary to secondary volts to obtain the resistance referred to primary. The voltage ratio is obtained by measurement of primary and secondary voltages at standstill with the slip rings open-circuited. Averages of several rotor positions are taken to avoid errors due to possible unbalance.

74. Stray-load Loss Tests.[1] Stray-load losses, W_s, are defined as the excess of the total measured losses above the sum of the friction and windage, core, and copper losses calculated for the conditions of load from the no-load tests described above. These extra losses are made up chiefly of high-frequency core losses and rotor I^2R losses caused by the pulsations of the leakage-reactance fluxes produced by load currents. While the stray-load losses may be determined by direct input-output tests with a dynamometer or calibrated driving motor, the result is a small difference between two large quantities and so accuracy is very difficult to obtain. Whenever such tests are made, it is desirable to repeat them with the direction of power flow reversed, so the measurement errors may be substantially canceled out.

There are several ways of determining the stray-load loss by separate loss measurements, but the procedure is fairly complex and must be carefully done if accurate results are to be obtained. These are described in the AIEE Test Code for Polyphase Induction Machine, as well as in several AIEE papers.[2]

[1] Koch, C. J. Measurement of Stray Load Loss in Polyphase Induction Motors; *Trans. AIEE*, 1933, Vol. 51, p. 756.

Morgan, T. H., and Narbutovskih, P. M. Stray Load Loss Test on Induction Machines; *Trans. AIEE*, 1934, Vol. 53, p. 286.

Morgan, T. H., and Siegfried, V. Stray Load Loss Tests on Induction Machines—II; *Trans. AIEE*, 1936, Vol. 55, p. 493.

Leader, C. C., and Phillips, F. D. Efficiency Tests of Induction Machines; *Trans. AIEE*, 1934, Vol. 53, p. 1628.

[2] Morgan, T. H., Brown, W. E., and Schumer, A. J. Reverse Rotation Test for the Determination of Stray Load Loss in Induction Machines; *Trans. AIEE*, 1939, Vol. 58, p. 319.

Ware, D. H. Measurement of Stray-load Loss in Induction Motors; *Trans. AIEE*, April, 1945, Vol. 64, p. 194.

75. Performance Calculations. From the foregoing tests, all the circuit constants may be determined, enabling the equivalent-circuit calculations as outlined in Par. **57** to be carried out. To facilitate this, the formulas for calculating the constants as defined in Form A, Par. **57**, are collected in Form B, below.

The procedure in making performance calculations based on test data is first to divide E_0 by the approximate expected value of normal current, an arbitrary value of R_2/s being thus obtained. With this value and the known circuit constants, calculations are carried through for one point, determining the actual value of I. By entering the test slip-current curve (Par. **73**), the true value of s is found, and from this and R_2/s, R_2 is calculated. All the circuit constants are then known, whence the efficiency, power factor, torque, etc., are determined. Additional points are calculated with different values of s, covering the desired range of loads, and the exact characteristics are taken off curves plotted from the calculated results.

If values of torque, current, etc., are desired for considerable overloads or throughout the accelerating range, the values of R_2 and X should be modified to allow for magnetic saturation and eddy currents. Curves of reactance against current obtained by locked-rotor tests over the desired range of values and values of R_{2e} and corresponding values of ΔX obtained by locked-rotor tests at different frequencies are desirable for this purpose, especially in cases of closed-slot or double squirrel-cage rotors.

Form B

Formulas for Calculating Circuit Constants from Test Data for 3-phase Motors

$$X = \frac{f}{f_t}\sqrt{\frac{V^2}{3I^2} - \left(\frac{W}{3I^2}\right)^2} \qquad \text{(Par. 72)}$$

$X_1 = X_2 = 0.5X$ for single squirrel-cage or wound-rotor motors

$X_1 = 0.4X$ and $X_2 = 0.6X$ for low-starting-current motors

$$W_H + W_F = W_{RL} - 3I_M{}^2R_1 \qquad \text{(Par. 71)}$$

W_s from Par. **74**

$$X_M = \frac{E_0}{I_M} - X_1$$

76. Temperature tests[1] are made to determine the temperature rise of insulated windings under load conditions. The USASI Standards specify a limiting temperature for continuous-rated machines of 50°C by thermometer or 60°C by either the resistance- or the embedded-detector method for Class A insulating materials and corresponding values of 70°C by thermometer and 80°C by resistance or embedded detector for Class B insulation. Usually, the temperature is measured by mercury thermometers or thermocouples applied to the hottest accessible parts of the core and windings in several different locations. A small amount of putty is used to shield thermometer bulbs from the surrounding air, and care is taken to avoid external air currents, varying ambient temperatures, or other factors which may introduce errors.

The preferred method of making a full-load temperature test is to maintain nameplate voltage, current, and frequency until the temperature becomes constant, readings being taken every half hour. When constant temperature is reached, the motor is stopped as quickly as possible and additional thermometers are applied to the rotating parts as soon as these have come to rest. The maximum permissible time of stopping is 1 min for machines of less than 50 kW rating, 2 min for 50 to 200 kW ratings, and 3 min for machines larger than 200 kW. The winding temperatures usually increase after

[1] SUMMERS, E. R. Determination of Temperature Rise of Induction Motors; *Trans. AIEE*, September, 1939, Vol. 58, p. 472.

POTTER, C. P. Measurement of Temperature in General-purpose Squirrel-cage Induction Motors; *Trans. AIEE*, September, 1939, Vol. 58, p. 468.

HILDEBRAND, L. E., CAIN, B. M., PHILLIPS, F. D., HOUGH, W. R., ROSSWOG, J. G., and POTTER, C. P. Investigation of Hot-spot Temperatures in Integral Horsepower Motors; *Trans. AIEE*, March, 1945, Vol. 64, p. 124.

shutdown; so readings must be recorded at frequent intervals until definitely falling temperatures are observed. The highest temperature reached at any time during the test is taken as the correct value. If the temperatures fall continuously after shutdown, a curve should be plotted of temperature vs. time and extrapolated back to the moment of shutdown.

For protected-type or totally enclosed machines, it is often preferable to determine the temperature by the rise-of-resistance method. In this case, the "cold" resistance of the winding is measured at a known temperature, usually after the machine has been standing overnight at a uniform room temperature; and the "hot" resistance is measured immediately after shutdown. The hot resistance is taken as the highest value obtained after shutdown or is extrapolated back to the moment of shutdown if the resistance falls continuously.

The temperature is then calculated from the following formula,

$$T = \frac{R_T(234.5 + t)}{R_t} - 234.5 \tag{18-49}$$

where T = winding temperature when R_T was measured, R_T = hot resistance, R_t = cold resistance, and t = winding temperature when R_t was measured.

CHARACTERISTICS OF POLYPHASE INDUCTION MOTORS

77. Types. All polyphase induction motors may be classified as (1) squirrel-cage and (2) wound-rotor. Squirrel-cage motors are further classified by the NEMA[1] as Designs A, B, C, D, and F, with characteristics described in Par. **78.** Both squirrel-cage and wound-rotor motors may be of the (*a*) single-speed or (*b*) multispeed type. Many interesting special designs of motors for different purposes are described in "Electric Motors," Vol. II, Polyphase Current (3d ed.), by H. M. Hobart, Sir Isaac Pitman & Sons, Ltd., 1923.

78. Squirrel-cage Motors. Design A motors are designed to withstand full-voltage starting and have the same locked-rotor torques and slips as Design B motors but have higher breakdown torques and proportionately higher locked-rotor currents than those shown in Table 18-9. Power-company rules may require current-reducing starters for Design A motors in some applications.

Design B[2] motors, the most popular of all types, are designed for and are usually used for full-voltage starting. They have locked-rotor and breakdown torques as tabulated in Par. **90,** which represents the upper limit of the range of application. Starting currents are in accordance with Table 18-9. Full-load slip is less than 5%, except for motors with 10 or more poles, which may have slightly more.

Design C motors are designed for full-voltage starting and are usually provided with double squirrel-cage rotor windings. Locked-rotor and breakdown torques are in accordance with the tabulation of Par. **90,** which represents the upper limit of the range of application. Locked-rotor currents are the same as for Design B motors (Table 18-9), and the slip is less than 5%.

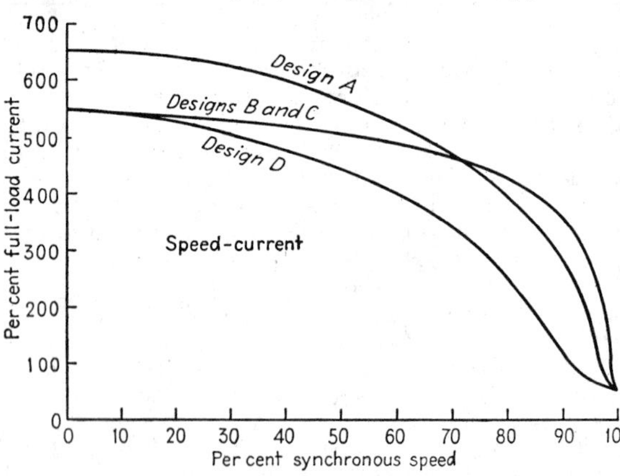

FIG. 18-27. Typical speed-torque and speed-current curves for squirrel-cage induction motors.

[1] NEMA Motor and Generator Standards; *Publ.* MG1-1955.
[2] KOCH, C. J. Line Start Induction Motors; *Trans. AIEE,* Vol. 48, p. 633.

Table 18-3. Induction-motor Application Outline

Application characteristics:							
Type of load	Constant	Varying	Constant	Varying	Varying	Varying	Varying
Starts	Seldom	Seldom	Seldom	Seldom	Frequently	Frequently	Frequently
Load peaks	Low	High	Low	High and frequent	High and of short duration	High	High
Starting torque	Normal	Normal	High	Normal	Normal	Normal to high	Very high
System WK^2	Low	Low	Low	Low	High	Low	
Examples	Majority of applications Centrifugal pumps Fans Unloaded compressors	Machine tools Lathes Saws Millers Etc.	Compressors Reciprocating pumps Loaded conveyors	High speed punch presses	Draw presses Bending brakes	Cranes Hoists	Extractors
Motor:							
Type	Design B	Design B	Design C	Design B	Design D	Design C or low slip D	Special
Time rating	Continuous	Continuous	Continuous	Continuous or intermittent	Intermittent	Intermittent	

Design D motors are designed for full-voltage starting and develop high locked-rotor torques of 275% of full-load torque, which represents the upper limit of the range of application. Locked-rotor currents are the same as for Design B (Table 18-9) and full-load slip is greater than 5%.

Design F motors have locked-rotor currents approximately 62% of those for Design B motors, but the resulting reduced locked-rotor and breakdown torques limit the application to certain well-defined loads with extremely well regulated voltage supply.

Typical speed-torque and speed-current curves for each of the four popular types of squirrel-cage motor are shown in Fig. 18-27.

79. Application of the various types of induction motor is shown in Table 18-3.

80. Wound-rotor motors are provided with an insulated phase winding (usually 3-phase) on the rotor, and the terminal of each phase is connected to a collector ring. Stationary brushes bear on the latter, and these are connected together through an adjustable resistance. Speed-torque and speed-current curves for a typical wound-rotor motor having various amounts of external resistance in the secondary winding are shown in Fig. 18-28, in which the numbers 100, 80, etc., refer to the percent external resistance; 100% resistance gives full-load torque at standstill.

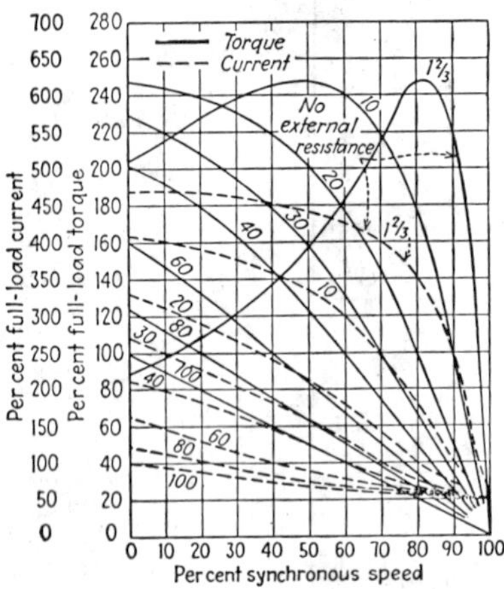

Fig. 18-28. Speed-torque and speed-current curves for typical wound-rotor induction motor.

Wound-rotor motors are normally started with a relatively high resistance in the secondary, and this resistance is short-circuited in one or more steps as the motor comes up to speed. This procedure allows the motor to deliver high-starting and accelerating torques and yet draw relatively little current from the line. Furthermore, most of the secondary losses which occur during acceleration are dissipated in the external resistor rather than in the motor itself.

The curves of Fig. 18-28 indicate that external resistance in the secondary circuit of a wound-rotor motor reduces the speed at which the motor will operate with a given load torque. The motor then has varying speed characteristics; i.e., any change in load results in considerable change in speed. The lower the operating speed, the more pronounced this effect, so that it is not usually feasible to operate at less than 50% of full speed by this method. Furthermore, the efficiency of a wound-rotor motor, including the losses in the external resistor, is reduced in direct proportion to the speed reduction obtained.

81. Multispeed squirrel-cage motors may be of the (1) single-winding or (2) two-winding type. The former have stator windings which can be connected to give either one of two speeds having a ratio of 2:1 (Fig. 18-21). The two-winding motor has two separate stator windings which can be wound for any number of poles so that any two synchronous speeds can be obtained. In addition one or both of the stator windings may be arranged for reconnection as in a single-winding motor, giving a total of three or four speeds, but the two speeds obtained on a single winding must have a ratio of 2:1. Thus a four-speed two-winding motor might have speeds of 1,800, 900, 1,200, and 600 r/min.

Multispeed motors may also be classified as (1) variable-torque motors, (2) constant-torque motors, and (3) constant-horsepower motors. Variable-torque motors have horsepower ratings at each speed proportional to the square of the speed, e.g., 20/5 hp, 1,200/600 r/min, and are used on loads, such as centrifugal pumps and fans, whose horsepower requirement decreases at least as rapidly as the square of the reduction in speed. Constant-torque motors have horsepower ratings at each speed directly proportional to the speed, for example, 20/10 hp, 1,200/600 r/min, and are used on conveyors, stokers, reciprocating compressors, printing presses, and other "constant-torque" loads. Constant-horsepower motors have the same horsepower rating at all speeds. They are used principally on machine tools such as lathes, boring mills, planers, and radial drills.

82. Multispeed wound-rotor motors are used occasionally. Where adjustable-speed operation (by varying the secondary resistance) is required on all synchronous-speed connections, only two speeds can be economically obtained, and these must have a speed ratio of 2:1. In some applications speed adjustment is essential on only one speed connection, in which case two, three, or four speeds can be provided.

83. Ratings. Standard horsepower ratings for polyphase induction motors are given in Table 18-4. Multispeed motors of the constant-torque or variable-torque type are usually given a standard horsepower rating at the top speed but may have odd horsepower ratings at the lower speeds, since the latter are fixed by the speed ratios.

Table 18-4. Standard Horsepower Ratings—Polyphase Induction Motors

1/8	3/4	5	25	75	250	500	1,000	2,250	4,500
1/6	1	7 1/2	30	100	300	600	1,250	2,500	5,000
1/4	1 1/2	10	40	125	350	700	1,500	3,000	
1/3	2	15	50	150	400	800	1,750	3,500	
1/2	3	20	60	200	450	900	2,000	4,000	

The synchronous (no-load) speed of a polyphase induction motor equals $(f \times 120)/P$, where f = frequency in cycles per second of the power supply and P = number of poles for which the motor is designed. The latter must be an even number. The full-load speeds obtained are slightly less than the synchronous speeds, the difference ranging from about 1% (on large high-speed motors) to 5% (on small low-speed machines). Exceptions are the Design D (high-slip) squirrel-cage motors, which are purposely designed to have full-load speeds ranging from 5 to 13% below their synchronous speeds.

Table 18-5. Standard Voltages—Polyphase Induction Motors

Voltage	Preferred HP Limits
110	No minimum; 15 hp maximum
208 and 220	No minimum; 200 hp maximum
440 and 550	1 hp minimum; 1,000 hp maximum
2,300	50 hp minimum; 6,000 hp maximum
4,000	100 hp minimum; 7,500 hp maximum
4,600	250 hp minimum; 8,000 hp maximum
6,600	400 hp minimum; no maximum

84. Voltage, Phases, and Frequency.[1] Standard voltages for polyphase a-c motors are given in Table 18-5. Many power systems, however, operate at other voltages. Thus the 3-phase 4-wire networks common in large cities usually provide 208 V. The table also gives reasonable horsepower limits of motors designed for the various voltages. Although motors can be designed in horsepower ratings above and below these limits, this usually involves an increase in cost or poorer operating characteristics. Motors are frequently provided with a "dual-voltage" feature; i.e., the windings are designed so that they can be connected for either one of two voltages such as 220 and 440 V. Sufficient terminals are provided so that either connection may be made external to the machine.

With few exceptions, polyphase distribution systems provide 3-phase power, although a few 2-phase systems exist. Two-winding multispeed motors are difficult to design 2-phase, so, where required, transformation to allow the use of 3-phase motors is often advisable.

A frequency of 60 c/s predominates in North America, but a few isolated systems operate in very limited areas at 50, 40, or 25 c. In Europe and some parts of South America 50 c is the general standard.

Certain special applications require higher speeds from induction motors than can be obtained at commercial frequencies. In such cases, induction frequency converters or special synchronous generators may be used to supply the higher frequency necessary.

[1] Miller, C. E. Voltage Rating versus Horsepower of Synchronous and Induction Motors; *Trans. AIEE*, 1952, Vol. 71, Pt. II, p. 306.

Table 18-6. General Effect of Voltage and Frequency Variation on Induction-motor Characteristics

Characteristic	Alternating-current (induction) motors			
	Voltage		Frequency	
	110%	90%	105%	95%
Torque:*				
Starting and maximum running	Increase 21%	Decrease 19%	Decrease 10%	Increase 11%
Speed:†				
Synchronous	No change	No change	Increase 5%	Decrease 5%
Full load	Increase 1%	Decrease 1.5%	Increase 5%	Decrease 5%
Per cent slip	Decrease 17%	Increase 23%	Little change	Little change
Efficiency:				
Full load	Increase 0.5 to 1 point	Decrease 2 points	Slight increase	Slight decrease
¾ load	Little change	Little change	Slight increase	Slight decrease
½ load	Decrease 1 to 2 points	Increase 1 to 2 points	Slight increase	Slight decrease
Power factor:				
Full load	Decrease 3 points	Increase 1 point	Slight increase	Slight decrease
¾ load	Decrease 4 points	Increase 2 to 3 points	Slight increase	Slight decrease
½ load	Decrease 5 to 6 points	Increase 4 to 5 points	Slight increase	Slight decrease
Current:				
Starting	Increase 10 to 12%	Decrease 10 to 12%	Decrease 5 to 6%	Increase 5 to 6%
Full load	Decrease 7%	Increase 11%	Slight decrease	Slight increase
Temperature rise	Decrease 3 to 4 C	Increase 6 to 7 C	Slight decrease	Slight increase
Maximum overload capacity	Increase 21%	Decrease 19%	Slight decrease	Slight increase
Magnetic noise	Slight increase	Slight decrease	Slight decrease	Slight increase

* The starting and maximum running torque of a-c induction motors will vary as the square of the voltage.

† The speed of a-c induction motors will vary directly with the frequency.

Typical frequencies are 90, 120, 180, 240, and 360 c, giving motor speeds of 5,400, 7,200, 10,800, 14,400, and 21,600, respectively, with two-pole motors. Motors for such operation must be of special design. Typical applications include textile machinery, woodworking machinery, and portable tools.

85. Temperature Rise. The standard temperature rises of the windings of squirrel-cage and wound-rotor induction motors are:

Size	Fractional hp				Integral hp					
Class of insulation	A		B		A		B		H	
Method of measurement	T	R	T	R	T	R	T	R	T	R
General-purpose motors	40	50	..	..	40	50				
Other open motors	50	60	70	80	50	60	70	80	110	125
Totally enclosed nonventilated	55	65	75	85	55	65	75	85	115	130
Totally enclosed fan-cooled	55	65	75	85	55	60	75	80	115	125
Smaller than 42-frame	..	65	..	85						

Where two methods of temperature measurement are listed, a temperature rise within the values listed in the table, measured by either the thermometer method (T) or the resistance method (R), demonstrates conformity with the standard. The values listed for the thermometer method (T) are usually used for nameplate marking, irrespective of the method of measurement actually used.

Very large motors (2,000 hp and above) are usually provided with resistance-type detector coils embedded in the windings for determining their temperature rise, in which case the allowable temperature rise is 10°C higher than that of the thermometer method.

General-purpose motors[1,2] rated 40°C rise are always given a "service factor"; e.g., the manufacturer guarantees successful operation at 1.15 times rated load, but the temperature rise may be higher, and the efficiency and power factor may be lower than normal.

Standard service factors[3] are 1.4 for motors rated 1⁄20 to 1⁄8 hp, 1.35 for 1⁄6 to 1⁄3 hp, 1.25 for 1⁄2 to 1 hp, 1.20 for 1 1⁄2 to 2 hp, and 1.15 for motors 3 hp and larger, for fractional horsepower, and Design A, B, C, and F integral-horsepower motors.

Open motors rated 40°C rise, and dripproof motors, if rated 40°C rise, by thermometer (or the equivalent) are usually given a "service factor" of 1.15; i.e., the manufacturer guarantees successful operation at 1.15 times rated load without injurious heating, although the efficiency and power factor may vary slightly from the rated load values.

86. Time Rating.[4] Most motors are rated on a "continuous" basis; i.e., they will carry their rated load continuously without exceeding the rated temperature rise. Certain special types of motor, however, are given "short-time" ratings such as 15 min, 1⁄2 h, or 1 h, indicating that they can carry their rated load for these periods of time, after which they must be allowed to cool to room temperature. In some cases a motor is given several different ratings, each for a different period of time. Short-time rated motors are used on loads of an intermittent nature, the time rating being chosen to give motor-heating equivalent to that produced by the actual duty cycle of the load.

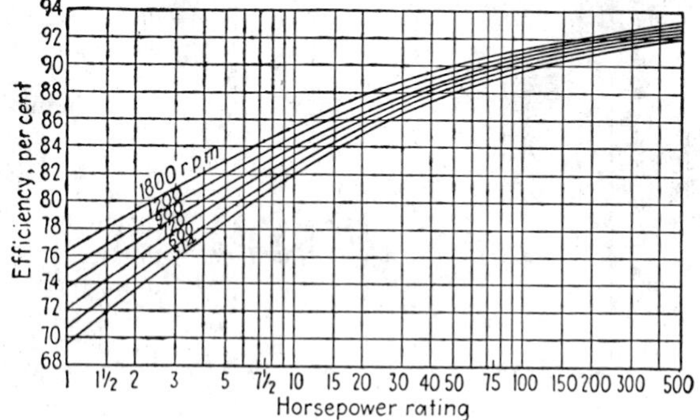

Fig. 18-29*a*. Full-load efficiencies of Design B squirrel-cage motors, 220- or 440-V 3-phase 60-c open type.

87. Efficiencies and Power Factors. Typical full-load efficiencies and power factors of standard Design B squirrel-cage induction motors are given in Figs. 18-29*a* and 18-29*b*, respectively. The variation of efficiency and power factor with load for a typical rating is shown in Fig. 18-31. The efficiencies of Design A motors are essentially the same, while those of Design C and wound-rotor motors are generally slightly lower, and those of Design D motors considerably lower. The power factors of Design A squirrel-cage induction motors are slightly higher, and those of Design C are slightly lower.

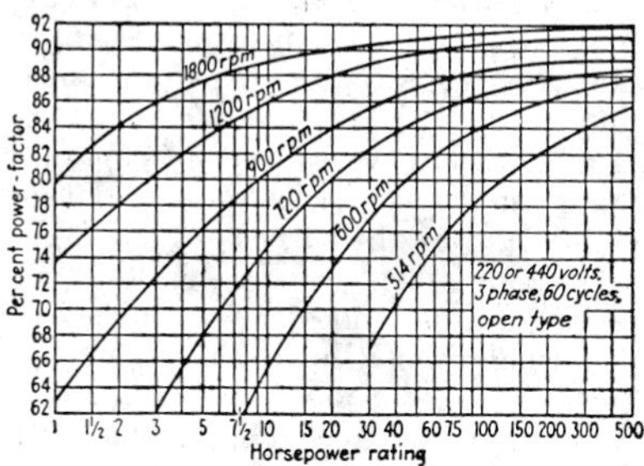

Fig. 18-29*b*. Full-load power factor of Design B squirrel-cage motors.

The efficiencies of all outputs will fall on a straight line on the nomograph of Fig. 18-30. Thus, if the efficiencies at any two loads are known, that of any other load is immediately obtained.

88. Full-load Current. With the efficiency and power factor of a 3-phase motor known, its full-load current may be calculated from the formula

$$\text{Full-load current} = \frac{746 \times \text{hp rating}}{1.73 \times \text{efficiency} \times \text{pf} \times \text{voltage}} \tag{18-50}$$

[1] NEMA Motor and Generator Standards; *Publ.* MG1-1955.

[2] Alger, P. I., and Johnson, T. C. Rating of General Purpose Induction Motors; *Trans. AIEE*, September, 1939, Vol. 58, p. 445.

[3] Hildebrand, L. E. Duty Cycles and Motor Rating; *Trans. AIEE*, September, 1939, Vol. 58, p. 478.

[4] Report on General Principles for Rating of Electrical Apparatus for Short-time, Intermittent or Varying Duty, September, 1941, AIEE Standard 1A.

where the efficiency and power factor are expressed as decimals. For 2-phase motors,

$$\text{Full-load current} = \frac{746 \times \text{hp rating}}{2 \times \text{efficiency} \times \text{pf} \times \text{voltage}} \tag{18-51}$$

Where 2-phase motors are supplied from a 3-wire circuit, the current in the common wire is 1.41 times the current in the other two wires.

89. Voltage and Frequency Variations; Voltage Unbalance.[1] Polyphase induction motors are designed to operate successfully under the following conditions of voltage and frequency variation but not necessarily in accordance with standards established for operation at normal rating:

1. Where the voltage variation does not exceed 10% above or below normal.
2. Where the frequency variation does not exceed 5% above or below normal.
3. Where the sum of the voltage and frequency variation does not exceed 10% (provided the frequency variation does not exceed 5%) above or below normal.

The effect of voltage and frequency variation on the characteristics of typical polyphase induction motors is given in Table 18-6.

The above statements presuppose the same line voltage in each phase. When line voltages applied to a polyphase induction motor are not exactly the same, unbalanced currents will flow in the stator winding, the magnitude depending upon the amount of unbalance. A small amount of voltage unbalance may increase the current an excessive amount.

Temperature rise will be increased considerably. Torques and full-load speed will decrease slightly. The locked-rotor current will be unbalanced to the same degree as the voltage unbalance, but the full-speed current will be greatly unbalanced, in the order of approximately 6 to 10 times the percentage voltage unbalance, making selection of overload protective devices difficult.

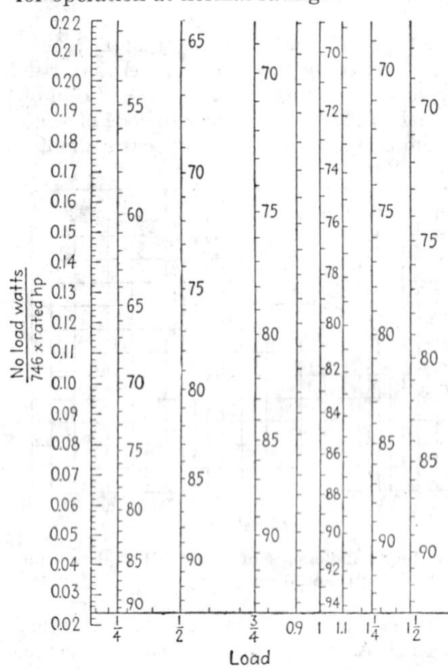

Fig. 18-30. Nomograph to evaluate fractional-load efficiencies for squirrel-cage induction motors.

90. Torques and Starting Currents. Starting and breakdown torques of common Design A, B, and C squirrel-cage induction motors are given in Table 18-8. Relative values for other classes of squirrel-cage motor are indicated by the curves of Fig. 18-27. The minimum breakdown torque for wound-rotor motors is 200% of full-load torque. As indicated by the curves of Fig. 18-28, the starting torque and starting current of wound-rotor motors vary with the amount of external resistance in the secondary circuit.

The starting kVA of a squirrel-cage motor is indicated by a code letter stamped on the nameplate. Table 18-7 gives the corresponding kVA for each code letter, and the locked-rotor current can be determined from

$$\text{Locked-rotor current} = \frac{\text{kVA/hp} \times \text{hp} \times 1{,}000}{k \times \text{line volts}}$$

where $k = 1$ for single-phase, $k = 2$ for 2-phase, and $k = 1.73$ for 3-phase.

[1] Reed, H. R., and Koopman, R. J. W. Induction Motors on Unbalanced Voltages; *Trans. AIEE,* November, 1936, Vol. 55, p. 1206.

Williams, J. E. Operation of 3-phase Induction Motors on Unbalanced Voltages; *Trans. AIEE,* Power Apparatus and Systems, April, 1954, No. 11, p. 125.

Table 18-7. Locked-rotor KVA for Code-letter Motors

Code Letter*	Kva per Hp, with Locked Rotor	Code Letter*	Kva per Hp, with Locked Rotor
A	0–3.14	L	9.0– 9.99
B	3.15–3.54	M	10.0–11.19
C	3.55–3.99	N	11.2–12.49
D	4.0 –4.49	P	12.5–13.99
E	4.5 –4.99	R	14.0–15.99
F	5.0 –5.59	S	16.0–17.99
G	5.6 –6.29	T	18.0–19.99
H	6.3 –7.09	U	20.0–22.39
J	7.1 –7.99	V	22.4 and up
K	8.0 –8.99		

* National Electrical Code.

Maximum locked-rotor current for Designs B, C, and D 3-phase motors has been standardized as shown in Table 18-9 for 220 V. The starting current for motors designed for other voltages is inversely proportional to the voltage.

91. Starting Methods. Wound-rotor motors are invariably started on full voltage but with external resistance in the secondary circuit. Ordinarily sufficient resistance is provided to give 100% torque at standstill, which means that 100% current will be drawn from the line. If a higher torque is required to start the load, less external resistance must be used, and the current drawn is proportionately higher. As the motor accelerates, the external secondary resistance is short-circuited in one or more steps.

The locked-rotor values in Table 18-9 are generally recognized as the minimum needed by motor designers to obtain the required torque characteristics for general-purpose motors. Squirrel-cage motors with these values are usually acceptable for full-voltage starting on power lines and also on combined light and power secondaries of 208 or 220 V, if manually controlled (infrequently started). In the case of automatically controlled (frequently started) equipment, with 208- or 220-V motors supplied from combined light and power secondaries, current-reducing starters to reduce the current to about 65% of these values may be required, unless consultation with the power company indicates that the available system capacity will permit use of full-voltage starting. In any case consultations with the power company for motor applications above 25 hp are advisable.

Autotransformer starters (compensators) are the most popular of any reduced-voltage type. They have the advantage that the ratio of torque developed by the motor to the current drawn from the line remains substantially the same as for full-

Table 18-8. Torques—Polyphase Induction Motors
(Per cent of full-load torque)

Rpm	3,600		1,800				1,200				900				720	
Torque	LR	BD	LR	LR	BD	BD	LR	LR	BD	BD	LR	LR	BD	BD	LR	BD
Design	AB	B	AB	C	B	C	AB	C	B	C	AB	C	B	C	AB	B
½ hp	...	...	...	...	...	...	...	...	...	...	150	...	250	...	150	200
¾ hp	...	...	...	...	...	...	175	...	275	...	150	...	250	...	150	200
1 hp	...	...	275	...	300	...	175	...	275	...	150	...	250	...	150	200
1½ hp	175	275	265	...	300	...	175	...	275	...	150	...	250	...	150	200
2 hp	175	250	250	...	275	...	175	...	250	...	150	...	225	...	145	200
3 hp	175	250	250	...	275	...	175	250	250	225	150	225	225	200	135	200
5 hp	150	225	185	250	225	200	160	250	225	200	130	225	225	200	130	200
7½ hp	150	215	175	250	215	190	150	225	215	190	125	200	215	190	120	200
10 hp	150	200	175	250	200	190	140	225	200	190	125	200	200	190	120	200
15 hp	150	200	165	225	200	190	135	200	200	190	125	200	200	190	120	200
20 hp	150	200	150	200	200	190	135	200	200	190	125	200	200	190	120	200
25 hp	150	200	150	200	200	190	135	200	200	190	125	200	200	190	120	200
30 hp	150	200	150	200	200	190	135	200	200	190	125	200	200	190	120	200
40–200 hp	*	200	*	200	200	190	*	200	200	190	125	200	200	190	120	200

NOTE: LR = locked-rotor torque.
BD = breakdown torque.
A, B, and C refer to Designs A, etc.
* Progressively lower values for these larger ratings.

voltage starting. The motor torque and the current drawn from the line (neglecting the magnetizing current of the autotransformer) are both reduced in proportion to the square of the voltage impressed on the motor. The magnetizing current of the autotransformer generally does not exceed 25% of motor full-load current. Normally, the motor accelerates nearly to full speed on the reduced-voltage connection and is then transferred to full voltage. Since the circuit to the motor is opened and then imme-

Table 18-9. Locked-rotor Current for 3-phase Motors at 220 Volts

Rated Horsepower	Classes B, C, D, Amperes	Rated Horsepower	Classes B, C, D, Amperes	Rated Horsepower	Classes B, C, D, Amperes	Rated Horsepower	Classes B, C, D, Amperes
1	24	7½	120	30	435	100	1,450
1½	35	10	150	40	580	125	1,815
2	45	15	220	50	725	150	2,170
3	60	20	290	60	870	200	2,900
5	90	25	365	75	1,085		

diately reclosed, a transient inrush of current occurs which may be of much greater magnitude than the current normally drawn by the motor at the speed at which the transfer is made. This transient inrush, however, is of such extremely short duration that it does not produce an objectionable voltage disturbance on the average power system. Standard autotransformer starters are provided with 65 and 80% voltage taps in sizes up to 50 hp and with 50, 65, and 80% voltage taps in the larger sizes.

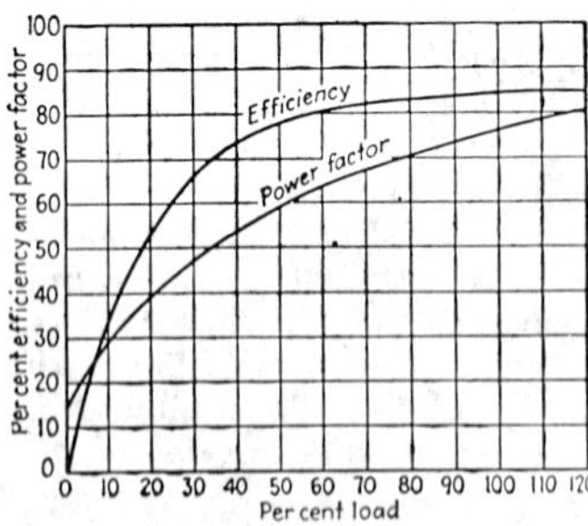

FIG. 18-31. Variation of efficiency and power factor with load for Design B squirrel-cage motors.

"Part-winding" starting is being more widely used for reducing starting current. This involves arranging the stator winding so that, by use of adequate control devices, one part of the stator winding is first energized and subsequently the remainder of the winding is energized in one or more steps. The purpose is to reduce the initial values of the starting current drawn and/or the starting torque developed by the motor. The usual arrangement involves energizing one-half the stator winding on the first step, resulting in approximately 50% of normal locked-rotor torque and approximately 60% of normal locked-rotor current. While this torque may be insufficient to start the motor in some applications, it permits drawing full-winding starting current from the system in two increments. Another method is to connect two-thirds of the winding on the first step, by using a 4-pole contactor, in which case the motor should accelerate promptly to full speed. The remaining third of the winding is then connected by closing a second contactor with only two poles.[1]

Resistor-type reduced-voltage starters are sometimes used. They have the disadvantage that the current drawn from the line is reduced in direct ratio to the impressed voltage, while the torque developed by the motor is reduced as the square of this voltage. The resistor is short-circuited, either all at once or in steps, when the motor comes up to speed. The circuit for the motor is not broken in transferring to full voltage, as is the case with the autotransformer starter. These features make the resistor-type starter adapted for use where "increment-type" starting-current restrictions exist. With the resistor-type starter, the contactors which short-circuit the resistors, as well as the line contactors, must carry the full current of the motor, whereas in part-winding starting, the contactors for the two parts of the winding each carry only half the total current.

[1] ALGER, P. L., WARD, H. C., JR., and WRIGHT, F. H. Split-winding Starting of 3-phase Motors; *Trans. AIEE*, 1951, Vol. 70, Pt. I, p. 867.

ALGER, P. L., and AGACINSKY, LORRAINE A New Method for Part-winding Starting of Polyphase Motors; *Trans. AIEE*, Power Apparatus and Systems, February, 1956, No. 22, p. 1455.

ALGER, P. L. Performance Calculations for Part-winding Starting of Three-phase Motors; *AIEE Conf. Paper* 56-515.

Reactor-type reduced-voltage starters are sometimes used on larger motors, most frequently on high-voltage motors (2,200 V or above), where oil circuit breakers are necessary to provide sufficient current-interrupting capacity. In such cases, the reactor and starting circuit breaker are placed in the neutral of the motor. The breaker can then be of low interrupting capacity, since the fault current at this point is limited by the reactance of the motor windings.

Wye-delta starting, though quite common abroad, is seldom used in the United States. This starter consists of a switching arrangement which transfers the motor winding from Y for starting to Δ for running. The current drawn and the torque developed by the motor are thus reduced to only one-third their full voltage values. This very low torque, the extra contactors required, and the current inrush when the circuit is reclosed on Δ make this scheme unattractive.

92. Construction. Polyphase induction motors are usually furnished with a shaft, two bearings, and mounting feet forming part of the stator frame. The bearings are usually carried by "end shields" which bolt to each end of the frame. Very large motors are often furnished with pedestal-type bearings. In this case a base may be provided which supports the stator and bearing pedestals, or individual soleplates may be provided for each bearing pedestal and for each side of the stator frame. Very small motors are sometimes furnished with a subbase that supports the motor by the end shields, usually through a resilient member to isolate vibration.

The smaller motors are sometimes furnished without mounting feet and are supported from the end shield on the drive end, the latter being provided with a machined flange or face for bolting to the driven machine. The "flange-type" and "face-type" end shields are also employed in wholly or partially supporting the driven machine by the motor.

93. Bearings. Grease-lubricated ball bearings are ordinarily used on integral-horsepower motors of all enclosures. Oil lubrication is used for the larger horizontal motors and for the thrust bearing of medium and large vertical motors. Oil-ring-lubricated sleeve bearings are sometimes used on all sizes and particularly on the very large high-speed motors.

Motors smaller than 1 hp at 1,800 are commonly built with oil-lubricated waste- (or felt-) packed sleeve bearings.

Some motors use prelubricated sealed or shielded bearings which are supported in simple cavities in the end shields, while others use open bearings with inner caps to complete the grease chambers. The former cannot be relubricated without complete disassembly of the motor, and sometimes the bearing itself, while the latter is usually arranged for relubrication from outside the motor. Present-day greases allow longer periods than formerly between relubrications.

94. Vertical motors are designed for operation with the shaft in a vertical position, usually with the shaft extension extending downward from the motor. Small motors may be supported by a machined face on the lower end shield, but the larger ratings commonly have an NEMA type D flange or type C face mounting for general applications or an NEMA type P base for water pumps, all of which are made integral with the lower end shield. The motor is designated "normal thrust" if the bearing is the same size as normally used in a horizontal motor of like rating or "high thrust" if a special thrust bearing is used. Ball bearings, either grease- or oil-lubricated, are generally used for both the thrust and the guide bearings. Vertical motors are also available in a "hollow-shaft" construction. These are used principally for driving deep-well pumps, the pump shaft being brought through the hollow motor shaft and coupled to the latter at the top of the motor. This arrangement allows convenient adjustment of the position of the pump impeller, and by the use of a special coupling, reversed rotation of the pump automatically disconnects the two shafts.

95. Motors for Belt Drive. Table 18-10 represents good practice (under normal operating conditions) for use of flat-belt or multiple V-belt drives on motors which are not provided with outboard bearings.

96. Gear motor designates the combination of a motor with an enclosed speed-reducing gear built as an integral unit. Gear motors are available in sizes up to approximately 75 hp and with output-shaft speeds ranging from 13.5 to 780 r/min. A high-speed (1,800 r/min at 60 c) motor is usually employed.

Table 18-10. Horsepower Limits for Flat-belt and V-belt Drives

Full load rpm of motor		Maximum hp rating	
Above	Including	Flat-belt drive	V-belt drive
3,000	3,600	25	20
1,800	3,000	30	40
1,200	1,800	40	100
900	1,200	75	150
720	900	125	200
560	720	200	300

Various types of gear-reduction unit are employed, including the internal planetary with concentric output shaft, the worm-and-wheel type with right-angle shaft, and the straight helical-gear reduction with parallel offset shaft.

The method of selecting gear motors for specific applications differs considerably from that of ordinary motors. The capacity of any gear is limited by the allowable stresses in the gear teeth, which must be kept within certain limits to obtain a reasonable life. Hence, peak torque, torque fluctuation and frequency, dynamic loading, number and severity of starts, hours of running per day, and expected years of life must all be considered in selecting a suitable gear.

To meet economically the requirements of the various applications, gear motors are offered having different "service factors," this term being defined as the ratio of rating of the gear employed to the rating of the motor itself. For example, one manufacturer offers three lines of gear motor having service factors of 1, 1.25 to 1.5, and 1.75 to 2, which are designated "general-purpose," "special-purpose," and "heavy-duty" gear motors, respectively. The classification of loads and corresponding service factors given below may be used as a general guide in the selection of gear motors.

Character of load	8–10 hr per day	24 hr per day
Uniform	1.00	1.25
Moderate shock	1.25	1.50
Heavy shock	1.75	2.00

The compactness, high efficiency, and safety features of gear motors have made them a very popular substitute for ordinary motors with separate speed reducers. Typical applications include conveyors, agitators, low-speed fans and blowers, mixers, screens, and line shafts.

97. Enclosing Features. A motor normally requires the circulation of air through its interior for removing the heat resulting from motor losses. If there is no restriction to the flow of ventilating air other than that necessitated by mechanical construction, the motor is designated "open type." The types of enclosure frequently used on polyphase induction motors are as follows:

a. Dripproof motors in which the ventilating openings are so constructed that drops of liquid or solid particles falling on the machine at any angle not greater than 15° from the vertical cannot enter the machine either directly or by striking and running along a horizontal or inwardly inclined surface.

b. Splashproof motors in which the ventilating openings are so constructed that drops of liquid or solid particles falling on the machine, or coming toward it in a straight line or at any angle not greater than 100° from the vertical, cannot enter the machine either directly or by striking and running along a surface.

c. Totally enclosed motors are so enclosed as to prevent the exchange of air between the inside and outside of the case but not sufficiently enclosed to be termed "airtight." Continuously rated motors in the totally enclosed (not fan-cooled) construction become uneconomic in sizes above approximately 15 hp.

d. Totally enclosed fan-cooled motors are totally enclosed machines equipped for exterior cooling by means of a fan or fans, integral with the machine but external to the enclosed parts. Totally enclosed construction is used where it is necessary to protect the motor from dirt, moisture, chemical fumes, or other harmful ingredients of the surrounding atmosphere. Totally enclosed fan-cooled motors are available in sizes up to approximately 1,000 hp (see Par. **68**).

e. "Pipe-ventilated" motors have openings for the admission and discharge of ventilating air, being further classified as "open" if only the inlet opening is arranged for connection of ducts or pipes and as "totally enclosed" if both inlet and outlet openings are so arranged. The air is normally circulated by means integral with the motor, but if the means is external to and not a part of the machine, these machines are known as "separately ventilated" or "forced-ventilated" machines. These constructions serve the same purpose as totally enclosed types in allowing air free from harmful ingredients to be brought to the motor and in some cases providing for removing the heat of the motor from the motor room. Separate, or forced, ventilation is usually more economical on low-speed machines whose rotor fans do not develop enough pressure to circulate sufficient ventilating air through the required duct length.

f. Totally enclosed machines with "closed systems of ventilation" are designated as "totally enclosed water-air-cooled" machines when they are cooled by circulating air which in turn is cooled by circulating water. Such machines are provided with a water-cooled heat exchanger for cooling the ventilating air and a fan or fans, integral with the rotor shaft or separate, for circulating the ventilating air. This type of machine is generally confined to rather large sizes. When the heat exchanger is of the air-to-air type rather than water-cooled, the machine is designated as a "totally enclosed air-to-air-cooled" machine.

Both the enclosed self-ventilated and the enclosed separately ventilated motors may be provided with a "closed system of ventilation"; i.e., the hot air discharged from the machine may be cooled and then recirculated through the machine. The cooling is usually accomplished by water circulated through surface air coolers. This system of ventilation is used only on rather large machines.

98. Motors for Hazardous Atmospheres.[1] In accordance with the degree of hazard involved, the NEC classifies as follows the locations in which hazardous gas or dust may be present:

Class I, Group A—Atmospheres containing acetylene.

Class I, Group B—Atmospheres containing hydrogen.

Class I, Group C—Atmospheres containing ethyl ether vapor.

Class I, Group D—Atmospheres containing gasoline, petroleum, naphtha, alcohol, acetone, lacquer-solvent vapor, and natural gas.

Class II, Group E—Atmospheres containing metal dust.

Class II, Group F—Atmospheres containing carbon-black coal or coke dust.

Class II, Group G—Atmospheres containing grain dust.

Since it is impractical to build a motor which is gastight, motors for Class I locations are constructed so as to withstand explosions within the motor, without ignition of the surrounding atmosphere. This is accomplished by (1) making the strength of the enclosing parts sufficient to withstand the maximum pressure caused by an ignition of the most flammable mixture of the gas involved and (2) providing long fits with close clearances at all joints and at the shaft opening to cool the escaping flame or gases to a point where ignition of the outside atmosphere is prevented. The motors may be of the totally enclosed or totally enclosed fan-cooled type and are termed "explosionproof."

Motors for Class II locations are built totally enclosed or totally enclosed fan-cooled and are sufficiently tight to exclude the specified dust. They are termed "dust-explosion-proof" when the enclosure is designed and constructed so as not to cause an ignition of an ambient atmosphere of the specified dust and also not to cause ignition of the dust on or around the machine.

Underwriters' Laboratories, Inc., a nonprofit organization, sponsored by the National

[1] NUCKOLLS, A. H. Inspection Tests of Explosionproof Motors; *Trans. AIEE*, February, 1936, Vol. 55, p. 151.

Board of Fire Underwriters, offers a testing and inspection service to manufacturers of equipment for hazardous locations. Motors which have been tested and inspected by them are provided with a label so indicating. Since most inspection authorities look to the Underwriters' Laboratories for guidance as to the suitability of equipment for hazardous locations, motors provided with this label are generally acceptable for use in the specified atmosphere anywhere in the United States. Complete lines of motors are available having Underwriters' Laboratories label for Class I, Group D or Class II, Group G locations in ratings up to approximately 1,000 hp.

In the large sizes where motors of conventional "explosionproof" construction are not available, enclosed motors arranged so they can be kept filled with inert gas have been used. Provision is made for keeping the gas (usually carbon dioxide) under a slight positive pressure at all times.

To keep the leakage (loss of gas) to a minimum, special oil seals at the shaft opening have been employed. Surface air coolers built integrally with the motor are usually used to dissipate the heat.

Motors for installation in mines come under the jurisdiction of the U.S. Bureau of Mines. This body does not issue approval of a motor and/or control but approves only a complete power-using assembly, consisting of a machine such as a pump, coal cutter, or conveyor, combined with the motor which drives it, the complete control equipment, and wiring. However, motors are available which have passed the inspection and test of the Bureau of Mines, indicating that they are eligible for ultimate approval, without further test, as part of a complete power-using assembly.

99. Shell-type motors consist of stators and rotors only, without shafts, end shields, bearings, or conventional frames (Fig. 18-32). The rotors are mounted directly on a shaft of the driven machine, which must also include a suitable support for the stator and a ventilating arrangement. The motors are built with relatively small outside diameters but may be slightly longer than standard machines. Furthermore, horsepower ratings over a rather wide range are built in each frame diameter, the ratings for the different diameters overlapping slightly. Although a great many of the motors used are for operation at standard commercial frequencies giving speeds up to 3,600 r/min (on 60 c), they are frequently supplied for operation at higher frequencies and correspondingly higher speeds. Frequencies up to 2,000 c with a corresponding two-pole motor speed of 120,000 r/min have been used, but the more common "high" frequencies range from 60 to 240 c, giving two-pole motor speeds up to 14,400 r/min.

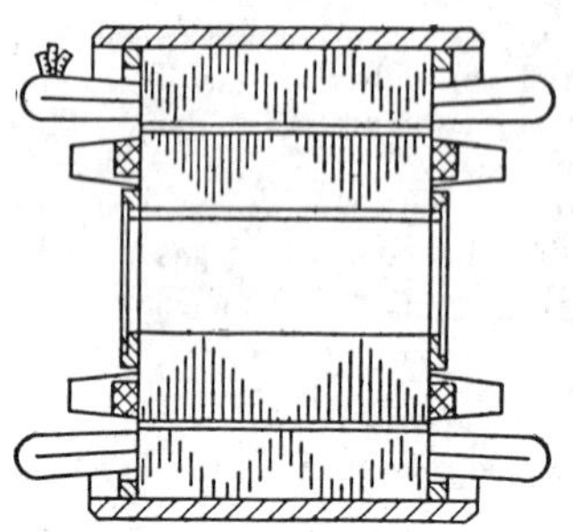

Fig. 18-32. Cross section of typical shell-type motor.

Shell-type motors are used principally on machine tools and woodworking machinery. Their relatively small physical size facilitates a compact design with maximum flexibility in arrangement of machine parts. The small diameter of the motors is of particular value, since it allows close spacing of spindle shafts.

The wide range of ratings available in each diameter reduces the cost of providing suitable mountings for the motors.

Motors of similar mechanical construction but with special insulation are used in hermetically sealed refrigeration and air-conditioning compressors, where the motor runs in an atmosphere of refrigerating gas. The insulation must neither harm nor be harmed by the refrigerant and, so that the refrigerant may be kept clean and dry, must not trap moisture or dirt.

100. Dimensions. NEMA has standardized mounting dimensions for various types of motor, those standardized for polyphase induction motors covering ratings from 1 to 125 hp (at 1,800 r/min). For convenience each set of standardized dimensions has been assigned a frame number, and the various ratings of motors have been assigned frame numbers from the series. Any motor offered by a manufacturer having a frame number from this series will have the corresponding standardized mounting dimensions. These are listed in NEMA Motor and Generator Standard, Publ. MG1-1955.

These NEMA frame dimensions along with a close counterpart in the metric system are included in the *Rept. on Dimensions* 72 issued by The International Electrotechnical Commission covering progress toward an international standard of motor dimensions.[1]

SINGLE-PHASE INDUCTION MOTORS

101. General Theory.[2] If one supply line to a polyphase induction motor is opened, the motor will not develop any starting torque, although if it is already operating, it will continue to run at slightly reduced speed, with a somewhat lower breakdown torque. The crux of the single-phase motor problem, therefore, is in providing auxiliary means for starting.

As explained in Par. **53**, the magnetic field of a single-phase winding carrying alternating current may be represented as a phasor stationary in space but alternating in time, or as the sum of two equal and oppositely revolving field phasors, which are constant in magnitude. In a polyphase motor, the backward-revolving field phasors of the several phases cancel each other, and the forward-revolving ones add directly, giving a uniform revolving field. In the single-phase motor, means are provided to reduce the backward field, but this field has always some remaining magnitude (except at one particular load in the case of certain capacitor-run motors), and consequently a single-phase induction motor always has extra losses and a double-frequency pulsating torque not possessed by a polyphase motor.

A simple way to visualize the effects of this backward field is to consider that the forward- and backward-revolving fields are separately produced by the same stator current; i.e., they are connected in series. Each field may then be treated as a separate polyphase induction motor, the forward field having a slip s with respect to the rotor, and the other a slip $2 - s$. At standstill, both values of slip are unity, and the two circuits are identical. At all times, the net torque developed is equal to the difference of the separate torques produced by the two fields. On this basis, the single-phase induction-motor equivalent circuit is given by Fig. 18-33.

The values of R_1, X_1, R_2, X_2, and X_M are the impedance constants derived by measurements across the single-phase terminals. Since half the total air-gap impedance at standstill is due to each field, the magnetizing and secondary impedance values are divided by 2 to obtain the values corresponding to the separate fields.

Inspection of this circuit reveals several interesting properties of the motor. At full speed, s is very small, and the backward field appears as an external series impedance of $R_2/4 + j(X_2/2)$. The corresponding loss $I^2R_2/4$ represents the power delivered to the rotor by the backward field. However, there is an equal loss due to the rotor's being driven forward at speed $1 - s$ against the backward-field torque; so the total loss caused by the backward field is $I^2R_2/2$, approximately. Since the backward-field rotor currents occur at double-line frequency, any double squirrel-cage or deep-bar rotor design which had an increased resistance at high frequency would greatly increase the power losses; and such designs, therefore, are seldom used for single-phase motors. The breakdown torque of a single-phase motor may be approximately calculated by Eq. (18-26) for a polyphase induction motor, if the impedance of the backward-revolving field is considered as a series impedance added in the primary circuit of the polyphase motor. Hence, any

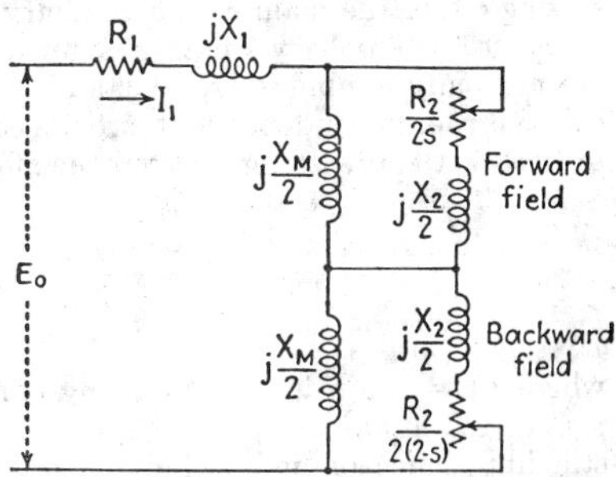

Fig. 18-33. Equivalent circuit of single-phase induction motor.

[1] This report is available through United States of America Standards Institute, 70 E. 45th St., New York 17.

[2] Lamme, B. G. Physical Conception of the Operation of the Single Phase Induction Motor; *Trans. AIEE*, April, 1918, Vol. 37, p. 627.

Kimball, A. L., and Alger, P. L. Single-phase Motor-torque Pulsations; *Trans. AIEE*, 1924, Vol. 43, p. 730.

Morrill, W. J. Characteristic Constants of Single-phase Induction Motors. 1—Air-gap Reactances; *Elec. Eng.*, March, 1937, Vol. 56, p. 333.

Veinott, C. G. Performance Calculations on Induction Motors; *Trans. AIEE*, Vol. 51, p. 743.

Suhr, F. W. Symmetrical Components as Applied to the Single-phase Induction Motor; *Trans. AIEE*, September, 1945, Vol. 64, p. 651.

increase in the secondary resistance of a single-phase motor actually reduces the breakdown torque, as well as lowering the speed at which breakdown occurs.

Another interesting characteristic is the double-frequency torque pulsation. The double-frequency current in the rotor reacting upon the slip-frequency forward magnetic field evidently produces a torque pulsation, even at no load. Physically, the no-load part of the pulsating torque provides the means for supplying and removing the magnetic field twice each cycle in the axis at right angles to the stator winding, and the additional part under load corresponds to the double-frequency pulsation of the single-phase power input to the rotor. To prevent objectionable transmitted vibration and noise from this cause, it is usual to mount single-phase machines on supports with torsional elasticity of some type, often rubber rings encircling the bearing housings in the case of fractional-horsepower motors.

102. Shaded-pole Motor.[1] The simplest way of providing a single-phase induction motor with starting torque is to place a permanently short-circuited winding of relatively high resistance in the stator at an electrical angle of 30 to 60° from the main winding. Usually this auxiliary winding, called a "shading coil," consists of an uninsulated copper strip encircling approximately one-third of a pole pitch. The current induced in the shading coil, by the portion of the main field linking it, reduces the magnitude of this flux and also causes it to lag in time phase. In consequence, the air-gap field has two components, an undamped alternating flux and a damped flux displaced 90° in space and $\theta°$ in time. Shaded-pole motors are used only in very small sizes normally below 50 W output. Principal applications are for desk fans and air circulators, where their simplicity, low torque, and low cost are well suited to the requirements.

The inherently high slip of a shaded-pole motor makes it convenient to obtain speed variation on a fan load by reducing the impressed voltage. It is common practice to provide multispeed fan operation by enclosing a small series reactor in the fan base, which can be switched in or out at will.

103. Resistance Split-phase Motors. A considerably greater starting torque can be obtained by providing a separate starting winding, or auxiliary phase, 90° displaced in space from the main winding of a single-phase induction motor. This extra winding is normally wound with fewer turns of a much smaller size of wire, so that it has a considerably greater resistance to reactance ratio than the main winding, and it is connected directly across the power supply, in parallel with the main winding. Just as in the case of the shaded-pole motor, the field of the auxiliary winding is displaced in time and space, so that its vectorial combination with the main field gives a much larger forward than backward field component. The motor can be reversed by reversing either the main or the auxiliary winding.

Since the auxiliary winding is normally located 90° from the main winding, the two are mutually noninductive at standstill and the standstill characteristics may be calculated from two independent circuits each like that of Fig. 18-33. By a similar analysis to that of the preceding section, the starting torque of a split-phase motor is

$$T = \frac{14.1aK}{N_S} I_M I_A R_2 \sin\theta \qquad \text{(ft·lb)} \qquad (18\text{-}52)$$

where I_M = main winding starting current in amperes; I_A = auxiliary winding starting current in amperes; N_S = synchronous speed; R_2 = resistance component of standstill impedance of main winding, less the primary resistance; a = ratio of effective number of turns in auxiliary winding to main-winding effective turns; K = an empirical coefficient which allows for nonfundamental rotor losses, usually equal to 0.9; and θ = angle of phase split between I_M and I_A.

Design limitations usually prevent θ from being greater than 30°, so that the starting torque per voltampere cannot exceed half that of a 2-phase motor built in the same

---

[1] Trickey, P. H. An Analysis of the Shaded Pole Motor; *Trans. AIEE*, September, 1936, Vol. 55, p. 1007.

Chang, S. S. L. Equivalent Circuits and Their Application in Designing Shaded Pole Motors; *Trans. AIEE*, 1951, Vol. 70, Pt. I, p. 690.

Kron, G. Equivalent Circuits of the Shaded-pole Motor with Space Harmonics; *Trans. AIEE*, 1950, Vol. 69 Pt. II, p. 735.

parts. Since, in addition, both I_M and I_A are drawn from a single phase of the power supply, the starting current is excessive, limiting the use of the resistance split-phase motor to sizes below ⅓ hp.

The auxiliary winding is opened automatically as the motor approaches full speed, as otherwise prohibitive losses would occur in it. Usually this is accomplished by means of a centrifugal switch or, in the case of hermetically sealed motors, by an electromagnetic relay. The high current density used to obtain an adequate resistance value makes the initial rate of temperature rise of the auxiliary winding very great, sometimes more than 50°C/s, so that these motors are not satisfactory for repeated starting or for inertia loads.

Earlier split-phase motor designs included motors with stationary external squirrel-cage members, with the primary windings on the rotor, receiving their power through slip rings. They are now normally built, however, with uniformly distributed partly closed stator slots, enameled-wire concentric stator windings, and a cast-aluminum or welded-copper squirrel cage on the rotor.

Typical characteristic curves for a ⅙-hp 60-c 1,725-r/min, resistance split-phase motor are shown in Table 18-13.

104. Repulsion-start Induction-run Motor.[1] As discussed in Par. **131,** a common way of obtaining single-phase-motor starting torque is to provide a d-c winding and commutator on the rotor, with a single pair of short-circuited brushes for starting and a centrifugal mechanism which short-circuits the entire commutator as the motor approaches full speed. This gives a pure repulsion-motor starting characteristic with very high torque per ampere and pure single-phase induction-motor operating characteristics. These motors are widely used in sizes up to about 5 hp. Typical characteristics of a 1-hp 60-c 1,800-r/min motor of this type are shown in Table 18-13.

105. Capacitor Motors.[2] Low-cost low-voltage capacitors have proved extremely useful in improving the performance of split-phase motors. By inserting an external series capacitor in the auxiliary winding circuit and making this winding with many more turns of much lower resistance, the angle of phase split θ can be increased to 90°, or even more, and the coincident increase in the turn ratio a permits a further decrease in the auxiliary winding current. Thus, the capacitor-start motor gives an adequate starting torque for a reasonable starting current and at the same time has so much greater thermal capacity than a resistance split-phase motor, by virtue of the reduced winding-current density, that it is satisfactory for nearly all industrial single-phase motor applications.

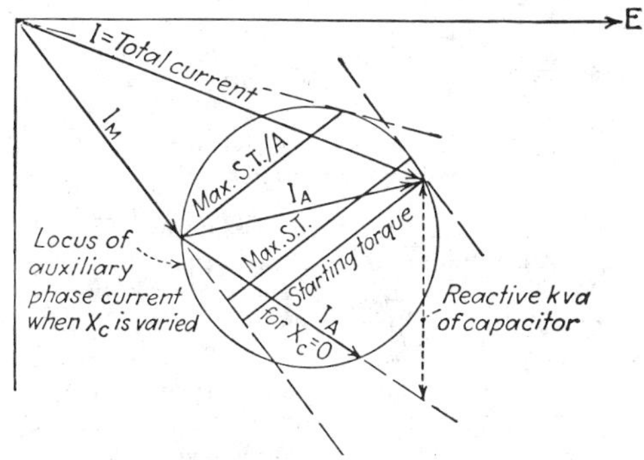

Fig. 18-34. Capacitor-motor starting-torque diagram.

Figure 18-34 illustrates a convenient method of determining the best size of capacitor to use with a given motor.[3] I_M represents the locked-rotor current in the main winding and I_A the current in the auxiliary winding. With no external capacitor, $X_C = 0$, and the motor becomes a plain resistance split type. As X_C is increased, I_A moves ahead in time phase, following a circular locus, increasing the torque and reducing the total current drawn from the line. Points of maximum torque, and maximum torque per ampere are indicated on the diagram.

---

[1] Hamilton, J. L. The Repulsion-start Induction Motor; *Trans. AIEE*, October, 1915, Vol. 34, p. 2443.

[2] Morrill, W. J. The Revolving-field Theory of the Capacitor Motor; *Trans. AIEE*, April, 1929, Vol. 48, p. 615.

Specht, H. C. The Fundamental Theory of the Capacitor Motor; *Trans. AIEE*, April, 1929, Vol. 48, p. 607.

Bailey, B. F. The Condenser Motor; *Trans. AIEE*, April, 1929, Vol. 48, p. 596.

[3] Morrill, W. J. The Revolving-field Theory of the Capacitor Motor; *Trans. AIEE*, April, 1929, Vol. 48, p. 615.

Usually low-voltage electrolytic capacitors are used for starting purposes, since these are economical in 110-V intermittent ratings. However, in cases of severe starting duty, higher-voltage motors; or where capacitors are retained in the circuit during operation; oil or Pyranol capacitors are desirable, and these are often connected through a transformer to secure the most economical capacitor voltage.

For most applications, the auxiliary winding is opened by a centrifugal switch or relay, as the motor approaches full speed, just as in the case of the resistance split-phase motor. Such motors are called "capacitor-start motors." In some cases of smaller-sized motors with low-starting-torque requirements, however, it is permissible to leave the capacitor permanently in circuit. These are called "permanent-split capacitor motors." The limitations of starting torque and motor size on this type are the result of the inherent tendency of the auxiliary-winding current to increase in magnitude and shift backward in time phase as the motor accelerates, so that unless the capacitor impedance is very high, the motor will have objectionable losses and large torque pulsations at full speed. However, the power loss in the capacitor circuit at speed is very much less than in a shading coil for a given starting torque, and so the permanent-split capacitor motor is finding increasing use for fan drive in sizes up to ¼ hp.

Speed variation of these motors is provided for by autotransformer or other adjustment of the voltage impressed on the main winding, leaving full voltage on the capacitor phase. Sometimes the voltage impressed on the entire motor is varied to change the speed. In other cases, a transformer is used to change the voltages of the main and capacitor phases inversely, causing a large phase unbalance as well as a reduction of the average excitation. This arrangement produces a large backward rotating-field component, which loads the motor and lowers its speed considerably for all values of externally applied, or shaft, load. Unusually stable operation at reduced speed is secured in this way.

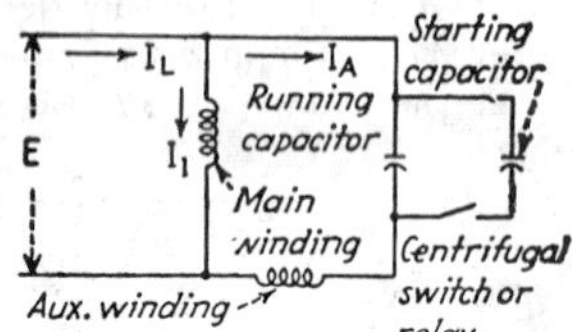

FIG. 18-35. Capacitor-run motor winding connections.

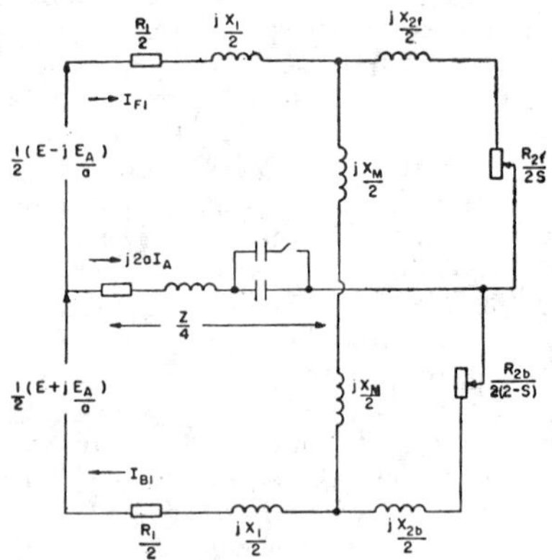

FIG. 18-36. Equivalent circuit of capacitor motor.

For the larger capacitor motors, in sizes of ½ hp and up, it is frequently economical to retain the auxiliary winding in circuit with a reduced capacitor size, to improve the operating characteristics. This is usually accomplished by providing a large electrolytic or highly stressed capacitor in parallel with a small oil- or Pyranol-type capacitor at starting and cutting the former out of circuit with a centrifugal switch or relay when the motor approaches full speed. Such motors are called "capacitor-run motors" and have winding connections as shown in Fig. 18-35. To appreciate the numerous considerations in the design of capacitor motors, it is desirable to understand the complete equivalent circuit,[1] given in Fig. 18-36. In this circuit

E = voltage impressed on main winding
E_A = voltage impressed on auxiliary winding (= E for usual case of Fig. 18-35)
a = ratio of effective turns, auxiliary to main winding

[1] KRON, GABRIEL Equivalent Circuit of the Capacitor Motor; *Gen. Elec. Rev.*, September, 1941, Vol. 44, p. 511.
VEINOTT, C. G. Moneca—A New Network Calculator for Motor Performance Calculations; *Trans. AIEE*, 1952, Vol. 71, Pt. III, p. 231.

$Z = R + jX$ = additional impedance of auxiliary winding, referred to main winding, where

$$R = \frac{R_{1A} + R_c}{a^2} - R_1$$

$$X = \frac{X_{1A} + X_c}{a^2} - X_1$$

R_{1A} = resistance of auxiliary stator winding (usually $\geqq a^2R_1$)
X_{1A} = reactance of auxiliary stator winding (usually $\leqq a^2X_1$)
$R_c - jX_c$ = impedance of capacitor in series with auxiliary winding
R_1, R_2, X_1, X_2, X_M = impedance constants of main winding (Fig. 18-33)
I_{F1} = forward-revolving component of main winding current $\times 2 = (I_1 - jaI_A)$
I_{B1} = backward-revolving component of main winding current $\times 2 = (I_1 + jaI_A)$
I_A = current in auxiliary winding

The main winding current $I_1 = \frac{1}{2}(I_{F1} + I_{B1})$, and $I_A = (j/2a)(I_{F1} - I_{B1})$. The line current is $I_L = I_1 + I_A$, if $E_A = E$.

If $Z = \infty$ (auxiliary winding open), the circuit becomes identical with Fig. 18-33 for the plain single-phase motor. In this case, $I_{F1} = I_{B1} = I_1$. If $Z = 0$, the circuit represents a 2-phase motor with unbalanced applied voltages.

The motor torque is the difference of the powers developed by the forward and backward fields, or

$$\text{Torque} = \frac{R_2}{2}\left(\frac{I_{F2}^2}{s} - \frac{I_{B2}^2}{2 - s}\right) \qquad \text{(synchronous W)}$$

106. Horsepower, Speed, and Voltage Ratings. Standard horsepower and speed ratings of single-phase motors are given in Table 18-11. Motors built in frames having a continuous rating of less than 1 hp, open type, at 1,700 to 1,800 r/min are designated "fractional-horsepower" motors, and those built in larger frames are called "integral-

Table 18-11. Standard Horsepower and Speed Ratings—Single-phase Constant-speed Motors

Standard horsepower ratings

1/20	1/6	1/2	1½	5	15
1/12	1/4	3/4	2	7½	20
1/8	1/3	1	3	10	25

Standard speed ratings

Rpm 60 cycles	Fractional hp	Integral hp
3,600	1/20–1	1½–25
1,800	1/20– 3/4	1 –25
1,200	1/20– 1/2	3/4–25
900	1/20– 1/8	1/2–25
Rpm 50 cycles		
3,000	1/20–1	1½–20
1,500	1/20– 3/4	1 –20
1,000	1/20– 1/2	3/4–20
750	....	1/2–20

horsepower" motors. Somewhat different standards of performance have been established for the two classes. USAS C-50 and NEMA Motor and Generator Standards, *Publ.* MG1-1955, include basic standards for both fractional- and integral-horsepower motors which are normally followed in specifications and testing. These publications may be obtained from the United States of America Standards Institute, 70 East 45th Street, New York City, and the National Electrical Manufacturers Association, 155 East 44th Street, New York City.

Both capacitor and split-phase motors are available in the multispeed as well as the single-speed type. They are used principally for belt and direct drive of centrifugal and propeller fans and are of the variable-torque class (Par. **81**). The multispeed motors for fan drive allow a change in fan speed without changing pulleys, which is essential where remote or automatic control of the rate of air delivery is required.

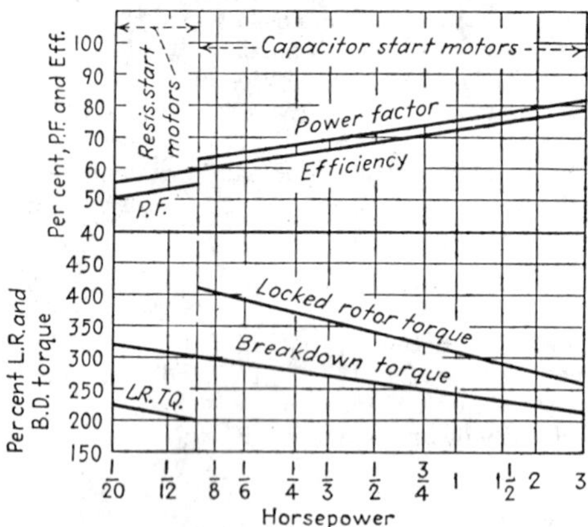

FIG. 18-37. Typical operating characteristics of 1,800-r/min single-phase motors.

The standard voltage ratings for single-phase motors are 115 and 230 V. However, many single-phase supply systems operate at other voltages, such as 110 and 220 or 120 and 240 V. Power companies place a limit on the size of motors that may be connected to single-phase lines. The limit usually falls between ½ and 1 hp for 110- to 120-V circuits and between 3 and 10 hp for 220- to 240-V circuits.

107. Temperature Rise. The standard temperature rises and service factors of single-phase motors are the same as for polyphase motors (Par. **85**).

108. Efficiencies and Power Factors. Typical efficiencies and power factors of

Table 18-12. Single-phase Motor Characteristics

	Hp	Approximate full load, amp		Locked rotor, amp				Breakdown torque (for defining hp ratings), oz-ft above line; lb-ft below line			
		115 volts	230 volts	115 volts		230 volts		3,600 rpm	1,800 rpm	1,200 rpm	900 rpm
Fractional hp	⅙	4.4	2.2	20		10		8.7–11.5	16.5–21.5	24.1–31.5	31.5–40.5
	¼	5.8	2.9	23		11½		11.5–16.5	21.5–31.5	31.5–44.0	40.5–58.0
	⅓	7.2	3.6	31		15½		16.5–21.5	31.5–40.5	44.0–58.0	58.0–77.0
	½	9.8	4.9	45		22½		21.5–31.5	40.5–58.0	58.0–82.5	
	¾	13.8	6.9	61		30½		31.5–44.0	58.0–82.5	5.16–6.9	
				Design		Design					
				L	M	L	M				
Integral hp	1	16	8	70	..	35	...	44.0–58.0	5.16–6.8	6.9–9.2	
	1½	20	10	..	..	50	40	3.6–4.6	6.8–10.1	9.2–13.8	
	2	24	12	..	..	65	50	4.6–6.0	10.1–13.0	13.8–18.0	
	3	34	17	..	..	90	70	6.0–8.6	13.0–19.0	18.0–25.8	
	5	56	28	..	..	135	100	8.6–13.5	19.0–30.0	25.8–40.5	
	7½	80	40	..	..	200	150	13.5–20.0	30.0–45.0	40.5–60.0	
	10	100	50	..	..	260	200	20.0–27.0	45.0–60.0		

the various types of induction motor that might be used to fill the requirements of the different ratings are shown in Fig. 18-37. Repulsion-start induction-run motors have about the same efficiencies and power factors except in the 1½- to 3-hp range, where they are lower. Repulsion-induction motors have roughly the same efficiencies but higher power factors.[1]

109. Full-load current of a single-phase motor is equal to

$$\frac{746 \times \text{hp}}{\text{Efficiency} \times \text{pf} \times \text{voltage}} \qquad (18\text{-}53)$$

where the efficiency and power factor are expressed as decimals. Approximate values of full-load current are given in Table 18-12. These are used for selecting wire and fuse sizes if no more accurate data are available.

110. Single-phase Motor Characteristics. 60-c four-pole 1,800-r/min synchronous speed. See Table 18-12.

111. Torques. The horsepower rating of a single-phase motor is defined by its breakdown torque. Thus, any 1,800 r/min motor with a breakdown torque between 31.5 and 40.5 oz·ft is, by definition, a ⅓-hp motor. The value used for definition is the minimum of the range of manufacturing variation for that particular design.

112. Starting Current.[2] Maximum values of locked-rotor current are established by NEMA for 60-c motors as shown in Table 18-12. In integral-horsepower sizes, NEMA has established two sets of locked-rotor values. The Design L motors include those types having inherently higher locked-rotor current than Design M motors. Power companies have developed a set of generally accepted starting current rules for single-phase motors (*EEI Publ.* Q-8).[3]

These rules specify limits for motor starting currents which will avoid excessive voltage dips, with resultant lighting flicker, and take into account the greater annoyances resulting from frequent motor starting during lighting hours as against the infrequent starting at any time. In general the rules permit 20 A at 115 V and 25 A at 230 V for automatically controlled single-phase motorized equipment. In the case of manually controlled (infrequently started) single-phase motorized equipment the allowable values are doubled for each voltage, and in all cases a tolerance of plus 15% is allowed in testing an individual motor. Locked-rotor currents in excess of these figures are permitted, upon approval of the electric utility, where available capacity permits, for example, in high-load-density areas.

113. Voltage and Frequency Variations. The allowable voltage and frequency variations for single-phase induction motors are the same as for the polyphase induction motors (see Par. **89**).

114. Construction. Single-phase motors are usually furnished with mounting feet either as an integral part of the stator or contained in a subbase. The subbase supports the motor by the end shields, usually through a resilient member in order to isolate pulsating-torque vibration. All sizes are available with a machined end-shield face (Par. **100**) on the drive end (with or without feet) for bolting to the driven machine.

Waste- or felt-packed sleeve bearings are standard for fractional-hp horizontal open motors and are available for the larger ratings. Ball bearings are standard for the larger horizontal machines of all enclosures as well as for vertical motors.

Standard single-phase motors are of the open or dripproof type, but all can be obtained in totally enclosed fan-cooled construction. Explosion-proof motors for Class I, Group D, conditions (Par. **98**) are available in sizes up to 10 hp.

Single-phase gear motors are available in sizes up to 5 hp with output shaft speeds of 13.5 to 780 r/min.

115. Motor Protection. Single-phase motors up to 2 hp often have an overload protective device integrally mounted with the motor, in which case the National

[1] Alger, P. L., and Johnson, T. C. Rating of General Purpose Induction Motors; *Trans. AIEE*, September, 1939, Vol. 58, p. 445.

[2] Williams, J. E. Design of Single-phase Motors to Minimize Voltage Dips; *Trans. AIEE*, 1953, Vol. 72, Pt. III, p. 484.

[3] *EEI Publ.* Q-8, single-phase motor starting current rules available at Edison Electric Institute, 420 Lexington Ave., New York, N.Y.

Electrical Code requires that the nameplate be permanently marked "thermal protection." The "thermal protector" used is an inherent overheating protective device responsive to motor current and temperature (except in the case of some small fractional-horsepower motors) and which when properly applied protects the motor against dangerous overheating due to overload or failure to start. The device can be either manually or automatically reset, although the latter type must not be used if an unexpected start of the motor could injure personnel or property.

OTHER TYPES OF INDUCTION MACHINES AND RELATED APPARATUS

116. Induction Generators. Any induction motor, if driven above its synchronous speed when connected to an a-c power source, will deliver power to the external circuit. This generator operation is easily visualized from the motor-circle diagram (Fig. 18-17) corresponding to the lower half of the circle in which the current vector is directed below the *OV* line. A unique feature is that the power factor of the output is fixed in value by the generator characteristics and is always leading, independent of the external circuit. The explanation is that the generator draws all its excitation from the system and so must receive a definite amount of lagging kilovoltamperes for a given voltage and load current. For this reason, induction generators alone cannot supply a power system but must always operate in parallel with synchronous machines or with capacitors.[1] They are, therefore, no more helpful in system stability than the addition of parallel reactors with a rating equal to the generator magnetizing reactance.

An induction generator delivers an instantaneous 3-phase short-circuit current equal to the terminal voltage divided by its standstill reactance, but its rate of decay is much faster than that of a synchronous generator of the same rating, and its sustained short-circuit current is zero.[2]

Since an induction generator must have a laminated rotor, to provide for the slip-frequency rotor magnetic field, its construction is not adapted to as high speeds as synchronous machines employing solid steel rotors. For these various reasons, induction generators have found few practical applications, their chief use being perhaps in variable-ratio frequency converter sets, where the induction end of the set operates as a motor or a generator depending on the direction of power flow through the set.[3]

117. Synchronous Induction Motors. In Europe, wound-rotor induction motors have often been provided with low-voltage d-c exciters that supply direct current to the rotor, making them operate as synchronous machines. With secondary rheostats for starting, such a motor gives the low starting current and high torque of the wound-rotor induction motor and an improved power factor under load. However, the extra cost, the small copper space available with a distributed rotor winding, and the necessary compromise between excessive ring voltage at start or excessive rotor current in operation are severe handicaps. Several different forms of these synchronous induction motors have been proposed, but they have not shown any net advantage over usual salient-pole synchronous or induction machines and are very seldom used in the United States. Reluctance motors (Par. **123**) also provide synchronous speed operation.[4]

The **Permasyn**[5] motor is a new type of permanent-magnet synchronous machine that promises to be widely useful in sizes up to about 5 hp. The Permasyn construction is the same as that of an ordinary squirrel-cage motor (either single or polyphase), except that the depth of rotor core below the squirrel-cage bars is very shallow, just enough to carry the rotor flux under locked-rotor conditions. The core may be interrupted by interpolar slots, to reduce pole-to-pole leakage. Inside this shallow rotor

[1] BASSETT, E. D., and POTTER, F. M. Capacitive Excitation for Induction Generators; *Trans. AIEE*, May, 1935, Vol. 54, p. 540.

WAGNER, C. F. Self-excitation of Induction Motors; *Trans. AIEE*, February, 1939, Vol. 58, p. 47.

[2] DOHERTY, R. E., and WILLIAMSON, E. T. Short Circuit Current of Induction Motors and Generators; *Trans. AIEE*, Vol. 40, p. 509.

[3] BERGER, M. J. La Machine asynchrone polyphasée et ses applications comme generatrice et comme transformatrice; *Intern. Elec. Congr.*, Paris, 1932, 3d section, *Rept.* 32.

[4] CROUSE, C. H. A Design Method for Polyphase Reluctance Synchronous Motors; *Trans. AIEE*, 1951, Vol. 70, Pt. I, p. 957.

CARR, L. H. A. Synchronous Induction Motor and Its Application; *Met.-Vick. Gaz.*, June, 1936, Vol. 16, p. 136.

CREEDY, F. Developments in Multispeed Cascade Induction Motors; *Jour. IEE*, 1921, Vol. 59, p. 511.

[5] MERRILL, F. W. Permanent-magnet Excited Synchronous Motors, *Trans. AIEE*, Power Apparatus and System, February, 1955, No. 16, p. 1754.

core is placed a permanent magnet, fully magnetized. The rotor core serves as a keeper, so that the rotor is not demagnetized by removing it from the stator. In starting, the rotor flux is small and is confined to the laminated core. As the speed rises, the rotor frequency decreases and the rotor flux builds up, creating a pulsating torque with the field of the magnet, as when a synchronous motor is being synchronized after the d-c field has been applied. As the Permasyn motor approaches full speed, therefore, the a-c impressed field locks into step with the field of the magnet and the machine runs as a synchronous motor. The absence of rotor I^2R loss, the synchronous speed operation, and the high efficiency and power factor make the Permasyn motor very attractive for special applications, such as high-frequency spinning motors. When many such motors are supplied from a high-frequency source, the kVA requirements are reduced to perhaps 50% of those needed for usual single-phase motor types, with consequent large savings. The high cost and low flux capacity of permanent-magnet materials have so far limited this new type of machine to relatively few uses, but its field is expected to grow (see also Par. **123**).

118. Induction-motor Speed Control. As mentioned in Par. **80,** the usual method of speed control for induction motors is to employ wound-rotor motors with adjustable secondary resistance, but this gives low efficiency at low speeds and poor speed regulation. If smooth speed control is needed, liquid rheostats are used but these make the cost high. Another method of speed control is to use multispeed motors, as discussed in Par. **81,** but this gives only a few definite speeds. Concatenation, or the connection of two or more wound-rotor motors electrically in series and coupling them mechanically also, gives speeds corresponding to the sum of the pole numbers, but this is comparable with, and more expensive than, the multispeed motor and is no longer used. Occasionally, the provision of an independent adjustable-frequency power supply may be justified, in which case any desired motor speed may be obtained by adjusting the supply frequency. This is likely to be very expensive, however.

Several methods of providing smooth, or "stepless," speed control of squirrel-cage motors have been used to a growing extent in recent years. One method is to provide a hydraulic coupling between the motor and the driven shaft. By control of the amount of liquid in the coupling, its slip can be adjusted over a wide range. The efficiency is low, as for a wound-rotor motor, because the motor must supply the load torque at full speed, the torque times slip energy being dissipated in the coupling, which, therefore, is usually water-cooled. This scheme has found favor for driving large blowers, as in power plants.

Another method utilizes cycloconverter-type frequency changers to generate adjustable frequency and voltage to power squirrel-cage induction motors. This is a means for providing efficient control of the motor torque over a wide speed range including stall. The principle of the cycloconverter, although known for years, is only now beginning to find wide application[1] as a result of the growing availability of high-current silicon-controlled rectifiers or thyristors.

119. Eddy-current Clutches. By constructing a squirrel-cage induction or synchronous motor with additional bearings, so that both primary and secondary members can revolve independently, a useful form of eddy-current clutch is obtained. Usually, such devices are designed with salient poles and d-c excitation, to simplify the control.[2] The d-c-excited member is generally fixed on the load shaft, and the squirrel-cage member (or frequently a solid iron or steel cylinder in which eddy currents are induced)

---

[1] Hamilton, R. A., and Lezan, G. R. Thyristor Adjustable Frequency Power Supplies for Hot Strip Mill Run-out Tables, *IEEE Conference on Rectifier Industrial Static Power Conversion*, November, 1965, p. 69.

Amato, C. J. Variable Speed with Controlled Slip Induction Motor, *IEEE Conference on Rectifier Industrial Static Power Conversion*, November, 1965, p. 181.

Usher, T. E., and Beck, C. D. Adjustable Speed with AC Motors and Adjustable Frequency Inverters, *AIEE Paper* No. DP 62-981.

Flairty, C. W. A 50 KVA Adjustable-frequency 24-phase Controlled Rectifier Inverter, *IRE Trans. on Industrial Electronics*, Vol. IE-9, No. 1, May, 1962, p. 56.

MacDonald, I. M. A Static Inverter, Wide Range, Adjustable Speed Drive, *IEEE International Convention Record*, Vol. 12, Part 4, 1964, p. 34.

Helmick, C. G., and Lipman, K. Adjustable-frequency Power Inverter Systems for Industry, *Westinghouse Engineer*, Vol. 23, No. 6, November, 1963, p. 167.

Mapham, N. W. The Classification of SCR Inverter Circuits, *IEEE International Convention Record*, Vol. 12, Part 4, 1964, p. 99.

Hulstrand, K. A. Adjustable Frequency—Static AC Power Supplies, *IEEE Paper* No. CP 64-571.

[2] Lory, M. R., Kilgore, L. A., and Baudry, R. A. Electric Couplings; *Trans. AIEE*, August, 1940, Vol. 59, p. 423.

is attached to the rotor of the driving motor. The motor is started up with the clutch unexcited, the load remaining at rest. By then supplying a graduated amount of direct current to the clutch, a magnetic field is produced which induces squirrel-cage currents and a torque that brings the load up to any desired speed. Of course, the entire power represented by the driving-motor torque, multiplied by the difference between the speed of the driving motor and the speed of the load shaft, is a power loss dissipated in the clutch; so the efficiency is no better than with a rheostatically controlled wound-rotor motor. However, the smooth adjustment of the clutch torque and, therefore, the load speed by control of the d-c excitation is an advantage that may justify the use of the clutch.

120. Selsyn Systems.[1] The word "selsyn" is an abbreviation of the term "self-synchronous," indicating machines which operate in space-phase alignment by virtue of electrical interconnection. If two wound-rotor induction motors are connected in parallel on their primary sides and to a common rheostat on their secondary sides, they will exert a synchronizing torque opposing any change in their relative speeds. The synchronizing torque falls off rapidly, however, as the external secondary resistance is reduced, and it is, of course, zero when both rotors are short-circuited. Hence, it is necessary to provide separate selsyn machines on the power-motor shafts if synchronism is to be maintained over normal ranges of speed and load. This characteristic is utilized in a variety of industrial applications, such as lift bridges and high-speed newspaper presses, where it is desired to have two or more motors operate in approximate synchronism during starting and load changes, without any mechanical tie between them.

The principle is also usefully applied in remote-indicating and follow-up mechanisms of many types.[2] Generally, there are a master selsyn and one or more receiver selsyns, the latter following exactly the angular position of the master. Each of the machines is essentially a small a-c-excited synchronous motor, usually with a single-phase primary and a 3-phase secondary winding. A common source of a-c excitation is connected to all the primary windings in parallel, and all the secondary windings are similarly connected in parallel, but not to any external source. If all the rotors are in the same angular position, all the secondary voltages will be in time phase; so no torque-producing secondary currents will flow. If one rotor is displaced through an angle θ, its secondary voltage will be $\theta°$ out of phase with all the

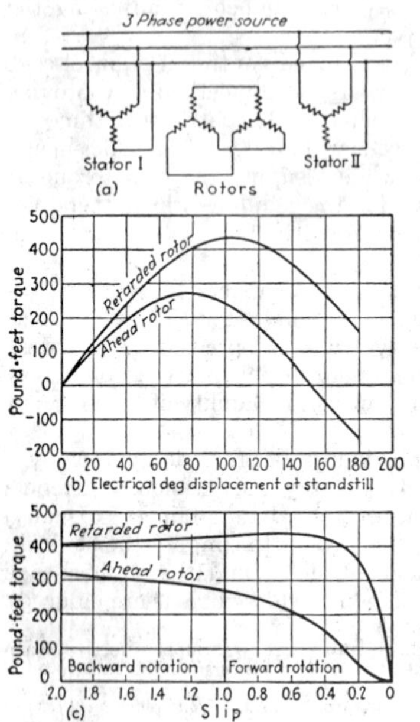

Fig. 18-38. Selsyn connections and characteristics.

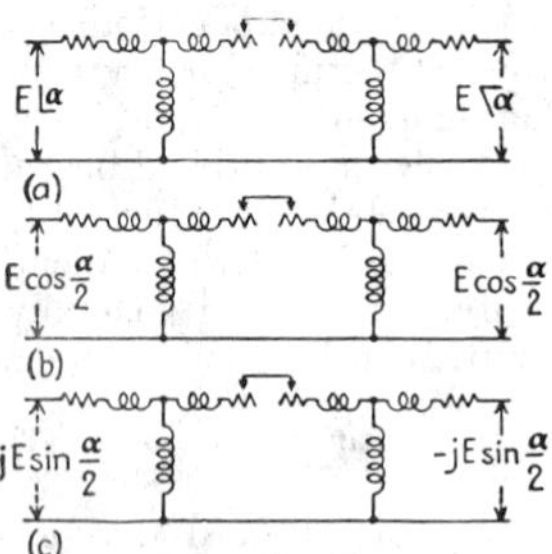

Fig. 18-39. Equivalent circuit for slip-ring motors paralleled on both stator and rotor. (*a*) When rotors are displaced by angle α; (*b*, *c*) two circuits equivalent to (*a*).

---

[1] Nowacki, L. M. Induction Motors as Selsyn Drives; *Trans. AIEE Suppl.*, 1934, Vol. 53, p. 1721.

[2] Linville, T. M., and Woodward, J. S. Selsyn Instruments for Pointer Systems; *Trans. AIEE*, 1934, Vol. 53, p. 953.

Johnson, T. C. Selsyn Design and Application; *Trans. AIEE*, October, 1945, Vol. 64, p. 703.

others and secondary currents will flow, creating torques that tend to restore the primaries to a common position. This positioning effect is present, whether the machines are rotating or standing still. For single-phase primaries, however, the synchronizing torque decreases as the speed is increased, with equal effects for either direction of rotation; while for 3-phase primaries, the synchronizing torque increases with speed in the direction against the phase rotation and decreases more rapidly than in the single-phase case, with increased speed in the direction of the phase rotation. At standstill, the single-phase primary has the greater torque, owing to the added reluctance torque of its salient poles. Small positioning selsyns are therefore usually built with single-phase primaries, while power selsyns are normally 3-phase and may be identical with wound-rotor induction motors.

Figure 18-38*a* to *c* shows the connection diagram, 3-phase standstill torque-angle characteristics, and 3-phase maximum torque vs. speed for two 25-hp 60-c 900-r/min, slip-ring motors connected in parallel on both sides. As diagram *a* indicates, no rotor current can flow unless the two rotors are displaced in space phase. Figure 18-39*a* shows the equivalent circuit for this condition, when the two rotors are displaced by an angle α. For calculation, it is convenient to consider the circuit as the result of superposing the two separate circuits shown in Fig. 18-39*b* and *c*. The in-phase component of the two voltages $E \cos (\alpha/2)$ is impressed on each side of circuit *b*, so that no rotor current flows. The opposite phase components of the voltages $jE \sin (\alpha/2)$ and $-jE \sin (\alpha/2)$ are impressed on the two sides of the circuit *c*, so that equal and opposite currents flow in the two rotors. These currents create induction-motor torques $T_0 \sin^2 (\alpha/2)$, tending to accelerate both rotors in the direction of the rotating magnetic field. Each of these currents also creates a synchronizing torque with the magnetic field, established by circuit *b*, tending to bring the two rotors together.

The synchronizing torque pulling forward the rotor that is behind (in the direction of the rotating field)

$$= \frac{6.09V(I - I_M)}{N_s} \sin \theta \sin \alpha + T_0 \sin^2 \frac{\alpha}{2} \qquad \text{(lb·ft)} \qquad (18\text{-}54)$$

The synchronizing torque pulling forward the rotor that is ahead

$$= \frac{6.09V(I - I_M)}{N_s} \sin \theta \sin \alpha - T_0 \sin^2 \frac{\alpha}{2} \qquad \text{(lb·ft)} \qquad (18\text{-}55)$$

where V = line voltage; I = induction-motor current for slip s when rotor is short-circuited; $\cos \theta$ = induction-motor power factor for slip s and load current I when rotor is short-circuited; T^0 = induction-motor torque for slip s when rotor is short-circuited; I_M = induction-motor magnetizing current; N_s = synchronous speed of induction motor; and α = electrical angle of displacement between two rotors, each revolving at slip s.

If the rotors are stationary and the transmitter is displaced ahead in the direction of the rotating field, the torque tending to hold the transmitter back is given by the lower curve in Fig. 18-38*b* and the torque pulling the receiver ahead is larger, as given by the upper curve. If the transmitter is moved backward, against the rotating field, the torque on the receiver will be less, following the lower curve, and the torque opposing the transmitter motion will follow the upper curve. Figure 18-38*c* shows the maximums of the corresponding torque-displacement curves plotted against speed. Evidently, the torque available to hold the receiver back, if the transmitter is retarded, is small when the rotor is operating in the neighborhood of synchronous speed. For this reason, it is preferable to connect the selsyns so that they rotate oppositely to their magnetic fields or with opposite phase rotation to the power motors that they are expected to hold in step.

Power-selsyn applications, therefore, consist of induction-driving motors, with separate direct-connected selsyns, often duplicates of the driving motors, all connected to the same power source. To bring the motors into alignment before starting the load, it is usual to apply single-phase voltage to the selsyn primaries for a brief period before applying 3-phase to all units.

121. High-frequency Motors. For high-speed tools and for spinning of rayon and other threads, a variety of interesting motor constructions have been developed. Normally these are two-pole 3-phase motors, with special high-frequency power supply of 90, 120, or 180 c, giving operating speeds between 5,000 and 10,500 r/min. In textile applications, the motors usually drive individual spinning buckets, which are subject to considerable unbalance due to uneven building up of thread, etc. The continual starting and stopping for loading and unloading the buckets requires the motors to carry unbalance reliably through the entire speed range, necessitating careful design of mounting flexibility and shaft stiffness. Most usual applications, however, are in woodworking and similar industries, where separate motor stators and rotors are supplied to the tool manufacturers for building into their particular devices (Par. **99**).

Three-phase 400-c power systems, used on large airplanes, have led to the development of 400-c motors with speeds of 12,000 and 24,000 r/min, having weights averaging 2 lb/hp for motors of 1 to 15 hp with 5-min ratings. These motors are open, with an external fan to force air over the windings.

122. Hysteresis Motors for Clocks and Phonographs. By constructing the secondary core of an induction motor of hardened magnet steel, in place of the usual annealed low-loss silicon-steel laminations, the secondary hysteresis loss can be greatly magnified, producing effective synchronous motor action. Such hysteresis motors, having smooth rotor surfaces without secondary teeth or windings, give extremely uniform torque, are practically noiseless, and give substantially the same torque from standstill all the way up to synchronous speed. A hysteresis motor is a true synchronous motor, with its load torque produced by an angular shift between the axis of rotating primary mmf and the axis of secondary magnetization. When the load torque exceeds the maximum hysteresis torque, the secondary magnetization axis slips on the rotor, giving the same effect as a friction brake set for a fixed torque.

Despite the interesting characteristics of this type of motor, it is limited to small sizes, because of the inherently small torque derivable from hysteresis losses.[1] Only moderate flux densities are practicable, owing to the excessive excitation losses required to produce high densities in hard magnet steel, and, therefore, about 20 W/lb of rotor magnet steel represents the maximum useful synchronous power on 60 c. Hysteresis motors have found an important use for phonograph-motor drive, their synchronous speed enabling a governor to be dispensed with and freedom from tone waver to be secured. A rubber-tired roller drive is sometimes used with those motors to secure quiet operation and easy adjustment of speed. The motor shaft is pressed against the roller by a spring, and the roller in turn runs on the interior of the cup-shaped rim of the phonograph turntable.

The Telechron motor, which is so widely used for operating electric clocks, also operates on the hysteresis-motor principle. In the Telechron motor, a two-pole rotating field is produced in a cylindrical air space, and into this space is introduced a sealed thin-metal cylinder containing a shaft carrying one or more hardened magnet-steel disks, driving a gear train. The 60-c magnetic field causes the steel disks to revolve at 3,600 r/min, driving through the gears a low-speed shaft, usually 1 r/min, which emerges from the sealed cylinder through a closely fitting bushing designed to minimize oil leakage. Although the magnetic field has to cross a very considerable air-gap length and pass through the thin walls of the metal cylinder, the power required to drive a well-designed clock is so small that ample output is obtained with only about 2 W input for ordinary household-clock sizes.

123. Reluctance and Induction Motors. If the rotor surface of a P-pole squirrel-cage motor is cut away at symmetrically spaced points, forming P salient poles, the motor will accelerate to full speed as an induction motor and then lock into step and operate as a synchronous motor. The synchronizing torque is due to the change in reluctance and, therefore, in stored magnetic energy, when the air-gap flux moves from the low- into the high-reluctance region. Such motors are often used in small-horsepower sizes, when synchronous operation is required, but they have inherently low pull-out torque and low power factor, and also poor efficiency, and therefore require larger frames than the same horsepower induction motor. The recently developed

[1] Teare, B. R. Theory of Hysteresis Motor Torque; *Trans. AIEE Suppl.*, 1940, Vol. 59, p. 907.

Permasyn motor has superior performance in every way, except possibly cost, and may replace the reluctance motor in the course of time.

If the number of rotor salients is nP, instead of P, and if the P-pole motor winding is arranged to also produce a field of $(n - 1)P$ or $(n + 1)P$ poles, the motor may lock into step at a subsynchronous speed and run as a subsynchronous motor. For the P-pole fundamental mmf, acting on the varying rotor permeance, will create $(n + 1)P$ and $(n - 1)P$ pole fields from this cause, and these will lock into step with the independently produced $(n - 1)P$- or $(n + 1)P$-pole field, when the rotor speed is such as to make the two harmonic fields turn at the same speed in the same direction.

It is difficult to provide much torque in such a subsynchronous motor, and their use is therefore limited to very small sizes, such as may be used in small timer or instrument motors.

ALTERNATING-CURRENT COMMUTATOR MOTORS

124. Classification.[1] As compared with the induction motor, the a-c commutator motor possesses two of the advantages of the d-c motor: a wide speed range without sacrifice of efficiency, and superior starting ability. In the induction motor, the starting torque is limited by the small space-phase displacement between the air-gap flux and the induced secondary current and by magnetic saturation of the flux paths. Besides limiting the torque-producing flux, such saturation has the additional effect of reducing the induced secondary current, by virtue of the increased diversion of primary current into the magnetizing-current path (Fig. 18-18). In the a-c commutator motor, on the other hand, the torque and current are held at the optimum space-phase displacement by proper location of the brush axis, and the secondary current is not limited by magnetic saturation, giving high torque per ampere at starting. Furthermore, the series commutator motor may be operated far above the induction-motor synchronous speed, giving high power output per unit of weight.

Alternating-current commutator motors may be grouped into two classes:

a. Those motors in which the resultant mmf providing the flux increases with the load. When operated from a source of constant voltage, the speed of such motors decreases with increasing load. They are termed "series motors" from the similarity of their characteristics to those of series-wound d-c motors. The speed at any given load may be varied by changing the applied voltage or, in some cases, by shifting the brushes.

b. Those motors in which the resultant mmf providing the flux is substantially constant irrespective of the load. For operation from a source of constant voltage, the speed of such motors is approximately constant. The speed may, however, be increased or decreased (independently of the load) by increasing or decreasing the voltage at the terminals of the motor, by brush shifting, or by the provision of suitably disposed and connected auxiliary coils. Such motors are termed "shunt motors."

Alternating-current commutator motors are either single-phase or polyphase.[2] A unique characteristic of all single-phase motors is a double line-frequency pulsation of the torque produced, corresponding to the sinusoidal variation twice each cycle of the single-phase power supplied. For any voltage V, current I, and power-factor angle θ, the instantaneous power is

$$\text{Power} = IV \sin \omega t \sin (\omega t - \theta)$$
$$= \frac{IV}{2} [\cos \theta - \cos (2\omega t - \theta)] \tag{18-56}$$

The second term of this equation represents the double-frequency torque pulsation,

[1] CREEDY, F. "Single-phase Commutator Motors"; Princeton, N.J., D. Van Nostrand Company, Inc., 1913.

OLLIVER, C. W. "The A-c Commutator Motor"; Princeton, N.J., D. Van Nostrand Company, Inc., 1927.

TEAGO, F. J. "The Commutator Motor"; New York, E. P. Dutton & Co., Inc., 1930.

GRABNER, A. On the Development of the Single-phase Commutator Motor and Its Commutation (in German); *Elek. u. Masch.*, Aug. 18, 1939, Vol. 57, p. 425.

[2] SMITH, S. P. Single and Three-phase Alternating-current Commutator Motors with Series and Shunt Characteristics; *Jour. IEE*, March, 1922, Vol. 60, p. 308.

which is partly transmitted to the load, causing small speed pulsations and necessitating special coupling and mounting designs to minimize vibration and fatigue stresses.

Polyphase commutator motors have the advantage of better inherent commutating ability, due in part to the need for shifting the rotor current only 60° in time phase at each brush stud for a 6-phase motor or 30° for 12 phases, as compared with 180° shift for a single-phase or d-c machine. Single-phase motors are generally limited to sizes below about 10 hp, except for railway applications.

125. Single-phase Straight Series Motor. An ordinary d-c series motor, if constructed with a well-laminated field circuit, will operate (although unsatisfactorily) if connected to a suitable source of single-phase alternating current. Since the armature is in series with the field, the periodic reversals of current in the armature will correspond with simultaneous reversals in the direction of the flux, and consequently the torque will always be in the same direction. But the inductance of the motor will be so great that the current will lag far behind the voltage, and the motor will have a very low power factor. The entire amount of armature flux produced along the brush axis generates a reactive voltage in the armature, which must be overcome by the applied voltage, without performing any useful function whatever. Such a motor is shown in Fig. 18-40, F representing the field and A the armature.

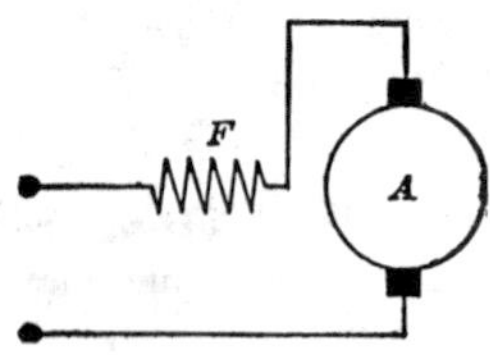

Fig. 18-40. Simple series-motor circuit diagram.

When the motor is first thrown on the circuit, and before the armature has moved from rest, the field constitutes the primary of a transformer and sends flux through the armature core. Those armature turns which at that instant are short-circuited under the brushes act as short-circuited secondary coils and are traversed by heavy currents which serve no useful purpose whatever and occasion serious heating. When the armature starts to revolve, these short-circuited turns are opened as they pass out from under the brushes and are replaced by other turns which are momentarily short-circuited and then opened. These interruptions of heavy currents are accompanied by serious sparking, since the heating is concentrated at the few segments on which the brushes rest. As soon, however, as a certain speed in acquired, the heating is distributed over all the segments and the conditions are ameliorated. This source of sparking is, then, most serious at the moment of starting. One way in which this difficulty has often been minimized is by the employment of leads of high resistance connecting the winding to the commutator segments.

The simple single-phase series motor has therefore two major faults, low power factor and poor commutation at low speeds, confining its use to fractional horsepower and very high speed applications.

126. Single-phase Compensated Series Motor. In all except the smallest sizes, it is usual to employ a compensating winding on the stator, in series with the armature and so arranged that its mmf as nearly as possible counteracts the armature mmf. A commutating winding is also frequently used, which somewhat overcompensates the armature reaction along the interpolar, or commutating-zone, axis and so provides a voltage to aid the current reversal, just as in a d-c motor. By these means, the flux along the brush axis is reduced to a small fraction of its uncompensated value, and the power factor of the motor is greatly improved. Further improvement of the power factor is secured by using a smaller air gap and correspondingly fewer field ampere-turns than in an uncompensated motor, thus reducing the reactive voltage in the series field to a minimum. Figure 18-41 illustrates the winding diagram of this motor.

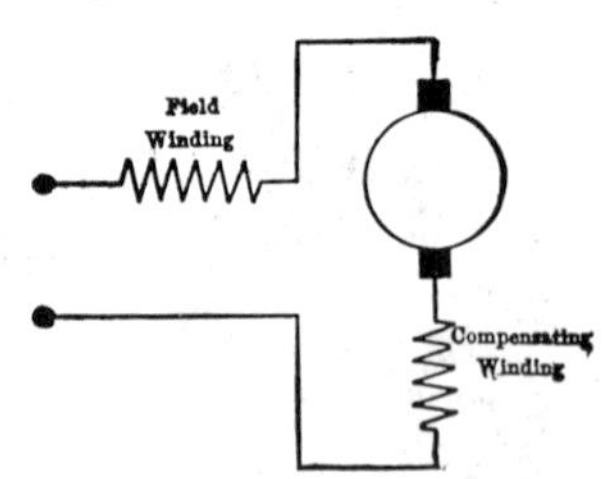

Fig. 18-41. Compensated series-motor circuit diagram.

127. Universal Motors.[1] Small series motors up to about ½-hp rating are commonly designed to operate on either direct current or alternating current and so are called "universal motors." Universal motors may be either compensated or uncompensated, the latter type being used for the higher speeds and smaller

[1] Packer, L. C. Universal Type Motors; *Trans. AIEE*, 1925, Vol. 44, p. 587.

ratings only. Owing to the reactance voltage drop, which is present on alternating current but absent on direct current, the motor speed is somewhat lower for the same load a-c operation, especially at high loads. On alternating current, however, the increased saturation of the field magnetic circuit at the crest of the sine wave of current may materially reduce the flux below the d-c value, and this tends to raise the a-c speed. It is possible, therefore, to design small universal motors to have approximately the same speed-torque performance over the operating range, for all frequencies from 0 to 60 c. On a typical compensated-type ¼-hp motor, rated at 3,400 r/min, the 60-c speed may be within 2% of the d-c speed at full-load torque but 15% or more lower at twice normal torque, while on an uncompensated motor the speed drop will be materially greater.

The commutation on alternating current is much poorer than on direct current, owing to the current induced in the short-circuited armature coils, and this provides a definite limitation on their size and usefulness. If wide brushes are used, the short-circuit currents are excessive and the motor-starting torque is reduced, while if narrow brushes are used, there may be excessive brush chatter at high speeds, causing short brush life. Good design, therefore, requires careful proportioning of commutator and brush rigging to meet conflicting electrical, mechanical, and thermal requirements. Universal motors are generally used for vacuum cleaners, portable tools, food mixers, and similar small devices operating at speeds of 3,000 to 10,000 r/min. These small series-wound motors are frequently furnished as motor parts, i.e., consisting of bare stators and rotors (with shaft) but without bearings or feet. They can then be compactly "built in" to the power-using devices.

Figure 18-42 illustrates a typical vacuum-cleaner motor's characteristics.

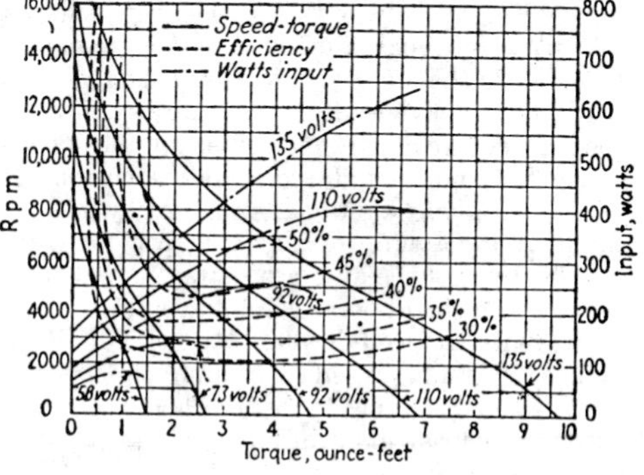

FIG. 18-42. Performance characteristics of universal motor.

128. Single-phase Railway Motor.[1] The series-motor characteristic with its very high starting torque and high light-load speed is ideal for traction purposes, so that single-phase series motors are used in the majority of a-c railway electrifications. The outstanding design problem in this case is to provide adequate commutating ability and thermal capacity to withstand the high circulating currents induced in the short-circuited armature coils at starting. With a sinusoidal flux wave, the rms voltage induced in each short-circuited turn at standstill is

$$\text{Transformer voltage/turn} = 1.11 \frac{f}{25} \phi_{max}$$

where f = supply frequency in cycles per second and ϕ_{max} = maximum value of the torque-producing flux per pole in megalines. A frequency of 25 c, or commonly 16⅔ cycles in Europe, has been adopted for railway electrifications, because of the serious limitations on higher-frequency motors imposed by this relationship.

Normal carbon brushes will glow if more than 2.5 V is continuously applied from heel to toe, so that the transformer voltage must be kept below this value in heavy starting duty if reasonable maintenance is to be secured. This requires that the 25-c motor flux be kept below about 2.5 megalines/pole. Additional torque can be secured only by increasing the number of poles and brush studs proportionately, so that modern 25-c railway motors of 200 to 600 hp usually have 12 to 18 poles. The low flux per

[1] PRITCHARD, F. H., and KONN, F. The Modern Single Phase Motor for Railroad Electrification; *Trans. AIEE*, March, 1931, Vol. 50, p. 263.
KONN, F. Single-phase Commutator Traction Motor; *Gen. Elec. Rev.*, April, May, and July, 1932, Vol. 35, pp. 206, 275, 397.

pole and one turn per armature coil requirement also limit the practicable voltage of 25-c railway motors to the range of 100 to 300 V. Speed control is therefore conveniently provided by tap changing on supply transformers.

129. Repulsion Motor.[1] If, instead of connecting the field and armature in series, the supply voltage is connected directly across the field and the armature brushes are short-circuited, the simple repulsion motor is obtained, as illustrated in Fig. 18-43. Actually, the field and compensating functions are performed by a single stator winding, and the brush axis is placed at an angle α with respect to the axis of this winding. The effective numbers of turns in the field and compensating stator axes are therefore equal to $N \sin \alpha$ and $N \cos \alpha$, respectively.

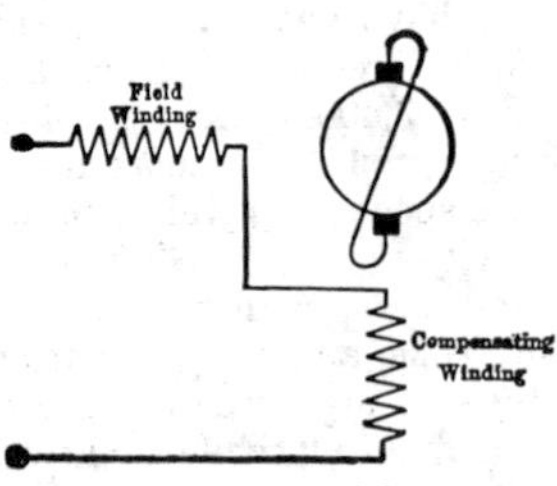

Fig. 18-43. Repulsion-motor circuit diagram.

If the brushes are shifted into line with the stator winding axis ($\alpha = 0$), the motor becomes a short-circuited transformer, drawing an excessive current and producing zero torque. As the brushes are shifted away from this position, called the "live neutral," a flux is produced in the cross axis and a torque is developed in the same direction as the shift. Further shift of the brushes results in a gradually decreasing current and a torque that passes through a maximum at an angle of 15 to 25°. Finally, at the false neutral ($\alpha = 90°$), the induced armature current is zero, and the torque is also zero.

As the motor speeds up, with any setting of α, the speed voltage due to the armature conductors cutting the field flux opposes the transformer voltage due to the compensating flux and the induced armature current also decreases, allowing a gradual increase in the flux along the brush axis. At synchronous speed, the fluxes in the two axes are equal, giving a pure rotating magnetic field.

The repulsion motor has better commutation than the series motor at speeds up to synchronism and poorer commutation at very high speeds, owing to the greater values of currents induced in the short-circuited coils under the latter conditions. These induced currents provide a braking effect which serves to limit the high no-load speed.

The repulsion motor has two important advantages over the series motor for small-power industrial applications. The armature, being isolated from the line, may be designed for any convenient low voltage, to ease the problem of commutation, and no brush-yoke insulation is needed. Also, the pure rotating magnetic field produced at synchronous speed gives exact neutralization of the voltage induced in the short-circuited coils and, therefore, gives good commutation at this speed.

Figure 18-44 illustrates the performance of a simple repulsion motor with fixed brushes, operated at different voltages, obtained by taps on a supply transformer. Such motors are used for fan drive and in variable-speed applications. Instead of varying the voltage, the brushes may be shifted, very similar performance curves being obtained (Fig. 18-45).

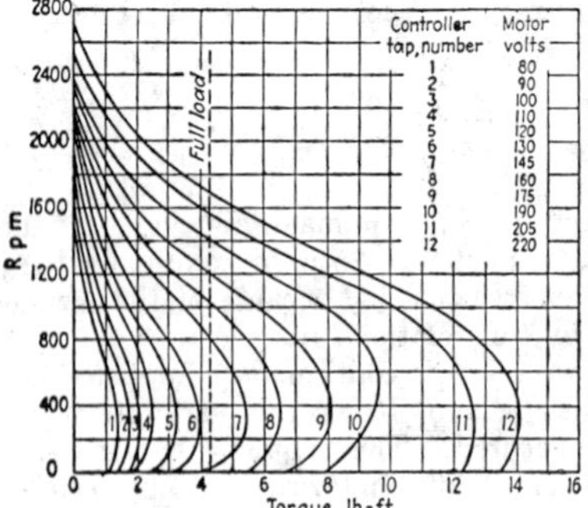

Fig. 18-44. Speed-torque curves of repulsion motor with varying applied voltage.

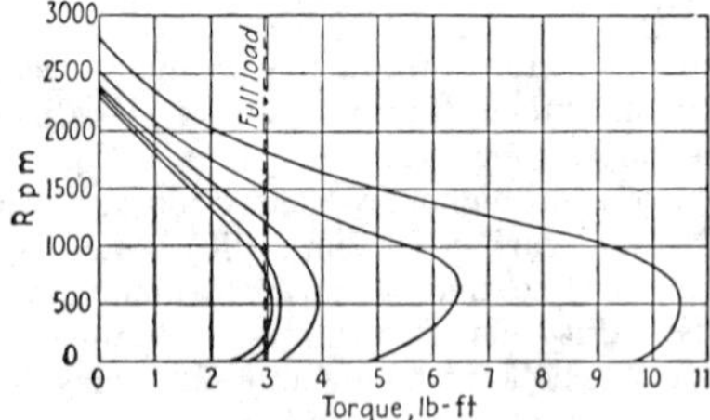

Fig. 18-45. Speed-torque curves of repulsion motor with brush shifting.

[1] West, H. R. The Cross Field Theory for A-c Machines; *Trans. AIEE*, February, 1926.

130. Armature-excited Repulsion Motors. Numerous varieties of armature-excited repulsion motors have been developed in past years, with the objects of power-factor correction and speed control. These motors have an extra set of brushes in the field axis, through which an exciting current is led from a variable-ratio transformer or other source (Fig. 18-46). The short-circuited brushes are then in line with the stator winding axis, and performance similar to a shunt d-c motor may be obtained, speed adjustment being secured by change in the voltage applied to the excitation brushes. However, for good commutation, commutating flux adjustments should be made in each brush axis for different speeds, and the overall result of this complication has been to restrict this form of motor to very limited uses.

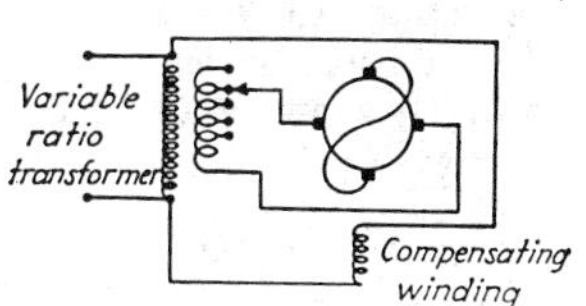

Fig. 18-46. Armature-excited repulsion-motor circuit diagram.

Instead of varying the excitation voltage, the two sets of brushes may be shifted together, varying the speed for a given load from a maximum when the short-circuited brushes are in the stator winding axis ($\alpha = 0°$) to zero when $\alpha = 90°$. This form of motor has a series speed characteristic, and it has been rather widely used for variable-speed spinning, printing-press drive, etc.

131. Repulsion-start Induction Motor. The desirable combination of the high starting torque and low starting current of the repulsion motor with the constant-speed characteristic of the induction motor has led to numerous designs which automatically shift from the repulsion to the induction characteristic at about two-thirds of synchronous speed. The most usual type has some form of centrifugal mechanism which short-circuits the commutator bars at a predetermined speed. At all higher speeds, therefore, the armature winding effectively becomes a simple squirrel cage, and the motor has normal single-phase induction-motor characteristics. Many such motors combine a brush-lifting mechanism with the short-circuiting device, so that the brush noise and wear are absent during normal load operation. The usual range of sizes is 1/4 to 5 hp at 1,800 r/min, with starting torques of 300 to 600% of full-load value. They are widely applied to drive compressors, pumps, and other constant-speed equipment (see also Pars. **104** and **114**).

132. Squirrel-cage Repulsion Motors.[1] Instead of employing a centrifugal switch, some designers provide a deeply embedded low-resistance squirrel-cage winding in the armature of the repulsion motor. The high reactance of the squirrel cage causes it to draw only a small additional current at starting, so that the starting characteristics are only slightly inferior to the pure repulsion motor. As the motor accelerates, the reduced rotor frequency enables more and more current to flow in the low-resistance squirrel cage, and at speed the induction-motor characteristic predominates. The no-load speed is slightly above synchronism and the no-load current and losses a little higher than at partial loads, owing to the induction-generator action of the squirrel cage in opposition to the repulsion-motor torque. The squirrel cage has also a strong

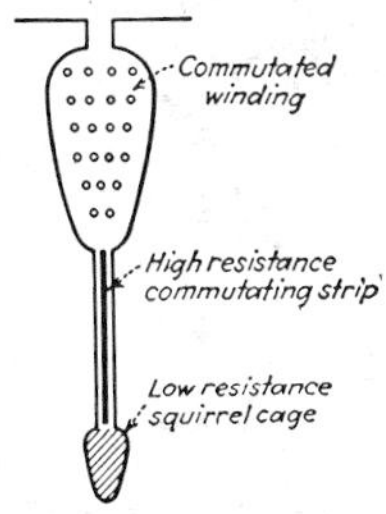

Fig. 18-47. Rotor slot of squirrel-cage repulsion motor.

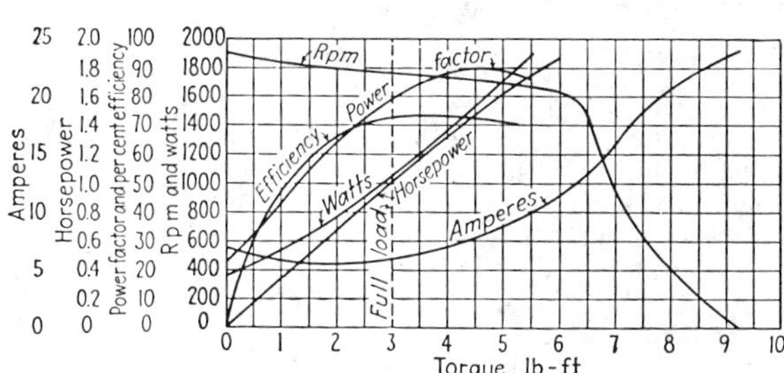

Fig. 18-48. Performance characteristics of normal-torque squirrel-cage repulsion motor.

[1] West, H. R. Theory and Calculation of the Squirrel Cage Repulsion Motor; *Trans. AIEE*, 1924, Vol. 43, p. 1048.
Bergman, S. R. A New Type of Single-phase Motor; *Trans. AIEE*, 1924, Vol. 43, p. 1039.

power-factor-correcting influence, giving these motors power factors 10% or more better than other repulsion-motor types.

Figure 18-47 indicates the slot arrangement of this motor. The brushes carry power current at all speeds, but especially good commutation is secured by resistance strips in the leakage slots of the squirrel cage. The resistance of these strips is chosen high enough to make them have a negligible effect on the starting performance but low enough for them to serve as an effective barrier to flux at the high frequency of commutation, so enabling the energy of any unreversed armature current to be inductively transferred to the squirrel cage instead of being dissipated in sparking at the brushes. Figure 18-48 illustrates typical performance characteristics of one of these motors.

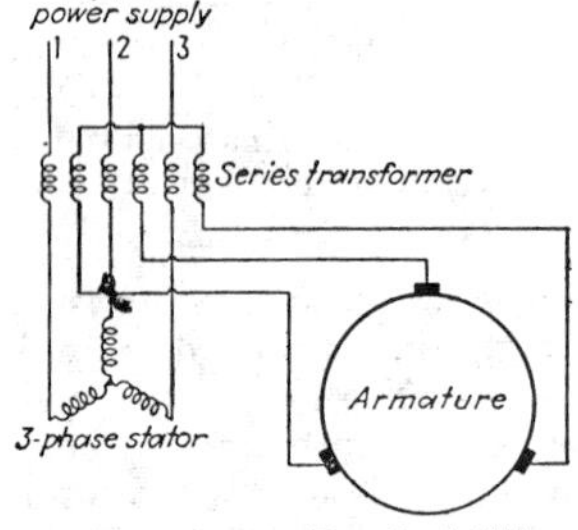

FIG. 18-49. Brush-shifting series-motor circuit diagram.

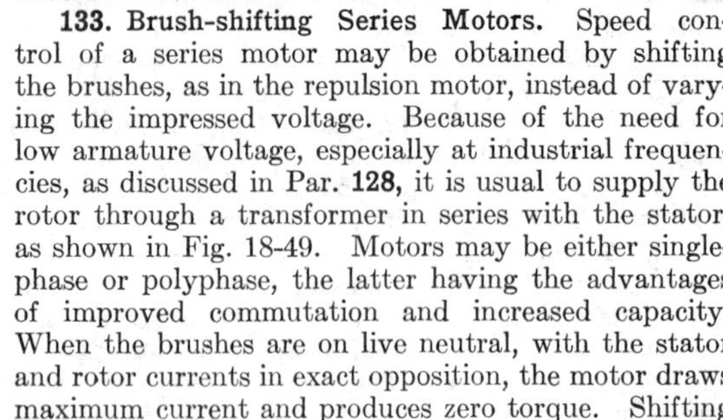

133. Brush-shifting Series Motors. Speed control of a series motor may be obtained by shifting the brushes, as in the repulsion motor, instead of varying the impressed voltage. Because of the need for low armature voltage, especially at industrial frequencies, as discussed in Par. **128**, it is usual to supply the rotor through a transformer in series with the stator, as shown in Fig. 18-49. Motors may be either single-phase or polyphase, the latter having the advantages of improved commutation and increased capacity. When the brushes are on live neutral, with the stator and rotor currents in exact opposition, the motor draws maximum current and produces zero torque. Shifting the brushes in either direction produces a torque in the same direction, the motor impedance rising with the angle of shift, until, at the dead neutral, 180°, the stator and rotor mmfs are additive, and maximum flux is produced with again zero torque.

The saturation limit of the transformer is conveniently used to hold the no-load speed down to a suitable value, usually about 150% of synchronous speed. This feature gives this type of motor a somewhat better speed regulation than other forms of series motors.[1]

134. Brush-shifting Polyphase Shunt Motor. The so-called "Schräge motor"[2] gives an adjustable constant-speed characteristic with good torque and efficiency. This motor consists of a polyphase induction motor with the primary winding on the rotor and insulated secondary winding on the stator, the latter being connected across two in-

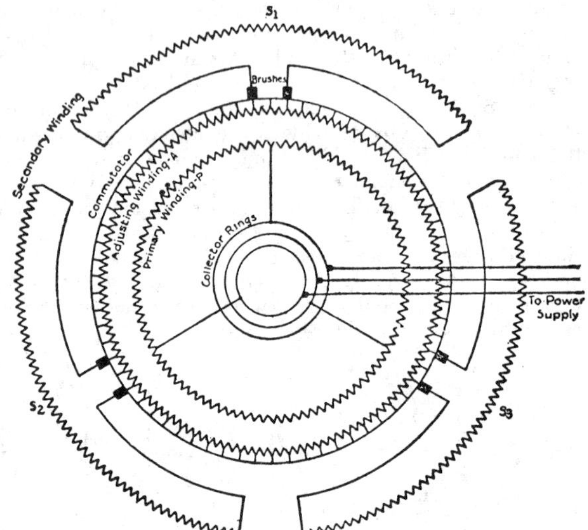

FIG. 18-50. Brush-shifting polyphase shunt-motor circuit diagram.

[1] JONES, R. A. Adjustable Varying Speed Alternating-current Commutator Motor; *Gen. Elec. Rev.*, November, 1921, Vol. 24, p. 921.

[2] LATOUR, M. C. A. Commutation in Alternating-current Machinery; *Trans. AIEE*, 1918, Vol. 37, p. 355.

SCHRÄGE, H. K. Ein neuer Drehstrom-Kommutatormotor mit Nebenschlussregulierung durch Bürstenverschiebung; *E.T.Z.*, Jan. 22, 1914, Vol. 35, p. 89.

ALTES, W. C. K. Polyphase Shunt Motor; *Trans. AIEE*, 1918, Vol. 37, p. 295.

UHL, H. C. Adjustable Speed A-c Motor with Shunt Characteristics; *Gen. Elec. Rev.*, April, 1925, Vol. 28, p. 248.

dependently movable sets of polyphase brushes on a commutator. The commutator is fed by an auxiliary adjusting winding, located in the outer parts of the rotor slots, as shown in Fig. 18-50. When the brushes connected to opposite ends of one phase of the stator winding are on the same commutator segment, they are short-circuited and the motor operates as a normal polyphase wound-rotor motor. When the two sets of brushes are shifted apart, a slip-frequency voltage is applied to the stator winding, thereby causing a change in speed. One direction of brush shift raises the speed; the other lowers it. A continuous speed range from full speed to standstill may be secured in this manner. Power-factor correction is also provided, by unsymmetrical spacing of the brush positions, introducing a secondary voltage component at right angles to the speed voltage.

To minimize the difficulties of commutation, these motors have large numbers of commutator bars and brushes, which together with the special windings result in relatively high costs. However, each motor is complete in itself, without external transformer or regulator, and they are suitable for direct connection to usual industrial voltages of 550 V and below.

One important application of these motors is to the drive of hosiery-knitting and ring-spinning machines, where an automatically controlled speed variation of 10 to 30% is often required. They are also employed for a wide range of industrial services, driving fans, pumps, conveyors, packaging machinery, paper mills, etc. Sizes up to several hundred horsepower have been built. They are normally rated on a constant-torque basis, the horsepower output varying in direct proportion to the speed.

The no-load speeds of these machines range from 7 to 10% above their full-load speeds with the brushes in the maximum-speed position and from 20 to 40% above the full-load speeds with the brushes in the minimum-speed position. Their temperature rise is 40°C when operating from maximum to 50% of maximum speed and 50°C when operating at lower speeds. Maximum running torques range from 140 to 250% depending on the rating and the brush position.

FRACTIONAL-HORSEPOWER-MOTOR APPLICATIONS

135. Scope of Fractional-horsepower Motors. Fractional-horsepower motor ratings are, with minor exceptions, from 1/20 to 3/4, inclusive. Motors of smaller ratings are classified as subfractional or miniature.

General-purpose motors are open 40°C motors offered in a wide variety of ratings and speeds for a large number of applications involving usual service conditions. **Definite-purpose motors** are designed for certain definite purposes involving known special characteristics (see Par. **145**). Use of a standardized general-purpose or definite-purpose motor will generally give the user maximum economy, greater ease in obtaining motors, and more all-round satisfaction. The first step in applying a motor is usually to select the type.

136. Selection of the Type. To aid in selecting the best type of motor, the principal characteristics of the major types of motor are given in Table 18-13. Each of these characteristics is a factor in selecting a particular type of motor. **Power supply** is important because only single-phase power is available where the majority of small motors are used; therefore, motors used on appliances built for general distribution are limited to single-phase types. Companion d-c and polyphase motors are usually available as required. Portable tools and office appliances often employ universal motors to permit use on either alternating or direct current. **Horsepower** and **speed** ratings required are important, because many types are not available in the whole range of ratings. However, the speed limitation may often be overcome by the use of a belt or a gear drive.

137. Speed classification required is another important factor. Most popular are **constant-speed motors,** which operate at a speed substantially constant over a wide range of loads. A **two-speed pole-changing motor** acts, on either speed connection, as a constant-speed motor. A **varying-speed motor** is one the speed of which varies with the load, ordinarily decreasing when the load increases, e.g., universal motors. An **adjustable varying-speed** motor is one the speed of which can be adjusted but which, when once adjusted for a given load, will vary in considerable degree with change

Table 18-13. Characteristics of

	Alternating				
	Single-phase				
	Split-phase types			Capacitor-start	Capacitor (1-value, or perm. split)
	General-purpose	High-torque	Two-speed, pole changing		
Schematic diagram of connections Arrangements shown are typical or representative; most of the types illustrated have numerous other arrangements which are also used.					
Characteristic speed-torque curves Ordinates are speed; 1 division = for all a-c motors, 20% of syn. rpm; for universal motors, 1000 rpm; for d-c motors, 20% of full-load rpm. Each abscissa division = 100% of full-load torque.					
Rotor construction	Squirrel-cage	Squirrel-cage	Squirrel-cage	Squirrel-cage	Squirrel-cage
Built-in automatic starting mechanism	Centrifugal switch	Centrifugal switch	Centrifugal switch	Centrifugal switch	None required
Horsepower ratings commonly available	$\frac{1}{20}$–$\frac{1}{3}$	$\frac{1}{6}$–$\frac{1}{3}$	$\frac{1}{8}$–$\frac{3}{4}$	$\frac{1}{8}$–$\frac{3}{4}$	$\frac{1}{20}$–$\frac{3}{4}$
Usual rated full-load speeds (for 60-cycle a-c motors; also d-c motors)	3450, 1725, 1140, 865	1725	1725/1140 1725/865	3450, 1725, 1140, 865	1620, 1080, 820
Speed classification	Constant	Constant	Two-speed	Constant	Constant, or adjustable varying
Means used for speed control	....	....	Two-speed switch	....	2-speed switch or autotransformer
Comparative torques: Locked-rotor	Moderate	High	Moderate	Very high	Low
Comparative torques: Breakdown	Moderate	High	Moderate	High	Moderate
Radio interference, running	None	None	None	None	None
During acceleration	One click	One click	Two clicks	One click	None
Approximate comparative costs between types, for same horsepower rating: Below $\frac{1}{20}$ hp	....	....	....	....	....
$\frac{1}{20}$–$\frac{1}{4}$ hp	100	75	210	125	140
$\frac{1}{3}$–$\frac{3}{4}$ hp	80	54	150	100	100–110
General remarks	For constant-speed operation, even under varying load conditions, where moderate torques are desirable or mandatory, this type is often used in preference to the more costly capacitor-start motor. Meets NEMA starting currents. Typical applications: blowers; centrifugal pumps; duplicating machines; refrigerators; oil burners; unit heaters.	High locked-rotor currents (in excess of NEMA) limit the use of this type on lighting circuits to applications where the motor starts only very infrequently, because of a tendency to cause flickering of the lights. Principal applications: washing and ironing machines; cellar-drainer pumps; tools for a home workshop.	Used where two definite speeds independent of load are required. Ratings above $\frac{1}{4}$ hp usually made capacitor-start. Motor shown always starts on high-speed connection; transfer to low speed made by starting switch. Common applications: belted blowers for warm-air furnaces or for other purposes; attic ventilators; air conditioning apparatus.	A general-purpose motor suitable for most applications requiring constant speed under varying loads, high starting and running torques, high overload capacity. Also available as two-speed pole-changing motor above $\frac{1}{4}$ hp. A few important applications are: refrigeration and air conditioning compressors; air compressors; stokers; gasoline pumps.	Primarily used for unit heaters, or for other shaft-mounted fans. Essentially a constant-speed motor, but by means of a two-speed switch, or by means of an autotransformer, other speeds can be obtained, *with fan loads*, if horsepower rating selected closely matches the fan load. Can also be made in intermittent ratings for plug-reversing service.

General remarks

Standard motors are ordinarily designed to operate in ambient temperatures from 10 C to 40 C (50 F to 104 F). Variations in line voltage of plus or minus 10%, or variations in frequency of plus or minus 5% are allowable.

Locked-rotor currents for single-phase motors, except split-phase high-torque and synchronous types, usually do not exceed the following limits established by NEMA:

Rating, hp	Amperes at	
	115 volts	230 volts
⅙ and smaller	20	10
¼	23	11½
⅓	31	15½
½	45	22½
¾	61	30½

Fractional horsepower motors are built for across-the-line starting.

The standard direction of rotation is counterclockwise facing the end opposite the shaft extension.

Fractional-horsepower Motors

Current				D-c or a-c (60 cycles or less), universal types		**Direct current**	
Motors		1, 2, or 3 phase	Polyphase				
Repulsion-start	Shaded-pole	Nonexcited synchronous (reluctance)	Squirrel-cage induction	Without governor	With governor	Shunt or compound	**Series**
		Stator winding may be: split-phase, capacitor-start, capacitor, polyphase			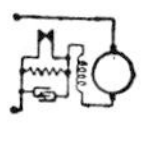	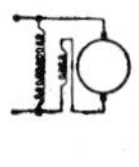	
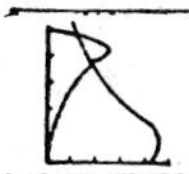				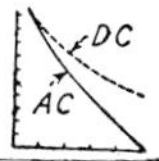 DC AC		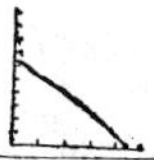	
Drum-wound; commutator	Squirrel-cage	Cage, with cutouts	Squirrel-cage	Drum-wound; commutator	Drum-wound; commutator	Drum-wound; commutator	**Drum-wound; commutator**
Short-circuiter	None	Depends on stator winding	None	None	None	None	None
$\frac{1}{8}$–$\frac{3}{4}$	$\frac{1}{2000}$–$\frac{1}{8}$	$\frac{1}{3000}$–$\frac{1}{3}$	$\frac{1}{6}$–$\frac{3}{4}$	$\frac{1}{150}$–1	$\frac{1}{50}$–$\frac{1}{20}$	$\frac{1}{20}$–$\frac{3}{4}$	$\frac{1}{25}$–$\frac{1}{30}$
3450, 1725, 1140, 865	1450–3000	3600, 1800, 1200, 900	3450, 1725, 1140, 865	3000–11,000	2000–4000	3450, 1725, 1140, 865	900–2000
Constant	Constant, or adjustable varying	Absolutely constant	Constant	Varying, or adjustable varying	Adjustable	Constant, or adjustable varying	Varying, or adjustable varying
....	Choke or resistor	....	....	Choke or resistor	Adjustable governor	Armature resistance	Resistor
Very high High	Low Low	Low Moderate	Very high Very high	Very high	Very high	Very high	Very high
None Continuous	None None	None	None None	Continuous Continuous	Continuous Continuous	Continuous Continuous	Continuous Continuous
.... 128 100	100	.... 200–400 275	.... 165–195 100	75 105–175	110 140–160	.... 175–225 120–140	185
A constant-speed motor suited to general-purpose applications requiring high starting torque, such as pumps and compressors. An associated type, the repulsion induction (buried cage) is used for door openers and other plug-reversing applications. Has been displaced for many applications by the capacitor-start motor.	For ratings below $\frac{1}{20}$ hp, this is a general-purpose motor. For fan applications, speed control is effected by use of a series choke or resistor. Applications: fans, unit heaters, humidifiers, hair driers, damper controllers.	Cutouts in rotor result in synchronous-speed characteristics. Curve shown is for split-phase stator. Pull-in ability is affected by inertia of connected load. Used for teleprinters, facsimile-picture transmitters, graphic instruments, etc. Clocks and timing devices usually use shaded-pole hysteresis motors rated at a few millionths of a horsepower.	Companion motor to capacitor-start motor with comparable torques and generally suited to same applications if polyphase power is available. Inherently plug-reversible and suitable for door openers, hoists, etc. High-frequency motors used for high-speed applications, as for woodworking machinery, rayon spinning, and portable tools.	Light weight for a given output, high speeds, varying-speed and universal characteristics make this type very popular for hand tools of all kinds, vacuum cleaners, etc. Ratings above $\frac{1}{4}$ hp usually compensated. Some speed control can be effected by a resistor or by use of a tapped field. Used with reduction gear for slower speed applications.	By means of a centrifugal governor, a constant-speed motor having the advantages of the universal motor is obtained. Governor may be single-speed or adjustable even while running. Speed is independent of applied voltage. Used in typewriters, calculating machines, food mixers, motion-picture cameras and projectors, etc.	A constant-speed companion motor for the capacitor-start or split-phase motor for use where only d-c power is available. For unit-heater service, armature resistance is used to obtain speed control. Not usually designed for field control.	Principally used as the d-c companion motor to the shaded-pole motor for fan applications. Used in these small ratings in place of shunt motors to avoid using extremely small wire.

in load. **Absolutely constant speed** is afforded only by the synchronous motor. **Adjustable-speed** motors are constant-speed motors the speed of which can be set to different predetermined values.

Other factors listed in the table are of varying importance, depending upon the particular application. Knowledge of the type used for the same or similar applications is often a helpful, but not infallible, guide. Much helpful information on theory and performance characteristics has been presented previously in this section.

138. Rating and Performance Standards. Standard voltage and frequency ratings and permissible variations of these are listed in Table 18-14.

Table 18-14. Approximate Full-load Inputs of 115-volt Motors

Item	Single phase	Polyphase	Universal	Direct current
Voltage ratings	115, 230	110, 220, 208 (60-cycle)	115, 230	115, 230, 32
Frequency ratings	25, 60	25, 60	0 to 60	
Allowable variation in circuit voltage, from rated, %	±10	±10	±6	±10
Allowable variation in frequency	±5	±5	0–60 cycles	
Allowable combined variation in voltage and frequency, %	±10	±10		

Standards of torques and efficiencies are given in this section; see also General Remarks, in lower left corner of Table 18-13.

139. Service Conditions. General-purpose motors are built to withstand all normal conditions. A few abnormal conditions requiring attention, and possibly special motors, are exposure to chemical fumes, explosive vapors, dust, dirt (particularly if gritty, combustible, or explosive), lint, steam, oil vapor, or salt air; exposure to weather, abnormal shock or vibration, ambients below 10 or above 40°C; prolonged operation at overloads; altitude of more than 3,300 ft; proximity of objects that block normal ventilation; dripping or splashing of liquids on the motor.

140. Thermal Protection. Many single-phase motors are now available with a built-in thermal protector which affords complete protection from burnout due to any type of overload, even stalled rotor. Most such devices are automatic-resetting, but some are manual-resetting. Motors so protected usually are marked externally in some way to indicate the fact.

141. Reversibility. In general, standard motors of the types listed in the table can be arranged by the user to start from rest in either direction of rotation. There are exceptions, however. Shaded-pole motors, unless of a special design, can be operated in only one direction of rotation. Small d-c and universal motors often have the brushes set off neutral, preventing satisfactory operation in the reverse direction. Single-phase motors which use a starting switch ordinarily cannot be reversed while running at normal operating speeds, because the starting winding, which determines the direction of rotation, is then open-circuited. By use of special relays this limitation of split-phase and capacitor-start motors can be overcome when necessary. Such motors are built for small hoists. High-torque intermittent-duty permanent-split capacitor motors; repulsion-induction (buried-cage) motors; and split-series d-c or universal motors are often built for plug-reversing service. Standard polyphase induction motors can be reversed while running, as can the smaller ratings of d-c motors; such applications should preferably be taken up with the motor manufacturer.

142. Mechanical Features. Rigid and rubber-mounted motors are commonly available. Sleeve and ball bearings are both standard. Sleeve-bearing motors are designed for operation with the shaft horizontal, but ball-bearing motors can be operated with the shaft in any position. For operation with the shaft vertical, sleeve-bearing motors may require a special design. Rubber mounting is widely used for quiet operation, because all single-phase motors have an inherent double-frequency torque pulsation. An effective and common arrangement uses rubber rings concentric with the shaft and so arranged as to provide appreciable freedom of torsional movement but little other freedom. Sometimes the driven member picks up the double-frequency torque pulsation and amplifies it to an objectionable noise, e.g., a fan with large blades mounted rigidly on the shaft. The cure for this difficulty is an elastic coupling between the shaft and the driven member; no amount of elastic suspension of the stator can help. Standard motors are generally open and of dripproof construction. Splashproof and totally enclosed motors are easily available.

143. Inputs of Small Single-phase motors. See Table 18-15.

Table 18-15. Voltage and Frequency Ratings, and Permissible Variations

Hp rating	3450 rpm		1725 rpm		1140 rpm		865 rpm	
	Amp	Watts	Amp	Watts	Amp	Watts	Amp	Watts
1/8	2.9	207	2.7	176	3.9	207	5.4	245
1/6	3.2	254	3.0	214	4.3	254	6.0	296
1/4	4.2	352	3.9	301	5.6	352	8.1	414
1/3	5.3	460	4.9	395	7.0	460	9.8	540
1/2	7.4	678	6.9	574	9.8	678		
3/4	10.6	981	9.9	835				

Full-load torque, in terms of horsepower and rated speed, is

$$\text{Full-load torque, oz}\cdot\text{ft} = \frac{84{,}000 \times \text{hp}}{\text{r/min}} \tag{18-57}$$

144. Application Tests. The primary object of any application test is to determine the power requirements of the appliance or device under various significant operating conditions. A convenient way of doing this is to use a motor of approximately the right horsepower rating and of predetermined efficiency at various outputs. Watts input are carefully measured under each condition. From the watts input observed (never use current as a measure of load except for d-c motors) and the known efficiency, the load is readily determined. Care should be taken in measuring the watts input to correct for the meter losses.

A second, and equally important, object of the test is to determine the actual locked-rotor and pull-up torques required by the appliance. The locked-rotor and pull-up torques of the test motor should be known or measured at rated voltage and frequency. (Locked-rotor torque often varies with slight changes in rotor position.) Using a transformer or induction regulator to obtain a variable voltage (do not use a resistance or choke for this purpose), measure the minimum voltage at which the motor will start the appliance and also the minimum voltage at which it will pull it up through switch-operating speed. Assuming that the pull-up and locked-rotor torques each vary as the square of the applied voltage, it is then a simple matter to determine the actual locked-rotor and pull-up torques required by the device. After a motor has been selected, it should be determined whether or not it can operate the device at 10% above and below normal rated voltage of the motor or over a wider range of voltage, if desired. If exceptional load conditions may occasionally be encountered, use of a motor equipped with inherent-overheating protection is often desirable.

145. Definite-purpose Motors. For a number of important applications, involving large quantities of motors, NEMA has developed standards to meet these special requirements effectively and economically. Motors built to these standards are usually more readily obtainable and economical than special motors tailored to one application. Highlights and distinguishing features are given in Table 18-16. More details can be obtained in NEMA Standards.

146. Miniature Synchronous Motors for Clocks and Timing Devices. *Shaded-pole hysteresis motors,* which operate at synchronous speed, are essentially the same as shaded-pole induction motors except that they use rotors of hardened-steel rings of a material having high hysteresis loss. Large quantities of such motors are built for clocks and timing devices. Clock motors have an input of 1.5 to 2 W and an output of a few millionths of a horsepower. Large motors with inputs up to 15 W are built for heavier duty applications. Rotor speeds are commonly 450, 600, and 3,600 r/min.

Table 18-16. NEMA Standards for Definite-purpose Motors

Application	Principal types	Distinguishing features
Universal motor	Universal: Salient-pole and distributed field	Dimensional standards; common practices utilizing parts
Hermetic motors	Split-phase, capacitor-start, polyphase	Parts only for hermetic refrigeration condensing units
Belt-drive refrigeration compressors	Capacitor-start. Repulsion-start, polyphase, d-c	Open; sleeve bearings, extended rear oiler; automatic-reset thermal overload protection
Jet-pump motors	Split-phase, capacitor-start, repulsion-start polyphase, d-c	3450 rpm; ball bearings; open; machined back end shield; automatic-reset overload protection
Motors for shaft-mounted fans and blowers	Split-phase, permanent-split capacitor, polyphase, d-c	Enclosed; horizontal; sleeve bearings; vertical, ball bearings; extended through bolts; capacitors on front end shield
Shaded-pole motors for shaft-mounted fans and blowers	Shaded-pole; two-speed, three-speed	Open or totally enclosed; sleeve bearings; high slips
Belted fans and blowers	Split-phase, capacitor-start, repulsion-start; two-speed split-phase and capacitor-start	Open; sleeve bearings; resilient mounting; automatic-reset overload protection; extended rear oiler
Stoker motors	Capacitor-start; repulsion start; polyphase, d-c	Totally enclosed recommended; automatic reset overload protection
Motors for cellar drainers and sump pumps	Split-phase, d-c	Vertical, dripproof, 50 C; two ball bearings, or one ball, one sleeve; mounts on support pipe; built-in float-operated line switch; overload protection
Gasoline-dispensing pumps	Capacitor-start, repulsion-start polyphase, d-c	Explosionproof; sleeve bearing; built-in line switch and capacitor; voltage-selector switch on single-phase
Oil-burner motors	Split-phase	Enclosed, face-mounted, round-frame; manual-reset overload protection; two line leads
Motors for home-laundry equipment	Split-phase, d-c	Low-cost, high starting current; open, 50 C; round-frame with ungrounded mounting rings; shaft extension with flat and hole for coupling
Motors for coolant pumps	Split-phase, capacitor-start, repulsion-start, polyphase, d-c	3450- and 1725-rpm; totally enclosed; ball bearings; machined back end shield
Submersible motors for deep-well pumps	Split-phase, capacitor, polyphase	3450-rpm; designed for operation totally submerged in water not over 25 C (77 F); use external relay for starting

Most of these motors are furnished with built-in reduction gears to give output speeds of 60 r/min to 1 r/month.

Reluctance motors, both self-starting and manual-starting types, are available for similar applications. Another type used is the synchronous-inductor motor, which is essentially an inductor alternator used as a motor; field excitation is furnished by a permanent magnet.

BIBLIOGRAPHY

147. References on Fractional-horsepower-motor Applications

VEINOTT, CYRIL G. "Fractional Horsepower Electric Motors," 2d ed.; New York, McGraw-Hill Book Company, 1948.

Standards for Fractional Horsepower Motors; *NEMA Publ.* MG2, 1951, New York, National Electrical Manufacturers Association.

MOTOR CONTROL

General

148. Industrial motor control is designed and built in accordance with rules and standards established by several organizations. Detailed design-construction and test information is contained in such publications as the National Electrical Code, National Board of Fire Underwriters; Standards for Industrial Control Apparatus, Institute of Electrical and Electronics Engineers; Industrial Control Standards, National Electrical Manufacturers Association; Standard for Industrial Control Equipment, Underwriters Laboratories; and Standard Rotation, Connections and Terminal Markings for Electric Power Apparatus, United States of America Standards Institute.

149. The essential functions of industrial motor control are the starting, speed regulating, stopping, and protecting of electric motors.

MOTOR-STARTING DEVICES

150. A contactor is a device, generally magnetically actuated, for repeatedly establishing and interrupting an electric power circuit.

Figure 18-51 illustrates a single-pole d-c contactor. Contactors of this type are rated on a continuous-current-carrying-capacity basis and are available in sizes of 15 to several thousand amperes. They are also rated on an intermittent-duty basis, at values depending on the duty cycle. The shunt operating coil is designed to withstand 110% of rated voltage continuously and to close the contactor successfully at 80% of rated voltage.

151. The magnetic blowout consists of a coil wound on a steel core and mounted between steel pole pieces. The blowout coil is generally connected in series with the contactor and carries motor current with the contactor closed. The current sets up a magnetic field through the core and pole pieces of the blowout structure and across the contact tips. When an arc is formed, the magnetic field of the arc and the magnetic field of the blowout repel each other and the arc is forced upward and away from the contacts. The extinguishing action, due to the lengthening of the arc, is extremely rapid and thereby greatly reduces the wear and burning of the contacts.

152. Performance of Contactors. To obtain trouble-free service and maximum contact life, the following items should be in accordance with the manufacturer's specifications: initial and final contact pressures, magnetic gap, arc gap, and wear allowance. The contact pressures can be measured by means of a spring balance, initial pressure with the contactor open, and final pressure with the contactor closed. The magnetic gap is the distance from the center line of the core to a corresponding point on the armature lever, and the arc gap is the distance between the arcing tips. Contact-wear allowance is the total thickness of material that may be worn away before the contact between the two surfaces becomes ineffective. The contacts should be renewed when worn so that the distance A (Fig. 18-51) between the back edges of the contacts, with the contactor closed, becomes less than the specified amount.

Current-carrying contacts should not be lubricated. Hinge pins and bearings should be lubricated with a light machine oil. The surfaces of the core and the armature, which seal when the contactor is closed, should be kept clean.

An a-c contactor is similar in construction to a d-c contactor, except that laminated iron structures are used. A shading coil is used at the core face to obtain quiet operation. Alternating-current contactors are available with two, three, or four main poles for interrupting all line circuits to single-phase, 3-phase, or 2-phase 4-wire motors (see Fig. 18-52 and Table 18-17).

153. Alternating-current across-the-line starters are simple in construction, easy to install and maintain, and inexpensive. A typical starter consists of a three-pole contactor with a thermal overload relay for protecting the motor. The starter con-

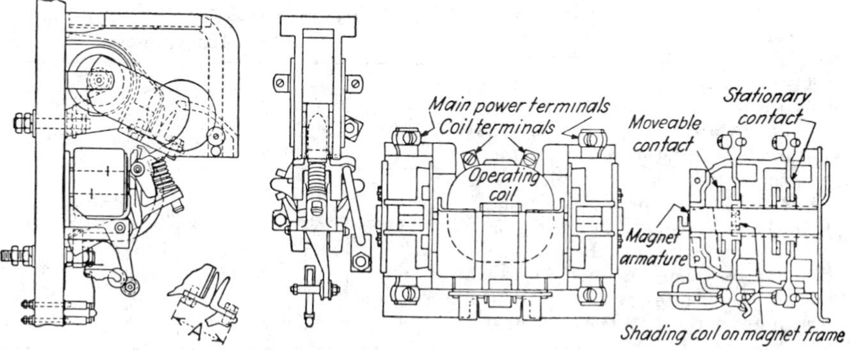

FIG. 18-51. Direct-current contactor. FIG. 18-52. Alternating-current contactor.

nects the motor directly to the line, impressing full voltage to the motor terminals. It is particularly suitable for squirrel-cage motors. Since these starters connect the motor directly to the supply lines, the motor will draw an inrush current of 6 to 10 times running current. In the majority of installations this is not objectionable and will not damage the motor or the driven machinery. When the starting inrush must be lower, some form of reduced-voltage starting must be used. The common types are autotransformer starters and primary-resistance starters.

154. Autotransformer starters have two autotransformers connected in open Δ to provide reduced-voltage starting. Three taps are supplied, as shown in Fig. 18-53, giving 50, 65, and 80% of full line voltage. The motor current varies directly as the voltage impressed on the motor terminals. The line current varies as the square of the impressed voltage and is therefore lower than with resistor-type starters. The torque also varies as the square of the impressed voltage. The 50% voltage tap will therefore provide 25% starting torque. Connections should be made to the lowest tap that will give the required starting torque.

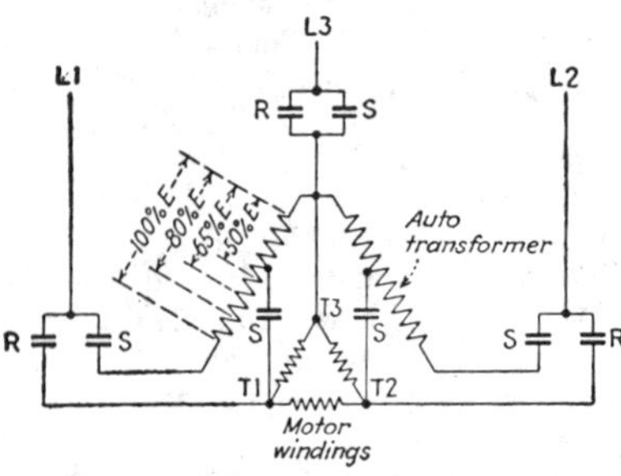

FIG. 18-53. Connections for autotransformer starter.

Characteristics of this type of starter are low line current, low power from the line, and a low power factor. Acceleration is not continuous, because the torque developed by the motor remains practically constant during the starting period, and the motor is momentarily disconnected from the line in transferring to line voltage.

Autotransformer starters are available in manual and automatic types. In the manual type the contacts are operated by means of a lever extending to one side of the enclosing case. The lever is equipped with a low-voltage release magnet. The automatic starter consists of a five-pole starting contactor *S* and a three-pole running contactor *R*. When the "run" button is pressed, the starting contactor closes, connecting the transformer to the line and the motor to the reduced-voltage taps. A timing relay is operated by the starting contactor. The motor accelerates, and, after a specified number of seconds, the timer contacts close, deenergizing the starting contactor and energizing the running contactor. The transformer is disconnected and the motor connected to line voltage.

155. Primary-resistor-type a-c starters connect the motor to the line through a resistor which is in series with the primary circuit. Reduced voltage at the motor is obtained because of the voltage drop across the resistor. As the motor accelerates, the current drawn from the line becomes less, and consequently the voltage drop

Table 18-17. Typical Ratings of A-c Contactors Used as Across-the-line Magnetic Starters with 3-phase Motors

Contactor size	Rating, amp	Horsepower at	
		220 volts	440 and 550 volts
00	10	1	1
0	15	2	3
1	25	5	7½
2	50	15	25
3	100	30	50
4	150	50	100
5	300	100	200
6	600	200	400
7	900	300	600
8	1350	450	900
9	2500	800	1600

Typical Ratings of D-c Contactors for Continuous-duty Applications

Contactor size	Rating, amp	Horsepower at		
		115 volts	230 volts	550 volts
1	25	3	5	
2	50	5	10	20
3	100	10	25	50
4	150	20	40	75
5	300	40	75	150
6	600	75	150	300
7	900	110	225	450
8	1350	175	350	700
9	2500	300	600	1200

across the resistor is lowered, and the voltage at the motor terminals is increased. The torque delivered by the motor is therefore constantly increased as the motor speed increases. After a definite interval, a timing device operated by the main contactor energizes the accelerating contactor, which short-circuits the resistor. There is no transfer period during which the motor may lose speed, and therefore smooth acceleration is obtained. In comparison with the autotransformer-type starter, the primary-resistor type takes more power from the line on starting but provides smoother acceleration, faster acceleration with a given initial torque, and higher power factor. In the smaller sizes the primary-resistor starter costs less than the autotransformer starter. For detail design of the primary resistor see "Control of Electric Motors" by P. B. Harwood, 3d ed., p. 399.

156. Slip-ring a-c motor starters consist of a contactor to connect the motor primary to the supply lines and a resistor and resistor commutating means for the secondary circuit. The starting torque depends on the ohmic value of resistance used, maximum torque being obtained when the resistance is selected for an inrush of approximately three times full-load current. Sufficient resistance is generally used to limit the inrush current to 150 or 200%. The resistor is cut out step by step as the motor accelerates, until the slip rings are short-circuited. The commutating means may be a faceplate controller, a drum, or a series of magnetic contactors controlled by current or time relays. High starting torque and low running slip can be obtained with a slip-ring motor. For detail design of a secondary resistor see Par. **173**.

157. Synchronous-motor starters of the full-voltage type connect the motor directly to the supply lines. The field winding is short-circuited through a discharge resistor during the starting period. The field is connected to the d-c lines when the motor is at a speed near synchronism. Reduced-voltage starters connect the motor to a reduced voltage for starting and transfer to full voltage at a speed just below synchronism. This transfer may be controlled by a time relay or a frequency relay. The field is energized either immediately before or immediately after the full-voltage switch closes.

Figure 18-54 is a simplified diagram of a synchronous motor controller arranged for starting the motor directly across the line and synchronizing by a relay operating at a selected frequency. Pressing the start button energizes relay *1CR* and contactor *M* to connect the motor stator windings to the a-c lines. Current at line frequency is induced in the rotor field winding and flows through the discharge resistor *FD* and the coil of relay *FR*. A small portion of this current flows through the reactor *X*, but the amount is limited, as the frequency is high. Relay *FR* closes rapidly, and the contacts on *FR* open the circuit to the *FS* field contactor. As the motor accelerates, the frequency of the induced current in the field winding decreases and an increasing portion of the current flows through reactor *X*. At a speed close to synchronism most of the current flows through *X*, and there will no longer be enough current flowing through the coil *FR* to keep the relay armature closed. Relay *FR* then opens, and the contacts on *FR* close to energize field contactor *FS*. Contactor *FS* connects the field to the d-c lines and opens the field discharge circuit through resistor *FD*, and the motor pulls into synchronism.

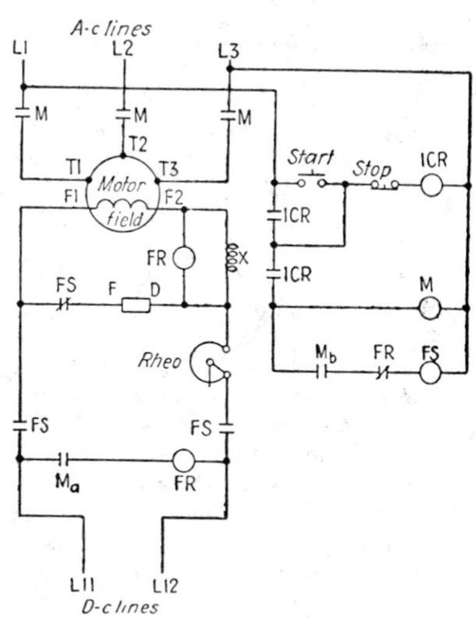

Fig. 18-54. Full-voltage controller with synchronization based on frequency.

Relay *FR* is polarized by a coil connected across the d-c lines through interlock contacts M_a. Polarizing the synchronizing relay provides a means for energizing the field contactor at a point in the a-c wave most favorable to synchronism.

158. Direct-current motors of small capacity may be started by connecting the motor directly to line voltage. Motors rated

2 hp or more generally require a reduced-voltage starter. The reduced voltage for starting is obtained by using resistance in series with the motor armature or by varying the armature supply voltage. Manual or magnetic control may be used.

159. Direct-current manual starters are satisfactory for applications that do not require frequent starting and stopping and where the starter can be mounted near the operator without requiring long motor leads. Across-the-line starters provide the simplest means of starting small d-c motors. Manually operated switches for this service are available in sizes up to 1.5 hp at 115 V and 2 hp at 230 V. For larger motors resistance is connected in series with the motor armature to limit the current inrush on starting. A manually operated means is then provided for removing the resistor from the circuit in a series of steps. Starters are available in the faceplate type, the multiple-switch type, and the drum type. The faceplate type is built for motors up to 35 hp, 115 V and 50 hp, 230 V. It consists of a movable lever and a series of stationary contact segments to which sections of resistor are connected. The resistor sections are short-circuited one at a time by moving the lever across the segments.

160. A drum switch consists of stationary contact fingers held by spring pressure against contact segments on the periphery of a rotating cylinder or sector. Drum controllers have many advantages over the faceplate and multiple-switch types. The mechanical construction is better, heavy contact pressures can be maintained, parts can be well insulated, blowout magnets and arc shields can be used, and the structure can easily be completely enclosed. Less space is required by the drum control, and it is easier to operate.

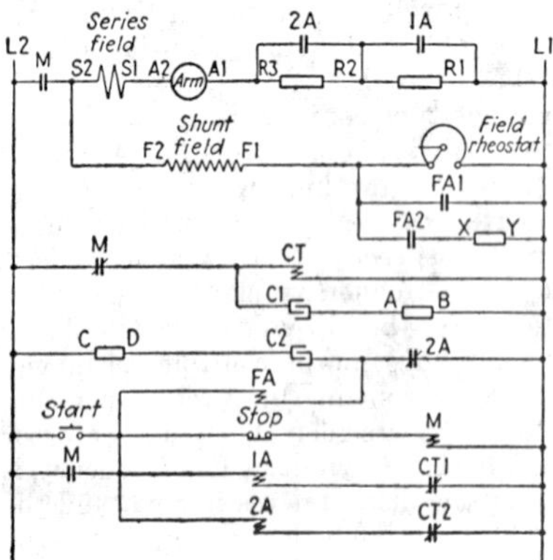

Fig. 18-55. Time-limit acceleration with definite-time relay.

161. Direct-current magnetic starters are used for applications where ease and convenience of operation are important; where the starter is operated frequently; where the motor is located at a distance from the operator; where automatic control by means of a pressure switch, limit switch, or similar device is desired; and for large motors which require the commutating of heavy currents. Resistance is connected in series with the motor armature to limit the initial current and is then short-circuited in one or more steps. Various devices are used to short-circuit the resistance steps automatically as the motor accelerates.

For larger motors a series of magnetic contactors is used, each of which cuts out a step of armature resistor. Figure 18-55 shows a type of **time-limit acceleration** where the operation of contactors, and therefore the rate of acceleration, is governed by a **magnetically operated definite time relay.** This time relay operates on the principle of discharging a capacitor, thus obtaining a definite time period which is unaffected by changes in temperature and load or by dust and dirt. With the motor at rest a circuit is obtained through a normally closed contact on *M* to energize the *CT* timing-relay coil and to charge capacitor *C*1. Contacts *CT*1 and *CT*2 on relay *CT* are open with the relay energized. Capacitor *C*2 is charged through the normally closed contact on the 2*A* contactor. Pressing the "start" button energizes the main contactor *M*, which maintains itself through a normally open interlock finger. Relay *FA* is energized, and its contact *FA*1 short-circuits the field rheostat. The motor accelerates from rest to a speed determined by the value of *R*1-*R*3 resistor. The circuit to timing relay *CT* is opened by the interlock on *M*, and capacitor *C*1 discharges through the *CT* coil and the *AB* resistor. Contacts *CT*1 and *CT*2 can be individually adjusted to close at any time during the capacitor discharge period. Closing *CT*1 energizes the 1*A* contactor, which short-circuits the *R*1-*R*2 resistor. The motor then accelerates to a speed determined by the value of the *R*2-*R*3 resistor step. Closing *CT*2 energizes 2*A* and connects the motor across the line, permitting it to accelerate to normal speed. Relay *FA* is deenergized when 2*A* closes. Contacts *FA*1 and *FA*2 remain closed for a definite time because of the discharge of capacitor *C*2 through the *FA* relay coil. When contact

*FA*1 opens, a resistance is inserted in the motor shunt field, equivalent to the field-rheostat resistance and *XY* resistance in parallel. Opening *FA*2 disconnects *XY*, and the motor runs at a speed determined by the setting of the field rheostat.

162. Inductive Time-limit Controller. By making use of the inductive effect of a coil wound on an iron core, the closing of accelerating contactors may be delayed. Two types of such controller are in general use. One makes use of an accelerating contactor having two coils, one for closing and one for holding out. The other type of controller uses small inductive relays, one with each accelerating contactor.

Figure 18-56 shows a typical reversing controller of the first type. The closing coils of the accelerating contactors are marked 1*A*, 2*A*, and 3*A*, and the holdout coils *HC*1*A*, *HC*2*A*, and *HC*3*A*. Moving the master controller to the forward position energizes contactors 1*F*, 2*F*, and *M*, which close to connect the motor across the line in series with resistors *R*1 to *R*4. Interlock contacts *M* and 1*F* close to energize the closing coils of the accelerating contactors. However, contactors 2*A* and 3*A* do not close immediately, as their holdout coils are connected across steps of the starting resistor and are therefore energized. Plugging contactor 1*A* closes without delay, as there is no countervoltage across the motor armature and holding coil *HC*1*A* is deenergized. Contactor 1*A*, when closed, short-circuits resistor step *R*1-*R*2 and also holding coil *HC*2*A*. Because of the inductance in the *HC*2*A* coil the current flowing in it drops to zero after a period of time. After the current in *HC*2*A* drops to a low value, contactor 2*A* closes to short-circuit resistor step *R*2-*R*3 and holding coil *HC*3*A*. Time is again required for the current in this coil to drop low enough to allow 3*A* to close. Closing 3*A* completes the accelerating cycle. The motor will be connected directly to the supply lines and will be running full speed in the forward direction.

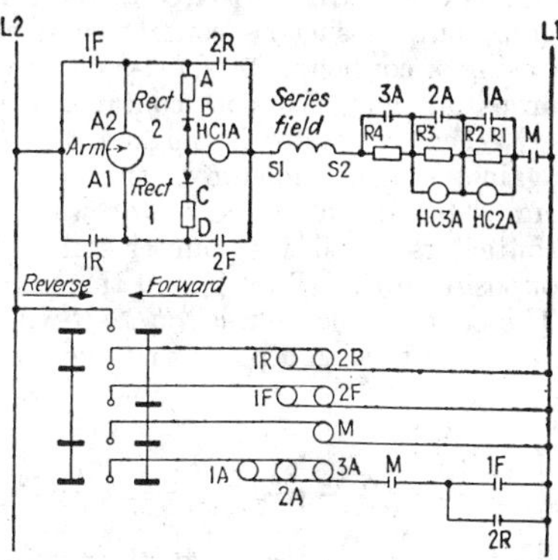

Fig. 18-56. Inductive time-limit controller.

Moving the master rapidly from the forward position to the reverse position will open contactors 1*F* and 2*F* and will close contactors 1*R* and 2*R*. Plugging contactor 1*A* and accelerating contactors 2*A* and 3*A* will be open. The armature countervoltage is high at the instant the motor is plugged, and current will flow through the lockout coil *HC*1*A* and one of the rectifiers. As the motor decelerates, the countervoltage decreases to a low value, permitting contactor 1*A* to close at or near zero speed. Time-limit acceleration is then obtained on contactors 2*A* and 3*A* to bring the motor up to speed in the reverse direction of rotation.

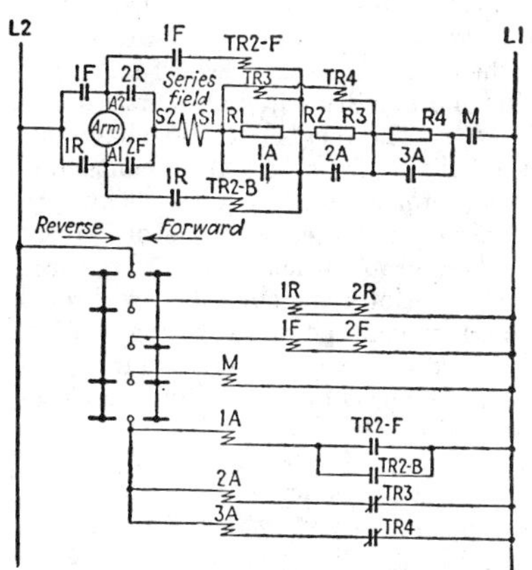

Fig. 18-57. Time acceleration with inductive relays.

163. Time acceleration with inductive relays is shown in Fig. 18-57. *TR*2-*F* and *TR*2-*B* are normally open relays the coils of which are connected in the main circuit and the contacts of which are connected in the circuit of the closing coil of the 1*A* contactor. *TR*3 and *TR*4 are normally closed inductive relays whose coils are connected across steps of resistance and whose contacts are connected to handle the coils of the 2*A* and 3*A* contactors, respectively. Time delay is obtained by the use of these relays, which have a magnetic characteristic such that, when their coils are short-circuited, a definite time is required for them to open.

Starting from rest, the operator moves the master handle to the forward direction, closing contactors 1*F*, 2*F*, and *M* and permitting current to flow through the motor and the resistor. A voltage equal to the *IR* drop across the resistance *R*1-*R*2 will immediately be impressed on the coil of the relay *TR*3, which will pick up its armature and open its contacts. Similarly, the relay *TR*4 will be opened. When contactor 1*F* closes, an interlock closes the coil circuit of relay *TR*2-*F*. The voltage impressed on this relay coil is equal to the algebraic sum of the *IR* drop across the resistance *R*1-*R*2 and the counter emf of the motor—a voltage sufficient to pick up the relay armature and close its contacts. The plugging contactor 1*A* closes immediately. The closing of this contactor short-circuits the coil of the relay *TR*3, which, after a definite time, drops out, closing its contacts and completing a circuit for the contactor 2*A*. The closing of contactor 2*A* short-circuits the coil of relay *TR*4, which, after a definite time, drops out and closes its contacts, completing a circuit for the coil of contactor 3*A*.

When the motor is plugged, the voltage impressed on the coil of relay *TR*2-*B* is again the algebraic sum of the *IR* drop across the resistance *R*1-*R*2 and the counter emf. Inasmuch as the motor was running in a forward direction when the reverse contactors closed, the counter emf is still in the same direction. This direction is now opposite to the *IR* drop, and therefore the voltage on the relay coil is practically zero. The contacts of the relay *TR*2-*B* will not close until the speed of the motor armature decreases to such a point that the difference between the *IR* drop and the counter emf is sufficient to close the relay. In this manner, time delay is obtained on the contactor 1*A* when the motor is plugged.

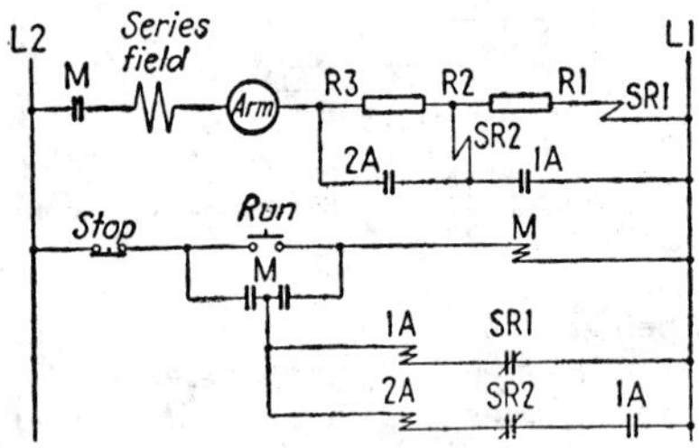

Fig. 18-58. Direct-current series-relay acceleration.

164. Direct-current starters with current-limit acceleration are designed to halt the starting operation whenever the required starting current exceeds an adjustable predetermined value, the starting operation being resumed when the current falls below this limit. With current-limit acceleration, the time required to accelerate will depend entirely upon the load. When the load is light, the motor will accelerate rapidly, and when it is heavy, the motor will require a longer time to accelerate. For this reason a current-limit starter is not so satisfactory as a time-limit starter for drives having varying loads. Time-limit starters are simpler in construction, accelerate a motor with lower current peaks, use less power during acceleration, and always accelerate the motor in the same time regardless of variations in load. Current-limit starters are desirable for motors driving high-inertia loads. A typical current-limit starter is shown in Fig. 18-58. The relays *SR*1-*SR*2 have normally closed contacts connected in series with the coils of the accelerating contactors. The coils of these relays are connected in the main motor circuit. The relays are provided with an adjustment so they can be set to close on a selected value of current.

Pressing the "run" button energizes the main contactor *M*, which closes and connects the motor to the line in series with the *R*1-*R*3 resistor. Motor current will flow through the *SR*1 coil, and its contacts will open rapidly and prevent 1*A* from closing. When the motor has accelerated enough to bring the line current down to the value for which *SR*1 is set, the relay contacts will close. A circuit is then provided for 1*A*, which closes, cutting out the first step of resistance and short-circuiting the *SR*1 coil. Current now flows through the *SR*2 coil and the 1*A* contacts. *SR*2 relay contacts open and prevent 2*A* from being energized. The motor accelerates again, and when the current falls to the value for which *SR*2 is set, its contacts close, energizing 2*A* and connecting the motor across the line.

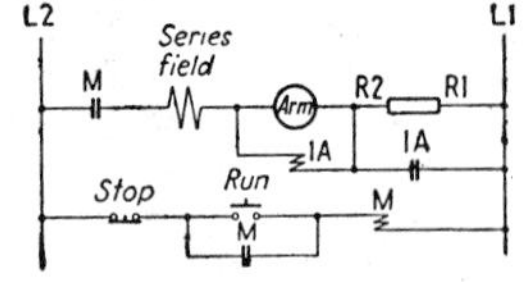

Fig. 18-59. Counter-emf acceleration.

165. Counter-EMF Acceleration. A controller using a form of current-limit acceleration known as counter-emf acceleration is illustrated in Fig. 18-59. Pressing the "run" button energizes main contactor *M* and connects the motor across the line, in series with resistor $R1$-$R2$. The accelerating contactor $1A$ has its closing coil connected across the armature and is designed to close on approximately half of normal voltage. The motor accelerates until the voltage across the armature is high enough to close $1A$. Contactor $1A$ then closes, connecting the motor to line voltage. This type of acceleration is seldom used for motors above 3 hp.

166. The design of starting-duty resistors requires the determination of the total ohms, the distribution of this resistance between the steps available, and the calculation of the current-carrying capacity. Standard resistors to meet various classes of service are designated by class numbers in accordance with the NEMA Table of Classification of Resistors. NEMA also publishes a resistor application table intended as a guide in specifying and designing resistors (see NEMA Industrial Control Standards). This table lists typical machines with the corresponding NEMA resistor-classification number. For example, a lathe should have a starter with resistor classification No. 115. This resistor has sufficient total ohms to limit the current inrush on starting to 150% of full-load current. It will be designed to have current-carrying capacity for an average accelerating current (rms value) of 125% full-load current, on the basis of starting once during each 80-s period and with an accelerating time of 5 s.

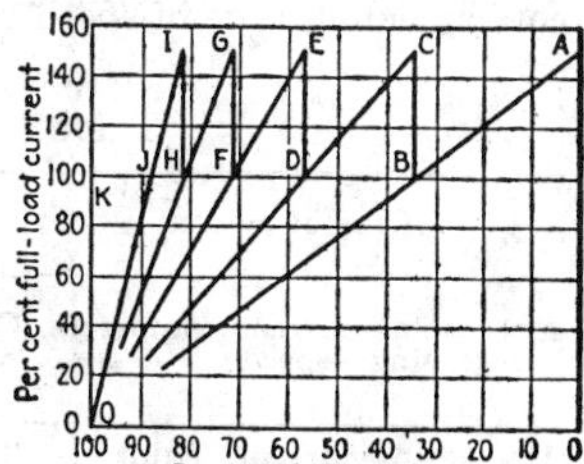

FIG. 18-60. Speed-torque curves of a d-c shunt-wound motor.

Figure 18-60 illustrates a graphical method for designing a resistor, as described above, for a d-c shunt-wound motor. The base line represents percentage of speed and the ordinates percentage of full-load current. With 150% inrush allowed, the total resistance, including motor and line, will be $R = E/1.5I$. Line OJI represents motor-characteristic curve, and KJ = drop in speed at full load due to the resistance in the armature and leads. The point A locates the first current inrush, and the motor accelerates along line AO. With full load the motor will not accelerate beyond point B. Point C locates the second current inrush, after which the motor accelerates to D, along line CO. Similarly points E, G, and I locate the third, fourth, and fifth inrushes, respectively. Line KB is the total resistance required to limit the inrush to 150%. The ohms required in the first step are $BD/BK \times E/1.5I$. The ohms required in the second step are $DF/BK \times E/1.5I$. By proceeding as above for each of the steps, a table as given below is obtained.

No. of step	Step ohms, % total	Time step in circuit, % total	Step ohms, % total	Time step in circuit, % total
1	40	40	37	37
2	28	68	25	62
3	19	87	18	80
4	13	100	12	92
5	100		8	100
			100	

For convenience, 100% ohms in the table represents the ohms in the controller, or BJ in Fig. 18-60. A five-step starter with No. 115 resistor will have 37% of the starter ohms in the first step. The first step of resistance will have a capacity for 37% of 5 s, or 1.85 s in each 80-s period, at 125% full-load current. Step 5 will have 8% of the resistor ohms, with a capacity for 100% time on, or 5 s in circuit for each 80-s period.

The example given refers directly to Class 115 resistors. Similar procedure can be followed to design a resistor for any classification. The fixed resistance in the armature and line can be assumed as 15% E/I for 1 to 3 hp, 11% E/I for 4 to 10 hp, 9% E/I for 15 to 35 hp, and 7% E/I for 40 to 75 hp. With this value determined, the only remaining variables in the graph (Fig. 18-60) are the allowable inrush and the number of accelerators. If the maximum inrush current is specified, the required number of accelerators can be determined. With the number of accelerators fixed, the current inrush can be determined and the resistance proportioned to hold the maximum inrush to a reasonable value.

SPEED-REGULATING DEVICES

167. Speed control of electric motors may be obtained by various means. The design of a speed-regulating controller is determined by the type of motor with which it will be used. Table 18-18 lists the various types of motor in general use and the corresponding type of speed control for each.

Table 18-18. Types of Motor-speed Control

Type of motor	Speed characteristic	Type of speed control
Alternating-current, squirrel-cage, single-speed	Constant; regulation about 4 to 15%	Primary resistor not practical. Varying line voltage and frequency sometimes used
Alternating-current, squirrel-cage, multispeed	Two or more constant speeds	Connecting separate motor windings. Reconnecting one motor winding to change number of poles
Alternating-current, synchronous	Constant; regulation determined by frequency	Constant speed determined by line frequency
Alternating-current, slip-ring	Regulation close at full speed	Secondary resistor
Direct-current, series-wound	Varying; wide speed regulation	Armature shunt and series resistors
Direct-current, shunt-wound	Constant at selected speed	Armature shunt and series resistors. Field weakening. Variable armature voltage
Direct-current, compound-wound	Regulation about 25%	Armature shunt and series resistors. Field weakening. Variable armature voltage

168. Control for multispeed a-c squirrel-cage motors may be of either the drum type or the magnetic type. Drum controllers are widely used, because the many changes in connections required to obtain different speeds can be readily accomplished. Drum controllers can be used with reconnected winding or separate-winding-type motors and with constant-torque, variable-torque, or constant-horsepower motors. Low-voltage and overload protection can be obtained by using a magnetic contactor and overload relay. When complete control by pushbuttons or other pilot devices is required, magnetic contactors are used to change the motor connections. Controllers of this type can be arranged to permit starting at any speed or to permit starting only at the slowest speed and changing to each higher speed in sequence.

Multispeed squirrel-cage motors are suitable for applications that require more than one operating speed but do not require speed regulation between these fixed speeds.

169. Alternating-current secondary-resistor speed regulators consist of a contactor to connect the primary of the motor to the supply lines and some form of resistance-commutating device for the secondary circuit. The commutating device may be a three-arm faceplate controller, a drum, or magnetic contactors. Regulating devices differ from starting devices in that the commutating means can remain continuously on any one of the resistor steps. The motor will therefore operate continuously at a reduced speed, as determined by the amount of resistance remaining in the motor circuit. The use of secondary resistance for speed regulation is not an efficient method because of the power loss in the resistor. The amount of speed reduction obtained will vary directly with the load on the motor. Speed regulators of this type are usually designed for 50% speed reduction. Under favorable conditions, however, motors can be operated at 75% speed reduction.

170. Control by Unbalance of Primary Voltage. When resistance is connected into the rotor circuit of a wound-rotor motor that is driving a positive load, satisfactory slow speeds are obtained. When the load is overhauling, the insertion of resistance in the rotor circuit of the motor will result in increased speeds. If the starter is connected for hoisting and a high resistance is inserted in the rotor circuit, the hoisting torque may be limited to a value that will not be enough to hoist the load but will be enough to retard its lowering speed. However, the speed-torque curves obtained by this method of lowering may be unstable.

If the voltages applied to the stator are unbalanced and resistance is inserted in the rotor circuit, speed-torque curves closely resembling those of a d-c dynamic-lowering hoist circuit are obtained. Figure 18-61 shows the circuit for an unbalanced-voltage control for a hoist, and Fig. 18-62 is a contactor-sequence chart for it.

On all hoisting points, contactors *M* and *H* are closed, and the stator is connected to the power lines in the conventional manner, the phase voltages being the same for all phases. On the first speed point, all rotor-circuit resistors are in circuit, and one rotor phase is opened by contactor 1*LA*. This results in a low hoisting speed. On the second point 1*LA* is closed, increasing the speed by completing the 3-phase rotor circuit. On succeeding points the speed is increased by short-circuiting the resistor, step by step, by the closing of contactors 1*A*, 2*A*, and 3*A*.

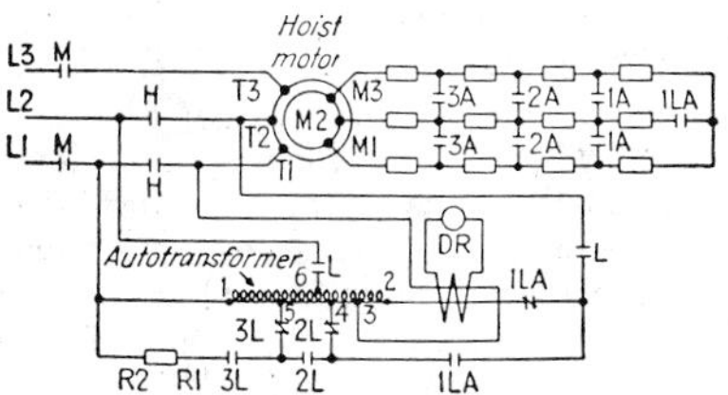

FIG. 18-61. Connections for a-c hoist controller using unbalanced primary voltage control.

SEQUENCE OF OPERATION											
CONTACTOR	HOIST					OFF	LOWER				
	5	4	3	2	1		1	2	3	4	5
M	O	O	O	O	O	⊖	O	O	O	O	O
H	O	O	O	O	O						
L						⊖	O	O	O	O	O
1LA	O	O	O	O				O	O	O	O
2L									O	O	O
3L										O	O
1A	O	O	O			O	O	O	O	O	
2A	O	O									
3A	O										

O CONTACTOR CLOSED
⊖ CONTROLLED BY DR RELAY

FIG. 18-62. Sequence chart of a-c hoist controller.

In lowering, the voltage unbalance is obtained by means of a multitap autotransformer. Contactors *M* and *L* are closed on all points. 1*A* is closed to establish a rotor-circuit resistance of suitable value, about 50% of $E/\sqrt{3}I$.

On the first point contactors *M* and *L* are closed, completing the stator circuit through the autotransformer. The stator voltages are unbalanced, being approximately 5, 110, and 115% of line voltage on terminals *T*1-*T*2, *T*1-*T*3, and *T*2-*T*3, respectively.

On the second point 1*LA* is closed. The corresponding voltages are now approximately 12, 110, and 104% of line voltage. Note that the phase rotation at the motor terminals is not reversed to provide power lowering.

On the third point 2*L* is closed, resulting in voltages of 30, 110, and 95% of line voltage.

On the fourth point 3*L* is closed, resulting in voltages of 120, 110, and 100% of line voltage.

On the fifth point, the voltages remain the same, but 1*A* is opened to increase the value of the rotor resistance to about 100% $E/\sqrt{3}I$.

When the master is moved to the "off" position, contactors *M* and *L* are under control of the *DR* relay. These contactors will remain closed, to maintain power on the motor, until the motor has decelerated to a low speed. A friction brake is applied to stop and hold the motor when power is removed from the motor.

Figure 18-63 shows the speed-torque curves obtained with a controller as described.

171. Direct-current speed regulators reduce motor speed below normal full field speed by using armature shunt and armature series resistors and increase the speed above normal by introducing resistance in series with the motor-shunt field. Figure 18-64 shows the connections for a d-c speed regulator which has series armature resistors *R*1-*R*2-*R*3, armature shunt resistor *R*4-*R*5, shunt-field resistor *B*-*C*, and field rheostat A-B. On the first speed point, contactors M and AS are energized. This connects resistors $R1$-$R2$-$R3$-$R4$-$R5$ directly across the line, with the armature of the motor connected across the $R4$-$R5$ resistor step. The resistors form a potentiometer from which a fairly stable low voltage can be obtained for the motor armature. Speed reduction to 20% of normal can be satisfactorily obtained. This connection should be used to obtain speeds below 50% of normal, particularly if a uniform slow speed with variations in load is desired. On the second speed point, the AS contactor is opened, and the motor is permitted to accelerate to the speed determined by the $R1$-$R2$-$R3$ resistor. On the third speed point, contactor $1A$ closes; and on the fourth speed point, contactor $2A$ closes, giving normal full field speed.

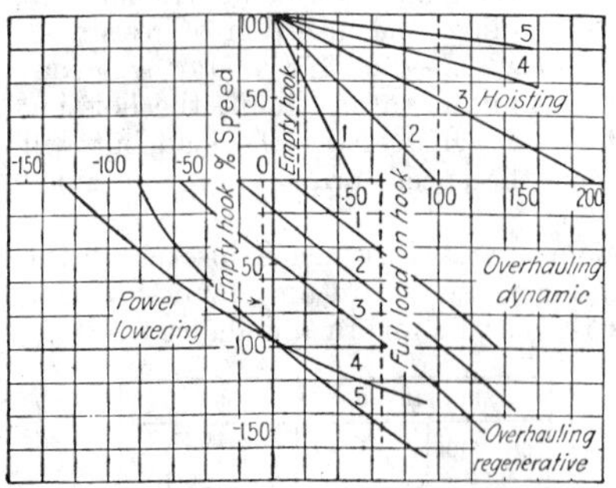

Fig. 18-63. Speed-torque curves obtained with a-c hoist controller of Figs. 18-61 and 18-62.

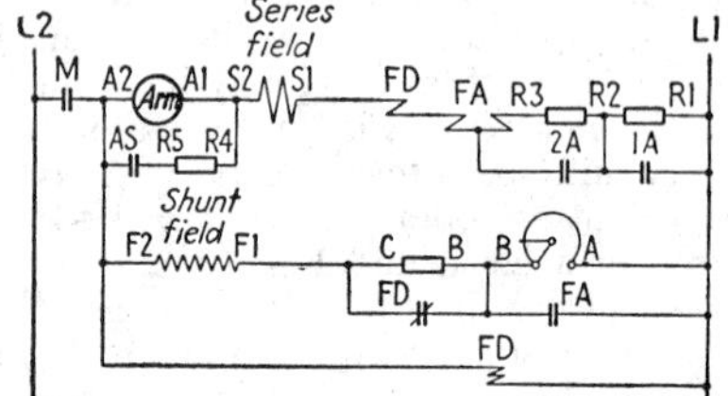

Fig. 18-64. Speed regulator for d-c motor.

The speed of the motor can then be increased above normal by means of rheostat AB. For motors having a speed increase by field greater than 25%, it is advisable to use an accelerating relay FA to limit the current inrushes on acceleration. A decelerating relay FD may also be used to limit the reverse currents obtained on deceleration. When the field rheostat is moved from A to B, the shunt-field strength and the countervoltage generated by the motor are reduced and an inrush of current to the motor armature is obtained. When the field rheostat is moved from B to A, the shunt field is strengthened and the countervoltage generated by the motor exceeds the line voltage. The current in the motor armature reverses, and an inrush is obtained from the motor to the line. Accelerating relay FA has two coils, both of which are in circuit at speeds below normal, which ensures that the relay will be closed and rheostat AB short-circuited to give full field for starting. When accelerator $2A$ closes, the starting coil in FA is short-circuited and the relay opens. The relay is usually set to close with 125% of maximum motor current through the running coil. The current inrush to the motor, when rheostat AB is inserted, is limited to the value for which relay FA is set. When the current reaches this value, FA closes, short-circuiting rheostat AB and thereby increasing the countervoltage and decreasing the current. The relay opens at 100% maximum motor current, at which point the rheostat is again inserted, and a second inrush obtained. This action continues until the motor has accelerated to full speed. The decelerating relay FD has a series coil in the armature circuit and a shunt coil connected across the line. During normal operation these coils are arranged to oppose each other. The contacts of the relay are normally closed and, when open, insert resistance BC into the field circuit. If the rheostat is moved rapidly from B to A, the motor armature current reverses. The action of the two coils on FD is then cumulative, and the relay closes. This inserts resistance BC and weakens the field. The relay will vibrate in the same manner as the accelerating relay until the motor speed has been reduced to the normal full field

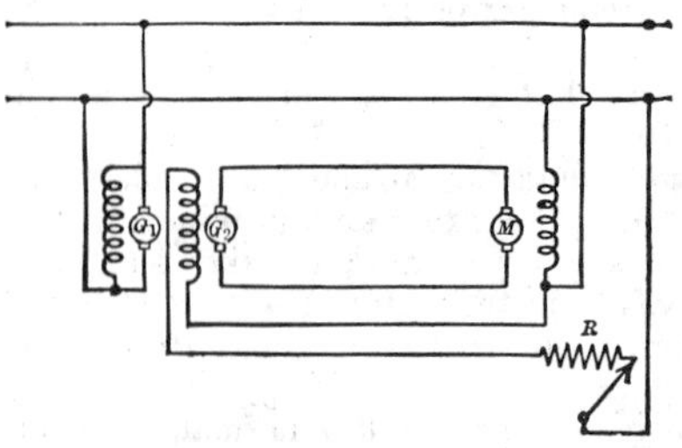

Fig. 18-65. Adjustable-voltage system of d-c control.

speed. A motor with a 3:1 speed range by shunt field, used with an armature resistance designed for 20% of normal speed, will give a total speed range of 15 to 1.

172. The adjustable-voltage system of d-c motor control provides a flexible and efficient means for obtaining a wide range of adjustable speeds. Figure 18-65 is a connection diagram for a typical adjustable-voltage system. G_1 is a self-excited small-capacity generator for supplying the excitation current to the fields of generator G_2 and motor M. The voltage of generator G_2 can be varied by means of rheostat R. By using a potentiometer rheostat for R, the voltage on the G_2 generator field can be changed from zero to maximum. The speed of motor M will vary, as the voltage of generator G_2 is varied, in accordance with the fundamental relation E = counter emf + IR. The IR drop remains constant, and therefore a change in the impressed voltage E must be compensated for by an equivalent change in countervoltage. The following table shows how the motor speed varies directly with countervoltage.

Line voltage	=	CEMF	+	IR	RPM
110	=	100	+	10	1000
60	=	50	+	10	500
35	=	25	+	10	250
20	=	10	+	10	100
10	=	0	+	10	Stalled

A change in the impressed voltage from 110 to 20, or approximately to 1/6, changes the motor speed from 1,000 to 100, or 1/10. The motor speed at a given voltage remains practically constant for wide variations in load. Speed control becomes unstable only when the IR drop is a large percentage of the generator voltage. Speed ranges of 10 to 1 are easily obtained. When used with a motor having a speed range by shunt field of 3 to 1, a total speed range of 30 to 1 is obtained.

This system may be thought of as an adjustable speed control from an a-c source of power, because generators G_1 and G_2 may be driven by an a-c motor. The first cost of the equipment is relatively high, as a motor-generator set is required in addition to the driving motor. However, the system has many advantages. It is possible to make the speed change per step extremely small, as 100 or more steps can be used in the generator field rheostat. The control can be very simple, as it is not necessary to open and close the main power circuits. Motor M can be reversed by reversing the field of generator G_2. The generator and motor form a closed isolated circuit the power output of which is limited by the capacity of the generator. The motor will produce the same torque at low as at high speed, since the field strength remains constant. Quick stopping, by means of regenerative braking, can be obtained by rapidly decreasing the field strength of generator G_2.

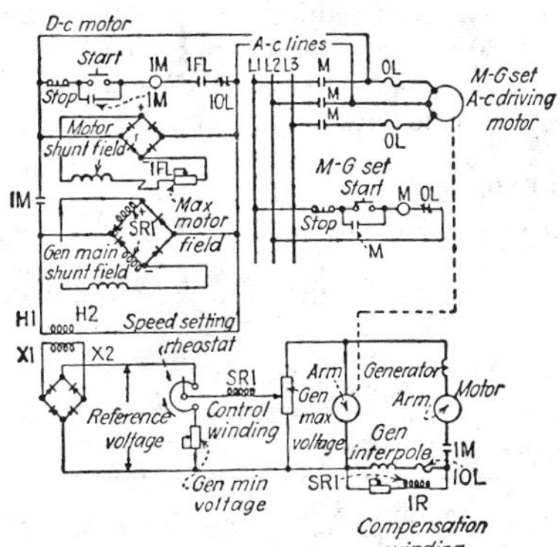

Fig. 18-66. Adjustable-voltage scheme using dry rectifiers and a magnetic amplifier for excitation.

The system is used extensively for elevators, paper calenders, planers, steel-mill drives, and textile range drives. For range-drive service, several motors are connected to a common generator. In order to obtain the same speed changes on all the motors, for a given change in terminal voltage, the motors must be selected to have approximately the same IR drop at full load. Small differences in speed can be adjusted by means of vernier, or dancer-roll, shunt field rheostats.

The packaged adjustable-voltage drive has the a-c driving motor, d-c generator, exciter, and control panel mounted in a single ventilated floor enclosure. The user need only wire the unit to the a-c supply and make the connections to the d-c motor and control station. Connections for a typical unit using magnetic-amplifier excitation are shown in Fig. 18-66.

A full-wave bridge type of selenium rectifier is used for the motor field supply with a series resistor for adjusting the maximum motor field. The generator field current

Table 18-19. Regulating-resistor Taper Table, Machine Load
(Resistor to give 50% speed reduction at 80% current)

Step	Amperes, % full load	Step ohms, % total	Amperes, % full load	Step ohms, % total
1	80	30	80	25.3
2	85	24	84.2	21.0
3	90	19	88.5	17.5
4	94	15	92	14.5
5	97	12	95.2	11.7
6			97.5	10.0

is supplied by the self-saturating magnetic amplifier *SR*1. A reference voltage is supplied for the speed-setting rheostat by a transformer and full-wave rectifier. A portion of this reference voltage, as determined by the position of the speed-setting rheostat, is matched through the magnetic-amplifier control winding with the generator armature voltage. If the generator voltage is low compared with the reference voltage set up by the rheostat, the current in the control winding will increase the output of the amplifier supplying the generator field. The increased generator field current will increase the armature voltage until it matches the reference voltage.

Better speed regulation and a wider speed range can be obtained by using a means to compensate for the *IR* drop of the motor circuit. *IR*-drop compensation may be obtained by connecting a magnetic-amplifier coil across the generator interpole and the overload relay coil as shown in Fig. 18-66. As the armature current increases, the voltage across this amplifier coil increases, which produces a higher magnetic-amplifier output. The result is an increase in generator armature voltage to compensate for the additional *IR* drop in the circuit.

173. Regulating-duty resistors are usually designed for continuous duty, in accordance with a NEMA classification number between 91 and 96 inclusive. The specific classification within this group is determined by the motor starting current allowed, with all resistance in circuit. The total ohms in the resistor are usually determined by the speed reduction required and not by the starting current. If the ohms used are too high to permit sufficient inrush for starting, a torque switch may be used to short-circuit some of the resistor during the initial accelerating period. For practical purposes the majority of machines can be considered to have one of two types of load characteristics: either the machine type of load or the fan type of load. The machine type of load requires approximately 80% of current at 50% speed reduction. The fan type of load requires approximately 40% of current at 50% speed reduction. For an a-c slip-ring motor with a 3-phase secondary and a star-connected secondary resistor, the ohms required for a given speed reduction can be calculated as follows,

$$R = \frac{E \times \%\text{ speed reduction}}{1.73 \times I \times \%\text{ current at reduced speed}} \tag{18-58}$$

where E = open-circuit voltage between rings with motor stalled and I = full-load secondary amperes per phase.

Table 18-20. Regulating-resistor Taper Table, Fan Load
(Resistor to give 55% speed reduction at 36% current)

Step	Amperes, % full load	Step ohms, % total	Amperes, % full load	Step ohms, % total
1	36	44	36	37
2	48	26	46	26
3	62	16	58	17
4	76	9	70	10
5	88	5	80	6
6			90	4

For 50% reduction, machine load

$$R = \frac{E \times 50}{1.73 \times I \times 80}$$

For 50% reduction, fan load

$$R = \frac{E \times 50}{1.73 \times I \times 40}$$

The corresponding formula for a d-c shunt-wound motor is

$$R = \frac{E \times \%\text{ speed reduction}}{I \times \%\text{ current at reduced speed}} \tag{18-59}$$

where E = line voltage and I = full-load amperes at normal speed.

Tables 18-19 and 18-20 show the distribution of the total ohms between the various steps in a speed regulator and also the current capacity required for each step. The machine-load taper will result in approximately equal speed changes per step. The fan-load taper table will not give equal speed changes per step but is designed to limit the current inrushes to a reasonable value at the higher speeds.

174. Field-regulating resistors should be designed on the basis of actual test data on the motor with which the control is to be used. The motor manufacturer will supply such data as the resistance of the shunt field or the maximum field amperes, the required rheostat resistance or the minimum field amperes. The total ohms in the rheostat will therefore be determined by the motor data. The number of steps in the rheostat must be large enough to limit the speed increase per step to an acceptable value. The character of the resistor material used may require a large number of steps, in order to keep the wattage per step on the rheostat within limits. The number of steps must be sufficiently large to keep the voltage drop at any step within specified limits to prevent arcing on the contacts. The arcing limits for sliding contacts vary from 200 V at 0.4 A to 50 V at 1 A and 25 V at 4 A. This voltage is the product of the step ohms and the amperes flowing before the step is inserted. The amount of resistor material required in the field rheostat is determined by the wattage which must be dissipated. Each step must be designed on the basis of the maximum current which it will have to carry. The total wattage capacity of the rheostat will be the summation of the step wattages. The summation watts can be calculated from the following formula:

$$\text{Summation watts} = \text{max. current} \times \text{min. current} \times \text{rheostat ohms} \tag{18-60}$$

The distribution of the resistor between the various steps can be determined on the basis of obtaining either equal speed changes per step or equal percentage speed changes per step. Common practice at present is to provide equal speed changes per step, and a rheostat having equal ohms per step will give this condition within practical limits. A rheostat for equal percentage speed changes has low ohms at the low-speed end and high ohms at the high-speed end. This gives a better wattage distribution in the rheostat and therefore requires less resistor material.

STOPPING DEVICES

175. Dynamic braking of a d-c shunt-wound motor can be obtained by disconnecting the motor from the line and shunting the armature with a resistor. The field must remain connected to the line during braking. For constant-speed motors the ohmic value of the braking resistor is $R = (E - I_aR_a)/I$. The value of I determines the amount of braking obtained and may vary from 150 to 300% of normal current. With shunt motors, having speed regulation by shunt-field control, the general practice is to strengthen the field during braking by short-circuiting the field rheostat. When this is done, the ohms in the braking resistor should be $R = (E - I_aR_a)/I \times$ speed range by field. The braking obtained decreases as the motor speed is reduced. The final stopping of the motor is due to friction, since at standstill no braking torque is obtained. For very quick stopping the braking resistance can be reduced in several steps as the speed decreases, thereby keeping the current at a high value. This type of braking is effective where it is necessary to stop a motor quickly. Mechanical brakes must be used to hold a load at a standstill.

176. Dynamic braking with an a-c induction motor can be obtained by disconnecting the motor from the power supply and applying direct current to one of the stator phases. With a squirrel-cage motor a direct current equal to 250% full-load current must be applied in order to obtain an average braking torque equal to 25% starting torque. The average braking torque can be increased, when a wound-rotor motor is used, by the addition of resistance in the rotor circuit. The maximum value of braking torque moves toward higher speeds as the rotor resistance is increased. A reasonable value of rotor resistance for good results is 40% of $E/\sqrt{3}I$. A transformer and full-wave selenium rectifier can be used for the d-c supply. For detailed discussion see R. W. Jones, "Electric Control Systems," 3d ed., p. 438.

177. Plugging is used with d-c motors to obtain very rapid reversing and is accomplished by connecting the motor to the line in the reverse direction while it is still rotating in the forward direction. The countervoltage of the armature is added to line voltage to force current through the armature and series resistor. The use of an additional step of resistor, with a contactor to short-circuit it when the motor reaches approximately zero speed, is described under Inductive Time-limit and Inductive-relay Controllers (see Pars. **162** and **163**). The total resistance in circuit for plugging should be $R_t = 2(E - I_aR_a)/\text{inrush current}$. With the armature drop assumed as 10%, the formula may be written $R_t = 1.8E/1.5I = 1.2E/I$, where E = line voltage and I = normal current. The plugging resistance is determined by subtracting the accelerating resistance from R_t.

178. Plugging is used with a-c motors to obtain a very quick stop. Three-phase squirrel-cage and wound-rotor motors may be plugged by reversing the line connections to any two of the stator terminals while the motor is running in the forward direction. In order to use plugging as a stopping means, a zero-speed switch is necessary to open the reverse contactor and prevent reversing the motor. A common form of zero-speed switch is the friction type, in which a contact is held closed by the friction of a small belt over a pulley driven by the motor. Any slight reversal of the motor will cause the contacts to open. Another type of zero-speed switch uses a disk which rotates in the magnetic field produced by Alnico magnets. Eddy currents are induced in the driven disk, and the magnetic reaction turns the magnet assembly to close the contacts. The contacts can be adjusted to open when the motor is near zero speed.

An electrical plugging relay is also used which remains closed until the current in the secondary decreases to the normal inrush current corresponding to the starting condition, or zero speed on the motor. This device has the advantage of not requiring a mechanical connection to the motor or machine.

179. Regenerative braking is used with adjustable-voltage control systems to obtain rapid stopping. A typical installation consists of a d-c generator, driven by an a-c synchronous motor, with a d-c motor connected electrically to the generator. In normal operation the generator voltage exceeds the countervoltage of the driven motor, and power is taken from the a-c lines to keep the driven machine rotating. If the field strength of the generator is decreased, the generator voltage becomes less than the countervoltage of the motor and the motor feeds power back to the generator and to the a-c lines. When the countervoltage exceeds the generator voltage, a heavy reverse current is obtained, as the value of this current is limited only by the low resistance of the loop circuit. A very rapid stop is obtained, since the voltage across the generator field can be reduced to zero in about 3 s.

180. Magnetic brakes are used to obtain quick, accurate stopping and to hold the load after stopping. Most brakes are electrically released and spring-set, so that braking will be obtained even though an electrical failure occurs. Disk-type brakes are supplied for mounting directly on the end bell of a motor and are available in ratings of 3 to 75 lb·ft. Shoe-type brakes are supplied with floor mounting bases and are available in sizes from 3 to 4,000 lb·ft.

181. Direct-current brakes are released by means of a solenoid or a direct-operating magnet. The coil in the operating device may be for either series or shunt connection and for continuous or intermittent duty. Series-wound brakes are operated by motor current and require 80% full motor current to release with a continuous-duty coil and about 40% full-load current with an intermittent-duty coil. The brake will be held released on about 10% full-load motor current. Intermittent-duty series brakes are rated as either ½-h duty or 1-h duty, to correspond to the rating of intermittent-duty series motors. A series-brake coil will carry full motor current continuously, for a period corresponding to its rating, without exceeding a temperature rise of 75°C. Shunt brakes may be for either continuous or intermittent duty. Intermittent duty is defined as 1 min on and 1 min off or the equivalent, the longest time on not to exceed 1 h. Shunt brakes will release at 80% of normal voltage, when adjusted for rated torque. The larger-sized brakes use partial-voltage coils and protecting resistors. Series brakes have a heavy wire coil, which is less likely to give trouble, are faster in operation, and will set whenever the armature circuit is open.

A d-c magnet-operated shoe brake is shown in Fig. 18-67, and data on a line of brakes of this type are given in Table 18-21. The magnet assembly consists of the inner and the outer magnet members *A* and *B* and a coil *C*, which is enclosed in a stainless-steel capsule and which is separate from either of the magnet members. The single torque spring *D* forces the inner armature toward the wheel and the outer armature away from the wheel. The force of the inner armature is transmitted directly to the inner shoe *E*, and the force of the outer armature is transmitted through the overhead pull rod *F* to the outer shoe *G*. When power is applied to the coil, the armature members are pulled together, thus moving the inner and outer shoes away from the surfaces of the wheel and releasing the brake. The braking torque can be adjusted by nut *H*.

Table 18-21. Direct-current Magnet-operated Shoe Brakes

Wheel diam., in.	Maximum torque, lb-ft			
	Shunt wound		Series wound	
	Int. duty	Cont. duty	½-hr duty	1-hr duty
8	100	75	100	65
10	200	150	200	130
13	550	400	550	365
16	1000	750	1000	650
19	2000	1500	2000	1300
23	4000	3000	4000	2600

NOTE: Int. = intermittent. Cont. = continuous.

182. Alternating-current brakes have three forms of operating mechanism: solenoid type, torque-motor type, and thrustor type. The smaller sizes of brakes are usually made in the solenoid type. On the larger sizes, a vertically mounted torque motor and antifriction ball jack provide a quiet, low-inrush-current operating means. A brake of this type is shown in Fig. 18-68. Upon application of power to the motor, the rotary motion of the armature is transformed into straight-line motion through the antifriction jack. With the brake fully released, the torque motor is stalled across the line. When the motor is disconnected from the line, the spring in the brake overhauls the motor mechanism and applies the brake. Data on a line of brakes of this type are given in Table 18-22.

The thrustor-type operating mechanism consists of a self-contained motor-driven centrifugal pump, oil chamber, and a piston which produces a straight-line movement to release the brake.

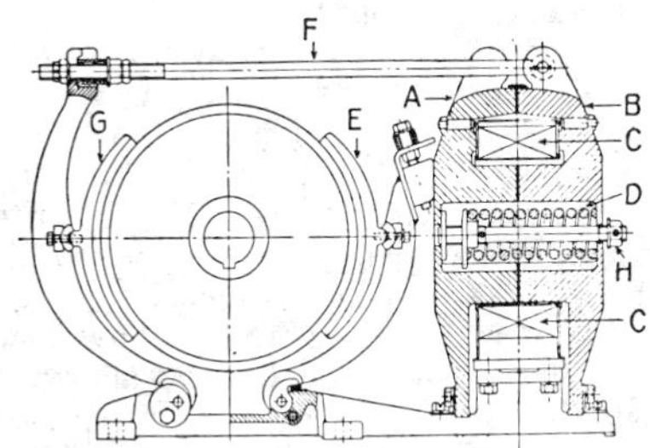

FIG. 18-67. Direct-current shoe brake. (*A*) Inner armature; (*B*) outer armature; (*C*) coil; (*D*) torque spring; (*E*) inner shoe; (*F*) pull rod; (*G*) outer shoe; (*H*) torque-adjusting nut.

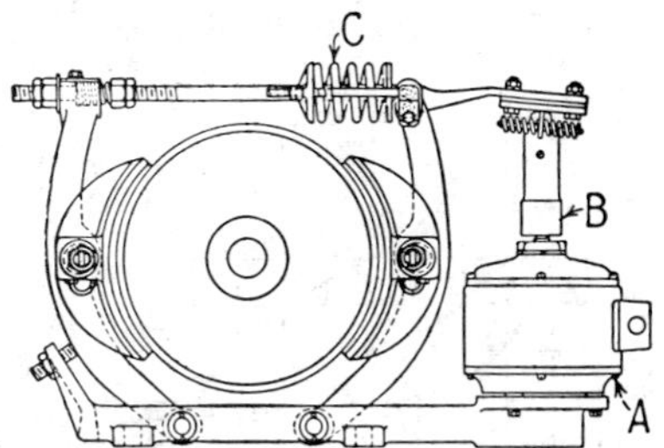

FIG. 18-68. Alternating-current torque-motor brake. (*A*) Torque motor; (*B*) antifriction jack; (*C*) operating spring.

Table 18-22. Alternating-current Torque-motor-operated Brakes

Wheel diameter, in.	Maximum torque, lb-ft		Volt-amperes		WR^2 of wheel	Safe maximum rpm	Weight of brake, lb
	Int. duty	Cont. duty	Int. duty	Cont. duty			
10	160	125	160	105	3.1	2015	150
13	400	325	210	140	12	1550	240
16	800	600	300	240	25	1260	370
20	1600	1200	1000	470	75	1012	750
25	3200	2400	1500	550	220	806	1210

Note: Int. = intermittent. Cont. = continuous.

183. **The brake size** for most applications can be determined by using the formula $T = 5.252 \times \text{hp}/\text{r/min}$, where T = full-load motor torque in pound-feet, hp = motor horsepower, r/min = speed of shaft on which brake wheel is mounted. A brake should be selected with a torque rating equal to or greater than the full-load motor torque T. In some cases the braking torque is determined by extreme operating conditions against which the brake must hold, e.g., heavy ice loads on bascule bridges or conditions of unbalance on skip hoists. In these cases the maximum load must be calculated and translated into pound-feet torque at the shaft on which the brake is mounted. Sufficient lining area is provided on all sizes of brake for the average application. However, a careful check as to lining area must be made when brakes are used for frequent stopping or for stopping high-inertia loads.

MOTOR-PROTECTING DEVICES

184. **Fuses** should be provided for motor circuits, in accordance with NEC. The current rating of the fuse must be considerably higher than the current rating of the motor, or the fuse will blow when the motor is started. For that reason fuses do not provide adequate overload protection. They furnish protection for the motor only in case of a short circuit or a very heavy overload. Their primary purpose is to protect the circuit rather than the motor.

185. **Magnetic-type overload relays** are operated by direct magnetic action of the motor current on a plunger. The relay consists of a series coil connected in the motor circuit and a plunger which is pulled up into the center of the coil when a certain value of current has been reached. When the plunger is lifted, a contact is tripped, opening the motor contactor-coil circuit and disconnecting the motor from the line. The tripping current can be varied by adjusting the initial position of the plunger with respect to the coil. Time delay in tripping is obtained by attaching a small oil dashpot to the plunger. The time delay can be adjusted so that the overload will not trip on the starting-current inrush but will trip on small sustained overloads.

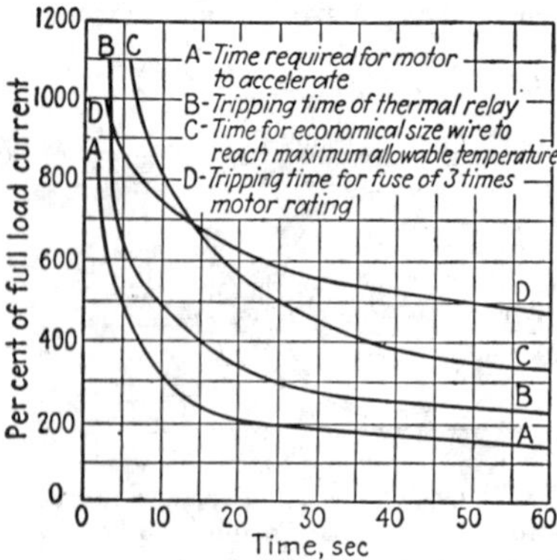

Fig. 18-69. Characteristics of thermal overload relays.

186. **Thermal overload relays** are available in the bimetallic type and the fusible-alloy type. The bimetallic type has two heaters in series with the circuit to be protected, and above these heaters are two strips of bimetallic material, which act as latches for the contact members. Bending of the bimetallic strips under heating of overload current will release the latches and allow the contacts to open. The fusible-alloy type has two heaters, each surrounding a thermal element consisting of a small tube, inside which is a loose-fitting shaft. The tube and shaft are rigidly joined by a special low-melting eutectic alloy. On overload, the increased current drawn melts the alloy, allowing the shaft to turn and the contacts to open.

Characteristics of a typical thermal overload are shown in Fig. 18-69. An inspection of these curves shows that the thermal overload adequately protects the wiring, that the fuse blows first on short-circuit current, and that the thermal relay allows the motor ample time to accelerate. A thermal overload has a tripping characteristic which corresponds closely to the heating characteristics of a motor and, therefore, provides an ideal protecting means. An overload coil should be selected so that the maximum permissible output can be obtained from the motor. A motor rated 40°C rise on the basis of 40°C ambient temperature will have a final safe temperature of approximately 95°C and will operate at 15% overload continuously without overheating. An overload coil should therefore be selected having an ultimate tripping current equivalent to 15% overload on the motor. A continuous overload of 15% would therefore ultimately trip the thermal relay. For overloads in excess of 15% the tripping time would be shorter than the time required for the motor to reach a dangerous temperature.

187. Low-voltage protection is the effect of a device, operative on the reduction or failure of voltage, to cause and maintain the interruption of power to the main circuit.

With magnetic controllers, this protection is obtained by using some form of 3-wire master switch. Should the line voltage drop to a low value or fail altogether, the main-line contactor will open and remain open, stopping the motor. To restart, it is necessary to push the "start" button. This type of control should always be used where the unexpected restarting of a motor after voltage failure may be dangerous to workmen or equipment.

188. Low-voltage release is the effect of a device, operative on the reduction or failure of voltage, to cause the interruption of power to the main circuit but not to prevent the reestablishment of the main circuit on return of voltage. Such protection is obtained when a 2-wire pilot device, e.g., a snap switch, float switch, or pressure switch, is used.

189. Phase-failure protection is the effect of a device, operative upon the failure of power in one wire of a polyphase circuit, to cause and maintain the interruption of power in all the wires of the circuit.

190. Phase-reversal protection is the effect of a device, operative on the reversal of the phase rotation in a polyphase circuit, to cause and maintain the interruption of power in all wires of the circuit. Protection of this type is necessary on elevators, where reversing of the phases would cause the car to start in a direction opposite to that in which the operator expects it to move.

191. Field-failure protection is usually provided in controllers for d-c shunt- and compound-wound motors. The coil of a relay is connected in series with the motor shunt field, and a normally open contact of the relay is connected in the stop circuit. If the field circuit is opened, the relay will be deenergized and the motor will be disconnected from the line. This prevents overspeeding the motor owing to an open circuit in the field. **A field protective relay** is used to insert resistance in series with the shunt field whenever the motor is not running. The coil of the relay is connected in parallel with the main switch coil, and a normally open relay contact is used to short-circuit a step of resistor in the field circuit. The resistor should be designed to reduce the voltage across the field to one-half line voltage. This reduces the field wattage to one-fourth the normal value and prevents overheating the field with the motor at standstill.

192. A field-discharge resistor should be provided for 230-V motors rated 7½ hp or more and for 550-V motors rated 5 hp or more whenever the shunt-field circuit must be opened. The ohmic value of a discharge resistor should be one to three times the ohms in the field. If a resistance of three times the field ohms is used, the induced voltage, when the circuit is opened, will be four times normal line voltage. This voltage, caused by the inductance of the field, must be limited to prevent damage to the insulation of the field windings. On nonreversing controllers without dynamic braking, the shunt field can be connected behind the main contactor and the field allowed to discharge through the motor armature.

BIBLIOGRAPHY

193. Books on Motor Control

HALL, G. H. "Motor and Control Applications"; New York, McGraw-Hill Book Company, 1937.

HARWOOD, P. B. "Control of Electric Motors," 3d ed.; New York, John Wiley & Sons, Inc., 1952.

JAMES, H. D., and MARKLE, L. E. "Controllers for Electric Motors"; New York, McGraw-Hill Book Company, 1952.

HEUMANN, G. W. "Magnetic Control of Electric Motors," 2d ed.; New York, John Wiley & Sons, Inc., 1954.

ROTERS, H. C. "Electromagnetic Devices"; New York, John Wiley & Sons, Inc., 1941.

SHOULTS, D. R., RIFE, C. J., and JOHNSON, T. C. "Electric Motors in Industry"; New York, John Wiley & Sons, Inc., 1942.

JONES, R. W. "Electric Control Systems," 3d ed.; New York, John Wiley & Sons, Inc., 1953.

CHESTNUT, HAROLD, and MAYER, R. W. "Servomechanisms and Regulating System Design"; New York, John Wiley & Sons, Inc., 1951.

Bibliography on Industrial Control; *AIEE Publ.* S39, September, 1950.

SYSTEMS OF MOTOR SPEED CONTROL USING AUXILIARY MACHINES

194. Adjustable-speed A-C Motors. Speed of an asynchronous or induction motor may be controlled by varying the secondary resistance, but this involves considerable losses and affords only subsynchronous speeds. Schemes for salvaging the slip energy, which is otherwise lost as heat, and either returning it to the line or putting it usefully to work as mechanical energy have been devised. These desirable results are accomplished by certain auxiliary apparatus, which may also in some cases permit operation above as well as below synchronous speeds.

195. Constant Horsepower and Constant Torque. These drives are often classified as either constant horsepower or constant torque; the designation refers to the inherent limitation in power based on full current and flux in the main machine. In the first scheme (*constant-horsepower drive*), the slip energy is converted into mechanical power and then returned to the main motor shaft. Since horsepower is a function of the product of torque and speed, such motors have high torque at low speeds and lower torque at higher speeds. In drives using this arrangement the auxiliary machine is mounted on or mechanically geared to the main motor shaft (Fig. 18-70*a*).

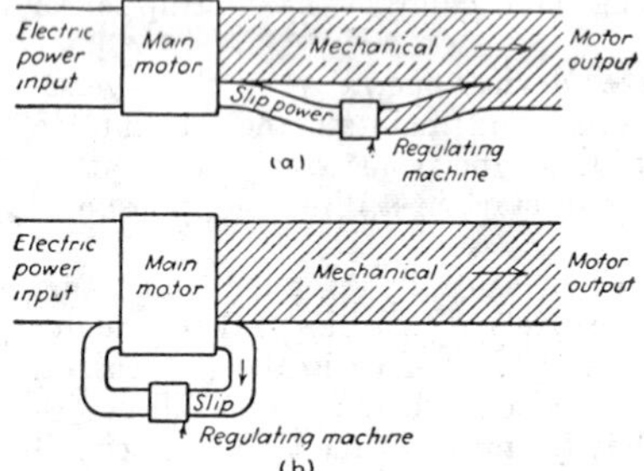

FIG. 18-70. (*a*) Constant-horsepower drive; regulating machine, coupled to main motor, returns power mechanically. (*b*) Constant-torque drive; regulating machine, mechanically separate from main motor, returns slip power electrically.

In the second scheme (*constant-torque drive*), the slip energy is converted into electrical power of the frequency and voltage of the supply circuit and is returned or fed back into the line. Since this power is not delivered to the main motor shaft, the auxiliary machine is not mechanically attached to the shaft but is separately driven. The limiting torque of the main motor being constant, the maximum horsepower output is proportional to the operating speed (Fig. 18-70).

196. Scherbius System. A common arrangement for such variable-speed control of either scheme is known as the Scherbius drive. In the *constant-horsepower* arrangement (Fig. 18-71) the regulating machine B is direct-connected to the main motor A and converts the slip energy of the main motor into mechanical energy. The regulating machine has a 3-phase stator winding which acts as a field winding and an armature with a commutator, but instead of two brush holders for each pair of poles these are spaced 120 elec deg apart. The field winding is connected through a transformer *C* to the slip rings of the main motor. The frequency at the commutator brushes is always the same as in the field, and the brush frequency is thus the same as on the slip rings of the induction motor, regardless of the motor speed. The voltage generated in machine *B* bucks the secondary voltage in the main motor, which tends to slow it down. By changing the taps on the transformer *C*, the value of the generated voltage and therefore the operating speed may be adjusted and, once adjusted, is practically constant regardless of the load. The regulating machine *B* performs as a shunt-wound d-c motor and converts the slip energy into mechanical power on the shaft.

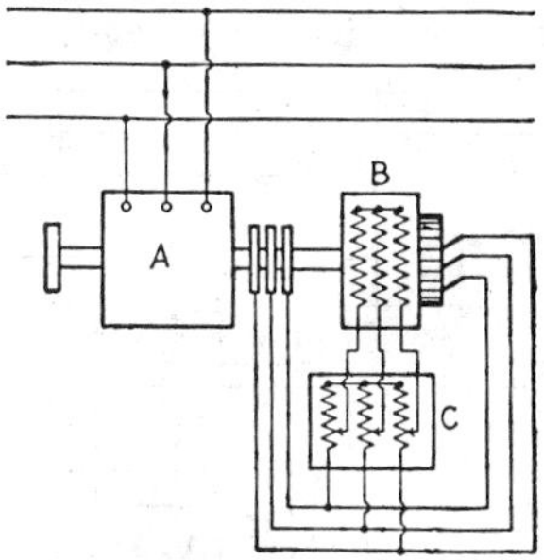

FIG. 18-71. Schematic diagram of principal connections for Scherbius drive below synchronous speeds at constant horsepower; regulating machine mechanically coupled to main motor. (*A*) Main motor. (*B*) Regulating machine. (*C*) Field-control transformer.

In the *constant-torque Scherbius* control, the slip energy is brought out to a separate motor-generator set. For operation above synchronous speed, power flows from the commutating machine to the slip rings when the main machine is running as a motor.

197. Speeds Both above and below Synchronous. In order to effect speed control not only below and above synchronous speed but to pass through synchronous speed, an additional device is necessary. When the main motor is running at synchronous speed, the slip-ring frequency and voltage are zero and there is thus no excitation in the regulating machine *B*; no voltage can therefore be generated in it, and there is thus no current flowing in the secondary winding of the motor that can increase the torque and force the speed above synchronism. A small exciter or frequency converter is therefore provided, mounted on the main shaft *E* (Fig. 18-72). This exciter consists merely of an armature similar to the armature of a rotary converter with an iron ring pressed on it to complete the magnetic circuit. The slip rings are connected to the line through transformer *F*. This armature acts to convert the line frequency to the frequency at the commutator brushes, which depends on the rotating speed of the main unit and which therefore is the same frequency as exists at the same time on the main-motor slip rings and the exciting circuit of the regulating machine. There is thus an added source of excitation to the machine *B*.

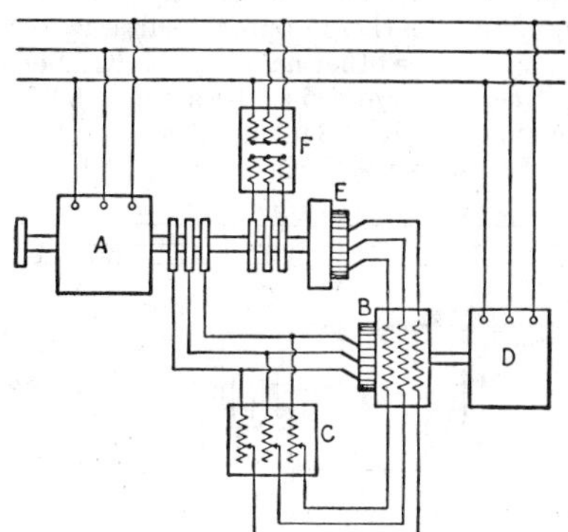

FIG. 18-72. Diagram of principal connections for Scherbius drive for speeds below, above, and at synchronism. (*A*) Main motor. (*B*) Regulating machine. (*C*) Field-control transformer. (*D*) Squirrel-cage motor (or generator). (*E*) Frequency converter. (*F*) Step-down transformer for *E*.

When the speed of the main unit passes through synchronism, the frequency of excitation is zero; i.e., direct current flows through the windings of the regulating machine and also the rotor of the main motor. But since voltage is generated in the regulating machine, the drive has full torque to pass through and above synchronous speed. When the exciter is at zero frequency, there is no reactance in the circuit and the exciter takes care only of the ohmic drop in the windings; hence the name "ohmic-drop exciter" is sometimes applied. The adjusting of speed is accomplished by tap changing on the exciter or field control transformer *C* or by moving brushes on the commutator of the frequency converter *E*.

198. Krämer System. In this system (Fig. 18-73) the slip energy of the main induction motor is transformed through a rotary converter into direct current, and this power is either returned to the main shaft (in a constant-horsepower machine) or further converted to a-c power and returned to the supply line for constant-torque drive. The basic difference between the Scherbius and Krämer systems is that, while the former employs a commutating machine to convert the slip energy, the latter accomplishes this by means of a rotary converter and motor. Speed regulation is accomplished by strengthening the field of the d-c motor, whereby the counter emf is increased, and this emf, acting through the rotary converter, bucks the secondary voltage of the main motor, and the motor is thus slowed down. The Krämer equipment thus lacks the characteristics of normal torque at or near synchronous speeds, but there are Krämer equipments in operation on drives of low-friction loads which operate above synchronous speeds.

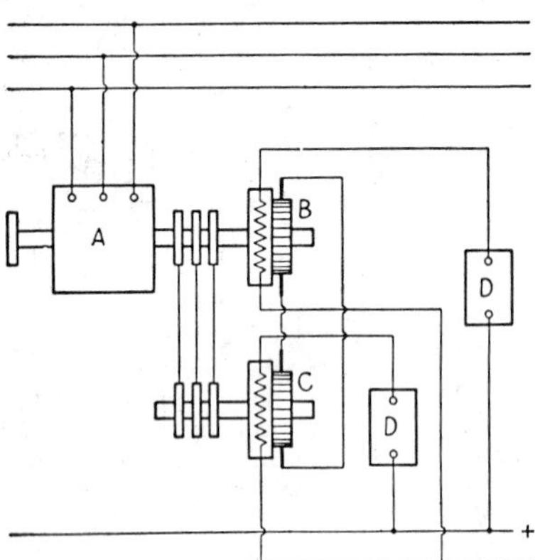

Fig. 18-73. Diagram of connections for Krämer drive. (*A*) Main motor. (*B*) Direct-current motor on main shaft. (*C*) Rotary converter. (*D*) Field rheostats for *B* and *C*.

The Scherbius and Krämer equipments have been developed for both 25- and 60-c service. However, the Scherbius commutator regulating machines inherently perform better on low-slip frequency, and the Scherbius equipment is therefore more suitable for 25-c service or for limited speed range in 60-c service. On the other hand, in the Krämer system, the rotary converter operates on the slip frequency; and on 60-c service, 50% speed reduction results in 30-c slip frequency, which can be handled on a standard 25-c converter, thus resulting in effective use of the converter capacity.

199. Modified Krämer Drive. A variation of the Krämer drive uses a synchronous motor and a d-c generator in place of the rotary converter and a constant-speed set feeding the slip power back into the line, as shown in Fig. 18-74. This drive has been used for a number of large wind-tunnel drives. It is particularly adapted to a wide range of speed control and to minimum disturbance on starting. For minimum starting kilovolt-amperes the constant-speed set is started first. The variable-speed set is then brought to synchronous speed, driving it from the d-c end. The slip rings are then energized by exciting the synchronous machines, and the main machine synchronized on closing the line circuit breaker, by using a synchroscope or automatic synchronizer. The top speed is limited by instability when the slip is below approximately 5 to 10% depending on the size and design characteristics.

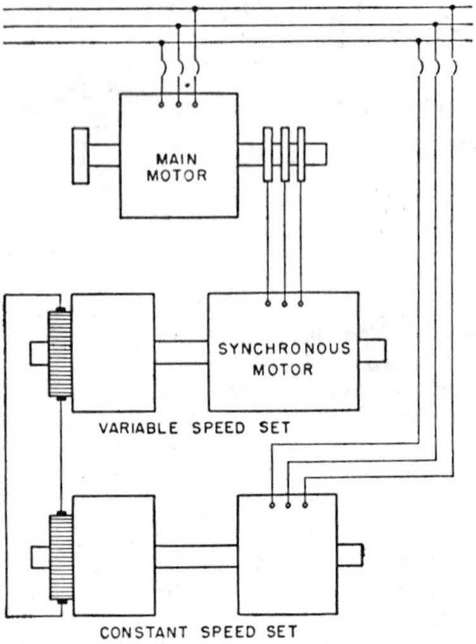

Fig. 18-74. Modified Krämer drive.

The product of the full torque of the variable-speed set and its synchronous speed will be approximately equivalent to the full horsepower of the drive. For a fan or pump type of load some advantage can be taken in the design of the d-c machine of the fact that the current drops off as the set speed increases. The power from the slip rings is equal to the input power times the per-unit slip, the losses being neglected. This means that the torque of the variable-speed set is proportional to that of the main drive.

200. Speed Control Using Electronic Converter. Several schemes for controlling the speed of a-c motors by electronic means have been tried, at least experimentally. One involves the substitution of a controlled rectifier in place of the rotary converter. Another is to supply a variable frequency to a synchronous or induction motor. The driving of an induction motor from a simple electronic inverter is complicated by the necessity of having large capacitors for furnishing all the kvar requirements, and these capacitors tend to cause instability at high slips.

A synchronous motor, once it has been brought up to one-sixth of synchronous speed or higher, may then be supplied from a simple electronic inverter or d-c link frequency changer.

201. Bibliography

Kincaid, C. W. Variable Ratio Frequency Changer Sets; *Elec. Jour.*, June, 1928, p. 279.

Laffoon, C. M., Dickey, D. A., and Kilgore, L. A. Variable Speed Drive for U.S. Army Wright Field; *Trans. AIEE*, 1942, pp. 126–130.

SECTION 19

ILLUMINATION

BY

WALTER STURROCK Illuminating Engineer, Lamp Division, General Electric Company; Past President, Illuminating Engineering Society; Fellow, Illuminating Engineering Society; Member, Society of American Military Engineers, Building Owners and Managers, Cleveland Engineering Society

KARL A. STALEY Lamp Division, General Electric Company; Member, Illuminating Engineering Society; Member, Society of Architectural Historians

REVISED BY

JACK F. PARSONS, P.E. Division Illuminating Engineer, Niagara Mohawk Power Corporation (retired)

CONTENTS

Numbers refer to paragraphs

SECTION 19

ILLUMINATION

RADIANT ENERGY AND LIGHT

1. For the principal purposes of illumination design, light is defined as **visually evaluated energy.** The visible energy radiated by light sources is found in a narrow band in the electromagnetic spectrum approximately from 4,000 to 7,000 angstroms (Å). By extension, the art and science of illumination also include the applications of ultraviolet and infrared radiation. The principles of measurement, methods of control, and fundamentals of lighting system and equipment design in these fields closely parallel those long established in lighting practice.

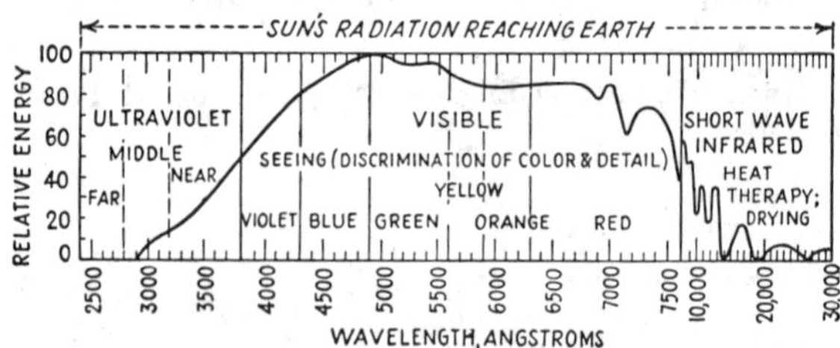

FIG. 19-1. Ultraviolet, visible, and short-wave infrared are the three principal bands of the electromagnetic spectrum with which illuminating engineering practice is concerned.

UNITS AND CONVERSION FACTORS

2. Photometric quantities in common use and their definitions are as follows:

Absorptance (absorption factor) is the ratio of the light absorbed to the incident light. The absorbed light is the difference between the incident light and the sum of the transmitted and the reflected light.

Blackbody is a temperature radiator which absorbs all the radiant energy that falls upon it. In practice it is a cavity with opaque walls at uniform temperature with a small opening for observation purposes.

Candlepower (cp). $I = df/dw$ cp is luminous intensity expressed in candelas.

Candela (cd) is the unit of luminous intensity. It is defined as one-sixtieth of the luminous intensity of one square centimeter of a blackbody radiator at the temperature of solidification of platinum (2046°K). An earlier definition was in terms of the average horizontal candlepower of a group of carbon-filament lamps. The original definition was in terms of the strength of a flame source, a standard candle.

Candelas per square inch (cd/in^2) is the unit of photometric brightness of a surface in a specified direction, similar in concept to the *footlambert.* Both express luminous intensity per unit area of surface.

In general, photometric brightness (average luminance) is useful only when it is reasonably uniform throughout a very wide angle of observation and over a large area of the surface considered. It can be computed for reflecting surfaces by multiplying the illumination (*footcandles*) by the *reflectance* of the surface. For a transmitting medium, the illumination is multiplied by the *transmittance* of the medium.

Color temperature of a nonblackbody is the temperature at which it is necessary to operate a blackbody so that the color of its light will match that of the source studied.

E-viton is the unit of erythemal flux corresponding to the quantity of radiant energy which produces as much reddening on untanned human skin as 10 μW of energy at 2,967 Å.

Fluoren[1] **(fn)** is the unit of ultraviolet flux from a black-light source equal to one milliwatt of energy emitted between 3,200 and 4,000 Å.

Fluoren per square foot[1] is the unit of incident flux density of surface one foot square on which there is a uniformly distributed flux of one fluoren.

Footcandle (fc) is the illumination on a surface of one foot square on which there is a uniformly distributed flux of one lumen.

Table 19-1. Conversion Factors for Lighting Terms

A. Illumination

1 footcandle = 1 lumen per square foot
1 lumen-hour = 60 lumen-minutes
1 lux = 1 lumen per square meter = 1 meter-candela
1 phot = 1 lumen per square centimeter

Number of → Multiplied by ↘ Equals number of ↓	Footcandles	Lux	Phots	Milliphots
Footcandles	1	0.0929	929	0.929
Lux	10.76	1	10,000	10
Phots	0.00108	0.0001	1	0.001
Milliphots	1.076	0.1	1,000	1

B. Brightness

1 footlambert = 1 lumen per square foot
1 millilambert = 0.001 lambert
1 lambert = 1 lumen per square centimeter
1 stilb = 1 candle per square centimeter
1 apostilb (international) = 0.1 millilambert

Number of → Multiplied by ↘ Equals number of ↓	Foot-lamberts	Lamberts	Milli-lamberts	Candelas per square inch	Candelas per square foot	Stilbs
Footlamberts	1	929	0.929	452	3.142	2,919
Lamberts	0.00108	1	0.001	0.487	0.0034	3.142
Millilamberts	1.076	1,000	1	487	3.382	3,142
Candelas per sq in.	0.00221	2.054	0.00205	1	0.00694	6.45
Candelas per sq ft	0.3183	295.7	0.2957	144	1	929
Stilbs	0.00034	0.3183	0.00032	0.155	0.00108	1

Footlambert (fL) is the unit of photometric brightness equal to that of a perfectly diffusing surface emitting or reflecting light at the rate of one lumen per square foot. The average photometric brightness of any reflecting surface in footlamberts is the product of the illumination in footcandles by the luminous reflectance of the surface.

Glow factor[1] **(fluorosity)** is the expression of the ability of a material to convert incident ultraviolet flux (fluorens) into visible light (lumens).

Lumen (lm) is the unit of luminous flux. It is equal to the flux through a solid angle from a point source of one candle or to the flux on a square foot of surface all points of which are one foot from a point source of one candle or to the flux on a square foot of surface all points of which are one foot from a point source of one candela. Light sources are rated in lumens.

Lumen hour (lm·h) is the unit of quantity of light. It is the luminous energy delivered in one hour by a flux of one lumen.

Luminaire is a complete lighting unit consisting of a light source, globe, reflector, refractor, and housing and the support, which is part of the housing.

Lux (lx) is the practical unit of illumination in the metric system and is equivalent to the *meter candle*. It is the illumination on a surface of one square meter on which is a uniformly distributed flux of one lumen.

Reflectance, or reflection factor, is the ratio of the light reflected to the incident light.

[1] Proposed term.

Regular, or specular, reflection is that in which the angle of reflection is equal to the angle of incidence. Diffuse reflection is that in which the light is reflected in all directions. The reflection from a body may be regular, diffused, or mixed. In most practical cases there is a superposition of regular and diffuse reflection.

Stilb (sb) is the unit of photometric brightness equal to one candela per square centimeter.

Transmittance (transmission factor) of a body is the ratio of the light transmitted to the incident light.

STANDARDS OF LUMINOUS INTENSITY

3. Primary Standards. The nominal primary standards of light in many countries have been flame lamps of various kinds, but a blackbody radiator developed at the National Bureau of Standards has recently been adopted as an international primary standard. Its essential element is a crucible containing pure platinum melted by electrical heating. While the molten platinum is solidifying, the temperature remains at 2046°K and the brightness of a small aperture from the crucible remains constant.

4. Reference Standards. Seasoned filament lamps held in the custody of government laboratories are in effect the official standards of light in all the more important countries. Groups of carbon-filament lamps kept in the laboratories of France, Great Britain, and the United States maintain the original International candela. Reference standards of vacuum-tungsten and gas-filled types (the latter for lumen output) have been established in the several countries. These are not exactly concordant, but the differences between countries using the International candela do not exceed 2%.

5. Working Standards. By comparison with the reference standards, other seasoned lamps are calibrated to serve as working standards. All electric-filament lamps change slowly with use; hence, lamps used as working standards should be verified periodically. For accurate measurements it is important to have a range of

Table 19-2. Brightness of Light Sources*

	Candelas per square inch	Footlamberts
Flame sources:		
Candle, plumber's	3.3	1,500
Kerosene lamp, flat wick	8.2	3,700
Welsbach mantle (gasoline lantern)	40	18,100
Natural sources:		
Sun	1,000,000	450,000,000
Moon	3.3	1,500
Sirius (brightest star)	9,500,000	
Clear sky	5.2	2,340
Overcast sky	1.4 or less	630 or less
Incandescent electric lamps:		
60-watt inside-frosted (center spot)	55	25,000
60-watt white, av. at 90 deg	19	8,600
100-watt inside-frosted (center spot)	94	42,300
500-watt inside-frosted (center spot)	270	122,000
500-watt clear bulb (filament)	7,500	3,390,000
150-watt PAR 38 spot *c*	740	334,000
150-watt PAR 38 flood *c*	240	108,000
300-watt R40 spot *c*	690	312,000
300-watt R40 flood *c*	120	54,200
500-watt R52 controlled spread *c*	120	54,200
Fluorescent lamps—cool white:		
20-watt T12, 380 mA	4.1	1,850
40-watt T12, 425 mA	5.2	2,350
75-watt T12, 425 mA	5.0	2,250
110-watt T12, 800 mA	7.3	3,300
215-watt T12, 1,500 mA	11.0	5,000
Electric arc lamps:		
High-intensity carbon arc	450,000	
Super-high-intensity carbon arc	580,000	
High-intensity mercury arc (H1)	900	4,070
High-pressure mercury arc (H4)	5,800	
Water- or air-cooled high-pressure mercury arc (H6)	194,000	

* "IES Lighting Handbook," Illuminating Engineering Society, New York. Forsythe-Smithsonian Physical Tables, 1954.
c = calculated average.

reference and working standards corresponding to the range of the lamps to be measured.

Gaseous-conductor lamps such as the fluorescent lamp may be calibrated as working standards against reference standards of the incandescent-filament type, provided that suitable filters for color correction are employed.

6. Brightness. The brightness of light sources and the luminous surfaces in an interior largely determine the visual comfort of the occupants, particularly when the luminaires in the room only partially shield lamps from view. The reflected images of lamps are specially significant as contributors to reflected glare.

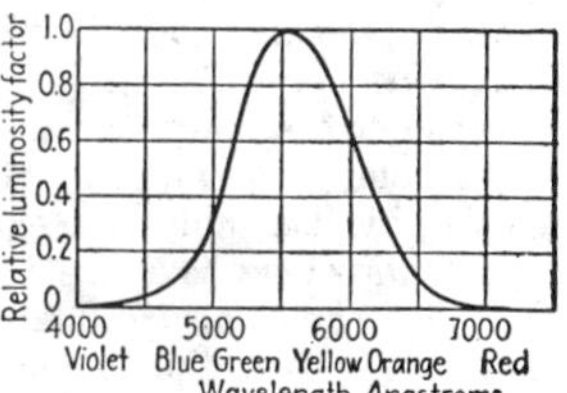

FIG. 19-2. Relative luminosity curve of normal human eye.

7. Luminous flux[1] is the time rate of flow of light. It is expressed in lumens. Radiant energy in the visible region of the spectrum varies in its ability to produce visual sensation, the variation depending upon the wavelength. The ratio of the luminous flux to the corresponding radiant flux is known as the *luminosity (visibility) factor* and is expressed in lumens per watt. This varies with wavelength, having a maximum at approximately 5,550 Å. The data are plotted in Fig. 19-2. At very low levels of illumination the position of the maximum gradually shifts to 5,100 Å as a result of greater use of rod vision.

From the foregoing it is apparent that two sources may radiate equal amounts of energy in the visible region of the spectrum but have different amounts of luminous flux emitted, depending upon the spectral distribution of the energy, the luminous flux being the integrated product of the energy per unit wavelength and the luminosity

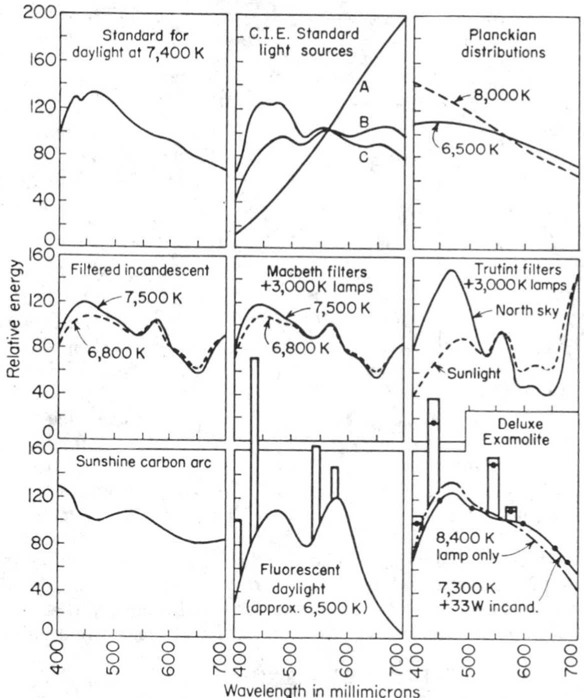

FIG. 19-3. Spectral-energy distribution curves for illuminants. Some are conventional substitutes for natural daylight. ("*IES Lighting Handbook*.")

[1] Illuminating Engineering Nomenclature and Photometric Standards, Illuminating Engineering Society.

factors. In comparing sources as to their efficacy in the production of light, we may consider either of two quantities: the luminous efficacy of the radiant energy emitted or the efficacy of the light source. The former does not take into account energy dissipated by the source by other than radiant means. The latter expresses the overall efficacy of the light source, establishing the ratio of the lumens secured to the watts expended regardless of how they are expended. For electric lamps this is expressed in lumens per watt, while for sources depending upon combustion it is expressed in lumens per thermal unit consumed per unit time.

8. Color. Within the visible spectrum, wavelengths are distinguished one from another by their ability to excite in the human eye various color sensations. Thus the shorter wavelengths excite the color known as violet; and as the wavelengths increase, the color sensation gradually changes through blue, green, yellow, and orange and finally to red at the longer wavelengths of the visible spectrum. The color of the sensation produced by light of a composite character is determined by its spectral energy distribution. Color is defined as that quality of visual sensation which is associated with the spectral distribution of light.

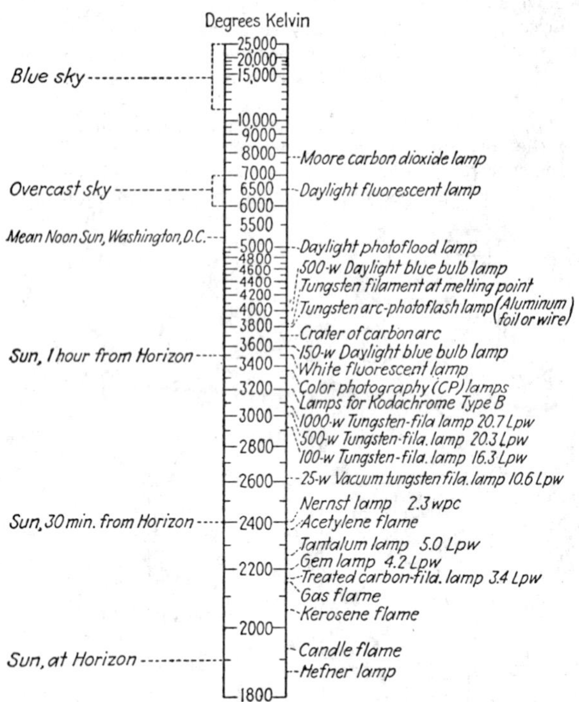

FIG. 19-4. Approximate color temperatures of flame, natural, and electric light sources.

Characteristic spectral energy distributions of a number of illuminants are shown in Fig. 19-3.

9. Color Matching. If spectrophotometric facilities are not available, color matches satisfactory for many purposes may usually be assured by the simple expedient of checking the match under each of two widely different illuminants of continuous energy distribution, red and green, for example, or yellow and blue. Satisfactory equipment for this purpose should permit two illuminants at extremes of below 2500°K and above 8000°K color temperature. For many simple color-matching problems a low-wattage incandescent lamp and a blue, daylight, or deluxe cool white lamp are adequate.

10. Color Temperature. Light from incandescent sources may be described further in terms of color temperature, which is the temperature at which a blackbody would emit luminous flux evoking the same color sensation and having substantially the same relative spectral energy distribution as the light under consideration. Macbeth[1] has shown graphically the relative positions of various natural and artificial light sources on a color temperature scale as in Fig. 19-4.

INCANDESCENT LAMPS

11. Tungsten-filament incandescent lamps were first produced in America in 1907. Tungsten has the highest boiling point of all metals, 8540°F. Its melting point is

[1] MACBETH Color Temperature Classification of Natural and Artificial Illuminants; *Trans. IES*, 1928, Vol. 23, p. 302.

6120°F, higher than that of all other elements, except carbon. The mechanical problems associated with tungsten filaments make the incandescent lamp an inherently compact, somewhat spherical structure. The filament's length and diameter limit its range of operation between 1.5 and 300 V. At 1.5 V, the filament is very short and thick, and it becomes difficult to heat it without excessively heating its support wires. The lamps in the low-voltage (6- to 12-V) class, however, are relatively rugged and will withstand the shocks of motor-vehicle and similar applications. At voltages near 300, the filament is very long and slender; it is fragile and difficult to support. As may be seen by comparing Tables 19-3 and 19-4, the high-voltage lamps are relatively inefficient. Of the so-called large lamps, 99% are designed for standard 120-V circuits. Several thousand types and sizes have been developed for hundreds of fields of application.

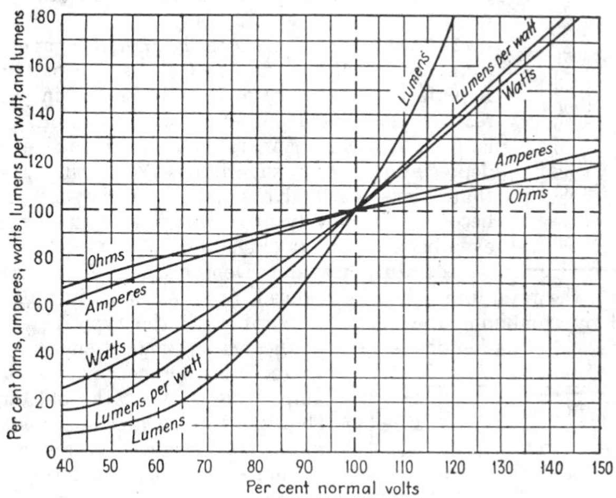

FIG. 19-5. Characteristic curves for large gas-filled lamps showing the effect of operating a lamp at other than its rated voltage. These characteristics are averages of many lamps. They enable the user and designer to predetermine lamp performance under varying conditions.

12. The performance of tungsten-filament lamps is affected by voltage, position of the bulb (if incorrect), size, construction, ambient temperature (if excessive), and quality of manufacture. The voltage characteristics[1] through a range of a few volts above and below design volts may be expressed as simple exponential equations in the following relationships, where capitals represent normal rated values:

$$\frac{\text{Life}}{\text{LIFE}} = \left(\frac{\text{LUMENS}}{\text{lumens}}\right)^a = \left(\frac{\text{LUMENS/WATT}}{\text{lumens/watt}}\right)^b = \left(\frac{\text{VOLTS}}{\text{volts}}\right)^d = \left(\frac{\text{AMPS}}{\text{amps}}\right)^u$$

$$\frac{\text{lumens}}{\text{LUMENS}} = \left(\frac{\text{volts}}{\text{VOLTS}}\right)^k = \left(\frac{\text{lumens/watt}}{\text{LUMENS/WATT}}\right)^h = \left(\frac{\text{watts}}{\text{WATTS}}\right)^s$$

$$= \left(\frac{\text{amps}}{\text{AMPS}}\right)^v = \left(\frac{\text{ohms}}{\text{OHMS}}\right)^z$$

$$\frac{\text{LUMENS/WATT}}{\text{lumens/watt}} = \left(\frac{\text{LUMENS}}{\text{lumens}}\right)^f = \left(\frac{\text{VOLTS}}{\text{volts}}\right)^g = \left(\frac{\text{AMPS}}{\text{amps}}\right)^j$$

$$\frac{\text{amps}}{\text{AMPS}} = \left(\frac{\text{volts}}{\text{VOLTS}}\right)^t \quad \text{and} \quad \left(\frac{\text{watts}}{\text{WATTS}}\right) = \left(\frac{\text{volts}}{\text{VOLTS}}\right)^n$$

[1] WEITZ, C. E. LD-1-Lamp Bulletin, General Electric Company, 1956.

Exponents d, k, and t are taken as fundamentals, and other exponents are derived from them. Values given apply to lamps operated at efficacies near normal and are accurate enough for calculations in the voltage range normally encountered.

The theoretical life of lamps calculated by the exponential relationship of life and voltage is seldom realized in practical installations in the case of excessive "under-voltage" burning, since handling, cleaning, vibration, etc., introduce breakage factors which tend to reduce lamp life.

Exponents

	Gas-filled	Vacuum
a	3.86	3.85
b	7.1	7.0
d	13.1	13.5
u	24.1	23.3
k	3.38	3.51
h	1.84	1.82
s	2.19	2.22
y	6.25	6.05
z	7.36	8.36
f	0.544	0.550
g	1.84	1.93
j	3.40	3.33
t	0.541	0.580
n	1.54	1.58

13. Filament Evaporation. Because of filament evaporation throughout life, the filament of a lamp becomes thinner and thus consumes less power. The light output decreases as the lamp progresses through life because of lowering filament temperature and bulb blackening. These curves apply to a 200-W general-service lamp on constant-voltage service (Fig. 19-6*a*).

14. Lamp life is necessarily based on averages obtained from life-testing hundreds of thousands of lamps annually. A perfect mortality record would be one in which all lamps reached their rated life and then burned out. (While some lamps fail earlier than rated life, others live longer.) The depreciation curve superimposed on this graph indicates that lamps which live substantially beyond rated life have become so relatively inefficient that they should be replaced even before burnout (Fig. 19-6*b*).

15. Mortality and Depreciation. The area under this curve represents the total lumen hours produced by an assumed installation of lamps. It is obtained by combining the mortality curve with the typical depreciation rate throughout life. The darker-shaded part indicates the logical smashing-point region where it is more economical to install new lamps than to keep the old ones in service. The light-shaded area represents the zone of group replacement—i.e., relamping the entire installation at one time before the normal rate of burnout reaches its peak (Fig. 19-6*c*).

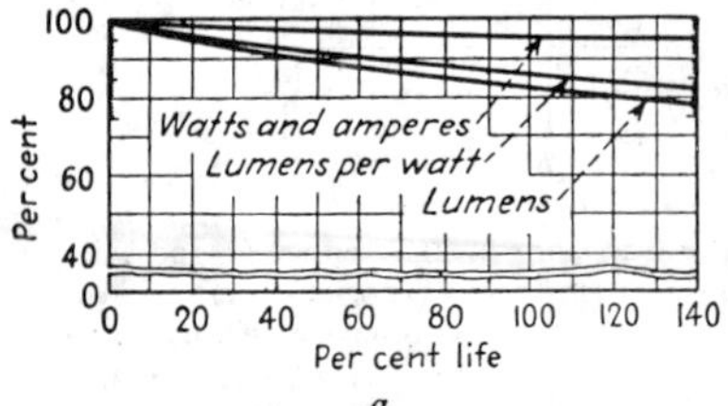

16. Burnouts. From the mortality curve the number of burnouts likely to occur within a given period can be computed for an in-

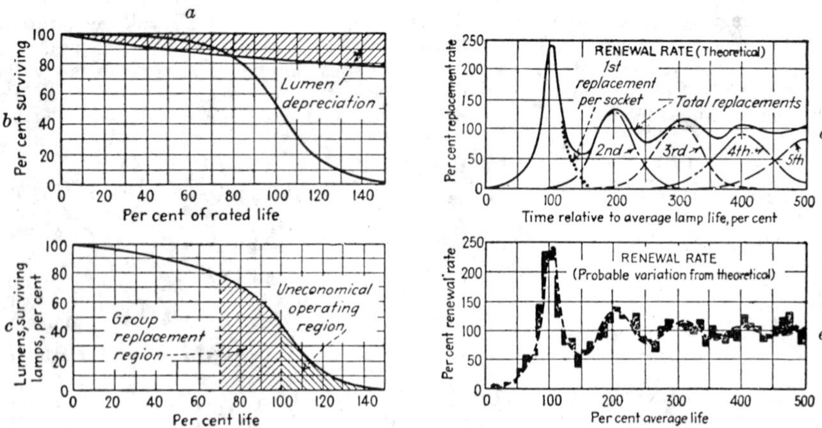

Fig. 19-6. Life characteristics and renewal rate.

finitely large installation. In a new installation few burnouts would be expected during the first several hundred hours. Approaching normal life, many burnouts would occur, necessitating frequent lamp replacement. Thus for a period of several lamp renewals per socket, the renewal rate first swings high, then low, finally settling down to a steady rate. The solid line represents the total replacements; the dotted curves the first, second, etc., replacement per socket (Fig. 19-6*d*).

17. Renewal Rate. The dotted curve showing the theoretical rate of renewals holds only for an infinitely large installation. Departures from this curve in practical installations will, by the law of probability, more likely be represented by the solid block-shaped pattern. The larger the installation, the more closely the two curves tend to coincide. Complaints on life are occasionally encountered during those periods when chance dictates that renewals run higher than average, even though a record of the actual number of renewals over an extended period of time would show average rated life had been obtained (Fig. 19-6*e*).

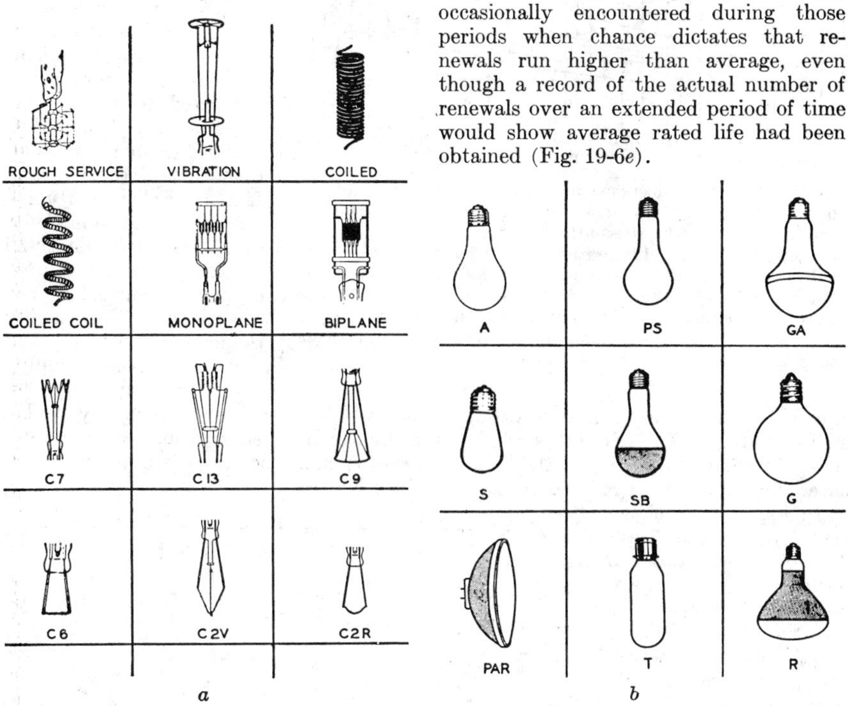

FIG. 19-7. Incandescent lamps. (*a*) Filament types. (*b*) Bulb shapes. Filaments: rough service, vibration, coil, coiled-coil, monoplane, biplane, C 7, C 13, C 9, C 6, C 2V, CZR. Bulbs: A, G, S, T, GA, PAR, R, PS, SB.

18. Transient Characteristics and Flicker.[1] The low cold resistance of a tungsten filament results in a marked inrush of current when the lamp is lighted. Though the current and the power consumption rise momentarily to values far above those of steady operation, the candlepower does not "overshoot" but rises gradually to its steady value. The relative magnitude of the initial surge of current and power depends to some extent upon the resistance and in greater degree upon the inductance of the circuit in which the lamp is connected. If the lamp is operated in an a-c circuit, the amount of the surge depends also upon the phase of the voltage wave at which the switch is closed. Oscillographic records of the starting of a 300-W, 120-V lamp, having a hot-to-cold resistance ratio 15.5, showed a maximum surge crest current of 47 A in a circuit having

[1] FORSYTHE and ADAMS Effect of Voltage Change on the Light Output of Tungsten-filament Incandescent Lamps; *Gen. Elec. Rev.*, 1936, Vol. 39, p. 497.
"The Visual Perception and Tolerance of Flicker"; Utilities Coordinated Research, Inc., 1937.
FORSYTHE, EASLEY, and HINMAN Time Constants of Incandescent Lamps; *Jour. Appl. Phys.*, 1938, Vol. 9, p. 209.

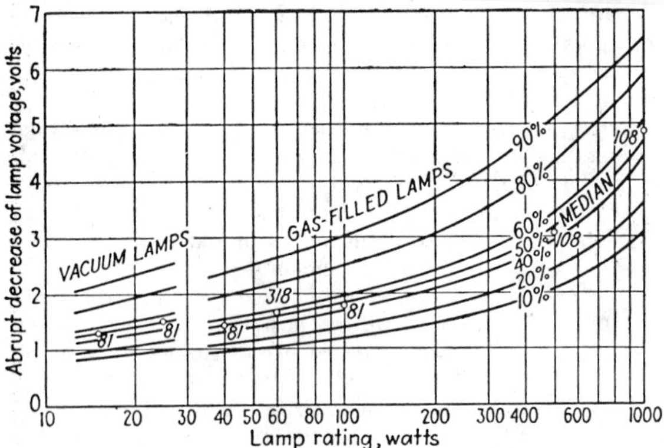

Fig. 19-8. Abrupt decrease of voltage at which flicker of 120-volt tungsten-filament lamp is perceptible. Lamps operating on 60-cycle alternating current; duration of voltage drop, 15 cycles; illumination on reading matter, 6.5 fc. Figures on curves denote percentages of observers expected to perceive flicker when abrupt voltage drops of indicated values are impressed on lighting circuits. Plotted points denote medians of observations, the number of which in each case is indicated by the adjacent figure.

0.1 Ω resistance and 0.04 Ω reactance at 60 c. This was reduced to 42.5 A when the resistance was increased to 1.0 Ω and to 36 A when the reactance was increased to 0.72 Ω.

When the voltage of a lighting circuit suffers an abrupt change, as when a heavy load is switched on or off, the light output of a tungsten-filament lamp connected to the circuit changes also, though this change is more gradual than that of the voltage. If a small change of voltage is spread uniformly over a short time interval, the rate of change of light output is reduced correspondingly. Where an incandescent lamp is operated from an a-c supply, the power input to the lamp and the light output pulsate with a frequency equal to twice the frequency of the applied voltage. Though the power input fluctuates between zero and a maximum value, the thermal capacity of the filament serves to limit the temperature swing and materially reduces the cyclic fluctuation of light output.[1]

In Fig. 19-8 is shown the relation of the threshold of perception of abrupt voltage change to the size of lamp and in Fig. 19-9 the relation of the threshold of perception of cyclic voltage pulsation to the frequency.

19. Rating. Tungsten-filament lamps for multiple circuits are rated in volts and watts. The practice is to maintain the watts of lamps at standard figures and to adjust the lumen output whenever improvements in lamp manufacture warrant increasing lamp efficacies.

The ratings of lamps of the more commonly used sizes are given in Tables 19-3 to 19-8 along with useful data from other sources.

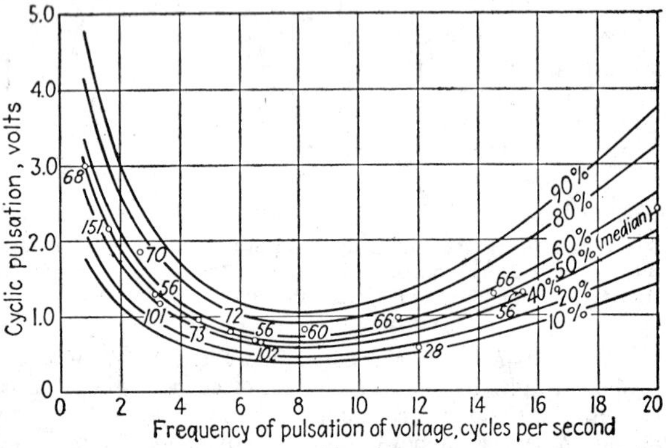

Fig. 19-9. Cyclic pulsation of voltage at which flicker of 115-volt tungsten-filament lamp is just perceptible. Derived from 1,104 observations by 95 persons in field tests of 25-, 40-, and 60-watt lamps conducted by Commonwealth Edison Co. Figures on curves denote percentages of observers expected to perceive flicker when cyclic voltage pulsations of indicated values and frequencies are impressed on lighting circuits. Plotted points denote medians of observations at various frequencies, the number of which in each case is indicated by the adjacent figure.

[1] Langmuir Flicker of Incandescent Lamps on Alternating-current Circuits; *Gen. Elec. Rev.*, 1914, Vol. 17, p. 294.

Table 19-3. General-service Lamps for 115-, 120-, and 125-volt Circuits
(Will operate in any position, but lumen maintenance is best for 40 to 1,500 W when burned vertically base up*)

Watts	Bulb	Base	Rated average life, h	Max. overall length, in	Average light-center length, in	Approx. initial lumens	Rated initial lumens per watt
10	S-14 inside frosted or clear	Medium	1,500	$3\frac{1}{2}$	$2\frac{1}{2}$	79	7.9
15	A-15 inside frosted	Medium	1,200	$3\frac{1}{2}$	$2\frac{3}{8}$	144	9.6
25	A-19 inside frosted	Medium	1,100	$3\frac{15}{16}$	$2\frac{1}{2}$	266	10.6
40	A-19 inside frosted	Medium	1,000	$4\frac{1}{4}$	$2\frac{15}{16}$	470	11.7
50	A-19 inside frosted	Medium	1,000	$4\frac{7}{16}$	$3\frac{1}{8}$	685	13.7
60	A-19 inside frosted	Medium	1,000	$4\frac{7}{16}$	$3\frac{1}{8}$	855	14.2
75	A-19 inside frosted	Medium	750	$4\frac{7}{16}$	$3\frac{1}{8}$	1,180	15.7
100	A-19 inside frosted	Medium	750	$4\frac{7}{16}$	$3\frac{1}{8}$	1,750	17.5
150	A-23 inside frosted	Medium	750	$6\frac{3}{16}$	$4\frac{5}{8}$	2,730	18.2
200	A-23 inside frosted or clear	Medium	750	$6\frac{3}{16}$	$4\frac{5}{8}$	3,940	19.7
300	PS-25 inside frosted or clear	Medium	750	$6\frac{15}{16}$	$5\frac{3}{16}$	6,300	21.0
300	PS-35 inside frosted or clear	Mogul	1,000	$9\frac{3}{8}$	7	5,750	19.2
500	PS-35 inside frosted or clear	Mogul	1,000	$9\frac{3}{8}$	7	10,500	21.0
750	PS-52 inside frosted or clear	Mogul	1,000	13	$9\frac{1}{2}$	16,700	22.2
1,000	PS-52 inside frosted or elear	Mogul	1,000	13	$9\frac{1}{2}$	23,300	23.3
1,500	PS-52 inside frosted or clear	Mogul	1,000	13	$7\frac{1}{2}$	33,000	22.0

* Consult manufacturers' technical literature for current data, as values change frequently.

20. Average Life. The rated average laboratory life, defined as the average of a representative group of lamps burned under correct operating conditions on a 60-c circuit, ranges from 750 to 1,500 h for the general-service types. As compared with life in the laboratory under controlled operating conditions, performance in service may differ widely. Lamp breakage and fluctuating line voltage tend to shorten life. Line-potential drop with resultant low-voltage operation often tends to lengthen life. It is customary to estimate that average life in service may be of the order of 75% of the rated average laboratory life. Extended-service lamps with a rated life of 2,500 h are available in a range of sizes from 15 to 1,000 W. They give an average of 20% less light than standard lamps under normal conditions. Such lamps are economically justified only when kilowatthour rates are very low and/or labor costs to replace lamps are very high.

21. Influence of Operating Conditions upon Lamp Performance. Tests show that ambient temperatures have little effect on performance characteristics. Very high temperatures, however, may cause mechanical difficulties.

On direct current, although the mortality rate is lower, the maintenance of light output is poorer than on alternating current.

Intermittent operation in general (not sign-flashing service) does not materially affect lamp performance. There is reason to believe that lamp life is shortened by voltage fluctuations, even though the voltage excess averaged over the life of the lamp is offset by an equal average voltage deficiency.

Table 19-4. Lamps for High-voltage Service—230 and 250 Volts
(Burned in any position)

Watts	Bulb	Base	Rated average life, h	Max. overall length, in	Average light-center length, in	Approx. initial lumens	Rated initial lumens per watt
25	A-19 inside frosted	Medium	1,000	$3\frac{7}{8}$	$2\frac{9}{16}$	220	8.8
50	A-19 inside frosted	Medium	1,000	$3\frac{7}{8}$	$2\frac{9}{16}$	480	9.6
100	A-21 inside frosted or clear	Medium	1,000	$5\frac{1}{4}$	$3\frac{13}{16}$	1,280	12.8
200	PS-30 inside frosted or clear	Medium	1,000	$8\frac{1}{16}$	6	3,040	15.2
300	PS-35 inside frosted or clear	Mogul	1,000	$9\frac{3}{8}$	7	4,800	16.0
500	PS-40 inside frosted or clear	Mogul	1,000	$9\frac{3}{4}$	7	9,100	18.2
750	PS-52 inside frosted or clear	Mogul	1,000	13	$9\frac{1}{2}$	13,600	18.2
1,000	PS-52 inside frosted or clear	Mogul	2,000	13	$9\frac{1}{2}$	18,600	18.6
1,500	PS-52 inside frosted or clear	Mogul	2,000	13	$9\frac{1}{2}$	27,000	18.0

Except in the case of lamps designed for a particular position of operation, operating position has little effect on lamp performance. Shock and vibration are likely to impair the performance of lamps with filaments of small diameter to a greater extent than in the case of lamps with filaments of large diameter. Special types of lamps are available for use in installations where vibration is likely to be encountered and others, known as "rough-service" lamps, for use where likely to be subjected to shock. Neither of these two lamps will function properly in place of the other.

Table 19-5. Basic Data on Standard Voltage Projector and Reflector Lamps

Watts	Bulb	Description	Base	Max. overall length, in	Approx. beam spread, deg[b]	Approx. beam lumens	Approx. total lumens	Approx. average candelas in central 10° cone
			Reflector Lamps for Spot Lighting and Floodlighting*					
30	R-20	Flood	Med.	3 15/16	85	140	200	290
75	R-30	Spot	Med.	5 3/8	50	400	710	1,800
75	R-30	Flood	Med.	5 3/8	130	610	710	430
150	R-40	Spot	Med.	6 1/2	40	810	1,765	6,000
150	R-40	Flood	Med.	6 1/2	110	1,500	1,765	1,250
300	R-40	Spot	Med.	6 1/2	35	1,800	3,550	13,500
300	R-40	Flood	Med.	6 1/2	115	2,800	3,550	2,500
300	R-40[a]	Spot	Med.	6 7/8	35	1,600	3,550	12,500
300	R-40[a]	Flood	Med.	6 7/8	115	2,700	3,550	2,550
300	R-40[a]	Spot	Mogul	7 1/4	35	1,600	3,550	12,500
300	R-40[a]	Flood	Mogul	7 1/4	115	2,700	3,550	2,550
500	R-40[a]	Spot	Mogul	7 1/4	35	3,100	6,100	22,000
500	R-40[a]	Flood	Mogul	7 1/4	115	5,400	6,100	5,200
			Reflector Lamps for General Lighting[c]					
500	R-52	Wide beam	Mogul	11 3/4	110	.....	7,550	
750	R-52	Wide beam	Mogul	11 3/4	110	.....	12,700	
1,000	R-52[a]	Wide beam	Mogul	11 3/4	110	.....	16,300	
1,000	RB-52	Wide beam	Mogul	12 3/4	130	.....	16,300	
550	R-52	Wide beam	Mogul	11 3/4	120	.....	8,100	
800	R-52	Wide beam	Mogul	11 3/4	120	.....	13,500	
550	R-57	Narrow beam	Mogul	12	70	.....	8,100	
800	R-57	Narrow beam	Mogul	12	70	.....	13,500	
1,000	R-57	Narrow beam	Mogul	12	70	.....	16,300	
			Projector Lamps for Spot Lighting and Floodlighting[d]					
75	PAR-38	Spot	Med. socket	5 5/16	30 × 30	465	730	4,800[e]
75	PAR-38	Flood	Med. socket	5 5/16	60 × 60	570	730	1,350[e]
150	PAR-38	Spot	Med. socket	5 5/16	30 × 30	1,100	1,730	10,500[e]
150	PAR-38	Flood	Med. socket	5 5/16	60 × 60	1,350	1,730	3,400[e]
150	PAR-38	Spot	Med. side prong	4 5/16	30 × 30	1,100	1,730	10,500[e]
150	PAR-38	Flood	Med. side prong	4 5/16	60 × 60	1,350	1,730	3,400[e]
200	PAR-46	Narrow spot	Med. side prong	4	17 × 23	1,200	2,350	30,000[e]
200	PAR-46	Med. flood	Med. side prong	4	20 × 40	1,300	2,350	11,000[e]
300	PAR-56	Narrow spot	Mogul end prong	5	15 × 20	1,800	3,650	70,000[e]
300	PAR-56	Med. flood	Mogul end prong	5	20 × 35	2,000	3,650	22,000[e]
300	PAR-56	Wide flood	Mogul end prong	5	30 × 60	2,100	3,650	10,000[e]
500	PAR-64	Narrow spot	Extended Mogul end Prong	6	13 × 20	3,000	6,000[f]	110,000[e]
500	PAR-64	Med. flood	Extended Mogul end Prong	6	20 × 35	3,400	6,000[f]	35,000[e]
500	PAR-64	Wide flood	Extended Mogul end Prong	6	35 × 65	3,500	6,000[f]	12,000[e]

* The rated average life of reflector (R) lamps is 2,000 h.
[a] Heat-resistant glass bulb.
[b] To 10% of maximum candlepower.
[c] Some of these types are also available for 230- to 260-V circuits.
[d] The rated average life of projector (PAR) lamps is 2,000 h. All PAR lamps have bulbs of molded heat-resistant glass.
[e] Central cone defined as 5° cone for all spots and 10° cone for all floods.
[f] Tentative value.

22. Mortality and Replacement of Incandescent Lamps. The lives of best-quality individual lamps, even when tested in the laboratory under controlled conditions, deviate from the average of the group in a manner similar to that shown in Fig. 19-10. Under service conditions the dispersion of individual lives is likely to be even greater. Therefore in a lighting installation started with new lamps, some of the lamps replaced as failures occur will have failed and will have been replaced before the entire initial installation becomes exhausted. This overlapping of failures causes a progressive smoothing out of the rate of lamp replacement until, after a period equal to about four or five times the average life, the rate of failure becomes relatively uniform. However, in the initial stages of operation of a new installation, the rate of lamp replacement is likely to fluctuate considerably, as indicated in Fig. 19-6.

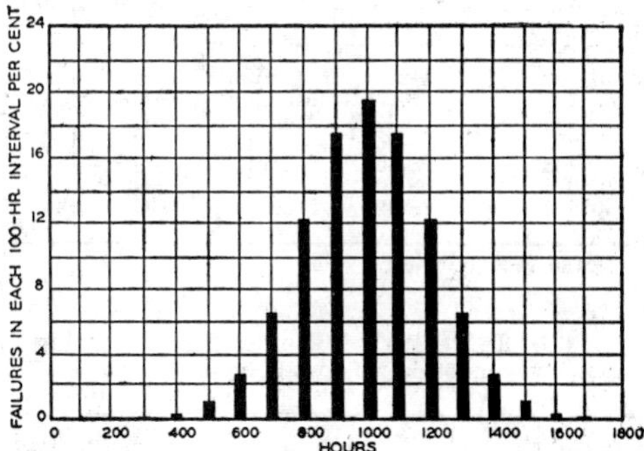

Fig. 19-10. Distribution of life of 60-watt lamps operated at rated voltage; $\bar{X} = 1{,}000$, $\sigma = 200$.

Table 19-6. Sealed-beam Lamps for Land Vehicles

Type of service	Trade number	Design watts	Design volts	Rated average laboratory life, h	Filament	Bulb: Type	Bulb: Max. diam.	Max. overall length, in	Base: SAE type	Base: Terminal	Base: Contacts
6-V Circuits											
Headlight....	5040S	50 40	6.1 6.2	75§ 150§	C-6 C-6	PAR-56	7.031	5¼	H-3	Lugs	3
Fog*........	4012	35	6.2	80§	C-6	PAR-46	5.70	3⅞	G-2	Screw	2
Spotlight....	4535	30	6.4	35§	C-6	PAR-46	5.70	3⅞	G-2	Screw	2
Spotlight....	4515	30	6.4	35§	C-6	PAR-36	4.46	2¾	G-2	Screw	2
Farm tractor.	4013	25	6.4	100§	C-6	PAR-46	5.70	3⅞	G-2	Screw	2
Farm tractor.	4019	35	6.2	80§	C-6	PAR-46	5.70	3⅞	G-2	Screw	2
Fog*........	4015	35	6.2	80§	C-6	PAR-36	4.46	2¾	G-2	Screw	2
12-V Circuits											
Headlight†...	4001	37.5	12.8	200‡	C-6	PAR-46	5.70	4	H-2	Lugs	2
Headlight†...	4002	50 37.5	12.8 12.8	320† 200†	C-6 C-6	PAR-46	5.70	4	H-3	Lugs	3
Headlight....	5400S	50 40	12.8 12.8	200† 320†	C-6 C-6	PAR-56	7.031	5¼	H-3	Lugs	3
Headlight....	5440	50 40	12.8 12.8	240† 400†	C-6 C-6	PAR-56	7.031	5¼	H-3	Lugs	3
Fog*........	4412	35	12.8	100†	C-6	PAR-46	5.70	3⅞	G-2	Screw	2
Fog*........	4415	35	12.8	100†	C-6	PAR-36	4.46	2¾	G-2	Screw	2

* Available with clear or amber lenses.
† Dual headlamp system (one unit of each type installed and used as a pair at each side on the front of vehicle).
‡ Rated life at 14.0 V in these types.
§ Rated life at 7.0 V in these types.

In certain classes of service where the burning hours are the same each day or where lamps are so located as to require special facilities for relamping, e.g., in street lighting, large stores, signs, and subways, the quality of maintenance may be improved by replacing all the lamps simultaneously at a fraction of rated life. The cost of such group replacement is a minimum when replacement intervals are such that the average daily relamping cost is equal to the cost of replacements on the last day before the whole group is replaced. It can be shown that the length of the interval is determined by the shape of the lamp-mortality curve and by the ratio of the unit cost of group replacement to the cost of individual replacement at failure. If this ratio is less than a certain

Table 19-7a. Multiple Street-lighting Lamps* (Group Replacement)
(Will operate in any position, but lumen maintenance best when burned vertically base up)

Watts	Bulb	Base	Class	Filament	Max. overall length, in	Light center length, in	Nominal lumens
92	A-23	Medium	C	C-9	5 15/16	4 7/16	1,000
189	PS-25	Medium	C	C-9	7 1/8	5 3/8	2,500
189	PS-25	Mogul	C	C-9	6 15/16	5 1/4	2,500
295	PS-35	Mogul	C	C-9	9 3/8	7	4,000
405	PS-40	Mogul	C	C-9	9 3/4	7	6,000
620	PS-40	Mogul	C	C-7A	9 3/4	7	10,000
860	PS-52	Mogul	C	C-7A	13	9 1/2	15,000

* 120 V; 3,000 h life.

Table 19-7b. Series Street-lighting Lamps* (Group Replacement)
(Except where noted, will operate in any position, but lumen maintenance best when burned vertically base up)

Rated initial lumens	Initial volts	Amperes	Clear bulb	Burning position	Class	Filament	Max. overall length	Light center length	Approx. hours life
1,000	10.0	6.6	PS-25	Any	C	C-8	7 1/8	5 3/8	3,000
1,000	8.8	7.5	PS-25	Any	C	C-8	7 1/8	5 3/8	3,000
2,500	22.0	6.6	PS-25	Base up	C	C-2V	7 1/8	5 3/8	3,000
2,500	22.4	6.6	PS-35	Any	C	C-2V	9 3/8	7	3,000
2,500	19.6	7.5	PS-35	Any	C	C-2V	9 3/8	7	3,000
4,000	34.2	6.6	PS-35	Any	C	C-2V	9 3/8	7	3,000
4,000	30.4	7.5	PS-35	Any	C	C-2V	9 3/8	7	3,000
4,000	14.6	15	PS-35	Base up	C	C-2V	9 3/8	7	3,000
4,000	15.1	15	PS-35	Base down	C	C-2V	9 3/8	6 1/4	3,000
6,000	50.2	6.6	PS-40	Any	C	C-2V	9 3/4	7	3,000
6,000	15.8	20	PS-40	Base up	C	C-2V	9 3/4	7	3,000
6,000	16.1	20	PS-40	Base down	C	C-2V	9 3/4	6 1/4	3,000
10,000	85.8	6.6	PS-40	Any	C	C-7A	9 3/4	7	3,000
10,000	27.3	20	PS-40	Base up	C	C-7	9 3/4	7	3,000
10,000	28.5	20	PS-40	Base down	C	C-7	9 3/4	6 1/4	3,000
15,000	37.5	20	PS-40	Base up	C	C-7	9 3/4	7	3,000

* All Mogul base, 3,000 h life.

fraction, determined by the shape of the lamp-mortality curve, group replacement at the most economical interval will cost less than individual relamping.

22a. Photoflood lamps produce approximately three times the photographic effectiveness of conventional lamps. They are high-efficiency short-life sources with a color temperature of 3400°K in clear bulbs and 4800°K in blue (daylight) bulbs. Smaller bulbs are used than with general-service types. Four of the lamps of Table 19-8 are of the reflector type for spot- or floodlighting service in the home or commercial photographic or television studio.

23. Reflector and Projector Lamps. Lamps embodying integral reflecting surfaces

Table 19-8. Incandescent-filament Photoflood or Superflood Lamps
(115–120 V)

Designation	Approx. watts	Rated life h	Approx. initial lumens	Approx. color tempera- ture, °K	Bulb	Base	Max. overall length, in	Amperes
No. 4 photoflood, PH/4	1,000	10	32,200	3,400	PS-35	Mog. Sc.	9 3/8	8.4
4B, B4 daylight blue, PH/B4	1,000	10	20,500	4,800	PS-35	Mog. Sc.	9 3/8	8.4
No. 2 photoflood, PH/2	500	6	17,000	3,400	PS-25	Med. Sc.	6 15/16	4.2
2B, B2 daylight blue, PH/B2	500	6	10,800	4,800	PS-25	Med. Sc.	6 15/16	4.2
RFL-2 R-2	500	6	4,500*	3,400	R-40	Med. Sc.	6 1/2	4.2
RSP-2	500	6	50,000*	3,400	R-40	Med. Sc.	6 1/2	4.2
No. 1 photoflood, PH/1	250	3	8,600	3,400	A-21	Med. Sc.	4 15/16	2.1
1B, B1 daylight blue, PH/B1	250	3	5,300	4,800	A-21	Med. Sc.	4 15/16	2.1
PH/300 R30	300	4	12,000*	3,400	R-30	Med. Sc.	5 3/16	2.5
PH/375/34R4	375	4	13,000*	3,400	R-40	Med. Sc.	6 1/2	3.1

* Mean candlepower in 10° cone.

are manufactured in a number of types. Bowl-silvered lamps are employed in direct-lighting equipment in which it is desired to shield the filament from view but principally in indirect equipment. Initial loss of light output due to the silvering is 6 to 10%; the rate of decline of light output is considerably greater than in clear-bulb lamps of corresponding sizes—60 to 80% greater in the case of 100- and 200-W lamps. However, a luminaire (of similar distribution) with an unprocessed lamp may produce less light because of poorer maintenance. In projector flood- and spot-light lamps (Table 19-5), the bulb is constructed of a molded bowl-shaped section of parabolic or other suitable profile, on the inner surface of which is a metal reflecting surface. This bowl is fused to a molded-glass cover plate, which may be clear or may consist of a pattern of lenses and prisms, depending upon the desired beam characteristics. Sealed-beam automobile lamps (Table 19-6) are of this construction. Reflector-type lamps are constructed with blown bulbs of suitable profiles (usually cylindrical for showcase lighting or parabolic for spot lighting) having parts of the inner surfaces covered with a reflecting metallic film. Their nominal life is usually 2,000 h.

A number of projector lamps are now available with dichroic filters (interference films) to control the spectral quality of the radiation in such a manner as to separate the heat from the light in the beam or to produce colored light without the usual losses due to absorption by filters. From 75 to 80% of the heat can be removed from the beam at a sacrifice of only 15 to 20% of the light. Colored dichroic lamps produce more deeply saturated colors with higher efficacy than is obtainable with color filters. Heat separating lamps are available in 150 and 300 W; colored lamps, in 150 W, in five colors.

24. Lamps for street lighting are available for series burning and multiple burning. The former are rated according to lumen output and the latter according to watts. In general the lumen maintenance of series lamps is better than that of multiple lamps. The latter are rated on a basis such that their average lumens throughout life are approximately equivalent to those of series lamps of corresponding sizes. For this reason, actual initial lumens of multiple lighting lamps are higher than nominal ratings (see Table 19-7 for performance data).

25. Iodine Quartz Incandescent Lamps. As the name suggests, the bulb of this type of lamp is made of quartz in tubular form and the iodine regeneration cycle is used to minimize bulb blackening. Since quartz has a melting point of 1650°C, it is ideal both for the iodine cycle and for smaller enclosures of high-wattage filaments. Ordinarily, tungsten evaporating from a hot filament is carried to the relatively cool bulb wall, where it accumulates and forms a black deposit. However, with temperatures of several hundred degrees centigrade, tungsten vapor and iodine vapor combine to form tungsten iodide, which does not adhere to the bulb wall but is borne by convection currents back to the filament, where it is reduced and the tungsten redeposited on the filament; the iodine vapor recirculates to continue the regenerative cycle. The bulb wall remains clear throughout life, and the filament has a higher efficacy and longer life.

Quartz lamps have optimum characteristics as an optical light source because of high brightness and compactness; the latter feature enables optics to be placed close

Table 19-9. Typical Iodine Quartz Lamps

Watts	Volts	Bulb	Rated initial lumens	Normal length, in	Burning position
300	120	T-4	5,000	3.2	Any
400	120	T-4	7,500	3.2	Any
500	120	T-3	10,500	4⅝	Within 4° of horizontal
1,000	120	T-6	19,000	5.6	Within 4° of horizontal
1,500	208 240 277	T-3	33,000	10	Within 4° of horizontal
250*	120	T-4	5,000	3.1	Any
500*	120	T-4	10,000	3.6	Any

Rated average life—2,000 h.
* Single-ended special screw base.

to the source for simplification and flexibility of optical design. The quartz envelope is physically rugged and impervious to shock.

Because of high temperature, the lamp should not be touched with bare hands for several minutes after it has been turned off. Portions of the bulb wall may reach 600°C; hence precautions must be taken to keep combustible materials away from the bulb. Oil, grease, or moisture from hands can cause early failure of the bulb. Because of the compactness of the light source, the temptation exists to reduce the size of luminaires to such a degree that heat dissipation is inadequate, resulting in excessive seal temperatures, with early lamp failure. Maximum recommended seal temperature is 650°F. The lamps are internally fused. In some types, successful operation of the iodine cycle requires operation in a position within 4° of horizontal. Typical iodine quartz lamps are listed in Table 19-9.

ARC LAMPS

26. Arcs.[1] Commercial arc-light sources may be divided into two major classifications: (1) those with consumable electrodes which operate in the open, with access of air, and (2) those which operate inside a sealed envelope of glass or quartz, in a gaseous atmosphere determined by the inclusion of fixed amounts of a gas, solid, or liquid, either singly or in combination.

Of the **open-arc type** all significant commercial examples are carbon arcs.[2] These are divided into flame arcs, low-intensity projector arcs, and high-intensity projector arcs. In the **flame arc** the light source is the entire arc stream made luminescent by the addition of flame materials. This arc is used in photography and related industrial photochemical processes, where, with a special coring of rare-earth salts, it produces an essentially continuous radiation more closely approximating natural sunshine than any other artificial source. A special form of the flame arc is the so-called "enclosed arc." It operates in a glass globe with limited access of air, the resulting nitrogen-rich atmosphere enhancing radiation in the violet and near-ultraviolet regions, which is quite effective in blueprinting and related copying processes. In the **low-intensity projector arc,** the principal light source is the incandescent tip of the positive carbon electrode, at or near its sublimation temperature. With a brightness of approximately 110,000 cp/in^2 uniformly generated over a considerable area, this source found early application in motion-picture, searchlight, and other projection systems where a concentrated source capable of producing a well-defined narrow beam is required. The **high-intensity carbon arc** is distinguished primarily from the low-intensity one by the inclusion of rare-earth materials in the core of the positive electrode, which volatilize into the arc stream as the electrode is consumed in burning. Thus, in addition to light from the incandescent carbon surface, there is a significant amount of light originating in the gaseous region immediately in front of the carbon as a result of the combination of a high current density and an atmosphere rich in flame materials. The color quality of the light from these sources[3] is well adapted to colored motion-picture photography and projection. Such arcs with brilliancies in the range from about 250,000 to 650,000 cd/in^2 have largely replaced the earlier low-intensity carbon arcs in motion-picture and searchlight projection, while the practicability of operation at brilliancies in excess of 1,300,000 cd/in^2 has been demonstrated in the laboratory.

27. Mercury Lamps.[4] All electric-discharge lamps operate by passing electric current through a gas or vapor. In carbon-arc lamps, light is produced by the incandescence of one or both electrodes. In most gaseous discharge lamps, however, the

[1] FOUND and FORNEY Hot Cathode Neon Arcs; *Trans. AIEE*, 1928, Vol. 47, p. 747.
COMPTON, K. T. The Electric Arc; *Trans. AIEE*, 1927, Vol. 46, p. 868.
FOUND, C. G. Fundamentals of Electric Discharge Lamps; *Trans. IES*, 1938, Vol. 33, p. 161.
DUSHMAN, S. The Search for High Efficiency Light Sources; *Jour. Opt. Soc. Am.*, 1937, Vol. 27, p. 1.
MACPHERSON, H. G. The Radiation from Carbon Arcs; *Jour. Appl. Phys.*, 1942, Vol. 13, p. 97.

[2] KALB, W. C. The Carbon Arc—A Valuable Industrial Tool; *Trans. AIEE*, 1942, Vol. 61, p. 581.

[3] BOWDITCH, F. T., and DOWNES, A. C. Spectral Distributions and Color Temperatures of the Radiant Energy from Carbon Arcs Used in the Motion Picture Industry; *Jour. Soc. Motion Picture Engrs.*, 1938, Vol. 30, p. 400.

[4] IES Lighting Handbook. BUTTOLPH, L. J. "Characteristics of Gaseous Conduction Lamps"; *Trans. IES*, February, 1935. FREEMAN, G. A. Short-arc Mercury Lamps, *Illum. Eng.*, April, 1950. NOEL, E. B. Radiation from High Pressure Mercury Arcs, *Illum. Eng.*, February, 1941. BEGGS, E. W. New Developments in Mercury Lamps and Their Applications; *Illum. Eng.*, April, 1947. BUTTOLPH, L. J. Mercury Lamps—Light Made to Order; *Gen. Elec. Rev.*, March, 1955.

Table 19-10. Direct-current Carbon Arc

	Low intensity	Nonrotating high intensity		Rotating high intensity					
	Application Number*								
	1	2	3	4	5	6	7	8	9
Type of carbon	Microscope	Projector	Projector	Projector	Projector	Projector	Projector	Searchlight	Studio "Yellow Flame"
Positive carbon:									
Diameter, mm	5	7	8	10	11	13.6	13.6	16	16
Length, in	8	12–14	12–14	20	20	22	22	22	22
Negative carbon:									
Diameter	6 mm	6 mm	7 mm	11/32 in	3/8 in	1/2 in	1/2 in	11mm	17/32 in
Length, in	4½	9	9	9	9	9	9	12	9
Arc current, A	5	50	70	105	120	150	180	150	225
Arc volts, d-c	59	40	42	59	68	78	74	78	70
Arc power, W	295	2,000	2,940	6,200	8,160	11,700	13,300	11,700	15,800
Burning rate, in/h:									
Positive carbon	4.5	11.6	13.6	21.5	20.5	17.0	21.5	8.9	20.2
Negative carbon	2.1	4.3	4.3	2.9	2.6	1.9	2.5	3.9	2.2
Approximate crater diameter, in	0.12	0.23	0.28	0.36	0.39	0.50	0.50	0.55	0.59
Maximum brightness of crater, candelas/cm²	15,000	55,000	83,000	90,000	85,000	87,500	95,000	65,000	68,000
Forward crater, candelas	975	10,500	22,000	36,000	44,000	63,000	78,000	68,000	99,000
Crater lumens†	3,100	36,800	77,000	126,000	154,000	221,000	273,000	250,000	347,000
Total lumens‡	3,100	55,000	115,000	189,000	231,000	368,000	410,000	374,000	521,000
Total lumens per arc watt	10.4	29.7	39.1	30.5	28.3	31.4	30.8	32.0	33.0
Color temperature, °Kelvin§	3,600	5,950	5,500–6,500	5,500–6,500	5,500–6,500	5,500–6,500	5,500–6,500	5,400	4,100

* Typical applications: 1, microscope illumination and projection; 2, 3, 4, 5, 6, and 7, motion-picture projection; 8, searchlight projection; 9, motion-picture-set lighting and motion-picture and television background projection.
† Includes light radiated in forward hemisphere.
‡ Includes light from crater and arc flame in forward hemisphere.
§ Crater radiation only.

Table 19-11. Typical Mercury Lamps for General Lighting, Street Lighting, and Black Light

Watts	Bulb	Base	USAS code	Description	Max. overall length	Light center length	Rated life hours†	Initial lumens
175	E-28	Mogul	H39-22KB/E	Gen. and street lighting	8¼	5	16,000+	7,300
			H39-22KC/C/E	Gen. and street lighting—color imp.	8¼	5	16,000+	7,100
250	E-28	Mogul	H37-5KB/E	General and street black light*	8¼	5	16,000+	11,000
			H37-5KC/C/E	General and street color imp.	8¼	5	16,000+	10,500
400	E-37	Mogul	H33-1CD/E	General and street black light*	11$\frac{5}{16}$	7	16,000+	20,500
			H33-1GL/C/E	General and street color imp.	11$\frac{5}{16}$	7	16,000+	20,000
1,000	BT-56	Mogul	H36-15GW/C	General—color imp.	15$\frac{1}{16}$	9⅜	16,000+	54,000
100	PAR-38	Admed. Skt.	H44-4GS	Black light (spot)*	5$\frac{7}{16}$	...	12,000	2,400
			H44-4JM	Black light (flood)*	5$\frac{7}{16}$	...	12,000	2,400
400	R-52	Mogul	H33-1DN/C	High bay—color imp.	11¾	...	16,000+	20,500

* Black light—use separate filter.
† Average lamp life at 5 or more burning hours per start.

luminosity results from the gas itself—neon, argon, krypton, or mercury or sodium vapors. The color of the light depends on the gas and its pressure, which may vary from a few microns to a hundred atmospheres or more.

Commercial mercury lamps include five types:

1. Low-pressure low-voltage lamps in long tubes, such as **fluorescent** and **germicidal** lamps.
2. High-pressure low-voltage lamps in short tubes, used principally for industrial lighting.
3. High-pressure high-voltage lamps in long tubes, for **photochemical** and **photocopying** purposes.
4. Very-high-pressure and very-high-voltage lamps in short tubes, such as the water-cooled 1,000-W mercury lamps for searchlight service.
5. Very-high-pressure very-low-voltage lamps in spherical-shaped bulbs.

The practical limit of a mercury-vapor lamp's current-carrying capacity is how high a temperature its enclosing tube can withstand without rupturing. By connecting an impedance in series with the lamp, the current is controlled. In most lamps about one-half the supply voltage is absorbed by a series ballasting device. A variety of ballasts are available for operating mercury lamps, singly or in pairs. Single lamp ballasts may have low (0.60 minimum) or high (0.90 minimum) power factor; two-lamp ballasts have inherently high power factor. The simplest lamp ballast is the reactor type used in series with the lamp when line voltage is sufficient for reliable starting. Voltage taps on the ballast match the ballast to the line voltage. A reactor-type ballast can be used when the line voltage is approximately twice the rated lamp voltage. The autotransformer-type ballast is used on circuits where the line voltage must be changed to suit the lamp requirements. The stabilizing ballast, sometimes referred to as a regulator or constant-wattage ballast, provides improved lamp-current

Table 19-12. Fluorescent Lamps* for Outdoor Applications

Lamp watts	105	150	200	105	150	200
Bulb†	T-10	T-10	T-10	T-14J	T-14J	T-14J
Length, in.	48	72	96	48	72	96
Initial lumens‡	5,900	9,500	13,000	5,800	9,400	12,900

* Lamp current 1,500 mA. Rated life 7,500 h at 3 h per start.

† Recommended lamps for open reflectors—jacketed outdoor fluorescents where temperature is likely to fall below 40°F; unjacketed outdoor fluorescents where temperature is usually about 40°F; 1,500-mA or 800-mA regular fluorescents where temperature is usually above 50°F. NOTE: If 1,500-mA, 800-mA, or unjacketed outdoor fluorescent lamps are operated in open equipment, low light output must be anticipated under adverse outdoor conditions.

Recommended lamps for closed reflectors—unjacketed outdoor fluorescents in single-lamp reflectors; 1,500-mA or 800-mA regular fluorescents in multiple-lamp reflectors.

‡ At peak light-output ambient.

operation and lower dropout voltage over an extended input-voltage range. Heavier wiring, oversized circuit breakers, and time-delay relays that may be required by the relatively high starting currents of conventional ballasts are eliminated with stabilizing ballasts, as the starting current is less than the rated operating current.

One of the limitations of the mercury lamp is the effect of power-supply interruptions. In the event of a power interruption or voltage dip lasting for more than 1 c, mercury lamps extinguish and do not restart for several minutes. The exact magnitude of the voltage drop to cause this condition depends on the ballast design. Regulator ballasts withstand a greater drop than other types. The delay in lamp restarting is caused by the high pressure which develops in the arc tube during operation. The ballast open-circuit voltage is not sufficient to restart the lamp until the lamp cools and the pressure decreases. In installations where this characteristic might be a safety hazard, the use of a few incandescent or fluorescent fixtures along with the mercury units assures emergency illumination until the mercury lamps restart.

Most mercury lamps of the industrial type (2) are constructed with two bulbs, an inner bulb, usually of quartz, which contains the arc, and an outer bulb, which shields the arc tube from changes in temperature and in some lamps acts as a filter to remove certain wavelengths of the arc's radiation. The inside of this bulb or tube in some types is coated with a phosphor to whiten the light. These lamps are called color-compensated, color-improved, or fluorescent-mercury. The phosphor adds fluorescent red to the reflected light if the lamp is in a reflector and also to the directly transmitted light. The transfer does not materially add to the efficiency, as in conventional fluorescent lamps, but it does make the color of emitted light more acceptable both in industrial and in street-lighting applications. The economy, efficacy, and long life of mercury lamps are definite advantages.

28. Mercury–Metallic Iodide Lamps. The 400-W lamp, an innovation still in the development stage, is similar in physical appearance and principle of operation to the mercury lamp. Added to the basic mercury ingredient are other metals in iodide form, which produce wavelengths in broad bands, approaching a continuous spectrum and giving a color rendition approaching that of fluorescent lamps. The addition of these metals and other innovations in design have resulted in a marked increase in efficacy, amounting to approximately 50% more lumens per watt—30,000 lm for the 400-W lamp. The smaller light source (the arc tube compared with the bulb surface of color-improved mercury lamps) permits more precise optical control and smaller control equipment. However, lumen maintenance is considerably lower; as a result, the economic life is only about 6,000 h, at which time the light output has fallen off to about 50% compared with the 16,000-h-life rating of mercury lamps. The most practical applications today are those where maximum light and good color rendition are required but total burning hours are short (3,000 to 4,000 h/year). For optimum operation, the lamp should be used with auxiliaries specifically designed for it. The lamp achieves full brightness in 4 min; restart time to full brightness is 20 min. Different lamps are required for base-up or base-down operation, the axis of the lamp to be within 75° of vertical.

29. Sun Lamps. All mercury-vapor lamps generate ultraviolet rays which cause sunburn or sun tan on human skin. The erythemal rays are transmitted (2,800 to 3,200 Å) through special glass bulbs of quartz type. Arc-type generators and bulbs which look like conventional reflector lamps are the two most common kinds. The RS type is a self-contained device with a 275-W inner quartz tube in an outer bulb (R-40) and a resistance ballast incorporated within the bulb. It also contains a self-activating switch, similar in action to thermal-type starters used with fluorescent lamps.

30. Glow Lamps. When sufficient voltage is applied to electrodes sealed within a bulb containing neon, argon, or helium, light is produced at the negative electrode. On direct current, one cathode glows; on alternating current, the reversal is so rapid, both electrodes appear to glow. The range of glow lamps is $\frac{1}{25}$ to 3 W (Table 19-13). Their useful life varies approximately as the inverse of the cube of the current. A glow lamp has a negative voltampere characteristic; hence a limiting resistance is used in series with it. In conventional screw-base types, the resistor is concealed in the base. The lamp life ranges between 3,000 and 25,000 h.

Glow lamps have wide use in electronic circuitry, where their action is that of a

Table 19-13. Typical Glow Lamps

Lamp No.*	NE-2	NE-51	NE-45	NE-48	NE-57	NE-17	AR-3	AR-4	NE-58	NE-30	NE-32	NE-56	NE-34	AR-1	NE-40
Nominal watts	1/25	1/25	¼	¼	¼	¼	¼	¼	½	1	1	1	2	2	3
Volts (circuit)	110–125	110–125	110–125	110–125	110–125	110–125	110–125§	110–125§	220–250	110–125	110–125	220–250	110–125	110–125	110–125
Bulb (clear)	T-2	T-3¼	T-4½	T-4½	T-4½	T-4½	T-4½	T-4½	T-4½	G-10 or S-11	G-10 or S-11	G-10 or S-11	S-14	S-14	S-14
Base†	Unbased wire terminal	S.C. Min. Bay.	Cand. Sc.	D.C. Bay. Cand.	Cand. Sc.	D.C. Bay. Cand.	Cand. Sc.	D.C. Bay.	Cand. Sc.	Med. Sc.	D.C. Bay. Cand.	Med. Sc.	Med. Sc.	Med. Sc.	Med. Sc.
Max. overall length, in	1 11/16	1 3/16	1 17/32	1½	1 17/32	1½	1 17/32	1½	1 17/32	2¼	2⅛	2¼	3½	3½	3½
Approx. starting voltage:															
A-c	65	65	65	65	55	55	80	80	65	60	60	60	60	65	60
D-c	90	90	90	90	75	75	115	115	90	85	85	85	85	90	85
Series resistance, Ω	200,000 external	200,000 external	30,000 internal	30,000 internal	30,000 internal	30,000 internal	15,000 internal	15,000 internal	100,000 internal	7,500 interna	7,500 external	33,000 internal	3,500 internal	3,500 internal	2,200 internal
Average useful life, h	Over 50,000	Over 15,000	Over 7,500	Over 7,500	Over 7,500	Over 7,500	150‡	150‡	Over 7,500	Over 10,000	Over 10,000	Over 10,000	Over 10,000	1,000‡	Over 10,000

* NE—neon gas type; AR—argon-nitrogen type.
† S.C. Min. Bay.—single contact, miniature bayonet; Cand. Sc.—candelabra screw; D.C. Bay. Cand.—double contact, bayonet candelabra; Skt.—skirted.
‡ As ultraviolet source.
§ A-c, may be operated on 135 V d-c.

practically instantaneous switch. At breakdown voltage, the lamp glows, and the switch is closed; at the extinguishing voltage, the lamp current drops to a fraction of its full value and may be considered as nonconducting, or open-circuit, in certain circumstances. This on-off characteristic suits the glow lamp to the dichotomy of binary arithmetic as used in computers and logic circuitry in general. Other glow-lamp applications in electronic circuitry include oscillators, pulse generators, voltage regulators, and coupling networks.

31. Electroluminescent lamps[1], which are essentially fluorescent lamps in sheet or panel form, are a recent development with a wide variety of applications.

This type of lamp is a thin-area source in which light is produced by a phosphor excited by a pulsating electrical field. In essence, the lamp is a plate capacitor with a phosphor embedded in its dielectric and with one or both of its plates transparent. Green, blue, yellow, or white light may be produced by choice of phosphor. The green phosphor has the highest brightness, with the others about 20% less for equivalent excitation. These lamps are available in sheet and ribbon form, flexible or with stiff backing, and are easily fabricated into simple or complex shapes. They have been used in decorative lighting, night lights, switchplates, instrument panels, clock faces, telephone dials, thermometers, and signs. Their application is limited to locations where the general illumination is low.

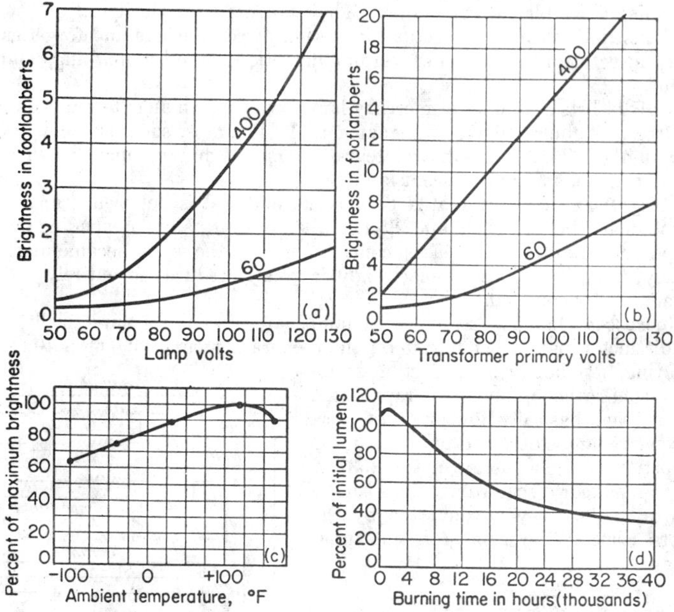

FIG. 19-11. Characteristics of electroluminescent lamps. (*a*) Brightness vs. voltage at 60 and 400 c/s; (*b*) same as (*a*), with 5:1 transformer ratio; (*c*) brightness vs. ambient temperature; (d) lumens vs. time.

Brightness varies with applied voltage, frequency (Fig. 19-11*a* and *b*) and temperature (Fig. 19-11*c*), as well as with the type of phosphor. At 120 V, 60 c, the brightness of the sheet form with the green phosphor is between 1 and 2 ftL; the brightness of the ribbon form may be as high as 5 ftL under these conditions or up to 30 ftL at 120 V, 400 c. With the sheet form at 600 V, 400 c, a brightness of 20 ftL has been achieved. These higher brightnesses are at the expense of useful lamp life.

Life is long (Fig. 19-11*d*) and power consumption low. There is no abrupt point at which the lamp fails; the time at which the brightness has fallen to 50% of initial

[1] *Sylvania Lighting Products Bull.* No. 0-255, 1963.

is sometimes used as a measure of useful life. For the sheet form, this is approximately 20,000 h at 120 V, 60 c. Approximate initial current and wattage values per square foot of lamp under these operating conditions are 65 mA and 3.5 W. Typical electroluminescent lamps are listed in Table 19-14.

32. Black-light Lamps.[1] Near-ultraviolet radiant energy (energy not visible to the human eye) causes certain materials to fluoresce or emit visible light. The normal human eye is sensitive only to radiant energy between 400 and 700 mμ in wavelength. Thus, lamps which produce primarily near-ultraviolet radiant energy in the 320- to 380-mμ range are popularly called "black" lights. This term is quite descriptive, since the ultraviolet energy from the "light" source cannot be seen by the human eye but the effects of the radiation on special materials can be visually dramatic.

When black light is directed at a fluorescent material, an energy conversion takes place. The material or chemical sensitive to ultraviolet energy absorbs the energy, then reradiates it at longer wavelengths (in the 400- to 700-mμ range) to which the eye is sensitive. The energy conversion is similar to that which takes place in fluorescent lamps; i.e., ultraviolet energy in the lamp activates the fluorescent phosphor coating to produce visible light in white or any other color.

Mercury lamps with filters to absorb the visible light and transmit the near-ultraviolet are used for fluorescent effects. Called black-light lamps, they are generally enclosed in a red-purple filter glass bulb that looks black. Many materials fluoresce when irradiated by black-light lamps. They are used for theatrical and advertising effects, industrial and food inspection, detection of counterfeits and forgeries, medical diagnosis, insect traps, crime and vermin detection, laundry marking, and copying equipment.

Table 19-11 lists a number of mercury lamps which with an external filter are black-light sources. Tubular sources designated as BLB lamps, such as the 15-W T-8 and the 20- and 40-W T-12 lamps, have integral filters and may be operated with the same ballasts as corresponding fluorescent lamps.

The brightness of an irradiated fluorescent material is between 1 and 5 ftL with printing inks and between 0.25 and 2.5 ftL with interior paints, depending on the color. The apparent brightness increases considerably as the eyes become dark-adapted. Conversely, the effectiveness of black light is greatly reduced or entirely negated by a small amount of visible light.

33. Fluorescent Lamps. Fluorescent lamps are available commercially principally in four distinct types, depending upon their operating circuits: (1) hot-cathode, preheat-starting, (2) hot-cathode, instant-starting, (3) cold-cathode, and (4) rapid-start lamps. All are basically low-pressure mercury-discharge lamps designed to emit a maximum portion of their energy in the 2537 Å line of the mercury spectrum. This short-wave ultraviolet energy is converted by the phosphors coating the inside of the tubes into light. The lamps are best adapted to operate on a-c circuits with reactance ballasts. Typical operating circuits are shown in Figs. 19-12 to 19-17.

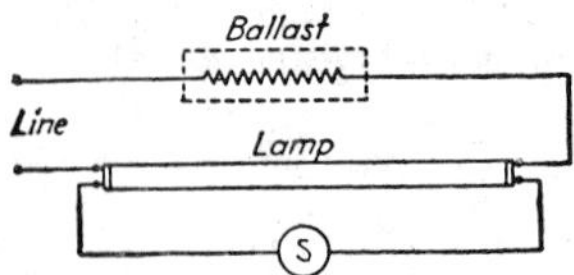

Fig. 19-12. Single-lamp ballast for 4- to 40-watt hot-cathode fluorescent lamp, preheat-starting. S = starting switch.

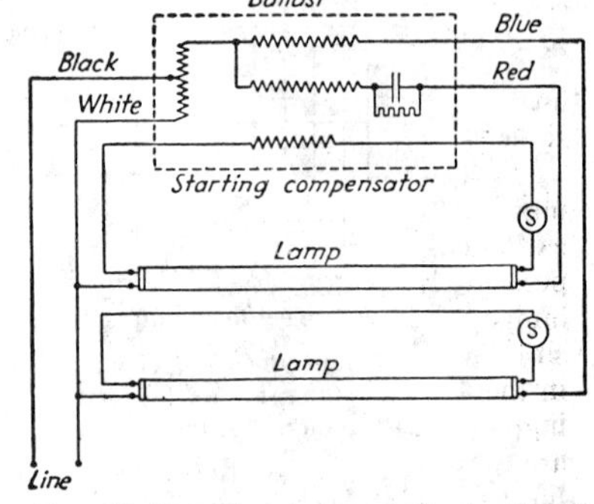

Fig. 19-13. Two-lamp ballast circuit for 30- and 40-watt hot-cathode fluorescent lamps, preheat-starting, showing built-in starting compensator.

[1] Kraehenbuehl, J. O., and Chanon, H. J. Technology of Brightness Production by Near-ultraviolet Radiation; *Trans. IES*, February, 1941.

Table 19-14. Typical Electroluminescent Lamps

Rated volts	Frequency, c/s	Current, mA/in²	Impedance, thousands of Ω/in²	Brightness, ftL	Typical applications
120	60	0.25 max.	350 min.	1.0–1.4	Any, on ordinary circuits
120	400	0.25–0.5	240–480	4–6	Aircraft
200	250	0.2–0.5	400–1,000	2.4–3.8	Automobiles
300	60	0.18–0.36	833–1,670	2.0–3.5	Increased brightness from ordinary circuits
600	60	0.25–0.5	1,200–2,400	5–7	Street signs
600	400	1.1–2.2	262–545	15–20	Present recommended maximum

The hot-cathode lamps (1) and (2) contain electrodes which are usually coiled-coil (or triple-coiled) tungsten filaments coated with one or more of the alkaline-earth oxides. By suitable circuit arrangements these cathodes can be heated to an electron-emitting temperature before the arc strikes, or they may be required to act momentarily as cold cathodes until they are heated by bombardment after the lamp has started. Lamps using these cathodes may be designed to carry currents of 1 to 2 A with low-voltage drop (18 V) at the electrodes.

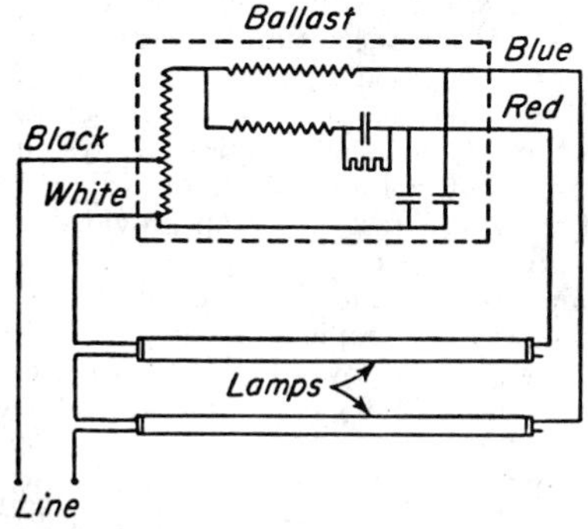

Fig. 19-14. Two-lamp ballast circuit for instant-starting, hot-cathode fluorescent lamps.

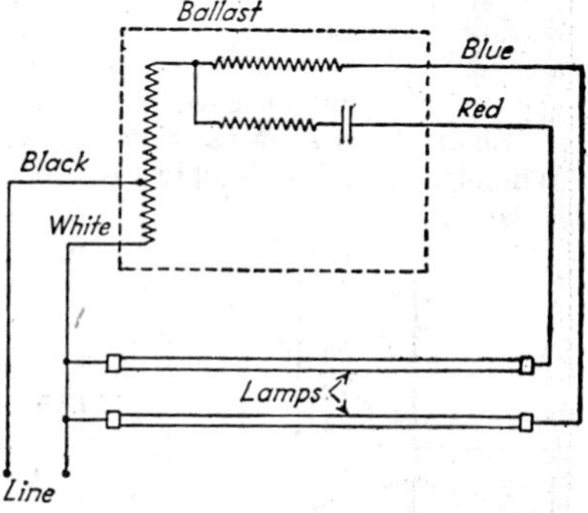

Fig. 19-15. Two-lamp ballast circuit for multiple operation of cold-cathode fluorescent lamps.

Unless the cathodes are kept at the proper temperature by means of an external heating circuit, they do not withstand frequent starting without correspondingly shorter life.

Cold-cathode lamps (3) are those which use electrodes of cylindrical form which may be coated on their inside surfaces with electron-emitting materials. These cathodes operate at temperatures which limit the lamps to low-current densities.[1] The electrode drop in these lamps is relatively high (105 V), but they are not subject to short life as a result of frequent instant starting.

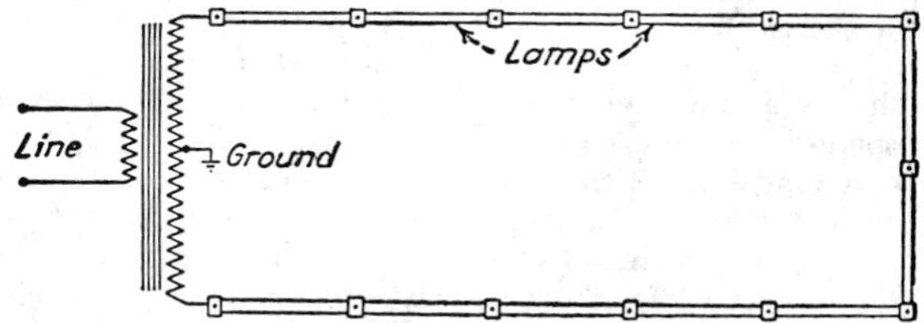

Fig. 19-16. Circuit for the series operation of cold-cathode fluorescent lamps from a leakage reactance transformer.

The only differences in the behavior of the two types of lamps are those imposed by the cathode characteristics outlined above. Otherwise their lumen rating and lumen maintenance are the same for the same tube size, phosphor, and loading in amperes per unit area of phosphor surface.

[1] IES Lighting Handbook.

Table 19-15. Commonly Used Preheat and Instant-start Fluorescent Lamps

Lamp watts	15	15	20	30	40	40	55	75
Type*	P	P	P	P	P	IS	IS	IS
Bulb†	T-8	T-12	T-12	T-8	T-12	T-12	T-12	T-12
Length, in‡	18	18	24	36	48	48	72	96
Watts—2 lamps and ballast	38	38	50	77	97	98	137	175
Lumens, initial:								
Standard cool white	780	740	1,170	2,000	3,150	2,900	4,250	6,100
De luxe cool white	612	520	850	1,520	2,180	2,020	3,080	4,300
Rated life, h§	7,500	7,500	7,500	7,500	¶	9,000	9,000	9,000

* P—preheat; IS—instant start.
† Diameter in eighths of an inch; T—tubular.
‡ Length includes two standard lamp holders.
§ Life at 3 burning hours per start. For 6-h basis, add 25%; for 12-h, add 60%; for continuous burning, average life is 2½ times rated value.
¶ Switch-start lead lamp—5,500 h. Switch-start lag lamp—6,500 h.

The rapid-start lamp (4), introduced in 1953, operates on a new principle—that of constant heating of the electrodes during both starting and operating phases. This minimizes loss of life due to starting and also reduces electrode drop several volts. These conditions mainly affect the ballast, which becomes smaller, lighter, and quieter. The lamp has a film coating and special cathodes, which, with other ballast-design features, produce a system which is much closer to the theoretical minimum volt-ampere requirements than any of the older instant-starting systems. This results principally from the effective utilization of cathode preheat and other starting aids.

Figure 19-17 shows a schematic diagram of a typical single-lamp rapid-start system. Windings J and G supply the proper cathode-preheat voltage simultaneously with the open-circuit voltage across the lamp provided by windings A and C. The C winding also provides the necessary impedance to limit lamp operating current. The characteristics of rapid-start lamp cathodes are of such a nature that the power loss of the cathode-preheat circuits is negligible even when no voltage cancellation is used after starting is effected. This accounts for an appreciable reduction in ballast complexity and cost. The rapid-start system is applied to single- and multiple-lamp applications, both high- and low-power-factor.

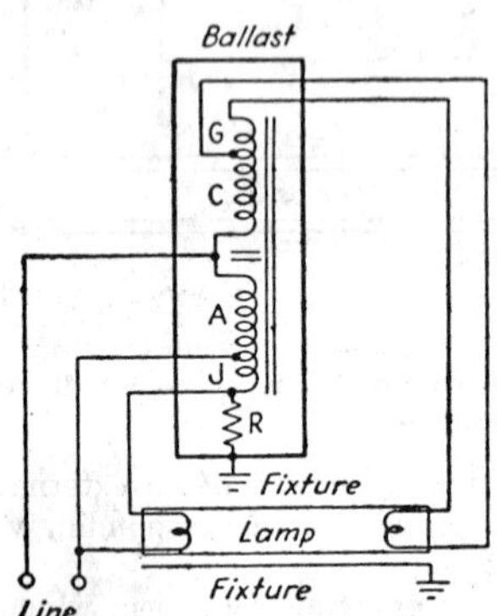

Fig. 19-17. Operating circuit for a single RS (rapid-start) fluorescent lamp.

When fluorescent lamps are operated on a-c circuits, the light executes cyclic pulsations of considerably greater amplitude than those of incandescent lamps of equivalent lumen rating. Though at commercial frequencies (50 and 60 c) this cyclic flicker[1] is not usually noticeable, it may produce unpleasant stroboscopic effects when moving objects come into the field of view. For this reason, good practice frequently dictates the operation of fluorescent lamps in a manner that will minimize this cyclic flicker. One method of accomplishing this object is to operate the lamps in pairs or threes on 2- or 3-phase circuits, respectively. Where only single-phase circuits are available, use may be made of the two-lamp ballast, for hot-cathode or cold-cathode lamps, in which leading current is supplied to one lamp and lagging current to the other, the phase difference being such that the light pulsations of the two lamps largely compensate each other. Present-day phosphors and ballast circuits minimize the flicker effect in all but the most critical applications.

[1] Brieger, Lawrence Effect of Voltage Dip Duration on Cyclic Light Flicker; *Elec. Eng.*, August, 1951, Vol. 70, No. 8, p. 685.
Davidson, G. E. Flicker in Lighting Systems; *Ontario Hydro Res. News*, October, 1952, Vol. 4, No. 4, p. 9.
Weise, W. R. Cyclic Flicker of Fluorescent Lamps; *Elec. World*, Oct. 27, 1945, Vol. 124, p. 80.
Eastman, A. A., and Campbell, J. H. Stroboscopic and Flicker Effects from Fluorescent Lamps; *Illum. Eng.*, January, 1952.

Table 19-16a. Commonly Used Rapid-start Lamps

Lamp watts	30	40	60	85	100*	110	110	160	215
Milliamperes	430	430	800	800	1,000	800	1,500	1,500	1,500
Bulb	T-12	T-12	T-12	T-12	T-12	T-12	T-12 PG-17	T-12 PG-17	T-12 PG-17
Length, in.	36	48	48	72	72	96	48	72	96
Watt—2 lamps and ballast	76	92	142	201	235	241	235	345	455
Lumens, initial:									
Standard cool white	2,250	3,150	3,950	6,300	6,950	8,900	6,900	10,900	15,500
De luxe cool white	1,530	2,230	2,750	4,300	4,600	6,100	5,050	8,000	11,000
Rated life, h	9,000	12,000	9,000	9,000	7,500	9,000	7,500	7,500	7,500

* Same as 85-W lamp, but operated at 1,000 mA. Used primarily in street lighting.

Table 19-16b. Circline Fluorescent Lamps

Lamp diam. and watts	Tube size	Tube diam., inches	Watts, lamp and ballast	Lumens, initial		Brightness, std. cool white, candles per square inch
				Std. cool white	Std. warm white and white	
8¼ in.–22	T9	1⅛	28.5	930	960	4.6
12 in.–32	T10	1¼	40.5	1550	1600	4.4
16 in.–40	T10	1¼	52.0	2200	2250	4.5

The fluorescent lamp, in itself, is inherently a high-power-factor circuit, but the reactive ballast normally used to stabilize the arc is inherently low-power-factor. Since in the usual circuit the voltage drop across the ballast is approximately equal to that across the lamp arc, the resulting power factor of a single-lamp reactive-ballast circuit is of the order of 50%. For many applications this low power factor is objectionable. In single-lamp ballasts, power-factor correction may be obtained by means of a capacitor shunted across the line connections or, where the lamp requires a higher voltage, by a capacitor across the transformer secondary. The two-lamp ballast, through displacement of the lamp currents, offers a ready means of power-factor correction and is usually designed to give a circuit power factor greater than 90%.

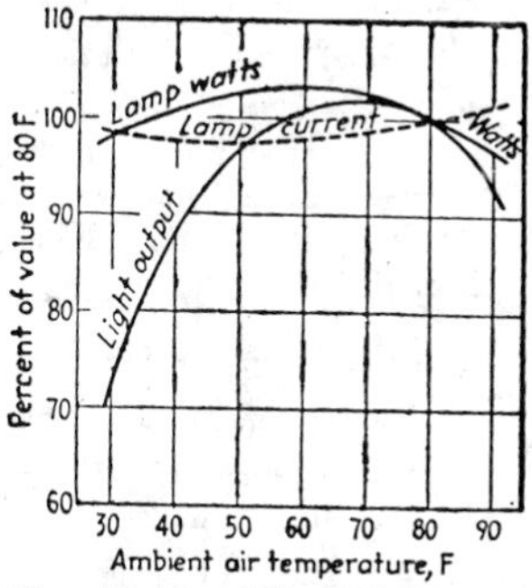

Fig. 19-18. Effect of air temperature on the characteristics of 40-watt hot-cathode fluorescent lamp.

All inductive fluorescent ballasts emit a certain amount of noise; the noise increases with the lamp current. A sound rating for ballasts has been developed by some manufacturers from A (quietest) to F (noisiest). The amount of cumulative ballast noise which is tolerable depends on two sets of principal factors:[1] (1) characteristics of the room and (2) characteristics of the luminaire.

A factory or a busy office, which has a high ambient sound level, does not require the same precautions as a television studio or a church. Obviously, the elimination of luminaire noise is not a requirement in a factory. A schoolroom or an office installation having high ambient sound levels during the day may be quiet in the evening. For this reason, luminaire noise which is inaudible during normal room usage may be quite audible during the quiet period.

As a result, the ballast-luminaire combination which is selected on the basis of average ambient sound level may be the source of a complaint if judged during the abnormally quiet period. Always select the ballast with the best sound rating available for the application. This will help to eliminate complaints which originate during quiet periods. Rooms which have sound-absorbent ceilings and perhaps carpeting,

[1] *Jefferson Electric Co. Bull.* FLB-1-65, 1965.

Table 19-17. Typical Cold-cathode Fluorescent Lamps

Lamp designation	2545			2569			2593			Hairpin (6-in diameter)		
Nominal length, in*		48			72			96			45	
Actual lamp length, in		45			69			93			¶	
Lamp base		Cap			Cap			Cap			Cap	
Lamp diameter, mm		25 (T-8)			25 (T-8)			25 (T-8)			25 (T-8)	
Starting volts—lamp type LP†		450			600			750			750	
Starting volts—lamp type HP†		600			750			835			835	
Lamp current, mA‡	120	150	200	120	150	200	120	150	200	120	150	200
Lamp voltage—lamp type LP	250	240		330	310		420	400		420	400	
Lamp voltage—lamp type HP	270	250	240	350	330	310	450	425	400	450	425	400
Lamp wattage—lamp type LP‡	26	30		34	40		42	49		42	49	
Lamp wattage—lamp type HP‡	28	33	40	37	43	52	46	54	65	46	54	65
Initial lumens—warm white§	1,100	1,300	1,600	1,700	2,000	2,350	2,300	2,700	3,400	2,300	2,700	3,400
Initial lumens—3500° white§	1,050	1,250	1,550	1,650	1,900	2,300	2,250	2,650	3,300	2,225	2,650	3,300
Initial lumens—4500° white§	1,000	1,200	1,500	1,600	1,850	2,200	2,200	2,600	3,200	2,200	2,600	3,200
Initial lumens—daylight§	950	1,150	1,450	1,550	1,800	2,150	2,150	2,550	3,100	2,150	2,550	3,100
Brightness, fL, warm white	1,180	1,400	1,710	1,270	1,500	1,780	1,310	1,530	1,930	1,310	1,530	1,930
Brightness, fL, 3500° white	1,150	1,340	1,660	1,240	1,470	1,730	1,280	1,500	1,880	1,280	1,500	1,880
Brightness, fL, 4500° white	1,100	1,280	1,600	1,210	1,440	1,650	1,250	1,480	1,700	1,250	1,480	1,700
Brightness, fL, daylight	1,090	1,230	1,550	1,190	1,410	1,600	1,230	1,450	1,650	1,230	1,450	1,650
Rated lamp life, thousands of h:												
Type LP	15	12.5		15	12.5		15	12.5		15	12.5	
Type HP	25	20	15	25	20	15	25	20	15	25	20	15

* Nominal length of lamps and sockets.
† Minimum ballast or transformer volts.
‡ Lamps can be operated over a wide current range. Lamp type LP can be operated at currents to 150. Lamp type HP can be operated at currents to 200.
§ Initial rating after 100 h for types LP and HP.
¶ Extended lamp length 93 in formed to U-shape with 180° 6-in arc.

heavy curtains, and furniture will help to deaden the ballast hum because sound will be absorbed rather than reflected. It should be remembered that such rooms may have a lower ambient noise level than the typical values shown in the sound-control chart (Fig. 19-19).

Typical applications	Typical average noise ambients, decibels*	Sound rating†
Study hall Residence Radio–TV broadcast studio Quiet office Church	20 to 24	A
Library Professional office Evening school classroom	25 to 30	B
Commercial buildings General office areas Stock rooms	31 to 36	C
Retail stores Noisy offices	37 to 42	D
Light manufacturing Outdoor lighting	43 to 48	E
Street lighting Factory	49 and up	F

* These ambients are typical at the time of normal room usage. Measure ambients with a standard sound meter and 40-dB weighing network.
† Where there is a choice of more than one ballast, it is recommended that the quietest one for the application be used.

Fig. 19-19. Sound-control chart for selection of fluorescent ballasts.

The construction of the luminaire should be such that it does not amplify the sound generated by the ballast. To be quiet, luminaires should be rigid and should be made of heavy-gage metal, and all broad expanses of metal should be formed or should have reinforcing ribs. Channels, louvers, strips, shields, and reflectors should not transmit or set up sympathetic vibration. These parts should be tightly secured. The luminaire mounting should properly support the luminaire but should not transmit luminaire vibration to the supporting structure. It may be necessary to isolate the luminaire with sound-absorbent materials. The ballast should be solidly mounted in the luminaire with good ballast-to-luminaire contact, since this not only will reduce noise but also will help to conduct normal operating heat away from the ballast. The sources of ballast hum increase as the number of luminaires in a room increases. So the greater the number of luminaires, the more urgent the need for ballasts with the best acoustical rating available.

For dimming hot-cathode fluorescent lamps, a number of arrangements are available using autotransformers or SCR devices. Depending on the dimming system, 50 to 90% of full lamp brightness can be achieved, with dimming control down to one three-hundredth of full brightness. Autotransformer systems are available to dim two or more lamps, while up to 1,200 lamps may be controlled on an SCR system. The lamps best suited to this application are the 40- and 30-W T-12 Rapid Start. The two wattages should not be used on the same circuit. For smooth operation, the lamps on any circuit should be made by the same manufacturer, at the same time, in the same color, and of the same age in use. Group replacement is the most satisfactory procedure. Lamps should be operated free from drafts at 50 to 80°F and should be seasoned 100 h at full brightness prior to dimming. Special dimming ballasts are required, and components must be securely grounded.

Commercial fluorescent lamps are designed to convert a maximum proportion of their input energy into radiation having a wavelength of 2,537 Å, and the fluorescent materials which coat the tubes are selected with a view to obtaining light of the desired color while having a maximum sensitivity to radiation of this wavelength. The efficiency of this primary energy conversion is very sensitive to changes in tube-wall temperatures, reaching a maximum value in the range 100 to 120°F, corresponding to the ambient temperature range, 65 to 85°F in still air.[1] The conversion efficiency and light output decline on both sides of the foregoing temperature range, rapidly at lower temperatures, more moderately on the high side. For this reason, lamps operated at ambient temperatures below 60°F should be enclosed to conserve their heat. Air movement over the lamp bulb also is effective in lowering the tube-wall temperature and, hence, has the effect of lowered ambient temperature. A typical example of the effect of air temperature on the performance of a fluorescent lamp is shown in Fig. 19-18.

Fluorescent lamps generally should be operated at voltages within ±10% of their designed operating points for best performance. Decreased life and uncertain starting may result from operation at lower voltages, and at higher voltages there is danger of overheating of the ballast or transformer as well as decreased lamp life. One exception to this is found in the series operation of cold-cathode lamps, where an adjustable voltage supply makes possible operation over a wide range of lighting intensities, i.e., dimmer operation such as that used in stage lighting.

34. Performance characteristics of fluorescent lamps are, to a considerable extent, dependent upon the characteristics of the ballast equipment. Typical of this is the effect of variations from rated line voltage on the conditions of lamp operation. Certified ballasts made in accordance with industry specifications and periodically field-checked by an independent laboratory are available for the more commonly used fluorescent lamps. They are to be distinguished by the letters CBM on the ballast case.

Failure of a hot-cathode fluorescent lamp usually results from loss of active material from the cathode or cathodes. This loss proceeds gradually throughout the life of the lamp and is accelerated by frequent starting. Depreciation of light output is caused principally by tube blackening and is rapid (as much as 10%) during the first 100 h

[1] MARDEN, J. W., BEESE, N. C., and MEISTER, G. Effect of Temperature on Fluorescent Lamps; *Trans. IES*, 1939, Vol. 34, p. 55.

but very gradual from that point on. For this reason the lamps are rated commercially on the basis of the lumen output after 100 h of operation.

Where direct current is available at circuit voltages comparable with the open-circuit voltages of the usual a-c ballast circuits, fluorescent lamps may be operated from these sources. For such operation resistance must be added to the usual series reactance ballast (transformer ballasts are not applicable) to limit the operating current to the designed value. This causes a marked reduction in the overall efficiency of the lamp and circuit combination over that obtained in a-c operation. Under d-c operation, lamps more than a few feet in length will promptly develop a concentration of the mercury vapor at the negative end of the lamp, with the result that only a fraction of the bulb will give off light. This condition can be overcome through a periodic (about once in 4 h) reversal of the polarity of the lines feeding the lamps. The life of lamps is likely to be shorter on direct than on alternating current.

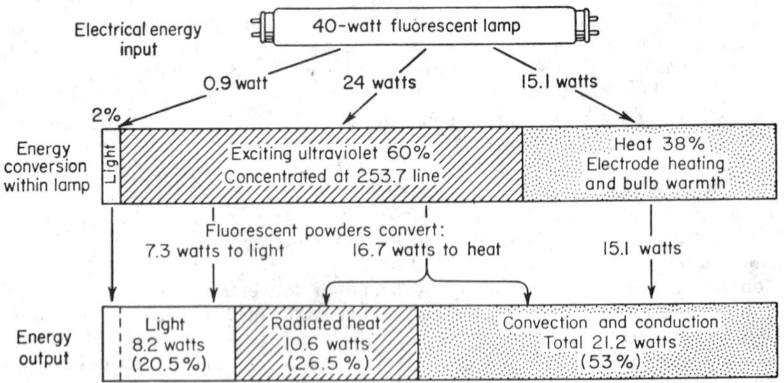

Fig. 19-20. Energy-conversion efficacy and distribution in a typical preheat-starting 40-W cool-white fluorescent lamp.

35. Energy Transformations.[1] The chart shown in Fig. 19-20 has been furnished by one of the manufacturers as typical of the energy transformations occurring in a 40-W hot-cathode fluorescent lamp.

DESIGN OF LIGHTING ACCESSORIES

36. Purposes commonly served by lighting accessories are: (1) redirection of light; (2) reduction of brightness and diffusion of light; (3) concealment of light source; (4) modification of color to light; (5) decoration; (6) protection of the light source; (7) polarization of light.

The extent to which redirection of light is accomplished depends upon the following factors: (1) proportion of total light flux intercepted by redirecting surface; (2) shape of redirecting surface; (3) properties of redirecting surface.

Surfaces as light-reflecting media are of metal, glass, plastic, and sometimes wood. Aluminum either with or without anodic coatings is much used. Steel or brass with chrome, nickel, silver, and rhodium when properly applied is a satisfactory reflecting surface of reasonable permanence. Synthetic coatings are being developed which are quite satisfactory, though not as a rule so permanent as the metallic coatings.

37. Enclosing translucent equipment is usually of glass or plastic. Glass of various degrees of transparency and translucency may be obtained. Plastics having the same general range of optical characteristics are available, though they are not capable of withstanding temperatures much in excess of 170°F.

Enclosing units are made in sizes up to 20 in in diameter. Cast and bent sections are satisfactorily used with fluorescent luminaires. They are shaped to produce diffusion and in some cases embody lens sections to control the light.

[1] Forsythe, Barnes, and Adams Fluorescence and Fluorescent Lamps; *Denison Univ. Bull., Jour. Sci. Lab.*, April, 1941, Vol. 36.

Polarizing multilayer diffusing panels for direct enclosed fluorescent luminaires are commercially available. These panels, when installed in suitable luminaires, vertically polarize a portion of the emitted light, the percentage of polarization depending on the angle of emission as shown in Fig. 19-21. Vertically polarized light, when reflected from nonmetallic surfaces, reduces veiling glare and improves the contrast between the critical detail of the seeing task and its immediate background as described in Par. **51,** Quality of Lighting. It is evident from Fig. 19-21 that polarization is most effective when the line of sight makes a large angle with the normal to the surface.

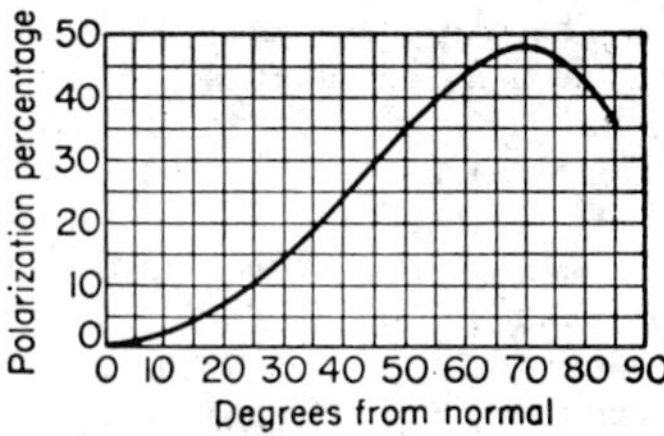

FIG. 19-21. Polarization efficiency curve for a four-lamp 2-ft-wide troffer.

38. Glass, porcelain enamel, plastics, etc., in their various forms may be employed as reflecting, transmitting, diffusing, or refracting media. Porcelain-enameled steel and mirrored glass combine the desirable characteristics of metals and glass.

Other materials such as fabrics, parchmentized papers, plastics, mica, marble, alabaster, pottery, and painted surfaces are generally characterized by high light absorption and low light transmission. In some cases their optical properties are subject to change. For these reasons their use is confined largely to decorative applications.

Reflection, transmission, and absorption data[1] on certain typical materials appear in Tables 19-18 and 19-19.

Table 19-18. Reflection Factors of New, Clean, Opaque Materials for Incidence of 20°

Surface	Reflection factor	Surface	Reflection factor
Aluminum, polished	0.67–0.73	Nickel	0.64
Aluminum, anodic	0.78–0.83	Silver, polished	0.92
Aluminum paint	0.55–0.70	Silvered mirror	0.85
Brass, polished	0.60	Stainless steel	0.55
Chromium, polished	0.65	White painted enamel	0.78–0.90
Rhodium, polished	0.75	White paint	0.75–0.90
		White porcelain enamel	0.75–0.86

Table 19-19. Reflection, Transmission, and Absorption Factors of Glasses for 20° Incidence

Glass	Reflection factor	Transmission factor	Absorption factor
Clear crystal	0.08	0.91	0.01
Rough or configurated crystal	0.14	0.83	0.03
White	0.74	0.12	0.14
Etched, single surface	0.15	0.82	0.03
Etched, double surface	0.28	0.69	0.03

LUMINAIRES

39. Luminaires are classified as:

Direct	0– 10% upward, 90–100% downward
Semi-direct	10– 40% upward, 60– 90% downward
General diffuse	40– 60% upward, 40– 60% downward
Semi-indirect	60– 90% upward, 10– 40% downward
Indirect	90–100% upward, 0– 10% downward

[1] LITTLE Modern Lighting Accessories; Illuminating Engineering Practice; Univ. Pa., IES, 1917, p. 195.

This classification applies to types of lighting such as industrial, commercial, and residential. Luminaires are designed to redirect light and to increase the effective light-source area, thus decreasing brightness while absorbing no more of the light flux than necessary.

40. Direct luminaires[1] generally are employed in industrial, commercial, and residential lighting. For industrial purposes the certified RLM standard units fabricated by a number of manufacturers comprise a complete line of incandescent and fluorescent industrial luminaires; angle reflectors for illumination of vertical surfaces; and dome, deep-bowl, and high-bay units for general and specific lighting purposes, together with industrial fluorescent units for space and bench lighting. Many types of certified industrial fluorescent luminaires are available for general and specific lighting purposes.

Built-in lighting units may be had for almost any type of direct-light distribution in either troffer form or downlight form.

41. Semidirect luminaires are used more generally in commercial and residential lighting where part of the light should fall upon the ceiling. These units are available in the form of enclosing glassware for incandescent lamps and either enclosing or louvered fluorescent units. In general they are suspended below the ceiling, though in some cases they are mounted upon or recessed in the ceiling.

42. General diffuse luminaires, used principally for commercial lighting in stores and offices, are usually of the enclosing types, where the inherent high brightness of the source is reduced through the use of diffusing glass, plastics, or louvers. For good lighting practice, brightness of luminaires in the field of vision should be less than 1½ cd/in^2. These units are usually suspended from the ceiling at a distance sufficient to produce reasonably uniform ceiling brightness, without unduly bright spots. Luminous ceilings are in this general class.

43. Semi-indirect luminaires are generally used for commercial lighting as in offices, stores, schools, auditoriums, etc., and also in residential lighting. Such luminaires must be suspended from the ceiling at a distance sufficient to produce reasonably uniform ceiling brightness. Approximately equal brightness of luminaire and ceiling is desirable, although luminaire brightnesses two to three times the average ceiling brightness are not objectionable. This type of lighting, produced by incandescent and fluorescent units of many types, is characterized by freedom from sharp shadows and specular glare.

44. Indirect luminaires are used principally for commercial purposes as in schools, stores, and offices, also for cove lighting in residences. They must be suspended from the ceiling at somewhat greater distance than is the practice in other types of lighting, since practically all the light is reflected from the ceiling, which should be of nearly uniform brightness. This type of lighting is free from sharp shadows and specular glare. It is good practice to design indirect luminaires with a sufficient downward component of light so that their brightness is approximately equal to the ceiling brightness; thus they do not become black silhouettes against a bright background. Luminaires of this kind are available for incandescent lamps and for fluorescent lamps.

45. Control of light distribution is usually accomplished through glass prisms and specularly reflecting surfaces. These surfaces may be of mirrored glass, chrome, rhodium, nickel, or anodic-processed aluminum coatings.

ARTIFICIAL-DAYLIGHT EQUIPMENT

46. Daylight varies in color, depending upon the time of day and the atmospheric conditions. Color temperature of direct sunlight varies from about 1900°K with the sun near the horizon to about 5300°K with the sun near the zenith. Clear blue sky ranges from about 11,000 to 50,000°K. Overcast sky ranges from 6000 to 7000°K. The color temperature of a perfectly white horizontal surface illuminated by overcast sky or sun and clear blue sky is of the order of 6500°K.[2]

Incandescent light in general is more yellow and less blue than daylight. The

[1] KETCH, J. M., and GIANINI, L. G. Engineering Aspects of Direct Lighting, Part II, Transmitting Systems; *Trans. IES*, 1939, pp. 411–426.

KETCH, J. M., and LA WALL, G. R. Engineering Aspects to Direct Lighting, Part I, Louvered Systems; *Trans. IES*, 1938, pp. 545–565.

[2] TAYLOR, A. H., and KERR, G. P. The Distribution of Energy in the Visible Spectrum of Daylight, *Jour. Opt. Soc. Am.*, January, 1941, Vol. 31, p. 3.

light from tungsten-filament vacuum lamps and from gas-filled tungsten lamps is 2350 to 3100°K. To correct the light from tungsten-filament lamps to a color temperature of, say, 6500°K involves the use of absorption filters which transmit only about one-tenth of the light. Where accurate equivalence of daylight is desired, this subtractive method of correcting tungsten-filament incandescent light is well worth while in spite of its physical inefficiency.

The fluorescent lamps that most nearly produce daylight-quality illumination are an acceptable substitute where critical matching of colors is not involved. The presence of the mercury lines in the light of fluorescent lamps operates to make them a less exact substitute for daylight than is desirable for critical work.

RESIDENCE LIGHTING

47. Residence Lighting. The technological, social, and economic changes since 1945 are influencing American living habits and altering residence architecture. Multiple uses of the decreasing space, brought about principally by economic changes, create a trend toward more open interior planning. These conditions require versatile and specifically planned lighting to meet the visual needs of all members of the family.

Reading and study, sewing, handicrafts, and household tasks are no less exacting than similar visual tasks in industry or in commercial establishments. The design of residential lighting systems agrees in basic principle with well-established techniques for public interiors. In addition, the home may be dramatized by lighting effects and good contemporary decorating similar in character to those successfully used in restaurants, shops, and theaters.

The Recommended Practice for Residential Lighting[1] is the principal authority on the subject. It outlines objectives of design, discusses brightness ratios, selection of luminaires and lamps, and lighting for specific visual tasks. The needs at specific furniture groups—desks, beds, sink, piano—and detailed studies of sewing, food preparation, and laundering are included. The 44-page brochure contains a complete résumé of lighting particulars for all rooms of the home. Another publication prepared by the same committee is Lighting Keyed to Today's Homes. The IES Residential Lighting Data Sheets provide detailed information on specific installations.

THE CANDLEPOWER DISTRIBUTION CURVE

48. Light Distribution. A candlepower distribution curve is a graphic presentation of the distribution of light intensity of a lamp or luminaire. Such presentations contribute valuable information to guide the engineer in determining the suitability of lighting equipment for application in various fields. As a background for using distribution curves it is first necessary to see how they are obtained.

The candlepower in any direction from a filament lamp equals the footcandles produced on a plane at right angles to the light rays times the square of the distance in feet from the lamp to the point of measurement ($\text{cp} = \text{fc} \times D^2$). For accurate measurements the distance should be at least five times the largest dimension of the source. (To simplify calculations 10 ft is often used in photometric laboratories.) If in this way the average candlepower around the axis of a filament lamp is determined for any angle from the vertical, say, 25°, the average value becomes one point which can be plotted to a convenient scale on polar-coordinate paper. Taking several measurements around the axis at the 25° angle from the vertical usually shows them all to be about the same. However, in laboratory photometry any slight differences that might be present because of filament structure or other variations are compensated by rotating the lamp so that one reading represents an average value.

To get sufficient data for a lamp or general lighting unit, 20 readings are usually taken at angles of 0, 5, 15, 25, 35, 45°, etc., up to 180° (Fig. 19-22), and the candlepowers are computed and plotted on polar-coordinate paper. A line connecting a series of such points forms the candlepower distribution curve. The value at 90° is the candlepower straight out from the unit, while that at 0° is directly below. For concentrated light sources such as searchlights and spotlights, photometric readings are often required 1 or 2° apart, rather than at 10° intervals.

---

[1] Prepared by Committee on Residence Lighting of the IES, New York.

49. Luminaire Output and Efficacy. The lumens of a light source are obtained by adding the lumens in the various zones. The zonal lumen values are readily computed by multiplying the candlepower at the center of the zone by the zonal constant. For example, at 55°, which is the center of the 50°–60° zone, the 300-cd reading multiplied by 0.90 gives 270 lm in that zone for the 200-W lamp (Fig. 19-22). Zonal constants for the usual 18 test zones as shown in Table 19-20 are simply numerical values of the square feet of area in each zone on the surface of a sphere having a radius of 1 ft. Since one candela generates one lumen of light per square foot of surface in the unit sphere, it follows that the lumens in each zone are equal to the square feet of sphere area in that zone multiplied by the candlepower at the center of the zone.

Luminaire Efficacy. The efficacy of a luminaire is expressed in terms of its lumens output divided by the lumens generated by the lamp. For example, if the 200-W lamp in Fig. 19-22 is placed in a white-glass enclosing globe whose candlepower distribution shows an output of 2,960 lm, the efficacy of the luminaire is 2,960 divided by 3,700, or 80%.

To determine the output of nonsymmetric or asymmetric luminaires, such as conventional fluorescent units, candlepower readings must be taken in a number of planes and their weighted average obtained for each zone. The zone candlepower multiplied by the zonal constant gives the zonal lumens. The sum of the zonal lumens equals the lumen output of the luminaire.

For a fluorescent luminaire, candlepower readings are usually taken in five planes, at 0, 22½, 45, 67½, and 90° from a plane through the luminaire axis. Candlepower values are measured in each plane at 10° intervals (5°, 15°, 25°, etc.). On assuming that the five planes are at the angles 0°, 22½°, 45°, 67½°, and 90° and that the candlepower readings at such intervals are designated as A, B, C, D, and E, then their weighted average is obtained by the formula

$$\text{cp} = \frac{A + 2B + 2C + 2D + E}{8}$$

In some laboratories, for similar tests, candlepower readings are taken in three planes only, crosswise (90°), lengthwise (0°), and at 45°. The candlepower values

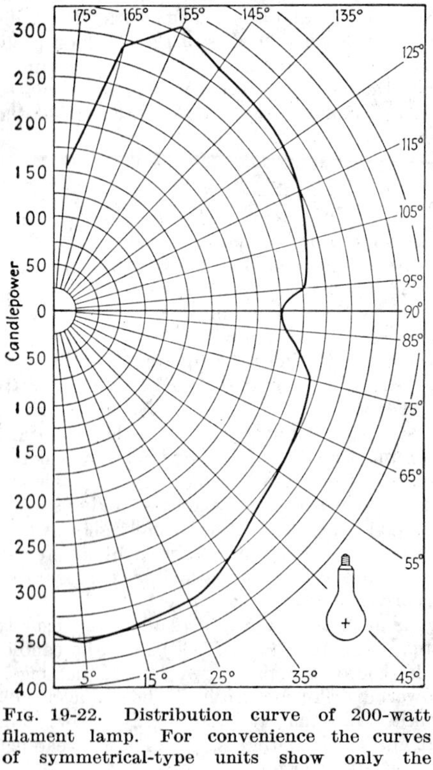

Fig. 19-22. Distribution curve of 200-watt filament lamp. For convenience the curves of symmetrical-type units show only the 0-180-deg half. Actually the distribution of light includes 360 deg. The two halves are symmetrical.

Table 19-20. Tabulation of Candelas and Lumens in the Various Zones in the Curve of Fig. 19-22

Computation of Lumens from Candelas

Angle	Candelas	Zone	Zonal constant	Zonal lumens
0	347			
5	352	0–10	0.10	35
15	348	10–20	0.28	97
25	342	20–30	0.46	157
35	326	30–40	0.63	205
45	307	40–50	0.77	236
55	300	50–60	0.90	270
65	288	60–70	0.99	285
75	285	70–80	1.06	302
85	259	80–90	1.09	282
90	257			
95	271	90–100	1.09	295
105	278	100–110	1.06	295
115	290	110–120	0.99	287
125	307	120–130	0.90	276
135	308	130–140	0.77	237
145	313	140–150	0.63	197
155	329	150–160	0.46	151
165	280	160–170	0.28	78
175	153	170–180	0.10	15
Total lumens	...	....	....	3,700

crosswise and lengthwise plus twice the values in 45° are added. The sum divided by 4 equals the average candlepower.

Similarly, nonsymmetric or asymmetric luminaires for filament lamps have such wide variations in candlepower at a given angle about the vertical that an average reading from which to compute zonal lumens cannot be obtained by rotating the unit. Candlepower distribution curves for such equipment are prepared from data obtained in specific planes, and in interpreting such curves one must be careful to observe the planes they represent.

QUANTITY AND QUALITY OF ILLUMINATION

50. Levels of Illumination. Many considerations are involved in specifying the illumination for various applications. In all cases, the illumination must meet the requirements for such seeing tasks as may be present, and in many cases it must also include lighting for aesthetic purposes and lighting to attract attention. In meeting these requirements, proper weight must be given such factors as brightness distribution, diffusion, direction, elimination of objectionable shadows, and color quality of the light, as well as footcandles.

Seeing is a function of a number of fundamentally important factors, some of which are the size of the object or detail, the contrast of the detail with its immediate background, the brightness of the detail, the time available to see it, and the brightness relation between the task and its surroundings.

In a store or sales area a pattern of brightness must be depended upon to capture the attention and interest of potential customers passing rapidly in vehicles or afoot, so that they will want to come into the place of business. Inside, as well, the distribution of brightness affects their movements, determines what they look at, and influences the numbers and kinds of things they buy and how often they return. Thus, the desirable illumination is often governed by more than the requirements of the particular seeing tasks.

The footcandle values found in present-day practice for various visual tasks are the result of several obvious factors which include the cost of lighting, methods of obtaining it, and habits in its use. Furthermore, lighting is always accompanied by a certain amount of heat and radiant energy, which in some instances is objectionable and which has retarded the approach to desirable illumination. The more favorable relation of light to radiant heat in the fluorescent lamp is one of the reasons why the newer source led so promptly to the use of higher footcandles and to the conclusion that air conditioning and proper lighting are economically compatible. Since the early 1900s, there has been a spectacular decrease in the cost of light. Lamps are ten times as efficient as they were then and cost about one-fifth as much. Electricity now costs about one-fourth as much as it did then. In 1906, for example, an employer's cost for providing one footcandle per worker amounted to 7% of the average worker's wages. In the 1960s, an employer can provide 200 fc at a cost of 3.5% of the average worker's wages.

Current recommended levels of illumination for a wide variety of visual tasks were adopted by the Illuminating Engineering Society in 1958. They were based on the results of research conducted at the Vision Research Laboratories of the University of Michigan by Dr. H. R. Blackwell under the sponsorship of the Illuminating Engineering Research Institute. Recommended levels of illumination for a number of visual tasks are given in Table 19-21. A more extended listing will be found in the IES Lighting Handbook.

51. Quality of lighting[1] comprises those aspects of the visual environment which make for comfortable, efficient seeing. It pertains to the distribution of brightness in the entire visual field. The requirements for quality of lighting may be concisely expressed for one specific class of installation, for only in this class are the guideposts sufficiently defined to achieve simplicity. This class of lighting installation may be defined according to use as follows: lighting in an area where critical seeing is done, over protracted periods of time, from fixed positions of the workers involving frequent viewing of the lighting equipment; specular reflections to be minimized; the work space

[1] These criteria for quality of lighting are taken from the IES Recommended Practice for Office Lighting, 1960, and the ASA Guide for School Lighting, 1962.

Table 19-21. Recommended Levels of Illumination

Area	Footcandles on Tasks*
Assembly:	
Rough easy seeing	30
Rough difficult seeing	50
Medium	100
Fine	500†
Extra fine	1,000†
Cloth products:	
Cloth inspection	2,000†
Cutting	300†
Sewing	500†
Pressing	300†
Inspection:	
Ordinary	50
Difficult	100
Highly difficult	200†
Very difficult	500†
Most difficult	1,000†
Machine shops:	
Rough bench- and machine work	50
Medium bench- and machine work, ordinary automatic machines, rough grinding, medium buffing and polishing	100
Fine bench- and machine work, fine automatic machines, medium grinding, fine buffing and polishing	500†
Extra-fine bench- and machine work, grinding, fine work	1,000†
Offices:	
Cartography, designing, detailed drafting	200
Accounting, auditing, tabulating, bookkeeping, business-machine operation, reading poor reproductions, rough layout drafting	150
Regular office work, reading good reproductions, reading or transcribing handwriting in hard pencil or on poor paper, active filing, index references, mail sorting	100
Reading or transcribing handwriting in ink or medium pencil on good-quality paper, intermittent filing	70
Reading high-contrast or well-printed material; tasks and areas not involving critical or prolonged seeing such as conferring, interviewing, inactive files, and washrooms	30
Corridors, elevators, escalators, stairways	20‡
Schools:§	
Reading printed material	30
Reading pencil writing	70
Spirit duplicated material:	
Good	30
Poor	100
Drafting, benchwork	100†
Lip reading, chalkboards, sewing	150†
Woodworking:	
Rough sawing and benchwork	30
Sizing, planing, rough sanding, medium-quality machine- and benchwork, gluing, veneering, cooperage	50
Fine bench- and machine work, fine sanding and finishing	100

* Minimum on the task at any time.

† Obtained with a combination of general lighting plus specialized supplementary lighting. Care should be taken to keep within the recommended brightness ratios. These seeing tasks generally involve the discrimination of fine detail for long periods of time and under conditions of poor contrast. To provide the required illumination, a combination of the general lighting indicated plus specialized supplementary lighting is necessary. The design and installation of the combination system must provide for not only a sufficient amount of light but also the proper direction of light, diffusion, and eye protection. As far as possible, it should eliminate direct and reflected glare as well as objectionable shadows.

‡ Or not less than one-fifth the level in adjacent areas.

§ Tasks are listed here, rather than areas.

to be a controlled environment (it is feasible to specify control of daylight, reflectances and textures, cleanliness, etc.). Fortunately, this class embraces office and school-classroom lighting, and it is in such areas that quality is of first importance.

1. *Control of Direct Glare.* Limit the average and maximum brightness of the luminaires. Specify the right reflectances and textures on architectural and work surfaces. Direct glare should not be a problem in school and office fluorescent lighting installations if ceiling, wall, and other reflectances comply with recommendations in Table 19-22 and if (1) the luminaires have crosswise and lengthwise average brightness distributions which fall entirely below any straight line(s) drawn through 250 fL at 75° lying between the two limiting solid lines in Fig. 19-23 and (2) the ratio of maximum to average luminaire brightness preferably does not exceed 3:1 and definitely does not exceed 5:1, as shown in Table 19-23. The maximum brightness must never be more than three times the value of the sloped limiting line of the graph.

2. *Control of Reflected Glare.* While control of direct glare is relatively simple and may be approached directly, practical methods for minimizing reflected glare are less specific. The presence of direct glare is usually obvious; reflected glare is much more

subtle in effect and often requires instrumentation to determine its presence. The losses in visibility due to reflected glare are real and measurable. The results of one study of reflected glare are shown in Table 19-24. These indicate the losses in contrast for two illumination levels and for tasks commonly found in the classroom. It has been shown that 1% loss in contrast requires a 15% increase in illumination to achieve equal visibility.[1] Hence, a small reduction in contrast has the effect of an appreciable reduction in the illumination level. It will be easily seen that the losses for glossy-background tasks are larger than those for mat-background tasks, but it must be remembered that even the latter are serious and hinder visual performance. It will also be seen that the losses decrease as the percentage of distribution of light upward from the

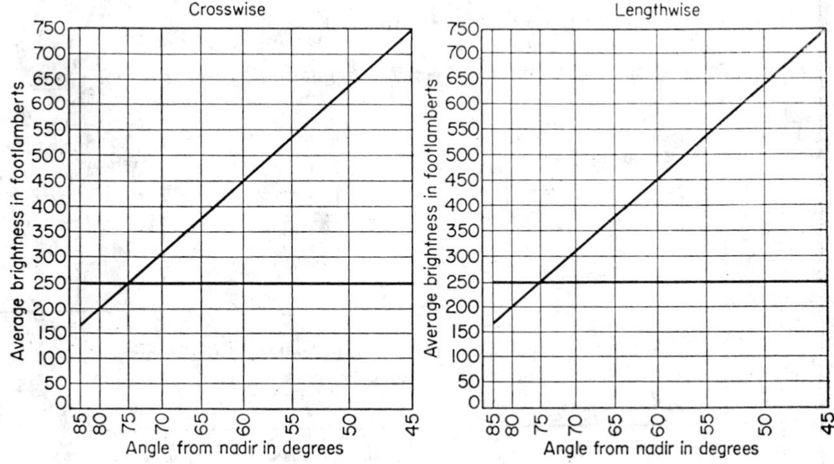

FIG. 19-23. Scissors-curve graph. Average and maximum brightnesses to be obtained from Photometric Report of the luminaire.

luminaires increases. The losses also decrease as the footcandle level of illumination is increased for the same number and location of luminaires.

Reflected glare can be minimized by controlling any or all of the factors that contribute to it. Specific recommendations include the following:

a. Use high-reflectance mat surfaces as in Table 19-22, and encourage the use of low-gloss inks instead of pencils.

b. Design the lighting system or locate the work areas so that the task will not be at the mirror angle with respect to the eye and any light sources (electric or daylight). Minimum loss of contrast from veiling reflections will usually occur when the light falls on the task in approximately the same direction as the observer's line of sight, but such conditions are difficult to achieve in multioccupant rooms. A more practical approach is to design the system so that most of the illumination on the task comes from favorable directions. This can be achieved by close spacing of luminaires and by avoiding layouts that locate the luminaires in the same vertical plane as the observer's line of sight.

c. Limit the luminance of light sources in the 0° to 45° zone.

Reflected glare can also be minimized by employing luminaires with a "batwing" distribution. Such luminaires have maximum candelas occurring in the 45° plane (instead of crosswise or lengthwise), and these candelas are at angles in the 35° to 45° zone from nadir, with the candela at nadir being ideally as low as possible.

Vertical polarization of the light can be helpful in reducing veiling reflections in certain horizontal tasks if they are viewed at angles of 40° or more from vertical. Studies of persons writing and reading have indicated, however, that the majority of continuous work is done at lower viewing angles, where polarization is of little advantage, if any.

[1] IES Lighting Handbook, Contrast vs. Brightness Curve.

3. *Acceptable Uniformity of Illumination.* Follow the spacing–mounting-height relations and distances from walls to rows of luminaires described in Par. **54.** Acceptable uniformity of illumination is usually accompanied by the elimination of undesirable shadows.

Table 19-22. Recommended Surface Reflectances*

Offices		Schools	
Ceiling	80–92%	Ceiling	70–90%
Walls	40–60%	Walls	40–60%
Furniture	26–44%	Chalkboards	Up to 20%
Office machines	26–44%	Desk	35–50%
Floor	21–39%	Floor	30–50%

* The textures of these surfaces should be mat and without gloss.

Table 19-23. Maximum to Average Luminaire Brightness Ratios

Crosswise					Lengthwise				
Angle	3 times value	Maximum	Average	Ratio max./avg.	Angle	3 times value	Maximum	Average	Ratio max./avg.
45°	2,250				45°	2,250			
55°	1,605				55°	1,605			
65°	1,125				65°	1,125			
75°	750				75°	750			
85°	495				85°	495			

Table 19-24. Loss in Contrast Due to Reflected Glare for Different Lighting Systems

55 Fc

Task*	Open-bottom-louver direct system, % loss	General diffuse system, % loss	Luminous indirect system, % loss
A-1	16.5	13.2	8.8
A-2	4.4	5.3	4.6
N-J	20.6	17.0	11.3
8-11	24.1	20.6	14.6
6-2	24.7	....	10.7
6-9	19.0	....	11.3

30 Fc

Task*	Open-bottom-louver direct system, % loss	General diffuse system, % loss	Luminous indirect system, % loss
A-1	16.6	17.6	9.0
A-2	8.2	7.7	7.2
N-J	21.6	20.2	10.8
8-11	30.1	23.6	15.7
6-2	25.3	19.4	9.9
6-9	22.9	21.3	10.8

* A-1 and A-2 are identical printed tasks except for the paper; A-1 is printed on glossy "coated book stock"; A-2 is printed on mat "offset book stock."

N-J, 8-11, 6-2, and 6-9 are pencil tasks using No. 2 pencil on ruled mat white paper having diffuse reflectance of 77%.

DESIGN OF LIGHTING SYSTEMS

52. For a discussion of design, lighting systems may be divided into two classes—**general** and **supplementary.** The general system supplies uniform illumination in the area; supplementary lighting supplies illumination of a specific nature, color, or distribution, usually to satisfy a local, specific requirement.

53. General Lighting. The design of general lighting systems is governed by room dimensions, structural features, reflection characteristics of walls and ceilings, mounting height of the luminaires, and the distribution and maintenance characteristics of the luminaire. The choice of the luminaire depends on the service to which it is to be put, which assumes a certain experience in selection, or other aids such as manufacturers' data, which assist the designer in making a selection appropriate from the standpoints of freedom from glare, efficiency, decorative value, and economy. The ultimate "brightness pattern" of the room is an important factor in the design.

The beginning concept of general lighting design is that of delivering a specified average footcandle level of illumination to a horizontal plane in a room. The light generated by the lamps in such a system is variously affected and considerably reduced by reflection, diffusion, and absorption as it impinges on reflectors and transmitting media in the luminaires and on ceilings, wall, floors, and on the objects in the room.

54. Luminaire Spacing and Mounting Height. The maximum permissible spacing for each type of luminaire is given in the Photometric Report provided by the manufacturer. These spacing limitations are related to the mounting height (usually above the floor but in some instances above the work plane) of direct, semidirect, and general-diffuse luminaires and to the ceiling height for indirect and semi-indirect systems. Observance of such limitations will ensure satisfactory uniformity of illumination throughout the major portion of the room so that all parts of the area will be equally suitable for the intended use. Peripheral areas may require special treatment, as indicated below. Uniformity of illumination is usually considered satisfactory if the maximum and minimum values are within plus or minus one-sixth of the average illumination in the area. Closer spacing than indicated by the spacing ratio will improve uniformity and reduce shadows. The spacing–mounting-height relations apply not only to individual luminaires but to the spacing between continuous sections, luminous panels, troffers, or sections of coves.

The distance between luminaires and the wall should not exceed one-half the distance between luminaires. Where desks or benches might be located along the wall, the distance between luminaires and the wall should not exceed 2½ ft. Likewise, the ends of continuous rows of fluorescent luminaires should preferably be within 6 to 12 in of the wall. Additional luminaires or luminaires having a greater number of lamps may be required adjacent to the walls, particularly where walls have low reflectance. Where direct and semidirect luminaires are used under such conditions, the perimeter luminaires should be carefully located to avoid shadows on the work from the worker.

A few typical layouts of luminaires are given in Fig. 19-24.

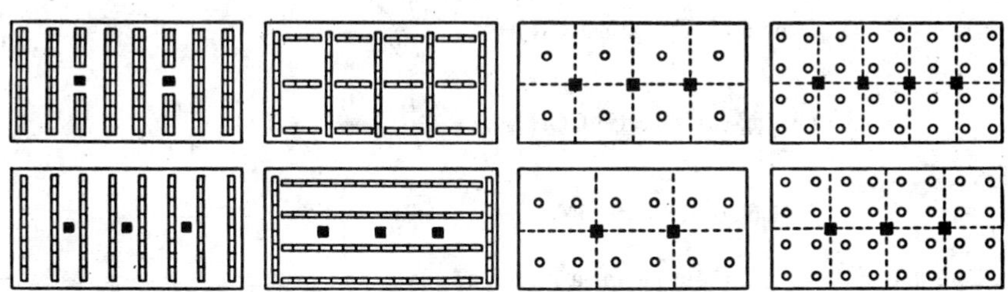

Fig. 19-24. Typical luminaire layouts in various interiors.

55. Zonal-cavity Method.[1] Over the last half century the art of predicting the illumination level which would be produced in a room from a lighting system has shown a history of steady improvement in accuracy and versatility. The basic illumination-calculation method was developed originally for use with coefficients of utilization values determined empirically. The basic formula still in current use is

$$\text{fc in service} = \frac{\text{rated lamp lumens} \times \text{CU} \times \text{MF}}{\text{area, ft}^2}$$

[1] The Zonal-cavity Method, Lighting Design Practice Committee of the IES, 1964.

where CU = coefficient of utilization, a decimal fraction =

$$\frac{\text{Lumens reaching the work plane}}{\text{Lumens generated by the lamps}}$$

and MF = maintenance factor, a decimal fraction =

$$\frac{\text{Illumination on a given area after a period of time}}{\text{Initial illumination on the same area}}$$

Later developments used methods of mathematical analysis in computing coefficient-of-utilization data for currently available lighting equipment. The newer mathematical methods of analysis of lighting distributions, taking into account the concept of interreflection of light, have led to progressively more accurate coefficient-of-utilization data, culminating in the IES-approved method of calculating illumination called the *zonal-cavity method.* Like all its predecessors, the method is based on the theory that average illumination is equal to lumens falling on the work plane divided by the area over which they are distributed. This present advance in the computation of coefficients of utilization features principally the introduction of means by which they may be calculated for many variable conditions which were formerly either ignored or established as fixed values or relationships.

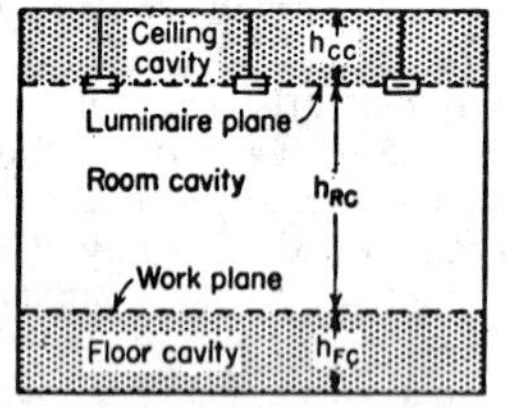

FIG. 19-25. Room cavities.

The new system considers the actual room as being made up of a ceiling cavity above the luminaires, a floor cavity beneath the work plane, and a room cavity located between the two (see Fig. 19-25).

In the general case, all these cavities are present. In the case of recessed or surface-mounted luminaires, the ceiling cavity is simply the ceiling. When the illumination on the floor is to be determined, the floor cavity becomes the floor.

It is now possible to calculate numerical relationships called "cavity ratios" which may be used to determine effective reflectance of the floor and ceiling and then to find the coefficient of utilization.

The basic steps in the calculation of any average illumination are as follows:

1. Determine *cavity ratios* for 3 cavities shown in Fig. 19-25 as follows:

$$\text{Room-cavity ratio RCR} = \frac{5h_{RC}(L + W)}{LW}$$

$$\text{Ceiling-cavity ratio CCR} = \frac{5h_{CC}(L + W)}{LW} = \text{RCR}\,\frac{h_{CC}}{h_{RC}}$$

$$\text{Floor-cavity ratio FCR} = \frac{5h_{FC}(L + W)}{LW} = \text{RCR}\,\frac{h_{FC}}{h_{RC}}$$

where h_{RC} = height of room between luminaire plane and work plane; h_{CC} = distance from luminaire plane to ceiling; h_{FC} = height of work plane above floor; L = room length; and W = room width.

Cavity ratios may also be found in Table 19-25 for typical-sized cavities which cover a wide range of room dimensions.

2. Obtain *effective ceiling-cavity reflectance* (ρ_{cc}) for combination of ceiling and wall reflectance to be employed, from Table 19-26. Note that, for surface-mounted or recessed luminaires, CCR = 0 and the ceiling reflectance may be used as the effective cavity reflectance. Unless *initial* illumination is to be calculated, ceiling and wall reflectances should be the expected *maintained* reflectances.

3. Obtain *effective floor-cavity reflectance* (ρ_{fc}) for combination of floor and wall reflectances to be employed, from Table 19-26. Unless *initial* illumination is to be calculated, floor and wall reflectances should be the expected *maintained* reflectances.

4. Obtain *coefficient of utilization* for the luminaire for 20% effective floor-cavity-reflectance condition, from Table 19-29a, interpolating between tabulated values as required to match room size and ceiling- and wall-reflectance combinations.

5. If *effective floor-cavity reflectance* (ρ_{fc}) obtained in step 3 differs significantly from 20%, obtain multiplier from Table 19-27 or Table 19-28. Multiply the coefficient of utilization by this multiplier.

6. Determine *average maintained illumination* by the following formula:

$$\text{fc} = \frac{(\text{rated lamp lumens/luminaire}) \times \text{CU} \times \text{MF}}{\text{area/luminaire}}$$

If *initial illumination* is desired, coefficient-of-utilization selection should be obtained by the above procedure except that initial ceiling and wall reflectances should be used in steps 2 and 3 and the maintenance factor should be omitted from the above formula.

When the desired average maintained illumination is known, the formula can more conveniently be expressed as

$$\text{Area/luminaire} = \frac{(\text{rated lamp lumens/luminaire}) \times \text{CU} \times \text{MF}}{\text{fc}}$$

This area divided by the luminaire length gives the approximate spacing between continuous rows, or it may be divided into the total room area to determine the number of luminaires required.

Example[1]

A typical classroom is 28 ft wide and 32 ft long and has a 12-ft 6-in ceiling height. Reflectances are: ceiling 80%; walls 50%; floor 10%. A semidirect two-lamp fluorescent luminaire is to be used on 2-ft 6-in stems. Work plane is 2 ft 0 in. Find the coefficient of utilization.

1. Calculate *cavity ratios* as follows, or look up in table of cavity ratios (Table 19-25).

$$\text{CCR} = \frac{(5)(2.5)(28 + 32)}{(28)(32)} = 0.84$$

$$\text{RCR} = \frac{(5)(8)(28 + 32)}{(28)(32)} = 2.7$$

$$\text{FCR} = \frac{(5)(2)(28 + 32)}{(28)(32)} = 0.67$$

2. In Table 19-26, look up *effective cavity reflectances* for ceiling and floor cavities. ρ_{cc} for the ceiling cavity will be 68%, while ρ_{fc} for the floor cavity will be 11%.

3. With the *room-cavity ratio* RCR known, it is now possible to find the coefficient of utilization for the luminaire in a room having an RCR of 2.7 and effective reflectances as follows:

$$\rho_{cc} = 68\% \qquad \rho_{w} = 50\% \qquad \rho_{fc} = 20\%$$

This CU = 0.56. Note that this is for an effective floor reflectance of 20%, while the actual effective reflectance of the floor ρ_{fc} is 11%. To correct for this, locate the appropriate multiplier in Table 19-27 for the RCR already calculated (2.7). It is 0.955 and is found by interpolating between the numbers for 60% and 70% ρ_{cc} and between RCR's of 2.0 and 3.0. Then

$$\text{CU final} = 0.56 \times 0.955 = 0.53$$

4. *Illumination level* can now be calculated if we know the area per luminaire, the lamp lumen rating, and the maintenance factor.

$$\text{fc in service} = \frac{(\text{rated lamp lumens/luminaire}) \times \text{CU} \times \text{MF}}{\text{area/luminaire}}$$

[1] From Holophane Co. Inc., "The Zonal-cavity Method," 1965.

Table 19-25. Cavity Ratios

Room dimensions		Cavity depth																			
Width	Length	1.0	1.5	2.0	2.5	3.0	3.5	4.0	5.0	6.0	7.0	8	9	10	11	12	14	16	20	25	30
8	8	1.2	1.9	2.5	3.1	3.7	4.4	5.0	6.2	7.5	8.8	10.0	11.2	12.5							
	10	1.1	1.7	2.2	2.8	3.4	3.9	4.5	5.6	6.7	7.9	9.0	10.1	11.3	12.4						
	14	1.0	1.5	2.0	2.5	3.0	3.4	3.9	4.9	5.9	6.9	7.8	8.8	9.7	10.7	11.7					
	20	0.9	1.3	1.7	2.2	2.6	3.1	3.5	4.4	5.2	6.1	7.0	7.9	8.8	9.6	10.5	12.2				
	30	0.8	1.2	1.6	2.0	2.4	2.8	3.2	4.0	4.7	5.5	6.3	7.1	7.9	8.7	9.5	11.0				
	40	0.7	1.1	1.5	1.9	2.3	2.6	3.0	3.7	4.5	5.3	5.9	6.5	7.4	8.1	8.8	10.3	11.8			
10	10	1.0	1.5	2.0	2.5	3.0	3.5	4.0	5.0	6.0	7.0	8.0	9.0	10.0	11.0	12.0					
	14	0.9	1.3	1.7	2.1	2.6	3.0	3.4	4.3	5.1	6.0	6.9	7.8	8.6	9.5	10.4	12.0				
	20	0.7	1.1	1.5	1.9	2.3	2.6	3.0	3.7	4.5	5.3	6.0	6.8	7.5	8.3	9.0	10.5	12.0			
	30	0.7	1.0	1.3	1.7	2.0	2.3	2.7	3.3	4.0	4.7	5.3	6.0	6.6	7.3	8.0	9.4	10.6			
	40	0.6	0.9	1.2	1.6	1.9	2.2	2.5	3.1	3.7	4.4	5.0	5.6	6.2	6.9	7.5	8.7	10.0	12.5		
	60	0.6	0.9	1.2	1.5	1.7	2.0	2.3	2.9	3.5	4.1	4.7	5.3	5.9	6.5	7.1	8.2	9.4	11.7		
12	12	0.8	1.2	1.7	2.1	2.5	2.9	3.3	4.2	5.0	5.8	6.7	7.5	8.4	9.2	10.0	11.7				
	16	0.7	1.1	1.5	1.8	2.2	2.5	2.9	3.6	4.4	5.1	5.8	6.5	7.2	8.0	8.7	10.2	11.6			
	24	0.6	0.9	1.2	1.6	1.9	2.2	2.5	3.1	3.7	4.4	5.0	5.6	6.2	6.9	7.5	8.7	10.0	12.5		
	36	0.6	0.8	1.1	1.4	1.7	1.9	2.2	2.8	3.3	3.9	4.4	5.0	5.5	6.0	6.6	7.8	8.8	11.0		
	50	0.5	0.8	1.0	1.3	1.5	1.8	2.1	2.6	3.1	3.6	4.1	4.6	5.1	5.6	6.2	7.2	8.2	10.2		
	70	0.5	0.7	1.0	1.2	1.5	1.7	2.0	2.4	2.9	3.4	3.9	4.4	4.9	5.4	5.8	6.8	7.8	9.7	12.2	
14	14	0.7	1.1	1.4	1.8	2.1	2.5	2.9	3.6	4.3	5.0	5.7	6.4	7.1	7.8	8.5	10.0	11.4			
	20	0.6	0.9	1.2	1.5	1.8	2.1	2.4	3.0	3.6	4.2	4.9	5.5	6.1	6.7	7.3	8.6	9.8	12.3		
	30	0.5	0.8	1.0	1.3	1.6	1.8	2.1	2.6	3.1	3.7	4.2	4.7	5.2	5.8	6.3	7.3	8.4	10.5		
	42	0.5	0.7	1.0	1.2	1.4	1.7	1.9	2.4	2.9	3.3	3.8	4.3	4.7	5.2	5.7	6.7	7.6	9.5	11.9	
	60	0.4	0.7	0.9	1.1	1.3	1.5	1.8	2.2	2.6	3.1	3.5	3.9	4.4	4.8	5.2	6.1	7.0	8.8	10.9	
	90	0.4	0.6	0.8	1.0	1.2	1.4	1.6	2.0	2.5	2.9	3.3	3.7	4.1	4.5	5.0	5.8	6.6	8.3	10.3	12.4
17	17	0.6	0.9	1.2	1.5	1.8	2.1	2.3	2.9	3.5	4.1	4.7	5.3	5.9	6.5	7.0	8.2	9.4	11.7		
	25	0.5	0.7	1.0	1.2	1.5	1.7	2.0	2.5	3.0	3.5	4.0	4.5	5.0	5.5	6.0	7.0	8.0	10.0	12.5	
	35	0.4	0.7	0.9	1.1	1.3	1.5	1.7	2.2	2.6	3.1	3.5	3.9	4.4	4.8	5.2	6.1	7.0	8.7	10.9	
	50	0.4	0.6	0.8	1.0	1.2	1.4	1.6	2.0	2.4	2.8	3.1	3.5	3.9	4.3	4.5	5.4	6.2	7.7	9.7	11.6
	80	0.4	0.5	0.7	0.9	1.1	1.2	1.4	1.8	2.1	2.5	2.9	3.3	3.6	4.0	4.3	5.1	5.8	7.2	9.0	10.9
	120	0.3	0.5	0.7	0.8	1.0	1.2	1.3	1.7	2.0	2.3	2.7	3.0	3.4	3.7	4.0	4.7	5.4	6.7	8.4	10.1
20	20	0.5	0.7	1.0	1.2	1.5	1.7	2.0	2.5	3.0	3.5	4.0	4.5	5.0	5.5	6.0	7.0	8.0	10.0	12.5	
	30	0.4	0.6	0.8	1.0	1.2	1.5	1.7	2.1	2.5	2.9	3.3	3.7	4.1	4.5	4.9	5.8	6.6	8.2	10.3	12.4
	45	0.4	0.5	0.7	0.9	1.1	1.3	1.4	1.8	2.2	2.5	2.9	3.3	3.6	4.0	4.3	5.1	5.8	7.2	9.1	10.9
	60	0.3	0.5	0.7	0.8	1.0	1.2	1.3	1.7	2.0	2.3	2.7	3.0	3.4	3.7	4.0	4.7	5.4	6.7	8.4	10.1
	90	0.3	0.5	0.6	0.8	0.9	1.1	1.2	1.5	1.8	2.1	2.4	2.7	3.0	3.3	3.6	4.2	4.8	6.0	7.5	9.0
	150	0.3	0.4	0.6	0.7	0.8	1.0	1.1	1.4	1.7	2.0	2.3	2.6	2.9	3.2	3.4	4.0	4.6	5.7	7.2	8.6

24	24	0.4	0.6	0.8	1.0	1.2	[illegible]	[illegible]	[illegible]	[illegible]	[illegible]	[illegible]	[illegible]	[illegible]	[illegible]	[illegible]	[illegible]	[illegible]	[illegible]	[illegible]	[illegible]
	32	0.4	0.5	0.7	0.9	1.1	1.3	1.5	1.8	2.2	2.6	2.9	3.3	3.6	4.0	4.3	5.1	5.8	7.2	9.0	11.0
	50	0.3	0.5	0.6	0.8	0.9	1.1	1.2	1.5	1.8	2.2	2.5	2.8	3.1	3.4	3.7	4.4	5.0	6.2	7.8	9.4
	70	0.3	0.4	0.6	0.7	0.8	1.0	1.1	1.4	1.7	2.0	2.2	2.5	2.8	3.0	3.3	3.8	4.4	5.5	6.9	8.2
	100	0.3	0.4	0.5	0.6	0.8	0.9	1.0	1.3	1.6	1.8	2.1	2.4	2.6	2.9	3.1	3.7	4.2	5.2	6.5	7.9
	160	0.2	0.4	0.5	0.6	0.7	0.8	1.0	1.2	1.4	1.7	1.9	2.1	2.4	2.6	2.8	3.3	3.8	4.7	5.9	7.1
30	30	0.3	0.5	0.7	0.8	1.0	1.2	1.3	1.7	2.0	2.3	2.7	3.0	3.3	3.7	4.0	4.7	5.4	6.7	8.4	10.0
	45	0.3	0.4	0.6	0.7	0.8	1.0	1.1	1.4	1.7	1.9	2.2	2.5	2.7	3.0	3.3	3.8	4.4	5.5	6.9	8.2
	60	0.3	0.4	0.5	0.6	0.7	0.9	1.0	1.2	1.5	1.7	2.0	2.2	2.5	2.7	3.0	3.5	4.0	5.0	6.2	7.4
	90	0.2	0.3	0.4	0.6	0.7	0.8	0.9	1.1	1.3	1.6	1.8	2.0	2.2	2.5	2.7	3.1	3.6	4.5	5.6	6.7
	150	0.2	0.3	0.4	0.5	0.6	0.7	0.8	1.0	1.2	1.4	1.6	1.8	2.0	2.2	2.4	2.8	3.2	4.0	5.0	5.9
	200	0.2	0.3	0.4	0.5	0.6	0.7	0.8	1.0	1.1	1.3	1.5	1.7	1.9	2.0	2.2	2.6	3.0	3.7	4.7	5.6
36	36	0.3	0.4	0.6	0.7	0.8	1.0	1.1	1.4	1.7	1.9	2.2	2.5	2.8	3.0	3.3	3.9	4.4	5.5	6.9	8.3
	50	0.2	0.4	0.5	0.6	0.7	0.8	1.0	1.2	1.4	1.7	1.9	2.1	2.5	2.6	2.9	3.3	3.8	4.8	5.9	7.2
	75	0.2	0.3	0.4	0.5	0.6	0.7	0.8	1.0	1.2	1.4	1.6	1.8	2.0	2.3	2.5	2.9	3.3	4.1	5.1	6.1
	100	0.2	0.3	0.4	0.5	0.6	0.7	0.8	0.9	1.1	1.3	1.5	1.7	1.9	2.1	2.3	2.6	3.0	3.8	4.7	5.7
	150	0.2	0.3	0.3	0.4	0.5	0.6	0.7	0.9	1.0	1.2	1.4	1.6	1.7	1.9	2.1	2.4	2.8	3.5	4.3	5.2
	200	0.2	0.2	0.3	0.4	0.5	0.6	0.7	0.8	1.0	1.1	1.3	1.5	1.6	1.8	2.0	2.3	2.6	3.3	4.1	4.9
42	42	0.2	0.4	0.5	0.6	0.7	0.8	1.0	1.2	1.4	1.6	1.9	2.1	2.4	2.6	2.8	3.3	3.8	4.7	5.9	7.1
	60	0.2	0.3	0.4	0.5	0.6	0.7	0.8	1.0	1.2	1.4	1.6	1.8	2.0	2.2	2.4	2.8	3.2	4.0	5.0	6.0
	90	0.2	0.3	0.3	0.4	0.5	0.6	0.7	0.9	1.0	1.2	1.4	1.6	1.7	1.9	2.1	2.4	2.8	3.5	4.4	5.2
	140	0.2	0.2	0.3	0.4	0.5	0.5	0.6	0.8	0.9	1.1	1.2	1.4	1.5	1.7	1.9	2.2	2.5	3.1	3.9	4.6
	200	0.1	0.2	0.3	0.4	0.4	0.5	0.6	0.7	0.9	1.0	1.1	1.3	1.4	1.6	1.7	2.0	2.3	2.9	3.6	4.3
	300	0.1	0.2	0.3	0.3	0.4	0.5	0.5	0.7	0.8	0.9	1.1	1.3	1.4	1.5	1.7	1.9	2.2	2.8	3.5	4.2
50	50	0.2	0.3	0.4	0.5	0.6	0.7	0.8	1.0	1.2	1.4	1.6	1.8	2.0	2.2	2.4	2.8	3.2	4.0	5.0	6.0
	70	0.2	0.3	0.3	0.4	0.5	0.6	0.7	0.9	1.0	1.2	1.4	1.5	1.7	1.9	2.0	2.4	2.7	3.4	4.3	5.1
	100	0.1	0.2	0.3	0.4	0.4	0.5	0.6	0.7	0.9	1.0	1.2	1.3	1.5	1.6	1.8	2.1	2.4	3.0	3.7	4.5
	150	0.1	0.2	0.3	0.3	0.4	0.5	0.5	0.7	0.8	0.9	1.1	1.2	1.3	1.5	1.6	1.9	2.1	2.7	3.3	4.0
	300	0.1	0.2	0.2	0.3	0.3	0.4	0.5	0.6	0.7	0.8	0.9	1.0	1.1	1.3	1.4	1.6	1.9	2.3	2.9	3.5
60	60	0.2	0.2	0.3	0.4	0.5	0.6	0.7	0.8	1.0	1.2	1.3	1.5	1.7	1.8	2.0	2.3	2.7	3.3	4.2	5.0
	100	0.1	0.2	0.3	0.3	0.4	0.5	0.5	0.7	0.8	0.9	1.1	1.2	1.3	1.5	1.6	1.9	2.1	2.7	3.3	4.0
	150	0.1	0.2	0.2	0.3	0.3	0.4	0.5	0.6	0.7	0.8	0.9	1.0	1.2	1.3	1.4	1.6	1.9	2.3	2.9	3.5
	300	0.1	0.1	0.2	0.2	0.3	0.3	0.4	0.5	0.6	0.7	0.8	0.9	1.0	1.1	1.2	1.4	1.6	2.0	2.5	3.0
75	75	0.1	0.2	0.3	0.3	0.4	0.5	0.5	0.7	0.8	0.9	1.1	1.2	1.3	1.5	1.6	1.9	2.1	2.7	3.3	4.0
	120	0.1	0.2	0.2	0.3	0.3	0.4	0.4	0.5	0.6	0.8	0.9	1.0	1.1	1.2	1.3	1.5	1.7	2.2	2.7	3.3
	200	0.1	0.1	0.2	0.2	0.3	0.3	0.4	0.5	0.5	0.6	0.7	0.8	0.9	1.0	1.1	1.3	1.5	1.8	2.3	2.7
	300	0.1	0.1	0.2	0.2	0.2	0.3	0.3	0.4	0.5	0.6	0.7	0.7	0.8	0.9	1.0	1.2	1.3	1.7	2.1	2.5
100	100	0.1	0.1	0.2	0.2	0.3	0.3	0.4	0.5	0.6	0.7	0.8	0.9	1.0	1.1	1.2	1.4	1.6	2.0	2.5	3.0
	200	0.1	0.1	0.1	0.2	0.2	0.3	0.3	0.4	0.4	0.5	0.6	0.7	0.7	0.8	0.9	1.0	1.2	1.5	1.9	2.2
	300	0.1	0.1	0.1	0.2	0.2	0.2	0.3	0.3	0.4	0.5	0.5	0.6	0.7	0.7	0.8	0.9	1.1	1.3	1.7	2.0
150	150	0.1	0.1	0.1	0.2	0.2	0.2	0.3	0.3	0.4	0.5	0.5	0.6	0.7	0.7	0.8	0.9	1.1	1.3	1.7	2.0
	300	....	0.1	0.1	0.1	0.1	0.2	0.2	0.2	0.3	0.3	0.4	0.5	0.5	0.6	0.6	0.7	0.8	1.0	1.2	1.5
200	200	....	0.1	0.1	0.1	0.1	0.2	0.2	0.2	0.3	0.3	0.4	0.5	0.5	0.6	0.6	0.7	0.8	1.0	1.2	1.5
	300	....	0.1	0.1	0.1	0.1	0.1	0.2	0.2	0.2	0.3	0.3	0.4	0.4	0.5	0.5	0.6	0.7	0.8	1.0	1.2
300	300	....	....	0.1	0.1	0.1	0.1	0.1	0.2	0.2	0.2	0.3	0.3	0.3	0.4	0.4	0.5	0.5	0.6	0.7	0.8
500	500	....	....	....	....	0.1	0.1	0.1	0.1	0.1	0.1	0.2	0.2	0.2	0.2	0.2	0.3	0.3	0.4	0.5	0.6

Table 19-26. Percent Effective Ceiling or Floor Cavity Reflectance for Various Reflectance Combinations

Ceiling or floor reflectance	90				80				70			50			30				10		
% wall reflectance	90	70	50	30	80	70	50	30	70	50	30	70	50	30	65	50	30	10	50	30	10
0	90	90	90	90	80	80	80	80	70	70	70	50	50	50	30	30	30	30	10	10	10
0.1	90	89	88	87	79	79	78	78	69	69	68	59	49	48	30	30	29	29	10	10	10
0.2	89	88	86	85	79	78	77	76	68	67	66	49	48	47	30	29	29	28	10	10	9
0.3	89	87	85	83	78	77	75	74	68	66	64	49	47	46	30	29	28	27	10	10	9
0.4	88	86	83	81	78	76	74	72	67	65	63	48	46	45	30	29	27	26	11	10	9
0.5	88	85	81	78	77	75	73	70	66	64	61	48	46	44	29	28	27	25	11	10	9
0.6	88	84	80	76	77	75	71	68	65	62	59	47	45	43	29	28	26	25	11	10	9
0.7	88	83	78	74	76	74	70	66	65	61	58	47	44	42	29	28	26	24	11	10	8
0.8	87	82	77	73	75	73	69	65	64	60	56	47	43	41	29	27	25	23	11	10	8
0.9	87	81	76	71	75	72	68	63	63	59	55	46	43	40	29	27	25	22	11	9	8
1.0	86	80	74	69	74	71	66	61	63	58	53	46	42	39	29	27	24	22	11	9	8
1.1	86	79	73	67	74	71	65	60	62	57	52	46	41	38	29	26	24	21	11	9	8
1.2	86	78	72	65	73	70	64	58	61	56	50	45	41	37	29	26	23	20	12	9	7
1.3	85	78	70	64	73	69	63	57	61	55	49	45	40	36	29	26	23	20	12	9	7
1.4	85	77	69	62	72	68	62	55	60	54	48	45	40	35	28	26	22	19	12	9	7
1.5	85	76	68	61	72	68	61	54	59	53	47	44	39	34	28	25	22	18	12	9	7
1.6	85	75	66	59	71	67	60	53	59	53	45	44	39	33	28	25	21	18	12	9	7
1.7	84	74	65	58	71	66	59	52	58	51	44	44	38	32	28	25	21	17	12	9	7
1.8	84	73	64	56	70	65	58	50	57	50	43	43	37	32	28	25	21	17	12	9	6
1.9	84	73	63	55	70	65	57	49	57	49	42	43	37	31	28	25	20	16	12	9	6
2.0	83	72	62	53	69	64	56	48	56	48	41	43	37	30	28	24	20	16	12	9	6
2.1	83	71	61	52	69	63	55	47	56	47	40	43	36	29	28	24	20	16	13	9	6
2.2	83	70	60	51	68	63	54	45	55	46	39	42	36	29	28	24	19	15	13	9	6
2.3	83	69	59	50	68	62	53	44	54	46	38	42	35	28	28	24	19	15	13	9	6
2.4	82	68	58	48	67	61	52	43	54	45	37	42	35	27	28	24	19	14	13	9	6
2.5	82	68	57	47	67	61	51	42	53	44	36	41	34	27	27	23	18	14	13	9	6
2.6	82	67	56	46	66	60	50	41	53	43	35	41	34	26	27	23	18	13	13	9	5
2.7	82	66	55	45	66	60	49	40	52	43	34	41	33	26	27	23	18	13	13	9	5
2.8	81	66	54	44	66	59	48	39	52	42	33	41	33	25	27	23	18	13	13	9	5
2.9	81	65	53	43	65	58	48	38	51	41	33	40	33	25	27	23	17	12	13	9	5
3.0	81	64	52	42	65	58	47	38	51	40	32	40	32	24	27	22	17	12	13	8	5
3.1	80	64	51	41	64	57	46	37	50	40	31	40	32	24	27	22	17	12	13	8	5
3.2	80	63	50	40	64	57	45	36	50	39	30	40	31	23	27	22	16	11	13	8	5
3.3	80	62	49	39	64	56	44	35	49	39	30	39	31	23	27	22	16	11	13	8	5
3.4	80	62	48	38	63	56	44	34	49	38	29	39	31	22	27	22	16	11	13	8	5
3.5	79	61	48	37	63	55	43	33	48	38	29	39	30	22	26	22	16	11	13	8	5

3.6	79	60	47	36	62	54	42	33	48	37	28	39	30	21	26	21	15	10	13	8	5
3.7	79	60	46	35	62	54	42	32	48	37	27	38	30	21	26	21	15	10	13	8	4
3.8	79	59	45	35	62	53	41	31	47	36	27	38	29	21	26	21	15	10	13	8	4
3.9	78	59	45	34	61	53	40	30	47	36	26	38	29	20	26	21	15	10	13	8	4
4.0	78	58	44	33	61	52	40	30	46	35	26	38	29	20	26	21	15	9	13	8	4
4.1	78	57	43	32	60	52	39	29	46	35	25	37	28	20	26	21	14	9	13	8	4
4.2	78	57	43	32	60	51	39	29	46	34	25	37	28	19	26	20	14	9	13	8	4
4.3	78	56	42	31	60	51	38	28	45	34	25	37	28	19	26	20	14	9	13	8	4
4.4	77	56	41	30	59	51	38	28	45	34	24	37	27	19	26	20	14	8	13	8	4
4.5	77	55	41	30	59	50	37	27	45	33	24	37	27	19	25	20	14	8	14	8	4
4.6	77	55	40	29	59	50	37	26	44	33	24	36	27	18	25	20	14	8	14	8	4
4.7	77	54	40	29	58	49	36	26	44	33	23	36	26	18	25	20	13	8	14	8	4
4.8	76	54	39	28	58	49	36	25	44	32	23	36	26	18	25	19	13	8	14	8	4
4.9	76	53	38	28	58	49	35	25	44	32	23	36	26	18	25	19	13	7	14	8	4
5.0	76	53	38	27	57	48	35	25	43	32	22	36	26	17	25	19	13	7	14	8	4

Table 19-27. Multiplying Factors for 10% Effective Floor Cavity Reflectance

(20% = 1.00)

Effective ceiling cavity reflectance, ρ_{cc}	80				70				50			30			10		
% wall reflectance, ρ_w	70	50	30	10	70	50	30	10	50	30	10	50	30	10	50	30	10
Room cavity ratio:																	
1	0.923	0.929	0.935	0.940	0.933	0.939	0.943	0.948	0.956	0.960	0.963	0.973	0.976	0.979	0.989	0.991	0.993
2	0.931	0.942	0.950	0.958	0.940	0.949	0.957	0.963	0.962	0.968	0.974	0.976	0.980	0.985	0.988	0.991	0.995
3	0.939	0.951	0.961	0.969	0.945	0.957	0.966	0.973	0.967	0.975	0.981	0.978	0.983	0.988	0.988	0.992	0.996
4	0.944	0.958	0.969	0.978	0.950	0.963	0.973	0.980	0.972	0.980	0.986	0.980	0.986	0.991	0.987	0.992	0.996
5	0.949	0.964	0.976	0.983	0.954	0.968	0.978	0.985	0.975	0.983	0.989	0.981	0.988	0.993	0.987	0.992	0.997
6	0.953	0.969	0.980	0.986	0.958	0.972	0.982	0.989	0.977	0.985	0.992	0.982	0.989	0.995	0.987	0.993	0.997
7	0.957	0.973	0.983	0.991	0.961	0.975	0.985	0.991	0.979	0.987	0.994	0.983	0.990	0.996	0.987	0.993	0.998
8	0.960	0.976	0.986	0.993	0.963	0.977	0.987	0.993	0.981	0.988	0.995	0.984	0.991	0.997	0.987	0.994	0.998
9	0.963	0.978	0.987	0.994	0.965	0.979	0.989	0.994	0.983	0.990	0.996	0.985	0.992	0.998	0.988	0.994	0.999
10	0.965	0.980	0.989	0.995	0.967	0.981	0.990	0.995	0.984	0.991	0.997	0.986	0.993	0.998	0.988	0.994	0.999

Table 19-28. Multiplying Factors for 30% Effective Floor Cavity Reflectance

(20% = 1.00)

% effective ceiling cavity reflectance, ρ_{cc}	80				70				50			30			10		
% wall reflectance, ρ_w	70	50	30	10	70	50	30	10	50	30	10	50	30	10	50	30	10
Room cavity ratio:																	
1	1.092	1.082	1.075	1.068	1.077	1.070	1.064	1.059	1.049	1.044	1.040	1.028	1.026	1.023	1.012	1.010	1.008
2	1.079	1.066	1.055	1.047	1.068	1.057	1.048	1.039	1.041	1.033	1.027	1.026	1.021	1.017	1.013	1.010	1.006
3	1.070	1.054	1.042	1.033	1.061	1.048	1.037	1.028	1.034	1.027	1.020	1.024	1.017	1.012	1.014	1.009	1.005
4	1.062	1.045	1.033	1.024	1.055	1.040	1.029	1.021	1.030	1.022	1.015	1.022	1.015	1.010	1.014	1.009	1.004
5	1.056	1.038	1.026	1.018	1.050	1.034	1.024	1.015	1.027	1.018	1.012	1.020	1.013	1.008	1.014	1.009	1.004
6	1.052	1.033	1.021	1.014	1.047	1.030	1.020	1.012	1.024	1.015	1.009	1.019	1.012	1.006	1.014	1.008	1.003
7	1.047	1.029	1.018	1.011	1.043	1.026	1.017	1.009	1.022	1.013	1.007	1.018	1.010	1.005	1.014	1.008	1.003
8	1.044	1.026	1.015	1.009	1.040	1.024	1.015	1.007	1.020	1.012	1.006	1.017	1.009	1.004	1.013	1.007	1.003
9	1.040	1.024	1.014	1.007	1.037	1.022	1.014	1.006	1.019	1.011	1.005	1.016	1.009	1.004	1.013	1.007	1.002
10	1.037	1.022	1.012	1.006	1.034	1.020	1.012	1.005	1.017	1.010	1.004	1.015	1.009	1.003	1.013	1.007	1.002

56. Supplementary Lighting. In those cases where the higher footcandle values are required for the more difficult seeing tasks in workplaces and for many home-lighting applications, it is often practical, economical, and desirable to install supplementary lighting equipment. Supplementary lighting, as the name implies, should be used in conjunction with the general lighting system, and when the combination is used for prolonged close work it should be planned to avoid too great a brightness ratio between the work areas and the surroundings. This is usually accomplished when the general lighting system provides illumination of the order of one-third to one-fifth the total footcandles. In merchandising displays a greater range is permissible in order to obtain maximum attention value. On the other hand a difference greater than about 2:1 between the illumination in a showcase and on the counter top may result in a less favorable appearance of goods when brought out for examination.

57. Calculations from Point Sources. The point-by-point method of calculating footcandles is employed for single luminaires when used alone or to supplement general lighting systems. This method is based on the application of the well-known inverse-

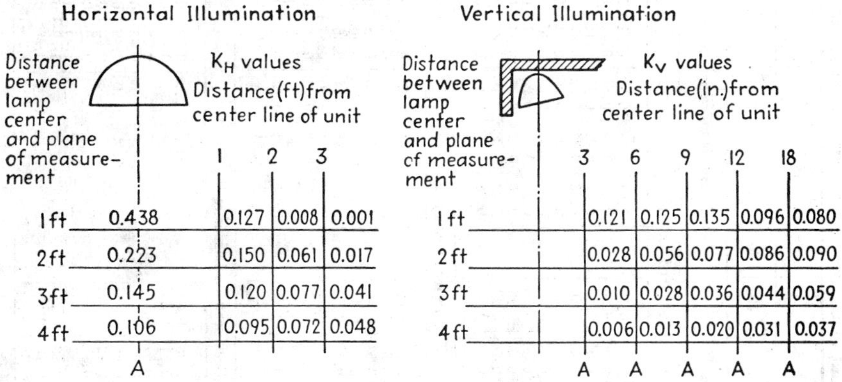

Horizontal Illumination

Distance between lamp center and plane of measurement	A	K_H values, Distance (ft) from center line of unit: 1	2	3
1 ft	0.438	0.127	0.008	0.001
2 ft	0.223	0.150	0.061	0.017
3 ft	0.145	0.120	0.077	0.041
4 ft	0.106	0.095	0.072	0.048

Vertical Illumination

Distance between lamp center and plane of measurement	K_V values, Distance (in.) from center line of unit: 3	6	9	12	18
1 ft	0.121	0.125	0.135	0.096	0.080
2 ft	0.028	0.056	0.077	0.086	0.090
3 ft	0.010	0.028	0.036	0.044	0.059
4 ft	0.006	0.013	0.020	0.031	0.037
	A	A	A	A	A

FIG. 19-26. Line-source data for typical broad- and narrow-distribution luminaires. For lumens-per-foot values, divide lumens by nominal lamp length. See Tables 16-17 to 16-19.

square law (fc = cp/D^2) to data obtained from the candlepower distribution curve for the luminaire. In applying this formula to a horizontal surface the footcandles at any point P are equal to the candlepower directed toward P multiplied by the cosine of the angle A and divided by the square of the distance D from the luminaire [fc = (cp × cos A)/D^2]. In this case A is the angle between the axis of the luminaire and a line from the light center to point P. In many applications the computed footcandles of most interest are those obtained where angle A is zero, and since its cosine then becomes unity, the formula is reduced to its simplest form: fc = (cp/D^2).

58. Calculations from Line Sources. Several technical papers have been published[1] in which the authors have discussed the calculation of footcandles at a given point from a line source. It has been shown that in the special case where one might assume a line source of infinite length the footcandles at a point P varies inversely as the distance D, not inversely as the square of the distance, which is the rule from a point source of light. The formula[2] for a line source of infinite length is fc = (fL × W)/2D, in which fL = brightness (footlamberts), W = width in feet of the line-light source, and D = distance in feet of the point from the light source.

At relatively short distances the approximate initial levels of illumination from line sources in two typical installations can be computed by multiplying the K values in Fig. 19-26 by the lamp lumens per foot.

[1] See IES Lighting Handbook for other formulas and graphical data.
[2] *Ibid.*

Table 19-29a. Coefficient of Utilization* for Six Typical Luminaires

Luminaires	Typical distribution	ρ_{cc}	80				70				50			30			10		
		ρ_w	70	50	30	10	70	50	30	10	50	30	10	50	30	10	50	30	10
		RCR	Coefficients of utilization, zonal-cavity method, ρ_{fc} = 20																
Coefficients for Luminaire 1																			
	67↑ 33↓	1	....	0.72	0.70	0.67	....	0.65	0.63	0.61	0.52	0.51	0.49						
		2	....	0.64	0.59	0.56	....	0.58	0.54	0.51	0.46	0.44	0.42						
		3	....	0.56	0.51	0.47	....	0.51	0.47	0.43	0.41	0.38	0.35						
		4	....	0.50	0.44	0.40	....	0.46	0.41	0.37	0.37	0.34	0.31						
		5	....	0.45	0.39	0.34	....	0.41	0.36	0.32	0.33	0.29	0.27						
		6	....	0.40	0.34	0.30	....	0.37	0.31	0.28	0.30	0.26	0.23						
		7	....	0.36	0.30	0.26	....	0.33	0.28	0.24	0.27	0.23	0.20						
		8	....	0.33	0.27	0.23	....	0.30	0.25	0.21	0.25	0.21	0.18						
		9	....	0.30	0.24	0.20	....	0.27	0.22	0.19	0.22	0.18	0.16						
		10	....	0.27	0.21	0.18	....	0.25	0.20	0.16	0.20	0.17	0.14						
Coefficients for Luminaire 2																			
	55↑ 45↓	1	....	0.74	0.71	0.69	....	0.67	0.65	0.63	0.56	0.54	0.53						
		2	....	0.65	0.61	0.57	....	0.60	0.56	0.53	0.50	0.47	0.45						
		3	....	0.58	0.53	0.49	....	0.53	0.49	0.45	0.45	0.41	0.39						
		4	....	0.52	0.46	0.42	....	0.48	0.43	0.39	0.40	0.36	0.34						
		5	....	0.47	0.40	0.36	....	0.43	0.38	0.34	0.36	0.32	0.29						
		6	....	0.42	0.36	0.31	....	0.39	0.33	0.30	0.33	0.29	0.26						
		7	....	0.38	0.32	0.27	....	0.35	0.30	0.26	0.29	0.25	0.22						
		8	....	0.34	0.28	0.24	....	0.31	0.26	0.23	0.27	0.23	0.20						
		9	....	0.31	0.25	0.21	....	0.28	0.23	0.20	0.24	0.20	0.17						
		10	....	0.28	0.23	0.19	....	0.26	0.21	0.18	0.22	0.18	0.16						
Coefficients for Luminaire 3																			
	25↑ 75↓	1	....	0.89	0.86	0.83	....	....	....	....	0.78	0.76	0.74	....	....	....	0.65	0.64	0.63
		2	....	0.79	0.74	0.69	....	....	....	....	0.69	0.66	0.63	....	....	....	0.58	0.56	0.54
		3	....	0.70	0.64	0.59	....	....	....	....	0.62	0.57	0.54	....	....	....	0.53	0.50	0.47
		4	....	0.62	0.56	0.51	....	....	....	....	0.55	0.50	0.46	....	....	....	0.47	0.44	0.41
		5	....	0.55	0.48	0.43	....	....	....	....	0.49	0.44	0.40	....	....	....	0.42	0.38	0.35
		6	....	0.50	0.42	0.37	....	....	....	....	0.44	0.39	0.34	....	....	....	0.38	0.34	0.31
		7	....	0.45	0.37	0.33	....	....	....	....	0.40	0.34	0.30	....	....	....	0.34	0.30	0.27
		8	....	0.40	0.33	0.28	....	....	....	....	0.36	0.30	0.26	....	....	....	0.31	0.27	0.24
		9	....	0.36	0.29	0.24	....	....	....	....	0.32	0.27	0.22	....	....	....	0.28	0.23	0.20
		10	....	0.33	0.26	0.21	....	....	....	....	0.29	0.24	0.20	....	....	....	0.25	0.21	0.18

Coefficients for Luminaire 4

		RCR																	
	0↑ 100↓	1	0.86	0.84	0.82	0.79	0.84	0.81	0.79	0.77	0.77	0.75	0.74	0.73	0.72	0.71	0.70	0.69	0.68
		2	0.81	0.77	0.73	0.70	0.79	0.75	0.71	0.69	0.71	0.69	0.66	0.68	0.66	0.64	0.65	0.63	0.62
		3	0.76	0.70	0.66	0.62	0.74	0.69	0.65	0.61	0.66	0.63	0.60	0.63	0.61	0.58	0.61	0.59	0.57
		4	0.71	0.64	0.59	0.56	0.69	0.63	0.59	0.55	0.61	0.57	0.54	0.58	0.55	0.52	0.56	0.54	0.51
		5	0.67	0.59	0.54	0.50	0.65	0.58	0.53	0.49	0.56	0.52	0.49	0.54	0.50	0.48	0.52	0.49	0.47
		6	0.63	0.55	0.49	0.45	0.61	0.54	0.49	0.45	0.52	0.47	0.44	0.50	0.46	0.44	0.49	0.45	0.43
		7	0.59	0.50	0.45	0.41	0.57	0.49	0.44	0.41	0.48	0.43	0.40	0.46	0.42	0.39	0.45	0.41	0.39
		8	0.55	0.46	0.41	0.37	0.54	0.45	0.40	0.37	0.44	0.40	0.36	0.43	0.39	0.36	0.41	0.38	0.35
		9	0.51	0.43	0.37	0.34	0.50	0.42	0.37	0.33	0.41	0.36	0.33	0.40	0.35	0.33	0.38	0.35	0.32
		10	0.47	0.38	0.32	0.29	0.46	0.37	0.32	0.29	0.36	0.31	0.28	0.35	0.31	0.28	0.34	0.30	0.27

Coefficients for Luminaire 5

		RCR																	
	0↑ 100↓	1	0.73	0.70	0.68	0.66	0.71	0.68	0.67	0.65	0.66	0.64	0.63	0.63	0.62	0.61	0.61	0.60	0.59
		2	0.67	0.63	0.59	0.56	0.66	0.62	0.58	0.56	0.59	0.57	0.54	0.57	0.55	0.53	0.55	0.54	0.52
		3	0.62	0.57	0.52	0.49	0.61	0.56	0.52	0.48	0.54	0.50	0.47	0.52	0.49	0.47	0.51	0.48	0.46
		4	0.58	0.51	0.46	0.43	0.57	0.50	0.46	0.42	0.49	0.45	0.42	0.47	0.44	0.41	0.46	0.44	0.41
		5	0.53	0.46	0.41	0.37	0.52	0.45	0.40	0.37	0.44	0.40	0.36	0.43	0.39	0.36	0.41	0.38	0.36
		6	0.50	0.42	0.36	0.33	0.48	0.41	0.36	0.32	0.40	0.35	0.32	0.39	0.35	0.32	0.38	0.34	0.32
		7	0.46	0.38	0.32	0.29	0.45	0.37	0.32	0.29	0.36	0.32	0.28	0.35	0.31	0.28	0.34	0.31	0.28
		8	0.42	0.34	0.29	0.25	0.41	0.33	0.28	0.25	0.32	0.28	0.25	0.32	0.28	0.25	0.31	0.27	0.24
		9	0.39	0.31	0.25	0.22	0.38	0.30	0.25	0.22	0.29	0.25	0.22	0.29	0.24	0.21	0.28	0.24	0.21
		10	0.36	0.28	0.23	0.19	0.36	0.27	0.23	0.19	0.27	0.22	0.19	0.26	0.22	0.19	0.25	0.22	0.19

Coefficients for Luminaire 6

		RCR																	
	0↑ 100↓	1	....	0.98	0.96	0.95	....	....	....	....	0.92	0.91	0.90	....	....	....	0.87	0.86	0.85
		2	....	0.94	0.91	0.89	....	....	....	....	0.89	0.87	0.86	....	....	....	0.85	0.84	0.83
		3	....	0.90	0.87	0.85	....	....	....	....	0.87	0.85	0.83	....	....	....	0.83	0.82	0.80
		4	....	0.87	0.83	0.81	....	....	....	....	0.84	0.81	0.80	....	....	....	0.81	0.79	0.78
		5	....	0.83	0.80	0.77	....	....	....	....	0.81	0.78	0.76	....	....	....	0.79	0.77	0.75
		6	....	0.81	0.77	0.75	....	....	....	....	0.79	0.76	0.74	....	....	....	0.77	0.75	0.73
		7	....	0.78	0.74	0.72	....	....	....	....	0.76	0.73	0.71	....	....	....	0.74	0.72	0.70
		8	....	0.75	0.72	0.69	....	....	....	....	0.74	0.71	0.69	....	....	....	0.72	0.70	0.68
		9	....	0.73	0.69	0.67	....	....	....	....	0.72	0.68	0.66	....	....	....	0.70	0.68	0.66
		10	....	0.70	0.67	0.64	....	....	....	....	0.69	0.66	0.64	....	....	....	0.68	0.66	0.64

* ρ_{cc}—percent effective ceiling-cavity reflectance.
ρ_w—percent wall reflectance.
ρ_{fc}—percent effective floor-cavity reflectance is 20 in above table. See Tables 19–27 and 19–28 for multiplying factors for ρ_{fc} at 10 and 30%.
RCR—room-cavity ratio.

Table 19-29b. Characteristics of Six Typical Luminaires of Table 19-29a

Luminaire	Description	Maintenance factors			Maximum* spacing
		Good	Med.	Poor	
1	Semi-indirect luminaire with two 1,500-mA lamps	0.70	0.60	0.50	1.2 × CH
2	General diffuse luminaire with two 430-, 800-, or 1,500-mA lamps; 45°C × 45°L shielding	0.75	0.70	0.65	1.1 × MH
3	Semidirect industrial luminaire with two 430-, 800-, or 1,500-mA lamps; 30° crosswise shielding	0.70	0.65	0.60	1.0 × MH
4	Direct industrial high-bay luminaire with aluminum or glass reflector, 400-W color-improved mercury lamp	0.65	0.60	0.55	1.0 × MH
5	Direct 24-in-wide troffer with two 430-mA lamps, prismatic low-brightness lens	0.70	0.65	0.55	0.9 × MH
6	Direct reflector-down light for R-40 flood lamp†	0.70	0.65	0.60	0.5 × MH

* MH—mounting height above floor.
CH—ceiling height.
† Coefficients of utilization are based on the lumen output of reflectorized lamps.

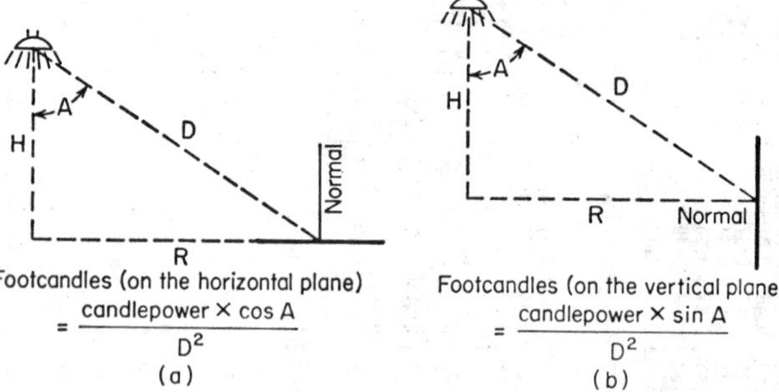

Fig. 19-27. Fundamental relationships for point-by-point calculations where the inverse-square law applies.

59. Structural Lighting. Lighting systems which form a substantial part of the structure of a building, as distinguished from individual or groups of luminaires suspended from the ceiling or bracketed to walls, are examples of structural lighting. The design of such systems was considerably stimulated by the advent of the fluorescent lamp in 1938, principally because of three of the lamp's characteristics: (1) Its tubular form made it relatively simple to incorporate flat lighting elements with heating and ventilating ducts, sprinkler systems, and other concealable services in the ceiling structure. (2) The relatively low lumens per fluorescent lamp necessitated more lamps to produce the lighting levels required; the resulting larger mass of the luminaires produced large-area low-brightness elements; these, when well handled, effected considerably improved illumination, especially in

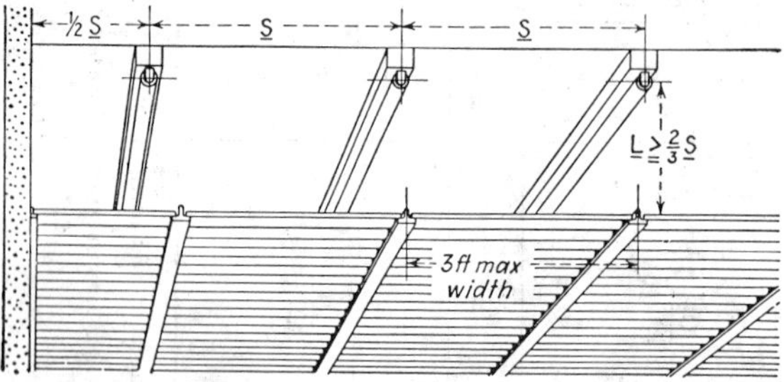

Fig. 19-28. Luminous ceiling.

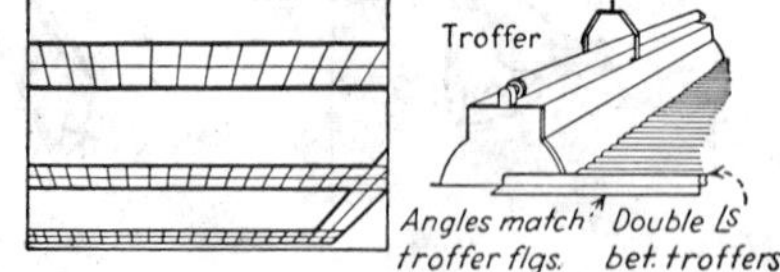

Fig. 19-29. Luminous ceiling with troffers.

the distribution of light over the area. (3) The low radiant heat of the fluorescent lamp made less demand per footcandle on cooling systems, and high levels (100 fc or more) of general lighting were economically possible.

The **luminous ceiling** of suspended plastic or other material is an outstanding form (Fig. 19-28). Other designs utilize horizontal sheets of corrugated or louvered plastic with vertical spacers forming acoustical baffles. In the sketch, recommended spacing of lamp rows and distance to diffusing media are shown. For a build-up of levels at desks, business machines, or other work surfaces, the combination of a **luminous ceiling with louvered troffers** (Fig. 19-29) is illustrated. The horizontal intensity produced by using this or similar combinations can be in hundreds of footcandles. An acceptable design of a **soffit** using fluorescent luminaires is shown in Fig. 19-30. The illustration shows its application at a laboratory sink; however, other horizontal surfaces near walls where more critical seeing is done could be similarly lighted. The **open canopy** (Fig. 19-30) produces about 20 fc of direct lighting, with an upward component for general space lighting. Other forms include additional downlights (incandescent) for highlighting furniture groups or merchandise displays. A commendable **cornice-and-cove** plan is shown in Fig. 19-32. For a public dining room or a residence with a wall having ceiling-to-floor draperies, it provides light to the ceiling and down on the draperies in a pleasing manner.

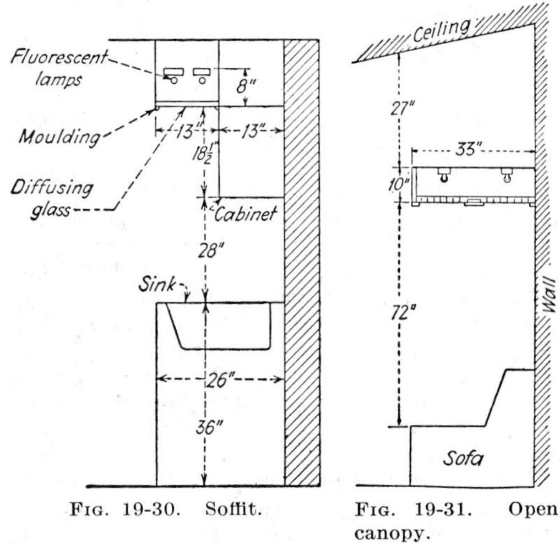

FIG. 19-30. Soffit.

FIG. 19-31. Open canopy.

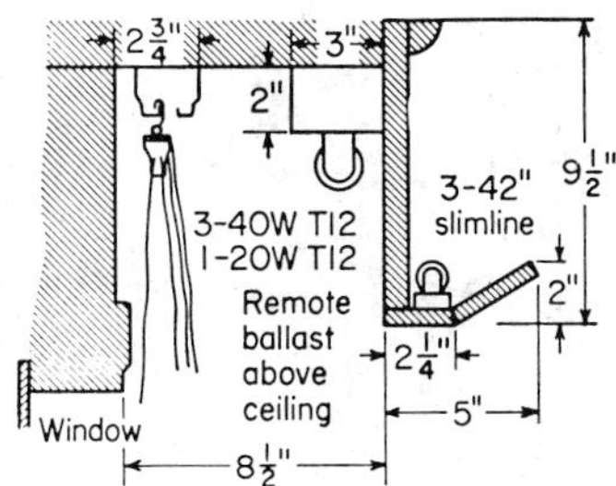

FIG. 19-32. Cornice and cove.

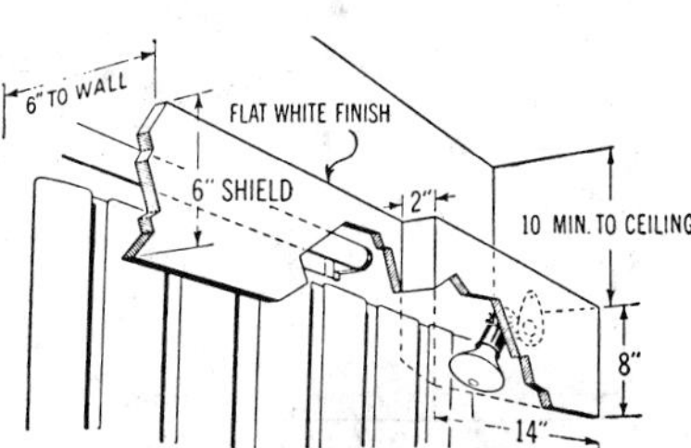

FIG. 19-33. Valance and spotlight.

Lamps of relatively low brightness are recommended in the cove (slim-line lamps operated at 120 mA). Reflecting surfaces are painted white. Other wall-lighting methods include the **valance-with-spotlight** plan as shown in Fig. 19-33. The spot-

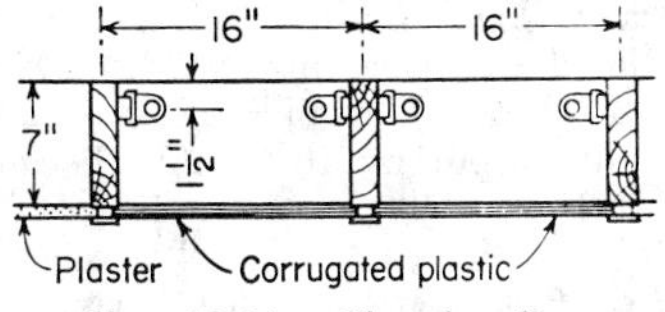

FIG. 19-34. Closed coffer.

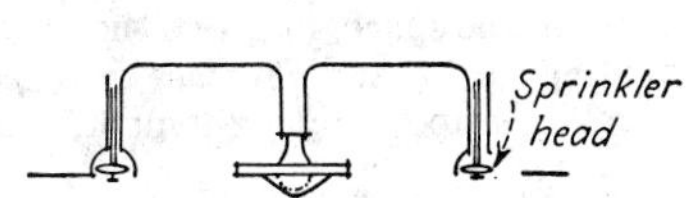

FIG. 19-35. Open coffer (incandescent).

lights in the end sections are separately controlled. They are used to add supplementary light to a sewing machine or similar point of critical visual demand. In the **closed coffer** (Fig. 19-34), the ceiling cavity is designed to fit the needs of the space to be lighted. The dimensions of the luminaire are 30 by 96 by 8 in for a room 7 by 11 ft. White corrugated vinyl plastic is the diffusing medium. Illuminated **open coffers** (Figs. 19-35 and 19-36) are equally acceptable in contemporary and classical interiors. Improved maintenance is effected by covering the concealed luminaire elements in the four sides with clear glass. Long, narrow coffers or free-form examples can be augmented by suspended incandescent luminaires for a combination of style, color effect, and sculptural distinction. The **shelf-valance** form (Fig. 19-37) is a space saver and is applicable to small desks, as in residences, or to longer wall-hung applications, as in a hospital corridor, where the staff would use it to complete reports, telephone, etc. If the shelf were of transparent material, a bulletin board above it would be illuminated. By using white glass for the valance board (instead of plywood, as shown), the surface could serve as the background for a departmental sign in a shop or supermarket. The **room divider** (Fig. 19-38) is representative of residence use in the contemporary interior. It is also found in foyers in public buildings and many other interiors where unobtrusive divisions of space are desirable. It can serve as a storage cabinet and planter base; the shelves are adaptable to a wide variety of uses. The fluorescent ballasts are concealed below the bottom shelf.

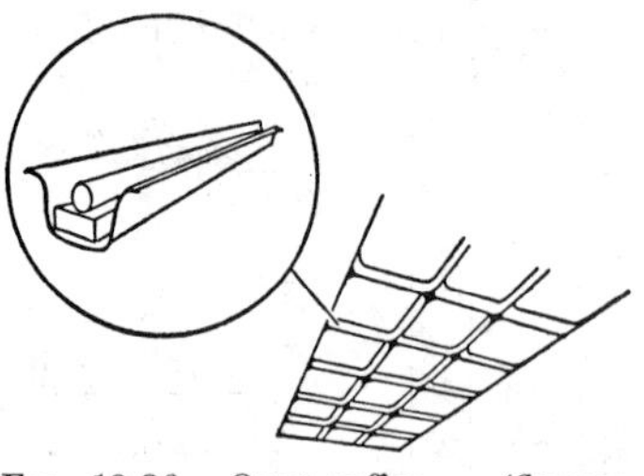

Fig. 19-36. Open-coffer (fluorescent).

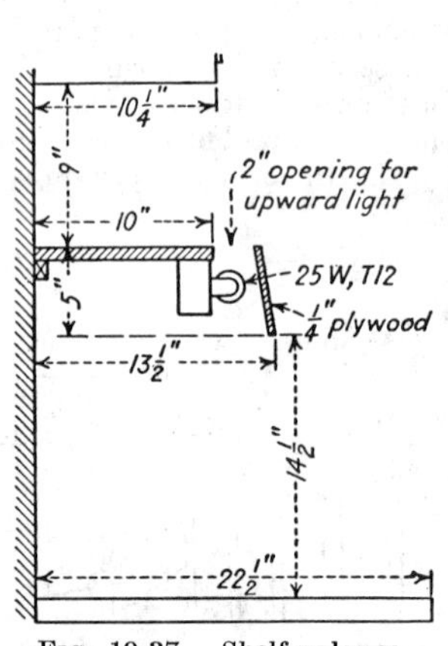

Fig. 19-37. Shelf valance.

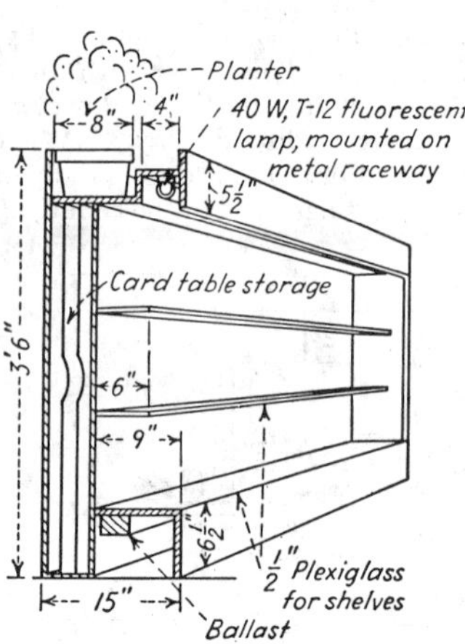

Fig. 19-38. Room divider.

60. Heat and Light.[1] Lighting at today's recommended levels has the capability of providing much of the heat needed in commercial buildings. Electrical space conditioning permits control of lighting heat, making it serve more usefully the varying heating needs of different parts of a building. At the same time, it provides for more efficient handling of the lighting heat during periods when cooling is needed.

Electrical space conditioning has the potential to lower both initial and operating costs for air handling compared with conventional methods, make available more usable and rentable space, improve the visual and thermal comfort of building occupants, reduce heating costs, and increase fluorescent light output by 10 to 15% because of more favorable lamp-operating temperatures.

[1] Fisher, W. S. "Electrical Space Conditioning"; Nela Park, Cleveland, Ohio, General Electric Co., 1964.

For an average multistory building, when the outside temperature is 0°F, the building heat loss is about 10 W/ft^2 of floor space. These losses can be offset by internal heat gains such as are produced by office occupants, who provide about 1 W/ft^2; electronic business machines, which provide about 1½ W/ft^2; and a modern 150-fc lighting system, which can provide about 8 W/ft^2. This adds up to slightly more than the building heat loss at 0°F. Only for lower temperatures or in periphery with outside wall and window exposures will supplementary heating be needed.

There are three essential factors which need to be considered if electrical space conditioning is to be successful.

First and foremost is a modern lighting system designed to current standards which will provide a high degree of visual comfort and good visual performance for building occupants. The heat generated by the higher lighting levels is energy that has already been paid for and that can be used to advantage.

The second consideration is the use of lighting components which can bring the lighting heat under control before it enters the occupied space. One of the most feasible methods of doing this is to return the room air through the luminaires, where 50 to 80% of the lighting heat is available for control. A major advantage of controlling the lighting heat at the luminaires is the reduction in air changes needed for an air-conditioned space. Since the air changes for a room are determined by the heat gains in the space, picking up 50% or more of the lighting heat before it enters the space can reduce supply-duct sizes and blower horsepower requirements, resulting in lower installation and operating costs for the air handling system.

The removal of heat from fluorescent luminaires increases light output and improves lamp efficiency. For example, a T-12 fluorescent lamp operating at 430 mA in still air at 100°F ambient produces 12% less light than the same lamp at 77°F. Since many modern office lighting systems produce ambients within the luminaire of 100°F and higher, the removal of this heat allows lamps to operate closer to their maximum luminous capabilities. Thermal comfort is improved by returning air through the lighting system, which reduces luminaire surface temperatures and, in turn, the radiant effects of these luminaires on the occupied space.

The third factor to be considered in electrical space conditioning is the installation of the proper heating and cooling devices to provide a coordinated system. Conventional heating and air-conditioning installations use major mechanical components such as an air-conditioning unit for cooling, fuel-fired furnace for heating, possibly facilities for fuel storage (depending on fuel), and a chimney to exhaust the combustion products. In addition, a cooling tower would be part of most building systems with central air conditioning. With electrical space conditioning, the furnace, fuel storage tank, and chimney are not required, a saving in floor space and capital investment.

In many situations a heat pump is feasible to move the heat from the center of the building to the periphery, or to storage, for later use. During the coldest winter days, when building heat losses may exceed the heat generated by the occupants and the lighting system, additional energy can be supplied through the heat pump by taking heat from a source such as well water or air in milder climates. In many locations during the early spring, late fall, and much of the winter, the building will have a net heat gain during occupied periods but will require some heating at night and over the weekends, when the building is unoccupied and the lighting is normally off. Under these conditions, a heat-pump system can transfer the excess heat supplied during the day to water storage, where it can be utilized at night and on weekends. An engineering study of each installation is required to determine the feasibility of a heat and light system and whether a heat pump or supplementary resistance heating in peripheral areas will prove more satisfactory and economical.

A diagrammatic representation of one such heat and light installation using air-handling troffers is shown in Fig. 19-39.

61. Street Lighting. Increasing motor-vehicle speeds and increasing congestion on streets and highways are bringing about requirements for higher levels of street and highway lighting.

Actuarial and safety experts studying the statistics have demonstrated that the provision of higher illumination on roadways is associated with the reduction in night accidents and thereby represents economy for the community, quite aside from the

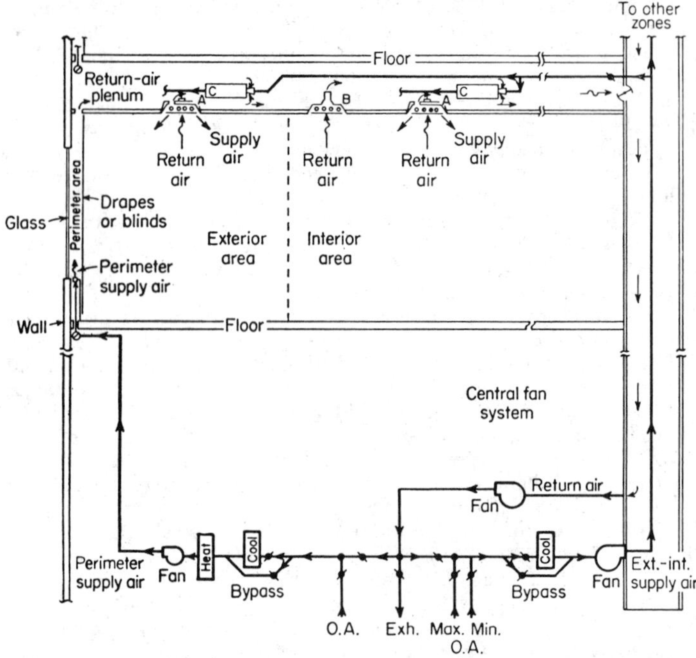

FIG. 19-39. Heat and light system. *A*, combination heat-extractor and supply-air light fixtures; *B*, heat-extractor light fixtures; *C*, venturi-type mixing units. (*From R. D. Schmidt, "Air Design Considerations for the Barber-Colman Heat-of-light Systems*; *Barber-Colman Co.*, 1963.)

humanitarian aspects of accident avoidance. As recognition of this fact brings about installations of higher illumination, knowledge of the impetus that they give toward traffic safety is spreading. Fixed lighting of streets and highways is therefore being modernized and extended at an accelerating rate, and many of the newer installations utilize mercury lamps (Table 19-11). The use of fluorescent luminaires is also expanding rapidly. Modern equipment design and application are improving seeing comfort as well as visibility.

The leading authority in this country on the technique of street and highway lighting is the Committee on Street and Highway Lighting of the IES. The USAS Standard Practice for Roadway Lighting, latest edition, is the modern guide in this field.

FLOODLIGHTING DESIGN AND PROCEDURE

62. Beam-lumens Method. All floodlighting-design methods include certain approximations, based on experience. In floodlighting systems containing a large number of luminaires, a detailed study of aiming diagrams and many calculations are usually required. A procedure which is useful for designing simpler systems is called the "beam-lumens" method. This method requires the solution of the two formulas (A and B), as discussed in the following paragraphs, and the coordination of the results.

In many locations in which floodlighting is proposed, there are some basic dimensions that can be assumed to be already fixed. For example, in ground-area floodlighting, the designer is usually able to locate points where the equipment should logically be placed, such as on nearby buildings, along high banks or fences, or on poles or towers. These locations establish the approximate perpendicular distance D from the floodlight to the plane of the surface to be lighted and the average aiming angles. They also guide the choice of floodlight type—narrow, medium, or broad-beam as listed in Table 19-31. In like manner the choice of equipment for lighting vertical surfaces can be

Table 19-30. Typical Recommended Illumination Levels for Floodlighting, with Recommended Coverage Factors

Area	Footcandles	Minimum coverage factor
Building:		3–4
General construction	10	
Excavation work		
Building exteriors and monuments:		2
Floodlighted		
Bright surroundings		
Light surfaces	15	
Medium-light surfaces	20	
Medium-dark surfaces	30	
Dark surfaces	50	
Dark surroundings		
Light surfaces	5	
Medium-light surfaces	10	
Medium-dark surfaces	15	
Dark surfaces	20	
Bulletin and poster boards:		1–2
Bright surroundings		
Light surfaces	50	
Dark surfaces	100	
Dark surroundings		
Light surfaces	20	
Dark surfaces	50	
Parking lots	5	2
Protective lighting:		2
Boundaries		
General lighting technique (nonisolated)	0.20	
Entrances		
Active (pedestrian and/or conveyance)	5	
Inactive (normally locked, infrequently used)	1	
Vital locations or structures	5	
Building surrounds	1	
Active shipping-area surrounds	5	
Storage areas—active	20	
Storage areas—inactive	1	
Loading and unloading platforms	20	
General inactive areas	0.20	
Service station (at grade)		3–4
Dark surrounding		
Approach	1.5	
Driveway	1.5	
Pump—island area	20	
Building faces (exclusive of glass)	10[r]	
Service areas	3	
Light surrounding		
Approach	3	
Driveway	5	
Pump—island area	30	
Building faces (exclusive of glass)	30[r]	
Service areas	7	

obtained by taking *D* as the horizontal distance from the luminaire to the plane in which the vertical surface is located.

The average aiming angle is measured from the perpendicular to the beam-axis line (Fig. 19-40). In a perimeter system in which the floodlights are mounted along or beyond the perimeter of an area, they will, of course, be aimed at various angles,

Table 19-31. Floodlight Beam Spread for Typical Areas

Type of area	Approx. distance to area	Beam spread
Buildings 2-3 story lighted from curb posts	10′–30′	Broad, also Fresnel type (wide spread)
Buildings lighted from across street:		
Areas 3,000 sq ft or less	50′–100′	Broad or medium
Areas 3,000 to 10,000 sq ft	50′–100′	Medium or narrow
Construction work:		
Parking spaces	At perimeter	Broad or medium
Baseball, football	Behind bleachers	Broad and medium

but the average aiming angle used in computation is measured between the perpendicular and the center line of the area to be lighted. When floodlights are on poles along the center line of an area, the average aiming angle is measured between the pole (perpendicular) and a point halfway to the boundary (one-fourth of the width of the total area).

Design Formulas and Procedure. Formula A

$$\text{Floodlights needed for coverage} = \frac{(\text{area lighted}) \times (\text{coverage factor})}{\text{beam-spot area}}$$

In this formula, *area* is the area to be lighted in square feet. This may be either a horizontal surface or a vertical one.

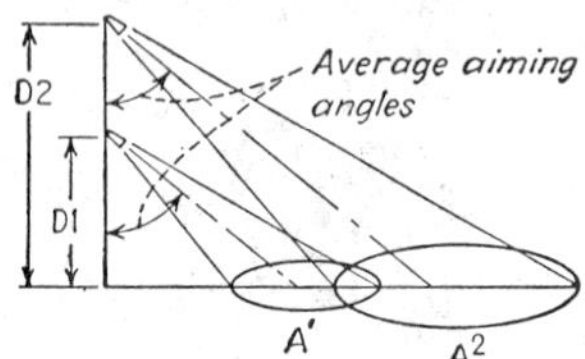

FIG. 19-40. Spot areas (for the same beam spread and aiming angle) vary as the square of the distance *D*. Spot length *L* and spot width *W* vary as the distance *D*. The spot area may be determined from $\pi LW/4$.

The **coverage factor** indicates the minimum number of directions from which each point in the area should be lighted, depending upon the use of the area. A coverage factor of 1 is acceptable in some applications, although in such systems one or two lamp burnouts might temporarily leave large, dark patches. Coverage factors greater than 1 therefore add desirable safety factors.

For example, a coverage factor of 2 is necessary for parking spaces and for protective lighting to reduce the effect of shadows between automobiles, rows of freight cars, piles of material, and similar bulky objects (see Table 19-30 for other recommended values).

The **beam-spot areas** at a 100-ft distance *D* in formula A are given in Table 19-33 for various beam spreads and aiming angles of usual equipment having symmetrical candlepower distribution. In this table, *D* is the perpendicular distance measured from the floodlight to the plane of the lighted surface. *L* and *W* are the lengths and widths of the ellipses formed when floodlights are aimed at an angle to the lighted surface. At 0° the area is assumed to be circular; at other angles, it is elliptical. At other distances and spreads and for similar beam spreads and aiming angles, the spot areas vary as the **square** of the distance *D*, while *L* and *W* values vary as the distance *D*. For example, if *D* is 80 ft and a 30° beam-spread floodlighting unit is aimed at a 50° angle, the elliptical spot area as computed from Table 19-33 will be $(80)^2/(100)^2 \times 9{,}978$, or 6,386 ft². Likewise, the length *L* of the ellipse will be $^{80}/_{100} \times$ 144.4, or 115.5 ft, and the width *W* will be $^{80}/_{100} \times 87.96 = 70.4$ ft.

Table 19-32. Data on Typical Floodlighting Equipment*

Floodlight watts	Lamp	Beam spread	Beam lumens
500	†	25°	3,900
		60°	5,300
		150° × 50°	5,300
1,000	†	15°	9,500
		55°	9,200
		120° × 25°	9,200
1,500	†	15°	12,300
		65°	17,200
		148°	19,600
400	BT-37‡	142°	13,900
1,000	BT-56‡	150° × 114°	28,300
500	T-3§	75° × 20°	4,500
	T-3§	90° × 75°	5,000
500	PAR56NS¶	15° × 30°	3,500
	PAR56WF¶	35° × 65°	5,600
1,000	PAR64NS¶	15° × 30°	8,500
	PAR64WF¶	45° × 70°	13,500

* See Table 19-5 for additional listing of reflector and projector lamps.
† Incandescent.
‡ Color-improved mercury.
§ Quartz-iodine, tubular.
¶ Quartz-iodine, projector.

Formula B

Floodlights needed for footcandles

$$= \frac{(\text{footcandles}) \times (\text{area lighted})}{(\text{maintenance factor}) \times (\text{utilization factor}) \times (\text{beam lumens})}$$

For this formula, typical **footcandle** recommendations are given in Table 19-30. The **maintenance factor** allows for dust and dirt and normal lamp depreciation. This is found under average conditions to be about 0.7. However, it may be as low as 0.3 for extremely dirty locations, where dust, dirt, and smoke are frequently suspended in the air.

The **utilization factor** is the ratio of the lumens effectively lighting an area to the beam lumens, and it can be estimated from the following conditions:

1. If half or more than half of the floodlights are aimed so that all their beam lumens fall within an area, the overall utilization factor will be about 0.75.
2. If one-quarter to one-half of the floodlights are aimed so that all their beam lumens fall within an area, the overall utilization factor will be about 0.60.
3. If fewer than one-quarter of the floodlights can be aimed so that their beam lumens fall within an area, the overall utilization factor is likely to be not more than 0.40.

Most floodlights and projector- and reflector-type lamps as listed in the manufacturers' catalogues are rated in beam lumens. These lumen ratings usually include only the light flux in that part of the beam in which the candlepower values are 10% or more of the maximum candlepower of the floodlight. Typical beam-lumen values are given in Table 19-32.

Table 19-33. Spot Areas for Narrow-, Medium-, and Broad-beam Floodlights at a 100-ft Distance*

Aiming angle	15-deg beam, narrow			30-deg beam, medium			50-deg beam, broad		
	Spot area	*L*	*W*	Spot area	*L*	*W*	Spot area	*L*	*W*
0	545	26.34	26.34	2,250	53.58	53.58	6,830	93.26	93.26
10	570	27.16	26.70	2,370	55.38	54.49	7,220	96.81	95.00
15	606	28.25	27.30	2,518	57.70	55.56	7,760	101.54	97.30
20	657	29.89	28.00	2,757	61.27	57.29	8,600	108.75	100.7
25	735	32.18	29.1	3,102	66.28	59.58	9,880	119.18	105.5
30	846	35.31	30.5	3,603	73.21	62.67	11,770	134.06	111.8
35	1,000	39.57	32.3	4,333	82.78	66.64	14,710	155.58	120.4
40	1,230	45.42	34.6	5,420	96.18	71.75	19,500	187.66	132.3
45	1,583	53.59	37.6	7,129	115.46	78.62	27,900	238.35	149.0
50	2,115	65.34	41.3	9,978	144.43	87.96	44,810	326.58	174.7
55	3,043	82.97	46.7	15,160	190.84	101.14	87,140	509.39	217.8
60	4,720	111.10	54.1	25,880	273.21	120.6	265,100	1072.99	314.6
65	8,165	160.19	64.9	54,480	447.94	154.85			
70	16,800	258.97	82.6	180,910	1000.2	230.3			

* The spot area for any other distance can be computed by multiplying the area in this table at the selected aiming angle by D^2 and dividing by 10,000.

As a general rule, it is wiser to design a system with a small number of large filament-lamp floodlights rather than a large number of small ones. This makes a simpler system to install, to control, and to maintain. Another reason is that in general the larger the lamp the more efficient it is. Also from a control-of-light point of view it is desirable to choose a floodlighting unit having as narrow a beam spread as can be used and still maintain the coverage-factor requirements.

The 1,500-W floodlight in Table 19-32 sets a practical upper limit to size. It should be remembered, however, that large floodlights are hard to conceal; this is important where their daytime appearance may be objectionable architecturally.

To solve formula B, after choosing the desired footcandles and estimating the maintenance and utilization factors, a size of floodlight is chosen for trial calculation and its beam lumens (Table 19-32) are substituted in the equation.

When the dimensions or shape of an area points to the use of several types of floodlights, with different beam spreads, it is customary to divide the area into sections and plan a system for each of them. Buildings with setbacks are typical examples, also very tall structures such as towers or monuments. In setback buildings, you would design one setback at a time, selecting the type of floodlight most suited for each. With towers or monuments, a similar approach is in order.

Floodlighting Design Problem. Assume that an area 200 by 200 ft is to be lighted to 5 fc. Also assume that it is located between two long buildings each 60 ft high, on top of which the floodlighting luminaires can be placed. By using a scale drawing to represent the luminaires placed 60 ft high and aimed toward the center line of the 200-ft-wide work space, a protractor will show an aiming angle of about 60°. For trial-computation purposes it is best to start with the assumption that a narrow-beam-spread floodlighting unit will serve the coverage requirements and then change to a wider beam if found desirable. Hence, for this work space the lighting-design procedure is as follows:

1. From Table 19-33 the area which can be lighted for a 15° floodlight mounted 60 ft high at an aiming angle of 60° is

$$\frac{(60)^2}{(100)^2} \times 4{,}720 = 1{,}700 \text{ ft}^2$$

2. Solving formula A,

$$\text{No. of 15° units needed for coverage} = \frac{(\text{area}) \times (\text{coverage factor})}{\text{spot area}}$$

$$= \frac{40{,}000 \times 3}{1{,}700} = 70$$

3. Solving formula B. A maintenance factor of 0.3 is assumed for this dirty location, and a utilization factor of 0.7 may be assumed for narrow-beam floodlights in this area. If a 1,000-W unit is selected for a trial computation, Table 19-32 shows that it may have as high as 9,500 beam lm. Then the number of

$$\text{1,000-W floodlights needed for footcandles} = \frac{(\text{footcandles}) \times (\text{area})}{\text{MF} \times \text{UF} \times \text{beam lumens}}$$

$$= \frac{5 \times 40{,}000}{0.3 \times 0.7 \times 9{,}500} = 100$$

4. Since the number of floodlights (100) in formula B is greater than the number needed for adequate coverage, it could be concluded that, if 50 floodlight units are conveniently spaced on top of each building along the work area, a satisfactory 5-fc lighting installation would be provided. On the other hand, if a minimum number of floodlights are desired, formula B can be re-solved on the basis of using 1,500-W luminaires having 12,300 beam lm (Table 19-32). In this case,

$$\text{Number needed} = \frac{5 \times 40{,}000}{0.3 \times 0.7 \times 12{,}300} = 77$$

Hence, this construction-work area can be satisfactorily lighted with 5 fc maintained in service if thirty-nine 1,500-W 15° floodlights are well distributed along each side of the work area and placed on top of the two 60-ft buildings.

As seen from the foregoing example, it is possible to use a fewer number of higher-wattage luminaires when the number of units (formula B) to provide the required footcandles is considerably greater than the number required for adequate coverage (formula A). On the other hand, when the trial computations show that the number of luminaires to provide the required footcandles is less than those for adequate coverage, it then becomes necessary to recalculate formula B by using beam lumens from smaller-sized units, until one is found which brings the answer equal to or greater than that for formula A. In other words, the answer to formula B should preferably never be less than that for formula A in order to provide adequate illumination as well as satisfactory coverage.

ECONOMICS OF LIGHTING

63. The Value of Lighting. The value of good lighting depends upon its use in the various fields of application.

Reports from a variety of industrial plants have indicated that better lighting created more satisfactory working conditions which in many instances meant an increase in production, reduced spoilage, fewer accidents, and less labor turnover. Likewise, in the store and other selling areas, good lighting properly applied has been recognized as a necessity, not only to permit the customer properly to inspect the merchandise but also to direct his attention to other items and thereby increase the volume of sales. Moreover, light and lighting are an integral part of many modern buildings where the architect has incorporated luminous elements for their functional use and aesthetic value.

For utilitarian installations which have resulted in an increase in factory production or an increase in sales in the store, a dollar-and-cents value can readily be given to better lighting. On the other hand, there is wise acceptance of light and lighting for its humanitarian and decorative aspects, which are of inestimable value.

64. Maintenance of Lighting. Good lighting maintenance is good economics; i.e., good maintenance assures the user that he actually gets the light he pays for. With a well-maintained lighting system, the user also gets the better seeing conditions that the system was designed to provide. These better seeing conditions contribute directly to higher productivity and improved morale in factories, offices, and schools. In stores, the better lighting that results from good maintenance helps to increase sales. In addition, all areas benefit from the more uniform appearance and fewer work interruptions that come with good maintenance.

65. Benefits of Cleaning. A lighting system should not only be cleaned properly; it should be cleaned at the proper time. This combination produces a cleaning program which is a profitable investment because three principal benefits can be obtained.

1. *More light delivered.* Dirt absorbs light. Cleaning removes the dirt and thus increases the lighting level. As a result, visibility is improved which pays off with benefits to the user. For example, in industrial and office areas, production may be increased and errors reduced; in merchandising areas, sales may be increased.

2. *Reduced maintenance costs.* A good maintenance program calls for cleaning the lighting system at the most economical time. It makes use of the most efficient methods and equipment. In this way the time and materials required are reduced; thus maintenance costs are lowered.

3. *Better appearance.* Clean lighting systems improve the appearance of the working or selling area. This is conducive to improved employee morale, better housekeeping, and increased customer comfort and satisfaction.

To further improve the appearance of the area being lighted, walls and ceilings should be periodically cleaned and repainted as part of the cleaning program.

66. Causes of Light Losses. Several factors contribute to light losses, and the effects of these factors vary with the kind of activity that takes place and the location of the establishment. For example, areas vary as to the amount and type of dirt in the air. Obviously, the amount of dirt in a foundry is greater than that in an air-conditioned office; and the amount and type of dirt in an office located near an industrial area are different from those in an office located in the country; the black type of dirt characteristic of steel mills is certainly unlike the relatively light-colored dirt of a bakery or woodworking shop. It is important to recognize these variations in considering the light losses which result from dirt on the lamps and lighting units and dirt on room surfaces. These two along with the unavoidable lamp-lumen depreciation are the principal factors which cause light losses in every lighting installation.

67. Relamping Benefits. The lamps in a lighting system can be replaced individually as they burn out, or the entire installation can be replaced before the lamps reach their average life. Individual replacement is usually called "spot replacement"; mass replacement is called "group relamping." The labor costs saved by group relamping in large installations and in many small ones more than compensate for the value of the depreciated lamps that are thrown away before they burn out. Other advantages which always accompany group relamping are more light, fewer work interruptions, better appearance of the lighting system, and less maintenance of auxiliary equipment.

Group relamping had its origin in the field of street lighting. It is still widely practiced in that field, because many street-lighting engineers realize that group relamping saves them money and reduces the number of unsafe dark areas. In factories, offices, stores, and schools, group relamping has grown in popularity in recent years; many examples can be cited where management has recognized the economic gains that result. Significant improvements during recent years in the length and uniformity of lamp life have helped make group relamping even more desirable.

Filament lamps and fluorescent lamps are both well suited to group-relamping programs. Mercury lamps, too, are often group-relamped; but because lamp costs are a high proportion of the total cost of lamp replacement, group relamping of mercury installations is not as widely practiced. The very-hard-to-reach locations of mercury lamps often make group relamping necessary to reduce the number of maintenance trips.

There are five principal advantages that group relamping offers to the user. The first three apply to all lighting systems, the last two chiefly to fluorescent lamps.

1. *Reduced Labor Costs Often Mean Net Savings.* Group relamping saves on labor costs, largely because much of the travel time and setup time required to change lamps individually is eliminated. The labor cost per lamp with group relamping generally ranges from one-fifth to one-tenth the spot replacement cost; it seldom exceeds 20 cents per lamp for fluorescent lamps. As labor cost of spot replacement increases, group relamping becomes a better buy because labor costs are a larger proportion of the total maintenance cost. When lamps are group-relamped after relatively few have failed, costs are usually reduced considerably.

2. *More Light Delivered.* All lamps depreciate in lumens continually as they burn. The earlier they are replaced, the higher the maintained illumination will be without adding to the cost of electric energy or the number of lighting fixtures used. This means that people will get the good seeing conditions that the system was designed to provide. When the higher illumination levels are assigned the value they deserve, in terms of work output or sales, shorter group-relamping intervals can be economically justified.

3. *Fewer Work Interruptions.* Group relamping can be done at a convenient time—during vacation shutdowns or after working hours, for example, when there will be no interruption of operations. The number of interruptions to report burnouts or to replace them is greatly reduced.

4. *Better Appearance of the Fluorescent Lighting System.* Black ends, color variations, and differences in brightness between adjacent old and new fluorescent lamps are common when spot replacement is used. With group relamping, all the lamps are the same age, and appearance is far more uniform. Since most of the lamps are replaced before they burn out, the distraction of blinking, flashing, or swirling lamps is minimized.

5. *Less Maintenance of Fluorescent Auxiliary Equipment.* Abnormal operating conditions that may occur at the end of fluorescent lamp life can damage starters and ballasts. When most of the lamps are replaced before they reach the end of life, auxiliary equipment lasts longer.

These five advantages of group relamping mean more satisfaction with the lighting system, as well as dollars-and-cents savings.

68. Group-lamping Costs. The cost of lamp replacement is made up of the cost of the lamp and the cost of labor required to replace the lamp. When the sum of these costs is reduced, of course, the total annual cost of operating the lighting system is reduced; you get more footcandles per dollar. It is difficult to assess the exact overall cost reduction without having all the other facts about the installation, but it should be kept in mind that economical lamp replacement means better overall lighting economics.

With spot replacement, the total replacement cost per lamp is equal to the cost of the lamp plus the labor cost of replacement. Group-relamping cost, to compare with this, is equal to the lamp cost plus group-relamping labor cost plus the cost of any interim spot replacements, divided by the group-relamping interval to put both sys-

Table 19-34. Cost-analysis Outline for Lighting Systems

General information	Lighting method 1	Lighting method 2
Installation data:		
Type of installation	....	....
No. of rows	....	....
Luminaires per row	....	....
Lamps per luminaire	....	....
No. of lamps	....	....
Watts per lamp (including accessories)	....	....
Total watts	....	....
Maintained footcandles	....	....
Calculation of complete expense		
Capital expense:		
Estimated cost of each luminaire installed	....	....
Estimated wiring cost per luminaire	....	....
Cost per luminaire (luminaire plus wiring)	....	....
Number of luminaires	....	....
Total cost	....	....
Assumed years life	....	....
Total cost per year of life	....	....
Interest on investment (per year)	....	....
Taxes (per year)	....	....
Insurance (per year)	....	....
Total capital expense per year	....	....
Energy expense:		
Total watts	....	....
Average hours used per year	....	....
kWh per year	....	....
Average rate per kWh	....	....
Total energy expense per year	....	....
Lamp-renewal expense:		
No. of lamps	....	....
Avg. hours used per year	....	....
Total lamp hours per year	....	....
Rated lamp life, h	....	....
Avg. lamp renewals per year	....	....
Net price each	....	....
Replacement expense each (labor)	....	....
Net price plus replacement expense each	....	....
Total lamp-renewal expense per year	....	....
Cleaning expense:		
No. of washings per year	....	....
Man-hours each (est.)	....	....
Man-hours for washing	....	....
No. of dustings per year	....	....
Man-hours each (est.)	....	....
Man-hours of dusting	....	....
Total man-hours	....	....
Expense per man-hour	....	....
Total cleaning expense per year	....	....
Repair expense:		
Repairs (based on experience, allocation of repair man's time, etc.)	....	....
Est. total repair expense per year	....	....
General information		
Recapitulation:		
Total capital expense per year	....	....
Total energy expense per year	....	....
Total lamp renewal expense per year	....	....
Total cleaning expense per year	....	....
Est. total repair expense per year	....	....
Complete lighting expense for year	....	....
Complete lighting expense per footcandle per year	....	....

tems on an equal time basis. These costs can be expressed in the formulas which are presented and discussed in other publications.[1]

69. Cost of Lighting. The overall items for comparing different lighting systems must include both the initial and the operating cost. While one of these may be a dominant factor in the final selection, it is usually desirable to combine the two into some type of "total cost" indicator.

The computation of initial, operating, and total annual cost for various systems considered for a given interior must be based on certain common assumptions, if the systems are to be fairly compared. Some of the important considerations are:

1. Equal illumination results—since different systems may not produce equal illumination levels in service, all costs should be equated to an equal maintained foot-candles basis.

2. Equal rates in amortizing the initial investment and allowing for interest, taxes, and insurance should be used.

3. Operating conditions, such as electrical-energy rate, burning hours per year, and starting frequency of the lamps, should be equal for the systems being considered.

4. Cleaning schedule should be appropriate to each type of system.

5. Uniform labor rates among systems should be used for estimating the cost of installations, cleaning, and relamping.

Table 19-34 tabulates the cost-analysis procedure for comparing the cost of two or more lighting systems and is self-explanatory.

RADIANT-ENERGY DETECTORS

70. Photovoltaic Cells. Selenium barrier-layer cells are of the self-generating type, with spectral sensitivity chiefly in the visible and near-ultraviolet regions, as indicated in Fig. 19-41. These cells are available in a variety of physical shapes and sizes, hermetically sealed and color-corrected to match the standard luminosity curve. Photovoltaic cells generate a potential difference, when irradiated, which bears a relationship to the intensity of illumination. These cells are used in photometers, light meters, color-temperature meters, exposure meters, reflectometers, goniophotometers, glossmeters, fluorometers, colorimeters, refractometers, and pyrometers.

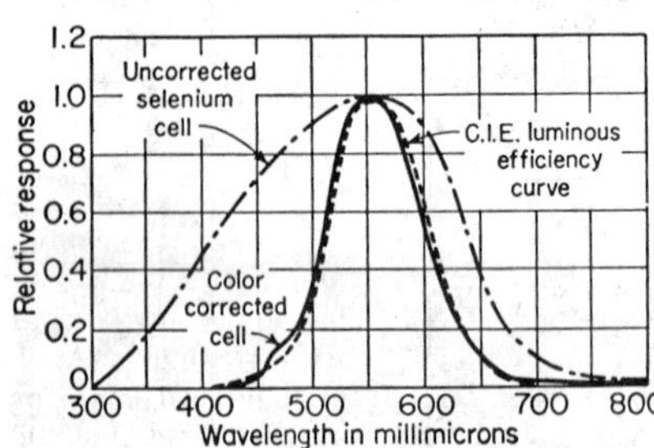

FIG. 19-41. Average spectral-sensitivity characteristics of selenium barrier-layer cells, compared with CIE spectral luminous efficiency curve.

71. Phototubes. Phototubes are photoemissive-type cells consisting of an anode and a cathode mounted in a transparent envelope which is either evacuated or gas-filled. When illuminated, the cathode emits electrons in proportion to the intensity of the illumination. An external voltage must be applied between the cathode and anode in order to sustain the electron flow from cathode to anode when the phototube is illuminated. Some of the materials used for the cathode in phototube construction are lithium, sodium, potassium, cesium, and rubidium.

The cells may be made in either the high-vacuum or the gas-filled type. The latter usually has greater sensitivity than the vacuum cells but is likely to give higher dark currents (current flow with the photocell shielded from all illumination).

Both high-vacuum and gas-filled photocells are used in photometric measurements, where a galvanometer and battery in series with the cell may suffice when illumination levels are high. For lower levels of illumination, the cells are used with vacuum-tube amplifier circuits in order to obtain high sensitivity. In extreme cases, where illumination intensities are very low, it is sometimes necessary to use multiplier tubes, which then make it possible to detect currents as low as 5×10^{-18} A, that is, 30 electrons/s.

The use of the multiplier phototube makes it possible to eliminate the vacuum-

[1] *Gen. Elec. Lamp Dept. Publ.* TP-105-R.

tube amplifier referred to above. These are high-vacuum photocells which incorporate, within the tube itself, a system for multiplying the photocurrent through secondary emission from successive electrodes (called "dynodes"). Such tubes usually require a total voltage of the order of 1,000 V but are capable of delivering current up to 1 mA without further amplification. They are readily adaptable to measurements at extremely low light intensities, when used with suitable amplifiers.

Phototubes are used in nephelometers, densitometers, recording microphotometers, color-transparency processing equipment, spectrophotometers, spectroradiometers, and the instruments listed under photovoltaic cells (Par. **70**).

Vacuum photocells usually give a linear response (constant microamperes per lumen) over a wide range of illumination intensity, once saturation has been reached. Such is not the case with gas-filled cells, which normally are linear for only small portions of their characteristic curve.

Barrier-layer cells for photometric use should be selected on the basis of linearity of response. Certain cells will be linear over ranges of 100:1 with an accuracy of 1% or better. Such cells must be used in a low-resistance circuit in order to take full advantage of their linearity. Special circuits have been devised to compensate for nonlinearity of barrier-layer cells.

PHOTOMETRY[1]

72. Visual Photometers. Because of its power of adaptation to different brightnesses, the eye cannot measure photometric properties with any accuracy. However, it is possible to detect very small differences in photometric brightness between two surfaces viewed simultaneously side by side, provided that they have nearly identical chromaticities (color). Visual photometers employ this principle and include a means for adjusting the photometric brightness of one or both fields, until the two appear equally bright.

73. Lummer-Brodhun Photometer. The Lummer-Brodhun cube consists of two identical 45° by 90° prisms, with a pattern etched in the hypotenuse face of one. The two hypotenuse faces are assembled together to make optical contact. Where the surfaces are in optical contact, light is transmitted, thus presenting an opportunity to compare two fields.

74. Flicker Photometer. The flicker photometer was developed as a means of comparing two sources of different colors. A single field is alternately illuminated by two sources so that the observer sees a flicker, which may be due to color difference, difference in photometric brightness, or both. Above a certain rate of alternation the color sensations blend, and the observer adjusts the sources to eliminate the remaining flicker due to differences in field brightness. When the flicker disappears, a photometric balance has been reached. Quite accurate determinations of intensities from light sources of different color characteristics are obtainable with this device.

75. Candlepower Distribution Apparatus. The radial distribution of candlepower normally is measured at a fixed distance (10 ft) from the center of the light source, though, in the case of street-lighting equipment, floodlights, and similar sources, longer distances are used. In order to economize on space for such measurements, mirrors are frequently used to reduce distances, relative data are obtained, and reliance is placed on the substitution method. To obtain the average candlepower, the light source usually is rotated about its own axis, though in the case of asymmetric sources this, of course, is not done.

76. Integrating Sphere Photometer. The integrating sphere photometer is the generally accepted method of determining the light output of light sources. Its proper use is subject to certain restrictions as to the reflection characteristics of the interior surface of the sphere, the dimensions of the test lamp or luminaire in relation to the diameter of the sphere, and the color quality of the light. Measurements of light output of luminaires usually call for a sphere 80 or 100 in in diameter, although in the photometry of incandescent lamps it is desirable to use a sphere at least 40 in in diameter. Errors are held to a minimum through the use of the substitution method with standard lamps of similar color to the test lamps.

[1] IES Lighting Handbook, 3d ed., 1959.

PORTABLE PHOTOMETERS[1]

77. Macbeth Illuminometer. One of the more widely used portable visual photometers is the Macbeth illuminometer shown in Fig. 19-42. It contains a Lummer-Brodhun cube, of which one field is a view of the test surface to be measured. The other field is lighted by a small comparison lamp mounted on a rack and pinion in a tube. The two fields are balanced by moving the comparison lamp back and forth in the tube. The inverse-square principle determines the result. The illuminometer may be equipped with a lens to restrict the test field and bring it into focus. The range can be extended over very wide limits by inserting calibrated neutral filters in either the test or the comparison lamp side. Colored filters may also be used to correct the color of the comparison lamp to that of a test surface.

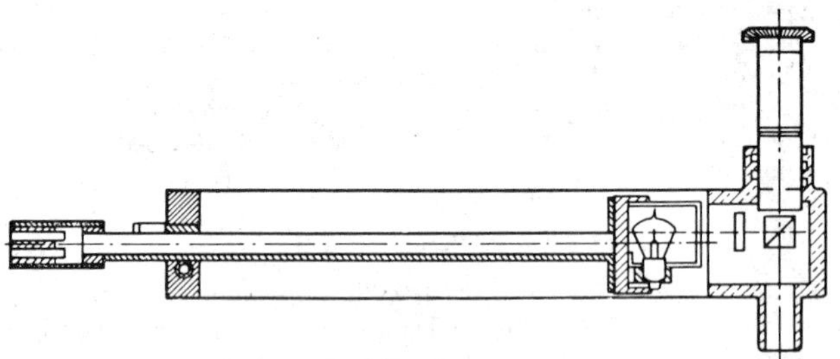

Fig. 19-42. Macbeth illuminometer.

78. Luckiesh-Taylor Brightness Meter. The Luckiesh-Taylor meter is a small, completely self-contained instrument for measurement of photometric brightness or illumination. A lens in the eyepiece brings the external test field into focus in the same plane as the comparison field, which is seen as two small trapezoids against the circular test field. Light for the comparison field is supplied by a small lamp. The brightness of the comparison field is adjusted to match the test field by rotating a photographic-film gradient. An illuminated scale, calibrated in footlamberts and candles per square inch, is seen through a second eyepiece. Neutral filters greatly extend the range of values that can be measured.

79. Photoelectric Photometers. Photoelectric photometers are convenient to use, eliminate personal judgment and variations among individual observers, and, under the best conditions, give very accurate results. Hence, these photometers have largely replaced visual devices for most routine measuring procedures. Photoelectric photometers may be divided into two classes: those employing barrier-layer cells; and those employing photoemissive tubes, which require additional equipment for operation.

Direct-reading photoelectric photometers for field use are available. One of these is the Pritchard photometer, which may be used to measure illumination, brightness, polarization, and chromaticity. Full-scale sensitivity is obtained through the use of calibrated filters for brightnesses of 1/10,000 to 100,000,000 fL. A variety of apertures provide fields subtending arcs over a range from 6′ to 2°.

Another instrument of this type is the Spectra Spot Brightness meter, available with ½° and 1½° fields with corresponding ranges of 1/10 to 1,000,000 fL and 1 to 10,000,000 fL.

80. Barrier-layer Cell Meters. Barrier-layer cells generate a voltage when radiant energy strikes the sensitive surface. However, the cells depart from linearity of response as the resistance of the circuit to which they are connected increases; and for precise results a current-balancing measuring instrument giving zero external resistance is necessary. The great majority of portable illumination meters in use today consist of a barrier-layer cell, or cells, connected to a meter calibrated directly in footcandles.

Barrier-layer cells have a number of limitations which must be taken into consideration:

1. The spectral response is different from that of the human eye; consequently, color-correcting filters should be used.

[1] IES Lighting Handbook, 3d ed., 1959.

2. Part of the light incident on the cell at wide angles from the normal are reflected from the cell surface and cover glass causing the response to be as much as 25% low. Corrective cover glasses giving so-called cosine correction are used with these cells to measure illumination.

3. The response of the cell varies with temperatures above and below 77°F. Prolonged temperatures above 120°F will permanently damage the cell.

4. The cell response when exposed to constant illumination decreases slightly over a period of time before reaching its final value. The cell should be well fatigued before beginning a series of measurements.

5. The microammeters used in portable meters are subject to certain inherent limitations of accuracy. Manufacturing tolerances alone may result in an overall uncertainty of reading at any point above one-quarter full scale of about ±8%. At lower scale readings, the percentage of error may be greater. The microammeter and cell should be checked frequently against a source of known illumination value.

6. When gaseous discharge sources are operated at frequencies above 60 c, precautions should be taken with regard to the effect of frequency on cell response.

81. Reflectometers. Measurements of the reflectance of surfaces are influenced by the nature and color of the surface and by the conditions of illuminating and viewing the specimens. A number of devices have been developed for these measurements, all employing the same general principles of operation but under slightly different geometric conditions. The method of observation may be either visual or photoelectric.

82. Reflectance Determination for Colored Materials. The same instruments employed for measuring reflectance of neutral materials may be used for measurements of colored surfaces if the observer, or the photocell, has a response identical with that of the CIE Standard, and the sphere coating and other parts are absolutely neutral. The reflectance is then determined for the particular light source employed. A far more satisfactory method in such cases is to use the recording spectrophotometer.

COLORIMETRY AND SPECTROPHOTOMETRY

83. Color may be specified in a variety of ways. Those in most general use are: (1) specification by trichromatic coefficients and, in addition, luminous intensity, total transmission, or total reflectance; (2) specification by dominant wavelength, purity and luminous intensity, total transmission, or total reflectance; (3) specification by comparison with material standards, e.g., the Munsell colors, Ostwald colors, and Maerz and Paul colors; (4) specification by color temperature in the case of incandescent lamps (see Par. **10**).

Trichromatic coefficients may be obtained by computation from the spectrophotometric data, by the following procedure in the case of transmitting or reflecting media. The spectral transmission or reflectance is determined. This is then multiplied by the spectral-energy distribution of the light source, and this product is multiplied by the tristimulus functions of the standard CIE observer, viz., the normal eye.[1] The areas under the three product curves so obtained are determined, and the ratios of the areas of the individual ones to the sum of the three areas are the trichromatic coefficients. Figure 19-43 has as coordinates the x and y trichromatic coefficients. The curved part of the solid line is the spectrum locus, i.e., the locus on such a diagram of the various pure spectral colors. All physically possible colors must lie on this locus or inside it. Methods for simplifying these calculations have been developed by Hardy[1] and others.

Computations of dominant wavelength usually are carried out for ICI[2] illuminants A, B, C or for a source of color temperature of 2360°K. CIE[2] illuminant A is defined as an incandescent lamp operating at a color temperature of 2854°K. CIE[2] illuminants B and C are defined in terms of illuminant A modified by filters designed to approximate color temperatures of 5000° and 6500°K, respectively.

Dominant wavelength is derived from the trichromatic coefficients through first

[1] Hardy, A. C. "Handbook of Colorimetry"; Cambridge, Mass., The Technology Press of the Massachusetts Institute of Technology, 1936.

[2] Commission Internationale de l'Eclairage (formerly International Commission on Illumination).

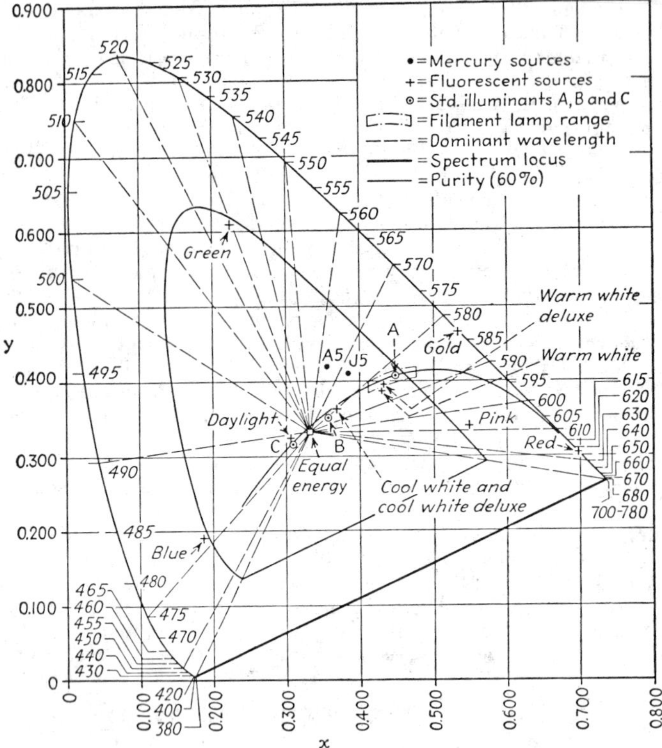

FIG. 19-43. Trichromatic coefficient diagram for various light sources.

plotting the point for the specimen on the trichromatic diagram, as shown in Fig. 19-43, together with a point representing the standard illuminant. A line is then drawn from the point for the standard illuminant through the point for the specimen and extended until it cuts the spectrum locus of the trichromatic diagram. If this intersection occurs in the upper curved portion of the diagram, the point of intersection represents, in trichromatic coefficients, the dominant wavelength. Should the intersection occur in the straight-line portion of the diagram (the line across the bottom), the line through the plotted points would be extended back to give an intersection with the curved portion of the diagram (spectrum locus); and this intersection then determines the complementary color. Purity is defined as the ratio a/b, in which a = distance from the point corresponding to the standard illuminant to the point representing the specimen and b = distance of the standard illuminant point from the spectrum locus measured along the line through the two points.

The Munsell and other systems consist of painted chips, dyed yarn samples, etc., and the color of the specimen is specified by comparison with these and identification on the appropriate scale, with or without interpolation. For extensive treatment of the subject, refer to the Report of the Colorimetry Committee of the Optical Society of America.[1]

84. A spectrophotometer consists of a photometric device combined with a spectrometer in such a manner as to permit of the comparison of two sources of light. The comparisons are made for spectral bands of varying widths depending upon the resolution of the spectrometer. Means are provided for varying the intensity of one or both

[1] Report of the Colorimetry Committee; *Jour. Opt. Soc. Am.*, 1943, Vol. 33, pp. 533, 534, 544; 1944, Vol. 34, pp. 183, 245, 633; 1945, Vol. 35, p. 1.

of the sources. Instruments are available using both visual and photoelectric methods of photometry. Because of the low intensities usually met, particular care must be taken to avoid stray light. It is also desirable that the slit widths be adjusted for the different spectral regions in order that the same wavelength-interval band be employed throughout. The CIE has standardized on 45° incidence and diffuse viewing or the reverse for the case of reflection surfaces. However, many spectrophotometers employ different conditions, an extremely important fact in that the conditions materially affect the results.

85. Photoelectric Spectrophotometers. Many spectrophotometers have been constructed employing photoelectric cells for measurement of transmission or reflection factors.

The most widely known of these is the General Electric recording spectrophotometer originally developed by A. C. Hardy, which measures both spectral-transmission factors and relative spectral-reflection factors.[1,2]

86. Colorimeters. In 1931 the CIE established tristimulus functions representing three primary color sensations: red, green, and blue. The color of a light source, a filter, or a reflecting surface may be specified by the trichromatic coefficients determined in accordance with the CIE method. In the case of photoelectric colorimeters the usual design involves the use of filter-photocell combinations whose respective spectral responses correspond to the tristimulus functions (two such combinations being required in general for the $\bar{x}$ function). With such an instrument the trichromatic coefficients of a light source may be determined and specified as in Fig. 19-43, or the design of a suitable light-source–photocell-filter combination will accomplish this for transparent media or reflecting surfaces.

87. The accuracy of photometry is of a lower order than that of most physical measurements owing to the fact that the quantity measured is not of a physical nature but is a sensation which is never exactly the same for two observers or for one observer on separate occasions. For this reason, a series of readings should be made and the average taken. The photoelectric cell has made possible, by elimination of idiosyncrasies of the human observer, much greater accuracy of reproduction of measurements. In some cases reproducibility is limited largely by the inconsistency of the light sources themselves. A distinction should be drawn between accuracy and reproducibility. Since the unit of luminous intensity is a statistical quantity expressing the sensation produced in the average observer by a given stimulus, photoelectric measurements can never exceed in precision the accuracy of establishment of the fundamental standard.

88. Bibliography. References on illumination may be found in texts on the subject of illuminating engineering and in appropriate journals. A few are listed below.

Texts:

Amick, C. L. "Fluorescent Lighting Manual," 2d ed.; New York, McGraw-Hill Book Company, 1947.

Barrows, W. E. "Light, Photometry, and Illuminating Engineering," 3d ed.; New York, McGraw-Hill Book Company, 1951.

Boast, W. B. "Illumination Engineering," 2d ed.; New York, McGraw-Hill Book Company, 1953.

Flynn, J. E., and Mills, S. M. "Architectural Lighting Graphics"; New York, Reinhold Publishing Corporation, 1962.

Hardy, A. C. "Handbook of Colorimetry"; Cambridge, Mass., The Technology Press.

IES Lighting Handbook; New York, Illuminating Engineering Society.

Jolly, L. B. W., Waldram, J. M., and Wilson, G. H. "Theory and Design of Illuminating Engineering Equipment"; New York, John Wiley & Sons, Inc.

Kraehenbuehl, J. O. "Electrical Illumination"; New York, John Wiley & Sons, Inc.

[1] Hardy, A. C. A New Photoelectric Spectrophotometer; *Jour. Opt. Soc. Am.*, 1935, Vol. 25, p. 305.
Hardy, A. C. History of the Design of the Recording Spectrophotometer; *Jour. Opt. Soc. Am.*, 1938, Vol. 28, p. 360.
Michaelson, J. L. Construction of the G.E. Recording Spectrophotometer; *Jour. Opt. Soc. Am.*, 1938, Vol. 28, p. 365.

[2] Pritchard, I. S. Design and Development of a New Recording Spectrophotometer; *Jour. Opt. Soc. Am.*, 1953, Vol. 43, p. 812.

MOON, P. "Scientific Basis of Illuminating Engineering"; New York, McGraw-Hill Book Company, 1936.

SHARP, HOWARD M. "Introduction to Illumination"; Englewood Cliffs, N.J., Prentice-Hall, Inc., 1951.

WALSH, J. W. T. "Photometry"; Princeton, N.J., D. Van Nostrand Company, Inc.

Journals:

Illum. Eng. (formerly *Trans. Illum. Eng. Soc.*) (N.Y.).

Trans. IES (Great Britain).

Jour. Opt. Soc. Am.

Light Magazine.

Lighting Handbook (Westinghouse Electric Corp., Bloomfield, N.J.).

Light and Lighting (Great Britain).

Phillips' Tech. Rev.

Proc. Intern. Comm. on Illumination (*Comm. Intern. de l'Eclairage*).

SECTION 20

INDUSTRIAL AND COMMERCIAL APPLICATIONS OF ELECTRIC POWER

BY

A. S. JUDD Electrical Engineer, Imperial Oil Enterprises Ltd.; Senior Member, Institute of Electrical and Electronics Engineers

L. L. QUINLAN Assistant General Manager, Operating Services, Inland Steel Company

F. T. WOLFORD Assistant Professor, Agriculture Machinery Department, Delaware Valley College of Science and Agriculture

C. T. MAIN II Associate, Chas. T. Main, Inc., Associate Member, American Textile Manufacturers Association, Inc.; Member, American Society of Mechanical Engineers; Registered Professional Engineer

J. F. HOWER Chief Electrical Engineer, Lehigh Portland Cement Company; Senior Member, Institute of Electrical and Electronics Engineers; Past Chairman, Cement Industry Committee, Institute of Electrical and Electronics Engineers; Registered Professional Engineer (Ohio)

DANIEL JACKSON, JR. Associate Editor, *Coal Age*

R. W. BERGMANN Manager, Electrical Engineering, Marion Power Shovel Company

HANS GARTMANN Chief Engineer (Retired), Pump and Compressor Department, De Laval Turbine, Inc.; Member, American Society of Mechanical Engineers; Member, Society of Naval Architects and Marine Engineers; Registered Professional Engineer

A. H. MYLES Chief Engineer, Heavy Industry Division, Square D Company; Fellow, Institute of Electrical and Electronics Engineers; Member, Association of Iron and Steel Engineers

JOHN LUSTI Chief Engineer, Otis Elevator Company; Member, Institute of Electrical and Electronics Engineers

S. W. HERWALD Vice-president, Electronic Specialty Products Group, Westinghouse Electric Corporation; Fellow, Institute of Electrical and Electronics Engineers

G. S. AXELBY Advisory Engineer, Aerospace Division, Westinghouse Electric Corporation

E. H. SPRECKELMEIER Chief Electrical Engineer, American Laundry Machinery Industries

R. C. HARDIE Manager, Mine Hoist Department, Process Machinery Division, Nordberg Manufacturing Company

J. A. KEELEY Manager, Operations Research, *The Miami Herald*

C. G. HELMICK Industry Engineering Department, Westinghouse Electric Manufacturing Co.; Associate Member, Institute of Electrical and Electronics Engineers; Registered Professional Engineer

CHARLES EMERSON *American Machinist*

JOHN DEB. SHEPARD Consulting Engineer, Greensboro, N.C.; Member, American Society of Heating, Refrigerating, and Air Conditioning Engineers; Member, American Society of Mechanical Engineers; Member, National Society of Professional Engineers; Registered Engineer (Alabama, Georgia, Maryland, North Carolina, South Carolina, Tennessee, and Virginia)

CONTENTS

Numbers refer to paragraphs

SECTION 20

THE PETROLEUM INDUSTRY

By A. S. Judd

WELL DRILLING

1. Cable Tool Drilling. A string of tools weighing between 2 and 5 tons is repeatedly raised and dropped by a cable. Cuttings and water are bailed out after the bit has been withdrawn. This method is used for depths up to 2,000 or 3,000 ft and will drill between 20 and 150 ft/day.

2. Rotary Drilling. A bit at the end of a hollow drill stem is rotated to cut or grind the rock. Drilling mud is pumped down the drill stem to cool the bit and to wash cuttings to the surface. The mud travels upward between the stem and the wall of the hole. It helps to prevent collapse of the wall and seals it against the flow of water and gas. The drilling mud also helps to prevent "blowing" of the well when a high-pressure stream is encountered.

Rotary rigs are used for depths greater than 3,000 ft and will drill 100 to 300 ft/day.

3. Power Required for Drilling. Diesel engines are generally used for drilling because of the requirements for portability. When electric drives are used, diesel-engine-driven generators are provided with the rig. Direct-current motors are used for the draw works and rotary table drives because of high starting-torque requirements, and a common motor may be used for both. The mud pump is often driven directly by the diesel engine.

OIL-WELL PUMPING

4. Methods of Forced Production. When there is insufficient natural pressure to force the crude oil to the surface, then some method of forced production is used. The most common methods are:

High-pressure Gas Lift. Gas is forced to the bottom of the well or to an intermediate point in the well. The gas mixes with the oil in the well and induces flow by decreasing the density of the fluid.

Water Flooding. Treated water is forced into the formation through nearby wells in order to increase the pressure in the formation and induce flow.

Bottom-hole Hydraulic Pump. High-pressure crude oil is carried down the well in tubing, and it is used to drive a reciprocating pump located at the bottom of the well.

Bottom-hole Centrifugal Pump. A special motor-driven multistage centrifugal pump is lowered to the bottom of the well. This method is used where large volumes of fluid must be pumped.

Sucker-rod Pump. A reciprocating single-acting pump is installed at the bottom of the well on the end of a tube inside the well casing.

5. Sucker-rod Pump Drives. The plunger is operated by a sucker rod from the surface. Various methods are available to drive the sucker-rod string, but generally a walking beam is used to provide the desired vertical motion.

6. Central Power Units. Central power units driven by electric motors are sometimes used to serve as many as 15 or 20 wells. Operating rods lead out to each pump to provide reciprocating motion to the walking beams.

7. Individual Engine or Motor Drives. These are more commonly used than multiple drives because they can be started and stopped individually and the pumps can be operated at different speeds. Electric motor drives are preferred because they can be started and stopped by a timer, they provide consistent, trouble-free performance regardless of weather conditions, and maintenance and investment costs are low. Also, it is easy to measure power demand and energy consumption of an electric motor. The well may be counterbalanced readily by an ammeter.

8. Motor Types. Torque requirements vary widely during the pumping cycle, and peaks occur when the sucker-rod string and fluid load are lifted and when the counterweight is lifted. NEMA design D motors, although relatively expensive, are well suited to this service, since they minimize current peaks and provide adequate torque under all service conditions, including automatic operation by time control. NEMA design C motors may be used where operating conditions are less severe. NEMA design B motors must be used with care in this service to avoid high cyclic current peaks, which may be objectionable on a small system, particularly if several wells should "get in step." The use of design B motors can also lead to oversizing of motors in an attempt to obtain sufficient starting torque. This results in the operation of the motor at a relatively low load factor, with consequent low power factor.

9. Double- or Triple-rated Motors. These are special motors developed for oil-well pumping. They are totally enclosed, fan-cooled NEMA design D motors that can be reconnected for 2- or 3-hp ratings at a common speed of 1,200 r/min. Typical horsepower ratings are 20/15/10 and 50/40/30. They provide flexibility in the field since they permit the selection of the horsepower rating at which the motor may be operated most efficiently. They also permit changing the pumping speed by changing the motor pulley and reconnecting the motor.

10. Single-phase Operation. If single-phase power only is available, then it is advisable to consider the use of single-phase to 3-phase converters and 3-phase motors. This avoids the use of large single-phase capacitor start motors, which are relatively expensive and contain a starting switch which could be a source of trouble due to failure or to the presence of flammable gas in the vicinity of the well.

11. Oil-well Control. A packaged control unit is available to control individual oil-well pumps. It contains, in a weatherproof enclosure, a combination magnetic starter, a time switch that can start and stop the motor according to a predetermined program, a timing relay that delays the start of the motor following a power failure, and lightning arresters. Pushbutton control is also provided.

12. Power-factor Correction. The induction motors used for oil-well pumping have high starting torques with relatively low power factor. Also, the average load on these motors is fairly low. Therefore it is advisable to consider the installation of capacitors to avoid paying the penalty imposed by most power companies for low power factor. They will be installed at the individual motors and switched with them, if voltage drop in the distribution system is to be corrected as well as power factor. Otherwise they may be installed in larger banks at the distribution center, if it is more economical to do so.

GAS-PROCESSING PLANTS

13. Natural Gas. Natural gas varies widely in composition and contains undesirable materials such as water and sulfur compounds, which must be removed before the gas enters the transmission pipeline. Various chemical processes are used, and plant capacity ranges from 5 to 1,000 million ft^3 of gas processed/day. Byproducts such as propane, butane, pentanes, and elemental sulfur are produced and marketed.

14. Power Supply. Purchased power is used where available. Local generation is by reciprocating gas engines or gas turbines.

Electrical installation practice in gas plants is similar to that followed in oil refineries, as described below.

OIL PIPELINES

15. Gathering Systems. These collect crude oil from the individual wells, or from tanks located near them, and carry it to tankage, where shipments may be accumulated.

16. Trunk Lines. These feed crude oil from gathering systems to the main crude pipeline pumping station.

17. Crude Lines. Crude lines are generally operated as common carriers. Since they handle crude oil for several companies and because crude-oil shipments vary greatly in quality and composition, storage tanks are necessary at stations along the pipeline so that batch shipments may be handled.

18. Products Lines. These convey products from a refinery to the market area.

Some products lines are operated as common carriers, while others are privately owned and operated.

19. Operation of Pipelines. Batches of crude oil or products are dispatched through a pipeline and are withdrawn to tankage at the end of the line or at intermediate points. If a batch is being drawn off at an intermediate point, then the downstream stations will operate at reduced flow.

Little mixing occurs at the interface between different batches. By careful scheduling and a knowledge of the pipeline it is possible to predict fairly accurately when an interface will arrive at a station. Interface detectors are installed also.

20. Pumping Stations. A schematic diagram of a typical pumping station is shown in Fig. 20-1.

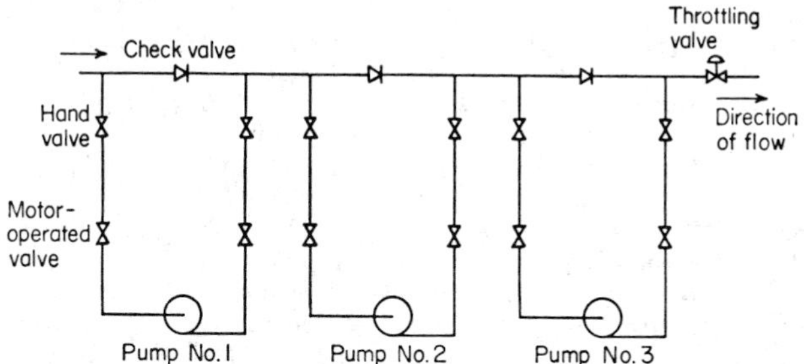

FIG. 20-1. Pipeline pumping station.

21. Arrangement of Pumping Stations. These are located at the head of the pipeline and at intervals along the line. Intermediate or booster stations must be capable of operating under varying conditions due to differences in liquid gravity, withdrawals at intermediate points, and the shutting down of other booster stations. Pumping stations often contain two or three pumps connected in series, with bypass arrangements using check valves across each pump. The pumps may all be of the same capacity, or one of them may be half size. By operating the pumps singly or together, a range of pumping capacities can be achieved.

Throttling of pump discharge may also be used to provide finer control and to permit operation when pump suction pressure may be inadequate for full flow operation.

22. Control of Pumping Station. Pumping stations are often unattended and may be remotely controlled by radio or telephone circuits.

23. Electrical System. Figure 20-2 is a typical electrical single-line diagram for a pumping station.

24. Motor Type for Main Pumps. The main pumps are driven by 3,600-r/min induction motors having NEMA design B characteristics. Full-voltage starting is used.

25. Motor Enclosure. Motor enclosures for outdoor use are NEMA weather-protected Type II, totally enclosed, fan-cooled, or dripproof with

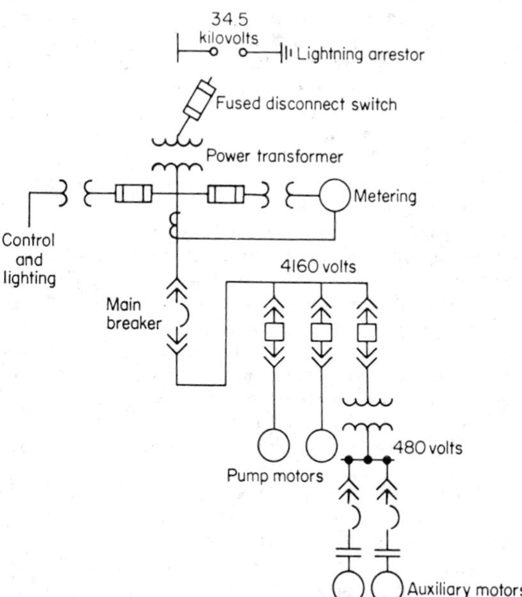

FIG. 20-2. Single-line diagram of a pipeline pumping station.

weather protection. Motors of the latter type are widely used. Not only are they less expensive than the other types, but they also have a service factor of 1.15. The above enclosure types are all suitable for the Class I, group D, division 2 classifications usually encountered.

If the pumps are located indoors, then a division 1 classification is likely to apply. Motors must be Class I, group D, explosion-proof, or they may be separately ventilated with clean outside air brought to the motor by fans. Auxiliary devices such as alarm contacts on the motor must be suitable for the area classification. The installed costs, overall efficiencies, and service factors associated with the enclosures that are available will influence the selection.

NATURAL-GAS PIPELINES

26. Natural-gas Pipelines. The operation of gas pipelines is similar to that of oil pipelines except that a single gaseous product is handled instead of batches of liquids. Compressors, either reciprocating or centrifugal, are used instead of pumps, and the pipe diameters are much larger than those used in oil pipelines.

27. Electrical Installation. The electrical installation and control systems are similar to those found in oil pipeline systems.

28. Compressor Drivers. These are usually reciprocating gas engines or gas turbines, to make use of the energy available in the pipeline. Electric motor drives use slow-speed synchronous motors for reciprocating compressors and four- or six-pole induction motors with gear increasers for high-speed centrifugal compressors. Motor voltages, types, and enclosures are selected as for oil pipeline pumps. Motors used with centrifugal compressors must develop sufficient torque at the voltage available under inrush conditions to accelerate the high inertia load. They must also have adequate thermal capacity for the long starting time required, which may be 20 or 30 s.

OIL REFINERIES

29. General. Oil refineries vary greatly in the variety and quantity of products and crude oil throughput. A basic refinery producing gasoline and other fuel products would include operations such as those described below. These are merely typical of the many processes in current use.

30. Crude-oil Desalting. Water is added to the crude oil to dissolve the unwanted salt. The mixture is passed through a vessel containing electrodes between which a potential of several kilovolts is maintained. The potential gradient causes the salt and water to coalesce and settle to the bottom of the vessel, where the mixture is drawn off. The desalted crude oil is discharged near the top of the vessel.

31. Crude-oil Distillation. The crude oil, which is a mixture of a large variety of hydrocarbons having different boiling points, is heated in a furnace to about 750°F and then enters a fractionating tower. The components are separated according to boiling range, since the lighter ones rise in the tower as gases and the heavier ones fall in the tower as liquids. Trays with specially designed openings in them are installed at intervals in the tower to ensure intimate mixing of the rising gases and the falling liquids and to provide places where liquids having certain boiling ranges may be drawn off the tower. The operation is first performed in a distillation tower in which the pressure is maintained somewhat above atmospheric pressure and again in another tower which is kept under vacuum in order to reduce the boiling ranges of the hydrocarbons and thereby prevent their destruction due to excessive heat. The combined unit is called an atmospheric and vacuum, or A & V, unit. It is also sometimes called a "two-stage pipe still." The main fractions produced are condensable gas, gasoline components (which are processed further), diesel fuel, heating oils, lubricating oil, and pitch.

32. Fluid Catalytic Cracking. To increase gasoline yield from a crude oil, heavy gas oil from the distillation unit is processed in a cracking unit. The heavy molecules are brought in contact with a catalyst under proper conditions of temperature and pressure and are converted into lighter molecules. Thus lighter products are formed which are suitable for use as gasoline components. The use of a catalyst promotes the cracking reaction at a lower temperature and pressure and produces larger quantities of products having more valuable qualities than is possible with straight thermal cracking.

The clay catalyst is in powder form, and it is handled as a fluid. The cracking reaction causes the formation of carbon deposits on the catalyst particles. These are removed by controlled burning in a regenerator vessel. The catalyst is continually being circulated through the reactor and the regenerator by means of air and gas flow.

33. Combustion Air Blower. Combustion air for the regenerator is provided by a large centrifugal air blower driven by an induction motor and gear increaser or by a steam or gas turbine.

34. Gas Compressor. Some of the products of the catalytic cracker are drawn off as gas. This gas is compressed, condensed, and fractionated to provide other fuel products and feedstocks for petrochemical processes. A centrifugal compressor is used, and it is driven by an induction motor through a gear increaser or by a directly coupled steam turbine or gas turbine.

35. Catalytic Re-forming. In this process relatively small hydrocarbon molecules are combined to form larger molecules which can be used as gasoline components. Hydrogen is produced in this process. Large reciprocating and centrifugal compressors are used to move the large volumes of gas involved in the process. Hydrogen is a byproduct of this process.

36. Hydrofining. This process uses hydrogen in the treatment of other products, such as lubricating oils, to improve quality and to remove sulfur.

37. Cooling-water Pumps. Large volumes of water are used for cooling-process streams and for condensers. Water may be conserved by the use of induced-draft cooling towers.

Cooling-water circulating pumps are driven by vertical motors. Standby pumps are driven by steam turbines.

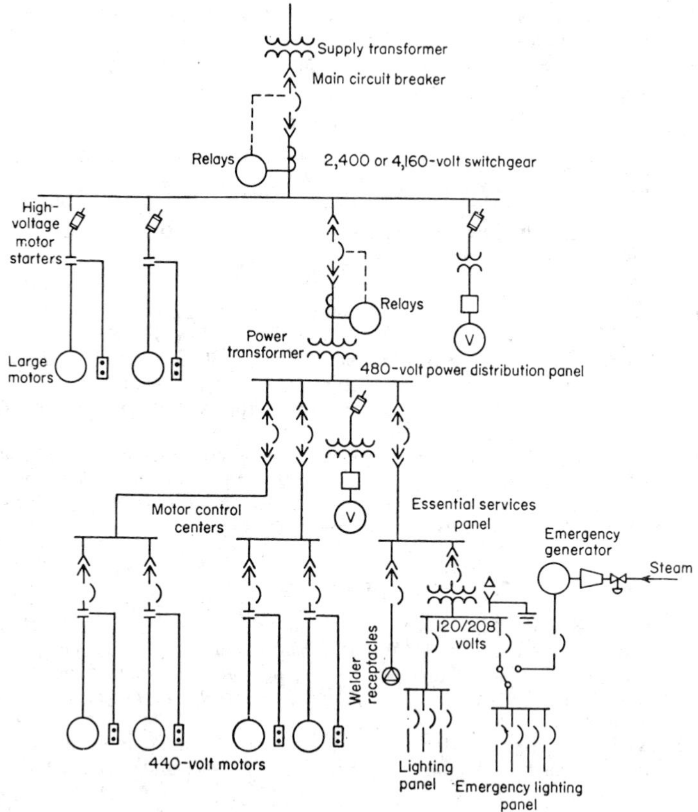

FIG. 20-3. Typical refinery process unit distribution system.

38. Power Supply. Refining operations are continuous processes, and uninterrupted runs of 1 or more years are expected between planned shutdowns for maintenance or turnarounds. Therefore it is essential that the power supply be extremely reliable.

Often duplicate full-capacity feeders are installed to the refinery, and sometimes these are run from different substations, for increased security.

39. Distribution Systems. Because refinery loads are often concentrated to a large extent in fairly well-defined process areas, it is common to install unit substations in the major process areas. These substations contain power transformers to provide 4,160- or 2,400-V power, 480-V power and lighting transformers, or provision for feeding lighting transformers. They also contain all associated switchgear, motor-control, and emergency generators.

40. Process-unit Distribution. A typical refinery process-unit distribution system is shown in Fig. 20-3.

41. Area Classification. Flammable gases and vapors are processed in oil refineries. Therefore it is necessary to classify the various locations according to the material that is present and also according to the degree of hazard expected.

Reference should be made to the applicable electrical code for requirements governing installations in classified areas.

The actual classification of areas is usually made by the electrical design engineer in consultation with persons who are familiar with the operation of the process.

The most common classification for refinery process units is Class I, group D, this class being that of hazardous gases and vapors, and the group comprising gasoline and many of the petroleum products.

Current practice is to classify outdoor, freely ventilated process areas as division 2. Indoor process areas that are not freely ventilated and places below grade level are classified as division 1. Areas in which a permanent ignition source is located, such as around a furnace, are not classified. Pressure-ventilated unit substations and control buildings are not classified. However, some companies follow the practice of classifying control rooms as division 2.

ELECTRIC MOTORS

42. Type of Motor. Two-pole induction motors having NEMA design B characteristics are used to drive the majority of refinery-process pumps. Motors operating at slower speeds are used for some applications, such as for driving reciprocating compressors or for driving centrifugal compressors through gear-speed increasers.

LIGHTING

43. Lighting Levels. Adequate lighting coverage must be provided for safe access by operating personnel to all parts of the process units. Local lighting is used extensively on tower platforms, in pump areas, etc. Footcandle levels follow general lighting practice. In some places, such as on towers, lighting levels are not critical, but the location of fixtures is important. For example, ladders and operating instruments must be illuminated.

44. Emergency Lighting. Emergency lighting fixtures give coverage of important accessways and tower ladders. They are also provided in the control building and the substation. Fixtures used in regular service are fed from an emergency lighting panel which is automatically transferred to the emergency power source when the main supply fails.

GROUNDING

45. Grounding System. An effective grounding system is necessary in refinery process units (*a*) to avoid having dangerous potentials on non-current-carrying equipment during electrical faults, (*b*) to ensure fast operation of electrical protective equipment under fault conditions, and (*c*) to dissipate electrical charges caused by atmospheric conditions.

46. Ground Loop. A ground loop consisting of buried cable and driven ground rods is installed around the process unit, and all major vessels and electrical equipment are connected to it.

EMERGENCY POWER SUPPLY

47. Emergency Generator. Emergency power is provided for emergency lighting and for the operation of electronic instruments. A small steam turbine-driven generator is provided at each process unit. The capacity ranges from 15 kW upward. Output voltage is 120/208 V 3-phase or 480 V 3-phase. The turbine is started automatically by means of a solenoid valve which controls the air supply to the pneumatic steam valve. Automatic switches transfer load to the generator output upon failure of the main power supply. For greater reliability, a dual-line generator may supply power for the instruments. A steam turbine is coupled to the motor-generator set, and upon failure of the motor supply voltage, the turbine picks up the load. Special speed control on the turbine maintains the desired standard of frequency. An automatic switch transfers the instrument load to the power-company system if the generator should happen to fail.

48. Batteries. Batteries may be used directly as a source of power for lighting and instruments.

CHEMICAL PLANTS

49. General. Electrical installations in chemical plants are similar to those found in oil refineries. However, some processes may involve materials that require special attention because of factors such as corrosivity and toxicity.

STEEL MILLS

By L. L. Quinlan

50. Power-distribution System. Integrated steel mills having blast furnaces and coke plants make use of the combustible gases from these processes by burning them to produce power and process steam. Many older steel plants produce and use power at 25 Hz, utilizing primary distribution voltages of 6.9 to 13.8 kV and secondary systems of 4,160 or 2,400 V and 480 V. Modern steel plants and modern parts of older plants utilize 60-Hz power exclusively, with primary distribution at 69 or 138 kV and secondary voltages of 13.8 kV, 4,160 or 2,400 V, and 480 V. Power can be supplied by public utilities, by in-plant generation, or by a combination of the two. Some plants having both 25- and 60-Hz systems have conversion equipment, typically large Scherbius sets or rectifier-inverter systems, for transfer of power from one frequency to the other. Where power is supplied from both a public utility and in-plant generation, there is usually some provision for controlling the maximum demand and improving the power factor of the portion of the load supplied by the utility to avoid high penalty charges.

The presence of numerous cranes and other equipment requiring d-c motors with some control over speed has led to the extensive use of 250-V constant-potential d-c shop circuits in most steel mills. In older plants the direct current is supplied by rotary converters, motor-generator sets, or mercury-arc rectifiers. In modern plants it is supplied by silicon diode or Thyristor rectifiers. The trend is toward the elimination of d-c shop circuits by the use of a-c cranes and package Thyristor power supplies for drives requiring variable speed.

51. Primary Production. The basic steelmaking areas of an integrated steel plant consist of *coke-oven batteries* for conversion of coal to coke, *blast furnaces* for conversion of iron ore to molten iron, and *steel-producing units* for refining molten iron and other alloy ingredients to steel. Once this basic steel has been produced in ingot (block) form or "continuous cast" into semifinished bars, it is ready for subsequent rolling into a usable size and shape.

Power consumption per ton of steel produced is low in the basic steelmaking areas because the products are handled in molten or bulk form, compared with the rolling mills, where reheated or cold steel is literally squeezed and stretched to the desired size and shape. Much of the electric power consumption in these primary producing areas is associated with auxiliary drives involved in material handling, water, air, and byproduct utilization and mobile equipment.

The processes of iron reduction and steel refining, with their many, sometimes elusive variables, do lend themselves to automatic and computer control. Open-loop computer systems have been applied to *blast-furnace* and *basic-oxygen-furnace* (BOF) operations. Raw-material handling and charging functions in *blast-furnace stock houses* and BOF have been automated extensively.

52. Rolling Mills. Rolling mills are classified either according to their construction or according to the material processed. Classified according to construction, mills are generally two-high or four-high, with a few existing three-high mills. Four-high mills consist of the usual two work rolls in contact with the product, with two additional "backup" rolls which are much larger and allow high rolling pressure without excessive deflection of the work rolls. A "universal" mill has vertical or edging rolls in tandem with the horizontal rolls. This permits a reduction of width or control of the edges of the product in the same stand where a reduction in thickness is taking place.

The principal types of mills, classified as to product rolled, are as follows:

a. Blooming mills roll ingots to blooms, or slabs. All material rolled in steel mills, except that which is direct continuous cast into slabs, blooms, or bars, first passes through this type of mill, or its equivalent, to be reduced to proper dimensions for handling in the finishing mills. These are generally single-stand, two-high reversing. Slabbing mills are a modification of the blooming mill. They are usually universal mills, which permits convenient rolling of wide slabs by eliminating frequent turning of the ingot. (See Fig. 20-4.) Blooming and slabbing mills may have such automatic features as preset of roll openings and speed synchronization between main and edger rolls. The preset information is sometimes stored on business-machine cards and read into the mill control system by a card reader, or it can be stored in the computer memory in computer-controlled installations. Blooming and slabbing mills are powered by large low-speed d-c motors operating from a variable-voltage system. A typical slabbing mill has a total of four 3,000-hp motors driving the horizontal rolls and two 2,000-hp motors driving the edger rolls at rated motor speeds of 40/80 r/min. The d-c power for blooming mills has been traditionally supplied by generators using the Ilgner system, but the present trend is toward the use of Thyristor power supplies connected in rectifier-inverter configurations.

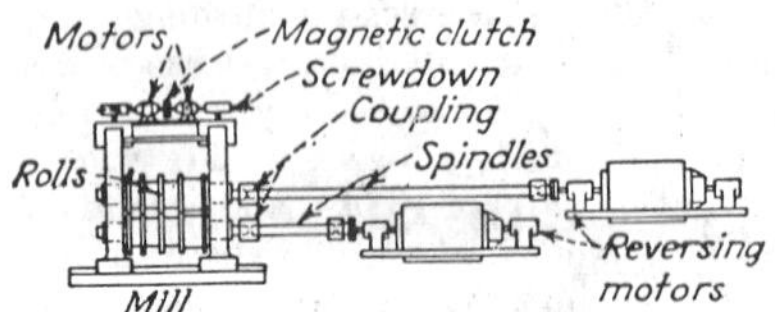

FIG. 20-4 Two-high reversing mill with twin roll-drive motors.

b. Hot-strip mills roll sheets, strip, and plate from heated slabs. These mills can be placed in two categories: continuous and semicontinuous. (These terms designate the type of rougher which precedes the finishing train.) The *continuous mill* has two to six mills in line which reduce the slab to a predetermined thickness for subsequent rolling through the finishing train (see Fig. 20-5).

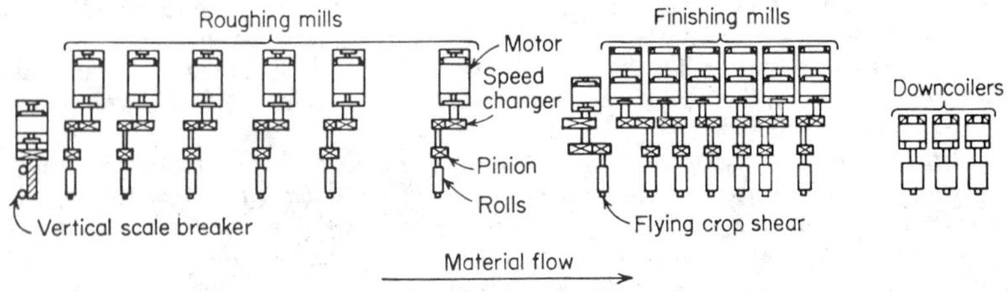

FIG. 20-5. Hot-strip mill.

The *semicontinuous mill* has a reversing rougher on which reduction is made by running the piece back and forth through the mill, which reduces the slab to a predetermined thickness for subsequent rolling through the finishing train.

The finishing train consists of five to seven stands, closely coupled and synchronized in speed, in which the piece is reduced to the desired gage. As a general rule, the piece is in all finishing stands simultaneously.

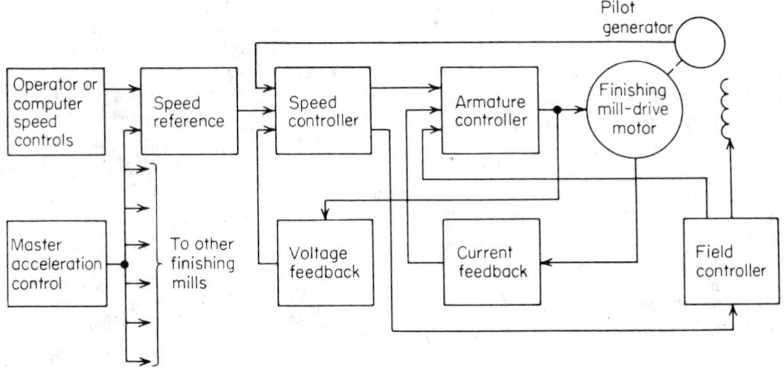

FIG. 20-6. Finishing-mill drive control.

Roughing stands of a continuous mill are usually driven by a-c motors, since speed synchronization with an adjacent stand is not required. Many existing roughing mills utilize wound-rotor induction motors with flywheels and water-slip regulators, but today synchronous motors predominate. The roughers on semicontinuous mills are driven by d-c motors.

Direct-current motors operating from a variable-voltage system are universally used for finishing stands. Tension between stands must be accurately controlled at a relatively low value, since the hot steel is in plastic form and excessive tension results in "necking," or breaking. Constant-tension "loopers" are used between stands for interstand tension regulation. The looper is pushed up by air, hydraulic pressure, or a torque motor, and the upward movement is restricted by the strip. The looper position is then fed back to the electrical control system, and the speed of individual finishing-stand motors is regulated to maintain the proper looper position.

Various measuring devices must always be in operation, measuring gage, screw position (roll opening), looper position, roll force, and width of strip. With the feedback from these devices, the positioning of the screw-downs, which control gage, and the edgers, which control width, can be regulated during the rolling of the strip. The newer mills make use of digital positioning to set up the mills prior to the entrance of the bar into the mill.

In Fig. 20-6, the thread-speed reference is set into the speed controller, and this speed is maintained by utilizing the pilot-generator feedback. When the strip enters the mill, the load increases sharply, bringing the current feedback to the armature controller into effect. By controlling the armature and field, rapid speed changes while complete control is maintained are possible, both with and without load. The acceleration control, when called upon, feeds into all the finishing-mill motors for uniform acceleration.

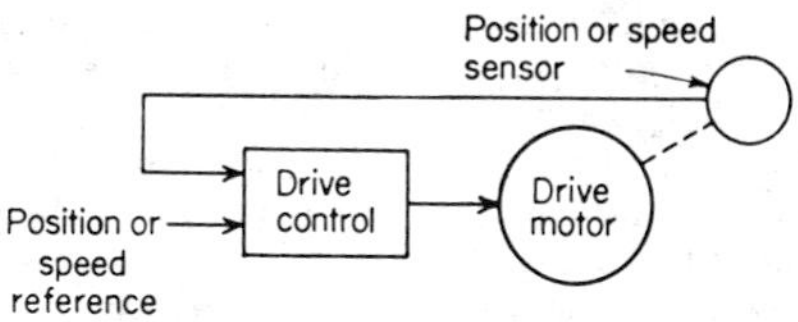

FIG. 20-7. Position regulator.

For setting up the mills in advance, digital position regulators are generally used; however, some of the newest mills have small digital computers controlling each stand. The principle of operation is basically the same in both systems. In Fig. 20-7, the desired position for the drive is set into the regulator, and then by taking the difference between this reference and the feedback signal from a position sensor, the drive is operated until this difference is zero. The thread speeds of the mills can be set up the same way except that, instead of moving to a preset position reference, the stands are accelerated or decelerated to a preset speed reference.

As the strip cools during rolling, it becomes harder and the output gage changes. Part of this change in gage can be controlled by accelerating the mill, but much of it must be controlled by changing the position of the screw-downs. The newer hot-strip mills use automatic gage control systems (AGC) to regulate screw-down position while the strip is in the finishing mill. There are two types of AGC in use, a constant-gage system and an absolute system. Both systems work basically the same, by taking a reference either from the head end of the strip in the constant-gage system or from preset switches in the absolute system and maintaining it throughout the length of the strip. This reference is compared with the actual gage of the strip, measured as a function of roll force and screw-down position, or by an x-ray gage, and the screw-downs are moved to bring both these signals to the same value. This is a continuous operation during the period of time the strip is in the finishing mills (see Fig. 20-8).

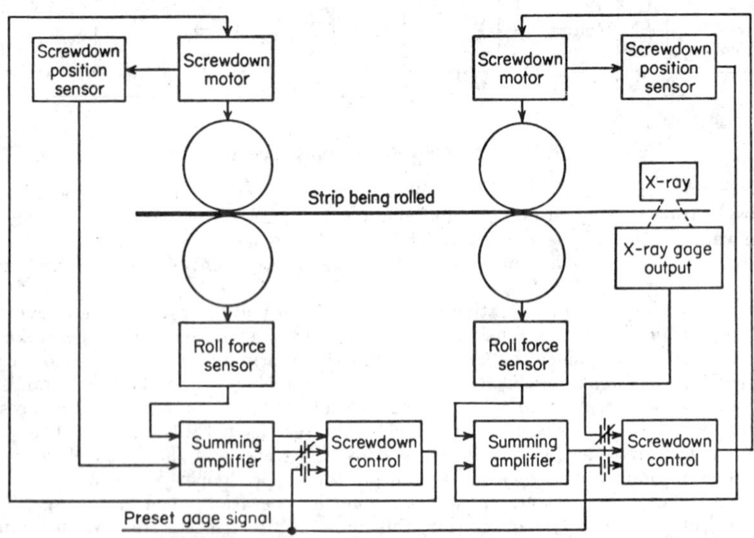

Fig. 20-8. Automatic-gage control system.

Modern hot-strip mills are built with partial or complete computer control options. Rolling in a completely automatic or computer-controlled state, the mills function as previously explained, except that the computer provides the position and speed references instead of operators. With a computer-controlled mill, these references can be constantly changed as the computer receives new information from the strip.

The handling of steel between roughing stands and between the finishing stands and the downcoilers is accomplished by roll "tables," which are a series of motor-driven rolls upon which the steel lies. In some installations, one large table motor drives a number of table rolls through mechanical gearing and shafting ("line-shaft drive"); in others, each roll is driven by one small motor. Permanent-magnet field d-c motors have been particularly successful for this arrangement. Table rolls are generally driven by d-c motors operating from a variable-voltage system, but there are a number of variable-frequency induction-motor systems in use. As Thyristor adjustable-frequency power supplies become more reliable and more available, this system could become standard.

c. Tandem cold-strip mills are used to cold-reduce previously rolled hot-strip mill products down to thicknesses as low as 0.002 in. Special "foil" mills have been built which can roll even lighter gages. A cold-strip mill is similar to the finishing stands of a hot-strip mill, except that tension between stands plays a much more important role in reducing the thickness of the steel. Modern cold-strip mills are built for finishing speeds in excess of 5,000 ft/min.

Cold-strip mills are generally three- to six-stand mills, four-high, with a coil box or payoff reel on the entry end and a tension reel at the delivery end. These mills are universally driven by d-c motors operating from a variable-voltage system. Usually the individual stands are voltage-regulated, and the operator establishes the motor field and, thereby, the stand speed according to the gage reduction and the strip tension desired. Load-cell tensiometers are used to indicate to the operator the interstand tensions.

Some of the most modern mills have each stand speed regulated, and all major drives incorporate full field acceleration to base motor speed for maximum torque. Above base motor speed, automatic field weakening with constant rated armature voltage is used to attain top motor speed. Payoff-reel tensions, interstand tensions, and winding-reel tensions are accurately controlled by using load-cell tensiometers to measure tension and provide feedback to tension regulators which operate on the appropriate drives (see Fig. 20-9).

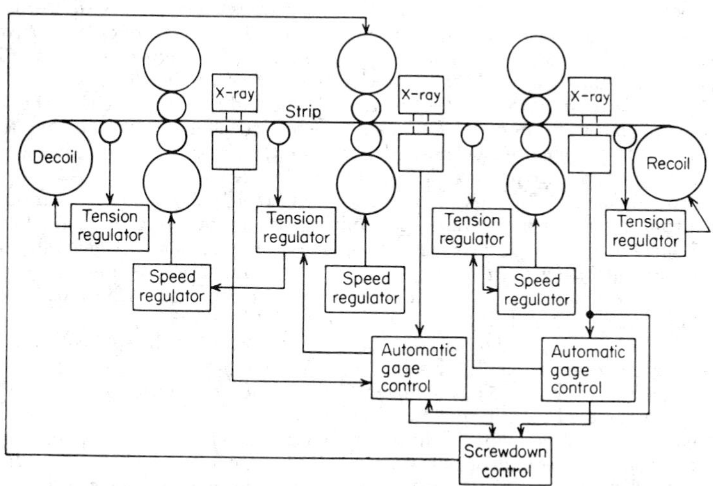

FIG. 20-9. Cold-strip mill.

Finished steel gage is controlled by utilizing x-ray gages to provide information to AGC regulators. Gage control may be accomplished by operating on interstand tensions only or by a combination of interstand tensions and work-roll openings. Work-roll openings ("screw-down positions") are controlled by digital position regulators which hold the screw-down position constant to within 0.0001 in. To help further in producing constant finished gage, the rolling forces on each stand are held constant by using speed-programmed digital regulators to hold constant rolling force at all speeds. Rolling force is measured and indicated by load cells placed under each stand housing.

d. Billet mills, used to roll blooms into billets, are frequently of a continuous type with rolling stands in tandem with several sets of passes in the rolls so that different-sized billets can be produced from a given bloom without changing rolls. This type of mill is used for producing only a limited range of sizes, which are further reduced in finishing mills.

e. Plate mills produce plates from slabs previously rolled by a blooming or slabbing mill. These mills are generally single-stand mills, either two- or four-high reversing or three-high running continuously in one direction, though some mills are provided with several stands in tandem, one serving as a rougher or breakdown stand. They are sometimes universal mills, being provided with vertical rolls for finishing the edges of the plates.

f. Structural mills are used for rolling beams, heavy angles, channels, etc., from blooms or billets. Such mills rolling the smaller structural sections and miscellaneous shapes are sometimes called "bar mills." They are frequently three-high, with more than one stand in line, and frequently have a separate rougher or breakdown stand. Some also have two-high reversing finishing mills with edger and vertical, as well as horizontal, rolls, to produce wide-flange (H) beams.

g. Rail mills are special mills for rolling this product only from blooms or billets, though rails are sometimes rolled on structural mills.

h. Merchant mills are used for rolling small angles, channels, rounds, squares, etc., from billets. This classification is generally applied to mills rolling the smaller sections used for miscellaneous purposes. Most are two-high, in tandem; however, some are arranged for "cross-country" rolling.

i. Rod mills are a specialized type of merchant mill for rolling small rounds, usually from No. 5 BWG and upward, which are later drawn into wire. Modern rod mills are arranged to roll a multiplicity of strands simultaneously.

j. Temper mills are used to produce steel strip of the desired temper, flatness, surface, and luster by using rolling pressure and tension. Reduction in thickness is incidental in the process and is normally very slight. Temper mills generally have one or two four-high stands and are similar to cold-strip mills, except that they are of lighter construction, have less powerful motors, and have simpler electrical control systems. Temper mills have been built and operated at speeds in excess of 7,000 ft/min.

k. Tube mills are used to produce tubes, e.g., pipe and conduit, by either the seamless or the butt- or lap-weld process. Seamless mills pierce a solid billet and then form the pierced billet into tubes of the desired size and thickness. Butt- or lap-weld mills form and weld previously prepared pipe skelp into tubes.

53. Process Lines. In most fully integrated steel mills, there are various types of process lines such as pickle lines, galvanizing and aluminizing lines, tin lines, continuous-annealing lines, and shear and slitting lines. Process lines are generally powered by d-c motors operating from a variable-voltage system. Tensiometers are sometimes used to indicate and even control tension at strategic places. The most important process lines are described below.

a. Pickle Lines. After steel is hot-rolled and before it can be cold-rolled, it is processed through a pickle line. In this process, all scale and impurities that have built up on the surface of the strip are removed. This is accomplished by running the strip through a tank of acid, which "pickles" the steel.

To maintain constant material flow through the pickling tank, a welding machine is used to weld the tail end of one strip to the head end of the following strip. To avoid stopping the line every time a weld is to be made, slack in the strip is allowed to develop immediately following the welder in a pit called a looping pit. Then, as the tail end of the strip comes into the welder, it is held there, and the slack in the looping pit is taken up to maintain constant material flow in the pickle tanks while the weld is made. There is another looping pit at the end of the line preceding the recoiler that performs the same function while a full coil is being removed and another started.

b. Galvanizing, or coating with zinc, and aluminizing can be described as the same process since the same line is used for both processes, with only the coating changed.

The steel must first be pickled and cold-rolled before coming to the galvanizing line. There, just as in the pickle line, the first process is welding, head to tail, of one strip to the next. In place of looping pits, however, traveling looping cars are used to absorb or supply strip while the entry and delivery ends of the line are stopped. From the entry loop car, the strip goes through a controlled atmosphere to a temperature near the melting point of the plating material. The actual plating then takes place as the strip is "dipped" or run through a tank of molten plating material. After plating, the strip passes through the delivery loop car and is coiled or cut to length.

c. Tin lines are used for coating strip steel with a protective layer of tin after it has been cold-rolled, annealed, and temper-rolled.

A welder and a looping tower are used at the beginning of the line. The looping tower stores strip in numerous vertical passes, as opposed to the horizontal looping car in the galvanizing line. Cleaning and rinsing are the next step in the process. The tin is deposited on the steel strip electrolytically, by using pure tin anodes and a plating solution which may be acid or alkaline depending upon the process. The low-voltage plating current is produced by plating generators or (on modern tin lines) by silicon-diode rectifiers. The current is applied directly to the tin anode and, after passing through the solution and the steel, is returned to its source through current-collector rolls in contact with the strip. After plating, the strip is put through a "reflow" process which heats it to the melting point of tin and gives it a shiny appearance and the proper iron-tin alloy bond. The reflow furnace may be gas-fired, or it may be a radio-frequency induction heating process.

After reflow the strip passes through the delivery looping tower and is then coiled or cut into predetermined lengths.

54. Instrumentation. More than any other factors, the rapid advances in instrumentation technology and techniques have made possible the revolutionary strides in automated process control and increased productivity in the steel industry.

Typical examples of steel-mill instrumentation are as follows:

a. X-ray and radioactive-isotope devices for (1) continuous measurement of moving strips and slabs from strips as thin as 0.002 in to slabs as thick as 2 in by using absorption techniques, (2) continuous measurement of coating thicknesses on steel strips by using fluorescence or reflectance techniques, (3) analysis of the chemical properties of steel by using spectrographic techniques, and (4) detection of refractory erosion in blast furnaces by using absorption techniques.

b. Infrared and ultraviolet devices for (1) continuous width measurements of moving hot-steel strips and slabs, optical scanning, and tracking of the edges, (2) measurement of hot-steel temperatures by using optical pyrometric techniques, and (3) detection of pinholes down to 0.001 in or less in diameter in moving strips, usually by using visible or near-ultraviolet light.

c. Ultrasonic devices for the detection of flaws in slabs.

d. Magnetic devices, which produce inferred measurements of physical or Rockwell hardness of moving strips by measurement of magnetic hardness.

e. Communications equipment, which is employed for (1) two-way voice communications by using either frequencies in the radio-frequency spectrum or frequencies near 100 kc for communications with overhead cranes by utilizing the hot rails as the information-transportation medium, (2) remote control of equipment such as overhead cranes, (3) closed-circuit television systems, which are especially useful for viewing high-temperature and remote areas, and (4) telemetry and supervisory control equipment, which allows operating personnel to remotely monitor and control the different processes involved in the production of steel.

f. Digital devices for (1) positioning of screw-downs, (2) control of machine tools, (3) measurement and control of elongation or stretching, length, and speed of steel strips, and (4) analog to digital conversion of process information for introduction into process control computers, data loggers, regulating systems, and telemetering systems.

ELECTRICITY IN AGRICULTURE

By Fred T. Wolford

FIELD OF APPLICATION

55. Extent of Electric Service. By June, 1963, about 3,505,000 farms, or approximately 98% of the total number of 3,580,000 farms reported by the U.S. Department of Agriculture for that year, were receiving central-station electric service. As of April, 1964, the U.S. Census Bureau listed the nation's farm population at 12,954,000. This is a reduction of 6,065,000 farm residents since 1954. Conversely, the average size of the nation's farms has increased from about 215 acres in 1950 to 302.5 acres in 1959. Farm production has increased by an estimated 17% in 1963 as compared with 1953 despite this 33% decrease in farm population. Farm output per man-hour has nearly doubled in the decade preceding 1964. Increased mechanization, electrification, chemical fertilizers, and better breeding and management practices are credited with this change.

56. Energy Use. Electricity supplies 10% of the purchased energy requirements of agriculture. Average annual use of electric energy on United States farms in 1965 was 8,592 kWh according to the Farm Electrification Council. About half this amount was consumed in the farm home. This amount represents a twofold increase since 1954, and it is expected nearly to double again by 1975 to 15,000 kWh.

FARM HOMES

57. Electric Home Heating. The electrical heating of homes has increased significantly in the 1960s. It is clean and conveniently controlled. The conversion of electric energy into heat energy is 100% efficient. According to the Edison Electric Institute, in 1965 there were 2.2 million residences using permanently installed electric heating units in the United States as a whole. This figure excludes various kinds of portable heaters. This is a threefold increase over the 750,000 installations in 1960. Twenty-four percent of these homes had been converted from other forms of energy, while the remaining number represents new construction. Although exact figures are not available, it has variously been estimated that 2.5 to 4% of farm residences have complete or partial electric home heat permanently installed. Average power consumption for a five-room, insulated farmhouse is 21,000 kWh/heating season, exclusive of portable heating devices.

58. Cooking. For the country at large, according to the Edison Electric Institute, the average connected load of the electric range is 12 kW, with an estimated annual consumption of 1,200 kWh. The demand has been calculated at 8 kW for a single-range installation according to the National Electric Code.

Tests made by the Home Economics Research Branch, U.S. Department of Agriculture, indicate that, for the complete comparable cooking job, electric ranges used half as much energy as bottled-gas ranges. In comparative cooking tests using the top of the range and oven, conducted over an 8-day period, the electric ranges used an average of 120,208 Btu and bottled-gas ranges used 248,706 Btu.

U.S. Department of Agriculture home economists' tests also indicated that there is no difference in speed of cooking on electric ranges or bottled-gas ranges.

59. Water Heating. It is extremely uneconomical to heat water on the surface burners of an electric range as is commonly done on the other types of farm range. This makes separate water heating an important consideration. For the temperatures usually found and required (a rise of about 100°F), 1 kWh will heat 4 gal of water. To this must be added all heat losses by radiation from tank or pipe. Storage-tank heaters with tanks of 60- to 120-gal capacity are sometimes used to obtain an off-peak rate when such a rate is available in the country. Under the more common policy of one rural rate for all energy consumed, where that rate is influenced by the customer's demand, a peak-limiting switch may be used to prevent the simultaneous operation of the water heater and the range. To be effective, instantaneous heaters usually have a demand of 3,000 to 5,000 W. Many tests under actual farm conditions in six northern states show an average monthly use of 289 kWh for household water heating.

60. Refrigeration. Freezing and refrigerated storage methods for processing and storing farm-produced food supplies have largely replaced such methods as pickling, brining, drying, and canning. Food freezers, home refrigerators, refrigerated storage rooms for cooling and preserving farm products, and use of community frozen-storage locker plants are popular in farm communities.

The larger sizes of home refrigerators with 10 to 12 ft^3 of storage space and an average energy consumption of 40 kWh/month are popular with farm families. Farmers prefer food freezers of the larger type, with 15 to 20 ft^3 of storage capacity. Average energy consumption varies from 3.56 to 6.27 kWh/(ft^3) (month) according to reports from Iowa State College.

FARM STRUCTURES

61. Water Systems. Water requirements for the farm household and farm enterprises, excluding irrigation, are frequently supplied by a single well. The water-supply equipment is usually an automatic hydropneumatic or air system having pumping capacity of 300-600 gal/h and using a ¼- to 1-hp motor, depending upon the total head in feet and the rate of pumping. Home water requirements average 50 gal/(person) (day). In addition, livestock requirements must be added: each horse, steer, or dry cow, 12 gal; each milk cow, 35 gal for drinking and washing equipment; each hog, 3 gal; each sheep, 2 gal; each 100 chickens, 8 gal. For yard fixtures, each ¾-in hose outlet requires 300 gal/h.

Where the source of supply is not more than 22 ft below the pump, a shallow-well system can be used. A jet-centrifugal pump has a practical lift limit of 80 to 100 ft, and piston-type pumps can go as deep as 800 ft with a suction lift below cylinder of 22 ft. This type is placed directly over the well and is generally recommended where pumping depths exceed 80 ft. Automatic pressure switches are usually set to start the pump when the pressure falls to 20 lb and stop it when 40 lb has been obtained. The energy requirement per 1,000 gal of water pumped rarely exceeds 2 kWh.

62. Heating Systems. Electrical heating of farmstead structures is generally confined to milk houses, individual pen-type areas for young livestock, and poultry brooders (see Par. **73,** Brooding). Electric heating in the milk house is ideal, as it is odorless, conveniently controlled and meets the high sanitary standards required. The milk-house temperature should not exceed 40°F. Several types of heaters have been successfully used: (1) the forced-air circulating type requiring 1,500 to 3,000 W, (2) batteries of 250-W infrared heat lamps directed toward working areas and water pipes, and (3) "heat-pump" systems, which utilize the heat removed in cooling the milk. In this type the ice-bank refrigeration system (either bulk or immersion coolers) extracts heat from the water in building up the ice, the heat thus being available for the milk-house. Electricity used in this indirect manner produces about three times as much heat as it would if directly used in a resistance heater. Only coolers with ½-hp or larger motors are recommended for this.

In the colder regions, the milk house must be insulated for the most economical cost of installation and operation. In these areas an electrically heated milk house needs at least a 1,500-W heater serviced by a 230-V line. Thermostats are usually attached to the heater unit, and operating consumption ranges from 1,000 to 3,000 kWh a season.

The need for infrared heat lamps during the first week of hog farrowing and sheep lambing has been proved. A 250-W lamp will heat an area 24 in in diameter when 3 ft above the floor. The lamps should be positioned at least 6 in above the animals and at least 30 in above the floor when bedding is used.

63. Ventilation Systems. Electrically powered mechanical ventilation of livestock structures provides low-cost positive control for the removal of excess animal body heat, objectionable odors, condensation, and temperature and humidity control. A full-grown cow will give off 3,000 Btu/h of body heat; 1,000 chickens about 800 Btu/h. Accurately controlled tests with dairy cows at the University of Missouri showed that temperatures above 75°F and relative humidities over 75% resulted in sharp declines in milk production and body weight.

In general, summer ventilation should maintain inside temperatures equal to or below the outside temperature, while in winter the reverse is true. Thermostatically controlled motor-driven fans are installed as required, with adequate fresh-air intakes to prevent excessive energy costs. Two speed fans, chosen to move the maximum air volumes required for various livestock, will permit air flow to be reduced in cold weather. Fan motors range from 1/20 to ½ hp and will consume 250 kWh/year and up, depending on usage. One kilowatthour of electricity will move about 1 million ft^3 of air.

64. Lighting. Next to the farm home, sufficient illumination is especially needed in poultry houses, milk houses, dairies, and livestock barns. Adequate lighting throughout the farmstead will stimulate production, prevent accidents, and expedite chores. For general lighting of farm structures, outlets should be provided at 10- to 12-ft intervals. The use of fluorescent fixtures is increasing, providing a softer, shadowless illumination at decreased cost. Dustproof fixtures are used in haylofts and feed-grinding and storage areas, and vaporproof fixtures are used where fire hazards exist. Nowhere on the farm is the presence of light more critical than in the poultry house. The intensity and duration of light stimulate maturity and subsequent egg production, govern the time of day eggs are laid, and determine feed consumption and rate of growth. Extending day lengths to 14 or 16 h can increase production 20 to 30% during the winter months. Sixty-watt incandescent lamps mounted in fixtures using 12-in shallow dome reflectors or 75-W lamps with built-in reflectors are used for intermittent lighting where morning or evening light is used to lengthen the day. Lights are spaced 10 ft apart and can be controlled by time switches. A dimmer is required for evening lights so that hens can find their way to the perches. Smaller wattage (12 to 25-W) lamps, sometimes colored, are used for all night lighting. Energy requirements per 100 birds range from 5 to 8 kWh/month during the lighting season.

DAIRY AND LIVESTOCK PRODUCTION

65. Automatic Feeding. Mechanical and automatic feeding of livestock and poultry is of growing importance. The flow of materials proceeds from individual storages to blenders, grinders, conveyers, and finally to distributors, as shown in Fig. 20-10.

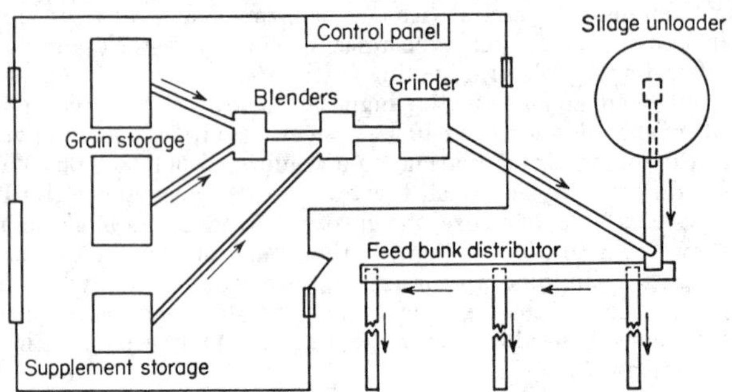

Fig. 20-10. Cattle-feeding system designed to feed silage in addition to grain and supplement.

In a typical automatic system feed is processed at a low rate over a maximum time period. This allows the use of small, low-capacity electric motors at each stage and does not require overhaul of the 100- and 200-A service found on most farms. Capacitor or repulsion start motors, totally enclosed and with continuous-duty rating, are recommended. Hammer or burr grinders with 2- or 3-hp motors are required for small grains (up to 5-hp motors for ear corn). Other motors are generally ½- to 1½-hp for the movement of the materials. In proportionally blending feed components from storages, belt, auger, or electric-vibrator feed meters are used. After this mixture is ground, it is moved by auger or pneumatic blower to distributor conveyers in the feedlot or barn which are of the endless-chain, belt, or auger types. In cattle-feeding systems silage may be automatically removed from the silo. For feeder cattle the grain-supplement mixture is then metered onto the silage and the complete ration conveyed to the feed bunk for distribution. Dairy cows usually receive the grain ration separately from the silage, either in the barn or in the milking parlor.

66. Controls are of the simple off-on switch type for manually controlled systems, but for fully automatic feeding the switching is accomplished by electric timers or sensing elements, which measure temperature, level of material, or pressures, at each point in the movement of feed from storage to final distribution. These sensing devices are used with current-carrying relays such as a magnetic starter. This starter may have auxiliary switches which control circuits to other magnetic starters. Thus, interlocking of each stage of feed transmission is accomplished. This is necessary where a sequence of operations requires overload protection for each motor. Since these systems depend on continuance of electric power, it is advisable to provide an emergency generator to handle at least a part of the demand in the event of a power failure.

Automatic feeding equipment introduces new hazards for operators. Starting by switch or time clock may be unexpected. Open conveyers should be shielded. Power must be disconnected while repairs are made. Wiring should be of the underground-feeder type, and all electrical equipment must be grounded.

67. Milk Handling. With the increase in numbers of cows being milked and the milk quality required to retain grade A fluid-milk markets, many producers have shifted to some degree of automation in milk handling. The completely automatic system is one whereby milk is taken from the cow, carried through a vacuum pipeline to a bulk cooling and storage tank, whence it is pumped into a bulk transporting truck for delivery to the processing plant, without having been lifted by, carried by, or exposed to human hands.

In the latest development, cows enter a central milking room, commonly called a milking parlor. Three or more are milked simultaneously by machine. The milk is moved by suction through glass or stainless-steel tubing to the bulk tank, located in a separate room. Bulk tanks are either of the vacuum or of the atmospheric type. The latter requires the addition of a releaser, since the tank itself is not under vacuum.

In conventional stanchion barns cows are milked by machine, the milk moving directly from the teat cups through a pipeline system to the bulk tank in a continuous flow. One variation in this system is where the milk is first collected in the milker pail, then poured into a centrally located receiver, to be moved through portable piping into the tank, either by vacuum or by a milk pump.

Where the above systems are not yet in use, the operator moves the machine from cow to cow, collecting the milk in cans, after which it is carried to the milk room for cooling.

68. Milking machines are operated by intermittent suction and pressure produced by a vacuum pump. The degree of vacuum is controlled by a relief valve and is usually maintained at 13 to 17 in Hg. The pulsations range from 45 to 55/min and may be regulated by an electromagnet energized through a distributor from a small generator driven by the pump motor, by the operation of the vacuum upon a piston, by a rotary or slide valve, or by the timed strokes of the pump itself.

Use of a milking machine saves about one-half the labor or one-fourth the cost of milking 20 to 25 cows. For larger herds, the savings increase. Milking technique has so improved that the machine is left on the cow only 2 to 4 min, and very little stripping is required. The power requirements depend upon the type and size of the system but should not exceed ¼ hp/milking unit. Average use of energy is below 2 kWh/(cow)(month).

69. Milk Cooling. Milk as drawn from the cow is at a temperature of approximately 98°F, a temperature favorable for bacterial growth. Standards for farm holding and/or cooling tanks formulated by the International Association of Milk and Food Sanitarians, the U.S. Public Health Service, and the Dairy Industry Committee require that milk be cooled to 50°F within the first hour after milking and to 40°F within the second hour. Furthermore, the milk temperature must not fluctuate more than ±2°F during the holding period. Milk-cooling standards such as these usually make it necessary for the producer to use mechanical refrigerated cooling equipment.

Where milk is collected and stored in cans, two types of milk coolers are used on the farm. They are commonly known as the immersion-tank type and the water-spray type. In the former the cans are stored in cooled water, and in the latter the cans are cooled by a cold-water spray. In each case the water is cooled by mechanical refrigeration. In the spray type a bank of ice is frequently frozen prior to cooling time in order to increase cooling capacity. Where the cans are cooled in the immersion type of cooler, a pump or agitator is used to circulate water for more rapid cooling.

70. Bulk handling and cooling of milk eliminate the necessity for using the conventional 10-gal milk can because the milk is cooled and stored in stainless-steel vats. Tank trucks pick up the milk and deliver it to processing plants. Two methods are used for cooling the vats. The first is commonly called the direct-expansion refrigeration type, where the Freon refrigerant is in direct contact with the milk tank liner, and the second is a water-spray cooling type, using an ice bank to cool the water.

Recent tests indicate that direct-expansion air-cooled refrigeration units require a motor rating of 1 hp for each 50 gal of milk to be cooled per milking. For water-spray types using an ice bank, power requirements are sharply reduced and are estimated at ⅓ hp for each 50 gal of milk cooled. Direct-expansion units are higher in first cost but cool milk more rapidly and at a slightly lower energy rate than water-spray types. Electrical energy per 100 lb of milk cooled averaged 3.5 kWh for bulk coolers and 2.9 kWh for can coolers according to University of Georgia reports.

71. Water Heating. A source of hot water is required by nearly all dairy milksheds for such purposes as sanitizing milking equipment, massaging udders before machine milking, veterinarian and insemination services, milk testing, etc. Two types of electric water heaters for the dairy are in common use, namely, the nonpressure, or "pour-in," and the pressure type. The latter is increasing in use.

The 10-gal pour-in type has been popular in small dairies. No water system or piping is required. To get a gallon of hot water, a gallon of cold water is poured in to force the hot water out. The heaters are insulated and have heating units in sizes ranging from 250 to 1,500 W.

The pressure type of electric water heater is the same as that used in the home. Where more than 15 gal of hot water is required daily, the pressure type of electric water heater is preferred by dairymen. Immersion-type heaters are also available for heating water in open pails. They are relatively low in efficiency and range in size from 500 to 1,500 W.

Warm water may be used for drinking by calves, other livestock, and poultry. Automatic, electrically heated stock waterers are used in areas where freezing is a possibility. Heating units of 150 to 200 W thermostatically controlled are usually adequate. Electric energy used averages from 15 to 20 kWh a month in the North Temperate zone.

72. Egg-handling Equipment. Automatic egg graders, washers or cleaners, and cooling and storage rooms are utilized in commercial egg production. Egg prices are based on quality and size. Mechanical, electrically driven graders have become standard equipment on large poultry farms because of possible labor savings. Some graders have a candler, which eliminates this separate operation. The larger units grade four sizes of eggs at the rate of 12 cases/h. Electric-energy requirement is about 1 kWh for each 100 cases of eggs graded. Dirty eggs are cleaned by automatic washing machines or by an abrasive wheel or buffer attached to the shaft of a small electric motor. Automatic machines are available that will dry-clean up to 1,500 eggs/h. Electrical-energy use is nominal, varying from 1/4 to 1/2 kWh/case of eggs dry-cleaned and 1 to 1 1/2 kWh for each 10 cases for the washing and drying models. Gathering eggs two or three times per day and cooling to about 55°F within 24 h help the poultry producer secure premium prices. Eggs held in refrigerated rooms at temperatures below 60°F and 75% relative humidity are becoming increasingly popular.

73. Brooding. Four different types of electrically heated chick brooding units are in common use. They are the (1) conventional hover type, (2) infrared lamp, (3) electrically heated slab brooder, and (4) underfloor hot water, heated electrically.

Conventional hover-type brooders are constructed for 300- to 500-chick capacity. The top and sides of these brooders are insulated in order to reduce heat loss. The resistance types of heating units, 750 to 1,000 W capacity, are used as sources of heat. Temperature is controlled thermostatically. The initial starting temperature is 90 to 95°F, with a gradually reduced rate of 5°F/week until 65°F is reached. Brooding periods vary from 6 to 10 weeks, depending on temperature and climatic conditions. Space requirements for chicks vary from 7 to 10 in^2 for Leghorns to 10 to 14 in^2/chick for large breeds. Energy consumption varies from 0.5 to 2 kWh/(chick) (brood), depending upon climatic conditions, condition of brooder house, and season of the year.

Brooding with infrared heat lamps in sizes from 125 to 375 W is common practice. A 250-W lamp provides heat for 50 to 100 chicks, depending upon brooding-room facilities and climatic conditions. Infrared heat lamps are used for farm-sized flocks and also for commercial chick brooding of 20,000 or more chicks in one house. In larger installations a part of the lamps may be thermostatically controlled. Reports from a Virginia experiment station give energy consumption at 2 to 3 kWh/chick for winter brooding.

Concrete slabs heated with embedded and thermostatically controlled heating cable are also being used for chick brooding. The cable is customarily installed about 1/2 in below the surface of the slab. From 30 to 50 W heating capacity/ft^2 of slab is required. Space requirements for chicks are similar to the hover-type brooder. Slab brooders are customarily covered with a hover. In some cases a combination of concrete-slab infrared-heat-lamp brooding system is used. In the event of power failure slab brooders tend to retain heat, and temperatures will drop only a few degrees in a 6- to 8-h power-off period. Energy use is 5 to 6 kWh/chick.

PLANT PRODUCTION

74. Irrigation Pumping. More electrical energy is used for irrigation pumping than for any other field operation. Proper design of an irrigating system will depend upon the following factors: (*a*) the acreage and kind of crop to be irrigated; (*b*) the amount of water that must be supplied; (*c*) the amount of underground water available; (*d*) the depth at which it is found.

Except where the water requirements are small and the depth to water great, plunger pumps are rarely used. The more common type is the centrifugal turbine pump, but where the lift is not more than 15 ft, the horizontal centrifugal pump is also used. The bowl of the turbine pump should be set below any expected drawdown in the well, and this will depend upon the porosity of the surrounding strata as well as upon the rate of pumping.

Vertical turbine pumps require vertical motors with either solid or hollow shafts and thrust bearings capable of carrying the pump load. Horizontal pumps should be connected to their motors through flexible couplings to avoid the use of belts. With average allowance for evaporation, irrigating an acre 1 ft deep requires 340,000 gal. The soil can be wet to a depth of 4 ft by using 4 to 6 in of water. From 10 to 20 in is required to produce the ordinary crops. With an overall efficiency of 50% for pump and motor, each acre-foot of water will require about 2 kWh of electricity for each foot of lift.

75. Methods of irrigation include overhead pipes, stationary spray plants, and portable sprinkler systems. In the overhead type the discharge pipes are supported on posts and are located about 50 ft apart in lengths up to 600 ft. The pipes are usually supported on rollers so that they can be oscillated by a type of water motor, and nozzles are spaced 2 ft or more apart. Sixty gal/min of water per acre at 30 lb pressure is satisfactory. Stationary spray plants can reduce spraying time in orchards by 50% or more compared with portable units. A central pumping station, mixing tanks, and symmetrically located discharge pipes complete the layout. The pumps are usually three- or four-cylinder, single-action, with capacities of 10 to 60 gal/min at pressures up to 600 lb or more, requiring motors of 5 to 30 hp. Outlets are located at regular intervals for attaching the spray hose. Spray nozzles discharge up to 8 gal/min depending on pressure and orifice size. Power required is usually under 10 kWh/(acre)(application). Portable systems utilize lightweight, quick-coupled pipes, with sprinklers attached. Laid on the ground, they require considerable labor to move, but the initial investment is less than with other types. Sprinklers operate at pressures of 20 to 50 lb/in^2 and cover circles 40 to 90 ft in diameter, delivering 3 to 30 gal/min. A motor as small as 2 hp will apply 1 in of water to 3 acres of land per week, although larger outfits are commonly used.

76. Grain Conditioning. Field harvesting and on-the-farm storage losses of small grains and ear corn can be materially reduced where mechanical crop-drying or conditioning equipment is utilized. Early harvest reduces field losses due to shattering or lodging of grain and shelling, which may occur during mechanical harvesting. Crops can be harvested when weather conditions are most favorable as soon as possible after they mature, thus reducing the chance of storm damage while the crop dries in the field.

77. Heated-air Crop Dryers. Equipment needed includes an oil burner, a power-driven fan, and a drying bin for the ear corn or small grain.

Most of the dryers are portable. Each unit consists of a power-driven fan, a heater, and safety controls. Such dryers have two characteristics that determine their performance in drying grain: (1) the rate at which heat is supplied (rate of fuel consumption per hour) and (2) the rate of air supply in ft^3/min. These dryers are normally equipped with oil burners that consume fuel at the rate of 3 to 14 gal/h and fans powered by 3- to 5-hp electric motors that deliver 9,000 to 15,000 ft^3/min of air. Usually 9,000 ft^3/min of 30°F air, with a relative humidity of 70%, can be heated to 70°F with an oil consumption of 3 gal/h used in a direct-heat dryer and 4.2 gal/h for the dryer if a heat exchanger is used. The U.S. Department of Agriculture reports that 1,000 bu of ear corn was dried from 30 to 13% moisture in 167 h.

Shelled corn, wheat, and oats can also be dried with heated air. Depth of grain in drying bins is 4 to 5 ft. Air flow must be uniform through grain, and temperatures of heated air should not exceed 110°F for seed corn and 140°F for wet milling. Temperatures up to 200 have been used without affecting feed value.

78. Unheated-air Crop Dryers. Wheat, oats, and barley are harvested in the summer, when atmospheric conditions are relatively favorable for grain drying with unheated air. Wheat combined at a moisture content as high as 20% can be successfully dried with unheated air. Minimum air flow is 3 ft^3/(min)(bu) with grain up to depths of 4 ft. With 16% moisture content air flow may be as low as 1 ft^3/(min)(bu) with wheat up to depths of 8 to 10 ft.

79. Forage Conditioning. Hay is finished in the mow by forcing air through it from ducts laid in the mow floor. The depth is limited to 6 to 8 ft for long hay or 4 to 6 ft for chopped or baled hay at one drying operation, but it is practicable to dry a total depth of 16 to 18 ft of long hay or 10 to 14 ft of chopped hay in the mow by repeated runs.

The hay should remain in the field until its moisture is reduced to 35 to 40%. Even then, 500 to 700 lb of water must be driven off to obtain 1 ton of finished hay containing 20% moisture or less. About 15 $ft^3/(min)(ft^2)$ of mow floor area is considered a desirable volume of air. Both centrifugal and propeller-type fans have been used satisfactorily, with motors of 3 to 7½ hp. Depending upon moisture in the hay, humidity of the air, and efficiency of the system, 25 to 65 kWh is used per ton of finished hay.

MATERIALS HANDLING

80. Conveyers and Elevators. Livestock and crop production requires much time and labor for loading, transporting, and unloading materials. Portable chain and flight conveyers, commonly called elevators, are available in lengths of 8 to 50 ft or more and in widths of 6 in to more than 20 in. They may be operated at angles up to 70°, depending on the material being handled, but care must be taken to prevent overturning or collapsing, particularly at the greater angles. The smaller sizes are generally used for moving loose, bulky materials such as small grains, chopped forage, and bedding and will require up to ¾-hp motors. Larger sizes are mounted on wheels and are used for baled, bagged, and packaged products, as well as other materials. Power requirements range from ¼ up to 5 hp, depending on the speed, angle of elevation, and weight of the material being handled. Vertical elevators for baled hay are mounted directly to the outside of barn walls. A 42-ft model will require a 2-hp motor.

81. *Auger conveyers* requiring fractional-horsepower motors are used for the horizontal and vertical moving of grains. Automatic feeding arrangements may employ 10-in-diameter forage augers in multiples of 5- or 10-ft sections up to 100 ft in length. Three-horsepower motors are required for lengths up to 90 ft, and 5 hp is needed for longer units. *Pneumatic conveyance* of grains and feed is increasingly popular on farms where the distance between storage or processing areas and feeding areas is considerable. This method is safe, has few moving parts, and is dust-free. The pipe can be placed in almost any path, above- or belowground. A University of Illinois report states that an air velocity of 4,000 ft/min is required for proper operation. A 5-in pipe will convey 4,500 lb of grain/h. This will require 2¾ hp for each 100 ft of length, according to the report.

82. Silo Unloaders. Mechanically operated silo unloaders remove the silage from the silo and deposit it at the foot. The operating mechanism of the top-unloading type is essentially a radial beam with scrapers or augers which collect the silage and bring it to the center of the silo, where it is picked up by a motor-driven air or mechanical device and delivered to the silo chute. Silage then falls down the silo chute, where it is collected for feeding.

There is also a bottom type of silage unloader. The operating mechanism consists of an endless chain mounted on a movable beam. The chain is equipped with scrapers which move the silage out of the silo as the chain revolves.

Unloaders eliminate the need for climbing the silo daily, reduce spoilage by removing silage at a uniform depth, and save up to 200 h/year of time. Results of Ohio State University tests indicate that top removal of grass silage at a rate of 1 ton/h requires 4.3 kWh and that 1.6 tons/h of corn silage requires 2.5 kWh. Three- to ten-hp motors operate the unloaders, and approximately 300 kWh is used annually.

83. Barn Cleaners. Electrically operated mechanical devices remove manure from poultry, dairy, and livestock barns. In poultry houses the cleaners may be placed under a slatted floor or in a wire-covered pit under tiers of mechanical feeders and waterers. In dairy barns they are installed in the gutters behind the cows. The dragline type uses a motor-driven drum to pull a belt or chain conveyer, equipped with cross flights, to an inclined elevator at the end of the barn, depositing the manure in a field spreader or pit.

The endless-chain type is well adapted to the larger stable where two rows of cows are housed. A single chain with wood or steel paddles travels around the gutters and up a short elevator, discharging the manure outside the stable. In this type of installation connecting or cross gutters must be installed at each end of the two rows of existing gutters so that an endless chain can be installed. The oscillating type uses a reciprocating bar with hinged paddle or auger conveyer. Portable types generally use a scoop steered by the operator and drawn along the gutter by a cable attached to a motor-driven drum. Cleaners are operated by electric motors of 2- to 5-hp capacity. They can be set to operate automatically for a predetermined cleaning period or can be switched on as need arises. Electric-energy use ranges from ½ to 1 kWh a month for each cow housed in the stable.

84. Feed Grinding. Two types of feed grinders are in common use for processing grains: the burr mill, in which grain is crushed between plates, and the hammer mill, in which rapidly revolving blades strike the grain, reducing its size and forcing it through a metal screen.

Feed-grinding and -handling units are used in conjunction with automatic feeding systems (see Par. **65**, on automatic feeding) and may also be equipped with conveyers or blowers to move the processed feed to storage bins.

Portable grinder-mixers are of growing importance among livestock feeders. With these units mixtures of small grains, ear corn, hay, concentrates, and home-grown feeds of all types may be fed to cattle, hogs, sheep, and poultry at different locations on the farm. They are of the hammer-mill type, are equipped with loading and unloading augers, and are pulled and powered by 25- to 35-hp tractors.

Several manufacturers market small burr and hammer mills that operate efficiently when powered by 3- to 7½-hp electric motors. Storage bins should be overhead so that grain will flow to the mills by gravity feed.

The most efficient speed for hammer mills is 3,300 to 4,000 r/min. Capacity and energy requirements vary widely with the kind and condition of grain and the fineness of grinding. In general, 1 kWh of energy will produce 35 lb of finely ground oats to 400 lb of coarsely cracked corn.

MAINTENANCE

85. Emergency Power. With increased dependency upon electric power for time-controlled mechanical feeding, pipeline milking systems, manure removal, etc., the added investment in emergency power units may be justified compared with the possible economic loss if regular power fails. Generators ranging from 3 to 15 kW and rated at 120/240 V are available in tractor power-takeoff (PTO) and engine-driven types. The latter may be manually or automatically started. Automatic generators must be of higher capacity, because peak-connected loads will be carried if power fails. Nonautomatic types should have a "power-off" alarm and must be power-takeoff-equipped with an overload circuit breaker. The tractor PTO-driven generators are least expensive to purchase, as the tractor engine serves as the generator drive. Output is controlled by an engine tachometer and/or voltmeter in the generator unit. Manufacturers claim a voltage rating within 2% of the normal supply voltage. Required generator capacity is obtained by totaling the power needs of essential loads, plus allowances for future loads and high starting currents of the motors. Double-throw switches must be used at the point of connection into the wiring, to prevent generator damage and power feedback into the supply line.

86. Arc Welders. A highly mechanized agriculture requires that many machinery and structural repairs be made by the farmer himself. A survey by the Kansas Farm Electrification Council covering the period 1958 to 1962 indicates that the number of dollars invested in electric welders was greater than for any other item of electric farm equipment. The electric arc welder is inexpensive, efficient, and an almost indispensable tool on modern farms. The 180-A transformer-type a-c welder is satisfactory for most farm shops. This machine can cut, hard-surface, and weld metals up to ½ in thickness. It requires a line voltage of 220 to 240 V single phase, 60 c. Current outputs from 30 to 180 A are possible. Duty cycle at maximum output is 20% with an open-circuit voltage of 25 V. A carbon-arc-torch attachment is used for brazing, soldering, and heating purposes. Larger generator-type units, either engine or tractor power-take-off-driven (hence portable), may be used as emergency power generators, supplying 5,000 W of 230- or 115-V single-phase 60-c power.

87. Phase Converters. Most farms have 100- or 200-A single-phase service, which limits them to the use of 7½- or 10-hp motors. Two types of phase converters are available which will convert single-phase to 3-phase current. By connecting the converters between the electric meter and the motor, they will permit the use of 3-phase motors up to 20 hp or more. In addition, the 1965 National Electric Code states that service entrances need be heavy enough to handle only the largest-power-demand equipment, plus a portion of all other equipment, rather than the total connected load as before. This is advantageous where irrigation pumps, grain dryers, large feed mills, etc., are in use.

88. Other shop equipment includes electrically powered air compressors, drill presses, grinders, hoists, lathes, saws, and paint sprayers. These generally require ¼- to ½-hp motors. Battery chargers drawing approximately 2 kWh/charge are popular.

89. Bibliography. Sources of additional information on the use of electricity in agriculture are as follows:

Agricultural experiment stations and Cooperative extension services, state colleges.

Edison Electric Institute, 750 Third Avenue, New York, N.Y. 10017.

Farm Electrification Council, 414 Professional Building, Oakbrook, Ill., 60523.

Farm Electrification Section, Agricultural Research Service, U.S. Department of Agriculture, Beltsville, Md.

National Rural Electric Cooperative Association, 2000 Florida Avenue N.W., Washington, D.C. 20009.

RICHEY, C. B., JACOBSON, P., and HALL, C. W. "Agricultural Engineers' Handbook"; New York, McGraw-Hill Book Company, 1961.

Rural Electrification Administration, Washington, D.C.

TEXTILE MILLS

BY C. T. MAIN II

90. Classification. Textile mills may be divided, generally, into the following classes: (1) cotton mills, (2) woolen and worsted mills, (3) dyeing and finishing plants, (4) knitting mills, (5) plants for processing rayon and other synthetic fibers or silk, jute, flax, and other fibers. Included are cordage and nonwoven plants.

Cotton mills for the most part produce and sell cloth and yarn in the "gray," or unfinished, state. Mills using dyed yarns in the fabrics that they weave either buy dyed yarns or operate a dyehouse.

Woolen mills practically always are complete units, starting with the raw material and producing finished fabrics.

Worsted mills may produce yarns only, may be confined to weaving and finishing, or may perform all operations from the grease wool to the finished cloth.

Cotton-finishing plants start with the "gray goods," which they bleach, dye, print, and finish as required. Woven rayon and some of the other synthetic fabrics are processed by some of the same machinery.

Knitting mills use cotton, wool, and synthetic and blended yarns to produce a variety of goods, including hosiery, underwear, jerseys, sweaters, sportswear, and rugs or carpets.

Rayon, nylon, and other synthetics are processed individually into fabrics or blended with cotton, wool, or other synthetics and processed by the regular cotton, woolen, or worsted machinery as the particular blend requires.

Jute and flax plants are only a small part of the industry and are highly specialized.

91. Process Steam. Worsted-yarn mills, without dyeing, use some process steam for wool scouring and combing.

Worsted-weaving plants, which do their own finishing, and all woolen mills use large quantities of process steam and hot water in the finishing work. This also would be true for certain knitting mills which do their own dyeing and finishing.

Cotton-finishing plants also use large quantities of process steam and hot water.

92. Source of Power. In general, mills using steam only for building heating and process in minor quantities can purchase power more cheaply than they can generate it,

Table 20-1. Power Requirements for Cotton-mill Machinery

Machines	Hp
Bale plucker, 6-bale cap	8.5–9.25
Automatic feeders	1–2
Axi-feed blender	1.5
Vertical opener	5–7.5
Lattice opener, S.L. No. 12, 3 hp, same driving No. 11 condenser	7.5
Opener, S.L. No. 15	1.0
Axi-flow cleaner A & B	3.0
Fans	5–20
Condensers	5.0
Distributors	1–3
Card and picker waste cleaner	5.0
Roving waste machine, hp/beater	5–7.5
Pickers, hp/beater	5.0
Pickers, automatic, complete, Trutzschler & Hergeth	19.25 & 21
Cards, cotton (up to 4 hp when cylinder is turning at 300 r/min)	1.5–4
Sliver lappers	1.5–3
Ribbon lappers	2.0
Comber, 8 head, one delivery	3.0–4.0
Comber, 12 head (6/side), two deliveries	5.0
Drawing frames, per delivery at 800–1,000 ft/min	1.5–2.0
Roving frames, cotton, 5–60 spindles per (consult mfr. for specific application)	1.0
Roving frames, synthetics, 5–25 spindles per (the larger the bobbin and the higher the speed, the fewer the spindles per hp)	1.0
Ring spinning frame, 12–35 spindles per	1.0
Ring twisters, 4–50 spindles per	1.0
Spoolers, automatic, 6 to 13 units per (connected hp)	1.0
Warpers, beam	7.5–25
Slashers	7.5–20
Winders, 10–25 spindles per	1.0
Automatic looms:	
Draper Corporation	
Up to 48 in each	0.75–1.0
50 to 90 in each	1.0
Above 90 in each	1.5
Crompton and Knowles Corporation	
C-4 60 to 72 in each	1.5
C-4 72 to 84 in each and above	2.0
C-5 up to 48, and usually 58 in each	1 0
C-6 up to 66 in each	1.0
C-8 up to 48 in each	1.5
C-8 up to 48 in each, with vacuum pump	2.0
C-8 50 to 72 in each	2.5
C-9 50 to 68 in each, with double-acting vacuum pump	3.0
C-9 72 to 88 in each, with single-acting vacuum pump	3.0
C-9 72 to 88 in each, with double-acting vacuum pump	4.0
C-9 over 88 in each	4.0

whereas in mills with finishing or dyeing departments, which use large quantities of low-pressure steam, the conditions are sometimes reversed. Then a mill may use an automatic extraction or a noncondensing turbine which will furnish power and process steam.

93. Motors. Textile machinery is generally driven electrically. Standard practice is individual motor drives rather than the "group drive." The most commonly used motor is the squirrel-cage induction motor. Because of lint and moist atmospheres, the totally enclosed nonventilated (TENV) type of motor or the totally enclosed fan-cooled motor of special textile design (TEFC-SC) is generally used. Where wool and other long-staple, not readily combustible fibers are processed which might clog the air passages or catch on the motor of the TEFC-SC type, a TENV or a motor with textile screens is used. A flywheel loom motor is used for better loom performance and reduced power consumption. Drip-proof motors with encapsulated windings are preferable in wet locations such as dyehouses. Where several motors are operated in a range as with finishing equipment, the trend is to use a separate packaged power unit of the static silicon-controlled rectifier (SCR) type for adjustable-voltage d-c controlled drive, but many a-c to d-c motor-generator sets are in use.

94. Starting equipment may be either manual or magnetic and of the full-voltage or reduced-voltage type, the choice depending upon the source and capacity of the power supply and on the requirements of the drive. Manual starters usually are limited in use to the control of infrequently started motors where remote control and undervoltage protection are not required. A manual loom switch in an NEMA type 12 enclosure has been developed for the industry. For the majority of cases a full-voltage magnetic

Table 20-2. Power Requirements for Woolen-mill Machinery

Machines	Hp
Wool- and waste-box duster	5–7.5
All-metal cone duster	20–30
48-in mixing pickers	15–30
Precarders, 48 in	10–15
Precarders, 60 in	20–25
Shredding pickers	20–40
Cards, three-cylinder sets, complete:	
60 × 48 in	15
60 × 60 in	17.5
60 × 72 in	22.5
60 × 84 in	25–30
Floor grinders	1.0
Mules, 55–70 spindles per	1.0
Ring frames, 8–12 spindles per	1.0
Ring frames, with collapsed balloon, 3–4 spindles per	1.0
Skein reels, 35–50 spindles per	1.0
Jack spoolers	1–2
Twisters, 200 spindles, 3¾-in gage	10
Twisters, 100 spindles, 10-in gage	25
Twisters, 120 spindles, 10-in gage	30–40
Winder, automatic filling, 10 spindles per	1.0
Dresser and beamer, high-speed	6–10.5
Looms, 82 in wide and up	1.5–2.0
Sulzer weaving machine	2.5–3.0

Table 20-3. Power Requirements for Worsted-mill Machinery

Machines	Hp	Machines	Hp
Wool-scouring machine, 2- to 4-bowl	10–25	Single-head double-ball gill reducer	3
Stock dryer	15–30	Single-head double-can gill reducer	2–3
Burr pickers	7.5–10	Single-head triple-can gill reducer	3
Worsted cards, one-cylinder:		Two-head four-can gill reducer	3–5
48 × 48 in	5–7.5	Two-head six-can gill reducer	5
54 × 54 in	7.5	Pin drafter, servo draft, auto. can del.	6
60 × 60 in	7.5–10	Pin drafter, two-head, two-can	3
Worsted card, two-cylinder, 30 × 60 in		Pin drafter, two-head, four-can	4
(compact)	15	80-spindle cone rover	5–7.5
Single-can or balling gill box	2	Roving frames, long draft, 72-spindle	5–7.5
Punch baller	0.5	Ring spinning frame, conventional, 200	
Noble comb	5	sp* 2¾ or 3-in ring	10–15
Intersecting gill box, can delivery, 4–6		Long draft, 240 sp, 4-in gage	15
head	7.5	Ring twisters, 200 sp, 3- or 3½-in ring	10–15
French or rectilinear comb	2–3	Hot-air slashers	10–40
Double ball per head, 4–6 heads	7.5	Looms, 72 in	1–1.5
Single-can gill reducer	3	Looms, 82 in	1.5–2.0

* sp = spindle.

starter with an NEMA type 12 enclosure is used. Usually it is of the combination type. Auxiliary control equipment such as control stations and limit switches located in lint atmospheres should be in lint-tight enclosures.

95. The power factor in cotton mills with mostly individually driven machinery probably will run about 75 to 80% without correction and the maximum billable demand factor 70 to 85% in full-production operation. A load factor of 90 to 95% would be expected for full-time operation. In woolen and worsted mills, the uncorrected power factor runs about 75% but can be as low as 65%. The maximum demand factor in these plants is about 70 to 75%. The load factor probably will run 70 to 80%. In finishing plants there is such a great variety of conditions that no general figures can

Table 20-4. Specific Recommended Levels of Illumination

(A portion of Table I in American Standard Practice for Industrial Lighting A11.1-1965 reproduced here by permission. Current recommended practice, footcandles in service)

Area	Footcandles on Task*	Area	Footcandles on Task*
Textile mills—cotton:		Spinning (frame):	
Opening, mixing, picking	30	White	50
Carding and drawing	50	Colored	100
Slubbing, roving, spinning, spooling	50	Spinning (mule):	
Beaming and splashing on comb:		White	50
Gray goods	50	Colored	100
Denims	150	Twisting, white	50
Inspection:		Winding:	
Gray goods (hand turning)	100	White	30
Denims (rapidly moving)	500†	Colored	50
Automatic tying-in	150	Warping:	
Weaving	100	White	100
Drawing-in by hand	200	White (at reed)	100
Textile mills—silk and synthetics:		Colored	100
Manufacturing:		Colored (at reed)	300†
Soaking, fugitive tinting, and conditioning or setting of twist	30	Weaving:	
		White	100
Winding, twisting, rewinding and coning, quilling, slashing:		Colored	200
		Gray-goods room:	
Light thread	50	Burling	150
Dark thread	200	Sewing	300†
Warping (silk or cotton system):		Folding	70
On creel, on running ends, on reel,		Wet finishing:	
on beam, on warp at beaming	100	Fulling	50
Drawing in on heddles and reed	200	Scouring	50
Weaving	100	Crabbing	50
Textile mills—woolen and worsted:		Drying	50
Opening, blending, picking	30	Dyeing	100†
Grading	100†	Dry finishing:	
Carding, combing, recombing and gilling	50	Napping	70
		Shearing	100
Drawing:		Conditioning	70
White	50	Pressing	70
Colored	100	Inspecting (perching)	2,000†
		Folding	70

* Minimum on the task at any time.

† Can be obtained with a combination of general lighting plus specialized supplementary lighting. Care should be taken to keep within the recommended brightness ratios. These seeing tasks generally involve the discrimination of fine detail for long periods of time and under conditions of poor contrast. The design and installation of the combination system must provide not only a sufficient amount of light but also the proper direction of light, diffusion, color, and eye protection. As far as possible it should eliminate direct and reflected glare as well as objectionable shadows.

be given. However, the figures usually would be no better than those for woolen and worsted mills.

96. Power requirements for textile machinery generally are not heavy, but they have been increasing with the trend for higher unit production. Tables 20-1 to 20-3 give, in most cases, connected horsepower. Individual machine running loads generally would be between 75 and 90% of the connected horsepower. Because of the great variety of dyeing, finishing, and knitting machinery and speeds and conditions under which these machines operate, a table of power requirements would be of little use, and has been omitted.

97. Air-conditioning systems to control temperature and humidity are extensively used in modern windowless plants. With the trend toward higher lighting intensities and increased power loads and a demand for more closely controlled atmospheric conditions within the production areas, refrigeration usually is required. The air-conditioning equipment adds a sizable demand to the plant power requirements. With refrigeration, it can increase the demand by as much as 50%. Controlling humidity also increases the boiler-plant requirements in the winter.

98. Lighting requirements in mills have become more exacting, and intensities continue to be greatly increased. Fluorescent lighting fixtures with high-power-factor ballasts generally are used. Ballasts with internal protection are preferred. Fixtures having an upward lighting component frequently are employed, with the trend toward general rather than individual lighting. Table 20-4 gives recommended footcandles for the various textile-mill departments.

99. Bibliography

Recommended Practice for Electric Installations on Textile Machinery, *IEEE Publ.* 77, February, 1965.

IES Lighting Handbook, 3d ed., 1962.

American Standard Practice for Industrial Lighting, A11.1-1965.

CEMENT MILLS

By John F. Hower

100. Chemical Composition. Portland cement is composed principally of the oxides of calcium (60 to 66%), silicon (20 to 24%), aluminum (4 to 6%), and iron (2 to 3%) in various mineral combinations. The first two of these form calcium silicates which are mainly responsible for its cementing qualities when mixed with water. The principal raw materials are limestone, marl, chalk, or marine shells together with clay, shale, or other alumina- and iron-bearing substances. Another raw-material source is cement rock, a natural formation so named for its nearly perfect proportions of the calcareous and argillaceous minerals.

101. Manufacturing Process. The manufacturing process consists in winning, grinding, and proportioning the raw materials, which are then sent to rotary cement kilns and brought to a white heat. The resulting clinker, formed by the exothermic and endothermic reactions, and a retarder (gypsum 3 to 5%) are jointly pulverized into the finished product, portland cement.

Either of two methods may be used for grinding, proportioning, storing, and conveying the raw mix to the kilns. In the dry process, these operations are conducted while the mixture is in the form of a dry powder. In the wet process, water is added to make a fluid slurry (about 35% moisture). Figure 20-11 shows two diagrammatic flow charts for typical dry-process and wet-process cement plants.

102. Power Requirements. Though varying in different plants with many factors, the following apply in general: connected motor load, 1.5 to 2.0 hp/daily bbl capacity; power consumption, 20 to 30 kWh/bbl; daily load factor, 90% average; yearly load factor, 75% average. (Cement-plant capacity historically has been calculated in barrels. A barrel of finished cement is four 94-lb bags, or 376 lb.)

103. Waste Heat vs. Purchased Power. Prior to World War II, many cement plants were equipped with waste-heat power systems which utilized the hot gases from the kilns for the generation of steam in specially designed boilers. Experience dictated

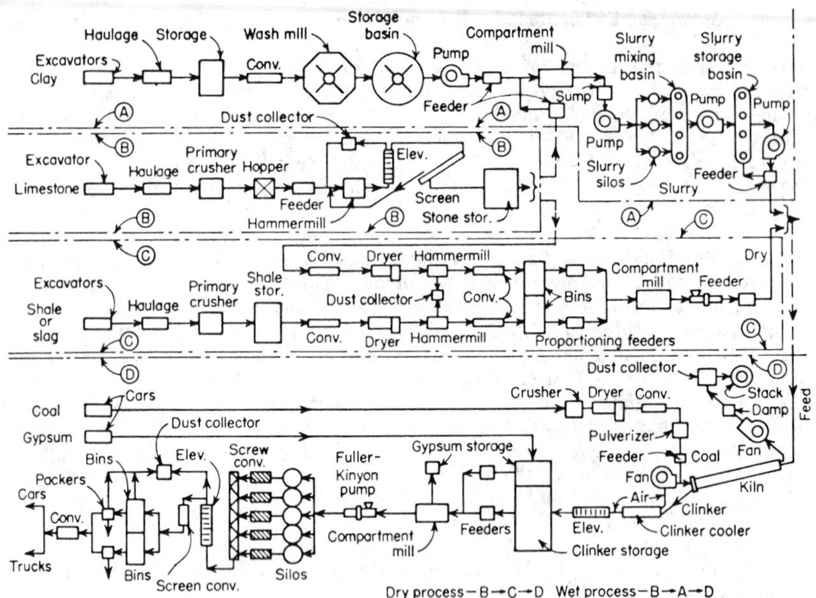

FIG. 20-11. Flow sheets of operations of a typical dry-process cement plant and of a typical wet-process cement plant.

that where the kiln exit gas temperatures were sufficiently high to justify the installation, the steam produced could supply electric power for the entire plant. Some type of standby power was provided either in auxiliary fuel-fired boilers or by a connection to the local utility.

Since World War II waste-heat power generation has all but been eliminated, since exit gas temperatures have been lowered with the advent of longer kilns. In addition, rising labor costs have made cement-plant power generation unattractive. The average 2.5- to 3-million-bbl cement plant has a purchased power connection of 15,000 kVA. Emergency standby generation is usually provided to turn the kilns, furnish air for slurry agitation in the wet-process plant, and provide other power for key drives and lighting in the event of a power outage.

104. Motors and Controls. Various types of motors and controls are employed depending upon the equipment being driven. The *squirrel-cage induction motor* is most commonly used, owing to its low cost, simplicity, and ruggedness where small- and moderate-sized high-speed constand-speed motors are required. NEMA Class B (normal torque, low starting current) and NEMA Class C (high torque, low starting current) meet the usual requirements. Control for squirrel-cage motors is ordinarily full-voltage magnetic control.

Wound-rotor induction motors are used to obtain high starting torque with low starting current on such applications as crushers and hammer mills and for some drives requiring continuous operation with speed control, e.g., kilns and draft fans. Drum controllers are often used for the smaller-sized motors, but magnetic control can be justified for the larger ones. A combination of magnetic primary and drum secondary control is popular, since overload and undervoltage protection can be obtained only through the use of the magnetic primary controller. Under average conditions, NEMA class 135 and 145 resistors are recommended for starting duty only. For continuous-speed regulating duty, the 90-series resistors are recommended.

Adjustable-speed d-c motor drives are used to obtain wide speed range and close speed adjustments in kilns and feeders. *Synchronous motors* with automatic field application control to ensure adequate synchronizing with minimum line disturbance are the preferred type in large tube and compartment mills. These synchronous-motor controllers should be provided with adequate field relays, preferably of the voltage and current type. In power systems supplied by a public utility, frequency relays are often employed to shut down the operation to prevent motor damage when power is restored after a momentary loss of voltage.

In general, motors are of open drip-proof construction with dustproof bearing features. Under extreme conditions where abrasive dusts or moisture from precipitation is present, totally enclosed (fan-cooled or nonventilated) or splash-proof enclosures may be required. In most cement-plant drive applications the trend is toward directly connected motors, gear motors, or enclosed speed reducers. In such drives the use of good, flexible couplings is necessary to secure the satisfactory service from both motors and mechanical equipment. Gear motors and speed reducers should be minimum AGMA Class II gearing for most applications.

In choosing enclosures for motors and controls, it is well to remember that cement dust is not explosive, although it may be abrasive. The dust which escapes into the plant atmosphere is made up of particles which are measured in microns. Accordingly, it will enter into the motor and control housings, coating the windings and contactors. Owing to its insulating qualities in coating contactor tips, control voltage under 250 V is not recommended unless special precautions are taken to keep the panels free from contamination. Centralized control rooms house the motor control centers and power centers for the various departments within the plant. These control rooms are normally pressurized with filtered air by a fan having a capacity to provide 25 to 30 air changes/h under a static pressure of ¾ to 1 in H_2O. Where control equipment is to be located outside the pressurized control room, dust-tight (NEMA type 4 or 5) or semi-dust-tight (NEMA 1 or 12 with gasket) enclosures are recommended.

The trend toward automating cement plants with the more sophisticated controls presents additional problems. Under this arrangement not only must the areas in the control room be kept dust-free, but comfortable temperatures and reasonable humidity have to be maintained. Transistors and SCR's to operate satisfactorily can be neither too hot nor too cold. As with human beings, they seem to perform their functions best under ideal climatic conditions.

105. Distribution System. The radial distribution system is satisfactory for cement-

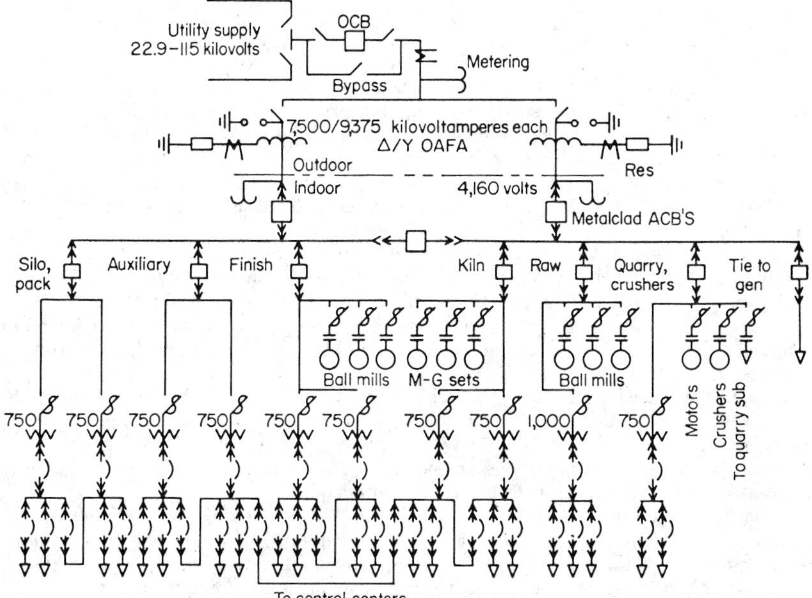

FIG. 20-12. Typical cement-plant distribution system.

plant service where department storage capacity allows independent departmental operation for a short time. Where an additional degree of continuity is desirable, loop feeds from other departments may be required. Each department is normally controlled by its own circuit breaker in the metal-clad switchgear lineup. Power-center transformers and motor control centers are recommended for each department. Power-center transformer sizes should be limited to the short-circuit capacity of the molded case circuit breakers or fuses in the motor control centers. Figure 20-12 shows a typical cement-plant distribution system.

The normal distribution voltage for plants below 25,000 kVA capacity is 4,160 V. Above 25,000 kVA the trend is toward the use of the 13.8-kV system or split-bus 4,160-V system.

The power distribution should be designed to limit voltage drop to 15% in starting large motors. The trend toward larger, tube-mill drive motors presents a problem in this regard, and at times the 15% value must be exceeded.

Protective relays should be chosen and set to selectively isolate faults in the distribution system. Only faulted circuits should be isolated.

Extra creepage insulators and bushings should be provided in exposed electrical equipment in outdoor substations. This practice will extend the time between cleaning periods and helps keep flashovers to a minimum. Silicone waxes and greases applied to the insulators and bushings will facilitate the removal of contaminants.

Special problems exist in the quarry that are not normally experienced in the plant proper. The use of portable equipment and the widely varying ground impedance necessitates grounding precautions to minimize shock hazards. The use of long lines and cables feeding concentrated loads gives rise to voltage regulation problems.

A Δ-Y isolating transformer of suitable size (1 kVA/connected hp) is recommended for the quarry. The neutral of the transformer should be grounded through a current-limiting resistor to limit ground-fault currents to low values (25 to 50 A).

106. Motor Applications. It is recognized that the foregoing cannot cover all applications within a cement plant. Accordingly, the following specific details are included to complement the above generalizations.

Gyratory crushers are commonly driven by belted wound-rotor motors. On small-sized crushers NEMA design B squirrel-cage motors are sometimes used. Synchronous motors have been used, particularly where power-factor correction is desirable (see Table 20-5 for typical motor-torque characteristics). Gyratory crushers are normally started unloaded; however, a power failure may occur during the crushing operation. To reduce the need for removing material from the crusher upon subsequent starting, many operators prefer higher-starting-torque motors.

Single-roll crushers are used extensively on some of the softer materials such as shale, cement rock, and certain limestones. Wound-rotor motors, either belted or driving through gear reducers, are the preferred type. Horsepower requirements vary considerably; however, ½ to 1 hp/(ton)(h) can be considered as the required amount for roll crushers.

Jaw crushers when used as primary crushers are usually driven by belted wound-rotor motors, although NEMA design D squirrel-cage motors are often used.

Hammer mills are usually directly connected to their drive motor through flexible couplings. Induction- and synchronous-motor drives are common. Reversing controllers should be considered for certain mills to obtain maximum service life from the hammers.

Rotary dryers used to dry raw materials have widely varying capacities and horsepower requirements, depending upon the type of material, amount of moisture, construction, and speed of rotations. Wound-rotor motors or squirrel-cage induction motors with eddy-current couplings are normally used when speed control is desirable; otherwise, constant-speed induction motors are adequate.

Wash mills are often used for the purification of raw materials, particularly clay, in a wet-process plant. The mill consists of a concrete tank in which a system of chains, drags, or rakes is rotated to mix the material thoroughly with water, thereby getting it into suspension and separating it from gravel or other foreign substances. The usual drive is by bevel gears through a belt or gear reducer from a wound-rotor motor.

Table 20-5. Summary of Drive Requirements or Crushers and Pulverizers

Crusher	Type drive	Typ. conn to load	Typ. hp	Typ. motor r/min	Relative WK^2 at motor shaft	Typical req'd torques			App. F.V. inrush times, normal	Typ. acc. time, s
						STG	P.I.	P.O.		
Compression types										
Jaw primary	Wound-rotor	Belt	150–300	720	Med.	150	...	250	Up to 3.5	15–60
Jaw secondary and clinker	Sq.-cage NEMA C	Belt	25–150	1,200	Med.	150	...	200	6	5–10
Gyratory, primary	Wound-rotor	Belt	50–1,000	600 720	Low	150	...	250	Up to 3.5	3–5
Cone	Sq.-cage NEMA B	Direct to pinion	100–500	600 514	Low	100	100	250	6	3–5
Single-roll, prim.	Wound-rotor	Belt	150–500	720	Med.	150	...	250	Up to 3.5	10–20
Double-roll, sec.	Sq.-cage NEMA C	Belt	10–200	1,200	Med.	150	...	200	6	5–10
Impact type										
Hammer mills	Wound-rotor	Direct to pinion	50–1,000	720 900	High	100	...	250	Up to 3.5	15–50 (all three)
Impactors	Sq.-cage NEMA B		50–1,000	720 900	High	100	...	250	6–8	
Impact brks.	Sync.		350–600	900	High	125	110	250	6–9	
Pulverizers										
Hercules mills	Sync.	Direct to pinion	350	400	Low	125	110	250	5–7.5	5–10
Babcock & Wilcox Raw and finish mills	Wound-rotor	Direct to pinion	300–800	900	Low	150	...	250	Up to 3.5	5–10 (all three)
	Sq.-cage NEMA C		300–800	900	Low	150	...	250	6–7.5	
	Sync.		300–800	900	Low	150	110	250	5-7.5	

Table 20-6. Typical Ball, Tube, and Compartment Mills
(Data courtesy F. L. Smidth & Company)

Mill diameter and length	Ball charge, lb	Mill use	Mill r/min	No. of compartments	Motor hp
8′6″ × 10′8″	45,000	Preliminary grinding or closed-circuit grinding	19.5	1	300
9′6″ × 10′8″	70,000		19.0	1	500
10′ × 15′6″	100,000		18.6	1	700
11′ × 17′	135,000		18.0	1	1,000
13′ × 25′	230,000		16.7	2	2,000
14′ × 24′	340,000		16.36	1	2,500
7′ × 30′	100,000	Finishing mill or compartment Unidan mill	20.00	2 or 3	600
8′ × 36′	130,000		19.5	2 or 3	800
8′6″ × 36′	145,000		19.0	2 or 3	1,000
9′6″ × 30′	170,000		18.5	2 or 3	1,250
10′ × 34′	200,000		18.0	2 or 3	1,500
12′ × 32′	300,000		17.3	2	2,000
12′ × 34′	310,000		17.5	2	2,500
13′ × 34′	400,000		16.6	2	3,000
12′ × 53′	550,000		17.3	3	4,000
13′ × 43′	560,000		16.7	2	4,400
15′ × 54′	890,000		15.7	2 or 3	6,660
15′ × 49′	800,000		15.7	2	6,600

Motors of 50 to 100 hp are usual; however, wash mills requiring 150 to 200 hp have been built.

Preliminary grinders are often employed ahead of the fine grinding ball or tube mills in both the raw and clinker grinding departments. Closed-circuit grinding is usually employed (1) to control the fineness of the finished product, (2) to increase the output for a given-sized mill, and (3) to lower the power requirements to produce a unit of finished product. Closed-circuit grinding is a system whereby the mill output of both coarse and fine particles is separated by means of screens, classifiers, air separators, or air currents within the mill. The oversized particles are returned to the mill feed for regrinding. Preliminary grinders are usually of the ball- or roller-mill type (see Table 20-6).

Ball, tube, and *compartment mills* are used for the fine grinding of both raw material and clinker. The compartment mill is a ball or tube mill having partitions separating the several sections, each section being loaded with different-sized grinding media. Closed circuit is normally employed in both the raw and clinker grinding departments.

Synchronous motors are usually employed to drive the respective grinding mills. Normally 0.8-pf machines are specified through the 2,500-hp size and unity machines above this size, since high power factor for plant power-factor correction is desirable. (They also provide reserve motor horsepower if required.) The mill motors are the largest within the plant, and the grinding power represents approximately 65 to 70% of the total plant load.

With the current emphasis on larger grinding mills, motors of 5,000 to 6,000 hp are not uncommon. The starting of these motors presents inrush problems even when the plant is supplied by a "stiff" utility system; accordingly, specially designed motors are commonly employed. The *synchronous induction motor* is one such design. It is started and brought up to full-load speed as a wound-rotor motor (approximately 98.5% synchronous speed). Subsequently, the d-c field excitation is applied at the most opportune angle for maximum synchronizing torque and minimum possibility of transient torques due to slipping poles. An alternate field application scheme has the field energized continuously and slips poles as the motor comes up to speed to "lock in." Resistance is kept in the rotor circuit to help cushion the transient torques incident to slipping poles at pull-in until after the motor has synchronized.

The speed of the driving pinions of mills ranges from 150 to 240 r/min. In the smaller mills (usually 1,000 hp and below) synchronous motors are coupled directly to the pinion shaft. In the larger mills synchronous motors at 600 or 720 r/min are coupled to intermediate gearing to drive the mill.

Average motor torque ratings for these drive motors are 180% starting, 140% pull-in, and 200% pull-out. Specific values should be obtained from the mill manufacturers because of the widely varying requirements.

Inching control for turning the mill at creeping speed is available in several types. Inching control is necessary to spot the mill for recharging balls and for repairs.

Conveyers including belt conveyers, screw conveyers, drag chains, bucket elevators, etc., are normally driven by directly connected gear motors. NEMA design B or C squirrel-cage motors are normally employed; however, on drives exceeding 100 hp wound-rotor motors are common where torque requirements are high or where cushioned starts are desirable.

Slurry pumps are generally of the centrifugal type and are directly connected to NEMA B squirrel-cage motors.

Agitators are used in a wet-process plant to mix the raw material and to keep it in suspension. Some agitators are entirely pneumatic in operation, while others combine air nozzles with rotating or traveling mechanical members. Mechanical agitating members are driven through gear reducers by squirrel-cage induction motors having a high starting torque.

Horsepower requirements vary; however, ¼ to ¾ hp/1,000 ft^3 of tank capacity is common for the mechanical operation and ½ to 1 hp for the same tank capacity for air compression.

Kilns. Rotary kilns are steel shells lined with high-temperature refractories and rotate on an axis inclined slightly from the horizontal. At the lower end of the kiln the mixture of atomized fuel and hot air is blown into the kiln to induce the required temperature. Raw material, in the form of either slurry or dry powder, is admitted at the high end. As it moves downhill owing to the rotation of the kiln, it is dried and calcined. In the burning zone the constituents combine, form a new set of compounds, and are transformed into hard nodules of portland-cement clinker.

Kiln-drive-motor horsepower requirements are dependent upon the size of the kiln, speed of rotation, and kind and quantity of raw feed (see Table 20-7). The trend is toward larger diameter and longer kilns to realize higher fuel economies. Approximately 900,000 Btu is required per barrel of clinker in a dry plant and about 1,000,000 Btu/bbl in a typical wet plant with kilns of equivalent size. Kiln torque requirements are practically constant throughout the entire speed range. Adjustable-speed drives are required ordinarily in the 3:1 or 4:1 range. Modern practice is to use adjustable-voltage d-c drive motors. The wound-rotor induction motor or the squirrel-cage induction motor with an eddy-current coupling is also used. Many operators specify a low "light-off" speed, usually one-sixth the normal operating speed. This requirement poses special heat-dissipating problems in the case of the drive motor.

Table 20-7. Power Requirements of Typical Kilns

(Data courtesy F. L. Smidth & Company)

Kiln size, diam., × length, ft	Approximate capacity, bbl/24 h		Motor hp
	Wet process	Dry process	
8′ × 80′	......	35	60
9′ × 150′	......	1,200	40
9′ × 323′	......	1,600	75
9′6″ × 320′	1,500	.....	75
10′ × 12′ × 350′	......	2,750	100
11′6″ × 400′	......	3,300	150
11′6″ × 400′	......	3,600	175
12′ × 450′	4,800	.....	200
13′6″ × 12′ × 14′ × 440′	5,850	.....	2 × 150
13′6″ × 15′ × 500′	6,700	.....	2 × 250
14′6″ × 16′6″ × 550′	......	8,500	500
16′ × 17′6″ × 570′	......	9,000	2 × 350
17′ × 18′ × 550′	9,500	.....	2 × 400
22′8″ × 21′ × 25′ × 760′	22,000	.....	2 × 1,200

The kiln motor is usually connected to the pinion gear through a suitable gear reducer. The drive should be arranged so as to "lift" the kiln. In kilns requiring more than 250 hp, parallel drives are normally used.

Kiln feeders are usually run at speeds proportional to the kiln speed. To accomplish this, the speed of the kiln is used as a reference for the kiln feeder power source.

Coolers. Portland-cement clinker from the kiln must be cooled before further handling or processing. Although other types of coolers exist, cooling is accomplished by (1) rotary coolers, (2) multicylinder units attached to the kiln shell, (3) the grate-type cooler, or (4) methods of handling which allow the material to disperse heat in transit. There are probably more rotary coolers in operation at the present time than any other type; however, the present trend is toward the grate cooler.

Heat thus recuperated may be used with the primary air for introducing the fuel, with the secondary combustion air, or for drying the fuel.

Rotary coolers are steel shells similar to kilns in construction and mounting but driven at constant speed by squirrel-cage motors.

Grate-type coolers are machines for cooling hot clinker by blowing ambient-temperature air through a perforated set of grates over which the hot material is moving in a uniform layer. The moving grates are usually driven by squirrel-cage motors in conjunction with eddy-current couplings to provide speed variation.

Coal Processing. Coal as it is received at the plant is normally crushed and then pulverized. Coal crushers are usually driven by squirrel-cage motors in the smaller sizes and by wound-rotor motors in the larger sizes. Coal pulverizers are generally of the bowl- or tube-mill type. High-torque squirrel-cage motors are normally used in both applications. The coal feeders supplying either type of mill are usually driven by variable-speed drives. The trend is toward using SCR power sources for these drives.

Fans. Fans find a universal use in cement plants. They are used for dust collection, inducing draft in the kilns, cooling clinker, pressurizing control rooms, air-sweeping mills, etc. Normally, standard-torque squirrel-cage motors are satisfactory. For high-inertia drives, special squirrel-cage or wound-rotor motors are required. Where speed control is desired, either wound-rotor motors with speed-regulating resistors or squirrel-cage motors with eddy-current couplings are used.

Fuller-Kinyon conveyer systems are used for transporting dry, finely divided materials. The principal element is a pump which, by the combined action of an impeller screw and compressed air jets, forces the material into a mixing chamber where air is injected. From this chamber the material is discharged into the conveying pipeline in an aerated and fluent state. Power requirements vary for both the pumping and the air requirements, depending upon the type and quantity of material handled and the distance transported. Normally, squirrel-cage motors, NEMA design B, are satisfactory for coupling directly to the pump and to the compressor.

Packing and Shipping. The larger part of portland cement is shipped in bulk by motor truck or railroad hopper cars. Approximately 20% is shipped in package; however, each year the volume of package shipments drops. Bulk cement is normally fed into the trucks or cars by gravity, requiring only electrically operated gates, valves, or feeders and possibly air-slide equipment to control the flow into the container. Packing machines are normally driven by directly coupled 1,200-r/min squirrel-cage motors, approximately 20 hp per three-valve machine and 25 hp for a four-valve packer.

107. Bibliography

Pit & Quarry Handbook, Chicago, *Pit & Quarry* Publications, 1964.

"Industrial Power Systems Handbook"; New York, McGraw-Hill Book Company, 1955.

"Standard Handbook for Electrical Engineers," 9th ed.; New York, McGraw-Hill Book Company, 1957.

Recommended Practices Paper, Distribution Working Group, Cement Industry Committee, IEEE, 1964.

Recommended Practices Paper, Drives Working Group, Cement Industry Committee, IEEE, 1965.

Annual Cement Issue, *Rock Products,* Chicago, Maclean-Hunter Publishing, 1965.

Annual Cement Issue, *Pit & Quarry,* Chicago, *Pit & Quarry* Publications, 1965.

Raney, A. K. Electric Equipment for Rock Crushing & Pulverizing, *Booklet* GET-2915, General Electric Company, 1959.

COAL MINES

By Daniel Jackson, Jr.

PRIMARY POWER

108. Purchase vs. Generation. A coal company's distribution system starts with a main substation or switchgear house where power is purchased. The main substation may reduce the voltage to the final levels of 220 or 440, or it may drop it only to 13,000, 7,200, 4,160, or 2,400 for transmission to load centers for final reduction. The number of voltage steps and the choice of voltage depend upon system load, transmission distance, safety, and limitations imposed by mine laws.

Few coal mines generate their own electric power today. The principal exception is new small- or medium-sized mines located in isolated areas where it is not economically possible to install power lines. Mines operating under these conditions install engine-generator sets in capacities up to 300 kW. Engines usually are diesel, and the generators or alternators normally produce 250 V d-c or 220 or 440 V a-c, 60-c, with 110 V in some instance. When demand exceeds the capacities of the largest standard "off-the-shelf" unit, twin or triple units are installed.

Sources of primary coal-mine power, aside from the privately owned public utilities, include, to a limited extent, REA-supported and local systems. The utilities make power available at 6,600—rarely less—to 69,000 V. Most commonly the voltage is 33,000. With one-step transformation, primary mine distribution usually is 4,160 or 7,200 V, nominal, with a trend toward 13,000 V. The 2,300-V level is being eliminated where it is economically feasible to do so.

In two-step transformation, the primary voltage usually is 13,000. Permanent transformer stations may employ either single- or 3-phase transformers, with a trend toward the latter.

109. Metering. Both central metering and metering at several load centers are employed by mining companies. Each case must be decided on its merits. In the past several years a number of companies have increased the primary voltage level and changed to central metering to reduce power cost and improve efficiency.

The electricity required to produce a ton of coal is approximately 13 kWh. Power cost per ton of coal has increased to about 15 cents, compared with 12 cents in 1955. The increased cost of power is reflected in not only the continued trend to mechanical mining but also the use of higher-capacity mining equipment as well. Power consumption, for example, by one of the largest coal companies in the United States, producing some 40 million tons of coal annually, consumes more than 48,000,000 kWh of power/month. The total installed transformer capacity required to supply this power is in excess of 245,000 kVA. In addition, transmission facilities include approximately 300 mi of overhead line and 90 mi of high-voltage cable.

110. Primary Distribution. Coal-mine power distribution systems generally fall into two classifications, including pole-mounted high lines and cable systems, with many using a combination of the two. The number and length of pole lines depend upon mine size, terrain, and so on, with one goal, the serving of secondary substations at load centers on the surface or underground. The stations are moved as the workings advance.

111. Strip Mines. At strip mines, pole-line practice is largely standardized; with a main line, pole-line laterals at intervals of 1,200 to 1,500 ft are run to the pit, terminating in switch houses which apply auxiliary transformers for low-voltage equipment and also supply the cables to the larger high-voltage equipment. As the pit moves across country, the laterals are shortened at intervals until the pit approaches the main line, which then is moved to restart the cycle. Cables on the equipment usually are 1,000 ft long. Thus, with a lateral spacing of 1,200 to 1,500 ft, equipment can operate freely between laterals with enough cable to spare to permit terminating laterals some distance back when shortening is necessary.

A large number of strip mines are now using the "ground-cable system" instead of the pole-line and cable combination. Otherwise, the basic plan is the same. A complete system consists of the main cable and cable laterals, the cable being sectionalized in 1,000- to 1,500-ft lengths as a rule, with connectors for termination in switch houses or for joining main cable lengths by junction boxes. Several types of cable may be employed, but the most common is type SHD with copper-shielding braid over each insulated conductor, to equalize surface stresses and eliminate static discharge, and grounding conductors in interstices. The latest improvement in this cable is the SHD-GC, with ground-check conductors for continuous cable monitoring. It is the safest and most widely used for high-voltage (up to 15,000 V) portable power applications at both strip and deep mines. Other SH cable types and also type W and G cables, without and with ground wires, respectively, may be used. A maximum rating of 5,000 V for W and G, however, usually results in such cables being limited to 2,300 V when employed in coal-mining work.

The largest power cable used in a strip operation to date is an SHD-GC type rated at 8,000 V, 880 A. It is 5⅜ in in diameter and weighs 24½ lb/ft. It consists of six 500,000-cmil conductors—two for each phase—one 750,000-cmil center ground conductor, and three No. 8 insulated ground-check conductors.

Another version of the ground-cable system for distributing primary power is to provide two separate cable circuits. This reduces the size of the individual cables and permits continuous operation in the event one cable fails.

112. Deep Mines. Underground mines may use a combination of surface pole-mounted high lines and cables. Where depth of cover and terrain make surface lines costly or impracticable, cable systems are used for primary distribution at underground mines. In this case the cables enter the mine at the main opening or through boreholes. Armored cable is employed in some instances, but the usual, as in strip mines, is a synthetic-jacketed type, usually type SHD, though other SH types or W or G may be employed. Cable manufacturers have designed a mine-power feeder cable for this purpose.

Where depth of cover and terrain favor pole-mounted high lines, the primary distribution system will include pole lines which follow the mine workings on the surface. At intervals boreholes are drilled and the power taken into the mine, rated voltage being kept near the working areas. The pole lines may serve d-c power-conversion units on the surface or underground or transformers near the working areas for reducing the voltage to 440 V for face utilization.

113. Power Factor. Without correction, power factor at the average coal mine would be 65 to 85%—sometimes less but seldom more. Corrective steps are necessary. Power contracts include rates which are based on classification, diversity factors, load power factor, demand power factor, monthly power factor, and cost of producing power. All contracts do not contain power-factor clauses. Whether this clause is included or not, it is to the advantage of the coal company to maintain a good power factor. Most power-factor clauses require the consumer to maintain an 85% power factor.

Corrective methods and equipment used to improve power factor include:

1. Synchronous motors for equipment requiring, say, 100 hp or more, e.g., pumps underground and in surface processing plants, air compressors, ventilating fans, and crushers.

2. Synchronous motors, 0.8 leading, in the motor-generator sets in large excavating units in strip mining and for supplying direct current underground.

3. Capacitors supplement other types of correction and bring power factor up to 90 to 95%, the usual goal, which normally takes the mine out of the penalty area. Theoretically, correction should be installed with each motor of any size, but a more practicable system is capacity for a group of motors near the center of such a group. The farthest back location, and the less preferable, is in the substation.

114. Primary Voltage Regulation. Increasing load density makes it more difficult to maintain economically adequate voltage at all points of the system. Increasing sophistication of automation equipment also requires more precise voltage, not wider voltage swings. Though voltage drop can be reduced by increasing the size of the conductors, shortening circuits, or reducing reactance with busway or differing cable configurations, these means can become uneconomical if carried to the extent sometimes required for optimum performance of extremely expensive and critical equipment.

In those cases where precise voltage is required at the load an automatic voltage regulator can provide accurately maintained voltage of the correct level at the point of utilization with minimum investment. The conductor in the feeder can be selected on the basis of load-carrying capability only, with the regulator compensating for any resultant voltage drops. The regulator can be selected to control a single feeder to a critical load, or it can be increased in size to handle the total output of a unit substation transformer. The latter alternative is selected when all the loads in a particular area, if supplied with best-quality voltage, will produce at a significantly improved performance level. Even though the utility supply is closely regulated, the load voltages will vary. Thus, it is rarely sufficient for an operator of a distribution system to rely solely upon primary power-supply regulation. Some of the newer automation equipment includes "built-in" voltage regulators, but in most cases the user must provide the well-regulated supply needed.

UNDERGROUND POWER

115. Alternating-current vs. Direct-current Power. Underground power systems have been in a state of change for several years. The "standard" voltage for underground coal-mine use was for many years 250 V d-c. Approximately 10 years ago the industry began using 440 V a-c at the working face. Practically all new mines installed today are using this voltage, with many of the older mines converting to the higher voltage.

Higher horsepower per machine, plus the fact that coal-seam thickness and size of opening limit the physical size of mining equipment, is generating more pressure for still higher voltages. The use of 440 V at the face has provided many advantages, including safety, that were not available with 250 V d-c. As loads continue to increase beyond the capabilities of 440 V, proposals are being made to consider the use of 4,160 V for large mining equipment such as continuous miners.

When 440 V was first introduced at the working face, there were many who objected to its use. Today, it is the standard. The proposal to use 4,160-V power in the face area is in the same position now as was the 440-V system a few years ago. The 250-V d-c system was developed and used to handle certain loads under set conditions. When these loads increased and mining methods changed, 440 V a-c was adopted because it could do the job more efficiently and economically. Now, with loads increasing beyond the capabilities of the 440-V system, there is a possibility that 4,160 will be adopted in the not too distant future. It will provide the same benefits on a larger scale as the 440-V system, even under more adverse conditions and greater loads. State and Federal mining laws will have to be changed before this voltage level can be applied in the working areas. At present, some state laws forbid anything over 250 V d-c or 220 V a-c in new mines. However, 550 V d-c and 440 V a-c may be used by special permission or because the mine was equipped for such voltage when the restrictions were adopted.

Alternating-current and direct-current loads encountered in underground coal mining fall into two categories:

1. Locomotive and shuttle-car haulage, calling for direct current.
2. All other applications for which either direct or alternating current may be employed.

Because direct current is so well suited to locomotive and shuttle-car operation, it is still widely used in mining. Underground power systems may consist of a combination of alternating and direct current, especially where locomotives and shuttle cars are used. The advantages of alternating current (440 V) have been proved to be great enough at the working face to provide another voltage system (250 V d-c) for coal haulage only. However, where belt conveyers are used, the mine may employ only alternating current.

Exceptions to the general rule of 250 and 550 V d-c and 220 and 440 V a-c underground include 2,300 and 4,160 V a-c for large slope hoists, pumps, compressors, and other equipment requiring motors of 100 or 125 hp and larger. Some 110-V a-c is used underground, but normally only for lighting and certain special applications.

116. Alternating-current Supply. An underground a-c power system consists of a surface substation, high-voltage cable couplers, portable underground switch houses, mine load centers (transformers), and distributions boxes. These circuits generally have a voltage rating of 4.16 or 6.9 kV, although there are a few instances where the

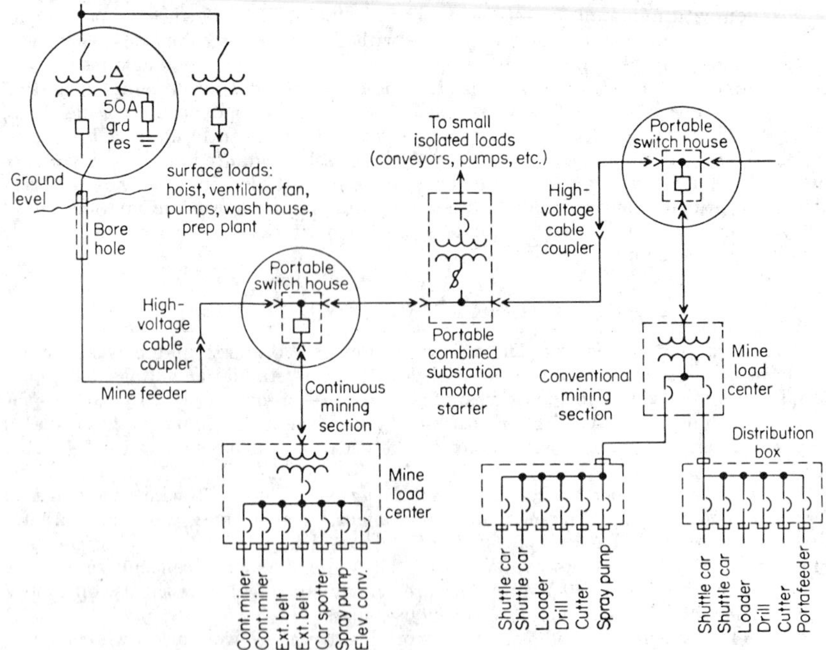

Fig. 20-13. Alternating-current power system in deep mine.

primary underground distribution is handled at levels as low as 2.4 kV and as high as 13.8 kV.

Figure 20-13 is a one-line diagram of a typical power system for an a-c mining operation. The equipment in the circles are the surface substation, where service voltage is stepped down to the primary distribution level of the mine, and the underground switch houses, which provide branch-circuit switching and relaying equipment for proper protection of the branch circuits and their connected apparatus.

All equipment for alternating current underground normally is portable to permit quick and easy transfer from one location to another. Transformers employed underground are of the dry type, along with sealed, nitrogen-filled units. A three-unit 150-kVA installation of this type, mounted on a skid, can be made with a total height of 25½ in and a total weight, including air circuit breaker and polarized plugs and receptacle potheads, of 2,600 lb. Capacitors on separate skids may be used with such substations—e.g., 45 kVA with a 225-kVA power center at another mine, which also provides skid-mounted 240-V distribution boxes with two circuit breakers, three feed receptacles, and ground-trip relays.

Underground voltage regulation is a major consideration in designing a power system for a-c mining. The need for good voltage regulation has generally resulted in the use of oversized cables selected primarily on the basis of allowable voltage drop in the system. Since the cable represents 50 to 60% of the initial investment for the mine power system, it is important to look for alternative ways to ensure adequate voltage at the face. An economic study was recently made of a proposed a-c power system for a new mine, with the object of comparing the cable cost of providing for voltage regulation in oversized cables or, as an alternative, through the use of a 10% buck-or-boost induction voltage regulator. The cable cost was $20,000 greater in the system employing cables selected primarily on the basis of voltage drop. These savings were based on the early operation of the mine when only one-third of the ultimate footage of high-voltage cable would be required. For the maximum distances expected after full development of the mine, the total savings in first costs for high-voltage cable was approximately $80,000.

The voltage regulator should be installed in the surface substation. This will eliminate additional costs for a special low-height unit or construction of a vault for an underground unit. The surface-substation location is ideal for the added reason that it can be applied to compensate for voltage drop in the system ahead of the substation and that, through line-drop compensation, it will adjust for the voltage drop in the mine feeder circuits.

117. Direct-current Supply. Although considerable rotating equipment (motor-generator sets and rotary converters) and mercury-arc rectifiers are still in service, the silicon diode is rapidly becoming the standard in purchases on new conversion equipment. Mercury-arc rectifiers are being converted to silicon-diode units because the latter offers high efficiency, long life, great reliability, simplicity, and a minimum of maintenance. The total annual saving for a 500-kW silicon-diode unit compared with a mercury-arc rectifier, reportedly, is approximately $1,800. Based on a price of $12 per kilowatt for the silicon equipment, a 500-kW unit will pay for itself in less than 4 years.

Most silicon-diode rectifiers are portable, with rail wheels for movement over mine track systems. Others are mounted on skids, and still others are mounted on rubber tires. Stationary units also are available.

Automatic operation is the rule, and portability is essential to keep down the cost of the more frequent moves necessitated by the rapid extraction with concentrated mining, particularly in the thinner seams, from which more and more of the tonnage is coming. Among its other advantages, the rectifier has reasonably good efficiency, even at one-quarter load, and also can withstand heavy overloads for relatively long periods of time.

Like motor size and connected horsepower, the rectifier is advancing in size. Some 15 years ago, 150 or 200 kW was a common size. Now, a conversion unit under 300 kW is seldom installed, and with larger face machines, and particularly with continuous miners, 500 kW is becoming fairly common. One reason is the increase in peak and steady loads. Now, momentary peaks of up to 2,000 A have been observed, along with average demands of 300 A or more.

118. Basic Distribution. Where the depth of the overburden permits frequent use of boreholes, or where, for example, working in a hilltop area permits frequent drift openings, it may be more economical to use pole lines for primary a-c service and distribute alternating or direct current from surface substations. Where underground stations are employed, the usual practice is to take primary alternating current in by cable—now usually type SHD, carrying 4.16 to 13.8 kV. Type W and G cables, normally rated 5,000 V maximum, also may be used, but usually only if the distribution voltage is not over 2,300.

In d-c mines, which normally rely on locomotive haulage, the basic distribution system is the trolley wire, with the track for return, supplemented by feeder and additional return circuits when the loads require them, which usually is the case. Some 4/0 trolley wire is employed, but the tendency is to use at least 6/0 and on up to No. 9 section, with supplementary feeder, where several large locomotives are used. Big locomotives also bring up the problem of not only placement and capacity of substations but also the effect on the remainder of the d-c system as the locomotive passes a given point. To reduce disturbances in face operation as a result of locomotive operation, substations may be equipped with automatic load distributors.

119. Sectionalization. The normal distribution system naturally includes circuit breakers and other basic protection devices. Overcurrent settings should conform to good practice. For reference to the latest recommendations for settings, see Bibliography (Par. **130**).

Aside from the basic facilities, a mark of a good distribution system is sectionalization with both manual and automatic equipment. Manual switches at intervals of 1,000 ft in trolley lines and feeders permit isolating sections or parts of the system in case of trouble or to allow change or repair, frequently without interruption in sections not experiencing trouble. Automatic reclosing circuit breakers protect stub-end trolley and feeder lines and the sections they serve from fire, annealing of trolley, and other difficulties resulting from short circuits. The breakers automatically cut out sections experiencing trouble and leave the remainder of the system free to continue operations.

120. Face Distribution. In the nature of things, mobile face equipment, whether alternating or direct current, must receive power through cables, with the exception of storage-battery-powered gathering locomotives and shuttle cars. Instead of connecting wire to wire or cable to cable, either solidly or by connectors, the practice today is to use junction and distribution boxes. Most such boxes, or circuit centers, also are designed to include circuit breakers and ground-protective equipment. Permissible types are available for use where permissible face equipment is required, usually where the U.S. Bureau of Mines or some other agency in authority finds an air sample containing 0.25% methane.

Main cables and, where the type of equipment, such as a stationary room conveyer, permits, distribution cables also are now being increasingly installed in short lengths—say, 150 ft—for convenience in shortening, lengthening, and handling. Permissible connectors are available and may also be obtained with pilot pins, corresponding to pilot wires in the cable, to trip the circuit breaker before the main pins are disengaged, thus making it impossible to open the connection under power.

121. Trailing Cables. The Federal Mine Safety Code and most state laws and regulations require that frames of electrically operated equipment be physically grounded to protect men in case of short circuit, either through the rail upon which the equipment rests, as a locomotive, or by a ground wire or cable. Where trailing cables serve equipment, the general practice is to include a grounding conductor or conductors in such cables.

For a-c service, type W and G round, multiple-conductor cables normally are employed. One conductor in 3-conductor type W may be used as a ground wire. Type G normally is made with three ground wires in the interstices between the cabled main conductors. These cables, plus heavy-duty drill cord, which is similar in general construction, are tough, flexible, and designed to take mining abuse.

In d-c locomotive service, single-conductor cables are the common type. Concentric mining machine cable finds limited use in gathering locomotives as well as in other equipment, but while diameters are small, this type is not designed to take the abuse cables must withstand in modern heavy-duty mining.

For practically all d-c service underground, except cable-reel locomotives, the flat-twin cable (type W without ground and type G with ground) normally is preferred. Advantages include smaller size; light weight; design for bending around small-diameter sheaves, cable drums, and guides; and good resistance to run-overs and physical abuse.

Normally, the maximum circuit voltage with single-conductor, 2-conductor concentric, 2-conductor round, and flat-twin cables should be limited to 1,000. Usual ratings for underground mining equipment are 600 to 1,000 V.

122. Protection. The solid wire or cable from machine frame to ground, if maintained, provides protection to men from short circuit but no protection for the machine. Consequently most mines employ equipment to open the circuit and thus reduce or eliminate damage as well as the hazard. The original and still widely used protective unit is the fuse, either in the junction or distribution box or in the "nip" on the end of the trailing cable used for hooking to the main feeder or trolley. In recent years, however, there has been a definite trend to circuit-breaker protection. With a-c service, the secondaries of the transformers are Y-connected, with neutrals grounded, and current transformers are included to trip the circuit breaker when a ground occurs. With Y-connection and a full 240 V at the face, the maximum potential from any conductor to ground is 139 V. One new a-c system includes doughnut-type current transformers around the 3 power conductors to detect imbalance in case of short circuit and to trip the circuit breaker in the same circuit center or distribution box.

In d-c service, a three-pole circuit breaker may be employed, with one pole in the ground circuit with a 5- or 10-A current-limiting relay in series with it to trip the circuit breaker in case of a fault in the machine or cable. In such cables, the regular power conductors are supplemented by a grounding conductor or conductors. Ground trips also may be used with two-pole circuit breakers, but the relay may be damaged if heavy current flows and the grounding circuit is not opened.

Proposed alternatives, particularly for such mobile units as shuttle cars, include the polarized relay, the polarized short-circuiting device, and certain electronic devices (see references in Par. **130** for fuller discussion).

123. Battery Service. The first major application of storage batteries in underground coal mining was in powering locomotives in gaseous mines. Next came the rubber-tired tractor-trailer unit and the shuttle car in the late 1930s, though the development of the cable-reel car soon cut the production of battery units to only a few a year. Now, aside from their use in locomotives, the principal use of batteries is in powering tractors pulling rubber-tired trail cars for hauling coal at many of the small mines in thin coal and also tractors for hauling supply cars and personnel carriers in larger trackless operations. The use of battery-powered equipment is gaining in popularity owing mainly to improved batteries and battery-powered equipment.

STRIPPING POWER

124. Operating Voltage and Loads. Fifteen years ago the largest stripping shovels and draglines had total connected loads on the order of 2,000. These machines drew peaks on the order of 3,000 and were considered difficult loads to supply at distances approaching 2 mi from the substation. Ten years ago the advent of the 60- and 70-yd^3 stripping shovels raised connected horsepowers to approximately 4,650 of main a-c driving motors with peak loads of 7,500 kW. Power at 7,200 V was required to drive these machines. The largest shovel in operation today includes a 140-yd^3 unit with more than 12,000 hp of main-drive a-c motors and peak loads on the order of 17,500 kW. Certain smaller units, such as drills and pumps, operate on 220 or 440 V, supplied by small service transformers usually in or near the pit where the equipment is being used. Some older and smaller machines operate on 2,300 or 4,160 V nominal. The latest increase in voltage level has been to 13,800 V to power a 65-yd^3 shovel.

Trailing cables as a rule are type SHD with ground-check conductors to facilitate ground protection, normally achieved by Y-connecting the transformer secondaries with the neutral grounded through a resistor. Resistor values are fixed so that maximum voltage to ground is limited to approximately 100 instead of, perhaps, 2,000. Quick tripping is necessary to reduce the time of possible shock and to limit damage to equipment. An extensively used method is installing a current transformer between the Y-neutral and earth ground at the substation. Or a special arrangement of three

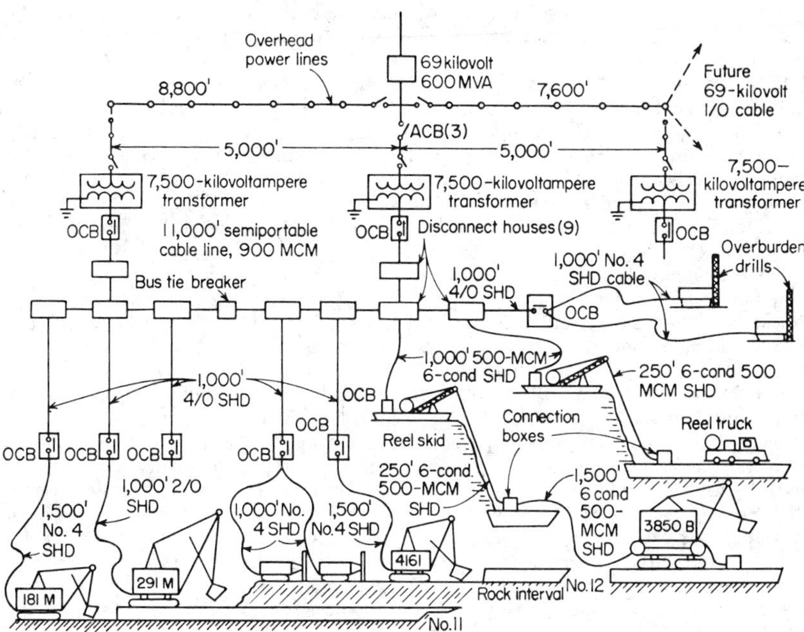

FIG. 20-14. Power system in strip mine.

current transformers on the 3 outgoing power conductors may be used to trip the feeder circuit breaker if a slight unbalance occurs, as when a conductor becomes grounded. Where cables are employed, a "doughnut"-type transformer may be employed to detect the imbalance.

Figure 20-14 shows the major items of electrical distribution apparatus for a strip mine, including the following:

1. The main substation, including equipment which transforms the incoming transmission-line voltage from the local utility to a primary distribution voltage for the mine.
2. A portable substation, including equipment which transforms the primary distribution voltage down to a lower distribution voltage for the smaller machines not suitable for operation on the primary distribution voltage.
3. Portable switch houses, including equipment which automatically protects the distribution circuits at the distribution voltage level(s) and enables flexible isolation and control of mine feeders.
4. Portable power centers, including equipment which transforms power from the distribution voltage to utilization voltages of 600 V and below for auxiliary equipment.
5. Transmission lines, cables, and cable couplers which carry power at the voltage used in the mine between the above electrical apparatus and eventually to the loads served by the distribution system.

The prime considerations in the choice of electrical apparatus used to perform these functions are safety, voltage regulation, reliability, and cost.

PLANT POWER

125. Voltage. Accepted voltage for most of the stationary motors in preparation plants is 440, leaving in most instances only the question of whether 2,300 or 4,160 V should be used for certain large units, such as pumps, crushers, and the like. A rough rule is that motors of 100 hp and larger should be powered by the higher voltages.

126. Transformer Location. Packaged substations with oil-filled transformers are available for outdoor service, with nonflammable units for indoor. If high-line voltage is over 10,000 and the reduction to 440 is made in one step, the substation should be outdoor. If the supply is under 10,000 V, then packaged indoor substations with nonflammable transformers are the general choice, principally because they can be placed closer to the load center.

127. Controls. Starters grouped in factory-assembled central cabinets are now standard for coal-preparation plants. One central cabinet is satisfactory for a small plant, but a large plant normally requires cabinets at several locations to keep the starters reasonably close to the motors. Draw-type starters which can be pulled out for quick replacement by a spare constitute the probable pattern for the future. Dust is a major problem with starters. The best method of eliminating it is to house cabinets in enclosed rooms with forced filtered-air ventilation to keep the pressure slightly higher than outside.

128. Capacitors. The induction-motor load of the preparation plant produces a low power factor, best offset by capacitors to bring the lagging power factor up to or near 100%. Theoretically, an appropriate capacitor should be connected to the motor terminals, but cost, space, and maintenance considerations normally dictate grouping the capacitors in the control room.

129. Motor Protection. The wide use of water and the difficulty of completely eliminating dust require some form of protection for many motors in preparation-plant service. This has led to growing use of special motors, such as drip-proof, splash-proof, and totally enclosed fan-cooled. The major type is the standard normal-starting-torque squirrel cage. Double-deck high-torque units are used where extra starting power is required, with wound-rotor units for extra-heavy starting. Control systems almost invariably provide for starting and stopping in sequence; some provide for locking out certain units when not needed or for maintenance. Plug-in-type bus provides for easy moves, changes, and additions and is being adopted by an increasing number of plants. Otherwise, conduit is standard, with armored cable as an alternative in some instances.

BIBLIOGRAPHY

130. References on Coal Mines

Power Systems, Deep and Strip Mines

LORDI, A. C. Trends in Open-pit Mine Power Distribution; *Coal Age*, January, 1961, pp. 96–103.

Efficient Power Designs; *Coal Age*, July, 1964, pp. 251–258.

Consol Power Systems . . . Deep and Strip; *Coal Age*, October, 1964, pp. 171–181.

Coal Division Power . . . AC and DC Service; *Coal Age*, October, 1962, pp. 165–169.

13.2 KV Distribution in Low Coal; *Coal Age*, June, 1964, pp. 84–88.

HAMILTON, D. E. Designs for Efficient, Safe, AC Power Systems; *Coal Age*, June, 1959, pp. 98–104.

WEICHEL, THOMAS R. Cables for AC Underground . . . Choice and Installation; *Coal Age*, May, 1959, pp. 90–98.

High Voltage at the Face; 4,160 vs. 440; *Coal Age*, April, 1965, pp. 102–117.

World's Largest Strip Cable . . . Design and Application; *Coal Age*, April, 1963, pp. 82–85.

Power Conversion Alternating to Direct Current

VIOLA, JAMES A. Rectifiers: Ignitron-to-Silicon Conversion; *Coal Age*, January, 1964, pp. 88–90.

CLOERN, CHARLES G. Silicon and Ignitron Rectifiers: Maintenance and Conversion; *Coal Age*, February, 1965, pp. 110–114.

Battery Power

HENSLER, J. F. The Case for Batteries; *Coal Age*, March, 1965, pp. 90–94.

Power Factor

HUGUS, FRANK R. Power Factor in AC Mines; *Coal Age*, March, 1962, pp. 75–85.

Voltage Regulation

PRIOR, ROGER About Automatic Voltage Regulation and Efficiency; *Coal Age*, June, 1964, pp. 78–80.

Circuit Protection

HARRISON, L. H. Attaining Simplicity in Ground-fault and Cable Protection; *Coal Age*, January, 1963, pp. 66–68.

About Silicon Diodes and DC Frame Grounds; *Coal Age*, August, 1964, pp. 98–100.

McDOWELL, STANLEY E. Selecting Protection for Intermediate Voltage Service; *Coal Age*, November, 1961, pp. 78–80.

BAKER, DONALD J., and BROWN, CLYDE L. What Overcurrent Settings for DC Feeder Circuit Breakers? *Coal Age*, November, 1953, pp. 86–89.

Safety

Federal Mine Safety Code for Bituminous Coal and Lignite Mines, revised, U.S. Bureau of Mines, 1954.

Federal Mine Safety Code for Anthracite Mines, U.S. Bureau of Mines, 1954.

Safety Rules for Installing and Using Electrical Equipment in and around Coal Mines, USASI.

EXCAVATING MACHINERY

BY ROBERT W. BERGMANN

TYPES OF EXCAVATORS

131. Shovels may be classified into three sizes: the small types used chiefly for highway construction and general excavation, the intermediate sizes used for heavy-duty quarry and bench mining, and the large sizes used in uncovering large deposits such as coal and iron ore. Small and intermediate-sized shovels are mounted on two crawlers, as shown in Fig. 20-15, and the large shovels, shown in Fig. 20-16, are generally mounted on eight crawlers.

Digging is accomplished by coordinating the three main motions: the "hoist," which pulls the *dipper* upward through the bank in digging and positions it vertically over the dumping position; the "crowd," which moves the dipper handle out or in to control

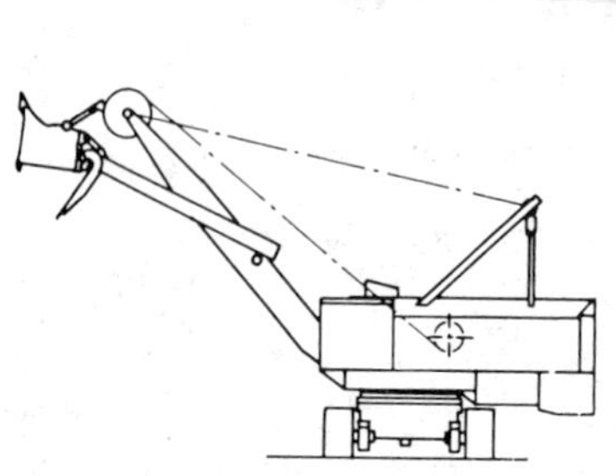

FIG. 20-15. Typical intermediate-sized shovel, mounted on two crawlers. (*Marion Power Shovel Company.*)

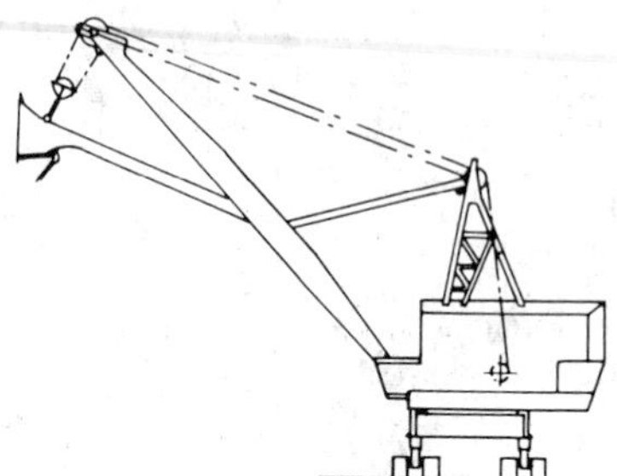

FIG. 20-16. Typical large shovel, mounted on eight crawlers. (*Marion Power Shovel Company.*)

the depth of cut and to position the dipper over the dumping position; and the "swing," which rotates the upper portion of the machine between the digging and dumping positions. A "propel" motion is also provided to move the machine from one digging position to another.

132. Draglines are classified into sizes in the same manner as shovels and are used for the same purposes except that draglines are not adaptable to handling rock or other "blocky" materials. Draglines are used where a great digging depth is required, where the material must be moved a great distance, or where the digging must be below the level of the machine. Small- and intermediate-sized draglines are mounted on two crawlers, as are shovels of the same sizes. Large, walking draglines of the type shown in Fig. 20-17 are mounted on large, circular bases to provide lower ground-bearing pressures.

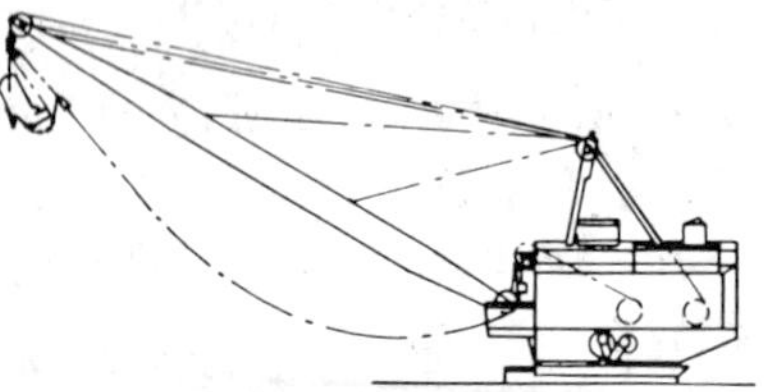

FIG. 20-17. Typical large dragline of the walking type. (*Marion Power Shovel Company.*)

Draglines generally dig below the elevation of their base and dump above it, whereas shovels generally dig above their base. Draglines also have three digging motions: the "drag," which pulls the *bucket* through the bank, toward the machine, and positions the bucket horizontally for dumping; the "hoist," which controls the depth of cut and raises the bucket to the dumping position; and the "swing," which is similar to that on a shovel. To "propel" a walking dragline, a large, rectangularly shaped shoe or pontoon on each side is lowered to the ground and used to raise the rear of the circular base. Resting on the two shoes and the front of the base, the machine slides backward several feet. The base is then lowered to the ground and the shoes moved back for another step.

133. Wheel excavators use a group of buckets mounted around the periphery of a wheel. By rotating the wheel, the buckets are filled in the bank and then dumped onto a conveyer belt, which carries the material to the dumping position. Digging is a nearly continuous operation, and thus the power fluctuations are not so great as for shovels and draglines.

Wheels are not adaptable to excavating blocky materials; so their use has been limited to areas where little or no rock is present.

DRIVE SYSTEMS

134. The mechanical-friction clutch drive is used almost exclusively on small shovels and draglines and in some intermediate-sized machines. Generally one gasoline or diesel engine is used to drive all motions through mechanical clutches with air or hydraulic control. Reversing is obtained by using separate clutches for each direction. Speed control is obtained by alternately engaging and disengaging the clutch or by partly engaging the clutch and allowing it to slip, both of which result in considerable wear and power loss (as heat). Braking is provided by mechanical brakes or by engaging the opposite clutch. The size of such machines is limited by the problems involved in providing adequate clutches and brakes.

135. The single electric motor drive is essentially the same as the engine drive described above except that the engine is replaced with a squirrel-cage a-c induction motor, a wound-rotor a-c motor with fixed rotor resistance, or a squirrel-cage motor with a torque converter. Such machines are used in relatively fixed locations, such as quarries, storage yards, etc., where the portability of engine power is not required.

136. The eddy-current clutch drive is used on some small- and intermediate-sized machines, the prime mover being either an internal-combustion engine or a squirrel-cage a-c induction motor. Two clutches are required to provide reversing, but, as a rule, only one is used on the hoist motion, the force of gravity being depended upon to accelerate the dipper, or bucket, and the machinery in the lowering direction. Speed control is obtained by only partly energizing the clutch coil, thus allowing it to slip; braking is accomplished by energizing the opposite clutch. The slippage involved in accelerating, in operating at reduced speed, and in braking causes considerable power loss, which is expended as heat in the clutch; lower operating speeds and elaborate ventilating systems are used to keep temperature rises within safe limits.

137. The d-c motor drive is used on some small machines, many intermediate-sized machines, and on all large shovels and draglines. Each motion is powered by one or more shunt-wound motors of the mill type, which provide the required features of low armature inertia, mechanical ruggedness, and high light-load speeds, overload capacity, and commutating ability. Armature voltages as high as 2.6 times rated and armature currents up to 200% of rated, with control of the shunt-field strength, make it possible to obtain light-load speeds of 3.4 to 4 times rated, stalled torques up to 2.4 times rated, and maximum intermittent horsepowers up to 250% of rated.

Generators driven by a-c induction or synchronous motors make it possible to use the Ward Leonard control system, in which the desired d-c motor torque and speed characteristics are obtained by controlling the excitation of the generator shunt field. Generator field-excitation control might be by means of contactors and resistors varying the current flow from a d-c constant-potential exciter, by rotating exciters such as the Rototrol (Westinghouse Electric Corporation) or the Amplidyne (General Electric Company), or by silicon-controlled rectifiers which vary the amount of rectified a-c current allowed to flow in the generator field in response to firing signals from magnetic amplifiers. These latter devices are used in regulating circuits to provide fine control of speeds and torques, even under severe shock-load conditions, thus reducing both mechanical and electrical stresses.

Reversing is obtained by reversing the direction of current flow in the generator field. Braking is accomplished by reversing or reducing the generator field current. Speed control is obtained by varying the generator field strength. The Ward Leonard system provides power conversion so that minimum power is used for accelerating and reduced-speed operation; braking actually regenerates or returns power to the source. Thus the d-c motor with Ward Leonard control provides the most efficient type of excavator drive.

138. Diesel-electric drives are used where portability is required but where it is desirable to retain some of the features of electric drives. In this case, engines are used to drive the generators. Some models use the d-c motor drive for only one or two motions, the other motions being clutch-driven.

POWER SUPPLY

139. Power-supply voltages used for excavators in the United States are 480, 2,400, 4,160, 7,200, and 14,400 V, for which the a-c motors are rated 440, 2,200, 3,810, 6,600, and 13,200 V, respectively. Table 20-8 indicates the maximum a-c motor horsepower recommended for use on these voltages.

140. Power-distribution systems for excavators must be designed to reduce voltage fluctuations, the long distances over which power is transmitted and the widely varying nature of the load being taken into account. Power is received at distribution voltages and transformed to the operating voltage by one or more transformer banks with the usual short-circuit and lightning protection in the primary and circuit breakers with

Table 20-8. Suggested Maximum Horsepower for Various Voltage Levels

Nominal power-supply voltage	Motor rated voltage	Suggested maximum horsepower
480	440	250
2,400	2,200	800
4,160	3,810	2,500
7,200	6,600	12,000
14,400	13,200	30,000

overload protection in the secondary circuits. Where mobility is required, substations are sled- or wheel-mounted. In large operations, secondary power is distributed by cable instead of overhead lines to take advantage of the low inductive reactance and thus lower voltage drops and also for mobility.

141. Trailing cables are generally type SH-D, which provide shielding over the phase conductors and ground conductors. Finely stranded conductors are used for flexibility; special outer jackets are used for mechanical protection. Where cables are to be moved less frequently and are not subject to mechanical damage, mine power cables may be used; these are similar to type SH-D, but with fewer strands and thinner jackets.

142. Safety grounding is essential to provide protection in the event of faults in

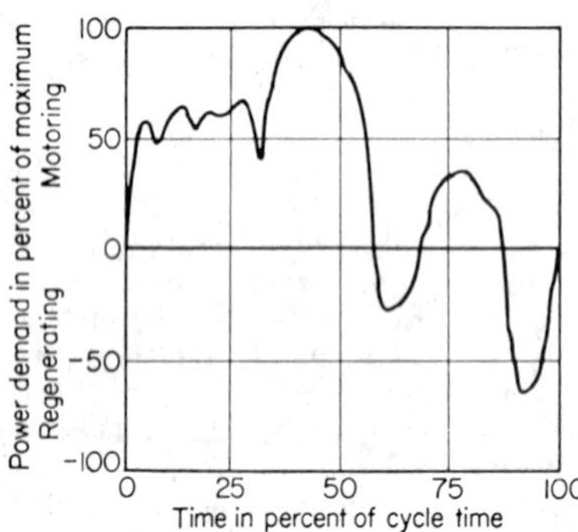

FIG. 20-18. Typical chart of the power demand for one digging cycle.

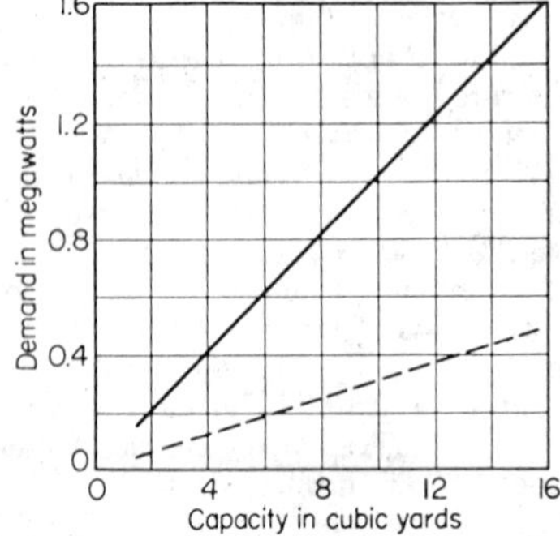

FIG. 20-19. Representative maximum demand (solid line) and 15-min demand (dotted line) values for small and intermediate-sized shovels and draglines.

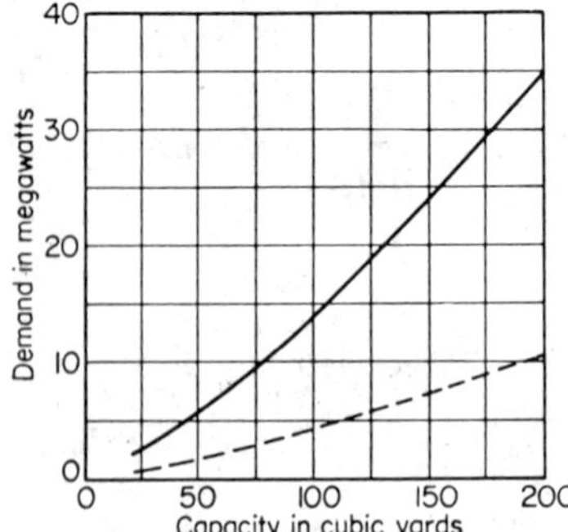

FIG. 20-20. Representative maximum demand (solid line) and 15-min demand (dotted line) values for large shovels.

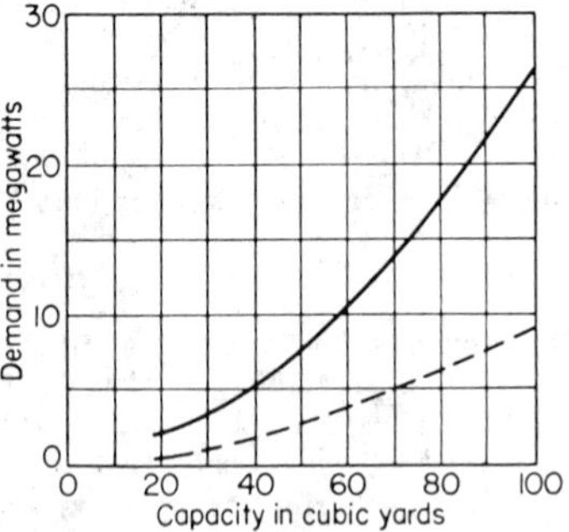

FIG. 20-21. Representative maximum demand (solid line) and 15-min demand (dotted line) values for large draglines.

the high-voltage system for personnel boarding or working around the excavator. A grounding resistor is connected between the system neutral and the cable grounding conductors to limit fault currents and thus the voltage between the excavator and earth. Any fault current flowing in this circuit is used to trip the circuit breaker. Several systems have been devised to continuously monitor the ground circuits and to open the circuit breaker in the event of an interruption.

POWER REQUIREMENTS

143. Starting is the most severe power requirement when the excavator has only one or two main a-c motors; when more than two or three main motors are used, the operating requirement is greater. Across-the-line starting is generally used, the inrush current being five to seven times rated; reduced inrush starting generally may be provided if required.

144. Operating power requirements are widely cyclical, as illustrated in Fig. 20-18; this cycle would be repeated every 20 to 60 s, depending upon the size of the machine. Typical maximum and 15-min demand values are given in Fig. 20-19 for small- and intermediate-sized shovels and draglines, in Fig. 20-20 for large shovels, and in Fig. 20-21 for large draglines. These demand values are given for machines with standard boom lengths and dipper, or bucket, capacities. When longer than standard booms (and thus smaller buckets, or dippers) are used, the demand will be higher than shown. These data may vary somewhat according to the manufacturer, but they may be considered representative; the manufacturer should be consulted if exact data are required.

145. Power-factor correction for machines with induction motors can be accomplished with capacitors, but the improvement is limited since it is impractical to vary the capacitance during the digging cycle and also because the maximum size of capacitor must be limited to prevent excessive transient torques and high voltages. When synchronous motors are used, the motor excitation can be varied in relation to the operator's motions, the d-c motor-armature voltages and currents, or the a-c input.

146. Energy consumption varies with the size of the excavator, the type of material handled, digging depth, dumping height, and the type of drive system used. For machines using the Ward Leonard d-c driving-motor system, the kilowatthours per cubic yard of material moved will vary from 0.25 to 0.4 for small- and intermediate-sized machines, 0.3 to 0.5 for large shovels, and 0.4 to 0.7 for large draglines. When either mechanical or eddy-current clutch drives are used, the energy consumption is 50 to 100% greater owing to the power lost in operating at low speeds, braking, etc.

BIBLIOGRAPHY

147. References on Excavators

Bennett, R. B., and Neslin, M. A. A Review of Open-pit Mine-power Distribution Practices, *IEEE Conf. Paper* 59–1213, October, 1959.

Cronquist, W. E., and Neslin, M. A. Excavator Controls; *Eng. Mining Jour.*, June, 1963, pp. 112–117, 128.

Hamilton, D. E. A Review of Grounding Practices for Mine Power Systems, *IEEE East Central and Allegheny–Ohio Valley Conf. Paper*, April, 1960.

McCrary, M. R. Voltage Regulation Problems Associated with Large Stripping Shovels; *Coal Age*, July, 1964.

Neslin, M. A. Predicting Excavator Dynamic Performance; *Mining Congr. Jour.*, June, 1962.

Rumfelt, Henry Application and Performance of Wheel Excavators; *Mining Congr. Jour.*, June, 1961.

Weis, J. F. Modern Electric Power Shovels, *IEEE Natl. Conf. Paper*, October, 1954.

Zeindler, R. W. Economic Comparison of Draglines vs. Shovels, *Paper* 25, November, 1964, Symposium on Opencast Mining, Institution of Mining and Metallurgy, 44 Portland Place, London W1.

Electric Power Distribution for Mines and Quarries, *Bull.* B-5447, Westinghouse Electric Corporation, East Pittsburgh, Pa., 1952.

DISPLACEMENT AND CENTRIFUGAL PUMPS

By Hans Gartmann

GENERAL

148. Classification of Power Pumps. Power pumps for all practical purposes can be divided into two general classes: displacement pumps and centrifugal pumps. Displacement pumps may be further divided into reciprocating pumps and rotary pumps, while centrifugal pumps may be divided into diffusing-vane and volute pumps. As will be shown later, the two classes of displacement pump differ radically in design and are really distinct classes, while the difference between diffusing-vane and volute pumps is a difference of construction detail rather than a difference of type.

149. Calculation of the horsepower required is based upon a consideration of the opposed head (expressed in feet or pounds per square inch), the pump delivery (expressed in gallons per minute or cubic feet per second), and the efficiency of pump and mechanical transmission. Pump efficiencies vary with the different types, sizes, and working conditions (see Pars. **154, 157, 170,** and **172**).

The head will include the static head pumped against and also the head of pipe friction. A calculation of required horsepower should always consider the latter component of total head.

Friction head for water may be calculated with reasonable accuracy by means of Eq. (17-11) for new clean pipes,

$$h_f = f \frac{L}{D} \frac{V^2}{2g} \tag{20-1}$$

where h_f = frictional resistance in feet of fluid, L = total length of pipe in feet, D = internal diameter of pipe in feet, V = velocity in pipe in feet per second, g = acceleration due to gravity, 32.17 ft/s², f = friction factor.

The value of f will vary approximately from 0.020 for 4-in pipes to 0.015 for 12-in pipes and larger. For accurate determination of friction losses in pipes and pipe fittings, pipe friction tables should be consulted.[1]

The horsepower (hp) required to pump a certain quantity of water against a given head is expressed as follows:

$$\text{hp} = \frac{G \times H}{3{,}960 \times \text{efficiency}} \tag{20-2}$$

where G = delivery in gallons per minute and H = sum of the static head and the friction head, both expressed in feet.

The above formula is based upon *cold* water of specific gravity equal to 1. If hot water or any other liquid of specific gravity different from 1 is pumped, it is necessary to multiply the horsepower obtained from the above formula by the specific gravity of the liquid. One cubic foot per second corresponds to 448.8 gal/min; also a pressure of 1 lb/in² corresponds to a head of 2.309 ft, for water weighing 62.34 lb/ft³ at 60°F.

DISPLACEMENT PUMPS

150. Slip. In all displacement pumps, the volume of liquid pumped is always less than the piston displacement or, in the case of the rotary pump (Par. **156**), the displacement of the rotating element. This difference is due partly to leakage past the piston (in the case of the reciprocating pump) or the rotating element (in the case of the rotary pump) and partly to leakage through the valves. The difference between the displacement of the moving part (piston or rotating element) and the volume of water discharged is called the slip and is usually expressed in percent of the total displacement. This slip may vary from 2% in a new pump to 50% in a badly worn pump. Generally speaking, anything under 5% may be considered fairly good performance for pumps that have been in service any length of time.

[1] See Hydraulic Institute, Pipe Friction Manual (Par. **172**).

151. Effect of Speed on Capacity of Displacement Pumps. Neglecting slip, the capacity of all types of displacement pump varies directly with the speed, regardless of the head pumped against. For constant head, the horsepower required varies almost directly as the speed and, therefore, as the capacity. This, of course, is not strictly true, for the pump efficiency will be slightly greater and the slip in percent slightly less as the pump approaches rated capacity. These characteristics and the service to be performed will determine the type of motor to be installed.

152. Displacement pumps under constant head require practically constant driving torque at all speeds and, therefore, at all capacities. This characteristic has a direct bearing on the type of motor to be selected in cases where it is required to operate the pump at variable capacity and, therefore, at variable speed. Variable speed of the pump is obtained and a saving of power is effected by using field control in the case of the d-c motor or a pole-changing device for an a-c motor or any similar device in which the torque of the motor remains constant.

A comparatively recent development in reciprocating pumps is the variable-stroke pump, by means of which the capacity may be varied from zero to maximum with the motor operating at constant speed, which results in high efficiency over a wide range. The stroke can be varied automatically to suit the demand of water, keeping the pressure or water level constant and thus making this type of pump well adapted for boiler feed service.

153. Starting Torque of Displacement Pumps. Displacement pumps operating at constant speed against a variable head will have a constant rate of discharge. The torque and horsepower input will vary (neglecting minor losses) directly with the head. In starting these pumps against the normal working pressure, the motor must exert full-load torque. As this requires very heavy starting currents, especially in the case of induction motors, the expedient of bypassing the pump in starting is usually employed wherever this is practicable. Where the bypass is installed, the head at starting, and therefore the starting torque, can be reduced to a negligible quantity. After the pump is up to speed, the by-pass is closed.

154. The efficiency of a reciprocating pump is a function of capacity, pressure, and type of pump. The curves in Fig. 20-22 show the efficiencies that can reasonably be expected of triplex pumps of good manufacture and in good condition. These curves show the efficiency of a series of pumps, each designed for a particular point on that curve or for a given capacity and head. The pumps are ordinarily designed to operate at a piston speed of 30 ft/min for pumps of 6-in stroke or less to 75 or 80 ft/min for pumps of 16-in stroke or over. These low piston speeds, in turn, require extremely low shaft speeds, as compared with the ordinary motor speeds. This condition is met by installing a slow-speed motor with a belt or silent-chain drive, by a double-gear reduction, or by a combination of the two. This reduction in speed, especially when accomplished by the double gears, necessarily involves a considerable loss of power, and this loss should be considered in comparing pump efficiencies.

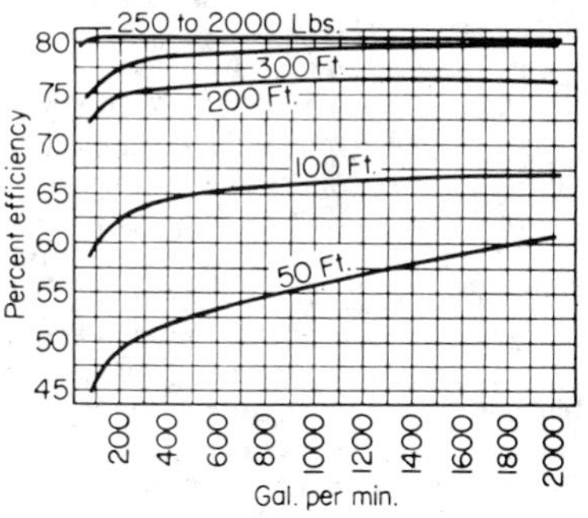

FIG. 20-22. Efficiency curves of triplex pumps under various loads.

155. Reciprocating-pump Types. Most displacement pumps are of the reciprocating type. In very small sizes, they are generally single-cylinder pumps, while in the larger sizes they are two- (duplex) and three-cylinder (triplex) pumps. All simplex, duplex, and triplex pumps are built in both the single-acting and double-acting pattern, i.e., to discharge only during one stroke of each revolution or during both strokes. It should be noted that, in the case of the **single-acting pump,** the simplex pump gives one impulse, the duplex two, and the triplex three impulses per revolution while, in the case of the **double-acting type,** the simplex gives two, the duplex four, and the triplex six impulses per revolution. Where uniformity of discharge and absence of pulsations are of primary importance, this feature will have a direct bearing on the type of pump selected.

156. Rotary-displacement pumps are built in several types, e.g., gear, guided- or swinging-vane, lobe, and screw pumps. They are used for any pressure up to 1000 psi or more and for capacities up to 3,500 gal/min. These pumps are self-priming and particularly suitable for handling liquid of high viscosity, e.g., heavy oils, tar, and sirup.

The speeds of rotary pumps are usually so low that motors of ordinary speed must be geared or belted to the pump shaft. However, some of the more recent types can be operated at comparatively high speed, making it possible in most cases to connect them directly without gears to the electric motor. Many rotary pumps require timing gears between the rotors to prevent them from making metallic contact. In the newer types, no timing gears are required, because the rotors are designed so that no power transmission takes place between the rotors, which obviously simplifies the design and diminishes operating difficulties.

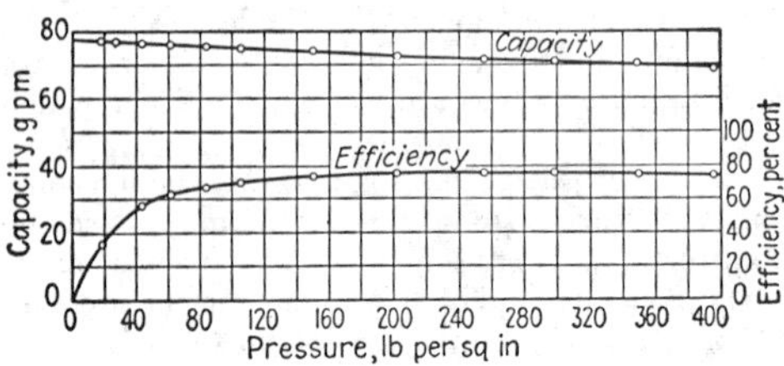

Fig. 20-23. Performance curves of a 3-in 1,150-rpm rotary pump handling lubricating oil, sp. gr. 0.923, at 67°F, 2,500 Saybolt Universal viscosity.

157. Efficiency of Rotary Pumps. Figure 20-23 represents the efficiency curve of a modern high-speed rotary pump. The efficiency is quite high for a pump of comparatively small size.

CENTRIFUGAL PUMPS

158. Classification. Centrifugal pumps may be classified in several ways: according to the number of complete elements in series, into **single-stage pumps** and **multistage pumps** (having more than one element or stage, Par. **166**); according to the arrangement of the suction inlet to the impeller, into **single-suction impeller,** in which the water enters only one side of the impeller, and **double-suction impeller,** admitting water to both sides of the impeller (Par. **159**); according to the construction of the casing, into **volute** and **diffusing-vane pumps** (Pars. **161** and **162**).

In recent years the **propeller pump** has found a wide application in dealing with large volumes of water at comparatively low heads, up to about 30 ft. The propeller pump may be operated at speeds more than twice as high as the centrifugal pump for the same head and capacity, resulting in a cheaper pump and driving motor. As the name implies, the impeller is similar to a ship's propeller. Such pumps are sometimes called "axial-flow pumps," the straight centrifugal pump being a radial-flow pump, as the water leaves the wheel in a radial direction.

Mixed-flow pump is a type in which the water leaves the impeller in a direction between axial and radial. This pump is also used for comparatively large volumes and for somewhat higher heads than the propeller pump, up to about 60 ft.

159. Relation of Suction to Hydraulic Thrust. A double-suction impeller is inherently balanced for hydraulic thrust; therefore the main purpose of a thrust bearing for such pumps is to locate the rotor centrally. A single-suction impeller is hydraulically unbalanced by an amount approximately equal to the head generated multiplied by the area of the entrance eye of the impeller. Single-suction pumps therefore have to be furnished with thrust bearings large enough to take care of the thrust, or special means for balancing the impeller, e.g., a balancing chamber enclosed by a ring on the back of the impeller, have to be provided. It is, accordingly, for the purpose of eliminating as much of the thrust as possible that multistage pumps sometimes are built of the double-suction type or with single-suction impellers placed back to back. It is also possible, however, to build multistage single-suction pumps with impellers facing in one direction with entirely automatic means for completely balancing the thrust so that thrust bearings are not required.

160. Volute and Turbine Casings. Single-stage pumps are almost invariably of the volute type, irrespective of head to be developed. Multistage pumps are of either the volute or the turbine type; the latter as a rule is designated as a diffusing-vane pump.

161. The volute pump, as its name implies, has the water passages built in the form of a volute or spiral of gradually increasing size toward the discharge opening. This is the cheapest type to manufacture, and it is, therefore, at present widely used also in multistage design for high pressures.

162. Turbine or Diffusing-vane Pumps. In a volute pump the velocity energy of the water leaving the impeller is converted into pressure in the volute and discharge passage of the casing. In the diffusing-vane pump the conversion takes place between adjacent vanes and passages between vanes surrounding the impeller. In reality a turbine or diffusing-vane pump may be classified as a multivolute pump, just as a volute pump may be referred to as a single diffusing-vane pump. It is possible to obtain smoother passages in a diffusing-vane pump than in a volute pump, and for very high heads per stage better efficiencies are, therefore, realized in the former.

163. The Importance of Suction Lift in the Selection of a Pump. One of the fundamental factors in the relative performance of displacement pumps and centrifugal pumps is that the former are capable of creating a partial vacuum in the suction connection without priming (Par. **164**), while the latter are not. Therefore, a centrifugal pump installed above its source of supply must be primed *before* starting; i.e., the suction piping and the pump casing must be filled with liquid, and all air expelled (see Par. **165**).

164. Priming. In large centrifugal pumps, priming is usually accomplished by actually "lifting" the water through the suction pipe and the pump from the source of supply by an auxiliary device. Where steam, high-pressure water, or compressed air is available, priming by ejector is a very satisfactory means. Where condensing apparatus is used, the water may be "lifted" into the pump by means of a connection from the vacuum chamber of the condenser to the pump casing.

If none of the foregoing means is available for priming the pump, a foot valve may be installed at the end of the suction piping, and the pump and piping filled with liquid from an outside source.

Another alternative is to employ a self-priming centrifugal pump; in this type a special air-pump element has been built in so as to make the pump self-priming.

165. Attainable Suction Lift. With all air leaks eliminated, almost any good centrifugal pump may be depended on for a suction lift of 15 ft (after it has been primed and started). Under favorable conditions, this may be increased to 20-ft suction lift. While there are centrifugal pumps operating under greater than 20-ft suction lifts, this requires the best of conditions, as to both pump and connections, and even then it must be understood that these same pumps would be much more reliable and dependable if the suction lift were reduced to 20 ft or less. This same characteristic makes a centrifugal pump much more sensitive to air leaks in the suction connection than a displacement pump. An air leak in the suction line of a centrifugal pump which is not even sufficient to cause a displacement pump to "knock" may cause a centrifugal pump to "lose its suction" or "priming."

Pumps handling water near the boiling point for the pressure involved must be installed below their source of supply. For instance, a pump handling condensate from a condenser should be installed about 3 ft below the bottom of the hot well, and a boiler-feed pump handling 212°F water should be located about 15 ft below the feedwater heater. Standards for obtainable suction lifts for various types of pumps and applications have been established by the Hydraulic Institute (Par. **172**).

166. Multistage Pumps. A pump designed to deliver 100 gal/min against a total head of 100 ft, with the supply flowing to the pump at zero pressure, will require exactly the same horsepower as a pump that delivers 100 gal/min against a 200-ft discharge head, with the supply flowing to the pump under a head of 100 ft. This characteristic is taken advantage of in designing pumps for heads higher than can be satisfactorily met by single-stage pumps, by using two or more pumps in series. These may be entirely separate and distinct pumps, the discharge of one connected into the suction of the other by pipe, or all the pumps or stages may be housed in one casing. For ordinary motor speeds, centrifugal pumps are usually built single stage for heads up to 300 ft, depending on the capacity of the pump and, therefore, the size of the impeller. As the peripheral speed of the impeller increases, either on account of larger diameter or because of higher rotative speed, the developed head per stage increases correspondingly. Multistage pumps are built with two to eight or more impellers in one casing. At a speed of 3,600 r/min, a head of 600 ft or more may be obtained from one impeller, so that a six-stage pump would develop a head of 3,600 ft or more, or in excess of 1,500 lb/in² when pumping cold water.

One very important characteristic of a centrifugal pump is that an impeller develops a certain number of *feet* of liquid head. It is extremely important to take this into account in dealing with liquids of different specific gravities, which occurs quite frequently. For instance, a centrifugal pump that develops 100 lb/in² (231-ft head) pumping cold water will *not* give 100 lb pressure when pumping gasoline. If the gasoline has a specific gravity of 0.70, the pressure developed by the pump will be only 70 lb. Therefore in order to obtain 100 lb pressure with gasoline the pump must be designed for a head of $231/0.70 = 330$ ft. The horsepower required to pump water against 100 lb pressure will be the same as pumping gasoline against 100 lb pressure.

167. The upkeep expense of centrifugal pumps is usually small, for the reason that the only moving parts in contact are the shaft and bearings. The clearance between the impeller rings and case rings is almost always greater than the permissible wear of the bearings. Even if they did make contact, this would probably be detected and the pump shut down before any damage was done.

High-head conditions require high impeller velocity, which gives high velocity to the water leaving the impeller. In order to obtain good efficiency these conditions demand high velocities in the diffusion chamber (if there is one) and in the volute. If the water contains much sand or other abrasive material, high velocity of the water results in rapid erosion of cast iron, the usual material used for volute pumps. For these conditions special materials must be selected for all wearing parts.

168. Piping Connections. From the principle of operation of centrifugal pumps, the velocity in the pump casing is very likely to be higher than good practice would permit for velocity of water in pipes; consequently the size of the pipe connection should be calculated on the basis of allowable velocity in the pipe rather than simply to make the connection of the size of the suction and discharge openings of the pump. To avoid shock due to sudden change of section of the pipe, this change in size of pipe should be made by means of a standard "increaser" or "reducer," and this fitting should be installed at the pump opening. Increasers bolted to the horizontal suction nozzle of the pump should be of the eccentric type in order to avoid an air pocket in the suction line. As already pointed out, the suction connections must be absolutely airtight if satisfactory operation is to be obtained. Incidentally, also, the packing glands must be drawn up tight but only tight enough to prevent leakage, care being taken to avoid excessive friction on the shaft at this point. Most pumps have a connection from the discharge chamber to the glands in order to keep them under a **water seal** at all times. When this connection is provided, the glands should be loose enough to permit constantly a slight leakage. If the pump is not furnished with a water-seal connection from the discharge chamber, such connection should be provided by the customer if the pump is operating with a suction lift. If the pump handles gritty water, the water-seal connection should be taken from a supply of clear water in order to avoid cutting of shaft or shaft sleeves. The sealing-water supply needs to come to the pump stuffing boxes under a pressure of only a few feet.

169. Electric Drive. The most common drive for centrifugal pumps is the electric motor. The centrifugal pump is ideally adapted for electric-motor drive in view of its requiring only low starting torque, resulting in comparatively small rush of current in starting, so that standard squirrel-cage motors can often be used, even of high horsepower.

The speed of centrifugal pumps is higher than can be used for displacement pumps; hence cheaper motors can be used, and there is an absence of gear reductions. In well-designed centrifugal pumps, there is very little danger of overloading the driving motor, so that motors can be selected close to the rating of the pump. This results in high efficiency and power factor of the motor.

For electrical distribution systems having a low power factor, the synchronous motor is an ideal drive for centrifugal pumps, as such motors can be operated to give a leading power factor and thus improve the power factor of the whole system. In view of the low starting torque, standard self-starting motors can as a rule be used for all centrifugal pumps.

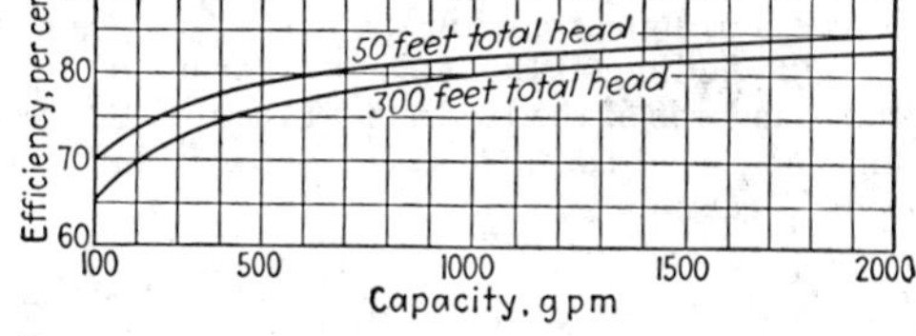

Fig. 20-24. Efficiencies of centrifugal pumps designed for various capacities and heads.

170. Characteristics of Centrifugal Pumps. Figure 20-24 shows typical curves of efficiency of standard centrifugal pumps designed for various heads and capacities. These curves may be compared with the curves of Fig. 20-22 of reciprocating pumps. Figure 20-25 shows the characteristic curves of a typical modern centrifugal pump. As may be noted, the head drops gradually from zero delivery or shutoff pressure, which may be anywhere from 10 to 50% above the designed head, depending upon the type of pump involved or its specific speed.

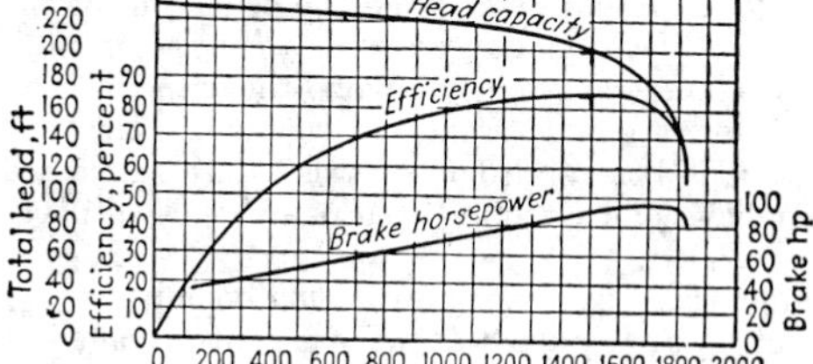

Fig. 20-25. Characteristic curves of a typical volute pump at constant speed.

For parallel operation of two or more pumps, it is important to have a head capacity curve that rises continually to shutoff pressure. If the curve has a "hump" so that the shutoff head is lower than the maximum head produced by the pump, unstable operation at low capacities will result.

The efficiency of centrifugal pumps is high and in well-designed pumps shows very little dropping off after comparatively long service. It is fairly safe to figure on efficiencies 85% and higher for pumps of sizes larger than 1,500 gal/min and heads in a single stage up to 200 ft. Large pumps for capacities over 10,000 gal/min will often give efficiencies higher than 88%. Pumps for 500 up to 1,000 gal/min will show efficiencies between 77 and 80%, and between 1,000 and 1,500 gal/min efficiencies from 82 to 85% should be obtained.

171. Performance under Variable Speed. It is important to bear this in mind in connection with pumps that are to operate for a considerable time at capacities and heads lower than normal. For instance, a circulating pump for a surface condenser may during the winter months be required to deliver only one-half of its normal capacity, and, in view of the fact that the head against which such a pump generally operates is composed only of pipe friction, the total head at one-half capacity would be only about one-quarter of normal head, so that these conditions would therefore be obtained at approximately one-half speed with one-eighth power consumption. Therefore, it pays in many cases to investigate the possibilities of variable-speed

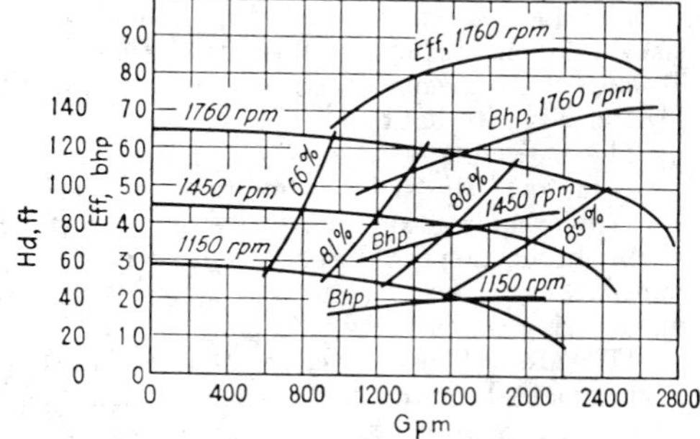

Fig. 20-26. Characteristics of a centrifugal pump at different speeds, with curves connecting points of equal efficiency.

wound-rotor induction motors, in spite of their low efficiencies at reduced speeds (see Fig. 20-26).

The power required at starting is very low, permitting the centrifugal pump to be started even against normal head without the use of a bypass and without requiring heavy starting currents in the motor—a most desirable feature.

172. Additional Pump Data. For more complete information on all types of pumps the Standards of the Hydraulic Institute, 122 East 42nd Street, New York, N.Y. 10017, may be consulted. These standards include reciprocating pumps, rotary pumps, and centrifugal pumps.

TRAVELING CRANES

By A. H. Myles

173. Overhead and gantry cranes have the same fundamental operating characteristics.

An overhead crane has a bridge traveling on an elevated runway and carrying one or more hoisting mechanisms either in fixed positions or on a movable trolley or trolleys.

A gantry crane is similar to an overhead crane except that the bridge is rigidly supported on two or more legs traveling on rails at ground level. There are several types of gantry cranes. A semigantry crane has one end of the bridge rigidly supported on one or more legs that run on a runway at ground level while the other end of the bridge runs on an elevated runway. In a cantilever gantry crane or a cantilever semigantry crane, the bridge girders or trusses extend transversely beyond the crane runway on one or both sides. Storage bridge cranes are gantry cranes of very long span ordinarily used for bulk storage of material. The bridge girders or trusses are supported on one or more legs. Such cranes often have one or more fixed or hinged cantilever ends.

174. Portal and tower cranes are characterized by having a rotating boom.

A portal crane has a traveling bridge structure similar to that of a gantry crane but has no trolley. Instead of a trolley, it has a pivoted boom attached to a revolving crane structure mounted on the bridge.

A tower crane is similar to the portal crane except that the revolving crane structure is mounted on a stationary or movable tower instead of on a traveling bridge.

175. Electrical equipment for cranes normally consists of the following: (1) a-c or d-c motors; (2) manual, electromagnetic, or static control gear with or without mechanical or electrical braking means; (3) one or more self-setting holding brakes; (4) a power-circuit or control-circuit overhoist limit switch; (5) a switch or circuit breaker of the enclosed type connected in the leads from the runway collectors; and (6) master switches, which may be located in the crane cab, suspended from the elevated structure, or positioned at a remote location.

176. Types of Motors. General-industry and steel-mill service cranes usually are powered by d-c series motors or a-c wound-rotor induction motors.

The d-c series motor has ideal inherent operating characteristics for crane service, since it adjusts its hoisting speed to the load—the greater the load, the slower the speed. This feature prevents excessive power demand with heavy loads and provides desirable high speeds with light loads.

The a-c wound-rotor or slip ring induction motor has less speed range than the d-c series motor. Under load conditions, its speed is less than its synchronous speed, and under overhauling conditions its speed should be controlled so as not to exceed 150% of synchronous speed. Speed control is obtained by varying secondary resistance or impedance, by varying the primary voltage, by varying the primary frequency, or by unbalancing the primary voltages.

177. Motor Size. Because crane motors are used intermittently and under different environmental conditions, the choice of motor size and type is determined by the service requirements. Any particular frame size of motor will, therefore, have different horsepower ratings for varying degrees of intermittent service and for different enclosures or ventilating systems. Steel-mill service cranes are usually supplied with totally enclosed motors. General-industry cranes may or may not require totally enclosed motors.

If the duty cycle is known, the horsepower rating must be sufficient to meet the duty-cycle requirements. When the duty cycle is not known, the motor horsepower is based on the 1-h rating of the motor, the formulas given below being used. The 1-h rated horsepower is defined as that horsepower which, if produced continuously for 1 h, will cause the motor to reach its specified temperature rise.

D-C Hoist Motors. A d-c series-wound hoist motor should be so selected that the 1-h rating will be not less than that given by the following formula,

$$\text{hp (1 h)} = \frac{0.75WV}{33{,}000E}$$

where W = weight of rated load plus the weight of the block and cable in pounds, V = specified hoisting speed in feet per minute, 0.75 = ratio of 1-h horsepower to ½-h horsepower rating of the motor, and E = combined efficiency of gears and sheaves.

For hoist service, experience has shown that d-c motors can be operated at a higher horsepower than their 1-h rated horsepower. This is taken into account in the foregoing formula so that selection can be based on the 1-h characteristic curve.

A-C Hoist Motors. An a-c wound-rotor motor should be so selected that the 1-h rating will be not less than that given by the following formula,

$$\text{hp (1 h)} = \frac{WV}{33{,}000E}$$

where W = weight of rated load plus the weight of the block and cable in pounds, V = specified speed in feet per minute, averaged between the maximum hoisting and lowering speeds for the maximum rated load (maximum hoisting and lowering speeds should be within ±25% of the specified speed), and E = combined efficiency of gears and sheaves.

Crane Bridge and Trolley Motors. The force required to drive the bridge or trolley is the sum of that necessary to overcome rolling friction and to accelerate or decelerate the drive. For bridge drives, the rolling friction is proportional to the total weight of the crane plus load; for trolley drives, the rolling friction is proportional to the trolley weight plus load. In either case, the friction is assumed to be constant at all speeds. Unless otherwise stated, an overall friction factor of 15 lb/ton is used for roller-bearing cranes and 24 lb/ton for sleeve-bearing cranes. Mechanical efficiencies are included in these factors.

D-C Motor Selection. The size of a d-c motor for the bridge or trolley should be not less than that computed from the following formula,

$$\text{hp (1 h)} = \frac{2{,}000WV}{33{,}000K} \times \frac{Nr}{Nf} \times \left(\frac{f}{2{,}000} + \frac{a}{32.2F}\right) \times SF$$

where W = weight of crane plus weight of load in tons for bridge-drive calculations or weight of trolley plus weight of load in tons for trolley-drive calculations, V = specified speed in feet per minute (determined from the speed-torque curve of the motor where the motor speed corresponds to free-running horsepower proportional to friction factor), K = ratio of maximum average accelerating torque (from zero to two-thirds specified speed) to rated torque, Nr = rated motor speed in revolutions per minute, Nf = motor speed in revolutions per minute at free running speed, f = rolling friction in pounds per ton, a = average acceleration in feet per second per second from zero to two-thirds specified speed, F = acceleration factor for mechanical efficiency and inertia of rotating parts including motor and brake wheel, which varies between 0.90 and 0.60 depending upon application, and SF = service factor (a thermal or heating factor which varies between 1.0 and 1.6 depending upon the application).

A-C Motor Selection. The size of an a-c motor for the bridge or trolley should be not less than that computed from the following formula,

$$\text{hp (1 h)} = \frac{2{,}000WV}{33{,}000K} \times \left(\frac{f}{2{,}000} + \frac{a}{32.2F}\right) \times SF$$

where W = weight of crane plus weight of load in tons for bridge-drive calculations or weight of trolley plus weight of load in tons for trolley-drive calculations, V = specified speed in feet per minute (determined from speed-torque curve of motor where motor speed corresponds to free-running horsepower proportional to friction factor), K = ratio of maximum average accelerating torque (from zero to two-thirds specified speed) to rated torque, f = rolling friction in pounds per ton, a = average acceleration in feet per second per second from zero to two-thirds specified speed, F = acceleration factor for mechanical efficiency and inertia of rotating parts including motor and brake wheel, which varies between 0.90 and 0.95 depending upon the application, and SF = service factor (a thermal or heating factor which varies between 1.0 and 1.6 depending upon the application).

178. Brakes. Each independent hoisting unit of a crane has at least one self-setting holding brake applied directly to the motor shaft or to some part of the gear train. The brake is usually released by energizing an electromagnetic operating coil to compress a spring which applies the braking force when the operating coil is deenergized.

On d-c-powered cranes, the hoist brake is usually series-wound, which is safer and faster than a shunt-wound brake. On a-c-powered cranes, a shunt-wound brake must be used; it may be a-c-operated or d-c-operated through rectifier equipment.

A holding brake, if provided on bridge and trolley drives, is automatically applied when electric power is removed. However, the control may be arranged to provide a drift point. Brakes designed to be applied by mechanical, electrical, pneumatic, or hydraulic means may also be used and are ordinarily controlled by a treadle operation.

179. Dynamic Lowering. This method of retarding the speed of the hoisting motor when lowering a load has been adopted almost universally on d-c cranes. In the lowering direction, the series motor is connected as a shunt motor, with its field in parallel with the armature; the parallel-connected field and armature have controlling series-connected power resistors. The motor thus becomes a shunt generator when the load is overhauling. The generated armature voltage will tend to rise above line voltage and will, in addition to delivering power to the field, actually deliver power back to the power system.

Alternating-current hoist controllers for slip-ring motors may also provide dynamic lowering. By applying unbalanced polyphase voltages to the primary of the motor and by inserting resistances of appropriate values in the secondary, it is possible to obtain dynamic lowering both below and above synchronous speed. In an extensively used controller of this type, two primary terminals of the motor are connected together, and single-phase voltage is applied between this common terminal and the third primary terminal of the motor to provide dynamic lowering.

180. Hoist Limit Switches. It is customary to provide a limit switch which is operable during hoisting to limit the upward travel of the hook block so that the hook block or moving sheaves will not be raised into the drum. The simplest arrangement is to gear a small control-circuit limit switch to the hoist mechanism so that as the hook block reaches the upper limit of travel the switch opens its contacts to deenergize the line contactors, thereby permitting the brake to stop the load. The preferred type of limit switch for crane hoists is one that directly interrupts the power circuits, as described later. It comprises a snap-action double-pole double-throw switch from which a resetting weight is suspended. The switch is so located with respect to the hoist mechanism that the suspended weight is directly over a platform attached to the hoist hook block. Raising the suspended weight permits a weighted lever to trip the switch, and lowering it permits the switch to reset.

D-C-POWERED CRANE HOIST CONTROL

181. Direct-current Hoist Control. The series motor is well suited to d-c hoists and, because of the simplicity of its associated control, is used universally. In the hoisting direction it is connected as a series motor, in the lowering direction as a shunt motor.

A five-point hoist power circuit is shown in Fig. 20-27. A power limit switch is connected in the motor circuit. The LS resistor is the dynamic braking resistor used with the limit switch. Four trolley bars and collector shoes are required. Nine normally open contactors H, $1L$, M, $1A$, $2A$, $3A$, $4A$, $2L$, and $3L$ and one normally closed contactor DB are provided. The H and $1L$ contactors are the hoist and lower positive line contactors, respectively; the $1A$, $2A$, $3A$, and $4A$ contactors are the accel-

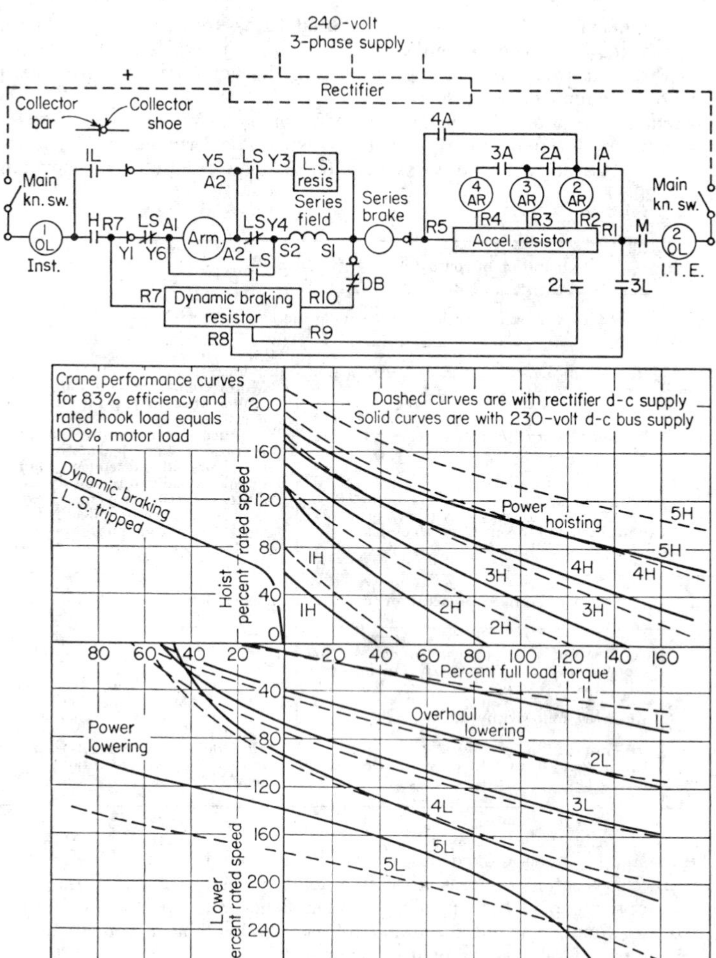

FIG. 20-27. Power-circuit and crane-hoist performance curves using a d-c motor operated from constant potential and from a rectified a-c power supply.

eration contactors for cutting out series resistance in the hoisting direction and for control of the field strength in the lowering direction. Contactors $2L$ and $3L$ are used for increasing the voltage across the armature in the lowering direction. The spring closing contactor DB is the lowering dynamic braking contactor connected in series with the dynamic braking resistor $R7$-$R10$.

If the master switch is moved from the "off" point to the fifth-point hoist, H and M close instantly, followed by $1A$ without time delay; contactor DB opens, and contactors $2A$, $3A$, and $4A$ close in timed sequence under the control of the time-current relays $2AR$, $3AR$, and $4AR$, respectively.

For some applications, slow first-point hoisting speeds are desirable to take up slack cable or to hoist light loads at slow speed. A motor shunt connection obtained by having contactor DB closed on first-point hoist will provide such results. In this circuit, additional protective equipment should be added to the control to assure setting of the brake when the power limit switch trips if the master switch is on first-point hoist.

Many schemes have been used to provide setting of the brake with a motor shunt connection. An extra collector bar can be used to assure brake setting, or a series or counter-emf relay. A counter-emf relay, in addition, provides a safe speed in lowering out of the limit switch by picking up to stop the hoist drive if excessive speed is reached before the limit switch resets.

In lowering, the motor is connected for operation as a shunt motor. The following point-by-point description of operation will best explain performance. The performance characteristic curves for a 230-V constant-potential power supply are shown in solid lines in Fig. 20-27.

Point	Operation of contactors	Purpose
	Lower—acceleration (moving the master slowly)	
1	*M*, 1*L*, 4*A*, 3*A*, and 2*A* close, *DB* remains closed	Provides strong current for brake release. Motor connected as a shunt motor. Dynamic braking provided in case load is overhauling
2	2*L* closes before *DB* opens	To reduce but maintain dynamic braking. Increase voltage to armature
3	3*A* and 4*A* open	To weaken field and reduce dynamic braking
4	2*A* opens	To weaken field and reduce dynamic braking
5	If the counter-emf of motor is about 50%, counter-emf relay picks up to permit 3*L* to close. 2*L* opens after 3*L* closes	Counter-emf relay prevents excessive current through armature. Field is further weakened and dynamic braking reduced. Voltage to armature is increased
	Lower—deceleration (moving the master slowly)	
5 to 4	Counter-emf relay opens, 2*L* closes. 3*L* opens after 2*L* closes. *DB* remains open	To maintain and strengthen dynamic braking
4 to 3	2*A* closes	Strengthens the field, and increases dynamic braking
3 to 2	3*A* closes, followed by 4*A*	Strengthens the field, and increases dynamic braking
2 to 1	*DB* closes; 2*L* opens	Maintains and increases dynamic braking
1 to off	1*L*, *M*, 2*A*, 3*A*, and 4*A* opens, *DB* remains closed	Dynamic braking maintained, and brake sets

If the master switch is moved quickly to the last point lower, 1*L*, *M*, and 4*A* close instantly, 2*L* closes after 1*L* closes with time delay of 2*AR*; after 2*L* closes, *DB* opens, followed by opening of 4*A* to provide acceleration. First point circuit for a time period gives maximum current for brake release. When the motor attains sufficient speed, a counter-emf relay picks up to permit closure of 3*L*; 2*L* opens after 3*L* closes.

If the master switch is moved quickly from the fifth point lower to the "off" point, 3*L* opens, *DB* closes, and 1*L* and *M* open in rapid sequence.

Entering the Limit Switch. With the master switch on the "off" point or on any hoisting point, the circuit established is such as to permit the brake to set when the limit switch trips. The counter emf of the motor is in a direction from *A*2 to *A*1, and the current will be maintained in the normal direction in the field from *S*2 to *S*1. The current reverses in the armature to provide dynamic braking, the motor acting as a series generator with the dynamic-braking resistor *LS* across its terminals.

Should the operator move the master switch to any of the lowering points at the instant of entering the limit switch, the motor will be "plugged" (power reversal while the motor continues to rotate in the direction previously established) as a series motor with line voltage in the same direction as the counter emf. The brake does not set, and hook travel is stopped by plugging.

The use of antifriction bearings and better trolley design results in relatively high light hook hoisting speeds, requiring an increase in the limit-switch resistance *LS* to obtain satisfactory motor commutation. The increase of the limit-switch resistance causes an increase in the stopping distance which, for some installations, cannot be allowed. The logical solution to this problem is to limit the maximum light hook hoisting speed, which in turn permits the use of lower limit-switch resistance. Both cause the stopping distance to decrease. The maximum hoist speed for most installations can be limited to 225% of rated motor speed by the addition of a permanent teaser field resistance (not shown) connected across limit-switch terminals *Y*6-*Y*4. The value of this resistance depends upon the crane hoist efficiency or percent of motor torque required to hoist an empty hook. For most cranes, 800% resistance having one-eighth full-load current capacity is satisfactory.

Lowering Out of Limit Switch. On all points in lowering out of the power limit switch, the motor is connected as a series motor, with the limit-switch resistor *LS* acting as a motor shunt. Under this condition, with a heavy overhauling load and with the brake released, there is no retarding torque. If the motor is not to reach an excessive speed, it is important that the limit switch reset in as short a travel distance as possible. If excessive speed is reached by the time the limit switch resets, then sparking at the motor commutator might be serious. Numerous means of limiting the motor speed have been used; for example, an armature shunt (not shown) of 40% resistance connected across terminals *Y*4-*Y*5 has been employed for this purpose, or a blocking rectifier can be used instead of the resistor. Another method has been the use of a counter-emf relay (as explained above) which opens the line contactors and sets the brake should the speed become excessive before the limit switch resets.

A-C-POWERED CRANE HOIST CONTROL

182. D-C Motors and Control. When the power supply to the crane is a-c, the use of d-c motors and control, described above, operated from a rectified power supply (Fig. 20-27) will give characteristics even more desirable than when power is obtained from a 230-V d-c bus.

When nominal 230-V 3-phase power is rectified by a set of six bridge-connected rectifiers, the d-c output voltage is approximately 300 V. This permits a 230-V d-c series motor to develop 1.3 times its 230-V horsepower rating, since, for a given load current and developed torque, the motor speed is directly proportional to the voltage impressed across the armature.

One feature of such a d-c dynamic lowering control is its ability to provide higher speed in lowering light loads while maintaining satisfactory motor stability in lowering heavy loads. When power is taken from a d-c bus, the maximum no-load lowering speed must be limited to approximately 140% of rated value, since with high armature current any further field weakening causes sufficient armature reaction to affect motor stability adversely. However, when d-c power is derived from a rectified source, the circuit constants are quite different. The performance curves of Fig. 20-27 for fifth-point lower compare the results obtained in lowering loads at maximum speed with power from a 230-V d-c line and with a rectified 300-V supply. Calculations for the data for the curves are on a per unit basis. Note particularly the improved stability and the higher no-load speed of the rectified d-c operation as compared with operation from a d-c bus.

183. A-C Dynamic Lowering Hoist Control. A typical a-c dynamic lowering hoist control power circuit with rectifier-operated d-c shunt magnetic brake and power limit switch is shown in Fig. 20-28. The motor, brake, and limit switch are mounted on the trolley. The brake relay *BR* and the rectifier are mounted on the control panel. Such an arrangement requires 10 collector rails, as shown.

In hoisting, contactors *H* and *M* and the brake relay *BR* close on the first point of the master switch to release the brake and apply hoisting power to the motor with all secondary resistance in the circuit. On the second point, 1*A* closes without delay; and on the third, fourth, and fifth points, 2*A*, 3*A*, and 4*A*, respectively, close in sequence under control of relays responsive to variation in rotor frequency or motor speed. The motor accelerates to a speed just under synchronous speed. On entering the limit switch, the two normally closed contacts *LS* open to remove power from the motor and to permit the brake to set.

On the first point lower, contactors *L*, *DB*, and 1*A* close, tying motor terminals *T*1 and *T*2 together and applying single-phase voltage across *T*2 and *T*3. The brake is also released, allowing the load to lower (if overhauling) against dynamic braking or retarding torque developed by the motor as it gains speed. The closure of 1*A* gives a secondary resistance value which provides a retarding torque of approx-

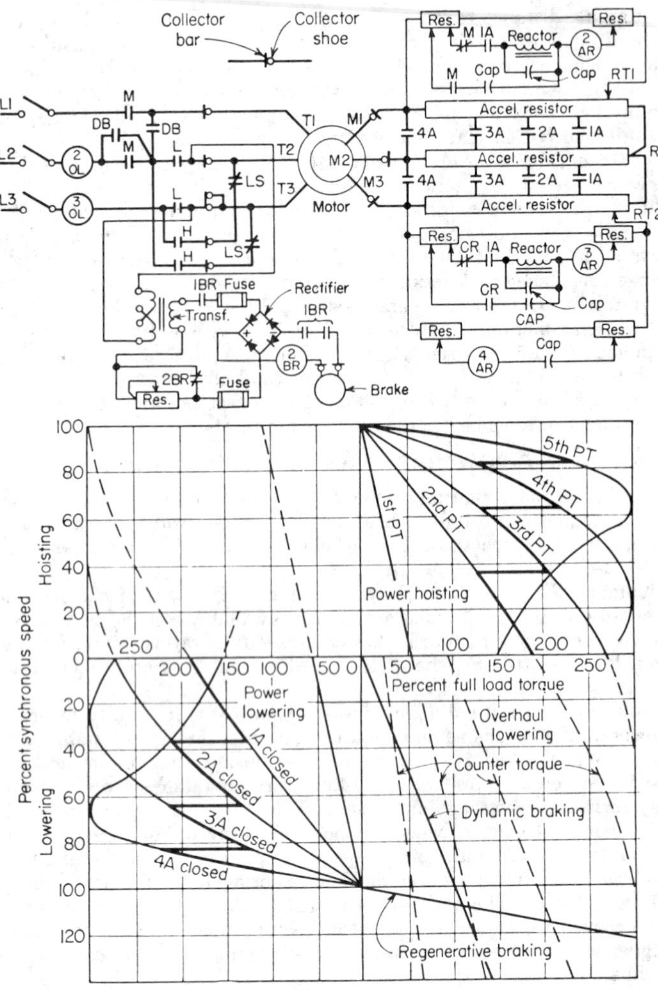

FIG. 20-28. Power-circuit and crane-hoist characteristic curves using an a-c wound-rotor motor and reversing dynamic lowering control.

imately full-load torque at synchronous speed. On the second point lower, *DB* opens and *M* closes to apply balanced 3-phase lowering power to the motor, and at the same time 1*A* opens to provide low torque. This point is used in "inching" a light load downward. If the master is held on the second point, 2*A*, 3*A*, and 4*A* will close in frequency-controlled sequence. If the master switch is moved to the third point from the "off" position, 1*A* closes instantly to provide higher driving torque in the lowering direction and 2*A*, 3*A*, and 4*A* follow in frequency-controlled sequence. With the secondary resistance short-circuited, and with an overhauling load, the motor runs at a speed somewhat in excess of synchronous speed, the speed being limited by regenerative braking.

On returning the master to the second point, 1*A* opens without any effect, but when the master is returned to the first point, *M* opens and *DB* and 1*A* close, while 2*A*, 3*A*, and 4*A* open to set up the dynamic braking connections. If the master switch is then moved to the "off" point, *DB*, *L*, and 1*A* remain closed and the brake sets. Dynamic braking assists the brake in bringing the motor to rest, at which time one of the frequency-responsive relays 2*AR* or 3*AR* opens its contacts to open *L*, *DB*, and 1*A*. This is referred to as "off-point dynamic braking." The retention of dynamic braking during the setting time of the magnetic brake prevents the load from dropping during this short time interval and greatly reduces brake-lining wear.

Figure 20-28 also shows the speed-torque curves of a dynamic braking a-c hoist controller built in accordance with the wiring diagram.

184. Alternating-current counter-torque hoist control is a control system in which the power to the motor during lowering is reversed to develop hoisting torque to decelerate, stop, or lower an overhauling load at stable speeds. The system is ideally suited for high-production cranes having severe duty cycles and with fixed overhauling loads such as lifting magnets and buckets. Motor heating and brake wear are a minimum.

185. Alternating-current static hoist control utilizing power magnetic amplifiers, or gate-controlled thyristors, provides the ultimate in fully regulated speed control. It is completely stepless in operation from zero to full speed for both hoisting or lowering, with the ability to float the load for precise positioning.

An elementary hoist circuit, utilizing the gate-controlled thyristor is shown in Fig. 20-29. The circuitry is similar when using power magnetic amplifiers. Either system of control is similar in performance and is for a wound rotor motor used to operate a hoist

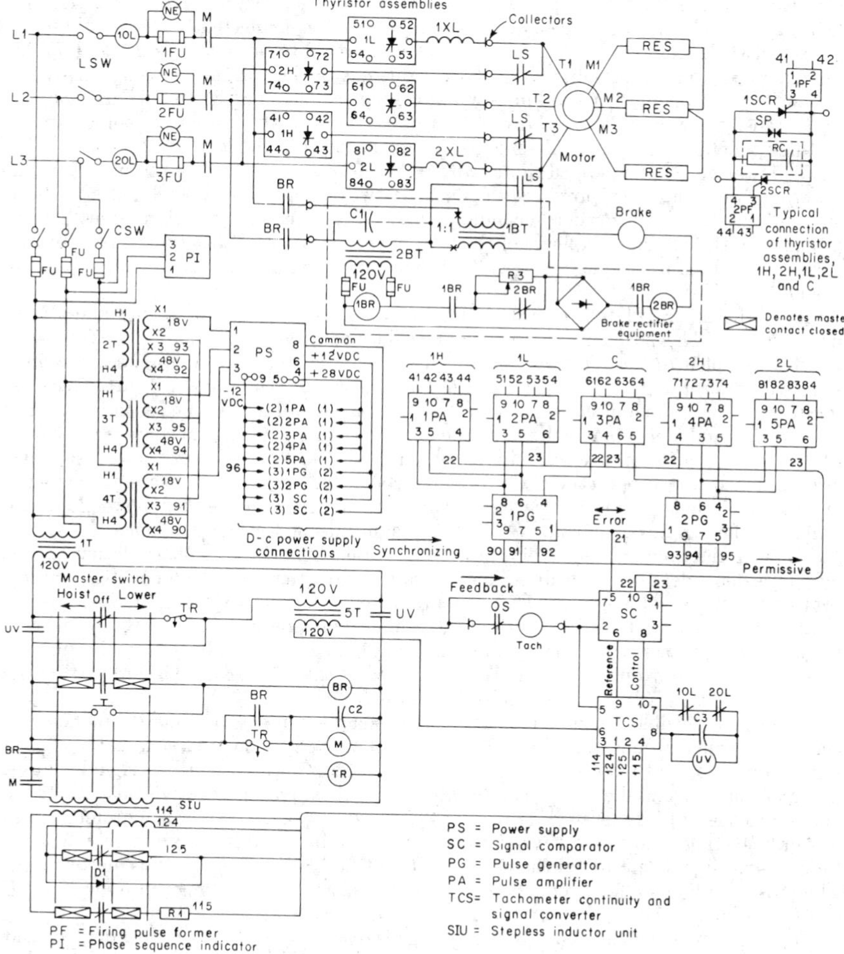

Fig. 20-29. Elementary circuit diagram for a static-power thyristor a-c hoist control.

having no mechanical or electrical load brake. The power circuit shown contains five gate-controlled thyristor assemblies, each being six-terminal units; two for the power circuit and four for the two gating circuits. The thyristor assemblies are arranged for phase reversal in the same manner in which contactors are normally used.

By simultaneously controlling the gating pulses of assemblies 1*H*, 2*H*, and *C*, hoist power of controlled voltage is applied to the motor terminals. By similarly controlling the assemblies 1*L*, 2*L*, and *C*, lowering power is applied to the motor.

Connected to the rotor circuit of the motor are slip resistors whose values are such as to produce pull-out torque at approximately 5% hoisting speed. Also shown in the power circuit is a standard power limit switch which limits the hook travel in the hoisting direction. Motor brake control includes rectification equipment for a high-speed d-c-operated brake from the a-c power.

The connection of the power limit switch and brake-control circuits are shown with the limit switch in its normal untripped condition. Hoisting power is supplied to the motor through the two normally closed limit-switch contacts. Power for the brake circuit is obtained from lines *L*1 and *L*2. Under normal operation, the primary and secondary windings of the 1*BT* transformer are connected in series by the normally closed limit-switch contact. The left side of the 1*BT* transformer primary winding is directly connected to the *L*1 phase of the supply voltage. By transformer action, the left side of the transformer secondary also appears to be connected to the *L*1 phase.

When the limit switch trips, its contacts reverse. Motor terminals *T*1 and *T*3 are disconnected from the supply voltage by the opening of the normally closed limit-switch contacts. The opening of the limit-switch contact connecting the two windings of the 1*BT* transformer in series also directly opens the supply voltage circuit for the brake rectifier panel. The brake is de-energized and sets to stop the load.

To lower out of the tripped limit switch, it is necessary to operate the master in the lowering direction. The previously open limit-switch contact is now closed and short-circuits the 1*BT* transformer secondary. The brake rectifier panel receives control power from the *L*2 and *L*3 phases of the supply voltage when the lowering thyristors are turned on. Balanced three-phase voltages with lowering phase rotation are applied to the motor.

Since this is a regulated control system, provision is made for a reference and a feedback signal. The speed reference signal is derived from an induction unit (*SIU*), which is coupled to the rotating element of a master switch. It comprises a magnetic vane which is rotated within the air gap of two transformer cores and is so arranged that master rotation in one direction produces an increasing voltage at the secondary of one transformer and a constant voltage at the other secondary, while the converse situation results with rotation in the opposite direction. These secondaries of the transformers are connected to bridge rectifiers in the signal converter, whose outputs are connected in opposition so that a net reference signal can be compared with the tachometer feedback signal at the signal comparator *SC*. The magnitude of the net reference signal is a function of the angular throw of the master switch handle while the polarity of this voltage is dependent upon the direction of master switch throw. Master switch contacts control the undervoltage relay *UV*, the brake relay *BR*, and time relay *TR*. When returning the master to OFF the time relay maintains power on the motor to provide motor braking to assist the magnetic brake in stopping. A pushbutton in the master-switch handle allows operation through zero speed at the off position without setting the magnetic brake. This provides load-floating operation.

The speed feedback signal is obtained from a permanent-magnet d-c tachometer generator which is mechanically coupled to the motor. The feedback voltage is in series opposition with the reference voltage and the net value of these signals causes an error signal from the signal comparator *SC* of positive or negative polarity to be applied to the pulse generators 1*PG* and 2*PG*. Polarity of the error signal determines whether hoisting or lowering thyristor assemblies will be gated for conduction. Magnitude of error signal determines the value of motor terminal voltage and thus motor speed. The pulse amplifiers *PA* include logic lock-out circuitry and the firing pulse formers *PF* provide improved gating signal reliability.

In a three-phase system each phase voltage is displaced by 120 electrical degrees and the instant of firing is different for each thyristor. Consequently there must be a separate pulse circuit for each thyristor.

Tachometer continuity equipment provides a continuity-sensing circuit by means of an alternating-current signal to check circuit continuity through the tachometer generator, the collectors, and the overspeed switch. It should be noted that the presence of this signal tends to improve the reliability of this circuit since an open circuit will immediately be followed by the presence of the a-c control voltage across the open which helps to keep the circuit continuous.

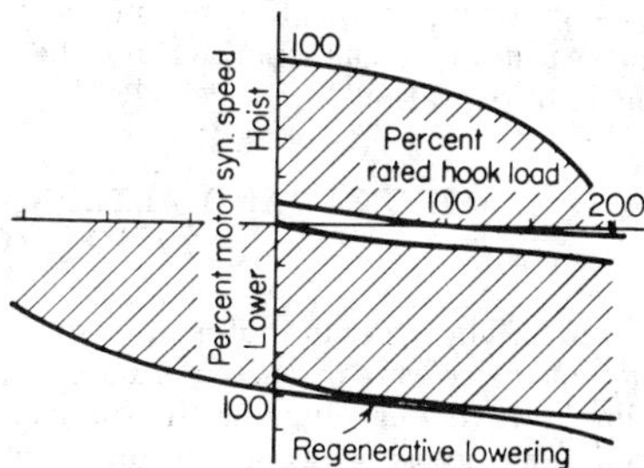

Fig. 20-30. **Crane performance curves for a-c static hoist control.**

The shaded envelopes of Fig. 20-30 illustrate the range of stepless control in both the hoisting and lowering quadrants. This area can be made continuous through zero speed

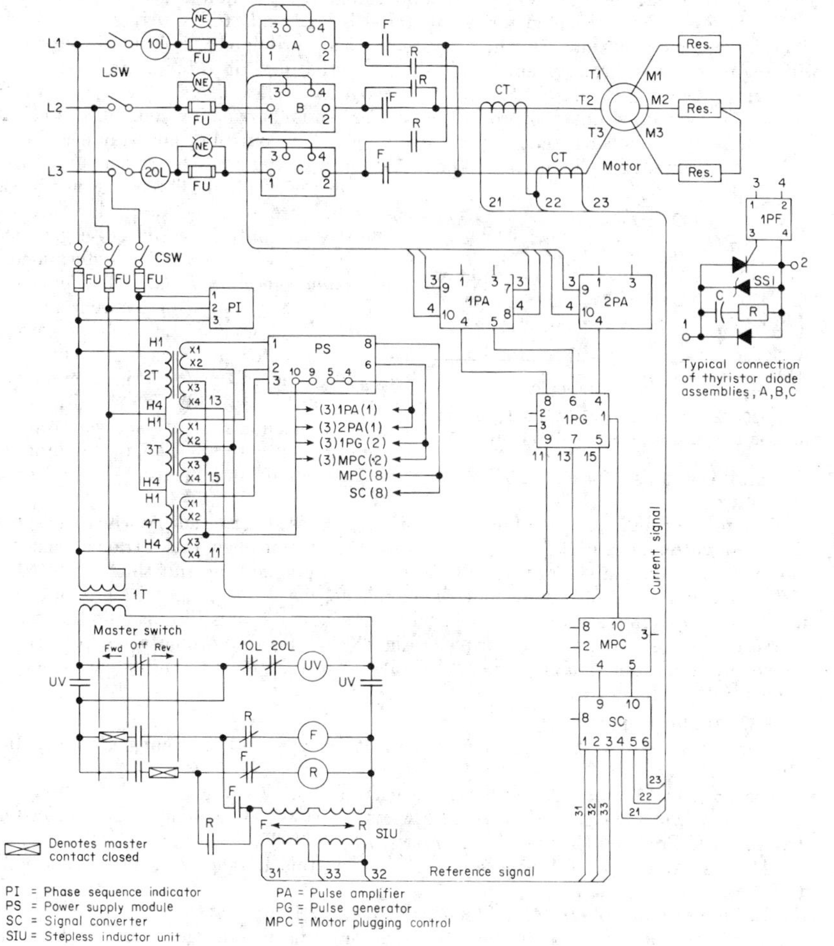

Denotes master contact closed

PI = Phase sequence indicator
PS = Power supply module
SC = Signal converter
SIU = Stepless inductor unit
PA = Pulse amplifier
PG = Pulse generator
MPC = Motor plugging control

Fig. 20-31. **Elementary circuit diagram for a static-power thyristor a-c bridge or trolley control.**

by depressing the pushbutton in the master-switch handle. The regenerative lowering curve is indicated as separate from the lowering control envelope since it is an unregulated mode of operation.

DIRECT- AND ALTERNATING-CURRENT CRANE BRIDGE AND TROLLEY CONTROL

186. The types of motors used on crane bridge and trolley drives and the computation of motor size were discussed in Pars. **176** and **177**. The most common type of control is a simple reversing-plugging full-magnetic controller. Although plugging is the most common means of bringing a bridge or trolley to rest, regenerative braking or dynamic braking is occasionally used to decelerate the motor in normal operation when d-c motors and control are used. The dynamic braking feature is referred to as "emergency dynamic braking" when applied upon power failure to slow down the drive. The emergency dynamic braking method, if used, is usually applied to the bridge motion of heavy steel-mill cranes and trolley drives of storage bridge cranes.

If static control is selected for a-c powered bridge and trolley drives, then several different arrangements of the control system can be provided. For full static control with regulated speed, the system is very similar to that described for static hoist control. However, since regulated speed control of bridge or trolley drives is not normally required, a somewhat simpler control scheme using primary reversing contactors and gate-controlled thyristors, or magnetic power amplifiers, for control of adjustable voltage to the motor terminals can be used. Such a system, utilizing gate-controlled thyristors, is shown in Fig. 20-31. Three power thyristor assemblies are required, and they are controlled by the output of the pulse generator. This pulse generator output is determined by the master reference signal. The pulses fire each of the thyristors for conduction of the proper portion of each half cycle of power to control motor torque and speed. Outputs of the pulse generators for bridge and trolley control need vary only in magnitude since direction of motor rotation is controlled by magnetic contactors. The pulse amplifiers *PA* and the pulse formers *PF* provide gating signal amplification and improved reliability.

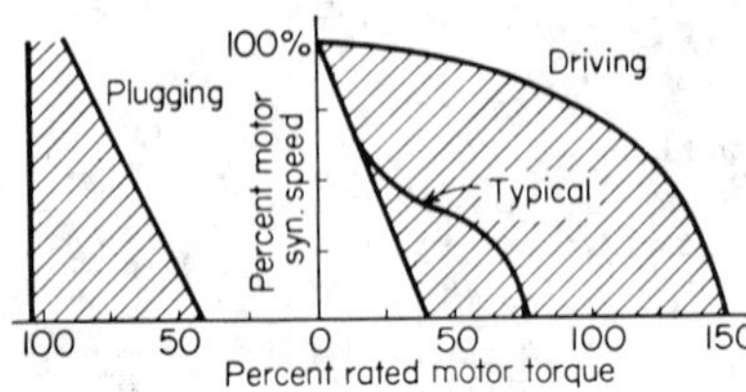

FIG. 20-32. Crane performance curves for a-c static bridge or trolley control.

The motor plugging control unit *MPC* incorporates static circuitry which accepts the signal converter output *SC*, compares it with the current signal, and translates it into a motoring or plugging signal for control of the pulse generator and thus motor torque. If the motor is in a plugging mode of operation, the *MPC* output is, proportional to master throw from OFF, and automatically increases at standstill, in a step function, for increased reverse acceleration torque. Fig. 20-32 shows representative performance. However, minimum plugging and acceleration torques may be adjusted to very low values if desired.

187. Bibliography

AISE Standard 6, Specifications for Electric Overhead Traveling Cranes, May 1, 1949.

USAS B30.2, Safety Code for Overhead and Gantry Cranes.

SRNKA, L. J., DAVIES, M. C., and MYLES, A. H. D-C Magnetic Crane Hoist Control for A-C Powered Cranes; *AIEE Trans. Paper* 60-30.

MYLES, A. H. Fast Response Amplifier Controls Large A-C Motor; *Control Eng.*, October, 1962.

Industrial Control, *NEMA Standards Publ.*, IC1-1965.

VAN NIEKERK, A. Single Phase Braking of 3-phase Induction Motors; *Elec. Jour.*, February, 1936.

SCHURR, C. A. Something Different in Reactor Crane Control; *Iron and Steel Engr.*, May, 1962.

ELECTRIC ELEVATORS

BY JOHN LUSTI

188. Introduction. The majority of elevators being installed today are electric-motor driven-traction type. Figure 20-33 shows in schematic form the principal hoistway and machine-room components of an elevator system.

189. Machines. Gearless machines have a grooved driving sheave and brake pulley directly mounted on the armature shaft of a shunt-type d-c driving motor.

Geared machines consist of a worm-gear reduction unit, with a driving sheave on the low-speed output shaft and a brake pulley on the high-speed input shaft, which is driven either by an a-c induction motor or by a d-c shunt motor. The gearing permits the use of a small high-speed drive motor such that the combination gear and motor cost less than a gearless machine.

To date, gearless machines are used almost exclusively for car speeds of 400 ft/min and up. Geared machines are rarely used beyond speeds of 350 ft/min because of excessive problems with noise and vibration and difficulties with gear wear and backlash brought about by frequent force reversals on the gearing during and due to the frequent starting and stopping of elevators.

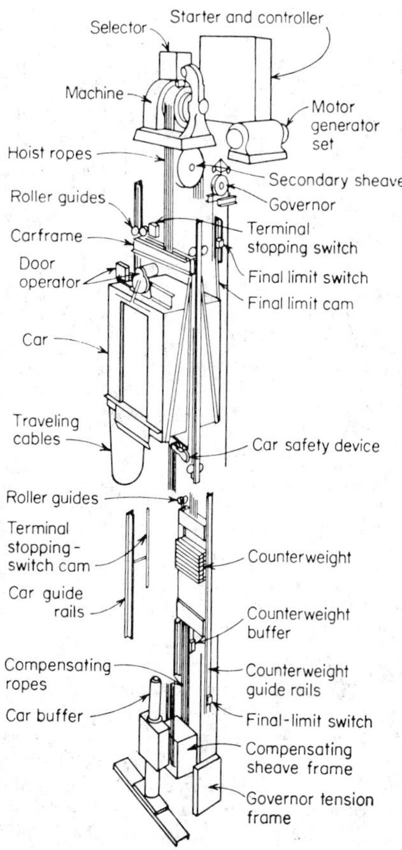

FIG. 20-33. Elevator machine room and hoistway equipment.

190. Brakes. Magnet-operated brakes are generally employed. They release electrically and apply through friction shoes held by springs against a cylindrical brake drum (pulley) on the machine driving shaft.

In the operation of gearless machines the brake function is to hold and not to slow down the elevator; therefore its size is determined solely by torque requirements. Its function is the same on higher-speed geared machines; therefore the brake mechanism, like the motor, is smaller on geared machines due to the torque reduction back through the gearing. On slow-speed a-c motor-driven machines, where the stopping is done solely by the brake, the heat generated by the absorption of the system kinetic energy becomes a major factor in determining brake size.

Direct-current magnets are used wherever possible, because they are quiet and can be readily controlled to give quick but smooth application. Alternating-current brakes may be magnet- or motor-operated and are usually provided with dashpots to regulate their application.

191. Traction. The elevator car is raised and lowered by hoisting ropes which pass over the machine drive sheave, the necessary traction being obtained by the friction between these ropes and the grooved surfaces of the sheave, and the pressure being applied by the weight of the elevator car and its load, the counterweight, and the weight of the ropes (and compensating chains or ropes, if used).

The traction-type elevator has an inherent safety feature in that, when either the car or the counterweight bottoms, the tension in the hoisting ropes is relieved and the driving sheave may rotate without moving the elevator owing to the loss in traction.

192. Ropes. Sets of three to eight steel hoisting ropes with diameters ranging from ½ to 1 in are used in parallel. The rope diameter used for the set determines the minimum sheave diameter which may be used. Sheaves that are too small will introduce excessive stresses in the ropes while flexing as they go over the sheave, causing shortened rope life. The sheave diameter is usually 40 or more times the rope diameter.

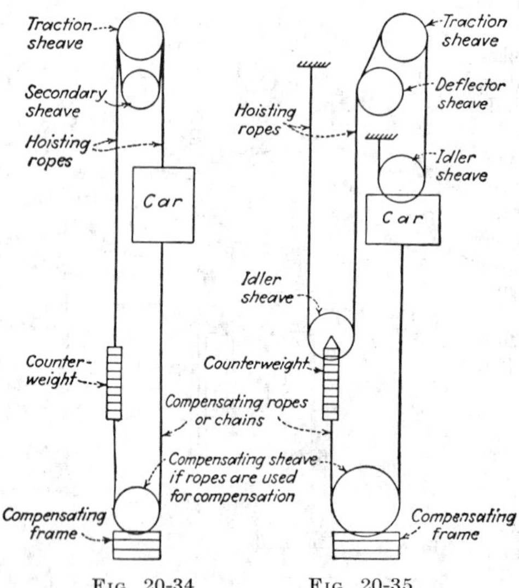

Fig. 20-34. One-to-one traction elevator with double wrap.
Fig. 20-35. Two-to-one rope-geared elevator with single wrap.

193. Roping Ratio (Figs. 20-34 and 20-35). Traction elevators usually have 1:1 or 2:1 ratios. The 2:1 rope gearing is employed on lower-speed gearless machines to reduce required machine size. The rope load on the drive sheave due to the suspended masses is halved, permitting lighter machine structure. The machine-torque requirements are halved owing to the ratio alone and are further reduced because the reduced load per rope permits smaller-diameter ropes and therefore smaller-diameter drive sheaves.

194. Roping Wrap (Figs. 20-34 and 20-35). The hoist ropes may pass over the driving sheave only once "single-wrap"; or they may be "double-wrap," wherein, after passing over the drive sheave, they go around a secondary idler sheave and over the driving sheave again in another set of grooves.

195. Roping Compensation. When the weight of the hoist ropes becomes appreciable, compensating ropes or chains are provided to keep the traction load on the machine from changing because of the weight of the hoist ropes as they pass from one side of the system to the other. The compensating sheave in the pit (used with ropes only) can be so arranged that it can take up slack by traveling in the down direction but cannot travel in the up direction; thus the car and the counterweight are constrained to move together, eliminating the possibility of getting appreciably slack hoist ropes owing to jumping of the car (counterweight) as a result of rapid retardation of a down-traveling counterweight (car). This arrangement is provided on higher-speed elevators primarily to ensure that all the moving masses will act together in the event of a mechanical safety or buffer operation.

196. Sheaves. Figure 20-36 shows the methods of grooving traction sheaves; the U-groove allows more load per rope than the others but requires double wrapping to obtain sufficient traction. The V, or undercut, usually gives enough traction with single wrap. The undercut type has the advantage over the V that the traction changes only slightly as the sheave wears.

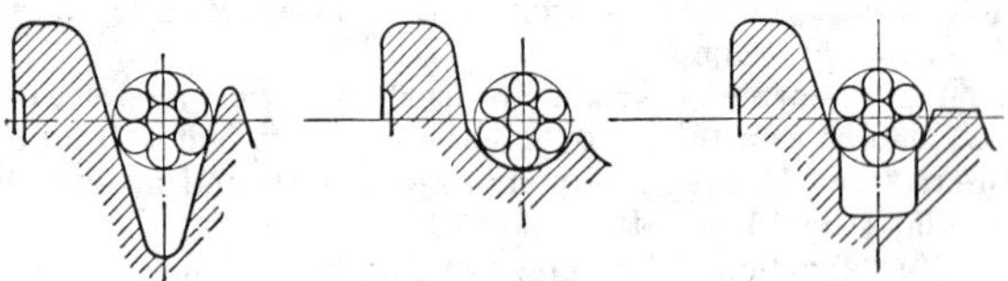

Fig. 20-36. Methods of grooving traction sheaves; V-groove, U-groove, undercut groove (left to right).

The machine-drive sheave is the only traction sheave; all other sheaves always employ the U-grooves.

197. Car. The car is the load-carrying unit, including its platform, enclosure, car frame, and car door. The car frame is the supporting structural frame to which the hoisting ropes or hoisting rope sheaves (2:1 roping), car guides, car safety and platform, and generally the door-operator machine are attached.

198. Counterweights. The function of the counterweight is to provide tension in the ropes on the opposite side of the drive sheave from the car to develop traction and to reduce the maximum net load that the driving machine must handle. The weight of the counterweight is generally equal to the weight of the car plus 40 to 45% of the duty load to be handled. This weight is generally a compromise between minimizing maximum machine traction load, minimizing average net load, and obtaining the lowest ratio between rope loads on the car vs. the counterweight side of the drive sheave (for both empty car and full load) to minimize traction-relation requirements.

199. Governor. A centrifugal governor is located at the top of the hoistway and driven by a governor rope attached to the safety operating mechanism on the car. In case of car overspeed (governor tripping speed) the mechanism holds the governor rope against the motion of the car, causing actuation of the car mechanical safety device. On moderate- and high-speed elevators, an electric contact on the governor trips open at somewhat above rated car speed and stops the car through the normal control circuits before it reaches the speed necessary to apply the safety. Additional contacts are sometimes furnished to aid in the control of the car.

200. Car Safety Devices. The car safety device consists of at least one mechanism mounted on each side of the car frame between or below the lower members of the car frame (safety plank). They stop the car by clamping the guide rails. On low-speed cars (to 150 ft/min) knurled rollers or cams are jammed between the safety block and the guide rails to bring the car to an abrupt stop. On high-speed cars smooth safety jaws clamp the rails with a controlled force to bring the car to a sliding stop. Maximum permissible retardation, by code, is gravity with full load in the car.

Car safeties are permitted to function only in the down direction, since retardation may exceed gravity, which would cause persons to leave the floor of the car on an up application. Excessive retardation must be avoided because of possible injurious effects on passengers.

Where there is occupied space under the pit or a used passageway, the counterweight is also provided with a mechanical safety.

201. Guides. The car and counterweight run on steel T-rail guides. Sliding guide shoes were formerly standard, but on many passenger elevators these are being replaced by roller guides, which give smoother riding and less friction. As they require no lubrication on the rail surfaces, it becomes easier to keep the hoistway clean and the fire hazard due to oil from the rails running down into the pit is eliminated.

202. Buffers. Electric elevators are provided with buffers in the pit under the car and counterweight. Spring buffers are used for slow-speed and oil buffers for high-speed elevators. On the latter, the counterweight oil buffer is frequently made a part of the counterweight structure.

203. Doors and Door Operators. Passenger-elevator car and hoistway doors are generally of the horizontal sliding type. The door panels have rollers at the top which ride on a steel track for support and guidance. The doors are guided at the bottom by shoes sliding in a machined self-cleaning slot. The door operator, mounted on supports to the car frame, is driven by an electric motor and coupled to the car door by belts, chains, or levers. The hoistway doors are automatically coupled to the car doors, when the elevator is at the landing, and operate in synchronism with them.

Freight elevators usually have a vertically sliding car gate and vertically sliding bi-parting hoistway doors. The car gate is counterweighted; the two halves of the bi-parting hoistway doors counterbalance each other. When the doors are power-operated, each door has an electrically powered door operator.

204. Selector. The function of a selector is to give car-position information to the control and operating systems so that automatic stops may be made at prescribed landings. Selectors located in the machine room may be mechanically coupled to the car by steel tape or wire, in which case they may be used to give all required positions down to the small fraction of an inch required for floor stopping accuracy. Some are coupled to the car by way of a sheave and the hoist ropes running over it. These are not sufficiently precise to permit accurate floor stops unless they are supplemented by devices, mounted on the car and interacting with cams fixed in relation to each landing, to take over position dictation when the car is within approximately 1 ft of the landing.

On slow-speed elevators switches in the hoistway handle the whole positioning, sometimes with an assist from a stepping selector in the machine room which is driven by pulses from the hoistway switches.

205. Controllers. Controllers contain the electrical elements which determine the "control" and "operation" of the elevator system. Strictly speaking, the "control" means the system for regulating the motion of the elevator, and "operation" refers to the logic system which determines what the elevator should do, i.e., start, go up or down, stop, etc., and which actuates the "control" accordingly.

206. Traveling Cables. Electric power and signals are carried to the car to operate the lights, door operators, etc., and call registration, door position, etc.; information is carried back to the machine room through traveling cables suspended from the bottom of the car at one end and attached to a junction box at the midpoint or top of the hoistway at the other end.

MOTORS, GENERATORS, AND CONTROLLERS

207. Special requirements of elevator motors and controllers are:

1. Utmost safety and reliability.
2. Automatic handling of negative or "overhauling" loads.
3. Ability to stand up and perform well under repeated starting and stopping. (A busy elevator may make up to 200 stops/h.)
4. Quietness and freedom from vibration.

208. Alternating-current Elevator Motors. Alternating-current elevator motors are practically all of the induction type. Most of them are single-speed, with squirrel-cage rotors and of comparatively small output—20 hp or less. Larger motors may be of the two-speed type and may have either squirrel-cage or wound slip-ring rotors. Squirrel-cage motors are more widely used than slip-ring because:

1. Their simpler construction makes for greater reliability and lower maintenance cost.
2. They give the designer more freedom in selection of number and shape of rotor slots, thus minimizing the magnetic-noise problem.
3. They do not require so much control complication as slip-ring motors.

Squirrel-cage elevator motors generally have high-resistance rotors to provide high starting torque and limited starting current, because a large proportion of the duty cycle consists of starts from rest. Usually the full-load slip of these motors is about 20%; thus a motor having a synchronous speed of 900 r/min may run at 720 r/min when the elevator is carrying full load in the up direction. Figure 20-37 shows typical characteristics for a squirrel-cage hoist motor.

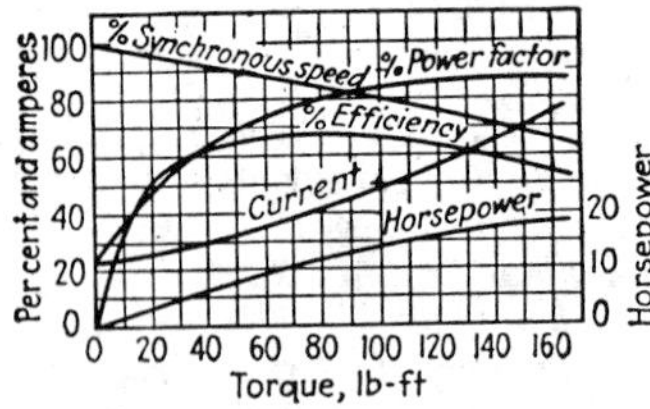

Fig. 20-37. Typical characteristics of squirrel-cage induction hoist motor for geared elevator.

Double-squirrel rotors are coming into more common use for elevators. They enable maintaining the high-starting-torque characteristic, but the full-load full-speed slip can be reduced to approximately 9%, giving better control and efficiency.

Frequent starting causes the outer high-resistance rotor bars to heat more rapidly than the inner bars. Care must be taken in the rotor design to prevent stresses due to thermal expansion causing mechanical failure of the double cage where the bars are interconnected.

209. Direct-current Elevator Motors. Practically all d-c motors now being built are the shunt type, with their speed controlled by varying the armature voltage.

Direct-current motors for worm-geared machines usually run from 10 to 75 hp, with speeds ranging from 1,400 r/min in the smaller units to 600 r/min in the larger, the speeds being limited by the requirements of the gear unit. For gearless machines the range is 20 to 150 hp and 60 to 150 r/min. These motors are required to deliver torque far in excess of nominal during acceleration and slowdown of the hoistway masses. The high currents produced during these periods increase armature reaction and reactive voltage effects. Extra-strong motor field excitation is used to overpower the armature reaction and prevent the loss of available torque due to its field-weakening effect. There is a greater need for interpoles, to buck out excessive armature reactive voltage, which causes commutator sparking, than there is in the more common continuous-duty motors.

Figures 20-38 and 20-39 show typical characteristics of d-c elevator motors.

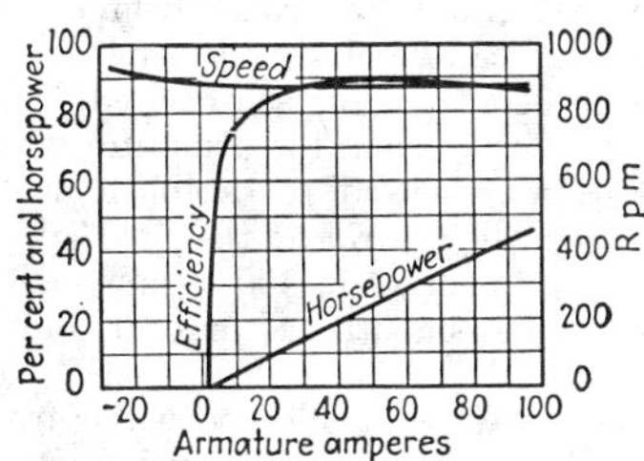

Fig. 20-38. Typical characteristic curves of d-c hoist motor for geared elevator.

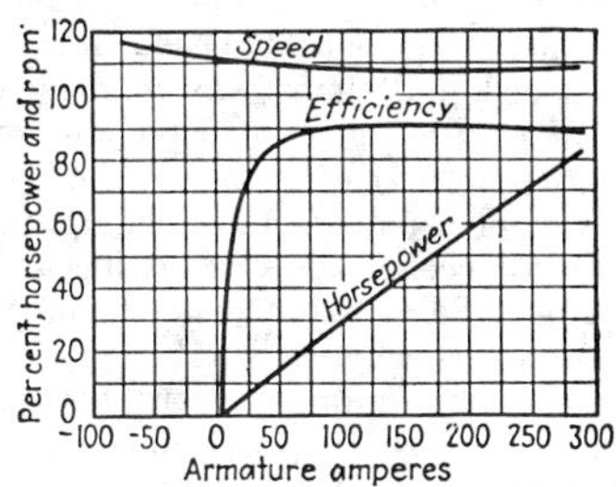

Fig. 20-39. Typical characteristic curves of gearless traction elevator motor.

210. Motor-Generators. The unit motor-generators used with elevators having voltage control are subject to varying loads and should be designed to be highly efficient and have low free-running losses, for the reason that they consume power while the elevator is stopped at the landings between runs.

These sets are usually of the two-bearing type, but the larger sizes (above 75 kW) may be three- or four-bearing. It is important that they be as compact as possible, to limit the space occupied. Elevator motor-generator sets are usually of 10 to 75 kW capacity (continuous rating).

On a-c power supply, low-slip induction driving motors are used. Sixty-cycle synchronous speeds are usually 1,800 r/min up to about 50 kW and 1,200 r/min above.

211. Ratings of Motor-Generator Sets and Motors. Ratings of motor-generator sets are based on a temperature rise of 50°C with continuous run at nameplate rating. Elevator motors are rated on the basis of horsepower required when the elevator is traveling at full speed in the up direction and carrying full-capacity load. As this occurs infrequently in actual operation, the rating for elevator motors is intermittent (¼-, ½-, or 1-h time basis). As rated by the manufacturers, they will usually operate in service without exceeding a 50°C rise.

Motor-generator sets are rated lower than the elevator motors but not less than the rms elevator-motor horsepower for the service duty cycle.

212. Alternating-current Rheostatic Control. Squirrel-cage motors up to about 6 hp driving slow-speed elevators (usually under 80 ft/min) are often started "across the line." For higher speeds and ratings one or more steps of accelerating resistance or reactance are used in the stator circuit for squirrel-cage motors and the rotor circuit for slip-ring motors to control starting torque.

Stopping is accomplished by disconnecting the motor and applying the brake. When two-speed motors are used, they start on the fast-speed winding (as above), and stopping is accomplished by first transferring to the slow-speed winding, one or more resistance steps being used to smooth out the transition, and then dropping the brake with the car running at slow speed. Single-speed motors are rarely used above 150 ft/min, because of poor stopping accuracy due to excessive changes in slide through the brake because of changes in net load. Two-speed motors are rarely used above 250 ft/min.

The resistance or reactance steps are ordinarily controlled by timed magnet switches, as, for example, a-c magnet switches timed by dashpots. Today it is more common to use capacitor-timed d-c switches, the d-c power for electromagnetic switch and brake operation being supplied by selenium rectifiers.

213. Voltage control, more popularly known as generator field control, provides a smoother and more refined control for elevators and is generally used for speeds above 150 ft/min. It employs an individual constant-running motor-generator set with a-c or d-c driving motor to supply variable-voltage d-c power to a d-c shunt-wound elevator hoist motor. Control of the hoist motor is obtained largely through control of the generator field excitation. This type of control is inherently smooth owing to the fact that the inductance of the generator field reduces fluctuations in the elevator-motor armature current. It has the further advantage that control devices may be much smaller because they handle only field currents instead of armature currents. Still another advantage is that the elevator may be more definitely controlled from full speed to zero speed, and almost any refinement of operation can be obtained.

There are two general categories of generator field control: self-excited, wherein the main power to the control field is derived from the generator armature which it controls, and separately excited. Separate excitation is used to establish direction and rate of acceleration on starting and to get accurate control at the final approach to the stop. Self-excited field control is employed up to speeds of 500 ft/min where the ultimate in performance is not required, because it gives smooth acceleration and slowdown with a minimum of control functions.

With separately excited field control the generator field controlled power is furnished by feedback amplifiers or by a constant-voltage d-c source in series with control resistances. Acceleration is time-based, and slowdown is distance-based. When feedback amplifiers are used, step and/or continuous signals at the amplifier input are used to control the speed. When the field power is regulated by resistance, the resistance is changed in discrete steps by switches. Some form of tachometric feedback for speed regulation is always used with feedback amplifier control.

Separately excited field control is always used when the utmost control is required and at speeds above 500 ft/min.

The equipment for starting and stopping the motor-generator set is of either the resistance or the Y-Δ-type. Y-Δ-starting is used on 3-phase a-c supplies and connects the drive motor in Y for starting and in Δ for running. Sometimes, in order to reduce standby losses, the drive motor is controlled so as to idle on the Y-connection when the elevator is standing at the main parking floor. Where service is intermittent, especially with automatic elevators, the standby losses are further reduced by employing a timing device to shut down the motor-generator set after the elevator has been idle for a predetermined time.

214. Leveling Control and Operation. In essence, leveling is moving the car in the vicinity of the landing toward the landing, with the motion so constrained that it can be safely done with the car and hoistway doors open.

The car position in which leveling is permissible and the direction of leveling motion are controlled through mechanical engagement of cams in the hoistway with switches on the car, through a change in magnetic reluctance when electromagnetic "inductor" devices on the car pass iron vanes in the hoistway or by similar leveling devices on positively driven selectors located in the elevator machine room. Special precautions are taken in the control systems of elevators whose normal running speed is greater than that permitted during leveling, to minimize the possibility of any faults causing excessive speed during leveling.

Today automatic leveling is used on almost all moderate- and high-speed elevators, and to a large extent on slow-speed elevators, especially those used for freight. It permits the door-opening operation to start while the elevator is approaching the landing so that the doors will be open far enough to permit passenger transfer as soon as the car is at floor level, thereby speeding up service. It also permits correcting for initial stopping inaccuracies and maintaining the car at floor level in loading or unloading if stretching or contracting of the hoist ropes should cause the car to move.

215. Continuous-pressure operation utilizes an up button and a down button in the car and at each landing which control the starting, stopping, and direction of travel of the elevator. The buttons must be held pressed to keep the car in motion. Automatic or manual means are provided to prevent interference from landing buttons when the car buttons are being used.

Continuous-pressure operation, owing to its simplified nature, is usually applied to freight elevators and limited to rheostatic control.

216. Car-switch operation utilizes a master switch in the car which gives the car attendant sole control of the starting, stopping, and direction of travel. On moderate- and high-speed elevators, the switch is usually provided with means to obtain one or more intermediate speeds. An annunciator or single-flash signal system is required so that the attendant may be informed of the location of landing calls.

Car-switch elevators have been installed up to 600 ft/min without leveling and 700 ft/min with leveling, but these speeds require skilled attendants and introduce complications when several elevators are operated as a group. For this reason, and because difficulties are also encountered in synchronizing waiting-passenger signals and power door operation with the position of the elevator, car-switch operation is rarely used today for high-speed passenger service and is largely confined to low- and moderate-speed freight service.

217. Single automatic pushbutton operation was the earliest type of automatic operation. It employs one button in the car for each floor served and one button at each landing. Momentary pressure of any button will cause the car to move toward and automatically stop at the corresponding floor. Once the car has started, pressure of other car or landing buttons will have no effect until the particular trip has been completed.

Since the single automatic pushbutton elevator serves only one person or group of persons at a time, it is usually applied only to small apartment houses or private residences or where the elevator is used for a special purpose.

218. Selective collective operation is automatic operation employing one button in the car for each floor served and up buttons and down buttons at floor landings. Car and landing calls may be registered by momentary pressure of the buttons at any time and in any sequence and remain registered until answered. Stops in response to the calls are made in the order in which the landings are reached. Up landing calls are answered on the up trip of the car, and down landing calls on the down trip, except that uppermost or lowermost calls are answered when reached irrespective of the direction of car travel.

The operation is sometimes modified to omit the up landing buttons, in which case it is termed "selective down collective" operation.

A "nonselective" type of collective operation is sometimes furnished. With this type of operation only a single button is employed at each landing, and all registered landing or car button stops are made in the order in which the landings are reached, irrespective of the direction of car travel.

Collective operation greatly extends the usefulness of an automatic elevator, as it can respond to several different calls on the same trip.

219. Group Automatic Operation. This is automatic operation of two or more nonattendant passenger elevators equipped with power-operated car and hoistway doors. The operation of the cars is coordinated by a supervisory control (dispatching) system.

In its simplest form two "selective collective" cars are operated from a common set of landing buttons. When there are no calls, one of the cars is generally parked at the lobby, the other car (the free car) at another landing. The lobby car will respond to lobby and basement hall calls; the free car will respond to all other hall calls. The supervisory system may be so arranged that the lobby car will start up to help out the free when calls appear behind the running free car or if the free car has not been able to clear up all calls within a given period of time. This system is commonly known as "duplex-collective" and is employed only where service demand is light and the utmost utilization of both cars is not required.

Group automatic operation for busy buildings where the utmost in car utilization is required has many special individual car as well as group supervisory dispatching features. A few of the more important individual car features will be mentioned here, and the supervisory features will be covered in Par. **220,** Dispatching. Of prime importance is the use of up and down hall lanterns with gongs to announce the arrival of the car. When a car is going to stop at a floor, the hall lantern and gong indicating in which direction the car will leave the floor are operated several seconds in advance of the arrival of the car to permit prospective passengers to move into position in front of the car to speed up passenger transfer. The lantern once operated, and the corresponding car direction, must not be changed for any reason until the doors have opened and reclosed. It is important, to obtain full utilization of the lanterns, that the passengers have complete confidence in them. It is especially important that the direction does not change while the doors are open, as a person might be entering the car at the time of the change, not be able to see the lantern change, and be carried in the wrong direction.

Load-measuring devices are used to bypass landing calls if there are too many people, approximately 75% of capacity, in the car to pick up further passengers efficiently. This has the effect of giving express service to a large number of people and reduces losses in passenger transfer time caused by shuffling around of people in the car to let people on and off at each stop, as well as eliminating stops when no further passengers could get on the car.

When the car stops for a call, the doors are opened and held open for a sufficient amount of time for normal passenger transfer. Various times are given as follows: for a stop to answer a car call only minimum time is allowed for passenger transfer; if a hall stop is answered, the time is increased because the prospective passenger is not quite as ready for the opening of the doors and generally hesitates, to allow for someone leaving the car; at a main landing where multiple transfers usually take place the time is further increased; and when many people are in the car, allowance may be made for the additional time required as a function of load. Where it is felt that appreciable time should be allowed, light-ray detectors with the beams intersecting the door are sometimes used to cancel the remaining time after a person has been detected crossing the threshold.

Of course not all passengers can transfer within the door-open times allotted; so automatic door-reversal devices are used to reopen the doors if they are closing while passengers are in the way. The light-ray device may be used for this purpose, but since it will cause door reversal any time a person crosses the threshold whether or not the doors are near enough to hit the passenger, it will sometimes cause unnecessary reversal, with consequent unnecessary delay in car operation. Less delay is experienced with a lightweight movable mechanical shoe extending slightly ahead of the door while it is closing, which when contacted will cause the doors to reverse, or with an electronic field device which will cause reversal before the doors can touch the passenger. With the latter device doors can be made to back away just far enough to keep from touching a person, eliminating the necessity of a full reopening.

If people should insist on holding up the doors in any manner, thereby delaying service after a period of time, the door-reopening operations may be cut out, at which time the doors are closed at slow speed to urge people away without danger of injury.

220. Dispatching. The object of a dispatching system is to direct the operation of a group of elevators so as to maximize their service capabilities.

Technically the service is *maximized* when the average person loses the least amount of time getting from one floor to another.

Actually the patience of the passengers must be taken into account in trying to achieve the *best* elevator service possible. When it is possible to reduce the average time lost in the system to a minimum only at the expense of having some people delayed excessively long times, it becomes necessary under *some* traffic conditions to sacrifice overall time lost in the system to bring the maximum time lost within acceptable limits.

In practice the service is optimized by programming the system for maximum handling capacity during peak loads and for minimum and equal waiting and traveling time for each passenger, individually, at other times.

The maximum number of passengers can be handled by a group of elevators, with a given total load capacity and hoistway heights, when each car is filled to load capacity in a single stop and emptied completely at the next stop made by the car. When the cars can operate well below load capacity, as is the rule during the greater part of the day, equal time separation (interval) between all cars in the group gives the fastest service and shortest travel time to the average passenger and lowest maximum waiting and travel times.

During heavy up peak traffic, such as occurs during the morning inrush in office buildings, practically all the demand for service will occur at the lobby; therefore cars are usually stored at the lobby during traffic lulls, selected for dispatch one by one, dispatched up upon demand one loading interval after the previous car, or sooner if

60 to 80% loaded as indicated by a load-measuring device. The cars go to the highest call and return to the lobby, answering any landing calls that may have been registered on the way. Additionally, when extreme traffic peaks are encountered, cars may be caused to bypass landing calls and may be assigned a limited number of floors to minimize total possible stops.

During down peaks cars may be time-dispatched from the upper landings or selected to serve zones in groups or individually. The zones may be in a preset pattern or as determined by traffic demand.

When the number of calls for service gets low enough, cars begin dropping out of service and park at random (the last stop), all at the same landing (generally the lobby) or individually at specific floors or zones according to anticipated traffic needs. At this time the cars are on call and go back into action as the traffic demands.

When there is sufficient traffic to keep all cars busy but not loaded, the dispatching system operates to keep the cars time-spaced. This may be done by time-dispatching cars from reversal points in accordance with the average interval between cars and the service demand as a function of direction or by assigning cars to load points when they become available (generally at reversal points).

The spacing of the cars is being continually disturbed because of the random nature of call registration. Once the spacing is disturbed, it tends to become more uneven. If a car is delayed, it has more chance to get more than its share of calls and so falls farther behind. As it falls behind, the car following it moves faster because it now has less chance to intercept calls. Ultimately the two cars get together and skip over one another, handling hall calls alternately. If there were no dispatching system to respace the cars, the whole group would be bunched most of the time.

221. Elevatoring Large Buildings. In determining the elevator equipment that will best serve an office or similar large building, it is necessary to take into account a considerable number of facts and particularly those which relate to the peaks of traffic.

The density of population in a single-purpose office building varies from one person per 90 ft^2 to one person per 120 ft^2 of net working area. In diversified-tenancy office buildings, the density may be one person per 100 to 150 ft^2.

Generally the 5-min peak of morning arrival varies from one-fifth the population in single-purpose buildings to one-ninth the population in diversified-tenancy buildings. The 5-min traffic peak during the luncheon period varies widely with occupancy.

In single-purpose office buildings the tenant usually limits the peak by staggering lunch hours; in diversified-tenancy buildings, it ordinarily runs from 11 to 14% of the population, combined up and down. The evening 5-min peak, when the building is being emptied, generally exceeds the morning peak. However, because of the more efficient passenger transfer during the down peak, the number of elevators required is usually determined by the morning arrival peak.

The handling capacity of the cars is calculated on the basis of peak loading of the cars, the probable number of stops based on loading and total possible stops, the time of passenger transfer as a function of load, and the round-trip running time based on the number of stops and probable distance traveled. To forecast the proper number, size, and speed of the elevators that will be required to handle the traffic in a particular building, it is necessary accurately to predict the peaks of traffic that will be experienced. This requires a large accumulation of data obtained from many different buildings of similar occupancy in similar locations.

222. Grouping and Arrangement of Elevators. In large buildings, it is usual to operate the elevators in local and express groups to gain maximum overall handling capacity and a reasonably short riding time. Tall buildings with large areas sometimes have as many as eight or more groups. Since a separation of the elevators into local and express groups reduces the number of elevators which serve a specific zone in the building, care must be exercised to ensure that the arrangement results in a satisfactory interval between cars.

Groups generally consist of 4 to 10 elevators, with half on each side of an 8- to 12-ft-wide hallway. The car capacity is usually 3,000 lb in large, diversified-tenancy office buildings and 3,500 to 4,000 lb in large, single-purpose buildings.

Car speeds are chosen in accordance with distance traveled typically: 12 stories, 500 ft/min; 24 stories, 800 ft/min; 48 stories, 1,200 ft/min.

PERFORMANCE

223. Overall Performance. The final measures of performance are the number of passengers or amount of material carried per unit of time, the speed and dispatch with which they are handled, and, for passengers, the waiting time for an elevator and the comfort of the ride. The performance depends on many factors such as speed, acceleration, and power-door operation. Even the shape of the elevator car and width of door opening influence the efficiency with which passengers can be loaded or unloaded.

An important performance index is the time required to run various distances, e.g., one floor or two floors. An 800-ft/min elevator can make a 12-ft run in about 4 or 5 s, start to stop, not including door-operation time.

224. Acceleration. This is limited by traction requirements and the comfort of the passengers.

To make the traction requirements reasonable, accelerations are generally limited to 5 ft/s^2 or below. However, where sufficient traction is available, acceleration as high as 10 ft/s^2 has been used.

Passenger comfort is affected by acceleration, rate of change of acceleration, and the length of time that they are experienced. Passenger feelings vary with the type and physical condition of the persons involved.

Accelerations of 5 ft/sec^2 or less and rates of change of acceleration of 10 ft/sec^3 are generally accepted as comfortable during the starting and stopping periods.

SAFETY FEATURES

225. Door Interlocks. The American Standard Safety Code for Elevators (now USASI) prohibits the starting of a car away from a floor unless the hoistway door is both closed and locked. This is usually accomplished by door interlock switches which open the control circuit when the door is either open or unlocked. Since these functions must be performed automatically on automatic elevators, a motor- or magnet-operated "retiring cam" is frequently furnished on the car. With power-operated doors, the interlock is usually combined with the operating mechanism.

226. Gate Contact. This opens the control circuit if the car door or gate is open.

227. Emergency Stop Switch. This is mounted in the car and when manually opened acts to cut off all power and apply the machine brake.

228. Slack-rope Switch. This is required for the hoist ropes of a drum-type elevator only. It is customary, however, to furnish a switch for the compensating sheave (Figs. 20-34 and 20-35) to stop the machine and apply the brake if the sheave moves above or below its normal position.

229. Terminal Devices. Electric elevators are usually provided with two types of electrical device for automatically limiting travel at either terminal. The "normal-terminal stopping device" slows down and stops the elevator when it is at or near the terminal landing and prevents further operation in the one particular direction. The "final-terminal stopping device" is arranged to function shortly after the normal device. It cuts off power to the elevator motor and brake and prevents further operation in either direction except by manual control from the machine room, which is necessary to return the elevator to its regular limits of travel.

Terminal stopping devices are usually mounted on the car or in the hoistway and are actuated by the movement of the car.

230. Door Clearances. On automatic-operation elevators, the landing doors must be placed substantially flush with the edge of the landing sill, and the space between these doors and a car door or gate must be kept small to avoid trapping a person, particularly a child, when the doors or gate is closed (see American Standard Safety Code for Elevators).

231. Runby and Clearance. The runby required for elevators of various speeds is usually fixed by law based on the American Standard Safety Code for Elevators. The clearance in the pit or between the top of the car and the overhead structure, when the car or counterweight bottoms, is ordinarily specified by law as 24 in minimum so that a man will be protected if he lies flat either in the pit or on top of the car.

232. Overloading. It requires extreme crowding with very tall people to overload seriously a passenger car conforming with the American Standard Safety Code for Elevators. This Code prescribes the minimum hoisting capacity as a function of inside net platform area. Typical figures are for 30 ft², 2,600 lb minimum; 5,000 lb for 50 ft²; 7,600 lb for 70 ft².

Additionally, the Code requires that the passenger elevator shall be designed to safely lower, stop, and hold loads up to 125% of rated load.

ENERGY DEMAND AND FEEDERS

233. The kinetic energy of elevators is important, as energy must be supplied for every run to start the moving masses. With voltage control this is partly recovered in stopping, but with rheostatic it is practically all lost. For high-speed elevators, operated by gearless hoisting machines, the principal kinetic energy is in the moving masses in the hoistway, and the kinetic energy of the motor and brake pulley is of comparatively small importance. For slow- and medium-speed elevators of the geared type, the rotor and the brake pulley have a large percentage of the kinetic energy of the elevator. When a-c motor-geared hoists are used, the kinetic energy of the rotor is considerably higher than for direct current, for the reason that the rotors are larger in diameter than the armature of d-c motors. For induction motors the kinetic energy per horsepower increases rapidly as the motors get larger (see Table 20-9 for computed kinetic energies of different parts of typical elevator equipments).

234. Energy consumption is generally given in kilowatthours per car-mile for a certain load and number of stops per car-mile. Figure 20-40 shows how the consumption varies with load and stops. The curves are plotted on the basis of average consumption for up and down trips; so the one for 1,000 lb load, which is about "balanced load," is the lowest of all. Usual values for stops per mile are 150 for local elevators, 75 or less for express elevators.

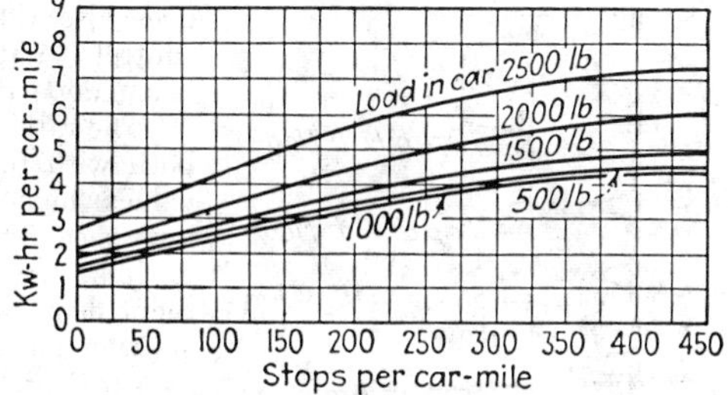

Fig. 20-40. Typical energy-consumption curves of gearless traction elevator; voltage control; a-c power; duty, 2,500 lb at 600 ft/min.

Note that curves such as those of Fig. 20-40 show only the energy consumed while the elevator is in motion. With voltage control the idling losses in the motor-generator, while the elevator is stopped at floors, must be added.

Considering all factors, with a moderate- or high-speed elevator that is reasonably busy, the total energy consumption with voltage control is usually less than with a-c rheostatic.

With car-switch or continuous-pressure operation some extra energy is used in inching to floors, especially if the control is rheostatic or there is no automatic leveling provided.

From 80 to 90% of the power taken from the input power lines for the operation of the elevators is dissipated in the elevator machine room in the form of heat. With

Table 20-9. Analysis of Kinetic Energy of Typical Elevators

Traction-machine type, hoist motor and control	Rated duty	Rise, ft	Kinetic energy ft-lb at rated load and speed in					
			Machine	Car	Counterweight	Ropes	Load	Total
Gearless 1:1, d-c motor, voltage control.	2,500 lb at 800 fpm	435	4180	15,250	18,200	10,000	6900	54,530
Gearless 2:1, d-c motor, voltage control.	3,000 lb at 500 fpm	200	2900	6,750	8,050	3,900	3250	24,850
Worm gear 1:1, d-c motor, voltage control.	3,500 lb at 250 fpm	125	6200	1,500	1,880	160	950	10,590
Worm gear 1:1, a-c motor, 1 speed, rheostatic control.	2,500 lb at 150 fpm	100	2450	460	560	50	245	3,765

gearless machines and generator field control, approximately two-thirds the heat released is from the motor-generator set, the remaining one-third being from the hoisting motor and control equipment. Since performance adjusted for normal temperatures becomes unsatisfactory at excessively high temperatures, it is the responsibility of the architect, consulting engineer, or owner to provide means, if necessary, to hold the ambient temperature of the machine room at a reasonable level. Spill air from the building air-conditioning system, separate air-conditioning units, or machine-room ventilating fans are often used for this purpose. The Btu released per hour by the elevators can readily be calculated from the kilowatthours per car-mile and the estimated car-miles per hour averaged on the basis of the working hours that the elevators remain in service.

235. Average Energy Consumption for Various Types of Installation (see Table 20-10). The table is presented only to give an idea of the quantities. Widely different figures may prevail on different installations of similar type, owing to differences in operating conditions.

236. Nature of Feeder Load. This is intermittent. Busy express elevators may actually run about 60% of the time they are in service. For other classes of service the time duty ranges down to 25% or even less. Figure 20-41 gives an idea of the duty cycle for a busy elevator. This is for voltage control and applies equally well to a-c or d-c supply.

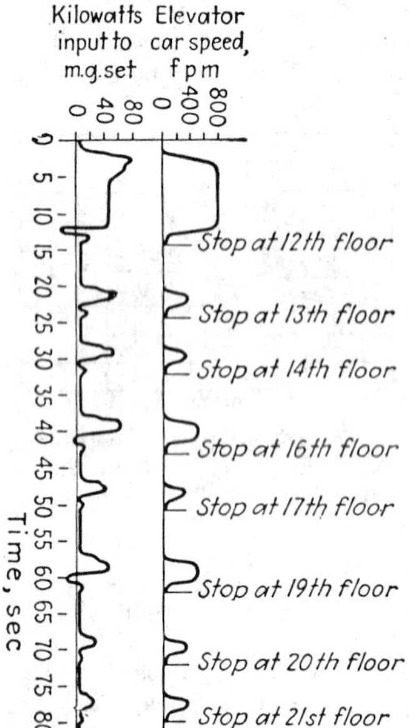

The running current varies with the load, as does also the peak starting current with voltage control. With rheostatic control the starting peak is less dependent on the load.

The ratio of peak-starting kVA to running kVA, both with full load up, has the following typical values: high-speed passenger, voltage control, 1.75; low-speed heavy-duty freight, voltage control, 1.4 or less; a-c rheostatic, 2.0 to 3.0 depending on refinement of control. The power factor of the starting peak is over 0.9 on voltage control and over 0.75 on rheostatic.

FIG. 20-41. Typical round-trip record of express elevator in morning peak of filling an office building; duty, 2,500 lb at 800 ft/min; travel—first to twelfth floor, express, twelfth to twenty-fourth, local; rise 261 ft.

237. Feeders for Single and Multiple Elevators. These should comply with the requirements of the National Electric Code (motor branch circuits) and any local regulations. The NEC bases feeder sizes on the nameplate current rating of the motor suitably modified for elevator service. In the case of motor-generator sets, the nameplate current is less than the full-load current owing to the continuous rating (see Par. **211**). However, the nameplate current rating is always equal to or greater than the rms of the current in the line during the duty cycle.

In addition to current capacity, voltage drop must be considered in the design of feeders. This should not exceed 3%.

When two or more elevators operate from the same set of feeders, it may happen that their current peaks coincide, in which case the rms line current becomes the arithmetic sum of the rms currents of all the cars. However, it is improbable that this will occur very frequently, so that advantage may be taken of the diversity of operation of the elevators in calculating the size of feeder.

For a group of n similar cars, each with an rms current of I, the line current will be nI when the peaks coincide and $\sqrt{n}I$ when the peaks are in the most random pattern. The line rms current falls somewhere between these two limits. The diversity factor (i.e., the ratio of the most probable line current to nI) varies within considerable limits and depends not only on the number of elevators using one set of feeders but on the type and intensity of the elevator service.

Table 20-10. Average Energy Consumption per Car-mile in Typical Elevator Installations

Traction-machine type and roping	Kind of supply and control	Rated duty	Rise, ft	Class of building	Kind of service	KWh per car-mile
Gearless 1:1.....	A-c voltage	2,500 lb at 800 ft/min	321	25-story office building	Passenger, express 1-25, no stops below 13	(30 mi in 10-h day) 4.4
Gearless 1:1.....	A-c voltage	2,400 lb at 600 ft/min	162	15-story office building	Passenger, local 1-15	(20 mi in 10-h day) 4.1
Gearless 2:1.....	A-c voltage	4,500 lb at 450 ft/min	135	12-story department store	Passenger, stop at each floor	(10 mi in 10-h day) 13.4
Worm gear 1:1..	A-c rheostatic	1,800 lb at 100 ft/min	61	7-story apartment house	Passenger	4.1
Gearless 1:1.....	A-c voltage	2,500 lb at 600 ft/min	267	18-story hotel	Passenger, express 1-18, no stops below 10	(12 mi in 15-h day) 5.8

238. Feeder Protection. If fuses are used for protection, they should be of a "lag" type which will not blow on short-time overloads. Fuses are affected by repeated heating and cooling and by heat conducted from poor connections at the clips, and so the best protection for feeders is by delayed overload circuit breakers.

239. Regeneration. The regenerative power of high-speed voltage-control elevators has been known to trip reverse power relays in the supply system. Such an elevator, fully loaded, running full speed down and stopping, may pump back about 20 and 90%, respectively, of its normal full-load-up running power. These figures are rough and subject to wide variation owing to differences in guide friction, etc. The problem of handling this negative power has required supplementary power-dissipating means when rectifiers are used to supply d-c elevator motors from a-c supply systems. Also, emergency power-supply systems have to be checked for the ability to handle the negative power.

240. Cost of Operating Elevators. This varies greatly with the service. In general, the higher class of equipment is the most economical for intensive-traffic elevators, since it gives the greatest service return per unit of power used. Power-consumption cost is only a part of elevator operating costs and should be secondary to the necessity of providing a good quantity and quality of elevator service.

In addition to power, operating costs include building costs chargeable to the elevators, such as the space occupied by the equipment, also attendants' wages, welfare, uniforms, etc. The elimination of the attendant costs with automatic elevators effects a considerable saving.

241. Hydroelectric elevators of the direct-lift plunger type are frequently installed for low-rise (up to 50 ft) low-speed (up to 150 ft/min) applications. The cars are generally not counterweighted, and safeties and governors are not required. The fluid (usually hydraulic oil) is pumped directly from a supply tank to the cylinder by an electric-motor-driven pump to move the car up; and gravity moves the car down.

Hydraulic motion-control valves operated by solenoid pilot valves control the motion of the car.

242. Moving Stairways (Escalators). These are extensively used in department stores; railway, elevated, and subway stations; and air-line buildings. They have also been applied to commercial buildings between the lower floors and to schools, race tracks, theaters, and other places of public assembly.

Moving stairways have the advantages that they can handle a large number of persons in a given time, there is no delay in getting on or off, the traffic is continuous, and even with very heavy traffic the transportation is comfortable.

Moving stairways are usually furnished in two sizes, 32 and 48 in wide between balustrades, with the nominal handling capacities in passengers per hour of 5,000 and 8,000, respectively, at a speed of 90 ft/min. A speed of 120 ft/min is sometimes used, which increases the handling capacity although not necessarily in proportion to the ratio of 120:90.

A number of large office buildings provide moving-stairway service to six or seven floors above the ground floor, which is reasonably close to the limit of traveling time that persons will accept in busy buildings. The use of moving stairways vs. elevators requires an economic consideration of such factors as space occupied, first cost, power consumption, maintenance, and operating personnel, if any.

Moving stairways are particularly desirable between the ground floor and basement or lower floors in office buildings. They enable basement and lower-floor space to be effectively used for merchandising, restaurants, banks, etc., which, under suitable conditions, return excellent rentals. With moving-stairway service, the lower-floor and basement space is almost as valuable as ground-floor space.

243. Moving-stairway Motors and Starters. On alternating current, squirrel-cage motors are generally used. They are designed for 6 to 10% slip, as the starting load is fairly heavy. A 32-in stairway with a rise of 20 ft requires about 10 hp when full-loaded.

On stairways that may run up or down, it is desirable to operate the a-c motors on reduced excitation for down motion, thus saving power and improving power factor. Y-Δ-starters are well adapted to such operation.

244. Moving-stairway Safety Devices. Moving stairways have magnetically released brakes to stop and hold them in case of failure of power. They also have switches to cut off power in case of slack chain and stop buttons at top and bottom landings.

BIBLIOGRAPHY

245. References on Electric Elevators

American Standard Safety Code for Elevators, New York, American Society of Mechanical Engineers, latest edition.

ANNETT, F. A. "Elevators"; New York, McGraw-Hill Book Company, 1960.

GLASER, WILLIAM F. "General Features of Electric Elevators"; Scranton, Pa., International Textbook Company, 1947.

ELECTRICAL-FEEDBACK CONTROL SYSTEMS

By S. W. HERWALD AND G. S. AXELBY

246. A feedback control system is a control system which tends to maintain a prescribed relationship between a controlled variable and a reference input by comparing functions of these quantities and using the difference as a means of control. Electrical-feedback control systems rely mainly on electrical energy for operation, as contrasted to thermal, hydraulic, pneumatic, mechanical, and other energy forms. Important characteristics usually specified of this type of control are (1) an **error-detecting device,** (2) an **amplifier,** and (3) an **error-correcting device.** Each element serves a functional purpose in matching the controlled variable to the reference input. The error-detecting device determines when the controlled variable is different from the reference input. It then sends out an error signal to the amplifier, which in turn supplies power to the error-correcting device. With this power, the error-correcting device changes the controlled variable so that it matches the reference input. The closed loop in Fig. 20-42, comprised of the error-detecting device, the amplifier, the error-correcting device, and the controlled variable, is characteristic of all feedback control systems.

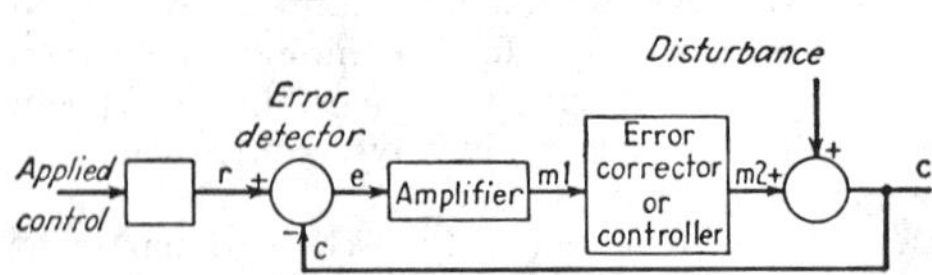

FIG. 20-42. Essential components of a feedback control system: r_1 measurement of reference input; c_1 measurement of controlled variable; e_1 error signal; m_1 output of amplifier; m_2 output of error corrector.

247. Reference and Controlled Variable. Any quantity, e.g., voltage, speed, temperature, position, direction, or torque, can be controlled. Any quantity also can be used as a reference input. The reference input need not have the same units as the controlled variable provided that the proper error-detecting device is employed.

When the controlled variable is mechanical position or any of its time derivatives, the feedback control system is commonly called a **servomechanism.**

248. The error-detecting device of Fig. 20-43 measures both the reference input and the controlled variable, as indicated by r and c, and detects when they are not similar. When they are not similar, an error signal, or actuating signal, represented by e is produced. The measuring units within the error-detecting device must be the same for both the reference and controlled quantities so that their difference may be obtained.

249. The amplifier raises the power level of signal from the error detector sufficiently so that the amplifier output m_1 can control the error corrector. Typical industrial devices such as generators, valves, relays, transistors, electronic tubes, operational amplifiers, and silicon-controlled rectifiers are commonly used amplifiers.

The error-correcting device of a feedback control system actually does the work of controlling against a disturbance by producing an output m_2 as shown in Fig. 20-42. The error-detecting device and amplifier serve only as a means of properly controlling the error-correcting device so as to make the controlled variable match the reference input or be a specified function of it. Typical error-correcting devices include electric motors, solenoids, gas engines, hydraulic motors and pistons, and fuel burners.

250. Applied control is the controlled variation of the reference input in order to obtain changes in the controlled variable. A common example of applied control to the reference input is the change in setting of a room thermostat, the reference temperature for the room being thus changed. Feedback control systems restore the balance between reference and controlled variables continuously and automatically to compensate for the effects of applied control and disturbances.

251. Forms of Error-detecting Device. The error-detecting device has many practical forms, and some representative types are given in Table 20-11. Not illustrated are more complex devices: space-angle error detectors such as a radar, or space-angle derivative detectors such as gyroscopes and accelerometers. Also not shown are various types of analog to digital converters which are ordinarily used to measure angles in digital form. Each error-detecting device, regardless of whether it is electrical, electronic, hydraulic, or mechanical, performs three distinct operations:

1. Measures the reference input r.
2. Measures the controlled variable c or a specified function of it.
3. Produces an error signal e sometimes called an actuating signal.

Of course, operations 1 and 2 may be made with different types of measuring devices.

252. The characteristics of an error-detecting device that are of the greatest importance to designers and engineers are (1) energy required to measure the reference input, (2) accuracy, (3) size, (4) reliability, (5) linearity, (6) noise, (7) signal level, and (8) resolution.

The greatest degree of design ingenuity can be exercised in the selection of an error-detecting device, since many special arrangements are possible to decrease operating forces or to increase accuracy.

The peculiar traits of mechanical, electrical, or hydraulic error-detecting devices can also be used to advantage. For example, magnetic saturation can be used to obtain a particular variation of error voltage with error. Similarly, a nonlinear mechanical linkage could be used.

253. Amplifiers. Some typical amplifiers are shown in Table 20-12. The broad similarity of all the amplifiers is apparent, since each contains a "gate" element G which controls power flow from the power source P to the load L. Each of these gates is a low-power or small-force element as compared with the power that they control. For example, the contact (No. 2 in Table 20-12) requires only ounce-inches to operate but can control many watts to the load. For high amplifications, single amplifiers can be cascaded (indicated in No. 1, Table 20-12). The difference in the power available as an error signal and power required to correct the regulated quantity determines the amplification necessary.

Many devices such as contacts, throttles, or valves have their power input determined by the time required for operation.

254. Error-correcting Devices. Two electrical error-correcting devices are shown in Table 20-13. Error correctors are power devices and include hydraulic motors

Table 20-11. Typical Electrical Error-detecting Devices and Their Characteristics

No. (Fig. 20-43)	Type	Main application	Operation	Operating features	Accuracy limited by	Features determining energy required to vary reference quantity measurement *r*	Amplifier of Table 20-12 frequently used with this device
1	D-c or a-c resistance bridge	Position control	Error voltage *e* appears when positions of moving arms of potentiometers *A* and *B* are not matched. Power source *E* is applied across both potentiometers. *A* measures reference position as voltage and *B* regulated position as voltage, their difference being *e*	*A* and *B* can be remote. Continuous rotation not possible	Potentiometer winding	Contact arm and bushing friction	2, 3, 4, 5
2	D-c tachometer bridge	Speed control	Error voltage *e* appears when speeds of tachometers *A* and *B* vary. *A* measures reference speed as a voltage and *B* regulated speed as a voltage, their difference being *e*	*A* and *B* can be remote. Top speed limited by commutator. *A* may be replaced by a potentiometer	Tachometer accuracy. Commutator resistance	Brush and bearing friction	2, 3, 4
3	A-c magnetic bridge	Position control, particularly for gyro pickups where very small forces prevail	Error voltage *e* appears when relative positions of rotor *A* and stator *B* do not match. Rotor *A* measures reference position magnetically and stator *B* regulated position magnetically. Voltage *E* across exciting coil *L* provides energy. When rotor covers unequal areas of each exposed stator pole (unbalanced magnetic bridge), pickup coils *M* and *N* have unequal voltages induced. Voltage difference is *e*	Limited rotation. Air gap usually small	Machining tolerance, magnetic fringing, and voltage-phase shift	Load taken from *e*. Bearing friction	2, 4
4	A-c synchro-system	Position control where continuous rotation is desired	Error voltage *e* appears whenever relative positions of rotors of synchrogenerator *A* and synchro-control transformer *B* are not matched. Reference position is measured by *A* as a magnetic flux pattern which is transmitted to the synchro-control transformer through interconnected stator windings. If rotor of *B* is not exactly 90° from transmitted flux pattern, *e* is produced	Unlimited rotation. Synchrogenerator and control transformer can be remote	Machining tolerance, accuracy of winding distribution	Distributed or nondistributed winding of control transformer rotor. Load taken from *e*. Bearing and slipping friction	2, 4

5	Frequency bridge	Frequency control	Error voltage e appears when reference and regulated frequencies differ. Tube channel A produces a filtered saw-tooth wave which gives a d-c voltage inversely proportional to the reference frequency. Tube channel B produces a similar voltage as a measure of regulated frequency. Difference of these d-c voltages is e	A and B can be remote. Tubes can be either gas or vacuum. Wide range of frequencies can be covered. Vacuum tubes should be used for high frequencies	Temperature and aging effects on tube and circuit elements	Input impedance	
6	Millivolt bridge	Temperature control	Error voltage e appears whenever regulated temperature differs from reference temperature. Regulated temperature is measured as a voltage by the thermoelectric effect of two dissimilar metals B. Reference temperature is represented as a voltage from battery-potentiometer source A. Difference in these voltages is e	A and B can be remote. Wide range of temperature can be covered	Ability to detect very low millivolt sign	Contact arm and bushing friction. If electronic voltage source A is used, input impedance	2, 4
7	Phototube bridge	Position control by intercepting a light beam	Error voltage e appears when movable shutter is in other than desired position. Light reaching phototube B measures shutter position. This light is measured as a voltage by the phototube-current variation. A reference position of shutter is represented by battery-potentiometer voltage. Difference of these voltages is e	A and B can be remote. Glass surfaces through which light travels must be kept clean	Continued accuracy of light source and phototube	Contact arm and bushing friction. If electronic voltage source A is used, input impedance	2, 4

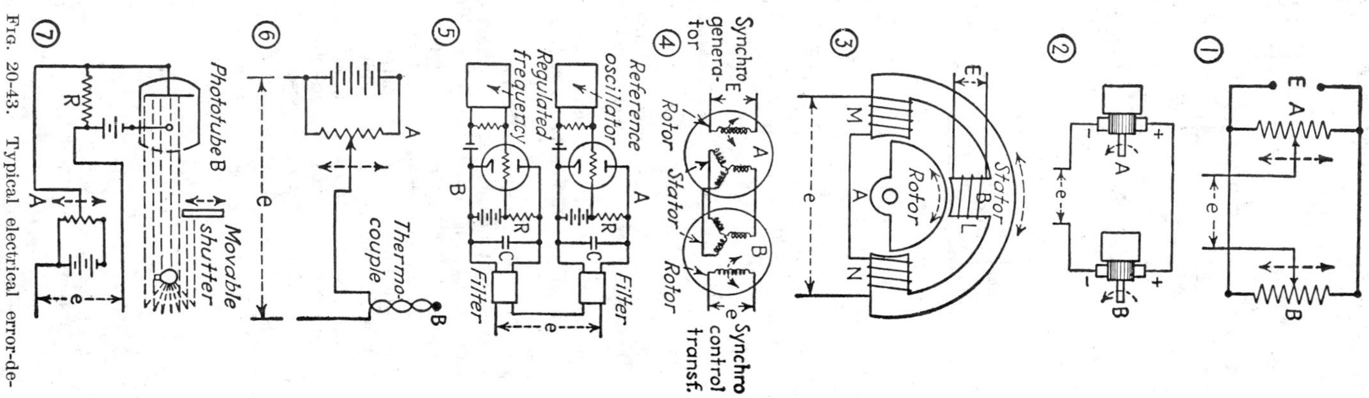

FIG. 20-43. Typical electrical error-detecting devices (see Table 20-11). (1) Resistance bridge (d-c or a-c); (2) direct-current tachometer bridge; (3) alternating-current magnetic bridge; (4) alternating-current synchro system; (5) frequency bridge; (6) millivolt bridge; (7) phototube bridge.

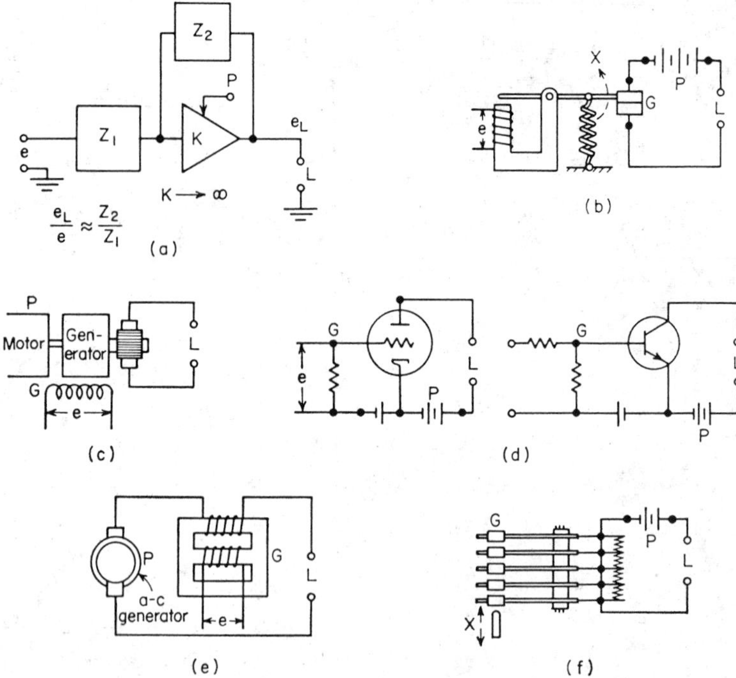

FIG. 20-44. Typical electrical amplifiers (see Table 20-12). (*a*) Operational amplifier; (*b*) relay; (*c*) generator; (*d*) electronic tube and transistor; (*e*) saturable reactor; (*f*) silverstat.

and pistons, steam and gas prime movers, and fuel burners as well as solenoids, and electric motors of various types such as d-c torque motors, a-c servo motors, and stepping motors.

255. The design of a feedback control system is usually separated into the same three basic functional elements as those of Fig. 20-42. The designer must select (1) an error-detecting device, (2) an amplifier, and (3) an error-correcting device that function compatibly with one another and perform the required job.

Aside from the usual cost and reliability factors, accuracy, including linearity, resolution, and noise, and operating force or power determine the selection of the error-detecting device. The amplifier is chosen on the basis of required amplification and the dynamic compensation, the forms of integral and derivative control, needed to achieve the required loop stability. The error-correcting device is determined by the nature of the controlled variable, the power required, and the ease of coupling to the selected amplifier.

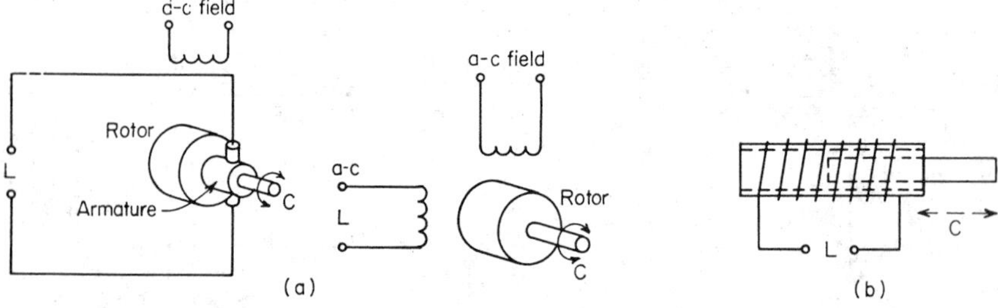

FIG. 20-45. Typical electrical error-correcting devices (see Table 20–13). (*a*) Electric motors, d-c and a-c; (*b*) solenoid.

Table 20-12. Typical Electrical Amplifiers and Their Characteristics

Schem (see Fig. 20-44)	Type	"Gate" element	Possible input units	Possible output units	Possible power-amplification factor	Devices represented by load L	Power control
(a)	Tubes or transistors	Grid or base	Microwatts	Watts	1×10^5	Relay motor Generator field Impedance Solenoid	Continuous
(b)	Relay	Contact	Watts	Watts or kw	1×10^3	Relay motor Generator field Impedance Solenoid	On-off
(c)	Generator	Field	Watts	Watts or kw	50	Motor impedance	Continuous
(d)	Electronic tube or transistor	Grid or base	Microwatts	Watts	1×10^5	Relay motor Generator field Impedance	Continuous
(e)	Saturable reactor	D-c coil	Milliwatts	Watts	3×10^2	Generator field Impedance	Continuous
(f)	Silverstat	Contacts	Grams	Watts	$1 \times 10^7 \times t$	Generator field Impedance	Stepped

Tables 20-11, 20-12, and 20-13 are helpful for devising feedback control systems. For example, suppose that it is desired to control angular position (a servomechanism). With relatively low accuracy, a d-c or a-c resistance bridge, a relay, and an electric motor could be used. Higher accuracy could be obtained by using an a-c synchro system with an electronic amplifier, a generator, and an electric motor. Each is a complete servomechanism, and each has its proper place in meeting desired performance requirements. The components can be chosen from the tables. The power of the error-correcting device is determined by the rate at which it is desired to vary the controlled variable. This rate of varying the regulated quantity also determines the transient errors and to a great extent the degree of stability, or the tendency of the entire servomechanism system to "hunt."

256. Hunting, or self-induced oscillation of the regulated quantity without change in the reference quantity, is the stability problem that complicates all quick-response, high-accuracy servomechanisms. Any closed-loop system, such as shown in Fig. 20-42, where the output feeds back into the input has a tendency to oscillate. However, the inherent damping in the simpler servomechanisms is usually great enough to overcome this tendency, and the system is stable. Where accuracy and response requirements are not too severe, a servomechanism of this self-damped type will perform satisfactorily. To meet the more exact accuracy requirements, special compensating circuits are used to give increased damping and the required degree of stability.

257. Feedback-control-system techniques are basic in the design for automation. A truly automatic factory, machine tool, aerospace vehicle, weapons control system, or process must have a means of comparing what it is producing with what it was asked

Table 20-13. Typical Electrical Error-correcting Devices

Schem (see Fig. 20-45)	Type	Input energy	Output energy	Approx output power range
(a)	Electric motor	Elec	Mech rotation	1×10^{-2} to 4×10^4 hp
(b)	Solenoid	Elec	Mech translation	1×10^{-3} to 15 hp

to produce and of being self-correcting if a discrepancy exists. These more sophisticated systems can be analyzed in much the same manner as the simple servomechanisms described.

BIBLIOGRAPHY

258. References

CHESTNUT, HAROLD, and MAYER, R. W. "Servomechanisms and Regulating System Design," 2d ed.; New York, John Wiley & Sons, Inc., 1959, Vol. 1.

JAMES, H. M., NICHOLS, N. B., and PHILLIPS, R. S. "Theory of Servomechanisms"; New York, McGraw-Hill Book Company, 1947.

AHRENDT, W. R., and TAPLIN, J. F. "Automatic Feedback Control"; New York, McGraw-Hill Book Company, 1951.

GIBSON, J. E., and TUTEUR, F. B. "Control System Components"; New York, McGraw-Hill Book Company, 1958.

GIBSON, J. E. "Nonlinear Automatic Control"; New York, McGraw-Hill Book Company, 1963.

HERWALD, S. W. Fundamentals of Servomechanisms: How to Select and Apply Them; *Product Eng.*, June, 1946, p. 464.

LAUNDRY AND DRY-CLEANING MACHINERY

BY E. H. SPRECKELMEIER

259. General. The laundry process requires the use of:

1. Water (hot and cold).
2. Detergents.
3. Power (electrical energy).
4. Fuel.
5. Heat.

260. Operations. Several operations are performed in laundering articles completely:

1. Identification—marking machinery.
2. Sorting—separating white articles from colored.
3. Washing—mechanical action with water and detergents in cylinder-type machines rotated in a forward and reverse direction at required speeds, special electric motors being used.
4. Extraction—removal of excess water in centrifugal-type machines using baskets or cylinders operated at high speeds between 500 and 1,300 r/min depending on the diameter of the basket or cylinder. Here again special motors are required.
5. Drying—use of heat in cylinder-type machines with high velocity of air moved by fans (motor-driven) and cylinders rotated by electric motors (reversing).
6. Flatwork ironing—use of ironers of chest type, steam-heated, having padded rolls which rotate and contact the chests, a roll for each chest. Variable roll speeds are required because of varying thicknesses of articles. Variable-speed drives are needed.
7. Folding of flatwork—done by automatic folding machines using special timers and conveyor drives with electric motors. The timing in many cases is electronic, digital, or analog.
8. Pressing—steam or electrically heated presses used, consisting of a stationary buck or table and a movable head.
9. Bundling and packaging—use of shirt folders and packaging machines.

261. The dry-cleaning process differs from the laundry process in that synthetic solvents and petroleum solvents are used instead of water and detergents. Aside from this, washing, extracting, and drying operations are similar. Since garments such as suits and overcoats, wearing apparel generally, and draperies constitute the dry-cleaning loads, flatwork ironers and folders are not used.

Synthetic solvents are most generally employed, and since they are not flammable, the electrical equipment need not be explosion-proof.

When petroleum solvents are used which have a flash point under 140°F temperature, explosion-proof Class I, group D electrical equipment is required.

The electrical equipment used with dry-cleaning machinery is important because the Underwriters' Laboratories, Inc., approval is usually required for insurance reasons, which results in the use of motors and control components that have previously been tested and examined by Underwriters.

262. Classification of Services. In the past, many different services were offered in the laundry field, such as damp wash, dry wash, rough dry, thrifty, and finished family.

At this writing, three services are most popular, soft finish, family finish, and the bachelor bundle.

In the dry-cleaning field, the usual processing involves cleaning and pressing.

The machines previously mentioned are used for laundering and dry cleaning.

263. Industrial Laundry Plants. The industrial laundry plants, which process wiping cloths and rags generally, are an important part of the laundry field.

The machinery used is essentially that used by commercial laundries with the exception that greater-capacity electric motors are required because of the severe and rugged duty imposed.

The cleaning solutions are special. Usually individual companies have their own formulas consisting of strong chemical solutions needed to attack greases and paints.

264. Linen-supply Plants. The linen-supply plants use machinery like that used by commercial laundries. Most machines are of the great-capacity type because of the large volume handled.

265. Electric Power Supplies. Since commercial, industrial, and linen-supply plants handle large-capacity loads, necessitating the use of integral-horsepower motors, the electrical equipment is furnished for polyphase power with voltages and frequencies as follows: 60-c 3-phase a-c at 208, 220, 240, 440, 480, 550, or 575 V for installations in the United States and Canada; 50-c 3-phase a-c at 220, 380, 400, 415, or 440 V for installations in Europe, England, Africa, South America, Asia, the U.S.S.R., and the Middle and Far East.

There are some laundry plants that generate their own electric power, using steam engines and boilers. This setup was popular in years past, because laundry processes require great quantities of live steam for drying, ironing, and pressing, and the exhaust steam is used for heating water, which resulted in high efficiency of boiler operation.

Since the power companies at this writing have attractive power rates for large power-consuming plants, the trend is to purchase power.

Dry-cleaning machinery, with some exceptions, is furnished for operation on 230-V 1-phase 60-c a-c power for United States and Canadian installations and 220-V 1-phase 50-c a-c for foreign installations.

The reason for single-phase power is because of the small-capacity machines used and because of the popularity of neighborhood dry-cleaning establishments, where only single-phase power is available.

Another reason is because of the popularity of coin-operated neighborhood establishments since 1960.

266. Water Supplies. Great quantities of water are required for laundering processes because of the many baths, and therefore experimenting is being carried on in so far as the reclamation of contaminated water is concerned, since the cost of water is an important item. This is particularly true in Europe.

The latest along this line is the reverse osmosis procedure, which involves chemical and some electric power. This process has reduced the cost of reclamation considerably according to the degree of purity required. The laundry process does not require a purity equal to that for drinking-water purposes.

267. Air Supplies. A considerable quantity of air is used in laundries for pneumatic operation of presses which perform the final drying and pressing of women's wearing apparel and men's shirts.

The air is supplied by large-capacity air compressors, electric-motor-driven, which constitutes a considerable electrical kilowatt consumption.

268. Power Required by Washers. In general, each 100 lb dry weight of laundry material requires 1.2 hp, based on a cylinder having a peripheral speed of 240 to 280 ft/min and reversing three to four times per minute.

The level of water in the tub affects the power requirements.

During the washing operation, the water levels are approximately one-half the levels used for rinsing since the washing and rinsing operations are intermittent. The rms horsepower mentioned above applies.

269. Listing of Machines Used

1. Identification machines (marking).
2. Washers—front loading, side loading (manual), side loading, automatic unloading.
3. Extractors.
4. Drying tumblers (including conditioning tumblers).
5. Ironers.
6. Feeders for ironers.
7. Folders.
8. Presses (laundry—collar, cuff, body). Presses (dry cleaning—suits, overcoats, top coats, dresses, draperies).
9. Starching equipment.
10. Combination washer and extractor (laundry).
11. Combination washer and extractor (dry cleaning).
12. Feather-cleaning machines.
13. Rug-cleansing machines.

Table 20-14. Water Steam and Energy for Laundering 100 Lb Dry-weight Load of Average Material

Processing	Water		Steam		Energy	
	Cold 62°F, gal	Hot 180°F, gal	Live at 100 lb, Btu × 1,000	Exhaust, Btu × 1,000	Hp·h	KWh
Washing	100	300	...	300	1.2	0.90
Extracting	...	...	...	...	0.4	0.30
Drying 30%	...	...	35	...	1.2	0.90
Ironing 70%	...	...	50	...	1.0	0.75
Miscellaneous	...	...	5	...	0.4	0.30
Total	100	300	90	300	4.2	3.15

270. Washers

Front-loading type (manual load and unload).

Side-loading type (manual load and unload).

Unloading type (side-loading manual—automatic unload accomplished by either lifting cylinder out of tub or raising both tub and cylinder to unload positions).

Industrial type for wiping-cloth and rag reclamation—require 50% greater-capacity motors than regular laundry washers.

Automatic-unloading washer. Loading is usually manual. Unloading is automatic by raising cylinder (with doors open) to position where load drops vertically downward out of open doors into containers of extractors, the containers then being conveyed by crane and hoist to extractor.

Side-loading type (hopper-loaded). For loading cylinder doors opened and positioned approximately 45° from vertical toward front of machine and load then dropped from floor above into chutes directed into cylinder-door openings.

Blanket type. Used for washing blankets at low speeds, usually one-half or one-third regular laundry-washer speeds (7 to 12 r/min). This is required to give gentle mechanical action during washing, to prevent felting.

Regular laundry washers can be equipped with two-speed motors of the two-winding type and used for blanket washing. A three-pole double-throw drum switch is used to transfer from high- to low-speed windings.

Dry-cleaning type. At this writing, dry-cleaning washers are not being used as they were in the past several years. When used, they operate similarly to laundry washers and differ in that cleaner's naphtha and petroleum-base solvents are used as the cleaning media. These solvents are flammable and therefore are a potential fire hazard. Because of this, special buildings are usually required, and in particular the electrical equipment has to meet National Electrical Code requirements which call for Class I, group D makeup. Motors and accessories must be explosion-proof. Magnetic controls if used in cleaning room must also be explosion-proof. If such controls are installed outside the cleaning room, they may be NEMA 1 type.

Table 20-15. Laundry-washer Data (**American Laundry Machinery**)

Type washers	Cylinder size		Capacity dry weight	Cylinder, r/min	Motor	
	Diameter, in	Length, in			Hp	Speed, r/min
American cascade......	36	36	110	25	1½	1,200
	36	54	165	25	3	1,200
	36	64	195	25	3	1,200
	42	36	150	23	2	1,200
	42	54	225	23	3	1,200
	42	64	265	23	4	1,200
	42	72	300	23	4	1,200
	42	84	350	23	4	1,200
	42	96	400	23	5	1,200
Mammoth cascade, Lo-door................	60	96	900	25	6(2)	1,800
	60	126	1,200	25	7½(2)	1,800
Front loading..........	30	15	25	30	½	1,800
	37	18	50	28	¾	1,800
	42	24	100	25	1½	1,800
Automatic-unloading cascade, cylinders....	42	54	225	23	3	1,200
	42	84	350	23	4	1,200
	42	96	400	23	5	1,200
Raise and lower, all sizes	..	...	.....	..	3	1,800

Motors (laundry). Motors are usually connected to machines by V-belts, chains, or gears and are of the reversing type with limited locked-rotor torque and current and with frame sizes sufficiently large to dissipate the heat due to frequent reversing (3 to 4 reversals/min).

A-c motors are of squirrel-cage induction type, with high-reactance-type rotors for across-the-line starting and reversing. Locked-rotor torques are as follows:

Four-pole motors—200% of full-load torque.

Six-pole motors—150% of full-load torque.

Locked-rotor current between 500 and 600% of full load.

Direct-current motors are compound-wound and operated at reduced voltage, starting and reversing by armature resistance in the control. All d-c motors have Class B insulation as standard.

General. All motors have ball bearings, grease-lubricated, some with grease fittings, others sealed. They are open drip-proof (unless otherwise required by special specifications), horizontal, with Class A insulation generally. Class B insulation is used in specific cases. Most motors have motor-mounted brakes.

Controls, general. Enclosures NEMA 1, sometimes NEMA 4 or 7, incorporate panel-mounted magnetic contactors, relays, overloads, and timers. Reversing is accomplished by a pair of reversing contactors and timer to give three to four reversals/min.

The timer has the feature of providing a variable drift period between reversals to allow the washer cylinder to reach rest before power is connected for reverse rotation.

Direct-current controls incorporate starting resistance to limit torque and current. This is accomplished by using counter-emf accelerators, current-limit accelerators, or inductive time accelerators. One-step acceleration is used for 2 hp and below; two steps for 3 to 10 hp. Acceleration time 3 to 4 s, coast period 2 s.

Alternating-current controls. Across-the-line type of reversing (full voltage), three to four reversals per minute.

Size contactors used for 220 V, 3-phase:

Size	Hp
2	7½–10
1	2–5
0	1–1½

Accessories. Magnetic brakes. Usually motor-mounted except for special applications. The torque of the brakes is that of the motor full-load torque rating or a little greater.

Pushbutton stations for start-stop-inching, etc. Have NEMA 1 and sometimes NEMA 5 enclosures.

Door interlocks, zero speed switches, temperature controls, electric solenoid valves, etc., are of NEMA 1 type except where otherwise required to meet specifications.

Automatic formula controls. These play an important part in washroom operation. The requirements are to control the closing of drain valve, the control of water level, the injection of supplies such as detergents, bleach, sour, blue, etc. As many as three or four washing operations with detergents and two or three rinses are required. Temperatures are also controlled by thermostats. When combination washer-extractor machines are involved, the extraction operation is also programmed. These controls usually are arranged with either a cylindrical device to which a formula chart is attached using electrical contact fingers which drop in slots to select the operation to be performed or a card device with raised and lowered hills and valleys to actuate microswitches or reed switches.

271. Laundry and Dry-cleaning Extractors. These are generally centrifugal machines which remove the excess water or solvent by centrifugal force.

Alternating-current motors are usually drip-proof, open type (laundry) or explosion-proof (dry cleaning), to develop approximately 300% locked-rotor torque, with locked-rotor currents approximately 600% of full-load values. Across-the-line starting. As the load decreases owing to removal of water and solvents, the final load on the motors is reduced to about 60% of motor-nameplate horsepower. For large-capacity loads, the motors are squirrel-cage induction types with frame sizes large enough to dissipate the heat due to a long period of acceleration. Motors are the high-resistance rotor type, with cast rotors for the small-capacity units and bar-wound for the higher-capacity units. In some cases the slip-ring type of motors is used. Class A insulation generally employed. Class B insulation is used when duty cycle requires it. In most applications, the braking from top speed to zero is accomplished through friction braking (band brake). In some applications the braking is accomplished by regenerative and plugging braking.

Direct-current motors are open, drip-proof, compound-wound, ball bearing, with Class B insulation.

Controls are across-the-line type, magnetic, a-c, NEMA 1. For d-c motors, reduced voltage, automatic controls are used for the armature resistance type, having current relays with magnetic contactors for acceleration for horsepower ratings 7½ and above. For smaller horsepower ratings, counter-emf acceleration is used. The following sizes of a-c contactors are used for the 208- to 240-V 3-phase horsepower ratings listed below:

Hp	Size Contactor
½–1	0
1½–5	1
7½–10	2
12½–15	3

Accessories. Pushbutton stations, cover and brake interlocks, are used of the NEMA 1 and 5 types for laundry extractors and of the NEMA 7 type for dry-cleaning extractors. Electric motor-driven timers are used with many applications to take care of the various machine functions such as lowering and opening cover, length of run, application of brake, etc.

Table 20-16. Laundry and Dry-cleaning Extractor Data
(American Laundry Machinery Industries)

Type machine	Dia. basket, in	Dry-weight capacity, lb	Basket speed, r/min	WK^2, lb-ft^2	Accelera-tion, s	Loads/h	Motor	
							Hp	60-c r/min
Monex	17	20	1,575	55	40	10	1	1,750
	30	85	1,320	524	90	6	3	1,750
Laundry (solid curb)	20	30	1,750	116	70	8	2	1,750
	26	50	1,600	286	75	8	3	1,750
	30	85	1,085	524	60	6	3	1,750
Laundry (open top)	40	150	875	2,300	110	4	5	1,750
	48	250	750	4,200	145	4	5	1,750
Laundry (Notrux)	50	320	800	11,270	250	5	12½	1,750
	54	400	750	14,775	280	5	12½	1,750
	54	400	875	14,775	90	8	12½–6¼	1,750
	60	500	700	25,300	340	5	15	1,750
Dry cleaning (solid curb)	20	25	1,375	72	30	8	1½	1,750
	20	20	1,050	67	25	10	1	1,750
	26	45	1,200	217	50	8	2	1,750
	26	35	900	202	35	10	1½	1,750
	30	75	1,125	450	60	6	3	1,750
	30	60	750	420	50	8	1½	1,750
Dry cleaning (open top)	40	115	875	2,020	95	4	5	1,750
	40	100	570	1,922	65	6	3	1,750
	48	195	750	3,485	120	4	5	1,750
	48	170	570	3,330	65	6	5	1,750

272. Drying Tumblers (Laundry). These consist of a perforated cylinder and housing in which damp load (removed from the extractor) is placed, usually to be finally dried. Heat is required, and fans (motor-driven) to draw heated air through the cylinder. Reversing motors are used in most cases to drive the cylinder (see tabulation below for data including electric motors).

Motors. For cylinder, the motors are of the reversing type, with characteristics the same as for washer motors.

For fan, the motors are of the general-purpose type.

Control. For cylinder, the same as for washers, with magnetic reversing contactors, relays, and timers used to accomplish three to four reversals of cylinder per minute. For fan, control consists of magnetic contactor and any related relays for interlocking.

Accessories. Pushbutton stations, interlock switches, magnetic brake to stop and hold cylinder.

Conditioning tumblers are used to partly dry damp work removed from extractors prior to being fed into ironers. These machines also break up a load removed from an extractor which is more or less compressed or caked. A rotating cylinder is involved which usually is reversed. Fans are also used. The electrical equipment is similar to that employed for washers and drying tumblers. Conveyors are used to convey work from extractors to finishing machines. Electrical equipment for conveyors is of the general-purpose type—open drip-proof motors and NEMA 1 control equipment.

273. Drying Tumblers (Dry Cleaning). These tumblers are similar to laundry tumblers except that, where solvents with flash point below 140°F are used, means must be provided to minimize explosions and fire hazards. Motors and accessories

Table 20-17. Drying-tumbler Data
(American Laundry Machinery Industries)

Type machine	Cylinder		Capacity dry load, lb	Drying time, min	Cylinder motor		Fan motor	
	Dia., in	Length, in			Hp	60-c, r/min	Hp	60-c, r/min
Open-end, 1 motor (nonreverse)....	36	18	20	22	...	.....	½	1,750
	36	24	30	24	...	.....	½	1,750
	36	30	40	28	...	.....	½	1,750
	44	42	110	24	...	.....	1½	1,750
Open-end, 2-motor (reversing)......	36	18	20	20	¼	1,750	½	1,750
	36	24	30	20	¼	1,750	½	1,750
	36	30	40	25	¼	1,750	½	1,750
	44	42	110	22	½	1,750	1	1,750
High-speed........................	42	40	65	30	1	1,200	3	1,750
	42	60	100	20	2	1,200	5	1,750
	42	90	135	25	2	1,200	7½	1,750
	42	120	200	25	3	1,200	10	1,750
Rotaire (conditioning).............	34	72	800/h	60	1½	1,750/125	½	1,750
	48	84	1,500/h	60	2	1,750/155	1½	1,750
American (thermatic).............	72	77	200-400	17	3	1,200	10	1,750

must be explosion-proof, Class I, group D. Controls, if installed outside the hazardous area, may be NEMA 1, non-explosion-proof. Grounding devices are used to protect against static electricity.

274. Flatwork ironers made up of friction load, which varies directly in proportion to speed. As mentioned, variable speed is required because different thicknesses of articles contain more water and therefore must be exposed to heated elements longer to accomplish final drying.

Table 20-18. Flatwork-ironer Data
(American Laundry Machinery Industries)

Type machine	Rolls		Delivery speed, ft/min	Motors		Type drive
	Number	Length		Hp	Speed	
Streamline............	2	100-120	7-22	1½	600-1,800	Multispeed motor
	4	100-120	16-48	3	600-1,800	Multispeed motor
	6	110-120	22-66	5	600-1,800	Multispeed motor
	8	120	28-80	7½	600-1,800	Multispeed or mech. speed changer, single speed for speed changer
Supersylon............	8	120	30-90	7½	600-1,800	Multispeed or mech. speed changer, single speed for speed changer
	8	120	37-110	10	600-1,800	Multispeed or mech. speed changer, single speed for speed changer
	12	120	30-90	10	600-1,800	Multispeed or mech. speed changer, single speed for speed changer
	12	120	37-110	15	600-1,800	Multispeed or mech. speed changer, single speed for speed changer
Hypro................	4	120		7½	1,800	Mech. speed changer
	6	120	40-125	15	1,800	Mech. speed changer
	8	120	40-125	15	1,800	Mech. speed changer
	6	120	50-150	15	1,800	Mech. speed changer
	8	120	50-150	20	1,800	Mech. speed changer
	8	120	58-175	25	1,800	Mech. speed changer
Model S..............	1	48	12	1/4	1,800	Constant-speed
	1	56	12	1/3	1,800	Constant-speed
Retron...............	1	85	15	3/4	1,800	Constant-speed
	1	100	15	3/4	1,800	Constant-speed

Drives. Many variable-speed drives are available. The least expensive is the multispeed a-c motor (four definite fixed speeds). The control for it is a drum-pole changing device (3:1 speed range).

The next speed-type device is a mechanical variable-pitch pulley driven by a constant-speed a-c motor (3:1 speed range, infinitely variable).

The third is an eddy-current clutch mechanism which has a speed range unlimited according to its construction. The control is electronic, with precise speed control.

The fourth, when direct current is available, is the shunt-wound adjustable-speed motor with shunt-field control from basic speed (either 3:1 or 4:1 speed range). Control is by a field rheostat with special automatic control (see Table 20-18 for drives and ironer data).

Exhausters. In the process of ironing sheets, etc. (flatwork), the padding of the rolls of the ironers becomes wet, and it is necessary to remove some of this moisture, which is accomplished by use of exhausters to keep padding dry. In the cases of Streamline and Supersylon ironers, the exhausters are usually driven by 1-hp 3,600-r/min motors. In the case of Hypro ironers, the exhausters are driven by 3-hp 3,600-r/min motors. The six-roll ironers require one exhauster. The eight-roll ironers require two exhausters.

Ventilating canopies. During ironer operations, moisture is evaporated from the material which is released in the air surrounding the ironer. In well-regulated plants, ventilating canopies are used to exhaust the moisture-laden air. The motors for these canopies are usually of ¾- or 1-hp size, 1,800 r/min, driving one or two fans.

275. Folding Machines. When the work comes off the ironer, it is folded either manually or automatically. In automatic folding, the work from the ironer is directed on folder conveyers driven by electric motors of ratings of ½ to ¾ hp, 1,800 r/min. The folding operations perform two types of folds, two transverse (full-width) folds, one-half length, and two cross folds. These operations are accomplished by mechanical or electrical means. In electrical folding, timers of the analog or digital type are used.

276. Presses. Other than flatwork material, wearing apparel and shirts are usually pressed on machines that are steam or electrically heated. Some of the machines are electrically motor-driven, but most are air-driven. Small, fractional-horsepower motors are used to rotate or cause functioning of the presses for positioning,

Air compressors. Small and large plants use air compressors to supply air for the operation of the heads which contact bucks. According to the size of the plant involved, motors driving air compressors are general-purpose motors of horsepowers of 3- to 30- to 40-hp units.

277. Miscellaneous. Blanket machines, which perform blanket drying, blanket brushing, etc., use steam for drying and rotating equipment for brushing operations. Electrical equipment consists of standard general-purpose motors for fans and rotating conveyers.

Rug dusters, cleaning and drying machines, and conveyors use general-purpose motors.

278. Washer-Extractors (Laundry). Instead of using separate washers and extractors, combination washer-extractors are used to eliminate the necessity of transferring the load from the washer to the extractor. These combination machines handle capacities from 25 lb (dry weight) to 800 lb.

Drives. Many types of drives are used. Some consist of a single motor of two-speed type with a variable-pitch sheave, resulting in the low-wash-speed operation of 30 to 45 r/min and the final extraction speed of 900 to 1,000 r/min for the smaller machines. Others of larger size, consisting of a single motor of two-speed type with a mechanical speed changer (variable-pitch sheave), result in low-speed wash operation of 25 to 30 r/min and the final extraction speed of 500 to 750 r/min. A third design consists of three motors, one for wash operation, one for distribution speed (intermediate between wash and extract), and a third for extraction speed.

Control. The control consists of forward and reverse contactors for reversing wash speeds (across the line), a magnetic contactor for the distributing speed motor, and a low- and a high-speed contactor for the extractor motor. The wash motor is a reversing-type drive, forward and reverse (usually four reversals per minute). The

Table 20-19. Washer-Extractor Data
(American Laundry Machinery Industries)

Type machine	Cylinder size		Capacity dry wt., lb	Cylinder, r/min			Motors			
							Wash		Extract	
	Dia., in	Length, in		Wash	Interm. extract	High extract	Hp	Speed	Hp	Speed
Econo 20	26	16	20	51	.....	420	1/3	1,800	1/3	1,800
Cascadex	30	15	25	43	.....	865	2/1	1,800/900	1	Motor
Cascadex	36	18	50	34	90	700	2/1	1,800/900	1	Motor
Cascadex	32	24	50	39	77	788	2/1	1,800/3,600	3	1,800
Cascadex	40	30	100	30	60	695	2.5/5	1,800/3,600	5	1,800
Cascadex	40	36	135	30	90	695	2	1,800	5	1,800
Cascadex	48	36	200	27.5	82.5	660	3	1,800	7½	1,800
Cascadex	60	36	350	24	72	570	4	1,800	10	1,800
Cascadex	46	74	400	21	41	540	5	1,200	5/10	1,800/900
Cascadex	46	110	600	21	41	540	7½	1,200	7.5/15	1,800/900
Cascadex	40	36	135	30.5	271	695	2/5	1,800/720	1	Motor
Cascadex	48	36	200	28	237	660	3/7½	1,800/720	1	Motor
Cascadex	60	36	350	24	200	570	4/10	1,800/720	1	Motor

distributing motor is a high-torque one-direction machine (across the line), and the extraction motor is a two-speed 900- to 1,800-r/min type. The two speeds are used for regenerative braking, with a mechanical-friction-type brake for final stopping.

MINE HOISTS

By R. C. Hardie

279. Single-drum Hoist. As a service or production hoist with cage or skip in balance with a counterweight, a single-drum hoist (Fig. 20-46) can efficiently service one or more levels, since the location of the counterweight at any time is not important. As a production hoist with skips in balance, the single-drum hoist is best for single-level hoisting.

280. Divided Single-drum Hoist. If multilayer winding is necessary, the single-drum hoist must have a divider (Fig. 20-47) to allow a separate compartment for each rope. If a counterweight is used with a divided drum, the counterweight rope can be wound on a smaller diameter. Consequently, the counterweight moves a lesser distance than the main conveyance, and rope-adjustment problems are reduced.

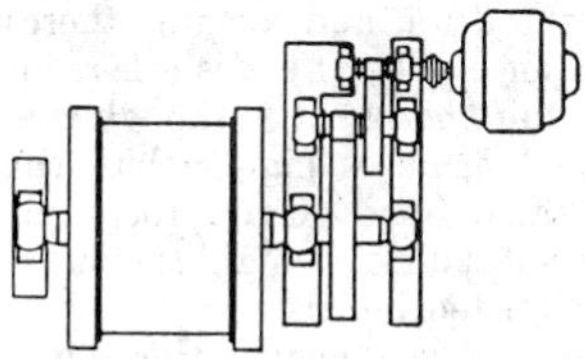

Fig. 20-46. Single-drum hoist.

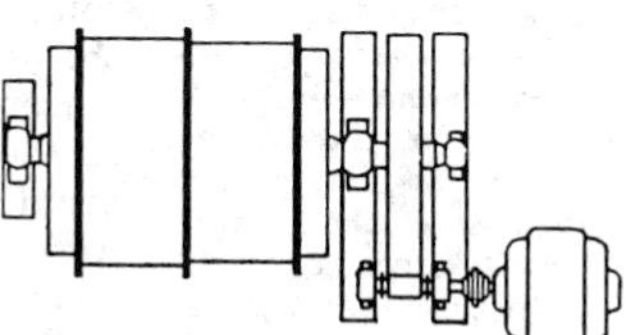

Fig. 20-47. Divided single-drum hoist.

281. Double-drum Hoist (One or Two Clutches). As a service hoist with cage and counterweight, this hoist (Fig. 20-48) can serve several levels efficiently, as the clutch facilitates quick adjustment of ropes to compensate for initial stretch. This hoist is also used occasionally as a production hoist with skips in balance for one-level hoisting.

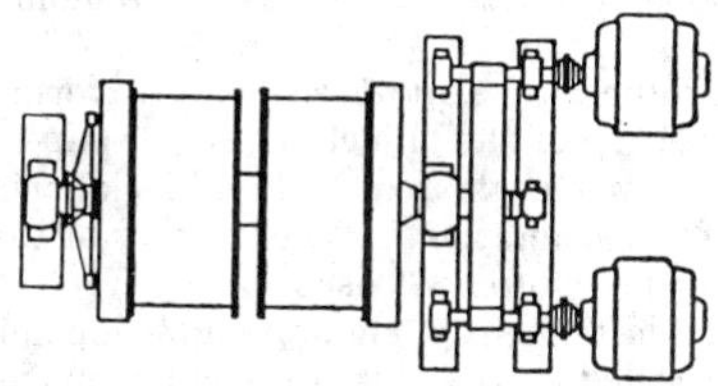

Fig. 20-48. Double-drum hoist, one or two clutches.

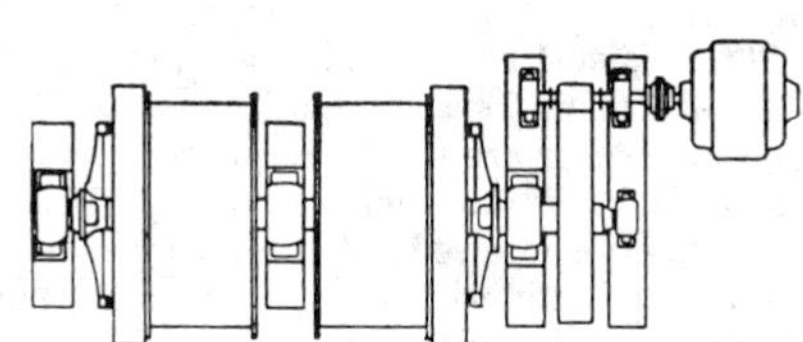

Fig. 20-49. Double-drum hoist, both drums clutched.

282. Double-drum Hoist (Both Drums Clutched). The main advantage claimed for this type (Fig. 20-49) is that, if something happens in one of the two compartments, the hoist can operate in the other compartment to raise and lower men and supplies. This hoist arrangement is favored if there is only one shaft entrance to the mine.

283. Single- and Multirope Koepe Hoist. Single- and multirope Koepe hoists (Fig. 20-50) can be efficient as service hoists with cage and counterweight for single- or multilevel hoisting or as production hoists with skip and counterweight for single- or multilevel hoisting. They can also be efficient as production hoists for single-level hoisting with skips in balance.

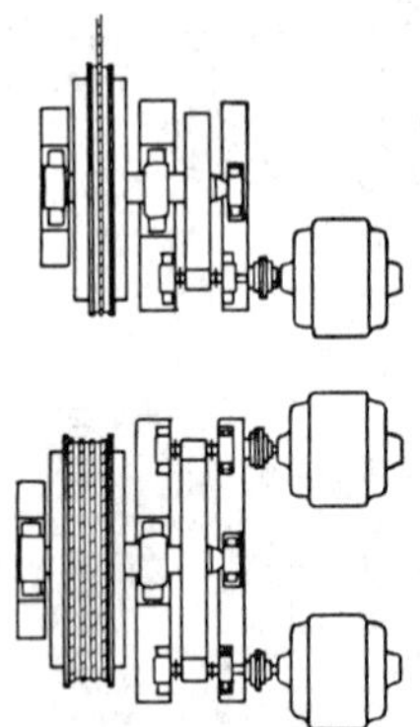

Fig. 20-50. Single and multi-rope Koepe hoist. The number of motors varies with the horsepower required. When the gear size for one motor is excessive, two motors are used. Generally with the gear set kept at a reasonable size (12 : 1 ratio or less) and the horsepower 1,500 or less, one motor is less expensive than two. If the ratio exceeds 12 : 1, double reduction gearing is usually employed.

The Koepe friction hoist has one or more ropes wrapped 180° around the wheel. The driving force from the wheel is transmitted to the ropes only through the amount of friction maintained between the rope and the material it rests upon. Thus the rope tension of the loaded side of the wheel cannot exceed the rope tension on the unloaded (or counterweight) side of the wheel by a ratio which is confined to a maximum of 1.5 to 1.6 with modern rope block liners. Also, a definite limit to the acceleration and retardation rates is necessary. On a drum hoist acceleration is limited only by motor and gearing.

284. Conical and Reel. Conical-shaped drums have been largely replaced by the Koepe friction hoist in deeper shafts and cylindrical drums in the shallow shafts, for horsepower savings. The cylindroconical type of drum accelerates the load on the small diameter with the empty conveyance on the large diameter. After acceleration of the loaded conveyance, the rope speed is increased further as the rope "climbs" the conical portion of the drum and remains there during the deceleration period. The effect is a leveling off of horsepower peak requirements, even though the inertia of the special-shaped drum is higher than that of a drum hoist for the same duty. Cylindroconical drums were suited for the short lifts at high speeds necessary when material was hoisted in cars.

The Koepe hoist effects nearly the same power saving because of lower inertia of the wheel and use of a tail rope, as only the net load is hoisted.

There are applications where a conical-shaped drum would be the most economical to operate with respect to kilowatts per ton hoisted. However, the original capital cost is greater than that for a Koepe or a cylindrical-drum hoist. Thus it becomes a matter of economics, because the money saved will not justify the capital expenditure.

The same can be said of reel hoists, but here it is rope cost that is a factor. The flat, hand-woven ropes are expensive and deteriorate fairly rapidly.

The horsepower requirements for conical and reel hoists are found by the moment method.

285. Hoist Drives. The a-c wound-rotor motor is the most economical form of mine hoist drive for a manually operated hoist. Some of the objections to the slip-ring induction motor are the higher-acceleration horsepower peaks required and lack of fineness of control for acceleration and particularly for deceleration. Where either of these is a problem, some refinements must be added or d-c equipment used.

For small induction motors which are operated infrequently, e.g., on an escape hoist, a drum controller with the hoist acceleration controlled by the operator's judgment may suffice. For larger horsepower (100 and more) and even for smaller-sized motors where frequency of operation warrants the additional cost, the secondary resistances are short-circuited by the secondary contactors under direction of time- or current-limit relays. If only time relays are used, a device for sensing when the motor has reached synchronous speed must be used or else on overhauling loads the motor could overspeed before the secondary resistance is completely short-circuited. The hoist operator can move the master switch to the full-speed position and the motor will accelerate uniformly in accordance with the setting of the relays.

With this manual operation the hoist is decelerated either by plugging the motor (applying reverse torque) or by gravity with the hoist brake. (Plugging the motor is wasteful of energy.) The hoist brakes must have sufficient capacity to stop the maximum downgoing load in less distance than normally required for deceleration and must be adequately sized for repeated stops under normal operating conditions.

Some of the usual accessory controls included are:

1. A maximum torque button to allow the motor to apply maximum torque at standstill. This is necessary to accelerate heavy loads, preventing dropping or roll-back when the brakes are released.

2. Overtravel-limit switches which remove power from the motor and set the brakes. This is to back up the safety controller, which is normally a Lilly controller. The Lilly controller removes power and sets the brakes if the full speed is exceeded or if the acceleration and deceleration rates (cam speeds) are exceeded.

3. Emergency stop button again removes power and sets brakes.

4. Back-out switches control direction of hoist travel after an overtravel such that the motor can rotate in the proper direction only.

When it is necessary or desirable to decelerate the hoist electrically as is done with automatic operation, some refinement must be added to the control. The a-c wound-rotor motor will not provide holdback torque at less than synchronous speed. Some alterations are used to overcome this:

1. Eddy-current brake added to provide adjustable torque load on the motor, applicable only to smaller-horsepower motors because of difficulty in dissipating heat in the brake.

2. Addition of d-c dynamic braking wherein the stator of the induction motor is cut off from the a-c source and excited from an adjustable d-c circuit. The motor is then an a-c generator, and the power must be dissipated in the secondary resistance. Dynamic braking has found application particularly on unbalanced hoists and on slopes where loads must be lowered at less than synchronous speed. Some hoists are controlled automatically, with dynamic braking used to decelerate the hoist in a closed-loop system just as with d-c equipment.

3. Instead of primary contactors connecting the a-c power supply to the motor primary, saturable reactors are used. Alternating-current motor torque varies as the square of the applied line voltage. The torque (voltage) is varied by increasing or decreasing the impedance of the saturable reactors, which consist of an a-c and d-c winding with a magnetic core. The a-c winding carries current to the motor, and the d-c winding is connected to an excitation source which varies the impedance from near zero to practically that of an open circuit by controlling the degree of saturation of the magnetic path. Saturable reactors have been used to automate hoists only in the smaller-horsepower motors used on service hoists, where d-c idle-time losses can be considerable.

4. As the hoist brakes must be sized to stop the maximum downgoing load repeatedly, some hoists are automated by controlling the brakes automatically with the secondary resistance, just as the hoist operator does when operating manually.

5. On slow-speed service hoists, automatic operation is easily effected by a two-speed squirrel-cage motor for low-horsepower applications. This is best adapted to cage control, where the cage rider replaces the hoist operator.

286. D-C Machines. On larger-horsepower applications where an a-c motor might have objectionable power peaks or where production hoist automatic operation is required, the d-c Ward Leonard control provides the finest control. The M-G set, usually synchronous motor-driven except in the smaller sizes, furnishes the d-c power to the hoist motor. Exact control of all speeds, including acceleration and deceleration, is effected by controlling the excitation of the generator, thus varying the voltage output. This allows close control over the speed of the driving motor, and the system is readily automated by making a closed loop between the d-c motor and generator by employing high-gain fast-response excitation devices—static or rotating—called regulators. The loop voltage can be made to follow the rate reference with a high degree of accuracy. The regulator compares a speed signal (tachometer generator voltage) from the hoist motor with that received from the rate reference and controls generator excitation accordingly. During acceleration the motor is under control of a current- (torque-) limit regulator for full loads and under control of the rate reference for light loads. The rate reference may be any device which accurately dictates the rates for acceleration, full speed, and deceleration, whereas the programmer follows the travel of the conveyance and initiates deceleration at the correct time. To do this with shaft switches with multiple levels is not practical; however, final stopping of the conveyance is by signal from a shaft switch. A programmer will not compensate for rope stretch caused by differences in loads.

A Koepe friction hoist requires a synchronizing device to drive the safety controller and programmer during conveyance travel. During a rest period, commonly at the collar or top level, this device drives the controller and programmer in the proper direction to correct for the distance the rope may have moved over the wheel. This resynchronizes the programmer and safety controller so that they again are properly oriented with respect to the conveyance in the shaft.

There are at least three modes of operation with d-c automatic hoists:

1. *Automatic* (production only), with the skip(s) correctly spotted; the cycle is initiated and will continue operating unattended until stopped.

2. *Semiautomatic,* with the conveyance (skip or cage) correctly spotted; the cycle is initiated by pushbutton. The conveyance goes to the selected level under control of the programmer and stops.

3. *Manual* control is from the master switch, with the programmer still overriding acceleration and deceleration rates. The hoist brakes are normally interlocked with the master switch and applied when the switch is moved to the zero-speed position.

To these may be added:

4. *Shaft or rope inspection speed*—speed is limited to 100 to 200 ft/min regardless of position of controller.

5. *Test*—a single trip of the conveyance initiated from the control panel.

At each level control panel a "jog up" "jog down" button provides creep speed within range of that level.

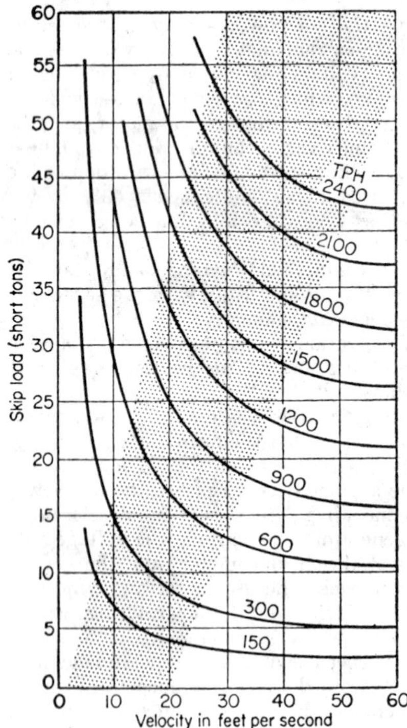

FIG. 20-51. Skip load vs. velocity curves for 1,600 ft depth.

287. Safety Features. The following safety precautions are normally included in the d-c closed-loop system. The loop contactor is opened and the hoist brakes are applied because of:

1. Overtravel or overspeed as detected by the safety controller.
2. Overcurrent protection normally timed.
3. Undervoltage a-c and d-c control power (may be timed).
4. Gear loss to programmer (and on some hoists the safety controller).
5. Loss of d-c generator excitation source.
6. M-G set power loss.
7. Generator circuit grounding.
8. M-G set and/or hoist bearing overheating.
9. Slack rope switch—drum hoist. Jammed conveyance detector—Koepe hoist.
10. M-G set overspeed.
11. Operating any emergency-stop pushbutton.

The main problem in selecting a hoist for a given capacity and depth is deciding on the proper skip load (pay load). A larger load hoisted at slower speeds requires less horsepower, but at the expense of increased rope diameter, which, in turn, increases the drum diameter, gearing, etc.

In selecting a skip size, therefore, it is helpful to know the relationship between skip load, speed, and capacity for the given depth. Such a relationship is shown in Fig. 20-51. These curves indicate that for any capacity, as skip load decreases, the speed

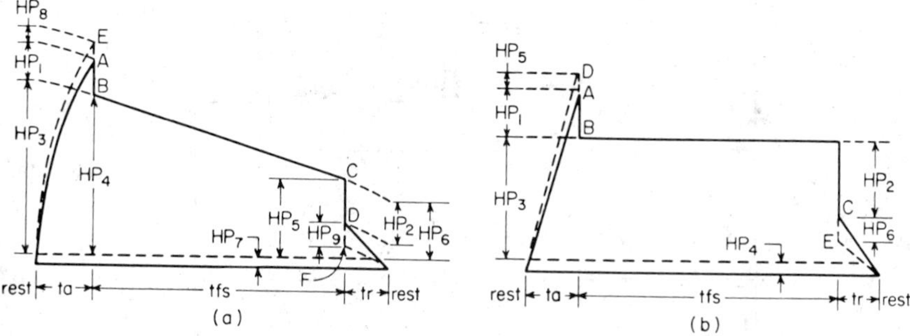

FIG. 20-52. Horsepower vs. time cycle for drum hoists (*a*) and friction or drum hoist with tail rope (*b*).

increases to the point where the cycle consists of only acceleration and retardation, with no full-speed time (approximately 63 ft/s at 1,600 ft in Fig. 20-51).

288. Skip Load. Use this equation:

$$SL \text{ (tons)} = \frac{\left\{\dfrac{\text{depth (ft)}}{\text{velocity (ft/s)}}\right\} + 0.4 \text{ velocity} + 12}{3{,}600/[\text{capacity (t/h)}]}$$

By substituting depth and capacity values, assuming different velocities, and solving for skip load, plot a skip-load–velocity curve similar to those in Fig. 20-51.

The next step is to select the proper skip load. The left side of the shaded area of Fig. 20-51 indicates the portions of the curves optimum for Koepe friction hoists; the right side shows portions optimum for drum hoists. (Experience has shown that the optimum skip load of a Koepe friction hoist usually is larger than that of a drum hoist, for the same tonnage and hoisting depth. By increasing the skip load, it is sometimes possible to jump to the next smallest motor size without greatly increasing the cost of the mechanical equipment. With a drum hoist, the cost of the mechanicals increases more rapidly than with a friction hoist—hence the different optimum areas.)

From the curve constructed, select the proper skip load. Multiply this skip load by 0.75 to find an approximate skip weight if it is not known. [SW (skip weight) = $0.75SL$.]

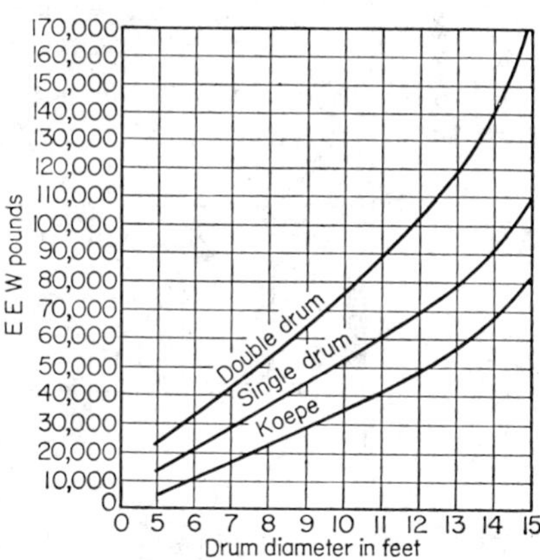

FIG. 20-53. Approximate equivalent effective weight reduced to rope center for different diameter drums.

289. Rope Size. Solve this equation for diameter:

$$\sqrt{\frac{[\text{skip load (tons)} + \text{skip weight (tons)}]/[(K_1/\text{factor of safety}) - K_2 \times \text{depth}]}{\text{number of ropes}}}$$

Rope type	K_1	K_2
Round-strand	41.8	0.00084
Flattened-strand	46.0	0.00090
Locked-coil	61.6	0.000122

Assume a D/d (drum to rope) ratio of 80 even though this may vary with the depth.

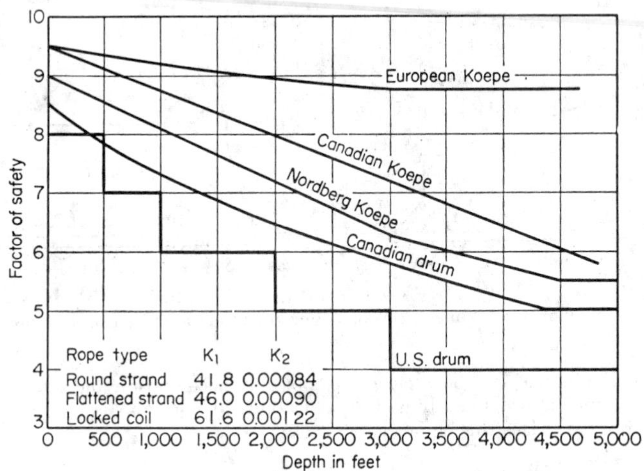

Fig. 20-54. Factors of safety.

290. Motor Horsepower (RMS Hp). Draw a horsepower/time cycle diagram similar to Fig. 20-52*a* or *b*, and label it as follows:

ta = acceleration time (s)
tfs = full-speed time (s)
tr = retardation time (s)

Creep time is not considered in acceleration or retard time and may be considered as rest time.

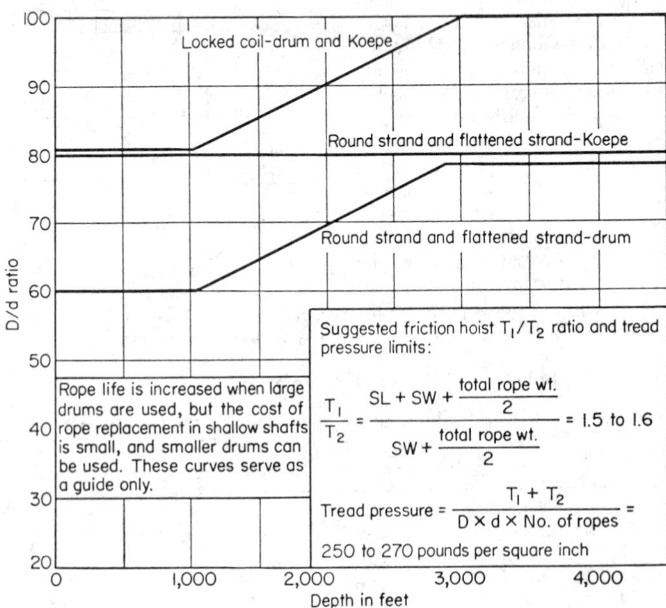

Locked coil-drum and Koepe
Round strand and flattened strand-Koepe
Round strand and flattened strand-drum
D/d ratio
Depth in feet
Rope life is increased when large drums are used, but the cost of rope replacement in shallow shafts is small, and smaller drums can be used. These curves serve as a guide only.
Suggested friction hoist T1/T2 ratio and tread pressure limits:
T1/T2 = (SL + SW + total rope wt./2) / (SW + total rope wt./2) = 1.5 to 1.6
Tread pressure = (T1 + T2) / (D × d × No. of ropes) =
250 to 270 pounds per square inch

Fig. 20-55. Suggested drum diameter (D) versus rope diameter (d) ratio.

From a skip-load velocity curve, select the full-speed velocity corresponding to the skip load.

Knowing the velocity and assuming 2.5 ft/s² for a and r rates, find ta and tr.

$$ta = tr = \frac{V}{2.5}$$

RMS Hp for Drum Hoist. Find the approximate equivalent effective weight (EEW) in Fig. 20-53.

Using the following steps, determine horsepower values corresponding to various points in the horsepower-time cycle (Fig. 20-52*a*). Note that hp_5 and hp_5 will be negative values.

Explanations of the abbreviations used in the drum-hoist horsepower equations follow:

R = depth × rope weight/ft
TSL = total suspended load = $EEW + SL + 2SW + 2R$
SLB = suspended load at bottom of shaft = $(SL + R) - (V \times ta \times \text{rope wt/ft})$
SLT = suspended load at top of shaft = $(SL - R) + (V \times tr \times \text{rope wt/ft})$

$$hp_1 = \frac{TSL \times V^2}{32.2 \times ta \times 550} \qquad hp_5 = \frac{SLT \times V}{550}$$

$$hp_2 = \frac{TSL \times V^2}{32.2 \times tr \times 550} \qquad hp_6 = \frac{(SL - R) \times V}{550}$$

$$hp_3 = \frac{(SL + R) \times V}{550} \qquad hp_7 = \frac{SL \times V}{550} \times 1.76^*$$

$$hp_4 = \frac{SLB \times V}{550}$$

Quantity hp_7 is the friction losses. These vary considerably with condition of shaft, skips, rope, etc. For inclined shafts, to this must be added rolling friction (2% of vertical component of skip and load) and rope friction (10% of vertical component of rope weight). These vary with degree of slope but are on the safe side.

$$A \text{ (peak accelerating hp)} = hp_1 + hp_7 + \frac{hp_4 + 2hp_3}{3}$$

$$B \text{ (full-speed hp at end of acceleration period)} = hp_4 + hp_7$$

$$C \text{ (full-speed hp at start of retard period)} = hp_5 + hp_7$$

$$D \text{ (retardation hp)} = hp_2 + hp_7 + \frac{hp_5 + 2hp_6}{3}$$

$$hp_8 \text{ (hp required to accelerate motor rotor)} = \frac{0.6A \times 1.2}{ta}$$

$$hp_9 \text{ (hp required to retard motor rotor)} = \frac{-0.6A \times 1.2}{tr}$$

$$E \text{ (total hp required to accelerate hoist and motor)} = A + hp_8$$

$$F \text{ (total hp required to retard hoist and motor)} = D + hp_9$$

Calculate the root-mean-square horsepower (rms hp) for d-c motors.

$$\text{rms hp} = \sqrt{\frac{E^2 \times ta + \frac{1}{3}(B^2 + C^2 + BC)tfs + F^2 \times tr}{0.75ta + tfs + 0.75tr + 0.5\ \text{rest}^1}}$$

$^*\ 0.176 = \dfrac{100 - E}{E}$, $E = 0.85$.

[1] Rest = creep time + idle time.

For induction motors,

$$\text{rms hp} = \sqrt{\frac{E^2 \times ta + \frac{1}{3}(B^2 + C^2 + BC)tfs + F^2 \times tr}{0.5ta + tfs + 0.5tr + 0.25 \text{ rest}}}$$

The procedure for unbalanced hoisting is the same except that to find the rms horsepower, the hp^2 divided by time for hoisting and lowering must be combined under the radical.

RMS Hp for Friction Hoist. Find the appropriate *EEW* in Fig. 20-53.

$$R = \text{depth} \times \text{rope weight/ft} \times 2 \times \text{number of ropes}$$

$$TSL = EEW + SL + 2SW + R$$

Using the following steps, determine horsepower values corresponding to various points in the horsepower-time cycle (Fig. 20-52*b*). Note that hp_2 and hp_6 will be negative values.

$$hp_1 = \frac{TSL \times V^2}{32.2 \times ta \times 550} \qquad hp_3 = \frac{SL \times V}{550}$$

$$hp_2 = \frac{TSL \times V^2}{32.2 \times tr \times 550} \qquad hp_4 = \frac{SL \times V}{550} \times 0.111^*$$

$$A \text{ (peak accelerating hp)} = hp_3 + hp_4 + hp_1$$
$$B \text{ (full-speed hp)} = hp_3 + hp_4$$
$$C \text{ (total retardation hp)} = hp_3 + hp_4 + hp_2$$

$$hp_5 \text{ (hp required to accelerate motor rotor)} = \frac{0.75A \times 1.2}{ta}$$

$$hp_6 \text{ (hp required to retard motor rotor)} = \frac{-0.75A \times 1.2}{tr}$$

$$D \text{ (total hp required to accelerate hoist and motor)} = A + hp_5$$
$$E \text{ (total hp required to retard hoist and motor)} = C + hp_6$$

Calculate root-mean-square horsepower (rms hp) for d-c motors.

$$\text{rms hp} = \sqrt{\frac{D^2 \times ta + B^2 \times tfs + E^2 \times tr}{0.75ta + tfs + 0.75tr + 0.5 \text{ rest}}}$$

For induction motors,

$$\text{rms hp} = \sqrt{\frac{D^2 \times ta + B^2 \times tfs + E^2 \times tr}{0.5ta + tfs + 0.5tr + 0.25 \text{ rest}}}$$

291. Example. As an example, assume a depth of 1,600 ft and a desired capacity of 300 t/h. From experience it is known that both a drum and a Koepe hoist would be applicable, but for this example assume that the cost of the Koepe is favorable.

From the skip-load velocity curve for 1,600 ft depth select a 13-ton load, as this is on the left side of the shaded area. This intersects the 300 t/h curve at a speed of approximately 12 ft/s or 720 ft/min.

As this is a friction hoist, assume the skip weight to be equal to the load, and solve for rope diameter.

$$\sqrt{\frac{[13 + 13]/[(K_1/\text{factor of safety}) - K_2 \times \text{depth}]}{\text{number of ropes}}}$$

$^*0.111 = \frac{100 - E}{E}$, $E = 0.90$.

Normally Koepe hoists use flattened-strand hoisting ropes. Round-strand ropes have been used, but this is the exception rather than the rule. Locked-coil ropes could also be considered.

From the table (Par. **289**) the factors for flattened-strand rope are

$$K_1 = 46 \qquad \text{and} \qquad K_2 = 0.00090$$

The factor of safety equals 7.5 for a Koepe (see Fig. 20-54). Then

$$d = \sqrt{\frac{[13 + 13]/[(46/7.5) - 0.00090 \times 1600]}{4}}$$

In selecting a Koepe hoist it is safe to assume four ropes. One can readily see from the value of d whether one, two, or six ropes would be more desirable.

$$d = \sqrt{\frac{26/(6.13 - 1.44)}{4}}$$
$$= 1.18$$

Pick a rope of 1¼ in diameter, and check the T_1/T_2 ratio from the formula in Fig. 20-55 (assume the tail ropes to be equal in weight).

$$\begin{aligned}
\text{Total rope weight} &= 3{,}200 \text{ ft} \times 2.81 \times 4 \\
&= 36{,}000 \text{ lb} \\
\text{Therefore} \quad T_1 &= SL + SW + 36{,}000/2 \\
&= 26{,}000 + 26{,}000 + 18{,}000 = 70{,}000 \text{ lb} \\
&= SW + 18{,}000 = 44{,}000 \text{ lb} \\
T_1/T_2 &= 70{,}000/44{,}000 = 1.59
\end{aligned}$$

This is on the high side because the rope size selected was somewhat larger than that found by the formula, but the ratio can be improved by adding weight to the skips. Adding 5,000 lb to each skip gives a T_1/T_2 ratio of 75,000/49,000 = 1.53.

Check the safety factor. The breaking strength of four 1¼-in F.S. ropes is 4 × 71, or 284 tons.

$$\text{F.S.} = \frac{284}{T_1} = \frac{284}{75{,}000/2{,}000} = 7.57$$

which is sufficient.

From Fig. 20-55 pick the wheel diameter.

$$80 \times 1.25 = 100\text{-in wheel}$$

Checking the tread pressure,

$$\text{Tread pressure} = \frac{T_1 + T_2}{D \times d \times \text{no. of ropes}} = \frac{75{,}000 + 49{,}000}{100 \times 1.25 \times 4} = 248 \text{ ft-lb/in}^2$$

A Koepe hoist necessary for 300 t/h from a 1,600-ft depth will be a 100-in-diameter wheel with four 1¼-in F.S. ropes, hoisting a 13-ton load in a 15½-ton skip at a speed of 720 ft/min.

To find the motor horsepower, from Fig. 20-53 select equivalent effective *EEW*, the inertia of the hoist at 25,000 lb.

$$R = 1{,}600 \times 2.81 \times 2 \times 4 = 36{,}000 \text{ lb}$$

$$TSL = 25{,}000 + 26{,}000 + 2 \times 31{,}000 + 36{,}000 = 149{,}000$$

Referring to Fig. 20-52*b*,

$$t_a = t_r = 12/2.5 = 4.8 \text{ s}$$

$$hp_1 = \frac{149{,}000 \times 12^2}{32.2 \times 4.8 \times 550} = 252 \text{ hp}$$

$$hp_2 = \frac{149{,}000 \times (12)^2}{32 \times 4.8 \times 550} = 252 \text{ hp}$$

$$hp_3 = \frac{26{,}000 \times 12}{550} = 567 \text{ hp}$$

$$hp_4 = \frac{26{,}000 \times 12 \times 0.111}{550} = 63 \text{ hp}$$

$$A = 567 + 63 + 252 = 882 \text{ hp}$$

$$B = 567 + 63 = 630 \text{ hp}$$

$$C = 567 + 63 - 252 = 378 \text{ hp}$$

$$hp_5 = \frac{0.75 \times 882 \times 1.2}{4.8} = 165 \text{ hp}$$

$$hp_6 = \frac{0.75 \times 882 \times 1.2}{4.8} = 165 \text{ hp}$$

$$D = 882 + 165 = 1{,}047 \text{ hp}$$

$$E = 378 - 165 = 213 \text{ hp}$$

To calculate the root-mean-square horsepowers (rms hp) the full-speed time (*tfs*) must be known.

$$t_a = t_r = \frac{12 \text{ ft/s}}{2.5 \text{ ft/s}^2} = 4.8 \text{ s}$$

The distance traveled during t_a and t_r is

$$\frac{12 \text{ ft/s} \times 4.8 \text{ s}}{2} \times 2 = 57.6 \text{ ft}$$

$$tfs = \frac{1{,}600 - 57.6}{12} = 128.5 \text{ s}$$

Checking,

$$\frac{882}{2} \times 4.8 + 630 \times 128.5 \times \frac{378}{2} \times 4.8 = \frac{26{,}000 \times 1{,}600}{0.90 \times 550} 83{,}979 \neq 84{,}040$$

which is sufficiently close. Therefore

$$\text{d-c rms hp motor} = \sqrt{\frac{(1{,}047)^2 \times 4.8 \times (630)^2 \times 128.5 + (213)^2 \times 4.8}{0.75 \times 4.8 + 128.5 + 0.75 \times 4.8 + 0.5 \times 12}}$$

$$= 631 \text{ hp}$$

ELECTRICITY IN THE GRAPHIC ARTS

By James A. Keeley

DESIGN CONSIDERATIONS

292. General. All producers of printed material live daily with the specter of a deadline. A deadline is necessitated by the perishable nature of information, by the demands of the customer, or by both. It defines a time space for production which does not allow for failure of equipment. Accordingly, the graphic arts industry expects the design engineer to place reliability above all else. Printing establishments almost always seek increased operating speed in new equipment but are never willing to attain this benefit through any diminution of reliability.

The electrical engineer faced with an equipment design assignment for a graphic arts customer should be guided, then, by the following principles, which collectively contribute to reliability.

293. Environment. The operating environment should be assessed realistically. The atmosphere of most pressrooms is permeated with ink mist and paper dust. Either of these agents can cause trouble in electric motors and control devices. Equipment to be mounted on or near presses must withstand long periods of severe vibration. Newspaper conveyors and mail-room equipment must also be designed to exclude paper dust from electrical components. Any device to be used in an engraving department must be designed in recognition of the corrosive effects of acid fumes.

294. Maintenance. All equipment should be designed in the belief that it will be maintained, rather than with the notion that it will not require maintenance. It is not the responsibility of the mechanical engineer alone to ensure that electrical components are readily accessible and can be replaced quickly. In complex circuit aggregates, such as press-drive controls, troubleshooting circuitry to pinpoint a malfunction quickly is a necessity. Where possible, and especially in electronic hardware, modular design should prevail.

295. Safety. Safety of operating personnel deserves much of the electrical engineer's attention. Much equipment found in a printing concern is massive and operates at high speed. Its potential danger can be minimized by application of automatic controls to halt machinery where certain conditions are met or specific protective devices are violated. Such controls should be of "fail-safe" design.

PRESS-DRIVE SYSTEMS

296. Group-drive Motors. If more than one unit or deck is driven by a single motor, the system is called a "group drive." This system is frequently employed on the smaller presses such as the deck type, unit-type tubular, and single-width semicylindrical. Group-drive motors are in the 50- to 200-hp range.

297. Unit-drive Motors. "Unit drive" refers to a press, the folder and each unit of which are powered by a motor. Unit-drive arrangements are used on large presses where flexibility and an even application of driving power are important. Flexibility is possible because individual motors can be used or not used depending upon the number of units required for a given run. Unit-drive motors are usually of 30, 40, or 50 hp.

298. Type of Current Used. Either alternating or direct current can be used in group-drive and unit-drive motors—the choice is the designer's. There are advantages and disadvantages to either choice and plenty of opinions supporting either. The important thing to recognize is that both alternating and direct current are used successfully in printing plants everywhere.

Alternating Current. Speed control of an a-c drive is obtained by use of a variable resistance in series with the rotor. Essentially, speed increases are stepped, the abruptness of steps being a function of the number of resistors employed. Smooth speed increases are desirable, because too great a step will cause a sudden speed change which will break the sheet being printed. In some a-c unit-drive systems the motors must

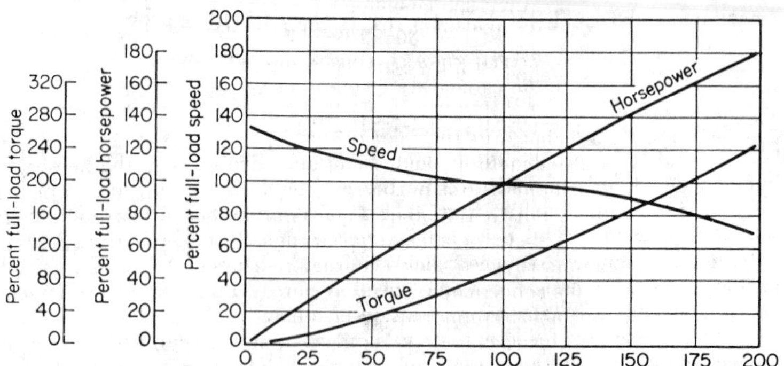

Fig. 20-56. Characteristics of compound-wound d-c press-drive motor (50 hp).

be synchronized to prevent an out-of-phase overheating. The problem is usually overcome in such designs by physically marking the drive-shaft clutch, so that it can be readily reengaged in the same position in which the motors were synchronized.

Both unit and group a-c drives utilize 3-phase wound-rotor induction motors. Because wound-rotor a-c motors undergo large speed changes with small changes in the torque required by the load, they are unsatisfactory for driving the press at threading speed. A small auxiliary motor, called the inching motor, is used. The inching motor is connected to the drive shaft through a slow-motion gear train and an overrunning clutch. As the press speed increases above threading range, the overrunning clutch disengages, the inching motor is deenergized, and the main-drive motors are energized.

Table 20-20. Power Required by Sheet-fed Presses

Sheet size group, in	Type feed	Platen		Cylinder flat bed		Offset		Cutting and creasing	
		Sheets/h	Hp	Sheets/h	Hp	Sheets/h	Hp	Sheets/h	Hp
8 × 12	Hand	2,800	⅓						
10 × 15	Hand	2,800	½						
	Auto	5,000	1½						
12 × 18	Hand	2,500	½	4,800	3				
	Auto	4,000	3	5,000	3				
14 × 20	Hand	2,000	1						
	Auto	2,600	2½	5,000	3				
	Color								
17 × 22	1	.....	...	5,000	5	7,000	3		
22 × 34	1	.....	...	4,500	5	7,000	5	4,500	3
	2	.....	...	.....	....	7,000	7½		
30 × 40	1	.....	...	3,000	5	6,800	7½	4,250	7½
	2	.....	...	3,750	10	6,800	15		
35 × 45	1	.....	...	2,800	7½	6,800	10	2,400	5
	2	.....	...	2,500	15	6,800	15		
38 × 56	1	.....	...	2,250	10	6,500	7½	2,200	5
	2	.....	...	2,250	15				
42 × 58	1	.....	...	.....	....	6,800	20	1,900	7½
	2	.....	...	1,500	15	6,800	25		
	3	.....	...	.....	....	6,800	40		
	4	.....	...	.....	....	6,800	50		
	5	.....	...	.....	....	6,800	50		
52 × 76	1	.....	...	.....	....	6,300	20	2,400	15
	2	.....	...	.....	....	6,300	30		
	3	.....	...	.....	....	6,300	40		
	4	.....	...	.....	....	6,300	50		
	5	.....	...	.....	....	6,300	60		

Notes: 1. Rotary letterpresses develop 90% of the maximum speed and require 60 to 80% of the power of an equivalent offset press.

2. Gravure presses develop 80% of the maximum speed and require approximately the same power as an equivalent offset press.

3. Platen presses tabulated by form size.

Direct Current. Direct-current drives were understandably popular in the days of d-c transmission networks. As a-c power transmission became the rule, the d-c drives waned in popularity. Control devices for power rectifiers and the convenience of solid-state rectification have caused a rejuvenation of direct current as press-drive power. A distinct advantage of a d-c system is the smooth acceleration of the press.

Both unit and group d-c drives employ compound-wound motors with separately excited shunt fields. Speed control of such motors is possible by varying armature current, field voltage, or both. In practice, both are varied. Up to 50% of speed the main field voltage is held constant and armature current is increased. Above 50% of speed armature current is maintained, and the main field voltage is decreased. Since d-c motors can be accurately controlled over their entire speed range, separate inching motors are not required.

Characteristics of a typical d-c press-drive motor are displayed graphically in Fig. 20-56.

299. Power Requirements. Power requirements of sheet-fed and newspaper presses are shown in Tables 20-20 and 20-21, respectively.

300. Press-drive Controls. The function of press-drive controls is simple: they regulate the power applied to the press in response to commands generated manually by remote pushbutton stations or automatically by condition sensors.

Manual Commands. Pushbutton stations on or near the press provide for inserting a safe condition (the press cannot be moved until the safe button is restored), for establishing a ready-to-run condition, for moving the press at a slow (inching or jogging) speed, for accelerating and decelerating the press, and for stopping the press.

Table 20-21. Power Required by Newspaper Presses

No. of straight-run pages	Factor for single width (note 1)	Press speed, thousands of papers/h							
		25	30	36	40	45	50	55	60
		Hp required, double-width press							
8	1.65	15	19	24	29	35	40	50	60
16	1.5	29	38	49	57	67	72	92	108
24	1.4	43	57	73	85	100	118	138	162
32	1.3	58	76	98	114	134	158	185	216
40	....	72	95	122	143	168	198	230	270
48	....	87	114	147	171	200	237	277	324
56	....	102	133	172	200	236	276	324	378
64	....	116	152	196	228	268	316	370	434

NOTES: 1. For single-width presses, multiply above power by factor in second column.
2. For each color cylinder, add one-third of eight-page power.

Automatic Commands. When a web of paper traveling at high speed breaks, it can cause severe press damage and possible personnel injury unless certain things happen instantly. A web-break detector, a shoelike device, which is supported by the sheet, falls as soon as a break passes beneath it to start the required sequence. A "red-button" stop is applied at once. In the time it takes a huge inertial mass like a press to stop, the broken web can wrap around a cylinder of the printing couple and smash plates and blankets and throw cylinders out of line. To prevent this, a web-severing device cuts the paper just ahead of the unit affected by the web break to lessen the amount of paper that can feed into the unit. Web-impeller devices to carry the broken web away from a printing couple by causing the web to wrap harmlessly on an idler roller are also used. Similar in function to the web-break detector is the folder-choke device, which initiates a red-button stop if the folder becomes jammed.

301. Press Brakes. Alternating-current-powered presses normally employ electromagnetically energized disk brakes or hydraulic brakes. Presses having d-c drive motors are customarily stopped by dynamic braking. Activation of the dynamic brake causes armature current to be cut off, but the field remains energized. The inertia of the press-drive train continues to turn the armature. Under these conditions, the d-c motor is acting like a d-c generator. The current generated is routed through the closed contacts of a relay to resistors, which convert the energy to heat. The magnetic force opposing rotation of the inertially powered armature is sufficient to stop large presses operating at top speed in about 8 to 10 s.

AUXILIARY PRESS CONTROLS

302. Insetting and Register Controls. Both insetting (the synchronous insertion of a preprinted web into the web array being printed) and automatic register control (ensuring that an image or superimposition of images laid down by one printing unit will be in register with an image to be superimposed by another unit) cope with the same general problem: control of a point on the web to enable its arrival at a certain location at a certain time.

Insetting. Insetting devices vary considerably in complexity and cost. One of the simplest is in the form of a stroboscope pulsed in synchronism with the impression cylinder. Its light is directed to illuminate preprinted marks on the sheet (spaced according to the repeat length used). If the mark appears to slip forward, web tension is increased; if it appears to slip backward, web tension is decreased. The tension corrections are applied not automatically, but by a pressman who reacts to what he sees.

In automatic insetting systems, a photoelectric cell scans the preprinted marks on the web being inset. Magnetic switches geared on a 1:1 basis with the impression cylinder are used in some systems to provide impression-cylinder location. It is possible also to use a photoelectric cell scanning marked gears or shafts to obtain the required impression-cylinder data. A special-purpose computer compares the time occurrence of the two pulses and, on the basis of the magnitude of the error and the rate of change of the error with respect to time, issues the commands appropriate to web-tension increase or decrease.

The principal difference among various lines of insetting equipment is how the tension changes are effected. Motor-positioned draw rollers, vacuum boxes, and belt tension devices are all used with success to apply web tension.

Register Controls. Automatic register controls are used in full color work to correct for variations that occur between units printing different colors. As in the case of insetting, a special-purpose computer generates corrections required to erase an error condition. Corrections are applied to motorized compensating rollers between the two units involved. Error sensing is accomplished by either scanning a mark printed by the first unit and comparing it with the position of the impression cylinder of the second unit or by comparing pulses generated by two photoelectric cells—one scanning a mark printed by the first unit and one scanning a mark printed by the second.

303. Automatic Paster Controls. High-speed web-fed presses become much more efficient machines if the tail of an expiring roll can be pasted to the head of a new roll without slowing or stopping the press. Modern automatic paster controls enable this to be done with a performance consistency well above 90%. The problem facing the designer of automatic pasting equipment is to sense the speed of the expiring web, bring the new roll up to this same speed within a narrow tolerance, press the tail to adhesive on the head of the new roll, and sever the expiring web from the core.

Although there are many differences in circuit design among manufacturers, the most noticeable difference among the various systems is in the method of bringing the new roll up to speed. Core drives, running tension belts, contact with the expiring web, and roller and friction belt predrives are all used. Although automatic paster devices cycle without manual intervention, they usually incorporate parallel manual controls to rotate the arms of the reel stand in either direction, to start the pasting cycle, and to cause pasting. In addition, visual signals are provided to tell the supervisors in the reel room and on the press deck when a paster cycle is in progress. The purpose of the signal is to alert the pressmen to pull the imperfect copies from the folder.

304. Ink-mist Controls. In letterpress operations, ink mist is a distinct nuisance. The mist is created at the outgoing nips of ink-transfer rollers by shearing of the tacky ink. The resultant force acting on the ink particles is tangential to the point of adjacent roller contact. Two different approaches to the problem have been developed:

1. Removal of ink mist from printing unit via ducting to a centrifugal filter.
2. Electrostatic repulsion of ink-mist particles.

Both methods are effective, but from an electrical engineer's point of view, electrostatic repulsion is more interesting. The theory of this system is that ink-mist particles ionized near their source will repel each other (hence retain minimum size) and will be attracted back to the uncharged rollers.

Alternating current is converted to high-voltage direct current (in the 15,000-V range) by a resonating coil circuit coupled to a step-up transformer. The ionization occurs in the region of a corona wire, or "comb," placed near each outgoing nip.

High-voltage d-c ink-mist ionization is not ideal for suppressing color ink mists. Some experimentation utilizing alternating current has been undertaken, but results have not been publicized as yet.

HOT-METAL OPERATIONS

305. Electricity is widely used in the graphic-arts industry to melt the lead-tin-antimony alloys used for the casting of type, base metal, and stereotype plates. Hot metal machinery includes linecasting machines, job furnaces, master remelt pots, plate-casting machines, monotype machines, and Ludlow machines.

Linecasting machinery requires fractional-horsepower constant-speed motors. Drive motor, blower motor, and possibly magazine-shifting motor and saw motor total a maximum of 1½ hp. Melting crucibles of linecasters require 1,800 to 3,200 W but operate at 1 kW average. Three or four heating elements are used.

Larger crucibles such as are found in remelt pots use larger elements and more of them. For example, an 8-ton-capacity plate remelt pot utilizes fourteen 17.5-kW elements to melt the plates and keep the molten metal in the 575 to 600° range.

AUTOMATION

306. The printing industry has made remarkable progress since 1960 but has really just scratched the surface of the possibilities of automation. The following partial listing points out areas that have received some attention.

307. Computers. Digital computers, both general and special-purpose, are gaining rapid acceptance as a tool for the hyphenation and justification of lines of type. In this capacity they provide paper tape which becomes input to automatic linecasting machines and photounits. Larger publishers also employ general-purpose computers for diverse business functions, such as payroll, billing, statistical analysis, and inventory control. In the future, computers will probably be more extensively used for process control, e.g., control of presses.

Special-purpose computers are integral parts of the circuitry of newspaper totalizing equipment, counterstacker machinery, and insetting and automatic register controls.

308. Computer Input Devices. Use of digital computers requires a means of communicating the data to be processed. Below is a list of electrically operated equipment now in use by the printing industry:

Card punches.
Paper-tape perforators.
Optical scanning equipment.
Digitizers of graphical coordinates.
On-line keyboards.
Light pencil equipment.
Magnetic tape and disk drives.
Computer console.

309. Phototypesetting Equipment. Photocomposition, the technique of setting text on film, is becoming increasingly popular. Light is directed against a grid or disk which is opaque except for the character being flashed. The transmitted light is focused against a photosensitive emulsion which records the character. Within limits, the projected image can be enlarged or reduced by the lens system employed. Input can be prepared manually or by computer.

At the present time, most photocomposition machines produce an output which is cut up in strips and hand-pasted where desired on the page layout. Subsequently, an offset plate is made from a photograph of the composed copy, or an engraving is prepared, depending upon the method of printing to be used.

Phototypesetting machines have recently been designed which utilize electronically stored characters. These are composed into desired groupings on a video screen placed next to the film. This technique enables high operating speed.

310. Materials Handling. Electromechanical conveyors are in use which transport rolls of newsprint, printing plates, copy, newspapers and magazines, and bundles of the finished product from place to place. Motors powering these conveyors vary in horsepower depending upon the weight of the transported load and length of conveyor. Generally, these are 3-phase a-c motors.

OTHER USES OF ELECTRICITY IN THE GRAPHIC ARTS INDUSTRY

311. Printing plants, of course, employ electricity for normal lighting, heating, and air-conditioning functions. In addition to these uses and the specific applications cited above, almost every tool of production is electric-motor-powered or uses electric current in its operations.

Examples follow:

Electroplating equipment.

Routers, shavers, and metal saws.

Engraving and photographic equipment.

Mat rolling presses.

Vacuum formers.

Collating machines, paper-cutting machines, stuffing machines, and paper-baling machines.

BIBLIOGRAPHY

312. References

Braasch, W. "Newspaper Presses"; ANPA Research Institute, 1958.

Hueman, M., Jr. "Comparative Analysis of Rotary Press Drives"; 1950.

MACHINE TOOLS

By C. G. Helmick

313. Individual drives are used on practically all modern machine tools. Many machines have motors for each of the various functions, thereby eliminating a considerable amount of mechanical shafting and gearing. For example, a separate motor may be used for the main drive, the feed drive, the traverse drive, and the lubricating pump. Frequently motors and control are built into the machines so that the only electrical work necessary during installation is connecting the power leads.

314. Standards for electrical equipment have been adopted by the National Machine Tool Builders Association. Open, semienclosed, and totally enclosed motors are used depending on service conditions. Ball-bearing motors are usually preferred, except by some grinder manufacturers who use sleeve-bearing motors. Most motors are supplied with continuous rating; however, 1 h and ½-h rated motors are used on many machine tools where the motor is not loaded continuously, such as traverse drives, clamping motors, and other intermittent applications.

The Joint Industry Conference (JIC) has developed standards for equipment to be supplied to the automotive and other mass-production or continuous-process industries. Automatic production lines and automation equipment require special emphasis on operator safety, minimum down time for maintenance, and standardization. NEMA Class XII control enclosure was developed specifically to meet JIC specifications. A single motor type has been standardized: foot-mounted only, totally enclosed, fan-cooled (or nonventilated in the smallest sizes), etc. For details see Combined JIC-NMTBA Electrical Standards cited in Par. **331.**

315. Classification of Drives. Most of the motors used on machine tools may be classified under the following types:

a. **Main drives** are those which do the main work of metal removal, shearing, or forming. Examples are the spindle drive on the lathe or the motor driving the cutter on a milling machine. Many of these drives must be adjustable in speed in order to operate under the most efficient conditions with various sizes of work or cutter. In general, constant horsepower is required regardless of the output speed. In many cases, however, the horsepower required at lower speeds is somewhat less than at higher speeds.

b. **Feed drives** are used to feed the work into the cutting tool or the cutting tool into the work. Wide speed ranges are required in order to meet varying conditions of operation. Most feed drives require approximately constant torque regardless of the speed. This is due to the fact that the load consists mainly of friction in the gearing and ways, the tool reaction being a minor quantity. On some machines, however, where extremely heavy cuts are taken at low speeds, the torque at low speed may be considerably higher than at high speed.

c. **Traverse drives** are used to position the cutting tool rapidly with respect to the work. In many cases the feed drive is used for traverse with the feed motor being run at its maximum speed. Where high traverse speeds are desired, however, a separate motor may be used for traversing. Constant-speed motors are used; and since traversing is usually done only at infrequent intervals, the motor may be intermittently rated.

d. **Accessories** such as lubricating pumps and coolant pumps are driven by constant-speed motors. These motors are small and are usually assembled with the pump as an integral unit. Clamping motors and gear-shifting motors are usually high-slip squirrel-cage torque motors or compound-wound d-c torque motors and are usually intermittently rated.

316. Squirrel-cage induction motors are used on most machine-tool applications. General-purpose squirrel-cage motors are used as the main drive on many machine tools with gear changes for adjusting the speed. Multi-speed motors with or without gear changes are also used for obtaining adjustable speed. High-slip squirrel-cage motors are used on punch presses, on machines requiring frequent reversing, and on numerous other applications.

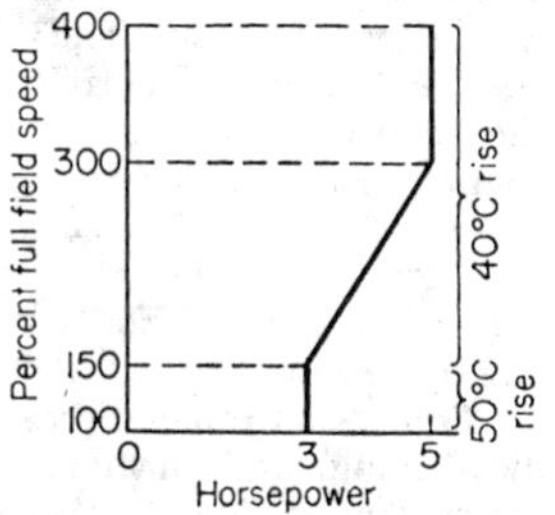

Fig. 20-57. Speed-horsepower characteristic of a tapered-horsepower adjustable-speed d-c motor.

317. Direct-current motors are used on many machine tools where a d-c power supply is available. Constant-speed motors are used where no speed adjustment is required. When speed adjustment is necessary, adjustable-speed motors are used. Adjustable-speed d-c motors may be rated as constant-horsepower, continuous; constant-horsepower, 1 h; or tapered-horsepower, continuous. Figure 20-57 illustrates the tapered rating. The same motor may be rated 3 hp, continuous, or 5 hp, 1 h; or a 3/5-hp tapered rating may be used. The tapered rating would be 3 hp from the minimum speed to 150% of minimum speed, the horsepower tapering from 3 hp at 150% speed to 5 hp at 300% speed, and the horsepower remaining constant at 5 from 300 to 400% speed. Special motors have been designated for 8-to-1 speed control by field weakening for some lathe drives requiring an exceptionally wide constant-horsepower range.

318. Direct-current adjustable-voltage drives are finding increased use as feed drives and main drives. The d-c adjustable-voltage drive generally consists of an a-c motor driving a d-c generator which supplies power for a d-c motor. Control of the d-c motor speed is obtained by adjusting the d-c generator voltage. Constant-torque characteristics are obtained over the armature-voltage control range. The speed of the d-c motor can also be adjusted by means of its field, and the characteristics over this speed range are constant horsepower. Figure 20-58 illustrates the horsepower and torque characteristics with respect to speed. The adjustable-voltage drive is well adapted for feed drives, as a wide speed range is available with constant-torque characteristics.

The entire speed range can often be covered without gear changes, although in some applications a single gear change is used. Speed ranges as high as 120 to 1 without gear changes have been obtained by using a Rototrol or similar speed regulator.

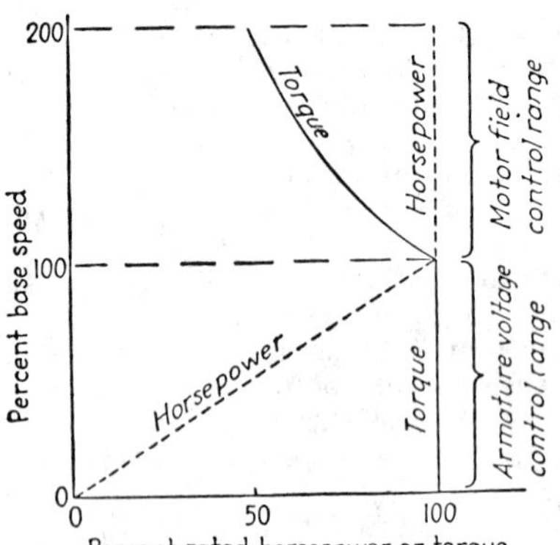

FIG. 20-58. Speed-torque and speed-horsepower characteristics of d-c adjustable-speed drives.

Direct-current adjustable-voltage drives have also found application as main drives. On these drives, part of the speed range is covered by motor-field control and part by armature-voltage control. An important advantage of the d-c adjustable-voltage drive is that the speed may be adjusted easily to a large number of values by means of the speed-control rheostat. The optimum speed for the job at hand may be obtained, and the speed may be set, while the machine is running. With a constant-speed drive and gear changes, a limited number of speeds are available, and the operator often must use a speed considerably lower than the optimum. Also, the speed usually must be selected while the machine is not running. Thus the adjustable-speed drive has a definite advantage from the standpoint of setup and production time.

319. Electronic adjustable-speed drives are also finding wide application to machine-tool drives, especially where fast response and extreme flexibility of control are required. The electronic drive consists of a d-c motor with electronic rectifiers to supply power for the armature and field. Speed control is obtained by adjusting either the armature voltage or the field current of the motor. This is done by phasing the firing angle of the electronic tubes to control the average output voltage. The horsepower and torque characteristics of this drive are the same as for the d-c adjustable voltage drive as shown in Fig. 20-58. The electronic drive is particularly well suited for feed drives, since a wide speed range with constant-torque characteristics is available. Excellent speed regulation with controlled accelerating current is obtained. The ease of control makes this drive especially adaptable for contouring and tracer work.

Tracer-controlled drives are being used more extensively, especially for milling, die sinking, and turning. The complexity of the drive depends on the number of dimensions to be controlled. A single-dimension tracer may be used on lathes, boring mills, etc., where the symmetry of the piece requires the control of only one feed motion. Profile milling or die sinking may utilize two- or three-dimension tracers. An electronic-tracer-controlled drive will consist of a d-c feed motor controlled by a high-gain position regulator. The regulator is actuated by an electrical signal from a stylus which is in contact with the master template and acts to control the feed motor to position the cutting tool accordingly.

Program-controlled drives are basically the same, differing only in the form of the controlling signal used. Machines have been developed which obtain the command signal from punched cards, punched tape, magnetic tape, etc., fed into a computer which then controls the feed drives. These drives are extremely flexible and are best suited for difficult work, especially in medium-volume quantities, where a specially designed machine is not economically practical.

320. Application of motors to machine tools is discussed below for the principal machines under four classifications.

A. SAWING AND SHEARING MACHINERY

a. **Power saws** include reciprocating hacksaws, rotating circular saws, and band saws having a linear motion. Hacksaws and circular saws are usually driven by constant-speed motors. Adjustable-speed drives are often used on band saws. Separate feed motors are sometimes used to move the saw into the material being cut or to move the material toward the saw (see Table 20-38).

Table 20-22. Values of Horsepower Required to Remove Metal

Material	Hp to remove 1 cu in. per min		
	Lathes	Drills	Milling
Aluminum	0.2 -0.4	0.2-0.4	0.2-0.4
Brass	0.25-0.5	0.3-0.4	0.3-0.6
Cast iron	0.25-0.5	0.4-0.7	0.4-0.7
Steel, mild	0.6 -0.8	0.8-1.0	1.0-1.3
Steel, medium	0.8 -1.0	1.0-1.5	1.5-2.0

Table 20-23. Bending and Straightening Rolls
(Wound-rotor induction motor or compound-wound d-c motor)

Capacity, in.	Length, in.	Hp	Capacity, in.	Length, in.	Hp
¼	62	5	½	122	15
¼	110	7½	⅝	74	10
½	62	7½	⅝	98	15
½	98	10	⅝	146	20

Table 20-24. Bolt and Nut Machinery
Motor A (Par. **323**)
(a) Bolt-forging machines

Size, in.	Hp	Size, in.	Hp
1¼	10	3	30
1½	15	4	60
2	20	5	75
2½	25	5½	75

(b) Bolt threaders

Size, in.	Hp	
	Single spindle	Double spindle
1	2	3
1½	3	5
2	5	7½
3	10	10

(c) Nut tappers

Size, in.	Hp	
	Six spindle	Eight spindle
1	7½ or 10	10 or 15
2	15	20

Table 20-25. Boring, Drilling, and Milling Machines
Motor A or B (Par. **323**)

Size of spindle, in.	Hp
3	10- 20
3½- 4½	15- 25
5 - 6	20- 50
8 -10	30-100

Table 20-26. Boring Mills, Vertical
Motor A or B (Par. **323**)

Size, in.	Hp	Size, ft	Hp
36	15-30	7- 9	40- 75
42	30-50	10-12	50-100
72	30-60		

Table 20-27. Bulldozers

Motor A (Par. **323**)

Size, in.	Stroke, in.	Hp	Size, in.	Stroke, in.	Hp
29 × 10	14	5	45 × 10	18	15
34 × 10	16	7½	62 × 16	20	20
39 × 10	16	10			

Table 20-28. Drilling Machines

Motor A or B (Par. **323**)

(a) Vertical drilling machines

Drill size, in.	Hp	
	Cast iron	Steel
⅜	½	1
½	¾	1
¾	1	2
1	1	3
1½	2	5
2	3	7½

(b) Radial drilling machines

Diam column in.	Hp	
	Normal duty	Heavy duty
11–15	7½–10	15
17–19	10	15–20
22–26	20 –40	50

(c) Multiple-spindle drilling machines

Size of drills, in.	No. spindles	Hp	
		Cast iron	Steel
1/32–1/4	6–10	3	7½
1/16–3/8	10	5	10
3/16–1/2	10	7½	15
1/4–3/4	10–50	10–50	20
3/8–1	10	10–15	30
2	4	10	25
2	6	15	40
2	8	20	50

Table 20-29. Gear-cutting Machines

Motor A or B (Par. **323**)

(a) Hobbers

Max. diam., in.	Pitch capacity, dp	Stroke, in.	Total hp
1	36	½	0.5
2 – 5	16–34	7	2.0
4½–10	6–10	12	7.25
4–14	3.5– 6	14	10.25
8	4.5	8	7.25
8	8	10	5.0
10	10	10	5.0

(b) Disk cutters

Size, in.	Hp
36	5– 7½
60	10–15

(c) Gear shapers

Max. diam., in.	Total hp
7– 9	1.0–1.5
12– 18	3.0
36– 40	5.0–7.5
72	5–20
100–120	7.5–25

(d) Bevel-gear generators

Max. diam., in.	Total hp
4.5	1.5
8.5–18	3.75–6.75
29	7.5
33–35.5	10.0
72	25.0

b. **Shears** may be built as squaring shears, rotary shears, or lever (alligator) shears. Rotary shears are equipped with general-purpose motors. Lever and squaring shears are usually equipped with flywheels to supply heavy momentary demands of power. The smaller shears operating at a high number of strokes per minute use general-purpose motors, and the larger shears operating at a low number of strokes per minute use high slip motors (see Par. **321;** also Table 20-40).

c. **Punch Presses.** See Par. **321;** also Table 20-37.

B. BENDING AND FORMING MACHINERY

a. **Bending rolls** require high-starting-torque intermittent-rated wound-rotor a-c motors or compound-wound d-c motors (see Table 20-23).

b. **Bulldozers** are driven by general-purpose constant-speed motors. On some applications high-slip motors are used (see Table 20-27).

c. **Forging hammers** or drop hammers are operated by general-purpose constant-speed motors. The motors should be mounted so as to minimize the mechanical shock upon them (see Table 20-31).

Table 20-30. Grinders

Motor A or B (Par. **323**)

(a) Plain grinders		(b) Universal grinders		(c) Surface grinders		(d) Centerless grinders	
Swing, in.	Total hp	Swing, in.	Total hp	Table width, in.	Total hp	Swing, in.	Total hp
3–4	1½– 4½	10	1 –3¾	6	1½– 4½	4¹⁵⁄₁₆	3
6	7⅓– 9	12	3¾–7½	8	1½– 4½	6¹⁵⁄₁₆–10⅝	9
10	12 –34	14–18	5¾–9¾	10	5– 7½	10¹⁵⁄₁₆	17½
14–16	12 –40			12	8–25	14–24	17½–23¾
20–36	26 –46			18–36	25–60	20–28	38½

C. CUTTING AND TURNING MACHINERY

a. **Horizontal and vertical boring mills** are usually driven by adjustable-speed d-c motors. Separate motors are used on the larger machines to drive the feed, traverse, and elevating motions (see Tables 20-25 and 20-26).

b. **Drilling machines** are driven by constant-speed squirrel-cage motors. On some machines only the motor parts are supplied, with the machine housing becoming the outside frame of the motor, and the rotor being mounted directly on the drill spindle (see Table 20-28).

c. **Engine lathes** are usually driven by constant-speed squirrel-cage motors. Multispeed a-c motors or adjustable-speed d-c motors are sometimes used (see Table 20-32).

d. **Axle lathes** are driven by adjustable-speed d-c motors or multispeed squirrel-cage motors. Wheel lathes are driven by adjustable-speed d-c motors.

e. **Milling machines** are usually driven by constant-speed squirrel-cage motors. Larger machines such as planer-type milling machines have separate motors for each milling head and separate motors for each of the feed motions (see Table 20-33).

f. **Planers** use d-c adjustable-voltage equipment to provide rapid reversal and wide speed range. Standard adjustable-voltage planer drives are available with a speed range of 40 to 1,200 r/min and a Rototrol, magnetic amplifier, or similar speed regulator to provide good speed regulation. Some planers, especially in the smaller sizes, use hydraulic drives. Some large planers for slow operation use constant-voltage adjustable-speed d-c motors (see Table 20-35).

g. **Shapers** and **slotters** employ constant-speed squirrel-cage motors. On certain large machines for special-purpose application, adjustable-voltage planer drives are used (see Tables 20-39 and 20-41).

h. **Broaching machines** utilize modified adjustable-voltage d-c planer drives. Hy-

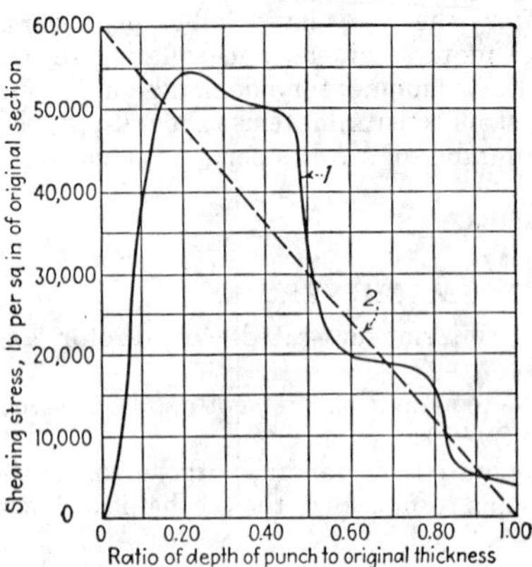

FIG. 20-59. Work diagram for punch press. Curve 1 shows work diagram for punching a hole in a steel plate. Curve 2 is the equivalent triangular work diagram, in which maximum pressure corresponds to maximum shearing stress of the plate.

draulic drives, and sometimes constant-speed d-c motor drives, are in use, especially in the smaller sizes (see Table 20-42).

D. GRINDING MACHINERY

a. **Grinding machinery** varies considerably in the design of individual machines. The general practice is to use adjustable-speed d-c motors for the headstock drive and either constant- or adjustable-speed motors for the wheel drive. Separate motors are often used to feed the wheel with respect to the work (see Table 20-30).

b. **Bench, pedestal, and centerless grinders** use constant-speed squirrel-cage motors with the grinding wheels mounted directly on the motor-shaft extension.

321. Punch presses, forging machines, shears, and similar machines are usually equipped with flywheels to supply momentary demands of power. Motors applied to these machines should have drooping speed-torque characteristics to allow the motor to slow down when heavy loads are encountered. The stored energy of the flywheel supplies the peak power while the motor is slowing down; and after the peak has passed, the motor accelerates the flywheel and restores its stored energy. High-slip squirrel-cage motors are used for this type of application. Machines that operate at a high number of strokes per minute use general-purpose squirrel-cage motors, as there is not enough time between strokes for the motor to slow down appreciably. The characteristic work diagram for a punch press is given in Fig. 20-59. Many large presses are hydraulic-operated, including some very large forging and extruding presses such as developed for aircraft structural sections. The high-pressure system is supplied by pumps driven by constant-speed a-c motors.

322. Power required to remove metal varies widely with the kind of metal, method of removal, shape of the tool, feed, speed, and other factors. The increased use of carbide tools has raised the cutting speeds and horsepower on many machine tools, so that general-purpose machines are designed to accommodate larger drive motors when such special tools are to be utilized. Typical values of horsepower to remove one cubic inch of metal per minute are listed in Table 20-22. These values are based on

Table 20-31. Hammers
Motor A (Par. **323**)

(a) **Board hammers**

Falling weight, lb	Hp	Falling weight, lb	Hp
1000	10	3000	30
1500	15	4000	40
2000	20	5000	50
2500	25		

(b) **Pneumatic hammers**

Falling weight, lb	Hp	Falling weight, lb	Hp
200	10	1000	50
300	15	1500	60
500	25	2000	75
750	40	3000	125
		5000	200

Table 20-32. Lathes

Motor A or B (Par. **323**)

(a) Engine lathes

Swing, in.	Hp	
	Normal duty	Heavy duty
12	3	7½
14–16	5– 7½	7½–20
18–25	10–20	15 –30
30–36	20–40	20 –40
42–48	30–40	50 –60
60–72	40–60	

(b) Turret lathes

Swing, in.	Hp	
	Normal duty	Heavy duty
12	3	5
14–16	5	7½
18–22	5–7½	10–15
24–26	15	40
28–30	25	75

(c) Automatic lathes, single spindle

Capacity, in.	Total hp
9/16	1¼– 3
1 1/16	2 – 5
1⅜	2¾– 7½
1½–2½	11
3¾	12½–17
5¾	17 –18½

(d) Automatic lathes, multispindle

Capacity, in.	Hp	
	4 spindle	6 spindle
1⅜	..	15
2	15	20
2½–3½	15	

Table 20-33. Milling Machines

Motor A (Par. **323**)

(a) Plain, universal, and vertical milling machines

Size	Longitudinal feed, in.	Hp
0	18	1
1	22	3– 5
2	28	5– 7½
3	34	10–20
4	42	15–30
5	50	25–50

(b) Plain or vertical milling machines for carbide steel milling

Size	Longitudinal feed, in.	Hp
2	28–34	20
3	34–42	30
5	42–50	50

(c) Planer-type milling machines

Size, in.	Stroke, ft	Hp per head
36–42	8–50	10– 25
48	8–50	10– 50
66	8–50	25– 50
78	10–50	25– 75
84–120	12–50	50–100

Table 20-34. Pipe Machinery

Motor A (Par. **323**)

(a) Pipe-threading and cutoff machines

Size of pipe, in.	Hp	Size of pipe, in.	Hp
¼–2	1	2½– 8	10
½–2	2	4½–13⅜	15
1 –4¾	5	8 –18	20
1 –6	7½	10 –20	25

(b) Pipe and nipple threading machines

Size of pipe, in.	Hp	
	Single spindle	Double spindle
⅛– ½	1½	3
⅛–1¼	3	5
¼–2	3	5
2½–4	7½	10

Table 20-35. Planers
Motor B (Par. **323**)

Width, in.	Hp
42	25– 35
48	35– 50
60	50– 75
72, 84, 100	50–100
120	75–100

Table 20-36. Presses, Horizontal Forcing
Motor A (Par. **323**)

Capacity, tons	Hp
150–200	20
300	25
400	30
500–600	40
750–800	60

Table 20-37. Punching Machines
Motor A (Par. **323**)

Diameter, in.	Thickness, in.	Hp
⅜	¼	1
½	½	2
⅝	⅝	3
⅞	¾	5
1	½	5
1¼	1	7½
1¾	1	10
2	1	15
2½	1½	20

Table 20-38. Saws, Cold
Motor A (Par. **323**)

Size, in.	Hp
20	10
26	10
32	15
36	20
42	25
48	30

Table 20-39. Shapers
Motor A or B (Par. **323**)

Stroke, in.	Hp
8	½
12–16	3– 7½
20–24	5– 7½
32–36	10–15

Table 20-40. Shears
Motor A (Par. **323**)
(a) Squaring shears

Thickness	Length, in.				
	32–42	50–60	72–96	120–144	168–192
	Hp for mild steel				
18 gage		1½	1½	2	
16 gage	1½	1½	2	3	
14 gage	1½	2	3	3	
⅛ in.	3	5	5	5 – 7½	7½
3/16 in.	5	5	7½	7½	10–15
¼ in.	5	5	7½	7½–10	15
⅜ in.	7½	10	10	10 –15	20
½ in.	10	15	15	15 –20	25
¾ in.	15	20	20	25 –30	40–50
1 in.	20	25	25	40	

(b) Alligator shears

Capacity, in.	Hp
1 × 1	5
2 × 2	7½
3 × 3	15
5 × 5	40
½ × 48 sheet	7½

(c) Rotary shears

Thickness	Max cutting speed, fpm	Hp
14 gage	22	¾
12 gage	22	¾
10 gage	33	1½
3/16 in.	10	2
¼ in.	16	3
⅜ in.	12	7½
½ in.	10	15
¾ in.	10	20
1	10	25

Table 21-41. Slotters and Key-seaters
Motor A or B (Par. **323**)

Stroke, in.	Hp
18	5
24	7½
30	10
48	20

Table 20-42. Broaching Machines
Motor A or B (Par. **323**)

Capacity, tons	Feet per min	Horsepower	
		Single ram	Duplex
5	30	10	15
6	30	10	15
10	30	15–20	20
15	30	20–30	25–30
20	30	25	30
25	20–25	30–40	40
50	15	40	

average machine efficiency and tool sharpness. For detailed information, see Boston and the ASTE handbook cited in Par. **331,** or contact the machine-tool builder directly.

323. Horsepower values given in Tables 20-22 to 20-42 are representative of the average practice of a large number of machine-tool builders. Very often these motors are overloaded as much as 50% for short times, e.g., for broaching strokes. This should be taken into account in selecting wiring, fuses, etc. The machine builder or agent should be consulted if accurate values are desired for a particular machine. Motor types are indicated as A or B: A, squirrel-cage induction motor; B, adjustable-speed d-c motor.

NUMERICAL CONTROL

By Charles Emerson

324. Numerical Control of Machine Tools. Where most standard machine tools move the work or tool at a given selected speed through a given distance to produce a single cut at a time, numerical control (NC) is a concept of keeping the part or tool motion under continuous, specific, and often varying control, producing a family of cuts. As originally developed for the aircraft and missile industry, numerical control permits moving a cutter through a previously designed path in the work, often a complex contour, to result in a shape of accurately defined dimensions. Most often an NC machine tool is controlled by an eight-channel punched tape that feeds through a tape reader and produces electrical signals that are processed by electronic mechanisms that operate motors and other power devices to operate the machine automatically. Under tape direction, the electronic control center selects machine-table and slide motions, speed of motions, rapid-traverse rate and direction, feed rate and direction, spindle speed, and auxiliary functions, such as tool changing.

325. Tape production for complex contouring requires the use of a computer, in most cases, to convert mathematical contour data to numerical data acceptable by the electronic control of the machine. But simple location via tape, as in drilling a number of holes in specific locations in a plate, can be organized manually and programmed by a technician, often in a short time. In either case, some note must be taken of the capacity of the machine to turn out work, reaction time of the positioning devices, limitations of power available, and other features of the mechanical system of the machine.

326. Accurate positioning of the machine table or slide under numerical control is much more demanding than for a standard machine tool as regards speed regulation, speed range, acceleration and deceleration, and other factors often considered of minor value, such as heat buildup, electrical losses, and friction effects. High rapid-traverse speeds are required for economical tool or table positioning, while accurate slow feeds are necessary for precision cutting. In addition, the rate of slowdown or reverse of motion is an important factor.

327. Position measurement is handled by linear or rotary servos or transducers, a linear unit mounted alongside or axially with a slide, or a rotary unit mounted on the end of a ball screw. These provide digital position data, often by a count of pulses developed in the electronic equipment, which is compared with a corresponding count of pulses dictated by the instructions punched in the tape. These pulse counts are compared electronically at the end of a machine stroke; if the feedback signal developed by the actual motion does not correspond to the command signal, the machine is automatically powered in the necessary direction to make up the difference. Thus the machine has a closed servo loop that provides accurate positioning.

328. Horsepower ratings for numerically controlled machine tools are not significantly different from those for standard machine tools; if anything, they tend to be lower. With refined machine design and the best available table supports and antifriction ways and bearings, friction losses are low. NC equipment tends to fall into the finishing category, though this rule does not apply in all cases. Many NC machines have individual table and spindle drives that range from 1 to 15 hp. For many spindles 5 hp is adequate; for many table and slide drives less than 1 hp is necessary. The control portion of an NC machine is of relatively high cost, often as much as or more than the machine it commands; so the machines themselves are generally designed and built to the highest standards of machine construction in an attempt to make the mechanical features parallel the control features.

329. Control Methods. Most often d-c shunt motors are applied to NC positioning devices, and they may be controlled by magnetic amplifiers, by thyratron or ignitron circuits, or by silicon-controlled rectifiers in electronic-amplifier circuits. Electronic circuits for pulsing, counting, and feedback functions, as well as for controlling motor speeds and directions, have generally been converted to transistors, and the newest trend is to integrated circuits. Such advances permit considerable reductions in the sizes of the control consoles necessary for the machine tools.

Machines to which numerical control has been applied include boring and milling machines, where a rotary multiblade cutter is operated in a fixed position and the workpiece is carried on a table with two or three motions. This arrangement allows a specific volume and shape of metal to be cut away from the workpiece by the cutter as the table is operated in several directions at once. In addition, extra functions can be provided by tape control: tool changing by means of a turret that retains several cutters, indexing a workpiece slide to one of several positions, and other powered motions. NC has also been applied to lathes, welders, grinding machines, and many types of drilling machines.

Numerical control generally divides into contouring systems and point-to-point systems. Contouring is the type of system that machines a mathematically defined shape on the workpiece, while a point-to-point system produces a straight cut or a series of specifically located holes in the work. The major differences in these systems are in the primary control devices, and not in the power units, though there may be more refinements in the contouring power system.

330. Servo Drives. A typical servo drive for a numerically controlled machine consists of a d-c shunt-wound motor with armature current controlled by two silicon-controlled rectifiers. An electronic circuit with negative feedback and accurate gain control fires the rectifiers in response to an applied control voltage, varying the motor speed to close accuracy and allowing instant reversal and controlled acceleration (and deceleration). Similar motor controls are applied to spindle drives, using fixed resistors and switching in place of gearing, which allows many more spindle speeds, and allowing the control tape to easily select the best speed for the job.

Other control methods are used, but to less extent. A stepping motor has been applied in some cases where feedback control of the motion is not considered necessary. With this system, pulses move the motor through incremental angular changes to drive a tool or slide through gearing.

Although numerical control has been added to standard machine tools in some cases, with the machines modified to accept special drives and measuring devices for feedback, the necessity for high rigidity in the machine, minimum friction, accurate speeds and feeds, reduction of backlash, and minimum reaction-mass effects has resulted in a new breed of machines that have, in turn, affected the design of standard machines. Many general-purpose machine tools are now designed to "NC standards" to improve their precision.

331. Bibliography. References to machine-tool drives and control have appeared in *Trans. ASME*, *American Machinist*, *Electrical Manufacturing*, *Machine Design*,

Machinery, Product Engineering, Tool Engineer, and other trade magazines. Specific articles worthy of reference are:

Combined JIC-NMTBA Electrical Standards; *Elec. Mfg.*, May, 1954, Vol. 53, No. 5.

Boston, O. W. A Bibliography of Cutting Metals, 1864–1943; published by ASME.

ASTE, Tool Engineers Handbook, 1st ed.; New York, McGraw-Hill Book Company, 1949.

Anderson, B. T. Three Applications of Electronic Drive to Lathes and Milling Machines; *Tool Eng.*, May, 1944, Vol. 13, No. 5, pp. 86–88.

Caldwell, G. A. Electronic Adjustable Speed Drives; *Tool Eng.*, May, 1944, Vol. 13, No. 5, pp. 83–85.

Rogers, G. L., and Dutcher, J. L. Practical Use of Tracer Controls; *Elec. Eng.*, August, 1954, Vol. 73, pp. 729–734.

Peaslee, L. R. Tape-controlled Machines; *Elec. Mfg.*, November, 1953.

Graves, B. P. Motor Drives and Electric Controls on Machine Tools; *Mech. Eng.*, October, 1938, Vol. 60, No. 10, pp. 729–734; discussion, March, 1939, Vol. 61, No. 3, pp. 245–246.

AIR CONDITIONING AND REFRIGERATION

By John deB. Shepard

AIR CONDITIONING

332. Air Conditioning Defined. The definition of air conditioning as approved by the American Society of Heating and Ventilating Engineers is as follows: "The simultaneous control of all or at least the first three of those factors affecting both the physical and chemical conditions of the atmosphere within any structure. These factors include temperature, humidity, motion, distribution, dust, bacteria, odors, toxic gases, and ionization, most of which affect in greater or less degree human health or comfort."

Accepted trade definitions consider air conditioning to be the control of temperature, humidity, motion, and cleanliness of the air within any structure, adding the qualifying prefix "summer" to those systems designed only to maintain temperature and humidity at a point below the normal outside summer condition and "winter" to those systems designed only to maintain elevated temperatures and humidities during cold weather. Systems designed to afford the stipulated control during all seasons of the year are usually designated as "year-round" air-conditioning systems.

333. Psychrometric Formula. A major step toward the development of air conditioning as a science was made by Willis H. Carrier with the derivation of the rational psychrometric formula (*Trans. ASME*, 1911) which established the relationship between dry-bulb and wet-bulb temperature, both easily ascertainable, and other important properties of the air such as density, relative and absolute humidity, vapor pressure, and total heat, the exact valuation of which is essential to the solution of problems involving the exchange of heat and humidity between air-conditioning spparatus and conditioned spaces and their contents. The rational psychrometric formula is the basis for numerous psychrometric charts.

334. Psychrometric Terms. Dry-bulb temperature is the commonly observed temperature of the air as indicated by a conventional thermometer not affected by the relative humidity of the air.

Wet-bulb temperature is the temperature of adiabatic saturation of air or the temperature at which air would normally saturate without any change in its heat content. Wet-bulb temperature is, therefore, also a measure of the total heat content of air; i.e., at a given wet-bulb temperature the total heat is constant regardless of changes in the dry-bulb temperature.

Relative humidity within the normal range is the percent ratio of the weight of water vapor actually contained in a unit of space to the weight of water vapor that the same unit of space would contain if fully saturated at the same dry-bulb temperature.

Absolute humidity is the actual weight of water vapor in air at any given condition and is usually expressed in grains per cubic foot or grains per pound of dry air.

Dew point is the temperature of saturation of air, or the point beyond which any further reduction in temperature would result in condensation.

Effective temperature is an arbitrary index of the degree of warmth or cold felt by the human body in response to temperature, humidity, and air motion. Experimentally determined, the scale has been fixed by the temperature of saturated air which reproduces the sensation of warmth felt at other combinations of conditions.

Sensible-heat effect is the exchange of heat between air and its surroundings or apparatus that changes the dry-bulb temperature without affecting the moisture content. It is convenient to consider sensible heat as dry heat.

Latent-heat effect is the exchange of moisture between air and its surroundings or apparatus by evaporation or condensation resulting in a change in the moisture content of the air with consequent change in the total latent heat of vaporization represented by the moisture in the air. It is convenient to consider latent heat as humidity.

335. Comfort Cooling. Investigations carried on under the auspices of the ASHVE indicate that effective temperatures varying from 70 to 73°F are acceptable inside conditions for peak summer weather depending upon length of occupancy and for normally clothed, normally active commercial workers. Table 20-43 gives approximate dry-bulb temperature and relative-humidity combinations within normal range which produce effective temperatures within the stated range with air motion 15 to 25 ft/min.

Table 20-43. Inside Design Conditions*

Effective temperature	Relative humidity		
	45 %	50 %	55 %
	Dry-bulb temperature, deg		
73	79½	79	78
72	78½	77½	77
71	77	76½	76
70	76	75	74½

* Data from ASHVE Comfort Chart; *Trans. ASHVE*, 1932, Vol. 38, p. 410.

Inside design conditions of 78 to 80°F dry bulb with 45 to 55% relative humidity, depending upon the character of occupancy, are accepted as standard inside design conditions against peak outside conditions for commercial comfort-cooling applications in almost all parts of the country.

It is seldom desirable to design for inside temperatures in excess of 80°F. It is generally impracticable to reduce the humidity sufficiently to produce effective temperatures within the desired range at dry-bulb temperatures much in excess of 80°F.

336. Industrial-process Air Conditioning. In industrial or process air conditioning, inside design conditions are frequently dictated entirely by the product requirements. In confectionery plants where chocolate is handled, it is usually not desirable to have temperatures in excess of 65°F (dry bulb), which is far below the comfort range. In chemical or pharmaceutical plants where deliquescent chemicals are handled, lower than average humidities may be required for satisfactory production. Many products can be satisfactorily processed at widely varying conditions as long as the condition is maintained constant and the process adjusted to it.

Where conditions are established by product requirement, design is usually directed toward maintenance of satisfactory working temperatures. Investigations to determine the limits for optimum health and efficiency of the industrial workers indicate that conditions should not exceed 80°F effective temperature (Pars. **334** and **335**), i.e., approximately 82.8°F d.b. at 80% RH, 84.4°F d.b. at 70% RH, 86.2°F d.b. at 60% RH, 88.2°F d.b. at 50% RH, and 90.3°F d.b. at 40% RH. Lower effective temperatures are required where perspiration must be prevented to avoid contamination or staining of products during handling.

There has been a marked increase in the use of complete air conditioning, including refrigeration, even in industrial operations where atmospheric control is not required by the process involved. An example is the trend toward larger manufacturing areas under single roofs, where windows cannot be depended upon for light or ventilation. In many cases it has proved less expensive in both first cost and operating cost to provide full air conditioning with refrigeration rather than ventilation.

Table 20-44. Maximum Summer Design Conditions*

City	D	W	City	D	W	City	D	W
Boston	88	74	Cleveland	89	75	Seattle	80	65
New York	93	76	Chicago	91	76	San Francisco	77	62
Philadelphia	90	77	St. Louis	94	78	Los Angeles	90	70
Washington, D.C.	92	77	Atlanta	92	77			
Jacksonville	94	79	Birmingham	94	78			
New Orleans	91	80	Dallas	99	78	Denver	90	64
Houston	94	80	Phoenix	106	76	Salt Lake City	94	66

* ASHRAE Guide and Data Book, 1965, Chap. 27, Table 1.

337. Winter Heating. Winter heating systems for normally occupied interiors are almost universally designed to maintain 70°F inside, with outside conditions assumed at 10°F above the lowest recorded temperature for the district in question.

338. Maximum outside design conditions are usually taken at values which will not exceed 2.5% of the total hours during June, July, August, and September. Accepted maximum design dry-bulb (D) and wet-bulb (W) temperatures in degrees Fahrenheit for larger cities representative of various climatic divisions of the United States are given in Table 20-44.

339. Calculation of the cooling load involves consideration of all possible internal as well as external heat gains not only for transmission and infiltration but also from sunlight radiation.

340. Sunlight radiation for 40°N lat on July 21 attains maximum values in Btu per square foot per hour transmitted by unshaded glass for variously oriented surfaces as follows: East, 200 at 8 A.M.; South, 91 at 12 noon; West, 200 at 4 P.M.; horizontal 258 at 12 noon.[1] Major consideration must be given to sun-exposed glass. Of the total impinging radiation, the amount passing through the net area of the window is approximately as follows: unshaded, 100%; half covered by buff shade, 70%; fully covered by light-finish venetian blind, 70%; equipped with canvas awning, 30%.[2] Although sunlight-radiation effect upon walls and roofs can be calculated, allowance is usually made for this effect by calculating heat transmission at a temperature above the design outside temperature. A substantial temperature head (usually 50°F) should be allowed in calculating transmission through the roof surface due to long hours of exposure to the sun. Heat from sunlight contributes only to the sensible-heat load.

341. Transmission of heat through the structural surfaces of buildings or spaces is directly proportional to the temperature difference between the two sides of the surface and to the thermal conductivity of the material or materials forming it. Overall coefficients for a number of common structural elements are given in Table 20-45. The coefficient U is expressed in Btu per hour per square foot per degree Fahrenheit temperature difference across the surface and is based upon an outside-surface coefficient for 15-mi/h wind velocity.

342. Total heat transmission through any exposed building surface may be computed by the formula

$$H = A \times U(t_o - t_i)$$

where H = sensible heat in Btu per hour, A = area of the surface element in square feet, U = coefficient of heat transmission of the surface, and t_o and t_i = respectively, the outside and inside temperatures in degrees Fahrenheit. This formula is used for computing heat loss from a space as well as heat gain to it.

343. Heat from Occupants. Every occupant of a conditioned space contributes both sensible and latent heat to the atmosphere, the relative proportions being dependent upon the dry-bulb temperature and the total quantity upon the state of activity of the individual (see Table 20-46).

[1] ASHRAE Guide and Data Book, 1965, Chap. 27, Table 12.

[2] Abstracted by permission from *Research Repts.* 975, *Trans.* ASHVE, 1934, Vol. 40; and 1180, *Trans.* ASHVE, 1941, Vol. 47.

Table 20-45. Coefficients of Heat Transmission U*
(Surface composition)

	U
Single window	1.13
Double window	0.56
12-in brick wall, no interior finish	0.35
12-in brick wall finished with ¾-in plaster on metal lath, furred	0.25
10-in hollow-tile wall, stucco exterior, no interior finish	0.33
10-in hollow-tile wall, stucco exterior finished with ¾-in plaster on metal lath, furred	0.24
10-in concrete wall, no interior finish	0.61
10-in concrete wall finished with ¾-in plaster on metal lath, furred	0.36
Wood siding, 1-in sheathing, studs, ¾-in plaster on metal lath	0.26
Wood siding, 1-in sheathing, studs, ¾-in plaster on metal lath, 3-in rockwool fill	0.07
Double partition, metal lath and ¾-in plaster on both sides of studding	0.39
4-in hollow-tile partition, plastered on both sides	0.37
Yellow-pine flooring (25/32 in) on joists, no ceiling	0.45
Yellow-pine flooring (25/32 in), ¾-in plaster on metal-lath ceiling below joists	0.31
6-in concrete slab floor, ¼-in asphalt tile, no ceiling	0.60
6-in concrete slab floor, ¼-in asphalt tile, ¾-in plaster on metal lath hung or furred ceiling	0.38
1-in wood flat roof, built-up roofing, no ceiling	0.48
1-in wood flat roof, built-up roofing, ¾-in plaster on metal-lath ceiling below joists	0.33
1-in wood flat roof, 1-in rigid insulation under built-up roofing, no ceiling	0.21
1-in wood flat roof, 1-in rigid insulation under built-up roofing, ¾-in plaster on metal-lath ceiling below joists	0.17
4-in concrete-slab flat roof, built-up roofing, no ceiling	0.70
4-in concrete-slab flat roof, 1-in rigid insulation under built-up roofing, no ceiling	0.24
4-in concrete-slab flat roof, built-up roofing, ¾-in plaster on metal-lath ceiling, hung or furred	0.41
4-in concrete-slab flat roof, 1-in rigid insulation below built-up roofing, ¾-in plaster on metal-lath ceiling, hung or furred	0.18

* ASHRAE Guide and Data Book, 1965, Chap. 24, Tables 5 to 18.

Only the sensible-heat fraction affects the temperature of the conditioned space, since latent heat from evaporation of body moisture appears as vapor at room temperature.

Table 20-46. Heat Loss from Average Men, in Btu per Hour as Sensible Heat SH and Latent Heat LH*

Condition of activity	Total heat loss, Btu/hr	Dry bulb temperature, deg F					
		70		75		80	
		Segregated heat loss, Btu/hr					
		SH	LH	SH	LH	SH	LH
Seated at rest	400	300	100	260	140	220	180
Light work	660	350	310	290	370	220	440
Moderate work	850	430	420	360	490	270	580

* Data from Charts of Research Report No. 908, *Trans. Amer. Soc. Heating and Ventilating Engrs.*, 1931, Vol. 37, p. 541.

344. Electrical Heat. The heat equivalent of the electrical input to all lights and electrical heating or power appliances used within a conditioned space contributes to the atmosphere heat which must be removed by the air-conditioning apparatus. In general practice, the approximate heat equivalent of 3400 Btu/(h)(kW) input to lights and appliances and 2500 Btu/(h)(hp) of motor capacity is used in calculating heat gain from this source. In considering heat from appliances and motors, load factor should be carefully investigated, since appliances or motors seldom run continuously at full load. Unless appliances are used for water heating or evaporation, the entire load from electrical sources will be sensible heat.

345. Process Heat. Heat from gas appliances is the most common form of process heat. Average heat input to common gas appliances in Btu per hour are as follows: 14-in coffee urn, 7500; 12-in coffee urn, 5000.[1] Heat output will be approximately

[1] ASHRAE Guide and Data Book, 1965, Chap. 27, Table 27.

75% sensible heat and 75% latent heat. Consideration should be given to load factor of appliances.

346. Ventilation and Infiltration. Air quantity for ventilation purposes is usually based upon the number of occupants of the conditioned space. Ventilation is required chiefly to overcome normal odors. Average accepted practice calls for the introduction of at least 10 ft³/min of outside air for each nonsmoking occupant and 20 ft³/min for each smoking occupant. For commercial interiors, such as restaurants and drugstores, where moderate smoking occurs, 15 ft³/min per person is usually allowed. In theaters, where ceiling heights are usually liberal, where there is no smoking and peak occupancy seldom occurs for long periods, an allowance of 7½ ft³/min per peak occupant is considered ample. Air positively exhausted from a space must be deducted from the positive ventilation supply in considering possible infiltration. If the exhaust exceeds the supply, infiltration will naturally be induced. Air for ventilation should be first brought to the conditioning apparatus and conditioned before introduction to the conditioned space.

Infiltration varies with the construction, exposure, and size of a conditioned area as well as with door traffic and location of doors. The following quantities are suggested as a general guide for buildings of average construction: cubical contents 0 to 5,000 ft³, one complete change of volume in 30 min; 5,000 to 50,000 ft³, 40 min; 50,000 to 100,000 ft³, 60 min; 100,000 to 200,000 ft³, 90 min; over 200,000 ft³, 120 min. Infiltration in cubic feet per minute for load calculation and comparison with positive supply in cubic feet per minute may be determined by dividing cubical contents of the space by the time factor for that space magnitude.

The sensible-heat component of ventilation or infiltration may be computed approximately from the formula

$$H_S = Q \times 1.07(t_o - t_i) \qquad \text{(Par. 348)}$$

where H_S = sensible heat in Btu per hour, Q = cubic feet per minute of air, and t_o and t_i = respectively, outside and inside temperatures in degrees Fahrenheit.

The latent-heat component of ventilation or infiltration may be computed approximately from the formula

$$H_L = Q \times 0.675(h_o - h_i) \qquad \text{(Par. 349)}$$

where H_L = latent heat in Btu per hour, Q = cubic feet per minute of air, and h_o and h_i = outside and inside absolute humidities, respectively, expressed in grains per pound of dry air (see a psychrometric chart).

Only infiltration air entering the conditioned space directly contributes to the internal heat load. Ventilation air conditioned before introduction to the space contributes to the load upon the conditioner and cooling apparatus only.

347. The internal sensible-heat load will be the summation of the sensible heat from transmission, sunlight radiation, occupants, lights, electrical devices, process heat, and infiltration. The **internal latent-heat load** will be the summation of the latent heat from occupants, process heat, and infiltration. A factor of safety of 10% is often added to these figures to allow for inaccuracies and, in the case of sensible-heat load, to cover the power input for air circulation. The **total heat load** on the conditioner or cooling apparatus will be the sum of the internal sensible- and latent-heat loads plus the sensible- and latent-heat load of ventilation air. Grand-total heat is usually expressed in tons of refrigeration effect (see Par. **362**).

348. Air Quantity for Sensible-heat Absorption. The sensible-heat load usually serves as the first guide to the selection of proper air quantity. The specific heat of air within the normal air-conditioning range is approximately 0.241 Btu/(lb) (°F). One Btu will raise 4.16 lb of air 1°F, or approximately 56 ft³ of air 1°F. The air quantity required to absorb a given sensible-heat quantity may be determined by the formula

$$Q = \frac{H_S}{60} \times \frac{56}{t_i - t_d} = \frac{H_S}{1.07(t_i - t_d)}$$

where Q = air quantity in cubic feet per minute, H_S = sensible heat in Btu per hour, t_i = space temperature in degrees Fahrenheit, t_d = delivery temperature of air into the conditioned space in degrees Fahrenheit, 60 = minutes per hour, and 56 = cubic feet per Btu per degree Fahrenheit temperature rise. Modern diffusion-type outlets will permit satisfactory delivery of air into spaces of 10- to 12-ft ceiling at 60°F. With higher ceiling heights and good conditions, air may be introduced at 55 to 50°F.

349. Latent-heat Absorption (Dehumidification). For the quantity of air selected to meet the sensible-heat load by the method just suggested, the moisture content at which the air must be introduced into the space to absorb the internal latent-heat load may be approximately calculated by the following formulas,

$$h_r \text{ (grains per pound)} = \frac{H_L}{Q \times 0.675}$$

and

$$h_d \text{ (grains per pound)} = h_i - h_r$$

where h_r = humidity deficiency in grains per pound of dry air; h_d and h_i = absolute humidity in grains per pound of dry air at delivery condition and within the room, respectively; H_L = internal latent heat in Btu per hour; Q = air delivery in cubic feet per minute; and the factor 0.675 is calculated from the average density of air, the average latent heat of moisture in Btu per grain, and rate correction. The delivery-moisture condition thus calculated, together with the delivery temperature determined as in Par. **348,** will permit graphical determination of the required delivery-air conditions from a psychrometric chart. **Dehumidification** is usually accomplished by condensation of moisture in the process of cooling the air.

350. Cooling-dehumidifying Apparatus. Two types of apparatus are used for cooling and dehumidifying: (1) the spray-type dehumidifier and (2) the surface cooling coil. In either, cooling and dehumidification may be accomplished by cold fluid at the proper temperature from any source, e.g., cold water from wells or ice tanks or water cooled directly by refrigeration apparatus or by the direct expansion of refrigerants in cooling coils.

Spray Type. In the spray-type dehumidifier, air to be conditioned is passed through a spray chamber into which a large volume of finely divided water is introduced. Spray-type dehumidifiers are of two general types: (*a*) those in which water cooled externally is introduced at a reduced temperature and (*b*) those in which water is sprayed over coils cooled by fluid from an external source.

Spray-type dehumidifiers have the advantage of permitting evaporative cooling (see Par. **353**), of cleaning the air of siliceous dust, and of absorbing some gases and odors.

Surface Cooling Coils. Surface cooling coils are widely used for cooling and dehumidification. Cooling of surface coils is usually accomplished by direct expansion of refrigerant, though frequently cold water or brine is used.

Common designs include (*a*) pierced-plate types in which the tubes pass through closely spaced thin metallic plates, (*b*) helical extended surface in which a ribbon of metal crimped or slit to accommodate curvature is wound in a helix about the central tube, and (*c*) the type in which fins are extruded directly from the tube body by rolling.

351. Heating Surface. Heating surface is generally similar to cooling surface in construction. Extended cast-iron surface is still used to some extent. Steam is an accepted heating medium, though use of hot water is increasing as it facilitates control.

352. Humidification and Humidifiers. In many comfort cooling applications and some industrial-process operations which have low winter heat loads and require only moderate humidification during the winter this is accomplished by small sprays which introduce finely divided water droplets into the air stream usually directly before or after the heating coil so that evaporation is stimulated. Pan-type humidifiers in which water is evaporated into the air stream by steam coils or direct steam sprays may also be employed. It should be noted, however, that these procedures add to the loading within the space and therefore should be employed only where there is a deficiency in heat gain or a heating requirement.

In large commercial or industrial operations employing spray-type air washers for summer cooling and dehumidification humidity control is usually achieved by maintaining a constant dew-point temperature and then allowing the air to rise to the dry-bulb temperature, resulting in the desired final relative humidity (Par. **334**) in absorption of heat gains within the space. It should be noted that the majority of industrial and many large commercial applications involve internal loading from power, lights, etc., that require the application of cooling during all seasons of the year when there is normal occupancy or operation. During winter this cooling may be obtained by outside air, as described subsequently for evaporative cooling, but there is no heating requirement during normal service.

The major shortcoming of the dew-point control method for humidity maintenance lies in the fact that there must be a fixed relationship between the maintained dew point and the final dry-bulb condition within the space to achieve the desired relative humidity. Thus there is a fixed relationship between the load and the temperature rise in supply air which may be taken to absorb this load, and this relationship may result in the requirement for excessively high circulating air quantities, particularly if high relative humidities must be maintained as is common in many textile-mill operations.

Accordingly, in such operations where high relative humidities must be maintained with high continuous internal heat loads common practice is to employ a *split system* in which the dew point of the circulating air is maintained at an arbitrarily low level which would normally result in insufficient humidification after absorption of the heat load, but the humidity is increased within the space by the direct application of water droplets from compressed-air or motor-operated mechanical atomizers. The atomized moisture introduced directly into the space is in water form and in the process of evaporation not only increases the relative humidity but absorbs sensible heat from the space in the process of being converted from water to vapor at the room condition. The moisture thus introduced does not add anything to the heat load within the space and further, by the process noted above, increases the sensible-heat-absorbing capacity of the primary supplied air, thus making it possible to perform the required cooling with a substantially lower volume of circulating air than would be needed for the straight dew-point method.

Atomizers are also frequently used as prime means for humidification in areas such as textile mills. Employed in this fashion or with the central-station system they also afford an excellent facility for direct humidity control in zones within a major area.

353. Evaporative Cooling. During intermediate seasons in normal climates or in climates such as prevail in dry mountainous country where temperatures are frequently excessive but humidities continuously low, evaporative cooling systems are very effective. Air is saturated adiabatically and thus assumes the wet-bulb temperature (see Par. **334**). Sensible cooling is thus done at the expense of increased humidity. Calculations for evaporative cooling systems are identical with those for other cooling loads except that 100% outdoor air is used.

Commercial evaporative coolers are similar to spray-type washers or humidifiers or spray-type dehumidifiers. The spray water is continuously recirculated by means of a pump.

354. Fans for Air Conditioning. Propeller-type fans used in cooling or heating units and normal ventilation service are usually driven by direct-connected motors, since their operating speed is within a suitable range. Most propeller-type fans require only fractional-horsepower motors, and capacitor-type single-phase motors, single- or multispeed, are usually provided as standard by manufacturers.

Axial flow fans offer good efficiency and pressure characteristics and are used extensively for *industrial* air-conditioning services in all capacity ranges. While their total sound output compares favorably with equivalent centrifugal fans, noise generation is at a high frequency and is relatively difficult to eliminate. Accordingly, they are not widely employed for *commercial* applications.

Centrifugal-type fans with volute housings are most commonly used in commercial air-conditioning work. Fan speeds are generally moderate to low, depending upon the air quantity to be circulated, and system resistance will vary, depending upon the design of duct work, apparatus, etc. Thus centrifugal fans are almost universally belt-driven, since this permits the use of economical a-c motor speeds, and since fan speed may be selected to fit the estimated requirement or readily changed on the job if necessary. Where single-phase a-c motors are used for fan drive, as on residential or small commercial air-conditioning applications, the capacitor-type motor is favored because of its quietness. Polyphase squirrel-cage motors are widely used for fan drive, since the average-size air-conditioning installation involves sufficient total motor load to justify polyphase supply. Regular squirrel-cage motors perform satisfactorily in most fan service. Two-speed single-phase or polyphase fan motors are sometimes used for year-round air-conditioning systems where circulation of air need not be so high in winter as in summer.

355. Filters. The most commonly used is the **unit-type viscous impingement filter,** which consists of a cell with a fibrous filler or filtering mat impregnated with an adhesive to which dust adheres upon impingement.

Dry-type unit filters are those in which the air is cleaned by straining action in passing through felt, cloth, or paper. Because of the density of the filtering media, large filtering area must be provided to keep resistance low.

Unless unit-type filters are carefully maintained, they may become so clogged as to restrict air circulation materially and affect adversely the operation of the system.

With the increase in size of systems in service and the increase in cost of maintenance labor, extensive development of automatic filtration equipment has occurred since 1960. One widely used design employs roll-type filter media which are fed clean onto a movable screen located across the air stream, are conveyed across the air stream, accumulating its dirt load, and then are taken up as a dirty roll on the opposite side of the filter elements. Since a substantial area of media can be provided in the roll, only occasional replacement of clean rolls and removal of dirty rolls are required. For normal dust loads a deep fibrous mat impregnated with adhesive is normally employed. For lint loads and similar dirt loads which will not penetrate such a mat and therefore cannot be stored in it, thin paper or glass-fabric media are generally employed. Electric drive is used to convey the filter media across the air stream, either operated on a fixed time cycle or arranged to automatically advance the filter media as the pressure drop across the unit increases owing to the accumulation of dirt. Other automatic filters employing fixed-media screens with electrically operated vacuum-cleaning devices which traverse the screen area are also available and find application, especially in areas where lint accumulation is the major problem.

The mechanical-filter types listed are efficient in removal of large dust particles which represent the greatest weight percentage of normal atmospheric dirt. Their inability to remove small particles which have little tendency to settle under gravitational action has resulted in the development of **electrostatic dust-precipitation equipment** which is efficient in removing even the smallest-sized particles such as those composing the visible elements of tobacco smoke.

The dirt accumulated on electrostatic filters must be periodically removed by cleaning. Large electrostatic units are often equipped with automatic washers to perform this operation, and one device employs collector plates mounted on conveying chains similar to the automatic filters described above which pass the plates continuously through an oil bath to clean them.

356. Automatic Control. Most control devices depend upon physical, thermal, or hygrostatic force for their operation. For example, thermostats frequently are operated by the distortion of bimetallic strips caused by the unequal expansion of the metals with increase in temperature or by the increase in pressure with rise in temperature of a liquid or gas contained in a closed element. Hygrostats generally depend upon the expansion or contraction of human-hair strands with changes in humidity to perform the basic control operation, though silk, paper, or wood is used in certain applications. The thermostatic and hygrostatic elements usually operate other devices to perform the control operation.

357. Air-conditioning control equipment may be divided into three separate classes: (*a*) self-contained, (*b*) pneumatic, and (*c*) electrical.

Self-contained devices are usually composed of thermostatic-fluid elements directly operating steam, water, or brine valves. These devices are not used where close control is necessary or where instantaneous tight shutoff is required. However, they are simple, cheap, and reliable and inherently give modulated control.

Pneumatic-control systems are those in which the thermostatic or hygrostatic element controls the rate of air leakage from a compressed-air system. The air pressure in the control system is used to operate diaphragm motors controlling valves and dampers. The spring loading of these motors can be varied so that within the given control-pressure range a whole sequence of separate valve and damper operations can be accomplished. Various types of relay and compensating arrangements are used to overcome inherent tendencies of such systems to override or "hunt" if set for high sensitivity or to give too wide a control range if set for low sensitivity.

Electrical Control. The most common electric thermostatic or hygrostatic controls are of the **snap action type;** i.e., the condition change simply serves to cause the device to make or break an electric circuit, which in turn controls the major element or elements to be operated. Because of contact difficulties, line-voltage controllers generally use mercury-tube contactors, which limit the accuracy of control. All wiring to the control must be run in conduit. Low-voltage controls (24 V a-c) which operate major motor drives or other elements through relays are more common; the light, open contacts permit accurate control, and open wiring (usually metal-clad) is allowed.

Electric controls employing thermocouples or resistance thermometers for sensing temperature conditions and various electrical leakage devices to determine humidity conditions are beginning to find extensive application for air-conditioning control. However, for large-system work in the actuation of valves and dampers it is hard to find a satisfactory substitute for the pneumatic or hydraulic operator, so that electric or electronic control circuits usually act through other relays to actuate the major system elements by means of such operators. Nevertheless, a tremendous increase in the use of electric and electronic equipment is anticipated, particularly because of the current trend to centralizing main control observation, recording, and actuating points as a result of the complexity and extent of the systems being installed in large commercial and industrial operations.

358. Air-conditioning Systems. Central systems are usually assemblies of air-conditioning apparatus designed to afford the necessary atmosphere control and may be arranged with (1) a single fan and single apparatus to care for one space, (2) two or more fans and a single apparatus to care for two or more zones, (3) a single fan blowing through warm and cool air chambers with mixing dampers delivering properly proportioned air to two or more zones, (4) a single apparatus and central fan delivering highly conditioned air to individual recirculating fans in two or more zones, (5) a single central apparatus and fan with supplementary cooling or heating equipment in the ducts leading to two or more zones.

The wide application of air conditioning to multiroom, multistory commercial structures presents a number of problems which have resulted in the development of a number of new systems and techniques. Among systems in common use are the following:

a. The fan-coil system which employs individual units served with hot or cold water from a central circulating system or zone.

b. The high-pressure induction unit system where highly conditioned air at a high velocity is used to aspirate the flow of room air through localized heat-transfer coils supplied with hot or cold water from an external source, the primary air caring for ventilation and dehumidification while the secondary coil cares for the major sensible-heat gain or loss within the zone.

c. The low-pressure hot and cold plenum system in which each zone is supplied from a central unit or apparatus where warm and cold air are mixed as required to meet the zone demand.

d. The high-pressure double-duct system in which warm and cold air from central apparatus served through parallel trunk supply systems are mixed at a local distributing point in each zone through a sound-deadening mixing-valve device subject to control from the immediate zone.

e. The single-effect high-pressure interior system for substantial areas with relatively constant load characteristics with control for various sections by manual or automatic volume regulation, etc.

359. The unit air conditioner is a factory assembly within a suitable casing of component apparatus necessary for atmosphere control which may include filters, sprays, cooling or heating coils, fan motor, dampers, etc. Unit air conditioners which include in the same casing the refrigerating system are defined as **self-contained unit air con-**

ditioners. **Direct units** are those designed for location within the conditioned space. **Indirect or remote units** are those which are designed for location outside the conditioned space. **Free-delivery units** are those with fixed fan characteristics designed for use without air ducts.

Self-contained room air conditioners designed for mounting in a window or special opening through an outside wall have been highly developed as a plug-in appliance requiring only electrical connection. These units consist of filter and cooling coil (see Par. **368**), conditioned air fan, and outlet and inlet grilles, together with refrigeration compressor, motor and drive, condenser air fan, and air-cooled condenser, these latter elements located outside the building wall.

These units are available with total heat-removal capacities in Btu per hour ranging from approximately 5000 to 24,000 and with nominal power input approximately 1 hp/10,000 Btu capacity. Units up through 8000 Btu capacity are generally suitable for operation on 115-V single-phase service. Larger units usually require 230-V single-phase service, but this is available for such use in most areas. Units of this type almost universally employ air-cooled condensers.

Self-contained commercial units with the refrigeration system built in as part of the factory assembly are manufactured in various sizes with nominal total heat-removal capacities ranging from 2 tons to in excess of 30 tons (see Par. **362**). Units in capacities of 2 to 10 tons are frequently employed as direct units within the space, and many employ water-cooled condensers. However, the current trend is to air-cooled condensers located external to the space. A number of designs are available in capacities ranging from 5 to 30 tons designed for roof mounting and complete with air-cooled condensers and in many instances with electric-resistance heating coils or automatic gas or oil-fuel burner systems for heating. Such units are sometimes used in multiple to take care of fairly substantial commercial or industrial areas in single-story construction.

Self-contained and factory-assembled condensing units complete with either water- or air-cooled condensers and suitable for connection to remote air-conditioning units are currently available in capacities of 2 to 75 tons.

360. Drives for Air-conditioning Systems. There is no satisfactory substitute for electric-motor drive for the power auxiliaries of air-conditioning systems. Compression-refrigeration systems which are used in conjunction with the large majority of air-conditioning installations are primarily designed for electric-motor drive (see Pars. **372** through **375**).

REFRIGERATION

361. Refrigeration Cycles. All practical refrigeration cycles produce controlled cooling effect or low-temperature heat removal by the free evaporation in an enclosed chamber (**evaporator**) of a liquid (**refrigerant**) under pressure conditions that produce the desired evaporation temperature. The refrigerant liquid absorbs its latent heat of vaporization from the medium being cooled and in this process is converted to vapor at the same pressure and temperature. This vapor is conveyed to another chamber (**condenser**) in which the pressure is maintained at a level sufficiently high to permit condensation of the refrigerant by water or air at normal temperatures. The heat quantity abstracted in the condenser is the latent heat of condensation (or vaporization reversible) of the refrigerant fluid, together with the heat that has been added to the refrigerant in the process of conveying it from the evaporator pressure level to the condenser pressure level. The condensed refrigerant (liquid) is allowed to flow from the condenser, through suitable throttling valves, back to the evaporator to repeat the cycle.

362. The standard unit of refrigeration is the ton, which is considered as heat removal at the rate of 200 Btu/min or 12,000 Btu/h or 288,000 Btu/24 h. This rate of cooling is approximately equivalent to the average cooling effect obtained by melting 1 ton of ice in 24 h at 43°F (latent heat of fusion of ice is 144 Btu/lb).

363. Refrigeration Systems. Practical refrigeration systems differ only in the method used to convey the refrigerant vapor from the evaporator, or low side, to the condenser, or high side. In closed refrigeration systems three methods are used to accomplish this transfer: (*a*) mechanical compression or pumping of the refrigeration vapor, (*b*) chemical absorption of the refrigerant vapor at low pressure with subsequent transfer to the high side as solute in a solution, and (*c*) physical absorption of the refrigerant vapor at low pressure with subsequent transfer to the high side as adsorbed vapor in a solid. Related to systems (*b*) and (*c*) are absorption and adsorption dehumidification systems used in air-conditioning processes requiring drying of air without cooling it.

364. Compression-Refrigeration Cycle. In the compression-refrigeration cycle a mechanical compressor or pump is used to convey the refrigerant vapor from the (low) evaporator pressure to the (high) condenser pressure. Vapor is drawn from the evaporator, where it has been evaporated in performing the cooling work, and is compressed and delivered to the condenser where it is liquefied, its latent heat of vaporization being absorbed by the condenser-cooling medium, which is usually water or air at normal temperature. The liquefied refrigerant is collected in the bottom of the condenser or in a separate container called a "receiver" and from there is fed back to the evaporator through suitable throttling valves.

There are three types of mechanical compressor in common use: (*a*) the reciprocating compressor, (*b*) the rotary compressor, and (*c*) the centrifugal compressor. To these must be added the jet compressor, which, while not usually considered as a mechanical compressor, is quite similar in operation. The **reciprocating and rotary compressors** are positive-displacement compressors; i.e., each cycle of the compressing member definitely opens a fixed volume into which fluid flows and then closes or occupies that same volume, forcibly expelling the fluid. The centrifugal or jet compressors depend for compression upon the kinetic or velocity energy imparted to the fluid.

The **reciprocating compressor** is suited to the compression of relatively small volumes of fluid through a high-pressure range. The **rotary compressor** is suited to the movement of moderate fluid volumes through moderate pressure ranges. The **centrifugal compressor** is suited to the movement of large volumes of fluid through small pressure ranges. Ammonia, carbon dioxide, Freon 12, methyl chloride, and sulfur dioxide are almost always used with reciprocating compressors. Freon 21 is used with rotary compressors, while Freon 11, methylene chloride, and water are used with centrifugal compressors.

365. Refrigerants which are used commonly enough to merit classification by the National Board of Fire Underwriters are listed in Table 20-47. The Freon group is now among the most widely used of all refrigerants. These halogenated hydrocarbons are chiefly derived from methane (CH_4) by the replacement of hydrogen molecules with chlorine and fluorine molecules and include F-11, F-12, and F-22.

Those listed as inflammable or in toxicity classification groups 1 to 4 (mildly to seriously toxic depending upon concentration, exposure, etc.) include ammonia, butane, dichlorethylene, ethane, ethyl bromide, ethyl chloride, methyl bromide, methyl chloride, methyl formate, propane, and sulfur dioxide.

Those listed as nonflammable (or practically so) and less toxic than group 4 (nontoxic in normally possible concentrations) include carbon dioxide, dichlorodifluoromethane (F-12), monochlorodifluoromethane (F-22), dichlorotetrafluoromethane (F-114), monofluorotrichloromethane (F-11), and methylene chloride (dichloromethane). Water is also used as a refrigerant and is, of course, considered as nontoxic and nonflammable.

366. Refrigerant Properties. Physical and thermodynamic properties of the nine most commonly used refrigerants and comparison of the performance of these refrigerants with the ideal (Carnot) cycle for standard conditions are listed in Table 20-47.

Ammonia is cheap and easily available; its efficiency is so high that it is the practical standard by which other refrigerants are rated. For these reasons it is widely used for ice manufacture and food processing, e.g., ice-cream manufacture, meat packing, cold storage, and brewery refrigeration.

Freon refrigerants, which have satisfactory characteristics with respect to toxicity and flammability as well as stable operation characteristics within the normal range, have largely replaced carbon dioxide for air conditioning, etc., where toxic refrigerants present public hazard.

Methyl chloride, sulfur dioxide, and Freon refrigerants find broad application in commercial and domestic refrigeration service.

367. Effect of Operating Conditions on Compressor Capacity and Power Require-

Table 20-47. Comparison of Refrigerants
("Refrigerating Data Book")

Refrigerants	Displacement cfm per ton	Work, hp per ton	Coefficient of performance	Pressure of saturated vapor, psia,* at saturation temperature			
				5 F	40 F	86 F	100 F
1. Anhydrous ammonia (NH_3)	3.44	0.989	4.76	34.27	73.32	169.2	211.9
2. Carbon dioxide (CO_2)	0.96	1.840	2.56	331.9	567.8	1043	
3. Dichlorodifluoromethane (CCl_2F_2)(F-12)	5.81	1.002	4.70	26.51	51.68	107.9	131.6
4. Monochlorodifluoromethane ($CHClF_2$) (F-22)	3.60	1.011	4.66	43.02	83.72	174.5	212.6
5. Monofluorotrichloromethane (CCl_3F)(F-11)	36.32	0.927	5.09	2.93	7.02	18.30	23.60
6. Methyl chloride (CH_3Cl)	5.95	0.962	4.90	20.80	42.60	95.50	119.0
7. Methylene chloride (CH_2Cl_2)	74.30	0.963	4.90	1.17	3.38	10.60	13.25
8. Sulfur dioxide (SO_2)	9.09	0.968	4.87	11.81	27.10	66.50	84.50
9. Water (H_2O)	476.70	1.125	4.10	......	0.25†	....	1.93†

* To obtain gage pressures deduct 14.7 psi atmospheric pressure.
† These pressures given in inches of mercury absolute.
NOTE: Figures are based on standard conditions and dry compression and represent the theoretical performance of the refrigerants for these conditions based upon their thermodynamic properties. Figures for carbon dioxide and water are unfavorable under standard conditions, since 86°F condensing temperature is close to the critical temperature of CO_2 and 5°F suction temperature is below the freezing point of water.

ment. The capacity of any refrigerating compressor depends upon the mass of refrigerant vapor that it can convey from the low- to the high-pressure condition. The power requirement of the compressor depends upon the mass of vapor conveyed and the pressure differential between the evaporator and the condenser. For this reason the capacity and power requirement of any given compressor operating at constant

Table 20-48. Characteristic Variation in Compressor Capacity with Changes in Suction and Discharge Condition*
(Freon-12, Constant Displacement)

Varying suction condition (discharge constant, 105 F, 126.2 psig)			
Saturated suction gas		Compressor capacity, Btu/hr/bhp	Gross capacity of compressor
Temp, deg F	Pressure, psig		
0	9.2	5,900	Increases 375% through this range
10	14.7	7,070	
20	21.1	8,460	
30	28.5	10,380	
40	37.0	12,760	
50	46.7	15,800	
Varying discharge condition (suction constant, 40 F, 37.0 psig)			
Saturated discharge gas		Compressor capacity, Btu/hr/bhp	Gross capacity of compressor
Temp, deg F	Pressure, psig		
82	87.0	22,400	Decreases 33% through this range
90	99.6	17,900	
98	113.3	14,700	
105	126.2	12,750	
112	140.1	11,050	

* Data calculated from typical compressor ratings.

displacement vary widely depending upon the temperature and pressure conditions of the suction and discharge vapor. The magnitude of this effect is illustrated in Table 20-48.

It should be noted that although the unit power requirement is much greater for low suction temperatures, the capacity of a constant-displacement machine is greatly reduced owing to reduced density of the vapor and reduced volumetric efficiency, so that the gross power requirement may be far below that at higher suction temperatures. For this reason it is general practice to operate belt-driven compressors at higher speeds when used at lower suction pressures to obtain greater displacement and capacity within the safe power rating of the machine. Care must be exercised in selecting equipment to be sure that safe power limitations will not be exceeded during starting-up periods when abnormally high temperatures may be encountered owing to high refrigerant temperatures. This is a very important consideration in the design of ultralow temperature systems, especially those employing multiple compressors in cascade arrangement. Within the normal range of condensing temperatures, the change in vapor density is not so pronounced, and the chief effect of departure above the design condensing temperature is a reduction in the output of the compressor.

368. Evaporators. The largest classification of evaporators for direct air cooling is surface cooling coils (see Par. **350**). For cooling water or brine, simple arrangements of submerged pipe coils or plate surface are sometimes used for small applications, especially where refrigeration storage is desirable. More widely used with reciprocating compression equipment is the so-called dry-expansion liquid cooler in which refrigerant is expanded into the tube surface within a shell and water passed under forced circulation across the tube surface as directed by baffles. Some larger reciprocating compressor installations and all large-capacity centrifugal refrigeration systems employ a conventional shell and tube surface, with forced circulation of water through the tubes, which are mounted within a pool of evaporating refrigerant within the shell surface. The majority of such applications employ integrally finned tubing (see Par. **350**).

369. Expansion Methods. Two general types of refrigerant expansion are employed: (*a*) dry expansion and (*b*) flooded expansion.

In the **dry-expansion system** the refrigerant is introduced into the evaporator through an expansion valve or pressure-reducing valve and makes one pass through the evaporating surface going to the compressor or absorber suction line. Since any liquid refrigerant which leaves the evaporator represents a loss of cooling effect, care is taken to assure only dry gas leaving the evaporator. The constant-pressure valve has been largely supplanted by the thermal-expansion valve, which through the use of the self-contained thermostatic-valve principle (see Par. **357**) controls the flow of refrigerant to give a constant superheat at the outlet of the evaporator regardless of the load. Dry-expansion systems are widely used with expensive refrigerants, since only a minimum refrigerant quantity is required to feed the evaporator properly. The dry-expansion system also permits feeding evaporators from the top so that any oil carried in the refrigerant will flow through the evaporator and be conveyed back to the compressor.

The flooded-expansion system maintains through float-valve control a constant level of refrigerant in the evaporator. In coil-type evaporators this is accomplished by means of a surge chamber external to the coil, usually located above the top of the coil. Refrigerant is fed into the surge chamber, which is connected with the top and bottom of the coil and to the suction line. As liquid refrigerant passes up through the evaporator coil, its density is reduced owing to the formation of vapor bubbles, so that the refrigerant circulates rapidly through the coil to the surge chamber, where the gas bubbles are released to the suction line. The major disadvantages of the flooded system are the relatively large refrigerant charge required and the necessity of providing means for oil removal when refrigerants that mix readily with oil are used.

370. Condensers. The function of the condenser is exactly the reverse of the evaporator, i.e., to afford a rapid transfer of heat from the condensing refrigerant to the cooling medium. For this reason there is a great similarity between condensing and evaporating equipment. Air-cooled condensers are frequently used with small compression systems which are located in well-ventilated places and where the difficulty of obtaining water supply and drain connections is not commensurate with the operating economies realized through the use of water cooling. Air-cooled condensers are similar in construction to surface-cooling coils.

Because of the effect that high condensing pressures have upon the economy of operation of refrigeration plants, water-cooled condensers are widely used. It is customary to assume the condensing requirement of the average compression plant to be 15,000 Btu/ton·h, of which 12,000 Btu represents the cooling work, and 3000 Btu the energy applied in heat pumping. This amount of heat can be absorbed by 1 gal/min of water rising through 30°F, 2 gal/min through 15°F, or 3 gal/min through 10°F. The design of the condenser and the amount of water required will depend upon the available water temperature and the desired condensing temperature and pressure.

371. Water Conservation. The increased use of tap water for condensing purposes in air-conditioning and refrigeration installations has seriously taxed the **water-supply** and **water-disposal** systems of many cities, with the result that restrictions in the form of increased water rates for condensing purposes, taxes upon water disposal, or outright prohibition of the use of water for condensing purposes have been effected by many municipalities. This has resulted in the development of equipment for the conservation of water. The familiar **cooling tower** has been redesigned to meet requirements of downtown buildings as to size and appearance, and the spray pond and atmospheric cooling tower find broader application in the industrial field. With spray ponds and atmospheric towers usually 5 gal/(min)(ton capacity) is allowed with water temperature entering the condenser at 10°F above the maximum outside wet-bulb temperature (see Par. **338**). With forced-draft towers usually 3 gal/(min)(ton) will suffice with the same temperature limits.

Thoroughly accepted in this field is the **evaporative condenser**, which consists of an extended-coil condenser (see Par. **367**) over which air is driven by fans and which is continually bathed in a water spray. Efficient condensing effect is obtained by the combined effect of air cooling and evaporation of water by the hot condenser coils. Small sizes for capacities up to 5 tons use propeller fans and direct tap-water connection for spray. Larger sizes (condensers in excess of 250 tons capacity have been built) are quite similar to unit air conditioners and employ centrifugal fans and a recirculating-spray system with a centrifugal pump.

Maintenance difficulties attendant upon the use of either cooling towers or evaporative condensers have resulted in a vast increase in the application of air-cooled condensers for reciprocating refrigeration systems in capacities up to 200 tons and possibly higher. While the use of air-cooled condensers results in a considerable operating-power penalty under maximum temperature condition, the apparatus employed is simple and relatively maintenance-free, factors which frequently justify the power penalty resulting from their use.

372. Refrigeration Compressors. The design of **reciprocating compressors** has been profoundly influenced by development of the production-line internal-combustion engine as offered today for automotive and aircraft service. Multicylinder in-line arrangements of 2 to 8 cylinders are thoroughly accepted, as are V and W arrangements of 4 to 16 cylinders and radial arrangements of 3, 5, and 7 cylinders. Practically all machines are single-acting with enclosed crankcases. With multicylinder design, large capacities are obtainable with good balance and relatively low piston speed, so that it is not unusual for machines of up to 120 tons capacity, with forced-feed lubrication, to operate at speeds as high as 1,750 r/min. Speeds below 500 r/min (except with very large machines) are the exception rather than the rule. Capacity variation is obtained by unloading groups of cylinders or by bypass or clearance-pocket arrangements.

The overall application of refrigerated air conditioning to large industrial and commercial buildings requires capacities far beyond the range of the largest ice-making or storage refrigeration installations. These requirements have been met adequately by development in the **centrifugal refrigeration** system, which finds current application in single units from 200 to 1,800 tons or over in capacity. Drive motors will vary from a low of 0.85 hp/ton for industrial applications to a high of 1.05 hp/ton for commercial installations.

The original designs for centrifugal systems employed multistage compressors with shaft seals and operating at speeds from 3,500 to 7,000 r/min. Standard four-pole or six-pole open motors with step-up gears were customarily used for compressor drives. Wound-rotor motors affording 25% speed regulation were frequently employed for capacity control. However, since 1955 design has largely been directed toward hermetic systems in which the drive motor is enclosed in the gas passage with the compressor, and with this arrangement the drive is naturally limited to the use of the squirrel-cage induction motor. Control of compressor capacity is afforded by regulating dampers in the compressor suction connection. Common designs employ either two-stage compressors direct-driven by two-pole motors at approximately 3,500 r/min or single-stage compressors operating at 11,000 r/min or higher and driven through step-up gears by enclosed two-pole motors. One design employs an external frequency converter providing 300-c current for operation of a two-pole induction motor for direct compressor drive at approximately 17,500 r/min. Centrifugal compressors are still offered without the enclosed motor and with the shaft extended through the compressor housing with a suitable seal; as such they can employ the open-motor or turbine drive. However, the general design and control arrangement of the system is generally the same for either hermetic or open drives for manufacturing standardization. Even with open units induction-motor drives are generally selected, for though motor sizes are large, average experience indicates that even where power-factor penalties are involved power-factor correction can be obtained more economically by the use of capacitors in connection with induction motors rather than through the use of synchronous motors. Where closed motors are employed for compressor drive, they are cooled either by water jacketing or by an arrangement employing a bypass of refrigerant or refrigerant gas for motor cooling.

373. Compressor Drives. While conventional open motors and V-belt drives are generally employed for large-capacity low-speed reciprocating compressors, which are still offered, the bulk of the production of reciprocating equipment is now of the direct-drive design generally employing four-pole 1,750-r/min motors. This design permits an extremely compact arrangement which is well suited to the unit-system combinations now widely offered by the industry.

In both the reciprocating and centrifugal compressor fields there is still considerable contention as to whether the use of standard open motors or sealed hermetic arrangements is most acceptable. The use of standard open-motor drives necessitates the shaft seal, which offers some disadvantages. The use of the hermetic arrangement requires careful design of the electrical components and close control of the refrigerant atmosphere, which must be kept absolutely moisture-free if electrical problems are to be avoided. Obviously, if failure occurs, the repair of the hermetic motor is a far more serious problem than the repair of the open motor. On the other hand, if properly designed and installed the fully enclosed motor operates in a controlled atmosphere and may be less likely to give trouble. However, with either arrangement the motor is generally a standard two-pole or four-pole induction motor as far as the basic electrical design is concerned.

Reciprocating systems generally employ means of unloading by holding suction valves open to afford capacity control in operation and are similarly arranged to start unloaded. Centrifugal systems employ dampers for capacity control, which are kept closed during the starting cycle so that these units also start unloaded. However, the characteristics of the reciprocating system are such that the starting-torque requirement is fairly high. Accordingly, line start for drive motors is preferred wherever utility-company regulations will permit, and since these units are usually to approximately 125-hp maximum size and are frequently installed in multiple units of smaller capacity, line start is generally permissible. Where reduced voltage starting is required, autotransformer or resistance-type reduced-voltage starters are generally employed, though frequently these may be effective only as a means of reducing inrush, as in many instances full line voltage must be applied before the compressor will actually pick up the load.

Centrifugal refrigeration systems generally represent larger-capacity units than reciprocating systems, with motor sizes ranging from 75 to 2,000 hp or occasionally higher. However, their starting-torque requirement is very moderate, and it is entirely practical to place the compressor in satisfactory operation by use of the Y-Δ starting system, in which the compressor drive motor is connected across the 3-phase source in a Y-arrangement during the starting interval and switched to the Δ-connection when the compressor has reached practically full speed and before the compressor damper is opened, permitting the unit to assume load. With this arrangement the current during the starting period can be limited to approximately 130% of the normal full-load running current, and this arrangement is generally entirely satisfactory to power companies. This starting arrangement is almost universally employed with centrifugal systems except for relatively small units, which may be within a range suitable for line start.

374. Motor Voltage. All standard equipment is based on the assumption that 60-c current is available, though some manufacturers will still list a capacity rating of their units of 60-c design when employed on 50-c service. All systems involving motors up through 5 hp are generally designed for 230-V 60-c/s 1-phase service but in sizes 3 hp and above are equally available for 208/220/440-V 60-c/s 3-phase service, which is generally available in most urban areas. Equipment for practically the entire available capacity range is designed for service on 208-V 60-c/s 3-phase current, which is commonly available as urban low-voltage distribution. However, it should be noted that because of the rigorous nature of compressor service it is usually inadvisable to attempt to use 220-V equipment for the compressor drive where only 208 V is supplied. Many feel that good design dictates the use of 200-V motors for such service. For larger commercial or industrial applications which are provided with primary service through independent substations, 440-V 60-c/s 3-phase equipment is generally preferred, as it permits a substantial reduction in wiring cost. Currently a number of systems employ 480/277-V 60-c/s 1-phase four-wire service, where the higher single-phase voltage may be satisfactorily used for fluorescent-lamp operation. This arrangement affords considerable economy in circuit design but presents complications with respect to the use of standard 115-V 60-c/s 1-phase conventional power or office equipment, for which separate transformers and circuiting must be provided. In the textile industry extensive use is made of 550-V 60-c/s 3-phase service, and equipment is generally available throughout the entire capacity range for this voltage characteristic. In larger plant operations employing plant primary service at either 2,300 or 4,160 V these voltages are frequently used for large compressor drives, and most lines make equipment available for this voltage in the range from 200 to 2,000 hp, though economics would generally indicate 500-hp drives as being the lowest acceptable size for this voltage.

375. Reverse-cycle or Heat Pump. The reverse-refrigeration cycle is being introduced for the purpose of heating interiors with electricity (see Pars. **361** and **364**). Heat absorbed at a low-temperature level is pumped to a level sufficient to permit satisfactory heating with the expenditure of only the amount of energy necessary to perform the pumping work, which is also reclaimed at the high-temperature level. Under certain circumstances this cycle is entirely practical. Many installations have been made where well water at moderate temperatures (50°F±) is available as a low-temperature heat source, and these installations have shown coefficients of performance up to 400%, that is, 5 Btu of heating effect for 1 Btu of energy applied. However, the performance of the refrigeration machine is sharply limited as the temperature head between the evaporator and the condenser is increased (see Par. **367**), so that the practical possibilities of this cycle are affected not only by the cost of electrical energy compared with other fuels but by the level of the low-temperature heat source and the temperature that must be maintained within the interior.

Growing application of air conditioning to residences, particularly in the South, has accelerated activity in the heat-pump field. Since the basic refrigeration equipment is being installed for summer conditioning, its utility is increased by arranging it to handle at least a part of the winter heating requirements. Most new developments are directed toward the use of outside air as a low-temperature heat source as well as a means for providing condensing effect when the system is on the cooling cycle. Heating capacity of such system will drop fairly rapidly as outside temperatures drop. However, there is a balance point between the requirement for summer cooling and available capacity for winter heating which will permit a rationally designed apparatus to care for the full heating requirements on the reverse-cycle basis for substantial periods during the heating season. Naturally, the more temperate the climate in which the apparatus is employed, the greater proportion of the total winter heat requirement can be handled entirely by the heat-pump principle. Since peak heating demand is experienced only for short periods, usually during early morning hours, the current procedure is to supplement the available output of the heat-pump device by direct resistance heating. This represents a substantial spot load, but its period of use is relatively short, and under most utility rate schedules the overall economics of the operation are reasonable. However, it presents problems in power distribution. A moderate-sized residence in the South Atlantic area with a basic 6-kW requirement for summer or heat-pump operation may be supplemented with as much as 12 kW of additional resistance heating load for use during peak heating periods. This imposes a maximum load of up to 18 kW per residential unit upon the lines of the utility, with little possible diversity.

On the other hand, this load occurs at a time when other urban loads are at a minimum and when outside ambient-temperature conditions are at a very low level, so that distribution circuits, transformers, etc., can handle a substantial overload with no difficulty. Accordingly, utility companies are finding that they can handle this load with no serious problems and feel that it is attractive.

A number of ingenious arrangements employing the heat-pump principle or other heat-recovery systems have been incorporated in recent designs of installations for large commercial buildings. It is evident that the core sections of these buildings are subject to cooling requirements throughout the entire year, as they are shielded from exterior exposure, but subject to heat gains from lighting and occupancy. A number of systems have been employed which use refrigeration for these interior zones throughout the entire year but during the wintertime take the effluent heat from the condenser of the refrigeration system and apply this as a means of heating the periphery of the building which is subject to normal winter exposure. Thus, in effect, the excess heat from the interior core of the building is transferred to the exterior for heating purposes while affording control of the heat load within the core of the building.

Other arrangements provide for direct cooling of lighting equipment either by circulating water or by exhaust ventilation and transfer the heat thus directly recovered to the periphery of the building for heating purposes. There are obviously many possibilities of this nature which must be studied for each individual project. However, where this means of heating is used, it is dependent upon a continued supply of electrical energy to the space as a primary heat source. In some instances arrangement is made to turn on interior lighting during cold periods for the purpose of heating the premises, or where lights are turned off, the equivalent amount of electrical energy may be applied through resistance heating, thus affording a heat source with no increase in electrical demand. These arrangements frequently make full electric heating entirely practical and economical for many large buildings.

GENERAL REFRIGERATION APPLICATIONS

376. Ice Making. Before the domestic electric refrigerator reached almost universal use, ice for domestic and commercial refrigeration was almost always intermittently produced can ice, frozen in conventional 300-lb blocks in cans submerged in a brine bath by reciprocating refrigeration plants using ammonia as refrigerant. Hundreds of such plants remain in service, but their output is usually converted to crushed ice for commercial storing or shipping use or to packaged cubed or crushed ice for various domestic uses. Most new plants manufacturing ice for industrial or commercial use produce ice continuously and automatically as cube, flake, or sheet ice, which is generally employed in loose form but which may be compressed into blocks if required. Many self-contained ice-making and storage units of fractional-tonnage capacity have been developed and are widely used by markets, restaurants, hotels, and motels to provide ice for packaging and beverage cooling use.

377. Ice-cream Manufacture. An efficient, well-operated ice-cream plant should produce ice cream on the can system at 0.4 kW/gal of product, including power for manufacturing, light, hardening, and storage; many plants investigated have reported even better performance.

378. Cold-storage Warehouses. Most cold-storage plants run at a much higher yearly load factor than other refrigerating plants. On a storage-space capacity and kilowatthour-use basis, investigation of a number of storage houses showed a performance of 0.68 to 0.8 kWh/(ft^3 space) (year operation). Yearly load factors showed a variation from 46 to 65.5%.

Because of the wide difference in temperature required for storage of different goods, many plants use the brine-circulation system, although small modern plants make extensive use of direct expansion. Brine temperatures for cold-storage work are carried from about −15 to +32°F.

379. Quick Freezing. If freezing is carried on at a rapid rate, the resulting ice crystals will be relatively small. This principle applies to the crystallization of many substances. It was, moreover, the key to the successful preservation of many vegetables by freezing. In normal slow-freezing processes large ice crystals are formed which pierce the normal cell walls of the product being frozen, so that upon subsequent thawing all natural juices are lost. The small crystals formed in the quick-freezing process do not break the cell walls, so the thawed product has all the characteristics of the fresh product.

Small vegetables such as peas and beans may be frozen on cold plates. Some products are immersed or sprayed with low-temperature sodium chloride brine. Certain fruits may be frozen in low-temperature solutions of sugars and water. In any case the process is sufficiently expensive to justify the handling of only the finest products brought direct from the field at the point of perfection.

380. Locker Plants. There are in service in the United States several thousand small plants for the processing, freezing, and storing of meats and vegetables on a commission or rental basis. Most of these plants have been located in farming communities and afford the farmer an opportunity to have his products (especially meats) processed and stored for his subsequent use. A typical locker plant will have a processing room maintained at 35 to 40°F, a sharp freezer held at −10 to −15°F, and a storage space containing 200 or more storage lockers held at 10 to 15°F. Lockers will average 4 ft^3 and will therefore hold a substantial quantity of wrapped cut meat or vegetables.

381. Precooling Plants for Shipments of Perishable Foods. The modern method of shipping perishable foods, e.g., fruit, vegetables, fish, meats, and poultry, includes provision for precooling either the carrier or the goods to be shipped or both before shipment in order to ensure safe delivery of the product. Precooling plants are usually located in districts of intensive fruit and vegetable farming or on the seaboard, the object being to put the product in the best possible condition for shipment by retarding those factors that cause decay and spoilage. Crops can be left to ripen before harvesting, thereby obtaining a much more desirable product with a greater market value.

382. Brewery Refrigeration. Refrigeration is essential in the modern brewery. In the process of fermentation, heat and carbon dioxide are evolved. Since the rate of fermentation, which affects the quality of the final product, is dependent upon temperature, the temperature of the fermenting wort and the fermentation cellars is controlled at the desired point, usually below 50°F, by the use of refrigeration. After fermentation has progressed to the desired point, the beer is cooled to approximately 33°F and stored for a period ranging from several weeks to months in aging cellars which are maintained at approximately 33°F. Following proper aging, the beer is filtered, finally cooled, and placed in barrels or bottles.

The steam requirements of breweries are high, but it has been demonstrated that they are not coincidental with the peak refrigeration or other power requirements,[1] so that the brewing process is not well adapted to heat-balance schemes involving the use of steam for refrigeration drive or power production. For this reason electric-motor drive supplied from central-station service is favored by progressive breweries.

383. Refrigeration Storage: Moderate Temperature. Many refrigeration applications having high peak loads for short periods of time can be most economically handled by the use of relatively small refrigeration machines which are operated over long periods of time to store refrigeration effect to meet the high short-period requirement. Refrigeration effect may be stored in bodies of brine or water, though a relatively large storage volume will be required, since the water or brine can only absorb or give

[1] Wilkes, Gilbert, Purchased Power in Breweries; *Elec. World*, Feb. 2, 1935.

up its sensible heat. A more efficient storage method involves the formation of ice upon submerged evaporating coils (see Par. **368**). Every pound of ice thus formed absorbs and can release upon melting at 32°F, its latent heat of fusion, 144 Btu/lb.

This principle has been frequently used to improve the load factor of refrigeration systems used in conjunction with air-conditioning systems having relatively short use periods (restaurants serving only one meal a day, theaters with only evening showings, and churches or infrequently used auditoriums).

384. Refrigeration Storage: Low Temperature. A solution of water and salt (or, in general, of any two substances) has a certain concentration which results in the lowest freezing temperature. A solution of this concentration is known as a **eutectic mixture.** A brine with a salt content lower than this concentration will start to freeze above this minimum temperature, and as cooling progresses, pure ice crystals will freeze out, thus increasing the concentration of the brine until the eutectic concentration is reached. At this point freezing will take place at a constant temperature until all latent heat is removed, the resulting crystals being a mechanical mixture of salt and frozen water. If the starting salt concentration of the mixture is greater than the eutectic concentration, salt crystals will first freeze out until the eutectic concentration is reached.

The use of eutectic mixture as a means of storing refrigerating effect, i.e., latent heat of fusion, at a temperature below 32°F (the freezing point of water ice) is receiving wide attention. The eutectic mixture of water 76.7% and sodium chloride (common salt) 23.3% has a freezing temperature of −6°F, and a latent heat of fusion of 101.5 Btu/lb is available at this temperature. Plate evaporators charged with eutectic mixtures are frequently used in refrigerator trucks which are cooled at night to produce sufficient stored cooling effect to meet the next day's operation.

385. Automatic ice-making machines produce ice continuously by the freezing of water on refrigerated surfaces with subsequent continuous or intermittent removal of the film by mechanical means. The final product closely resembles crushed ice and may be used directly for packing purposes or may be compressed into blocks or briquets or extruded in desired shapes.

386. Industrial Refrigeration. Refrigeration has always found extensive application in the chemical and petroleum fields in the many special applications where the maintenance and control of temperature at low levels are required. Refrigeration is used to control the temperature of cutting oils in certain machine-tool operations, typically the operation of high-speed grinders and lappers. In this application, its use permits higher operating speeds and notably improves finish and accuracy of product dimension owing to constant temperature maintenance. It is used in the storage of aluminum rivets, which unless held at low temperatures after annealing will harden at normal room temperatures. It is used in the heat treating of certain special alloys whose characteristics are improved by reduction to low temperature. In this connection, the use of refrigeration holds promise of eliminating the necessity of long aging which is required for certain metal parts during processing at normal temperatures. It has been extensively used in cooling the electrodes of spot-welding machines.

387. Reference on Air Conditioning and Refrigeration

ASHRAE Guide and Data Book (biannual), New York, American Society of Heating, Refrigeration & Air Conditioning Engineers.

SECTION 21

ELECTRICITY IN TRANSPORTATION

BY

T. B. OWEN Missile & Space Systems Division, Douglas Aircraft Company, Inc.; Senior Member, Institute of Electrical and Electronics Engineers

GERALD E. RIGSBY Senior Design Engineer, Delco-Remy Division, General Motors Corporation; Member, Society of Automotive Engineers

W. E. JACOBSEN Industrial Sales Division, General Electric Company; Fellow, Institute of Electrical and Electronics Engineers; Member, Society of Naval Architects and Marine Engineers; Civil Member, American Society of Naval Engineers; Registered Professional Engineer (New York)

F. N. HOUSER Mechanical Editor, *Railway Age*; Managing Editor, *Railway Locomotives and Cars*; Member, American Society of Mechanical Engineers; Registered Professional Engineer (Pennsylvania)

CONTENTS

Numbers refer to paragraphs

SECTION 21

ELECTRICITY IN TRANSPORTATION

ELECTRICITY IN AEROSPACE

By T. B. Owen

AEROSPACE ELECTRIC SUBSYSTEMS

1. Electric Power in Aerospace Vehicles. From the first aircraft, electric power has been in use in some form; and with the accelerated development of both commercial and military aircraft, such use has expanded greatly. Then, with the development of the missile and its extension into space vehicles, the use of electric power has increased rapidly. Basically, the same criteria apply for aircraft, missile, and space-vehicle electric power generation and utilization; the difference occurs in the means of power generation and its application. In aircraft, a ready means of power generation has been available—the aircraft engine; in the missile and spacecraft, no such means generally exist. Again, on aircraft, the use of electric power was quickly applied to passenger or crew comfort items. This was not a requirement on the missile but has again become a prime item on manned spacecraft.

In this section, aircraft uses and development will be treated first, and then the divergencies due to missiles and spacecraft will be given, plus new concepts in this latter field.

AIRCRAFT ELECTRIC SUBSYSTEMS

2. Uses of Electric Power. Electric power is used in an airplane to increase the effectiveness or utility of the airplane over that provided by the basic airfoil, airframe, and propulsion means. Most of the functions performed by electricity which contribute to the improved effectiveness of aircraft may be included in the following major categories:

1. Lighting.
2. Instruments (flight, engine, position, etc.).
3. Actuators (flaps, landing gear, bomb doors, etc.).
4. Control systems (engine, propeller, warning, etc.).
5. Pumps and blowers.
6. Heating (deicing, cooking, special equipment, etc.).
7. Communication and navigation.
8. Offensive and defensive armament (turrets, radar, computers, etc.).
9. Flight control.

The many uses of electricity in present-day aircraft have required a capacity of 270 kW (360 kVA) on large bombers.

3. Types of Electric Power Subsystems. There are three principal types of power subsystems in current use:

1. For many years, the most extensively used system for both military and commercial aircraft was 28 V d-c. While inadequate for primary power on larger aircraft, 28-V d-c power is still required in current designs, particularly in executive and other small aircraft.

2. The 400-c 3-phase 200-Y/115-V system is that preferred, with some application

of higher voltages for generation and primary distribution. This system usually operates with paralleled generators.

3. Variable-frequency (usually 400 to 800 c) single-phase 115-V or 3-phase 200-Y/115-V systems are often used to supplement a primary 28-V d-c system. They have been considered for primary power, but the usage is not general. Load switching and lack of continuity of power make this system less desirable than the constant-frequency system.

Among other systems which have been used to a more limited extent are (1) 120-V d-c (115-V d-c is used widely by the British), (2) 12-V d-c for small private aircraft, (3) 400-c Δ-connected 115-V, and (4) higher frequencies such as 800, 1,600, or 2,000 c, single-phase or 3-phase.

Since practically all modern aircraft have metallic structures, one side of d-c and single-phase a-c and 1 phase of Δ-connected 115-V systems are grounded to save cable weight. The 200-Y/115-V system has 3-phase conductors, with the neutral grounded to permit application of single-phase loads between any phase conductor and ground.

4. Voltage and Frequency Ratings. The voltage ratings of aircraft electric apparatus are shown in Table 21-1. These ratings are taken from the latest revision of IEEE Standard 127 and represent general usage. Frequency ratings are shown in Table 21-2 and are taken from the same source.

5. Operating Conditions. Aircraft equipment must operate satisfactorily under the following conditions: (1) wide temperature range, (2) wide altitude range, (3) exposure to sand and dust, (4) wide range in humidity, (5) exposure to conditions promoting corrosion and fungus growth, (6) exposure to oil and other vapors, (7) explosion hazard, (8) vibration, (9) accelerations from landing shocks and maneuvers, and (10) changes in flight altitude.

The high degree of airplane mobility makes the full range of operating conditions possible on a single flight. Furthermore, this range is continually increasing, particularly for temperature, altitude, and vibration. The problem of equipment cooling has been aggravated by the ram-air-temperature rise encountered in supersonic aircraft

Table 21-1. Aircraft-apparatus Voltage Ratings

	Direct current‡			Alternating current		
Nominal system designation (volts)	14	28	120	115	115/200	230/400
Generators:						
Rated voltages	15	30	125	120	120/208	240/416
Voltage adjustment range, %	+0 to −15	+0 to −15	+0 to −15	115 ± 5	115/200 ± 5	230/400 ± 5
Continuous-duty devices:*						
Rated voltages	13	27	115	115	115/200	230/400
Voltage range, %	±10	±10	±10	±5	±5	±5
Intermittent-duty devices:*						
Rated voltages	12.5	26	115†	115	115/200	230/400
Voltage range, %	±10	±10	±10	+5 to −10†	+5 to −10	+5 to −10
Battery-operated devices (devices that must operate while generators are not in use):						
Rated voltages	11.5	23				
Voltage range, %	±25	25				
Emergency-duty devices:						
Rated voltages	....	....	....	115	115/200	230/400
Voltage range, %	....	....	....	+5 to −15	+5 to −15	+5 to −15
Dielectric tests (rms volts for 1 min at 60 c or 120% of value for 5 s):§						
Factory-test volts	1,500	1,500	1,500	1,500	1,500	1,800
Field test or retest before and after use (clean and dry only):§						
(75% of factory-test volts)	1,125	1,125	1,125	1,125	1,125	1,260

Apparatus is to function satisfactorily over the voltage and/or frequency ranges given (simultaneously), but with performance not necessarily in accordance with guarantees at rated voltage.

* For operation from a d-c voltage-regulated system. If operation is required from battery alone, use values for "battery-operated" devices.

† It is assumed that most 115-V wiring will be applied on the basis of thermal rating and that provisions for wider tolerances for voltage regulation on intermittent loads have not been made.

‡ Direct-current devices such as starters, etc., and items operated during starting must operate under conditions of much wider voltage ranges. These devices are not covered by these standards.

§ In accordance with recommendations of AIEE Standard 803 (now IEEE Standard 135), Proposed Test Code for Aircraft Equipment Insulation, July, 1957.

Table 21-2. Aerospace Equipment Steady-state Frequency Ratings (Alternating Current)

Nominal system frequency, c/s	400	2,000	3,200
Type I, standard	±5%	±5%	±5%
Type II, optimum	±1%	±1%	±1%
Type III, precision	±0.01%	±0.01%	±0.01%
Type IV, broad	±25%	±25%	±25%

Type I. Standard frequency is in accordance with MIL-STD-704, Military Standard, Electric Power, Aircraft, Characteristics and Utilization of, Oct. 6, 1959.
Type II. Optimum frequency represents current state-of-the-art capability.
Type III. Precision frequency represents close tolerance power for systems with high accuracy requirements which cannot be met by using power with wider frequency tolerances.
Type IV. Broad frequency represents an a-c generating system with broad frequency tolerances such as results from direct, or simple, coupling of an a-c generator to a prime mover which is not regulated for frequency control.
Unless otherwise stipulated, type I, standard frequency, is assumed to be applicable.

and the need for more compact designs. Generators and other equipment have been developed to accept 120°C cooling air, and liquid cooling is being used in a few cases.

6. Importance of Weight. Weight is important for all air-borne equipment. Each additional pound of equipment requires 5 to 10 lb additional in airplane structure, propulsion engines, and fuel. Stated otherwise, decreased equipment weight increases pay load. Authorities differ on the exact value of weight saving in equipment, but it appears that an increased investment of $50 can be justified for each pound of aircraft weight saved. Thus, each pound of equipment weight may be worth as much as $500.

The high value of weight also justifies a sacrifice in equipment life for achievement of lighter weight provided that reliability is not sacrificed. Typical required life is 1,000 h for military aircraft and 5,000 h or longer for commercial applications. For engine-mounted equipment, the no-maintenance interval may be matched with the engine-overhaul interval.

7. Parts of an Aircraft Electric Subsystem. The aircraft electric system consists of three well-defined parts: (1) generation (or power supply); (2) distribution; (3) utilization.

POWER-SUPPLY SYSTEM

8. Generator Prime Movers. Practically all aircraft generators which make up the main power-supply system receive their energy from the main engines either through mechanical shaft power or through compressor bleed air taken from gas-turbine-type engines. Figure 21-1 shows the various methods employed to extract accessory power from a typical jet engine. Auxiliary generators for ground or emergency operation may be driven by ram air or an auxiliary engine.

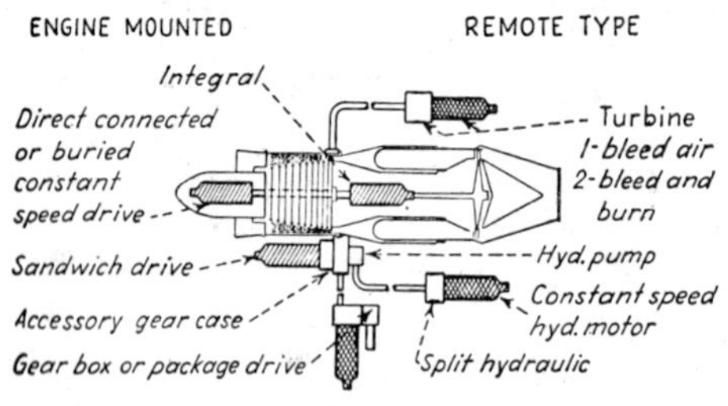

Fig. 21-1. Possible methods of extracting power from a jet engine.

a. Direct Engine Drive. Generators mounted directly on pads on the accessory gear case of reciprocating engines are usually limited in power because of limitations on overhung moment imposed by vibration. Larger units may be mounted directly on turbine engine pads or separately mounted and shaft-driven. Direct-driven generators often operate over the entire engine-speed range from idle to military power. This range averages 3 or 3.5 to 1 in reciprocating and older single-rotor jet engines. However, some newer jet engines and the high-pressure rotor of dual rotor engines have speed ranges as low as 1.5 to 1. Turboprop engines often have only 10% variation for all flight conditions, although ground idle may be as low as 50% of maximum speed.

b. Variable-ratio Transmission. In order to obtain constant speed from the main engines for constant-frequency a-c systems, continuously variable-ratio hydraulic transmissions are used. A differential-type transmission has been widely adopted, since only an amount of power proportional to the difference between input and output speeds is handled hydraulically. Thus, space and weight are saved. Transmissions are operated either on engine oil or on a separate oil system including an air-oil or fuel-oil cooler. Governors hold steady-state speed within ±¼ to ½% and are provided with a means of sensing difference in real load between paralleled generators and of dividing it within 10%. Transient recovery times of the order of 0.5 to 2.0 s are obtained.

c. Engine-compressor-bleed Turbine Drive. The pneumatic turbine is used where sufficient compressor bleed air can be obtained and where either (1) insufficient engine pads or space is available or (2) an overall performance study of various accessory systems shows it to be superior for a particular application. Among its advantages are (1) suitability for constant-frequency systems and parallel operation, (2) flexibility as to location and number of units, and (3) ability to operate all generators from a common duct system after an engine failure. Turbines generally operate at much higher speeds than the generators; thus a speed-reducing gearbox is required. Automatic controls maintain constant speed and divide real load among paralleled generators. Methods of speed control employed are (1) throttling, (2) partial arc admission, and (3) variable-area nozzle, in order of increasing complexity and improved performance.

It is generally necessary to make a rather detailed application study before using air turbines for generator drives. The large power-transmission weight associated with high-temperature air ducts may be partly offset by use of the same ducts for cabin pressurization, anti-icing, and pneumatic engine starting. Also, engine thrust and specific fuel-consumption penalties for compressor bleed vary widely with different engines and operating conditions. For these reasons, turbine-drive weights cannot be compared directly with hydraulic-drive weights. In general, the dry weight of turbine, gear, and control will be of the order of 2 lb/kW of generator rating. A further development which permits a reduction in bleed-air requirements is the "bleed-burn" turbine in which fuel is used to increase the available energy of the air consumed.

d. Auxiliary Engine Drive. Small reciprocating-engine generator sets have been used for many years to (1) make aircraft self-sufficient, i.e., provide starting power in absence of ground equipment; (2) provide power for servicing, lighting, etc., without operating the main engines (particularly in water-based aircraft); and (3) provide an emergency source of power under low-altitude flight conditions, including take-off and landing. Recent developments in small turbine engines have resulted in increasing use of this type in larger ratings and with specific weight greatly reduced to about 4 to 5 lb/kW for the prime mover.

e. Ram-air Turbine Drive. The wind-driven generator of early aviation days is again finding application, this time as an emergency source of power which is particularly useful in single-engine aircraft to permit restarting after a jet-engine flame-out or to supply instruments and controls for an emergency landing. It also has an advantage as an emergency source when the main generating system fails in that a battery often has only sufficient capacity for a few minutes of operation whereas the turbine within its load rating is unlimited in this respect. Although such units are relatively heavy, they are more than competitive with a battery plus a small d-c to a-c inverter as an emergency source.

9. Direct-current Generator Characteristics. Multiengined aircraft designed before the advent of the jet engine generally featured 28-V d-c systems paralleled to provide reliability. Many such aircraft are still in service and probably will be so for some years to come. In the meanwhile, a new series of d-c generators has been developed for the small airplane market, and newly designed large aircraft feature a-c systems. This section will deal briefly with the older d-c systems, with some detail on the newer developments.

1. *Conventional D-C Generators and Controls.* The conventional aircraft d-c generators are usually self-excited by a shunt-field winding. Commutating and compensating series windings are nearly always used, in some cases with differential series fields. Usually, the series field is provided with a tap which gives 2-V to ground (aircraft structure) at full load. This is for paralleling purposes. With the exception

Table 21-3. Characteristics of Typical Aircraft 30-V D-C Generators (Conventional)
(Data from General Electric Co.)

Continuous rated current, amp	200	200	200	300	300	400	400	400*
Base speed, rpm	2200	4400	3000	4550	3000	6000	4500	3000
Maximum rated speed, rpm	4500	8000	8000	8000	8000	8500	8000	8000
Rated air flow, sea level, cfm†	....	60	63	70	120	120	120	130
Weight, lb	48	34	44	50	60	51	61	73
Lb per kw	8.0	5.7	7.3	5.5	6.7	4.2	5.1	6.1
Frame diam., in	....	6	6	6	6½	6½	6½	7$\frac{3}{16}$
Length including radial blast cap, in	....	10$\frac{13}{16}$	12$\frac{3}{16}$	13¾	15$\frac{1}{16}$	13$\frac{9}{16}$	15$\frac{1}{16}$	17⅜

* This machine is a jet-engine starter-generator capable of delivering 93 lb·ft of torque with 1,000-A current.
† Based on pressure drop of 6 in of water across the machine and inlet air temperature of 40°C at sea level. Some recently developed machines are designed for 120°C inlet air temperature.

of a few recent models for use with constant-speed drives, standard generators are designed for variable-speed operation as dictated by main-engine fixed-ratio gear drive.

Direct-current generators are classified with respect to

1. Operating voltage (15, 30, 125 V).
2. Current (50 to 400 A).
3. Speed range (see Table 21-3 for typical values).
4. Cooling air (40, 80, or 120°C at sea level with a pressure drop of 6 in of water).

In Table 21-3 are listed a number of standard generators in the 30-V 40°C class.

Generated armature voltage and field-resistance curves resemble those of a commercial generator at the base operating speed. As operating speed increases, the generated voltage is increased. The available output likewise increases (Fig. 21-2), as does also the electrical rate of response. The high available output at high speed provides a higher order of power-stability margin but entails possible thermal destruction of the generator in a short time if the high load is maintained. The high rate of electrical response is advantageous in quickly recovering from a shock-load addition but decreases the voltage-regulation stability margin. Generators usually have overload capacity of 150% of rated load for 2 min at 120% of base speed, starting from 40°C average winding temperature.

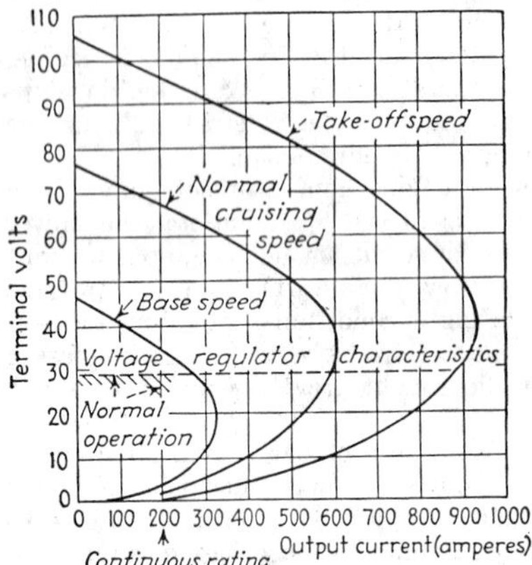

FIG. 21-2. Aircraft d-c generator external volt-ampere characteristics at various speeds.

2. *Brushless D-C Generators and Controls.* With the advent of the brushless a-c generator, we have also the brushless d-c generator. Both were made necessary by the brush-wear problems of commutators or slip rings at altitude. A pronounced benefit has been the lengthening of the time between overhauls. Essentially, the brushless d-c generator is an a-c machine with the commutator replaced by high-reliability silicon rectifiers. Operating voltages and currents are the same as those for conventional machines. The present trend is to make the brushless machine a direct replacement for the conventional brush machine.

The brushless d-c machine development was accompanied by a more radical one, that which replaced the not-too-reliable carbon-pile voltage regulator with a transistorized version and the heavy control panel with a lighter, solid-state version. These, too, are designed as direct replacements of the conventional designs.

Some of the features of the brushless d-c generator lend themselves to a simplification of the system. For example, since the brushless machine is an a-c generator with rectifiers, there is no reason to use a cutout relay between the generator and the bus. The rectifiers serve to isolate it except when the machine is up to speed and is correctly excited.

The machine parameters resemble those of the brush machine in most cases. Table 21-3 may be used as a guide for these machines also.

10. Direct-current Generator Controls. Variable-speed aircraft generators do not inherently produce a sufficiently uniform voltage to serve the constant-potential power-supply system, and various failures must be protected against. Thus, controls are provided to accomplish the following:

1. Maintain voltage at the bus within specification limits.
2. Divide load current among the several generators in parallel.
3. Switch on the generator when its voltage is suitable.
4. Switch off the generator when its voltage is abnormal.
5. Provide short-circuit protection.
6. Provide means of manual supervision:
   a. Indication of performance.
   b. Manual switching.

A typical 28-V generating system is shown in Fig. 21-3, using a conventional-type generator, a carbon-pile regulator, and standard components for protection. A brushless type 28-V generator, transistor regulator, and solid-state protection and control system are shown in Fig. 21-4.

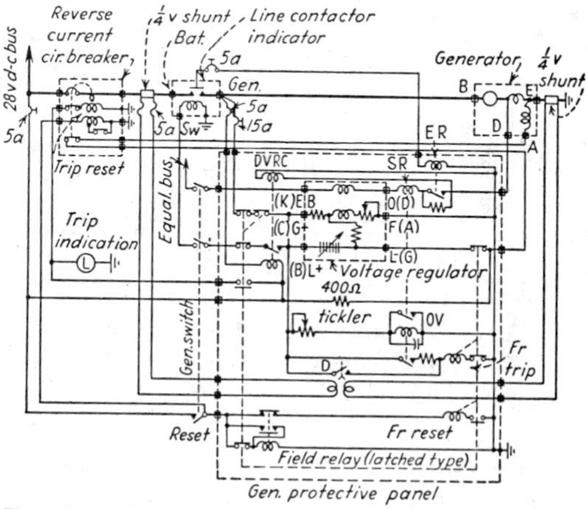

FIG. 21-3. Typical generator circuit arrangement for 28-V d-c system.

a. Voltage Regulators. The generator voltage regulator serves the dual purpose of automatically maintaining system voltage constant and balancing load current among paralleled generators. The usual voltage regulator consists of a solenoid operating against spring action on a carbon pile in series with the shunt field of the generator; but a transistorized voltage regulator has been developed which is quicker in action than the carbon-pile regulator. It will become increasingly encountered as the brushless d-c generator systems develop.

Where several generators are operated in parallel, an equalizer circuit of some type is provided on the voltage regulator for obtaining equal division of current among the several units. Current unbalance among paralleled generators is caused by error voltages, e.g., those due to regulator inaccuracies in setting or droop or nonsymmetrical generator circuits. Methods of paralleling are directed at creating voltages to oppose these difference voltages and thus reduce the unbalance. The regulator-equalizer method described corrects unbalance without adding to voltage regulation at the bus. In fact, since bus voltage is the average of all the regulating points, bus-voltage regulation tends to be reduced by equalizer action.

b. Generator Switching. Generator switching is accomplished by means of a main power-switching contactor in combination with appropriate pilot relays and controlled by them to accomplish automatically the following:

1. Close the contactor when the associated generator voltage reaches normal value and is of normal polarity.

2. Open the contactor when the associated generator voltage becomes subnormal.

Provision is made also to permit opening the contactor manually from a remote position. Where the generator is used also as a starter, provision is made for bypassing the pilot relay to close the contactor and operate the machine as a motor.

One type of pilot relay is polarized with a permanent magnet and has two coils. One coil senses the difference between generator and bus voltage, and the other senses generator current. When generator voltage exceeds bus voltage, the pilot relay closes

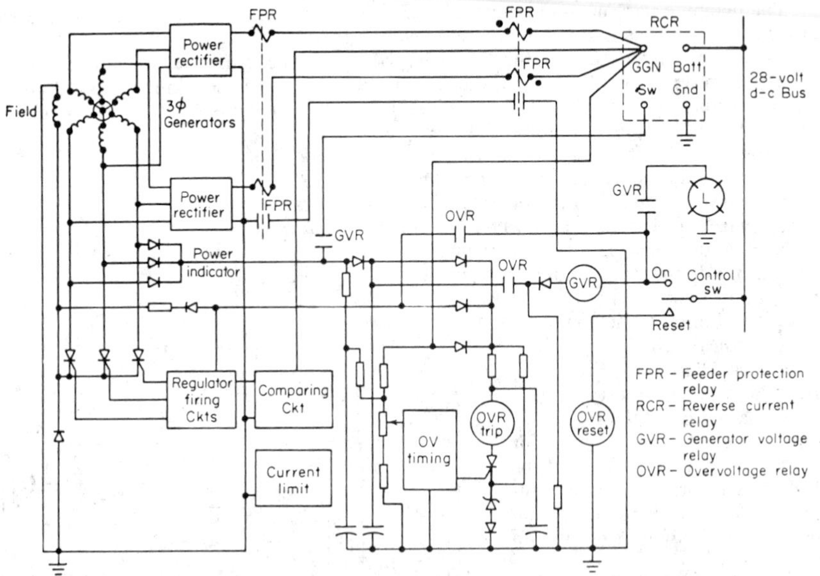

Fig.21-4. Brushless d-c generator with transistor regulator and solid-state control system.

and completes the circuit to the contactor. When the generator voltage drops below normal, reverse current flows through the current coil and opens the pilot relay removing the generator from the bus.

The "generator reverse-current cutout" consists of the line contactor and pilot relay mounted in one enclosure. In other system arrangements, the pilot relay is mounted in a separate control box (which often includes the voltage regulator) and senses reverse current through voltage drop across the generator series windings. The contactors used may either have a nominal 28-V interrupting rating or be designed with sufficient capacity to interrupt the maximum output voltage and current of a wide-speed-range generator under full field, as could occur in case of regulator failure.

c. Short-circuit Protection. The generator relay provides some measure of protection to the system for generator and generator-feeder faults. However, a circuit breaker with a reverse-current trip having sufficient time delay to override normal switching transients is often used to provide (1) higher interrupting capacity (up to 12,000 A), (2) backup protection for a welded contactor or pilot relay failure, (3) auxiliary contacts to deenergize the generator (open the field) and prevent its continuing to feed the fault, and (4) interrupting capacity for overvoltage conditions using a shunt trip coil actuated by an overvoltage relay (see the following section). Forward current tripping is not provided, since it is accepted practice to leave an overloaded generator on the line, although an overtemperature warning may be provided to enable the crew to take corrective action. In a few cases, differential-current or balanced-current protection has been used to remove a generator and deenergize it without time delay in order to reduce the hazard of fire and smoke.

d. Overvoltage Protection. The capacity of a wide-speed-range generator for producing high voltages under full field is illustrated in Fig. 21-2. The potential damage to load equipment is so great that it is customary to provide fast-acting overvoltage relays to deenergize and remove the faulty machine from the bus. Several schemes have been employed, but all depend upon sensing either generator field voltage or bus voltage. The former is inherently selective, since the field voltage in normal machines is always below rated owing to the regulator series resistance, but it does not protect for a partial failure at high speed where the field voltage may still be below rated yet sufficient to produce above-rated terminal voltage. Bus-voltage sensing

Table 21-4. Characteristics of Typical Aircraft 3-phase A-C Generators with Integral Exciters

(Data from General Electric Company)

Continuous rating, kva	10	20*	20	30	30	30	40	60
Rated power factor	0.75	0.75	0.75	0.9	0.75	0.75	0.75	0.75
Min. rated speed, rpm	5700	7600	5700	4000	4800	5700	5700	5700
Max. rated speed, rpm	6300	8400	6300	8000	7200	6300	6300	6300
Frequency	380/420	380/420	380/420	400/800	320/480	380/420	380/420	380/420
Cooling air, S.L. temp., deg C	80	120	80	40	40	80	80	80
Cooling air, rated cfm	140	220		300	330	330	330	330
Frame diam., in	7	8¼	11	11	11	11	11	11
Length with axial blast cap, in	16 15/32	14 5/16	15⅝	17⅝	16⅞	16⅞	17⅝	19 1/32
Weight, lb	45	38*	62	99	87	75	87	115
Lb per kva	4.5	1.9*	3.0	3.5	3.1	2.7	2.3	2.0
Max. harmonic and % of fundamental	5–1.90		5–1.45	7–1.40	5–0.75	5–0.75	5–1.4	5–0.44
Phase voltage unbalance with ⅓ current unbalance, max. %	3.7		3.6	3.9	3.8	3.0	2.9	2.9
3-phase S.C. current at base speed, p.u	5.0		3.6	3.3	3.5	3.0	3.3	3.5
Reactances at 400 cycles, p.u.:†								
Synchronous, direct axis (xd)	120	2.02	2.13	1.12	2.51	2.14	1.99	1.61
Synchronous, quad. axis (xq)	0.68	1.16	1.25	0.65	1.35	1.20	1.12	0.79
Transient, direct axis ($x'd$)	0.20	0.38	0.42	0.22	0.37	0.37	0.34	0.23
Subtransient, direct axis ($x''d$)	0.14	0.28	0.24	0.14	0.21	0.20	0.19	0.12
Negative sequence (x_2)	0.18	0.36	0.34	0.16	0.23	0.21	0.20	0.12
Zero sequence (x_0)	0.28	0.34	0.012			0.17	0.011	0.15
Negative sequence resistance (r_2)	0.035		0.057			0.042	0.039	0.025
Zero sequence resistance (r_0)	0.026	0.32	0.039			0.026	0.024	0.020
Efficiency, full load, 6000 rpm	77	85	77	80	78	83	83	

* This machine is designed for static excitation and has no integral exciter.
† Calculated values.

protects for all overvoltages above the relay setting but requires a more complex means of selecting the faulty machine. Either biasing of the overvoltage relay or the use of a separate selector relay acting on the directional unbalance in regulator equalizer circuit current is effective for this purpose. Such circuits are included in Figs. 21-3 and 21-4.

11. Storage Batteries. The lead-acid battery was used almost exclusively in the past, but recently the nickel-cadmium battery has been replacing it in new design, and in some aircraft the battery has been deleted entirely. The nickel-cadmium battery is much lighter than the lead-acid but does introduce some complications. There is no method of reliably indicating the state of charge. The battery should be completely discharged frequently. Battery capacity is adversely affected by incomplete discharge cycles. Twenty-four-volt batteries operating in conjunction with transformer rectifiers in flight and providing power for lights, control, and possibly starting on the ground as well as for emergency flight operation are still found on most a-c aircraft.

12. Alternating-current Generators. Alternating-current generators found on aircraft are almost exclusively of the synchronous type. They usually have an integral exciter which is so designed that it will build up on self-excitation, making the a-c system independent of other power sources. Some recent designs have a small overhung permanent-magnet pilot exciter for build-up and exciter control power which may also be used for other control functions. The most recent trend is to the brushless type of machine, which eliminates the brushes, commutator, and slip rings. Virtually all new design is brushless, and many of the brush-type machines are being replaced with the brushless. This type of machine rectifies the output of the exciter and applies the result directly to the rotating field, eliminating the need for the commutator and slip rings and enhancing machine reliability.

Another important development is the variable-speed constant-frequency generator (VSCF). This machine delivers its constant-frequency output without the necessity for a hydromechanical constant-speed drive. The usual method of accomplishing this result is to use a 3-phase generator delivering 3,000 to 4,000 c/s, depending upon engine speed. This is synthesized into a 400-c 120-Y/208 output voltage from the high-frequency supply, which is then filtered to produce a sine wave. The frequency synthesizer is a solid-state device operating silicon-controlled rectifiers in the proper sequence to provide the desired output, utilizing an accurate frequency reference to

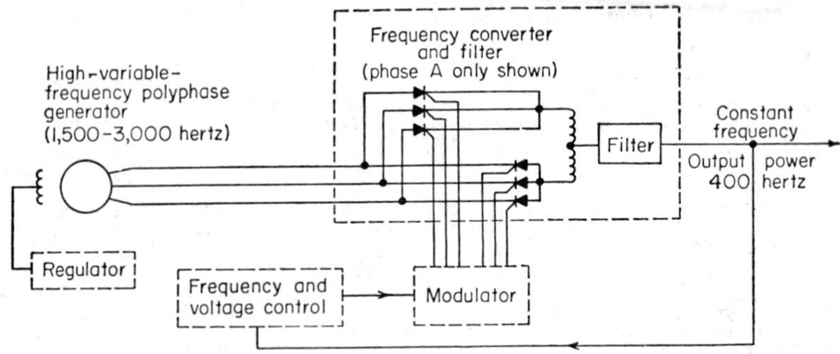

FIG. 21-5. Block diagram of variable-speed constant-frequency generator.

maintain the output frequency constant. A block diagram of the VSCF system is shown in Fig. 21-5.

Alternating-current generator characteristics are established by the requirements of Specification MIL-G-6099. VSCF system requirements are established by Specification MIL-E-23001, for a 40-kVA system; this follows MIL-G-6099 in most requirements and will be followed by a more definite specification as the VSCF is developed.

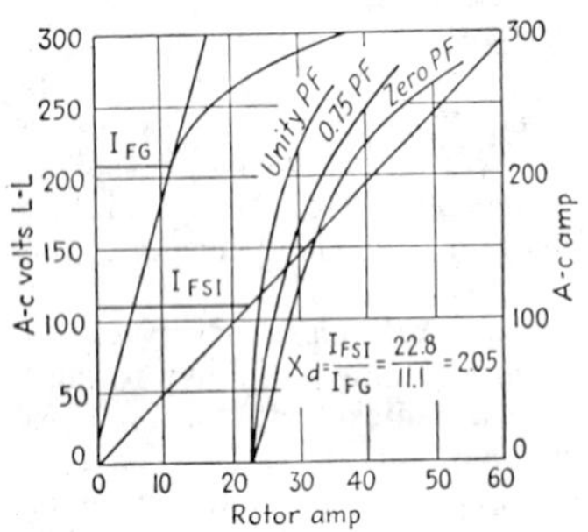

FIG. 21-6. Typical aircraft a-c generator, saturation and short-circuit characteristics.

Typical characteristics of a-c generators of the conventional and brushless types are listed in Table 21-4. Typical saturation and short-circuit characteristics are shown in Fig. 21-6. Alternating-current generators are classified with respect to

1. Voltage (120 or 208-Y/120).
2. kVA (10 to 90).
3. Power factor (0.9 or 0.75).
4. Speed range and frequency (see Table 21-4 for typical values).
5. Cooling air (40, 80, or 120°C at sea level with 6 or 11 in water-pressure drop).

Generators are provided with a mounting flange for attachment to standard engine drive pads. For stiffness and to reduce overhung moment, the a-c generator field is located next to the flange-end bearing on the rotor. The exciter armature is next followed by the commutator, slip rings, and opposite-end bearing. Salient-pole designs have been used for speeds up to 12,000 r/min. Heavy amortisseur circuits are used in both direct and quadrature axes to reduce transient disturbances, to reduce phase-voltage unbalance, and to provide greater stability in parallel operation. A flexible drive shaft is normally used to reduce torsional oscillations from the engine drive. The shaft is provided with a friction damper to prevent breakage at resonant frequencies. The flexible shaft also helps in reducing the transient peak torques imposed on the drive during short circuits and synchronizing.

13. Alternating-current Generator Control. The majority of newer a-c systems have been constant-voltage and constant-frequency, and many have been parallel systems. Controls are supplied to accomplish the following:

a. Maintain voltage at the bus within specification limits (Par. **14**).

b. Divide real and reactive load among the several generators in a parallel system (Par. **14**).

c. Switch the generator on the bus when frequency and phase voltages (and rotor angle in parallel systems) are suitable (Par. **15**).

d. Switch the generator off when it is the cause of abnormal frequency or any abnormal phase voltage (Pars. **16** and **17**).

e. Switch the generator off the bus when it is the cause of unequal generator current distribution in a parallel system (Par. **18**).

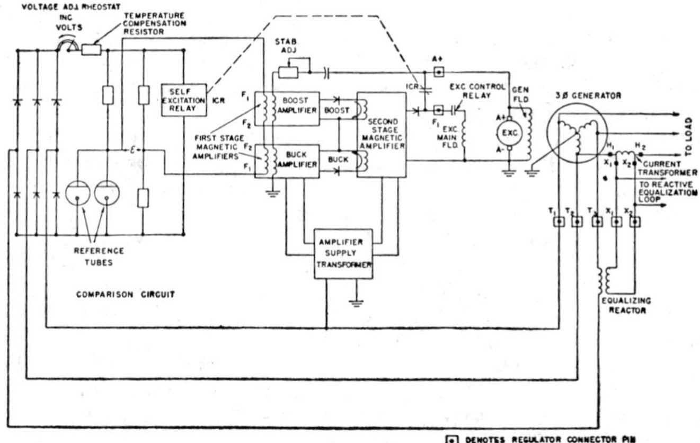

FIG. 21-7. Diagram of magnetic-amplifier type of a-c generator voltage regulator.

f. Provide protection against short circuits in the generating system (Par. **19**).

g. Provide manual and/or automatic means of supervision.

In many modern installations, equipment is provided to accomplish the foregoing automatically to simplify aircraft operation. In some systems, no manual supervision is required or provided for other than a reset switch to permit placing the automatic start-up sequence in effect after a generator has been tripped.

14. Voltage Regulators. Voltage regulators are generally designed to sense and respond to the approximate average of the magnitudes of the three line-to-line voltages on 3-phase systems furnished with high-phase take-over if one phase exceeds the average. Reactive division has been obtained by feeding differential current signals from current transformers in one generator line into a discriminator circuit which biases the sensing circuit of the regulator according to the magnitude and sense of the difference between the reactive flow of the particular machine and the average flow of all machines. This produces a system in which bus voltage is the average of all the regulator operating points and no added regulation is caused by the reactive division circuitry.

Regulators in general use are the magnetic-amplifier type and the transistorized type. The transistor regulator has made possible more accurate voltage control and better transient control. Virtually all new design is transistorized. A schematic diagram of a magnetic-amplifier regulator is shown in Fig. 21-7, that of a transistorized regulator in Fig. 21-8.

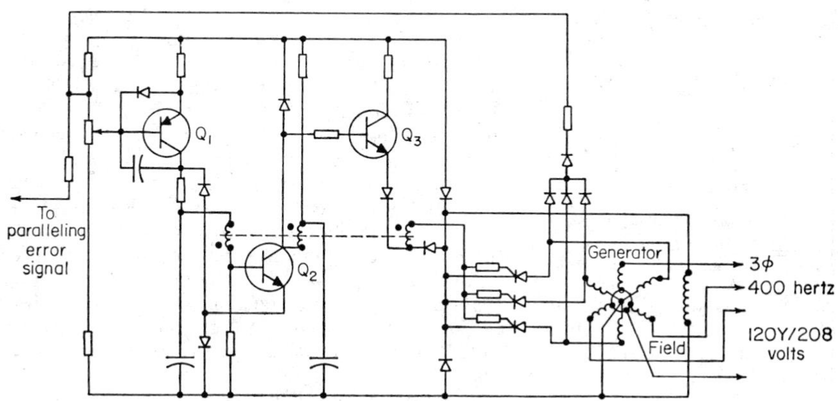

FIG. 21-8. Brushless a-c generator with transistor regulator.

SWITCHING AND PROTECTION

15. Switching Devices. A latch-type generator control relay is used to open and close the exciter field for energizing and deenergizing the machine. The protective equipment shuts the machine down by tripping this relay.

Both latch-type and magnetic hold-in contactors are used to connect the machine to the bus. They switch all 3 phases simultaneously and must be capable of interrupting the maximum fault current that may flow through them. Protective equipment isolates a faulty machine by tripping this contactor.

Generator switches are used for resetting, line-contactor switching, and, in some cases, deenergizing the machine. One type has three positions: "off," "on," and a spring-loaded "reset" position. The "off" position opens the line contactor; "on" closes the line contactor or initiates automatic paralleling; the "reset" position will close the generator control relay.

16. Frequency-sensing Devices. Induction-motor and transformer loads may overheat, or load-protective equipment may be operated owing to underfrequency operation. For this reason, frequency-sensing devices are used to open the line contactor when frequency is below permissible values.

Underfrequency relays operating off bus voltage have been used widely. A tuned circuit or a high- or low-pass filter is used as the sensing element. Many recent applications have used a mechanical underspeed switch on the drive to open the line contactor.

If a drive overspeeds for any reason, both the drive and the machine may be destroyed. For this reason an overspeed device is used to shift into full underdrive in hydraulic transmissions or to remove the source of energy from turbine drives. Selectivity in a parallel system is obtained by an overrunning clutch which prevents large speed increases in normal drives; thus, only the faulty drive control is actuated by the protection. The machine may be removed from the bus by underspeed and/or difference current protection.

Very accurate frequency control and real load division are incorporated as a part of the hydromechanical constant-speed drive. Some type of frequency standard is used, and all drives are compared against this standard and the speed regulated accordingly. Also, current transformers in one phase of each paralleled machine are connected in series. A sensing circuit coupled with a discriminator controls the individual drives to divide the real (kilowatts) load.

17. Voltage Sensing. An inverse time type of relay using an air dashpot is used for overvoltage protection. Its voltage-time trip curve is designed to override normal system transients without tripping, but overvoltage transients in excess of normal will cause a trip and remove the machine from the system. These relays usually sense the same quantity as the regulator, i.e., the approximate average magnitude of the three line-to-line voltages.

If the bus voltage falls below specifications, rotating loads may overheat, and static load performance is affected. Undervoltage relays sensing the approximate average of the three line-to-line voltages have been used to trip the line contactor under these conditions after a suitable delay.

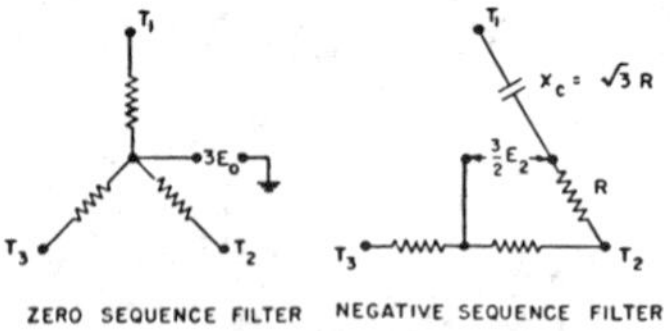

ZERO SEQUENCE FILTER NEGATIVE SEQUENCE FILTER

FIG. 21-9. Sequence filters used for unbalanced voltage sensing.

Unbalanced phase voltages resulting from large unbalanced loads, unbalanced faults, and open conductors may seriously affect system performance owing to possible overvoltages in some phases and unbalanced currents that may be extremely large in some instances. Phase-voltage unbalanced relays are used to remove a machine with large voltage unbalance after a suitable delay to permit fault-protection coordination. These relays may be operated from sequence filters such as shown in Fig. 21-9. The negative-sequence relay has been used also to hold a generator off the line when it has reversed phase sequence due to maintenance error.

18. Unbalanced Generator Currents. If sizable differences in generator currents exist in a parallel system, total generating capacity is reduced. Furthermore, since load-division circuitry is designed to keep such unbalance within specifications for any normal conditions, excessive current differences are a direct indication of a failure within the generating system. A relay connected to a differential current-transformer loop senses the difference between one generator current and the average of all generator currents. It may be used to trip bus tie contactors and split a parallel system, thereby removing the unbalance. Time delay sufficient to override normal transient unbalances must be incorporated.

If a machine in a parallel system is over- or underexcited for a given load condition, differential reactive current will flow, bus voltage will be changed but not always sufficiently to actuate abnormal voltage relays, and, in some cases, out-of-step operation may result. One general approach has been to open bus tie contactors on difference current as described and allow over- and undervoltage protection to trip the faulty machine. The more commonly accepted method is to use differential reactive current to bias the over- and undervoltage relays. Thus, the abnormal machine voltage relays see a greater abnormality in voltage than do normal machine relays and trip the abnormal machine only.

If a feeder or machine phase is opened on a parallel system, phase voltages will be kept in good balance by the normal machines and loads are little affected. If, however, the phase in which load-division current transformers are located is opened, very severe unbalanced currents and neutral currents will flow in all machines and overheating in both normal and abnormal machines will result. Relays sensing neutral current have been used to trip bus tie contactors and split the system so that a larger unbalance voltage sufficient to operate unbalance voltage relays will exist in the abnormal machine and it will be tripped. Time delay sufficient to clear the worst fault involving neutral must be incorporated to avoid splitting the system on load- or distribution-system faults. This arrangement has been used in some cases to isolate a bus with a ground fault and to trip the associated machine.

19. Short-circuit Protection. As noted in the preceding, in case of a short circuit in a parallel system, unbalanced voltage protection may be used to isolate a machine and its associated bus and to trip the machine after a time delay of the order of 5 s.

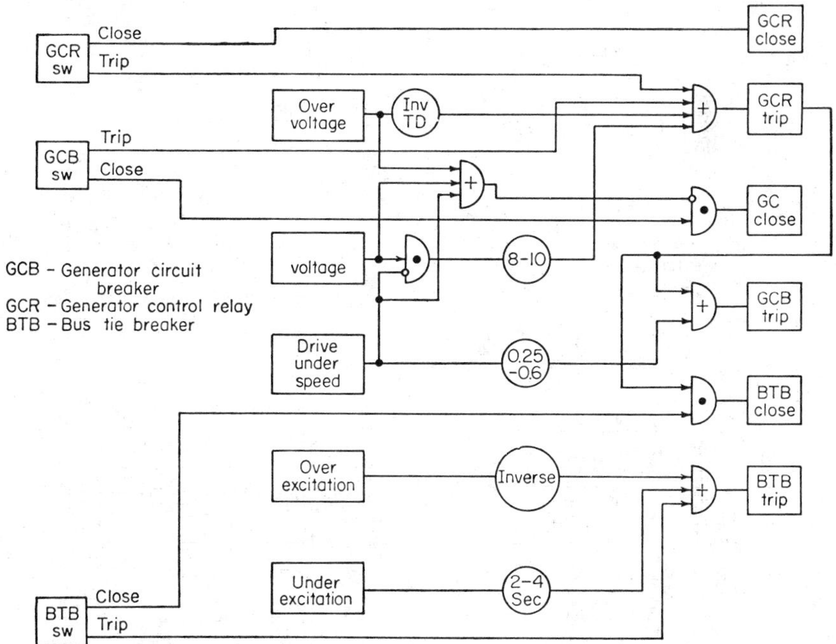

FIG. 21-10. Block diagram: a-c generator control system.

The most commonly used method is to protect the machine and feeders with differential fault protection and rely on unbalanced and undervoltage relays for bus protection. This offers the advantages of high speed and fully selective operation, and the bus associated with the faulted machine or feeders need not be isolated from the system. The usual arrangement is to use one current transformer in each phase located at or near the feeder connection to the bus and one located on the neutral side of each machine phase. A difference in excess of 20 to 40 A between the two transformers on any phase is used to operate a relay which trips the machine.

20. Generator-circuit Arrangement. Figure 21-10 is a logic diagram of a generator system incorporating the above protective features. This is for a single generator of a multigenerator system and includes automatic paralleling, as well as features discussed previously. Refer to Fig. 21-11*c* for the distribution circuit configuration to which this logic circuit would be applicable. The logic circuit may be implemented either by relays or by solid-state logic. Both are in present use, with the trend toward solid-state logic.

DISTRIBUTION

21. The distribution system is comprised of a number of buses, interconnecting cables, and associated protective devices.

22. Buses. A bus is a junction of three or more circuits. Since damage of this point can make more than one circuit inoperative, the objective is usually to keep the area as small as possible and to give it good mechanical protection. The bus is usually a copper bar mounted on insulating material (aluminum has been used in a few installations). The bus and associated protective equipment are often enclosed in a metal housing or, in some cases, a Fiberglas housing in order to reduce the probability of short circuits.

Small aircraft may have only one bus, but some large aircraft have had more than 15 buses. Buses may be classified as to service: (*a*) source bus, (*b*) load bus. Source buses often need to be considered a part of the generating system in that generator protective equipment is designed to provide source-bus protection as well in many applications.

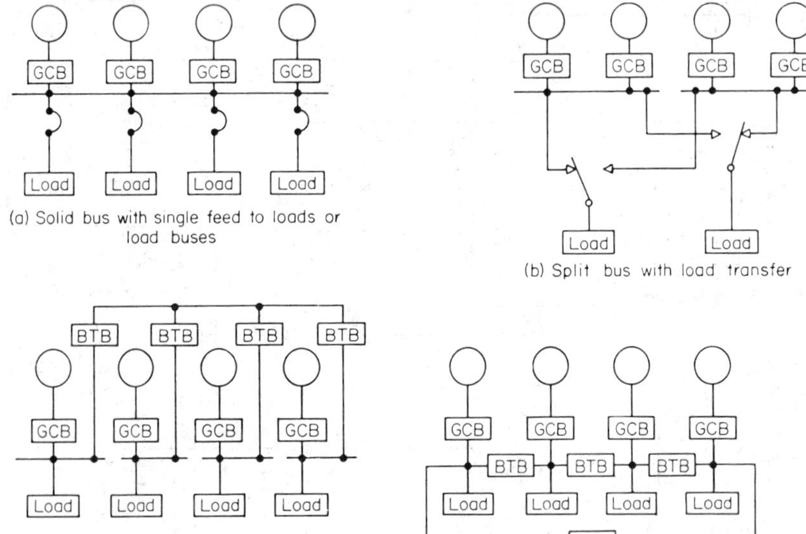

Fig. 21-11. Distribution-circuit configurations commonly used in aircraft.

23. Distribution Circuits. Figure 21-11 shows several system configurations in use today in both a-c and d-c systems. Each of these has advantages and disadvantages; the choice is dictated by the application and requirements of the particular airplane. Generally, larger airplanes tend toward a split or synchronizing source-bus arrangement with alternate feeds to vital load buses, whereas smaller, high-performance aircraft tend toward the simpler arrangements. However, a trend toward high-speed, fully selective protective equipment is discernible, since, for some loads, transient power disturbances of long duration are as bad as (or even worse than) complete loss of power to the load. As this trend continues, emphasis is expected to shift more toward the split-bus or ring-bus configurations at the expense of more complex protective equipment.

Load-transfer contactors as shown in Fig. 21-11 have been used widely to transfer loads from a dead bus over to a live one. These devices will transfer on loss of voltage or upon the presence of sizable voltage unbalance. Time delay sufficient to override clearing any fault beyond the contactor must be incorporated to prevent chattering with a fault on the load side. Recently, consideration has been given to deletion of these in favor of automatic reclosing of bus tie contactors and/or fully selective generating-system protection. If a source bus is faulted or lost through sequential protection, the loads connected to that bus are lost.

24. Circuit Protection. Circuit-protective devices are provided in the circuits at the point of junction with a bus. These devices both detect and interrupt abnormally large currents. Limiters are used in generation, bus tie, and distribution circuits;

Table 21-5. Aircraft-cable Characteristics

Wire or cable size		Max.* d-c resistance, ohms per 1000 ft at 20 C	400-cycle positive† (and negative) sequence impedance, ohms per 1000 ft	400-cycle zero† sequence impedance, ohms per 1000 ft	Nominal weight* of finished wire, lb per 1000 ft	Continuous-duty current, amp*	
Aluminum	Copper					Single wire in free air	Wires and cables in conduit or bundles
	AN-22	16.45	....	....	6.3	...	5
	AN-20	10.25	$10.25 + j0.304$	$10.44 + j2.20$	8.4	11	7.5
	AN-18	6.44	$6.44 + j0.289$	$6.33 + j1.93$	11.6	16	10
	AN-16	4.76	$4.76 + j0.278$	$4.95 + j1.88$	14.3	22	13
	AN-14	2.99	$2.99 + j0.264$	$3.18 + j1.80$	21.0	32	17
	AN-12	1.88	$1.88 + j0.247$	$2.07 + j1.72$	32.0	41	23
	AN-10	1.10	$1.10 + j0.228$	$1.29 + j1.64$	47.5	55	33
	AN-8	0.700	$0.700 + j0.226$	$0.792 + j1.53$	76.5	73	46
	AN-6	0.436	$0.440 + j0.221$	$0.632 + j1.42$	115	101	60
	AN-4	0.274	$0.282 + j0.212$	$0.474 + j1.33$	176	135	80
	AN-2	0.179	$0.188 + j0.206$	$0.380 + j1.24$	278	181	100
	AN-1	0.146	....	....	322	211	125
	AN-0	0.114	$0.127 + j0.204$	$0.319 + j1.15$	415	245	150
	AN-00	0.090	....	....	525	283	175
	AN-000	0.072	....	....	657	328	200
	AN-0000	0.057	....	....	820	380	225
AL-8	....	1.093	$1.093 + j0.226$	$1.28 + j1.53$	32	60	36
AL-6	....	0.641	$0.648 + j0.221$	$0.840 + j1.42$	55	83	50
AL-4	....	0.427	$0.439 + j0.212$	$0.631 + j1.33$	74	108	66
AL-2	....	0.268	$0.282 + j0.206$	$0.474 + j1.24$	107	152	82
AL-1	....	0.214	....	....	133	174	105
AL-0	....	0.169	$0.188 + j0.204$	$0.370 + j1.15$	167	202	123
AL-00	....	0.133	....	....	203	235	145
AL-000	....	0.109	....	....	242	266	162
AL-0000	....	0.085	....	....	296	303	190

NOTE: Impedance data are based upon tight, equilateral bundle. Positive sequence data are nearly independent of skin distance†; zero sequence data are based upon 2-in skin distance.

* Data from Military Specification MIL-W-5086, Wire, Electrical, 600-volt, Copper, Aircraft, Nov. 30, 1950; Military Specification MIL-W-5088-A (ASG), Wiring, Aircraft, Installation of, Oct. 11, 1954; Military Specification MIL-W-7072, Wire, Electrical, 600-volt, Aluminum, Aircraft, Nov. 30, 1950.

† Data from Impedance Data for 400-cycle Aircraft Distribution Systems, by D. W. Exner and G. H. Singer, Jr., *AIEE Tech. Paper* 52-322, October, 1952. Data for copper taken directly from this paper; data for aluminum extrapolated by the authors from this paper. Extrapolation method was spot-checked with data in Experimental Determination of 400-cycle Impedance of Wire in Aircraft Power Distribution Circuits, by J. D. Andrew, *Trans. AIEE Paper* 54–374, June 14, 1954. Positive sequence calculations checked within 5% and zero sequence within 17% of the data for the two aluminum conductors in this paper.

fuses and thermal or magnetic circuit breakers are used in load circuits. Generally, limiters are used for short-circuit protection; fuses and circuit breakers are used for cable overload protection as well as fault clearing. In a-c systems, relaying is used widely for generator, feeder, and bus circuit protection.

Most 3-phase loads cannot operate satisfactorily or may even be a hazard if 1 phase is opened to clear a fault. Recently, 3-phase devices which interrupt all 3 phases with overcurrent in any 1 phase have been marketed and are expected to come into use for 3-phase load protection. No satisfactory 3-phase device suitable for main-distribution-circuit protection has been marketed other than a 3-phase contactor actuated by thermal current-sensing devices in each phase. Two or more separated conductors per phase have been used to reduce the possibility of opening a phase because of a fault.

Breakers, limiters, and fuses have inverse-time characteristics, some of which are shown in Fig. 21-12. Some magnetic circuit breakers for load protection offer the added feature of nearly instantaneous operation on large fault currents. This is an advantage in reducing system disturbances due to faults.

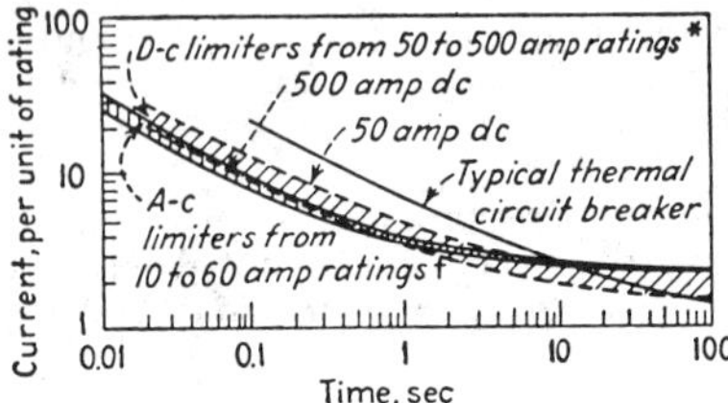

Fig. 21-12. Typical aircraft circuit-breaker and current-limiter characteristics. (**Plotted from curves supplied by Bussman Manufacturing Company; †plotted from curves supplied by Burndy Engineering Company.*)

25. Aircraft Cable. Aircraft cable is stranded of fine wire to obtain flexibility. The insulation withstands abrasion, has flame-retarding characteristics, and resists heat, fuel, and lubricants under the wide range of environmental conditions found in aircraft. Both copper and aluminum conductors are used in aircraft, but aluminum used more widely in heavy-current circuits, since its weight is only 55 to 70% of that of copper of equal conductivity in the larger cable sizes. Both copper and aluminum cables are usually terminated in pressure-applied lugs with hand tools or power presses. Cable characteristics for both a-c and d-c use are shown in Table 21-5.

ELECTRICALLY OPERATED EQUIPMENT

26. Motor Application. The largest aircraft have used as many as several hundred electric motors per airplane. The characteristics of these motors vary to suit a wide variety of applications, and sizes range from a few watts for control applications to large intermittent-duty units of 15 to 25 hp for actuation of landing gear and major aerodynamic surfaces. Large actuators are often of the electrohydraulic type. Many intermittent-duty applications require electric brakes, which are supplied as part of the motor. Continuous-duty motors are usually not required in sizes larger than 5 to 10 hp. These larger sizes are commonly used for such applications as cabin air conditioning. Little standardization of aircraft motors has been achieved up to the present time.

27. Motor Construction. Motors are of either totally enclosed or open-ventilated construction. In a few cases, externally supplied air blast may be provided, but most motors have to operate in an enclosed area where no external air is available. In addition, there is increasing demand for motors which will pass explosion-proof and oilproof (½ pt/h drip) tests. For these reasons, most d-c motors are totally enclosed because of the commutator and brushes. Continuous-duty, totally enclosed motors may be provided with an external fan which forces air through a shroud over the outside of the frame to improve heat transfer to the surrounding air. Open-ventilated a-c induction motors can often pass the required environmental tests and thus are used in the larger sizes for continuous duty because of substantial weight savings.

Motors are flange-mounted with the exception of a few cases in which they are built into an actuator or gear unit and then may be of single-bearing construction. Standard flange sizes are used. Square flanges with four boltholes are used in smaller sizes, and round flanges with additional boltholes to prevent distortion are used in larger sizes. Keyed shafts or splines are used for coupling to the load. Where leads are AN-10 cable or larger, terminal boards are generally used, whereas the leads are brought out directly in the smaller sizes. Three-phase a-c motors are 4-wire with four terminals, and continuous-duty d-c motors require only two terminals. Intermittent-duty d-c motors for reversing service have three terminals for split-series and four terminals for compound-wound types.

Brush problems encountered on d-c motors at altitude have accelerated the development of brushless d-c motors. A short treatise on these motors is included in the section on spacecraft electric subsystems.

28. Motor Characteristics. Motor-design objective is the attainment of adequate performance with minimum weight and acceptable efficiency. To accomplish this, speed is usually high to reduce size and weight. Commonly used speeds for 400-c a-c motors are 7,300 and 11,000 r/min, corresponding to six-pole and four-pole designs, respectively. Direct-current motor speeds vary widely, but the trend is toward use of the a-c motor speeds stated to increase the degree of interchangeability. Reduction gears are used frequently to obtain an acceptable speed for the driven equipment.

Aircraft motors are rated for either continuous or intermittent duty. The intermittent-duty cycles vary for the application, although 1 min on and 9 min off is typical. Temperature and altitude limit the rating and must be specified. In addition, where requirements for high-performance aircraft are especially severe, it is necessary to specify the maximum mounting-pad temperature and the velocity of the air surrounding the motor. Since an important consideration in these applications is also the total heat rejection which must be disposed of in the aircraft, it is desirable to achieve high efficiency in continuous-rated motors. Over 75% efficiency is usual for units larger than 1 hp, with somewhat lower values for intermittent-duty motors.

Direct-current motors may be shunt-, series-, or compound-wound, depending on the application. The split-series motor has been used extensively for reversing service because of the simplicity of the reversing control, which needs only to switch the input from one end of a tapped series winding to the other to reverse direction of rotation. Speed-torque curves for typical aircraft motors fall within the limits shown in Fig. 21-13. Starting currents are in the order of 400% for series motors and 1,000% for shunt motors.

Alternating-current motors are usually of the 3-phase squirrel-cage induction type, although single-phase capacitor motors are used where no 3-phase power is available. Peak torque, locked-rotor torque, and power factor vary over wide limits, depending on the application requirements, although starting current is limited to 800% by military specifications.

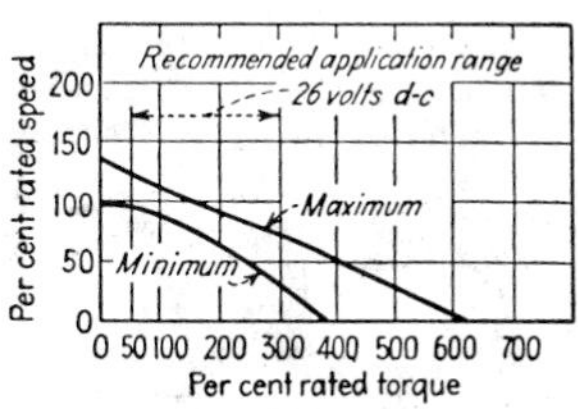

Fig. 21-13. Range of aircraft a-c motor speed-torque characteristics.

29. Conversion Equipment. *a. Application.* Inverters have been used on practically all aircraft with d-c generating systems to supply a-c loads such as instruments, fluorescent lighting, radio and radar, and control systems such as automatic pilots and supercharger regulators. With increasing use of electronic equipment and other a-c loads, and because rotary-inverter efficiencies are in the neighborhood of 50%, it is not unusual for 20 to 25% of the d-c power to be used for inverters. The poor weight economy resulting from this situation has been a principal factor in the increasing use of a-c power generation and the development of the static inverter, which has full-load efficiencies in the region of 75 to 80%. In aircraft having only a-c power generation, it is often necessary to convert 10% or more to 28 V d-c by means of transformer-rectifiers. However, the penalty is not so severe because of the higher efficiency (80 to 90%) and lower specific weight of the conversion units and the lower specific weight of the a-c generating system. Of course, transformers are used for the conversion of voltage in aircraft as elsewhere. Step-up transformers are used principally in electronics for plate supplies, and step-down transformers are required in various load equipments such as lighting, certain low-voltage instruments, and very small motors.

b. Aircraft Inverters. Inverters have been built to supply frequencies of 60 to 3,200 c, although 400 c is most generally used. The rotary inverter is the most prevalent type in older aircraft but may be found as the emergency a-c supply in more modern aircraft. Basically, it is a motor-generator set with control of frequency by d-c motor speed control, of voltage by a-c generator field control. Both single- and 3-phase inverters are found, and output voltages are usually 26, 115, and 115-Y/200. Automatic speed control is required to meet the close frequency tolerances required by today's aircraft equipment. The older inverters have carbon-pile regulators, but in order to meet both frequency and voltage requirements, it has been necessary to resort to solid-state regulators.

The trend to solid-state apparatus found in regulators is also found in the inverter field. In many cases, individual apparatus will incorporate a small inverter sufficient to supply the single load at high efficiency. The static inverter to replace the rotary inverter is characterized by lighter weight, higher efficiency, and freedom from mechanical faults, since it has no rotary armature, commutator, or slip rings and does away with the mechanical regulator. There are significant differences between the rotary and static inverters which must be taken into account when replacement is contemplated. Some are:

1. The rotary contains its own fan as a matter of course; the static may require special cooling.

2. Voltage regulation of the rotary inverter is by regulation of the field for all 3 phases at once; the static regulation is commonly for each phase.

3. Usually, almost full inverter output may be taken from a single-phase or a 3-phase rotary inverter; only normal load may be so taken in the static inverter.

4. A static inverter is much more easily designed for a fixed load, and many are so designed. Care should be taken that such an inverter is not installed where varying loads occur. A possible arrangement for a static inverter is shown in Fig. 21-14. A series regulator controls the frequency of a 3-phase square-wave oscillator; the series regulator is controlled from the master frequency control. The output of the square-wave oscillator is fed into the bases of a transistor amplifier, and the output, which is a quasi-square wave, is transformed to the proper voltage and filtered to a sine wave. A feedback controls another series regulator to give the correct voltage output. This is only one method of producing an acceptable output; many others are in use. This one has the merit of suppressing the third harmonics.

5. The number of reactive voltamperes supplied by the static inverter is fixed by the filter capacitors installed. This influences the types of loads to be supplied, the voltage regulation on surge loads, and the current that may be supplied on faults.

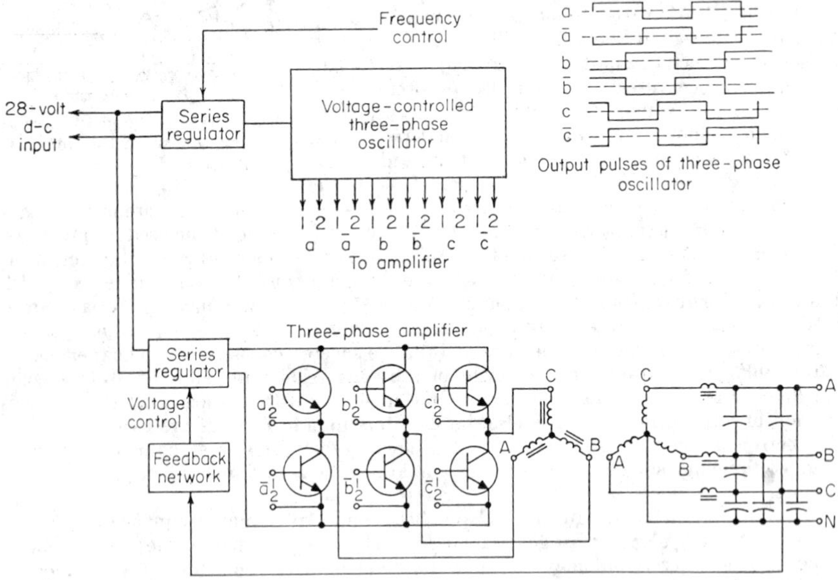

FIG. 21-14. Static inverter, 3-phase block diagram.

c. Transformer-Rectifiers. Ratings of aircraft transformer-rectifiers to supply 28-V d-c power vary from 1 to 200 A. Three-phase units are customary because of the excessive ripple in the output of the single-phase type. They are built in both regulated and unregulated types. Voltage variation of an unregulated unit over the input voltage range is shown in Fig. 21-15. Such units can be operated in parallel from the same input power supply if consideration is given to the increased rectifier voltage drop with age. That is, a new unit may not divide load with an older unit within acceptable limits. Principal elements of the unregulated type are a transformer to reduce the voltage to about 35 V a-c, a silicon rectifier stack, and a suitable enclosure. A fan is often used for cooling the stack, and a filter for smoothing the output ripple. The transformer usually has a Y-Δ-secondary to obtain 6-phase output for reducing ripple.

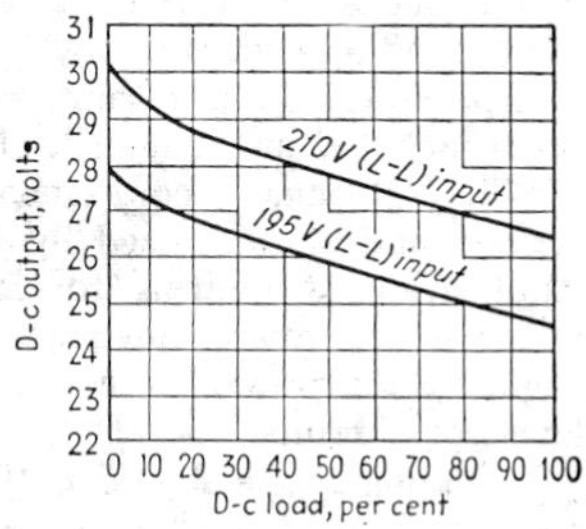

FIG. 21-15. Unregulated transformer, rectifier voltampere characteristics. (*Data from General Electric Company.*)

While unregulated transformer-rectifiers have been used widely, there are many applications in which the loads require the narrower voltage limits obtained with a standard d-c generator system. Thus, regulating saturable reactors have been added to the basic transformer and rectifier. Effectively, regulation is accomplished by varying the a-c voltage drop across the saturable reactors to maintain fixed output voltage. This voltage drop is controlled by a voltage regulator utilizing a cold-cathode reference tube and a control magnetic amplifier. A load-equalizing circuit similar to that used with d-c generators is also provided, acting through an additional control winding on the magnetic amplifier. The d-c voltage proportional to load current for this purpose is obtained from the rectified output of a current transformer in one of the a-c lines. By using the same equalizing voltage per ampere of load, it is possible to parallel and divide load with d-c generators.

Improvements in rectifiers in recent years permit operation at plate temperatures up to 140°C, with life in aircraft use in excess of 1,000 h. With transformers, magnetic amplifiers, and saturable reactors using Class H insulation, operation in ambients of 120°C is possible. In some cases, power-factor-correction capacitors are used.

30. Electric-load Analysis. Owing to the variety of the types and characteristics of the electric loads used in aircraft, it has been found convenient to employ a standardized approximation method of determining the required generating capacity on any particular airplane. The principal element of this method is the load-analysis chart, shown for d-c systems in Table 21-6, by means of which the total average amperes is determined for specified time intervals during the various aircraft operating conditions. Typical conditions for military-combat-type aircraft are loading and anchor, start and warm-up, taxi, take-off and climb, cruise, cruise-combat, and landing. The chart headings are self-explanatory, only two operating conditions being shown. Under each condition, the "Amp" column is obtained by multiplying the number of units operating simultaneously during this condition by the amperes per unit. Average amperes for each of the specified intervals is the "Amp" column multiplied by the actual operating time accumulated during the interval divided by the interval. Totals are obtained for each interval column.

The a-c load-analysis chart is similar, taking into account additional data required, such as power factor, watts and vars per phase, and allowable frequency limits. Time intervals generally used are 5 s, 5 min, and 15 min (30 for cruise and cruise-combat) instead of the 30 s, 2 min, and 15 min used for direct current.

Once the "total average amperes" has been obtained, it is then necessary to determine the smallest generator rating which will meet these requirements. Aircraft generators reach ultimate stabilized temperature in 15 to 30 min; so these intervals usually are equivalent to continuous generator rating. However, for any given operating condition the generator continuous rating will depend upon engine speed (or drive limitations) and available cooling. The allowance for spare generating capacity may vary for different manufacturers and particular aircraft requirements, but general rules for military types are 50% spare capacity on single-engine aircraft and 100% spare capacity on multiengine aircraft. For commercial types, all essential loads must be supplied with one-half the engines inoperative. Of course, it is necessary to match the short-time load requirements as well as the continuous values with the corresponding generator ratings.

Space limitations permit brief discussion of only a few of the more important electric loads included in the load-analysis chart.

31. Lighting. Requirements for some lighting affecting safety are rigidly prescribed and standardized by FAA regulations, as are position and navigation lights. Application of lighting for instruments and cockpits requires careful consideration in order to provide adequate illumination without distraction, glare, or eye fatigue. Lighting for passenger comfort and convenience is dictating the adoption of higher levels of illumination. Both incandescent and fluorescent lighting are used.

Landing lights should be so installed as to reduce to a minimum the veiling glare resulting when light in the beam is reflected back from fog, moisture, and dust particles in the air.

32. Gyroscopic Instruments. Gyroscopes provide a reference axis. The gyroscope is usually a 3-phase or single-phase 400-c motor rated either 115 or 26 V. The rotor of the motor comprises the gyroscopic mass. Electrical gyroscopes are used in gyro-horizon instruments, drift meters (gyrostabilized), electric turn-and-bank indicators, remote-indicating earth-inductor compasses (gyrostabilized), automatic pilots, gunsights, and radar.

Table 21-6. Aircraft D-C Electric-load Analysis Chart Showing Typical Figures for Two of the Several Aircraft Operating Conditions
(Data from U.S. Military Specification MIL-E-7017, Feb. 28, 1951)

A	B	C	D	E	F-1					F-6				
					Loading and anchor					Cruise-Combat				
Equipment	Part designation	Number of units	Amp per unit	Operating time, min	Amp	Average amp				Amp	Average amp			
						0.5 min	2.0 min	15.0 min	Min.		0.5 min	2.0 min	30.0 min	Min.
A—Armament and bombing:														
Turret (stand-by)		2	40.00	30.00						80.0	80.0	80.0	80.0	
Turret (max.)			100.00							60.0	60.0	60.0	60.0	
Control, bomb arming		24	0.50	30.00						12.0	12.0	12.0	12.0	
Motor, bomb door		1	150.00	0.25	150.0	75.0	18.8	5.0		150.0	75.0	18.8	2.5	
Relay, bomb door		2	2.00	0.25	2.0	1.0	0.3			2.0	1.0	0.3		
Rocket firing		4	10.00	0.03						Neg.				
C—Flight controls:														
Motor, elevator tab		1	4.00	0.10										
Motor, trim tab		2	3.00	0.10										

33. Remote Indication. Self-synchronous devices provide one means for lightweight, reliable, and accurate indication from remote stations. Both d-c and a-c types are used in aircraft, sometimes with as many as four indications incorporated on one dial weighing less than 1 lb. Individual transmitters weigh approximately 0.3 lb. Liquid-level indication has also been provided by means of variation in the capacitance of a tank unit as the dielectric changes with the level of the fluid in the tank. The signal is amplified to operate the visual indicator. Indication of engine speed is accomplished with a permanent-magnet a-c tachometer generator driving a magnetic drag indicator. Mass rate of fuel flow is measured by means of a transmitter in which a synchronous-motor-driven impeller imparts a fixed rotational velocity to the fuel passing through it. The torque produced in then removing this velocity is proportional to the momentum, which is, in turn, proportional to the mass rate of fuel flow.

34. Remote and Automatic Controls. Remote controls are very common and include solenoid values, interlocks, control surface positioning, and crew and passenger comfort controls. Much of this is accomplished by servomechanisms on closed-circuit controllers. These vary from a simple on-off heater to a highly complex attitude control system. In military aircraft, gun or rocket pointing at speeds of one or more Mach numbers involves pointing the whole airframe, in many cases with complex computers on board both the aircraft and the projectile. Development of the inertial platform for nuclear submarines has led to parallel development of such systems for both aircraft and missiles, relieving the pilot of many of his duties on long-range flights.

35. Communications and Radar. Communication systems for modern aircraft have become integrated with position-sensing and display devices; Loran, Tacan, and various others have been developed and often serve for the dual purpose of communication and distance measuring. The complete communication system has in the past taken a large portion of the generated electric power of an airplane, but the introduction of transistorized equipment has tended to cause a decrease in total load. On the other hand, the miniaturization possible with transistors has made feasible the use of more equipment so that the total load has decreased very little.

A very marked change has been brought about by the use of radar. Virtually all commercial aircraft are equipped with weather-avoidance radar and many with collision avoidance. A popular version of radar is height-measuring equipment, as well as modifications of the Doppler effect for distance measurement.

In military aircraft, an additional requirement is for jamming enemy radar. This may require large amounts of power.

SPACECRAFT ELECTRIC SUBSYSTEMS

36. General. The whole subject of electric subsystems for spacecraft is in a continual state of change, and an authoritative statement today will be outmoded tomorrow. There are, however, some basics which can be explored, such as energy sources and possible means of conversion, typical distribution schemes, and exotic utilization schemes that are being pursued.

37. Electric Power in Spacecraft. In general, the usages specified in Par. **42** on electric systems in aircraft serve here, but with some modification.

a. Unmanned Spacecraft

1. Control systems (attitude, orbital readjustment, computers, etc.).
2. Communication (telemetry, radar, etc.).
3. Pyrotechnics (separation squibs, retro and ullage rockets, engine starting, etc.).
4. Pumps (fuel and oxidizer pumps; cooling).
5. Precision timing devices (programmers).
6. Navigation (gyros, inertial platforms, etc.).
7. Camera controls.
8. Actuators.
9. Instrumentation.

b. Manned Spacecraft

1. More complex communication requirements.
2. Life-support systems (air regeneration, waste disposal or conversion).
3. Body-temperature control. (The temperature of an unmanned spacecraft may be rigidly controlled, as many components will not operate at either low or high temperature. Thus, body-temperature control is only an extension of the normal control systems which may be installed on unmanned spacecraft.)

38. Types of Electric-power Subsystems. Early spacecraft electric-power subsystems were at first extensions of aircraft practices, and this influence still lingers.

1. The most extensive use is found for the 28-V d-c systems. The exact voltage used will vary considerably according to local need. The supply voltage for telemetry transducers is 5 V, and there are many instances of 56-V systems (two 28-V systems in series) in use for power. There is no real standardization; the missile designer has tended to consider each case as unique. Additionally, there are many voltage levels for the logic systems in use; usually these are internally generated in the system itself.

2. The a-c subsystem is a follow-on from the aircraft 115-Y/200-V system, but the general trend now is to use a d-c to a-c converter at the place of need rather than installing a central supply. This may change in the future when ion-propulsion systems are sufficiently developed.

39. Operating Conditions. The operating conditions experienced in spacecraft are much more severe than those in aircraft, as follows:

a. Extreme temperature ranges, extending to very near absolute zero.

b. Pressure differential from atmospheric to that of space.

c. Possible high humidity in the spacecraft interior on the ground.

d. High vibration levels while being boosted into space.

e. High acceleration levels.

f. Zero-gravity (0-g) conditions in orbit.

g. In space, no protective atmospheric coating over materials, and oils, greases, etc., sublime off. Two bare surfaces may form a "weld" upon contact.

Any or all of the above may be operative at the same time. The spacecraft-electric-subsystem designer is required constantly to evaluate combinations of new conditions for his designs.

At this point, it should be emphasized that every watt of power used in a spacecraft must be dissipated to space as heat. That is, in a spacecraft no infinite heat sink is available such as that provided by the earth and its atmosphere. The watts loss due to inefficiency and the watts put into a device both remain aboard the spacecraft. This fact may complicate the spacecraft design immensely. The generation of large quantities of power must be associated with large radiators to radiate the generator inefficiency to space at a relatively high temperature. The generated wattage used on board must be radiated to space at a much lower temperature. Various methods to accomplish these purposes have been designed. Heat dissipation from the spacecraft is not a subject for this section. It is mentioned because it is a serious restraint upon and a primary consideration of the electrical-system design.

40. The Total Complex. The spacecraft is only a part of a system and may well be only a small part. The spacecraft must be put into orbit by a booster rocket. This, in turn, must be supported by a large ground-support installation. The components of this system are shown in Fig. 21-16, giving the electrical interfaces.

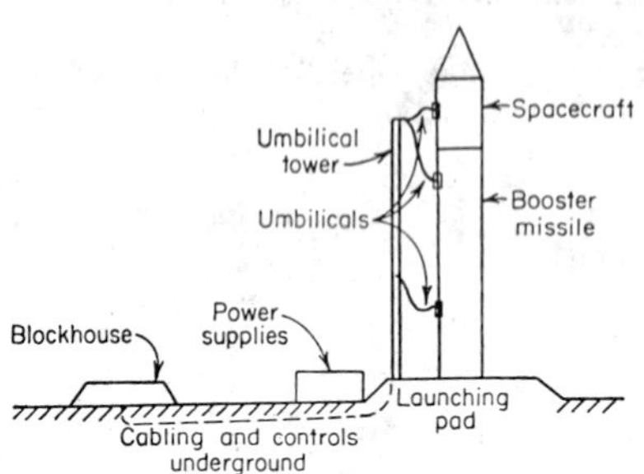

Fig. 21-16. Booster rocket and spacecraft on launch pad. Not shown is the gantry, which surrounds the booster and spacecraft prior to launch, and which serves as both protection and work platform. It is mobile and is moved away a short while before launch.

1. The spacecraft sits on top of the booster, and there are interstage connectors between the two. Guidance and control components are commonly located in the spacecraft, and separation functions upon burnout of the booster may be controlled from there.

2. The umbilical connectors bring power to the spacecraft and booster prior to lift-off to prevent depletion of their power supplies. A number of functions are monitored through the umbilicals. These connections are severed, generally, by explosive charges just prior to lift-off.

3. The power supplies for both the booster and the spacecraft are located close to the pad for minimization of voltage drop.

4. All signals monitored through the umbilicals are brought to the control center. In addition, telemetry transmitters aboard the booster and spacecraft send intelligence which is monitored. The items monitored may include various critical voltages and currents, performance of the propulsion system, vibration about the spacecraft during launch, and, in manned spacecraft, the physiological condition of the astronauts. Firing controls, destruct controls, etc., are located in the control center.

Since the ground-support items are not flight hardware, there is no need for the miniaturization that exists in items installed in space vehicles. However, ground-support circuitry becomes quite sophisticated in most cases, since automated check-out of the vehicle systems is incorporated.

41. Miniaturization and Weight Saving. Every ounce of weight in the spacecraft must be accelerated to orbital or earth-escape velocity, and this requires the expenditure of propellant. Thus, there has been an increasing push for weight savings. A 5-A switch in aerospace would probably be rated at less than 1 A in commercial use. At the same time, such equipment must operate under environmental conditions that a commercial component will never experience. The result has been an increased cost, but much more reliable components. The limiting criterion in almost all cases is the generation of heat and its dissipation. A parallel development with miniaturization has been the reduction of wattage dissipation. This has been quite successful. Relays have more efficient magnetic circuits; contacts are specially designed to be low-resistance; transistors have replaced vacuum tubes; and at present writing, integrated circuitry is replacing discrete components. The state of the art is changing so rapidly that it is difficult to predict the trend of the future except to say that extreme miniaturization will continue, solid-state logic will replace relaying, and all functions possible will be made automatic, but the automatic systems will incorporate manual backup where manpower is available.

42. The Spacecraft Electric System. In common with the aircraft electric system, the spacecraft electric system consists of three well-defined parts: (1) generation (or power supply), (2) distribution, and (3) utilization.

43. Spacecraft Electric-power-supply Methods. The spacecraft must carry its total power supply with it; refueling stations do not exist. However, orbiting refueling or resupply stations are being proposed and may be used in the near future. Power-supply methods are generally divided into two classifications: (1) direct conversion, for example, converting the photon energy of the sun directly into electric energy, and (2) indirect conversion, converting nuclear fission or chemical energy to heat and then using a heat-conversion apparatus for conversion to electric energy. Greater efficiencies are usually attainable with the first class, since Carnot efficiencies are not involved, but at this writing, larger amounts of power are available by the second. Each class will be treated separately.

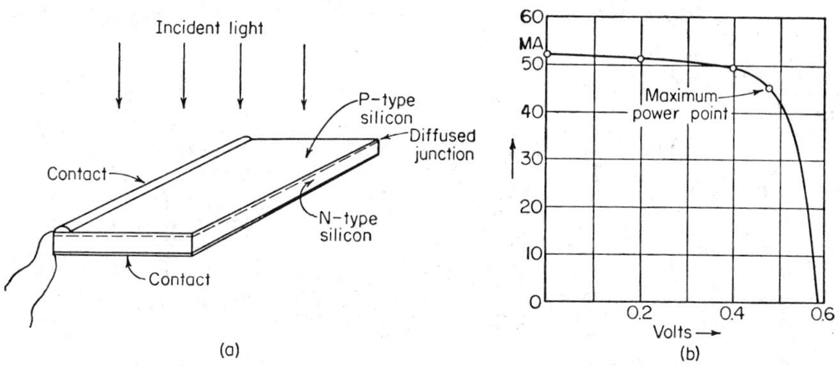

FIG. 21-17. Solar-cell construction (*a*); characteristic curve of solar cell (*b*).

44. Direct-conversion Methods (see also Sec. **12**). 1. *The Solar Cell.* The solar cell is probably the device most often used in energy conversion today, particularly for unmanned spacecraft. It is reliable and in the proper quantities produces enough power for many applications. It operates on the solar photon flux, which, in the vicinity of the earth, is about 1,400 W/m^2. This value increases as one nears the sun, being about 120% of this value in the vicinity of Venus and 80% in the vicinity of Mars. In Fig. 21-17*a*, a thin piece of silicon is mounted on a metal backing. This silicon wafer is made with an *n-p* junction near its surface. This may be either *n* on *p* or *p* on *n*; *p* on *n* is more radiation-resistant. The *n* layer contains an excess of electrons, the *p* layer an excess of holes; at the junction thermal diffusion has caused the electrons to migrate to the *p* layer and the holes to migrate to the *n* layer, leaving a thin layer of hole and electron-free silicon. A photon with the requisite energy, i.e., of a wavelength less than 12,000 Å, will be absorbed by one of the silicon atoms in the equilibrium area, creating an electron-hole pair. The field at the junction will carry the electron into the *n* layer and the hole into the *p* layer, causing a voltage to form between the *n* and *p* layers.

Other types of solar cells are being developed, such as thin-film or dendritic, and semiconductors other than silicon look promising. Present efficiencies may be surpassed by such developments.

Only a portion of the solar photons are of the energy necessary to create an electron-hole pair, and some of the pairs recombine before they can migrate to the proper side of the junction to be effective. The individual solar cell is small, usually 1×2 cm, and thus cannot supply much power. It is necessary to join thousands of the individual cells in series-parallel array or combination to obtain the voltage and current necessary to supply spacecraft loads. A typical voltampere curve of a solar cell is given in Fig. 21-17*b*.

Owing to the very large number of cells required for supplying even a small load, the solar cell is best suited to loads of under 1 kW, but larger installations have been recorded. Overall cell efficiency approaches 12%, with a probable limit of 16%. The cell is readily damaged by space radiation, and very thin covers are supplied for protection against such damage.

Maximum efficiency is not obtained unless the cell face is normal to the sun's rays, and various cell configurations are employed to accomplish this end, such as large paddles with the solar cells upon them and means for orienting the vehicle for maximum effect. The greatest use, however, is in vehicles which are spinning at a low rate. The outside of the vehicle is covered with cells so that at least a portion are exposed at all times. Such an arrangement gives an efficiency of 36% of the maximum possible but does have the virtue of not having to be oriented. Solar concentrators are used, but these require proper orientation to attain their potentialities.

2. *The Fuel Cell.* Fuel cells are devices by which fuels and oxidants are chemically combined to produce electrical energy. In this respect, the fuel cell is the exact opposite of the ordinary electrochemical reaction, where a current is passed through a chemical solution and hydrogen and oxygen are liberated. There are three general areas of fuel-cell development:

a. Hydrogen fuel cells.
b. Hydrocarbon fuel cells.
c. Biochemical fuel cells.

The hydrocarbon fuel cell is an attempt to use some hydrocarbon, such as hydrazine, directly as the fuel element, and it has attained some success. The biochemical fuel cell uses the fundamental electrochemistry of enzymes and bacteria in a life situation. Possibilities here include the conversion of waste products to useful products and the simultaneous generation of electricity. However, the hydrogen-oxygen fuel cell has seen the greatest development and is at present in use.

The hydrogen-oxygen fuel cell depends for its operation upon the ionization of gaseous hydrogen and oxygen at the cathode, with electron flow as a consequence. The chemical formulas underlying one such reaction are

Anode: $$2H_2 \rightarrow 4H^+ + 4e^-$$

Cathode: $$4e^- + 4H^+ + O^2 \rightarrow 2H_2O$$

The electrolyte here is acid, and the water is produced at the cathode. In the so-called "Bacon" cell, the electrolyte is a base, and the water is formed at the anode. It is mechanization of this process which has required development. The electrodes must be nonconsumable, and there must be a means of ionizing the hydrogen. The ion-exchange membrane is one such development, but other catalysts are also used.

The theoretical efficiency of the fuel cell is high compared with heat engines, running from 80 to 90%. This efficiency is not obtained in practice, since there is some heat of reaction generated, and there are circuit losses; around 55% is claimed. In addition, there is the voltage-regulation problem; the cell must be supplied with gaseous hydrogen and oxygen as a function of load, and the water formed must be carried away. There are various solutions to this problem, depending upon the particular development.

There have been several difficulties associated with the development of the fuel cell. These are listed below:

a. The pure bulk of the fuel and oxidizer in the gaseous form shows that it must be stored in the liquid form, and this, of course, at cryogenic temperatures. Further, there must be means of bringing the fuel and oxidizer to gaseous form before use. This storage and means for usage consume considerable space and weight and account for some of the complexity that accompanies a fuel-cell installation.

b. The fuel and oxidizer must be supplied in accordance with the load on the cell. Since hydrogen and oxygen are stored at cryogenic temperatures, there must be a means for conversion to usable pressures and volumes. To supply the necessary amperes capacity, it may be necessary to parallel several fuel-cell series banks.

The advantages of the fuel cell are so great that their development will undoubtedly be very rapid, and the advances in spacecraft power supplies will be reflected in the cells developed for industrial use. Here, fuels will be drawn from the hydrocarbons, while the oxidizer will be air. The utilization means are more complex and the efficiencies lower. The block diagram of a fuel-cell system is shown in Fig. 21-18.

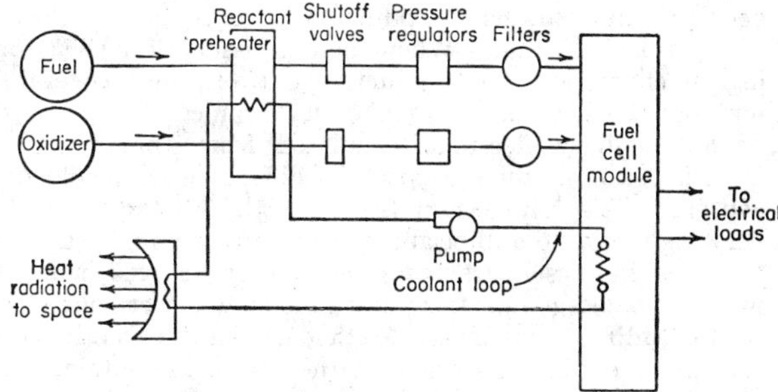

FIG. 21-18. Fuel-cell system.

3. *The Magnetohydrodynamic (MHD) Generator.* The conventional d-c generator consists of a conductor forced through a magnetic field. In the MHD generator, the conductor becomes a plasma, forced through a magnetic field. The plasma is contacted by conductors and the power drawn off. A simplified diagram of an MHD generator is shown in Fig. 21-19.

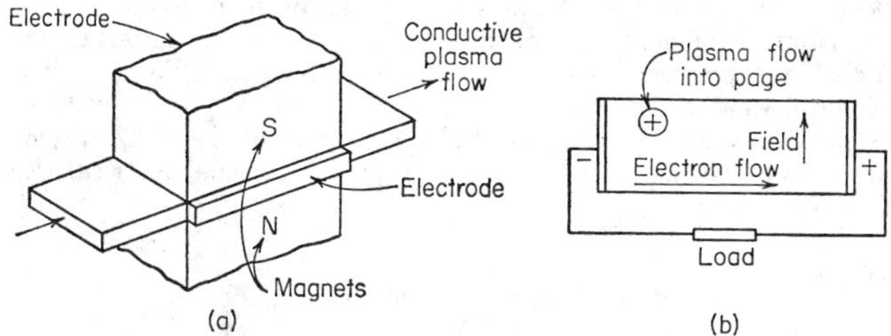

FIG. 21-19. Schematic of magnetohydrodynamic generator.

The conductive plasma flow, the electron flow, and the field are at right angles to one another; the voltage and power density per unit of volume flow are related thus:

$$E = (U \times B)n$$

$$PD = \sigma U^2 B^2 n(1 - n)$$

where U is the plasma velocity, B is the field strength, n is the ratio of the load resistance to the sum of the internal and load resistances, and σ is the conductivity.

Several major problems have appeared which have been only partly solved:

a. Conductivity of the plasma. Unless the plasma is at an elevated temperature, it may not be ionized and conductive. Further, at such elevated temperatures, the plasma is likely to erode the magnet faces and the electrodes. Methods have been employed to make the plasma conductive, such as by seeding with readily dissociated molecules.

b. High field strength is exceptionally difficult; for to obtain a large amount of power, the "channel" must also be large. A solution here is the development of superconductive materials. In outer space, with cryogenic temperatures, it will be possible to obtain a field strength of 100 kG over a large area.

c. The high temperatures involved have made the development of new materials necessary. Research is currently advancing the state of the art here.

The MHD generator would seem to be well suited to the generation of electricity where a continuous plasma source is available, as in a rocket engine, and large amounts of power are to be extracted. The reverse process, where power is supplied and motor action results, is in use for pumping liquid metals.

4. *Atomic Fission*. When there is fission of a U^{235} atom, the result is two or more highly charged decay products, neutrons, and several beta particles. Fissionable material has been used to make an "atomic battery" by capture of the beta particles on an auxiliary electrode, insulated from that carrying the fissionable material. Over a period of time, a potential difference will build up as a result of beta-particle capture. The voltage may become high, but the internal impedance of the battery is very high, and no appreciable power output is realized.

Since the fission fragments are highly charged, some attempts are being made to isolate the charged particles and thus have direct conversion to electrical power.

45. Indirect-conversion Methods. All the methods to be described utilize the conversion of heat to electricity, the heat being obtained in a number of ways. These include:

a. Solar energy.

b. Nuclear fission.

1. Nuclear reactors.
2. Radioisotopes.

The methods of obtaining heat are not a part of this write-up. It will suffice to say that a variety of types of solar-ray concentrators are used to obtain the necessary high temperatures, and the reactor power-conversion schemes are too numerous to cover in a short paragraph. There are many types of radioisotopes with different lengths of lives and different heat-generating capabilities. Which one is used will be governed by the individual requirements.

1. *Thermoelectricity*. This generation device has been in use for many years, and the chief contribution of aerospace technology has been the materials improvement, giving a much higher power density and efficiency than have been available with the usual thermocouple materials.

Thermoelectric devices are characterized by a figure of merit, Z, which is the ratio of the square of the Seebeck coefficient to the product of the electrical resistivity and the thermal conductivity.

$$Z = \frac{S^2}{Pk} \qquad \text{per } ^\circ\text{K}$$

where S = Seebeck coefficient, a measure of the open-circuit voltage generated per degree centigrade temperature difference of the junction, P = electrical conductivity, ohm-centimeters, and k = thermoelectric conductivity, watt-centimeters per square centimeter per degree centigrade.

This formula says that the Seebeck coefficient should be high and the product of the electrical and thermoelectric conductivities should be low. In an electric generator, the conductivity should be high to prevent ohmic losses; so the thermoelectric conductivity must be very low. These requirements eliminate metals and insulators and leave only the semiconductors, and it is in this area that the greatest development has occurred.

Two materials in particular have been developed which are useful as thermoelectric generators, lead telluride (PbTe) and germanium silicon (GeSi). The peak Carnot efficiency with these materials is around 30%, and the peak temperature of the hot junction is 800°C, but the realizable efficiency is no more than 3 to 4% owing to heat conduction, contact resistance, etc.

Thermoelectric-development problems have centered on (1) making successful contact between the materials, (2) the fragility of the material, (3) and heat flow across the materials. Successful thermoelectric generators have been constructed, but in general they are less efficient than the solar cell, and since high temperatures are required, some sort of focusing collector is required for sunlight conversion.

2. *The Thermionic Generator.* Another device requiring high temperatures for operation is the thermionic generator shown schematically in Fig. 21-20. Cesium is extruded through the pores of a coarse cathode material and, having a low work function, is ionized at the surface. Both cesium positive ions and electrons form about the cathodes, and dependent upon the elimination of the space charge provided by the cesium positive ions, the electrons travel to the anode. If a load is connected between cathode and anode, current will flow through it. The power generated is proportional to the difference of the Fermi levels at the cathode and anode. With cathode temperatures of 2200°K, a cesium thermionic-converter cell can provide nearly 1 V with 5 to 8 W/cm² of cathode surface and at an efficiency of over 15%. Theoretical efficiencies are above 40% but have not been realized in practice.

No thermionic converters have been installed in spacecraft at this writing. It appears probable that their most successful use will be when nuclear reactors are successfully adopted for space use. The thermionic converter is, of course, limited by Carnot efficiency. Its virtue is that the operating temperature is higher than that of other heat devices, and so the Carnot efficiency is higher.

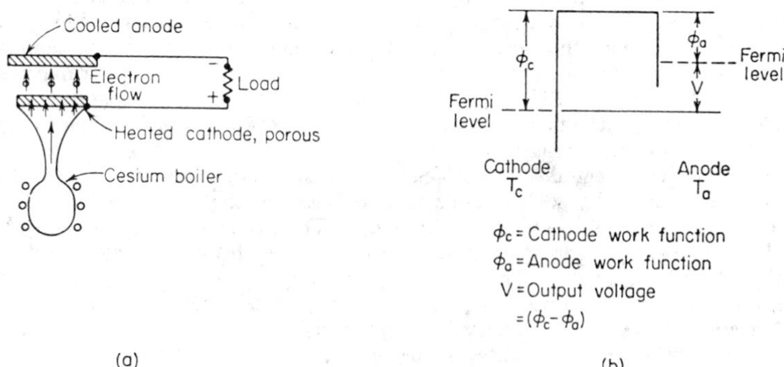

FIG. 21-20. Schematic of thermionic generator (*a*); thermionic-generator work-function diagram (*b*).

3. *Rotary Devices.* Rotary devices are subject to the same ills that plague earthbound motors and generators, bearings, seals, and the like. In addition, a rotating drive acts as a gyroscope and may inhibit maneuvering of a spacecraft. Nevertheless, the rotary device is proposed for use in cases where the loads are very high, and the state of the art in other conversion devices cannot supply the demand. Generally, the heat source is a nuclear reactor, and the problem is to transfer the heat to a generator located some distance from the reactor, so as not to be affected by nuclear radiation. In most cases, two series heat-exchange loops are used. Reference is made to the literature (Par. **49**) for further information.

Two general types of heat engines are used:

a. The turbine.

b. The piston engine—Brayton cycle, Stirling cycle, and Rankine cycle.

4. *The Electrostatic Generator.* Proposed, but not built, this generator holds promise for a high power generator in space. It would operate in the space environment, which is a far better vacuum than can be produced on earth and which is necessary for the production of very high voltages at very low currents. This technology is an extension of the old "electrical-influence" machines used in early electrical experimentation. A well-designed machine here has at least as high an efficiency as the electromagnetic machine.

5. *Radioisotopes.* The use of radioisotopes is for heat sources only and is discussed only because their characteristics affect the design of the conversion means. The

Table 21-7

Radioisotope	Major radiation	Half-life, years	Power density, W/cm³
Cs^{137}	α, β, and γ	30	0.21
Sr^{90}	β and γ	28	1.4
Ce^{144}	β and γ	0.78	30.0
Po^{210}	α	0.38	75.0

salient features of radioisotopes are:

1. Their half-life, i.e., the period in which they will decay to one-half their original potency. If a generator is to deliver 10 W for 0.38 year with the radioisotope polonium 210, the initial design must supply at least 20 W, for 0.38 year is the half-life of polonium 210 (see Table 21-7 for the characteristics of some representative radioisotopes).

2. Their radiation characteristics, which influence the shielding requirements, since much spacecraft electronic equipment will not withstand nuclear radiation.

3. Safety, first, on handling prior to launch; and, second, the degree of burnup and atmospheric contamination to be expected upon reentry.

4. Reactions with conversion apparatus. Since good thermoelectric devices use semiconductors and semiconductors tend to have their lattice structure destroyed by some types of radiation, the radioisotope and the semiconductor must be compatible.

46. Primary and Secondary Batteries. Batteries have been the primary source of power in all the early satellites and are the usual source for the spacecraft booster rocket. In the case of the booster rocket, the requirement is only for sufficient capacity to supply the loads until it burns out, and for this source a class of small, dense, high-energy batteries have been developed. Satellite or spacecraft service requires a battery which may be charged by solar cells or thermoelectric or thermionic devices to replace the normal usage of the satellite or spacecraft. The primary battery is one which is not rechargeable and, in aerospace use, is activated by addition of an electrolyte, either manually or automatically. The secondary battery is rechargeable; i.e., after discharge, the electrochemical process may be reversed by application of a voltage.

The lead-acid type of secondary battery once was widely used in aircraft. This battery is not applicable in satellites or spacecraft, for a number of reasons:

a. Very corrosive electrolyte.

b. Cannot be hermetically sealed.

c. Weight.

d. Poor voltage regulation.

Both the primary and the secondary battery are employed in aerospace applications. The primary battery is particularly useful in booster-rocket applications, where the service is for a single shot; recharging and long life are not important. One variety of the primary battery is the quick-energized battery. The battery is charged and dry until a special mechanism forces the electrolyte into the plates. Such a mechanism may be a lanyard or an electrical signal activating an explosive device that shatters a diaphragm and releases gas to force the electrolyte into the proper area.

A feature of almost all the high-energy batteries is "peroxide voltage." That is, the battery when charged will have an open-circuit voltage somewhat higher than normal owing to the formation of a peroxide rather than an oxide. This voltage disappears after a very short interval of use and is only a small percentage of the battery capacity. A 28-V battery may read around 35 V on open circuit; the current transient occurring when the load is first connected may be damaging to some electronic equipment. Special precautions are often taken to get rid of this peroxide voltage by a small preload applied prior to the application of the actual load. This is necessary since much aerospace equipment is designed to operate within narrow limits and might well be damaged.

Special charging means must be provided where secondary batteries are used, as the special spacecraft batteries are much less tolerant of incorrect methods of charging than are the familiar lead-acid batteries.

Table 21-8 provides a comparison of the energy densities to be expected from the most common primary and secondary batteries. This is by no means complete; as

Table 21-8

Battery type	Power density, Wh/lb	
	Primary	Secondary
Zinc-silver oxide (An/KOH/AgO)	80	40
Silver-cadmium (AgO/KOH/Cd)	33	30
Nickel-cadmium (NiOOH/KOH/Cd)	17	10

various manufacturers claim special advantages for their particular batteries, only a reasonable estimate is given.

Where very long life is expected, the two batteries showing the poorest power density, the silver-cadmium and the nickel-cadmium, are the only ones which have been able to operate in space for a year or more and then only when seriously derated. In any event, every battery installation in a rocket or spacecraft is so special that very little standardization has occurred. It is not possible to present tables of dimensions, weights, etc., as with the airplane lead-acid battery. Among the items that make such a table impossible for spacecraft are:

1. Quick-energized batteries are heavier and bulkier by reason of their actuation mechanisms.

2. The differences in life requirements in various missions change the battery size. A battery designed for low drain and charging rates will be very different from one that is to be quickly expended.

3. Good voltage regulation requires a heavier battery.

4. Sealed batteries designed for a long period of usage may have a stainless-steel noncorrosive case; a lighter battery without such requirements may have a magnesium case.

A very special type of battery is the thermal battery. This has been a classified item and not much has been available for publication, except that activation is by thermal means. When the internal temperature rises sufficiently, the chemical ingredients liquefy and the battery is usable. Life of the battery is very short, often being in seconds, but never over a few minutes. These batteries are used when large amounts of energy are needed for very short periods.

47. Special Electric Utilization Equipment. 1. Missile and spacecraft environmental and functional requirements have created some entirely new equipment which will be briefly described here. No treatment will be given to special types of usual equipment such as radars, etc.

2. *Brushless D-C Motors.* Altitude has always been a problem in d-c generators and motors because of the commutator problem. This is accentuated in vehicles that operate above the atmosphere permanently, and the brushless d-c motor has been developed to eliminate it. The a-c motor is not directly applicable, since the majority of the power-generating devices in spacecraft are d-c and conversion to a-c power involves another inefficiency, that of the inverter. Two general types of brushless motors are in use, the inverter–induction motor and a d-c motor with an electronic commutator.

a. The inverter–induction motor uses an inverter which is specially designed to work with a specific induction motor and which uses the motor windings as the usual filter. The operation is square-wave, and the combined efficiencies of the inverter and induction motor are at least as high as for a d-c motor alone. The best of such combinations operate 3-phase, but 2-phase is also successful. In both cases, either the motor must be designed for low starting current or the inverter must be designed to saturate so that starting current is limited; otherwise the transistors or silicon-controlled rectifiers in the inverter will be overloaded.

b. To synchronize the electronic commutator with the armature of the motor, some position-sensing device is necessary. This device controls the operation of the transistors feeding the armature, assuring that the magnitude and phase of the current supplied are correct.

3. *The Electroexplosive Device (EED).* The EED is simply a resistance wire embedded in an explosive substance; it is used to initiate various sequences in the spacecraft, such as stage separation, automatically activated batteries, or the ignition of the main engines of a booster or sustainer rocket. Ideally, it requires only a small amount of current to actuate and must not be sensitive to extraneous impulses, such as coupled r-f energy. The EED has become a very specialized device. In most cases, a large part of the load on a booster or sustainer missile will be the EEDs.

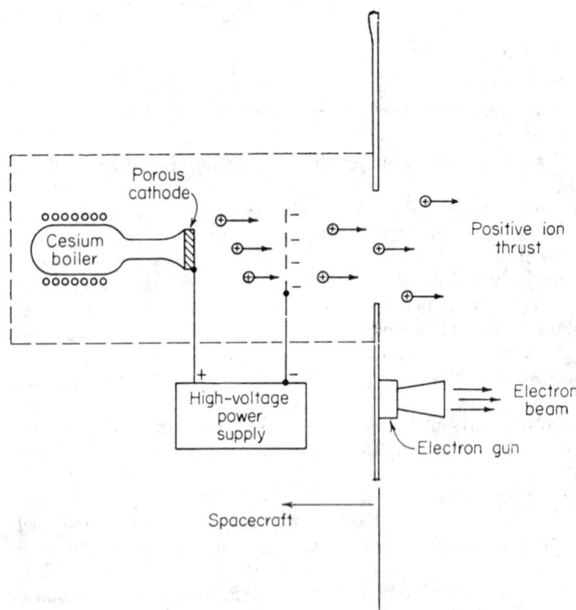

Fig. 21-21. Ion-propulsion engine for spacecraft.

4. *The Ion Engine.* After the spacecraft is free from the earth and is in space, very little acceleration is required to bring it to extremely high velocities. Small accelerations are not obtained efficiently with normal thrust-generating devices, and so the ion engine has been proposed for this usage. This engine accelerates relatively heavy ions to very high velocities and directs them so as to accelerate the spacecraft.

Figure 21-21 gives a schematic diagram of a simple ion engine. Cesium is ionized at the cathode, and the positive ions are accelerated toward a negatively charged grid. The grid being open-meshed, most of the ions pass through and, leaving the spacecraft, act to accelerate it. There are other grids in the engine designed to control the exit velocity of the ions. The ions, being positive, leave the vehicle with a net negative charge. This could build up to such proportions that no positive ions would escape and thus no thrust would be provided. The electron gun shown is intended to keep the spacecraft at or near zero potential.

48. Distribution and Control Systems. 1. The booster or ballistic-missile electric distribution and control system is quite simple, generally, being only a battery or batteries feeding loads, no battery charging or regulation systems being needed. The life of such systems is measured in minutes, at the most, and the emphasis is on reliability for the severe environment to be encountered.

When the spacecraft, manned or unmanned, is considered, the problem is compounded by the need for voltage regulation. A schematic of a power distribution system for a typical unmanned earth satellite is shown in Fig. 21-22, and the salient features will be discussed briefly:

a. The power source is generalized and may be any of the devices discussed. Two are shown for reliability; blocking rectifiers prevent reverse current flow.

b. Each power source charges a battery through a regulator, since spacecraft batteries require careful charging to prevent damage. Again, a blocking rectifier is used to permit battery charging but prevent the battery from feeding back into the regulator.

c. The bypass regulator is used to divert the power source current directly to the load bus to avoid charging rates in excess of those which the battery charge regulator can handle. The unregulated bus is kept at a maximum of 32 V d-c; nominal is 28 V d-c.

d. All loads may be fed from either No. 1 or No. 2 distribution bus. Blocking rectifiers are used to decouple the distribution buses from one another.

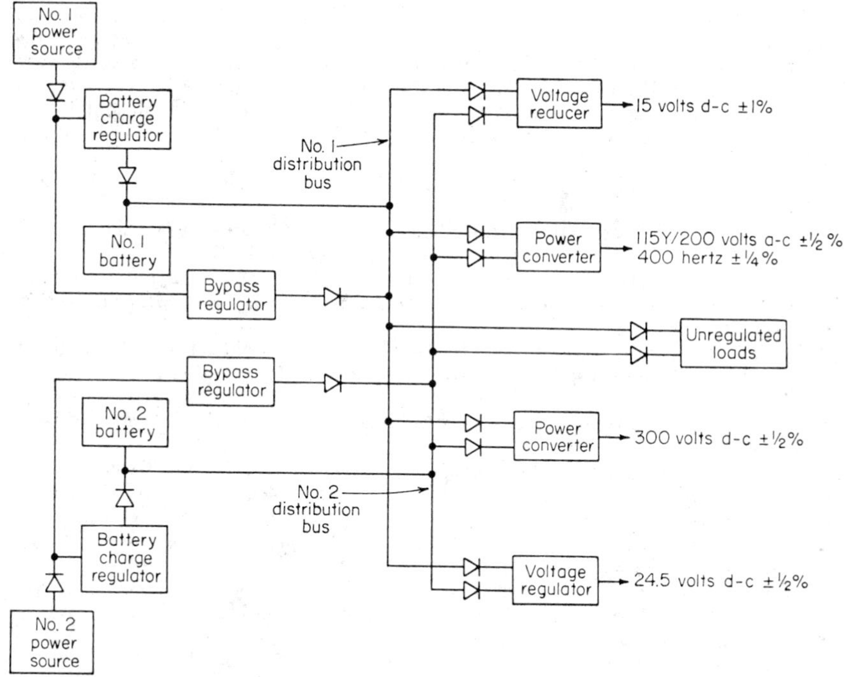

FIG. 21-22. Distribution system for spacecraft.

2. *Conversion Devices.* Conversion devices used in missiles and spacecraft are generally similar to those used elsewhere but differ in some respects. The differences are noted briefly:

a. Voltage regulators may be complex, depending upon the regulation requirements. They are invariably solid-state.

b. The voltage reducer may be some form of lossy device or may be a d-c to d-c converter. This will be treated in more detail in the next paragraph.

c. The generalized power converter, or d-c to d-c converter, is shown in the schematic of Fig. 21-23. A d-c supply powers an oscillator, the output of which is a square wave. The duty on the transistors is very light, since they are always operating in the switching mode. The voltage output is dependent upon the transformer ratio, and since the output is a square wave at a relatively high frequency, rectification and filtering are not difficult. Where voltage regulation is required, a control of the base drive of the transistors will suffice in most cases. If more control is needed, a series regulator in the output will be required. Overall efficiency will be 80 to 90%, and this type of power converter is thus superior to the usual series of shunt type of control for voltage reducers.

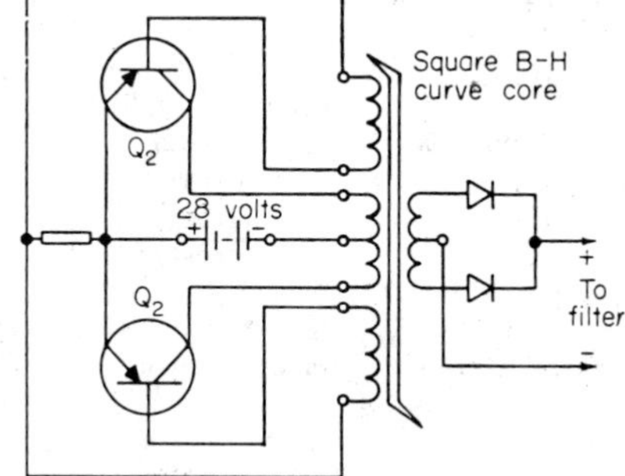

FIG. 21-23. Generalized power converter.

Power converters for conversion to precision alternating current are of the same type described in connection with aircraft electrical systems, solid-state devices, generally precision-frequency-controlled. They may be single or 3-phase, as the requirements dictate.

49. Bibliography

1. *Aircraft*

ANDERSON, H. C., CRARY, S. B., and SCHULTZ, N. R. Present D-c Aircraft Electric Supply Systems; *Trans. AIEE*, 1944, Vol. 63, pp. 265–272.

CARLSON, K. W., and SHERRARD, G. S. Distribution System Reliability of 28 Volt D-c Aircraft Electronic Systems; *AIEE Paper* 52-132; *Trans. AIEE*, Vol. 71, Pt. II, pp. 113–117.

FINISON, H. J., and KAUFMANN, R. H. "Dc Power Systems for Aircraft"; New York, John Wiley & Sons, Inc., 1952.

CALDWELL, S. C., and WOOD, A. J. The Effects of Abnormal Conditions on Aircraft, Parallel A-c Power Systems; *Trans. AIEE*, 1953, Vol. 72, Pt. II, pp. 379–384.

ANDREW, J. D. Aircraft Power Distribution Circuitry, Experimental Determination of 400 Cycle Impedance of Wire On; *AIEE Paper* 54-374, *Trans. AIEE*, Vol. 73, Pt. II, pp. 469–478.

KAHLE, H. A., and McCONNELL, H. M. Analogue Computer Methods Applied to steady-state A-C System Problems; *AIEE Paper* 56-766, *Trans. AIEE*, Vol. 75, Pt. II, pp. 279–282.

SMITH, R. E. A-C Generator: A Brushless Air-cooled Aircraft; *AIEE Paper* 57-468, *Trans. AIEE*, Vol. 76, Pt. II, pp. 189–191.

RIAZ Energy Conversion Properties of Induction Machines on Variable Speed Constant-frequency Generating Systems; *AIEE Paper* 58-917, *Trans. AIEE*, Vol. 78, Pt. II, pp. 25–30.

HAMER, W. J. Aircraft Storage Batteries; *AIEE Paper* 60-849, *Trans. AIEE*, Vol. 79, Pt. II, pp. 277–287.

IEEE, Aircraft and Missiles Electric System Guide 750, January, 1960.

GAYEK, H. W. Transfer Characteristics of Brushless Aircraft Generator Systems, *IEEE Trans. on Aerospace*, April, 1964, Vol. AS-2, No. 2, pp. 913–928.

PLETTE, D. L., and CARLSON, H. G. Performance of a Variable Speed Constant Frequency Electrical System; *IEEE Trans. on Aerospace*, April, 1964, Vol. AS-2, pp. 957–970.

2. *Spacecraft*

ABRAHAMSON, L. T. Silver-Zinc Batteries or Source-primary Electric Power for Pilotless Aircraft; *AIEE Paper* 57-491, *Trans. AIEE*, Vol. 76, Pt. II, pp. 297–300.

SCHUH, N. F., and TALLENT, R. J. Solar Powered Thermoelectric Generator Design Considerations; *AIEE Paper* 59–847, *Trans. AIEE*, Vol. 78, Pt. II, pp. 345–352.

BECKMAN, C., BEDROSIAN, S. D., BEKOWITZ, R. S., and CHEN, T. C. Research Guidelines for Digital Computer—Controlled Launch Control and Checkout of Operational Satellite System, *IEEE Trans. on Aerospace*; August, 1963, Vol. AS-1, No. 2, pp. 1065, 1073.

LEVIN, H., and ASAM, A. R. MHD Generator Relationships; *Jour. AIAA*, Feb. 13, 1963.

KLAVAN, L. S., and YOCHELEY, S. B. Computer Controlled Launch Control and Checkout of Operational Satellite System; *IEEE Trans. on Aerospace*, August, 1963, Vol. AS-1, No. 2, pp. 1249–1261.

SZEGO, G. C., and COHN, G. M. Fuel Cells for Aerospace Applications; *Astronautics and Aeronautics*, May, 1963.

KAHN, B., and GOURDINE, M. C. Electrodynamic Power Generation; *Jour. AIAA*, August, 1964.

KOTNIK, J. T., and SATER, B. L. Power Conditioning Requirements of In Rockets; *IEEE Trans. on Aerospace*, April, 1964, Vol. AS-2, No. 2, pp. 497–504.

GINGRICH, J. E. Radioisotope Fueled Thermionic Space Power System; *IEEE Trans. on Aerospace*, April, 1964, Vol. AS-2, No. 2, pp. 669–674.

GOULD, C. L. Solar Cell Power Systems for Space Stations; *IEEE Trans. on Aerospace*, April, 1964, Vol. AS-2, No. 2, pp. 759–768.

PIERRO, J. J., and PHILLIPS, J. E. Investigation of High Frequency Power Conversion and Generator Techniques; *IEEE Trans. on Aerospace*, June, 1965, Vol. AS-3, No. 2, pp. 411–422.

Holt, W. T., and Lamonte, R. J. Monitoring and Recording of Physiological Data of the Manned Space Program; *IEEE Trans. on Aerospace*, June, 1965, Vol. AS-2, No. 2, pp. 341–344.

ELECTRICAL EQUIPMENT FOR AUTOMOTIVE VEHICLES

By Gerald E. Rigsby

50. Automotive electrical systems are in reality a complex system of systems working together to achieve the high degree of satisfactory performance, durability, and reliability necessary for their specific operating requirements. There are five basic systems to be considered: the starting system, the ignition system, the charging system, the lighting system, and the accessory system. Each must be designed to meet its individual performance requirement as well as that of the total electrical system. Passenger cars in almost all cases use 12-V batteries with working system voltages of 12 to 15 V. Passenger cars have very complex electrical systems, whereas the farm tractor may feature only starting, ignition, and charging equipment. Trucks, buses, and special vehicles have equipment designed to meet their specific operating requirements.

51. Electrical equipment for passenger cars consists of battery, generator (diode-rectified), voltage regulator, cranking motor, ignition system, lights (for driving, turn and brake warning signals, instruments, indicators, and interior), accessory motors for heating and air-conditioning blowers, actuators, clocks, windshield wipers, radio, horns, and other automatically and remotely operated controls and accessories. Manual and vacuum switches, solenoids, and relays control the equipment (see Fig. 21-24 for a simplified schematic wiring diagram).

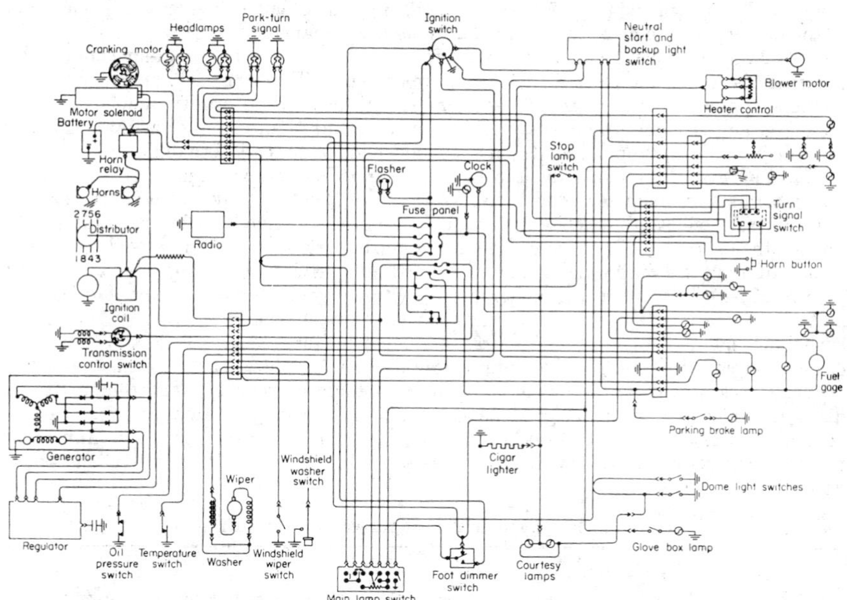

Fig. 21-24. Typical passenger-car wiring diagram.

52. The battery is the reserve power source for the automotive electrical system. It supplies the current to crank the engine, to supplement the generator when the load exceeds its output, and to operate the accessories when the engine is not running.

A battery is an electrochemical device and, therefore, depends on chemical reactions to convert chemical energy to electrical energy. The efficiency of this conversion is very dependent upon temperature. Unfortunately the temperature drop which causes loss of battery energy also increases the cranking load (engine friction). Because of this combination of circumstances, cold cranking conditions are generally used as criteria to determine whether or not battery performance will be satisfactory. Figure 21-25 shows the temperature effect on battery output.

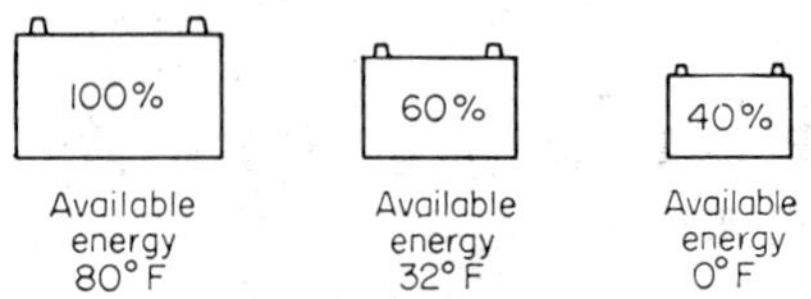

Fig. 21-25. Temperature effect on battery output.

Batteries for automotive vehicles are usually either 6- or 12-V, the 12-V being used for passenger cars, farm equipment, and light-duty trucks. The 6-V batteries are of the heavy-duty type, used in series-parallel arrangements on heavy-duty trucks, buses, and off-highway equipment. Sometimes additional switching is done to provide 24 V for starting, while 12 V is retained for lighting and ignition.

Batteries are made up of cells, each producing an open-circuit voltage of approximately 2 V. Each cell is made up of positive and negative plates insulated from one another by a "separator" usually made of microporous materials. These microporous materials are usually rubber, plastic-impregnated paper, or paper made from plastic-impregnated wood fibers. The number of plates in a cell varies depending upon the performance required. The cells are connected in series by means of a lead-alloy connecting strap. Cases are molded of hard rubber or bituminous compounds, with compartments for each cell. The electrolyte consists of a solution of sulfuric acid with a specific gravity of 1.260 to 1.280 (at 80°F) when the battery is fully charged. The specific gravity will decrease to about 1.140 (at 80°F) when the battery is fully discharged. Recent improvements in battery construction are the one-piece molded cover and through-the-partition cell connectors. The latter reduces the *IR* drop in the battery during cranking, thus improving performance.

53. Data on typical electrical loads for air-conditioned and non-air-conditioned automobiles with 12-V electrical systems are shown in Table 21-9. Typical operation of the automobile does not make necessary the inclusion of intermittent loads such as cranking, actuator motors, horns, and turn signals and requires only partial consideration for periodically applied loads such as windshield wipers and brake lights.

54. The generator normally is capable of supplying the continuous electrical load under high-speed operating conditions. Its performance and drive-ratio characteristics must be selected so that system output capabilities at high speeds are sufficient to replace the energy removed from the battery during cranking and low-speed operation. This requires a thorough knowledge of typical city driving characteristics such as the percent time spent under various traffic conditions; generator, regulator, and battery ambient operating temperatures; relationships between car speed and engine speed with both manual and automatic transmissions; battery charge, discharge, and efficiency characteristics, etc. To obtain desirable battery life, it is necessary to select performances and drive ratios in a manner to limit battery cycling from excessive discharge at idle to a full charge state after prolonged driving. Manufacturers' ratings generally represent only high-speed operating conditions; however, it is also important to consider low-speed performance characteristics in making automotive-vehicle applications.

Table 21-9. Typical Passenger-car Electrical Loads

	Load, A at 14.0 V	
	Non-air-cond.	Air-cond.
Driving lights, front and rear, high beam	15.0	15.0
Ignition	2.0	2.0
Instrument lights and gages	2.5	2.5
Radio	1.5	1.5
Heater or air conditioning (clutch and high blower)	7.0	21.0
Total	28.0	42.0
Generator rating used	37	55

Generators usually are driven by the engine through V-belts at ratios of 2:1 to 3.25:1. When high engine speeds are combined with high drive ratios, excellent bearings and lubrication, a well-balanced rotor, and concentric slip rings are required for satisfactory life and performance. Passenger-car generators usually are of the 3-phase diode-rectified Lundell type, have between 8 and 14 poles, and are between 5.00 and 6.6 in in diameter, with lengths closely approximating these diameters. The output of these machines is d-c current suitable for battery charging.

55. Special-duty generators are required for trucks, buses, taxis, police cars, emergency vehicles, and off-highway equipment. These generators must be custom-tailored to satisfy the specific needs of output and environment. Some motor coaches have very high loads at low and high speeds to supply air conditioning and fluorescent lights. A generator that uses the engine oil to cool the windings and rectifying diodes is employed to satisfy these requirements. With the oil-cooling feature, these generators are capable of supplying up to 300 A at full output. Typical output curves for various types of automotive generators are shown in Fig. 21-26. Some off-highway-equipment applications use an integral generator-regulator. The transistor regulator is built into and connected internally with the generator.

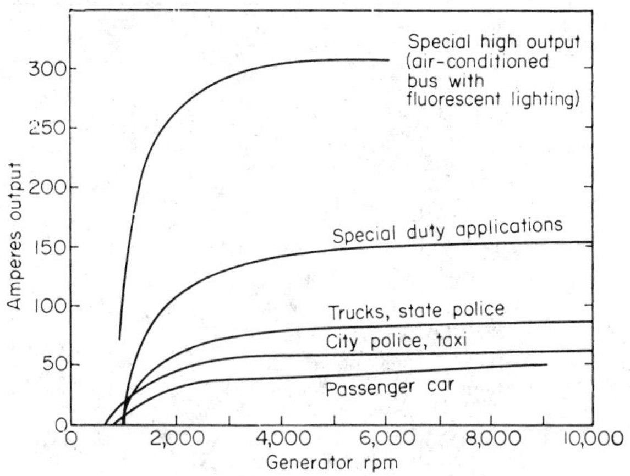

FIG. 21-26. Generator output curves.

56. Parallel operation of two generators is sometimes desired for multiengine operation. Because diode-rectified generators designed for automotive use generally are self-regulating for safe maximum current, they may be paralleled directly into the same battery system without any special considerations. Sometimes two generators are connected to the same high-capacity voltage regulator to operate in parallel.

57. Regulators for diode-rectified-generator-equipped electrical systems require only a voltage regulator or limiter since these generators control their own current output. The purpose of the voltage regulator is to protect the battery, lamp bulbs, and accessories by limiting the voltage developed by the generator. It consists of a voltage-sensitive relay with its contacts in series with the generator field coil. The vibrating contacts open and close the field-coil circuit to limit the amount of field current and, in turn, control the strength of the magnetic field in the rotor. This limits the voltage to the desired maximum value.

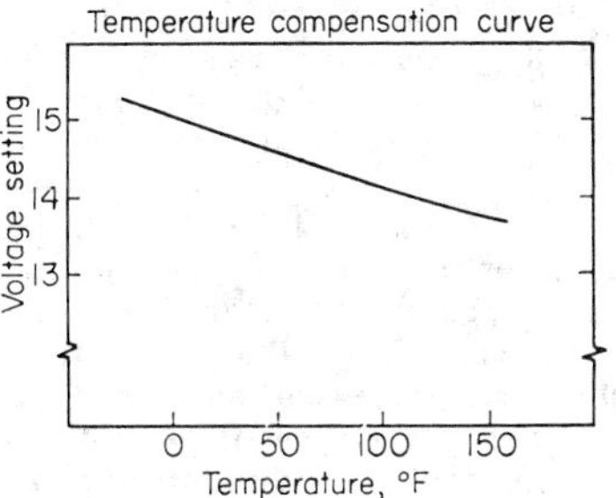

FIG. 21-27. Voltage-regulator temperature-compensation curve.

In order to match the charging characteristics, driving schedules, and requirements of the battery for passenger cars and some special-duty equipment, it is necessary to have the regulated voltage temperature compensated. A typical temperature-compensation curve is shown in Fig. 21-27. To obtain the maximum benefit of temperature compensation, it is necessary to mount the regulator in an area that will have the same ambient temperature as the battery.

An important consideration for improved battery life is the point in the circuit where the voltage regulator is connected. The regulator should sense, and thus control, the voltage at the battery terminals.

58. Transistor Voltage Regulators. The method of voltage control is similar to that of the vibrating-contact regulators, the difference being that a transistor circuit opens and closes the generator field circuit to control the field current. It is possible to obtain very accurate voltage control with a transistor regulator if so desired. Many special-duty applications require transistor voltage regulators designed to have accurate voltage control, long life, and satisfactory performance under extreme vibration conditions.

59. The cranking motor is a small, high-powered, intermittent-duty motor operated by the storage battery. It must rotate the engine at a speed sufficient to establish starting conditions. Passenger-car motors are about 4½ in in diameter and are usually series-wound four-pole machines. The addition of a shunt field is sometimes necessary to limit the no-load top speed. Owing to the intermittent nature of the starting operation, watts output can be very high in a unit of small overall size. Armatures are bar-wound, series-field coils are of strap copper, and brushes are of high copper content to keep resistance low enough to permit high current flow to develop the required output watts. A typical motor used with a 12-V passenger-car system will have about 0.010 Ω total resistance.

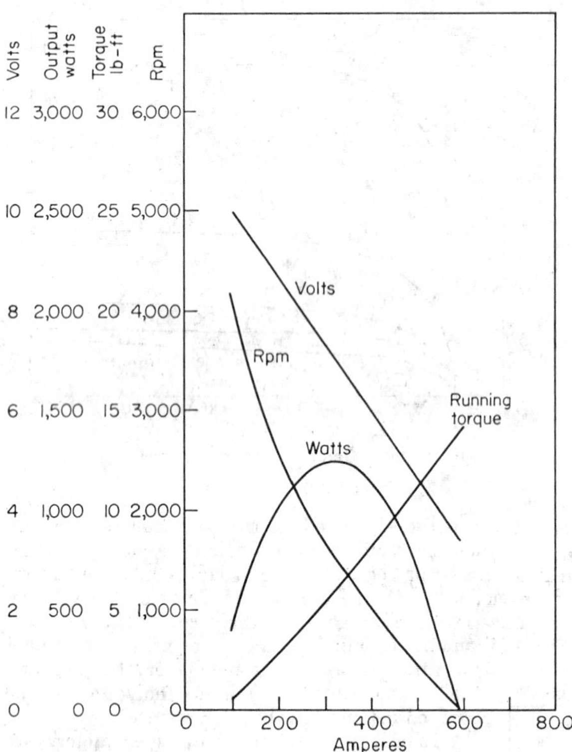

FIG. 21-28. Typical passenger-car cranking-motor performance curve.

60. Drive pinions on the cranking motor shaft apply torque to the engine when meshed with the engine ring gear. The gearing ratios of 12:1 to 20:1 result in good mechanical advantage for the motor and allow it to rotate at reasonable speed during cranking. Pinions are usually meshed to the ring gear electrically with a solenoid or other electrical-mechanical arrangement. After the engine starts, the motor is protected from being driven at high speed by use of an overrunning clutch.

61. The size of a cranking motor is determined by the torque and speed required to crank a particular engine. The low-temperature condition, when lubricating oil is stiff and when battery characteristics provide the least energy for the motor, is the condition used for motor design. Good engine characteristics in terms of carburetion, manifolding, ignition, and valve timing are important in reducing size and cost of the cranking motor. Typical 12-V passenger-motor performance curves are shown in Fig. 21-28.

62. Low-temperature cranking of a 400-in³ passenger-car engine at −10°F will require about 400 A of a "12-V" motor. Actual battery voltage may drop to about 6 V with this current draw. Torque required varies with condition of bearings, oil viscosity, piston displacement, and compression ratio but may be in the vicinity of 160 lb·ft to provide 40 engine r/min for starting. The same engine when warm may require only 150 to 200 A for cranking. The largest single variable for low-temperature cranking is oil viscosity. The force required to shear the oil film increases with viscosity, and viscosity increases with decreasing temperature. Figure 21-29 shows engine cranking speed vs. temperature for three different grades of engine oils. Although gasoline engines vary somewhat in starting characteristics, generally speaking, rotation at about 40 r/min is needed for good consistent firing and starting. It may be seen from the curve that one grade of oil makes a difference of about 10 to 15° in minimum starting temperature.

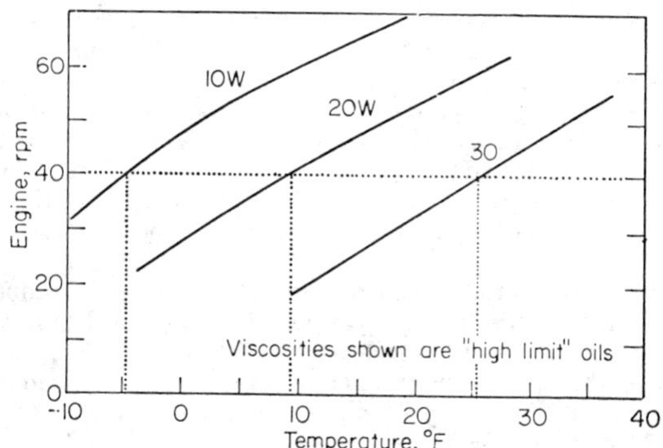

FIG. 21-29. Effect of oil viscosity on cranking speed.

Another factor that must be considered for modern high-compression-ratio passenger-car engines is the cranking requirement when the engine is hot. There is a tendency toward preignition with a hot engine. This preignition opposes the cranking motor, and the net result is too low a torque to crank the engine. To crank these engines, a larger battery and higher-horsepower cranking motor are required.

63. The ignition system produces and distributes the high-voltage spark to ignite the combustible fuel mixture in the engine cylinder. A schematic diagram of an ignition system is shown in Fig. 21-30.

64. The ignition coil consists of a primary and secondary winding mounted upon a common magnetic circuit with considerable air gap. In a 12-V system, the character-

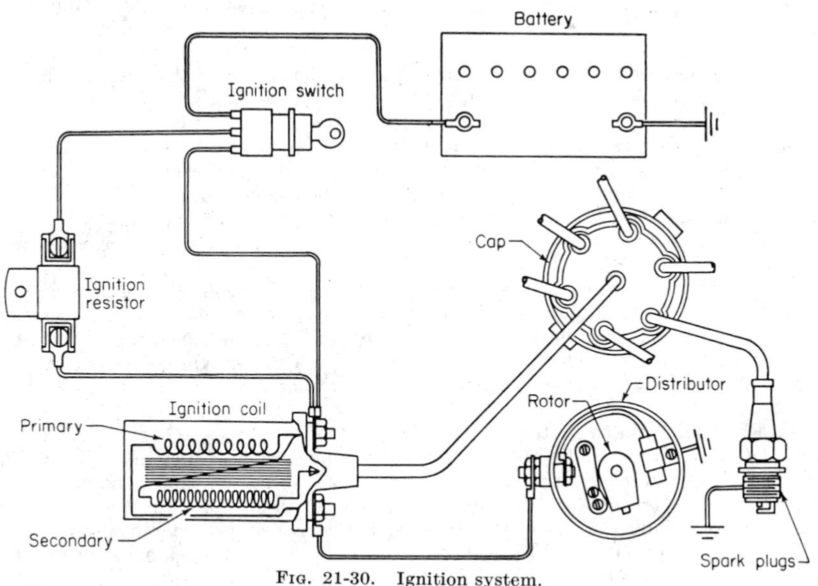

FIG. 21-30. Ignition system.

istics of the coil are as follows: Primary resistance 3.5 to 7 Ω (includes about 1.8 Ω external resistance to reduce coil heating), primary inductance 8 to 20 mH. Primary-to-secondary turns ratio is usually 1:60 to 1:110, the secondary having 14,000 to 28,000 turns of 40- or 42-gage enameled wire. The circuit-breaker cam usually has as many lobes as engine cylinders and in four-stroke-cycle engines is driven by the engine at one-half engine speed. Whenever a spark is to be produced, the cam separates two contacts normally held together by a spring. Contacts are usually made of pure tungsten, and a capacitor of 0.15 to 0.30 μF is used across them to reduce arcing and to ensure a sudden interruption of the current. When the primary current is broken, the sudden decay of flux in the magnetic circuit of the ignition coil produces a high-voltage pulse in the secondary of the coil which is carried through the rotor, usually driven by the circuit-breaker cam, and thence distributed to the proper spark plug. Since the voltage developed depends upon the current broken in the primary circuit, it is desirable to keep the current value as high as possible without injuring circuit-breaker contacts and to keep the time during which contacts are together as long as possible. On a six-lobe cam, six-cylinder ignition distributor, at 60-mi/h car speed, the contacts are together about 0.004 s each time they close.

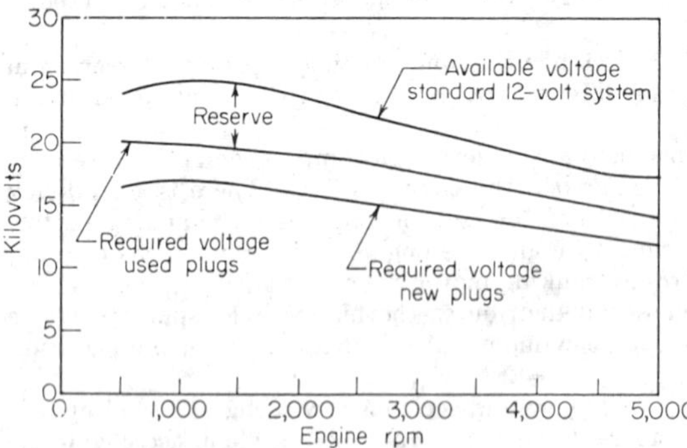

Fig. 21-31. Ignition performance and requirement curves.

The trend toward higher compression ratios in modern engines has required a substantial increase in ignition-system performance. The breakdown voltage of the spark-plug gap is a function of the gas pressure around the plug; so plug-firing voltages have increased with compression pressures. The reserve value of voltage available over voltage required is shown in Fig. 21-31.

65. Automatic Spark Advance. Since the combustion rate of the mixture in the engine cylinder is not proportional to engine speed, the time at which ignition should occur with relation to the engine crankshaft should normally advance with increased speed. To care for this automatically, most ignition distributors are equipped with centrifugally operated weights which advance the circuit-breaker cam with respect to the engine drive, with increase in speed. The amount of advance desirable for best performance at each speed is controlled by a number of conditions involved in the engine characteristics and hence must be determined experimentally for each engine design. The amount of advance obtained by centrifugal means is usually a compromise between maximum power and detonation.

66. Vacuum advance control is incorporated in many ignition distributors, in addition to the centrifugal automatic advance, to improve economy at part-load operation where timing of the spark needs to be advanced beyond the advance supplied by the centrifugal mechanism. This control is actuated by a diaphragm connected in such a way to the engine intake manifold that the spark timing is advanced during part-load operation.

67. Electronic ignition systems are available as options for those applications which warrant them. There are two basic types of electronic systems—the inductive system and the capacitor discharge system, the difference in the systems being the manner in which the energy is stored. The inductive system stores its energy in the inductance of the primary winding of the ignition coil, while the capacitor-discharge system stores its energy in a capacitor. With either of these electronic systems a semiconductor circuit is required to control the timing of the discharge of energy.

68. Lighting for automotive vehicles consists primarily of head, tail, and instrument-panel lights for night driving, stop and turn signal lights for warning other drivers, and indicator lights for information. Trucks and buses are required to have additional warning and marker lights. Increasing traffic loads on highways have required much improvement in the amount of light produced by headlights without increasing the glare to oncoming cars. The use of four headlights makes this possible, since each lens can be designed optically for each beam, with fewer compromises.

Because of safety considerations the lamps, their controls, and wiring must be designed to provide trouble-free service. Head lamps maintain high light output over a long period of time and are designed so that they can be mechanically aimed, if necessary, or replaced if burned out, in a short period of time.

69. Controls. Complex electrical systems require many forms of control. Switches are used for ignition, starting, lighting, turn signals, door and compartment lights, indicators of oil pressure, windows, heating and air conditioning, windshield wipers, transmission control, electric antennas, rear-seat speakers, and horns. Solenoids are used as remote actuators and for valve control. Relays are used to control high-current circuits without high wiring losses and to break the field circuit for diode-rectified generators to eliminate battery drain when the engine is shut off.

70. The wiring harness is the "lifeline" of today's modern automotive electrical system. It joins each electrical unit to its control and must be designed for adequate current-carrying capacity, minimum voltage drop, and mechanical strength. Its terminals must make positive low-loss connections in spite of vibration and corrosion from weathering and from ice-melting solutions used on roads. It is usually made up of vinyl-covered cable; however, there is a growing trend toward using printed-circuit harness behind the instrument panel and flat-wire harness under the floor covering to the rear of the car. Many heavy-duty systems use synthetic-rubber-covered cable for added protection in long-life, more rugged applications.

71. Accessories for today's automotive electrical system present many electrical design problems. Electric windshield wipers, heaters, defrosters, and actuator motors for seats, windows, antennas, and convertible tops require special small-motor designs for their application as well as the associated controls. Air conditioning for cars and buses poses the problem of high blower loads at low engine speeds. The refrigerant systems have required solenoid and magnetic controls of valves, compressor clutches, and throttle controls. Electric horns are tone-matched to produce a brilliant, pleasing sound as well as an authoritative warning signal. They are usually mounted in pairs as standard equipment, with trios and quartets optional. Recent developments in the field of electronics have made possible transistor radios—AM, FM, and stereo FM—automatic headlight-dimming controls, automatic speed control, and automatic turning on and off of headlights.

BIBLIOGRAPHY

72. References on Automotive Electrical Equipment

1. Onksen, G. W. Why Cars Have Four Headlights; *Gen. Motors Eng. Jour.*, January–February–March, 1958, Vol. 5.

2. Norris, J. C. Delcotronic Transistor-controlled Magnetic Pulse-type Ignition System; *SAE Paper* 617B, Jan. 14, 1963.

3. Redick, D. C. Selecting the Proper Electrical System for Farm, Construction, and Industrial Equipment; *SAE Paper* 734D, Sept. 9, 1963.

4. Cameron, G. L., Pettit, C. W., and Rowls, G. A. Cold Cranking Team: Battery, Cables, Cranking Motor, Engine Oil; *SAE Paper* 894B, Sept. 14, 1964.

5. Vinal, G. W. "Storage Batteries," 4th ed.; New York, John Wiley & Sons, Inc., 1955.

MARINE POWER APPLICATIONS

By William E. Jacobsen

GENERAL MARINE CONSIDERATIONS

73. Marine service conditions differ from industrial or land service primarily because of the physical environmental conditions under which the equipment is operated and because of the greater importance of reliability in shipboard service. Equipment must be suitable for service under rolling and pitching of the vessel and for possible list of the vessel caused by hull damage or unbalanced loading. In addition, the equipment must be able to withstand the adverse effects of moist, salt-or oil-laden atmospheres, as well as the vibrations incident to operation of other machinery and the pounding of the vessel in a heavy sea.

74. The safety of a ship at sea is dependent on the maintenance of propulsion, steering, lighting, communication, navigational, and other auxiliary services, which are, in turn, dependent upon the ship's-service power system. Vital services are safeguarded by duplicate units, alternate circuits, and emergency sources of supply.

75. United States Coast Guard. The U.S. Coast Guard is responsible by law for the enforcement of rules and regulations intended to safeguard life and property on American ships. Electrical engineering regulations cover Coast Guard requirements for electrical equipment for the entire range of vessels coming under its cognizance. Compliance with these regulations, as applicable, is mandatory.

76. American Bureau of Shipping. The American Bureau of Shipping is a private, nonprofit marine classification society which establishes rules and requirements as to design, construction, and inspection of ships and ships' machinery. This bureau classifies ships in accordance with their rules and thus provides assurance that industry-recognized good practices are incorporated in the vessel and in its equipment.

77. IEEE Marine Rules. The Institute of Electrical and Electronics Engineers' Committee on Marine Transportation is the author of IEEE Standard 45, Recommended Practice for Electric Installations on Shipboard. This publication, generally referred to as IEEE Marine Rules, is the basis for marine electrical engineering practices in this country. This widely accepted standard, in itself, is not a mandatory marine requirement, but both the U.S. Coast Guard and the American Bureau of Shipping use it as the basis of their rules and regulations.

78. Marine. The term "marine" is commonly associated with merchant marine ships and not with United States naval vessels. The following practices are applicable to merchant marine vessels.

SHIP'S-SERVICE POWER SYSTEMS

79. Use and Capacity. The ship's-service generation capacity now employed on new vessels ranges from 1,500 to 2,500 kW on cargo vessels to about 4,000 kW on medium-sized passenger or passenger-cargo vessels. Very large passenger liners may have generating sets totaling 10,000 kW or more. The usual practice is to employ two generating sets of equal capacity on cargo vessels, while three or more sets will generally be used on passenger vessels.

80. Standard Voltages. The voltages indicated in the following table have been accepted as standard on merchant vessels built in the United States:

	Direct current, volts	Alternating current, volts
Lighting	115	115
Power	115 and 230	115–220–440
Generators	120 and 240	120–230–450

81. Standard Frequency. A frequency of 60 c/s is recognized as standard for a-c lighting and power systems.

82. Selection of Voltage and Distribution System, Direct Current. For vessels having little power apparatus, 120-V generators are recommended with 115-V light and power distribution systems.

Where an appreciable amount of power apparatus is to be considered, 240/120-V 3-wire generators, 230-V power distribution system, and 230/115-V 3-conductor lighting feeders should be selected. Branch circuits from lighting panelboards should be 115-V 2-wire.

83. Selection of Voltage and Distribution System, Alternating Current. For small vessels having little power apparatus, 120-V 3-phase generators may be used, with 115-V 3-phase distribution for power and lighting. As an alternative, lighting feeders may be single-phase, balanced at the generator and distribution switchboard to provide approximately equal load on the 3 phases.

Table 21-10. Ship's-service Loads—Typical Passenger Vessel

Load group	Total installed load, kW	Normal sea load, kW	Normal port load, kW
Engine and boiler rooms	2,607	444	275
Deck auxiliaries	1,688	90	280
Navigation and communication	24	6	3
Hotel	1,900	940	940
General auxiliaries	1,759	1,547	1,547

NOTE: Four 1,100-kW generating sets installed.

For vessels with considerable power apparatus (particularly for intermediate-sized vessels), generators may be 230-V 3-phase, the power distribution 220-V 3-phase, and lighting distribution 115-V 3-phase, derived from transformers. As an alternative, a 120/208-V 3-phase 4-wire system may be used for lighting and power distribution.

For all vessels of a size and type to require a dual voltage system, first consideration should be given to the application of 450-V generators with power distribution at 440 V 3-phase and lighting distribution at 115 V 3-phase, 3-wire or 120/208 V 3-phase 4-wire, derived from transformers.

As the use of a-c power increases, the trend is clearly toward the use of 440-V systems. Power is generated at 450 V, and most of the utilization is at 440 V. Lower-voltage power for lighting and small auxiliaries is obtained as necessary from transformers.

84. Trend toward Use of Alternating Current—Advantages. Alternating-current ship's-service power systems are now selected for virtually all new American merchant ships, and in fact almost all oceangoing vessels built in this country since World War II have used a-c ship's-service power. This type of electric power has been found advantageous on tankers, bulk carriers, general cargo vessels, and passenger ships.

The main advantage of the a-c system comes through the use of squirrel-cage induction motors and across-the-line motor starters. This rugged and reliable motor, requiring extremely low maintenance, is used for literally hundreds of drives in sizes ranging from fractional horsepower to several hundred horsepower. The absence of commutators and brushes and the ability to start these motors directly across the line in all except the largest sizes make them exceedingly attractive for shipboard application. Most applications employ single-speed motors, and where the service requires it, two-, three-, or four-speed motors are utilized.

Another major advantage of a-c power is the ability conveniently to employ a higher-voltage 3-phase distribution system. Generation and most of the distribution and utilization are carried out at 440 V, with only a relatively small percentage of the power transformed to 115 V for lights and small devices.

Weight and first cost are lower for a-c systems, with the possible exception of those ships in which d-c power demands for cargo handling are an extremely large part of the total load.

85. Classification of Auxiliaries. Shipboard electric auxiliaries (Tables 21-10 to 21-14) may be grouped and described generally as follows:

a. Engine- and Boiler-room Auxiliaries. These include pumps, fans, air compressors, etc., of the same general type as used on land. Special features provided for these auxiliaries are those having to do with the motion of the vessel and the corrosive atmosphere.

Table 21-11. Ship's-service Loads—Typical Cargo Vessel

Load group	Total installed load, kW	Normal sea load, kW	Normal port load, kW
Engine and boiler rooms	975	275	110
Deck auxiliaries	2,050	.....	350
Navigation and communication	20	10	5
Hotel	350	180	180
General auxiliaries	300	240	240

NOTE: Two 1,000-kW generating sets installed.

Table 21-12. Ship's-service Loads—Typical Tanker

Load group	Total installed load, kW	Normal sea load, kW	Normal port load, kW
Engine and boiler rooms	1,000	350	110
Deck auxiliaries	100		
Navigation and communication	20	10	5
Hotel	300	150	150
General auxiliaries	175	100	100

NOTE: Two 750-kW generating sets installed.

b. Deck Auxiliaries. These auxiliaries include steering gear, anchor windlass, capstans, boom topping, and lifeboat, deck, and cargo winches. These are specialized ship applications and with the exception of the steering gear are deck-mounted and directly exposed to sea water.

The steering gear has been classed as a deck auxiliary because it has been placed under the jurisdiction of deck officers. As the steering gear is very vital to the safe operation of the ship, much attention has been given to the rugged character of both motor and control units in order to provide maximum reliability.

With the exception of the main cargo winches, practically all deck machinery is powered by a-c motors, which are frequently of the two-speed type. Cargo winches, almost without exception, employ d-c motors. The winches are normally used in pairs in the burtoning method, and the power for each pair of winch motors is derived from a single associated motor-generator set.

c. Navigation and Communication Equipment. This includes gyrocompass and pilot, loran and radio direction-finding equipment, radar, echo-sounding machines, radio, intercommunication equipment, and fire-alarm system.

d. General Auxiliaries. Items such as ventilating blowers, fans, and air-conditioning equipment are employed in great numbers on modern ships.

Table 21-13. Major Motor Applications—Modern Cargo Vessel

Application	No. of units	Hp rating	No. of speeds
Engine and boiler rooms:			
Forced-draft fan	2	150/66/39/17	4
Main circulating pump	1	150/38	2
Auxiliary circulating pump	2	20	1
Main condensate pump	2	25	1
Auxiliary condensate pump	2	5	1
Fuel-oil service pump	2	20/10	2
Fuel-oil transfer pump	2	75/37.5	2
Lubricating-oil service pump	2	30	1
Fire pump	1	100	1
Fire and bilge pump	1	50	1
Deck auxiliaries:			
Cargo winch motor-generator sets	12	60	1
Capstan	2	35/35	2
Boat winches	2	20	1
Boom topping winches	14	5	1
Boom topping winches	10	15	1
Steering-gear pump	2	50	1
Anchor windlass	1	100	1
General auxiliaries:			
Engine-room supply	4	15/3.7	2
Engine-room exhaust	4	10/1.25	2
Quarters supply	3	4/1	2
Quarters supply	1	10	1
Quarters exhaust	2	4/1	2
Quarters exhaust	16	1/2	1
Hold supply	2	2	1
Hold exhaust	5	3	1
Service refrigeration	2	25	1
Cargo refrigeration	2	7.5	1
Air compressor	2	25	1
Air conditioning	2	60	1

NOTE: All motors full-voltage-starting squirrel-cage type.

Table 21-14. Major Motor Applications—Modern Tanker

Application	No. of units	Hp rating	No. of speeds
Engine and boiler rooms:			
Forced-draft fan*	2	175/78	2
Main circulating pump*	1	150/38	2
Auxiliary circulating pump	1	40	1
Main condensate pump	2	20	1
Auxiliary condensate pump	1	20	1
Fuel-oil service pump	2	15/7.5	2
Fuel-oil transfer pump	1	75/37.5	2
General service pumps*	1	75	1
Deck auxiliaries:			
Cargo stripping pump*	3	125/67.5	2
Steering-gear pumps	2	40	1
General auxiliaries:			
Engine-room ventilation supply	2	10/5.6	2
Engine-room ventilation exhaust	2	10/5.6	2
Quarters ventilation supply	2	3	1
Quarters ventilation supply	1	2	1
Boiler-room ventilation supply	2	15/8.4	2
Air compressor	2	25	1
Air conditioning	2	50	1
Refrigeration	2	7.5	1

* Reduced-voltage-starting. All others full-voltage-starting. Squirrel-cage type.

e. Hotel Load. Electrical load so characterized is to be found principally on passenger vessels and includes electric cooking and heating units of all kinds, elevators, special lighting, refrigeration, etc.

f. Special auxiliaries for particular services are to be found on such vessels as oil tankers having large pumps installed for handling cargo; self-unloaders, e.g., the coal and limestone carriers on the United States Great Lakes; and vessels equipped to carry railroad cars.

86. Typical Ship's-service Loads. The data given in Tables 21-10 through 21-14 show the nature and size of ship's-service loads on modern oceangoing ships. Tables 21-10, 21-11, and 21-12 show the combined ship's-service loads for various types of oceangoing vessels, and Tables 21-13 and 21-14 show the type of major motors comprising the bulk of the connected motor load.

87. Generating Sets. Two or more generating sets are required for ship's-service power. Rules require that the number and rating of sets be such as to provide one spare set at all times. Thus, most cargo vessels use two sets of equal rating, each capable of supplying the full-operating-load requirements of the vessel.

On steam-driven ships, the ship's-service generating sets are invariably turbine-driven. High-speed turbines operating at the full-steam conditions of the propulsion plant drive generators at acceptable speed through reduction gears. On diesel-driven vessels, ship's-service sets are diesel-driven. Emergency power sets, where used, are usually diesel-driven, although gas-turbine-driven generating sets are beginning to find acceptance.

88. Switchboards. Switchboards, generally of the dead-front type, are provided for the control, metering, and handling of the ship's-service power system. Generators are paralleled and voltages and load adjusted from devices located at the main switchboard. Power feeders emanate from this board to the various major loads or to feeder panels and distribution boards arranged for supplying and switching many individual load circuits. A simplified one-line diagram of a typical cargo vessel is shown in Fig. 21-32.

89. Continuity of Service. Continuity of service and adequacy of interrupting capacity are of paramount importance in ship's-service electric power systems. Figure 21-32 is a simplified diagram of an electric power system for a typical modern cargo vessel.

All vital auxiliaries are connected to the main 450-V bus through individual circuit breakers, each of which is capable of interrupting the maximum fault current available at that particular point. The majority of vital services are accommodated with molded-case circuit breakers either fused or unfused—depending upon the system short-circuit capacity. The relatively few very large auxiliaries are provided with open-frame air circuit breakers. Prospective short-circuit currents are calculated at the main buses and at the ends of each of the major feeder cables such as those connecting the main switchboard bus to the group controls. Appropriately rated circuit breakers are then selected for each location. The selection is based upon (*a*) the maximum prospective short-circuit current at the point of installation, (*b*) the degree of aysmmetry of the circuit up to that point, and (*c*) the continuous current-carrying ability required by the load circuit.

90. Group Control. The combining of groups of related motor controllers into one or more free-standing mechanical assemblies known as group controls is finding increased favor with ship designers in view of the present trends toward centralization of control functions. Appearance is improved, servicing is made easier, and shipyard installation costs are usually diminished by the use of group controls. They have been successfully applied in the case of engine-room auxiliaries and deck-machinery auxiliaries such as vang and topping winch controls.

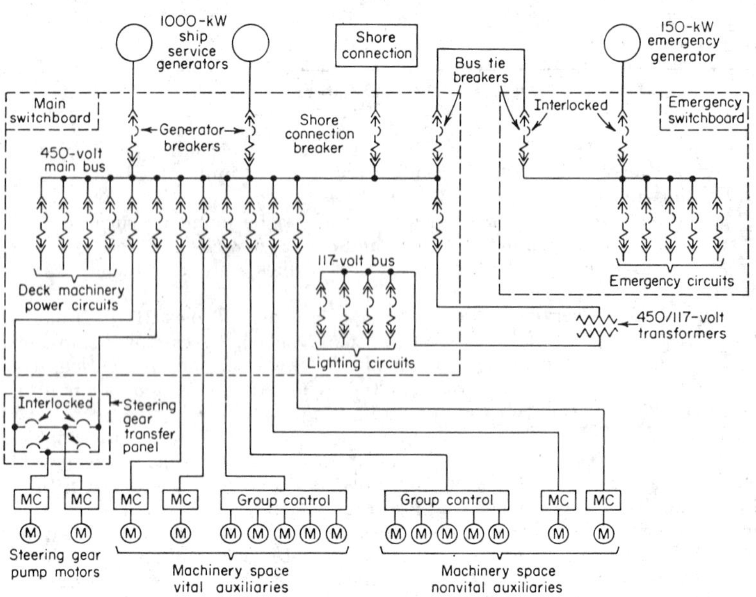

FIG. 21-32. Ship's-service electric power system—modern cargo vessel.

ELECTRIC PROPULSION

91. Electric Drives. The term "electric drive," when used in conjunction with ship propulsion, refers to the interposing of an electric generator and motor between the prime mover and propeller shaft. The justification for its use lies in its ability to afford a satisfactory speed reduction between a high-speed, efficient, lightweight prime mover and a slow-speed, efficient propeller; to make use of a unidirectional prime mover; to allow freedom of relative location of prime mover and propeller shaft; and to afford rapid maneuvering control by means of a switching of the electrical connections. Two forms of electric drive have been widely used: (1) the turbine-electric drive; (2) the diesel-electric drive.

Both turbine-electric drive and diesel-electric drive may be further classified according to the type of electric transmission used, e.g., alternating current and direct current. The a-c transmission is broadly a fixed-speed-ratio reversible drive; the d-c transmission, a variable-speed-ratio reversible drive.

Electric couplings used with internal-combustion engines may be considered as a third form of electric transmission of fixed speed ratio and nonreversible.

92. Advantages. Electric drive offers a number of advantages, which include the following:

a. Flexibility of Arrangement. Electric drive increases the flexibility of vessel design because the generating sets may be located without reference to the propulsion motor. Likewise, the motor may be located so as to result in a short propeller shaft.

b. Economy of Operation and Maintenance. In vessels having multiple generators, part of the generating plant may be shut down when operating at sufficiently reduced power, resulting in better reduced-speed economy. Electric drive has demonstrated that its maintenance costs are low.

c. Maneuvering. With electric drive, it is possible to provide full power for reversing by electrical reversal of the motor while allowing the prime mover to continue rotation in one direction. Manipulation of controls for electric drive is exceedingly simple and lends itself to pilothouse control where desired.

d. Auxiliary Power from Propulsion Generator. On certain vessels, advantage can be taken of the propulsion generator as a supplementary source of power for cargo handling or other special power service while in port, with consequent reduction in size of the ship's-service power-system generating sets.

93. Disadvantages. *a. Transmission Losses.* Losses in the electrical system will range from 7 to 15% of the transmitted power. Some of this loss is offset by the better prime-mover performance, but a net loss must be considered.

b. First Cost. Electric drive costs more than straight mechanical drives. Whether or not such additional extra investment is justified depends on many complex factors in any specific case.

94. Economic Considerations. The use of electric drive must be based on sound economics in each specific case. The overall economy must be studied on the basis of ship speeds between ports of call, fuel costs, turn-around time, type of cargo, cost of investment, operating maintenance costs, etc.

95. Present Application. Electric drive finds present use in medium- and small-size vessels in which maneuvering ability is very important or where dual use can be made of the propulsion generator sets. Vessels such as tugs, ferries, fireboats, dredges, and icebreakers are in this category. No large oceangoing electric-drive vessels have been built in this country since World War II.

TURBINE-ELECTRIC DRIVE

96. Electrical System. Turbine-electric d-c drive has been used only infrequently and for special-service vessels. It is similar to the diesel-electric d-c drive except that the turbine runs at substantially constant speed.

In the turbine-electric a-c drive, both synchronous motors with an amortisseur starting winding and induction motors have been used. The induction motor is easily arranged for pole changing and is advantageous if two basic operating-speed ratios are desired. The induction motor with external rotor resistance is applied where auxiliary electric power is to be supplied from the propulsion turbine generator while delivering propulsion power over a wide range of propeller speeds. The lightest and most efficient drive results with a synchronous motor and hence is used in almost all modern installations. A turbine-electric drive therefore normally consists of the following elements:

a. A single-unit condensing steam turbine, direct-connected to a 3-phase cylindrical-rotor generator.

b. A propulsion motor of the synchronous type with an amortisseur starting winding directly coupled to the propeller shaft.

c. An exciter for supplying electric power to the turbine-generator and propulsion-motor fields.

d. A control cubicle which contains the operating levers for controlling the direction of rotation of the motor, application of field excitation, and turbine speed.

97. Types of Ship to Which Adapted. Turbine-electric drive is particularly suitable for ships of large power, above about 6,000 hp/shaft, e.g., large passenger and passenger-cargo ships having two or more propellers and multiple turbine generators, because of the reduced-speed economy and the economical auxiliary power obtained from the propulsion generators for such purposes as refrigeration of cargo. Likewise, tankers find it advantageous to use power from propulsion generators for cargo pumping, and self-unloading bulk-cargo vessels for cargo-handling conveyer systems.

98. Generators and Motors. Generators are normally of the cylindrical-rotor type, 3-phase, Y-connected, generating 2,300 to 4,000 V. The generators are direct-connected to the driving turbine and run at speeds ranging from 3,600 to 5,500 r/min, depending upon the generator rating and the number of motor poles required to produce the desired propeller speed.

Propulsion motors are of the salient-pole synchronous type and are normally direct-connected to the propeller shaft. Normal propeller speeds on merchant ships are about 100 r/min. Motors must be equipped with heavy-duty amortisseur windings for starting and reversing operations.

Both motors and generators are normally totally enclosed with surface air coolers transferring ventilating air heat to circulated sea water. Ventilating air may be circulated by rotor fans or by separate motor-driven blowers. The lubricating system must be arranged to operate properly for the degree of pitch and roll to which the vessel may be subjected. This is usually taken as 5° pitch, 30° roll, and 15° permanent list for merchant vessels.

99. Propulsion Control. The control is of the dead-front type and normally employs cam-operated air-break contactors. These contactors are capable of opening rated load at rated voltage, but normal switching is done with the generators deenergized.

The scheme of operation is quite simple, as will be seen by reference to Fig. 21-33.

With the propulsion turbine-generator set operating at about 20% speed and the

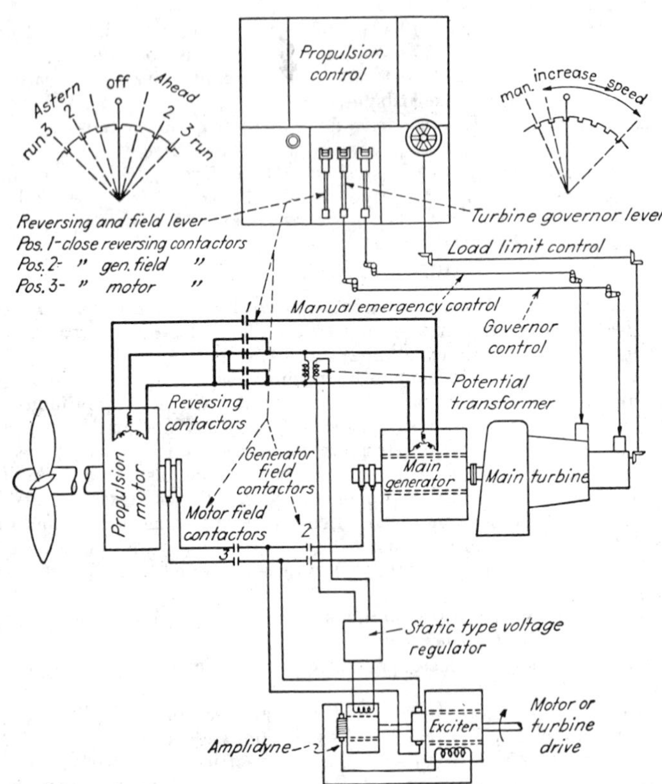

Fig. 21-33. Schematic diagram of a-c propulsion control.

motor and generator fields deenergized, the sequence of operation for starting ahead is as follows:

a. The direction contactors are closed in the ahead position (reversing contactors).

b. The generator-field contactors are closed, and overexcitation applied to the generator field. The motor now starts as an induction motor, requiring three to five times normal current, until it comes up to a low value of slip from which it can be synchronized.

c. The motor-field contactors are closed, and the motor pulls into synchronism.

d. The motor and generator excitations are immediately reduced to normal, either automatically or manually.

e. By operating the turbine-governor lever, the speed desired is set.

Note that the propeller speed is now proportional to the turbine speed just as though it were geared mechanically. To deenergize the propulsion drive, the turbine speed lever is put in the maneuvering position, the field lever is moved to the "off" position, and the reversing lever is moved to the "off" position, thus allowing the turbine-generator to reduce to maneuvering speed and no load.

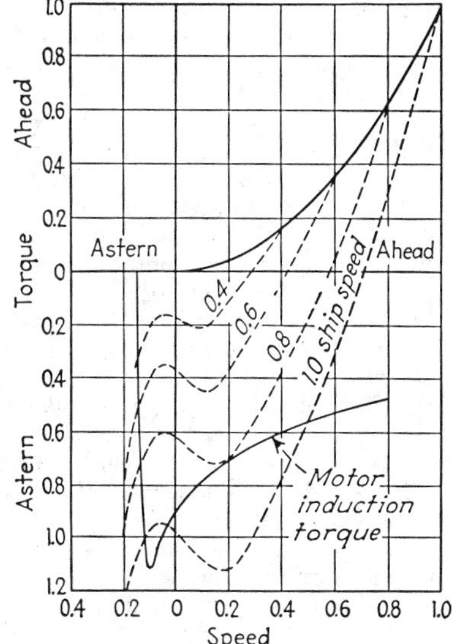

FIG. 21-34. Propeller torque vs. speed at various ship speeds in per unit of full-ahead speed and torque.

In reversing the propeller when the ship has forward headway, the same sequence of starting operation is involved, but because of the increased motor torque on the propeller produced by ahead motion of the vessel, the time required for each step is changed. Typical curves of propeller torque vs. speed at various ship speeds are shown in Fig. 21-34. Also shown is a curve of the torque developed by the motor operating as an induction motor with the connections reversed and with the generator overexcited and running at 15% speed.

100. Excitation System. The excitation system must be capable of applying excitation to the generator, the motor, or both, in proper sequence and magnitude during maneuvering or steady-state operation. Under steady-state operation, sufficient excitation must be provided to motor and generator to preserve synchronism. The necessary value varies with the roughness of the sea, hull cleanliness, ship loading, and other factors which dictate the actual motor torque demand. The excitation may be adjusted manually, in which case it must be set high enough to maintain synchronism during the maximums of propeller torque swings. On modern ships, an exciter regulator is provided which holds approximately constant volts per cycle on the generator. The rapid response characteristic of the regulator results in the excitation being changed from moment to moment as required. The consequent reduction in average excitation produces less field heating, and a reduction in size of motor and generator is possible.

A typical excitation system of great rapidity of response and simplicity makes use of a pilot exciter of the Amplidyne type. Because of the high amplification of the input signal by this exciter, control circuits are greatly simplified and have static elements. The arrangement shown in Fig. 21-33 is of this type.

DIESEL-ELECTRIC DRIVE, DIRECT CURRENT

101. Electrical System. With but few exceptions, the type of electric transmission that has been used between diesel engines and propellers has been of the d-c type. This has developed because the power-speed characteristics of the diesel engine may be used fully for a wide range of speed-torque requirements associated with harbor and coastal craft. In addition, the system has less complication than an a-c system when, as is frequently the case, a multiplicity of diesel generating sets is electrically connected to each propeller shaft through one or more propulsion motors. The system consists of the following elements:

a. Diesel engines directly connected to d-c separately excited generators.

b. A separately excited propulsion motor or motors directly coupled or geared to each propeller shaft.

c. An excitation system that may obtain its power from an exciter directly connected to the propulsion generator or the ship's-service electric power system or both. Alternatively, a motor-generator set may provide excitation power.

d. A control cubicle in the engine room, selective with one or more remote deck stations (optional but usually required) through which engine speed, generator excitation, and sometimes propulsion-motor excitation are controlled, and hence the speed and direction of the vessel.

102. Special Advantages. In addition to the general advantages listed in Par. **92** for electric drives in general, diesel-electric d-c drives offer the following additional particular advantages:

a. Ease of arranging for coupling electrically a multiplicity of units to a single propeller and the greater reliability associated therewith.

b. Better operating economy and higher maintained speed of the vessel under different load and sea conditions, resulting from the adjustable-ratio speed reduction between high-speed diesels and slower-speed propellers.

c. Better adaptation of motor speed to actual propeller requirements after installation without sacrificing power or compromising diesel-engine operation.

d. Possibility of using duplicate electrical machines for generators and mechanically geared motors, resulting in interchangeability and reduction in spares.

103. Types of Ships to Which Adapted. Diesel-electric d-c transmission is particularly adaptable to ships of moderate power—4,000 hp or less/shaft—used for special purposes. Many coastal and harbor craft are of this type and can advantageously employ diesel-electric d-c drive in the following services:

a. Tugboats, which must operate over a wide range of speeds and powers involving both running-light and towing conditions.

b. Ferryboats, which require exceptional reliability and maneuverability to operate efficiently in congested harbors.

c. Fireboats, dredges, fishing boats, harbor tankers, etc., in which high propulsion power demands alternate with high power demands for special functional services.

d. Special naval auxiliary craft of similar characteristics to those above.

104. Generators. For diesel-electric drive the d-c generators are of the separately excited type. They operate as variable-voltage units connected to the propelling system and at constant voltage when supplying the auxiliary-power bus. The voltage rating is usually 250 or 500 V. Generators are available in ratings as high as 1,200 kW at speeds up to 800 r/min and 1,600 kW at lower speeds. Bearings are generally of the self-aligning sleeve type, forced-feed lubricated. The lubrication system is especially designed for sea conditions of pitch and roll. Heaters are provided to prevent moisture condensation on electrical parts when shut down.

105. Propulsion motors are very similar to the generators regarding constructions and enclosure. They may be of single- or double-armature types. The location within the hull influences the selection in moderate ratings, the double-armature motor being of smaller diameter and longer for the same power. · High-speed motors are connected to the propeller shaft through gears; low-speed motors, rigidly coupled. The bearings are generally self-aligning, flood- or disk-lubricated. The propeller-shaft thrust bearing may be incorporated in one of the motor-bearing housings if the motor is direct-connected. The designed voltage of a single armature should not exceed 1,000 V.

106. Propulsion control in diesel-electric d-c drive is generally of the dead-front type using air-break cam-operated switches. The equipment on a typical generator panel is a handwheel for operating the generator-field and -armature setup switches, a control switch for actuating the generator-field contactor, a small handwheel for operating a rheostat in the generator-field circuit when the generator is supplying power to the auxiliary-power bus, an ammeter, and a voltmeter. The motor panel contains a handwheel for controlling the speed and direction of rotation of the propeller, a transfer switch for shifting the control to either pilothouse or engine room, a handwheel for operating the motor-field rheostat, an engine-speed-control switch, and an excitation-control switch. There are also ammeters and voltmeters for the propulsion-power and excitation circuits. Protection against short circuit in the armature loop is usually provided by an instantaneous overcurrent relay.

It is common for the engines to have an idling and maneuvering speed of 50% of maximum speed, although this varies. If this is the case, the first points on the speed controller vary the generator-field excitation until about 50% of rated propeller speed is attained. Moving the speed lever further increases engine speed and, hence, output power and propeller speed. The pilothouse control works on the same principle. The propulsion-power and excitation circuits are shown schematically in Fig. 21-35 for a single generator and motor arrangement.

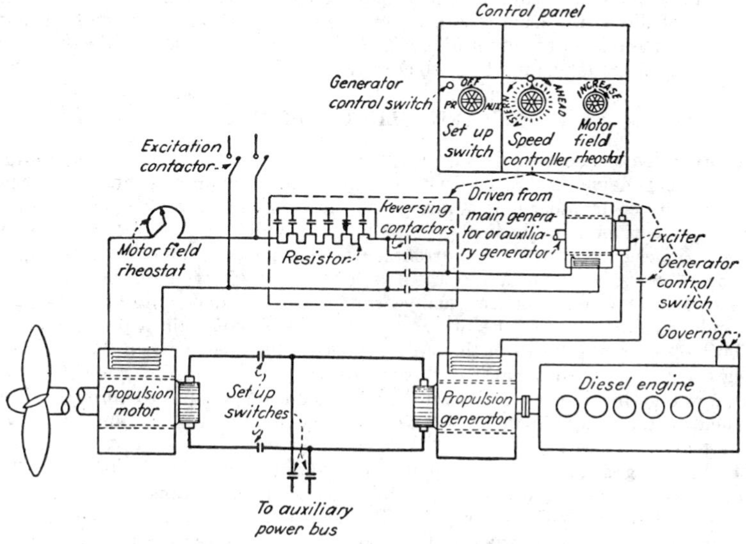

FIG. 21-35. Schematic diagram of d-c propulsion control.

107. Excitation System. In the diesel-electric d-c drive system, the generator field may be controlled by varying resistance in series with it. In this case, excitation current is supplied from a constant 125-V d-c ship's-service bus. Although the system is simple, it results in comparatively large energy loss in the resistors and somewhat larger control equipment. The pilothouse control is large unless a servo system is used to operate the rheostat.

Another system which is more generally used has a variable-voltage generator exciter, usually directly coupled or belt-driven from its generator. The losses are small, as the exciter-field currents are small. A number of remote-control stations are therefore feasible, located on the afterdeck, pilothouse, etc.

If three or more generators are connected in a series loop, it is necessary to limit the load on any generator in the event that its engine, for any reason, fails to produce the torque required. The remaining generators force current through the armature of the generator attached to the weak engine and tend to stop and reverse it. To prevent reversal, the excitation of the generator must be removed.

DIESEL-ELECTRIC DRIVE, ALTERNATING CURRENT

108. Differences from Turbine-Electric A-C Drive. The major differences between the turbine-electric and diesel-electric a-c drive are brought about by the necessity of operating a number of generators in parallel per propeller and by the torque characteristics of the diesel engine. Because of the slower prime-mover speed, the generators are of salient-pole construction, with amortisseur windings. The methods of speed control and reversal are the same fundamentally as for the turbine-electric drive.

109. Use. Primary advantages of using a-c generators and motors with diesel engines are to obtain reduced weight and greater power per shaft. Alternating-current motors are more suitable, particularly in large horsepowers, as higher voltages may be used and commutation is not a factor. Nonreversal of diesel engines is also important. The speed ratio of the generators and motor is fixed by the ratio of poles; therefore, all speed changes are made by changing engine speed. The lowest propeller speed obtainable as a synchronous drive is limited by the minimum engine-operating speed (maneuvering speed), which must be of the order of 25 or 30% of full speed. Some subsynchronous operating speeds can be obtained by induction-motor-type operation on the motor amortisseur winding if such reduced speeds are necessary.

The use of diesel a-c drive is largely for ships whose power is above that suitable for d-c drive, which is on the order of 5,000 hp/shaft, and whose service does not require the high order of maneuverability of the d-c system.

ELECTRIC SLIP COUPLINGS

110. Function. The electric slip coupling is a device for transmitting torque by means of electromagnetic forces in which there is no mechanical contact between the driving and driven member. It is interposed between each diesel engine and its associated single-reduction mechanical-gear pinion, the bull gear being connected to the propeller. Poles excited by direct current are usually mounted on one rotating member and an armature winding, usually of the double squirrel-cage type, on the other rotating member. The two members are usually arranged so that one containing the squirrel-cage winding rotates within the other, separated by a small air gap of about ¼ in. Relative motion or slip between the two members induces currents in the squirrel-cage winding which react with the air-gap flux to produce torque, thus holding the two members closely together in speed. Direct-current excitation is supplied through slip rings to the field poles. Adjustment of the excitation current controls the amount of torque that can be transmitted.

111. Advantages and Disadvantages. The slip coupling is a torsionally flexible member between the engine and gear and hence limits the transmission of torsional vibrations and propeller-shock torques. It allows operation with a small amount of misalignment between the two coupling halves and provides a ready means of disconnecting a disabled engine. A ship with two or more engines per propeller may be maneuvered without frequent reversal of the diesel engines by operating one engine with reversed rotation and one with normal rotation and energizing the proper coupling for ahead or reversed propeller, as desired.

Slip couplings have a maximum torque of about 150% full-load torque and about 75% torque up to 140% slip. The coupling is designed to provide the necessary torque for reversing duty and to absorb the associated loss energy. It does not have the flexible gear ratio of the d-c electric drive, and thus the engine-coupling combination cannot deliver full power at reduced speeds.

112. Use. The electric coupling, because of its lower cost, space, and weight compared with a complete electric drive, has made possible to some degree certain advantages of the electric drive on cargo ships having geared propellers and two moderate-speed diesel engines per shaft. It has also been applied on a small number of tugs and ferries. However, for vessels requiring full power over a speed range, and for vessels requiring dual use of the generating plant for special functional services, the slip coupling has little to recommend it when compared with full diesel-electric drive.

SHIP AUTOMATION

113. Status. Automation of shipboard systems, particularly propulsion and ship's-service power systems, has made very rapid progress in recent years. Practically all major new-construction ships are now automated to some degree, and programs are under way for modifying existing vessels by the addition of automation equipment.

114. Purpose. The term "automation" is very broad and means many things to different people. However, in the marine industry, "automation" is used to describe systems or arrangements of sensing, monitoring, indicating, and actuating devices that get work done more economically and with improved reliability. Usually this means operating with fewer men, and automation systems are therefore aimed at reducing the work necessarily done by men and making the necessary work more easily accomplished. These systems are commonly referred to as *central operation systems* (COS) or *central engineering operating station* (CEOS). A schematic block diagram of such a system is shown in Fig. 21-36.

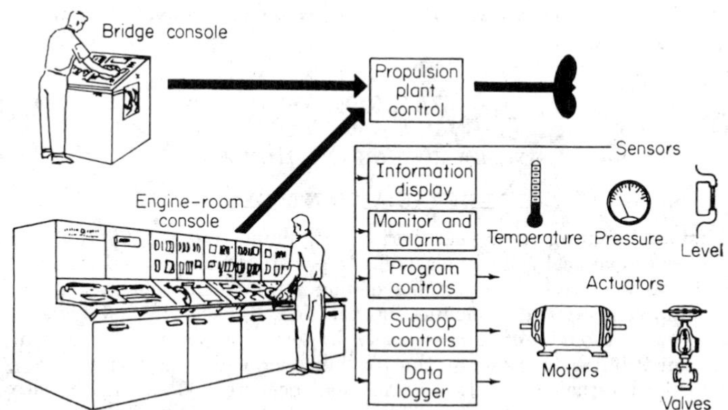

FIG. 21-36. Block diagram of central operation system (COS).

115. Advantages. Shipowners who install automation equipment are endeavoring to reduce operating costs by the following:

a. Reduction in Operating Personnel. Ships with well-automated power plants can be operated with one-man watches instead of the three- and four-man watches on non-automated vessels. Direct propeller speed and direction control are accomplished from the bridge.

b. Improved Operating Efficiencies. Improved and readily available data, as well as self-regulating subsystems, produce lower fuel costs and help reduce maintenance requirements.

c. Better Vessel Utilization. Reduced operating personnel reduces the required crew living spaces and support services, leaving more space for cargo.

BIBLIOGRAPHY

116. Reference Literature on Marine-power Applications

COLEMAN, HARRY C. "Marine Engineering"; New York, Society of Naval Architects and Marine Engineers, 1944, Vol. II, Chap. IX, Electric Propulsion.

FOX, BENJAMIN, and COLEMAN, HARRY C. Alternating Current for Auxiliary Plants of Merchant Vessels; *Trans. Soc. Naval Architects and Marine Engrs.*, 1946, Vol. 54.

GEARY, ELMER A. N.S. Savannah—Electrical System in First Merchant Nuclear Reactor System; *Trans. AIEE*, 1962, Vol. 81, Pt. II, Applications and Industry.

HALL, WILLIAM A. Determination of Shipboard Electric Short-circuit Currents; *Trans. AIEE*, 1962, Vol. 81, Pt. II, Applications and Industry.

HANSEN, H. H. Voltage Regulation of D-C Power Supply with Suddenly Applied Load; *Trans. AIEE*, 1962, Vol. 81, Pt. II, Applications and Industry.

HARVEY, H. FRANKLIN, JR. "Marine Engineering"; New York, Society of Naval Architects and Marine Engineers, Vol. II, Chap. X, Electric Plant.

IEEE Committee on Marine Transportation, Recommended Practice for Electric Installations on Shipboard, IEEE Standard 45, New York, Institute of Electrical and Electronics Engineers.

JACOBSEN, WILLIAM E., and KOCH, RICHARD L. Diesel-electric Propulsion for Polaris Submarine Tender; *Naval Engrs. Jour.*, August, 1962, Vol. 74, No. 3.

SMITH, FRANK V. Electric Ship Propulsion, in "Marine Engineers' Handbook"; New York, McGraw-Hill Book Company, 1945.

THORNBURY, J. W. Current Electrical Systems and Equipment for Warships; *Trans. Soc. Naval Architects and Marine Engrs.*, 1961, Vol. 69.

WASMUND, JAMES A. Propulsion Equipment for the U.S.S. Glacier; *Westinghouse Engr.*, September, 1955, Vol. 15.

WASMUND, JAMES A. Series vs. Parallel Connected Generators for Multiple Engine, Direct Current Diesel-Electric Ship Propulsion Systems; *Trans. AIEE*, 1954, Vol. 73, Pt. II, Applications and Industry.

RAILROADS

BY FREDERICK N. HOUSER

GENERAL ADVANTAGES

117. History. During a 40-year period, beginning about 1895, the engineering and economics of classical railway electrification were developed and refined. In this period electrification overcame a number of major railroad problems, making possible numerous improvements and economies in operation. During its first 30 years railroad electrification was thought of almost solely in terms of power generated at central stations and distributed to trains by third rail or overhead trolley wire.

By 1925 the internal-combustion engine had been refined so that its size, rating, and reliability allowed it to be considered as a prime mover for railway service. After early attempts at producing mechanical transmissions had proved generally unsuccessful, electrical transmissions were found satisfactory as torque converters for gasoline and diesel engines. During the fourth decade of the twentieth century the internal-combustion engine became a significant factor in railway motive power. The following 25 years saw the diesel locomotive with electric transmission eliminate completely the steam locomotive in many parts of the world. The diesel-electric at times also displaced short stretches of conventional electrification which had been installed to overcome specific operating problems on otherwise all-steam lines.

118. Reasons for Electrification. Electric traction has been applied for a variety of reasons since its initial use in 1895. The successful utilization in locomotives of diesel engines with electric (and, more recently, mechanical) transmissions has tended to cloud what were once clear-cut distinctions between the characteristics of operations based on central-station electrification and operations based on steam locomotives.

Smoke abatement was once a major advantage of electrification, especially in thickly settled districts near terminals and in lengthy tunnels. The first main-line electrification—put in service on the Baltimore & Ohio through a tunnel in Baltimore, Md., in 1895—was installed to overcome a smoke problem. Diesel-electric locomotives eliminated this electrification after more than 50 years of service and proceeded then to eliminate most of the other short, isolated stretches of electrification which had been installed specifically for smoke-abatement purposes.

Mountain grades often presented major operating problems with steam-locomotive operation which could be alleviated by electrifying. Because the slow-speed traction characteristics of electric and diesel-electric locomotives are practically identical, the diesel has readily supplanted full electrifications installed solely because of the high torque which can be developed in the series traction motor at low speeds.

Dynamic or regenerative braking, possible with locomotives and cars fitted with traction motors, has many operating advantages on grades and even in level territories where frequent stops must be made. Dynamic braking involves dissipation of energy in resistor banks mounted on the vehicle where the kinetic energy of the moving train is converted into electrical energy by operating the traction motors as generators. In regenerative braking, possible only with full electrification, the current produced by the traction motors is fed to the trolley wire so that a train descending a grade may actually be producing the electrical energy to power a train ascending a hill elsewhere on the same line. While regeneration is relatively simple in d-c electrifications, the complication of phase matching and other problems have made it much less frequently used on a-c electrifications. The advantage of concentrating the retarding effort on the motive-power unit so that the air-brake system need not be utilized is so great that many fully electrified lines are completely satisfied with dynamic braking where there is absolutely no feedback into the power system. In all cases there must be an air-brake system to take over in the event of a failure of the electric braking system and to bring the vehicles to a complete stop and hold them; the effect of electric braking decreases as the armatures rotate at low speeds and stops completely when they cease to rotate.

Suburban service in metropolitan areas places demands of service frequency and reliability on railroad plants which at one time could be met only by electrified operation. The rapid acceleration made possible by motors distributed throughout the train permits high schedule speeds without increasing the maximum speeds. The result of such operating characteristics is maximum capacity of crowded tracks during rush hours. Diesel-powered rail-car trains with individually powered units can approach the accelerations which are possible on an electrified line. The possibility of using lightweight, high-performance gas turbines with short-time characteristics even more comparable with those of electric traction motors has been under investigation. In the case of transit lines with subway operation, the possibility of utilizing anything but full electrification appears remote.

Transit and suburban operations impose requirements of operating flexibility which once could be achieved only with electric multiple-unit cars. The ability to change train consists rapidly and to reverse direction with no switching was once possible only with electric cars. Individually powered rail diesel cars have the same flexibility as electric multiple-unit cars. A recent development is push-pull operation, where conventional unpowered passenger cars are fitted with the multiconductor train line which is used for diesel-locomotive control and where certain unpowered cars are fitted with operator's compartments from which a trailing diesel locomotive can be controlled. Trains operating into stub-end terminals under push-pull operation have the diesel locomotive on the outbound end. In the outbound, or "pull," operation, the train is operated in the conventional manner from the locomotive cab. On the inbound, or "push," operation, the train is controlled from the cab in the leading passenger car, with the locomotive operated remotely at the far end of the train. By fitting all, or certain, intermediate coaches with cab controls, train consists can be varied almost as readily as with regular electric multiple-unit or rail-diesel-car operation. All types of multiple-unit operation involving cars, locomotive units, or cars and locomotive units are based on the pioneering work of the traction pioneer Frank B. Sprague, who developed the concept before the turn of the century.

119. Fuel Saving. Full electrification is most popular in regular line-haul operations in countries where hydroelectric power is cheap and plentiful or where fuel oil is scarce or expensive or must be imported. Many early electrifications were justified economically on fuel-cost savings as compared with steam locomotives. Because the thermal efficiency of the diesel locomotive is four to five times greater than that of its steam predecessor, the economic advantage of full electrification is much less in many areas than was formerly the case.

120. Capital Charges. Installation of the power-transmission and -distribution system is a heavy charge against electrification. The maintenance of these facilities also is costly, and in some circumstances taxes are increased. If density of traffic is high, these costs may be divided by large train mileage, and the burden will not be great, but if density of traffic is low, this may be a serious obstacle to economic electrification. If, for instance, the maintenance costs and interest, depreciation, and tax charges of the power-distribution facilities in a zone 100 mi long are $500,000 per year, and if there are 200 trains per day over the section, or 7,300,000 train mi annually, the cost of maintaining the power-distribution facilities is 6.85 cents per train-mile—not a high figure if there are offsetting advantages in the electrification. In general, it may be said that straight electrification is most advantageous if there are:

a. Dense traffic.

b. Traffic evenly distributed throughout the day and year.

121. Maintenance Costs. Because full electric locomotives and cars have no prime movers, their maintenance costs are lower than for diesel or turbine-powered units. This must be countered, of course, by the maintenance costs of the transmission, conversion, and distribution facilities that are involved in full electrification. Experience has shown that, despite the higher maintenance costs of diesel-powered equipment, there is about the same availability, due largely to the standardization of servicing procedures, preventive maintenance practices, and standardized parts, which make rapid repair possible.

Ownership costs of electrically powered motive power must obviously be lower than those of independently powered units if complete electrification is to be justified and if there is no significant difference in operating costs. Experience has shown that electric locomotives are capable of economical operation for 30 years or more. The economic life of diesel-electric units has been approximately 10 to 15 years for road power and 20 to 25 years for switching locomotives.

A major advantage of electric motive power is its intrinsic overload capacity for short periods. This serves to increase greatly the effective rating of diesel-electric locomotives at low speeds and of electric locomotives throughout their entire speed range. As an example, a short grade may require capacity for, say, 15 min, which may not be required for longer periods and for which the overload capacity of traction motors and transformers may well be suited. Starting from stops or accelerating after slowdowns may also utilize overload capacity to advantage in maintaining fast schedules with heavy and high-speed trains. In preparing for full electrification, it is vital to know the characteristics of the traffic to be handled.

GENERAL PRINCIPLES OF TRAIN OPERATION

122. Capacity of motive power to perform a given service depends upon the profile of the line, curves, weights of trains, frequency of stops, and schedule speed required.

123. Frequency of stops is especially important, since the more stops for the same schedule speed, the more time is consumed and the greater the need for acceleration, hence the more horsepower required.

124. Power of the locomotive is applied at the drawbar, and the weight of the locomotive for a given drawbar pull must be such that friction to prevent slipping will exist between the driving wheels and the rail.

125. Tractive effort, which is usually expressed in pounds, has a definite relation to the torque of the motors, dependent upon the gear ratio, etc.

The tractive effort TE may be expressed as follows:

$$\mathrm{TE} = T \times G \times E \times R \tag{21-1}$$

where T = motor torque in foot-pounds, R = driving-wheel radius in feet, G = gear ratio, E = efficiency of mechanical transmission.

126. Train Resistance. The tractive effort is used in overcoming certain factors, i.e., normal train resistance, grade resistance, curve resistance, wind resistance, air resistance, and acceleration, etc.

Basic **train-resistance figures** may be used to advantage. Values calculated will provide a basis for determining locomotive characteristics to meet particular speed requirements, designing the distribution system and substations, and determining energy consumption and power demand. A table was prepared by W. J. Davis, Jr., of the General Electric Company, some years ago, but the figures are somewhat high for modern track with heavy rail.[1]

In addition to the values derived from these tabulated data there must be included under certain conditions the following: *grade resistance, curve resistance, wind resistance,* and *resistance due to acceleration.*

[1] *Gen. Elec. Rev.*, October, 1926.

127. Speed-Time Curve. Movement of a train can be represented graphically by what is known as a speed-time curve. After each stop the train must accelerate—at first perhaps rapidly, then more slowly until a balanced speed is reached. This balanced speed is maintained until a point is reached when the power is cut off and the train coasts until the brakes are applied to bring it to a standstill. The cycle can be indicated graphically (see Fig. 21-37). Acceleration is shown by the line *AB*; constant speed by the line *BC*; coasting, *CD*; braking, *DE*; and stop, *EF*.

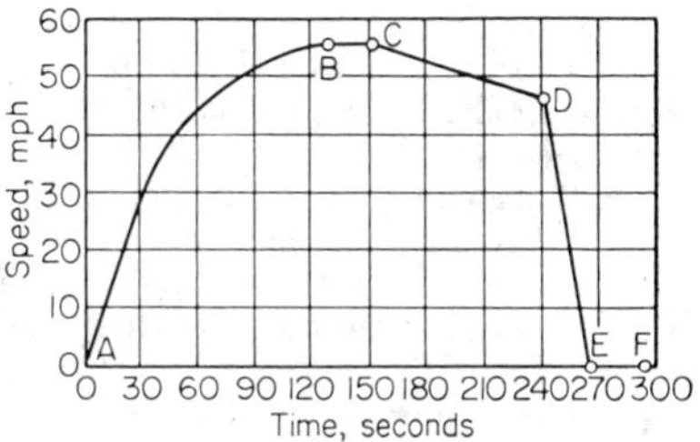

FIG. 21-37. Typical speed-time curve.

128. Distance-Time Curve. Distance is the first integral of speed with respect to time; hence the area included beneath the speed-time curve is a measure of the distance covered up to any point. The distance-time data may be constructed from the speed-time curve by a process of integration or finding the area under the curve by a planimeter. The data may be calculated readily, and this is often the method adopted.

129. Current-time curve can then be plotted for each run. From the motor-characteristic curve, the current for each value of tractive effort and speed can be transferred to the speed-time curve, and the current requirements obtained. The power input at any point is measured by the current times the voltage, which may be assumed constant for this analysis. (In a-c power systems in figuring current requirements this must be divided by the power factor.) **A power-time curve** may be thus plotted.

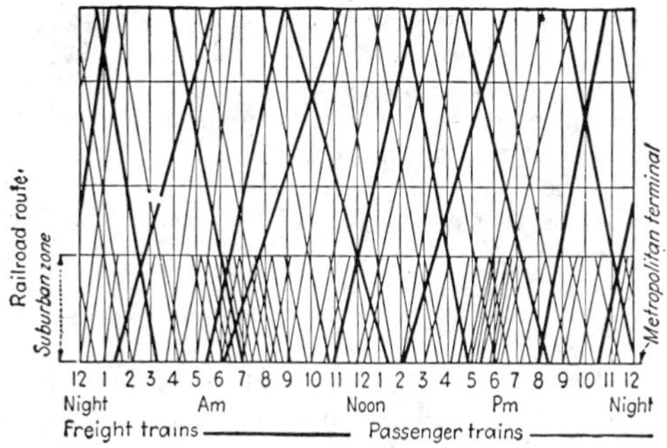

FIG. 21-38. Typical graphic train sheet for suburban and through passenger and freight trains.

130. A graphic train diagram is built up, usually plotted with time as a base on a 24-h day and geographical distances as ordinates (see Fig. 21-38). This will show at a glance how many trains are on the line at any given instant and therefore, roughly, the power demand at that instant.

131. Total power requirements in an electrified section for the system may be determined by adding the simultaneous values of all the power-time curves of trains on the system at one time. Average power per train can be determined by integrating the power-time curve over the total time and dividing by the same period. For ordinary purposes the average power may be multiplied by the total number of trains in order to arrive at the demand. The maximum possible demand will occur if all trains are starting simultaneously, which may be the case if schedules are disarranged for any reason. The energy input to the train is determined by integration of the power-time curve, and the summation of the energy input to all the trains gives the total energy. Such analysis of the demands and energy requirements is essential in designing a power plant or in negotiating for purchase of power. In figuring power-supply requirements, losses in the motive power and losses in power transmission, substations, and distribution to the moving trains must be taken into account.

From these data also may be calculated the characteristics of the transmission and distribution systems, most economical locations and sizes of substations, etc. The substations must be of sufficient capacity and number to supply the loads imposed. The transmission, substation, and distribution analysis is similar to that for any other type of power distribution, except that in the railroad the load center moves about. A balance must be maintained between cost of power losses and the capital and operating charges in the system, just as in any other power network.

Substations are located naturally in the vicinity of crossover, or interlocking, facilities, where trains may change from one track to another. At these points it is logical to sectionalize the distribution circuits in order to segregate electric troubles which may occur.

132. Momentum Grade Operation. When a train is operated over a rolling profile, it will normally lose some speed while ascending grades. This loss of speed represents a portion of the kinetic energy of the moving train which is converted into some of the work of moving the train up the grade. This kinetic energy was produced by the work of the locomotive as the train approached the ascending grade, assisted possibly by a descending approach grade. In railroad circles this is commonly known as momentum operation.

In many cases, advantage may be taken of this form of operation to increase the maximum tonnage rating of a locomotive over a given profile to a value greater than that which would obtain if the locomotive were required to start the train on any grade or to maintain some specified minimum speed on any grade without benefit of momentum. The amount of increase depends upon a number of factors, including the type of locomotive, the length and steepness of the ascending grades, the speed at which the train can approach the grades, the average weight of car in the train, and the maximum permissible value or tractive effort. The latter may be determined by adhesion limits or, in some cases where electric propulsion is used, by the necessity of protecting the electrical apparatus from overheating. The minimum speed corresponding to this maximum tractive effort may be determined from the characteristic curves of the locomotive.

The relationships of speed, tractive effort, train resistance, and variations in profile are too involved to permit the development of a simple formula for taking account of this momentum effect. Solution is usually a step-by-step process.

If the longest grade on the profile is also the steepest, it will determine the maximum tonnage rating. This will be the tonnage which the locomotive can haul up the grade at the minimum speed corresponding to the maximum permissible tractive effort. The step-by-step calculation is then made, starting at the foot of the grade with the train moving at the speed it can be expected to have at that point. If the speed does not decrease to the permissible minimum before the summit is reached, the calculation is repeated, with a somewhat greater tonnage. If the longest grade is not the steepest, the computation is somewhat lengthier.

A greater gain in tonnage rate by this method can be expected for electric locomotives than for any type of prime mover locomotive. This is because of the short-time overload capacity of electric locomotives and their ability to draw power greater than their continuous rating from the contact system.

SYSTEMS OF ELECTRIC TRACTION

133. Systems. Electrification of railroads has been accomplished by means of a number of different methods mainly characterized by the system of distributing the power to the locomotives or motor cars. The principal systems may be designated as follows:

134. Alternating-current Single-phase System. This system utilizes an overhead contact wire which usually in this country carries single-phase voltage of nominally 11,000 V of 25 c/s and generally abroad of 15 or 16⅔ c/s. The power return is in the running rails.

Commercial frequency (50 c) in the trolley wire has been adopted to a considerable extent in Europe and Asia, and there are several thousand miles of track now in operation in this system in France, Germany, Britain, and Japan. The trolley voltage in these systems varies from 6,600 to 25,000 V. The motive power in these systems has converters, rectifiers, or straight single-phase series motors. Many of these railroads are equipped for regenerating braking.

Commercial-frequency (50- or 60-c/s) high-voltage (25,000-V) distribution seems destined to be used on all future main-line electrifications. The light, single-phase contact system with its necessary circuit breakers represents the lowest-cost fixed installation which has ever been possible in the railway industry. Step-down transformers, supplied by the utility and covered in the power-rate structure, can be fed from the central power stations that serve commercial and residential loads. In all today's highly industrialized nations the load represented by even the heaviest-traffic rail lines is relatively small. Where once there was concern about imposing a main-line railroad or rapid-transit load on a public-utility network, today's rail loads are being handled readily.

Locomotives taking power from the single-phase distribution systems are, in general, four types: (1) step-down transformer, with single-phase series-commutator traction motors; (2) phase converter, either synchronous or asynchronous, taking in single-phase power and delivering 3-phase power to induction-type traction motors; (3) motor-generator sets, where single-phase power in a synchronous motor drives a d-c generator furnishing power to series-commutator traction motors; and (4) static rectifiers to change the alternating current to direct current for delivery to standard d-c traction motors.

Alternating-current, 3-phase System. This system originally utilized *two* overhead contact wires over each track, the third leg of the 3-phase circuit being connected to the rail, at ground potential. While this arrangement has disappeared completely, some work has been done in applying a 3-conductor 3-phase third rail for rapid-transit power supply.

135. Direct-current System. This system comprises distribution by means of overhead contact wire or third rail, depending upon the impressed voltage, which is usually nominally 600, 1,200, 1,500, 2,400, or 3,000 V, the power return being through the running rails, as in the single-phase system.

As much as 750 V d-c may be satisfactorily carried in a third rail, but higher voltages are usually carried in overhead-trolley wires.

Rapid Transit Systems. The distribution is usually at a nominal voltage of 600 V. Rapid-transit systems are nearly always operated with third rail at 600 V. The traction motors are usually series-commutator type.

136. System Comparison. The commercial-frequency (50- or 60-c/s) high-voltage (25,000-V) a-c system of railway electrification seems destined to be used for all major extensions. Only relatively short extensions of existing systems will utilize direct current or lower-voltage lower-frequency alternating current. Modern electronic techniques make practicable dual-, triple-, or even quadruple-voltage motive power, which can operate efficiently over routes where current must be collected from distribution systems at different voltages and frequencies and where even alternating and direct current must be utilized.

Commercial-frequency high-voltage a-c electrification makes possible smaller support structures, lighter-weight and simplified overhead wiring, and fewer substations. When such installations are properly proportioned, the cost of changes to the communication circuits will not be large and will produce a modern plant with greatly reduced maintenance and interference from the weather. Signal changes are necessary with a-c electrification, but recent developments also make it possible for the costs of such changes to be minimized. Only in areas of very low overhead clearances is the cost of securing adequate electrical clearance for high-voltage systems an important factor. Commercial-frequency a-c systems eliminate problems of electrolytic corrosion.

POWER SUPPLY

137. The high-power controlled semiconductor rectifiers have opened a new era in railway electrification. The potential for the Thyristor in power engineering can be compared with the revolution brought about by the transistor in communication electronics in the 1950s. Wherever electric power is used and the associated devices and machines must be switched and controlled, electronic components prove to be inertia-free, require no maintenance, and can be operated in any position; being unaffected by vibration, they are superior in every way to conventional devices employing contacts. In some cases, as true with railway electrification, they make solutions possible to problems which could not previously be overcome technically or economically. Semiconductors are altering many of the traditional approaches to railway electrification, and even to the electrical systems which diesel-electric locomotives carry.

138. The single-phase system of distribution is standard in the United States, at a frequency of 25 c/s, with nominal trolley voltage of 11,000 V. In Europe the frequency was generally at 16⅔ c, though on an occasional railroad 25 or 15 c had been adopted. The trolley voltage in Europe was generally 15,000 V. Since 1960, European railway systems have almost universally adopted 50-c/s 25,000-V distribution.

139. Direct-current-system distribution is at voltage ranging from 600 V nominal potential to 1,200, 1,500, 2,400, and 3,000 V. The higher voltages are always carried in overhead trolley wires. It is obvious that the higher voltages will require relatively less copper or equivalent in the distribution circuits and that substations may be more widely spaced with higher-voltage circuits, all else being equal.

140. Relatively greater voltage fluctuations may be tolerated in railroad distribution circuits than in commercial circuits, where dips may be troublesome to the users. The chief consideration, within limits of proper operation, lies, as in the design of commercial transmission lines, in the economic balance between capital charges of the power distribution system and fuel charges in heating the conducting materials. It is true that a serious drop in voltage will result in slowing the trains, and a relatively long period may affect the schedule; but if the voltage dip does not last long, the schedule is not seriously slowed. Voltage drops on the order of 10 to 15%, with occasional dips to 20%, are generally not excessive, depending upon the type of service operated.

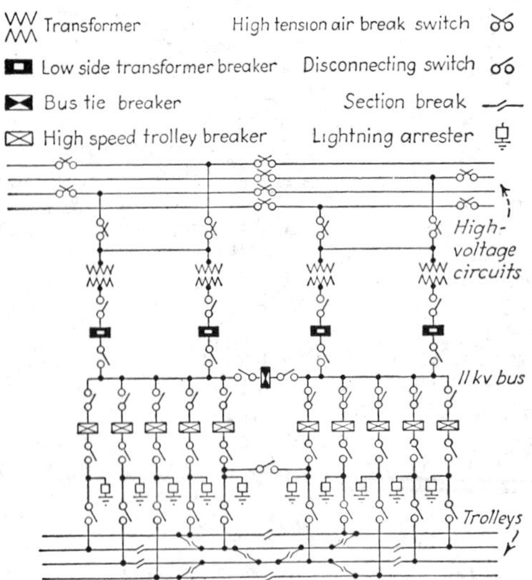

FIG. 21-39. Typical step-down substation, 132/11-kV, 2-wire, single-phase.

141. The voltage drop may be computed as in the case of commercial circuits from the load data assumed and the design of the distribution system which gives resistance (in d-c systems) or impedance (in a-c systems). The voltage is maintanied at the desired point by means of feeders operated in connection with the trolley conductors.

142. Feeder Arrangement in Single-phase Systems. A number of feeder arrangements have been used. The voltage can be stepped up and down readily by means of transformers. A common arrangement is to carry one or more single-phase transmission circuits on the catenary supporting structures, of the voltage desired—usually 44,000, 88,000, or 132,000 V—stepping down at intervals as desired at substations, to the trolley-rail voltage (see Fig. 21-39). This is known as the *2-wire system.*

143. Three-wire Systems. An alternative scheme is known as the *3-wire system* (see Fig. 21-40). The voltage of the auxiliary feeder may be any desired voltage as conditions may indicate—usually an integral multiple of the trolley voltage; i.e., if the trolley is 11,000 V, the feeder-rail voltage may be 11,000, 22,000, or 44,000 V. The transmission and distribution arrangement is designed to produce proper voltage regulation and also to provide satisfactory inductive coordination conditions with parallel communication circuits. The transformers are known as **balancing transformers.**

144. Feeder Arrangement. In d-c systems the feeders are usually installed in parallel with the contact conductors. The simplest arrangement is, of course, with no feeders, a stub-end trolley carrying the load. There may be a single feeder in parallel, with each section of the contact-conductor circuit connected into it at intervals. This arrangement is, in effect, merely an enlargement of the contact-conductor cross section. The cross sections of the feeders may be a function of the loads and of the distances of taps into the trolley from the substation, in order to equalize voltage drops and maintain the contact-conductor voltages as uniform as is practicable.

145. Trolley Sectionalizing. In a-c or d-c systems, the sections into which the contact conductor is divided may vary from a few hundred yards to several miles, depending upon the physical conditions and the importance of the service. The sectionalizing points are generally located at crossover points in order to enable trains to cross over from one track to another and run around a defective section.

146. Direct Suspension of Trolley Wire. Voltages above 600 V are carried in overhead conductors, and where speeds are low, the trolley is directly suspended above the tracks. Where high speeds are operated, the trolley wire is suspended by what has come to be known as a "catenary system."

147. Catenary Construction. If the power is distributed to high-speed trains by means of overhead conductors, it is usually catenary construction. The *contact wire* is supported from a so-called "messenger" by means of hangers of varying lengths (Fig. 21-44) so designed that the contact wire normally lies nearly parallel to the rails, the sag being taken in the supporting messenger.

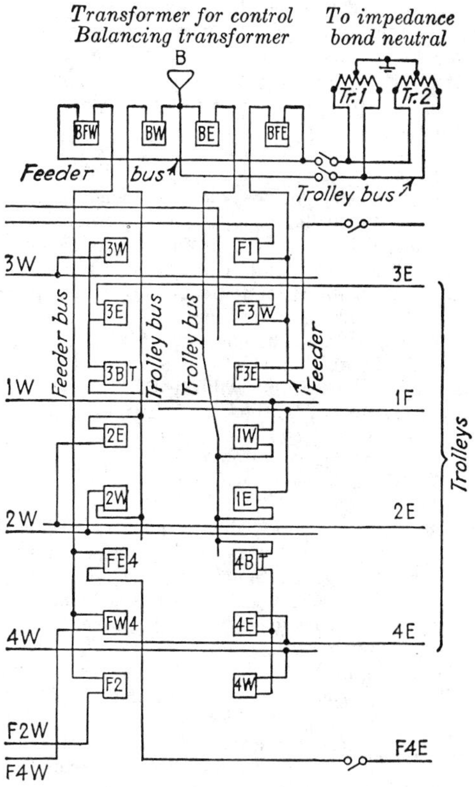

FIG. 21-40. Typical railway substation, 3-wire, single-phase.

148. Catenary and Parabolic Curves. The term *catenary* is derived from the Latin *catena,* meaning "chain," and is the curve assumed by a completely flexible material hanging freely between two supports and loaded *uniformly throughout its length.* The formula for a catenary curve is somewhat complicated; so all design of overhead construction of the so-called "catenary type" is based upon the *parabola,* the formula for which is relatively simple. This procedure is justified because a flexible cable supported at two points and loaded *uniformly throughout the horizontal projection of its length* assumes a parabolic curve. The actual curve of the messenger supporting a contact wire lies between a parabola and a catenary, and errors are relatively small within the limits of the sags and stresses adopted in this type of work, with the accuracy that is economically possible in installing the construction also considered.

149. Catenary Design. The principle of the track catenary design is the same basically as that of designs of transmission lines, but certain additional factors must be considered.

The parabola formula may be expressed thus,

$$S = \frac{w \times d^2}{8 \times T} \tag{21-2}$$

where T = horizontal component of tension in messenger in pounds; S = sag of messenger in feet; d = length of span in feet; w = weight of the system messenger, trolley wires, hangers, etc., in pounds per foot of span.

From this formula, with the weight per foot of the system (including messenger strand, trolley wire, and clips) and the permissible stress in the messenger known, the sag can be determined for any given span. The total tension in the messenger is greater than the horizontal component as expressed in the formula, the vertical component being only the dead weight of the system. Nevertheless, the vertical component is relatively so small as compared with the horizontal component (generally less than 2%) that the error may be safely ignored. A not uncommon sag is about 5 ft in a 300-ft span.

150. Ice and Wind Loads. In the design of the system, the assumed weights must include not only the dead weight of the material of the construction itself but, in addition, the weight of possible *glaze* which may be deposited upon the wires and amounts perhaps to a cylinder about each member, assumed at ½ in radial thickness in heavy loading districts and ¼ in medium as defined in the National Electrical Safety Code, U.S. Bureau of Standards—and also the horizontal *wind* load, which is assumed generally at 8 lb/ft² on two-thirds of projected area on each wire and hanger. These conditions are computed for maximum stress analysis, and the sags are so fixed that the resulting stress in the messenger will be safe.

151. Deflection due to wind across the tracks is of importance because, if this is too great, the pantographs will ride off the wire in a bad storm. This deflection is measured in the same way as the sag of the messenger.

$$D = \frac{P \times d^2}{8T} \tag{21-3}$$

where D = deflection due to wind, d = span length, T = combined tension of messenger and contact wires, P = wind pressure per linear foot.

152. Temperature Effects. An additional consideration is the temperature effect upon the messenger. Obviously, with the span fixed, the stress in the messenger will depend upon the temperature, all other things being equal, and the sag in any given distance must be so chosen that at the lowest possible temperature, as well as with the maximum assumed wind and ice loads, the stresses will not rise above a safe point. This analysis is somewhat involved, and reference is made in the Bibliography (Par. **259**) to data describing it in detail.

153. Messenger Sag. After the weights and stresses are determined, the sags are computed on the preceding formula—usually based upon the average temperature encountered (60°F in temperate latitudes)—and the hanger lengths determined on the basis that each hanger length is equivalent to the sag of the messenger at a span equal to twice the distance of that hanger from the low point of the span, plus a constant representing the shortest hanger in the span (see Fig. 21-41).

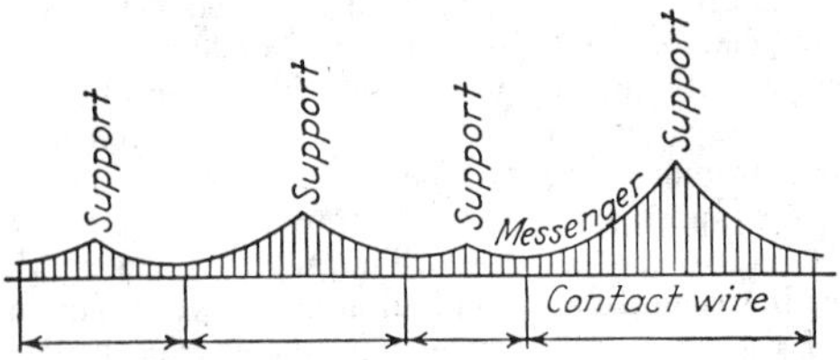

Fig. 21-41. Varying lengths of catenary spans symmetrical about points of supports.

154. Span Lengths and Hangers. The catenary spans usually cannot be of constant lengths but must be chosen to conform to local requirements, e.g., track curves and overhead-highway-bridge structures with limited clearance. As will be noted in the formula, the sag of any span is a function of the square of the span. The stress in the messenger is thus (at an assumed temperature) approximately constant in adjacent spans regardless of the span length. It is thus convenient in laying out the catenary construction so to design it that the hanger sets or groups and messenger sags are symmetrical about each point of support. Since the messenger is horizontal at the low point in the sag or span, it is immaterial whether the span is symmetrical about the low point or is composed of two half spans of varying lengths (see Fig. 21-41). This method permits all hangers to be cut in bundles of standard lengths, varying only in number of uniformly spaced hangers chosen per span.

The foregoing analysis is based upon tangent-track construction where all hangers are normally vertical.

155. Chord Construction on Curves. On curves it is sometimes the procedure to install the catenary around the curve in a series of chords, pulled off at intervals so spaced that the middle ordinate in any chord is not too great to permit safe passage of the collecting pantograph shoe (Fig. 21-42). On this basis the profile of the catenary is the same as that in tangent track. The spacing of pull-offs is dependent upon the sharpness of the curve. In general, the midordinate should not be greater than one-quarter the length either side of the center of the collecting shoe, which is usually about 4 ft long.

156. Pull-off Chord Construction. The maximum length between pull-off points of a chord for a curve of a given radius, if at the point of pull-off the contact wire is 1 ft toward this to the outside of the curve and at the center 1 ft inside the center line, is represented by

$$d = 4\sqrt{r} \qquad (21\text{-}4)$$

where d = span and r = radius of curve.

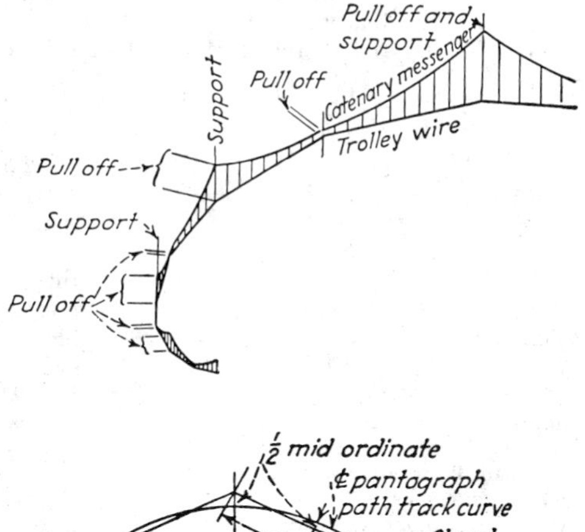

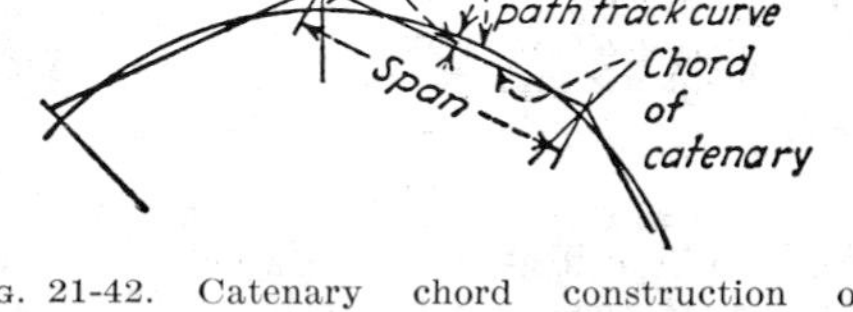

FIG. 21-42. Catenary chord construction on curves. Perspective view (above); plan view (below).

157. Inclined Catenary. It is common practice to install the catenary on curves with inclined hangers, each hanger taking a position that is a resultant of the vertical load of the contact wire and the horizontal load due to curve pull. This is known as **inclined-catenary** construction. The point of support of the messenger is offset toward the outside of the curve to a point that permits a horizontal as well as a vertical sag in the messenger (Fig. 21-43). This construction is very satisfactory for curves up to 4°. For sharper curves the horizontal pull is so great that the hangers take a position too close to the horizontal, the offset of the messenger support is too great, and the longest hangers are too long to be satisfactory. Furthermore, not much distance is saved in pull-off spacing, so the inclined catenary construction is not much used in curves sharper than about 4°.

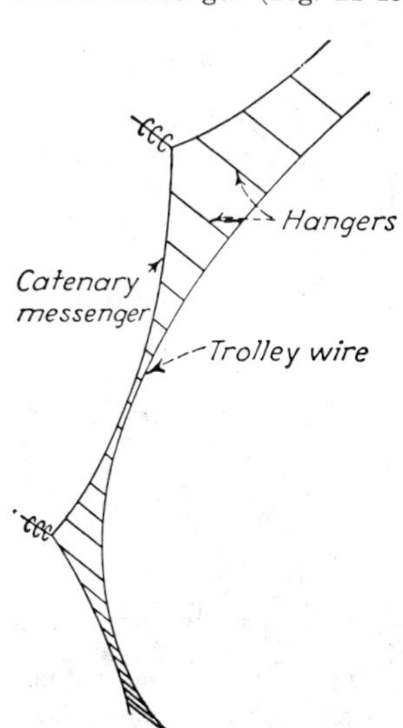

FIG. 21-43. Inclined-catenary construction on curves, perspective view.

158. Calculating Inclined Catenary. In inclined-catenary construction the computations are, in general, of the same type as with tangent construction, the horizontal component of the sag being computed from the curve pull per foot, which is based on the tension of the trolley wire. With the degree of curvature of the track known, the radius and the required pull per foot of the curve are readily found,

$$F = \frac{T \times d}{R} \qquad (21\text{-}5)$$

where F = horizontal force normal to the wire in pounds, T = tension in wires in pounds, d = span in feet, R = radius of curve in feet; w = horizontal force per foot normal to wire, $w = T/R$.

159. Horizontal and Vertical Sags. From these data the horizontal sag may be obtained just as the vertical sag was. The resultant sag follows automatically, as does the length of the inclined hangers.

160. Span Lengths and Sags. The span length most popular in this country is on the order of 300 ft between supports, though in many cases the limit is 275 ft. The limitations are the practicable length of hangers and sags of the messenger, which, as has been indicated, varies as the square of the span. Large sags mean supports higher above the rail and therefore higher structures and greater deflection due to transverse wind loads.

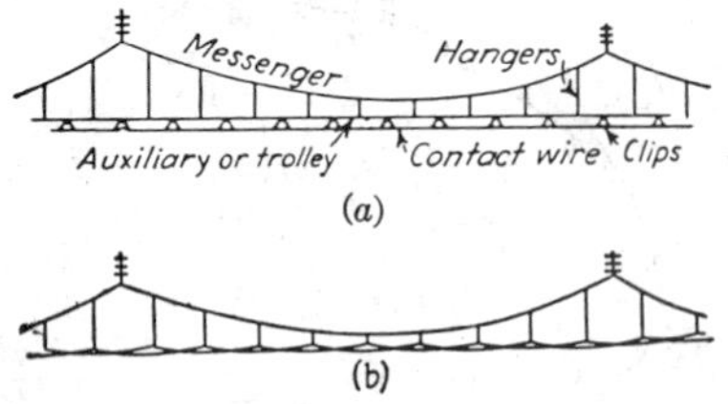

Fig. 21-44. Typical arrangements for double trolley wire. (*a*) One serving as contact wire and supported by clips between hangers from the other wire, which serves as an auxiliary messenger. (*b*) The two trolley wires supported in the same horizontal plane by alternate hangers from the catenary messenger.

161. Contact Wire below Trolley. On many railroads, a contact wire is installed immediately below the auxiliary or trolley wire, which thus becomes an auxiliary messenger. This design performs a double duty, in that the contact wire takes the wear of the sliding pantograph shoes and the supporting clips between hangers provide a flexible support without "hard spots," particularly undesirable in high-speed operation (see Fig. 21-44).

Twin Trolley Wires. In some designs two copper contact wires are installed side by side supported by alternate hangers (see Fig. 21-44*b*). This provides double contact surface with the pantograph shoe and is of advantage when large currents are collected.

162. Materials, Messenger. The material of the catenary construction depends upon local conditions. If electric conductivity is of importance, as in low-voltage distribution systems, the messenger may be of copper. Copper, however, has relatively low tensile strength. Bronze strand is sometimes used as a messenger on account of relatively higher tensile strength and satisfactory noncorrodible characteristics. A bronze alloy (silicon or cadmium) with conductivity on the order of 40 to 60% and occasionally 90% of that of copper is commonly used. Copper-covered steel strand is popular as a messenger material on account of its relatively high tensile strength and satisfactory corrosion resistance. It has the advantage of high melting point, so is not so likely as some other materials to burn upon arcing contact. Galvanized heat-treated alloy steel of high tensile strength is sometimes employed for messenger satisfactorily, though it is subject to serious corrosion under certain atmospheric conditions. "Stainless steel," which resists corrosion and has high tensile strength, is occasionally used but is costly.

163. Materials, Trolley and Contact Wires. *Hard-drawn* copper trolley wire, of sizes between Nos. 2/0 and 4/0 and sometimes 6/0, is usually employed for distribution at all voltages. It is sometimes used as a contact wire but with steel pantograph shoes wears relatively rapidly. The contact wire installed below the copper wire is usually of bronze of an alloy possessing about 40% conductivity.

164. Materials, Hangers. The hangers may be of cold-rolled galvanized steel, copper-covered steel rods, or bronze rods. Steel is relatively less expensive than other material but is subject to serious corrosion unless continuously painted.

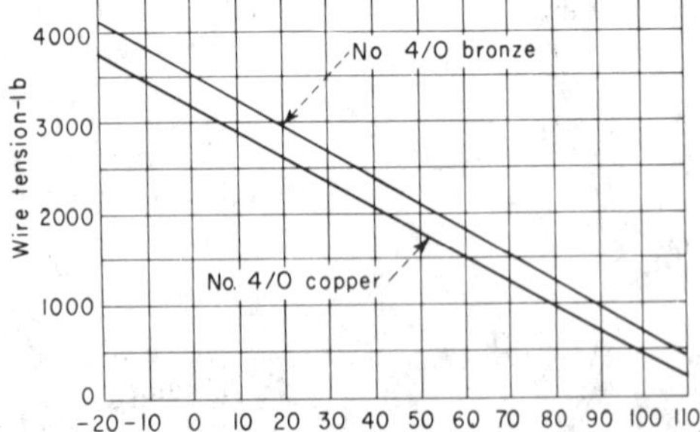

Fig. 21-45. Stresses in bronze and copper trolley wire at various temperatures.

165. Anchoring Trolley and Contact Wires. Trolley wire or auxiliary messenger and the contact wire in the United States are usually anchored rigidly at each end of a section perhaps 4 or 5 mi long. Tension at installation must be so chosen that the material neither is overstressed at very low temperatures nor hangs too slackly at high temperatures. The temperature-stress curve is practically a straight line; so if the two limits, maximum and minimum stress, are correctly fixed, the wire will perform satisfactorily at all temperatures (see Fig. 21-45).

166. Constant-tension Trolley Wire. In railroads abroad the contact wire is commonly installed upon what might be called a "constant-tension basis," and the dead end at the far end of each section is hung over a pulley by weights which rise and fall with temperature changes.

167. Types of Catenary Design. In addition to the simple catenary construction outlined above, other forms of catenary have been employed as follows:

Double Catenary. This consists of twin messengers each installed with both vertical and horizontal sag, supporting the contact wires by means of triangular hangers (see Fig. 21-46). The design has horizontal stability but is expensive to install. On curves, the construction is based on chords with pull-offs.

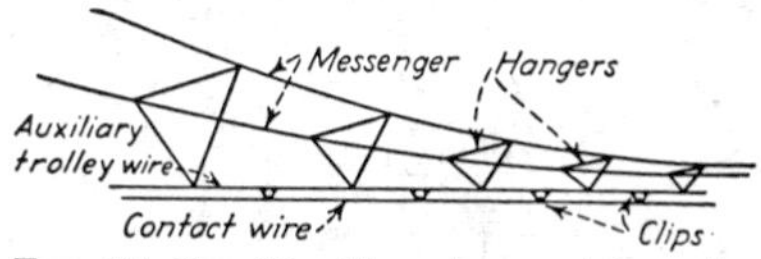

FIG. 21-46. Double catenary triangular hangers, perspective view.

Compound Catenary. This consists of heavy messengers hung from the supporting structures and in turn supporting the single or simple catenary members from intermediate points (see Fig. 21-47). Typical of modern practice is the compound catenary used on Japan's high-speed New Tokaido Line, on which trains operate at speeds up to 150 mi/h. The copper trolley wire has a cross section of 110 mm^2; the auxiliary messenger is made of stranded cadmium copper wires of 60 mm^2; the messenger, also stranded cadmium copper, is 80 mm^2. Total area of the overhead is 250 mm^2. Other overhead contact-line facilities are a negative feeder of aluminum and booster transformers for prevention of induction on communication lines. Section breaks in this 25,000-V 60 c/s distribution system consist of a dead section with a 10-Ω resistor at each end to prevent wear of the pantograph slider and trolley wire due to the breaking of large currents being drawn by accelerating trains.

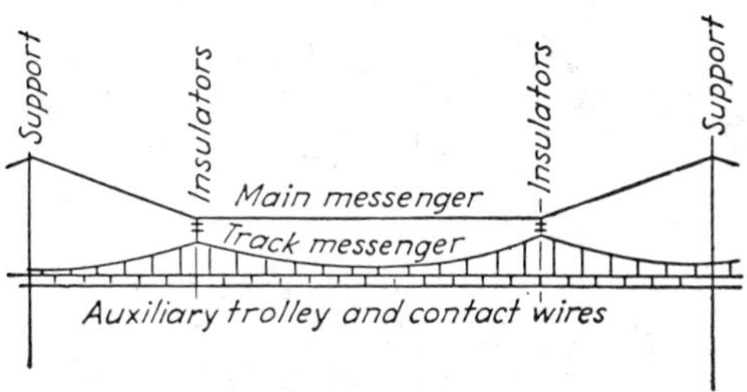

FIG. 21-47. Compound catenary.

168. Low-inertia Problems. The light overhead installation possible with commercial-frequency electrification does simplify construction and maintenance and makes possible low-cost supporting structures. At higher speeds, however, there has developed the problem of maintaining continuous contact between the trolley and pantograph because of the low inertia of the overhead. Springs and dampers are used in the Tokaido Line's contact system, adding considerably to initial and maintenance costs. Without proper damping, in either the contact wire or the pantograph, or both, serious resonant oscillation can develop, with the possibility of the pantograph shoe separating from the contact wire while drawing currents of 8 to 10 mW. A solution more economical than the constant-tension arrangement used in Japan has been the introduction of dampers in the pantographs and control of the level and stiffness of the contact wire to give the smoothest possible pantograph trajectory. Problems of satisfactory pantograph operation can be complicated when locomotives or cars with pantographs are operated in multiple unit and follow each other closely under the contact wire. Because the contact wire cannot be kept at a constant height above the rail, particularly when existing routes are being electrified, its gradients must be carefully designed to avoid pantograph bounce. At high speeds a maximum contact-wire gradient of 1 in 500 has been found satisfactory. Actually the constant inertia, which the compound catenary was designed to achieve, has been reproduced by the British Railways with a lower-cost simple catenary system where the trolley is allowed to sag at midspan. Figure 21-48 shows not only typical catenary construction for commercial-frequency systems but the contemporary supporting structures for such a system.

169. Supporting Structures. Catenary supporting structures are of varying designs. In many cases, especially when multiple tracks are involved, light, fabricated trusses or rolled sections, channels, or H-beams are installed over the tracks to support the messengers. This construction has the advantage of also supporting track signals as required and also possessing a certain degree of stability along tracks in the event of messenger failure. The cross member, or bridge, is supported by posts, which may be latticed or of a rolled section. The posts may be self-supporting, or the entire unit may be of the so-called **portal type**, the overturning moment being taken care of by the corners, so that shear is imposed only upon the top of the foundation.

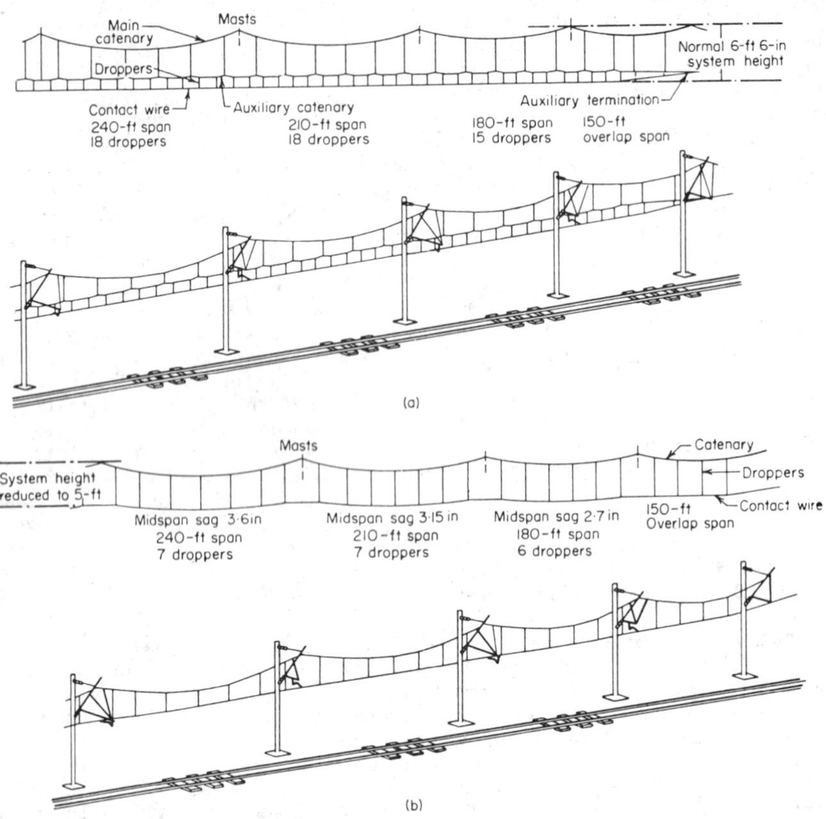

Fig. 21-48. Lightweight overload possible with commercial frequency simplifies supporting structures. (*a*) Compound catenary; (*b*) sagged simple catenary.

170. Cross Catenary. A common form of construction is the cross catenary, supported usually by guyed poles, when there is sufficient width of right-of-way. From this cross span is hung the track-messenger construction (see Fig. 21-49). This type of construction is economically adapted to multiple-track electrification.

171. Pole-and-bracket Construction. For single- or double-track electrification, so-called pole-and-bracket construction is often employed (see Fig. 21-49). Sometimes the poles are located between tracks with double brackets. In some cases independent poles are used for each track. This has the advantage of complete independence of each track construction from the others—an advantage especially in the event of derailment of a train.

172. Feeder Supports. The catenary supporting structures are available for carrying the traction-power feeder and auxiliary circuits, e.g., signal-power-supply circuits.

173. Clearance of the overhead trolley varies from a normal minimum of 16 ft above the top of the rail to a maximum of 24 to 25 ft on certain railroads.

174. Third-rail construction is of two general types: (1) overrunning and (2) underrunning.

In the **overrunning** design the rail of low-carbon steel is supported on insulators, generally either treated wood blocks or porcelain, which are in turn supported on long track ties, generally every sixth tie. Rail joints are loosely installed to permit temperature expansion and contraction, and the joints are bonded to provide conductivity for large currents. Anchors are provided at intervals to prevent creepage. The rail is sometimes protected by a wood or composition shield to protect it from accidental contact (see Fig. 21-50). In these cases the shoe is horizontal and slips under the protective covering. In some instances when this arrangement is not practicable, the rail is protected by a strip of wood at one or both sides, the top being left open, directly under the third-rail shoe.

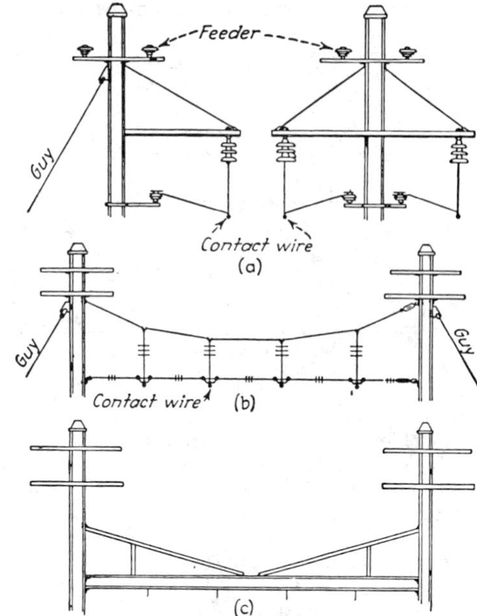

FIG. 21-49. (a) Pole and bracket. (b) Cross catenary. (c) Light catenary bridge.

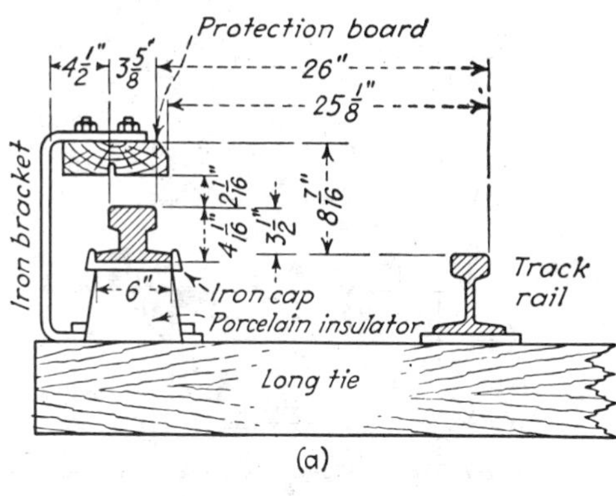

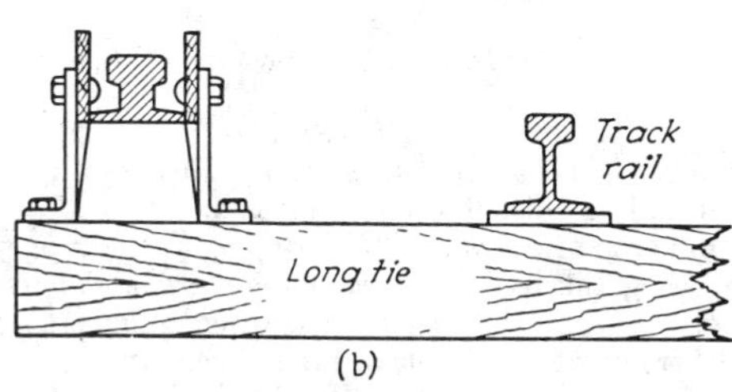

FIG. 21-50. Overrunning third rail. (a) Top protection. (b) Side protection.

Underrunning third rails have the contact on the lower surface and are supported by gooseneck, or curved, brackets. The rail is protected by wood or composition trunking on three sides, the bottom only being left bare for the contact shoe which projects horizontally from the truck frame of the equipment (see Fig. 21-51). This arrangement is free from difficulty from ice and relatively free from snow trouble.

Location of the third rail is determined by the rolling stock passing along it. Sufficient clearance must be allowed to permit cars and locomotives of maximum dimensions to clear the rail; usually on steam railroads the Association of American Railroad Clearance diagrams govern the location. On

American railroads the distance from the track gage to the center line of the third rail varies usually from 20 to 28 in, and the height above the top of the track rail, from zero to 7½ in.

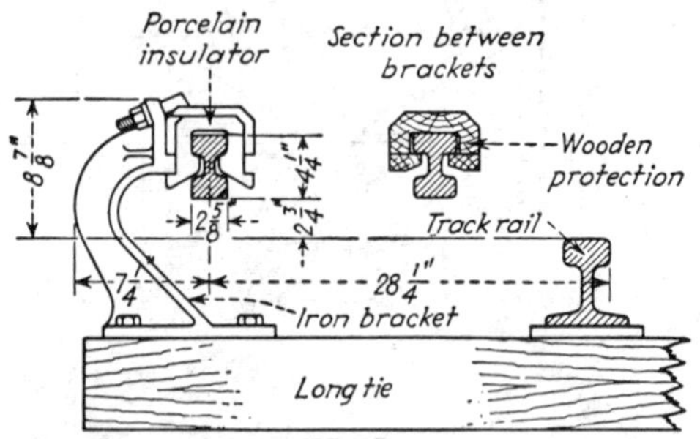

FIG. 21-51. Underrunning third rail.

175. Rail Bonds. Third-rail and track-rail bonds are (1) mechanically attached, (2) welded.

Mechanically attached bonds may be attached to the head or the web of the rail in holes drilled or punched and reamed in the rail. The bonds are usually stranded copper swaged or welded into terminals with cylindrical or tapered heads at right angles to the length. These heads are compressed into the holes in the rail by expansion, hammering, hydraulic compression, or expanding by means of a pin forced into the terminal.

Welded bonds are attached by gas or electric welding usually to the head of the rail.

Impedance bonds consist of a flat iron core surrounded by two coils, with the center point led out for a terminal. This provides high impedance between rails and low impedance for the power-return current, which passes through each coil in balanced amount and, through the neutral, to the neighboring impedance bond or to cross bonding (see Fig. 21-52*a* and *b*). Multiple tracks are cross-bonded at intervals to permit full advantage to be taken of all track conductivity possible. Since most electrified sections of steam railroads are equipped with signal track circuits, the cross bonding must be taken from the neutral points of impedance bonds, as the running rails must be insulated from each other by the ties in order to preserve the integrity of the track circuits.

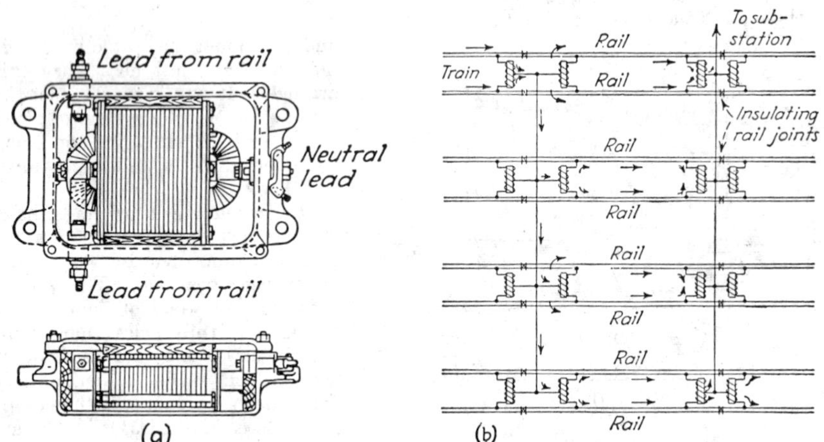

FIG. 21-52. (*a*) Single impedance bond. (*b*) Cross bonding, power return distributed over four tracks.

176. Substations. In single-phase electrification the substations are usually of two general types: (1) supply substation and (2) distribution substation.

Unless the power is to be used at commercial frequencies (50 or 60 c/s), the supply substation usually has facilities for changing the frequency from the commercial frequency to that of the railroad system. In some instances the power is generated directly at the required frequency, so the frequency changers are not necessary.

177. Distribution substations for single-phase railroads are generally transformer stations with oil or air circuit breakers to control the trolley and feeder circuits, which are usually sectionalized at the substation. All this apparatus is usually out of doors.

The circuit breakers are generally single-pole. Since the substations are often located at so-called "interlockings" on the railroad, where trains can be crossed over from one track to another, attendance is often available to operate the circuit breakers, though in many cases the circuit breakers are remotely controlled by the load dispatcher from a central point. No attendance is required except for occasional inspection and cleaning. Typical a-c substations are shown in Figs. 21-39 and 21-40.

178. Direct-current Substations. In d-c electrification three types of substation are used: manual, automatic, and supervisory-controlled. Each type may use for conversion the synchronous converter, the motor-generator set, the mercury arc, or the solid rectifier. **Manually operated substations** require attendants to start and stop the apparatus, open and close switches, and supervise other operations. **Automatic substations** replace the attendants with relays which automatically control the various operations in response to the load on the system or the voltage. The character and functioning of the automatic devices depend on the type of conversion apparatus and the conditions to be met. Design of the automatic apparatus is such that manual control is possible under abnormal conditions when desired.

Supervisory control substitutes for the manual operations relay-actuated movements, controlled from a central point. Indications of the condition of load, etc., are made at the load dispatcher's control board, and he can energize circuits which perform the necessary actions at the substation. By this means the load dispatching for an entire system may be concentrated at one point, under the supervision of a load dispatcher who is thus at all times informed of the status of the load. This is similar to such operations in commercial power practice.

Apparatus of d-c substations usually includes the following, which are common to all types: (1) incoming a-c 3-phase feeders with lightning arresters; (2) high-tension circuit breakers; (3) step-down transformers; (4) switchboard, hand, automatic, or remotely operated, for control of incoming a-c feeders and d-c outgoing feeders; (5) conversion apparatus, details of which will vary with the types of converting apparatus; (6) outgoing feeders and switches.

179. Synchronous converters for changing 3-phase alternating current to 600-V direct current were formerly standard equipment. The **compound-wound converter** is used especially with widely fluctuating output, e.g., suburban or rapid-transit systems in which heavy units accelerate rapidly. The series field winding is usually adjusted for flat compounding at 600 V throughout the range of the load.

180. Motor-generator sets have been used on some high-voltage d-c systems instead of synchronous converters as it is difficult to design a converter running on a 60-c circuit for more than 750 V, whereas a generator may be built for 1,500 V. Two 750-V machines may be run in series for a 1,500-V electrification, but it is not considered practicable to use four in series for a 3,000-V line. The efficiency of motor-generators is less than that of synchronous converters, but the former are better adapted for power-factor correction, and in some instances where the voltage of the primary feeder is low, step-down transformers can be omitted by winding the synchronous motor for the voltage of the power supply.

181. Electronic rectifiers are replacing other conversion equipment for the supply of direct current, both for 600-V city systems and for high-voltage electrifications. The connections resemble those for synchronous converters, because step-down transformers are required to obtain the correct ratio of voltages between alternating and direct current. Three-phase supply is suitable at any frequency, and the rectifier itself can be designed for any desirable number of phases, connections of the transformers changing the 3-phase power accordingly. Rectifiers can be built to have shunt or compound regulation in a manner somewhat similar to that with synchronous converters and are adaptable for regeneration of power.

Advantages of rectifiers are high efficiency, particularly at high voltage; light weight; small floor space; and ease of automatic control (see Fig. 21-53).

182. Operation of Automatic Substations. Substations without attendants may be made to start automatically in various ways, but for the most part they function through remote control or automatically by means of a voltage relay, which closes its contacts at a particular value of reduced contact-line voltage. The machine remains in operation until cut out of service by a light-load control. It is necessary to provide a sequence of circuits to start the converter, bring it to synchronous speed, connect it

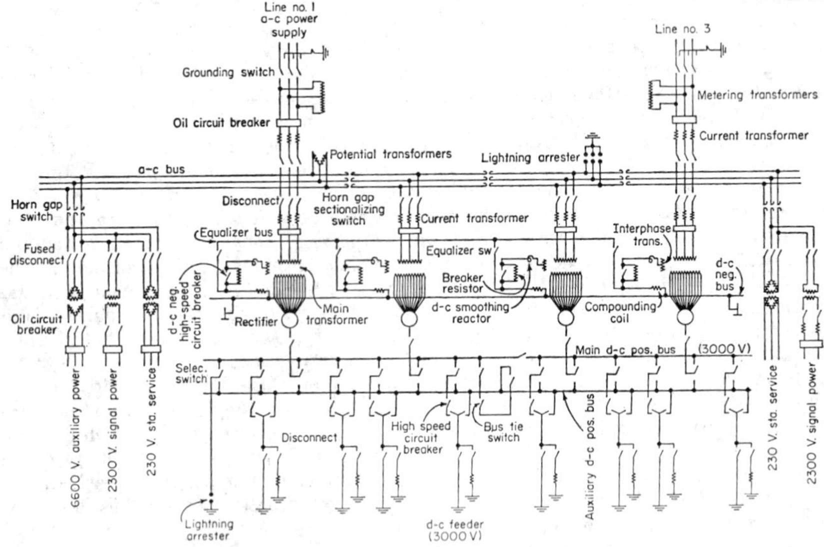

Fig. 21-53. Mercury-arc-substation diagram.

to the d-c circuit, and transfer to the several machines in the station the proper proportions of the total load. Operation of mercury-arc-rectifier automatic substations is even simpler.

ELECTROLYTIC AND INDUCTIVE COORDINATION

183. Earth Currents. Practically all railroad-electrification power distribution systems utilize the track rails to complete the traction circuit. Since the track rails are normally supported by wood ties on the ground, there is relatively low resistance between the rails and the ground, and much of the return current is likely to leak from the rails into the earth. In the case of d-c systems this gives rise to electrolysis problems and in a-c systems to inductive problems. Coordination is necessary with other agencies in either case to prevent interference from the grounded railroad circuits.

184. Conditions Contributing to Electrolysis. The stray currents from the rail-return circuit may flow into the earth and into underground structures, e.g., water pipes, gas pipes, and electric cable sheaths, returning to the rails or negative feeder taps in the vicinity of the substations or power plant. There is a neutral point where the rails are at earth potential. Between the substation and that point, the rails generally are negative with respect to the earth, and current tends to flow from earth or underground structures to the rail, whereas from the neutral point to points farther removed from the substation, the rails are positive with respect to the earth and current tends to flow from the rails into the earth.

185. Electrolytic Damage. Electrolysis occurs where the current leaves metal underground structures to flow in the earth, i.e., where the structures are positive to the earth. The extent of electrolytic damage is a function of the character of the metal and the amount of current and of time and soil conditions. One ampere flowing continuously for a year may dissolve 13 to 20 lb of iron or 75 lb of lead at the points where the current leaves the metal.

186. Electrolysis-preventive methods may be divided into four classes:

1. **Insulating rails** from the ground so far as possible and installing insulating joints in pipelines or cable sheaths to increase the resistance of the current-return path through the earth, though care must be taken to ensure that stray currents shunting into the earth around the joints do not produce electrolytic corrosion in the positive side of the joint.
2. Ensuring that **rail bonding** is efficiently installed and maintained.

3. **Installing drainage** by bonding the metallic structures directly to the rails at points where the current tends to leave the structures to enter the earth and return to the rails.

4. Separate **return feeders,** insulated from the ground, which reduce the voltage drop in the rails, such feeders to be connected to the rails at various points, so chosen that the voltage at each location is about the same under normal conditions of load.

Electrolysis sometimes occurs in reinforced concrete. The oxides of iron thus formed may occupy more space than the original reinforcement and cause cracking of the concrete. It is a safe precaution to install insulating joints in all pipes and cable sheaths leading to or from reinforced concrete structures in the earth.

187. Electrolysis effects due to alternating currents are less than 1% of those produced by direct current and thus have been found to be of little importance in practice.

188. Electrolytic self-corrosion is often noticed when metallic structures are embedded in soil by varying properties, owing to local galvanic action and not to stray currents. For example, a piece of coke in contact with a pipe may cause a current to flow from the pipe to the coke and return elsewhere from the earth to the pipe, causing rapid local corrosion. It is very desirable on this account that pipes or metallic sheath cable not be buried in cinders, even though no stray direct current is present in the vicinity.

189. Inductive coordination problems between traction power circuits and neighboring communication facilities are similar in many respects to those of commercial power and communication circuits. The principles are the same, and insofar as the traction power-transmission circuits are concerned, the problems are practically identical, as these are metallic circuits and may be transposed as desired and there is no earth current.

In the **traction power distribution systems,** the detail problems are somewhat different, because the ground must be used to some extent by the power circuit which returns to the substation or power plant through the rails, part of it unavoidably passing into the earth. This constitutes a loop between the trolley wire and the rail which may be as much as 24 ft apart and which it is, of course, impossible to transpose. Insofar as the ground-return circuit is concerned, the loop is roughly considered as twice the distance between the trolley and the rail, the earth current being considered as concentrated about an imaginary plane a distance below the ground equal to that of the trolley wire above ground, although this distance depends much on the nature of the ground at any point.

190. Special Measures in Coordination. Since it is impossible to transpose the traction-distribution circuits, other means are necessary to avoid inductive disturbance in paralleling communication circuits. It is often not possible to separate geographically the power and communication facilities, as the communication facilities are usually necessary for railroad operation and the right-of-way is usually restricted in width.

191. Principles and Practices. In addition to methods adopted for power circuits in general, a number of methods of coordination have been developed between traction power and communication circuits.

MOTIVE POWER

General Classifications

192. Electric locomotives may be grouped as follows:

1. Locomotives receiving all their **power from outside sources** by either overhead wire or third rail.

2. **Internal-combustion** locomotives: diesel, gasoline, or gas-turbine engines with electric drive or power transmission to traction motors.

3. **Combination** of internal-combustion engine with collecting device and control to take power from an outside source. These last two classes are not, strictly speaking, "electric locomotives" but contain much electric equipment.

The locomotives in group 1 may be classified as a-c or d-c equipment, dependent upon whether they take power from a single-phase or a d-c distribution system.

Alternating-current locomotives may be further divided as follows:

1. **Straight single-phase** with step-down transformer supplying power to series-commutator traction motors.
2. **Phase converter** with a rotating machine which changes single-phase power to 3-phase for delivery to 3-phase induction-type traction motors.
3. **Motor-generator** taking single-phase power into a synchronous motor which drives a d-c generator which furnishes power to the traction motors.
4. **Rectifier** delivering power for single-phase trolley system to rectifiers for supplying standard d-c traction motors.

Direct-current locomotives are classified on the basis of voltage at the collecting shoe, usually 600, 1,200, 1,500, 2,400, 3,000, or 4,000 V.

Locomotives have been divided generally as follows:

1. Road passenger.
2. Road freight.
3. Road *freight* and *passenger*.
4. Road *switcher*.
5. Yard switcher.

193. Classification of Driving and Trailer Axle or Wheel Arrangement. Starting at the front end of the locomotive designed for single-end operation or at either end of locomotives built for double-end operation, the axles and truck connections are designated in their consecutive order. Letters A, B, C, D represent the number of driving axles in a rigid wheel base; numerals 1, 2, 3, the number of guiding or trailing axles; and plus or minus signs, the absence or presence of connection between trucks.

194. Rating of locomotives is now usually expressed in terms of horsepower. This is more important than maximum tractive effort in hauling a train. Diesel-electric locomotives are usually rated in terms of horsepower at the diesel-engine shaft driving the main generator. To convert this figure to horsepower at the driving wheels, it is customary to multiply the rated horsepower by 82%. A great advantage of electric locomotives is their ability to draw on central power plants for short periods of time and accelerate rapidly or maintain speed in short grades. The diesel engine itself has little overload capacity. Gas turbines are somewhat more flexible in this regard.

The continuous rating is commonly based on temperature rise of transformer or traction motors of 120 to 130° and sometimes as high as 140° above the surrounding air, with Class B insulation, consisting of mica, glass, or asbestos, as indicated in USAS C35.1-1943, and AIEE Standard 11 for railway motors.

The 1-h rating is based on similar temperature rise reached in 60 min, starting at normal temperature; overloads of shorter periods are similarly figured and are indicative of loads the locomotive can haul at given speeds for the periods indicated without overheating.

The starting rating is the sustained tractive effort that can be exerted during acceleration and may correspond to the 25% adhesion coefficient for a short time in high-speed passenger service and up to 10 min in freight service.

195. Similarity of Electric and Diesel-Electric Systems. Electrical and electronic developments since 1950 have served to cause electric and diesel-electric locomotives to have more and more in common. The successful application of ignitron rectifiers, followed by selenium and then silicon rectifiers, on a-c electric locomotives has made it possible to drive them with the same d-c traction motors used on diesel locomotives. The series d-c motor has torque characteristics well suited for rail traction applications. The same rectifier developments have also made it possible, since 1965, to have the engine on diesel locomotives drive an alternator rather than the d-c main generator, which had always been used prior to that time. Alternator output is rectified for delivery to conventional d-c traction motors. The drive for higher diesel-electric locomotive ratings was, until then, being restricted by the physical dimensions and maintenance requirements of the d-c main generator necessary to transmit the horsepower which improved diesel engines were capable of producing. A suitable d-c generator for engine ratings of 3,000 hp, or more, was of such size that it could not readily be fitted within railway clearance restrictions. The smaller-diameter lighter alternator without brushes fitted readily in a high-horsepower diesel locomotive and simultaneously cut maintenance requirements.

Developments in electronics are making possible new traction concepts. Along with more sophisticated control, the static converter, now feasible because of Thyristors, can supply multiphase variable-frequency current. This ultimately could make possible the use of the asynchronous traction motor. The ability to supply power at variable frequency overcomes shortcomings that formerly ruled out the asynchronous motor for traction: lack of flexibility for speed regulation and low starting torque. The asynchronous motor offers the advantages of simplicity and ruggedness; it costs less than the commutator-type motor and requires little maintenance. Because higher rotating speeds are possible, it makes possible small, light units with high ratings.

While Thyristor, or cycloconverter, design and operation are relatively simple with a-c feed, the operation with a d-c supply is relatively complicated because there is no following half wave to turn the static unit off. This is overcome with oscillating circuits incorporating capacitance, but as circuit complexity increases, there is a sacrifice of efficiency. The Thyristor, however, can make possible the elimination of the conventional speed-control rheostat used on d-c locomotives, using instead a converter with smooth output and no loss of power in speed-regulating resistances.

196. Power Sources. Electric locomotives, with no prime movers, collect current from a trolley or third rail. Diesel-electric and gas-turbine-electric locomotives generate current in a main generator.

197. Pantographs are mounted on the roof of the equipment and usually carry a sliding shoe for contact with the overhead trolley wire. They are always used for high-speed operation and when large amounts of current are to be collected. They consist of a jointed frame usually of steel tubing (Fig. 21-54). The contact shoes are usually about 4 ft long, and these may be a single shoe or two shoes on each pantograph. The shoes may be straight throughout their length or cambered slightly or (especially abroad) may be in the form of an auxiliary bow. In this country the

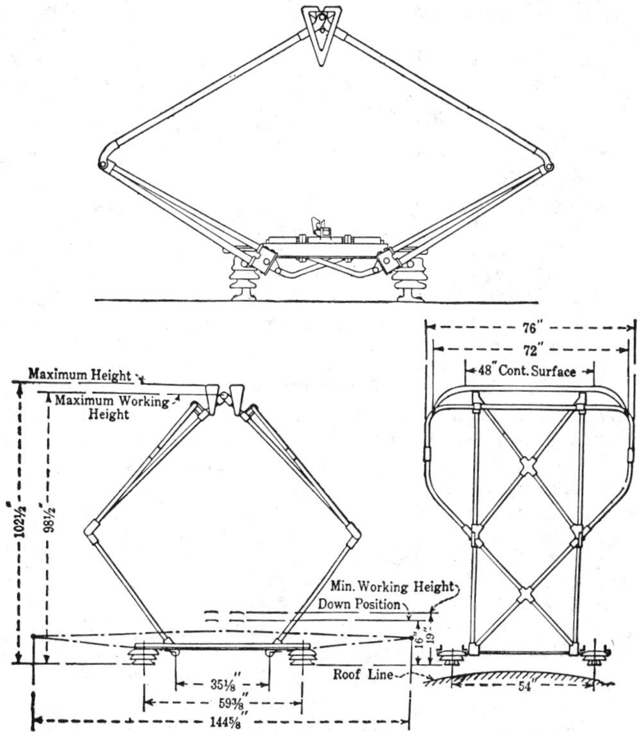

FIG. 21-54. Pantographs. Single (above); double shoes (below).

material is often steel with, sometimes, wearing plates of copper or bronze inserted, whereas aluminum or an alloy is often used abroad. The pressure varies from 20 to 35 lb in this country, but pressures from 10 to 20 lb are common abroad. The pantograph is usually held up by springs and lowered by air pressure, but this may be reversed, especially abroad, where the pantographs are often raised by air and dropped by gravity.

Since the trolley wire is placed above the tracks a distance that may vary between 15 ft 3 in and 25 ft, the pantographs must be so constructed that they will operate satisfactorily between these limits, maintaining continuous contact with the overhead wire at reasonably constant pressure at any height. They must be light enough so that they will follow the wire as it changes height above the rails to pass under overhead crossings, etc., with minimum inertia effects even at high speeds. They must be sufficiently strong and rigid to resist air pressures, both head on, due to speed of the train, and transverse, due to wind; also they must resist stresses due to the sway of the locomotive at high speeds and, to some extent, stresses due to blows from the overhead system at deflectors or turnouts, etc., and from striking birds. If, however, there is serious trouble on the overhead system, the pantograph must be sufficiently pliable so that it will be damaged rather than pulling down the overhead structure.

Matching the development of the lightweight overhead construction in high-voltage electrifications has been the development of lightweight pantographs. The Faivlie (French) and the AMBR (British) designs consist essentially of an elbow-shaped frame which is really half the traditional design illustrated. The low inertia of this assembly enables it to have a smooth trajectory with low shoe wear.

198. Third-rail Shoes. There are two general types of third rail: the *overrunning*, in which the contact is made on the upper surface, and the *underrunning*, where the contact is made on the under surface. In electrified main-line railroads in this country it is considered necessary to protect the rail effectively from accidental contact from trespassers, so the collecting shoes are in these cases horizontal whether the type is overrunning or underrunning.

In rapid-transit systems, where the right-of-way is fenced or is underground or elevated, more simple protection is adequate, and here often the shoes are merely supported directly over the rail, sliding over it.

In general, where the electric motive power operates solely from the third-rail system, it is sufficient that the shoes automatically slide to the surface of the third rail, on one side of the locomotive or the other.

There are generally four shoes on each locomotive, two on each side of the equipment and at both ends. The power is taken from the shoes to the locomotive by means of cable into a fuse box. Generally each shoe is connected to a separate fuse with magnetic-blowout arrangement. In case a fuse does not operate properly under conditions of short circuit, sometimes a grounding device is installed to make a dead short circuit in the third rail so that the substation circuit breaker will open and protect the locomotive in emergency.

Transformers

199. Single-phase transformers in electric locomotives are similar to those in stationary practice. It has been the practice in this country to use air-cooled or (more recently) *Inerteen-* or *Pyranol*-cooled (*Askarel*) transformers, although abroad oil-cooled transformers have been common.

In **air-cooled transformers** the air is taken generally from the cab, as it is for the motor cooling. The cab forms a sort of settling chamber for air from outside. Careful screening through a filter is necessary in winter to prevent snow from getting into the cooling system and wetting the coils of transformer or motor. This is especially important with multiple-unit cars when the transformers are generally mounted under the car in a position to pick up dust and moisture, unless care is used in the design of air intake to provide screen and baffles.

200. Voltage control is generally on the secondary side through switch groups, as described. It is sometimes foreign practice to make the voltage taps on the high-voltage side of the transformers. This has the advantage of less copper in the taps and smaller switches, which also may be all immersed in the oil used for cooling the transformers, but has the limitation of requiring greater kilovoltampere capacity in the transformer windings.

On equipment using d-c supply the circuit is obviously necessarily grounded, because the rail is used for the negative power return to the substation. In a-c systems the ground on the secondary winding of the main transformers is usually effected by grounding the midpoint of a reactor or an autotransformer connected to two of the secondary taps. This connection limits fault currents in case of ground faults.

Diesel-electric locomotives, through their first four decades, were propelled by direct current developed primarily in commutating-pole shunt-wound generators coupled to the engines. Output in such a system is regulated by varying engine speed and controlling generator excitation (Fig. 21-55). The load regulator and engine governor serve to maintain a constant-horsepower output at any throttle setting. If the engine demands more fuel than called for at that throttle setting, the load regulator reduces field excitation to avoid overloading the engine. In addition to the shunt and battery (separately excited) fields, the main generator is usually fitted with a starting field which is energized from the battery to motorize the generator and start the diesel engine. There are actually several types of diesel engine control, but all conform to this general principle.

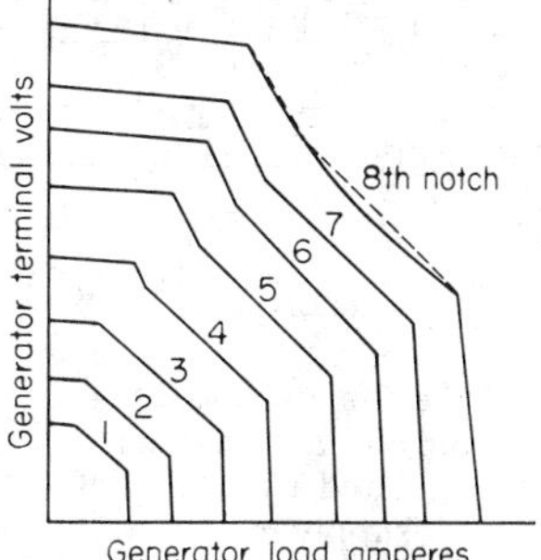

FIG. 21-55. Generator demand for eight throttle positions.

When about 1965 diesel-engine ratings went above 2,500 hp, the trend to alternators began. The small, lightweight alternator used on United States diesel-electrics for production of propulsion current is a rotating-field machine which develops 3-phase power in the stator windings. Because of the high ratings possible with alternators, it was practicable to eliminate some steps of transition (alternation of the series and parallel relation of the traction motors on a unit) and to reduce the amount of motor shunting over a locomotive's speed range. A separate starting motor is necessary.

201. Locomotive Rectifiers. When current is not collected or generated as direct current on units which have d-c traction motors, it is necessary that they be fitted with suitable converter or rectifier equipment. In the case of diesel-electric locomotives, semiconductor rectifiers were already perfected before the alternator came into use. Because the alternator is connected to a variable-speed diesel engine, the input current is of widely varying frequency. Rectifier banks on electric locomotives are fed with constant-frequency power. The same rectifiers are capable of being used in both types of locomotives. Typical silicon diodes have an average rated current of 250 A, a peak current of 500 A during each rectifier half wave, and a maximum peak reverse voltage of 1,200 V, corresponding to a reverse current of less than 6 mA under cold conditions and less than 20 mA under hot conditions. The forward drop is always in the range of 1 to 1.2 V/diode at the rated current.

202. Diode Arrangement. Rated values lead to a theoretical number of diodes mounted in series and parallel, according to the circuit design. In this arrangement, however, the following must be considered:

1. Permissible overload current, which affects the number of diodes in parallel.
2. Possible surge voltage, which affects the number of diodes in series.
3. Cooling or forced cooling, which determines the physical arrangement of the rectifier assembly.
4. Protective or fault-detection devices and indicating devices for diode failure.

Prior to the adoption of "dry" rectifiers, electric locomotives and multiple-unit cars were fitted with ignitron (single-anode-tube) rectifiers. These units are considerably less rugged than selenium and silicon rectifiers and also require liquid cooling and antifreeze protection. Silicon diodes have been substituted for ignitrons in some electric locomotives, and a comparison of the installations is shown in Fig. 21-56.

203. Motor-Generator Locomotives. The motor-generator set usually consists of a single-phase synchronous motor driving one or two d-c generators and an exciter

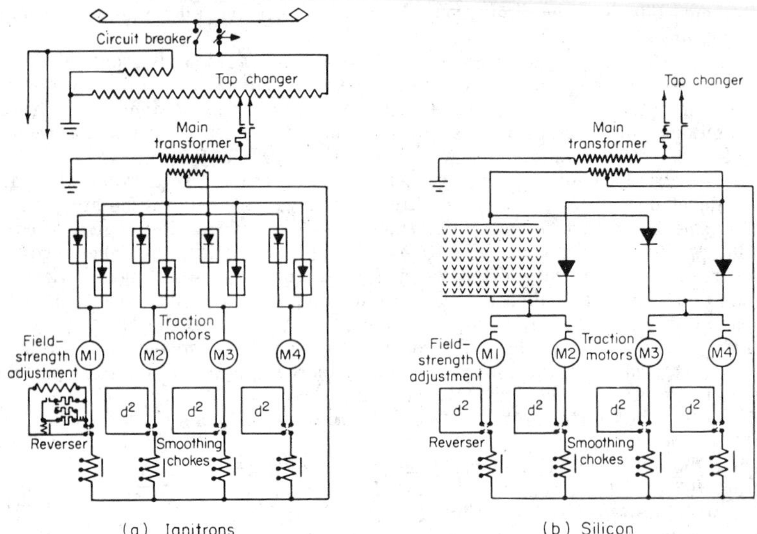

Fig. 21-56. Comparison of ignitron and silicon-diode rectifier installations on comparable electric locomotives.

on the same shaft. The set is usually started and brought up to speed by a starting motor and automatically synchronized when it is up to speed. The generator voltage is regulated by means of field control from the exciter; so resistance in the traction-motor supply circuit is not required, nor is it necessary to provide series-parallel control to obtain economical running speeds or save rheostat losses in the traction-motor control. A series-parallel arrangement is sometimes used to permit economy in the size and weight of the generator.

Control Circuits

204. Control-battery Circuits. The switch control of electric and diesel-electric locomotives and multiple-unit cars is usually by means of a storage-battery supply (64 V or other convenient voltage) actuating electromagnetic or electropneumatic switches or contactors which control the power circuits. The controller is usually of the drum type, where a cylinder which may be revolved through any desired arc is equipped with insulated segments which press upon stationary fingers. Any desired circuit or combination of circuits can be set up by properly connecting the segments for contact as the controller handle is manipulated (see Fig. 21-57).

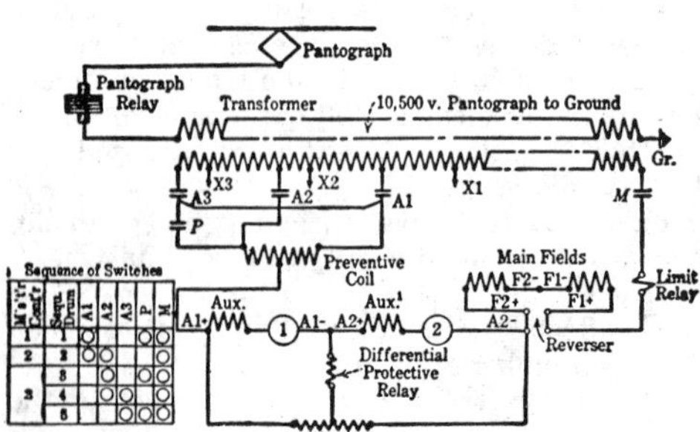

Fig. 21-57. Control diagram, main circuit, two single-phase motors on 11,000 volts.

205. Multiple-unit Control. As the individual control circuits are set up by the controller contacts, they may be carried to one set of contactors, or switches, or by parallel arrangement may be carried by means of multiple-conductor cable to another motive-power unit, which is thus controlled in a manner identical with the heat unit in the train. A number of locomotive cabs or motor cars through a train can thus be simultaneously controlled by one man. This is known as multiple-unit control. It is desirable to incorporate interlocks or relays to ensure that the connections will take place in the proper sequence and to protect against overloads, failures of line voltage or motors, overspeed due to slipping of drivers, etc.

206. Automatic Control. In certain types of service, such as rapid-transit trains or suburban motor-car trains, automatic control is used. This is accomplished by inserting current-limiting relays, with current coils, in the motor circuits which operate the control circuits when the current reaches a predetermined value and permit the next sequence to operate. By using this scheme, the engineman merely closes the main control, and the acceleration is automatically accomplished with no further manual operation on his part.

207. Switch Groups. The 64-V circuits from the controller operate contactors, or switches, which may be electromagnetic or electropneumatic. The electropneumatic system of switches, or contactors, is the more common. Compressed air is generally used to close the switches. The control is by small electromagnetic valves operated by the battery circuit from the main controllers or auxiliary switches. The switches are usually assembled in groups of 6 to 12. The air is generally piped into the end of each group—usually at about 70 to 90 psi pressure—and fills a pipe reservoir running the length of the group. When the engineman operates the controller, the battery current flows through the proper conductor to the proper magnet, opening the proper valve, which permits the air pressure to enter the proper cylinder, compress a spring, and close the switch contacts, establishing the desired main-circuit connection. The switch will remain in contact as long as the magnet is energized and air pressure is available. When the circuit is disconnected, the air is exhausted and the switch is rapidly opened by pressure of the spring. Arc chutes with blowout coils are usually provided. On motor cars there are various devices to make contact on a rotating drum. Some are electromechanical, some electropneumatic. In the process of acceleration these contacts may operate an electric power circuit or cam switch or energize a magnet valve.

208. Auxiliary Control. The 64-V control circuits are used also to operate pantographs, third-rail shoes, and, in the case of rapid-transit or urban electric cars, sliding doors, etc., throughout the train, as well as electromagnetic air brakes when that type of brake is used.

Miscellaneous Equipment

209. Auxiliary apparatus on the locomotives is dependent upon the system of power supply. If the locomotive is operated from a single-phase trolley, a tap from the main transformer, generally at the lower voltages, is taken for the local supply to single-phase auxiliary motors, e.g., blower, compressor, oil and water pumps for train-heating boilers, etc. If the supply is 600 V d-c, the auxiliary motors are usually designed for 600-V operation. If the supply is a-c at a higher voltage, generally a motor-generator is installed which provides auxiliary power at usually 110 V for compressors, blowers, lights, etc. Diesel-electric locomotive control is usually at 64 V. Motor-generators are generally used to produce power at 32 V for control; the storage battery, a necessary source which must be independent of power supply, floats on this circuit.

210. Electric Braking. One of the important advantages in electrification or diesel-electric operation is in electric braking or in using the traction motors as generators and either feeding power back into the power distribution system or absorbing it in heating resistance on the locomotive or in some other manner. Since the energy of a moving train is measured by the mass and the square of the speed, braking at high speeds must absorb a very considerable amount of energy. The potential energy of a heavy train descending a long grade is also very considerable and, if the train is to be operated safely, must be absorbed in some manner. If this mechanical energy in either form is absorbed through brake shoes, great care must be exercised to avoid overheating the shoes or wheels.

211. Regenerative Braking. In order to feed power back into the supply system, the traction motors must generate power at a voltage higher than the supply voltage and at a reasonably constant voltage. In d-c systems this condition is met by the traction motors acting as generators which will return the energy to the line with a slight increase above normal no-load speed. With compound-wound motors the increase in speed to obtain regeneration is somewhat greater. The speeds at which regeneration takes place are reduced if it is possible to increase the field somewhat during the braking period. Since braking is effected by the absorption of kinetic energy of the moving train, the field strength must be increased as the train slows down if retardation is to be uniform.

212. Dynamic braking is used in some instances instead of regenerative braking. When this is done with either a-c or d-c series motors, they are short-circuited through suitable resistances. The residual magnetism in the poles will cause the machine to generate an emf which sends current through the local circuit. This current, passing through the field coils, builds up magnetism which in turn increases the voltage, causing a current to flow in amount limited by the resistance. The heating of the resistance absorbs the energy generated by the motor, and the train slows down. As the motors reduce speed, the voltage drops, and to obtain uniform retardation, the resistance must be reduced. The starting resistance grids are ordinarily also used in the braking. The chief limitation to dynamic braking lies in the relatively small amount of energy which can be dissipated by the amount of resistance it is possible conveniently to install in the locomotive.

Traction-motor Characteristics

213. Direct-current Series Motors. The railroad types are usually insulated for operation on 600- to 3,000-V circuits, depending on distribution voltage. On 3,000 V usually either two or three motors are operated permanently in series, and on some cars it is found desirable on account of improved commutation to wind motors of the small sizes for 300 V, placing two permanently in series for use with 600-V supply.

Since the same current in a series motor flows through armature and field circuits, the field strength varies with the load on the motor. The speed is reduced as the current increases on account of the higher flux density in the magnetic circuit.

214. Field Excitation. Flexibility in performance of the series motor may be obtained by a change in the relation between the armature current and the field strength. The field current may be reduced by shunting the field coils with resistance, or there may be field coil taps. Either method will weaken the field flux and produce higher armature speed for any given armature current. This provides to some extent a method of controlling the speed.

215. Terminal Voltage. Reduction in the voltage at the motor terminals of a series motor will cause it to run slower at any given current value, since the armature does not have to develop so much counter emf. The speed is thus very roughly proportional to the terminal voltage. Reduction in terminal voltage is obtained by placing resistance in series with the motor or by reducing the supply voltage by transformer taps in a-c circuits or by placing the traction motors two or more in series.

216. The compound-wound traction motor has an additional set of field coils fed directly from the line, forming a shunt circuit. This shunt winding supplies a considerably smaller proportion of the field ampere-turns than is customary in stationary practice and therefore does not give so flat a motor-speed characteristic as it does in its application to the industrial compound-wound motor. The compound winding tends to limit excessive speeds in descending grades, a characteristic of value in electric braking.

217. Single-phase commutator series motors of several types have been used in railroad service. Their general construction is similar in many respects to that of d-c series motors. Since the change in direction of the field and armature currents takes place simultaneously, the torque remains in the same direction. There are necessary, however, certain modifications in the single-phase motor as compared with the d-c motor. The field coils can no longer be simple windings around a pole piece, because such windings would offer too high an impedance to the alternating current, so that fewer turns and more poles with a lower flux density are used. Likewise, the impedance of the armature, or rotor windings, must be designed with particular care to make the reactance low, and special compensating windings may be introduced in the field to effect this. In the voltage equation of the series a-c motor, the resistance drop of the d-c equation is displaced by a term for the impedance drop which may be sufficiently large to play a part in the output characteristics and thus make them somewhat different from those of the d-c motor. This type of motor operated at 15, 16⅔, 25, 50, and 60 c.

218. Single-phase Motor Design. The current-heating limitations may be more rigid in a-c motors than with d-c motors because of the transformer action in the armature coils, which are short-circuited by the commutator brushes. Heavy short-circuit currents are induced in these coils, which act as secondary windings of a transformer of which the field coils are the primary. The magnitude of these currents can be limited by insertion of resistances in the armature leads or by keeping the induced voltage low by having a small number of turns in the primary or field coils. **Commutating interpoles** (Fig. 21-59) can be used to offset this short-circuit current during running if their flux is such as to induce a counter emf in the windings, opposite to that induced by the field flux in this transformer action. The conditions for overheating due to this action may be especially serious at starting, when a single armature coil may be short-circuited for some time.

219. Commutation limitations of a-c motors may tend to be greater than those of d-c motors because of the need to commutate this short-circuit current as well as the main motor current. A variety of methods of connections and the addition of compensating windings have been devised to give satisfactory commutation over a wide range of speeds. The repulsion motor with the armature current supplied by induction rather than by direct connection is a step in this direction. Its output characteristics are practically identical with those of the simple series motor. It has better commutating characteristics in the neighborhood of the synchronous speed than the straight series motor, though at starting it has the same limitations. By double feeding of the armature by both induction and conduction, the commutating properties of the series motor can be improved over those of a simple connection, because the compensating field can be made to induce a voltage in the short-circuited windings to compensate for the transformer-action voltage (see Fig. 21-58).

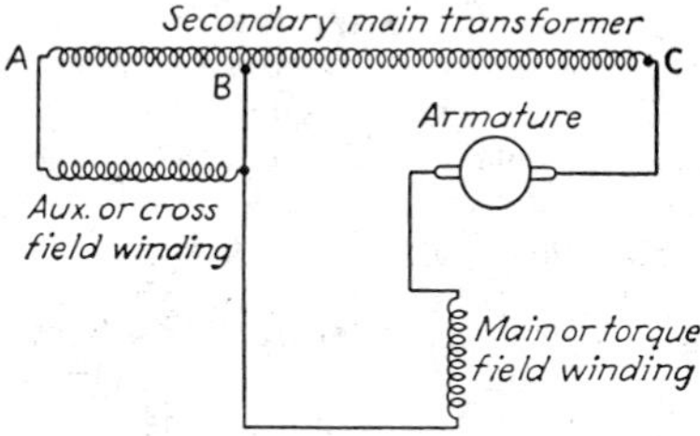

FIG. 21-58. Double-fed a-c motor connection.

220. Power factor of the a-c commutating motor is dependent upon the reactance in the armature and field circuits, and when nothing is done to compensate for these reactances, the power factor is likely to be low. With the series motor, the armature reactance is likely to be the major item. Since the reactance of the armature is a function of the flux leakages of the conductors, the installation of a compensating winding with a flux in the opposite direction can be used to offset this. With the repulsion motor, the power factor is largely a function of the exciting-field reactance. This is compensated for by the use of extra brushes supplying current to the armature to produce the compensating field.

221. Interpoles are placed between the main field poles and consist generally of small field poles having windings connected in series with the armature.

222. In the **doubly fed** motor the resistance leads in the armature are omitted and the armature and fields are not connected permanently in series. The motor starts as a repulsion motor with the armature short-circuited and then operates as a doubly fed series motor. A diagram of connections is shown in Fig. 21-58. The motor has an auxiliary field winding, or "cross-field winding," as it is sometimes called, which has twice as many effective turns as the armature. The action of this winding and the armature winding is similar to that of a transformer. With the magnetizing current neglected, the total ampere-turns in each winding are equal and opposite; consequently the current in the auxiliary winding is only one-half that in the armature. This is of value in control.

223. The a-c interpole motor has three fields: the *main* field, *auxiliary* field, and *interpole* field. The interpole fields are located between the main fields and are connected to give alternate positive and negative poles, as in any motor. This type has an advantage over the double-fed motor in that it has the interpoles in addition to the distributed auxiliary windings. These interpoles accomplish the same result as the interpoles in d-c motors, i.e., an improvement in commutation which results in a reduction of weight for a given horsepower. At the same time, interpoles applied to single-phase motors improve the power factor. The design of the interpole may be varied over a wide range to accomplish or meet any operating condition. It is possible with a fixed interpole coil to vary the field strength and its effect on motor operation by shunting the interpole coils with resistance or inductance. Because of this interpole flexibility, the motor lends itself to regeneration.

224. Polyphase induction motors are operable from a 3-phase supply circuit.

The polyphase-induction motor may be supplied from a single-phase trolley circuit by use of a **phase converter,** or **phase splitter,** mounted in the locomotive cab. This device consists of a motor with a 2-phase wound stator and a wound rotor. One stator phase is fed from the main transformer, and proportional voltage is generated in the other, with a phase displacement of about 90° electrically from the first. One terminal of the generated phase is connected to the midpoint of the transformer secondary, resulting in a Scott connection to give 3-phase power. The phase converter may be either synchronous or asynchronous.

The induction-type, or asynchronous, **polyphase traction motor** has a primary winding usually on the stator, which receives current from the supply circuit, and a secondary winding usually on the rotor, which is short-circuited on itself either directly or through resistances. This permits rugged construction with high weight efficiency. The characteristics of the induction motor are "constant speed," similar to those of the d-c shunt motor, the speed falling off but slightly from a fixed no-load value. The reduction of speed is 5 to 10% from no load up to maximum load developed. This means that, as long as the locomotive is on the line, the train speed is nearly constant regardless of grade of track or weight of trains.

The induction-motor secondary is built with a 3-phase winding, the leads being brought out to slip rings on the shaft. The secondary circuit may be completed through external resistances which reduce the speed at starting in a manner somewhat similar to the insertion of resistances in the armature circuit of a shunt motor. The resistance may be varied to give a certain limited amount of speed control and short-circuited for full-speed running.

Another method of speed control is varying the number of poles in the induction motors.

Control of Traction Motors

225. Motor Control. If a motor is connected across the supply circuit when the vehicle is standing still, the current it would take would be limited only by the resistance or inductance of the circuit and would far exceed the safe carrying capacity of the windings. The torque would also probably cause slipping of the driving wheels. Some means, therefore, is necessary to limit the starting current. It is also desirable to provide more than one economical running speed.

With a **d-c series motor,** the current and torque produced at a standstill may be reduced by strengthening the field or lowering the terminal voltage or both. Motors may be placed in series, reducing the terminal voltage of each without loss in external resistance. External resistance may be placed in series with the motors to limit the starting current to any desired amount, and by varying the resistance the current may be kept constant during acceleration as desired, as the counter emf is being built up. Since maximum torque while starting demands full field strength, any shunts or reduced field connections are usually thrown out of action in starting.

With the **d-c compound-wound motor** the start may be made with full armature current in the series-field coils and maximum current in the shunt-field coils. A starting resistance inserted in the armature circuit is reduced in steps until the armature and series field are connected across the line. Further speed increase is effected by reducing the shunt-field current in steps to the point where the shunt winding is disconnected, and the action is then identical with that of a plain series motor.

226. In the **control of the double-feed motor** (Fig. 21-58), by assuming that the voltage across AC, the secondary of the main transformer, is 300 V, that 1,100 A is flowing through the armature, and that transformer tap B is at the same point as C, then the voltage, due to the transformer relation between this winding and the armature winding, would be 300, and the voltage induced in the armature and torque-field winding would be 150 V. Since the ampere-turns must be equal, there will be flowing through the auxiliary winding 550 A in the opposite direction to the current in the torque field, so that tap *B* is carrying only 550 A. If transformer tap *B* were at the same point as tap *A*, then 300 V is connected across the armature and torque field, and under the same local conditions of 1,100 A flowing through the armature, there would be 550 A in the opposite direction as before, flowing through the auxiliary windings, resulting in 550 A in tap *B*. By moving the lead of tap *B* gradually from a point on the transformer winding near *C* toward *A*, the effective motor voltage can be changed from approximately 150 to approximately 300 V. This provides a means of controlling the speed of the motor by switching only one-half the armature current.

227. Terminal voltage with single-phase motors is usually varied, not by inserting external resistance, but by voltage taps brought out from the supply transformer; so there is no rheostatic loss with a-c motors as there is with d-c motors. There are important exceptions to this. In some cases it has been found economical to provide only three transformer taps and then cut out resistance to acceleration. (Fig. 21-59). The series, series-parallel, and parallel positions of the motors may be obtained with either a-c or d-c motors (Fig. 21-60).

Some a-c electric locomotives are equipped with **transformer voltage taps** which can raise the motor terminal supply voltage at high speeds to counteract the speed-horsepower drooping characteristic inherent in the series motor. This gives the locomotive a great horsepower at high speeds, a valuable characteristic in hauling heavy trains on fast schedules.

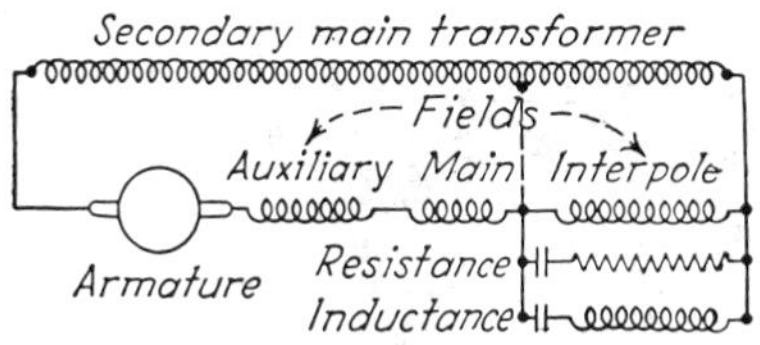

FIG. 21-59. Single-phase interpole railway motor.

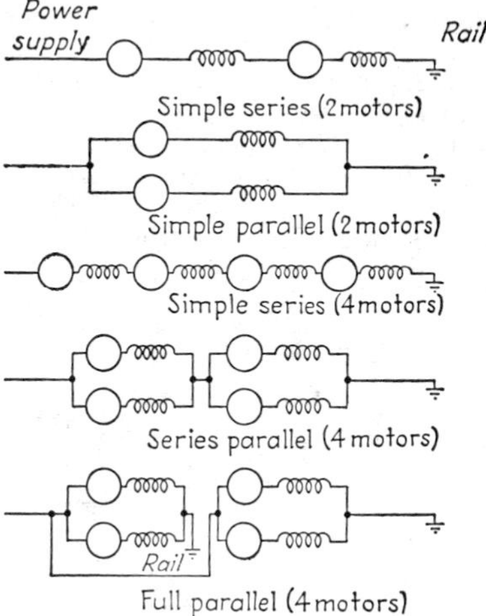

FIG. 21-60. Arrangements of traction motors.

228. Series-parallel operation of traction motors is usually adopted to economize on weight and produce good efficiency at low speeds and high tractive efforts.

The motors of a locomotive or car may be grouped in various combinations depending upon the number of motors in the vehicle.

With two motors, only the simple series and parallel positions of the motors are possible (see Fig. 21-60). With four or six motors or motor groups, series, series-parallel, and full-parallel positions give considerable flexibility. With supply voltages of 600 V d-c, it is customary to use each motor individually in these combinations, though often two motors are connected permanently in series. With higher line voltages (1,500 to 3,000) it is customary to couple two motors permanently in series. With a-c supply the transformer secondary voltage may be varied as desired, and these limitations do not exist.

229. Transition from one connection combination to another must be made with as little variation in torque as possible. One scheme commonly used permits one of the motors at a time to change with resistance temporarily in its circuit.

230. Resistance grids are usually of cast iron, assembled in frames, usually mounted on steel tie rods, insulated with mica, and supported at each end by steel frames bolted securely to the locomotive or car structure. Edgewise-wound resistors may be used, wound from a noncorrodible, unbreakable strip of special resistance alloy with low temperature coefficient

231. Preventive Coils. With alternating current, since the voltage supply is taken from transformer taps, the supply voltage is flexible. In starting, the voltage applied is low, and all points may be continuous running points. The problem of maintaining continuous torque in changing taps is an important one, just as in changing motor combinations. Two transformer taps cannot be tied together during the instant of change without some method of limiting short-circuit currents in the transformer windings involved. This is done by the use of preventive coils or impedance shunted across the taps during transition.

232. Pulsating-current Motors. While the pulsation factor of d-c traction motors has generally been suppressed to about 25%, motors capable of successful operation at a current pulsation factor of 50% have been developed. Such motors require lamination of certain internal components in a manner duplicating the construction of a-c motors. Communications interference must be carefully controlled when current with such pulsation is involved. Motors capable of operating on pulsating d-c current simplify considerably the circuitry associated with rectifiers. They also offer the possibility of another semiconductor system of control—the d-c "chopper" concept.

233. Chopper Control. The replacement of the conventional rheostat of d-c locomotives by a static converter incorporating Thyristors is attractive even though oscillating d-c circuits must be designed to turn the Thyristors off after they have been fired. Rather than control voltage by rheostat control, the "chopper" can vary voltage or current by impulses set up in Thyristors in series in the circuit.

The mean voltage supplied to the motor is then determined by the time of opening of the circuit, or its frequency. Two means of varying the mean voltage are then

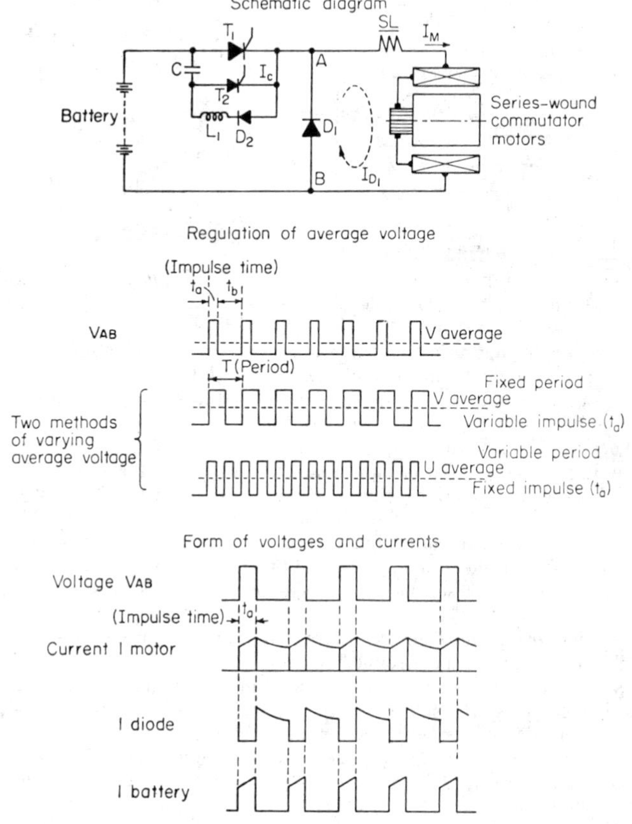

FIG. 21-61. Feed for a commutator motor, impulse control.

possible (Fig. 21-61):

1. At a fixed frequency, the value of opening impulses is varied.
2. With impulses fixed, their frequency is varied.

Method of Drive

234. Single-reduction Gearing. Diesel-electric locomotives and multiple-unit cars, along with recent electric locomotives also having d-c traction motors, are usually powered with one motor per axle. The d-c motor is normally "nose-suspended" with a single-reduction spur gear drive. This installation places the motor axis parallel with the axle. The motor frame incorporates bearings that are supported by the axle; the other end (nose) of the motor is supported resiliently by the truck frame. The pinion mounted on the armature shaft engages the drive gear, which is mounted by press fit on the axle just inboard of one of the wheels. This nose suspension maintains a fixed relation between the armature shaft and the axle. The motor is free to move through only a small arc, with the axle as the center, when track irregularities are encountered and when torque is being exerted. Such a motor, with most of its weight carried on the unsprung axle, must be extremely rugged because it is subjected to extreme shock and vibration as rail irregularities occur.

235. Double-reduction Gearing. Direct-current traction motors on some locomotives and cars are body-hung at right angles to the axles, with the torque from each transmitted to one or more axles by Cardan shafts which drive through right-angle gearboxes. Such a drive may incorporate double-reduction gearing. Double-reduction gearing, regardless of the motor location, permits smaller, higher-speed motors and has found wide application in transit cars. On many transit cars these light, economical motors are truck-mounted, usually supported entirely from the truck frame rather than being axle-suspended. The size can be reduced to such an extent that the motor is part of the gear unit, rather than the gear unit being part of the motor. Small, high-speed motors may safely be self-ventilating.

The driving of both axles in a truck by a single truck-mounted motor incorporating double-reduction gearing is growing in popularity. A single motor having the same capacity as two conventional traction motors is lighter and less costly than the two. The coupling of the two axles through the gear drive significantly reduces wheel slipping. The ability to incorporate gear shifting in a double-reduction drive makes possible the building of a "universal" type of locomotive which can exert its horsepower over a much wider speed range.

Conventional a-c locomotives have been fitted with a variety of drives, including bipolar and quill types. Recent developments indicate that the costly, complicated drives once necessary with a-c motors can be avoided in all future models.

SIGNALING

236. Purpose. The purpose of railroad signals is to keep trains apart but permit spacing as close as safety permits.

237. Wayside signals are installed along the right-of-way where they may be observed by the engineers of passing trains. The changes of indication may be made by the movement of mechanical parts or the change in the color or position of one or more electric lamps or both. So-called color light signals use lamps that are brilliant enough to be seen in the brightest daylight. With semaphores or movable arms a lamp is usually so placed that, as the arm moves to one or two more positions, colored lenses are shifted in front of the lamp to give color indications to the passing engine crew to correspond to the indication of the arms.

238. Signal Indications. There are many indications in signals. In the position of the semaphore arm, commonly, vertical is proceed; horizontal, stop; and 45°, an approach indication. In light signals, green is usually proceed; red, stop; and yellow, usually caution. Some railroads use light-position indications, two or more lights arranged horizontally being stop; vertically, proceed; and at 45°, caution. Additional indications are obtained by the use of two or more arms or sets of lights at a given signal location.

239. Block Systems. A block is a length of track which is controlled or protected by a signal at the entering end. A block system is a series of consecutive blocks.

Block signals are operated generally on one of the following principles:

1. *Manual.* The signals are set by hand by the operator on orders or information received by him, usually by telephone or telegraph or by timetable instructions. In this system when a train has entered the block, the signalman, or operator, sets the signal to the stop position. On single-track railroads, in order to provide for changes in schedule or for extra trains, the dispatcher issues train orders providing meeting points and train movements, and the signal operators are governed accordingly.

2. *Automatic Signals.* The signals are operated automatically by the presence of a train in the block. This type is common on busy railroads. Automatic signals may employ either battery-current or a-c supply. The track circuits are energized by power between 2 and 8 V, either a-c or d-c. The rails at each end of the block are cut and connected by insulated joints. There is sufficient insulation in the wood ties so that the rails constitute conductors in a closed circuit with signal power supply at one end and a relay at the other to actuate the signal (see Fig. 21-62). When a train passes the insulating joints, the axles short-circuit the rails and cause the relay to drop. An advantage of this arrangement is that a broken rail or a failure of power will also cause the relay to drop and give a stop-and-proceed indication to the signal. In Fig. 21-62, *B* is the source of energy, alternating or direct current; *J*, the insulated rail joints; *T*, the track rails; and *R*, the relay which holds contact *A* closed when the relay is energized and allows it to open by gravity if for any reason the relay is deenergized. If a-c power is used, the relay is a frequency type, usually 60 or 100 c; so if it should be energized at a lower frequency than that for which it is designed to operate, it will open its contacts, causing the signal to go to a stop position. The source of energy is of as low voltage as practicable, to prevent operation in the event the short circuit by the train is not complete—and there should be stray currents present. In railroad practice, relays are employed having 2 to 16 Ω resistance in track circuits varying from 60 to 6,000 ft. A minimum air gap between the relay armature and pole pieces must exist to prevent undesirable effects from residual magnetism. As this air gap is increased when the armature falls away under the force of gravity, it is evident that more current through the relay is required to lift the armature than to hold it in position once it has been raised. The usual 4-Ω track relay requires about 70 mA to pick up and will drop when the current is reduced to about 35 mA.

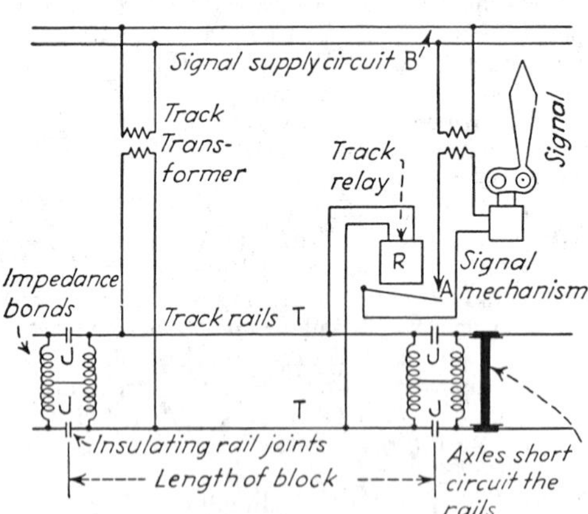

FIG. 21-62. Simplified diagram, automatic-signal track circuits.

240. Impedance Bonds. In electrified railroads, whether a-c or d-c, the traction-rail return power must be permitted to flow around the insulated rail joints with low resistance or impedance. The ends of each track section are thus equipped with impedance bonds, consisting of a few turns of heavy copper wound about a laminated core, the terminals of which are connected each to a rail, and the midpoint or neutral brought out for connection to a similar bond beyond the insulated rail joints. These bonds are so designed that the traction power (direct or alternating current), traveling in nearly equal amounts in each rail, passes through the coils of the impedance bond, which is so wound that the impedance is very low, because the magnetization of each half balances the other half for currents thus flowing. The impedance between rails for flow from one rail to the other for the higher-frequency signal power is high (see Fig. 21-52). If there is unbalance in the rails which permits more traction power to flow in one rail than in the other, there is a tendency to saturate the iron core and thereby reduce the impedance of the bond. This unbalancing effect is limited by introducing an air gap into this magnetic circuit. The bonds are usually designed to withstand 20% unbalance with a decrease of not more than 10% impedance. The size of the bond is, of course, dependent upon the fault current that it is expected to carry, the impedance for which it must be wound, and the unbalance to be accounted for. When good track bonding is maintained, unbalance may be assumed to be low and a smaller bond can be employed (see Par. **174**).

241. Cross Bonding. In multiple tracks, cross bonding is installed between the neutrals of the impedance bonds in adjacent tracks to permit advantage to be taken of power return using the conductivity in all tracks. Care must be taken in this connection not to have too frequent cross bonding, with possible danger of permitting, through leakage, faulty signal operation.

AUTOMATIC TRAIN CONTROL, CAB SIGNALS

242. Purpose of automatic train control was set forth in a report of the Signal Section, ARA, September, 1924, and AREA, March, 1925, to prevent that class of accidents due to failure of employees to observe or obey signal indications.

243. Wayside and Locomotive Apparatus. The automatic train-control or train-stop system consists of two parts—that on the wayside and that on the locomotive. These, working together, transmit to the locomotive the indications necessary for operation and that may be translated into controlling the locomotive independently of any action of the engineman if he makes a mistake in running by a signal with restrictive indication without acknowledging that he is doing so.

244. Systems of Control. Train-control devices are classified into two groups: (1) **intermittent** and (2) **continuous.** The intermittent group may be divided into **contact** and **noncontact** types. The contact type depends upon physical contact between the train element and the wayside element; the noncontact type, upon an electric or magnetic impulse without physical contact between the roadside and train elements.

In American railroads the contact systems are little used because of possible disturbance from weather, trespassers, etc., but such installations in the rapid-transit field are not uncommon, the right-of-way being generally less accessible to trespassers and more intensively patrolled.

245. Code System. In the noncontact system the most common installation is the **continuous inductive scheme,** often installed without a train-stop feature but with cab-signal indications continuously in front of the engine crew in the cab. This has done much to speed traffic. One common type is known as the code system.

246. Code Transmitter. For the cab-signal control, an interrupter or code transmitter is usually employed; this produces and impresses on the track rails groups of pulsating currents. These groups are applied at such rates as 75, 120, and 180 pulsations/min to produce a set of distinctive codes, one for each degree of proceed cab-signal indication. The most restrictive indication is produced by the absence of any code.

Power supply in this instance is usually at a frequency of 100 c/s in order to eliminate interference from 25- or 60-c frequencies commonly used for industrial or traction purposes. In one instance where there were two types of propulsion circuit in the vicinity, one 25-c a-c in an overhead trolley and the other 600 V d-c in third rail, the a-c locomotives picked up some d-c power, which resulted in harmonics in their transformers. In this case the frequency of the cab-signal power supply was changed to 91⅔ c from 100 c in order to avoid this fourth harmonic of the 25-c power.

247. Frequency of Supply and Code. Energy generally at 100 c (except as indicated above) is fed into the far, or leaving, end of each block through the medium of a code transmitter, which produces a distinctive code in accordance with existing track and block conditions. For example, the code for a *clear*, or *proceed*, indication is 180; for approach-medium speed, 120; for approach, 75 interruptions/min; and for stop or restricting or slow speed, no code. The device that produces the codes, or the code transmitter, may be a small motor-operated mechanism which has three cams; and as the shaft rotates, the track circuit is alternately made and broken at the proper code frequency. A synchronous motor is used which ensures uniform operation.

More recent installations employ a spring-hung pendulum operated by electromagnetic action. There are decoding apparatus operated from the coded track relay to operate other relays that provide wayside signal indications as required. The same rail circuit that operates the coded track relay can also be used to energize the locomotive receiver coil for cab signals or train control.

248. Receiver. The electric coded impulses fed into the track rails are picked up inductively by two receiver coils, one mounted centrally about 6 in above each rail ahead of the forward truck wheels of the locomotive. The code impulse in the rails

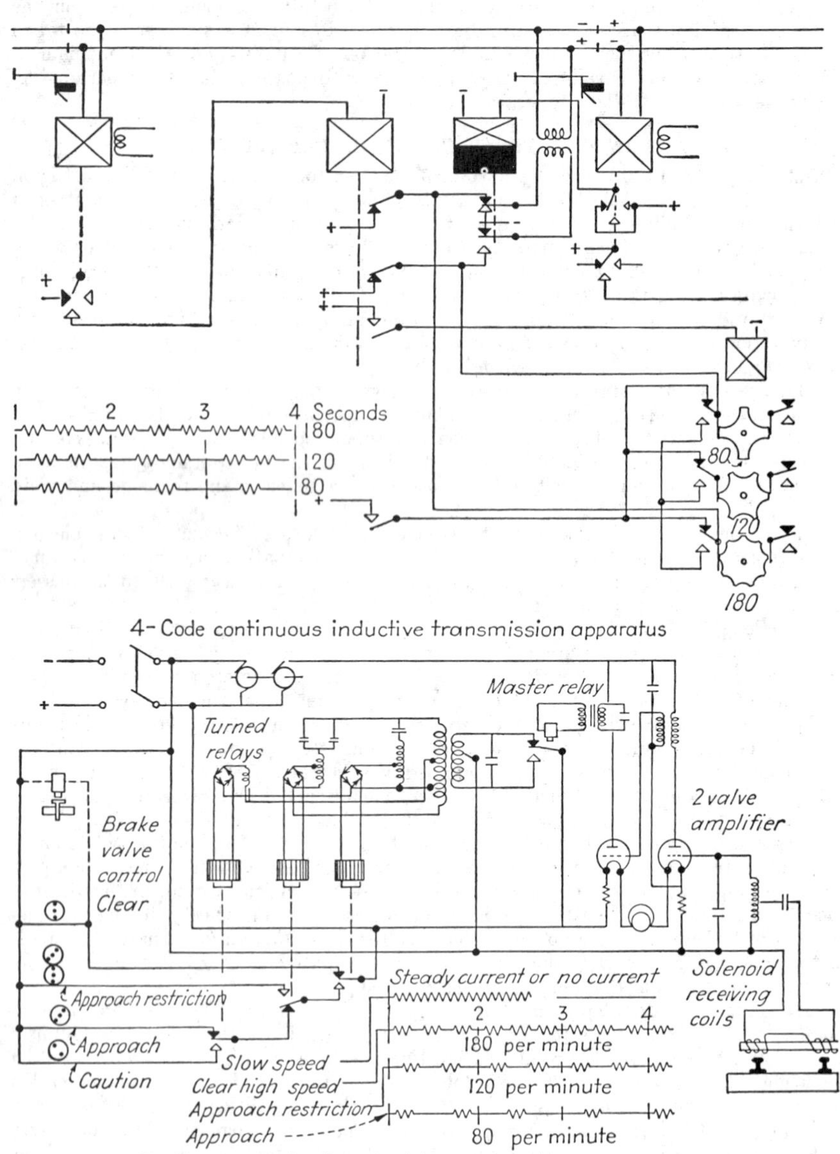

FIG. 21-63. Cab signal code circuits, transmitting and receiving.

energizes the receiver coils and produces corresponding impulses in them, the track circuit being completed through short circuit of the rails by the wheels and axles of the train (see Fig. 21-63). The energy picked up by the receiver coils first passes through an amplifier and filter, where all except the 100-c frequencies are filtered out, and where the voltage is amplified sufficiently to operate a master relay. The rate of operation of the master relay is identical with the code frequency in the track.

249. Decoding. The frequency is then changed by the locomotive generator to a d-c code at 32-V supply. A decoding transformer tunes the decoding relays to their proper frequencies. The tuning of these circuits is such that, when the clear code frequency of, say, 180 c/min is present, the circuits to the *clear* and *approach* decoding relays are energized; when the *approach-medium* code frequency of 120 c/min is present, the *approach* and *approach-restriction* relays are energized. When the *approach* code frequency of 75 c/min is present, only the *approach* decoding relay is energized. When there is no power flowing, or when there is no code, none of the decoding relays is energized. Cab signals thus controlled by the decoding relays indicate the track conditions in the block by means of either color-light or position-light aspects. The cab signals are so arranged in the locomotive that both engineman and helper can see them.

250. The acknowledging switch is located within convenient reach of the engineman. A whistle sounds when a signal changes to a more restrictive indication, but this whistle is stopped by the engineman's acknowledgment. Automatic brake application may be added to stop the train automatically if the signal change is not properly acknowledged. Speed-control governors are also sometimes employed, but this is not usually felt to be necessary.

251. The intermittent type of train stop is used on many railroads instead of the continuous type (the latter including, as has been said, cab signals). This type, like the continuous type, may be superimposed on the wayside-signal facilities. The scheme consists of two parts, one in the locomotive and the other attached to the track in the rear of the block signal with which it operates. There is no provision between the wayside-signal locations to actuate the apparatus on the locomotive. The part attached to the locomotive is called the **receiver** and is usually attached to the outside of the truck frame on the locomotive or tender. It consists of an inverted U-shaped magnet with laminated core, large pole pieces, and two coils. The primary coil of this receiver is continuously energized at 32 V, producing a strong magnetic field. The secondary coil is connected with the same circuit and also with a relay through which a small amount of current normally flows (see Fig. 21-64).

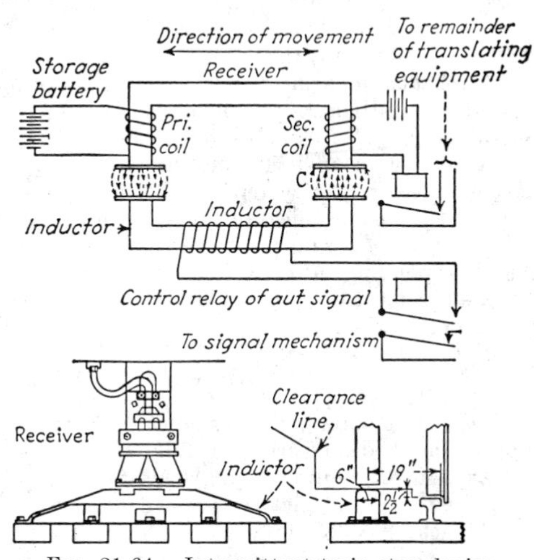

FIG. 21-64. Intermittent train-stop device.

252. Wayside Inductor. The track element is called the **inductor** and is very similar to the receiver. It is attached to the ties immediately outside the rail, with its pole face about 2½ in above the top of the rail and its inner face about 17¼ in from the rail. It is protected by a nonmagnetic housing. The inductor coils are so controlled automatically that, when the signals with which they are connected are clear, the coils are closed automatically on themselves, or short-circuited. When the receiver on a passing locomotive approaches an inductor having an open circuit, i.e., when the signal indication is restrictive, a surge of magnetic flux builds up in the secondary coil, producing a variation of current on the engine, which causes the relay to open and apply the brakes through an electromagnetic device attached to the engineman's brake valve.

253. Receiver. In passing a clear wayside signal, no action on the part of the engineman is necessary, but if the signal is less favorable than clear, indicating that speed should be reduced, the engineman must operate a forestalling level as the locomotive passes over the inductor. This device cuts out the automatic-brake application until the next signal is reached. If a restrictive signal indication is not forestalled, the brakes are applied automatically and the train is stopped. It is necessary for the engine crew to break a seal to release them and proceed.

254. Centralized traffic control is a development in railroad operation that has made possible important operating economies. Under this system all signals and switches at crossovers, passing sidings, or junctions on a given section of railroad are controlled from a central point. The trains are operated entirely by signal indications, and so no train orders are issued. All track switches thus installed in the section are remotely controlled, and the signals are either automatic or remotely controlled. Such a system consists of three parts: (1) control office, (2) line control circuits, and (3) wayside apparatus. Two general schemes have been developed, one employing multiple conductors with a circuit to each unit of apparatus, and the other, a code-selector system of circuits. Microwave is extensively utilized.

255. The code-selector system employs 2, 3, or 4 wires for the control circuits from the central office to the various wayside locations. Coded impulses sent out on the line circuits select the desired location and execute the desired control at that point. The system is analogous to the supervisory control of a power substation. The control machine is usually provided with a track model with lights which help the dispatcher to visualize the operation under his control. Telephone or radio provides channels through which information may be conveyed to the dispatcher in the event of an emergency from points on the line.

256. Highway-crossing signals and electrically operated crossing gates are usually controlled automatically by track circuits in a manner generally similar to automatic block signals. An interlocking relay is ordinarily employed, so that the crossing signal ceases operation when the rear of the train passes the crossing. Audio-frequency control circuits, now utilized, can eliminate insulated joints.

257. Train Identification. When train headway is very close, as in suburban service or on urban rapid-transit systems, a system of train identification may be used for the guidance of signal-tower operators. In this system each train to be identified is equipped with an inert coil without power connection but tuned to a specific frequency, each frequency representing one train classification. Along the right-of-way at points where identification is to be made, there is a "bridge" tuned to each identifying frequency. As a train passes the identification point, the coil on the train unbalances the corresponding bridge and activates equipment to identify the train. The system may be arranged to operate the proper switches automatically to route the train to its proper destination.

AUTOMATIC TRAIN CONTROL

258. The mobile and wayside equipment described previously makes possible, when properly applied and oriented, fully automatic train control (ATC). The preciseness of train control required in high-speed close-headway rapid-transit systems is generally recognized as being beyond the capabilities of even the most skilled operators. The advantages of ATC apply also to main-line railroads. Automatic train control produces a consistency of operation using optimum performance as the criterion. Because less is required of an operator in terms of skill, a larger number of persons are capable of performing his function. Human errors in judgment are eliminated, and operating requirements are rigidly enforced. In the case of electrification, ATC can reduce power consumption on a per train and an overall basis.

The proper approach to building any automatic train-control system is analogous to building a control system for a process industry. In a major transit system there are three classifications of control function in the overall ATC concept:

1. Automatic train operation (ATO) is that portion of the ATC system which substitutes for the work done by the engineman or motorman and produces the traction and braking responses in the train.

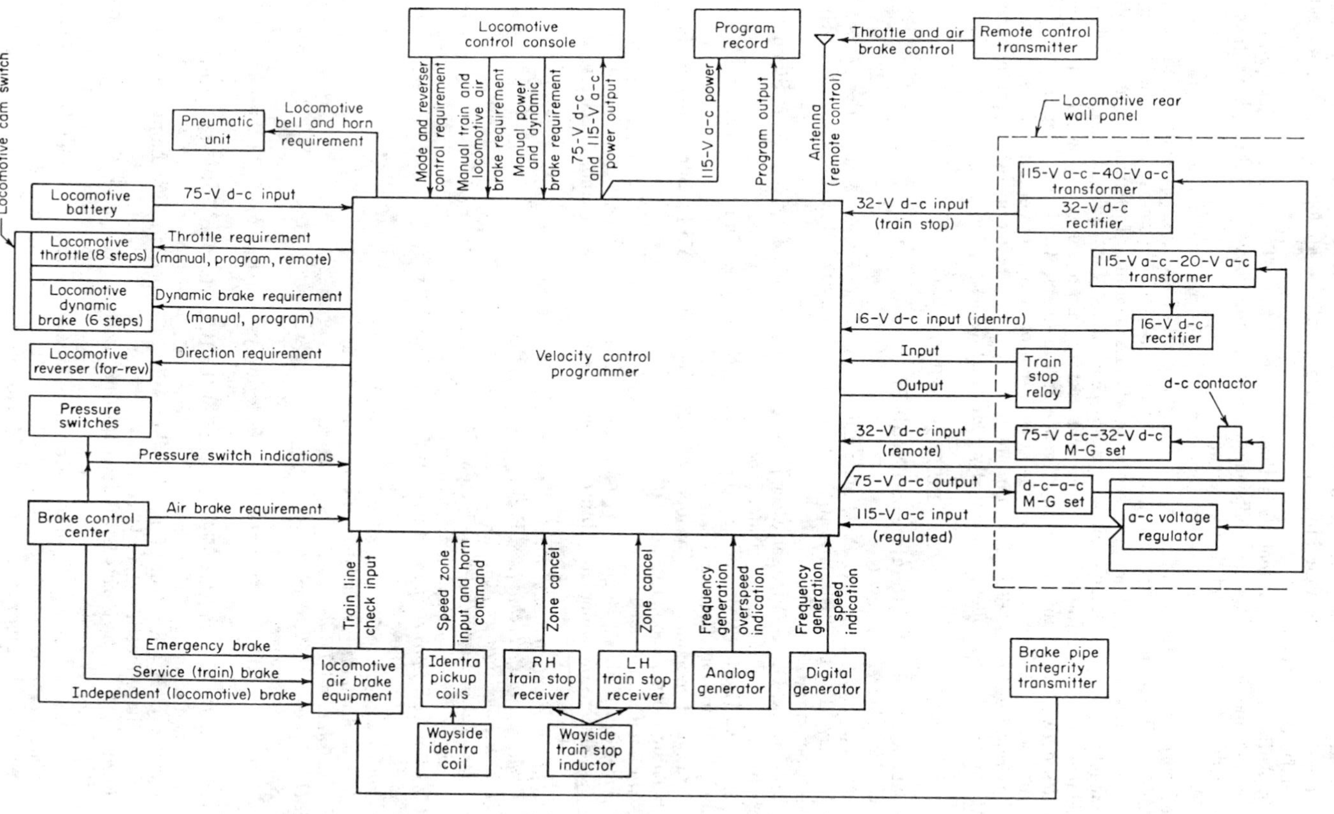

FIG. 21-65. Automatic train operation (ATO), system block diagram.

2. Automatic line supervision is that portion of the ATC system which enforces schedules and routes of individual trains and generally is analogous to a high-speed automatic CTC system.

3. Automatic train protection is that portion of the ATC system which protects controlled operations against collision, overspeed, and movement through switches, those things traditionally done by a signal system.

The step-function multiple-unit control and the programmed acceleration and braking incorporated in cars and locomotives are typical of the developments which make possible full automatic train control. Automatic block systems and interlocking, centralized traffic control, and cab signaling with its capability for transmitting wayside intelligence to the moving trains have also contributed to the technology which is producing ATC. Automatic train control does require a vastly expanded communications network to transmit the intelligence and feedback which are essential for successful operation. An example of the automation of heavy conventional freight trains is illustrated in Fig. 21-65. The equipment is capable of being operated in fully automatic, remotely controlled, or manually controlled modes.

BIBLIOGRAPHY

259. References to Literature on Electric Railways

Selected Reference Books

Dorer, A. T. "Electric Traction"; London, Sir Isaac Pitman & Sons Ltd., 1954.

Healey, K. T. "Electrification of Steam Railroads"; New York, McGraw-Hill Book Company, 1929.

McGowan, G. F. "Diesel Electric Locomotive Handbook"; New York, Simmons-Boardman Publishing Corporation, 1953.

Sillcox, L. K. "Mastering Momentum"; New York, Simmons-Boardman Publishing Corporation, 1941.

"Car and Locomotive Cyclopedia," 21st ed.; New York, Simmons-Boardman Publishing Corporation, 1966.

Current Publications

American Railway Engineering Association, Chicago.

Association of American Railroads—Engineering Division and Mechanical Division, Chicago.

Bulletin of International Railway Congress Association, Brussels.

French Railway Techniques, Paris.

International Railway Gazette, London.

International Railway Journal, New York.

Japanese Railway Engineering, Tokyo.

Mechanical Engineering and *ASME Transactions*, New York.

Railway Age, New York.

Railway Locomotives and Cars, New York.

Railway Signaling and Communications, New York.

Spectrum and *Trans. IEEE*, New York.

UTIP Journal, Vienna.

Specific References to Articles

Broughall, J. A. High-voltage A-C Railways; *Intern. Ry. Gaz.*, Apr. 15, 1966, p. 307.

Brown, H. F. Railway Electrification in the U.S., Has It a Future? Paper at New York Section meeting, AIEE, 1961.

Curtis, L. B. Electrification—Devil or Angel? Paper at AIEE Land Transportation Committee meeting, 1962.

Forray, R. The Series BB Silicon Rectifier Locomotives; *French Ry. Tech.*, 1965, No. 2, p. 95.

Macefert-Tassin, Y. Silicon Rectifier Developments in Europe; *IEEE Trans. on Industry and General Applications*, October, 1963.

New Tokaido Line; *Hitachi Rev.*, 1964, No. 8.

New Tokaido Line; *Japan. Ry. Eng.*, December, 1964.

NOUVION, F. Evolution of Motive Power Units for Electric Traction; *French Ry. Tech.*, 1965, No. 3, p. 1.

SELL, R. G. British Railways Research on Current Collection; *Intern. Ry. Gaz.*, April, 1966, p. 312.

Thyristor Control of Traction Power; *Intern. Ry. Gaz.*, Jan. 7, 1966, p. 15.

Diesel Electric Drive without Commutators; *Bull. Intern. Ry. Congr.*, October, 1965, p. 431.

SECTION 22

ELECTRIC HEATING AND WELDING

BY

JOHN F. CACHAT Vice President and General Manager, TOCCO Division, Park-Ohio Industries, Inc.; Senior Member, Institute of Electrical and Electronics Engineers and Illuminating Engineering Society

CONTENTS

Numbers refer to paragraphs

SECTION 22

ELECTRIC HEATING AND WELDING

ELECTRIC HEATING

PRINCIPLES OF HEATING

1. Material is heated for processing purposes either (*a*) in the open air or (*b*) protected from the air in a vacuum or a special gas chamber. In comparative terms: In open air the rate of heating is fast and the time short. With enclosures the rate of heating depends upon the source. It may be fast; generally it is slow, however, and the time long. The basic consideration in each case is the correlation of temperature and the time, dictated by the material to be heated and the results desired.

2. Electric heating has certain unique characteristics, viz:

a. The precision of electric control is extended to the transfer of heat. Uniformity of temperature within a relatively narrow limit is readily attained.

b. Its development does not involve combustion.

c. There is no upper limit to the temperature obtainable except the ability of the materials to withstand heat.

The collateral merits of electric heat vary in value with the conditions in each case. The most important of these are:

a. Application at the precise point needed.

b. Flexibility; includes easy subdivision, freedom of location, and general adaptability.

c. Good working conditions, cleanliness, quietness, ambient temperature is little affected, etc.

d. Fast response.

e. Safety.

3. Heat-process Engineering. The heating method for a particular process may be selected after consideration of two basic principles of industrial heating, (1) temperature and (2) mode of heat transfer. This narrows considerably the field from which a choice of the heating method should be made. Generally there is more than one possible method. Economic factors must then be taken into account. This, in addition to the relative capital, maintenance, and energy cost, includes a study of the collateral merits of methods that always have a bearing on the overall cost of production.

4. A convenient temperature classification is:

a. Low temperature—up to 400°C (752°F).

b. Medium temperatures—400 to 1150°C (752 to 2102°F).

c. High temperatures—beyond 1150°C (2102°F).

5. Heat energy is transferred by conduction, radiation, convection, high-density electromagnetic force, and concentrated electrostatic fields. The transfer can be obtained with a combination of these methods.

6. Conduction of heat can take place in all three states of matter, i.e., in solids, in liquids, and in gases. Thermal conductivity is defined as the quantity of heat which flows in unit time through unit area of a plate of unit thickness having unit difference of temperature between its faces. There is no generally accepted combination of units for expressing thermal conductivity. Some of the more common combinations of units for this value are:

k = Btu/(ft^2) (h) for 1-in length of path/°F (British units)

k = cal/(cm^2) (sec) for 1-cm length of path/°C (cgs units)

k = W/(in^2) (1-in length of path) (°C)

Values of k for various materials as published are not based on a common method of determination. Hence, agreement in thermal-conductivity data for a given material cannot be expected.

The thermal conductivity of a material is affected by temperature, in some cases increasing, in other cases decreasing with rising temperature. Values representative of the general order of the thermal conductivities of ordinary temperatures of the three states of matter are given in Table 22-1.

Table 22-1. Thermal Conductivities

Solids at or near Room Temperature	k*
Silver	2824
Copper	2661
Aluminum	1463
Zinc	769
Iron	467
Nickel	412
Marble	21
Ice	15
Porcelain	7
Chalk	7
Glass	5
Liquids at Temperatures Noted	
Water, 20°C	4.15
Glycerin, 9–15°C	1.85
Ethyl alcohol, 20°C	1.23
Petroleum, 13°C	1.03
Gases at Temperatures Noted	
Hydrogen, 100°C	1.00
Methane, 7–8°C	0.188
Air, 0°C	0.165
Ammonia gas, 0°C	0.130
Carbon dioxide, 0°C	0.090

* British units; divide by 2,903 to obtain corresponding cgs values of k.

An arbitrary standard by which to judge the value of a material as a nonconductor of heat (heat insulator) is $k = 1$ (British unit).

The law of thermal conduction for a constant-temperature gradient is expressed thus:

$$h_c = \frac{Ak_m t'}{l} \tag{22-1}$$

where h_c = rate of heat transfer by conduction, A = area of cross section of path of heat flow, t' = temperature gradient, l = length of path, k_m = mean value of thermal conductivity of material in which the temperature gradient is established for a given range of temperature.

7. Shape Factor. Equation (22-1) as written is applicable only to paths of uniform cross section. If the area of the cross section of the path of heat flow varies along its length, the equation is applicable if the logarithmic mean area of the path is used.

For boundary surfaces that are not smooth

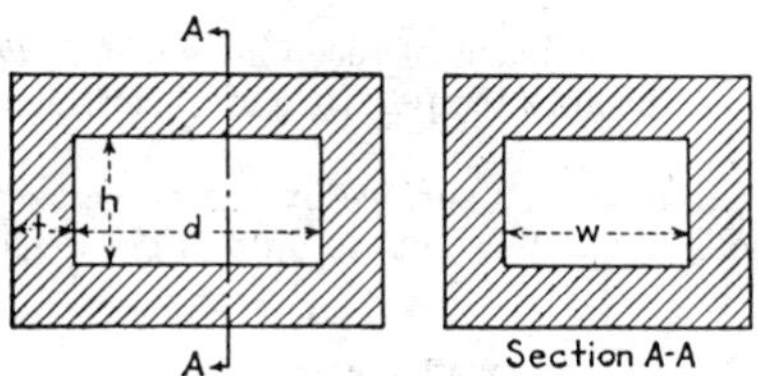

FIG. 22-1. Shape factor—rectangular enclosures.

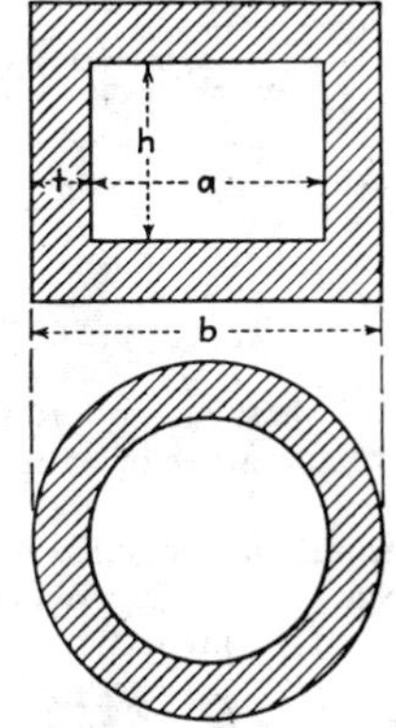

FIG. 22-2. Shape factor—cylindrical enclosures.

curves, a shape factor S must be substituted for the term A/l in Eq. (22-1), thus:

$$h_c = Sk_m t' \qquad (22\text{-}1a)$$

Two of the equations for the shape factor S are

a. Rectangular enclosures (Fig. 22-1),

$$S = \frac{A}{t} + 0.54\Sigma l + 1.20t \qquad (22\text{-}2)$$

where A = inside surface area, t = thickness of wall, Σl = sum of length of all inside edges.

b. Cylindrical enclosures (Fig. 22-2),

$$S = \frac{A}{t} + 0.54\Sigma l + \frac{2\pi h}{2.30 \log (b/a)} \qquad (22\text{-}3)$$

where A = inside area of top and bottom surfaces, t = thickness of wall, Σl = sum of lengths of inside top and bottom edges, h = inside height, b = outside diameter, a = inside diameter.

8. Radiation is a surface phenomenon. Surfaces emit and absorb radiant energy in various degrees, depending on the nature of the surface.

Average values of radiation and absorption coefficients are given in Table 22-2.

Solids for the most part are opaque to radiation; a notable exception is quartz. Liquids transmit radiation to some extent. Only two gases, water vapor and carbon dioxide, absorb radiation to any marked degree.

Table 22-2. Average Values of Radiation (and Absorption) Coefficients

Material	ϵ, radiant-energy coefficient
Blackbody	1.0
Lampblack	0.95
Asbestos board	0.93
Steel, oxidized	0.79
Copper, oxidized	0.72
Lead, oxidized	0.63
Cast iron:	
Oxidized	0.62
Bright	0.22
Brass:	
Oxidized	0.60
Polished	0.10
Nickel, oxidized	0.42
Zinc, oxidized	0.11
Silver, polished	0.03
Aluminum:	
Polished	0.04
Oxidized	0.11
Paper	0.93
Aluminum paint	0.30
White enamel	0.64
Black lacquer	0.87

The rate of heat transfer by radiation under the conditions usual in heating practice is

$$h_r = \epsilon \times 3.68(T_0^4 - T_a^4) \times 10^{-11} \quad \text{(centigrade scale)} \qquad (22\text{-}4a)$$

$$h_r = \epsilon \times 0.35(T_0^4 - T_a^4) \times 10^{-11} \quad \text{(Fahrenheit scale)} \qquad (22\text{-}4b)$$

where h_r = watts per square inch, ϵ = absorption coefficient of receiving surface, T_0 = absolute temperature of emitting surface, T_a = absolute temperature of absorbing surface.

9. Natural convection is a transfer of heat to or from a surface by the movement of a fluid or gas when this movement is caused solely by a difference in fluid or gas density.

For natural convection in air

$$h_b = 11.70 \times t \times 1.25 \times 10^{-4} \quad \text{(centigrade scale)} \qquad (22\text{-}5a)$$

$$h_b = 6.50 \times t \times 1.25 \times 10^{-4} \quad \text{(Fahrenheit scale)} \qquad (22\text{-}5b)$$

where h_b = watts per square inch, t = temperature gradient. These equations apply to vertical planes more than 12 in high. For heights less than 12 in multiply the coefficient in Eq. (22-5*a*) or (22-5*b*) by the constant noted below.

Height, In	Constant
8	1.35
6	1.53
4	1.76
2	2.70

The rate of heat transfer by natural convection from a horizontal surface facing upward is about 35% greater and a rate from a horizontal surface facing downward is about 30% less than from a vertical surface. Hence, for average conditions, and on the sides of the body exposed, the average rate of heat transfer by natural convection is well represented by Eq. (22-5).

If the natural-convection heat transfer is to a surface, the portion affected is the reverse of that stated for transfer from a surface. This does not affect the average rate of heat transfer.

The segregation of the components of heat transfer by radiation (ϵ = 0.90) and natural convection for various temperature gradients is given in Fig. 22-3. The total rates of heat transfer (radiation + natural convection) for various surfaces are given in Fig. 22-4.

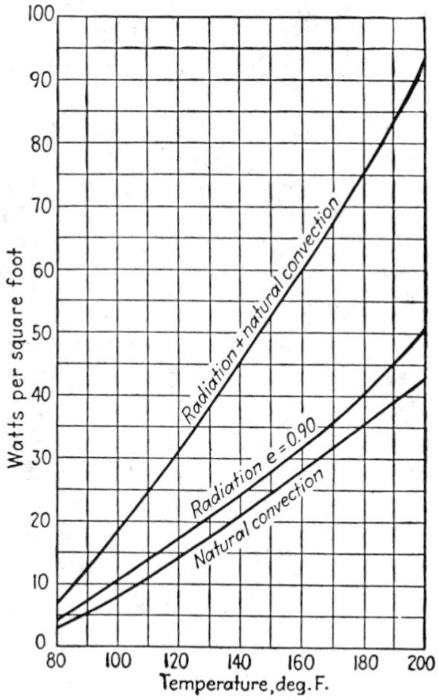

FIG. 22-3. Segregation of components of heat transfer from a surface to surroundings at 65°F.

10. Forced Convection. The velocity of the gas is a dominant factor in the transfer of heat by forced convection. The flexibility of its control, together with the penetrating property of gases, makes forced convection an effective method of heat transfer in many cases. Forced convection can also be applied to fluids.

In electric-heating practice the main limitation on the use of forced convection on gases is the upper limit of operating temperature of fans, about 650°C (1202°F). In some cases the gas can be heated after leaving the fan, and thereby higher temperatures than that noted above can be used.

There is in each use of forced convection a requirement of defined passages for the movement of the gas. This method also entails an investment charge for the fans, ducts, etc., and requires an expenditure of additional energy.

Some of the more common arrangements of surfaces with various conditions of forced convection have been studied, and equations for the corresponding heat-transfer coefficients have been developed. Two of these equations of general interest relate to (*a*) gases flowing in pipes and (*b*) gases flowing past plane surfaces.

11. Gas in Pipes. Turbulent flow and steady state of heat transfer. Straight pipe,

$$k' = 0.027 c_p \mu^{0.20} \frac{G^{0.80}}{d^{0.20}} \qquad (22\text{-}6)$$

where k' = Btu per hour per square foot of internal surface area per degree Fahrenheit, c_p = mean specific heat of the gas, μ = its viscosity in pounds per hour per foot, d = internal diameter of pipe in feet, G = weight velocity of the gas in pounds per hour per square foot of cross-sectional area of the pipe.

For coiled pipes multiply the value of k' for a straight pipe by the coefficient

$$a = 1 + 3.6 \frac{d}{d_c} \qquad (22\text{-}7)$$

where d = internal diameter of pipe and d_c = diameter of the coil.

The heat-transfer coefficients obtained with forced convection in air, nitrogen, oxygen, and carbon monoxide are practically the same. Carbon dioxide is about 7% and hydrogen about 30% better than air for the transfer of heat by forced convection.

12. Weight Velocity. Variations in temperature and pressure will cause changes in the linear velocity and the density of the gas. The weight-velocity—product of linear velocity and density of the gas—is independent of temperature and pressure variations. The relations are

$$G = v_p = \frac{W}{A} \tag{22-8}$$

where G = pounds of gas per hour per square foot of cross-sectional area of the passage, v = linear velocity of the gas in feet per hour, p = density of the gas in pounds per cubic foot, W = flow of gas in pounds per hour, A = cross-sectional area of the passage in square feet.

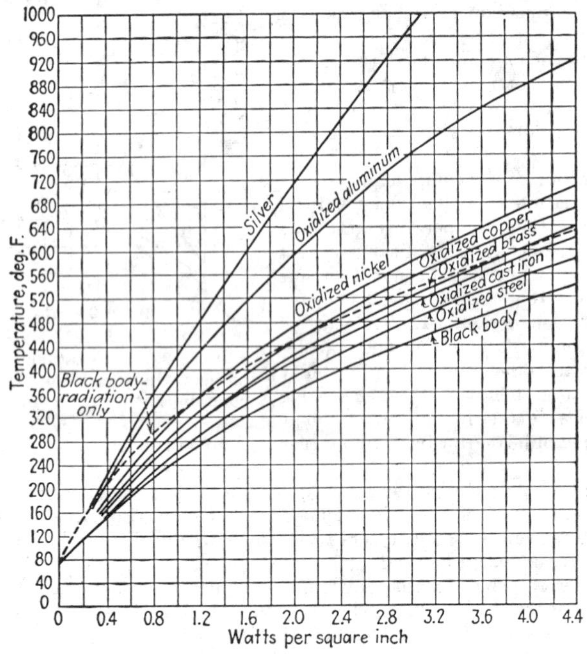

FIG. 22-4. Heat transfer from surfaces to surroundings at 70°F.

13. Critical Velocity. That velocity which marks the transition from streamline flow to turbulent flow in a pipe is given by the equation

$$\frac{v_c dp}{\mu} = 2{,}100 \tag{22-9}$$

where v_c = critical velocity of the gas in feet per second, d = internal diameter of the pipe in feet, p = density of the gas in pounds per cubic foot, μ = viscosity*,† of gas in pounds per second per foot, 2,100 = Reynolds number.

For values of the left-hand member of this equation below 2,100 the flow is streamline; above, turbulent.

* Centipoises × 2.42 = viscosity in pounds per hour per foot. 1 centipoise = 0.01 poise, the cgs unit.
† Centipoises × 0.000672 = viscosity in pounds per second per foot.

Example 1. Air at 325°F flowing through a 2-in pipe.

$\mu = 0.024$ centipoise

$p = 0.0576$ lb/ft³

$d = \frac{1}{6}$ ft

$$v_c = \frac{2{,}100 \times (0.024 \times 0.000672) \times 6}{0.0576}$$

$= 3.5$ ft/sec

With streamline flow in a pipe the average velocity of the gas is one-half the maximum velocity. With turbulent flow this ratio is about 0.80; thus the advantage of turbulent flow for forced convection.

14. Hydraulic Diameter. For pipes of noncircular cross section the hydraulic diameter is used in computation:

$$\text{Hydraulic diameter} = \frac{4 \times (\text{cross-sectional area})}{\text{perimeter of cross section}} \tag{22-10}$$

Example 2. Air flowing through a straight metal pipe [Eq. (22-6)].

Temperature of pipe = 180°F
Mean temperature of air in pipe = 1000°F

$G = 6{,}000$ lb/(h)(ft²)

$c_p = 0.2430$

$\mu = 0.037$ centipoise

$d = 0.166$ ft

$$k' = \frac{0.027 \times 0.2430 \times (0.037 \times 2.42)^{0.20} \times (6{,}000)^{0.80}}{(0.166)^{0.20}}$$

$= 6$ Btu/(h)(ft²)(°F)

15. Plane Surfaces. The heat-transfer coefficient of forced convection in gases flowing past plane surfaces does not differ greatly from the value of that coefficient obtained for gases in turbulent flow through pipes 12 in in diameter and larger.

Average values of the coefficient k' for the transfer of heat between a plane surface and a gas flowing past that surface are given by Fig. 22-5. The equation of that curve is

$$k' = 0.50(v)^{0.78} \tag{22-11a}$$

where k' = Btu per square foot per degree Fahrenheit temperature difference, v = velocity of the gas in feet per second, or

$$k' = 0.1465(v)^{0.78} \quad [\text{W/(ft}^2\text{)(°F temp. diff.)}] \tag{22-11b}$$

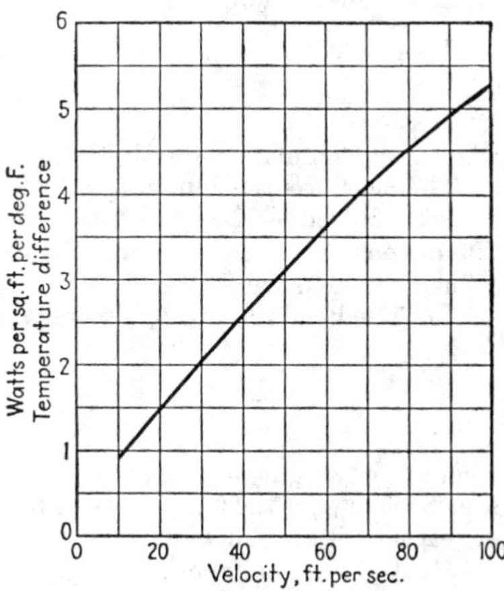

Fig. 22-5. Heat transfer by forced convection from a surface.

16. Rate of Heat Transfer. The product of the heat-transfer coefficient and the temperature gradient gives the rate of heat transfer per unit area. With forced convection the temperature difference between the surface and the gas varies along the length of the surface. In such cases the logarithmic mean difference should be used:

$$\theta' = \frac{\theta_1 - \theta_2}{2.30 \log (\theta_1/\theta_2)} \tag{22-12}$$

where θ' = logarithmic mean temperature difference, θ_1 = larger temperature difference, θ_2 = smaller temperature difference.

If the ratio of the larger and smaller temperature difference is not greater than 2:1, the arithmetic mean temperature difference can be used.

17. Finned Surface. If the heat-transfer coefficient of a surface is less than the heat-transfer coefficient which is the measure of the rate of flow of heat through the body, to or from the surface, the addition of fins or ribs to the surface will increase the rate of heat flow to or from the surface.

As a general rule a spacing and dimensioning of fins that increase the surface area seven to nine times is good practice. The height of the fins should not exceed three to four times the spacing.

The principle of increasing the surface area per unit length of the surface by the addition of fins is equally applicable to the transfer of heat by radiation.

Table 22-3. Diffusivity of Materials

(At room temperature except where otherwise noted)

Material	*n*, CGS Units
Silver	1.737
Copper	1.133
Aluminum	0.826
Zinc	0.402
Steel	0.173
Nickel	0.152
Ice	0.0112
Marble	0.0090
Fireclay brick (0–800°C)	0.0074
Glass	0.0057
Water	0.0014
Air (0°C)	0.179

18. The rate of heat absorption for given external thermal conditions is determined by the diffusivity coefficient of the substance. This coefficient is defined as the ratio of the thermal conductivity to its heat storage capacity per unit volume; thus,

$$n = \frac{k}{p_c} \tag{22-13}$$

where k = thermal conductivity, p = density, c = specific heat.

This coefficient is but little affected by temperature. Some values for comparison are given in Table 22-3. All other conditions being the same, the rate of heating or the rate of cooling for the two materials will be the factor for the material having the higher diffusivity coefficient.

19. The heat absorbed by a material for a given temperature rise is

$$q = \frac{w \times c_m \times t'}{3.412} + L + R \qquad \text{(Wh)} \tag{22-14}$$

where w = weight in pounds; t' = temperature rise in degrees Fahrenheit; c_m = mean specific heat over the temperature range; L = latent heat, if any, in watthours; R = heat of reaction, if any, in watthours.

Heat of reaction here means the quantity of heat that must be added to the charge or subtracted from the charge, depending on whether the reaction is endothermic or exothermic.

Example 3. Heat absorption of lead, 70 to 750°F.

$W = 300$ lb

Melting point = 621°F

Latent heat of fusion $L = 3$ Wh/lb

c_m = 70 to 621°F = 0.031

c_m = 621 to 750°F = 0.04

$$q = \frac{[300 \times 0.031 \times (621 - 70)] + [300 \times 0.04 \times (750 - 621)]}{3.412} + (300 \times 3)$$

$$= 2{,}580 \text{ Wh}$$

The heat contents of the common metals and gases are given in Figs. 22-6 and 22-7 and in Table 22-4. Horizontal sections in the graphs show latent-heat quantities. Latent heats of alloys must be determined for each alloy.

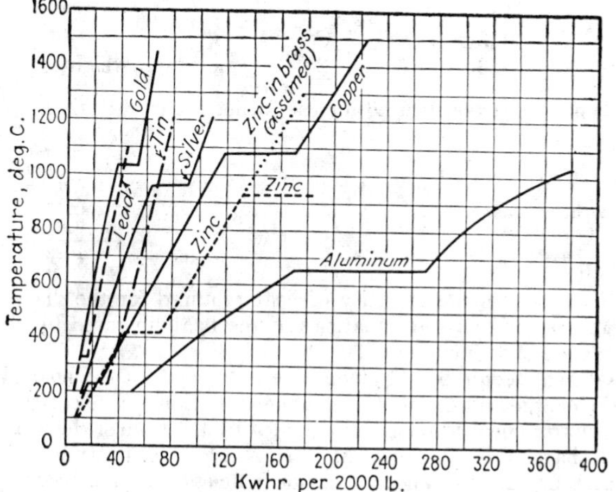

Fig. 22-6. Heat contents of metals.

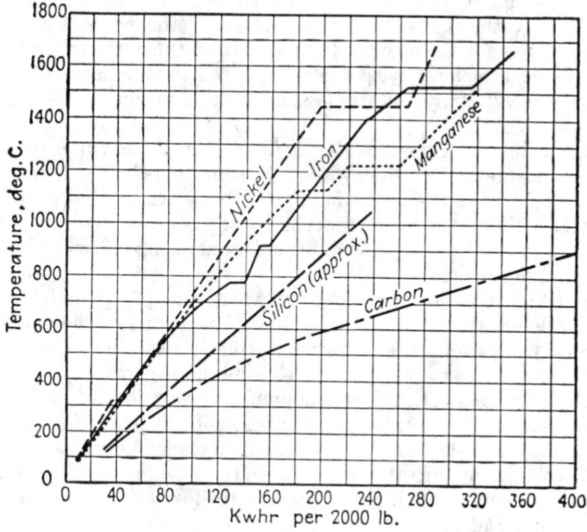

Fig. 22-7. Heat contents of metals.

Table 22-4. Heat Content of Gases, Watthours per Pound

°C	°F	H_2	O_2	N_2	CO	CO_2	CH_4	H_2O	Air
100	212	182	12	13	13	11	29	23	13
200	392	365	24	26	26	23	62	46	26
300	572	549	36	40	40	36	100	72	38
400	752	734	49	53	54	50	143	98	51
500	932	920	61	67	68	63	189	124	64
600	1112	1,107	75	81	82	78	240	152	77
700	1292	1,296	88	96	97	93	293	180	90
800	1472	1,487	102	111	112	109	350	209	103
900	1652	1,677	115	125	127	124	410	239	116
1000	1832	1,873	129	141	142	141	473	269	130
1110	2012	2,069	143	156	157	157	538	300	143
1200	2192	2,267	157	171	173	174	605	332	157

Example 4. The production of calcium carbide from calcium oxide (lime) and carbon (coke).

$$\begin{array}{cccc} CaO + 3C \rightarrow CaC_2 + CO \\ 56 \quad 36 \quad 64 \quad 28 \quad \text{(molecular weights)} \end{array} \tag{22-15}$$

Thus two-thirds of the carbon is used for the formation of calcium carbide.

	Heat Absorption, KWh/Lb
Heat of reaction	0.77
Sensible heat	0.45
Total	1.22

The heat of reaction is the quantity of heat required for the disassociation of the calcium oxide minus some of the quantities of the heat liberated by the formation of the carbon monoxide and calcium oxide.

20. The total heat required in a given case is the sum of the heat absorbed by the material, i.e., the charge and the heat loss of the operation. The corresponding average power input is the total heat in watthours divided by the time of the heat cycle in hours. The power-time characteristic is individual for each class of heating surface.

21. Thermal Efficiencies. The conversion efficiency of the heating device is the ratio of the heat absorbed by the material, i.e., the charge for a given temperature rise and the corresponding heat input to the device. If the charge is under cover, the heat absorbed by the material of the enclosure (the heating chamber) must be taken into account, along with the heat dissipated from the outer surface of the chamber. With increase in length of heat cycle the heat stored in the materials of the enclosure affects the conversion efficiency less and less, and with continuous operation it is usually negligible.

The operating efficiency expresses the combination of the conversion efficiency and the efficiency of the method of operation. If containers or removable supports for the charge are used, it is necessary to distinguish between net and gross operating efficiencies; the latter includes the heat absorbed by the containers or supports. A much-used rule of thumb for preliminary estimates is an operating efficiency of 50% for many heat applications under cover.

22. Thermal efficiencies are expressed as watthours per pound, pounds per kilowatthour, kilowatthours per ton, etc., according to the scale of the operation.

23. High-density Electromagnetic Force. Usually utilized to achieve fast heating times in air, vacuum, or gas atmospheres, this method of heating is more commonly known as induction heating. The electromagnetic force is developed at some frequency through the utilization of a suitable conducting coil. The coil is generally made of copper, often it is liquid-cooled, and it is designed to encircle or to be in close proximity to the material (nominally metal) to be heated. The high-density electromagnetic force coupled to the metal load induces current to flow in the load, and this current directly heats the material.

24. Concentrated Electrostatic Fields. Usually applied to fast heating times in air, vacuum, or gas atmospheres, this method of heating is more commonly known as dielectric heating. The electrostatic field is developed at a very high frequency. Through the utilization of metallic plates (often copper) placed above and below or on either side of the material to be heated (nominally a nonconductor) the electrostatic field causes a rapid reorientation of the molecules in the load, and heat is thereby produced internally.

25. Industrial heating is classified as follows:

a. The heating of materials to temperatures below the temperature at which a change of state occurs.

b. The heating of materials to cause a change of state, i.e., evaporation and melting.

c. The heating of materials to bring about a chemical reaction.

METHODS OF ELECTRIC HEATING

26. The methods of electric heating are:

a. Resistance heating = I^2R effect.

b. Electric arc.

c. Induction heating = special application of I^2R.

d. Dielectric heating.

The resistance method and the induction method are widely employed. The use of the electric arc is generally confined to high-temperature applications. Dielectric equipment is widely used for the heating and welding of nonconductors.

Alternating current is used in most applications of electric heating. Direct current is required for a thermal process in which electrolysis is involved.

27. Resistance Heating. The I^2R effect is expressed by the equations

$$q = i^2rt = \frac{e^2}{r}t \qquad \text{(Wh)} \tag{22-16}$$

where r = resistance of the path of the current in the load; i = the current in amperes, for varying current, the rms value; e = applied voltage, across load; t = time, in hours, of the flow in current.

The resistivity-temperature coefficient of metals is positive; that of nonmetals, negative. An exception among nonmetals is graphite.

28. The relation between applied voltage and change of resistance—a general case in heat applications—for constant power is

$$e_2 = e_1 \frac{z_2}{z_1} \sqrt{\frac{r_1}{r_2}} \tag{22-17a}$$

or if the power factor of the circuit is unity,

$$e_2 = e_1 \sqrt{\frac{r_2}{r_1}} \tag{22-17b}$$

where e_1 = initial voltage, r_1 = initial resistance, z_1 = initial impedance, r_2 = new resistance, z_2 = new impedance, e_2 = new voltage.

29. The rate of heat development, i.e., watts, is obtained by dividing the quantity obtained by Eq. (22-16) by the time period in hours or frequencies of an hour.

Example 5

$$q = 60{,}000 \text{ Wh}$$

$$t = 1 \text{ h, } 60 \text{ kW}$$

$$t = 10 \text{ min, } \tfrac{1}{6} \text{ h, } 360 \text{ kW}$$

$$t = 1 \text{ min, } \tfrac{1}{60} \text{ h, } 3{,}600 \text{ kW}$$

30. Indirect Resistance Heating. The heat is developed in resistors (heating units), a circuit apart from the charge, and is transferred from the surface of the resistors to the surface of the charge by one or more of the modes of heat transfer, viz., conduction, radiation, and convection (natural or forced).

If the heat transfer is by conduction, the resistor must be in contact with the charge.

An enclosure for the charge (heating chamber) is required for heat transfer by radiation and convection.

31. Resistors. The materials in general use for resistors for medium- and low-temperature surfaces are the two alloys B82 and B83 designated by the ASTM (see Sec. **4** under Electrical-resistance Alloys).

B82 alloy. Additional data on round wires and rectangular sections in Table 22-5. Approximate maximum operating temperature 1160°C (2100°F).

B83 alloy. See the footnotes, Table 22-5. Approximate maximum operating temperature 900°C (1652°F).

The melting ranges of these alloys are narrow. They do not soften if the temperature of the resistor is kept at a reasonable margin below the melting range. The coefficients of expansion are low, and the change of resistance with the change of temperature is small. The resistivities permit the use of reasonable cross sections and lengths at standard low-distribution voltages.

Table 22-5. B82-alloy (Ni-Cr) Average Values

No. B & S	Diameter, in	Surface area, in²/lin. ft	Ω/ft straight round wire, at 20°C*	Ft/lb bare round wire†
1	0.289	10.91	0.00778	4.18
3	0.229	8.65	0.0124	6.66
5	0.182	6.86	0.0196	10.5
7	0.144	5.44	0.0314	16.8
9	0.114	4.31	0.0500	26.9
11	0.091	3.42	0.0785	42.2
13	0.072	2.71	0.1250	67.4
15	0.057	2.15	0.200	107
17	0.045	1.71	0.321	172
19	0.036	1.35	0.502	26
21	0.0285	1.073	0.800	430
23	0.0226	0.851	1.27	684
25	0.0179	0.675	2.03	1,090
27	0.0142	0.535	3.22	1,730
29	0.0113	0.424	5.09	2,730
31	0.0089	0.337	8.21	4,420
33	0.0071	0.267	12.90	6,930
35	0.0056	0.212	20.70	11,150
37	0.0045	0.168	32.10	17,200
39	0.0035	0.133	53.10	28,400

* For B-83 alloy multiply by 1.037.
† For B-83 alloy multiply by 1.02.

These alloys have the prime requisite of resistance to oxidation and scaling when operated in air. However, this protection is not perfect, and it decreases in value with increases of resistor temperature. The alloys are susceptible to harm from compounds of sulfur and are affected to some extent by carbon monoxide. Unexpected causes of chemical action on resistors of these alloys are foreign gases or gases evolved from the material being heated.

In addition to the B-83 alloy, several other similar alloys are listed by manufacturers for low-temperature service.

It is often desirable for low-temperature heating applications to use resistors known as metal-sheath heaters. These devices provide a metal sheath, usually a nickel alloy or a stainless steel, as a protective cover over the resistance device and insulated from it by a magnesia oxide. Sheath heaters are ideal for direct immersion and conductive heating, as the sheath tends to assume the temperature of the resistance element. Direct contamination of the resistance element is thus avoided.

32. Resistors for operating temperatures above 1150°C (2100°F) are made of silicon carbide, molybdenum, tungsten, and graphite.

33. Silicon carbide is the basis of a resistor material for operating in air for temperatures up to about 1500°C (2732°F). The material is formed into rods of diameters and lengths for combination into circuits of required electrical rating. The range between

Table 22-6. B82-alloy Ribbons for Ovens and Furnaces, Average Values

Width, in	Thickness, in	Surface area, in²/lin ft	Ω/ft at 20°C	Ft/lb
0.75	0.125	21.0	0.00544	2.92
0.75	0.090	20.16	0.00756	4.06
0.75	0.070	19.68	0.00971	5.22
0.75	0.050	19.20	0.0136	7.32
0.75	0.030	18.72	0.0228	12.20
1.00	0.100	26.40	0.00510	2.74
1.00	0.080	25.92	0.00638	3.42
1.00	0.060	25.44	0.00850	4.57
1.00	0.040	24.96	0.0127	6.85
1.50	0.125	39.0	0.00272	1.46
1.50	0.09	38.16	0.00378	2.03
1.50	0.07	37.68	0.00486	2.61
1.50	0.05	37.20	0.00680	3.66
1.50	0.03	36.72	0.0113	6.10

the cold and hot resistances is small enough to permit single-voltage operation. The resistance gradually increases with use, and an autotransformer with taps to compensate for this increase of resistance is desirable.

34. Molybdenum resistors are suitable for temperatures up to 1650°C (3002°F). This metal is ductile enough at room temperature for drawing into wire for resistor windings. The resistivity-temperature characteristic is shown in Fig. 22-8. The wide range between the cold and hot resistances makes multiple-voltage operation necessary. The supports (insulators) of the windings should be of magnesia or zirconia. Molybdenum resistors cannot be operated in air and also must be protected against reactions with silicon and carbon. The metal is immune from reactions with sulfur, nitrogen, hydrogen, and water vapor. A hydrogen atmosphere is ordinarily used for the protection of these resistors. Molybdenum is not suitable for resistors of vacuum furnaces because of its high vapor pressure.

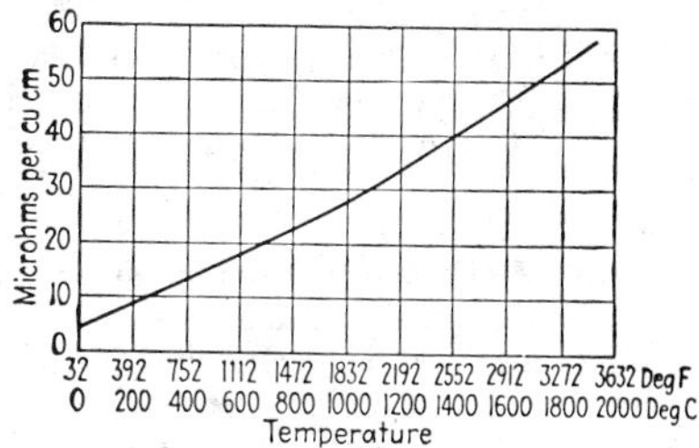

Fig. 22-8. Resistivity-temperature characteristic of molybdenum.

35. Tungsten resistors can be used for temperatures up to 2000°C (3632°F). The maximum temperature is limited by the refractory supports of the resistor. The metal must be heated for drawing into shapes. The resistivity-temperature characteristic is practically the same as that of molybdenum. The low vapor pressure of tungsten makes it useful for resistors of vacuum furnaces.

36. Graphite resistors are suitable for any temperature that can be used. The resistors must be protected against oxidation above about 600°C (1112°F). Also because of the chemical activity of carbon, special consideration must be given to the atmospheres surrounding the resistors. The material is available in a wide range of shapes and dimensions. See Sec. **4**.

37. The useful life of a resistor (barring accidents) is dependent mainly on the temperature at which it is operated. That temperature must be high because of the required rate of heat transfer per unit area of the surface. Thus the operating temperature of a resistor is a function of the surface power density (watts per unit area) of the resistor. For a resistor of given shape and dimensions the operating temperature will vary with the watts input, since the surface area remains the same. Hence, watts per unit area is a first consideration in resistor design. This value in practice varies according to the conditions of use. Some examples are given later.

A rectangular shape for the resistor section gives a greater surface area for a given resistance than the round shape. Hence, the frequent use of the ribbon, or strip, form of resistor, particularly for the upper range of temperature. The cast resistor for furnaces also embodies this principle.

38. The maximum voltage of resistor circuit is limited by electrical insulation and elevated temperatures and by safety considerations to 600 V. This value may be exceeded in special cases. Standard distribution voltages and standard frequencies are used. Resistor windings may be in one, two, or more circuits and may be connected either single-phase or polyphase.

39. As a general rule the heat-equalizing effect of the heat-storage capacity of the charge into the surrounding enclosure (if any) makes permissible voltage fluctuations within reasonable limits. However, the average voltage must be equal to the normal voltage to maintain the normal heating rate. The relation between heating rate and voltage is shown in Fig. 22-9.

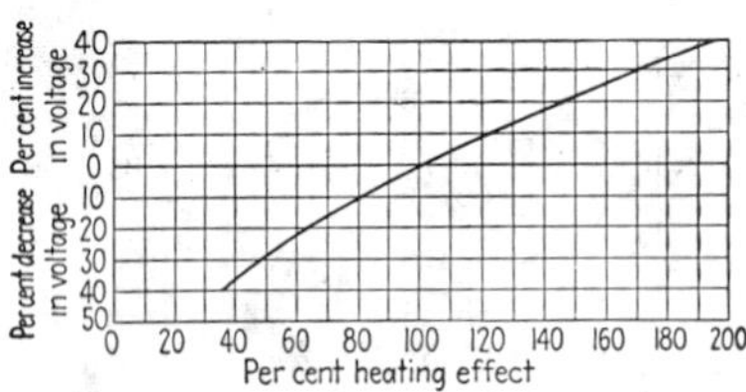

FIG. 22-9. Variation of heating effect with changes of voltage.

40. The power input to resistors supplied from constant-voltage circuits can be varied by the use of multiple circuits and series-multiple-switching combinations. Another method is the use of an autotransformer or transformer with taps. With 3-phase resistor circuits the change from delta connection to Y-connection, and vice versa, is available. In each case the power input to the resistor is proportional to the square of the voltage. For example, with the change from delta to Y the power input is reduced to one-third its value with the delta connection.

Other devices are available for power controls such as: saturable reactors, regulated or unregulated; variable inductances; and stepless variable autotransformers.

41. Heating Chambers. The term is usually applied to the enclosure of a charge which is heated by radiation or convection or by the joint effects of these modes of heat transfer. Aside from mechanical considerations the design of the chamber is related primarily to temperature and the major mode of heat transfer to be used. The essential features are: the enclosing members, the resistors, and the atmosphere within the chamber.

The functions of a heating chamber are:

a. The control of the distribution of the heat within the chamber.

b. If required, the control of the cooling rate of the charge.

c. The conservation of as much of the heat supply as may be practical and economical.

d. In many cases the confinement of the atmosphere around the charge.

A cylinder with an inside diameter equal to its height represents the ideal shape in proportions of the heating chamber, considering only the utilization of material and the rate of heat loss from the external surface. This ideal is seldom attained because of the many practical considerations in heating service. Electric heat permits wide latitude in the proportions and shape of heat chambers, and factory planning takes this into account in developing the sequence of movements of material through manufacturing operations.

Heating chambers are of two types, viz.:

a. Batch chambers. The charge is stationary during the heat application. The cycle may include cooling the charge in the chamber.

b. Continuous chambers. The charge is heated as it moves through the chamber. In some cases the chamber is extended for more or less cooling of the charges before it leaves the chamber.

The application of the batch-type heating chamber is defined by the following conditions:

a. Intermittent production.

b. Long periods of heating and in some cases slow cooling.

c. For material that cannot be handled on conveyers.

d. Heating service beyond the range of the capacity of conveying mechanisms.

e. Supplementary heating service.

A continuous-type heating chamber is indicated where the flow of material is reasonably uniform and continuous, i.e., under mass-production conditions. In some cases batch heating chambers with automatic charging and discharging are essentially of the continuous type.

A resistor for a heating chamber is a sinuous winding located on the interior surface of the chamber. Muffles except in special cases are not required. The distribution of the resistor winding is designed with reference to the proportions of the chamber, the character of the charge, the major mode of heat transfer, and the openings of the chamber.

SELF-CONTAINED UNITS, WATER HEATING, STEAM GENERATORS

42. Resistor heating equipment includes many small devices such as electrically heated tools, appliances of various kinds, self-contained units, ranges, water heaters, stills, sterilizers, and baths. The larger units include ovens, kilns, lehrs, furnaces, etc. Each of these descriptive terms relates to a field—some more general than others—of electric heating service.

43. Small Devices. As a rule the best design of a resistor for any small device is obtainable only by trial. The space is generally limited, and the conditions which affect heat transfer vary widely. For many devices a useful guide is 20 W/in^2 of resistor surface. Both higher and lower values are used according to the conditions in each case. The data in Table 22-7 represent average practice.

Metal-sheath heaters can be obtained for operation at many of the wattages listed in Table 22-7, to be operated at specified voltages of 115 or 230 V. These are further described in Par. **44.**

Table 22-7. Cold Resistances (75°F) for Common Wattages at Operating Temperature, 110- to 120-V Circuits

(Multiply resistances by 4 for 220- to 240-V circuits)
(From Driver-Harris Company Data Book)

Watts	Nichrome V,* Ω	Nichrome,† Ω	Recommended B&S sizes	
			110–120 V Max Min	220–440 V Max Min
100	123.52	118.10	26–30	29–33
200	61.76	59.05	25–29	28–32
300	41.17	39.37	24–28	27–31
400	30.88	29.53	22–26	25–29
500	24.70	23.62	20–24	23–27
600	20.59	19.68	19–23	22–26
700	17.65	16.87	18–22	21–25
800	15.44	14.76	18–22	21–25
900	13.72	13.12	17–21	20–24
1,000	12.35	11.81	16–20	19–23
1,100	11.23	10.74	16–20	19–23
1,200	10.29	9.84	15–19	18–22
1,300	9.50	9.08	14–18	17–21
1,400	8.82	8.44	13–17	16–20
1,500	8.23	7.87	12–16	15–19
2,000	6.18	5.91	10–14	13–17
2,500	4.94	4.72	9–13	12–16
3,000	4.12	3.94	8–12	11–15

* Corresponds to ASTM Designation B82 alloy.
† Corresponds to ASTM Designation B83 alloy.

Example 6*a* (refer to Table 22-7). 1,000 W, 110 V, single-phase. Cold resistance 11.81 Ω.

No. 18 Brown and Sharpe B83 alloy wire
Cold = unit resistance = 0.406 × 1.037 = 0.421 Ω/ft

$$\text{Length of wire} = \frac{11.81}{0.421} = 28 \text{ ft}$$

$$\text{W/in}^2 \text{ of resistor surface} = \frac{1{,}000}{1.52 \times 28} = 23.50$$

Example 6*b*. 1,000 W, 220 V, single-phase.

Cold resistance $= 11.81 \times 4 = 47.24\ \Omega$
No. 23 Brown and Sharpe B82 alloy wire
Cold unit resistance 1.27 Ω/ft

$$\text{Length of wire} = \frac{47.24}{1.27} = 37.2 \text{ ft}$$

$$\text{W/in}^2 \text{ of resistor surface} = \frac{1{,}000}{0.851 \times 37.2} = 31.5$$

44. Self-contained units meet manifold needs for localized heat in small quantities within the lower temperature range and extending up to about 815°C (1500°F).

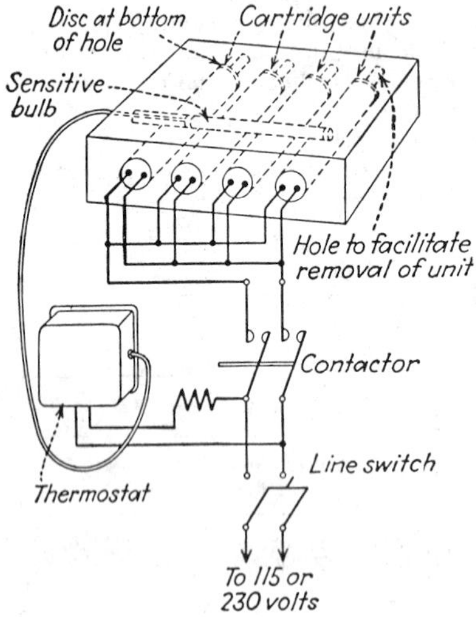

The construction of these units consists of a resistor of the 80 Ni–20 Cr alloy embedded in a heat-resisting material such as powdered fused magnesia and enclosed in a cylindrical metal sheath. In many cases they are applied for direct-contact heating, i.e., for heat transfer by conduction.

FIG. 22-10. Application of insertion (cartridge) units.

45. The forms of the self-contained heating units include: insertion (or cartridge) units; straight or curved tubular units in a shape to conform to the method of use; space (or strip) heaters; immersion units for heating liquids; fin-type units for air heating; cast-in units; and a special construction for melting soft metals.

Some typical uses of the first four forms named are indicated by Figs. 22-10 to 22-13.

Standard voltages of self-contained units are 115, 230, and 460 V. Space restrictions of the terminals do not permit higher voltages except for special designs. The ratings of these units range from 30 W to 10 kW.

46. Space heaters are made in two types:

a. With steel sheath; specific rating 10 W/in² of surface; maximum operation temperature 399°C (750°F).

b. With porcelain-enameled steel or alloy-steel sheath; specific rating 15 W/in² of surface; maximum temperature 650°C (1200°F).

Both types are used for ovens, warming tables, process machinery, space heating, etc.

47. The rating in watts per square inch of affected area of the sheath of the unit is considered in relation to its application: (1) the resistance to oxidation or other chemical action; (2) the rate of heat absorption by the surrounding medium. Representative temperature values and the corresponding watts per square inch of sheath surface area for operation in open air are noted below.

Sheath material	Maximum temperature in open air	W/in²
Plain steel	400°C (725°F)	9
Nickel silver	540°C (1004°F)	17
Calorized steel	650°C (1202°F)	24
Alloy steel	815°C (1500°F)	30

Immersion units for heating water have comparatively high ratings, 40 to 50 W/in². However, these units should always be installed so as to permit free circulation of

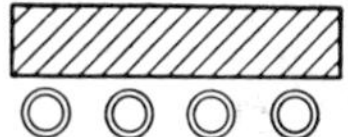

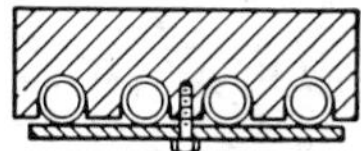

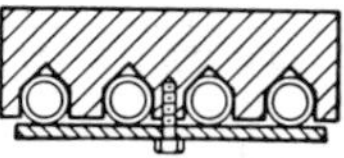

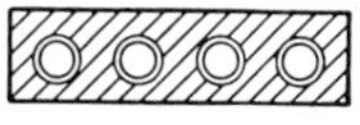

FIG. 22-11. Applications of straight tubular units.

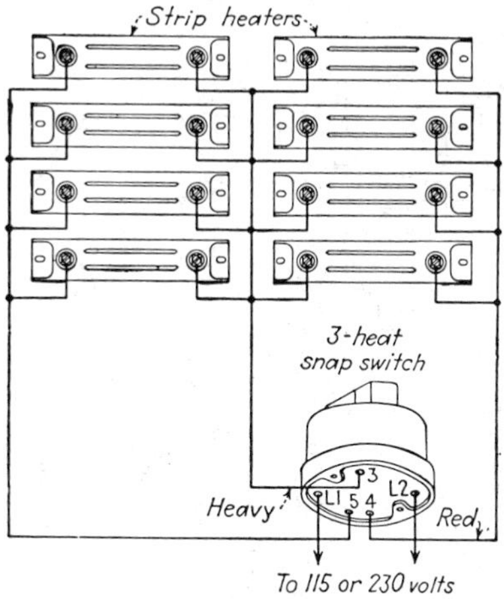

FIG. 22-12. Space (or strip) heating units.

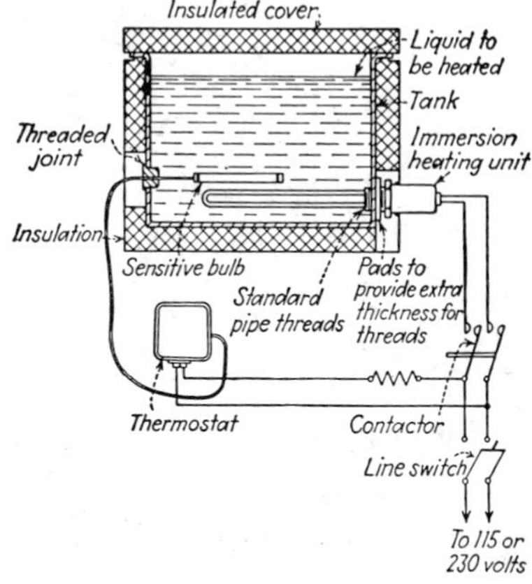

FIG. 22-13. Application of immersion unit.

water, and they should not be operated unless completely submerged.

Oils in general do not absorb heat as rapidly as water. Also, immersion units in oil accumulate carbon deposits on the sheath. Hence, units for this service have a much lower rating—20 to 25 W/in². Steel is a satisfactory sheath material for mineral oils. Other oils must be given individual consideration with reference to the chemical action on the sheath material. The same consideration applies to heating chemical materials in liquid form; e.g., a unit with a lead sheath is used for nickel and copper solutions.

48. The temperature-watts characteristic of the fin-type units (Fig. 22-14) is given in Fig. 22-15. Its application for air heating by forced convection is shown by the chart in Fig. 22-16.

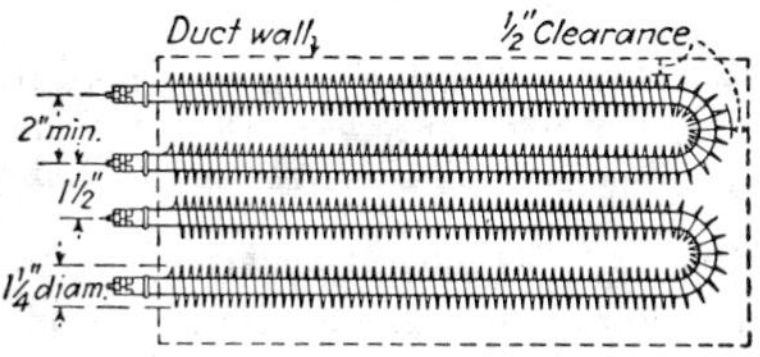

FIG. 22-14. Fin-type heating units.

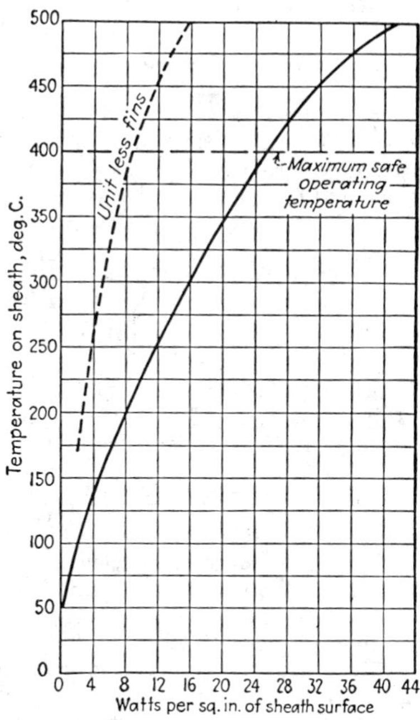

FIG. 22-15. Temperature-watts characteristic of fin-type unit in still air.

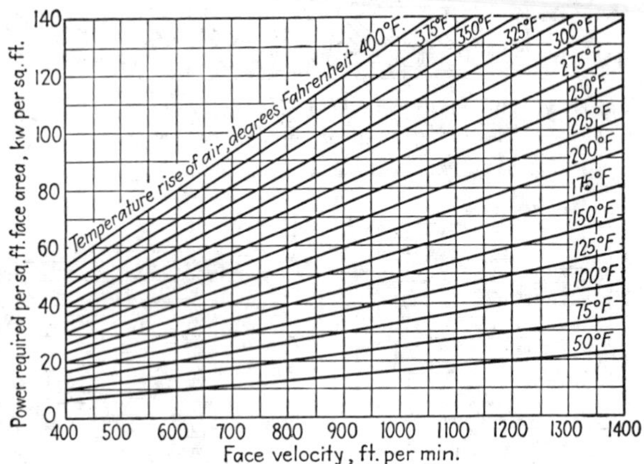

FIG. 22-16. Application of the fin-type unit for forced-convection heating.

Example 7. Air heated with fin-type unit.

Face area of duct	4 ft²
Air flow	2,000 ft³/min at 70°F
Ratings of units	60 kW
Face velocity	500 ft/min
KW/ft² face area	15
Temperature rise of air (Fig. 22-16)	90°F
Outgoing air temperature (70 + 90)	160°F

49. Room Heating. The space heater is well adapted to heating rooms, radiation and natural convection being used for the heat transfer.

An approximate equation for the rate of heat loss from structures of ordinary building construction is

$$\text{kW} = \frac{(0.02NV) + (1.13G) + KA}{3{,}412}(T_r - T_0) \qquad (22\text{-}18)$$

where N = number of air changes per hour (usually two or three); V = room volume in cubic feet; G = area of windows in square feet; A = area of ceiling and exposed walls in square feet; $T_r - T_0$ = temperature difference, in degrees Fahrenheit, between inside and outside; K = heat-transfer coefficient in Btu per hour per square foot per degree Fahrenheit.

Average Values of k

Wood construction	0.25
8-in brick wall	0.50
12-in brick wall	0.36
6-in concrete wall	0.79
10-in concrete wall	0.62

Example 8. Corner room, one side opening to cold hall. Room dimensions 12 × 16 × 10 ft high. Two windows each 8 × 5 ft, wood construction. Outside temperature 0°F; inside, 70°F. Two air changes per hour.

$$\text{kW} = \frac{(0.02 \times 2 \times 1{,}920) + (1.13 \times 80) + (0.025 \times 552)}{3{,}412} \times 70 = 6\ \text{kW}$$

The heat-transfer coefficient can be reduced by the addition of proper insulating material in the walls and ceiling of the room. Other heat sources must be considered in determining the actual heating-equipment requirements, such as lighting, appliances, machinery, and people.

50. A heating cable with a lead sheath for temperatures up to 74°C (165°F) has a variety of uses, e.g., warming soil to promote plant growth, prevention of freezing of liquids in pipes, and keeping gutters and downspouts open.

51. Water Heating. An example illustrates the use of the charts of Figs. 22-17 and 22-18.

Example 9. An open-top hot water tank.

Inside dimensions: length 3 ft; width 2 ft; depth 2 ft. Depth of water 1 ft 6 in. Quantity of water in tank 67.5 gal. Tank insulation: see Fig. 22-18. Required: (*a*) to heat this body of water from 50 to 150°F in 2 h and (*b*) to continue the heating at the rate of 4 gal/h.

From Fig. 22-17, 4.2 gal/kWh,

$$\frac{67.5}{4.2 \times 2} = 8 \text{ kW}$$

From Fig. 22-18, rate of heat loss at 150°F = 1.70 kW.

a. Power required = 8 + 1.7 = 9.7 kW.

b. Heating water at the rate of 4 gal/h.

The power required is 1 + 1.7 = 2.7 kW.

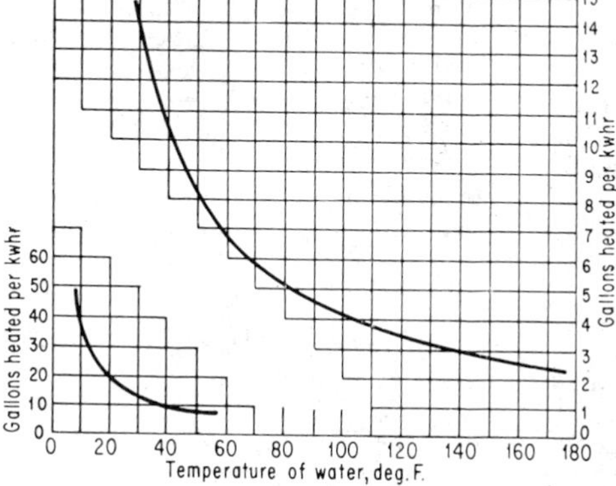

Fig. 22-17. Water heating, gallons per kilowatt-hour.

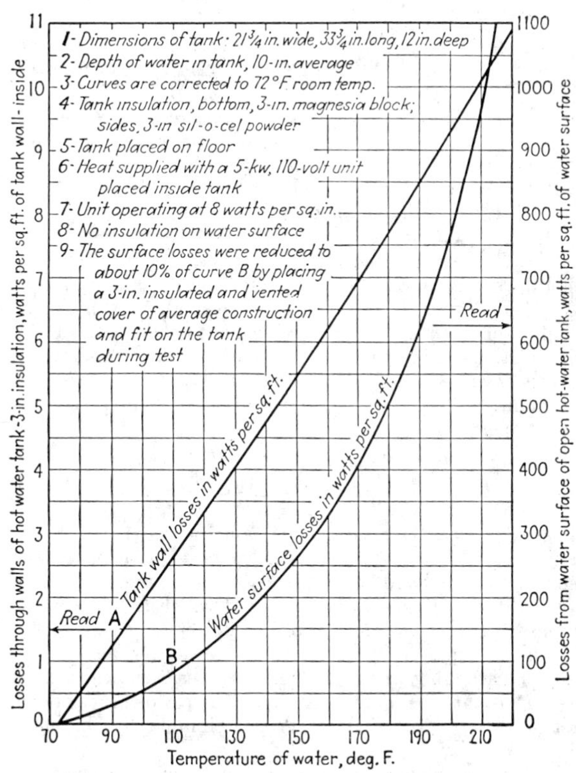

Fig. 22-18. Heat loss from hot-water tanks.

The installation would be a 10-kW three-heat water-immersion heating unit, a switch, and a thermostatic temperature regulator.

52. Electric steam generators are made in three types:

a. With immersion heating units up to 300 kW, single-phase, standard low voltages.

b. With 80 Ni–20 Co alloy resistors mounted in the tubes of a fire-tube boiler. Sizes up to 1,000 kW, single-phase or 3-phase, standard low voltages.

c. Electrode type. Large units 2,000 kW and above. All 3-phase, either 2,300 or 6,600 V.

The heat loss from a generator in operation is the sum of the loss through the heat insulation and the loss by conduction through the pipe connections. The time of heating is very short and the total loss within a given time is small compared with the heat output in the steam. Hence, the insulation of the shell can be moderate in amount for a high conversion efficiency—95 to 97% and higher. This efficiency is usually given as 3 lb of steam/kWh. The boiler-horsepower rating is one-tenth the kilowatt rating.

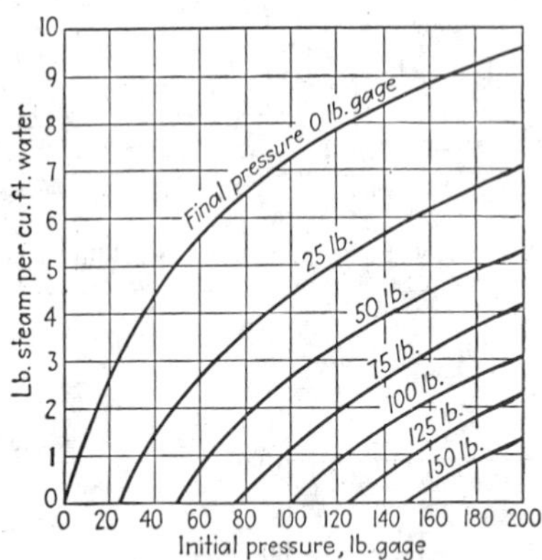

Fig. 22-19. Flashing capacity of steam generators.

There is but little storage capacity in a steam generator, and the power demand cannot be reduced by storage as in water heaters. When a quantity of steam beyond the rating of the generator is required, it cannot be obtained by releasing the steam through a pressure-reducing valve. The amount of steam thus available per cubic foot of water in a generator can be read from Fig. 22-19.

Example 10

Generator pressure 90 lb gage.

Outlet pressure (through reducing valve) 25 lb gage.

Steam output 4 lb of steam/ft^3 of water.

The cubic feet of water available for the above method of use is the volume between the maximum and the minimum levels.

The auxiliary equipment of electric steam generators is the same as that of fuel-fired steam generators. However, full automatic control is more frequently used with the electric steam generator. The main use of the small sizes of electric steam generators is for the supply of steam at points which cannot conveniently be reached by steam pipelines and in locations where the fuel-fired unit is not permissible.

The larger units have been applied mainly for utilization of the excess water capacity of hydroelectric plants and by industrial plants which require large quantities of processed steam, e.g., paper mills.

RESISTOR OVENS

The two types of resistor ovens are convection and radiation.

53. An oven is a low-temperature heating chamber with provision for ventilation.

54. Oven service is drying and baking. Drying is the removal of liquids by evaporation. The rate of drying is limited by the rate at which the liquid will move from the interior of the mass. This is a limitation on the power input. The effect of a too high power value is skin drying.

55. Baking is a hardening process which involves chemical and physical changes in the material. If a material contains a solvent, much of that component is removed during the first stage of baking. Air drying preliminary to baking may be advantageous.

The technique of baking is a correlation of temperature and time. The general rule is: The higher the temperature, the shorter must be the heat cycle. The temperature-time relation for each material is individual and must be determined by trial.

56. The temperature ranges for baking some materials are:

Insulating materials, organic	65–150°C (150–300°F)
Silicone resins	150–300°C (300–570°F)
Wire enamels	300–500°C (570–930°F)
Solventless varnish	Maximum 100°C (212°F)
Foundry cores	90–260°C (200–500°F)
Food products	90–120°C (200–250°F)

57. Heat Absorption. The chart of Fig. 22-20 is useful in baking. The power values are for sensible heat only.

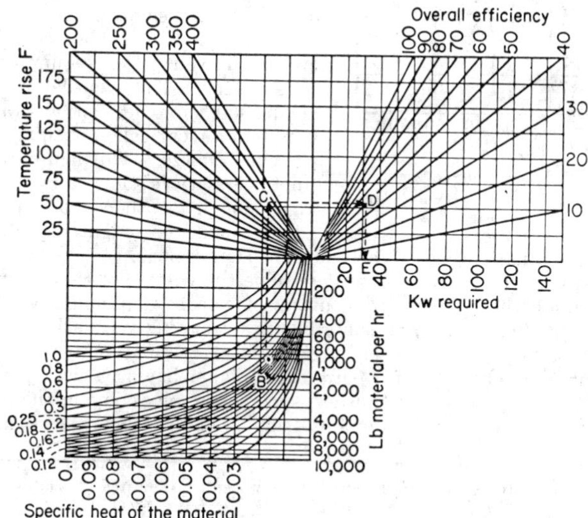

FIG. 22-20. Heat-absorption chart for oven service.

58. Ventilation. As a rule, the ventilation of oven chambers is forced by means of fans. The removal of inflammable vapors is a special procedure. The following paragraphs are quoted by permission from *Pamphlet* 74, Associated Factory Mutual Fire Insurance Companies, Boston, Mass.

"**Oven Ventilation.** The amount of ventilation obtainable by natural draft will vary between very wide limits depending upon oven temperature and atmospheric conditions outdoors. If sufficient ventilation is allowed for safety under the maximum demand, a great deal of heat is wasted at other times. The least ventilation is obtained at the start of the bake, where the need is greatest; and the most ventilation is obtained at baking temperature, when it is least needed.

"Attempts to correct the short-comings of natural draft by the use of manually operated dampers have many times resulted in their use for the purpose of reducing the time required for heating the oven. This is a dangerous practice and is the cause of many oven explosions. The motor driven blower, preferably direct connected, exhausting only a sufficient volume of air for safety, is a much more satisfactory system. The expense of such an installation, except on the ovens of the very smallest size, is more than offset by the saving due to heating smaller quantities of air.

"**Ventilation in Box-type Ovens.** It has been found that, in typical one hour bakes on sheet-metal parts, practically all the thinner is evaporated at the end of 20 minutes and that the evaporation rate during the period from 4 to 8 minutes after loading is about three times the 20 minute average (see Figure 22-21). In figuring ventilation for intermittent ovens it is proper to take these facts into consideration in order that there will be no time during the baking period when explosive conditions will exist.

"For example, it was found in the case of one oven investigated that during the course of a one hour bake the work lost in weight 9 pounds (equal to 1.5 gal) of thinner. Since the vapor from one gallon of thinner requires approximately 2,300 cubic feet of air to barely keep the mixture below the lower explosive limit, it might at first seem that 1.5 × 2,300 or 3,450 cubic feet of air per hour (58 cubic feet per minute) would be sufficient. Based on the evaporation time and rate mentioned above, the necessary ventilation was (3,450/20) × 3.0 = 518 cubic feet per minute in order to be safe under the worst conditions occurring during the first 8 minutes.

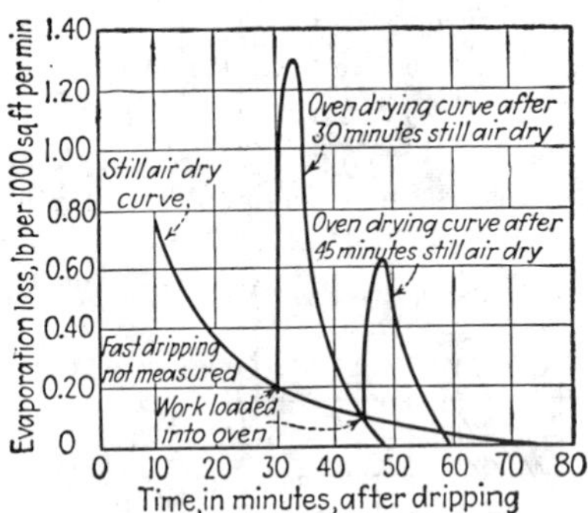

Fig. 22-21. Evaporation curve for typical japan baking cycle. Oven temperature, 325°F.

"The foregoing example is representative of conditions usually found in Japan ovens and consequently can be used as a guide in laying out ventilation requirements for box-type ovens. The ventilation required is about 350 cubic feet of air (at 70°F) per minute for every gallon of thinner evaporated per oven charge. This figure may be higher than necessary in some cases and may not be high enough in others; consequently, expert advice should be sought regarding the minimum allowable ventilation, particularly for new installations.

"Sometimes it is not convenient to determine the volume of thinner used per oven charge by weighing the representative pieces of work. Consequently, the figures in the next table are offered to serve as a guide in estimating ventilation. These figures are taken from tests on a moderately pigmented enamel of average dipping viscosity containing approximately 55% of a petroleum thinner having a flash point of around 110°F. For straight-asphalt-base finishes, such as the black japan, the figures will be slightly less, due to the faster flowing and dripping of these less viscous materials.

Dripping time, min at 70°F	Thinner left on work, gal/1,000 ft	Indicated ventilation, $(ft^3/min)/1{,}000\ ft^2$
10	2.0	700
20	1.5	525
30	1.1	385
40	0.9	320
50	0.7	250
60	0.6	210

"Note: Freshly sprayed work will carry approximately the same amount of thinner as dipped work at the end of a 20-min dripping and air-drying period.

"While the amount of ventilation recommended for box-type ovens will take care of the extreme conditions at work during the first 20 minutes, it is realized that the ventilation will be much more than is necessary for safety after this 20 minute period and would be extravagent to maintain throughout the baking period. Where the size of the oven warrants it, the ventilating equipment may be arranged so that the amount of air supplied can be reduced during the latter part of the bake to the necessary proportion.

"One method of accomplishing this is to provide a ventilating fan to exhaust a constant volume from the oven and an auxiliary fan which will give the required increase of ventilation during the first 20 minutes. Another method of producing the desired results is to use a single fan displacing the calculated amount of air but with the exhaust piping so arranged that a portion of the discharged air can be recirculated automatically after the 20 minute period is over. A third suggestion for cutting down the circulation of fresh air during the latter half of the bake is the automatic operation of a balanced damper by a thermostat which is actuated by the temperature of the gases in the exhaust pipe.

"In practically all cases the time required for the oven to reach baking temperature is more than sufficient to take care of the bulk of the thinner evaporation. Consequently, if the thermostatic mechanism is adjusted to partly close the exhaust damper or to recirculate the gases when the exhaust temperature reaches a point almost up to baking, a decided improvement in heating economy should result. The damper in such an arrangement should be of the balanced type, counterweighted to give full ventilation if at anytime the thermostat control should fail.

"**Ventilation in Conveyor-type Ovens.** In continuous ovens of the conveyor type, the solvent evaporation takes place at nearly constant rate, and, therefore, it is necessary to allow only sufficient ventilation to give a good margin of safety over the exact requirements necessary to prevent an explosive mixture. However, when calculating ventilation for conveyor ovens it should not be forgotten that air which has been drawn through the baking compartment and then out through the centrally located exhaust will not be effective in reducing flammable vapor concentration in the draining end of the oven. The amount of air required to dilute the vapor from one gallon of evaporated thinner to the minimum explosive condition is approximately 2,300 cubic feet, but, on account of the fact that a considerable portion of the air may pass through the oven without traversing the zone in which thinner vapors are given off and also on account of the lack of uniform distribution of the ventilation air currents, it is not safe to allow less than four times this amount or 9,000 cubic feet (at 70°F) for each gallon of evaporated thinner."

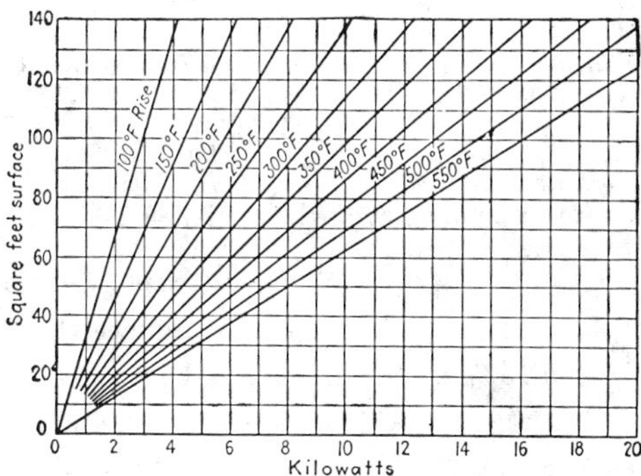

FIG. 22-22. Heat loss through oven walls for 1-in thickness. Divide by thickness for approximate losses with other wall thicknesses.

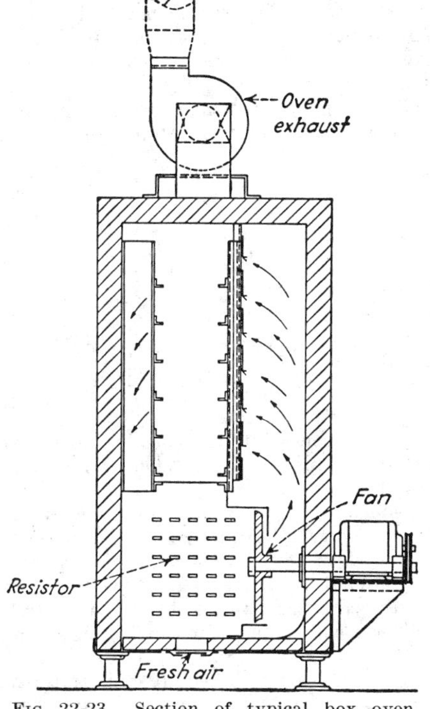

FIG. 22-23. Section of typical box oven with forced-convection heating.

CONVECTION OVENS

59. Convection Ovens. The transfer of heat is by forced convection in gases, mainly air. Steel panels are used for the enclosure of the heating chamber. A double wall with inner heat insulation is general practice. The rates of heat loss through walls of this construction are shown in Fig. 22-22. The heat-storage capacity of these enclosures is small.

60. Box ovens—batch or stationary heating—are of two designs, viz.:

a. Heating chambers with hooks, racks, or shelves for the support of the charge.

b. Heating chambers arranged to receive drawers, trucks, or similar removable supports for the charge.

The section drawing for Fig. 22-23 illustrates a simple box oven arranged for forced-convection heating, recirculation, and forced ventilation.

Example 11. Calculation for box oven.

Inside dimensions, width 2 ft, height 3 ft, depth 3 ft.

Charge 200 lb steel parts.

Heat cycle 45 min.

Ventilation, four changes per charge, 72 ft^3 of air; weight at 70°F, 5.76 lb.

Initial temperature 70°F; end temperature 250°F.

Heat absorption, see Eq. (22-14).

a. Steel

$$q = \frac{200 \times 0.12 \times (250 - 70)}{3{,}412}$$
$$= 1.27 \text{ kWh}$$

b. Air

$$q = \frac{5.76 \times 0.237 \times (250 - 70)}{3{,}412}$$
$$= 0.072 \text{ kWh}$$

Total 1.34 kWh.
Corresponding average power 1.80 kW.
Losses:

Outer surface of oven, 2-in wall = 54 ft^2

Surface (Fig. 22-22) = 1.45 kW

Add 15% for door loss and leakage and about 25% for temperature regulation.

$$\text{kW rate} = (1.80 + 1.45) \times 1.40 = 4.5 \text{ kW}$$

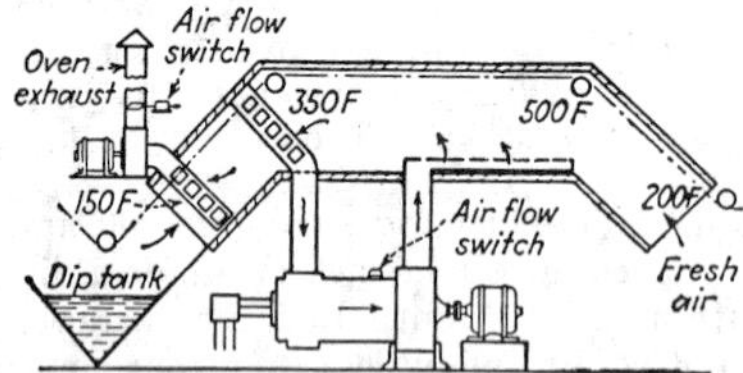

Fig. 22-24. A form of continuous convection oven.

61. Continuous, or air, ovens are flexible in design. The structure may be horizontal, with or without inclined entrance and exit, or vertical. The travel of the charge may be one-way or a U-shape movement. The latter permits the absorption of heat from the outgoing material by the incoming material. The chain conveyer is applicable generally. An arrangement of a continuous-convection oven is shown in Fig. 22-24.

62. Operating efficiencies of convection ovens vary widely. Representative values for baking common finishes are:

Oven	Lb/kWh
Box:	
Rack type	6–8
Truck type	10–12
Semicontinuous	10–18
Continuous	25–3

RADIATION OVENS

63. Radiation Ovens.[1] The transfer of heat is by radiation. The heating units are tungsten-filament lamps (heating lamps) with self-contained or external reflectors.

64. Standard heat lamps are 125, 250, 375, 500, and 1,000 W, 115 V. The self-contained reflector lamps of the first three ratings are most generally used. The filament temperature around 2,500 is low enough to give a comparatively long life, average 5,000 h or more. The large part of the radiation from filament lamps is in the infrared region of wavelengths; hence the term "infrared heating."

65. Construction of the radiation oven is similar to that of the convection oven. The main difference is the detail incident to the lamp-type heating units. The general arrangement is a bank of heat lamps which surrounds or partially surrounds the charge. In a continuous oven this bank forms a tunnel through which the charge travels. This type of oven is frequently built with metal-sheath heaters as a source of infrared energy. High temperatures are now available through the use of quartz-infrared units, used particularly in higher wattage ratings.

66. General service is drying and baking processes which require temperatures below 315°C (600°F). The method is best adapted to continuous horizontal ovens for heating charges, which present a large surface area in proportion to the mass, e.g., sheet metal, textiles, paper, some plastics, etc., and require only surface heating.

[1] Superior numbers in this section refer to the Bibliography, Par. **241**.

Table 22-8. Typical Power Densities of Radiation Heating

Lamp wattage	Spacing of lamps, in	Watts/in²
At 120 V:		
250	12 12	1.094
250	9 9	2.094
250	6 6	5.062
375	9 9	3.187
375	6 6	7.125
500	12 12	2.187
1,000	12 12	4.37
At 240 V:		
1,600T3	12 between ctrs.	4.82
1,600T3	6 between ctrs.	9.45
3,800T3	12 between ctrs.	1.30
3,800T3	6 between ctrs.	2.85*

* Higher voltages permit operation at a higher power density.

There is a trend today to go to higher oven temperatures and higher processing temperatures by utilizing the high-temperature quartz-infrared units that are now available.

67. Absorption coefficients of the surface of charges (see Par. **8**) should be reasonably high. Correspondingly, the inner wall surface of the heating chamber should have a low absorption coefficient. Aluminum linings are used in many cases.

68. Concentration of radiation (power density) on the charge surface depends on the wattage and spacing of the lamps; the efficiency of the lamp reflector units; and, to a limited extent (within the range of practice), the distance between the lamps and the surface of the charge. The concentration of typical arrangements is given in Table 22-8.

69. Charge temperature that can be obtained depends on:

a. Time of exposure.
b. Power concentration.
c. Absorption coefficient of surface of charge.

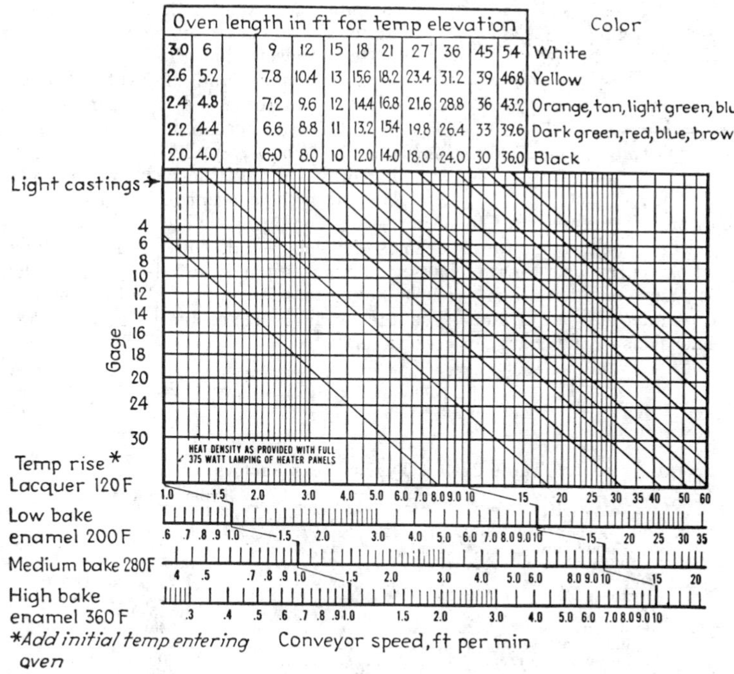

FIG. 22-25. Selection chart for continuous radiation ovens.

d. Mass and thermal properties of the charge.

e. Rate of heat loss from the charge by reradiation and by convection as governed by the interior wall temperature, in velocity, and the temperature of the atmosphere in the heating chamber.

The effect of these factors on the continuous heating of steel is illustrated by Fig. 22-25. This chart is based on a typical oven construction—the power density is 7 W/in² and the interior wall and air temperatures are approximately 80% of the specified charge temperature. Processes that require holding the charge at the end temperature for a specified period of time will require an additional length of oven in direct relation to the conveyer speed.

The chart of Fig. 22-25 does not take into account the heat required for solvent evaporation, which is usually negligible, or the heat necessary for the evaporation of water if present on or in the charge. If evaporation of water is required, the temperature of the charge usually will not exceed 160°F as long as free evaporation takes place.

70. In tall ovens, air stratification may be expected unless provision for air circulation is provided. Air circulation is particularly important for drying ovens. Heat locks or air curtains should be provided at entrance and exit openings of ovens for temperatures above 250°F. The byproduct convection heat is thus preserved.

71. Spot Heating. In addition to oven service there are many special applications of radiation heating in open air. An example is spot or localized heating for various purposes.

DIELECTRIC HEATING

72. Dielectric Heating. The term relates to dielectric loss (see Sec. **4**). The method is used for heating to moderate temperatures certain materials that have low thermal conductivity, e.g., rubber for vulcanizing, wood for drying and setting glue in plywood manufacturing, textiles for drying, and plastics for softening, and for core baking in foundries.

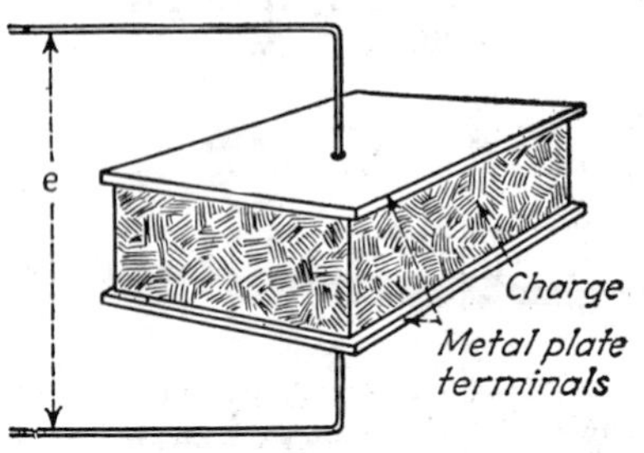

FIG. 22-26. Assembly for dielectric heating.

73. Construction. The simplest assembly is indicated by Fig. 22-26. A slab of the material to be heated is placed between metal plates or electrodes. The equivalent circuit of this assembly and the explanation thereof are given in Sec. **4**. The opening paragraphs of that section, pertaining to insulating materials, are equally applicable to dielectric heating provided that the viewpoint is shifted to heat from the concept of insulation.

74. Power. If the material to be heated is homogeneous and the alternating (or varying) electric field is uniform, heat is developed uniformly and simultaneously throughout the entire mass of the charge. The rate of this heat development depends on (*a*) the nature of the material, (*b*) the field intensity, and (*c*) the frequency. This rate, with parallel-plate electrodes in contact with the charge in a sine-wave form, is:

For dimensions of the charge in centimeters,

$$P\ \text{cm}^3 = \frac{1}{1.80} f \left(\frac{e}{t}\right)^2 k \cos\varphi \times 10^{-12} \qquad (\text{W/cm}^3) \qquad (22\text{-}19a)$$

Table 22-9. Typical Loss Factors

Material	Typical Values of Loss Factors
Spruce wood, dry	0.011
6% moisture	0.175
10% moisture	0.29
Fiber	0.26
Textolite	0.62
Hard rubber	0.015
Porcelain	0.044
Cellulose acetate	0.25
Cellulose nitrate	0.50
Phenol formaldehyde	0.45
Urea formaldehyde	0.16

For dimensions of the charge in inches,

$$P \text{ in}^3 = 1.41 f \left(\frac{e}{t}\right)^2 k \cos \varphi \times 10^{-12} \qquad (\text{W/in}^3) \qquad (22\text{-}19b)$$

where f = frequency, e = applied voltage, t = distance between electrodes (thickness of the charge), k = dielectric constant of the material, $\cos \varphi$ = power factor of the material (see Fig. 22-27).

75. Values of loss factors vary as widely as the nature of the materials. Furthermore, the loss factor of the material is affected by both frequency and temperature. Some typical values of loss factors are given in Table 22-9. In general, the better the insulating property of the material, the lower its loss factor. The best result in each case is obtained by predetermination of the loss factors with different frequencies over the temperature range under consideration.

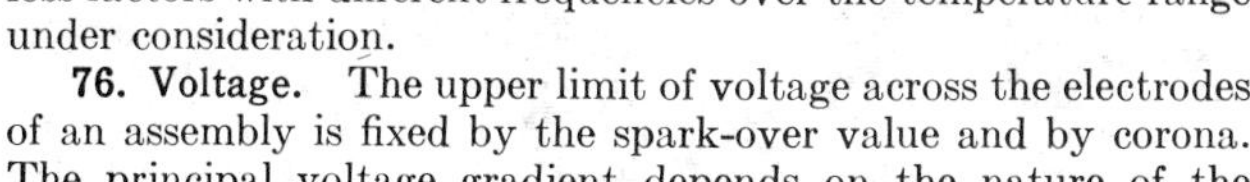

FIG. 22-27. Phase diagram of the dielectric heating circuit.

76. Voltage. The upper limit of voltage across the electrodes of an assembly is fixed by the spark-over value and by corona. The principal voltage gradient depends on the nature of the material of the charge. Values within the range of 2,000 to 5,000 V/in² are found in practice. As a rule the voltage across the electrodes should not exceed 15,000.

77. Frequency. The general rule is that the higher the frequency the lower the voltage required for a given rate of heat development. This rule is modified by the effect of frequency on the loss factor. This effect alone is usually not sufficient to be a determining factor in the selection of frequency for a given material. However, the variations of the loss factor of different materials necessitate the selection of frequency with reference to each material.

The relations expressed by Eq. (22-19) are such that frequencies required for dielectric heating are within the megacycle range, usually 2 to 40 Mc. The most general practice is within the range 10 to 30 Mc. Today there is a trend toward using frequencies of the order of 10 Gc/s. Tubes are at present available in suitable ratings for operation in this frequency range, and larger power ratings will soon be available

In using frequencies in this area special consideration is given to the electrode employed. This may be in the form of a waveguide or a cavity.

The upper technical limit on frequency, charge alone being considered, is the acceptable variation of voltage over the electrode area. (Nonuniformity of this voltage distribution means a corresponding nonuniformity of heat development in the charge.) For a given electrode-surface dimension, the higher the frequency the greater the probability of an undesirable nonuniform voltage distribution due to the standing-wave effect. Some voltage variation is permissible; 10% is deemed reasonable. That value is the basis of Fig. 22-28.

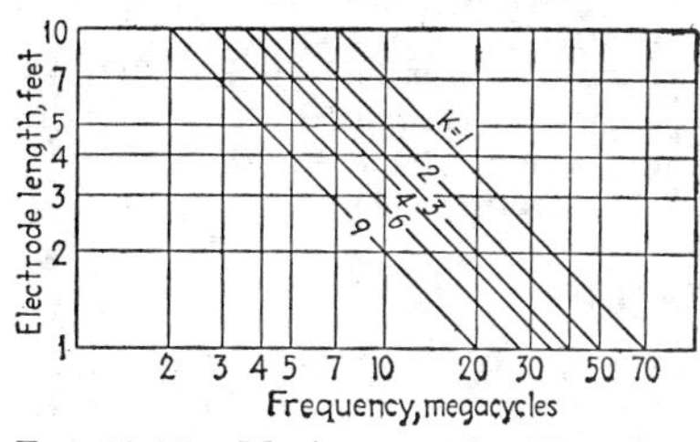

FIG. 22-28. Maximum tolerable electrode lengths for 10% voltage variation. Connection to the shorter edge of electrode.

78. Charge Capacitance

$$c = \frac{8.85 \times k \times A}{t} \times 10^{-14} \qquad (\text{farads}) \qquad (22\text{-}20)$$

where A = surface area of the electrode in square centimeters and t = thickness of charge in centimeters.

79. Power Supply. The common practice is the use of the self-excited oscillator (see Sec. **28**) for each assembly. The output of an oscillator tube decreases with increase of frequency. Therefore, in the choice of frequency it may be necessary to consider magnetrons and klystrons to operate efficiently in the gigacycle range.

For the charge to absorb energy, the circuit must be tuned to approximately resonance. This involves the capacitance of the charge [see Eq. (22-20)]. Inasmuch as the charge capacitance is inversely proportional to the charge thickness, that dimension is a factor in the selection of frequency. Thus, in general, the highest frequencies are used with thin charges.

80. Electrodes. The dimensions of the electrodes have been referred to in Par. **77**. Flat metal electrodes in contact with the charge can be used if the material is reasonably homogeneous and of uniform thickness. Contact is not necessary, but separation increases the voltage required. If the thickness of the charge is not uniform, a corresponding nonuniform spacing of the electrodes is required; this is indicated in Fig. 22-29. Otherwise there will be nonuniformity of heat development in the charge.

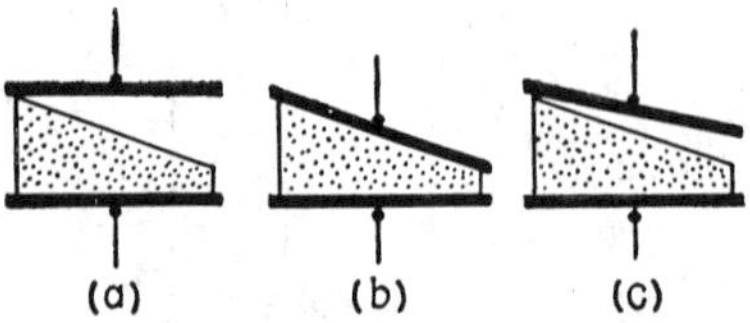

(a) (b) (c)

FIG. 22-29. Electrode spacings for charges of uneven thicknesses (see Taylor in Bibliography, Par. **241**); (*a*) and (*b*) incorrect, (*c*) correct.

If the electrodes are in contact with the charge, there is a flow of heat from the charge to the cold electrodes. The result is a temperature gradient in the charge adjacent to the electrodes. A layer of heat insulation between each electrode and the charge may be advisable, particularly if the electrodes are massive. An air gap, where this can be used, serves the same purpose.

In continuous heating, an air gap between one of the electrodes and the charge or between each of the electrodes and the charge, according to operating conditions, is necessary.

81. Shielding. Interference with communication channels may necessitate the shielding of dielectric equipment. Reference to rulings of the Federal Communications Commission should be made. Safety of operating personnel is the main consideration.

82. Time. The length of the heat cycle is determined by consideration of:

a. Usable values of voltage and frequency.

b. Physical and chemical conditions pertaining to the heat process.

c. Operating conditions.

d. Economics.

The thermal conductivity of the material of the charge is a negligible factor. As a rule, the heat cycle is a comparatively short period.

83. Efficiency. The heat absorption of the charge is calculated by Eq. (22-14). Some allowance for heat loss by dissipation from charge surfaces must be made; usually 10 to 15% is a reasonable estimate. To this loss must be added the loss of the individual electrical apparatus. An overall efficiency of 50% is representative.

INDUCTION HEATING

84. Induction Heating. The material to be heated is exposed to an alternating (or varying) magnetic flux. The material to be heated must be a conductor so that currents are thus induced in the material (or load) and flow wholly within its mass. Generally this type of heating is produced by an electromagnetic field caused by an alternating-current source. It is possible, however, to produce similar results, but at lower current densities, with a high-flux permanent magnet moved at high speed across conductive materials. The term "eddy-current heating" is sometimes applied to the method.

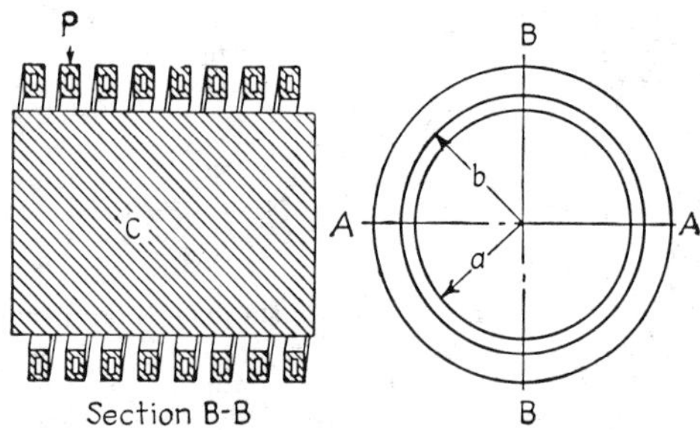

FIG. 22-30. Assembly for induction heating.

85. Assemblies. A typical assembly is shown in Fig. 22-30. In all cases the assembly is a form of transformer. The coil around, or adjacent to, the charge forms the primary circuit. The secondary circuit is the path of the current in the load. Figure 22-30 shows, in section *B-B*, *P* the coil and *C* the load. Note that the ID of the coil is larger than the OD of the load. These two items should not touch. General practice calls for the use of an insulator between the coil and the load. The insulator may be air, but it can be a refractory or other nonconductor, depending on the temperatures and other factors involved in the application.

86. Shape of Load. The ideal load shape is a cylinder. It may be a solid, as in Fig. 22-30, or a tube. Shape alone does not affect the principle involved. It may introduce some complications in geometry. Any shape load can be heated by the inductive method, but not always to obtain the desired results. Special techniques are required for some unusual shapes.

87. Basic Relations. For a cylindrical load and other loads where the coil encompasses the load, the directions of the magnetic flux are parallel to the longitudinal axis of the load and hence to its lateral surface. The factors which, for a given intensity of longitudinal magnetic flux at the lateral surface of the load, determine the rate of heat development in the load are:

a. The electric and magnetic properties of the material.

b. The frequency.

c. The radius, or one-half the thickness of the charge.

The relations of these factors are given in Eqs. (22-21) to (22-23), which follow.

88. Reference Depth (see Fig. 22-31)

$$\text{Depth} = \sqrt{\frac{\rho \times 10^9}{2\pi\sqrt{2}}} \times \sqrt{\mu f} \qquad \text{(cm)} \qquad (22\text{-}21)$$

or

$$D = 3{,}560 \times \sqrt{\frac{\rho}{\mu f}} \qquad \text{(cm)} \qquad (22\text{-}21a)$$

where D = reference depth in centimeters; ρ = resistivity of the material in ohm-centimeters cubed; μ = permeability of the material. For nonmagnetic materials, μ = 1.0; f = frequency in cycles per second.

This reference depth may also be found in inches by a similar, short formula:

$$D = 3{,}160 \times \sqrt{\frac{\rho}{\mu f}} \qquad \text{(in)} \qquad (22\text{-}21b)$$

where D = reference depth in inches; ρ = resistivity in ohm inches; μ = permeability; f = frequency in cycles per second.

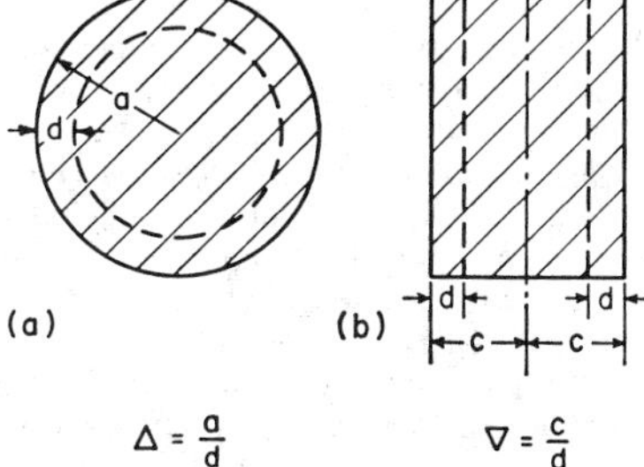

FIG. 22-31. Index ratio diagrams.

89. Index Ratio (see Fig. 22-31). This ratio is expressed as

Radius = r over reference depth d.

a. For a cylindrical charge

$$\Delta = \frac{r}{d} \qquad (22\text{-}22)$$

where the radius is in the same units of measure as the reference depth. Ideally $\Delta = 4$.

b. For a rectangular section

$$\Delta = \frac{c}{d} \qquad (22\text{-}23)$$

where c = one-half the thickness of the charge at its smallest section and d = reference depth [see Eq. (22-21)]. Ideally, $\Delta = 4$.

When the index ratio is less than 2, efficiency can be improved by increasing the frequency. If greater than 4, it may be desirable to decrease the frequency.

90. Heat Concentration. In considering the solid load illustrated in Fig. 22-30, it is necessary to visualize the current subject to skin effect induced to flow in circular paths parallel to the circumference of the load. Each circuit possesses inductance, and there is a subsequent progressive decrease in current strength radially. The rate of decrease changes with frequency.

The current distribution results in a concentration of heat development in the circumferential zone of the charge, as indicated for a given charge and the frequencies in Figs. 22-32 and 22-33. The data for Fig. 22-33 are the same except that the frequency is four times that of Fig. 22-32. The ordinate at radius $r = d/2$ is 100%.

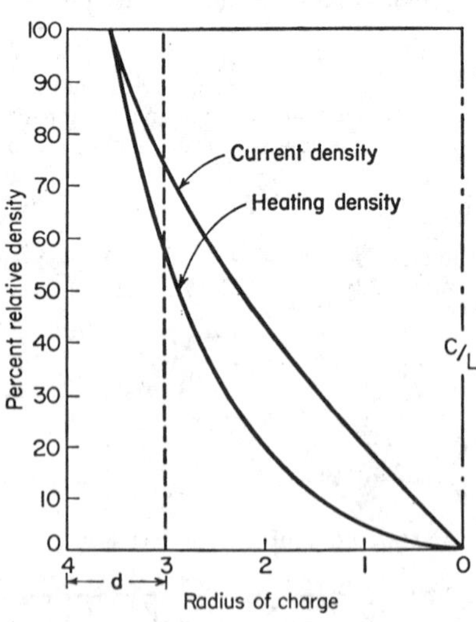

Fig. 22-32. Heat concentration in a solid cylindrical charge. Index ratio is 4.0.

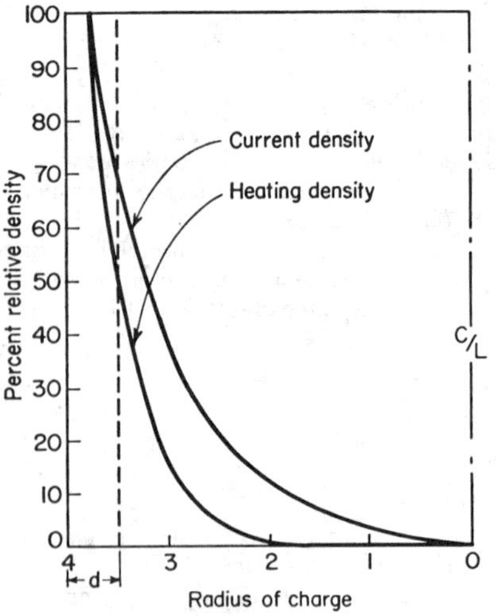

Fig. 22-33. Heat concentration in a solid cylindrical charge. Index ratio is 8.0.

Figure 22-32 and 22-33 show the relative relations of heat and current density for a given frequency. Time has not been considered, but it is extremely important. The curves are typical of high-density short-time conditions. With longer times the heat density will conduct toward the center, showing a heat-density curve to the right of the current-density curve.

91. The Rate of Heat Development. The general equation for the rate of heat development by symmetrical currents in a long load (length much greater than its diameter) of any shape for a longitudinal and sinusoidal alternating magnetic flux at its lateral surface is

$$P = H^2 sL \times \sqrt{\mu f} \times \sqrt{\rho_a \times 10^9} \times 10^{-7} \qquad \text{(W)} \qquad (22\text{-}24)$$

$$H = \frac{4\pi N \times \sqrt{2}i}{10L} \qquad \text{(Oe)} \qquad (22\text{-}25)$$

where ρ_a = resistivity of the material of the load in ohm centimeters cubed, μ = permeability of material of the load, s = shape factor of the load, L = length of the load in centimeters, N = number of turns in the coil around the load, i = current in the coil in amperes.

Note: For other than the long assemblies, Eq. (22-24) is multiplied by a constant less than unity to correct for shortness.

92. Shape. Round, square, or rectangular bars or sections can be heated equally well. The frequency used and the cross section to be heated play an important part in the results obtained. The reference-depth consideration is essential. Any shape can be heated efficiently with the proper frequency when the dimensions of the load are at least four times the reference depth.

It is always wise to have a load that is several times longer than its cross-sectional dimensions. This result can often be obtained by stacking or abutting loads together.

The efficiency of the heating of the load decreases if the length is not substantially greater than the cross section.

93. Magnetic Charges. The a-c excitation curve of the magnetic material more or less coincides with the d-c magnetization curve as saturation is approached. Hence, the latter curve (usually more readily available) can be used as a reference in induction heating. For values of H higher than that required for saturation, the value of μ corresponds to the saturation values of p and H as used in Eq. (22-24).

94. Permeability is dependent on frequency up to around 10 kc/s. Above that frequency, permeability decreases to some extent with increase of frequency, but this effect is not well defined.

95. Hysteresis. In induction-heating practice, the rate of heat development by hysteresis in a magnetic material at room temperature is a small percentage of the eddy-current watts. The hysteresis watts decrease rapidly with rise of temperature. Hence, hysteresis is a negligible factor in the inductive method of heating.

96. Primary Coils. Single-layer coils are standard practice. Tubular conductors—round, oval, or rectangular—provide for water cooling of the coil.

The number of turns is selected with reference to the desired relation of amperes and volts for a given service. The restrictions on this choice are the space required for conductor insulation and the requirement of water cooling.

The space factor of the coil is

With round conductor:
$$s = \frac{N\pi h}{4L} \qquad (22\text{-}26)$$

With rectangular conductor:
$$s = \frac{N \times h}{L} \qquad (22\text{-}27)$$

where N = number of turns, h = axial dimension of the conductor, L = length of coil over turns.

The space factor of a coil with an oval conductor must be estimated on the basis of the axial dimension of the conductor.

The radial dimension of a copper-coil conductor on the basis of 80°C (176°F) operating temperature should be

$$r = \frac{11.5}{\sqrt{f}} \quad \text{(cm)} \qquad (22\text{-}28)$$

This dimension for a tubular conductor is the thickness of the inner wall.

The i^2r loss in the coil increases rapidly with the decrease of the radial dimension below the value given by Eq. (22-28). Little is gained by the use of a greater dimension.

If the assumed operating temperature differs from 80°C, the constant in Eq. (22-28) should be multiplied by the ratio

$$\frac{\sqrt{p_b \times 10^3}}{1.47} \qquad (22\text{-}29)$$

where p_b = resistivity of copper at the assumed operating temperature in ohm centimeters cubed.

97. Air-gap ratio is the ratio b/a (Fig. 22-30), or the ratio of the corresponding dimensions of the assemblies with other than cylindrical charges.

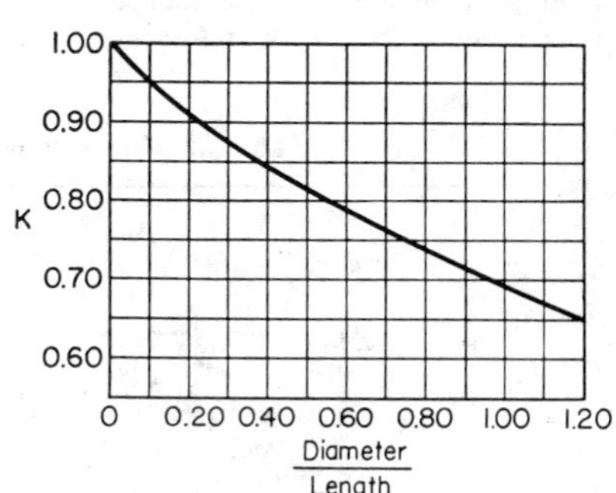

FIG. 22-34. The constant K in Eq. (22-31).

98. Circuit Equations. The assumptions of the equations of the circuits of an assembly that follow are:

a. A solid cylindrical charge of a nonmagnetic material.
b. A single-layer primary coil with its conductor in accordance with Eq. (22-28).
c. Length of charge equal to length of primary coil.
d. A sinusoidal alternating current.

Primary resistance

$$r_b = \frac{1.83 \times \sqrt{f} N^2 b \times 10^{-6}}{\sqrt{sL}} \qquad \text{(ohms)} \qquad (22\text{-}30)$$

where b = the internal radius and L = length, both in centimeters, and s = space factor of the primary coil.

With a full-length charge in place, the a-c resistance of the coil is approximately 91% of the value given by Eq. (22-30).

For a conductor temperature other than 80°C multiply Eq. (22-30) by the ratio of Eq. (22-29).

Primary reactance,

$$x_b = \frac{8\pi^3 f N^2 (b')^2 K}{1} \times 10^{-9} \qquad \text{(ohms)} \qquad (22\text{-}31)$$

$$b' = b + \frac{5}{\sqrt{f}} \qquad \text{(cm)} \qquad (22\text{-}32)$$

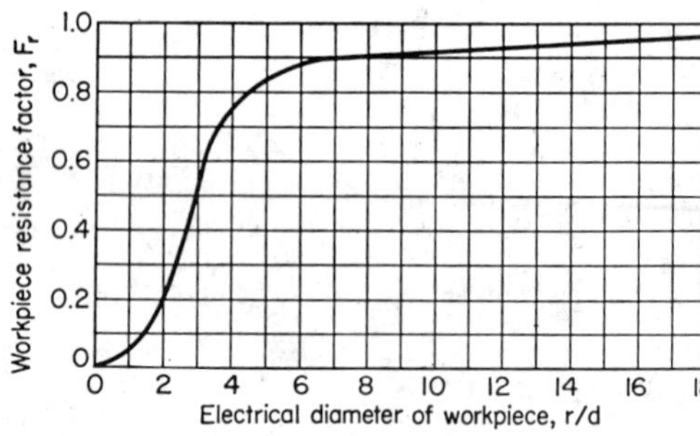

Fig. 22-35. Resistance factor F_r.

where K = a constant for the proportion of the coil from Fig. 22-34 (for a complete table of these constants see Table 10, Circ. C74 of the National Bureau of Standards). The value of this constant is based on the ratio $2b'/l$.

Secondary resistance,

$$r_a = \frac{2\pi p_a N^2}{l} \Delta F_r \qquad \text{(ohms)} \qquad (22\text{-}33)$$

where r_a = the resistance of the load, p_a = the resistivity of the load, N = the number of turns in the coil, l = the length of the coil parallel to the axis of the load, Δ = the index ratio, and F_r = the resistance constant shown in Fig. 22-35.

Table 22-10. Correction Y_r for Eq. (22-33)

Length/dia.	Air-gap ratios			
	1.10	1.30	1.50	1.70
0.50	0.67	0.46	0.35	0.30
1.0	0.73	0.55	0.47	0.43
2.0	0.77	0.63	0.59	0.57
3.0	0.83	0.76	0.72	0.71
4.0	0.88	0.83	0.79	0.78
5.0	0.91	0.87	0.84	0.83

Table 22-11. Correction Y_x for Eq. (22-34)

Length/dia.	Air-gap ratios			
	1.10	1.30	1.50	1.70
0.50	0.36	0.24	0.19	0.16
1.0	0.52	0.40	0.33	0.30
2.0	0.64	0.56	0.51	0.49
3.0	0.73	0.68	0.65	0.63
4.0	0.80	0.75	0.72	0.70
5.0	0.84	0.80	0.77	0.76

Secondary reactance,

$$x_a = \frac{8\pi^3 f N^2 a^2 Fx}{l} \times 10^{-9} \quad \text{(ohms)} \tag{22-34}$$

$$F_x = \text{function}\left(1 - \frac{2\theta}{\Delta}\right) \tag{22-35}$$

$$\theta = \frac{(ber\ \Delta\ bei') - (bei\ \Delta\ ber')}{(ber\ \Delta)^2 - (bei\ \Delta)^2} \tag{22-36}$$

Values of θ are given in Fig. 22-36.

99. Shortness Correction. Equations (22-33) and (22-34) are based on the long-charge proportion, i.e., a length much greater than the diameter. The term cannot be made definite, because its application varies with the air-gap ratio of the load. However, in general, Eqs. (22-33) and (22-34) must be corrected for charge shortness if the ratio of the length and diameter is less than 6.0.

Representative values of these corrections—Y_r for Eq. (22-33) and Y_x for Eq. (22-34)—are given in Tables 22-10 and 22-11.

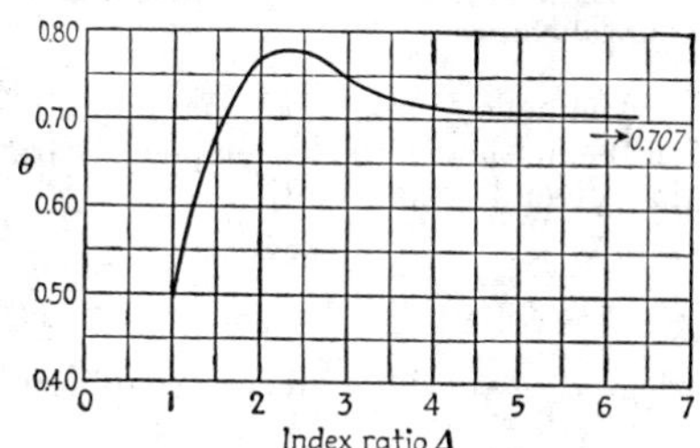

Fig. 22-36. The factor θ in Eq. (22-35).

100. Effective Values

Resistance:

$$R = r_b + r_a \quad \text{(ohms)} \tag{22-37}$$

Reactance (for long-coil assembly):

$$X = \frac{8\pi^3 f N^2 X(b^2 - a^2)}{l} \times 10^{-9} + \frac{r_a\theta}{f_r} \quad \text{(ohms)} \tag{22-38}$$

Impedance:

$$Z = (R^2 + X^2)^{1/2} \quad \text{(ohms)} \tag{22-39}$$

Power factor:

$$\cos\varphi = \frac{R}{Z} \tag{22-40}$$

Electrical efficiency:

$$N = \frac{r_a}{r_b + r_a} = \frac{r_a}{R} \tag{22-41}$$

Power:

$$P = \frac{e^2 R}{Z^2} \quad \text{(watts)} \tag{22-42}$$

where e = applied voltage to the coil; r_a = resistance of the load; r_b = resistance of the coil; in all the foregoing b = internal diameter of the coil.

101. Power Density. Equation (22-42) divided by the area of the lateral surface of the load gives the gross power density. That value multiplied by the electrical efficiency is the net, or effective, power density.

102. Frequency. The technical criterion for the selection of frequency for induction heating—other than for metal-melting service, to be considered later— is the electrical efficiency of the assembly.

The relations of electrical efficiency for the two basic shapes of charge are shown in Figs. 22-37 and 22-38.

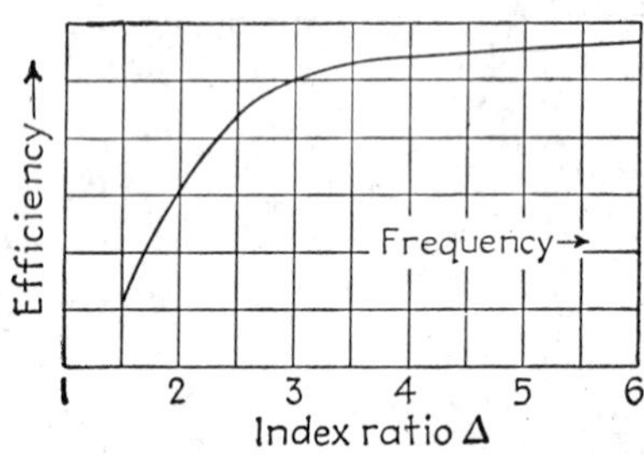

FIG. 22-37. Frequency-efficiency relations of an assembly with a solid cylindrical charge.

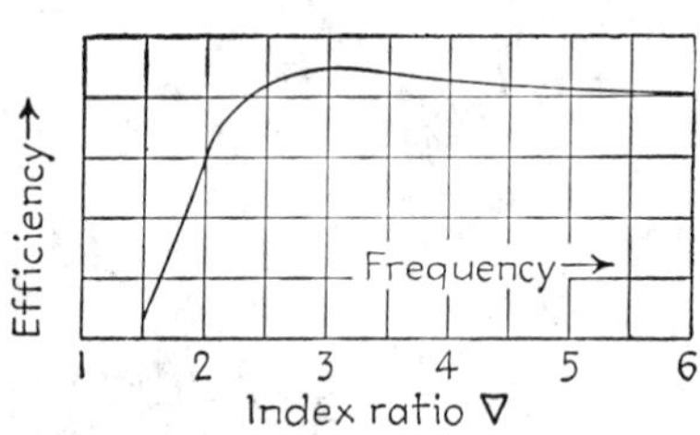

FIG. 22-38. Frequency-efficiency relations of an assembly that has a rectangular section.

The hollow cylindrical charge (tube or pipe) is a special case. The maximum electrical efficiency with this shape of charge is obtained with a frequency somewhat higher than given by the equation

$$f_0 = \frac{44 \times pa \times 10^6}{a(a - t)} \qquad \text{(c/s)} \qquad (22\text{-}43)$$

where a = outside radius in centimeters, t = wall thickness in centimeters, pa = resistivity of the material at the temperature specified for the heat application, in ohm centimeters cubed.

The shape of the efficiency-frequency characteristic of an assembly with a hollow charge is the same as that of the charge with a rectangular section (see Fig. 22-38).

NOTE: If the wall thickness varies, the minimum thickness should be used in Eq. (22-43).

103. Standard Frequencies. Economics usually dictates the use of a standard frequency nearest the technical choice. The standard frequencies (so-called because of usage) are: line-frequency equipment; magnetic and solid-state multipliers at

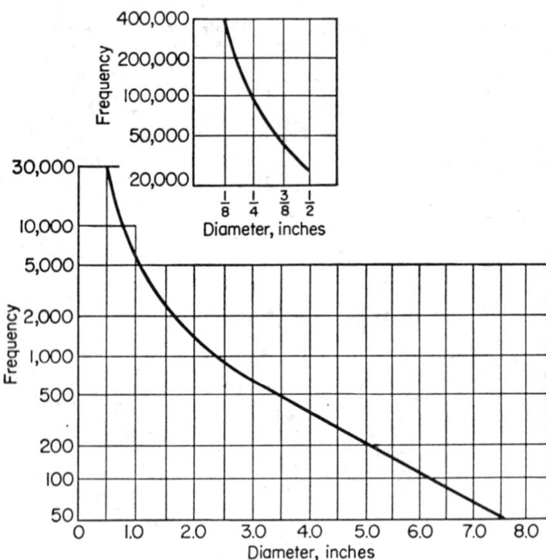

FIG. 22-39. Frequency-dimension relation of solid cylindrical charges. Charge resistivity 125 μΩ·cm³.

frequencies of 180 c, rotating apparatus at frequencies of 180, 360, 400, 960, 3,000, and 9,600 c (based on a 60-c drive); solid-state multipliers, operating at 1,500 and 3,000 c; spark-gap apparatus, 20 to 40 kc/s; electronic apparatus, around 300 to 450 kc/s, and with some frequencies going to 4 and 8 Mc/s.

104. Selection of Frequency. The conditions outlined in the preceding paragraphs are usually met as follows.

Solid Cylindrical Charge. A standard frequency to give a rated index ratio within the range 2.50 to 6.0. The upper limit is preferred, particularly for short assemblies.

Charge with Rectangular Section. A standard frequency to give a rated index ratio within the range 1.70 to 5.0. For short assemblies the rated index ratio should be near 3.0.

Hollow Cylindrical Charge. A standard frequency near the frequency given by Eq. (22-43).

The graphs of Fig. 22-39 show reference values of frequency and the corresponding diameters of solid cylindrical charges on the basis of the charge resistivity 125 $\mu\Omega \cdot cm^3$ (steel at forging temperature). Unit permeability; rated index ratio 2.50.

The graphs of Fig. 22-40 are for the data of Fig. 22-39, with the exception that the charge resistivity is 10 $\mu\Omega \cdot cm^3$ (aluminum at forging temperature).

In the case of a magnetic material heated to a temperature above the Curie point, the initial boosting effect of the magnetic property on the rate of heat development can be taken into account in the selection of frequency. For example, 9,600 c with a good coil design is effective for carbon-steel charges down to about ½ in in diameter.

105. Plant Economics. The economic factors of plant operation, aside from the use of standard frequency, relate to the diversity of charges. Generally the standard frequency that is effective for the entire range of work is desirable. An example from practice is the heating of steel bar stock in sizes ranging from ½ to 3 in with 9,600 c.

The investment charge for electrical apparatus is a matter for individual study in each case. For example, the use of electronic apparatus may lead to the use of a frequency much higher than the technical choice.

106. Methods of Heating. Temperatures below the melting points of metals:

a. Mass heating, e.g., for forging and annealing.

b. Surface or localized heating, e.g., heating steel for hardening.

107. Heating Rates. As a rule, metal charges can be heated faster than is sanctioned by either metallurgy or economics. The best practice is the heating rate at which the flow of parts is continuous during the working period. The heat cycles of the inductive method are short, being measured in seconds or minutes according to the class of service and scale of operation.

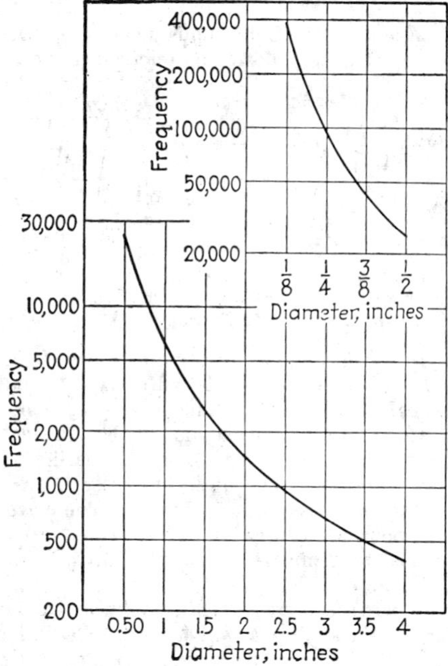

Fig. 22-40. Frequency-dimension relation of solid cylindrical charges. Charge resistivity 125 $\mu\Omega \cdot cm^3$. Index ratio 2.50.

108. Overall Efficiency. The number of pounds of material heated per hour varies with the material, temperature, rate of heating, electrical efficiency of the coil, and efficiency of the special electrical apparatus. A reference value for steel heated for forging with 9,600 c and below is 4 to 6 lb/kWh.

Example 12. Solid cylindrical charge, SAE 1045 steel.

Diameter 1 in; length 11 in; $N = 75$; $F = 9{,}600$ c; air-gap ratio, 1.50 (measured values).

Resistance of primary coil, empty, 0.115 Ω.

Reactance of primary coil, 1.98 Ω.

$$p_a = 121\ \mu\Omega\cdot\text{cm}^3$$
$$\mu = 1.0$$
$$\text{Ohms, rated index ratio} = 3.175$$
$$r_a = 0.21\ (0.235)$$
$$R = 0.315\ (0.34)$$
$$Z = 1.58\ (1.70)$$
$$\text{Power factor} = 20\%\ (20\%)$$
$$\text{Electrical efficiency} = 68\%\ (67\%)$$

Example 13. Solid cylindrical steel charge.

Diameter 2 in; length 12¼ in; weight 11⅞ lb. Heated to 1427°C (2600°F) in 110 s.

Frequency = 9,600 c
Rated index ratio = 5.80
Power average = 70 kW
Heat content (Fig. 22-7) = 1,425 Wh
Heat absorption, average rate = 46.6 kW
Heating rate = 388 lb/h
Overall efficiency = 5.5 lb/kWh

Example 14. Continuous heating service. Charge: round bar stock, SAE 1045 steel. Heating followed by quenching and reheating as indicated by Fig. 22-41, 9,600 c.

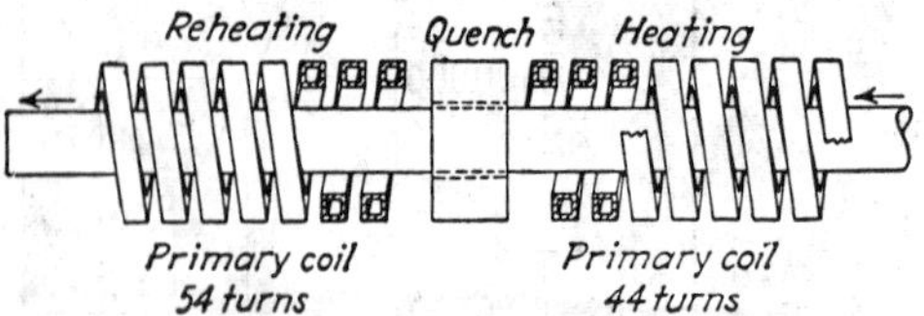

Fig. 22-41. Arrangement of an assembly for the continuous-operation heating, quenching, and reheating of steel bars (Example 14).

Heating station.
End temperature 871°C (1600°F).
Diameter of charge 1 in.
Air-gap ratio 1.75.
Rated index ratio 4.0.
Speed of charge 1.56 ft/min.
Heating rate 250 lb/h.
Power 32 kW.
Overall efficiency 7.8 lb/kWh.

A typical performance of a group of eight of the units in Example 14 in the surface load, with one operator, is 2,000 lb/h of 1-in-diameter SAE 1045 steel bar stock hardened to 64-63 Rockwell C and drawn to 31-30 Rockwell C.

109. Applications. The largest uses of induction heating for temperatures below the melting points of metals are heating steel for hardening and heating metals for forging. The use of this method for heating metal parts inside glass bulbs and tubes for degassing during evacuation is general. A more recent development is its application for making glass-to-metal joints. There are many applications for brazing and soldering, both stationary and continuous service. This type of equipment can also be applied to many welding applications. This will be covered further in the section on welding.

There are many rules of thumb that can be used for quick evaluations or requirements for specific applications. For example, in heat treating, power densities of the order of magnitude of 5 to 10 kW/in² of load surface should be used in general. For heating for forging the power density related to the square inches of load surface should be of the order of magnitude of ½ to 1 kW/in² of surface. A general rule of thumb that gives some indication of the power required for brazing is that 1 kW of energy should be used for each lineal inch of brazed joint. All the above rules of thumb apply to the usual short-cycle induction-heating times.

TERMINAL RESISTANCE HEATING

110. Terminal Resistance Heating. The heating of bodies of uniform cross section, e.g., metal rods, tubes, and wire, in the open air, by the passage of current through the body. The method for stationary work requires clamp- or pressure-type terminals; for continuous work, some form of sliding contact, contact wheels, or rolls suited to the form of the work is needed. The requirement for terminals, together with the low voltage and high current densities, limits the use of the method to comparatively small dimensions. However, there have been several installations in Europe made where bars 2 to 4 in in diameter and 20 m long have been heated in this manner.

Typical uses are: heating rivets, heating bar stock for forging (stationary applications), and the continuous heat-treatment of steel wire.

End-to-end terminal heating is illustrated by Fig. 22-42. The expansion of the piece serves to indicate its temperature and causes the closing of contacts in a control circuit for temperature limitation. This arrangement has been applied to heating steel bars of 1 in cross section and up to 30-ft lengths for air hardening. It also has been applied in conjunction with induction heating for the heat treating of such items as torsion bars where the cross section is not uniform.

An automatic machine for heating bar stock—¼ to 1¼ in diameter in lengths up to 24 in—with photoelectric control is shown in Fig. 22-43. Similar manually operated machines extend the size range to 2¼ in in diameter.

An example of this class of work is the end heating of steel bars, 4- to 6½-in lengths of ¾-in 0.20°C steel, to 2100°F for forging. The time required is 21 s for the 4-in length and 24 s for the 6½-in length.

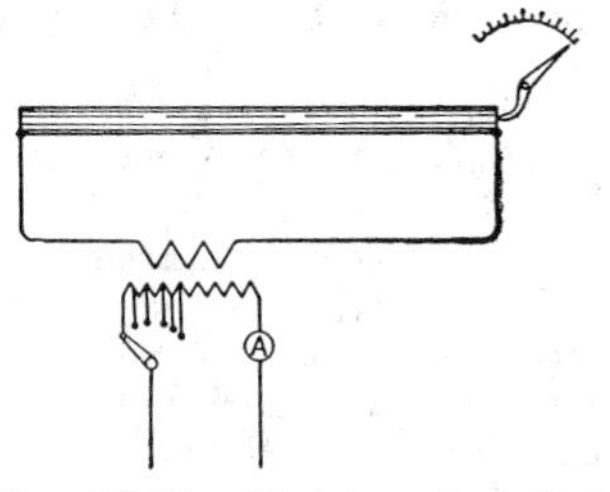

FIG. 22-42. End-to-end method of direct-resistance heating.

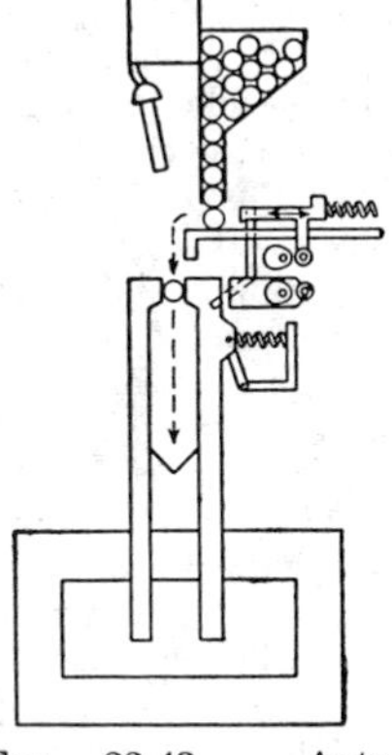

FIG. 22-43. Automatic direct-resistance heating machine.

With clamp terminals the metal near the terminals does not attain the temperature of the main portion of the stock because of the cooling effect of the terminals. The cold ends are not always wasted, depending on the use to be made of the material; e.g., in the air hardening of hacksaw blades it is desired that the ends of the blades be left with normal hardness.

Standard frequencies are employed which within the range of diameters used give uniform current densities and uniform heating with heating periods of 5 min or less.

The heat loss is the sum of the heat flow from the terminals plus the loss from the surface of the work by radiation and natural convection. The operating efficiency for a given metal thus depends on the rate of heating; values obtained in practice range from 60 to 90%.

RESISTOR FURNACES, LEAD BATHS, SALT BATHS

111. Resistor Furnaces. The term is applied arbitrarily to indirect-resistance heating chambers for temperatures of the heat-treat process as applied to steel; above 700°C (1292°F) except for tempering below that temperature.

112. Classification of resistor furnaces is the same as for resistor ovens, i.e., batch furnaces and continuous furnaces.

Standard types of batch furnaces are:

a. Box furnaces. Simple enclosures for manual operation in handling work into and out of the heating chamber.

b. Car-bottom furnaces. The hearth of the furnace is mounted on a car, useful for heavy loads of castings and slabs.

c. Elevator furnace. The heating chamber is elevated. The hearth is mounted on an elevator for floor-level loading and unloading; an effective type for annealing steel sheets and similar charges.

d. Bell furnaces. The heating chamber is a cylindrical hood or bell which is lowered over the charge on a stationary hearth. The design is primarily for annealing coils of steel strip and like charges. One hood can serve in rotation a number of hearths. Dummy insulated hoods replace the chamber for cooling of the charge under cover.

113. Continuous Furnaces. Practically any type of material-conveying mechanism can be incorporated in the design of a resistor furnace. The most common types are noted below:

a. Conveyor furnaces. The moving mechanism is a unit built into the heating chamber. One form is a mesh-belt conveyer (made of a heat-resisting alloy) for light-weight articles. The overhead chain conveyer is adapted to articles that can be hung on hooks, e.g., sheet-metal parts.

b. The rotary hearth furnace is designed for charges which require bottom support during heat-treatment. The particular feature is the return of the charge to a point adjacent to the starting point of the heat cycle.

c. Roller-hearth furnaces have power-driven rolls for conveying heavy parts, castings, and the like, through the heating chamber. The articles may be loaded on trays as vehicles for the travel of the charge.

d. Pusher furnaces are similar to roller-hearth furnaces except that the straight-line movement is obtained by a pushing mechanism.

e. The walking-beam furnace is similar to the two types noted in preceding paragraphs except for the propelling mechanism.

The selection of the type of conveying mechanism for a furnace is a material-handling problem around which the heating chamber, including the method of heating and cooling, is designed.

114. The heating chamber of a resistor furnace is an enclosure with a refractory lining; a surrounding layer of heat insulation; and an outer casing of steel plate, brick, or tile. The hearth of a batch-type furnace is made of a heat-resisting alloy and usually is in sections to prevent warping and to facilitate removal. The conveyer forms the hearth of some types of continuous furnaces; in others, a separate hearth is required.

The inside proportions of the heating chamber are made to suit the character of the charge and the type of furnace. The only space requirement for the source of heat is sufficient room for the desired distribution of the resistor windings, which are carried on supports attached to the refractory lining.

It is general practice to use insulating firebrick—a semirefractory material—for the inner lining of the heating chamber. This material has thermal and physical properties intermediate between those of fire-clay brick and heat-insulating materials. The lining of a semirefractory material has less heat-storage capacity than a fire-clay brick lining, and its use accordingly decreases the time periods of heating and cooling the chamber and also decreases the stored-heat loss for a given cycle of operation. These savings are important mainly in batch furnaces. Other advantages of the semirefractory lining are its heat-insulating value and the decrease in weight of the furnace construction.

The maximum temperature of the inner face of the layer of the insulation of the heating chamber determines the nature of the material required for the insulation. This temperature for furnace service requires an inorganic material and the use of diatomite (the porous siliceous residuum left by the decay of minute marine plants, diatoms) as thermal insulation is general in resistor-furnace construction.

The economic thickness of a layer of heat insulation is determined by the factors of diminishing decrease in the rate of heat loss with increasing thickness of the insulation, cost of energy, and investment of charge of the insulating material in place.

The construction of a heating chamber on a resistor furnace as shown in Fig. 22-44, with a 4½-in-thick semirefractory lining and a layer of heat insulation 9 to 13 in thick, represents general practice for heating-chamber temperatures up to 1150°C (2102°F).

115. Resistors. The major mode of heat transfer of furnace temperatures is radiation. Heat distribution by radiation is obtained by a combination of direct radiation from the resistors and reradiation from the interior boundary surface of the chamber. This is the general guide in the location of the resistors in the chamber to provide for radiation to all surfaces of the charge.

The temperature gradient between the resistor and the heating-chamber space ranges in practice from 50 to 100°C (90 to 180°F) for heating-chamber temperatures of

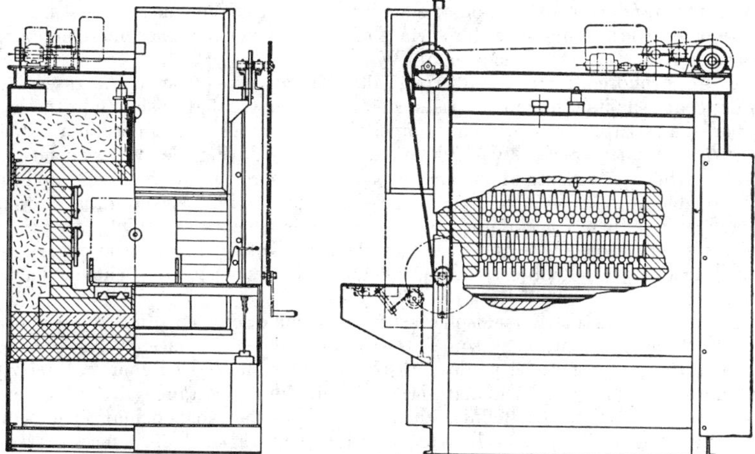

Fig. 22-44. Typical construction of a box-type resistor furnace.

800 to 1200°C (1472 to 2172°F). Correspondingly higher temperature gradients are used for lower chamber temperatures.

General guides in the design of nickel-chromium alloy resistors for the upper values of heating-chamber temperatures are a power density of 10 to 15 W/in² of resistor surface and 1½ to 4 kW/ft² of chamber surface area occupied by the resistor winding.

Example 15. Design of resistor. B82 alloy.

Rating 45 kW, 220 V.

Connection 3-phase, Y-connection.

Arrangement 15 kW on each side wall, 15 kW on hearth.

Temperature of heating chamber 1000°C (1832°F).

Temperature gradient 50°C (90°F).

Resistor temperature 1050°C (1922°F).

Amperes per phase 118.

Resistance per phase 1.08 ohm.

On the assumption of 12 W/in² of resistor surface, the surface area required per phase is 1,250 in².

The surface area of 0.75- by 0.040-in. ribbon is 18.96 in²/ft.

$$\text{Length of ribbon/phase} = \frac{1{,}250}{18.96} = 60 \text{ ft}$$

$$\text{Resistance/phase at operating temperature} = 0.017 \times 1.073 \times 60 = 1.09\ \Omega$$

In the design of a resistor for a given kilowatts rating and to fit a given space, there are a number of choices as to width and thickness of ribbon, spacing, length of loops, etc. The thicker ribbons give longer life and are preferred for the higher temperatures. Short loops are better than long loops, and close spacing of the winding is undesirable.

115a. The distribution of the resistor winding must conform to the variations and rates of heat transfer through the enclosure at different points. For example, a door introduces a heat loss which requires the concentration of resistor capacity at that point. Hence, the distribution of the resistor surface is not uniform but must be fitted to the design constants of the chamber.

The resistor of the example given would be suitable for a heating chamber with inside dimensions as in the tabulation.

Length, in	54
Width	24
Height	20½

With the standard construction of enclosures and a 9-in thickness of heat insulation, the rate of heat loss from the outer surface of the chamber for a heating-chamber temperature of 1000°C (1832°F) is 9 to 10 kW.

116. Atmospheres, Natural and Artificial. The mixture of air in the gases evolved from the charge as its temperature rises constitutes a natural atmosphere in the heating chamber of a resistor furnace. The composition of such an atmosphere in a batch furnace during a heating cycle is variable and at any time during the cycle depends on the nature, weight, and volume of the charge, on its temperature at that time, and on the degree to which the heating chamber is sealed against the infiltration of air. A natural atmosphere in a heating chamber of a continuous furnace is mainly air. Natural atmospheres are used where the extent of the action of oxygen on the charge during the heating cycle is not objectionable and for processes where that chemical action is desired.

The basis of an artificial atmosphere is an exclusion of oxygen (air) from the heating chamber by the substitution of some other gas or mixture of gases. The selection of the substitute gas or gases is made with reference to the desired chemical activity of the artificial atmosphere on the material of the charge or the range of the temperature of the heat application. A definite chemical action other than oxidation may be required, e.g., the addition of carbon to a steel (carburization); or it may be required that the artificial atmosphere be chemically inactive.

Thus artificial atmospheres are classified as (*a*) active, or process, atmospheres and (*b*) inactive, or protective, atmospheres. However, in many cases it is necessary that a protective atmosphere have some degree of chemical activity in addition to its protective function, and the term "protective atmosphere" is modified to that extent.

The primary gases in protective atmospheres are hydrogen, carbon monoxide, and nitrogen. Nitrogen serves three purposes:

a. To reduce the velocities of the reactions of the active gases present.

b. To lower the potential pressure of each active gas.

c. To reduce the inflammability of the active gas or gases. As a rule, nitrogen constitutes a large part of the volume of a protective atmosphere.

The main uses of protective atmospheres are:

a. The prevention of the formation of oxides on the materials of the charge or, conversely, the reduction of any oxides that may be present.

b. The prevention of a change in the carbon content of a steel undergoing a heat-treatment.

Each of these uses denotes a chemical system in which the action is reversible.

The chemical systems relating to metallic oxides are:

(*A*) Oxide plus hydrogen = metal plus water vapor.

(*B*) Oxide plus carbon monoxide = metal plus carbon dioxide.

The chemical systems with reference to carbon in steel are:

(*E*) Methane = hydrogen plus carbon.

(*F*) Carbon monoxide = carbon dioxide plus carbon.

Equilibrium-volume ratios for these four chemical systems for carbon steel over the usual range of temperature of heat-treatment processes are shown in Fig. 22-45. Oxidation of iron is active down to about 600°C (1112°F). There is but little tendency toward a change in the carbon content of a steel below the critical range.

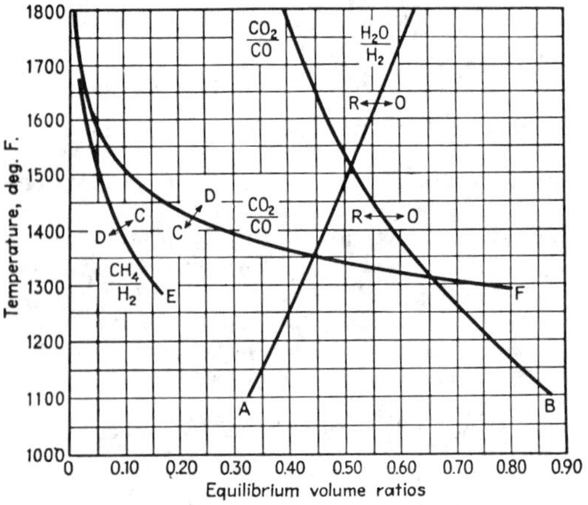

FIG. 22-45. Equilibrium-volume ratios of artificial atmospheres. (*C*) Carburizing; (*D*) decarburizing; (*O*) oxidizing; (*R*) reducing.

The technique in the use of each of the chemical systems is to maintain the volume ratio of the two gases to correspond to the desired direction of the reaction, if no chemical action is desired, to maintain that volume ratio at (or near) its equilibrium ratio for the temperature of the heat application.

In the case of the hydrogen–iron oxide reaction (curve *A*), the water-vapor content of the mixture of gases at equilibrium decreases with a decrease in temperature. Hence, if a steel should be cooled in a protective atmosphere of this kind, the permissible water-vapor content of the controlled atmosphere is dictated by the lowest temperature of the operation. The reverse is true of the carbon monoxide–iron oxide reaction (curve *B*). Thus, if at a given temperature the carbon dioxide content of the mixture of carbon monoxide and carbon dioxide is less than the volume for equilibrium at that temperature, it will be less than the volume for equilibrium at any lower temperatures and the steel can be cooled at that atmosphere without oxidation.

With a mixture of carbon monoxide, carbon dioxide, hydrogen, and water vapor (chemical systems *A* and *B*), these gases will adjust the proportions of the mixture to conform to the equilibrium ratio of each system for the particular temperature.

Curves *E* and *F* in Fig. 22-45 show the volume ratios of systems *E* and *F* for equilibrium with graphite. The equilibrium-volume ratios of these two chemical systems for carbon and solid solution in steel (austenite) depend in each case on the carbon content of the steel. For the methane–hydrogen–carbon system (*B*), the volume ratio of the two gases at equilibrium with carbon in an unsaturated steel at a given temperature is less than the value shown by curve *B*. For the carbon monoxide–carbon dioxide–carbon system (*F*), the volume ratio of the two gases at equilibrium with a carbon in low- and medium-carbon steel at a given temperature is somewhat greater than the value shown by curve *F*; for high-carbon steels the equilibrium-volume ratios approach the values of curve *F*.

The terms "oxidation," "reduction," "carburization," and "decarburization" refer to the chemical activity of a particular atmosphere and not to the extent of its effect on a charge. In all cases the concentration of the active gas or gases, time, temperature, in the case of steel the gas pressure, and the catalytic effect of hot surfaces within the heating chamber are factors in the results obtained.

The first consideration in the selection of a protective atmosphere is the air absorbed on the surface of the charge and the infiltration of air into the heating chamber. The oxygen thus admitted to the chamber increases the volume ratio H_2O/H_2 and CO_2/CO, and this must be taken into account in the analysis of the protective atmosphere as it enters the heating chamber.

The effects, if any, of artificial atmosphere on the resistors as noted under materials for resistors (Par. **31**) should be considered, along with the dimension and the composition of the atmosphere for a given heat-treatment process.

The available data concerning protective atmospheres for the protection of alloy steels during the heat-treatment process indicates that the technique for alloy steels is much the same as for carbon steels; i.e., an atmosphere suitable for carbon steel would in general be suitable for an alloy steel of the same carbon content.

In the heat-treatment of nonferrous metals and alloys the use of either chemical system *A* or chemical system *B* requires for each oxide a knowledge of the equilibrium-volume ratios of the chemical system used over the range of the operating temperature. Individual problems may arise. For example, copper can be bright-annealed in an atmosphere of dry steam, an inactive gas for this application, but the resultant staining of the copper during cooling may be objectionable. Also, usually copper contains a small percentage of oxide, and in annealing such copper in an atmosphere containing a reducing gas the temperature of the metal must be kept below about 400°C (752°F); otherwise the oxide will be reduced and the copper made brittle.

117. Sources of Gases for Protective Atmospheres. *Nitrogen.* The role of this gas in protective atmospheres has been stated. Commercial nitrogen contains more or less oxygen, water vapor, and carbon dioxide—small amounts but sufficient to cause slight oxidation. The addition of a small amount of a reducing gas to neutralize these impurities is generally necessary. Any tendency toward decarburization of steel (chemical system *E* or *F*) can be corrected by the addition of a small amount of methane.

Ammonia Dissociation. This process is carried out in a resistor-type heating unit at about 870°C (1598°F) and yields a gas mixture 75 H_2 and 25 N_2, with traces of ammonia and water vapor. It is usually desirable to burn a part of the hydrogen to increase the nitrogen content of the mixture. The resulting water vapor and the impurities noted above are removed by purification. A mixture largely nitrogen can thus be obtained. This is the most expensive source, but the gas mixture is uniform and thus dependable for the most exacting service, e.g., the bright annealing of chromium-bearing steels.

Charcoal Gas. Air passed through granular carbon at an elevated temperature yields a mixture of hydrogen, carbon monoxide, carbon dioxide, and nitrogen. For a given temperature the CO_2/CO volume ratio depends on the nature of the carbon. The hydrogen content depends on the mixture contents of the air and carbon. The source of carbon is usually charcoal which has been given a prior heat application to remove volatile matter and moisture.

Partly Burned Gases. A hydrocarbon gas such as natural gas, coke-oven gas, city gas, butane, or propane is partly burned to yield a mixture of carbon monoxide, carbon dioxide, hydrogen, methane, water vapor, and nitrogen. The composition of the gas product varies with the fuel, the air-gas ratio, and the conditions under which it is burned. This composition can be varied over a wide range by adjusting the air-gas ratio.

Purification. The volume ratios of the component gases obtained by combustion must be adjusted to the equilibrium ratios of the chemical systems A and B for all metals and also for the systems E and F for steel, corresponding to the temperature of the heat application. This adjustment is made by removal of carbon dioxide by an absorption process and water vapor by refrigeration prior to the admission of the gas to the heating chamber.

The extent of the purification process and the degree of concentration of the active gas or gases required for a given protective atmosphere depend on the nature of the metal of the charge, the temperature of the heat application, the cooling of the charge in the heat chamber, if practiced, and the particular degree and kind of chemical activity required. Methane as a corrective agent for decarburization of steel as noted with reference to a nitrogen atmosphere is equally applicable to protective atmospheres produced by partial combustion.

The flowsheet of the process of partial combustion of the fuel gas is shown in Fig. 22-46. The removal of water vapor in the last stage down to about 3% when the gas is cooled to 75°F is sufficient for many uses of protective atmospheres. If further drying is required, the gas can be passed through a refrigerated cooler to remove moisture to a saturation point of 40°F and then through activated alumina to reach a saturation point of −76°F, the "bone-dry" condition.

In the cooling of steel in a protective atmosphere made by partial combustion the rate of cooling with reference to chemical system A determines the extent of the re-

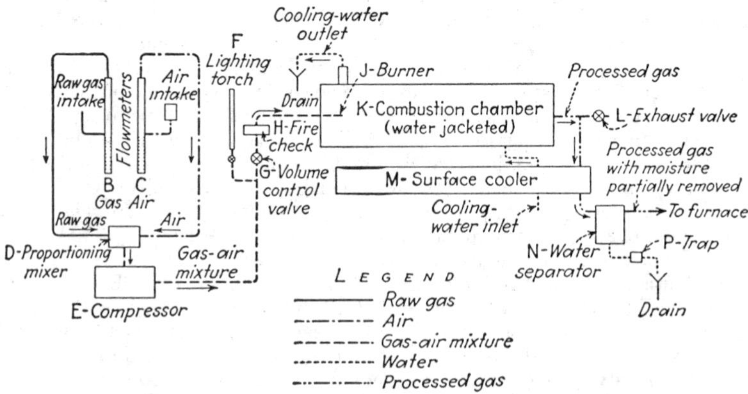

FIG. 22-46. Flowsheet of partial combustion of a fuel gas.

moval of water vapor necessary. A light charge which cools rapidly through the oxidizing range, down to about 600°C (1112°F), may not require a refrigerated gas, whereas it would be required for the comparatively slow cooling of a heavy charge.

In the heat-treatment of high-carbon steels without decarburization—which is desirable for scale-free hardening and bright annealing—both carbon dioxide and water vapor must be completely removed.

Some hydrocarbon gases contain sulfur in small quantities. In all cases this must be removed in the purification process.

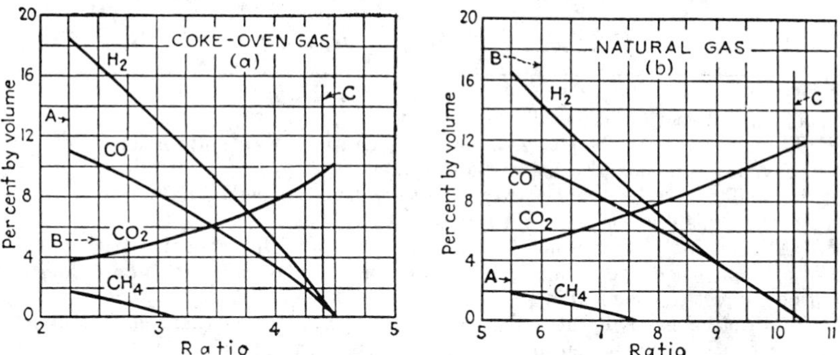

FIG. 22-47. Constituents of gas mixtures other than nitrogen obtained by partial combustion. (*A*) Reducing mixture for brazing steel; (*B*) reducing mixture for bright-annealing steel; (*C*) inactive mixture. (*A*) and (*B*) are inflammable; (*C*) is noninflammable.

The charts of Fig. 22-47*a* and *b* show the range of the analysis of the gas products obtainable by the partial combustion of the coke-oven gas and a natural gas, respectively. The legends of the diagrams indicate the composition and typical use of these gases as protective atmospheres.

118. Effectiveness of Gases. At about 800°C (1472°F) hydrogen and carbon monoxide are equally effective as reducing agents. At lower temperatures carbon monoxide is better; above that temperature hydrogen is the more effective agent.

119. Quantity of Active Gas. The quantity of an active gas required for the reduction of a given quantity of metallic oxide is calculated from the molecular weights of the oxide and the gas product of the reaction.

Example 16. The reduction of iron oxide by hydrogen. The reaction is

$$\begin{array}{ccccl} Fe_3O_4 + & 4\,H_2 \rightarrow & 3\,Fe + & 4\,H_2O & \\ 232 & 8 & 168 & 72 & \text{(molecular weights)} \end{array}$$

Thus, per pound of oxide $(1 \times 72)/232 = 0.31$ lb of water vapor is produced. Hydrogen has one-ninth the weight of water vapor; hence, there is required $0.31/9 = 0.034$ lb of hydrogen/lb of iron oxide.

120. Selection of Atmospheres. A natural atmosphere in a sealed heating chamber with a charge that fills the chamber is to a large degree a protective atmosphere, inasmuch as the amount of oxygen present is small and the surface area of the charge is large. For other conditions and for a definite chemical service (other than oxidation) the selection of an artificial atmosphere requires consideration of a number of factors, viz.:

a. The metal to be heated.
b. The method of cooling the charge.
c. The range of temperature.
d. The time period of the cycle.
e. The initial chemical nature of the surface of the charge.
f. The surface effect—or effects—to be prevented.
g. The surface effect—if any—to be produced.
h. The internal effect that will be produced in the metal.
i. The gases evolved from the metal and from the surface of the heating chamber.

Table 22-12. Average Operating Efficiencies of Resistor Furnaces

Process	Type of furnace	Lb/kWh
Annealing copper and brass	Batch	15–40
Annealing steel and cast iron	Batch	7–11
	Continuous	10–25
Hardening steel	Batch	7–11
Tempering steel	Batch	20–35
Galvanizing	Batch	15–22
Porcelain enameling	Batch	3–6 (gross)
Sheets	Continuous	15–20 (gross)
	Batch	4–6 (gross)
Carburizing	Continuous	8–11 (gross)

j. The volatilization of metals.

k. The impurities in the charge, and the infiltration of air and foreign gases.

Other and more general considerations are:

a. The effects of the artificial atmosphere on the material of the resistors, as noted in the paragraphs relating to resistor materials.

b. The selection of the refractory material in the inner lining of the heating chamber must be made with care to avoid impurities in the material, e.g., iron oxide, which would be subject to chemical attack by the atmosphere.

c. The active gases hydrogen, carbon monoxide, and methane are combustible. Safety precautions should ensure that inflammable mixtures of these gases with air will not be present by accident or otherwise.

d. Ventilation to ensure against toxic effects of the gases used. Heating chambers with artificial atmospheres are operated under positive pressure, and leaks may develop.

121. Maintenance. The establishment and maintenance of the desired chemical nature of an artificial atmosphere require first the purging of the heating chamber to remove all air from the chamber and afterward the continuous flow of the selected gas mixture through the chamber during the cycle of operation.

122. The field of the resistor furnace is the heat-treatment of metals or alloys, annealing glass, porcelain enameling, brazing of metal parts, etc.

The resistor furnace can be placed anywhere that electric power is available. It has no adverse effect on working conditions in the plant. Its adaptability is the same as that of motor-driven machine tools. The present trend is to include the furnace in the line of manufacturing operations.

The majority of resistor furnaces are within the size range 1 to 100 kW; sizes up to 500 kW are common, and a considerable number within the range 500 to 1,000 kW are in service. The only upper limit on size is the practicability of utilization.

Average operating efficiencies are shown in Table 22-12.

123. Data for typical batch (or box) resistor furnaces for general heat-treatment service are given in Table 22-13.

Typical uses of resistor furnaces are represented by the examples which follow.

Table 22-13. Batch Resistor Furnaces, 1000°C (1832°F) Class

Item	kW rating	Heating chamber						
		Dimensions, in						Lb of steel/h to 815°C (1500°F)
		Inside			Outside			
		Width	Depth	Height	Width	Depth	Height	
A	29	18	36	18	64	89	97	300
B	45	24	54	21	70	108	101	500
C	60	30	63	23	97	125	98	650
D	72	36	72	23	94	135	104	750

124. Annealing Brass.

Example 17. Pieces of miscellaneous sizes loaded in steel pans. Heating-chamber dimensions: width 6 ft 6 in; height 2 ft 6 in; depth 12 ft.

Weight of brass per charge, 8,000 lb.

Weight of steel pans, 850 lb.

Temperature 1100°F.

Heat cycle 1½ h.

$$\text{Heat absorbed by charge} = \frac{8{,}000 \times 0.105 \times 1030}{3{,}412} = 252 \text{ kWh}$$

$$\text{Heat absorbed by pans} = \frac{850 \times 0.14 \times 1{,}030}{3{,}412} = 36 \text{ kWh}$$

Rate of heat loss from furnace at 1100°F = 41 kWs

Heat loss per cycle = 62 kWh

Total energy per charge = 252 + 36 + 62 = 350 kWh

Operating efficiency for continuous operation = 8,000/350 = 23 lb brass/kWh

Av. power input = 350/1.5 = 233 kW

Rating of furnace = 300 kW

The average performance from the log of five heats with this furnace is given in the table:

Weight of brass, lb	5,962
Weight of pans, lb	726
Temperature	1,095°F
Heat cycle, min	66
KWh	256
Operating efficiency, lb of brass/kWh	23.2

125. Tempering Steel. The reheating of steel after hardening to a temperature below the critical range, followed by any desired rate of cooling. Electrically heated oil baths and salt baths are used for tempering many kinds of small parts. Another type of tempering equipment is a vertical resistor oven with a removable inner metal cylindrical basket to contain the charge and to provide an angular passageway for the circulation of air (by a fan mounted on the oven) over the resistors and then through the charge—an application of forced convection heating.

126. Bath Heating. The use of a bath of molten lead or a molten salt bath for heating a metal part by immersion is in effect the use of an artificial atmosphere.

The rate of heat transfer from the liquid to the charge is by conduction and liquid convection and is much faster than the transfer by radiation for the same temperature conditions.

127. Lead bath has a working temperature range of 340 to 900°C (644 to 1652°F). A typical use is heating edge tools for hardening. Lead is inactive with respect to steel, is stable, and does not absorb moisture. It oxidizes readily, and the surface of the bath should be covered with a layer of protective material such as crushed charcoal. This layer also serves as heat insulation for the top surface of the bath.

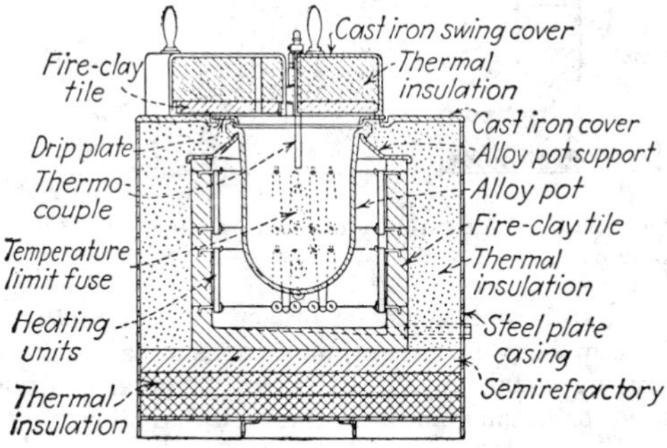

FIG. 22-48. Lead melting pot.

The design and construction of the lead bath are similar to melting pots with external heating units (see Fig. 22-48).

128. The salt bath is used for heat-treatment processes within the temperature range up to 1315°C (2400°F). The salts are nitrate mixtures, chloride mixtures, and cyanide mixtures, selected to suit the need in each case. The cyanide mixtures are used for cyanide hardening and carburizing. The rate of heat transfer to the charge is somewhat slower than in the lead bath.

Three types of salt bath are manufactured:
a. Pots with immersion heating units for nitrate mixtures up to 593°C (1100°F).
b. Pots with external heating units (see Fig. 22-49).
c. Pots with one or more pairs of electrodes (see Fig. 22-50).

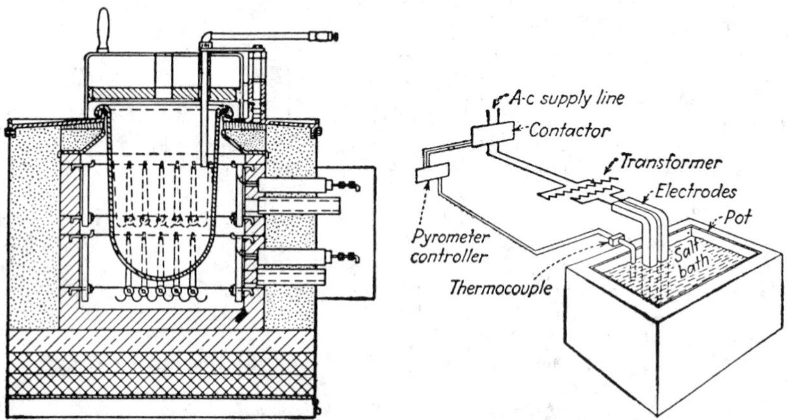

Fig. 22-49. Salt bath with external heating units.

Fig. 22-50. Electrode-type salt bath.

129. Electric Furnace Brazing. A method of brazing that has wide applicability to light assemblies such as automotive parts. The manufacture of the fin-type heating unit of Fig. 22-14 is an example of this method.

The parts *A* and *B* to be joined can be like or unlike metals. The brazing metal *C* is selected for a lower melting point than that of either metal *A* or metal *B*, to form an alloy with metal *A* or *B* and, when molten, to wet the surface of parts *A* and *B*.

The assembly is prepared in some such manner as indicated by Fig. 22-51, with a snug fit to give a capillary joint space. The heating chamber contains a protective atmosphere reducing in nature and is held at a temperature to melt the brazing metal quickly.

The gas in the chamber serves as a flux to reduce any oxide present and to give chemically clean surfaces for joining. The limitation of the gas flux is the dissociation pressure of the oxide on the joint surface. If this pressure is very low, a gas may not be effective. Assemblies of this kind are oxides or steels containing appreciable additions of chromium, aluminum, or silicone. A solvent flux is necessary in such cases.

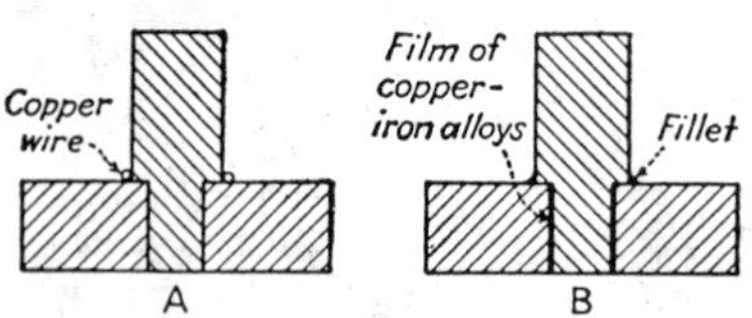

Fig. 22-51. Assembly of parts for brazing.

The molten braze metal is drawn by capillary action into the joint space and, in cooling, forms an alloy bond between the parts *A* and *B*. No pressure on the joint is required.

Metals which can be brazed by this method include practically all steels, cast iron, malleable iron, and nonferrous metals and alloys in general. Copper is the general brazing metal for iron and steel. Nonferrous alloys with suitable melting ranges are used for brazing nonferrous metals and alloys. Brazing alloys in common use are given in the next table.

Brazing Metals

Metal	Temperature Range
Copper	1093–1149°C (2000–2100°F)
Brass, 60 copper, 40 zinc	953–1093°C (1750–2000°F)
Silver alloys	648–926°C (1200–1700°F)

A protective atmosphere obtained by partial combustion, with a composition of

20% hydrogen, 12% carbon monoxide, 3% carbon dioxide, remainder nitrogen, is typical for this service.

Assemblies must be made self-supporting at the temperature of the brazing operation by riveting, tack welding, etc., according to the type of assembly.

The time required for the brazing action is short. The overall time depends on the mass of the parts; 1 to 4 min is representative for light work.

The types of resistor furnace most used for brazing are batch (or box) furnaces, continuous furnaces with mesh-belt conveyers, and roller-hearth furnaces. All have water-cooled cooling chambers.

The batch furnace is adapted to brazing assemblies of any weight and for intermittent and miscellaneous production. Also, a furnace of this kind is available for other heat-treatment processes.

The mesh-belt conveyer furnace is designed for light work in mass production. The conveyer loadings range from 4 to 7.5 lb/ft^2 uniformly distributed.

The roller-hearth brazing furnace is not limited as to size or length—a design particularly suited to large-scale production of assemblies of all kinds.

129a. Porcelain Enameling. The final operation in the production of porcelain enamel is fusing of the glass on the metal surface. Both batch resistor furnaces and continuous resistor furnaces are used for this service.

Temperature and time are closely related in the fusing operation (burning or firing). The glass must be brought to the temperature at which it will flow to an even surface. Close temperature regulation is essential. The on-and-off method is general practice. The time period depends on the mass of the base metal.

The burning (or firing) temperatures for the ground coat of sheet-steel enamels (wet process) vary from 800 to 1000°C (1472 to 1832°F). The corresponding temperatures for the cover coats are somewhat lower depending on the combustion. The burning temperatures for cast-iron enamels (dry process) range from 815 to 980°C (1500 to 1800°F).

The atmosphere in the heating chamber of enameling furnaces is clean air. Cleanliness is essential.

A typical batch resistor furnace for porcelain enameling has the inside dimensions: width 4 ft; height 18 in; length 12 ft; rating 200 kW, 3-phase, 220 V. Heat loss at 830°C (1526°F), 21 kW. Average operating efficiency for sheet-steel enamels: 8 lb net/kWh; 3.2 ft^2/kWh, 1 coat.

The U-shape continuous resistor furnace with an overhead conveyer (Fig. 22-51) is particularly well adapted to burning sheet-steel enamels. As a rule the heating chamber is elevated to provide a heat seal at the incoming and outgoing openings. With this arrangement the outgoing work gives up part of its heat to the incoming work.

The loss of heat from the U-shape continuous furnace is the sum of (*a*) the loss through the enclosure, (*b*) the door loss, (*c*) the conveyer loss, and (*d*) the loss incurred in reheating the supports attached to the conveyer. The operating efficiency varies with the class of work, temperature, method of loading, speed of conveyer, idle time, etc. For average conditions the operating efficiencies of burning enamel on sheet-metal parts range from 4 to 12 net lb of ware/kWh.

130. Annealing glass is a heat-treatment process (involving both heating and cooling) for strain relief only; there is no change in the structure of the glass. Strains in glass are relieved by heat. This being done, the cooling of the glass must be controlled so as to prevent the return of the strained condition.

The annealing range of a glass refers to the range of temperature between the temperature at which strain relief takes place slowly (lower annealing temperature) and the temperature at which it anneals rapidly (upper annealing temperature). This temperature range is related closely to the composition of the glass. The upper annealing temperature for soda-lime glass usually ranges from 490°C (914°F) to 610°C (1130°F).

Heating Rates. The heating of a body of glass places the outer layers under compression and the inner portion under tension. The heating rate is important only in that the glass must be safeguarded against temperature gradients which would cause stress in the glass beyond its compression strength and thereby cause fracture. With indirect resistance heating, the maximum temperature to which glass can be heated is fixed by the fixed temperature of the heating chamber, and this is protection against a too rapid rate of heating.

The essential point in heating glass is uniformity of temperature within the mass and the attainment of the required temperature. The time required for strain relief at the usual annealing temperature is a short period. This time period should not be confused with the annealing time, i.e., total time required for strain relief (see Table 22-14).

The second stage of annealing is the control of the cooling rates so as to prevent the reappearance of strains. This is accomplished by progressively lowering the temperature to which the glass is subjected, slowly at first, accelerating as the glass approaches rigidity. With batch heating the temperature in the chamber is lowered by decreasing the power input. Zone cooling is used with continuous chambers. The latter are termed "lehrs."

The design of electric lehrs is similar to resistor heating equipment generally. Glass can be handled in only one layer. Hence, the proportion of the inside dimensions of the heating chamber is a width much greater than the height. Examples of the designs of electric lehrs follow:

a. Lehr for bottles, etc. Inside dimensions: width 4 ft; length 100 ft; rating 60 to 90 kW, depending on the class of ware. Four zones, each with individual temperature regulation.

b. Lehr for plate glass. Inside dimensions: width 14 ft 6 in; length 730 ft; rating 1,590 kW. The total length of the zones of heat distribution is 400 ft. The first three sections contain 164 kW each; the following sections, 122 kW each. Each zone with temperature regulation. Operating efficiency 30 to 60 kWh/ton (2,000 lb) of glass.

The conveyer of a lehr is a part of the cooling system. It serves as a carrier of heat, and the rate of this heat transfer is under control by the control of the speed of the conveyor. The operating speed varies from 3 to 12 in/min, depending on the class of glass being annealed and the loading of the conveyer.

The on-and-off method of temperature regulation for each zone is general practice. The atmosphere in the heating chamber for annealing glass is clean air; cleanliness is essential.

131. Galvanizing. The production of zinc-iron alloy coatings on the surfaces of iron and steel parts by immersion of the articles in a bath of molten zinc—the hot-dipping process.

The containers for the molten zinc—pots or tanks—are made of low-carbon steel plate and in shapes and sizes to suit the article to be heated. External heating units (resistors) are mounted on the side walls; the distribution is 3 to 4.5 kW/ft^2 of side-wall area. The construction of the enclosure—side walls, ends, and bottom—conforms to resistor-furnace practices. The ratings of galvanized tanks in service vary from 25 to 630 kW.

Materials galvanized vary widely in shape, size, and analysis. The operating temperature varies accordingly; the average value is about 454°C (850°F). This temperature is critical, and close regulation is essential.

The rate of heat loss from electrically heated galvanizing tanks with exposed zinc surfaces is about 1.5 kW/ft^2 of the top surface of the bath at the above noted operating

Table 22-14. Average Operating Efficiencies of Annealing Glassware in Electric Lehrs

Ware	Weight, oz	kWH/gross	Lb/kWh
Ovals	11⅞	2.67	40
Jars	11¼	1.97	52
Milk bottles	27¼	4.19	59
Jars	22	2.86	69
Jars	20¼	2.54	72
Jars	12	1.29	84
Jars	19½	1.37	128
Soda bottles	16	1.10	131

temperature. The operating efficiency ranges from 10 to 20 lb of work/kWh, depending on the size of the tank, character of the work, rate of working, etc.

During idle periods it is necessary to hold the zinc in a molten state. As the larger part of the heat loss is from the exposed surface of the molten metal, the idle-period loss can be reduced by an insulated cover for the tank.

132. Malleable iron is made by heating white cast iron to a temperature above the critical range—820 to 950°C (1508 to 1742°F)—and holding the charge at that temperature for a period of time, followed by cooling at a controlled rate. This heat-treatment process converts the combined carbon of the white iron into graphite (temper carbon) and thus renders the metal ductile.

The type of furnace for this service is selected with reference to the size or sizes of castings and the rate of production required. With batch furnaces the gases evolved from the charge, largely carbon monoxide, form a protective atmosphere in the chamber. As a rule a protective atmosphere slightly reducing in nature must be introduced and maintained in the chambers of continuous furnaces.

Example 18. An example of the use of a roller-hearth resistor furnace for this service is as follows:

Charge, lightweight castings handled in trays.

Production, 35 tons (2,000 lb)/day of 24 h.

Heating-cooling cycle: heat to 1750°F for 4 h; cool rapidly to 1450°F for 1 h; slow cooling from 1450 to 1200°F for 7 to 8 h. Total time in furnace, 13 h. Protective atmosphere slightly reducing.

The design of the furnace includes a heating chamber, rapid-cooling chamber with air- or water-cooling coils, and the slow-cooling chamber. The chambers are separated by vertical sliding doors. Double inlet and outlet doors are provided. All doors are interlocked for sequence of movement of the loaded trays. The rolls and doors are motor-driven. Overall dimensions: width 62 in; height 16 in; length 130 ft. Electrical ratings: 750 kW, 3-phase, 60 c, 220 V. Gas consumption 1,500 ft^3/h. Operating efficiency about 300 kWh/ton (2,000 lb).

TEMPERATURE REGULATION

133. Temperature regulation of ovens and furnaces means maintaining the temperature at one or more selected points in the heating chamber within certain limits. Uniformity of distribution of temperature within a heating chamber is a part of the design of the chamber with reference to the character of the charge and the major mode of heat transfer to be used.

The resistors of the larger units are divided into two or more independent circuits, each with individual temperature regulation. This subdivision of resistors is also used for zone heating—and zone cooling where needed—in continuous ovens and furnaces.

Automatic temperature regulation is a feature added to temperature measurement (see Sec. **3**, Pyrometry). Contacts for the regulating devices are mounted on the temperature instrument. A multipoint instrument may serve to record temperatures at different points in a heating chamber, but an individual instrument is required for each point of temperature regulation.

The control and regulation of temperatures on induction equipment must be accomplished by fitting a cycle to a regulating or control device that will either alter the power-supply voltage or turn the equipment off and on. This type of regulation is occasionally used for special applications involving induction-heating equipment.

134. Temperature Response. Thermostatic devices such as bimetallic springs and fluid columns are used for temperature response up to about 400°C (752°F). The thermocouple is required for the temperature range of the resistor furnace and is used to some extent for lower temperatures.

The bimetallic spring responds to changes of temperature by expansion and contraction. This movement is transmitted to a contact device in the power circuit. Its use is limited mainly to water heaters, portable devices, etc.

The fluid-column device consists of a bulb containing a liquid or vapor, a small-bore connecting tube, and a bellows or hollow spring. The expansion of the fluid in the bulb is transmitted to the bellows or spring, which in turn actuates the snap-action switch opposed by a spring.

The basic considerations which determine the selection of thermocouples for temperature regulation of heating chambers are (*a*) the maximum temperature and (*b*) the chemical nature of the atmosphere within the chamber.

Infrared sensing devices are used for controlling certain temperature conditions, and occasionally radiation pyrometers can be used effectively for temperature control.

135. Representative thermocouple practice is:

a. Natural atmospheres. The iron-constantan thermocouple is used for temperatures up to about 760°C (1400°F); for higher temperatures up to about 1,000°C (1832°F), the chromel-alumel thermocouple is employed.

b. Protective atmosphere. The iron-alumel thermocouple with a protective tube is used to some extent for temperatures up to about 1150°C (2102°F). More generally employed in practice is the radiation thermocouple, which, being outside the heating chamber, is not subject to chemical action by the chamber atmosphere. This type of thermocouple is often classified as a radiation pyrometer.

c. The 90 platinum–10 rhodium thermocouple is applicable to both natural and protective atmospheres and for temperatures up to about 1500°C (2732°F). Its use is limited mainly to temperatures above the range of base-metal thermocouples.

The degree of the regulating temperature is dependent largely on the accuracy of the temperature-response element. Periodical checking and replacement when needed are essential in this service.

136. The on-and-off (**current-on and current-off**) **method** of temperature regulation is most generally used in practice. This method utilizes the thermoequalizing effect of the heat-storage capacity of the inner lining of the heating chamber, together with the heat-storage capacity of the charge; the value of the latter may be small or large according to the mass of the charge and its diffusivity. The main reliance, particularly with furnaces, is on the heat-storage capacity of the inner lining of the chamber and, more specifically, on that of the first few inches of the thickness of the lining.

The on-and-off method is applicable only with resistors or materials that have small resistivity-temperature coefficients, e.g., the 80 Ni–20 Cr alloy. Also, it is more effective where the major mode of heat transfer is by radiation.

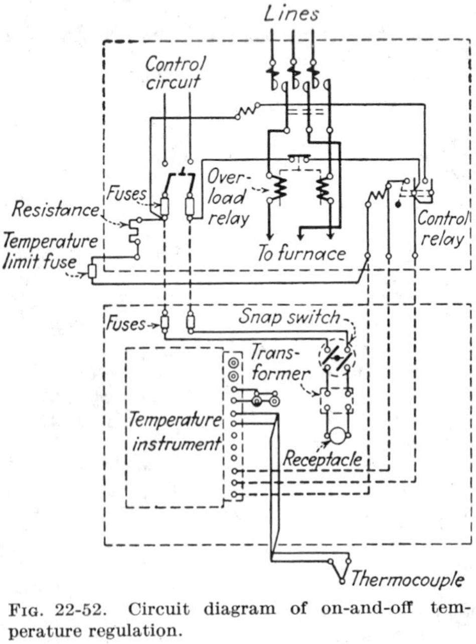

The on-and-off method is also used for self-contained devices, generally by utilizing the heat-storage capacity adjacent to the resistors and relying on conduction for the thermoequalizing effect.

The circuit diagram (Fig. 22-52) of an on-off temperature regulator is typical. The relay functions to operate contactors in the power circuit in response to the closing and opening of the low and high contacts mounted in the temperature instrument. The diagram indicates the use of a thermocouple; it is equally applicable to fluid-column devices. The instrument may be either a millivoltmeter or a potentiometer. A kilowatts capacity beyond that required for the average rate of heating must be included in the design of the resistors for the functioning of the on-and-off method of temperature regulation. This method does not affect the power control of the furnace load.

FIG. 22-52. Circuit diagram of on-and-off temperature regulation.

137. The second method of temperature regulation is by adjustments of the voltage supply to the resistor circuit. This method is not so dependent on heat storage. It is particularly applicable to the transfer of heat by forced convection in oven practice. A closer regulation can usually be obtained, and the trend is toward the use of this method for the larger oven and furnace installations.

Voltage adjustments for temperature regulation can be obtained by (*a*) voltage taps in a transformer or autotransformer, (*b*) an induction regulator, (*c*) a saturable-core reactor, and (*d*) a regulated saturable-core reactor. In each case the device is arranged to function with a temperature instrument for automatic temperature regulation.

With the saturable-core-reactor method of temperature regulation (Fig. 22-53), the voltage of a potentiometer-type temperature instrument is applied to the grid of the rectifier, which supplies direct current to the reactor winding.

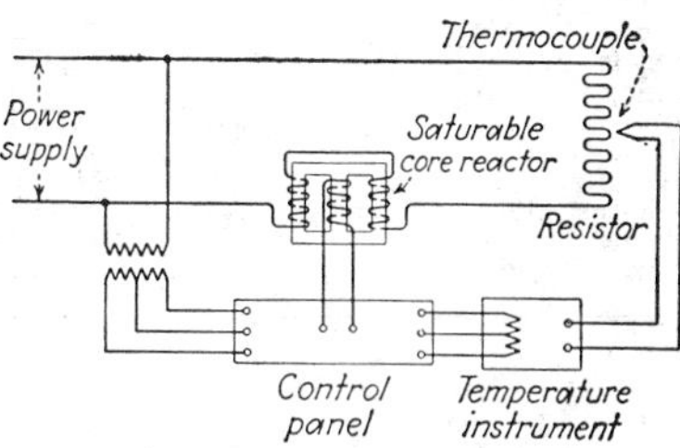

Fig. 22-53. Temperature regulation by voltage adjustment. Saturable-core-reactor type.

The effect of the saturable-core reactor on the power factor of the furnace load depends on the kilowatts rating of the resistor circuit with reference to the average power required by the charge. The average performance of a resistor oven or furnace for this method of regulation with reference to the power factor of the furnace circuit is about the same as that obtained with general-purpose induction motors.

The regulated saturable reactor, the saturable reactor, or the induction regulator has been and can be used in controlling power to radio-frequency types of induction heating equipment. This is not necessarily tied in with temperature regulation.

138. Temperature protection for resistor heating equipment is obtained by a temperature fuse (gold alloy) mounted in the heating chamber and connected in the control circuit to the temperature regulator.

MELTING METALS

139. Melting includes, in addition to the change of state, the further heating of the metal to a specified temperature, known as the "pouring temperature" if the metal is to be poured into a mold or the "working temperature" if the metal is to be used for coating, as in galvanizing, or as a liquid heating bath.

Table 22-15 refers to melting service. An alloy belongs in a group of this table to which its major component belongs. The melting-temperature range of an alloy must be obtained from the constitution diagram of a system to which the alloy belongs.

Table 22-15. Melting Points of Metals

Metal	Melting point	
	°C	°F
Group 1:		
Tin	232	450
Bismuth	271	520
Cadmium	321	610
Lead	327	621
Zinc	420	788
Antimony	630	1166
Magnesium	651	1204
Aluminum	659	1218
Group 2:		
Silver	961	1761
Copper	1083	1981
Nickel	1452	2646
Cobalt	1480	2696
Iron	1530	2786

140. The pouring temperatures of nonferrous metals and alloys range from 100 to 200°C (180 to 360°F) above the melting point. They vary for each metal or alloy with the type of load and the size and type of casting according to the purpose for which the casting is to be used. Typical values are given in Table 22-16.

Table 22-16. Pouring Temperatures of Metals

Metal or alloy	Melting temperature		Approximate pouring temperature	
	°C	°F	°C	°F
Tin	232	450	316	600
Lead	327	621	454	850
Zinc	419	786	449	840
Aluminum	659	1218	800	1472
Type metal	238	460	343	650
Babbitt metal	235–260	455–500	460	860
Die-cast alloy	395	743	427	800
Solder ½ and ½	182–232	360–450	343	650
Solder 2 and 1	185	365	271	520

141. Volatilization. In melting metals there is a certain loss by volatilization. This loss of metal is usually negligible except for charges which contain a high percentage of zinc. Another consideration is the poisonous nature of zinc fumes.

142. Stirring an alloy while it is in the molten state is often necessary to prevent segregation. Stirring is also an aid in bringing about equalization of temperature and is a safeguard against overheating in the surface of the molten mass.

Automatic stirring can be obtained with suitable induction coils or by the use of an induction furnace.

143. The melting of the soft metals, or group 1 (see Table 22-15), and their alloys is within the range of the 80 Ni–20 Cr alloy-resistor heating equipment. Containers for these materials are designated as melting pots, solder pots, lead pots, galvanizing kettles, tanks, etc.

144. Melting pots are made of cast iron or steel and of various nonferrous alloys. The selection of the material for the pot depends on the temperature of the molten metal and the possible reactions of that metal with the material of the pot.

Open-top pots are used in which the molten metal is removed by dipping or pumping and into which metal is dipped for coating. A greater depth than is necessary for the service is desirable as an aid to temperature recovery when cold metal is added.

Closed-top pots are advantageous in reducing the heat loss from the molten-metal surface and in reducing oxidation. Also, space can be provided for a protective atmosphere. Discharge of the molten metal by gas pressure, e.g., steam, is incorporated in some designs.

The heat insulation of a metal melting pot should be sufficient to limit the outside surface temperature to a safe value. Otherwise, insulation is a matter of economy. As a rule, a layer of refractory material is unnecessary for temperatures below 600°C (1112°F). A refractory layer is useful as an aid in maintaining an even metal temperature.

145. Heat Loss. The rates of heat dissipation from open-top melting pots of typical construction are shown in Fig. 22-54. This loss can be reduced by a layer of insulating material in granular form, e.g., charcoal or diatomite on the surface of the metal. This layer also reduces oxidation of the metal.

146. Cast-in Unit. A self-contained heating unit embedded by casting in a mass of gray iron. This unit is designed for metal-melting services up to 510°C (950°F). A typical method of use is shown in Fig. 22-55. The unit is applicable to the melting of all the soft metals with the exception of aluminum and zinc, exceptions because of their alloying action with iron. The rating of the unit should be on the open-air basis.

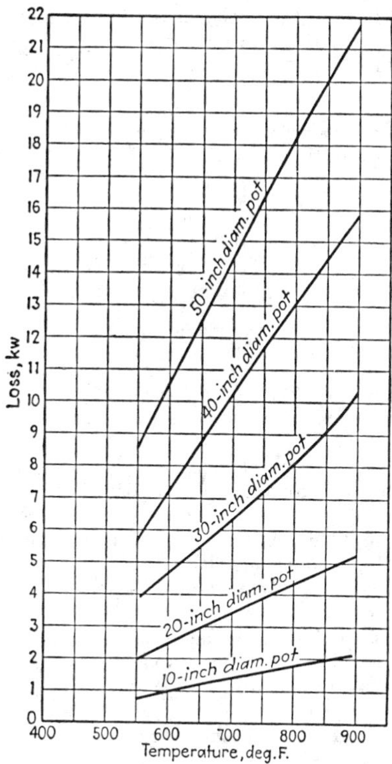

FIG. 22-54. Heat loss from open-top metal-melting pots.

147. External heating units supplied to a melting pot are shown in Fig. 22-48. The rating should be on the basis of 20 to 30 W/in² of the side surface of the container. This design of melting pot is suitable for molten-metal temperatures up to 898°C (1650°F).

148. The temperature-regulating equipment of melting pots is the same as used for resistor ovens and furnaces. As a rule, small melting pots—say, 5 kW and below—have only manual control by a two- or three-point switch.

149. The melting rate of a melting pot is measured by the time required for the charge of molten metal to regain the pouring or working temperature after the addition of a quantity of cold metal. For example, if 100 lb of cold metal is added and the metal in the pot regains its temperature within 10 min, the melting rate for that metal is 100 × 6 = 600 lb/h.

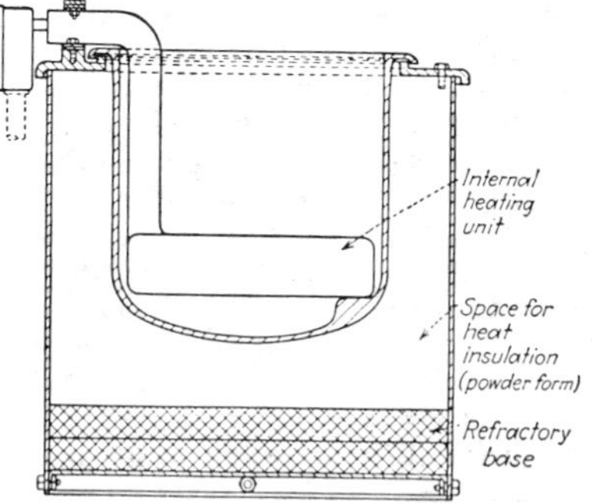

FIG. 22-55. Soft-metal melting pot with internal heating unit.

150. The kilowatts rating of a melting pot is based on the required rate of melting a given metal or alloy. For melting quantities of metal, e.g., for castings, the kilowatts rating should be no less than the rate of heat input to the metal plus the rate of heat loss. For melting for coating work, e.g., galvanizing, the kilowatts capacity needed is the sum of the capacities required for melting, for heating the base material, and for the rate of heat loss. As a rule, some additional capacity is installed to accelerate heating up and to prevent too large a drop in temperature when cold metal is added. The operating efficiency of the melting range is not affected by its kilowatts rating.

Example 19 (continuation of Example 3). Melting lead and heating to 750°F.

Wh/lb from Example 3 = 8.3.

20-in-diameter open-top melting pot.

Melting rate 1,000 lb/h.

Rate of heat loss (Fig. 18-54) 4 kW.

kW input while melting (8.3 + 4) = 12.3 kW.

An addition of about 3 kW would be sufficient for average conditions which would give a rating of 15 or 16 kW.

151. Metals of group 2 (see Table 22-15) and their alloys require either the arc furnace or the induction furnace for melting. Brass and steel are the major alloys of this group.

Brass. The heat contents of some of the common copper-zinc alloys are given in the following table. The total heat required for melting brass varies with the composition of the alloy, the extent of the impurities in the charge, the size of the furnace, and the degree of continuity of its use. Average figures for electric furnaces are:

Yellow brass, 220 to 350 kWh/ton. Red brass, 250 to 400 kWh/ton.

Brass	Temperature		Heat content, kWh/ton
	°C	°F	
61 Cu, 36 Zn, 3 Pb	1000	1832	146
65 Cu, 35 Zn	1000	1832	136
66⅔ Cu, 33⅓ Zn	1000	1832	138
80 Cu, 20 Zn	1100	2012	161

Steel. Melting and refining service refers to the manufacture of steels of a specified analysis. Some charges are made up of steel scrap, more or less pig iron, and alloy additions as required.

Acid process and basic process refer to methods of removing impurities from the bath of molten steel in the furnace by a blanket of slag—an acid slag for the acid process, a basic slag for the basic process.

An acid slag does not remove sulfur and phosphorus; hence, this type of slag is limited to refining charges free from these impurities. All common impurities can be removed by the basic process. Furnaces for the acid process have acid linings (silica). Furnaces for the basic process have basic linings (magnesia).

Melting service is mainly the production of castings from scrap-metal-foundry service. Scrap metal carries more or less scale, oil, dirt, etc.; and in most cases some refining is necessary. Where scrap free from sulfur and phosphorus is available, the acid process is commonly used. Also, the acid lining furnace is used for the production of cast iron. It is necessary to use basic lining furnaces for melting some alloy steels, e.g., manganese steels.

The heat content of steel at pouring temperatures is approximately 360 kWh/ton (2,000 lb). Average values of the total heat required for electric-furnace production are: 500 to 600 kWh/ton in melting service; 500 to 900 kWh/ton in melting refining service, depending on the size of the furnace and length of the refining period.

ARC FURNACES

152. The two types of arc furnace in common use are (*a*) the 3-phase furnace and (*b*) the single-phase furnace. The general field of the 3-phase furnace is the production of alloy steels; that of the single-phase furnace, the production of nonferrous alloys. Both types of furnaces can be used for the manufacture of high-quality gray-iron castings. However, the use of large-size induction coreless and core-type furnaces may be preferable.

153. Three-phase Arc Furnace. Standard sizes of these furnaces range from 250

Table 22-17. Representative Sizes of Three-phase Arc Furnaces in General Use

Shell diameter	Normal charge, lb	Transformer rating, kVA	Lb/h single slag heats
4 ft	800-1,000	250-350	500
4 ft 6 in	1,200-1,500	350-500	900
5 ft	1,500-2,000	500-750	1,300
6 ft	3,000-4,000	750-1,000	2,000
7 ft	5,000-6,000	1,000-1,500	3,000
8 ft	7,000-9,000	1,500-2,000	4,500
9 ft	10,000-12,000	2,000-3,000	6,000
10 ft	16,000-20,000	2,500-3,000	10,000

to 80,000 kVA; loading range, 500 lb to 250 tons. Sizes 1,000 to 5,000 kVA predominate.

The design of the 3-phase furnace is shown in Fig. 22-56. The chamber is a steel bowl with a refractory lining. The hearth is a shallow bowl formed in the bottom lining. The roof is a removable dome-shape refractory structure carried on a steel roof ring. The roof has three round ports in equilateral triangular arrangement through which vertical carbon or graphite electrodes travel. Each electrode is carried on a winch-and-rope system, motor-driven. The supports for the electrode mechanism may or may not be attached to the furnace shell. The structure of the furnace is mounted on a tilting mechanism for pouring the molten metal through the door opening in the side of the shell.

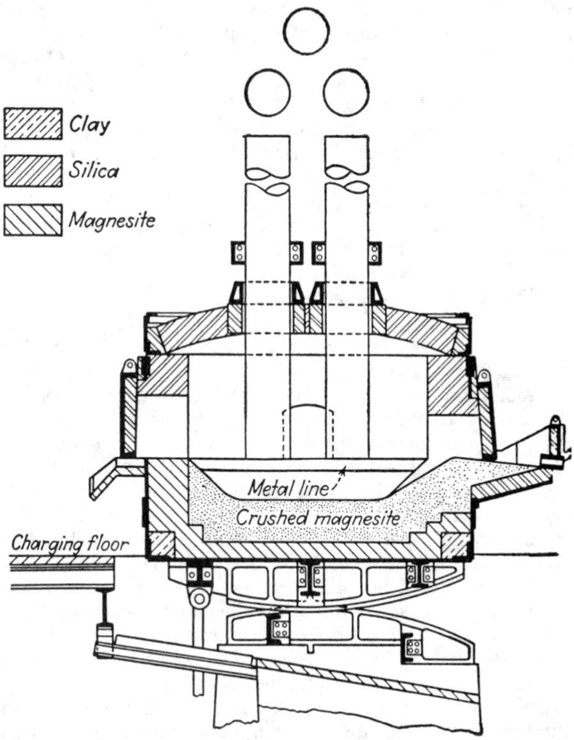

Fig. 22-56. General design of three-phase arc furnace. Basic lining.

Refractories. The chemical nature of the slag, acid or basic, determines the required chemical nature of the lining of the hearth and side wall of the chamber up to a few inches above the top surface line of the slag, i.e., an acid refractory (silica) for acid slags, and a basic refractory (magnesia) for basic slags. The roof is usually made of silica brick. Silica has a tendency to spoil during heating and cooling, and furnaces which are in intermittent use often have roofs made of fire-clay brick. The two types of refractory lining are shown more clearly in Fig. 22-57.

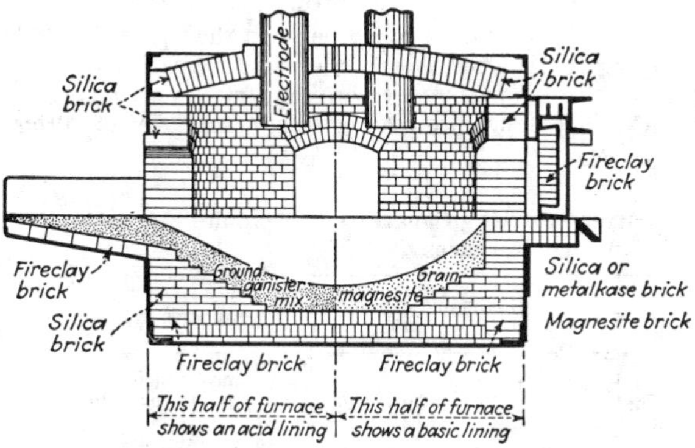

Fig. 22-57. Refractory linings of three-phase arc furnace. (*Harbison-Walker Refractories Company*.)

Temperature. The operating temperature of the chamber is limited by the softening point of the refractory, particularly that of the roof, where there is a concentration of heat. A refractory material can be operated with the temperature of its inner face close to its softening point provided that the outer face is exposed to the open air, thus permitting a flow of heat through the refractory body. A temperature gradient is thus established in the refractory so that if its thickness is correctly related to its thermal conductivity the mean temperature of the refractory body will not be high enough to impair its strength materially.

The temperature of molten steels is around 1600°C (2912°F). The melting point of silica is 1713°C (3115°F), but the softening point of a silica refractory is somewhat lower because of impurities in the refractory body. Hence, in a steel melting furnace the temperature of the inner face of the refractory lining is too high to permit the use of heat insulation. Even a thick coat of dust on the roof of the melting furnace is undesirable.

154. The designation for a 3-phase arc furnace may be given in terms of the holding capacity, the shell diameter, the pouring capacity, the melting rate, or a combination of these. A given diameter of shell can be attached to a range of ratings by varying the thickness of the refractory linings. Sizes are given in Table 22-17.

The method of charging is as follows:

Sizes, 4 and 4 ft 6 in	Hand
Sizes, 5 and 6 ft	Hand or chute
Larger sizes	Hand, chute, or top charging

155. Charges. The 3-phase arc furnace is primarily a scrap-metal-conversion unit. The two types of furnace with respect to the method of charging are (*a*) the door-charge type and (*b*) the top-charge type. Depending upon the character of the scrap, hand charging and chute charging are the usual methods for small furnaces. Large furnace installations are often equipped with side-door-charging machines. Top charging is growing in favor for medium-size furnaces. In this method the roof of the furnace is removed, and a complete charge is placed in the chamber by a drop bucket handled by an overhead crane. This is both a timesaving and a laborsaving method. The charging time is only a few minutes, e.g., a reduction from 30 to 5 min. Top charging has the other advantage of a full chamber and a lower heat loss during the charging.

Some 3-phase arc furnaces are used for refining service only. Molten metal from an open-hearth furnace, Bessemer converter, or cupola is the charge.

The weight of scrap metal varies with the degree of its subdivision (see the following table). The weight of charge that can be placed in a given furnace depends on the kind of scrap. If sufficient scrap metal cannot be placed in the furnace initially to form the weight of molten metal desired, additional quantities can be added later in the heat cycle. This practice affects adversely to some extent both the operating efficiency and the consumption of electrodes.

Approximate Weight of Scrap Iron and Steel

Kinds of Scrap	Lb/ft^3
Sprues, gates, and risers	70–90
Borings and turnings	100–150
Small miscellaneous scrap	200
Large scrap	300

156. Electrodes. The arc in each phase is maintained between the tip of the electrode of that phase and the charge (bath after the molten state is reached). The charge thus serves as a common electrode for the three arcs and makes a connection of the 3-phase circuit at that point. The designation "direct-arc furnace" refers to this arrangement.

Table 22-18. Approximate Current-carrying Capacities of Graphite Electrodes for Arc Furnaces

Nominal diameter, in	Amperes	Nominal diameter, in	Amperes
2	600–1,000	9	6,400–10,800
2½	800–1,500	10	7,800–12,500
3	1,200–2,100	12	11,300–17,000
4	1,800–3,000	14	15,400–21,500
5½	2,300–4,100	16	20,100–26,100
6	3,100–5,400	17	22,700–27,400
7	4,200–6,900	18	25,500–30,500
8	5,500–9,000	20	28,300–34,600

Table 22-19. Approximate Current-carrying Capacities of Carbon Electrodes for Arc Furnaces

Nominal diameter, in	Amperes	Nominal diameter, in	Amperes
8	2,000–3,000	20	11,000–17,300
10	3,000–4,800	24	15,800–24,800
12	4,500–6,800	30	24,700–35,300
14	5,400–8,500	35	28,800–38,400
17	7,900–12,500	40	37,700–50,200

The trend is toward the general use of graphite electrodes. Carbon electrodes are preferred in some cases. Standard sizes in the corresponding current ratings are given in Tables 22-18 and 22-19.

The consumption of electrodes is caused largely by volatilization and burning. There is some breakage. Graphite begins to oxidize at about 600°C; carbon, at 400°C. Under average conditions the consumption of graphite electrodes is about one-half that of the carbon electrodes. Average values for melting service, pounds of electrodes per ton of metal melted, are: graphite 4 to 10; carbon 8 to 15. The corresponding consumption in melting refining service is about 10 lb for graphite and 18 lb for carbon.

157. Selection of the size of furnace for foundry service is based on several factors, viz., average production, maximum and minimum production, casting facilities, weights of castings, power supply, rate schedule. For continuous use, a large unit is more efficient than two or more small units. However, operating a large unit for small production is not economical. Two sizes of furnaces may be the better arrangement. Similar considerations prevail in the size selection of the furnace for ingot production, although the operating conditions are somewhat different.

The more exact procedure possible with the comparatively small charges of arc furnaces is responsible to a considerable extent for the continuous growth of that furnace for making alloy steels, both castings and ingots. The term "electric steel" is the accepted designation of a uniform and high-quality product.

158. A typical performance of a medium-size furnace in foundry practice (acid process) is given in the table.

Three-phase Arc Furnace

Diameter of shell	6 ft
Range of loading	2–4 tons
Electrical rating	1,000 kVA
Weight per heat	4,000 lb
Time of first heat	2–2½ h
Average time of succeeding heats	1¼–1½ h
Number of heats per 10-h day	6–7
Average kWh/ton of molten steel	540
Tons of molten metal/day	12–14

159. The voltampere characteristic of the arc is negative (curve *A*, Fig. 22-58), and a stabilizing element (curve *B*, reactors for an a-c arc) is necessary for circuit stability (curve *C*). Reactance also serves to limit the current in the circuit when an electrode touches the charge. This reactance is a total reactance of the circuit from the furnace terminals to the point in the power system where the voltage is held constant. Thus a furnace at the end of a long feeder is a different problem from a furnace installed adjacent to a large substation.

The operation of an arc furnace is dependent on the stabilizing element of the circuit only to the extent of ensuring continuity of operation. The limitation of current fluctuations is a problem of power service and is individual for each location. The chart in Fig. 22-59 is based on reactance only. The resistance of the circuit is also a factor, and the actual value of the short-circuit current will be less than that indicated.

160. Circuit Characteristics. The arc-furnace circuit (containing resistance and reactance) is operated at constant voltage and supplies the unity power-factor load, the arc or arcs. The characteristics of this type of circuit for a given applied voltage are shown in Fig. 22-60. The maximum power of the circuit occurs at 0.707 power factor. The maximum power in the arc occurs at a higher power factor of the circuit, a value dependent on the constant of the circuit.

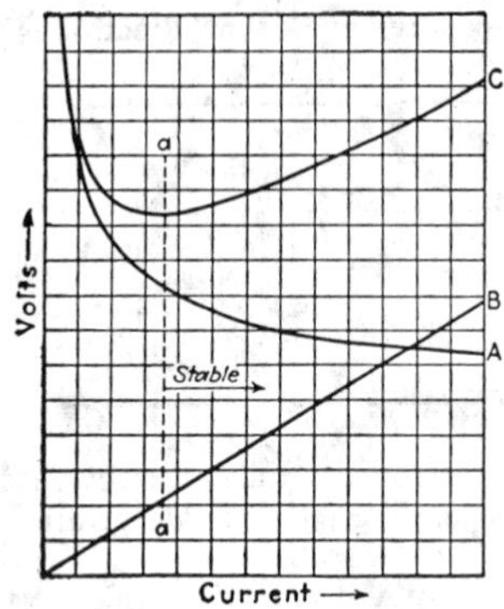

FIG. 22-58. Volt-ampere characteristic of arc in air.

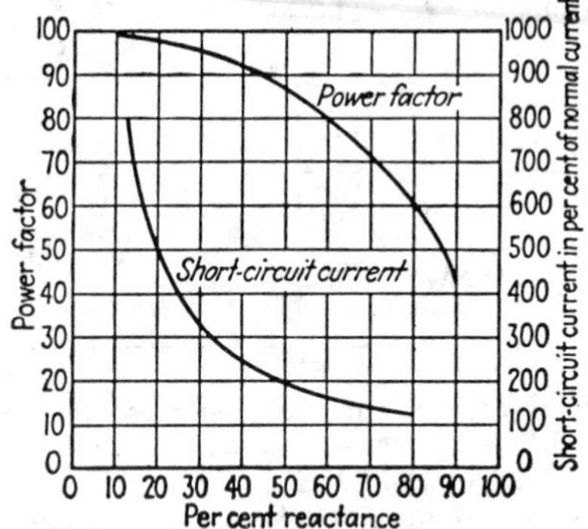

FIG. 22-59. Effect of series reactance with straight-line characteristic on the power factor and short-circuit current of an arc-furnace circuit.

The optimum current value is

$$i_{\text{opt}} = \frac{e}{\sqrt{2x}}\sqrt{l - \frac{r}{z}} \qquad \text{(amperes)} \tag{22-44}$$

The corresponding value of the maximum power in the three arcs of a 3-phase furnace is

$$p_{\max} = -\frac{3e^2}{2(z+r)10^3} \qquad \text{(kW)} \tag{22-45}$$

The power factor of the circuit corresponding to the optimum current is

$$\cos\varphi = 0.707\sqrt{l + \frac{r}{z}} \tag{22-46}$$

where e = applied voltage, r = resistance of the circuit in ohms, x = reactance of the circuit in ohms, z = corresponding impedance. All are phase-to-neutral values, counting from the point in the power system where the voltage is held constant.

There is an individual optimum current value for each value of voltage applied to a given circuit as shown in Fig. 22-61.

The slope of the power-current characteristic on each side of the optimum current value is small; the term means practically a range of current values below and above the actual optimum current. This value of current relates only to the circuit charac-

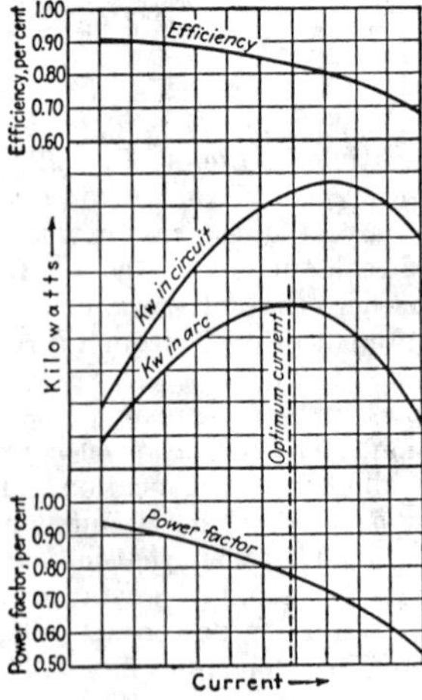

FIG. 22-60. Characteristics of an arc-furnace circuit.

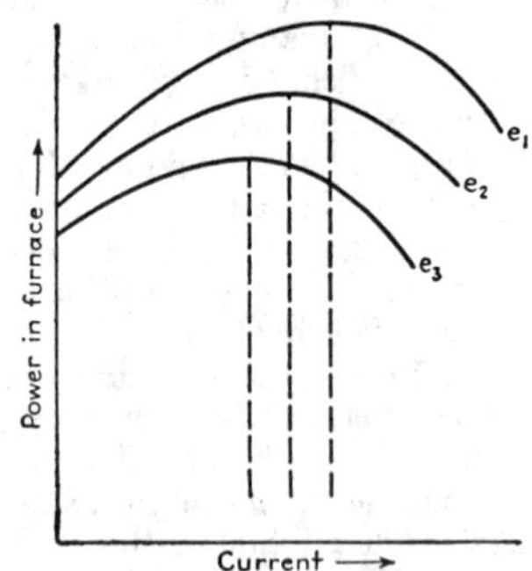

FIG. 22-61. Relation of optimum-current values and voltage.

teristics. Some value of current lower than the optimum value for a given applied voltage may give the desired power in the arcs. A common error in the operation of furnaces is the use of current values higher than the optimum value.

The values of optimum current and maximum power for a given voltage can usually be determined by trial toward the end of a heat when the circuit is stable and balanced.

161. Electrical Apparatus. The rating of the electrical equipment of a 3-phase arc-furnace installation varies for a given size furnace for the class of service and in some cases according to the power-service conditions. The electrical equipment includes:

a. A variable-ratio power transformer.

b. Reactors if required.

c. An automatic current regulator.

d. A control panel for the operator.

e. Electrode motors and tilting motors.

f. A main-line circuit breaker and disconnecting switches.

162. Transformers. The features which distinguish the arc-furnace transformer from the conventional power transformer are:

a. Individual service.

b. No requirement of regulation.

c. A wide range of comparatively low secondary voltages and correspondingly high secondary currents.

The power-time relation of a melting furnace—batch operation—is a declining characteristic corresponding to the decreasing temperature gradient within the chamber as a charge of metal passes from the solid to the molten state. At the end of the melting period the power required is the practically constant value of the rate of heat loss from the furnace. This continues until the metal is poured.

The input of power to the furnace is proportional to the square of the applied voltage. Hence, the applied voltage should be reduced as the heating cycle progresses, to follow the decreasing temperature gradient. This ideal procedure is approached in practice by a multiple-voltage operation. Practice during the past few years for new installations has been four operating voltages. The trend is to increase this number. There is considerable variation in arc-furnace service. Hence, furnace transformers have a range of voltage taps for the selection of the operating voltages found to be best suited in each case.

The maximum secondary voltage (line-to-line open-circuit voltage) of 3-phase arc furnaces seldom exceeds 275 V; this limit is fixed because of insulation and safety considerations. A maximum voltage within the range 200 to 250 is common practice.

A typical specification for a 3-phase arc-furnace transformer includes an extended primary winding with two taps therein with a secondary voltage range 235/220/205/190/175/160 V, with the primary winding connected in delta. This voltage range is extended by changing the connection of the primary winding from delta to Y to get 58% voltage from each tap.

An example of operating voltages from the range of taps cited is 235/205/175/118 V. The last-named voltage is obtained from the 205-V tap by using the Y-connection in the primary winding.

The rating of a variable-ratio transformer is proportional to the product of the maximum secondary voltage (open-circuit value) and the maximum secondary current. Full rated capacity at the three highest secondary voltages is usually sufficient. With the primary windings changed to the Y-connection, the kVA rating of each voltage tap is 58% of its rating for the delta connection.

Depending on the size of the transformer, there may be one, two, or more secondary coils per phase. The bar leads from these coils extend through the transformer tank, plus and minus arrangement, for the completion of the 3-phase connections outside the tank. The delta connection is standard practice for the secondary circuit.

The 3-phase water-cooled transformer is a preferred type. Self-cooled units and forced-air-cooled units are used to some extent. Space and weight limitations in some cases make three single-phase units necessary.

163. Reactance. The performance of arcs in metal-melting furnaces is illustrated by oscillograms by Clark.[7] At the start of a heat the charge is cold, and a carbon metal arc is erratic. Within a short time the conditions are much improved by the entrance of metal vapor into the arc stream, and the circuit becomes stable to a degree dependent on its reactance.

The performance of a furnace circuit during the initial period of a heat can be improved by the use of one of the lower operating voltages during the starting period.

There are no criteria for stability in current limitations in arc-furnace circuits and hence no standard values of reactance in these circuits. As a rule, 40 to 60% reactance is satisfactory.

The inherent reactance in the circuit of a large furnace—5,000 kVA and larger—may be, and usually is, sufficient for the need. As the secondary voltage is fixed by conditions other than the kVA rating of the circuit, the smaller installations require more or less supplemental reactance. The values of supplemental reactance given in the next table represent average practice for 60-c circuits. (Percent reactance denotes the percent reactive-voltage drop in the circuit with rated current.) This reactance is added by reactors in the primary circuit.

Transformer Rating, kVA	Supplemental Reactance, % (Rating of Reactors)
Up to 1,000	35–40
1,001–2,000	30–35
2,001–3,000	25–30
3,001–4,000	20–25
4,001–5,000	10–20

The normal reactance of 60-c furnace transformers ranges from 5 to 7%. Reactance values higher or lower in each case, within certain limits, than the range noted can be obtained by design but may entail a sacrifice one way or another in the design of the transformer. Hence it is considered better practice to use a normal design of transformer and to add supplemental reactance, if needed, by reactors.

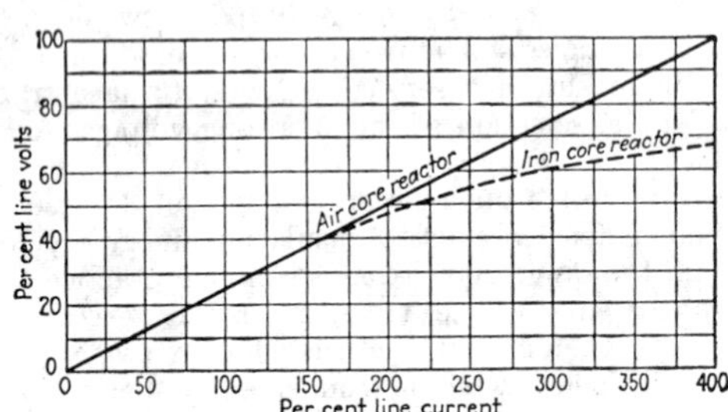

FIG. 22-62. Volt-ampere characteristic of a 25% iron-core reactor and corresponding air-core reactor.

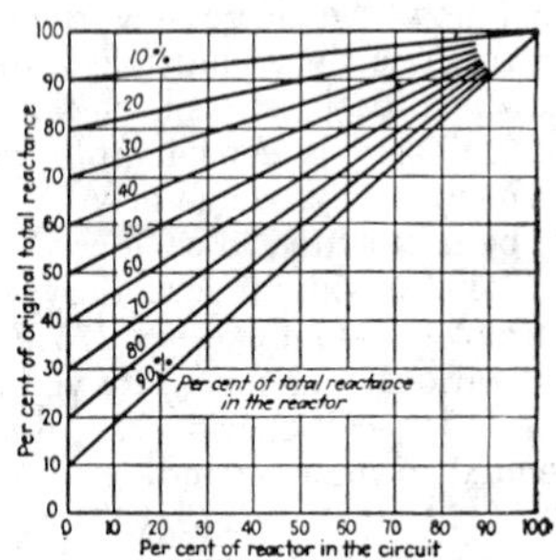

FIG. 22-63. Percentage of reactor to be left in the circuit for a given percentage of the original total reactance.

Iron-core reactors mounted inside the transformer tank for supplemental reactors, if required, are standard practice. The characteristic of this type of reactor for this service, along with the straight-line characteristic of air-core reactor for comparison, is shown in Fig. 22-62.

Four-voltage Operation of a Three-phase Arc Furnace

Switch position	Connection of primary winding	Reactor taps	Voltage taps
I	Delta	30%	220
II	Delta	20%	180
III	Y	10%	235/136
IV	Y	0%	200/116

Reactors should have a number of taps for adjustment after installation. The chart of Fig. 22-63 gives the percentage of the reactor to be left in the circuit for a given reduction in the total reactance of the circuit. For example, with a 50% reactor, to reduce the total reactance of the circuit to 70% of its original value, 40% of the reactor winding is retained in the circuit.

The transformer taps and reactor taps are connected into a common terminal board so arranged that any combination of transformer taps and reactor taps can be made for each of the selected operating voltages.

The diagram of Fig. 22-64 illustrates a tap arrangement and switching arrangement for operating voltages and the use of reactor windings for both the delta and the Y-connection.

An example is the connection of one installation given in the table.

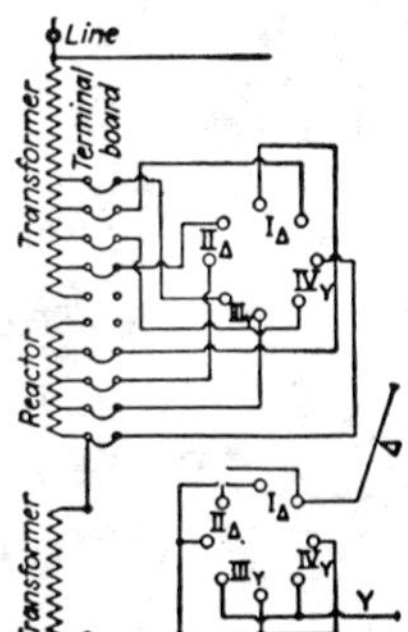

FIG. 22-64. Diagram for four operating voltages. One phase only of a 3-phase transformer.

164. Automatic Current Regulator. By reference to Fig. 22-60 it is noted that a change in current causes a change in the power of an arc-furnace circuit. Within the limits of circuit stability the current with a given applied voltage can be changed by changing the length of the arc. This is the principle of the power regulation of arc furnaces.

The intermittent-type current regulator, which operates within preset limits of current variation, has been superseded almost entirely by the continuous-type current regulator.

A simplified diagram, one phase only of the continuous-type regulator, is shown in Fig. 22-65. The principle of operation is the opposition of a voltage B derived from the circuit of the arc by a reference voltage A. The resultant value of these two voltages determines the polarity of the generator that drives the electrode motor. Thus the length of the arc—and correspondingly the current in the arc—is maintained at a predetermined value.

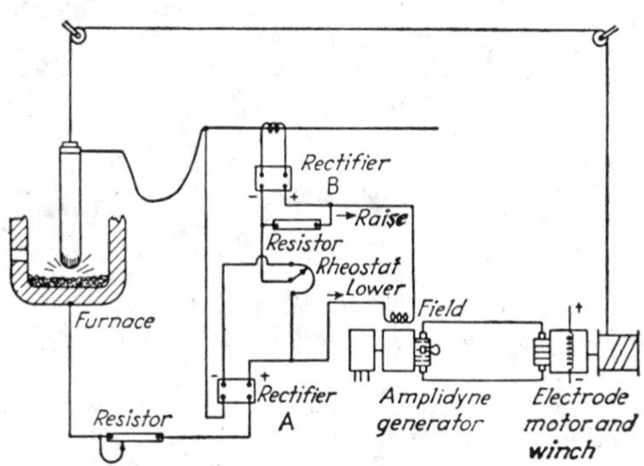

FIG. 22-65. Simplified circuit diagram of continuous-type automatic current regulator for 3-phase arc furnace. One phase only.

Each phase in the 3-phase circuit is regulated independently. However, because of the common electrode, the charge, or both, the three elements of a 3-phase regulator work together for the maintenance of equal currents in the three circuits of the power system.

The automatic regulator performs three other functions:

a. The feeding of the electrodes at the rate of consumption.

b. The removal of partial short circuits caused by the electrodes coming into contact with the charge.

c. The protection of the equipment in case of a failure of the power supply.

165. Operators Panel. The standard equipment consists of three ammeters, a polyphase wattmeter, a voltmeter, and the necessary rheostats and switches for the operation of the furnace.

166. The electrode drive is a reversing service—rapid at certain times—and a motor with a low WR^2 effect is desirable.

167. The main-line circuit breaker serves both as a protective device and as a switch. The switching service rate is many times per day. In normal operation the arc circuit is opened by raising the electrodes so that the circuit breaker opens the magnetizing part of the power transformer.

All changing of taps is done with a main-line circuit breaker open (no-load tap changing). The tap-changing switch is interlocked with the circuit breaker to prevent incorrect operation. As a rule the tap-changing switch is mounted inside the tank in 3-phase transformers and outside when three single-phase units are used. The tap-changing switch can be motor-operated or hand-operated; the former is the more general practice.

168. Single-phase Arc Furnace. The most common single-phase arc furnace is the automatic rocking furnace (see Fig. 22-66). This furnace is used extensively for melting both ferrous and nonferrous metals and alloys. Standard sizes extend up to and include 600-kW ratings for melting 4,000 lb of cold steel scrap in 90 min.

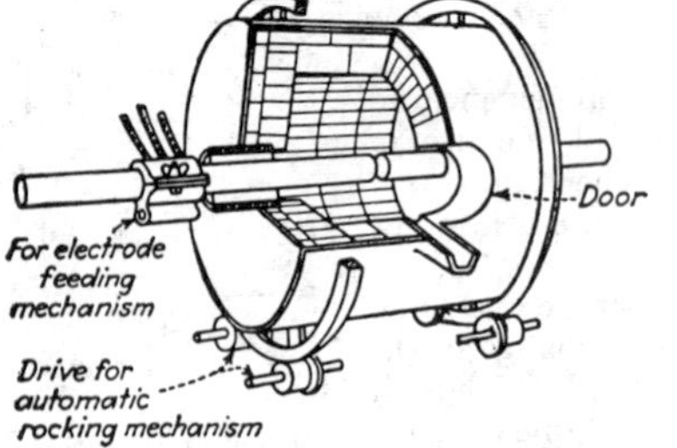

Fig. 22-66. General design of single-phase rocking arc furnace.

169. Typical performance of the rocking arc furnaces in brass-foundry service is represented by the following record of melting:

Charges 1,000 lb red brass.
Operation 8 to 9 h/day.
Melting time, average, 30 min.
Pouring temperature 1200 to 1230°C (2192 to 2246°F).
Production, 8 to 9 h, 12,000 lb.
Operating efficiency 290 kWh/ton.
Electrode consumption 3¼ lb/ton.
Metal loss 0.5%.

170. The load characteristics of a single-phase arc furnace are similar to those of 3-phase arc furnaces. However, as there is no arc between an electrode and the charge, the initial performance of the single-phase furnace is somewhat better than the initial performance of the 3-phase arc furnace. The average power factor of the single-phase furnace is 70 to 80%.

The accompanying data pertaining to the rocking arc furnace are from the paper by Clark,[7] which contains oscillograms of the operation of this furnace.

Transformer kVA	Holding capacity, scrap iron	Maximum swing, kVA
190	350 lb	400 at 0.40 pf
525	½ ton	1,000 at 0.40 pf
875	1½ tons	1,500 at 0.40 pf

It should be evident from an analysis of a single-phase arc furnace that a balanced 3-phase load can be obtained by a proper capacitor-reactor relationship in the other phase.

171. Electrical equipment for single-phase arc furnaces is similar to that for 3-phase arc furnaces. Usually only one operating voltage is used.

172. Gray Iron. Gray iron with uniform structure and high engineering properties, viz., tensile, bending, shearing, and impact strength, is produced in arc furnaces. The method of production involves superheating the iron after melting to about 1600°C (2912°F) and holding it at the elevated temperature for a brief period of time. Gray irons with the tensile strength of 40,000 lb/in² are produced regularly by this method.

With a comparatively large and continuous production of gray iron, say, 25 tons/day and more, it is practicable to combine melting in a cupola and heat-treating the molten metal in an arc furnace, i.e., duplexing. When the production rate is too low to warrant duplexing, the arc furnace can be used for either continuous cold melting (periodical charging of cold metal and tapping) or batch melting. Induction furnaces can also be used in this type of operation and to the extent that they automatically stir the melt are more advantageous than the arc furnace.

Heat absorbed by the iron for melting only is about 270 kWh/ton. The total heat-absorption values for temperatures above the melting point are given in Fig. 22-67.

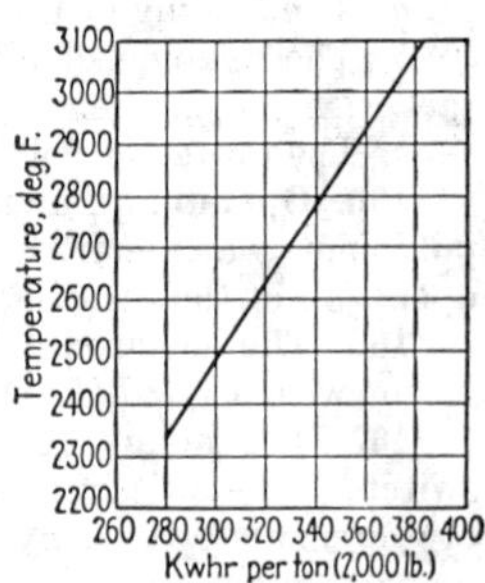

Fig. 22-67. Heat content of molten gray iron.

A record of an arc furnace in the production of high-quality gray iron by batch melting is given in the adjoining table.

Heat No.	Time per heat (includes time for charging and pouring)	KWh/ton
1	2 h 35 min	670
2	2 h 30 min	550
3	2 h 25 min	520
4	2 h 20 min	520
5	2 h 20 min	520
Average.	. . . .	556

NOTE: The higher rates of the first two heats were caused by the absorption of heat by the refractory lining of the furnace.

In the duplexing method the amount of energy required for the second stage of the process, i.e., superheating, depends on the entering and leaving temperatures of the molten metal, the conversion efficiency of the furnace, the length of the holding period, etc. The value ranges in practice are 50 to 150 kWh/ton; a fair average is 100 kWh/ton.

The energy added for the superheating of molten gray iron serves the same purpose of betterment of the engineering properties of the iron as alloy additions in the production of alloy cast iron.

INDUCTION FURNACES

173. Two types of metal-melting furnace that embody the induction principle are:

a. The coreless furnace. This furnace can be used in operation from frequencies of main-line 60 c up to many hundreds of kilocycles. The frequency generally depends upon the size of the furnace.

b. The core-type furnace, most generally operated from main-line frequencies, principally 60 c.

174. The Coreless Induction Furnace. The general design of this furnace is shown in Fig. 22-68. The assembly consists of three main parts: (*a*) the primary coil, (*b*) the refractory container, and (*c*) the frame, which includes supports and tilting mechanism.

The distinctive feature of this furnace in common with other assemblies for induction heating is the absence of a continuous iron path for the magnetic flux. However, iron laminations are frequently used on the larger sizes, and particularly for line-frequency furnaces, to reduce the reactance for the flux on the back or outside of the furnace. Another feature for comparison with other types of melting furnace is the small quantity of refractory material in the construction.

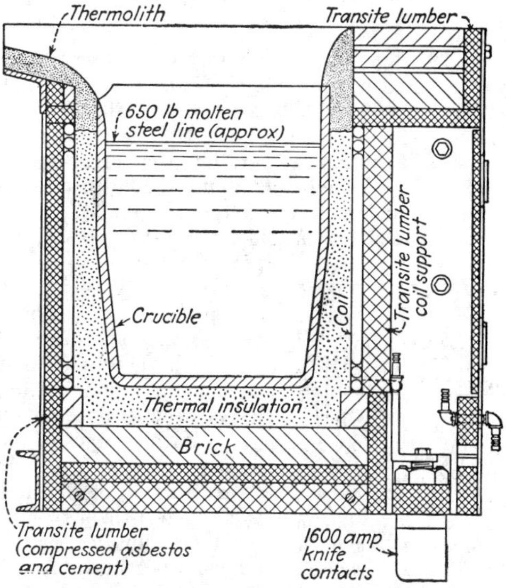

Standard preformed crucibles are used for the smaller furnaces, up to about 500 lb holding capacity. Over 500 lb capacity rammed linings are employed rather than preformed crucibles. The preformed crucibles are generally used for nonferrous melting. The base and the wall around the crucible are made by ramming a granular refractory material. The top of the wall is sealed with a refractory cement.

FIG. 22-68. Coreless induction furnace.

The containers in the larger furnaces which are made in place have rammed linings, the procedure being the same as for the smaller furnaces, except that a hollow collapsible form is substituted for the crucible to form the receptacle. The refractory practice is conventional. Acid or basic materials are used as required. In some instances the receptacle is melted in place to frit the inner wall of the refractory.

While in use, the inner refractory material of the wall and base becomes sintered to a certain depth—3/8 in or more. The outer portion of the refractory remains in granular form and serves as a support for the inner sintered lining, as thermal insulation, and as a barrier against the leakage of molten metal in case of a crack in the inner lining.

For floor-level mounting, the electrical connections are made generally by flexible cables or in some instances for small furnaces by knife contacts. If it is desirable to handle the furnace as a ladle, particularly in the small sizes, the coil is entirely separated from the crucible and the crucible is held on a small platform consisting of sand, a mounting base to hold the coil, and the crucible. The coil is removed by lifting it up, and the crucible can then be handled as a pouring ladle. In general for platform mounting the electrical connections are made with flexible cables. With this arrangement, power can be left on in the furnace during pouring—a feature often desirable when a number of castings are poured from one heat.

175. Standard sizes of coreless induction furnaces for melting nonferrous metals and alloys range from 10 to many thousands of pounds.

Table 22-20. Operating Data of Core-type Furnaces

Charge	KWh/ton	KWh to hold overnight	Net metal loss, %
Red brass	252	11	0.40
Yellow brass	195	8	0.30–0.60
Alloy: 75 Cu, 20 Sn, 3 Pb, 2 Zn	200	8	0.50 Zn
Nickel-silver	275	12	0.60 Zn
Alloy: 87 Cu, 6 Sn, 5 Zn, 2 Pb	260	12	0.50 Zn
Bronze	285	12	0.70 gross
Copper	310	13	
Zinc	90	4.5	0.25

176. Standard sizes for steel melting furnaces are 50 up to 60,000 lb. These are often listed in the ratings of 100 lb capacity, 200, 300, 600, 1,000, 1,500, 2,000, 4,000, 8,000, 10,000, 15,000, 20,000, and 40,000 lb capacity, plus the various intermediate sizes.

177. Features of operation peculiar to this type of melting furnace are that:

a. The refractory container makes necessary a large air gap (loose coupling), with consequent low power factor, 20 to 30%.

b. The charge is cold scrap metal. Thus, initially, the secondary circuit is a current path through a variety of shapes and dimensions of pieces, and contact resistance is a large part of the total resistance. The charge becomes homogeneous as the metal melts, shrinks in height, and thus decreases the coupling of the circuit. At this time, cold metal may or may not be added to the charge. It is a common practice after the first melt to maintain a molten heel in the furnace to expedite the melting of the second load.

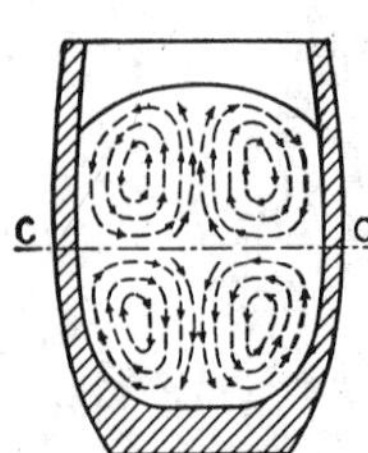

Fig. 22-69. The stirring effect in a mass of molten metal being heated by induced currents.

c. With charges of magnetic material the effect of the magnetic property is pronounced at the start of the heat cycle.

d. Stirring of the molten metal, as indicated by Fig. 22-69, is characteristic of this furnace. This movement of the bath has much metallurgical significance in the production of homogeneous alloys. The movement is often desirable and can be regulated to a certain extent by the frequency selected for the size furnace involved.

178. The characteristics of charges vary widely, but the general characteristic of the circuit during the melting of the charge is shown in Fig. 22-70. During the run from which this information was obtained, the power factor was partly corrected. Thus, the power-factor graph relates only to operation with a fixed capacitance. The initial input may be either higher or lower than the input a few minutes or seconds after the circuit is closed.

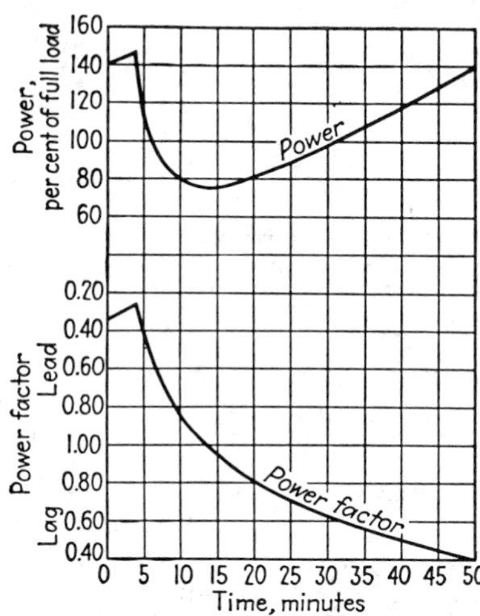

FIG. 22-70. Circuit characteristics of a coreless induction furnace.

The general practice is to vary the capacitance on the circuit during the heat cycle to maintain approximately unity power factor. The circuit diagram of Fig. 22-71 shows the arrangement for switching capacitors for this purpose.

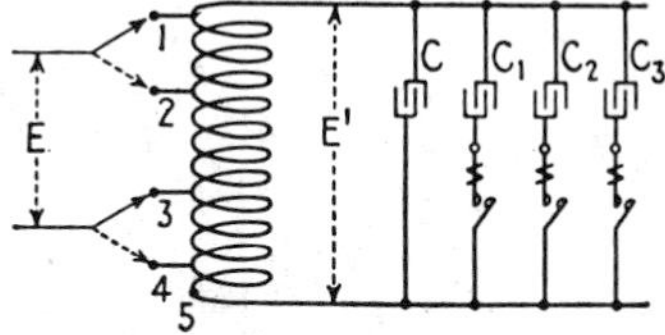

FIG. 22-71. Circuit diagram of a coreless induction furnace.

179. Frequency. The primary technical factor in the selection of frequency for a metal-melting furnace is the desired degree of stirring of the molten metal for the size of furnace required. This stirring effect is proportional to the square of the ampere-turns and inversely proportional to the frequency. For a given power value the current decreases and the voltage increases with increase of frequency.

A second consideration is the fineness of the scrap metal of the charge. If the pieces are very small, there may be difficulty in starting the melting of a cold charge. The frequency must be high enough in a given case to give an electrical efficiency high enough for starting (see Fig. 22-37). These conditions and the economics of the service early led to the adoption in this country of 960 c for steel-melting furnaces, 100 kW and above, and 3,000 c for smaller furnaces. As a rule these frequencies are also suitable for melting nonferrous charges. Various frequencies are used for laboratory furnaces. In general industrial practice today the trend is to use 60 c where large-tonnage furnaces are being used.

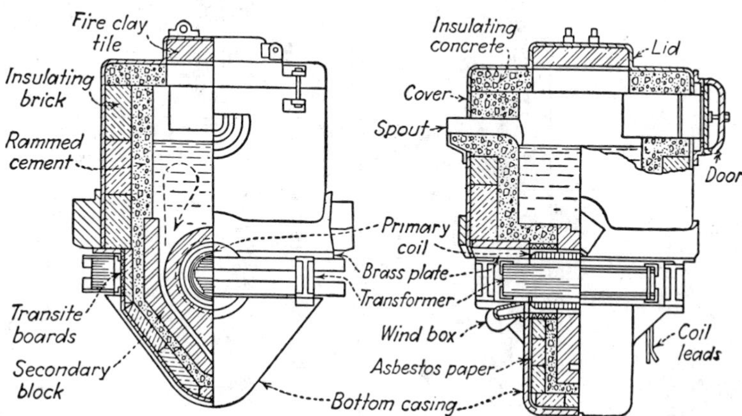

FIG. 22-72. Submerged resistor furnace for melting heavy nonferrous metals and their alloys.

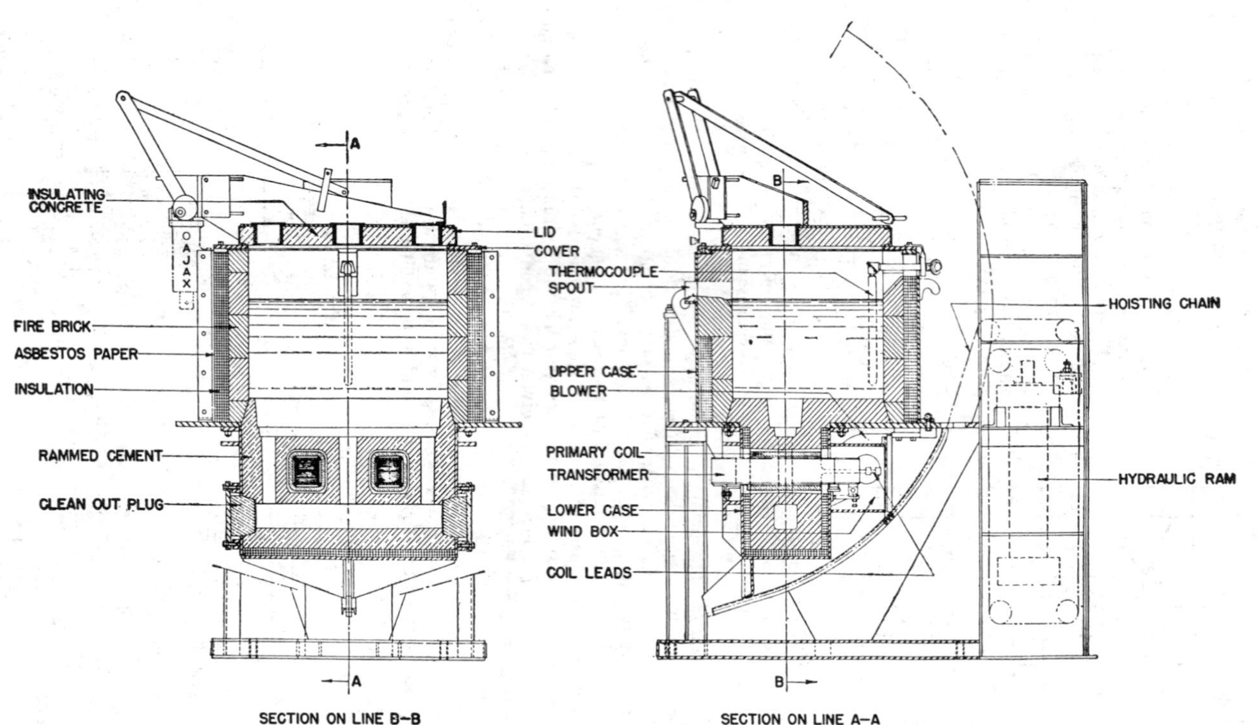

Fig. 22-73. Submerged resistor furnace for melting aluminum and its alloys.

180. Service. The coreless induction furnace is primarily a metal-melting unit. An important use of this furnace is the production of carbon ferrous alloys. The refining of steel, i.e., the removal of phosphorus and sulfur, in this furnace has not been developed to a fine art. The deoxidation of the melt, e.g., by the addition of aluminum just before pouring, is not here classed as refining. Various special services are: duplexing steel, vacuum melting, heating of charges of nonconducting material (with or without melting) by the use of conducting crucibles, etc.

181. Performance. The melting rate of a furnace, and consequently the power input, is determined in each case by the rate at which the molten metal can be used. Heat cycles of ½ h for small furnaces, 1 h for medium-sized furnaces, and 1½ to 2 h or more for large furnaces represent typical practice.

182. Energy consumption varies for a given material with the size of the furnace, the melting rate, and the idle time between heats. Representative values are 375 kWh/ton (2,000 lb) for red brass and 600 kWh/ton (2,000 lb) for steel. These figures are lower with larger furnaces in continuous operation and somewhat higher with small units and infrequent service.

183. The Core-type Furnace. The general design of this type of furnace for melting nonferrous metals and alloys is shown in Fig. 22-72. The basic feature is a single-turn loop of molten metal below and connected to the bath serving as a secondary circuit. A conventional laminated steel core and primary winding complete the transformer feature.

The heat is developed in the loop of molten metal below the body of the charge. Electrodynamic action causes motion of the metal in the loop as indicated. This movement serves to transfer the heat and to stir the molten metal in the chamber above the loop.

184. Starting the furnace requires sufficient molten metal to close the secondary circuit. In changing from one metal or alloy to another the furnace must be emptied. For day-by-day melting of the same metal or alloy—the usual practice—the charge, or a portion of the charge, is held molten overnight by using a below-normal voltage.

185. The V-shaped loop furnace (Fig. 22-73) is designed for melting the heavy nonferrous metals and alloys. It is used widely for melting brass. With these metals the slags, or nonmetallic particles, in the melt are much lighter than the metal, so that they tend to float on the surface of the bath and do not interfere with the circulation of the metal in the loop of the secondary circuit.

Standard sizes on these furnaces range from 60 to 1,000 kW on single-phase, 60 c in this country, and for standard voltages up to 600 V inclusive.

186. A design of the core-type furnace for melting aluminum and its alloys embodies two or more vertical channels below the bath and connected to the bottom by horizontal channels of larger cross section, as shown in Fig. 22-73. The slag particles of the lighter metals have about the same specific gravity or are heavier than the molten metal. Such particles tend to accumulate in the channel of the secondary circuit. The design of the channel noted above provides for its easy cleaning from time to time while the furnace is in operation.

187. Power and power factor of the core-type furnace with constant applied voltage vary with the resistivity of the metal in the loop that forms the secondary circuit. The graph of these characteristics is shown in Fig. 22-74.

188. Operating Data of Core-type Furnace (see Table 22-20).

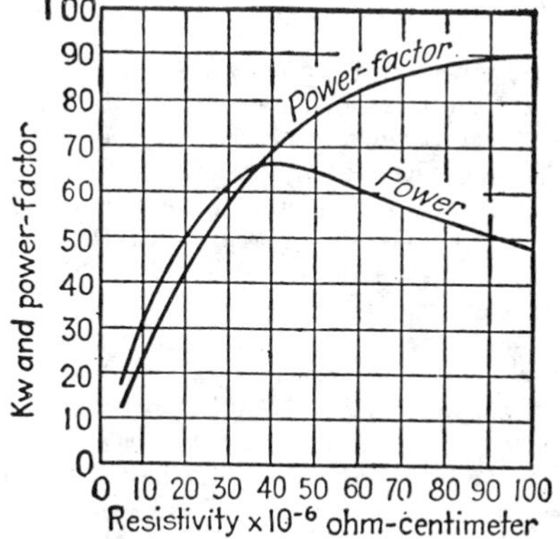

FIG. 22-74. Circuit characteristics of a submerged resistor furnace.

RESISTANCE FURNACES

189. Resistance Furnaces. Temperature range 1500°C (2732°F) and above. This type of furnace is generally an open-top heating chamber with electrodes—movable or fixed—buried in the charge. Variations are furnaces with closed tops, furnaces with a resistor buried in the charge, etc.

The general service is heating charges of a refractory nature to bring about chemical reactions or changes in the physical structure of a material. Where the product is obtained by a chemical reaction, e.g., the reduction of a metallic oxide, the term "smelting furnace" applies.

The length and cross-sectional area of the path of the current through the charge are proportioned to suit the power characteristic of the material to be heated. Where the buried resistor is used, its design is based on similar considerations. In all cases the load is resistance, i.e., the I^2R effect. In some cases where the charge forms a resistor there may be arcing along the path of the current.

There are no standard designs of resistance furnaces. Each application is an individual undertaking, and the furnace is assembled in place. The simplicity of the construction permits a wide range of designs and much latitude in dimensions.

Resistance furnaces are unity-power-factor loads, and the circuit is stable; i.e., reactance is not needed as with other furnaces. However, there are reactances and resistances in the furnace conductors, and the circuit characteristics are the same as for arc-furnace circuits. The voltage is low, generally less than 250 V. The corresponding high-current values introduce problems of inductance in the leads to the furnace, particularly with 60-c power supply.

190. Furnaces with movable electrodes (see Fig. 22-75) are stationary, vertical, cylindrical, or rectangular structures, single- or 3-phase. The latter is more common.

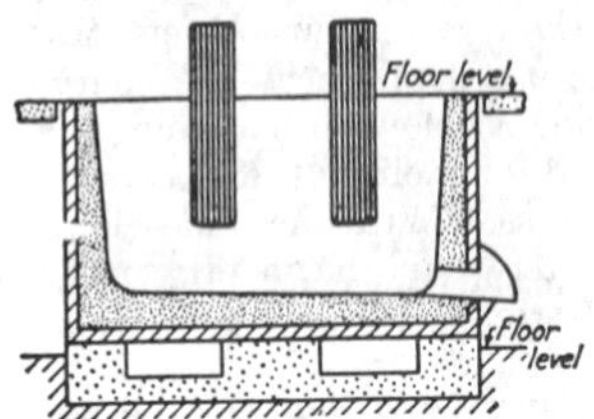

Fig. 22-75. Open-type resistance furnace with movable electrodes (three electrodes for 3-phase furnace).

The material of the charge, granular in form, is fed, more or less continuously, into open-top furnaces or through chutes into closed furnaces. The product, if molten, is discharged by tapping through a side, usually near the bottom. Vapor products pass out at the top or through the side openings in the chamber. If the product is solid, e.g., fused magnesia, the batch method is used and the furnace is disassembled for removal of the product.

The regulation of movable-electrode furnaces is accomplished by electrode movements to hold constant current, as with arc furnaces. The load is fairly uniform, and the regulating duty is light. Either d-c motors or a-c motors can be used to drive the electrodes. Direct-current motors are more often employed because of the ease of speed adjustment and the superiority of dynamic braking.

Continuous operation at one voltage is general practice. Transformers for these furnaces have a number of taps in the primary windings for the selection of the secondary voltage to suit operating conditions, which may change from time to time.

191. Batch Furnaces with Fixed Electrodes. A horizontal rectangular chamber with an electrode at each end is typical construction where the current is passed through the charge.

Refractory materials have negative resistivity vs. temperature coefficients. Hence, to maintain constant power in a batch furnace as the temperature of the charge rises, the applied voltage must be varied as expressed by Eq. (22-17*a*) or (22-17*b*). Accordingly the regulation of furnaces of this type is by voltage adjustments, either load-ratio control or no-load tap changing, the latter with or without an induction regulator for intermediate voltages. Generally manual operation is sufficient. Constant power is not the rule. In many cases the power input is varied to correspond to a prescribed heat cycle.

Some of the more common uses of resistance furnaces are noted in the following paragraphs.

192. Calcium carbide furnaces are open-top vertical chambers with movable carbon electrodes (round or rectangular), one electrode for single-phase furnaces, three for 3-phase furnaces. The 3-phase furnace is illustrated by Fig. 22-75. Both solid and self-baking electrodes are used.

The raw materials are calcium oxide (lime) and carbon (coke or anthracite coal). The molten carbide is discharged by tapping and cast into shapes for crushing and sizing when cold. The chemical reaction is

$$\begin{array}{cccc} CaO + & 3C \rightarrow & CaC_2 + & CO \\ 56 & 36 & 64 & 28 \end{array} \quad \text{(molecular weights)} \qquad (22\text{-}47)$$

The carbon monoxide product of the reaction is usually a waste product, being burned as it escapes from the furnace.

The sizes of calcium carbide furnaces range from 1,000 to 10,000 kVA, electrical rating. Transformers for furnaces above 5,000 kVA are often single-phase units. A typical secondary-voltage range for large 3-phase furnaces is 150/144/138/132/126/120/114 V, line-to-line open-circuit values. The wide range of voltages is provided for power adjustments as it rises. No-load tap changing is general practice.

These furnaces can be operated over a considerable range of voltages and are thus adaptable to load equalization of power systems.

The total heat required per ton (2,000 lb) of calcium carbide depends on the purity of the raw materials, size of the furnace, and the rate of working. The heat absorbed (see Example 4) is 1.22 × 2,000 = 2,440 kWh/ton. The average operating efficiency of large furnaces may be 75%, that is, 3,250 kWh/ton. The operating efficiency of smaller furnaces is about as follows:

Sizes, kVA	KWh/Ton (2,000 Lb)
1,000	5,500–7,500
3,000	4,000–5,000
5,000	3,500–4,000

The electrode consumption is 50 to 75 lb/ton of calcium chloride.

193. Ferroalloys consist of iron combined with another element, e.g., silicon, manganese, chromium, or tungsten, and with a low carbon content. Some ferroalloys with a low percentage of the alloying element are used as scavengers in the refining of steel. Examples are ferrosilicon, a deoxidation agent; and ferromanganese for deoxidation and for removal of sulfur. This class of ferroalloys is usually made in the blast furnace. Ferroalloys with a large percentage of the alloying element are electric-furnace products. These are used as vehicles for introducing the desired element or elements in the manufacture of alloy steels.

The making of a ferroalloy is by the reduction or smelting process. The charge is a mixture of the ore (oxide) of the selected metal, scrap iron, and a reducing agent, usually carbon. A fluxing agent is required when the ore is not self-fluxing. A considerable quantity of coke to furnish the carbon is required, e.g., about 1,400 lb of coke/ton of 50% ferrosilicon. Another reducing agent, e.g., aluminum or ferrosilicon, is required for some low-carbon-content ferroalloys. In some cases the low-carbon ferroalloy is made by remelting a high-carbon alloy under oxidizing conditions, thus removing a large part of the carbon.

194. Ferroalloy furnaces are of the general design shown in Fig. 22-75. The method of operation is much the same as with calcium carbide furnaces. Carbon linings are used for ferrosilicon furnaces; the linings of other ferroalloyed furnaces are usually magnesite.

Sizes range from 2,500 to 10,000 kVA. The electrodes are carbon except for some low-carbon alloys for which the graphite electrodes are used. A representative range of voltages is 140/130/120/110/100 V, line-to-line open-circuit values.

195. Ferrosilicon is the largest item in the list of ferroalloys. The standard grade is 50% silicon content. Other grades are 75 to 80% silicon and in small quantities 90 to 95% silicon. Above 95% the product is designated as metallic silicon. The charge consists of silica, coke, and steel scrap. The energy consumption range is

% Silicon	KWh/Ton (2,000 Lb)
50	5,000–6,000
75	1,050–12,000

The electrode consumption runs from 60 to 70 lb/ton of product.

196. Ferromanganese. The charge consists of manganese ore, coke, and steel scrap. The standard grade is 80% manganese. The energy consumption is 6,000 to 7,000 kWh/ton (2,000 lb) and the electrode consumption around 75 lb/ton.

197. Silicomanganese is a ferromanganese in which a part of the manganese is replaced by silicon. The composition varies within wide limitations. A typical analysis is: manganese 55, silicon 25, and iron 20%.

198. Ferrochromium is made from charges of chrome-iron ore (chromite), coke, lime, and fluorospar. The chromium content of the alloys runs from 60 to 70%. From 6,000 to 7,000 kWh/ton of alloy is required. The electrode consumption is around 100 lb/ton. For the manufacture of low-carbon stainless steel the normal carbon content of the alloy (4 to 5%) is brought down to less than 1% by refining in an arc furnace.

199. Ferrotungsten is produced in batch-type furnaces. The ores mostly used are ferberite, $FeWo_4$, and wolframite, $(Fe, Mn)Wo_4$. The alloy is resmelted to lower the carbon and phosphorus contents. The total energy required for 65 to 70% alloy is 8,000 to 10,000 kWh/ton.

200. Ferromolybdenum is a product of the batch-type furnace with 50 to 85% molybdenum. The ore is molybdenite, MoS_2, which is roasted prior to smelting. The energy consumption is 8,000 to 9,000 kWh/ton.

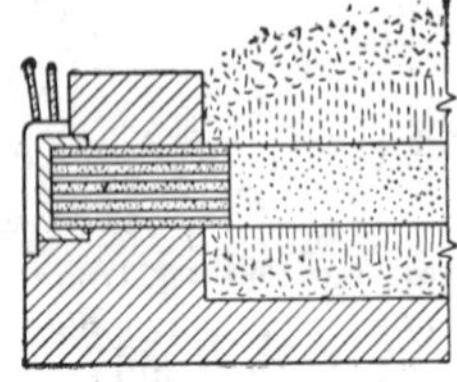

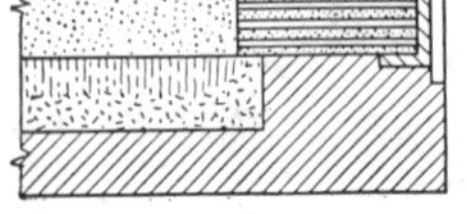

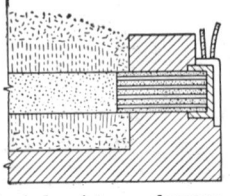

Fig. 22-76. Open-top buried-resistor furnace (silicon carbide furnace).

201. Various other ferroalloys, e.g., ferrovanadium, ferrotitanium, and ferronickel, are also electric-furnace products.

202. Silicon carbide furnaces are of the buried-resistor type (see Fig. 22-76). The charge is a mixture of silica (sand), carbon (coke or anthracite coal), and small quantities of sawdust and salt.

The mixture of raw materials is packed around a core of granulated coke which serves as a resistor for heating the charge. Previously used core material which has been converted to graphite is also used to form the core. The reaction is

$$\begin{array}{cccc} SiO_2 + & 3C \rightarrow & SiC + & 2CO \\ 60 & 36 & 40 & 56 \end{array} \quad \text{(molecular weights)} \qquad (22\text{-}48)$$

The sawdust serves to make the charge porous for the escape of the gases. It removes impurities as chlorides.

Close temperature regulation is essential, as there is a comparatively narrow region within which silicon carbide forms and above which it is decomposed.

The electrical resistance of the furnace portion of the circuit decreases with the rise of temperature, the range from initial to final temperature being 4 or 5:1. This requires either a variable-ratio transformer or induction regulator to follow the change of resistance.

Upon completion of the heat cycle the furnace is partly dismantled for removal of the product, which is a solid with a crystalline structure.

The energy required for the production of silicon carbide varies with the size of the furnace, rate of working, etc. The usual range is 7,000 to 9,000 kWh/ton (2,000 lb).

203. Silicone oxycarbides have varied uses as refractories and for heat insulation. A furnace for the production of Si_2C_2O (also a byproduct of the silicon carbide furnace) is similar to the silicon carbide furnace but has two or more resistors in parallel.

204. Graphitizing Furnaces. Artificial graphite (graphitic carbon) is made by heating amorphous carbon while shielded from air to a temperature of around 2500°C (4532°F). The presence of a metallic oxide in the charge is thought by some to aid the conversion of carbon from the amorphous to the graphitic form. The output of the furnace is:

a. Bulk graphite. Used for dry cells, electrotyping, paint pigment, lubrication, etc.

b. Graphitized carbon articles, e.g., anodes, electrodes, and brushes.

The raw material for making bulk graphite constitutes both the charge and its protective layer. Graphite shapes are made by first forming the amorphous carbon into the shape desired. These are embedded (between electrodes) in the material used for making bulk graphite and in some such manner as indicated by the sketch in Fig. 22-77 of the furnace for making graphite electrodes. Upon completion of the graphitizing process, for which several days are required, the furnace is partially dismantled for removal of the product.

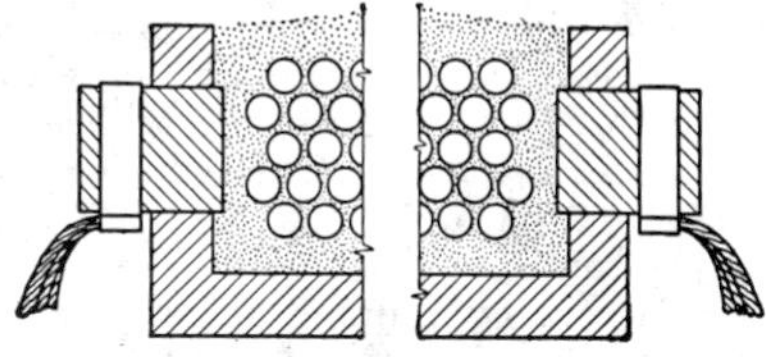

FIG. 22-77. Graphitizing furnace.

The details of the design of graphitizing furnaces vary widely. The general design is an open-top rectangular chamber with brick walls and with fixed electrodes at each end for single-phase operation.

The resistance of the charge at the end of a heat cycle is much lower than at the beginning and may be one-tenth or even a smaller fraction of the initial resistance. During the run the power input is adjusted from time to time to conform to a specified program based on the character of the charge.

The program of the heat cycle requires a transformer with a wide range of secondary voltages, e.g., 200 to 40 V, with closely spaced taps. Usual practice is to employ two groups of secondary coils, in series for the maximum voltage with each group delivering one-half that voltage. Taps are placed in the primary winding for the intermediate voltages with each of the secondary-coil connections. Load-ratio control is general practice.

The sizes of graphitizing furnaces vary over a wide range; there is no standard size. However, all are alike in requiring a wide range of voltage. Transformer sizes 1,500 to 3,500 kVA are representative for the larger furnaces. It is common practice to use one transformer to serve in rotation several furnaces. For this purpose the transformer is mounted on a truck for moving along the line of furnaces.

The power factor of the furnace circuit declines during the heat cycle as the voltage is lowered and the current increased. With large furnaces, particularly on a 60-c circuit, the power factor may reach low values, 25 to 30% at the end of the heat cycle. A method of power-factor correction used for this circuit is indicated by the diagram in Fig. 22-78. The correction of the series capacitor is automatic and relieves the power transformer of the kilovars.

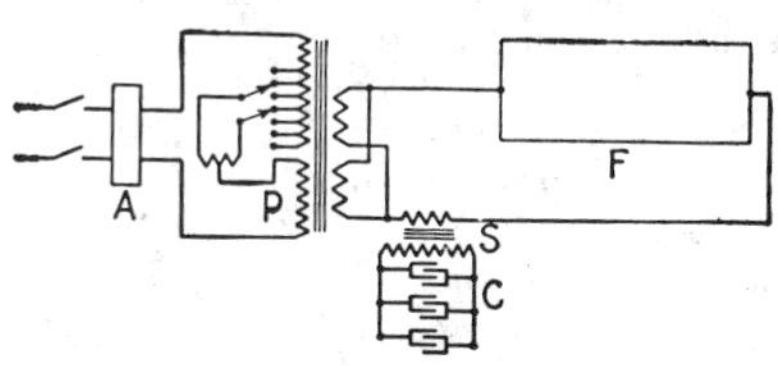

FIG. 22-78. Application of series transformer with capacitors for correction of power factor of resistance furnaces.

The energy consumption in graphitizing varies widely, from 3,000 to 8,000 kWh/ton of product, depending on the size of the furnace, the character of the charge, the length of the heat cycle, etc.

205. Fused alumina, an abrasive and a high-temperature refractory, is made by fusing calcined bauxite together with carbon (coke) and iron (usually borings and turnings); the latter two components serve to remove impurities which settle to the bottom of the furnace.

The furnace chamber is an open-top vertical cylindrical-sheet-steel shell with the hearth of carbon blocks and with two movable carbide electrodes for single-phase operation. The carbon hearth serves to complete the circuit for starting. The materials of the charge are fed into the chamber, and the electrodes are raised gradually as fusion of the charge takes place. This is continued until the chamber is filled. The part of the charge surrounding the core serves as a refractory lining. At the end of the cycle the furnace is allowed to cool, and the core solidifies.

206. Fused magnesia, a high-temperature refractory, is made in a manner similar to that for fused alumina.

207. Fused silica is produced by melting silica (quartz sand) in a furnace with carbide resistors embedded in the charge. The fused mass is blown out of the furnace by compressed air into molds to form the articles desired, a glassmaking operation. The transparent form of fused silica is made by melting selected quartz crystals in a graphite crucible placed in a vacuum furnace with a graphite resistor.

208. Zinc Oxide. A continuous fixed-electrode vertical-resistance furnace for the last stage of the production of zinc oxide from zinc sulfate ores is described by Wheaton.

209. Pig Iron. Economic conditions in the United States do not favor the use of electric heat for smelting iron ores, as is the practice in Norway and Sweden.

Smelting furnaces for iron ores are the same in principle as ferroalloy furnaces. In both, shaft furnaces with inclined electrodes and open-top furnaces with vertical electrodes are used. The reducing agent is carbon obtained from charcoal or coke; the former is preferred. About 750 lb of charcoal or 850 lb of coke/ton (2,240 lb) of pig iron is required.

The energy required ranges from 2,300 to 3,000 kWh/long ton of pig iron. The electrode consumption is 20 to 30 lb of carbon electrodes/long ton of iron. These values depend on the silicon content of the iron, kind and grade of ore, size of furnace, and operating conditions.

ELECTRIC WELDING

RESISTANCE WELDING

210. Resistance welding includes spot welding, projection welding, seam welding, and butt welding. All are based alike on the principle of resistance heating but differ in the details of application.

211. Spot welding is applied to overlapping sheets. These are clamped between water-cooled electrodes (Fig. 22-79), pressure is applied, and a current impulse is passed through the assembly. The material in the zone of pressure is heated to fusion, and the joint thus made is cooled under pressure.

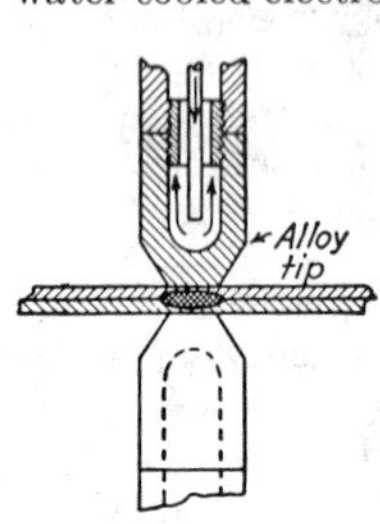

FIG. 22-79. Spot-welding electrodes and cross-sectional view of heated area.

Spot welding is largely employed on sheets 0.025 to 0.125 in thick, although welds on thicker sheets and plates up to 1 in thick are made regularly.

The area of the spot weld is small; weld diameters of 0.25 to 0.625 in are the usual range. The maximum diameter is limited by the area of uniform contact between sheets that can be obtained with practical pressure.

Rapid heating by using high current values is necessary to bring the metal that is to form the joint to the required temperature without more than normal heating of the adjacent metal. Current values range from 5,000 A upward. The voltage between the electrodes is usually less than 2 V. The open-circuit voltage is generally less than 12 V. The time period of current flow varies widely, depending on the thickness of the sheets, kinds of material, etc. For thin steel sheets this period is about 1 c ($\frac{1}{60}$ s) for each 0.010 in of the total thickness of the two sheets to be joined. The pressure, current value, and time are variables that must be correlated in each case. A spot-welding machine is shown in Fig. 22-80.

Spot welding with a single current impulse is limited to sheets not thicker than about 0.125 in. Because of the heating of the electrodes, and consequent short life, with heavier material, thick sheets are welded by using a number of current impulses, e.g., 4 c on, 2 c off (Fig. 22-81), repeated as required. The no-current intervals, i.e., cycles off, prevent the overheating of the electrodes.

The conditions which are met in spot welding vary widely; the following examples are illustrative:

a. Two 0.025-in stainless-steel sheets. Single current impulses, 2 c ($\frac{1}{30}$ s), 6,000 A. Weld diameter 0.25 in.

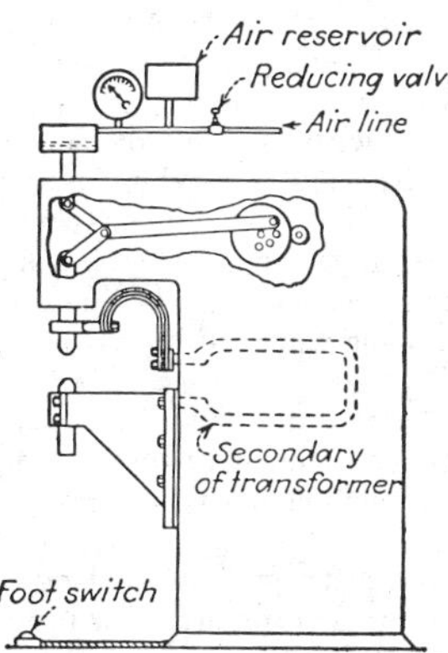

FIG. 22-80. Spot-welding machine with high-speed movement and air cushion.

b. Two 0.25-in carbon-steel plates. Eleven cycles on, 8 c off, repeated four times. Total time of current flow 44/60 s, 25,000 A, rms of cycle. Weld diameter 0.625 in.

c. Two 1-in carbon-steel plates. Twenty-one cycles on, 51 c off, repeated 34 times. Total time of current flow 11.9 s, 73,000 A, rms of cycle. Weld diameter 1 in.

The list below gives average values of measurements:

Number of spot welds 180.

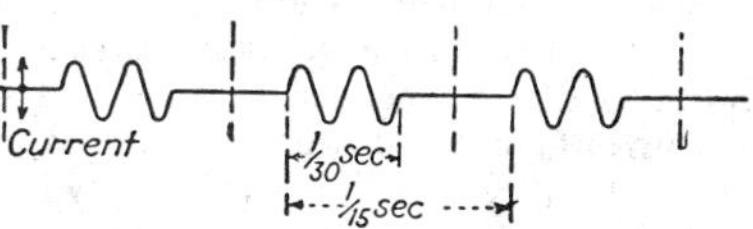

FIG. 22-81. Timing of resistance welding by thyratron control.

Time cycle, average, 0.50 s.
kVA/weld, average, 31.4.
Power factor, average, 50.7%.
kW demand/weld, average, 16.
Wh/weld, average, 2.12.

212. Projection welding differs from spot welding only in the use of the buttonlike projection on one of the sheets to define the path of the current. A number of adjacent welds can be made simultaneously without interference. A useful method for assembly work.

213. Seam welding can be made by putting a series of spot welds on thin sheets by passing the sheets between two pressure electrodes (Fig. 22-82) and with continuously interrupted current. Speeds range from 10 to 400 in/min, approximately 60 in/min being the most common speed. Twelve welds per inch with 6- to 10 in-diameter electrodes is representative practice. Examples of this type of seam welding are:

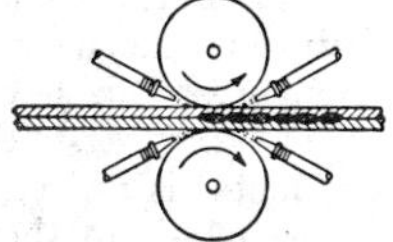

FIG. 22-82. Seam-welding electrodes with water cooling.

a. Two 0.025-in stainless-steel sheets. Speed 72 in/min, 2 c on, 2 c off; current impulses 1/30 s; 6,300 A, rms of cycle. Voltage across electrodes 0.60 V; open circuit 2.34 V.

b. Two 0.032-in carbon-steel sheets. Speed 100 in/min, 2 c on, 2 c off; current impulses 1/30 s; 13,420 A, rms of cycle. Voltage across the electrodes 0.85 V; open circuit 4.25 V.

Sometimes classed as seam welding but otherwise known as edge welding is the type of welding accomplished on continuous tube mills or continuous butt-welding installations where the edges of metal are brought together and welded as a seam. In this type of welding pulsed currents are not used to obtain the weld, but rather a continuous a-c or d-c supply fed through rotating electrodes to the abutting edges of the seam produces the electrical energy required for the weld.

214. The variables in these three methods of welding which must be correlated are:

a. Electrode diameter, material of electrodes, contact area, cooling, and pressure.

b. Nature and surface of metal or metals to be welded.

c. Timing of current flow.

d. Current value and waveform.

215. Timing circuits for resistance welding are either (*a*) nonsynchronous or (*b*) synchronous (relating to the welding current zero).

Nonsynchronous timers are of various types, ranging from the simple application of foot pressure to close and open the welding circuit, according to the judgment and skill of the operator, to the combinations of cams, limit switches, etc., the functions of which are independent of the operator.

The synchronous timers are electronic, using thyratrons. The most simple of these are those which are adjustable, as regards to current impulse, in 1-c steps. A complete electronic control with an ignitron contactor is indicated by Fig. 22-83. When used with an a-c supply, the ignitron will pass current each half cycle that the anode is positive. A second ignitron is connected inversely to pass current during each of the other half cycles.

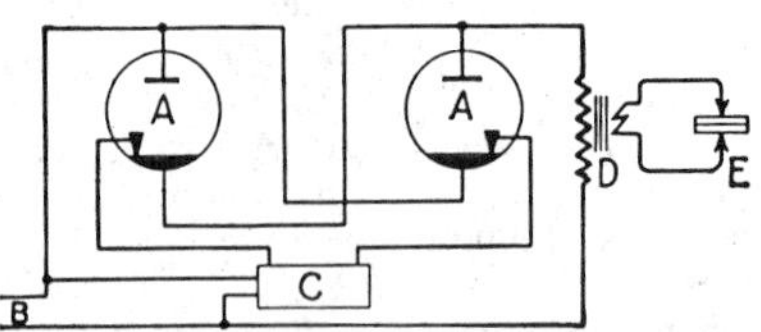

FIG. 22-83. Electronic control for resistance welding.

Electronic control for resistance-welding circuits is inherently a precision device. The choice between the two types of control is primarily a matter of the value of exactness in timing, which becomes more important as the welding time is decreased (see Fig. 22-83).

216. Current control in welding circuits, i.e., the control of the rate of heat development, is obtained by taps in the primary winding of the welding transformer or by phase control of the welding current, retarding the point at which current starts to flow in each half cycle beyond the power-factor angle. A combination of the two methods is often used.

217. Percussion welding is a self-timing spot-welding method. A current impulse is obtained by the discharge from a capacitor or from a magnetic field. (This method is usually classified under d-c resistance welding.) The capacitor arrangement for percussion welding is used extensively in the manufacture of lamp filaments.

218. Butt-welding Upset Method. End-to-end welds, lap welds, and butt welds are included here. The faces of the parts should be prepared for even contact. The parts are clamped and brought together by hydraulic pressure, thus closing the circuit. Plasticity occurs simultaneously over the entire contact area. The joint is cooled under pressure. The method has a wide variety of applications, large and small. In continuous pipe welding, the current is conducted to the work by a pair of rotating electrodes. This, as previously mentioned, is sometimes referred to as the seam-welding technique. Butt welding can be obtained by utilizing induction-heating equipment; how this can be accomplished will be described later.

219. Flash Welding. No preparation on the face is necessary. One face is stationary; the other movable. Voltage is applied through pressure contacts. A momentary contact establishes current flow and flashing (the melting and vaporization of the minute contact points). The movable face is fed forward continuously at the rate at which the metal is melted. The pressure of the gases developed between faces prevents the access of air to the hot faces and blows out the molten metal. The metal adjacent to the faces becomes plastic, and the joint is made by fusing the parts together by hydraulic pressure. The circuit is opened by a jam relay as the faces make contact. However, if a considerable amount of flashing is required to remove impurities, the circuit is held closed for a predetermined time and may be opened just prior to the application of the last pressure value.

An example of flash welding is the production of continuous rails. The length is limited only by the facilities for handling, for transportation, and for track laying. Lengths of ½ mi or more are in service. A 131-lb rail, 13 in² of cross-sectional area, requires current impulses of around 40,000 A and about 60,000 A for a few cycles when the faces are brought together for forging the joint. The open-circuit voltage is 12 to 18 V for 60-c circuits. The time period for welding operation is 1½ to 2 min.

Other items that are commonly flash-butt-welded both for economic reasons and for strength of the component parts are aircraft-landing-strut assemblies and heavy-duty crankshaft sections. It is possible not only to flash-butt-weld these types of products but also to anneal the weld by a lower-value current being allowed to pass through the welded joint after the weld is completed. Welds in this type of product may take anywhere from 2 to 5 min to complete.

ARC WELDING

220. Welded joint is a union of metal parts made by localized heating, without pressure in arc welding and supplemented by pressure in resistance welding. The technique of welding is built around the methods of concentration of heat and is based on the metallurgy of welds in the mechanics of joints.

221. Arc Welding; Manual Operation. Methods of arc welding are classified as follows:

1. Direct-current arcs, one electrode—metal or carbon.
   a. Unshielded arc.
   b. Shielded arc.
2. Alternating-current arcs, metal electrodes.
   a. Unshielded arc, one electrode.
   b. Shielded arc, one electrode.
   c. Shielded arc, two electrodes (atomic-hydrogen process).

Joints. The five principal types of arc-welded joints are butt joints, corner joints, edge joints, lapped joints, and T-joints. Any length of joint can be made by travel of the electrode; one or more passes may be required.

Joint Positions. The position of the joint (see Fig. 22-84), especially for multipass work, is an important factor in the economics of welding. The maximum welding speeds obtained with the flat position are about twice as fast as those with the horizontal position and about four times faster than the maximum speeds obtained with the overhead and vertical positions. Welding in the flat and horizontal positions is much easier for the operator than overhead welding—so much so that the arc can be kept in operation about twice as long per hour of overall time. Likewise, vertical welding is less tiresome than overhead welding and represents an increase of about 50% in welding time without an increase in the fatigue of the operator.

The weldability of metals and alloys can be stated as easy, intermediate, and difficult to weld. This with reference to carbon steel relates to carbon content. Carbon steels up to about 0.30% C are easy to weld. That class of steels is used in a large part of the fabrication of materials by welding. Practically all metals and alloys can be electrically welded, but in many cases an individual study of the metal or alloy, dimensions, and position of the weld is required to develop the particular technique necessary.

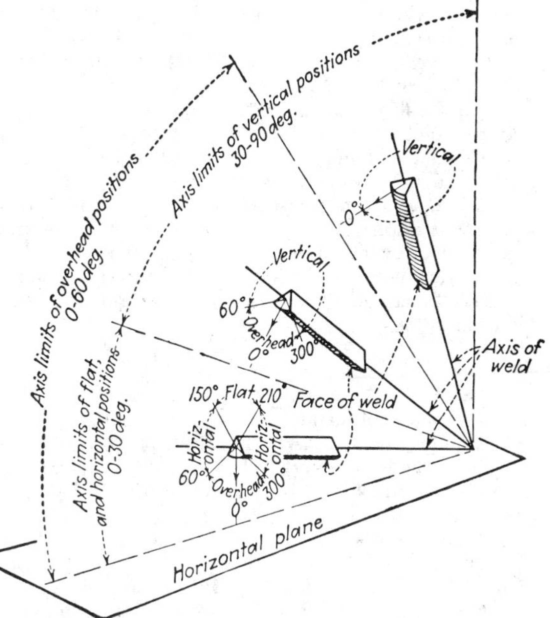

FIG. 22-84. Position of welds—arc welding.

A characteristic feature of arc welding is that symmetrical sections (I-beams and T-beams) can be used.

222. Single-electrode Welding. The arc is maintained between the electrode and the work. Round steel electrodes, with carbon content suited to the class of work, supply the metal (filler metal) for the joint.

The phenomenon of the transport of the molten metal from the electrode to the work is dependent on the position of the joint. It is, however, a factor in the selection of the electrode for each joint position. The making of a sound weld requires that the filler metal be deposited only on molten-base metal. The depth of the molten-base metal is termed "penetration."

223. Metal electrodes are of two types:

a. Bare electrodes (or lightly coated electrodes).

b. Coated electrodes.

The bare electrode is covered with a film or its oxide to aid in stabilizing the arc circuit. This type of electrode is used mainly for joints which carry only static loads.

The coated electrode has a jacket made of either organic (combustible) or inorganic material or a combination of the two types. This coating serves several purposes:

a. The organic material provides the gaseous shield to restrict the access of oxygen and nitrogen (air) to the molten metal.

b. The inorganic material makes a slag to refine the molten metal. (The principle is the same as the refining of steel and the manufacture of steel in an arc furnace.) This slag also serves as an insulating blanket to retard the rate of cooling of the weld metal.

c. The gas envelope and the slag aid materially in the stabilization of the arc circuit.

d. In some cases the coating is used to add alloying elements to the weld metal.

Electrode coatings are of different types, corresponding to the class of weld service (kind of base metal, joint position, etc.). The selection of electrode size, carbon content, and type of coating is a major consideration in welding practice.

Powdered-metal Electrodes. Some electrodes are available which have powdered iron, which will make up a good portion of the metal to be deposited in the joint, mixed in with the flux coating. This makes the electrodes much larger on the inside for the same-diameter core wire as standard flux-coated electrodes. These electrodes are well suited to down-hand welding and are said to deposit more metal in a given period of time than standard flux-covered electrodes.

224. Low-hydrogen-type Electrodes. Some types of steel welding rods have special treatment during manufacture to reduce the amount of hydrogen in the coating. The reduction of hydrogen in the coating produces a weld with little or no absorbed hydrogen in the base metal and reduces the likelihood of underbead cracking.

225. Electrode Quantities. The weight of electrode metal required for a given joint is

$$W = \frac{\text{weight of metal deposited}}{l - \text{electrode loss}} \qquad (22\text{-}49)$$

The weight of metal deposited is calculated from the volume required to fill the joint plus reinforcement, if any. The electrode loss is the sum of the scrap end loss, the spatter loss, and the coating loss. The scrap end loss varies with the diameter and length of the electrode. With 14-in electrodes the loss varies from 10 to 20%. The average value of this loss is about 17%.

The other losses depend on the type and size of the electrode, joint position, welding current, arc voltage, and skill of the operator. With bare electrodes these losses vary from 8 to 15%, average 13%; with coated electrodes, from 15 to 35%, average 27%.

Example 20. Normal fillet weld, right-angle joint, ¼-in bare electrode.

In^3 of steel deposited/lin ft, 0.375 = 0.106 lb.

Weight of electrode consumed, 0.151 lb (30% total loss).

Weight of coated electrode, 0.189 lb (44% total loss).

Example 21. Positioned fillet weld 45°.

Angle joint, ⅜-in coated electrode.

In^3 of steel deposited/lin ft, 0.96 = 0.272 lb.

Weight of electrode consumed, 0.486 lb (44% total loss).

226. Carbon (graphitic carbon) electrodes are used with direct current for arc welding of nonferrous metals and alloys and with the addition of external shielding (gases from material placed adjacent to the arc) for welding steel. The weld metal is supplied by melting a portion of the base metal. If additional metal is required, it is supplied by a filler rod. Carbon electrodes are not suitable for vertical and overhead welding. The use of a carbon electrode is declining.

227. The d-c arc is adapted to practically any class of welding service. The arc voltage varies with the type of electrode, ranging from 15 to 40 V. Both straight polarity (work positive, electrode negative) and reverse polarity (work negative, electrode positive) are used, depending on the character of the work. (Polarity here relates to the difference in the rates of heat development at the anode and cathode of the d-c arc.) Metal electrodes should be selected with reference to the polarity arrangement. Carbon electrodes can be used with either arrangement; straight polarity is more common practice.

The rotating d-c-generator single-operator unit, with a drooping voltampere characteristic, driven by any type of prime mover, is made in sizes ranging from 159 to 600 A, usual load-voltage rating 40 V, all sizes. A wide range of adjustments of welding current is provided in each size. The larger part of d-c welding is done with this type of equipment.

Multiple-operator equipment and an adjustable series resistor for each arc circuit for operation from constant distribution systems are used to some extent. The special field for this equipment is where the service of each arc-welding circuit is infrequently in use, say, 25% of the time or less. In such cases the loss in the series resistor is less than the no-load loss of the single-operator unit for the same service.

Another type of d-c-welding power source is a dry-type rectifier used in conjunction with a multiphase high-leakage-reactance transformer. Many of these rectifier-type welders use selenium rectifiers which are forced-air-cooled.

Rectifier-type welders are said to combine some of the desirable arcing characteristics of d-c welding, such as easy arc starting, with those of welding transformers, such as reduced no-load losses. The same range of rating is available in rectifier-type welders used in rotating d-c welders.

228. Alternating-current Arc Welding. The field of a-c arc welding is practically the same as that of d-c arc welding.

The standard equipment for single-operator a-c arc welding is a single-phase standard frequency transformer, sometimes with a number of voltage taps, and an adjustable reactance. Generally, capacitors are included to correct the power factor of the circuit to about unity between one-half and three-fourths load. The welding leads are limited to about 300 ft radius because of voltage drop.

Standard equipment includes ratings within the range of 150 to 1,000 A, 30 to 40 V.

229. Atomic-hydrogen Arc Welding. This method uses an a-c arc between the tungsten electrodes (not including the work) in an atmosphere of hydrogen (shielded arc). Either commercial hydrogen or disassociated ammonia can be used. The weld metal is supplied by the base metal or, if additional metal is required, by a filler rod.

The method is very flexible. Practically any metal or alloy, ferrous and nonferrous, can be welded. Its greatest usefulness is for the fusion welding of certain ferrous alloys, e.g., chrome-nickel steels, aluminum, and duralumin, but it must use flux to weld some of these materials. The welding of thin sheets, the production of tubing, and the repairing of expensive tools and dies are some of the common uses of the method.

230. Inert-gas-shielded Arc Welding. The method comprises the use of a single tungsten electrode, a filler rod, and argon or helium as a gas shield. The particular field for the method is the welding of the light metals. No flux is required. The technique varies with the type of work. The following procedures are typical.

Welding Aluminum and Aluminum Alloys. Alternating current and argon.

Welding Magnesium and Magnesium Alloys. Either alternating current and argon or direct current, reverse polarity, and either argon or helium.

Welding Stainless Steel, Mild Steel, Copper, and Copper Alloys. Direct current, electrode negative, and either argon or helium or alternating current, high frequency, stabilized with argon or helium.

231. Consumable-electrode Gas-shielded Welding. This type of welding uses small, bare electrodes in coil forms. The welding is done in gas atmospheres. The 0.020- to 0.125-in-diameter electrodes are fed mechanically into a hand gun for manual welding at feed rates up to 1,000 in/unit. The voltampere arc characteristics of this type of welding are rising (Fig. 22-85). Standard d-c welding generators do not give the best results. Special low-voltage d-c welders give better results. The optimum characteristic is one which is rising, to match the arc characteristic.

Argon, argon and 1% oxygen, helium, and carbon dioxide or mixtures of these gases are used. Argon is used for welding aluminum, magnesium, copper alloys, and titanium. Argon-oxygen or carbon dioxide is used for mild-steel welding. Almost all welding is done with the electrode positive.

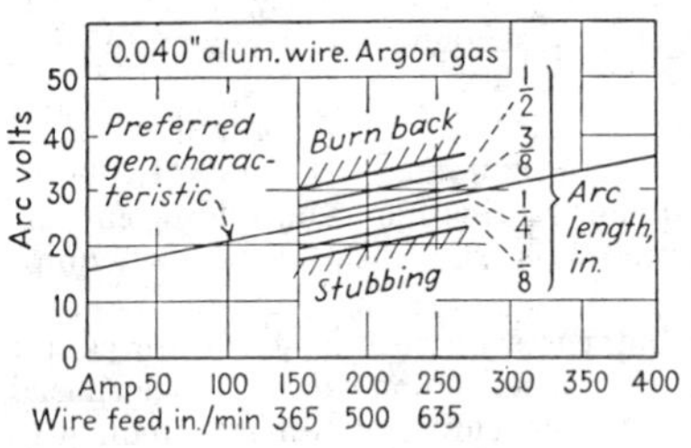

Fig. 22-85. Typical volt-ampere curves, consumable electrode gas-shielded arc.

The automatic consumable-electrode gas-shielded equipment does not require elaborate control systems. With the low-voltage generators the current flow automatically regulates itself to burn off whatever amount of wire is being fed into the arc.

As a result of this type of welding the generator produces the current to make the burn-off of wire rate equal to the wire feed speed.

232. Direct vs. Alternating Current. The considerations that affect the choice between direct current and alternating current are mainly the following:

a. Efficiency. Alternating current is higher (85%). Rotating direct current and rectifier direct current are about 65% under load. Also, no-load losses for rectifier direct current and alternating current are less than the no-load loss of the single-operator d-c equipment. The saving is 25 to 30%. A larger saving may be made compared with multiple-operator d-c equipment.

b. Power factor of alternating current and rectifier direct current is low (about 40% lagging) and can be corrected by the use of capacitors.

c. The magnetic deflection of the arc (magnetic blowout) is not so pronounced with the a-c arc. This phenomenon is often troublesome in working with d-c arcs of high current values. The a-c arc permits higher current and larger electrodes in such cases, with a consequent higher welding speed.

d. The comparative cost of electrodes.

e. The cost of equipment comparison.

f. Where the range of work in a shop is widely varied, the use of both types of arc welding may be found desirable.

233. Automatic arc welding, either d-c or a-c, is confined mainly to simple joints which can be placed in the flat position for welding. The schematic diagram of Fig. 22-86 illustrates a typical arrangement of automatic equipment.

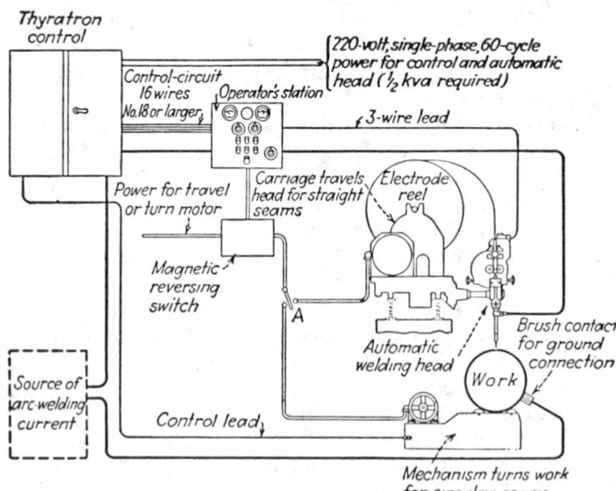

Fig. 22-86. Schematic diagram of automatic arc-welding equipment.

The electrode for direct current may be bare or coated. Coated electrodes for alternating current is usual practice except when the flux (in granular form) is applied to the work (submerged-arc method). Coated electrode wire in long lengths for feeding from a reel has a slot in the coating for electrical contact. Coated electrodes in short lengths with end contact are also used by adapting the feeding mechanism for automatic return to the starting position when the electrode is consumed.

The power-supply equipment, d-c or a-c as the case may be, for automatic arc welding is the same as for the corresponding manual operation.

The automatic regulator controls the rate of feeding of the electrode to match the rate of melting of the metal electrode by controlling the speed of the electrode motor in response to slight changes in the arc voltage so as to maintain a practically constant average arc voltage.

One type of automatic regulator in wide use is an electric device which contains the value and the direction of armature coil of a d-c electrode motor. This control uses two thyratron tubes in the armature circuit of the electrode motor, passing current from an a-c source in one direction or the other, as required, to feed the electrode up or down. The magnitude of the current passed is controlled by grid-phase shift of the thyratron tubes in response to changes in the arc voltage in comparison with the preset reference voltage.

Automatic arc welding, a method independent of the human element in the operation, has many merits where it is applicable. The most important of these are uniform welds, high production rate, and low electrode loss.

INDUCTION WELDING

234. Induction Resistance Welding. This method of welding incorporates the use of induction heating principles along with the application of high-frequency energy, usually in a radio-frequency range. It is particularly useful as a method of welding in tube mills. Here energy can be induced into the slit-formed tubing with a single-turn inductor and current caused to flow across the gap between pressure rolls where resistance welding is caused by the I^2R drop across the joint between the weld rolls. This is illustrated in Fig. 22-87 and Fig. 22-88, where the physical relationship between the inductor and tubing is shown in Fig. 22-87 and the current paths in the formed tubing are shown in Fig. 22-88.

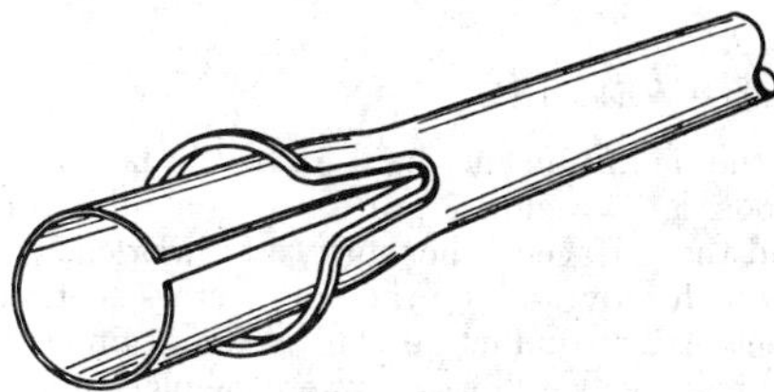

FIG. 22-87. Basic design of high-frequency induction resistance-welding inductor with concentrating nose.

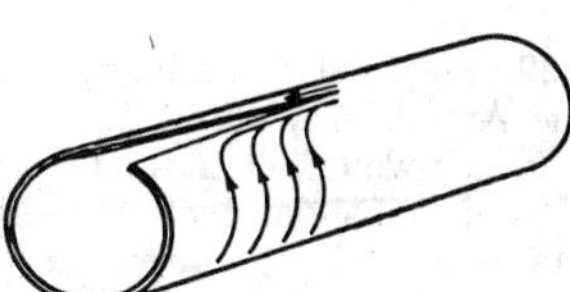

FIG. 22-88. Current path in tube.

This method of welding has been and is being used effectively for the welding of carbon steels, aluminum, copper, titanium, and mechanical stainless steel. This technique has been applied to portable weld units where it is possible to weld tubing and pipe on site from skelp.

The radio frequencies normally used for this type of welding are in the neighborhood of 200 to 450 kc/s. It is possible to weld heavier wall material in this manner, frequencies being used in the neighborhood of 10,000 c or even 3,000 c, although these frequencies will not give as efficient an operating installation as the radio-frequency installation.

235. High-frequency Resistance Welding. In recent years the application of contacts to line pipe utilizing radio-frequency energy has been successfully accomplished, and a weld similar to that obtained with induction resistance welding is thus secured. In this type of welding installation, high-frequency, preferably radio-frequency energy is passed through sliding contacts to the V-shaped edges of the tubing, as illustrated in Fig. 22-87. This current is then caused to flow across the contact between the pressure rolls, giving the I^2R heating effect of resistance welding at this point.

This method of welding is also effective on nonferrous materials, as well as carbon steels.

In the last few years several installations have been made where this method of welding has been adapted to structural shapes and the finned-tube shapes.

236. Induction Butt Welding. Two properly faced tubular sections or rectangular sections butted together can be induction-heated with a suitable frequency. This work is dependent upon the depth of penetration and the size of the part. The work is so heated that when the parts are brought to temperature they can be welded by the application of a suitable pressure. One of the members can be fixed and the other moved as in the case of resistance welding described in Par. **220.**

This method of welding has been particularly suitable for the welding of high-pressure boiler tubing, for the welding of railroad rails in the field, and for the butt welding of certain types of large cylindrical objects. Equipment has been furnished to weld transmission pipe in the field. A weld made in this manner is usually much faster than a weld made by a conventional electric process.

237. Longitudinal Butt Welding. The longitudinal butt weld can be accomplished by induction on such mechanical equipment as a tube or pipe mill where hot-rolled skelp can be formed and welded continuously without the necessity of edge cleaning such as is required for conventional resistance welding. Of course, it is also possible to perform this type of welding by either induction resistance or high-frequency resistance welding.

ELECTRON-BEAM WELDING

238. Electron-beam Welding. In recent years a method of welding, particularly in vacuum, has been developed which incorporates the utilization of an electron beam or beams for the welding of very thin and very narrow and unusual metals. Lately this technique has been adapted to thin sections of conventional materials where it is desirable to obtain a weld by heating a very narrow section of the parent metals and obtaining a bond as strong as the parent metal. Commercial units for this type of welding are available, and the units are applicable to certain production operations as well as to laboratory welding.

ELECTROSLAG WELDING

239. Electroslag welding according to the definition in Chap. 55 of the latest issue of the American Welding Society Handbook is "a welding process which relies on a molten slag which melts the filler metal and the surface of the work to be welded. The weld pool is shielded by this molten slag which moves along the entire cross section of the joint as welding progresses. The electrically conductive slag is maintained in a molten condition by its resistance to the current which flows between the electrode and the work."

This type of welding appears particularly suitable to replace multipass welding techniques. The process is especially applicable for welding plates ranging in thickness from 1¼ to 18 in.

This method may require welding current of the order of 550 to 650 A for ⅛-in-diameter electrodes; however, the range can vary depending on the material to be welded from 400 to 900 A.

The welding voltage required is normally in the neighborhood of 40 to 55 V for ⅛-in-diameter electrodes.

The electroslag method has been widely used in Europe for some time. Its use should be considered when the welding application involves large structural members, pressure parts in nuclear reactors, and high-pressure boiler downcomer tubes. It could be used for forged crankshaft welding. It would be advisable in some instances to use it on press frames. It is suitable for cylinders for hydraulic presses, for turbines and alternator shafts, for rolling-mill frames, and for ship hulls.

240. Plasma is a field for electric heating now becoming commercially usable. There are basically two types of plasma heating, (*a*) arc plasma, and (*b*) induction plasma.

Plasma heating involves the ionization of gases and develops extremely high temperatures.

Arc plasma heating has been used primarily for rapid melting of small quantities of material, both metals and refractories. It can be employed for cutting metals and is being used experimentally in investigations involving extremely high temperatures. A d-c arc is generally used in obtaining the plasma.

Induction plasma heating is developed generally at frequencies of about 4 Mc. Experimentally frequencies of 8 to 10 Mc have been utilized, and investigations are proceeding at many other frequencies. It is reported that plasma temperatures up to 20,000°F have been reached. Various gases are required, depending on the results desired. Gases commonly used are hydrogen, helium, argon, and nitrogen, and some work has involved the use of air.

BIBLIOGRAPHY

241. Specific Reference on Electric Heating and Welding

1. Goodell, P. H. *Trans. AIEE*, 1941, Vol. 60, p. 464.
2*a*. Scott, G. W. *Trans. AIEE*, 1945, Vol. 64, p. 558.
2*b*. Taylor, H. P. *Trans. ASME*, 1943, Vol. 65, p. 201.
3. Slowter, E. E., and Gonser, B. W. *Metals and Alloys*, 1937, Vol. 8, pp. 159, 195.
4. Adam, W. *Trans. Electro-Chem. Soc.*, 1936, Vol. 70, p. 143.
5*a*. Farlow, V. R., and Myer, E. R. *Trans. ASM*, 1932, Vol. 30, p. 516.
5*b*. Vaughn, J. T., and Williamson, J. W. *Trans. AIEE*, 1945, Vol. 64, p. 587.
6. Peters, F. P. *Metals and Alloys*, 1937, Vol. 7, p. 281.
7. Clark, L. W. *Trans. AIEE*, 1935, Vol. 54, p. 1173.
8. Clamer, G. H. *Metals and Alloys*, 1934, Vol. 5, p. 242.
9. Tuthill, R. W. *Welding Jour.*, February, 1954, p. 128.
10. Osborn, H. B., Jr. Metallurgical Aspects of Induction Heating; *Ind. Heating Mag.*, June–July, 1957.
11. Osborn, H. B., Jr. Surface Hardening by Induction Heat; *Metal Progr.*, December, 1955.
12. Osborn, H. B., Jr. High Frequency Welding Longitudinal Seam and Butt; *Tooling and Production Mag.*, May, 1963.
13. Osborn, H. B., Jr. High Frequency Welding of Pipe and Tubing; *Welding Jour.*, July, 1963.
14. Wheaton, G. F. *Trans. AIME*, 1936, Vol. 121, p. 599.
15. Osborn, H. B., Jr. Metallurgical Aspects of Induction Heating; *Ind. Heating Mag.*, June–July, 1957.
16. Osborn, H. B., Jr. Surface Hardening by Induction Heat; *Metal Progr.*, December, 1955.
17. American Welding Society, Welding Handbook, Chap. 55.

General References on Electric Heating and Welding

McAdams, W. H. "Heat Transmission," 3d ed.; New York, McGraw-Hill Book Company, 1954.

American Society of Metals, "Induction Heating"; Metals Park, Ohio, 1946.

Cable, J. W. "Induction and Dielectric Heating"; New York, Reinhold Publishing Corporation, 1954.

Gröber, H., Erk, S., and Grigull "Fundamentals of Heat Transfer," 3d ed.; New York, McGraw-Hill Book Company, 1961.

Clarke, J. A. "Theory and Fundamental Research in Heat Transfer"; New York, The Macmillan Company, 1963.

Irvine, T. F., Jr. "Advances in Heat Transfer"; New York, Academic Press, Inc., 1964, Vols. 1 and 2.

Kutateladze, S. S. "Fundamentals of Heat Transfer"; New York, Academic Press, Inc., 1963.

Simpson, P. G. "Induction Heating Coil and System Design"; New York, McGraw-Hill Book Company, 1960.

Tudbury, C. A. "Basics of Induction Heating"; New York, John F. Rider, Publishers, Inc., 1960.

Sacks, R. J. "Theory and Practice of Arc Welding," 2d ed.; Princeton, N.J., D. Van Nostrand Company, Inc., 1960.

American Welding Society, Welding Handbook, New York.

"Procedure Handbook of Arc Welding Design and Practice," 11th ed.; Cleveland, Ohio, Lincoln Electric Co., 1957.

ASM Committee on Furnace Atmospheres, American Society of Metals, "Furnace Atmospheres and Carbon Control"; Metals Park, Ohio.

Paschkis, V., and Persson, J. "Industrial Electric Furnaces and Appliances," 2d ed.; New York, Interscience Publishers, Inc., 1960.

Barnshah, R. F. "Vacuum Metallurgy"; New York, Reinhold Publishing Corporation, 1958.

Drummond, J. E. "Plasma Physics"; New York, McGraw-Hill Book Company, 1961.

U.S. Steel Corporation, "Making, Shaping, and Treating of Steel," 8th ed.

SECTION 23

ELECTROCHEMISTRY AND ELECTROMETALLURGY

BY

FREDERICK A. LOWENHEIM Research Coordinator, M & T Chemicals, Inc.; Chairman, Electrodeposition Division, Electrochemical Society (1943–1946); Chairman, Metropolitan Section Electrochemical Society (1948); Member, American Electroplaters' Society; Member, Institute of Metal Finishing; Member, American Chemical Society; Fellow, American Association for the Advancement of Science; Member, Chemists' Club (N.Y.); Member, Committee B-8, American Society for Testing and Materials; Technical Editor, *Plating*

CONTENTS

Numbers refer to paragraphs

SECTION 23

ELECTROCHEMISTRY AND ELECTROMETALLURGY

ABSTRACT PRINCIPLES

INTRODUCTION

1. Definition. **Electrochemistry** is the study of the interconversion of chemical and electrical energy. In practice, when the conversion is from chemical to electrical energy, the result is a primary or secondary battery (see Sec. **24**); when from electrical to chemical, it deals with electrolytics, electrothermics, or gaseous discharge. **Electrolytics** is concerned with the effect of the electric current upon solutions (usually aqueous) or upon fused salts.

2. Nomenclature. The passage of current through a solution is associated with the movement through it of charged particles called **ions**; the terminals leading the current into the solution are **electrodes,** the positive pole being the **anode** and the negative the **cathode**; the ions that move, or **migrate**, toward the anode are **anions,** and toward the cathode **cations**; the solution itself is the **electrolyte**; that surrounding the cathode is the **catholyte,** and that surrounding the anode the **anolyte.** (NOTE: Although most authorities use the term "electrolyte" in the sense here given, it is also frequently used in the sense of the solute, e.g., a solution of an electrolyte.) The over-all process of decomposition due to the passage of the current is called **electrolysis.**

3. Conductors. The electric current may be conducted through a medium in two principal ways. When the conduction is **metallic** or **electronic,** it is not accompanied by any movement of material through the conductor or by chemical reaction; metals and alloys and a few closely related substances such as carbon and graphite conduct in this manner. In the class known as **electrolytic conductors,** conduction is accompanied by the movement of matter through the medium and by chemical reaction at the electrodes; this class includes solutions of acids, bases, and salts; fused salts; a few solid substances; and hot gases. A few intermediate cases also exist.

The passage of electricity from one type of conductor to the other is always accompanied by chemical reaction.

4. Electrolytic Conductance. There is no evidence for the existence of molecules of true salts, either in the solid state or in solution. Their behavior can be explained entirely on the basis of their **complete dissociation** into ions, and deviations from ideal behavior are traceable to the electrical effects between particles of unlike charge. Many ions are hydrated in solution. Besides the salts, other types of compound give rise to ions in solution: acids like HCl, which is a molecular compound, ionize completely in water according to

$$HCl + H_2O \rightarrow H_3O^+ + Cl^-$$

and weak acids and bases such as acetic acid and other organic acids,

$$CH_3COOH + H_2O \rightleftharpoons H_3O^+ + CH_3COO^-$$

in which case the dissociation is not complete. The ion H_3O^+ is commonly called the **hydrogen ion**, although the actual hydrogen ion H^+, which would be a bare proton, does not exist in aqueous solution.

Under the influence of an applied voltage, the ions move toward the electrodes as explained in Par. **2**. In very dilute solutions, the forces between the ions may be neglected, and the conductivity of the electrolyte is the sum of the conductivities of its constituent ions; these individual ionic conductivities are called **mobilities.** In general, the cation and anion do not have the same mobility, so that more current is carried by one than by the other. The proportion of the current carried by a given ion is its **transference number.** In more concentrated solutions, many complications enter, due to interionic forces; the mathematical formulation of these deviations from simple behavior is one of the principal aims of theoretical electrochemistry.

In water solution, hydrogen ion H_3O^+ and hydroxyl ion OH^- have much higher mobilities than any other ions, probably because their migration involves a different mechanism. For this reason, solutions of strong acids and bases are much better conductors than any other solutions.

CURRENT RELATIONS

5. Faraday's laws state that (1) the amount of chemical change produced by an electric current is proportional to the quantity of electricity and (2) the amounts of different substances liberated by a given quantity of electricity are proportional to their chemical equivalent weights. (Equivalent weight = atomic weight divided by valence change.) These laws are fundamental; apparent deviations from them are in all cases traceable to side reactions or, exceptionally, to a partially nonelectrolytic nature of the conduction process.

6. The Faraday. By definition, the **coulomb** is the quantity of electricity that will liberate 0.001118 g of silver. The quantity of electricity required to liberate 1 g-equivalent weight of silver, and therefore by Faraday's law of any other element, is 107.88/0.001118, or 96,500[1] C; this quantity is called the **faraday.** Faraday's laws may now be stated in the form

$$G = Iet/96{,}500 \tag{23-1}$$

where G = grams of substance reacting, e = equivalent weight, I = current in amperes, and t = time in seconds.

If the reaction in question takes place at unit voltage, the energy consumed in liberating one equivalent weight will be 96,500 W·s, or J. One joule = 0.2389 g·cal; so 96,500 J = 23,050 cal. This quantity is known as the **thermochemical equivalent of the volt.**

7. Electrochemical Equivalents. In Eq. (23-1), when $I = 1$ A and $t = 1$ s, $1{,}000 \times G$ = mg/C; if $t = 3{,}600$ s, G = g/Ah; obviously, G may be obtained in any other convenient units, e.g., kilograms or pounds per ampere-hour or per ampere-day, by use of the proper conversion factors. G is called the **electrochemical equivalent** of the element. Some values of G for the more important elements are given in Table 23-1, along with a generalized formula for calculating G for any element.

Although the word "element" has been used, it should be understood that Faraday's laws hold for compounds, complex radicals, and any other units entering into electrochemical change; the calculations are analogous.

If the voltage of the cell reaction is known, such data as kilograms per kilowatthour may be calculated by recalling that ampere-hours × volts = watthours.

8. Use of the Table of Electrochemical Equivalents. When a metal exhibits more than one valence, it is important to select the correct value. Otherwise, the use of Table 23-1 should be obvious. Other values may be obtained from those in the table by applying the following formulas, where E = voltage of the reaction, and G = electrochemical equivalent in grams per ampere-hour (column 7):

$$\begin{aligned} \text{kg/hp·hr} &= 0.746G/E \\ \text{hp·hr/kg} &= 1.341E/G \\ \text{lb/hp·hr} &= 1.645G/E \\ \text{hp·hr/lb} &= 0.6080E/G \\ \text{kg/hp·year} &= 6{,}535G/E \\ \text{lb/hp·year} &= 14{,}407G/E \end{aligned}$$

9. Electrochemical Equivalents of Elements. See Table 23-1.

[1] For highest accuracy the figure 96,490 should be used.

Table 23-1. Electrochemical Equivalents of Elements

Element (1)	Symbol (2)	Atomic weight (3)	Valence or change of valence (4)	Mg/C (5)	C/mg (6)	G/Ah (7)	Ah/g (8)	Lb/1,000 Ah (9)	Ah/lb (10)
Any	....	A	z	$0.0103638A/z$	$96.49z/A$	$0.0373097A/z$	$26.8028z/A$	$0.0822537A/z$	$12{,}157.6z/A$
Aluminum	Al	26.98	3	0.09320	10.729	0.33554	2.9802	0.73973	1,351.8
Antimony	Sb	121.75	3	0.42059	2.3776	1.5141	0.66045	3.3381	299.58
Arsenic	As	74.92	3	0.25882	3.8637	0.93174	1.0733	2.0541	486.83
Beryllium	Be	9.012	2	0.04670	21.414	0.16812	5.9483	0.37064	2,698.1
Bismuth	Bi	208.98	3	0.72194	1.3851	2.5990	0.38475	5.7298	174.52
Cadmium	Cd	112.40	2	0.58245	1.7169	2.0968	0.47693	4.6227	216.33
Calcium	Ca	40.08	2	0.20769	4.8149	0.74769	1.3375	1.6484	606.66
Chlorine	Cl	35.457	1	0.36747	2.7213	1.3229	0.75592	2.9165	342.88
Chromium	Cr	52.00	3	0.17964	5.5667	0.64669	1.5463	1.4257	701.40
			6	0.08982	11.133	0.32335	3.0925	0.71287	1,402.7
Cobalt	Co	58.93	2	0.30537	3.2748	1.0993	0.90966	2.4236	412.61
Copper	Cu	63.54	1	0.65851	1.5186	2.3707	0.42181	5.2264	191.34
			2	0.32926	3.0371	1.1853	0.84364	2.6132	382.67
Fluorine	F	19.00	1	0.19691	5.0784	0.70888	1.4107	1.5628	639.88
Gold	Au	196.97	1	2.0414	0.48987	7.3489	0.13608	16.202	61.723
			3	0.68046	1.4696	2.4496	0.40823	5.4005	185.17
Hydrogen	H	1.008	1	0.01045	95.724	0.03761	26.590	0.082912	12,061.
Iron	Fe	55.85	1	0.57882	1.7277	2.0837	0.47990	4.5939	217.68
			2	0.28941	3.4553	1.0419	0.95981	2.2969	435.36
			3	0.19294	5.1830	0.69459	1.4397	1.5313	653.05
Lead	Pb	207.19	2	1.0736	0.93142	3.8651	0.25873	8.5211	117.36
Magnesium	Mg	24.31	2	0.12597	7.9383	0.45350	2.2051	0.99979	1,000.2
Manganese	Mn	54.938	2	0.28468	3.5127	1.0249	0.97576	2.2594	442.60
Mercury	Hg	200.59	1	2.0789	0.48103	7.4840	0.13362	16.499	60.61
			2	1.0394	0.96206	3.7420	0.26724	8.2496	121.21
Nickel	Ni	58.71	2	0.30423	3.2870	1.0952	0.91306	2.4146	414.16
Oxygen	O	16.00	2	0.08291	12.061	0.29848	3.3504	0.65803	1,519.7
Palladium	Pd	106.4	2	0.55135	1.8137	1.9849	0.50381	4.3759	228.52
Platinum	Pt	195.09	2	1.0109	0.98922	3.6394	0.27478	8.0234	124.64
			4	0.50547	1.9873	1.8197	0.54954	4.0118	249.27
Potassium	K	39.102	1	0.40525	2.4676	1.4589	0.68545	3.2163	310.92
Rhenium	Re	186.2	7	0.27568	3.6274	0.99244	1.0076	2.1879	457.05
Rhodium	Rh	102.91	2	0.53327	1.8752	1.9198	0.52089	4.2324	236.27
			3	0.35551	2.8129	1.2798	0.78136	2.8215	354.42
			4	0.26664	3.7505	0.95990	1.0418	2.1162	472.55
Selenium	Se	78.96	4	0.20458	4.8881	0.73649	1.3578	1.6237	615.89
Silver	Ag	107.870	1	1.11800*	0.89450	4.0246	0.24847	8.8727	112.71
Sodium	Na	22.99	1	0.23826	4.1970	0.85775	1.1658	1.8910	528.82
Tellurium	Te	127.60	4	0.33061	3.0248	1.1902	0.84021	2.6239	381.12
Tin	Sn	118.69	2	0.61504	1.6260	2.2141	0.45165	4.8813	204.87
			4	0.30753	3.2518	1.1071	0.90328	2.4407	409.72
Titanium	Ti	47.90	4	0.12411	8.0576	0.44678	2.2382	0.98499	1,015.2
Tungsten	W	183.25	6	0.31757	3.1490	1.1432	0.87471	2.5204	396.76
Zinc	Zn	65.37	2	0.33874	2.9521	1.2195	0.82003	2.6885	371.96
Zirconium	Zr	91.22	4	0.23635	4.2311	0.85085	1.1753	1.8758	533.11

* By definition.

10. The Coulometer. If conditions can be arranged to ensure that the electrode reaction will not be complicated by any side reactions, the gain in weight of a cathode, the volume of gas liberated, or the quantity of titratable iodine set free, etc., can be used as measures of quantity of electricity. Such an apparatus is a **coulometer**, also called a **voltameter**.

11. Current Efficiency. It is often found that the amount of chemical change produced by the passage of a given quantity of electricity is less than that calculated from the table of electrochemical equivalents. This is due not to any failure of Faraday's laws but to the failure to take into account *all* the electrode reactions, e.g., liberation of hydrogen at the cathode or oxygen at the anode. If these are included, it will be found that Faraday's laws are exactly satisfied. Since attention is usually focused upon one of the reactions to the exclusion of others, an important characteristic of the electrolytic process is given by the **current efficiency**, defined as

$$\frac{\text{Amount of desired chemical change}}{\text{Theoretical amount from Faraday's laws}} \times 100$$

Cathode efficiency is the foregoing ratio applied to the cathode process, and **anode efficiency** is the ratio for the anode process.

12. Current Density. Although the total current flowing through a cell will permit calculation of gross product, the thickness of metal deposited is often of greater importance. This depends on the current density, defined as the current per unit electrode area. In general, the current density will vary from point to point in any system and the value calculated from the amperes divided by the electrode area is an average.

Most electrolytes exhibit a range of current densities over which the reactions are satisfactory and outside which they are not. This range for any electrolyte depends upon the temperature, degree of agitation, and solution concentration and often upon specific factors. The current density often affects the nature of the electrode reactions.

At the anode also, the current density is important. Too high a current density may cause passivity: the anode fails to dissolve; oxygen is evolved instead; and the anode does not fulfill its function of replenishing the metal content of the solution.

13. Current Distribution. As mentioned above, the current density obtained by dividing the total amperes by total electrode area gives an average figure. Actually the current distribution over the face of an electrode depends upon the geometry of the system; and except for the simplest cases, rarely met with in practice, the mathematics of this distribution is complicated. Where, as in electroplating, considerable uniformity of metal, and therefore current, distribution is desired, anode design often requires considerable care and ingenuity; this is of greater or less consequence depending on whether the throwing power of the solution is low or high.

14. Effect of Temperature. The resistance of electrolytes decreases with increasing temperature, so that the heating effect of the current tends to become cumulative provided the voltage is kept constant. Since temperature affects the properties of the electrolyte, its close control is usually of considerable importance.

VOLTAGE RELATIONS

15. Electrode Potentials. When an electrode is in contact with a solution of its ions, a potential difference is set up between the two. Its magnitude is given by

$$E = E_0 + RT/nF \ln a \qquad (23\text{-}2)$$

where E_0 = a constant, R = gas constant, T = absolute temperature, F = faradays, n = valence change, and a = metal-ion activity. Where approximate treatment is sufficient, we may use, instead of a, c, the concentration of metal ions: $c = C\alpha$, where C = stoichiometric concentration and α = apparent degree of dissociation as given by the conductance ratio.

When a (or c) is 1, $\ln a = 0$, so that $E = E_0$. E_0 is called the **standard electrode potential** and is the potential of an electrode in contact with a solution of its ions of unit activity. By definition the standard potential of the hydrogen electrode is zero at all temperatures.

A tabulation of E_0 for the elements results in the familiar **electromotive series** (see Table 23-2). In this table the elements having the greatest negative potentials have the greatest tendency to pass into solution: they are the "base" metals; those with positive potentials tend to be inert and are the "noble" metals.

16. Electrode-potential Series (see Table 23-2)

Table 23-2. Electrode-potential Series

(Molal electrode potentials referred to the hydrogen electrode as zero)

Electrodes	Potential, volts	Electrodes	Potential, volts	Electrodes	Potential, volts
$Li \rightleftharpoons Li^{+}$	−3.045	$Cr \rightleftharpoons Cr^{++}$	−0.56	$Sb \rightleftharpoons Sb^{+++}$	+0.15
$Rb \rightleftharpoons Rb^{+}$	−2.93	$Fe \rightleftharpoons Fe^{++}$	−0.441	$Bi \rightleftharpoons Bi^{+++}$	+0.2
$K \rightleftharpoons K^{+}$	−2.924	$Cd \rightleftharpoons Cd^{++}$	−0.402	$As \rightleftharpoons As^{+++}$	+0.3
$Ba \rightleftharpoons Ba^{++}$	−2.90	$In \rightleftharpoons In^{+++}$	−0.34	$Cu \rightleftharpoons Cu^{++}$	+0.34
$Sr \rightleftharpoons Sr^{++}$	−2.90	$Tl \rightleftharpoons Tl^{+}$	−0.336	$Pt/OH^{-} \rightleftharpoons O_2$	+0.40
$Ca \rightleftharpoons Ca^{++}$	−2.87	$Co \rightleftharpoons Co^{++}$	−0.277	$Cu \rightleftharpoons Cu^{+}$	+0.52
$Na \rightleftharpoons Na^{+}$	−2.715	$Ni \rightleftharpoons Ni^{++}$	−0.250	$Hg \rightleftharpoons Hg_2^{++}$	+0.789
$Mg \rightleftharpoons Mg^{++}$	−2.37	$Sn \rightleftharpoons Sn^{++}$	−0.136	$Ag \rightleftharpoons Ag^{+}$	+0.799
$Al \rightleftharpoons Al^{+++}$	−1.67	$Pb \rightleftharpoons Pb^{++}$	−0.126	$Pd \rightleftharpoons Pd^{++}$	+0.987
$Mn \rightleftharpoons Mn^{++}$	−1.18	$Fe \rightleftharpoons Fe^{+++}$	−0.04	$Au \rightleftharpoons Au^{+++}$	+1.50
$Zn \rightleftharpoons Zn^{++}$	−0.762	$Pt/H_2 \rightleftharpoons H^{+}$	0.0000	$Au \rightleftharpoons Au^{+}$	+1.68
$Cr \rightleftharpoons Cr^{+++}$	−0.74				

17. Potentials Due to Mechanical Forces. The potentials given in Table 23-2 are for metals in the so-called **standard state.** The previous history of an electrode often has a great effect upon its potential: a metal under strain is positive to the same metal unstrained; large crystals have a different potential from small, etc. This subject is of importance in explaining some types of galvanic corrosion.

18. Concentration Cells. In Eq. (23-2), if numerical values are substituted for R and F, if T is taken as 25°C (298°K), and if logarithms are taken to the base 10 instead of e, we have

$$E = E_0 + 0.059/n \log a \qquad \text{(or } \log c \text{ approx.)} \tag{23-3}$$

Since E_0 and n are constant for a given electrode reaction, it is seen that there is a potential difference between two identical electrodes if they are dipping into solutions of different concentrations. For a tenfold change in the concentration of a univalent metal ion, the potential will change by 59 mV; by 29.5 mV for a bivalent ion; etc.

19. Voltage Applied to Electrolytic Cell. The total voltage registering on a voltmeter across an electrolytic cell while current is flowing may be regarded as being made up of three components: the voltage required to overcome the ohmic resistance of the electrolyte, the decomposition voltages, and the polarization voltages,

$$E = E_c + E_d + E_p \tag{23-4}$$

E_c, the conduction voltage, follows Ohm's law and may be calculated in the usual way. E_d, the decomposition voltage, can be calculated from the thermochemistry of the reaction: when the cell is reversible and the anode reaction is the exact opposite of that at the cathode, this term disappears. E_p is the so-called "polarization emf" and includes as well as true polarization the overvoltages and other factors (see Par. **26**).

20. Voltage Delivered by an Electrogenetic Cell. In this case, the available voltage is generated by the cell itself, so that before a useful voltage can be delivered the cell first has to overcome its internal resistance and any polarization effects involved. The available voltage is then

$$E = E_s - E_c - E_p \tag{23-5}$$

E in this case is termed the "terminal" or "working voltage" of the cell; E_s is the "static voltage" and corresponds to the decomposition voltage of an electrolytic cell.

21. The Normal Voltage of a Cell. The normal voltage of a cell may be defined as the voltage required to do the fundamental work of the cell, disregarding any side reactions. If no current is passing, the normal voltage is the E_s of the cell. Since the fundamental work of the cell necessarily involves the passage of a current through the cell, the normal voltage of a cell in operation must take into consideration the E_c of the cell. The normal voltage of an electrolytic cell is then $E_s + E_c$; and of an electrogenetic cell, $E_s - E_c$.

22. Voltage Efficiency. The voltage theoretically necessary to carry out the cell reaction is E_d, the decomposition voltage; but in any actual case E_c, the conduction voltage, always enters, and E_p, the polarization voltage, usually does. The voltage efficiency of an electrolytic cell is E_d/E; of an electrogenetic cell, E_s/E. This is never 100% but may approach that figure when the current passing is very small.

23. Voltage of Conduction. The voltage of conduction for any given cell E_c can be calculated if one knows the distance between the electrodes, the cross section of the conducting area, and the specific conductance of the electrolyte, since

$$R = \rho l/a = l/\kappa a \tag{23-6}$$

where R = resistance of the cell in ohms, ρ = specific resistance of the electrolyte, κ = specific conductance of the electrolyte, l = distance between the plates, a = area of the conducting section; l and a must be in the same system of units.

24. Decomposition Voltage. The voltage of decomposition of any compound is the voltage that must be applied to the electrolysis in order to effect the decomposition of the compound in question. This voltage is readily calculated using the value for the thermochemical equivalent of the volt, 23,050 cal. The voltage required for any particular reaction will be determined by the ratio between the total heat required for the decomposition of one chemical equivalent and 23,050 cal. A reaction that liberates instead of absorbs heat will generate a voltage, and, properly connected, such a combination would serve as an electrogenetic cell. There is no difference, however, in the method of calculation:

$$E_d = Q/(n \times 23{,}050) = 0.000{,}043{,}377Q/n \tag{23-7}$$

where E_d = decomposition voltage, Q = heat balance of the reaction in question, n = number of chemical equivalents involved in the reaction. This is known as **Thomson's rule.** Since the heats of formation used in the calculation are determined from and at 0°C, the values obtained are strictly correct only at that temperature. For any other temperature, the values will vary from those calculated, owing to the fact that the heat contents of the starting materials and the final products are different.

25. Polarization Voltage. As has been shown in Par. **24**, the static value of the emf of any cell may be calculated from the thermochemical reactions taking place during the operation of the cell. When, for any reason, the emf required by an electrolytic cell is greater than the calculated normal value, or that delivered by an electrogenetic cell is less than the calculated normal value, the cell is said to be polarized. This variation may be brought about by one or more of three causes:

1. Concentration effects.
2. Ohmic resistance of the products of the electrolysis on or in the vicinity of the electrodes.
3. Electrode effects.

Concentration effects and ohmic resistance due to accumulated products of electrolysis may be encountered accompanying any kind of electrochemical reaction but are usually not of importance except in reactions where the products at one or both electrodes are gaseous; in that case the ohmic resistance and its accompanying electrode effects are invariably causes of polarization to a greater or lesser degree.

26. Concentration Polarization. When a cell has been in operation for a short time, the concentrations of metal ions about the anode and cathode will be different owing to the fact that metal is being removed near the cathode and dissolved at the anode. In effect, a concentration cell is set up whose emf opposes the applied voltage (see Par. **18**). This concentration polarization is opposed by the forces of diffusion and migration of the ions; nevertheless, it is often the factor limiting the current density that may be applied.

27. Ohmic Resistance Due to Products of Electrolysis. A very considerable proportion of the total polarization voltage is in many cases merely additional potential required to force the cell current through an increased ohmic resistance in the cell, due to the accumulation on and around the electrodes of some of the products of electrolysis. This is particularly true of all electrochemical reactions in which a gas is liberated but also applies in many cases where the product is a solid. An extreme case of the latter is found in the electrolysis with aluminum anodes, where a thin film of Al_2O_3 is formed over the entire surface. This film is of high resistance, and a considerable additional voltage is required to force the current through it. In this and similar cases the polarization voltage may build up to many times the normal voltage of the cell.

In every electrochemical reaction in which a gas is liberated at one of the electrodes, polarization due to ohmic resistance is quite marked, since it requires only an extremely thin film of the gas to interpose a fairly material resistance to the passage of the current. Of course, before the gas accumulates to any great thickness it forms into bubbles and escapes, but the presence of a film of gas 0.001 mm in thickness could account for a polarization voltage considerably greater than any ordinarily encountered.

In addition, as it frees itself from the electrode in the form of bubbles, the electrolyte is pushed back from the surface of the electrode; and for a distance of several millimeters, instead of presenting a cross section that is 100% electrolyte to the passing current, a considerable percentage of the space is occupied by gas bubbles, with a proportionate increase in resistance.

28. Electrode effect is any effect that may be made manifest in the polarization through the agency of electrode material. The character and magnitude of the electrode effect are determined by the composition of the electrode and the nature of its surface. Electrode effects are usually not manifest except in reactions in which one or more of the products are gases. With gases they may reach the order of several tenths of a volt. Just what enters into the electrode effect is not definitely known. Some of the possible contributing factors are the formation of chemical compounds between the electrode and the gas and the solution, occlusion, or absorption of the gas in the electrode material or of compounds of the gas with the electrode material.

In the case of electrode effect there is the possibility of a **negative polarization.** For example, with the deposition of hydrogen on a palladium cathode, the palladium takes up the hydrogen rapidly, presumably with formation of a hydride, whose heat of formation, converted into emf, assists the applied emf in forcing the current through the cell, with the result that the normal voltage of the cell is lowered instead of raised.

29. Overvoltage. Any departure of an electrode from equilibrium conditions regardless of the cause is called "polarization." The special polarization associated with ionic discharge at the electrode is known as overvoltage; it is the difference in volts between the actual deposition potential and the reversible electrode potential in the same solution. Although some metallic ions, e.g., iron, nickel, and cobalt, exhibit overvoltages, most such phenomena are associated with gases. Hydrogen and oxygen have been most thoroughly studied.

From the relation $E = E_0 + 0.059 \log a$, it would be expected that hydrogen ($E_0 = 0$) would be discharged from an HCl solution of unit activity at a cathode potential of 0 V. Actually, hydrogen bubbles do not form until the potential becomes considerably more negative. In practice, then,

$$E = E_0 + 0.059/n \log a - V$$

where V = overvoltage.

The prime determining factor in overvoltage is the particular combination of electrode and gas that is brought together. The smoother the surface of the electrode, the higher the overvoltage. Overvoltage increases with increasing current density and in most cases increases with time, at least up to a certain limit. In some cases the values vary considerably with the character of the electrolyte, particularly between acid and alkaline electrolytes. The temperature and pressure of the electrolyte also have an effect.

The values of overvoltage as determined by different investigators vary considerably. Table 23-3 gives some maximum and minimum values recorded by various investigators; this shows the general character and magnitude of a number of typical overvoltages.

30. Overvoltage of Gases at Different Electrodes. See Table 23-3.

31. Counter Emf. If we break the circuit in an ordinary electrolytic cell with a voltmeter connected across its terminals, it will frequently be noticed that the voltmeter readings continue momentarily after the current has ceased to flow and then gradually drop off to zero. This, being opposed to the applied emf of the cell, has

Table 23-3. Overvoltage of Gases at Different Electrodes

Overvoltage of hydrogen		Overvoltage of oxygen	
Electrode	Volts	Electrode	Volts
Pt, platinized	0.00 –0.07	Ni, spongy	0.05 –0.76
Pd	0.00 –1.16	Ni, smooth	0.12 –0.87
Graphite	0.002–1.22	Co	0.13
Pt, smooth	0.02 –0.77	Pt, platinized	0.24 –0.77
Au	0.02 –0.95	Fe	0.24 –0.83
Te	0.05 –0.66	Pb (PbO_2)	0.28 –1.12
Ag	0.10 –1.08	Cu	0.42 –0.84
Carbon	0.14 –1.23	Pd	0.42 –1.00
Ni, smooth	0.14 –1.25	Pt, smooth	0.44 –1.38
Duriron	0.17 –1.14	Au	0.52 –1.66
Monel	0.19 –1.21	Graphite	0.53 –1.28
Fe	0.20 –1.29	Ag	0.58 –1.14
Cu	0.23 –1.27	Ir	0.65 –1.19
Sn	0.24 –1.23		
Hg	0.28 –1.30		
Brass	0.38 –1.25	Overvoltage of chlorine	
Bi	0.39 –1.29		
Pb	0.40 –1.29		
Cd	0.47 –1.26		
Zn	0.48 –1.24	Pt, platinized	0.006–0.10
Al	0.30 –1.29	Pt, smooth	0.008–0.26
		Graphite	0.19 –0.54
		Magnetite	0.60 –1.85

been called "counter emf." If there is any concentration polarization in the cell, it will be manifest as a counter emf on breaking the circuit; but most of the counter emf is due to the accumulation around the electrodes of some of the products of electrolysis in a reactable form, so that as soon as the electrolyzing current is no longer acting, these products on the electrodes set up a recombination, which of course is in the reverse direction to the applied emf, since the direction of the reaction is reversed.

32. Depolarization. As its name indicates, depolarization is the process of more or less completely removing or neutralizing in a cell the polarizing effect of one or more of those particular polarizing factors which are operating in the cell. Since the character and magnitude of the polarization in different cells depend on which of the four contributing causes enter into it, the methods of accomplishing depolarization vary accordingly.

Since concentration polarization is caused by the normal diffusion of the electrolyte failing to keep pace with the changes of concentration brought about by the action of the current, depolarization can be effected by any process which will increase the rate of diffusion. Agitation of the solution or movement of the cathode relative to the electrolyte is widely used, with the limitation that the stirring up of mud, slimes, etc., is sometimes objectionable. Agitation by means of ultrasonic waves is very effective where applicable and economically feasible; it may also modify the character of the deposits obtained. Increasing the temperature also helps.

Increased temperature and mechanical stirring are also effective to a certain degree in combating resistance polarization due to gas accumulation at the electrodes, but the chief agency is the addition to the electrolysis of a substance that can chemically combine with the film accumulated on the surface of the electrode and thus remove it more or less completely.

Almost the only way to reduce the polarization voltage due to the electrode effect is to change the material or surface of the electrode. Selection of a material with low overvoltage will care for both these last two factors to some extent, since low overvoltage means low degree of polarization due to the combination of the resistance and electrode factors. The electrode polarization can also be reduced by using a rough, or "mat," surface instead of a smooth, or polished, surface, since overvoltage increases with the smoothness of the electrode surface.

Depolarization is particularly important in electrogenetic cells and in many types of electrolytic cells used in the synthesis of certain types of compound—in fact, in any cell in which the electrochemical reaction is accompanied by the liberation of oxygen or hydrogen. As carried out in practice, the depolarization of reactions liberating either gas is usually accomplished by the use of a chemical compound that will not interfere with the main reaction and is easily reducible for use at the cathode or easily oxidizable for use at the anode. The depolarizer may be used either as an insoluble solid, packed in close contact with the electrode, or as a soluble salt dissolved in the electrolyte itself.

Further features of depolarization are discussed in Pars. **39** and **43**.

ENERGY RELATIONS

33. Energy efficiency of an electrochemical reaction is determined by combination of the current and voltage efficiencies of the reaction, i.e.,

$$(\text{Current efficiency}) \times (\text{voltage efficiency}) = \text{energy efficiency} \qquad (23\text{-}8)$$

Current efficiency may reach 100% but usually does not; voltage efficiency is always considerably below 100% owing to the conduction voltage and is still further reduced if the cell is subject to polarization. For these reasons, the energy efficiency of an electrochemical reaction is never very high; cells with little or no polarization usually come within the range of 60 to 80%, and as the polarization increases, this may be cut to half or to even less in extreme cases.

34. Energy Output. From Par. **22**, the voltage efficiency of a cell $= E_d/E$; but in some cells—those where the anode reaction is the exact opposite of the cathode reaction—$E_d = 0$. In order to have a basis for comparing cells of this type, it is common to specify the amount of product per unit of energy, e.g., pounds per kilowatthour.

35. External Work. When any reaction is accompanied by the liberation of a gas, the expansion of that gas against any pressure other than atmospheric requires the doing of external work and the energy represented by this work may have to be taken into consideration in the calculation of the energy required to carry on the operation. In the calculation of the energy requirement of any reaction, allowance must be made for external work if the final condition of the products is different in volume or pressure from the original condition of the substances entering into the formation of the compound whose heat of formation is being used in the calculation. If the volume of gases produced is greater than the volume in the original condition, or if the pressure against which they must expand on formation is greater than atmospheric, the energy consumption of the reaction will be proportionately increased; but if the volume of the gases produced is less or their pressure less than atmospheric, the energy consumption will be less and proper deduction must be made.

36. Internal Work. The theoretical energy requirements of a reaction under standard conditions may sometimes be changed by the fact that some product of the reaction is in a different state of aggregation from the corresponding raw material or different from the normal state of the substance at standard conditions of temperature and pressure. Proper allowances must then be made for the internal work involved, corresponding to the latent heat of the change of state involved, usually the heat of fusion or the heat of vaporization. The heat involved in this change of state will proportionately change the thermal balance of the reaction as calculated from data for standard conditions. For example, if the normal state at standard conditions of a product of the reaction is liquid but it is actually produced and removed from the reaction in the gaseous state, the latent heat of vaporization must be added to the normal heat balance of the reaction in applying Thomson's rule; or a corresponding deduction must be made if the conditions are reversed. It should be noted that allowances for internal work are required in a calculation only when the changes of state involved are not normal to standard conditions.

ELECTRODE REACTIONS

37. Electrode Reactions. The reactions taking place at the electrodes during an electrolysis may be classified in various ways, depending on the point of view. Probably as satisfactory as any scheme of complete classification is the following:

1. At the cathode:
   a. Deposition of metal.
   b. Evolution of hydrogen.
   c. Electrolytic reduction.

2. At the anode:
   a. Solution of metal.
   b. Evolution of oxygen or some other anion.
   c. Electrolytic oxidation.

38. Secondary Reactions. In most cases the concept of so-called secondary reactions represents an unnecessary complication in electrolytic theory. In a dilute aqueous solution containing Na^+ and OH^-, it is gratuitous to suppose that Na^+ is first discharged, then reacts with water to form H_2. Since hydrogen ions are always present, and since, even allowing for the large differences in concentration and for overvoltage, they deposit at a lower potential than sodium, the reaction is simply the cathodic discharge of hydrogen. This is not to deny that there are well-authenticated cases of secondary electrode reactions, but they are somewhat exceptional.

39. Deposition and Solution of a Metal. The deposition or solution of a metal is usually a reversible reaction, within the limits of our practical definition of a reversible reaction. Such polarization as is present is chiefly concentration polarization but usually to such a slight extent that for practical purposes the reaction may be considered reversible. There are some exceptions to this statement, particularly iron, nickel, and cobalt, all of which show a material amount of polarization. A satisfactory explanation of these exceptions has not yet been reached. Similarly, these same metals, when acting as anode, may, under certain conditions, be subjected to such a decided increase in polarization that solution practically ceases. This phenomenon is known as **passivity** (see Par. **43**).

When the conditions of electrolysis are such that the deposited metal can readily alloy with the electrode or with a second metal being deposited at the same time, the energy generated by this combination will tend to lower the potential required for the deposition. This may cause a material lowering of the polarization or, with little or no polarization present, may even result in a material lowering of the normal voltage of the reaction. This is of importance particularly with a liquid cathode, e.g., mercury, or in the simultaneous deposition of two metals, where the alloying may proceed without the necessity of much diffusion, owing to the intimacy of the mixture that is being deposited. The simultaneous deposition of two metals in the form of an alloy is discussed in Par. **70**.

40. Simultaneous Deposition of Metal and Hydrogen. Provided that no polarization is involved, a metal will begin to deposit on a cathode when the cathodic potential reaches the value given by $E = E_0 + 0.059/n \log c$ (at 25°C and by neglecting the approximation involved in writing c for a, the activity). But since hydrogen ions are always present in any aqueous solution, there is always the chance that hydrogen will deposit instead of, or in addition to, the metal desired. In the case of gases, however, there is usually an overvoltage to be taken into account. Thus, zinc can be deposited from its aqueous solution in spite of the fact that its standard potential is much more negative than that of hydrogen; for when overvoltage is considered, a sample calculation shows that in a solution 1 molar in zinc and 0.1 molar in hydrogen ions, for instance, the deposition potentials are

For zinc: $E = -0.76 + 0.059/2 \log 1 = -0.76$ V
For hydrogen: $E = 0 + 0.059 \log 0.1 - 0.7 = -0.759$ V

Thus, the deposition potentials under these conditions are almost identical, and both zinc and hydrogen will deposit. By increasing the current density somewhat, the hydrogen overvoltage will be increased, and the efficiency of zinc deposition will rise.

Metals more negative than manganese have not been deposited from aqueous solutions.

Both the hydrogen overvoltage and the cathode potential depend to a great extent on the current density; the overvoltage also depends on the temperature. The metal-ion concentration and the hydrogen-ion concentration (acidity) are also obviously important; thus, in a cyanide solution, where the concentration of copper ions is very small owing to complex formation, some hydrogen is always given off at the cathode. At the very beginning of electrolysis, the nature of the cathode material plays a part, since the hydrogen overvoltage differs from one material to the next; once a film of metal has been deposited, this factor drops out.

Usually the simultaneous deposition of hydrogen has no ill effects except the obvious loss of current efficiency. In a few cases where the hydrogen tends to embrittle the deposit or the basis metal, its effects may have to be watched more closely. The embrittling of steel during the electrodeposition of zinc has been the subject of much study.

41. Evolution of Hydrogen and Oxygen. The deposition and evolution of hydrogen or oxygen at an electrode are always accompanied by polarization and therefore can never be reversible. The conditions involved have already been discussed in connection with polarization and overvoltage (see Pars. **25** and **29**).

As with the deposition of a metal, we may have the deposited gas dissolving in or combining with some other material but with a much broader field of possibilities. The gas may be adsorbed into the electrode (Par. **28**); it may chemically combine with the electrode or the electrolyte, giving a secondary reaction with its accompanying oxidation or reduction; or it may combine with some purposely added depolarizer.

42. Oxidation and Reduction. Strictly speaking, every cathode reaction is a reduction, and every anode reaction is an oxidation. When speaking of electrolytic oxidation or reduction, however, it is usually customary to apply the terms to the particular reaction that the cell in question is designed to perform. The deposition of copper from a copper sulfate solution, for the sake of the recovery of the copper, is considered a reducing reaction, although there is at the anode an equivalent amount of oxidation. Most electrolytic processes are designed to make particular use of either the oxidizing or the reducing action of the current. It is important, in a cell in which reduction is desired, to minimize the effect of the accompanying oxidizing action as much as possible.

In some cases the electrode reactions are such that they do not interfere with each other, even though no effort is made to isolate them. For example, the electrolysis of a copper sulfate solution with platinum electrodes would result in the deposition of copper at the cathode and the formation of H_2SO_4 and O_2 at the anode. Although one of these is a reducing action and the other oxidizing, they do not interfere with each other, for there is no appreciable tendency for either H_2SO_4 or O_2 to combine with the deposited copper, which can be recovered at practically 100% current efficiency. A good example of the opposite extreme may be had by replacing the copper sulfate solution with a mixture of ferrous and ferric sulfates. This could be electrolyzed indefinitely without any appreciable change in the solution, for the oxidizing action of the anode on the ferrous ion would be balanced by the reducing action of the cathode on the ferric ion, and the solution as a whole would remain unchanged. But if the solution surrounding the two electrodes is separated by a porous diaphragm and the action of the two electrodes thus isolated from each other, the operation would eventually result in the oxidation of all the ferrous salt in the anode compartment and the reduction of all the ferric salt in the cathode compartment.

Where the character of the two reactions or the products is such as to require isolation to prevent losses, this separation may be accomplished in one of several different ways:

1. By **keeping the anode and cathode far enough apart** to separate the reactions and their products.
2. By inserting a **porous diaphragm** between the two electrodes.
3. By making the **physical conditions** such as to promote a high efficiency of the reaction desired, but a low efficiency of the reaction not desired.

Method 1 is generally utilized only in cases where the tendency to reaction is already small, as with the copper electrolysis cited above.

Porous diaphragms are extensively used, but they increase the resistance and complicate the cell construction and operation materially and hence are avoided, if possible.

The simplest device and the one most often used is the **control of the physical conditions** to favor the desired reaction. Of these, that which lends itself best to manipulation is the **current density.** In general, a low current density favors high efficiency in oxidation or reduction, and increasing the current density cuts down the efficiency very rapidly. If a reducing action is desired, a low cathode density is used, but a high anode density; and the reverse for an oxidizing action.

43. Passivity. Most metals when dissolving as anode show a degree of polarization, which in some cases rises to the point where the anode reaction changes—metal solution being partly or wholly replaced by oxygen discharge. Such an anode is said to be "passive." Often a visible film forms on the anode, e.g., an insoluble oxide, cyanide, or other compound of the metal, which mechanically inhibits solution; this is **mechanical passivity.** In other cases, sometimes called **electrochemical passivity,** there is no visible film, and yet the passive behavior is noted. The cause is probably a very thin invisible coat of oxide, tough and adherent enough to prevent metal solution. Iron, cobalt, and nickel are particularly subject to this type of passivity.

In a few cases, e.g., lead anodes in a chromium plating solution and in many types of electrowinning solution, passivity at the anode is actually desired. Where it is to be prevented, solution compositions are adjusted so as to contain a high concentration of an ion that will react with the film, e.g., excess cyanide in silver plating to dissolve the AgCN, which would otherwise coat the anode. Current density is kept at sufficiently low values, often by increase of anode area; heating and agitation may be employed; in the case of the iron-group metals, chlorides help prevent passivity. Often the conditions finally arrived at represent a compromise between factors favorable for the anode and those favorable for the cathode process; these do not necessarily coincide.

44. Acidity and pH. Although water dissociates into ions only slightly, the following equilibrium always exists in aqueous solutions or in pure water:

$$H_2O + H_2O \rightleftharpoons H_3O^+ + OH^-$$

The foregoing undoubtedly comes nearer the truth than the more usual formulation, $H_2O \rightleftharpoons H^+ + OH^-$, since H^+ would be a bare proton and must be hydrated. If it is remembered, however, that the symbol H^+ is employed as a convenient shorthand, we can write for the reversible reaction: $(H^+)(OH^-)/(H_2O)$ = a constant, where ( ) means "the concentration of." Since (H_2O) is very nearly constant in fairly dilute aqueous solutions, this reduces to $(H^+)(OH^-) = K$. At ordinary temperatures, K is very nearly 1×10^{-14}. A solution in which $(H^+) = (OH^-) = 10^{-7}$ is a neutral solution; when $(H^+) > 10^{-7}$, the solution is acid; when less, it is alkaline. **The negative logarithm of the hydrogen-ion concentration** is called the **pH** of the solution; at neutrality, pH = 7.0. When pH < 7, the solution is acid; when >7, alkaline.

The pH is a very important property of a solution, and its close control is often a necessity in electrochemical processes. When pH is too low, quite often hydrogen is deposited at the cathode instead of metal; when too high, basic salts may be precipitated in the solution, or with the deposited metal, causing unsatisfactory plates. It has an effect, too, upon the quality of the deposit and many other factors governing the success of electrolytic reactions.

pH is sometimes controlled by the use of **buffer solutions,** which are solutions relatively insensitive to additions of acids or bases. They usually contain a weak acid and a salt of that acid or a weak base and a salt of the base; e.g., a solution containing both acetic acid and sodium acetate would be buffered.

The measurement of pH is carried out either potentiometrically or by means of indicators; one of the most convenient, though not necessarily most accurate, methods is by the use of the **glass electrode.**

This very important subject cannot be covered within the scope of this volume. See, for example, in addition to the books mentioned in the Bibliography (Par. **48**):

Bates, R. G. "Determination of pH: Theory and Practice"; New York, John Wiley & Sons, Inc., 1964.

45. Reduction. Electrolytic reduction may involve any of the following operations:

1. Changing a cation from a higher to a lower valence.
2. Changing a cation to the elementary state.
3. Adding hydrogen to the composition of a compound.
4. Abstracting oxygen or some other anion from the composition of the compound.

Reactions coming under headings 1 and 2 are usually carried on with only such slight amounts of polarization as can be accounted for by concentration polarization; but in a few cases, particularly of type 1, the polarization is considerable. Reactions coming under headings 3 and 4 are often secondary reactions, resulting from the reaction of the primary hydrogen on the adjacent electrolyte. Since the absorption of these gases is practically never complete, and since more or less gas is liberated at the electrode, a considerable amount of gas polarization always accompanies reactions of these two types.

46. Oxidation. Electrolytic oxidation may also be divided into four types, similar to the corresponding subdivision of reduction:

1. Changing a metal from its elementary state to an ion.
2. Changing an ion from a lower to a higher valence.
3. Adding oxygen or some other anion to the composition of the compound.
4. Abstracting hydrogen from the composition of the compound.

Polarization is much more pronounced in all types than is the case with reduction.

Electrolytic oxidation is more powerful and more readily controlled than any chemical agency for doing the same work, and through its use many new compounds have been produced that cannot be made in any other way.

47. Effect of Polarization on Reduction and Oxidation. The magnitude of gas polarization varies widely with different electrodes and different physical conditions surrounding the electrolysis, and the character of the resulting reaction is in many cases determined by the magnitude of the polarization. In Par. **40** it is pointed out how the magnitude of the electrode potential, on the one hand, and the sum of the electrode potential and the hydrogen overvoltage, on the other, determined the relative deposition of metal or hydrogen. In a similar manner, cathodic reduction reactions require a certain critical voltage. If the electrode potential of the hydrogen, plus its overvoltage, is greater than this critical voltage, the reaction will proceed; if less, the reaction will not proceed. It is thus possible to carry on many electrolytic reductions using a lead cathode, having a high hydrogen overvoltage, that are impossible with a platinum cathode, with its lower hydrogen overvoltage. Or if the reaction is one in which there are several possible stages in the reduction, an electrode with a high hydrogen overvoltage will carry the reduction to a farther stage than an electrode with a lower hydrogen overvoltage.

Since oxygen overvoltages are in general much smaller than the corresponding values for hydrogen, there is not quite so much latitude in oxidation as reduction. In other respects the two opposing reactions are quite similar in their behavior.

While polarization is the prime determining factor in reduction and oxidation reactions, it is subject to control only within certain limits. It cannot be done away with entirely, and it cannot be increased beyond 1 or 2 V; but within those limits it can be manipulated quite readily by modifying the various physical factors which affect the polarization. Foremost among these are current density, temperature, concentration, and electrode material. In addition, the results may be still further modified by the presence of certain catalysts in the solution.

48. Bibliography. For more detailed consideration of basic electrochemical theory, the following are recommended:

Glasstone, S. "An Introduction to Electrochemistry"; Princeton, N.J., D. Van Nostrand Company, Inc., 1942.

Creighton, H. J., and Koehler, W. A. "Principles and Applications of Electrochemistry," Vol. 1, Principles, 4th ed., 1943, Vol. 2, Applications, 2d ed.; New York, John Wiley & Sons, Inc., 1944.

Kortüm, G., and Bockris, J. O'M. "Textbook of Electrochemistry" (2 vols.); Houston, Tex., Elsevier Press, Inc., 1951.

Delahay, P., and Tobias, C. (eds.) "Advances in Electrochemistry and Electrochemical Engineering"; New York, John Wiley & Sons, Inc., 1961–1965.

Potter, E. C. "Electrochemistry"; New York, The Macmillan Company, 1956.

ELECTRODEPOSITION

ELECTROPLATING

49. General Principles of Electroplating. The object that is to receive a coating is made cathodic in a solution of a salt of the coating metal. The anode may either be soluble and consist of the same metal, or it may be inert. The aim of the former arrangement is to dissolve as much metal from the anode as is deposited on the cathode during the plating process. When an inert anode is used, it is necessary to regenerate the solution either chemically or by electrolytic solution in a separate tank.

In an electroplating solution in which the anode consists of the same metal as that which is plated out on the cathode, the electrochemical reaction at the cathode is the reverse of that at the anode, so that no emf is required to bring this reaction about (if we neglect the small concentration polarization). The emf applied at the terminals of the cell is therefore used up in overcoming the ohmic resistance and exceeds only slightly the product of current multiplied by internal resistance of the cell. In case an inert anode is used, the anodic reaction is not the reverse of the cathodic reaction and a certain emf is required, viz., the decomposition voltage of the solution. The ohmic loss of voltage should always be made as small as possible, and the decomposition voltage of any solution is never more than a few volts. Thus the voltage at the electrodes of a plating cell must be relatively low; it is in practice generally between a fraction of 1 V and 6 V. In all cases a voltmeter and ammeter should be used by the electroplater so that he may know exactly the electrical condition of his cells and regulate them accordingly. Some kind of rheostat is needed for this purpose.

50. The only way to reduce the voltage at the electrodes of a plating tank is to reduce the internal resistance. This may be accomplished by suitable additions of high-conductivity electrolytes. It may also be accomplished to a certain extent by placing the anodes very near the articles to be plated. But this should not be overdone, since it is important to obtain a uniform deposit over the whole surface, and this will be the more nearly accomplished the further the anode is away from the article to be plated.

51. In case a great many small articles are plated simultaneously in one plating tank, it is usual to have the articles in continuous motion, so that they tumble over each other and present successively all parts of their surfaces to the electroplating action of the current. This is called **barrel plating.**

52. Objectives in Electroplating. The most familiar aspects of electroplating are its uses in decorating or protecting the underlying metal. Plating may also be used as a manufacturing operation or in repair and reconstruction of worn parts. Examples of these uses will be found in Table 23-4 (see Par. **68**).

53. Preparation for Plating. The preparation of an object for plating may involve any or all of the following operations:

1. The removal of oil, grease, or other organic material.
2. The removal of rust, scale, oxides, or other inorganic coatings adhering to the metal.
3. The mechanical preparation of the surface of the metal to receive the deposited metal, by polishing, etc.

For the first, soaps, hot alkali solutions, or organic solvents such as chlorinated hydrocarbons are used; for the second, various acids, alkali and salt solutions, mechanical abrasion, and electrolytic cleaning; for the third, mechanical abrasion and polishing.

54. Cleaning Methods. Electrolytic cleaning in hot alkaline solutions supplements the chemical action of the solution by the mechanical effect of the vigorous gas evolution at the surface. Electrolytic pickling is sometimes used for the same reason. Ultrasonic agitation also helps to dislodge foreign material. Degreasing by means of vapors of organic solvents, which condense upon the work and run back into the tank, has the advantage over merely dipping in the solvent in that a process amounting to constant distillation is being carried out so that clean, fresh solvent is always in contact with the work. Diphase and emulsion cleaners have certain advantages. In general, the formulation and operation of cleaners are a highly proprietary field, and vendors of these materials should be consulted.

55. The Plating Solution. The various constituents of the plating solution usually fall into one of the following classes, with respect to the reason for their use:

1. To furnish the metal content.
2. To increase the conductivity of the solution.
3. To increase or decrease the metal-ion concentration.
4. To increase the anode corrosion.
5. To regulate the pH.
6. To modify the character of the deposit.

56. Variations in Operating Conditions. In addition to the composition of the solution itself, variations in operating conditions affect the character of the metal deposit. Those of particular importance are current density, metal concentration, agitation, temperature, conductivity, metal-ion concentration, hydrogen-ion concentration, addition agents, structure of the basis metal, and metal distribution.

57. Current Density. The effect of current density is important, since it is desirable to deposit the metal in the shortest possible time to maintain maximum output. In general, increasing the current density decreases the crystal size in the deposited metal, but this is true only to a certain point, beyond which the crystal size increases. This limiting current density depends on several factors, e.g., temperature, degree of agitation, and solution composition. Most plating solutions exhibit a range of current densities over which the deposit is satisfactory in brightness, adhesion, etc., and outside which the deposit is unsatisfactory for any one of a number of reasons. In any practical plating electrolyte, this range must not be too narrow, for it is not feasible to control current densities too rigidly, particularly upon cathodes or irregular shapes. Deposits produced at too high a current density so that they are dark, loosely adherent, or powdery are said to be **burned.**

Current density is also important at the anode, since too high current density may cause passivity; and in some cases, e.g., stannate tin, too low a current density causes the anode to dissolve in an undesired valence state (in the foregoing case, as stannite rather than stannate).

58. Conductivity. The use of a solution of good conductivity is important from the standpoint of economy in power consumption and also because it reduces the tendency to form trees and rough deposits.

59. Temperature. A low temperature of the solution favors formation of small crystals of metal; and a high temperature, large crystals. In some cases this is very marked, a difference of only 15°C resulting in a 50% decrease in strength of the metal deposited. On the other hand, high temperature may give beneficial results due to (*a*) increased solubility of the salts, permitting greater metal concentration and higher current densities; (*b*) increased conductivity, which also permits higher current densities and reduces the tendency to form trees; (*c*) decreased occlusion of hydrogen in the deposited metal, which in many cases is the cause of bad deposits. Since both (*a*) and (*b*) tend to decrease crystal size, they may in some cases counteract the tendency of temperature alone to increase the crystal size.

60. Metal Concentration. Bad deposits that are due to insufficient metal at the cathode may be remedied by providing an increased metal supply, unless the solution is already saturated with the salt in question. This, however, is not always practical from an economic standpoint.

61. Agitation. Since the limiting current density depends to a great extent upon the speed with which metal ions are brought to the cathode surface to be discharged, agitation is often helpful in permitting a higher plating speed. Its usefulness is limited by the facts that it tends to stir up sediment and slimes and it increases the absorption of atmospheric carbon dioxide. Where the latter effect is not harmful, air agitation is often employed. In the cases of automatic plating machines and continuous-strip and -wire plating, the effect of agitation is obtained by the movement of the cathode through the solution. In still-tank plating, agitation may be accomplished by mechanical reciprocating motion of the cathode bar; standard apparatus is available for this purpose. Agitation is often combined with continuous filtration by circulating the electrolyte from the tank, through a filter, and back to the tank.

62. Metal-ion Concentration. The most favorable conditions for fine-grained deposits include a very low metal-ion concentration. It is not practicable to realize this by the use of very dilute solutions, however, since in such a case the metal around the cathode would quickly be exhausted. The two desiderata of a high metal concentration and low metal-ion concentration can be attained by the use of complex ions in which the metal is rather firmly bound. At equilibrium,

$$xM^+ + yR^- \rightleftharpoons (M_xR_y)^\pm \tag{23-9}$$

where M = a metal ion and R = a complex-forming ion like cyanide, CN^-. The equilibrium lies far to the right; but as soon as metal is discharged at the cathode, the equilibrium is disturbed and more M^+ ionizes out of the complex. The most familiar examples of the use of complex ions are the various cyanide electrolytes, but many others are also of value in electroplating.

63. Hydrogen-ion Concentration (pH). Deposition of hydrogen, as of a metal, depends on its standard potential, its concentration, and its overvoltage (see Pars. **15, 29,** and **40**). Regulation of the pH is important, particularly in the plating of those metals above hydrogen in the electromotive series. Often the pH must be kept within narrow limits, in order to avoid, on the one hand, the excessive evolution of hydrogen and, on the other, the precipitation of basic salts or hydroxides in the electrolyte; additionally, some metals, e.g., nickel, vary in their physical properties depending upon the pH of the solution from which they are deposited. Close regulation of the pH is generally accomplished by the use of buffered solutions (see Par. **44**).

64. Metal Distribution and Throwing Power. In the plating of an irregularly shaped object, some parts of the cathode receive higher current densities than others and consequently are more heavily plated. Since specifications for platings often call for a minimum thickness of metal, the excess plated upon the high-current-density parts represents a waste of metal, current, and time. In some cases conforming anodes are resorted to, to correct this condition; but fortunately many solutions have the property of good **throwing power,** which automatically mitigates it to a greater or lesser extent. Throwing power may be defined as the improvement in metal distribution over the current distribution and ranges from excellent for the stannate tin solution to practically nil for the chromic acid electrolyte. Although there are many devices for measuring throwing power, none is free of objections and the term is used in a rather qualitative sense.

The **covering power** of a solution is related to the throwing power and to the lowest current density at which metal can be deposited from the solution; it determines whether or not areas of very low current density will be covered at all.

65. Addition agents, in general, are substances added to the bath to promote formation of small crystals and smooth deposits, thus permitting the use of higher current densities. These substances are usually colloidal in character and include such materials as glue, gelatin, albumen, glucose, dextrose, dextrin, phenol, cresol, carbon bisulfide, gum tragacanth, glycerin, tannic acid, and a host of others. The exact mechanism of the action of these colloidal materials is not well understood, but the general effect is to prevent the growth of individual crystals and produce a larger number of smaller crystals.

One function of addition agents—the ability of some types to produce a **bright plate**—is of increasing importance. Since the labor of polishing and buffing is usually a larger item of cost than the purely electrochemical aspects of plating, the production of a deposit that is bright as plated has many attractions. Addition agents have been developed that make this possible for some metals, notably nickel. Sometimes deposited alloys are brighter than either metal alone, as with the nickel-cobalt "bright nickel." These processes are almost without exception patented, and reference should be had to the patent literature and the booklets of the various plating-supply firms.

66. Structure of the basis metal on which another metal is plated has a decided effect on the structure of the deposited metal; undesirable conditions in the former may cause defects in the latter. This relation is so close that copper deposited on a clean cast- or rolled-copper base consists of crystals that are extensions of those in the basis metal.

67. Metals Used in Electroplating. Table 23-4 lists the metals most commonly electroplated and gives for each the type of solution used, the principal constituents of the solution, and some of the average conditions of plating, e.g., the current density, the temperature, the voltage, and the cathode efficiency. Throwing power of the solution is indicated qualitatively, and the principal applications of the metal are listed; the final column indicates the average time to deposit 0.001 in (a mil) of plate.

68. Summary of Electroplating Practice (see Table 23-4).

69. Notes on the Use of Table 23-4. It will be realized that the information contained in Table 23-4 represents a very much condensed summary of a subject that for detailed treatment requires a complete volume: the data should be looked upon as a bird's-eye view, not a set of working directions. For example, almost all the

Table 23-4. Summary of Electroplating Practice
(Average operating conditions)

Metal (1)	Principal uses (2)	Type of solution (3)	Principal ingredients (4)	Temp, deg F (5)	CD, A/ft² (6)	Volts (7)	Cathode efficiency, % (8)	Throwing power (9)	Time to deposit 0.001 in. (10)
Cadmium	Protection	Cyanide	CdO, NaCN, brighteners	70–95	15–45	1–4	90	Good	20 min
Chromium	Decorative Engineering (hard) Cylinder liners (porous)	Chromic acid	CrO_3, H_2SO_4	120	250	6–8	15	Poor	2 hr
Copper	Electroforming Undercoat for other metals Stop-off in casehardening, etc.	Acid	$CuSO_4.5H_2O$, H_2SO_4	75–120	15–40	1–2	100	Fair	35 min
		Cyanide	CuCN, NaCN, Na_2CO_3	75–100	5–15	1.5–3	50	Good	90 min
		Rochelle	Above + rochelle salts	140–160	20–60	2–3	60	Good	45 min
		Many other types, *e.g.*, fluoborate, pyrophosphate, all-potassium cyanide		....	....	.....	...	.....	......
Gold	Decorative Electronics	Cyanide (Solutions vary considerably, depending on application)	$KAu(CN)_2$, K_2CO_3, KCN	120–160	5–15	2–6	80	Good	......
Indium	Bearing surfaces	Cyanide	$InCl_3$, NaCN, addition agent	Room	10–150	.....	40	Good	......
		Sulfate	$In_2(SO_4)_3$, Na_2SO_4	Room	20	.....	75	Poor	......
		Fluoborate	$In(BF_4)_3$, H_3BO_3, NH_4BF_4	70–90	50–100	.....	50	Good	......
Iron	Electroforming Repair	Chloride	$FeCl_2$, $CaCl_2$	190	60	.....	95	.....	20 min
		Sulfate	$FeSO_4(NH_4)_2SO_4$	Room	20	.....	95	.....	1 hr
Lead	Protection Bearing surfaces	Fluoborate	$Pb(BF_4)_2$, HBF_4, glue	Room	10–80	0.5	100	Good	40 min
Nickel	Protection Decorative Electroforming Undercoat for Cr, etc.	Sulfate-chloride Sulfamate Fluoborate	$NiCl_2$, $NiSO_4$, NH_4 ion, H_3BO_3 (Formulations differ widely, depending on purpose) Ni sulfamate, sulfamic acid $Ni(BF_4)_2$, HBF_4, addition agents	75–100	Varies greatly	0.5–3	95	Fair	30 min
Rhodium	Decorative Optical	Sulfate Phosphate	Prepared salts	110–120	10–80	2.5–5	15	.....	......
Silver	Decorative Protective Bearing surfaces	Cyanide	AgCN, KCN, K_2CO_3, CS_2 (Or Na in place of K)	80	5–15	1	100	Good	......

Tin	Protection Food and dairy Bearings Electrical To enable easy soldering	Sulfate	$SnSO_4$, H_2SO_4, addition agents	Room	40	1–3	90	Fair	15 min
		Fluoborate	$Sn(BF_4)_2$, HBF_4, addition agents	75–100	50	.....	100	Good	10 min
		Other acid electrolytes		....	....	.....	...	.....	......
		Stannate	Na_2- or $K_2Sn(OH)_6$, Na- or KOH	150–190	40	4–8	80	Excel-lent	30 min
Zinc	Protection	Sulfate	$ZnSO_4$, NH_4Cl, addition agents	75–100	15–400	.....	99	Fair	10 min
		Cyanide	$Zn(CN)_2$, NaCN, NaOH, brighteners	100	10–50	.....	85	Good	40 min
			Alloys						
Brass	Rubber-bonding Decorative	Cyanide	CuCN, $Zn(CN)_2$, NaCN, Na_2CO_3	75–100	3–10	2–3	75	Good	......
Bronze	Decorative Undercoat for chromium Stop-off for steel	Cyanide-stannate	CuCN, KCN, KOH, $K_2Sn(OH)_6$, rochelle salt	155	20–100	3–6	70	Excel-lent	30 min
Lead-tin	Bearings Solderability Electrotyping	Fluoborate	$Sn(BF_4)_2$, $Pb(BF_4)_2$, HBF_4, addition agents	Room	60	1–2	100	Good	......
Tin-zinc	Solderability	Cyanide-stannate	$Zn(CN)_2$, KCN, KOH, $K_2Sn(OH)_6$	150	10–75	4–5	80–95	Excel-lent	30 min
Tin-nickel	Printed circuits	Chloride-fluoride	$NiCl_2$, $SnCl_2$, NH_4NF_2, HF	150	25	1–2	98	Excel-lent	30 min

metals can be plated by proprietary processes, for each of which various advantages over the standard solutions are claimed. Which of several competing processes is preferred depends upon the circumstances in each case.

Usually the combination of circumstances will indicate a choice of one solution or another, for there is no such thing as a single electrolyte that combines all the desired properties. No acid copper solution can be employed to plate directly on steel, because of the formation of an **immersion** deposit (chemical replacement: $Fe + Cu^{++} \rightarrow Fe^{++} + Cu$). Consequently, a cyanide-type solution is used; as soon as a thin deposit is formed, the work can be transferred to an acid solution, and the advantages of faster plating in the latter can be obtained. Such a preliminary thin deposit is referred to as a **strike.**

In silver plating, a strike solution lower in metal and higher in cyanide than the principal electrolyte is used to prevent the immersion deposit. The reason for these precautions is that immersion deposits are generally of inferior properties as compared with good electroplates.

Other conditions may dictate the choice of a certain solution. In tin plating, the acid solutions offer faster plating and economy of power but are relatively poor in throwing power, and thus for plating complicated shapes the alkaline stannate solutions are much preferred.

Alkaline solutions, including the cyanide, have in general better throwing power than acid. Other factors that must be taken into consideration are ease of control, physical properties of the deposit, cost per gallon of solution, etc.

70. Alloy Plating. In order to codeposit two metals, it is necessary that their deposition potentials be brought close together, if they are not so already. Thus, copper and zinc cannot be codeposited from an acid electrolyte; but when they are in the form of their complex cyanides, the copper-ion concentration is so greatly reduced that its potential is made more nearly equal to that of zinc and brass can be deposited. Tin bronzes are plated from mixed cyanide-stannate electrolytes, as well as from proprietary baths. They are superior to copper in corrosion resistance and are used both as an undercoat and as a final decorative finish. Tin-zinc alloy (80% tin), plated from similar types of solutions, has some applications in electronics because of its excellent solderability. Tin-nickel (65% tin), plated from chloride-fluoride solutions, finds uses in printed circuitry and as a decorative finish. Tin-lead alloys are deposited from fluoborate solutions.

There are many gold-alloy processes in use, almost all of a proprietary nature. Gold plating has expanded from purely decorative applications in jewelry to assume major importance in electronics and aerospace; the codeposition along with the gold of small amounts of alloying elements confers on the deposit various specific properties including hardness, electrical conductivity and contact resistance, and wear resistance.

"Thin films" for magnetic memory devices make use of alloy plating, depositing various alloys of the iron-group metals: iron, cobalt, and nickel.[1]

71. Continuous Plating. The tin plating of continuous coils of sheet steel in wholly mechanized equipment was given great impetus during World War II by the need to conserve tin; at present all but a very minor proportion of tin plate is produced electrolytically, and in terms of tonnage of product strip electrotinning undoubtedly represents the largest single application of electroplating. On a somewhat smaller scale, other metals are continuously plated on steel strip, steel and copper wire, electrical terminals, and other substrates.

72. Selection of Plated Coatings. Table 23-5 is intended as a rough guide to the selection of plated coatings, indicating those coatings which should be considered for specific applications. Since individual requirements will vary widely, the table will have obvious limitations and cannot take the place of expert judgment; nor are the many highly specialized uses of plated coatings covered.

73. Plating on Aluminum. Electroplates on aluminum, e.g., bright chromium or nickel, add nothing to the resistance of aluminum to corrosion; and although they are perfectly satisfactory where the corrosion conditions are not too severe, they may not be suitable for continuous outdoor exposure.

[1] Many references are cited by ZENTNER, V. *Plating*, 1965, Vol. 52, p. 868.

The principal problem in the electroplating of aluminum is to secure a coating which will show satisfactory adhesion under service conditions, particularly in the presence of moisture. Satisfactory electroplates can be applied if the proper technique is employed in preparing the surface before plating.

74. Anodic Coating of Aluminum. Aluminum is protected by the oxide film which always forms on the metal when it is exposed to the air. This film is very thin, however—about 0.5×10^{-6} in in thickness. By making aluminum or aluminum alloys anode in suitable electrolytes, e.g., dilute sulfuric acid, oxalic acid, or chromic acid, oxide coatings of substantially greater thickness, say, 0.0001 to 0.001 in, can be formed. These anodically formed oxide coatings offer good protection against corrosion, good resistance to abrasion, and good dielectric strength and are decorative in appearance. As formed, they are porous and permeable and can be colored by the absorption of suitable dyes. They can also be sealed so as to become nearly impermeable and nonabsorbent without changing the appearance.

Table 23-5. Selection of Plated Coatings

Application	Platings Generally Useful
Corrosion prevention	Zinc, cadmium, lead
Decoration-protection: general	Copper + nickel + chromium, bronze + chromium
Decoration-protection: special colors and effects	Bronze, brass, precious metals, tin-nickel alloy
Light reflectance	Rhodium, silver
Solderability	Tin-zinc alloy, tin, tin-lead alloy, cadmium, silver, gold
Bearing surfaces	Tin, tin-lead alloy, chromium, indium
Building up worn parts	Nickel, iron (with or without copper)
Hard surfacing	Chromium
Electrical contacts	Gold, silver, platinum group
Foods and beverages	Tin
Rubber bonding to metals	Brass
Stop-off in nitriding	Bronze, copper, tin

75. Surface Treatments for Magnesium. Most of the surface treatments for magnesium and its alloys are directed toward providing a base for subsequent painting or lacquering; the coatings themselves do not possess particularly good corrosion resistance. Magnesium can also be electroplated, by means of preparatory treatments somewhat similar to those used for aluminum, with the conditions even more critical.

76. Conversion Coatings on Zinc. The corrosion resistance and appearance of zinc and cadmium electrodeposits can be improved by treatment in solutions that will form on the surface an adherent and impervious coating of a metal chromate. Several proprietary processes are used for this purpose.

77. Phosphate Coatings on Steel. By appropriate chemical treatment the corrosion resistance of plain steel can be somewhat enhanced by processes known as Bonderizing and Parkerizing. These have their principal application as a base for subsequent painting or lacquering.

78. Bright Dipping. The appearance of zinc and cadmium deposits can be greatly enhanced without significant removal of metal by a dip in a solution containing H_2SO_4 and H_2O_2. Copper and copper alloys can be given a good surface finish by dipping in H_2SO_4-HNO_3 mixtures (see G. Soderberg, *Trans. Electrochem. Soc.*, 1954, Vol. 88, p. 297).

79. Electropolishing. Closely allied to bright dipping is electropolishing, which utilizes anodic treatment in specially formulated electrolytes to bring up a polished surface on such metals as stainless steel. Electropolishing is also useful as a tool in preparing metallic surfaces for microscopic examination. Both bright dipping and electropolishing depend on the more rapid eating away by the solution of microprojections on the metal, so that a smoother surface results, although the exact mechanism is a matter for debate.

80. Stripping of Deposits. Rejected and imperfect electroplates can usually be stripped, and the part replated. Methods, some purely chemical and some involving making the part anode, are to be found in the Metal Finishing Guidebook-Directory (see Bibliography, Par. **86**).

81. Testing of Deposits. As the specifications for the performance of electroplates become more and more rigorous, more attention is directed toward testing methods.

The principal properties subjected to test include:

1. *Corrosion Resistance.* Although many accelerated weathering tests have been devised and used, some assurance that results of these tests will correlate with actual performance in the expected service should be sought before placing reliance on them. Thus the neutral salt spray,[1] although appearing in many specifications, is often of questionable validity. For the specific case of automotive brightwork plated with copper-nickel-chromium, the so-called CASS and Corrodkote[2] tests have gained wide acceptance and appear to predict service performance reasonably well. Outdoor exposure tests on stationary racks and observation of parts in actual service, although time-consuming, are still required to confirm indications from accelerated tests.

2. *Porosity.* The presence of pores in a metallic coating may have much or little bearing upon its protective value depending on whether the coating is cathodic or anodic to steel. Pores in a zinc coating do not greatly detract from its protective value, because the zinc protects the steel "sacrificially"; i.e., in the galvanic cell which is set up under corrosive conditions it is the zinc and not the steel that corrodes. If the metal is cathodic to steel, the opposite condition obtains and the steel may corrode more seriously about a pore than if it were not coated at all. The salt-spray test, noted under (1) above, is under some conditions a test of porosity. Pores may also be detected by the **hot-water** test and the **ferroxyl** test; in the first, each pore shows as a rust spot and, in the second, as a blue coloration.

3. *Thickness.* Other things being equal, the performance of an electroplate correlates fairly well with its thickness. For nonmagnetic coatings on steel, a widely used test method is the Magne-gage, which can also be used for nickel coatings on steel or on nonmagnetic basis metals. Other instruments based on similar principles are also available. The thickness of a coating will affect the strength of x-rays, beta rays, or other electromagnetic radiation reflected from the substrate, and several instruments and methods are based on this principle. Chemical tests rely on the chemical solution of the coating by a standard reagent; the procedure is standardized, so that the time necessary to penetrate to the basis metal gives a measure of the thickness. A piece of known area may be weighed before and after stripping the coating; or the stripping solution may be analyzed for the coating metal (tin plate is usually tested in this way). The Mesle chord method and microscopic examination are also used. An automatic instrument is available based on anodic solution of the coating at controlled current. For details see Chap. 32 in "Modern Electroplating" (Bibliography, Par. **86**).

4. *Adhesion.* The quantitative measure of the force with which an electroplate adheres to the basis metal has been the subject of considerable study. However, most tests in actual use are rather qualitative and depend on the observer's experience. A coating is rated as either sufficiently adherent for the purpose or not, and no attempt is made to express the adhesion quantitatively.

5. *Other properties* sometimes subjected to test are hardness, abrasion resistance, solderability, lacquer or paint adhesion, or such other characteristic as may be important in a particular application. Anodic coatings on aluminum may be tested by measurement of their electrical resistance.

ELECTROFORMING AND OTHER PLATING PROCESSES

82. Electroforming. The production or reproduction of articles by electrodeposition upon a mandrel or mold is known as electroforming. The mandrel is usually subsequently removed from the deposit, which then becomes the finished electroformed article; but sometimes the combination of mandrel plus deposit is left intact.[3]

The oldest and most used variety of electroforming is electrotyping (see Par. **83**), but the process is becoming of increasing importance as a means of manufacturing components of unusual shape or to close tolerances. Typical of such articles are venturi nozzles, pitot tubes, waveguides, and reflectors. A well-established use, and one that illustrates the extremely fine detail that electroforming is capable of, is the production of masters and stampers for phonograph records.

Although any metal that can be electroplated can be electroformed, the principal

[1] ASTM Designation B-117.
[2] ASTM Designations B-368 and B-380.
[3] MISSEL, L., SHAHEEN, M. E., and TAYLOR, R. *Plating*, 1965, Vol. 52, p. 34.

emphasis has been on nickel and copper. Molds or mandrels may be expendable (used once) or permanent and may be metallic or nonmetallic.[1] The surface of a metallic mold is treated to prevent the adhesion of the deposited metal, so that the finished object can be removed; a nonmetallic mold is treated to render its surface conductive.[2]

83. Electrotyping. The object of electrotyping is to reproduce printing, setup type, engravings, medals, etc. A mold of the object to be reproduced is first made, e.g., of wax, by impressing the object in wax. If the mold is a nonconductor of electricity, as in the case of wax, its surface is made conducting by giving it, with a brush, a coating of graphite. Instead of a graphite coating, the mold may receive a metallic coating of copper by pouring copper sulfate solution over the surface of the mold and dusting on it from a pepper box very finely divided iron filings, brushing the mixture over the surface; in other cases the wax is coated with metal by interposing it between a cathode and an anode of the metal to be plated and passing a high-tension discharge. By suitable electrical connections of clamps or wire to the surface of the mold the latter is then made a cathode in an acid copper electroplating bath. In the case of reproducing type matter, two cases containing prepared molds are always suspended back to back between two large copper anodes so that the conducting surfaces of the molds directly face the anodes. The copper shell is then separated from the mold on which it is deposited, and in order to give it the necessary strength for further use, it is backed with type metal.

84. Other Plating Processes. The uses of electroplating and allied processes are many. The following deserve passing mention:

a. Electroanalysis. The complete stripping of a metallic constituent from an electrolyte by plating on a weighed cathode is a common and convenient analytical tool (see standard books on quantitative analysis).

b. Immersion Plating. Chemical replacement methods for applying coatings without the use of an electric current include immersion and contact processes and "electroless" plating (see process *g* below). Immersion processes have rather specialized uses, but where applicable they can be cheaper and simpler than standard electroplating techniques. Tin on copper and its alloys, tin-copper alloys on steel wire, and tin on aluminum-alloy pistons are prominent among the major uses for immersion processes.

c. Stripping Processes. The process of detinning tin scrap by making it anode in a caustic solution has been superseded in this country by other nonelectrolytic methods but is still apparently used to some extent abroad.

d. Brush Plating. Plating can sometimes be carried out in the field by using a brush wet with electrolyte; the technique is also useful for plating localized areas without the need to stop off the remainder of the piece.[3]

e. Plating on plastics and other nonconductors has enjoyed a revival of interest. The art is very old and until recently had limited use; but the development of specially formulated plastics, particularly of the ABS (acrylonitrile-butadiene-styrene) type, and processes suitable for these materials has brought plated plastic parts into some prominence, for both decorative and functional applications.[4]

f. Electrophoretic processes are used for the deposition of nonmetallic substances like rubber and metal oxides and for some metals that cannot be deposited electrolytically from aqueous solution. "Electrocoating" is used to deposit paints (at present principally primer coats) electrophoretically on automobile bodies.[5]

g. Electroless Plating. Considerable development has resulted from the work of Brenner and co-workers[6] on the plating of nickel by chemical reduction rather than electrolytically. Advantages of the technique include (*a*) almost perfect throwing power and distribution of deposit; (*b*) ability to plate complicated shapes, in fact, anywhere that solution can penetrate; (*c*) good deposit characteristics. The deposit is not pure nickel but contains about 7% phosphorus. Offsetting disadvantages are principally economic, since the chemical reducing agents used are expensive.

---

[1] ASTM Committee B-8, Sub VII, *Plating*, 1964, Vol. 51, p. 1075.
[2] Symposium on Electroforming, *ASTM Spec. Tech. Publ.* 318, 1962.
[3] Rubinstein, M. *Materials and Methods*, 1954, Vol. 40, No. 6, p. 98.
[4] Saubestre, E. B., Durney, L. J., and Washburn, E. B. *Metal Finishing*, November, 1964, Vol. 62, p. 52.
[5] Hutchinson, C. O. *Plating*, 1965, Vol. 52, p. 1133.
[6] Symposium on Electroless Nickel Plating; *ASTM Spec. Tech. Publ.* 265, 1959.

Electroless processes are also available for copper, cobalt, gold, and some other metals and alloys.

85. Plating-room Layout. In designing a plating room, attention should be given to placing of the generators or rectifiers in relatively dust-free locations, with due consideration to economy of bus bars. Proper floor drainage, ventilation, materials for tanks, and heating coils are other factors that require consideration. These matters are discussed in Graham's book (see Par. **86**).

86. Bibliography on Electroplating. Where no other reference is given, it may be assumed that further details, and references to the original literature, may be found in:

LOWENHEIM, F. A. (ed.) "Modern Electroplating," 2d ed.; New York, John Wiley & Sons, Inc., 1963.

GRAHAM, A. K. (ed.) "Electroplating Engineering Handbook," 2d ed.; New York, Reinhold Publishing Corporation, 1962.

Other useful general references are:

BLUM, W., and HOGABOOM, G. B. "Principles of Electroplating and Electroforming," 3d ed.; New York, McGraw-Hill Book Company, 1949.

Metal Finishing Guidebook-Directory; Westwood, N.J., Metals & Plastics Publications, issued annually.

Files of *Trans.* and *Jour. Electrochem. Soc.; Trans. Inst. Metal Finishing; Plating; Proc. Am. Electroplaters' Soc.; Metal Finishing.*

ELECTROLYTIC REFINING

87. Fundamental Principles. In electrolytic refining of metals the starting material is a highly concentrated alloy or mixture of metals, and the purpose is to remove the last impurities and to recover not only the principal metal in pure form but also the foreign metals, especially the precious metals. The impure metal is made the anode, and the fundamental principle of the process is that by the electrolytic action the metal to be refined is dissolved from the anode and passes into the electrolyte, carrying with it all less noble metals that are present; any more noble metals present in the anode are left undissolved; at the cathode the prime metal is deposited alone, since the less noble metals require more energy to deposit. This cannot be satisfactorily accomplished, except with a comparatively pure, high-grade anode; in American copper refining the impure copper anode is generally 98 to 99.5% pure.

88. Economic Considerations in Electrolytic Refining. The economic factors concerned in the use of electrolytic refining are as follows:

a. Higher price for the pure metal.

b. Value of byproducts recovered, particularly precious metals.

c. Lack of any other satisfactory method of refining.

d. Comparative cost of electrolytic refining and other methods.

e. Need for highly pure metals in advanced technology, where cost is not a factor.

The premium on the pure metal is seldom sufficient to pay the cost of the operation, so that the other factors, particularly the second, are of great importance. The values of the precious metals recovered are usually more than sufficient to pay the cost in copper refining. For many years no other satisfactory method of refining was available for the removal of bismuth from lead, and this alone accounted for the electrolytic refining of that metal, although the recovery of the bismuth helped to pay the cost. In most cases the cost of electrolytic refining is greater than by other methods, so that without the other factors there would be little possibility for the use of this method. In the case of zinc, the price margin is low; there are no valuable byproducts; and where refining is required, redistillation is satisfactory and less costly than electrolysis. Hence, no zinc is electrolytically refined, but it is reduced electrolytically, direct from the ore (see Par. **101**).

89. Copper Refining. The electrolyte is a copper sulfate solution containing free sulfuric acid. The copper content usually is between 3 and 3.75%, and the free sulfuric acid is between 15 and 18%. (Usually a very small amount of a soluble chloride, like NaCl, is added to precipitate as chloride any silver which may dissolve and to slime antimony as oxychloride.) An anode of typical composition contains 99 to 99.3% copper, up to 110 oz silver/ton, up to 7 oz gold/ton, and up to 0.07% arsenic. The cathode copper is exceedingly pure, usually running about 99.98% copper, with hydrogen as the chief impurity. In order to have high electric conductivity, the copper must be free from arsenic and antimony (the amount should be but a few thousandths of a percent); in order to prevent brittleness, the cathode copper must be free from tellurium and lead. Since the electric conductivity of copper very delicately indicates its purity, it is commonly used as a measure of the purity (see Sec. **4**, Par. **16**).

It is customary to circulate the electrolyte from tank to tank, and this is more important the higher the current density. The current density varies in American refineries usually between 16 and 28 A/ft^2; the question of what current density to carry is largely one of energy cost. An elevated temperature of the electrolyte is used. Soluble sulfates of impurities in the anode pass into the solution, which therefore needs purification at intervals. This is usually done by working up a certain quantity of electrolyte regularly into bluestone and adding fresh acid to the electrolyte.

90. Arrangement of Electrodes in Copper Refining. With respect to the arrangement of the electrodes, two different systems are in use. In the parallel or multiple system all cathodes in a tank are in parallel, and all anodes are in parallel; the arrangement is shown in Fig. 23-1. In the series, or Hayden, system, diagrammatically shown in Fig. 23-2, only the first and last electrodes are connected to the electric circuit; the first, which is the anode, is of impure copper; the last, the cathode, of pure copper. All intermediate electrodes are made of the copper to be refined and act as bipolar electrodes; copper is dissolved from the side acting as anode and pure copper simultaneously deposited on the side acting as cathode. In both systems enough tanks are connected in series to give circuit voltages of 100 to 200 V. The main points of difference between the multiple system and the series are in energy cost, compactness, and cost of preparing anodes. The power in the series system is about 70% of that in the multiple system. In the multiple tank close attention must be paid to the contacts. The series tank has no contacts or conducting bars, and the electrodes are very close together, the anodes being thin, even plates. Such anodes must either be rolled or specially hand-cast, and the grade of material used must be good. The interest on the metal tied up in process and the investment in plant are less in the series system. The series system requires no starting sheets but much closer supervision to keep the quality of the cathodes up. As lead-lined tanks cannot be used in series work, owing to the relatively high voltages used, tank maintenance becomes an important item. Most plants use the multiple system.

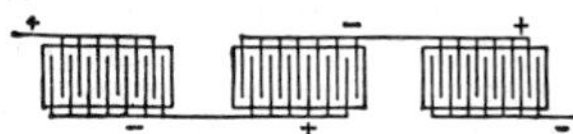

FIG. 23-1. Multiple system.

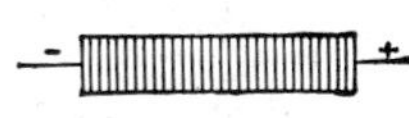

FIG. 23-2. Series system.

91. Silver Refining. The raw material for the production of pure silver consists largely of anode slimes from copper refining and those from the electrolysis of lead, nickel, and zinc; the concentrates obtained from the desilverization of lead; and silver-gold bullion of several classes.

The anode slimes are allowed to settle in order to remove the excess electrolyte; they are then washed, filtered, roasted, and leached in sulfuric acid to remove the copper. After again filtering and washing, the slimes are melted and refined in a small reverberatory furnace—doré furnace. The doré metal is cast into anodes for electrolytic parting of the silver and gold by the Balbach or Moebius system.

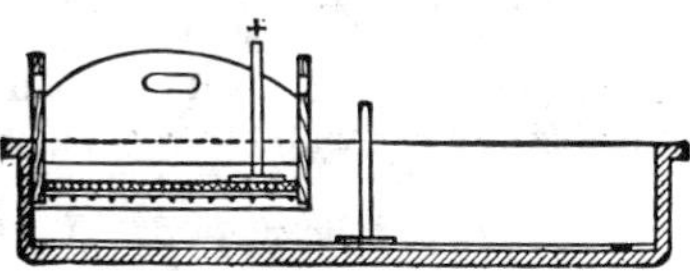

FIG. 23-3. Balbach cell.

In the **Balbach[1] process** the cells are mastic-lined concrete tanks with graphite slabs fitted to the bottom to form the cathode as shown in Fig. 23-3. The doré-metal anodes are supported in wooden frames above the cathode, and the frames enclosed in a cloth case. During electrolysis the silver is deposited on the cathode and the gold collected as a slime in the cloth cases. A current density of 40 to 50 A/ft^2 of anode is used. At this density the voltages range from 3 to 3.8 V/tank, and a current efficiency of 93 to 95% is obtained. The energy requirements are about 31.5 Wh/fine oz of silver produced.

[1] Also known as the Thum process.

92. Moebius Process. In the Moebius system the anodes and cathodes are arranged vertically in much the same manner as in a copper refinery. The doré-metal anodes are encased in a cloth bag to collect the gold slimes as shown in Fig. 23-4. The cathodes are thin sheets of stainless steel. Mechanically operated wooden scrapers brush the silver crystals from the cathodes into trays. A current density of about 50 A/ft^2 of anode surface is employed. At this density the voltage is about 2.7 per tank, and a current efficiency of 98% is obtained.

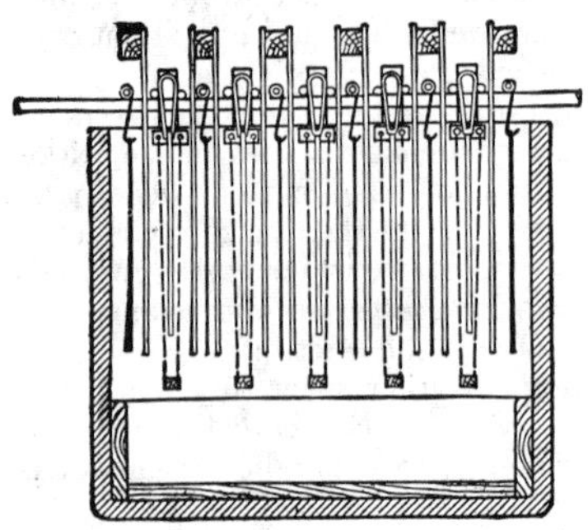

FIG. 23-4. Moebius cell.

93. Gold Refining. The Wohlwill process of gold refining with recovery of platinum and palladium employs a hot acid solution of gold chloride (7 to 8% gold, 10% hydrochloric acid) at a temperature of 65 to 70°C. A slight modification of the Wohlwill process (55°C, purer anodes, and lower current density) is in use in the United States mints. The applicability of the Wohlwill process to alloys richer in silver has been rendered possible by the employment of a pulsating current (obtained by superposing an alternating current on the direct current) instead of a purely direct current.

The gold cells are much smaller than the silver cells—sometimes no larger than 12 by 16 in—and 12 in deep. The anodes are cast slabs, suspended from the anode supports by C- or S-shaped hangers through a hole in the top of the anode. The current density is 50 to 70 A/ft^2 of cathode surface. The cathodes are strips of fine gold rolled to a thickness of 0.01 in and are removed for melting when they weigh 150 to 200 oz.

94. Lead Refining. The Betts process employs a solution of lead fluosilicate, containing an excess of hydrofluosilicic acid (7 to 9% Pb and 5 to 8% free acid) with a very small addition of gelatin or glue, depositing lead in dense, coherent form and free from bismuth. The voltage per tank is 0.35 to 0.6 including losses, the temperature 40°C, the energy consumption from 95 to 110 kWh/ton of lead. The efficiency is about 90%, and the current density 16 to 22 A/ft^2.

95. Nickel Refining. Crude metal anodes contain about 95% nickel, 2.5% copper, and other impurities such as iron, cobalt, and sulfur. They are produced by smelting of nickel oxide, which in turn originates from nickel sulfide separated from ore. Since the impurities also dissolve in the electrolyte and would redeposit at the cathode, diaphragm cells are used and the anolyte and catholyte systems are kept separate, the anolyte being chemically purified before being returned to the system as catholyte.

In a recent innovation introduced by the International Nickel Company, much of the pyrometallurgical part of the process is avoided by using sulfide anodes, thus bypassing the conversion of the sulfide to oxide and subsequent reduction of the oxide to metal. These matte anodes contain 72% nickel, 23% sulfur, plus copper, iron, and other impurities. Since all the sulfide is anodically oxidized to elemental sulfur, anode slimes are voluminous and anodes are bagged.

Typical electrolyte composition is nickel (as sulfate) 60 g/l, sodium chloride 60 g/l, boric acid 15 g/l. Electrolytic cathodes may be 99.9% pure nickel.

96. Solder. At the plant of the U.S. Metals Refining Company at Carteret, N.J., solder is produced from white metal scrap by using an electrolyte containing $SnSiF_6$, $PbSiF_6$, and free H_2SiF_6. The anodes contain 50 to 85% lead, 48 to 13% tin, and up to 10% impurities. No attempt is made to control the proportion of lead to tin in the cathode deposit; blending to specification is done afterward.[1]

97. Other Metals. Bismuth, indium, and tin are electrorefined to some extent. These and other metals are in some demand in a highly pure state for specialized uses, and small-scale electrorefining plants— sometimes hardly larger than laboratory setups—operate to fill this demand.

98. Bibliography on Refining. References to the older literature as well as a good

[1] HERMSDORF, R. P. E., and HEBERLEIN, M. F. W. *Trans. AIME*, 1936, Vol. 121, p. 289.

summary of recent practice will be found in:

MANTELL, C. L. "Industrial Electrochemistry," 4th ed.; New York, McGraw-Hill Book Company, 1960.

Articles on individual metals will be found in:

HAMPEL, C. A. "The Encyclopedia of Electrochemistry"; New York, Reinhold Publishing Corporation, 1964.

Older but still valuable is:

LIDELL, D. M. "Handbook of Nonferrous Metallurgy," 2d ed.; New York, McGraw-Hill Book Company, 1945, Vol. II.

Current developments and statistics are covered in the annual volumes of the Mineral Year Book of the U.S. Bureau of Mines.

ELECTROWINNING

99. Fundamental Principles. Electrowinning differs from electrorefining principally in that the metal is introduced into the solution chemically instead of electrochemically and the anodes are insoluble under the conditions of electrolysis. Because, unlike electrorefining, the reaction at the anode is not substantially the reverse of that at the cathode, the cell voltage is higher. The product released at the anode is predominantly oxygen. Anodes last for several years but must eventually be replaced, because either they are slowly attacked chemically or periodic cleaning has gradually worn them down.

Feed to the process may be an ore, a concentrate, or a metallurgical byproduct from which the desired metal has been chemically leached. The resulting solution is purified and clarified before being introduced into the cells.

Metals which have been or are produced by electrowinning from aqueous solution include antimony, cadmium, chromium, cobalt, copper, indium, manganese, thallium, and zinc. Electrowinning is also conducted from fused salts to produce aluminum, magnesium, sodium and others; these processes are considered in Pars. **130** to **136**.

100. Copper. About 10% of copper production results from the electrowinning process, the rest being produced by pyrometallurgy and refined electrolytically. The final products are identical. In electrowinning of copper, the leaching solution is sulfuric acid; since the electrolytic stripping of the leach solution regenerates the acid used in leaching, the process essentially consumes no acid, but of course there are inevitable processing losses. The feed electrolyte to the cells contains 3 to 5% copper and 2.5 to 3% free acid; it is electrolyzed until the copper content has been reduced to about 1% and the acid has increased to 7 or 8%, when it is withdrawn and used for leaching another batch of ore. Cell voltage is about 1.9 V, temperature 30 to 50°C, and current density 5 to 15 A/ft^2.

One of the principal keys to the economic success of copper electrowinning was the development of a satisfactory anode material. Lead containing 6 to 15% antimony is used in most plants; at Chuquicamata, Chile, the so-called Chilex anode is a copper-silicon alloy. In addition to this plant, copper electrowinning is carried out at Inspiration, Ariz., another plant in Chile, and several in Africa.

101. Zinc. Ores used for electrolytic reduction are complex zinc-lead-copper-iron sulfides with impurities of gold, silver, cadmium, and other metals. They are concentrated and roasted in preparation for electrolysis. The operation is cyclic, in that zinc is added to the circulating electrolyte in one department of the plant and deposited out in another.

Spent electrolyte from the tank house (high in acid and low in zinc) is used to leach calcined ore. The zinc oxides and sulfates dissolve in the solution, which becomes richer in zinc and lower in acid. The solution is purified and piped to the electrolytic tank house. Chemical lead or lead with 1% silver is used for the insoluble anodes, and aluminum for the cathode starting sheets. The electrodes are assembled in tanks as in the multiple system of copper refining.

The two processes in use in the United States differ principally in the acidities and current densities employed. The processes begin with solutions containing 8 to 22% zinc and no free acid. At the Anaconda Copper Mining Company plants, the acidity is allowed to increase only to 10 to 12% free sulfuric acid, whereas at the Sullivan Mining

Company this increases to 27%. Current densities at Anaconda are about 30 A/ft^2 of cathode, and at Sullivan 80 to 100 A.

102. Cadmium occurs as an impurity in many zinc, lead, and copper ores. It is recovered from dusts collected in the bag houses of lead and copper furnaces and from purification of electrolytes in zinc-reduction plants.

The raw materials are leached and chemically purified and the solution freed from heavy metals by precipitation with zinc. The resultant solution containing zinc, cadmium, and free sulfuric acid is electrolyzed in a similar manner to electrolytic zinc except on a much smaller scale. Insoluble lead anodes and aluminum cathodes are employed.

At a current density of 3.5 to 12.1 A/ft^2 of cathode the voltage per tank is 2.5 to 3.8 V, and 0.6 to 1 kWh is required to produce 1 lb of cadmium.

103. Manganese. The roasted ore is leached in a solution containing ammonium sulfate and sulfuric acid. Since manganese is less noble than any other plateable metal, almost all the impurities present will deposit more easily, so that careful purification of the solution before electrolysis is very important. Electrolysis takes place in a diaphragm cell, with stainless-steel or titanium cathodes and anodes of a lead–1% silver alloy; at about 5 V, current density 40 to 60 A/ft^2, and a current efficiency of 60 to 65%, power consumption is 3.5 to 4 kWh/lb of manganese.

104. Bibliography on Electrowinning (see Par. **98**).

ALKALI CHLORIDE ELECTROLYSIS

105. Products of Alkali Chloride Electrolysis. Electrolysis of an aqueous solution of alkali chloride, such as sodium chloride, produces caustic soda (NaOH), hydrogen, and chlorine, if the anodic and cathodic products are not allowed to interact to form other compounds—sodium hypochlorite, bleaching powder (NaOCl), and hydrogen when the solution is cold, or sodium chlorate ($NaClO_3$) and hydrogen when the solution is hot. If potassium chloride is used instead of sodium chloride, the products are caustic potash (KOH) and chlorine.

106. Chlorine and Caustic Soda. The chlor-alkali industry produces the two greatest tonnage products of the electrochemical industry (if electric furnace steel is neglected as not being strictly electrochemical). They are coproducts; i.e., they are produced by a single process in a fixed ratio. This poses a problem, since the demand for chlorine is rising much faster than that for caustic soda—although the situation may change in the future as it has in the past. The electrolysis of waste hydrochloric acid can produce chlorine without the accompanying caustic, and this process is being actively developed.

More than 2% of all the electrical energy consumed in the United States is used by the chlor-alkali industry. The electrolysis of sodium chloride brines accounts for more than 90% of the approximately 5.5 million tons of chlorine produced in the United States. Diaphragm cells account for 74%, mercury cells for 20%, sodium cells 5% (see Par. **131**), and miscellaneous nonelectrolytic processes less than 1% of the total.

Separation of the anodic product (chlorine) from the cathodic (caustic and hydrogen) is obtained by (*a*) preventing the formation of hydroxyl ions and hydrogen by use of a mercury cathode or (*b*) use of a porous diaphragm. Diaphragm cells are dominant in the United States, but new installations tend to use the mercury cell, which is accordingly gaining. In other parts of the world the mercury cell is favored.

107. Diaphragm Cells. Nearly saturated sodium (or potassium) chloride brine at a temperature of 60 to 70°C is fed into the anolyte compartment; it flows continuously through the diaphragm into the catholyte compartment, flow being maintained by a differential head. Only a portion of the alkali chloride entering the cell is electrolyzed, the unreacted portion leaving with the hydroxide solution from the cathode compartment.

Graphite anodes are used, and the principal anode reaction is the formation of chlorine from chloride ion; this takes place at an efficiency of about 97%; side reactions include the formation of hypochlorite and the oxidation of the anode itself to carbon dioxide. At the steel cathodes the reaction is the evolution of hydrogen at practically 100% efficiency, leaving an equivalent amount of hydroxyl ions behind as caustic alkali.

Cell designs differ in detail, but are all similar in principle; they all incorporate vertical graphite anodes, steel-screen cathodes, and deposited asbestos diaphragms. Principal types are the Hooker, Diamond, and Dow cells.

108. Mercury Cells. In the mercury cell, the anode process is the same as in the diaphragm cell, the evolution of chlorine; but the cathode reaction is the deposition of sodium into the mercury cathode to form sodium amalgam (or potassium amalgam in the electrolysis of potassium chloride). The amalgam is then transferred to another vessel, called a denuder or decomposer, where it reacts with water to form alkali hydroxide and regenerate the mercury. (In practice the amalgam is not completely denuded of its alkali metal.)

Advantages of the mercury cell include the greater purity of the caustic produced—or alternatively the greater ease and consequently cheaper plant needed to produce it—since it does not have to be separated from the original salt as in the diaphragm process, which produces a fairly dilute salt-caustic solution. On the other hand, energy requirements are somewhat higher, and the cells themselves more expensive because of the investment in expensive mercury. As stated, present preference in new installations seems to favor the mercury cell.

Mercury cells, like diaphragm cells, differ in detail but not in principle. Horizontal denuder types include the Krebs B.A.S.F., Solvay, and Uhde; vertical denuder types are known as the de Nora and Olin-Mathieson.

109. Hypochlorite (Bleaching Liquor). While for the production of caustic and chlorine by electrolysis of sodium chloride the anodic and cathodic products must be kept separate, the reverse requirement must be fulfilled for the electrolytic production of hypochlorites (bleaching liquors) by electrolysis of sodium chloride. Sodium hypochlorite is the result of the reaction of chlorine on caustic soda. To obtain the hypochlorite in the electrolytic cell itself, the electrodes are placed near together, and the electrolyte is maintained in steady motion in order to mix the anodic and cathodic products together.

110. Chlorates. The production of sodium and potassium chlorates by electrolysis of sodium or potassium chloride requires interaction between caustic soda or potash and chlorine under conditions of moderately high temperature (30 to 50°C) and absence of reducing conditions. For the latter purpose the addition of chromate is used. No diaphragms are used.

111. Fluorine. The need for fluorine for the production of uranium hexafluoride stimulated development of fluorine cells. The Union Carbide Nuclear Company's cell uses an electrolyte (anhydrous) of potassium bifluoride (KF 60%, HF 40%) at a temperature of about 90°C. Anodes are carbon and cathodes of steel; the cell itself is constructed of monel-lined steel. Current density (anode) is 90 to 140 A/ft^2, current efficiency about 96%, and cell voltage about 10 V.

112. Bibliography on Alkali Cells. In addition to the references cited in Par. **98**, see:

Sconce, J. S. (ed.) "Chlorine"; New York, Reinhold Publishing Corporation, 1962.

ELECTROLYSIS OF WATER

113. General Theory. In the electrolytic decomposition of water, as carried out on an industrial scale for the production of oxygen and hydrogen gases, instead of pure water, which has too low an electrical conductivity, a 25% solution of caustic soda or its equivalent, caustic potash, is used as electrolyte. Sulfuric acid is no longer used. With caustic soda, NaOH, as electrolyte, the cathode reaction is

$$2Na^+ + 2H_2O + 2\epsilon \rightarrow 2NaOH + H_2 \qquad (23\text{-}10)$$

and the anodic reaction is

$$2OH^- \rightarrow H_2O + \tfrac{1}{2}O_2 + 2\epsilon \qquad (23\text{-}11)$$

where $\epsilon = 1$ faraday; so that H_2 is set free at the cathode and O_2 at the anode, water disappearing, while the quantity of caustic soda remains constant, since as much is formed at the cathode as is decomposed by the current. It is, therefore, necessary to add new water to the solution from time to time.

114. Types of Cell. Hydrogen-oxygen cells are generally of four kinds: bell, unit, filter-press, and pressure. At 2 to 2.2 V/cell, each produces 6.5 to 7.5 ft^3 of hydrogen and 3.25 to 3.75 ft^3 of oxygen (atmospheric pressure and 32°F)/kWh. Electrodes are usually iron, anodes being nickel-plated to reduce oxygen overvoltage.

115. The Knowles cell is a typical example of the **bell type.** It consists of a steel tank containing several iron or steel gas-collecting bells, or hoods, of rectangular cross section. Inside each hood and insulated from it is an iron electrode. Alternate hoods carry asbestos extensions which act as porous diaphragms between an anode in one hood and a cathode in the adjacent hood. Several sets of anodes and cathodes are connected in parallel in a single tank. Hydrogen gas collects in cathode hoods, and oxygen gas in anode hoods, which are connected to gas-collecting pipes on top of the cells.

116. The Levin cell is typical of the **unit type.** It is a thin, vertical steel tank of rectangular cross section. Two asbestos diaphragms divide the tank vertically into three narrow compartments, the middle one containing a cathode, and the outer two containing anodes. The cathode is cobalt-plated to reduce hydrogen overvoltage. Generated gases pass upward in these narrow compartments through sight indicators to gas-collecting pipes or manifolds.

117. Filter-press cells are an expansion of the unit principle, with several units assembled in series as in a filter press. Each electrode, except the two end ones, is bipolar, acting as cathode on one side and anode on the other. Porous diaphragms of asbestos separate the electrodes and divide the space between them into anode and cathode compartments. Rubber gaskets along the edges of the diaphragms insulate the electrodes from each other and through bolts hold the entire assembly together. Holes in the top or the side near the top of the electrodes permit discharge of hydrogen and oxygen gases, and holes in the bottom permit introduction of water and any make-up alkali that is required.

118. Pressure Cells. For several years European manufacturers have been developing the use of filter-press cells at high pressures instead of pressures only slightly above atmospheric conditions. Such cells have been operated at 3,000 lb/in². Theoretically, reduced electrolyte resistance and reduced polarization voltage more than offset the small increase in energy necessary to produce gases at higher pressures.

ELECTROLYTIC CORROSION

119. Galvanic Corrosion. Chemical attack on metals usually forms reaction products which retard further attack unless the product is liquid, volatile, or porous or flakes off or cracks. Probably the most destructive corrosion results from indirect electrochemical attack, frequently thought of as chemical in nature.

The best-known galvanic or electrochemical corrosion occurs when two dissimilar metals are in contact in the presence of an electrolyte. In this case a galvanic couple is created and is short-circuited on itself through the electrolyte. The metal of higher potential will become the anode, tend to go into solution in the electrolyte, and, therefore, corrode. An almost identical condition is obtained in an alloy which is not perfectly homogeneous or in a metal of which different parts have been subjected to different heat treatments or mechanical stresses. Under these conditions, certain parts will have a higher potential than others, and in the presence of an electrolyte, a galvanic couple is formed. The part of higher potential will tend to go into solution and corrode. The electrolyte need only be rain water with impurities dissolved from the air or from the surface of the metal itself.

When different parts of a metal have unequal access to oxygen, the parts more greatly exposed acquire on their surfaces thin films of oxide or hydroxide, ennobling these portions. The part protected from oxygen then has a higher potential, tends to become an anode, and corrodes in the presence of an electrolyte. An example of this is the corrosion of a steel sheet or bar partially immersed, vertically, in brine. Corrosion occurs much more rapidly on the lowest portion than on the part immediately below the surface of the liquid where oxygen has greater access to the metal. Because of the lesser oxygen content at that level, the lowest part becomes anodic and corrodes and the portion just below the liquid level becomes cathodic and is protected. Another example is the corrosive behavior of a drop of brine on the surface of a piece of steel or iron. Corrosion is much more pronounced under the center of the drop, where oxygen is at a minimum, than around the edges.

120. Methods of Retarding Corrosion. The environment is of primary importance in any consideration of methods of retarding corrosion. Metals which perform well in one type of environment may be entirely unsuited to another. Those which perform well at the seacoast, for example, may be unsuited to an industrial environment, or vice versa, since the agents causing attack are quite different in the two cases. The principal method of minimizing corrosion is to isolate the metal from its environment by means of an envelope of more corrosion-resistant material: paint, organic coatings, or electrodeposits. Some metals, as aluminum and chromium, form corrosion-resistant oxide coatings which retard further attack, but in any medium which reacts with these oxide films the metals are not corrosion-resistant.

Paints and organic coatings must be free of pores and discontinuities in order to fulfill their function. This is also true of electrodeposits of metals more noble than the basis metal, such as copper-nickel-chromium on steel. If the deposited metal is considerably less noble than the basis—zinc on steel, for example—pores or discontinuities are of less moment, for the zinc protects the steel galvanically, by so-called sacrificial action. It becomes the anode and the steel the cathode of the galvanic cell, and the steel does not corrode until all the zinc in its immediate neighborhood is consumed.

Zinc plates have for many years been placed in marine boilers in order to make the iron cathodic and thus retard corrosion. Zinc corrodes and must be replaced. In some instances, corrosion in boilers and condensers has been retarded by application of an external voltage so as to make the iron a cathode.

121. Corrosion by Stray Currents. Underground pipes and cables are corroded by electrolytic action from unidirectional electric currents in the ground. These currents may result from galvanic couples in the ground, track returns in street-railway systems and electrified steam railroads, or a variety of other causes.

122. Methods of Retardation—Cathodic Protection. Stray current from electric traction equipment may be retarded by reducing the electric resistance of the track and increasing the resistance of the ground circuit. The former is accomplished by welding sections of track together and by welding flexible connections across track joints which cannot otherwise be made electrically continuous. The latter is helped by maintaining good, dry ballast below and between the ties.

The method of widest use is cathodic protection in which, by the application of an external source of d-c voltage, the underground material to be protected (pipeline, cable, etc.) becomes lower in potential than the surrounding soil. The buried metal is thus made a cathode instead of an anode. In practice it has been found that all usual forms of corrosion are prevented when the cathode protection makes the pipe or other metallic structure 0.25 to 0.30 V negative to the soil or liquid surrounding the pipe.

This voltage is obtained by using small motor generators, engine generators, batteries, or rectifiers. The negative lead is connected to the metal structure to be protected, and the positive lead to a ground some distance away. Or magnesium or zinc rods, externally connected to the pipe, are sunk into the ground and protect the pipe in a "sacrificial" manner, since they are anodic to it.

It is not possible to calculate the number of these units to use for a specific installation, because the length of pipeline which one unit can protect varies with the resistance of the soil, size of pipe, condition and extent of protective coating on the pipe, and many other factors. In pipeline practice, it has been found that one unit may protect anywhere from as little as ¼ mi of bare pipe to as much as 20 mi or more of well-coated line. It is necessary, therefore, to install temporary units at several places along the pipeline and, by checking the voltages between pipe and ground at different points for several values of generator output, determine the number and rating of d-c units to employ.

The positive lead of the cathodic unit must have a low-resistance connection to ground. This is accomplished by making use of a specially constructed ground bed 500 to 700 ft away from the buried metal. The most usual ground bed is made by burying several tons of junk cast iron 6 or 8 ft in the ground. Another form is obtained by sinking 6- or 8-in pipes vertically in the ground. Several of these pipes (six or more) are joined together at the top with a pipe header or suitably insulated cable. Very successful ground beds have been made by burying a continuous length of junk pipe parallel to the pipeline or cable being protected. All such beds are improved by the addition of large quantities of sodium or calcium chloride to obtain better conductivity between ground bed and soil.

123. References to Corrosion

UHLIG, H. H. (ed.) "Corrosion Handbook"; New York, John Wiley & Sons, Inc., 1948.

SPELLER, F. N. "Corrosion, Causes and Prevention," 3d ed.; New York, McGraw-Hill Book Company, 1951.

EVANS, U. R. "The Corrosion and Oxidation of Metals"; London, Edward Arnold & Co., 1960.

CAMPION, F. A. "Corrosion Testing Procedures"; New York, John Wiley & Sons, Inc., 1952.

LAQUE, F. L., and COPSON, H. R. "Corrosion Resistance of Metals and Alloys," 2d ed.; New York, Reinhold Publishing Corporation, 1963.

UHLIG, H. H. "Corrosion and Corrosion Control"; New York, John Wiley & Sons, Inc., 1963.

Also files of *Corrosion* (published by National Association of Corrosion Engineers); *Jour.* and *Trans. Electrochem. Soc.*

OTHER AQUEOUS ELECTROLYTIC PROCESSES

124. Inorganic Synthesis. A large variety of inorganic compounds may be synthesized by electrolytic action. In some cases, this method is merely of scientific interest, since the material may be produced more cheaply by other methods; in other cases, the electrolytic method has replaced other methods of production, while in still other cases, materials have been produced by electrolytic methods that could not be produced by other means. Among the compounds that may be produced electrolytically are hypochlorites and hyposulfites, perchlorates and persulfates, bromates, potassium ferricyanide, potassium permanganate, and hydrogen peroxide.

125. Organic Synthesis. For many years use of electrolytic methods in organic synthesis was confined almost entirely to the laboratory and had little commercial importance. The past few years, however, have seen a great change in this respect, and several new and important electroorganic processes have been commercialized: in fact one such process won the Kirkpatrick Chemical Engineering Achievement Award in 1965, and another won a merit award in the same year. The first was Monsanto's process for the "hydrodimerization" of acrylonitrile to adiponitrile by electrolysis; the product is a key intermediate in the manufacture of nylon. The second was Nalco's electrolytic route to tetraethyl lead (or tetramethyl lead) used in antiknock gasolines (see *Chem. Eng.*, June 21, 1965, p. 102; Nov. 8, 1965, p. 235). A large number of fluorocarbons are produced electrolytically. Electrolytic reduction of sugars to sugar-alcohols, formerly a commercial process, has been superseded by nonelectrolytic methods.

For references see entries on electroorganic chemistry and fluorocarbons in Hampel, Par. **98**.

MATERIALS FOR CELL CONSTRUCTION

126. Insoluble Anodes. Probably the most troublesome problem in the development of commercial electrolytic processes has been that of securing satisfactory insoluble anodes. In the early days platinum was comparatively cheap and was much used; where a very cheap material was required, coke or retort carbon was used, and soon artificial graphite was produced and largely supplanted them. As more processes were developed and requirements broadened, other materials were added to the list.

For hypochlorites, platinum, platinum-iridium alloys, and graphite are chiefly employed; for chlorates, platinum; and in alkali and chlorine cells, the platinum and amorphous carbon of early days have been replaced by artificial graphite.

In electrolysis of water for production of hydrogen and oxygen, nickel-plated iron is used.

In electrolysis of fused salts, graphite is employed almost exclusively.

In the direct electroreduction of copper and zinc, lead is mostly used, as also in most processes where sulfuric acid is the corrosive agent to be resisted. The oxygen liberated at the anode gradually covers it with a layer of lead peroxide, so that it is in effect an anode of lead peroxide. If operating conditions are such that the coating of peroxide is solid and adherent, the anode lasts almost indefinitely; but if the peroxide is loose and porous, the peroxidizing action soon progresses clear through the lead plate and it disintegrates. This action may be slowed up, and the anode's life extended, by alloying with the lead another metal to reduce the rate of attack; small amounts of silver, thallium, and tin have been found very effective for this purpose. The same result may also be obtained by the presence of small amounts of manganese in the solution; the MnO_2 that deposits on the anode with the formation of the PbO_2 materially reduces the disintegration.

At Chuquicamata, Chile, the presence of nitrates in the copper solution made use of lead impossible, and anodes cast from fused magnetite were first employed. When these could no longer be obtained, an iron-silicon alloy with about 13% of silicon was used. This was not satisfactory, since it slowly dissolved and contaminated the solution with iron, and a systematic research was inaugurated to develop a new anode. This resulted in a copper-silicon-iron alloy and a lead-antimony alloy for use as anode material.

127. Cathodes. Since the cathode is protected by action of the current and by the metal deposited on it, it is much less subject to corrosion difficulties than is the anode. In cells for hypochlorites, chlorates, and electrolysis of water, the cathode is usually the same as the anode. In alkali-chlorine cells it is usually of iron. In cells in which a metal is deposited, the cathode is immediately covered with this metal, so that in effect the cathode is the metal being deposited, even though sometimes something else is used as a starter. In the electroreduction of zinc, sheets of aluminum are used as cathode, from which the deposited metal may be easily stripped.

128. Diaphragms. Second only to the problem of the insoluble anode is the difficulty of securing a satisfactory diaphragm. This must not only resist the corrosive action of the solution but must also be sufficiently impervious to diffusion currents in the cell and sufficiently pervious to the electric current. These requirements are so contradictory that the result is usually a compromise, but the best results are obtained with a material having the maximum number of very fine pores. The greater the porosity, the less resistance will be interposed in the cell circuit; and the finer the pores, the less will be the diffusion through the pores. The materials used in various processes include paper, cloth, clay, fused alumina, bonded silica, asbestos cloth, asbestos sheet, asbestos sheet impregnated with sulfur, porous rubber, and synthetic plastics. The many new materials being made available by the plastics industry are being actively investigated both as diaphragms and for the cell itself.

129. Cell Construction. The most common materials used for cell construction are wood, steel, and concrete, with or without a covering of protective material, usually asphalt or sheet lead. For large cells, concrete protected by asphalt or mastic is much used. Small cells for chloride electrolysis are sometimes made of slate or soapstone, and for gold and silver refining and similar operations, chemical stoneware is frequently used.

COMMERCIAL PRODUCTS OF THE ELECTROLYTIC FURNACE—ELECTROLYSIS OF FUSED SALTS

130. Aluminum. The most important electrolytic-furnace operation is the production of aluminum by the electrolysis of a solution of pure alumina, Al_2O_3, in fused cryolite, $AlF_3 \cdot 3NaF$. This process was developed simultaneously and independ-

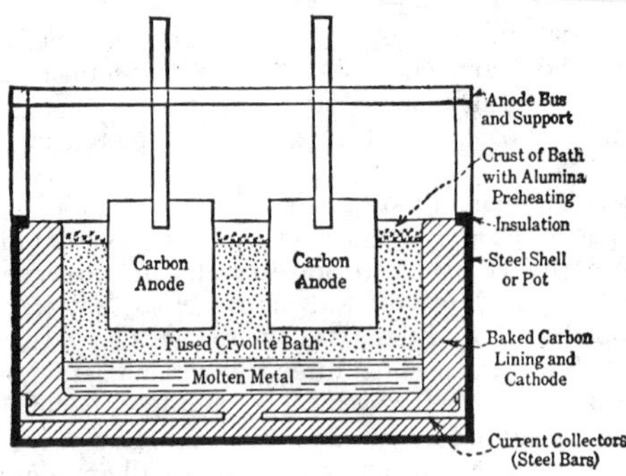

FIG. 23-5. Hall-Héroult aluminum furnace.

ently by Charles M. Hall in the United States and P. L. T. Héroult in France and is generally known as the **Hall process** in the United States and the **Héroult process** in Europe. The fused bath carries about 5% alumina and is electrolyzed at a temperature of about 950°C, with carbon blocks used as anodes and the carbon lining of the furnace and the accumulated metal as cathode (see Fig. 23-5). The liquid metal collects at the bottom of the bath and is periodically tapped off. As the alumina is used up, fresh material is added to the top of the bath. The heat required to maintain the bath in the liquid condition is furnished entirely by the current used in the electrolysis. About 7-8 kWh is required for the production of 1 lb of aluminum.

The electrolytic furnace has also been used for electrolytic refining of aluminum by the **Hoopes process.** This process uses an anode of copper-aluminum alloy and an electrolyte of cryolite and barium fluoride. The gravities of the anode and electrolyte are so adjusted by controlling their composition that the anode lies in the bottom of the cell, and the pure metal deposited at the cathode floats on top of the electrolyte. The cell operates at about 6 V, with 20,000 A; metal with a purity as high as 99.999 can be made by this process.

131. Sodium. The Castner process, using caustic soda as raw material, is now obsolete. Metallic sodium is now produced by the electrolysis of a fused sodium chloride–calcium chloride mixture, and an equivalent quantity of chlorine is recovered as byproduct; the principal reactions are

$$2NaCl \rightarrow 2Na + Cl_2 \tag{23-12}$$

$$2Na + CaCl_2 \rightleftharpoons Ca + 2NaCl \tag{23-13}$$

The resulting raw sodium metal thus contains about 5% calcium, but this is removed by filtration outside the cell leaving a final product containing less than 0.04% Ca.

The cell is composed of special firebrick with or without an iron shell, a graphite anode, and iron cathodes. Chlorine gas is withdrawn from the anode compartment and liquid metallic sodium from the cathode compartment. Great care must be exercised in handling the liquid sodium, because it is very fluid and finds its way through the smallest cracks and, if exposed to oxygen, will burn very briskly.

Energy consumption is about 5 kWh/lb of sodium and the 1.5 lb of chlorine liberated at the same time. Cell circuits are operated at about 300 V d-c and from 20,000 to 40,000 A. Operating temperature is about 600°C, and current efficiency varies between 40% (for cells of older design) up to 80% or more for modern cells.

See, in addition to general references:

SITTIG, M. "Sodium, Its Manufacture, Properties and Uses"; New York, Reinhold Publishing Corporation, 1956.

132. Magnesium. Magnesium is made by electrolysis of a melt containing anhydrous magnesium chloride. The only process of present importance in the United States utilizes sea water as raw material. Magnesium hydrate is precipitated from the water by means of lime; the hydrate is converted to chloride by hydrochloric acid. The chloride is then dried and finally partly dehydrated—the last is a critical step.

Electrolytic decomposition is accomplished in large, rectangular, cast-steel pots externally heated to maintain proper cell temperature and to reduce consumption of electric energy. The anode consists of large graphite bars, and the pot itself serves as cathode. The electrolyte is molten magnesium chloride, to which some sodium chloride is added to reduce the melting point and increase the conductivity. The cells or pots operate on direct current at about 70,000 A—8 to 13 kWh being required to produce 1 lb of magnesium. Chlorine gas is liberated at the anode. Metallic magnesium formed at the cathode is lighter than the bath and, therefore, floats to the surface, from which it is removed daily, and the process continued. The cell bath is maintained at a nearly constant level by either a continuous or an intermittent feed of dehydrated magnesium chloride.[1]

133. Calcium is made by the **contact-electrode process** by electrolysis of fused calcium chloride, the cathode being continually and slowly raised during the process of electrolysis so that its end always just touches the surface of the bath.[2]

134. Beryllium. Metallic beryllium is obtained in America by the **chloride process,** in which beryllium chloride, $BeCl_2$, is electrolyzed in a fused bath of sodium chloride. The pot or cell is made of nichrome or chromium-iron alloy and acts as the cathode. The carbon anode is inserted into the pot through a dome-shaped cover which also permits the introduction of inert gas and the controlled discharge of gases formed during electrolysis. At the conclusion of electrolysis the beryllium, in large, bright crystalline plates and spangles, is found adhering to the sides of the pot.

The **fluoride process** electrolyzes a fused mixture of sodium-beryllium fluoride and barium-beryllium fluoride in a graphite crucible which also serves as an anode. The cathode is a special iron alloy and must be water-cooled. Since the principal use for beryllium is in alloys with copper, it is considered preferable to electrolyze it directly into a copper cathode (see B. R. F. Kjellgren, *Trans. Electrochem. Soc.*, 1946, Vol. 89, p. 247; H. C. Kawecki, *ibid.*, p. 229).

135. Cerium. The first production of cerium was by electrolysis of fused cerium chloride and later from a mixture of cerium chloride with some sodium chloride. Most of the so-called "cerium" metal of commerce is a mixture of cerium with other metals of the same group, particularly lanthanum, neodymium, and praseodymium, and is usually known as "misch metal." It is prepared by the electrolysis of the fused mixed chlorides.

136. Other Metals. Electrolysis of fused salts is used or projected for the winning or refining—or both—of many metals which cannot, because of their electrochemical activity, be deposited from aqueous solution or reduced by carbon or hydrogen. These include titanium, zirconium, lithium, potassium, cesium, and rubidium. Tantalum, molybdenum, tungsten, and some other "refractory" metals have recently been both electroformed and electroplated from fused salt electrolytes.[3]

137. Phosphoric acid is made by smelting a mixture of phosphate rock, silica, and carbon. The reaction in the furnace is

$$(CaO)_3P_2O_5 + SiO_2 + 5C \rightarrow P_2 + 5CO + (CaO)_3(SiO_2)_2 \qquad (23\text{-}14)$$

The phosphorus vapor and the carbon monoxide products of the reaction are burned at the outlet of the furnace chamber to form phosphorus pentoxide, P_2O_5, and carbon dioxide. The pentoxide after cooling is hydrated with an atomizing spray to phosphoric acid, H_3PO_4. Some of the acid is collected in cooling towers; the larger part is recovered by Cottrell precipitators. The slag is tapped intermittently from the bottom of the furnace.

Electric furnaces for this service have closed chambers and are 3-phase with movable electrodes. The energy required is about 2.30 kWh/lb of phosphorus pentoxide.

138. Carbon bisulfide is produced in a shaft-type resistance furnace with fixed electrodes. Charcoal or coke is fed in at the top and sulfur through channels in the walls of the hearth and of the shaft.

The carbon reacts with the vaporized sulfur to form carbon bisulfide, CS_2, which passes up through the incoming charge of charcoal and is discharged in vapor form from an outlet at the top of the shaft. The sulfur vapor is condensed under water.

[1] For further details, see Hunter, R. M. *Trans. Electrochem. Soc.*, 1944, Vol. 86, p. 21; Schambra, W. P. *Trans. AIChE*, 1945, Vol. 41, p. 35.

[2] Mantell, C. L., and Hardy, C. "Calcium"; New York, Reinhold Publishing Corporation, 1945.

[3] Mellors, G. W., and Senderoff, S. *Jour. Electrochem. Soc.*, 1965, Vol. 112, p. 266; Senderoff, S., Mellors, G. W., and Reinhart, W. J. *Ibid.*, p. 840.

The furnace is 2-phase with four electrodes, 240 to 330 kW at 60 V. The energy consumption is about 1,000 kWh/ton of product.

139. References on the Production of Metals by the Electrolysis of Fused Electrolytes. Further details on the electrolysis of fused electrolytes will be found in the following:

MANTELL, C. L. "Industrial Electrochemistry," 3d ed.; McGraw-Hill Book Company, 1960; EDWARDS, J. D., FRARY, F. C., and JEFFRIES, Z. "The Aluminum Industry"; McGraw-Hill Book Company, 1930.

SECTION 24

BATTERIES

BY

GEORGE W. HEDDERSON Electrical Engineer (Retired), Exide Industrial Division, ESB Incorporated

SAMUEL RUBEN Consultant, P. R. Mallory & Company, Inc.

J. J. COLEMAN Vice President, Engineering & Research, Burgess Battery Company, Division of Servel, Inc.

PAUL L. HOWARD President, P. L. Howard Associates, Inc.

F. B. PIPAL Manager, Battery Engineering Department, Consumer Products Division, Union Carbide Corporation

J. J. LANDER Director of Electrochemical Research, Delco-Remy Division, General Motors Corporation

E. F. ERRICO P. R. Mallory & Company, Inc.

J. F. DONAHUE Manager, Product Development, Alkaline Battery Division, Gould-National Batteries, Inc.

CONTENTS

Numbers refer to paragraphs

SECTION 24

BATTERIES

LEAD-ACID AND NICKEL-IRON BATTERIES

By George W. Hedderson

1. Definitions. A storage battery is one in which the electrochemical action of the cells is reversible; i.e., following their output of electric current on discharge, they can be recharged to their original state by passing current through them in the opposite direction. They are also known as "secondary" cells and, in England, as "electric accumulators."

The basic unit of any battery is the **cell,** consisting essentially of positive plates, negative plates, and electrolyte. One or more cells connected together for a given purpose constitute a **battery.** The cells are usually connected in series, but parallel or series-parallel connections may also be used.

The open-circuit voltage, or emf, of a cell is a fundamental characteristic of the elements which constitute it. Almost any two dissimilar metals or elements in a conducting electrolyte will produce some voltage. Only a few such combinations, however, have any commercial value. The voltage of a battery is the additive of all cells connected in series.

The capacity, usually expressed as ampere-hours, is basically dependent on the size (volume and surface area) of the plates. At any particular time, however, it may vary with such circumstances as temperature and discharge rate, etc. The ampere-hour capacity is the additive of all cells connected in parallel.

2. Classification. The **lead–sulfuric acid** type (using lead-antimony grids) is the most common and versatile one in use. It is applicable to almost every battery requirement.

The **nickel–iron–potassium hydroxide** type, developed by Thomas Edison, is also in considerable industrial use. Its primary applications are in cycled or semicycled service such as electric industrial trucks and railway cars.

The **nickel–cadmium–potassium hydroxide** type, developed in Europe, is receiving increased use mostly in standby or other light service.

In addition to these there are various "special" types developed for scientific and military purposes, but which are not satisfactory or economical for common uses.

3. Application. A storage battery's outstanding attributes are these:

1. It provides a portable source of electric power. This power is available in considerable quantity for use on moving equipment or where no power lines are accessible. It is unfettered by cords or cables.

2. It is capable of delivering large quantities of power for short periods and being recharged at low rates over an extended time. Thus heavy surges of power are available when required, without heavy demands on a power system or equipment.

3. It provides the most reliable source known of emergency power, instantaneously, when normal power fails.

4. It provides a source of "pure" direct current for laboratory and other specific purposes, either as a separate and independent supply or by acting as a "filter" in a normal supply system.

The applications of storage batteries include the following:

Automobiles.
Aircraft.
Electric power and substations.
Emergency light and power.
Electric industrial trucks.
Various portable uses.
Telephone service.
Railway diesel engines.
Railway-car lighting and air conditioning.
Marine applications.
Submarines.

LEAD-ACID BATTERIES

4. Theory. In a lead-acid type of cell, lead and lead oxides are acted upon electrochemically by a solution of dilute sulfuric acid (H_2SO_4). When the battery is fully charged, the active material of the positive plate is lead peroxide (dioxide) (PbO_2); the negative plate is sponge lead (Pb). As the cell is discharged, the electrolyte (H_2SO_4) divides into H_2 and SO_4. The H_2 combines with some of the oxygen formed at the positive plate to produce water (H_2O). The SO_4 combines with the lead (Pb) of both plates, forming lead sulfate ($PbSO_4$).

When the cell is charged, this action is reversed and the lead sulfate ($PbSO_4$) on the positive and negative plates is reconverted to lead peroxide (PbO_2) and sponge lead (Pb), respectively. The strength of the electrolyte increases as the SO_4 from the plates combines with hydrogen from the water to form H_2SO_4.

The chemical formula is therefore

$$PbO_2 + Pb + 2H_2SO_4 \underset{\text{charge}}{\overset{\text{discharge}}{\rightleftarrows}} 2PbSO_4 + 2H_2O$$

In a fully charged battery all the active material of the positive plates is lead peroxide, and that of the negative plates is pure sponge lead. All the acid is in the electrolyte, and the specific gravity is at maximum. As the battery discharges, the specific gravity of the electrolyte gradually decreases, because the proportion of acid is decreasing and the water is increasing.

When a battery is placed on charge, the reverse action takes place. The acid in the sulfated active material of the plates is driven out and back into the electrolyte. The specific gravity continues to rise until all the acid is driven out of the plates. Additional charging will not raise the gravity any higher. All the acid in the cells is in the electrolyte, and the battery is said to be fully charged. On discharge the plates absorb acid, and on charge they return the absorbed acid back to the electrolyte.

In a charged cell the positive plates are of a dark brown color and the negative plates a medium gray with a metallic sheen. When normally discharged, both groups of plates are somewhat lighter in color and the negative plates lose their sheen.

5. Physical Construction and Assembly (see Fig. 24-1). The active materials alone have no rigid mechanical form or strength and, particularly the positive, are poor conductors of electricity. It is necessary, therefore, to mount them in some sort of frame, or "grid," to maintain physical shape and to conduct the current to all parts of the material. This grid usually takes one of these forms:

a. A flat lead-antimony latticework into which is pasted finely powdered lead oxides which are later electrochemically "formed" into the desired active material. These are known as "flat-plate" or "pasted-plate" types. This is the most common plate construction and is adaptable to both positive and negative plates.

b. A series of vertical lead-alloy spines, or "core rods," cast integral with a "top bar." Each core rod is surrounded by a "tube," which may be of slotted or otherwise perforated rubber or plastic, or braided from plastic or glass thread, or combinations of these. The active material, again in the form of finely powdered lead oxides, is packed firmly between the core rods and the retaining tubes. The bottoms of the tubes are sealed and joined together to form a semirigid plate. This is known as a "tubular" plate (or Ironclad) and is used only in positive plates. It is used largely in heavily worked or cycled batteries.

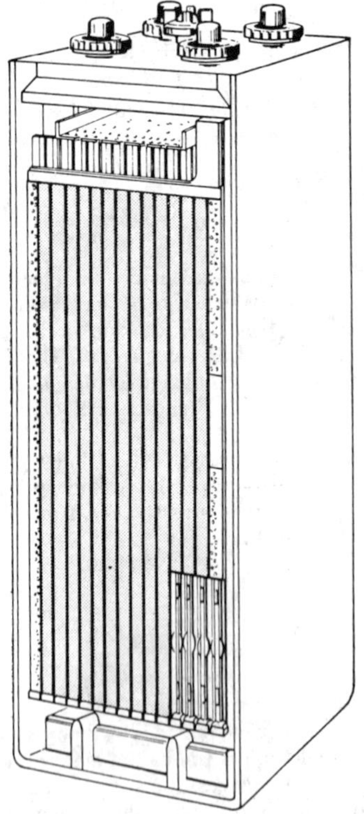

FIG. 24-1. Cutaway view of lead-acid tubular-type cell.

c. A pure lead plate cast with complex ridges or grooves, or mechanically furrowed (Tudor types), or a lead-antimony grid into which corrugated lead ribbon, rolled into spiral "buttons," has been inserted (Manchester type). Both these are known as Planté plates as the active material is formed electrochemically from the lead of the plate itself. They are used only as positive plates and have long life in standby service.

The addition of antimony to the lead grid gives it greater physical strength and rigidity and offers greater resistance to "formation" (corrosion) by the electrolytic action with the acid.

The **plates** are arranged parallel to each other, alternately positive and negative. All the positive plates are joined and thus connected by an alloy **strap**, and likewise the negative. This strap, through its **post**, leads to the external circuit.

The length, width, thickness, and number of plates in a cell are determined by the capacity required for the desired application. It is common practice to have a negative plate at each end of the element, thus making one more negative than positive plate in the cell. Thus a 15-plate cell has 7 positive and 8 negative plates. The 2 outside negative plates are frequently thinner, as the outer surface gets little use.

The positive and negative plates must not come in contact with each other and are prevented from doing so by a **separator.** Separators are usually in sheet form and are commonly made of wood, rubber, glass, or plastic. They must be microporous in structure to permit the electrolyte to permeate them.

The **element,** consisting of the positive and negative **groups** and separators, is placed in a **jar** or multicell **container** which holds the dilute sulfuric acid electrolyte. A **cover** is placed over the element and sealed to the top of the jar to exclude dirt or foreign material and reduce the evaporation of water from the electrolyte. The cover has a **vent plug,** which has small holes for the escape of gas and which can be removed for adding water and taking hydrometer readings. The materials used for the jar or container and cover vary considerably with the economics involved and with the type of application. For example, any "portable" service requires sturdy materials not easily broken, whereas stationary battery installations are not so critical.

Some typical assemblies are:

1. Automotive batteries—usually in 3- or 6-cell integral containers of asphalt "composition."

2. Aircraft—6- or 12-cell containers of hard rubber or plastic, sometimes with an external aluminum case.

3. Electric industrial trucks—hard rubber jars and covers assembled in steel trays.

4. Railway diesel-engine and railway car batteries—hard-rubber multicell containers.

5. Stationary services—most of these are in transparent plastic multicell containers or individual jars. Those of any appreciable size are mounted on steel "racks" in one or more "steps," or tiers, for convenience in servicing. (Glass jars were used for many years until suitable plastics became available.)

All the types are of nominally "sealed" construction, in that each cell is completely sealed against dirt, moisture, etc., except for the small vent in the plug which allows the escape of gas during charge. For certain cells which are *completely* sealed, airtight, and spill-proof, see Pars. **25** and **26.**

6. Voltage. The lead-acid cell has the highest voltage (per cell) of any commercial type. It is generally referred to as having a nominal voltage of 2 V, although this varies with the specific gravity and whether or not it may be discharging or on charge at the time. Thus a 3-cell battery is usually referred to as a 6-V battery, a 60-cell battery as a 120-V battery, etc.

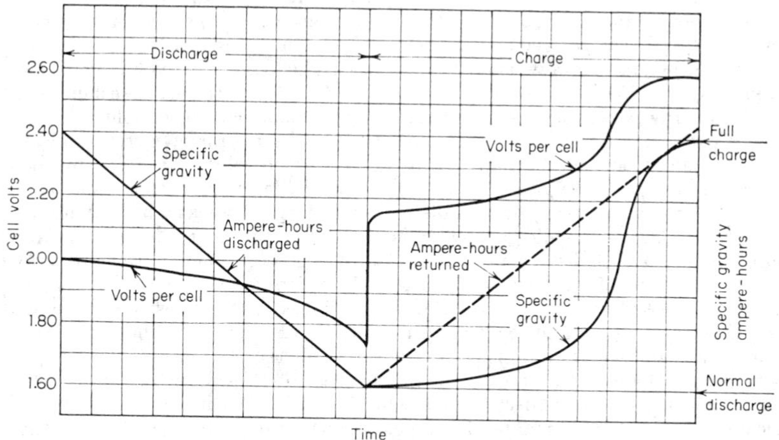

FIG. 24-2. Typical voltage and gravity characteristics during a constant-rate discharge and recharge.

The voltage on open circuit (with no current flowing in either direction, and after sufficient time for the voltage to stabilize) is a direct function of the specific gravity and is represented closely by the formula V = specific gravity + 0.84. Thus the open-circuit voltage of a cell with a specific gravity of 1.210 will be 2.05 V; one with a gravity of 1.280 will be 2.12 V.

As soon as a cell starts to discharge, there is a decrease in voltage due to the effective internal resistance of the cell. Also, at a continuous given rate of discharge, the voltage gradually becomes lower as the discharge progresses until, as the cell nears exhaustion, the voltage drops rapidly to and below a value where it is no longer effective for the application to which it is put. The value at this point is the "final" voltage. It varies with the rate of discharge, being lower with higher ampere rates. It may be as high as 1.85 V for low rates of discharge or as low as 1.0 V at extremely high rates. A representative value of 1.75 V is, however, commonly used for a large proportion of typical battery applications (see Figs. 24-2 and 24-3 for representative discharge curves).

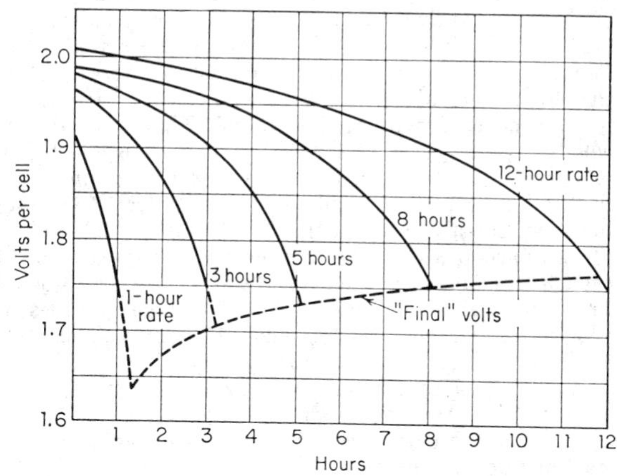

FIG. 24-3. Time-voltage curves to "final" volts.

When a discharged battery is placed on charge, its voltage immediately rises. With commonly used rates the voltage will rise within minutes to 2.10 or 2.15 V and then increase gradually until the charge is three-quarters complete. Near that point the voltage rises more sharply and then levels off at a maximum when the battery is fully charged. The voltage at this point is about 2.6 V/cell at the normally used "finish rate" of charge (see Fig. 24-2).

7. Electrolyte and Specific Gravity. The electrolyte is a dilute solution of sulfuric acid and water, being known by the trade as "battery grade." On a volume basis it can be supplied in any desired specific gravity, but in the comparatively small quantities required by industrial users for the replacement of spillage or leakage it is usually supplied at a gravity of 1.400 and diluted further as necessary.

The specific-gravity value is its relation in weight to an equal volume of water. In speaking of it, 1.250 is referred to as "twelve fifty," 1.400 as "fourteen hundred," etc.

Specific gravity is measured by a **hydrometer,** which floats in the solution at a depth varying with the gravity. A calibrated scale on the hydrometer stem indicates the gravity value at the point where it emerges from the solution. For convenience in reading the gravity of typical "closed" cells, the hydrometer itself is mounted inside the glass barrel of a syringe, the complete device being known as a hydrometer syringe.

Another form of specific-gravity indicator is known as "pilot balls." These are small wax or plastic balls, usually ¼ to ⅜ in in diameter, made in several "gravities" by using materials of appropriate densities. They are installed in a vertical raceway, with the highest-gravity ball at the lowest position and the lowest-gravity one at the top. Their gravities are so selected with respect to the gravity range of the cell that all are floating when the cell is fully charged. As the cell gravity decreases during discharge, the balls sink successively as the electrolyte gravity becomes lower than their own. Thus the number of balls which are still floating gives an approximate indication of the battery's state of charge. Either two or three balls are customarily used, made in different colors for easy identification. They can be used only in cells with transparent containers. While by no means so accurate as a hydrometer, they do indicate at a glance the approximate state of charge.

It is never necessary to add electrolyte to a battery except to replace loss due to spillage or leakage in some form.

Additives. Various special compounds, both powders and liquids, are frequently offered as "additives" to the battery electrolyte, with claims of larger capacity, longer life, etc. While some of these may not be actually harmful, their use has not been found beneficial by battery manufacturers or other informed sources. Their use, therefore, is not recommended.

8. Specific Gravity in Operation. The value of the specific gravity of a battery when fully charged is a matter of design and is affected by many factors. The gravity must be high enough for the electrolyte to contain a sufficient amount of sulfuric acid to fulfill the chemical requirements of the cell. On the other hand, if the gravity is too high, the acid content may be strong enough to have a direct chemical effect on certain parts of the cell. Between these two extremes there are other factors, such as capacity, temperautre, and battery life, etc., which dictate the particular gravity best suited to a given purpose. The full-charge gravities most commonly used, usually expressed as a range of ±10 points (1 "point" equals 0.001), and certain representative applications are:

1. 1.275. Heavily worked, or "cycled," batteries such as electric industrial trucks.
2. 1.260. Automotive and aircraft service.
3. 1.245. Partly cycled batteries such as railway-car lighting and large engine-starting batteries, etc.
4. 1.210. Batteries in stationary standby or emergency service.

In tropical climates or where the temperature averages 90°F or higher, it is customary to reduce these values by 30 or 40 points.

The decrease in specific gravity on discharge is proportional to the amount in ampere-hours discharged. This is shown by the straight line in Fig. 24-2, which indicates the approximate gravity readings taken during a discharge. Thus at any time the gravity is an approximate indication of the state of charge of the battery. This is determined by comparing the gravity as read with the full-charge value and the published "specific-gravity drop," which is the decrease from full charge to nominal discharge.

Example. A certain type of cell reads 1.195, corrected for temperature and level. Assume this type has an average full-charge gravity of 1.245 and a gravity drop of 125 points. It is then 50 points below full charge and is therefore about $^{50}\!/_{125}$, or 40%, discharged.

On recharge, however, as shown by the curved line, the rise in specific gravity as indicated by a hydrometer is not uniform or proportional to the amount of charge (in ampere-hours). During the early part of the charge there is nothing to mix or stir the electrolyte, and some of the heavier acid coming from the plates does not rise to the top of the cell and cannot be read by the hydrometer. During this part of the charge, therefore, the hydrometer reading does not represent the true state of charge of the battery. Later in the charge, when gassing begins, all the electrolyte becomes more completely mixed and the gravity, as read at the top of the cell, rises rapidly to its full-charge value. The broken line in Fig. 24-2 indicates ampere-hours returned and (up to where it nears the gravity line) what the gravity *would read* if it were possible to keep the electrolyte completely mixed throughout the charge. This "lag" in gravity rise at the top of the cell does not mean that the battery is not taking the charge and does not reduce the discharge capacity available.

Specific-gravity Corrections. The specific gravity varies, however, with changes in temperature. This is due, not to any characteristic of the battery, but merely to the fact that the electrolyte expands as the temperature rises, resulting in a lower gravity reading. Conversely, as the temperature is lowered the gravity rises. This change is equal to one point (0.001) in gravity for every 3°F change in temperature. Likewise the gravity will vary as the electrolyte level falls and rises with the use and addition of water. As water is "consumed" by gassing and evaporation, the level falls and the remaining electrolyte contains a greater proportion of acid; thus the specific gravity is higher. After water is added and becomes mixed, it will return to its previous value. (In a certain type of cell, for example, the gravity may rise 15 points with each ½-in drop in level.)

To compare accurately specific-gravity readings taken at different times and at different temperatures and electrolyte levels, such readings are "corrected" to the normal reference temperature of 77°F and the normal level. Such "corrected specific-gravity" readings indicate what the gravity *would be* if the temperature and level were at the above-normal values. To make this correction on the above type of cell: (1) Add one point of gravity for each 3°F above 77°F, or subtract one point of gravity for each 3°F below 77°F. (2) Subtract 15 points of gravity for each ½ in below normal level, or add 15 points for each ½ in above normal level.

Example. Specific gravity of cell reads 1.235 at 89°F and ½ in low level. $89 - 77 = 12°F \div 3 = 4$ points to be added. ½ in low level = 15 points to be subtracted. Net result, subtract 11 points; corrected gravity is 1.224.

The full-charge specific gravity will decrease slightly in value as the battery ages owing to minute losses of acid from spray, etc. No definite value can be given, but this decrease is small, usually not over a few points per year.

9. Capacity. The capacity of a storage battery—its ability to deliver energy—is usually expressed in ampere-hours, which is simply the product of the discharge in amperes over a number of hours. However, a capacity of, say, 200 Ah has little significance unless it is qualified by the many factors which influence a battery's capacity, the principal ones being discharge rate, temperature, specific gravity, and final voltage.

Discharge Rate. The higher the discharge rate in amperes, the fewer total ampere-hours a battery will deliver under otherwise similar conditions. This relationship varies somewhat with different types of plate and cell construction; Fig. 24-4 shows a nominal relationship of typical commercial cells.

This effect is due to two causes. The primary one is the lack of immediate diffusion of the electrolyte. During discharge, the only portion of the electrolyte which is "useful" is that in the pores of the plates in actual contact with the active material. As the acid in this portion becomes depleted, the electrolyte must diffuse or circulate to bring more acid to the active material, where it is needed. As the rate increases, however, this circulation does not increase proportionately, with the result that the cell voltage decreases more rapidly, thus limiting the total capacity.

Another result of higher current rates is to increase the voltage drop *within the cell.*

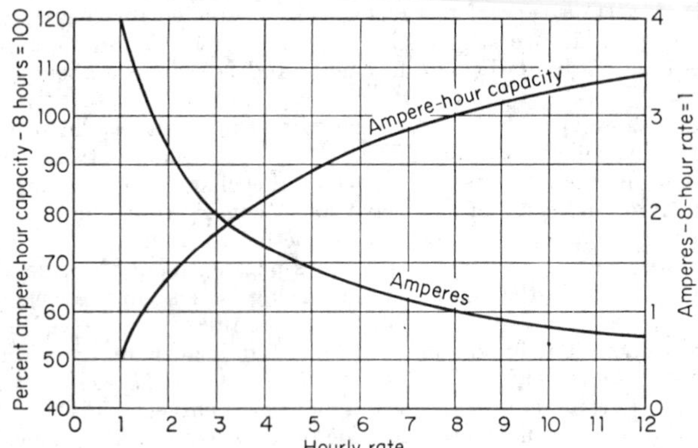

FIG. 24-4. Capacity-rate curve based on 8-h rate.

All cells have a certain internal ohmic resistance (see Par. **19**). The higher the current, the greater the voltage drop or loss in this resistance within the cell, thus reducing its external or useful voltage which supplies the load.

The rate most commonly used as a standard is the 8-h rate. There are exceptions, however, the principal ones being automotive batteries, which are customarily rated at the 20-h rate and motive power (industrial truck) types rated on a 6-h basis. Any correct rating is proper, so long as it is correctly specified. Manufacturers usually list several hourly ratings, nearly always including the 8-h, for the convenience of users in making comparisons.

Temperature. Many chemical reactions are accelerated at higher temperatures, and this is true of those in a storage battery. Also, the resistance and viscosity of the electrolyte are reduced at higher temperatures, thus reducing the voltage drop and energy loss within the cell. These combine to increase the battery's capacity at higher temperatures and reduce it at lower temperatures. Because of the effect of temperature on the resistance of the electrolyte, this difference in capacity varies with the rate of discharge as well. Figure 24-5 shows an approximate relationship for typical cell types. The "standard" reference temperature almost universally used is 77°F (25°C).

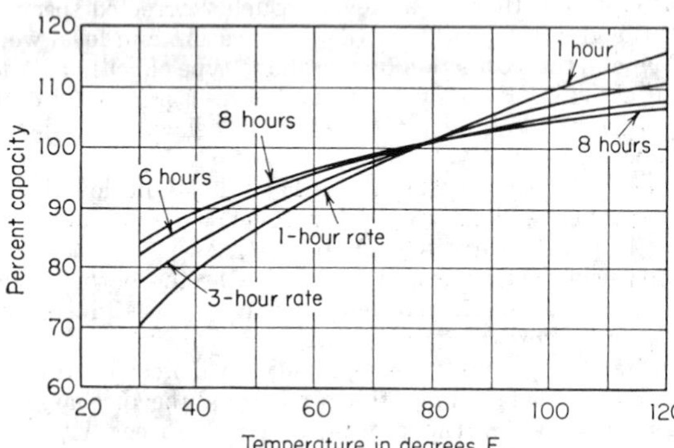

FIG. 24-5. Effect of temperature on capacity (77°F = 100%).

Specific Gravity. This affects cell capacity, as electrolytes of higher gravity have more acid in contact with the active material and available for the chemical reactions than do electrolytes of lower gravity. Also, the higher-gravity electrolyte has a lower electrical resistance, which better maintains the terminal voltage of the cell.

The degree to which full-charge specific gravity affects cell capacity varies considerably with different designs, but a rule of thumb frequently applied is that a difference of 25 points in gravity will change the capacity 8 to 10%. For example, a cell having a capacity of 100 Ah with a full-charge gravity of 1.275 will have 90 to 92 Ah capacity if the full-charge gravity is reduced to 1.250.

The choice of full-charge gravity is a compromise of many factors. Some effects of difference in gravity (in varying degree), other features being the same, are:

Higher Gravity	Lower Gravity
More capacity	Less capacity
Shorter life	Longer life
Less space required	More space required
Higher momentary discharge rates	Lower momentary discharge rates
Less adaptable to "floating" operation	More adaptable to "floating" operation
More standing loss	Less standing loss

Final Voltage. This term is used to designate the minimum useful and accepted voltage at various rates of discharge (see Par. **6**). The lower the permissible voltage, the longer the discharge and the more ampere-hours obtained.

10. Discharge Characteristics. In general, a battery may be discharged without harm at any rate of current it will deliver, but the discharge should not be continued beyond the point where the cells approach exhaustion or where the voltage falls below a useful value.

Discharging at a constant current value, the initial voltage depends on the rate of discharge and the normal characteristics of the cell. As the discharge continues, the cell voltage will slowly decrease during perhaps the first 70 to 80% of the total time period. It will then fall more rapidly, passing over the "knee" of the curve to the "final" voltage as full time and capacity are reached.

The total ampere-hours available varies with the rate of discharge, being lower at higher rates. This lower ampere-hour capacity does not, however, represent any specific loss of energy—it simply means that the cell voltage falls to its minimum useful value in a shorter period of time at that rate. If the rate is then decreased, the voltage will "recover," or rise, and further capacity (ampere-hours) can be obtained before the voltage again falls to the same minimum value. It is generally true, therefore, that, where *decreasing current rates* are used, approximately the total ampere-hours for a given hourly rate can be obtained over that period of time, regardless of higher rates during the early portion of the discharge. Figure 24-6 shows the approximate effect of discharging a cell at successively lower rates, carrying each one to the same final voltage. This result is not obtained when the higher rates are at the latter part of the discharge period, as there is then no opportunity for sufficient diffusion of the electrolyte to maintain the cell voltage.

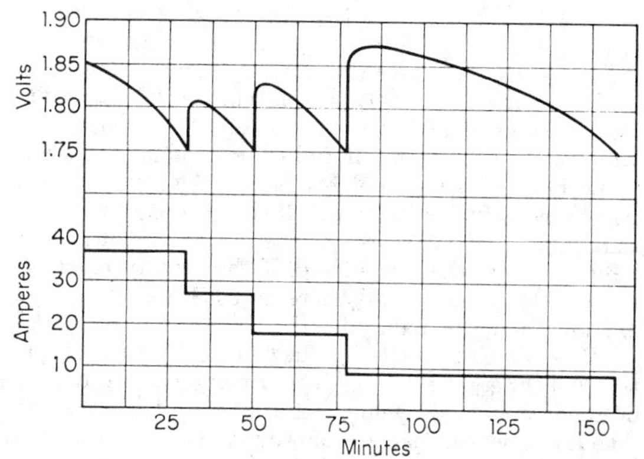

FIG. 24-6. Effect of decreasing discharge rates on cell voltage.

During discharge there is a small rise in battery temperature, perhaps 5 to 10°F, depending on the type of battery assembly from the standpoint of heat dissipation. The actual chemical reactions on discharge absorb a small amount of heat, which would tend to cool the battery slightly, but the heat due to the internal resistance (I^2R) of the cell is greater so that the net result is an increase in temperature.

A battery should not be discharged beyond the point where the cells approach exhaustion. This is referred to as "overdischarging" and can have harmful results, particularly if the battery is not promptly recharged.

11. Charging Characteristics. In general, a battery may be charged at any rate in amperes that will not produce excessive gassing or temperatures above 110°F (125°F for short periods). Another index is that any rate is safe which does not result in a cell voltage of more than 2.4 V while the current is above the "normal" or "finish rate" of charge. The manufacturer usually determines and publishes such a "normal" or "finish rate" in amperes for each type and size of cell made. This rate is the current value which can safely be used at any time when charging is required and which can be continued to the completion of the charge without causing excessive gassing or high temperature. This finish rate is usually between 4 and 10 A/100 Ah of the battery's capacity (8 h) depending on the type of cell assembly. When a number of high-capacity cells are assembled as a compact mass, the available surface for heat dissipation is much less than for separate individual cells and comparatively lower finish rates must be used to avoid high temperatures.

A battery which is partly or completely discharged can safely absorb much higher currents than the finish rate, up to possibly 10 times that value, but as it approaches full charge, the current must be reduced, either gradually or in one or more steps, to the finish rate or less. In practical applications, it is seldom necessary to use currents of more than 4 or 5 times the finish rate to charge in the time available. When the charge is complete, it should be stopped or reduced to a low value.

In any type of service, a battery should receive the "correct" amount of charge, sufficient to charge it fully or maintain it in that condition, but no more. Undercharge or overcharge should be avoided to whatever extent is practical under the conditions of use.

An insufficient amount of charge (even to a small degree, if continued) will cause gradual sulfation of the negative plates, with eventual loss of capacity and reduction of battery life.

An excessive amount of charge will tend to "form up" (corrode) the grids of the positive plates into lead peroxide, thus weakening them physically and increasing their electrical resistance. If the overcharging is at comparatively high rates, the gassing will be excessive and this tends to wash the positive active material from the plates. All these results reduce the capacity and shorten the life of the battery.

There are reasonably simple checks to determine whether or not the amount of charge is correct. If the proper amount of charge is being given, the specific gravity will reach its approximate full-charge value at the end of a recharge or remain at that value in floating or similar service. Also, the amount of water required by the cells will be a normal minimum. If the gravity does not reach full-charge value or tends to show a continual decrease, the battery is not getting enough charge and more should be given. On the other hand, if the gravity reaches or remains at full charge and an excessive amount of water is required, the battery is receiving too much charge and the amount should be reduced.

It is difficult to specify the "normal" water requirements, as they vary with batteries of different full-charge gravities and with the type of service from the standpoint of the amount of cycling (charge and discharge) which the battery receives. The manufacturer can usually, however, specify an approximate value which will serve as a guide for a given application.

12. Charging Methods. For charging equipment see Pars. **43** to **54**.

Cycle Charging. This refers to the complete recharge of a battery after it has been partly or normally discharged. For this charging, a time of 8 h is usually assumed, as it cannot be accomplished safely in a much shorter period under practical conditions, although it may, of course, be extended over a longer time if available (see Fig. 24-7). A nominal assumption is that about 110% of the ampere-hours discharged must be returned to charge the battery fully. This requires a starting rate of 20 to 25 A/100 Ah of the battery's rated capacity. This current is either "tapered" (reduced gradually) to the finish rate at the end of charge or reduced in one "step" (referred to as "two-rate" charging) when the charge is 80 to 85% complete. When full charge is reached, the charge should be stopped; this is normally done automatically.

The most typical application of daily cycle charging is that of the batteries in electric-industrial-truck service. There are usually a number of these at each location, and the charging is frequently done from a single piece of equipment by what is known as "modified constant-potential" charging (Par. **48**).

"Boost" and "Quick" Charging. As applied to industrial batteries, particularly those in electric industrial trucks, "boost" charging is additional charging during short periods, such as lunchtime, etc., to extend the battery's capacity to complete the day's work. The current (in amperes) used should not exceed the value of the ampere-hours

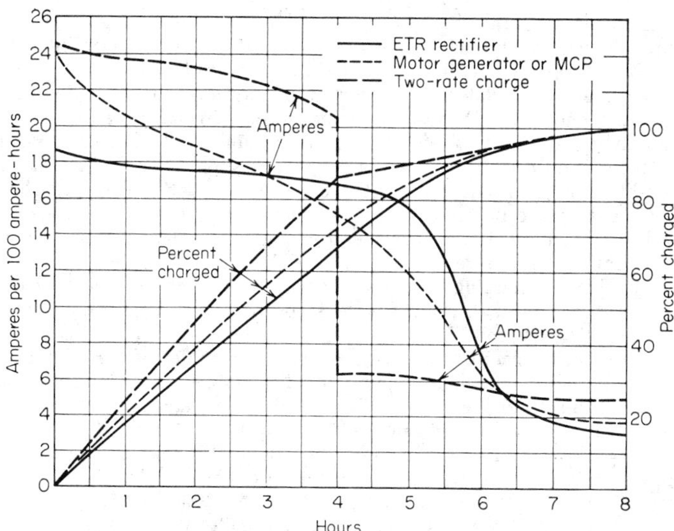

FIG. 24-7. Typical "tapered" and "two-rate" recharging of cycled batteries in 8 h.

still remaining "out" of the battery at the end of the charging period. In practice, however, with modern charging equipment, the battery may simply be placed on charge in the usual way to obtain satisfactory results.

"Quick" charging as applied to automobile batteries should preferably be continued only long enough to make the battery serviceable, the completion of the charge being accomplished by the car generator in the usual way. To avoid prolonged charging at the high current rates commonly used, most of these chargers have an automatic cutoff which stops the charge when the battery reaches a predetermined temperature.

Floating Charge. This type of charging is one where the battery is continuously connected to an electrical system, including a charger and load. The charger is designed to maintain a constant voltage throughout its load range. The value of this voltage is such that it supplies to the battery sufficient current to overcome its internal losses and keep it fully charged, but without appreciable overcharge. Following any discharge, the battery will automatically draw a higher current, which will decrease, as full charge approaches, until it is again reduced to the low maintenance value. It is merely necessary to keep the voltage of the charger at the correct value to operate the battery properly under all normal conditions of system load and temperature. The battery may be said to be "balanced" across the system.

A typical example of floating operation is the switchgear control battery used in utility and industrial electrical generating plants and substations. The battery, charger, and load are connected in parallel to a common bus. There is usually a small continuous load consisting of pilot lamps, relays, etc., and momentary or short-time loads of comparatively high values, such as those from circuit-breaker operation and the emergency operation of lights and other vital equipment. The continuous load is carried by the charger at the normal floating voltage, and the battery draws its own maintenance current at the same time. Any load which exceeds the charger capacity will lower its voltage slightly, to the point where the battery discharges to supply the remainder. If there should be a complete failure of normal power, the battery supplies the entire load until power is restored. It then is automatically recharged.

In the majority of floating-battery installations the plastic-jar type of assembly is used with "low" full-charge gravity of 1.210. Under these conditions the charger and battery are usually operated at a voltage of 2.15 V/cell—129 V for a typical 60-cell battery. The current drawn by the battery itself, when fully charged, will vary somewhat with its age and condition but will be in the general range of 50 to 100 mA/100 Ah of its rated capacity (8-h rate).

Where its use is practical, this type of operation is considered the most nearly ideal, as it requires the least maintenance and results in the longest battery life.

Partial Float (partial cycle). This is similar to "full float" but applies where the charging current is available only a portion of the time. For example, the battery of a railway passenger car is charged from a generator driven by the car axle and therefore can supply current only when the car is in motion. When the car is standing at stations or before and after a run, the battery must supply the entire load of lighting and air conditioning, etc. In motion, the car generator must carry the load on the system and recharge the battery so that it will be ready for the next demand. The floating voltage must be carried considerably higher so that the battery will draw sufficient current to become recharged in the time available. On a "local run" with frequent stops, higher charging currents and therefore higher voltage will be required than on a long cross-country run, where the stops are few. A generally representative floating voltage is 2.37 V/cell.

Automotive service, small boats, etc., are other examples of this partial-float type of operation. While the engine is running at sufficient speed, the battery must receive sufficient charge to supply the loads of lights and radios, etc., during traffic stops and when parked. The usual voltage value is 2.40 V/cell.

The battery of a railway diesel locomotive falls in this category, although it sometimes approaches full float in the case of switching engines, which are in virtually continuous use 24 h a day. The commonly used voltage is 2.30 to 2.32 V/cell.

Two-rate Charging. This has been used considerably in the stationary-battery field. It is usually a second choice after floating operation, as the equipment is less expensive. While less desirable than floating, it frequently is satisfactory. Its normal application is on systems where there are a small continuous load and heavier short-time or momentary loads. A continuous "low" charging rate is adjusted to the value of the steady load. A much higher rate is initiated either by a periodic device such as a time switch or by some function of the related equipment. The high rate is stopped by a voltage relay, which responds to the rising voltage as the battery approaches full charge. Thus the length of the high rate is adapted to the total charge requirements.

One disadvantage of this method is that the equipment requires readjustment whenever appreciable changes occur in the steady load and certain definite relationships must exist between the current output of the charger and the battery size.

Continuous Average Current. In some applications, while the load may vary widely from time to time, the total over a given period of, say, 24 h can be reasonably well estimated. For example, on a railway signal circuit, the load may vary from 0.3 to 6 A at different times, but the daily total may be reasonably close to 18 Ah. A continuous charge of 0.75 A (plus the necessary amount to maintain the battery itself) will therefore meet the average demands and the charger is adjusted to that value. Whenever the load is more than 0.75 A, the battery is discharging, and whenever it is less, the battery is receiving charge, thus effecting a "balance" over the 24-h period.

This operation, employing inexpensive equipment, is adaptable to various small battery systems which have a routine use. They should be checked frequently, however, as some unanticipated load may discharge the battery, necessitating a manual recharge.

Trickle Charge. This is a continuous constant-current charge given to a battery to maintain it in a fully charged condition, with no external load connected to it. This may be used for batteries in storage or in standby service, where their only use is in an emergency such as failure of the normal power supply. In setting up such a trickle charge, a current value of 50 to 100 mA/100 Ah of battery capacity is a good trial value. However, the voltage should be checked and the current readjusted until an average voltage of 2.15 to 2.17 V/cell is maintained.

Only comparatively low-gravity batteries are suitable for "floating" or the various other methods of continuous charging. High-gravity batteries in standby service or storage should be kept charged by periodic recharges.

Current Requirements. The current drawn by batteries which are floated from a constant-voltage source varies with:

1. The voltage of the charging source. The current will approximately double

for each increase of 0.05 V/cell (2.15 to 2.20, etc.) and, of course, will be reduced to one-half for a similar decrease.

2. Temperature. An increase of 15°F in temperature will also approximately double the current drawn, and a similar decrease will reduce it to one-half.

3. Battery age. Owing to the increase of the internal losses in a cell, the current which it requires to maintain full charge also increases. No accurate values can be given, but here also the requirements probably double during the battery's life. The value of 50 to 100 mA/100 Ah given above may be considered a fair average—a new battery requiring less and one near the end of its life requiring more.

Equalizing Charge. A battery must be brought to a state of full charge to avoid excess sulfation; yet appreciable overcharge must also be avoided. To accomplish this, it is common practice to stop daily or other frequent recharges when the battery is nominally but not completely recharged and then give a periodic "equalizing charge." This is simply a continuation of a regular charge until a complete state of charge is attained. Such a charge should be continued until successive readings of gravity and voltage show no increase over a period of several hours. In practice it is usually done by continuing the charge by time clock for a certain period which experience has shown to be adequate. The frequency of these equalizing charges varies with the service application—from weekly in industrial cycle service to several months in standby floating service.

GENERAL SUBJECTS

13. Efficiency. The efficiency of a battery—energy input vs. energy output—can vary widely depending upon use. A small amount of energy is required to maintain it, even without any use, so that the greater the amount of proper use, the higher the efficiency. Nominally the relation between a normal discharge and the necessary recharge is the basis on which efficiency is considered.

Efficiency may be expressed in two ways—as the ampere-hour efficiency or the watthour efficiency. In ampere-hours, it is usually considered that the recharge should equal 110% of the discharge, giving an efficiency of about 91%. However, the average voltage on charge is considerably higher than on discharge, in an approximate proportion of 17 to 18%, giving a voltage efficiency of 85%. Combining these two (91 × 0.85) results in a watthour (or total energy) efficiency of 77 to 78%, which can be considered as a representative figure.

14. Local Action. This refers to the internal losses of a battery standing on open circuit or when on float charge, and without considering any losses incidental to any discharge. It is due to the local chemical action between component parts of the plates and is almost entirely in the negative plates. For example, the negative active material—pure lead—and the antimony of the grid and any other constituents of the alloy react with the electrolyte as a "cell." If it were practical to use a pure lead grid and eliminate every trace of impurity in the cell, there would be virtually no local action.

The degree of local action may be expressed either as the percentage loss in capacity per month on open circuit or by the amount of current required on float or trickle charge to overcome it and keep the battery fully charged. This varies with temperature, being greater at high temperatures and less with low. A general estimate may be obtained from the approximate float current mentioned under Floating Charge in Par. **12**, Charging Methods.

15. Gassing. A battery cell cannot absorb all the energy from the charging current toward the end of charge, and the excess energy dissociates water by electrolysis into its component gases hydrogen and oxygen. The oxygen is liberated at the positive plates and the hydrogen at the negatives. When a battery is completely charged, all the energy, except the small resistance loss, is consumed in this electrolysis.

During a recharge, gassing is first noticed when the cell voltage reaches 2.30 to 2.35 V/cell and increases as the charge progresses. At full charge, when most of the energy goes into gas, the amount of hydrogen liberated is about 1 ft^3/cell for each 63 Ah input. Inasmuch as a 4% content of hydrogen in the air may be hazardous, the above value may be used to relate the maximum amount from a given battery to the size of the room in which it is located.

16. Mossing. This describes the possible deposition of a spongelike layer of lead on the negative plates or strap. This material was originally shed from the plates (mostly the positives) in fine particles and circulated throughout the cell by gassing, falling on both the positive and the negative plates. When in contact with either plate, it is changed to the active material of that plate. That on the positive is "loose" and noncohesive and washes off again from the gassing of the cell. Material on the negative plate, however, is cohesive and adheres to and builds up on the top edge and possibly along the side edges of the plate. It can accumulate to such an extent that it bridges over or around the separators, touching an adjacent positive plate and causing a partial short circuit.

The accumulation of any appreciable amount of moss is usually an indication of overcharging in ampere-hours or high charging currents in amperes, either of which should be corrected.

17. Sediment. There is a tendency for some of the active material on the surface of the plates to separate from the main body of material and settle to the bottom of the container. This is counteracted in various ways. The material may contain a "binding agent," or it may be held in place by the various types of tubular construction or on flat plates by perforated glass, rubber, or plastic sheets or mats known as "retainers."

Despite these means, a small amount of such material may fall from the plates, most of it usually from the positives, and a certain space in the bottom of the container, below the plates, is usually reserved for this "sediment."

With proper floating types of operation, this sediment is entirely negligible and after years of operation may amount to hardly more than a layer of dust. In active cycle service, an appreciable quantity may accumulate after years of use, but the size of the sediment space is designed to accommodate all that will fall during the battery's life. Thus it should never be necessary to remove or "clean" the sediment from a battery.

Certain types of abuse, primarily excessive overcharging or charging at high rates, will accelerate the accumulation of sediment and shorten the useful life of the battery. If the sediment increases to the point where it reaches the bottom of the plates, it will partly short-circuit them and cause failure.

18. Temperature. The operating temperature of a battery should preferably be in the "normal" range of 60 to 80°F. Higher temperatures give some additional capacity at the time but will reduce total battery life. High temperatures (125°F and higher) can actually damage some of the battery components and cause early failure.

Low temperatures reduce capacity but will prolong battery life under floating operation or in storage. Low temperature may freeze the electrolyte, but only if the battery is discharged (low in specific gravity) at the time. At the temperatures shown in the following table, the electrolyte will not freeze unless the specific gravity is lower than indicated:

Temperature, °F	Specific gravity (approx.)	
	At same temperature	Corrected to 77°F
+20	1.100	1.080
+10	1.150	1.130
0	1.185	1.160
−10	1.210	1.180
−20	1.235	1.200
−30	1.250	1.215
−40	1.265	1.225

19. Internal Resistance. The actual ohmic value of the internal resistance of a cell is sometimes requested. It is low, and no general answer can be given, as it varies with the state of charge, specific gravity, cell size in ampere-hours, temperature, physical construction, and the condition of the cell or the degree to which it is worn out. While it can be estimated for a given set of conditions, this resistance value has little practical value in the application or operation of a battery. The voltage and current characteristics on discharge and charge can always be used to solve any practical problems.

20. Testing. The testing of battery capacity can be done only by conducting a discharge under controlled and recorded conditions. Test discharges are conducted in the following manner:

The battery is first properly and completely charged. Temperature and specific gravity must be at their normal or standard values or corrections applied to allow for any difference. A discharge rate is selected, depending on the time and load equipment available. Usually a rate between the 3- and 8-h discharge rates is chosen. The discharge rate in amperes is held constant at the chosen value and the total battery voltage read and recorded at frequent intervals. The test is concluded when the battery voltage falls to the preselected final voltage value. The capacity is expressed as the percentage of time at which final voltage was reached.

Example. A 32-cell battery rated at 275 Ah at the 5-h rate is to be tested at that rate to a final voltage of 1.75 V/cell. The 5-h rate is 275/5, or 55 A, and the final battery voltage is 32×1.75, or 56.0 V. On test the total voltage fell to 56.0 V in 4 h 36 min, or 4.6 h. 4.6 h/5 h = 0.92, or 92% capacity obtained.

21. Storage. If a battery will be out of service for several months or longer, it should be given a thorough equalizing charge and stored in a cool, dry location, but where the temperature is not low enough to permit freezing. Check the gravity at intervals of 2 months, and recharge the battery when the gravity drop has reached 25 to 35% of the specified amount. Experience may indicate that the checking interval may be longer, depending on temperature, full-charge gravity, and battery condition.

Once a battery has been filled with electrolyte, *it must not be "dumped" and allowed to remain "dry"* or with the plates simply moist. It will be quickly harmed under such a condition.

22. "Dry-charged" Batteries. When it is desired to keep or store **new** batteries for a considerable time before they are required, they are frequently manufactured or prepared in a "dry-charged" condition. This consists essentially in charging and drying the plates, before assembly, in an atmosphere devoid of air or oxygen. All elements of the assembled cell are completely dry, and the cell is partly or completely sealed to keep out any moisture. Such batteries must be stored in a cool, dry location until ready for use, and under these conditions the plates will retain most of their charge for as long as perhaps 2 years.

In placing these batteries in service, they should be filled with electrolyte of a specific gravity about 10 points lower than the nominal full-charge gravity of the battery. For example, for a battery with a rated full-charge gravity of 1.275, fill with 1.265 electrolyte.

The battery may be used shortly thereafter, although it may not have full capacity, depending upon the length and condition of storage. It is preferable first to give it a thorough equalizing charge after filling. Once it has been properly prepared, its capacity, characteristics, and life are the same as a new "wet" battery.

23. Maintenance. The routine maintenance of storage batteries varies widely with the type of battery and its use. A battery of "stationary"-type cells in an electric power substation may require no specific attention for many months, whereas an industrial-truck battery in a busy plant will require weekly watering and cleaning, etc. Only general rules can be given.

Proper charging is the most important factor in battery service and life, and the proper method for each application should be carefully followed. A battery in frequent cycle service need not be completely recharged each time but should be given a proper equalizing charge weekly. A battery in floating or standby service should be kept fully charged, or as nearly so as conditions will permit.

Water should be added at necessary intervals to keep the electrolyte level between its normal upper and lower limits. The plates must not be allowed to become dry. Water must be of "approved" quality. The battery manufacturer will supply specifications if desired, but most "city" water sources are satisfactory.

Batteries must be kept clean and dry to the extent that no corrosion, dust, or moisture offers a conducting path to partly short-circuit the cell or contact "ground."

Lead-acid batteries do not require any routine overhaul or solution changes during their entire life except as a result of accidental damage.

24. Counter Cells. These cells, properly known as counter-electromotive-force (cemf) cells, are similar in some respects to battery cells but are used as constant-voltage resistors. They consist, physically, of two groups of plates, but the plates are all alike and are simply "grids" without any active material. Neither group of plates is essentially either positive or negative, although when direct current is passed through them, they take on polarity, just as a resistor does.

Their advantage over resistors is that they maintain a nominally constant voltage drop of approximately 2 V over a wide range of currents, whereas the voltage drop in a resistor is directly proportional to the current. They lower the voltage of a d-c source, and their usual application is to maintain a constant voltage across a battery-supplied load when the battery voltage is necessarily raised for charging. As an example (see Fig. 24-8), assume a telephone installation with a nominal system voltage of 48 V and a permissible range of 46 to 50 V; 24 cells are necessary in order to supply sufficient voltage during an emergency discharge, and yet the floating voltage for 24 cells is 51.5 V and, for recharging, at least 56 V. Therefore three counter cells would be used, one in the circuit during normal floating operation, and all during a recharge. A voltage-sensitive device normally accomplishes this automatically and cuts the counter cells in or out successively as required, to maintain a nearly constant load voltage.

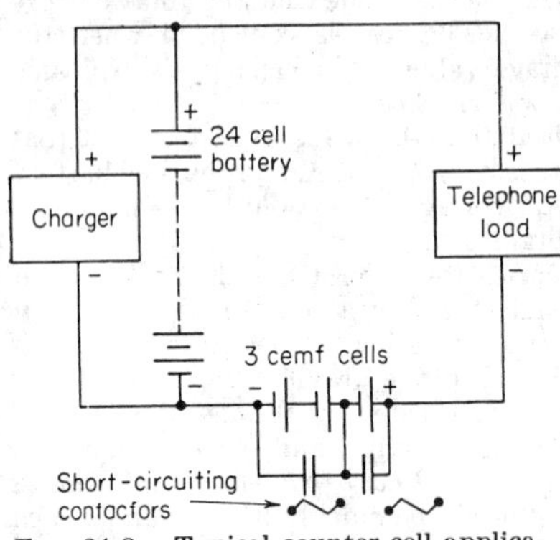

FIG. 24-8. Typical counter-cell application.

As shown in Fig. 24-8, counter cells are *connected in the load circuit, not the charge circuit,* as is sometimes assumed. By short-circuiting them, rather than "cutting them out," there is no break or other interference in the load circuit.

The earlier counter cells were groups of lead-antimony plate grids without any active material and using sulfuric acid electrolyte. The plate which acts as the positive scales within a comparatively short time, however, and also builds up a certain amount of surface capacity, which may cause arcing at the switch used to short-circuit it.

In recent years, all counter cells are made with either nickel or stainless-steel plates and with an electrolyte of sodium hydroxide (NaOH). This cell does not build up any significant capacity, and the plates have an almost indefinite life. A ½-in layer of oil is floated on top of the electrolyte to reduce the spray from gassing and to minimize contact between the electrolyte and the air. The electrolyte of these cells does require renewal at long intervals, depending on the amount of use.

As all counter cells are simply "water resistors," they require considerable addition of water, in direct proportion to the ampere-hours passed through them.

25. Lead-Calcium Cells. The lead-calcium battery is a lead-acid battery whose lead parts have been hardened by a very small amount of calcium instead of the antimony normally used. This change has reduced the corrosion rate, increased the battery's life, and improved the operation in emergency standby service. Generally it is not used in frequent cycling service, especially where time between charge and discharge is short such as in motive power service (industrial truck, mine locomotive, golf carts, etc.). The charge current, when fully charged and when floating on a voltage regulated bus, is very small so that water is infrequently needed. It has been found that, with the lead-calcium battery floating on voltage regulated bus between 2.20 and 2.25 volts per cell, equalizing charges are not required. The battery, therefore, requires very little maintenance.

The lead-calcium battery has become increasingly popular in standby service as an emergency power source since it was first introduced on a production basis about 1950. In spite of several unrecognized problems in design, manufacture, and operation, many of the lead-calcium batteries made in the early years are still in service and operating satisfactorily. Lead-calcium batteries of later design and manufacture are expected to show further increase in life and performance.

26. Maintenance-free Batteries. These are batteries of the lead-calcium type which have a vent which retains electrolyte but releases a small amount of gas formed near the end of the charge. Such cells, if operated with a properly designed, current-limited, constant-potential rectifier, decompose very little water so that maintenance is not required and, therefore, last for several years. Chargers used to charge or boost automotive batteries should not be used since excessive decomposition of the water in the electrolyte may occur, thereby shortening the life of the battery.

Batteries of this type have found a place in portable equipment where electrolyte spillage must be avoided and in service where water additions are inconvenient or readily forgotten. At present such batteries are available in the smaller sizes only and are used for powering hand tools, such as electric drills, hedge shears, portable televisions, and emergency lighting units.

NICKEL-IRON BATTERIES

27. Theory. The nickel-iron battery is unique in many respects, both its chemistry and physical properties being different from those of other types. It is often known as an "alkaline" battery from the electrolyte used.

In a fully charged cell the positive active material is nickel oxyhydrate (NiOOH), the negative is metallic iron sponge (Fe), and the electrolyte potassium hydroxide (KOH) and water with the addition of lithium hydroxide. The chemical reactions during discharge produce intermediate transitory compounds, but the end result is a transfer of oxygen from the positive to the negative, leaving the positive as nickel hydroxide, $Ni(OH)_2$, and the negative as iron hydroxide, $Fe(OH)_2$. Neglecting the intermediate reactions, the end formula is

$$2NiOOH \cdot H_2O + Fe \underset{\text{charge}}{\overset{\text{discharge}}{\rightleftharpoons}} 2Ni(OH)_2 + Fe(OH)_2$$

The electrolyte takes no part in the chemical reaction; its composition is unchanged. It does, however, act as a transfer agent by "accepting" oxygen at the positive during discharge and "depositing" it at the negative. On charge, conversely, the action is reversed, and the positive and negative materials revert to their original charged state. The active material is finely ground powder and requires a "container" to give it a physical form and provide a conductor to the external circuit.

The positive material is encased in perforated nickel-plated steel tubes, made by spirally winding a flat strip into tubular form. The ends of these tubes are closed by crimping, and the tubes themselves are reinforced at intervals by annular steel rings. Since the active material is a poor electrical conductor, it is interspersed with alternate layers of metallic nickel flake to increase its conductivity to acceptable levels.

The negative material is encased in flat, perforated nickel-plated steel pockets to accomplish a similar result. No additional conductive material is required.

The "plate" is formed by grouping the desired number of positive tubes or negative pockets and crimping or welding them together or to a nickel-plated sheet-steel frame.

28. Physical Construction (see Fig. 24-9). The **plates** have an extension at the top known as a **lug,** or **ear,** in which there is a hole to receive a connecting **rod.** The desired number of plate lugs, with intermediate spacing washers to separate the plates by the required distance, are placed over the rod and retained by nuts at either end. Also on the rod, at the center, is the lower part of the **post,** which extends upward as the external connection to the circuit. This assembly of plates, rod, and post is known as a **group.**

The plates of the positive and negative groups are intermeshed between each other and prevented from coming in contact by round, hard rubber or plastic **pins,** or **insulators.** This assembly of positive and negative groups and insulators is known as an **element.**

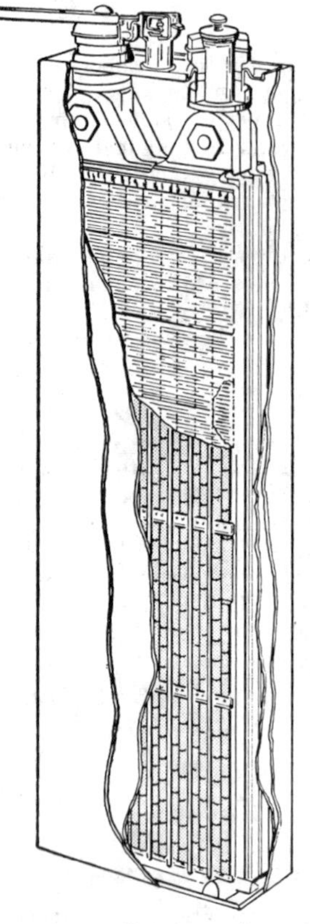

FIG. 24-9. Cutaway view of nickel-iron industrial cell.

The **jar,** or container, is of nickel-plated sheet steel. Inasmuch as this is an electrical conductor, it must be insulated internally from the element, and also from the posts where they pass through the top or cover. The insulation from the element is by hard rubber or plastic sheet, and hard rubber bushings, or **gland caps,** encircle the posts. Inasmuch as the cell containers are metal, each cell must be insulated from the adjoining ones and from any other electrical conductor. This is done in either of two ways: The sides of each container may be equipped with small, round extensions, known as **bosses,** which project into rubber **buttons** in holes in the sides of a wood crate, or **tray.** The cell is suspended from these bosses, thus insulating it from adjoining cells. This also permits the bottom of the tray to be open for ample ventilation, this being necessary in alkaline-type cells to dispose of the heat generated on both charge and discharge. As additional insulation and protection from moisture, the entire tray and the sides and bottom of the cells are covered with a black insulating paint known as **Esbalite.** The top of the cells is covered with a resinous, varnishlike coating known as **Esbaline.**

Another method is to assemble the cells in steel trays or boxes. The containers are not equipped with bosses but rest on individual rubber pads on the bottom of the tray. They are coated with an insulating plastic film and separated from each other by vertically corrugated plastic sheets. This type of more compact assembly is, however, usually considered less satisfactory from the standpoint of both electrical and heat insulation.

29. Voltage. The voltage of nickel-iron cells is usually referred to as being 1.2 V; thus a 10-cell battery is nominally considered a 12-V battery, etc. The actual voltage at any time, however, depends on whether it is on open circuit, on discharge, or on charge and, in the latter cases, on the current rate and the state of charge. The open-circuit voltage varies between 1.25 and 1.35 V, but its relation to the state of charge is not sufficiently definite to be used as an indication of that condition.

On discharge, the voltage will decrease to a value where it is no longer effective for the application to which it is put. This is referred to as the "final" voltage. It varies with the rate of discharge, being lower at higher rates and vice versa. The most commonly used value is 1.0 V/cell, although either higher or lower ones may be more suitable for particular uses (see Fig. 24-12 for discharge curves at various rates).

Conversely, on charge, the voltage rises and follows a distinct pattern throughout the charge, to a maximum value varying with the charging rate. At the rates most commonly used, this maximum voltage is 1.7 to 1.8 V/cell (see Fig. 24-14 for typical charge-voltage curves).

30. Electrolyte. The electrolyte is a solution of potassium hydroxide (KOH) and water, with the addition of lithium hydroxide. While necessary to the chemical reaction, it is not significantly changed by it throughout the normal range of charge and discharge.

The specific gravity of the electrolyte of a new cell is about 1.210 to 1.215 at the normal electrolyte level at 77°F, this one value being used for nearly all types. This gravity slowly decreases over months or years owing to loss from spray during charge and possible spillage. The electrolyte also tends to carbonate because of contact with the air. When the gravity decreases to about 1.160, it definitely affects the capacity, and the entire solution requires renewal (see Par. **41**). This may be necessary several times during the life of the battery.

The only time that there is any significant change in the specific gravity is if the discharge is carried beyond its normal limits, down to or near zero voltage, as when cells are put in storage. This drives much of the lithium from the plates into the electrolyte, and its gravity may rise 25 to 35 points (0.025 to 0.035). This action is reversed when the cells are again charged.

Specific-gravity Corrections. The correction for temperature is the same as for lead cells—one point (0.001) in gravity for each 3°F. The change with the electrolyte level varies with the physical design of the cell, but 4 to 6 points/¼ in change may be considered average.

To compare accurately specific-gravity readings taken under different conditions, they must be corrected to a common standard, this being the normal or specified electrolyte height of the cell, and a temperature of 77°F. The "corrected gravity" is what it *would read* under these standard conditions.

1. Add one point (0.001) of gravity for every 3°F above 77°, or subtract one point for every 3° below 77°.

2. Subtract five points of gravity (assuming the cell has a five-point change per ¼ in) for every ¼ in that the level is below its normal position, or add five points for every ¼ in above normal.

Example. Specific gravity reads 1.220 at 89°F and ½ in low level. 89°F − 77°F = 12/3 = 4 points to be added. ½ in low level = 10 points to be subtracted. Net result, subtract 6 points; "corrected" gravity is 1.214.

Under normal operating conditions there is little danger of the electrolyte freezing, but at low temperatures and with the specific gravity reduced below normal, there is that possibility. At the temperatures shown in the following table, the electrolyte will not freeze unless the gravity is lower than indicated.

Temperature, °F	Specific gravity	
	At same temp.	Corr. to 77°F
+20	1.095	1.075
+10	1.145	1.120
0	1.180	1.150
−10	1.210	1.180
−20	1.235	1.205
−30	1.255	1.225
−40	1.280	1.240

Alkaline electrolyte for nickel-iron cells is usually furnished in two specific gravities: (*a*) 1.200 to 1.210, known as **refill electrolyte,** and (*b*) 1.240 to 1.250, known as **renewal electrolyte.**

The former is used primarily for replacing accidental spillage loss and the latter for complete replacement when required by low gravity (see Par. **41**).

31. Capacity. The capacity of the alkaline storage battery is expressed in ampere-hours, the product of the discharge in amperes over a number of hours. However, a figure such as 100 Ah has little significance unless it is qualified by the various factors which influence the capacity. The principal ones which must be considered are: (1) discharge rate, (2) temperature, and (3) final voltage.

Discharge Rate. The higher the discharge rate in amperes, the fewer total ampere-hours a battery will deliver under otherwise similar conditions. This relationship varies slightly with different cell types but is generally represented by the graph in Fig. 24-10.

The primary reason for this is the internal resistance of the cell, which causes a voltage drop or loss within the cell itself. The higher the current being discharged, the greater is this internal voltage drop and, therefore, the less voltage is available for the external load. The discharge rate most commonly used as a standard reference for alkaline batteries has been the 5-h rate, although a 6-h basis is now frequently used. The difference between these two is only a couple of percent. Other hourly ratings can be supplied to cover applications requiring them. All such ratings are equally correct, as long as they are properly specified.

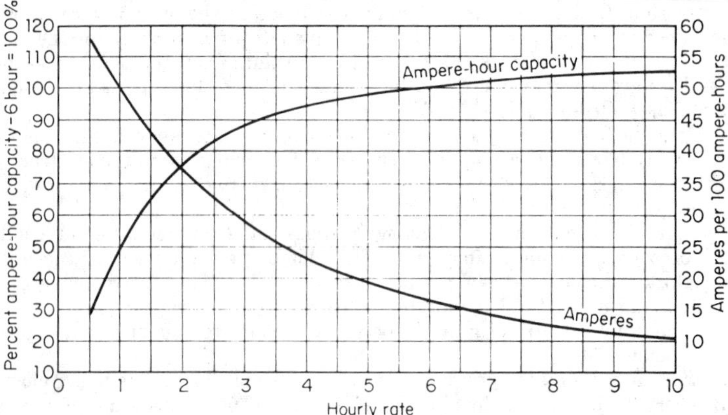

FIG. 24-10. Capacity-rate curves to a "final" voltage of 1.0 V/cell.

Inasmuch as the electrolyte does not change during normal charge and discharge there is no lack of diffusion to affect the capacity.

Temperature. The reactions of a storage battery are accelerated at higher temperatures, and the resistance and viscosity of the electrolyte are reduced, it having a negative coefficient of resistance. Thus the voltage drop within the cell is less, and the resultant capacity is greater. Conversely, at lower temperatures the capacity is reduced. Figure 24-11 gives a general indication of this temperature effect at various discharge rates. A "standard" reference temperature of 77°F (25°C) is most commonly used and is assumed unless otherwise specified.

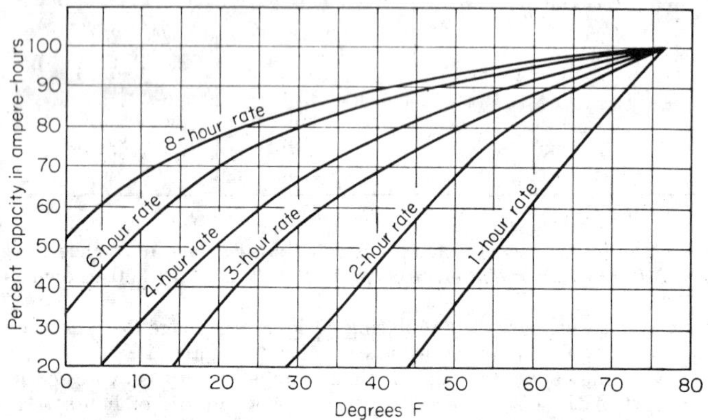

FIG. 24-11. Effect of temperature on capacity (77°F = 100%).

Final Voltage. This term defines the voltage commonly accepted as the practical minimum for a given application. It varies with the use to which the battery is put, being higher for low rates of discharge and lower for high rates. A final voltage of 1.0 V/cell is most commonly used for applications of alkaline batteries.

32. Discharge Characteristics. These batteries may be discharged at any rate of current they will deliver, but the discharge should not be continued beyond the point where the cell nears exhaustion. They are well adapted for low or moderate rates of discharge, but not for extremely high rates such as large-engine starting.

At any constant rate of discharge, the cell voltage decreases throughout the discharge until the appropriate final voltage is reached. Figure 24-12 illustrates the approximate relation between cell voltage and discharge rates for typical alkaline cells.

The lower capacity at high rates does not, however, represent a comparable loss of energy; it means that the voltage falls to its final value in a shorter period of time.

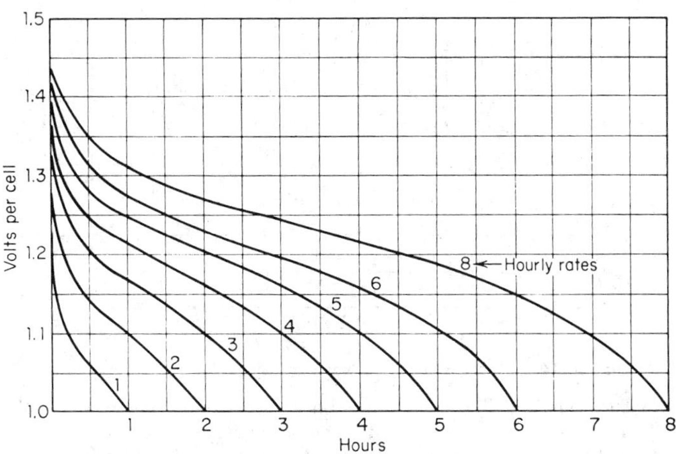

FIG. 24-12. Time-voltage curves to a "final" voltage of 1.0 V/cell.

If, after a high-rate discharge to a given voltage, the current is reduced, the cell voltage will recover and further capacity can be obtained at the lower rate before the voltage again drops to the same value. It is generally true that, when *decreasing current rates* are used, the total ampere-hour capacity of a battery at a given hourly rate (or nearly that amount) may be obtained during that time period, regardless of higher than average rates at the beginning. This does not hold, of course, if the high rates are at the end of the discharge (see Fig. 24-13).

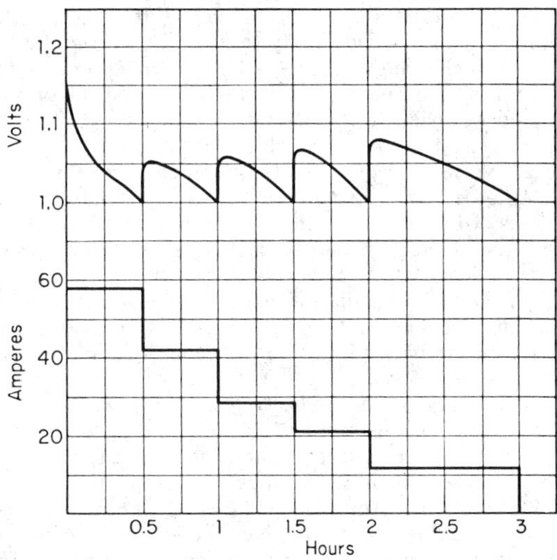

FIG. 24-13. Effect of decreasing discharge rates on cell voltage.

During discharge there is an appreciable rise in temperature due to the heat losses (I^2R) in the cell, this increasing with an increase in current rates. During a typical discharge of 6 to 8 h, this may cause a rise of 20 to 30°F or more if good ventilation is lacking. Where practical, it is well to limit the discharge to 80% of capacity, particularly if moderately high rates are used, to minimize this temperature rise.

The state of charge of an alkaline battery cannot be determined as readily as with other types. The specific gravity does not change appreciably during discharge or recharge, and a gravity reading, therefore, has no significance. The only practical means is to read the voltage of a pilot cell under a momentary high-rate discharge. A device known as a "test fork" is available for this purpose and consists of a resistor drawing a high current and a voltmeter to measure the cell voltage at the time. By a chart or table, the voltage reading can be related to the state of discharge of the battery. While not at all a precise means of determining the state of charge, it does meet most practical requirements.

33. Charging Characteristics (for charging equipment see Pars. **43** to **54**). A battery may be charged at any current rate which does not produce excessive gassing or temperature rise. The maximum temperature should not exceed 115°F. Any cell voltage

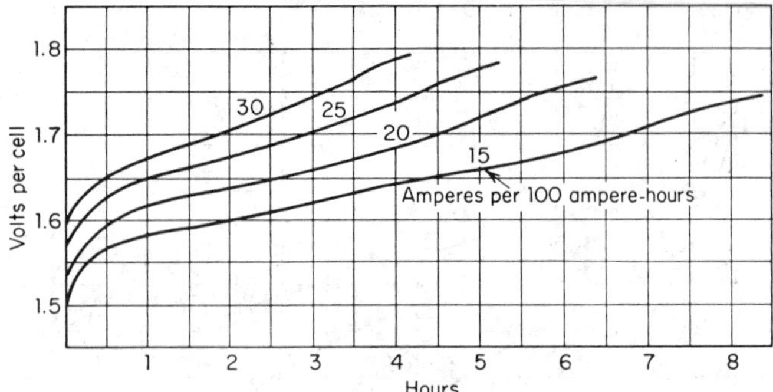

FIG. 24-14. Typical charging voltages at various current rates.

which does not exceed 1.70 V/cell will keep both these conditions within limits, on the assumption that the charge is terminated when complete.

Each type and size cell is assigned a suggested "charge rate," which is usually between 15 and 20 A/100 Ah of rated capacity. For recharging, this charge rate can be used throughout the entire charge, which would thus require a 6- to 8-h time period. There is, however, considerable latitude permissible in the charge rate to be used, a starting rate of twice the specified value not being prohibitive. The current at the end of charge should not exceed the specified rate but should preferably be not less than one-third of it. (This does not apply to floated batteries, or similar operating methods.)

For recharging, an ampere-hour input approximately 25% greater than the previous output is necessary to return the battery to the previous state of charge. Every effort should be made to give the correct amount. Too little charge will result in reduced capacity, and too much, while not harmful *as such*, will cause high temperature and excessive water requirements, which can bring internal damage to the cells.

The detection of improper amounts of charge is primarily a matter of checking the causes of low capacity, higher than normal temperatures, or water requirements. Current and voltage values indicated above are representative for normal ambient temperatures of 70 to 80°F and batteries in good condition.

While lack of sufficient charge will reduce the capacity of alkaline batteries, no permanent harm results. One or two discharge and recharge cycles will usually return them to full capacity. With proper operation, no routine "equalizing" or special charges are required.

34. Charging Methods. *Cycle Charging.* This represents a routine recharge following a partial or complete discharge and is best illustrated by industrial truck service. The current rate is usually one which has an average value about equal to the specified charge rate. With some types of equipment this will remain at a nominally constant value throughout the entire charge—with other types it will "taper" somewhat, starting higher and finishing lower than the specified value. For this charging, an 8-h period is usually assumed to be available, and an average value of the specified rate will complete the charge in that time or less. It should be stopped automatically.

Floating Charge. The use of nickel-iron batteries in "full-float" service is comparatively small. Where applied, however, a voltage value of 1.50 to 1.55 V/cell is commonly used. To maintain the battery in a fully charged condition, approximately 4 to 6 mA/Ah of battery capacity will be drawn at the midpoint of this range.

Partial Float (Partial Cycle). This method of operation is similar to "floating" except that the charging source is not continuously available and during such periods the battery must carry the load. The most common application of this is railway passenger-car operation for lights and air conditioning. Charging current is available only when the car is in motion, as the generator is driven from the car axle, and the battery must supply the load during station stops.

The floating voltage must be carried at a comparatively high value to replace the discharge during periods of operation. The value of this floating voltage may range anywhere from 1.55 to 1.70 V/cell, 1.65 being a representative figure. Train schedules and ambient temperature conditions are the principal determining factors.

Nickel-iron batteries find comparatively little application in types of service where two-rate charging, continuous average current, and trickle charging are commonly used. If, however, it is desired to use one of these methods, they can be applied in a manner similar to that used with lead batteries (see Par. **12**). However, the reduction or termination of the charge by a voltage relay is not so accurate or reliable as might be desired.

35. Efficiency. The efficiency of a battery—energy input vs. energy output—will vary widely depending on the circumstances of use. Nominally the relation between a normal discharge and recharge is the basis on which efficiency is considered. This may be expressed in two ways, as the ampere-hour efficiency or the watthour efficiency. In ampere-hours, it is usually considered that the recharge should equal 125% of the discharge, giving an efficiency of 80%. However, the average voltage on charge is considerably higher than on discharge, in the approximate proportion of 35%, giving a voltage efficiency of 75%. Combining these two (80×0.75) results in a watthour (total energy) efficiency of about 60%, which can be considered a representative figure.

36. Local Action. This indicates the internal losses of a cell, standing either on open circuit or on float charge. On open circuit this may be represented by the loss in capacity over lengthening periods of time. With any type of continuous charge it is usually measured by the amount of current required to maintain it in a fully charged condition (see Par. **34**). These losses increase at higher temperatures and with the age of the cell.

37. Gassing. During charge there is a certain amount of energy which the cells cannot absorb, and this is dissipated as "gas." This gas actually consists of oxygen evolved at the positive plate and hydrogen at the negative. In nickel-iron cells these gases are evolved during the entire charge, the amount increasing, however, as the charge progresses. The proportion varies during the charge, but when the cell is fully charged, it represents the relation in which they exist in the water which is being disassociated—two parts of hydrogen to one of oxygen.

The positives "retain" a certain amount of oxygen during the charge, and this continues to evolve for some time after the charge has stopped. This is "useful" in that a nickel-iron battery will deliver an appreciably greater capacity if discharged within several hours after a thorough recharge than if allowed to stand until it has stopped gassing.

When cells are fully charged, at which time virtually all the energy is dissipated as gas, the amount of hydrogen is about 1 ft^3/cell for each 63 Ah of excess charge. When hydrogen in the air reaches a proportion of about 4%, it can constitute an explosive mixture. The above value can therefore be used to relate the approximate amount evolved from a given battery to the size of the room or compartment in which it may be located.

38. Ventilation. Proper ventilation of nickel-iron batteries is important to avoid high temperatures, as the temperature factor can be important if the battery is in cycled service.

In most assemblies of these batteries the cells are separated by an air space of ½ in to permit free air circulation between them. They are suspended from the sides of the trays, rather than resting on the bottoms, so that the bottoms may be entirely open.

However, the compartment or space in which the battery is located must permit airflow through it. For example, an industrial truck battery thus constructed, but placed in an enclosed, nonventilated compartment, cannot usually be kept at a reasonable temperature. These cells can be made in compact assemblies for use where space requirements are critical, but special means may be required to avoid high temperatures unless the discharge rates are low.

Ventilation is equally important during charge. When cycled batteries are charged out of a truck or other enclosed compartment, the charging rack or bench should be of open grille construction for free airflow. Fans may also be desirable to achieve the best results.

39. Internal Resistance. Any battery cell has its own internal ohmic resistance consisting of the path through the active materials, plate frames, electrolyte, and cell posts. While this resistance is usually low in comparison with that of the external load, it can be significant at high current rates of discharge.

The actual value is seldom used or required in practical battery applications, as the known voltage and current characteristics usually suffice. A general approximation, however, is a value of about 0.005 Ω for a fully charged 100-Ah cell at normal temperature. During the latter half of discharge this will increase perhaps 50%. For cells of other ampere-hour capacities, the value would vary inversely.

40. Tests. The actual testing of battery condition or capacity is done by a discharge under controlled conditions. To indicate how capacity is thus determined, such discharge tests are conducted in the following manner:

The battery is completely recharged. A discharge rate is selected, depending on the load equipment and time available, the 5-h rate being most commonly used to a final voltage of 1.0/cell. If the test cannot be run at about the normal temperature of 77°F, corrections must be made. During the test the current must be continuously maintained at the selected rate, and the proportion of time until final voltage is reached, compared with the rated time, represents the battery capacity.

Example. On the basis of normal temperature and specific gravity it is desired to test a 24-cell 450-Ah battery at the 5-h rate. Current is 450/5 = 90 A. Final voltage is 24 × 1.0 V = 24 V. During the test the total battery voltage decreases to 24.0 V in 4.4 h. This is 88% of the rated time of 5 h; therefore, the battery has 88% capacity.

41. Maintenance. While a battery requires comparatively little maintenance, what is required is necessary to proper operation and long life.

Proper Charging. This is the most important factor. Every effort should be made to give the correct amount of charge at proper rates and with adequate ventilation.

Water Addition. Approved water must be added at necessary intervals to maintain the electrolyte level between the tops of the plates and its specified height. *Do not overfill,* as this may cause overflowing during charge.

Cleaning. The battery must be kept reasonably clean and dry to avoid any paths of electrical "leakage" between adjoining cells or from cells to conducting trays or boxes. Such leakage will cause electrolytic action, resulting in corrosion of the steel containers.

These batteries should preferably be cleaned with steam or hot water and blown dry with compressed air. They should be raised off the floor when this is done in order that any dirt will be completely flushed away. Always keep filler caps closed during such cleaning.

After cleaning renew any of the protective coatings as necessary.

Solution Change. Over long periods of time the specific gravity of the electrolyte will decrease owing to small amounts lost through spray, etc. When the corrected value falls to about 1.160, the entire contents of the cells should be replaced. This should be done as follows: Obtain a sufficient amount of "renewal solution," which has a specific gravity of 1.245 at 77°F. Discharge the battery completely, and then short-circuit it for a number of hours. Dump the old solution and replace with the new, followed by a recharge of about twice the normal length of time. This solution renewal is usually necessary only perhaps twice during the battery's life.

If this type of battery is in light use for a considerable period of time or is not given sufficient charge, it may develop low capacity. This can usually be remedied by one or two "cycles" of discharge and recharge. Such a cycle consists in discharging the battery down to about ½ V/cell and then short-circuiting it. This short circuiting should be done individually or in groups of just a few cells. After being short-circuited for at least a few hours, the short-circuiting leads should be removed and the battery recharged at the specified rate for twice the normal length of time (equivalent to 2.5 or 3 times the rated ampere-hour capacity of the battery). The battery should then be restored to full capacity.

42. Storage. If these batteries are to remain out of service for several months or longer, they should preferably be completely discharged, short-circuited, and left in that condition during the storage period. If this is not done, several "cycles" may be required to restore capacity. First, of course, add water to normal level if necessary, to ensure that plates remain covered. Also, keep filling caps closed.

To prepare for use, add water if necessary, remove short-circuiting jumpers, and give a recharge at specified rate for about twice the usual length of time (equivalent to three times the ampere-hour rating). Once a battery is filled with electrolyte, *it must not be dumped and allowed to remain empty.*

CHARGING EQUIPMENT AND CONTROLS

43. Basic Requirements. The basic requirements of a charger from the standpoint of the battery are:

1. A safe value of charging current throughout the entire charge.
2. Completion of the charge in a time period compatible with the service involved.
3. Accurate termination of the charge when complete, or reduction of the current to a safe maintenance value.

In addition to these the charger itself must have: (1) maximum reliability, (2) automatic operation to the degree practical in the application, (3) simplicity in design and construction, (4) good efficiency and power factor, and (5) reasonable cost.

44. Definitions

1. Owing to the many types and sizes of batteries involved, most of the values given are at the "common denominator" of one 100-Ah cell.
2. The term "charging time" refers to the time to recharge a nominally (100%) discharged battery.
3. "Constant voltage" assumes a tolerance of about 2% and "constant current" perhaps 5 to 10%.
4. The prime power source is assumed to have a nominally constant voltage.
5. Ambient temperatures are assumed to be 70 to 80°F and the equipment "warm."
6. Only basic components are discussed; all equipment is assumed to have the necessary switches, fuses or circuit breaker, relays, meters, etc.

Inasmuch as equipment for lead-acid and nickel-iron batteries is covered, nickel-iron batteries, because of their different cell voltages, will require a greater number of cells than lead-acid ones for the same application. This is usually in the proportion of 5:3. A charger designed for either a lead or a nickel battery is not suitable "as is" for the other type. An attempt to "convert" a charger from one type to the other may be simple or impractical, depending on the individual circumstances.

45. Basic Characteristics and Uses. A battery on charge is not a fixed or static load. It has *a voltage of its own*, and is connected to the charger so that the two voltages oppose each other—positive of battery to positive of charger, and negative to negative. Thus the current which flows is the result of the *difference between the two voltages* and the low ohmic resistance of the battery. The voltage of the battery itself rises during the charge, further opposing the flow of current as the charge progresses. If a discharged battery is connected directly to a d-c voltage high enough completely to charge it in a reasonable time, the initial current flow would be extremely high owing to the low battery voltage at that time. While this would not, in most cases, harm the battery, the charger required to deliver this current would be prohibitively large and expensive. Likewise, if the battery were connected to a d-c voltage just sufficient to produce a normal starting rate, the rise in battery voltage would soon almost reach the impressed voltage, producing a "balance" which would reduce the current to a negligible value.

Therefore, the elementary requirements of charger design are to limit the starting current to a value that will keep the equipment within reasonable size and cost, yet maintain a sufficiently high current throughout the entire charge to complete it in a time period acceptable to the application. There are three basic inherent charger characteristics (Fig. 24-15).

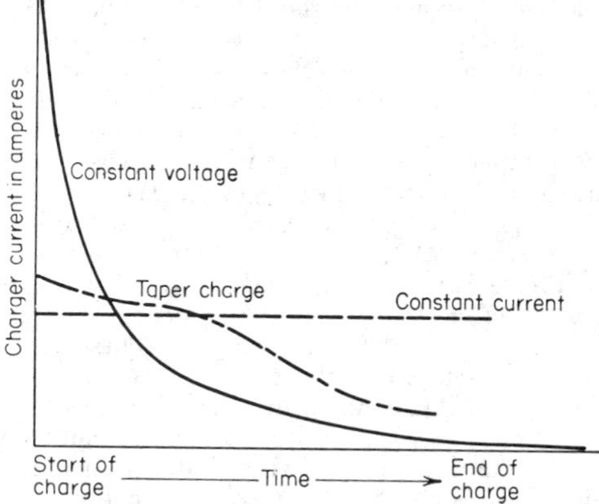

FIG. 24-15. Currents produced by basic charging characteristics.

a. Constant Voltage. The d-c output voltage remains constant throughout its load range. This is used for "floating" methods of operation and also for "modified-constant-potential"

(MCP) charging (Pars. **12** and **34**). If this is not modified the result is extremely high starting rates, requiring prohibitively large and costly equipment. In floating service it is modified in the charger itself to limit the current to or near its rated value. In MCP the charger voltage remains constant, but a resistor in each battery circuit limits the current (Par. **48**).

b. Constant Current. The current is limited to a nearly constant value. This is used in many low-current applications of lead-acid batteries and also in nearly all cycle charging of nickel-iron batteries. Of course, it can be used for any lead-battery charging if the current does not exceed the "finish rate," but this frequently requires too much time. Two-rate charging employs the same basic characteristic, except that the current value is reduced during the charge by some special means.

c. Taper Charging. This is a compromise between the other two characteristics and relates to methods in which the current starts at a moderately high value but is gradually reduced, or "tapered," to a low rate as the battery voltage rises. Its principal application is in the cycle charging of lead-acid batteries (particularly electric industrial trucks).

The degree of taper (the relation between the current at the start and end of charge) can be varied over a wide range by appropriate design. In the cycle charging of lead batteries mentioned above the ratio is usually around 4 or 5:1, that is, a starting current of 18 to 25 A/100 Ah of battery capacity, finishing at 4 or 5 A/100 Ah. For nickel-battery types, a slight taper is used with an average value of 15 to 20 A/100 Ah of battery capacity.

46. Basic Equipment. All charging must be done by direct current. Inasmuch as d-c utility or industrial systems are no longer common, the prime power source is invariably alternating current, rectified either by a motor-generator set or by a static rectifier. Either can provide any of the basic charging characteristics.

Considering all types of service, the majority are single-battery installations and employ, therefore, a "single-circuit" charger. The outstanding exception is electric industrial-truck operation, where groups of several or many are at the same location. Single-circuit chargers may still be used if the number is small, but for large groups one or more large pieces of equipment are usually employed (see modified-constant-potential charging, Par. **48**).

For many years motor-generators were used almost exclusively for all except the smallest industrial batteries. Since the 1940s static rectifiers have been progressively developed and can now be applied to any size installation. While both types of equipment are still available, the trend is toward rectifiers rather than rotating apparatus.

47. Motor-Generators. Single-circuit chargers of this type are usually comparatively small in capacity (seldom more than 10 kW) and use a typical polyphase induction motor. Practically all generators are basically "shunt" machines, these being machines in which the shunt field winding is the predominating source of the magnetic flux. Most of them are modified, however, with small series windings on the main poles or on interpoles, or both. These windings can be combined in various ways, including a "differential," or reversed, connection, to produce, along with other features of the machine, such as air gap, brush location, etc., a characteristic ranging from a steep taper to constant current. This can be made inherent in the generator, the battery being connected directly to the generator without any other modifying components. Both horizontal and vertical motor-generator sets are commonly used, the trend in recent years being toward the vertical construction for cycle chargers.

Single-circuit motor-generators for constant-voltage floating service are nearly always of horizontal design. Some years ago they were usually "straight" shunt machines, with little modification. These vary in voltage somewhat with changes in load and tend to overload themselves during heavy demands or when the battery is discharged. A unique generator, known as a "split-pole" type, has been developed to overcome these deficiencies. In this machine the main pole pieces are split or divided, with certain portions of the windings on only one section of the pole. The overall effect of this construction is to provide a precisely constant voltage throughout the load range of the generator and at the same time prevent overload by a decrease in voltage when the output tends to exceed its rating. This generator has been widely used in recent years wherever an accurate d-c voltage is required.

Multicircuit generators for the charging of groups of batteries are typical d-c generators with a nominally constant voltage, "flat-compounded" by the appropriate relation of their shunt and series windings. However, each battery must have its individual panel with proper resistances and controls (Par. **48**).

48. Multicircuit Equipment. This equipment will be considered in its most common application—recharging industrial truck batteries.

The system consists of one or more chargers, which supply direct current to a common "bus," and a number of panels, each designed to charge an individual battery from that source. It is usually referred to as "modified-constant-potential" charging.

The source of direct current can be either a motor-generator or a rectifier, the former being more common, as motor-generators were developed in large sizes much earlier than comparable rectifiers. Whichever is used, the characteristics must be the same—a nominally constant voltage throughout its load range ($\pm 2\%$ or less) and a current capacity equal to the sum of the currents drawn by all the individual panels. While there is no limit to the number of batteries which can be charged from one machine, it is common practice to use several chargers in parallel on one d-c bus or to separate the system into several groups if more than perhaps 20 or 30 batteries are to be charged.

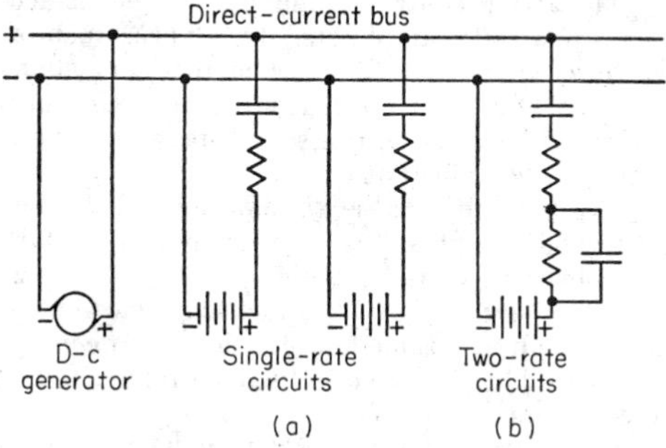

FIG. 24-16. Basic circuit for modified-constant-potential charging.

In using a single d-c bus, all batteries should consist of the same number of cells and be of the same general type, either lead-acid or nickel-iron. The ampere-hour capacity, however, may vary. The basic circuit simply consists of a ballast resistor in series with the battery across the d-c bus (see Fig. 24-16). However, there must be the proper relation among (1) bus voltage, (2) number of battery cells, (3) battery ampere-hour capacity, and (4) value of ballast resistor to charge at proper current rates for the battery, and in the desired length of time.

The actual circuits are not so simple as shown in Fig. 24-16. The motor-generator set or rectifier must be protected against overload and reversed current, and the individual panels must have similar protection, as well as means for automatically terminating the charge. Also, each requires necessary meters to indicate conditions at all times. Such a system requires proper overall design by the battery or equipment manufacturer.

49. Rectifiers. While large rectifiers for multicircuit charging have been developed and some are in use, most are of the single-circuit variety. In a rectifier the a-c and d-c portions of the circuit are electrically interconnected. The most elementary type of rectifier consists of a transformer to change the a-c supply voltage to a value appropriate for the battery being charged and a rectifying element to act as a valve permitting the current to flow in only one direction. This would constitute a half-wave rectifier. For greater efficiency and a better waveform (more uniform current), the rectifying elements are usually arranged to provide full-wave rectification in either single-phase or 3-phase systems (see Fig. 24-17). Early rectifiers used copper oxide "stacks" to achieve the one-way valve effect, but these were later superseded by selenium stacks or silicon diodes. Both these latter types are still in common use, each having its own particular advantages. The silicon type is the more costly, but this is outweighed by smaller space requirements, longer life, and the absence of aging characteristics.

The valve effect, which permits current to flow in only one direction, is an inherent characteristic of the "junction" between the metal and its compound, which together constitute the rectifying element.

A basic rectifier (transformer and rectifying element only) has a drooping, or tapering, characteristic similar to a shunt generator, the degree of taper being a function of the "impedance" of its component parts. This fundamental characteristic can be modified to obtain almost any desired one by the use of:

Special transformers.

Reactors.

Ballast resistors.

Transistorized "control units."

Magnetic amplifiers.

For small applications, particularly where cost is a factor, selenium stack rectifiers are satisfactory. Their controls usually consist of transformer taps or resistors. In larger applications where requirements are not particularly critical, saturable reactors alone can provide basically good characteristics. For more precise performance patterns and regulation, however, magnetic amplifiers with transistorized control units are highly desirable. The basic components are the same, all the characteristics being determined and controlled by the control unit. The circuitry of these units cannot be covered here.

The actual control of the load current is accomplished by varying its *value* by reactors or magnetic amplifiers, etc. These open or close their magnetic "gates" to permit the desired current flow. The rectifiers all use a simple rectifying element, silicon diodes. These permit current to flow in the one desired direction at any time, as determined by the impressed voltage and the impedance of the circuit as a whole, at that particular instant.

More recently these simple diodes have been replaced by, or used with, "silicon-controlled rectifiers," usually referred to as SCR's. These SCR's by an auxiliary "gate control circuit," can be made to *limit the time period* of current transmission (during each half cycle) by a "signal" from a sensing component. Thus the desired effective average load current is obtained by increasing or decreasing the time during which it is permitted to flow. This eliminates the need for reactors or magnetic amplifiers and reduces the components of the system.

With any type of charger the voltage values and current requirements are, of course, determined by the battery and load to which it is connected.

50. Current Form—Measurements and Effects. *General.* In measuring the current to batteries on charge, both the type of meter used and the "waveform" of the current must be taken into consideration. While any charging current is necessarily unidirectional, its waveform can vary considerably, depending on the charging source.

If charging were to be done from another battery, the current would be "pure direct current," but from almost any other source there is a certain amount of "ripple" in it which may affect its measurement if the proper meter is not used.

Current from a d-c generator has little ripple and can be considered as pure direct current.

Current from rectifiers, however, will vary in waveform, depending on:

1. Half-wave or full-wave rectification.
2. Single-phase or polyphase circuit.
3. Relative values of charger and battery voltages.
4. Load on the rectifier.
5. Characteristics of the particular rectifier.

See Fig. 24-17, which shows certain waveforms under the conditions indicated, a sine wave being assumed. This indicates the differences in ripple between single and polyphase and half- and full-wave. However, the wave is usually distorted (more peaked, more flat, or irregular) by the particular characteristics of the rectifier. It must also be remembered that charging current flows only during that part of the cycle in which the rectified voltage is higher than the countervoltage of the battery (not where shown by broken lines in the single-phase diagrams).

Meters. From the standpoint of battery charging (ampere-hours input), the "average" value of the current is the important factor and therefore the desired indication. This is what is measured by the so-called "standard" or moving-coil type of d-c meter, and this type will give a correct reading in practically any commercial charging circuit.

The tong-type, or clip-on, d-c meter does not, however, measure the "average" value, but rather the "effective," or rms, value. For pure direct current the average and rms are the same, but for any other waveform the rms is always higher by an

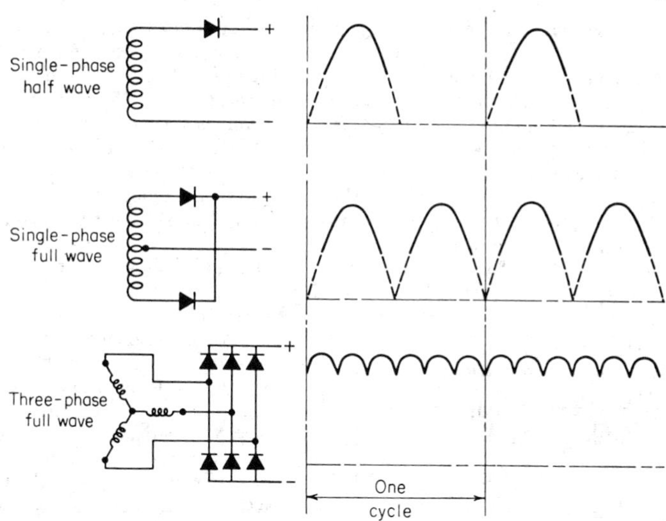

FIG. 24-17. Basic rectifier circuits and waveforms produced.

amount which increases with the degree of ripple. Thus a tong meter will read higher than the average, or correct, value from the standpoint of current input to the battery.

The relation between the average and rms values is known as the "form factor" and is equal to the rms divided by the average.

Cable Heating. The form factor also has an effect on the heating value of the current during charge in all conducting portions of the circuit including cables, fuses, and the battery itself. A pulsating current causes more heating in a conductor than a steady current of the same average value. The heating effect increases as the square of the form factor. This is why single-phase rectifiers require heavier cables and other current-carrying components than 3-phase ones of the same average current output.

51. Efficiency. For battery efficiencies see Pars. **13** and **35**. *Charger Efficiency.* There are certain energy losses, depending on equipment type. In motor-generators these are friction, windage, "copper" losses in the windings, and "iron" losses in the magnetic circuits. In rectifiers there are the copper losses in transformers, reactors, etc., losses in any ballast resistors, iron losses, and the rectification losses in diodes or rectifier stacks.

It is impractical to try to evaluate each of these in all the various chargers or charging systems. Also, the efficiency of any charger varies with the load upon it, usually being at a maximum at or near full-rated load and less at lighter loads. Any efficiency value, when specified, is therefore assumed to be at full load.

In most well-designed commercial equipment an efficiency of perhaps 65 to 75% can be expected, irrespective of its type, whether motor-generator or rectifier. Greater efficiency can be obtained at higher initial cost, but this does not necessarily represent a better overall value.

Overall Conversion Efficiency. It is sometimes desired to estimate the overall efficiency from the a-c input to the battery output, i.e., the actual cost of power to the apparatus which the system supplies. This can be done by combining the values for the charger and the battery. The result is about 40 to 45% for nickel batteries and 50 to 55% for lead types, and these values are frequently used as a rule of thumb for such purposes, on the basis, of course, of cycle charging.

At first thought this may seem like a low figure, representing a wasteful use of energy. However, when the actual cost is compared with the utility of the system as a whole, its real economy is apparent.

52. Charging Controls and Termination. *General.* Thus far, only the proper *charging currents* and the equipment to obtain them have been considered. An important function is the accurate termination of the charge when complete. (This applies, of course, only to cycle charging, as in floating and other similar methods the charge is not actually stopped, but only reduced to a low maintenance value.) In theory, a charge is complete when the battery voltage reaches a maximum value and/or the current falls to a minimum. In lead batteries there is also the fact of the specific gravity rising to a maximum. None of these, however, is an actual fixed value. They vary with changing current rates, temperature, and the age and condition of the battery. At present, no means has been commercially developed to sense a minimum or maximum value *as such,* although certain methods are under consideration. It is therefore necessary to use some other practical means of determining when the battery is fully charged, to stop the charge at that time.

The Voltage-Time Method. During charge, under given circumstances, the voltage of a battery rises in a well-established pattern. This includes a rise, or "hump," toward the latter part of the charge, the voltage at this point varying less with battery age and condition than the final charge voltage. Inasmuch as this rise occurs at approximately a certain state of charge, it indicates how much longer the charge should continue for completion. While this is evident in both battery types, it is more distinct (and its application more accurate) in lead batteries than in nickel types. These characteristics permit an automatic method of charge termination known as the "voltage-time" method. A voltage relay is adjusted to operate at the point where the distinct voltage rise occurs, and this starts a time switch which runs through a preset time interval and then stops the charge. Inasmuch as the voltage of a battery on charge is affected by its temperature, the voltage relay must be "compensated" to vary its operating point with ambient temperature changes, always to function at the same state of charge (see Fig. 24-18, which illustrates the temperature compensation of a typical relay used, when calibrated for either lead-acid or nickel-iron cells).

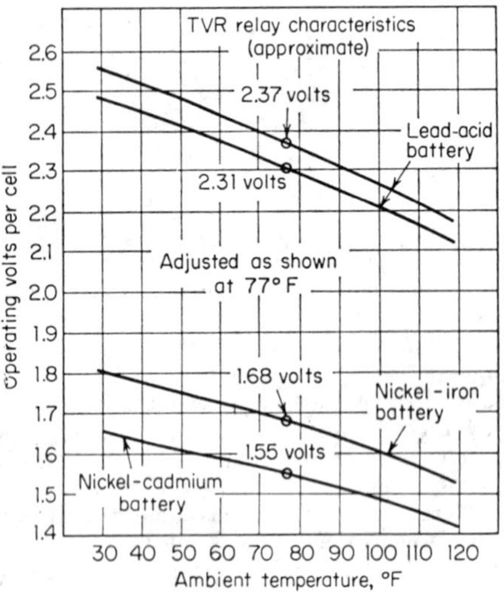

FIG. 24-18. Approximate voltage-temperature characteristics and typical settings of voltage relays for charging control.

While the precise adjustment of this equipment may vary with particular circumstances, the commonly used values in industrial truck service, based on 77°F, are these:

1. Lead-acid batteries—a relay setting of 2.37 V/cell, followed by a 3-h time period.
2. Nickel-iron batteries—a relay setting of 1.68 V/cell, followed by a 2½-h time period.

Inasmuch as the charging current usually bears a certain relation to the ampere-hour capacity of the cell, this capacity has no bearing on the above voltage or time.

The voltage-time method *inherently* senses the state of charge of the battery and thus provides accurate, automatic charge termination.

Most "relays" used at the present time are of the electromechanical type, designed for accurate operation ($\pm 1\%$) at the desired voltage and compensated for temperature as described above. Recently, however, static transistorized relays have been developed and used, and the proportion of this type will probably increase.

Ampere-hour Meter Control. An ampere-hour meter may be used to measure both the discharge and recharge. This meter is an integrating type with a single pointer, operating through a scale range at least equal to the capacity of the battery. On discharge the pointer moves clockwise, indicating the ampere-hours discharged. On charge, with the current flowing in the opposite direction, it moves counterclockwise, registering the amount of charge. When it returns to its starting point, it indicates that the battery is again in a full-charged condition.

The additional ampere-hours of charge required (see Pars. **12** and **34**) compared with those discharged are given by designing the meter to run slow, by the desired percentage, in the charge direction.

For the meter to operate as indicated above, it must be in the battery circuit on both discharge and charge, which is impractical in many types of service. Also, the meter cannot register, or compensate for, the internal losses in the battery and does not respond accurately to short-time current peaks. Because of these disadvantages and the fact that such meters require occasional recalibration and repair, their use has largely been superseded, in recent years, by the voltage-time method.

Other Methods. There are various manual or semiautomatic means of terminating a recharge, but these depend upon manual attention or estimates of the charge required rather than an accurate, automatic termination.

For example, the required charge for a lead battery may be estimated from a specific-gravity reading or that for a nickel-iron type from the indication of a "test fork." In either case the charge may then be stopped by a preset time switch or ampere-hour meter. Both these methods of estimating are, however, subject to certain errors or "corrections" which make them much less reliable than the voltage-time method, which inherently senses the state of charge.

53. "System-governed" Batteries. This term is frequently applied to applications where the battery and charger are components of a complete "system," such as those on automobiles, railway passenger cars, railway diesel engines, boats, etc. These fall into the partial-float or partial-cycle category, inasmuch as the charger can supply current only when the vehicle (or its engine) is operating above a certain speed. The complete system must be so designed, therefore, that the anticipated discharge, and any load required during operation, can be met by the generator output during the time it is available.

The charger in most of these systems is a d-c generator or an a-c "alternator," the latter having a built-in rectifier to change the alternating to direct current. Any such equipment, however, operates at varying speeds, and its output voltage must be controlled to adapt it to the comparatively constant voltage required by the system and battery. This is accomplished by a "regulator," which is fundamentally the same for all the above applications, although, of course, larger, more accurate, and more sophisticated in design on industrial equipment than on automobiles. Such a regulator normally consists of:

1. An automatic switch, or "cutout," which closes the circuit between the generator and the remainder of the system whenever the generator voltage is sufficiently high for charging and which opens the circuit when its voltage falls below a useful value.

2. A voltage regulator that maintains a constant voltage at the generator terminals whenever the generator speed is sufficient to power the existing load. Whenever this portion of the regulator is functioning, the charger is said to be "on voltage regulation."

3. A current regulator that has a "series" winding which carries the full output of the charger. It protects the generator against overload. Whenever the current tends to exceed its rated value, it reduces the field strength and keeps it within safe limits. When this item is functioning, the regulator is said to be "on current regulation." Figure 24-19 illustrates the results of this operation during a recharge.

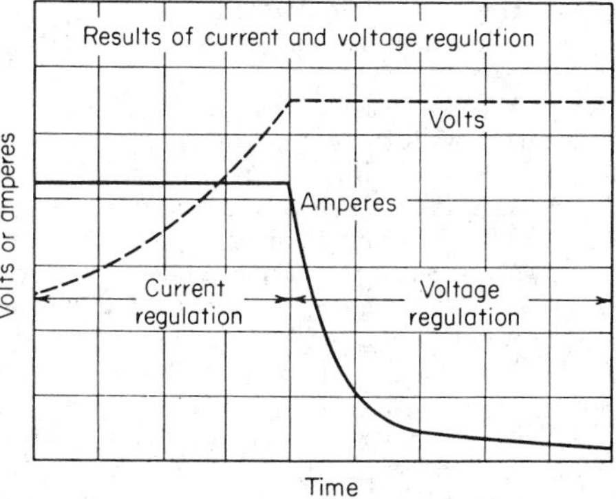

Fig. 24-19. Effects of current and voltage regulation during a recharge.

The actual voltage values will depend on the particular circumstances involved, but these are representative:

Type of Service	Regulator Voltage, V/Cell
●Lead-acid batteries:	
Railway diesel locomotives	2.30
Railway passenger cars	2.37
Automobiles and small boats	2.40
●Nickel-ircn batteries:	
Railway passenger cars	1.65

54. Sequence Control. This is an accessory available for many single-circuit cycle chargers. It consists of additional controls (relays, etc.) which permit it automatically to charge two batteries in sequence. Each battery is initially connected to a separate charging cable. One is charged first in the usual manner, and, when complete, the charging circuit is transferred to the second one. This permits the charger to do double duty without any manual attention, assuming, of course, that sufficient time is available for both batteries to receive their charge.

These sequence controls have no effect on the normal characteristics of the charger. They merely disconnect the first battery and connect the second at the appropriate time.

SEALED ZINC-MERCURY CELLS

By Samuel Ruben

55. General. The type of battery suitable for a given application is no longer a matter of cell geometry, but one of an electrochemical system that will operate most effectively. The size of the cell is related more to the total capacity or energy developed for the load range, and not as limited in respect to the rate of discharge. This was highlighted in 1941 by the large-scale use of transceivers and miniaturized equipment where more exacting requirements were encountered. There were three requirements: maximum electrical capacity per unit volume, maximum storage life under varying ambient temperature, and as constant a discharge voltage as possible.

In 1941, the only commercially available dry-cell system for portable communication equipment was the Leclanché cell, commonly known as the zinc-carbon cell. The development of the sealed zinc–mercuric oxide cell[1*] provided a dry-battery system that could meet the requirements unfilled by the Leclanché system. A number of fundamental factors were found necessary to produce a sealed zinc–mercuric oxide alkaline cell and overcome the limitations met with in the early attempts to use HgO as a cathode element:

a. High-density cathode.

b. Anodes having an adequate ratio of effective electrode areas to cathode coulombic capacity.

c. Alkali-resistant and -absorptive barrier spacer of a submicroporous material in contact with the cathode.

d. A compressible electrolyte-retaining spacer.

e. An alkaline electrolyte containing a saturated zincate-ion component.

f. Quantitative balance between anode and cathode coulombic capacities.

g. An anode contactor of nonelectrolytically consumable but amalgamatable metal insulated from the cell container by an elastomeric plastic grommet.

h. A cell container composed of a passive metal in contact with the cathode.

i. A sealed structure provided by crimping the container against the insulated anode contactor.

The combination of these factors provides a cell having internal chemical stability within an airtight seal, with freedom from atmospheric effects, pressures, or humidity.

The characteristics of the mercury cell are high capacity per unit volume of 40 to 50 Wh/lb compared with 10 to 20 Wh/lb for the Leclanché zinc-carbon system, a flat voltage discharge at a relatively constant impedance allowing maximum performance up to the exhaustion of its capacity, and ability to withstand temperature variations and high gravitational forces. These characteristics have made it desirable for hearing

---

* Superior numbers correspond with the numbered references in Par. **66**.

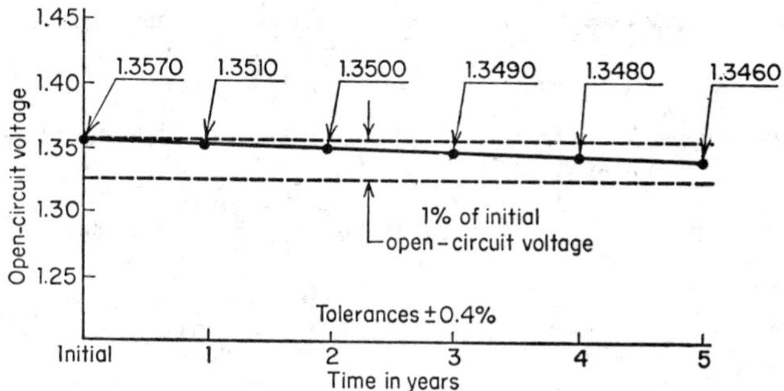

Fig. 24-20. Open-circuit voltage of RM 12R mercury cells vs. years of storage at 70°F (21.11°C).

aids, medical electronics heart-beat stimulators as pacemaker, space and aviation, blood-testing equipment, electronic thermometers, missiles and satellites, ballistic telemetry, high-altitude balloons, sonobuoys and marine devices, small-craft fathometers and tachometers, undersea communications, instruments and computers, radiation detectors, oil-well survey instruments, laboratory meters, pH meters, reference voltage source, air-sea rescue units, radio beacons, electronic wrist watches, and clocks.

Since the Zn-HgO cell does not depend upon cathode activity coefficients controlled by crystal structures, impurities, stoichiometric variations, or the physicochemical constitution of its discharge products, it can provide reproducible potentials within ±0.001 V between cells. The chart of Fig. 24-20 illustrates the change in cell potentials over a 5-year storage period at 21.1°C of 1.357 to 1.346 V, with a useful stability up to 10 years, depending upon the composition of the microporous spacer.

56. Reactions at the Electrodes. The electrochemical reaction of the system is as shown in Fig. 24-21.

Material: Zn/KOH, ZnO, H_2O/HgO
After solution: Zn/KOH, K_2ZnO_2, H_2O/HgO
Ionization product of electrolyte: $K^+ + OH^-$, $2K^+ + ZnO_2^{--}$, $H^+ + OH^-$

	Anode	Cathode
Reactions in producing electricity	$Zn - 2e^- \rightarrow Zn^{++}$ $Zn^{++} + 2OH^- \rightarrow Zn(OH)_2$	$HgO + H_2O \rightarrow Hg(OH)_2 \rightarrow Hg^{++} + 2OH^-$ $Hg^{++} + 2e^- \rightarrow Hg\ 2OH^- + 2H^+ \rightarrow 2H_2O$
Since electrolyte is saturated with ZnO	$Zn(OH)_2 \rightarrow ZnO + H_2O$	
Electrode end products	ZnO	Hg

Fig. 24-21. Theory of zinc–mercuric oxide system. Since the basic electrode reactions are the oxidation of the $Zn - 2e^- \rightarrow Zn^{++}$ and the reduction of the $Hg^{++} + 2e^-$ at the cathode, and since water appears at both electrodes, there is no significant change in KOH or H_2O concentration. The overall chemical reaction for producing 2 faradays/g mole of anodic zinc and cathodic mercuric oxide is $Zn + HgO \rightarrow ZnO + Hg$.

It has been reported[2] that the discharge reaction of the mercuric oxide electrode may be less direct than indicated, such as an electrochemical reduction through an intermediate product $(Hg_2O)_n$ or $(HgOH)_n$, and decomposed by a disproportionation reaction to HgO + Hg. Chemical analysis and x-ray diffraction patterns indicate a deposition in crystalline form of zinc oxide (zincite) on anode and Hg on cathode as end products.

57. Electrode Potentials. The half-cell potential in reference to the hydrogen electrode, expressed by the Nernst equation, is

$$E = E^0 - (RT/nF) \ln Q$$

where Q is the product of the activities of the resulting substances divided by the activities of the reacting substances, each raised to that power whose exponent is the coefficient of the substances in the chemical equation. Applying this:

Zn Anode:

$$E_{Zn;Zn^{++}} = E^0{}_{Zn} - \frac{RT}{nF} \ln a_{Zn^{++}}$$

Since activity product $(a_{Zn^{++}})(a_{OH^-})^2$ is 4.5×10^{-17}, OH^- molality is 7.7 with activity coefficient of 3.75[3] and $n = 2$. At 25°C and $\log_{10}$

$$E_{Zn;Zn^{++}} = 0.76 - 0.0296 \log \frac{4.5 \times 10^{-17}}{(28.9)^2} = 1.332 \text{ V}$$

HgO Cathode:

$$E^0{}_B = Hg + 2OH^- \rightarrow HgO + H_2O + 2e^- = -0.0984$$

$$E_{HgO;OH^-} = E^0{}_B + \frac{RT}{nF} \log a^2{}_{OH^-} \cdot E_{HgO;OH^-}$$

Activity coefficient HgO/OH = 1, $n = 2$.

$$E_{HgO;OH^-} = E^0{}_B + \frac{RT}{F} \log a_{OH^-}$$

$$= -0.0984 + 0.592 \log 28.9$$

$$= -0.012 \text{ V}$$

$$E_{cell} = E_{Zn;Zn^{++}} - E_{HgO;OH^-}$$

$$= 1.332 - (-0.012)$$

$$= 1.344 \text{ V}$$

which is in close agreement with commercial cell potentials.

58. Balanced Cells. The basic principle of the balanced cell is a requirement for sealed alkaline cells. This necessitates that the coulometric capacity of the mercuric oxide cathode and the zinc anode be equal so that on completion of the cell's discharge there is no residual unoxidized zinc. This is accomplished by a nonconsumable anode contactor which is the top closure for the hermetically sealed cell. The cell top is made of amalgamatable metal such as copper or tinned steel, to avoid local couple action, the latter being used commercially except where nonmagnetic materials are required, in which case copper is preferred.

Without this balanced relation between the coulometric capacities of the electrodes gas would be generated at the cathode when a discharge is continued after it is exhausted. Since the reversible deposition potential of hydrogen in the electrolyte is

$$E_{H_2}H^+ = 0.000 - 0.052 \log \frac{10^{-4}}{7.7 \times 2} = 0.898 \text{ V}$$

and the hydrogen overvoltage of the steel container is 0.15 V, the sum of both is less than the zinc potential of 1.332 V. Therefore hydrogen gas will be generated as long as there is zinc available in unbalanced cells. An external potential of 0.28 V is observable, which is the differential potential derived from $E_{Zn;Zn^{++}} - E_{H_2;H^+}$. The balanced-cell principle has been applied to other sealed alkaline cells, silver and manganese oxide cathodes being used. For balance, 1.295 g Zn = 1 Ah = 4.07 g HgO.

59. Cell Structures. The sealed zinc-mercuric oxide alkaline cell is manufactured in three basic structures:

a. The wound-anode type.

b. The flat-pressed powdered cathode and anode type.

c. The cylindrical-pressed powdered electrode type.

These three types of structures are shown, respectively, in Figs. 24-22, 24-23, and 24-24. The wound-anode type (Fig. 24-22) has its anode composed of a corrugated zinc strip with a paper absorbent wound in an offset manner so that it protrudes at one end and the zinc protrudes at the other end. The zinc is amalgamated with mercury (10%), and the paper impregnated with the electrolyte, which causes it to swell and produce a positive contact pressure. The other necessary details of construction are apparent in Fig. 24-22. In the pressed powder cells (Figs. 24-23 and 24-24), the zinc powder is preamalgamated prior to pressing into shape, and its porosity allows electrolyte impregnation with oxidation in depth when current is discharged. A double can structure is used in the larger-sized cells, the space between the inner and outer containers providing passage for any gas generated in a cell not properly balanced in relation to the stoichiometric requirements of anode to cathode weights, or the presence of impurities in the anode material. The construction is such that if excessive gas pressures are experienced the compression of the upper part of the grommet by internal pressure allows the gas to escape into the space between the two cans. A paper tube surrounds the inner can so that any liquid carried by discharging gas will be absorbed, maintaining a leak-resistant structure. Release of excessive gas pressure automatically reseals the cell. Adjacent to Figs. 24-22, 24-23, and 24-24 are the mechanical and electrical specifications of representative cells embodying the respective structures. The type designation RM* is used in the industry to identify Zn-HgO cells.

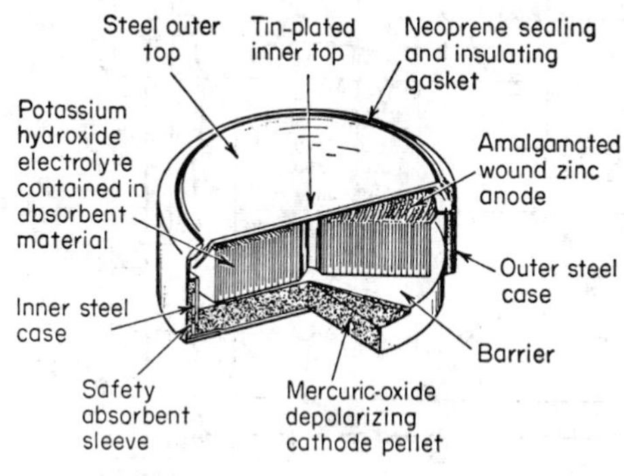

Type No.	Max. diam., cm	Height, cm	Weight, g	Rated capacity, mAh
RM 640	1.587	0.965	9.68	360
RM 3	2.498	1.37	22.56	1,500
RM 1438	3.71	1.003	36.22	2,700
RM 1450	3.71	1.36	51.80	4,500
RM 2550	6.58	1.394	165.20	13,000

FIG. 24-22. Relation of geometry and weight of cell to capacity, wound-anode cells.

Type No.	Max. diam., cm	Height, cm	Weight, g	Rated capacity, mAh
RM 312	0.87	0.34	0.56	36
RM 575	1.143	0.33	1.4	100
RM 675	1.143	0.54	2.24	160
RM 630	1.549	0.58	4.76	350
RM 640	1.574	1.104	7.84	500
RM 4R	3.02	1.66	40.88	3,400

FIG. 24-23. Relation of geometry and weight of cell to capacity, flat-pellet structure.

Type No.	Max. diam., cm	Height, cm	Weight, g	Rated capacity, mAh
RM 24	1.0	4.396	14.0	900
RM 601	1.59	2.857	34.16	1,800
RM 3R	2.489	1.651	28.56	2,200
RM 502	1.358	4.90	30.44	2,400
RM 401	1.133	2.844	11.20	800
RM 1R	1.579	1.638	12.04	1,000
RM 12R	1.519	4.959	30.88	3,600
RM 42R	2.922	6.032	148.33	14,000

FIG. 24-24. Relation of geometry and weight of cell to capacity, cylindrical structure.

* RM is a contraction of the names Ruben and Mallory, the inventor and the manufacturer, respectively.

60. Cell Discharge Characteristics. The general characteristic of the Zn-HgO system is noted in Fig. 24-25. This figure illustrates an equilibrium period T_2 representing practically 97% of the cell life when the current drain is relatively low; e.g., for an RM 1 cell at 1 mA drain, voltage is constant (at 98% of no-load potential) within less than ½% during this period. At higher level drains such as 20 mA, the potential will vary by ±2% from 90% of no-load level, and the equilibrium period is approximately 67% of cell life. Larger cells, or groups of smaller cells in parallel, would show longer equilibrium-period characteristics for equivalent rates of drain.

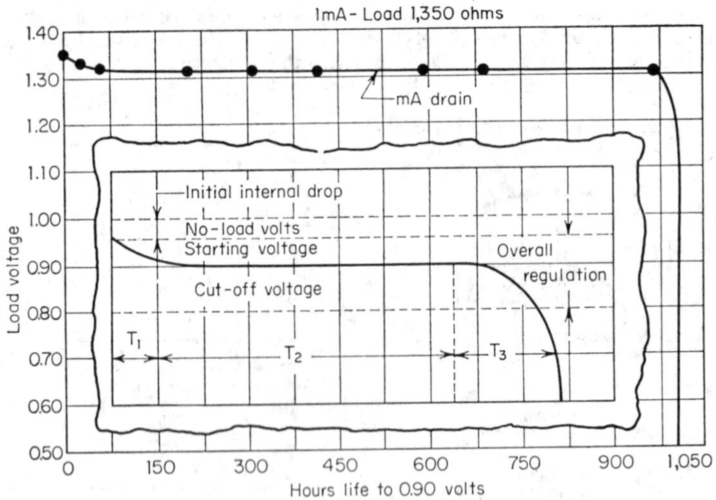

FIG. 24-25. Load voltage vs. hours of life for RM 1 cells. General characteristic of RM-cell system:
T_1, initial polarization time. T_2, equilibrium period—best regulations. T_3, end of life—decay period.

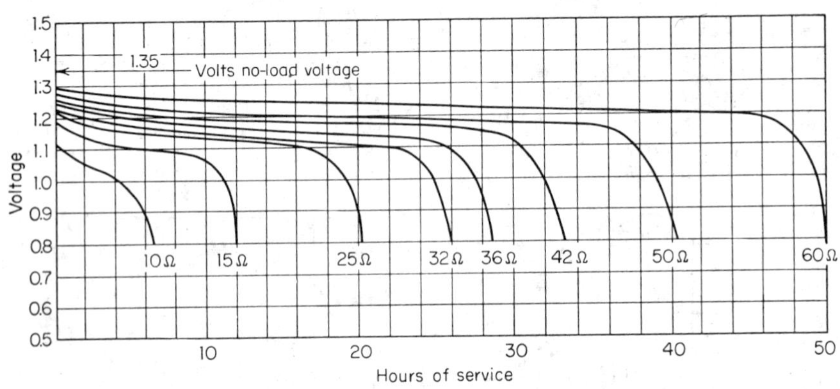

FIG. 24-26. Voltage discharge curves for RM 1R cell. Resistance values and equivalent current drains at 1.25 V:

Ohms	Milliamperes	Ohms	Milliamperes
10	125	36	35
15	83	42	30
25	50	50	25
32	40	60	20

Since the current density per unit of electrode area determines the equilibrium period, in the proper cell design the voltage may be as constant as desired (for any predetermined equilibrium period) to meet the demands of any particular circuit. The electrical and mechanical specifications for the RM 1 cell are 1.580 cm diameter, 1.638 cm height, 1.960 cm³ volume, 12.191 g weight, and 1,000 mAh capacity. The voltage discharge curves for the RM 1 cell at various loads vs. time are shown in Fig. 24-26.

61. Internal Resistance of the Zn-HgO Cell during Discharge. The maintenance of low internal resistance or impedance is an important characteristic of the mercuric oxide cell. The factors which are responsible for this desirable characteristic are:

a. Cathode mix of HgO plus graphite compressed at high pressures such as 20,000 lb/in².

b. Direct reduction of the semiconductor HgO to the metal Hg, maintaining a low cathode resistance.

c. Anodes designed with enough effective area to allow complete oxidation in depth.

d. Electrolyte composition and conductivity remaining relatively constant with use of the cell.

e. Low heat of formation of HgO.

f. Rapid ion diffusion at electrodes, which does not impose a rigid limit of discharge current density.

This last factor is one of the important ones in the mercuric oxide cell. The ions immediately adjacent to the cathode must be discharged for current to flow through the circuit. When the electrons are produced by ionization of the zinc anode ($Zn \rightarrow Zn^{++} + 2e^-$) and discharged to the cathode ($2e^- + Hg^{++} \rightarrow Hg^0$) through the circuit, direct reduction takes place. Since the number of electrons in 1 C is 6.28×10^{18}, a rapid reduction without an appreciable ionic gradient is necessary for maintaining the cell potential.

The relation of cell impedance to discharge time is shown in Fig. 24-27, and while it refers to an RM 625 cell, the specifications of which are 1,549 cm diameter, 0.592 cm height, 1.115 cm³ volume, 3.969 g weight, and 350 mAh capacity, it is the characteristic discharge form of all Zn-HgO cells. The rapid rise in a-c resistance at the end of discharge is indicative of the complete oxidation of the Zn anode. Maximum allowable current density in an electrolytic discharge is related to the expression for the limiting current density,

$$i = \frac{DnF}{\delta t} c \times 10^{-3} \text{ A/cm}^2$$

where D is the diffusion coefficient for the ion being reduced, n and F have their usual significance, δ is the depth of the ionic gradient of the electrolyte next to the electrode surface (0.05 cm), t is the transport number of the ions in solution, and c is the concentration of the diffusing ion in moles per liter.

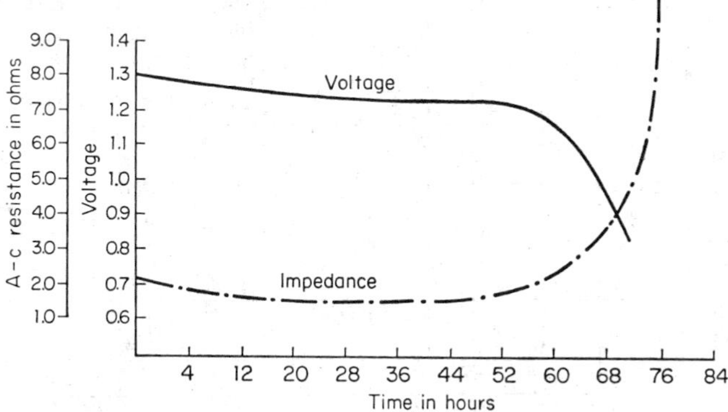

Fig. 24-27. Cell impedance vs. electrical discharge. This resistance is taken at the indicated drain. This curve is representative over the range of current from 1 to 20 mA. Internal a-c resistance at 1 kc, 70°F (21.11°C). RM 625 under 250-Ω load (5 mA).

While there is no indication of a critical limiting current density in Zn-HgO cells, it has been found that current densities below 1.5 A/cm² of cathode surface are conducive to obtaining maximum cell efficiency and high terminal voltages, since current densities of such magnitude do not deplete the concentration of HgO at the effective cathode surface. The value of 1.5 A/cm² was observed with cells designed for general use; however, this value increases with electrodes having higher porosities.

The growth of the deposited ZnO on the anode is responsible for a decreased rate of ion discharge at low temperatures. The relation of discharge characteristics of cells having anode areas designed specifically for low temperatures is shown in Fig. 24-28.

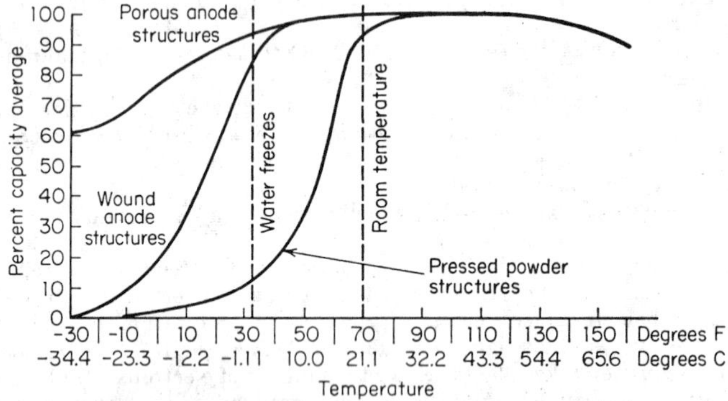

FIG. 24-28. Extended performance of mercury batteries at low temperatures.

62. Mercury Voltage Reference Cell. The unusually fixed and reproducible potential of the Zn-HgO cell has led to its development as a voltage reference battery. As shown in Fig. 24-29, this type of voltage reference is stable over a wide range of temperature (−50 to +50°C) and has the advantage of being able to withstand accidental overloads which may be experienced in a measurements laboratory.

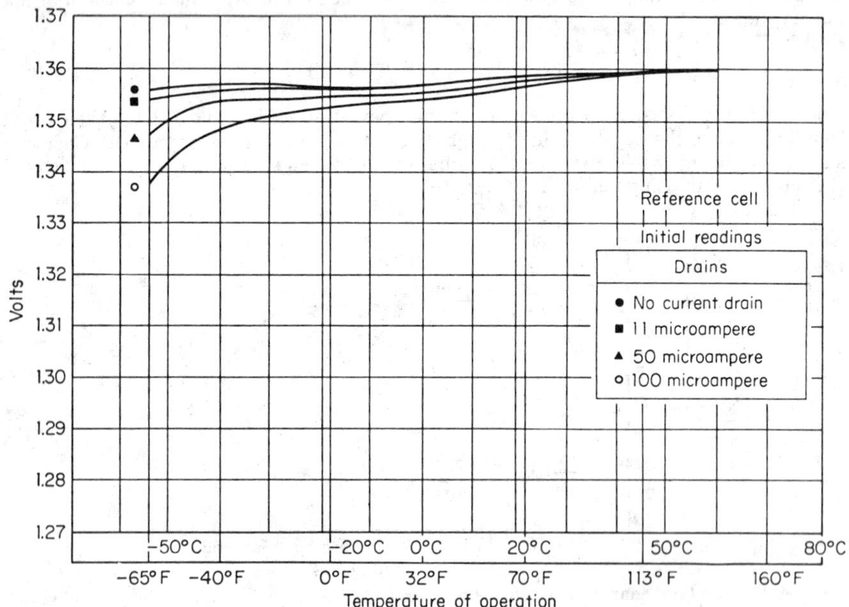

FIG. 24-29. Voltage stability of mercury reference cell.

63. Rechargeable HgO Cells.[5] The problem of mercury conservation has necessitated the development of a rechargeable Zn-HgO cell. Since the electrochemical reactions are completely reversible, $Zn + HgO \leftrightarrows ZnO + Hg$, new requirements were introduced to allow electrodes and separators to remain in an operable condition during cycling. The cathode requirements were met when silver powder was substituted for the graphite content in the semiconductive mercuric oxide cathode. This provided means for immobilizing in place the Hg formed during discharge, forming an amalgam and preventing the formation of large mobile mercury globules. It maintains wide distribution of, and continuous contact to, the reduced mercury. This function is twofold. The first is to allow reconversion in depth of the mercury to mercuric oxide. The second is that the silver allows a readily controllable means for avoiding overcharging. The recharging potential of 1.35 V is maintained until all the cathode's Hg component is oxidized.

If the charge is continued beyond this point, the silver component becomes oxidized to Ag_2O, causing a rapid rise in potential from 1.35 to 1.6 V. If continued even further, the potential would rise to 1.82 V as the Ag_2O is converted to AgO. If charging were to continue even further, oxygen gas would be produced. The differential potential (0.25 V) allows a relay or semiconductive device to disconnect the charger, thereby preventing the formation of colloidal silver or cathode gas. The cyclic performance of a rechargeable Zn-HgO cell is related to the conductivity of the cathode barrier or spacer, which must be a more oxidation-resistant microporous material than that applied to primary cells and must have filter action for colloidal silver.

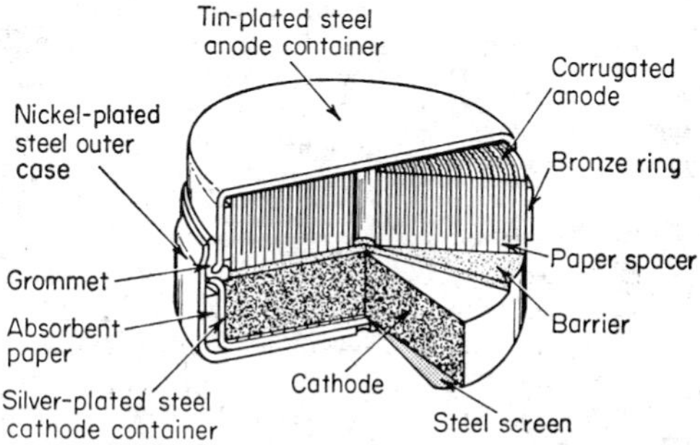

Laboratory experience has shown that a ceramic barrier produced by the compression and sintering of a disk of magnesium oxide gives the greatest number of cycles without change in cell characteristics. The structure of the rechargeable cell anode differs from that of the primary cell, as shown in Fig. 24-30. It provides a continuous contact to the zinc oxide formed at the anode so as to allow reduction to zinc or deposition of plated zinc at the anode surface. In the wound-anode structure the anode consists of a bronze strip electroplated with an adequate amount of zinc to provide a balanced cell. When the pressed-powder anode-type structure is used, silver powder is added to the zinc amalgam. The silver powder amalgamates and serves as a continuous contacting network for contact to the zinc oxide produced at the anode. The mechanical and electrical specifications for several of the rechargeable Zn-HgO cells are shown in Table 24-1.

Fig. 24-30. Structure of rechargeable mercury cell.

Table 24-1. Specifications for Rechargeable Cells

Type	Diameter, cm	Height, cm	Weight, cm	Capacity, mAh	Drain, mA
625S	1.549	0.591	4.44	70	10
640S	1.574	0.977	7.84	120	15
1420S	3.715	0.591	25.48	500	250

64. Zinc Mercuric Dioxysulfate Cell.[6] Another system has been developed for low current requirements which has some of the desirable flat voltage discharge characteristics of the zinc mercuric oxide alkaline cell, while allowing the use of the less expensive conventional dry-cell construction. In the past, attempts have been made to use mercury compounds in the cathode of a nonalkaline dry cell, but these have failed because of the inherent limitations of the materials used. In the new cell, which might be termed the mercury-zinc-carbon cell, the cathode reactant is the basic mercuric sulfate, or mercuric dioxysulfate, and the electrolyte is a zinc sulfate solution in which the cathode is stable. The anode is amalgamated zinc. The electrochemical system in the presence of an aqueous solution of zinc sulfate can be expressed as $Zn/ZnSO_4/HgSO_4 \cdot 2HgO + C$, with an overall reaction on discharge as $3Zn + HgSO_4 \cdot 2HgO \rightarrow ZnSO_4 + 2ZnO + 3Hg$. The potential of the cell is 1.36 V. The theoretical capacity of $HgSO_4 \cdot 2HgO$ is 0.2204 Ah/g, and the practical capacity of the carbon mix is 0.15 Ah/g.

65. Cell Structures. While this system can be used in a standard dry-cell structure, one of the important characteristics of the zinc–mercuric dioxysulfate electrochemical system is that it allows the use of stainless-steel containers for contact to the cathode. This has permitted production of thin wafer-type cells shown in Fig. 24-31. The case is a shallow stainless-steel cup and the cathode a pressed disk of mercuric dioxysulfate, Shawinigan carbon, and zinc sulfate solution containing 1% potassium dichromate as an inhibitor, in contact with the bottom of stainless steel. The spacer is a laminate of paper and cellophane which separates the cathode from the amalgamated zinc disk anode. The anode has a polyethylene grommet around its edge, insulating it from the cathode container, which is crimped against it for sealing of the cell. The electrical and dimensional specifications of the WD5 wafer-type cell are 2.54 cm diameter, 0.272 cm height, 1.376 cm^3 volume, and 4.9 g weight. Initial flash currents are on the order of 0.5 A, and capacity is 230 mAh, with a 1,000-c impedance of 10 Ω.

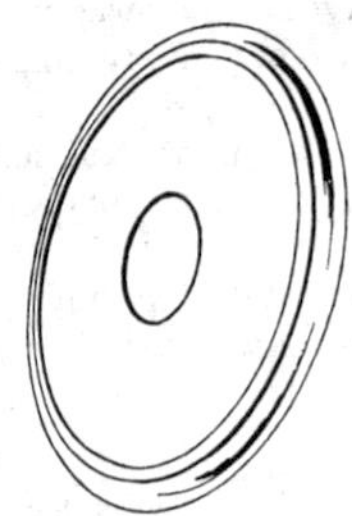

Fig. 24-31. Zinc mercuric dioxysulfate cell.

66. References

1. Friedman, M., and McCauley, C. E. The Ruben Cell, A New Alkaline Primary Dry Cell Battery; *Trans. Electrochem. Soc.*, 1947, Vol. 92, pp. 195–215. Vinal, G. W. "Primary Batteries," New York, John Wiley & Sons, Inc., 1950, pp. 304–321. Potter, E. C. "Electro-chemistry"; New York, The Macmillan Company, 1956, pp. 375–377. Garrett, A. G. "Batteries of Today"; Dayton, Ohio, Research Press, 1957, pp. 84–87. Milazzo, G. "Electro-chemistry"; Amsterdam, Elsevier Publishing Company, 1964.

2. Yoshizawa, Shiro, and Takihara, Zenichiro *J. Electrochem. Soc. Japan*, 1962, Vol. 30, No. E93; 1966, Vol. 30, pp. 93–99.

3. Harned, H.S., and Owen, B.B. "Physical Chemistry of Electrolytic Solutions"; New York, Reinhold Publishing Corporation, 1958, p. 730. Latimer, W. M. "Oxidation Potentials," 2d ed.; New York, Prentice-Hall, Inc., 1953.

4. Ruben, S. Balanced Alkaline Dry Cells; *Trans. Electrochem. Soc.*, 1947, Vol. 92, pp. 183–193.

5. Hampel, C. A. "Encyclopedia of Electrochemistry"; New York, Reinhold Publishing Corporation, 1964, pp. 74–75, 83–84.

6. Ruben, S. Zinc Mercuric Dioxysulfate Dry Cell; *J. Electrochem. Soc.*, 1959, Vol. 2, pp. 77–78.

SILVER CHLORIDE–MAGNESIUM AND CUPROUS CHLORIDE–MAGNESIUM BATTERIES

By J. J. Coleman

67. Operating Characteristics. The operational characteristics of commercial silver chloride–magnesium batteries and cuprous chloride–magnesium batteries differ from those of conventional batteries (carbon-zinc and lead-acid) in several important respects:

a. They can be efficiently discharged completely in 1 h or less; i.e., they operate efficiently at high rates.

b. After electrolyte has been added, rapid deterioration sets in and most of these batteries are exhausted by self-discharge in a period of less than 10 days and some in less than 1 day.

c. As a result of (*b*), these batteries are sold dry, and electrolyte is added a short time before battery discharge is initiated; thus, batteries free of electrolyte and protected from high humidities can be stored for years with little or no deterioration.

Because of (*b*), these batteries are called "one-shot"; because of (*c*), "reserve" batteries.

A comparison of silver chloride batteries with cuprous chloride batteries can be briefly put as follows:

a. Silver chloride batteries permit efficient operation at higher rates.

b. In the electrolyte-free condition they need no special packaging to resist high-humidity storage (90 days at 90°F, 90% RH is possible).

c. They are more expensive.

d. They are manufactured in a larger range of sizes. Both appear in fractional kilowatthour sizes, and silver chloride batteries to supply vehicular power for torpedoes may be as large as 60 kWh.

e. Silver chloride batteries require saline water for activation, while cuprous batteries normally contain some sodium chloride so that fresh water can be used.

68. Constructional Characteristics. *a. Cuprous Chloride–Magnesium Batteries.* Cathodes are made by compacting CuCl and graphite plus some plastic binder such as polystyrene. NaCl may be included to give additional conductivity to the electrolyte. Other agents may be added for special purpose. The anode is thin magnesium sheet. The separator is some soft absorbent material that will soak up and hold the needed electrolyte.

b. Silver Chloride–Magnesium Batteries. Cathodes are made from sheet silver chloride—a malleable material which is easily rolled into thin strips. The surface is usually reduced to a porous conductive film of metallic silver. The anodes are thin magnesium sheets. Many of these batteries are dropped into the sea, and provision is made for rapid entry of electrolyte (sea water). Thus the separators used are variously plastic filaments, small "dots" of plastic film, glass beads, and the like, to provide an open structure between electrodes.

69. Typical Uses

Silver Chloride–Magnesium

Application field	Device	Function in device
Air-sea warfare Underwater communication	Sonobuoys	Power electronic gear Fire squibs
Air-sea warfare and undersea warfare	Torpedoes	Primary vehicular power Power for heat-engine ignition Guidance gear
Antisubmarine warfare	Bathythermograph	Power electronic gear
Meteorology	Radiosonde Meteorological balloon	Power electronic gear Power incandescent lamp
Rescue	Rescue beacons Life rafts Submarine marker beacon Pyrotechnic marker	Power electronic gear Power location lights Power electronic gear Power ignition

Cuprous Chloride–Magnesium

Application field	Device	Function in device
Meteorology	Radiosonde Meteorological balloon	Power electronic gear Power incandescent lamp

70. Performance Characteristics. Batteries employing different electrochemical systems can be compared on the basis of their capacity in watthours per pound or watthours per cubic inch at various rates. This comparison is crude because these factors for any particular battery depend on the style of terminals and containers employed and on the space required within the battery for components and assemblies sufficiently rugged for the service required. The following figures are applicable to commercial silver chloride–magnesium batteries:

Wh/lb	Wh/in³	Rate, h
25	1.1	½
30	1.3	¾
50	2.5	3
56	2.8	8

Cuprous chloride–magnesium batteries would show one-half to three-fourths the capacity of a corresponding silver chloride battery.

71. References

AgCl-Mg

Vinal, G. W. "Primary Batteries"; New York, John Wiley & Sons, Inc., 1950, pp. 274–281.

Kirk, R. E., and Othmer, D. F. "Encyclopedia of Chemical Technology"; New York, McGraw-Hill Book Company, 1960, Vol. 3, pp. 579–580.

Condon, E. U., and Odishaw, H. "Handbook of Physics," 1st ed.; New York, McGraw-Hill Book Company, 1958, pp. 4–155 to 4–158.

Hampel, C. A. "The Encyclopedia of Electrochemistry"; New York, Reinhold Publishing Corporation, 1964, pp. 76–77.

CuCl-Mg

Condon, E. U., and Odishaw, H. "Handbook of Physics," 1st ed.; New York, McGraw-Hill Book Company, 1958, pp. 4–155 to 4–158.

Hampel, C. A. "The Encyclopedia of Electrochemistry"; New York, Reinhold Publishing Corporation, 1964, pp. 76–77.

SILVER OXIDE–CADMIUM BATTERIES

By Paul L. Howard

72. General. The silver oxide–cadmium secondary battery is the newest system giving a higher energy density than either the lead-acid or nickel-cadmium batteries. The cell in the charged condition consists of silver oxide positives and cadmium negatives separated by a semipermeable membrane with a potassium hydroxide solution as the electrolyte. The simplified chemical reaction is

$$AgO + Cd \underset{\text{charge}}{\overset{\text{discharge}}{\rightleftarrows}} Ag + CdO$$

This battery has found use in various applications where high energy density and long cycle life are required. Depending on the rate of discharge the system gives 20 to 30 Wh/lb. Recharging efficiency ranges from 85 to 95% based on rates of recharge; i.e., the faster the charge, the lower the efficiency.

There are two types of cells in use, i.e., vented and sealed. The sealed types are designed to withstand some overcharge and give maximum oxygen recombination. The vented type requires some maintenance. For best service the vented type is stored dry and filled and formed in accordance with manufacturer's instructions at the time of use. The sealed types must be filled, formed, and sealed at the factory; so they are shipped in the wet discharged condition. For best life, delivery should be scheduled to coincide with use.

Cells of various capacities of both types are manufactured by Yardney Electric Corporation, Gulton Industries, Inc., and Electrochimica Corporation. These may be

Table 24-2. General Characteristics of Silver Oxide–Cadmium Batteries

Type cell	Nom. Ah		Cell dimensions, in			Vol., in³	Weight, oz	
	Volts	Cap.	H	W	L		Vented	Sealed
YS-01	1.4	0.1	1.38	0.63	0.22	0.19	0.18	0.19
YS-05	1.4	0.5	1.55	1.08	0.54	0.90	0.75	0.78
YS-1	1.4	1.0	2.02	1.08	0.54	1.18	1.20	1.3
YS-2	1.4	2.0	2.53	1.72	0.59	2.57	2.30	2.5
YS-3	1.4	3.0	2.86	1.72	0.59	2.92	3.20	3.4
YS-5	1.4	5.0	2.91	2.08	0.79	4.78	5.00	5.3
YS-10	1.4	10.0	4.81	2.32	0.74	8.25	9.20	9.7
YS-18	1.4	18.0	7.00	2.13	0.81	12.10	13.00	13.9
YS-20	1.4	20.0	4.28	1.73	2.05	15.2	15.10	16.1
YS-40	1.4	40.0	7.05	3.25	0.99	22.7	26.30	27.4
YS-60	1.4	60.0	4.50	2.73	2.36	29.0	42.50	44.1
YS-70	1.4	70.0	6.25	3.64	1.41	32.1	42.00	44.1
YS-100	1.4	100.0	4.81	2.78	3.44	46.0	53.00	55.3
YS-300	1.4	300.0	17.5	4.19	1.78	131.0	183.00	193.0
BD3/4	3.3	4.0	3.75	1.87	1.47	8.1	9.6	
BD3/2	4.0	2.0	2.63	1.87	1.47	5.7	5.6	
AG 4.6	1.4	4.6	2.6	2.10	0.84	4.58	5.3	7.4
AG 9.0	1.4	9.0	3.87	2.1	0.84	6.84	8.8	11.5
AG 15.0	1.4	15.0	4.35	2.31	1.06	10.0	14.1	16.6
AG 22.0	1.4	22.0	5.75	2.31	1.06	14.0	17.6	23.0
AG 39.0	1.4	39.0	6.75	3.17	1.07	22.9	31.8	38.4
AG 69.0	1.4	69.0	10.17	3.16	1.07	34.4	52.8	62.0
SCR-2	1.4	2.0	2.63	1.58	0.64	2.7	2.0	
SCR-5	1.4	5.0	3.00	1.76	0.71	3.8	4.9	
SCR-10	1.4	10.0	3.94	2.20	0.89	7.7	9.3	
SCR-20	1.4	20.0	4.90	2.81	1.13	15.6	19.4	
SCR-45	1.4	45.0	6.38	3.65	1.47	34.2	43.5	
SCR-150	1.4	150.0	7.47	5.43	1.63	66.1	99.0	
S-250	1.1	0.25	0.225	0.985 Dia.	....	0.18	......	0.35

YS manufactured by Yardney Electric Corporation.
AG manufactured by Gulton Industries.
SCR manufactured by Electrochimica Corporation.

assembled into batteries of various voltages and capacities for the particular end use. The general characteristics of these cells are shown in Table 24-2. Typical discharge curves for a range of rates are given in Fig. 24-32 and charge voltage in Fig. 24-33.

Charging may be by either the modified-constant-potential method, where the maximum initial surge current is limited to about two times the rated capacity of the cell, or constant current with a voltage cutoff. The constant potential or voltage cutoff for the vented type is about 1.6 times the number of cells and for the sealed between 1.5 and 1.6 times the number of cells as specified by the manufacturer. As seen in

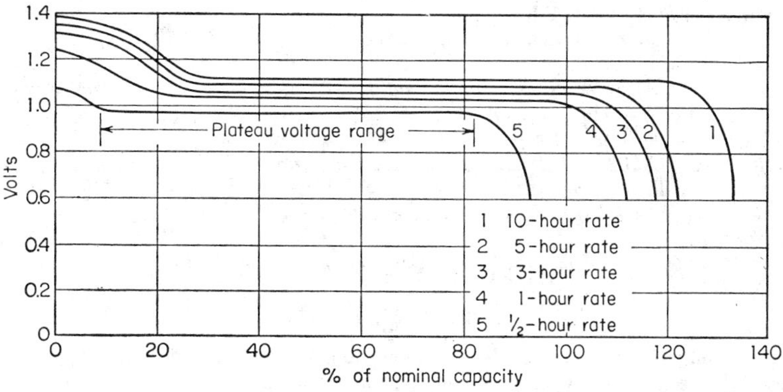

FIG. 24-32. Discharge curves of a silver-cadmium cell at various rates.

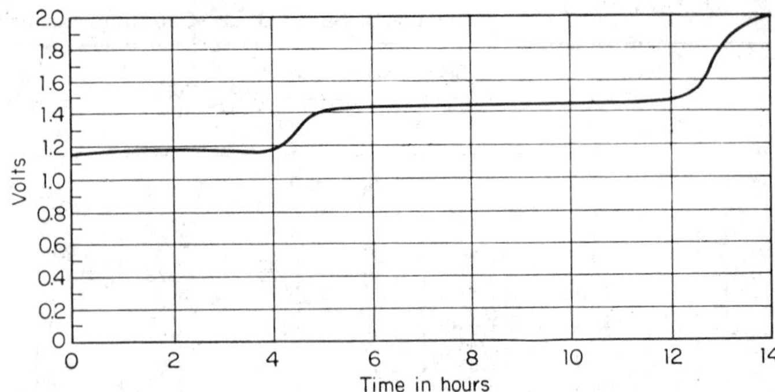

FIG. 24-33. Typical charge curve of a silver-cadmium cell (cutoff voltages 1.55 to 1.60 for sealed cells, 1.60 to 1.65 for vented cells).

Fig. 24-33, the voltage rises rapidly at the end of charge, which causes the current to drop on the constant-potential charge to a low value. This feature reduces the problem of serious overcharge.

As in other types of batteries performance drops off as the temperature is lowered. Low rates of discharge are less affected than high rates at low temperatures. Thus a temperature range of −20 to +60°C is satisfactory for low rates, while 0°C is minimum for high rates. In the latter case the voltage dips at the beginning and rises as the cell warms up. When voltage is critical, the battery should be maintained at normal temperatures by external heaters.

Cycle life is dependent on the depth of discharge and rates of recharge. There are many regimes ranging from a partial discharge of 10 to 50% with recharges in a short time interval with up to 20,000 c to complete discharge on every cycle with normal recharge with up to 1,000 c at low rates and 300 to 500 c at higher rates of discharge. During the life of the battery it is advisable to schedule reconditioning cycles of special charge-discharge as specified by the manufacturer to restore the capacity lost during repeated cycling.

This system is specifically suited for applications where high energy density, maximum cycle life, and long overall wet-stand life are required. The vented type is used in all forms of portable electric appliances, communication equipment, standby, and other, allied equipment. At present the sealed type is mainly used for space and satellite applications. It is applicable to use in general commercial areas as they develop.

73. References

VINAL, GEORGE W. "Storage Batteries," 4th ed.; New York, John Wiley & Sons, Inc.

HOWARD, PAUL L. The Sealed Silver Cadmium Battery; *Electrochem. Technol.*, September–October, 1963, 1-No. 9–10.

FRINK, A. M., JR. Sealed Cadmium Silver Oxide Batteries: *Proc. 17th Ann. Power Sources Conf.*, May, 1963.

SHAIR, R. C., LISKA, JOHN, and RAMPEL, GUY Silver Cadmium Batteries; *Proc. 18th Ann. Power Sources Conf.*, May, 1964.

"Energy Data Book"; Yardney Electric Corporation.

Bull. C-102, C-103, B-201, Electrochimica Corporation.

HENNIGAN, T. J., and APELT, A. O. Use of Sealed Silver Cadmium Battery on Explorer XII; *NASA Tech. Note* D-1543, January, 1963.

ZINC-CARBON BATTERIES

BY F. B. PIPAL

74. Introduction. The standard zinc-carbon Leclanché type primary battery is widely used because of low cost, reliable performance, and ready availability.

Cells of many sizes and characteristics have been developed to meet the requirements of a large variety of applications. Manufacturers' lines include hundreds of battery types, sizes, and terminal arrangements. Data covering standard dimensions, performance, and methods of testing the various commercial sizes and types of zinc-carbon cells and batteries are given in USA Standard Specifications for Dry Cells and Batteries, C18.1-1965.

75. Definition. Zinc-carbon primary batteries use cells which are a modification of the Leclanché wet cell. Zinc-carbon batteries are dry, since the electrolyte is made nonspillable by combination with absorbent materials: they are primary batteries inasmuch as they are not designed to be recharged (Par. **85**).

76. Composition and Construction. Zinc is the anode, or negative electrode, in a zinc-carbon primary cell. It gives up electrons easily to an external circuit and accordingly becomes a source of positively charged ions.

Manganese dioxide is the cathode, or positive electrode. In a standard type of cylindrical cell the manganese dioxide is mixed with acetylene black and solid ammonium chloride, wet with a zinc chloride–ammonium chloride electrolyte, and is in the form of a bobbin. The acetylene black is used to improve conductivity of the mix and to retain moisture. The manganese dioxide is a depolarizer and a source of about one-half the energy output of the cell. As a zinc-carbon cell delivers current, the manganese dioxide loses oxygen, while the zinc is oxidized. As the available oxygen from the manganese dioxide becomes depleted, the manganese dioxide becomes less and less active as the cathode.

The basic zinc-carbon cell is made in many shapes and sizes. The two general categories, however, are cylindrical and flat cells. The chemical ingredients are the same for both, the difference being mostly physical. In any one specific size the cell ingredients and formulas may be varied to give certain performance qualities for different applications.

In the usual cylindrical cell a carbon rod is inserted into the bobbin. This rod, with a metal cap, serves as a current collector and is chemically inert. It is also porous enough to permit the escape of gases which accumulate in the cell, while, at the same time, it does not permit leakage of electrolyte.

The separator, as, for example, cereal paste or plastic film, is an electrolyte-containing layer that physically separates the positive and negative electrodes while permitting ion transfer between electrodes through the electrolyte.

The zinc-carbon cell must have vents for gases which are generated in the cell during reactions and while idle. Consequently, the cell cannot be hermetically sealed. Pitch,

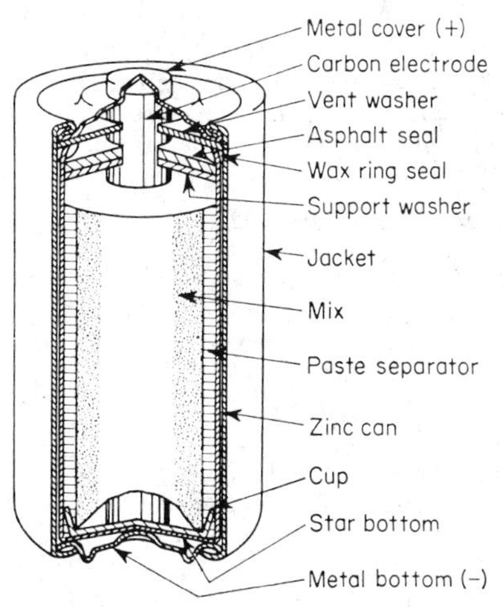

Fig. 24-34. Cross section of standard round zinc-carbon cell.

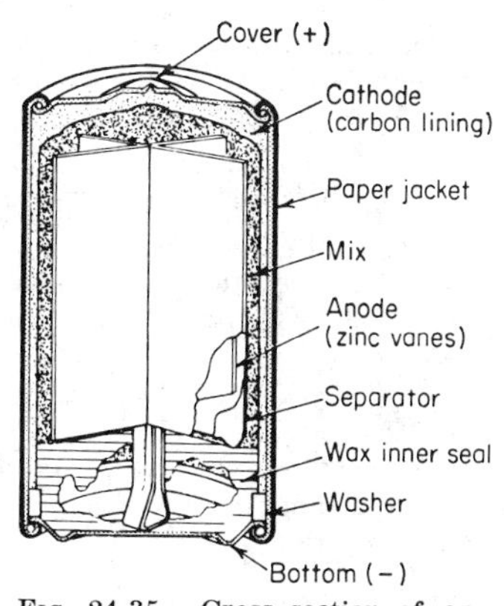

Fig. 24-35. Cross section of external-cathode, or "inside-out," round cell.

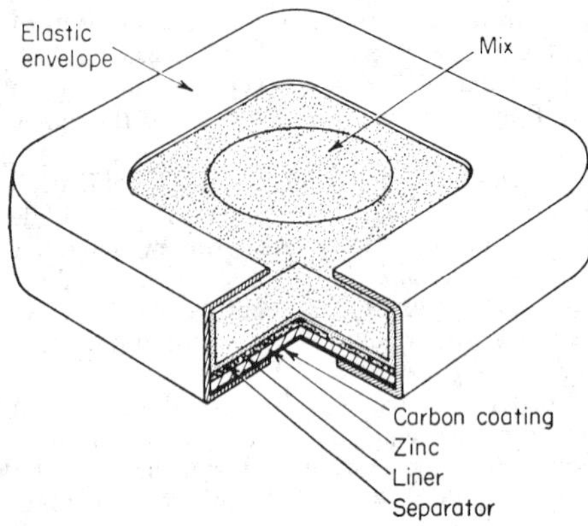

Fig. 24-36. Cross section of "mini-max" flat cell.

rosin, wax, plastic, waterproofed cardboard, and insulated metal covers are seals used which minimize air access and moisture loss by evaporation and thus contribute to service maintenance. Air spaces left between the depolarizing mix and the seal provide for expansion of the cell contents as the cell is used.

A cylindrical or round cell of usual or standard construction is illustrated in Fig. 24-34.

In addition to the standard round cell, there is an external-cathode or "inside-out" construction which is shown in Fig. 24-35. In this cell an injection-molded impervious inert carbon wall is the container of the cell and also functions as the current collector. Zinc vanes are inside the cell and are surrounded by the cathode mix. This ensures efficient zinc consumption and, since zinc is not used as a container, a high degree of leakage resistance.

A flat-cell construction is pictured in Fig. 24-36. In this cell, carbon is coated on a zinc plate to form a duplex electrode. This electrode is the combination of the zinc of one cell and the carbon of the adjacent one. There is no expansion chamber or carbon rod. Flat cells are used only in multicell batteries.

77. Chemical Reactions. The chemical reactions occurring in a zinc-carbon cell, if considered completely, would be found very complicated. Familiarity with these is not necessary to an understanding of the behavior of the cell. There are, however, certain major fundamental chemical reactions.

One of the most direct methods of presenting the reactions in the dry cell is to start with the electrolyte, since it is the part of the cell which contacts both electrodes and fills the space between them. The electrolyte consists of ammonium chloride and zinc chloride dissolved in water. The ammonium chloride is the electrolyte salt and dissociates to form ammonium cations and chloride anions as shown by the following equation:

$$NH_4Cl \rightleftharpoons NH_4^+ + Cl^- \tag{24-1}$$

It is convenient to present the reactions at the individual electrodes in the cell to clarify their functions. The manganese dioxide is the positive electrode, or cathode, of the cell, and the carbon rod acts as an electrical conductor. The manganese dioxide reacts electrochemically with the ammonium cations, only when current is drawn, by one or both of the following reactions to form the by-products shown:

$$2MnO_2 + 2NH_4^+ + 2e \rightarrow 2MnOOH + 2NH_3 \tag{24-2}$$

$$2MnO_2 + 2NH_4^+ + 2e + ZnCl_2 \rightarrow ZnO\cdot Mn_2O_3 + 2NH_4Cl \tag{24-3}$$

The conditions of current drain, temperature, and several other factors determine which of the two above reactions predominate in any given dry-cell discharge. The ammonia formed in Eq. (24-2) reacts chemically with zinc chloride as follows:

$$ZnCl_2 + 2NH_3 \rightarrow ZnCl_2\cdot 2NH_3 \tag{24-4}$$

The zinc chloride in the electrolyte may be considered as the buffer salt to form the diamino zinc chloride shown above, since it functions to maintain the electrolyte acidity in the desired range. At the same time the zinc anode reacts electrochemically with chloride anions thus,

$$Zn + 2Cl^- \rightarrow ZnCl_2 + 2e \tag{24-5}$$

to form zinc chloride.

The cathode reaction [Eq. (24-2)], the chemical reaction [Eq. (24-4)], and the anode reaction [Eq. (24-5)] can be added to give an overall cell reaction,

$$2MnO_2 + 2NH_4Cl + Zn \rightarrow 2MnOOH + ZnCl_2 \cdot 2NH_3 \qquad (24\text{-}6)$$

When the cathode reaction [Eq. (24-3)] and the anode reaction [Eq. (24-5)] are added, the overall cell reaction becomes

$$2MnO_2 + Zn \rightarrow ZnO \cdot Mn_2O_3 \qquad (24\text{-}7)$$

The underlying principle of the battery operation is that zinc metal at the anode is consumed by oxidation, using a part of the oxygen in the manganese dioxide cathode, to provide electricity. When zinc is oxidized in a battery, the energy is produced as electricity and usually at a high efficiency.

78. Voltage. With a zinc negative electrode, a manganese dioxide positive electrode, and an electrolyte of ammonium chloride–zinc chloride the initial open-circuit voltage is referred to as a nominal 1.5 V. The actual initial voltage, however, may vary from about 1.5 to over 1.6 V, depending upon the type of cell and the type of manganese dioxide and on slight differences in electrolyte formulation. The open-circuit voltage of zinc-carbon cells using identical materials is the same regardless of size. Voltmeter readings, however, will show slightly higher values as the cell size increases, because of meter load and the difference in impedance between the cells. A higher-voltage battery is not necessarily better than one of lower voltage.

The open-circuit voltage of zinc-carbon cells on the shelf declines a few hundredths of a volt per year until the battery reaches a condition of final exhaustion, at which time the voltage begins to drop rapidly.

The terminal voltage of a cell under discharge is called the working voltage, or voltage under load. It is more important than the open-circuit voltage. It is lower than the open-circuit voltage by an amount which depends upon the current passing through the cell and the internal resistance of the cell. The working voltage of a zinc-carbon cell falls as it is discharged, as shown in Fig. 24-37. The service hours delivered by the cell or battery are greater as the cutoff or end-point voltage is lower.

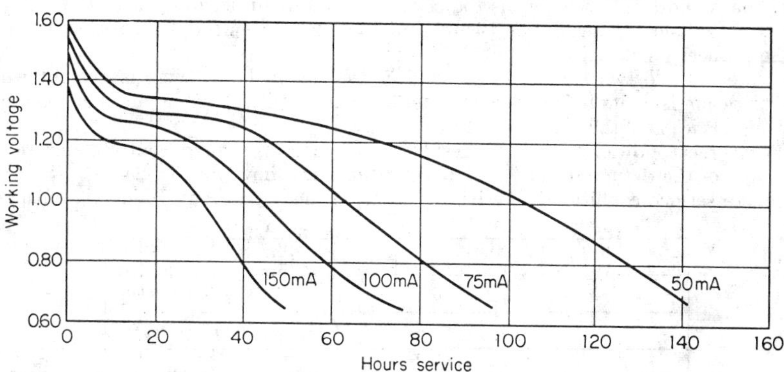

FIG. 24-37. Voltage discharge characteristic of fresh zinc-carbon size D cell discharged 2 h a day at 70°F.

Zinc-carbon primary batteries of the following nominal voltages are in common use: 1.5, 3, 4.5, 6, 7.5, 9, 12, 15, 22.5, 30, 45, 67.5, and 90. Other voltage batteries, less readily available commercially, are 33, 75, 225, 240, 300, and 510. Cells and batteries may, of course, be connected in series to obtain any desired voltage.

79. Service Capacity. The service capacity of a zinc-carbon battery cannot be stated as a single value. The amount of electrical energy which can be withdrawn varies under different conditions, because the cells function at different degrees of efficiency depending upon the load. The point of maximum service efficiency of a cell varies with the make and type of cell and the conditions of its use—i.e., whether the current drain is continuous or intermittent, whether the temperature is high or low, and whether the cutoff voltage is high or low.

In more detail, factors which affect the service capacity are:

Current drain. The capacity is greater as the discharge current is less, down to a certain point beyond which the service efficiency decreases because the spontaneous reactions within the cells become an increasingly important factor.

Operating schedule. This is the relative time of discharge and recuperation. The performance is normally better when the discharge is intermittent.

Cutoff voltage. The capacity is greater as the cutoff voltage is lower.

Temperature. This is discussed in Par. **82,** Effect of Temperature.

Storage period of the battery prior to use. See Par. **83,** Shelf Life and Storage.

If the load is too heavy and energy is withdrawn too rapidly from the cell, the reaction rate of both electrodes can be exceeded. Reaction products which are generated

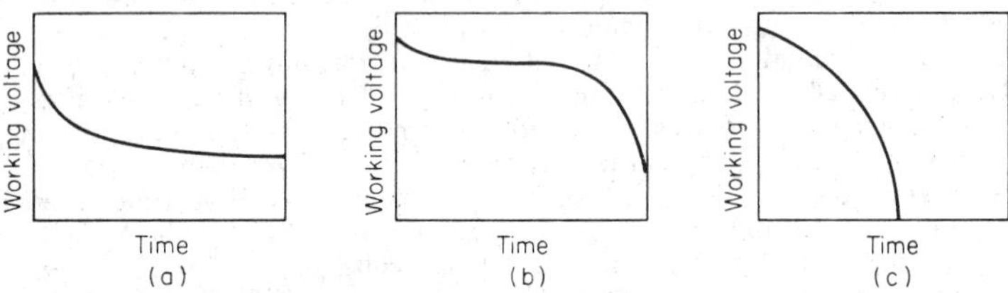

FIG. 24-38. Performance of No. 6 general-purpose cells. (*a*) Heavy withdrawal of energy, incomplete depolarization, elapsed time of a few hours; (*b*) highly efficient performance, elapsed time 6 to 12 months; (*c*) light service, reduction in energy output due to shelf depreciation, elapsed time 2 to 3 years.

in the cell cannot diffuse away fast enough; thus the working voltage drops off, and the cell does not function efficiently. If the energy is withdrawn more slowly, the depolarizer functions more effectively and diffusion can take place so that the cell maintains its current and working voltage holds up well.

If the withdrawal of energy is too slow, the time required to exhaust the cell is so extended that shelf deterioration is sufficient to detract from the cell's available energy and so reduces its electrical output.

Between these two conditions of service, heavy and light, there is a point where depolarization is at its best and shelf deterioration is still negligible, and here the most efficient service is obtained.

The three conditions described result in differences in energy output and also in the shape of the discharge curve. The conditions are shown in Fig. 24-38.

Severe service conditions may be due to either heavy current drains or continuous

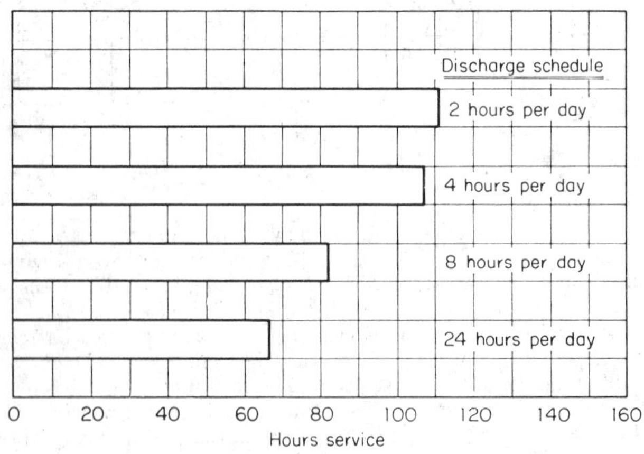

FIG. 24-39. Hours of service vs. discharge schedule : size D zinc-carbon cell, 50 mA drain to 0.9 V cutoff at 70°F.

or long discharge periods. Continuous use is not necessarily inefficient if the drain is light. A relatively heavy drain may be permissible if the discharge periods are short and the rest periods constitute a large proportion of the time.

Zinc-carbon batteries are intended primarily for intermittent service. The total hours of life obtained is normally greater for a discharge schedule of 2 h/day, for example, than for continuous discharge. Intermittent service allows sufficient time for the back diffusion of reaction products and other reactions in the recuperation process. Service life as a function of discharge schedule is depicted in Fig. 24-39.

Approximate service capacity for typical cylindrical and flat zinc-carbon cells is given in Table 24-3 for three different current drains. The values given are for fresh cells at 70°F. The discharge schedule is 2 h/day. The cutoff voltage is 1.0 V/1.5 V cell for the first 10 cells listed and 0.8 V/1.5 V cell for the remaining cells. This is in accordance with typical cutoff voltages associated with their use. The data are based on starting current drains and constant-resistance tests.

Table 24-3. Approximate Service Capacity of United States of America Standards Institute Sizes of Cylindrical and Flat Zinc-Carbon Cells at Various Current Drains

USASI cell size	Starting drain, mA	Service capacity, h	USASI cell size	Starting drain, mA	Service capacity, h	USASI cell size	Starting drain, mA	Service capacity, h
N	1.5	275	G	15	820	F24	1	475
	7.5	52		75	150		5	150
	15	24		150	65		10	72
AAA	2	290	6	50	700	F30	1.3	275
	10	45		250	150		6.5	40
	20	17		500	70		13	16
AA	3	350	F15	0.4	210	F40	1.3	450
	15	40		2	30		6.5	108
	30	15		4	8		13	52
B	5	420	F12	0.5	435	F60	2	190
	25	65		2.5	103		10	40
	50	25		5	51		20	15
C	5	430	F17	0.6	710	F70	3	550
	25	100		3	155		15	150
	50	40		6	75		30	65
D	10	500	F20	0.7	210	F80	3	600
	50	105		3.5	35		15	165
	100	45		7	12		30	72
E	15	400	F22	0.8	475	F90	3	770
	75	70		4	98		15	200
	150	30		8	49		30	90
F	15	520	F25	1	500	F100	5	1,000
	75	105		5	105		25	260
	150	45		10	45		50	110

Service capacity of individual cells ranges from less than 100 mAh to about 30 Ah. When assembled into batteries in various cell arrangements, the capacities of currently available battery types range to over 100 Ah.

80. Cutoff Voltage. The cutoff, or end-point, voltage is the closed-circuit voltage at the end of a useful discharge. It is the cell voltage below which battery-powered equipment will not operate satisfactorily.

Typical cutoff voltages for carbon-zinc cells range from 0.6 to 1.2 V/1.5 V cell, depending upon the application. The cutoff voltage should be made as low as possible so as to use the available energy in the battery. This is sometimes done, on the assumption that the equipment can tolerate it without causing failure, by using a slightly higher voltage than the application normally requires. The cutoff voltage per cell is then lower, which ensures efficient use of the battery. The curves of Fig. 24-40 show the advantage of a low cutoff voltage if a cell is subjected to moderate or heavy use.

81. Internal Resistance. When battery-operated devices require a high current for

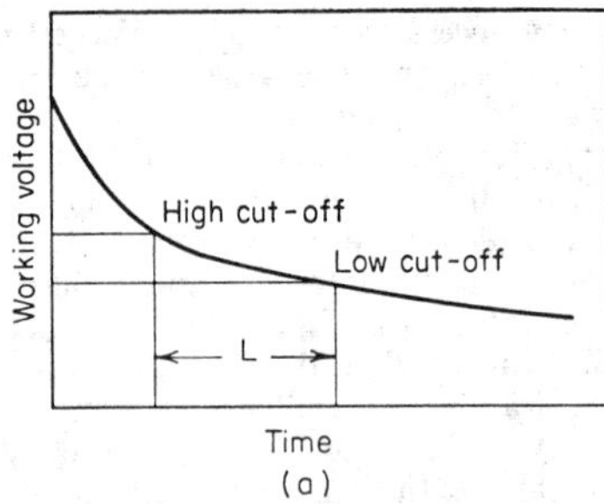

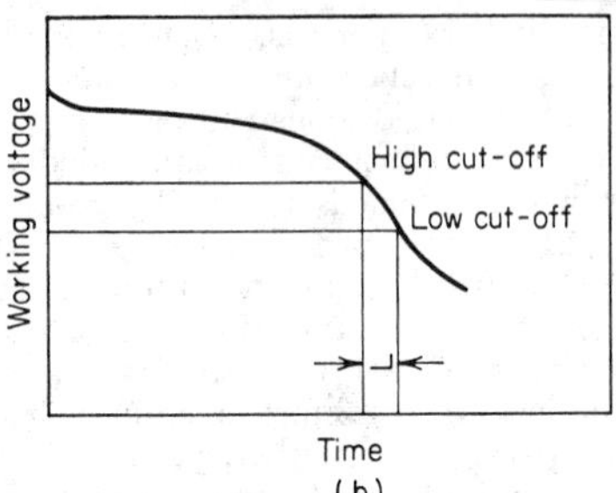

FIG. 24-40. Cutoff voltage characteristics. (*a*) Advantage of lower cutoff when use is moderate or heavy; (*b*) lesser advantage of lower cutoff when use is light (*L* = life advantage).

short periods of time, the internal resistance of cells becomes important. If the internal resistance of a single cell is too high to provide the current, two or more cells may be connected in parallel or a larger cell may be used.

The internal resistance of a zinc-carbon cell is a combination of solid conductor and electrochemical and contact resistance. The internal resistance of unused cells is low and is usually negligible in most applications.

Internal resistance may be measured by flash current (short-circuit amperage). Amperage readings are not necessarily related to service capacity. Amperage is usually higher in large cells. In cells of any size, amperage may vary with different grades and different manufacturers.

The internal resistance increases with storage time, use, and decreasing temperature. The cell dries out with age. During discharge, some of the MnO_2 is converted to Mn_2O_3, which has higher resistance. Coatings over the reactants also reduce currents.

Table 24-4 lists the approximate internal resistance, as determined by the flash current, of several typical round cells of the general-purpose type.

Table 24-4. Approximate Internal Resistance of Fresh Round Cells

USASI cell size	Average flash current, A	Approximate internal resistance, Ω
N	2.4	0.69
AAA	3.5	0.44
AA	5.4	0.29
C	3.3	0.47
D	5.8	0.27
F	9.0	0.17
G	11.0	0.11
6	38.0	0.03

82. Effect of Temperature. Most zinc-carbon cells are designed to operate at room temperature. High operating or storage temperature is an abusive condition. The higher the operating temperature, the greater the energy output of the cell. Voltage and depolarization are increased by high temperature, but this is offset by rapid shelf deterioration, caused by rapid loss of moisture and accelerated chemical action.

High temperature may cause early breakdown of a cell. Any temperature above 70°F reduces the normal shelf life of the cell. Prolonged exposure to temperatures above 125°F may cause erratic operation and sudden failure.

Low temperatures have a marked effect on the output of zinc-carbon cells. They become practically inoperative at a temperature of −21°C (−6°F). Special low-temperature cells will operate at a limited capacity down to temperatures of −40°C (−40°F). These cells will perform best on light current drains, where the effect of internal resistance is not so marked.

The effect of temperature on zinc-carbon cells is covered in the following from the viewpoint of voltage, service capacity, internal resistance, and flash current.

The open-circuit voltage of a cell decreases on the average 0.0004 V/°C when the temperature is decreased from 25°C (77°F) to −20°C (−4°F). Practically the actual working voltage at the terminals of the cell or battery is of more significance than the open-circuit voltage. Figure 24-41 shows the different voltage characteristics of a size D zinc-carbon cell discharged at three different temperatures. In each case the cells are discharged with a continuous current drain which has an initial value, at full cell voltage, of 667 mA (this simulates a ½-A lamp).

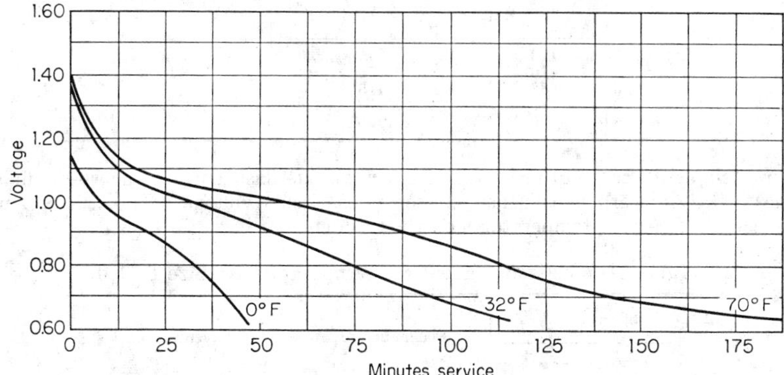

Fig. 24-41. Voltage characteristics of size D zinc-carbon cell discharged at three different temperatures with a continuous current drain starting at 667 mA.

Table 24-5 shows the initial voltages of a size D cell for different temperatures as the current drain is varied.

Table 24-5. Initial Terminal Voltage of Size D Cell for Varying Conditions of Temperature and Current Drain

Temperature		Open circuit	Terminal voltage at drains shown, mA			
°C	°F		25	50	75	150
45	113	1.58	1.57	1.56	1.55	1.53
21	70	1.57	1.56	1.54	1.52	1.50
0	32	1.57	1.54	1.52	1.49	1.46
−18	0	1.56	1.53	1.51	1.48	1.44
−29	−20	1.55	1.48	1.45	1.33	1.25
−40	−40	0.75	0.06	0	0	0

The effect of temperature on service capacity may be expressed as a percentage based upon the capacity at 70°F. Table 24-6 shows the effect of temperature on the capacity of a size D cell for both a light and a heavy current drain.

Table 24-6. Effect of Temperature on Service Capacity of Size D Cell at Two Different Current Drains

Temperature		Light current drain (30 mA starting-continuous to 0.75 V), %	Heavy current drain (375 mA starting-continuous to 0.75 V), %
°C	°F		
38	100	130	125
27	80	120	110
21	70	100	100
16	60	90	83
4	40	70	63
−7	20	49	48
−18	0	26	28
−29	−20	3	0

The internal resistance of zinc-carbon batteries increases at decreased temperatures. This is shown in Table 24-7 for a size D cell.

Table 24-7. Variation with Temperature of Internal Resistance (Flash-current Method) of Size D Cell

Temperature		Average resistance, Ω	Percent of resistance at 70°F
°C	°F		
45	113	0.245	89
21	70	0.276	100
0	32	0.415	150
−18	0	0.478	173
−29	−20	1.93	700

Low temperatures produce a greater effect on the flash current (short-circuit amperage) of a zinc-carbon cell than on the open-circuit voltage. Table 24-8 depicts the effect of decreasing temperatures on the flash current of a size D cell.

Table 24-8. Flash Current vs. Temperature for Size D Cell

Temperature		Flash current, A	Percentage of flash current at 70°F
°C	°F		
45	113	6.2	113
21	70	5.5	100
0	32	3.7	67
−18	0	3.2	58
−29	−20	0.8	14
−40	−40	0	0

83. Shelf Life and Storage. As zinc-carbon cells become older, they gradually deteriorate, either in use or while idle. The rate of deterioration is greater for partially discharged than for unused cells. The deterioration while the cell is idle is the result of zinc corrosion, very slow side chemical reactions, and moisture loss from the cells. These changes gradually reduce the service output.

Shelf life is a measure of the lasting quality of cells prior to their being put in service. It is the period of time, at a storage temperature of 70°F, after which a battery retains a specified percentage (usually 90%) of its original energy content. Zinc-carbon cells are normally designed to operate at room temperature. High temperatures reduce shelf life. The shelf life of a battery stored at 90°F is only about one-third that of one stored at 70°F. Prolonged exposure to temperatures above 125°F causes very rapid disintegration of the battery.

The rate of deterioration is retarded by low temperatures, and shelf life is improved. Low temperatures or even freezing are not harmful to zinc-carbon cells as long as there is not repeated cycling from low to higher temperatures. A storage temperature of 40 to 50°F is very effective, and the major benefit is derived by avoiding exposure to temperatures in excess of 70°F.

In connection with cold storage of zinc-carbon batteries, it is necessary to observe the following precautions:

1. Handling of the batteries should be minimized after the batteries are chilled.
2. When batteries are removed from low-temperature storage, they should be allowed to reach room temperature in their original packing so as to avoid condensation of moisture, which may cause electrical leakage and destruction of the jackets.

84. Methods of Testing. The variety of uses for zinc-carbon cells has led to the adoption of a considerable number of standard tests. Service tests have been developed by a Sectional Committee of the USASI acting under the sponsorship of the National Bureau of Standards. These tests are described in USA Standard Specifications for Dry Cells and Batteries, C18.1-1965. They have been universally adopted by the battery industry.

There is no simple or rapid method of determining the service capacity of a zinc-carbon battery. Tests must be run which closely duplicate the class of service for which the battery is intended. The schedule of operation is very important. The service capacity of a battery used 2 h/day is considerably different from that of the same battery used 12 h/day. Quick, continuous tests are not related to longer intermittent uses and are not indicative of cell behavior in actual service.

The output of a battery on a service test is usually expressed as the time that the battery can maintain a closed-circuit voltage above a specified cutoff voltage that marks the end of the test. This cutoff voltage depends upon the characteristics of the equipment which the battery powers.

Relative service capacities of batteries cannot be determined by short-circuit amperage readings. The short-circuit amperage of a cell may be adjusted over a wide range by varying the carbon content of the depolarizing mix. Carbon contributes nothing to the cell energy and is used primarily to control cell resistance. As carbon is added to a cell, depolarizer must be removed so that energy content is reduced.

Zinc-carbon batteries can be meter-tested to check their condition. A loaded voltmeter provides the best spot check. Open-circuit voltmeter readings give no indication of internal impedance. A short-circuit amperage reading is damaging, especially to higher-voltage batteries utilizing flat cells.

Even a loaded voltmeter test will give no indication of remaining service capacity unless the exact history of the battery is known and can be compared, on a service vs. meter-reading basis, with other batteries tested in similar service.

85. Rechargeability. It has been known for years that the Leclanché type of zinc-carbon cell is rechargeable to some degree if the discharge and charge cycles are controlled with precision. On this matter the National Bureau of Standards (*Letter Circ.* LC965) makes the following comments:

"From time to time attention has been turned to the problem of recharging dry cells. Although the dry cell is nominally considered a primary battery it may be recharged for a limited number of cycles under certain conditions. Briefly these are:

"1. The operating voltage on discharge should not be below 1.0 volt per cell when battery is removed from service for charging.

"2. The battery should be placed on charge soon after removal from service.

"3. The ampere-hours of recharge should be 120%-180% of the discharge.

"4. Charging rate should be low enough to distribute recharge over 12-16 hours.

"5. Cells must be put into service soon after charging as the recharged cells have poor shelf life.

Recharging of dry cells may be economically feasible only when quantities of dry cells are used under controlled conditions with a system of exchange of used cells for new ones already in practice, and with equipment available to provide direct current for charging. Such a system would not be practical for home use."

By reversing the flow of current through the battery, both the anode and the cathode can, with proper controls, be restored to a near-original condition through the process of electroplating. The efficiency of these replating operations determines how useful the system is for rechargeable batteries. This efficiency is affected by (1) electrolyte conductivity, (2) the nature of the reaction products, and (3) the type of battery separator that is used.

In the zinc-carbon system, zinc dissolves in the electrolyte during use and often forms reaction products in combination with the manganese dioxide. Upon recharging, the zinc ion must travel from the electrolyte and redeposit on the anode. To produce a smooth plating operation, it is necessary that a good portion of the original zinc remain intact and that current distribution with the cell be very uniform. Conditions existing in the ordinary dry cell quickly lead to unevenness in the plating after successive charge and discharge cycles. Zinc depositing more heavily in certain areas of the anode causes the formation of dendrites, or treelike growths, which penetrate the separator, touch the cathode, and cause internal short circuiting of the battery.

During discharge of the cathode, the manganese dioxide is reduced to one of the lower-valent oxides. The reoxidation or re-forming of the manganese dioxide during recharge may not proceed smoothly if substantial insoluble reaction products prevent even distribution of the current within the cell.

Deep discharge employs battery materials nonuniformly. The low conductivity of the electrolyte used in carbon-zinc batteries, when compared with those used in rechargeable systems, also limits the rate of discharge and the rate of charge acceptance to values that are lower than normally useful in rechargeable systems.

Recharging cells which are not specifically designed for charging can be dangerous. Excessive amounts of gassing, which may result from too high a value of charging current, may cause a tightly sealed cell to explode, with consequent personal injury or damage to equipment.

86. Selecting a Battery. To specify a battery, at least the following minimum information must be known about the application:

Maximum starting voltage.

Initial current drain at full battery voltage.

Cutoff voltage below which the device will not operate properly.

Operating schedule.

Service life desired.

Other factors which must also be considered are temperature during operation of the battery, size, weight, environment, and type of terminals.

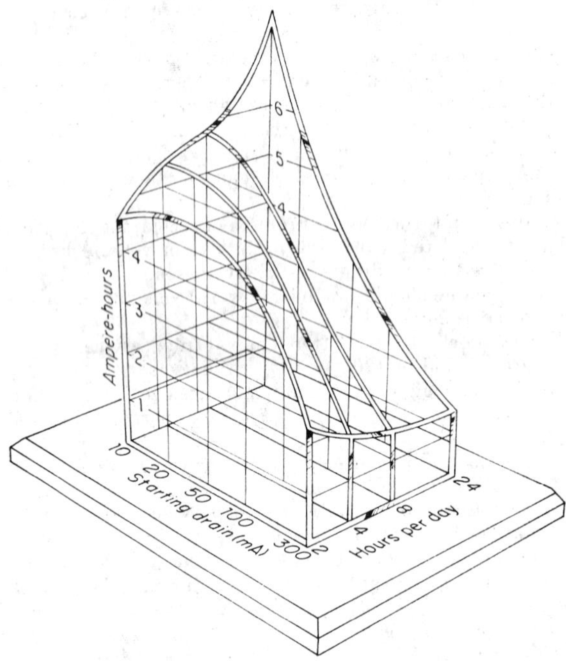

Fig. 24-42. Battery service life as a function of the initial current drain and operating schedule for a D-size zinc-carbon cell.

Service information as a function of current drain, operating or discharge schedule, and cutoff voltage is shown in the battery manufacturer's literature. The efficiency of zinc-carbon batteries improves as current density in the battery decreases. Thus, use as large a battery as possible in the application. This will give a lower cost per hour of operation and longer shelf life. Over a certain range of current density, service life may be tripled by cutting the current drain in half. This is equivalent to using a larger battery for a given application and so reducing the current density within the cells. This is true down to a certain point beyond which shelf deterioration becomes important.

All the ideal characteristics of any given application cannot usually be found in one battery. Thus, optimum performance of a battery can usually best be achieved by meeting the critical needs of the application and subordinating the others.

Service capacity for zinc-carbon batteries depends on the relative time of discharge and recuperation periods. Figure 24-42 shows service life as a function of initial current drain and operating schedule.

Service temperatures must be known, because they will affect battery life. If the battery is to be stored for any period of time before use, the length of time and the temperature should be considered for the effect on the operation of the battery.

Allowable size and weight will sometimes determine which battery is selected in spite of other requirements. Bulky and heavy batteries can be reliable and economical if their size can be accommodated. A final selection is based on a balance of size, weight, and operating cost. The choice should also consider availability. If necessary, batteries with other chemical systems should be examined.

SILVER-ZINC CELLS

J. J. Lander

87. Electrochemical Reactions. The active materials of the silver-zinc cell consist of divalent silver oxide (AgO) at the positive plate and metallic zinc at the negative plate. Plate groups are immersed in an aqueous solution of potassium hydroxide (30 to 45% by weight). During discharge the positive plate is electrochemically reduced to monovalent silver oxide (Ag_2O) and to metallic silver. At sufficiently low rates the discharge proceeds in a stepwise fashion at two distinct voltage levels corresponding to the two discharge reactions at the positive plate. The zinc negative plate is oxidized to zinc oxide and zincate ion.

Successful utilization of this couple as a secondary cell is enabled by mechanical separation of the positive and negative plates with semipermeable membrane materials which imbibe the electrolyte to provide the necessary ionic conductivity and which slow down migration of silver toward the zinc plate and growth of metallic zinc toward the positive plate. Cellophane and modifications thereof are in general use.

88. Plate Construction. Positive plates can be made by compacting silver or silver oxide powder onto a silver grid of screen or expanded metal. A preferred technique for mechanical stability is to sinter the metallic powder or the oxide after compaction. This may be followed by a subsequent pressing operation for sizing.

Negative plates can be made by compacting zinc dust or zinc oxide powder mixed with small percentages of mercuric oxide onto a silver (or copper) grid. Some mechanical stability is conferred by wrapping the plate in a thin, porous, nonwoven cellulosic paper and dampening with water before compaction. The mercuric oxide is added to obtain low self-discharge rates. Current-collecting wires are welded to the grids.

89. Cell and Battery Assembly. The positive plates are completely wrapped or encased in a number of layers of cellophane or other semipermeable-membrane material. The negative plates may or may not be similarly wrapped, depending on the requirements of the design or application. A thin layer of porous, chemically inert plastic sheet is frequently used between the positive plate and the first cellophane layer, to prolong separator life.

After wrapping, the plate groups are intermeshed and the total element is inserted into a tightly fitting plastic case. The lead wires from the plates are connected to their terminals in the cell cover by soldering, and the cover is sealed to the case by heat sealing or chemical solvent sealing, depending on the case and cover material used. Cells are grouped in a metal frame or box to form a battery unit. Cells are activated by adding the electrolyte through a filling vent after sealing on the cover.

Either preformed plates or uncharged plates may be used in cell assembly. If uncharged, cells or batteries are given a formation charge at a low rate before putting into service.

90. Discharge Characteristics. At low rates, cells discharge for roughly one-quarter their ampere-hour capacity at close to 1.8 V, after which the balance of the capacity is yielded at about 1.55 V. At higher rates the upper voltage is not obtained except briefly, and useful discharge voltage can be as low as 1.1 to 1.2 V, depending on the rate of discharge, plate thickness, separator thickness, electrolyte concentration, and temperature.

Voltage, faradaic capacity, and total energy yield all fall off approximately linearly as a function of the discharge current. Practical designs yield 50 to 70% of their low-rate energy capability at high rates of discharge at room temperature. Energy yield falls off, roughly linearly as temperature is reduced, to about 10% of room temperature capability at −40°F.

91. Charge Characteristics. Cells may be charged efficiently by constant-current, constant-voltage, or taper-current methods. It is desirable to limit the finishing voltage on charge to 2.00 to 2.10 V/cell, to limit the amount of overcharge, for excessive overcharge shortens life owing to more rapid growth of metallic zinc paths through the separator. Eighty to 100% charge can be returned from a full discharge in 8 to 20 h at room temperature at ampere-hour efficiencies of 98% or better. Charge-time requirements increase rapidly below 40°F, and recharge can scarcely be accomplished below about 10°F.

92. Life. The cycle-life capability of the silver-zinc cell varies from a few thousand cycles at low depths of discharge (~20%) to a few hundred cycles, or less, when full-capacity discharges are given. Life depends on the cycling regime given, and in terms of elapsed time it is somewhat dependent on the thickness and quality of the separation used, because cellulosic-type separation is subject to deterioration in the cell environment. As many as 300 deep discharges over a 22-month period have been reported.

Twenty months' stand time in the charged condition is readily achieved, although subsequent life on a cycling regime would be expected to be sharply reduced. Batteries have been floated for up to 3 years. Shelf life in a protected environment in the unactivated condition (i.e., electrolyte not added) is unknown but is expected to be many years.

93. Applications. Silver-zinc cells are commercially available in a wide range of ampere-hour capacities. Extensive commercial exploitation of this high-energy-yield cell (40 to 70 Wh/lb, 2 to 4 Wh/in^3) is prevented by its high cost and short life, relative to other secondary-battery types. Typical uses are for portable photographic and communication equipment, model boats and airplanes, instrumentation equipment, helicopter aircraft, and torpedo propulsion.

In space applications, where weight is a prime consideration, this cell is likely to find increased usage; it has already been employed in a variety of space-probe vehicles. Current research and development programs on separator materials and plate formulations may be expected to result in substantial performance improvement.

Automatically activated primary silver-zinc batteries have found wide usage in a variety of missiles.

94. Bibliography

VINAL, G. W. "Storage Batteries," 4th ed.; New York, John Wiley & Sons, Inc., 1955.

"Characteristics of Separators for Alkaline Silver Oxide–Zinc Batteries"; Wright-Patterson AFB, Ohio, Air Force Aero Propulsion Laboratory, August, 1964, ed. by J. E. Cooper and Arthur Fleischer.

SOLOMON, F., and WORK, G. W. Present-day Long Life Silver–Zinc Secondary Batteries; *Proc. 4th Intern. Symp. Research and Develop. in Non-Mechanical Electrical Power Sources*, September, 1964, Brighton, England, pp. 463–472.

CLARK, W. W., et al. Alkaline Battery Evaluation; *Rept.* APL-TDR-64-76, Air Force Contract AF 33(616)-7529, June, 1964.

KERELLA, J. A., et al. Applied Research Investigation of Sealed Silver–Zinc Batteries; final report, Contract AF 33(615)-1583, June 30, 1966, and references therein quoted.

Sealed Silver–Zinc Batteries, J. J. Lander, in "Space Power Systems Engineering"; New York, Academic Press, Inc., 1966, Vol. 16, pp. 1101–1110.

ZINC–ALKALINE–MANGANESE DIOXIDE BATTERIES

BY E. F. ERRICO

95. Definition. One of the most recent battery systems to reach extensive commercial acceptance is the zinc–alkaline–manganese dioxide dry battery. This electrochemical system utilizes a negative electrode of compressed amalgamated zinc granules and a depolarizing positive electrode of compacted synthetic-type manganese dioxide made conductive with graphite. Both electrodes are in the presence of highly alkaline potassium hydroxide electrolyte. Since the electrolyte is immobilized in cellulosic-type separators, the term "dry battery" applies to this system.

96. Construction. Alkaline-manganese cells are available in two structural types, namely, cylindrical construction and flat, or wafer, construction. Cylindrical types include the commonly used AAA, AA, N, C, and D cells, while the newer wafer designs are available for specialized equipment. The cylindrical construction is shown in Fig. 24-43.

Metal case parts are fabricated to close tolerances and have appropriate electroplated coatings to resist corrosion and to provide the electrochemically compatible surfaces required. Particular emphasis is placed on chemical purity of the electrode and electrolyte constituents during production operations, to minimize contamination. This is especially important with the zinc anode material because of a galvanic corrosion reaction between zinc and alkaline electrolyte when small amounts of certain contaminants are present. Further to reduce any wasteful corrosion of the zinc while the cell is idle, zinc oxide is added to the electrolyte.

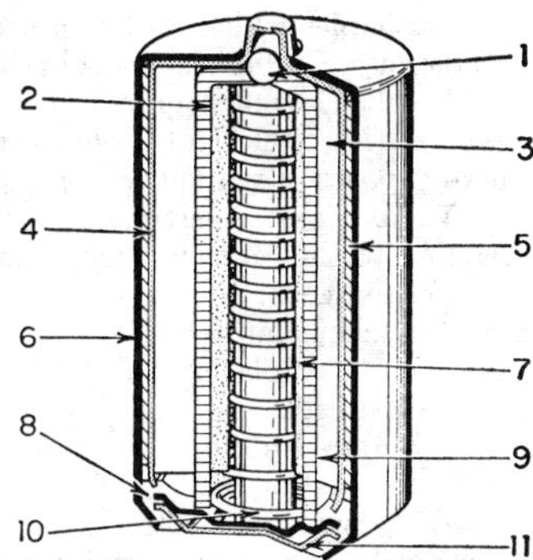

FIG. 24-43. Zinc–alkaline–manganese dioxide battery of cylindrical construction showing (1) glass insulator, (2) electrolyte absorbent, (3) depolarizer, (4) gold-plated inner can, (5) adapted sleeve, (6) outer can, (7) zinc anode, (8) molded grommet, (9) barrier, (10) spring contact, and (11) double-top assembly.

These cells have a low internal resistance and a capability for high current delivery. Also, the system is structurally stable and exhibits excellent shelf-life characteristics. Energy up to 30 Wh/lb and 4 Wh/in^3 is obtainable with high efficiency over a range of discharge currents and temperatures. The cylindrical cells available offer a discharge capacity range from 10 Ah for the D size to 0.58 Ah for the N size.

97. Reactions and Voltage. The overall discharge reaction of the zinc–alkaline–manganese dioxide cell is complex because of the reaction at the positive electrode. In strongly alkaline electrolytes manganese dioxide can be electrolytically reduced to divalent manganese. Based on a valence change of 2, the simplified combined reactions at the positive and negative electrodes are

$$Zn + MnO_2 + 2H_2O \rightarrow Zn(OH)_2 + Mn(OH)_2 \text{ with } Zn(OH)_2 \rightleftarrows ZnO + H_2O$$

The closed-circuit voltage is influenced by the discharge current and temperature. High current and low temperature requirements decrease operating voltage and service life.

Open-circuit voltage is dependent on the type of synthetic manganese dioxide used in the depolarizing cathode, and cells from various sources will have an open-circuit voltage range of 1.47 to 1.59 V; consequently, a load-voltage measurement is more reliable for determining cell condition.

98. Applications. Alkaline-manganese cells are especially useful in relatively heavy or continuous drain applications. In photographic uses they provide high amperage and several times the energy for photoflash and are ideal for powering the motor drive in motion-picture cameras. This system is also suitable for tape recorders, toys, model airplanes and boats, electric shavers and toothbrushes, and both heavy-duty and novel lighting devices.

NICKEL-CADMIUM BATTERIES

BY J. F. DONAHUE

99. Introduction. *Types.* There are two general classifications of nickel-cadmium alkaline batteries: vented and sealed. The vented nickel-cadmium system came into being during the 1890s, through the inventive efforts of W. Junger, a Swedish chemist. However, it is only since World War II that the nickel-cadmium system has gained widespread commercial acceptance in the United States. A major factor behind this market growth was the invention of the sealed nickel-cadmium battery. Until the early 1950s all commercially produced secondary batteries, whether alkaline or acid, were of the vented type. Gases generated within these cells during charge or discharge required access to the atmosphere to avoid a rupture of the cell case through buildup of internal gas pressure. In the 1940s Neumann and Gottesman of France discovered the principles which lead to a practical sealed nickel-cadmium battery. Provision was made for recombination within the sealed cell of gases produced during normal operation. Such cells do not require the addition of water, since no water can be lost through evaporation or electrolysis of the electrolyte during charging. They also can be operated in any position without danger of electrolyte leakage.

Various means have been utilized to fabricate the positive and negative electrodes, leading to the following subclassifications of nickel-cadmium batteries:

- Vented cells:
  - Pocket plate.
  - Sintered plate.
- Sealed cells:
  - Sintered plate.
  - Pressed plate.

Each type of electrode offers certain advantages and disadvantages; selection of the most appropriate type can be made, therefore, only after evaluating performance and cost criteria of the particular application.

100. Cell Components. The cell of a nickel-cadmium battery is made up of six components: (1) $Ni(OH)_3$-$Ni(OH)_2$ positive plate; (2) Cd-$Cd(OH)_2$ negative plate; (3) separator to insulate the positive from the negative plate; (4) electrolyte consisting of a solution of potassium hydroxide; (5) plastic or metal container and cover; and (6) suitable electrical leads to conduct current from the plates to the external connections.

During discharge of a cell, trivalent nickel hydroxide is converted to divalent hydroxide at the positive plate; the reverse process occurs during charge. The charged state of the negative plate is metallic cadmium, which converts to hydroxide during discharge and back to metallic cadmium when it is recharged. Specific gravity of the potassium hydroxide electrolyte does not change during charge or discharge. Thus, specific-gravity measurements of a nickel-cadmium cell's electrolyte cannot be used as an indication of the state of charge.

VENTED POCKET-PLATE BATTERIES

101. Applications. Pocket-plate batteries are primarily used for standby service. In such service the batteries are normally connected to a power source and trickle-charged or floated at a voltage maintained slightly above the battery's open-circuit voltage. The battery is thus kept in a fully charged condition, and overcharge is held to a minimum. Float service is applicable to requirements such as engine starting, emergency lighting, switchgear tripping, marine power systems, alarm systems, and railway-car lighting. The long life under float conditions and relative low cost of pocket-plate batteries compared with sintered-plate nickel-cadmium batteries make them ideally suited for such applications. Cells are available in capacities ranging from a few ampere-hours to over 500 Ah.

102. Construction. The first rechargeable nickel-cadmium battery used powdered active materials contained in perforated steel pockets. This principle of construction remains unchanged to the present time. Nickel hydroxide, usually mixed with graphite or flake nickel to improve conductivity, is used for the positive electrodes, and cadmium oxide is used for the negative electrodes. These active materials are encased in rectangular pockets formed from thin, perforated, nickel-plated steel ribbon. The many perforations permit access by the electrolyte, but they are small enough to retain the active material particles. The pockets, generally ½ in wide and in a thickness range of 0.060 to 0.250 in, are welded into nickel-plated steel frames to form electrodes of various sizes and capacities needed to meet demands of specific applications.

The positive plates of the cell are bolted together by a rod extending through a hole in the frame of each plate. A washer is placed between the plates to allow for interleaving with the negative plates, which are similarly connected. Insulation between plates is provided by a corrugated and perforated plastic sheet or vertical plastic rods. The assembled plates are inserted into a plastic or nickel-plated steel container.

The cell is filled with a potassium hydroxide electrolyte usually having a concentration of about 20% (1.2 specific gravity). Somewhat higher concentrations may be used to decrease internal resistance for low-temperature applications. However, a decrease in operating life results when higher concentrations are used.

The cells are series-connected by nickel-plated copper straps of a size suitable for the current-carrying capability required. Cells are normally mounted in wooden trays, with an air space around each cell to provide both insulation between the metal-cell containers and air convection to dissipate the heat generated during charging or discharging.

A variation of the pocket-plate construction is found in some nickel-cadmium batteries that use a tubular positive, which better resists deformation stresses that develop during cycling. Layers of nickel hydroxide are alternated with layers of thin nickel flakes within a perforated nickel-plated steel tube. The nickel flakes serve as electrical conducting paths to the tube walls. The tubular construction gives better cycle life than the pocket type, but at an increase in cost. Few tubular-type batteries are built at the present time.

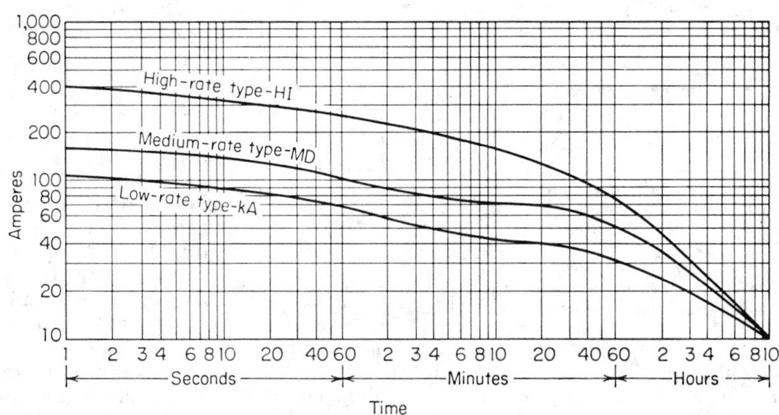

FIG. 24-44. Discharge capacity to 1.14 V for 100-Ah pocket-plate cells.

103. Discharge Characteristics. The discharge characteristics of any battery system depend heavily on the internal resistance of the battery. The battery manufacturer can vary the internal resistance, within limits, by slightly modifying the cell element. Where discharges of high current and short duration are required, the internal resistance can be minimized by using as thin a plate as possible, to maximize surface area and hence decrease current density during discharge. When lower currents for longer periods are required, the battery manufacturer will use thicker plates, fewer in number, thereby decreasing the cost of the battery.

Pocket-plate batteries are now available in three different plate thicknesses. Figure 24-44 compares the discharge capability of three types of 100-Ah cells designed for high-, medium-, and low-rate service. The plot shows the time when the discharge voltage will be

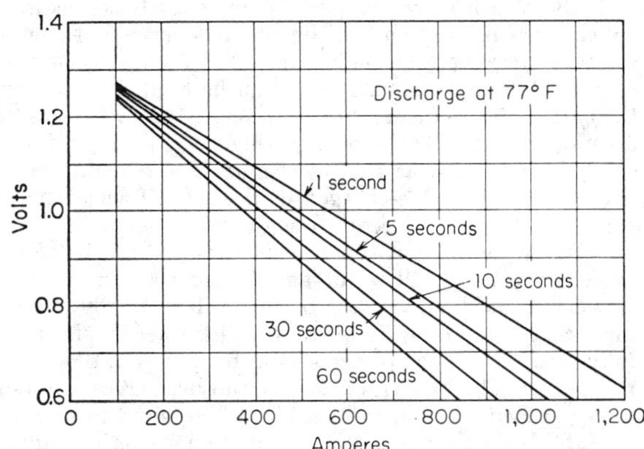

FIG. 24-45. Typical voltage-current characteristics for 80-Ah engine-cranking pocket-plate cell (discharged at 77°F).

above 1.14 V at various current drains for each of the three types of cells. For instance, at a discharge rate of 100 A the low-rate design will stay above 1.14 V for only 3 s, while the medium-rate design will be above this voltage for 1.0 min and the high-rate design for 40 min.

In selecting the proper-sized cell for a particular application, the temperature extremes under which the battery will have to operate are of considerable importance. Figures 24-45 and 24-46 illustrate the effect of a temperature drop from 77 to 32°F on the high-rate discharge capability of a pocket-plate-engine cranking battery. The individual cell voltages which result from discharging at high rates are shown in these figures for five different discharge periods for an 80-Ah size of cell. For instance, the curves show that, at 77°F in discharging at 800 A, the 1-s voltage will be 0.83 V, the 5-s voltage 0.80 V, the 10-s voltage 0.78 V, etc.

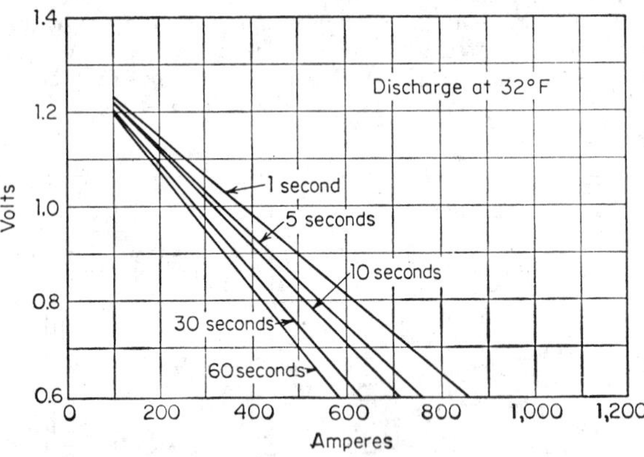

FIG. 24-46. Typical voltage-current characteristics for 80-Ah engine-cranking pocket-plate cell (discharged at 32°F).

104. Charging. Pocket-plate batteries may be charged either by constant-current or by constant-potential methods. Since the normal application for pocket-plate batteries is standby service, they must be maintained in the fully charged condition for extended periods of time. This may be accomplished by either trickle charging or float charging.

Constant-potential Charging. This is the normal method for engine-starting batteries. The generator voltage regulator should be set to hold the battery charging voltage between 1.45 and 1.50 V, multiplied by the number of cells in the battery. Readings should be taken at the battery terminal posts with the voltage regulator at proper working temperature and only after the battery voltage has ceased to rise.

Constant-current Charging. This method consists in charging at a constant current, not for a definite length of time or to a definite end voltage, but until the battery voltage ceases to rise, indicating a fully charged battery. A d-c line voltage equal to 1.40 V/cell, times the number of cells in the battery, is necessary at the beginning of the charge, and 1.85 V/cell at the end of the charge. It is necessary to insert a variable resistor between the battery and the line so that a constant charging current can be maintained as the battery voltage builds up. The 7-h rate is normally recommended for constant-current charging. Higher rates can be used; however, care must be exercised so that the battery temperature does not exceed 140°F. Rates lower than the 7-h are perfectly acceptable and will minimize gassing.

Trickle Charging and Float Charging. To maintain batteries at full charge, they may be either trickle-charged or floated. Trickle charging consists in maintaining a constant charging current through the battery; the charging current is calculated to be slightly more than the self-discharge of the battery. If too high a rate is used, excessive gassing will result and frequent watering of the battery will be necessary. If the current is too low, the battery will gradually lose capacity. Floating the battery involves maintaining a constant voltage across the battery terminals to maintain it fully charged. The voltage should be maintained at a value equal to 1.40 V times the number of cells in series. Where the discharges, although momentary, are relatively heavy or frequent, the voltage should be raised to 1.45 V/cell.

105. Maintenance. Nickel-cadmium batteries require little maintenance. Unlike lead-acid batteries, a nickel-cadmium battery may be stored for long periods in either the charged or the discharged condition without danger of damage to the battery. The electrolyte must be maintained at the proper level; too high a level may result in electrolyte being pumped out of the cell during overcharge, and too low a level will permit the upper portion of the plates to dry out and become inactive. The batteries' cell covers and intercell connectors should be maintained clean and dry to prevent any ground currents developing.

VENTED SINTERED-PLATE BATTERIES

106. Applications. Sintered-plate nickel-cadmium batteries are more expensive per ampere-hour than pocket-plate batteries. However, they give better performance at high rates and suffer a smaller capacity loss at lower temperatures. These improvements in performance are primarily due to the lower internal resistance of the sintered-plate cells. Sintered-plate battery applications have mainly resulted where high discharge rates are required over a wide temperature range and where the size and weight of the battery must be kept to a minimum. Principal uses have been in the military area for missiles, rockets, and aircraft and similar auxiliary power applications. The commercial market has been limited owing to cost. However, costs have been decreasing and should continue to do so, so that new commercial market areas will become feasible. Sintered-plate batteries have been used to a limited extent for emergency lighting, emergency controls and alarms, engine starting, and switch gear tripping.

107. Construction. Prior to World War II all nickel-cadmium batteries were either the pocket or the tubular type. But during the 1930s sintered-plate positive and negative electrodes were developed in Germany. Sintered electrodes could be fabricated much thinner (as thin as 0.020 in) than was possible with either the pocket or the tubular designs. It thus became practical to have more plates, and also less space between plates in a given-sized cell. The resulting internal resistance was about one-half that obtainable from pocket-plate batteries. A sintered plate is fabricated from three components: a metal grid, which serves as a current collector; a porous sintered structure surrounding this grid; and the active material in the porous structure. Pure nickel, or nickel-plated steel in the form of woven screen, expanded metal, or perforated sheet has been used as grid materials.

To prepare the porous sintered plaque, a fine nickel powder is sintered onto the grid. The resulting plaque has a porosity of about 80%. The active materials—a nickel salt in the positive and a cadmium salt in the negative—are impregnated into the pores of the plaque.

The cell element is then assembled, alternating positive and negative plates with a separator between each. Separators are usually woven or nonwoven nylon cloths with a thin, porous film such as cellophane. The positive plate leads are brought together and welded to a nickel-plated steel terminal. The negative plate leads are similarly welded. The element is then placed in a container, which is usually a plastic material such as nylon or high-impact polystyrene. Nickel-plated steel containers have been used on occasion when the battery is likely to be exposed to high temperatures.

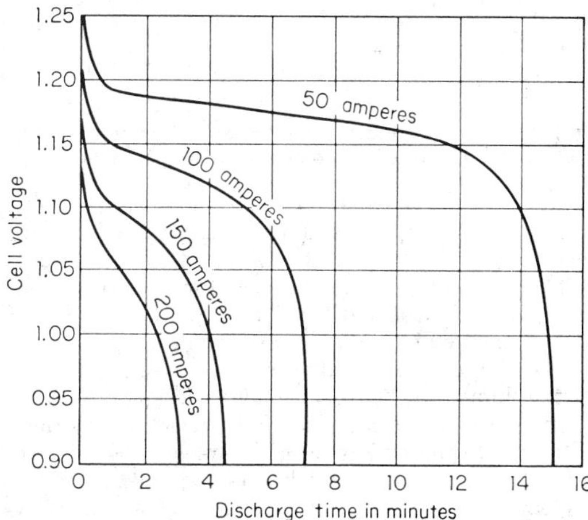

FIG. 24-47. High-rate constant-current discharges for 12.5-Ah vented sintered-plate cell.

108. Discharge Characteristics. The sintered-plate battery's low internal resistance makes it ideally suited for applications requiring long battery life and high current drains over a wide temperature range. There are other alkaline systems (silver-cadmium and silver-zinc) which give more watthours per unit weight and volume, but at a higher cost and with a much shorter life. Figure

24-47 shows the voltage obtained from a 12.5-Ah cell when subjected to high discharge rates. Continuous drains as high as 200 A can be obtained from this small cell, which weighs only 1.3 lb and occupies a volume of 19 in³. Figure 24-48 shows the discharge characteristics of the same cell at lower current drains. Generally, vented sintered-plate nickel-cadmium batteries will deliver approximately 13 Wh/lb and 1.3 Wh/in³.

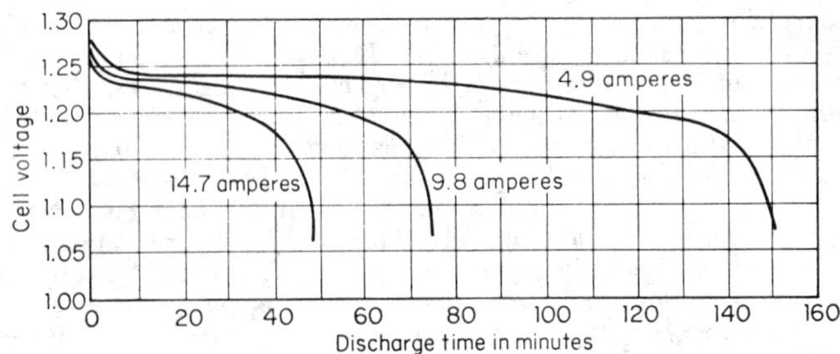

Fig. 24-48. Constant-current discharges of a 12.5-Ah vented sintered-plate cell.

The watthours per cubic inch will be somewhat less in the smaller sizes, since the cell terminals occupy a disproportionate percentage of the total volume in these cells. Figure 24-49 shows the effect of reduced temperature on a 90-Ah 6-V battery designed for engine starting. At 300 A the capacity is reduced by only 20% when the temperature is lowered from +77 to −40°F.

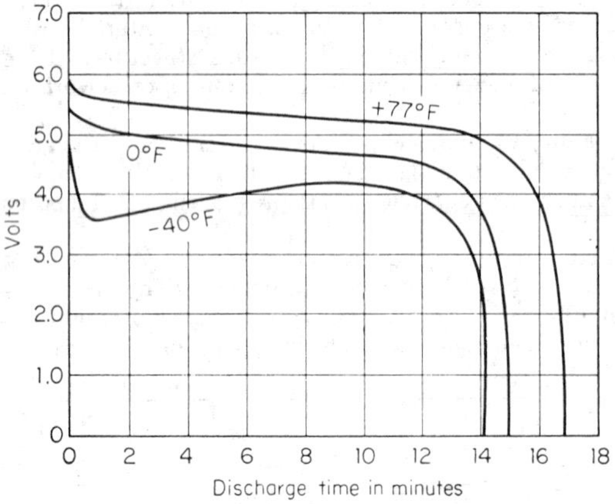

Fig. 24-49. Temperature discharge characteristics at 300 A for a 90-Ah 6-V vented sintered-plate battery.

Individual cells are available in sizes ranging from a fraction of an ampere-hour to several hundred ampere-hours. These may be connected in series to obtain any desired voltage.

109. Charging. As with the pocket-plate-type batteries, sintered-plate nickel-cadmium batteries can be charged by applying either a constant-potential or a constant-current charge to the battery.

Constant-potential Charging. Constant-potential charging is the easiest and most rapid method of charging for most applications. However, the charging equipment is usually more expensive than for the constant-current type. A fully discharged battery can be almost completely recharged by the constant-potential method in 1 h, although it is preferred that the charge be continued for 2 or 3 h until the current stabilizes for three consecutive half-hour readings. The voltage of the charger should be set at 1.55 V, times the number of cells in the battery. Figure 24-50 shows the current-time relationship for a typical charge at 1.55 V for a single cell.

Some types of installations use constant-potential charging, with the maximum current limited by charger capacity, design, or automatic regulation. This is the type of system used for aircraft. The generator or other charging source is controlled by a voltage regulator having a preset value of 1.45 to 1.55 V/cell. The current is constant initially until a predetermined voltage is reached, at which time the current decreases in much the same manner as shown in the lower part of the curve in Fig. 24-50.

Constant-current Charging. To charge at a constant current, the d-c charging source must be capable of operating at 1.80 V, times the number of cells in the battery. The

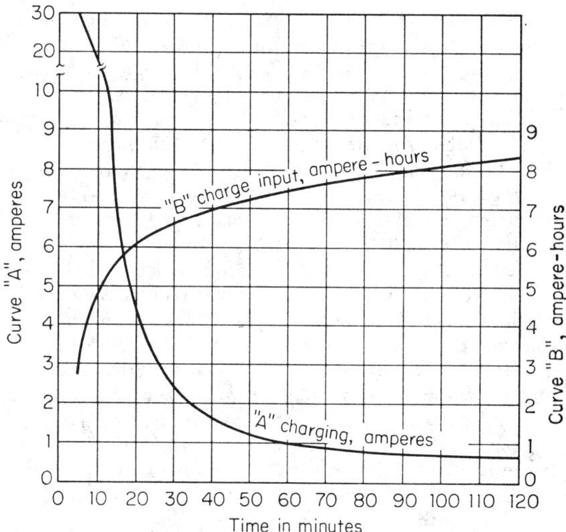

FIG. 24-50. Constant-potential charge curve at 1.55 V for 5-Ah vented sintered-plate cell.

charger must have an ammeter and rheostat of suitable capacity and range. The charging current is usually selected to supply a full charge in 5 to 8 h. It should be noted that it is always necessary to replace about 125 to 150% of the ampere-hours removed on the previous discharge. Thus, to charge completely a 10-Ah battery a current of 3 A could be applied for 5 h or 2 A for 7.5 h. A typical charging curve at the 5-h rate is shown in Fig. 24-51. If the extent of the previous discharge is not known, the battery may be charged at a constant current until the battery voltage ceases to rise. Reasonable overcharge is not harmful as long as the electrolyte level is maintained over the tops of the plates and the electrolyte temperature is not permitted to exceed 125°F. Care must be exercised to terminate a constant-current charge at the proper time to preclude overheating and damage to the battery.

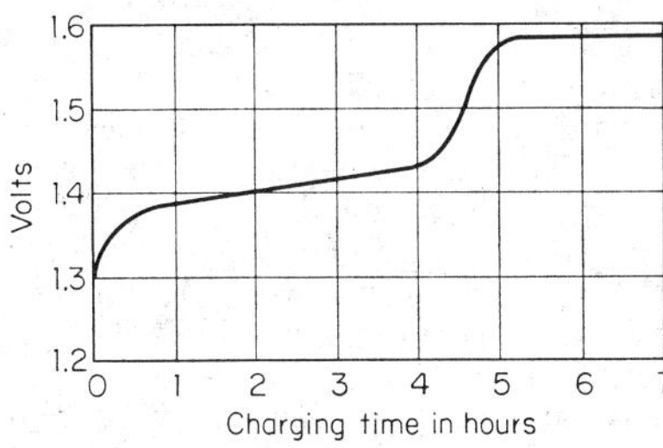

FIG. 24-51. Typical charging curve at constant current for a fully discharged vented sintered-plate cell.

Trickle Charging and Float Charging. Batteries which are used on standby service such as switch tripping or control applications are maintained in the fully charged condition either by a trickle charge or by a float charge. The procedure is identical to that described under Vented Pocket-plate Batteries, with one exception. The float voltage for sintered-plate batteries is 1.36 to 1.38 V times the number of cells in the battery.

110. Maintenance. Little maintenance is required for sintered-plate batteries. The major points are identical to those described in Par. 105.

SEALED NICKEL-CADMIUM BATTERIES

111. Applications. Though sealed cells were first manufactured in the 1950s, it was not until 1960 with the introduction of the first cordless razor by the Remington Shaver Division of Sperry Rand that sealed cells were produced in large volume. Since that time the industry has grown rapidly. A great variety of cordless appliances have subsequently been introduced. However, the cost of sealed cells has limited most of their large-volume applications to low-wattage appliances. The electric toothbrush, knife, razor, etc., use batteries which will deliver about 1 to 10 Wh.

Large-volume production of sealed cells has allowed extensive investment in automated facilities and a resultant reduction of price. Prices have been reduced by over one-third in the 7-year period from 1960 to 1967. Sharply improved electrical performance has also been obtained. Higher charge rates and more capacity per unit weight and volume have been achieved. As further reductions in cost and improvements in performance are achieved, new sealed-cell markets will become practical. Appliances which require more energy, such as vacuum cleaners and power tools, and which are now produced in only limited quantities will become more attractive to the public.

Sealed cells have been used extensively in both military and space programs. Target drones, torpedoes, missiles, and rockets are a few of the common military applications. Sealed cells, in conjunction with solar cells, continue for at least the present to be the most reliable power source used in satellites.

112. Construction. The major problem associated with sealing any type of secondary battery is preventing the buildup of pressure caused by gassing of the electrodes during charging. In vented cells, oxygen is given off by the positive electrode and hydrogen at the negative as they reach full charge. The prevention of pressure buildup in sealed nickel-cadmium batteries was achieved by incorporating the following two fundamental features.

Excess Negative Capacity. In a sealed cell the negative plate capacity is about 1.5 times the capacity of the positive. Thus, when a cell is charged, the positive plate reaches the fully charged state before the negative. As the positive approaches full charge, it starts to evolve oxygen. The oxygen passes through the separator to the negative plate and oxidizes (discharges) the negative. On the assumption that the proper charge current is maintained, the gas is consumed at the negative plate at the same rate at which it is generated at the positive plate. The charged condition of the negative plate also reaches a steady-state condition whereby it is reduced by the charging current and oxidized by the evolved gas. This results in the negative plate never reaching a point where it evolves hydrogen, even though the cell itself remains on charge indefinitely.

Limited Electrolyte. For the oxygen to pass from the positive to the negative plate, it is necessary to restrict the quantity of electrolyte in the cell to as low a level as practical to provide for ionic conductivity. There is no free electrolyte in a sealed cell, as it is all absorbed by the plates and separators.

Practically all sealed nickel-cadmium batteries are manufactured using sintered-type electrodes. Construction of these electrodes is identical to that of vented cells. Sealed cells are manufactured in three different shapes: rectangular, cylindrical, and button.

Rectangular cells are available in sizes ranging from about 1 to 25 Ah. Their construction is almost identical to that previously described for the vented sintered-plate cells; the only real differences are in the plate separator material and a lesser quantity of electrolyte in the cell. The sealed cell utilizes a thin separator material, which is usually either a woven or nonwoven nylon or polypropylene material. A film-type separator such as that used in vented cells is not employed, as it would inhibit the passage of oxygen from positive to negative plate during overcharge.

Cylindrical cells are used most commonly. They are available in sizes ranging from 0.5 to 7 Ah. Larger experimental cells have been built but have not been mass-produced as yet. Present production of cylindrical cells is almost entirely of the sintered-type electrode. Each cell contains one positive plate, one negative plate, and one separator; the separator is interleaved between the positive and negative. The assembly is coiled in a "jellyroll" configuration, and this cylindrical coil is then inserted into a nickel-plated steel can. A tab from the negative plate is welded to the can. The lead from the positive is welded to the cover. An insulating gasket is placed between the can and cover, and the can rim is crimped against the gasket. The cover usually contains a pressure-release mechanism that will vent excessive gas pressure if the cell malfunctions, thus avoiding the possibility of a violent rupture of the cover from the can.

The cost of cylindrical cells per ampere-hour is less than that of rectangular cells. This is due primarily to cylindrical cells having only one positive and one negative handled per cell during manufacture, whereas in the manufacture of rectangular cells a multiplicity of positives and negatives must be handled simultaneously.

Button cells are built in capacities ranging from about 0.02 to 2.0 Ah capacity. They are made either with sintered plates or with pressed plates. The pressed plates are made by pressing the powdered active materials into a disk and then tightly enclosing the disk in wire screen. The sintered or pressed-disk positive and negative are interleaved with a separator and placed in a nickel-plated steel cup, which serves as the cell container. A tab from the positive electrode is welded to the cup, and the negative is welded to the cover. Cell closure is similar to that employed for cylindrical cells.

113. Discharge Characteristics. Sintered-plate sealed cells give reasonable performance at both high and low discharge rates over a wide temperature range. Figure 24-52 shows the effect of discharge rate on the voltage of a 1.2-Ah cell. The variation of capacity and voltage with discharge rate shown on this curve is typical of what can be expected from sealed cells. The effect of temperature on discharge capacity is shown in Fig. 24-53 for two moderate discharge rates. Improvement in discharge performance at low temperature can be obtained by increasing the concentration of the potassium hydroxide electrolyte from the usual 25% to as high as 40%. Care must be exercised in charging such cells, as recombination of the gases evolved on overcharge does not proceed so rapidly as in cells with the lower electrolyte concentration.

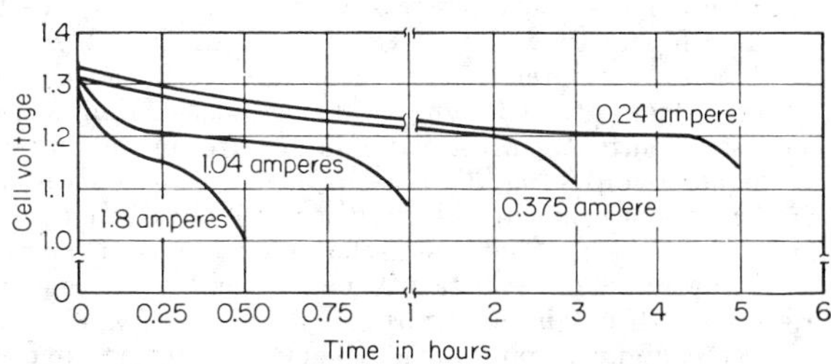

Fig. 24-52. Effect of discharge rate on operating voltage of a 1.2-Ah sealed cell.

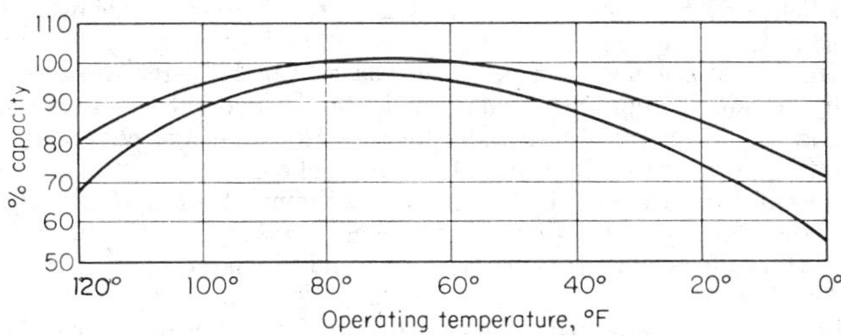

Fig. 24-53. Effect of temperature on discharge performance of sealed cells: 10-h rate (top); 5-h rate (bottom).

Sealed cells deliver about 14 Wh/lb and 1.4 Wh/in³ at the 5-h discharge rate. Table 24-9 shows a few electrical and physical characteristics of sealed cells having the same dimensions as the familiar Leclanché dry cells:

Table 24-9. Characteristics of Typical Nickel-Cadmium Cells

Dry-cell type	Ni-Cd type	Weight, oz	Ah rating	Internal impedance*
AA	500 SC	0.8	0.5	0.020
C	1.8 SC	2.4	1.8	0.014
D	4.0 SC	5.2	4.0	0.006
F	7.0 SC	7.6	7.0	0.005

* Values are for a fully charged cell at 60 c/s. For a semidischarged cell impedance increases 20% and 50% for a fully discharged cell.

In discharging a battery composed of many sealed cells in series, the discharge should be stopped before any cell in the battery drops below zero voltage. If the discharge is not terminated, the remaining cells in the battery will be charging the low cell in the reverse direction, which can severely shorten the low cell's life. Fortunately, most battery-operated devices will no longer operate when one cell of a battery has dropped to zero. However, the more cells there are in a battery, the greater the possibility that the using equipment's function will not be seriously impaired by the dropping of one cell, so that the device continues to function and the low cell of the battery is driven into reverse.

The electrical characteristics of sealed pressed-plate cells are similar to those given above for sintered-plate cells, with two exceptions. Pressed-plate cells give much less capacity at discharge rates higher than $C/1$, where C is equal to the capacity of the cell in ampere-hours. The pressed-plate cells should not be left on continuous overcharge at rates higher than $C/50$. Overcharge at rates higher than this, though not at all detrimental to sintered-plate cells, will result in shorter life for pressed-plate cells.

114. Charging. Charging of sealed nickel-cadmium cells is normally performed at a constant current. Most manufacturers recommend that this be performed at currents no higher than $C/10$. To obtain full capacity, it is necessary to charge about 140% of the previous output. Thus, a completely discharged 10-Ah cell would be charged at 1.0 A for 14 h. The end-of-charge voltage is in the range of 1.42 to 1.50 V at room temperature. The voltage will be less at higher temperatures (for example, 1.39 to 1.44 V at 110°F) and higher at lower temperatures (for example, 1.52 to 1.58 V, 32°F).

Overcharging at rates higher than $C/10$ can result in cell overheating, or if the oxygen does not recombine as rapidly as it is generated, the seal may be broken. Both conditions will reduce cell life. Charging at temperatures above 100°F will result in somewhat less than rated capacity on the subsequent discharge. For instance, a cell charged at 110°F will exhibit about 15 to 20% less capacity than a cell charged at room temperature. Continual overcharge at the $C/10$ rate will have no harmful effect on the cells at temperature above 50°F; however, care must be exercised in overcharging a cell at temperatures below 50°F. Recombination of the evolved oxygen is slower at the lower temperatures, so that reduced charging rates must be used. For special applications, where rapid recharge is a necessity and the charge can be carefully controlled, it is possible to charge cells to about 90% capacity in about 1 h at the $C/1$ rate. To prevent overcharging at this high rate, care must be exercised that the charge is terminated at the proper time; otherwise, the cell seal will be broken.

At the present time there is considerable development work underway to devise a foolproof method for rapid charging of sealed cells. Several approaches offer promise:

1. Using a third electrode either to signal the end of charge or to recombine oxygen at a rapid rate.
2. Using a diode in parallel with the cell to shunt the current as the cell reaches full charge.
3. Employing electronic circuitry to terminate the charge at the proper voltage levels.

115. Maintenance. Sealed nickel-cadmium cells require no maintenance beyond normal charging.

116. Bibliography

BRIGGS, G. W. D., JONES, E., and WYNNE-JONES, W. F. K. The Nickel Oxide Electrode—Part 1; *Trans. Faraday Soc.*, 1955, Vol. 51, p. 1433; The Nickel Oxide Electrode—Parts 2 and 3; 1956, *ibid.*, Vol. 52, pp. 1260, 1272.

CARSON, W. N., JR., and MCQUADE, J. M. The Use of Auxiliary Electrodes in Sealed Cells, *Extended Abstr. Battery Div. Electrochem. Soc. Meeting*, Sept. 30–Oct. 3, 1963, Vol. 8, p. 32.

CONWAY, B. E., and BOURGAULT, P. L. The Electrochemical Behavior of the Nickel–Nickel Oxide Electrode—Part I. Kinetics of Self-discharge; 1959, *Can. J. Chem.*, Vol. 37, p. 292; The Electro-chemical Behavior of the Nickel–Oxide Electrode—Part II. Quasi-equilibrium Behavior; 1960, *ibid.*, Vol. 38, p. 1557.

FLEISCHER, A. Characteristics of Nickel Cadmium Batteries; *AIEE Winter Conf. Paper* CP 60-322, 1960.

RÜETSCHI, P., and OCKERMAN, J. B. Sealed Batteries with Auxiliary Electrodes, *Extended Abstr. Battery Div. Electrochem. Soc. Meeting*, Sept. 30–Oct. 3, 1963, Vol. 8, p. 29.

VINAL, G. W. "Storage Batteries," 4th ed.; New York, John Wiley & Sons, Inc., 1955.

SECTION 25

TELEPHONE, TELEGRAPH, AND DATA SYSTEMS

BY

C. R. WILLIAMSON Assistant Vice President—Engineering, American Telephone and Telegraph Company; Member, IEEE

CONTENTS

Numbers refer to paragraphs

SECTION 25

TELEPHONE, TELEGRAPH, AND DATA SYSTEMS

TELEPHONE INSTRUMENTS

1. General Description. The requirements for a telephone instrument are:

a. Acoustical-to-electrical energy converter.

b. Electrical-to-acoustical energy converter.

c. Induction coil or hybrid circuit.

d. Means for signaling.

e. Source of power.

f. Transmission medium.

A typical telephone instrument of today consists of a carbon transmitter, electromagnetic receiver, anti-side-tone induction-coil hybrid circuit, dial and switchhook to signal the switching system, and a ringer to signal the user. Power is usually derived from the switching system over the telephone line that also serves as the transmission medium.

Much of the efficiency of modern telephones is due to the use of the anti-side-tone circuit as a basis for telephone instrument design. One of the common anti-side-tone circuits is shown in Fig. 25-1.

Incoming alternating currents from the line cause voltages to be induced in the three windings of the induction coil. The network impedance is high relative to the receiver and transmitter impedances so that little power is wasted in the network. Alternating currents originating in the transmitter induce voltages in the three windings so as to cause the energy to be divided between the network and line impedances, with little, if any, available to the receiver. This circuit accomplishes two results.

1. The fraction of power delivered to the receiver from the line can be traded against the fraction of power delivered to the line from the transmitter by choosing the proper impedances and turns ratios,

$$\frac{P_{\text{rec}}}{P_{\text{line}}} + \frac{P_{\text{line}}}{P_{\text{trans}}} = \frac{m}{(n+1)(m+n)} + \frac{n(m+n+1)}{(n+1)(m+n)} = 1 \qquad (25\text{-}1)$$

2. The amount of side tone (transmitter power delivered to the receiver) can be controlled by choosing the impedances and turns ratios to satisfy the conditions of biconjugacy,

$$\frac{Z_{\text{net}}}{Z_{\text{line}}} = \frac{mn}{m+n+1} \qquad (25\text{-}2)$$

$$\frac{Z_{\text{rec}}}{Z_{\text{trans}}} = \frac{m(m+n-1)}{n} \qquad (25\text{-}3)$$

3. For maximum efficiency the impedances and turns ratios should be such as to make Z_{set} equal Z_{line}.

$$Z_{\text{set}} = Z_{\text{line}} \qquad \text{when } \frac{Z_{\text{rec}}}{Z_{\text{line}}} = \frac{m(m+n)}{n+1} \qquad (25\text{-}4)$$

The choice of impedances and turns ratios in practice represents a compromise between side-tone balance, set efficiency, and requirements of other parts of the telephone system.

2. Types of Telephone Instruments. There are two basic types of telephone instruments, magneto (local-battery) sets and common-battery sets.

Magneto (local-battery) telephone sets provide a battery at the station-set location to supply direct current to the transmitter. Signaling to the operator is accomplished by a hand generator (magneto) that sends alternating current over the line.

Common-battery telephone sets receive transmitter direct current from the central office (CO) over the telephone line. Signaling the CO is accomplished by the switchhook, which causes a d-c change in the line that in turn operates a supervisory relay in the CO.

The Bell System's 500-type telephone is representative of the typical common-battery telephone instrument in use today.

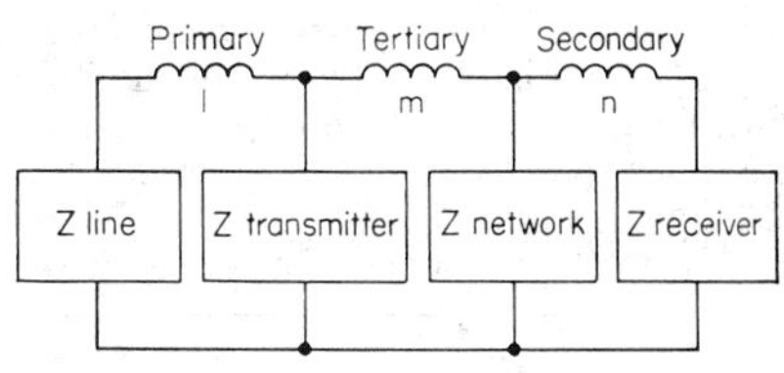

Fig. 25-1. Common anti-side-tone circuit.

In Fig. 25-2, the circuit of this telephone is drawn to resemble the anti-side-tone circuit shown in Fig. 25-1.

The ringer is shown connected for individual or bridged ringing. Party ringing is accomplished by connecting the ringer from one side of the line to ground. The switchhook breaks both sides of the line when the set is on-hook and closes both sides of the line when the set is in use. A break contact mutes the clicks generated by the switchhook in the receiver. The rotary-dial contacts make and break one side of the line to generate a dial pulse.

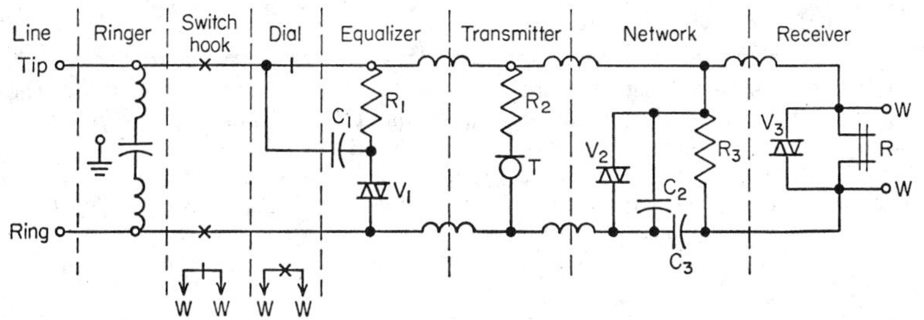

Fig. 25-2. Typical common-battery telephone instrument.

The dial "make" contact mutes the dial clicks in the receiver. Capacitor C_1 suppresses dial pulse transients to prevent the dial from causing radio interference. The resistor R_1 and the varistor V_1 form the line equalizer that acts as a shunt on short loops to limit the power transmitted by the set to the line, and vice versa. Too high a level might create crosstalk problems. The varistor decreases in resistance as the voltage applied to the set increases with decreasing loop length. This effect is shown in Fig. 25-3*a*.

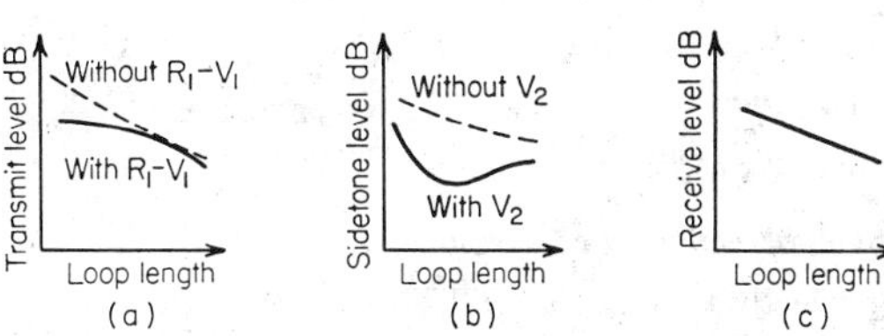

Fig. 25-3. Effect of line equalizer (*a*), anti-side-tone network (*b*), and overall receiver power-loop length characteristic (*c*).

The primary winding of the induction coil is split to balance the impedance of each side of the line. An unbalanced circuit is more susceptible to inductive noise, since it allows a difference in noise potential which causes higher noise currents. The carbon transmitter T and resistor R_2 make up the transmitter impedance of the anti-side-tone circuit. Capacitor C_2, varistor V_2, capacitor C_3, and resistor R_3 make up the anti-side-tone network impedance. Varistor V_2 compensates for changes in line impedance so that conditions approximating biconjugacy can be maintained to control side-tone levels (Fig. 25-3*b*). The receiver R makes up the impedance of the receiver leg of the circuit. Varistor V_3 limits the output level of the receiver to levels well below those which might be objectionable to the user. The receiver power varies little with loop length (Fig. 25-3*c*).

3. Common Versions. There are many versions of the standard telephone instrument to accommodate particular use requirements. A partial listing is as follows: (1) desk, (2) wall, (3) hang-up, (4) panel, (5) coin, and (6) key.

In addition, each of these may be equipped with special equipment such as:

1. Amplifier-equipped handsets.
2. Hands-free speaker systems.

4. Requirements. The telephone instrument must be able to function with the various switching systems (PBX, panel, step-by-step, crossbar, electronic switching) over varying loop lengths (typically 0 to 15,000 ft of 26-gage cable) and to provide the desired levels of transmit, receive, and side tone under these conditions. Certain compromises must be made in an optimum design, because there may be conflicting requirements between the telephone instrument and other parts of the telephone plant. For instance, the maximum d-c resistance of the set is constrained by the minimum current required to operate supervisory relays in the CO even though on long loops the set would be more efficient with a higher resistance.

Typical requirements for a telephone instrument are tabulated below:

	Short loop	Long loop
Line current	200	20 mA
A-c impedance	600	900 Ω
Transmit levels	−10	−20 dBV
Receive levels	95	86 dB(RAP)
Side tone	95	85 dB(RAP)
Rotary dial	10 + 0.5 pps	
Make-break ratio	61 ± 3%	

5. Ringing Systems. Ringers are classified in two groups according to the d-c resistance of the coil; high-impedance ringers have a resistance of 2,500 Ω or more, and low-impedance ringers have less than 2,500 Ω coil resistance. Ringing bridges are classified as capacitor type (using a capacitor in series with the ringer) or tube type (using a cold-cathode electron tube in series with the ringer). Ringing systems are further classified as polarized or nonpolarized.

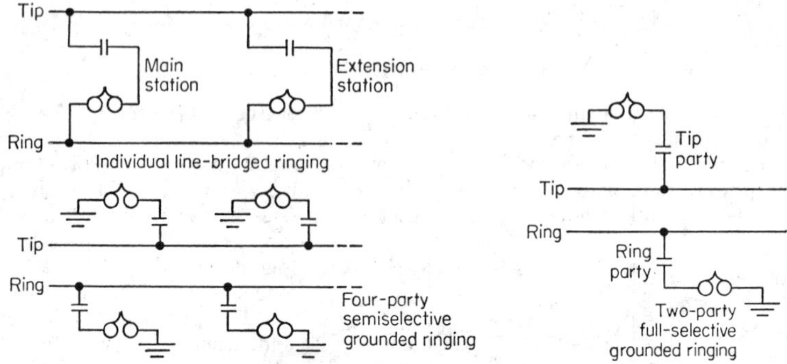

Fig. 25-4. Nonpolarized ringing connections.

Any ringing system where ringing selectivity is not provided or is accomplished by means other than polarized ringing current (such as frequency-selective ringing) is classified as nonpolarized. Examples of ringing connections are shown in Fig. 25-4. Semiselective ringing uses code ringing to differentiate between stations on the same side of the line.

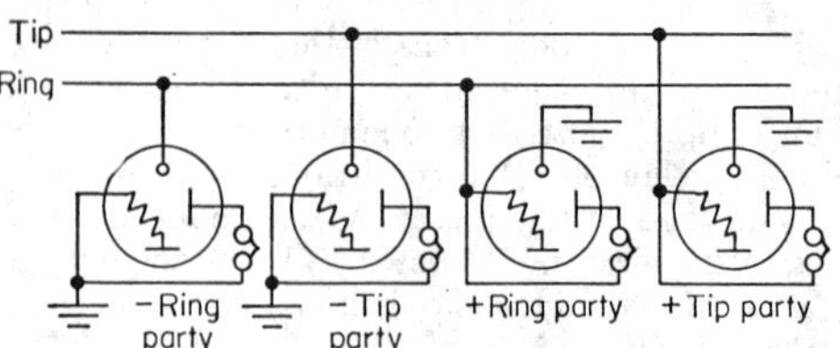

Fig. 25-5. Polarized ringing connections.

Polarized ringing denotes any ringing system which accomplishes ringing selectivity through bridges which are connected, solid, and biased to operate on a particular polarity of ringing current. Four-party full-selective or 8-party semiselective ringing is accomplished (Fig. 25-5) by using sets employing a cold-cathode electron tube. The tube and ringer at the station are poled to operate only on a positive or negative superimposed ringing signal. Magneto ringing and frequency-selective ringing are also used in full-selecting ringing systems.

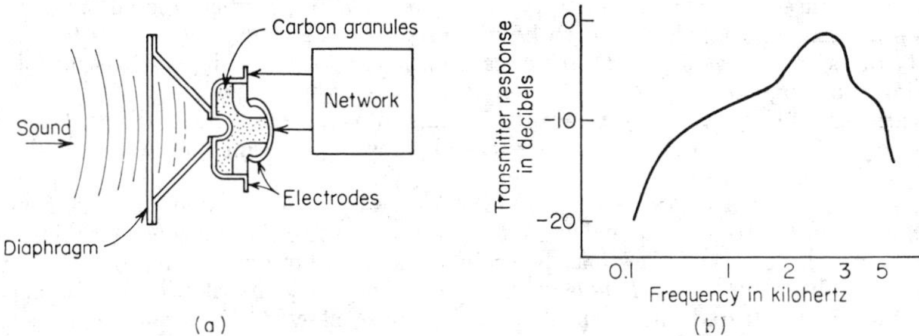

FIG. 25-6. Carbon-granule transmitter (*a*) and its response (*b*).

6. Transmitters. The most common telephone transmitter (see Fig. 25-6*a*) in use today is the carbon-granule type. Two concentric electrodes, one attached to a backplate, the other to the diaphragm, enclose the carbon granules. Vibration of the diaphragm causes the resistance of the carbon to vary inversely with pressure. This change in resistance acts to convert the sound energy into electrical energy. It is an amplifier in that it uses the energy supplied by the voice to control the energy supplied by the battery. A gain of 20 to 30 dB can be realized in this manner. The efficiency of the carbon transmitter is directly proportional to the direct current through it. A typical response of a carbon transmitter is shown in Fig. 25-6*b*.

Some use is made of electromagnetic transmitters in some telephone instruments. This type of transmitter is less efficient than the carbon type and must be supplemented with an amplifier.

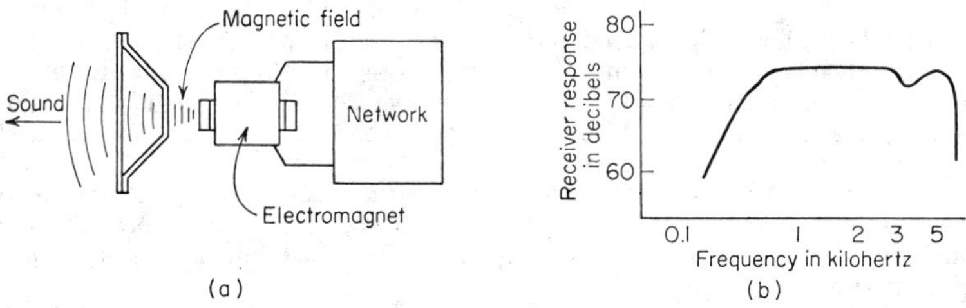

FIG. 25-7. Typical telephone receiver (*a*) and its response (*b*).

7. Receivers. The basic components of the typical telephone receiver (Fig. 25-7*a*) are the diaphragm, the magnetic core, and the coil. Voice currents in the coil produce a change in the magnetic field that causes the diaphragm to follow the fluctuations of current. The coil usually has a low impedance on the order of 100 to 200 Ω. The same structure, with a higher-impedance winding, can be used as an electromagnetic transmitter, but it usually requires some additional gain. A typical receiver response is shown in Fig. 25-7*b*.

8. Handsets. The handset of the telephone actuates the switchhook and places the receiver and the transmitter in the proper relationship to the mouth and ear for maximum efficiency. It must be heavy enough to actuate the switchhook but light enough to be held by the user for extended periods. The critical dimensions are the center-to-center distance from R to T (modal distance)

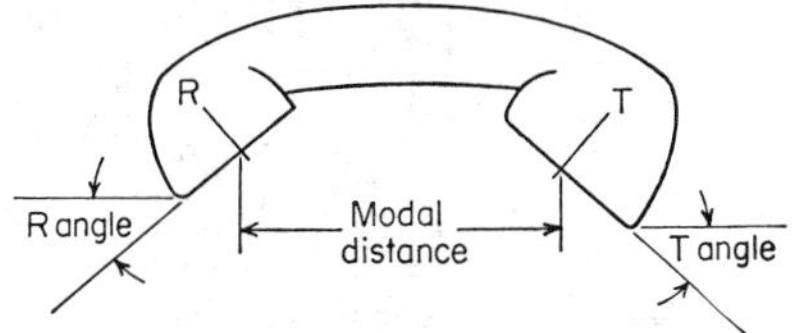

FIG. 25-8. Telephone handset.

and the *R* angle and *T* angle (Fig. 25-8), which must be adjusted to compensate for the efficiencies of the transmitter and receiver types used in the design.

9. Ringers. There are two basic types of ringers, those which are tuned to respond to a certain frequency of ringing signal to provide selectivity and straight-line ringers, which are polarized to obtain selectivity.

In both cases, the ringing signal, typically 85 V at 20 c/s superimposed on 48 V d-c, causes a changing magnetic field to be generated by the ringer coil. This field alternately pulls and releases the armature, causing it to swing the clapper back and forth, striking the gongs in the process and generating the audible signal.

Variations of the bell ringer include buzzers and horns. A recent innovation is a tone ringer that makes use of tones generated by the ringing signal, amplified, and made audible through the use of a loudspeaker.

The important acoustic frequencies generated by the ringer are in the 1,000 to 2,500-c/s band. Higher frequencies are not so effective for persons with impaired hearing, and lower frequencies may be unduly masked by room noise.

The required sound level depends upon the frequency distribution of the energy. A weaker concentrated energy peak may be more effective than more total energy spread over a wider band.

10. Induction Coils. Induction coils are used in most modern telephone instruments to form the basic hybrid of the anti-side-tone circuit. The coil core is commonly made of a laminated silicon iron core and has three or more coil windings. The turns ratios of the windings are calculated so as to implement the conditions of biconjugacy necessary in the anti-side-tone design. Adjustments can be made to accommodate transmitters and receivers of various impedances and efficiencies by varying the turns ratio.

11. Keys. Telephone sets are often equipped with keys that are used with key systems to provide certain features not available with single-line sets. A key is used to provide local switching functions so that one telephone set may be used with two or more lines. A typical six-button key telephone might provide the following features:

1. Ability to pick up one of three central-office lines.
2. Ability to hold one or more lines and use another.
3. A key to signal an assistant.
4. A key to connect to an "intercom" system.

In addition, the keys are often used as a signal source by use of various lamp codes to indicate when a line is being used, on hold, or has an incoming call.

12. Switchhooks. The switchhook is used to signal the switching system when a call is originated or terminated. It is usually operated by the weight of the handset on a spring pile up that opens "tip" and "ring" when the set is on-hook, and closes the "tip" and "ring" when the set is off-hook.

13. Touch-tone Dial. The "Touch-tone" dial is a dual-frequency oscillator capable of the generation of any one of 16 combinations of two frequencies under the control of a push-button-actuated mechanism. Operation of a push button causes the generation of a two-frequency signal.

Although the dial may be used in a 4 × 4 array, the most common versions have a 3 × 4 array and use one of each group of frequencies to dial one of the digits, 0 through 9, as illustrated in Fig. 25-9*a*.

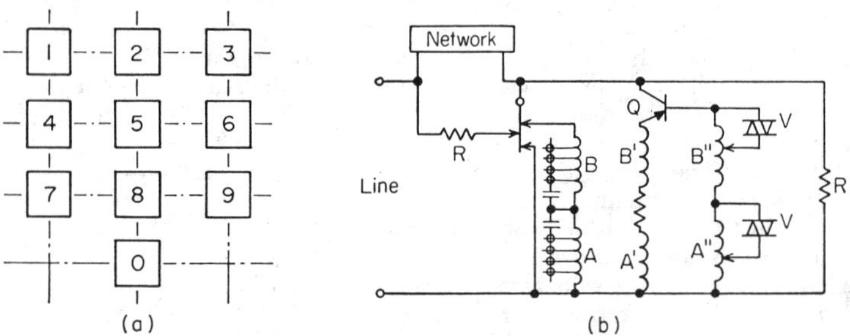

Fig. 25-9. Touch-tone dial array (*a*) and circuit for two-frequency operation (*b*).

Two-frequency operation is accomplished by employing two completely independent tuned transformers *A* and *B*, with their respective windings in series. Simultaneous oscillation at two frequencies is possible because the transistor is operated linearly and amplitudes are limited by other elements.

The circled points on the transformers in Fig. 25-9*b* represent frequency-selecting contacts, one of which is closed on each winding when a button is depressed. The tuned windings normally have a direct current through them. Pressing the button interrupts this direct current and shock excites the oscillator.

14. Rotary Dial. Dial speed is controlled to 10±0.5 pps, which plays an important part in achieving the close control of the make and break time. By controlling the make and the break time to close limits, appreciable extension of dialing range is possible.

Modern dials are of the single-lobe cam-and-pawl arrangement, driven by a motor spring (wound by the user for each digit) and controlled by a governor to maintain a constant speed.

15. Dialers. Several mechanical or electromechanical dialer designs are in use that will dial the desired number, which has been precoded into the dialer, with the depression of a start button. Rotary and touch-tone dialing can be provided in this manner.

The coding is done by recording the number in memory media such as cards, tapes, or drums. Sensing devices interpret the code and cause pulsing contacts to send the proper digit out on the line.

KEY TELEPHONE SYSTEMS

16. Definition and Application. A key telephone system consists of multibutton station sets or keys, connecting cables, and relay switching apparatus on central-office, PBX, or private lines. Whether associated directly with central-office lines or as supplements to manual or dial PBX station circuits, key-system switching does not require the services of a special operator or attendant to handle calls incoming to or outgoing from the system.

In many of the more complex systems, the line-relay apparatus and power plant required for its operation are contained in apparatus boxes or cabinets or on large

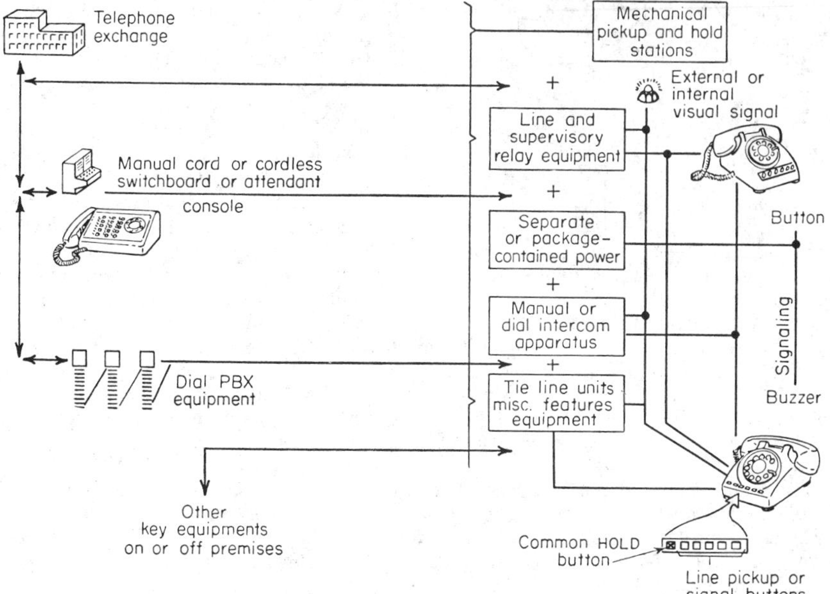

FIG. 25-10. Correlation of key-telephone-system arrangements—broad concepts.

floor-supported mounting frames. Subject to station location and grouping on the premises, such equipment is most often centrally situated for ease of cabling, maintenance, change, or rearrangement—and to take full advantage of the relatively restricted operating range of the equipment. At these locations, also, may be found the equipment needed to provide two-way manual or dial selective intercommunicating between stations in the system, tie-line apparatus interconnecting separate systems, etc. The broad concept is diagrammed in Fig. 25-10.

17. Equipment and Major Services. Station instrumentalities used in key systems range from single-line pickup desk or hang-up instruments—with or without separately mounted lever or push-button keys—through (turn-button) two-line pickup and (multibutton) four- and six-line sets, to large-capacity (10, 18, 20, 30, and upward) lever-type or push-button key boxes, turrets, and "consoles." These sets are generally arranged for 2-wire line, common battery operation, although increasing use is being made of sets arranged to handle 2- or 4-wire line terminations.

The multibutton stations are often arranged with plastic keys, which may be illuminated as status indicators of line usage. In some instances visual signals are provided separately, in the form of one, two, three, or more "beehive"-type indicators.

The ringer in most multiline pickup sets is electrically associated with one of the lines appearing at the station, i.e., as a line ringer operating whenever an incoming call is made to that line. On the other hand, this same ringer can be connected as a common ringer—activated from local circuitry to signify whenever calls are incoming on any (or all) line(s) appearing at the station.

Sometimes switchhook plunger buttons or other control keys integral to the telephone set are provided to allow manual switching control of subordinate stations or for bell transfer or cut off, etc. Automatic supervisory relay control of comparable functions can be provided optionally, as desired.

Fig. 25-11. Typical station apparatus.

Most telephone-set mounting cords are dressed with spade-tipped conductors for termination with connecting line wires beneath screw terminals on multipost connecting blocks. In larger-capacity multiline-set categories, however, the trend is to provide one or more multicontact plugs (up to 50 contacts, generally) which allow for rapid installation or change in conjunction with mating cable terminating hardware. Figure 25-11 illustrates typical station apparatus.

When key-system relay equipments are used to afford line pickup and hold services, and to control auxiliary visual and audible signals, the multibutton station set usually features a common "hold" button, regardless of the number of lines appearing at the station. If line-relay apparatus is not furnished, line pickup and hold functions are handled mechanically at or within the station itself. A resistance holding bridge applied mechanically at a station can be retired or released from the line only by the station which applied it, whereas, in a relay-actuated line pickup and hold system, any station which selects and goes off-hook on a previously held line will immediately release the holding bridge, regardless of its location in the system.

Basically, two different types of line-circuit configurations have been employed in telephony to allow station control of line pickup and hold relays in key telephone systems. One (Fig. 25-12) uses the central-office or PBX line current for the application and release of the holding bridge. Owing to its unbalanced transmission characteristics, a "balance" lead and a "hold" lead in addition to the tip and ring "talking leads" are needed per line. When needed for auxiliary service functions, additional pairs of leads are required in providing local visual and audible signals, etc.

The second type of line circuit uses a station-controlled auxiliary conductor (A-lead) per line, plus the tip and ring leads, for line pickup and hold switching functions. The A-lead circuits require local power at the key-system site for proper relay operation. Figure 25-13 illustrates this type of arrangement.

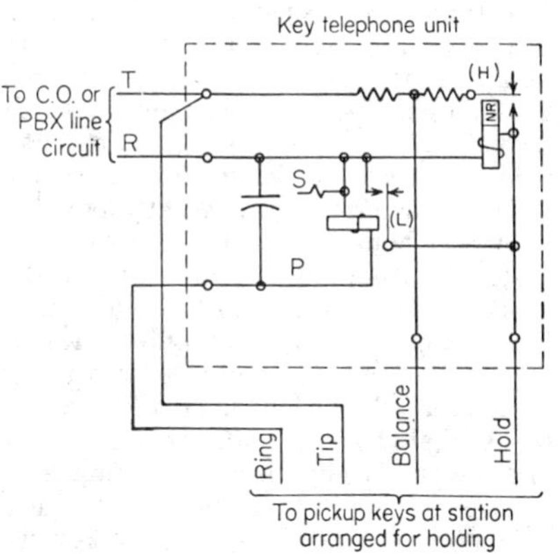

Fig. 25-12. Line pickup and hold circuit. (Operates and releases on line current.)

Many multibutton station sets and key equipment relays are arranged to afford visual and audible signals to aid in user interpretation of line-circuit status in the system. A rhythmical (60 ipm) flashing signal indicates the presence of an incoming call on a line. This can be augmented by interrupted audible signals, if needed. A steadily burning lamp signifies a line in use or a held-line condition. To differentiate visually between these latter two signals, a third visual indication—an arrhythmic "wink," or flutter, interruption—is often used to mark the line-held condition specifically.

These interrupted signals are derived from flashing (A/B) relay combinations, from electromechanical cam-actuated devices, as well as from solid-state drivers. Such equipment forms part of the common array of materials serving an installation—as contrasted to per station, or per line, or per line per station switching gear also required.

The visual and audible signal controls are arranged to lock in for specific timing intervals, so that the station user is aware of the calling traffic situation within his system. Release of locked-in circuits can be accomplished manually—via cutoff key or toggle switch—or automatically, via thermal- (bimetallic-) element relay release, or by R/C timing arrangements—including transistor detection. Locked-in circuit release can be accomplished on a system basis or on a per line basis, adjusted to meet optimum service needs.

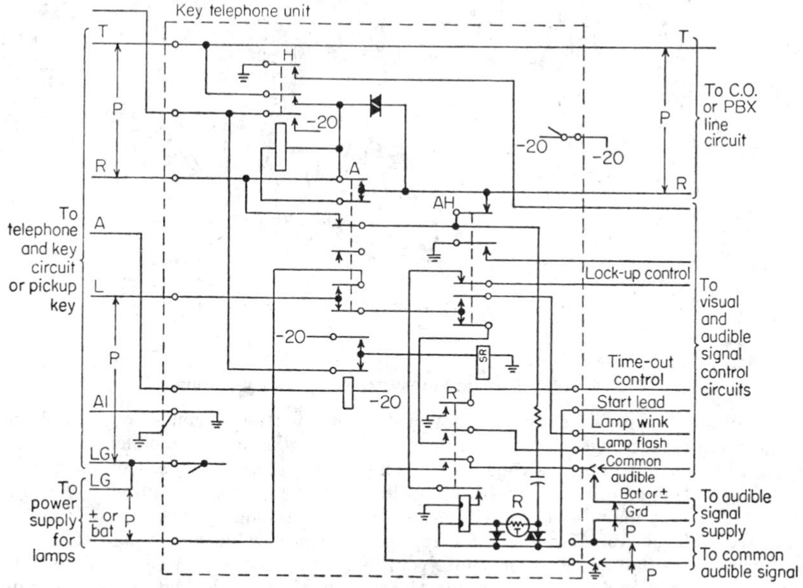

Fig. 25-13. A-lead-type central-office or PBX line circuit.

18. Characteristics of Design and Convenience. Line pickup and hold are the most common features furnished at key-system installations. These two services permit the station user to select one of several lines available to him for making outgoing calls or to answer incoming calls, while permitting him to leave the conversation momentarily at any time (and maintaining the circuit during this interval) to consult privately, use another line, etc., then return to the held line to complete the call at will. To provide this service, only a small mounting plate equipped with relays is needed to perform the necessary pickup and hold switching functions. It can be efficiently engineered in compact form and made at modest cost. Incrementally, several or many of these "key telephone units" (KTU's) can be interrelated in building-block fashion at a key-system installation.

Similarly, other feature service building-block KTU's can be interassociated with line-circuit units to control visual and audible signals, provide "intercom" or tie-line service, permit relay-controlled cutoff or privacy features, etc.

As the numbers of lines and stations increases and as the complexity of services to be offered compounds itself, the incremental building-block installation approach loses favor in comparison with factory-wired assemblies, "packages," of switching apparatus contained in cabinets, on relay racks, and the like.

Power needed to operate these key-system arrangements involves small transformers or rectifiers—or combination transformer/rectifiers—which provide relay operating power (generally 24 V direct current), audible signaling power (low-voltage alternating or direct current for buttons and buzzers, as well as 105-V 20-c/s or 115-V 30-c/s a-c ringing current), and lamp-operating power (usually 10 V a-c) to care for visual-signal requirements. In the majority of situations, these power plants are connected to the customer's commercial power-distribution system. However, some key-system installations are powered from building floater batteries, from PBX power plants, or from central-office cable pairs.

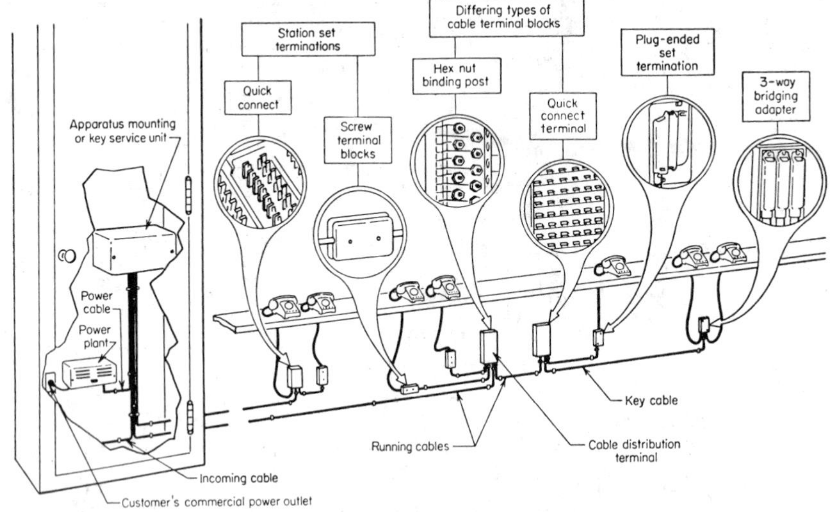

FIG. 25-14. Typical key-system installation layouts.

Figure 25-14 illustrates typical key-system installation layouts, oversimplifying the more complex arrangements, and considerably complicating the smaller-sized jobs—particularly as regards extensiveness of cabling, hardware and apparatus used, interconnecting station equipment involved, and line-circuit equipment required. All in all, however, the basic ingredients remain quite consistent among systems.

19. Other Types of Key-system Circuits. A considerable percentage of installations incorporate one or more local intercommunicating paths between stations. Depending upon the degree of communications sophistication desired, these intercommunicating paths range in complexity from the communal "farm-line" circuit (with separate push-button and buzzer signaling) to more elaborate dial-selective intercommunicating facilities having one or more talking links, options for visual and machine ringing features—plus the ability to conference or tie in any station on the "intercom" with an "outside" PBX or central-office line.

The key-system "intercom" circuit is of value in relieving the traffic load and call holding times of manual or dial PBX equipment serving the stations.

Ring-down and automatic signaling tie-line circuits are also designed for specific use within and among key systems. They do not require PBX or central-office switch trains interposed between tie-line stations, although many operate off premises with respect to the key equipment location. Other circuits available include manual and automatic privacy control over subordinate stations in a system, secretarial-type circuits, and other configurations to embellish further the basic line pickup and hold service as the customer wishes.

PRIVATE-BRANCH-EXCHANGE (PBX) SYSTEMS

20. General. Private-branch-exchange (PBX) systems are designed to provide telephone communications service for business organizations of all types and sizes. Communications service within the business (intra-PBX) and to other businesses, both nearby and far away, may be provided by manual, electromechanical, dial, or electronic-dial switching equipment.

The equipment required for PBX service is usually located on the business customer's premises and involves an agreement between the customer and the telephone company regarding the location of the equipment and the amount of space to be assigned to the service facility. Each installation is engineered on the basis of the customer's particular needs for quantities of elements such as telephone stations, central-office trunks, foreign-exchange trunks, tie trunks, and miscellaneous trunks for serving code call and paging systems, and telephone-dial dictation systems. Other features that may be provided include conference circuits, night-service arrangements, power-failure transfer or reserve power facilities, and many more to provide comprehensive telephone service.

21. Manual Systems. In manual systems, the switching work on all calls is handled by the customer's attendant. Manual systems offer two types of PBX attendant switchboards, cord and cordless, which may be either nonmultiple or multiple boards.

A manual cord PBX (Fig. 25-15) consists of one or more sections of framework containing a jack field on which are terminated station lines, central-office trunks, and tie lines of various types, a quantity of double-ended cord circuits for connecting the jacks together, a telephone and dial circuit which connects the attendant to the cord circuits by keys, and miscellaneous common, auxiliary, and alarm circuits.

The cord circuits are arranged so that the front cords are used for answering central-office trunk calls and the rear cords are used for answering extension or tie trunk calls. Lamps are associated with jacks to provide a visual signal on incoming calls.

Each of the cord circuits provides means for the attendant to ring, talk, and supervise a connection. This is accomplished by providing talking, splitting, listening, and ringing keys on the front and rear cords. A key is associated with each cord so that the attendant may set up a night connection

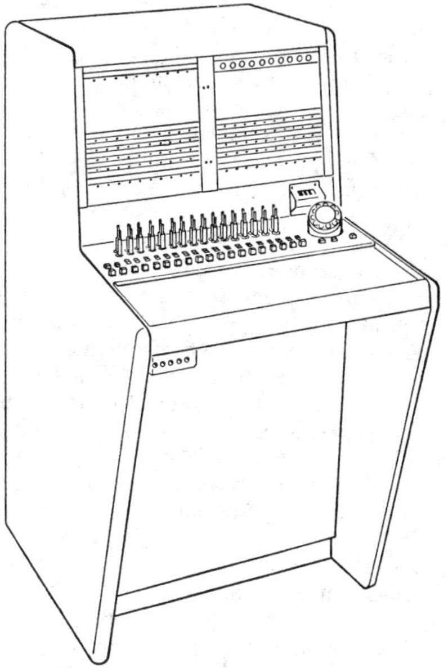

Fig. 25-15. Manual cord PBX.

or permit a station user to dial over a central-office trunk. Provision is also made for attendant recall by a flashing lamp when the station user is in need of operator assistance.

A talking battery for transmission on station-to-station (intra-PBX) calls is provided through a retard coil in the cord circuit. A central-office battery is provided for transmission when one end of the cord is plugged into a central-office trunk jack. Ringing current may be supplied either over cable pairs from the central office or from the ringing power plant of the associated dial switching equipment.

Fig. 25-16. Nonmultiple manual cordless PBX.

In general, nonmultiple manual switchboards operate on 16 to 24 V d-c, while multiple manual and both nonmultiple and multiple switchboards used with dial systems operate on 48 V d-c.

A nonmultiple manual cord PBX (Fig. 25-15) consists of one or two individual attendant's positions in which each line or trunk has only one appearance. The capacity of a typical two-panel single-position nonmanual cord PBX is 300 extensions (also called station lines) and a total of 60 to 120 central-office, tie, miscellaneous, and attendant trunks and 15-cord circuits.

A nonmultiple manual cordless PBX (Fig. 25-16) is housed in a compact metal cabinet arranged with a sloping top. Each cabinet is a self-contained unit and is connected to the station and central office over cable pairs. These units are designed to be placed on top of an ordinary desk and are available in sizes up to 20 extensions, with an appropriate number of trunks.

Fig. 25-17. Multiple manual cord PBX.

Multiple manual cord PBX's (Fig. 25-17) consist of two or more switchboard positions, similar to the nonmultiple type, except that each station line and trunk appears on each of the three positions. Multiple manual cord PBX's are available with capacity for over 1,000 stations and over 100 trunks.

22. Dial Systems. In dial systems most of the switching work on intercommunicating calls and outgoing calls is accomplished by switching machines. An attendant is required to receive and place certain types of calls, but even these calls are generally switched by the machine.

The switchboard or attendant's console is usually located in a separate room adjacent to the room containing the dial switching equipment. The separate room prevents the noise from the operating switches and relays from interfering with the attendant and also prevents dust and other foreign objects from getting into the equipment to cause malfunctions.

23. Step-by-step (S × S) PBX. 1. *Equipment.* The step-by-step dial PBX (Fig. 25-18) consists of an assembly of step-by-step electromechanical switches which perform various functions to connect one station to another or to trunks and tie lines on the basis of pulses or signals received from the dial of the telephone instrument. Line relays connect the extension station set to the equipment. Trunk units connect trunks of various kinds to the equipment. A manual switchboard or attendant's console, power equipment, and miscellaneous alarm and register circuits are part of the complete system.

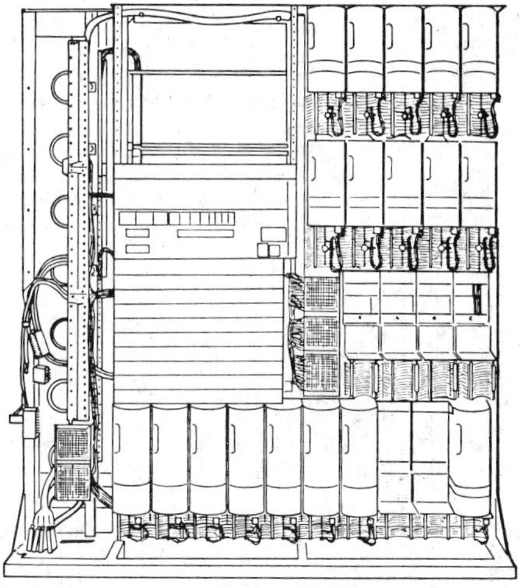

FIG. 25-18. Step-by-step dial PBX switching equipment.

2. *Operation.* To place a call through this equipment, the calling party lifts the receiver from the switchhook, at which time an idle line finder (LF) connects to that line. Each line finder is also connected to a first selector switch (1ST SEL.) (see Fig. 25-19), after which the dial tone is heard, signifying the system is ready to accept dialing digits. Assume that the PBX is arranged for four-digit dialing for station-to-station service. When the first digit is dialed, the first selector steps from the bottom level up to the level corresponding to the first digit and then connects to a second selector switch (2ND SEL.). The second dialed digit causes the second selector to step up to the level of the dialed number and then connect to the connector switch (CONN.). The third number causes the connector to step up to the level of the dialed number and after receipt of the fourth dialed digit rotates to make the final connection to the called station through the line relay and the main distribution frame (MDF).

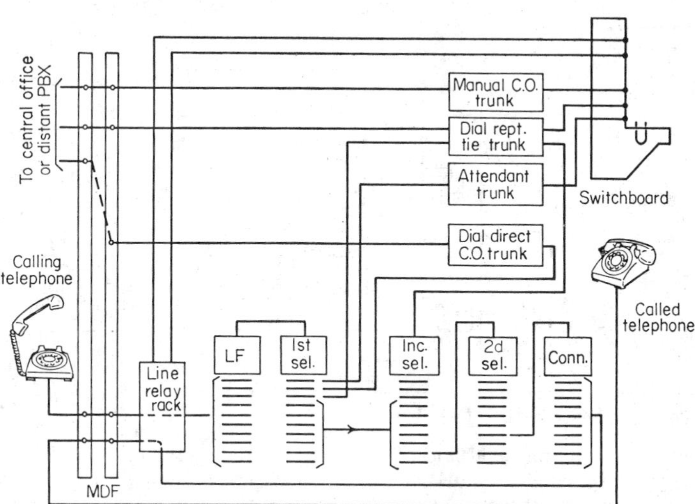

FIG. 25-19. Step-by-step dial private branch exchange.

If the PBX is on a three-digit dialing basis, a second selector is not furnished. If two-digit dialing is used, a selector-connector switch is used in place of the first selector and connector. Then the selector-connector operates as a selector on the upper levels of the switch and as a connector on the lower levels.

When the PBX has dial repeating tie trunks, they connect to an incoming selector (INC. SEL.). Access to the attendant trunk, dial direct central-office trunk (DIAL DIRECT C.O. TRUNK), and dial repeating tie trunk (DIAL REPT. TIE TRUNK) is achieved by dialing the proper digit as the first number dialed.

24. Crossbar (XB) PBX. 1. *Equipment.* This type PBX is designed primarily to operate with telephone consoles, although a cord-type switchboard is sometimes employed for the use of the attendant. The equipment is generally furnished as a self-contained unit (Fig. 25-20), quiet in operation, and finished in paint colors to harmonize with most reception area or general office decors. A separate equipment room is generally unnecessary.

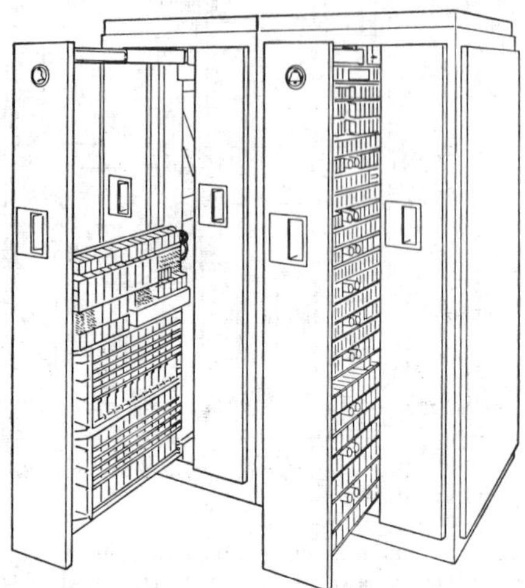

Fig. 25-20. Crossbar dial PBX switching equipment.

The switching plan of crossbar PBX's is designed to provide for the interconnections between station lines, central-office trunks, tie trunks, attendant trunks, and miscellaneous trunks for special services. Crossbar switches (space division), under control of a common group of circuits, are the basic switching element of the system, which is sometimes called a wired logic common control system. The common group control circuits are designed with the marker as the basic element. As can be seen in Fig. 25-21, the marker has interconnections with all the other circuits. The functions involved in completing calls are accomplished partly by the marker and partly by the other circuits.

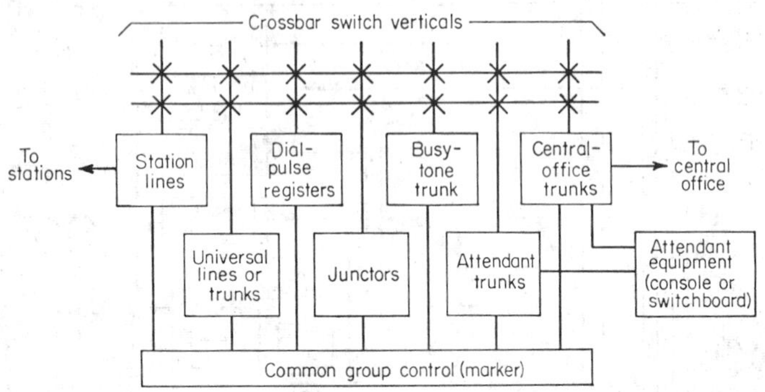

Fig. 25-21. Block diagram of crossbar dial PBX system.

The station lines, universal lines or trunks, central-office trunks, attendant trunks, busy-tone trunk, junctors, and dial pulse registers use relays and are terminated on the verticals of crossbar switches.

2. *Operation.* When a station originates a call, a line relay operates and signals the identity of the calling line to the marker. The marker then selects an idle dial pulse register and establishes a connection from the calling station to this register through an idle horizontal path in the crossbar switches, and the register sends the dial tone to the calling station.

As the calling station dials, the register stores the dialed pulses. When the dial pulse register has stored all the dialed information, it calls for a marker and passes information to the marker as to the type of call being placed (station-to-station, tie trunk, central-office, trunk, etc.), the number called, and whether the calling station is denied the ability to make any type of call. The marker then completes the call through an idle junctor circuit and links to the calling and called stations. When the called station answers, the junctor disconnects ringing current and completes the transmission path.

This is a time-sharing system in that the marker operates on only one call at a time. While the marker is setting up a call, all other calls are locked out to prevent interference with the call in progress. However, the marker operation is relatively fast so that under normal operating conditions no appreciable delay is experienced in handling all the calls placed on the system.

A station-to-station call requires two links, one from the calling station to the originating end of a junctor, and one from the terminating end of the same junctor to the called station line. The marker sets up this connection and the junctor applies ringing and provides supervision and talking battery.

A dialed tie trunk-to-station call also uses two links and a junctor, and the call is completed as for a station-to-station call. Only one link and no junctor is required to complete calls from a station to a trunk, from a trunk to a station, or from a central-office trunk to a tie trunk.

When an incoming trunk to the PBX is seized at the central office, an indication of the seizure appears at the attendant's equipment. The attendant answers the call and completes it to the station requested by the calling party by dialing from the attendant's position. When the attendant starts this completion, the trunk requests the marker and identifies itself. The marker then connects the trunk to an idle dial pulse register in the same manner as for a station line.

Calls encountering busy stations, all trunks or all paths busy, are routed to a busy-tone trunk, which returns a busy indication to the originator of the call. If the busy-tone trunk is busy, the dial pulse register returns the busy indication.

25. Centralized Electronic PBX System. The transistor, other solid-state devices, printed wiring, miniaturized components, integrated and thin-film circuitry, and magnetic-memory devices are among the factors contributing to the existence of electronic switching systems.

1. *Equipment.* In a centralized electronic PBX system stored program control guides and directs all the system actions in much the same way the common group control functions in crossbar PBX's. A major difference is that semipermanent magnets and transistor-resistor logic circuitry are used in the electronic system, as opposed to the wired logic of the crossbar system. Wiring changes are necessary to effect a different sequence of operations in crossbar systems, whereas changing the state of magnetization of some of the semipermanent magnets in the program control is generally all that is required in the stored program electronic systems.

Fig. 25-22. Line switch unit of an electronic switching system.

A quantity of PBX customer equipments, called switch units (Fig. 25-22), are controlled by the common equipment, called the control unit, located in the central office. This arrangement minimizes the amount of equipment located on the customer's premises while maximizing control economy by concentrating as much of the control function as possible in the common equipment. Supervisory and control signals at voice frequencies are transmitted between the control unit and switch units over cable pairs called data links and digit trunks (Fig. 25-23).

In the Bell System No. 101 electronic system the control unit has a capacity up to 3,200 PBX station lines, approximately 800 central-office, tie, and foreign exchange trunks, distributed among as many as 32 switch units. Switch units are available in different arrangements to serve 60 up to 3,000 lines.

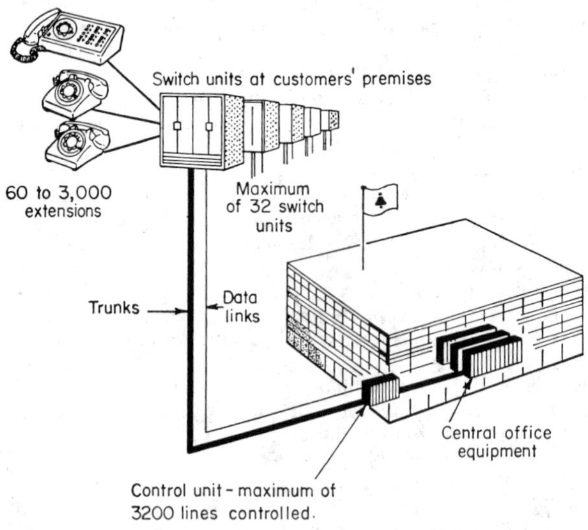

FIG. 25-23. Electronic-switching-system layout.

2. *Operation.* The control unit (Fig. 25-24), guided by instructions stored in a semipermanent magnet memory, called the program control store, sets up and supervises all call processing activities of the system. Inputs to the control unit are indications of changes in the supervisory states of terminal circuits and dialed digits from PBX telephone stations served by the switch units. Outputs consist of switching orders to the switch units and outpulsed digits to the central-office or tie lines through a sender feature of the control unit. The digit and data store assembles both input and output information in a scratch-pad temporary memory element. Input information is read out sequentially and processed by program control. Output information is transmitted to switch units or the central office.

Supervisory equipment in the switch unit (Fig. 25-25) continuously scans the terminal circuits and informs the control unit of their states. A change in state from one scan to the next requires call-processing activity by the system.

Call processing involves all the memory elements in a continuous interplay. The identity of a line, trunk, or other source of input signals is referred to the call-status

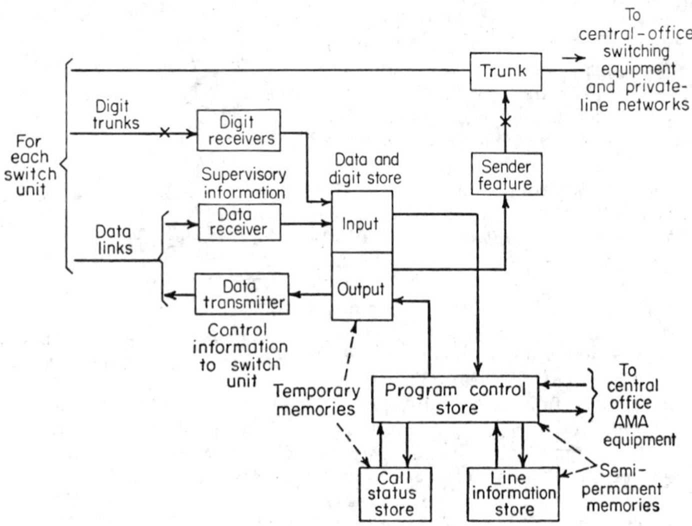

FIG. 25-24. Control unit of an electronic switching system.

store; the resulting status information influences the output information required to advance the call. Output information may be determined directly in the program store or after consulting the line-information store. The control unit processes one call at a time on a time-sharing basis, but system actions are so fast that no calls are delayed.

The call-status store is also searched to determine the busy or idle state of the called line. If a called line is busy, the line-information store is consulted to determine whether or not the line is in a hunt group and what other lines are in the group. The call-status store is then searched to determine whether or not a connection can be made to an idle line in the group.

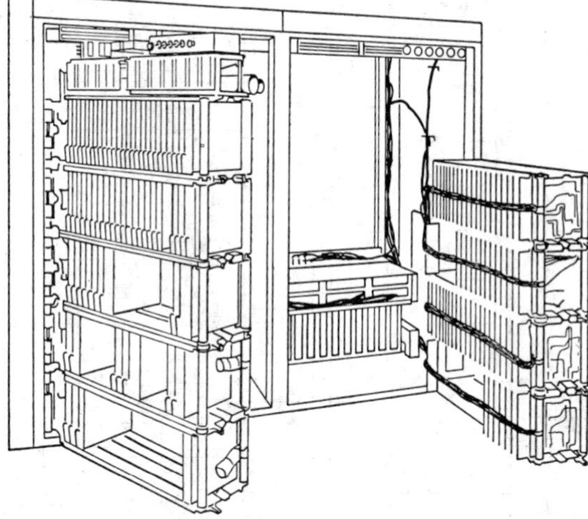

FIG. 25-25. Self-contained electronic dial PBX.

This stored-program concept of operation takes full advantage of the microsecond response speed of the system and provides routine diagnostic maintenance activities as well as traffic-data collection activities sandwiched in between call-processing routines. And since it is relatively simple to change the instructions in the program control store, the system has almost limitless flexibility and adaptability for accommodating customer service features of the future.

26. Self-contained Electronic PBX. 1. *Equipment.* Dial electronic PBX's are also available in the up to several hundred station line size to satisfy customer needs for modern service features in a small, attractive package which will blend into the decor of general office areas. They are generally self-contained wired program electronic PBX's employing solid-state devices and ferreed or diode space-division switching networks (Fig. 25-25). A cordless console or cord switchboard provides attendant functions.

2. *Operation.* When a PBX station user lifts the handset to make a call to another PBX station, the common control circuit (Fig. 25-26) recognizes the off-hook condition, goes into the line dial-tone mode of operation, and scans for the station line and an idle register, marking them when found. The network control then sets up a path through the switching network from the station-line-side appearance to a register on the trunk-side appearance. After a path has been selected, the register sends the dial tone to the station, both register and line marks are removed, and the common control and network control return to the idle condition to await the appearance of another service request.

The dialed digits are received by the register, and when dialing is complete the register bids for common control. This service request from the register causes the common control to go into the read register mode and select the register requesting service. The register passes the called station number to the common control, causing the common control to seize and mark the idle called station line and scan for an idle intercom trunk. When the intercom trunk is seized, the terminating end is marked for connection to the trunk side of the switching network.

The path through the switching network is established with the common control held to complete the originating end of the call, and a signal from the network control to the common control causes the originating end of the intercom trunk to be marked. The common control instructs the register to call back the calling station line for identification.

After the calling station is called back and marked for connection to the line side of the switching network, the path through the switching network is established, the

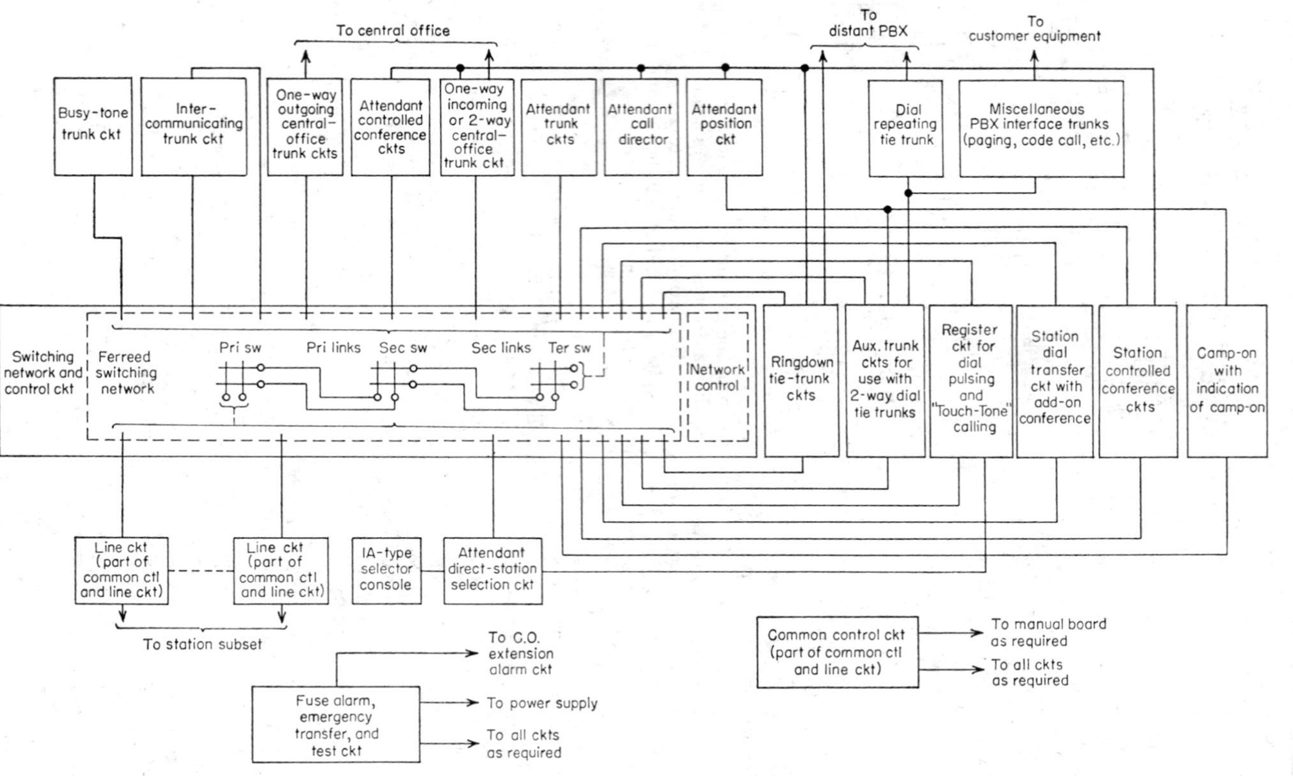

FIG. 25-26. Block diagram of self-contained electronic dial PBX.

common control is released, and the calling station is connected to the intercom trunk. When the common control releases, the register releases.

The intercom trunk supplies ringing current to the called station and ring-back to the calling station. When the called station is answered, a d-c path is completed back to the intercom trunk, which removes the trunk mark. This mark removal inhibits ringing current and ring-back, provides talking battery voltage, and completes the transmission path.

When the calling station goes on-hook, the intercom trunk restores to the idle condition, which releases the originating and terminating connections.

27. Attendant Equipment. Attendant equipment provided with a dial PBX gives the attendant the ability to answer, originate, and intercept calls and to give assistance to subscribers. Attendant equipment consists of either a switchboard or a console.

1. *Switchboard.* Switchboard equipment has been discussed under manual systems.

2. *Console.* The console (Fig. 25-27) is a small, attractive, key-operated "switchboard" which may be placed on a desk or table top. The attendant uses push buttons and a dial to answer and complete calls. Actual switching occurs in the associated dial switching equipment. Tones, illuminated panels, and illuminated buttons provide the signaling and supervisory information needed to process calls.

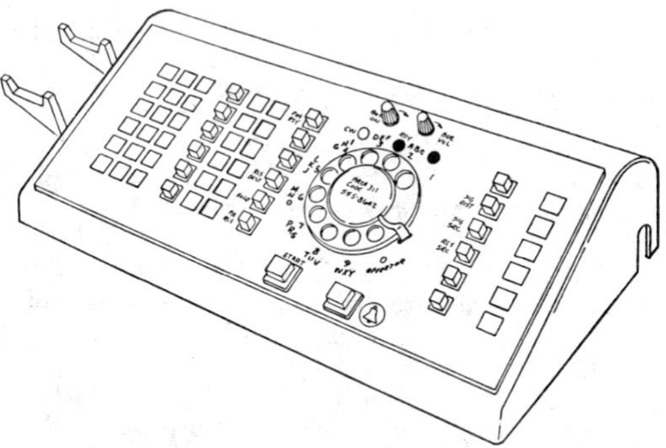

FIG. 25-27. Universal attendant console.

28. Special Systems. Special systems or arrangements have been developed for the use of particular customers to serve specific functions. Examples and brief descriptions of these follow.

1. *Hotel-Motel Systems.* The needs of hotels and motels are unique in that the guests prefer to dial their calls and the billing information must be immediately available to hotel management. Means are provided for registering local calls automatically and for a message waiting feature.

2. *Telephone Answering Systems.* Specially designed switchboards are available for telephone-answering service bureaus to provide secretarial service for their customers.

3. *Emergency Reporting.* A special PBX reporting telephone system is available for use as a fire-reporting system, police-reporting system, industrial-plant emergency-reporting system, industrial-plant routine reporting system, as a monitor of plant supervisory devices, and as a reporting system that combines fire, police, and other emergency reporting by the general public. "Call boxes" of the outdoor type, dial cord switchboard, and associated switching equipment comprise the system (Fig. 25-28).

Call boxes are strategically located on street corners, near schools, and near hospitals, where the general public has easy access to them.

In calls originated by the public, removal of the handset from the call box establishes a connection to the fire headquarters switchboard, where it is recognized and answered by the switchboard attendant. A light on the switchboard serves to identify the location of the call box. Oral information about the emergency is given to the attendant so he can arrange to dispatch fire apparatus, police assistance, or emergency personnel.

Other special features can be included which automatically record the time of the call, the call-box number, and details of the oral report and attendant's reply.

4. *Four-wire PBX.* Four-wire switching systems have been developed primarily for use by the military. Several features of these PBX's are:

a. Priority control of calls, including override of established connections.

b. Large-capacity conference capability.

c. Push-button operation and automatic identification of established connections.

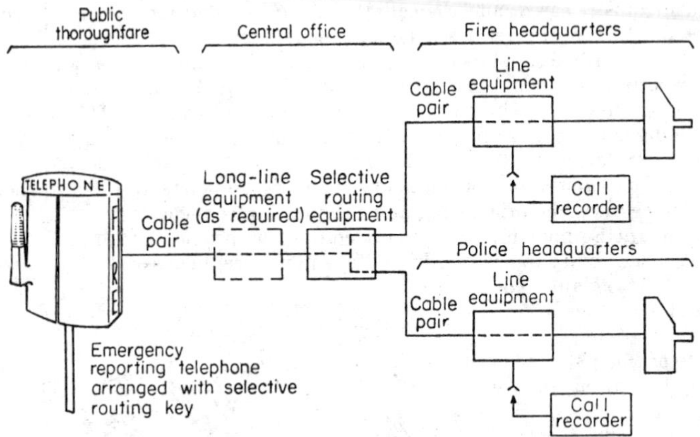

FIG. 25-28. Major equipment elements of a combined fire and police reporting system.

5. *Centrex.* Centrex is a PBX service offering designed to improve the service features available to business customers. In addition to the basic PBX features of station hunting, restrictions, dial intercom, and night connections the following features are offered:

a. Direct outdialing (DOD) from all unrestricted PBX stations, with identification of the station provided for billing purposes.

b. Direct indialing (DID) to all unrestricted PBX stations without the necessity for attendant completion (except for listed-number calls).

c. Transfer of calls by station or attendant.

d. Consultation-hold permits a station user to hold an existing incoming trunk call, originate a call to another station for consultation, and then return to and complete the original call.

e. Conference-calling to permit a station user to establish a conference with up to five conferees plus the originator without attendant aid.

f. Trunk-answer-any-station permits any station to answer a call to the attendant when night service is activated, by dialing a special code.

DATA AND TELEGRAPH SERVICES

29. Definition. The definition of data, for the purposes of this section, is information in electrical form which is to be transmitted from one place to another.

30. Data Signals. Multilevel signals are those having more than one state, such as binary (two levels or states), ternary (three levels or states), quaternary (four levels or states), etc. Each signaling element can convey any one of a number of conditions (e.g., any one of four conditions for a quaternary signal). Analog (continuously variable) signals may be thought of as having an infinite number of states or levels. As the number of states or levels is increased, the amount of information per signal element is correspondingly increased, but so is the difficulty of electrically separating one level from another at the receiving end of a transmission system. Noise entering the system tends to make the electrical level of a given signal less distinct and accurate identification more difficult.

Signals ready for transmission are usually in binary form but may also be of higher order or in analog form. Binary signals are preferred, since two-state devices are relatively easy to manufacture, are quite reliable, and are generally inexpensive. In addition, the receiving end of the transmission system must decide only whether the signal is in state one or state two. It need not discriminate among many incremental levels. Analog transmission usually requires system response to direct current and poses problems both in maintaining an accurate d-c zero or reference point and in maintaining linearity of the reproduced signal.

Data signals can be either synchronous or asynchronous. Synchronous data signals are those where the data pulses occur at an unvarying rate, such as in step with a clock or trigger signal. Asynchronous signals are those which have no such strict relationship with time and often begin each character with a start pulse (such as start-stop teletypewriter signals).

31. Bits per Second vs. Bauds. The bit (a contraction of the words "binary digit") is a single binary decision or the equivalent amount of information. Bits per second is the amount of information handled per second. The baud (named for Émile Baudot) is the signaling rate in code elements per second,

$$\text{Signaling speed in bauds} = \frac{1}{\text{length of shortest signaling element in seconds}} \qquad (25\text{-}5)$$

Example. A modulator receives a binary data signal at 2,000 bits/s. The incoming information is grouped into pairs of bits and used to phase-modulate the carrier frequency. Depending on the bit grouping (00, 01, 10, or 11) the carrier is shifted in phase a predetermined amount. The line signaling rate is then only 1,000 bauds, since each signaling element represents 2 bits of information. This system uses 4-phase, or quaternary - phase, modulation.

Bits per second and baud are therefore two different parameters and are numerically equal only under certain conditions.

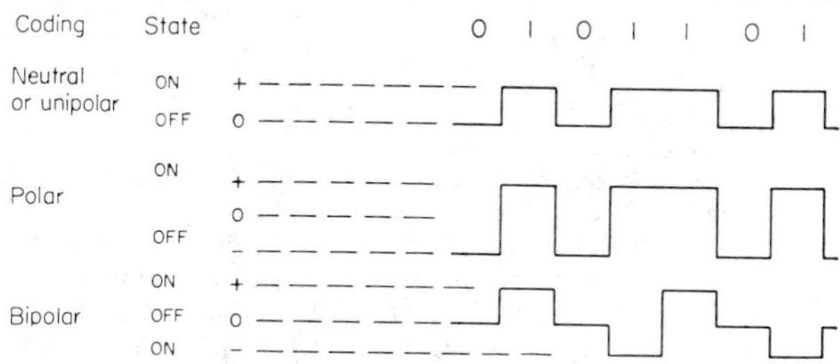

FIG. 25-29. Form of binary signals. Neutral or unipolar—current or voltage represents one condition; no current or voltage represents other condition. Polar—current or voltage of one polarity represents one condition; current or voltage of opposite polarity represents other condition. Bipolar—each "on" bit has opposite polarity from previous one. A neutral signal is converted to this form to reduce the d-c component on the communication channel.

Binary signals may be either serial (serial by bit and serial by character, where each character is sent binary bit by binary bit) or parallel (parallel by bit and serial by character, where all the bits of a given character are sent at the same time on parallel channels but the characters are sent one after another). These signals may be generated by mechanical contacts (e.g., a switch) or by the operation of electronic circuitry (e.g., a semiconductor flip-flop) (see Fig. 25-29).

32. Data-set Interface. The data-set interface is the point of interchange of information between the data set (or "modem") and the associated apparatus. To provide for interchangeability of different types and different manufacturers of terminal equipment, a number of different organizations have written a variety of standards defining this interface for a variety of purposes. Examples are the serial currents which conventional teletypewriters deliver or expect to receive and which are on-off currents of, generally, 62.5 mA or 20 mA. In electronic or voltage interfaces there are standards such as MIL-STD-188B, EIA Standard RS-232-B, and other less common ones. The following table compares the two common teletypewriter current interface standards:

Teletypewriter standard	Unit	Neutral operation
62.5 mA	Mark	62.5 mA current
	Space	0 current
20 mA	Mark	20 mA current
	Space	0 current

The widely accepted EIA Standard RS-232-B* was developed to permit interchangeability of serial data transmission equipment. The older, military standard MIL-STD-188B† was a similar attempt to provide universal operation of military data equipment. The following table compares some points of EIA RS-232-B and MIL-STD-188B (standard interface—low level).

	RS-232	188B
Data signal transmitted:		
Mark	Volts −5 to −25	+6 ± 1
Space	Volts +5 to +25	−6 ± 1
Control signal transmitted:		
ON	Volts +5 to +25	+6 ± 1
OFF	Volts −5 to −25	−6 ± 1
Signal receiving sensitivity—minimum	Volts +3	+0.5
Undefined signal	Volts +3 to 0 to −3	+0.5 to 0 to −0.5
Minimum load impedance	Ohms 3,000	5,000
Maximum source impedance	Ohms not directly specified	100

33. Codes and Coding. Many codes have been developed to transmit information. Each code consists of a series of 1s and 0s arranged in an order such that each character of the code is uniquely defined.

1. *Morse Telegraph.* The earliest telegraph system that is still in limited use is the manual Morse telegraph system. The system is used to communicate between two or more locations, an electrical circuit being employed to transmit long and short pulses, commonly known as dots and dashes. Hand-operated sending keys are ordinarily used for the transmitter, while an audible sounder is used by the operator to detect the signals at the receiver. The most commonly used format is the International Morse Code shown in Table 25-1.

Table 25-1. International and American Morse Codes

	International	American		International	American
A	·–	·–	R	·–·	· ··
B	–···	–···	S	···	···
C	–·–·	·· ·	T	–	–
D	–··	–··	U	··–	··–
E	·	·	V	···–	···–
F	··–·	·–·	W	·––	·––
G	––·	––·	X	–··–	·–··
H	····	····	Y	–·––	·· ··
I	··	··	Z	––··	··· ·
J	·–––	–·–·			
K	–·–	–·–	1	·––––	·––·
L	·–··	——	2	··–––	··–··
M	––	––	3	···––	···–·
N	–·	–·	4	····–	····–
O	–––	· ·	5	·····	–––
P	·––·	·····	6	–····	······
Q	––·–	··–·	7	––···	––··
Period	·· ·· ··	··––··	8	–––··	–····
Comma	·–·–·–	·–·–	9	––––·	–··–
Interrogation	··––··	–··–·	0	–––––	———

2. *Five-level Telegraph Codes.* About 1880, in a successful attempt to derive a multichannel telegraph system from a common circuit, Baudot designed a code with a constant 5-bit character length, and the present CCITT international alphabet 1 is based upon it. This arrangement required a trained operator to send the code using a special five-key transmitter. Fifteen years later Murray developed a different 5-bit code to be used with a typewriterlike device to punch holes in paper tape for later transmission. The latter code is the basis of the present CCITT international telegraph

* EIA Standard RS-232-B, obtainable from Electronic Industries Association, 2001 Eye St. N.W., Washington, D.C. 20036 (9 pages).

† MIL-STD-188B, obtainable from Naval Supply Depot, 5801 Tabor Avenue, Philadelphia 20, Pa. (179 pages).

alphabet 2. It is alphabet 2, rather than the Baudot, which is used almost universally as the communication code of 5-bit teleprinter equipment. (The 5-bit code is also called the five-level or five-unit code.)

Each 5-bit code has a capability of $2^5 = 32$ discrete combinations. This is not sufficient to send all plain language characters (26 alphabetic + 10 numeric + punctuation + special characters ≥ 32). It is necessary to use two combinations to permit each one remaining to have two meanings (FIGURES SHIFT and LETTERS SHIFT). Allowing for the SHIFT characters, the all 1's (ERASE or NULL), and all 0's (BLANK) leaves $2 \times (32 - 4) = 56$ characters total. The CCITT alphabets 1 and 2 are shown in Table 25-2.

Table 25-2. Five-level Telegraph Alphabets

Lower case[1]	Baudot or International No. 1						Murray or International No. 2					
	Rare					Upper case[1]	Common[3]					Upper case[1,2]
	1	2	3	4	5		1	2	3	4	5	
A	0					1	0	0				
B			0	0		8	0			0	0	?
C	0		0	0		9		0	0	0		:
D	0	0	0	0		Ø	0			0		
E		0				2	0					3
F		0	0	0			0		0	0		
G		0		0		7		0		0	0	
H	0	0		0		+			0		0	
I		0	0					0	0			8
J	0			0		6	0	0		0		Bell
K	0			0	0	(	0	0	0	0		(
L	0	0		0	0	=		0			0	)
M		0		0	0	)			0	0	0	.
N		0	0	0	0				0	0		,
O	0	0	0			5				0	0	9
P	0	0	0	0	0	%		0	0		0	Ø
Q	0		0	0	0	/	0	0	0		0	1
R			0	0	0	−		0		0		4
S			0		0	.	0		0			′
T	0		0		0						0	5
U	0		0			4	0	0	0			7
V	0	0	0		0	′		0	0	0	0	=
W		0	0		0	?	0	0			0	2
X		0			0	,	0		0	0	0	/
Y			0			3	0		0		0	6
Z	0	0			0	:	0				0	+
Space	4								0			Space
Letters					0	Letters	0	0	0	0	0	Letters
Figures				0		Figures	0	0		0	0	Figures
Car. return	0	0				Car. ret.				0		Car. ret.
Line feed	0				0	Line feed		0				Line feed
Blank	5											
Error				0	0	Error						

NOTES: 1. Upper case is obtained by preceding group of characters with *figures*. Lower case is obtained by preceding group of characters with *letters*.
2. Blanks in column are available for national (local) use. Some of the international characters are also changed for specific usage symbols such as for the U.S. Weather network.
3. This is now the standard five-unit teleprinter code subject to Note 2.
4. The Baudot code uses the *letters* and *figures* characters as the *space*.
5. The rest or idle condition of the Baudot system is a continuous spacing signal.

KEY

Sign	Baudot polar	Murray	
		Neutral	Polar*
0 Mark Space	Negative current Positive current	Positive current No current	Positive current Negative current

* Polar signals are not widely used in North America for teleprinter transmission, and the polarity is essentially undefined. Polarities shown are those recommended by CCITT and in use in Europe. See "Teleprinter Switching" by E. A. Rossberg and H. E. Korta (Van Nostrand). Also Chap. VI, Article 16, Telegraph Regulations (Geneva revision, 1958) published by the International Telecommunications Union, Geneva.

Table 25-3. Proposed Revised American Standard Code for Information Interchange

b_7 b_6 b_5 →					000	001	010	011	100	101	110	111
b_4	b_3	b_2	b_1	Column → / Row ↓	0	1	2	3	4	5	6	7
0	0	0	0	0	NUL	DLE	SP	0	`	P	@	p
0	0	0	1	1	SOH	DC1	!	1	A	Q	a	q
0	0	1	0	2	STX	DC2	"	2	B	R	b	r
0	0	1	1	3	ETX	DC3	#	3	C	S	c	s
0	1	0	0	4	EOT	DC4	$	4	D	T	d	t
0	1	0	1	5	ENQ	NAK	%	5	E	U	e	u
0	1	1	0	6	ACK	SYN	&	6	F	V	f	v
0	1	1	1	7	BEL	ETB	'	7	G	W	g	w
1	0	0	0	8	BS	CAN	(	8	H	X	h	x
1	0	0	1	9	HT	EM	)	9	I	Y	i	y
1	0	1	0	10	LF	SS	*	:	J	Z	j	z
1	0	1	1	11	VT	ESC	+	;	K	[	k	(
1	1	0	0	12	FF	FS	,	<	L	~	l	¬
1	1	0	1	13	CR	GS	—	=	M	]	m	)
1	1	1	0	14	SO	RS	.	>	N	^	n	|
1	1	1	1	15	SI	US	/	?	O	_	o	DEL

3. *The Seven-level ASCII.* The limitations of the 5-bit code are overcome in the new ASCII,* which is a 7-bit code with a capability of $2^7 = 128$ discrete characters and which is based on a straightforward binary progression. This code is intended for all devices requiring electrical interchange of information with other devices. The International Organization for Standardization (ISO) is preparing to recommend a 7-bit code very similar to ASCII for worldwide communication and data-processing purposes. The ASCII alphabet is outlined in Table 25-3. Many other codes have been developed, such as the U.S. Military Fieldata, the Binary-Coded-Decimal, and others, but these tend to be more specialized in application than the ones mentioned.

34. Printing Telegraph Systems. These systems have replaced nearly all Morse telegraph systems because of their inherently greater speed, written record, relative ease of operation, and availability of automatic features. The two types of systems in use in the United States employ either "start-stop" or "synchronous" modes of transmission. Speeds range from 60 to 150 words/min, with signaling speeds up to 150 bauds. Start-stop operation is used for most commercial applications, while synchronous operation is used mainly for government encrypted traffic.

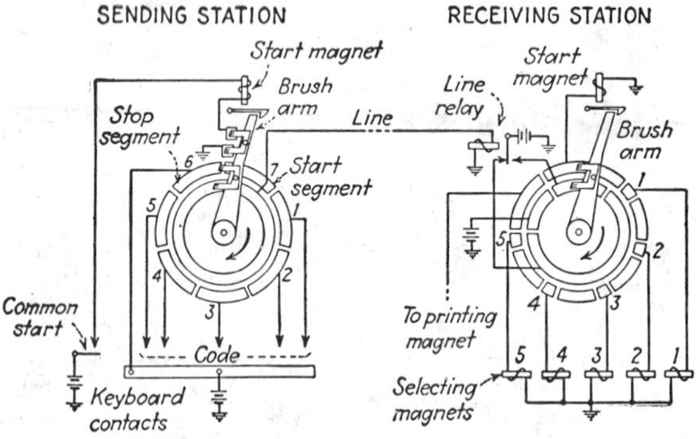

Fig. 25-30. Simplified diagram of start-stop systems.

Start-stop systems are asynchronous in nature. Each character consists of a unity-length start pulse (always spacing), five to eight unity-length information bits (marking or spacing), and a stop pulse (marking) of unity or longer length. An electromechanical version of a five-level system using segmented distributors and revolving brushes using a simplified, all-metallic facility is illustrated in Figure 25-30. The brush arms at both the transmitter and receiver are at rest between characters, although constant torque is applied through friction clutches connected to the respective electric motors. When a character is to be sent by either starting a paper tape through a tape reader or depressing a key on the keyboard, the start magnet at the transmitter is activated, releasing the brush arm, which rotates off the stop segment and passes across the start segment. This opens

* USA Standard Code for Information Interchange X3.4-1963, obtainable from United States of America Standards Institute, 10 E. 40th St., New York, N.Y. 10016 (11 pages).

the previously closed line, releases the start magnet at the receiver, and both transmitter and receiver brushes rotate together to transmit and detect the pulses that make up the code for that character. A typical line signal for the letter F sent in five-level code at 100 words/min is shown in Figure 25-31.

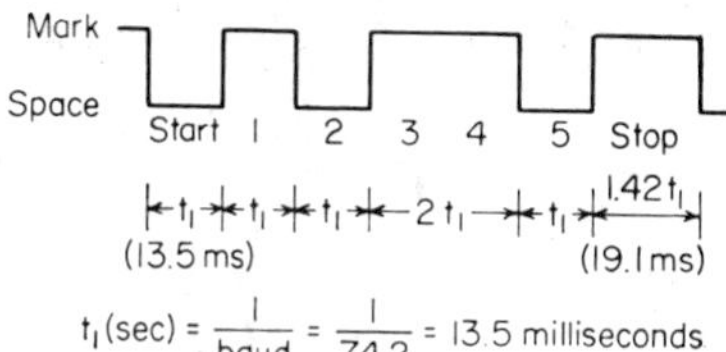

FIG. 25-31. Letter F in five-level alphabet code at 75 words/min.

Historically, start-stop operation was an economical choice, since timing starts afresh with the start pulse of each character and speed regulation of motors in electromechanical machines or frequency precision of oscillators in electronic devices is not critical. Also, the propagation time from a transmitter to any of several receivers may be disregarded. Table 25-4 lists the nominal speeds and pulse lengths of several start-stop printing telegraph codes.

Table 25-4. Start-Stop Printing Telegraph Codes (Nominal speeds and pulse lengths)

Baud rate	Unit code	Words/min	Characters/sec	Unit pulse, ms	Stop pulse, ms
45.5	7.42	61.3	6.13	22.0	31.2
50.0	7.50	66.7	6.67	20.0	30.0
56.9	7.42	76.7	7.67	17.6	25.0
74.2	7.42	100	10.0	13.5	19.1
75.0	7.50	100	10.0	13.3	20.0
110	11.0	100	10.0	9.09	18.18
150	10.0	150	15	6.67	6.67

35. Data Input/Output Devices. Devices providing input to and using the output from a data-transmission system may have any complexity, from a simple alarm unit, to a teletypewriter, or even to a computer. Each must be capable of providing the appropriate control signals to the data terminal apparatus. Figure 25-32 shows a typical arrangement of control logic for a piece of terminal equipment such as a tape unit or computer for use in controlling a data set where the interface corresponds to EIA RS-232-B. This control arrangement is based on operation over the public telephone network, but an equivalent arrangement is necessary on any facility.

In Fig. 25-32 the data terminal is activated by either the Ring-Indicator (lead *CE*) signal from the data terminal in the case of an incoming call or by the push button Start in the case of an outgoing or originating call. Either signal sets a flip-flop, which in turn activates the data terminal and returns a signal to the data set on the Data-Terminal-Ready (*CD*) lead. When the data set senses this signal, it turns the Data-Set-Ready (*CC*) lead ON and goes into either the sending or the receiving condition, depending on the status of the Request-to-Send (*CA*) lead as presented by the data terminal. If the Request-to-Send lead is not turned ON, a signal returned from the data set on the Data-Carrier-Detector (*CF*) lead tells the data terminal to prepare to receive data. At the end of data transmission the data terminal must tell the data set that transmission is complete, and this is accomplished by turning off the Data-Terminal-Ready (*CD*) lead. The method of turning this lead off depends on the logical arrangement of the data terminal, but it is essential that the data terminal be arranged to recognize a signal such as an End-of-Transmission character when it is in the receive mode or a similar signal internally generated if it is in the transmit mode.

A data input device may be one which mechanically reads holes in punched paper tape and by a simple mechanical arrangement translates this information into a series of serial pulses. The device may also be much more sophisticated and perform considerable code translation and error-control functions before transmission. Similarly, a data output device is a device which accepts data from a transmission system and makes use of it in some manner such as punching holes in a paper tape or passing in-

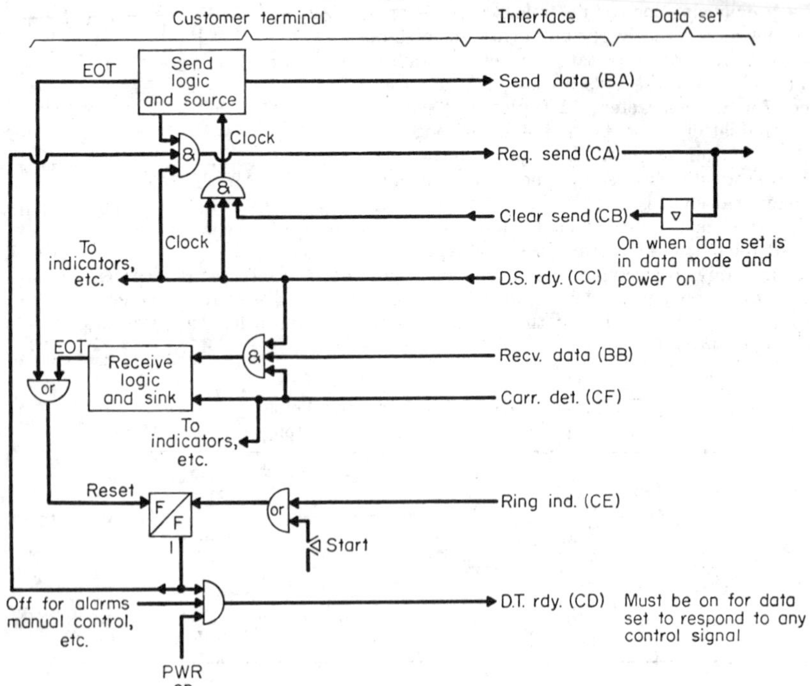

FIG. 25-32. Typical control and logic requirements in a customer terminal.

formation on to a computer. The data output device may also be arranged to check for errors and make corrections where necessary, the complexity being a function of the purpose of the device. A wide variety of data input-output devices is available; even a partial listing is beyond the scope of this section.

36. Data Sets. Direct currents and signals having frequencies below about 300 Hz and above about 3,000 Hz normally cannot be transmitted through ordinary voice communication circuits. Rapid transitions of signals such as square waves couple into adjacent circuits and cause interference (crosstalk or noise). For these reasons data signals are fed into data sets (modems) and there are caused to modulate carrier

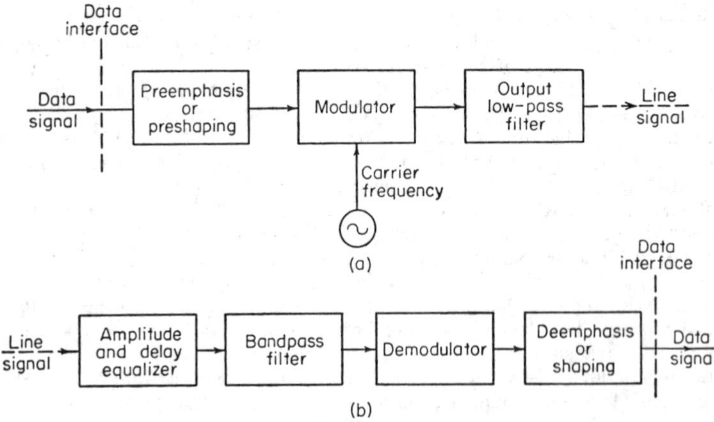

FIG. 25-33. Typical arrangements for data modulator (*a*) and demodulator (*b*).

signals. Figure 25-33 illustrates typical data sets in block form. The frequency and amplitude of the carrier, the type of modulation, and the equalization, if any, must be carefully chosen to provide accurate data transmission and at the same time to avoid any interference to or from adjacent circuits, pilot tones, signal tones, etc., which are also a part of the transmission medium.* Frequency and phase modulation are commonly used on the switched telephone network and other facilities where wide variations in circuit loss occur from call to call and where amplitude-sensitive devices such as compandors are in use. Amplitude modulation has been used frequently on private (point-to-point) facilities where the circuit loss can be held reasonably constant.†

The following table compares some features of the public switched telephone network and private facilities for data transmission:

Public Switched Network	Nonswitched Point-to-point Lines
Reaches all business and home telephone	Reaches only specified points
Automatic rerouting in case of trouble	Usually out of service in case of trouble
Circuit characteristics highly variable from call to call (owing to use of different dialed routes)	Circuit characteristics tailored to needs and quite constant
Bandwidth limited	Bandwidth available up to full video capabilities
Charged for as used	Charged for full time

37. Errors and Error Rates. Error rates vary widely with time in any one transmission medium, owing primarily to interference effects (noise, fading, etc.). Long-run error rates on a voice channel should be 1 in 10^5 for data above 600 bauds. Slower speeds may approach 1 in 10^6. Some typical error rates experienced in the United States telephone network in 1958 are shown in Fig. 25-34. Quite similar error rates are experienced in wideband digital systems.

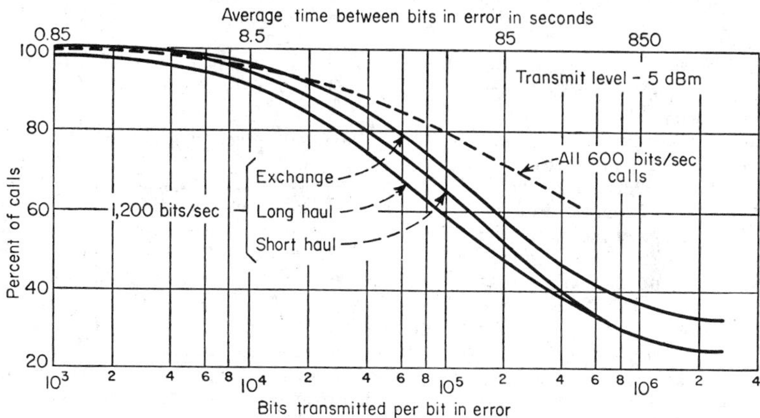

FIG. 25-34. Error rates typical in telephone network.

Statistically, in a practical system, errors occur in bursts. This is due mainly to the interfering effect of such things as relay chatter, transient bounce, etc.

It is usual to experience lower block-of-data-in-error rates than would be expected from the bits-in-error statistics. The reason for this is that digital data transmission is usually encoded in rather long blocks compared with a single bit interval. Short error bursts, therefore, often affect only one block.

The effective rate of transmission of usable data is often considerably less than the line rate because of start pulses, redundant pulses for error control, etc. This is especially true in systems where each block of data is inspected for errors prior to transmitting a succeeding block. Transmission time and backward-acting error-correcting

* For a description of typical characteristics see Capabilities of the Telephone Network for Data Transmission by A. A. Alexander, R. M. Gryb, and D. W. Nast, in the *Bell System Tech. Jour.*, May, 1960, Vol. 39, pp. 431–476.

† A description of data-set design is given in "Data Transmission" by W. R. Bennett and J. R. Davey, McGraw-Hill Book Company, 1965.

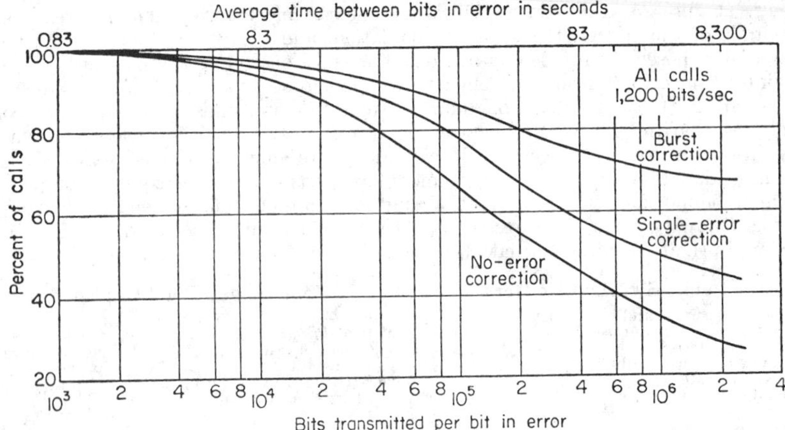

Fig. 25-35. Error-rate improvement with error correction.

schemes also reduce the effective rate of transmission. It is this "effective rate" which defines the "throughput" of a data system.

Error control can improve performance. Many applications require only sensing of an error and retransmission of the block of data containing the character in error (e.g., simple parity check). Other systems use forward-acting codes that detect and correct errors. Figure 25-35 shows the improvement that can be expected from one such scheme.

Some systems use redundancy in transmission to control errors. For example, a message may be sent three times, and in case of a discrepancy the 2 bits out of 3 that agree are accepted.*

38. Coupling to the Telephone Network. Acoustic, inductive, and electric connections are the three methods used.

Acoustic and inductive coupling arrangements are employed where it is desired to couple energy into and out of the telephone network through the telephone instrument itself. It is impossible to predict within close limits the signal level on the telephone line for a given acoustical output from a data transmitter. This comes about because of the nature of the carbon-button telephone transmitter itself and the variation of current through it as line resistance to the central office varies with distance. Resonances in the acoustic coupling chamber and harmonic and intermodulation distortion in the carbon-button transmitter can greatly affect the character of the transmitted signals. Transmitter coupling efficiency can vary widely from telephone to telephone owing to carbon-button characteristics and can even vary considerably during a given data transmission. A high acoustic input to the telephone transmitter can overcome the possibility of weak signals but will then normally result in a high line-signal level, overloading telephone plant and causing severe crosstalk and noise. Crosstalk of this nature interferes seriously with other users of the telephone system and in most areas is considered illegal.

Receiving data by acoustic coupling to the telephone receiver is perhaps the most difficult acoustic operation. The signal level is often quite low owing to circuit and coupling losses, causing problems in reliable detection which are made more difficult by any local ambient acoustic noise.

Inductive coupling is more unreliable than acoustic coupling, since the parameters which permit this form of coupling are a byproduct of telephone design and not a main part of it. Some telephones couple quite well, while others will not couple inductively at all. This is a good form of coupling to avoid!

Direct electrical connection to the telephone line is the preferred method of coupling. Transmission is then a function only of the telephone channel itself and not of the

* For a much more complete discussion of error control see Coding for Error Control: An Examination of Techniques by Albert G. Franco and Mark E. Wall, *Electronics*, Dec. 27, 1965, Vol. 38, No. 26.

Table 25-5. Typical Data Sets Available from United States Telephone Companies for Use on the Public Telephone Network

Data set	Max. bit rate, bits/s	Interface*	Use
103A	300	EIA	Teletypewriter and similar
201A	2,000†	EIA	Tape and computer
202	1,200	EIA	Tape and computer
401	20 ch./s	Cont.	Low-speed punched cards and remote metering
402	75 ch./s	Cont.	8-channel paper tape
602	....	Anal.	0 to 7 V facsimile dc to 900 Hz
603	....	Anal.	±2 V electrocardiogram dc to 100 Hz
801	....	ACU	Automatic dialing

* EIA—meets EIA RS-232-B Specifications. Cont.—contact closure interface. Anal.—analog interface. ACU—automatic calling unit.
† Operates synchronously at 2,000 bits/s.

telephone instrument. In addition such features as answering and dialing without human intervention are possible. Telephone companies insist that they provide all direct connections to the switched network. This is the only means by which they can assure that the connected equipment will perform as desired and, even more important, that telephone plant and other customers and services will not be adversely affected.

Table 25-5 is representative of data sets available from the Bell system, the general system, and many independent United States telephone companies for data transmission over the switched network.

39. Bandwidth-selectable Service. Operation is restricted to voice bandwidths only when the public telephone network is involved. Western Union offers a switched bandwidth-selectable service between certain cities.* In this service a subscriber voice-data handset allows the user to place a call to another subscriber and, at the same time, to select the 2-, 4-, 8-, 16-, or 48-kHz channel desired. All circuits are "4-wire"; i.e., the two directions of transmission are completely separate from end to end. Communication common carriers also offer a variety of private-line rvoice-bandwidth and wider-than-voice-bandwidth data services up to a complete television channel. Many equipment manufacturers offer data sets for use on appropriate private channels. Table 25-6 is a partial list of available digital data sets which illustrate the data-handling capabilities of private-line voice-bandwidth circuits.

Table 25-6. Partial List of Available Medium-speed Digital Data Sets
(Characteristics are those published by the manufacturers)

Type	Avail. from	Bit rate, bits/s	Max.*	Synch.*	Channel required	Modulation†
201B	Telephone Company	2,400		×	Note	Phase
202	Telephone Company	1,800	×		Note	F.S.
SC-309	Stromberg Carlson	2,000	×		850 Hz	VSB
		2,400		×	850 Hz	
HC-270	Hughes Aircraft	600, 1200,			Note	Phase
		2,400,				
		4,800		×	600–3,600 Hz	
26C	Lenkurt	1,200,	×		Note	F.S.
		600, 1,200,		×	Note	F.S.
		2,400‡				
Sebit 24B	Rixon	600, 1200,		×	Note	VSB
		2,400,				
		4,800		×	625–3,250 Hz	VSB
TE-216	Collins	600, 1,200,		×	Note	Phase and AM
		2,400, 3,600				
		4,800, 6,200				

NOTE: Specially equalized voice circuit.
* Max.—bit rate is asynchronous up to this maximum. Synch.—bit rate is synchronous at this value.
† F.S.—Frequency shift. VSB—vestigial sideband amplitude.
‡ Uses "duobinary" coding.

* For a more complete description see *Western Union Tech. Rev.*, July, 1962, Vol. 16, No. 3, pp. 98, 120; also October, 1965, Vol. 19, No. 4, pp. 142–147.

In addition to voice-bandwidth services there are a variety of lesser-bandwidth services available. Table 25-7 lists a number of services available at teletypewriter speeds—the supplier usually provides the teletypewriter as part of the service. Some common carriers also will provide services of even less capability for signaling and for low-speed telemetering (usually digital).

Table 25-7. Services Using Less than Voice Bandwidth

Service	Description	Speed, words/min	Notes	Supplier*
Telex:				
Domestic.........	Exchange TTY service	66.7	1	WU
Overseas.........	Exchange TTY service	66.7	1, 2	I.T.&T., R.C.A.C., W.U.I.
TWX..............	Exchange TTY service	100	2, 3	Bell
		60	2, 3	Bell
Private TTY.......	May be two-point or multipoint	60–150		Bell
				W.U.

Notes: 1. Domestic and overseas Telex can interconnect.
2. TWX and overseas Telex can interconnect.
3. 60- and 100-speed TWX can intercommunicate.
* Supplier: Bell = Bell System. I.T.&T. = International Telephone and Telegraph. R.C.A.C. = RCA Communications. W.U. = Western Union. W.U.I. = Western Union International.

Table 25-8 lists a number of Western Electric Company channel terminals which require a channel bandwidth larger than that required for voice. Each of these terminals on the subscriber's premises must be associated with frequency-translating or -modulating equipment located at the telephone company's toll office and any necessary line amplifiers in between.

Table 25-8. Western Electric High-speed Channel Terminals

Type	Data-set capability			Channel req'd. nominal B.W.	Modulation†
	Bit rate, kbits/s	Min.‡	Synch.‡		
301B	40.8	..	×	48 kHz	Phase
303B*	52 μsec element length	×		24 kHz	VSB
303B	19.2	..	×	24 kHz	VSB
303C*	20 μsec element length	×		48 kHz	R.P.
303C	50	..	×	48 kHz	R.P.
303D*	4.3 μsec element length	×		250 kHz	R.P.
303D	230.4	..	×	250 kHz	R.P.

* For binary facsimile signals.
† VSB—vestigial sideband amplitude. R.P.—restored polar line signal.
‡ Min.—minimum (set is asynchronous). Synch.—synchronous.

CENTRAL-OFFICE SWITCHING SYSTEMS

40. General. For individual telephones to be useful, they must have access to an instrumentality that is capable of establishing a connection between any two stations. Where only one instrumentality is involved, it may be manual or automatic and in either case may be one of many types. Where several instrumentalities are associated in switching longer-haul calls, they may be connected in any combination. In the manual systems one basic division is magneto or common battery. The automatic systems may be classed as step-by-step, all-relay, panel, crossbar, and electronic and also as common control or noncommon control. The latter terms refer to the degree of control exercised by the customer's dial over the actual switching mechanisms. For instance, in the crossbar systems the dialed digits are stored and do not directly control any switching operations, while in step-by-step systems the dial directly controls the movement of the switches. The manual switching systems are being replaced rapidly by automatic systems, and some automatic systems are being replaced by the latest, or electronic, type. All switching systems have certain appurtenances such as power plant, distributing frames, and test bureaus.

41. Power Plant. Power plants for telephone central offices produce electrical energy at voltages suitable for use in the various operations of telephone equipment. Mainly these plants convert commercial 60-c alternating current to direct current at the nominal voltages of 24, 48, and ±130. The two lower voltages are used for transmission battery, for the operation of electromechanical apparatus, and for vacuum-tube heater supplies. The higher voltage is used for coin-handling functions and for the plate-voltage supply in vacuum-tube circuits. Almost universally when only one polarity of a particular voltage is used, the positive end of the central-office battery is connected to ground. Other voltages have been used, but those above are now the most common; however, the advent of electronic switching systems is forcing the introduction of new voltages. The generators supplying direct current are backed up by storage batteries charged on a full float basis, usually. The batteries not only aid in voltage regulation but also provide a source of emergency power if the commercial power fails. In addition, diesel-engine alternators are provided to supply basic 60-c power in cases of extended commercial power failure.

Ringing energy can be obtained in several ways, (*a*) from a motor-generator driven by the commercial supply or by a storage battery, (*b*) from a ringing dynamotor supplied on the primary side by a storage battery, (*c*) from a vibrating pole changer operating from a storage battery, (*d*) from a subharmonic generator driven from the commercial supply, and (*e*) from a hand-driven magneto or from the commercial supply. These generators supply alternating current at fixed frequencies ranging from 16 to 66 c/s and at voltages ranging from 75 to 105. One widely used combination is 20 c at 86 V. The type of supply used depends somewhat on the output required. This may vary from 25 W for a small office to 1,000 W to serve a large building housing several offices. The ringing-distribution leads are fused, and also a resistance lamp is inserted in series to prevent excessive current in case of inadvertent short circuits. In most applications the alternating current is superimposed on a d-c component in order to provide for tripping (cessation of ring upon answer).

The usual ringing cycle is 2 s ring and 4 s silence. These intervals are used on most individual, two-party, and four-party full selective lines. The cycle is obtained by feeding the continuous ringing voltage through slowly revolving drums, sections of which are segmented to provide the stated intervals. Other types of configurations are used for semiselective and code ringing patterns.

Harmonic ringing is a type of selective ringing and employs a ringing source which generates several frequencies. Each subscriber set on a particular line must then be tuned to a different frequency in order to provide the selective feature, and the switching machine must determine and apply the proper frequency to ring a particular number.

42. Distributing Frames. Distributing frames or cross-connection facilities are installed in all types of central offices. These frames take two shapes, single-sided and double-sided. If single-sided, the protectors or terminal strips are often mounted vertically, although there may be a vertical and a horizontal section. If double-sided, one side is the vertical and the other the horizontal. The purpose of these frames is always to provide flexibility in the association of one type of equipment with another. A common type of main frame provides facilities for terminating incoming cable pairs on the vertical side and central-office equipment on the horizontal side. This permits any cable pair to be connected to any line number or trunk or other equipment by means of a twisted wire pair called a cross connection. An intermediate distributing frame provides for connecting any line number to any particular line equipment; likewise there is a trunk distributing frame, etc. The protectors spoken of above provide protection against high voltages or excessive current on the cable pairs.

43. Test Bureaus. Facilities must be provided to test the cable pairs and all items of outside plant. Such facilities are concentrated in a test bureau, which consists of several test positions together with records of subscriber lines, trunks, cables, etc.

These positions enable the tester automatically to reach each subscriber line and apply certain electrical tests to it. If he feels that a different type of test is required or if he is dealing with a trunk circuit, he may have a test shoe manually connected to the cable pair at the main-frame protector.

44. Cabling. Central-office equipment is most usually connected by switchboard cable. This type of cable is made up of various sizes of annealed tinned copper wire, which is sometimes enameled also. Twenty-two-gage wire has been in common use, but the recent tendency is to 24- and 26-gage conductors. Likewise, the insulation was commonly a silk and cotton wrapping or braid but is now more often polyvinyl chloride. Wires carrying signaling currents may be single, but wires carrying voice currents are twisted in pairs to avoid crosstalk.

45. Manual Switching Equipment. 1. *Magneto.* The magneto switchboard is suitable only for serving a small number of telephones and even in that application is rapidly being replaced. Its basic characteristics are the requirement for local batteries at each telephone set and the use of ringing current for all signaling.

2. *Common-battery.* Common-battery manual switchboards are suitable for use in all sizes of offices. As the name implies, batteries are not required at each local station, because one battery in the central office is used in common by all stations. As a result of this change the signaling, except for ringing on an incoming call, is done by direct current. The latest types of these manual boards provide certain features of the automatic systems such as automatic ringing and message register operation.

46. Automatic Switching Equipment. 1. *Step-by-step.* The first type of automatic switching equipment to be developed was the step-by-step system. This system derives its name from the fact that the switches are controlled directly by the customer's dial and step, or move, one unit for each impulse generated. The original device and the type in most common use today is a switch with two degrees of motion, vertical and rotary; i.e., when a switch is seized, it first steps vertically in accordance with the number of pulses (1 to 10) contained in the first digit it receives. It then rotates and may stop on any one of 10 terminals. The amount of rotation may be governed by dial pulses, as in a connector, or by the busy or idle state of the trunks appearing on the terminals, as in a selector. If the rotation is governed by the busy state of the terminals, the maximum number of terminals which can be covered is 10, because the rotation and seizure of the succeeding switch must be accomplished within the interdigital interval. There is an exception to this in a system based on a type of switch called a motor switch. This is a rotary switch with 100 sets of terminals which can store incoming digits while hunting for an idle path outgoing, and therefore its search time can be extended beyond the interdigital interval and its search field beyond 10 terminals. There is also another step-by-step or direct control system in which the motion is planar. This system is likewise limited to 10 steps, but they are taken in each of two directions in the horizontal plane.

2. *All Relay.* This type of office is classed as a step-by-step type, although the dial pulses cause relay operation only. It is used in small community dial office applications in sizes under 1,000 lines. The offices operate on a three- or four-digit basis, and in the smallest the line finders are connected directly to the connectors. In the larger offices a link is used to permit any line to have access to any connector hundred.

3. *Panel.* The panel system was the first venture into a common, as opposed to a direct, control system. It derives its name from the fact that the sets of three or four terminals per line of trunk are arranged in flat banks or panels with corresponding terminals aligned vertically. The banks are composed mainly of 40 or 100 sets each. A selector rod which can move vertically is mounted in front of each set of terminals and carries brushes which contact the terminals before it. Since some terminals in the bank are used for control purposes, an individual selector will have access to 90 outputs in a 100-point bank. These 90 outputs are arranged in eight groups of 10 and two groups of 5. The groups may be used individually or combined so that a group size may be 10, 20, or 30 terminals, and so on. The two sets of 5 terminals are always located at the top of the bank and usually are not combined with any groups below. The banks are mounted on frames, with each frame mounting 60 selectors. The 100-point banks are mounted 5 to a frame; the 40-point, 10 to a frame. The selec-

tors are motor-driven by means of a clutch mechanism mounted at the bottom of the frame before continuously revolving cork-surfaced rolls.

When a call is initiated, dialed digits are stored in a sender and the first three are translated in a decoder—this translation permits arbitrary routing in selecting a path out of the office. The last four digits of a seven-digit call are also translated, but according to a fixed pattern so that the routing on the incoming side of the office is not arbitrary. After translation the sender controls setup of the call, which is sequential, as in the step-by-step system; i.e., as each selector chooses the subsequent idle piece of equipment, the path thus set up is used first for control purposes. When this function is finished, the individual pieces of equipment advance to a cut-through position for talking. Thus the conversation path is built up piece by piece.

4. *Crossbar.* The crossbar systems are a further development of the common control principle. They also derive their names from the type of switch employed. This switch contains sets of contacts, called crosspoints, with three to six individual contacts per set, and the switch size may be either 100-point or 200-point. That is, each switch has 10 horizontal, or select, magnets and either 10 or 20 vertical, or hold, magnets. Any individual crosspoint, or set of contacts, in the field of 100 or 200 can be closed by the operation of one select and one hold magnet.

These switches are mounted on frames called link frames. Each link frame usually has a set of 10 primary and secondary switches each, which are connected in various patterns called cross-net patterns and which provide each primary switch access to every secondary switch, or vice versa.

Access between frames is provided by junctors. The junctors contain the transmission and supervisory apparatus and are connected in specific junctor patterns, depending on the size of the frame and the number of frames to be interconnected. The patterns provide specific numbers of paths from one frame to any succeeding frame and are used to obtain an equitable distribution of traffic over the office.

As in the panel system the customer dials the desired number into a sender or register. This piece of equipment seizes a common control unit known as a marker and transmits to it the first three dialed digits and other information. The marker then translates this code to determine which link frame carries trunks outgoing to the called office, selects an idle channel from the line to the selected idle trunk, and then causes simultaneous closure of the crosspoints to establish the conversation path.

5. *Electronic.* The concept of an electronic switching system is a radical departure from the concept of an electromechanical switching system in so far as control of the system is concerned. The plan of the switching network resembles a crossbar switching network, and relays are employed for certain functions, but the system derives its name from the fact that the information storage, call-processing, and control equipment is electronic rather than electromechanical.

As in all common control systems the customer-dialed number is registered, translated, and then processed. The electronic system performs individual logic functions in microseconds rather than milliseconds and will complete setup of an intraoffice call within 1 s of completion of dialing.

The basic concept of an electronic switching system is that of an electronic central processor operating with a stored program to control the actions of the switching network on a time-sharing basis. All information processing is handled by a central processor consisting of central control and the temporary and semipermanent memories. The temporary memory is used for storage of the transient information required in the processing of calls, such as dialed digits and the busy or idle state of lines and trunks. The semipermanent memory contains the stored program (method of operation) and the translation information.

Scanning points are provided in each line and trunk circuit and in the signal receivers. Electronic scanners periodically examine the electrical condition of these points, and information on their state is transmitted to the central processor. This state is compared with the state recorded on the last previous scan, and if there has been a change, the central processor recognizes that there may be need for action. Subsequent scans will determine whether the change was transient or bona fide, and the processor will then decide whether or not to proceed with a programmed response.

Analogous to the scanning points are distributing points. These are the points by which the central processor initiates action in the various equipment units and is thus enabled to establish and disconnect calls.

47. Switching Facilities. 1. *Step-by-step.* The step-by-step system is primarily a local switching system. As such, it uses base-metal contacts in the transmission path and likewise employs only one pair of wires for both directions of conversation. It has been arranged for local automatic message accounting, but this is employed in only a minority of the offices. Recently it has also been arranged for a limited number of common control features, but the application of these is not general.

The step-by-step system can also be used for tandem switching. In this application it often has associated with it centralized automatic-message-accounting equipment.

2. *Panel.* The panel system is also primarily a local switching system, although it can be used for tandem switching. In neither application does it have automatic-message-accounting equipment. It is also a base-metal contact system.

3. *Crossbar.* The crossbar is the most versatile electromechanical switching system and is used for local, tandem, and toll switching. In any of these applications it may have automatic-message-accounting facilities, either local or centralized, and for toll switching it uses two pairs of wires, one for each direction of transmission. All crossbar systems use precious-metal contacts in the transmission path to minimize noise.

4. *Electronic.* The electronic switching system is even more versatile than the crossbar system. It also may be used for local, tandem, or toll switching and may include automatic-message-accounting facilities in any application. It employs precious-metal contacts in the transmission path, and in the toll application the system uses separate pairs for the two directions of transmission.

48. Switching Features. The foregoing discussion is concerned with basic automatic telephone switching which can be accomplished by means of a rotary dial at the station and a switch which will follow pulses generated by it. However, an improvement in service can be achieved by adding various features to the systems. The first five lines of Table 25-9 indicate the switching features available in the various systems and those added as the state of the art has progressed.

Table 25-9. Features of Central-office Switching Systems

	S × S	Panel	X-bar	Electronic
Memory	No[a]	Yes	Yes	Yes
Translation	No[a]	Yes	Yes	Yes
Second trial	No	No	Yes	Yes
Alternate route	No[a]	No	Yes	Yes
Message register	No[b]	Yes	Yes[d]	No[f]
Class-of-service	No	Few	More	Many
ANI	Yes	Yes	Yes[d]	No[f]
AMA	No[b]	No	Yes[e]	Yes
Centrex	Yes[c]	No	Yes[e]	Yes
Custom calling	No	No	Yes[e]	Yes

[a] Available when office is arranged for common control.
[b] Available but used only on a minority of lines.
[c] Requires unusual equipment arrangements.
[d] Primarily Bell System No. 1 crossbar.
[e] Primarily Bell System No. 5 crossbar.
[f] Not required; LAMA facilities eliminate need for this.

In addition to switching telephone calls, the equipment must also provide for developing the charge information. Lines 5, 7, and 8 of Table 25-9 indicate the methods used in the various systems for doing this.

Finally, the recent tremendous growth in the telephone population and the advent of the electronic age have spurred the development of new services. The last two lines of Table 25-9 indicate that these can be furnished now only by the most recently designed systems.

The headings are defined briefly below.

1. *Memory.* The ability of a switching system to store dialed digits and other information and take action appropriate to the situation.

2. *Translation.* The ability of a switching system to refer dialed digits and class-of-service or other input information to recorded information to determine how to process a call.

3. *Second Trial.* The ability of a switching system to recognize that there has been an equipment irregularity and try again.

4. *Alternate Route.* The ability of a switching system to recognize that a route or routes are blocked and choose an alternative.

5. *Class-of-service.* Distinctions provided in a switching system to indicate the type of line or type of treatment to be accorded a particular line.

6. *Message Registration.* A method of counting the number of completed charged calls originated from a particular line. In local areas there is one scoring per call, but in zone registration areas calls between zones will cause more than one registration per call.

7. *Automatic Number Identification* (ANI). An equipment interrogation method of determining the calling line number.

8. *Automatic Message Accounting* (AMA). A means for recording call details and processing them for billing purposes. This may be LAMA, wherein the recording equipment is located in the local office, or CAMA, wherein this equipment is located at a central switching point. The ANI equipment is used in association with CAMA but not with LAMA.

9. *Centrex.* A service provided to PBX's whereby calls to extensions are terminated at the extension without intervention by the attendant and outward toll calls are identified by the extension number.

10. *Custom Calling Services.* Proposed new services which are under trial and development. Those most nearly ready for commercial application are speed calling and call waiting. The former provides for originating calls to any of a limited series of specific 7- or 10-digit numbers on a 2-digit basis. Call waiting is a service that provides an indication to a customer whose line is in use that someone is endeavoring to reach him.

The rotary dial, which generates pulses of direct current, is now the general means for a customer to transmit the desired number to his serving central office. A new method, called by various trade names, which generates combinations of audible frequencies is being applied to all types of local switching systems and is supplanting the rotary dial. Rotary-dial pulses, being interruptions of direct current, cannot, in general, be transmitted over a completed telephone connection. However, audible frequencies can be so transmitted, and providing a source of these frequencies to all types of customers opens prospects of a whole new field of use for switched, business, and nonbusiness telephone connections for "machine-to-machine" conversations.

49. Tandem and Toll Switching. Tandem and toll switching systems are logical outgrowths of the local switching systems. Basically, the same equipment is used in comparable types of local and toll offices. The differences lie in the fact that tandem and toll machines do not serve customers directly but switch incoming trunks to outgoing trunks; they provide more elaborate alternative routing, translation, and transmission arrangements and often use kinds of machine language other than dial pulsing (panel call indicator, revertive, and multifrequency) as input and output information. Many of these offices are equipped with CAMA facilities.

WIRE-LINE AND CARRIER TRANSMISSION

50. Definitions. The word "transmission" describes the conveyance of information. When this word is applied to the field of telecommunications, it includes communications for entities as well as human beings. In its most basic form, the transmission system consists of the seven parts shown in Fig. 25-36.

The information usually exists in either of two forms—digital or analog. Because of this, there are four system configurations which arise, depending upon the mode of

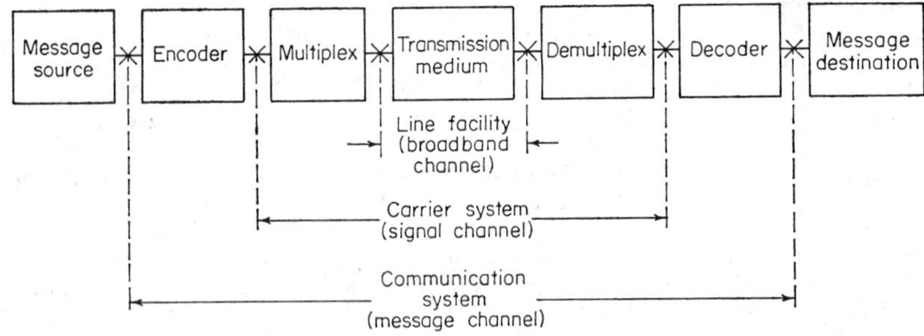

FIG. 25-36. Basic form of a transmission system.

the information coming in and the mode of the transmission medium. Figure 25-37 displays the four possibilities divided into analog input-output (I/O) and digital I/O systems.

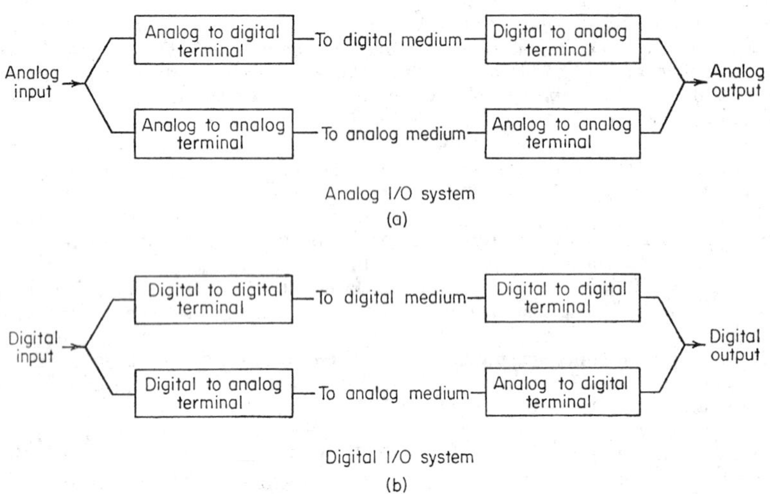

FIG. 25-37. Possible system configurations: analog input/output (*a*) and digital input/output (*b*).

51. Carrier Systems. The actual facilities (open-wire or cable) represent most of the cost in a circuit. Carrier systems permit many circuits to share the same facility and thus reduce the facility cost per circuit.

52. Major Components. Figure 25-38 shows a carrier system in its most elementary form and identifies the major components of a system. The line facility may be an open-wire pair, one or more cable pairs, or coaxial tubes, or a channel of a radio system. It is shared by all the circuits on the system, which are called derived circuits or channels. The multiplex terminals at each end process the

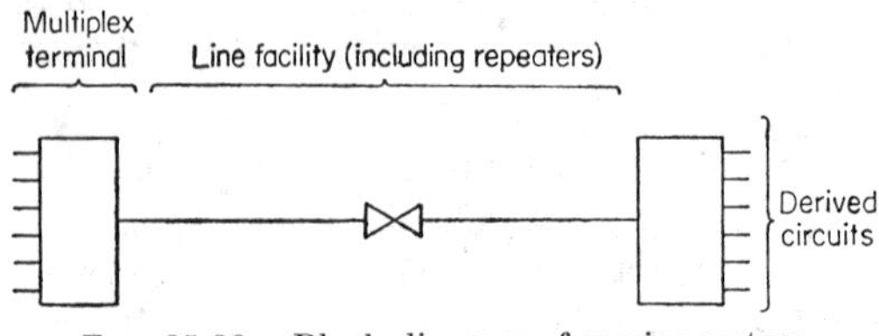

FIG. 25-38. Block diagram of carrier system.

derived circuits, converting or modulating them into a form that permits them to share the line facilities.

53. Line Facilities. The line facility generally includes amplifiers spaced at regular intervals to offset the attenuation of the open-wire or cable pairs used. (Radio systems are a somewhat different case and will be discussed later.) In many cable systems, power for these amplifiers (which may be located along the cable route in manholes or on poles) is transmitted over the cable pairs from a nearby office where power is available.

Since the attenuation of an open-wire pair or a cable pair is a function of frequency, it generally is necessary to use equalizers to produce a uniform sum of actual loss plus equalizer loss at all frequencies of interest. However, some systems use amplifiers whose gain is a function of frequency, complementary to the attenuation of the pair, thus combining the functions of amplification and equalization.

It also is generally necessary to provide some means for automatically adjusting the amplification and equalization to offset changes in the attenuation of the open-wire or cable pair, chiefly due to temperature. This function is called regulation. The composite device that performs these three functions—amplification, equalization, and (if needed) regulation—is called a repeater.

54. Terminals. The principal function of the terminal is the processing of signals from each circuit to form the line signal. However, there generally are other functions, such as mop-up regulation and equalization or feeding power to remote repeaters.

The signal-processing function may be accomplished by either of two different techniques: frequency-division multiplex (FDM) or time-division multiplex (TDM). In an FDM system the transmitting terminal modulates each circuit to a particular frequency assignment, each channel being assigned to a unique part of the frequency spectrum. The modulation technique may be amplitude modulation (AM), either double side band (DSB) or single side band (SSB), or frequency modulation (FM). In some systems, several successive stages of modulation are used.

In a TDM system each circuit is sampled in sequence and a signal transmitted describing the value of the sample. Most systems use pulse code modulation (PCM), in which the value of each sample is translated into a numerical code, usually seven or eight binary digits. However, it is also possible to transmit only the difference between the sample in question and the previous sample; this is called delta modulation (DM). If PCM is used, it is necessary to sample at a rate at least double the highest voice frequency to be transmitted; an 8-kc/s sample rate is common. If DM is used, a single binary digit is sufficient to describe a sample (larger than or smaller than the previous sample), but much higher sampling rates are needed.

55. System Design. The design of a particular carrier system is strongly influenced by its proposed field of use, particularly the distance between terminals. A system designed for long-haul use (a long distance between terminals) can economically use elaborate terminal arrangements to derive as many channels from the line facility as possible. On a short-haul system, however, it is important to keep terminal costs low, even at the expense of less efficient use of the line facilities.

56. Short-haul Systems. Short-haul carrier systems are available which make use of DSB, FM, or PCM; some systems, intended for intermediate lengths, use SSB.

57. Compandors. Short-haul systems often use compandors to reduce the effects of line noise (and crosstalk) on speech. A compandor is a device for compressing the volume range of a speech signal in the transmitting terminal and expanding the compressed signal to its original range in the receiving terminal. It effectively reduces noise during the silent intervals, when it is most objectionable.

58. Survey of Systems. A survey of available systems is shown in Tables 25-10 and 25-11.

Table 25-10. Wire-line Carrier Frequency Allocations
(Open-wire carrier system)

Manufacture	Type	Frequency allocations
Budelman	34 PLC Models A, B, & C	12, 24, 36, 48, 60, 72, 84, 96
Budelman	16, 26, 37, 38, 46, 49 Models A, B, & C	10, 20, 30, 40, 50, 60, 70, 80, 90, 100, 110, 120, 130, 140, 150, 160, 170, 180, 190, 200
	"A" allocation	20, 30, 40, 50, 60, 70, 80, 90, 100, 110, 120, 130, 140, 150, 160, 170, 180, 190, 200, 210
	"B" allocation	30, 40, 50, 60, 70, 80, 90, 100, 110, 120, 130, 140, 150, 160, 170, 180, 190, 200
Stromberg-Carlson	661–665	12, 24, 32, 40, 48, 60, 72, 84, 100, 112, 124, 136, 152, 168, 184, 200, 224, 240, 256, 272, 296, 312, 328, 344
Stromberg-Carlson	561	27, 39, 51, 63, 75, 87, 99, 111, 123, 135
Stromberg-Carlson	565	27, 39, 51, 63, 75, 87, 99, 111, 123, 135, 156, 172, 188, 204, 220, 236
Kellogg	K31–K32	14, 24, 32, 40, 48, 60, 72, 84, 108, 124, 140, 156, 172, 188, 204, 220
Kellogg	K31–K32	24, 36, 48, 60, 72, 84, 96, 108, 124, 140, 156, 172, 188, 204, 220, 236
Western Electric Lenkurt Lynch Lynch	"O" 45C B-510 B-500	2, 18, 20, 36, 40, 56, 60, 76, 80, 96, 100, 116, 120, 136, 140, 156
Western Electric Lenkurt Lynch Collins G.E. of Canada	"J" 45A B-60J WMX 106 TY SPO 1026	36–88, 92–150: Allocations for these carrier systems fall within this frequency range

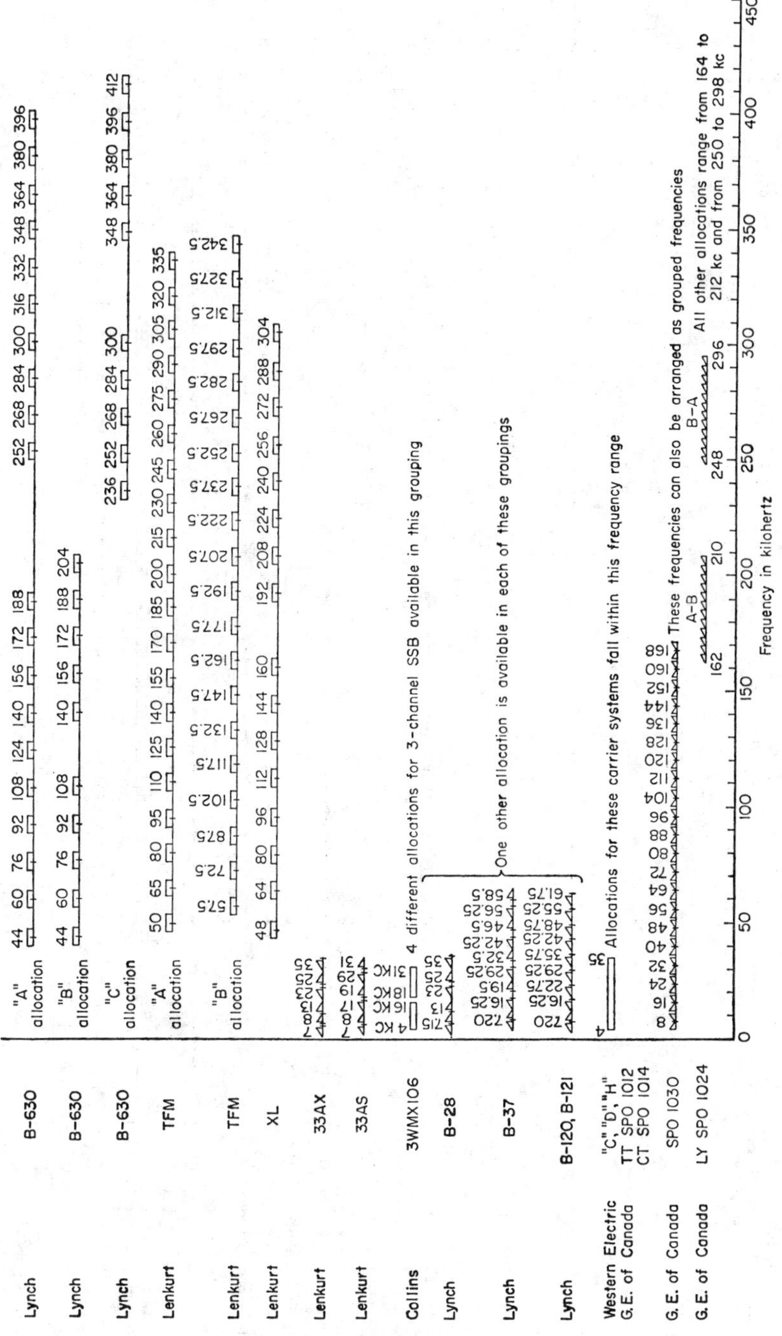

Lynch B-630 "A" allocation
Lynch B-630 "B" allocation
Lynch B-630 "C" allocation
Lenkurt TFM "A" allocation
Lenkurt TFM "B" allocation
Lenkurt XL
Lenkurt 33AX
Lenkurt 33AS
Collins 3WMX106
Lynch B-28
Lynch B-37
Lynch B-120, B-121
Western Electric G.E. of Canada "C", "D", "H" TT SPO 1012 CT SPO 1014
G.E. of Canada SPO 1030
G.E. of Canada LY SPO 1024
4 different allocations for 3-channel SSB available in this grouping
One other allocation is available in each of these groupings
Allocations for these carrier systems fall within this frequency range
These frequencies can also be arranged as grouped frequencies
All other allocations range from 164 to 212 kc and from 250 to 298 kc
A-B
B-A
Frequency in kilohertz

Table 25-11. Wire-line Carrier Frequency Allocations
(Cable carrier system)

Manufacture	Type	Frequencies
Western Electric Lynch Lenkurt Stromberg-Carlson	N-1, N-2 B-410 LN-1, LN-2 SC N-2	40 48 56 64 72 80 88 96 104 112 120 128 136 — 168 176 184 192 200 208 216 224 232 240 248 256 264
Kellogg (trunk and subscriber)	K-24	68 84 100 116 132 148 164 180 196 212 228 244 — 292 308 324 340 356 372 388 404 420 436 452 468
Lenkurt	81A	24.5 31.5 38.5 45.5 52.5 59.5 66.5 73.5 80.5 87.5 94.5 101.5 108.5 115.5 122.5 129.5 136.5 143.5 150.5 157.5 164.5 171.5 178.5 185.5 192.5 199.5 206.5 213.5 220.5 227.5 234.5 241.5 248.5 255.5 262.5 269.5 276.5 283.5 290.5 297.5 304.5 311.5 318.5 325.5 332.5 339.5 346.5 359.5 367.1 381.5 388.5 402.1
Lenkurt (formerly panhandle) (trunk and subscriber)	x	16 32 48 64 80 96 112 128 144 160 — 192 208 224 240 256 272 288 304 320 336
Lynch	B-630	44 60 76 92 108 124 140 156 172 188 204 220 236 252 268 284 300 316 332 348 364 380 396 412
Western Electric	K	12 — 60

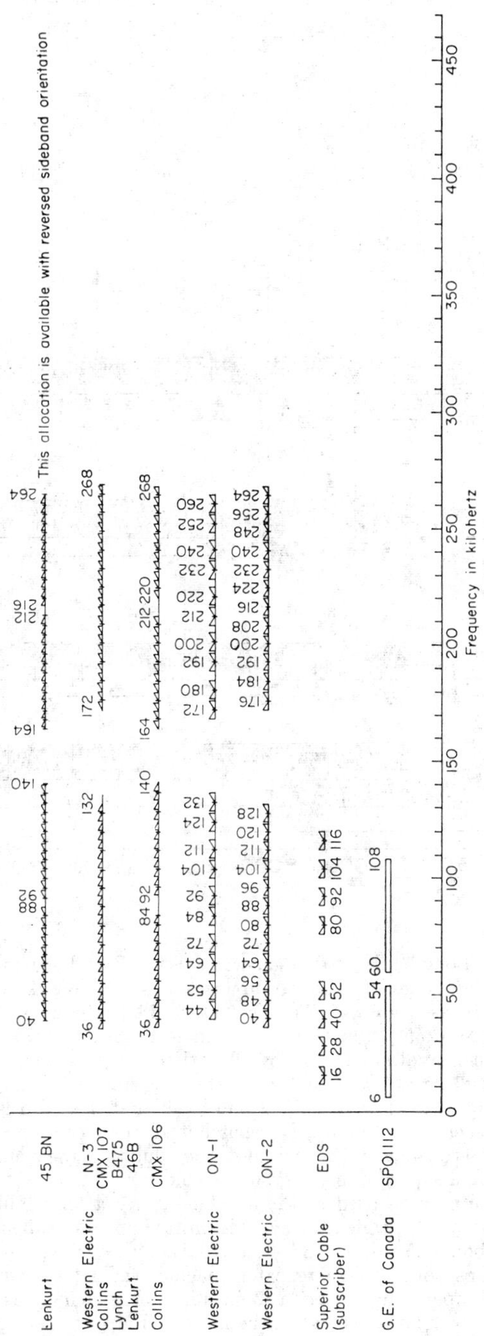
Lenkurt
45 BN
Western Electric
Collins
Lynch
Lenkurt
N-3
CMX 107
B475
46B
Collins
CMX 106
Western Electric
ON-1
Western Electric
ON-2
Superior Cable (subscriber)
EDS
G.E. of Canada
SPO1112
This allocation is available with reversed sideband orientation
Frequency in kilohertz

Table 25-12. Broad-band
(Typical Bell and

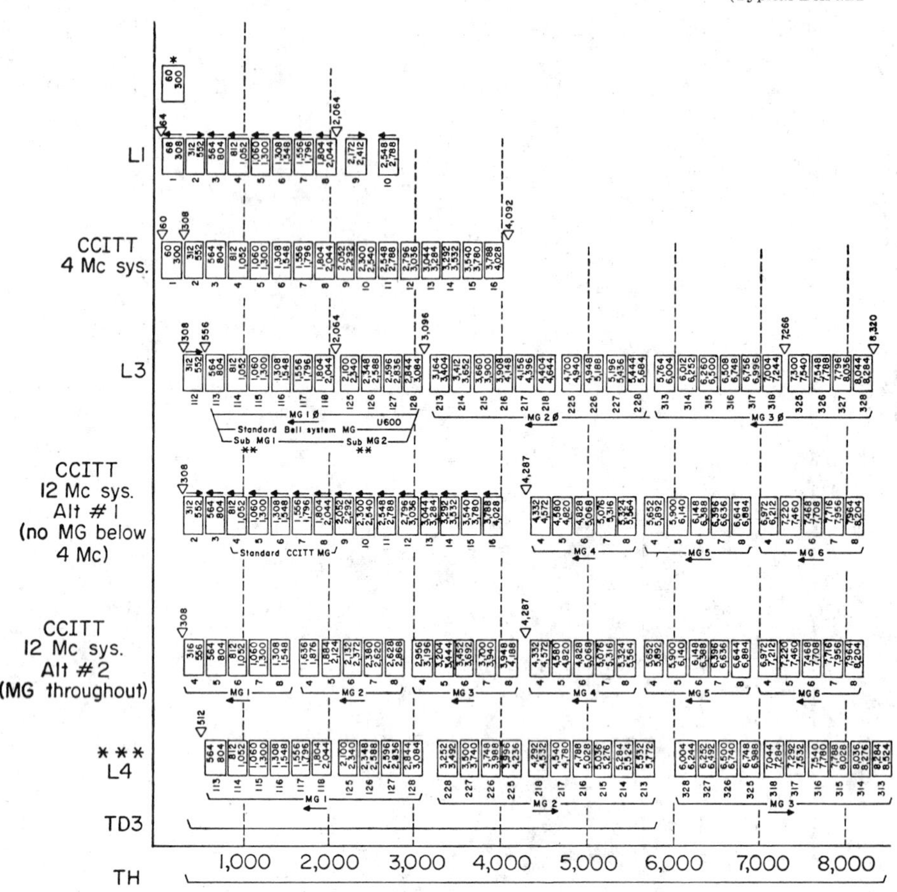

59. Long-haul Carrier Systems. Long-haul systems generally use SSB and several successive stages of modulation. One stage of modulation translates 12 circuits to the 60- to 108-kc/s range, referred to as a group. The next stage translates five of these groups (60 circuits) to the 312- to 552-kc/s range, referred to as a supergroup. Supergroups are then translated to their final line frequency allocation.

Figure 25-39 gives these frequency allocations. As shown, the first stage of modulation is called the channel bank, the second is called the group bank, and the third is called the supergroup bank. As is seen in the figure, it is possible to connect 60 circuits (or 12 circuits) through an intermediate office, without demodulating them all the way to voice frequencies, by using a supergroup (or group) connector.

60. Coaxial Cable. Coaxial cable may be used as the line facility for a long-haul system of this type. The first coaxial system, called the L1, transmitted up to 10 supergroups with a repeater spacing of about 7.5 mi on 0.375-in coaxials. A later system, called L3, transmitted 31 supergroups with a 4-mi repeater spacing; the L4 system (promised for 1967) transmits 60 supergroups with a 2-mi repeater spacing. In L3 and L4 systems a fourth stage of modulation is added, the master group, consisting

Carrier Frequency Allocations
CCITT systems)

→ Channel sidebands are erect (spectrum frequency increases with increasing audio frequency)

← Channel sidebands are inverted (spectrum frequency increases with decreasing audio frequency)

▽ Pilot

No._ No._ (box) Top number indicates lower-band edge of super group. Bottom number indicates upper-band edge of super group

No. (box) Number outside of box indicates super group

* Alternate assignment of SGI in the LI which matches CCITT standard

** There is no submaster stage of modulation on the solid-state multiplex (LMX-2)

*** MG's 1 and 2 will be used for TD-3 and MG's 1, 2, and 3 will be used for TH assignments

Ø This MG designation not compatible with new designation as on L4. It is proposed to change these, in the future, from 1, 2 and 3 to A, B and C

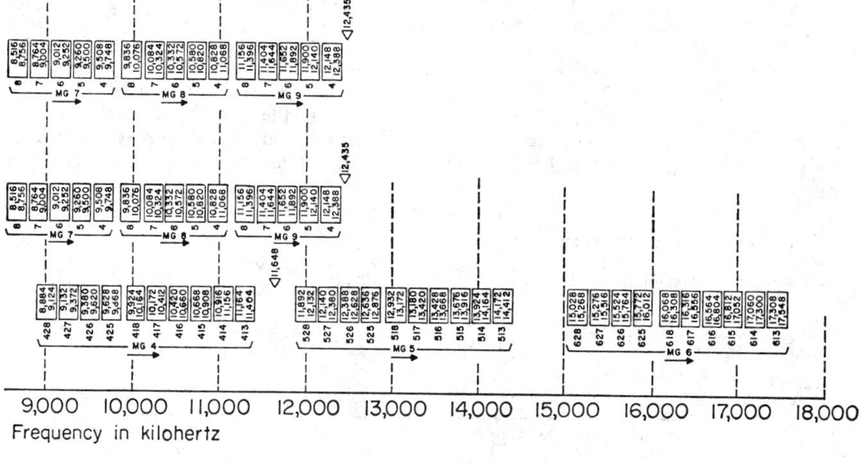

of 10 supergroups translated as a group to the final line-frequency assignment. Table 25-12 shows typical Bell and CCITT frequency allocations.

61. Submarine Cables. Modern transoceanic submarine-cable telephone systems represent an important extension of coaxial-cable-system design techniques. The transmission problems posed by the submarine cable systems are essentially similar to those posed by land-based systems except that the ocean-bottom environment introduces some new problems. For example, submarine cables and associated repeater housings must withstand hydraulic pressures up to about 5,000 lb/in² or more for long periods of time. Also, since the nature of the installation precludes, for practical purposes, any extensive adjustment or check of the cable or repeaters after placing, high reliability and long service life become essential requirements to be provided by system design and manufacturing control techniques. Early systems used two one-way coaxial-type cables—one for each direction of transmission. The cable designs were adapted from telegraph experience and depended on a layer of armor wires outside the cable for protection and strength. Recent cable designs have used single cables without armor but with a core of high-strength steel strands inside the coaxial center conductor for strength. In early systems, the repeater elements were arranged axially along the cable in elongated flexible housings to facilitate handling with the cable-laying machinery available at the time. In recent systems, use has been made of rigid repeater housings. This is practicable with the cable-laying machinery currently available. Derivation of voice channels is accomplished by using supergroup and group modulation stages, as described above. However, because of the high cost of the line facility, more efficient use is generally made in submarine cable systems of the available bandwidth by the use of a 3-kc/s spacing plan for voice channels instead of the 4-kc/s plan normally used for domestic systems. In addition, the number of channels provided by a given cable system is frequently increased (doubled in some cases) by the application of TASI (time assignment speech interpolation). Table 25-13 compares the principal designs for transoceanic submarine cable telephone systems currently in use.

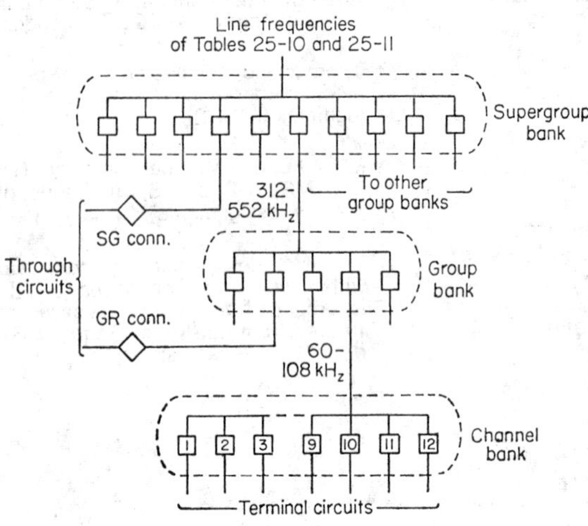

Fig. 25-39. Arrangement of banks and frequency allocation in a carrier system.

Table 25-13

System designation	Coaxial size inside diameter of outer conductor, in*	Nominal repeater spacing, nmi	Circuit capacity without TASI, 3 channels
SB	0.62	37	48
SD	1.00	20	128
SF	1.50	10	720

* SB system requires two cables for two-way operation; SD and SF systems require only one cable.

62. Voice Transmission. The subject of voice transmission covers the transmitting of signals of voice-bandwidth grade from a source to a receiver. In past years this consisted in providing a talking path to transmit information between two or more human beings. It has now become necessary to add the capability for data transmission between machines and for both voice and data transmission between a machine and a person. Any circuit must be able to work in any one of these three modes at any one time. The transmission field has been a dynamic one through the years of its evolution. The electron tube made possible a good long-distance communication facility. Later, the transistor provided economies in space and money, and its low power requirements made the use of amplifiers feasible in places where the cost of powering electron tubes had been prohibitive.

63. Quality. The quality of communications is determined by many factors. The objectives for these various factors are outlined in subsequent paragraphs, and examples of transmission parameters, objectives, and design criteria are discussed. The transmission performance of a telephone connection depends on a number of parameters, mainly loss, noise, distortion, and echo. How well each person hears depends not only upon these parameters of the connection but also upon the talker and upon his own hearing. The loudness of the speech that reaches the listener depends upon the loudness at the talker's lips and how well he directs his words into the transmitter. How well the listener hears depends to some extent upon the sensitivity of his ear and upon the margin by which the speech overrides the total telephone noise. Figure 25-40 shows the speech power levels produced by an average talker throughout a typical telephone connection. The abrupt rise at the left and the abrupt fall at the right are the result of plotting both the acoustic field strength and electrical power on the same numerical ordinate scale but in different fundamental units. There is an actual power gain through the transmitter and a loss through the receiver, but not necessarily of the amounts shown in the figure. Note that the received speech is at about a 38-dB higher level than the total telephone noise. A difference in this general range or greater is considered good, but as differences fall below about 30 dB, the noise starts to interfere with intelligibility.

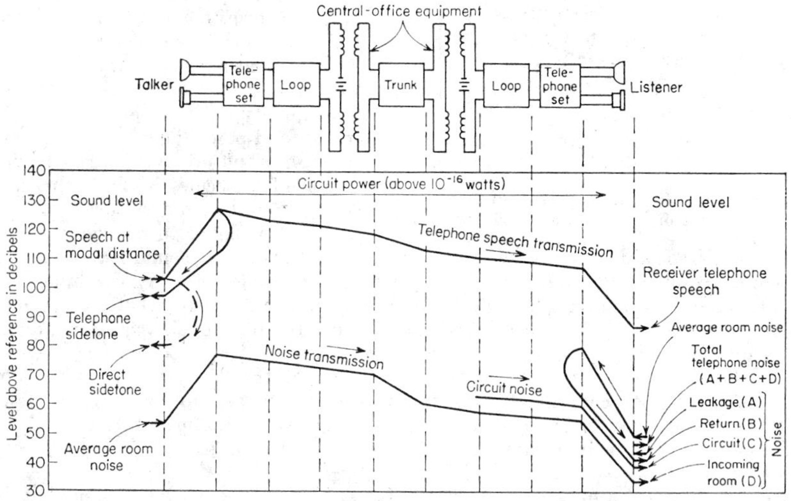

FIG. 25-40. Power-level diagram of typical telephone connections.

64. General Transmission Objectives. 1. *Overall transmission performance objectives* for voice telephone communication are primarily determined by opinion appraisals.

The general problem in establishing transmission objectives for the interlocal and toll switching networks is to design each circuit element (set, loop, switching office, and trunks) so that the complete system will please the customers when these parts are combined in each way permitted by a particular switching network. Specific objectives will not be discussed here, as they change through the years with changes in customer preference and advances in the state of the art. Performance objectives for voice transmission have been established for the parameters discussed in Par. **65.**

2. *The provision of good transmission* requires a compromise among: (*a*) the need for sufficiently low loss to provide satisfactorily high received volumes, (*b*) minimum contrast in received volumes on different calls, and (*c*) the need for sufficiently high losses to ensure adequate performance from the standpoint of talker echo, noise, and near singing.

3. *Objectives for transmission of the voice* are based on various subjective tests. Volume, noise, frequency response, echo, crosstalk, and side tone are adjusted to obtain customer preference curves. Since customer reaction and patterns change throughout the years, these tests must be made periodically and objectives updated accordingly.

4. *Objectives for the transmission of data* cannot be determined in a subjective manner. Since data-receiving devices have little or no intelligence built into them, the various parameters must be controlled within limits determined by the expected error performance. Objectives for data transmission are determined by the required error performance and are discussed in the paragraphs on digital I/O systems.

65. Parameters of Importance. 1. *Attenuation* is a general term used to denote a loss in power when energy is transmitted through some medium. It is usually expressed in decibels, and in this case it is ten times the logarithmic ratio of the transmitted power to the received power. Attenuation is measured with the facility terminated in its characteristic impedance at both ends.

2. *Insertion loss* is the loss measured when the network or circuit element is terminated at both ends, not in its characteristic impedance, but in some other specified impedance.

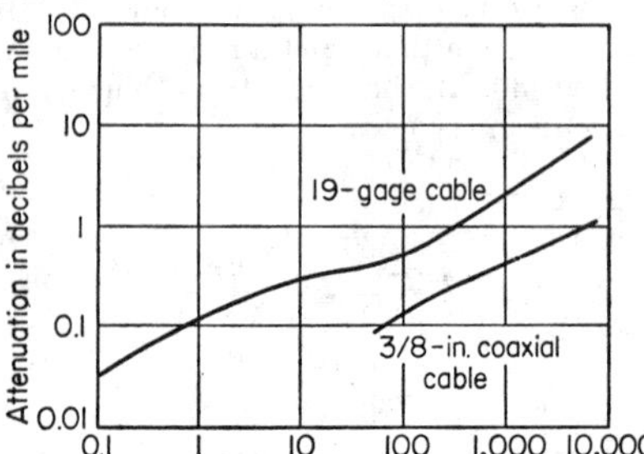

Fig. 25-41. Attenuation vs. frequency characteristics of various wire facilities.

3. *Frequency response* is used to describe the loss-vs.-frequency characteristics of a transmission medium. Since attenuation increases with increasing frequency, the loss-vs.-frequency curve will tilt upward, as shown in Fig. 25-41.

4. *Noise* may generally be defined as any unwanted signal associated with the desired signal. In telephone transmission the major types of noise are gaussian, impulse, and power hum. Gaussian noise is random in nature and is caused by the collision of molecules in the transmission system. The sound resulting from gaussian noise is similar to rain on a tin roof. Impulse noise is a disturbance of short duration. The resulting sound is a click or pop and may result from contact closures, switches operating, etc. Power hum is the result of a transmission medium being in close proximity to a source of power such as a 60-c power system. The energy finds its way into the transmission path by inductive and capacitive coupling.

5. *Crosstalk* means, in general, the unwanted signal that finds its way from its own channel into other channels through inductive and capacitive coupling. Crosstalk of voice signals may be intelligible if it is loud enough, or it may be inherently unintelligible because of frequency shift or frequency inversion in carrier terminal equipment.

6. *Return loss* is a measure of the reflection at a point where a transmission medium changes character. It is generally expressed as 10 log (P_i/P_r), where P_r is the power reflected from the junction and P_i is the power that would arrive at the junction if the medium beyond the junction were replaced with a nonreflecting termination. The smaller the reflection, the higher the return loss. The phenomena associated with return loss are "instability" (singing), "hollowness" (near singing), and "echo," defined below.

7. *Echo* is the attenuated and delayed sound heard distinct from, and in addition to, the original transmission by a talker or listener as a result of distant reflection in the transmission path. By extension, it is also attenuated and delayed data signals, distinct from, and in addition to, the original signals.

8. *Side tone* is a term used to indicate that portion of the talker's signal appearing at his own receiver without appreciable delay. The effect of excessive side tone is to cause the talker to lower his voice, thereby reducing the volume which the listener receives.

9. *The linearity* of a circuit is a measure of its ability to reproduce a signal without distortion.

66. Voice-frequency Transmission Loops. Telephone loops are the facilities, usually cable pairs, that provide the electrical connections (transmission paths) between the station equipment at the various customer locations and the local switching offices that serve them.

1. *Resistance Design.* This is the most commonly used method of designing loops. Resistance design is used to determine the permissible lengths of two different gages of wire that can be used for a loop of specified length to obtain a certain value of total conductor loop resistance. The value of total conductor loop resistance is usually the external conductor-loop-resistance signaling limit for the type of switching office in which these loops will be terminated. For economic reasons, the design will normally consist of complementary lengths of adjacent gages, i.e., 26 and 24, 24 and 22, or 22

and 19 gage, with the finer-gage cable located next to the central office. The following formula can be used to determine this design,

$$R = r_1(L - x) + r_2 x \tag{25-6}$$

in which R is the total conductor-loop-resistance signaling limit in ohms, r_1 and r_2 are the resistance per unit length for the finer and coarser gages, respectively, in ohms per kilofoot, L is the total loop length in kilofeet, and x is the length of the coarser gage in kilofeet. Where frequent application is required, graphical layouts of resistance vs. loop length can be used to solve these problems more readily. In order to provide a satisfactory distribution of transmission losses for loops using cable facilities designed to a resistance signaling limit, it is necessary to apply inductive loading, usually H88 (88-mH coils spaced at 6,000-ft intervals), to all loops over 18 kft in length, limit bridged tap (any portion of the cable facility that is not part of the direct transmission path) to a maximum of 6 kft, and, on loops longer than 10 kft, use only the Bell System 500-type or equivalent telephones.

2. *Telephones.* The present design of loops is based upon the use of the Bell System 500-type telephone or equivalent. These telephones provide about 5 dB improvement, both transmitting and receiving, over the Bell System 300-type telephones when used on facilities having loop resistance of about 900 Ω or more and a 48-V central-office battery supply. It is the increased efficiency of the 500-type set on the longer loops that is used to control the losses of loops over 10 kft in length, as mentioned in (1) above. The older types of telephones, such as the Bell System 300-type, may be used on loops less than 10 kft in length. The transmitting and receiving efficiencies of the 300-type set on short loops are about as great as can be used and still adequately control crosstalk and noise. Universal use of the more efficient 500-type set for both short and long loops is made possible by the equalizer in the 500-type set. Figure 25-42 indicates the relative transmitting and receiving response vs. length of 26-gage loop for a 300-type set and 500-type sets with and without the equalizer. These curves show the effect of the equalizer in the 500-type set on the transmitting and receiving transmission characteristics.

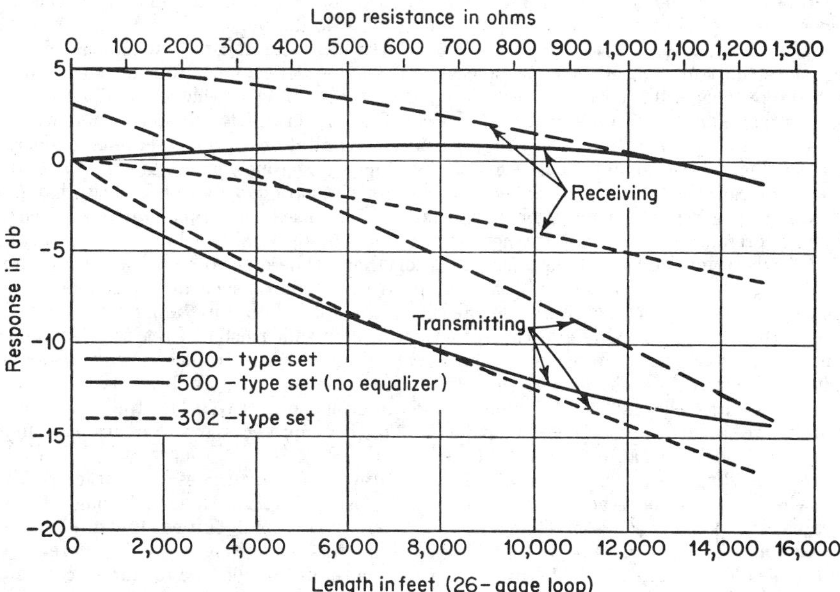

FIG. 25-42. Relative transmitting and receiving response vs. length for various loop configurations.

The most notable effects are the almost constant receiving response for the 500-type set with equalizer for loop length up to about 12,000 ft and a transmitting response that is comparable with the 300-type set for loop lengths up to about 10,000 ft. Beyond 10,000 ft the effect of the equalizer in the 500-type set diminishes, and the total increased efficiency of the 500- versus 300-type set becomes available to provide a better grade of transmission for the longer loops than could be obtained with the 300-type sets. The 500-type set also contains an improved side-tone balancing network, which is required to control side tone that would otherwise be excessive with the more efficient transmitter and receiver used in the 500-type sets.

67. Trunks and Switching Network Design. A trunk is a communication circuit extending between two switching offices. Trunks may be provided over either 2-wire or 4-wire facilities. Two-wire trunks transmit over the same pair of wires in both directions, while 4-wire trunks transmit over separate pairs in the two directions. Trunks may be established on open-wire, cable, or radio facilities. They may operate at either carrier or voice frequencies on open-wire lines or cables, but only at carrier frequencies on radio channels. Many factors enter into the choice of facilities for minimizing the long-range cost of a trunk group: transmission requirements, sizes of group, expected rate of growth, geography (over-water spans limit the choice to submarine cable or radio), etc.

An interlocal switching network for a city or metropolitan area involves a minimum of switching between local and tandem offices. Usually only one to three trunks are switched together over relatively short distances to complete a connection. The transmission requirements for such trunks are therefore less restrictive than for trunks in the toll switching network. Performance objectives for interlocal trunks are specified for all the important circuit parameters (i.e., loss, noise, etc.) with the exception of echo. Echo is not a significant impairment owing to the short round-trip delays encountered in interlocal networks, and therefore minimum loss objectives are dependent only on the need to prevent singing and near-singing conditions.

The design of trunks for the toll switching network involves tighter limits on all circuit parameters. In addition, the concept of via net loss (VNL) design is applied to control echo. This is necessary because the toll switching network plant involves a hierarchy of toll switching offices (Classes 1, 2, 3, and 4) and automatic alternative routing. A possible maximum of nine trunks can be connected in tandem in a continental United States call, resulting in significant echo delays.

The design loss of toll connecting trunks (between local and toll switching offices) and intertoll trunks (between toll offices) is primarily determined by the requirement that talker echo will be satisfactorily low on more than 99% of the connections which encounter the maximum delay. VNL is defined as the lowest loss in decibels at which it is desirable to operate a toll trunk, considering limitations of echo, crosstalk, noise, and singing. It is calculated as VNL = (VNL factor) × (one-way distance) + 0.4 dB, where specific VNL factors have been derived for the various types of facilities having differing velocities of propagation. The factors are based on subjective tests of customers' preferences, and they further assume that no appreciable echo paths are introduced where intertoll trunks are switched together. There is no echo path at 4-wire switching offices, and balance requirements have been established at 2-wire switching offices. These balance requirements effectively control echo by providing a high return loss in the echo path. One additional method of controlling echo in the toll switching network is the application of echo-suppressor equipment on all trunks between regional centers (Class 1 to Class 1) and on all interregional high-usage trunks. Based on the above design considerations, intertoll trunks are designed at VNL, and toll connecting trunks are designed at VNL + 2.5 dB (the 2.5 dB being necessary to assure stability from near singing in an overall condition).

68. Loading. Loading is a system for improving the transmission characteristics of wire facilities for certain purposes without using amplification. Although there are several possible ways of changing the characteristics of facilities, the most practicable and effective method is to add series inductors, called loading coils, at regular intervals along the line. While preserving the continuity of the line for direct current, these coils increase the line's impedance and thereby reduce its attenuation. In addition, they tend to make both impedance and attenuation more uniform throughout the "passband" of the loading system. These advantages, illustrated in Fig. 25-41, are attained at the sacrifice of all transmission capabilities at frequencies above the passband.

The impedance, the attenuation, and the passband of a system are roughly approximated by the following formulas:

$$\text{Impedance} = L/C \qquad \text{(ohms)} \tag{25-7}$$

$$\text{Attenuation} = R/2\sqrt{C/L} + G/2\sqrt{L/C} \qquad \text{(Np/loading section)} \tag{25-8}$$

$$\text{Passband} = 0 \text{ to } f_c \qquad \text{where } f_c = \frac{1}{2\pi\sqrt{LC}} \tag{25-9}$$

where L = inductance (henrys) of each loading coil, C = mutual capacitance (farads) of the facility between successive loading coils, R = series resistance (ohms) of the facility between successive loading coils, G = shunt conductance (mhos) of the facility between successive loading coils, and f_c = cutoff frequency in cycles per second.

Table 25-14. Cable Loading Systems in General Use

Designation*	Nominal cutoff frequency f_c, c/s	Use
H88	3,500–4,000	Message trunks and long subscriber lines
H44	5,000–5,600	Certain data circuits
B22	10,000–11,200	Program networks

* The letter designates the spacing,
H = 6,000 ft B = 3,000 ft
The number gives the inductance of each loading coil in millihenrys.

Economics dictates that the loading coils be spaced not too close together, and the need for a reasonably wide passband puts an upper limit upon L. The loading systems that have evolved for voice-frequency cable facilities under these constraints have employed coils ranging from a few millihenrys to about ¼ H and spacings ranging from about 1,000 to 9,000 ft. The trend for loading on regular telephone trunks today is to use coils of 88 mH at a spacing of about 6,000 ft. For this system, f_c ranges from 3,500 to about 4,000 c/s, depending upon the cable capacitance per unit length. For wider-band facilities, the inductance, the spacing, or both must be reduced. For 15-kc/s program facilities, for example, coils of 7.5 mH at a spacing of 1,500 ft have been used.

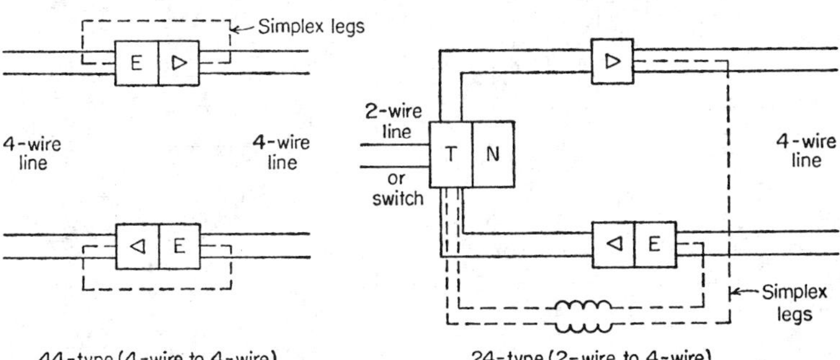

FIG. 25-43. Amplifier-type repeaters: 4-wire to 4-wire (*a*) and 2-wire to 4-wire (*b*).
◁—amplifier; triangle points in direction of transmission.
E—equalizer; compensates for unequal propagation at different frequencies in the line section leading up to it.
T—terminating set; accomplishes transition from 2-wire to 4-wire.
N—balancing network; balances 2-wire line to minimize power fed from incoming 4-wire line to outgoing 4-wire line, in order to limit echo to acceptable levels.

Table 25-15. Electrical Characteristics of Exchange-type Cables at 68°F

Gage (AWG)	Primary constants[1] (at 1,000 c/s) C, μF/mi	R, Ω/mi	L, mH/mi	G, mhos/mi		Secondary constants (at 1,000 c/s) Attenuation Db/mi	Attenuation Np/mi	Phase shift Rad/mi	Phase shift Deg/mi	Characteristic impedance R, Ω	Characteristic impedance X, Ω	Midsection image impedance R, Ω	Midsection image impedance X, Ω	Cutoff frequency, c/s	Speed of propagation, mi/s
19	0.084	86	0.886	1.219	Nonloaded	1.27	0.146	0.156	8.9	296	−276				40,000
					H88-loaded	0.42	0.049	0.519	29.8			1,013	−93	3,440	12,000
22	0.082	173	0.870	1.190	Nonloaded	1.81	0.208	0.214	12.3	417	−403				29,000
					H88-loaded	0.79	0.091	0.519	29.7			1,035	−180	3,480	12,000
24	0.084	274	0.950	1.219	Nonloaded	2.31	0.266	0.272	15.6	516	−503				23,000
					H88-loaded	1.21	0.140	0.536	30.7			1,045	−272	3,440	12,000
26	0.079	440	0.995	1.146	Nonloaded	2.85	0.329	0.332	19.0	671	−660				19,000
					H88-loaded	1.79	0.206	0.542	31.0			1,121	−425	3,540	12,000

Electrical Characteristics of Open Wire at 68°F
(Non-pole-pair physical—12 in spacing)

Gage, mils	Type conductor	Primary constants (at 1,000 c/s) C, μF/mi	R, Ω/mi	L, mH/mi	G, mhos/mi	Secondary constants (at 1,000 c/s) Attenuation Db/mi	Attenuation Np/mi	Phase shift Rad/mi	Phase shift Deg/mi	Impedance R, Ω	Impedance X, Ω	Speed of propagation, mi/s
104	Copper (dry)	0.00846	10.4	3.66	0.052	0.067	0.0077	0.036	2.1	674	−144	176,000
	Copper (wet)	0.00876	10.4	3.66	1.75	0.073	0.0084	0.036	2.1	664	−132	173,000
104	Copper-steel[2] (dry)	0.00858	26.0	3.72	0.11	0.15	0.0177	0.040	2.3	737	−327	158,000
	Copper-steel (wet)	0.00894	26.0	3.72	3.26	0.17	0.0192	0.040	2.3	729	−300	157,000
109	Steel (dry)	0.00835	85.0	10.4	0.23	0.29	0.0333	0.067	3.9	1,283	−629	94,000
	Steel (wet)	0.00873	85.0	10.4	3.36	0.31	0.0360	0.068	3.9	1,271	−579	93,000

[1] The figures given take account of the effects of both wires in each mile of cable pair.
[2] 40% conductivity of same gage copper.

Many years ago, before amplifiers came into general use, loading was used on open-wire lines. With the advent of amplifiers, however, this use was discontinued, since it increased to an undesirable extent the line characteristics' sensitivity to rain and sleet.

The cable loading systems in general use today are shown in Table 25-14.

69. Transmission-line Characteristics. Table 25-15 lists some of the more important characteristics of some of the more widely used transmission facilities.

70. Repeaters. A repeater is a device that boosts the energy levels on a communication circuit in both directions of transmission. The earliest telephone repeater was a person who acted as an intermediary between customers too far apart to hear each other clearly. He listened to one customer and then repeated what he heard to the other customer. Early electrical repeaters depended upon telephone receivers mechanically coupled to transmitters for stepping up the energy levels. For many years, however, repeaters have been entirely electrical and now use either vacuum tubes or transistors to boost energy levels. Today's repeaters are of two general kinds, the amplifier type and the negative-impedance type. The amplifier type includes two amplifiers, one for each direction, while the negative-impedance (NI) type employs only one energy source. The amplifier type, without auxiliary equipment, does not pass direct current or signals below the usual frequency range of the voice; the NI type does. Figures 25-43 and 25-44 show schematically the ways in which the two types of repeaters are used.

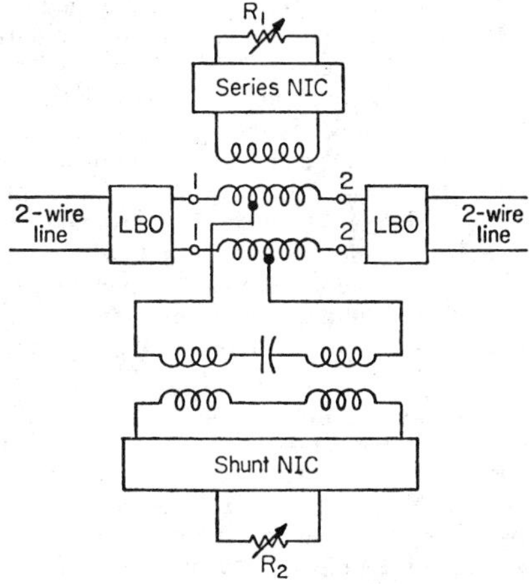

FIG. 25-44. Negative-impedance repeater.

For 2-wire circuits, in which the same wire facilities are used for both directions of transmission, either the negative-impedance type or the amplifier type can be used. Since the former is cheaper and needs no signaling bypass equipment, it is usually preferred. In the amplifier type a transition is made to 4-wire operation at each repeater point, and a one-way amplifier is inserted in each side of the 4-wire part, one for each direction of operation. Most of the amplifier types have greater gain capability, but crosstalk restrictions usually prohibit large gains. The main purpose of using a 4-wire instead of a 2-wire circuit is to prevent reflections that occur between the ends of the circuit from reaching either end as echoes. Amplifier-type repeaters are essential in such circuits. Each amplifier transmits in one direction only and blocks all reflections. Negative-impedance repeaters cannot be used in 4-wire circuits, since they do not block reflections.

71. Reference. Transmission Systems for Communications, Bell Telephone Laboratories, Incorporated, 1964.

72. Television Transmission. The present standard bandwidth required for television is 4.25 MHz, which is the FCC standard. The required bandwidth is determined by the horizontal resolution required.

73. Spectrum. The scanning process determines the basic distribution of energy in the signal band. Line scanning of picture information concentrates the signal energy into harmonics of the line frequency. In addition, modulation of the line-frequency harmonics by the 60-Hz field scan gives rise to 60-Hz sidebands on each line-frequency harmonic. The television signal, therefore, consists of a number of fixed frequencies which vary in phase and amplitude at a slow rate only as a result of motion in the picture. Each of these frequencies can be considered as a carrier and the effect of motion as adding sidebands around the carrier. The net result is a signal-frequency composition similar to that shown in Fig. 25-45.

Figure 25-45*a* illustrates the entire 4-MHz bandwidth, indicating the levels of line-frequency harmonics for a typical signal. Nine-tenths of the harmonics have not been drawn in, and the 60-Hz components near zero frequency have been omitted for clarity.

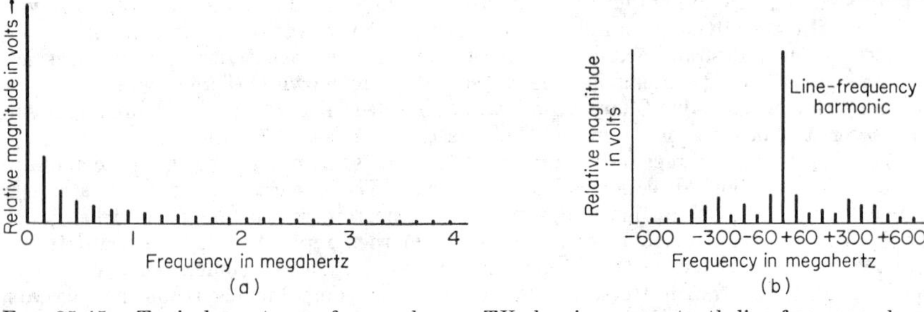

Fig. 25-45. Typical spectrum of monochrome TV showing every tenth line-frequency harmonic (*a*) and typical sidebands around each line-frequency harmonic (*b*).

A small section of Fig. 25-45*a* magnified to illustrate the presence of the 60-Hz sidebands that cluster about each line-frequency harmonic is shown in Fig. 25-45*b*.

74. Color Signal. The National Television Systems Committee (NTSC) color television system is based on the principle that color may be adequately defined in terms of three characteristics: luminance, hue, and saturation. Luminance is defined as intensity or brightness and is the basis on which the present monochrome system operates. Hue is in terms of whether a color is red, blue, green, yellow, etc. Saturation is the degree to which the hue is mixed with white. For example, pink is a low-saturation red. A high-saturation red would be a brilliant crimson.

The color signal, therefore, must contain information as to these three characteristics. The color system uses the same type of signal to transmit luminance information as is used in the monochrome or black-and-white system. To this is added the saturation and hue information which is the basic difference between the monochrome signal and a color signal. The necessity of transmitting three pieces of information instead of one, simultaneously and without interaction or distortion, imposes new requirements on the transmission facilities.

The phase relationship of the color subcarrier to a reference synchronizing signal of the same frequency (color burst) determines the hue of the color. The color burst consists of approximately 9 Hz of the color carrier frequency placed on the back porch of the horizontal blanking signal, as shown in Fig. 25-46*a*.

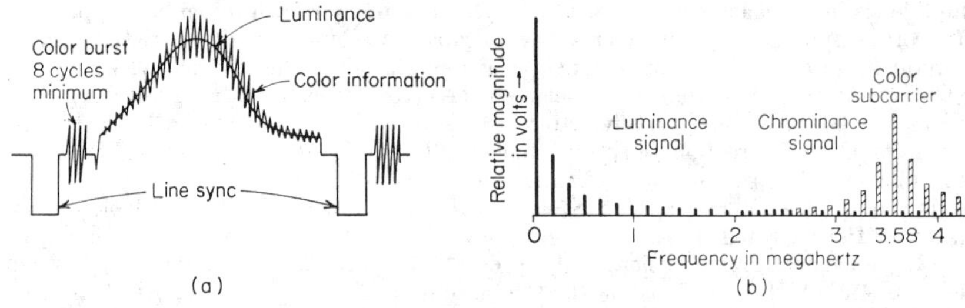

Fig. 25-46. Color signal waveform (*a*) and NTSC color TV spectrum (*b*).

Distortions of hue will occur if the phase shift of the transmission system at color carrier frequency is a function of the amplitude of the luminance signal. This variation in color carrier phase shift caused by variations in amplitude of the luminance signal is called differential phase. The presence of differential phase in a system used to transmit color television results in a distortion in the hue of the colors. The frequency of the color subcarrier, 3.579545 MHz, is an odd multiple of half-line frequency (7.867 Hz for color television). The frequency composition of a typical NTSC color signal is shown in Fig. 25-46*b*.

75. Local Transmission. Wire facilities (A2A facilities) for broadcast television transmission involve the use of base-band video transmission system over video pairs.

Television wire connecting circuits are used for a variety of purposes in network operations. Some examples of these uses are shown in Fig. 25-47a. Network broadcasters usually have their studios, master control, and broadcast transmitter distributed among several buildings in various locations in a city. The wire facilities furnish broadband tie lines to interconnect these facilities. Two-way connections between the master control location and the studio are often required for programming purposes.

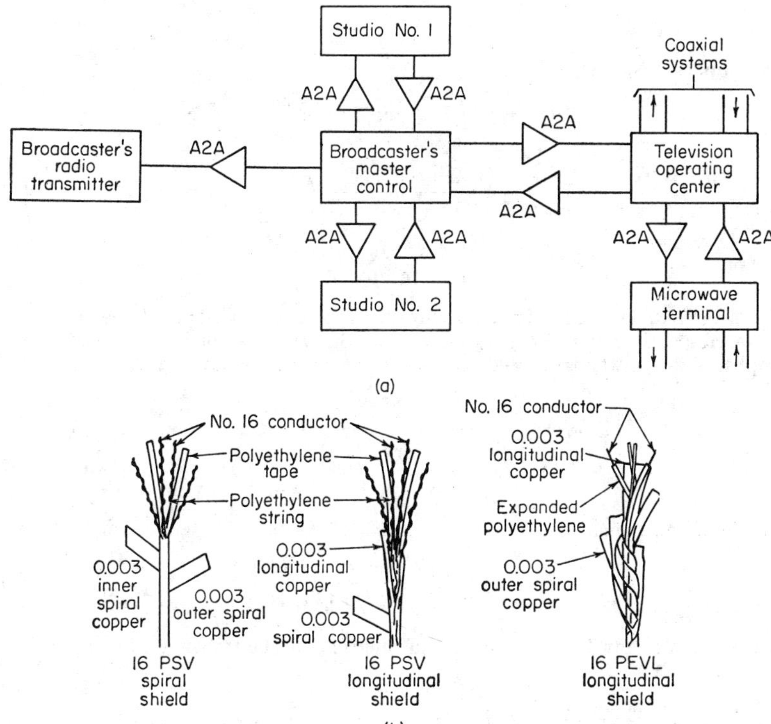

FIG. 25-47. Video transmission facilities (a) and cables used (b), A2A system.

For network operation, connecting circuits are required between the master control and the central switching point where connections to the intercity microwave radio systems are made. From the television operating center as this switching point is called, wire circuits are frequently used to send or to receive from a microwave terminal which may be located outside the city.

The A2A system is functionally made up of three main parts in addition to the video cable facilities. A transmitting terminal is provided at the originating end of the video channel, to which is delivered the video signal from the customer or connecting equipment. A receiving terminal is provided to terminate the video channel at the receiving end; it will deliver either an unbalanced 75-Ω or a balanced 124-Ω signal to the customer or connecting equipment. Repeater equipment is available to provide additional gain and equalization along the video cable facilities. The A2A system was designed primarily for use underground in urban areas. This system can be designed for uses up to 10 to 15 mi.

Figure 25-47b gives the construction of video cable facilities.

76. Data Transmission. In 1928, Nyquist set forth three requirements for the suppression of intersymbol interference. They state:

1. Signaling without intersymbol interference is possible only at rates up to twice the bandwidth of the channel.

$$\omega_1/\pi = 2f_1 \qquad (f_1 = \text{cutoff}) \tag{25-10}$$

2. Intersymbol interference should be removed at instances halfway between signal impulses (at the sampling instant). The required transmittance function is

$$Y(\omega) = \begin{cases} \cos \dfrac{\pi\omega}{2\omega_1} & 0 < \omega < \omega_1 \\ 0 & \omega > \omega_1 \end{cases} \tag{25-11}$$

The impulse response is

$$\begin{aligned} g(t) &= \frac{1}{\pi} \int_0^{\omega_1} \cos \omega t \cos \frac{\pi\omega}{2\omega_1}\, d\omega \\ &= \frac{2\omega_1 \cos \omega_1 t}{\pi^2(1 - 4\omega^2 t^2/\pi^2)} \\ &= \frac{4f_1 \cos 2\pi f_1 t}{\pi(1 - 16 f_1^2 t^2)} \end{aligned} \tag{25-12}$$

3. The area under the received wave during a signal time unit should be proportional to the corresponding impressed signal value. Such a function, whose transmittance accomplishes this, is the truncated reciprocal $(\sin x)/x$ function,

$$Y(\omega) = \begin{cases} \dfrac{\pi\omega}{2\omega_1 \sin (\pi\omega/2\omega_1)} & 0 < \omega < \omega_1 \\ 0 & \omega > \omega_1 \end{cases} \tag{25-13}$$

The impulse response is

$$g(t) = \frac{1}{\pi} \int_0^{\omega_1} \frac{\pi\omega \cos \omega t\, d\omega}{2\omega_1 \sin (\pi\omega/2\omega_1)} \tag{25-14}$$

Over any given bandwidth, the number of bits that can be transmitted is

$$C = W \log_2 n \qquad \text{bits/s} \tag{25-15}$$

where W = the usable bandwidth and n = the number of recognizable states. This expression can also be written as

$$C = W \log_2 \left(1 + \frac{S}{N}\right) \qquad \text{bits/s} \tag{25-16}$$

where S = the average signal power and N = the average noise power in the band. The latter formula is normally referred to as the "Shannon formula."

Various systems of modulation are often used to transmit digital information. Usually the low-frequency response does not permit the transmission of the base-band signal. The common types of systems used are AM, FM, frequency shift keying (FSK), PM, and AM VSB. Amplitude-modulation systems normally do not use envelope detection, as the envelope of the waveform is quite often destroyed by the transmission medium. Comparisons of commonly used systems are shown in Table 25-16 and in Figs. 25-48 and 25-49.

77. Signal processing is used to increase capacity and minimize effects of transmission impairments.

78. Techniques Used. Some of the more important schemes also include multiphase and multilevel inputs. Parallel inputs, where the entire channel is subdivided into several narrower slots, are often used to send all the bits of a character simultaneously. This often simplifies the terminal equipment.

79. Multiphase. Multiphase transmission schemes permit encoding of digital information by means of discrete changes in the phase of a transmitted carrier. Popular systems are 4- and 8-phase. The average number of bits of information conveyed

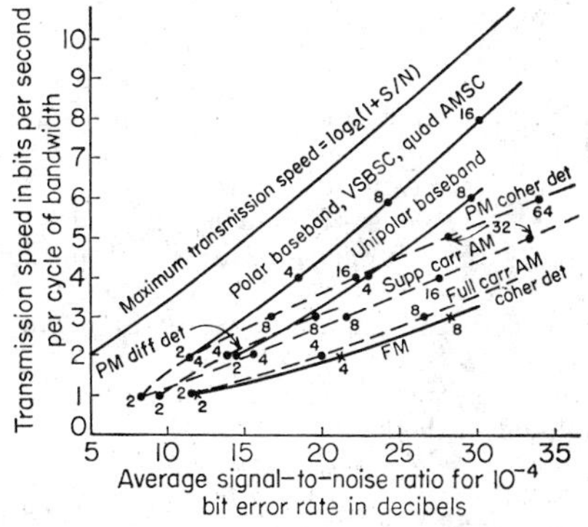

FIG. 25-48. Average signal-to-noise ratio (*a*) and maximum steady-state power to noise power (*b*) for 10^{-4}-bit error rate. (*From W. R. Bennett and J. R. Davey, "Data Transmission"; New York, McGraw-Hill Book Company*, 1965, *p*. 228.)

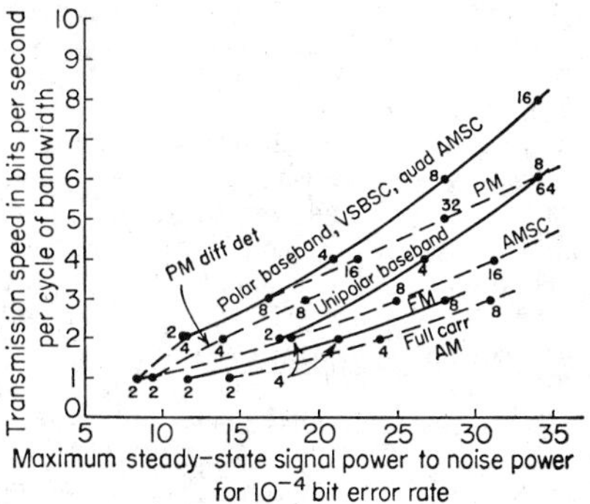

FIG. 25-49. Average time between bits in error (*a*) and expected improvement with error-reduction schemes (*b*). (*From W. R. Bennett and J. R. Davey, "Data Transmission"; McGraw-Hill Book Company*, 1965, *p*. 229.)

Table 25-16. Comparisons of Various Data Systems under Limitations of Average and Maximum Steady-state Signal Power*

System	Number of states	Speed in bits/(s)(c) of bandwidth	Signal-to-noise ratio for 10^{-4} bit error rate: Average signal power	Signal-to-noise ratio for 10^{-4} bit error rate: Maximum steady-state signal power
Unipolar baseband	2	2	14.4	17.4
	4	4	22.8	26.9
Bipolar baseband	2	2	14.4	17.4
Polar baseband	2	2	11.4	11.4
	4	4	18.3	20.8
	8	6	24.3	28.0
	16	8	30.2	34.4
Full carrier AM:				
Envelope detection	2	1	11.9	14.9
Coherent detection	2	1	11.4	14.4
	4	2	19.8	23.8
	8	3	26.5	31.0
Suppressed carrier AM, coherent detection	2	1	8.4	8.4
	4	2	15.3	17.8
	8	3	21.3	25.0
	16	4	27.2	31.4
PM, coherent detection	2	1	8.4	8.4
	4	2	11.4	11.4
	8	3	16.5	16.5
	16	4	22.1	22.1
	32	5	28.1	28.1
	64	6	34.1	34.1
PM, diff. detection	2	1	9.3	9.3
	4	2	13.7	13.7
	8	3	19.5	19.5
FM	2	1	11.7	11.7
	4	2	21.1	21.1
	8	3	28.3	28.3
VSB, 50% modulation	2	2	16.2	17.9

VSB, suppressed carrier, coherent detection (see polar baseband)
Quadrature AM, suppressed carrier, coherent detection (see polar baseband)
* From W. R. Bennett and J. R. Davey, "Data Transmission," McGraw-Hill Book Company, 1965, p. 227.

per phase sample is equal to

$$H = \log_2 \phi \tag{25-17}$$

where ϕ is the number of possible phase conditions. Thus, for 4-phase transmission, 2 bits (a dibit) are transmitted per phase sample.

Compared with FM or AM, phase modulation is relatively insensitive to amplitude distortions and has improved S/N performance. Its constant line energy permits simple AGC techniques to be used. It also permits direct bit synchronization from the modulated base-band signal.

80. Multilevel. Multilevel transmission schemes are employed mainly to permit increased signaling rates.

Nyquist's familiar relationship among the rate of transmission C, the number of different states (levels) per sample N, and the bandwidth W is

$$C = W \log_{2N} \tag{25-18}$$

Nyquist also proved that the maximum number of samples which can be transmitted per second is equal to twice the bandwidth W,

$$\therefore S = 2W \tag{25-19}$$

Shannon's relationship among signaling rate, bandwidth, and noise,

$$C = W \log_2 (1 + S/N) \tag{25-20}$$

therefore becomes

$$C = S/2 \log_2 (1 + S/N) \tag{25-21}$$

or

$$C = S \log_2 \sqrt{1 + S/N} \tag{25-22}$$

Therefore, the average number of bits per sampling interval is

$$H = \log_2 \sqrt{1 + S/N} \tag{25-23}$$

While this deviation is theoretical, it does illustrate the utility of multilevel transmission.

81. Parallel. One method of overcoming low S/N ratios in a frequency-multiplexed carrier system is to employ parallel channel transmission rather than a single wideband channel. The advantage lies in the fact that single-frequency energy is limiting and by spreading energy among a number of channels higher total signal energies can be realized.

A secondary advantage of parallel transmission is the resulting reduction of envelope delay distortion requirements due to the narrower individual bands of frequency.

82. Restored Polar. Restored polar is a method of digital transmission employing signal differentiation at the transmitter and low-frequency restoration at the receiver. This results in less stringent transmission requirements at the extreme low end of the base-band spectrum.

83. Bipolar. Bipolar transmission is another example of a technique to reduce the requirement for transmitting very low base-band frequencies. Here the scheme is to encode alternate binary 1s into opposite polarity pulses. Transmission of 0s is at ground reference, midway between the positive and negative levels of alternate 1s.

The Fourier transform of a random bipolar signal has zero energy at direct current.

84. Randomizing. Randomizing of synchronous digital signals is necessary to reduce the interference effect set up by repetitive patterns. Such patterns occur most frequently during idle periods or when certain unique data combinations are transmitted. Randomizing reduces the probability of discrete interfering tones from being transmitted and causing interference in other channels of frequency-multiplexed systems.

85. Pulse Shaping (Raised Cosine). A method for reducing signal energy beyond the first zero axis crossing (reciprocal of polar pulse width) is to shape the repetitive pulses, in the frequency domain, with a low-pass filter. The result is a reduction in intersymbol interference, with a corresponding improvement in S/N ratio.

A host of filtering schemes are available, the most popular characteristics being $(\sin x)/x$, gaussian, and raised cosine. Of the three, raised cosine is usually preferred, since reduction of intersymbol interference is affected in a minimum of bandwidth in the frequency domain.

86. Advantages of Synchronous Transmission. An outstanding advantage of synchronous transmission is the ability to regenerate and retime a signal distorted by noise and transmission irregularities. Provided that a minimum S/N ratio is met, depending on the system, essentially perfectly timed signals can be regenerated from a distorted input. With regeneration and retiming, distortions are not cumulative.

87. Need for Asynchronous Operation. One disadvantage of synchronous operation is the resulting time quantization of input signals displaying random time transitions. A typical example is black-and-white (two-level) facsimile. Here asynchronous operation is preferred, to eliminate the effects of line jitter due to quantization.

88. Coherent and Noncoherent Modulation Techniques. Examples of coherent modulation techniques are double-sideband AM, carrier-transmitted vestigial sideband AM, FM, and PM. Some examples of noncoherent modulation are single-sideband suppressed-carrier AM and vestigial-sideband suppressed-carrier AM.

Coherent techniques permit precise recovery of the modulated base-band information, both in amplitude and in phase. Noncoherent systems using local carrier reinsertion for detection often display both frequency and phase wander owing to the differences between the original carrier and the reinserted (nonrelated) carrier.

For certain kinds of digital-data transmission, noncoherent systems can produce detected signals having inverted outputs (i.e., marks become spaces, and vice versa).

89. Transmission Parameters That Affect Data. *Loss.* Excessive loss will cause the signal to fall below the sensitivity of the receiver. Sensitivities down to levels of 40 to 50 dBm are common for data sets operating over large switched networks such as the public telephone network.

Delay Distortion. Differences in velocity of propagation of various frequencies cause relative phase shifts that spread out a pulse in time (when demodulated). This spreading, if it is great enough, can cause errors. Delay distortion is measured by estimating the derivative of the phase shift vs. frequency curve at particular frequencies of interest. This measurement is called envelope delay and normally carries the unit of microseconds. Figure 25-50 shows an estimate of the spread in the public telephone network in 1958.

Attenuation Distortion. Frequencies are attenuated differently over a band of interest. Often this adds a bias toward mark or space in certain modulation schemes. Figure 25-51 and Table 25-17 show the attenuation distortion in the public telephone network in 1958.

Noise. There are two general types of noise to consider. They are message circuit (gaussian) and impulse noise. Many of the commonly encountered facilities have been designed for voice. The human ear is more critical of message circuit noise than most data sets. A 20-dB S/N ratio will yield error rates of the order of 1 in 10^5 for many common types of modulation schemes. Impulse noise is more serious. It is generated by switches and other man-made electrical disturbances or by band-limiting high-frequency natural phenomena. These disturbances are coupled magnetically or electrostatically or directly through common feeds (battery supply, etc.). They are often short in duration (approximately 1 ms) and, as such, may be inaudible. Some characteristics are shown in Fig. 25-52.

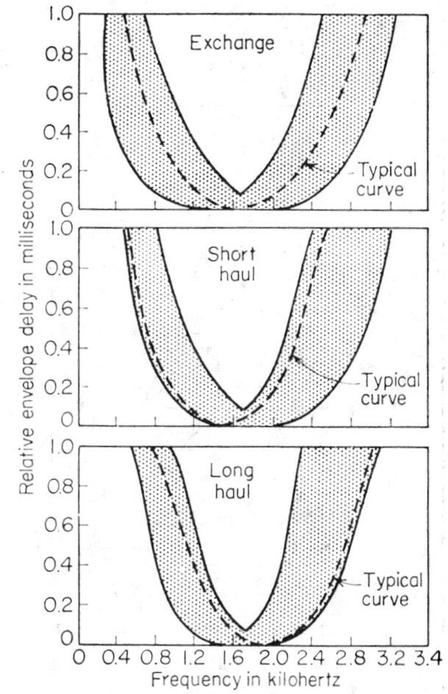

Fig. 25-50. Typical envelope delay in the public telephone network.

Return Loss. Impedance irregularities on a 2-wire transmission line cause an incident signal to be partly reflected. The reflection itself can encounter a second im-

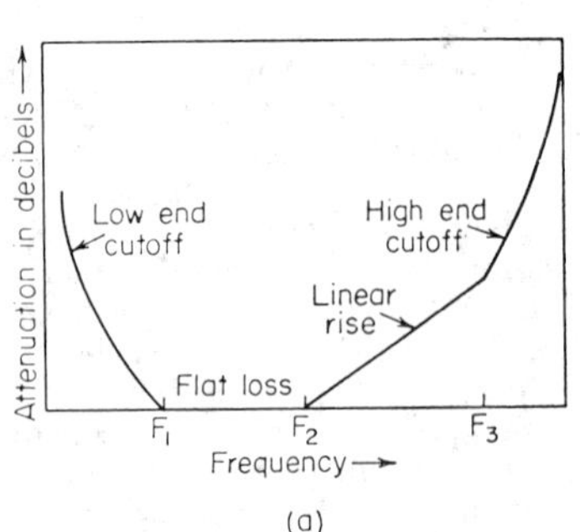

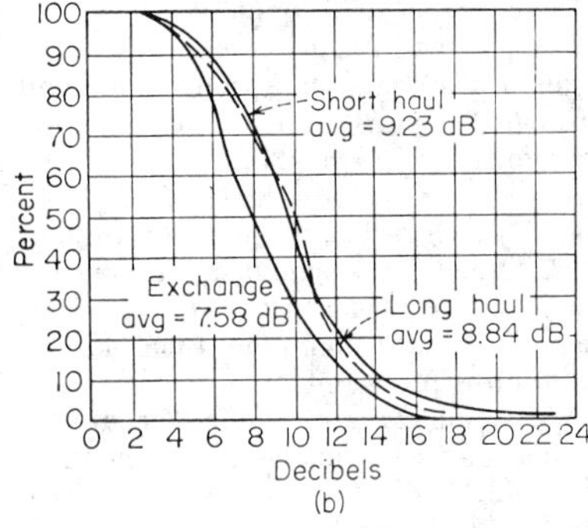

FIG. 25-51. Attenuation distortion as function of frequency (*a*) and level (*b*).

Table 25-17

	Roll-off below f_1, dB/octave	f_1, c/s	f_2, c/s	f_3, c/s	Roll-off above f_3, dB/octave
Exchange	15	240	1,100	3,000	80
Long distance:					
Short-haul	24	300	1,075	2,950	90
Long-haul	27	280	1,150	2,850	80

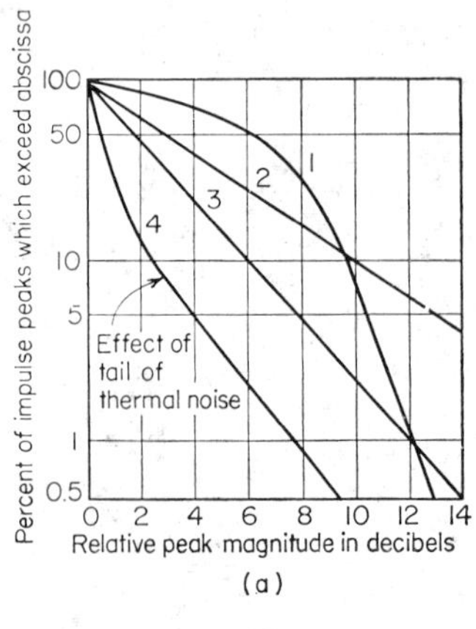

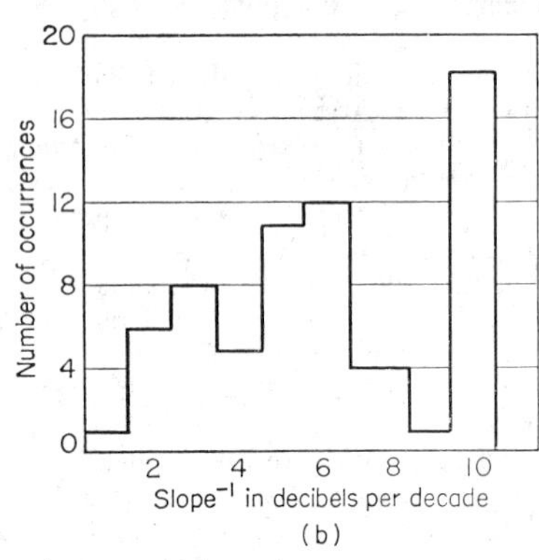

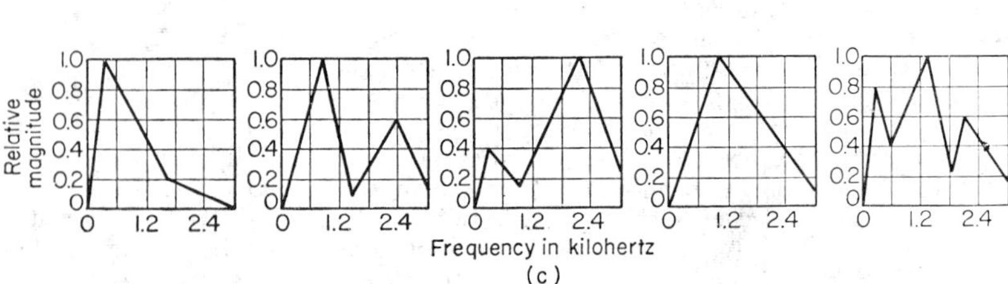

FIG. 25-52. Cumulative amplitude distribution of noise (*a*), histograms of inverse slopes (*b*), and families of impulse noise (*c*).

pedance irregularity and be partly reflected. This second reflection, if great enough in amplitude, will cause errors when it reaches the demodulator of the data set. Many data sets employ AGC circuitry and will neglect reflections if they are 12 dB or less than the incident signal.

Frequency Offset. If a frequency f is transmitted into a transmission line, $f \pm \Delta f$ may be received if suppressed carrier multiplexing is used. Normally this shift may be regulated to small values (0 c/s in 1 s), but in some older and poorly regulated systems as much as 25 to 50 c/s can occur. This translation may cause errors, particularly in narrowband systems.

Nonlinearities and Instantaneous Effects. Some clipping and compression are likely to occur in many telephone grade circuits, especially when compandors are encountered. In addition, there are intermodulation products that create nonlinearities. Amplitude-phase distortions and FM disturbances are also likely to be encountered. Most of the effects are minor but are likely to cause trouble in higher-speed and more refined systems.

In addition, short interruptions (dropouts) and instantaneous changes in amplitude or phase (hits) can occur. If only an occasional error is caused, this may be acceptable. If loss of sync occurs, this can be a more serious problem.

90. Telegraph Transmission. In widespread cities low-speed data circuits using private-line telegraph facilities are often composed of several stations. The circuit consists of "loops" from the stations to their serving telegraph offices, one or several "trunks" in tandem between offices, and "hubs" wherever two or more trunks or loops on the same circuit are connected at a serving telegraph office. A typical example is seen in Fig. 25-53. A telegraph test board would normally be located at the hub locations.

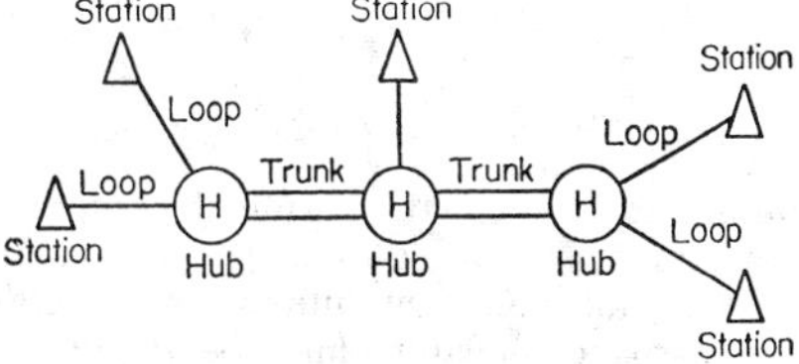

FIG. 25-53. Typical telegraph network.

91. Loops. Several loop transmission schemes are generally in use. The choice depends on transmission speed, loop facility length and makeup, and the amount of allowable distortion.

Transmission at Speeds Up to 75 *Bauds.* Neutral line currents (mark = full line current, space = no line current) of either 20 or 62.5 mA obtained from keying voltage of +130 and −130, +130 and −48, or +130 and ground may be used. Better transmission may be obtained by using polar rather than neutral line currents; however, this is less common in the United States today, because the cost of purchasing and maintaining polar relays and repeaters is greater than the equipment used for neutral signal systems. Extremely long loops or loops requiring abnormally low distortion may be served over frequency-shift tone-carrier channels right to the station location.

Transmission at Speeds Up to 180 *Bauds.* Telegraph loops operating at signaling rates above 75 bauds may not use 60- or even 20-mA line currents if they utilize pairs in communication cables carrying voice data or program-grade channels, because the line current pulses may cause interference in these other services. Transmission systems using about 20 V and low currents of the order of 6-mA balanced neutral or 3-mA polar signals can be economically used on 2-wire facilities for half-duplex circuits up to about 2,000 Ω loop resistance at these speeds. Full-duplex and long half-duplex circuits also may be served on 2-wire facilities by using single-channel frequency-shift carrier schemes.

92. Trunks. Interoffice trunk facilities are typically provided on frequency-shift multiplex carrier systems which derive from one 4-wire voice-grade facility up to 23 telegraph channels spaced 120 c/s apart for 60-baud circuits, 17 telegraph channels spaced 170 c/s apart for 75-baud circuits, or 8 channels spaced 340 c/s apart for 150-baud circuits. Back-to-back start-stop telegraph distortion on such systems is about 5%.

This same type of carrier multiplex system can be economically used on loops as well as trunks, especially where the stations are served by communication cables from nontelegraph offices as shown in Fig. 25-54. Here, the only equipment in the nontelegraph office need be a passive branching filter which effectively permits "dropping off" certain telegraph-grade channels along the route. The station also must contain a single-channel terminal to demodulate its particular carrier signal to the signals re-

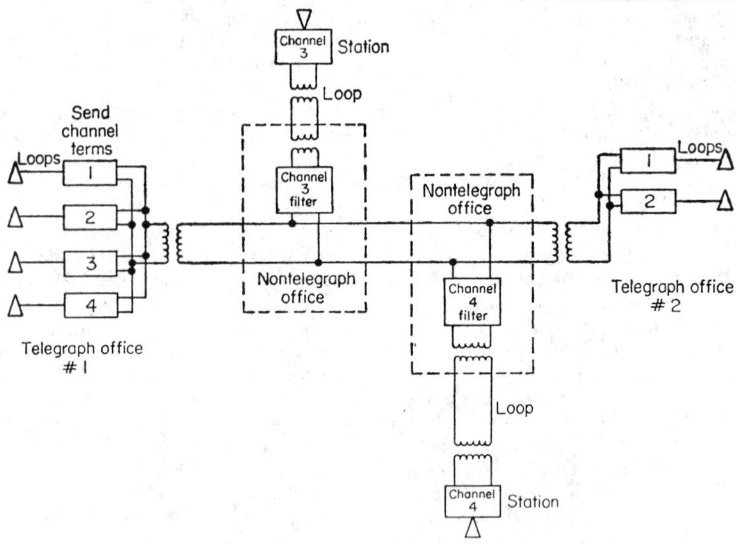

Fig. 25-54. Telegraph carrier branching arrangements.

quired to operate the terminal equipment. Table 25-18 indicates typical frequency assignments for voice-frequency carrier telegraph multiplex equipment.

On multipoint circuits where several telegraph trunks in tandem are assigned, regenerative repeaters may be required. They ensure that the signal distortion on the circuit does not exceed the point where the receiver can reliably detect the correct mark and space pulses. Most electromechanical devices, such as teletypewriters, will transmit a signal with less than 8% distortion and be able to receive signals reliably with up to 35% distortion. Electronic devices generally have more margin in that they can transmit with 5% distortion or less and receive signals with up to 40% distortion. The connecting facility will introduce varying amounts of distortion, depending on its physical characteristics, length, and the transmission (baud) rate.

Table 25-18. Typical Voice Frequency Telegraph Carrier Multiplex Channel Assignments and Signaling Rates

120-c spacing ±30-c frequency shift (signaling rates to 60 bauds)		170-c spacing ±35-c frequency shift (signaling rates to 75 bauds)		340-c spacing ±70-c frequency shift (signaling rates to 150 bauds)	
Channel No.	Midband freq.	Channel No.	Midband freq.	Channel No.	Midband freq.
1	420	1	425	1	680
2	340	2	597	2	1,020
3	660	3	765	3	1,360
4	780	4	935	4	1,700
5	900	5	1,105	5	2,040
6	1,020	6	1,275	6	2,380
7	1,140	7	1,445	7	2,720
8	1,260	8	1,615	8	3,060
9	1,380	9	1,785		
10	1,500	10	1,955		
11	1,620	11	2,125		
12	1,740	12	2,295		
13	1,860	13	2,465		
14	1,980	14	2,635		
15	2,100	15	2,805		
16	2,220	16	2,975		
17	2,340	17	3,145		
18	2,460				
19	2,580				
20	2,700				
21	2,820				
22	2,940				
23	3,060				

Table 25-19. Nominal Telegraph Distortion in Various Facilities

System	% distortion at baud rate		
	45.5	56.9	74.2
I. 75 baud telegraph multiplex carrier systems (170 c/s channel spacing)			
a. 17-channel telegraph carrier on:			
1. 4-wire cable <500 mi	8.8	10.7	13.4
2. 4-wire cable 500–1,500 mi	9.8	12.2	14.5
b. 17-channel telegraph carrier on:			
1. Short-haul voice carrier (N, ON carrier) <200 mi	8.8	10.7	13.4
2. Long-haul voice carrier (K or L carrier):			
<500 mi	8.8	10.7	13.4
500–1,500 mi	9.8	12.2	14.5
>1,500 mi	11.4	13.4	15.6
II. Metallic telegraph trunks			
1. 4-wire on equalized bare pairs <100 mi	11.4	13.4	15.6
2. 2-wire neutral:			
1–10 mi	8.0	9.1	11.3
10–20 mi	9.8	11.3	13.9
III. Metallic telegraph loops			
1. Using neutral telegraph carrier or electron tube repeater:			
1–10 mi	5.6	6.7	8.0
10–20 mi	8.0	8.8	10.4
IV. 76–150 bauds—use 340 c/s spaced carrier, and use distortion values for (I) at 45.5 bauds			

Table 25-19 summarizes the approximate amount of distortion to be expected in several types of facilities. To ensure adequate working margins, a regenerative repeater would normally be placed in a circuit at each telegraph office where the nominal cumulative distortion is expected to reach about 25%.

PLANT MAINTENANCE AND COORDINATION

93. Testing of Transmission Paths, Purpose. Testing from a maintenance standpoint is done for several reasons, (1) to determine whether a circuit, or portion thereof, is suitable for service when it is first established, rearranged, or restored to service following repair or readjustment, (2) on a periodic basis to determine whether it is still suitable for service or to detect incipient troubles, and (3) to locate a source or sources of trouble. Such testing includes transmission measurements, signaling tests, and d-c tests of the conductors which form all or part of the path of the circuit.

94. Scope. 1. *Transmission Tests.* Transmission tests include measurements of loss or gain at one or more frequencies, noise, crosstalk, delay distortion, and return loss.

2. *Signaling Tests.* Signaling tests include determination of the ability of a circuit to transmit and/or receive address and supervisory information and a measurement of the margin available before failure.

3. *D-c Tests.* Direct-current tests of conductors include tests for open circuits, short circuits, resistance, capacitance, resistance and capacitance unbalance, crosses, and insulation resistance.

95. Transmission Loss. Transmission loss or gain is measured as shown in Fig. 25-55 between resistances whose values are specified in advance. It is a measurement of the insertion loss or gain of the circuit; however, for practical reasons, the same physical resistor is not necessarily used in the reference and measuring condition. Transmission loss is the ratio, expressed in decibels, of the power dissipated in the termination in the reference condition to that dissipated in the termination in the measuring condition. If the value of this ratio in decibels is negative, a gain is measured.

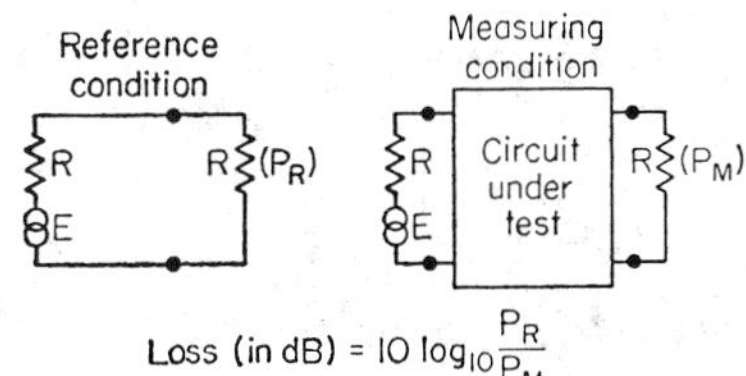

Fig. 25-55. Measuring transmission loss or gain.

96. Noise Testing. Noise tests consist of the measurement of the power dissipated in a resistance termination of specified value connected to one end of a circuit with no test

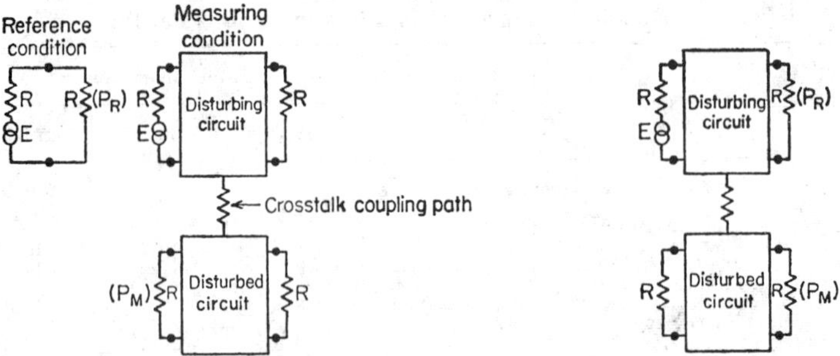

FIG. 25-56. Crosstalk measurement: near end (*a*), far end (*b*).

power applied to the circuit. The noise power at each frequency may be weighted equally in the measurement, or it may be weighted according to its relative interfering effect on the human ear when listening through a specified type of telephone set. The present weighting for telephone circuits carrying voice message service is called C-message weighting and assumes that the listener is using a 500-type telephone set. Noise is expressed in decibels above reference noise (abbreviated dBrn). Reference noise has been established at −90 dB below 1 mW. Noise measurements are of two types, (1) with a meter whose dynamic characteristics integrate or average the noise power over a specified length of time (about 200 ms) or (2) with a device which counts the number of noise peaks exceeding a specified magnitude in a specified length of time. The first measurement relates the noise power to its interfering effect on the human ear, while the second relates it to its effect on the transmission of digital signals. In the latter case, it is called impulse noise.

97. Crosstalk Measurement. Crosstalk is measured as shown in Fig. 25-56. There are two types—near-end (Fig. 25-56*a*) and far-end (Fig. 25-56*b*). For near-end crosstalk, the coupling is the ratio (expressed in decibels) of the power dissipated in a termination connected to the source of the disturbance to the power dissipated in a termination connected to the source of the disturbed circuit. For far-end crosstalk, the coupling is the ratio (expressed in decibels) of the power dissipated in a termination connected to the disturbing circuit to the power dissipated in the termination connected to the disturbed circuit, as shown in Fig. 25-56*b*. In the far-end case, both measurements are made at the same level points on the disturbed and disturbing circuits; so this coupling is called "equal-level" crosstalk coupling. Crosstalk coupling has been expressed in dBx. Crosstalk can be the result of coupling between the separate physical circuits or of intermodulation between two channels in a single carrier circuit.

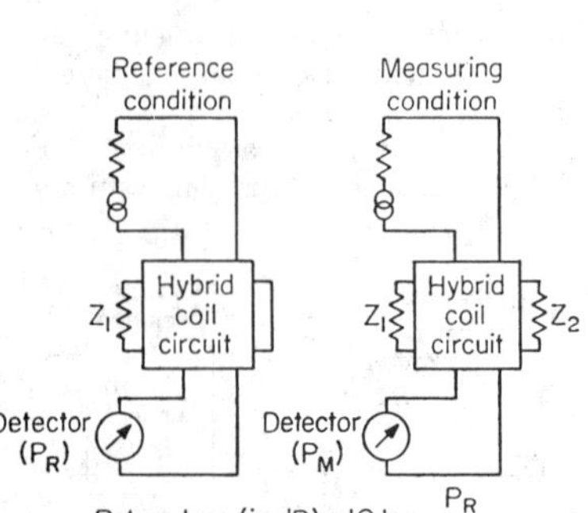

FIG. 25-57. Measurement of return loss.

98. Delay Distortion. Delay distortion is the difference in propagation time of two frequencies which are transmitted simultaneously over a circuit. It is expressed in seconds or fractions thereof. Measurements are made by special equipment described in the references in Par. **100**.

99. Return Loss. Return loss is the relation between the impedance looking into one end of a circuit and a specified value of impedance when the opposite end of the circuit is terminated in the same or another value of impedance. The relation is as follows:

$$20 \log \frac{Z_1 + Z_2}{Z_1 - Z_2} \tag{25-24}$$

One method of measuring return loss is by the use of a hybrid coil circuit as shown in Fig. 25-57. The return loss, expressed in decibels, is the loss measured between the two 4-wire parts of the hybrid, less the sum of the losses between each 4-wire part and the 2-wire part.

100. References

Dunbar, F. C. A Delay Distortion Simulation Set; *Trans. AIEE*, 1960, Vol. 79, Pt. II, p. 183.

Codd, W. A. A Test Set for Measurement of Envelope Delay Distortion at Audio Frequencies with 1-microsecond Precision; *Trans. AIEE*, 1960, Vol. 79, Pt. I, p. 241.

Schroeder, A. C. A Sweep Method for Measuring Envelope Delay; LB-883, Radio Corporation of America, RCA Laboratories Division, Industry Service Laboratory, Princeton, N.J.

101. Protective Devices, Basic Objectives. Voltages and currents used in the normal operation of a communication system are generally considered to be safe to persons, property, and equipment. The dielectric strength and current-carrying capacity of the instruments handled by people, together with the plant and equipment, are designed to be more than adequate for the operating voltages and currents. Power supplies for communication systems are, when necessary, equipped with devices for limiting the current within safe values or with interrupting devices for clearing abnormal currents in the event of fault conditions. High voltages are required to operate some long-haul facilities such as submarine and coaxial cables, and under these situations special protection and operating maintenance procedures are followed to safeguard personnel and equipment.

It is also necessary to protect communication facilities from disturbances from voltages and currents resulting from direct electrical contact with an electrical power system, lightning stroke, or induction from either of these sources. It is usually not practical to avoid exposure to lightning; therefore, protective measures are generally designed to reduce the amount of damage. The principal means of preventing accidental contacts with power circuits is to maintain adequately constructed plants from the standpoint of both strength and clearances. This requires cooperative planning by all wire- and cable-using organizations and adherence to accepted standards of the National Safety Code for pole-line construction.

When adequate protection measures have been applied to the communication system, basic objectives will be met, namely:

1. *Safety to Persons.* Safety to persons—i.e., the user, the workman, and the general public.

2. *Safety to Plant.* Safety to plant—i.e., the communication system and the property where the communication system is used.

3. *Reliability.* Ensuring the reliability of the communication service. Generally, all serial communication facilities are exposed to some electrical disturbance and will therefore require protection. In some cases, facilities placed underground may be exposed to electrical disturbances such as lightning or induction from power lines and may also require protection.

102. Principles. The various protective devices are used to perform specific functions as follows:

1. *Voltage Limiting.* Voltage-limiting or -equalizing devices—(1) protector gaps discharging in air or gaps enclosed in a tube containing gases and (2) semiconductor devices neutralizing or isolating (insulating) transformers.

2. *Current Limiting.* Thermal elements, such as ballast lamps.

3. *Current Interrupting.* Fuses, fine-gage wire or cable conductors; thermal devices, heat coils.

All these devices are designed to be sufficiently sensitive to provide the required degree of protection but not sensitive so as to cause unnecessary service interruptions.

103. Protector Gaps. This device is used to protect equipment and plant facilities and consists of a mounting for two carbon blocks.

Variations and uses for 3-mil, 6-mil, 10-mil, 20-mil, or 30-mil carbon blocks.

In one design, one block is entirely of carbon and the other is of porcelain with a small carbon insert. The carbon insert is held in place by cement and is depressed slightly below the bearing surface of the porcelain to provide the gap spacing. Different gap spacings are available: the 3-mil with spark-over d-c volts of 500 V, the 6-mil gap operating at 800 V, and the 10-mil at 1,250 V. On steady-state discharges of considerable duration the cement will melt from heat produced by the arc. This permits the carbon insert, under the influence of spring pressure, to move into direct contact with the solid carbon block. When this occurs, the gap has permanently operated and the blocks must be replaced before the circuit again will be operative. Protectors of this type have a continuous 60-Hz current rating of the order of 10 to 30 A rms; however, these ratings vary depending upon different manufacturers.

When this device is used to protect station installations, in accordance with fire-protection practice, a gap spacing of 3 mils is used and it is necessary that a fuse (7 A) or fine-gage conductors (24-gage) be placed ahead of the device to prevent overheating in the event of long-duration contacts which exceed the current-carrying capacity of the protector mounting.

The 3-mil gap protector is also used to protect central office equipment, and fine-gage conductors are placed between the exposed outside plant and the central office in order to prevent overheating of the protector mounting. Generally, heat coils are placed in series with the office wiring to protect office equipment against sustained current when voltages below the spark-over of the protector gap are impressed on the circuit.

The 6-mil gap protector is used to equalize potentials at junctions where facilities of different dielectric strengths are connected together, such as open-wire cable junctions, or to provide a parallel path around equipment, such as a repeating coil in an open-wire circuit.

The 10-mil gap protector is used to reduce the electrical stress placed on the 6-mil gap protectors in heavy-lightning areas, such as the placement of a 10-mil gap in an open-wire circuit approximately ½ mi ahead of a 6-mil gap protector at an open-wire cable junction.

In another design of the carbon-block protector, the blocks are entirely of carbon with a permanently fixed gap of either 20 or 30 mils. These gaps spark-over at 2,000 and 3,000 V peak ~60 Hz, respectively. The protectors are usually designed with three blocks to provide protection to a pair of wire conductors exposed to high-voltage (over 5,000 V) power circuits.

104. Gas-tube Protector. This device consists primarily of a discharge gap (or gaps) enclosed in a tube containing an inert gas or combination of gases at reduced pressure. It is essentially a surge-protection device and generally is incapable of discharging large amounts of steady-state (60 Hz) current. In order to obtain the fail-safe feature usually designed into the carbon-block gap, it is necessary to use a bimetal mounting for the gas-tube protector, which will establish metal-to-metal contact in the event of sustained discharge of abnormal currents. Various ranges of spark-over voltages for the gas-tube protectors are available and are comparable with those of the commonly used carbon-gap protectors.

The principal advantage of the gas tube as compared with the carbon-block protector is its ability to withstand repeated operations without appreciably lowering the gap impedance. Smudging of carbon blocks, with resulting lower leakage path, may occur after approximately 40 operations. A corresponding degradation does not occur in the gas-tube protector until approximately 2,000 operations, thus substantially improving service reliability and maintenance costs of circuits so equipped.

If gas-tube protectors are used as station protectors, a fail-safe feature must be provided. This may be accomplished by placing a gas tube in parallel with a carbon-block protector and coordinating the spark-over of the two gaps so that the gas tube will discharge first; or a special fail-safe mounting may be provided.

105. Fuses. Fuses are designed to protect against excessive current and are employed in various ratings from 1 to about 15 A. The rating of the fuse, extensively used for station protection, is 7 A; ratings of 5, 10, and 12 A are also employed. Fuses of the enclosed type are invariably used; if the enclosing tubes are slotted to permit escape of hot gases, the slots should be turned toward the mounting surface and the latter protected with an asbestos mat or the noncombustible base of the fuse mounting.

106. Heat Coils. Heat coils (or sneak-current fuses) are, essentially, fuses of special construction designed to overcome the unreliability of ordinary fuses for very small currents. For example, a heat coil can be constructed to operate on a current of 0.2 A in 30 s or a variety of other ratings. This device comprises a small winding of fine wire, which develops sufficient heat to melt a small piece of fusible metal and release a spring, which grounds the line circuit; the resistance of the winding usually is of the order of several ohms. The function of heat coils is to protect telephone apparatus wound with fine-gage wire (relays, drops, etc.) against damage from foreign currents insufficient to ground protector blocks. The proper location for the heat coils is between the protectors and the apparatus.

107. Semiconductor Devices. Semiconductor devices (*p-n* or combined *n-p-n* junctions) have reverse conduction characteristics which make them exceptionally fine voltage limiters. Such devices, when conducting in the reverse direction, present a high resistance at applied voltages below a critical surge breakdown value and a well-defined region of substantially constant voltage above the critical breakdown value.

There are two general types of junction diodes at present available for protection purposes: (1) the silicon-alloy and (2) the silicon diffused junction. Diffused junction diodes are at present available having effective junction dimensions of about 0.125 by 0.125 in, as compared with the alloy types, which have a maximum junction diameter of only about 0.025 in. For protection applications where the higher capacitance of the larger junction area can be tolerated, diffused junction diodes offer distinct advantages because of their greater current-carrying ability and less increase in terminal voltage after breakdown.

Semiconductor diodes can be arranged to provide metallic or longitudinal protection. However, they are chiefly employed for metallic protection, since, in most cases, the dielectric strength of equipment, with respect to ground, exceeds 600 peak volts and adequate voltage limitation may be obtained with carbon-block or gas-tube protectors.

The response time of diodes is extremely fast, of the order of nanoseconds. In use in communication circuits for protection purposes, their operating time may be considered negligible. It may be assumed that the inverse-voltage breakdown point will be the same on surge as it is on d-c potentials.

The forward voltage characteristic of a semiconductor diode may also be employed to provide limitation in a low-voltage range. The drop across a silicon diffused junction when conducting significant current in the forward direction is of the order of 0.5 V and will increase only a small amount above this value as the current increases. The transition between the nonconducting and the conducting state is not so sharp as breakdown in the inverse direction, but it still exhibits reasonably good limiting performance. Units are available having two junctions in parallel and oppositely poled to provide bipolar operation. If voltage limitation is needed in excess of 0.5 V (for example, at 1.5 V), the units may be used in series to meet the requirement (for example, three units).

108. Ground Wires. Protector ground wires are a very important link in the protection system. The conductor should be continuous (without splices) and not smaller than No. 14 AWG solid copper. The size should be increased depending upon current-carrying requirements. The wire should be run in the most direct route between the protector and ground electrode and located so as not to be subject to mechanical damage. Connections to the protector and grounding electrode should, where possible, be exposed for inspection purposes, and all connections should be securely made. It is most important for the grounding electrode to have a good electrical connection with the earth, one which is substantial and reliable.

A public metallic piping system (water or gas) provides the best and preferred grounding electrode. However, when a water pipe is not available, a connection may be made (in order of preference) to the neutral of a multigrounded neutral power system, a private water system having at least 10 ft of metallic pipe buried in the ground, other buried metallic structure, and pipes or rods. Ground rods or pipes should be selected so as to have long life and not corrode in the environment in which they are placed. They should, if possible, be driven into moist earth to a depth of at least 5 ft. The driven ground may have resistances in the range of 25 to 100 Ω and so should never be used if a water system or neutral of the multigrounded neutral power system is available.

Where possible, communication grounds should be bonded to the power-system ground in order to minimize voltage differences between the two systems. This practice should always be followed at station protector locations.

109. Neutralizing Transformers. Neutralizing transformers are devices which provide a means of introducing a voltage into a circuit to neutralize an unwanted voltage. It has a one-to-one ratio winding and consists of a single primary winding and secondary windings for the pairs to be neutralized. These transformers are commonly designed to neutralize voltages from 2,000 V to as high as 16,000 V and are designed for single-pair or multiple-pair communication facilities.

When the interfering voltage is caused by the ground-potential rise at a power substation, one side of the primary of the neutralizing transformer is connected to the substation ground and the other side to a remote ground (a ground outside the influence of the rise of potential of the substation ground). The communication pair to be neutralized (the telephone central-office side, for example) is connected to one side of the secondary, and the equipment at the substation is connected to the other side.

During normal conditions at the substation, when large fault current is flowing into the ground, the ground potential rises. A portion of the fault current also flows through the primary of the neutralizing transformer, which induces an opposite and equal voltage in the secondary winding. Communication equipment on the substation side of the neutralizing transformer is at the high potential of the substation, and communication equipment (cable or wire pairs, for example) on the other side is at remote ground potential and therefore not subject to the possible damaging voltages of the substation area.

110. Protective or Insulating Transformers. Protective or insulating transformers are employed to protect telephone sets, switchboards, or terminals connected to lines occupying the same structures as the high-tension high-energy transmission circuit. Another application is to protect communication equipment from abnormal voltages resulting from rise in ground potential (such as might occur at power substations). An insulating transformer for this purpose, made by one of the leading manufacturers, will withstand a 25,000-V test between windings for 1 min. The transmission loss is about 1 dB. The transformer is mounted in a weatherproof case for outdoor mounting; the casing should always be thoroughly grounded. It should be noted that such transformers introduce additional losses in both transmission and ringing, which should be taken into account in arranging circuit layouts.

Special protectors are usually employed between the line and the insulating transformers. The protector consists of extra long fuses in the line circuit, with spark gaps bridged to ground; sometimes the fuses are so mounted as to be an integral part of an air-break disconnecting switch.

For the protection of power-company personnel engaged in patrolling transmission lines, insulating stools are often used at telephones connected to communication circuits exposed to the transmission line. An ordinary four-legged hardwood stool, with the legs inserted in inverted porcelain line insulators, will serve to isolate the telephone user from earth, thereby eliminating a shock hazard. Insulating mats are sometimes used at dry or interior locations in place of stools but, in general, are less effective.

111. References.

Tech. Rept. 6, Electrical Protection of Tactical Communication Systems, Final Report—Task BQ-CD, reissue date December, 1963, Contract DA36-039 SC73089, U.S. Signal Research and Development Laboratory, Fort Monmouth, N.J.

Cable Protection; *Bell System Tech. Jour.*, 1961, Vol. 40, pp. 547–576.

Lemieux, J. E. R. Field Experience with Gas-filled Protectors on Communications Lines; *Trans. IEEE, Paper* 63-236, 1963.

Bodle, D. W., and Hayes, J. B. Electrical Protection for Transistorized Equipment; *Trans. AIEE, Paper* 59-71, 1959.

COLEMAN, O. K. Why Ground; *Trans. AIEE,* May, 1956, Vol. 75.
National Electrical Code, 1965 edition.
BONNESSEN, E. J. The Application of Neutralizing Transformers for Rise in Station Ground Protection; *AIEE Relay Committee Rept.* S-117, Application and Protection of Pilot-wire Circuits for Protective Relaying, July, 1960.

112. Corrosion of Telephone Cable Sheaths. Stray-current corrosion of lead-sheath telephone cables may occur when direct current flows from the sheath through an electrolyte consisting of soil or water surrounding the cable. A current of 1 A flowing from lead through an electrolyte can cause a weight loss as great as 74.7 lb/year, according to Faraday's law. Sources of stray direct current are electric railways, elevators, local power systems, foreign cathodic protection systems, etc.

Corrosion may also be caused by chemically developed potentials or corrosion cells. Galvanic action involves two dissimilar metals in contact with an electrolyte such as lead and copper grounding systems or lead and galvanized manhole hardware. Differential aeration involves a variation in the concentration of dissolved oxygen in the electrolyte contacting the cable. Local action involves local cells set up between two adjacent areas of cable sheath such as differences in surface conditions caused by abrasion.

113. Typical Corrosion Mitigative Measures. 1. *Control of the stray current* by minimizing pickup and returning it to the source over a metallic drain wire.

2. *Use of rectifiers or sacrificial anodes* to keep the cable sheath at a negative potential to earth.

3. *Isolation of the metallic shield* from earth contacts by using an insulated covering. Most of the present-day telephone cables have a polyethylene outer sheath over the metallic shield. Corrosion of the shield is not a problem if the cable is installed and maintained so that the polyethylene jacket is free of voids and breaks.

Mitigative measures in built-up areas that cause interference to other underground structures are uneconomical and undesirable. Many local correlating committees and sections of the National Association of Corrosion Engineers have been organized to encourage the cooperative solution of common corrosion problems.

114. References

UHLIG, HERBERT H. "Corrosion and Corrosion Control"; New York, John Wiley & Sons, Inc.
SUNDE, ERLING D. "Earth Conduction Effects in Transmission Systems"; Princeton, N.J., D. Van Nostrand Company, Inc.
UHLIG, HERBERT H. "Corrosion Handbook"; New York, John Wiley & Sons, Inc.

115. Inductive Coordination of Communication and Power Systems. Inductive Fields. Power lines have electric and magnetic fields surrounding them that induce voltages in other wires passing through these fields. For convenience, the effects of these fields are classed as electric and magnetic induction and are analyzed separately.

116. Electric induction into an exposed telephone line is due to the voltage on the power circuit. As shown in Fig. 25-58*a*, the voltage of the power wire to ground E_P

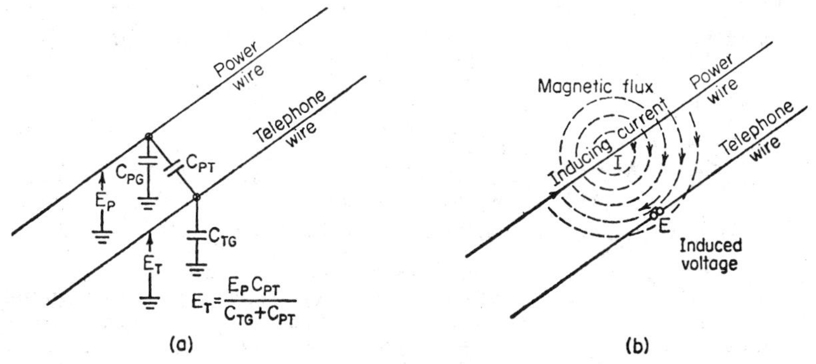

FIG. 25-58. Electric induction (*a*) and magnetic induction (*b*).

divides over the capacitances C_{PT} and C_{TG} so that the voltage on the telephone wire

$$E_T = \frac{E_P C_{PT}}{C_{TG} + C_{PT}} \tag{25-25}$$

117. Magnetic induction into an exposed telephone line is due to the current on the power circuit which sets up an alternating magnetic field. As shown in Fig. 25-58*b*, a voltage is induced along the telephone wire which is proportional to the rate of change of the magnetic flux.

118. Noise Frequency Induction. Every noise induction problem involves consideration of the inductive influence of the power line, the susceptiveness of the telephone circuit, and the coupling between them. Reducing any one of the factors to zero would eliminate the noise problem.

119. Power-line influence is dependent on the magnitude and frequencies of the harmonics on the power system and its degree of balance. The approximate relative interfering effects of telephone-circuit noise currents are shown in Fig. 25-59.

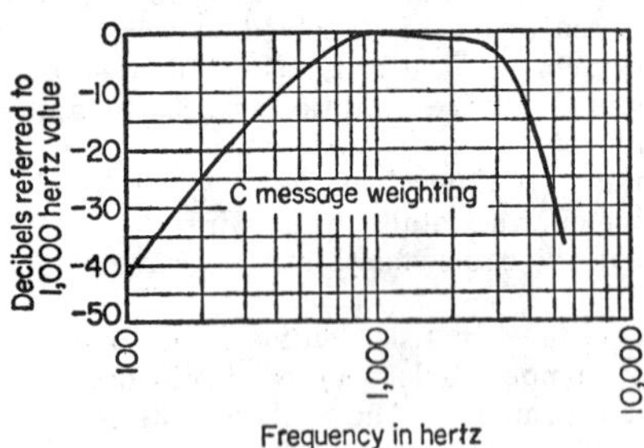

FIG. 25-59. Appropriate relative interfering effects of telephone circuit noise.

A useful overall index for the waveshape of the voltage or current is a quantity called telephone influence factor (TIF). The TIF of a voltage or current wave is the ratio of the square root of the sum of the squares of the weighted rms values of all sine-wave components to the rms value (unweighted) of the entire wave. The 1960 TIF weighting factors are shown in Fig. 25-60.

An important factor in the balance of a power system is the magnitude of residual harmonic currents and voltages. Since the triple-harmonic currents in a grounded Y-system are in phase, they are inherently residual components.

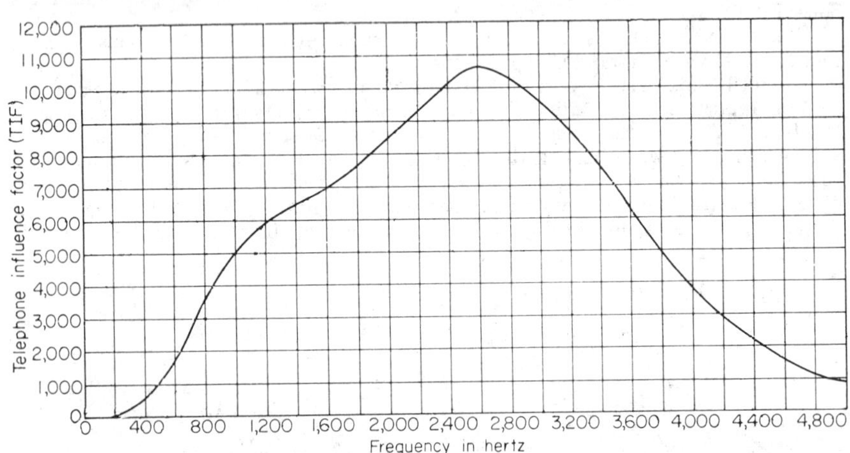

FIG. 25-60. Telephone-influence-factor weightings.

120. Coupling factors can be controlled by separation and transpositions in the power and communications circuits and shielding. Grounded metallic-sheath telephone cable provides nearly perfect shielding for electric induction. For magnetic shielding, a cable sheath must be electrically continuous and effectively grounded at each end. Susceptiveness of a telephone system is the characteristic which determines the extent to which it is capable of being adversely affected by being in electric and magnetic fields. Designing and maintaining a telephone system that is well balanced, i.e., symmetrical with respect to earth, is the most effective means of improving the susceptiveness.

121. Low-frequency Induction. High voltages are sometimes induced in paralleling communication circuits due to large residual currents or voltages on the power line at the time of a fault. If the earth resistivity is 100 Ω·m, a telephone line 50 ft from a 60-c/s power line would have approximately 90 V induced/kft exposure for a 1,000-A fault to ground. An effectively grounded cable sheath might reduce the voltage induced on the conductors within the cable by about 50%. Cooperative planning in advance of construction is important in avoiding low-frequency induction problems.

122. References

Engineering Reports of the Joint Subcommittee on Development and Research of the Edison Electric Institute and the Bell Telephone System: The Telephone Influence Factor of Supply System Voltages and Currents; *Eng. Rept.* 33. Supplement to *Eng. Rept.* 33, The Telephone Influence Factor of Supply System Voltages and Currents; *EEI Publ.* 60-68. Inductive Coordination of Rural Power and Telephone Systems; *Eng. Rept.* 40. Wave Shape of Multi-phase Rectifiers; *Eng. Rept.* 49. Inductive Coordination Aspects of Power Factor Correction Capacitor Installations; *Eng. Rept.* 50. Coupling Factors for Ground Return Circuits—General Considerations and Methods of Calculation; *Eng. Rept.* 14. Shielding of Ground Return Circuits at Low Frequencies; *Eng. Rept.* 26. Low Frequency Shielding in Telephone Cables; *Eng. Rept.* 48.

RADIO TRANSMISSION SYSTEMS

123. General. Radio systems have been used for many years for the transmission of telephone and telegraph services. In recent years they have been used for the transmission of hundreds of voice bandwidth channels. Engineering descriptions and considerations for the use of radio systems are contained in the following paragraphs.

124. Mobile radio provides a convenient and reliable means of communicating with vehicles, ships, trains, aircraft, and pedestrians. Recent advances in the state of the art employing integrated circuitry, solid-state techniques, ferrite devices, etc., make possible compact, lightweight, low-power-drain radio equipment which readily lends itself to mobile or portable use.

125. Vehicular Mobile Service. Frequency modulation, being less susceptible to interference than amplitude modulation, is generally employed in vehicular mobile communication systems. Amplitude modulation is used in certain instances where low cost is a controlling factor and interference is not a problem.

Frequencies assigned to the various vehicular mobile services generally fall within three broad frequency ranges: v-h-f low-band, 25 to 50 MHz; v-h-f high-band, 144 to 174 MHz; and u-h-f band, 450 to 470 MHz. Each of these bands exhibits certain transmission characteristics. The v-h-f low-band frequencies provide somewhat greater range than the higher frequencies, but exhibit greater susceptibility to man-made electrical interference and long-distance skip interference. The latter is due to reflection of radio signals from the upper layer of the ionosphere, caused by the rise of sunspot activity. During this period radio services separated by thousands of miles and normally well out of range of each other may experience mutual interference. It is estimated that future sunspot peaks will occur about 1968 to 1970 and 1979 to 1981 and at approximately 11-year intervals thereafter. The v-h-f high-band, which generally exhibits the most desirable characteristics, is the most extensively employed in the vehicular service.

The u-h-f 450- to 470-MHz frequencies exhibit greater path and obstruction losses than the v-h-f frequencies. However, compensating for these increased losses are two important factors: (1) lower ambient noise at u-h-f frequencies, permitting lower received signal levels, and (2) low cost of additional antenna gain at u-h-f frequencies. Thus the range obtained for a given transmitter power at ultrahigh frequency is approximately the same as for very-high frequency except in mountainous or heavily wooded areas. The cost of u-h-f mobile equipment is somewhat higher than that of v-h-f equipment of comparable power.

126. Marine Mobile Service. Mobile radiotelephone service to ships is provided along the United States coasts and on the Great Lakes and the Mississippi River.

Three frequency bands are employed; 2 to 3 MHz (AM), 4 to 8 MHz (AM—inland waterways), and 156 to 174 MHz (FM). The m-f and h-f bands are used to provide communication between ships, ships and Coast Guard stations, and also ships and land stations connecting to the public telephone network. The v-h-f band frequencies are similarly employed. In addition companies and organizations having marine activity may also communicate with v-h-f maritime mobile stations for business and operational purposes.

The m-f and h-f bands provide communication up to several hundred miles during the daytime and much farther at night depending upon frequency, transmitter power, antenna efficiency, season of year, atmospheric noise, etc. The v-h-f band provides line-of-sight propagation, and the range is primarily dependent upon antenna height and whether transmission is over fresh or sea water. Fresh-water path losses with ordinary antenna heights are practically the same as those over land. A typical range over fresh water from a 25-W ship transmitter to a land station with 100-ft-high antennas would be on the order of 30 to 35 mi.

127. Aeronautical Mobile Service. Segments of many frequency bands are assigned to the aviation services for navigational purposes, aircraft operation, and aircraft control. These allocations cover a wide range of frequencies from l-f and m-f ranges for radio beacons and radio ranges up to s-h-f microwave frequencies for radar, radio altimeters, and radio navigation.

The radiotelephone employing very-high frequency (108 to 132 MHz civil use), ultrahigh frequency (225 to 400 MHz military use), and high frequency (2 to 25 MHz) is generally employed for aircraft operation and control. Amplitude modulation is generally employed, permitting instant break-in or override in event of emergency. High-frequency systems employing single-sideband modulation are generally used for long-range overwater communication.

Tropospheric scatter techniques are employed in certain long-range overwater aeronautical communication systems. By the use of high-power high-gain antennas and highly sensitive low-noise receiving equipment, v-h-f ranges have been extended beyond the point of grazing to distances over sea upward of 300 to 400 mi (statute).

128. Point-to-point Radio Systems. Long-range radio communication without intermediate repeaters and communication to and between ships on the high seas is best accomplished by using frequencies in the h-f part of the spectrum, 3 to 30 MHz. Radio emissions from the earth at these frequencies are bent by the interaction of the propagated wave and the ionosphere. With the proper selection of frequency for a given set of ionospheric conditions the bending can return the signal to the earth, often at a great distance from the transmitter. Since the signal is also reflected upon striking the earth, the process can be repeated two or more times for communication between stations half a world apart.

Radiotelephone service provided by this means is usually restricted to services that cannot be provided by other means for technical or economic reasons. This limitation results from the very small total bandwidth usable with this type of propagation. Furthermore, only part of the total is usable at a given time, because of the vagaries of the ionosphere, whose structure varies with the time of day, season, and sunspot activity.

High-frequency radio is a feasible method for communication with or between ships at sea and with airplanes at great distances. High frequency also provides an economical means of providing small numbers of direct telephone or telegraph circuits between widely separated countries. Submarine cables and communication satellites are better suited for this service wherever the number of circuits required can justify the large investments for these facilities.

The preferred method of modulation in this band is single sideband (SSB) with reduced or suppressed carrier. This permits maximum frequency conservation in this very important band, even though SSB equipment is more expensive than double-sideband (DSB) equipment. Transmitter outputs range in power from about 100 W to 10 kW, with higher powers available for special applications.

Since propagation varies continuously, special control-terminal equipment is usually applied. A control terminal typically includes some or all of the following: noise suppressors; voice-operated devices permitting one direction of transmission at a time, to prevent echoes; variable-gain amplifiers that present essentially the same input level to the transmitter with loud and soft talkers and that restore the spread in level at the receiver; and squelch devices, to prevent very high noise levels from reaching the user's ear. Since it is relatively easy for anyone with a "short-wave" receiver to eavesdrop on these frequencies, commercial h-f radiotelephone circuits are equipped with privacy devices to prevent unauthorized reception.

129. Beyond-horizon tropospheric radio-relay communications systems are suitable for remote regions and where there are geographical obstacles, such as large bodies of water, precluding the use of intermediate repeaters. A band of frequencies in the u-h-f spectrum is generally used for this type of system. Path lengths of 100 to 500 mi, depending upon transmitter power and antenna gain, are usually satisfactory.

The variability of tropospheric scatter phenomena requires the use of dual or quadruple diversity to ensure good reliability. This is achieved by combinations of frequency and space diversity.

Transmitter powers up to 50 kW and large antennas having gains of 40 dB or more may be required to meet transmission objectives over multihop systems.

130. Microwave radio systems, operating over line-of-sight paths, are used to provide transmission trunks between switching offices. A number of frequency bands in the u-h-f and s-h-f catagories have been allocated for fixed government and public service. For commercial service the most useful bands are 3,700 to 4,200 GHz, 5,925 to 6,425 GHz, and 10,750 to 11,750 GHz. These allocations are further subdivided into a number of radio channels, each of which is used to transmit several hundred voice-bandwidth circuits.

The selection of a particular radio system depends upon the type of service for which the trunk is intended. The more important considerations are trunk length, number of voice-bandwidth channels, type of service, spacing of multiplex terminals, transmission requirements, and reliability. The following paragraphs describe two general types of systems and discuss their characteristics and uses.

131. Short-haul radio systems are used for multihop trunk service over relatively short distances. Typical systems are designed to meet transmission and reliability objectives for about 10 repeater sections in tandem, providing trunks up to about 250 mi in length.

Short-haul systems are usually designed to demodulate the multiplex signal to base-band at each repeater but, in the process, add distortions inherent in the demodulator and base-band amplifier portions of the repeater. However, these systems are particularly adaptable to service where it is desirable to drop and add circuits at a number of localities along the route.

In a typical short-haul radio repeater the incoming frequency-modulated radio signal is selected from all other signals by a bandpass filter and then passes to the receiving converter. Here it is heterodyned by a locally generated frequency to produce an intermediate frequency. Amplification to make up for the loss of the previous air path is provided by a low-noise i-f preamplifier and i-f amplifier. Since the path loss varies owing to propagation anomalies, the i-f amplifiers are compensated by automatic gain control (AGC) to maintain a constant output power. Following the i-f amplifiers there is a limiter to remove amplitude modulation produced by transmission irregularities, a discriminator, and a base-band amplifier.

At this point the signal is a close reproduction of the original multiplex signal, modified only by the noise and distortion introduced by the radio system and any extraneous radio signals not completely eliminated by the bandpass characteristic of the system.

The base-band signal may be connected to multiplex equipment to drop and add groups of circuits as the trunking facilities require. The resulting signal complex is amplified by a base-band amplifier to frequency-modulate the radio transmitter. The signal is then beamed to the receiver of the following repeater.

The most extensive development for commercial short-haul systems has been in the 5,925- to 6,425-GHz and 10,750- to 11,750-GHz allocations. The choice of the band to use depends upon equipment availability, interference considerations from existing systems, and propagation characteristics affecting reliability, as discussed in later paragraphs.

132. Long-haul radio systems are designed to meet transmission and reliability requirements for trunk service 4,000 mi or more in length. Unlike short-haul systems, the signal is not demodulated to base band at every repeater, and distortions that would be introduced in the process are avoided. Instead the signal is demodulated to intermediate frequency, and terminal and multiplex equipment is introduced at widely separated localities.

Commercial long-haul systems have been developed mainly in the 3,700- to 4,200-GHz and 5,925- to 6,425-GHz frequency bands. Rain attenuation of signals in the higher-frequency bands has been a deterrent to development of long-haul systems in the 10,750- to 11,750-GHz allocation.

In a typical long-haul system repeater the receiver function from antenna through the i-f main amplifier is the same as in a short-haul system. The transmitter, however, consists of a modulator to combine the receiver output i-f signal with a locally generated signal to produce the assigned radio frequency. The signal is then further amplified for transmission to the next repeater. Thus the signal is in a frequency-modulated state from receiver input to transmitter output.

The interface between the multiplex equipment and the first radio transmitter is a terminal transmitter that converts the multichannel base-band signal to an i-f frequency-modulated signal. At the receiving end of a trunk section a terminal receiver performs the reverse function.

133. Radio-system engineering of both short- and long-haul systems requires close attention to parameters affecting transmission objectives and reliability. These are controlled to a large extent by the circuit and equipment designs produced by the manufacturers, but the application and route design must match the equipment capability to achieve the overall results that are desired.

Manufacturers of radio equipment specify thermal-noise performance based upon a transmitter output power and a normal received signal level. The radio-system engineer must select antennas and transmission lines to optimize gains and losses and must select repeater sites and repeater spacing to optimize path losses and minimize interference from other routes.

On the assumption of free-space transmission, the received signal level can be calculated from the following expression,

$$P_r = P_t + G_t + G_r - L_{tl} - P_l \qquad (25\text{-}26)$$

where P_r = received signal level, dBm
P_t = transmitter output, dBm
G_t = gain of transmitting antenna, dB
G_r = gain of receiving antenna, dB
L_{tl} = sum of transmission-line losses, dB
P_l = path loss, dB = $20 \log (4\pi d/\lambda)$ (d = distance between antennas and λ = wavelength in same units as d)

Figure 25-61 is a nomograph of path loss as a function of frequency and distance between repeater sites and may be used to determine P_l in expression (25-26). Antenna gains are usually expressed as a power ratio in decibels of the field strength on the axis of the main lobe compared with the field strength from an isotropic radiator. Transmission-line losses include the loss of all networks, cables, and waveguide between the transmitter-equipment point at which the output power is specified and the transmitting antenna and likewise between the receiving antenna and the receiver equipment.

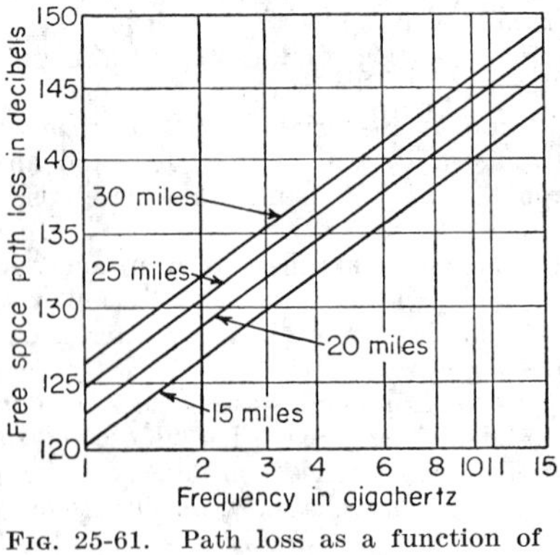

FIG. 25-61. Path loss as a function of frequency.

In addition to thermal noise there are other noise components introduced by the radio system caused by intermodulation products due to transmission imperfections, noise added by repeater local oscillators, and tones produced by beats between the desired radio carrier frequency and interfering frequencies.

Reliability is usually expressed as the

percentage of time a trunk facility is not available for service or does not meet transmission objectives. The main factors affecting reliability are equipment failures, power failure, down time for maintenance, and low received signal level during periods of radio fading. Outages due to equipment failures and maintenance activity can be greatly reduced by automatically or manually transferring the service to duplicate equipment. Outage due to power failure can be reduced by providing stand-by battery or emergency power facilities at each radio station. The effects of propagation anomalies are usually reduced by diversity techniques discussed in subsequent paragraphs.

134. The propagation paths of radio waves between two points are the sky wave, direct wave, ground reflected wave, and surface wave. The surface wave is useful below about 1.5 MHz. Sky waves are used for long-distance transmission in the h-f spectrum. Little use is made of ground reflected waves in commercial radio applications, except in tandem with sky waves over very long distances. The direct wave is the principal mode of reception in point-to-point radio systems operating in the s-h-f bands.

During periods of normal propagation the received signal of the direct wave is steady and, if the path has been engineered with adequate ground clearance and to avoid ground reflections, the received level will agree closely with calculated free-space values. During abnormal propagation the path loss may be much greater than free-space calculations would indicate. These variations in propagation are referred to as fading.

Fading results in a decrease in received signal level and consequently an increase in thermal noise. The effect on transmission performance must be considered, and if the service to be carried by the radio facility requires a high degree of reliability, a means of diversity protection must be included in the system design. The two most commonly used methods of protecting against frequency-selective fading are space diversity and frequency diversity.

135. Space diversity is achieved in multihop line-of-sight systems by utilizing two receiving antennas with a vertical separation on the order of 100 to 200 wavelengths. The output of the associated receivers having the stronger signal is connected to the following transmitter.

Usually it is desirable to provide a working stand-by transmitter to protect against equipment failures in the regular transmitter. The stand-by is also switched into service to permit maintenance of the regular transmitter. An economic disadvantage of space-diversity systems is the need for taller and stronger towers. Since the lower antenna of the pair should have clearance for free-space transmission, additional tower height is required for the second antenna.

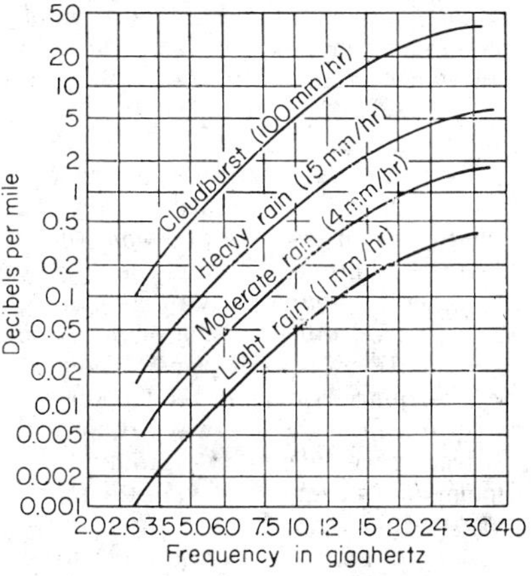

FIG. 25-62. Estimated atmospheric absorption.

136. Frequency-diversity systems employ two radio channels with sufficient frequency separation so that fading is essentially uncorrelated. A simple form of frequency diversity in common use for short-haul systems uses two radio channels transmitting the same signal information. The signal outputs of the two corresponding receivers are monitored and the one with the stronger signal selected for service. This type of diversity also gives protection against equipment failures.

Frequency-diversity systems of the type described above are not so efficient in the use of the available frequency spectrum as are space-diversity systems. More efficient systems are generally used for long-haul routes in which one stand-by channel protects against fading and equipment failures of several working channels. The switchover to protection at the two ends of the switching section must be coordinated to maintain circuit continuity.

137. The effects of rain attenuation of radio signals must be considered in selecting equipment for a particular route. Radio energy is scattered and absorbed by raindrops, and the effect is more pronounced at higher frequencies, where the wavelength approaches the dimension of the raindrops. The effect of rain on propagation in the 4- and 6-GHz bands is relatively small. At 11 GHz, however, the path loss can be quite high.

Figure 25-62 shows the estimated path loss vs. frequency for various intensities of rain. It can be seen that in areas subjected to widespread rain of cloudburst proportions the path loss may increase 5 dB/mi or more in the 11-GHz band. Since all radio channels in the band will be attenuated to approximately the same degree, neither space diversity nor in-band frequency diversity will be effective. Crossband frequency diversity utilizing a 4- or 6-GHz radio channel paired with an 11-GHz channel will, however, mitigate outage due to fading, rain attenuation, and equipment failures.

ANTENNAS FOR RADIO COMMUNICATION

138. V-H-F and U-H-F Mobile Communication Antennas. Antennas for mobile or portable use should be small and of light weight, rugged construction, pleasing appearance, and low cost. For base-station use, high gain, rugged construction, desired radiation characteristics, lightning protection, and reasonable cost are considered as being of major importance.

139. Base-station Antennas. Since uniform circular coverage from a base station is generally desired, it is quite natural that the resonant vertical radiator, which exhibits such characteristics, is the basic radiating element found in most mobile base-station installations. The simplest vertical radiator consists of a quarter-wave whip or monopole operating above a ground plane. The ground plane is usually simulated by four elements extending horizontally from the base of the monopole like the spokes of a wheel. The resonant-frequency radiation resistance of such an antenna is in the range of 35 to 37 Ω, which can, in most cases, be adequately matched by conventional 50-Ω coaxial transmission lines.

By folding the vertical radiator back to the base a so-called folded unipole is formed. A d-c path is thus established to ground, providing increased protection against static or lightning surges. In addition, the radiation resistance is raised by a factor of about 4, and the antenna exhibits increased broadband characteristics as compared with the single whip antenna. By making the grounded side of the loop smaller in diameter the radiation resistance can be made to approximate 50 Ω, thus providing a better match to conventional coaxial transmission lines.

The quarter-wave ground plane or folded-monopole antenna is generally regarded as equivalent to a half-wave dipole, although under theoretically idealized conditions (in free space with infinite ground plane) the quarter-wave whip is 3 dB poorer than a dipole. However, such conditions are rarely realized in practice, and for engineering purposes 0-dB gain is generally assumed.

A more streamlined and structurally advantageous radiator is found in the coaxial antenna. The ground-plane radials are replaced by a tubular skirt extending downward from the point of feed. The coaxial transmission line is then fed up through the antenna support centered inside the skirt. This makes a solid, neat-appearing structure having radiating characteristics comparable with those of the ground-plane antenna. This and similarly constructed collinear gain antennas are probably the most widely used base-station antennas in the mobile-communication services.

140. Mobile Antennas. With vertical polarization generally employed at base stations, it follows that for optimum signal strength vertical polarization is also a requirement at the mobile units.

The quarter-wave whip antenna operating above a ground plane most effectively meets the basic requirements of a mobile antenna and is generally employed in the mobile services. The modern metal car top provides an excellent ground plane and readily lends itself to the installation of 150- and 450-MHz quarter-wave whip antennas. A fairly uniform omnidirectional radiation pattern is generally obtained with such installations. In those situations where rooftop installation at 150 MHz and 450 MHz is not possible or convenient, use of a coaxial-type antenna mounted on the cowl, rear deck, or rear fender provides a good compromise.

141. Rhombic antennas are used for long-range radio communication systems operating in the h-f portion of the frequency spectrum. Each installation is usually for a specific purpose, and the design depends upon the desired elevation angle, beamwidth, gain, and bandwidth. These parameters are controlled by the length and angle of the rhomboid sides and the height above ground. Combinations of rhomboids stacked vertically or horizontally are used to gain certain advantages. Design considerations for rhombic antennas are covered in detail in the McGraw-Hill "Antenna Engineering Handbook" (edited by Henry Jasik, 1961).

142. Parabolic antenna systems are generally used for line-of-sight microwave radio transmission. A highly directive beam is formed by the parabolic reflector in the same way a searchlight concentrates the light from a source located at the reflector focal point.

Parabolic antenna systems are available in several configurations, each having advantages and disadvantages to be considered in selecting the proper arrangement for a particular route. Direct radiating reflectors are mounted on towers and buildings and aimed directly at the antenna at the opposite end of the path. Periscope systems utilize a parabolic antenna directed at a plane or slightly curved reflector which redirects the signal energy to the distant receptor. Another form of parabolic antenna is the horn reflector utilizing a sector of a large paraboloid of revolution. A rectangular or conical horn with apex at the focal point extends to intercept the parabolic sector.

143. Direct radiating antennas are capable of high directivity and good discrimination against unwanted signals (relative to periscope systems). This is an important consideration particularly in areas where radio systems are widely used. The discrimination characteristics, particularly front-to-back ratio, of parabolic direct radiating systems permit the use of a two-frequency plan for transmitting and receiving in two directions, and this conserves the frequency spectrum. Antennas of this type are adaptable for both short-haul and long-haul radio systems.

Direct radiators mounted on tall towers have the disadvantage that long transmission lines are usually required for connection to the radio equipment. The attenuation of long lines is an important factor in the overall system design, for it reduces the strength of the received signal. Even with the use of a waveguide for the transmission line the loss may be an appreciable factor.

Impedance mismatches at each end of a transmission line cause delayed echo signals due to reflections of a portion of the signal energy from each impedance discontinuity. The amplitude and delay distortions associated with the echo signals produce spurious noise and intermodulation products in the transmitted message information. Since the magnitude of the effect is a function of the period of delay as well as the amplitude, the transmission-line length must be considered in the overall allocation of noise parameters in the system design.

144. Periscope antennas are used mainly with short-haul radio systems and where it is feasible to use four-frequency plans for two directions of transmission. With the parabolic reflector mounted near the radio equipment, frequently on the roof of a repeater hut, long transmission lines are avoided; consequently, the losses and distortions associated with long lines are avoided to a large degree.

In selecting components for periscope systems the important transmission parameters to consider are: the relative diameters of the exciter antenna and projected area of the reflector, the separation of the exciter and reflector, the height of the reflector with respect to the path terrain, and whether a plane or curved reflector is required. A more complete discussion of periscope antenna systems may be found in the McGraw-Hill Antenna Engineering Handbook, Chap. 22.

145. Horn-reflector antennas have high efficiency, good directivity, and good impedance characteristics over a wide band of frequencies. They are easily matched to waveguide to transmit or receive signals in several frequency bands simultaneously. For these reasons horn reflectors are particularly suitable for long-haul high-density radio routes. Table 25-20 gives the performance characteristics of a typical horn-reflector antenna.

146. Passive reflectors are occasionally used in microwave relay systems to redirect the radio beam in much the same way as a plane mirror can be used to reflect a light beam around a corner. They are used to change the direction of the radio beam to

Table 25-20. Horn-reflector Antenna Characteristics*

Frequency	4 GHz		6 GHz		11 GHz	
Polarization	Vert.	Hor.	Vert.	Hor.	Vert.	Hor.
Midband gain, dB	39.6	39.4	43.2	43.0	48.0	47.4
Front-to-back ratio, dB	71	77	71	71	78	71
Beam width (azimuth), deg	2.5	1.6	1.5	1.25	1.0	0.8
Beam width (elevation), deg	2.0	2.13	1.25	1.38	0.75	0.88
Sidelobes, dB below main beam	49	54	49	57	54	61
Cross-polarization discrimination, dB	50	46	51	51	57	51
Side-to-side coupling, dB	81	89	120	122	94	112
Back-to-back coupling, dB	140	122	140	127	139	140

* These characteristics are for a particular pair of antennas without any waveguide system attached.

get over or around an obstruction, to reduce the number of repeaters, or to place them at a more accessible location.

A single passive reflector is generally used for transfer angles greater than about 40°. When the angle is less than this, the size of the reflector required for good efficiency becomes too large. In such cases a double reflector is more practical.

The total loss for a path using a passive reflector is given by the expression

$$10 \log 1.23 \times 10^{17} (D_1 D_2 / A)^2 \tag{25-27}$$

where D_1, D_2 = distance in miles from reflector to transmitting and receiving antennas and A = projected area of reflector in square feet.

The above expression takes into account the "gain" of the reflector as well as the loss of both sections of the path, and the result is not dependent upon frequency.

147. References

1. Bullington, K. Radio Propagation Fundamentals; *Bell System Tech. Jour.*, May, 1957, p. 593; "Antenna Engineering Handbook"; New York, McGraw-Hill Book Company, 1961, Chap. 33.

2. Bullington, K. Radio Propagation Variations at VHF and UHF; *Proc. IRE*, January, 1950, p. 27.

3. The Propagation Characteristics of the Frequency Band 152–162 Mc/s Which Is Available for Marine Radio Communications; Radio Technical Commission for Marine Services, *Rept. Spec. Comm.* 19, Appendix F, February, 1956.

4. Marine Radiotelephone Service in the 2-Mc Band; Radio Technical Commission for Marine Services, *Rept. Spec. Comm.* 11, Dec. 14, 1949.

5. Kirby, Herbstreit, and Norton Service Range for Air-to-ground and Air-to-air Communication at Frequencies above 50 Mc; *Proc. IRE*, May, 1952.

6. Development of Long Distance VHF Communication System; *Federal Aviation Agency Rept. Project* 116-10D, September, 1963.

7. Transmission Systems for Communications; Bell Telephone Laboratories.

SECTION 26

ELECTROPHYSICS

BY

J. H. HAGENGUTH Manager, High Voltage Research, General Electric Company; Fellow, Institute of Electrical and Electronics Engineers

J. G. ANDERSON Technical Director, Project EHV, General Electric Company; Fellow, Institute of Electrical and Electronics Engineers

E. J. ROGERS Brookhaven National Laboratory

H. E. SEEMANN Physicist, Research Laboratories, Eastman Kodak Company

CONTENTS

Numbers refer to paragraphs

SECTION 26

ELECTROPHYSICS

LIGHTNING AND THE PROTECTION OF LINES AND STRUCTURES FROM LIGHTNING

BY J. H. HAGENGUTH AND J. G. ANDERSON

PART I. THEORY OF LIGHTNING

1. Origin of Thundercloud Charges. The average number of days per year on which thunderstorms occur in various regions of the United States of America is shown on Fig. 26-1.

Lightning storms seem to have certain characteristics which are peculiar to a given storm. As an illustration, some storms consist mostly of cloud-to-cloud strokes, while in other storms cloud-to-ground strokes predominate. Furthermore, it has been shown*[1] that some storms produce strokes having a higher percentage of successive current peaks within the stroke than other storms. These habits are apparently due to differences in the location of centers of charges within the cloud; the successive-current-peak (multiple-stroke) type being the result of the discharge of various centers in sequence.

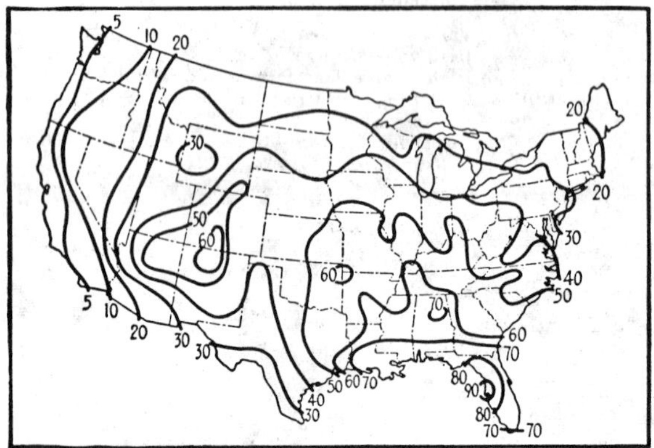

FIG. 26-1. Average number of thunderstorm days for the 40-year period 1904 to 1943. (*From "Thunderstorm Rainfall." Hydrometeorological Report 5, U.S. Weather Bureau in cooperation with Corps of Engineers, U.S. War Department,* 1946.)

The exact process is not known through which a cloud builds up electrical charges to such a magnitude as to cause a lightning stroke. Several theories[2,3,4] have been advanced to account for the accumulation of these charges, but the problem is complex and not subject to easy proof or determination through the construction of models. Investigations[5,6] seem to indicate that one very likely source of generation of charges is the freezing process in the upper portions of the cumulus clouds, which seems to extend beyond 20,000 ft before electrical discharges occur. In laboratory experiments[5] potentials of the order of several hundred volts were produced at the water-ice interface during the freezing process.

In the theory of the process of production of the charged cloud, a separation of charges must take place. A considerable number of data[7,8] indicate that the ground

* For all numbered references, see Bibliography (Par. **38**).

end of at least 95% of cloud-to-transmission-line strokes are positive. Thus, it seems certain that for most strokes the lower surface or cloud base is negatively charged, at least over most of its area. There is evidence that regions of positive charge can exist in the base of the cloud.[5,9,10] That the more remote or upper portions of the cloud are positively charged is evidenced by oscillograms[11] which show a change from negative to positive after some time has elapsed. The action of convection currents in breaking up drops of water which have coalesced, and raising these droplets to the upper portions of the cloud with a subsequent reuniting and falling, is regarded as an important part of the mechanism of separating electrical charges in the cloud. It is quite obvious that the air currents must do the work of separating the charges and keeping them separated within the cloud. When these convection currents die out, the charges will come together and the cloud returns to approximately its original condition.

Mechanism of Lightning Discharge

2. Conditions existing on the earth, prior to a stroke between cloud and earth, must be recognized in order that one may be familiar with the stroke process. This is necessary for a proper understanding of the protective measures taken, and may help to explain why protection is not always attained.

As a negative charge builds up in the cloud base, a corresponding positive charge appears in the earth underneath by induction—the negative earth charges being driven away from under the negative cloud. Thus, as a storm cloud is carried along above the surface of the earth, its counterpart of opposite polarity is carried along the surface of the earth underneath. These positive charges, being attracted by the negative charges in the cloud, move up transmission towers, along overhead ground wires, up church steeples, flagpoles, trees, and other conducting or semiconducting objects. When the gradient is high, a corona discharge or St. Elmo's fire may be seen. Thus it is clear that, with a cloud of considerable height, say 5,000 ft, a very large area of earth is involved underneath.

3. Stroke Mechanism. Through the use of photographs taken of lightning strokes, where the lens has been moved rapidly with respect to the photographic film, Schonland[12,13] found that the lightning stroke is initiated by a streamer from the cloud which progresses toward the earth in a series of steps. The time taken for this initiating step leader to reach the earth was in some cases of the order of 0.01 s. It was found that upon reaching the earth a "return stroke" took place, progressing from the earth to the cloud. At times, it was found that this return stroke did not get all the way back to the cloud.

That more than one discharge could take place through the same path between cloud and ground has been known for many years. In 1905, Larson[14] published a photograph of 40 such discharges occurring in a stroke which persisted for 0.624 s, as shown by the moving camera which he used.

Schonland's work showed that discharges, or current peaks which followed the path determined by the step leader discharge, were in general preceded by a downward leader called a continuous leader. Upon making contact with the earth, a return discharge occurred from the earth just as in the case of the initial step leader. The velocity of the step leader is of the order of 50 m/μs, with time intervals of about 100 μs between steps. On the other hand, the continuous leaders have a velocity of about 1 to 23 m/μs. The return stroke from the earth moves up the channel with a velocity of from about 20 to 140 m/μs, which on the average is considerably less than half the speed of light.

McEachron[11] has shown that at least 80% of the strokes to the Empire State Building are initiated by the building, rather than by the cloud. Apparently the stress at the top of the building, due to its great height (1,275 ft to top of lightning-collecting system) and its configuration, is such that a step leader begins at the building and progresses toward the cloud. The velocities are of the same order of magnitude that Schonland found for the downward step leader. No return stroke was found from the cloud, but instead a continuing current flowed in one case as long as 0.625 s. Following the initial step leader, successive discharges may occur, and these, where leaders have been photographed, were always downward followed by the return discharge from the building. It appears, therefore, that the mechanism involved in the discharges, after the first, is the same to a tall structure as to the open country or to low objects (see Fig. 26-2). The propagation velocities of both the continuous leader and the return discharge are of the same order whether the stroke was initiated by the cloud or by a tall object.

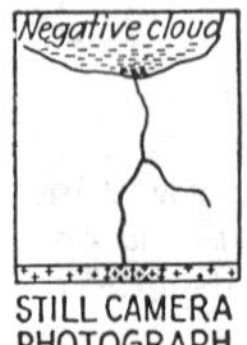

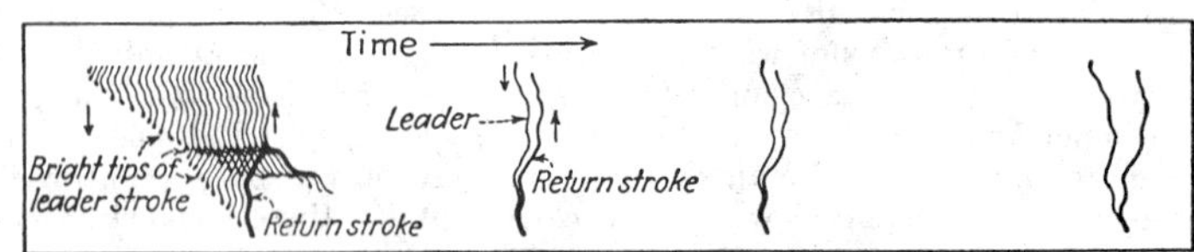

STILL CAMERA PHOTOGRAPH OF STROKE

Initial downward stepped leader Subsequent discharges with downward continuous leaders

HIGH SPEED BOYS CAMERA PHOTOGRAPH OF THE SAME STROKE

CLOUD TO EARTH

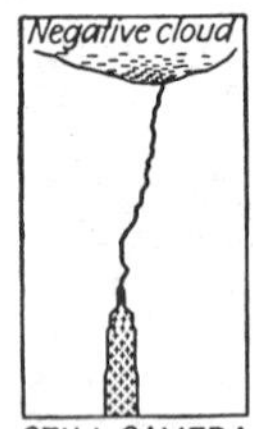

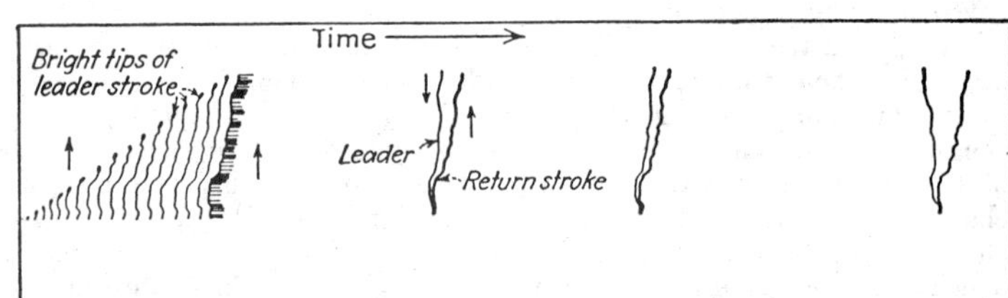

STILL CAMERA PHOTOGRAPH OF STROKE

Initial upward stepped leader Subsequent discharges with downward continuous leaders

HIGH SPEED BOYS CAMERA PHOTOGRAPH OF THE SAME STROKE

CLOUD TO TALL CONDUCTING STRUCTURE

Fig. 26-2. Schematic diagram of mechanism of lightning discharge. (*From K. B. McEachron, Jour. Franklin Inst.*, 1939, *Vol.* 227, *p.* 149.)

Unpublished data and experience of other investigators[15] seem to bear out the expectation that as the heights of structures become less the earth-initiated stroke becomes less frequent. Possibly it could occur from a transmission or radio tower located at the top of a high rocky pinnacle, but, in general, it is not expected that transmission towers will be high enough to initiate lightning strokes.

Many photographs[11,13] taken both in this country and abroad show continuing illumination between successive current peaks as shown by the moving camera or as seen sometimes when the path is blown along by the wind. Oscillograms, e.g., Fig. 26-3, taken by McEachron, beginning in 1937, show that the change in density of such a photograph can be closely correlated with the current in the stroke[16] (except where the film is overexposed, which is not likely to be the case with the continuing part of the stroke). Thus, the photographs become good evidence to indicate that a stroke of lightning is in reality a d-c arc between the cloud and the earth with superimposed current peaks which may occur in almost any manner, depending upon the construction of the cloud and location of concentrations of charge. Figure 26-3 is a typical oscillogram showing a stroke to the Empire State Building starting with an upward leader. Downward leader strokes will, no doubt, begin with a current peak, but complete data are lacking.

It seems certain now that the direction of branching is also the direction of propagation of the initial step leader, so that still photographs which show branching will reveal information on direction of propagation. As a rule, branching[11,13] occurs only at the time of the initial discharge; succeeding discharges, if present, seldom show any branching.

It also seems clear that direction of propagation is determined by the configuration of the electrodes rather than by polarity as was thought to be the case for some years.

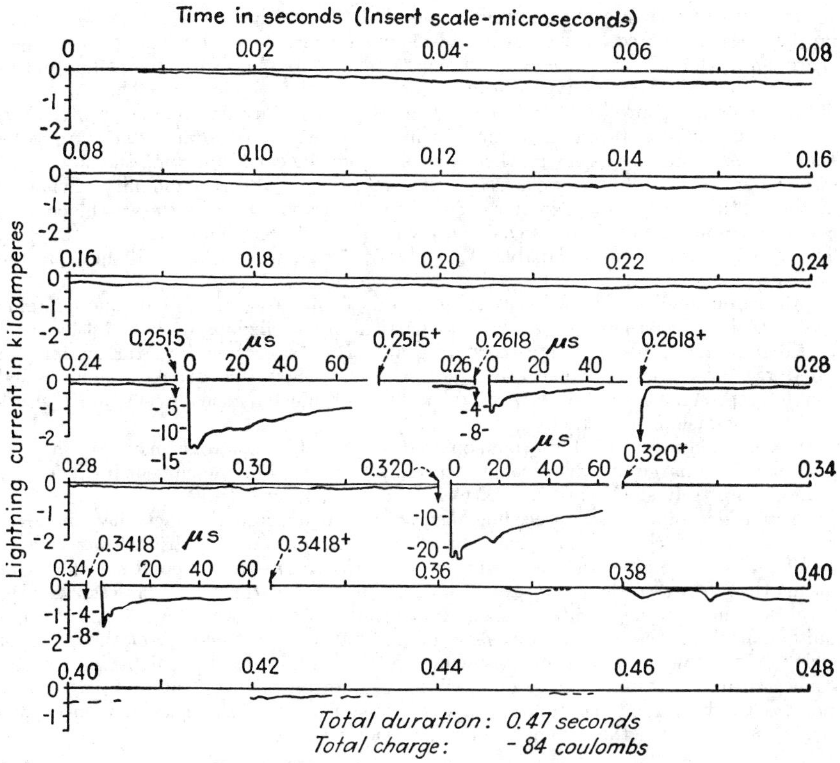

Time in seconds (Insert scale-microseconds)

Lightning current in kiloamperes

Total duration: 0.47 seconds
Total charge: - 84 coulombs

FIG. 26-3. Replot of low-speed cathode-ray oscillogram of stroke 3 with inserts of high-speed cathode-ray oscillograms showing current peaks. Empire State Building, New York City, 1940.

As the stress in the cloud becomes higher and higher, a condition is finally reached where a leader starts on its way to the earth, picking the best path from instant to instant, which is the reason why the discharge does not follow the electrostatic field. As it progresses, it in effect lays down a negative[17] space charge in the area surrounding the path, and carries with it a high concentration of charge in the progressing end of the leader. The leader is connected to the cloud at all times, since it requires a supply of charges from the cloud to keep it going. From the point of view of the electrostatic field, it is as though a conductor had been let down from the cloud, and this will cause a redistribution of charge on the earth beneath. As the leader comes closer and closer, charges[18] in the earth move in to the point underneath the down-coming leader until the stress on the earth's surface may become so great that the air is broken down and streamers form from the earth's surface. In Fig. 26-4 is a drawing of a photograph taken on a beach in New Jersey, and it shows the existence of the streamers. It should be noticed that the earth streamers are branched in an upward direction, and that there are downward streamers on the stroke itself. At the time the step leader formed the downward streamers, there was no connecting stroke to the ground, but the potential gradient was sufficiently high so that three upward streamers formed and the stroke made contact with the longest of the three.

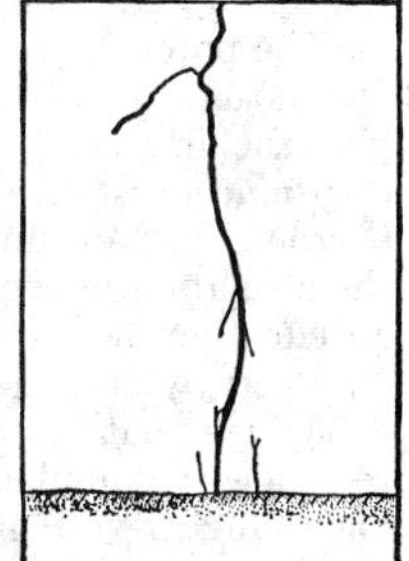

FIG. 26-4. Lightning stroke to beach showing upward earth streamers.

The return-discharge phenomenon means that the positive earth charge* is moving up the channel created by the downward-moving leader. A field exists between these charges and those in the space surrounding the channel, and also in the cloud, and in any branches which may have been formed by the initial step leader. The movement of these charges up into the channel constitutes the flow of current, and here, apparently, is the reason why there was no return discharge following the upward leader from the Empire State Building, although there was one from the earth following the downward step leader. The mobility of charges in the cloud is low, while the mobility of charges in the earth is very much greater. So when the leader reached the earth, charges could move quickly up into its channel, giving rise to a relatively high current peak. The cloud, however, is not capable of supplying charges so quickly, and therefore the current peak cannot develop; instead a slow discharge results.

Measurements[15] are available that substantiate the large amount of photographic evidence that strokes to the earth, and particularly to regions of low resistance of considerable area, begin with a current peak. It is suggested by McEachron and McMorris[18] that the movement of charges in to the point to be struck, prior to actual contact, represents the wavefront, and the tail is determined by the movement of charges into the channel after contact.

Goodlet, however, is of the opinion that "as the leader stroke moves toward its goal, point discharge and displacement currents will flow, but the current in the object struck is probably small until the start of the main or return stroke."

A consideration of the preceding discussion will disclose the fact that, in high-resistance areas, charges may have to come from considerable distances, with the result that considerable voltage is required to get the charges to the point struck. This means that in high-resistance areas, although the stroke current may be less, the area involved will be much greater than if the ground resistance were low. This effect is quite apart from the effect of tower-footing resistance upon flashover of the insulator string on the tower, or from unsatisfactory protection afforded by a lightning arrester connected to a high-resistance ground without interconnection. These are local situations, dependent upon the action of a local resistance, and have little influence upon stroke current or whether or not troubles are widespread.

Electrical Characteristics of Lightning

4. Voltage. The potential between cloud and ground, just prior to a stroke of lightning, has been variously estimated from 100,000,000 to 1,000,000,000 V.[19,20] The protection engineer is concerned principally with the potential which appears on the line conductor. If the current in a wave traveling along a line is known, a satisfactory value of voltage is obtained by multiplying the current by the surge impedance Z of the conductor, which is frequently taken as 500 ohms.

The potential which can appear upon apparatus is limited only by either protective apparatus or flashover of insulating structures, such as bushings or line insulators, plus the impulse strength of wood or other insulating support, including the effect of ground resistance. Direct strokes must be given consideration on all circuits carried overhead and on underground circuits connected to overhead conductors. Increasing the insulation between conductors, and between conductors and ground, is not of itself an effective means of preventing flashover due to lightning. Some reduction can be obtained by this means, and on low-voltage circuits the spacing may be made great enough so that system current does not ordinarily follow the lightning flashover, thus reducing the number of interruptions to service.[21] However, increasing the insulation allows higher-voltage traveling waves to reach stations.

The potential allowed by line insulation is determined by flashover values resulting from laboratory tests. The flashover potential is dependent upon the shape of the wave applied, as well as its crest value. As the time to flashover becomes shorter, the flashover voltage becomes higher, and may be of the general order of twice the 60-cycle

* When reference is made to the movement of positive charges along a conductor, it is to be understood that this refers to the movement of the electric field and the positive ends of the field terminating on the conductor. Thus the word "charges" is used to signify the terminal points of the electric field, at either conductors or other bodies. It is also to be remembered that the movement of a positive charge from left to right is equivalent to the movement of a negative charge in the opposite direction.

crest flashover when the rate of voltage rise is approximately 1,000 kV/μs, and flashover is on the front of the wave.

Lightning may cause voltages in conductors by induction, as well as by direct stroke to the conductor. The electrostatic field from a charged cloud induces in the earth beneath charges of the opposite sign, and these will be found also on transmission-line conductors within the cloud field. Charges, having the same sign as the cloud, will be driven off to remote parts of the line, or over the surface of insulators to ground as a slow leakage. Since any charge bound on the conductors will be released as the field is reduced in strength and will travel away from under the cloud along the conductor at the speed of light, the potential developed as the result of a sudden change of cloud field will depend not only on the change of field gradient but on the rate of change.

$$V = agh \tag{26-1}$$

where V = induced voltage; a = a factor less than unity, dependent upon rate of change of the cloud field and the distribution of bound (charge); g = actual gradient, in volts per foot, where line is located; h = height of line in feet.

Peek[20] gives a maximum value of 100 kV/ft gradient, while Norinder[22] has measured values of ag up to 80 kV/ft.

Wagner and McCann have calculated the induced potential on the basis of the streamer and return-stroke mechanism of the lightning discharge.[23]

In a traveling wave, it is necessary that the energy be equally divided between the electromagnetic and the electrostatic fields. When the released bound charge has fully developed into a traveling wave, it will be found that the voltage of the traveling wave is half of that induced, since the electromagnetic field is developed at the expense of the electrostatic.

In general, experience seems to show, in light of the available data, that lines insulated for a system voltage of more than 69 kV will not be troubled much by induced voltages, but as the insulation becomes less with lower voltage circuits, the voltages induced by lightning are of greater importance.

Direct measurements of induced voltages on circuits are not available; but it is known that with a negative cloud base the polarity of the traveling wave, due to the release of the bound charge, would be positive. Some indication of the magnitude of voltages induced in this manner may be obtained from the results of field studies of currents through distribution[24] and station[25,26] type lightning arresters.

5. Current Amplitudes. The most comprehensive data have been collected from measurements with the magnetic link[27] arranged to read current in tower legs.[29,30] In some cases current was measured in all four legs. In other cases, measurements were made in one leg and extrapolated. The stroke currents were obtained from adding together currents in all towers which seemed to be involved in a particular stroke. It is to be noted that 50% of the stroke currents are in excess of 10,000 A (see Table 26-1). However, only 10% of the tower currents are in excess of 32,000 amp. About 90% of the currents recorded were of negative polarity.

Experience curves are given showing lightning currents through distribution[24] (Fig. 26-5) and station lightning arresters[25,26] (see Fig. 26-6). The original data show, among other characteristics, a much greater percentage of positive records for distribution circuits than were recorded on transmission lines. It seems likely that many of these are the result of the release of bound charges from negative clouds.

Table 26-1. Crest Values of Current Peaks

Per cent of cases exceeding value given in succeeding columns	Direct strokes, ka		Transmission-line towers, ka	
	Empire[28] State Bldg.	McCann[15]	Lewis[29] and Foust	Waldorf[30]
50	10	5.5	9.5	11
20	18	15	25	23
10	26	25	34	32
Maximum	58	160	132	114

Walter[31] suggested in 1905 that the continuing illumination observed on moving-camera photographs of multiple lightning strokes was a continuing current. It remained for McEachron[11] to show in 1939 that the continuing illumination did represent current, through the use of rotating lens or film cameras of the Boys[32] type in conjunction with an oscillograph capable of recording over 1 s of time. Broadly speaking, the lightning stroke consists of one or more current peaks superimposed on a more or less continuous current flow. In one case[11] a record was obtained of a continuous discharge of 250 A lasting for more than 0.25 s, and having the characteristics of a d-c arc being fed continuously from the cloud. One high-current peak of 13,000 A was recorded, with a reversal to positive polarity occurring at almost the end of the discharge.

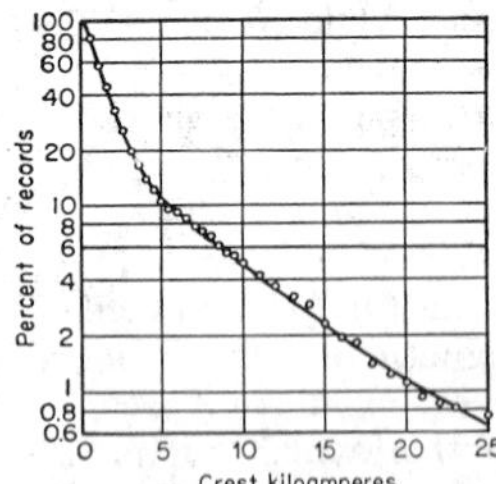

Fig. 26-5. Lightning currents through distribution arresters.[24] Of 1,608 records obtained, 1,011 were negative and 597 were positive.

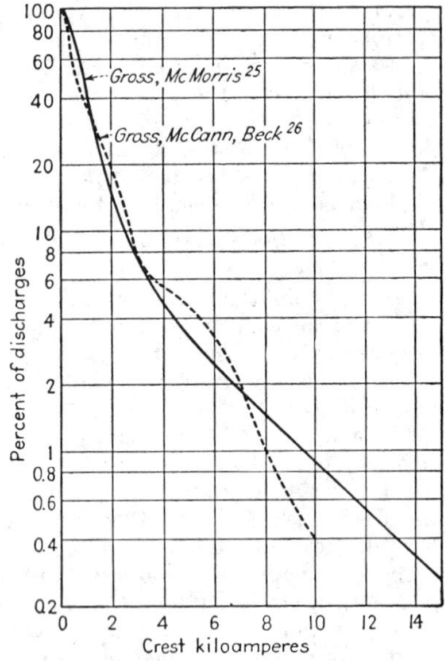

Fig. 26-6. Lightning currents through station arresters.

It is important to recognize the existence of the continuing type of discharge, since it is no doubt responsible for burning of cable sheaths and conductors, and it will play some part in the blowing of fuses on distribution circuits.[33] It may also have a bearing on the length of time it is necessary to keep the circuit deenergized when using automatic reclosing.[34] The duration of many strokes, including the total time of all discharges in the same path, has been determined by camera, oscillograph, fulchronograph, or photorecorder.[28,15] The maximum time recorded was 1.5 s (see Table 26-3).

In a number of cases the total stroke current as it flowed through a single path has been measured by means of oscillograph or fulchronograph. The maximum current

Table 26-2. Waveshapes of Current Peaks

Per cent of cases exceeding value given in succeeding columns	Rate of rise, ka per μsec			Direct strokes			
	Transmission-line towers, Berger[36]	Direct strokes		Front, μsec to crest		Tail, μsec to half value	
		Empire[28] State Bldg.	McCann[15]	Empire State Bldg.	McCann	Empire State Bldg.	McCann
50	11	12.5	9	1	2.5	34	42
20	21.5	26	17	2.7	4.2	68	58
10	31	33	35	5.1	5.5	97	70
Maximum		50	45	8.3*	10.0	120	90

* Maximum normally encountered—two unusual peaks exceeded this value.

recorded was 160,000 A. Fifty percent of the records were in excess of 7,000 A (see Table 26-1). The majority of all direct-stroke current peaks measured were of negative polarity.

6. Current Waveshapes. Data are available showing waveshapes of lightning currents measured directly. The wavefronts, wave tail, and rate of rise obtained by various investigators[15,28] with oscillographs, fulchronographs, and other devices are shown in Tables 26-1 and 26-2. Data obtained by other investigators[17,35,36,37] give results falling within the limits indicated in the tables. In 1938, McEachron[38] secured oscillograms of successive current peaks in strokes to the Empire State Building. In one case as many as 12 current peaks were measured in 0.28 s. The waveshapes of all were similar. Three of them reached a crest current of 5,000 A. In another case two peaks, 89,200 μs apart, were secured and show considerable similarity in waveshape (see Fig. 26-7).

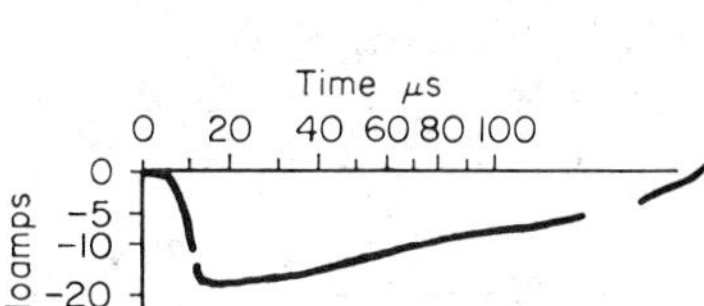

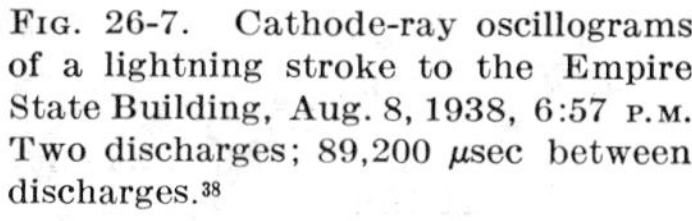

FIG. 26-7. Cathode-ray oscillograms of a lightning stroke to the Empire State Building, Aug. 8, 1938, 6:57 P.M. Two discharges; 89,200 μsec between discharges.[38]

It cannot be said that lightning has a particular waveshape; it can be expressed only between limits, and in terms of the frequency of occurrence, as indicated in the curves and tables.

A more recent lightning investigation[79] showed that the average lightning-stroke current was 35 to 40 kA. Berger's results on Monte Salvatore near Lugano[80] indicate occasional current peaks of positive polarity and as high as 180 kA.

The American Institute of Electrical Engineers has produced three bibliographies[81] on lightning and related phenomena covering the years 1918 to 1960, providing a rather complete coverage technically as well as historically.

Rate of Rise. It should be pointed out that rates indicated in Table 26-2 apply only at the stricken point. The wavefront on a transmission line will be only half the rates indicated, unless the stroke occurs at the very end of the line, and these will decrease rapidly with travel.

7. Effects on Objects Struck. Nonconductors are often shattered by lightning, while conductors may be considerably burned or eliminated entirely. The high current peak may shatter trees or poles without setting fire to them, while a succession of current peaks with continuing current may well cause a fire. High-current short-time discharges passed through a No. 14 rubber-covered wire can eliminate the wire but leave the rubber apparently undamaged. Such cases have been reported from the field, and similar results obtained in the laboratory.[39,40] Hollow or flat conductors are often crushed by high lightning currents owing to both heating and magnetic effects.

The pressure effects developed inside chimneys or inside dwellings as a result of the pressure in the spark often blow such structures apart violently. The effects upon apparatus will not be discussed in this section. It seems likely, in view of the rather large number of coulombs found in lightning discharges, that some conductor burning thought to have been due to power current may be the result of lightning itself.

8. Charge. The charge IT in the current peaks of a stroke should be considered separately from the total charge in a stroke, since it often happens that the continuing low-current portion will have a relatively large charge on account of its long duration (see Table 26-3).

9. Voltage Waveshape on Lines. Waveshape data are available for transmission lines only. They were obtained with cathode-ray oscillographs coupled to the overhead conductors. In most cases the location of the stroke was not known, and the waveshapes measured have been the waveshapes of traveling waves, and as such have no doubt undergone considerable change due to travel. These waveshapes are, however,

Table 26-3. Lightning Stroke Characteristics

Per cent of cases exceeding value given in succeeding columns	Total stroke duration, sec		Total stroke charge, coulombs		Charge in individual current peaks, coulombs,* Empire State Bldg.
	Empire[28] State Bldg.	McCann[15]	Empire State Bldg.	McCann	
50	0.27	0.08	19	9	0.15
20	0.44	0.3	56	39	0.70
10	0.55	0.5	81	88	1.36
Maximum	1.5	1.3	164	100	4.9

* Coulomb values based on time to half value.

typical of those that might reach apparatus to be protected. They are voltage waveshapes, but the time to crest and the time to half value are the same as for corresponding current waveshapes. As a matter of fact, the voltage waveshape is important from the insulation point of view, while the current waveshape is important in deter-

Table 26-4. Voltage-waveshape Characteristics

(A) Measured at end of transmission line

43 cases	Front	44 cases	Tail
%	Reached crest in less than stated μsec	%	Reached half value in less than stated μsec
86	20	86	50
75	10	75	40
51	5	20	20
Times to crest varied from (0–0.2) to (80–90) μsec		Times to half value varied from (3–4) to (150–200) μsec	

(B) Measured between ends of transmission line

17 cases	Front	17 cases	Tail
%	Reached crest in less than stated μsec	%	Reached half value in less than stated μsec
94	20	83	30
82	10	65	10
71	5		

mining the performance of protective devices. In a traveling wave on a transmission line, they are related through the surge impedance, so, in general, knowing either one is sufficient to determine at least approximately the other. It is not definitely known, however, how to translate the current measured in a direct stroke into current which would be flowing in transmission-line conductors away from a stricken point. Fortescue[41] and Bewley[42] have used the direct stroke as though it acted as a vertical conductor, using stroke surge impedances of 200 and 400 Ω, respectively. On the other hand, Lewis and Foust have assumed that the numerical sum of the currents in the various conducting paths should equal the current in the stroke as measured at the top of the tower. It is quite certain, as a result of the discovery of the leader- and return-stroke mechanism, that the stroke channel cannot be considered strictly as a conductor connected to an impulse generator which is the cloud. However, considering the lack of great accuracy in the fundamental data concerning lightning itself, either method may be used, since the result obtained by either method is well within the accuracy of the data available. The only known voltage oscillogram taken close to a direct stroke was obtained on a 220-kV system.[43]

Waveshapes change with travel, in general, the fronts becoming less steep and the tails longer. The crest value also decreases with travel. Data relating to change of waveshape with travel are given in papers by various investigators.[44–49]

Investigators agree that corona plays an important role in changing waveshapes with travel. Additional changes not explained by corona are explained by Bewley's[50] multivelocity theory, which is based on the varying depths of penetration of ground current with earth resistance and the increased effective diameter of the conductors due to corona.

A replot of oscillograms is shown in Fig. 26-8 to show the change of waveshape with travel for positive and negative surges both full wave and chopped. The effect of travel on the wave-crest potential is given in Fig. 26-9. It should be noted from these results that the chopped waves attenuate more rapidly than do the full waves. In Bewley's multivelocity theory this results from a lesser depth of penetration in the earth, as well as the separation into multivelocity waves.

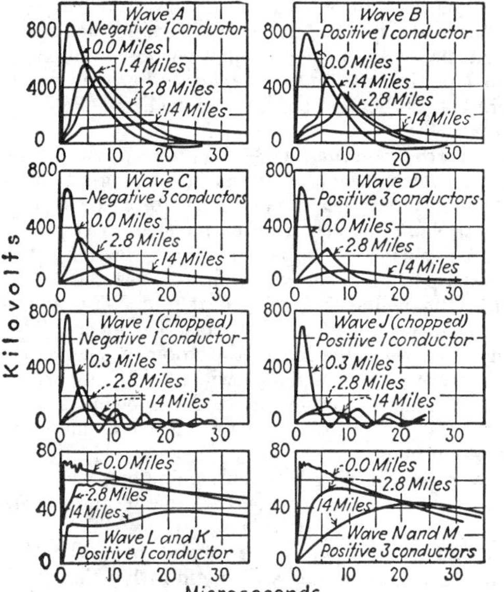

FIG. 26-8. Replot of oscillograms to show change of waveshape with travel.[44]

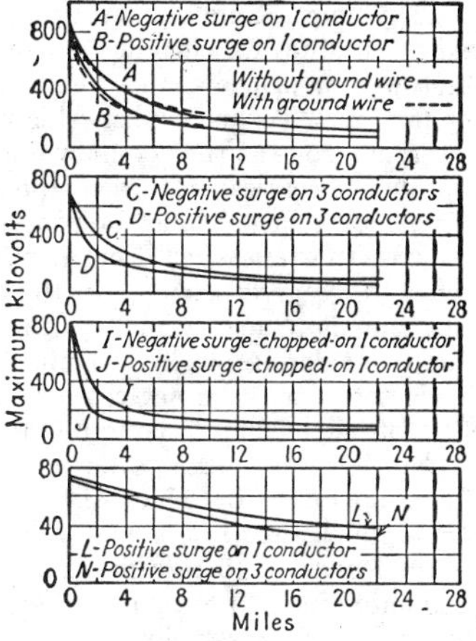

FIG. 26-9. Change of crest of voltage wave with wave travel.[44]

For calculations of attenuation the Foust-Menger[47] formula is ordinarily used, where

$$e = \frac{e_0}{kse_0 + 1} \tag{26-2}$$

and

$$A = -ke^2 \tag{26-3}$$

where e_0 = initial surge voltage at point of inception; k = an empirical proportionality factor; s = distance, in miles, that wave had traveled; e = voltage, in kilovolts, at distance s; A = attenuation, in kilovolts per mile, at any point where surge voltage is e.

According to Bewley and Rudge,[51]

$$\begin{aligned} k &= \text{(approx.) } 0.0006 \text{ for chopped waves } (2\ \mu s) \\ &= \text{(approx.) } 0.0003 \text{ for short waves } (5\ \mu s) \\ &= \text{(approx.) } 0.00016 \text{ for long waves } (20\ \mu s) \end{aligned}$$

More recent investigations, obtaining more detailed oscillograms on the front of the waves, indicate that corona is the main influence affecting the attenuation by shifting the wavefront. This shift, interacting with the tail of the wave, increases attenuation the shorter the wave tail.[82] Tests with steep-fronted waves but fixed tails indicate similar phenomena for conductors up to 2 in in diameter and 1,800 kV.[83] In general, attenuation is beneficial.

10. Voltage-waveshape Characteristics.[7] See Table 26-4.

11. Typical Voltage-waveshape Characteristics.[52] The typical waveshape measured on transmission lines for types of waves classified as oscillatory was 9 μs to crest and 18 μs to half value (15 cases); for unidirectional waves, 3.5 μs to crest and 15 μs to half value (33 cases).

PART II. THE PROTECTION OF LINES

12. General. The protection of transmission lines naturally divides into two general methods. The first seeks to keep the stroke off the line conductors; and the second allows the line conductor to be struck but prevents the power follow current from creating an interruption to service. The first method makes use of overhead ground wires, while the second uses such devices as protector tubes, ground-fault neutralizers (Petersen coils), or automatic reclosing breakers.

13. The principle of operation of the **overhead ground wire** is to intercept the stroke and to conduct its current to ground without sufficient potential developing either at the tower or in the span to cause a flashover between the ground wire or tower, and conductors. To accomplish this result a sufficient clearance at midspan is required, so that a stroke at that point will not develop flashover potential before reflections from adjacent towers can return to the stricken point and relieve the stress. At the tower, sufficient insulation must be provided so that voltages developed at that point, as the result of the resistance drop at the base of the tower, will not cause flashover. This means, for a given current in the lightning stroke, the higher the tower-footing resistance the more insulation required. Tower-footing resistance will also be important when considering strokes to midspan. For usual tower heights the impedance drop in the tower may be neglected. In the case of high towers, as at river crossings, the effect of height may have to be taken into account.

Another factor also to be considered is the coupling between conductors, which, in effect, operates to reduce the potential between the stricken conductor and others within its electrostatic field.

14. Physical Requirements. Overhead ground wires are placed over the line conductors in such a manner that lightning makes contact with them rather than with the line conductors. From the point of view of lightning, the ground wires can be made of any suitable material such as steel, copper, aluminum, or copper-covered steel. The conductor size will, in general, be determined by mechanical considerations—if too small it might be burned seriously by the lightning current. The largest conductor definitely known to have been burned completely through, according to the author's records, is a No. 4 solid copper conductor. It is probable that in most cases the overhead ground wire should not be smaller than No. 1/0. When considering the design of overhead ground wires, it is to be remembered that more clearance between the ground wires and conductors is required at midspan than at the point of support.

15. Location and Effect of Overhead Ground Wires. Overhead ground wires should be placed so as to intercept any lightning strokes that otherwise might reach the line conductors. The basic concept for the application of these ground wires has usually been the "cone-of-protection" idea. This idea states that any grounded object throws a protective "shadow" over any object below it and that lightning will not usually enter this shadow zone. Laboratory tests[57–60] and field experience indicate that, for the case of a ground wire, this protective shadow zone is not a cone, but a wedge or tent with the ground wire as the apex and a horizontal dimension of about 2 for each vertical dimension of 1. However, improvement in protection can be obtained if a 1-to-1 wedge, as has been substantiated in practice, is used. If two overhead ground wires are used, they are placed so that no other conductor is outside the overlapping wedges.

It is often possible to locate the ground wires substantially over the outermost line conductors, thus improving the shielding. The increased cost must be balanced against improved expected performance.

A relationship to approximate the number of strokes expected to contact typical transmission lines per mile per year, as determined by the protective shadow created by either one or two ground wires, is as follows,

$$S = \frac{KI[4h_t - 2.67(h_t - h_g) + b]}{5{,}280} \tag{26-4}$$

where S = number of strokes to line per mile per year, K = a constant (0.2 to 0.5), I = isokeraunic level in thunderstorm days per year, b = distance between adjacent ground wires in feet, ($b = 0$ for one ground wire only), h_t = height of ground wires at tower in feet, h_g = height of ground wires above earth at midspan in feet.

The constant K represents the number of strokes per square mile per year per unit isokeraunic level. It has often been assumed that this constant is approximately 0.5, but recent operating experience on completely unshielded lines indicates a value of 0.26 as more realistic.

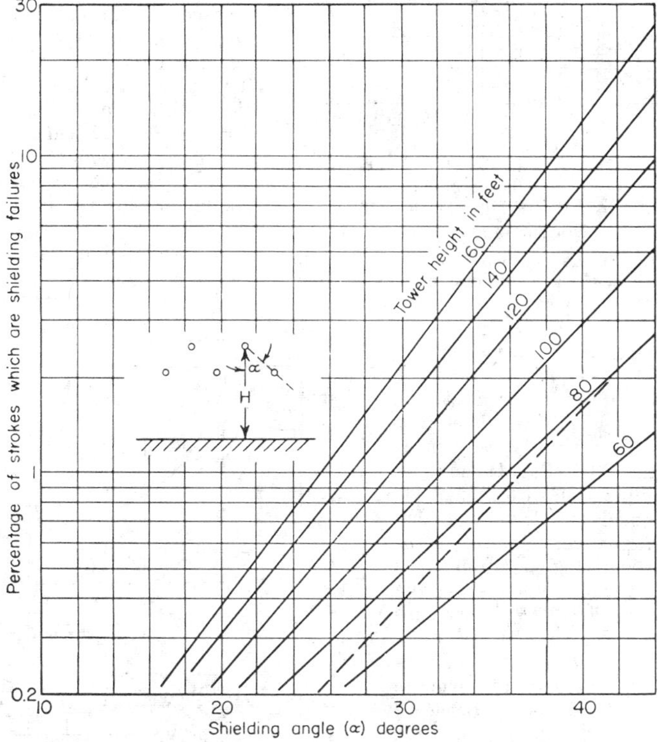

Fig. 26-10. Transmission-line shielding failure experience. [*From Kostenko*[87] *(solid lines) and Burgsdorf*[88] *(dotted line).*]

For strokes from Eq. (26-4) calculated to strike the line, there is a certain probability that any one of them can bypass the ground wire and strike one of the phase conductors directly. This condition is called a shielding failure. The probability of a shielding failure has been calculated from theoretical conditions,[85,86] but greatly simplifying assumptions have to be used. Two authors[87,88] have reported the actual transmission-line shielding failure experience of Fig. 26-10. This figure shows that, as tower heights are increased, the probability of shielding failure increases, and smaller shield angles must be used. A shield angle of 20° or less is usually very effective.

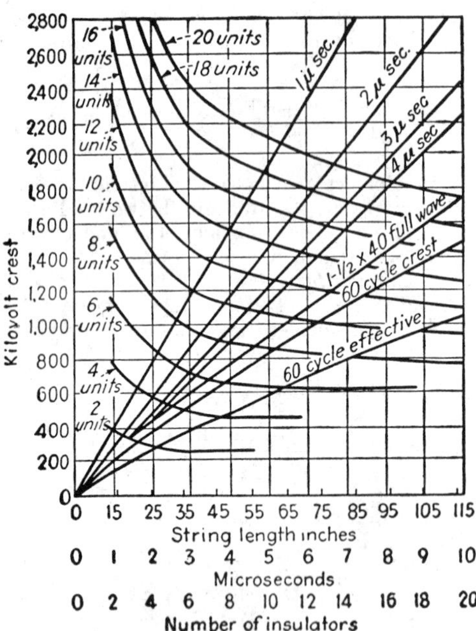

FIG. 26-11. Flashover characteristics of 10-in.-diameter suspension insulators having a 5¾-in. spacing. Positive polarity, 1.5 × 40 wave.

16. Insulator and Gap Flashovers. In designing an overhead ground-wire system, it is necessary to use flashover data for suspension insulators and rod gaps.[62,63]

The curves in Figs. 26-11 and 26-12 are tentative average values from several laboratories (unpublished material of EEI-NEMA Subcommittee on Correlation of Laboratory Data) and are not standard values. (The values are for 0.6085 in Hg vapor pressure, 25°C temperature, 76 cm Hg barometric pressure.) The 60-c and 1.5 × 40 wave values are correct to within ±8%. The flashover voltages for times of less than 6 μs are subject to large variations depending on the exact waveshape.[64] The figures can be used in connection with the estimating charts given in Pars. **17** and **20.** The rod-gap flashover values represent the voltage required to flash between conductors for the indicated spacing. Tests[64] have extended insulator and rod-gap flashover data to 240 in of spacing.

Positive flashover values are given because they are usually lower than the negative values. Thus used they will result in a margin of safety, since most impulses on transmission lines are negative. For other types of insulators, Table 26-5 (Par. **19**) may be used in conjunction with Fig. 26-11.

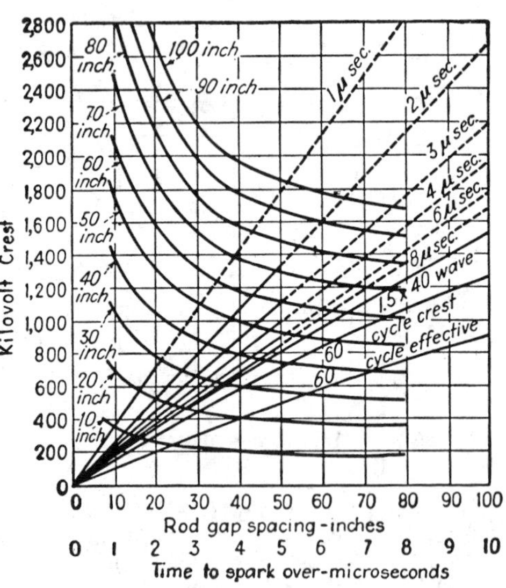

FIG. 26-12. Spark-over characteristics of rod gaps. Positive polarity, 1.5 × 40 wave.

17. Methods of Calculating Lightning Performance. Several methods of calculating lightning performance of transmission lines have been devised.[61,89,90,67,91] The subject is complicated and not amenable to complete handbook development. One very approximate method was developed by Bewley,[61] in which he prepared an estimating chart based on the relation

$$I = \frac{V}{(1 - C_n)\alpha R'} \qquad (26\text{-}5)$$

where I = lightning current that the stroke could deliver to a zero-resistance ground (which is equal to twice the traveling-wave current in the stroke; the stroke is assumed to have a surge impedance of 400 Ω); V = voltage across the insulation (insulators or insulators plus wood); C_n = coupling factor (depends upon the number, configuration, and spacing of the conductors and is modified by corona); α = crest factor (depends upon the length of spans or the distance between protector tubes, tower-footing resistances, lightning wavefront, and conductor surge impedance); R' = effective resistance (depends upon the grounding resistance and the surge impedance of the lightning stroke and of the stricken conductors).

The chart (Fig. 26-13) is based on the assumption of a wave rising at a uniform rate to crest in 2 μs, on a crest factor α which takes into account reflections from adjacent towers, and on the time to flashover t, based on average tower-footing resistances. The effect of system voltages has been neglected.

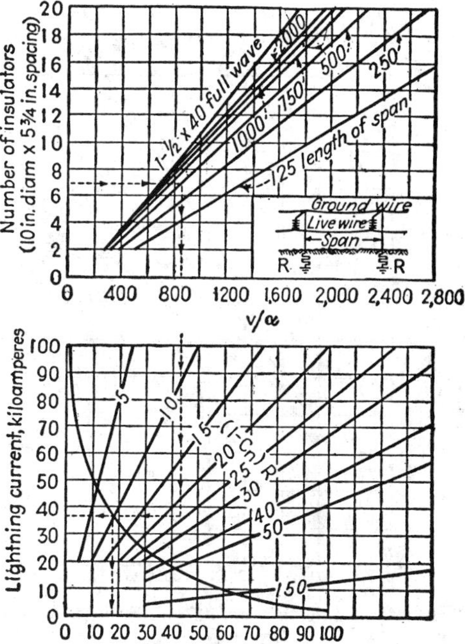

FIG. 26-13. Chart for estimating percentage of insulator flashovers, based on a 2-μsec wavefront.[61]

The upper part of the chart is a plot of V/α (horizontal scale), with the length of span as parameter. A value of R = 20 was assumed in taking α and t as functions of span length only. The lower portion of the chart contains the Lewis and Foust lightning-current experience curve and a sheaf of straight lines representing the equation

$$\frac{V}{\alpha} = (1 - C_n)R'I$$

with $(1 - C_n)R'$ as parameter.

The use of the chart, which was prepared to include strokes within one-fourth span, may be illustrated by an example, whose solution is shown by the dotted line. A line has seven insulator units per string, two ground wires at an elevation of 60 ft, average clearance of 15 ft, and 500-ft spans with tower-footing resistances of 60 Ω. What percentage flashovers can be expected at or close to the towers?

It is first necessary to find a value for $(1 - C_n)R'$, which is obtained from Figs. 26-14 and 26-15. Using R = 60 and two ground wires, R' from Fig. 26-14 equals 39. The configuration of overhead ground wires and line conductors is such that $(b/\alpha)_{\text{mean}}$ (see Fig. 26-15) is 7.7 and C_n = 0.39. Thus $(1 - C_n)R'$ = 23.8.

Referring now to Fig. 26-13, follow horizontally the seven-disk line until it intersects the spacing line of 500 ft, thence downward to the $(1 - C_n)R'$ = 23.8 line, and read horizontally a current of 37,000 A, which corresponds, from the Lewis and Foust experience curve, to 18% flashovers.

If it is desired to reduce the flashovers to 5%, this may be done by increasing insulation or decreasing the tower-footing resistance; as a rule, the latter is preferable. From the lower part of Fig. 26-13 it will be noted that 67,000 A lightning current corresponds to 5% flashovers, and if the tower-footing resistance is kept at the original 60 Ω so that $(1 - C_n)R'$ = 23.8, it will be found that V/α is 1,540 and 13 disks will be required. On the other hand, if the line insulation is not increased, R must be decreased so that

$(1 - C_n)R' = 0.61R' = 12.8$, from which $R' = 21$; from Fig. 26-14, R is found to be 25 Ω. If wood arms are used on a steel tower, or if wood poles are used, the equivalent number of insulators may be determined,[65] or, in case the flashover value from the conductor to the tower or down wire is controlling, this is to be converted into terms of insulator units.

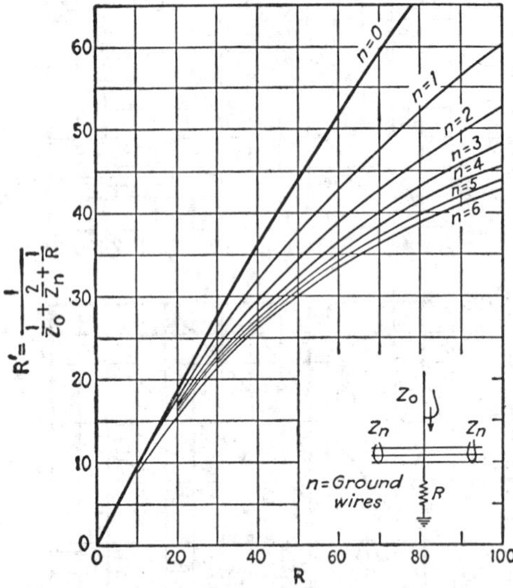

FIG. 26-14. Effective resistance at the stricken point.[61]

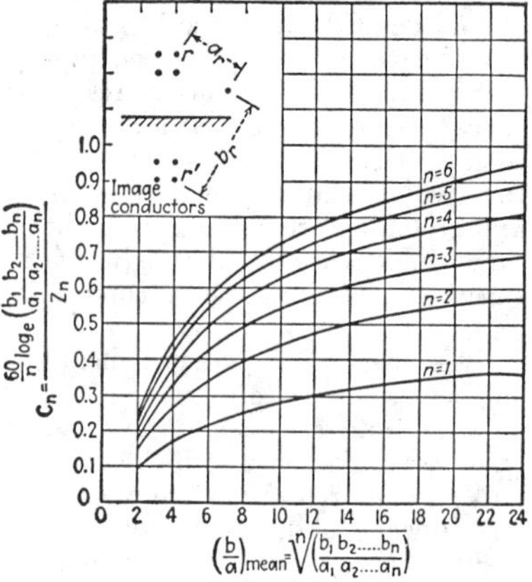

FIG. 26-15. Coupling factor between a group of conductors and a single isolated conductor.[61]

If lightning should fail to be intercepted by the ground wires and should make contact with a line conductor, the lightning current to cause flashover may be obtained by the relation E/R', where E is the flashover voltage and R' the effective resistance, which, for the condition assumed, will be equal to 154 Ω,

$$R' = \left(\frac{1}{Z_o} + \frac{2}{Z_n} + \frac{1}{R}\right)^{-1} = \left(\frac{1}{400} + \frac{2}{500} + \frac{1}{\infty}\right)^{-1} = 154$$

Z_o being the stroke surge impedance, and Z_n being the conductor surge impedance as in Fig. 26-14.

Thus, if the flashover of seven insulator units is 1,300 kV at a time of 2 μs, a current in the stroke of 8,400 A in 2 μs would cause a flashover. After flashover, the tower-footing resistance and also the overhead ground wire would be effective. In this case, determination of the current required to cause a second conductor to flash over could be obtained approximately by the use of the estimating chart of Fig. 26-13, on assuming the stricken conductor to act as an additional ground wire. This would not take account of the fact that negative reflections would not be available from adjacent towers for the stricken conductor, although they would be available, of course, for the ground wires.

Another approach to calculating lightning performance of transmission lines is presented in a 1950 AIEE Committee Report.[67] This report also takes into account the inductance of the tower, which may have an appreciable effect on insulator flashover, particularly for the higher rates of rise of lightning strokes or for tall towers. Although this method is more inclusive than the approximate method developed by Bewley, it has failed to predict outage rates on tall-tower 345-kV lines by over an order of magnitude.[79] This failure may have been due to improper considerations of probability of shielding failure.

A modification of this 1950 AIEE method was developed by Clayton and Young,[92] and the generation of lightning voltages explored in some detail by Wagner and Hileman.[93] A "Monte Carlo" digital-computer method of computing lightning performance was devised by Anderson,[91] and has been used extensively to calculate performance of extra-high-voltage (EHV) lines in the United States. The observed lightning performance of transmission lines varies over a wide range. For EHV lines, outage rates of 0.2 to 2 tripouts/(100 mi)(year) are common, and tall-tower single-ground-wire lines have experienced rates as high as 8 tripouts/(100 mi)(year).

18. Strokes to Midspan. The performance of the overhead ground wire can be estimated roughly for strokes to midspan, midspan being considered as covering a quarter-span length on either side of the middle of the span. The flashover voltage between conductors is taken as being equal to the flashover of a rod gap of the same spacing. The reduction of voltage is accomplished by returning reflections from adjacent towers, and it becomes a race between the rising potential at the stricken point and the arrival of these reflections.

However, a simplified approach is useful in preliminary design procedures, and using this simplified approach, Bewley has prepared a convenient estimating chart, which is reproduced as Fig. 26-18 together with Figs. 26-16 and 26-17. Quoting from Bewley,[61] "Equation (26-5) (Par. **17**) is solved in Fig. 26-18 by the intersection of two sets of curves:

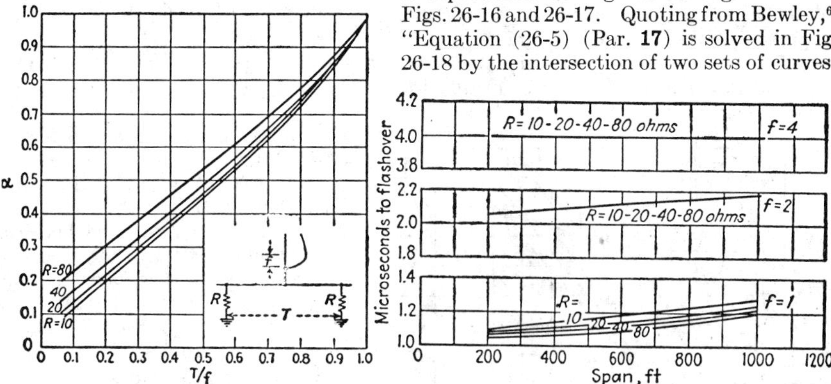

FIG. 26-16. Crest factor at midspan for a stroke at midspan.[61]

FIG. 26-17. Time to flashover at midspan for a stroke to midspan.[61]

the flashover voltages V (as function of the midspan clearance s and the time to flash-

Table 26-5. Tabulation of String Lengths for Suspension Insulators

No. of units	4¾-in. spacing		5¼-in. spacing		5¾-in. spacing
	String length	Equiv. No. 5¾-in.	String length	Equiv. No. 5¾-in.	String length
1	8	1	8	1	8
2	12¾	1⅞	13⅛	2	13¾
3	17½	2¾	18¼	2⅞	19½
4	22¼	3½	23⅜	3¾	25¼
5	27	4⅜	28½	4⅝	31
6	31¾	5¼	33⅝	5½	36¾
7	36½	6	38¾	6⅜	42½
8	41¼	6⅞	43⅞	7¼	48¼
9	46	7¾	49	8⅛	54
10	50¾	8½	54⅛	9	59¾
11	55½	9⅜	59¼	10	65½
12	60¼	10⅛	64⅜	10⅞	71¼
13	65	11	69½	11¾	77
14	69¾	11¾	74⅝	12⅝	82¾
15	74½	12⅝	79¾	13½	88½
16	79¼	13½	84⅞	14⅜	94¼
17	84	14¼	90	15¼	100
18	88¾	15	95⅛	16⅛	105¾
19	93½	15⅞	100¼	17	111½
20	98¼	16¾	105⅜	18	117¼

over t), and the curves representing the value of V in terms of the other variables from Eq. (26-5). The use of Fig. 26-18 is as follows: (*a*) find α from Fig. 26-16, and t from Fig. 26-17; (*b*) for a given midspan clearance s find in Fig. 26-18 the vertical intersection with the appropriate t curve, and interpolate the value of I; (*c*) for these values of αI and α find the lightning current I and the corresponding percent flashovers from the inset curves in Fig. 26-18. This procedure can be reversed, of course, to determine the proper midspan clearance to prevent the flashovers from exceeding a specified number."[61]

As an example, assume a ground resistance of 40 Ω, a clearance s of 15 ft, and a span length T = 800 ft. Assume also a wavefront f of 1 μs or 1,000 ft. T/f = 0.8; from Fig. 26-16, α = 0.76; and from Fig. 26-17, t = 1.17. From Fig. 26-18, αI = 55,000, and I = 72,000 A, with the expectancy equal to 3%. If the span length is decreased to 500 ft, I becomes 110,000 A, with expectancy about 1%. On the other hand, if the span length is kept at 800 ft and the wavefront is increased to 2 μs or 2,000 ft, a current of about 110,000 A is required to cause flashover. These results indicate that, if wavefronts are of the order of 2 μs or longer, spacings much in excess of 15 ft will not be required unless spans are comparatively long.

Wagner and Hileman[94] have pointed out that, where the action of corona discharges is included, this concept must be modified because of changes both in waveshape and time of flashover.

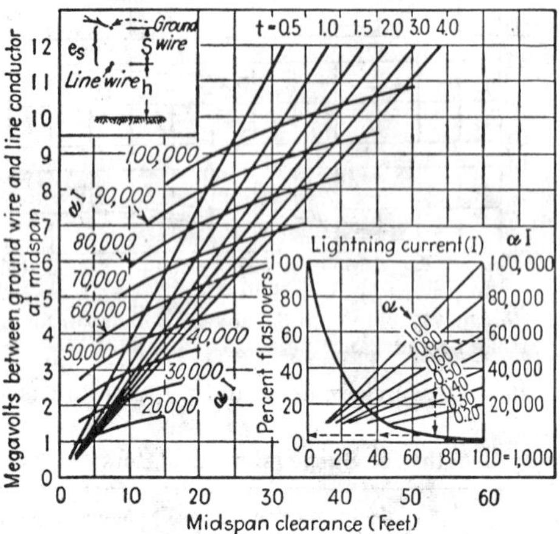

FIG. 26-18. Chart for estimating midspan flashovers.[61]

Experience indicates that midspan flashover is a very uncommon occurrence. Wagner and Hileman[94] have suggested that the heavy discharge of corona currents between a stricken ground wire and a nearby phase conductor greatly reduces the overvoltage between the two conductors during the initiation of the breakdown process, thereby increasing the time required for complete breakdown to occur to several microseconds (Fig. 26-12). This increase in breakdown time becomes longer than the time required for reflections to arrive from nearby towers to neutralize most of the voltage across the gap. The result is that many strokes that would produce midspan flashover according to classical computations are actually prevented from doing so. Laboratory tests[95] support this idea. In any case, a midspan flashover to a phase conductor is likely to create a surge voltage on the phase conductor of sufficient magnitude to flash over the insulator to ground on the nearest tower. If the power-frequency follow current flows through this new path across the insulator, it can cause insulator burns and suppress any arcs in midspan.

19. Tabulation of String Lengths for Suspension Insulators. See Table 26-5.

20. Tower Grounding. Two methods are in general use for reducing tower-footing resistances. One is through the use of **driven grounds**, and the other through the use of the buried **counterpoise.** Frequently, to get the desired lightning performance, it is necessary to reduce the resistance to 10 Ω or less.

While calculations[68] may be made of ground resistance under assumed conditions, it is common practice to determine the adequacy through the use of measurements. Where grounding conditions are particularly poor, very long ground rods may be driven[69,70] with apparently quite satisfactory results, provided, of course, that the soil is of such a character that long rods can be driven.

The **counterpoise** is of two types: (*a*) radial, (*b*) continuous. With either type, the conductor is buried a depth of 1 to 3 ft and may be connected solidly to each tower or through a small gap.[71] The conductor material of the counterpoise is not important, except that it should be mechanically strong and resistant to corrosion. The effect of conductor size is not known, but experience seems to indicate that a size which is suitable mechanically will be sufficient from the point of view of conductivity. Copper conductors suffer the disadvantage of being likely to be stolen.

It has been shown theoretically and by tests[72–74] that the counterpoise has an initial transient surge impedance of 150 to 200 Ω which decays to its final value of resistance, the transition time being the time required for the first reflection to return, traveling at one-third of the speed of light. Bewley[61] has prepared charts showing how the counterpoise length may be calculated when of the radial type. It is the usual practice to bury the radial counterpoise parallel with the transmission line conductors and within the right of way. Such a counterpoise may add something less than 10% to the overhead ground wire coupling.

Bewley points out that, for a 3-μs wavefront and 20 Ω/1,000 ft, the counterpoise length should be at least 600 ft and the impedance 33 Ω. If the length is reduced to 400 ft, the counterpoise impedance is increased to 50 Ω. For wavefronts as short as 1 μs, the counterpoise need not be longer than 200 to 300 ft. Bewley concludes, "In any event, it is more effective to utilize a given length of wire as several radial counterpoises 200 ft long than as a long continuous counterpoise."

Effectiveness of the radial counterpoise in reducing tower resistance and line outages has been demonstrated.[30]

For a comparison between the performance of the radial and continuous types on a 220-kV line, Bell[75] concludes that "buried counterpoise when used in conjunction with overhead ground wires provides multiple paths to ground for lightning, and distributes such current over long distances. Continuous counterpoises seem ideal, and a 2.6-mi section has had a perfect 10-year record. Noncontinuous or crowfoot cables have had an excellent but not quite perfect 9-year record in preventing insulator flashovers." During the 10-year period, the towers flashed per mile per year were zero in the continuous-counterpoise section, and 0.015/(mi)(year) in the radial counterpoise section. (These varied in length from 50 to 250 ft.)

The continuous counterpoise may have some advantage in keeping lightning trouble localized, especially when areas of low resistance are more or less insulated from other areas by rock or similar nonconducting material. One practical method of arriving at type and number of counterpoises to be installed is given by E. Hansson.[76]

Lewis and Foust[29] found that, with reference to currents measured in counterpoises and ground rods, particularly in territory in which a considerable depth of sand lies on the surface, driven rods offer a very effective means of reducing tower resistance. The rods conducted about 80% of the current, and the counterpoise lying in the high-resistance sand about 20%.

21. Wood-pole Lines. Wood[65] may be used to augment the lightning strength of porcelain supports. Wood is not suitable without porcelain in series with it. The impulse strength of wood depends upon its moisture content, and, according to Lusignan and Miller,[77] reducing the moisture content from an average greenwood value of about 35% to an average moisture equilibrium value of about 15% may nearly double the insulation strength of creosoted pine structures, and raise that of fir structures by about one-fourth. These authors also point out that, during a rain storm, structures using wood as supplementary insulation may suffer reduction to as low as 65% of dry flashover in the first 2 or 3 min of rain, without much further reduction on account of additional wetting.

Lusignan and Miller give several useful curves which may be used for estimation purposes. These data are based on laboratory tests.

When designing wood structures for lightning, it is usually important to adjust the flashover, either through gaps or otherwise, so that flashover does not take place over wood surfaces, thus subjecting them to possible damage.

From the lightning point of view, the design of a successful line is a problem of insulation and spacing of conductors, ground wires, and downwires to utilize the wood insulation to the greatest degree and yet meet the conditions already discussed in sections on protection.

An interesting use of the principle of diversion of lightning strokes by overhead ground wires, without allowing the lightning current to flow through the tower-footing resistance, is the 154-kV wood-pole line of the TVA in the Tennessee Valley described by Evans and Daniels.[78] This line is an H-frame line using wood spar arms with nine insulators. There are no grounded wires on either the spar arms or the poles. Lightning diverter wires on each side of the line are supported on separate poles set 14 ft from the outer conductors. The diverter wires are 57 ft above the ground and approximately 15 ft above the line conductors. According to Evans and Daniels, this H-frame line, which is 96 mi in length, has had a perfect operating record over three seasons. With such a construction, it is necessary only that the tower-footing resistance be low enough to prevent flashover between the down lead on the diverter poles and the line conductor—in this case a normal distance of about 13.5 ft.

22. Protector tubes are operated in parallel with the insulator they are intended to protect. The tubes are always installed with an external air gap. The tube itself consists of a fiber tube of suitable dimensions within which is arranged an air gap between electrodes. The impulse discharge voltage of the tube with its external gap must be less than the flashover of the insulator at all wavefronts.

Protector tubes were used widely in the 1930s.[81] Largely because of the rapid loss of life of the tubes and the great increase in and diversity of short-circuit current, which eroded the current-quenching fibers, protection has been changed to valve-type lightning arresters.

23. Ground-fault Neutralizer (Petersen Coil).* The ground-fault neutralizer is a reactance connected between the system neutral and ground. Its reactance is so proportioned with reference to the capacitance of the system that the lagging current of the neutralizer is equal in magnitude and opposite in phase to the charging current at the fault. Thus, the fault current becomes zero, and the fault is cleared.[81] While the neutralizers, when properly tuned, will prevent outages from line-to-ground faults, they have not become popular in the United States, primarily because of the rapid increase in interconnections within systems and between systems. This makes it most difficult to obtain proper tuning and therefore reduces the capability of preventing outages in which they excelled under simpler system configurations.

24. Automatic Reclosing. Successful automatic reclosing of circuit breakers following a lightning flashover requires that the line be deenergized long enough so that the restoration of system potential does not continue the arc at the point of original fault.

Recent field and laboratory investigations have extended knowledge of deionization time needed for successful reclosure up to 345-kV systems.[84] It was found that a dead time of 19 c would result in successful reclosures in about 92% of flashovers due to lightning. More information can be found elsewhere.[81]

25. Masts and Ground Wires over Stations. Either masts or overhead ground wires[51] may be used to prevent direct strokes of lightning from reaching conductors in stations. While one tall mast may be used, so located that all system conductors such as disconnecting switches are within the cone of protection, yet there is some evidence that a larger number of lesser height are likely to be more effective. This is partly because the ratio of base to height of the cone increases as the height becomes less, and also experience seems to show that the tall mast has its cone of protection violated somewhat more often than does a low mast.

If several masts are used, their intersecting cones will determine the area of protection. The heights and locations of the masts are varied until all the points requiring protection are within the cones. The masts are usually a part of the steel work and are grounded through the station steel to the station ground.

In some cases overhead ground wires are located over the station, their spacing and height being such as to give proper clearance to live conductors underneath and yet so arranged that the intersecting wedges of protection cover all live parts. The calculations for ground wires with strokes to midspan (Fig. 26-18, Par. **18**) are conveniently made, the clearance to nearest conducting point underneath being equal to the clearance indicated between ground wire and line conductor at midspan.

Frequently, it is possible to bring the overhead ground wire or wires from a trans-

* See also Sec. **13**.

mission line over the station. This method will give satisfactory shielding if the station is not too extensive. In that case additional ground wires or masts may be required.

If overhead ground-wire protection for lines is used, it is mutually advantageous to both line and station to have the overhead ground wires carried in and tied into the station ground. The chief advantages are (*a*) protection of the last span from direct strokes resulting in severe duty on station apparatus and (*b*) reduction of ground resistance through paralleling of ground connections.

PART III. NEED AND METHODS OF LIGHTNING PROTECTION

26. Protection of buildings and other similar structures is the subject of a booklet issued by the National Fire Protection Association as NFPA No. 78, 1965. Only a very brief summary is given here, including some of the more important aspects.

27. Casualties from Lightning. Approximately 400 persons are killed and 1,000 injured in the United States annually by lightning. Nine-tenths of these accidents occur in rural districts.

28. Considerations for Protection. Whether or not protection is justified will depend upon the value and nature of the building and its contents, the frequency of occurrence of lightning storms, and the degree of shielding offered by other structures as well as the availability of fire-fighting apparatus. In some states, a reduction in insurance rates can be obtained as a result of protecting buildings against lightning. The hazard to human beings is practically eliminated if they are within a properly protected structure.

29. Operation of lightning-rod system depends upon the principle of intercepting the lightning stroke before it reaches the structure to be protected, and discharging the lightning current to ground through a sufficiently low resistance so that dangerous voltages do not result on account of the *IR* drop. Tall trees cannot be depended upon to protect nearby structures; if provided with lightning rods, however, they may act as other similar tall grounded structures to protect nearby objects if they are within the cone of protection.

Protection of Ordinary Buildings and Miscellaneous Property

30. In placing a lightning-rod system on any structure, **air terminals** should be provided for the purpose of keeping the earth end of the lightning arc away from the building, and also to provide a cone of protection so that the building itself is completely protected. Air terminals should be spaced at intervals not exceeding 25 ft. NFPA No. 78, 1965, should be consulted for details.

31. Roof conductors and down conductors should be coursed in such a way as to join each air terminal to all the rest. At least two paths to ground should be provided. "Dead ends" should not exceed 16 ft.

32. Roof conductors and down conductors may be made of copper, copper-clad steel, galvanized steel, or a metal alloy which is as resistant to corrosion as copper. Copper cables should weigh not less than 3 oz/lin ft. Thus a round, solid conductor of at least ¼ in in diameter is required. If steel or galvanized-steel conductors are used, they must have a net weight of steel of not less than 320 lb/1,000 ft.

33. Metal-roofed buildings should be treated as any other building from the point of view of lightning rods and down conductors, unless the metal sheets of which the roof is composed are made electrically continuous by means of suitable bonding or interlocking. Such a roof should be provided with air terminals to receive the lightning arc and should be connected to the earth at two points, at least preferably on opposite corners.

34. Grounds are very important, connections being made to water pipes where available. In some cases it will be necessary to provide an extensive buried system of wires to obtain a suitably low ground resistance. The treatment of grounds depends of course upon the character of the soil and its conductivity.

35. Bonding and Grounding of Metal Bodies. Metal bodies extending through the structure to be protected or wholly exterior to the structure should be bonded to the nearest lightning conductor, and under some conditions may require an additional separate grounding. Metal bodies situated wholly within the structure, and those projecting through the sides of the building below the second floor, which come within 6 ft of a lightning conductor or a metal body connected thereto, should be bonded to the nearest lightning conductor and, if of considerable length, should be grounded at the lowest or farthest extremity.

36. Protection of Areas. In general, areas may be most satisfactorily protected by the use of masts with interconnected wires from their tops, all suitably grounded to an interconnected ground system. This arrangement gives a subdivision of the lightning currents down any one mast, and provides for further subdivisions of currents in the ground. The area to be protected should be wholly within a cone of protection which will have a base radius of approximately two times the height of the axis of the cone. For a more detailed discussion of the cone of protection reference should be made to NFPA No. 78.

37. Protection of Structures Containing Inflammable or Explosive Materials. In this connection extra precautions must be taken to prevent all sparks, no matter how minute, from occurring within the protected area. Since these sparks can result from induction, it is essential that all metal objects be bonded together and frequently connected to ground. In some cases, it may be desirable to build a complete Faraday cage, which of course represents the most certain form of protection. Even in completely enclosed steel tanks, however, if internal structural members come close to the roof and the space contains an explosive mixture, special precautions must be taken. In other cases where the areas are particularly large the use of overhead ground wires or grounded mesh may be used. Before undertaking the protection of such structures one should consult those experienced in this field to be sure that all the proper precautions are taken. Consult also Part III of NFPA No. 78.

BIBLIOGRAPHY

38. Literature on Lightning and Protection from Lightning

1. McEachron, K. B. *Trans. AIEE*, 1934, Vol. 53, p. 1633.
2. Simpson, G. C. *Trans. Roy. Soc. (London)*, 1909, Ser. A, Vol. 209, p. 379.
3. Wilson, C. T. R. *Jour. Franklin Inst.*, July, 1929, Vol. 208, p. 1.
4. Geitel, H. *Physik. Z.*, 1916, Vol. 17, p. 455.
5. Byers, H. R., and Braham, R. R. The Thunderstorm, *U.S. Dept. Commerce Publ.*, June, 1949.
6. Workman, E. J., and Reynolds, S. E. *Phys. Rev.*, 1950, Ser. 2, Vol. 78, p. 254.
7. Lewis, W. W., and Foust, C. M. *Gen. Elec. Rev.*, 1936, p. 543.
8. Grunewald, H. *Paper* 323, CIGRE, 1939.
9. Jensen, J. C. *Jour. Franklin Inst.*, 1933, Vol. 216, p. 707.
10. Simpson, G. C., and Scrase, F. J. *Proc. Royal Soc. (London)*, August, 1937, p. 309.
11. McEachron, K. B. *Jour. Franklin Inst.*, 1939, Vol. 227, p. 149.
12. Schonland, B. F. J., and Collens, H. *Proc. Roy. Soc. (London)*, 1934, Ser. A, No. 849, p. 654.
13. Schonland, B. F. J., Malan, D. J., and Collens, H. *Proc. Roy. Soc. (London)*, 1935, Ser. A, Vol. 152, No. A-877, p. 595.
14. Larson, A. *Ann. Rept. Smithsonian Inst.*, 1905, p. 119.
15. McCann, G. D. *Trans. AIEE*, 1944, Vol. 63, p. 1157.
16. Flowers, J. W. *Gen. Elec. Rev.*, 1944, Vol. 47, p. 9.
17. Goodlet, B. L. *Jour. IEEE*, 1937, p. 1; 1938, p. 209.
18. McEachron, K. B., and McMorris, W. A. *Gen. Elec. Rev.*, 1936, p. 487.
19. Wilson, C. T. R. *Trans. Roy. Soc. (London)*, 1920, Ser. A, Vol. 221, p. 73.
20. Peek, F. W., Jr. *Trans. AIEE*, 1931, Vol. 53, p. 1077.
21. Andrews, F. E. *Elec. World*, Apr. 25, 1931, p. 780.
22. Norinder, Harald *Jour. Franklin Inst.*, 1934, Vol. 218, p. 717.
23. Wagner, C. F., and McCann, G. D. *Trans. AIEE*, 1942, Vol. 61, p. 916.
24. McEachron, K. B., and McMorris, W. A. *Trans. AIEE*, 1938, Vol. 57, p. 307.
25. Gross, I. W., and McMorris, W. A. *Trans. AIEE*, 1940, p. 417.
26. Gross, I. W., McCann, G. D., and Beck, E. *Trans. AIEE*, 1942, Vol. 61, p. 266.
27. Foust, C. M., and Gardner, G. F. *Gen. Elec. Rev.*, 1934, Vol. 37, p. 324.

28. Hagenguth, J. H., and Anderson, J. G. *Trans. AIEE,* 1952, Vol. 71, Pt. III, p. 641.
29. Lewis, W. W., and Foust, C. M. *Elec. Eng.,* March, 1945, Vol. 64, p. 107.
30. Hansson, E., and Waldorf, S. K. *Trans. AIEE,* 1944, Vol. 63, p. 251.
31. Walter, B. *Ann. Physik,* 1905, Vol. 18, p. 863.
32. Hagenguth, J. H. *Gen. Elec. Rev.,* 1940, Vol. 43, pp. 195, 248.
33. Bergvall, R. C., and Beck, E. *Trans. AIEE,* 1940, Vol. 59, p. 442.
34. McEachron, K. B. *Trans. AIEE,* 1939, Vol. 58, p. 635.
35. Berger, K. *Assoc. Suisse Elec. Bull.,* 1936, Vol. 27, p. 145.
36. Norinder, H. *Jour. Franklin Inst.,* 1935, Vol. 220, No. 1, p. 69.
37. Sokolnikow, J. S. *Elektrichestvo,* 1937, Vol. 58, No. 2, p. 1; abstract, *Elektrotech. Z.,* Aug. 12, 1937, Vol. 58, p. 32.
38. McEachron, K. B. *Elec. World,* 1940, Vol. 113, No. 6, p. 428.
39. McEachron, K. B., and Thomason, J. L. *Gen. Elec. Rev.,* 1935, Vol. 38, No. 3, p. 126.
40. Bellaschi, P. L. *Trans. AIEE,* 1935, Vol. 54, p. 837.
41. Fortescue, C. L., Atherton, A. L., and Cox, J. H. *Trans. AIEE,* 1929, Vol. 48, p. 449.
42. Bewley, L. V. *Gen. Elec. Rev.,* 1933, Vol. 36, p. 516.
43. Bell, Edgar, and Price, A. L. *Trans. AIEE,* 1931, Vol. 50, p. 1101.
44. Brune, O., and Eaton, J. R. *Trans. AIEE,* 1931, Vol. 50, p. 1132.
45. Conwell, R. N., and Fortescue, C. L. *Trans. AIEE,* 1930, Vol. 49, p. 872.
46. Dowell, J. C. *Trans. AIEE,* 1931, Vol. 50, p. 551.
47. Lewis, W. W., and Foust, C. M. *Trans. AIEE,* 1930, Vol. 49, p. 917.
48. McEachron, K. B., Hemstreet, J. G., and Rudge, W. J. *Trans. AIEE,* 1930, Vol. 49, p. 885.
49. Wagner, C. F., Gross, I. W., and Lloyd, B. L. *Trans. AIEE,* 1954, Vol. 73, Pt. III, p. 196.
50. Bewley, L. V. *Trans. AIEE,* 1934, Vol. 53, p. 1749.
51. Bewley, L. V., and Rudge, W. J. *Gen. Elec. Rev.,* 1937, Vol. 40, p. 363.
52. Lewis, W. W. *Trans. AIEE,* 1930, Vol. 49, p. 1501.
53. Cox, J. H., and Beck, E. *Trans. AIEE,* 1930, Vol. 49, p. 857.
54. George, R. H., and Eaton, J. R. *Trans. AIEE,* 1930, Vol. 49, p. 877.
55. Smeloff, Nicholas N., and Price, A. L. *Trans. AIEE,* 1930, Vol. 49, p. 895.
56. Sporn, Philip, and Lloyd, W. L., Jr. *Trans. AIEE,* 1930, Vol. 49, p. 905.
57. Peek, F. W., Jr. "Dielectric Phenomena in High Voltage Engineering"; New York, McGraw-Hill Book Company, 1929.
58. Matthias, A., and Burkhardtsmaier, W. *Elektrotech. Z.,* 1939, pp. 681, 720.
59. Schwaiger, A., and Ziegler, H. *Mitt.* Rosenthal-Isolatoren, GmbH, 1939, No. 23.
60. Wagner, C. F., McCann, G. D., and MacLane, G. L. *Trans. AIEE,* 1941, Vol. 60, p. 313.
61. Bewley, L. V. *Gen. Elec. Rev.,* 1937, Vol. 40, pp. 180, 236.
Bewley, L. V. Book "Traveling Waves on Transmission Systems"; New York, John Wiley & Sons, Inc., 1933.
62. Dowell, J. C., and Foust, C. M. *Gen. Elec. Rev.,* 1937, Vol. 40, p. 141.
63. McAuley, P. H. *Elec. Jour.,* July, 1938, p. 273.
64. Hagenguth, J. H., Rohlfs, A. F., and Degnan, W. J. *Trans. AIEE,* 1952, Vol. 71, Part III, p. 455.
65. AIEE Committee Report, *Trans. AIEE,* 1956, Vol. 75, Pt. III, p. 16.
66. Committee Report, *Trans. AIEE,* 1946, Vol. 65, p. 70.
67. Committee Report, *Trans. AIEE,* 1950, Vol. 69, p. 1187.
68. AIEE Committee Report, *Trans. AIEE,* 1960, Vol. 79, Pt. III, p. 52.
69. Schahfer, R. M., and Knutz, W. H. *Elec. Light and Power,* 1938, p. 30.
70. Hemstreet, J. G., Lewis, W. W., and Foust, C. M. *Trans. AIEE,* 1942, Vol. 61, p. 628.
71. Cozzens, B. *Trans. AIEE,* 1939, Vol. 58, p. 140.
72. Bewley, L. V. *Trans. AIEE,* 1934, Vol. 53, p. 1163.
73. Fortescue, C. L., and Fielder, F. C. *Trans. AIEE,* 1934, Vol. 53, p. 1116.

74. Sunde, E. D. *Trans. AIEE*, 1940, Vol. 59, p. 987.

75. Bell, Edgar. *Trans. AIEE*, 1940, Vol. 59, p. 822.

76. Hansson, E. *Elec. Jour.*, 1936, Vol. 33, No. 6, p. 281.

77. Lusignan, J. T., Jr., and Miller, C. J., Jr. *Trans. AIEE*, 1940, Vol. 59, p. 534.

78. Evans, Llewellyn, and Daniels, H. C. *Elec. World*, 1939, Vol. 111, p. 1434.

79. Schlomann, R. H., Johnson, I. B., Price, W. S., and Anderson, J. G. *Trans. AIEE*, 1957, Vol. 76, Pt. III, p. 1447.

80. Berger, K. Paper 27, *Proc. Intern. Conf.* Central Electricity Research Laboratories, Leatherhead, Surrey, England, May 7–11, 1962. [Book edited by J. S. Forrest, P. R. Howard, and D. J. Littler, Butterworth & Co. (Publishers), Ltd., London, 1962. Also *Bull. Swiss Inst. Elec. Engrs.*, 1955 and 1965.]

81. AIEE Lightning Reference Book (1918–1935).

AIEE Lightning Reference Bibliography (1936–1949), *AIEE Special Publ.* S-37, April, 1950.

Lightning Reference Bibliography, 1950–1960, AIEE Committee Report, *Trans. AIEE*, 1962, Vol. 81, Pt. III, p. 944.

82. Bockmann, M., Hylten-Cavallius, N., and Rusck, S. *Paper* 314, *Conf. Intern. Grands Réseaux Elec. Haute Tension*, 1950.

83. Wagner, C. F., Gross, I. W., and Lloyd, B. L. *Trans. AIEE*, 1954, Vol. 73, Pt. IIIA, p. 196.

84. Johnson, I. B., et al. *Paper* 307, *Conf. Intern. Grands Réseaux Elec. Haute Tension*, 1962.

85. Golde, R. H. *Trans. AIEE*, 1945, Vol. 64, p. 902.

86. Young, F. S., Clayton, J. M., and Hileman, A. R. *Trans. IEEE*, Power Apparatus and Systems, Spec. Suppl., 1963, p. 132.

87. Kostenko, M. V., Polovoi, I. F., and Rozenfeld, A. N. *Elektrichestvo*, 1961, No. 4, p. 20.

88. Burgsdorf, V. V. *Paper* 326, *Conf. Intern. Grands Réseaux Elec. Haute Tension*, 1958.

89. Wagner, C. F. *Trans. AIEE*, 1956, Vol. 75, Pt. III, p. 1233.

90. Hagenguth, J. H., and Anderson, J. G. *Trans. AIEE*, 1957, Vol. 76, Pt. III, p. 1379.

91. Anderson, J. G. *Trans. AIEE*, 1961, Vol. 80, Pt. III, p. 414.

92. Clayton, J. M., and Young, F. S. *IEEE Conf. Paper* 63-1116, 1963.

93. Wagner, C. F., and Hileman, A. R. *Trans. AIEE*, 1960, Vol. 79, Pt. III, p. 589.

94. Wagner, C. F., and Hileman, A. R. *Trans. IEEE*, Power Apparatus and Systems, 1963, Vol. 82, p. 117.

95. Wagner, C. F. *Trans. IEEE*, Power Apparatus and Systems, September, 1964, Vol. 83, No. 9, p. 931.

HIGH-VOLTAGE GENERATORS AND PARTICLE ACCELERATORS

By E. J. Rogers

CONSTANT-POTENTIAL ELECTROSTATIC GENERATORS

39. Van de Graaff Generator. The electrostatic generator invented by R. J. Van de Graaff affords d-c voltages from several hundred kilovolts to 10 million V and steady currents up to a few milliamperes. It is used for many purposes, such as the production of controllable monoenergetic beams of positive ions for precision nuclear research and the acceleration of electrons and their conversion into penetrating X-rays for treatment of deep-seated malignancies and radiography of heavy metal sections and for the simulation of micrometeoroids in space-vehicle development.

40. Operating Principle. The Van de Graaff electrostatic belt generator consists essentially of a high-voltage electrode supported on an insulating column and of a rapidly moving insulating-belt system (Fig. 26-19) which carries electric charge to the terminal. At the ground end of the machine, electric charge is sprayed on the belt from corona points directed at the pulley and supplied with controlled potential from a small transformer-rectifier. The resulting ionization of the gas in the vicinity causes a transfer toward the pulley of electric charge which is carried off by the intervening insulating belt. At the terminal end of the belt, a similar means is used to transfer charge to the terminal.

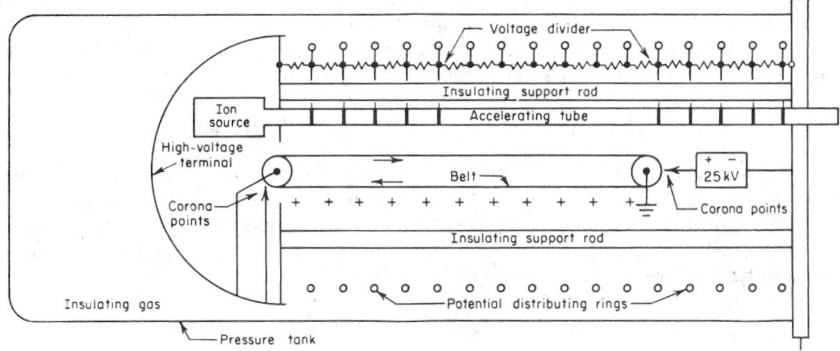

FIG. 26-19. Van de Graaff accelerator (for positive-ion acceleration).

Compressed-gas insulation is used to attain high voltages in compact apparatus. A typical pressure-insulated generator designed for operation at 2,000 kV in 27 atm of air has a tank with inside dimensions of 36 in diameter and 60 in high. The 0.5-mA charge-conveying system is a single 6-in-wide rubber-fabric belt running at 5,000 ft/min. The insulating column is built up of spaced metallic equipotential planes separated by solid insulators. Each plane is surrounded by a metal ring which serves as a corona shield and potential-distributing electrode. The total voltage is divided by resistors or corona gaps to distribute the potential evenly along the column. Both voltage rating and current capacity of a given electrostatic generator are increased by improving the dielectric strength of the insulating medium through either an increase in pressure or use of superior insulating gases or both. CCl_2F_2 (Freon) and SF_6 with saturation pressures at normal temperature of 85 lb/in² and 300 lb/in², respectively, have several times the voltage-insulating strength of air or nitrogen at the same pressure. Higher total pressures may be attained by adding nitrogen.

41. Tandem Van de Graaff. In a more advanced version of the Van de Graaff accelerator the high-voltage terminal is located in the center of the tank (Fig. 26-20). Negatively charged ions, injected into the accelerating tube at one end of the tank, are accelerated to the potential of the positively charged terminal. Inside the terminal they pass through a stripping foil or gas canal in which they lose two electrons (in the case of H^- ions). The ions thus become positively charged and are accelerated away from the terminal to emerge from the exit end of the tube with an energy equal to twice the potential of the high-voltage terminal.

The concept of multistage operation can be extended further to provide three stages of acceleration using two cascaded accelerators. In this type of operation the first machine (called the *injector*) has within its high-voltage terminal a negative-ion source.

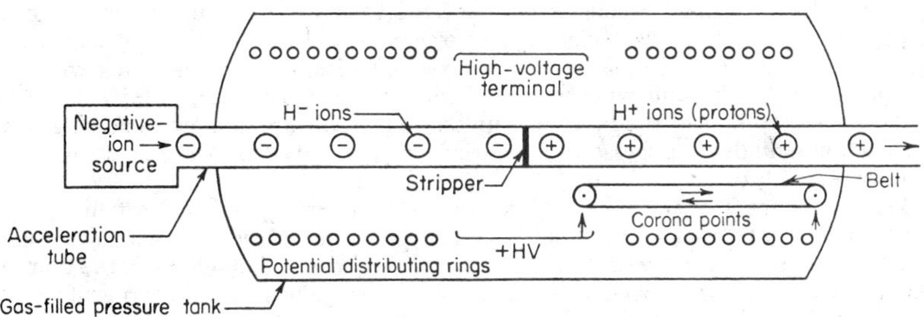

FIG. 26-20. Tandem Van de Graaff accelerator.

The negative ions emerging from the injector after a single stage of acceleration enter the second machine, which employs the tandem principle to provide two more stages of acceleration, thus producing an output beam of protons with an energy of three times the voltage of a single machine.

Very large three-stage accelerating systems designed to produce monoenergetic proton beams of 30 MeV are presently under construction. The accelerators are housed in tanks approximately 80 ft long and 20 ft in diameter. It is anticipated that ions of heavier elements which can be ionized to higher charge states may be accelerated to energies in excess of 100 MeV with these machines.

HIGH-VOLTAGE SURGE GENERATORS

42. Impulse Generators. Some of the highest voltages now in use are the transient impulses produced by surge generators. Ten-million-volt surges have been produced between two 5-million-V generators charged to opposite polarities.

Impulse waveforms generally have a rapid rise followed by a less rapid decline to zero, the voltage-time characteristic being given by the equation

$$v = A(\epsilon^{-mt} - \epsilon^{-nt}) \qquad (26\text{-}6)$$

Three common laboratory impulse-testing waves are the 0.5-5, the 1-10, and the 1.5-40. The first figure refers to the time in microseconds to the crest of the wave and the second figure to the time in microseconds to one-half the crest value on the tail of the wave. Impulse voltages are used to test the strength of power equipment against lightning and switching surges.

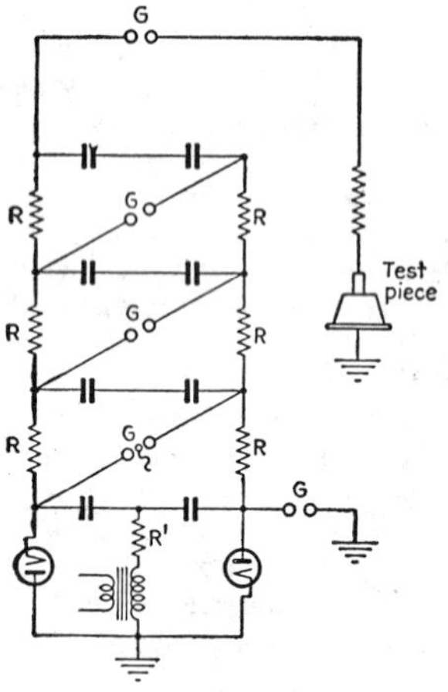

43. The Marx surge-generator circuit multiplies the voltage of the power source by charging capacitors in parallel and discharging them in series. In the typical Marx surge-generator circuit (four stages in Fig. 26-21) vacuum-tube-rectified a-c power is used, and one terminal of the test piece and one terminal of the transformer are grounded. The capacitors are first charged in parallel through the charging resistors R and then connected in series and discharged through the test piece by the simultaneous spark-over across the sphere gaps G. The impulse follows when an initiating voltage sufficient to precipitate breakdown is placed on the middle electrode of the three-electrode gap between the first and second condenser banks. Cumulative insulation of the condenser banks is usually met by assembling the units on a stairlike frame or an ascending helical frame of porcelain and wood.

FIG. 26-21. Typical fourstage Marx surge-generator circuit.

HIGH-VOLTAGE TRANSFORMERS AND RECTIFIERS

44. Cascade transformers are used primarily for very high a-c voltage applications which require little power, such as high-potential tests; for the measurement of high a-c potentials; and for the acceleration of electrons and positive ions to high energy. Each transformer unit, except the last, is provided with a tertiary winding which supplies power to the unit following it. Transformers beyond the first are insulated from ground for the voltage of the preceding units. The units are mounted in a single stack on porcelain cylinders, progressive insulation being supplied by the transformer tanks themselves.

45. Resonance Transformers. Excessive reactive input power in very-high-voltage transformers can be avoided by designing the capacitance of the secondary winding and high-voltage terminal to neutralize the inductance. Up to 2,000 kV for x-ray production has been attained by such air-cored gas-insulated resonance transformers.

46. Rectified A-C Voltages. In the production of x-rays and in nuclear research, the use of high-d-c potentials is preferable to a-c potentials. Unless the Van de Graaff generator is used, this calls for a form of transformer-rectifier circuit.

The simplest transformer-rectifier circuit is the half-wave circuit (see Fig. 26-22*a*). Full-wave rectification is obtained by combining two half-wave units as shown in Fig. 26-22*b*. These circuits are widely used for low-power applications requiring d-c voltages up to 200 kV.

47. Voltage Doubler. A common circuit for obtaining twice the transformer voltage is the Greinacher half-wave voltage doubler (see Fig. 26-22*c*). During each half cycle, one of the two capacitors is charged to full transformer voltage so that twice transformer voltage is available at the load. The Graetz bridge circuit (Fig. 26-22*d*) is a full-wave rectifier circuit which delivers a maximum voltage approximately equal to the a-c peak value. This circuit is commonly used in x-ray diagnostic and 200-kV therapy equipment.

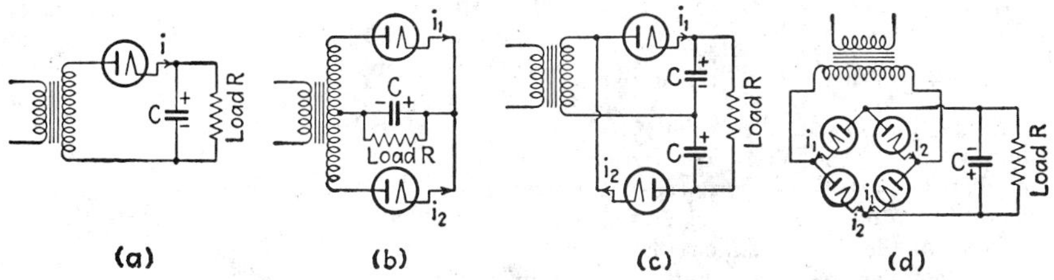

FIG. 26-22. Schematic connections: (*a*) half-wave transformer-rectifier circuit; (*b*) full-wave transformer-rectifier circuit; (*c*) half-wave voltage-doubler circuit (Greinacher); (*d*) full-wave bridge circuit (Graetz).

48. Chain Rectifier Circuit. Greinacher also suggested a cascaded high-voltage generator (Fig. 26-23) which has been used to produce 1.5 MV to earth. The circuit $aa'b'$ is a simple half-wave rectifier. Capacitor C_1' is therefore charged to the peak transformer voltage E. The potential b' and the voltage across rectifier ab' alternate between the value $2E$ above ground and zero. This voltage acts upon the half-wave rectifier circuit $ab'b$ to charge capacitor C_1 to a potential $2E$. Similarly, all the capacitors are charged to $2E$ V except C_1'. The potential of point e is therefore $8E$ V above earth. The capacitors are at progressively higher potentials in two columns. The last unit is surmounted by a well-rounded high-voltage terminal.

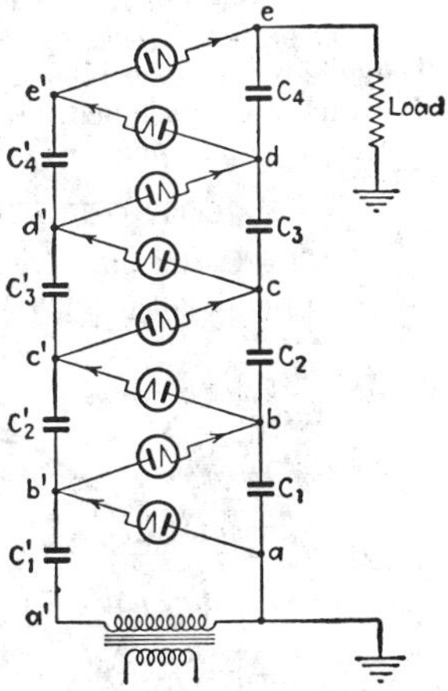

FIG. 26-23. Chain rectifier circuit (Greinacher, Bouwers, Cockroft).

49. Insulating Core Transformer. The insulating core transformer (ICT) was conceived by R. J. Van de Graaff, inventor of the electrostatic generator which bears his name.

It is not feasible to construct a transformer of conventional type for voltages above a few hundred kilovolts because of the difficulty of insulating the high-voltage secondary winding from the core. In the ICT (Fig. 26-24) this problem is circumvented by separating the magnetic core into segments which are insulated from each other and connecting a secondary-winding section and a high-voltage rectifier and filter capacitor to each core segment. The core segments with their associated components are stacked to form the high-voltage column, which, like that of the Van de Graaff accelerator, is surrounded by potential-distributing rings to ensure a uniform voltage gradient along the column.

In practical form the ICT is a 3-phase device with the core consisting of three segmented legs with massive flux-return rings at each end of the system. Primary excitation is provided by three primary windings at the bases of the core legs, and the top end of the stack is capped with a shield to form the high-voltage terminal.

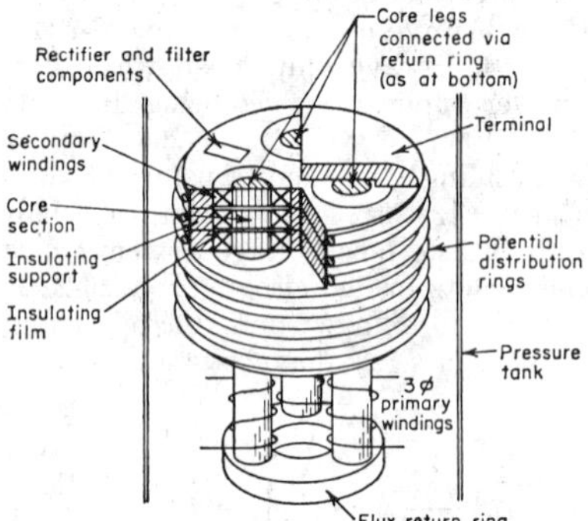

Fig. 26-24. Insulating core transformer.

The ICT can produce d-c outputs of millions of volts at currents of many milliamperes.

50. Dynamitron Accelerator. A relatively new addition to the accelerator family is the Dynamitron (Fig. 26-25), which converts low-voltage a-c power to high-voltage d-c power by means of a cascaded rectifier system driven in parallel from an r-f oscillator whose frequency is of the order of 100 kc/s.

This machine, structurally similar to the Van de Graaff, is insulated by compressed gas and contains a high-voltage electrode which is supported by an insulating column and connected to ground through a series string of high-voltage rectifier tubes. The column, as in the Van de Graaff, is surrounded by a set of potential-distributing rings, but unlike those of the Van de Graaff, the rings are split into semicircular pairs.

The semicircular rings and all their enclosed components are positioned between a pair of semicylindrical electrodes. These electrodes form the tuning capacitance for the r-f oscillator circuit and couple power to the high-voltage rectifiers via their capacitance to the rings.

Dynamitrons have been constructed which can accelerate particles to energies of 4 MeV. They share with the ICT the advantage that they can produce much larger currents than electrostatic machines.

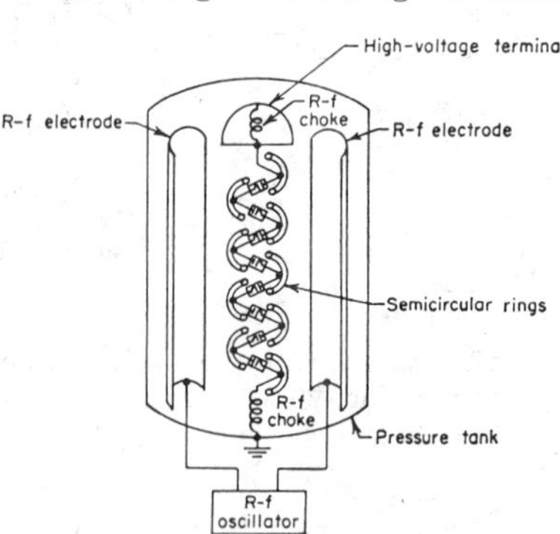

Fig. 26-25. Dynamitron accelerator.

CYCLOTRON, SYNCHROCYCLOTRON, AND SYNCHROTRON

51. The Cyclotron, invented by E. O. Lawrence, is a magnetic resonance device for the production of high-energy positive ions for nuclear research (Fig. 26-26). It uses the principle that, below relativistic energies, a charged particle in a uniform transverse magnetic field moves in a circular orbit at constant angular velocity. The D's are the two halves of a pillbox-shaped copper chamber between which a high-frequency voltage is applied. They are mounted within a squat, cylindrical brass vacuum chamber which lies between the pole faces of a large d-c electromagnet. Positive ions from a source near the center are accelerated by the electric field to one of the D's, moving in a circular path because of the magnetic field. The magnetic-field strength and the oscillator frequency are adjusted so that the ions always traverse the gap between the D's in an accelerating field.

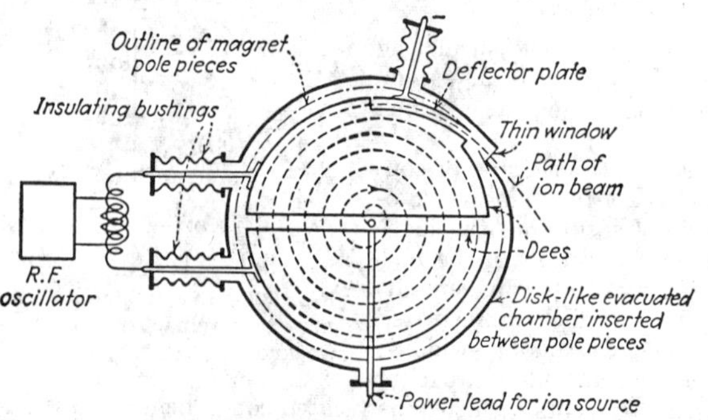

Fig. 26-26. Diagram of essential elements of a cyclotron.

The repeated accelerations cause the ions to move in progressively larger semicircles. As the ions approach the periphery, they pass through a region where the deflector plate at high negative potential directs them toward the portal or to a target in the rim of the vacuum chamber. The final kinetic energy in electronvolts of the ion is $B^2eR^2/2m$, and the applied frequency is $Be/2\pi m$ c/s, where B is the magnetic flux in webers per square meter, e is the charge in coulombs, R is the maximum orbit radius in meters, and m is the particle mass in kilograms. A typical cyclotron has a 70-ton d-c magnet with a 42-in-diameter pole face and produces 8-MeV protons or 16-MeV deuterons, using a maximum flux of 1.8 $\mathrm{Wb/m^2}$, an orbit radius up to 19 in, and a peak r-f voltage of 140 kV at 13.5 MHz.

At ion energies producing more than 1% increase in relativistic mass, synchronism is lost because of the decrease in angular velocity. In the synchrocyclotron this energy limitation is removed by cyclically diminishing the r-f frequency to match the ion-rotation frequency. Far fewer positive ions now emerge in the discrete pulses of this acceleration method, but a tenfold increase in particle energy is possible.

52. Sector-focusing Cyclotron. A recent development in the cyclotron based upon a principle first described by L. H. Thomas in 1938 permits ions to be accelerated to relativistic energies without the disadvantages of frequency modulation and pulsed operation characteristic of the synchrocyclotron.

In the normal cyclotron relativistic particles would remain in synchronism with a constant-frequency accelerating voltage if the strength of the magnetic field were made to increase radially so that as the particles increased in relativistic mass their radius would increase more slowly than otherwise, thus maintaining constant angular velocity. Unfortunately, however, a magnetic-field gradient of this type is defocusing to the beam.

In the sector-focused cyclotron, advantage is taken of the fact that dividing the magnetic field into sectors of alternating strength will produce a focusing effect. It is possible to design a sector-focusing magnet which will have a net focusing effect on the beam even though the average radial field gradient is positive.

The sector-focusing effect is enhanced if the magnetic field is divided into spiral sections of alternating strength rather than into simple sectors. In practice this is accomplished by using magnet pole tips which have spiral-shaped ridges instead of flat surfaces.

Several "spiral-ridge" cyclotrons are at present in use. An 88-in machine at Oak Ridge National Laboratory produces protons of 60 MeV energy, with beam currents in the vicinity of 100 μA.

53. Synchrotrons, suggested independently by McMillan and Veksler, are designed either for positive-ion or for electron acceleration to the 100- to 10,000-MeV energy range by the repeated application of a radio-frequency field to particles constrained to a stable circular orbit within an evacuated doughnut chamber in an increasing magnetic field. The Brookhaven proton synchrotron utilizes many C-shaped core sections to produce an annular ring of magnetic field of 30 ft radius and produces pulses of 3-BeV protons for basic nuclear studies. The protons are injected in pulses from a 3.6-MeV Van de Graaff. The frequency of the r-f accelerator must be carefully matched to the increasing orbit frequency of the particles, which execute approximately 3×10^6 r.

54. Alternating-gradient Synchrotrons. In a synchrotron the condition for vertical focusing of the beam is that the field index, $n = (-R\, dH/H\, dR)$, where R is the machine radius and H is the magnetic-field strength, must be greater than zero, whereas for horizontal focusing n must be less than 1. For $1 > n > 0$, which is the case for the conventional synchrotron, focusing occurs in both planes but the focusing forces are weak, resulting in large oscillations of the particles around the equilibrium orbit. As the diameter of the machine is increased to obtain higher energies, the oscillation amplitudes increase. The size of the magnet aperture required to contain the oscillating beam imposes a practical energy limit of about 10 BeV on the conventional synchrotron.

A new focusing principle called alternating gradient or "strong" focusing was described by Courant, Livingston, and Snyder at Brookhaven National Laboratory in 1952 and independently two years earlier by Christofilos, a Greek electrical engineer.

In the alternating-gradient synchrotron (AGS) the particles pass alternately through sectors of the magnet which have large negative and positive values of n. It can be shown that, as the beam passes through the large focusing and defocusing fields, it experiences a net focusing effect. This statement holds both in the horizontal plane, where the large negative values of n are strongly focusing and the large positive values of n are strongly defocusing, and in the horizontal plane, where the converse situation exists.

The net focusing forces in both planes, although much weaker than the alternating focusing and defocusing forces in the individual sectors, are still far stronger than those in the conventional, or "weak," focusing synchrotron.

The AGS principle, by greatly reducing the beam-oscillation amplitudes, allows the use of a small-aperture magnet, making it economically feasible to construct machines of much larger radius and consequently higher energy than the conventional synchrotrons.

A 33-BeV AGS is in operation at Brookhaven National Laboratory and a 29-BeV machine in the CERN Laboratory at Geneva. A 70-BeV AGS is under construction in Russia, and design studies are under way in the United States for an AGS to provide protons of 200 BeV and one to provide 750- to 1,000-BeV protons.

55. The Betatron. In the betatron developed by D. W. Kerst, electrons moving in a circular orbit of fixed radius are accelerated to high energies by the time increase in magnetic flux linking this orbit. The apparatus utilizes an evacuated "doughnut" tube between the poles of a magnet fed by low-frequency power. Acceleration takes place during the repeated quarter-cycle intervals of increasing flux. By shaping the poles for radially diminishing field strength, a stable circular orbit for all electron energies is established at that radius at which the induced potential is a minimum. Both radial and axial focusing are obtained if the field strength H obeys the relation $H \sim 1/R^n$, where R = distance from the pole axis and n lies between 0 and 1, usually about 0.75. Electrons are injected into the orbit for about 10 μs early in the rising flux cycle and, after some hundred thousand revolutions, are caused to impinge against an internal target by pulsing an auxiliary coil to shrink the orbit. The laminated silicon-steel magnet of a 20-MeV betatron weighs 3½ tons, is fed at 16.5 kV with 180-c/s power, and has a 15-in-diameter equilibrium orbit. The 100-MeV betatron at Schenectady has a 76-in pole diameter, weighs 130 tons, requires 200 kW of 60-c power, produces a maximum flux density of 4,000 G at the orbit, and imparts about 4,000 eV/r.

LINEAR ACCELERATORS

56. The linear accelerator (linac) is a device for accelerating charged particles to high energies by high-frequency r-f pulses applied to a periodic array of electrodes or to an evacuated waveguide designed for the required phase velocity.

The main advantages of the linac for the acceleration of protons (and heavy ions) to high energies are its ability to produce large peak beam currents with high duty factors and the absence of beam-extraction problems. Machines have been built to produce protons with energies over 50 MeV and peak beam currents as high as 80 mA. The accelerating voltage of the order of 200 Mc is supplied by high-frequency power triodes in resonant cavity oscillators.

57. Electron Linac. In any orbital accelerator a charged particle radiates energy due to the centripetal acceleration inherent in such a machine.

The radiated energy increases rapidly as the particle approaches the velocity of light, c. Electrons, being some 2,000 times lighter than protons, approach c so closely at energies above a few hundred million electronvolts that the energy loss due to radiation imposes excessive power requirements on the r-f accelerating system. The linac thus provides the only feasible means for accelerating electrons into the BeV energy range.

An electron linac under construction at Stanford will produce 20-BeV electrons at peak currents of 25 to 50 mA and average currents of 15 to 30 μA. The machine, 2 mi long, will use two hundred and forty 3,000-Mc klystrons for acceleration.

58. References

Constant-potential Electrostatic Generators

Van de Graaff, R. J., Trump, J. G., and Beuchner, W. W. Electrostatic Generators for the Acceleration of Charged Particles; *Rept. Progr. Phys.*, 1948, Vol. 11, p. 1.

ROGERS, E. J., and TURNER, C. M. Injection System. Part I. Van de Graaff Accelerator; *Rev. Sci. Instr.*, 1953, Vol. 24, pp. 816–820.

ROGERS, E. J. Van de Graaff Proton Source Receives 110 Kilovolt Boosts; *Electronics*, 1962, Vol. 13, pp. 58, 59.

VAN DE GRAAFF, R. J. Tandem Electrostatic Generators; *Nucl. Instr. Methods*, 1960, Vol. 8, pp. 195–202.

ROSE, P. H. The Three-stage Tandem Accelerator; *Nucl. Instr. Methods*, 1961, Vol. 11, pp. 49–62.

High-voltage Surge Generators

ALLIBONE, T. E., and PERRY, F. R. Standardization of Impulse-voltage Testing; *Inst. Elec. Engrs.* (*London*), 1936, Vol. 78, p. 257.

MARX, EDWIN Investigations in the Testing of Insulators with Impact Voltages; *Elektrotech. Z.*, 1924, Vol. 45, p. 652.

High-voltage Transformers and Rectifiers

BOUWERS, A. Modern X-ray Development; *Brit. Jour. Radiol.*, 1934, Vol. 7, p. 21.

CAMILLI, G. Cascade-type Potential Transformers; *Gen. Elec. Rev.*, 1936, Vol. 39, p. 95.

CHARLTON, E. E., WESTENDORP, W. F., and DEMPSTER, L. E. New Million-volt X-ray Outfit; *Jour. Appl. Phys.*, 1939, Vol. 10, p. 374.

SARSFIELD, L. "Electrical Engineering in Radiology"; London, Chapman & Hall, Ltd., 1936.

Insulating-core Transformer

BENN, T. R., and SCATURRO, A. J. A New Mobile 600 kV Test Set, *Paper* 31TP65-123, IEEE Winter Power Meeting, Jan. 31–Feb. 5, 1965.

The Dynamitron

CLELAND, M. R., and MORGANSTERN, K. H. A New High-power Electron Accelerator; *IRE Trans. on Ind. Electronics*, 1960, Vol. IE-7, No. 2.

CLELAND, M. R., and FARRELL, P. Dynamitrons of the Future; *IEEE Trans. on Nucl. Sci.*, 1965, Vol. NS-12, No. 3.

Cyclotrons and Synchrotrons

LIVINGOOD, J. J. "Cyclic Particle Accelerators"; Princeton, N.J., D. Van Nostrand Company, Inc., 1960.

LIVINGSTON, M. S., and BLEWETT, J. P. "Particle Accelerators"; New York, McGraw-Hill Book Company, 1962.

ROGERS, E. J. Method for Damping Phase Oscillations in a Synchrotron; *Rev. Sci. Instr.*, 1958, Vol. 29, No. 3, pp. 215–217.

Sector-focused Cyclotrons, *Rept.* 26, *Nucl. Sci. Ser.*, National Academy of Sciences–National Research Council, 1959.

The Betatron

KERST, D. W. Acceleration of Electrons by Magnetic Induction; *Phys. Rev.*, 1941, Vol. 60, p. 47.

KERST, D. W. A 20-million Electron-volt Betatron or Induction Accelerator; *Rev. Sci. Instr.*, 1942, Vol. 13, p. 387.

WESTENDORF, W. F., and CHARLTON, E. E. A 100-million Volt Induction Accelerator; *Jour. Appl. Phys.*, 1945, Vol. 10, p. 581.

The Linear Accelerator

ALVAREZ, L. The Proton Linear Accelerator; *USAEC Document AECU*-120, No. 30, 1948.

MULLETT, L. B. Design Study for a 600-MeV Proton Linear Accelerator, *Atomic Energy Res. Document* G/M151, April, 1953.

DEMOS, P. T., et al. MIT Linear Electron Accelerator; *Jour. Appl. Phys.*, 1952, Vol. 23, pp. 53–65.

LIVINGOOD, J. J. "Cyclic Particle Accelerators"; Princeton, N. J., D. Van Nostrand Company, Inc., 1960.

LIVINGSTON, M. S., and BLEWETT, J. P. "Particle Accelerators"; New York, McGraw-Hill Book Company, 1962.

X-RAYS

By Herman E. Seemann

59. General Properties. X-rays occupy a region in the spectrum having (in practical applications) a wavelength range from 0.0006 to 2 Å. (One angstrom = 10^{-8} cm.)

Because of high-voltage developments, x-rays may now be produced with wavelengths shorter than the gamma rays emitted by radium. Their range extends to the extreme ultraviolet.

The x-radiation penetrates materials opaque to ordinary light and causes certain chemical changes. It may be reflected, refracted, and diffracted, although not to the same degree as light—only for very small angles and with very special experimental procedures. Its action on photographic emulsions is much the same as that of light. When passed through a gas, x-radiation produces ions, thus making the gas electrically conducting. This phenomenon is utilized in ionization chambers for measuring x-ray intensity.

60. Production. X-rays are produced by allowing a cathode-ray beam to impinge upon a body of metal within a highly evacuated tube. The electron (or cathode) stream is produced by a hot-filament cathode. The wavelength range of the continuous x-ray spectrum depends upon the potential applied between the cathode and anode or target (see Par. **61**), and in a single tube a wide range of wavelengths is always produced simultaneously.

61. X-ray Spectrum. X-rays emanating from a target present a continuous spectrum having a definite spectral energy distribution. Above certain critical voltages (depending upon the material of the target) there appear superposed on the continuous spectrum definite spectral lines of outstanding intensity which are characteristic of the elements composing the target. Conversely, from the wavelengths of these spectrum lines from the target, the chemical elements present in the target may be determined. Such spectral lines are called the **characteristic radiation** of the target. The spectrum is obtained by diffracting the beam by means of a suitable crystal (calcite, for example) and measuring the spectral intensity of this diffracted beam, photographically or by means of an ionization chamber. The actual wavelengths are determined from the relation $n\lambda = 2d \sin \theta$, where n = order of the spectrum, λ = wavelength, θ = angle through which the beam is diffracted, and d = interatomic spacing of the crystal (d for calcite = 3.029×10^{-8} cm).

62. Secondary Radiation. When a body is irradiated by x-rays, secondary radiation is reemitted. This radiation is composed of three parts: *scattered radiation* of the same wavelengths (or nearly so) as the incident radiation, *characteristic radiation* of the material irradiated, and *corpuscular radiation* or electrons. The characteristic radiation appears when the primary radiation is produced at a certain minimum voltage which depends upon the irradiated substance.

63. X-ray Absorption, Opacity of Materials. All materials absorb x-rays to a degree which depends on the material and the quality of the x-rays. The law of absorption for *monochromatic* x-rays is given by $I = I_0\epsilon^{-\mu d}$, where I = intensity (energy per second) of the radiation incident on the absorber, I_0 = transmitted intensity, ϵ = natural logarithm base, d = thickness of the absorbing material, and μ = absorption coefficient. Within a certain range μ is approximately proportional to the cube of the wavelength, for a given material. The approximate half-absorption value of thickness x for several metals, when using tungsten characteristic radiation of wavelength 0.209×10^{-8} cm (characteristic line $k_{\alpha 1}$), is given in the following table:

Element	Atomic number	μ	x, cm
Aluminum	13	0.624	1.11
Iron (steel)	26	8.46	0.082
Copper (brass)	29	13.62	.051
Zinc	30	12.58	.055
Silver	47	60.40	.011
Lead	82	52.20	.013

64. X-ray Quality. The quality of an x-ray beam is determined by the spectral wavelengths contained in the radiation. As a practical method of expressing the quality, the half-absorption-value layer is often used; this is defined as the thickness of a given medium (usually copper) which absorbs one-half the incident radiation (see Par. **63**). For a given peak voltage, there is a minimum wavelength to the x-rays produced. This is given by the relation $\lambda_0 = 12.34/V$, where λ_0 = minimum x-ray wavelength in angstroms and V = peak voltage in kilovolts.

65. Modern X-ray Tubes. Gas x-ray tubes are obsolete. Tubes are of the Coolidge type or a modification thereof, the vacuum in them being sufficiently high so that the gas content plays no part. The source of bombarding electrons (or cathode rays) is a hot tungsten filament, which permits the passage of a large electron stream between cathode and anode and insures steadiness of operation. The space current is regulated in general by the filament current, increasing with the filament temperature.

The conversion of the energy of the electron stream into x-rays is a process of low efficiency. The efficiency is approximately proportional to the kilovoltage and reaches a value of about 1.1% at 200-kV peak for a tungsten anode. Almost the entire input of the x-ray tube is therefore wasted as heat. Cooling is necessary in order to dissipate the large amount of heat generated at the focal spot where the electron stream impinges.

The problem is complicated by several variables, viz., watts to be dissipated, focal-spot size, time of exposure, and features of design. The manufacturer's rating chart should be consulted for specific conditions.

66. Generating Apparatus. Early forms of high-voltage generators have been supplanted by the transformer. A transformer with high leakage reactance, which for other purposes would have poor operating characteristics, is not undesirable since it prevents the accidental passage of too heavy currents through a tube. It is usually desirable to ground the center or one end of the secondary. While milliammeters for measuring the space current may be placed at the grounded part of the secondary, it is preferable that they be placed in the high-tension circuit. A sphere spark gap, with spheres 125 mm in diameter placed across the tube terminals, gives a sufficiently accurate voltage measurement in terms of centimeter spark equivalent. A point gap may be used but is not preferable; electrostatic voltmeters of the attracted-disk type may be used satisfactorily but are cumbersome. The most suitable method is to use a shielded high resistance[1] (100 to 150 MΩ) in series with a suitable milliammeter, thus giving in effect a high-potential voltmeter which draws only about 1 ma of current.

Synchronous rotary switches in the high-tension circuit for rectification are obsolete and are replaced by circuits employing electronic rectifying tubes. The latter are used in a wide variety of circuits, embodying half- or full-wave rectification. By inserting an electrical filter of proper capacitance and inductance in the high-tension circuit, a potential may be produced in which the fluctuations are as low as 0.1%, that is, practically constant potential. Three-phase systems are satisfactory though somewhat more complicated (see Sec. **9**).

Improvements in x-ray tubes have extended the use of self-rectifying x-ray tubes into the high-voltage region. In this case, only half-wave current passes through the tube. The equipment is less complicated than where a separate rectifying system is used.

67. Intensity Measurement. The direct measurement of x-ray energy is impractical outside the physical laboratory. Until recently, the output of a tube was expressed simply in terms of the milliamperes space current and kilovolts potential applied. The intensity is now expressed more exactly by means of the ionization produced by an x-ray beam at the point in question. Various portable instruments called *dosage meters* are made which can be calibrated against a standard[2] in terms of the ionization in a cubic centimeter of air, the unit being 1 R.[3]

68. X-ray Diffraction (Crystal-structure Analysis). Diffraction of x-rays by means

[1] TAYLOR, L. S. Apparatus for the Measurement of High Constant or Rippled Voltages; *NBS Jour. Res.*, 1930, Vol. 5, p. 609.

[2] TAYLOR, L. S. The Precise Measurement of X-ray Dosage; *NBS Jour. Res.*, 1929, Vol. 2, p. 771; also 1930, Vol. 5, p. 507.

[3] "The roentgen shall be the quantity of x- or gamma-radiation such that the associated corpuscular emission per 0.001293 gram of air produces, in air, ions carrying 1 esu of quantity of electricity of either sign." Adopted by Fifth International Congress of Radiology.

of a crystal was first conceived by Laue in 1912. Atoms in crystals have a systematic arrangement which affects x-rays as a ruled grating does light. A beam of monochromatic x-rays passing through a crystal is diffracted and recorded photographically as spots, bands, or rings, depending upon the material and the technique used. By such means the atomic arrangement of most diffracting materials may be studied. This has very important industrial as well as physical applications in the study of materials and control of processes. The most common commercial method (Hull's) uses a monochromatic beam of x-rays passing through the crystalline sample in finely powdered form. The diffracted beams then appear as lines. The angular displacements of these lines indicate the crystal structure and atomic spacings of the powdered material.

69. Medical Radiography.[1] For medical diagnostic purposes, voltages ranging from 40 to 130 kV peak are generally used, with exposures varying in length from 1/60 to 10 or 15 s and tube currents ranging from 10 to 1,000 mA. The exact combination depends upon the nature of the body part being radiographed. Thick parts require higher voltage for adequate penetration than thin parts. However, this principle cannot be applied too generally because the x-ray transmissions of bone and tissue, for example, become more and more similar for the shorter wavelengths generated at the higher voltages. A radiograph taken at too high a voltage is said to lack *contrast.* If a trial indicates sufficient contrast but not enough exposure, the milliamperage or the time should be increased since this may be done without affecting the quality of the x-rays.

By using fluorescent intensifying screens in medical radiography the decrease in exposure time amounts to a factor of from 10 to 30, depending on conditions. X-ray film for use with intensifying screens is referred to as "screen film." X-ray film with greater sensitivity for the direct radiation is also available.

Secondary radiation, as well as the primary image-forming radiation, is recorded in all radiographs and may be a serious detriment when thick parts are radiographed. The use of the Potter-Bucky diaphragm between subject and film eliminates a large proportion of the secondary radiation and yields a correspondingly improved radiograph, by increasing the visibility of detail.

70. Fluoroscopy and Photoradiography. The image produced by the x-rays on a fluorescent screen may be viewed directly if sufficient x-ray intensity is transmitted by the specimen and the observer is dark-adapted. This procedure is commonly used in the medical field during the reduction of fractures or for the observation of movement of internal organs. Contrast may be enhanced by ingestion or injection of radiopaque substances.

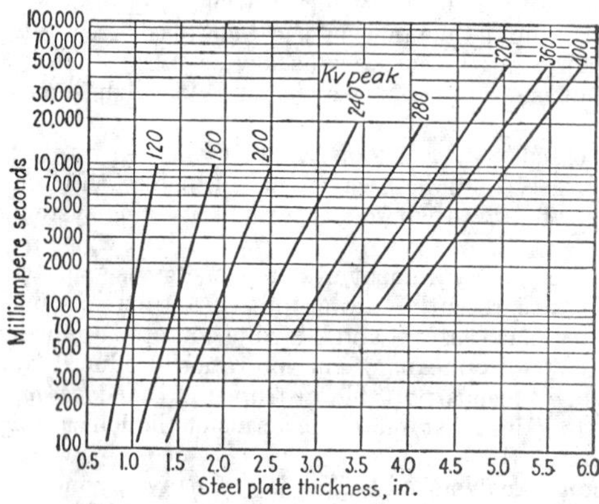

Fig. 26-27. Exposure chart for 36-in target-film distance with intensifying screens. (*General Electric Company, X-ray Dept.*)

In industry, castings of the light metals may be inspected in a similar way. The method does not yield such good detail as the radiograph but is less expensive. The operator is protected by lead glass or by viewing indirectly in a mirror.

The fluorescent screen image may be photographed with a camera having a fast lens. The equipment and method have been worked out for use in mass medical surveys.

71. Industrial Radiography. It is common practice to radiograph important castings and welds with the purpose of rejecting those pieces showing serious defects or to improve the technique of manufacture. The principles of the process are the same as in medical radiography

[1] Radiography for medical or surgical diagnosis should not be undertaken by electricians or photographers, as the correct interpretation of the shadows is difficult and requires special training.

(Par. **69**), but longer exposures and higher voltages are common because of the greater opacity of the heavier elements.

Both kinds of film are used, but the direct-exposure type may usually be used advantageously with lead foil about 0.005 in thick on both sides of the film. This acts as an intensifier to some extent (principally because of electron action) and reduces the effect of secondary radiation. Lead screens do not shorten exposures nearly as much as fluorescent screens and are not of much advantage beyond 1-in steel unless more than 200 kV are available.

Aluminum is several times more transparent to x-rays than steel. It is seldom necessary to employ kilovoltages higher than 150. Brass is more opaque than steel. For a given technique, the thickness of steel penetrated is about 50% greater than the corresponding thickness of brass. The graphs of Figs. 26-27, 26-28, and 26-29 serve as a guide to exposure technique but may require slight modification by the operator to meet local conditions.

The development of supervoltage units has made possible the radiography of very thick steel. The best known are the General Electric 1,000,000- and 2,000,000-V resonance transformer units, the 20,000,000-V Kerst betatron, and the 2,000,000-V Van de Graaff electrostatic generator (see Par. **39**). At these high voltages the relative exposures for different materials of the same thickness depend primarily on their densities rather than on their atomic numbers.

A Westinghouse unit[1] makes x-ray exposures in 1/1,000,000 s at high tube currents so that fast motion may be "stopped." Such equipment has been used in ballistic studies and is suitable for the study of rapidly moving machinery. One-inch steel may be penetrated at 300 kV.

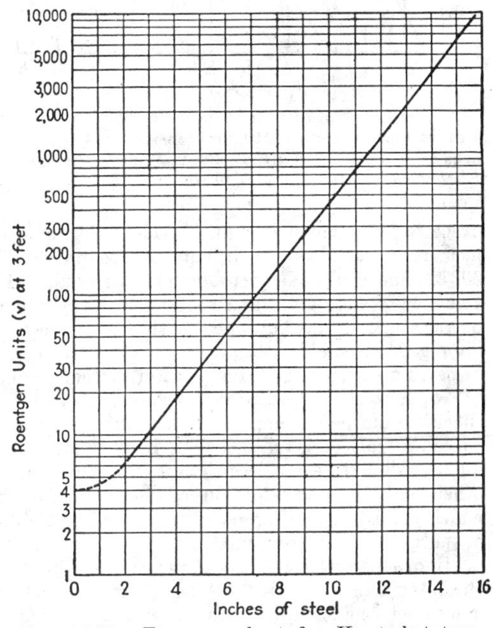

FIG. 26-28. Exposure chart for Kerst betatron, 20-million-volt x-rays; standard development D-19b; density =2; 0.04-in front lead screen; fine-grain film. For approximate exposure time in minutes divide ordinate by 100. (*Dr. G. D. Adams.*)

72. X-ray Therapy. X-rays are used for the treatment of various diseases, particularly those of a cancerous nature, and including skin diseases. For superficial treatment, voltages from 10 to 100 kV are used, and the rays are filtered by passing through thin sheets of suitable metals to eliminate the longer wavelengths. For deep application, voltages up to 2,000 kV are used, also with proper filtration. To prevent burns, it is imperative that the administered dose be carefully measured (see Pars. **67** and **74**).

73. X-ray Burns. The so-called "x-ray burns" may be acute, i.e., developed from a single prolonged exposure; or they may be chronic, developing gradually as a result of repeated small exposures which individually would produce no noticeable effect. The acute burn, which seldom appears until some days after the exposure, is evidenced by the reddening, blistering, and sloughing observed in ordinary burns. All x-ray injuries are especially slow in healing. They should receive immediate treatment by a competent specialist.

[1] SLACK, C. M., and THILO, E. R. Field Emission Applied to Ultra-speed X-ray Technique; *Proc. First Natl. Electronics Conf.*, October, 1944.

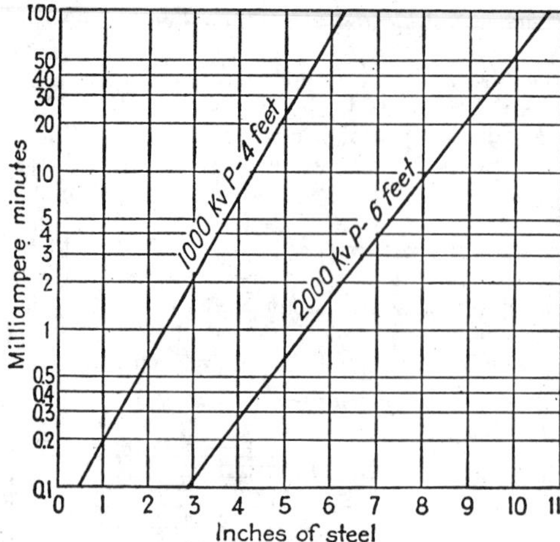

FIG. 26-29. Exposure chart for 1,000- and 2,000-kV x-rays; lead screens 0.010 in; density = 1.5 (approx.); typical fine-grain x-ray films. (*General Electric Company, X-ray Dept.*)

74. X-ray Protection. In the ordinary application of x-rays for diagnosis or treatment, the danger to the patient is small, but physicians working daily around x-ray tubes must observe caution to escape injury. To increase protection of the patient, filters of copper from 0.3 to 1.5 mm thick or of aluminum 0.5 to 5 mm thick are interposed in the main beam to cut out the rays of low penetration (long wavelength) which would otherwise be absorbed by the skin. X-ray protection[1] is afforded by placing sheet lead of suitable thickness between the operator and all sources of radiation, i.e., direct and scattered from air and solid objects. Concrete and solid concrete blocks have been used with some success.[2]

The protective thickness of lead or concrete depends upon the voltage and current used and the distance to the nearest occupied space. For details, the latest publications on this subject from the National Bureau of Standards should be consulted (e.g., Handbooks 50, 51, and 60). X-ray rooms may be equipped with lead-glass windows for the convenience of the operator. Pipes (e.g., ventilating ducts) that pass through protective barriers should be equipped with lead elbows. Tubes should be enclosed in lead-impregnated shields. Aprons and gloves composed of rubber impregnated with lead or barium oxides afford the operator further protection in medical fluoroscopy.

75. Electrical Protection. Protection for patient and operator against high-voltage shock should be accomplished by making all high-tension parts completely inaccessible while the voltage is on. Spacing between floor and exposed high-tension overhead conductors should be not less than 9 ft. Well-grounded metal guards should enclose any other high- and low-tension conductors. The best protection is afforded by units which completely oil-immerse the x-ray tube, transformers, and all conductors in one or more grounded containers.

In order to protect other electrical apparatus, surge arresters should be placed across the transformer primary. A simple surge arrester can be made by placing a 10,000-Ω noninductive resistance, with its center grounded, across the line. Quick-acting double-pole circuit breakers should also be used in the primary line.

76. Bibliography

Symposium on Radiography, Philadelphia, ASTM, 1943.

CLARK, G. L. "Applied X-rays," 4th ed.; New York, McGraw-Hill Book Company, 1955.

COMPTON, A. H., and ALLISON, S. K. "X-rays in Theory and Experiment"; Princeton, N.J., D. Van Nostrand Company, Inc., 1935.

"Radiography in Modern Industry"; Rochester, N.Y., Eastman Kodak Co., 1946.

SPROULL, W. T. "X-rays in Practice"; New York, McGraw-Hill Book Company, 1946.

MORGAN, R. H. "Handbook of Radiology"; Chicago, Year Book Publishers, Inc., 1955.

WEYL, C., WARREN, S. R., JR., and O'NEILL, D. B. "Radiologic Physics"; Springfield, Ill., Charles C Thomas, Publisher, 1941.

[1] NBS Handbook 76: X-ray Protection.
SINGER, G., TAYLOR, L. S., and CHARLTON, A. L. Concrete as a Protective Material against High-voltage X-rays; *Radiology*, 1939, Vol. 33, p. 68.
BRAESTRUP, C. B. Industrial Radiation Hazards; *Radiology*, 1944, Vol. 43, p. 286.
Also consult American Standards Association, 70 East 45th St., New York 17, N.Y.

[2] TAYLOR, L. S. Industrial X-ray Protection; *ASTM Bull.* 99, August, 1939, p. 23.

Lonsdale, K. "Crystals and X-rays"; Princeton, N.J., D. Van Nostrand Company, Inc., 1949.

Tentative Recommended Practice for Radiographic Testing E94-52T, Philadelphia, American Society for Testing Materials.

Industrial Applications of the X-ray, General Electric X-ray Corporation, 1936.

Glasser, O. "Wilhelm Conrad Röntgen"; Springfield, Ill., Charles C Thomas, Publisher, 1934.

Bragg, W. L. "The Crystalline State"; New York, The Macmillan Company, 1934, Vol. 1, A General Survey.

Zirkle, R. E. "Biological Effects of External X and Gamma Radiation"; New York, McGraw-Hill Book Company, 1954.

ULTRAVIOLET ENERGY

77. Ultraviolet radiant energy extends over a spectral range from the short-wave limit of the visible spectrum, λ4,000, to an indefinite limit in the Roentgen region beyond λ1,000. It causes a variety of uncounted chemical, physical, and biological effects; but the relative effectiveness*[1] of energy of various wavelengths has been determined for only a few of these. Air absorbs ultraviolet energy shorter than λ1,850, and the production of ozone is one result of this absorption. The maximum germicidal effectiveness is in the region of λ2,575, but living organisms are destroyed by energy of wavelengths longer than λ3,000. Maximal erythemal (sunburning), tanning, and antirachitic effectiveness is in the region of λ2,967. Measurements of ultraviolet energy are meaningless unless weighted according to the effectiveness of energy of various wavelengths in producing a given effect. This can be done only for those effects for which the spectral effectiveness of ultraviolet energy has been determined.

78. Erythemal effectiveness has been determined for average untanned skin; and inasmuch as erythema is an important effect of exposing the human skin, it provides a practical means of rating ultraviolet sources, intensities, and exposures where human beings are involved. The latest determinations[2] for wavelengths corresponding to the mercury spectral lines are presented in Table 26-6.

Table 26-6. Relative Erythemal Effectiveness

λ2400	0.95	λ2654	0.30	λ2804	0.06	λ3022	0.50
2482	0.90	2675	0.20	2894	0.25	3130	0.02
2537	0.80	2700	0.15	2925	0.70	3342	0.004
2576	0.70	2753	0.05	2967	1.00	3663	0.001

The **spectral effectiveness** of ultraviolet energy in tanning (white) human skin is practically the same as for producing erythema for wavelengths longer than λ2,950. In general, erythema appears shortly after an adequate exposure, and tan is a more belated result. However, energy longer than λ3,300 can produce tan shortly after, or even during, exposure without a characteristic erythema. The relative tanning effectiveness is the same as for erythemal effectiveness in the last column of the table excepting for λ3,342 and λ3,663. The values for tanning for these two wavelengths are in the neighborhood of 0.004 and 0.001 instead of zero.

A systematic nomenclature[3] has been developed for this complex realm. An **E-viton** per square centimeter is a unit of erythemal effectiveness equivalent to 10 μW of radiant energy per square centimeter at λ2,967, where the erythemal effectiveness is maximal and is relatively equal to unity. A minimum perceptible erythema is produced on average untanned (white) skin by an exposure of 40 E-viton min (or ⅔ E-viton h)/cm^2 of skin.

79. Ultraviolet Outdoors. The spectra of sunlight and skylight at the earth's surface extend nearly to λ2,990. The intensity and short-wave limit of ultraviolet energy reaching the earth outdoors varies with latitude, solar altitude, cloudiness, atmospheric conditions, etc. Erythemal effectiveness is important in itself and is generally a fair measure of antirachitic, and a rough measure of germicidal, effective-

* For all numbered references see Bibliography (Par. **89**).

ness. The results[4] of an hourly record over 4 years of the erythemal effectiveness of ultraviolet energy received by a horizontal surface from the sun and entire sky are presented in Table 26-7 as obtained in a suburb of Cleveland, Ohio. The hourly records were summarized for months and averaged for the 4 years.

Table 26-7. Erythemal Ultraviolet Energy on a Horizontal Surface Exposed to Sun and Entire Sky

Month	Hours of sunshine		% of possible	E-viton hr/sq cm	% of total for year
	Possible	Actual			
Jan.	295	84	28	51	1.9
Feb.	297	97	33	65	2.4
Mar.	370	185	50	182	6.8
Apr.	401	201	50	265	9.8
May	452	303	67	432	16.1
June	455	304	67	452	16.8
July	461	352	76	435	16.2
Aug.	429	285	66	346	12.9
Sept.	374	239	64	250	9.3
Oct.	343	185	54	130	4.8
Nov.	295	106	36	53	2.0
Dec.	285	71	25	27	1.0

The possible hours of sunshine and the percentages of actual sunshine are given for each month in order to allow for cloudiness. The effect of solar altitude seasonally is seen to be similar to the daily variation. The total E-viton hours per year varied from a low of 2,300 to a high of 2,900. About 80% of the total erythemal energy for the year is received in the 6-month period April to September. Only 7% of the year's total is received in the 4-month winter period November to February. The highest intensity of erythemal ultraviolet energy recorded in the 4 years was 4.2 E-vitons/cm². This is equivalent to 42 μW of energy/cm² of maximal erythemal effectiveness, λ2,967. A 10-min exposure of average untanned skin to this intensity will produce a barely perceptible erythema. Generally, on clear days at noon in summer (maximum solar altitude 73°) the average contributions of erythemal energy were about equal for sun and sky. On clear days the total contribution by the entire sky, upon a horizontal surface, exceeds that from the sun excepting occasionally at noon.

80. Artificial Sunlight. Many sources of ultraviolet energy are now available including arcs, mercury-vapor lamps, and filament lamps. The ultraviolet output of arcs using impregnated carbons can be varied so widely and readily that data should be obtained from the manufacturers. The data presented in Table 26-8 include the present sources of a fairly constant nature. The intensities of ultraviolet energy of these spectral ranges are given in microwatts per square centimeter at a distance of 1 m perpendicular to the axis of the bare lamp. The limit λ2,800 is used for sun lamps because this approximates the limit of the spectrum of daylight to which human beings are acclimated.

81. The ultraviolet output of the constant and controllable sources included in Table 26-8 may be compared with the ultraviolet energy contributed on the earth's surface by the sun on a clear midsummer day at noon. This value varies with latitude and atmospheric conditions, but the total energy shorter than λ3,130 is 90 to 140 μW/cm². Weighted according to erythemal effectiveness, the average maximal value is about 25 μW/cm². Artificial sunlight can be supplied to produce a minimum perceptible erythema on average untanned skin in any given time by supplying the proper intensity of erythemal flux or erythemally weighted energy in E-vitons per square centimeter. The time required to produce a minimum perceptible erythema on average untanned skin is indicated in Table 26-9 for various sources in an efficient reflector.

82. Germicidal effectiveness of radiant energy is maximal in the region of λ2,575, and it decreases rapidly with increasing wavelength. The relative germicidal effectiveness as determined with *Bacillus coli* for energy of various wavelengths is as follows: λ2,537, 100; λ2,700, 87; λ2,800, 60; λ2,900, 30; λ3,000, 6; λ3,100, 1.3; λ4,000, 0.01.

Visible energy from λ4,000 to λ7,000 is only one-thousandth to one ten-thousandth as effective as energy of λ2,537. Energy near the short-wave limit of the solar spectrum is relatively weak in germicidal effectiveness. Notwithstanding the high intensities of energy at the earth's surface due to the sun and sky, new germicidal sources provide far greater intensities of lethal energy. This has opened many new fields of disinfection by means of germicidal energy.

Table 26-8. Approximate Ultraviolet Energy and E-vitons Emitted by Various Sources

Source	Total watts	Watts radiated in spectral regions			E-vitons output
		Below λ2800	λ2800–λ3165	λ3165–λ3800	
Sunlamps:					
S-1 (776 glass)	400	0.01	2.7	5.0	68,000
S-4 (776 glass)	100	0.01	0.9	3.6	50,000
RS-4 (776 glass)	100	0.01	1.6	3.3	25,000
RS (776 glass)	275	0.01	1.6	3.3	25,000
Germicidal:					
8-watt (9741 glass)	8	1.5	0.03	0.03	84,000
15-watt (9741 glass)	15	2.9	0.06	0.05	160,000
30-watt (9741 glass)	30	7.3	0.16	0.13	400,000
High-pressure mercury-arcs:					
AH-6 (774 glass)	1000	0.5	40.0	110.0	1,000,000
AH-6 (quartz)	1000	21.0	87.0	124.0	3,500,000
Fluorescent or F sunlamps:					
40-watt (9821 glass)	40	0.01	3.1	7.2	100,000

Table 26-9. Approximate Duration of Exposure to Produce a Minimum Perceptible Erythema (MPE) on Average Untanned Skin by Average Midday Midsummer Sunlight and by Various Artificial Sources in Efficient Reflectors at a Distance of 24 In. from the Skin

Source	Min	Source	Min
Summer sunlight	20	Tungsten CX, 500 watt	50
Type S-1, 400 watt	4	Germicidal, 4 watt	8
Type S-4, 100 watt	5	Germicidal, 8 watt	5
Type RS-4, 100 watt	18	Germicidal, 15 watt	2.7
Type RS, 275 watt	6	Germicidal, 30 watt	0.7
Quartz mercury arc, 360 watt	3	Type AH-6, 774 glass, 1000 watt	0.2
Sunshine carbon arc, 6 mm	10	Type AH-6, quartz, 1000 watt	0.05

Microorganisms vary greatly in their resistivity to germicidal energy. *Bacillus coli* is fairly representative of many bacteria, but some of these are ten times as resistant. Exposures are the products of intensity and time. The resistivity of various organisms including fungi, spores, yeast cells, algae, and protozoa varies over a range of at least 50 to 1 compared with *B. coli*. The wavelength of energy of maximal lethal effectiveness may vary somewhat for these various organisms, but this is a minor matter compared with the variation in resistivity.

The new germicidal sources consisting of low-pressure mercury-discharge sources have germicidal efficiencies so much greater than sources heretofore available that only they are considered here. Basically they are similar to fluorescent lamps but have a tube or bulb of special glass which is not coated with phosphors. The 30-, 15-, and 8-W sources emit, respectively, about 7, 3, and 1.5 W in the region of λ2,537. A 4-W source with a single base is also available. The output of germicidal flux depends upon the composition and thickness of the glass.

About 1 ft below a 30-W germicidal source in a proper reflector, an intensity of germicidal flux of 600 μW/cm^2 is readily obtained. This is 100 times that at the earth's surface during midday on clear days in midsummer.[5]

83. Disinfection of air is practicable for reducing the spread of airborne disease and the contamination of products and processes. Air of such depths as are encountered indoors does not appreciably absorb energy of λ2,537. Therefore, the practical aim is to project this energy through large volumes of air before it reaches surfaces that absorb it. The germicidal efficiency of a given installation indoors increases with the ceiling height. The germicidal sources are located at about the 7-ft level, and the upper stratum of air is irradiated. Natural or artificial circulation of air dilutes the air in the occupied zone so that the concentration of airborne organisms is decreased. The practice is so new and involves so many variables that specifications cannot be discussed in the space available here. Germicidal sources may be installed in air ducts or in other confined spaces if necessary. In any case, the intensity of germicidal flux reaching human eyes and skin should be limited to a small value. Tentatively, this is 0.5 μW/cm^2 for persons exposed for 8 h.[6]

84. Disinfection of water by means of germicidal sources is now far more practicable[7] than in the past with quartz mercury arcs. Water absorbs energy of λ2,537, which makes it necessary to limit the depth. The depth of water that absorbs 90% of the incident germicidal energy varies from a few inches to a few feet for waters from supply systems of various cities and other sources. If the organisms have a resistivity comparable with that of *Bacillus coli*, hundreds of gallons of water per hour can be disinfected to a high degree by a 30-W germicidal source.

85. Disinfection of other materials[1] depends upon the absorption coefficient. Surfaces can be disinfected as well as the surrounding air. Energy of λ2,537 does not penetrate deeply into most substances, but if adequate time is available this means of disinfection and sterilization may be practicable. The depths that absorb 90% of the incident energy are as follows: milk, 0.003 in; wines, 0.03 to 0.1 in; beers, 0.05 in; colorless vinegar, 0.6 in; brown vinegar, 0.1 to 0.2 in.

86. Reflecting Media. In using ultraviolet energy, materials having high reflectances are generally desirable. In using germicidal energy in an occupied room, the upper walls and ceiling should have low reflectances. In Table 26-10 are presented the reflectances for three different wavelengths of ultraviolet energy.[8]

Table 26-10. Reflectances of Materials

Material	% reflectance			Material	% reflectance		
	λ2652	λ2967	λ3650		λ2652	λ2967	λ3650
Aluminum:							
Paints	55–75	56–77	58–79	Magnesium oxide, pressed	77	86	87
Etched and brightened	88	89	91	smoked	93	93	94
Mill finish	56	56	57	Calcium carbonate, pressed	78	83	86
Vaporized on glass	87	88	88	Wall plaster, white	46	65	76
Foil	72	80	84	White water paints	10–33	14–41	40–76
Chromium	44	54	61	Porcelain enamel, white	4.7	5.4	63
Nickel	37	38	45	Paper, white	20–40	30–50	45–70
Stainless steel	20	24	31	Wallpapers	15–30	20–40	30–50
Silver	24	17	77				
Zinc oxide, pressed	2.5	2.5	4				
Titanium oxide	6	6	31				

87. Transmitting Media. The approximate short-wave limit of transparency of ordinary glasses varies with the composition, but in general it may be considered to be at λ3,200. It is not difficult to make glass transparent farther into the ultraviolet region, and a variety of glasses is now available particularly for bulbs and tubes of artificial sources which efficiently transmit energy near the short-wave limit of sunlight, λ2,900 to λ3,000, and even the highly germicidal energy of λ2,537. One practical difficulty is to prevent solarization, an increasing opacity of the glass due to exposure to ultraviolet energy. Heating commonly corrects and prevents solarization of glasses at temperatures far below the melting point of the glass. The transparency of quartz, both natural and fused, varies considerably. The short-wave limit for fused quartz is commonly near λ1,850, and that for natural crystals is sometimes at λ1,250. Fluorite ranks with quartz crystals.

The increasing number of plastics extends the possibilities of transmitting media. Thin sheets commonly transmit somewhat farther into the ultraviolet than ordinary glass. Clear films 0.001 in in thickness were found to transmit 60 to 75% of energy of λ2,967. Thin gelatin is transparent to about λ2,000. However, the deterioration of such materials under exposure to short-wave ultraviolet must be considered. Pure water is transparent to the neighborhood of λ2,000.

88. Various Uses. Inasmuch as photochemical reactions in which ultraviolet energy plays a part are almost numberless, there are too many applications and potential uses to be discussed here.[9] Sterilization of water and other materials, paint and dye testing for permanency, curing and aging processes, chlorination, drying of oils and varnishes, polymerization, vitalizing of food products, and photographic processes are among the present prominent uses of ultraviolet.

During recent years the significance of solar energy outdoors as a powerful environmental factor during the evolution of life has been receiving attention. It is definitely known that even suberythemal dosages of ultraviolet from λ2,800 to λ3,200 are effective in the prevention and cure of rickets. Solar energy outdoors also appears to be influential upon various abnormal conditions, such as tetany, chronic anemias, skin diseases, and extrapulmonary tuberculosis. Through its production of vitamin D in the skin, it aids in the growth of sounder bones and teeth.

Supplying ultraviolet for health along with light for seeing is likely to be the eventual service of light sources alone or combined with special sources of ultraviolet energy. Various practical sources of light and of ultraviolet energy are now available for this purpose. E-vitons are delivered to the occupied area in sufficient intensity just as lumens are in lighting practice. It is necessary to know only that from 15 to 50 E-viton min/cm² produces a minimum perceptible erythema on untanned skin. If the duration of exposure is fixed, the intensity in E-vitons per square centimeter is readily computed.

89. Bibliography

1. LUCKIESH, MATTHEW "Applications of Germicidal, Erythemal and Infrared Energy"; Princeton, N.J., D. Van Nostrand Company, Inc., 1946.

2. LUCKIESH, MATTHEW, and TAYLOR, A. H. Erythemal and Tanning Effectiveness of Ultraviolet Energy; *Gen. Elec. Rev.*, 1939, Vol. 42, p. 274.

3. LUCKIESH, M., and HOLLADAY, L. L. Fundamental Units and Terms for Biologically Effective Radiation; *Jour. Opt. Soc. Am.*, 1933, Vol. 23, p. 197.

LUCKIESH, M., and HOLLADAY, L. L. Nomenclature and Standards for Biologically Effective Ultraviolet Radiation; *Jour. Opt. Soc. Am.*, 1931, Vol. 21, p. 420.

4. LUCKIESH, MATTHEW, TAYLOR, A. H., and KERR, G. P. A Four-year Record of Ultraviolet Energy in Daylight; *Jour. Franklin Inst.*, 1939, Vol. 228, p. 425.

LUCKIESH, MATTHEW, TAYLOR, A. H., and KERR, G. P. Seasonal Variations of Ultraviolet Energy in Daylight; *Jour. Franklin Inst.*, 1944, Vol. 238, p. 1.

5. FORSYTHE, W. E., BARNES, B. T., and EASLEY, M. A. Ultraviolet Sources and Their Radiation; *Jour. Opt. Soc. Am.*, 1934, Vol. 24, p. 178.

6. WELLS, W. F. Bactericidal Irradiation of Air; *Jour. Franklin Inst.*, 1940, Vol. 220, p. 347.

LUCKIESH, MATTHEW, and HOLLADAY, L. L. Tests and Data on Disinfection of Air with Germicidal Lamps; *Gen. Elec. Rev.*, 1942, Vol. 45, p. 223.

LUCKIESH, MATTHEW, and HOLLADAY, L. L. Designing Installations of Germicidal Lamps for Occupied Rooms; *Gen. Elec. Rev.*, 1942, Vol. 45, p. 343.

7. LUCKIESH, MATTHEW, and HOLLADAY, L. L. Disinfecting Water by Means of Germicidal Lamps; *Gen. Elec. Rev.*, 1944, Vol. 47, p. 45.

8. LUCKIESH, MATTHEW, and TAYLOR, A. H. Transmittance and Reflectance of Germicidal Energy; *Jour. Opt. Soc. Am.*, 1946, Vol. 36, p. 227.

9. ELLIS, C., and WELLS, A. A. "The Chemical Action of Ultraviolet Rays"; New York, Chemical Catalog Company, Inc.

SECTION 27

ELECTRONIC DATA PROCESSING

BY

JOHN M. CARROLL Associate Professor of Industrial Engineering, Lehigh University; Senior Member, Institute of Electrical and Electronics Engineers; Registered Professional Engineer (Pennsylvania)

CLAUDE M. SMITH Process Computer Business Section, Information System Division, General Electric Company

CONTENTS

Numbers refer to paragraphs

SECTION 27

ELECTRONIC DATA PROCESSING

By John M. Carroll

INTRODUCTION

1. Definitions. *Electronic data processing* (EDP) consists in (1) organizing data into a form suitable for processing, (2) manipulating new data and previously prepared files according to some plan, and (3) reporting the resulting information in a desired format and at a specified time and place through the use of electronic equipment.

Data are unstructured collections of facts concerning a given process.

Information is useful reports concerning a given process. Information is derived by structuring data according to a specified format.

Origination of data consists in (1) collecting facts when and where they are available, preferably in machine-readable form, (2) converting data to the medium most efficient for subsequent processing, and (3) verifying data to achieve a desired degree of accuracy.

Manipulation of data involves (1) classifying input data by type, (2) ordering them into sequence, (3) updating files, (4) looking up tabulated facts, and (5) performing computations.

Reporting information entails producing output documents, reports, and instructions for various purposes.

Machine-readable form means a type of input capable of producing the desired electrical responses in processing equipment.

EDP equipment consists of an electronic computer and all associated peripheral equipment required to perform the desired operations. In this section, an *electronic computer* will be understood to mean an automatic general-purpose stored-program digital computer whose active elements are vacuum tubes, transistors, or integrated circuit packages (ICP's).

2. Analog Computers. A full discussion of analog computers is beyond the scope of this section, but they will be discussed briefly to show how they differ from digital computers.

An analog computer manipulates one or more continuous variables to produce an output which is some desired function of the input variable or variables. The manipulation is performed by mechanical, electrical, or electronic devices. The slide rule is a rudimentary mechanical analog computer.

1. *Mechanical Analogs.* *a.* Addition. A differential gear (Fig. 27-1*a*) produces rotation of the left axle y and the right axle x, the sum of which, in radians, is equal to the rotation of the drive shaft z,

$$z = x + y$$

b. Multiplication by a constant. Two bevel gears of unequal radii (Fig. 27-1*b*) produce the functional relationship

$$z = kx$$

c. Integration. A ball-disk integrator (Fig. 27-1*c*) consists of a disk rotated by input x. A ball located a distance y from the center of the disk turns an output shaft whose rotation z follows the relationship

$$z = \int y \, dx$$

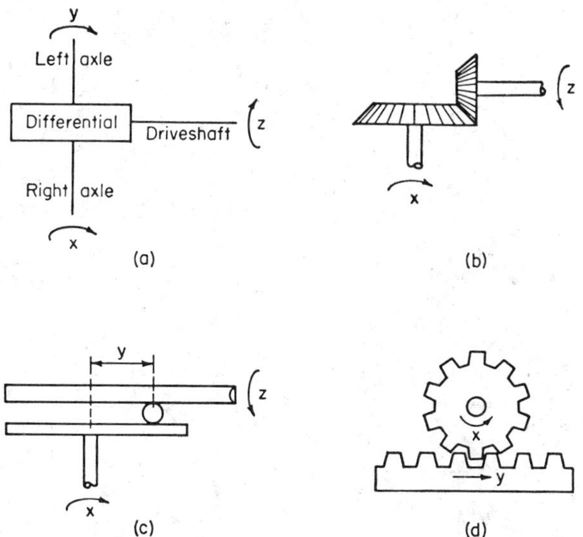

FIG. 27-1. Mechanical analogs. (*a*) Differential gear; (*b*) bevel gears; (*c*) ball-disk integrator; (*d*) rack and pinion.

d. Interconversion. A rack-and-pinion gear (Fig. 27-1*d*) converts a linear displacement y to a shaft rotation x, or *vice versa.*

Mechanical analogs are used in special-purpose computers such as gunsights. They are capable of accuracies of 1 part in 10,000.

2. *Electromechanical Analogs.* *a.* Multiplication by a constant. A linear potentiometer (Fig. 27-2*a*) can multiply a voltage by a constant k, between 0 and 1. The parameter can be varied through 0 by grounding the midpoint of the potentiometer and using an electronic amplifier to produce signal inversion (Fig. 27-2*b*).

b. Function generation. A nonlinear potentiometer can be used to generate any desired function. A device called a *resolver* (Fig. 27-2*c*) uses a potentiometer whose resistance wire is wound on a sinusoid, which is then bent to form a cylinder. Such a device can resolve a vector of magnitude R into its components

$$x = R \cos \theta$$

$$y = R \sin \theta$$

An arbitrary function can be generated by a *servo table* (Fig. 27-2*d*). A wire is bent into a desired function $f(x)$, and a linear potentiometer is driven across the table by a servo constrained to keep the distance x proportional to the input voltage. The voltage obtained at the end of the wire y will be proportional to the amount of resistance tapped off,

$$y = f(x)$$

c. Multiplication. The error-reducing property of a servomechanism is used to produce an electromechanical multiplier (Fig. 27-2*e*). One input y is applied to a servomechanism, and the other input to the servo is constrained to take on the same value by moving the tap on a potentiometer. The tap on a second potentiometer is ganged to the first, and the second input x is applied to the second potentiometer. The output from the tap of the second potentiometer is proportional to the desired product,

$$E_o = xy/100$$

Since division is the inverse of multiplication, the latter operation can be performed by similar arrangements.

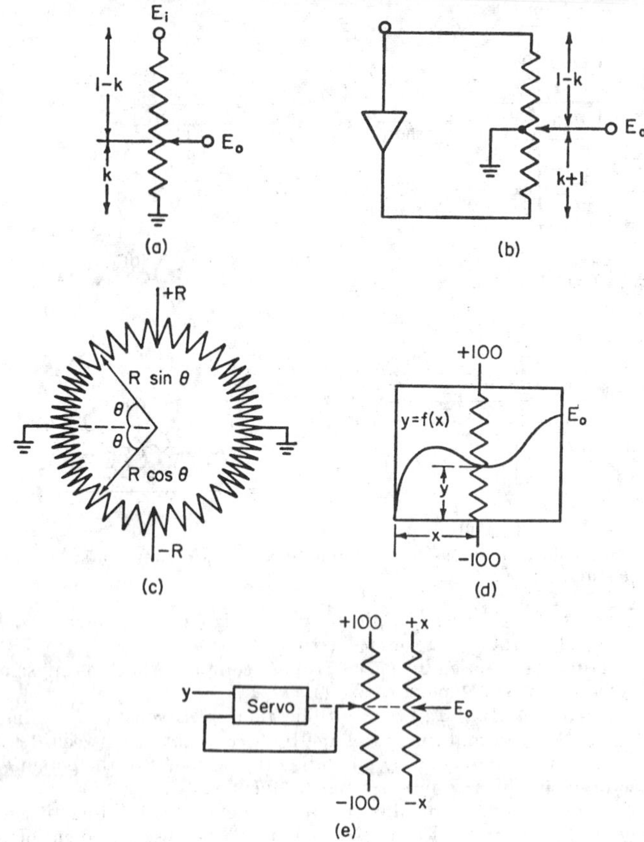

FIG. 27-2. Electromechanical analogs. (*a*) Linear potentiometer; (*b*) linear potentiometer that varies parameter through zero; (*c*) resolver; (*d*) arbitrary function generator; (*e*) servo multiplier.

3. *Electronic Analogs.* The operational amplifier (Fig. 27-3*a*) is the basis of most electronic analog computers. It consists of a three-stage d-c amplifier of high gain having appropriate *RC* circuits in the input and feedback. The input and output voltages have a nominal range of 100 to −100 V. Operational amplifiers are frequently constructed with vacuum tubes.

Analysis of an operational amplifier is performed in terms of its *transfer function.* Since the impedance of the first grid is very high, essentially all the current passing through R_i passes through R_f; hence,

$$\frac{E_a - E_i}{R_i} = \frac{E_o - E_a}{R_f}$$

The ratio of output voltage to the voltage at the first grid is determined by the gain μ of the amplifier,

$$E_o = \mu E_a$$

Eliminating E_a from the equation of the transfer function,

$$\frac{-R_f E_i}{R_i} = E_o - \frac{E_o}{\mu} - \frac{R_f E_o}{R_i \mu}$$

The error in setting E_o equal to $-R_f E_i / R_i$ is proportional to $(1 + R_f/R_i)/\mu$. Since E_o is about 100 V and μ is typically 1,000,000, the approximation is very good and it can be assumed that the voltage at the first grid is zero. The overall accuracy of electronic analog computation ranges from 0.01 to 2% and is largely dependent upon maintaining high gain in the operational amplifiers.

a. Addition. If several inputs (Fig. 27-3*b*) are applied to an operational amplifier,

$$E_o = -\sum_i \frac{R_f}{R_i} E_i$$

Thus the operational amplifier can act as an adder.

b. Multiplication by a constant. The relationship between R_f and R_i can be used to multiply any of the inputs to an operational amplifier by a constant.

c. Integration. If a capacitor is placed in series with the feedback resistor (Fig. 27-3*c*), the output becomes

$$E_o = -\frac{1}{R_i C_f} \int_0^t E_i \, dt + E_{ic}$$

where E_{ic} is the constant of integration (initial condition) and represents the voltage across C_f at time $t = 0$. Hence, the operational amplifier can integrate, simultaneously add several inputs, and multiply each by an appropriate constant.

d. Multiplication. Several schemes have been used to accomplish the multiplication of variables. The most common technique is the *quarter-square* principle,

$$xy = \tfrac{1}{4}[(x + y)^2 - (x - y)^2]$$

The addition, two subtractions, and multiplication by $\frac{1}{4}$ are performed by operational amplifiers. Diode networks are frequently employed to perform the squaring function.

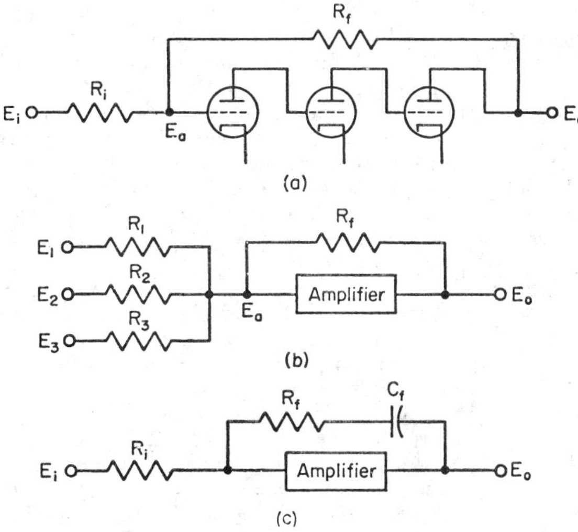

FIG. 27-3. Electronic analogs. (*a*) Operational amplifier; (*b*) addition of variables and their multiplication by constants; (*c*) integration.

4. *Applications.* The general-purpose analog computer is often called a *differential analyzer.* It consists of a collection of operational amplifiers, power supplies, other components, and input/output devices provided with convenient plug-in connections. A frequently used output device is the *xy table,* or plotting board, that produces a graphical representation of the output function.

Special-purpose analog computers such as those used in gunsights or industrial-process controllers have interconnections permanently wired in. There is also a large variety of *network analyzers* in which the flow of electricity is made analogous to the flow of water in pipes, electricity in transmission lines, or petroleum in oil-field deposits.

Occasionally a general-purpose analog computer will be interconnected with a digital computer to produce a *hybrid computer.* These arrangements are usually employed in the simulation of dynamic systems such as aerospace vehicle controls.

3. Digital Calculators. Digital computation deals with discrete rather than continuous variables. The most rudimentary form of digital computing device is the abacus. There are a great many other digital computing devices which are semiautomatic or even automatic in nature and which cannot truly be called computers. These devices, which lack the power, speed, and flexibility of computers, are called *calculators.*

1. *Adding machines* perform addition of pairs of decimal digits and propagation of carries by number wheels which are advanced by gear segments, which may be driven manually or by motor. The terminal positions of the number wheels are determined by stops which are set when the operator depresses number keys (see Fig. 27-4). The wheel is returned to zero by spring tension after a rachet mechanism disengages it from the gear segment. Subtraction is usually accomplished by a reversal of machine drive.

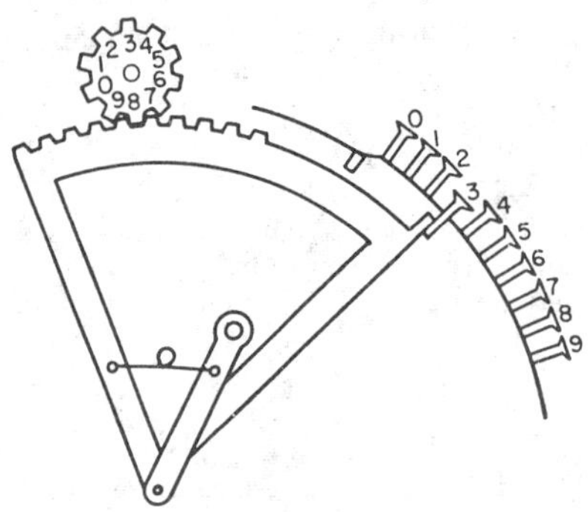

FIG. 27-4. Mechanical addition.

The keyboard may have a column of 10 keys for each digit (*full keyboard*) or only 10 number keys. In the latter, the function of the full keyboard is performed by a register which is set by successive key depressions (10-key machine).

In a machine having *dial display*, the registers are composed of number wheels visible to the operator (Fig. 27-4). Printing, or *listing,* machines cause the contents of registers to be printed on paper tape.

2. *Desk calculators* perform multiplication and division as well as addition and subtraction. Multiplication by a decimal digit m is accomplished by entering the multiplicand in the keyboard and repeating the addition operation m times. Multiplication by 10 is accomplished by shifting each digit of the multiplicand to the left. Division may be performed by repeatedly subtracting the divisor from the dividend and counting the number of operations required to reduce the remainder to less than the divisor. These operations are mechanically programmed into desk calculators. More complex operations such as square root may also be programmed.

3. *Accounting machines* (also called bookkeeping machines) are printing calculators programmed in an operation sequence by sets of program pins attached to a movable carriage. Some features available include: keyboard entry to and printing from several accumulators; transfers between accumulators with or without reset. Print-out can be made on two rolls of printing paper, and a complete alphabetic keyboard is often made available. Multiplication and division are not usually provided, although multiplication by a constant can be programmed.

4. *Tabulators* (such as the IBM 407) accept input on punched cards and print out listings or totals on 12-in-wide paper (120 characters to a line). The tabulator, also called an *electric accounting machine,* will handle 110 to 125 cards a minute (120 machine cycles). It prints both final and subtotals, which it holds in accumulators called *counters*. The number of counters is expandable. The operation of the carriage can be programmed by a loop of punched paper tape, but most of the machine operations are programmed by rear-panel plugboard wiring.

5. *Electromechanical calculators* are of two types, nonprogrammed and programmed. The nonprogrammed machines perform a single arithmetic operation on data from a single card. One model of *multiplying punch* will multiply factors read from a single punched card and punch the result into the same card or print it on paper tape.

Plugboard programmed calculators can execute a program consisting of up to 12 steps. They are equipped with a dozen or more storage registers and accumulators. Typical machine speeds are 200 to 220 machine c/min.

4. Electronic calculators, sometimes confused with electronic computers, have electronic units for storage, calculation, and program control. They operate at relatively high speeds: 100 to 150 cards a minute. Typically addition requires 10 to 50 msec. A typical program can be executed in 325 msec. There may be 100 or more program steps. From 10 to 80 registers are available to store program constants and intermediate results. However, these calculators are considerably less powerful than general-purpose electronic computers because:

a. They are programmed by back-panel plugboard wiring, not by the same medium used to feed in data, which is the way an electronic computer is programmed.

b. Although they can perform all four arithmetic operations and make logical decisions, they can modify their programs in response to a logical decision only by skipping parts of the program or going back to the start. As will be shown, an electronic computer can execute any number of internal program loops.

c. Speed is much slower and storage capacity much smaller than in the case of electronic computers.

5. Digital Computers. The following are characteristics of modern digital computers:

a. Digital. They deal with discrete quantities rather than continuous variables.

b. Automatic. Program control is switched automatically through an unlimited number of program steps depending upon the results of logical decisions made by the machine. Program control may skip forward or loop backward to any desired program step including START, depending upon the program.

c. Stored Program. Program instructions are entered into the machine in the same manner as data to be processed (e.g., on punched cards or magnetic tape). Instructions are stored in the same type of electronic storage locations as are program constants and intermediate results.

d. Electronic. The active elements of the computer are vacuum tubes (now seldom used), transistors, or integrated circuit packages.

e. High-speed. Machine cycle times (access) of early digital computers were in the millisecond range. Later machines operated in the microsecond range. Modern machines have memory access times that are measured in nanoseconds.

f. High-capacity. A large number of electronic storage locations is available. The number of locations is expandable in multiples of 4,000; 8,000 is a typical basic machine configuration. These locations include only those immediately accessible by the machine's control circuitry.

g. Instruction Suite. Most digital computers provide the ability to add and subtract (with both single and double precision), multiply, divide, make logical decisions (determine whether one quantity is less than, equal to, or greater than another), and manipulate data (including shifting right and left).

6. General-purpose Machines. General-purpose computers can be programmed (instructed) to perform a large number of tasks. They may differ in configuration depending upon the applications for which they are intended.

a. Scientific and Engineering. These machines typically perform a large number of calculations on a limited quantity of input data, e.g., computing stress in structural members.

b. Business and Commercial. These machines typically perform a small number of calculations on great quantities of input data, e.g., preparing payroll checks for factory employees.

In early machines there were marked differences in speed, storage capacity, and instruction suite between the two types of digital computer, the scientific machines having greater capability in all three areas. However, today it is customary to use the same basic machine for either application. The major difference today is that business computers typically have a larger and more diverse suite of peripheral equipment, e.g., tape handlers, magnetic character-recognition equipment, optical character readers, etc.

7. Special-purpose Computers. A special-purpose digital computer is designed to solve a specific problem or class of problems in the most efficient manner but has extremely limited capability for solving other types of problems. They are said to be "dedicated" to a specific application and have certain instructions and routines permanently wired in. A few representative applications for special-purpose computers are:

a. Industrial-process control computers are designed to accept continuous inputs from industrial processes, convert them to digital form, compare them with stored program constants, compute control signals, and transmit them to the process under control (see Par. **114**).

b. Digital differential analyzers solve differential equations, using finite differencing techniques.

c. Guidance computers are installed aboard aircraft, missiles, and nuclear submarines and at missile sites. They accept inputs from orthogonal accelerometers aboard the vehicle, compute the distances the vehicle has moved in each direction from a known reference point, compare the vehicle's position with its expected position, and compute corrective signals for the vehicle's control mechanisms.

d. Fire-control computers, used in artillery batteries, accept information on speed, velocity, drift, etc., and produce data for altering range settings, elevation, and azimuthal angles.

e. Message-switching computers (communications terminals) are located at the terminals and switching centers of a communications system and control its switching. The largest such computer-controlled system is the U.S. Air Force Autodin (automatic digital network) system, in which five regional switching centers connect to terminals in each region and to each other. The system has a capacity of 500 terminals. Computers at the regional centers switch messages on the basis of priority and security classification, record all messages, and perform statistical analyses.

8. Computer Organization. A computer is a system that consists of five subsystems (Fig. 27-5).

a. Input. The input subsystem accepts data in machine-readable form (e.g., punched cards, perforated paper tape, magnetic tape, magnetic-ink characters, etc.), translates them into computer code, and transmits them to the processing system for storage in the computer's memory.

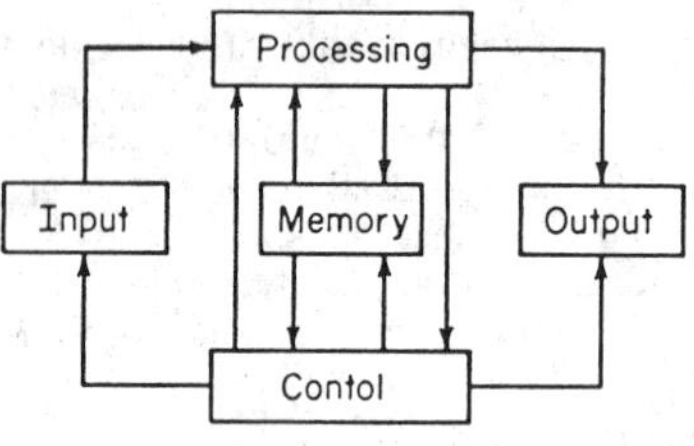

Fig. 27-5. General organization of a digital computer.

b. Memory. The memory, or storage, system stores data and instructions in binary form, using a large bank of two-state electronic devices (most commonly magnetic cores). An *address* provides a unique identification for each memory location. The *immediate-access memory* is backed up for longer-term storage by *addressable bulk storage* (e.g., magnetic-disk storage units) and by *nonaddressable bulk storage* (e.g., magnetic-tape units) which properly form part of the input system.

The immediate access memory is subdivided to provide a *scratch-pad memory.* It is usually part of the main core memory although scratch-pad memories have been constructed as dots of thin magnetic film deposited on glass substrates. The scratch-pad memory helps speed up computer operations, but it is not usually accessible to the programmer.

Another specialized storage subsystem is the *read-only memory.* This memory is not usually accessible to the programmer. It is commonly programmed to add special subroutines to the machine's instruction suite. One manufacturer's read-only memory consists of a deck of copper-plated Mylar punched cards that function as a type of three-dimensional plugboard.

c. Control. The control subsystem originates commands that specify exactly what each part of the computer system is to do and when each operation is to begin.

d. Processing. The processing subsystem is also called the *arithmetic unit.* It performs the four basic arithmetic operations, makes logical comparisons, and edits data to alter their form.

e. Output. The output subsystem accepts binary data from the computer's memory and translates them into the form required by output media. Output media include punched cards, perforated paper tape, high-speed printers, magnetic tape, and video displays.

DATA FOR PROCESSING

9. Number Systems. A number consists of an *absolute value a* selected from the set of values belonging to the number system used and a *positional value* which represents a power of the *base,* or *radix b,* of the number system. Positional values greater than unity are counted to the left of the *number-system point*; positional values less than unity are counted to the right of the point. The radix is the *modulus* of the set of admissible absolute values. For example,

$$a_n \times b^{n-1} + \cdots + a_3 \times b^2 + a_2 \times b^1 + a_1 \times b^0 \cdot a_{1'} \times b^{-1'} + a_{2'} \times b^{-2'} + \cdots + a_m \times b^{-m'} \tag{27-1}$$

is expressed as

$$a_n\, a_{n-1} \cdots a_2\, a_1 \cdot a_{1'}\, a_{2'} \cdots a_{m'}$$

Absolute value a may assume any value from 0 to $b - 1$.

10. Decimal Numbers. In the decimal system the base b equals 10 and the absolute value a may assume any of the values 0, 1, 2, 3, 4, 5, 6, 7, 8, 9. In expressing a number, the power of ten may be written above each position. For example,

millions	hundred thousands	ten thousands	thousands	hundreds	tens	ones
2	5	7	7	4	5	0

or, using the representation of Eq. (27-1),

$$
\begin{aligned}
2 \times 10^6 &= 2{,}000{,}000 \\
5 \times 10^5 &= 500{,}000 \\
7 \times 10^4 &= 70{,}000 \\
7 \times 10^3 &= 7{,}000 \\
4 \times 10^2 &= 400 \\
5 \times 10^1 &= 50 \\
0 \times 10^0 &= 0 \\
\hline
&\quad 2{,}577{,}450
\end{aligned}
$$

Addition. To add decimal digits, it is necessary to remember the 100 combinations of the digits 0 through 9,

+	0	1	2	3	4	5	6	7	8	9
0	0	1	2	3	4	5	6	7	8	9
1	1	2	3	4	5	6	7	8	9	10
2	2	3	4	5	6	7	8	9	10	11
3	3	4	5	6	7	8	9	10	11	12
4	4	5	6	7	8	9	10	11	12	13
5	5	6	7	8	9	10	11	12	13	14
6	6	7	8	9	10	11	12	13	14	15
7	7	8	9	10	11	12	13	14	15	16
8	8	9	10	11	12	13	14	15	16	17
9	9	10	11	12	13	14	15	16	17	18

Wherever a sum exceeds 9, a 1 digit is carried to the next column to the right; all accumulated digits are carried.

Subtraction is the reverse of addition. It is performed by borrowing from the next column to the left whenever a difference falls below 0. The result is checked by adding the *difference* to the *subtrahend* to produce the *minuend.*

Multiplication is repeated addition, but the cumbersome process of adding the *multiplicand* first to a zero and then to the successive sums as many times as the number of the *multiplier* may be avoided by consulting the decimal multiplication table,

×	0	1	2	3	4	5	6	7	8	9
0	0	0	0	0	0	0	0	0	0	0
1	0	1	2	3	4	5	6	7	8	9
2	0	2	4	6	8	10	12	14	16	18
3	0	3	6	9	12	15	18	21	24	27
4	0	4	8	12	16	20	24	28	32	36
5	0	5	10	15	20	25	30	35	40	45
6	0	6	12	18	24	30	36	42	48	54
7	0	7	14	21	28	35	42	49	56	63
8	0	8	16	24	32	40	48	56	64	72
9	0	9	18	27	36	45	54	63	72	81

Division consists in repeatedly subtracting the *divisor* from the *dividend* until a difference less than the divisor is obtained. This difference is the *remainder*. The number of subtraction operations performed is the *quotient*.

11. Binary Numbers. In the binary-number system the base b equals 2, and the admissible values of a are 0 and 1. In expressing a number, the power of two may be written above each position,

thirty-twos	sixteens	eights	fours	twos	ones
1	0	1	1	1	0

Using the representation of Eq. (27-1),

$$\begin{aligned}1 \times 2^5 &= 32\\ 0 \times 2^4 &= \ \ 0\\ 1 \times 2^3 &= \ \ 8\\ 1 \times 2^2 &= \ \ 4\\ 1 \times 2^1 &= \ \ 2\\ 0 \times 2^0 &= \ \ 0\\ &\ \ \overline{46}\end{aligned}$$

12. Binary Arithmetic. *a. Addition.* The binary-addition table comes from the upper left-hand corner of the decimal-addition table,

+	0	1
0	0	1
1	1	*

The asterisk indicates that there is no digit 2 in the binary system. Instead, 2 is represented by 10, which means 0 and carry 1. Therefore,

+	0	1	
0	0	1	
1	1	0	and carry 1

Example. Add 101, 111, 001, 110.

```
   1   }
1111   }Carry digits
 101    Augend
 111    Augend
 001    Augend
 110    Addend
-----
10011   Sum
```

b. Subtraction. Subtraction is the reverse of addition.

Example. Subtract 1101, 1001, 1111, and 0010 from 1100110. Check by addition.

```
1100110  Minuend
   1101  Subtrahend
   1001  Subtrahend
   1111  Subtrahend
   0010  Subtrahend
111111
  11111  Borrow digits
    1
-------
 111111  Difference
```

Check.

```
    1101
    1001
    1111
    0010
  ------
  100111   Sum of subtrahends
 +111111   Difference
 -------
 1100110   Minuend
```

c. Multiplication. The binary multiplication table comes from the upper left-hand corner of the decimal multiplication table,

×	0	1
0	0	0
1	0	1

In computers, binary multiplication is accomplished by forming the product of the multiplicand by each digit of the multiplier, shifting the position of each digit to agree with the position of the digit, and adding the product.

Example. Multiply 101100101 by 1011.

```
101100101       Multiplicand
1011            Multiplier
---------
101100101       Step 1. Multiplicand is reproduced
                  by leftmost digit of multiplier
 000000000      Step 2
----------
1011001010      Sum
  101100101     Step 3
-----------
11011111001     Sum
   101100101    Step 4
------------
111101010111    Sum (=product)
```

Note that the addend of each addition is shifted one place to the right of the previous addend.

d. Division. The divisor may be held in a fixed position and the dividend shifted one place to the left for each step. In each step, the digits of the dividend are compared with those of the divisor. When the divisor digits are equal to or smaller than the dividend digits, a 1 is written in the quotient, the divisor is subtracted from the dividend, and the difference written as a new dividend. This dividend is shifted to the left, and the process repeats. When the divisor digits are larger than the dividend digits, a zero is written in the quotient and the dividend is shifted left.

Example. Divide 111101010111 by 1011.

```
                                        Quotient Digits
    111101010111   Dividend
1011               Divisor (larger)           0
   111101010111    Dividend shifted
1011               Divisor (larger)           0

  111101010111     Dividend shifted
1011               Divisor (larger)           0

111101010111       Dividend shifted
1011               Divisor (smaller)          1
-----------
 10001010111       New dividend

10001010111        Dividend shifted
1011               Divisor (larger)           0

10001010111        Dividend shifted
 1011              Divisor (smaller)          1
-----------
  110010111        New dividend
```

110010111 1011	Dividend shifted Divisor (smaller)	1
110111	New dividend	
110111 1011	Dividend shifted Divisor (larger)	0
110111 1011	Dividend shifted Divisor (larger)	0
110111 1011	Dividend shifted Divisor (smaller)	1
1011	New dividend	
1011 1011	Dividend shifted Divisor (larger)	0
1011 1011	Dividend shifted Divisor (equal)	1
R = 0000	(no remainder)	

The resulting quotient digits 101100101 provide a check on the multiplication operation of the previous example.

Thus all the operations of binary arithmetic can be accomplished by addition, subtraction, shifting right, shifting left, and comparing two numbers to see whether one is equal to, smaller than, or larger than the other (logical decisions).

13. Decimal-to-binary Conversion. To convert a decimal number to a binary number, successively divide the decimal number by 2. Record a 1 if the division leaves a remainder; record a 0 if there is no remainder. Start the binary number with the *last* binary digit recorded.

Example. Convert 181 to binary.

	Remainder
2)181	
2)90	1
2)45	0
2)22	1
2)11	0
2)5	1
2)2	1
2)1	0
0	1

Thus the binary equivalent of 181 is 10110101.

14. Binary-to-decimal Conversion. A binary number may be converted to decimal by adding the products of the absolute and positional values.

Example. Convert 101100101 to decimal.

$$
\begin{aligned}
1 \times 2^8 &= 256\\
0 \times 2^7 &= 0\\
1 \times 2^6 &= 64\\
1 \times 2^5 &= 32\\
0 \times 2^4 &= 0\\
0 \times 2^3 &= 0\\
1 \times 2^2 &= 4\\
0 \times 2^1 &= 0\\
1 \times 2^0 &= 1\\
\hline
&\ 357
\end{aligned}
$$

Thus the decimal equivalent of 101100101 is 357.

15. Octal Numbers. Octal numbers are sometimes used as an intermediate conversion between binary and decimal numbers. In some computers, memory maps, or dumps, which are helpful in debugging programs, are printed out in octal. In the octal-number system the base b is equal to 8, and the admissible values for a are 0, 1, 2, 3, 4, 5, 6, 7.

16. Binary-to-octal Conversion. Binary numbers can be converted to octal by inspection. Group the binary digits in triads starting at the binary point, and write the decimal value of each triad to obtain the corresponding octal digits.

Example. Convert 111101010111 to octal.

7	5	2	7
111	101	010	111

Thus the octal equivalent of 111101010111 is 7527.

17. Octal-to-decimal Conversion. An octal number may be converted to decimal by adding the products of the absolute and positional values.

Example. Convert octal 7527 to decimal.

$$\begin{aligned} 7 \times 8^3 &= 3584 \\ 5 \times 8^2 &= 320 \\ 2 \times 8^1 &= 16 \\ 7 \times 8^0 &= 7 \\ &\overline{3927} \end{aligned}$$

Thus the decimal equivalent of octal 7527 is 3927.

18. Decimal-to-octal Conversion. It is sometimes necessary to enter corrections to programs in octal form. A decimal number may be converted to octal by making successive divisions by 8 and tallying the remainders.

Example. Convert 8,000 to octal.

	Remainder
8)8,000	
8)1,000	0
8) 125	0
8) 15	5
8) 1	7
0	1

Thus the octal equivalent of 8,000 is 17,500.

19. Binary-coded Decimal Numbers. Each decimal digit may be represented by a group of four binary digits in a form known as binary-coded decimal numbers. This form is often used in computers programmed for commercial applications, since the conversion between the binary and decimal systems is greatly simplified.

Decimal Equivalent	Binary Bit Set 8	4	2	1
0	0	0	0	0
1	0	0	0	1
2	0	0	1	0
3	0	0	1	1
4	0	1	0	0
5	0	1	0	1
6	0	1	1	0
7	0	1	1	1
8	1	0	0	0
9	1	0	0	1

20. Complementation. Subtraction may be performed in computers by complement addition.

a. Nines Complement. The nines complement of a decimal number is formed by subtracting each digit from 9. Nines-complement subtraction is performed by forming the nines complement of the subtrahend and adding it to the minuend with an *end-around carry.*

Example. Subtract 346 from 799.

The nines complement of 346 is 653.

```
 799
+653
----
1452
 ↪1   End-around carry
----
 453
```

b. Ones Complement. The ones complement of a binary number is obtained by making each 0 into a 1 and each 1 into a 0. Ones-complement subtraction is performed by adding the ones complement of the subtrahend to the minuend with an end-around carry.

Example. Subtract 1010 from 1111.

The ones complement of 1010 is 0101.

```
  1111
 +0101
 -----
 10100
    ↪1   End-around carry
 -----
   101
```

21. Excess-3 Code. The process of forming the natural ones complement of binary-coded-decimal (BCD) numbers can result in numbers unacceptable to the system. For example, the complement of 0010 (decimal 2) is 1101 (decimal 13), which is not a member of the set of BCD numbers. However, by adding binary 0011 (decimal 3) to each BCD binary bit set, the numbers are so coded that the nines complement of the equivalent decimal digit can be obtained by forming the ones complement of the binary bit set.

Excess-3 Code

Decimal equivalent	Natural BCD	Excess-3	Nines complement
0	0000	0011	1100
1	0001	0100	1011
2	0010	0101	1010
3	0011	0110	1001
4	0100	0111	1000
5	0101	1000	0111
6	0110	1001	0110
7	0111	1010	0101
8	1000	1011	0100
9	1001	1100	0011

Example. Subtract 346 from 799.

```
 799 =   1010   1100   1100   Excess-3
-346 =   1001   1000   0110   Nines complement
         ----   ----   ----
         0011   0100   0010
            1      1      1
         ----   ----   ----
 453 =   0100   0101   0011   Natural BCD
```

The excess-3 code is a commonly used representation of decimal digits in computer codes.

22. Gray Code. The Gray code (also called the *cyclically permuted*, or *reflected binary*, code) is used in analog-to-digital converters that make use of a coding disk to obtain a digital representation of shaft rotation. In the Gray code, in proceeding from any digit to the next larger, there is always a change of one and only one digit, either from a 0 to a 1 or from a 1 to a 0. In analog-to-digital conversion, use of the Gray code holds errors, caused by reading part of the digits in one sector of the code wheel and part in another, to a maximum of 1 in the least significant digit.

Gray Code

Decimal Equivalent	Gray Code
0	0000
1	0001
2	0011
3	0010
4	0110
5	0111
6	0101
7	0100
8	1100
9	1101

Gray-code representations may be converted to decimal numbers by use of the following formula:

$$x = \sum_{i=0}^{k-1} \pm a_i(b^{i+1} - 1)$$

Example. Convert the Gray representation 1011 to a decimal number.

Starting with the leftmost binary digit, taking it as positive, and according each succeeding *nonzero* quantity an alternate sign (a equals 1 or 0 and b equals 2; i depends upon position),

$$\begin{aligned} +1(2^{3+1} - 1) &= 15 \\ 0(2^{2+1} - 1) &= 0 \\ -1(2^{1+1} - 1) &= -3 \\ +1(2^{0+1} - 1) &= \underline{\ 1} \\ & \quad 13 \end{aligned}$$

23. Hexadecimal Numbers. The hexadecimal numbering system has been adopted to represent characters and addresses in several third-generation computers. In the hexadecimal system, the number base b is 16, and there are 15 admissable values for a, including the numbers 0 to 9 and the alphabetic characters A to F.

Hexadecimal Marking System

Decimal (base 10)	Hexadecimal (base 16)	Binary (base 2)
0	0	0000
1	1	0001
2	2	0010
3	3	0011
4	4	0100
5	5	0101
6	6	0110
7	7	0111
8	8	1000
9	9	1001
10	A	1010
11	B	1011
12	C	1100
13	D	1101
14	E	1110
15	F	1111

Note that the hexadecimal system makes full use of the four-binary-bit set, which is only partly used in the BCD system.

24. Binary-to-hexadecimal Conversion. Binary numbers can be converted to hexadecimal by inspection. Group the binary digits by fours, starting at the binary point, and evaluate each group in the hexadecimal system:

Example. Convert 1101101111 to hexadecimal.

$$0011 \quad 0110 \quad 1111 = 36F$$

25. Hexadecimal-to-decimal Conversion. A hexadecimal number may be converted to decimal by adding the products of the absolute and positional values.

Example. Convert hexadecimal 36F to decimal.

$$\begin{aligned} 3 \times (16)^2 &= 768 \\ 6 \times (16)^1 &= 96 \\ F = 15 \times (16)^0 &= \underline{\ 15} \\ & \quad 879 \end{aligned}$$

26. Decimal-to-hexadecimal Conversion. A decimal number may be converted to hexadecimal by making successive divisions by 16 and tallying the remainders.

Example. Convert 16,000 to hexadecimal.

```
                    Remainder
16)16,000
16) 1,000              0
16    ) 62             8
16     ) 3          14 = E
         0             3
```

Thus decimal 16,000 equals hexadecimal 3E80.

27. IBM Punched Cards. Punched cards are the principal input to digital computers. The IBM card measures $7\frac{3}{8}$ by $3\frac{1}{4}$ in. One corner, usually the upper-left-hand one, is cut diagonally. The card's format is a 12×80 matrix. The rows, starting with the top, have the values 12, 11, 0, 1, 2, 3, 4, 5, 6, 7, 8, 9. Each column represents a digit. Values are recorded by making one or more rectangular punches in the desired column. Punches in rows 12, 11, and 0 are called *zone punches*; punches in rows 1 to 9 are called *digit punches*. The *Hollerith code* used with IBM cards provides for representing the numerals 0 to 9, alphabetical characters A to Z, and several special characters by combinations of zone and digit punches.

Hollerith Code

Character	Zone punches	Digit punches
0	0	
1	..	1
2	..	2
3	..	3
4	..	4
5	..	5
6	..	6
7	..	7
8	..	8
9	..	9
A	12	1
B	12	2
C	12	3
D	12	4
E	12	5
F	12	6
G	12	7
H	12	8
I	12	9
J	11	1
K	11	2
L	11	3
M	11	4
N	11	5
O	11	6
P	11	7
Q	11	8
R	11	9
S	0	2
T	0	3
U	0	4
V	0	5
W	0	6
X	0	7
Y	0	8
Z	0	9
&	12	
#	..	3-8
,	0	3-8
-	11	
/	0	1
@	0	4-8

28. Remington Rand Punched Cards. The Remington Rand card is said to contain 90 columns. However, it is actually a 12×45 matrix divided horizontally into two fields. Within each field, six rows represent values and the 45 columns represent characters. Thus the card can contain up to 90 characters. The rows have the values 0, 1 or 2, 3 or 4, 5 or 6, 7 or 8, and 9. In the *Powers code* used with Remington Rand cards, the presence of a nine punch indicates that the dual-valued punch or punches for that character assume their higher values.

Powers Code

Character	Values	Actual punches
0	0	0
1	1	1/2
2	2	1/2-9
3	3	3/4
4	4	3/4-9
5	5	5/6
6	6	5/6-9
7	7	7/8
8	8	7/8-9
9	9	9
A	2-6	1/25/6-9
B	1-5	1/2-5/6
C	0-7	0-7/8
D	0-3-5	0-3/4-5/6
E	0-3	0-3/4
F	2-8	1/2-7/8-9
G	5-7	5/6-7/8
H	3-7	3/4-7/8
I	3-5	3/4-5/6
J	1-3-5	1/2-3/4-5/6
K	4-6	3/4-5/6-9
L	0-9	0-9
M	0-5	0-5/6
N	0-6	0-5/6-9
O	1-3	1/2-3/4
P	1-3-7	1/2-3/4-7/8
Q	3-5-7	3/4-5/6-7/8
R	1-7	1/2-7/8
S	1-5-7	1/2-5/6-7/8
T	4-8	3/4-7/8-9
U	0-5-7	0-5/6-7/8
V	0-4	0-3/4-9
W	0-3-7	0-3/4-7/8
X	0-8	0-7/8-9
Y	2-4	1/2-3/4-9
Z	5-7	5/6-7/8-9

29. Six-bit Code. Since the four-bit BCD code is inadequate to represent the complete character set desired for computer print-out, the *zone-bit* technique is combined with a basic BCD notation to produce a six-bit code. There are four possible zone-bit combinations: 00, 11, 10, and 01. Combined with the *digit bit* groups (in excess-3 or BCD code), the six-bit code can represent a character set of 40 (or more). The table below illustrates how a complete computer character set can be represented by the six-bit code:

Alphanumerical Character

Zone bits	Numeric bits													
	0000	0001	0010	0011	0100	0101	0110	0111	1000	1001	1011	1100	1101	1110
00	0	1	2	3	4	5	6	7	8	9	#	@	–	=
01	+	A	B	C	D	E	F	G	H	I	.	..	..	..
10	–	J	K	L	M	N	O	P	Q	R	$	*	..	..
11	Space	/	S	T	U	V	W	X	Y	Z	,	%	(	)

The foregoing code is representative of six-bit computer codes, but there are wide differences among the codes used by different manufacturers. If fact, different codes may be used within a single computer system, for example, in core storage and on magnetic tape.

30. Multimode Code. Within a given computer either pure binary or binary-coded-decimal (six-bit) representation may be used. For example, consider a system using a 20-bit word.

Bit number:	0	1	2	3	4	5	6	7	8	9	10	11	12	13	14	15	16	17	18	19
Binary	1	1	1	1	1	1	1	1	1	1	1	1	1	1	1	1	1	1	1	1
		+									262,143									
BCD	0	0	Z 1		Digit 1				Z 2		Digit 2				Z 3		Digit 3			

Bit number 0 is a code indicating whether the *recording mode* is binary (1) or BCD (0). Bit number 1 is a sign bit in the binary mode, 1 indicating positive and 0 indicating negative (in BCD, the sign would be a distinct character). The remaining 18 bits represent either an 18-bit binary number or three BCD characters in 6-bit alphanumeric code.

31. Extended-binary Code. Computer users have become dissatisfied with the conventional uppercase print-out, especially when the computer is used to prepare reports for management.

In third-generation computers, the basic BCD bit set has been extended from six to eight bits. The set of eight bits (or six, as the case may be) is known as a *byte*. The eight-bit code allows an extended set of characters which can include lower case alphabetic characters and several control characters such as "tab," "backspace," "line feed," etc. The extended binary-coded-decimal interchange code (EBCDIC) is shown in Table 27-1.

Table 27-1. Extended Binary-coded-decimal Interchange Code (EBCDIC)

BIT POSITIONS → 76, 54

3210	00				01				10				11			
	00	01	10	11	00	01	10	11	00	01	10	11	00	01	10	11
0000	NULL				SP	&	–									0
0001							/		a	j			A	J		1
0010									b	k	s		B	K	S	2
0011									c	l	t		C	L	T	3
0100	PF	RES	BYP	PN					d	m	u		D	M	U	4
0101	HT	NL	LF	RS					e	n	v		E	N	V	5
0110	LC	BS	EOB	UC					f	o	w		F	O	W	6
0111	DEL	IL	PR	EOT					g	p	x		G	P	X	7
1000									h	q	y		H	Q	Y	8
1001									i	r	z		I	R	Z	9
1010			SM		¢	!										
1011					.	$	,	#								
1100					<	*	%	@								
1101					(	)	_	'								
1110					+	;	>	=								
1111					|	¬	?	"								

BIT POSITIONS: 76543210

Control characters:

NULL — All zero-bits
PF — Punch off
HT — Horizontal tab
LC — Lower case
DEL — Delete
RES — Restore
NL — New line
BS — Backspace
IL — Idle
BYP — Bypass
LF — Line feed
EOB — End of block
PR — Prefix
PN — Punch on
RS — Reader stop
UC — Upper case
EOT — End of transmission
SM — Set mode
SP — Space

Note:
Chart is read by order of significance as designated by "BIT POSITIONS," i.e., 7 is 2^7 bit, 6 is 2^6 bit. . .etc.

For example:
E is 11 00 0101

32. USA Standard Code. Computer systems differ widely in their codes. This fact has led to chaotic situations when several computers of different types were linked together into networks. Furthermore, there has been an increasing requirement for computers to accept data directly from communications lines without complex interfacing systems for code-to-code conversions. An answer to these problems is the USA Standard Code for Information Interchange (ASCII, pronounced *Ash-Key*). Most modern computers will accept either the EBCDIC or the ASCII code as specified by the program. The ASCII code utilizes an eight-bit byte. (See Table 27-2 for the code and Sec. **26** for a discussion of the problems of transmitting computer data over communications circuits.)

33. Parity Bits. A frequent machine error is dropping a bit, because of dirt or some flaw in the magnetic storage medium, or picking up an extraneous bit, because of incomplete erasure of previously recorded data. These errors may be detected by a technique known as parity checking. A bit position is added to each computer word

Table 27-2. USA Standard Code for Information Interchange (Extended to Eight bits)

BIT POSITIONS → 76, 54

3210	00 (76) 00 (54)	00 01	00 10	00 11	01 00	01 01	01 10	01 11	10 00	10 01	10 10	10 11	11 00	11 01	11 10	11 11
0000	NULL	DLE			SP	0					—	P			@	p
0001	SOH	DC1			!	1					A	Q			a	q
0010	STX	DC2			"	2					B	R			b	r
0011	ETX	DC3			#	3					C	S			c	s
0100	EOT	DC4			$	4					D	T			d	t
0101	ENQ	NACK			%	5					E	U			e	u
0110	ACK	SYNC			&	6					F	V			f	v
0111	BELL	ETB			'	7					G	W			g	w
1000	BS	CNCL			(	8					H	X			h	x
1001	HT	EM			)	9					I	Y			i	y
1010	LF	SS			*	:					J	Z			j	z
1011	VT	ESC			+	;					K	[			k	{
1100	FF	FS			,	<					L	\			l	|
1101	CR	GS			—	=					M	]			m	}
1110	SO	RS			.	>					N	^			n	¬
1111	SI	US			/	?					O	`			o	DEL

BIT POSITIONS: 76543210

Control characters:

NULL — Null/idle
SOH — Start of heading (CC)
STX — Start of text (CC)
ETX — End of text (CC)
EOT — End of transmission (CC)
ENQ — Enquiry (CC)
ACK — Acknowledge (CC)
BELL — Audible or attention signal
BS — Backspace (FE)
HT — Horizontal tabulation (punch card skip) (FE)
LF — Line feed (FE)
VT — Vertical tabulation (FE)
FF — Form feed (FE)
CR — Carriage return (FE)
SO — Shift out
SI — Shift in

DLE — Data link escape (CC)
DC1, DC2, DC3 — Device controls
DC4 — Device control (stop)
NACK — Negative acknowledge (CC)
SYNC — Synchronous idle (CC)
ETB — End of transmission block (CC)
CNCL — Cancel
EM — End of medium
SS — Start of special sequence
ESC — Escape
FS — File separator (IS)
GS — Group separator (IS)
RS — Record separator (IS)
US — Unit separator (IS)
DEL — Delete

Note:
Chart is read by order of significance as designated by "BIT POSITIONS," i.e., 7 is 2^7 bit, 6 is 2^6 bit...etc.
For example:
E is 10 10 0101

or to each byte on magnetic tape so that the lateral bit count of ones is always even (*even parity*) or odd (*odd parity*) depending upon the design of the machine.

Example. Even parity checking in core storage.

Consider a 20-bit computer word consisting of the BCD characters AND:

Bit number:	0	1	2	3	4	5	6	7	8	9	10	11	12	13	14	15	16	17	18	19	20
	0	0	0	1	0	0	0	1	1	0	0	1	0	1	0	1	0	1	0	0	1
					A						N						D				P

There are seven ones in the first 20 bit positions; so a one is added in the parity *P* position to make the bit count *even.*

Example. Odd parity checking in magnetic-tape storage. Consider a six-bit byte recorded laterally on magnetic tape which consists of the BCD character A:

Bit number:	0	1	2	3	4	5	6
	0	1	0	0	0	1	1
				A			P

Since there are two ones in the first six bit positions, a one is added in the parity position to make the bit count *odd.* This representation is referred to as *seven-channel* magnetic tape.

Parity bits are computed each time a word is transferred to or from storage and each time a byte is read from or written onto magnetic tape. The computed bit is compared with the bit carried along. If an error is detected, the parity-bit computation will be repeated one or more times (depending upon the design of the machine). If recomputation confirms that a parity error exists, the computer will stop or go to a recovery routine depending upon its programming.

34. Error Correction. Some computers have a wired-in facility not only for detecting but also for correcting parity errors. In addition to the lateral parity bit, a *longitudinal* parity byte is computed for each block of, say, 10 data bytes. This system permits locating the erroneous digit and rewriting it correctly without the machine having to stop on parity error or go to a programmed recovery routine.

Example. Consider a block of 10 bytes on seven-channel magnetic tape with *orthogonal* odd parity check for error correction.

Decimal representation	Zone bits	Digit bits	Parity
1	00	0001	0
2	00	0010	0
3	00	0011	1
4	00	0100	0
5 (error)	00	0111	1 ×
6	00	0110	1
7	00	0111	0
8	00	1000	0
9	00	1001	1
0	00	0000	1
Parity	11	1110 ×	0

The lateral parity check will discover an error in the fifth byte of this block. The longitudinal parity check will discover an error in the fifth channel. The machine will then erase the one in the fifth channel of the fifth byte, thus eliminating the extraneous bit and correcting the block of data. Note that this system will correct only a single error in a block.

35. Floating Point. In floating-point arithmetic, the computer converts numbers to a form consisting of a mantissa and an exponent, or characteristic. The characteristic is the power of ten (or other radix) required to change the mantissa into its original decimal form. In normal form, the mantissa has one place to the left of the decimal point. The standard form of the mantissa for 1,358.8 is therefore 1.3578 with an exponent of 3. To avoid negative exponents, the exponent is often written in excess form, such as excess-50, in which 50 is added to the exponent.

Example. Express 0.00181 in excess-50 floating point,

$$0.00181 = 1.81E\,47$$

Express 1,357.8 in excess-50 floating point.

$$1{,}357.8 = 1.357E53$$

Example. Express 0.000332 in normal floating point.

$$0.000332 = 3.32 - 04$$

Express 841.925 in normal floating point.

$$841{,}925 = 8.41925 + 05$$

PROCESSING AND STORAGE

36. Boolean Algebra. Boolean, or logical, algebra provides a mathematical technique for designing and analyzing the switching circuits used in digital computers.

a. It permits convenient calculation of switching circuits.

b. It provides basic theorems for simplifying or reducing a given switching-circuit network.

c. It is a straightforward way of describing computer circuitry.

Admissible States. Two states are admissible in boolean algebra: 1 and 0. The states may represent the *truth* or *fallacy* of a proposition, the *closed* or *open* states of a switch, or the presence or absence of a pulse.

Basic Operators. There are three basic boolean logic operators: *addition* (also called *disjunction, logical sum, inclusive or, union*), *multiplication* (also called *conjunction, logical product, and,* or *intersection*), and *negation* (also called *complementation* or *not*).

a. Logical addition. The rules of logical addition are

$$0 + 0 = 0$$
$$0 + 1 = 1$$
$$1 + 0 = 1$$
$$1 + 1 = 1$$

b. Logical multiplication. The rules of logical multiplication are

$$0 \cdot 0 = 0$$
$$0 \cdot 1 = 0$$
$$1 \cdot 0 = 0$$
$$1 \cdot 1 = 1$$

c. Negation. The rules for negation are

$$\overline{0} = 1$$
$$\overline{1} = 0$$

These basic operators, and three derived operations, *equivalence, exclusive or,* and *implication,* are summarized in Table 27-3.

Table 27-3. Important Boolean Operations

Symbol	Interpretation	Logical translation	Terminology
A	Not A	$\overline{A}$	Negation
A+B	A or B	A+B	Disjunction, logical sum, inclusive OR
A · B	A and B	A·B	Conjunction, logical product
A=B	A if and only if B	$A \cdot B + \overline{A} \cdot \overline{B}$	Equivalence
A≠B	Either A or B	$A \cdot \overline{B} + \overline{A} \cdot B$	Exclusive OR
A→B	If A, then B	$\overline{A}$+B	Implication

The rules for applying the operators are summarized in the form of *truth tables* (Table 27-4).

Table 27-4. Boolean Truth Tables

A	0	1	0	1
B	0	0	1	1
$\overline{A}$	1	0	1	0
A+B	0	1	1	1
A · B	0	0	0	1
A=B	1	0	0	1
A≠B	0	1	1	0
A→B	1	0	1	1

Logical operations are conveniently explained in terms of *set theory*. A set is a collection of objects sharing a common characteristic. Operations with sets are represented by a *Venn diagram* (Fig. 27-6). The members of a set are enclosed by a continuous line. All objects in the *universe* under consideration are enclosed in the rectangle. This is called the *entire set*. The members of a set are said to assume the truth value 1 or 0. In Fig. 27-6*a*, the shaded area represents objects belonging to the set A; the unshaded area represents objects of the entire set not belonging to the set A, that is, $\overline{A}$ or the logical complement of A.

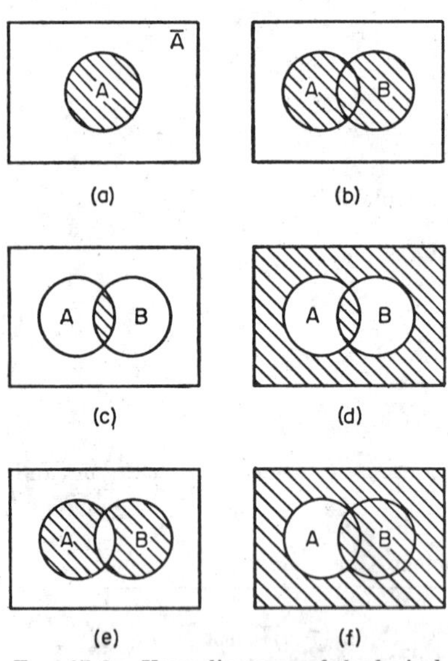

Fig. 27-6. Venn diagrams of the logical operations. (*a*) Complementation; (*b*) inclusive or; (*c*) product; (*d*) equivalence; (*e*) exclusive or; (*f*) implication.

In Fig. 27-6*b*, the shaded area represents objects belonging to set A, set B, or both sets A and B. This is the logical sum of A and B, or A + B. This operator is called the *inclusive or*, meaning and/or.

In Fig. 27-6*c*, the shaded area represents objects belonging to both sets A and B, that is, the logical product of A and B, or A·B.

The objects in the shaded area of Fig. 27-6*d* represent objects that are *equivalent* with respect to the characteristics of sets A and B; they either share these characteristics or possess none of them.

In Fig. 27-6*e*, the shaded areas represent objects that possess the characteristics of either set A or B but not both. This operator is called the *exclusive or*, that is, *or* but not *and*.

Figure 27-6*f* illustrates *implication*. The shaded area represents the set of objects for which A implies B. If A is true and A implies that B is true, then B must be true; but if A implies that B is true and B is true, A may or may not be true.

37. Boolean Theorems. A list of boolean theorems is given in Table 27-5. These relationships are useful in reducing logical expressions, an essential step in the design of computer circuitry.

Table 27-5. Theorems of Boolean Algebra

Proposition	Name
1. $A+A=A$	Absorption rule for addition
2. $A\cdot A=A$	Absorption rule for products
3. $A+B=B+A$	Commutative rule for addition
4. $A\cdot B=B\cdot A$	Commutative rule for products
5. $(A+B)+C=A+(B+C)=A+B+C$	Associative rule for addition
6. $(A\cdot B)\cdot C=A\cdot(B\cdot C)=A\cdot B\cdot C$	Associative rule for products
7. $A\cdot(B+C)=A\cdot B+A\cdot C$	Distributive rule for products
8. $A+B\cdot C=(A+B)\cdot(A+C)$	Distributive rule for addition
9. $A+I=I$	Idempotent rule for addition
10. $A\cdot I=I$	Idempotent rule for products
11. $0+A=A$	Nilpotent rule for addition
12. $0\cdot A=0$	Nilpotent rule for products
13. $\overline{A}+A=I$	Definition of the entire set
14. $A\cdot\overline{A}=O$	Definition of the empty set
15. $\overline{A\cdot B}=\overline{A}+\overline{B}$	De Morgan's rule for products
16. $\overline{A+B}=\overline{A}\cdot\overline{B}$	De Morgan's rule for addition
17. $\overline{\overline{A}}=A$	Second complement rule
18. $A+A\cdot B=A$	
19. $A+\overline{A}\cdot B=A+B$	
20. $A\cdot(A+B)=A$	
21. $A\cdot B+B\cdot C+C\cdot A=(A+B)\cdot(B+C)\cdot(C+A)$	
22. $A\cdot\overline{B}+A\cdot C+B\cdot C=A\cdot\overline{B}+B\cdot C$	

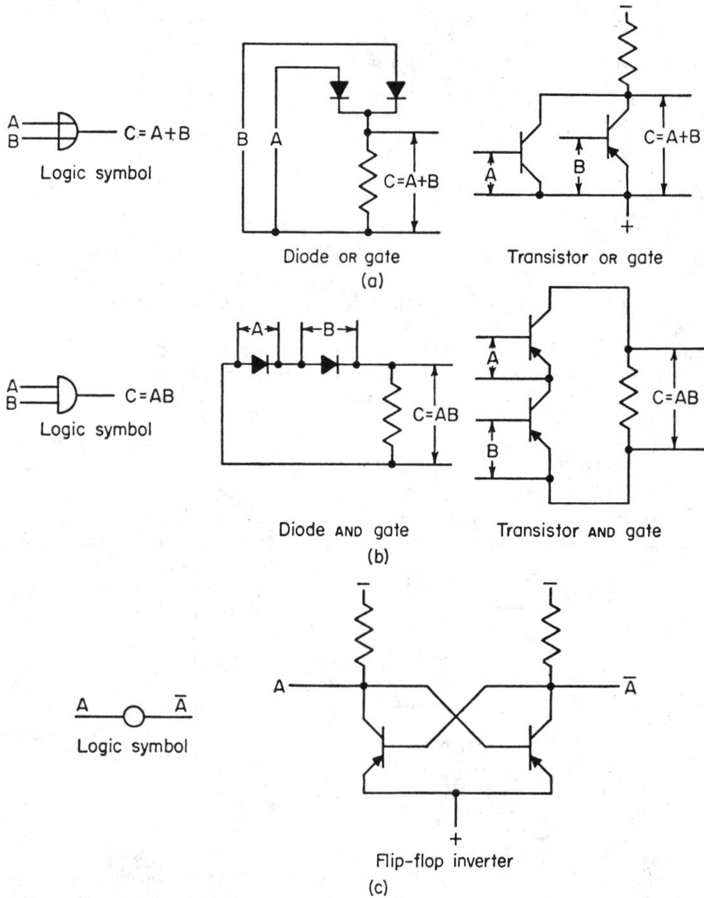

FIG. 27-7. Logic symbols and realization of boolean functions. (*a*) *or*; (*b*) *and*; (*c*) *not*.

38. Logical Design. The logical designer starts with two known sets of values: (1) the states that the inputs to the logical network can assume and (2) the outputs desired for each input condition. There are two basic approaches: (1) the sum of products and (2) the product of sums.

The inputs and desired outputs are written down in the form of a truth table. The sum terms are formed by writing down the variables corresponding to zeros in the truth table and complementing the variables corresponding to ones in the truth table. The product terms are formed by writing the variables corresponding to ones in the truth table and complementing the variables corresponding to zeros in the truth table.

Inputs		Output C	Sum terms	Product terms
A	B			
0	0	1	(A+B)	$\bar{A}\bar{B}$
0	1	0	$(A+\bar{B})$	$\bar{A}B$
1	0	0	$(\bar{A}+B)$	$A\bar{B}$
1	1	1	$(\bar{A}+\bar{B})$	AB

The *product of sums* is formed by choosing the terms for which the output is *zero* and multiplying them together. The *sum of products* is formed by selecting the terms for which the output is *one* and adding them together,

$$(A + \overline{B})(\overline{A} + B) = C$$

$$AB + \overline{A}\,\overline{B} = C$$

or

$$\overline{A}\,\overline{B} + AB = C$$

thus demonstrating that the two techniques are equivalent.

39. Or Gates. The function of boolean addition can be realized by an electronic circuit that *mixes* the inputs and *limits* the output to a standardized value. One such circuit is the diode-or gate; the function can also be realized by transistor circuitry (Fig. 27-7*a*).

40. And Gates. The function of boolean multiplication can be realized by an electronic circuit that *gates* the inputs and *limits* the output to a standardized value. One such circuit is the diode-and gate; the function can also be realized by transistor circuitry (Fig. 27-7*b*).

41. Inverters. The function of boolean complementation can be realized by an electronic circuit that *inverts* its input. Vacuum-tube and transistor amplifiers perform this function. However, a *flip-flop* is commonly employed (Fig. 27-7*c*).

42. Flip-Flops. Figure 27-8 is a basic flip-flop circuit utilizing a pair of *p-n-p* transistors in the grounded-base connection. It will be analyzed in terms of negative-pulse logic. Assume that transistor *Q*2 is saturated and *Q*1 is cut off. The collector of *Q*2 will be close to ground potential, while the collector of *Q*1 will be at negative potential. The voltage-divider action of resistors *R*4 and *R*5 will maintain a positive potential at the base of *Q*1, which will keep it cut off. The flip-flop is thus held in one of its two stable states. This is the *zero state*, since there is a negative potential at the collector of *Q*1.

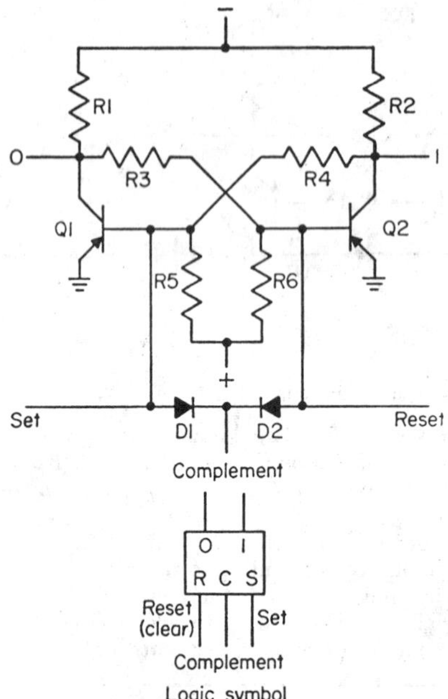

Fig. 27-8. Rudimentary flip-flop circuit and logic diagram.

a. Set. If a negative pulse is applied to the *set line,* the negative potential on the base of *Q*1 will bring the transistor out of cutoff. As *Q*1 begins to conduct, the current through *R*1 will increase and the voltage-divider action of *R*1, *R*3, and *R*6 will make the base of *Q*2 less negative. The current through *Q*2 will decrease. When the collector of *Q*2 approaches ground potential, *R*3 and *R*6 will hold the base of *Q*2 positive with respect to the emitter and will keep it cut off. The flip-flop will now be in its *ONE state* and there will be a negative potential at the collector of *Q*2.

b. Reset. If a negative pulse is now applied to the *reset line, Q*2 will be brought out of cutoff and driven toward saturation. As current through *R*2 increases, voltage-divider action of *R*2, *R*4, and *R*5 will make the base of *Q*1 less negative and the current through *Q*1 will decrease. When the collector of *Q*1 approaches ground potential, *R*4 and *R*5 will hold its base positive and keep it cut off. The flip-flop will now have been returned to its ZERO state, or *cleared,* and there will again be a negative potential at the collector of *Q*1.

c. Complement. A negative pulse applied to the *complement line* will be *steered* by diodes *D*1 and *D*2 to both the set and the reset inputs. The negative pulse will have no effect if a transistor is already saturated, but if it is cut off, the pulse will bring it out of cutoff and drive it toward saturation. Thus if a flip-flop is in the ONE state, a pulse on the complement line will reset, or clear, the flip-flop. If the flip-flop is in the ZERO state, a pulse on the complement line will set the flip-flop.

43. Quarter Adders. Adders are boolean circuits that perform the function of arithmetic addition. The quarter adder has two inputs A and B and produces their sum S at the output. (The carry output is not present in this configuration.) The truth table corresponds to the boolean exclusive-or function.

Inputs		Output S	Sum terms
A	B		
0	0	0	$(A+B)$
1	0	1	$(\overline{A}+B)$
0	1	1	$(A+\overline{B})$
1	1	0	$(\overline{A}+\overline{B})$

$$S = (A + B)(\overline{A} + \overline{B}) = A\overline{B} + \overline{A}B$$

This function can be realized by two inverters, two and gates and one or gate (see Fig. 27-9*a*).

44. Half Adder. A half adder accepts inputs A and B and produces both sum S and carry C outputs.

Inputs		Outputs		Sum terms	Product terms
A	B	S	C		
0	0	0	0	$(A+B)$	AB
1	0	1	0	$(\overline{A}+B)$	AB
0	1	1	0	$(A+\overline{B})$	AB
1	1	0	1	$(\overline{A}+\overline{B})$	AB

The product-of-sums expression for S is the same as before,

$$S = A\overline{B} + \overline{A}B$$

The sum-of-products expression for C is

$$C = AB$$

Thus a half adder is realized by paralleling the quarter adder with an and gate (see Fig. 27-9*b*).

45. Full Adder. A full adder accepts as inputs A, B, and a carry from a preceding stage C_i. It produces a sum output S and an output carry C_o. The truth table is as follows,

Inputs			Outputs	
A	B	C	S	C_o
0	0	0	0	0
0	0	1	1	0
0	1	0	1	0
0	1	1	0	1
1	0	0	1	0
1	0	1	0	1
1	1	0	0	1
1	1	1	1	1

which results in the following boolean expressions:

$$S = \overline{A}\,\overline{B}C_i + \overline{A}B\overline{C}_i + A\overline{B}\,\overline{C}_i + ABC_i$$

$$C_o = \overline{A}BC_i + A\overline{B}C_i + AB\overline{C}_i + ABC_i$$

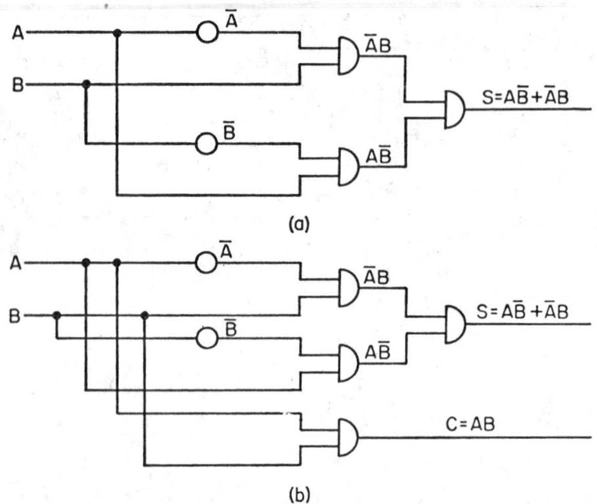

FIG. 27-9. Logic diagrams of (*a*) quarter adder and (*b*) half adder.

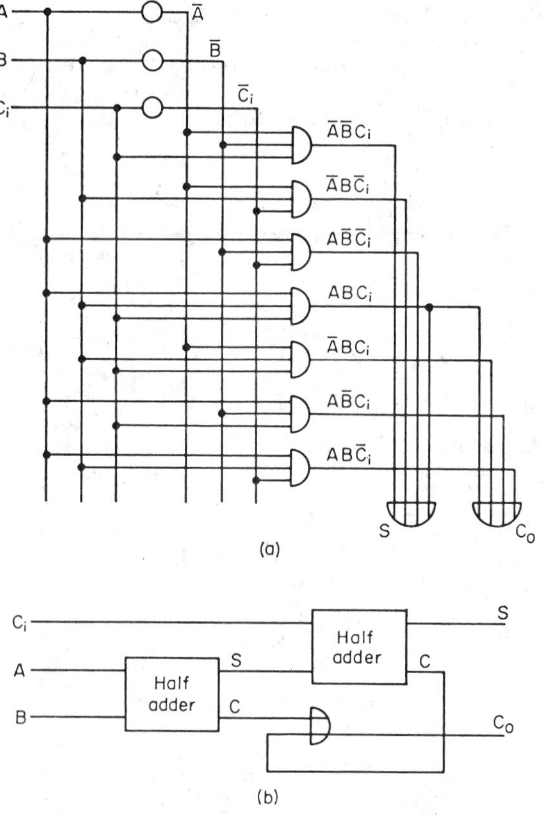

FIG. 27-10. Realization of a full adder (*a*) by direct implementation of the boolean function and (*b*) by cascading half adders.

Note that one and gate is redundant. The boolean expressions can be realized directly by logic circuitry (Fig. 27-10*a*) or by using two half adders and an or gate (Fig. 27-10*b*). The first approach requires three inverters, seven and gates, and two or gates. The second approach requires four inverters, six and gates, and three or gates. The logic designer will base his choice of approach on the relative cost of the basic circuits and upon whether his system is *modularized* at the gate or half-adder level.

46. Series vs. Parallel Machines. In a *bit*-serial machine bits enter and leave the arithmetic unit over single lines in the form of pulse trains. They are said to be *clocked* in and out under control of a *master clock* or highly stable pulse generator.

In a *bit-parallel* machine a separate line is provided for each bit of a computer word. Information is transferred a word at a time rather than a bit at a time.

A serial organization requires less circuitry than parallel organization, but computer operations are slower in a serial machine. The serial organization is used in small computers and some special-purpose machines. Most large general-purpose machines are parallel.

There are hybrid approaches which are employed in some machines, such as bit-parallel, byte-serial, etc.

A serial machine is most compatible with input-output equipment such as electric typewriters, paper-tape reader-punches, teleprinters, and other communications gear. A parallel machine is most compatible with peripheral equipment such as magnetic-tape handlers and high-speed printers.

When a parallel machine must work with serial input-output equipment, serial capability is provided in the form of *buffer storage.*

47. Adder Circuits. Full adders can be designed for serial operation or for parallel operation.

a. Serial Adder. The inputs to a serial adder consist of two pulse trains which are clocked into a single adder. A one-bit-time delay line is provided so that the carries arrive at the input in proper time phase. The output will be another pulse train. The least significant bits of a number are the rightmost pulses of the train (see Fig. 27-11*a*).

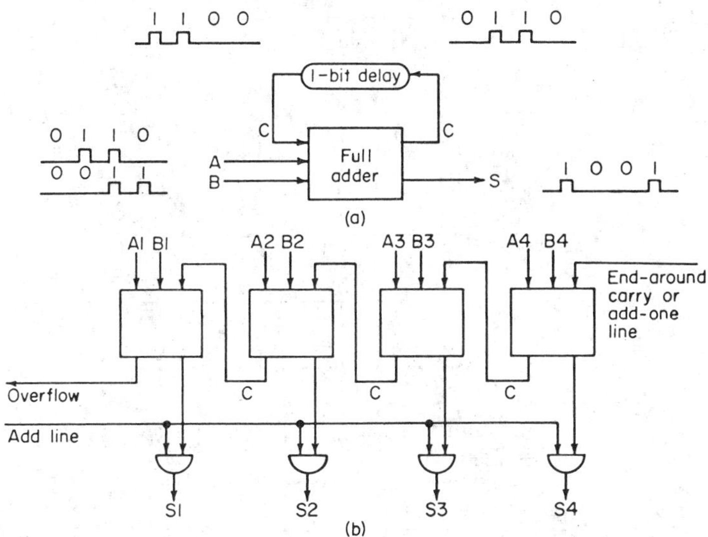

FIG. 27-11. (*a*) Serial full adder and (*b*) parallel full adder.

b. Parallel Adder. A four-bit parallel adder is shown in Fig. 27-11*b*. There are four lines for the bits of the addend, four lines for the bits of the augend, and four lines for the sum bits. The *overflow line* transmits carry bits that exceed the most significant bit of the *register* (in this case, the four flip-flops). The *add line* transmits an *enabling pulse* from the computer's control circuitry. The extreme rightmost input line may accept carries from the overflow line or an *add*-1 signal from the computer's control circuitry.

48. Register Operations. A register is a group of flip-flops each one of which stores a binary bit of the number stored in the register. (Other types of registers, e.g., magnetic cores, have been used, but flip-flops are a common realization of this function.)

a. Clear Register. In this operation a *clear pulse* at the input removes any information stored in the flip-flops, returning them all to the zero state (see Fig. 27-12*a*). (A three-bit register is used to illustrate these functions; in practice, there will be one flip-flop for each bit of the computer word.)

b. Complement. A pulse on the *complement line* changes all ones stored in the flip-flops to zeros and all zeros to ones (see Fig. 27-12*b*).

c. Transfer. The contents of the lower register are transferred to the upper register, regardless of the state of the upper register flip-flops, upon receipt of a *transfer pulse.* This is called *double-line transfer* (see Fig. 27-12*c*). Transfer may be accomplished with half as many and gates if the control circuitry (or program) is required to clear the upper register before transfer is initiated. In this case, the 1 output only is utilized.

d. Shift. In a *shift register,* the contents are shifted one place to the right (left) each time a pulse is applied on the shift input line. The register shown in Fig. 27-12*d* is loaded serially. Shifting may also be accomplished by a transfer operation between registers so aligned as to produce the desired bit shift. When the output of a shift register is tied back to the input with appropriate logic and delay circuitry, the result is a *circulating register,* often used for dynamic storage in machines or to provide right (left) shifting without loss of information.

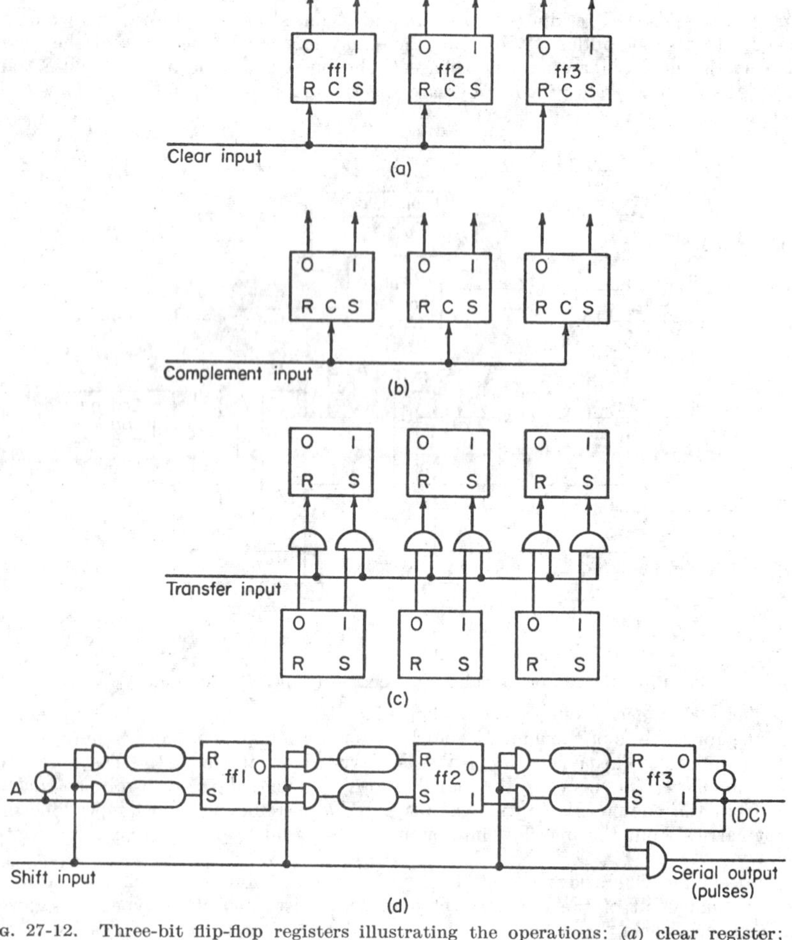

FIG. 27-12. Three-bit flip-flop registers illustrating the operations: (*a*) clear register; (*b*) complement; (*c*) double-line transfer; (*d*) shift right.

49. Counters. A counter records the number of events that have occurred. Counters are used to keep track of elapsed time and operations. Figure 27-13*a* shows a four-stage binary counter. It consists of a chain of flip-flops. The *clear line* sets each flip-flop to zero to start a new count cycle. The input line accepts pulses representing the events to be counted.

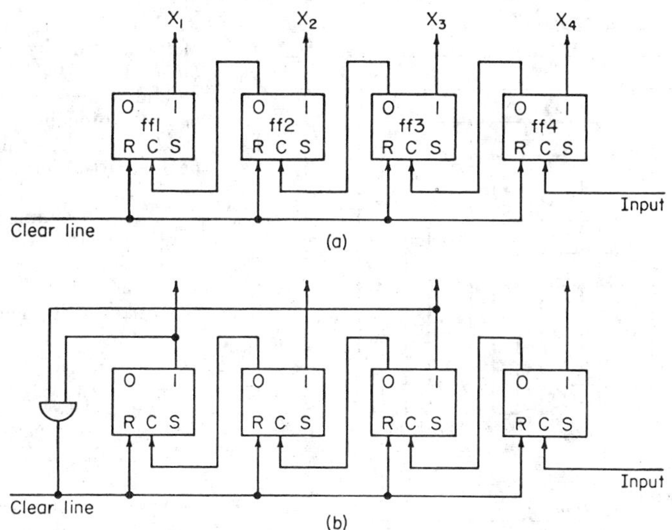

FIG. 27-13. (*a*) Four-stage binary counter and (*b*) decimal counter using 8-4-2-1 code.

a. Binary Counter. The total number of states a binary counter can have is 2^n, where n is the number of flip-flops. The counter shown counts up to 15 and resets on the sixteenth pulse. Each flip-flop stage effectively divides its input by 2. The sixteenth pulse complements FF4, causing FF3, FF2, and FF1 to be complemented in turn.

b. Decimal Counter. The four-stage binary counter may be modified to count on a scale of 10 by connecting the 1 outputs of FF3 and FF1 to the clear line through an and gate. Thus, when the count of 10 is achieved, the entire counter will be reset. The output of a four-stage decimal-connected binary counter will be in the 8-4-2-1 BCD code (see Fig. 27-13*b*).

50. Encoders and Decoders. These are special logic circuits that perform code-to-code conversion in computing machinery.

a. Encoders. An encoder translates data from decimal form to binary or binary-coded-decimal form. The encoder shown in Fig. 27-14*a* assumes that there is a separate line for each decimal digit. This could be realized by use of a key set. There is one output line for each binary position; thus the output is bit-parallel, byte-serial.

b. Decoders. The decoder is the reverse of the encoder. There is one input line for each BCD position and one output line for each decimal digit (see Fig. 27-14*b*).

51. Immediate-access Storage. The storage of data within a computer's central processing unit (CPU) is called immediate-access storage. It is also known as the computer's *memory*. Storage of data outside the CPU is called *bulk storage* and is properly considered as part of the input and output subsystems. Immediate-access storage is realized by means of flip-flop registers, core memories, magnetic drums, or thin-film memories. The last type of memory is functionally similar to core storage although potentially capable of faster operation than either cores or drums.

Core storage is relatively expensive and affords *access times* in the microsecond and high-nanosecond ranges. Drum storage is less expensive, usually provides greater capacity than core storage, and typically affords access times in the millisecond range. In using drum storage, the efficiency of computer operation is often dependent upon programming techniques that take advantage of the rotation of the drum. These techniques are collectively known as *minimum-latency coding.*

Most modern computers utilize core storage. Drums are occasionally used to store programs in control computers and general-purpose computers having *multiprogramming* capabilities.

52. Core Storage. A magnetic core is a toroidal-shaped component made from ferromagnetic ceramic material. (Some cores for special applications are made by

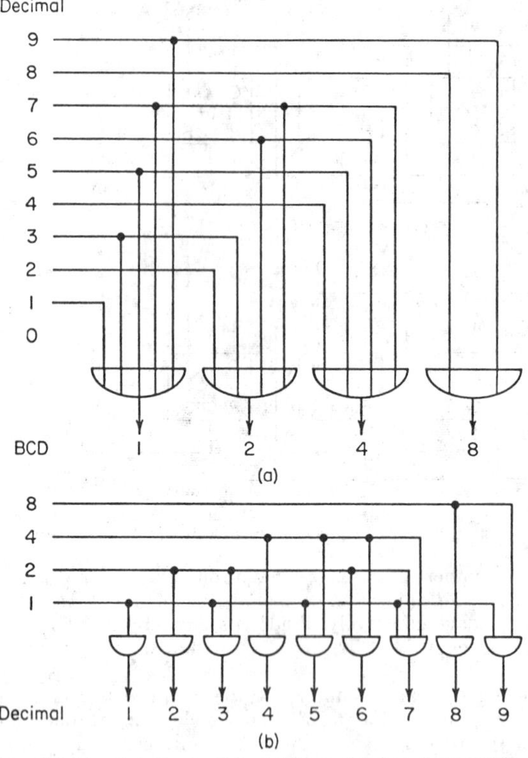

FIG. 27-14. (*a*) Decimal-to-BCD encoder and (*b*) BCD-to-decimal decoder.

winding thin ferromagnetic metal around a ceramic bobbin.) If a wire is wound around this core and current passed through this *input winding,* the core will become magnetized in either a clockwise or a counterclockwise direction, depending upon the direction of the current. The retentivity of the material is such that it will retain a large part of its magnetizing flux when the current is removed. This property is utilized to produce a bistable storage device. A second, or *sense, winding* is used to determine whether a core is magnetized in its *one* or *zero* state. Suppose that a pulse of current is passed through the core of sufficient magnitude and in the proper direction to magnetize the core in its one state. If the core is already in its one state, the current induced in the sense winding (which is proportional to the rate of change of flux in the core) will be small. If the core is in its zero state, however, the rate of change of flux will be high and a much larger pulse will be induced in the sense winding.

53. Magnetic-core Planes. A core plane having only nine cores is shown in Fig. 27-15 to illustrate the operations of *reading, writing,* and *sensing.* Each core has two *read-write windings.*

Suppose that all nine cores are in the zero state and it is desired to switch core 2, 2 to the one state. To do this, a current pulse of magnitude I_m must be passed through core 2, 2. To the select core 2, 2, *half-select pulses* of magnitude $+I_m/2$ are impressed on line $X2$ and on line $Y2$. The currents are *coincident* in only core 2, 2, and consequently only this core is magnetized in the one state. This operation is called *writing.*

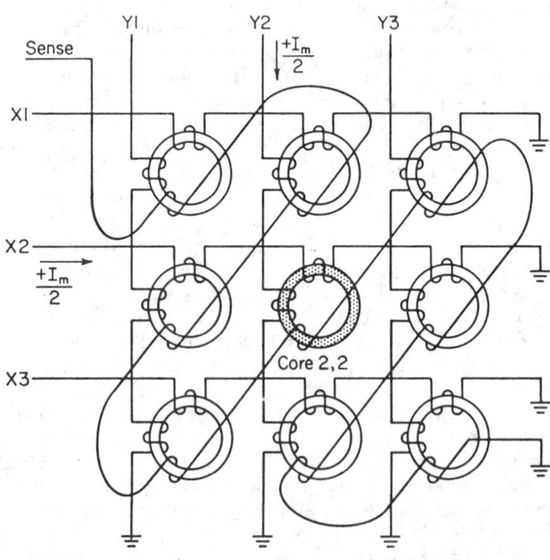

FIG. 27-15. Nine-core memory showing read-write and sense windings.

To read the contents of core 2, 2, *half-select* read pulses of magnitude $-I_m/2$ are impressed on lines $X2$ and $Y2$. The currents are coincident in core 2, 2, and the rate of change of flux in this core is sufficient to register an output pulse on the *sense* line, which links all cores in a diagonal fashion to minimize interaction with the read-write windings.

In practice, there are 64 X lines and 64 Y lines and a core plane thus consists of 4,096 cores. There is a single sense winding. This configuration comprises the basic module of a core memory. Each core represents one bit of a computer word. Thus a module of core memory is said to contain 4K computer words.

54. Magnetic-core Stacks. There is one core plane for each bit of a computer word. If a computer word consists of, say, 20 bits, there will be 21 core planes—one for each bit and one for the parity bit. The planes are stacked one atop the other to form a 4K module.

The windings of a given X or Y read-write line link all the cores belonging to each word on the line. In a computer having a 20-bit word, this would be $64 \times 21 = 1{,}344$ cores. Thus the write operation would switch all cores belonging to the selected word into the one state.

To enable writing zeros as well as ones, a second diagonal winding (inhibit) links all the cores of each plane. During the write operation, a half-select pulse of magnitude $-I_m/2$ is impressed on the *inhibit windings* of the core planes corresponding to the bit positions that are desired to be left in the zero state.

Example. Suppose that it is desired to write the computer word 10101111 into core-memory position 256 of a computer having an eight-bit word. Half-select write pulses of magnitude $+I_m/2$ would be impressed on read-write lines $X64$ and $Y4$, and half-select inhibit pulses of magnitude $-I_m/2$ would be placed on the inhibit windings of core planes $P2$ and $P4$.

To read this word out of memory, half-select read pulses of magnitude $-I_m/2$ would be impressed on read-write lines $X64$ and $Y4$, and output pulses would be registered on the sense windings of core planes $P1$, $P3$, $P5$, $P6$, $P7$, and $P8$.

Note that the read operation is *destructive*. The operation destroys the information written into the cores. To restore this information, the pulses on the sense windings are stored in flip-flops as well as being transmitted to the registers in which information is required. A write cycle is initiated following the read operation, and the information stored in flip-flops is once again written into its proper place in the core stack. Furthermore, core memory is said to be *volatile* inasmuch as information is lost if system power is dropped during computer operations.

Switching current in core memories is supplied by current amplifiers called *core drivers* that operate typically at ampere levels.

The ferrite core memory described is representative of word-addressed memories, the most common kind. In addition, there are data-addressed memories (DAM). In these memories, it is possible sequentially to compare any desired bit sequence of all the words in memory with a bit sequence stored in a register. When a match is achieved, the XY coordinates of the word containing the desired bit sequence are made available for further processing.

55. Magnetic Drums. A magnetic drum is a rotating cylinder coated with a thin layer of magnetic material having the same properties as the material used in magnetic cores (i.e., a square *hysteresis loop*). They rotate at speeds of 120 to 75,000 r/min. Information is recorded by magnetizing small areas of the drum surface. This is accomplished by *recording heads* (see Fig. 27-16*a*). The heads may be used to read and write, or separate read and write heads may be provided. Drums may have as few as 15 or as many as 400 heads. The heads are mounted along the surface of the drum (see Fig. 27-16*b*). The area which passes under a given head as the drum rotates is called a *track*. Each track is subdivided into *cells*, each capable of storing one bit. All the cells that are under a set of heads at the same time are called a *slot*. One track has a series of *timing pulses* permanently recorded on it to provide timing for the drum. The timing track determines the location of each slot of storage cells. Information is packed on drums with densities that vary from 200 to 1,000 bits/in.

a. Parallel Operation. When a drum is operated in the parallel mode, all the bits of a computer word are written or read simultaneously using as many read-write heads (i.e., tracks) as there are bits in the computer word. To locate a particular *address* on the drum, it is necessary to select the proper set of read-write heads, then utilize a counter to count the timing pulses until the desired slot passes under the heads.

b. Serial Operation. When a drum is operated in the serial mode, all the bits of a computer word are recorded on a single track. The track is subdivided into *sectors,* each capable of storing one word. A vacant cell may be left between sectors. To address a specific word in the serial mode, it is necessary to select the proper read-write amplifier for the track and to count timing pulses until the start of the desired sector is attained, then read the word out a bit at a time.

c. Serial-Parallel Operation. When BCD information is recorded on a drum in serial fashion, the 8-4-2-1 channels of the BCD code may be recorded in parallel on four adjacent tracks. Such a group of tracks is called a *band.*

d. Dynamic Registers. The registers of a computer may be realized by a magnetic drum. Such an arrangement requires separate read and write heads spaced 180 or 90° apart around the periphery of the drum. Pulses are continually read on and off the drum. The separation between the reading and writing heads is made to correspond to one-word time, thus achieving the delay function required of *circulating* or dynamic registers (see Fig. 27-16*c*).

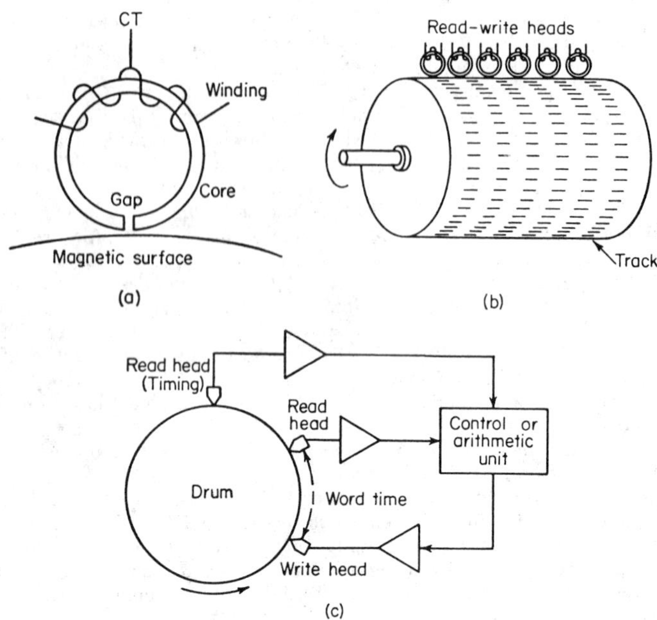

Fig. 27-16. (*a*) Read-write head, (*b*) magnetic drum, and (*c*) drum used as a dynamic register.

56. Central Processing Unit. The central processing unit of a computer consists of the arithmetic, control, and memory subsystems. Figure 27-17 shows a typical CPU. It consists of the core memory, arithmetic unit, and nine registers.

a. M *register.* The M register receives all information entering or leaving the computer and all information entering or leaving the core memory.

While a word is in the M register, the parity bit is recomputed and checked.

b. B *register.* All words going to either the arithmetic or the control circuitry pass through the B register. It holds the addend in addition, the subtrahend in subtraction, the multiplicand in multiplication, and the divisor in division.

c. I *Register.* The *I* register is part of the control subsystem and holds the instruction the computer is executing.

d. P *Register.* The P register is a counter that keeps track of program steps and holds the address of the next instruction.

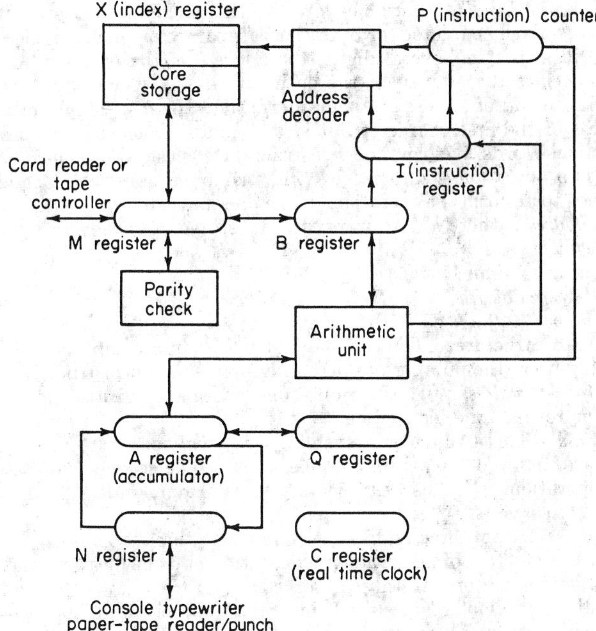

FIG. 27-17. Typical central processing unit.

e. A *Register.* The A register is the accumulator. It holds computer words prior to storage or directed movement to other registers. It holds the augend in addition and receives the sum; it holds the minuend in subtraction and receives the difference; it receives the most significant half of the product in multiplication; and it holds the most significant half of the dividend in division and receives the remainder. Data can be entered into the accumulator by means of console switches.

f. Q *Register.* The Q register holds the multiplier in multiplication and receives the least significant half of the product; it holds the least significant half of the dividend in division and receives the quotient. It functions as an extension of the accumulator in double-precision addition and subtraction and in receiving the accumulator overflow in shift-right operations.

g. N *Register.* The N register acts as a one-character buffer when transferring information from the accumulator to the console typewriter or paper-tape reader-punch (and *vice versa*).

h. X *Register.* The X registers are index registers. Physically they are special locations in core storage. They are used in *address modification.*

i. C *Register.* The C register is a counter that functions as a *real-time clock* to provide a time base for the computer's log (which is kept on the console typewriter).

57. Data Words. A data word may be fixed-length or variable-length.

a. Fixed-length Words. In a fixed-word-length computer, the memory locations store a specified number of digits. The entire block is transferred each time data are moved. The registers, arithmetic circuitry, etc., are all designed to handle a block of characters. Fixed word-length machines thus have parallel transfer of information and parallel arithmetic. The machine discussed in Par. **56** is a fixed-word-length machine. Some late-model machines have a fixed word length of, say, 36 bits that represent four bytes plus parity in the EBCDIC or ASCII code, with each byte separately addressable.

The entire word must be used even if only one digit is to be stored in memory. To circumvent this waste of memory several data items may be *packed* in one word. When any part of the data is needed, it is unpacked by using *editing* instructions.

A fixed-word-length machine has a fixed precision for arithmetic. This difficulty is overcome by use of wired-in *double-precision* arithmetic and *floating-point* arithmetic.

b. Variable Word Length. Variable-word-length machines use serial transfer and serial arithmetic. Each character (or byte) has a separate memory position, or *address*. Since a word will consist of several digits, the address of the leftmost (or rightmost) digit location specifies the address of the entire word. A special bit called a *flag bit* is used to mark the end of a word. As an alternative, the instruction may specify not only the address of the first character but also the number of characters in the word.

58. Instruction Words. A computer instruction consists of an *operation code* and an *operand*. The operation code specifies what is to be done; the operand specifies what is to be used to perform the operation. The operand can be:

a. Memory address, where a data word is to be stored or retrieved.

b. An addressable register.

c. Memory address of an instruction.

d. An input-output device.

e. Literal numbers to be used in the program.

58a. Command Structure. The command structure of a computer can take several forms, depending upon the number of operand addresses in an instruction word.

a. Single- (one-) address. There is only one operand, and a complete arithmetic cycle requires more than one instruction.

b. Two-address. The two-address system uses two operands, which specify the locations of both factors of an arithmetic operation. The second address receives the result of the operation and thus destroys the original data unless programming provisions are made to save it.

c. Three-address. The three-address system uses three operands, which specify the locations of both factors of an arithmetic operation and the location where the result is to be stored. Both factors are thus preserved.

Since not all instructions make use of more than one operand, the two- and three-address systems waste digits for instructions, although they reduce the number of instructions needed in the program. Variable-word-length computers often provide for instructions with a variable number of operands.

d. Four-address. In a system using four addresses, the fourth address specifies the memory address of the next instruction in the computer program.

e. Modified Two-address. The modified two-address system is essentially a single-address system in which the second operand specifies the memory location of the next instruction.

f. Modified Three-address. The modified three-address system is essentially a two-address system in which the third operand specifies the memory location of the next instruction.

59. Memory Cycles. The basic timing mechanism of a computer is a crystal-controlled oscillator, or *master clock*. The frequency of this oscillator determines the operating speed of the computer. Large computers of the second generation had clock rates on the order of 1 Mc/s. Medium-sized computers had clock rates on the order of 500 kc/s. Small computers were frequently slower. Large third-generation computers have clock rates of 10 Mc/s or more.

A memory cycle, or *word time,* is the time it takes to retrieve a word from memory, store it in one of the registers, and rewrite the word in memory. It may take, for example, eight clock pulses to execute a memory cycle.

The implementation of a program step in the single-address system typically requires one word time to fetch the instruction from memory and store it in the instruction register and another word time to fetch the operand (i.e., the *contents* of the operand address) and execute the operation. Operations such as double-precision arithmetic, multiplication, and division require several word times for execution.

60. Arithmetic Operations. The instruction suite of a large, general-purpose computer may contain several hundred instructions. Among these are arithmetic instructions. A few of these will be described in the context of a single-address machine:

a. Add. The contents of a specified memory location are moved to the B register and added to the contents of the accumulator. The sum is received in the accumulator. The operation takes two word times.

b. Subtract. The contents of a specified memory location are moved to the B register and subtracted from the contents of the accumulator. The result is stored in the accumulator. Time: three word times.

c. Double-precision Add. The contents of a specified location and the next highest-numbered location are added to the contents of the accumulator and the Q register, which receive the sum (three word times).

d. Double-precision Subtract. The contents of a specified location and the next highest-numbered location are subtracted from the accumulator and the Q register, which receive the difference (five word times).

e. Multiply. The contents of a specified location are multiplied by the contents of the Q register. The most significant half of the product is stored in the accumulator, the least significant half in the Q register (9 to 23 word times).

f. Divide. The contents of the accumulator and the Q register are divided by the contents of a specified memory location. The quotient is stored in the accumulator, the remainder in the Q register (26 to 29 word times).

61. Branching. An important advantage of an automatic computer is that it is not required to execute a specific sequence of instructions. It can *jump* (i.e., skip over several instructions to a predetermined point later in the program sequence), or it can *loop* (i.e., return to an earlier point in the program and repeat one or more operations). Jumping and looping are accomplished by branching instructions. Usually the computer tests for some condition of a register or unit of input-output equipment and transfers control to a designated program step if the condition tested for is found to hold.

a. Unconditional. Control is unconditionally transferred to the named program step.

b. Parity Error. Branch to a recovery routine (or stop) if a parity error is discovered in the M register.

c. I/O *Ready* (*Not Ready*). Branch to an information-transfer routine if a designated input or output device is ready to transmit or receive. The condition given in parentheses is the converse condition. Generally test and branch instructions can be stated either way at the convenience of the programmer.

d. Accumulator Zero (*Nonzero*)

e. Accumulator Odd (*Even*)

f. Accumulator Minus (*Plus*)

g. Accumulator Overflow. Branch to a routine to cope with this condition.

h. X *Register Less than a Constant*

i. X *Register Greater than a Constant* (*Equal*)

61a. Address Modification. The property that enables a computer to process a large number of records, search a table, or perform various routines according to sequence is called address modification. Address modification may be carried out either arithmetically or by use of index registers.

Arithmetic Address Modification. Essentially, one or more program steps will be brought into the accumulator, and constants will be added to or subtracted from its operand portion.

Example. Suppose that it is desired to find the sum of 30 numbers. The numbers will be read into core from cards and stored in 30 sequential memory locations. The first part of the program would be a *load* (accumulator), *add,* and *store* routine. The load instruction would have as its operand the first memory location in which data are stored. After operating upon the first of the 30 numbers, the load instruction itself would be brought into the accumulator by a second load instruction and its operand augmented by one. The load, add, and store would now be repeated on the second data number.

To get the computer out of this *loop,* an *exit routine* employing a *count box* is set up. This is a memory location in which the constant *thirty* is stored and a *one* subtracted each time the load, add, and store routine is accomplished. A *test-and-branch* routine is the last portion of each *iteration* and determines when the count box is reduced to zero, at which time control of the program is switched to a designated program step.

Index Registers. Index registers are special core locations used in automatic address modification. When an instruction is *tagged* to an index register, its operand address is automatically incremented by the contents of the specified index register. Instructions are available to load the contents of index registers into the accumulator, increment them by a constant, or store them in a specified memory location. In addition, the number of each index register serves as its operand address.

Example. In the previous example, the first load instruction is tagged to one of the index registers which has previously been set to zero (by loading zeros into the accumulator and storing the contents of the accumulator in the index register). Another index register is used as the count box. After each iteration, the contents of both registers are incremented by one. The contents of the second index register are tested for equality to thirty and, depending upon the outcome of the test, control is switched either to the load instruction or out of the loop (when contents of X2 = 30).

INPUT-OUTPUT SYSTEM

62. Magnetic Tape. Magnetic tape is the most popular form of bulk storage of computer data. It consists of a layer of iron oxide affixed to a Mylar or acetate base using either hard or soft binder (see Fig. 27-18*a*). Steel tapes have also been employed, but their use is not common at this time. Tape ranges from ½ to 3 in wide; ½-in tape is the most common. It is usually wound on 10½-in-diameter reels. Each reel contains about 2,400 ft of tape.

Characters, or bytes, are written laterally across the tape. There used to be seven channels: two for zone bits, four for BCD bits, and parity (see Fig. 27-12*b* and *d*). The ASCII and extended binary codes use eight channels plus parity. Longitudinal parity by computer word is usually added.

Binary information is written by beginning with sign and identifying code bits, then recording the binary number in groups of six or eight bits (see Fig. 27-18*e*).

The end of a magnetic-tape record is denoted by an interrecord gap (IRG). It is usually ¾ in wide. The end-of file is denoted by an end-of-file gap (EOF). It is

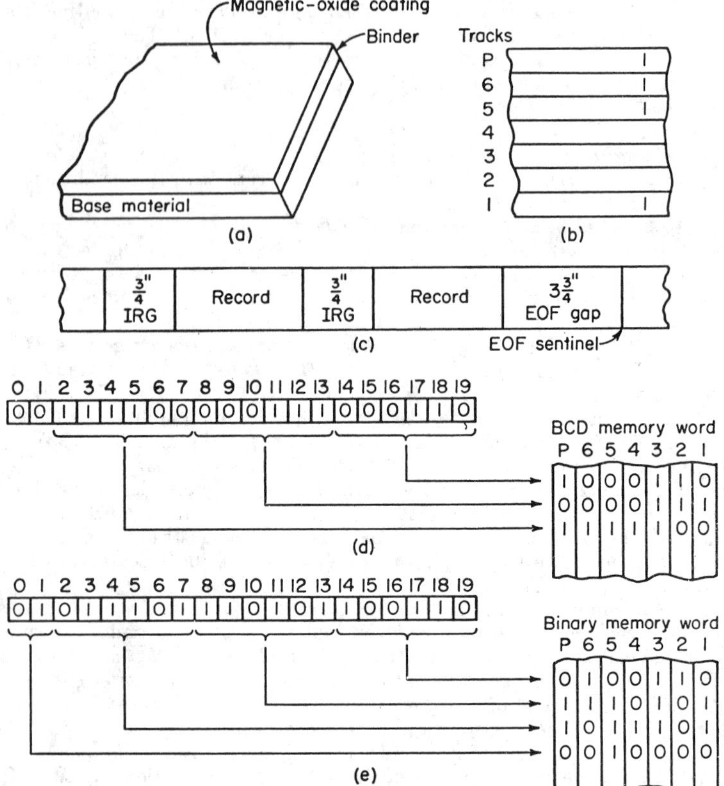

FIG. 27-18. Magnetic tape. (*a*) Cross section; (*b*) tracks for 6-bit computer code; (*c*) tape gaps; (*d*) BCD representation of 20-bit computer word; (*e*) binary representation.

usually 3¾ in wide, followed by a special character (e.g., octal 17) known as the EOF sentinel (see Fig. 27-18*c*). The beginning and end of tape (EOT) are denoted by aluminum-foil strips $\frac{3}{16}$ by 1 in, embedded in the tape 15 and 20 ft before the physical end. These are sensed by photoelectric cells in the tape handlers to prevent runout of the tape reel.

Because writing automatically destroys previous data written on tape, some tape handlers have file-protection rings ("write rings") which must be in place before writing can occur. Without the ring in place, only reading can take place.

Several tape densities are in use. Low-density tape has 200 characters (bytes)/in. High-density tape used to mean 555 characters/in, but now tape systems recording 800 and even 1,600 characters/in are available. Systems capable of recording 3,000 characters/in are under development.

Several systems of recording data on tape are in use. In the conventional return-to-zero technique (RZ), no current flows through the write head except when a 1 or 0 is to be written. If a 1 is to be written, a pulse of positive polarity is applied to the write head. If a zero is to be written, a negative pulse is applied (see Fig. 27-19*a*).

An alternative RZ technique impresses a steady-state value of negative direct

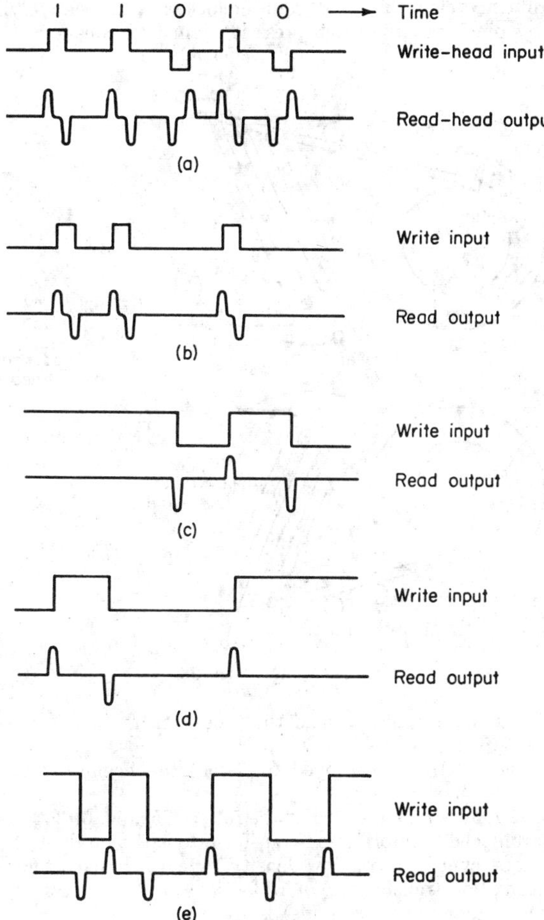

FIG. 27-19. Tape recording techniques: (*a*) and (*b*) return to zero; (*c*), (*d*), and (*e*) non-return to zero.

current to the write-head windings and applies a positive pulse only when a 1 is to be written (Fig. 27-19*b*).

Non-return-to-zero (NRZ) recording techniques are used to conserve tape and depend upon change of state rather than the direction of current in the write-head windings. In one NRZ technique, head current is negative for the entire bit time when a 0 is recorded and positive for the entire bit time when a 1 is recorded (Fig. 27-19*c*).

In the modified NRZ technique the polarity of the current in the write winding is reversed each time a 1 is recorded and remains constant when a 0 is recorded (Fig. 27-19*d*).

A third NRZ technique, sometimes called a phase-modulated system, records a 0 as a ½-bit-time positive pulse followed by a ½-bit-time negative pulse and a 1 as a ½-bit-time negative pulse followed by a ½-bit-time positive pulse (Fig. 27-19*e*).

63. Tape Handlers. Also called tape transports or tape stations, these units drive the tape past separate reading and writing heads. The tape handler contains two reels: the supply reel and the take-up reel. Some tape stations collect the tape in supply and take-up bins. The tape is threaded around tape guides, which are mechanically suspended to relieve the shock of rapid accelerations and decelerations and between capstans and pinch rollers. There are a forward and a reverse capstan both of which are constantly rotating clockwise and counterclockwise, respectively. The tape is caused to move by pinch rollers, which press it against the desired capstan (see Fig. 27-20).

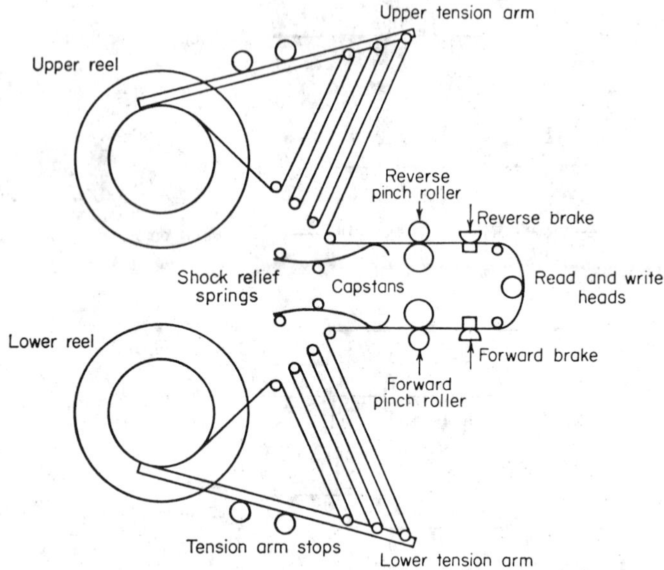

Fig. 27-20. Tape handler.

Pressure pads or vacuum pockets hold the tape even and smooth as it passes the read and write heads.

Two photoelectric cells are positioned to detect the beginning and end of tape markers.

The tape handler moves tape forward at the rate of 75, 150, or 300 in/s and rewinds at twice the forward speed. Information-transfer rate is the product of recording density and forward tape speed. Thus low-density tape (200 characters/in) moving at 75 in/s has an information transfer rate of 15 kc/s.

The time to pass a tape file (in seconds) is given by

$$T = \frac{NRW}{K} + (R - 1)S$$

where N = number of tape characters (bytes) per computer word, R = number of records in the file, W = number of computer words per record, K = information transfer rate in cycles per second, and S = tape start-stop time in seconds.

The length of a tape file (in inches) is given by

$$L = \frac{NR(C + W)}{B} + (R - 1)g + G$$

where B = number of characters per inch, C = number of nondata words (format) per record, W = number of data words per record, g = length of interrecord gap in inches, and G = length of end-of-file gap in inches.

64. Tape Controllers. The tape controller comes between a tape handler and the central processing unit. It usually contains a clock which governs the information transfer rate to and from the tape handlers. (In some systems, timing pulses are recorded on the tape itself.)

It contains a buffer known as the N register, which usually contains several tape characters, typically a computer word. This word is clocked in or out of the M register in the CPU by the master clock of the CPU. It is read from or written onto tape under control of the clock of the tape controller. The tape controller may contain additional buffer registers for high-density interrupt-mode operations.

The tape controller contains logical circuits to:

1. Count words and characters transferred to and from its N register.
2. Perform lateral and longitudinal parity checking.
3. Accomplish code-to-code conversion when the tape character code differs from that used in the CPU.

65. Tape Writers. Magnetic tapes are usually prepared by first punching cards or paper tape, then running this through a computer, tape controlle, and tape handler to create the desired tape record. However, keyboard devices arer available that will create records directly on magnetic tape. Operationally they resemble keypunches. Provision is also made for verification of the tape record.

66. Disk Storage Units. One point of dissatisfaction with magnetic tape is that it is impossible to address a tape record directly without usually spending considerable time passing or rewinding the tape reel. For many applications *addressable* bulk storage is desired.

Disk storage units (DSU) store data on steel disks about 16 in in diameter and coated on both sides with magnetic oxide, similar to that used to coat tape. The disks are stacked on shaft and rotate continually at uniform speed (e.g., 1,200 r/min) when power is applied. Mechanically the arrangement is not unlike a coin-operated phonograph.

Access to any record is gained by one or more pneumatically driven arms which position reading and writing heads over the desired record. A typical disk storage unit can store 6 million words or more.

Figure 27-21*a* shows the layout of one disk of a typical DSU. (A unit may have 16 or more such disks.) The disk is divided into an inner and an outer zone. The outer zone is divided into 16 sectors; the inner zone, into 8 sectors. Each sector contains portions of 128 concentric recording tracks (corresponding to grooves on a phonograph record). Each track sector contains one disk record (Fig. 27-21*b*). Each record contains 3 control words and 64 data words. The control words are the header, end-of-record (EOR) word, and a longitudinal parity check containing a parity bit position for each bit position of the 64 data words. Thus 3,072 records containing 196,608 words can be stored on one side and 98,304 records or 6,291,456 words in one DSU.

There are eight sets of read and write heads on each arm (Fig. 27-21*c*). The heads access both sides of the disk simultaneously, half accessing the inner zone and half accessing the outer zone. Thus 96 records can be read or written without moving the arm. There are 64 discrete positions of the arm that enable all records on both sides of a disk to be accessed. The average positioning time for the arm is 199 ms, and latency due to disk rotation for this particular unit is 26 ms.

Thus, to specify a location on a DSU, it is necessary to (1) address the DSU subsystem, (2) select the proper DSU (four or more units may be available with some systems), (3) select the proper disk, (4) head position, and (5) record.

67. Disk Controllers. The DSU controller can handle one or several disk storage units. It provides an interface with the CPU. Its buffer register stores one computer word.

68. Disk Packs. To make disk storage units more flexible in application, some manufacturers have made it possible conveniently to remove disks and store them in much the same way in which reels of magnetic tape are stored. The disks are usually handled in packs of four.

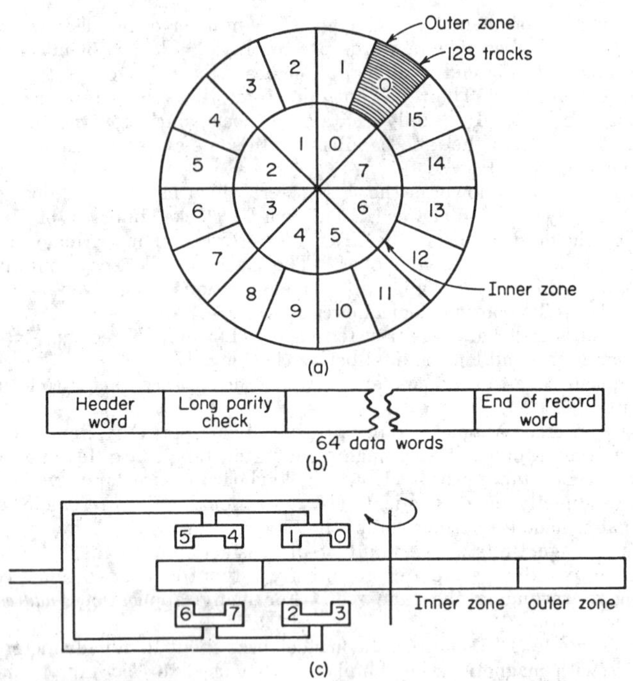

Fig. 27-21. Disk storage unit showing (*a*) layout of disk, (*b*) disk record image, and (*c*) read-write head arrangement.

69. Magnetic Cards. Several other types of addressable bulk storage have been developed. One system uses thin metal or plastic cards measuring about 9 by 12 in which are coated with magnetic oxide and contain identifying notches on their top side. These magnetic card systems include CRAM (card random-access memory) and random-access card equipment. In the latter system, the cards are suspended in a long tub file. When it is desired to access a particular record, a hydraulically driven arm selects the desired card and wraps it around a drum. The drum quickly accelerates to operating speed, and data are transferred as in the case of a magnetic drum. Access time is comparable with that of a DSU, but storage capacity is much greater.

70. Data Cell Drives. Other addressable bulk storage systems store information on strips of magnetic tape which are automatically selected and read when addressed.

71. Keypunches. By far the most popular basic input medium to computers is the Hollerith card. These cards are prepared on a machine known as a keypunch. It has a keyboard not unlike that of a typewriter but containing special control characters not found on a typewriter. A proficient operator can punch about 200 cards an hour (10,000 key strokes). Three basic models are available: the IBM 024 is the basic unit; the 026 interprets as well as punches, printing the contents of the card along the top edge; the 029 is designed to function with IBM 360 computers and incorporates special automatic features.

In addition to punching, the keypunch can duplicate the information in one card on the cards following it through the feed. All keypunches can be equipped with program drums that instruct them to skip columns or duplicate according to instructions punched in a special *program card* which is wrapped around the drum.

The 026 keypunch can interpret cards at the rate of 12/min.

Other means for putting data on cards include Porta-punch, in which special cards having partly punched holes are placed in a frame so that the chad can be knocked out with a needlelike tool, and mark-sense cards, in which data are entered by marking appropriate rows and columns with an electrographic pencil.

72. Verifier. To ensure against the keypunch operator introducing errors, punched card decks must be verified. The deck is fed into an IBM 56 verifier, and the operator enters the same information as previously punched. If a card is error-free, the verifier notches it adjacent to the eightieth column. If an error is detected, an alarm signal is given and the operator must repeat the work three times. If all three tries are in error, the card is notched on the 12 edge above the column in which the error occurs.

Remington Rand cards are verified on the keypunch, which makes the round holes oblong if a punch is correct. Incorrectly punched cards are detected by a fast machine which scans a verified deck to discover round holes.

73. Reproducer. Since mark-sense cards are likely to become smudged and Porta-punch cards may be incompletely punched, it is recommended that decks of this nature be reproduced. The IBM 514 reproducer can copy 100 cards a minute. Features offered include:

1. Exact reproduction of a deck fed into the read feed (also called 80-80 reproduction).
2. Transposition of columns, accomplished by plugboard wiring of the reproducer.
3. Gang punching, or multiple reproduction of uniquely identified master cards.
4. Digit emission, punching specified digits (e.g., date or department number) into card decks.
5. Mark-sense punching.

74. Interpreting. Card output from a computer is usually not interpreted as delivered. This can be done at the rate of 60 cards a minute on the IBM 552 interpreter. However, this machine prints a line only 60 characters long, and to interpret a full 80-column card requires two passes.

75. Sorting. Inasmuch as sorting by computer is often time-consuming, card files of a few hundred cards may frequently be sorted more conveniently by electromechanical equipment. The IBM 082 sorter has one brush and sorts cards, one column at a time, into one of 13 pockets, i.e., reject (no punch), 12, 11, 0, and 1 to 9. It sorts 650 cards a minute; the model 083 can sort 1,000 cards a minute; and the 084 sorts 2,000 cards a minute. It is customary to sort proceeding from the rightmost digit. In the case of alphabetic data, a numeric sort is performed first on a given column, followed by a zone sort. The model 084 sorter has limited capability for one-pass alphabetic sorting.

76. Printing. Where it is not desired to tie up the high-speed printer of a computer in order to list program decks for proofreading, it may be economical to install an IBM 402 (407 or 408) tabulator or electric accounting machine. The machine executes 120 c/min, which is not that much slower than a slow-speed computer card reader. The tabulator has one print wheel for each of 120 character positions. The wheels, which carry about 50 different characters, are positioned independently, and the entire line is printed by a platen pressing interleaved paper and carbon rolls against the type roll. Up to 14 copies can be produced at a time.

77. Card Readers. The card reader, card punch, and associated circuitry in the CPU constitute a computer's direct input-output subsystem.

Card readers accept 400 to 2,000 cards a minute. Reading is accomplished one character at a time, and data are converted to the computer code and accumulated in the direct input-output buffer until an entire word can be transferred to the threshold (M) register.

Inasmuch as card reading is extremely slow in contrast with other computer operations, provisions are usually made to have the computer perform other operations while card reading is being accomplished.

Punches in cards can be read according to the standard Hollerith code or in 10- or 12-row binary notation. In the latter notations, the presence of a hole is read as a binary 1, and each card column is read as a 10- or 12-bit binary number (see Fig. 27-22).

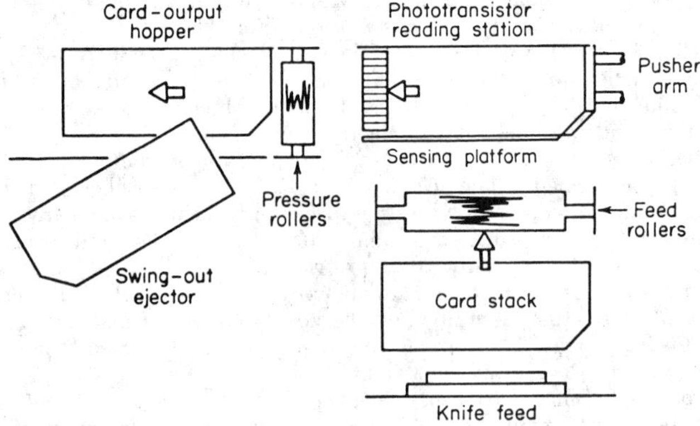

Fig. 27-22. Card reader.

78. Card Punches. The card punch can produce 100 to 300 cards/min in Hollerith or binary notation. Usually cards are not interpreted. The cards are punched one row (80 columns) at a time, 12 row first. Punching takes place under control of a card-punch controller, which is part of the computer's direct input-output subsystem. Data are transferred from the M, or threshold, register to an 80-bit position buffer before the punching dies, under control of punch logic, are activated. The card punch may have the optional feature of checking punched cards for blank or double-punched columns.

79. Paper-tape Reader-Punches. Paper tape is an important computer input when a data-processing system is integrated with a communications network. Standard paper tape comes in a roll 8 in in diameter, 800 ft long, 0.004 in thick and in the widths shown in Fig. 27-23. Note that "channel S" refers to the sprocket hole used for tape drive.

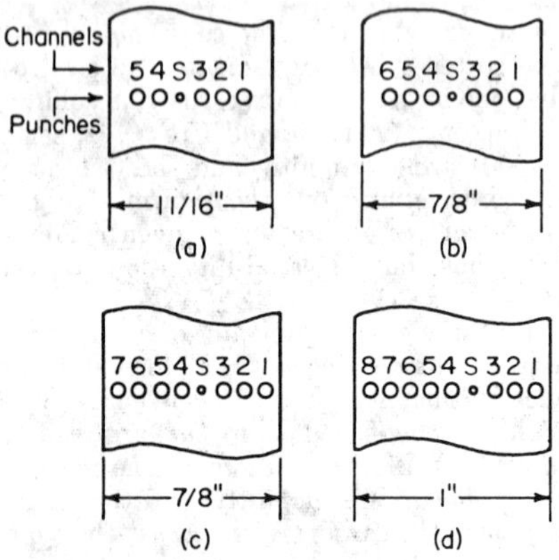

Basically, the five-channel code (Fig. 27-23*a*) is the Baudot telegraph code (see Sec. **26**); the six-channel code (Fig. 27-23*b*) is a computer code in which channels 1-4 represent the 1-2-4-8 positions of BCD code and channels 5 and 6 are zone bits; the seven-channel code (Fig. 27-23*c*) adds a lateral parity bit (even) in channel 5, pushing the zone bits to channels 6 and 7; the eight-channel code adds an end-of-line (EL) bit in channel 8. The eight-channel code is shown in Table 27-6.

Fig. 27-23. Perforated paper tape. (*a*) Five-channel; (*b*) six-channel; (*c*) seven-channel; (*d*) eight-channel.

Paper-tape readers use nine photocells and a light source to read all information holes and the sprocket hole. Tape readers operate 200 to 1,000 characters/s. On some systems the spooler cannot be used at high speeds. Each time a character is read, its parity is checked (on 7- and 8-channel tape only), and it is converted to computer code and placed in the N register of the CPU. The contents of the single-character N register are clocked into the accumulator a character at a time for further transfer on a word basis.

Table 27-6. Eight-channel (Level) Paper Tape Code

Character	Tape image (by channel)								
	1	2	4	Feed hole	8	Parity	0	×	End of line
TAPE FEED	○	○	○	●	○	○	○	○	
A	○			●			○	○	
B		○		●			○	○	
C	○	○		●		○	○	○	
D			○	●			○	○	
E	○		○	●		○	○	○	
F		○	○	●		○	○	○	
G	○	○	○	●			○	○	
H				●	○		○	○	
I	○			●	○	○	○	○	
J	○			●		○		○	
K		○		●		○		○	
L	○	○		●				○	
M			○	●		○		○	
N	○		○	●				○	
O		○	○	●				○	
P	○	○	○	●		○		○	
Q				●	○	○		○	
R	○			●	○			○	
S		○		●		○	○		
T	○	○		●			○		
U			○	●		○	○		
V	○		○	●			○		
W		○	○	●			○		
X	○	○	○	●		○	○		
Y				●	○	○	○		
Z	○			●	○		○		
0				●			○		
1	○			●					
2		○		●					
3	○	○		●		○			
4			○	●					
5	○		○	●		○			
6		○	○	●		○			
7	○	○	○	●		○			
8				●	○				
9	○			●	○	○			
SPACE				●		○			
–				●				○	
/	○			●		○	○		
&				●		○	○	○	
$	○	○		●	○	○		○	
,	○	○		●	○	○	○		
.	○	○		●	○		○	○	
@			○	●	○	○			
%			○	●	○		○		
*			○	●	○			○	
□			○	●	○	○	○	○	
#	○	○		●	○				
P11		○		●	○	○			
P12		○		●	○			○	
P13		○		●	○		○		
P14	○		○	●	○	○	○		
P15	○		○	●	○		○	○	
P16	○		○	●	○	○		○	
P17	○		○	●	○				
EL				●					○
SKIP		○	○	●	○	○	○		
CORR	○	○	○	●	○	○			
ERR	○	○	○	●	○			○	
SP1		○		●	○	○	○	○	
SP2		○	○	●	○		○	○	
CR		○	○	●	○	○		○	
EC1		○	○	●	○				
EC2	○	○	○	●	○		○		

Characters are punched 10 to the inch at the rate of 110/s. Characters must be transferred from the A to the N register before punching can take place. Parity is calculated and code conversion accomplished before punching occurs.

Computer words can be punched in binary form by using three channels of paper tape to denote the octal digits represented by the successive triads of binary bit positions of a computer word. For a computer having a 20-bit word length, the binary representation of a word would consist of an octal number using only channels 1 and 2, followed by six octal numbers utilizing channels 1, 2, and 3. The first octal digit connotes the word break.

80. Factory Data Collection. In certain applications, computer data are collected on line by a network of key sets which communicate with a central computer. Often these sets require insertion of a prepunched plastic card to validate the identity of the user. The system provides a central clock.

81. Tag Readers. Special equipment has been devised for reading needle-punched tags attached to merchandise and converting this input into machine-readable form, usually punched paper tape.

82. Magnetic-ink Character Readers. This equipment, known as MICR, is widely used in banks to read the bank identification and account numbers on checks. It reads and sorts up to 1,200 magnetically encoded source documents a minute. It sends information to the CPU for further processing while sorting the documents into 12 pockets. It accepts documents ranging in size from $2\frac{1}{2}$ by $5\frac{1}{4}$ to $3\frac{3}{4}$ by $8\frac{3}{4}$ in. It reads information imprinted in the E13B font in magnetic ink (see Fig. 27-24*a*).

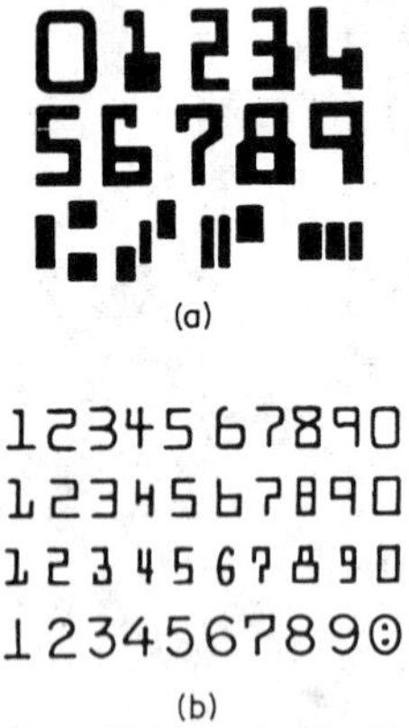

01234
56789

(a)

1234567890
1234567890
1234567890
1234567890

(b)

Fig. 27-24. (*a*) E13B type font used with magnetic-ink character-recognition equipment, and (*b*) typical type fonts used with optical scanners.

83. Optical Character Readers. These devices accept input documents printed or typed in ordinary carbon ink. The usual character readers now employed read characters imprinted from credit cards and convert each customer's number to machine-readable form. The type font must be the print on the document. An imprinter accomplishes this. Optical character readers have televisionlike pickups that scan each character and convert it to machine-readable form. The Post Office Department has done experimental work reading and coarse-sorting mail. Although optical character readers exist that can read alphanumeric page copy and even handwriting to a limited extent, the machines are expensive and not yet considered adequate for on-line production use. They are easily confused by broken or smudged characters and occasionally find it impossible to locate the beginning of a line of type in the vertical dimension.

Figure 27-25 illustrates one realization of an optical character reader. A 5-by-7 grid is superimposed on each successive character. A photocell-light-beam system scans the grid vertically. A binary 1 is recorded for a particular cell if it is predominantly dark and a binary 0 if it is predominantly light. With this system each printed character can be converted to a 36-bit binary number (12-digit octal).

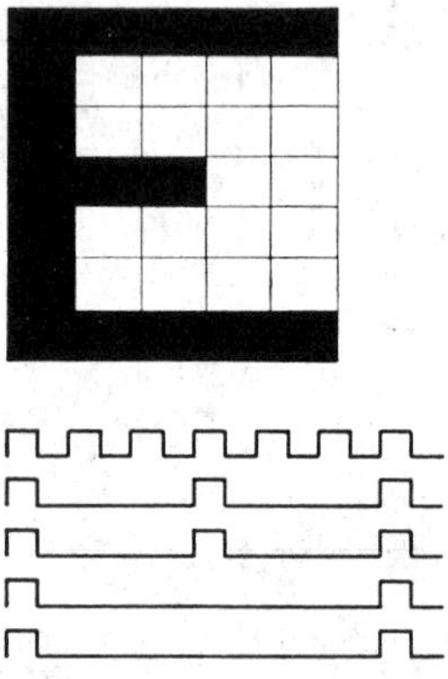

84. Console Typewriters. An electric typewriter is part of nearly every computer system. It communicates with the accumulator through a character buffer (N register). Although information may be entered by the typewriter, it is most frequently used to communicate with the operator. The computer prints out on the typewriter under program control. The medium is used for error messages, instructions to the operator (e.g., "Replace reel on tape handler number 1"), and to keep a log showing the beginning and end of each program.

85. High-speed Printers. The usual method of acquiring printed output (hard copy) from a computer is by a high-speed

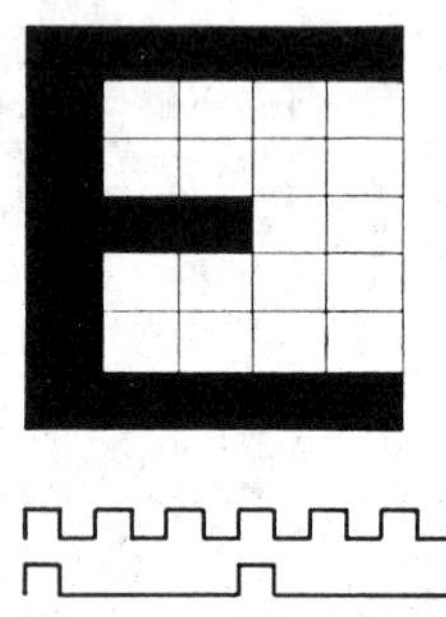

011,111,111,001,001,100,100,110,000,011,000,001
3 7 7 1 1 4 4 6 0 3 0 1

Fig. 27-25. Conversion of printed characters into binary and octal code by optical scanning.

printer. The common types of printers are mechanical and produce 300 to 1,000 lines/min. Some manufacturers have furnished electrical printers that produce 3,500 lines/min or more.

Printer output is obtained on a continuous roll of paper 12 in wide perforated in sheets 11 in long. A printed line contains 120 or 160 characters.

In a mechanical printer, type characters move continuously at high speed. A roll of carbon paper is interposed between the tape and the paper. Individual hammers are fired for each character position in the printed line under control of the printer controller and press the paper and carbon against the type. The type may be in the form of a *type roll,* consisting of 120 or 160 type wheels permanently welded together. There are usually 50 different characters on a type wheel. Alternatively, the type may be carried on an endless *type chain,* containing 128 different characters.

In electrical printers, each character is represented by a matrix of 35 electrodes, which are selectively energized by the printer controller to form the desired characters, then pulsed to form their images. *Electrographic printers* emit sparks from the electrodes to a platen and thus selectively darken special electrographic paper.

Electrostatic printers create an electrical charge pattern on ordinary paper, which is then automatically dusted with carbon particles in a binder, and the resulting printed information fixed by application of heat.

The printer controller contains a one-word data buffer which receives information from the computer's output register. Timing circuits in the printer controller then clock the data into a 120- or 160-character print-line image buffer, which controls the firing of hammers or the setup of electrode matrices. The printer controller may also contain a one-word format buffer to receive programmed format instructions for setting up a line of print.

86. Graphic Applications. Computers have been used to prepare lists for printing (see Sec. **20**). The basic output mechanism is a special, long-persistence cathode-ray tube with a matrix grid electrode containing the shapes of characters in the computer's character set. Under control of the graphics controller the cathode-ray-tube beam passes through the selected shaped holes of the grid matrix and composes a line of type on the face of the tube. Each successive line is then captured on silver halide film to provide a negative suitable for preparing offset printing plates. The special cathode-ray tubes are known by various names, such as Charactron or Typotron.

87. Teletypewriters. Remote access to a computer may be gained by teletypewriters (see Sec. **26**). These are keyboarded machines that transmit characters in the form of d-c pulses called *bauds* which are transmitted in a bit serial mode in five-, six-, seven-, or eight-level code. Teletypewriter communication is asynchronous, with special start and stop bauds denoting the beginning and end of each character. There are three general types: automatic send receive (ASR), keyboard send receive (KSR), and receive only (RO). Teletypewriter lines are classified as full duplex (send and receive simultaneously), half duplex (send and receive alternately), and simplex (send or receive only). Standard teletypewriter rates in use are 45, 50, 56, 66, 75, 110, and 150 bits/s.

88. Data Sets. At data rates of 300 to 3,000 bits/s, it becomes difficult to distinguish d-c pulses from line noise. At these rates, communication is effected by data sets which make use of an interrupted tone or two tones (one for mark, the other for space) to transmit intelligence over the line. In the asynchronous mode, standard bit rates are 300, 600, 1,200, 1,800, 2,000, 2,400, and 3,000 bits/s. Communication is in five-, six-, seven-, or eight-level code with start and stop bauds.

Special synchronous data sets are sometimes used for direct computer-to-computer communications at rates of 2,000 and 2,400 bits/s. Code level may be from 5- to 16-bit codes without start-stop bauds.

Most standard data sets have provision for dialing up telephone stations and for voice transmission to aid in establishing communications.

A rough rule of thumb for selecting line facilities is to use a bandwidth of twice the maximum expected data rate to be transmitted (see Sec. **26**).

89. Communications Buffers. Inasmuch as data transmission is accomplished in the bit-serial mode and most computers operate bit parallel and character or word serial and also since data communications rates are much slower than computer word times, it is necessary to provide buffering between remote lines and the computer's CPU. A communications buffer usually contains a timing unit as well as the required buffer registers.

A typical unit may scan 15 communications lines sequentially. In the receive mode, the bits will be accumulated in separate line buffer registers, parity computed and checked for seven- and eight-level code inputs, and the necessary conversion accomplished to convert the input to computer code. These operations and transmission of data to the lines are performed under control of the buffer timing unit. When a line buffer register is filled, data are transmitted word or character parallel to the input register of the CPU under control of the computer's timing unit.

90. Data Terminals. Some communications buffers are so complex as to constitute free-standing computers in their own right. A typical unit might be a single-address stored-program special-purpose digital computer capable of processing both alphanumeric and binary information. It may perform arithmetic operations as well as acting as central control of a communications system. It will usually accommodate any standard transmission rate ranging from 45 to 3,000 bits/s and will control transmission over common-carrier facilities to another data terminal or one or more standard teletypewriters. Such a unit would incorporate a magnetic-core memory to store program instructions and data. The memory might contain 4,096, 8,192, or 16,384 words. Communications terminals have memory-cycle times in the microsecond range, comparable with those of general-purpose computers.

Terminals typically can scan as many as 128 communications lines. Each line will be assigned a buffer, where data will be accumulated prior to passing it on to the working registers of the communications terminal. Data from teletypewriters will be accumulated in bit buffers; data from character-oriented data sets, in character buffers; and data from word-oriented sets, in word buffers. Data from buffers will be passed on a bit, character, or word basis to the input register of the terminal, although data from a high-speed computer peripheral may pass directly into the terminal's memory. From the input register (or memory) data pass to other working registers (for computation or processing) or to storage for later transmission to the CPU at machine speed.

Typical arithmetic operations include binary addition, logical functions (e.g., and, or, exclusive or), shift operations (e.g., right, left, circular), bit change, and address modification.

Program input to the terminal is by punched paper tape over a special channel provided for that purpose. The basic program cycle of a terminal is to:

a. Receive bits.
b. Assemble bits into characters.
c. Assemble characters into words.
d. Assemble blocks into messages.
e. Assign message routing.
f. Disassemble blocks into words for transmission.
g. Disassemble words into characters.
h. Put the character in the buffer for transmission.

The terminal also checks input parity of seven- or eight-bit code and parity of words coming from the computer or high-speed peripherals, performs code-to-code conversions, and can be programmed to perform a limited amount of data processing.

The terminal includes interface units which look toward the associated computer and its high-speed peripherals such as tape handlers, disk storage units, or high-speed printers. Data are transferred in a word-serial mode from the terminal's computer interface data register, and an assigned memory address in the CPU is also transferred in parallel from a corresponding address unit in the terminal's computer interface unit. This transfer takes place under control of the CPU timing unit. Similar provisions are made for transferring data to and from high-speed peripherals (see Fig. 27-26).

91. Video Displays. A video display system consists of a display controller and one or more terminal memory units each feeding one or more display terminals.

A *display terminal* consists of a television display screen with a typewriterlike keyboard. Here is how a typical unit functions: The screen presents a fixed-format display (46-character lines) consisting of alphabetics, numerics, punctuation, and special symbols that allow presentation of charts, tables, and diagrams. Each character entered on the keyboard (local mode) is displayed on the screen so that a message can be verified and corrected before it is sent to the CPU. In its receive mode, the terminal receives computer messages via the communications line. A teletypewriter may be attached to the display terminal to furnish hard copy.

The *display controller* has its own master timing circuits to provide clock drive for its multiplexer and display terminals. Operating the keyboard generates control and data characters in six-level code. A character generator converts character codes into digital video signals for display.

Each memory unit can store up to 1,472 characters. The controllers are capable of half- or full-duplex operation, synchronous or asynchronous, at rates of 1,200, 2,000, and 2,400 bits/s.

Video display units generally communicate with the CPU *via* data sets and a communications buffer.

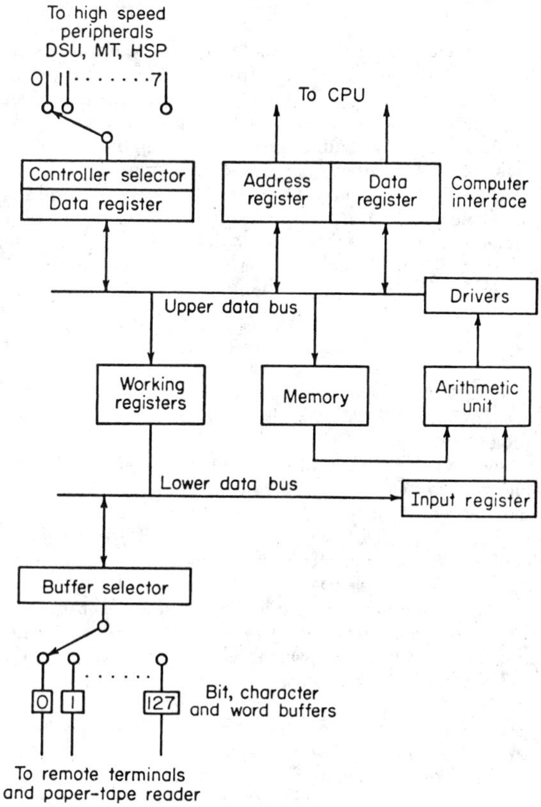

FIG. 27-26. Typical communications terminal computer.

PROGRAMMING

92. Machine Language. Machine language is completely numeric; it is the lowest level of programming language. Each instruction in the computer's repertoire is denoted by a binary number, as is each location in memory.

Example. Suppose that we desire to add two numbers A and B and store their sum C. Suppose further that numbers A and B already reside in memory. To carry out this operation, we require three instructions, whose octal representations are

01 = load accumulator
02 = store accumulator
10 = add *operand* to contents of accumulator

Assume that our computer has a modified two-address command system and that we are reserving memory locations 101, 102, and 103 for data storage and locations 201, 202, 203, and 204 for program storage. In machine language, our program would be

01 0101 0202

which means load the accumulator with the contents of memory location 101, and go to location 202 to find the next instruction.

Then

10 0102 0203

which means add the contents of location 102 to the contents of the accumulator, and go to location 203 to find the next instruction. Then

02 0103 0204

which means store the contents of the accumulator in memory location 103, and go to location 204 to find the next instruction in the program, which completes our example.

Thus, in machine-language programming the programmer must keep track of a great many details such as assignment of *absolute* memory locations.

93. Assembly Languages. An assembler is a machine-oriented language which makes use of a stored *symbol table* to convert *mnemonic instructions* to numerical instruction codes and to convert *symbolic addresses* to absolute addresses. Use of an assembler frees the programmer from having to keep track of address allocation and having to remember the numerical instruction code. The programmer can, within certain limits, devise his own symbolic address representations.

Example. In assembly language, the program segment of Par. **92** would appear as

```
LDA  A
ADD  B
STA  C
```

Assignment of instruction locations is taken care of without concern on the part of the programmer.

Many assemblers offer the feature of *relative addressing,* or specifying memory locations with reference to an operand. For example, the operand C + 3 would denote memory location 0106. Use of relative addressing can simplify making insertions in a program. It also gives the programmer some control over the machine's assignment of memory locations to instructions as, for example, to specify the location of a given segment of a program.

The assembler is loaded into the computer from cards or tape as is any other program. The *source program* written in assembly language is fed in as data would be, usually on punched cards. The assembler produces a deck of binary cards which contains the *object program* assembled in machine language.

If the program will not assemble properly because of operator errors (e.g., an undefined program step to which control is switched), appropriate *error messages* are printed out in a *listing* of the source program.

94. Problem-oriented Languages. The first step toward problem-oriented language is the use of *macroinstructions* (macros). A *macro* is a miniature *program generator* which produces the coding necessary to accomplish a common function in computer operations. For example, the macro

ADD A TO B GIVING C.

would automatically generate the coding described in Pars. **93** and **92.** Other macros might include

MULTIPLY A BY B GIVING C.
DIVIDE A INTO B GIVING C.
READ
WRITE, etc.

At a higher level of complexity and more specialized in nature are *subroutines.* An *open subroutine* is merely a part of the main program. A *closed subroutine* is part of a library of subroutines that are called as needed, entered, and executed, control being returned to the main program afterward. Typical closed subroutines may compute square root, compute randomized addresses on a disk storage unit, or perform housekeeping chores at the end of a file such as rewinding tapes and writing appropriate tape labels.

Problem-oriented languages consist of libraries of macros, subroutines, and *pseudocodes,* i.e., instructions which do not result in the generation of coding but rather call in subroutines as required by the main program. The main program itself may be nothing but a *calling sequence,* with jumps to appropriate subroutines.

95. Interpretive Routines. The earliest problem-oriented languages were macro-instruction translators or interpretive routines. The library of macros and subroutines resided in memory, and the source program was executed without producing an object program, i.e., translation and execution proceeded simultaneously. These routines had the disadvantages of wasting memory and requiring translation each time a given source program was run.

96. Compilers. Modern problem-oriented languages are known as compilers. The compiler and the source program are both loaded into the computer, and an object program is produced in binary-coded machine language. All the steps that go toward producing the object program are known collectively as the *compiling run*; when the object program is run with data, the *execution run* is said to have occurred. Sometimes a program in assembly language is produced as an intermediate step. In some computer centers, the object program is not normally punched out but rather written on a *scratch tape*. In this mode of operation, the compiler is externally indistinguishable from an interpretive routine. A compiler consists of two phases: translation and generation. The translation phase produces a symbol table to translate compiler words to machine instruction codes and subroutine calls, and names of variables to storage locations. The generation phase of the program produces machine coding for macros and subroutines.

There are three general types of compilers: scientifically oriented, commercially oriented, and special-purpose. The best-known scientifically oriented compiler is FORTRAN (formula translation). Other scientific compilers include: ALGOL (algorithmic language), MAD (Michigan algorithmic decoder), JOVIAL (Jules's own version of the international algebraic language), BALGOL (basic ALGOL), BASIC, and WIZ.

The most widely used commercially oriented compiler is COBOL (common business-oriented language); other commercial compilers include: GECOM (general compiler) and Commercial Translator. The NPL (new programming language) compiler is a combined commercial and algebraic language. (A more recent version is PL-1).

Specialized compilers are available for list processing (e.g., LISP) and for compiling digital simulations (e.g., SIMSCRIPT, GASP, and GPSS).

97. FORTRAN. The development of FORTRAN has been an evolutionary process, and several versions exist: FORTRANSIT, FORTRAN I, FORTRAN II, and FORTRAN IV. FORTRAN II can be regarded as the basic language at the present stage of development. However, it requires compilation of an object deck. It is being superseded by FORTRAN IV, which needs no compilation run (i.e., *load-and-go*).

Typical FORTRAN II programming conventions:

Variable names must begin with an alphabetic character, may contain no special characters, and are limited to six characters.

Integer variables start with one of the letters I, J, K, L, M, or N. (These are ordinarily used for subscripts and exponents.)

Floating-point variables start with the letters A through H and O through Z. Integer and floating-point quantities cannot be mixed on the same side of the "equals" sign.

Constants are formed by the decimal digits. *Integer constants* have no decimal point and are limited to 4 to about 10 digits. *Floating-point constants* contain a decimal point and up to 8 digits. Larger (or smaller) values can be realized by using exponential notation (that is, the letter E followed by a signed or unsigned integer not greater than 38 in most systems).

Operational Characters. The following characters are used in FORTRAN arithmetic statements:

Character	For card punch without FORTRAN	Meaning
(	%	Left parenthesis
)	□	Right parenthesis
+	&	Addition
=	#	Equivalence
−	−	Subtraction
/	/	Division
*	*	Multiplication
**	**	Exponentiation

Formatting. The following conventions are used in format statements:

Letter	Meaning
F	Floating point
E	Exponent of floating point
I	Integer
X	Space fill
H	Hollerith output

Program Preparation. Programs are usually punched into cards according to the following format:

Card Column	Contents
1	Comments; precede with C in col. 1
2–5	Statement number
6	Continuation; use a zero in col. 6
7–72	Statement
73–80	Card identification only

Statements. FORTRAN statements may direct input and output operations; perform arithmetic operations; change the order of program execution; cause a set of statements to be executed a specified number of times; and direct the compiler translating process.

Input-Output. These statements include READ, PRINT, PUNCH, and FORMAT.

1. READ *n*, *list*: *n* is the corresponding format statement; *list* is the list of variables on the card, separated by commas.
2. PRINT *n*, *list*.
3. PUNCH *n*, *list*.
4. *n* FORMAT (list of specifications separated by commas).

Following are types of specifications where w = field length and d = number of positions to the right of the decimal point:

Fw.d
Ew.d
Iw
wX
wH

Arithmetic. In addition to the operations previously defined, FORTRAN performs the following mathematical functions:

Name	Meaning
SQRTF ( )	Take square root of the expression in parentheses
ABSF ( )	Take absolute value
LOGF ( )	Take natural log
EXPF ( )	Exponentiate e to the power given in parentheses
SINF ( )	Take sine of angle in radians
COSF ( )	Take cosine of angle in radians
ATANF ( )	State the value in radians of the angle whose tangent is enclosed in parentheses

Transfer of Control. Transfers may be unconditional or conditional:

GO TO *n*: unconditional transfer of control to statement *n*.

IF (expression) *l*, *m*, *n*: conditional transfer depending upon the value of the expression in parentheses: negative, go to statement *l*; zero, go to *m*; positive, go to *n*.

STOP or STOP *n* stops the computer without provision for restarting except by going to the beginning of the program; STOP *n* displays *n* on console typewriter when stop occurs.

PAUSE or *PAUSE n* halts the computer with possibility of restarting where stopped.

Looping. Looping is accomplished by the DO LOOP, which, in turn, requires use of *subscripted variables*. The DIMENSION statement reserves space for the various values a subscripted variable may assume:

DIMENSION $a(k)$, $b(k, l)$, $c(k, l, m)$: reserves space for values of subscripted variables *a*, *b*, and *c* having *k* rows, *l* columns, and *m* levels, respectively.

DO $i\ j = k, l, m$, where *i* is the number of the last statement in the loop; *j* is the index variable, which must be an integer; *k* is the initial value of the loop variable; *l* is the terminal value of the loop variable; and *m* is the incremental value by which the loop variable is modified.

In English the DO LOOP statement reads: "perform the operations down to and including *i* while varying *j* from *k* to *l* by *m*'s."

Compile. The statement END signals the compiler that there are no more statements.

98. COBOL. The COBOL language was put together by a committee of computer manufacturers and users who met first in 1959 at the behest of the Unites States government. COBOL has evolved as did FORTRAN. The first compiler, described in a committee report issued in 1960, was called COBOL 60; a subsequent version was called COBOL 61.

Inasmuch as the Federal government, the largest single user of computers, has served notice that it will not buy or lease machines that cannot accept COBOL programs, computer manufacturers have responded actively to make their products COBOL-compatible. Frequently a double compile is required. The first pass translates a COBOL program to the manufacturer's own commercially oriented compiler (e.g., GECOM). Subsequent passes produce a program first in assembly language, then in binary machine language (object program). A COBOL program consists of four divisions: identification, environment, data, and procedure.

Program Preparations. COBOL source programs are punched in the following format:

Card Column	Contents
1–6	Card sequence number
7	Continuation indicator (hyphen)
8–11	Names of divisions, sections, and paragraphs (beginning)
12–72	Text area
72–80	Program identification

Identification Division. The only required parts of this division are the heading and the first paragraph. The *program name* fulfills the requirements of *data names* given in the data division.

IDENTIFICATION DIVISION.
PROGRAM-ID. program-name
The following paragraphs may be added to document the program:
AUTHOR
INSTALLATION
DATE-WRITTEN
DATE-COMPILED
SECURITY
REMARKS

Environment Division. This division describes the computers to be used for the compile and execution runs and assigns files to appropriate peripherals. The *hardware names* used in this division are members of the set of 800 words making up the compiler's vocabulary or set of *reserved words.* The *file-names* and *special-names* fulfill the requirements of *data-names* given in the data division.

ENVIRONMENT DIVISION.
CONFIGURATION SECTION.
SOURCE-COMPUTER. hardware-name.
OBJECT-COMPUTER. hardware-name.
SPECIAL NAMES: hardware-name IS special-name.
INPUT-OUTPUT SECTION.
I-O-CONTROL. RERUN EVERY *n* RECORDS.
FILE-CONTROL. SELECT file-name ASSIGN TO hardware-name.

Data Division. This division sets up all memory locations needed in running the program. These are referred to by *data names.* Data names may refer to files, records, or data elements. A data name may use any combination of 30 or fewer alphabetic or numeric characters (one must be a letter). Each data location is described by a *picture* which uses the following conventions:

Symbol	Meaning
9	Numeral
A	Letter
X	Alphanumeric
B	Space
V	Assumed decimal point (input only)
Z	Numeral with leading zeros suppressed (output only)

In addition the special characters period, comma, and $ may be used in output. In shorthand notation a string of, say, 10 numerals can be denoted as 9(10).

The data division consists of three sections: file, working storage, and constant. The *file section* describes output and input files. The *working-storage section* sets up program work areas. The *constant section* describes program constants.

There are several levels of data denoted as follows:

Level	Meaning
FD	File
01	Record
02	Group item (data element)
03	Elementary item (data element)
77	Independent work location
88	Condition name describing a data element

Level numbers may range from 01 to 49 inclusive depending upon the hierarchical structure of data.

Constants include *numeric literals* (1 to 18 numerals) *alphanumeric literals* (1 to 120 letters or numerals), *figurative constants* (e.g., ZEROS, SPACES, ONES, etc.); *integer constants*; and *special names* which relate to conditions of hardware (e.g., SWITCH-1-ON).

Subscripts are data names which appear in parentheses following the data names which they identify. A subscripted variable must be followed by an "OCCURS *n* times" clause setting up the dimensions of the resulting table.

File descriptions contain several items of information in addition to the file name. Some of these are:

1. Recording mode (if other than BCD).
2. File size—number of records.
3. Blocking factor in words.
4. Record size in characters.
5. Label records, whether standard or omitted. [COBOL automatically writes identifying records at the beginning and end of each file and each tape. The beginning records (BTL or BFL) identify tapes and files by name and number. The ending records (ETL or EFL) provide record and block counts.] This is a required clause for tape files.
6. Value of identification is the file name to be used on the BFL (in some systems this must be a nine-character literal). This clause is required with standard label records.
7. Data record names comprising the file. This is a required clause.
8. Record sequencing keys.

File Section. This section contains a section heading, file descriptions, and record descriptions. Output files may have to precede input files.

```
FILE SECTION.
FD file-name-1, LABEL RECORDS ARE STANDARD, VALUE OF IDEN-
   TIFICATION IS literal, DATA RECORDS ARE record-name-1, record-name-2.
  01 record-name-1
       02 group-item-1
            03 data-element-1, PICTURE IS 9(10).
            03 data-element-2, PICTURE IS A(10).
       02 data-element-3, PICTURE IS 9.
            88 condition-name-1 VALUE IS 1.
            88 condition-name-2 VALUE IS 2.
  01 record-name-2.
FD file-name-2, . . .
```

Working-storage Section. This section is composed only of record description entries following the section header.

```
WORKING-STORAGE SECTION.
  77 data-name-1...
  01 data-name-2...
       02 data-name-3...
```

Constant Section. This section is set up similarly to the previous section.

```
CONSTANT SECTION.
  77 data-name-1...
  01 data-name-2...
       02 data-name-3...
```

Procedure Division. The procedure division carries out the work of the program. In illustrating the format of procedure statements we shall use the following generalized terms:

Term	Meaning
File name	As used in the Data Division
Record name	As used in the Data Division
Data name	As used in the Data Division
Hardware name	As used in the Environment Division
Imperative statement	COBOL verb, e.g., GO TO . . .
Procedure name	Identifying name of a paragraph in the procedure division formed according to the rules for forming date names
Conditional statement	Statement of the form (data-name) (condition) (data-name or constant), e.g., (PAY) (GREATER THAN) (DEDUCTION)

The following conditions are acceptable in conditional statements:

GREATER THAN	POSITIVE
LESS THAN	NEGATIVE
EQUAL TO	IS ZERO

Input and Output Verbs

READ file-name.

READ file-name AT END imperative-statement.

WRITE record-name.

ACCEPT data-name FROM hardware-name.

DISPLAY data-name ON hardware-name.

Data Transfer Verb

MOVE data-name-1 to data-name-2.

Arithmetic Verbs. (In some cases, numeric literals may be substituted for data-names.)

ADD data-name-1 TO data-name-2 GIVING data-name-3.

ADD data-name-1 TO data-name-2.

SUBTRACT data-name-1 FROM data-name-2 GIVING data-name-3.

SUBTRACT data-name-1 FROM data-name-2.

MULTIPLY data-name-1 BY data-name-2 GIVING data-name-3.

DIVIDE data-name-1 INTO data-name-2 GIVING data-name-3.

ON SIZE ERROR imperative-statement.

ROUNDED. (Rightmost digit retained is increased by 1 if the leftmost of trailing digits discarded is 5 or greater.)

Sequence-control Verbs

GO TO procedure-name.

GO TO procedure-name-1, . . . procedure-name-*n* DEPENDING UPON data-name-1.

IF conditional-expression THEN imperative-statement-1 OTHERWISE imperative-statement-2.

PERFORM procedure-name-1. (Unconditional transfer of control for one paragraph only, after which control returns to statement following.)

PERFORM procedure-name UNTIL conditional-expression.

PERFORM procedure-name VARYING data-name-1 FROM data-name-2 BY data-name-3 UNTIL conditional-expression.

Processor-control Verbs

OPEN INPUT file-name.

CLOSE file-name.

STOP RUN.

END PROGRAM.

In addition, some COBOL systems admit the verbs ENTER and EXIT, which permits the programmer to enter a lower-order compiler or assembly language to obtain greater facility, e.g., for use of mathematical functions or bit manipulation.

99. Decision Tables. Use of decision tables facilitates programming logical operations on the computer. They have been used to design products such as transformers, meters, and telephone switching equipment. Wherever a program involves a large number of business or scientific decisions, a tabular chart is often easier to draw and understand than a textual or flow-chart representation. Decision tables are also known as *decision structure tables* or *tabular system charts.*

The basic format of a decision table is shown in Fig. 27-27*a*. The two sets of double lines divide the table into four parts. Above the horizontal double lines are the *primary,* or *name, rows.* Below are the *secondary,* or *value, rows.* To the left of the vertical double line are *condition blocks.* To the right are *action blocks.*

Associate the word IF with the first vertical line on the left, AND with other vertical single lines, THEN with the vertical double line, and PERIOD with the last vertical line.

Condition names are generally *data names* as defined in Par. **98.** Action names are generally *verbs,* e.g., GO and PERFORM. Condition names and values make up *conditional statements.* Condition values generally include a *condition* (see Par. **98**) followed by a *data name* or *constant.*

Action values are frequently *procedure names,* although they may also be made up of conditions and data names or constants. As each statement is executed, control drops to the following statement when transferred by an action.

Example

IF TEMPERATURE (condition-name-1) is GREATER THAN LIMIT (condition-value-1: condition + data-name) *AND* THERMOCOUPLE (condition-value-2) is EQUAL TO UNBROKEN (condition-value-2: condition + data name) *THEN* WRITE (action-name-1) ALARM (action-value-1: a constant) *AND* GO TO (action-name-2) EMERGENCY-PROCEDURE (action-value-2: a procedure-name).

See also Fig. 27-27*b*.

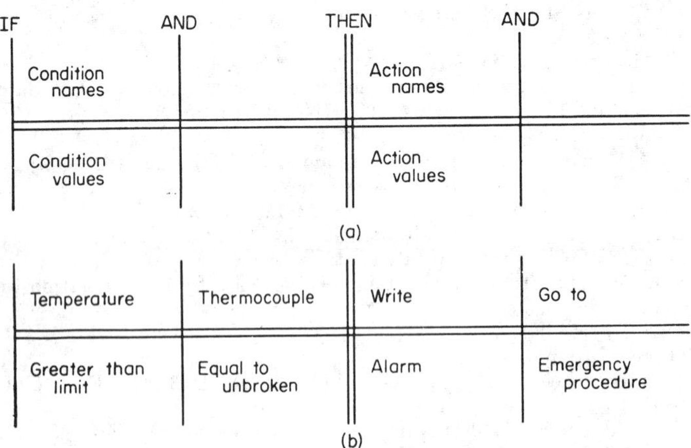

Fig. 27-27. (*a*) Decision table format and (*b*) example.

100. Report Writers. A report-writer routine is an extension of a commercially oriented compiler which provides business reports in conventional format to facilitate understanding and rapid reading.

Basically the routines provide a high degree of format control for the high-speed printer. Following are typical functions performed by a report-writer routine.

1. Print report headings at the beginning.
2. Print report footings at the end.
3. Maintain page control by line count, skipping to a new page when directed by the program.
4. Maintain line spacing.
5. Print page headings, footings, and numbers.
6. Accumulate totals.
7. Detect control breaks, print totals and control headings.
8. Count detail lines.
9. Edit data by suppressing leading zeros, inserting commas and dollar signs.

101. Sort-Merge Generators. These belong to a class of programming aids known as generators or packaged routines. They automatically compile machine-language programs for performing certain specific tasks common to a large number of data-processing operations.

Basically, these generators produce programs which prepare *strings* of records by sequentially ordering a block of records taken into the computer's high-speed memory (*tournament routine*). The block is replenished by replacing each tournament winner with a new record from input if this record is equal to or larger than the current winner. Otherwise it provides the seed of the next block. The resulting strings are accumulated on two or more tape handlers in the *presort* subprogram. The tapes are then collated by a *rotary-merge* subprogram.

The efficiency of a sort routine is determined by computer-memory cycle time, tape information-transfer rate, size of core memory, number of tape handlers available, and the machine's facility for moving data about.

To compile a sort-merge program using a generator, the programmer punches parameter cards describing the file to be sorted and the sorting operations required. The generator then compiles a sort program which is subsequently loaded into the computer when the program is to be executed.

102. Executive Routines. Executive routines (also known as *supervisory routines*, or *monitors*) set up, monitor, and control the operation of other programs.

As illustrated in Fig. 27-28, the operating system consists of the basic executive monitor and groups of executive routines concerned with execution control and control of translators and generators.

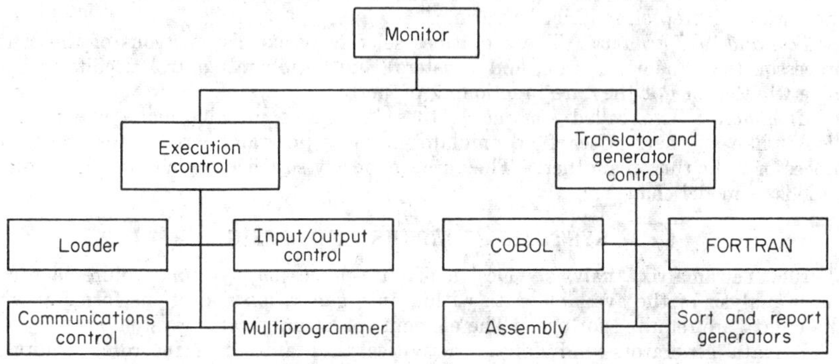

FIG. 27-28. Major components of an executive program.

Loader. This routine initializes storage by placing zeros or blanks in work areas and loads the program into storage.

Input-Output Control. This portion of the operating packages handles input-output system details such as:

1. Swapping and rewinding tape units.
2. Blocking and unblocking records.
3. Making three attempts to read after parity error and branching to an appropriate recovery routine if unsuccessful.
4. Checking tape labels.
5. Providing for halt or recovery routine after processing errors such as: exceeding storage capacity, wrong-length record, or unequal address comparison.
6. Execution of certain macros such as OPEN and CLOSE (files), GET and PUT (records), READ, PUNCH, and STACK (cards), and SEEK and SCAN (disk files).

Multiprogramming. This type of executive routine controls the simultaneous execution of multiple object programs stored in core. Segments of the programs are executed on a time-sharing basis (i.e., time-division multiplex). The multiprogramming routine handles the storage of register contents relevant to a particular program when its execution is interrupted and restores the contents when control is returned.

Communications Control. These routines are required in systems with a number of remote input-output units. The executive routine handles incoming messages, controls the processing of each output, and routes outgoing messages. The routines observe priorities and provide for storing messages until the required channels are available.

Compile and Run. This type of routine handles the steps of compilation, assembly, and execution sequentially, without manual intervention. More complex systems can accept several source languages (e.g., COBOL, ALGOL, FORTRAN, assembly language), calling up appropriate translators from library storage (tape or disk).

103. Utility Routines. These programs handle common housekeeping tasks in an information-processing system. Representative programs include:

1. Card to tape.
2. Card to printer.
3. Tape to printer.
4. Clear (zero) disk.
5. Load disk.
6. Disk to tape (or card).
7. Copy disk (or tape).
8. Disk to printer, etc.

104. Test Routines. These routines are used to "debug" programs and to diagnose hardware troubles. Representative programs include:

Memory Dump. Prints the contents of the immediate-access storage unit, usually in octal notation.

Trace Routine. Prints location of instructions and operands and contents of registers for each step of an object program.

Test and Maintenance. These routines selectively exercise portions of the central processor (e.g., memory, logic, and registers) and peripheral control circuits to determine whether or not they are functioning properly.

Emulators. Also called *simulators*, these programs literally make one processor "look" like another. Use of an emulator permits programs written for one central processor to be run on another. They are frequently supplied by manufacturers during periods of model changeover.

SYSTEMS ANALYSIS AND DESIGN

105. Tetramatrix Analysis. Design of data-processing systems requires a preliminary analysis of the available data within an organization and of the forms in which these data are required for use. The elements of an information system are:

a. Data generators—individuals, activities, and automatic recording equipment that originate data.

b. Data users—individuals, activities, and control devices which require reports or other information.

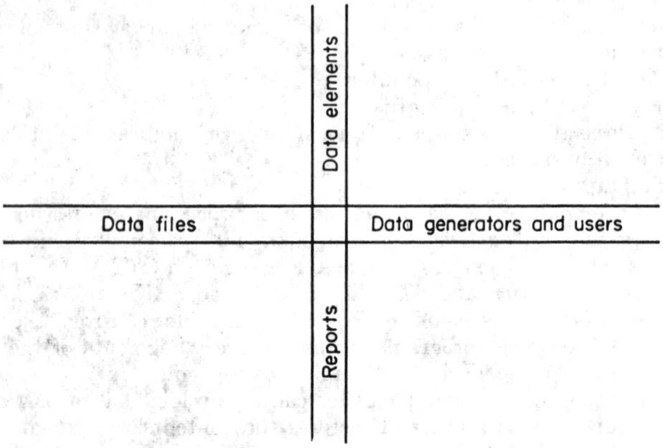

Fig. 27-29. Layout of a tetramatrix for information-systems analysis.

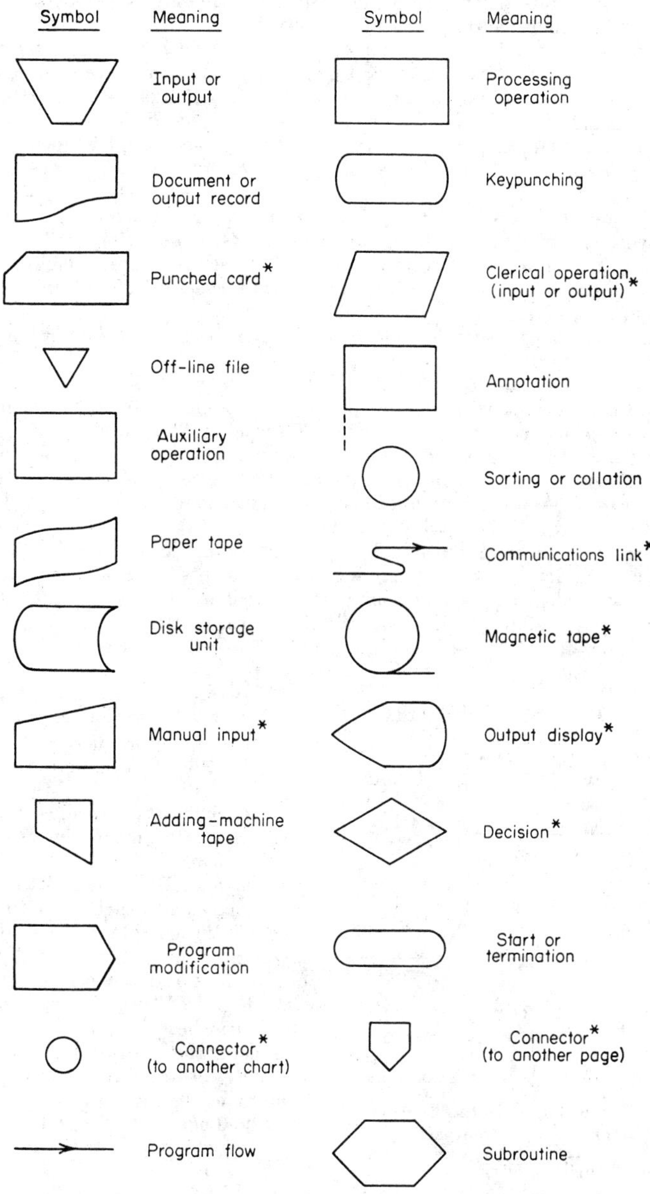
Symbol
Meaning
Input or output
Document or output record
Punched card*
Off-line file
Auxiliary operation
Paper tape
Disk storage unit
Manual input*
Adding-machine tape
Program modification
Connector* (to another chart)
Program flow
Processing operation
Keypunching
Clerical operation (input or output)*
Annotation
Sorting or collation
Communications link*
Magnetic tape*
Output display*
Decision*
Start or termination
Connector* (to another page)
Subroutine
* = Proposed USA Standard symbols

FIG. 27-30. **Symbols used in program documentation.**

c. Data elements—facts constituting the inputs to the system (e.g., names, dates, costs, quantities, etc.).

d. Data files—collections of data from which reports may be generated.

e. Reports—structured data delivered to users.

These factors are used as row or column headings in a four-part (tetra-) matrix, as shown in Fig. 27-29. The analyst first makes an exhaustive list of all data generators and users in the system and enters them on the right horizontal axis. He then lists all reports needed by the various users and enters them on the lower vertical axis after first verifying that these reports are indeed needed and used by the recipients and enters ones in appropriate cells of the lower right-hand matrix associating reports with data users.

He next breaks out all the data elements used on all reports, identifying which basic data elements enter into computation of report items. He verifies which of the data elements are in fact required by the users. These elements are entered on the upper vertical axis. The ones are then entered in appropriate cells of the upper right-hand matrix associating data generators with data elements.

The analyst now coalesces data elements into files based on the following principles:

a. Nonduplication.

b. Convenience of data collection.

c. Association of the greatest possible number of quasi-static data elements into *data banks* (e.g., master personnel file, master inventory file, etc.).

d. Association of the greatest number of conveniently collected data elements into *transaction files* (e.g., employee time cards, materials issue cards, etc.).

106. Record Definition. Each data element must be defined in terms of:

a. Maximum expected length.

b. Representation: numeric, alphabetic, or alphanumeric.

c. Special characters required: assumed decimal points, decimal points, dollar signs, zero suppression, floating decimal points or dollar signs, check protection, or commas. (This step implies defining media, e.g., punched card, printer, etc.).

Higher levels such as records and groups must be defined in terms of data elements and spacing (filler).

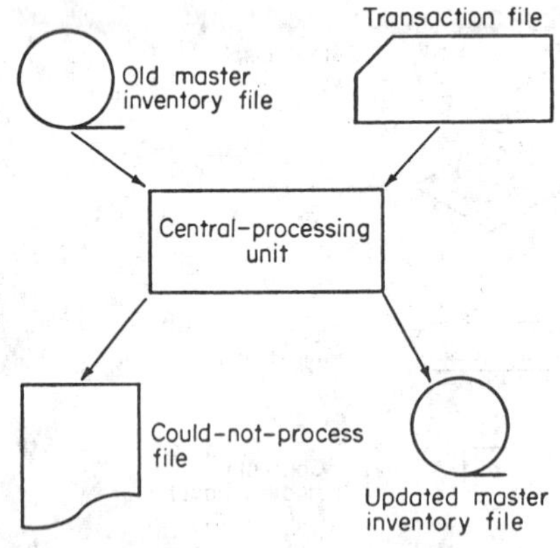

FIG. 27-31. Run diagram of a program for updating a materials inventory.

Files are defined in terms of records and media. This step in systems analysis furnishes the basic data needed to complete the file section of a COBOL data division.

107. Symbols. Symbols to be used in program documentation are illustrated in Fig. 27-30.

108. Run Diagrams. After files have been described, each run to be made through the computer's central processor is diagrammed, showing the input and output files and the peripherals involved (see Fig. 27-31 for an example). This step furnishes the basic data needed to complete the file control section of a COBOL data division and, by implication, describes the object computer.

109. Logic-flow Diagram. The logic-flow diagram makes use of the symbols of Par. **107** and illustrates:

a. Opening, closing, and accessing of all relevant files.

b. End-of-file procedures.

c. All logical tests and consequent branching.

d. Program loops.

e. Processing operations, including subroutine calls.

f. Error procedures.

g. Termination procedure.

Example. The logic-flow diagram for updating a materials inventory is shown in Fig. 27-32. The following files are involved:

a. Old master inventory file (OM), which consists of the following data elements: quantity available (AA); amount on hand (OH); amount on order (OO); and amount reserved for future jobs (RE). There is one master record for each class of material, identified by inventory number (IN).

b. Transaction file (T), which consists of the following data elements: orders placed (O); reservations of stock (R); shipments received (S); and amount issued (I). There is one transaction file for each class of material for which there has been activity, identified by transaction number (TN). The transaction file is made up of several *unit records* (i.e., punch cards), representing transactions occurring over a given period of time. Before the program is executed, these transaction records are sorted by number and written on magnetic tape.

c. Updated master inventory file (UDM), which consists of the old records brought up to date to reflect transactions occurring in the period in question.

d. Could-not-process file (CNP), which consists of records out of order which will be processed manually.

The basic processing equations derive from the relationship

$$OH + OO - RE = AA$$

Logic-flow charts are written conventionally from top to bottom and from left to right. The procedures shown on the chart can be translated directly to COBOL statements, and the resulting set of statements, with appropriate procedure names added where required for reference, constitute the COBOL procedure division.

A program of this general type may be thought of as an operation in *records management.* The operation is conducted *off-line* inasmuch as there is no physical connection between the data-collection mechanisms and the computer. It may be characterized as a *batch-processing* operation, since transactions are accumulated over a period of time rather than being processed as they occur.

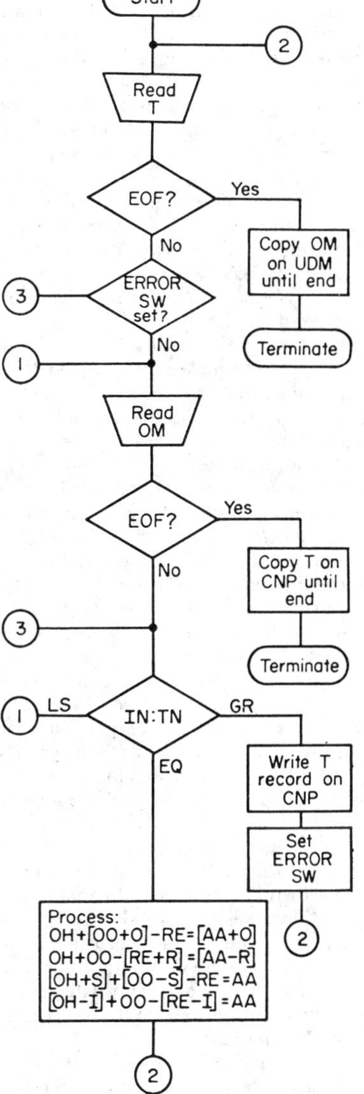

FIG. 27-32. Logic-flow diagram of a program for updating a materials inventory.

110. Information Retrieval. Information retrieval may be looked upon as one of the three basic modes of EDP system operation, along with *scientific computation* and *records management.* There are, in turn, at least five modes in which an information-retrieval system may operate:

a. Library Automation. These applications are essentially records-management operations performed within a library or information-center environment. Detailed programs exist for:

i. *Serials control.* Subscribing to publications, renewing subscriptions, requesting missing issues, keeping track of copies during binding or other processing operations, and generating a periodically updated list of serials holdings.

ii. *Circulation control.* Use of punched cards in check-out and check-in of circulating holdings, generating a periodically updated list of overdue books and borrowers holding them.

iii. *Book catalogs.* Recording bibliographical (catalog-card) descriptions of accessions and holdings and generating a periodically updated catalog of holdings in book-form for use by library patrons.

iv. *Interlibrary loans.* Determining nonduplicating holdings of two or more libraries to facilitate interchange of information.

b. General Dissemination of Information. These systems entail periodically printing and circulating lists of new accessions, usually with catalog-card information. These lists may result as by-products of book-catalog programs.

c. Retrospective Search. This mode of operation requires that center holdings be indexed in depth by using a *coordinate indexing* (or *concept coordination*) system in which several (usually 10 to 14) standardized *indexing terms* (or *descriptors*) are manually or automatically assigned to each document. The user's request for information is framed in terms of boolean functions of a set of descriptors (for example, $A \cdot B$, $A + B$, $A \cdot B \cdot \bar{C}$, etc.). Machine-stored *thesauri* (or *word-authority lists*) may be employed to standardize the set of terms submitted by a requester.

The searching operations may involve use of an *inverted file* in which the *items* (document numbers) are listed under their *characteristics* (indexing terms) rather than conversely, as is conventional practice.

d. Selective Dissemination of Information. In these systems an *interest profile* of indexing terms is developed for each user from a questionnaire, his past information-seeking habits, or both. Periodically, the descriptor sets of recent accessions are matched with user-interest profiles, usually on a probabilistic basis, and notification is sent to users of accessions in their specific areas of interest only.

e. Demand Retrieval. The basic technique is similar to that used in retrospective search, with the following differences:

i. No attempt is made to search the file exhaustively; effort is concentrated in areas where successful search is most probable (e.g., recent accessions).

ii. User requests and system responses are usually in an *on-line* mode (i.e., input and output equipment are electrically connected to the central processor).

iii. There is often an *adaptive-programming* feature connected with the search program which permits the user to reach the information he desires in several steps by progressively modifying his request based upon system response.

Information retrieval is not limited to retrieval of documents as such. It has been successfully employed to identify individuals having specified combinations of skills, cutting tools having desired properties, etc. Most *management information systems* operate in one or more of the information-retrieval modes.

111. On-line Computing. It has long been evident that the bulk of available central processor time has been wasted waiting on slower-speed peripheral equipment. Modern information-processing systems are capitalizing on this available time to extend computer capabilities.

Definitions

a. In-line processing. Processing data in random order in which they arrive at the computer without prior editing or sorting.

b. On-line processing. Operation of input-output devices under direct control of the central processor; usually implies that transactions update files directly at the time of data recording at the source.

c. Real-time operation. Performance of computation during the actual time when the related physical process occurs so that computation results are useful in guiding the physical process.

d. Time sharing. A mode of using a computer facility to shorten solution which has the following characteristics:

i. On-line, remote consoles.

ii. Response in real time (in context of application and human reaction time).

iii. Multiple users with direct access.

iv. Users react independently (protected from each other's mistakes).

v. Users can state arbitrary procedures.

vi. File system with protected storage and controlled access.

Problems in time-sharing include dynamic scheduling; priority and memory interaction; communications; file organization and protection; equipment and program interaction; queueing at hardware units; reliability and switchover of hardware; fallback and recovery after failure; programmer coordination; and program testing in the face of interaction effects, multiple input complexity, and inability to repeat errors.

e. Multiprogramming. Multiple-object programs reside in the core memory of a single system. They are executed alternately to improve throughput by simultaneous input, output, and processor operation.

f. Multiprocessing. Two or more processors reside in a common core memory. One processor usually acts as control, and the other is subordinate.

g. Multicomputing. Requires two or more system processors each of which operates as an independent system but has direct communication between respective memories.

112. Process Control. A process control system consists of a computer and associated hardware and software capable of receiving analog and digital input signals from a process (e.g., steel mill, chemical plant, atomic reactor, etc.) and providing outputs to the process for controlling the operation.

Process control systems operate in real time, typically seconds and milliseconds but ranging from microseconds to hours in special instances.

The systems are on-line inasmuch as the computer is intimately associated and connected with the particular process.

Reliability is a prime consideration in process control, with a requirement for at least 99.5% availability with no scheduled downtime. As a consequence *worst-case design* is employed for the system components so that the circuitry will perform accord-

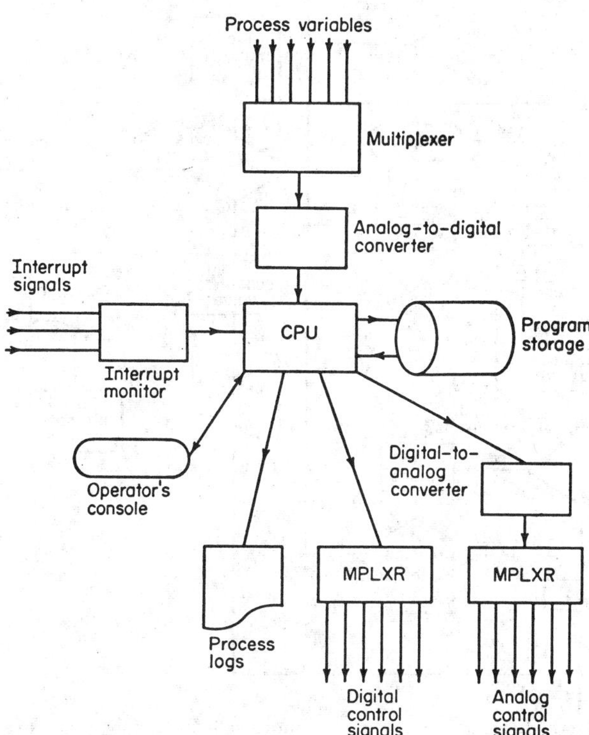

FIG. 27-33. Process-control computer showing the usual suite of peripheral equipment.

ing to specifications when all parameters are simultaneously at their worst conditions in relation to all other parameters.

Ruggedness is another requirement imposed upon control computers and not generally imposed upon general-purpose computers. Typical specifications include:

a. Temperature 32 to 131°F.

b. Humidity 5 to 95% RH.

c. Vibration 5 to 33 c/s, with 15-mil displacement.

d. Power-supply variation of ±10% in voltage; ±1 c/s in line frequency.

e. Ability to filter out dust particles down to 10 microns in diameter.

Central Processor and Memory. Modern process-control computers operate in the microsecond range. Their instruction suite is smaller than for general-purpose computers but includes special instructions for bit manipulation. Control computers must have a high order of interrupt capability: 15 interrupt lines are common. Immediate-

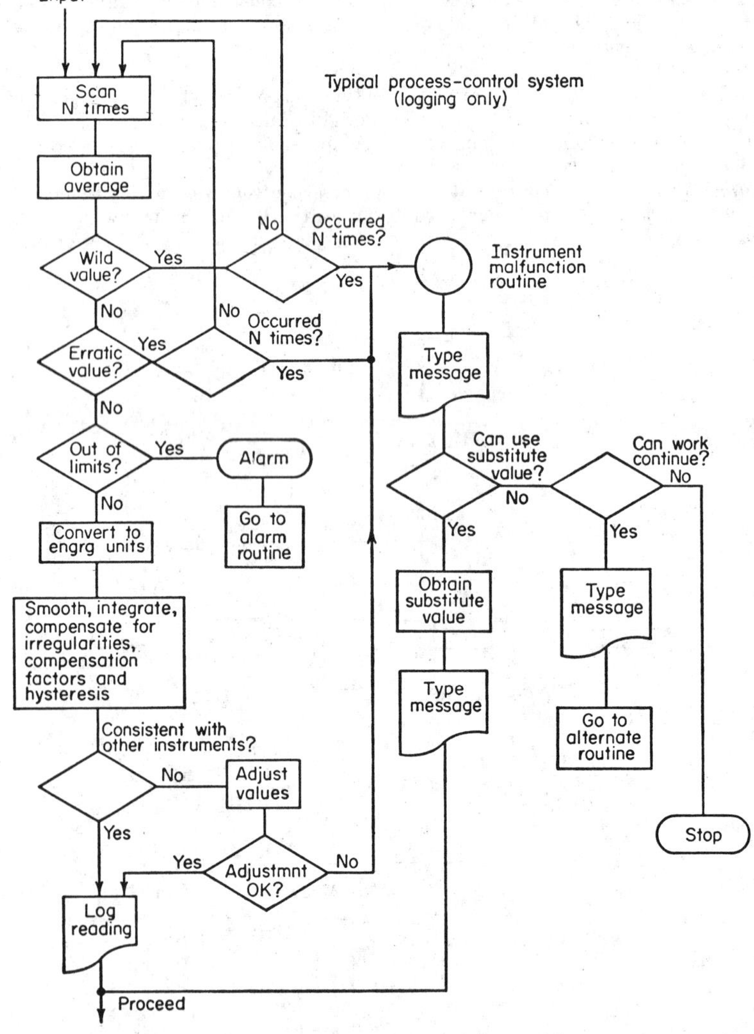

Fig. 27-34. Logic-flow chart of program to validate and log process variables.

access storage requirements may exceed those of general-purpose computers: core memories are about the same size, but a high-speed drum may be employed for program storage. Little use is made of tapes or drums except to store backup data for use in general-purpose data processing during *free time* (i.e., between process inputs).

The common peripherals of a process-control computer are shown in Fig. 27-33. They include:

a. Process sensors, transducers, actuators, and controllers.
b. Manual input console.
c. Paper-tape reader-punch (program loading).
d. Analog-digital and digital-analog converters (50 to 200 inputs).
e. Signal conditioners for analog and digital lines.
f. Analog and digital input controllers.
g. Output controllers.

Process-control computers are characterized by a heavy emphasis on error detection and correction. Figure 27-34 is the flow chart of a logging operation showing how the computer relies on reasonableness checks, limit checks, and self-consistency and consistency among inputs to discriminate between instrument malfunctions and process irregularities.

113. Bibliography on Electronic Data Processing

Gregory, Robert H., and Van Horn, Richard L. "Automatic Data-processing Systems: Principles and Procedures," 2d ed.; Belmont, Calif., Wadsworth Publishing Company, Inc., 1965.

Goode, Harry H., and Machol, Robert E. "System Engineering"; New York, McGraw-Hill Book Company, 1957.

Brooks, Frederick P., Jr., and Iverson, Kenneth E. "Automatic Data Processing"; New York, John Wiley & Sons, Inc., 1963.

Davis, Gordon B. "An Introduction to Electronic Computers"; New York, McGraw-Hill Book Company, 1966.

Fink, Donald G. "Computers and the Human Mind"; Garden City, New York, Anchor Books, Doubleday & Company, Inc., 1966.

Bartee, Thomas E. "Digital Computer Fundamentals," 1st ed.; McGraw-Hill Book Company, 1960.

Williams, William F. "Principles of Automated Information Retrieval"; Elmhurst, Ill., The Business Press, 1965.

Ledley, Robert S. "Programming and Utilizing Digital Computers"; New York, McGraw-Hill Book Company, 1962.

Van Ness, Robert G. "Principles of Punched Card Data Processing"; Elmhurst, Ill., The Business Press, 1962.

Savas, E. S. "Computer Control of Industrial Processes"; New York, McGraw-Hill Book Company, 1965.

COMPUTER CONTROL OF POWER SYSTEMS

By Claude M. Smith

114. Introduction. The power industry pioneered the use of process-control computers. Now, computers have become a vital, necessary factor in helping to solve the complex problems of the industry as it expands to meet the needs of the population explosion.

The process-control computer can be economically justified on the basis of (1) greater safety, (2) fuel savings, (3) reduced outages, and (4) manpower reduction. However, experience has shown that through the use of real-time data processing, mathematical modeling, and control techniques, efficiencies in nearly every aspect of everyday production and distribution can be realized.

115. Functional Organization. The commercial, engineering, and operating functions of an electric utility obtain the greatest benefit from the application of computers. Scientific and business digital computers for many years have been used by these functions to handle their responsibilities more efficiently in a before-the-fact or after-the-fact method.

New off-line computer applications are continually being developed. The commercial applications include the ever-increasing task of maintaining up-to-date records and processing customer billing. The engineering function has found many applications in the design and construction of the power system. To mention a few:

- Mechanical engineering:
  - Power plant design.
    - Mechanical design of piping.
    - Mechanical design of buses.
    - Heat balance for steam stations.
  - Transmission and substation structures.
    - Transmission-tower spotting.
    - Broken-conductor calculations.
    - Sag and tension calculations on transmission lines.
- Electrical engineering (design and system planning):
  - Load-flow calculations.
  - Short-circuit current calculations.
  - Transient-stability calculations.
  - Generation expansion.
  - Primary-circuit impedance calculation.
  - Radial primary-circuit protection studies.
  - Secondary-network load flow.

The economical operation of a power plant requires continual improvement in procedures for control and maintenance. The operating function whose responsibilities and problems primarily concern the everyday operation of the system provided limited applications for the scientific- or business-type computer. However, with the advent of the on-line process computer with its ability to rapidly accumulate data and immediately compute unit and component performance, the operating function had a powerful tool with which to meet the growing complexities of the everyday normal and abnormal operational problems. Some of the applications now being implemented include:

- Steam-electric power plants:
  - Data logging.
  - Performance calculations.
  - Monitoring and performance.
  - Sequence monitoring—operator guides.
  - Turbine-generator control.
  - Boiler optimization.
  - Automatic start-up and shutdown.
- Electrical power dispatching:
  - Load and frequency control.
  - Economic dispatching.
  - Hydrothermal optimization.
  - Interchange billing.
  - System and unit production statistics.
  - Interchange transactions.
  - Unit commitment.
  - Selection of fuels.
  - Maintenance scheduling.
- Nuclear power plants:
  - Rod-position monitoring.
  - In-core ion-chamber monitoring.
  - Core-performance calculations.
  - Control-rod pattern calculations.
  - Fuel burnup optimization.
  - In-core chamber calibration.

116. Process-control Computers. A process-control computer is one which can communicate with process instrumentation and control equipment. It can be analog, digital, or a hybrid system. Analog systems are used where a low-cost equipment solution is required for simple single-control loop applications involving few variables. The analog systems are often termed "wired programs" and as such are inflexible.

Digital process-control computers can communicate with (1) process instrumentation, (2) control equipment, (3) human operators, and (4) other computers. Process communications refers to the continuous scanning of any instrument, sensor, or transducer, with or without an electrical output, thereby producing analog or digital inputs. Analog or digital output signals (latched relays, timed relays, printed logs, displays, etc.) are sent out as required. Figure 27-35 is a block diagram of a process-computer application.

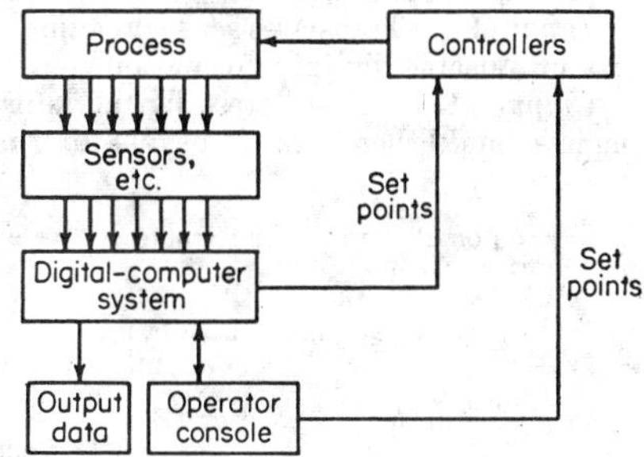

FIG. 27-35. Process-computer application block diagram.

117. Parts of a Process Computer. *Memory.* Data and instructions are electronically stored.

Arithmetic and Control Unit. Performs calculations, logical operations, sequences, and distributes data throughout the system. It supplies and receives information from the memory, the automatic priority interrupts and I/O buffer, and the peripheral buffer.

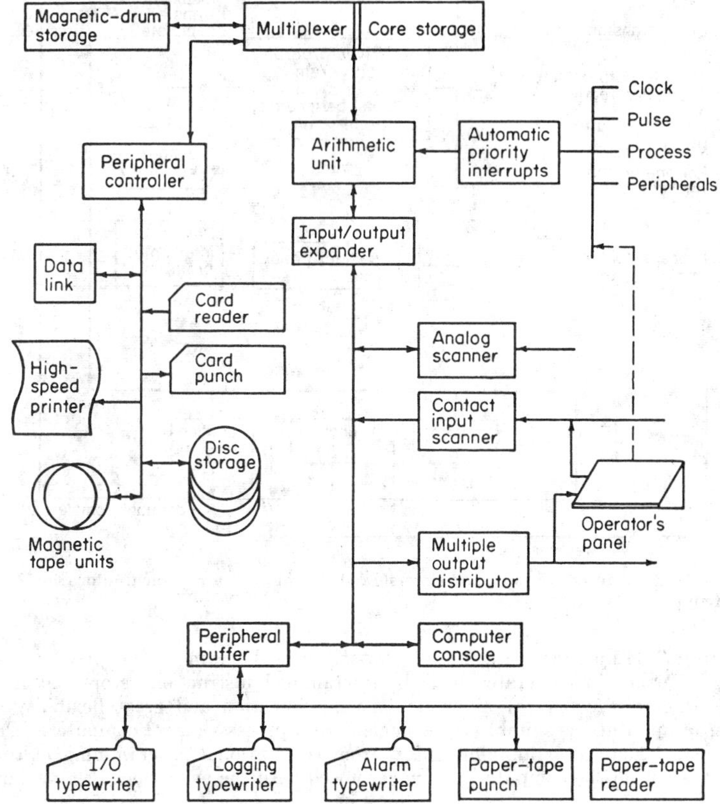

FIG. 27-36. Process-computer block diagram.

The automatic priority interrupt detects and locates ready and complete signals from devices that require testing at long time intervals and also provides instantaneous signals when process conditions change.

Input (I). Translates sensor readings, contact closures, manual switches, and other data into usable digital information.

Output (O). Translates digital information into logs, analog set points, lighted displays, and other forms of usable information external to the computer.

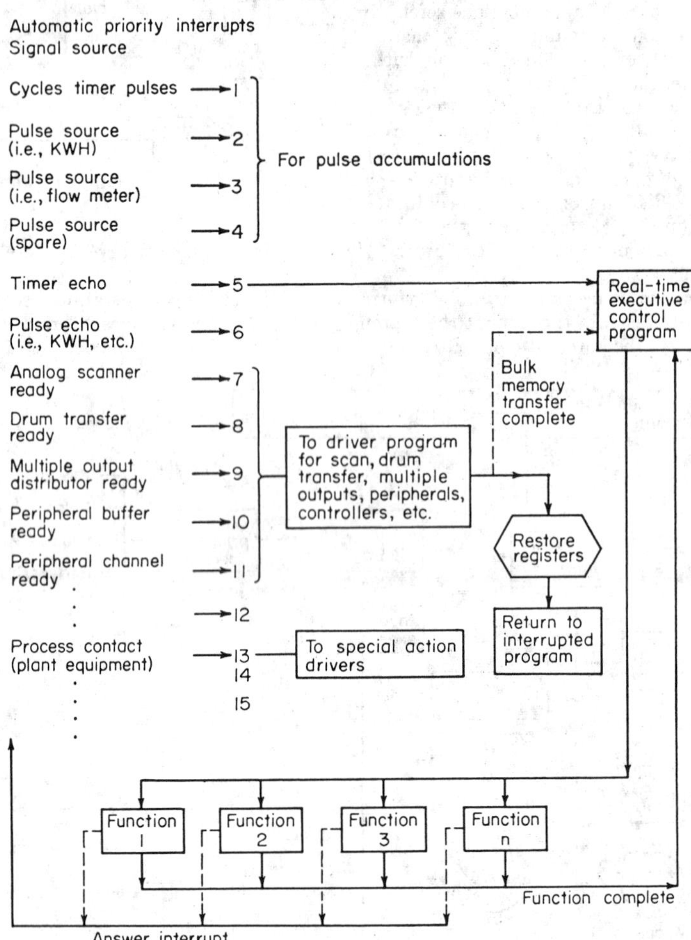

Fig. 27-37. Multiprogramming for time-shared operation via a typical allocation of inputs to interrupt.

Figure 27-36 illustrates a computer-system block diagram.

Digital computers operate through programmed instructions stored in memory. Stored programs (software) allow for easy modification and great flexibility in the number of applications suitable for a single digital process-control computer. Process-control (real-time) programming differs from conventional programming in that the operations within the computer are synchronized with the events occurring external to the computers. At any time the computer system must be able to stop its programmed sequence of operations in response to a signal from a higher-priority function, complete the actions required by the higher-priority functions, and then complete the interrupted function.

The *real-time operating system* provides the basis for a process-computer system. It accomplishes the timing and scheduling operation through the *real-time executive program* and the basic input and output (I/O) functions through the *automatic priority interrupt system.*

Interrupt capability is provided through unique hardware communication. What to do in response to a process signal interrupt is defined by user specifications and is implemented through software logic. It is through the interrupt feature that the computer can stay synchronized with the process and respond with action in microseconds. Figure 27-37 is a block diagram showing multiprogramming for time-shared operation via typical allocation of inputs to interrupt.

The ability to communicate in synchronization (real time) with the process and the human operators established the role of the digital process-control computer. Modular construction is the key word in designing process-control hardware and software. The following modules or building blocks are some of those used in developing the typical process-computer application in the power industry. For a variety of detailed computerized power systems see the references (Par. **123**).

118. Scan, Log, and Alarm Functions. With the multiprogramming real-time executive programs that keep the software system in order ignored, the heart of any process computer system is: (1) scanning the instrumentation, (2) testing the measured values against predetermined limits and alarming through printed messages, lights, or bells if the values are out of limits, and (3) logging desired data periodically or on operator demand. Detection of incipient component failure substantially reduces the probability of damage and prolonged outage.

Breakpoint 0 — Initiate start-up
Preliminary conditions
Plant auxiliaries
Boiler prestart conditions
Turbine prestart conditions
Breakpoint 5 — Boiler ready to light off
Boiler warm-up
Feedwater and condensate
Steam seal and vacuum
Breakpoint 10 — Turbine ready for roll-off
Turbine acceleration
Boiler warm-up
Breakpoint 12 — Ready to synchronize
Boiler warm-up
Turbine generator loading
Turbine transfer to partial arc
Breakpoint 16 — Transfer to partial arc completed
Unit at 18 megawatts
Minimum load
Final conditions checked
Breakpoint 18 — Start-up complete

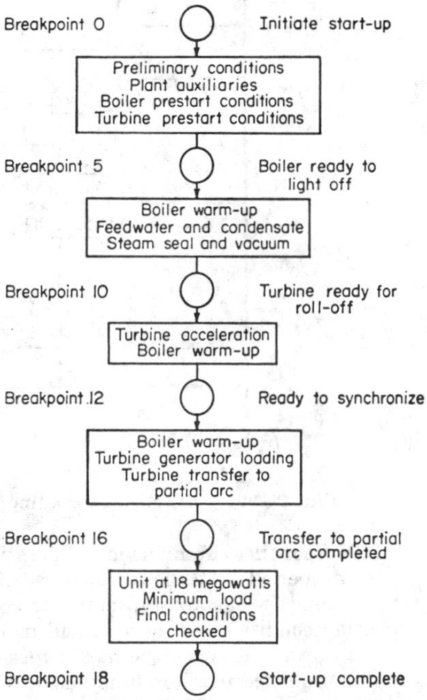

FIG. 27-38. Basic start-up program.

119. Performance Calculations. Performance-calculation programs, when added, provide quick detection of changes in equipment performance and efficiency, informing the operator immediately and allowing the operator to adjust controls or put equipment in or out of service in an effort to improve performance or prevent damage.

Procedures for calculating power-station component performance criteria are, in most cases, described by ASME test codes or recommended by user-manufacturer groups, based on favorable application experience.

Performance calculations for a thermal power plant usually include:

1. Boiler efficiency.
2. Unit heat rate.
3. Turbine-cycle heat rate.
4. Turbine-section efficiencies.
5. Feedwater heating monitoring.

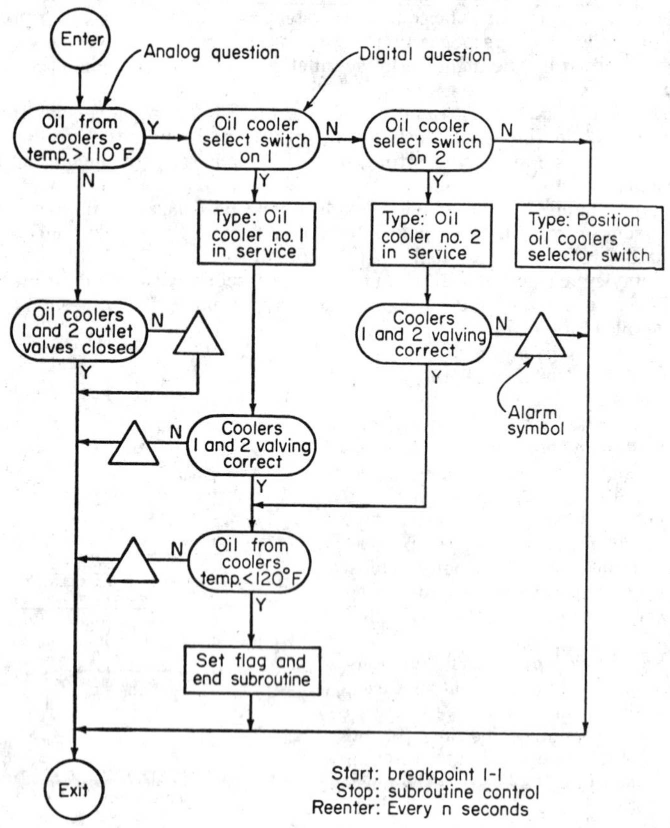

FIG. 27-39. Oil-cooler subroutine with typical analog and digital questions.

120. Sequence Monitoring. Operating errors, occurring during the start-up and shutdown periods, are the major cause of plant and equipment damage.[4]* Boiler combustion tends to instability, metal stresses and metal expansions near their safe limits, operating conditions are severe, and monitoring all the equipment in order to start up or shut down taxes even the most experienced operator.

The sequence monitor function provides a consistent approach for each operating circumstance, thereby aiding the experienced operator and lessening the difference in performance among operators. New operators may be added without extensive and tedious training, as all operators are presented a limited amount of guidance information at any one time, with defined break points displayed when significant steps have been completed. Figure 27-38 is a simplified flow diagram of a basic start-up program involving several break points. Before implementing an operator guide function, experts in the operation of the specific pieces of equipment and the system must define and document to the smallest detail the procedure for each circumstance. It is important to remember that *man must instruct the computer* through programmed sequence logic instructions which the computer follows implicitly. The logic instructions are usually a series of analog or digital questions (Fig. 27-39) followed by instructions which provide paths of computer action depending upon whether or not the answer to the question is acceptable.

---

* Superior numbers correspond with the references in Par. **123**.

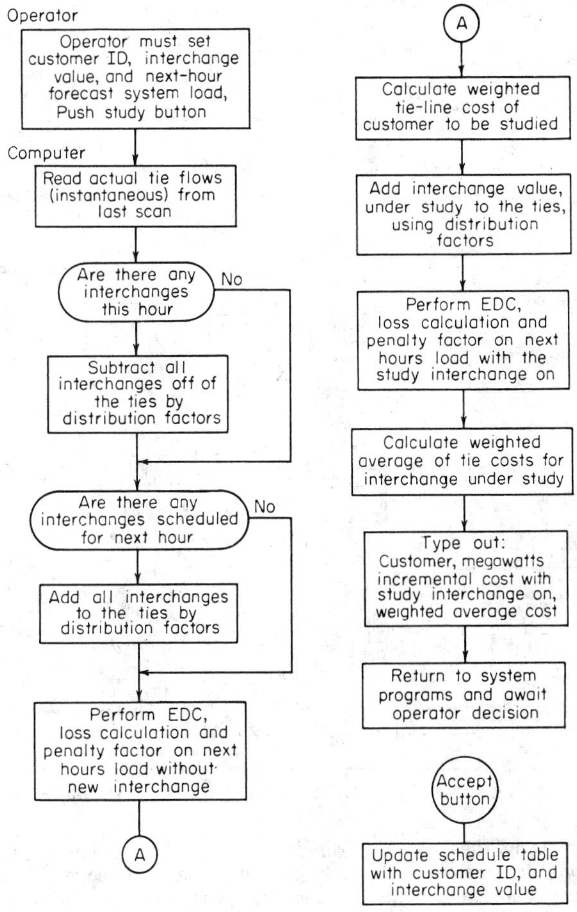

FIG. 27-40. Flow chart of an interchange transaction study.

121. Economic Dispatching. The ability of the digital process computer to contain a mathematical model of the power system provides a convenient method for the solution to many operating and engineering planning problems. The economic-dispatch program is a mathematical model of the system for operation with minimum fuel costs, with equal incremental loading of units, recognizing transmission losses. In addition to controlling the load on each generating unit, the resulting data from the economic dispatching calculations (EDC) can be used in the following programs:

1. Unit commitment. Determines the most economical starting and stopping time for all generating units of the system.
2. Interchange negotiations. Economic evaluation of future interchange transactions (purchase or sale of power).
3. Load-duration costing. Economic evaluation for having generating units out of service for maintenance.

To describe briefly one of the preceding: An interchange transaction program (Fig. 27-40) is started by the operator pressing the *interchange study* button on the operator's

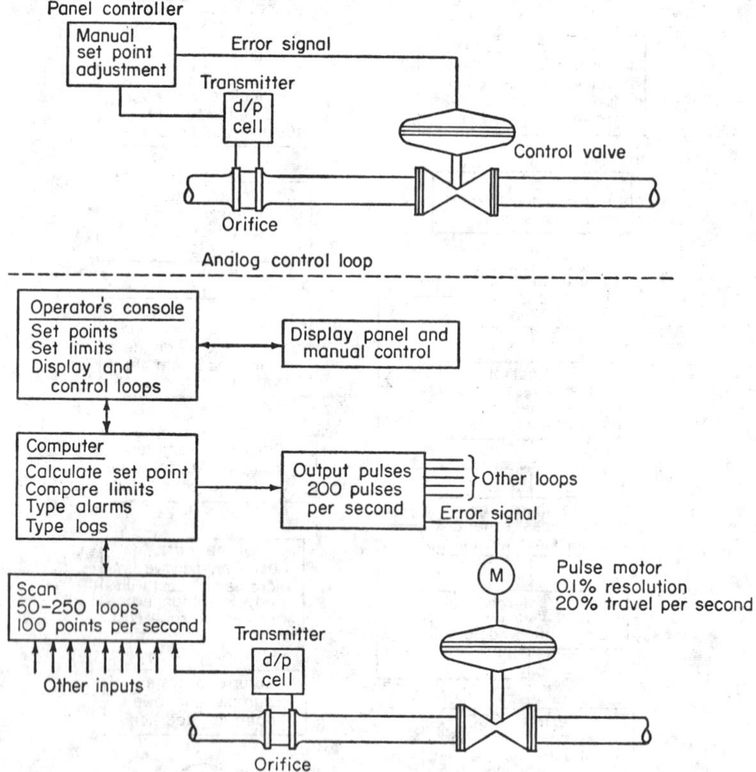

FIG. 27-41. Direct digital control vs. analog control loops.

console and entering the proposed transaction by setting dials. Daily-load forecast curves are stored in memory. Incremental costs are calculated with and without the transaction, by using tie-line distribution factors. The typewritten results provide the dispatcher with the necessary cost information for making a decision. If the transaction is made, an *Accept* pushbutton is pressed, and this now becomes the base condition for the next transaction.

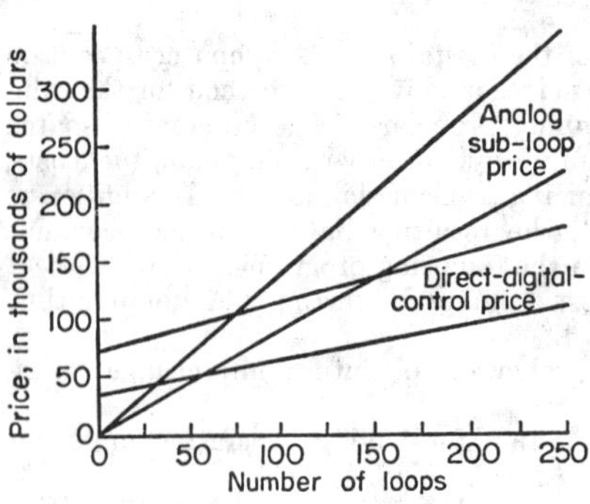

FIG. 27-42. Economic evaluation of direct digital control.

122. Closed-loop Control. To date the control loop has been closed in very few functions. In steam-generating plants the loops have been closed on *automatic start-up* and *automatic shutdown,* which includes boiler warm-up and turbine acceleration. Even these loops open when abnormal events occur, and then the human operator must assist or take over completely.

Only when more and better instrumentation is available, when more is known concerning the process and its control, and when man is able to instruct the computer more completely—only then can the control loops which lead to complete computer control be closed. On- and off-

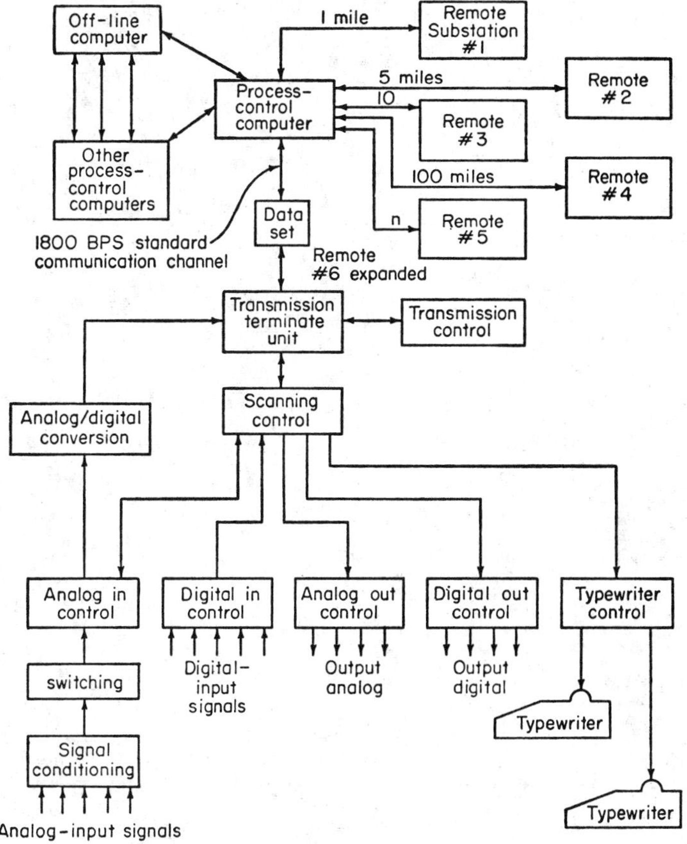

FIG. 27-43. Integrated computer system with remotes.

line computer studies[2,3] are adding to process knowledge. Two major areas seem to be the immediate future of process-computer control.

1. Direct digital control (Figs. 27-41 and 27-42).
   a. Eliminates analog subloops, thereby reducing costs and improving accuracy.
   b. Automatic calculation of set points.
   c. Computer calibration of sensors—improved accuracy.
2. Integrated computer systems (Fig. 27-43).

123. Bibliography on Computer Control of Power Systems

1. CHADWICK, W. L., and GOULD, W. R. Computer Shakedown at Huntington Beach Consolidated Experience; *Elec. World,* Nov. 19, 1962.

2. *Proc. Power 2nd Computer Appl. Conf.,* 1963, IEEE.

3. *Proc. Power 2nd Computer Appl. Conf.,* 1965, IEEE.

4. LIND, K. C., and SAEZ, R. Sequence Monitoring: A Power Plant Operation Guide Function; *Proc. Power 2nd Computer Appl. Conf.,* 1965, p. 333, Puerto Rico Water Resources Authority.

SECTION 28

INDUSTRIAL ELECTRONICS

BY

VIN ZELUFF Electronics consultant

CONTENTS

Numbers refer to paragraphs

SECTION 28

INDUSTRIAL ELECTRONICS

1. General Applications. The uses of electronics in industry can be classified in three groups: measurement, processing, and supervision. Tables 28-1, 28-2, and 28-3 list typical applications in each of these groups and the devices and components used to perform the various functions. As shown in the tables, virtually all industrial electronic equipment is composed of three elements: (1) an input transducer, which converts from the quantity measured, controlled, or supervised, to an electrical quantity, usually a variable voltage; (2) a conversion device, usually containing tubes or transistors, which amplifies or otherwise modifies the electrical output of the input transducer; and (3) an indicator, control element, or output transducer, which indicates and/or controls in terms of the initial quantity applied to the input transducer.

ELECTRON-TUBE CHARACTERISTICS

2. Thermionic Vacuum Tubes (see also Sec. **5**, Pars. **198** to **216**). Electron tubes which employ thermionic means of obtaining electron emission and in which no gaseous ionization effects take place (high-vacuum conditions) are known as thermionic vacuum tubes. Thermionic tubes are classed according to the number of electrodes they contain. Examples are diodes, triodes, tetrodes, pentodes, hexodes, etc. Such tubes are commonly used in communications technology and hence are often called "radio tubes." They are also widely used in industrial applications, and some, in fact, have been redesigned for the more stringent requirements of industry, particularly for use in equipment where long life and high reliability are needed.

Vacuum tubes having a power-handling capability ranging from a few milliwatts up to 10 or 20 W are usually called "receiving tubes." Higher-power types are usually classed as "transmitting tubes," but there is no sharp line of demarcation. High-power transmitting vacuum tubes range in power output up to 500,000 W. Although there are upward of 1,000 tube types, they all break down into the basic types to be described, diode, triode, tetrode, pentode, and so on, or combinations of these.

3. The diode vacuum tube is a two-element vacuum tube employed for rectification of alternating currents or for detection of modulated waves. The essential elements of the diode are the negative electrode, or cathode, which serves as the emitter of electrons; and the positive electrode (anode or plate), which collects the electrons (see also Fig. 5-113). The rectification function results from the unidirectional character of the current flow, the result of the fact that only one electrode is an emitter. The detection characteristic is related to rectification and is dependent also on the curvature of the curve relating applied voltage to the resulting current.

4. Dynamic Plate Resistance. The principal operating parameter of the diode tube is the dynamic plate resistance, which is the slope of the curve between applied voltage and resulting current. Specifically the dynamic plate resistance is

$$r_p = de_p/di_p \qquad \text{(ohms)} \tag{28-1}$$

where de_p = an elementary change in voltage between cathode and anode and di_p = corresponding elementary change in current between cathode and anode. As shown

Table 28-1. Measurement and Control of Physical Quantities

Quantity measured or controlled	Name of device	Input transducer	Conversion device	Indicator or output transducer
Acoustic:				
Pitch	Sound analyzer	Microphone	Audio amplifier and filters	Meter
Volume	Sound analyzer	Microphone	A-f amplifier and filters	Meter
Noise	Noise meter	Microphone	A-f amplifier, filter, rectifier	Meter
Chemical:				
Conductivity	Fluid-conductivity control	Bridge	Null detector	Meter
Gas content	Gas analyzers	Vacuum chamber	Mass spectrograph	C-r oscilloscope
Humidity	Humidity control, moisture indicator	Hair control	Contact amplifier	Meter
Ion concentration	pH meter/control	Titration electrode	Potentiometer	Meter
Radioactivity	Geiger counter ionization chamber	Geiger tube	Amplifier counter	Meter
Turbidity	Turbidity indicator	Phototube	Amplifier	Meter
Electrical:				
Current	Electrometer	Low-grid-current triode	D-c amplifier	Meter
	Current regulator	Triode	D-c amplifier	Meter
Flux	Flux meter	Pickup coil	A-f amplifier	Meter
Frequency	Frequency meter	Bridge	Null detector	Meter
Circuit parameters	Various bridges	Bridge	Null detector	Meter
Power	Electronic wattmeter	Triode-tetrode		Meter
Voltage	Voltage regulator	Triode	Amplifier	
	Electronic voltmeter	Diode-triode	D-c amplifier	Meter
Mechanical:				
Angle	Selsyn or synchro	Selsyn transmitter		Selsyn receiver
Level	Level-indicator control	Float contact	Electronic switch	Meter or relay
Liquid flow	Flow-meter control	Flow turbine	Generator	Meter or relay
Position	Positioning control	Pantograph	Potentiometers	Motor drive
Pressure	Pressure-indicator control	Bridge gage	Null indicator	Meter or relay
Speed	Speed-indicator control	Tachometer generator	Voltmeter	Meter or relay
Strain	Strain gage	Crystal pickup	Amplifier	C-r oscilloscope
Tension	Slack regulator	Tension contact	Amplidyne	Motor
Torque	Torque-indicator control	Magnetic brake	Amplifier	Meter or relay
Vacuum	Vacuum (ionization) gage	Ionization gage	D-c amplifier	Meter
Vibration	Vibration meter	Crystal pickup	A-f amplifier	Meter
Hardness	Hardness tester	Hammer and gage	Acoustic pickup	C-r oscilloscope
Viscosity	Viscosity meter	Paddle	Magnetic brake	C-r oscilloscope
Optical:				
Brightness	Illuminometer, photometer	Phototube	Amplifier	Meter
Color	Spectrometer, colorimeter	Phototube and filters	Amplifier	Meter
Density, opacity	Densitometer	Phototube	Amplifier	Meter
Infrared	Infrared-detector meter	Infrared phototube	Amplifier	Meter
Reflection	Reflectometer, reflection meter	Phototube	Amplifier	Meter

Table 28-1. Measurement and Control of Physical Quantities.—*Concluded*

Quantity measured or controlled	Name of device	Input transducer	Conversion device	Indicator or output transducer
Ultraviolet	Ultraviolet-detector meter	Ultraviolet phototube	Amplifier	Meter
Thermal:				
Conductivity	Thermal-conductivity meter	Temperature-difference bridge	Null indicator	Meter
Temperature	Thermometer-temperature control	Resistance-thermometer thermocouple	Amplifier	Meter or relay
	Pyrometer (optical)	Phototube and filters	Amplifier	Meter
Time:				
Standard	Primary frequency standard	Quartz crystal	Frequency divider	Clock
Interval	Interval timer	Capacitor charge	Amplifier	Relay
Delay	Delay timer	Multivibrator	Amplifier-pulse selector	Meter or relay
Sequence	Sequence timer	Trigger circuit	Amplifier-pulse selector	Meter or relay

in Fig. 5-115, r_p varies with the absolute values of e_p and i_p. When a resistance is connected in series with the anode and cathode (for the purpose of developing the rectified voltage), the operating point may be found by the "load-line" method. (This technique is illustrated in Fig. 28-3.)

5. The triode vacuum tube is a three-element tube used for amplification and the associated functions of modulation and demodulation. In addition to the cathode and anode (which serve the same purposes as in the diode), the grid electrode is situated as shown in Fig. 5-114. Variations in voltage applied between the grid and cathode produce changes in the plate current by modifying the space potential at the surface of the cathode.

6. Triode Operating Characteristics. Since there are two applied voltages e_p and e_g and a resulting current i_p, the possible operating points of the triode are expressed as a family of curves, typical examples of which are given in Fig. 5-116. Ordinarily, the e_g value is negative with respect to the cathode. If e_g is positive, another current i_g, flowing from cathode to grid, must be considered. The sum of i_p and i_g equals the current i_k taken from the cathode.

7. The amplification factor of a triode expresses the relative ability of the grid and the plate voltages to cause variations in the plate current. It is measured by the ratio of the plate-voltage change to the grid-voltage change, which has an equal and opposite effect on the plate current, thus:

$$\mu = -de_p/de_g \qquad (ip \text{ const}) \qquad (28\text{-}2)$$

The amplification factor of a triode depends on the geometry and dimensions of the electrode structure. In general, the closer the grid to the cathode, relative to the cathode-anode spacing, the higher the value of μ. The use of a fine wire and close spacing between the wires also produces a high μ value. The values of amplification factor for triodes vary from 5 to 100, with 30 a representative value in modern tubes.

8. Grid-plate Transconductance. The effect of the grid voltage on the plate current is expressed by the grid-plate transconductance g_m (or "mutual" conductance), which is the ratio of an elementary change in the plate current to the elementary change in grid voltage causing it,

$$g_m = di_p/de_g \qquad (\text{A/V or mhos}) \qquad (28\text{-}3)$$

with e_p held constant. Values of g_m for triodes vary from 1,000 to 10,000 μmhos (0.001 to 0.01 A/V).

Table 28-2. Processing and Process Control

Process	Name of device	Transducer or processor	Conversion device	Indicator or control element
Coating	Paint-depth control, etc.	Capacitance pickup	Bridge oscillator	Magnetic valve
Combustion	Combustion control	Phototube	Amplifier	Relay
Counting	Counter	Phototube	Amplifier	Relay
Food processing	Food-processing control	R-f oscillator		
Gas analysis	Gas analyzer	Ionization gage	....	Meter
Drying	Electronic dryer	R-f oscillator		
Gluing	Dielectric heating	R-f oscillator		
Grading	See Sorting			
Heating:				
Dielectric	Dielectric heater	R-f oscillator		
Induction	Induction heater	R-f oscillator		
Mixing ingredients	Color control	Phototube	Amplifier	Relay
Packing and wrapping	Wrapping control	Phototube	Amplifier	Relay
Photographic exposure	Exposure control	Phototube	Amplifier	Relay
Power conversion	Rectifier	Phanotron-thyratron	Rectifier circuit	
	Invertor	Thyratron	Inverter circuit	
	Frequency changer	Thyratron	Rectifier inverter	
Precipitation	Electrostatic precipitator	Kenotron		
Preheating	Dielectric heater	R-f oscillator		
	Induction heater	R-f oscillator		
Registering	Register control	Phototube	Amplifier	Relay
Surface treatment	Surface tester	Crystal pickup	Amplifier	C-r oscilloscope
Sorting:				
By size	Size sorter	Phototube	Amplifier	Meter, relay
By color	Color sorter	Phototube and filter	Amplifier	Meter, relay
By electrical properties	Resistance sorter, etc.	Bridge	Null indicator	Meter or relay
Vibration	Ultrasonic vibrator	Quartz crystal		
Weighing	Weight control	Phototube or capacity	Amplifier oscillator	Relay
Welding:				
Arc	Arc-welding control	Thyratron-ignitron	Controlled rectifier	
Resistance	Resistance-welding control	Thyratron-ignitron	Cycle timer and electronic switch	

9. Triode Parameters. The triode parameters (dynamic plate resistance r_p, amplification factor μ, and grid-plate transconductance g_m) express the tube's operating capabilities at a given operating point. Each parameter is the slope of the operating curve at the operating point (the curve in question is that plotted in terms of the particular variational quantities expressed in the ratio; for example, g_m is the slope of the $i_p - e_g$ curves shown in Fig. 5-116). The parameters are measured without resistance or impedance in the external circuits and hence are representative of the tube itself. The three quantities are related by

$$gm = \mu/r_p \tag{28-4}$$

In general r_p and g_m vary with the absolute magnitudes of the applied voltages and currents, but μ is dependent primarily on the structure. The load-line method (Fig. 28-3) may be used to determine the operating point when a resistance is connected in series with the plate.

10. The tetrode vacuum tube is a four-element vacuum tube used for the same purposes as the triode but having a higher amplification factor (and greater freedom

Table 28-3. Detection, Inspection, Protection, Supervision of Processes, Personnel, or Property

Function	Name of device	Transducer	Conversion device	Indicator or control element
Fume detection	Fume detector	Phototube	Amplifier	Relay
Elevator leveling	Elevator leveler	Pickup coil and vane	Thyratron	Relay-brake
Door opening	Door control	Phototube	Amplifier	Relay-pneumatic plunger
Intrusion detection ("burglar alarms")	Photoelectric alarm	Phototube	Amplifier	Relay-alarm
	Capacity-operated alarm	Capacitance	Amplifier	Relay-alarm
Hazard protection	Machine control-safety control	Phototube	Amplifier	Relay-guard
Illumination supervision	Illumination control	Phototube	Amplifier	Relay
Flaw detection in:				
Metals	Metal detector	Capacitance	R-f oscillator	Relay
Fabrics	Weft straightener	Phototube	Amplifier	Amplidyne
Sheet materials	Hole detector	Phototube	Amplifier	Relay alarm
Solid materials	X-ray inspector	X-ray tube	Photograph	
Load supervision in power systems	Carrier-current control	Carrier generator and modulator	....	Meter, relay
Metal detection in:				
Processing	Metal detector	Inductance	R-f oscillator	Headphones
Personnel	Gun detector	Inductance	R-f oscillator	Headphones
Prospecting	Geophysical apparatus	Seismic receptors	Amplifier	Photographic record
Fire detection	Flame detector	Phototube	Amplifier	Relay
Remote inspection (telemetering)	Telemetering apparatus	Gage or meter	R-f transmitter and receiver	Recording relay
Smoke detection	Smoke detector	Phototube	Amplifier	Meter or relay
Traffic supervision	Traffic control	Phototube or contact	Amplifier	Relay
Power-system supervision	Carrier-current indicator	Carrier generator and modulator	....	Meter or relay
Voltage supervision	Voltage regulator	Triode-pentode	Amplifier	
Microscopic inspection	Micrometer	Capacitance	Amplifier	Meter or relay
	Electron microscope	Electron stream	Phosphor or film	Photographic record
Medical detection	X-radiation detector	Ionization chamber	Amplifier	Meter
Mercury processing	Mercury-vapor detector	Vapor tube	Amplifier	Meter
Synchronization	Synchronizer	Triodes-pentodes	Amplifier	Meter and relay
Germicidal protection	Sterilamp, etc.	Gas-discharge lamp		
High-speed inspection	Stroboscope	Gas-discharge lamp	....	Camera or visual
High-speed photography	Impulse light source	Gas-discharge lamp	....	Camera
Slack regulation	Slack control	Tension contacts	Amplifier	Amplidyne

from interaction between grid and plate circuits) by virtue of a screen grid between the control grid and the plate (see Fig. 5-121). The screen grid is maintained at a fixed d-c positive potential to shield the plate from the remainder of the tube structure. The presence of the shield makes the plate voltage ineffective in controlling the plate current and thereby produces a high value of amplification factor, together with a correspondingly high value of dynamic plate resistance.

Since the operating electrodes are the cathode, grid, and plate, the tetrode is described by the triode parameters μ, r_p, and g_m in the usual sense. The only difference in operation is the steady direct current collected by the positive screen grid. The tetrode is used principally as a voltage amplifier because of its high amplification factor. It has been replaced in most modern applications by the pentode.

11. Pentode Vacuum Tube. The pentode is a five-element vacuum tube similar to a tetrode but employing an additional grid, between the screen grid and plate, for the purpose of suppressing secondary emission from the plate. The suppressor grid is maintained usually at ground (or cathode) potential and collects a negligibly small current. The pentode construction permits wide variations in the plate voltage without incurring the difficulties arising from negative internal plate resistance that are present in the tetrode. The suppressor also adds further to the shielding effect of the screen grid and hence increases the amplification factor and the dynamic plate resistance. The pentode is widely used for voltage as well as power amplification over the entire range of frequencies employed in electrical communication. Its operation is described by the triode parameters μ, r_p, and g_m at the operating point.

12. The beam-power vacuum tube is a tetrode structure which has the operating characteristics of a pentode (see Fig. 5-124). The electrodes are so arranged as to direct the electron current to regions of zero potential between the screen grid and plate, which regions take the place of the suppressor grid. The suppressing and shielding action is smoother and more complete than in the pentode, permitting higher plate-voltage swings without excessive effect on the grid circuit. The power sensitivity (power output in the plate circuit per volt applied to the grid circuit) is correspondingly high, higher than is available in pentodes, tetrodes, or triodes of similar proportions

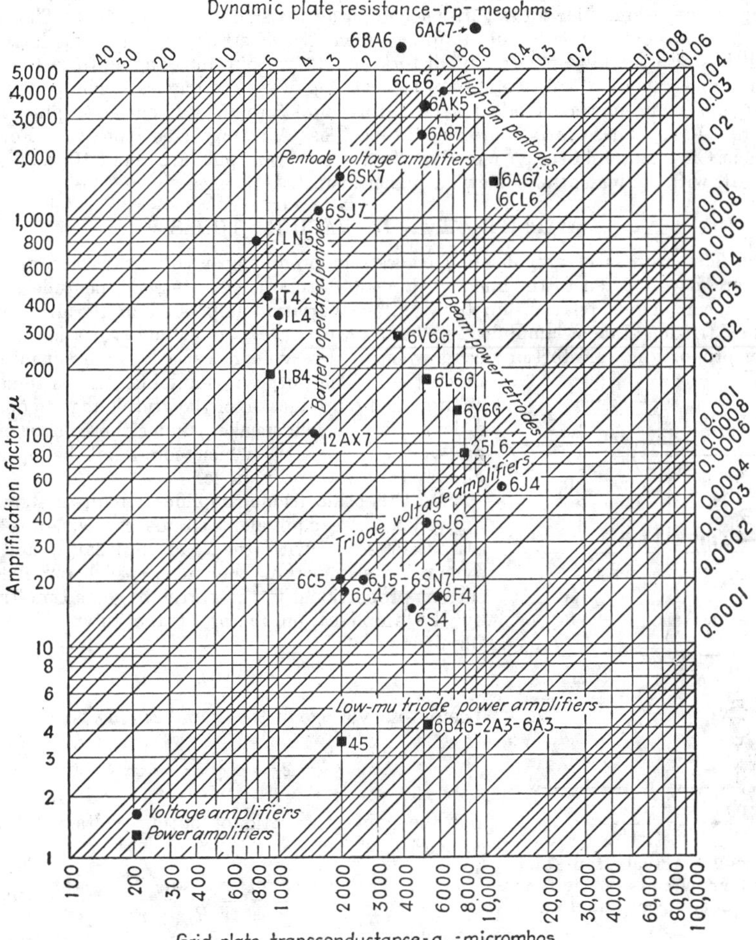

FIG. 28-1. Nominal values of μ, r_p, and g_m associated with typical tubes.

13. Multielement Vacuum Tubes. Tubes having six and seven elements are available with several grids which can exert effects simultaneously on the plate current and hence are useful for mixing two communication signals together. In addition there are many types of tubes which are simply mechanical combinations of two or more structures within one envelope.

14. Cathode Structure. There are two widely used forms of cathode: the filamentary and the equipotential. Barium–strontium oxide coatings are universally used in small tubes. The emission efficiencies vary from 50 to 100 mA emission current/W of heating power.

15. Classification by Triode Parameters. Figure 28-1 shows the triode parameters μ, r_p, and g_m in triple coordinates for various voltage-amplifier and power-amplifier tubes, triodes, tetrodes, and pentodes. In general the tubes with the highest figures of merit have positions nearest the upper right-hand corner of the chart.

16. Typical Receiving Tubes and Important Operating Characteristics (see Table 5-17)

17. Typical Transmitting Tubes and Important Operating Characteristics (see Table 5-19)

18. Classification According to Power Output and Distortion. The **power-output rating** of power-amplifier tubes varies according to the structure employed. The triode tube has the lowest power output for a given size of structure and cathode rating, but it has also the lowest harmonic distortion. Power outputs ranging from 1 to 5 W maximum at about 5% distortion are typical of power-type triodes. The power pentodes (and beam-power tetrodes) have distortion of 5 to 15% at power-output ratings of 2 to 10 W. These ratings are single-tube Class A ratings (see Par. **24**). For two tubes in Class B amplification (see Par. **26**) the power outputs are about 10 to 20 times as great, with equivalent or somewhat higher distortion ratings.

AMPLIFICATION BY VACUUM TUBES

19. The process of amplification consists in producing a voltage variation in the load circuit connected to the anode of a vacuum-tube triode (tetrode, pentode) which bears a continuous relationship to a voltage variation impressed in the grid circuit of that triode. Linear amplification occurs when the two voltage variations are in direct proportion. Distortion occurs when the two voltage variations are not in direct proportion. The purpose of amplification is to secure an amplified reproduction of the grid voltage to operate a device, e.g., a loudspeaker, under the control of the grid-voltage variation, which could not be operated directly from the grid voltage.

20. The basic triode amplifier circuit is shown in Fig. 28-2. The equivalent circuit consists of a generator, in series with the load impedance and the dynamic resistance of the tube, which produces a current proportional to the grid voltage and multiplied by the amplification factor of the tube. Thus, by Ohm's law,

$$\mu e_g = i_p(R_L + r_p) \tag{28-5}$$

The output voltage of the amplifier i_pR_L is

$$e_L = \frac{\mu e_g R_L}{R_L + r_p}$$

and the gain (the ratio of output voltage to input voltage) is

$$G = \frac{e_L}{e_g} = \frac{\mu R_L}{R_L + r_p} \tag{28-6}$$

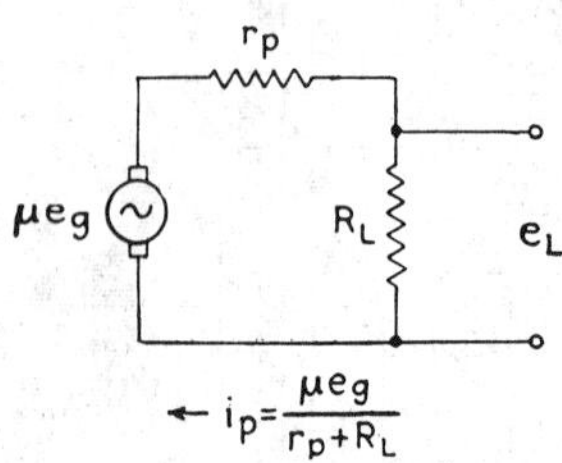

Fig. 28-2. Actual (top) and equivalent circuits of the fundamental electronic amplifier.

These relationships assume that the amplifier operates as a linear translator, i.e., that r_p is sensibly constant over the operating range of e_g, an assumption that is valid for small amplitudes.

21. Dynamic Characteristic by the Load-line Method. A more comprehensive analysis of amplifier action must take into account the succession of operation points on the tube characteristics over which the electrode voltages and currents pass during operation. The locus of points thus traced out is known as the dynamic characteristic of the circuit. The dynamic characteristic of an amplifier with a resistance in the anode circuit is a single, substantially straight line. When the load circuit contains reactance as well, the dynamic characteristic is an ellipse, due to the time-phase displacement between the current and voltage across the load impedance.

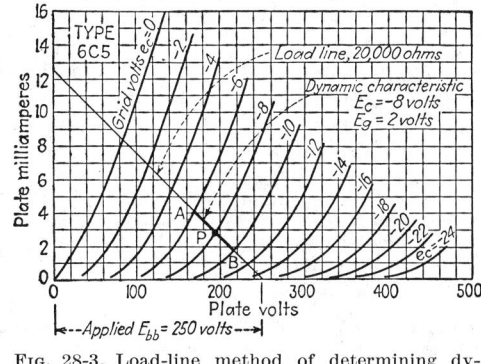

FIG. 28-3. Load-line method of determining dynamic characteristic of Class A amplifier.

A simple method of determining the dynamic characteristic in resistance-loaded amplifiers involves the load line, as shown in Fig. 28-3. The load line is laid out on the anode-voltage vs. anode-current family as a line having a slope equal to the inverse of the load-resistance value and intersecting the anode-voltage scale at the value of applied anode voltage. The dynamic characteristic is that portion of the load line over which the grid voltage actually swings, as in Fig. 28-3. The dynamic characteristic so determined may also be plotted on the grid-voltage vs. plate-current curves, but the load line cannot be determined so simply in this case. Figure 28-4 shows the dynamic characteristic on the $e_g - i_p$ curves.

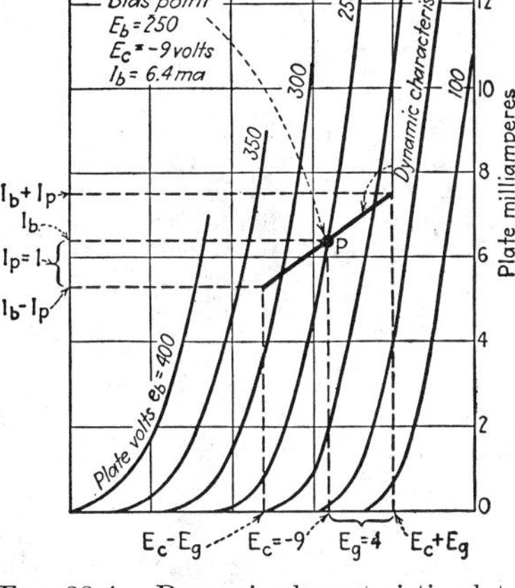

FIG. 28-4. Dynamic characteristic plotted on e_g–i_p coordinates.

22. Tetrode and Pentode Amplification. Tetrodes and pentodes are used in amplifier circuits in the same manner as triodes, and the gain and dynamic characteristics are determined in the same way. The only assumption in using the triode equations is that the screen and/or suppressor grids remain at fixed potentials, which is usually sufficiently true to allow calculations of engineering accuracy. The primary difference in the values of the circuit constants arises from the difference in the μ and r_p values of the tubes.

23. Grid-bias Voltage. It is customary in most amplifiers to maintain the average potential applied to the grid at a value negative with respect to the cathode. This assures that the grid impedance will remain high (since no grid current is collected). Further, it assures that the circuit shall operate over a more closely linear region of its dynamic characteristic, thus avoiding distortion. The negative grid-bias voltage is commonly obtained in one of two ways. The first is a separate source of voltage, either a dry battery or a separate rectified and filtered power-supply source (or a portion of such a separate source). The second and more widely used method is the self-biasing circuit, in which a shunt combination resistance and capacitance is connected in series with the cathode, as shown in Fig. 28-5. In general the resistance R_c is chosen such that

$$R_c = E_c/I_b \tag{28-7}$$

where E_c = required value of negative grid voltage and I_b = value of plate current which flows when the grid voltage has that value (and when the anode voltage has a specified value). The capacitance is chosen such that its reactance X_c, at the lowest frequency $f_{\min}$ under consideration, is at least 20% of the value of R_c; thus,

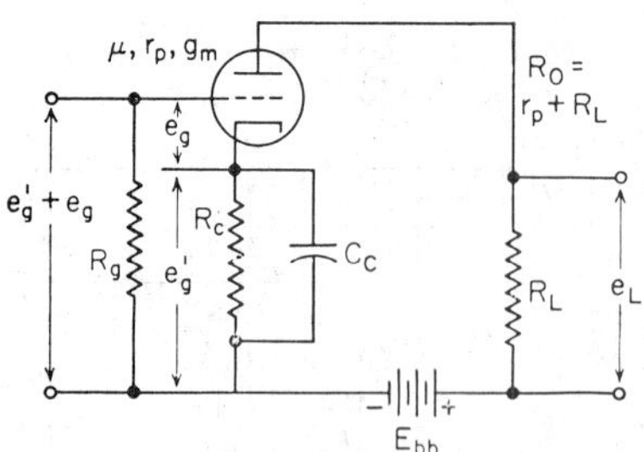

FIG. 28-5. Circuit of amplifier utilizing cathode self-biasing resistor to maintain grid at negative potential with respect to cathode.

$$X_c = \frac{1}{2\pi f_{\min} C_c} \geqq 0.2R_c \tag{28-8}$$

The greater value of C_c, the lower the frequency $f_{\min}$ at which the amplifier will maintain its gain. At frequencies below $f_{\min}$, the gain falls off because the resistance R_c feeds a voltage into the grid circuit which is out of phase with the applied signal voltage.

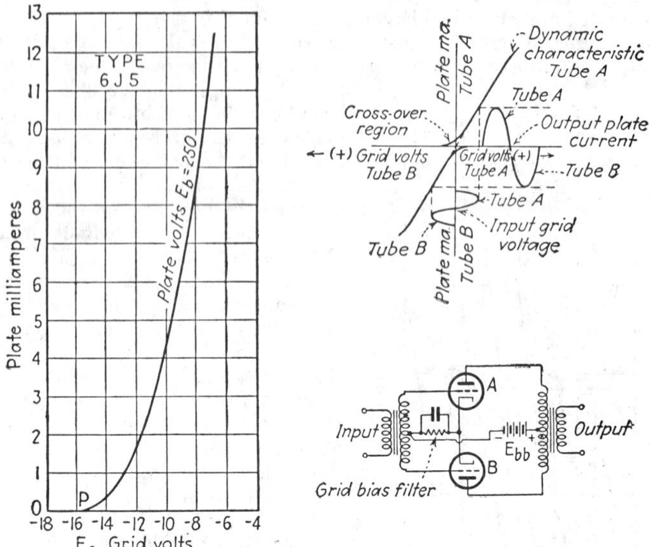

FIG. 28-6. Static and dynamic characteristic employed in push-pull Class B amplification and circuit diagram of push-pull amplifier.

24. Class A Amplification. A Class A amplifier is one whose grid-bias voltage is fixed at a value which permits the plate current to flow at all times. In the Class A1 amplifier, the excursion of the grid voltage over the dynamic characteristic is restricted to negative values, whereas in the Class A2 the grid-voltage excursion is wider and grid current flows during part of each signal cycle. Class A1 is used when the distortion introduced by grid-current flow (and by attendant nonlinearity of the dynamic characteristic) must be avoided. The characteristics of Class A amplification are low distortion (high fidelity of reproduction) and low power efficiency. The latter effect results from the fact that the continuous flow of plate current introduces a steady d-c power loss which goes into heating the plate.

25. Power Output of the Class A Amplifier. On the assumption of a linear dynamic characteristic, the power output of the Class A amplifier may be deduced from inspection of the dynamic characteristic (see Fig. 28-3). The maximum and minimum values of plate current and voltage are those occurring at the ends of the dynamic characteristic. The rms values of the voltage and current amplitudes are $(E_{max} - E_{min})/(2\sqrt{2})$ and $(I_{max} - I_{min})/(2\sqrt{2})$, and the power output is the product of these rms values,

$$P_o = \frac{(E_{p\,max} - E_{p\,min})(I_{p\,max} - I_{p\,min})}{8} \qquad (28\text{-}9)$$

26. Class B Amplification. A Class B amplifier is one whose grid-bias voltage is fixed at the plate-current cutoff point, i.e., at point P in Fig. 28-6. Accordingly plate current flows only during those half cycles of the signal voltage which make the plate positive; i.e., the reproduction is of the "half-wave" variety. The distortion associated with this mode of operation is very high unless two tubes are operated in push-pull, as shown in Fig. 28-6. In this case the positive and negative half cycles are amplified alternately by the two tubes, and the distortion is greatly reduced, but not to the low values possible in the Class A amplifier. The advantage of the Class B amplifier is that the average plate current is small and there is accordingly much less power lost in heating the plate than in Class A operation. The full advantage of the increase in efficiency of the Class B amplifier is obtained, however, only when the signal excursion is large enough to cause grid current to flow (Class B2). This is the usual mode of operation. The input signal circuit to a Class B2 amplifier must be designed to supply the grid-current flow without distortion. A pair of tubes in Class B is ordinarily capable of supplying about 10 to 20 times the power output of a single tube of the same type operating in Class A.

27. Class C Amplification. A Class C amplifier operates with a grid-bias voltage value considerably more negative than the plate-current cutoff point (see Fig. 28-7). The plate current accordingly flows for less than ½ c of the signal alternation. The efficiency in the plate circuit is thereby increased greatly, but the distortion becomes so great that Class C operation cannot be used except in circuits employing tuned circuits (r-f amplifiers) which emphasize the fundamental frequency and attenuate the harmonic-distortion components. The power output of a Class C stage is about 50 to 500% greater than that of a stage operating in Class B, other factors being equal.

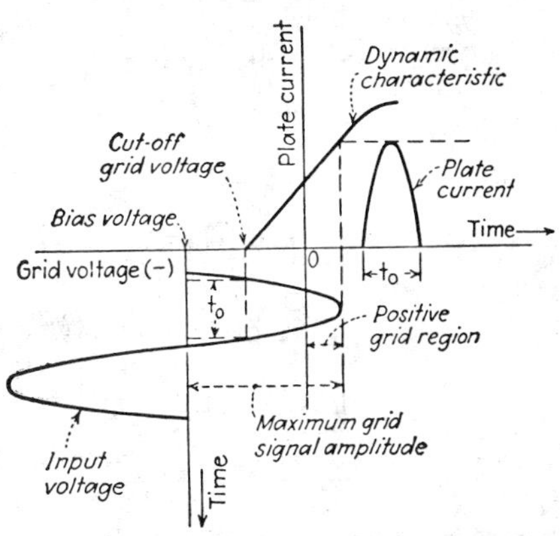

FIG. 28-7. Dynamic operation of a Class C amplifier.

28. Oscillation. An electronic oscillator circuit is essentially an amplifier, part of the output of which is coupled back to the input. The coupling connection permits the generation of sustained variations or oscillations in the plate current without the need of an external signal source, since the oscillator output acts in itself as the signal source. Several forms of oscillator circuit are shown in Fig. 28-8. The tuned-circuit oscillators are most widely used. In this type, a tuned circuit is employed in the grid or plate circuit, or in both, to act as a storage agent for the energy of the oscillations. In this type of oscillator the output current has very nearly a pure sine-wave shape, provided that the circuit is sharply tuned. In the relaxation oscillator, no tuned circuit is employed, and the output waveshape is usually of irregular outline, varying from a square wave to a transient discharge wave, depending on the circuit constants.

The feedback connection between output and input may be inductive (self- or mutual), capacitive, or a combination of the two. The plate-to-grid interelectrode capacitance may serve the purpose when the frequency of operation is high enough to permit the coupling of a sufficiently large portion of the output to the input voltage. The high values of plate-to-grid capacitance in triode tubes make them susceptible to this type of oscillation. When oscillation is not desired, special circuit precautions must be taken to balance out ("neutralize") the voltage fed back. The tetrode and pentode tubes, having much smaller values of grid-to-plate capacitance, are comparatively free of the feedback tendency and hence do not require such careful neutralizing in circuits designed to be free of oscillations. The oscillator frequency may have any value from several cycles per second to several hundred millions of cycles per second, depending primarily on the inductance and capacitance values employed in the tuned circuits.

29. Relaxation Oscillation. The oscillators described in previous paragraphs produce oscillations of sinusoidal shape. When more steeply sided oscillations are required, e.g., in pulse and rectangular-wave generators, relaxation oscillators are usually employed. The two most widely used relaxation oscillators are the multivibrator and the blocking oscillator.

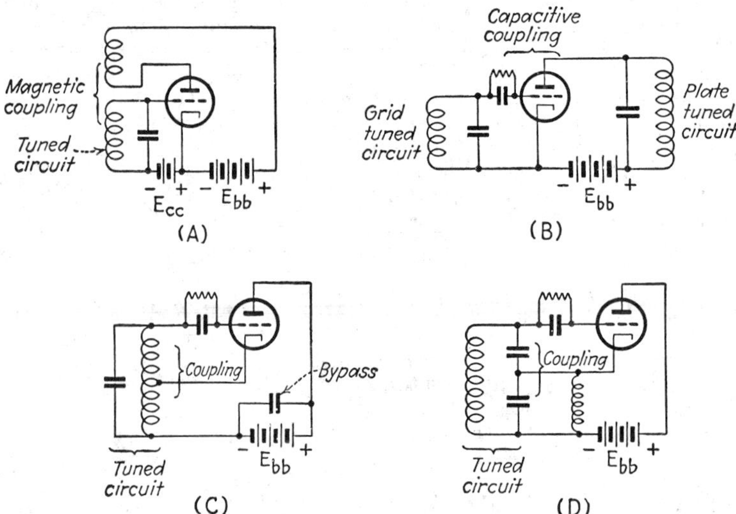

FIG. 28-8. Types of oscillator circuits: (*A*) tuned grid, mutually coupled; (*B*) tuned grid, tuned plate; (*C*) Hartley; (*D*) Colpitts.

30. The multivibrator is essentially two stages of resistance-capacitance-coupled amplification with output connected to input. Usually the coupling is between the plates and the grids, but cathode coupling may also be used. Any casual disturbance in the potential of the first grid (tube 1) is amplified and reversed in polarity by the two stages. The two reversals of polarity produce, at the first grid, a signal of large magnitude and the same polarity as that initially assumed. Consequently the potential of the first grid is rapidly shifted, by cumulative action, in the direction of the first disturbance. If the disturbance was negative, the first stage is cut off (becomes nonconducting), usually in a few millionths or thousandths of a second. In this case, the plate of the first stage is raised in potential, driving the grid of the second stage sharply positive and holding the second stage at full conduction. If the initial disturbance was positive, the first stage reaches full conduction at once, while the second stage is cut off.

During the ensuing period, charge trapped on the cutoff grid leaks off (the grid "relaxes") at a rate determined by the capacitance and resistance connected to it, usually in a time comparable with RC s, where R = grid resistance to ground in ohms and C = coupling capacitance in farads. When the cutoff grid is raised in potential to the conduction level, the continuing rise in the potential becomes a positive disturbance in the sense first assumed, and the first stage is thereby rapidly driven to full conduction, while the second stage, depressed by the fall in the first-stage plate potential, is cut off. In this manner the two stages exchange roles, one being cut off while the other is fully conducting, the exchanges being separated by a time approximately equal to the time constant of the coupling connection.

The waveform taken from either plate has a rectangular shape in the upper, positive portions and an exponential shape in the lower, negative portions. If both coupling circuits have the same time constant, the duration of positive and negative portions is the same (symmetrical multivibrator). By suitable choice of the time constants, it is possible to vary the relative duration of the two portions and, in effect, to produce short, sharp pulses separated by longer quiescent intervals.

31. The blocking oscillator operates in similar fashion, but the feedback from the plate to grid is accomplished by a transformer, and so a single tube suffices. The blocking transformer is so connected that any change in grid potential is fed back to the grid in the same polarity, with amplified magnitude. Consequently, any initially assumed negative disturbance on the grid is increased in cumulative fashion, and the tube is cut off. The grid recovers as its charge is carried away by the grid resistor; and when conduction begins, the positive disturbance thereby introduced is rapidly increased by the cumulative feedback connection. Consequently, the tube is driven sharply to full conduction, and the first half of a sinusoidal oscillation begins. This half wave is abruptly terminated, however, as the grid is driven to cutoff by the increasing potential on the plate. The circuit thus produces sharp pulses of half-wave sinusoidal shape at the plate, separated by longer quiescent intervals during which the grid recovers from cutoff.

32. Tuned-circuit Amplification. Amplifiers containing tuned circuits in grid or plate circuits are intended to amplify a comparatively narrow band of frequencies, compared with the audio and video (television) amplifiers, which amplify a wide range of frequencies without discrimination. The tuned amplifier is commonly used to increase the amplitude of an alternating current, e.g., one generated in an oscillator. Tuned amplifiers may also serve to amplify a modulated alternating current, which is constituted of the carrier frequency and a cluster of symmetrically disposed side-band components.

GAS-FILLED ELECTRON TUBES

33. Types of Gas-filled Tubes. Some electron tubes contain a gas or vapor at low pressure and make use of ionization effects. These gas-filled tubes are classed as thermionic gas-filled diodes (phanotrons), triodes (thyratrons), and tetrodes (shield-grid thyratrons) and as pool-type diodes (tank rectifiers) and triodes (ignitrons). They are characterized by much higher current-carrying capacity, lower voltage drop between cathode and anode, and more limited control (in the case of triodes and tetrodes) than the corresponding high-vacuum tubes. They find use in a-c rectification and in industrial control of power sources.

34. Gas-filled Diode (Phanotron). The thermionic gas-filled diode employs a cathode (usually of the barium–strontium oxide type) and an anode immersed in gas or vapor. Mercury vapor, obtained from liquid mercury enclosed in the tube, is commonly employed. For applications where variations with temperature are troublesome, the inert gases neon, argon, or helium may be used. The tube serves as a rectifier of alternating current. The full conduction current must be available thermionically from the cathode, and the anode must not emit electrons appreciably, to avoid reverse conduction (arc-back). The primary advantage of the gas-filled diode relative to the vacuum diode is the greater efficiency which results from the low voltage drop between anode and cathode during conduction.

35. Operating Characteristics of Typical Gas-filled Diode Rectifier Tubes (Phanotrons) (see Table 5-20).

36. Gas-filled Triode (Thyratron). A grid electrode may be inserted in a gas-filled tube to obtain a limited control over the conduction. The gas-filled triodes employing grids are commonly called "thyratrons." They are rated very similarly to the phanotrons as to forward and inverse voltage, average and peak anode current, deionization, and cathode-heating times. In addition a characteristic curve is necessary to specify the grid voltage at which conduction is initiated, specified for various values of the anode potential. A typical control characteristic is shown in Fig. 28-9. The tendency of the control grid to collect grid current (electronic when positive, ionic when negative) is usually expressed in an empirical curve. The grid current often reaches values as high as several hundred microamperes, and this fact must be taken into account in designing the control circuit. Two types of thyratron are available: positive grid-voltage and negative grid-voltage types, in which the grid-control range extends over voltages positive and negative with respect to the cathode, respectively.

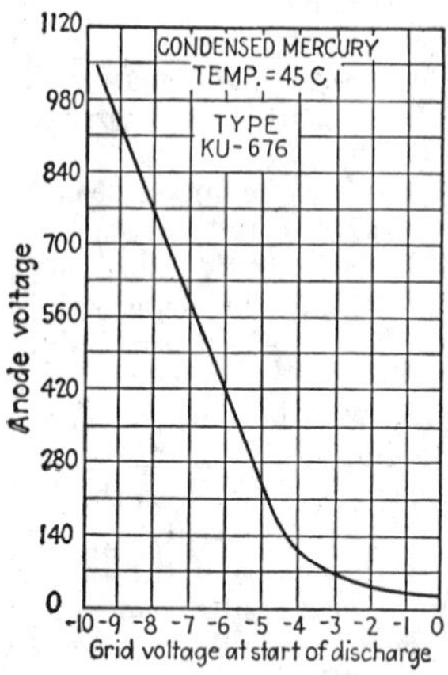

Fig. 28-9. Typical thyratron grid-control starting characteristic.

37. Gas-filled Tetrode. The shield-grid thyratron is a gas-filled tetrode with an extra grid structure partly surrounding the control grid and anode. This arrangement permits a smaller structure to be used as the control grid and thus lowers the grid current considerably. Also, the shield-grid potential (usually within 5 V plus or minus of the cathode) may be used to shift the control characteristic, as shown in Fig. 28-10. This shift may be used as an operating variation, but ordinarily it is employed simply to compensate for variations in the control characteristic as tubes age or as they are replaced.

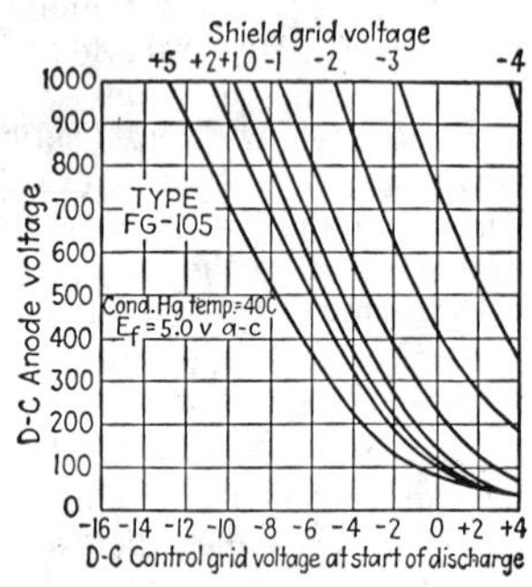

Fig. 28-10. Typical control-grid starting characteristics of a shield-grid thyratron.

38. Pool-type Triode (Ignitron). A gas-filled triode electron tube of very high anode current rating is the ignitron, which employs a pool of mercury as a cathode. A luminous spot on the surface of the pool acts as the source of emitted electrons (by field emission primarily, since the temperature and bombarding energy are not high enough to account for the high emission by thermionic or secondary means). A single spot is capable of emitting about 15 A; but if more current is required, several cathode spots are formed. Consequently the mercury pool has a very high overload capacity, in contrast to the thermionic cathode. The power rating is determined by the available area for cooling rather than by electronic considerations.

Conduction is initiated in the ignitron tube by applying power to the ignitor electrode, a nonmetallic rod which dips into the cathode pool. The potential gradient built up along the rod succeeds in initiating the cathode spot, which thereafter maintains itself until the anode potential is removed. The ignitor circuit consumes a maximum of 10 to 50 A at 100 to 300 V, but the duration of the ignition cycle is about 50 μs, so that the average power consumed (in 25- or 60-c/s service) is very small. Thyratron tubes, capacitor discharge circuits, or saturated reactors carrying alternating current as well as direct current are used as sources of the ignition voltage. The starting-time characteristic of a typical ignitron is shown in Fig. 28-11.

The ratings of ignitrons are similar to those of thyratrons; but the values of anode current are higher and those of inverse anode voltage lower.

39. Operating Characteristics of Typical Triode and Tetrode Thyratron Tubes (see Table 5-20).

40. Ratings of Typical Ignitron Tubes (see Table 5-20).

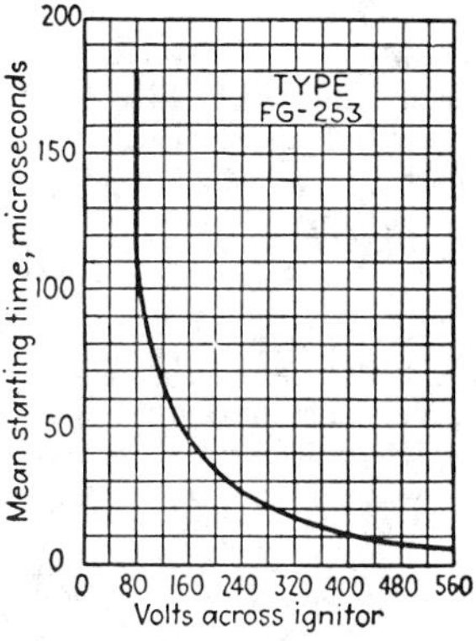

Fig. 28-11. Starting time vs. ignitor voltage of a typical ignitron.

RECTIFICATION

41. Action in Tube. The process of converting an alternating voltage or current into a unidirectional voltage or current is called rectification. Rectification is carried out by applying the alternating voltage to the cathode and anode of a diode (triode or tetrode in controlled rectification) electron tube. The tube permits the anode current to pass only during that portion of the a-c cycle which makes the anode sufficiently positive to draw anode current. The anode current is zero when the anode is negative. The unidirectional anode current is passed through a load circuit (resistance or impedance) across which a corresponding unidirectional voltage appears. If the unidirectional voltage is to be converted to direct voltage, it is passed through a smoothing filter of the low-pass variety.

42. The basic rectifier circuit is shown in Fig. 28-12. The source of alternating voltage, the diode, and the load circuit are in series, and the instantaneous sum of the voltages across the elements is zero at all times. The graphical method illustrated in Fig. 28-12 may be carried out most simply if the load circuit is a pure resistance (typical of many applications). A simple approximate method consists in assuming that the voltage drop across the cathode and anode, during conduction, is a constant equal to the voltage abscissa corresponding to 0.75 of the peak anode current. This voltage drop is then subtracted from the applied voltage to give the voltage across the load resistance during each positive-anode half cycle. The average voltage across the load is

$$E_{avg} = 0.318R_{peak} \qquad (28\text{-}10)$$

where E_{peak} is the peak value of the voltage across the load. The average current is determined from E_{avg} by Ohm's law.

When the rectifier diode is of the gas-filled kind, the same type of graphical and approximate solutions applies, except that the voltage drop across the tube during conduction is then fixed at a value between 5 and 25 V, depending on the tube characteristic, operating temperature, etc., but independent of the value of the anode current.

When the load circuit is a low-pass filter terminated in a resistance, the solution is best attacked empirically. Curves of output voltage vs. current for various values of applied a-c voltage and typical filter elements, calculated for 60-c/s conditions, are available from the manufacturers for several types of vacuum diode rectifier.

The choice of tubes for rectifier circuits depends fundamentally on the average values of

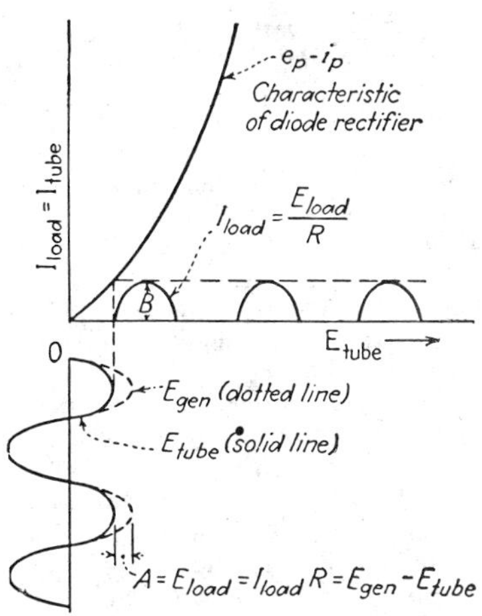

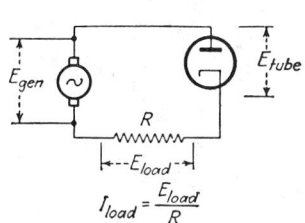

Fig. 28-12. Basic diode circuit and graphical solution. Successive approximation is applied until the voltage across the load (at A) equals the resulting load current (at B) times the load resistance. Usually E_{tube} is small compared with E_{load}.

current and voltage required in the load circuit. From these average values, corresponding peak values can be calculated by using Eq. (28-10). The average and peak values of load current are then compared with those of available tubes. Upon assuming ideal rectification (no voltage drop across the tube), the required rms alternating-voltage input may be calculated, and from this the inverse-peak voltage, which is

$$E_{\text{inv}} = 1.41 E_{\text{rms}} \tag{28-11}$$

This inverse-peak value must also be compared with the ratings of available tubes.

43. Other Direct Rectifier Circuits. The simple circuit shown in Fig. 28-12 is usually elaborated in engineering practice. Typical circuits, shown in Fig. 28-13, are the full-wave rectifier, the voltage doubler (which charges two capacitors in series), and various multiphase arrangements.

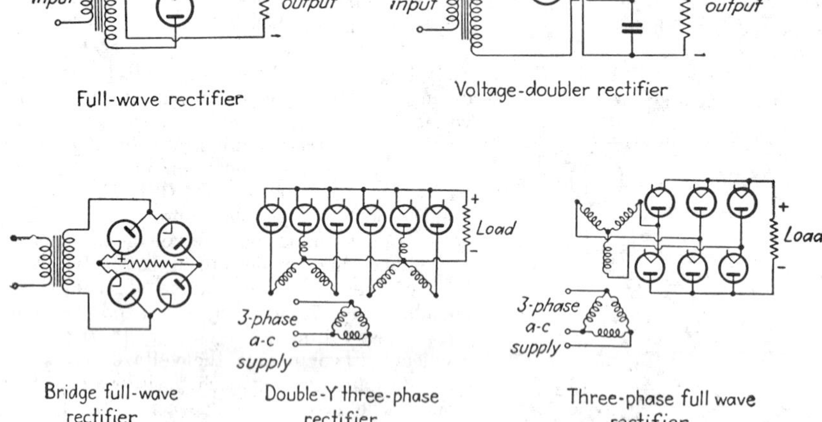

Fig. 28-13. Rectifier circuits.

44. Grid-controlled Rectification. Grid-controlled rectification, in which the magnitude of the anode current is controlled by varying a direct or alternating potential applied to the grid of a triode tube, may be carried out in vacuum or in gas-triode circuits. However, in power practice, the greater efficiency of the gas-filled tube has given it universal preference. Controlled rectification with vacuum tubes is employed in high-frequency pulse-forming and shaping circuits.

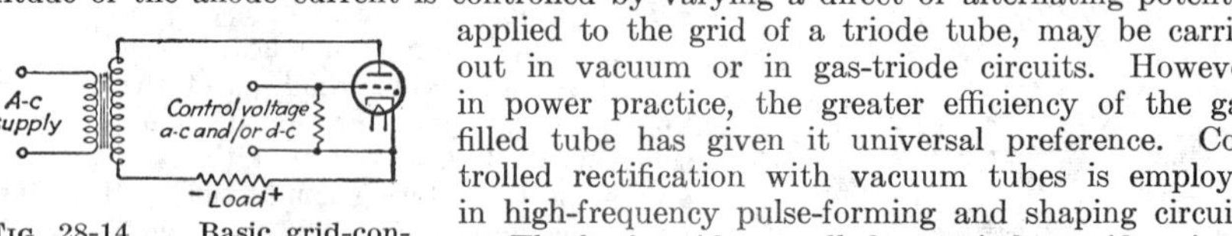

Fig. 28-14. Basic grid-controlled rectifier circuit.

The basic grid-controlled gas-triode rectifier circuit is shown in Fig. 28-14. To determine the instant at which conduction begins in each positive-anode half cycle, the control characteristic of the thyratron must be known. When the values of grid and anode potential simultaneously meet or exceed the values necessary for conduction, conduction begins; and it continues until the anode voltage at the end of the half cycle falls below the point at which conduction ceases. The graphical method of obtaining the dynamic grid-control characteristic from the static control characteristic is shown in Fig. 28-15. Whenever the grid voltage intersects this dynamic control curve, conduction begins and continues thereafter to (approximately) the end of the cycle. The applied grid voltage may be caused to intersect with the dynamic control characteristic in several ways. If the grid voltage is direct voltage, its amplitude must simply be made more positive until intersection occurs and the point of intersection determines the instant at which conduction begins. If the grid voltage is alternating, its amplitude may be increased, with its phase fixed, to obtain intersection; or the phase may be shifted with the amplitude fixed. Typical circuits for each type of control are shown in Fig. 28-16.

PHOTOTUBES AND PHOTOCELLS

45. There are three general classes of photoelectric "cells" in use: **photoemissive, photoconductive, and photovoltaic.** The photoemissive type has been widely used in

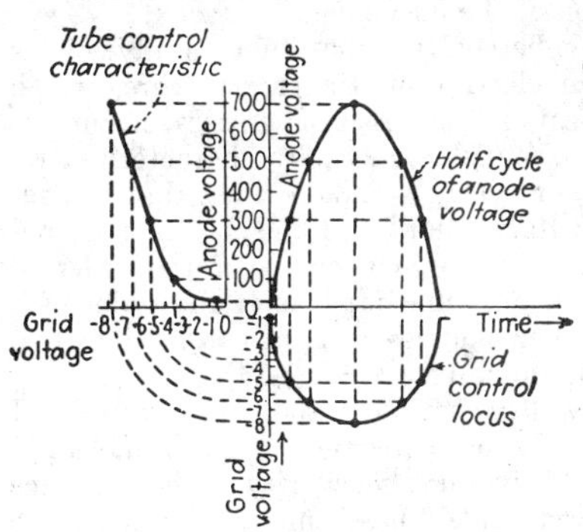

Fig. 28-15. Development of grid-controlled locus from tube-control characteristic.

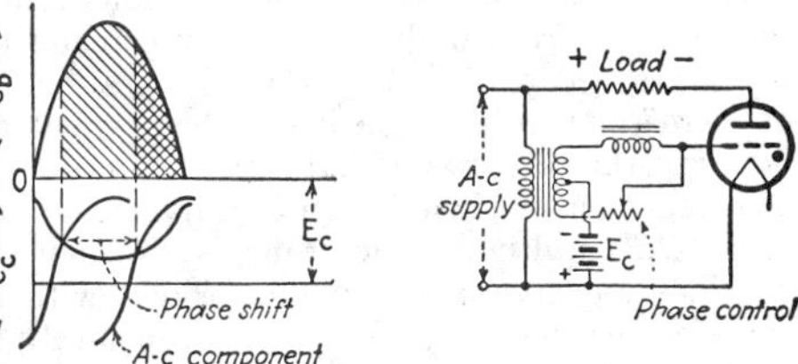

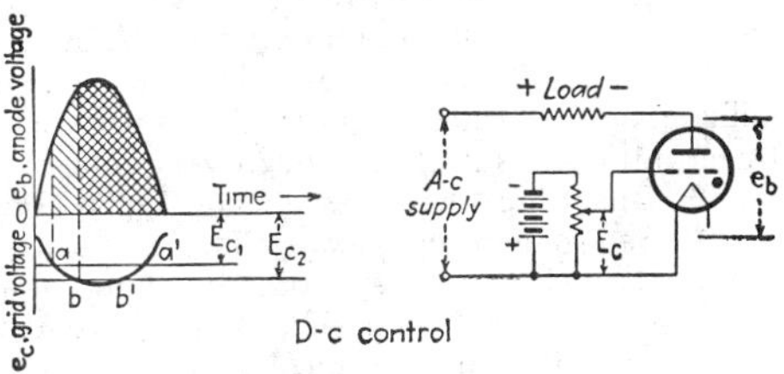

Fig. 28-16. Alternating-current amplitude, a-c phase shift, and d-c circuits for grid-controlled rectification. Shaded areas represent conduction.

industry for many years as a glass-enclosed device, the phototube. Phototubes are classed as vacuum types or gas-filled types as well as by the material used as the photoelectric emitter and by their mechanical construction. The photoconductive cell is a semiconductor device whose electrical resistance, or conductivity, changes when it is exposed to light. The third main type, the photovoltaic cell, generates a voltage of about ½ V when exposed to light.

46. Light-Current Characteristic. The saturation current flowing through a phototube (Fig. 28-17) is the product of the total radiant flux falling on the cathode and the luminous sensitivity of the cathode. If the illuminated area of the cathode is constant, the current is also proportional to the illumination in footcandles. The maximum value of current which can be withdrawn without injury to the cathode is limited to 20 to 50 μA. The amount of current passed depends also on the voltage applied between cathode and anode; so in general a family of curves is needed to represent the operating region of the phototube. Usually at the operating point, the voltage is sufficient to draw the full saturation value of the photoelec-

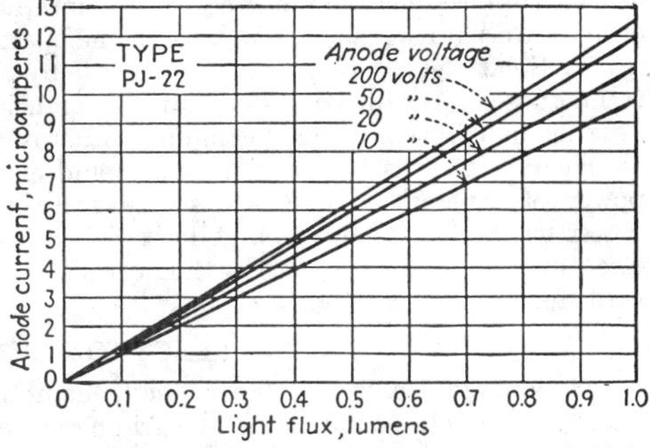

Fig. 28-17. Light vs. current in a typical high-vacuum phototube.

tric emission; and in this case, the current output is linearly related to the incident light flux.

47. Voltage-Current Relationship. For a given value of incident light flux, the current through a phototube depends on the three-halves power of the applied voltage until saturation is reached, and thereafter the current is very nearly independent of applied voltage. When a resistance is connected in series with the anode (to develop the output voltage), the point of operation on the curves is found by the load-line method in a manner exactly analogous to that employed with thermionic diodes.

48. Gas Amplification Ratio. The purpose of the gas in gas-filled phototubes is to enhance the amount of emission from the cathode by employing the effects of electron regeneration. However, if any appreciable gas discharge occurs, the severe bombardment of the cathode by the positive ions injures its luminous sensitivity. Hence no more than 90 V is usually applied to the anode, and the degree of ionization is small—not enough to secure full electron regeneration. The gas amplification ratio expresses the increase in current relative to that available from the same tube structure in vacuo. Values of the ratio vary from 3 to 10 times. Since the ionization effect is cumulative, the current output is not a strictly linear function of the incident radiant flux.

49. The multiplier phototube is another member of the phototube family. This combines a conventional photocathode with an electron-multiplier structure, which increases the photocurrent by electron multiplication (successive increases of current by secondary emission). The multiplier types (see Table 5-24) are characterized by very high luminous sensitivity, from 2 to 6 A/l, or nearly a million times that of conventional vacuum-type phototubes. A high-voltage supply (1,250 V) is required for the multiplier structure. Since the maximum output current is of the order of milliamperes, the maximum light flux that can be accommodated is of the order of a thousandth of a lumen, although very much smaller amounts of light can be detected.

50. Operating Characteristics of Typical Phototubes (see Table 5-24).

51. Self-generating Photocells. A device widely used for the measurement of illumination is the self-generating or photovoltaic photocell (barrier-layer cell, blocking-layer cell, exposure-meter cell, etc.). The seat of the photosensitivity is the interface between a metal and a semiconductor. Two forms widely used employ cuprous oxide of copper and selenium on iron, respectively. The interface possesses a degree of unilateral conductivity, so that electrons, released from the interface by the incident light, are trapped by the barrier layer and find a return path through the external circuit, provided that the external resistance is low compared with that of the interface. The process is a direct transformation of light energy into electrical energy. Since no external battery or power supply is required, photocells, combined with portable microammeters, have found wide use as photographic exposure meters and illumination meters. The luminous sensitivity varies from 100 to 500 μA/lm, and the spectral response of commonly used forms is not markedly different from that of the human eye. The maximum voltage generated on open circuit is of the order of several hundred millivolts; and maximum output current of typical commercial forms, into a short circuit, is in the neighborhood of 500 μA. The current is not strictly proportional to the incident flux because of space-charge effects at the interface, and the light-current curve depends on the external resistance as well as on the operating temperature.

52. Solar Cells. In recent years a photovoltaic cell made of silicon processed by techniques similar to those used in manufacturing transistors has had wide application in space exploration and is beginning to be used in industrial equipment. Thousands of silicon solar cells arranged in panels and exposed to sunlight can provide electrical power of several hundred watts. Several satellites have been so equipped. The power has been used to recharge batteries in the satellite and, in some cases, to power electronic equipment directly. Single cells are useful in "reading" punched holes in cards for data processing.

ELECTRONIC RELAYS

53. Voltage Control Relay. The general function of electronic relays involves translating a voltage impulse in the grid circuit of a gas-filled or vacuum triode (tetrode or pentode) into a current impulse in the anode circuit. The current associated with the grid voltage may be very small (enough only to supply the leakage and/or grid current, usually of the order of microamperes), and the power in the grid circuit is correspondingly small. The voltage associated with the anode current, on the other hand, may be very large, limited only by the available anode voltage supply and insulation requirements. Consequently the power available in the anode circuit is large. The relay function thus involves the transformation of a weak power impulse into a much stronger one. The transformation may be instantaneous (direct relaying) or delayed (delayed relaying) depending on the circuit configuration employed in the grid circuit. With vacuum tubes the delay is limited only by the unavoidable time constant of the grid-circuit resistance and the distributed capacitance. Usually the delay is 1 μs or less. In gas-filled tube relay circuits the shortest delay is the ionization time, usually not more than 100 μs.

If a capacitance is employed in the grid circuit (usually in shunt with the grid and cathode), the time constant may be built up to any value required. In this case the delay is equal to RC s (where R = grid resistance in megohms and C = capacitance in microfarads) when the control level of grid voltage is reached at 63% of the voltage applied to the grid circuit. Typical direct and delayed relay circuits are shown in Fig. 28-18.

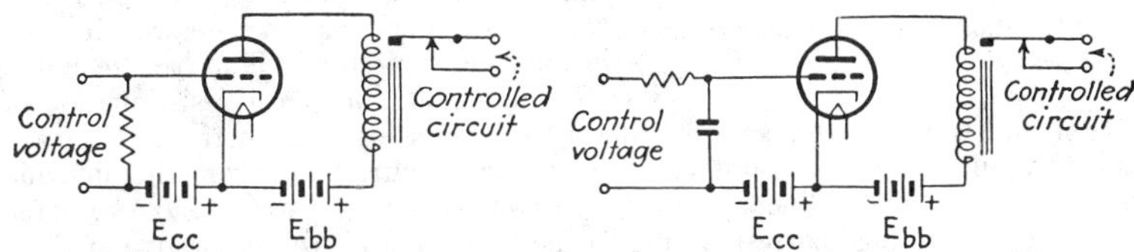

FIG. 28-18. Basic electronic-relay circuit (left) and time delay using resistance-capacitance combination.

Another, more accurate circuit capable of introducing time delay is the **relaxation oscillator.** The most widely used form is the multivibrator, two stages of resistance-capacitance-coupled amplification, with the output of the second stage connected to the input of the first. Any negative change in the grid potential of the first stage is thereby amplified and reapplied in the same polarity to the first grid. The grid is thereby driven sharply to cutoff, while the second stage remains at full conduction. This condition persists until the charge trapped on the first grid leaks off through a resistor to ground, whereupon the first stage suddenly regains full conduction, and the second stage is cut off. These rapid changes of potential at the first grid (and on the other electrodes of both tubes) continue periodically. The interval between potential shifts can be varied by changing the resistance or capacitance values in the coupling between stages. Adjustable and controlled time delays of a few microseconds to several minutes may be achieved in circuits of this type.

In ordinary electronic relaying practice an electromechanical relay is connected in the anode circuit and is operated by the change in anode current which results from the change in control grid voltage. The ratings of the coil of such electromechanical relays must be chosen in accordance with the values of anode current available from the relay tube. The required contact ratings of the relay depend, of course, on the equipment to be controlled.

In gas-filled tubes, the anode current begins suddenly when the grid and anode voltages simultaneously reach the combination of values required for conduction. Moreover, in gas-filled tube circuits, the voltage drop across the anode and cathode of the tube is small and constant during conduction; and, in consequence, substantially the full value of the anode supply voltage is applied to the coil of the electromechanical relay, instantaneously, when conduction conditions obtain. This makes for positive relay action.

When direct voltage is applied to the anode of the gas-filled relay circuit, provision must be made for open-circuiting the anode circuit before the grid can regain control for the next cycle. One simple method of so doing makes use of an extra pair of contacts on the electromechanical relay in series with the anode circuit. These contacts must open, however, only after the relay armature has pulled up far enough to exercise its control of the circuit to which the relay contacts are connected, and the auxiliary contacts must, of course, reclose when the relay armature falls back, ready for the next control cycle. For simple manual resetting, a pushbutton may be provided to open the anode circuit momentarily. The use of alternating current in the anode circuit is the simplest means of assuring that the grid circuit shall regain control when the grid voltage falls below the control level.

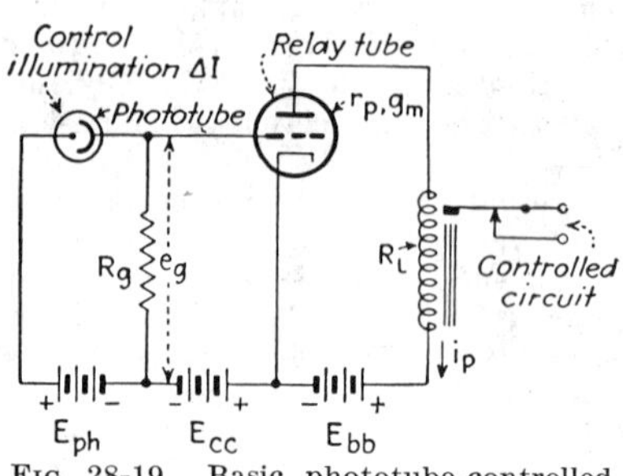

FIG. 28-19. Basic phototube-controlled electronic relay.

Because of its low values of grid current, which suit it to work from high-impedance sources, the vacuum tube is better adapted for relaying from extremely weak control sources. This is true, for example, when a phototube is the source of the control voltage. The vacuum tube is also better suited when the control level must be precisely set and when variations due to changes in the tube-control characteristic cannot be tolerated.

54. A typical phototube relay circuit is shown in Fig. 28-19, which illustrates the factors underlying the design of vacuum-tube relay circuits. The transconductance G_m and dynamic plate resistance r_p of the relay tube must be known, as well as the resistance of the electromechanical relay coil and the pull-up and drop-out values of the relay current. From this information, and on the assumption of linear operation of the relay tube, it is possible to determine the range of grid voltage corresponding to the pull-up and drop-out range of the relay, as follows:

$$\Delta e_g = \Delta i_p \frac{R_L/r_p + 1}{g_m} \tag{28-12}$$

where Δe_g = grid-voltage range corresponding to the relay-current range Δi_p; and R_L, r_p, and g_m are shown in Fig. 28-19.

If the phototube is operated with sufficiently high anode voltage to secure saturation emission, the range of light flux corresponding to the pull-up to drop-out range of relay coil current may be determined from

$$\Delta I = \frac{\Delta e_g}{sR_g} \tag{28-13}$$

where ΔI = illumination range in lumens falling on the phototube cathode, s = luminous sensitivity in microamperes per lumen of the cathode to the illumination used, and R_g = value of the grid resistance in megohms.

SPECIAL TUBE TYPES

55. Cathode-ray Tubes (see also Sec. **5**, Par. **213**). The cathode-ray tube is essentially a device for converting the energy of an electron beam into light energy and utilizing the light as an indication of the motion of the electron beam. It finds use as a measuring instrument, as an oscillograph for examining alternating waveforms, as an image-reproducing device in television reception, and in radar equipment. The elements of the cathode-ray tube include an electron gun for forming a beam of electrons, a deflecting system (employing electric or magnetic deflection forces), and a screen of luminescent material on which the electron beam impinges, forming a spot of light. The brilliance of the spot may be controlled by a grid element in the electron gun and its position on the screen by the two-coordinate deflection system. For oscillography, a horizontal deflection proportional to time and a vertical deflection proportional to the measured quantity are employed to form the waveform on orthogonal coordinates.

Another application of the cathode-ray tube is in radar. The cathode-ray tube registers the reception of a reflected radio signal and indicates the time between the transmission of the wave and its reception, a time interval proportional to the distance of the reflecting object from the radar. Various presentation systems have been devised to indicate not only the presence and distance of the target (aircraft or ship detected by the radar) but also its angular bearings in azimuth and elevation.

56. Electron Microscopes. The electron microscope is an electron tube which employs electron trajectories to form an enlarged image of a surface or of a translucent substance. The advantage of the electron microscope over the conventional optical microscope lies in its great resolving power. The resolving power of an electron ray is determined by the wave properties of the electron, which in turn depend on the electron velocity. By employing high-velocity electrons, wavelengths shorter than those of visible light (or ultraviolet) may be obtained and correspondingly higher resolving power obtained. Resolutions corresponding to a useful magnification of some 200,000 diameters have been reported by using the shadow-type electron microscope, whereas 6,000 diameters is the limit of the immersion-type optical microscope using ultraviolet radiation.

Two general types of electron microscope have been developed. One is useful only for the study of electron-emissive surfaces. In this type, electrons emitted from a surface are caused to take divergent paths and finally to impinge on a photographic plate or fluorescent screen, there producing an enlarged image of the electron distribution at the emitting surface. The more widely useful type of electron microscope employs beams of electrons which are passed through thin sections of the substance under examination. The relative opacity of the various portions of the substance to the passage of the electrons forms a "shadow" image, which is recorded on a photographic plate or fluorescent screen placed beyond the section under examination. The necessary divergence and convergence of the electron paths to form the enlarged image are usually produced by the action of magnetic focusing coils placed externally to the chamber through which the electrons pass. Very high accelerating voltage (above 100,000 V) is customarily employed to secure the highest possible resolution.

ELECTRONIC MOTOR CONTROL

57. Machine operations of every description have been simplified by electronic control of electric motors, and the quality and quantity of processed materials have been considerably improved, with reduced first costs and operating expense. Electron tubes with appropriate circuits are used to control practically any electric motor function, such as speed, acceleration and deceleration, starting torque and current, and overloads.

Methods for controlling some of the more common motor characteristics follow.

58. D-C Shunt Motor Control. The most commonly used motor-control setup is built around a d-c shunt-wound motor fed from a tube rectifying system which is in turn fed from standard a-c distribution lines.

Basically d-c motor-control setups are comprised of three main parts: the motor itself, with its armature and field windings; and two tube-rectifier units, one for supplying the necessary field power, and the other for rectifying the necessary armature power. The control circuits are usually incorporated as integral parts of the rectifying systems.

The circuit for the shunt-wound d-c motor is shown in Fig. 28-20. The field coil consists of a large number of turns of relatively small wire wound on radial poles around the periphery of the motor. As d-c current is forced through these field coils by the application of a voltage across them, a magnetic field is set up along the axis of the poles around which the coils are wound.

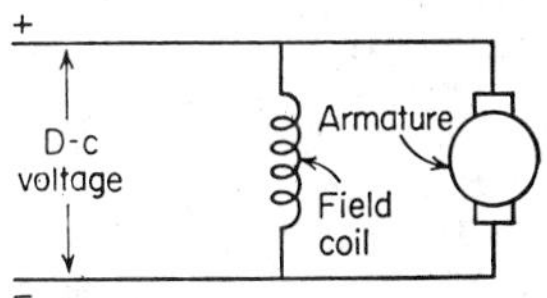

FIG. 28-20. Diagram of a shunt-connected d-c motor, showing the field coil and armature.

The armature winding consists of several turns of relatively heavy wire or strap wound on the rotatable member of the motor, and as d-c current is forced through this winding, another magnetic field is produced. These two fields are of such magnitude and direction that they cause the armature to rotate about its axis.

The armature winding offers very low resistance to the flow of current while the motor is standing still. When it begins to revolve and the segments of the armature winding begin cutting the magnetic lines of force set up by the field coils, a voltage is induced in the armature winding which opposes the original voltage applied to the armature and acts to cancel out the effect of a portion of that applied voltage. This opposing voltage, or "counter-emf," limits the current through the armature by reducing the effective voltage that causes the flow of current.

The speed of the motor is determined by the balance point at which the effective armature voltage (which is approximately the applied voltage minus the induced voltage) is just sufficient to cause enough armature current to flow to drive the mechanical load and make up for friction losses in the motor and transmission system.

59. Speed Control. If the magnetic field set up by the field winding is decreased, the induced canceling voltage in the armature winding will be less. Thus the effective armature voltage will be greater, the armature current will increase, and the motor will speed up.

On the other hand, if the voltage applied to the armature is reduced, the effective armature voltage is decreased and less current flows. The motor will thus slow down.

These two speed-changing phenomena are the basis for motor speed control. Where it is desired to have a motor run at some speed slower than its normal speed, control of the armature current may be used. Where the speed is to be increased, field current control may be used. Combinations of both types of control are commonly employed to obtain speed regulation over very large ranges.

60. Typical Speed-control Circuit. A greatly simplified motor speed-control circuit is shown in Fig. 28-21. Here the d-c voltages for the field and armature are supplied by two separate rectifier systems. The voltage applied to each may be varied by the addition of resistance in series with the windings. Special power rheostats are available for this type of circuit, but these components are usually bulky and expensive. They are also difficult to maintain because of the large amounts of power they must dissipate.

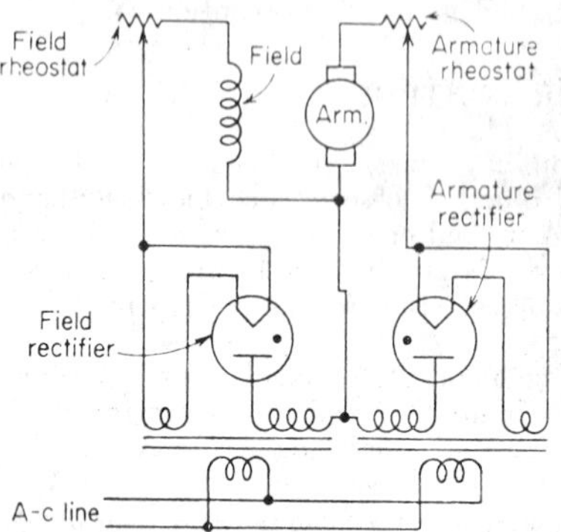

Fig. 28-21. Schematic for controlling the speed of a d-c motor by using gas rectifiers to convert a-c line power to direct current.

By substituting thyratrons for the diode rectifiers shown in Fig. 28-21 and adding suitable electronic circuits to control the rectifying action of the thyratrons, armature and field currents can be adjusted at less expense and with numerous other advantages.

Figure 28-22 shows a basic electronic control system, again greatly simplified. The thyratron grids are connected to phase-shifting devices. The phase of the grid voltage is made variable so that the thyratrons may be caused to conduct for different portions of the half cycle during which the thyratron anodes are positive. Note that, when diode rectifiers were used to supply direct current to the field and armature, they conducted current for the full positive half cycle, affording no control.

If a thyratron grid is allowed to swing positive at a time when the plate of the tube is just going positive, the tube will conduct over the entire positive half cycle. Any load connected to that tube will draw power during these positive half cycles. If the grid voltage is retarded, however, so that the grid is in some portion of its negative half cycle at the time the plate voltage starts its positive half cycle, the firing of the tube will be delayed until the grid voltage swings to a value which will permit the tube to conduct. Thus the tube conducts during only a portion of its positive half cycle, and current in the load will flow for shorter intervals of time. The average current, i.e., the total of the instantaneous current divided by time, will decrease. If it is the armature current which is decreased in this manner, the motor speed will decrease.

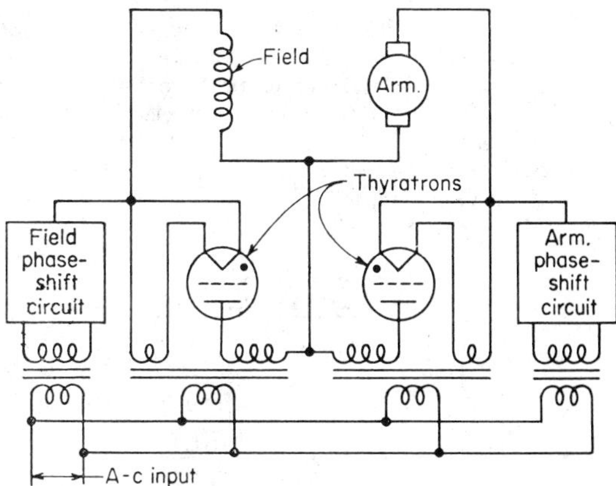

FIG. 28-22. Simple circuit showing the use of grid-controlled gas tubes to vary voltages applied to the motor field and armature.

It is worth noting that, if a single thyratron is used to control the armature or field, current cannot flow more than half the time, because the tube will not conduct during the half cycles of the applied a-c voltage which make the anode negative with respect to the cathode. This, then, is the condition which will produce the maximum average current through the tube and the load. With two tubes in a full-wave rectifier circuit it is possible to force current through the load at all times if desired.

61. Phase-shifting Methods. The armature portion of an electronic control system which uses the phase-shift principle is shown in Fig. 28-23. Here the d-c armature current is obtained from the a-c distribution lines by means of a single thyratron rectifier circuit. The grid of the thyratron is connected to a transformer whose primary is connected across the output terminals of a special phase-shift bridge circuit.

In such a circuit, as the value of the variable reactance is increased, the a-c grid voltage is retarded in phase. Thus instead of being positive when the plate becomes positive, the grid voltage lags behind by a certain amount. The tube conducts, not as soon as the plate goes positive, but at some time later in the positive plate excursion as determined by the time when the grid voltage goes positive. This latter time is determined by the tube characteristics. The result is that the thyratron will conduct for shorter and shorter periods of time and the average current will be less as the phase lag increases.

If the armature were rotating in a fixed field, its speed of rotation would be decreased by increasing the amount of reactance in the phase-shift bridge.

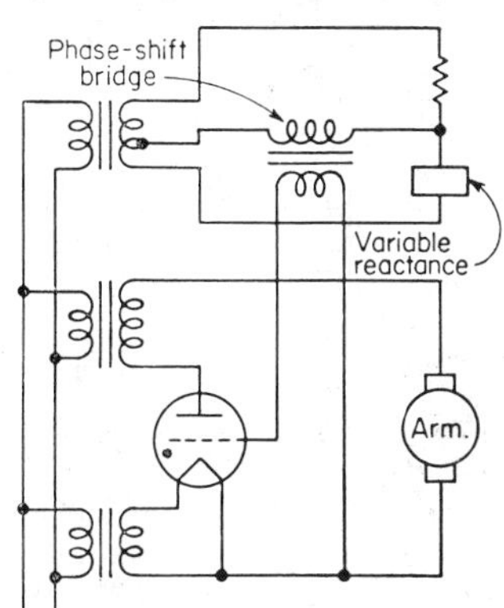

FIG. 28-23. Diagram of means for retarding the phase of thyratron grid voltage for varying the current supplied to the motor armature.

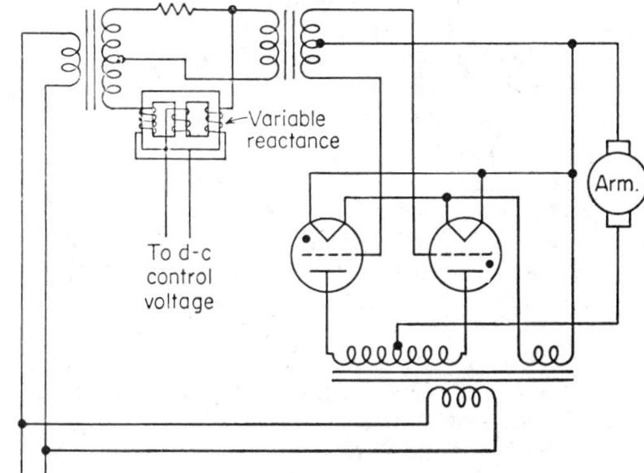

FIG. 28-24. The use of saturable reactors greatly increases, or amplifies, the amount of control possible.

Actually the circuit shown in Fig. 28-23 is applicable only to small motors, since only half-wave rectification is employed. To supply adequate armature power for larger motors, high values of peak rectified current would be required. The circuit in Fig. 28-24 employs a full-wave rectifier with exactly the same type of phase-shift arrangement but with a more practical type of variable reactance in the bridge.

62. Saturable-reactor Phase Shifter. The variable reactance shown in Fig. 28-24 is a saturable reactor. The inductive reactance introduced into the circuit by the saturable reactor is determined by the amount of d-c current flowing in the d-c winding.

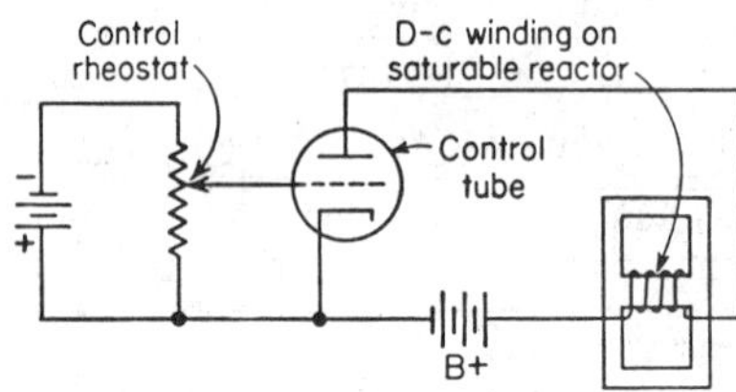

Fig. 28-25. Simple circuit for varying the direct current applied to the d-c winding of a saturable reactor.

The advantage of the saturable reactor is that only a small change in d-c power is required to cause a large change in its inductance. The small d-c change can be achieved by a potentiometer connected across a source of d-c voltage, or it may be derived electronically by a circuit such as the one shown in Fig. 28-25, where a potentiometer is used to determine the amount of bias on a control tube. As the arm of the control rheostat is moved up, making the grid more negative with respect to its cathode, the plate current of the tube will decrease, reducing the current which flows through the d-c winding on the reactor. The a-c side of the reactor will present more inductance in a grid phase-shift circuit, causing less armature current to flow and the motor speed to decrease.

Thus we have effectively added resistance in the armature circuit. However, instead of requiring a large power rheostat with a great power loss, we have accomplished the same effect with a small potentiometer having almost negligible power loss.

63. Variation of Resistance. In explaining the phase-shift bridge used to control the armature current in the previous section, it was shown that the phase of the grid voltage was changed by varying the inductance of one of the arms of the bridge. Phase-shift variation also may be accomplished by varying the resistance in the arm containing the reactance.

This system is illustrated in Fig. 28-26. The circuit is shown connected to the field winding of a motor. The basic phase-shift bridge is still present, but in this case the reactance used for phase shift is a capacitor, and the element used to vary the phase shift is actually a variable resistance or rheostat in series with the capacitor.

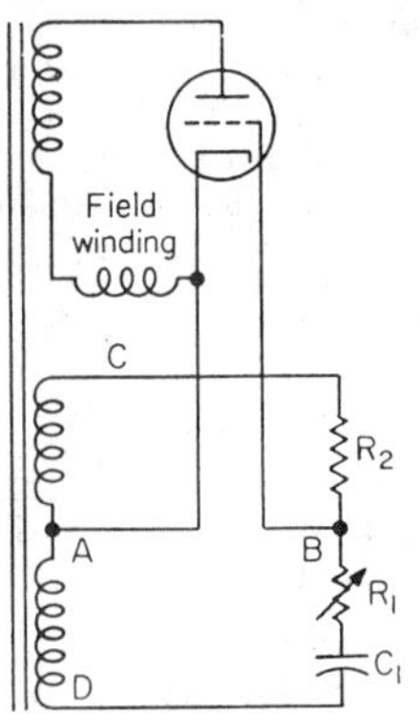

Fig. 28-26. Basic phase shift circuit shows a bridge configuration.

The voltage across a capacitor is 90° out of phase with the current flowing through it. Therefore, since the voltage across a resistance is in phase with the current flowing through that resistance, the voltage across the capacitor and the voltage across a resistor in series with it are 90° out of phase with each other.

Thus, if the value of the resistor R_1 in Fig. 28-26 is chosen so that its effect is very much greater than that of the capacitor C_1, the voltage across that arm of the bridge will be practically in phase with the current flowing through it and consequently in phase with the voltage across the other arm of the bridge, which is made up of R_2. The voltage between A and B is then nearly in phase with the voltage across points C and D, since little phase shift occurs around the bridge.

If now the value of R_1 is decreased, so that its effect is quite small as compared with the reactive effect furnished by C_1, then the voltage across that arm of the bridge will be made up primarily of the voltage across the capacitor, and it will be out of phase with the voltage across the other arm of the bridge, which is still caused by the pure resistance R_2. Thus decreasing the value of R_1 effects a phase shift in the voltage output taken from points A and B. So, by decreasing the resistance in the capacitor arm of the phase-shift bridge, the field current can be decreased and the motor speed increased.

64. Continuous Control. We have seen that the speed of a motor can be increased by lowering the field voltage and decreased by lowering the armature voltage. In many cases a given d-c motor with a certain rated speed is to be used in a system where speed control above and below the rated speed is desirable. It is possible to maintain smooth control over the entire range of speeds with a single speed-changing knob by using the rheostat shown in Fig. 28-27.

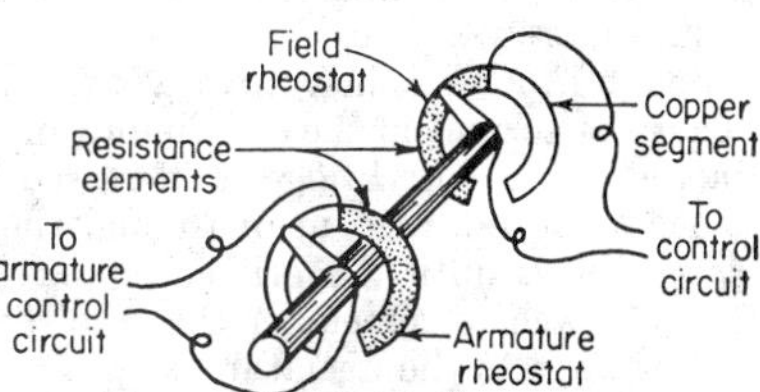

FIG. 28-27. Special dual potentiometer provides continuous speed control with a single knob for low-power applications.

If the motor is designed to run at its rated speed when the full voltage is applied to both the armature and the field, it will operate under this condition when the arms of the rheostats are straight up and no resistance is in either circuit.

Moving the knob counterclockwise places some resistance in the field circuit but at the same time makes the arm of the armature rheostat move onto a copper segment which provides a direct path to the power line. The armature circuit will continue to have full voltage applied to it. However, the resistance added to the field circuit reduces the field current, and the motor speeds up.

To decrease the motor speed the knob is rotated clockwise, thereby taking resistance out of the field circuit. When the rated speed point is passed, resistance is added in the armature circuit, while the field current remains unchanged.

65. Speed Regulation. We have seen that motor speed can be controlled by a d-c voltage. We have also seen that the speed of rotation of a motor determines the amount of induced voltage in the armature and thus the effective voltage across the armature. It would seem that we should be able to devise a loop arrangement such that if the motor speed were reduced, due to some external load change for instance, the voltage change in the armature winding caused by the speed change could be used to restore the motor to its original speed, thereby providing a degree of regulation.

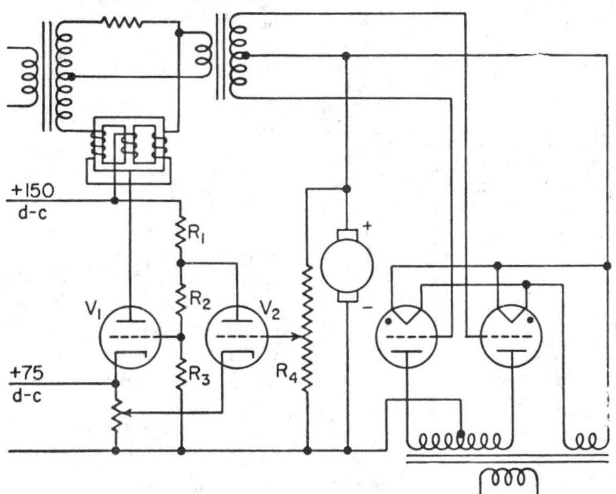

FIG. 28-28. Circuit providing regulation of the speed of rotation of a d-c motor.

A circuit capable of doing this is shown in Fig. 28-28. The voltages shown applied at the left side of the diagram are assumed to be provided by some regulated source. They serve as reference voltages in the speed-regulation circuit, as will be explained shortly.

Tube V_1 is a triode connected across the upper half of the regulated voltage supply, with the d-c winding of a phase-shift saturable reactor in its plate circuit. The grid voltage of V_1 is determined by the three-resistor voltage divider which is connected across the entire regulated voltage supply. For the moment assume that V_1 is conducting, because its grid voltage is relatively positive with respect to its cathode. The current flowing through the d-c winding of the saturable reactor reduces its a-c reactance, and the armature rectifier thyratrons conduct over a maximum portion of the positive half anode cycle. Suppose that the motor tends to speed up, as might happen if its load were decreased. As the armature accelerates, the induced voltage increases, causing the total voltage across the armature to increase. A portion of this armature voltage increase is applied to the grid of V_2 through R_4 in such a way that the grid becomes more positive. As the grid of V_2 becomes more positive, that tube conducts more current. The V_2 plate current flowing through R_1 lowers the voltage at the grid of V_1, causing that tube to conduct less current. Thus the current in the d-c winding of the saturable reactor decreases, and the reactance in the phase-shift bridge is increased, causing the grid voltage of the rectifier thyratrons to be retarded in phase and the armature current to be decreased. The motor speed drops until the original speed (before the load decrease occurred) is attained.

It is well to remember that the speed regulation is accomplished by comparing the voltage across the armature with a known fixed voltage. To have perfect regulation, the armature voltage would necessarily have to vary exactly linearly with the armature speed. This is actually not the case, because the total armature voltage is determined by the applied voltage, the opposing induced voltage, and the IR drop caused by the current flowing through the armature resistance. This third factor, the IR drop, is directly proportional to the armature current, since the resistance is constant. It is the factor which prevents the armature voltage from deviating in direct proportion to motor speed.

66. *IR*-drop Compensation. The discrepancy in speed regulation caused by the armature IR drop may be compensated by the circuit shown in Fig. 28-29. Another transformer, T_1, has been introduced. Its primary windings are connected in the armature circuit in such a way that the voltage across its secondary increases as the armature current increases. The secondary transformer voltage is rectified by V_1, and the portion of the resulting d-c voltage appearing across the lower portion of R_1 is applied in series with the armature voltage divider in the speed-regulation circuit. By proper adjustment of R_1, the voltage added in series to this circuit will be exactly equal and opposite to the voltage discrepancy caused by the armature IR drop (since both are proportional to armature current). The voltage tapped off for the regulator circuit will be exactly proportional to machine speed.

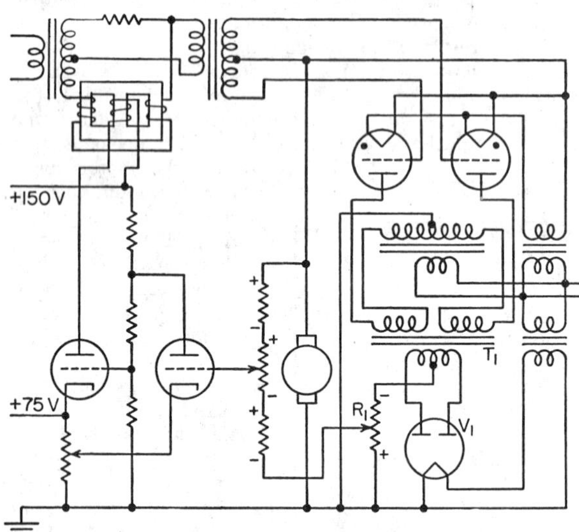

Fig. 28-29. Speed regulation and IR-drop compensation are provided by this circuit.

This IR-drop compensation circuit also serves as an armature current-regulating device, since a large armature current will cause the addition of large negative voltage into the speed-regulator voltage divider across the armature. The result will be a reduction in armature current through the speed-regulation circuit.

67. An armature current-regulating circuit operating on the above principle is shown in Fig. 28-30. Here the armature current flows through the regulation transformer primary as before. However, instead of using the rectified regulating voltage in series with the armature voltage divider, as in the IR-drop compensation circuit, here the voltage is used directly to control the firing time of the armature thyratrons through V_1 and the saturable reactor connected in its plate circuit. The circuit of Fig. 28-30 shows the field of the motor being supplied by the same anode transformer, but with straight diode rectifiers and a series potentiometer for controlling field current.

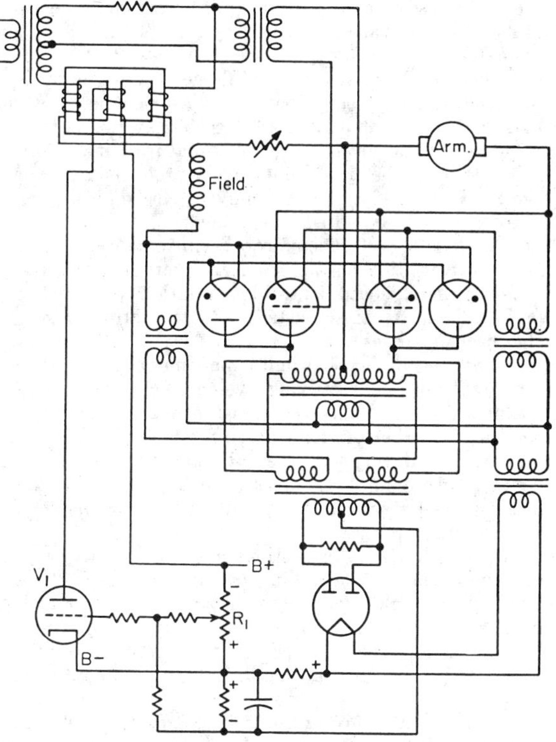

Fig. 28-30. Circuit providing adjustable speed control with *IR*-drop compensation and armature-current regulation.

68. Motor Reversing. If the voltage applied to either the field or the armature of a d-c motor is reversed, the direction of rotation will be reversed. However, this reversal cannot be accomplished instantaneously.

For instance, quickly reversing the polarity of the armature voltage would cause the armature current immediately to rise to a dangerously high value. The reason for this high current is that the applied voltage (having been reversed) would now add to rather than oppose the induced voltage in the armature. This induced voltage depends upon the direction of rotation, and the armature would not immediately reverse.

Several methods for avoiding this heavy current have been devised. Probably the commonest and simplest method makes use of a large magnetic contactor which disconnects the armature voltage and connects a resistance across the armature, giving the effect of dynamic braking. Once the armature has come to rest, the same, or another, contactor may be operated to apply the reverse polarity voltage to the armature circuit, thereby causing the motor to begin accelerating in the opposite direction. Where more rapid direction changing is required, a regenerative system is employed whereby the kinetic energy stored up in rotation of the armature is forced back into the power lines. This system is found more frequently in nonelectronic setups; however, it becomes quite complicated in electronic versions because of the fact that electrons can flow in only one direction in a tube rectifier.

69. Refinements. In commercial control units, additional refinements are necessary to provide protection against overloads and make it possible for inexperienced personnel to operate the motors without damage to the equipment. In many cases thermally operated overload relays are integral parts of electronic control units. They are seldom necessary with fractional-horsepower motors, since usually these motors have sufficient internal impedance to limit the starting and accelerating currents to values which will not damage the motor or paralyze the d-c power supply.

70. Starting-current limiting devices may be placed in the circuit in a number of different ways. Some setups use a starting button which is depressed and held until the motor reaches a certain speed, at which time the button is released. With this type the limiting device is connected in the circuit while the start button is depressed. An automatic system involves a centrifugal switch or some other means for cutting the starting-current limiting device out of the circuit when the motor reaches some safe speed. These starting-current limiting aids may be separate units, such as a resistance in series with the armature, or they may be part of the speed-control or speed-regulation systems.

When we say that the speed range of a d-c motor is extended by the use of current-reducing devices in series with the motor elements, we must understand that the rated horsepower available for rated speed may not be available when the motor is slowed down. At very low speeds the armature current must be reduced below its normal value, because the air cooling usually produced by armature rotation is reduced. However, this would be true not only of electronically controlled motors but of all types running at reduced speeds.

The twisting effort, or torque, will remain constant in a motor whose speed is regulated by variation of the armature voltage alone, but the horsepower varies. If, on the other hand, speed control by variation of the field is employed, the horsepower output available remains constant, but the torque decreases as the field is weakened and the speed increases.

It will be noticed that in d-c machinery circuits the armatures and fields are supplied directly from the rectifying system. No d-c smoothing reactors or other external means are used to absorb the instantaneous differences between the output voltage of the tube and the counter voltage produced in the armature. Thus pulses of armature current flow during the interval when the instantaneous applied voltage exceeds the counter voltage in the armature. The armature winding reactance affords some filtering of these pulses. Some circuits incorporate a separate reactance in the armature circuit to smooth the pulses.

In some cases filter sections are required to prevent hunting. Electronic circuits are almost instantaneous in operation, whereas the equipments they control usually have finite time delays before they respond to control. These circuits are often found between the grids and cathodes of rectifying thyratrons.

This discussion is limited to single-phase applications, in the interests of simplicity. The circuits described are applicable to larger multiphase equipment with correspondingly more electronic components.

71. Reversible Motor Control. Many types of industrial machinery, such as milling machines, shapers, lathes, and variable-depth boring machines, require motors which move work repeatedly between two points—in other words in which the drive motor is continually reversing.

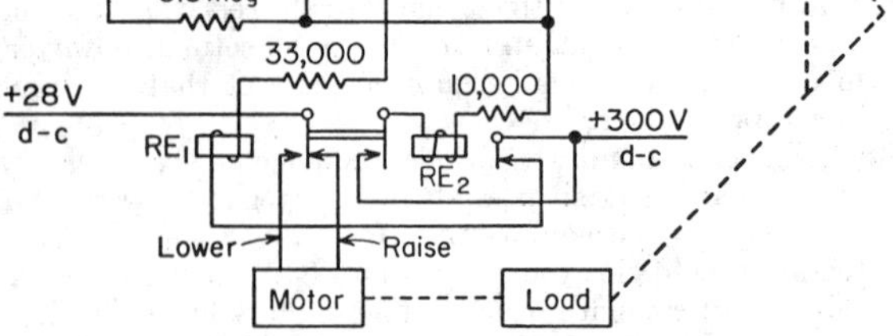

The electronic circuit shown in Fig. 28-31 is capable of reversing a motor when the load on the motor reaches a preset limit. Thus it qualifies for the job described above. Either limit can be set without regard to the other by means of potentiometers R_1 and R_2. Potentiometer R_3 is geared to the turning mechanism of the load so that its full resistance is traversed with the complete range.

Fig. 28-31. Circuit for rapid reversal of the direction of rotation of a motor.

Relay 1 controls the direction of travel of the motor. The load will move one way when it is energized and return the other way when it is deenergized. With the potentiometers in the positions shown, RE_1 will initially be deenergized, and the motor will move the load to such a position that R_3 picks off a voltage equal to that set on R_1. Relay RE_1 will then be energized by the firing of V_1, and the motor will reverse. The other extreme reversal will occur when the load moves to a point where the voltage tapped by R_3 equals that set at R_2, causing V_2 to conduct and RE_1 to be deenergized again. The values shown were chosen to reverse the direction of an automatically rotated radio dial, but by proper alteration of components it could easily be adapted to heavier operations—in fact any reversible motor setup which can be controlled by the action of a relay can be used in conjunction with such a circuit.

72. Tachometer Speed Regulation. Figure 28-32 shows a circuit containing a tachometer generator mechanically coupled to the motor which is being electronically controlled. The output of such a generator is a d-c voltage which is directly proportional in magnitude to the speed of rotation of the motor.

This type of device affords a simple means for controlling and regulating motor speed. The voltage may be applied to the circuit supplying motor current in such a way that a decrease in the tachometer output, such as might be caused by a decrease in motor speed, would cause an increase in armature current so that the motor will regain its original speed.

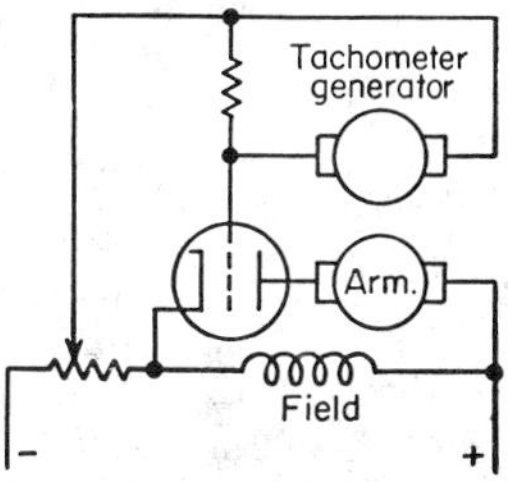

The setup shown in Fig. 28-32 would have extremely limited usefulness because of the small power-handling capabilities of a tube which might be controlled by the voltage of the tachometer generator. Actually, the voltage derived from the tachometer would be used in one of the circuits discussed previously to control the motor.

FIG. 28-32. Tachometer generator provides the reference signal for this speed-regulation circuit.

73. Machine Tools. Among the types of machines to which electronic drives have been successfully applied are grinders, milling machines, toolroom lathes, turret lathes, and thread mills. In addition, these drives have been supplied for automatic welding machines and for special equipment such as testers for magnetos, airplane-propeller governors, and instrument tachometers.

A particularly interesting application has been made for driving the headstock on grinders. The wide speed range obtainable and the full and constant torque provided at slow speeds make it possible to provide the proper speed for every type of grind just by turning a small control knob. In several instances the new control has made possible a simplification in the design of the headstock itself, through elimination of the gears and pulleys formerly used.

The electronic motor drive for machine tools has several important advantages. An infinite number of speeds are available, instead of just a certain number of steps provided by gear changes. Speed changes are made without stopping, simply by turning a small knob, and the motor itself is the only moving element. On a gear-change mechanism, the machine must be stopped to change speed; this requires considerable time and makes a mark on the work. Smooth automatic acceleration is provided. Finally, the electronic motor drive is ideally suited for setting up an automatic sequence of operation where it is necessary to change the feed rate frequently as the sequence of operation changes, as well as for automatically regulating the feed rate to maintain a given load on the cutter motor.

A reduction in finishing time from 13½ h to 5 min in the machining of aluminum spar beams for airplane wings has been accomplished at one aircraft plant with the help of an electronic motor-control system installed on a large automatic contour milling machine. Spar beams are long, one-piece structural channels that run lengthwise through the wing, from fuselage to wing tip. The spar must be machined accurately to permit perfect joining of wing ribs and cap strips, and it must also be contoured exactly to conform with the irregular shape of the wing itself.

The carriage of the milling machine houses four cutter-motor assemblies, which finish the long spar in a single setup despite the wide variety of cuts required. Flexible speed control of the carriage motor by the electronic drive assures that the cutters are fed to the work at all times in proper relation to the changing contours of the spar beam.

In one pass over the table, the depth of cut may increase and decrease several times, while the number of cutters entering the work may change from one to four. Such varying conditions require a change of feed to avoid overloading of the cutter motors. Moreover, a fast skip speed is essential to save time when no cutting at all is necessary.

The field of usefulness of an ordinary drill press is greatly increased by the addition of electronic speed control for the motor. Operating speeds can be adjusted from 25 r/min to 1,750 r/min simply by rotating a knob, and the direction of rotation can be changed by pressing a button. Independent speed adjustments are provided for both tapping and backing out. The range of speeds makes it possible to use the drill press for a wide variety of hard-to-drill materials, from molded compounds to the hardest steel pieces, without changing pulleys.

An electronic drill-tool alarm responds to the increase in torque load on the spindle of a drill when cutting edges have dulled so much that the torque load exceeds a predetermined amount, usually between a 1 and 10% increase. The drill is pulled out automatically when dull, reducing breakage and speeding up the drilling of deep holes. Where danger of breakage is negligible and automatic retraction is unnecessary, the electronic control can simply be connected to operate a signal light.

74. Balancing Machines. In machines for balancing large rotating parts such as flywheels or armatures, rapid acceleration and deceleration present a difficult problem, since couplings and other parts must be light in order to avoid distortion of results. Torque must be smoothly applied and limited in amount so that such parts will withstand the strain; yet the acceleration and deceleration must be done quickly to keep total test time at a minimum.

With electronic motor control, balancing machines with drives rated up to 200 hp can bring a huge wheel up to speed in the shortest possible time without increasing the torque above the safe operating value for the machine. As one example, automatic electronic control of the accelerating rate is used in bringing 17-ft-diameter marine propulsion gears up to full speed of 200 r/min for dynamic balancing. The 50-ton gear to be tested is supported by bearings that permit it to vibrate freely in a horizontal direction. Pickup coils convert this movement into electrical impulses that energize a wattmeter in a dynamic balancing circuit which indicates where correction of unbalance is required. Electronic control equipment was chosen because both squirrel-cage and wound-rotor induction motors heated excessively during the required long acceleration time of 10 min.

Reduction of time in balancing submarine electric-motor armatures from 20 to 5 man-h is accomplished by using electronic control on the balancing machine. The time for one adjustment cycle, starting at the instant the stop button is pushed and including dynamic braking to a stop, readjustment of balancing weights, starting the motor again, and bringing it up to critical speed, is only 37 s with electronic control, as compared with 2 to 4 min required for free-wheeling stop and manual starting.

ELECTRONIC HEATING

75. Electronic Heating. Basically there are two types of high-frequency heating: induction heating and dielectric heating. Both methods share a common principle: power in the form of alternating current is transferred from an electrical circuit, where it originates, into the material, where it is converted into heat by the resistance of the material and raises the temperature of all or part of that material. The heat actually originates inside the body of the material being heated, but the energy which causes this heating is furnished externally.

There are two basic differences between these methods. Only conductors of electricity can be heated by induction heating, whereas dielectric heating applies only to nonconducting materials. The second difference lies in the apparatus used to transmit the energy from the electric circuit into the material. In the case of induction heating, the heating element is a coil of copper wire or tubing through which the alternating current flows. Energy is transferred by induction—thus the name induction heating.

In dielectric heating, on the other hand, the material to be heated is placed between two metallic plates, which are alternately charged to positive and negative voltages with respect to each other. The entire apparatus resembles a capacitor, with the material acting as the dielectric—thus the term dielectric heating.

76. Induction Heating. In induction-heating apparatus, the material to be heated is placed in an alternating electromagnetic field. The nature of the field required depends on the size, shape, and material of the work.

Figure 28-33*a* shows a typical case. A metallic bar is to be heated for purposes of hardening. The coil surrounding the bar is referred to as the work coil. Its function is analogous to that of the primary winding of a transformer. The work, or charge, being a conductor, acts as a short-circuited secondary. When the primary, or work coil, is energized, a current is induced in the secondary, or work. The magnitude of this induced current will depend on the amplitude and frequency of the alternating current flowing in the work coil. It is the induced current that causes the heating of the material.

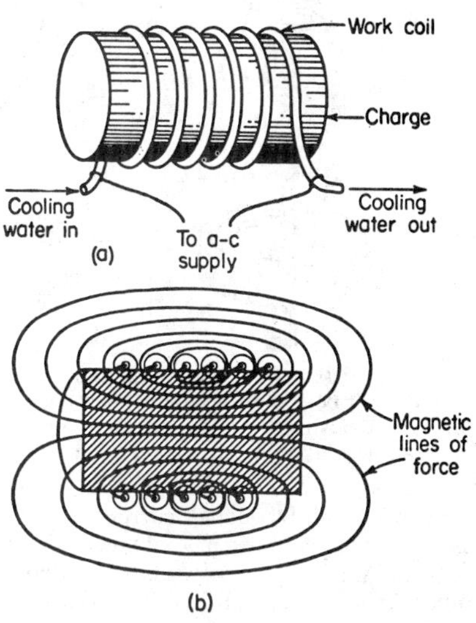

FIG. 28-33. Basic induction-heating setup. High-frequency alternating current passed through the coil heats the metallic charge. The depth of heating depends on the frequency of the heater current.

The lines of magnetic flux which are set up in the charge by the current-carrying coil are parallel to the charge axis in the case cited, as shown by the cutaway drawing (Fig. 28-33*b*). The distribution of this flux (and thus the distribution of heating) within the work will depend on several factors: (1) the electric and magnetic properties of the material, (2) the dimensional cross section of the work itself, and (3) the frequency of the applied alternating current. The first two of these factors are usually fixed by the job at hand. The remaining variable is the frequency of the alternating current which is made to flow in the work coil and which determines the rate at which the magnetic field within the charge builds up and collapses and then builds up in the other direction.

The selection of frequency for a particular job depends on the application. If, for instance, a piece of metal is to be melted, it may be placed in a suitable crucible, which is surrounded by a fairly low-frequency work coil. As the charge experiences the regularly reversing magnetic field, it begins to heat uniformly and continues until the desired temperature is attained.

If the frequency is raised, skin effect (in which the current tends to flow only in a thin shell on the charge surface) comes into play and the secondary currents induced in the charge tend to remain on the outside. Since more heating occurs in the vicinity of the high-flux density, only the outside is heated by induction. This sort of arrangement is advantageous where the outside of the material is to be heat-treated for high surface hardness and the interior is to remain soft. Uniform hardening would cause the piece to be brittle; no hardening at all would cause the piece to be easily damaged. Such a piece might be used in the bearing of a crankshaft. By the same means, the cutting tip of a lathe tool may be hardened, while the main part is left soft and strong.

The induction-heating effect may be noticed when frequencies as low as standard 60-c power line are used, but higher frequencies almost always are employed in actual practice. As might be expected, the production of relatively large amounts of power at frequencies of hundreds of kilocycles or several megacycles presents a job for electron tubes. In some processes where only a few hundred cycles per

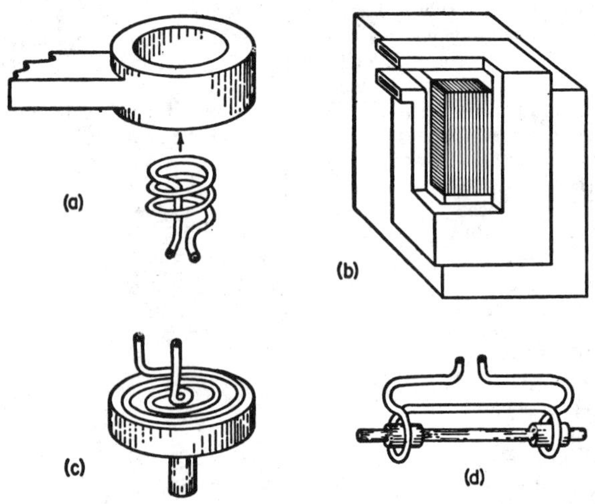

FIG. 28-34. Many special induction-heating coil forms are used. In *a*, the inside of the connecting-rod bearing is to be heated. Drawings *b*, *c*, and *d* show other special cases for localized heating.

second are required, frequency-multiplying rotating machinery operating from regular 60-c input can be used to better advantage.

Induction-heating work coils may take on a number of shapes. The example shown in Fig. 28-33 is typical but by no means a standard form. The technique of designing a work coil that will produce just the right amount of heat in just the right portion or portions of the charge is a highly specialized science. A few typical configurations are illustrated in Fig. 28-34. In *a* the induction-heating coil passes inside the bearing shown. Thus the inner surface is heated by the flow of high-frequency current within the coil. In *b* the coil is concentrated around the corner of the piece of tool steel. Thus only that corner is heated. In *c* the surface of the piece immediately under the spiral-wound work coil is heated. A system for applying heat in two different portions of the piece simultaneously is shown in *d*. Here the induction-heating current flows through two coils, and each coil imparts a certain amount of heating in the desired locations.

Very special work coils are sometimes required. In applications where induction heating is used in soldering or brazing (Fig. 28-35), it is usually important that only the portions which actually come in contact with the solder or brazing material be heated. Otherwise distortion of the assembly may occur. In cases like these the work coil may take the form of a knife-edged single-turn loop, which fits snugly around but does not touch the immediate vicinity of the proposed joint. Since the work coil is also part of an electrical circuit, its inductance and resistance must be considered in choosing coil shape and material. Low resistance is advantageous, since less power will be wasted in the work coil and the overall operation will be more efficient. Where moderately high frequencies are involved, the diameter of the wire or tubing used in the work coil should be large because of the skin effect. In most high-powered induction-heating equipment, provisions are made for circulating cooling water through the work-coil tubing. Care is taken to see that this does not hinder the operation of the device by short-circuiting out the ends of the coil where the water enters and leaves.

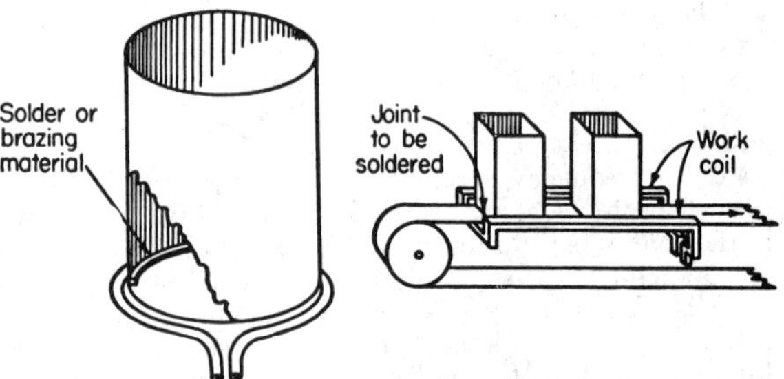

FIG. 28-35. When induction heating is used for soldering or brazing, the heat produced must be concentrated in a small area surrounding the proposed joint to prevent distortion of the work.

Frequently in the production of strip materials where continuous processes are employed, the material is passed through an induction-heating coil, where it is brought up to the proper temperature. In regulating and controlling this type of process the speed of travel through the coil may determine the amount of heating.

77. Dielectric Heating. In dielectric heating, nonconducting materials are placed between a pair of conducting plates which are electrically charged by high-frequency a-c potential. The basic setup is illustrated in Fig. 28-36.

When the molecules of a nonconducting material are placed in an electric field, they tend to line up physically with the lines of force of that field. If the field is reversed, these particles tend to reverse themselves to line up with the field. If the field is periodically reversed at some fairly high rate, these particles will be constantly realigning themselves. In doing so, heat is created, and the temperature of the material rises.

The examples illustrated in Fig. 28-36 are typical of the dielectric heating work. Between the two plates is a homogeneous piece of the nonconducting material. At any one time, when a certain voltage exists between the two plates, a share of the voltage difference is assumed by each minute section of the material. In other words, if the top plate is positive with respect to the bottom, the bottom half of each particle will be slightly more positive than the top half.

When the material being heated is not homogeneous, a different situation exists. Depending on the electrical characteristics of the different materials, a larger or smaller share of the voltage difference will be assumed by each material. The voltage that each material is responsible for determines the amount of heating that will occur in that material.

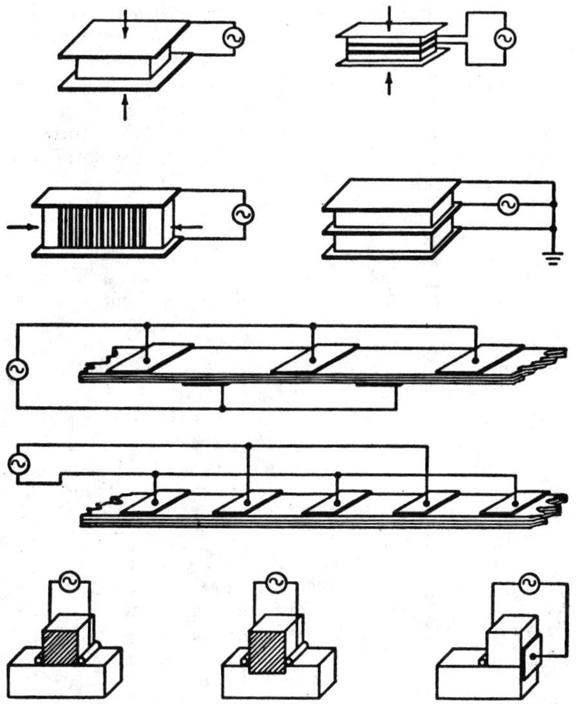

Fig. 28-36. Dielectric heating apparatus may take many forms. Basically, all these contain two metallic plates that are alternately charged at a high rate. Here glued joints are heated.

Figure 28-36 shows a system for applying even heat to speed the drying of a glued joint. The voltage change is gradual throughout the thickness of the material, since there is no change in the type of material between any two opposite points on the electrodes.

Where other than rectangular work is to be heated by the dielectric method, air gaps can be permitted, as long as their share of the voltage drop is taken into consideration and enough voltage is applied between the plates to allow sufficient drop across the material. Air acts just like another material (which it is) and must be treated as such.

The first commercial use of dielectric heating had to do with introducing artificial fevers into various parts of the human body (diathermy). The most desirable characteristic of the artificial fevers thus produced was that the fever spread uniformly and extended well below the surface of the body, which was not the case with simple heat-radiation devices. Dielectric heating is applicable to almost all nonconducting materials, such as wood, plastic, certain liquids, etc. One of the most common applications of dielectric heating is the preheating of plastic materials. By this means much time is saved in the plastics industry. Originally the material had to be heated very slowly in order that it be heated uniformly. By dielectric heating, only a few moments are required, instead of hours.

Dielectric heating is also widely used in the processing of both natural and synthetic rubber products. Again, even heating is required—and obtained—by this method. Where heat is required for drying, such as in the printing and dyeing industry, dielectric heating serves very well. In the lumber industry and its associated fields dielectric heating is often used to process lumber and to speed and perfect the drying of glued joints.

The electrical power required for both types of high-frequency heating is usually provided by some sort of oscillator, although the power could be obtained from an amplifier driven by an oscillator. What is needed in dielectric heating is a high voltage. This is provided at the terminals of the capacitor in the tuned circuit of an oscillator. In inductive heating, the need is for high current at some desired frequency. This high current is provided by the current flowing through the inductance of the tuned circuit of an oscillator. Therefore, a tube oscillator provides both kinds of electrical energy for high-frequency heating. The major engineering is in the design of the oscillator circuit to provide the proper energy and in the design of the applicators by which the energy is imparted to the workpiece to be heated.

Heating and cooking of many foods can be done between metal plates on a production basis by dielectric heating. Cookies can be baked in a few seconds. Roasting of meat that would take several hours by conventional heating can be done in a half hour.

78. Microwave Heating. Another method of heating nonconductive material is that of literally spraying radio-frequency energy onto the object. For this purpose, a magnetron tube is used. The magnetron is a tube that generates power at frequencies above 1,000 Mc for modern radar transmitters. At these frequencies, hollow pipes called waveguides are better conductors than wire, and an empty funnel, a horn antenna, is a good radiator of microwave energy when mounted at the end of the waveguide. Heating of nonconductive objects or materials can be done by inserting the object at certain spots in the waveguide or in front of the horn antenna. Cooking of food is done by microwave heating, and it is much faster than conventional heating methods, although it is also more expensive. Restaurants have been using microwave cookers for some years. In some cases the food is precooked and stored. When an order is received, the meat and vegetables are prepared on a dish, inserted in the oven and exposed to the radio-frequency energy from the horn antenna, and heated in a minute or two to a temperature ready for serving. Models for home use were introduced a few years ago. Airlines are beginning to install units on jet airplanes. The magnetron usually is operated at a frequency of 2,450 Mc for microwave heating. The power input is about 2,000 W from the line; roughly half of this is available for heating or cooking.

ELECTRONIC WELDING CONTROL

79. Resistance Welding. Heat caused by the flow of current through materials is usually considered to be power lost and is, of course, undesirable. Many modern-day industrial processes depend on just such heating for the accomplishment of useful and productive work. Such an application is resistance welding.

Resistance welding is the process of joining two pieces of metal together by passing a heavy current through them in such a way that intense heat is generated at exactly the spot where the desired joint is to be formed. This heat causes a small portion of each piece to melt. The molten portions flow together, and when the current is turned off and the metal cools and solidifies, a permanent bond, or weld, is formed.

All resistance-welding systems consist of three general components—a power source, a system of electrodes for applying power to the pieces being welded, and a control system for timing the application of power. The design of any of these components depends on the materials and type of weld.

Spot welding is a method of joining two metals by concentrating welding current on a relatively small area, small electrodes being used. Such welding can be accomplished with portable tools, such as gun welders, in which the electrodes are connected to the power source by a pair of heavy conductors. There are many types of spot welders. A number of spot-welder electrodes may be connected in series to form a series weld.

Seam welding is similar to spot welding except that current-carrying wheels replace the small spot-welder electrodes. The two pieces (usually sheet materials) are overlapped and fed through the welding electrodes at speeds varying from 2 to 100 ft/min. Seam welding is illustrated in Fig. 28-37. The wheels are pulsed regularly with welding current; so the seam weld resembles an even series of closely spaced spot welds.

A third type of resistance welding is the **projection method.** This is of two general kinds—butt and flash welding. In butt welding, two pieces to be joined are held together firmly, and current passes through the joint. When the two pieces are sufficiently heated, they are forced together under high pressure, giving a solid joint. In flash welding, the pieces are held together only lightly while current is passed through them. Owing to irregularities of the metal surfaces considerable flashing occurs, and the surfaces are "burned" smooth, as well as heated. They are then joined under high pressure.

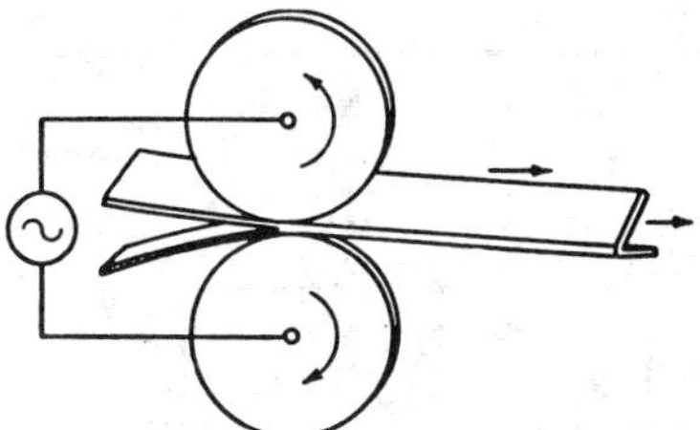

Fig. 28-37. Seam welding is accomplished by passing the two materials to be welded through a pair of alternately charged wheels, which serve to apply welding pressure as well as welding current.

The heat generated in any material may be determined by the formula $W = I^2Rt$, where W is the total amount of heat, I is the current flowing, R is the resistance of the material, and t the length of time the current flows. Since the current in this equation is squared, any change in current in a welding process will cause even more of a change in the heating. For instance, if the current is doubled, the heat generated will be four times as great.

In welding two pieces of metal, the resistance factor in the heating formula is very low—two pieces of metal brought together under pressure usually form a good contact. The time element is usually quite short, as determined by other considerations; so the remaining factor of the equation (the current) must be large if enough heat is to be generated to cause portions of each piece of metal to melt and run together. In practice, welding currents run into hundreds of amperes, though the voltage required to cause such large currents need not be high—a fraction of a volt to 15 or 20 V being typical values. The current is usually applied through a transformer which has a voltage step-down ratio from some readily available voltage to the desired welding voltage. The electrodes for connecting this low-voltage high-current power source must make extremely good contact with the materials to be joined. If more resistance appears at the electrodes than between the two pieces of material, more heating will take place at the electrodes and they will melt and fuse before the proposed weld forms.

Mechanical pressure plays an important part in the operation. To be effective, the application of pressure must be timed just right with respect to the application of the welding current. In a properly executed welding cycle, full electrode pressure must be exerted before the welding current is turned on. This pressure must be maintained until after the current is interrupted and the molten "nugget" has cooled to a temperature at which it regains a large portion of the inherent strength of the metal from which the weld is formed. This delay in releasing pressure is called the hold, and it helps to prevent cracks and faulty welds. The period before welding current is applied, while the electrodes are approaching the proper pressure, is called the squeeze time.

Welder Control. Accurate timing of the squeeze, weld, and hold times is very important. Squeeze time must be sufficient to allow sufficient mechanical pressure to be built up, but not unnecessarily long, for economic reasons. The weld time must be just enough to form the desired nugget, and the hold time must be right to prevent too rapid cooling and resulting flaws in the weld.

80. Resistance-welding control circuits are of two general types. The first type puts the equipment through its proper cycle at random times as initiated by the operator. In other words, as the operator positions the materials to his satisfaction, he actuates a device which initiates the welding cycle. The second type of control runs through squeeze, weld, and hold intervals and then, after a fixed period of "off" time, automatically initiates the next welding cycle. In this case some positive means for positioning the proper areas between the welding electrodes at the proper time is a necessity. In seam welding the second type of control is used, and the electrode wheels are arranged to move the materials a fixed amount during each off period.

Welding control circuits are interval timing circuits. Figure 28-38 shows a typical example. The time interval is started by closing Sw_1, which connects the thyratron circuit across the a-c supply line. The thyratron grid voltage at that instant assumes a value depending on the position of the tap on the resistor R across the line. The thyratron conducts on portions of the a-c cycle when its anode is positive with respect to its cathode, and a small amount of cathode current flows to the thyratron grid, which is slightly more positive than the cathode. This current flow charges the grid capacitor with a voltage of the polarity shown. At a certain time, depending on the values of grid resistor and capacitor, the voltage across the capacitor will make the grid sufficiently negative so that, on the next positive excursion of the a-c supply voltage, the thyratron will no longer conduct, thereby ending the interval. The smaller the grid capacitor, the shorter the charging time required for the grid to regain control, and vice versa.

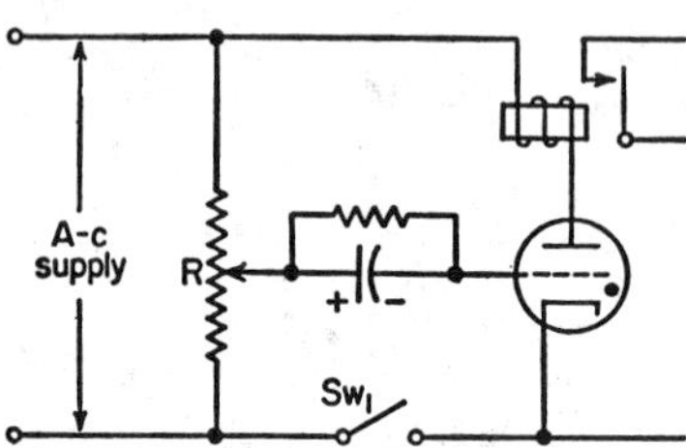

FIG. 28-38. Basic circuit of typical welder time-interval control.

The length of interval can easily be adjusted by setting the tap on the resistance network, which determines the voltage at the supply side of the grid capacitor. After the thyratron ceases to conduct on positive half cycles of the supply voltage, there is no longer any charging current in the grid circuit, and the capacitor loses its charge by forcing current through the parallel resistor, which dissipates the charge in the form of heat. The time required for this discharging process depends on the value of the resistor—a high ohmic value offering more resistance to the discharging current and thus taking a longer time.

If the switch is still depressed when the capacitor discharges sufficiently to again allow thyratron conduction during positive excursions of supply voltage, the circuit will again operate, and so on. Relays in the anode circuit of the thyratron may be used to control the primary circuits of welding transformers, or, as in the usual case, these smaller tubes are used to control the conduction in ignitrons, which in turn control the extremely heavy welding currents.

81. Ignitron Circuit. A typical circuit employing thyratrons as firing tubes for the larger ignitrons is shown in Fig. 28-39. The ignitrons are connected so that they will conduct current in both directions through the primary of the welding transformer. Each ignitron will conduct on half the a-c cycle when its anode is positive with respect to its cathode, if the thyratron connected to its grid is conducting. The firing time of the thyratrons is easily controlled by circuits of the type discussed in the section on motor control. Some smaller types of welders use what might be called half-wave controls, which employ only one control tube. This tube conducts current only on every other half of the a-c cycle.

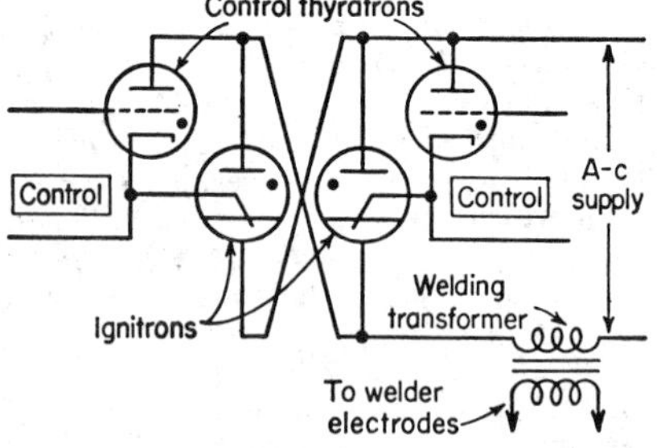

FIG. 28-39. Very large values of welding current can be controlled by this circuit. Actual control is applied to the primary of the welding transformer by a combination of thyratron firing tubes and ignitrons.

82. Synchronous and Asynchronous Control. Resistance-welder circuits do not present a pure resistance load to their power sources; the current in the line does not reach its maximum value at the same instant that the applied voltage does. This is the result of a finite amount of reactance being in the circuit.

If welding current is turned on at any random portion of the a-c cycle, certain undesirable effects may take place. Most serious of these is the production of unwanted voltage or current surges. These transients can be eliminated by making sure that the welding current is always applied at the proper instant in the a-c cycle. The current should rise from zero to maximum at a definite time relationship with the voltage wave. This relationship depends on the amount of reactance in the welding circuit. Synchronous welder timers take this condition into consideration. In addition to controlling the amount of heat applied to the weld, the control system delays the application of the current (after the switch is closed) until the voltage wave is at the proper point in its cycle. From that point at which the welder current is applied the delay circuit loses control, and the interval timer determines how long the current will flow.

The elimination of these transients is most important in large-sized welders where, for economy, equipment must run very close to its limit. Another factor that is taken into consideration in designing control circuits is the fact that the core material in the welder transformer will become saturated if a predominance of current in a certain direction flows through its windings. This effect is overcome by alternating the polarity of the first current surge applied to the transformer.

83. Capacitor Discharge Welder. Most resistance welds are accomplished in a relatively short period of time. The shortness presents a rather difficult design problem. Whenever large amounts of power must be delivered, certain precautions must be taken to prevent equipment failure. In resistance welding, tremendous amounts of power may be required to flow during a period which may be no longer than a fraction of a second. The average power is low, but the equipment must be designed to withstand the peak power.

The capacitor storage type of welder bypasses this difficulty and is suitable for a great many industrial welding applications. A typical capacitor storage circuit is shown in Fig. 28-40. There are two phases to the welding operation. First, the switch is thrown (either automatically or manually) to the left, where it connects a capacitor, a d-c voltage source, and a resistor in series. The capacitor receives a charge from the power source through the resistor, which limits the current flow to a reasonable amount. In the welding step the switch is thrown to the right, and the capacitor dumps its slowly accumulated charge into the weld almost instantaneously.

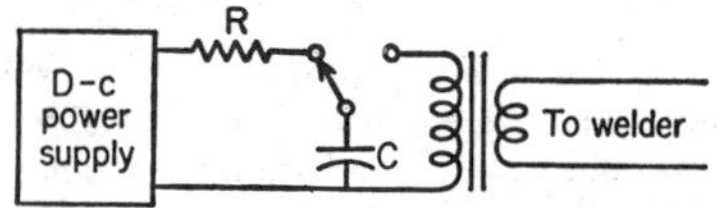

FIG. 28-40. Principle of capacitor-discharge welding equipment. A capacitor is charged by a d-c power source and discharged through the welding transformer.

The voltage source need supply only enough power to charge the capacitor slowly, while the weld still receives its extremely high power surge over the desired short period of time.

Electronics enters the welding picture chiefly through the control of welding time. A vast amount of engineering design has gone into this phase of the applications of tubes to industrial operations, and as a consequence the circuits developed may be highly complex.

84. Welding by Induction Heating. Soldering or welding of small metal parts is often done by induction heating, with transmitting-type vacuum tubes used to produce the heat in the parts to be joined. An interesting installation of this kind employs, in addition, a phototube to "see" the temperature of the metal involved, a receiving tube, and thyratron tubes to operate relays. The workpieces are themselves metal parts for cathode-ray tubes.

In the manufacture of television picture tubes, a vacuum-tight weld is required in the exhaust-tubulation assembly to join the copper tubing to the sealing sleeve.

Figure 28-41 is a circuit diagram of the arrangement devised to utilize the drop in temperature to control the r-f welding generator. The generator induces about 5 kW of power into the assembly. Radiation from the assembly is reflected by the prism and focused by the lens into the phototube. The drop in photocurrent, passing through the amplifier as a voltage, is reversed and appears as a rising wavefront at the differ-

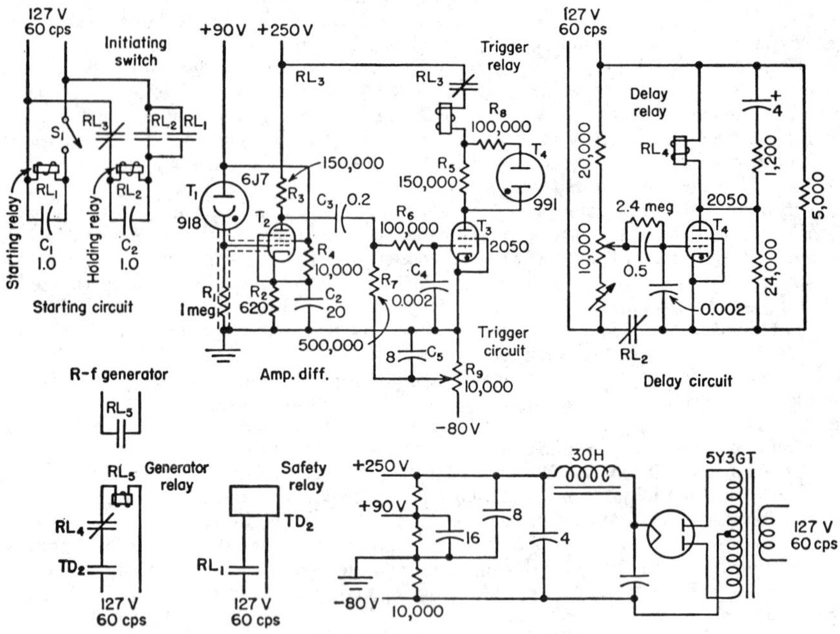

FIG. 28-41. Complete circuit of welding control system.

entiator. Upon differentiation the wave becomes a positive pulse. This trips a thyratron, which in turn starts an electronic delay stage.

The delay stage produces a delay of about ½ s. This arbitrary delay period ensures that the copper has melted around the entire circumference of the weld. At the end of the delay period a relay shuts off the radio-frequency generator. The initiating switch, in addition to starting the generator, also triggers a safety relay. This safety relay is set for a delay of about 12 s, which is greater than the time required for the longest weld. The relay shuts off power in case of a faulty weld or a failure of the electronic circuit.

The initiating switch S is a foot switch by means of which relay coil RL_1 is energized. Capacitor C_1 serves to quench the resultant transient so that it does not affect the thyratrons in another part of the circuit. Holding relay coil RL_2 is energized by the momentary current through the contacts of RL_1; RL_2 is held by its own contacts. Contacts of RL_1 also start the timing of safety relay TD_2, which, in turn, energizes RL_5, thus starting the radio-frequency generator. Holding relay RL_2 permits the operator to remove the foot from the initiating switch during the weld.

The type 918 phototube is used because of its infrared-sensitive S-1 surface. The fixed bias on the 6J7 tube, operated at cutoff, is about −6.0 V. At the first peak of 1.7 V, the anode voltage of the 6J7 drops to about 150 V. The drop in signal of 1.0 V at the time of the copper fusion causes the anode voltage to rise to about 220 V. A change of 70 V is realized.

The temperature drop of the weld is rapid (0.1 s), and the output of the differentiator circuit C_3R_7 is a positive pulse of 44 V magnitude, more than enough to fire T_3. When T_3 fires, the anode current energizes relay RL_3 and one set of contacts interrupts the anode current. If the grid is still sufficiently positive, the 2050 tube will reignite and then again be interrupted in the fashion of a relaxation oscillator.

Potentiometer R_9 controls the sensitivity of the tube. When T_3 fires, the grid current during conduction lowers the terminal grid voltage to a value less than the bias. Capacitor C_3 holds this less negative value over into the period when the contacts of RL_3 reapply anode voltage. Thus, after the bias is decreased to the value at which the circuit starts to oscillate, a large increase in bias at the potentiometer is necessary to stop the oscillation.

Specifically, with R_6 short-circuited out, the 2050 tube starts to oscillate at a bias of −3 V and stops oscillating at −22 V. The addition of R_6, however, reduces this lower limit to −10 V, which is satisfactory. Resistor R_6 isolates the differentiator from the grid current of the thyratron.

Capacitor C_4 stabilizes the operation of T_3 by bypassing any transient pickup. The neon tube T_4 indicates when this 2050 tube fires.

Relay RL_3 has a second set of normally closed contacts in series with the coil of RL_2. Because the latter is a holding relay, one operation of RL_3 causes RL_2 to deenergize and remain deenergized. Relay RL_2, therefore, may be energized by the initiating switch S_1 and deenergized by a drop in intensity of radiation on the phototube.

It has been found desirable to allow the generator to remain on for a short period after the temperature drop occurs. This additional time allows the copper to flow around the entire periphery, making a tight seal. A delay of 0.6 s has been determined to be optimum for the purpose. This delay is produced by a commercial thyratron time-delay relay consisting of thyratron T_4 and associated circuit. At the end of the delay period, the current through T_4 energizes relay RL_4, which in turn shuts off the generator through RL_5.

TRANSISTOR APPLICATIONS

85. The **transistor** is an electronic amplifier in which the amplified current is emitted and transported within a semiconducting solid. In the transistor, the current carriers are emitted over a range of temperature which includes room temperature, whereas the cathode of an electron tube must be heated to many hundreds of degrees above room temperature. Moreover, the applied voltage needed to cause conduction in the transistor is small compared with the voltage (B supply voltage) required by electron tubes. The absence of cathode-heating power and the low operating voltage make the transistor a highly efficient device, particularly for amplifying weak signals, and permit large aggregations of transistors to be operated in a small space without generating excessive heat. For information on semiconductor materials. See Sec. **4,** Pars. **589** to **596.** Transistors are treated in Sec. **5,** Pars. **183** to **186.**

86. Basic Transistor Circuits. The mechanism by which transistors amplify is illustrated in Fig. 28-42. The transistor shown consists of a thin slab of germanium (or silicon), with electrodes on opposite faces of the slab. Connections are made to each electrode and to the slab itself. One electrode is known as the emitter, corresponding to the cathode of an electron tube. The opposed electrode is the collector (anode or plate in a vacuum tube). The semiconductor itself is known as the base and corresponds to (although not strictly analogous to) the control grid of a vacuum tube. The arrow on the emitter symbol indicates the conventional direction of current flow (direction of positive charges, opposite to direction of negative charges) in the emitter circuit.

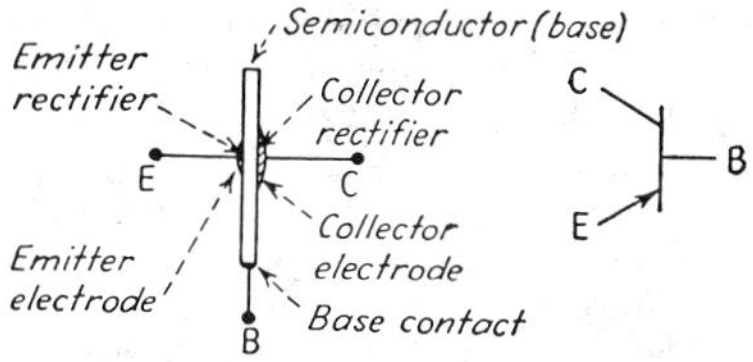

FIG. 28-42. Typical bipolar transistor (left) and circuit symbol.

The transistor consists essentially of two rectifiers (one formed by the emitter and base, the other by the collector and base), which are biased by external batteries so as to present widely dissimilar resistances. Power amplification occurs, as a result of the disparity in the emitter and collector resistances, when an input signal is applied between two of the electrodes and the output signal is taken between one of these and the remaining electrode. The three circuit arrangements possible with these connections are designated, by the electrode which is common to input and output (Fig. 28-43), as the common-base, common-emitter, and common-collector circuits.

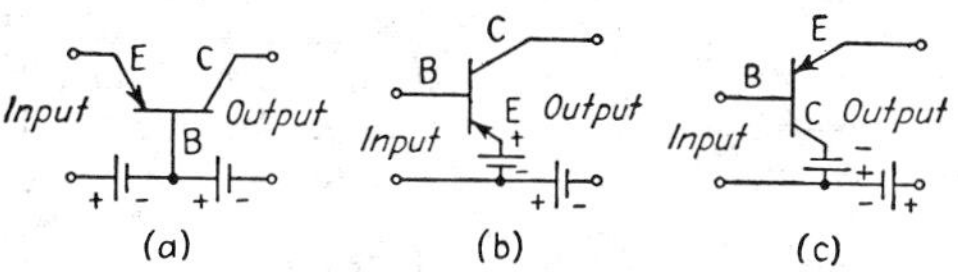

FIG. 28-43. Transistor amplifier connections: (*a*) common base; (*b*) common emitter; (*c*) common collector.

The **common-base transistor circuit** corresponds to the grounded-grid vacuum-tube circuit. It has a current gain less than unity, low input impedance, and high output impedance; it introduces no phase reversal between input and output; and it is comparatively stable against changes in the transistor parameters with temperature.

The **common-emitter circuit** corresponds to the grounded-cathode electron-tube circuit and (like the latter) is the most widely used form. It has a current gain above unity, a power gain generally higher than the common-base circuit, higher input impedance and comparatively lower output impedance than the common-base connection, and it produces a phase reversal between input and output.

The **common-collector circuit** corresponds to the grounded-anode (cathode-follower) vacuum-tube circuit. It has substantial current gain, voltage gain close to but not equal to unity, and input impedance substantially higher than its output impedance.

AMPLIFICATION BY TRANSISTORS

87. Transistor Amplification. The basic principle of the transistor amplifier is best illustrated by the **common-base connection** (Fig. 28-43). The equivalent circuit at low frequencies, shown in Fig. 28-44, consists of three resistances, arranged in the form of a T. One of the resistances is associated with a current generator. The **emitter resistance** r_e is related to, but smaller than, the "forward" resistance of the rectifier formed by the emitter and the base. The polarity of the emitter bias battery V_{ee} is so chosen that current flows through the emitter rectifier in the low-resistance direction, and this establishes a low resistance r_e through which the variational input signal current i_i passes. A typical value of r_e is about 25 Ω.

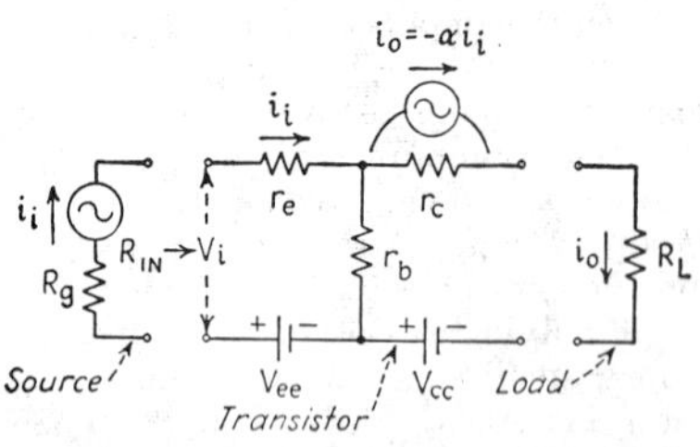

FIG. 28-44. Equivalent circuit, common-base connection.

The **collector resistance** r_c is related to, but smaller than, the resistance of the rectifier formed by the collector and the base, but in this case the polarity of the collector bias battery V_{cc} is such as to cause current to flow in the reverse direction through the rectifier, thereby establishing a high resistance within which the variational output signal current i_o is generated. The power needed to generate the additional output power is taken from the collector battery. A typical value of r_c is about 1 MΩ.

The **base resistance** r_b is the sum of the so-called "spreading resistance" and an approximately equal resistance determined by the geometry and current gain of the transistor. It is ohmic resistance; a typical value of r_b is about 200 Ω.

When the input signal voltage v_i is applied between emitter and base, a current i_i of minority carriers is injected by the emitter into the base semiconductor. These carriers pass through the base (by a process of diffusion) to the collector. On the assumption that the lifetime of the carriers is sufficiently high, only a small number of the carriers are removed by recombination, and a large fraction of the input current reaches the collector, where it emerges as the output current. To account for the passage of this slightly diminished current through the high collector resistance r_c, the collector resistance must be considered to be associated with a current generator, as shown. The fraction of the input emitter current reaching the collector is defined as the common-base current gain α, that is, $i_o = -\alpha i_i$. In the foregoing discussion, α has been taken as less than unity. This discussion does not apply to the point-contact transistor, the first type to be developed. It has an alpha value greater than unity, but it is now obsolete. It had two inherent shortcomings, high noise level and circuit instability.

It should be emphasized that the transistor amplifier is unlike a vacuum-tube amplifier in that it is not a voltage-operated device and its voltage gain, while of incidental interest, is not a fundamental quality. Rather, since both the current and voltage at the input are significant, the transistor is essentially a power amplifier. The power gain is the ratio of the power in the load $i_o^2 R_L$ to the power delivered to the transistor $i_i^2 R_{in}$.

88. Common-base Circuit. On the assumption that the load resistance R_L is small compared with the collector resistance r_c, simple expressions for **current gain, input**

resistance, voltage gain, and power gain of the **common-base circuit** are as follows:

Current gain: $$G_i = -\alpha \tag{28-14}$$

where the negative sign signifies that the output and input currents are in phase.

Power gain: $$G_p = \frac{\alpha^2 R_L}{r_{\text{in}}} \tag{28-15}$$

Voltage gain: $$G_v = \frac{G_p}{G_i} = \frac{\alpha R_L}{r_{\text{in}}} \tag{28-16}$$

Input resistance: $$r_{\text{in}} = r_e + r_b(1 - \alpha) \tag{28-17}$$

89. Common-emitter Circuit. The low-frequency equivalent circuit of the common-emitter connection is shown in Fig. 28-45. Here the input current to the amplifier is the base current, which is the difference between the emitter and collector currents. Consequently, i_i in the common-emitter amplifier is much smaller than in the common-base connection. Therefore, for a given value of collector current, the current gain is correspondingly high. With the emitter current i_e taken as a reference, the collector current is αi_e (having the same significance as in the common-base circuit), and the base current is $(1 - \alpha)i_e$. As the common-base current gain approaches unity, therefore, the common-emitter-circuit current gain approaches infinity. With $\alpha = 0.98$, as in the previous example the current gain of the common-emitter amplifier is $0.98/0.02 = 49$.

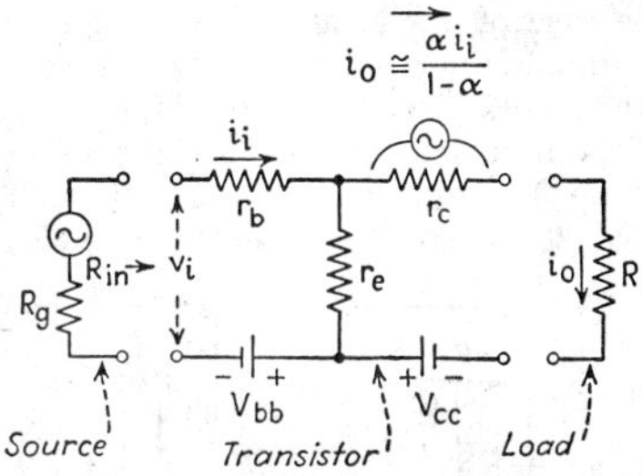

FIG. 28-45. Equivalent circuit, common-emitter connection.

When the load resistance R_L is small compared with $r_c(1 - \alpha)$, the following approximate expressions apply to the common-emitter circuit:

Current gain: $$G_i = \frac{\alpha}{1 - \alpha} \tag{28-18}$$

Power gain: $$G_p = \left(\frac{\alpha}{1 - \alpha}\right)^2 \frac{R_L}{r_{\text{in}}} \tag{28-19}$$

Voltage gain: $$G_v = \frac{\alpha}{1 - \alpha} \frac{R_L}{r_{\text{in}}} \tag{28-20}$$

Input resistance: $$r_{\text{in}} = \frac{r_b + r_e}{1 - \alpha} \tag{28-21}$$

An important advantage of the common-emitter circuit appears in the expression for power gain [Eq. (28-19)] when the load resistance R_L is made equal to the input resistance r_{in}. This is the so-called direct-cascade circuit, in which no impedance transformation (no transformer) is used between stages. In the common-emitter connection, the power gain for such an impedance match is approximately $\alpha^2/(1 - \alpha)^2$, which has a high value for values of α near unity (power gain 2,400 or 33.8 dB for $\alpha = 0.98$). Under the same condition, the direct-cascade power gain of the common-base circuit is α^2, which is less than unity.

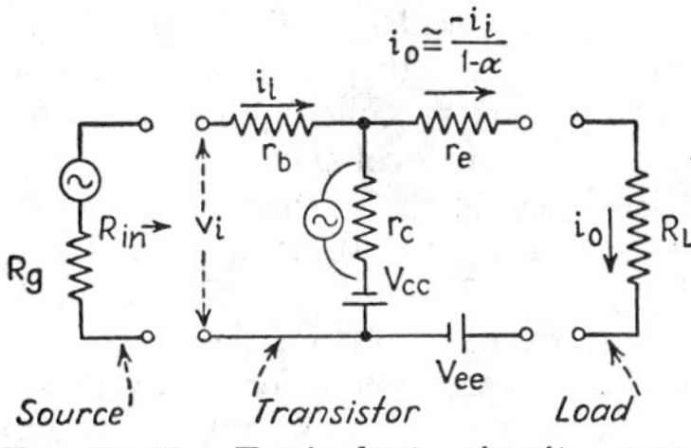

FIG. 28-46. Equivalent circuit, common-collector connection.

90. Common-collector Circuit. The low-frequency equivalent circuit of the common-collector connection is shown in Fig. 28-46.

This circuit, like the common-base connection, does not reverse the polarity of the signal passing through it. The current gain and power gain are equal; i.e., the voltage gain is approximately unity. When the load resistance R_L is small compared with $r_c(1 - \alpha)$, the approximate expressions for the common-collector circuit are

Current gain: $$G_i = \frac{1}{1 - \alpha} \tag{28-22}$$

Power gain: $$G_p = \frac{1}{1 - \alpha} \tag{28-23}$$

Voltage gain: $$G_v = -1 \tag{28-24}$$

Input resistance: $$r_{\text{in}} = \frac{R_L}{1 - \alpha} \tag{28-25}$$

91. Operating Point and Dynamic Characteristic of Transistor Amplifiers. As in the case of vacuum-tube amplifiers, transistor operation can be specified in terms of characteristic curves. These are customarily plotted in terms of the collector current i_c and voltage v_c (corresponding to the i_p and e_p curves of a triode) for discrete values of the input current (i_e in the common-base connection, i_l in the common-emitter connection). Typical examples of such curves for the two connections are plotted in Figs. 28-47 and 28-48 respectively.

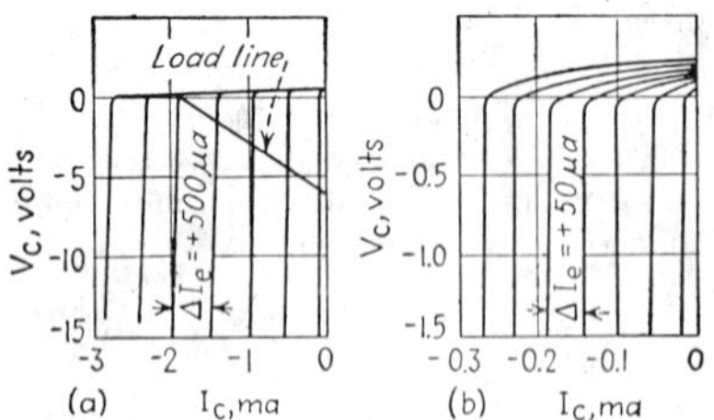

Fig. 28-47. Collector characteristics and typical load line, common-base connection.

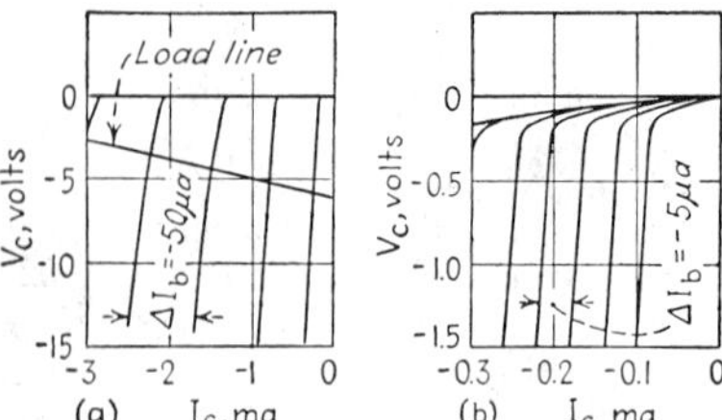

Fig. 28-48. Collector characteristics and typical load line, common-emitter connection.

The dynamic characteristic of the amplifier falls on the load line, a line having a slope in volts per ampere equal to the load resistance R_L and intersecting the v_c axis ($i_c = 0$) at the value of the collector bias battery voltage V_{cc}. The operating point of the amplifier is set by the quiescent (zero-input-signal) value of the input current, and the active region of the dynamic characteristic is contained within the maximum and minimum values of the input current corresponding to the superposition of the input signal on the operating point. Typical dynamic characteristics are shown in the figures.

An attribute of the $v_c - i_c$ curves of the transistor, of outstanding importance in comparing transistors with vacuum tubes, is the fact that the curves lie parallel to one another and equally spaced (thus indicating normal, distortion-free amplification) at very low values of v_c and i_c typically down to $v_c = 0.1$ V and $i_c = 0.05$ mA. This means that the transistor amplifier is capable of maintaining high power amplification (30 to 45 dB/stage) with very low applied bias voltage in the collector circuit, with corresponding high power efficiency in the amplification of low-level signals.

The operating point of a transistor amplifier is subject to drift from a number of causes, the most important of which is the change in collector saturation current I_{co} which occurs as the temperature of the transistor increases. The amplification depends also on the stability of the current gain α, which may decrease during the life of the transistor, particularly if the transistor is subject to storage or operation at excessive temperatures. The common-emitter and common-collector circuits, whose power gain depends on the factor $1 - \alpha$, are particularly sensitive to alpha instability. The common-base circuit is much less so.

At high frequencies, the power gain of a transistor amplifier is limited by two mechanisms: (1) the fact that the current gain of the transistor (fraction of the emitter current reaching the collector) decreases as the quarter period (90 elec deg) of the signal oscillation approaches the transit time of the minority carriers, and (2) the shunting effect of the unavoidable capacitance c_c associated with the base-collector rectifier. (A similar capacitance of larger magnitude is associated with the base-emitter rectifier, but it is shunted across a very low resistance r_e; so its effect is not felt.) The first effect is expressed by stating the frequency f_α at which the common-base current gain α falls to 0.707 of its low-frequency value (3 dB loss of current gain).

92. Transistor Circuits. The general functions of electron-tube circuits can be performed by corresponding transistor circuits. These include amplifiers of various types, demodulators, sinusoidal and relaxation oscillators, switching and counting circuits, and pulse generators. Figure 28-49 shows a number of basic transistor circuits.

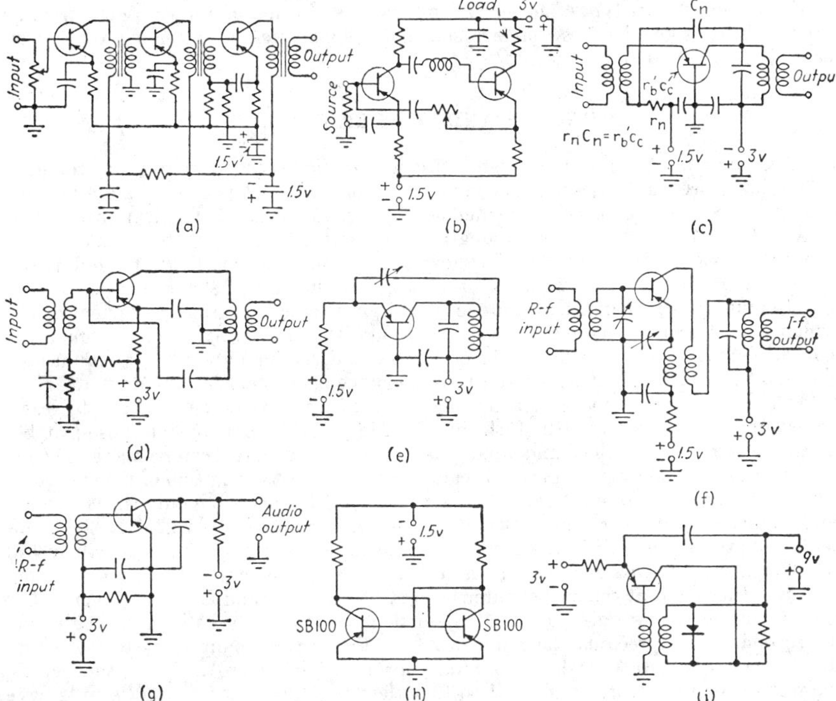

FIG. 28-49. Typical transistor circuits: (*a*) hearing aid, audio amplifier; (*b*) video amplifier, variable bandwidth; (*c*) bandpass r-f amplifier, common-base; (*d*) bandpass r-f amplifier, common-emitter; (*e*) r-f oscillator; (*f*) frequency converter (self-oscillating mixer); (*g*) amplitude-modulation detector; (*h*) direct-coupled switching circuit; (*i*) pulse generator.

The load resistance can be increased to obtain higher power output, provided that the collector bias voltage is large enough to draw the maximum allowable value of collector current with the high value of R_L. The collector voltage cannot be increased without limit, since at low values of collector current (during the low-current half cycles of the output signal) the reverse voltage across the collector rectifier may become high enough to exceed the inverse-voltage rating of the rectifier, with consequent distortion and possible destruction of the transistor.

As in vacuum tubes, the requirements for high power tend to be opposed to those for high-frequency operation. Power output levels available from a single transistor at frequencies above 100 Mc/s are limited to date to 10 W or less, whereas up to about 50 W can be generated by a pair of power transistors in a Class B stage operating over the audio-frequency spectrum. Higher power levels, of the order of tens of watts, are possible in very-low-frequency operation, as in the servomechanisms of an autopilot.

The power ratings of transistors are variously described. The factor most generally applicable, as in a vacuum tube, is the rated collector dissipation in watts or milliwatts. The allowable dissipation in small-signal transistors ranges from about 10 to about 150 mW, depending on the size of the emitter and collector electrodes. The maximum power output is 50% of the collector dissipation in Class A service.

93. Types of Transistor Structure. Many variations of the manufacturing processes of transistors are possible and are in use today. They are classified into types by process or by geometry of construction. There are rate grown, alloy junction, grown junction, fused junction, alloy diffused, double diffused, grown diffused, mesa, epitaxial, microalloy, and planar types, and even combinations of some of these. Most transistors are "small-signal" types that are similar in application to receiving tubes (see Sec. **5**, Pars. **185** and **186**).

COUNTERS AND DIVIDER CIRCUITS

94. Counters. Tube and transistor circuits for high-speed counting and for controlling high-speed processes are now very important tools. These circuits are in common use in computers, in radiation detectors, in communication applications, and in many industrial processes where timing and synchronization of events are important. Basically an electronic counter is the same in function as its mechanical counterpart. It is a storage device. It is quiescent until it gets an impulse of some sort; it reacts to this stimulus, and it maintains whatever condition it is left in after the arrival of the stimulus until another impulse is received. But it is inherently faster in operation than a mechanical counter since it has no inertia; and, having no weight, it cannot load the circuit to which it is attached and thus affect the operation of that circuit.

An electronic counter acts as a slowing-down device. It can be arranged to react, for example, to every 2nd, 4th, 16th, or 100th impulse so that, by interposing it between the events to be tallied and a mechanical counter, the latter can be made to count events occurring 100 times as fast as it can handle without the addition of the electronic circuit. If the arm of a mechanical counter is tripped 23 times, the number 23 appears on its indicator and will remain until another event is counted. It will store this number indefinitely. Of course, there is a limit to the number of impulses it will record before it starts counting over again. At this point a second counter comes into action, recording the fact that the first counter has reached its limit and is starting over. The second counter is now quiescent until the first has again reached its limit, whereupon the second counter records the number 2. In the decade counting system the first counter can count up to 9, the next impulse bringing up 0 (zero). When this occurs, the second counter records number 1 so that the two will indicate that 10 events have taken place. Two counters can tally 99 impulses before they repeat.

95. Electronic Counters. In electronic counters the actuating impulse is a voltage or other electrical stimulus and not a mechanical movement. Because the impulse is electrical, wires can connect the counter with apparatus at the place where the events occur, and a huge number of events per second can be recorded at almost any distance. An electronic counter is usually much more complex than a simple mechanical system, but it is faster and just as accurate. For each count on a mechanical counter, a mechanical force must be applied to the arm of the counter. Many processes do not have sufficient physical force to trip the counter arm, thereby prohibiting the use of mechanical counters.

Electronic counters are built up from a basic unit which has only two degrees of freedom, like an on-off switch. If two such switches are mechanically coupled so that the second moves only when the first is turned on (and does not move when the first is turned off), then the second switch moves only half as often as the first. The counting rate has been cut in half. Addition of more switches can divide the counting rate still further.

Before any phenomenon can be counted by electronic means it must be converted into or cause a pulse of electrical energy for actuating the counter circuits. Many ingenious methods have been devised for converting mechanical processes into electrical pulses. A few are shown in Fig. 28-50. In *a*, a system is shown where small pieces are allowed to fall upon a plate which is attached to a microphone. The impact of the piece upon the plate causes a sound, which, by means of the microphone, is converted into an electrical voltage. In *b*, the assembly-line motion is sufficient to operate a switch as each piece passes a counting point. In *c* and *d*, a phototube and light-source setup is used. In *c*, the light beam is interrupted as objects pass the inspection point, whereas, in

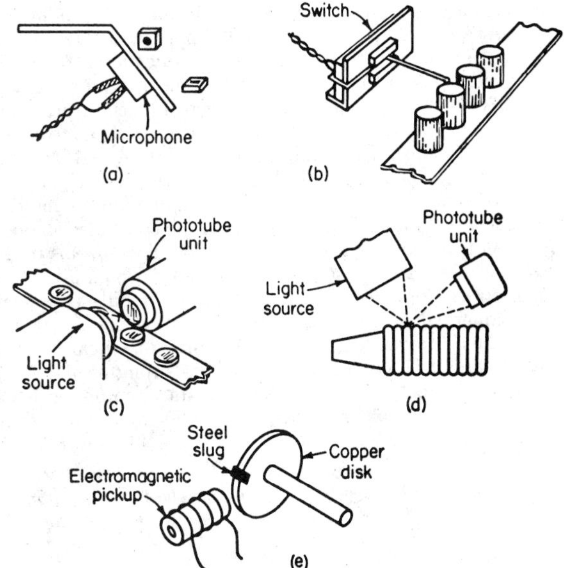

Fig. 28-50. A microphone or switch may be arranged to convert a mechanical phenomenon into an electrical signal for counting purposes, as shown in *a* and *b*. In *c* and *d* a phototube circuit detects movement and creates the counting signal. A means for obtaining electrical signals for counting revolutions of rotating machinery is shown in *e*.

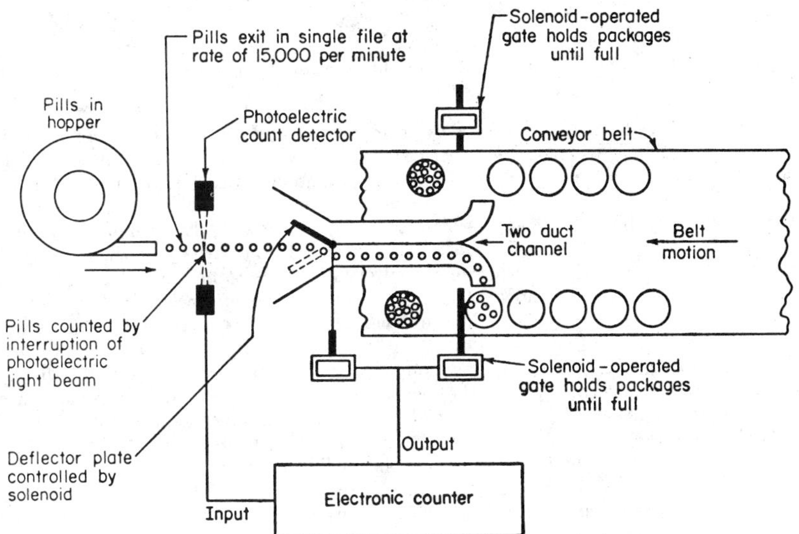

Fig. 28-51. Photoelectric setup for counting pills and portioning them into lots of a predetermined number.

d, a small amount of light is reflected onto the phototube as each paper drinking cup passes a certain point. A means for counting revolutions of a rotating disk is shown in *e*. The disk can be attached to any piece of rotating equipment. Each time the steel slug passes the electromagnetic pickup, an electric signal is formed.

96. Counter Applications. Electronic counting, timing, and control instruments are mainly used in industry to measure and control such things as quantity, length, time, sequence, frequency, and revolution. A typical application is illustrated in Fig. 28-51. Here pills are being counted, but any object large enough to block a portion of the light in the phototube system could be handled in the same manner. When the desired number of pills falls into the container on one side, the counter sends a signal to the deflection door, which diverts the flow of pills to the other side of the moving belt. At the same time a solenoid is operated which allows the filled container to pass to a subsequent operation. Thus the flow of pills is not interrupted, and the passage of containers along the belt is continuous.

The system of Fig. 28-52 is often used to measure linear footage in a continuous process. The friction wheel or roller is brought in contact with the moving material, and its diameter is chosen so that its circumference bears some integral relationship to the desired linear dimension of the material. Thus the counting (measuring) of these dimensions is simple.

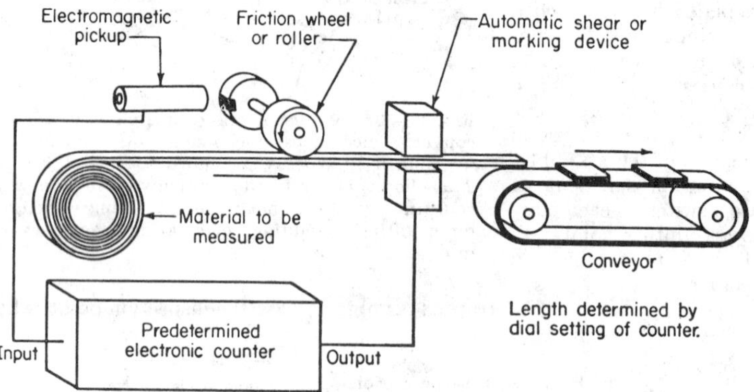

Fig. 28-52. Electromagnetic pickup for counting the revolutions of the wheel geared to the roller riding on the strip to actuate the cutter.

Time and speed can be measured with extreme accuracy by the use of counters. Very high speeds (thousands of feet per second) and very short lengths of time (down to a few microseconds) can be measured as readily as low speeds and longer lengths of time.

In measuring time intervals, the counter circuit is ordinarily used in conjunction with an extremely stable high-frequency oscillator. If the output of the oscillator is applied to the input of the counter for the interval of time to be measured and the frequency of the oscillator is known, then, by counting, the number of pulses or cycles the oscillator produces during the time interval can be determined. This is the system used to measure high velocities. For example, a shell is fired past two points. As it passes the first point, a pickup arrangement starts a high-frequency oscillator. As the shell passes the second point, the oscillator is turned off. All that is needed for accurate determination of shell velocity is an accurate measure of the distance between the two points, the frequency of the oscillator, and an accurate count of the number of cycles that occurred during the time it took the shell to pass from one point to the other.

97. Dividers. There are many industrial applications in which the basic counter circuits can be employed where the job of actually counting something does not occur at all. In other words, the counting circuits have uses other than in counting. Basically they are "dividing" circuits; they give output impulses which are some integral submultiple of the input impulses.

The basic circuit is like an electrical switch in that it has two degrees of freedom; it can count up to 2 before it repeats. That is, it divides the number of incoming impulses by 2, thus cutting in half the rate at which the final tally must be made. In Fig. 28-53*a*, tube *A* is either conducting, or it is not. When it conducts, tube *B* does not. If something makes tube *B* conduct, then tube *A* goes out (ceases to conduct). Suppose that before the first impulse comes along tube *A* is conducting, the lamp in its cathode circuit is lighted, and tube *B* is not conducting. Now the first impulse can be arranged to make tube *B* conduct, whereupon tube *A* and its light go out. The next impulse will make tube *A* go on and tube *B* go out. Thus the lamp in the cathode circuit of tube *A* indicates every other impulse and actually notes the arrival of the even-numbered events, i.e., the second, fourth, sixth, etc.

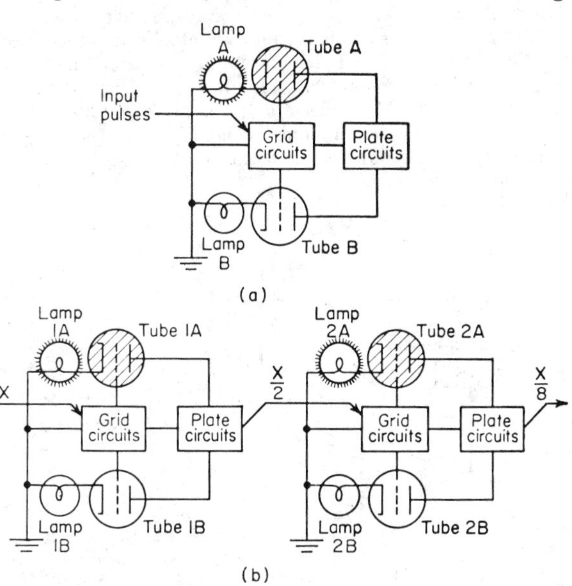

FIG. 28-53. Basic flip-flop circuits, showing counter indication method and cascading of basic circuits for counting higher than the basic 2.

Now a second set of two tubes may be connected in place of the lamp in the circuit of tube *A* and arranged so that a stimulus is passed onto the second set of tubes only when tube *A* of the first set is changed from a nonconducting to a conducting condition. The important thing in these circuits is not actual conduction or lack of it but the shift from one state to the other.

Refer to the four-tube system of Fig. 28-53*b*. Consider V_{1B} only as a switch to turn on or off tube V_{1A}. V_{1A} then is the critical counting tube. The second set of tubes is connected to tube V_{1A} and registers only when this tube is changed from a nonconducting to a conducting condition. Nothing happens in the second set of tubes when V_{1A} goes out, only when it goes back on. Under these conditions V_{2A} will register the arrival at V_{1A} of every fourth impulse. And if a third set of tubes is connected to V_{2A} or V_{2B}, it will record every eighth stimulus that has caused something to happen in the circuits of the first set of tubes.

In Fig. 28-53*b* the initial condition is that tubes V_{1A} and V_{2A} are conducting. The first impulse causes tube V_{1B} to conduct and V_{1A} to go out. But since V_{2A} receives an input impulse only when tube V_{1A} reverts to its original conducting state, the arrival of the second pulse at the input of the first two-tube circuit causes no change in the condition of the second set of tubes.

Tubes

Impulse	V_{1A}	V_{1B}	V_{2A}	V_{2B}
0	On	Off	On	Off
1	Off	On	On	Off
2	On	Off	Off	On
3	Off	On	Off	On
4	On	Off	On	Off

The second impulse cuts off V_{1B} and fires V_{1A}, which fires V_{2B} and cuts off V_{2A}. A counter in this tube circuit will not indicate until this tube again conducts—which it will do upon the arrival of the fourth impulse at the input to the first set of tubes.

The manner in which tube V_{2A} counts every fourth impulse can be seen from the table. The counter in V_{2A} circuit is set at zero at the beginning when tubes V_{1A} and V_{2A} are conducting. Nothing happens to the counter when tube V_{2A} cuts off; it indicates only when it goes back on. A third counter circuit, connected to receive a pulse every time V_{2B} returns to its original state, would return to *its* original condition after every eighth pulse, and so on.

98. Eccles-Jordan Pair. A simple circuit of this type is shown in Fig. 28-54. It is known as the Eccles-Jordan circuit and consists of two vacuum triodes. The circuit has two completely stable sets of operating conditions, and it has the further property that it can be switched from one condition to the other by the application of an electrical pulse. The switching action is almost instantaneous. A detailed study of this circuit will be made.

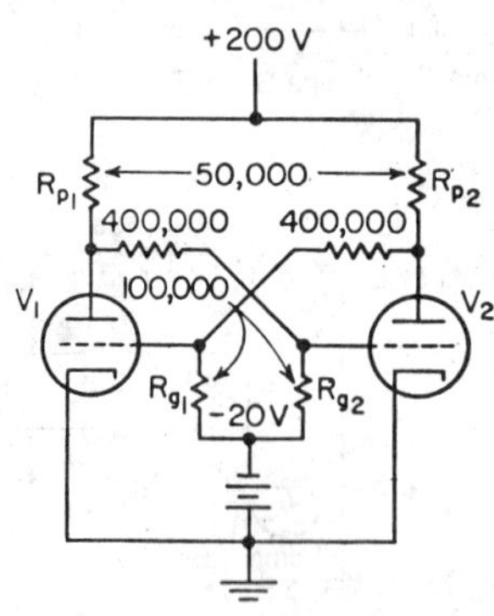

FIG. 28-54. Basic Eccles–Jordan pair has two completely stable sets of operating conditions.

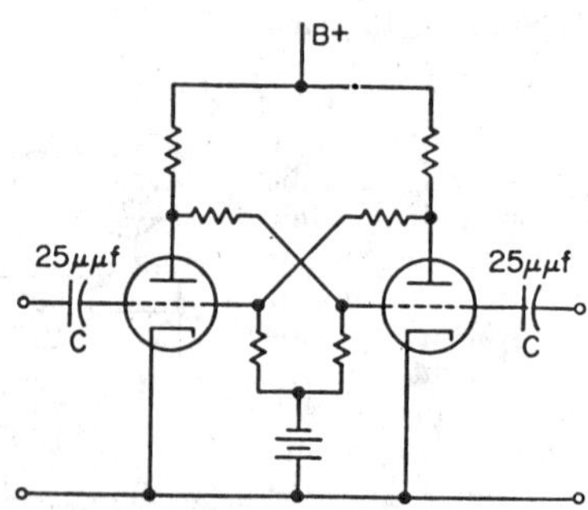

FIG. 28-55. Eccles–Jordan pair with double-controlled triggering connections for changing from one stable condition to the other.

The circuit shown in Fig. 28-55 is symmetrical; the components in the circuit of one tube are identical to those in the other circuit. At the beginning, assume that a 2-mA current is flowing through each triode. The resulting voltage drop in each plate resistor will be 100 V, and the voltage at each plate will be 100 V positive with respect to ground, or cathode, potential.

Suppose that the transconductance of each tube is 1,000 μmhos (for each volt change in grid voltage the plate current will change 1 mA). If perfect circuit symmetry existed, the currents in the tubes would remain equal. However, if some change takes place that causes the plate current of V_1 to increase slightly, its plate voltage will drop accordingly. Say the plate voltage of V_1 drops 0.1 V. The grid of V_2 experiences a drop of 0.02 V, because it is connected on the tap of a voltage divider circuit formed by two resistors (400,000 and 100,000 Ω) in such a way that it experiences one-fifth of any changes at the plate of V_1.

The plate current in V_2 will then drop, because its grid becomes more negative, $0.02 \times 1{,}000 \times 10^{-6}$, or 0.02 mA. Then the plate voltage of V_2 will rise $0.02 \times 50{,}000$, or 1 V. As before, because of voltage divider action this time involving R_{P2} and R_{G1}, one-fifth of this voltage, or 0.02 V, rise will be felt at the grid of V_1. This voltage rise causes the plate current of V_1 to rise by 0.2 mA. The resulting plate voltage drop is $0.2 \times 50{,}000$, or 10 V. The original change has thus been augmented.

The current in V_1 thus continues to increase (and that of V_2 to decrease) until stability occurs. Assume the full conducting plate current to be 3 mA. Then the plate voltage of V_1 will be 50 V, and there will be 70 V across the 400,000- and 100,000-Ω resistors, allowing 0.14 mA to flow through the combination. The grid voltage with respect to the cathode voltage is then $20 - 0.14 \times 10^{-3} \times 100{,}000$, or -6 V. Current through the plate-load resistor of V_2 is found by dividing 220 V (200 + 20) by 550,000 Ω (50,000 + 400,000 + 100,000) to be 0.4 mA, on the assumption that the tube is completely cut off. This establishes the plate of V_2 at a potential of 180 V.

99. Triggering. To flip the circuit to its other set of stable operating conditions, the original balance condition must be reached and then a little more voltage added so that the circuit behaves in the reverse of the example given above. This change-initiating process is known as triggering, and it may be accomplished in a number of ways. For example, by applying strong enough negative voltage pulses on the plate of the conducting tube, triggering could be accomplished. However, because of the large magnitude of the required pulse, this system is seldom used. The generally accepted method of triggering the Eccles-Jordan circuit is to apply a pulse of voltage to one or both of the grids, negative to the conducting grid or positive to the grid of the cutoff tube. This could be accomplished by short-circuiting the grids to the cathode. The circuit diagram for double-controlled triggering is shown in Fig. 28-55. Either side may be triggered by the application of a pulse of the proper polarity and sufficient amplitude.

A circuit for single-controlled triggering is shown in Fig. 28-56. The only modification to the basic Eccles-Jordan circuit is the addition of C_A and C_B, called commutating capacitors. Assume the circuit to be in a stable condition, with V_1 conducting and V_2 cut off. If a negative pulse is applied to the grids of the tubes, it will have no effect on V_2, since that tube is already cut off. The negative pulse will, however, decrease the plate current flowing through V_1, and its plate voltage will rise in the form of a positive pulse. This entire positive pulse is applied to the grid of V_2, since the voltage across C_A cannot change instantaneously. The positive pulse thereby delivered to the grid of V_2 is great enough to overcome the original negative input pulse; so V_2 will start conducting. When V_2 conducts, the voltage at its plate decreases rapidly. This decrease is felt at the grid of V_1 as a negative pulse which is sufficiently strong to cut that tube off. Both negative and positive pulses will trigger this circuit if they are large enough. Positive pulses, however, must be larger than negative pulses for proper triggering action.

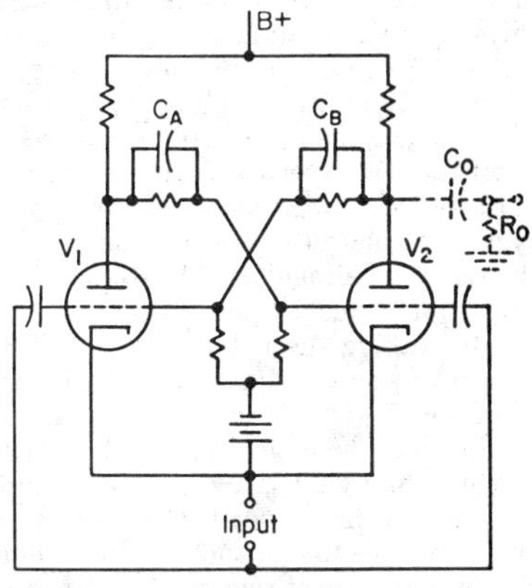

FIG. 28-56. Addition of commutating capacitors C_A and C_B permits the use of a single triggering input.

100. Failure to Trigger. When a circuit fails to trigger upon the application of a pulse of the proper polarity, several factors may be responsible. The pulse may have insufficient amplitude to push the circuit past its balanced condition. The pulse time must be less than the time constant of the input *RC* combination, or the capacitor will charge and the grid will not receive enough triggering voltage.

The size of the input capacitor is subject to limitations. If it is small compared with the input capacitance of the tube, the pulse voltage is divided and may be too small at the tube. Also, the smaller the input capacitor, the steeper the required pulse front edge. In other words, the trigger voltage must rise to the required value almost instantaneously. Steep-front pulses are often hard to produce and are inconvenient to handle.

If the input capacitor is made too large, it will tend to prevent changes in the grid voltage of one tube when a pulse is applied to the other. In other words, assume V_1 conducting and V_2 cut off and the coupling capacitor of V_1 too large. A positive pulse of sufficient amplitude applied to the grid of V_2 will cause it to conduct, and its plate voltage will drop. This drop would normally be enough to decrease the value of grid voltage at V_1 to a point where it would cease to conduct, but because of the large input capacitor the grid voltage is held fixed, and when the pulse to V_2 ends, the circuit reverts to its original untriggered condition.

There is also a limit to the closeness of adjacent pulses, but in most industrial applications this limitation is of little consequence. By proper choice of components most of these circuits can be made to operate from input repetition rates of 100,000/s and higher.

101. Counter Output. There are several methods of taking output signals from these counter circuits. The method employed depends on the purpose for which the output is used. If the circuit is to be followed by another counter circuit, it is usually desirable to have the output in the form of sharp pulses, one pulse for every two input pulses. For other purposes it might be advantageous to take out a square wave, i.e., a voltage which fluctuates between two definite values, one while the circuit is in one condition and the other when in the opposite state.

Figure 28-56 shows a method for taking either kind of output—the constants C_0 and R_0 determining the nature of the output wave. The plate voltage of V_2 is high when V_2 is cut off and low when V_2 is conducting. If C_0 is quite large, it will tend to maintain a constant voltage drop across itself, and when the plate voltage remains at this raised value for any length of time, the capacitor will eventually charge up (assume a voltage drop) to that value, and charging current will cease to flow. When this happens, the output voltage becomes zero. The output signal has thus suddenly gone positive and then gradually decreased to zero.

If the plate of V_2 stays positive for a brief time only, as would be the case if the input pulses were applied at a fast rate, the capacitor would not have time to charge on each cycle and the output would be almost the same as the plate voltage of V_2. That is, the output voltage would be high while V_2 is cut off and low when V_2 is conducting. This constitutes a square wave.

By making the value of C_0 very small, the charging time may be reduced to a very low value; so the output has the form of short pulses of positive voltage occurring whenever V_2 is cut off. When V_2 starts conducting again, its plate voltage decreases rapidly but the capacitor is still charged to the high voltage. When the capacitor discharges through R_0, a negative output pulse occurs. Also, the choice of R_0 will have an effect on the shape of the output signal. A smaller value of R_0 will allow C_0 to charge and discharge faster and will provide a sharper pulse.

Thus for every two input pulses to the circuit of Fig. 28-56 we have a single output pulse which occurs when V_2 returns to its original condition. Actually, there will be two pulses: a negative pulse when V_2 conducts and the desired positive pulse when it is again cut off. However, either of these pulses can be removed from the output. Subsequent counter stages may be used that are sensitive only to pulses of one type, and in this way either the positive or the negative pulses may be counted. This circuit actually registers every other pulse; it is a device for slowing down the necessary counting rate by two. A second two-tube circuit connected similarly will count every fourth input pulse, and so the number of counts per second can be increased geometrically, while the number of tubes goes up arithmetically. Since both triodes can be included in one envelope (and usually are employed in this way), the number of counts actually possible is 2^N, where N is the number of double-diode tubes employed. Thus five such tubes will count every thirty-second pulse.

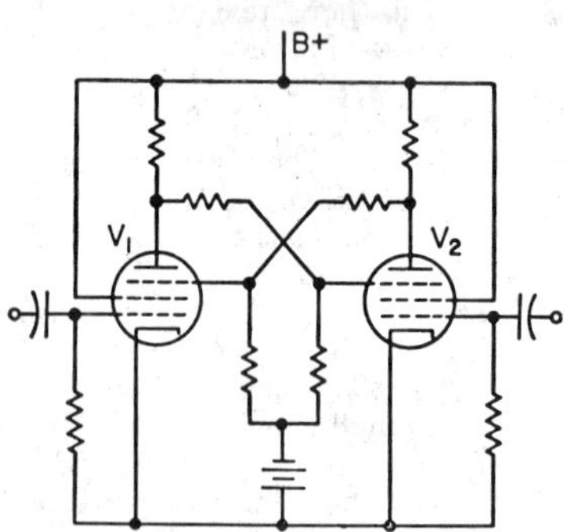

Fig. 28-57. Pentode trigger circuit is sensitive only to negative trigger pulses.

102. Pentode Trigger Circuit. The circuit shown in Fig. 28-57 is sensitive only to negative pulses because, even though a positive pulse is applied to the control grid of the cutoff tube, the suppressor will hold the tube cut off. Interchanging the control and suppressor grid connections makes the circuit sensitive to positive pulses. Changing the cutoff voltage of the tube is accomplished by changing the suppressor voltage. With this circuit, single input triggering is possible as in the triode circuit, and triggering will occur with as little as ½ V.

103. Thyratron Counters. The circuit shown in Fig. 28-58 employs thyratrons and will count up to 4 million pulses/h, or about 1,000/s. Counters of this type are frequently used between high-speed recurring phenomena and mechanical counters which are too slow-moving or require too much energy to be actuated directly by the phenomena being counted. In each unit the steady bias potential applied to both grids is slightly more negative than the critical negative potential required to prevent thyratrons from firing.

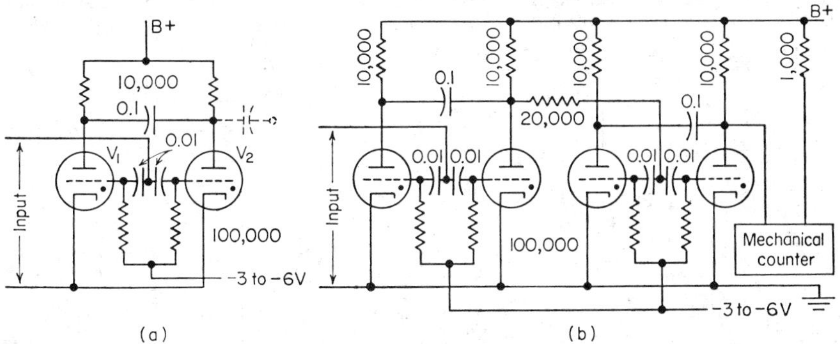

Fig. 28-58. Thyratron counter circuits. In *b* is shown how one thyratron counter circuit may be connected to a following stage.

In explaining the circuit operation, assume that V_1 is conducting. As V_2 is fired by the application of a positive pulse at the input, the plate voltage of V_2 drops. This resulting voltage drop is transmitted through a capacitor to the anode of V_1, stopping the current flow in that tube. Since each thyratron responds to every other input pulse, scale-of-two operation is effected. Two tubes per stage are required. The coupling between stages is accomplished as shown in Fig. 28-58*b*. One of the thyratrons in each stage is coupled to the preceding unit by connecting the input side of the grid capacitors to the plates of one of the thyratrons through high resistances.

104. Ring Counters. Divisions greater than 2 may be achieved by the use of several basic counters in what is known as a ring-counter circuit. A typical ring-counter block diagram is shown in Fig. 28-59, along with an operating schedule for a complete cycle of operation. In this circuit, division by 5 is possible.

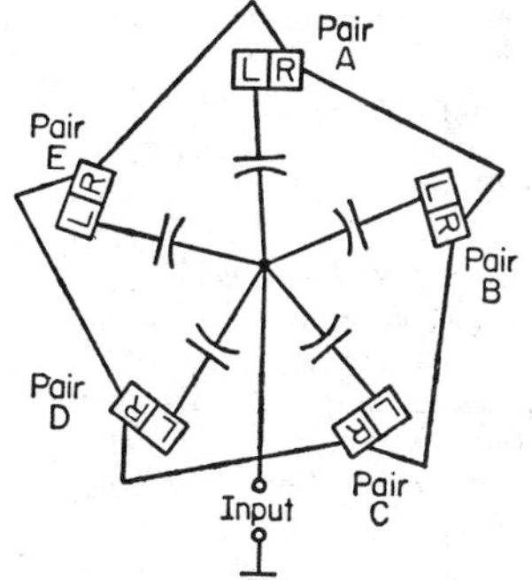

Input pulse number	Pair A		Pair B		Pair C		Pair D		Pair E	
	L	R	L	R	L	R	L	R	L	R
0	X			X		X		X		X
1		X	X			X		X		X
2		X		X	X			X		X
3		X		X		X	X			X
4		X		X		X		X	X	
5	X			X		X		X		X
6		X	X			X		X		X

X indicates conduction

Fig. 28-59. Block diagram of ring counter. The chart shows the condition of each tube section during progression through six counts. The counter shown is a ring of five.

In the initial condition the left side of pair *A* is conducting, while the left sides of the other four pairs are cut off. Upon receipt of the first negative pulse at the input terminals, the left side of pair *A* is cut off, and the voltage rise at its plate causes the right side to conduct. The voltage drop at the plate of the right side, which is coupled to the right side of pair *B*, causes that circuit to flip. The next pulse applied to the input terminal will return pair *B* to its original cutoff condition and will flip pair *C*. Subsequent pulses will pass this flipped condition around the ring so that it is present at each tube once for every five input pulses. Thus an output taken from any one of the five stages will be one-fifth that of the input. Output may be taken from the plate of either of the paired tubes, depending on the polarity of the desired output.

105. The Binary Decade. The circuit shown in Fig. 28-60 has found wide application in industrial electronics because of the ease with which it may be incorporated into such an extensive variety of counting problems.

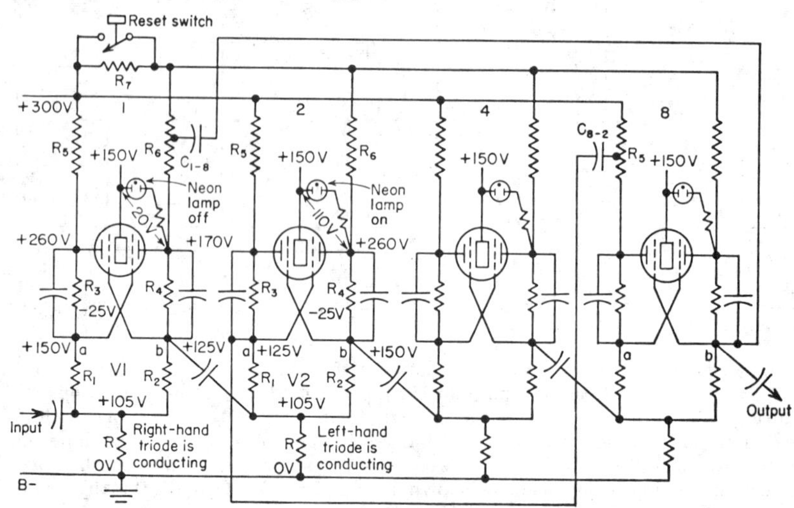

FIG. 28-60. Binary-decade-counter circuit used commercially in many applications. This system provides decade (10) counting with binary indication.

In presenting a number, four neon bulbs are required for each digit of the number. The binary progression 1-2-4-8 is used. This system of number presentation is shown in Fig. 28-61. The numbers corresponding to the lighted neon bulbs are added vertically to get the indicated number. The appropriate neon bulb or bulbs are ignited in their proper order, since two pulses from one circuit cause one pulse to be passed to the succeeding circuit.

The counter decade consists of four stages, each employing a dual-triode vacuum tube (such as the 12AT7 tube) arranged in an Eccles-Jordan trigger circuit in conjunction with an indicating neon bulb. The basic functioning of all stages is alike and is dependent on shifting of operating potentials of each stage back and forth between the two sets of values indicated in stages 1 and 2 of Fig. 28-60.

Considering V_1, note that a fixed potential of +150 V is applied to the cathode of the tube at all times by the power supply. Grid and plate voltages are obtained from the resistance network included in the basic circuit and depend on relative currents through the two branches of the network $R_1R_3R_5$ and $R_2R_4R_6$.

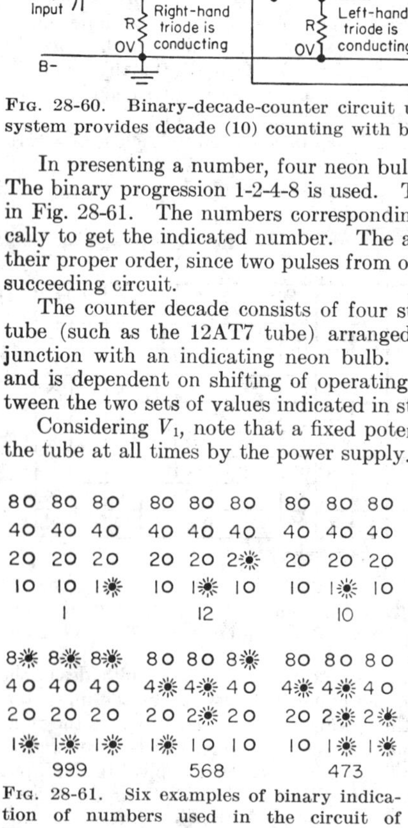

80	80	80	80	80	80	80	80	80
40	40	40	40	40	40	40	40	40
20	20	20	20	20	2✹	20	20	20
10	10	1✹	10	1✹	10	10	1✹	10
	1			12			10	
8✹	8✹	8✹	80	80	8✹	80	80	80
40	40	40	4✹	4✹	40	4✹	4✹	40
20	20	20	20	2✹	20	20	2✹	2✹
1✹	1✹	1✹	1✹	10	10	10	1✹	1✹
	999			568			473	

FIG. 28-61. Six examples of binary indication of numbers used in the circuit of Fig. 28-60.

To explain circuit operation, assume that the right-hand side of V_1 is essentially at cathode potential (zero bias) and that no signal pulses have arrived. At this zero value of bias the right-hand side of V_1 will conduct strongly, and its plate voltage will drop to a low value, about 20 V, lower than the extinction voltage of the neon lamp connected across this particular section. For this condition, then, the neon bulb for the first stage is out.

As the right-hand side of V_1 conducts,

the voltage drop across R_6 is greatly increased. This leaves less voltage for R_4 and R_2, and hence point b is driven negative with respect to the cathode. A shift of 25 V is produced by the resistance values shown.

The grid of the right-hand side of V_1 is connected to point a, which is at cathode potential, as assumed for this explanation and required for full conduction of that portion of the tube. The grid of the left-hand portion of V_1 goes to point b, which is at +125 V with respect to ground and −25 V with respect to the cathode. Thus the cutoff bias is applied to the left-hand side of V_1.

The circuit can remain in this condition indefinitely until a negative pulse is applied to the input circuit. Any positive pulse has no effect, because it is dissipated in the low impedance of the conducting triode section. When a negative pulse arrives at the input circuit, however, both grids swing in a negative direction. There is no direct effect on the cutoff side (the left-hand side), but the negative grid swing on the right-hand grid stops the plate current of that side. The resulting redistribution of the voltage drops across R_6, R_4, and R_2 places point b at cathode potential, as indicated in stage 2. The grid of the left-hand side of V_1 swings positive to zero bias, and that side starts conducting. Point a goes 25 V negative with respect to the cathode, keeping the grid of the right-hand triode negative once it is driven negative by the negative input pulse. Thus the circuit conditions in stage 2 are reversed.

With the right-hand triode of stage 2 cut off, its plate voltage goes up to about 110 V above the striking value of the neon lamp. Thus the lamp glows.

The arrival of a second negative pulse again triggers the stage, the lamp goes out, and the succeeding lamp is lit, since the voltage drop at the plate of the right-hand portion of V_2 appears at the input of the third stage in the form of a negative pulse.

106. Count Indication. A complete counter decade consists of four identical stages with neon lamps as shown in the circuit diagram (Fig. 28-61). These stages are numbered 1, 2, 4, and 8 according to the binary system; the number assigned to each stage is equal to the number of negative pulses required at the input of the decade to make the neon lamp in the stage come on the first time.

In preparation for a count, a momentary displacement of the reset switch inserts a common dropping resistor R_7 in the voltage supply line to all right-hand triodes. This drives the grids of all left-hand sections negative, so that the decade is preset, with all right-hand triodes conducting and all neon lamps extinguished.

When the tenth pulse is being counted, a negative pulse goes to the succeeding decade and at the same time resets the zero condition, where all lamps in the decade are out. In the binary system, however, lamps 2 and 8 should be on, and this is undesirable in indicating counts in the decimal system. It being remembered that pulse 9 was indicated by lamps 1 and 8 being lit, for the count of 10 in the decimal system, lamp 8 must be extinguished and lamp 2 prevented from lighting.

After the ninth pulse the left-hand sections are conducting in both stages 1 and 8. When stage 1 is reversed by the tenth pulse, a negative pulse obtained from a tap on R_6 in stage 1 is fed through C_9 to stage 8, causing point b of stage 8 to swing negative. This drives the grid of the left-hand side of stage 8 negative, causing the flip of stage 8 and extinguishing the lamp corresponding to the number 8. The negative pulse taken from the tap on R_6 must be large enough to cause switching, but not too large, because at the eighth count it must be overpowered by the negative pulse that comes from stage 4 and correctly triggers stage 8 to turn its lamp on.

In a similar manner a positive pulse from R_5 of stage 8 fed through C_{10} to point a of stage 2 overwhelms the negative pulse sent to it from stage 1. Since the lamp corresponding to stage 2 was out for the count of 9, it stays out. This type of counter operates reliably at pulse rates as high as 100,000/s. It is simply necessary to supply negative pulses of approximately the proper shape and amplitude.

107. Another Decade Unit. Another arrangement which produces an output pulse for every 10 pulses it receives is shown in Fig. 28-62. As the diagram indicates, this circuit employs a ring-type circuit, but it will be noticed that a ring-of-5 circuit is used for division by tens. This is accomplished by following the ring-of-5 circuit with a scale-of-2 circuit, so that for every two output pulses from the scale-of-5 circuit only one is applied to the output on the succeeding stages. Thus division by 10 is accom-

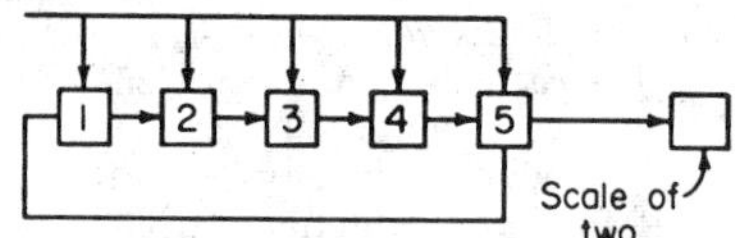

FIG. 28-62. Block diagram of a simple ring-of-five scale-of-two decade counter.

plished. This circuit has the advantage of requiring fewer tubes for decade division. For a straight ring-of-10 circuit, 10 tubes are required. Another advantage of this type of circuit is the fact that neon lights (NE 51) may be arranged to read to an accrued number of input pulses directly, and several such decades may be arranged to indicate counts of very high numbers.

In the circuit diagram of the complete unit shown in Fig. 28-63, the cathode resistor R_2 is chosen to produce the desired voltage on the common right-hand cathode bus with only one right-hand section conducting. Resistor R_1 produces the same voltage drop with four left-hand sections conducting. The left- and right-hand cathodes are split in this manner to assure that the odd condition will exist in only one element at a time.

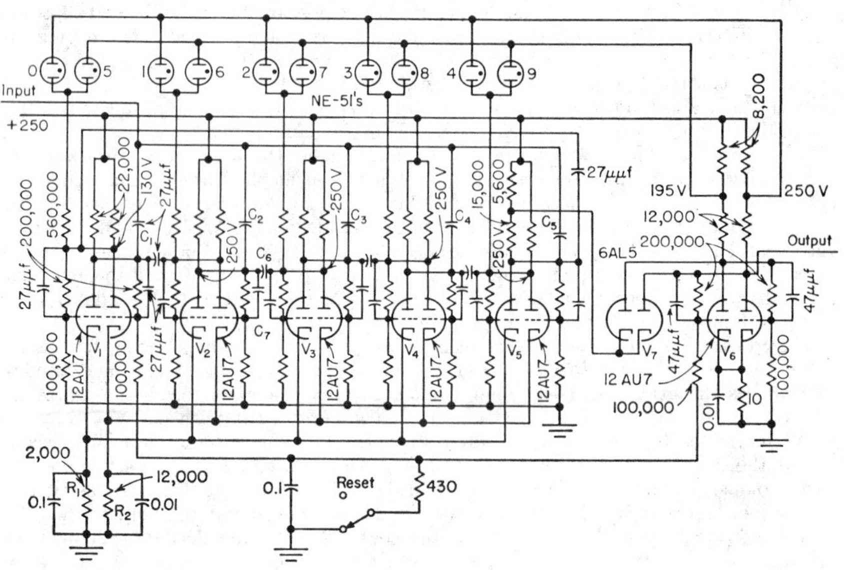

FIG. 28-63. Schematic of the counter whose block diagram is shown in Fig. 28-62.

Assume that V_1 is the odd tube. (By "odd" we mean that the right-hand section of the tube is conducting and the left side is cut off. The tubes are said to be in their normal condition when the reverse is true.) It is desired to advance the odd condition to V_2. This can be done by a negative pulse either at the right side of V_1 or at the left side of V_2. The negative input pulse through C_1 appears at both places, initiating the desired change in both V_1 and V_2. The flip-flop action in both tubes aids the input pulse in this respect. Since the cathodes are bypassed, input pulses through C_2, C_3, C_4, and C_5 are effectively short-circuited out through the respective left-hand plates of V_2, V_3, V_4, and V_5 before the transition, while during transition the pulse through C_5 is still short-circuited out by the left-hand plate of V_3. Any tendency for the input pulse at the left-hand grid of V_3 to appear via C_2, C_6, and C_7 to cut off the left-hand side of V_3 is overcome by the input pulse at the right-hand grid of V_3, in addition to being overridden by the positive pulse from the left-hand plate of V_2.

As long as the input-pulse rise time (before or after the pulse is differentiated by C_2 and C_3 through the left-hand side of V_2 and V_3) is short enough in comparison with the transition time and the pulse is of sufficient amplitude, wide variations may be permitted in pulse height and shape as well as in the values of interstage and input coupling capacitors.

The scale of two represented by V_6 is coupled to the ring by the diode V_7. The diode cathode will switch from +250 V to 195 V after the fifth count and back to 250 V after the tenth count. The circuit was developed for use with radiation counting devices, but it is completely adaptable to industrial applications where high-speed counting is required.

108. Ring-of-three Counter. A counter circuit designed to give one output pulse for every three applied to its input is shown in Fig. 28-64. Only one tube is conducting at any one time. The triggering pulses are applied across R_k. Each tube has associated with it a potentiometer network $R_2 - R_3$ to repeat a fraction of the plate voltage suitable for the following tube grid. The grids are connected into the plate circuits of the preceding tubes by resistors R_4. (All similarly labeled components have the same values for each stage.)

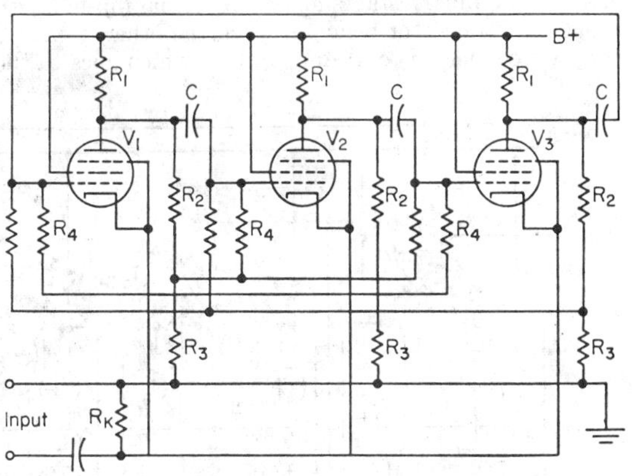

FIG. 28-64. Ring-of-three counter produces one output pulse for every three input pulses.

All the circuits have identical component values. Therefore, when the circuit is first turned on, any one of the stages can start conducting and the remaining stages must remain cut off until the stable operating condition of the circuit is interrupted by the application of an input pulse. Assume that V_1 is conducting initially. All the stages share resistor R_k in their cathode circuits. The value of this resistor is chosen so that, when any one of the tubes conducts current, the voltage drop across the resistor will be sufficient to hold the remaining tubes cut off. The voltage drop across R_k is, of course, positive at the cathode end of each tube, thus making the grid negative with respect to the cathode.

If, then, at a particular instant V_1 is conducting, V_2 and V_3 are cut off. The voltage at the plate of V_1 is less than that at the other tubes; so we may say that the voltage at the plate of V_1 is negative with respect to the voltage at the plates of V_2 and V_3. The grid of V_2 is connected by R_4 to the positive (relatively speaking) plate of V_3 and the negative plate of V_1. The grid of V_3 is connected to one positive plate (V_2) and the negative plate of V_1. The grid of V_1 is connected to the positive plate of V_2 and to the positive plate of V_3 and is positive with respect to the other two grids by an amount sufficient to ensure that V_1 continues to conduct.

To change this initial condition and pass the count on to a subsequent stage, something must be done to cut off V_1 and make V_2 conduct, while V_3 experiences no change. A positive pulse applied at the input terminals (which are connected directly to the cathodes of all tubes) will cause V_1 to be cut off momentarily because its cathode becomes positive with respect to its grid. As current stops flowing through the load resistance of V_1, its plate voltage will rise abruptly and a positive pulse will be sent via R_2 to the grids of V_2 and V_3. But the grid of V_2 receives a greater share of the V_1 plate-voltage rise because of the capacitor between these two points. When the input pulse subsides, the grid of V_2 is left momentarily more positive than any of the other grids. Thus V_2 conducts, and the voltage drop across R_k prevents either of the other tubes from conducting.

Subsequent pulses pass the conducting condition first to V_3 and then back to V_1, and so on, so that each tube conducts once for every three input pulses. Output may be taken from the plate of any one of the stages by coupling to the plate of the tube comprising that stage.

A ring of greater number than three can be obtained by simply adding more stages before feeding back to the first stage. The only limitation to the number of stages that can be incorporated in a circuit of this kind is the fact that the plate current of one tube must be large enough to hold all the other tubes cut off (the plate current must cause cutoff voltage to appear across R_k when one tube is conducting).

In practice, rings of five are usually sufficient. They are used in conjunction with scale-of-two circuits for decade counting. Many different combinations are, of course, possible.

109. Decade Counter Using Transistors. For comparison with the tube counter, the circuit of a decade scaler with transistors is shown in Fig. 28-65. Silicon *n-p-n* transistors are used with diode clamps. The transistors must have collector- to base-emitter breakdown voltages of the order of 90 V or better because a 90-V supply is used. The binary stages are conventional flip-flop circuits with two transistors in each stage, one transistor being on when the other is off. The bases of TR_2, TR_4, TR_6, and TR_8 are connected to a common bus, which goes to the reset switch.

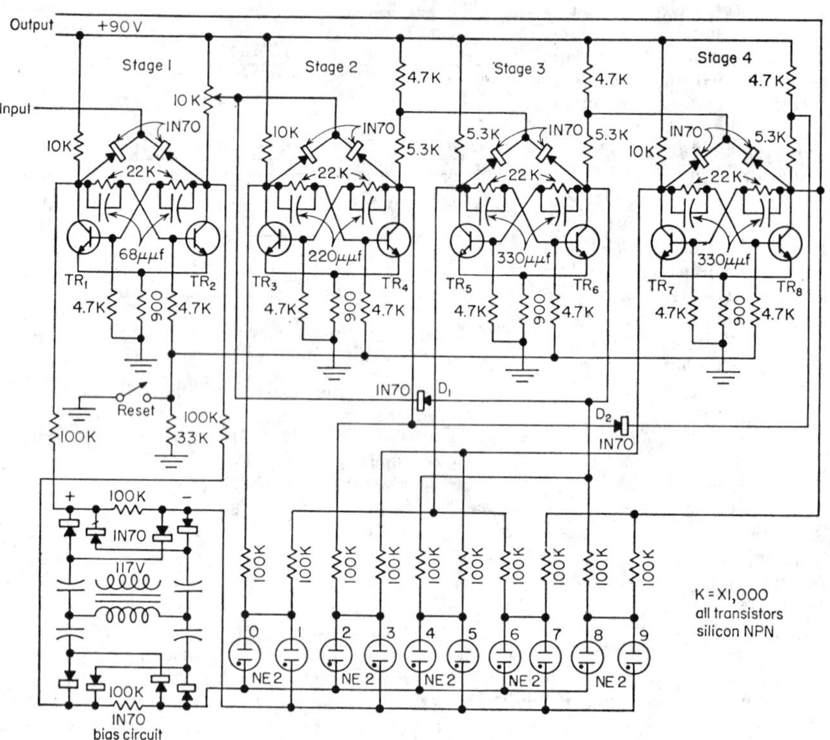

FIG. 28-65. Circuit of 10-lamp scaler using silicon transistors.

Clamping diode D_1 is connected between the collector of TR_2 (stage 1) and the collector of TR_6 (stage 3). Diode D_2 is connected from the collector of TR_8 (stage 4) to the collector of TR_4 (stage 2). The up-down voltage patterns for this scaler are given in Fig. 28-66. When the switch is in the reset position, the ground return is opened and a 33,000-Ω resistor placed in series with the bases as shown in Fig. 28-65. This makes the bases more positive and causes TR_2, TR_4, TR_6, and TR_8 all to conduct regardless of count.

All even-numbered lamps are connected to a common bus, which goes through the bias supply to the collector of TR_2. All the odd-numbered lamps connect to a common bus, which goes through the bias supply to the collector of transistor TR_1.

Since stage 1 flips on every count and stages 2, 3, and 4 have the same pattern for two consecutive counts, one side of the odd-numbered lamps can be connected to the collector of TR_1 and one side of the even-numbered lamps can be connected to the TR_2 collector. The other side of lamps 0 and 1, 2 and 3, 4 and 5, 6 and 7, and 8 and 9 can be connected to the other collectors in the manner shown. To light the neon lamps reliably, the bias circuit is used. The capacitors block direct current from one collector circuit to another. Thus the collector swings are superimposed on rectified alternating current.

It is found that a transistor with an alpha of 0.92 or better will operate satisfactorily. Transistors TR_4 and TR_6 require an h_{22} parameter 0.3 to 0.4 $\times$ 10^{-6} mho for best operation. All the other transistors must have like parameters except that the h_{22} parameter may be 0.5 to 1.0 $\times$ 10^{-6} mho. Typical transistor parameters are h_{11}, 50 to 150 Ω; alpha, 0.94; h_{22}, 0.3 to 0.4 $\times$ 10^{-6} mho for TR_4 and TR_6 and 0.5 to 1.0 $\times$ 10^{-6} mho for all others; h_{12}, 1.0 to 4.0 $\times$ 10^{-4}; peak inverse voltage, 90 V.

Count by units	Transistor designation: TR_1	TR_2	TR_3	TR_4	TR_5	TR_6	TR_7	TR_8
0	60	10	60	10	60	10	60	10
1	10	60	60	10	60	10	60	10
2	60	10	10	60	60	10	60	10
3	10	60	10	60	60	10	60	10
4	60	10	60	10	10	60	60	10
5	10	60	60	10	10	60	60	10
6	60	10	10	60	60	10	10	60
7	10	60	10	60	60	10	10	60
8	60	10	60	10	10	60	10	60
9	10	60	60	10	10	60	10	60
0	60	10	60	10	60	10	60	10

Collector potential in volts

Fig. 28-66. Collector potentials for 10-lamp scaler.

Performance figures of the scaler are: frequency range up to 200 kc/s; minimum input voltage −10-V pulse; supply voltage +90 V; supply current 40 mA; collector dissipation 40 to 50 mW; total dissipation 3.6 W; operating temperature 20 to 70°C.

APPLICATIONS OF PHOTOELECTRIC DEVICES

110. Photoelectric Circuits. A beam of light has many advantages in industrial work. It is weightless, is instantaneous in its action, has no inertia, and is easy to control by reflection, refraction, focusing, or filtering by wavelength. Applications of light-sensitive phototubes and photocells fall into two classes: those which use light simply as an on-off switch, a tool for counting, grading, detecting the presence of objects, protecting workers at dangerous machines, and the like; and those applications in which the photoelectric device responds to an optical quantity, such as illumination, brightness, color, or color temperature.

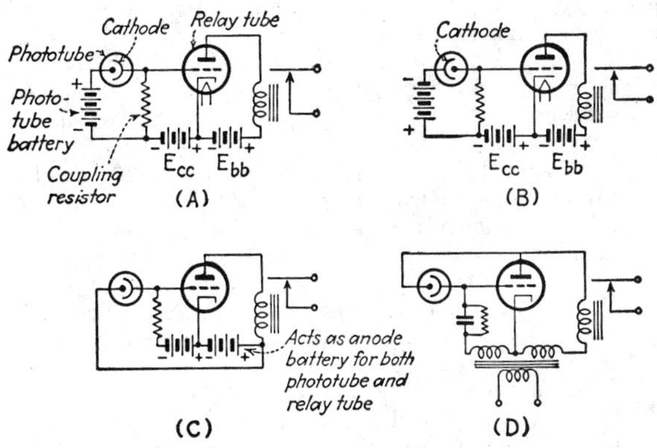

Fig. 28-67. Typical phototube relay circuits. (*A*) Separate phototube battery, positive control; (*B*) same, negative control; (*C*) use of common battery for phototube and anode; (*D*) a-c-operated phototube relay.

111. Light Relay. Several forms of photoelectric relay are shown in Fig. 28-67. Adjustment of the grid resistance, cathode-bias resistor, or applied voltage allows the pull-up of relay current to be set at various values of illumination of the phototube cathode. The marginal relay has two relays which operate to close and open an indicating circuit when the illumination falls outside a given range. A plate-current meter may be used as an illumination indicator, if the relay is calibrated against a known source of light.

112. Illumination Control. The simple light relay (on-off or marginal type) may be used to control the illumination of a room. The relay contacts turn on lamps as the illumination falling on a control surface decreases. Lighting in classrooms and factories may be controlled in this manner; in some elaborate installations several marginal relays are employed to turn on successive banks of lamps as the outside illumination decreases. The simple on-off phototube relay is used to turn on electric signs snd street lamps with the coming of darkness (or during a storm) without supervision. In precise work, the illumination supplied to a photometer may be controlled within several parts in a thousand by amplifying the output-relay current and using the amplified current to control the input to the lamp.

113. Color Analyzers. Color analysis by phototubes implies the use of a tube having good response over the entire significant spectral range covered by the colors under test. However, equal response to all colors is not required if simple comparative measurements are made. Trichromatic analysis falls in the latter class. In this method, three filters, whose spectral-transmission characteristics slightly overlap and cover the visible range, e.g., red, green-blue, and violet, are placed successively between the sample and the phototube (or self-generating photocell). The anode current is then recorded for each filter, and these values of current may be compared with those found similarly for the standard color composition. Care must be taken, of course, to see that the source of light used has the same spectral distribution for all measurements, which is possible only when the lamp is operated at fixed voltage and is known to be free from aging effects.

A complete spectrophotometric analysis of a color sample may be performed on a null basis by the color analyzer developed by Hardy. The device forms a widely dispersed spectrum from the light of an incandescent lamp and presents different portions of this spectrum in succession to the sample and standard white surface. The reflected (or transmitted) light is compared by a rotating optical system which presents both illuminations to a single phototube. The output of the phototube in turn controls the effective transmission through the optical system so as to make the two illuminations equal. The amount of control necessary to produce equal light is used as a measure of the spectral composition of the instrument. The analyzer proceeds through the spectrum in this manner automatically and traces the spectrophotometric curve of the sample on tracing paper.

114. Illumination and Exposure Meters. The self-generating photocell is widely used as a portable exposure meter, and its range and accuracy, as used with the associated microammeter, are sufficiently good. The more accurate and sensitive phototube-and-amplifier meters are used for scientific purposes.

115. Sorting by Color and Shade. Automatic methods of sorting objects on the basis of color make use of a phototube relay fitted with a light filter which admits sufficient light to actuate the relay only when the given color is present. More elaborate means of sorting on the basis of shades of color may be provided by employing two phototubes fitted with limiting filters, as shown in Fig. 28-68. The relay is actuated in this case only if the light admitted to the two phototubes contains the necessary two components associated with the controlling shade of color. The same idea may be extended indefinitely to as many shades as are desired, provided only that suitable optical filters are available. The two-tube system also has the advantage of being independent of changes in brightness of the illumination, provided that such changes in brightness are not accompanied by color changes.

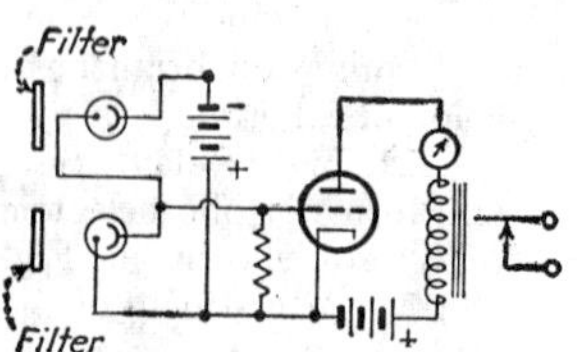

Fig. 28-68. Circuit for discriminating between colors by use of filters.

116. Interruption Relay. The phototube relay may be used as a counting and detecting device if it is so arranged that the light entering the phototube may be interrupted by the object to be counted or sorted. An obvious application is the photoelectric burglar alarm. For this purpose infrared light and the infrared-sensitive cells are ordinarily used. The light beam may be reflected back and forth by mirrors to protect an extended area or enclosure from intrusion. The phototube relay counter employs a magnetic relay counter in the plate circuit of the relay tube; it is used for counting objects on conveyor belts and similar applications. Objects may also be sorted according to the degree to which they interrupt the light beam. Several relays may be arranged with individual phototubes to sort objects into several grades on a marginal basis. Considerable precision may be achieved if optical magnification is used.

The interruption of a light beam has been widely used for automatic door opening, especially in restaurants and railroad stations, where traffic is heavy. A pneumatic plunger is commonly used to move the door. A magnetic solenoid, controlled by the phototube relay, actuates the plunger. Two relays are often employed in series, one whose beam is directly in front of the door entrance, the other with a beam shining obliquely through the doorway. The first beam acts to open the door, and the second to keep the door open while anyone is passing through it.

The protection of operators of machinery is also a common use of the interruption relay. A curtain of light, formed by several lamps and cylindrical lenses, is thrown in front of the machine area to be protected, e.g., the area immediately in front of the jaws of a punch press. Several phototubes, likewise fitted with cylindrical lenses, accept the light. If any part of the beam is interrupted, the phototube which receives the decreased light increases its resistance. All the phototubes are connected in series, so that the action of any phototube is sufficient to operate the relay. The relay in turn is connected to the electric control circuits of the press, to prevent its operation when the curtain of light is in any way interrupted.

The timing of athletic events, in particular horse races, is also performed by phototube interruption relays. Several lamps throw parallel beams of light across the track into phototubes placed opposite. The interruption of any of the beams actuates the relay. The relay contact in turn may control a chronometer or, in the case of horse racing, may actuate an automatic camera which photographs the finish of the race within a few milliseconds of the interruption of the beam.

The interruption principle is also used to supervise all manner of electrical measurements by the meter-limit system. In this application, the pointer of a meter is caused to interrupt a narrow beam of light, which shines through the face of the instrument at a point on the scale corresponding to the limit above which the phototube is to act. Two phototubes may be used with two such beams to form an upper and lower limit of supervision. A similar method makes use of a small mirror glued to the meter pointer, which reflects light from a source in front of the meter to a phototube likewise in front of the meter but slightly to one side. Marginal control may be obtained from two or more phototubes, using the single beam of light.

The meter-limit system is ordinarily applied to low-voltage control applications where the voltage output applied to the meter is smaller than can be conveniently amplified electronically. Thus, for example, in temperature measurements employing the thermocouple, sufficient voltage is not available, but several microamperes or milliamperes of current are available, which are sufficient to actuate a moving-pointer meter.

117. Optical Transmission Relay and Meter. The light transmission of any substance may be measured and monitored readily by employing a phototube relay with a meter in series with the relay coil. The applications in photography are obvious. In one such application, the timing of printing exposures is determined automatically by integration of the light entering the phototube through a portion of the negative. The turbidity of sewage, density of smoke, etc., may all be measured and monitored similarly. In smoke control, marginal control may be desired, in which case two mechanical relays are used in the relay-tube plate circuit.

118. Photoelectric Calculation Device. The photoelectric integraph is a device for integrating a function under a curve (or a combination of such functions). The curve is plotted as a boundary between black and white on a transparent base. The image of a straight-line filament lamp is passed through the plot to a phototube and the plot shifted along its ordinate scale. Two such plots may be integrated simultaneously in any relationship (including a relationship variable with one of the quantities in the curves) by passing the light through two plots in succession.

119. Register Control. Phototube relays are used in continuous processing (in wrapping machinery, high-speed printing of paper and fabrics, fabric weaving, etc.) to keep a cutting or printing operation in register with control marks printed on the edge of the material being processed. A source of light (passed through the material if it is transparent enough; otherwise, reflected from it) enters the phototube after being intercepted by the control marks. The phototube in turn controls a registering motor whose field circuit, for example, may be fed from a grid-controlled rectifier under the control of the phototube. A similar method is used for the inspection for pinholes in sheet metals undergoing continuous rolling or finishing.

INSPECTION TECHNIQUES

120. Inspection Techniques. One reason why electronic devices have received so much attention in industrial plants is their ability to combine with ordinary instruments or machines to do jobs previously considered impossible. Thus by combining the glowing pattern of a graph line or curve on the face of a cathode-ray tube with the ability of a phototube, we have an eye and a brain that can inspect an object for visible flaws and automatically accept or reject the object under test. Foods such as beans, seed corn, peanuts, potatoes, and coffee are inspected for faults by electronic equipment using phototubes and cathode-ray tubes.

In a typical fault-finding machine of this type, the beans slide down a chute to a rotary feed that passes them one at a time through an optical system consisting of a lamp, a focusing lens, two mirrors, and two phototubes. Rays from the light source are reflected by the bean through a partial mirror. It reflects 50% of the light through a red color filter to the first phototube and transmits the rest to a mirror that reflects it through a green color filter to the second phototube.

The amplified outputs of the two phototubes are applied to a cathode-ray tube whose face is covered with a mask cut in such a way that for acceptable beans the glowing pattern is entirely covered. When an off-color bean passes through the optical system, the pattern traced by the electron beam moves outside the marked area and the glow is picked up by a third phototube whose amplified output actuates an ejector mechanism that trips the bean into the reject bin.

121. Roundness Inspection. Automatic electronic inspection of the trueness of periphery and the width of gap of piston rings is much faster than hand checking methods, and the production rate is determined by the speed at which the rings are presented to the gage. The typical inspection cycle per piece is less than 5 s. The piston ring to be checked is inserted inside a master ring, which is placed on the instrument table and rotated by a power-driven roller. The gaging functions are performed by scanning beams of light directed onto phototubes that energize electronic circuits to illuminate three signal lights.

As the ring revolves, one beam of light is projected on the periphery of the piston ring. Clearance between it and the master ring results from an out-of-round condition of the piston ring and permits a portion of the light beam to fall on the phototube. A red rejection signal flashes when excessive light indicates that the piston ring is out of round beyond an acceptable point. A green signal flashes at the end of one complete revolution if the width of gap is within tolerance. Another beam of light actuates a yellow signal should the gap be undersize. A third beam of light energizes another circuit to illuminate the red rejection signal should the width of the gap be oversize.

122. Meter Testing. Electric utility companies have used phototubes for watthour meter testing. The revolutions of the disk are counted by projecting a beam of light through the anticreep holes of the disk to a photocell. The resulting pulses can be used to actuate a telephone-type rotary switch to start and stop another meter used as a standard for a predetermined number of revolutions of the meter being tested. Also, a meter can be quickly adjusted to match the speed of the standard by reflecting light from the serrated edges of the disks to photocells whose outputs are compared.

123. Phototransistor Card Reading. One application of the phototransistor is shown in Fig. 28-69. It is used in a 118-channel card reader for automatic handling of toll telephone calls. The phototransistor is illuminated by a light beam modulated at 400 c/s when a card punch hole passes over the reader. The light acts as the emitter of the phototransistor, which has a collector impedance of about 10,000 Ω. This impedance is reduced to approximately 3,000 Ω by the illumination. The a-c signal from the phototransistor is applied to the transistor amplifier. The amplifier is a conventional common-base circuit having a voltage gain of 40 to 100. The amplifier output triggers a cold-cathode gas tube that operates a relay.

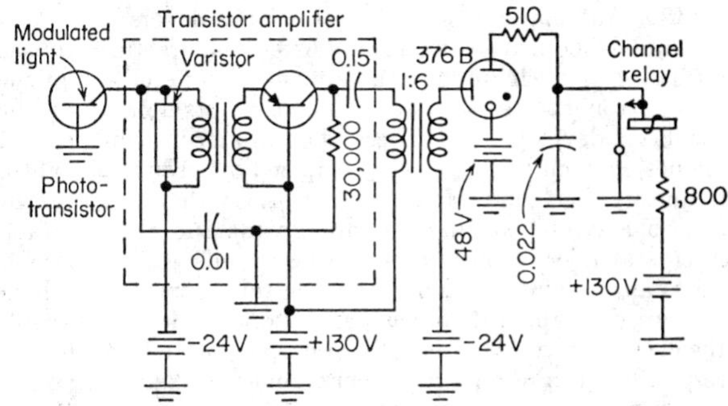

Fig. 28-69. Transistor amplifier for boosting signal from phototransistor to trigger cold-cathode tube.

MISCELLANEOUS APPLICATIONS

124. Telemetering. An industrial application of amplification and communications techniques involves the transmission of meter indications to a distant point (telemetering) by wire or radio. The techniques employed vary with the particular application. A simple, stable system involves controlling the output of an oscillator (modulation) and using the meter current, after amplification, as the modulating agent. At the receiving end of the circuit, the modulated current is demodulated, and the d-c component of the output is applied to the remote-indicating meter.

Telemetry has been used for many years for remote metering, monitoring, and supervisory instrumentation of power plants by utility companies. Wire lines were adequate for the purpose, and speed was not an important factor, as it is today. Considerable expansion and development of techniques took place during World War II to meet the need to measure stresses and deformations of aircraft structures during flight tests. For this purpose, frequency-modulation (FM) transmitters modulated by the output of strain gages are used and oscillographic records made of the corresponding output of receivers. Several gages can be sampled in time sequence by electronic switching mechanisms for commutating. Telemetering systems are also used to measure air-blast and water-shock levels in atomic-bomb tests and for measuring the performance of many components and systems in vehicles of all kinds, including missiles, satellites, and spacecraft. Many systems have been developed to handle information from many sources, such as transducers, in the vehicle. In most systems, signals from each source are sampled in succession, and then they modulate the radio transmitter.

An example of a modern telemetering system and the extent of its sophistication is the equipment used in the Gemini space capsule. Electrical signals are produced by 269 sources in the capsule. Measurements of these signals are made sequentially, and 6,400 measurements each second are sampled, encoded, and transmitted to the ground. The sampling rate for individual signal is as high as 640 samples/s (for the astronauts' heartbeats). Slowly varying conditions, such as the internal temperature of the capsule, are sampled at rates as low as once in 2.4 s. During the 2.4 s, each of the 269 sources is sampled at least once. The measurements are transmitted to ground by pulse code modulation of solid-state transmitters.

At the receiving point of a telemetering system, there can be many pieces of electronic equipment. Oscillographs, pen recorders, magnetic-tape recorders, readout devices such as meters, analog and digital computers: the complexity depends on the job to be done. Some of these are employed in industry, in field testing vehicles, in measuring the mechanical strain of moving or rotating equipment, in detecting flaws in metals during production, and in measuring critical conditions, such as temperature and vibration, in locations where no human observer can be stationed.

125. Automatic Curve Tracer. Transducers such as loudspeakers operate over the audio-frequency spectrum. After manufacture, it is necessary to check the performance of such units at a number of different frequencies, a time-consuming job. An electronic tester that automatically plots a curve on the face of a cathode-ray tube has been developed. Devices that produce sound, such as handset receivers, headphones, and loudspeakers, are energized by a sound signal that is produced by a variable-frequency electronic signal generator, so that the resulting sound changes gradually from a low-pitched buzz on through the audible tone range to a high-pitched squeal. A crystal microphone picks up the sounds. The microphone output is amplified by tubes and is made to move the electron beam of a cathode-ray tube vertically on the screen by an amount that is proportional to loudness, while mechanical linkage sweeps the electron beam horizontally in proportion to frequency. The result is the frequency-response curve of the instrument under test, traced as a glowing fluorescent curve or graph on the face of the cathode-ray tube.

For testing microphones, the procedure is reversed. The signal generator feeds a loudspeaker that serves as the artificial sound source having known characteristics, and the test microphone is arranged to "listen" to the sounds. The curve obtained in this way tells how well the microphone is able to "hear" sounds.

By shading the face of the cathode-ray tube with a wax pencil or paint above and below two curves that represent permissible limits, even unskilled operators can tell at a glance whether a unit passes its test. For good units, the entire response curve will glow and be visible, while for bad units a part or all of the electronically traced curve will be behind the shaded areas.

126. Crack Detector. Determination of the quality of a metal product has frequently been a laborious process in which it has been necessary to rely on fatigue tests and the study of microsections of samples. Such procedures are slow and costly and conducive to waste in production. One solution to this problem of high-speed production testing for cracks in metal wire, tubing, and bars is based on the electrical characteristics of the metal under test.

In the instrument, a coil is arranged in an electronic oscillator circuit so that it induces circulating electrical currents in the sample being tested. These currents are strongest when there are no cracks to break up the current path. The frequency of the oscillator changes when a crack enters the field of the coil. The frequency change shows on an indicating meter. To avoid worker fatigue and to ensure against missing short flicks of the meter pointer, a neon lamp flashes at the same time.

The fact that cracks are rarely uniform in depth for more than a fraction of an inch or that they may be full of oxide has not been found to be a serious hindrance. The instrument gives indications of cracks 0.0005 to 0.25 in deep. To make the operation entirely automatic, a relay in the anode circuit of one electronic tube operates a small compressed-air paint sprayer, which marks the faulty material over the crack.

127. Magnetic Inspection. Bolts and other parts made of magnetic material have been inspected by an electromagnetic method. The equipment used for this purpose shows up flaws and cracks and can handle part sizes ranging from small bolts to reasonably large castings.

When electrical current is passed through a magnetic substance, such as a bolt, a north and south pole are produced at each crack in the material. When finely divided iron oxide in solution is poured over the bolt, the oxide particles adhere at the cracks and make them readily visible. Such electromagnetic inspection in the production of aircraft has long been considered a vital necessity as a safety measure.

128. Utility Uses. In power-generating plants, turbine supervisory instruments are used to ensure safe operation of turbine generators, particularly during starting and stopping periods. Of the various observations normally taken, two, at least, involve electronic devices. An eccentricity recorder measures the eccentricity of the front end of the turbine shaft by means of a detector coil placed near the shaft. An audio tone applied to the coil is modulated by changes in the air gap. The modulation component is separated out, amplified, and rectified for use to actuate a recording instrument which, if desired, can be equipped with alarm contacts to warn of any abnormal reading.

129. Prospecting. Electrical measurements made at the surface of the earth for investigating the various electrical properties of the upper strata often call for electronic equipment for increasing the sensitivity of the measuring instruments. These instruments are used in making measurements of the natural earth potential, polarization effects, radio field strength, and others. Correct interpretation of the measurements indicates the location of ore masses and oil-containing faults. Thousands of square miles of inaccessible territory and offshore areas are explored by air by modern prospectors searching for mineral deposits, using the airborne electronic magnetometer equipment developed for detecting submerged submarines from the air. Shoran, special mapping cameras, and other electronic devices have been combined with the supersensitive electronic magnetometer, which is built into a bomblike casing suspended from an airplane or attached to the wing tip to free it from the magnetic influences of the airplane.

For intensive surveys over a small area, the reflection-seismograph method of geophysical exploration has proved invaluable in locating petroleum-bearing properties. By means of seismic pickups, high-gain electronic amplifiers, and photographic recording galvanometers, the seismic effects produced by controlled dynamite explosions are recorded and analyzed. The depths to the reflecting strata are determined from the total time required for the seismic vibrations to travel from the explosion to the reflecting layer and back to the seismic pickups used as detectors on the surface of the earth.

When the seismic waves arrive at the earth's surface, their acoustic energy is converted to electrical energy by means of seismic pickups, which are electrically similar to microphones. This energy is fed to electronic amplifiers connected to recording galvanometers that produce a photographic record, or seismogram. Timing marks at intervals of 0.01 s and the instant of explosion of the dynamite charge are recorded on this seismogram, together with the seismic refracted and reflected waves. When such a seismogram is made with a large number of seismic pickups placed on the surface of the earth in line with the shot point at intervals of around 200 ft, it is possible to distinguish between the direct, the refracted, and the reflected waves. The time from the instant of explosion of the dynamite charge to some definite reflected or refracted characteristic of this event is counted to the nearest 1/1,000 s, and the depth to this reflecting horizon is then computed. By this means, contour maps can be drawn of the subsurface strata at depths of a few hundred to around 20,000 ft. When the survey is made over a large area, subsurface irregularities where petroleum might exist are detected.

Subsurface information is also obtained by recording or logging the various underground layers through which the shaft has been driven. Within the last decade an electrical log has been widely adopted. This consists of an electrode suspended from a cable that is moved up and down in the shaft. Wires connect the log to recording equipment on the surface that makes a continuous record of the electrical resistivity of the earth formations in terms of depth.

Interpretation of the readings produced by electrical logs indicates the locations of such underground conditions as sands and shales. In many fields, new wells can be drilled between producing wells with the help only of electrical logs. In wildcat or new-field wells, the electrical log sometimes shows up sands which are good bets as oil sands and which would otherwise go unnoticed in ordinary drilling operations.

Many old fields are known to have oil-bearing sands in the upper formations, sands that were not found when the wells were first drilled and cased in steel. If these sands can be logged through the steel casing, oil production can be increased by perforating the casing with a special gun.

Radioactive elements like radium, the thorium series, and potassium often exist in oil-bearing formations and give off gamma rays that can be detected through steel pipe. The first gamma-ray detector used for this purpose was an ionization chamber filled with nitrogen at a pressure of about 500 lb/in^2. The chamber, made of drill pipe, was capable of withstanding the pressure of several thousand pounds per square inch encountered in wells. Housed in a compartment adjacent to the ionization chamber was an electronic detecting circuit with batteries. The detector unit transmitted pulses that were automatically amplified at the surface and read on a galvanometer. Measurements were made at various points in the well. In more recent work the gamma-ray technique has been adapted to commercial logging equipment. Power is fed down the cable from the surface to the vacuum tube in the chamber at the same time that pulses are transmitted up to the surface. An interesting application of the gamma-ray logging technique is in tracing the cement that has been forced out around the outside of a casing under great pressure through holes made in the casing by bullets fired from a gun. If cement is used that has been made radioactive by addition of carnotite ore, it is easily located by gamma-ray measurements. The cement may travel a fairly large distance, as much as 70 ft, up the outside of the pipe. Squeeze jobs are ordinarily used in sealing off objectionable water and gas sands from the producing sand, so that it is important at times to know the disposition of the cement.

130. Vibration Analysis. The sound-reproduction technique is used for analysis of vibration and noise in industrial applications. A microphone or vibration pickup element is employed to produce a current proportional to the amplitude of vibration. After amplification, the corresponding voltage waveform vs. time may be examined on the screen of a cathode-ray oscilloscope, or the waveform may be analyzed into its harmonic components in a harmonic analyzer. Study of the waveform and its component harmonics often leads to the source (and eventually to the control) of the noise or vibration.

Vibration-amplitude recorders show the extent of transverse vibration occurring at each of the main bearings of a turbine-generator set. A detector unit, consisting of a permanent magnet and a moving coil mounted at each bearing, produces voltage proportional to vibration velocity. This voltage is integrated, amplified, and applied to the recorder.

131. Metal Detection. The presence of metal in the field of an inductance carrying a high-frequency current and connected in a tuned circuit may be revealed in one of two ways. If the metal is ferromagnetic, its presence may have a decided effect on the inductance displayed by the coil and hence may vary the resonant frequency. Or if the metal acts as a short-circuited turn, the eddy currents induced in it will reflect a resistive component into the impedance of which the coil is a part. This change in resistive component has a similar (though less marked) effect on the resonant frequency, especially if the resistive component is appreciable compared with the internal resistance of the oscillator tube. Usually the tuned circuit forms the output circuit of an oscillator. The exploring coil then acts to change the loading on the oscillator as the metal enters the field, and this change in loading may be detected by the change in the d-c plate current of the oscillator. In the more sensitive forms of detector, the frequency of the oscillator is compared with the standard output frequency of a beat-frequency oscillator and the beat-note frequency between the two sources detected and observed aurally or directly on a frequency meter.

132. Capacitance-operated Relay. If an oscillator is arranged so that it is highly sensitive to variations in the capacitance of its tuned circuit, it may be made to measure (or to control other quantities in terms of) the capacitance variation. The capacitance-operated relay has a wide variety of applications, e.g., burglar detection, the automatic operation of display devices, and the detection of moisture content and thickness in continuous processing.

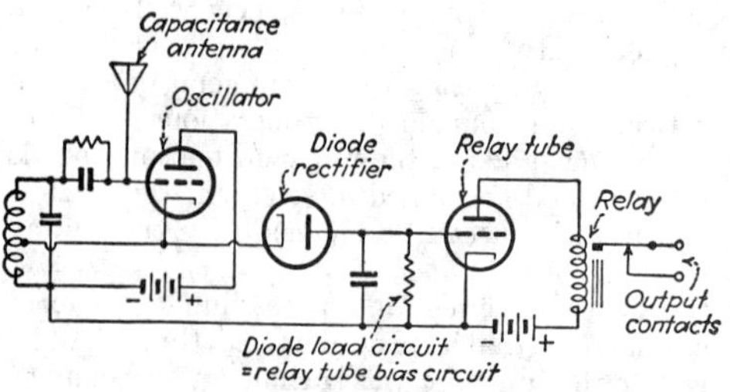

Fig. 28-70. Typical capacitance-operated relay.

A typical form of capacitance-operated relay is shown in Fig. 28-70. A Hartley-type oscillator is fitted with the capacitance-detecting element (antenna) connected to its grid. Variations in this capacitance against ground have the effect of changing the time constant of the *RC* filter in series with the grid, which determines the amplitude of oscillation. A portion of the oscillating voltage existing across the tuned circuit is rectified and the output filtered to form the d-c control voltage applied to the relay tube. This relay can be made extremely sensitive by employing a small value of capacitance in the *RC* grid filter, so that the antenna capacitance becomes a corresponding large part of the time-constant product. The motion of one's hand 3 or 4 ft from the antenna is sufficient to actuate the relay when set at maximum sensitivity, but a lower sensitivity is desirable for continued and unattended operation.

133. Control of Illumination. The grid-controlled rectifier serves many specialized applications. One such application is in the precise control of lighting in theaters. Alternating current to the lamp load is passed through one winding on a saturable-core reactor, while rectified direct current from the grid-controlled rectifier passes through another winding. The variation in d-c magnetization resulting from the controlled rectifier permits smooth and efficient control of the lamp load from full brilliance to extinction. Voltage regulation may also be incorporated by a feedback connection between the lamp circuit and the control grid of the rectifier. The use of thyratron rectifiers in conjunction with saturable-core reactors in similar applications is widespread.

134. Gas-triode Relay-tube Applications. The gas-triode voltage-operated relay serves wherever a large current flow is to be initiated by a weak voltage impulse. The advantages of the gas-triode relay over mechanically operated switches are the speed of response; freedom from external sparking (hence, value in explosive atmospheres); and the ease with which, and range over which, the control level (corresponding to the pull-up of a mechanical relay) may be adjusted. One typical example is in the control of stroboscopic light sources. A gas-discharge lamp is connected to a bank of capacitors, in series with a group of thyratrons. The discharge of the capacitors through the lamp is controlled cyclically by the thyratron relay tubes, at rates varying from several to several thousand cycles per second. Furthermore, the duration of the conduction period of each cycle may be controlled readily, thus varying the amount of light energy produced in each flash.

SECTION 29

CODES AND STANDARD PRACTICES

BY

LESLIE D. PRICE Consultant, Formerly Manager, Engineering and Safety Regulations Department, National Electrical Manufacturers Association. Fellow, Institute of Electrical and Electronics Engineers. Fellow, Standards Engineers Society

CONTENTS

Numbers refer to paragraphs

SECTION 29

CODES AND STANDARD PRACTICES

BY LESLIE D. PRICE

RESUSCITATION

1. General Procedure. Artificial respiration should be started without delay in all cases where normal breathing has been stopped or is weak, spasmodic, or in any way inadequate as a consequence of electric shock, gas asphyxia, drowning, or any other form of suffocation. It should be continued uninterruptedly until recovery or until there are positive evidences of death. Records[1*] show success after periods of one to several hours of continuous artificial respiration, the longest record of complete success being 10 h before the patient, a victim of electric shock, was breathing without assistance.

2. Chances of recovery are very good if artificial respiration is administered promptly (within 2 min) but decrease rapidly if resuscitation is delayed. After 4 min delay there is little chance of success; however, artificial respiration should be administered, since in exceptional cases persons have been revived after a delay of ½ h or more.[2,15]

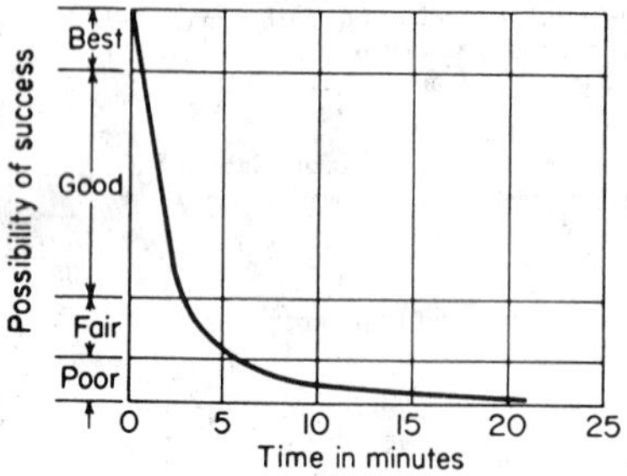

FIG. 29-1. Curve showing possibility of success plotted against elapsed time before start of artificial respiration.

Figure 29-1 shows the possibility of successful revival by artificial respiration for each minute of delay. There must be no delay in loosening clothing, warming the victim, getting him down from the pole, or moving him to a more comfortable position. However, an immediate check of the victim's mouth should be made by a quick pass of the fingers through the mouth to pull the tongue forward and remove false teeth, tobacco, chewing gum, etc. After resuscitation is started, the victim's belt, collar, and other clothing may be loosened, provided that this does not interfere with the resuscitation process.

3. Effects of Shock on the Human Body. The effect of electric shock on a human being is rather unpredictable and may manifest itself in a number of ways.

a. Asphyxia. Electric shock may cause a cessation of respiration (asphyxia). Current passing through the body may temporarily paralyze either the nerves or the area of the brain which controls respiration.

b. Burns—Contact and Flash. Contact burns are a common result of electric current passing through the body; they vary in severity, the same as thermal burns. The burns can normally be seen at the points where the current entered and left the body; however, internal tissues along the path of the current are also damaged. Therefore, the seriousness of electric burns may not be immediately evident, because the appearance does not indicate the depth or extent of the internal injury.

* Superscript numbers refer to the Bibliography, Par. 17.

In some accidents there is a flash, or electric, arc, the rays and heat from which may damage the eyes or result in thermal burns to exposed parts of the body.

c. Heart Stoppage and Fibrillation. Electric shock may disturb the natural rhythm of the heartbeat. When this happens, the heart muscles may stop completely, or they may be thrown into a twitching or trembling state known as "ventricular fibrillation," in which the actions of the individual muscle fibers are no longer coordinated. In either case (complete stoppage or ventricular fibrillation) the pulse disappears and circulation ceases.

When this occurs, the brain, which is the most highly developed part of man, may suffer irreparable damage after about 4 min.

d. Muscle Spasm. A series of erratic movements of a limb or limbs may occur owing to alternating contraction and relaxation of the muscles. This muscle-spasm action on the muscles of respiration may be a factor in the stoppage of breathing.

4. Choice of Method. In cases of electric shock there is no single ideal method of artificial respiration. The preferred manual method is mouth to mouth (insufflation), because it assures positive movement of air into the lungs. However, the victim's injuries may make any particular method unsuitable; therefore the choice must depend on conditions. On a pole or elevated structure, resuscitation (preferably mouth to mouth) should be started immediately without waiting to lower the victim.

5. General Instructions. *a.* As soon as artificial respiration has been started and while it is being continued, an assistant (if available) should loosen any tight clothing about the victim's neck, chest, or waist (Fig. 29-2). Liquids are not to be given by mouth until the victim is fully conscious.

b. Resuscitation should be carried on at the nearest possible location to where the victim was injured. He should not be moved from this location until he is breathing normally, and then only upon his own volition and while lying down. Should it be necessary, because of extreme weather conditions, to move the victim before he is breathing normally, resuscitation should be carried on while he is being moved.

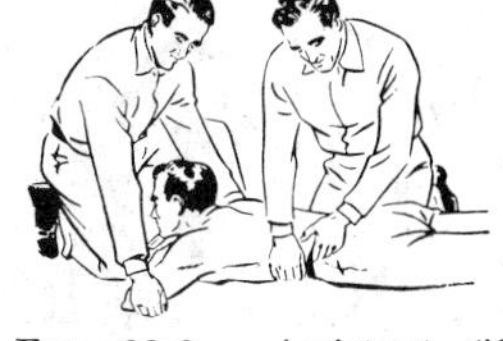

FIG. 29-2. Assistant (if available) loosens tight clothing about victim's neck, chest, or waist.

c. A brief return of normal breathing does not necessarily indicate that resuscitation should be discontinued. Not infrequently the victim, after temporary recovery, stops breathing again. He must be watched, and if normal breathing stops, artificial respiration must be resumed at once.

d. Artificial respiration must be continued (4 h or longer, if necessary) until natural breathing is restored or rigor mortis sets in.

e. To avoid strain on the victim's heart when he revives, he should be kept lying down and not be allowed to sit up or stand. If he revives before the doctor arrives, he should be given a stimulant, such as ammonia by inhalation, or a hot drink, such as coffee or tea. *The victim should be kept warm.* However, when heating devices are applied to an unconscious person, great care must be taken to prevent possible burns. The heating devices should be tested on one's own body before use and, if too hot, should be wrapped in a towel or other suitable insulation or allowed to cool to the proper temperature.

If it becomes necessary to change operators, this change should be made without losing the rhythm of respiration. By this procedure no confusion results at the time of change and a regular rhythm is kept up.

6. Mouth-to-mouth (Mouth-to-nose) Method. *a.* The victim should be placed with his head as far back as possible so that his neck is extended. On the ground, a "prop" (rolled-up jacket, seat cushion, etc.) under his shoulders helps straighten the windpipe by allowing the head to fall farther back. On a slope, the body should be placed with the head slightly downhill.

b. The operator places one hand on the victim's forehead and the other under the victim's neck and forces the head into an extended position as far back as possible (Fig. 29-3).

1. For mouth-to-mouth breathing, he then releases the forehead and pinches the nostrils (Fig. 29-4).

2. The operator uses one hand to elevate the victim's jaw by inserting a thumb between the victim's teeth (Fig. 29-5).

3. For mouth-to-nose breathing, he removes his hand from under the victim's neck and pushes on the chin, lifting it upward and, at the same time, closing the mouth (Fig. 29-6).

c. An alternative method is to lift the jaw forcibly upward with both hands, placing the fingers on both sides of the jaw (near the ear lobes) (Fig. 29-7).

1. For mouth-to-mouth breathing, the operator closes the victim's nose by pinching the nostrils.

2. For mouth-to-nose breathing, he closes the victim's mouth by pushing on the lower lip with the thumbs.

FIG. 29-3. Extend head.

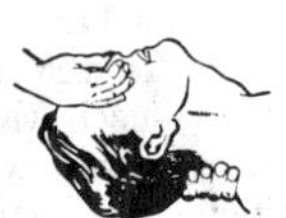

FIG. 29-4. Pinch nostrils for mouth-to mouth breathing.

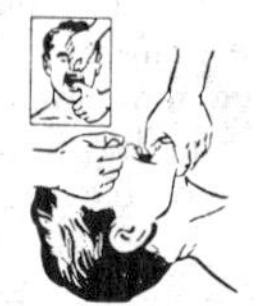

FIG. 29-5. Elevate jaw.

FIG. 29-6. Close victim's mouth for mouth-to-nose breathing.

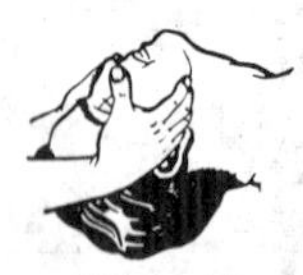

FIG. 29-7. Alternative method of closing victim's mouth.

FIG. 29-8. Mouth-to-mouth breathing.

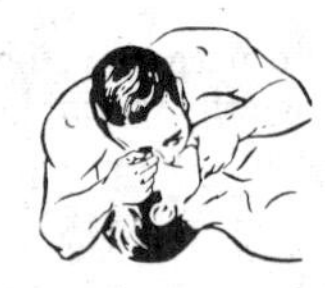

FIG. 29-9. Mouth-to-mouth breathing.

FIG. 29-10. Mouth-to-nose breathing.

d. After taking a deep breath, the operator places his mouth over the victim's mouth (or nose), being sure to make an airtight contact. In mouth-to-mouth resuscitation, the victim's mouth should not be held open too wide, for it must be completely covered by the operator's lips.

(With an infant, the operator's mouth should be placed over the child's *mouth and nose.*)

e. The operator then breathes or blows into the victim's mouth or nose, forcibly for adults and gently for children (Figs. 29-8, 29-9, and 29-10).

f. The jaw must be held in an elevated position on both the inspiration and the expiration phases.

g. If the chest does not rise, the position of the head (paragraph *a*) and jaw (paragraph *b*) should be improved and the blowing done more forcibly. If the victim's lungs are still not ventilated, the airway may be obstructed. He should be placed in a face-down, head-down position, with the tongue pulled forward, and patted firmly on the back to dislodge any foreign object.

h. The cycle of inflation and exhalation should be repeated 12 times/min for adults and 20 times/min for infants and small children.

i. If the victim's stomach swells during resuscitation, air may be entering it. This may be corrected by the operator gently pressing on the victim's stomach during exhalation.

7. Back-pressure–Arm-lift Method. Follow These Instructions, Even If the Patient Appears Dead. As soon as possible feel with your fingers in the patient's mouth and throat and remove any foreign body (tobacco, false teeth, etc.). If the mouth is tight shut, pay no more attention to it until later. Do not stop to loosen the patient's clothing, but immediately begin actual resuscitation. Every moment of delay is serious. Proceed with the following standard technique:

a. Position of the Subject. Place the subject in the face-down position. Bend his elbows and place the hands one upon the other. Turn his face to one side, placing his cheek upon his hands (see Fig. 29-11).

b. Position of the Operator. Kneel on either the right or left knee at the head of the subject, facing him. Place the knee at the side of the subject's head close to the forearm. Place the opposite foot outside the elbow. If it is more comfortable, kneel on both knees, one on either side of the subject's head. Place your hands upon the flat of the subject's back in such a way that the heels of your hands lie just below a line running between the armpits. With the tips of the thumbs just touching, spread the fingers downward and outward (see Fig. 29-12).

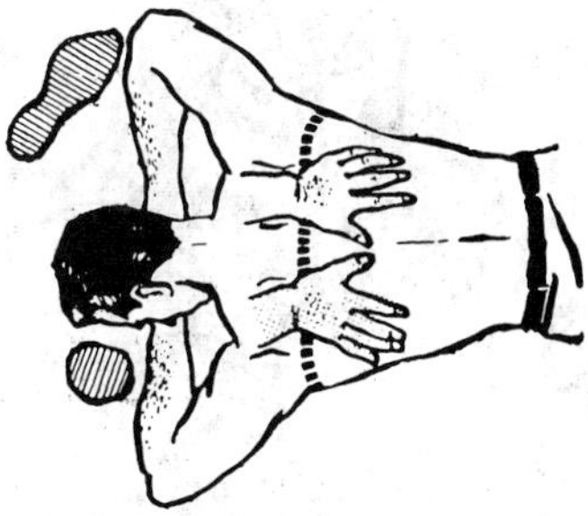

Lay victim in prone position

FIG. 29-11. Holger Nielsen's Back Pressure–Arm Lift Method. Lay victim in prone position. Elbows bent, one hand on the other. Head on hands, face to one side. Kneel at victim's head, on either knee.

c. Compression Phase. Rock forward until the arms are approximately vertical, and allow the weight of the upper part of your body to exert slow, steady, even pressure downward upon the hands. This forces air out of the lungs. Your elbows should be kept straight and the pressure exerted almost directly downward on the back (see Fig. 29-13).

d. Position, Expansion Phase. Release the pressure, avoiding a final thrust, and commence to rock slowly backward. Place your hands upon the subject's arms just above the elbows (see Fig. 29-14).

e. Expansion Phase. Draw the subject's arms upward and toward you. Apply just enough lift to feel resistance and tension at his shoulders. Do not bend your elbows, and as you rock backward, his arms will be drawn toward you. Then lower the arms to the ground. This completes the full cycle. The arm lift expands the chest by pulling on chest muscles, arching the back, and relieving the weight on the chest (see Fig. 29-15).

f. Cycle. The cycle should be repeated 12 times/min at a steady, uniform rate.

Place hands

FIG. 29-12. Place hands with fingers spread, thumbs touching, heels of hands placed just below a line connecting the armpits.

Apply pressure

FIG. 29-13. Apply pressure. Rock forward slowly until the arms are vertical. Keep elbows straight. Exert steady, even pressure down.

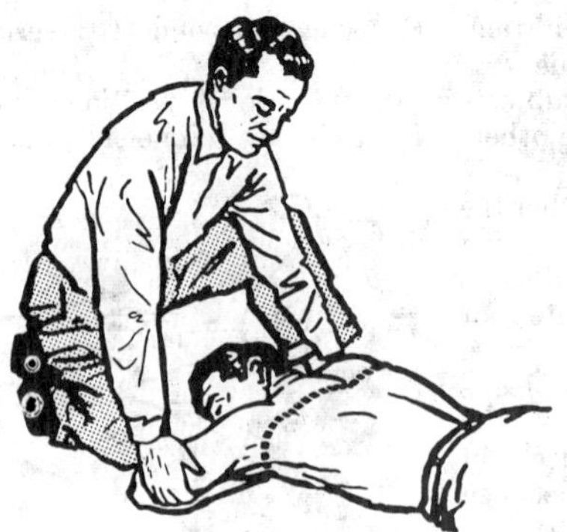

FIG. 29-14. Release pressure. Rock back slowly. Grasp victim's arms just above the elbows. Continue backward.

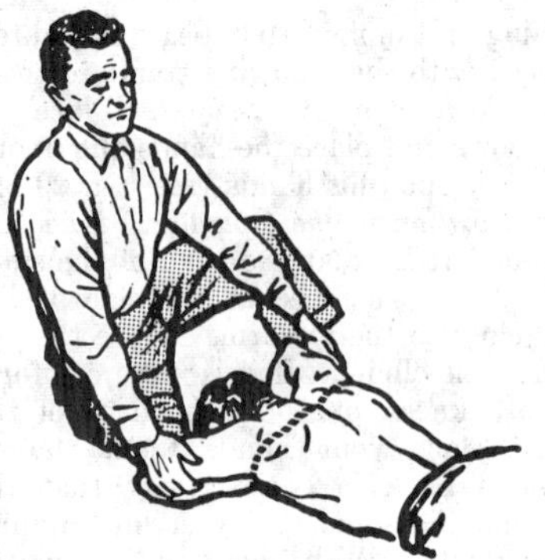

FIG. 29-15. Lift arms. Raise victim's arms upward and backward until tension is felt, then lower them, completing the cycle.

The compression and expansion phases should occupy about equal times, the release periods being of minimum duration.

g. Change of Operators. The relief operator kneels beside the operator, facing him. The latter falls away upon the completion of the lift; the relief operator merely pivots into place and starts with the compression phase. He may count with the operator just before relieving him, thereby establishing his timing.

Additional Directions. There should be a slight inclination of the body of the subject in such a way that fluid drains better from the respiratory passage. The head should be extended, not flexed forward, and the chin should not sag lest obstruction of the respiratory passages occur. These aspects can be cared for in placing the subject into position or shortly thereafter, between cycles. A smooth rhythm in performing artificial respiration is desirable, but split-second timing is not essential.

8. Other Approved Methods. In keeping with the recommendation of the National Research Council, the other methods acceptable under special circumstances are the hip-lift–back-pressure and arm-lift–chest-pressure methods. NOTE: In these and all alternative methods the cycle is 10 to 12 times/min.

The Hip-lift–Back-pressure Method. This method is useful in instances where it is impossible or undesirable to move the victim's arms. This method compares favorably with the back-pressure–arm-lift method in providing ventilation; however, the operator may become fatigued during sustained performance, particularly if the subject is heavy. (Some operators prefer to roll the hip instead of lifting it, the so-called hip-roll method.)

a. The subject is placed in the prone position with the face turned to the side and resting on the back of one hand. The other hand is extended above the head.

b. The operator kneels on either knee at the level of the subject's hips straddling the subject; he places his other foot near the opposite hip.

c. The operator places his hands beneath the subject's hips (at the bony pelvis, not at the waist). The fingers should be just below the hip bones where they touch the floor. The operator lifts upward from 4 to 6 in, keeping his arms straight. The hips are then replaced. The operator must not drop the subject.

d. The operator moves his hands to the subject's midback at the lower tips of the shoulder blades.

e. With the fingers spread and the thumbs an inch or two on either side of the spine, the operator rocks forward and exerts moderate pressure. This forces air out of the lungs. The elbows should be kept straight and the pressure exerted almost directly downward.

f. In changing operators, the relief operator kneels on the opposite side of the operator. After the operator designates which lift phase he is leaving on, the relief operator begins the press phase from where he is as the operator falls away on his side of the subject. The relief operator simultaneously straddles the subject and assumes the proper position for the next lift phase.

The Arm-lift–Chest-pressure Method (Silvester). If the subject is pinioned on his back or for any other reason must lie on his back, the arm-lift–chest-pressure method should be used. With this procedure, there is some danger that the tongue will relax backward and obstruct the airway. A helper can see that the tongue is pressed into normal position; if no helper is present, the operator himself should insert his index and middle fingers into the subject's mouth and press the tongue into place as soon as possible and should check occasionally to see that it remains there. Sometimes it will remain in position if once so placed provided that the lower jaw does not sag and the head is turned somewhat to one side.

a. A small pad should be placed under the subject's back, just below the neck, so that the head is slightly extended.

b. The operator places himself at the head end, kneeling on one or both knees.

c. With the subject on his back and his arms lying at his sides, the operator grasps the wrists.

d. The operator then draws the upper extremities vertically and headward, in an arc that brings the wrists finally to the floor beyond the subject's head.

e. The operator then brings the arms back, placing them on the lower part of the subject's chest, immediately adjacent to each other. The subject's hands are so placed on his chest that the base of the little finger of one hand is approximately adjacent to the base of the thumb of the other hand. Continuing to hold the wrists, the operator exerts slow and steady pressure upon the wrists and thus upon the chest.

f. When a second rescuer is available, he may "take over" the arm-lift–chest-pressure method with practically no break in rhythm. To change operators, the first operator moves off to one side while his replacement comes in from the other side. When the replacement is ready, the first operator releases the subject's arms at the stretch phase. The new operator grasps them and continues the application.

Schafer Prone-pressure Method. The prone-pressure method of resuscitation was first advanced by Sir Edward S. Schafer in 1903 and is applicable in those cases where injuries or other circumstances make the use of other methods inadvisable. It is an efficient and safe procedure. It can be applied for a considerable length of time by one person, and unless the operator exerts undue pressure on the victim's back, no damage results from its use.

a. The victim should be laid on his stomach with one arm extended straight beyond his head, the other arm bent at the elbow and the face turned *slightly* outward, with the side of the chin resting on the hand or forearm so that the nose and mouth are free for breathing (Fig. 29-16). *Do not turn the head too much as this may constrict the windpipe.* Air flow through the windpipe is dependent upon an unobstructed passage, which can be promoted by bending the head backward and pulling the jaw forward. If there is a slope, the body should be placed so that the head is slightly downhill.

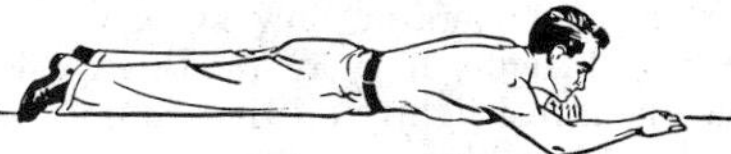

FIG. 29-16. Lay victim on his stomach.

b. The operator then kneels, straddling the victim's thighs, with his knees placed at the distance from the hip bones, which will allow him to assume the position shown in Fig. 29-17.

If the victim is large, the operator will perform just as efficiently and may be more comfortable if he straddles one thigh (Fig. 29-18). This should be the thigh on the side toward which the face is turned; this will give a better view of the face and permit seeing that the victim's nose and mouth remain free from obstruction.

The palms should be placed on the small of the victim's back, with the fingers resting on his ribs, the little finger just touching the lowest rib. The thumb should be alongside the fingers and the tips of the fingers just out of sight (Fig. 29-19).

c. With the arms held straight the operator swings forward slowly, so that the weight of his body is gradually brought to bear upon the victim. The shoulders should be almost over the heels of the hands at the end of the forward swing. The elbows should

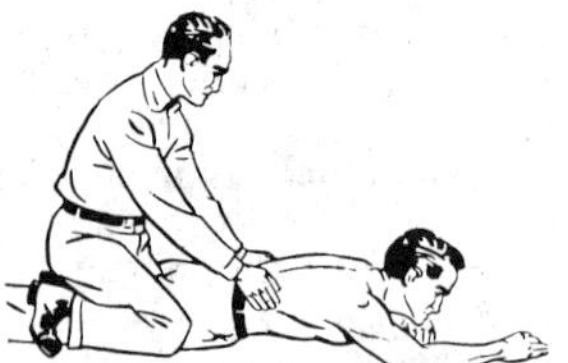

Fig. 29-17. Straddle victim's thighs.

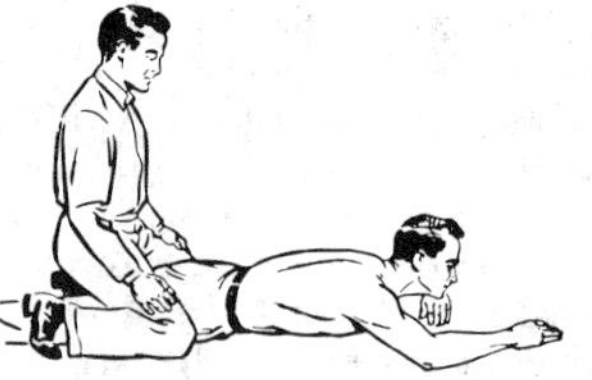

Fig. 29-18. Straddle one thigh if victim is large.

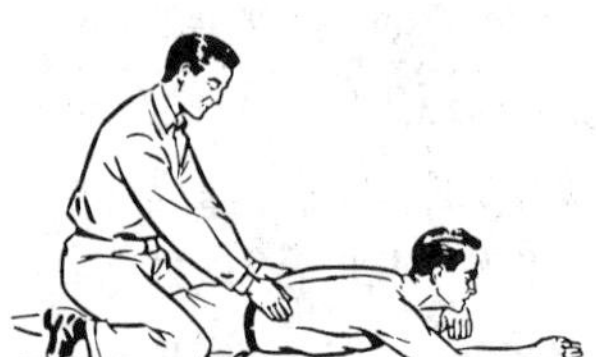

Fig. 29-19. Place palms on small of victim's back.

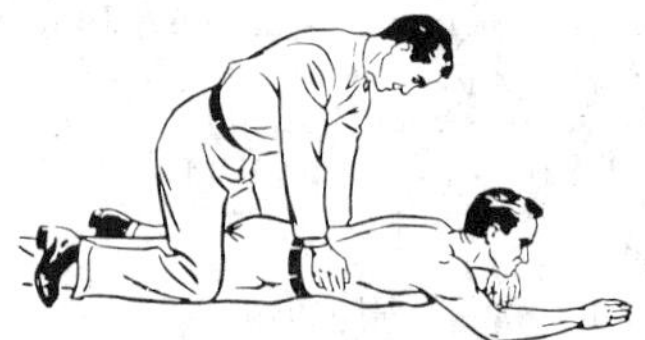

Fig. 29-20. Apply 40 to 50 lb of pressure.

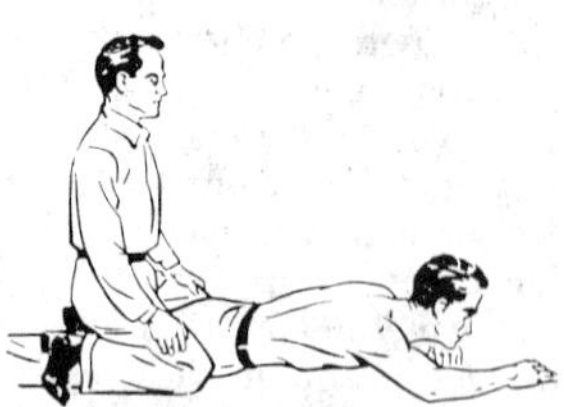

Fig. 29-21. Swing backward to remove pressure.

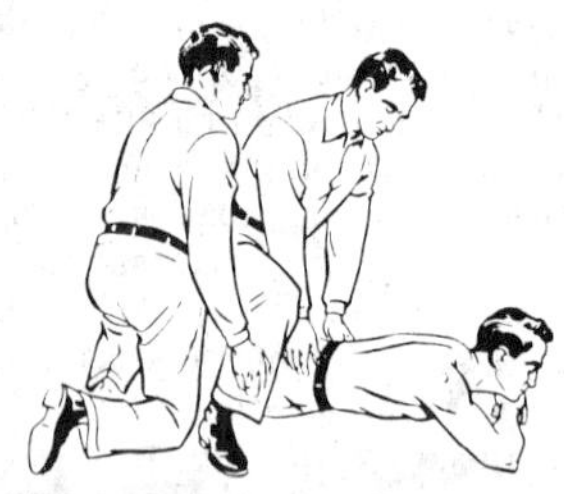

Fig. 29-22. Proper positions of operators immediately before change.

not be bent (Fig. 29-20). About 40 to 50 lb pressure should be applied to the normal adult. This operation should take about 2 s.

d. The operator now swings backward to remove the pressure completely. To relieve fatigue, he may remove his hands completely and rest them on his knees or let them rest lightly on the victim's back (Fig. 29-19 or 29-21).

e. After about 2 s, the forward swing is begun again. The double-movement cycle of compression and release should be repeated unhurriedly approximately 12 times/min. This results in a complete respiration about every 5 s.

f. To continue resuscitation for some time, it may be necessary to change the operator. This change must be made without losing the rhythm of respiration. Figure 29-22 illustrates the proper positions of the operators immediately preceding the change. By this procedure no confusion results at the time of change, and a regular rhythm is maintained.

9. Pole-top Resuscitation. As with manual resuscitation on the ground, there are a number of ways of performing pole-top resuscitation, and each has certain advantages and disadvantages.

After examining conditions, the rescuer climbs to the victim as quickly as possible. First he removes the victim from electrical contact (if necessary) and lets him hang by his safety strap. Then the rescuer determines whether artificial respiration is needed, how it can safely be applied, and which method to use.

1. *Mouth-to-mouth (Mouth-to-nose) Method.* This is the preferred manual method because it assures positive movement of air into the lungs.

a. The rescuer positions himself alongside and slightly above the victim.

b. He tilts the victim's head back and pulls the jaw forward to ensure an open air passageway.

c. The rescuer places his mouth over the victim's mouth, being sure to make an airtight contact. The rescuer's cheek should rest against the victim's nose to prevent leakage of air, or the rescuer places his mouth over the victim's nose and with his hand closes the victim's mouth (Fig. 29-23).

d. The rescuer then breathes into the victim's mouth or nose forcibly until the chest rises.

e. The first few breaths should be rapid, to inflate the lungs fully. The rescuer then continues at the normal breathing rate of about 12 times/min.

f. Watch the victim's chest. If it does not rise, the lungs are not being ventilated. Check the mouth for foreign matter which may be obstructing breathing. Clean the mouth, then retilt the head, and try again.

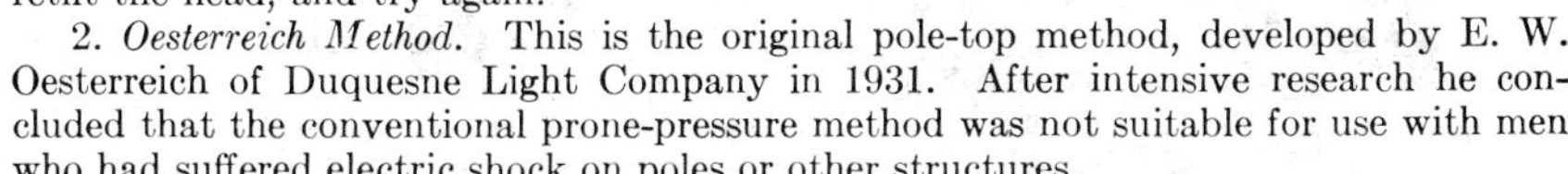

FIG. 29-23. Pole-top resuscitation.

2. *Oesterreich Method.* This is the original pole-top method, developed by E. W. Oesterreich of Duquesne Light Company in 1931. After intensive research he concluded that the conventional prone-pressure method was not suitable for use with men who had suffered electric shock on poles or other structures.

Research at Johns Hopkins University revealed that the Oesterreich method of artificial respiration is, for the purpose intended, an adequate and valuable procedure. It assures adequate lung ventilation and eliminates the time required to lower a man from a pole or other structure.

a. Putting his safety strap around the pole or structure below the victim, the rescuer climbs upward, keeping his strap between the victim's legs until the victim is straddling the strap.

b. Next, he encircles the victim's waist with his arms, placing both hands on the abdomen, thumbs below the lower ribs, fingers touching (Fig. 29-24).

c. With his arms and hands, he compresses the victim's abdomen in an upward motion. At the finish of the stroke his hands are cupped, with the fingers depressing the abdomen under the rib margin. The pressure is then quickly released and the procedure repeated at the rate of approximately 12 times/min.

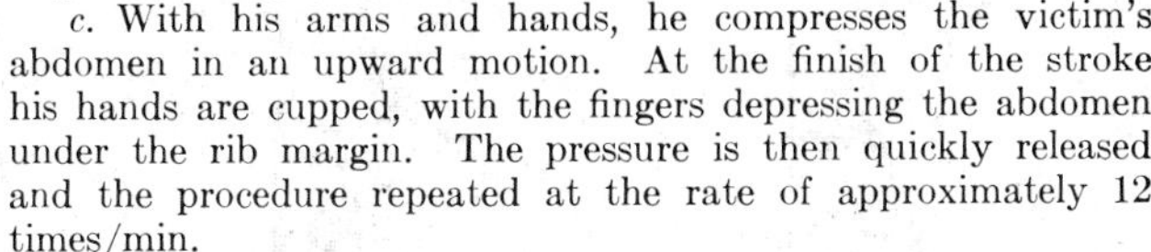

FIG. 29-24. Encircle victim's waist with arm.

The victim's head should be kept back so that the airway is open.

3. *Double-rock Method.* *a.* The rescuer gets into position with the victim astride his safety strap, as described in the Oesterreich method.

b. Next, he encircles the victim's waist and compresses the abdomen in an upward motion. The pressure is then quickly released, and this expiration phase is augmented by rocking the upper part of his body backward with the victim (Fig. 29-25).

FIG. 29-25. Double-rock method—backward phase.

FIG. 29-26. Double-rock method—forward phase.

FIG. 29-27. Backward rock, drawing victim's arms upward.

c. Then the rescuer rocks forward and moves his hands up over the victim's chest (Fig. 29-26).

d. He then rocks backward again, simultaneously drawing his arms upward and backward (Fig. 29-27). This lifts and stretches the victim's arms, thereby expanding the chest and producing active respiration.

The complete cycle is repeated approximately 12 times/min. As in the Oesterreich method, the victim's head should be kept back.

Fig. 29-28. Bear-hug method—squeeze.

Fig. 29-29. Bear-hug method—discontinue pressure.

4. *Bear-hug Method.* *a.* As in the double-rock method, the rescuer places the victim astride his safety strap, then encircles his chest and squeezes him in a firm "bear hug" (Fig. 29-28).

b. The pressure on the chest is discontinued, and the rescuer spreads his arms slightly and moves them upward, lifting the elbows of the victim to about a horizontal position (Fig. 29-29). The hugging and arm-lifting procedure is then repeated at a frequency of approximately 12 times/min. To keep the victim's airway open, the head should be kept back against the rescuer's shoulder.

5. *Arm-grasp Method.* *a.* This method is similar to the double-rock and bear-hug methods. However, at the beginning, the rescuer grasps the victim's wrists with his hands and retains his hold during the entire procedure (Fig. 29-30). (He can grasp the victim's right and left hands with his own, respectively, or he can grasp the left one with his right hand and the right one with his left hand.)

Fig. 29-30. Arm-grasp method.

Fig. 29-31. Raise victim's wrists up.

Fig. 29-32. Raise victim's wrists backward.

b. He applies compression by pressing the victim's hands against the abdomen. Active inspiration is obtained by raising the victim's wrists up and backward (Figs. 29-31 and 29-32). The use of this method assures that the victim's arms will not flail around and contact nearby energized equipment.

10. Artificial Circulation (External Heart Compression). Electric shock may disturb the natural rhythm of the heartbeat. When this occurs, the ventricle muscles may fibrillate or stop entirely; in either case, no blood is pumped, and the victim's brain will be irreparably damaged after about 4 min.

1. *How to Determine Heart Stoppage.* When the heart stops (or fibrillates), the following symptoms can be noticed:

a. Victim is unconscious.

b. Breathing is stopped.

c. No pulse is detectable.

d. The pupils of the eyes are greatly enlarged.

e. The skin usually turns a gray-blue color.

Since (*a*) and (*b*) may occur for other reasons than heart stoppage and (*e*) takes a few minutes to develop, the most accurate indication is lack of pulse. The best way to determine this is to place the pads of the fingers (which are more sensitive than the finger tips) alongside the victim's Adam's apple and check for pulse. If no pulse can be detected, the rescuer should also check the victim's pupils. If they are enlarged and do not narrow down in response to bright light when the lid is raised, the heart has stopped or is fibrillating. (Observing the pupils should be a countercheck on the pulse, because some eye defects cause the same effect.)

2. *External Heart Compression.* If all indications are that the heart has stopped or is fibrillating (no blood is being pumped), the victim should be placed on his back on a hard surface, such as the ground, the bed of a truck, the floor of a building, etc. The rescuer then kneels at the victim's side, gives the victim half a dozen quick breaths (see mouth-to-mouth method), and starts heart compression.

To do this, he places the heel of one hand on the lower third of the victim's breastbone (sternum) and the other hand on top of the first. A simple way to find this spot is to put a finger on the xiphoid (bottom end of the breastbone) and place the hand alongside it (Fig. 29-33). With the fingers extended so that no pressure is applied to the ribs, he presses down firmly and quickly, so that the breastbone is depressed about 1½ to 2 in. (With children, the pressure from one hand is sufficient; with babies, two fingers suffice.) He then releases the pressure. This cycle is repeated once a second.

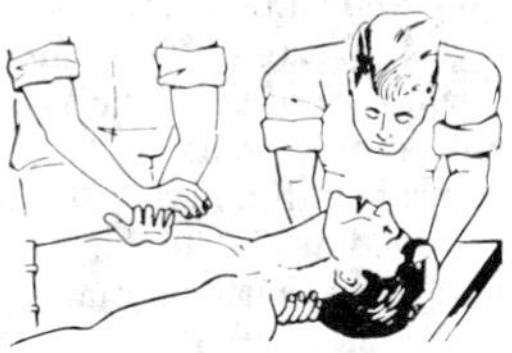

FIG. 29-33. Heart compression.

If the rescuer is working alone, he must interrupt heart compression about every 15 or 20 strokes to give two or three breaths of air to the victim. If another rescuer is available, one rescuer should apply heart compression (60 c/min) while the other gives mouth-to-mouth respiration (12 c/min).

3. *When to Stop.* Heart compression (and mouth-to-mouth respiration) should be continued until:

a. The victim revives.

b. Medical help arrives (or the victim is taken to a hospital or doctor's office).

c. Rigor mortis sets in.

Note that, even after the victim revives, he must be closely watched, because he may have a relapse and require resuscitation again. Also, note that both heart compression and respiration *must be continued in the ambulance and hospital room until a doctor "takes over."*

4. *When Heart Compression Should Not Be Done.* Heart compression should *not be done* when the victim's pulse can be felt, or when the ribs are broken, or when the pupils do not remain widely dilated.

11. Artificial Circulation (Portable Defibrillator). Under the direction of Dr. W. B. Kouwenhoven of Johns Hopkins University, a portable-type defibrillator has been developed to give an electric countershock which is effective in arresting ventricular fibrillation. The apparatus provides for a simulated a-c countershock produced by discharging two capacitors in sequence, the first positive and the second negative. The electrodes are placed above and below the victim's heart, and the discharge is controlled through insulated hand-held electrodes that are actuated through spring contacts. Such defibrillators are most effective if immediately available. Otherwise, the external heart-compression technique must be started immediately.

The use of the countershock defibrillator is obviously much safer than the open-chest method and appears to be as effective in cases of ventricular fibrillation. The open-chest method involves a major surgical procedure, either for countershock or direct heart massage.

ELECTRIC SHOCK

12. Breaking the Contact. The victim must be freed from contact with the live conductor as promptly as possible. Whenever possible, try to have the power turned off. If you decide to attempt rescue, you should be on a dry surface (nonconductor) and then use a dry stick (or a known nonconductor) to remove the victim from the contact or the contact from the victim. Extreme care must be taken to avoid contact with a live conductor regardless of what it is.

13. Action of Electric Current. In electric shock, the current may pass through the breathing center at the base of the brain and cause this center to stop sending out the nerve impulses which act upon the muscles responsible for breathing. As a consequence, breathing stops abruptly. In such cases, starting artificial respiration immediately substitutes for the natural respiration of the victim. If the shock has not been severe, the breathing center recovers after a time and resumes its vital function; but the current may so paralyze the breathing center as to require several hours for recovery, and artificial respiration must be continued unceasingly until there is recovery or positive evidence of death.

Under the foregoing conditions, victims of electric shock are unconscious, but heart action and blood circulation continue. Recovery depends upon prompt and effective lung ventilation until normal respiration is restored.

14. Current Tolerance of the Human Body. Alternating currents of high frequency produce little sensation, compared with alternating currents of low frequency and equal strength. Tests have been made to determine the "tolerance current" of typical individuals at several frequencies. The tolerance current was arbitrarily assumed as the limiting current strength which the subject could take through his arms and body, without marked discomfort or distress.

A man can tolerate only about 30 mA at 11,000 c/s but can tolerate nearly ½ A at 100,000 c/s. Although the tolerance current was found to increase very rapidly above 11,000 c/s, the increase between 60 and 11,000 c/s was much less rapid, ranging from about 5 mA at 60 c/s to 30 mA at 11,000 c/s.

An investigation made by F. M. Farmer of the amount of 60 c/s current which will produce mild shock disclosed that the average current at which 42 men first observed the sensation of shock upon sudden contact was 1.2 mA and that the average of the maximum currents which the men could withstand without serious discomfort was 8.0 mA.[23]

Voltages as low as 12 V, 25 to 60 c/s alternating current, have been shown by test to be all that some individuals could withstand under conditions of good contact with the circuit. Other tests have shown that, when dry electrodes were held in both hands, a-c voltages ranging from 20 to 40 were all that were required to produce the maximum current that individuals could withstand for a short time and still have voluntary control of their muscles. In these tests the currents ranged from 6 to 10 mA. Tests with direct current indicated that slightly higher current values could be withstood for a short time until a hot spot occurred at the point of contact.

The comparative immunity to injury in handling circuits up to 120 V with bare hands is due to the high resistance of the dry, uninjured human skin. This resistance is greatly lowered by any thorough wetting of the skin, especially by perspiration and many chemicals, and also by cuts and blisters. Under such conditions 110 to 120 V alternating current can produce a change in heart action which may prove fatal. Furthermore, it requires only a few seconds' contact with a 50-V standard-frequency a-c circuit to produce blisters at the point of contact, thus destroying the dry-skin high-resistance protection.[11,15]

15. Other First Aid. Electric-shock victims often require first aid in addition to artificial resuscitation.

1. *Calling a Doctor and Ambulance.* Sometimes it is very difficult to decide when a doctor's services will be required or when an injured man should be moved by ambulance. Certain companies have fixed policies in this respect, and in most of the severe injuries where there is a possibility of fracture, when there is bleeding of any significance, or when the injured man has lost consciousness (even for a short time), it is obvious that the victim should be seen by a doctor.

In other situations, where a man has not been so severely injured or there has been no apparent injury, a rough rule of thumb can be applied: if the man has lost consciousness for any reason (trauma, shock, diabetic coma, epileptic seizure, etc.) or has suffered any change in the vital signs (rate of pulse, rate of breathing, fullness of pulse, etc.), a doctor should see him.

An ambulance should be called to transport any victim who (*a*) has had more than a momentary loss of consciousness, (*b*) has sustained an actual or suspected fracture of the lower extremities or back, or (*c*) has had some injury which may have damaged internal organs (such as a blow to the kidney region or under the ribs, where the spleen might be ruptured). Victims who are ill without any apparent injury should also be transported by ambulance, particularly if their illness has been sufficiently severe so that there have been changes in pulse rate, breathing, or condition of the skin (cool and clammy instead of dry and warm).

2. *Treatment of Traumatic Shock.* Traumatic shock is the relative collapse of the blood-vessel system (lowered blood pressure, rapid and weak pulse, cold and moist skin). It may result from fear, pain, or other nervous stimulation without any loss of blood. The initial treatment consists in keeping the victim warm and quiet (be careful of excessive heat, which will increase the condition of shock) and keeping him lying down, preferably prone with the legs elevated slightly.

Shock may also occur because of loss of blood brought about by burns, severe injury, or hemorrhage. This is not so easily counteracted by the first-aider. Wounds should be treated and bleeding stopped as well as possible under the circumstances, but every effort should be made to transport the victim to a hospital or doctor's office as soon as it can be done properly.

3. *Treatment of Other Injuries*

a. Minor injuries

1. Burns, cuts, abrasions, scratches. Small first-degree burns (redness of skin only) and small cuts, abrasions, and scratches should be treated by gently cleansing the skin with soap and water and then covering the area with a sterile gauze dressing. If the injury is on the face, cleansing alone may be sufficient, although sterile petrolatum may be applied to burns to keep the burned skin from cracking.

b. Major injuries

1. Bleeding. Simultaneously with the restoration of breathing, serious bleeding or hemorrhaging must be controlled before any other treatment is given. This is best done by the pressure of a sterile dressing directly on the wound. Tourniquets should be used *only* if firm pressure on the wound fails to stop the bleeding.
2. Burns. Second-degree burns (where blister formation is expected) and third-degree burns (where skin is burned through) should be covered with sterile dressings without washing. No antiseptic, grease, petrolatum, "burn ointment," or similar substance should be applied, because these will have to be removed subsequently (sometimes painfully) and may contaminate the relatively sterile burned areas.
3. Fractures. All fractures should be seen by a doctor. Pending the doctor's arrival or transportation of the victim to the doctor, the fractured part should be immobilized with splints or with improvised splints made of rolled-up newspapers, sticks, tool shafts, etc. The injured part should not be moved until it has been so immobilized. The splinting material should be padded if it is firm (like a stick or ax handle) where it rests against bony prominences. It is particularly important that injured persons who are suspected of having a fracture of the spine anywhere from the head on down to the sacrum not be moved until the whole spine has been immobilized. If a fracture of the neck is suspected, the head should be pulled gently and slightly extended in the axis of the body proper and held while the rest of the body is being manipulated. If the injured person is conscious and complains of numbness and inability to move his body below the neck, *he should not be moved at all until professional help is on hand.*

MECHANICAL RESUSCITATION

16. Mechanical resuscitators are in use in hospitals, by fire and police departments and other rescue squads, in various branches of the armed services, and by some industrial concerns. One type is designed for intermittent positive-pressure breathing. Another type of mechanical resuscitator supplies oxygen at positive pressure and then aids exhalation by applying negative pressure.

Another type is the respirator, or "iron lung," used in hospitals and other first-aid centers.

Prompt application of resuscitation is essential; consequently, manual resuscitation should be applied immediately. If mechanical resuscitators are used as "second aid," they should be applied only by persons trained in their use or under the direction of a competent physician.

Modern mechanical resuscitators for the application of artificial respiration in the field are of either the "blow" or the "suck-and-blow" type. They are as safe as the manual methods (except when used in conjunction with external heart compression, as indicated in the next paragraph), provided that the pressures exerted by them are within safe limits and they are operated by experts especially trained in their use.

All "pressure-cycled" mechanical resuscitators are ineffective when used simultaneously with external heart compression, for the device which controls the "pacing" of the resuscitator is affected by the compression of the victim's chest in such a way as to cause the machine to operate too rapidly, resulting in inadequate air exchange.

In any case in which external heart compression must be applied to provide artificial circulation, no "pressure-cycled" mechanical resuscitator should be used to ventilate the lungs. Instead, mouth-to-mouth (mouth-to-nose) artificial respiration should be applied to provide adequate lung ventilation.

This statement parallels one contained in a statement* of policy approved by the American Heart Association, Inc., American National Red Cross, Industrial Medical Association, and the U.S. Public Health Service.

In a case requiring lung ventilation only, a proper mechanical resuscitator may be used if available. Remember, however, that no mechanical resuscitator is superior to correctly performed manual artificial respiration.

Manual artificial respiration should never be delayed for any purpose, including waiting for a mechanical resuscitator to arrive.

BIBLIOGRAPHY

17. References on Resuscitation

1. Maclachlan, W. *Jour. Ind. Hyg.*, 1930, Vol. 12, p. 291.

2. Bates, Gordon, M.B., Gavy, R. E., M.D., and Maclachlan, W., B.A.Sc. The Need for Prolonged Artificial Respiration in Drowning, Asphyxiation, and Electric Shock; *Can. Med. Assoc. Jour.*, August, 1938.

3. Gordon, A. S., Raymon, F., Sadove, M., and Ivy, A. C. Manual Artificial Respiration; *Jour. Am. Med. Assoc.*, 1950, Vol. 144, pp. 1447–1452.

4. Gordon, A. S., Fainer, D. C., and Ivy, A. C. Artificial Respiration; *Jour. Am. Med. Assoc.*, 1950, Vol. 144, pp. 1455–1464.

5. Gordon, A. S., Sadove, M., Raymon, F., and Ivy, A. C. Evaluation of Manual Methods of Resuscitation; *Elec. Eng.*, January, 1952, Vol. 71, pp. 38–40.

6. Gordon, A. S., Frye, C., and Sadove, M. S. Comparative Studies of New Push-Pull Methods for Pole-top Resuscitation; *Elec. Eng.*, February, 1953, Vol. 72, pp. 132–137.

7. Oesterreich, E. W. Discussion of article on Comparative Studies of New Push-pull Methods for Pole-top Resuscitation; *Elec. Eng.*, February, 1953, Vol. 72, pp. 137–140.

8. Drinker, Cecil K., M.D. The Use of Drugs in Resuscitation from Electric Shock; *Jour. Am. Med. Assoc.*, June 30, 1945.

9. Talbott, S. A., M.D. (Johns Hopkins Hospital) Practical Effects of Electricity on the Heart; *AIEE Conf. Paper* 55–239, February, 1955.

10. Kouwenhoven, W. B., and Milnor, W. T. (Johns Hopkins University) Electric Defibrillation; *AIEE Tech. Paper* 55–95, February, 1955.

11. Report of Investigation of Electric Shock as It Pertains to the Electric Fence; *Underwriters' Lab., Inc., Subject* 855, September, 1939.

12. Dalziel, Charles F., Lagen, John B., and Thurston, Joe L. (University of California) Electric Shock; *AIEE Tech. Paper* 41–7, November, 1940.

13. Dalziel, Charles F., Ogden, Eric, and Abbott, Curtis E. (University of California) Effect of Frequency on Let-go Currents; *AIEE Tech. Paper* 43–134, July, 1943.

14. Dalziel, Charles F. Effect of Wave Form on Let-go Currents; *AIEE Tech. Paper* 43–135, July, 1943.

* *Circulation*, May, 1965, Vol. 31, No. 5.

15. DALZIEL, CHARLES F. (University of California) Dangerous Electric Currents; *AIEE Tech. Paper* 46-112, May, 1946.

16. KARPOVICH, PETER V. "Adventures in Artificial Respiration"; New York, Association Press, 1953.

17. "Resuscitation in Gas Asphyxia, Electric Shock and Drowning"; American Gas Association, 1955.

18. "American Red Cross First Aid Textbook"; rev. 1945 (and resuscitation supplement, ARC 1086).

19. MOTLEY, H. L., COURNAND, A., WERKO, L., DRESDALE, D. T., HIMMELSTEIN, A., and RICHARDS, D. W., JR. Intermittent Positive Pressure Breathing, a Means of Administering Artificial Respiration in Man; *Jour. Am. Med. Assoc.*, 1948, Vol. 137, p. 370.

20. HANDFORD, S. W., et al. Positive and Alternate Positive-negative Pressure Resuscitation from Curare Poisoning; *Jour. Appl. Physiol.*, March, 1951.

21. MALONEY, J. V., M.D. (Johns Hopkins Hospital) Importance of Negative Pressure Phase in Mechanical Respirations; *Jour. Am. Med. Assoc.*, May 16, 1953.

22. OESTERREICH, E. W. Pole Top Resuscitation; *EEI Bull.*, March, 1939.

23. *Elec. World*, Apr. 27, 1929, p. 831.

24. GEIGES, K. S. Electric Shock Hazard Analysis; *AIEE Conf. Paper*, October, 1956.

25. GORDON, ARCHER S., M.D., and FRYE, CHARLES, M.D. A Comparison of Mouth-to-mouth and Manual Artificial Respiration Techniques; *AIEE Conf. Paper*, October, 1957.

26. KOUWENHOVEN, W. B., and MILNOR, W. R., M.D. Field Treatment of Electric Shock Cases, 1; *AIEE Trans. Paper*, January, 1957.

27. PRICE, L. D., and MICHENER, HOWARD P. Electrical Safety Regulations in the United States, Canada and Europe; *AIEE Conf. Paper*, June, 1959.

28. DALZIEL, CHARLES F. Temporary Paralysis Following "Freezing" to a Wire; *AIEE Trans. Paper*, February, 1960.

29. DALZIEL, CHARLES F. Threshold 60-cycle Fibrillating Currents; *AIEE Conf. Paper*, February, 1960.

30. KOUWENHOVEN, W. B., KNICKERBOCK, G. G., MILNOR, W. R., and JUDE, J. R. Field Treatment in Electric Shock Cases, II; *AIEE Trans. Paper*, February, 1960.

31. DALZIEL, CHARLES F. Improvements in Electrical Safety; *AIEE Conf. Paper*, February, 1961.

32. RORDEN, HAROLD L., and LANE, JOHN E. Safety Aspects of Working by the Live Line Barehand Method; *IEEE Conf. Paper*, January, 1963.

33. KOUWENHOVEN, W. B., and KNICKERBOCKER, G. G. Status of Closed Chest Cardiac Massage; *IEEE Conf. Paper*, October, 1964.

34. Edison Electric Institute, Resuscitation Manual, June, 1965.

COORDINATION OF ELECTRIC SYSTEMS

GENERAL CONSIDERATIONS

18. Need for Coordination. It is well recognized that electric-power, transportation, and communication services are essential elements of modern civilization. The various wire-using organizations render their separate and distinct services to the same communities and usually to the same individuals, so that it frequently is necessary that the instrumentalities which they use be closely interrelated. For example, the distribution lines of the telephone company, the distribution lines of the power company, the transit system, the fire and police alarm systems, etc., must all be on the same city streets and alleys. The long-distance lines of the telephone companies, the intercity circuits of the telegraph companies, the transmission lines of the power companies, the rights-of-way of main-line railroads, etc., all run between the same communities and must of necessity be on routes which are more or less generally parallel.

The fact that the structures and circuits used in these various services must frequently be in proximity requires that they be "coordinated" to avoid structural interference, electrical interference resulting from contacts, and inductive interference resulting from the interrelation of the circuits through the electric and magnetic fields inherent in the operation of all electric circuits.

19. Principles and practices have been developed and adopted by many organizations in order to promote the orderly handling of coordination matters. They all are based upon the principle that the public interest is paramount, that all the services are essential to the public, and that, therefore, the wire-using organizations must find ways and means to avoid interference between the different services.

The following groups have adopted "principles and practices": Edison Electric Institute and Bell Telephone System,[25] American Railway Association and Bell Telephone System,[26] Edison Electric Institute and Western Union Telegraph Company,[27] Association of American Railroads and Edison Electric Institute.[28,29] These generally accepted principles and practices are used as guides by the operating organizations in these groups and have proved to be very advantageous, particularly since the areas in which these organizations operate are widespread and are not coextensive.

As an aid in the application of these principles and practices, many cooperative studies have been carried on, the most extensive being those conducted by the Joint Subcommittee on Development and Research of the Edison Electric Institute and the Bell System. The results of these studies form the principal technical background for this discussion of inductive and structural coordination.

20. Wire-using systems involved in coordination problems can be grouped under two general heads: one comprising those systems which are characterized by relatively low energy level and various degrees of **susceptiveness** (see definition, Par. **28**), and the other embracing the systems which are characterized by relatively high energy level and various degrees of **influence** (see definition, Par. **27**. The first group consists of systems which can be classed generally under the heading of communication and signal type, and the second group comprises the systems which can be classed generally under the head of power and railroad type as follows:

Communication and Signal Type	Power and Railroad Type
Wire telephone and telegraph	Commercial power and light
Carrier telephone and telegraph	Surface, elevated, and interurban railroads
Block signal and train control	Electrified trunk railroads
Centralized traffic control	Customer utilization equipment
Supervisory and remote control	Industrial electric systems
Fire and police alarm systems	Isolated private systems

DEFINITIONS

21. Inductive Coordination. The location, design, construction, operation, and maintenance of supply and communication systems in conformity with harmoniously adjusted methods which will prevent inductive interference.

22. Structural Coordination. The location, design, construction, operation, and maintenance of supply and communication systems in conformity with harmoniously adjusted methods which will prevent structural interference.

23. General Coordinative Methods. Those methods reasonably available for general application to supply or communication systems which contribute to structural or inductive coordination without specific consideration to the requirements for individual situations.

24. Specific Coordinative Methods. Those additional methods applicable to specific situations where general coordinative methods are inadequate.

25. Inductive Exposure. A situation involving supply and communication circuits where the conditions are such that inductive coordination must be considered.

26. Inductive Interference. An effect due to the inductive influence of a supply system, the inductive susceptiveness of a communication system, and the inductive coupling between the two systems, of such character and magnitude as to prevent the communication system from rendering satisfactory and economical service.

27. Inductive Influence. Those characteristics of a supply circuit with its associated apparatus that determine the character and intensity of the inductive field which it produces.

28. Inductive Susceptiveness. Those characteristics of a communication circuit with its associated apparatus which determine the extent to which its operation may be affected by inductive influence.

29. Coupling. The interrelation of electrical circuits by electric or magnetic induction or both, or by conduction through a common earth path, or by combinations thereof.

30. Structural Exposure. A situation involving supply and communication facilities where the conditions are such that structural coordination must be considered.

31. Structural Interference. A condition arising from the structural relationship of supply and communication facilities which by reason of the possibility of contacts or conduction between the respective facilities, or by reason of their proximity, prevents the safe and economical operation of either system.

32. Conflicts or Conflicting Construction. Situations where two separate pole lines parallel each other in close proximity, under conditions defined more specifically in the National Electrical Safety Code.[3]

33. Discontinuity. A point at which there is an abrupt change in the physical relations of supply and communication circuits or in electrical constants of either circuit.

Transpositions are not rated as discontinuities, although technically included in the definition, because of their application to coordination.

34. Transposition. An interchange of position of conductors of a circuit between successive lengths.

35. Coordinated Transpositions. Transpositions which are installed in either supply or communication circuits or in both for the purpose of reducing inductive coupling and which are located effectively with respect to the discontinuities in both the supply and communication circuits.

36. Power Level. The level of the electrical power flowing in a communication circuit. At any point the power level depends on the conditions of input and of losses between the point of input and the designated point.

37. Abnormal Operating Conditions. Electrical operating conditions resulting when operating arrangements other than normal are established.

38. Fault Conditions. Conditions resulting when a fault to ground or a short circuit occurs on an electrical supply circuit.

COOPERATIVE PROCEDURE

39. Cooperative arrangements between the various wire-using organizations serving the same general territory are necessary in order to carry out coordination work. In most localities there is direct interchange of information and discussion between the engineers of these organizations regarding matters of mutual interest. This can be most effectively carried out where a specific individual or position is designated in each organization to be primarily responsible for coordination matters and where these individuals maintain close contacts with each other.

40. Cooperative planning greatly simplifies coordination. It involves joint review by the engineers of the organizations involved of plans relating to the type and location of lines and major equipment before definite commitments are made.

41. Cooperative studies of specific coordination problems are made by engineers of the organizations involved to determine the facts as to whether coordinative measures are needed and, if so, what measures should be applied.

LINE LOCATION

42. Coordination of line location is an important factor in cooperative planning. It involves consideration of either or both of the following: whether **joint use** or **separate lines** should be employed, and where separate lines are to be used, their location with respect to each other so that unnecessary crossings, conflicts, and inductive exposures will be avoided and the application of coordinative measures will be facilitated.

In urban areas, lines of the various classes must be on the same streets and alleys, and joint use usually provides greater safety and is more economical than separate lines. A more detailed discussion of joint use is given in Pars. **46** and **48.**

Along highways in rural areas separate lines have been generally employed for power and communication circuits, although recent developments increase the applicability of joint use in rural areas. Where separate lines are to be used for power and communication circuits along a highway, one side of the highway should be employed for communication circuits and the other side for power circuits. This sometimes involves the common use of poles by two or more communication companies and, in some instances, by two or more power companies. Crossings from side to side should be avoided even where only one line is in place or is being constructed.

The same private right-of-way is seldom utilized by two wire-using organizations for important facilities except where railroad right-of-way is involved. The occupancy of a railroad right-of-way by circuits of another organization ordinarily involves a contractual relationship between the other organization and the railroad.

In the use of generally paralleling rights-of-way for major transmission lines and trunk communication lines, cooperative advance consideration should be given to various possible routes so that full advantage can be taken of separation along with other measures to obtain the best and most economical coordination.

STRUCTURAL COORDINATION

43. Coordinated methods of construction have been worked out by the various wire-using organizations in order to minimize the probability of contacts between circuits of the power or railroad type and circuits of the communication or signal type and to provide safe working conditions. These methods include the establishment of relative levels for lines of various classes, strength of construction, clearances, and other arrangements which will minimize the probability of structure and conductor failure from mechanical or electrical causes. For example, it is recognized that the higher-voltage lines (which usually employ the larger and stronger wires) should generally occupy the upper positions. It is also recognized that climatic loading (ice, wind, and temperature) on conductors and structures varies widely over the United States—and even between localities not far apart—and that these conditions must be taken into account. The National Electrical Safety Code[3] contains minimum requirements for clearances, strengths of conductors, strengths of supporting structures, etc., under different climatic loading conditions, and its provisions are generally accepted as a guide to practice throughout the United States in connection with crossings, conflicts, and joint use.

In some cases, the coordinated methods of construction have been incorporated in jointly prepared specifications.[1,2,29]

44. The coordination of electrical protection of the circuits involved in situations of proximity is of great importance in limiting the adverse effects of contacts between circuits. In the power system, such coordinative measures usually consist of relaying or fusing systems which will ensure prompt deenergization of the circuit in the event of a contact. In the communication system, these measures usually consist of arrangements and protective devices which will provide a safe path to ground for the power currents until the power circuit is de-energized.

45. Crossings and conflicts between various classes of circuits can be kept at a minimum by cooperative planning. This not only reduces the probability of contacts but frequently results in overall economies by reducing the amount of special construction required.

At crossings, the probability of contacts can be minimized by adequate clearances and strengths and by coordination of structure location. The use of a common crossing pole may be advantageous and facilitates coordination of protection.

46. Joint use of poles between power circuits of the 2,300- and 4,000-V class, as well as the utilization voltage circuits and communication circuits, has been general for over thirty years. Since about 1940 there has been a decided trend toward increasing power-distribution voltages up to 13 kV and even to 33 kV in a few cases. At first, joint use of communication systems with these higher voltages was confined to cable plant, but later adequate structural and electrical coordination was adopted which permitted joint use with open-wire communication circuits.[13,23,76]

Specifications for construction and maintenance of jointly used wood pole lines carrying supply and telephone circuits were revised in 1945.[1] Information on joint-use construction methods will also be found in Sec. **13.**

Joint use between rural power (usually of the 13-kV class) and communication circuits has become quite general since. Experience has shown the improved methods of coordinated electrical protection to be quite effective.[22] Specifications for long-span joint construction were issued in 1946 by the EEI–Bell System Joint Committee on Plant Coordination.

47. Joint-use agreements are advantageous in order to provide definite understandings with regard to standards of construction, arrangements for carrying out cooperative advance planning, procedures when joint use is to be entered into or abandoned, rental rates (or other methods of handling the costs involved), and other pertinent matters. The tendency in recent years has been toward the type of agreement which covers a substantial area (such as a city or an operating area) rather than the type which covers only a single pole or a small group of poles.

48. Joint use of wire facilities by applying carrier on power-distribution circuits to extend telephone service in rural areas has been employed to a limited extent. Experience indicates that the electrical protection and structural coordination methods for ordinary joint use will provide a suitable guide on poles carrying telephone-carrier terminal equipment.

INDUCTIVE COORDINATION

49. Induction problems are divided into **noise-frequency problems,** i.e., problems relating to the harmonics of the power system which manifest themselves as noise in the telephone system; and **low-frequency problems,** i.e., problems relating to the fundamental frequency of the power system. These low-frequency problems are further subdivided into **normal** and **abnormal induction.** Normal induction relates to induction which occurs during the normal operation of the power system and is of a more or less continuous nature. When sufficiently severe, it manifests itself in disturbances to signaling in telephone circuits, or distortion of signal current in telegraph circuits. Abnormal induction relates to discrete occurrences at the time of faults to ground or other abnormal conditions in the power circuit. If sufficiently severe, it manifests itself in the communication circuit as high voltages which may result in protector operation, breakdown of insulation, or hazard to workmen.

In dealing with a specific inductive coordination problem it has been found advisable to study separately the three basic factors, namely, **inductive influence** (see Par. **27**), **inductive coupling** (see Par. **29**), and **inductive susceptiveness** (see Par. **28**). The net effect of an inductive exposure depends on all three of these factors so that, if any one of them could be reduced to zero, there would be no inductive effect, and if any one is unusually high, at least one of the others must be unusually low to maintain the same overall effect. Since it is impracticable, in most cases, to reduce any one of the three to zero, the practical problem of coordination is to so control all of them that adequate results are secured at reasonable overall cost.

COUPLING

50. Coupling. There are three types of coupling between power and communication circuits:

Magnetic—that due to the magnetic field set up by the currents in the power conductors which links the communication conductors and induces a voltage along them (see Fig. 29-34). The magnitude of the **voltage induced** is dependent upon the length of parallel, separation between the circuits, the configurations of the circuits (particularly whether the metallic or ground return circuit is being considered), earth resistivity, shielding, and the magnitudes and frequencies of the power currents. The **actual current** which will flow in the communication circuit is dependent primarily upon the induced voltage and the communication-circuit impedance since the coupling between the two circuits is, with rare exceptions, sufficiently loose so that the reaction in the power circuit of this current is negligible.

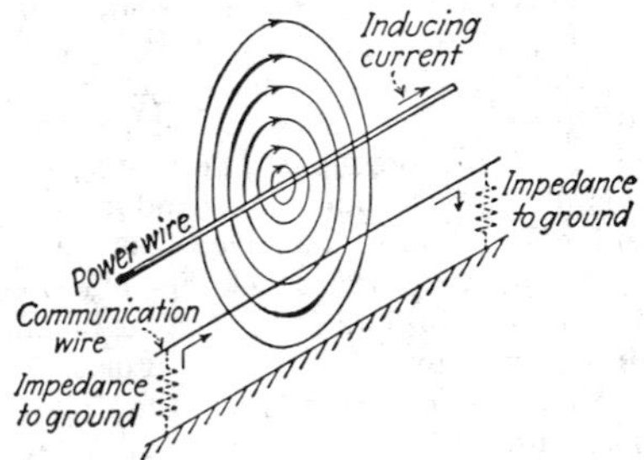

FIG. 29-34. Fundamentals of magnetic induction.

Electric—that due to the electrical field caused by the voltages on a power line and within which the communication circuit lies. The fundamental equivalent circuit involved in this situation is illustrated in Fig. 29-35. It will be noted that this circuit is equivalent to a capacitance between the power line and ground (C_{PG}), a capacitance between the power line and the communication line (C_{PT}), and a capacitance from the communication line to ground (C_{TG}). Since the capacitance between the power line and ground is of relatively little significance as far as coupling is concerned, and if the admittances to ground on the communication circuit are neglected, this whole arrangement may be considered with a fair degree of approximation as a capacitor potentiometer and the magnitude of the voltage on the communication line is dependent on the **relative magnitudes** of the two capacitances (C_{PT} and C_{TG}) and not on their **absolute magnitudes.** In other words, the actual magnitude of the **electric voltage induced** is governed only by the configuration of the parallel and the voltage on the power line, and not by the length of the parallel or the frequency of the power voltage.

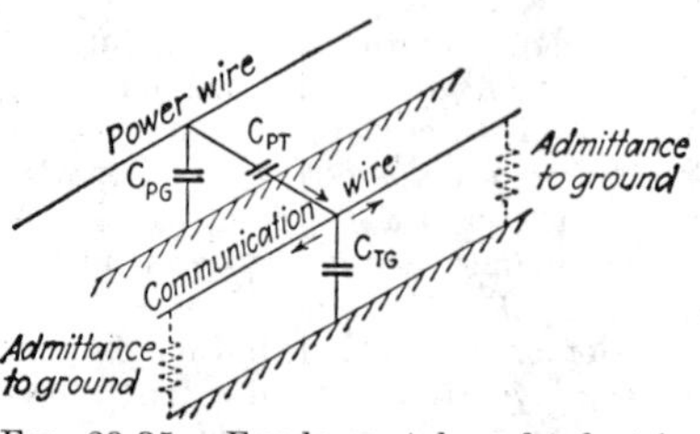

FIG. 29-35. Fundamentals of electric induction.

In the actual field case, of course, the communication line usually extends beyond the exposure and is connected to equipment and hence has admittances to ground in addition to the capacitances within the exposure. (These admittances are shown dotted in the figure.) As a rule these admittances are large compared with those within the exposure, and in this event the communication-line voltage is determined by the current which can be transferred to the communication line and the total admittance of this line to ground. This current is proportional (approximately) to the magnitude of the capacitance between the power and communication lines and to the magnitude and frequency of the power-line voltage. The magnitude of the capacitance is proportional to the length of the exposure, so that the net result of electric coupling is (like magnetic coupling) proportional to exposure length and frequency.

Conductive—that due to power-line current flowing through an impedance which is common to both the power and communication circuits; usually a common ground resistance. Conductive coupling is usually unimportant except in the vicinity of a grounding point on the power system and where the conditions are such that all or part of the power-system ground potential (fundamental frequency) may be impressed on the communication circuit.

51. Noise-frequency coupling coefficients for both electric and magnetic induction for exposures at 20 ft or greater separations have been published.[41,42] Data are given on the eight coupling coefficients used in noise-frequency work; i.e., direct metallic-circuit and longitudinal-circuit induction due to balanced and residual voltages; and direct metallic- and longitudinal-circuit induction due to balanced and ground-return currents. For this discussion these terms have the following meanings:

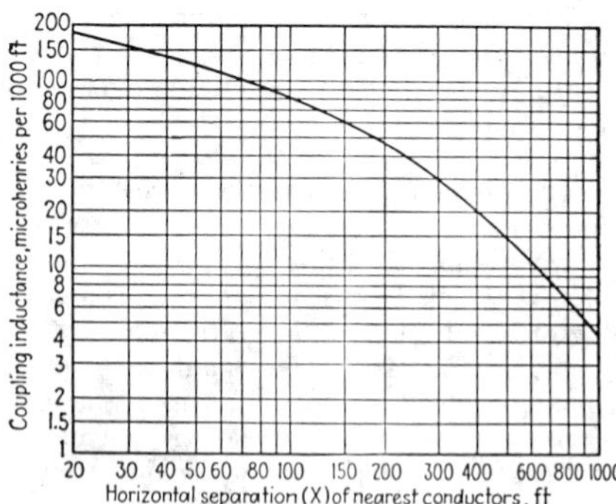

FIG. 29-36. Longitudinal-circuit induction from ground-return current, any configuration. Do not use this chart for estimating induced voltages at fundamental frequency.

Direct metallic-circuit induction means the difference in the induced voltage on or along the two sides of a communication circuit; i.e., it is the induction acting in the metallic circuit.

Longitudinal-circuit induction means the voltage induced **along** the wires of a communication circuit (or line) in parallel by magnetic induction or **between** the wires and ground by electric induction.

Balanced voltages or currents are those components of the power-circuit voltages or currents which add vectorially to zero.

Residual voltages and currents are, respectively, the vector sum of the voltages to ground of the power wires and the vector sum of the currents in the power-phase wires.

Ground-return current is the vector sum of the currents in all the wires of a power line—including the neutral or ground wire, if present. If there is no neutral or ground wire, the **residual** and **ground-return** currents are the same.

In many noise-frequency induction problems the longitudinal induction from residual voltages and from ground-return power currents are the controlling components. Curves for these two quantities are given in Figs. 29-36 and 29-37. The electric-induction curve is given in terms of volts induced on the communication line per kilovolt residual voltage, and it is, as discussed above, independent of frequency and length of exposure. The magnetic-coupling curve is given in terms of microhenrys per kilofoot of exposure.

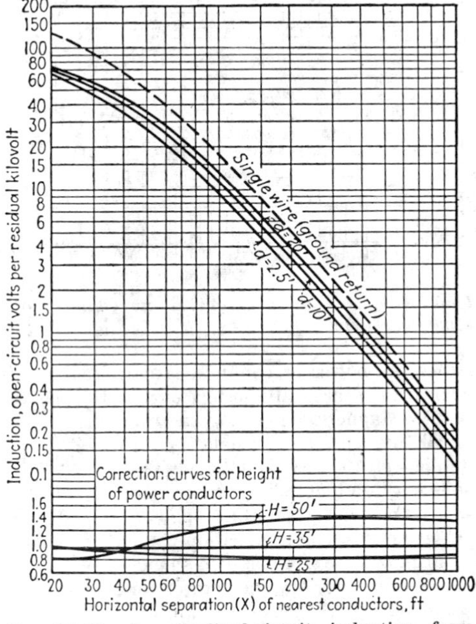

FIG. 29-37. Longitudinal-circuit induction from residual voltage of 3- or single-phase circuit, any configuration. For single-phase circuits multiply ordinates by 1.2.

For joint-use situations these coefficients cannot be used, and special methods are required.[34,38]

The coupling coefficients give what might be called the "raw" coupling, which is only one of many factors which determine the net voltages and currents in the communication circuit. These factors are discussed later in connection with shielding, influence, and susceptiveness.

52. Low-frequency Coupling Coefficients. Low-frequency induction is most commonly concerned with magnetic induction from ground-return power-line currents into the longitudinal communication circuit. In cases involving ground-return telegraph circuits at close separation, magnetic induction from balanced components frequently is important. The coupling at the fundamental frequency of the power circuit is generally calculated on the basis of theories developed by Carson and others and experimentally verified by the Joint Subcommittee on Development and Research. Figure 29-38 shows coupling curves for 60 c for three values of earth resistivity, viz., 10, 100, and 1,000 m·Ω, based on this theory.[6,39]

In the use of these coupling curves it is necessary to know the earth resistivity at the particular location. While, in some cases, approximations of the earth resistivity can be reached, based on other measurements made in the vicinity (particularly if the geological conditions are similar), coupling tests are almost always required if reasonably accurate results are desired, particularly for wide separations.

Coupling can be determined by any of several methods, of which the following are the more common:

a. In overall coupling tests, the power circuit is grounded at one end and energized at low voltage to ground at the other. The longitudinal voltage in the communication line is measured, and this voltage divided by the power-line current is the coupling.

b. In probe-wire tests, longitudinal current is caused to flow in a communication or power circuit and the voltage induced in short wires, laid on the ground parallel to this primary circuit and at suitable distances from it, is measured. From these measurements and coupling curves the effective earth resistivity can be determined. The overall coupling can be determined by extrapolation or by computation using the effective earth resistivity as determined by the tests.

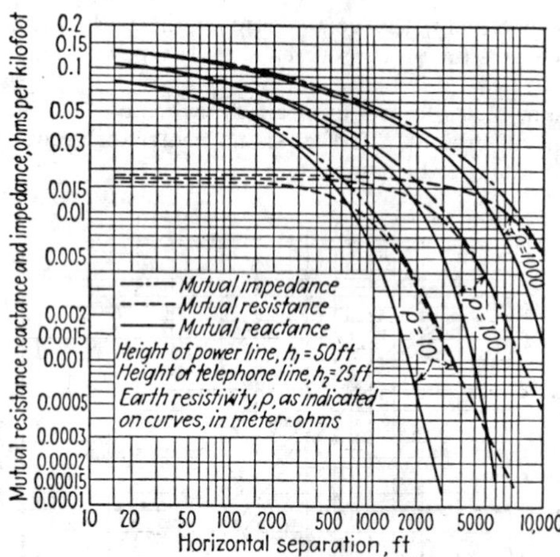

FIG. 29-38. Variation of mutual impedance with separation (60 c).

c. In d-c tests, the effective earth resistivity is determined from measurements of the mutual resistance between ground-return circuits of variable length. The overall coupling is then computed from the earth resistivity thus determined.

53. Electric shielding is provided by any metallic conductor grounded at one or more points. The most common and effective example of electric shielding is that provided by a cable sheath around either the power or communication circuits. The electric shielding provided by such a sheath is practically perfect even with relatively high-resistance grounds.

54. Magnetic shielding is provided by currents circulating in ground-return circuits.[17,51,73] The shielding conductors may be ground wires on the power line, power-cable sheaths, communication-cable sheaths, or specially installed shielding conductors on the power or communication line. In addition, paralleling railway tracks, pipelines, etc., afford some shielding. Also, conductors on which protectors have operated offer shielding, and in cable circuits working conductors add to the shielding because of their more or less effective grounding through terminal apparatus and through the distributed capacitance of the cable. The magnitude of this latter effect depends on the length and makeup of the cable and on the type of terminal apparatus employed.

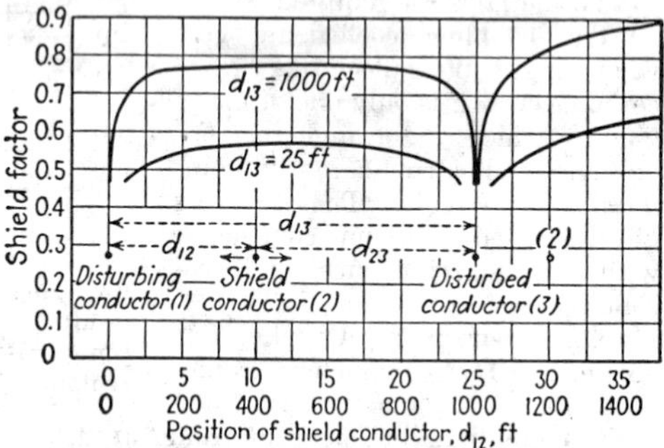

FIG. 29-39. Variation of shield factor with location of shielding conductor. Frequency 60 c. Earth resistivity 100 m·Ω. Shielding conductor No. 1/0 copper wire; perfect ground connections; location variable in the horizontal plane (elevation 40 ft) of the disturbing and disturbed conductors.

The term **shield factor** is frequently employed as a measure of shielding. This term is used for the ratio of the induced voltage which exists in the presence of the shielding to that which would exist if the shielding conductor were removed, on the assumption that the current in the disturbing conductor is unchanged. It depends on the relation of the mutual impedance between the shielding conductor and the disturbing and disturbed conductors to the self-impedance of the shielding conductor with earth return. Figure 29-39 shows the variation of shield factor with location of the shielding conductor with respect to the disturbing or disturbed conductors. Figure 29-40 shows the change in shield factor with frequency and grounding resistance for a 2/0 copper shield wire 3 ft from the disturbed conductor and for a cable sheath.

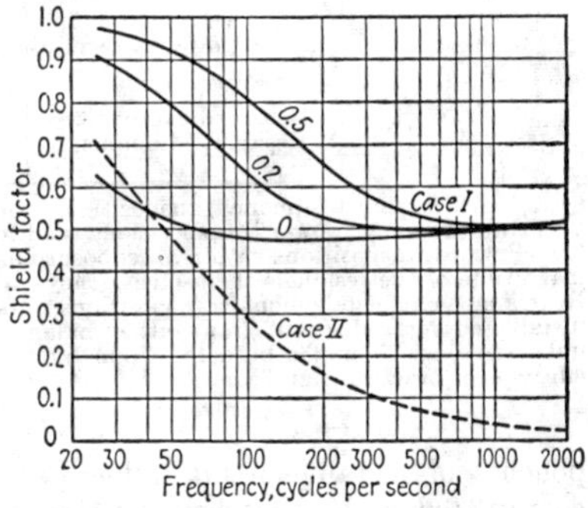

Fig. 29-40. Variation of shield factor with frequency. *Case* 1: No. 2/0 copper shield wire located 3 ft from disturbing conductor; grounding resistance in ohms per kilofoot of shielding conductor as indicated on curves (full lines). *Case* 2: Lead cable sheath surrounding disturbing conductor; resistance 0.11 Ω/kft; grounding resistance zero (dashed curve). Earth resistivity 100 m·Ω for both cases.

55. Transpositions. A communication circuit is said to be transposed when the two sides of the circuit reverse their respective pin positions on the line. A metallic-circuit voltage induced from an external source in a section of the circuit on one side of a transposition tends to be neutralized by the corresponding voltage in a section of the same length on the other side of the transposition, since the voltage in the second section is reversed with respect to the first by the transposition. Telephone transpositions have no effect on longitudinal-circuit induction or induction to ground, because this type of induction takes place in the circuit composed of the communication wires in parallel as one side and the earth as the other.

The neutralizing effect of power transpositions results from the change in phase of the electric and magnetic fields of the **balanced voltages and currents,** respectively, accompanying a change in the relative positions of the power wires. In a single-phase circuit this change is 180°, analogous to the change effected by a transposition in a communication pair. In a 3-phase circuit, where a transposition is made by changing the positions of each of the 3-phase wires, the phase change is 120°. In this case, in order to obtain maximum neutralization of the inductive fields from balanced components, two transpositions of identical type are necessary. Corresponding changes in the phase of the induction in the exposed communication circuits are dependent upon uniformity of exposure conditions. A section of power line, within which each phase conductor occupies each of the respective phase-conductor positions for approximately equal distances, is often termed a "barrel." Power transpositions have no effect on the induction from residual or ground-return components.

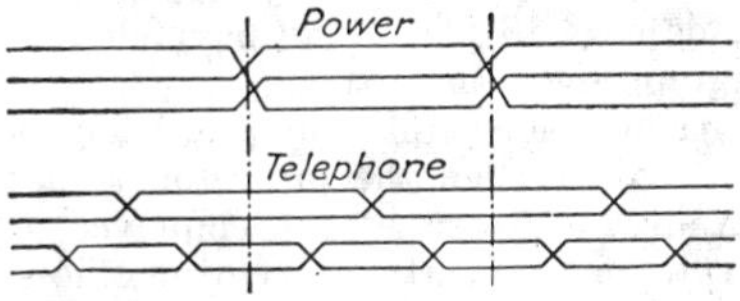

Fig. 29-41. Schematic layout of coordinated transpositions.

Transpositions are usually used on multicircuit telephone lines throughout their length to limit crosstalk. Various transposition systems have been developed for this purpose (see Sec. **24**). Some of them are particularly suitable for use in inductive exposures. When power and telephone transpositions are both employed in an inductive exposure, they should be "coordinated." The principle of coordinated transpositions is shown in Fig. 29-41.

56. Common functions of power and telephone transpositions[61] within exposures are summarized in the following table. In this table the heading "Direct metallic-circuit induction" refers to the voltages directly induced in the metallic telephone circuit within the exposure. The heading "Indirect metallic-circuit induction" refers to the voltages arising within the metallic telephone circuit due to the action of longitudinal-circuit induction or voltages to ground on telephone-circuit unbalances.

This table shows that transpositions in telephone circuits in exposed sections should be thought of from the coordination standpoint as a means of reducing the direct metallic-circuit induction from either balanced or residual currents or voltages of the power circuit. Similarly, power-circuit transpositions should be thought of as directed toward reducing only the longitudinal-circuit induction (and hence only the indirect com-

Functions of Transpositions within Exposures

Source of induction in power system	Direct metallic-circuit induction	Indirect metallic-circuit induction
Balanced components	*T*	*P*
Residual components	*T*	*

T indicates that telephone transpositions tend to reduce this component of induction.
P indicates that power transpositions tend to reduce this component of induction.
* Power transpositions within an exposure may tend to reduce this component only if they reduce the magnitude of the residuals themselves; they do not affect the induction from given amounts of residuals. Rearrangement of telephone transpositions outside of exposure sections may tend to reduce the indirect metallic-circuit induction (from either balanced or residual power-circuit components) if they reduce unbalances which might be acted upon by longitudinal-circuit induction. Ordinarily, neither of these effects is of practical significance.

ponents of metallic-circuit induction) from **balanced** currents and voltages. When coordinated with the telephone-circuit transpositions, the effect of the power-circuit transpositions on direct metallic-circuit induction will be small on the average.

Transpositions in power **distribution circuits** are usually of little value, because of the relatively large ratio of residual to balanced components inherent in such systems.[65]

57. Power Transpositions outside Exposure Sections. Theoretical considerations show that transpositions in power circuits outside exposure sections would reduce the magnitude of residual charging-current components at nontriple-harmonic frequencies. Experience shows that other factors are usually controlling in most coordination problems and that transpositions for this purpose are not warranted.

LOW-FREQUENCY INDUCTION—ABNORMALS

58. Low-frequency inductive coordination problems from **abnormals** on power lines generally involve discrete occurrences associated with faults to ground on the power system. Such situations are most important, of course, where the influence of the power system, combined with the coupling between the power and communication system, is sufficiently great to cause operation of protectors in the communication system. The disturbances which may occur in the communication system include service interruption due to protector operation, damage to plant, the possibility of acoustic or electric shock, false signaling, and telegraph-signal distortion.[17]

Interruption to communication service may result from permanent grounding of protectors, which may occur if the discharge current through the protector is not promptly extinguished, or if this current is very high. Protectors are employed as required for the protection of persons, equipment, and property at central offices, subscriber stations, repeater stations, etc.

59. Influence of the power system depends principally on the magnitude and duration of the abnormal current to ground which might flow through a parallel at the time of fault-to-ground and on the frequency of such occurrences. Calculations of fault current are usually made by the symmetrical-component method described in Sec. **15.** As a result of extensive field investigations it has been found advisable in inductive-coordination calculations to add a **fault resistance** of 20 Ω for a fault occurring along the line and 5 Ω for a fault occurring at a substation.

Frequency of occurrence of faults is an important factor in the evaluation of the influence of the power system in a particular exposure. This can be estimated from the general behavior of the particular system involved or proposed and also from the location of the exposure on the system.

Duration of fault current on a power system likewise is an important element in the determination of its influence, since the energy transferred to a communication system for a fault of short duration is much less than for one which lasts a long time. The duration is usually controlled by relay and circuit-breaker time except for self-clearing faults through protector tubes.

Remedial measures applicable to reduction of the influence of a power circuit[17] under abnormal conditions naturally involve measures associated with the reduction of the magnitude of the fault current (e.g., neutral impedance), the duration of the fault current (e.g., relays and protector tubes), and reduction in the number of faults (e.g., line clearances, construction, and insulation). It is evident that, with the possible exception of neutral impedance, these measures are the same as those commonly used to improve power-system operation. Even in the case of neutral impedance, its applicability and limitations are inextricably interwoven with power-system operation. For this reason, no attempt is made to discuss the measures in detail here, and the reader is referred to Sec. **15**.

60. Coupling can be determined by calculations or tests as described in Pars. **50** to **52**. In connection with the determination of coupling, full account must be taken of all shielding which is present in the parallel under consideration. Increased separation between power and communication circuits is, of course, the most obvious way of reducing coupling in a particular situation. When new power transmission or long-distance communication lines are contemplated, it frequently is practicable through cooperative advance planning to secure sufficient separation to be very helpful even where it is not practical to reduce the coupling sufficiently to eliminate the problem. For existing lines increased separation involves relocation of lines which, in many cases, would be unduly expensive. Coupling can in some cases be reduced by shielding associated with the power circuit or the communication circuit (see Par. **54**).

61. Susceptiveness of communication circuits to abnormal induction, as stated above, is generally associated with the breakdown of the protectors applied to the circuit. Figure 29-42 is a simplified diagram illustrating the application of such protectors. In Fig. 29-42*a* is illustrated a circuit open wire or cable between two terminals. In such case the protectors are applied at the terminals. In Fig. 29-42*b* is illustrated a combination circuit of open-wire and entrance cable. Here, in addition to the protectors at the terminals, protectors having a larger air gap are applied at the junction of the open wire and the entrance cable.

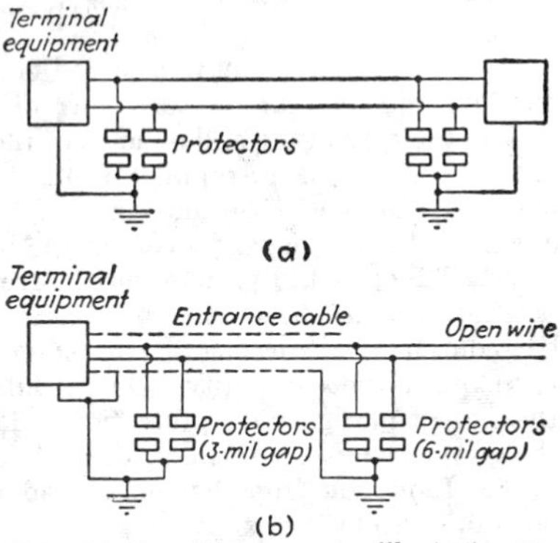

Relay protectors are devices consisting essentially of a relay with rugged contacts to short-circuit carbon block protectors almost instantly following the initial discharge of the protector and thus to protect them against permanent grounding. These may be either of the unit type for the protection of a single communication circuit or of the multigrounding type arranged so as to ground all wires on the line simultaneously.[66]

Neutralizing transformers installed along a communication line provide counter emfs which largely neutralize induced longitudinal voltage and permit the line to operate normally through inductive disturbances.[69] They have not been developed for carrier telephone circuits.

Fig. 29-42. Diagrams to illustrate application of telephone open-space protectors.

A unit-type neutralizing transformer of the three-winding type may be used to

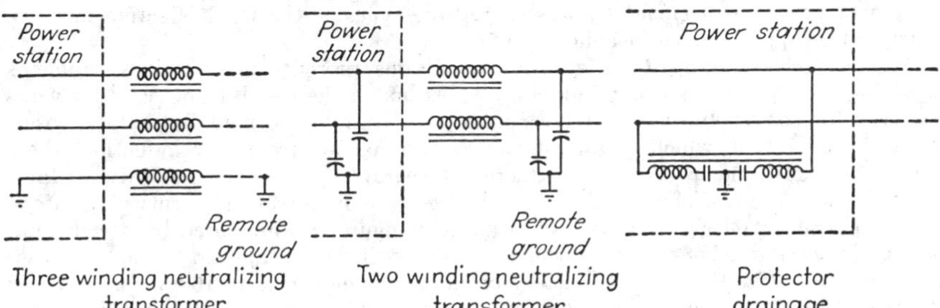

Fig. 29-43. Principles involved in protective arrangements to guard against interruptions to communication circuits supplying power stations.

maintain uninterrupted communication or signaling service to power stations during power disturbances which result in large induced voltages or power-station ground-potential rise.[11] Where metallic-circuit operation only is needed, two-winding transformers or a form of protector drainage can sometimes be used for the same purpose.[20] Figure 29-43 shows the general features of these three forms of protection. In an actual installation, these arrangements would be supplemented as necessary by additional protective features to prevent damage to the equipment due to lightning or to contacts between power circuits and the communication or signaling circuits.

Acoustic shock reducers of the copper-oxide rectifier type are the most frequently used means for reducing acoustic disturbance in operators' telephone receivers.[54] They are also effectively used on power-line dispatchers' circuits.

LOW-FREQUENCY INDUCTION—NORMALS

62. Low-frequency inductive coordination problems under **normal** operating conditions are generally associated with ground-return fundamental-frequency or third-harmonic currents.

63. Influence. In transmission lines ground-return current is generally due to induction into the ground wire caused by the load currents. The magnitude of this current is dependent on the magnitude of the load current, the coupling between the various phase wires and the ground wire, and the impedance of the ground wire and the resistance of the ground-wire grounds. This may be calculated from the configuration of the wires and the impedance of the ground-wire circuit.

The magnitude of the third-harmonic ground-return current on transmission lines is dependent on the number, size, and excitation of the transformers and their connections.

In 3-phase 4-wire multigrounded neutral distribution circuits the unbalanced fundamental-frequency earth-return current is dependent on the resistance of the neutral and on the magnitude and balance of the load currents and, of course, where all 3 phases are not present, on the number of phases. The third-harmonic component is dependent upon the number, size, and excitation of the various transformers connected to the circuit and is not dependent upon the balance of transformers between phases.

64. Effect of the ground-wire current in long exposures can be reduced by transpositions in the phase wires to balance the induction in the ground wire and can also be reduced by increasing the resistance of the ground wire or inserting insulators in it to break the continuity. The insulators can be bridged by gaps so as to provide the necessary lightning protection. The latter measures reduce the effectiveness of the ground wire as a shield against abnormal induction.

65. Induction from balanced load current can be reduced by coordinated transpositions within the exposure.

66. Coupling for normal conditions is determined by the same factors as for abnormal conditions.

67. Susceptiveness. The only type of telephone plant generally involved in problems due to normal fundamental-frequency induction is that using grounded ringers or other grounded signaling equipment. In most cases readjustment of the signaling equipment or its replacement by less susceptive types is the most effective measure which can be applied in the telephone plant.

68. The telegraph plant under some conditions presents more serious problems regarding normal low-frequency induction.[18] Although the trend in the past few years has been for conversion of major lines from ground-return to metallic-circuit carrier-frequency operation, which is not affected by normal low-frequency induction, there remain a large number of local commercial telegraph circuits in the United States which involve the use of ground-return circuits employing several forms of printers.

In the printer telegraph system each letter or figure is represented by a code consisting of several units of time. During each time unit there is applied to the line either positive or negative battery, the battery ordinarily being of 100 to 160 V. A measure of the speed of the circuit is the "dot frequency" of the circuit, which is the reciprocal of the time required for two consecutive units of the code character. For the higher-speed telegraph circuits the dot frequency may exceed the fundamental power frequency. Such circuits have a high susceptiveness to induced 25- or 60-c current. In the higher-speed multiplex circuits the 180-c component may also become important.

69. Multiwire neutralizing transformers—associated, of course, with a pilot wire through the exposure—may be used for neutralizing induced voltage (see Fig. 29-44). The transformers add impedance in each telegraph wire and also increase the cross fire between wires. These effects may be controlled by holding to a low value the resistance of the primary circuit. The limitation on resistance of the primary sets a limit on the length of exposure in which multiwire neutralizing transformers can be used economically on high-speed telegraph circuits.

By double tuning of the primary, neutralizing transformers can also be used for two frequencies, as shown in Fig. 29-45.

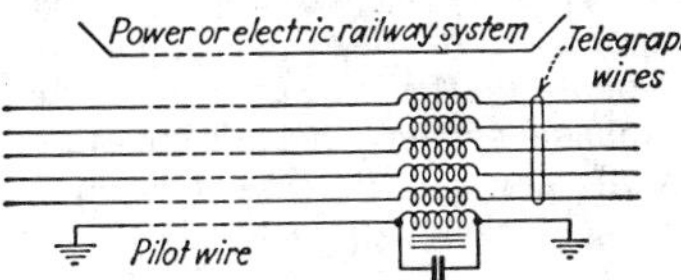

FIG. 29-44. Multiwire neutralizing transformer.

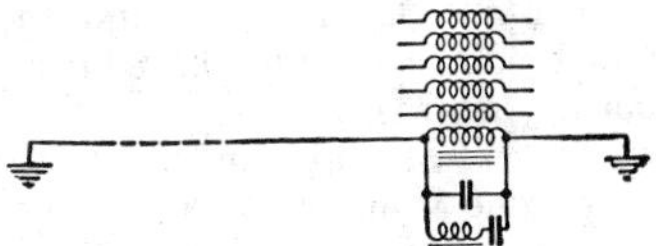

FIG. 29-45. Neutralizing transformer for correction of two frequencies.

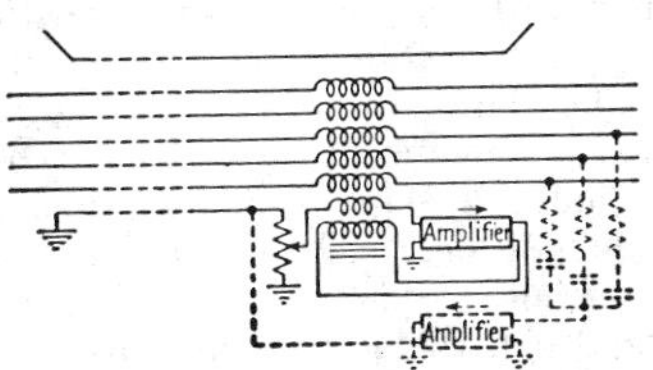

FIG. 29-46. Neutralizing amplifying system.

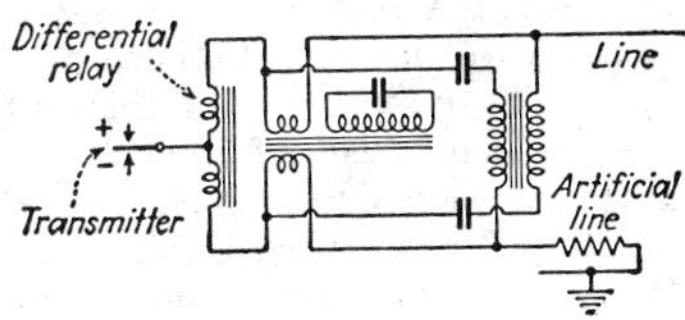

FIG. 29-47. Balanced double resonant network.

70. The neutralizing amplifying system is illustrated in Fig. 29-46. A transformer serves to impress on each telegraph wire a voltage to counteract the disturbing voltage induced in line wires. In this case the transformer is energized by a vacuum-tube amplifier. The voltage induced in the pilot wire is also opposed by the voltage in one coil of the transformer, and the residual is impressed on the amplifier. The ratio of the potentiometer is not the same as the turn ratio of the opposing coils with reference to line coils. Through adjustment of each of these ratios it is possible to obtain step-up action which provides neutralization approaching 97%.

71. A balanced double resonant network has been developed, as shown in Fig. 29-47, which is effective in removing most of the induction with relatively little effect on the telegraph circuits. This device may be used where the dot frequency of the telegraph signal approaches nine-tenths of the frequency of the induced current. Its effectiveness, of course, is dependent upon the power frequency remaining nearly constant.

72. Neutralizing-wire operation generally uses an additional wire for each working facility. This wire is connected to a winding on the relay in the working circuit which is adjusted so that voltages due to induction, ground potential, etc., impressed on both wires are balanced out in the relay. This is essentially metallic-circuit operation.

73. Discriminating networks usually consisting of one or more circuits resonant at the interfering frequency may be employed to reduce the amount of the interfering current flowing in the receiving relay. These networks are, of course, not applicable where the dot frequency is close to that of the interfering current.

NOISE-FREQUENCY INDUCTION

74. Noise-frequency coordination problems arise owing to induction from the harmonic voltages and currents in the power circuit, which manifests itself as noise in the telephone receivers. Noise in carrier circuits is in this general classification but is not discussed here, since very few such problems have been encountered.

75. Influence. The two principal factors in the power system which control its influence are the **waveshape** of the power currents and voltages and the **balance** of the power lines and equipment to ground. Sine wave, 60 c, is the primary product manufactured by the power companies and used by their customers. However, it is impossible practically to generate and transmit a pure sine wave because of the generation of harmonics in the generators, transformers, and utilization equipment.

The harmonics of the power-system voltages and currents fall into two general classes: (*a*) the triple components and (*b*) the nontriple components. The nontriple components such as the 5th harmonic (300 c), the 7th harmonic (420 c), the 11th harmonic (660 c), etc., are 120° apart on a 3-phase system, the same as the fundamental voltage, and, therefore, they appear on the lines primarily as balanced components. Owing to unbalance of lines and equipment, they may also appear as residual components. On the other hand, the triple harmonics, e.g., the 3d (180 c), 9th (540 c), and 15th (900 c), are in phase on the wires of a 3-phase system and appear on the lines primarily as residual components. On a **single-phase system** there is no difference in the action of triple and nontriple harmonics. In any specific case it is important to determine whether the components are **balanced** or **residual,** as the treatment for their control may be entirely different.

In general, the generating equipment and transformer equipment (where the three transformers in a bank are symmetrical) are reasonably well balanced, both at the fundamental and at harmonic frequencies except that, with certain types of connections, triple harmonics may be impressed on the lines. Untransposed lines are inherently unbalanced to some extent, both as to their reactance and as to their capacitance to ground. The importance of this unbalance, which is frequently negligible at 60 c, increases with increase in frequency. Transmission lines can be reasonably well balanced at the lower harmonic frequencies by means of transpositions. Three-conductor cable circuits are inherently well balanced, both in their reactance and in their capacitance to ground. In addition, where metallic sheaths are used, there are practically perfect electric shielding and considerable magnetic shielding. Noise-frequency inductive coordination problems involving such cables are rare.

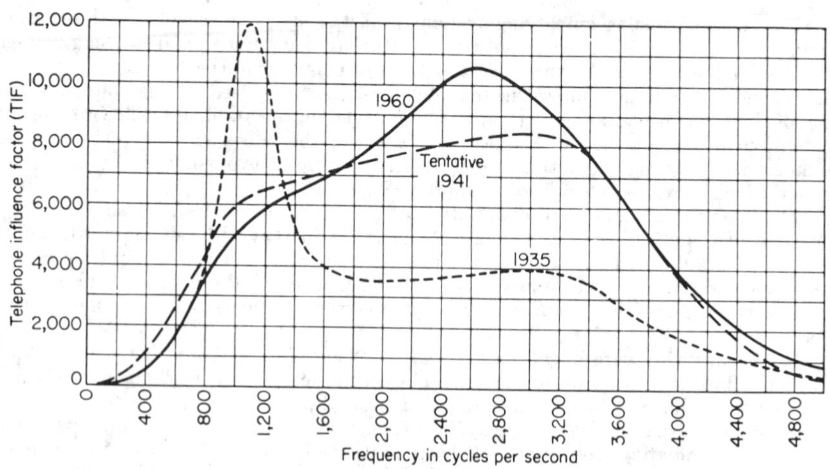

Fig. 29-48. Frequency-weighting characteristic for TIF measurements.

76. The effect of harmonics depends on the frequency response of the telephone system combined with that of the human ear.[57] Since the coupling between the power and telephone system is proportional to frequency, the influence of a given harmonic current or voltage on the power system is proportional to the frequency response of the telephone system multiplied by a factor proportional to its frequency.[58] Thus the **telephone influence factor** (TIF) for currents or voltages of any frequency are derived and are shown as the "1935 curve" in Fig. 29-48. The "Tentative 1941 curve" shows values of a proposed weighting which was under study by the EEI–Bell System Joint Subcommittee on Development and Research and adopted tentatively in 1941 to recognize changes in characteristics of telephone equipment since the 1935 TIF weighting was adopted. Further developments in telephone-equipment characteristics resulted in adoption of the 1960 values.[77] The influence of a given **harmonic current** or voltage component is proportional to the **magnitude** of the current in amperes times its TIF (called the I-T product) or the magnitude of the voltage in kilovolts times its TIF (called the kV-T product). The I-T or kV-T product for a current or voltage wave containing a number of harmonics is proportional to the square root of the sum of the squares of the I-T or kV-T products for *all* the component harmonics. The *overall* TIF of such a current or voltage is the I-T or kV-T product thus obtained divided by the rms value of the current or voltage as measured with an ordinary ammeter or voltmeter. Meters are available for measuring TIFs and I-T and kV-T products directly.

77. Upper limits of TIF for balanced components on synchronous machines, at no-load open-circuit rated voltage, are adopted by the National Electrical Manufacturers Association. These values for balanced TIF are as follows:

Size of Machine, kVA	Balanced TIF
62.5–299	300
300–699	200
700–999	150
1,000–2,499	125
2,500–9,999	60
10,000 and above	50

The balanced TIF is measured by connecting the TIF meter to the secondary of a potential transformer connected directly across the terminals of the machine on open circuit at rated voltage and frequency.

78. For residual-component TIF a limit of 30 was adopted at the same time for machines rated 5,000 kVA and above. For smaller machines, special consideration should be given to each individual problem. In situations where coordination problems exist or are anticipated from the residual components of the generator waveshape, special consideration should be given by the utility and the manufacturer to machines where the residual-component TIF is limited to 2.5. Residual-component TIF is, of course, important only for generators with grounded neutral which are connected directly or through Y-Y-connected transformers to lines.[37]

The residual-component telephone influence factor of a 3-phase machine may be obtained by connecting the machine in delta with one corner open and with the machine operating at normal voltage and speed, no load. One-third of the current in microamperes in a TIF meter placed directly in the open corner divided by the phase-to-neutral rms voltages gives the residual-component TIF.

79. Frequency-selective devices are of course applicable for reducing any harmonics which may be present in a generator waveshape.[59] Shunt devices have been used only on the smaller machines.

Wave traps for reducing the effects of residual components have been applied in a number of instances. These consist of reactors and capacitors in parallel, connected between the neutral of the machine and ground to insert a high impedance in the residual circuit for the particular troublesome frequency or frequencies.[59]

80. Power and distribution transformers inherently require harmonics in connection with the magnetization of the iron in their cores. The magnitude of these harmonics is dependent upon the size, design, and excitation current of the transformers. In general, transformer design is adjusted so as to result in a practical minimum distortion of waveshape. Exciting current I·T usually increases rapidly as the operating voltage is increased above rated voltage.

In single-phase transformers, the harmonic exciting current of the transformer is supplied from the lines connected to the primary and secondary in inverse proportion to the relative impedances of the circuits on the two sides of the transformer.

In 3-phase transformer banks connected **delta-delta,** all the triple-harmonic exciting current is confined to the windings of the three transformers. In transformers connected **delta-Y,** most of the triple-harmonic current is supplied by the delta winding. In transformers connected **Y-Y** with grounded neutral, the triple-harmonic currents are supplied as residuals circulating through the transformer from line to neutral. The addition of a tertiary delta winding to such a transformer bank tends to provide a low-impedance path for triple-harmonic exciting current, with a consequent reduction in the triple-harmonic residual current. The effectiveness of the tertiary delta is dependent upon its impedance, compared with the impedance of the primary and secondary windings and of the lines connected to them.

81. Rectifiers used for the conversion of alternating current to direct current and other similar devices inherently produce harmonics in both the d-c and a-c circuits with which they are associated.[21] In a 6-phase rectifier the 6th, 12th, 18th, 24th, etc., harmonics are present on the d-c side; and in a 12-phase rectifier corresponding multiples of 12 are present. Selective devices may be used to reduce these harmonics.[46]

The power-system waveshape on the a-c side is also distorted by mercury-arc rectifiers. For a 6-phase rectifier the prominent harmonics are the 5th, 7th, 11th, 13th, etc., and for a 12-phase rectifier the prominent harmonics are the 11th, 13th, 23rd, 25th, etc. In a specific case the effect of the rectifier in distorting power-system waveshape is dependent upon the load on the rectifier, the number of phases, and the impedance of the power system supplying it. Frequency-selective devices have been used on the a-c side of rectifiers and usually consist of resonant shunts or series reactors with shunt capacitors.[59]

In a-c railroad electrifications, there has been a recent trend toward the use of rectifiers on locomotives and multiple-unit cars so that d-c propulsion motors may be used. The installation of filters near the rectifiers has been found effective in reducing the influence of the a-c trolley and feeder system from the inductive-coordination standpoint.[24]

82. Phase-shifting arrangements have been used with multiunit rectifier installation to provide operation of the whole installation at a relatively large number of phases.[15,74] In this way, the frequency of the lowest pair of harmonics (equal to the number of phases ± 1) and the frequency interval between successive pairs (equal to the number of phases) are increased. Installations having 24, 30, 36, 60, 72, and 108 phases have been made in the United States.

83. Shunt capacitors used either by the power company on distribution circuits or by customers, for power-factor correction, may lead to an increase in harmonic currents.[75] In most cases where this increase in harmonic currents has been important, it has been due to the resonance between the capacitors and the supply line, or the associated transformers. Resonance due to shunt capacitors may be destroyed by the insertion of reactors (or reactors and resistors in parallel) in series with the capacitors, or by relocation or change in size of the capacitor banks. Under some conditions, shunt capacitors may destroy resonance or propagation effects and result in reduction of harmonics.

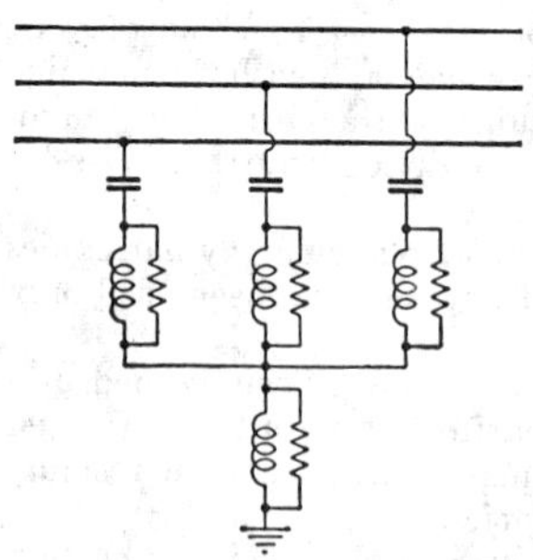

FIG. 29-49. Nonresonant shunt or terminating network for 3-phase line.

Shunt capacitors may also be used in a network to provide a means of loading the balanced and residual circuits of the line with an impedance having a large resistance component at harmonic frequencies without consuming appreciable energy at the fundamental frequency. This arrangement is applicable for use as a shunting network near the supply point to prevent resonance between the line capacitance and supply-system inductance or as a terminating network toward the end of an electrically long line (see Fig. 29-49).

84. Single-frequency control tones on power systems have been increasing in use lately. These include tones for water-heater control, street-light control, and time-clock control. These systems with control frequencies of about 360 to 3,500 c are generally installed on power-distribution lines, although many of the time-clock control systems are confined to lower voltages within a building or manufacturing site. In most cases the duration of the signal varies from about 5 to 40 s, and the number of operations per day has ranged from 1 to more than 12.

85. Transmission Lines. In general the loads on 3-phase transmission lines are reasonably well balanced.

Transmission lines with delta-delta or Y-delta-transformers only connected to them generally do not have large residual harmonic currents, and the triple harmonics on such lines are usually small. Where Y-Y-transformers are used, the residuals are likely to be much larger and the triple harmonic currents may be controlling. In some cases the paralleling of a Y-Y-bank with the Y-side of a Y-delta-bank in the same substation with neutrals grounded will substantially reduce the transmitted triple-harmonic currents.

86. Distribution systems differ from transmission systems in that the loads, and frequently much of the line mileage, are single-phase. In **delta** and **unigrounded neutral** circuits, the **load currents** are confined to the line wires although charging currents may be earth return. In **multigrounded neutral circuits,** there is residual current due to load unbalances, part of which becomes earth return. Control of this current component can be obtained by maintaining reasonable load balance on the 3 phases. Triple-harmonic exciting currents are, of course, residual and are not affected by load balance.

87. Urban distribution circuits are in general relatively short and have high load densities. Under these conditions, charging currents can usually be neglected, and the influence generally is controlled by transformer exciting currents and load currents.[12]

88. Rural distribution circuits differ from urban circuits in that usually they are longer and the load density is smaller. Particularly for the higher-voltage circuits, harmonic **charging currents** often accentuated by **resonance** or **propagation effects** may be more important than load or transformer exciting currents.[14,19,65] These charging currents can flow to ground regardless of the condition of neutral grounding, and the inductive influence of delta and unigrounded neutral circuits may, therefore, be as high as or higher than that of multigrounded neutral circuits. It has been found practicable, in some cases, to use **capacitors** or other devices at the supply points of rural circuits to improve the waveshape and reduce the influence. **Resonance** or **propagation effects** can also be reduced by nonresonant devices as shown in Fig. 29-50.

89. Susceptiveness of telephone circuits at noise frequencies depends primarily on the **balance** of the circuits to ground and the **power level** of the telephonic currents. Unbalances may be of the **shunt type** (difference in the admittances of the two sides to ground) or the **series type** (differences in the series impedances of the two sides) and may occur in the central-office equipment, the station sets, or the lines.

90. Unbalances in central-office equipment usually are due to (1) relays, retard coils, repeating coils, etc., connected from one side to ground or presenting different impedances between the two sides and ground; or (2) relays in one side of the line only, or differences in the impedances of capacitors or other apparatus connected in series with the two sides. Toll central-office equipment is usually designed to be balanced, and such unbalances as occur are due to manufacturing tolerances. Some types of local central-office equipment, however, are inherently unbalanced.

91. Station sets used in party-line service and in certain other types of service frequently have ringers or other signaling apparatus connected to ground in such a way as to result in shunt unbalances.[71] The effect of such connections depends on the impedance of the ringer or other apparatus connected to ground and on the characteristics of the station set. One type of set employs a gas-tube relay in the grounded ringer circuit which eliminates the ringer unbalance during the talking condition.

92. Drainage is sometimes used on telephone circuits to reduce voltages to ground which might act on shunt unbalances in central offices or station sets. It usually consists of a well-balanced retard coil connected across the circuit and with the midpoint grounded.[65] To provide d-c continuity, the connections to the line wires are frequently made through well-balanced capacitors.

93. Longitudinal chokes may be used to reduce the longitudinal currents acting on series unbalances.[65] A retard coil with two equal windings is used, one winding being connected in series with each line wire. The poling is such that the coil is noninductive to the metallic circuit and offers a high impedance to the longitudinal circuit.

94. Telephone cable circuits are extensively used in urban areas. In the construction of cables the wires are paired, and each pair is individually twisted. This results in close spacing and frequent transposition of the pair so that there is practically no direct metallic-circuit induction. Induced voltages act on the shunt unbalance of the pair to other pairs and sheath, on station sets and central-office unbalances, and on the series unbalances. The lead cable sheath acts as a highly effective electric shield, and therefore electric induction need not be considered for circuits carried in lead sheath cable. Grounding of the telephone-cable sheath at both ends results in substantial magnetic shielding, the effectiveness of which is dependent on frequency and on the resistance of the sheath and ground connections.

Low-resistance ground connections for shielding purposes at the central-office end of a cable sheath can be generally obtained by connection of the aerial cable to the underground cable or to the central-office ground. If an unsatisfactory electrolysis condition results, this can be avoided by installing an insulating joint between the aerial and underground cables and connecting the aerial cable to a power-circuit multigrounded neutral conductor. Such a conductor has been found to provide an effective ground in non-water-pipe areas as well as where extensive public water-pipe systems exist.[68]

95. Open-wire lines are more susceptive than cables, since they are subject to direct metallic-circuit induction and more liable to be unbalanced. Transpositions of open-wire lines are effective in reducing direct metallic-circuit induction and in reducing the capacitance unbalance to ground and to external circuits. Rolled joints practically eliminate unbalances due to joint resistance. Shunt unbalances may be important. The more common types of such unbalances are (*a*) leakage unbalance due to inadequate or missing insulators, or contacts with trees or brush, and (*b*) capacitance unbalances due to inadequate transpositions, or split quads or pairs in entrance cables.

96. Toll lines in either cable or open wire are designed to have a high degree of balance. In addition, the power level can be controlled to a certain degree by the telephone repeaters. Crosstalk requirements frequently limit the power level which can be used on multicircuit lines. On private lines with only a few circuits, high-level transmission has been used as a specific noise coordination measure.

97. Carrier telephone systems as a rule are not affected by power-system harmonics. Coordination with power-line carrier systems may be required in some cases.

BIBLIOGRAPHY

98. Miscellaneous

1. Joint Pole Practices for Supply and Communication Circuits, A Report of the Joint Committee on Plant Coordination of the Edison Electric Institute and the Bell Telephone System; *EEI Publ.* M12, October, 1945.

2. Specification for Communication Lines Crossing the Tracks of Railroads; Specification 1B1, Telephone and Telegraph Section, AAR.

3. National Electrical Safety Code, 5th ed.; Government Printing Office.

4. Generator Wave Shape; *Natl. Elec. Light Assoc. Publ.* 239, August, 1932.

5. Moore, L. M. Rural Phase-inverters Cut 'Phone Induction; *Elec. World,* Jan. 6, 1945.

99. American Institute of Electrical Engineers Papers

6. Gilkeson, C. L., and Bowen, A. E. Mutual Impedances of Ground-return Circuits; *Trans. AIEE,* 1930, pp. 1370–1379.

7. Gilkeson, C. L., and Hanks, A. J. Iron Armored Aerial Communication Cable; *Trans. AIEE,* 1934, pp. 890–895.

8. Gilkeson, C. L., and Jeanne, P. A. Overvoltages on Transmission Lines; *Trans. AIEE,* 1934, pp. 1301–1309.

9. Blye, P. W., Barstow, J. M., and Kent, H. E. Measurement of Telephone Noise and Power Wave Shape; *Trans. AIEE,* 1935, pp. 1307–1315.

10. Power System Faults to Ground; *Trans. AIEE,* 1937, pp. 421–433. Part 1, Characteristics, by C. L. Gilkeson, P. A. Jeanne, and J. C. Davenport. Part 2, Fault Resistance, by C. L. Gilkeson, P. A. Jeanne, and E. F. Vaage.

11. GEORGE, E. E., HONAMAN, R. K., LOCKROW, L. L., and SCHWARTZ, E. L. Neutralizing Transformers to Protect Power Station Communication; *Trans. AIEE*, 1036, pp. 524–529.

12. COLEMAN, J. O'R., and DAVIS, R. F. Inductive Coordination of Common-neutral Power-distribution Systems and Telephone Circuits; *Trans. AIEE*, 1937, pp. 17–26 and 189.

13. COLEMAN, J. O'R., and SCHIRMER, A. H. Protection Features for the Joint Use of Wood Poles Carrying Communication Circuits and Power-distribution Circuits above 5,000 Volts; *Trans. AIEE*, 1938, pp. 131–140.

14. TAYLOR, T. A., and WAHLQUIST, H. W. Noise Coordination of Rural Power and Telephone Systems; *Trans. AIEE*, 1938, pp. 613–620.

15. MARTI, O. K., and TAYLOR, T. A. Wave Shape of 30- and 60-phase Rectifier Groups; *Trans. AIEE*, 1940, pp. 218–224.

16. KENT, H. E., and BLYE, P. W. Inductive Coordination with Series Sodium Highway-lighting Circuits; *Trans. AIEE*, 1939, pp. 325–332.

17. COLEMAN, J. O'R., and TRUEBLOOD, H. M. Coordination of Power and Communication Circuits for Low-frequency Induction; *Trans. AIEE*, July, 1940, Vol. 59, pp. 403–411.

18. MILNOR, J. W. Control of Inductive Interference to Telegraph Systems; *Trans. AIEE*, 1940, pp. 469–474.

19. PLUCKNETT, K. J., SMITH, W. T., and TAYLOR, T. A. Inductive Coordination of R.E.A. Distribution Systems and Telephone Systems; *Trans. AIEE*, 1941, p. 586.

20. KILLEN, R. B., and LAW, G. G. Protection of Pilot Wires from Induced Potentials; *Trans. AIEE*, 1946, pp. 267–270.

21. Reports of AIEE Electronic Power Conversion Committee:

a. Inductive Coordination Aspects of Rectifier Installations; *Trans. AIEE*, 1946, pp. 417–436.

b. Inductive Coordination Aspects of D-c Systems Supplied by Rectifiers; *Trans. AIEE*, 1951, pp. 1034–1054.

22. CAMPBELL, J. W., HILL, L. M., MOORE, L. M., and SCHOLZ, H. J. Joint Use of Pole Lines for Rural Power and Telephone Services; *Trans. AIEE*, 1947, pp. 519–524.

23. BLOECKER, W. E., and BULLARD, W. R. Some Aspects of Joint Use of Wood Pole Lines with Increasing Line Voltage; *Trans. AIEE*, 1953, Pt. I, pp. 709–712.

24. KING, E. B., GORDON, K. H., and HIBBARD, L. J. Rectifier Motive Power—Inductive Coordination Considerations; *Trans. AIEE*, 1954, Pt. II, pp. 107–114.

100. Principles and Practices

25. Reports of Joint General Committee of Edison Electric Institute and Bell Telephone System, reissued July, 1945, EEI *Publ.* M5, including:

a. Principles and Practices for Inductive Coordination.

b. Allocation of Costs for Inductive Coordination.

c. Principles and Practices for Joint Use of Wood Poles.

26. Reports of Joint General Committee of American Railway Association and Bell Telephone System on The Inductive Coordination of Railway Electrical Supply Facilities and the Communication Facilities of the Bell System, Apr. 15, 1931, and Sept. 1, 1932.

27. Reports of Joint General Committee of Edison Electric Institute and The Western Union Telegraph Company on The Physical and Inductive Coordination of Electrical Supply and Communication Systems, July 22, 1935.

28. Reports of Joint General Committee of Association of American Railroads and Edison Electric Institute on The Inductive Coordination of Electric Supply and Communications Systems, Oct. 7, 1936.

29. Reports of Joint Engineering Committee of Association of American Railroads and Edison Electric Institute on Principles and Practices, and Specifications, for Crossings of Electrical Supply Lines and Facilities of Steam and Electrified Railroads, August, 1946, *EEI Publ.* O-12.

101. Reports of Joint Subcommittee on Development and Research of EEI-Bell System

30. Athenia-Passaic Ground Potential and Induction Investigation; *Eng. Rept.* 5, May 29, 1929.

31. Minneapolis Joint Use Investigation; *Eng. Rept.* 6, May 29, 1929.

32. Train Control Tests Made on the Delaware, Lackawanna and Western Railroad Near Horseheads, N.Y., during 1926–1928; *Eng. Rept.* 7, May 29, 1929.

33. Termination of Isolated Exposure Sections to Obtain Normal Metallic Circuit Currents; *Eng. Rept.* 8, Oct. 31, 1929.

34. Short-Cut Methods for Calculating Noise in Local Telephone Subscriber Circuits in Cable Due to Exposures to Power Distribution Circuits; *Eng. Rept.* 9, Apr. 15, 1930.

35. A Method of Measuring the Balance of Open-wire Telephone Circuits with Respect to Longitudinal-circuit Induction; *Eng. Rept.* 10, Apr. 22, 1930.

36. Methods of Measuring Noise on Open-wire Toll Circuits; *Eng. Rept.* 11, Feb. 16, 1931.

37. The Effects on Inductive Coordination of Generators Feeding Directly on the Line and Operating with Grounded Neutrals; *Eng. Rept.* 12, Feb. 16, 1931.

38. Short-cut Methods for Calculating Noise in Open-wire Subscriber Circuits Due to Joint-use Exposures to Power Distribution Circuits; *Eng. Rept.* 13, Feb. 16, 1931.

39. Coupling Factors for Ground-return Circuits—General Considerations and Methods of Calculation; *Eng. Rept.* 14, May 14, 1931.

40. Wave-shape Survey on Operating Power Systems; *Eng. Rept.* 15, Feb. 3, 1932.

41. A Method of Estimating Noise in Open-wire Toll Circuits; *Eng. Rept.* 16, Aug. 15, 1932.

42. Short-cut Methods for Calculating Noise in Open-wire Subscribers' Circuits Exposed to Power Lines at Roadway Separations; *Eng. Rept.* 17, Aug. 15, 1932.

43. Longitudinal-circuit Propagation along Open-wire Lines at Voice Frequencies; *Eng. Rept.* 18, Feb. 10, 1933.

44. Direct Capacitances and Capacitance Unbalances in Open-wire Telephone and Joint-use Lines; *Eng. Rept.* 19, Feb. 10, 1933.

45. Exploring Wire Methods for Toll and Local Circuit Noise Studies; *Eng. Rept.* 20, Aug. 25, 1933.

46. Selective Devices for Reducing Harmonies in Power Systems, Pt. I, D-c Systems; *Eng. Rept.* 21, Aug. 25, 1933.

47. Wave-shape Characteristics of Power Circuits Supplying Rectifiers; *Eng. Rept.* 22, Aug. 25, 1933.

48. The Impedances of A-c Supply Systems at Harmonic Frequencies to Balanced Components; *Eng. Rept.* 23, Aug. 25, 1933.

49. The Impedances of A-c Supply Systems at Harmonic Frequencies to Residual Components; *Eng. Rept.* 24, Mar. 16, 1934.

50. Articulation Studies on the Effects of Noise; *Eng. Rept.* 25, Mar. 16, 1934.

51. Shielding of Ground Return Circuits at Low Frequencies; *Eng. Rept.* 26, Mar. 16, 1934.

52. General Considerations in the Use of Neutral Impedance; *Eng. Rept.* 27, Mar. 16, 1934.

53. The Effect of Vacuum Tube Amplifiers and Telephone Repeaters on Acoustic Shock; *Eng. Rept.* 28, Nov. 2, 1934.

54. Use of Copper Oxide Varistors in Reducing Acoustic Shock; *Eng. Rept.* 29, Nov. 2, 1934.

55. Overvoltages on Transmission Lines Due to Ground Faults as Affected by Neutral Impedance; *Eng. Rept.* 30, Nov. 15, 1935.

56. Low Resistance Grounds; *Eng. Rept.* 31, June 7, 1935.

57. Relative Interfering Effects of Different Single-frequency Noises in Telephone Circuits; *Eng. Rept.* 32, June 7, 1935.

58. The Telephone Influence Factor of Supply System Voltages and Currents; *Eng. Rept.* 33, Oct. 24, 1935.

59. Selective Devices for Reducing Harmonics in Power Systems, Part II, A-c Systems; *Eng. Rept.* 34, Jan. 21, 1936.

60. Study of Observations from Low-frequency Inductive Exposures; *Eng. Rept.* 35, Jan. 21, 1936.

61. The Use of Power and Telephone Transpositions within Exposure Sections; *Eng. Rept.* 36, July 22, 1936.

62. Computation of Zero-sequence Impedances of Power Lines and Cables; *Eng. Rept.* 37, July 22, 1936.

63. Power Level and Sensitivity of Telephone Circuits; *Eng. Rept.* 38, July 22, 1936.

64. Characteristics of Power System Faults to Ground; *Eng. Rept.* 39, *EEI Publ.* E9, November, 1937.

65. Inductive Coordination of Rural Power and Telephone Systems; *Eng. Rept.* 40, *EEI Publ.* F2, January, 1938.

66. Short-circuiting Relay Protectors; *Eng. Rept.* 41, *EEI Publ.* F17, August, 1938.

67. Inductive Coordination Aspects of Petersen Coil Grounding; *Eng. Rept.* 42, *EEI Publ.* H3, March, 1940.

68. Effects of Bonding Aerial Telephone Cable Sheath to Multi-grounded Power Neutrals; *Eng. Rept.* 43, *EEI Publ.* H10, August, 1940.

69. Neutralizing Transformers; *Eng. Rept.* 44, *EEI Publ.* H12, August, 1940.

70. Frequency Weighting for Message Circuit Noise; *Eng. Rept.* 45, *EEI Publ.* I-7, July, 1941.

71. Susceptiveness of Subscriber Station Sets; *Eng. Rept.* 46, *EEI Publ.* I-8, July, 1941.

72. Positive Disconnection of Distribution Circuits during Faults to Ground; *Eng. Rept.* 47, *EEI Publ.* J-1, July, 1941.

73. Low-frequency Shielding in Telephone Cables; *Eng. Rept.* 48, *EEI Publ.* J-4, July, 1941.

74. Wave Shape of Multi-phase Rectifiers; *Eng. Rept.* 49, *EEI Publ.* J-6, April, 1942.

75. Inductive Coordination Aspects of Power Factor Correction Capacitor Installations; *Eng. Rept.* 50, *EEI Publ.* 51-16, October, 1951.

76. Joint Use of Poles, Telephone Circuits and 6.6- and 13.2-kv Power Circuits, Safety Features; *Prov. Rept.* 19, *EEI Publ.* E-4, January, 1937.

77. The Telephone Influence Factor of Supply System Voltages and Currents; Supplement to *Eng. Rept.* 33, Edison Electric Institute–American Telephone and Telegraph Company, Sept. 2, 1960.

78. *NEMA Standards Publ.*, Motors and Generators, MG1-1963, TIF-MG1-22.44.

CLEARANCES FROM ENERGIZED EQUIPMENT AND LINES

WORKING DISTANCES

102. Minimum clearances have been established for the distances to be maintained between unguarded "live" parts and men performing nonelectrical operations in the vicinity or men working on electric equipment or playing hose streams on energized electric lines and equipment; also clearance between energized apparatus and fog or spray nozzles has been established (see Table 29-1).

103. Guarding Live Parts.[1] In industrial plants, mercantile establishments, and similar places, electric utilization equipment should be located or guarded to prevent unqualified employees or visitors venturing too close to live parts. Some parts require frequent inspection and servicing, and these should be accessible only to authorized or qualified employees.

Maintenance men, repair men, operators, and others must work frequently near exposed or partly exposed live parts. Having met the required qualifications for adjusting or handling energized parts, they are considered as "qualified" persons to engage in such work. For these individuals a lesser degree of protection is sufficient than for unauthorized and unqualified persons.

In generating stations and substations where electric equipment is accessible only to qualified attendants, general use of guards and complete isolation is allowable to a less extent than is allowable for electric utilization equipment accessible to unqualified persons, such as in industrial plants and similar places.

[1] For further details, see National Electrical Safety Code (National Bureau of Standards Handbook H30, March, 1948), Pt. 1, Rules for the Installation and Maintenance of Electrical Supply Stations and Equipment; also Pt. 3, Rules for the Operation of Electric Equipment and Lines.

Table 29-1. Safe Distances for Working and for Using Fire-hose Streams and Spray Nozzles on or near Electric Equipment

Voltage						Min clearances§ from unguarded live parts; for nonelectrical work in vicinity of elec equipment				Clearance‖ from live parts; min approach for working on elec equipment, ft	Safe distance,¶ ft, when using a solid stream of fresh water, with a nozzle pressure of 50 lb		Clearance,¶ in., between fog or spray nozzles and live elec apparatus
Operating, line to line	Spark-over*												
	Sphere			½-in. rods									
	Diam cm	Spacing†		Spacing‡		Vertical ft in.		Horizontal ft in.			1⅛-in. nozzle	1½-in. nozzle	
		Cm	In	Cm	In	ft	in.	ft	in.				
120	.....	.....	.....	.....	.....	..	..	..	..	..	½	½	
300–7,500**	.....	.....	.....	.....	.....	7	6	3	0				
550††	.....	.....	.....	.....	.....	..	..	..	..	..	4	4	
600	.....	.....	.....	.....	.....	7	8	3	2				
750–7,500**	.....	.....	.....	.....	.....	..	..	..	..	½			
2,300	.....	.....	.....	.....	.....	7	9	3	3	..	11	16	
6,600	.....	.....	.....	.....	.....	7	10	3	4	..	19	29	
7,500	.....	.....	.....	.....	.....	8	6	3	5	1	..	..	6
11,000	.....	.....	.....	.....	.....	9	0	3	6	2	20	30	
22,000	6.25	0.67	0.268	1.6	0.64	9	3	3	9	..	25	33	8
33,000	6.25	1.06	0.424	2.6	1.04	9	6	4	0	..	30	40	12
44,000	6.25	1.47	0.588	3.7	1.48	9	10	4	4				
50,000	6.25	1.72	0.688	4.4	1.56	..	..	..	..	3	..	..	23
66,000	6.25	2.38	0.852	6.8	2.72	10	5	4	11				
70,000	6.25	2.56	1.024	7.6	3.04	..	..	..	..	5			
88,000	12.5	3.19	1.276	11.6	4.64	11	0	5	6	..	..	..	30
110,000	12.5	4.28	1.672	16.8	6.72	11	7	6	1	..	..	..	37
132,000	12.5	5.3	2.12	21.7	8.68	12	2	6	8	..	..	..	44
166,000	12.5	7.65	3.06	28.8	11.52	..	..	..	..	..	..	..	52
220,000	25	8.93	3.57	39.0	15.63	..	..	..	..	..	..	..	76
330,000	25	11.81	6.72	59.6	23.46	..	..	..	..	..	..	..	120

* Values interpolated from American Standard for Measurements of Test Voltage in Dielectric Tests, C68.1-1953, American Standards Association. Also see Sec. **3** (negative impulses for 6.25-cm sphere with 0.5- to 2.5-cm sphere-gap spacing, 12.5-cm sphere with 3.0- to 8.0-cm spacing, and 25-cm sphere with 2.5- to 25.0-cm spacing).

† Spacing using sphere-gap spark-over voltages for the listed diameters of spheres; 60-c and negative-impulse crest; one sphere grounded; at 760 mm barometric pressure and 25°C.

‡ Spacing using rod-gap spark-over voltages (crest); at 25°C, 760 mm barometric pressure, and 1.55 cm vapor pressure.

§ Natl. Elec. Safety Code, Part 1, 1948.

‖ Natl. Elec. Safety Code, Part 4, 1948.

¶ Natl. Board Fire Underwriters *Bull.* 91, 1963 (interpolated values).

** Volts to ground.

†† Trolley service direct current.

104. Minimum clearances and safe distances for working conditions with respective operating voltages are given in Table 29-1. There is included for comparison a list of sphere-gap and rod-gap spacings for the various operating voltages. Three voltage listings in the table are for line to ground; all other listings are for line to line.

At all times electric equipment should be considered as being alive, unless it is *positively* known to be dead.

FIRE-HOSE STREAMS

105. Fire-hose Streams—Conductivity.[1] Tests have shown that a stream of water carries current as long as it remains unbroken, but from the point where the stream begins to break into spray, the resistance of the electric circuit increases rapidly. Beyond the point at which the stream breaks, no appreciable current could be detected.

Certain facts have been deduced relating to conductivity of fire-hose streams in extinguishing fires. The amount of current conducted to the nozzle by the stream will depend on the following conditions:

a. The voltage of the wire or device.

[1] Fire Streams and Electric Circuits, *Natl. Board Fire Underwriters Bull.* 91, December, 1963. ALLEN, H. L. Insulator Washer Solves Contamination Problem; *Elec. World*, Nov. 17, 1952.

b. The distance from the nozzle to the electrically charged line or device.

c. The diameter of the stream.

d. Whether the stream is solid or broken.

e. The conductivity of the water.

For voltages up to 600, classed by NEC as the limit of low potential, there is usually no danger unless the nozzle is brought very close to the line or device. This is particularly true when a fireman wears rubber boots.

For voltages higher than 600, the nozzle must be held at greater distances (see Table 29-1).

106. Safe distances for using 1⅛- and 1½-in nozzles in the vicinity of electric lines, apparatus, or devices are given in Table 29-1. These distances apply when using streams of fresh water.

Sea water, salty water, or the discharge from a soda-acid extinguisher may have such high conductivity that no rule can be applied as to the safe distance for a solid stream. If it becomes necessary to use water known to have high conductivity, then an unbroken stream should not be directed at energized parts.

107. Safe distances for using a small hose[1] to extinguish fires in electric equipment, and allowing for a liberal factor of safety against current being conducted by the stream, are based (1) on tests that have been run on the conductivity of ¼- and ⅜-in streams (up to 70 lb pressure) at 440 V and (2) on the fact that higher voltages generated in powerhouses would cause only slight changes in these distances because of the high air resistance between the drops after the stream breaks up. The following distances should be maintained:

Small hose can be used in powerhouses provided that the operator stands with the nozzle of ¼-in streams at least 15 ft away from any "live" electric apparatus which may be struck accidentally or 20 ft away with ⅜-in streams.

For complete safety the circuit should be opened to "kill" electric equipment before turning water on it.

108. The spraying of a stream may be an important factor in electrical fires, especially those involving transformers, oil-cooled switches, burning insulation and wood, or other combustible material near electric equipment. Tests indicate that the spray or "fog" nozzle provides protection for the holder and also that such a nozzle is very efficient in cooling the combustion gases and vapors and extinguishing the fire. For clearances, see Table 29-1.

109. Various portable manually operated fire extinguishers are listed by Underwriters' Laboratories, Inc. The units suitable for electrical (Class C) fires are those which discharge carbon dioxide and those with a carbon tetrachloride base. Dry-powder extinguishers are also suitable.

Foam, soda acid, and hand-pump tanks with antifreeze solutions are not suitable for extinguishing fires in electric equipment; see *Natl. Board Fire Underwriters Bull.* 91, December, 1963; also Factory Mutual Engineering Division's *Loss Prevention Data Sheet* 4-3, October, 1950.

INDUSTRY CODES AND STANDARDS

110. Industry standards are a necessity for the successful and safe operation of today's electric power systems. Systems are generally interconnected and may be supplied by a number of generators built by different manufacturers in several generating stations. Transmission systems and distribution facilities involve large numbers of components which, though produced by many manufacturers, must function together harmoniously. Utilization equipment, its wiring, protection, and control, from steel rolling mills to toasters, requires standard supply facilities and standard utilization voltages and frequencies. The requirements for standards for telecommunication systems are similar in nature.

Industry standards are voluntary. Recognition of the need for such standards and the economic advantages of mass production and interchangeability accruing from their use result in a remarkable degree of acceptance and use.

[1] Conductivity of Small Hose Streams; *Factory Mutual Record*, January, 1933.

In addition, industry standards, employed as they are as instruments of commerce, particularly in the preparation of purchase specifications, both nationally and internationally, must use a common language, to minimize or eliminate misunderstandings between the buyer and the seller, and to assist the purchaser in selecting and obtaining the proper products for his particular needs.

Standards are essential to the electrical industry in the following broad categories:

1. Fundamental units of mass, time, nd length.
2. Fundamental and derived units of electricity—current, potential difference, freqency, magnetic strength, dielectric strength, etc.
3. Definitions of technical terms.
4. Physical standards to determine uniformly the magniude of these units.
5. Standards of instrumentation and methods of measurement.
6. Product standards defining a product, process, or procedure and relating to nomenclature, composition, construction, dimensions, tolerances, safety, operating characteristics, performance, quality, rating, testing, and application.
7. International standards.

Industry standards are developed by a large number of engineering societies, manufacturers' trade associations, departments of the Federal government, national standards organizations of countries throughout the world, and international standards organizations. While the list of standardizing organizations in the United States is imposing, each has its place in the whole picture related to the competence and responsibilities of the separate organizations in their specific disciplines. Coordination is effected through the facilities of the United States of America Standards Institute.

111. Specifications are the adaptation of nationally recognized standards to purchase specifications, which include, in addition to the reference to the technical content of product standards, methods of test, nomenclature, etc., and the commercial agreements necessary for acceptance and purchase. These latter commercial agreements are matters of negotiation between the purchaser and supplier and are not appropriately subjects of technical standards.

For some products of a complex nature, standards for specifications are developed as guides to purchasers in the development of their own specifications. Adherence to the standards for specifications again assists the supplier and the purchaser in determining the exact product which meets the need of the user.

112. Codes are essentially standards relating to safety, public health, or welfare. Codes may be given the force of law through legislation, which may incorporate the text of the code, adopt the code by reference, or effectuate the code through the prima facie rule-of-evidence method of reference. Examples of the various methods are given in USAS Z56, Nationally Recognized Standards in State Laws and Local Ordinances.

The two safety codes most important to the electrical industry are the National Electrical Code (USAS C1) sponsored by the National Fire Protection Association, and the National Electrical Safety Code (USAS C2) sponsored by the National Bureau of Standards. The National Electrical Code represents the consensus of expert opinion as to the practical methods and materials of installation to provide safeguarding of persons and property in the use of electrical equipment. The National Electrical Safety Code is a standard governing the safe installation, maintenance, and operation of electrical supply systems.

Standards for safety for electrical equipment are formulated by the Underwriters' Laboratories, Inc., based on the safeguarding of persons and property from fire and accident hazards arising from the use of electrical equipment when installed in accordance with the provisions of the National Electrical Code.

113. The United States of America Standards Institute, 10 East 40th St., New York, N.Y. 10016, serves as the national clearinghouse and coordinating agency for the establishment of a uniform set of national standards on all subjects for which standards exist or may be prepared nationally in the United States or internationally by the organizations (International Electrotechnical Commission, International Organization for Standardization, Pan-American Standards Commission) in which the United States is represented through the institute. In addition, the USASI has on file standards of all organized national standardizing bodies in the world. Published in the language of country of origin (not translated), these are available for use in the United States and may be obtained from the institute.

114. Organizations involved in standards activities of primary concern to electrical engineers, together with their identifying initials, are listed below. In general, the names of the organizations give some indication of their areas of standards activities and responsibilities. However, in all cases, inquiry to the USASI will result in information as to the appropriate source of standards information if it is not immediately available from the association.

Continual revision of standards makes it impracticable to list titles and identifying numbers of existing standards. Catalogues of standards may be obtained from the USASI and from the various organizations listed.

AA —Aluminum Association, 420 Lexington Avenue, New York, N.Y. 10017.

AAR —Association of American Railroads, Transportation Bldg., Washington, D.C. 20006.

AEIC —Association of Edison Illuminating Companies, 51 East 42d St., New York, N.Y. 10017.

AFBMA —Anti-Friction Bearing Mfrs. Asso., Inc., 60 East 42d St., Room 764, New York, N.Y. 10017.

AFMF —Associated Factory Mutual Fire Insurance Companies, Factory Mutual Engineering Division, 1151 Boston-Providence Turnpike, Norwood, Mass.

AGA —American Gas Association, 605 Third Avenue, New York, N.Y. 10016.

AGMA —American Gear Manufacturers Association, 1 Thomas Circle N.W , Washington, D.C. 20005.

AIA —American Institute of Architects, The Octagon, 1735 New York Avenue N.W., Washington, D.C. 6.

AIA —American Insurance Association (successor to National Board of Fire Underwriters), 85 John St., New York, N.Y. 10038.

AIF —Atomic Industrial Forum, Inc., 850 Third Avenue, New York, N.Y. 10022.

AISE —Association of Iron and Steel Engineers, 1010 Empire Bldg., Pittsburgh, Pa. 22.

AISI —American Iron & Steel Institute, 150 East 42d St., New York, N.Y. 10017.

AMC —American Mining Congress, 1102 Ring Bldg., Washington, D.C. 6.

AMCA —Air Moving & Conditioning Association, Inc., 205 West Touhy Avenue, Park Ridge, Ill.

API —American Petroleum Institute, 1271 Avenue of the Americas, 7th Floor, New York, N.Y. 10020.

ARI —Air-Conditioning & Refrigeration Institute, 1815 North Fort Myer Drive, Arlington, Va. 22209.

ASCE —American Society of Civil Engineers, 345 East 47th St., New York, N.Y. 10017.

ASHRAE—American Society of Heating, Refrigerating and Air Conditioning Engineers, Inc. (successor to American Society of Heating and Air Conditioning Engineers, and American Society of Refrigerating Engineers), 345 East 47th St., New York, N.Y. 10017.

ASME —American Society of Mechanical Engineers, 345 East 47th St., New York, N.Y. 10017.

ASTM —American Society for Testing & Materials, 1916 Race St., Philadelphia, Pa. 19103.

ASTME —American Society of Tool and Manufacturing Engineers, 20501 Ford Road, Dearborn, Mich. 48128.

AWS —American Welding Society, 345 East 47th St., New York, N.Y. 10017.

AWWA —American Water Works Association, Inc., 2 Park Avenue, New York, N.Y. 10016.

CAGI —Compressed Air & Gas Institute, 122 East 42d St., New York, N.Y. 10017.

CBM —Certified Ballast Manufacturers, 2120 Keith Bldg., Cleveland, Ohio 44115.

CDA —Copper Development Association, Inc., 405 Lexington Avenue, New York, N.Y. 10017.

CEE —International Commission on Rules for the Approval of Electrical Equipment, c/o United States of America Standards Institute, 10 East 40th St., New York, N.Y. 10016.

CEMA —Canadian Electrical Manufacturers Association, 10 Price St., Toronto 5, Canada.
CGA —Compressed Gas Association, Inc., 500 Fifth Avenue, New York, N.Y. 10036.
CS —U.S. Department of Commerce, Commodity Standards Division, National Bureau of Standards, Washington, D.C. 20234.
CSA —Canadian Standards Association, 235 Montreal Road, Ottawa 7, Canada.
DOD —U.S. Department of Defense, Defense Supply Agency, Cameron Station, Alexandria, Va.
EEI —Edison Electric Institute, 750 Third Avenue, New York, N.Y. 10017.
EIA —Electronic Industries Association, 2001 Eye St., Washington, D.C.
GAMA —Gas Appliance Manufacturers Association, 60 East 42d St., New York, N.Y. 10017.
GSA —General Services Administration, Federal Supply Service, Washington, D.C. 20407.
IEC —International Electrotechnical Commission, U.S. National Committee, c/o United States of America Standards Institute, 10 East 40th St., New York, N.Y. 10016.
IEEE —Institute of Electrical & Electronics Engineers (successor to AIEE and IRE), 345 East 47th St., New York, N.Y. 10017.
IES —Illuminating Engineering Society, 345 East 47th St., New York, N.Y. 10017.
IFI —Industrial Fasteners Institute, 1517 Terminal Tower, Cleveland, Ohio 44113.
IPCEA —Insulated Power Cable Engineers Association, 283 Valley Road, Montclair, N.J. 07042.
ISO —International Organization for Standardization, c/o United States of America Standards Institute, 10 East 40th St., New York, N.Y. 10016.
ITA —Industrial Truck Association, 250 Gateway Towers, Pittsburgh, Pa. 15222.
JEDEC —Joint Electronic Devices Engineering Council, c/o Electronic Industries Association, 2001 Eye St., Washington, D.C.
NBFU —National Board of Fire Underwriters; see American Insurance Association.
NBS —U.S. Department of Commerce, National Bureau of Standards, Washington, D.C. 20234.
NEMA —National Electrical Manufacturers Association, 155 East 44th St., New York, N.Y. 10017.
NEMI —National Elevator Manufacturing Industry, Inc., 101 Park Avenue, New York, N.Y. 10017.
NFPA —National Fire Protection Association, 60 Batterymarch St., Boston, Mass. 02110.
NMTBA —National Machine Tool Builders Association, 2139 Wisconsin Avenue N.W., Washington, D.C. 20007.
NSC —National Safety Council, 425 Michigan Avenue, Chicago, Ill. 60611.
PASC —Pan-American Standards Commission, c/o United States of America Standards Institute, 10 East 40th St., New York, N.Y. 10016.
RMA —Rubber Manufacturers Association, Inc., 444 Madison Avenue, New York, N.Y. 10022.
RWMA —Resistance Welder Manufacturers Association, 1900 Arch St., Philadelphia, Pa. 3.
SAE —Society of Automotive Engineers, Inc., 485 Lexington Avenue, New York, N.Y. 10017.
SAMA —Scientific Apparatus Makers Association, 20 North Wacker Drive, Chicago, Ill. 60606.
SMPTE —Society of Motion Picture & Television Engineers, 9 East 41st St., New York, N.Y. 10017.
SPE —Society of Plastics Engineers, Inc., 65 Prospect St., Stamford, Conn.
SPI —Society Plastics Industry, Inc., 250 Park Avenue, New York, N.Y. 10017.
UL —Underwriters' Laboratories, Inc., 207 East Ohio St., Chicago, Ill. 60611.
USASI —United States of America Standards Institute, 10 East 40th St., New York, N.Y. 10016.

An expert is largely
a man who knows where and
how to find the answers.

INDEX

Numbers in **boldface** refer to sections; all other numbers refer to paragraphs. Page numbers are not used in this index.

(References are to sections and paragraphs—not pages)

INDEX

(References are to sections and paragraphs—not pages)

(References are to sections and paragraphs—not pages)

INDEX

(References are to sections and paragraphs—not pages)

(References are to sections and paragraphs—not pages)

(References are to sections and paragraphs—not pages)

INDEX

(References are to sections and paragraphs—not pages)

(References are to sections and paragraphs—not pages)

(References are to sections and paragraphs—not pages)

INDEX

(References are to sections and paragraphs—not pages)

(References are to sections and paragraphs—not pages)

(References are to sections and paragraphs—not pages)

(References are to sections and paragraphs—not pages)

Modulus—Noise

(References are to sections and paragraphs—not pages)

(References are to sections and paragraphs—not pages)

INDEX

(References are to sections and paragraphs—not pages)

(References are to sections and paragraphs—not pages)

INDEX

(References are to sections and paragraphs—not pages)

(References are to sections and paragraphs—not pages)

INDEX

(References are to sections and paragraphs—not pages)

(References are to sections and paragraphs—not pages)

INDEX

(References are to sections and paragraphs—not pages)

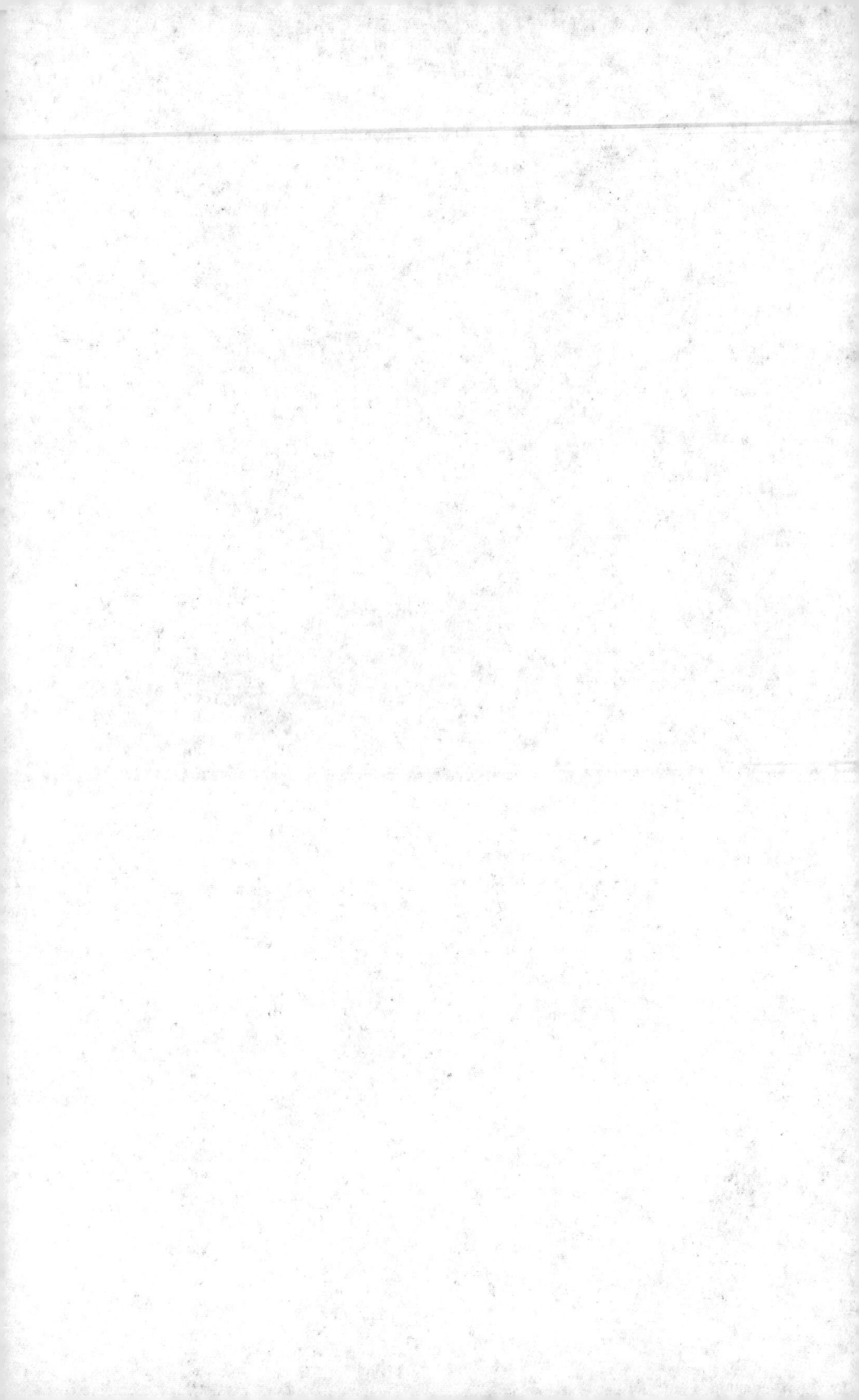

INDEX TO SECTIONS FOR QUICK REFERENCE